MILLS'S HISTOLOGY

for

PATHOLOGISTS

MILLS'S HISTOLOGY *for* PATHOLOGISTS

SIXTH EDITION

Rhonda K. Yantiss, MD
Clinical Professor
Vice Chair of Faculty Development and Mentoring
Department of Pathology and Laboratory Medicine
University of Miami
Miller School of Medicine
Miami, Florida

Wolters Kluwer

Philadelphia • Baltimore • New York • London
Buenos Aires • Hong Kong • Sydney • Tokyo

Acquisitions Editor: Nicole Dernoski
Senior Development Editor: Ariel S. Winter
Editorial Coordinator: Erin E. Hernandez
Marketing Manager: Kirsten Watrud
Senior Production Specialist: Bridgett Dougherty
Manager, Graphic Arts & Design: Stephen Druding
Manufacturing Coordinator: Lisa Bowling
Prepress Vendor: Aptara, Inc.

Printed in Mexico

Library of Congress Cataloging-in-Publication data available upon request from Publisher.

ISBN-13: 978-1-9752-3429-4

shop.lww.com

QUADM0925

Contributors

Sylvia L. Asa, MD, PhD
Professor
Department of Pathology
University Hospitals
Case Western Reserve University
Cleveland, Ohio

Leomar Y. Ballester, MD, PhD
Assistant Professor
Department of Pathology
University Hospitals
Case Western Reserve University
Cleveland, Ohio

Phoenix D. Bell, MD, MSc
Assistant Professor
Department of Pathology
University of Pittsburgh Medical Center
Pittsburgh, Pennsylvania

Jennifer A. Bennett, MD
Associate Pathologist
Department of Pathology
University of Chicago
Chicago, Illinois

Gerald J. Berry, MD
The Richard L. Kempson Endowed Professor in Surgical Pathology
Department of Pathology
Stanford University
Stanford, California

Justin A. Bishop, MD
Professor
Department of Pathology
UT Southwestern Medical Center
Dallas, Texas

Baris Boyraz, MD, PhD
Assistant Professor
Department of Pathology
Weill Cornell Medicine
New York, New York

Yasmeen M. Butt, MD
Consultant
Department of Laboratory Medicine and Pathology
Mayo Clinic
Scottsdale, Arizona

Alison L. Cheah, MBBS
Anatomical Pathologist
Department of Anatomical Pathology
Douglass Hanly Moir Pathology
Sydney, NSW, Australia

Elizaveta Chernetsova, MD, FRCPSC
Assistant Professor
University of Ottawa
Pathologist
Department of Pathology and Laboratory Medicine
The Ottawa Hospital
Ottawa, Ontario, Canada

Benjamin B. Cho, MD
Clinical Assistant Professor
Department of Pathology and Laboratory Medicine
University of North Carolina at Chapel Hill
Chapel Hill, North Carolina

Andrew Churg, MD
Professor
Department of Pathology
University of British Columbia
Vancouver, British Columbia, Canada

Laura C. Collins, MBBS
Vice Chair of Oncologic Pathology and Chief of Breast Pathology
Department of Pathology and Laboratory Medicine
Professor of Pathology (appointment pending)
Weill Cornell Medicine, New York Presbyterian Hospital, New York, New York

Antonio L. Cubilla, MD
Director
Department of Pathology
Instituto de Patología e Investigación
Asuncion, Central, Paraguay

Thomas J. Cummings, MD
Professor of Pathology
Department of Pathology
Duke University Medical Center
Durham, North Carolina

Ronald A. DeLellis, MD*
Emeritus Professor
Department of Pathology
Warren Alpert Medical School
Brown University
Providence, Rhode Island

Lori Erickson, MD
Professor
Department of Laboratory Medicine and Pathology
Mayo Clinic
Rochester, Minnesota

William Faquin, MD, PhD
Professor of Pathology
Department of Pathology
Massachusetts General Hospital
Harvard Medical School
Boston, Massachusetts

Yuri Fedoriw, MD
Distinguished Professor
Department of Pathology and Laboratory Medicine
University of North Carolina at Chapel Hill
Chapel Hill, North Carolina

Samson W. Fine, MD
Attending Pathologist
Department of Pathology and Laboratory Medicine
Memorial Sloan Kettering Cancer Center
New York, New York

Karen Fritchie, MD
Pathologist
Department of Anatomic Pathology
Cleveland Clinic
Cleveland, Ohio

Gregory N. Fuller, MD, PhD
Professor
Department of Anatomic Pathology and Neuroradiology
The University of Texas MD Anderson Cancer Center
Houston, Texas

Monica Garcia-Buitrago, MD
Professor of Pathology
Department of Pathology
University of Miami
Miami, Florida

C. Blake Gilks, MD
Consultant in Pathology
Department of Pathology and Laboratory Medicine
Vancouver General Hospital
Vancouver, British Columbia, Canada

Carolyn H. Glass, MD, PhD
Associate Professor
Division Chief of Cardiovascular Pathology
Department of Pathology
Duke University
Durham, North Carolina

Rondell P. Graham, MBBS
GI/Liver and Molecular Pathologist
Department of Laboratory Medicine and Pathology
Mayo Clinic
Rochester, Minnesota

Joel K. Greenson, MD
Professor of Pathology
Department of Pathology
University of Michigan Medical School
Ann Arbor, Michigan

Maria A. Gubbiotti, MD, PhD
Assistant Professor of Neuropathology
Anatomic Pathology
MD Anderson Cancer Center
Houston, Texas

Krisztina Z. Hanley, MD
Professor of Pathology
Department of Pathology and Laboratory Medicine
Emory University Hospital
Atlanta, Georgia

Seung-Mo Hong, MD, PhD
Professor
Department of Pathology
Asan Medical Center
University of Ulsan
College of Medicine
Songpa-gu, Seoul, Republic of Korea

Jutta Huvila, MD, PhD
Clinical Lecturer
Department of Biomedicine
University of Turku
Turku, Finland

Muhammad T. Idrees, MD
Professor
Department of Pathology and Laboratory Medicine
Indiana University School of Medicine
Indianapolis, Indiana

Patricia Iorfino, MD
Senior Pathologist
CBLPATH (A Sonic Healthcare Anatomic Pathology Practice)
Rye Brook, New York

*Deceased.

Susanne K. Jeffus, MD
Associate Professor
Department of Pathology
University of Arkansas for Medical Sciences
Little Rock, Arkansas

Ivy John, MD
Associate Professor of Pathology
Director of Bone and Soft Tissue Pathology Center of Excellence
Department of Pathology
University of Pittsburgh
Pittsburgh, Pennsylvania

Andrew Kanik, MD
Director of Dermatopathology
Department of Dermatopathology
CBLPATH (A Sonic Healthcare Anatomic Pathology Practice)
Rye Brook, New York

Staci Keene, MD
Assistant Professor
Department of Pathology and Laboratory Medicine
University of North Carolina at Chapel Hill
Chapel Hill, North Carolina

Darcy A. Kerr, MD
Associate Professor of Pathology
Department of Pathology and Laboratory Medicine
Dartmouth Health
Lebanon, New Hampshire

S.H. Kroft, MD
Professor and Chair
Department of Pathology
Medical College of Wisconsin
Milwaukee, Wisconsin

Oleksandr N. Kryvenko, MD
Professor
Department of Pathology and Laboratory Medicine
Desai Sethi Urology Institute
Radiation Oncology
Sylvester Comprehensive Cancer Center
University of Miami, Miller School of Medicine
Miami, Florida

Elizabeth M. Kurian, MD
Associate Professor
Department of Pathology
University of Texas Southwestern Medical Center
Dallas, Texas

Laura W. Lamps, MD
Godfrey D. Stobbe Professor and Director of Gastrointestinal Pathology
Department of Pathology
University of Michigan
Ann Arbor, Michigan

Steven H. Lewis, MD, FACOG, FCAP
Clinical Professor of Pathology and Faculty Associate of Bioethics and Humanities
Department of Pathology and Bioethics and Humanities
University of Colorado School of Medicine
CUHealth
Aurora, Colorado

Matthew Leong, MD
Resident
Department of Pathology and Laboratory Medicine
Cedars-Sinai Medical Center
Los Angeles, California

M. Beatriz S. Lopes, MD, PhD
Professor of Pathology and Neurological Surgery
Department of Pathology
University of Virginia School of Medicine
Charlottesville, Virginia

Fiona Maclean, MBBS, BAppSc
Consultant Histopathologist
Department of Anatomical Pathology
Douglass Hanly Moir Pathology
Sydney, New South Wales, Australia

Shamlal Mangray, MB, BS
Chief
Department of Pathology and Laboratory Medicine
Nationwide Children's Hospital
Columbus, Ohio

Jesse K. McKenney, MD
Staff Pathologist
Department of Pathology
Cleveland Clinic
Cleveland, Ohio

Ozgur Mete, MD, FRCPC
Professor of Pathology and Consultant Endocrine Pathologist
Department of Pathology and Laboratory Medicine
University Health Network
Toronto, Ontario, Canada

Elizabeth A. Montgomery, MD
Professor of Pathology
Department of Pathology
University of Miami, Miller School of Medicine
Miami, Florida

Julia R. Naso, MD, PhD
Anatomical Pathologist
Department of Pathology and Laboratory Medicine
University of British Columbia
Vancouver, British Columbia, Canada

Attilio Orazi, MD, FRCPath
Professor and Chairman
Department of Pathology
Paul L. Foster School of Medicine
Texas Tech University Health Science Center
El Paso, Texas

Doreen N. Palsgrove, MD
Assistant Professor
Department of Pathology
University of Texas Southwestern Medical Center
Dallas, Texas

Nicole C. Panarelli, MD
Professor of Pathology and Laboratory Medicine
Vice Chair of Education and Faculty Development
Weill Cornell Medicine
New York, New York

Liron Pantanowitz, MD, PhD, MHA
Chair and Professor of Pathology
Department of Pathology
University of Pittsburgh
Pittsburgh, Pennsylvania

Julio C. Poveda, MD
Assistant Professor of Clinical Pathology
Department of Pathology and Laboratory Medicine
University of Miami
Jackson Health System
Miami, Florida

Alan D. Proia, MD, PhD
Professor
Department of Pathology
Duke University School of Medicine
Durham, North Carolina
Professor and Clinical Chair
Department of Pathology
Campbell University Jerry M. Wallace School of Osteopathic Medicine
Lillington, North Carolina

Peter Pytel, MD
Professor
Department of Pathology
University of Chicago
Chicago, Illinois

Bradley J. Quade, MD, PhD
Chief of Perinatal Pathology
Department of Pathology
Brigham and Women's Hospital
Boston, Massachusetts

Charles M. Quick, MD
Professor of Pathology
Department of Pathology
University of Arkansas for Medical Sciences
Little Rock, Arkansas

Robert Riddell, MD, FRCPC, FRCPath
Pathologist
Department of Pathology
Mount Sinai Hospital
Toronto, Ontario, Canada

Drucilla J. Roberts, MD
Pathologist
Department of Pathology
Massachusetts General Hospital
Boston, Massachusetts

Lisa M. Rooper, MD
Associate Professor
Department of Pathology
Johns Hopkins Hospital
Baltimore, Maryland

Andrew E. Rosenberg, MD
Vice Chair; Professor of Pathology
Director of Anatomic Pathology
Director of Bone and Soft Tissue Pathology
Department of Pathology
University of Miami, Miller School of Medicine
Miami, Florida

Steven P. Salvatore, MD
Associate Professor of Clinical Pathology
Department of Pathology and Laboratory Medicine
Weill Cornell Medicine
New York, New York

Stuart J. Schnitt, MD
Chief of Breast Oncologic Pathology
Dana-Farber/Brigham Cancer Center
Department of Pathology
Brigham and Women's Hospital
Boston, Massachusetts

Theresa Scognamiglio, MD
Associate Professor of Clinical Pathology and Laboratory Medicine
Department of Pathology
Weill Cornell Medicine
New York, New York

Camila C. Simoes, MD
Assistant Professor
Department of Pathology
University of Arkansas for Medical Sciences
Little Rock, Arkansas

Edward B. Stelow, MD
Professor
Department of Pathology
University of Virginia
Charlottesville, Virginia

David Suster, MD
Pathologist
Department of Pathology
Rutgers University Hospital
Wayne, New Jersey

Saul Suster, MD
Professor and Chairman Emeritus
Department of Pathology and Laboratory Medicine
Medical College of Wisconsin
Milwaukee, Wisconsin

Henry D. Tazelaar, MD
Consultant and Geraldine Colby Zeiler Professor of Cytopathology
Department of Laboratory Medicine and Pathology
Mayo Clinic in Arizona
Scottsdale, Arizona

Curtis T. Thompson, MD
Dermatopathologist
CTA Pathology
Portland, Oregon

Arthur S. Tischler, MD
Professor of Pathology
Department of Pathology and Laboratory Medicine
Tufts Medical Center
Tufts University School of Medicine
Boston, Massachusetts

J. Han J.M. van Krieken, MD, PhD
Professor of Pathology
Department of Pathology
RadboudUMC
Nijmegen, The Netherlands

Elsa F. Velazquez, MD
Director of Dermatopathology
Department of Pathology and Dermatopathology
Massachusetts General Hospital
MGPO Dermatopathology
Boston, Massachusetts

Hannes Vogel, MD
Professor, Director of Neuropathology
Department of Pathology
Stanford University
Stanford, California

Kevin Waters, MD, PhD
Associate Professor
Department of Pathology and Laboratory Medicine
Cedars-Sinai Medical Center
Los Angeles, California

Bruce M. Wenig, MD
Chair and Senior Member
Department of Pathology
Moffitt Cancer Center
Tampa, Florida

Maria Westerhoff, MD
Professor
Department of Pathology
University of Michigan
Ann Arbor, Michigan

Shawn Winer, MD, PhD, FRCPC
Staff Pathologist and Assistant Professor
Department of Laboratory Medicine and Pathobiology
Mount Sinai Hospital and the University of Toronto
Toronto, Ontario, Canada

Evgeny Yakirevich, MD, DSc
Professor
Department of Pathology
Lifespan Academic Medical Center
Providence, Rhode Island

Rhonda K. Yantiss, MD
Clinical Professor
Vice Chair of Faculty Development and Mentoring
Department of Pathology and Laboratory Medicine
University of Miami, Miller School of Medicine
Miami, Florida

Preface

The sixth edition of *Mills's Histology for Pathologists* arrives just five years after publication of the fifth edition. Although normal histology has not changed during this interval, our understanding of human development and the spectrum of normal findings continues to grow. Advances in immunohistochemical and molecular techniques have improved our ability to study normal and disease states, providing insights into pathologic processes arising from, or differentiating toward, various tissue types. Increased awareness of the spectrum of normal findings and widespread use of immunohistochemistry have enabled us to recognize variants of normal that cause diagnostic confusion. Accordingly, the sixth edition brings incremental but valuable improvements in our perceptions of human histology.

This new edition also brings quite a few new authors and their fresh perspectives. The chapters on Nail, Adipose Tissue, Blood Vessels, Peripheral Nerve, Larynx and Pharynx, Major Salivary Glands, Lungs, Serous Membranes, Stomach, Small Intestine, Appendix, Liver, Pancreas, Lymph Nodes, Urinary Bladder, Ureter, and Renal Pelvis, Vagina, Uterus and Fallopian Tubes, Placenta, Thyroid, Adrenal, and Neuroendocrine are all written by new senior authors. Many more chapters include new junior authors who bring their own ideas.

As with prior editions, this text aims to bridge the gap between normal histologic findings and pathologic alterations. Although the text emphasizes normal histology and normal features that may be confused with pathologic conditions, prepathologic conditions and pathologic processes confused with normal are briefly discussed in most chapters. It is this pathologic perspective that continues to set *Mills's Histology for Pathologists* apart from standard histology texts written by anatomists. Considerable effort has been expended to improve and update the illustrations, adding new ones whenever appropriate.

We believe that the sixth edition of this text is the best yet and that it will continue to provide valuable aid to both the neophyte pathology trainee and the experienced anatomic pathologist.

Rhonda K. Yantiss, MD

Preface to the First Edition

Histology textbooks exist in abundance. Some are classics of their kind and have gone through innumerable editions over many years. They have served pathologists well, for the most part, especially in terms of strict tissue and cell histology. There is, however, a borderline between histology and pathology in which information for the pathologist is often lacking.

With this textbook we made an attempt to fill the gap. The significance and function of many histologic structures in terms of pathologic interpretation is often absent or obscure. In particular, variations of the norm related to such variables as age, sex, and race are often not clarified in conventional textbooks. For example, the chapter on Paraganglia notes that the connective tissue between the lobules in the carotid body increases with age. Another example related to age is in the pediatric kidney chapter, where it is noted that the glomeruli of fetuses are disproportionately large and are rarely seen in a state of histologic "immaturity." While the chapter on the myofibroblast details the location, staining, ultrastructure, and cytoskeletal protein composition of this unusual cell, we also learn of its importance in the desmoplastic reaction in cancerous tissue and, most importantly, that it is not found in carcinomas which are still in situ.

Some gross observations occasionally will be found as lagniappe, such as the notation that in patients with congenital absence of a kidney, the ipsilateral adrenal will be round rather than angulated. Another example would be that there is a crease in the earlobe associated with coronary artery disease.

Variations in staining reactions are considered, such as the failure of factor VIII to stain renal glomerular vessels. One finds that intestinal endocrine cells can be detected with hematoxylin and eosin (sic) stains by the infranuclear location of the granules. Uncommonly known fixation artifacts are uncovered; for example, the prickle-cell layer (with the so-called intercellular bridges) is actually a retraction artifact of the plasma membranes with the desmosomes remaining relatively fixed.

In most chapters, "prepathologic" considerations are emphasized, while in others the developed pathologic alterations related to the norm represent the major thrust of the chapter.

Some comments will be perceived as gratuitous, such as the remark in the Penis chapter to the effect that "the prepuce could be a mistake of nature." Furthermore, we learn that the "collagen fibers are wavy in the flaccid state and become straight during erection."

The pathology neophyte as well as the many esteemed and experienced pathologists will find helpful information in this book.

Stephen S. Sternberg, MD

Acknowledgments

I am deeply indebted to Dr. Stacey E. Mills for entrusting me with production of *Mills's Histology for Pathologists*. In this edition, we built upon his efforts and the foundational works of prior authors to create a comprehensive and up-to-date textbook that reflects the latest developments in the field and affords rising stars in our field new publishing opportunities. I am grateful to both new and seasoned chapter authors who contributed their best work to create this vital resource for pathologists. I would also like to thank my mentors Robin Young, Harvey Goldman, Donald Antonioli, and Daniel Knowles who put me on the right path; Laura Lamps, Laura Collins, Nicole Panarelli, Jose Jessurun, and countless others who kept me on that path; and all my colleagues at the University of Miami who continue to teach me how to be a better pathologist and person. My own contributions would not have been possible without the support of my family—Andrew, Zach, and Maddie, you are simply the best.

Rhonda K. Yantiss, MD

Contents

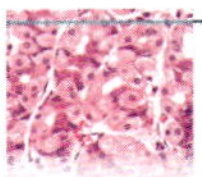

SECTION VII Alimentary Tract

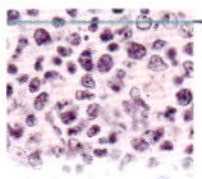

SECTION VIII Hematopoietic System

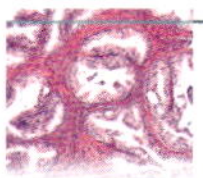

SECTION IX Genitourinary Tract

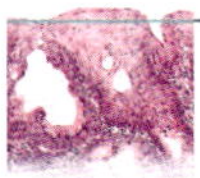

SECTION X Female Genital System

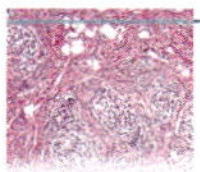

SECTION XI Endocrine

SECTION

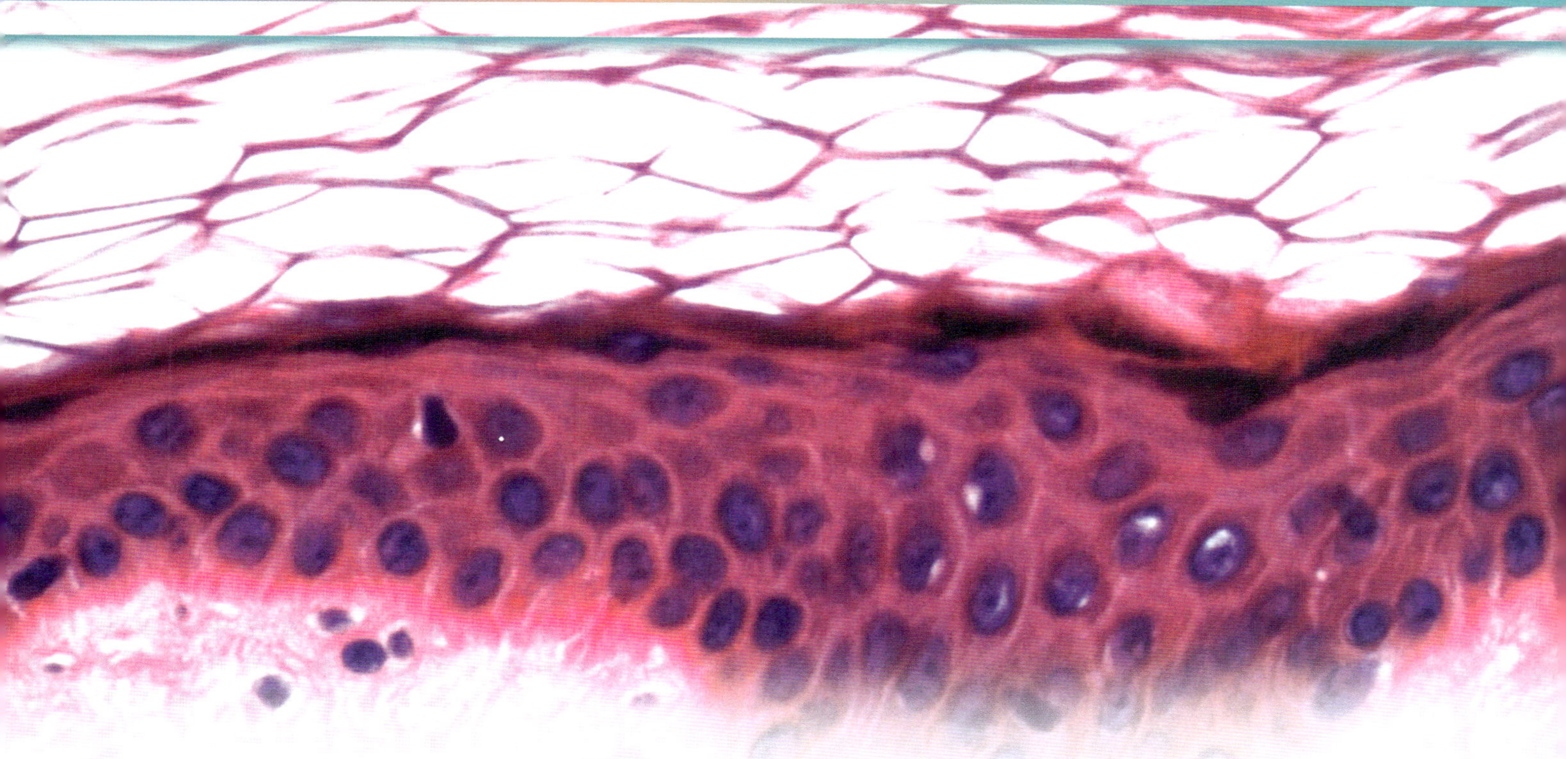

Cutaneous Tissue

1

Skin

Andrew Kanik ■ Patricia Iorfino

The skin accounts for approximately 15% of the total body weight and is the largest organ of the body. It is composed of three layers: (a) epidermis, (b) dermis, and (c) subcutaneous adipose tissue. Each component has its unique and complex structure and function (1–3), with variations according to age, gender, race, and anatomic location. Functions of the skin are extremely diverse. It serves as a mechanical barrier against external physical, chemical, and biologic noxious substances, and as an immunologic organ. It participates in body temperature and electrolyte regulation. It is an important organ of sensuality and psychological well-being. In addition, it is a vehicle that expresses primary cutaneous conditions as well as diseases of the internal organs. An understanding of normal skin histology is essential to the understanding of its pathologic conditions.

EMBRYOLOGY

Epidermis

The ectoderm gives rise to epidermis and its appendages. The embryo is initially covered by a single layer of ectodermal cells that differentiates into two layers by the 6th to 8th week of development: the basal layer and an overlying periderm. Developmental abnormalities in the ectoderm produce among others a variety of syndromes grouped under the umbrella term of ectodermal dysplasias (4). The basal layer is the germinative layer, creating additional rows of cells that form a multilayer between the ectoderm and periderm (5). Keratinization takes place in the upper stratum by the 23rd gestational week, at which time the cells of the periderm have already been shed (5,6,7). The CD30 antigen, which is considered to be restricted to tumor cells of Hodgkin disease and anaplastic large cell lymphoma, participates in the terminal differentiation of the fetal skin (8).

Cell junction proteins are expressed in the early two-layered embryonic epidermis and as early as the 8th week of estimated gestational age (9). The dermoepidermal junction and its components are ultrastructurally similar to that of mature skin by the end of the first trimester (10). Keratinocytes constitute 90% to 95% of the cells in the epidermis, and the rest of its cells are melanocytes, Langerhans cells, and Merkel cells. Nonkeratinocytes are seen in the epidermis of 8- to 10-week-old embryos. The precursor cells of melanocytes migrate from the neural crest to the dermis and then to the epidermis, where they differentiate into melanocytes during the first 3 months of development. During this migration, melanocytes can reside in other organs and tissues. Ultrastructurally recognizable melanosomes may be seen in melanocytes of the fetal epidermis at 8 to 10 weeks of gestational age (11).

Langerhans cells are derived from the CD34-positive hematopoietic precursor cells of the bone marrow. The characteristic ultrastructural marker, the Birbeck granule, is seen in the cytoplasm of Langerhans cells in 10-week-old embryos (12). Immunohistochemical expression of CD1a is completed by 12 to 13 weeks of estimated gestational age (12,13).

Merkel cells can also be seen in the epidermis of the 8- to 10-week-old embryo. The origin of Merkel cells is debatable. Some have suggested a neural crest derivation (14), whereas others suggest epidermal origin through a process of differentiation from neighboring keratinocytes (15–17). Merkel cells in the epidermis are initially numerous, but diminish with increasing gestational age (18).

Dermis

The mesoderm provides the mesenchymal elements of the dermis and subcutaneous fat (4). The dermis is derived from the primitive mesenchyme underlying the surface ectoderm. The papillary and reticular dermis are recognized by 15 weeks of intrauterine life (19,20).

As described by Breathnach (19), three types of dermal mesenchymal cells are recognized in 6- to 14-week-old embryos. Type I cells are stellate-dendritic cells with long, slender processes. These are the most numerous primitive mesenchymal cells and probably give rise to endothelial cells and pericytes. Type II cells have less extensive cell processes; their nuclei are round and their cytoplasm contains large vacuoles. They are classified as phagocytic macrophages of yolk-sac origin, and are rarely seen after week 14 of development. Type III cells are round with little or no membranous extension. They contain numerous vesicles, some with an internal content suggestive of granule secretory-type cells. These cells could be melanoblasts on their way to the epidermis, or they could be precursors of mast cells. Schwann cells associated with neuroaxons are also identified during this period, but they lack basal lamina. Macrophages and mast cells are also frequently seen during this time.

Fibroblasts are numerous and active from 14 to 21 weeks of development. They are elongated spindle cells with abundant rough endoplasmic reticulum. These are the fundamental cells of the dermis and synthesize all types of fibers and ground substance (1). Type III collagen fibers are abundantly present in the matrix of the fetal dermis, whereas type I collagen fibers are more prominent in adult skin (20). Elastic fibers appear in the dermis during the 22nd week of gestational age and a well-developed network of elastic fibers is formed by week 32.

The dermis is initially organized into somites, but this segmental organization soon gives way to development of dermatomes along the segmental nerves as they form (21). Adipocytes develop in the subcutaneous tissue from the primitive mesenchymal cells and accumulate from the 24th week of gestation until term.

Epithelial Skin Appendages

Most epithelial cells of skin appendages derive from follicular epithelial stem cells located in the basal layer of epidermis at the prominent bulge region of the developing fetal hair follicles. These multipotent stem cells may represent the ultimate epidermal stem cell (22). Mesenchymal cells of the developing dermis interact with epidermal basal cells in 10-week-old embryos. These epidermal cells grow both downward to the dermis and upward through the epidermis to form the opening of the hair canal. As the growing epithelial cells reach the subcutaneous fat, the lower portion becomes bulbous and partially encloses the mesenchymal cells, descending with them to form the dermal papillae of the hair follicle; this structure plays an important role in the future processes of hair follicle regeneration (23). The descending epidermal cells around the dermal papillae constitute the matrix cells from which the hair layers and inner root sheath will develop. The outer root sheath derives from downward growth of the epidermis. The first hairs appear by the end of the 3rd gestational month as lanugo hair around the eyebrow and the upper lip. The lanugo hair is shed around the time of birth. The developing hair follicle gives rise to the sebaceous and apocrine glands.

The sebaceous glands originate as epithelial buds from the outer root sheath of the hair follicles and are developed during the 13th to 15th gestational weeks (24). Differentiated sebaceous glands with hairs protruding through the skin surface are present at the 18th week of gestation (25). They respond to maternal hormones and are well developed at the time of birth.

The apocrine glands also develop as epithelial buds from the outer sheath of the hair follicles in 5- to 6-month-old fetuses (21). They continue to grow and increase in number into late embryonic life as long as new hair follicles develop.

The eccrine glands develop from the fetal epidermis independent of the hair follicles (21). They are initially seen as regularly spaced undulations of the basal layer. At 14 to 15 weeks, the tips of the primordial eccrine glands have reached the deep dermis, forming the eccrine coils (26). At the same time, the eccrine epithelium grows upward into the epidermis. The primordial eccrine epithelium acquires a lumen by the 7th to 8th fetal month, and, thus, the first eccrine unit is formed. Both ducts and secretory portions are lined by two layers of cells. The two layers in the secretory segment undergo further differentiation. The luminal cells become tall columnar secretory cells, and the basal layer transforms into secretory cells or myoepithelial cells. The first glands are formed on the palms and soles by the 4th gestational month, then in the axillae in the 5th month, and finally on the rest of the hair-bearing skin (27).

HISTOMORPHOLOGY

Epidermis

The epidermis is a stratified and keratinizing squamous epithelium that dynamically renews itself, maintaining its normal thickness by the process of desquamation. The cells in the epidermis include (a) keratinocytes, (b) melanocytes, (c) Langerhans cells, (d) Toker cells (in certain anatomic locations), and (e) Merkel cells. In addition, the epidermis contains the openings for the eccrine ducts (acrosyringium) and hair follicles. Recent immunohistochemical studies have demonstrated that the epidermis contains free nerve axons in association with Langerhans cells (28).

Keratinocytes

The keratinocytes of the epidermis are stratified in four orderly layers from bottom to top: (a) the basal layer (i.e., stratum basale, germinativum), (b) the squamous layer (i.e., prickle cell layer or stratum spinosum), (c) the granular layer (i.e., stratum granulosum), and (d) the cornified or horny layer (i.e., stratum corneum). The dermoepidermal junction has an irregular contour because of the upward extension of the papillary dermis to form the dermal papillae. The regions of the epidermis separating the dermal papillae are the rete ridges (Fig. 1.1). The transcription factor p63 plays an important role in maintaining this orderly arrangement and the continuous development of the pre- and postnatal skin (29).

› ***The Basal Layer*** Basal cells are the mitotically active cells that give rise to keratinocytes. They are seen as a single layer of cells above the basement membrane and show minor variations in size, shape, and melanin content. Basal cells are columnar or cuboidal with basophilic cytoplasm. The nucleus is round or oval with coarse chromatin and an indistinct nucleolus. Basal cells contain melanin in their cytoplasm as a result of pigment transfer from neighboring melanocytes. Basal cells are connected to each other and to keratinocytes by desmosomes located in the cell membranes. They are aligned perpendicular to the subepidermal basement membrane and attached to it by modified desmosomes (i.e., hemidesmosomes).

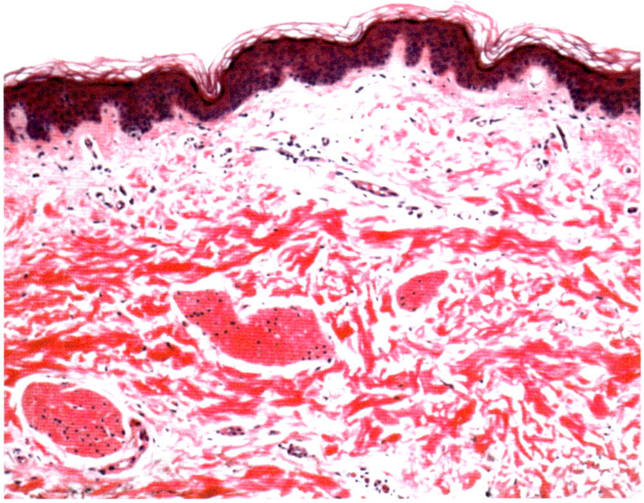

FIGURE 1.1 Normal skin showing stratified epidermis with rete ridges, papillary dermis, and reticular dermis (H&E).

› ***The Squamous Layer*** The squamous layers are composed of 5 to 10 layers of keratinocytes that are larger than the basal cells. They contain cytoplasmic intermediate filament proteins composed of keratin protein pair types, which change as maturation occurs from the basal layer toward the stratum corneum (30). The suprabasal keratinocytes are polyhedral with somewhat basophilic cytoplasm and a round nucleus. Melanin is variably present in many of these keratinocytes, and provides protection from the damaging effect of ultraviolet light. The more superficial cells are larger, flattened, eosinophilic, and oriented parallel to the surface. The keratinocytes contain one or two conspicuous nucleoli and tonofilaments within the cytoplasm.

The main squamous layer is also called the spinous or prickle cell layer because of the characteristic short projections that extend from cell to cell. These projections reflect retraction of the plasma membrane during tissue processing, whereas the desmosomes remain relatively fixed and correlate with intercellular bridges.

Desmosomes are composed of a variety of polypeptides. Desmogleins and desmocollins are the transmembrane constituents and desmoplakin, plakoglobin, and plakophilin are cytoplasmic components. Other intercellular junctions, namely gap junctions and adherens junctions, are distinct from desmosomes in composition and distribution; they provide alternative cell-to-cell adhesion mechanisms (31). An intercellular space of constant dimension is present around each cell; acid and neutral mucopolysaccharides are present in the intercellular spaces as indicated by special stains. The pemphigus antigens are localized in the cell membranes (32) or in the desmosomes of these cells (33).

Toker cells are occasionally seen in the epidermis of the nipple (34), accessory nipple (35,36), the pubic regions, and in the milk-line distribution (37). These cells contain clear or pale cytoplasm and are detected in men and women. They have been considered to be nonneoplastic ductal epithelial cells, although some authors hypothesize that they might be the precursors of mammary or extramammary Paget diseases (36,38). Toker cells outside of the nipple are considered to result from either abnormal keratinization or aberrant derivatives of eccrine or apocrine sweat gland epithelial cells (39–41). They are benign clear cells that have a pyknotic nucleus surrounded by a clear halo and a narrow rim of clear cytoplasm (Fig. 1.2). They lack the pleomorphism, nuclear morphology, and intensity of the chromatin staining seen in Paget cells (Fig. 1.3). The immunohistochemical staining pattern of benign clear cells

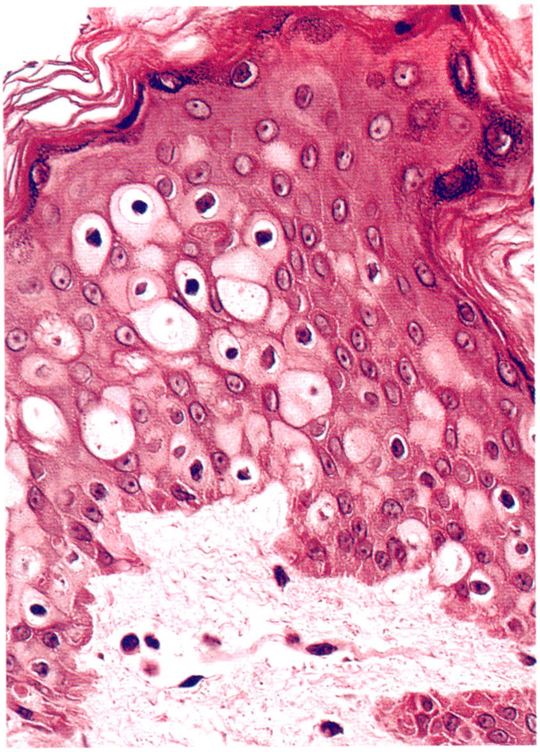

FIGURE 1.2 Clear cells of the nipple epidermis.

may resemble that of Paget cells in that they react with the cytokeratin 7 (CK7) but, unlike Paget cells, they are usually negative for GCDFP-15.

Common inflammatory changes seen in the squamous layer include (a) spongiosis, or intercellular edema, as may occur in patients with allergic contact dermatitis, (b) acanthosis, or thickening of the epidermis, typical of psoriasis, (c) atrophy of the epidermis, as is commonly present in discoid lupus erythematosus, (d) acantholysis, or detachment of keratinocytes because of changes involving intercellular junctions, such as may occur in blistering disorders, and (e) dyskeratosis, or abnormal keratinization of keratinocytes.

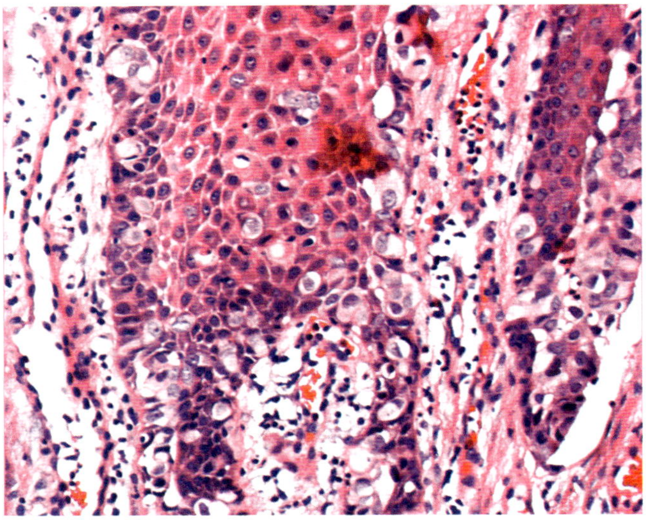

FIGURE 1.3 Paget cells in extramammary Paget disease.

› *THE GRANULAR LAYER* The granular layer is composed of one to three layers of flattened cells lying parallel to the skin surface. Their cytoplasm contains intensely basophilic-stained keratohyalin granules. In contrast, trichohyalin granules are produced by the inner root sheath of hair follicles and are stained red on routine hematoxylin and eosin (H&E)-stained sections. The keratohyalin granules are histidine rich and are precursors to the protein filaggrin, which promotes aggregation of keratin filaments in the cornified layer.

Keratinocytes located between the squamous layer and the granular layer contain small membrane-coating granules known as lamellar granules (i.e., Odland bodies or keratinosomes). They are composed of the acid hydrolase and neutral sugars conjugated with proteins and lipids. These granules are present both intra- and extracellularly. They are approximately 300 nm in diameter and are not visible by light microscopy. Their functions are to provide epidermal lipids, increase the barrier property of the cornified layer against water loss, and aid in the desquamation process. This interface between the squamous and the granular layer is also the site of the synthesis and storage of cholesterol (42).

› *THE CORNIFIED LAYER* The cornified layer (i.e., stratum corneum) is composed of multiple layers of polyhedral eosinophilic keratinocytes that lack a nucleus and cytoplasmic organelles. These cells are essentially dead, and represent the most differentiated cells of the keratinization system. They are composed entirely of high–molecular-weight keratin filaments and, in formalin-fixed sections, are arranged in a basket-weave pattern (Fig. 1.4). These cells eventually shed (desquamate) from the surface of the skin. The process of keratinization takes 20 to 45 days.

Tissue sections taken from the skin of the palms and soles reveal a homogenous eosinophilic zone, the stratum lucidum,

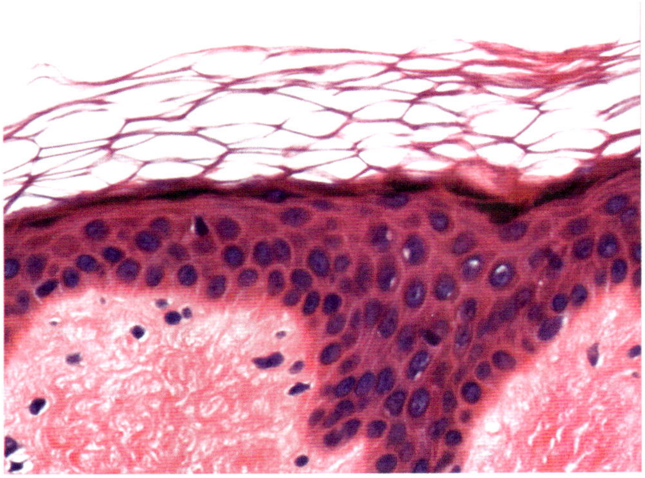

FIGURE 1.4 Basket-weave pattern of the cornified layer (also in Fig. 1.1).

in the lowest portion of the cornified layer just above the granular layer. This additional layer is rich in extracellular elements such as energetic enzymes and sulfhydryl groups adding to the normal functional barrier of the skin (43).

Common abnormalities of the cornified layer include (a) increased thickness in the cornified layer (i.e., hyperkeratosis), as may occur in patients with ichthyosis; (b) parakeratosis, which features nuclei in the cornified layer, and can be seen in chronically irritated skin and actinic keratosis; and (c) presence of fungal organisms, or superficial dermatophytosis.

› BASEMENT MEMBRANE ZONE The basement membrane zone separates the epidermal basal layer from the dermis. It appears as a continuous, undulating, and thin periodic acid–Schiff (PAS)-stained layer (Fig. 1.5). The basal cells are attached to the basal lamina by hemidesmosomes. Ultrastructurally, the basement membrane zone is composed of four distinct structures, from top to bottom (44). First, the plasma membrane of the basal cells contains the hemidesmosomes. Bullous pemphigoid antigen 1 is localized in the intracellular component of hemidesmosomes. Second, the lamina lucida is an electron-lucent area with anchoring filaments containing various laminin isoforms (45). Bullous pemphigoid antigen 2 (type XVII collagen) is associated with the transmembrane component of hemidesmosome-anchoring filament complexes in the lamina lucida. It is also the site of the blister in dermatitis herpetiformis (46). Third, the lamina densa is an electron-dense area composed of mainly type IV collagen. Finally, the sublamina densa zone, or pars fibroreticularis, contains mainly the anchoring fibrils (47) (type VII collagen) that attach the basal lamina to the connective tissue of the dermis. Antibodies against epidermolysis bullosa acquisita react with the carboxy terminus of type VII collagen (48,49). Inflammatory conditions of the basement membrane can result in thickening (e.g., discoid lupus erythematosus) or formation of subepidermal vesicles (e.g., bullous pemphigoid).

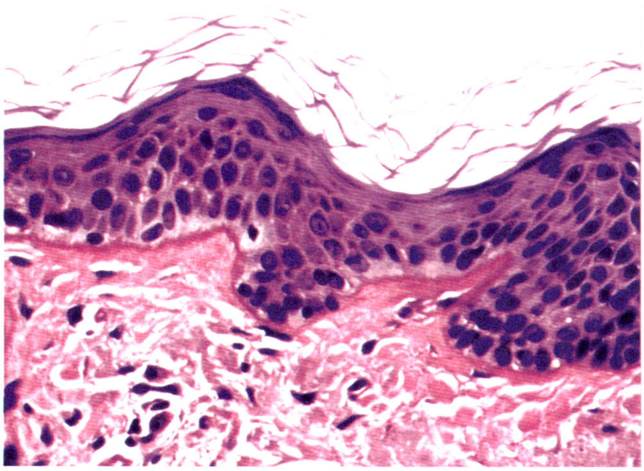

FIGURE 1.5 PAS-positive basement membrane.

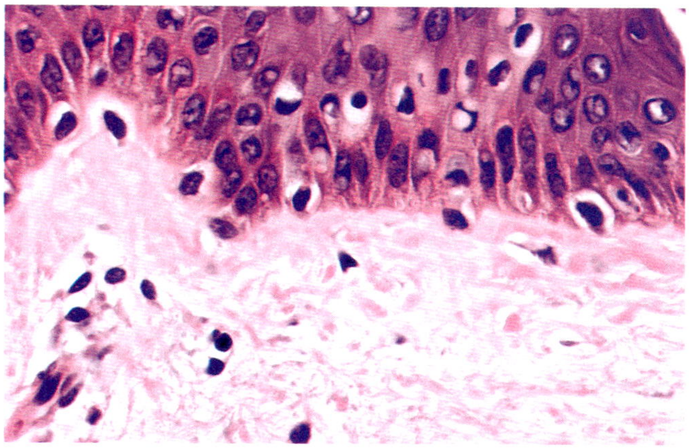

FIGURE 1.6 Melanocytes in the basal layer, composed of ovoid nuclei within a clear space.

Melanocytes

Melanocytes are dendritic cells derived from the neural crest. Melanocytes are localized in the basal layer of the epidermis, although their dendritic processes extend in all directions. The dendritic nature of normal melanocytes is usually not evident in routine H&E-stained sections. Rather, melanocytes contain elongated or ovoid nuclei surrounded by a clear space (Fig. 1.6). They are usually smaller than neighboring basal keratinocytes. Melanocytes do not contain tonofilaments and do not attach to basal cells with desmosomes. However, anchoring filaments extend from their plasma membranes to the basal lamina. Laminin-5, a component of anchoring filaments, may be a ligand for melanocyte attachment to the basement membrane in vivo (50). Melanocytes that are close to the basal lamina have structures resembling hemidesmosomes of basal keratinocytes (51).

Melanocytes produce and secrete melanin, which can be red (pheomelanin) or yellow-black (eumelanin). The most important function of melanin is to protect against the injurious effects of nonionizing ultraviolet irradiation. Melanin is formed through a complex metabolic process in which tyrosinase is the main catabolic enzyme, using tyrosine as substrate. The synthesis of melanin takes place in melanosomes, lysosome-related organelles. In the early stages of development, melanosomes are membrane-limited vesicles located in the Golgi-associated endoplasmic reticulum. The maturation of melanosomes undergoes four stages. Stage I melanosomes are round without melanin. These are seen in balloon-cell melanomas. Stage II through stage IV melanosomes are ellipsoidal with numerous longitudinal filaments. Melanin deposits start at stage II. In stage III, melanin deposits are prominent. Stage IV melanosomes are fully packed with melanin, obscuring the internal structures.

The developing melanosomes, with their content of melanin, are transferred to the neighboring basal keratinocytes and hair follicular cells. The mechanism of melanin transfer is a complex process (52,53), with the end result being phagocytosis of the tip of melanocytic dendrites by the keratinocytes through pigment donation (54).

The number of melanocytes in normal skin is constant in all races with 1 melanocyte for every 4 to 10 basal keratinocytes; alteration of the ratio of is important when diagnosing pigmented lesions and vitiligo. The color of the skin is determined by the number and size of melanosomes present both in keratinocytes and melanocytes, not by the number of melanocytes. The number of melanocytes remains constant from newborn to adulthood but decreases in older age (55). As a result, the availability of melanin to keratinocytes diminishes in aged skin; the skin becomes lighter in color and the incidence of skin cancer increases due to a lack of protection that melanin provides.

Melanin is both argentaffin and argyrophilic. It can be recognized by Fontana–Masson silver stains. In addition, melanocytes and their dendritic processes are identified by the dopa reaction in histologic slides prepared from frozen sections and in paraffin-embedded sections with immunohistochemical stains with SOX-10 and S100 protein. The latter is highly sensitive but not specific for cells of melanocytic lineage; it can be detected in Langerhans cells, Schwann cells, eccrine, and apocrine gland cells, among others. Melanocytes can also be identified with monoclonal antibodies directed against Melan-A/MART-1 (Melanoma Antigens Recognized by T cells-1), a melanocytic differentiation marker. The Melan-A/MART-1 antigen is expressed in normal melanocytes, common nevi, Spitz nevi, and malignant melanoma. Under normal conditions, the melanoma-associated antigen HMB-45 does not react with adult melanocytes (56). It is expressed in embryonic melanocytes, hair bulb melanocytes, and activated melanocytes (57). It is usually seen reacting with most melanoma cells and can be seen in the junctional portion of a Spitz nevi, common nevi, and dysplastic nevi.

An absence or significant decrease in the number of melanocytes is seen in vitiligo. Albinism results from a defect in the synthesis of melanin but the number of melanocytes is normal in a skin biopsy. Melanocytic hyperplasia is seen in lentigo, benign, and malignant melanocytic neoplasms, and as a reaction pattern in a variety of neoplastic and nonneoplastic conditions (e.g., dermatofibroma and chronic photoactivation). In a freckle, there is an increase in pigment donation to adjacent keratinocytes rather than melanocytic hyperplasia.

Langerhans Cells

Langerhans cells are mobile, dendritic, antigen-presenting cells present in all stratified epithelium, predominantly in the mid to upper parts of the squamous layer. They appear to lie within lacunae of the skin and have darkly stained nuclei with an indented, reniform shape (Fig. 1.7). Langerhans cells can be recognized by histoenzymatic stains for adenosine triphosphatase (ATPase); they can also be detected in formalin-fixed, paraffin-embedded tissue using S100 protein immunohistochemistry and, more specifically, the antibodies to the CD1a antigen or langerin (Fig. 1.8). Histoenzymatic and immunohistochemical stains demonstrate the extensive dendritic nature of Langerhans cells.

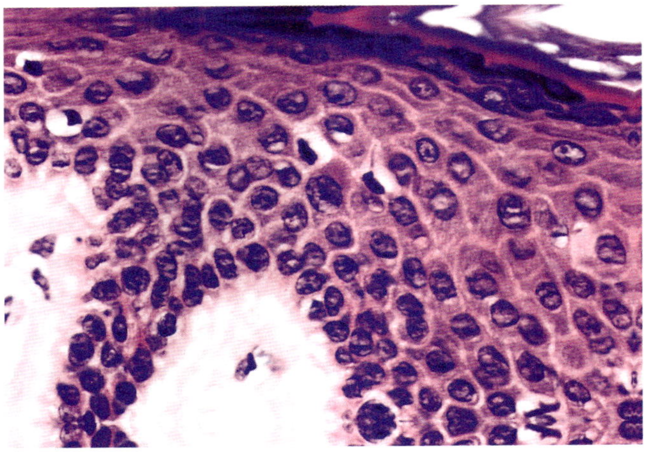

FIGURE 1.7 H&E section of possible Langerhans cells composed of elongated nuclei surrounded by a clear space in the mid epidermis.

Langerhans cells lack desmosomes, tonofilaments, and melanosomes by electron microscopy. They contain small vesicles, multivesicular bodies, lysosomes, and the characteristic Birbeck granule (58), a rod-shaped organelle varying in size from 100 nm to 1 μm (59). Birbeck granules contain a centrally striated density and an occasional bulb at one end with a unique tennis-racket appearance. Langerhans cells are also present in other types of epithelia, lymphoid organs, and dermis and are increased in the skin in a variety of inflammatory conditions, such as contact dermatitis, where they can be seen as minute nodular aggregates in the epidermis.

Merkel Cells

Merkel cells were first described by F.S. Merkel in 1875. They are usually scattered and irregularly distributed in the basal cell layer in the epidermis. They may group together as clusters within the epidermis where they are coupled with enlarged terminal sensory nerve fibers to form slowly adapting mechanoreceptors that mediate tactile sensation

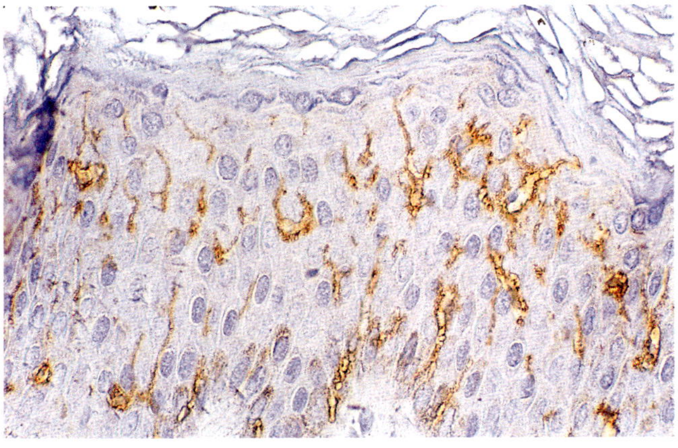

FIGURE 1.8 CD1a-specific reaction of Langerhans cells. Note the dendritic processes.

(60–62). They are located in higher concentration in the glabrous skin of the digits, lips, and oral cavity, in the outer root sheath of hair follicles (63), and in the tactile hair disks (64).

Merkel cells are not recognized in routine histologic preparations. Electron microscopy and immunostaining are required for their identification. Merkel cells are ultrastructurally attached to adjacent keratinocytes by desmosomes. They have scant cytoplasm, invaginated nuclei, a parallel array of cytokeratin filaments in the paranuclear zone, and the characteristic membrane-bound dense core granules that are often, but not always, related to unmyelinated neurites. Normal and neoplastic Merkel cells may express neuron-specific enolase, chromogranin, synaptophysin, neural cell adhesion molecule, and various neuropeptides and other substances by immunohistochemistry (65–67). However, the expression of these substances is heterogeneous and variable. The constant pattern seen in Merkel cells is the presence of paranuclear aggregates of keratins (15,67,68), which include low–molecular-weight keratins 8, 18, 19, and 20. The most specific keratin is CK20 because it is expressed in simple epithelial cells but not in adjacent keratinocytes (69,70) (Fig. 1.9).

Pilar Unit

The pilar unit is composed of the hair follicle, sebaceous gland, arrector pili muscle, and (when present) eccrine and apocrine glands.

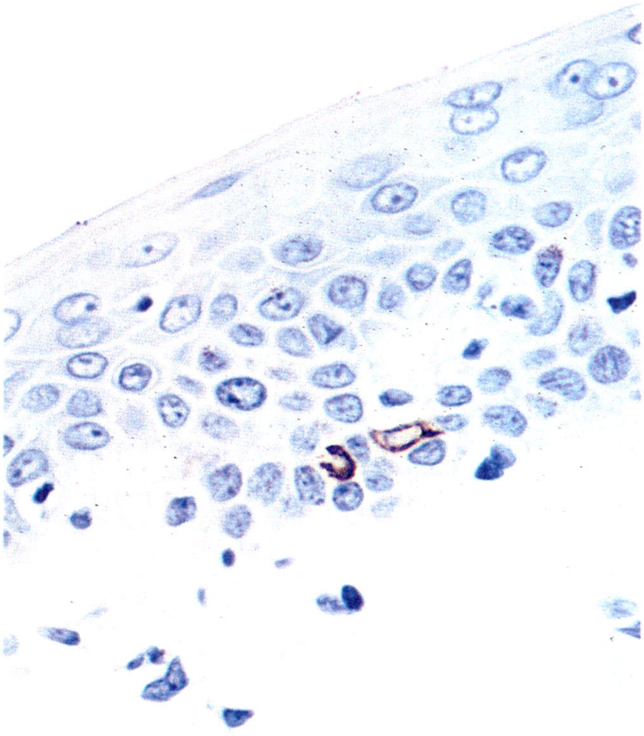

FIGURE 1.9 Cytokeratin 20 staining a Merkel cell in the basal layer of the epidermis.

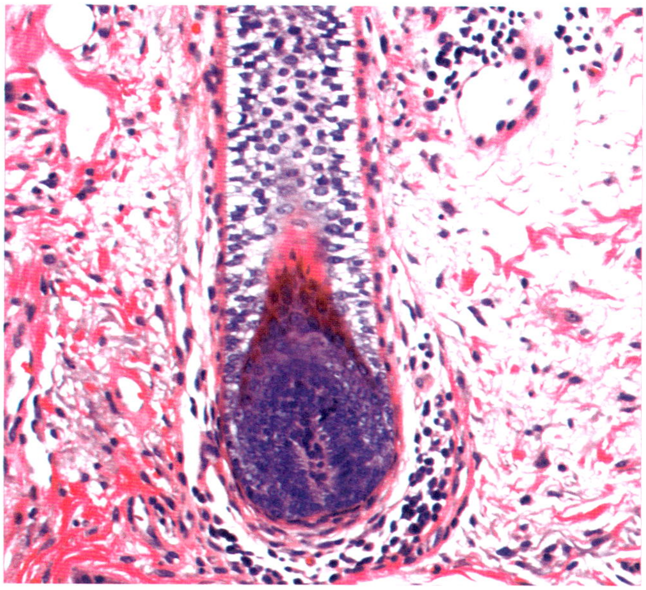

FIGURE 1.10 Inferior segment of the hair follicle, showing the hair papilla.

› ***Hair Follicle*** The hair follicle is divided into three segments from top to bottom: (a) the infundibulum, which extends from the opening of the hair follicle in the epidermis to the opening of the sebaceous duct; (b) the isthmus, which extends from the opening of the sebaceous duct to the insertion of the arrector pili muscle; and (c) the inferior segment, which extends to the base of the follicle. The inferior segment is bulbous and encloses a vascularized component of the dermis referred to as follicular (dermal) papilla of the hair follicle (Fig. 1.10).

The microanatomy and function of the hair follicle are very complex. The cells of the hair matrix differentiate along six cell linings. Beginning from the innermost layer, they are (a) the hair medulla; (b) the hair cortex; (c) the hair cuticle; and (d) three concentric layers of the inner root sheath that form the cuticle of the inner root sheath, Huxley layer, and Henle layer.

The inner root sheath of the hair follicle is surrounded by the outer root sheath (Fig. 1.11), which is composed of clear cells. These glycogen-rich cells are seen in some of the neoplasms that show hair follicular differentiation (e.g., trichilemmoma). A PAS-positive basement membrane separates the outer root sheath from the surrounding connective tissue. Thus, the hair shaft is formed from the bulb region that occupies the hair follicular canal.

Dendritic melanocytes are present only in the upper half of the bulb, whereas inactive (amelanotic) melanocytes are present in the outer root sheath. These melanocytes can become active after injury, migrating into the upper portion of the outer root sheath and to the regenerating epidermis.

The cells of the inner root sheath disintegrate and disappear at the level of the isthmus, whereas the cells of the outer root sheath begin an abrupt sequence of keratinization termed trichilemmal keratinization (71). Trichohyalin granules are red in routine H&E-stained sections, as

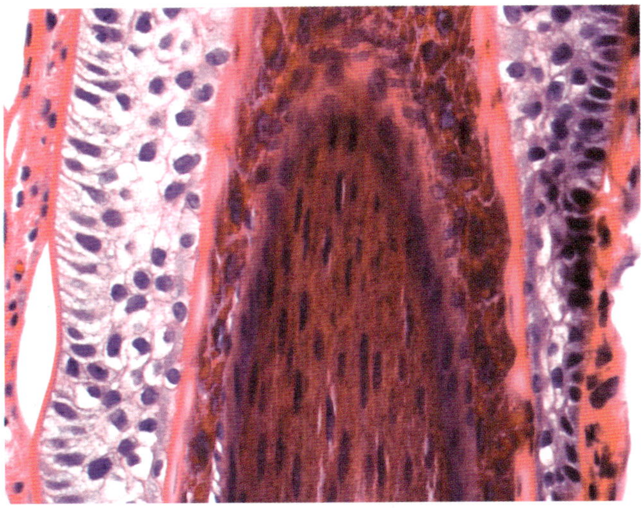

FIGURE 1.11 Hair follicle showing the hair shaft (**center**) surrounded by the inner root sheath, which contains trichohyalin granules. The outer root sheath is composed of clear cells.

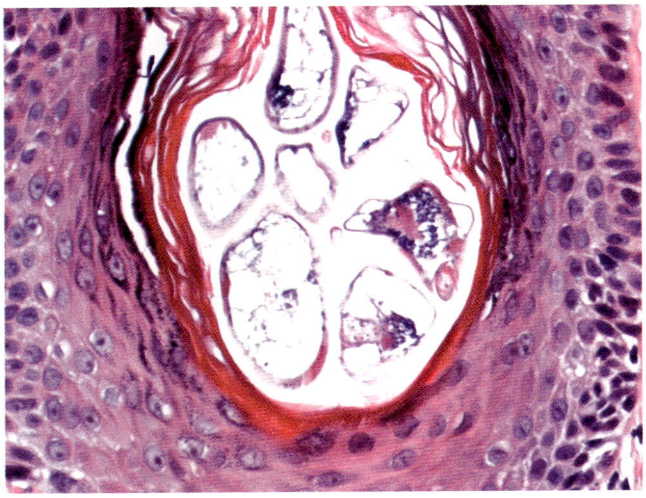

FIGURE 1.13 *Demodex folliculorum* mites in the follicular infundibulum.

opposed to the blue granules of the keratohyalin of epidermal keratinization and of the epithelium of the follicular infundibulum of the hair follicle. The staining features of these granules permit neoplasms and cysts to be distinguished from either pilar or epidermal origin. Under normal circumstances, microorganisms like *Staphylococcus epidermidis*, *Pityrosporum* yeast (Fig. 1.12), and the *Demodex folliculorum* mites (Fig. 1.13) are encountered in the follicular infundibulum.

The mantle hair of Pinkus is a hair follicle in which proliferation of basaloid epithelioid cells emanating from the infundibulum is seen. Sebaceous proliferation is present in those cords (Fig. 1.14). The significance of this hair follicle is not known (72).

The hair growth is in lifelong cyclic transformation. Hormones and their receptors play prominent roles in hair cycle regulation (73). Three phases are recognized: (a) anagen—active growth phase; (b) catagen—involuting phase (apoptosis-driven regression); and (c) telogen—relative resting phase. The histologic features previously described correspond to the anagen hair.

During the catagen phase, mitosis and melanin synthesis cease at the level of the hair bulb. The hair bulb is progressively replaced by a cornified sac formed by retraction of the outer root sheath around the hair bulb, and a club hair is formed. A thick glassy basement membrane surrounds the hair follicle. Apoptosis of single cells in the outer root sheath is a characteristic finding during the catagen phase.

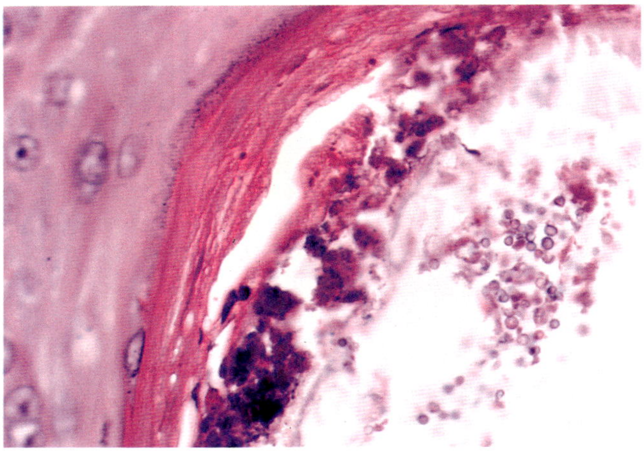

FIGURE 1.12 Yeasts of *Pityrosporum* in the follicular infundibulum.

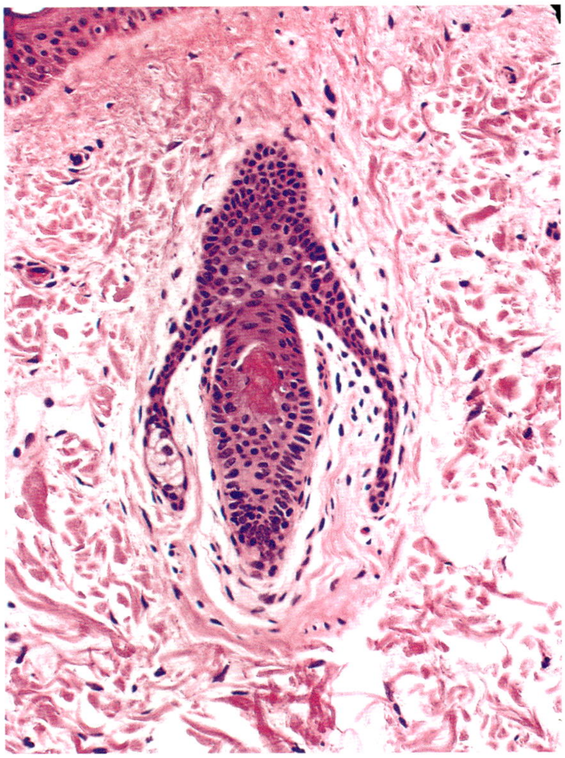

FIGURE 1.14 Mantle hair of Pinkus with lateral extensions containing sebaceous cells.

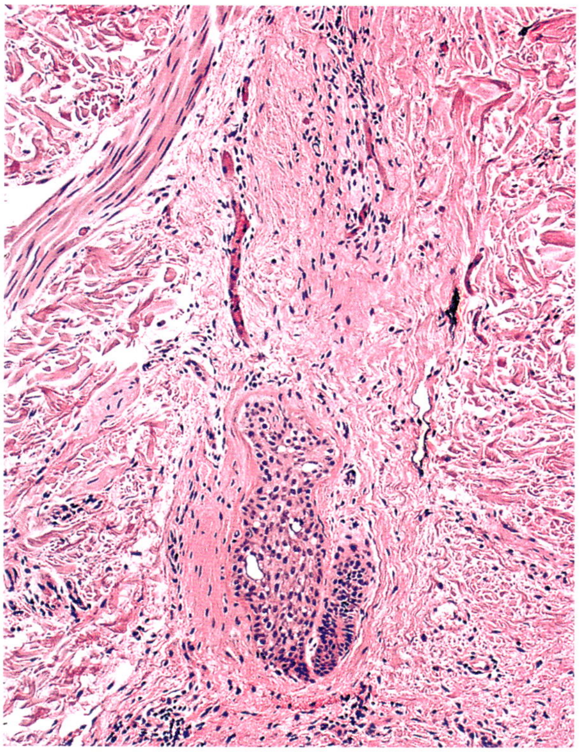

FIGURE 1.15 Catagen–telogen hair follicle located entirely within the dermis.

During the telogen phase, the club hair and its cornified sac retract even further to the insertion of the arrector pili muscle, leaving behind the dermal papilla, which is connected to the retracted hair follicle by a fibrous tract (Fig. 1.15) (21). When the cycle is complete, a new anagen phase begins with the formation of a new hair matrix.

The duration of the normal hair cycle varies. The anagen phase is measured in years for the scalp, but it is measured in shorter periods in other regions of the body. The length of the hair is also related to the amount of the anagen hair. More than 80% of the hair present in normal scalp is anagen hair. The catagen phase takes 2 to 3 weeks and the telogen phase may last a few months.

The color of normal hair depends on the amount and distribution of the melanin in the hair shaft (21). Normal human epidermal melanocytes may synthesize both eumelanin and pheomelanin (72). The melanins in black hair are eumelanin, which is present in ellipsoidal eumelanosomes, while those in red hair are mainly pheomelanin, which reside in spherical pheomelanosomes (74,75). Fewer melanosomes are produced in the bulbar melanocytes of blond hair. A relative absence of melanin and fewer melanosomes are seen in gray hair. Multiple internal or external regulatory factors are involved in hair pigmentation. There might be some correlation between tryptophan content and tyrosinase expression with hair color (76,77).

Another structure related to the pilar unit is the hair or pilar disk (i.e., the Haarscheibe). The Haarscheibe is a specialized spot in close vicinity to hairs. This structure is usually not recognized on routine histologic sections. It may present as an acanthotic elevation of the epidermis, limited by two elongated rete ridges laterally (1). The epidermis in this area contains more Merkel cells in the basal layer, and the dermal component is well vascularized with myelinized nerve fibers in contact with Merkel cells (21,65). The Haarscheibe is considered to represent a highly sensitive, slowly adapting mechanoreceptor (1,78).

› ***Sebaceous Glands*** The sebaceous glands are holocrine glands typically associated with hair follicles. They are lobulated structures composed of multiple acini in some locations like the head and neck; in other sites, such as the chest, they are composed of a single acinus. The periphery of the lobules contains the germinative cells, which are cuboidal and flat with large nucleoli and basophilic cytoplasm without lipid droplets. As differentiation occurs, several inner layers show lipid accumulation in the cytoplasm that fills the cell.

The more differentiated cells have a characteristic multivacuolated cytoplasm and are termed sebocytes (Fig. 1.16). Their nuclei are centrally located and scalloped due to lipid imprints. Fully differentiated cells disintegrate and discharge cellular debris (sebum) into the excretory duct, which opens into the hair follicle in the lower portion of the infundibulum. The excretory duct is short, shared by several lobules, and lined by keratinized squamous epithelium.

Sebaceous glands are prominent in facial skin. They are also seen in hairless areas including the buccal mucosa, vermilion of the lip, prepuce, and labia minora (i.e., Fordyce spots), on the areola surrounding the nipple (i.e., Montgomery tubercles), and even at times, in the parotid gland. Sebaceous heterotopias may also occur in the upper third of the esophagus. The palms and soles are the only cutaneous regions devoid of sebaceous glands.

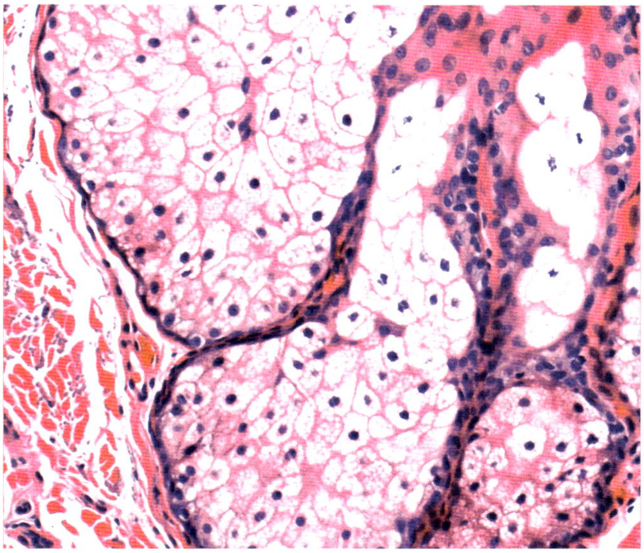

FIGURE 1.16 Sebaceous glands with peripheral germinative cells and, toward the center, the differentiated vacuolated cells.

The sebaceous glands develop in the 13th to 14th week of fetal life and contribute to the production of vernix caseosa, which is a protective waxy substance that coats the embryonic skin. Sebaceous glands diminish postnatally until puberty when elevated hormonal levels, particularly androgens, cause their enlargement and ramp up sebum production (79). The number of glands remains essentially the same through adult life, although the glands increase in size with age. Sebum levels drop after menopause in women but do not change much until the eighth decade in men (80).

Within sebaceous glands, the germinative cells express appreciable quantities of keratins. Mature sebocytes demonstrate cytoplasmic reactivity for high–molecular-weight keratins and epithelial membrane antigen. Adipophilin shows membranous and vesicular reactivity of lipids.

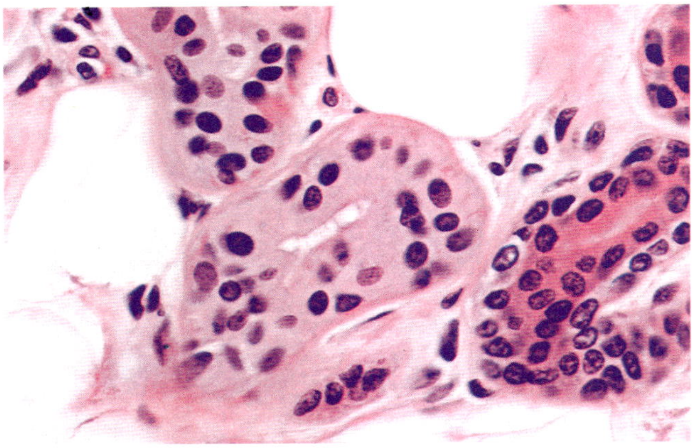

FIGURE 1.18 Clear cells of the eccrine glands.

› ***Eccrine Glands*** The eccrine glands are true sweat glands responsible for thermoregulation. They are found in higher concentration on the palms, soles, forehead, and axillae and have dual secretory and excretory functions. Eccrine glands begin to form around the fourth gestational month in palmoplantar skin and around the fifth across the remaining body. They are morphologically mature before birth and their number is fixed throughout life (81). Sweat gland density decreases with skin surface area expansion as an individual grows and, thus, their density is inversely proportional to body surface area.

The secretory portion of an eccrine gland is a convoluted tube located in the deep dermis at the interface with the subcutaneous tissue, and rarely, within the subcutaneous tissue. Several glandular structures with a central lumen form the secretory coils. These appear as lobular structures often surrounded by fat even when located within the dermis (Fig. 1.17).

The eccrine coil contains three cell types: clear cells, dark cells, and myoepithelial cells. The clear cells rest directly on the basement membrane and on the myoepithelial cells (Fig. 1.18). They are composed of pale or finely granular cytoplasm with round nuclei in the center of the cells. Deep invaginations of the luminal membranes form intercellular canaliculi lined with microvilli (Fig. 1.19) (82). These intercellular canaliculi are often present in neoplasms derived from eccrine glands. The clear cells contain abundant mitochondria and variable amounts of PAS-positive, diastase-labile glycogen. The dark cells border the lumen of the glands. They contain abundant secretory granules that have glycogen-staining characteristics, as well as sialomucins, PASD-positive mucopolysaccharides, and high concentrations of proteins (83). The dark cells are difficult to identify in routine H&E-stained sections but can be highlighted with acid-fast and PAS-diastase (PAS-D) stains. Immunostains for S100 protein highlight the granularity of the cells (Fig. 1.20). The myoepithelial cells are contractile spindle cells that surround the secretory coil (Fig. 1.21). They are supported by a PAS-positive basement membrane, and associated with elastic fibers, fat, and small nerves in the adjacent stroma.

The excretory component of the eccrine gland is composed of three segments: (a) a convoluted duct in close association with the secretory unit (Fig. 1.22), (b) a straight dermal component, and (c) a spiral intraepidermal portion, the acrosyringium, which opens onto the skin surface (Fig. 1.23). The transition between the secretory and the excretory component is abrupt. Both convoluted and straight dermal ducts are histologically identical. They are narrow

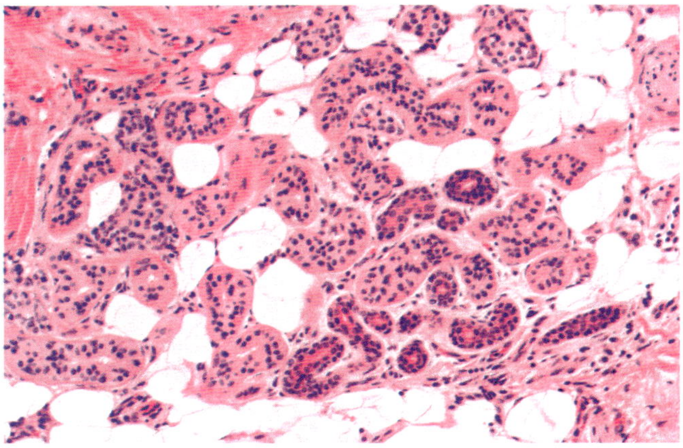

FIGURE 1.17 Eccrine lobule containing fat, glands, and ducts.

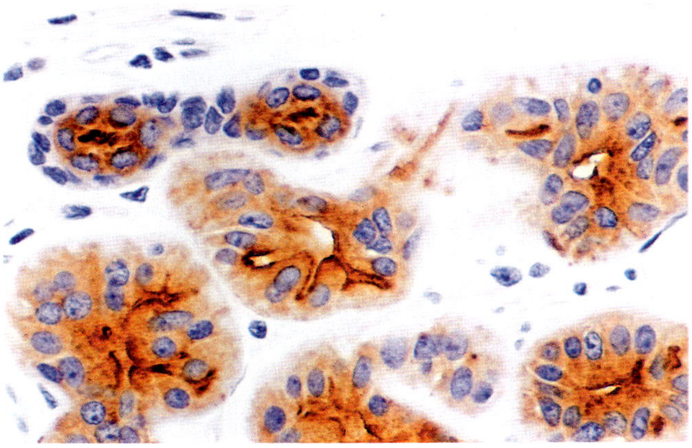

FIGURE 1.19 Intercellular canaliculi (anti-CEA).

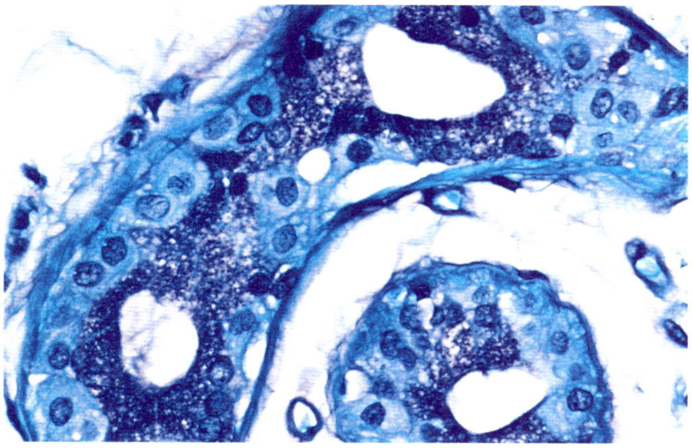

FIGURE 1.20 Dark cells with granular cytoplasm (acid-fast stain).

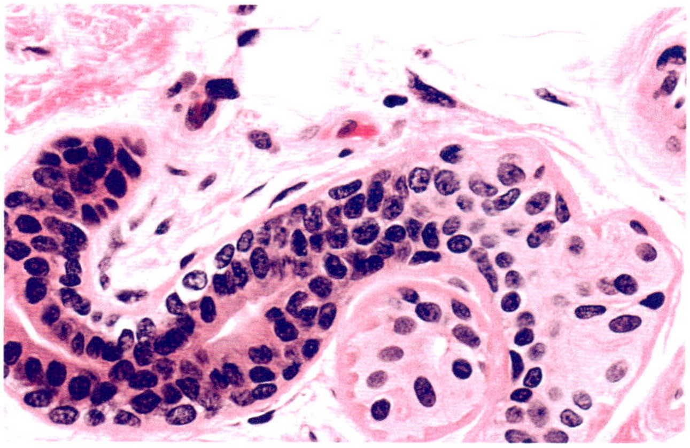

FIGURE 1.22 Eccrine duct. Note the abrupt transition from the secretory portion.

tubes with slit-like lumina lined by double layers of cuboidal cells. The luminal cells contain more granular eosinophilic cytoplasm and a larger nucleus than the peripheral row of cells, whereas the latter are rich in mitochondria. Luminal cells produce a layer of PASD-positive tonofilaments near the luminal membrane that is often referred to as "the cuticular border." This cuticular border often persists in eccrine neoplasms. There are no myoepithelial cells or peripheral basement membrane in the eccrine ducts.

The intraepidermal segment of the eccrine duct, known as acrosyringium, has a unique symmetrical and helicoidal course in the epidermis; its length roughly correlates to the thickness of the epidermis (41). The acrosyringium consists of a single layer of luminal cells and two or three rows of concentrically oriented outer cells. The presence of keratohyalin granules in acrosyringium in the lower levels of the squamous layer indicates that they keratinized independently. The intraepidermal lumen is lined by an acellular eosinophilic cuticle before keratinization (3,21). Melanin granules are absent.

› ***Apocrine Glands*** Apocrine glands are mostly located in the axillary, anogenital areas, mammary region, eyelids (Moll glands), and external ear canal (ceruminous glands), and their presence is characteristic in Nevus Sebaceus of Jadassohn. These specialized sweat glands are present at birth but do not become active until puberty (84). The apocrine gland (Fig. 1.24) has a coiled secretory portion and an excretory (ductal) component. The secretory portion is much longer than its eccrine counterpart and may reach 200 μm in diameter, compared with 20 μm for the eccrine glands. The secretory glands are located in the subcutaneous fat or deep dermis. They are lined by one layer of cuboidal, columnar, or flat luminal cells, and an outer layer of myoepithelial cells, which is surrounded by a PAS-positive basement membrane. The luminal cells contain eosinophilic cytoplasm filled with lipid, iron, lipofuscin, and PASD-positive granules, and a large nucleus located near the base of the cell. Detached fragments of apical cytoplasm are found in the gland lumina. Secretions from apocrine glands release secretory materials accompanied by loss of part of cytoplasm (83). Other forms of secretion, including merocrine (i.e., granular contents within numerous vesicles are released without loss of cytoplasm) and holocrine

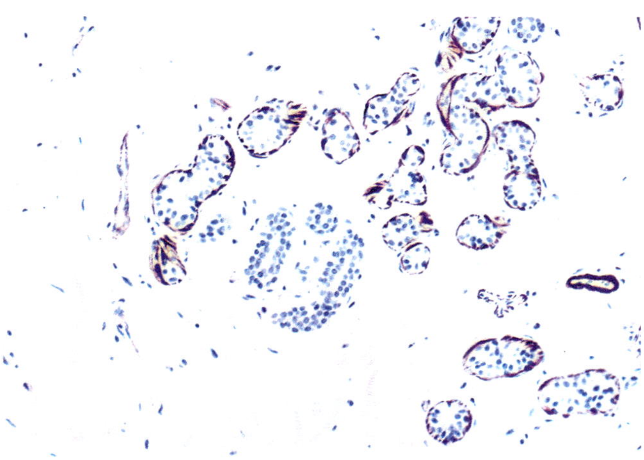

FIGURE 1.21 Glands, but not the ducts, are surrounded by myoepithelial cells (anti-HHF35).

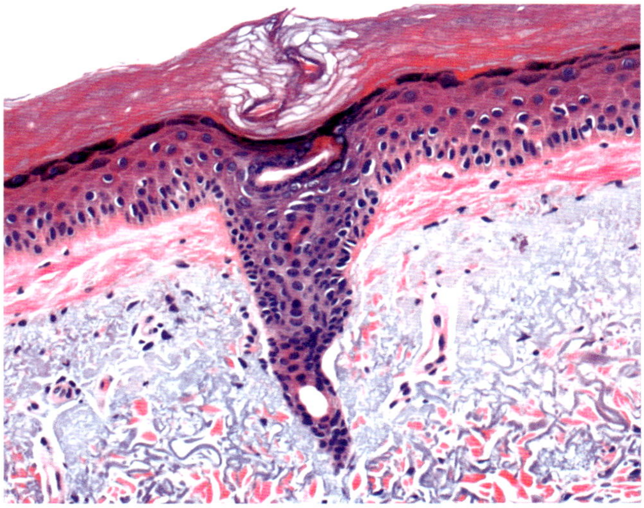

FIGURE 1.23 Acrosyringium.

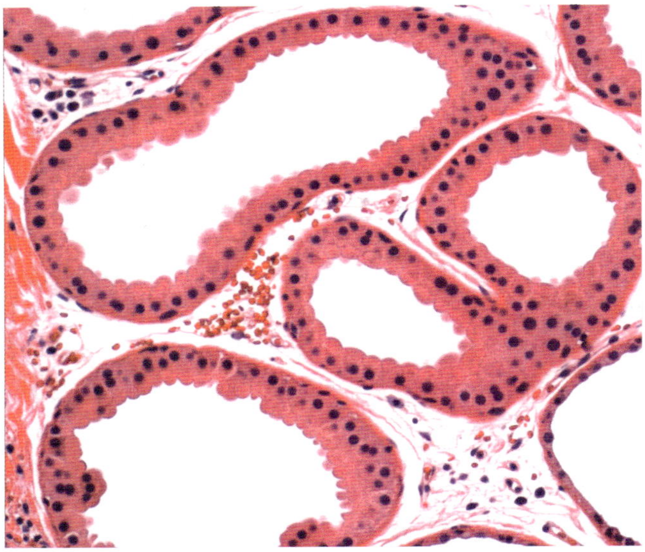

FIGURE 1.24 Secretory apocrine glands.

(i.e., the entire cell is secreted into the glandular lumen) have been observed (83).

Similar to the eccrine duct, the excretory (ductal) component of the apocrine gland has a double layer of cuboidal cells. Microvilli are identified on the surfaces of the luminal cells and keratin filaments fill the cytoplasm, imparting an eosinophilic hyaline appearance to the inner lining of the duct. No myoepithelial cells or peripheral basement membrane are identified in the excretory duct. Apocrine glands are always connected to a pilosebaceous follicle. The intrafollicular or intraepidermal portion of the apocrine duct is straight, rather than spiraled as seen in acrosyringium.

A third type of sweat gland, the so-called "apoeccrine glands" of the human axillae (85), is composed of a dilated secretory portion that, by electron microscopy, is indistinguishable from the apocrine glands. However, these glands retain the intercellular canaliculi and dark cells of the eccrine glands. The duct opens onto the epidermis rather than the hair follicle. These glands, which develop from eccrine glands during puberty, account for up to 45% of all axillary sweat glands in a young person. Obstruction of intraepidermal apoeccrine sweat ducts by apoeccrine secretory cells might be the cause of Fox–Fordyce disease (86).

Dermis

The dermis is a dynamic, supportive connective tissue harboring adnexal structures, mesenchymal cells, fibrous tissue, and ground substances, as well as vascular and nerve plexuses (1). The dermis (Fig. 1.25) consists of two zones, the papillary and the reticular dermis. The adventitial dermis (87) combines the papillary and the periadnexal dermis.

The papillary and periadnexal dermis can be recognized by the presence of a loose meshwork of thin, poorly organized type III collagen and lesser amounts of type 1 collagen accompanied by a delicate branching network of fine elastic fibers (88–91). The papillary dermis also contains abundant ground substance, fibroblasts, and the capillaries of the superficial arterial and venous plexuses.

The reticular dermis is thicker than the papillary dermis and is composed of multiple layers of well-organized thick bundles of collagens, predominantly type I collagen, arranged parallel to the surface. These layers are built from overlapping individual fibers of uniform size. The plates are oriented randomly in different directions (92). There are also thick elastic fibers with a fragmented appearance that can be detected by special elastic tissue stains (Fig. 1.26). Some ground substance and vessels of the deep plexuses are also present in the reticular dermis.

Most of the resident cells in the dermis are dermal dendritic cells, fibroblasts, mast cells, and macrophages. Dermal dendritic cells represent a group of immunophenotypically and functionally heterogeneous cells with dendritic morphology (88). At least three types of dermal dendritic cells are recognized (93,94). Factor XIIIa–positive dermal dendrocytes are distributed around vessels and sweat glands in the papillary dermis. These dermal dendrocytes, also known as dendrophages (95), express some markers of mononuclear macrophages (96) and have phagocytic function (97). CD34-positive dendritic cells are present in the mid and deep dermis around adnexae (93). The third group of cells have a Langerhans cell–like phenotype and are HLA-DR and CD1a-positive. They are present in in a perivascular distribution and are involved antigen presentation but differ from true Langerhans cells in that they lack Birbeck granules (94,98,99). Fibroblasts are the dynamic and fundamental cells of the dermis, synthesizing all types of fibers and ground substances. They appear as spindle-shaped or stellate cells that contain prominent, well-developed, rough endoplasmic reticulum by electron microscopy. Mast cells are derived from CD34-positive bone marrow progenitor cells and are sparsely distributed

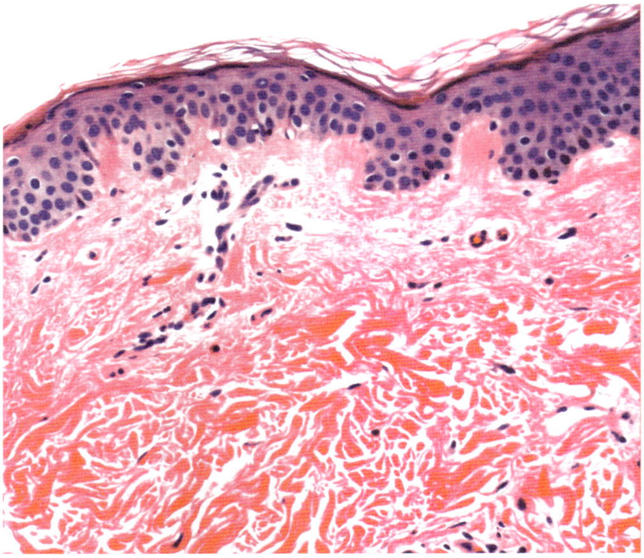

FIGURE 1.25 Papillary and reticular dermis.

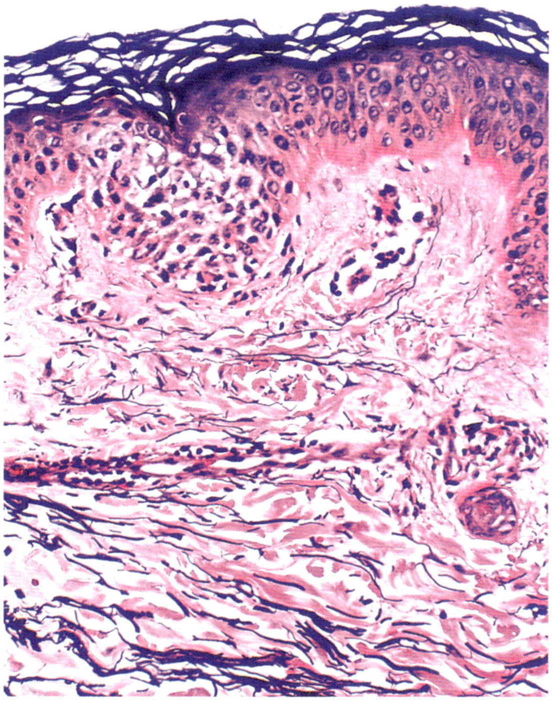

FIGURE 1.26 Distribution of elastic fibers. Elastic fibers are thin and branching in the papillary dermis and thick and fragmented in the reticular dermis.

in the perivascular and periadnexal dermis. They contain darkly stained ovoid nuclei and granular, basophilic cytoplasm that is highlighted by Giemsa and toluidine blue stains. Mast cells are immunopositive with tryptase and CD117 stains (100–103). Macrophages are present in the normal dermis but become more evident when pigments or other ingested material are present in their cytoplasm.

The dermis also contains amorphous ground substance filling the spaces between fibers and dermal cells. It mainly consists of glycosaminoglycans or acid mucopolysaccharides. The latter are mostly nonsulfated acid mucopolysaccharides, predominantly hyaluronic acid and, to a lesser degree, sulfated acid mucopolysaccharides such as chondroitin sulfate. The ground substance is present in small amounts and is seen as empty spaces between collagen bundles in routine H&E-stained sections; it is also easily identified with Alcian blue and toluidine blue special stains. In pathologic conditions, such as lupus erythematosus, granuloma annulare, and dermal mucinosis, the excessive quantity of ground substance appears as strings of bluish material that can be seen without the aid of special stains.

Subcutaneous Tissue

Subcutaneous tissue, also called subcutis or hypodermis, is crucial in thermoregulation, insulation, provision of energy, and protection from mechanical injuries. It is composed of mature adipose tissue arranged in lobules. The mature adipocytes are round and rich in cytoplasmic lipids that compress the nucleus. Adipocytes express S100 protein, adipophilin, and vimentin with immunohistochemical stains. Lobules of mature adipocytes are separated by thin bands of dermal connective tissue that constitute the interlobular septa (Fig. 1.27). Thus, inflammatory changes involving the subcutaneous tissue can be divided into septal panniculitis (e.g., erythema nodosum) and lobular panniculitis (e.g., panniculitis associated with pancreatitis).

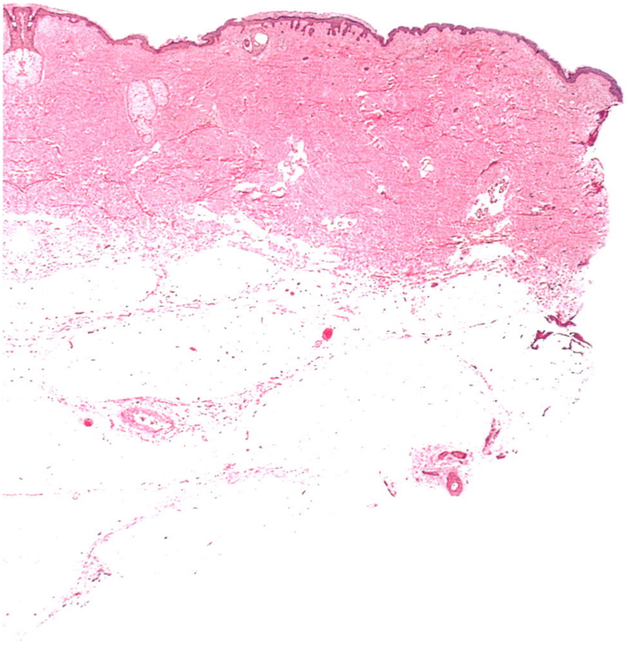

FIGURE 1.27 Septa and lobules of subcutaneous fat.

Blood Vessels, Lymphatics, Nerves, and Muscle

The large arteries that supply the skin are located in the subcutaneous tissue, usually within the interlobular septa and are accompanied by large veins. Smaller arteries, venules, and capillaries constitute the main vasculature seen in the dermis and within the lobules of the subcutaneous fat. A superficial plexus composed of these smaller vessels is located in the papillary dermis and a deep plexus is present in the reticular dermis. The superficial vascular plexus separates the papillary dermis from the reticular dermis, whereas the deep vascular plexuses define the boundary between the reticular dermis and subcutaneous tissue. The division of superficial and deep plexuses is important in the classification and recognition of many inflammatory diseases of the skin in which characteristic infiltrates are located around the superficial, deep, or superficial and deep plexuses. The cutaneous lymphatic vessels (104) accompany the venules and are located in the deep and superficial plexuses. Unless valves are seen within these vessels, their recognition in routine sections is impossible. They are normally surrounded by a cuff of elastic fibers.

Special anastomosing arteriovenous structures known as glomera are present in the reticular dermis of the acral

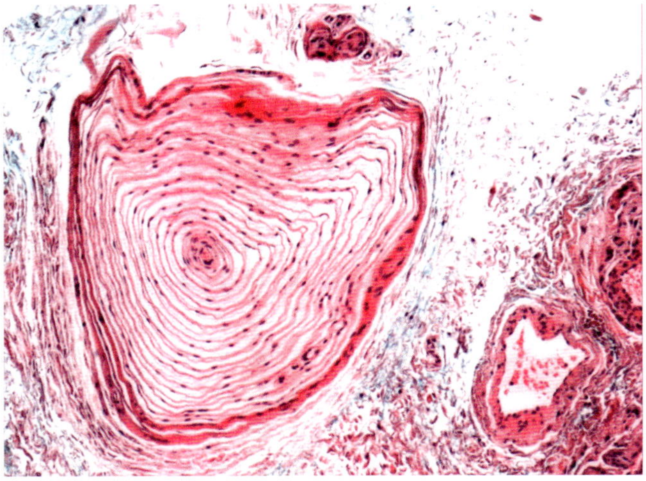

FIGURE 1.28 Vater–Pacini corpuscle.

skin. Each glomus is composed of an arterial segment (the Sucquet–Hoyer canals) connected directly with venous segments. Each Sucquet–Hoyer canal is surrounded by four to six layers of glomus cells, which are modified vascular smooth muscle cells that serve as a sphincter. The glomera appears to be involved in thermoregulation.

Small nerve fibers are present throughout the skin and can extend into the papillary dermis. Larger nerve bundles are seen in the subcutaneous fat and in the deep reticular dermis (105). Some sensory nerves form nerve ending organs in the palms and soles. Meissner corpuscles of the papillary dermis are composed of several parallel layers of Schwann cells containing an axon; they function as rapid mechanical receptors for touch sensation. The Vater–Pacini corpuscles occur in weight-bearing areas. They consist of concentrically arranged Schwann cells around an axon and are located in the deep dermis and/or subcutaneous fat. They serve as receptors for sense of deep pressure and vibration (Fig. 1.28).

Aggregates of smooth muscle cells form the arrector pili muscles, which arise in the connective tissue of the dermis and insert into hair follicles below the sebaceous glands. Melanocytes of congenital nevi are often seen within the arrector pili muscle. Smooth muscle is also seen in the skin of external genitalia (tunica dartos) and in the areolae. Strands of striated muscle are found in the skin of the neck, and face, particularly the eyelids.

HISTOLOGIC DIFFERENCES OF SKIN WITH AGE

Newborns and Children

With the exception of acral skin, the epidermis of newborns and children is of the same thickness as in adults. However, it contains a greater density of melanocytes and Langerhans cells. The dermis is more cellular than that of the adult and contains a higher concentration of ground substance. The density of eccrine glands is higher at birth, while apocrine glands are not well developed until after puberty (106). The sebaceous glands are developed in children, but sebaceous secretion begins at puberty under the influence of androgen stimulation (107).

The adipocytes of the subcutaneous tissue in newborns and children are thin walled and larger than adult adipocytes. Infants also possess more brown fat; it comprises up to 5% of body weight at birth but diminishes with age and virtually disappears by adulthood. Brown adipocytes contain centrally located nuclei and are rich in mitochondria; they contain multiple lipid droplets of varying size in the cytoplasm. Lobules of brown fat contain an abundance of capillaries and are of particular importance in neonates because its fat molecules can be degraded into fatty acids and produce heat (108,109).

Elderly

The histologic changes seen in the skin of elderly patients are mainly due to atrophy and reduction of most cutaneous elements (110,111). The cells of the epidermis are arranged haphazardly because proliferation of the basal cell layer becomes progressively aberrant, which may predispose to the development of neoplasms (112). There is a marked decrease in the number of melanocytes and in the number of melanosomes, leading to reduced pigmentation (113,114) and greater exposure to ultraviolet light. The Langerhans cells also decrease in number and function with advanced age. These changes diminish resistance to the damaging effects of contactants and contribute to age-associated deterioration of immune function (115).

The dermis of aged skin is thin, relatively acellular, and avascular with decreased amounts of collagen, elastin, and ground substance (110,114). Fibroblasts, dendritic cells, and mast cells are reduced in number. Reduction in the cutaneous vascular supply leads to a decrease in inflammatory response, absorption, and cutaneous clearance (112). Elastic fibers show structural and biochemical alterations that diminish skin elasticity, while collagen bundles are thicker and stiffer. The net effect is that age-associated alterations make the dermis less stretchable, less resilient, and prone to wrinkling (116,117). Both eccrine and apocrine glands are reduced in number and function, resulting in diminished secretions in elderly individuals. Sebaceous glands increase in size and manifest clinically as sebaceous hyperplasia, although their secretory output is lessened by decreased activity (109,118). The number and rate of growth of hair follicles also decreases with age. Vellus hairs develop into terminal hairs in unusual sites, such as the ear, nose, and nostrils, resulting in possible cosmetic problems. There is also a decreased functioning of Meissner and Vater–Pacini corpuscles (119). Finally, atrophy of subcutaneous tissue occurs in the face, shins, hands, and feet, but it increases in other areas, particularly the abdomen in men and the thighs in women (105). The pathologic hallmark of extrinsic aging is solar elastosis (Fig. 1.29) (120).

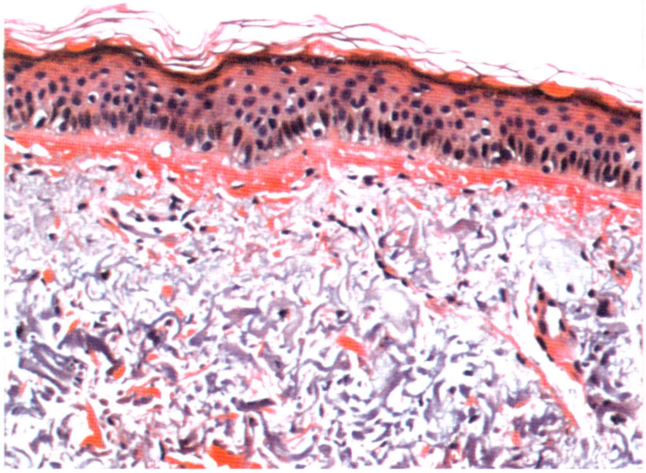

FIGURE 1.29 Solar elastosis in the dermis.

HISTOLOGIC VARIATIONS ACCORDING TO ANATOMIC SITES

Regional variations of the normal histomorphology are important to recognize so as to avoid the misinterpretation of variation as abnormality. The normal scalp and other dense hair-bearing regions show hair follicles extending through the dermis into the subcutaneous fat (Fig. 1.30). This is not typically seen in areas with less dense hair growth. Abundant vellus hairs are present in the skin of the ear, whereas facial skin characteristically contains numerous pilosebaceous units (Fig. 1.31). Sebaceous glands are particularly large on the nose. The eyelid epidermis is thin and composed of two to three layers of cells and basaloid epithelial buds. Modified apocrine glands (Moll glands) and vellus hairs are seen in the dermis.

Sections taken from the skin of the trunk, especially the back, show a thicker reticular dermis than that seen at other sites (Fig. 1.32). Unawareness of this normal variation may lead to the erroneous diagnosis of scleroderma or other fibrosing process. The skin around the umbilicus also shows thick and fibrotic dermis (Fig. 1.33).

The palms and soles contain stratum lucidum with a thick, compact cornified layer and loss of the characteristic basket-weave pattern (Fig. 1.34). The dermis in these areas contains numerous eccrine units, nerve end organs, and glomus structures. There are no pilosebaceous units (glabrous skin) and friction ridges (i.e., epithelial primary ridges, dermatoglyphs, fingerprints) are present (121). Sections of lower leg skin may contain thicker blood vessels in the papillary dermis as a result of gravity and stasis (Fig. 1.35). Smooth muscle fibers are seen in the dermis

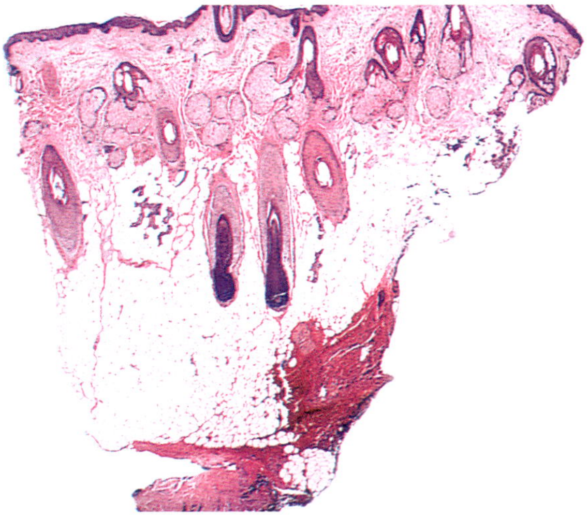

FIGURE 1.30 Section of scalp showing hair follicles extending into the subcutaneous tissue.

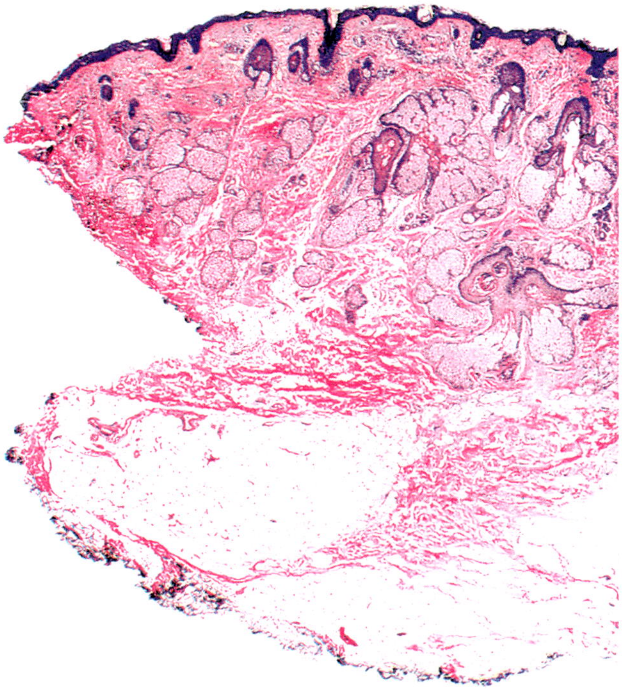

FIGURE 1.31 Skin of face with pilosebaceous units.

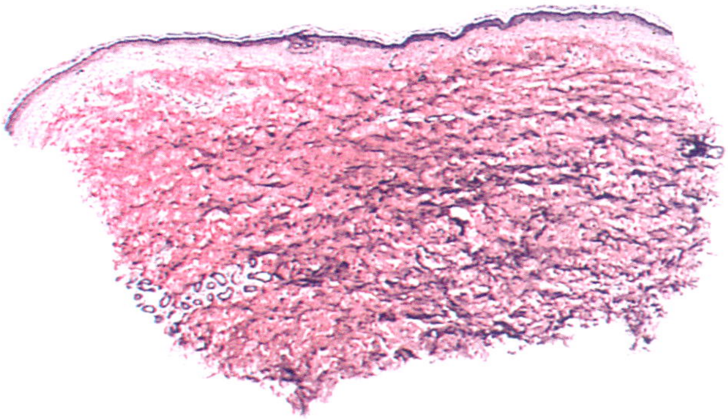

FIGURE 1.32 Section of skin of the back showing the normal reticular dermis.

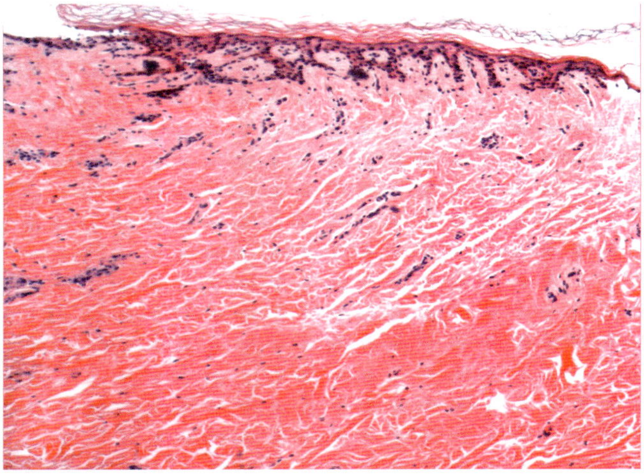

FIGURE 1.33 Umbilicus with dermal fibrosis.

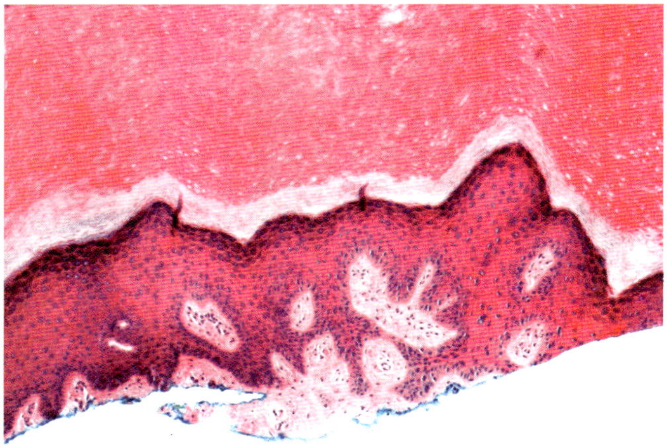

FIGURE 1.34 Histologic section of the palm with compact cornified layers and stratum lucidum.

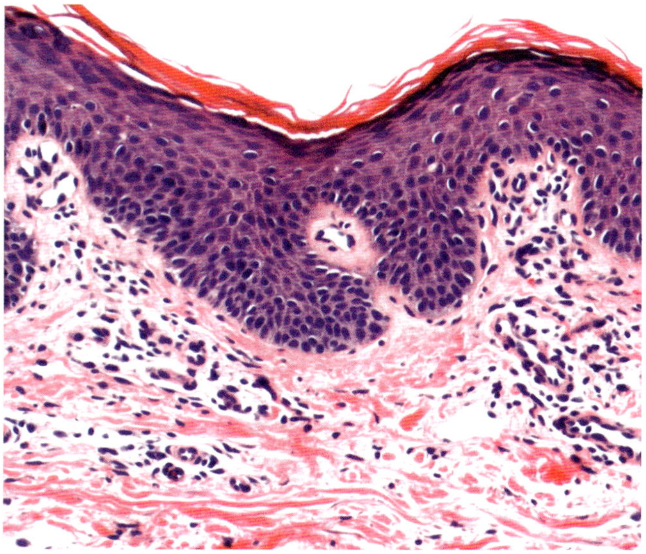

FIGURE 1.35 Skin of the leg showing a proliferation of small thickened blood vessels secondary to stasis.

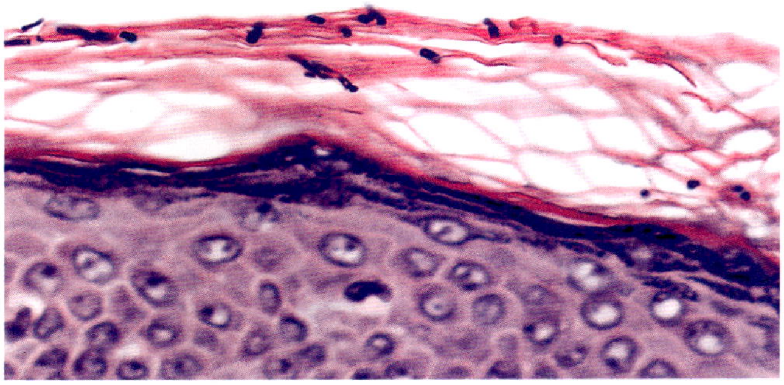

FIGURE 1.36 Superficial dermatophytosis (PAS stains).

of the skin of external genitalia and areola of the nipple. Cutaneous–mucosal junctions may lack granular and cornified layers, and cells of the squamous layers are larger, with higher glycogen content.

PATHOLOGIC CHANGES FOUND IN BIOPSIES AND INTERPRETED AS "NORMAL SKIN"

Biopsies taken from clinically abnormal skin lesions may be interpreted histologically as normal because of the presence of subtle changes in a variety of situations. The following are some examples:

- Dermatophytosis is seen in the cornified layer (Fig. 1.36) of an otherwise normal skin.
- A thick or absent granular layer may indicate an abnormal process of keratinization like psoriasis or an ichthyosiform dermatosis.
- Vitiligo (Figs. 1.37 and 1.38) may give the histologic impression of normal skin unless one searches for melanocytes and melanin.
- Macular and lichen amyloidosis (Fig. 1.39) may be overlooked because the pink globules of amyloid are mistaken for normal dermis.

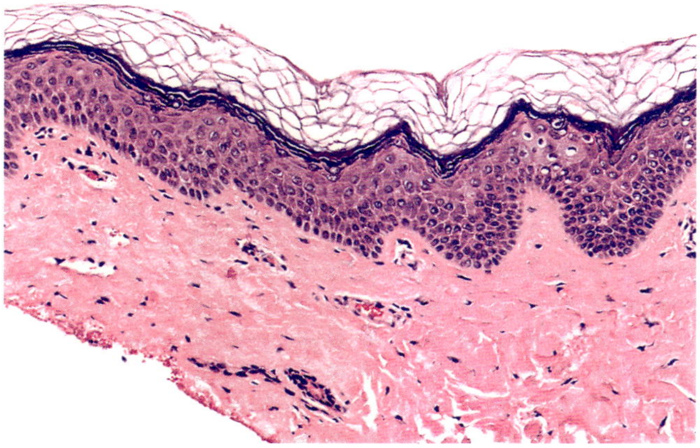

FIGURE 1.37 Vitiligo. Note the absence of basal melanocytes.

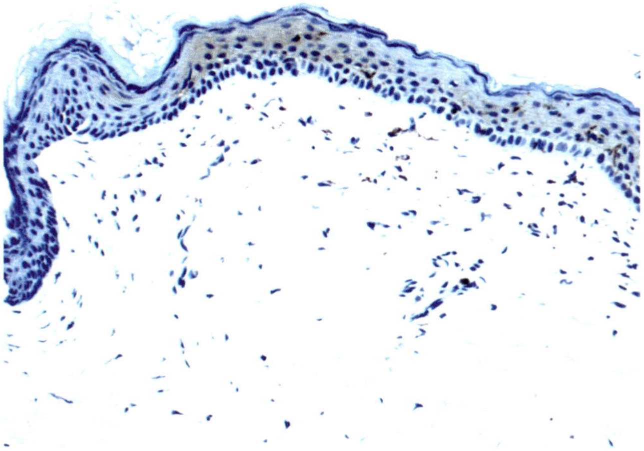

FIGURE 1.38 Vitiligo. S100 protein stains show the absence of basal melanocytes.

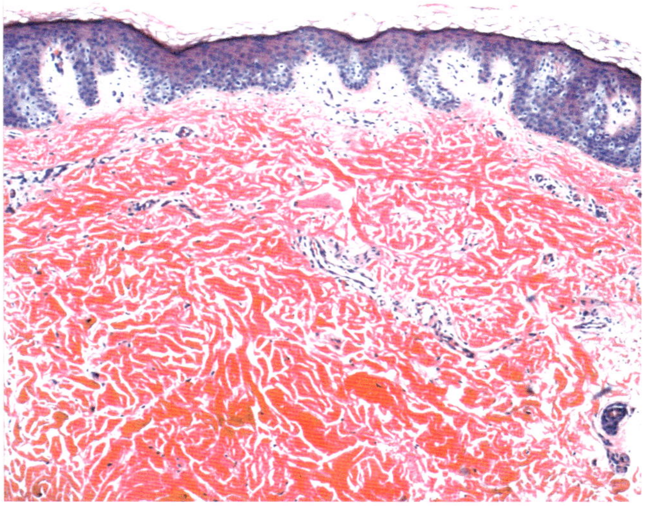

FIGURE 1.40 Urticaria shows only dermal edema.

- Urticaria (Fig. 1.40) may produce only edema, which in routine sections is seen as separation of the collagen bundles in the dermis. Similar dermal changes are noted in the case of dermal mucinosis, in which deposition of mucinous material may be inconspicuous in routine sections. Special stains for mucin such as Alcian blue pH 2.5 or colloidal iron will be helpful.
- The changes of telangiectasia macularis eruptiva perstans, a subtype of cutaneous mastocytosis, are subtle and include dilated blood vessels in the upper dermis with a scant infiltrate of mast cells. The infiltrate must be confirmed with appropriate stains for mast cells such as CD117.
- Trichotillomania is a hair-pulling habit resulting in areas of alopecia. Although histologic changes can be numerous (21), at times hair follicles devoid of hair are the only changes seen.
- Some degenerative diseases of the skin, such as anetoderma, can show only a partial loss of elastic fibers in the dermis, which will be demonstrated by special stains of elastic tissue.
- The "connective tissue nevus" represents a hamartoma with an overproduction of collagen bundles and increased, normal, or decreased elastic tissue in the dermis.

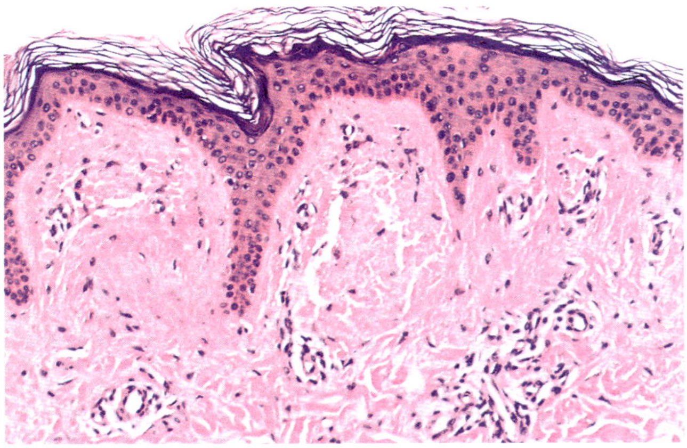

FIGURE 1.39 Lichen amyloidosis composed of pink globules in the papillary dermis.

Other conditions that might be interpreted to represent "normal skin" include café au lait spots, cutis laxa (elastolysis), myxedema, and scleromyxedema. Therefore, the combination of clinical information and careful histologic examination, special stains, or immunohistochemical studies of the biopsy material is crucial to avoid misinterpretation.

SPECIMEN HANDLING

Biopsy samples should be placed in formalin fixative immediately after procurement for the purpose of routine histologic examination. Specimens obtained for direct immunofluorescent study should be placed in Michel medium or, alternatively, placed on saline-moistened gauze if it is going to be processed within 24 hours. Specimens required for flow cytometry, molecular studies, and electron microscopy are sent fresh in saline-soaked gauze or in the appropriate transport media; they should be processed as soon as possible. Excisional biopsy samples or and larger specimens require proper sampling with consideration that histology has priority if no prior diagnosis exists for that particular patient. Punch and shave biopsies are described grossly and either embedded intact or sectioned, depending on the size of the biopsy. Specimens should be embedded on "edge" (vertical). Excisional biopsies and surgical specimens obtained for neoplasia should be described and the entire deep and lateral margins of the specimen inked before sectioning. The margins can be evaluated by cutting

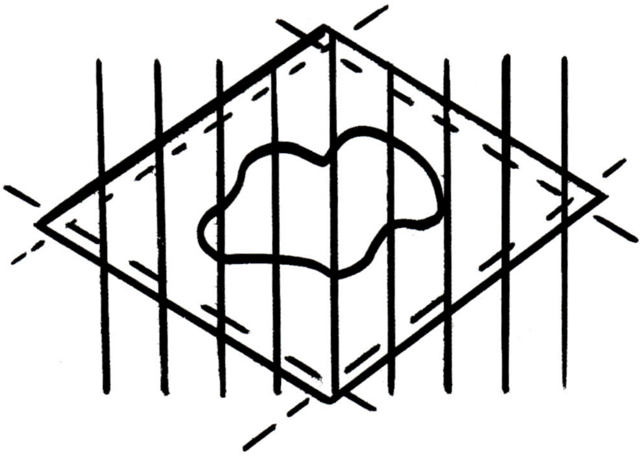

FIGURE 1.41 The entire neoplasm in the center of the lesion is examined by "bread loafing" the specimen; the deep margin is also evaluated. The lateral margins are included in each section submitted for histologic evaluation, or they can be submitted separately by cutting them along the depicted interrupted lines.

along all margins or, most commonly, by serially sectioning the specimen and submitting it entirely (Fig. 1.41).

ARTIFACTS

Poor histologic preparation can result in artifacts that hamper evaluation of slides by the pathologist. These artifacts can be the result of various factors.

1. Artifacts produced at the time of excision, such as cautery (Fig. 1.42) and excessive squeezing of the specimen.
2. Specimens stored at low temperatures, resulting in freezing artifacts (Fig. 1.43).

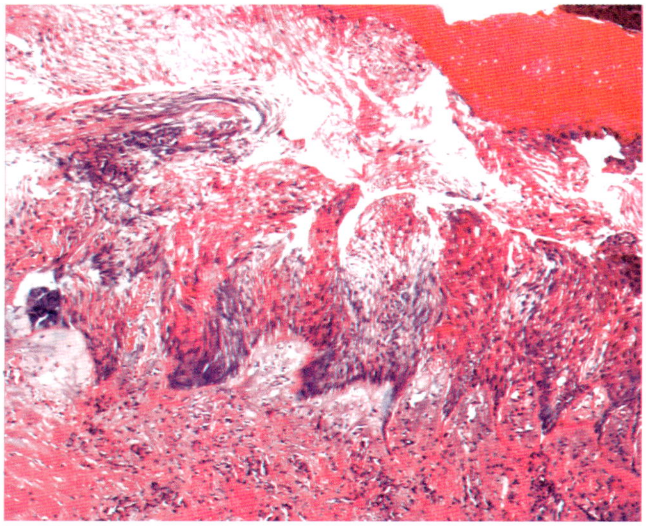

FIGURE 1.42 Cautery effect with vertical elongation of keratinocytes; such a sample is difficult to evaluate.

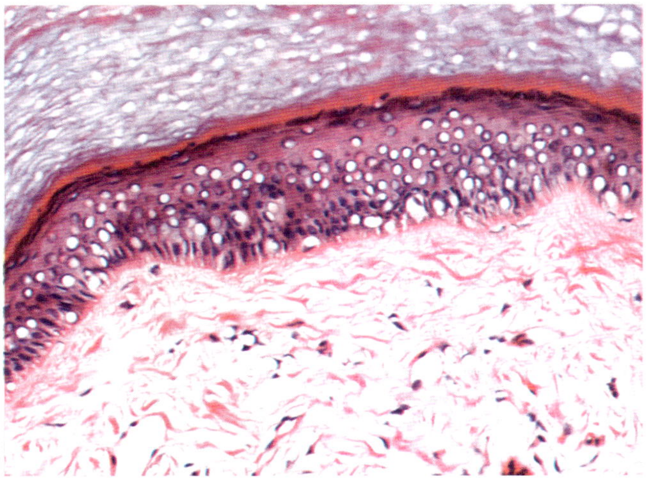

FIGURE 1.43 Freezing artifacts of a specimen with vacuolar changes in the epidermis.

3. Fixation problems that include poor, or no, fixation of the specimen before cutting, old solutions, insufficient fixation time, or inadequate volume of fixative; the specimen should be placed in a solution 15 to 20 times the volume of the specimen (122).
4. Improper monitoring of steps involved in the preparation of a slide, such as cutting, temperature of the water bath, freshness of the staining solutions employed, and other factors.
5. Cutting artifacts caused by a dull knife, calcifications, or other hard substance.
6. Artifacts characteristically seen in some conditions, such as tissue holes in basal cell carcinoma (Fig. 1.44) and the lack of epidermis in sections from toxic epidermal necrolysis.
7. Artifacts by exogenous materials such as gelfoam, sutures, cotton/gauze fibers, nail polish, tattoo ink.

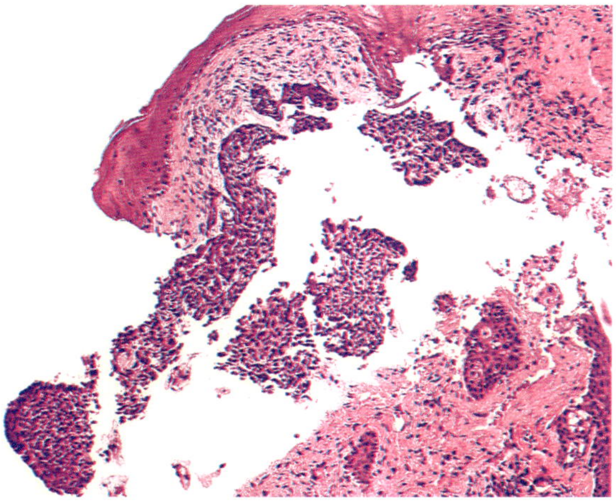

FIGURE 1.44 Tissue defects in a basal cell carcinoma, which appear in the spaces after multiple sections were performed.

STAINING METHODS

The majority of the skin lesions can be diagnosed with well-prepared H&E-stained sections. However, they will not provide an adequate answer in all cases. A comprehensive review of "special stains" is beyond the scope of this chapter because every case is different and may require a specific approach. The following are the most common stains used in our laboratory in the study of cutaneous tissue.

Histochemical Stains

1. PAS: To study the thickness of the basement membrane, for glycogen (diastase labile) and fungal organisms (diastase resistant).
2. Gomori methenamine silver (GMS): For fungal organisms and cutaneous *Pneumocystis carinii.*
3. Ziehl–Neelsen and Fite stains: For acid-fast organisms.
4. Gram stains: For bacteria.
5. Steiner and Warthin–Starry stains: In cases of bacillary angiomatosis and for spirochetes, although the latter immunohistochemical method is preferable.
6. Giemsa: For mast cells and *Leishmania.*
7. Mucicarmine: For epithelial complex acid mucin as well as capsule of *Cryptococcus.*
8. Alcian blue: For acid mucopolysaccharides (pH 2.5), sulfated mucopolysaccharides (pH 1.0), and acid complex connective tissue mucin (pH 0.5).
9. Congo red: For amyloid.
10. Elastic van Gieson (EVG): For elastic fibers.
11. Fontana–Masson: For melanin.
12. von Kossa: For calcium.
13. Perls iron: For stainable iron.
14. Trichrome: For collagen and muscle.

Immunofluorescence

Immunofluorescence plays an important role in the diagnosis and evaluation of skin disorders such as lupus erythematosus, vasculitis, and autoimmune blistering diseases. Specimens are provided to the laboratory either immediately following biopsy on saline-soaked gauze or in Michel transport medium. Michel transport medium is not a fixative. It is a solution that stabilizes proteins for immunofluorescence. It is made up of citric acid, ammonium sulfate, N-ethylmaleimide, and magnesium sulfate and maintains at room temperature the isotonicity and pH of the tissue (7.0 to 7.2). Specimens that have been fixed in formalin are not amenable to immunofluorescence studies. Specimens that have been placed in Michel transport medium and then processed for routine H&E staining show a recognizable artifact in which there is a loss of nuclear detail. Immunofluorescent studies are either direct, using the patient's tissue, or indirect, using the serum of patients and a control tissue containing the relevant target molecules (e.g., human skin, monkey esophagus). Direct immunofluorescence studies are performed on frozen sections of skin using fluorescein isothiocyanate (FITC)–conjugated antisera to immunoglobulins A, G, and M, as well as fibrinogen and complement (C3). Normal nonspecific staining includes autofluorescence of dermal elastic fibers, internal elastic laminae surrounding small- to medium-sized arteries, and necrotic keratinocytes (colloid bodies). Basement membrane material may show weak autofluorescence with IgG and needs to be distinguished from true positive staining noted in disorders such as pemphigoid and lupus. Many nonimmunobullous and nonautoimmune disorders such as actinic keratosis, rosacea, solar elastosis, and polymorphous light eruption can show weak, often discontinuous, linear or granular staining that must be distinguished from true positive basement membrane deposition. Nonspecific staining can also be seen in serum crust, spongiotic epidermis, and linear zones surrounding blood vessels that have increased vascular permeability. These staining patterns tend to involve all FITC-labeled immunoglobulins rather than specific disease-associated ones.

Immunohistochemical Stains/Molecular Studies

Immunohistochemistry detects cellular and fibrillar antigens in tissue sections. The antigens may be cytoplasmic, membranous, or nuclear and show cellular events such as apoptosis or altered proliferation. Most commercially available antibodies can be used on formalin-fixed tissue. Most immunohistochemical studies utilize either a DAB detection kit ("brown stain") or an alkaline phosphatase detection kit ("red stain") (Fig. 1.45), the latter which is especially

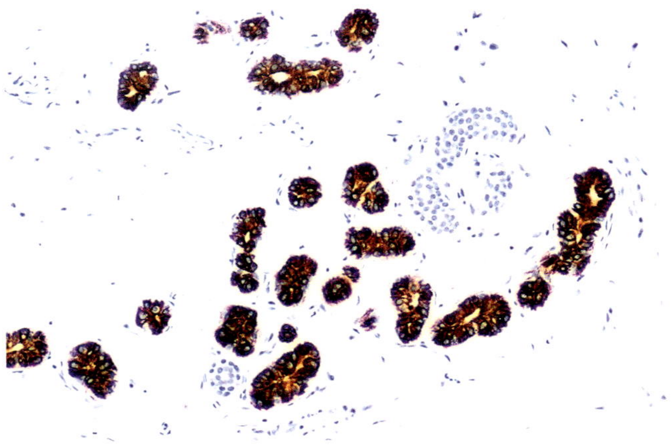

FIGURE 1.45 CAM5.2 immunostaining. The secretory glands but not the ducts are stained.

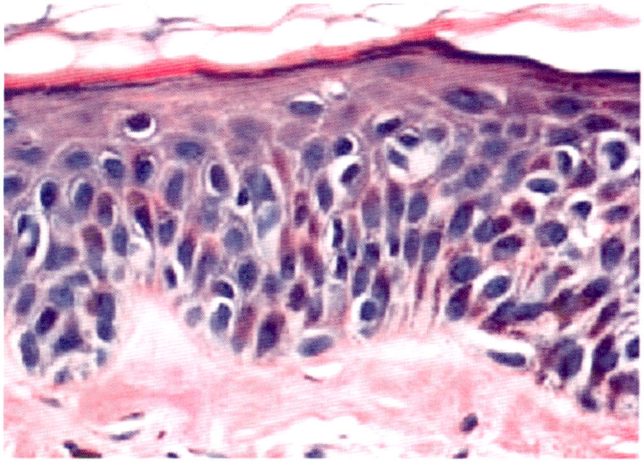

FIGURE 1.46 Normal melanocytes with brown melanin pigment. This coloration is identical to that of the diaminobenzidine chromogen often used with immunohistochemistry, making distinction of a positive immunostain versus background melanin difficult.

helpful in the diagnosis of pigmented melanocytic lesions (Figs. 1.46 and 1.47). Positive or negative reactivity of the antibodies tested often depends on variables such as quality and duration of fixation as well as the use of additional steps such as protease digestion and antigen retrieval. Although a separate positive control slide is standard for each immunohistochemical stain, this does not guarantee that the stain has worked properly for the patient's tissue. Interpretation of immunohistochemical studies relies on knowledge of normal skin components that can serve as internal positive controls (Table 1.1).

Immunohistochemistry is now widely used as an aid in the diagnosis of difficult cases where there is a defined differential. Examples include pigmented actinic keratosis and melanoma in situ, reactive and neoplastic T-cell infiltrates, benign cutaneous lymphoid hyperplasia (pseudolymphoma) and B-cell lymphoid neoplasms, atypical melanocytic proliferations and melanoma, basal cell carcinoma, squamous cell carcinoma, intraepidermal pagetoid proliferations, and poorly differentiated epithelioid and spindle cell tumors. Some commonly used panels in our laboratory are shown in Tables 1.2 and 1.3.

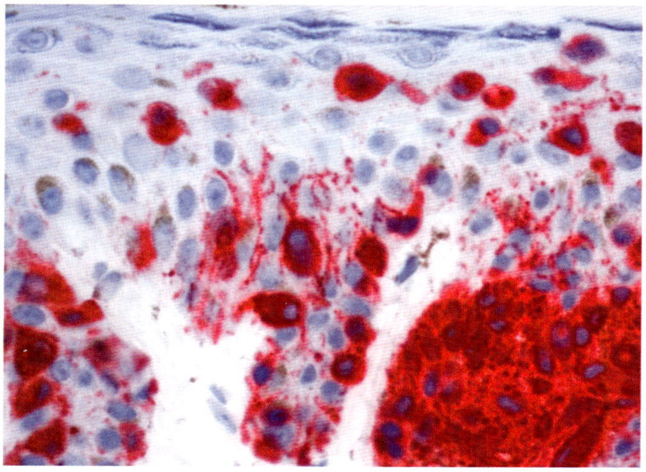

FIGURE 1.47 Proliferating melanocytes decorated with MART-1 and alkaline phosphate detection.

Numerous keratin immunostains are available ranging from cocktails to individual keratins. High–molecular-weight keratins 34be12 and 5/6 are reactive with epidermal and follicular keratinocytes along with eccrine and apocrine sweat glands. As cytokeratin 5/6 is found primarily in mesothelium and other lining-type epithelia such as epidermis and squamous mucosa, it is a particularly useful antibody in differentiating metastases from primary cutaneous tumors (123). We have also found keratins to be helpful in the diagnosis/detection of poorly differentiated squamous cells and infiltrating/morpheaform basal cell carcinomas. Cytokeratin 20 is a heavy-weight keratin that shows a specific perinuclear (Golgi) dot-like staining pattern in cutaneous neuroendocrine carcinomas. CK7 shows similar dot-like perinuclear reactivity with extracutaneous (metastatic) neuroendocrine carcinomas. Low–molecular-weight keratins such as CAM5.2 and CK7 are reactive with secretory glandular cells found in sebaceous apocrine and eccrine glands, and are valuable in the diagnosis of Paget disease along with adnexal carcinomas and metastatic lesions.

Other antibodies useful in the diagnosis of epithelial tumors include BerEp4, p63, p40, adipophilin, and FXIIIa (AC-1A1). BerEp4 is an epithelial antibody used in differentiating basal cells from squamous cell carcinomas, as well as aiding in differentiating microcystic adnexal carcinoma from basal cell carcinomas (124,125). Antibodies that detect squamous proliferation, namely p63 and p40, are helpful in classifying many cutaneous malignancies. While both are particularly helpful in the diagnosis of spindle cell/sarcomatoid squamous cell carcinomas (126), p40 is as sensitive as, and more specific than, p63 (127). Adipophilin is an antibody to a surface protein of intracellular lipid material. A membranous and cytoplasmic vesicular pattern staining of adipophilin is useful in distinguishing sebaceous tumors, particularly intraepithelial sebaceous carcinomas, from other pagetoid neoplasms. Of note, granular adipophilin staining is nonspecific and can be noted in pagetoid squamous cell carcinoma (128). FXIIIa (AC-1A1), a nuclear marker for sebaceous differentiation, has recently been reported as an even more sensitive and specific antibody for the differentiation of sebaceous neoplasms from squamous and other clear cell neoplasms (129).

Melanocytic antibodies include the highly sensitive S100 protein and SOX-10 immunostains along with antibodies specific to melanocytes, such as Melan-A/MART-1, HMB-45, and tyrosinase. Melan-A/MART-1 and tyrosinase are antibodies associated with premelanosomes and are present in both benign and malignant neoplasms. HMB-45 reactivity is limited to activated melanocytes and melanocytes with abundant melanin (blue nevus cells, type A nevus cells). This latter fact is utilized in differentiating atypical

TABLE 1.1 Epithelial, Melanocytic, and Mesenchymal Antibodies

Epithelial Antibodies		
Cytokeratins	AE1:AE3	Appendage and epidermal keratinocytes
	34be12	Epidermal and follicular keratinocytes
	CAM5.2	Apocrine, eccrine, and sebaceous glands
	Cytokeratin 5/6	Lining epithelium of skin and squamous mucosa
	Cytokeratin 7	Apocrine, eccrine, and sebaceous glands
	Cytokeratin 19	Apocrine and eccrine glands
	Cytokeratin 20	Normal epidermal basal layer Merkel cells
p63		Basal and suprabasal epidermal keratinocytes, matrix and outer root sheath of hair follicles, sebaceous gland basal cells, and myoepithelial cells of sweat glands
p40		Basal and suprabasal epidermal keratinocytes, matrix and outer root sheath of hair follicles, sebaceous gland basal cells, and myoepithelial cells of sweat glands
EMA		Perineural cells, sometimes epidermis, sebocytes, luminal surfaces, and lateral borders of sweat glands
BerEp4		Matrical and outer root sheath follicular keratinocytes and apocrine/eccrine glands
CEA		Eccrine and apocrine gland luminal surfaces
Androgen receptor		Sebaceous glands
Adipophilin		Sebaceous glands, adipose tissue
FXIIIa (AC-1A1)		Nuclei of sebocytes
Melanocyte Antibodies		
S100		Nevomelanocytes, nerves, adipose tissue, Langerhans cells, some dermal histiocytes/dendrocytes, and myoepithelial cells
SOX10		Schwann cells, melanocytes along with myoepithelial cells in breast and salivary gland tissue; normal mast cells also can be positive
Melan-A/MART-1		Nevomelanocytes
HMB-45		Activated melanocytes and pigmented melanocytes (blue nevus, type A nevus cells)
Tyrosinase		Nevomelanocytes
MITF		Nevomelanocytes, lymphocytes, smooth muscle cells, mast cells, fibroblasts, and Schwann cells
p16		Benign nevomelanocytes
PRAME		Nuclear staining in some single melanocytes/sun-damaged (older individuals) sebaceous glands, apocrine glands, and hair germs (cytoplasmic)
Mesenchymal Antibodies		
Vascular endothelium	CD31	Vascular endothelium
	CD34	Vascular endothelium
	FVIII	Vascular endothelium and serum
	FLI-1	Vascular endothelium and small lymphocytes
	D2-40	Lymphatic endothelium
	ERG	Vascular Endothelium
Muscle	SMA (alpha smooth muscle)	Myofibroblasts, smooth muscle Myofibroblasts, smooth muscle
	HHF-35 (muscle common actin)	Smooth muscle Smooth muscle
	Desmin	Smooth muscle
	Caldesmon	
Neural/neuroendocrine	CD56	Small dermal nerves
	Neurofilament	Nerves
	S100	Nevomelanocytes, nerves, adipose tissue, Langerhans cells, some dermal histiocytes/dendrocytes, and myoepithelial cells
	Cytokeratin 20	Merkel cells
	Chromogranin	Merkel cells and nerves and some cells in eccrine/apocrine glands
	Synaptophysin	Merkel cells, nerves, and some cells in eccrine/apocrine glands
Fibroblast	CD10	Fibroblast proliferation and follicular center lymphocytes
	CD34	Mesenchymal and dendritic interstitial cells in dermis and around nerves, vascular endothelium, hematopoietic stem cells
	FXIIIa	Dermal dendrocytes
CD117		Mast cells, immature Langerhans cells, basal layer keratinocytes, and melanocytes

EMA, epithelial membrane antigen; CEA, carcinoembryonic antigen; MITF, microphthalmia-associated transcription factor.

TABLE 1.2 Basal Cell Carcinoma Versus Squamous Cell Carcinoma

	34βe12	EMA	BerEp4
Basal cell carcinoma	+	–	+
Squamous cell carcinoma	+	+	–

nonneoplastic dermal nevomelanocytic proliferations from melanoma. With the exception of a blue nevus, most nonneoplastic nevomelanocytic proliferations show an absence or loss of HMB-45 reactivity upon descent into the dermis. Conversely, most melanomas show no loss of HMB-45 reactivity (130,131). Other antibodies useful in the diagnosis of malignant melanocytic lesions are the nuclear proliferation antibody Ki-67 (132) and p16 (133). Similar to HMB-45, benign nevomelanocytic lesions show low to absent Ki-67 positivity, while many melanocytic malignancies show an increased Ki-67 staining pattern (133). On the other hand, p16 shows loss of staining in most malignant melanocytic lesions (134). However, not all melanomas show increased Ki-67 reactivity and/or loss of p16 reactivity (135,136).

Melanocytic antibodies are also useful in the classification of atypical dermal spindle cell malignancies and atypical epidermal proliferations. Desmoplastic/spindle cell melanomas show variable reactivity with MART-1/Melan-A and are always negative for HMB-45. S100 protein and SOX-10 are often the only melanocytic antibodies that are positive, with SOX-10 showing greater sensitivity for spindle melanocytic proliferation than S100 protein (137). Of note, SOX-10 is not specific for neural/melanocytic lesions, with positivity noted in both breast carcinomas and salivary gland lesions as well as mast cells (138). MART-1/Melan-A and other melanocytic immunostains are useful in differentiating pigmented keratinocytic proliferations from atypical junctional melanocytic proliferations (139). Two newer immunohistochemical antibodies, PRAME (Preferentially Expressed Antigen in Melanoma) and BRAF help in the distinction of melanoma from Spitz nevi and other benign melanocytic proliferations, although some melanomas can be negative for either marker. In our experience, as well as recently reported, PRAME, has also shown positive staining in dysplastic nevi (most often high-grade lesions), congenital nevi, as well as rare melanocytic proliferations that would be otherwise classified as low-grade dysplastic nevi (140). In these low-grade lesions, the staining intensity is less than that of the high-grade and malignant melanocytic lesions. This phenomenon of weak staining has also been reported in hematolymphoid cells within sentinel lymph nodes (141). PRAME has also been reported to be reactive in sebaceous carcinomas (142), basal cell carcinomas, atypical fibroxanthoma (140), and some scars (143). Other tumors that can be PRAME-positive include synovial sarcomas, myxoid liposarcoma (144), malignant peripheral nerve sheath tumors (145), and a handful of pulmonary, salivary, ovarian, and mammary tumors (146,147). Two other immunohistochemical antibodies which have been used to further evaluate Spitzoid/epithelioid melanocytic proliferation are BAP-1 (BRCA1-associated protein-1), and BRAF. BAP-1 negative staining has been used to subclassify some Spitzoid melanocytic tumors as BAP-1 inactivated melanocytic tumors that show an indolent clinical course. There are also reports of BAP-1 negative melanomas (148). BRAF positivity (mutation) has been reported as a finding noted in Spitzoid (and non-Spitzoid) melanomas; its absence a reassuring finding in Spitzoid lesions (149). BRAF positivity is also used to qualify patients for some melanoma chemotherapeutic regimens such as Vemurafenib (Zelboraf), dabrafenib (Tafinlar), and encorafenib (Braftovi).

TABLE 1.3 Intraepidermal Atypical Pagetoid Cells

	MART-1	34βe12	CK7	p63	CAM5.2
Pagetoid Bowen disease	–	+	–	+	–
Melanoma	+	–	–	+/–	–
Paget/ extramammary Paget	–	+/–	+	–	+
Sebaceous carcinoma	–	+/–	+	+	+

Antibodies that can be reactive with varied normal mesenchymal elements within the dermis and subcutis are useful when identifying their neoplastic counterparts. Vascular neoplasms, both benign and malignant, may be reactive with endothelial antibodies directed against CD31, CD34, FVIII, FLI-1, ERG, and D2-40. Of these, CD31 and ERG are the most specific with FLI-1 showing more specificity for epithelioid angiosarcoma. FVIII often shows nonspecific staining of perivascular tissue; it is associated with von Willebrand factor and is present not only in lining endothelial cells but in the surrounding extravascular plasma. CD34 is the least specific, as it is reactive with mesenchymal and dendritic interstitial cells in the dermis as well as hematolymphoid cells. D2-40 is specifically reactive with lymphatic endothelium and can be used to confirm metastatic tumor within lymphatic vasculature. Although not reactive with any normal skin component, human herpes virus type 8 (HHV8) antibody is useful in the diagnosis of Kaposi sarcoma (150). However, it can be seen in non-Kaposi sarcoma vascular and fibroblastic proliferations in some populations where the virus is endemic (151). Smooth muscle antibodies, alpha smooth muscle actin (SMA), and HHF-35 are reactive with both vascular myoepithelium and arrector pili muscles, whereas desmin is limited to smooth muscle. Dermatofibroma and dermatofibrosarcoma protuberans may be differentiated with a combination of FXIIIa and CD34 (152). CD10 is reactive with lymphoid tissue and mesenchymal and dendritic interstitial cells in the dermis, as well as atypical fibroxanthoma. Neural tumors show reactivity with S100 protein along with NKI/C3. In addition, S100 protein is positive in both cartilaginous and adipose tumors.

Hematologic antibodies are used to define both the cell type in an inflammatory infiltrate and determine whether the infiltrate is reactive or malignant. Lymphoid markers include all available CD antibodies for the characterization of T-cell and B-cells. Macrophage-directed antibodies include CD68 and CD163 for non–Langerhans cell histiocytes and CD1a, langerin (CD207), and S100 protein for Langerhans cells and tumors. CD117 is positive in mast cells, immature Langerhans cells, basal layer keratinocytes, and melanocytes. Neuroendocrine cells are limited in normal skin to Merkel cells and eccrine sweat glands and can be demonstrated with chromogranin and synaptophysin. As mentioned above, Merkel cells are positive with cytokeratin 20.

Antibodies that demonstrate cell proliferation and apoptosis such as Ki-67 and p53 are valuable in differentiating actively proliferating malignancies and dysplasias from the histologic atypical-appearing benign counterparts. Ki-67 is useful in the distinction of melanomas and follicular center lymphomas from atypical nevi and benign lymphoid proliferations, respectively (153–155). Ki-67, in combination with p53, is also helpful in the diagnosis of keratinocytic dysplasia in sun-damaged skin (156).

Immunostains and in situ hybridization are valuable in diagnosing or confirming infectious agents. There are immunostain antibodies specific for herpes simplex virus (I and II), varicella-zoster virus, herpes virus type 8, cytomegalovirus, spirochetes, and *Mycobacteria*. Of note, the immunostain for spirochetes, while sensitive for *Treponema pallidum*, does show cross reactivity with other spirochete organisms (157). In situ hybridization is currently used for the diagnosis of high- and low-risk human papillomavirus (HPV) condyloma accuminata (Fig. 1.48).

Comparative genomic hybridization and fluorescent in situ hybridization have been utilized to help distinguish melanoma from atypical melanocytic proliferations and reactive or nonneoplastic melanoma mimickers. Comparative genomic hybridization distinguishes melanoma from cellular and atypical blue nevi (158). Isolated chromosome number copy abnormalities (DNA) can be seen in benign melanocytic proliferation as well as atypical melanocytic lesions (159). Fluorescent in situ hybridization testing for DNA copy number abnormalities in chromosome regions associated with melanoma are utilized as an ancillary diagnostic test for differentiating benign melanocytic lesions from melanomas. Gains in *RREB1* at 6p25 and *CCND1* at 11q13, as well as abnormalities in *MYB* at 6q23 and *CEP6* (centromere 6) are highly sensitive and specific for malignant melanoma. However, 10% of metastatic melanomas are reported to show negative results by in situ hybridization and, conversely positive results have been reported in benign melanocytic lesions (160–162). Spitz nevi have 11 p gains, alterations at 9p21 and 8q34, and a variety of fusions (ROS1, NTRK1, NTRK3, ALK, BRAF, Met, and RET), some of which can be assessed using immunohistochemical techniques (163,164). Proliferative nodules in congenital nevi often show complete loss or gain of entire chromosomes in contrast to melanomas which harbor partial segments (165). Overall concordance between these methodologies is high, although false positive and false negative results do occur (164). There are many other settings in cutaneous pathology where genetic loci can be tested to establish a diagnosis such as dermatofibrosarcoma protuberans (COL1a1::PDGFB) and nodular fasciitis MYH9-USP6 (USP6) (166,167).

Molecular studies in combination with immunohistochemical phenotyping have also become useful in aiding the differentiation of reactive and neoplastic T-cell and B-cell proliferations, although false-negative gene rearrangements (pseudoclonality) do occur. T-cell pseudoclonality can be seen in inflammatory dermatoses, such as lichen planus, pityriasis lichenoides, and lichen sclerosus along with histologic neoplastic mimickers such as lymphomatoid drug eruption and nickel contact dermatitis (168). B-cell pseudoclonality has been reported in some insect bite reactions and drug reactions (169). Newer genetic techniques utilizing next generation sequencing are becoming increasingly common in the diagnosis of lymphomas of the skin as well as in several other dermatologic applications.

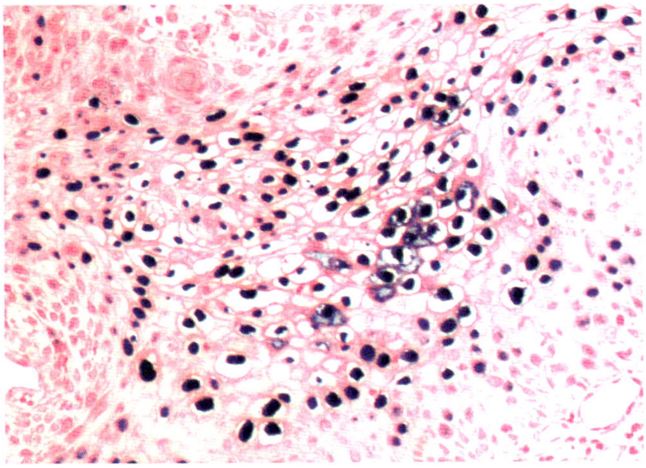

FIGURE 1.48 In situ hybridization with low-risk HPV 6 in a warty lesion.

REFERENCES

1. Kanitakis J. Anatomy, histology and immunohistochemistry of normal human skin. *Eur J Dermatol*. 2002;12:390–399.
2. Montagna W, Parakkal PF. *The Structure and Function of the Skin*. 3rd ed. Academic Press; 1974.
3. Montagna W, Freedberg IM. Cutaneous biology 1950–1975. *J Invest Dermatol*. 1976;67:1–230.
4. Visinoni AF, Lisboa-Costa T, Pagnan NAB, Chautard-Freire-Maia EA. Ectodermal dysplasias: clinical and molecularreview. *Am J Med Genet A*. 2009;149A(9):1980–2002.

5. Murphy GF. Histology of the skin. In: Elder DE, Elenitsas R, Johnson BL Jr, Murphy GF, eds. *Lever's Histopathology of the Skin*. 9th ed. Lippincott Williams & Wilkins; 2005.
6. Breathnach AS. The Herman Beerman lecture: embryology of human skin, a review of ultrastructural studies. *J Invest Dermatol*. 1971;57:133–143.
7. Holbrook KA, Odland GF. The fine structure of developing human epidermis: Light, scanning and transmission electron microscopy of the periderm. *J Invest Dermatol*. 1975;65:16–38.
8. Tamiolakis D, Papadopoulos N, Lambropoulou M, et al. Ber-H2 (CD30) Immunohistochemical staining of human fetal tissue. *Int J Biol Sci*. 2005;1:135–140.
9. Hentula M, Peltonen J, Peltonen S. Expression profiles of cell-cell and cell-matrix junction proteins in developing human epidermis. *Arch Dermatol Res*. 2001;293:259–267.
10. Smith LT, Sakai LY, Burgeson RE, Holbrook KA. Ontogeny of structural component at the dermal–epidermal junction in human embryonic and fetal skin: the appearance of anchoring fibrils and type VII collagen. *J Invest Dermatol*. 1988;90:480–485.
11. Sagebiel RW, Rorsman H. Ultrastructural identification of melanocytes in early human embryos [abstract]. *J Invest Dermatol*. 1970;54:96.
12. Foster CA, Holbrook KA, Farr AG. Ontogeny of Langerhans cells in human embryonic and fetal skin: expression of HLA-DR and OKT-6 determinants. *J Invest Dermatol*. 1986;86:240–243.
13. Foster CA, Holbrook KA. Ontogeny of Langerhans cells in human embryonic and fetal skin: cell densities and phenotypic expression relative to epidermal growth. *Am J Anat*. 1989;184:157–164.
14. Winkelmann RK, Breathnach AS. The Merkel cell. *J Invest Dermatol*. 1973;60:2–15.
15. Moll R, Moll I, Franke WW. Identification of Merkel cells in human skin by specific cytokeratin antibodies: changes of cell density and distribution in fetal and adult plantar epidermis. *Differentiation*. 1984;28:136–154.
16. Moll I, Lane AT, Franke WW, Moll R. Intraepidermal formation of Merkel cells in xenografts of human fetal skin. *J Invest Dermatol*. 1990;94:359–364.
17. Polakovicoca S, Seidenberg H, Mikusova R, Polak S, Pospisilova V. Merkel cells—review on developmental, functional and clinical aspects. *Bratisl Lek Listy*. 2011;112:80–87.
18. Boot PM, Rowden G, Walsh N. The distribution of Merkel cells in human fetal and adult skin. *Am J Dermatopathol*. 1992;14:391–396.
19. Breathnach AS. Development and differentiation of dermal cells in man. *J Invest Dermatol*. 1978;71:2–8.
20. Smith LT, Holbrook KA, Madri JA. Collagen types I, III, and V in human embryonic and fetal skin. *Am J Anat*. 1986;175:507–521.
21. Mehregan AH, Hashimoto K, Mehregan DA, et al. Normal structure of the skin. In: Mehregan AH, Hashimoto K, Mehregan DA, Mehregan DR, eds. *Pinkus' Guide to Dermatohistopathology*. 6th ed. Appleton & Lange; 1995.
22. Lavker RM, Sun TT, Oshima H, et al. Hair follicle stem cells. *J Investig Dermatol Symp Proc*. 2003;8:28–38.
23. Matsuzaki T, Yoshizato K. Role of hair papilla on induction and regeneration processes of hair follicles. *Wound Repair Regen*. 1998;6:524–530.
24. Downig DT, Stewart ME, Strauss JJ. Biology of sebaceous glands. In: Fitzpatrick TB, Eisen AZ, Wolff K, Freedberg IM, Austen KF, eds. *Dermatology in General Medicine*. Vol 1. 3rd ed. McGraw-Hill; 1987:185–190.
25. Muller M, Jasmin JR, Monteil RA, Loubiere R. Embryology of the hair follicle. *Early Hum Dev*. 1991;26:159–166.
26. Hashimoto K, Gross BG, Lever WF. The ultrastructure of the skin of human embryos. I. The intraepidermal eccrine sweat duct. *J Invest Dermatol*. 1965;45:139–151.
27. Montagna W. Embryology and anatomy of the cutaneous adnexa. *J Cutan Pathol*. 1984;11:350–351.
28. Hosoi J, Murphy GF, Egan CL, et al. Regulation of Langerhans cell function by nerves containing calcitonin gene-related peptide. *Nature*. 1993;363:159–163.
29. Koster MI. p63 in skin development and ectodermal dysplasias. *J Invest Dermatol*. 2010;130:2352–2358.
30. Kirfel J, Magin T, Reichelt J. Keratins: a structural scaffold with emerging functions. *Cell Mol Life Sci*. 2003;60:56–71.
31. Ishiko A, Matsunaga Y, Masunaga T, Aiso S, Nishikawa T, Shimizu H. Immunomolecular mapping of adherens junction and desmosomal components in normal human epidermis. *Exp Dermatol*. 2003;12:747–754.
32. Wolff K, Schreiner E. Ultrastructural localization of pemphigus autoantibodies within the epidermis. *Nature*. 1971; 229:59–61.
33. Stanley JR, Klaus-Kovtun V, Sampaio SA. Antigenic specificity of fogo selvagem autoantibodies is similar to North American pemphigus foliaceus and distinct from pemphigus vulgaris autoantibodies. *J Invest Dermatol*. 1986;87: 197–201.
34. Val-Bernal JF, Diego C, Rodriquez-Villar D, Garijo MF. The nipple-areola complex epidermis: a prospective systematic study in adult autopsies. *Am J Dermatopathol*. 2010;32:787–793.
35. Toker C. Clear cells of the nipple epidermis. *Cancer*. 1970;25:601–610.
36. Willman JH, Golitz LE, Fitzpatrick JE. Clear cells of Toker in accessory nipples. *J Cutan Pathol*. 2003;30:256–260.
37. Kumarasinghe SPW, Chin GY, Kumarasinghe MP. Clear cell papulosis of the skin: a case report from Singapore. *Arch Pathol Lab Med*. 2004;128:e149–e152.
38. Marucci G, Betts CM, Golouh R, Peterse JL, Foschini MP, Eusebi V. Toker cells are probably precursors of Paget cell carcinoma: a morphological and ultrastructural description. *Virchows Arch*. 2002;441(2):117–123.
39. Kuo TT, Chan HL, Hsueh S. Clear cell papulosis of the skin. A new entity with histogenetic implications for cutaneous Paget's disease. *Am J Surg Pathol*. 1987;11:827–834.
40. Tschen JA, McGavran MH, Kettler AH. Pagetoid dyskeratosis: a selective keratinocytic response. *J Am Acad Dermatol*. 1988;19:891–894.
41. Kim YC, Mehregan DA, Bang D. Clear cell papulosis: an immunohistochemical study to determine histogenesis. *J Cutan Pathol*. 2002;29:11–14.
42. Elias PM. Epidermal lipids, barrier function, and desquamation. *J Invest Dermatol*. 1983;80(suppl):44s–49s.
43. Zirra AM. The functional significance of the skin's stratum lucidum. *Morphol Embryol (Bucur)*. 1976;22:9–12.
44. Katz SI. The epidermal basement membrane zone—structure, ontogeny, and role in disease. *J Am Acad Dermatol*. 1984;11:1025–1037.

45. Foidart JM, Bere EW Jr, Yaar M, et al. Distribution and immunoelectron microscopic localization of laminin, a non-collagenous basement membrane glycoprotein. *Lab Invest.* 1980;42:336–342.
46. Smith JB, Taylor TB, Zone JJ. The site of blister formation in dermatitis herpetiformis is within the lamina lucida. *J Am Acad Dermatol.* 1992;27:209–213.
47. Leblond CP, Inoue S. Structure, composition, and assembly of basement membrane. *Am J Anat.* 1989;185:367–390.
48. Woodley DT, Burgeson RE, Lunstrum G, Bruckner-Tuderman L, Reese MJ, Briggaman RA. Epidermolysis bullosa acquisita antigen is the globular carboxyl terminus of type VII procollagen. *J Clin Invest.* 1988;81:683–687.
49. Shimizu H, McDonald JN, Gunner DB, et al. Epidermolysis bullosa acquisita antigen and the carboxy terminus of type VII collagen have a common immunolocalization to anchoring fibrils and lamina densa of basement membrane. *Br J Dermatol.* 1990;122:577–585.
50. Scott GA, Cassidy L, Tran H, Rao SK, Marinkovich MP. Melanocytes adhere to and synthesize laminin-5 in vitro. *Exp Dermatol.* 1999;8:212–221.
51. Tarnowski WM. Ultrastructure of the epidermal melanocyte dense plate. *J Invest Dermatol.* 1970;55:265–268.
52. Seiberg M. Keratinocyte–melanocyte interactions during melanosome transfer. *Pigment Cell Res.* 2001;14:236–242.
53. Barral DC, Seabra MC. The melanosome as a model to study organelle motility in mammals. *Pigment Cell Res.* 2004;17:111–118.
54. Murphy GF. Structure, function and reaction patterns. In: Murphy GF, ed. *Dermatopathology*. WB Saunders; 1995.
55. Glimcher ME, Kostick RM, Szabo G. The epidermal melanocyte system in newborn human skin. A quantitative histologic study. *J Invest Dermatol.* 1973;61(6):344–347.
56. Gown AM, Vogel AM, Hoak D, Gough F, McNutt MA. Monoclonal antibodies specific for melanocytic tumors distinguish subpopulations of melanocytes. *Am J Pathol.* 1986;123:195–203.
57. Kanitakis J. Immunohistochemistry of normal skin. In: Kanitakis J, Vassileva S, Woodley D, eds. *Diagnostic Immunohistochemistry of the Skin. An Illustrated Text*. Chapman & Hall Medical; 1998:38–51.
58. Birbeck NS, Breathnach AS, Everall JD. An electron microscope study of basal melanocytes and high-level clear cells (Langerhans cells) in vitiligo. *J Invest Dermatol.* 1961;37:51–64.
59. Niebauer G, Krawczyk W, Wilgram GF. The Langerhans cell organelle in Letterer Siwe's disease. *Arch Klin Exp Dermatol.* 1970;239:125–137.
60. Halata Z, Grim M, Baumann KI. The Merkel cell: morphology, developmental origin, function. *Cas Lek Cesk.* 2003;142:4–9.
61. Halata Z, Grim M, Bauman KI. Friedrich Sigmund Merkel and his "Merkel cell," morphology, development, and physiology: review and new results. *Anat Rec A Discov Mol Cell Evol Biol.* 2003;271:225–239.
62. Tachibana T, Nawa T. Recent progress in studies on Merkel cell biology. *Anat Sci Int.* 2002;77:26–33.
63. Santa Cruz DJ, Bauer EA. Merkel cells in the outer follicular sheath. *Ultrastruct Pathol.* 1982;3:59–63.
64. Camisa C, Weissmann A. Friedrich Sigmund Merkel. Part II. The cell. *Am J Dermatopathol.* 1982;4:527–535.
65. Gu J, Polak JM, Van Noorden S, Pearse AG, Marangos PJ, Azzopardi JG. Immunostaining of neuron-specific enolase as a diagnostic tool for Merkel cell tumors. *Cancer.* 1983;52:1039–1043.
66. Leff EL, Brooks JS, Trojanowski JQ. Expression of neurofilament and neuron-specific enolase in small cell tumors of skin using immunohistochemistry. *Cancer.* 1985;56:625–631.
67. Rosen ST, Gould VE, Salwen HR, et al. Establishment and characterization of a neuroendocrine skin carcinoma cell line. *Lab Invest.* 1987;56:302–312.
68. van Muijen GN, Ruiter DJ, Warnaar SO. Intermediate filaments in Merkel cell tumors. *Hum Pathol.* 1985;16:590–595.
69. Wang NP, Zee S, Zarbo RJ, Bacchi CE, Gown AM. Coordinate expression of cytokeratins 7 and 20 defines unique subsets of carcinomas. *Appl Immunohistochem.* 1995;3:99–107.
70. Moll I, Kuhn C, Moll R. Cytokeratin 20 is a general marker of cutaneous Merkel cells while certain neuronal proteins are absent. *J Invest Dermatol.* 1995;104:910–915.
71. Headington JT. Transverse microscopic anatomy of the human scalp. A basis for a morphometric approach to disorders of the hair follicle. *Arch Dermatol.* 1984;120:449–456.
72. de Viragh PA. The 'mantle hair of Pinkus.' A review on the occasion of its centennial. *Dermatology.* 1995;191:82–87.
73. Alonso LC, Rosenfield RL. Molecular genetics and endocrine mechanisms of hair growth. *Horm Res.* 2003;60:1–13.
74. Nakagawa H, Imokawa G. Characterization of melanogenesis in normal human epidermal melanocytes by chemical and ultrastructural analysis. *Pigment Cell Res.* 1996;9:175–178.
75. Jimbow K, Ishida O, Ito S, Hori Y, Witkop CJ Jr, King RA. Combined chemical and electron microscopic studies of pheomelanosomes in human red hair. *J Invest Dermatol.* 1983;81:506–511.
76. Biasiolo M, Bertazzo A, Costa CV, Allegri G. Correlation between tryptophan and hair pigmentation in human hair. *Adv Exp Med Biol.* 1999;467:653–657.
77. Burchill SA, Ito S, Thody AJ. Effects of melanocyte-stimulating hormone on tyrosinase expression and melanin synthesis in hair follicular melanocytes of the mouse. *J Endocrinol.* 1993;137:189–195.
78. Smith KR Jr. The Haarscheibe. *J Invest Dermatol.* 1977;69:68–74.
79. Niemann C, Horsley V. Development and homeostasis of the sebaceous gland. *Semin Cell Dev Biol.* 2012;23(8):928–936.
80. Zouboulis CC, Boschnakow A. Chronological ageing and photoageing of the human sebaceous gland. *Clin Exp Dermatol.* 2001;26(7):600–607.
81. Cui CY, Schlessinger D. Eccrine sweat gland development and sweat secretion. *Exp Dermatol.* 2015;24(9):644–650.
82. Baron DA, Briggman JV, Spicer SS. Tubulocisternal endoplasmic reticulum in human eccrine sweat glands. *Lab Invest.* 1984;51:233–243.
83. Sbarbati A, Osculati A, Morroni M, Carboni V, Cinti S. Electron spectroscopic imaging of secretory granules in human eccrine sweat glands. *Eur J Histochem.* 1994;38:327–330.
84. Baker LB. Physiology of sweat gland function: the roles of sweating and sweat composition in human health. *Temperature (Austin).* 2019;6(3):211–259.
85. Schaumburg-Lever G, Lever WF. Secretion from human apocrine glands: an electron microscopic study. *J Invest Dermatol.* 1975;64:38–41.

86. Sato K, Leidal R, Sato F. Morphology and development of an apoeccrine sweat gland in human axillae. *Am J Physiol*. 1987;252:R166–R180.
87. Kamada A, Saga K, Jimbow K. Apoeccrine sweat duct obstruction as a cause for Fox-Fordyce disease. *J Am Acad Dermatol*. 2003;48:453–455.
88. Reed RJ, Ackerman AB. Pathology of the adventitial dermis. Anatomic observations and biologic speculations. *Hum Pathol*. 1973;4:207–217.
89. Meigel WN, Gay S, Weber L. Dermal architecture and collagen type distribution. *Arch Dermatol Res (1975)*. 1977; 259:1–10.
90. Junqueira LC, Montes GS, Martins JE, Joazeiro PP. Dermal collagen distribution. A histochemical and ultrastructural study. *Histochemistry*. 1983;79:397–403.
91. Sorrell JM, Caplan AI. Fibroblast heterogeneity: more than skin deep. *J Cell Sci*. 2004;117(pt 5):667–675.
92. McNeal JE. Scleroderma and the structural basis of skin compliance. *Arch Dermatol*. 1973;107:699–705.
93. Nestle FO, Nickoloff BJ. A fresh morphological and functional look at dermal dendritic cells. *J Cutan Pathol*. 1995; 22:385–393.
94. Narvaez D, Kanitakis J, Faure M, Claudy A. Immunohistochemical study of CD34-positive dendritic cells of human dermis. *Am J Dermatopathol*. 1996;18:283–288.
95. Kanitakis J. Immunohistochemistry of normal human skin. *Eur J Dermatol*. 1998;8:539–547.
96. Nickoloff BJ, Griffiths CE. Not all spindled-shaped cells embedded in a collagenous stroma are fibroblasts: recognition of the “collagen-associated dendrophage”. *J Cutan Pathol*. 1990;17:252–254.
97. Headington JT. The dermal dendrocyte. *Adv Dermatol*. 1986;1:159–171.
98. Headington JT, Cerio R. Dendritic cells and the dermis: 1990. *Am J Dermatopathol*. 1990;12:217–220.
99. Sepulveda-Merrill C, Mayall S, Hamblin AS, Breathnach SM. Antigen-presenting capacity in normal human dermis is mainly subserved by CD1a+ cells. *Br J Dermatol*. 1994; 131:15–22.
100. Meunier L, Gonzalez-Ramos A, Cooper KD. Heterogeneous populations of class II MHC+ cells in human dermal cell suspensions. Identification of a small subset responsible for potent dermal antigen-presenting cell activity with features analogous to Langerhans cells. *J Immunol*. 1993;151: 4067–4080.
101. Walls AF, Jones DB, Williams JH, Church MK, Holgate ST. Immunohistochemical identification of mast cells in formaldehyde-fixed tissue using monoclonal antibodies specific for tryptase. *J Pathol*. 1990;162:119–126.
102. Arber DA, Tamayo R, Weiss LM. Paraffin section detection of the c-kit gene product (CD117) in human tissues: value in the diagnosis of mast cell disorders. *Hum Pathol*. 1998;29:498–504.
103. Metcalf DD, Akin C. Mastocytosis: molecular mechanisms and clinical disease heterogeneity. *Leuk Res*. 2001;25: 603–625.
104. Ryan TJ, Mortimer PS, Jones RL. Lymphatics of the skin. Neglected but important. *Int J Dermatol*. 1986;25:411–419.
105. Glatte P, Buchmann SJ, Hijazi MM, Illigens BM-W, Siepmann T. Architecture of the cutaneous autonomic nervous system. *Front Neurol*. 2019;10:970.
106. Johnson BL, Honig PJ, Jaworsky C, eds. *Pediatric Dermatopathology*. Butterworth-Heinemann; 1994.
107. Pochi PE, Strauss JS, Downing DT. Age-related changes in sebaceous gland activity. *J Invest Dermatol*. 1979;73:108–111.
108. Klaus S. Functional differentiation of white and brown adipocytes. *Bioessays*. 1997;19:215–223.
109. Klaus S. Adipose tissue as a regulator of energy balance. *Curr Drug Targets*. 2004;5:241–250.
110. Fenske NA, Lober CW. Structural and functional changes of normal aging skin. *J Am Acad Dermatol*. 1986;15(pt 1): 571–585.
111. Smith L. Histopathologic characteristics and ultrastructure of aging skin. *Cutis*. 1989;43:414–424.
112. Patterson JAK. Structural and physiologic changes in the skin with age. In: Patterson JAK, ed. *Aging and Clinical Practice: Skin Disorders, Diagnosis and Treatment*. Igaku-Shoin; 1989.
113. Kurban RS, Bhawan J. Histologic changes in skin associated with aging. *J Dermatol Surg Oncol*. 1990;16:908–914.
114. Montagna W, Carlisle K. Structural changes in ageing skin. *Br J Dermatol*. 1990;122(suppl 35):61–70.
115. Sauder DN. Effect of age on epidermal immune function. *Dermatol Clin*. 1986;4:447–454.
116. Lavker RM, Zheng PS, Dong G. Morphology of aged skin. *Clin Geriatr Med*. 1989;5:53–67.
117. Balin AK, Pratt LA. Physiological consequences of human skin aging. *Cutis*. 1989;43:431–436.
118. Bolognia JL. Aging skin. *Am J Med*. 1995;98:99S–103S.
119. Cerimele D, Celleno L, Serri F. Physiological changes in ageing skin. *Br J Dermatol*. 1990;122(Suppl 35):13–20.
120. Glover JD, Sudderick ZR, Shih BB, et al. The developmental basis of fingerprint pattern formation and variation. *Cell*. 2023;186(5):940–956.e20.
121. Rongioletti F, Rebora A. Fibroelastolytic patterns of intrinsic skin aging: pseudoxanthoma-elasticum-like papillary dermal elastolysis and white fibrous papulosis of the neck. *Dermatology*. 1995;191:19–24.
122. Mondragon G, Nygaard F. Routine and special procedures for processing biopsy specimens of lesions suspected to be malignant melanomas. *Am J Dermatopathol*. 1981; 3:265–272.
123. Plumb SJ, Argenyi ZB, Stone MS, De Young BR. Cytokeratin 5/6 immunostaining in cutaneous adnexal neoplasms and metastatic adenocarcinoma. *Am J Dermatopathol*. 2004; 26(6):447–451.
124. Tellechea O, Reis JP, Domingues JC, Baptista AP. Monoclonal antibody Ber EP4 distinguishes basal-cell carcinoma from squamous-cell carcinoma of the skin. *Am J Dermatopathol*. 1993;15(5):452–455.
125. Krahl D, Sellheyer K. Monoclonal antibody Ber-EP4 reliably discriminates between microcystic adnexal carcinoma and basal cell carcinoma. *J Cutan Pathol*. 2007;34(10):782–787.
126. Dotto JE, Glusac EJ. p63 is a useful marker for cutaneous spindle cell squamous cell carcinoma. *J Cutan Pathol*. 2006;33(6):413–417.
127. Alomari AK, Glusac EJ, McNiff JM. p40 is a more specific marker than p63 for cutaneous poorly differentiated squamous cell carcinoma. *J Cutan Pathol*. 2014;41(11):839–845.
128. Ostler DA, Prieto VG, Reed JA, Deavers MT, Lazar AJ, Ivan D. Adipophilin expression in sebaceous tumors and other cutaneous lesions with clear cell histology: an

immunohistochemical study of 117 cases. *Mod Pathol.* 2010;23(4):567–573.
129. Tjarks BJ, Pownell BR, Evans C, et al. Evaluation and comparison of staining patterns of factor XIIIa (AC-1A1), adipophilin and GATA3 in sebaceous neoplasia. *J Cutan Pathol.* 2018;45:1–7.
130. Schaumburg-Lever G, Metzler G, Kaiserling E. Ultrastructural localization of HMB-45 binding site. *J Cutan Pathol.* 1991;18(6):432–435.
131. Magro CM, Crowson AN, Mihm MC. Unusual variants of malignant melanoma. *Mod Pathol.* 2006;19:S41–S70.
132. Soyer HP. Ki 67 immunostaining in melanocytic skin tumors. Correlation with histologic parameters. *J Cutan Pathol.* 1991; 18(4):264–272.
133. Nasr MR, El-Zammar O. Comparison of pHH3, Ki-67, and survivin immunoreactivity in benign and malignant melanocytic lesions. *Am J Dermatopathol.* 2008;39(2):117–122.
134. Al Dhaybi R, Agoumi M, Gagné I, McCuaig C, Powell J, Kokta V. p16 expression: a marker of differentiation between childhood malignant melanomas and Spitz nevi. *J Am Acad Dermatol.* 2011;64(2):357–363.
135. Blokhin E, Pulitzer M, Busam KJ. Immunohistochemical expression of P16 in desmoplastic melanoma. *J Cutan Pathol.* 2013;40(9):796–800.
136. Mason A, Wititsuwannakul J, Klump VR, Lott J, Lazova R. Expression of p16 alone does not differentiate between Spitz nevi and Spitzoid melanomas. *J Cutan Pathol.* 2012; 39(12):1062–1074.
137. Karamchandani JR, Nielsen TO, van de Rijn M, West RB. Sox10 and S100 in the diagnosis of soft-tissue neoplasms. *Appl Immunohistochem Mol Morphol.* 2012;20(5):445–450.
138. Miettinen M, McCue PA, Sarlomo-Rikala M, et al. Sox10—a marker for not only Schwannian and melanocytic neoplasms but also myoepithelial cell tumors of soft tissue: a systematic analysis of 5134 tumors. *Am J of Surg Pathol.* 2015;39(6):826–835.
139. Helm K, Findeis-Hosey J. Immunohistochemistry of pigmented actinic keratoses, actinic keratoses, melanomas in situ and solar lentigines with Melan-A. *J Cutan Pathol.* 2008;35:931–934.
140. Turner N, Ko CJ, McNiff JM, Galan A. Pitfalls of PRAME immunohistochemistry in a large series of melanocytic and nonmelanocytic lesions with literature review. *Am J Dermatopathol.* 2024;46(1):21–30.
141. Lezcano C, Pulitzer M, Moy AP, Hollmann TJ, Jungbluth AA, Busam KJ. Immunohistochemistry for PRAME in the distinction of nodal nevi from metastatic melanoma. *Am J Surg Pathol.* 2020;44(4):503–508.
142. Cazzato G, Colagrande A, Ingravallo G, et al. PRAME immuno-expression in cutaneous sebaceous carcinoma: a single institutional experience. *J Clin Med.* 2022;11(23):6936.
143. Plotzke JM, Zoumberos NA, Hrycaj SM, Harms PW, Bresler SC, Chan MP. PRAME expression is similar in scar and desmoplastic melanoma. *J Cutan Pathol.* 2022;49(9):829–832.
144. Lezcano C, Müller AM, Frosina D, et al. Immunohistochemical detection of cancer-testis antigen PRAME. *Int J Surg Pathol.* 2021;29(8):826–835.
145. Cadwell CR, Yuksek GE, Hirbe AC, et al. Preferentially expressed antigen in melanoma (PRAME) expression in malignant, but not benign, peripheral nerve sheath tumors. *J Neuropathol Exp Neurol.* 2021;80(4):384–386.
146. Elsensohn A, Hanson J, Ferringer T. Preferentially expressed antigen in melanoma expression in nonmelanoma skin cancers and melanocytes in surrounding skin. *J Cutan Pathol.* 2021;48(9):1150–1155.
147. Xu B, Jungbluth AA, Frosina D, Alzumaili B, et al. The immune microenvironment and expression of PD-L1, PD-1, PRAME and MHC I in salivary duct carcinoma. *Histopathology.* 2019;75(5):672–682.
148. Donati, M, Martinek, P, Steiner, P, et al. Novel insights into the *BAP1*-inactivated melanocytic tumor. *Mod Pathol.* 2022;35:664–675.
149. Cheng TW, Ahern MC, Giubellino A. The spectrum of Spitz melanocytic lesions: from morphologic diagnosis to molecular classification. *Front Oncol.* 2022;12:88923.
150. Pantanowitz L, Pinkus GS, Dezube BJ, Tahan SR. HHV8 is not limited to Kaposi's sarcoma. *Mod Pathol.* 2005;18:1148–1150.
151. Patel RM, Goldblum JR, Hsi ED. Immunohistochemical detection of human herpes virus-8 latent nuclear antigen-1 is useful in the diagnosis of Kaposi sarcoma. *Mod Pathol.* 2004;17:456–460.
152. Altman DA, Nickoloff BJ, Fivenson DP. Differential expression of factor XIIIa and CD34 in cutaneous mesenchymal tumors. *J Cutan Pathol.* 1993;20:154–158.
153. Scholzen T, Gerdes J. The Ki-67 protein: from the known and the unknown. *J Cell Physiol.* 2000;182:311–322.
154. Li LX, Crotty KA, McCarthy SW, Palmer AA, Kril JJ. A zonal comparison of MIB1-Ki67 immunoreactivity in benign and malignant melanocytic lesions. *Am J Dermatopathol.* 2000;22:489–495.
155. Abdelsayed RA, Guijarro-Rojas M, Ibrahim NA, Sangueza OP. Immunohistochemical evaluation of basal cell carcinoma and trichoepithelioma using Bcl-2, Ki67, PCNA and P53. *J Cutan Pathol.* 2000;27:169–175.
156. Talghini S, Halimi M, Baybordi H. Expression of P27, Ki67 and P53 in squamous cell carcinoma, actinic keratosis and Bowen disease. *Pak J Biol Sci.* 2009;12(12):929–933.
157. Pavlidskey P, Seminario-Vidal L, McKay KM, Sami N. Spirochete immunostaining is not just for syphilis: diagnostic utility in borreliosis. *J Cutan Pathol.* 2015;42:370–372.
158. Chan MP, Andea AA, Harms PW, et al. Genomic copy number analysis of a spectrum of blue nevi identifies recurrent aberrations of entire chromosomal arms in melanoma ex blue nevus. *Mod Pathol.* 2016;29:227–239.
159. Miedema J, Andea AA. Through the looking glass and what you find there: making sense of comparative genomic hybridization and fluorescence in situ hybridization for melanoma diagnosis. *Mod Pathol.* 2020;33(7):1318–1330.
160. Hindi Z, Sidiropouos M, Al Habeeb A, et al. Fluorescence in situ hybridization (FISH) copy number abnormalities at 6p (REBI), 6q (MYB) and 11q (CCND1) reliably distinguish metastatic versus benign melanocytic lesions. *J Dermatol Res Ther.* 2016;2:017.
161. Gerami P, Li G, Pouryazdanparast P, et al. A highly specific and discriminatory FISH assay for distinguishing between benign and malignant melanocytic neoplasms. *Am J Surg Pathol.* 2012;36:808–817.
162. North JP, Garrido MC, Kolaitis NA, LeBoit PE, McCalmont TH, Bastian BC. Fluorescence in situ hybridization as

an ancillary tool in the diagnosis of ambiguous melanocytic neoplasms: a review of 804 cases. *Am J Surg Pathol.* 2014;38:824–831.

163. Yeh I, de la Fouchardiere A, Pissaloux D, et al. Clinical, histopathologic, and genomic features of Spitz tumors with ALK fusions. *Am J Surg Pathol.* 2015;39:581–591.
164. Gammon B, Beilfuss B, Guitart J, Gerami P. Enhanced detection of spitzoid melanomas using fluorescence in situ hybridization with 9p21 as an adjunctive probe. *Am J Surg Pathol.* 2012;36:81–88.
165. Nguyen TL, Theos A, Kelly DR, et al. Mitotically active proliferative nodule arising in a giant congenital melanocytic nevus: a diagnostic pitfall. *Am J Dermatopathol.* 2013;35:e16–e21.
166. Minoletti F, Miozzo M, Pedeutour F, et al. Involvement of chromosomes 17 and 22 in dermatofibrosarcoma protuberans. *Genes Chromosomes Cancer.* 1995;13(1):62–65.
167. Erickson-Johnson MR, Chou MM, Evers BR, et al. Nodular fasciitis: a novel model of transient neoplasia induced by MYH9-USP6 gene fusion. *Lab Invest.* 2011; 91(10):1427–1433.
168. Alessi E, Coggi A, Venegoni L, Merlo V, Gianotti R. The usefulness of clonality for the detection of cases clinically and/or histopathologically not recognized as cutaneous T-cell lymphoma. *Br J Dermatol.* 2005;153:368–371.
169. Boer A, Tirumalae R, Bresch M, Falk TM. Pseudoclonality in cutaneous pseudolymphomas: a pitfall in interpretation of rearrangement studies. *Br J Dermatol.* 2008;159:394–402.

2

The Normal Nail Unit

Curtis T. Thompson

Fingernails represent an important epithelial miniorgan system of the hand with a complex anatomical structure. The human nail represents an evolutionary advantage beyond claws and has two different functions: it forms a protective plate and enhances sensation of the fingertip. The protective function of the fingernail is commonly known, but sensation is equally important. The fingertip contains many nerve endings, allowing us to receive volumes of information about objects we touch. The nail acts as a counterforce to the fingertip providing even more sensory input when an object is touched. Sophisticated sensory detection of the paronychium allows for the identification of directional movement when it comes in contact with external items (1). The nail unit is also useful in hygiene and defense.

Although most pathologic specimens from the nail show well-known changes such as fungal infection, psoriasis, lichen planus, and characteristic malignant tumors, a broad spectrum of other diseases may also be found (2,3). For the pathologist, knowledge of normal histology and its more common variations is important when establishing a correct diagnosis. Unfortunately, much of the literature can be confusing because concepts and nomenclature have changed over the years, as has our understanding of embryology, physiology, genetics, and immunohistochemistry. The purpose of this chapter is to provide the general pathologist and dermatopathologist with a comprehensive discussion of the anatomy and unique histology of the nail area.

HISTORY

Historical interest in the nail structure can be traced to the works of Galen in the second-century BCE when he noticed the nail's resemblance to the hair shaft. In-depth evaluation of the nail unit did not begin, however, until the end of the 19th century with studies by Zander (4), Kolliker (5), and Unna (6). These early works focused on nail embryology and comparative anatomy of birds and other primates (4,7,8). In the early 20th century, researchers increased the understanding of nail development and anatomy, although technical limitations led them to incorrectly conclude that the nail plate was formed only by nail unit matrix independent of the nail bed and adjacent structures (9,10). It wasn't until the 1950s that Lewis (11) proposed his concept of a nail unit composed of dorsal, intermediate, and ventral nail plates that could be differentiated using a silver-proteinate stain. In 1963, Zaias (12) expanded the concept of an anatomically and histologically complex nail unit consisting of the proximal nail fold, nail matrix, nail bed, and hyponychium, all of which contribute to the formation of a functional nail.

Modern studies have focused on the molecular composition of the nail, especially different types of keratins, and on ultrastructural analysis. Characterization of the specialized dermis in the nail unit, termed "onychodermis," has been of particular interest, since the onychodermis has a distinct architecture and cell-type not seen in the dermis of the skin elsewhere (13). The emergence of clinical and surgical expertise specific to the nail unit has allowed for more comprehensive descriptions of a variety of neoplasms that are specific to the nail unit (3). Immunostaining has shown utility in the evaluation of these neoplasms, particularly nevi and melanoma (14). Molecular analysis using polymerase chain reaction (PCR) is also an important tool used to classify fungal infections (onychomycosis) of the nail unit, allowing for more sensitive and specific identification of tinea (dermatophytes), mold, and yeast infections (15). PCR tests are now able to identify nucleic acid mutations that promote terbinafine resistance, a primary medication used to treat onychomycosis. Thus, treatment failures with terbinafine may be better evaluated to guide further management (16).

EMBRYOLOGY

Our understanding of embryonic nail unit development initially relied on imaging modalities, particularly light and scanning electron microscopy. In recent decades, however, our understanding has advanced through the use of molecular studies focused on expression of different types of keratins, as well as larger sets of genes. Comparative studies among humans, primates, and other animals have further elucidated mechanisms underlying embryogenesis of nail unit, such that we now have a sophisticated understanding of its embryonic development and function after birth (17).

Nail unit development commences by day 14 of gestation when gastrulation leads to formation of three germ layers: the endoderm, mesoderm, and ectoderm. The mesoderm and ectoderm give rise to the limbs and, ultimately, the nail unit. A limb bud is recognizable by day 30. The mesoderm subsequently differentiates during days 14 to 56 such that the lateral plate mesoderm and ectoderm produce recognizable limbs and digits. The initial limb field forms when cells in the lateral plate mesoderm express specific combinations of Hox transcription factors that bind DNA and regulate mRNA transcription (18). Fore- and hindlimb (i.e., arm and leg) determination is controlled by T-box genes (19). Limb bud outgrowth results from a feedback loop mediated by fibroblast growth factor 8, which is expressed in the apical ectodermal ridge (20). This cluster of ectodermal cells proliferates to drive limb bud extension, the distal portion of which is termed the "autopod" and forms the hand, foot, and associated digits (21–28). Nail unit formation occurs on the dorsal surface of the distal tip of each digit. Localization to the dorsal surface is dependent upon Wnt7a expression in the dorsal ectoderm and development is driven by Msx1 and Msx2 homeobox genes (29). Msx1 maintains epithelial cells in an undifferentiated, proliferative state, while Msx2 expression leads to differentiation unique to the nail unit. Specific regulation of nail unit development in different animals (i.e., claw vs. nail) is not well understood, but some authors believe it may be in response to the phenotype of the digit, namely size and shape (30).

The differences in nail unit structure among primates and other animals largely result from regulatory mechanisms rather than sequence differences in genes important to embryogenesis; the genetic sequences governing nail development are largely conserved across different species. The regulatory mechanism surrounding these important genes is termed the "cis-regulatory hypothesis" (31,32). Cis-regulatory elements are present near important genes and function as promoters, enhancers, or silencers of transcription. In this way, fine-tuning gene function leads to differences in the size and shape of the nail that suit the needs of humans, nonhuman primates, and other animals.

Scanning electron microscopy has been used to evaluate embryonic development of the nail plate, although this tool is less useful when studying the mature nail because it is difficult to image the hard plate of a fully-formed nail unit (33–42). Embryonic development of the nail progresses through five stages: plate phase, fibrillar phase, granular phase, squamous phase, and definitive nail phase (34) (Table 2.1). The earliest recognizable fingers are seen in 42- to 45-day-old embryos (16-mm crown rump), while the toes are visible at 52 to 54 days of age (18.5-mm crown rump) (43).

The plaque phase occurs during the 7th to 10th weeks of gestation. It is characterized by the accumulation of proliferative cells that show abundant mitotic activity accompanied by apoptotic debris and T-lymphocytes in the primary nail base (Fig. 2.1). The combination of proliferation and apoptosis allows for architectural remodeling of the nail unit and invagination of the nail folds.

TABLE 2.1 Stages of Nail Development of the Embryo

Nail Unit[a]	Embryonic/Fetal Skin[b]	Development
Plaque phase	Earliest recognizable digits	Day 42+ gestation
Fibrillar phase	Primary nail field appears	Day 70+
Granular phase	Keratohyalin granules develop	Day 84+
Squamous phase	Distinct keratin plate identifiable	Day 120+
Definitive-nail phase	Matrix produces true nail plate	Day 150+

[a]Suchard R. Des modifications des cellules de la matrice et du lit de l'ongle dans quelques cas pathologiques. *Arch Physiol. (Paris)* 1882;2:445.

[b]Holbrook KA, Odland GF. The fine structure of the developing human epidermis: light, scanning, and transmission electron microscopy of the periderm. *J Invest Dermatol.* 1975;65: 16–38.

The fibrillar phase occurs at 70 to 84 days gestation and is notable for the large size of the primitive nail unit which occupies the distal one-third of the digit (Fig. 2.1B). At this point, the primitive plate is attached to surrounding tissue with peri–ungual-fixing filaments. The nail unit first appears as a smooth, shiny quadrangular surface delineated by continuous shallow grooves, and corresponds to the "primary nail base" by Zander (4) as well as the "primary nail field" by Zaias (12). Primitive epithelium in the primary nail field covers the dorsal distal digit and contains two clusters of epithelial cells, one at each end. The proximal cluster of epithelial cells proliferates backward and downward at an angle into the subjacent mesenchymal tissue, which is highly cellular and contains abundant ground substance. The superficial part of this wedge of epithelial cells forms the nail matrix and proximal nail groove, the latter of which ultimately becomes the proximal nail fold. The deeper part will eventually become the dorsal and intermediate nail matrices. There is a crease of cornified cells at the junction between the superficial and deeper parts that will become the cuticle of the fully developed nail. The cluster of epithelial cells in the distal primary nail field forms the transverse distal fold which ultimately becomes the hyponychium. By day 77, the proximal and lateral nail folds have also formed.

During the granular phase (85 to 120 days), the nail unit has identifiable nail folds and is surfaced by a primitive epithelium with granular-appearing keratohyalin vesicles (Fig. 2.1C). The epithelium has a basal layer and three to

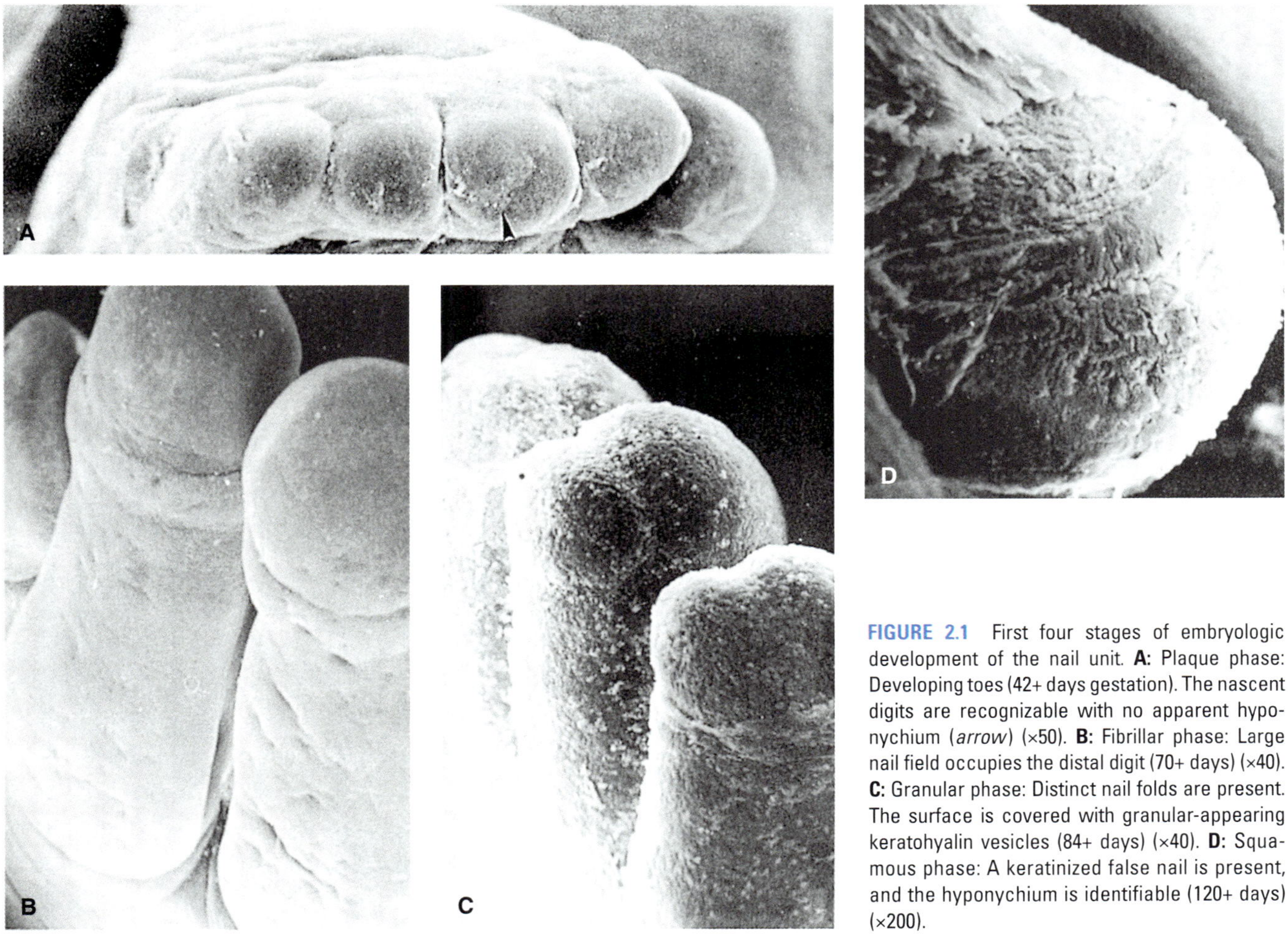

FIGURE 2.1 First four stages of embryologic development of the nail unit. **A:** Plaque phase: Developing toes (42+ days gestation). The nascent digits are recognizable with no apparent hyponychium (*arrow*) (×50). **B:** Fibrillar phase: Large nail field occupies the distal digit (70+ days) (×40). **C:** Granular phase: Distinct nail folds are present. The surface is covered with granular-appearing keratohyalin vesicles (84+ days) (×40). **D:** Squamous phase: A keratinized false nail is present, and the hyponychium is identifiable (120+ days) (×200).

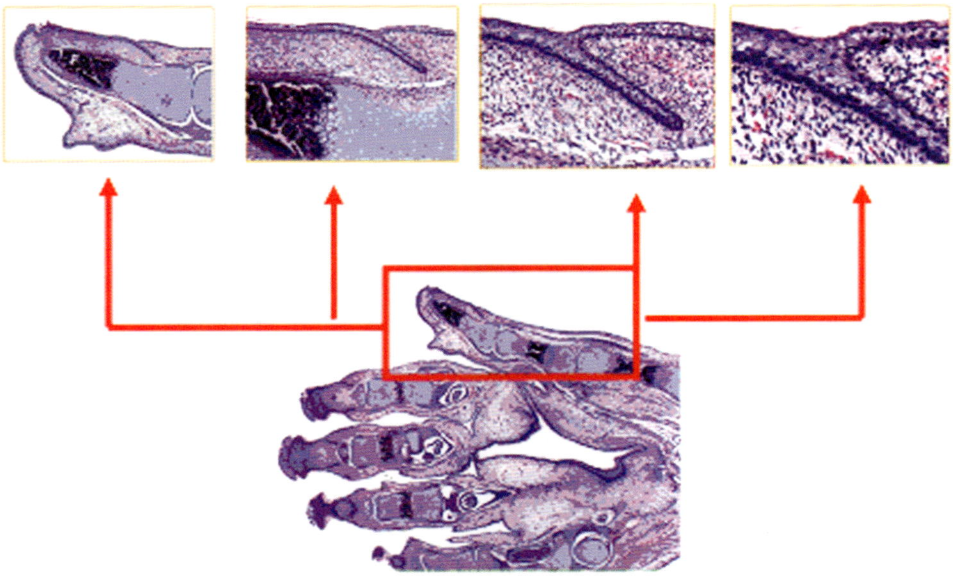

FIGURE 2.2 Sagittal section of a fetal hand and digit (~112 days gestation). Matrix epithelium is easily identifiable and extends into the onychodermis. Partially calcified bone and cartilage is present.

four superficial cell layers. These layers are composed of primitive keratinocytes with clear cytoplasm, a thin, acellular layer at the surface, and a well-formed matrix that projects into the nascent onychodermis (Fig. 2.2). Keratinization is observed in the transverse distal fold at 3.5 months of gestation (11,12,33). Indeed, this focus of keratinization is the first site of keratinization in the entire developing embryo (44).

The squamous phase features formation of a keratin plate that covers the entire distal digit. This early plate has been termed the "false nail," since true nail plate production has not yet begun (Fig. 2.1D) (11,12). The definitive nail phase begins with the production of true nail plate, which is characterized by the presence of sulfhydryl radicals and cornified cells. The true nail plate gradually appears proximal-to-distal and extends to the hyponychium at birth, after which clipping of the distal plate is required (42,45).

STRUCTURES OF THE NAIL UNIT

Pathologic assessment of normal digits is most easily performed in the autopsy suite since digits and nail unit specimens received as pathology specimens are usually altered by disease. The primary structures of the nail unit include: (1) nail plate; (2) lunula; (3) nail matrix; (4) proximal nail fold (including cuticle); (5) lateral nail folds; (6) nail bed; and (7) hyponychium (Fig. 2.3). The proximal nail fold has a hyperkeratotic, acellular cuticle that is closely opposed to the proximal nail plate (Fig. 2.4). The nail plate is generated by matrix epithelium, and it emerges below the proximal nail fold with its acellular, hyperkeratotic cuticle. The distal plate is tightly adherent to the subjacent nail bed epithelium, which moves distally with the plate. The onychodermis extends beneath the entire length of the nail matrix and nail bed. The onychodermis contains specialized fibroblasts that have been of great interest in recent years, details of which will be offered later in this chapter. Finally, the distal interphalangeal bone, though not considered to be a component of the nail unit, appears to be important in helping maintain a healthy nail unit with continuous nail plate production.

Nail Plate

The nail plate is flat or slightly convex from the proximal to distal end (Fig. 2.5), and there is significant convexity

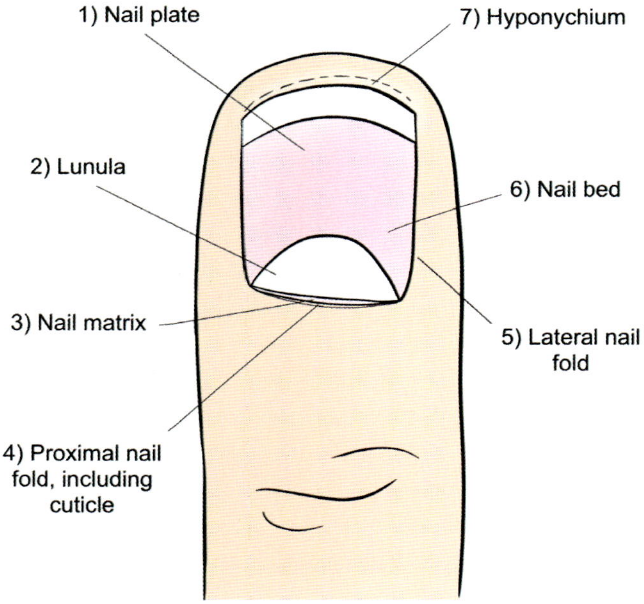

FIGURE 2.3 Grossly visible structures of the nail unit. (1) Nail plate; (2) lunula; (3) nail matrix; (4) proximal nail fold, including cuticle; (5) lateral nail folds; (6) nail bed; and (7) hyponychium.

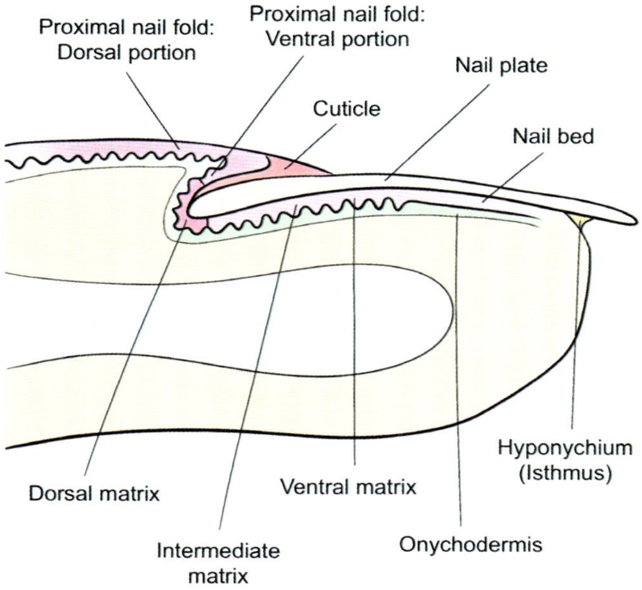

FIGURE 2.4 Cross-sectional structures of the nail unit. The close relationship of the proximal nail fold with its hyperkeratotic, acellular cuticle to the nail matrix, nail bed, hyponychium, and subjacent onychodermis is evident. The plate is attached proximally to the nail fold generating nail matrix epithelium and the distal nail bed to which it is tightly attached. The subjacent onychodermis extends below the nail matrix and nail bed. The distal interphalangeal bone, though not considered to be a component of the nail unit is positioned closely to the overlying nail unit.

from medial-to-lateral as it dives into the lateral nail folds (Fig. 2.6). The nail plate is overall rectangular in shape, and its size varies with each digit, ranging from 1 × 1 cm to 2 × 3 cm in the adult depending on the size of the digit. The proximal plate is hidden under the proximal nail fold with its cuticle. The visible plate occupies 25% to 50% of the dorsal surface of each finger distal to the dorsal interphalangeal joint, whereas the plate occupies approximately 75% of the dorsal surface of the distal big toe. The nail plate is approximately 0.5 mm thick in women and 0.6 mm thick in men. Its thickness varies with age. It becomes thicker with increasing maturity, but often thins with advanced age. Age-related nail thinning may become problematic because the nail plate can be damaged more easily.

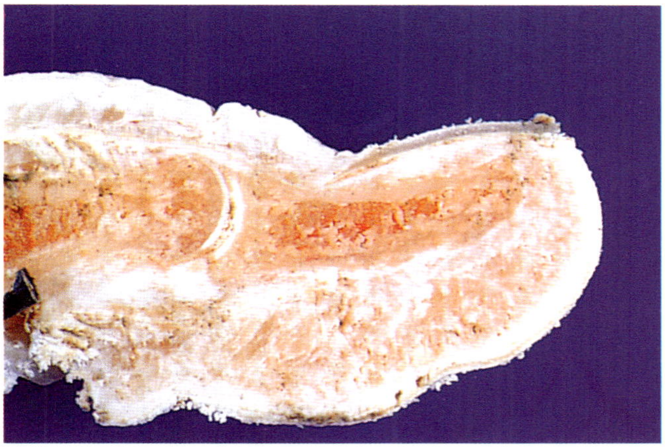

FIGURE 2.5 Gross sagittal section of an adult distal thumb. The slightly darker-colored nail plate is convex. The two vertical *white lines* represent the interphalangeal joint. The proximity of the distal interphalangeal bone to the structure of the nail unit is evident.

FIGURE 2.6 Gross cross-section of an adult distal finger. The slightly green, very convex nail plate dives into both lateral nail fold sulci. The lateral nail folds obscure the lateral nail plate.

Nail color varies with skin type, reflecting the density of melanocytes present in the nail unit. A healthy nail is translucent, and much of the color comes from the subjacent vascular network in the onychodermis. Changes in the translucency generally reflect a disease involving the nail unit, which could be dermatitis, infection, or neoplasm. Under normal circumstances, the nail plate is mostly resistant to ultraviolet light exposure; it blocks UV-B entirely and only permits a small amount of UV-A. As a result, the nail bed does not develop solar elastotic fibers with chronic sun exposure, a feature seen elsewhere on the skin. This characteristic also makes it difficult to treat diseases of the nail using UV light. Psoralen-UV-A light treatment (PUVA) for psoriasis, for example, is largely ineffective due to the minimal penetration of UV light (46).

The nail plate is composed of a dorsal (surface) layer, intermediate layer, and ventral layer (37–40,47). The dorsal and ventral layers are structurally similar and, thus, will be described together. The intermediate layer is essentially the nail matrix and will be described subsequently.

The dorsal and ventral layers consist of closely packed, interdigitating corneocytes lacking nuclei and organelles (Fig. 2.7). The corneocytes are composed mostly of hard α-keratin (80%) and soft α-keratin (20%). They have tight intercellular desmosomal junctions and their size increases with age (35). In the dorsal layer, the corneocytes are oriented in a slanted fashion that imparts a smooth, shiny surface to the nail. The ventral layer attaches tightly to the nail bed epithelium via a series of striking longitudinal ridges and furrows on the deep nail plate surface. The α-keratin in

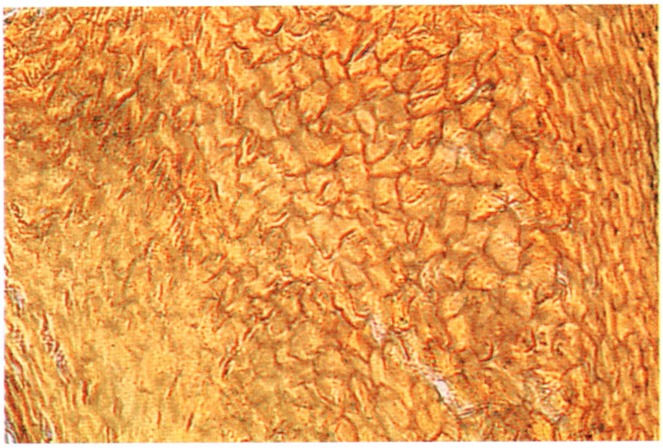

FIGURE 2.7 Dorsal surface of the nail plate. The anucleate corneocytes are polyhedral (VVG stain, ×200).

the dorsal layer is arranged in a haphazard, overlapping pattern that creates a hard, rigid layer, and is similarly oriented in the ventral layer (Fig. 2.8).

The dorsal and ventral layers also contain lipid bilayers (49-Å thick) oriented parallel to the nail surface. These lipid bilayers normally fill the ampullar dilations of the dorsal plate and intercellular spaces in the ventral plate, and their loss may underlie age-related thinning of the plate (48). The lipid bilayers also play an important role in preventing water loss through the nail plate. Water normally comprises

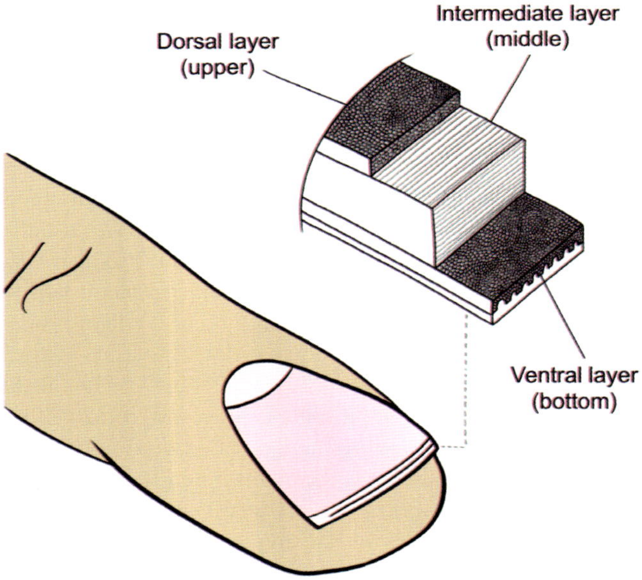

FIGURE 2.8 Schematic showing the three layers of the nail plate: dorsal, intermediate, and ventral. The three layers, though quite similar in composition, are generated from different regions of the nail matrix epithelium and have different functions. The arrangement of the keratin in each layer is also shown, with the dorsal and ventral layers having a cobblestone appearance. The ventral layers interdigitate with the nail plate with parallel, longitudinal ridges. The keratin filaments in the intermediate layer are parallel to the direction of nail plate growth.

18% of the nail plate but, like other charged molecules, can move easily through the nail plate (49–51). Decreased lipid in the nail can lead to excess water loss, termed "transonychial water loss" (TOWL), and produce brittle nails (52). Loss of lipid and water can also contribute to pathologic states, such as psoriasis and onychomycosis, wherein the nail plate is altered and usually acanthotic (53,54).

The intermediate layer comprises two-thirds of the nail plate mass. Unlike the dorsal and ventral layers, the intermediate layer contains corneocytes with nuclear remnants. Importantly, the α-keratin fibers (81 Å in diameter) in the intermediate layer are oriented transversely to the direction of nail growth, thereby providing significant strength to the nail plate rather than the hard, rigid texture present in the other two layers (Fig. 2.8). This arrangement of the α-keratin fibers produces a plate that is much more difficult to cut transversely than longitudinally. Of note, the nail plate shape changes when force is applied during grasping and similar activities. It is likely that this change in shape makes the nail function more powerfully, however, the biomechanics of this temporary structural change are not fully understood (50,55).

The biochemical composition of the nail plate has been extensively studied. Calcium is an important component of the dorsal and ventral layers of the nail plate (56). Calcium comprises only 0.1% of the weight of the nail plate, but its concentration is 10 times greater than that of a hair shaft (57). Some authors have suggested that calcium does not contribute significantly to nail plate hardness. However, it is mostly present in the dorsal and ventral layers, which are the hardest of the plate, and decalcification of the nail during histologic processing makes the tissue easier to manage (58,59). Other metals such as copper, manganese, zinc, and iron are also found in small quantities in the nail plate (59). Like calcium, these elements could be absorbed from the blood, or introduced exogenously through soaps, moisturizers, chemicals, and other exposures. Some diseases are associated with increased organic and metallic components of the nail plate. The nail contains increased urea nitrogen and uric acid in patients with gout (60); creatinine accumulates in chronic renal failure (61); patients with cystic fibrosis have increased sodium content (62); calcium increases with advanced age (57,63); copper accumulates in patients with Wilson disease (64); environmental exposures can lead to increased arsenic content (65); and medications and their metabolites, such as morphine, 6-acetylmorphine, and cocaine, can be detected in the nails of patients taking these drugs (66). Detection of these agents is a feature useful in forensic analysis.

Lunula

The lunula appears as a white crescent in the proximal portion of the nail plate (Fig. 2.9). It represents the most distal portion of the nail matrix and is composed of the keratinocytes that produce the nail plate. The size of the lunula correlates with the amount of nail matrix present in the nail, most of which is not visible grossly (67). The lunula produces the deepest portion of the nail plate and is most visible on the thumbs. The

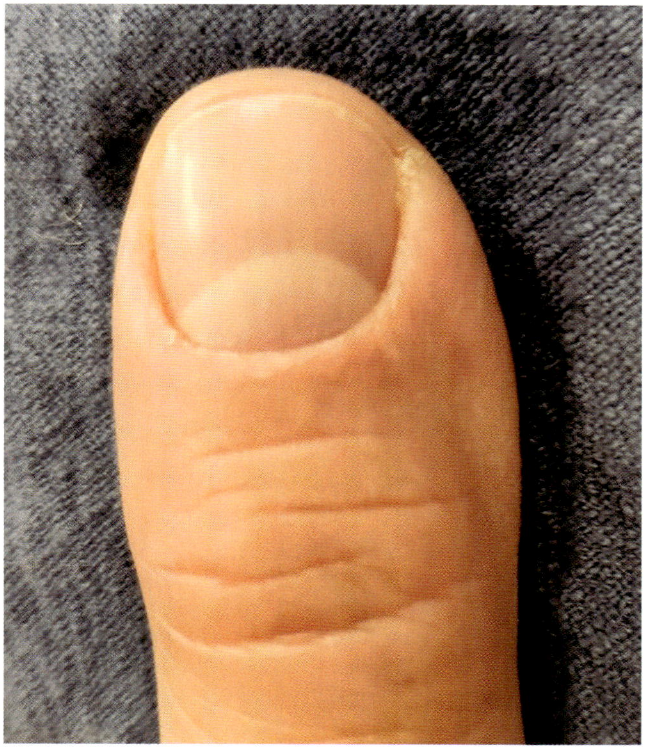

FIGURE 2.9 Human thumb showing a prominent lunula in the proximal portion. The distal edge of the lunula is the residence of the stems cells for the nail epithelium; they produce epithelial cells that attach to the underside of the nail plate and travel distally with the plate as it grows.

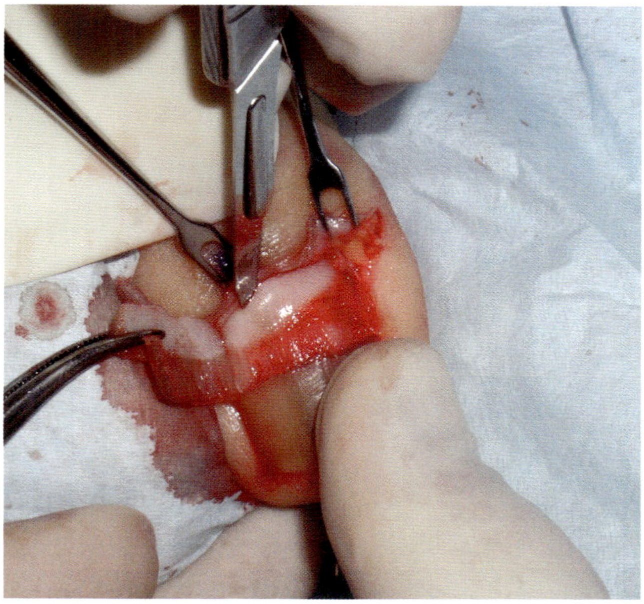

FIGURE 2.10 Nail surgery showing reflection of the proximal nail fold. The white nail matrix extends proximally very near the distal interphalangeal joint. (Photo provided courtesy of Dr. Nathaniel J. Jelinek.)

lunula may not be visible when the proximal nail fold completely covers the distal matrix. The color of the lunula is due, in part, to the effect of light scattered by the nucleated cells within the keratogenous zone of the matrix, as well as the presence of a thick layer of epithelial cells in the matrix (68,69). The nail plate over the lunula is thinner than the distal nail plate, since the region is still undergoing nail plate formation (70). Some authors have speculated that the white hue reflects fewer vessels within the subjacent onychodermis as well as a different collagen bundle arrangement (71–73). The distal edge of the lunula contains the stem cells for the nail bed. As these stem cells divide, the cells attached to the underside of the nail plate move distally with the plate (Fig. 2.9).

Nail Matrix

The nail matrix, including the lunula, is mostly obscured by the proximal nail fold. Thus, it is important to remember that the nail unit extends proximally under the proximal nail fold to within a few millimeters of the distal interphalangeal joint (Fig. 2.10). Because surgeons sometimes lack this understanding, nail unit excisions may be incomplete, leaving behind residual nail matrix that continues to produce an irregular nail plate and necessitating an additional surgery to achieve a complete excision. The close proximity of the distal interphalangeal joint to the nail unit may also lead to leakage of synovial fluid from the joint into the nail unit, producing a digital myxoid/mucous cyst (Fig. 2.11). Synovial fluid can accumulate in any part of the nail unit, though it is usually limited to the interstitium.

Like the nail plate, the nail matrix has been divided into three parts: ventral, intermediate, and dorsal (11,12). The ventral, or deepest portion of the nail matrix, produces the ventral nail plate, and its epithelium merges with that of the proximal nail bed. The intermediate portion of the nail matrix produces the bulk of the nail plate, and the dorsal portion contributes to the shiny, visible smooth surface of the nail plate. For simplicity, others have limited the division of the nail matrix to proximal and distal areas (74).

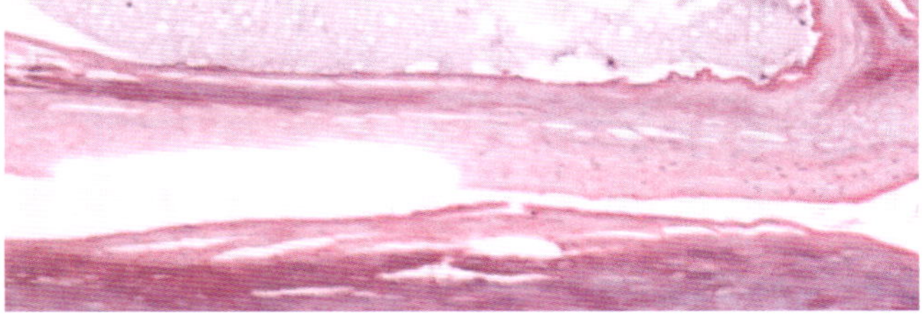

FIGURE 2.11 Digital myxoid/mucous cyst. Purple-staining mucinous synovial fluid has leaked into the nail plate (H&E, ×40).

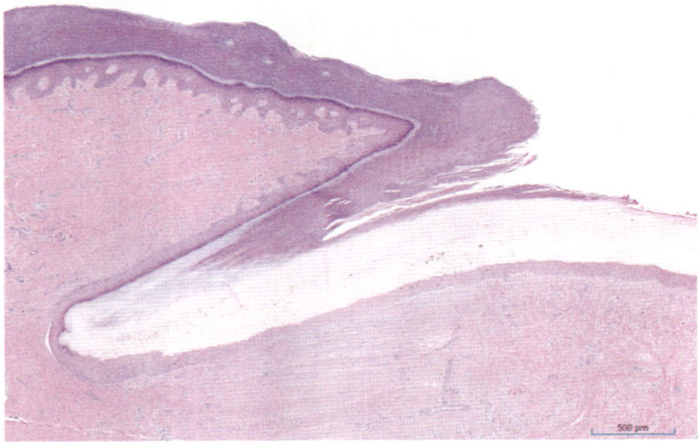

FIGURE 2.12 Proximal nail unit. The proximal nail fold, including the cuticle, and the nail matrix are evident (H&E, ×20).

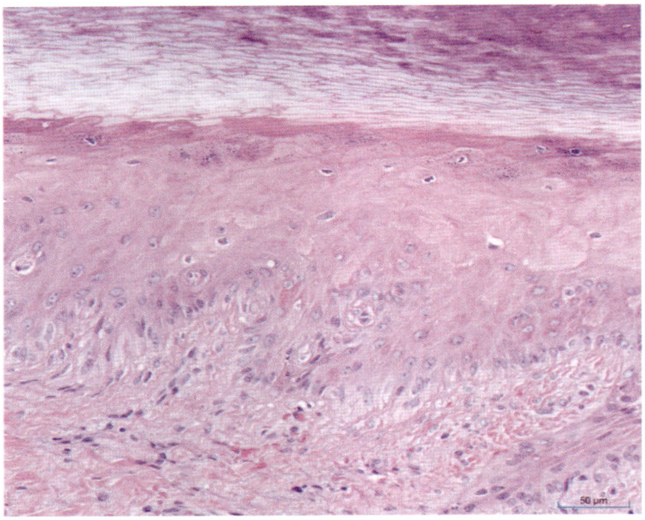

FIGURE 2.13 Nail matrix epithelium. The undulating basalis shows abrupt keratinization forming the nascent nail plate (H&E, ×100).

The nail matrix located immediately below the proximal nail fold is a slightly acanthotic squamous epithelium with a variable thickness, measuring 80 to 150 μm (Fig. 2.12). The epithelium contains interspersed melanocytes, Langerhans cells, and Merkel cells. It has an undulating basilar layer and abrupt keratinization forming the nascent nail plate (Fig. 2.13). Epithelial cells in the basilar layer are mitotically active along the entire dorsal, proximal, and ventral portions. The proliferative rate in this region is approximately 20%, compared to only 1% in the remainder of the nail bed (75). The basilar cells attach to the onychodermis with finger-like elongations that interdigitate between the micropapillae of the papillary dermis (37–40). Suprabasilar epithelial cells are also mitotically active with a proximal-to-distal orientation parallel to the direction of nail plate growth. These cells have abundant desmosomal attachments to each other, conferring early strength in the nascent nail plate (Fig. 2.14). As the nail matrix epithelial cells keratinize and terminally differentiate into flat corneocytes, the

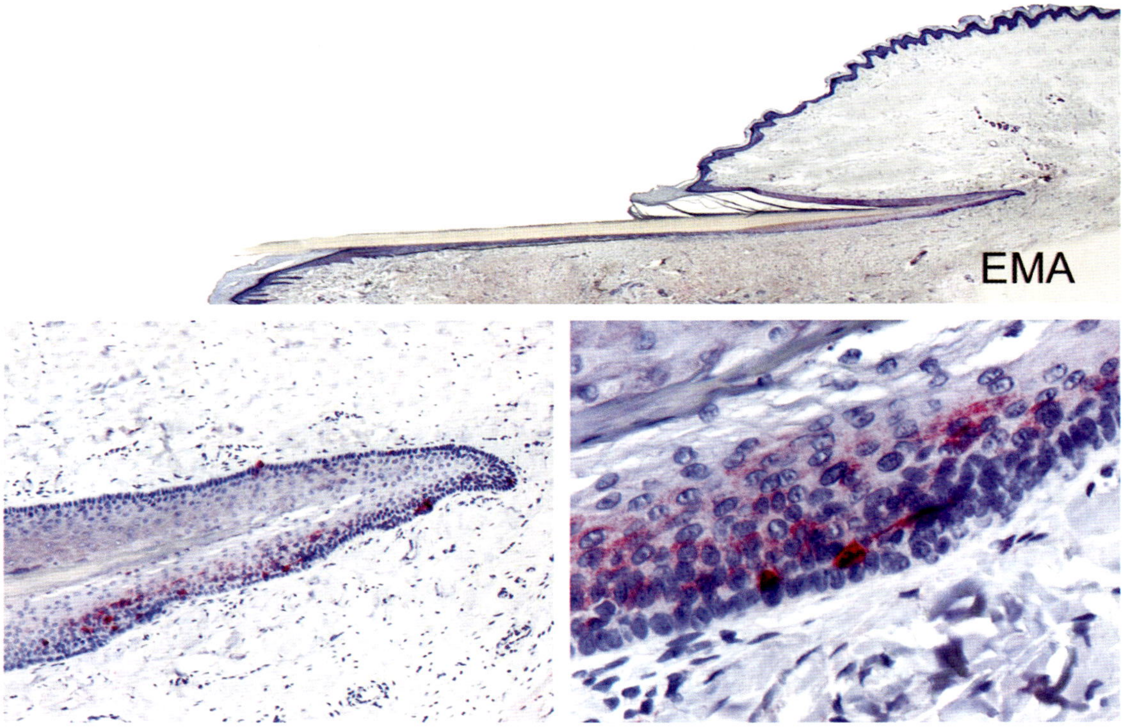

FIGURE 2.14 Suprabasilar nail matrix. Abundant desmosomal junctions are present in the suprabasal layer of the nail matrix epithelium. These desmosomal junctions are bigger and more abundant than in the epidermis elsewhere in the skin (EM, ×12,000).

nuclei and organelles disappear, and tonofibrils fill the terminally differentiated nail matrix cells. The entire process of cornification occurs in only three to four cellular layers. Expression of integrin a2b1 and integrin a3b1 promotes strong cohesion within the nail matrix (76).

Proximal and Lateral Nail Folds

The nail plate is adjacent to one proximal and two lateral nail folds (Fig. 2.3). The proximal nail fold covers 25% of the nail plate and serves to protect the subjacent nail matrix and nascent proximal nail plate. It is contiguous with the epidermis of the dorsal digit. Of note, the terminal tendon of the digital extensor is immediately subjacent to the proximal nail fold and can be easily injured during surgery or as a result of trauma (77). As the dorsal digit epidermis merges with the dorsal component of the proximal nail fold epithelium, hair follicles, and other adnexal structures are notably absent (Fig. 2.15).

The distal edge of the proximal nail fold is slightly hyperkeratotic, forming the cuticle. The cuticle seals and protects the blind space under the proximal nail fold, thereby preventing contamination by external agents. Manipulation of the cuticle disrupts this seal and can lead to inflammation and infection (paronychia). The proximal nail fold epithelium invaginates into the ventral component of the proximal nail fold, and its granular layer disappears as it merges into the nail matrix epithelium (Fig. 2.16) (12,69,78). This transition zone between the proximal nail fold and the nail matrix epithelium is important to the creation of the smooth, shiny surface of the dorsal nail plate (68).

The lateral nail folds do not contribute to nail plate formation, but they do promote the even distal movement of the nail plate. The lateral nail folds also represent a transition zone from the nail bed epithelium to the cutaneous epithelium of the digit. Like the proximal nail fold, the lateral nail folds are hairless and devoid of a dermatoglyphic pattern (fingerprints). The epithelium of the proximal portion of the lateral nail fold merges with both the proximal nail fold epithelium and the matrix epithelium. Interestingly, the lateral nail folds contain a specialized sensory ability allowing the identification of directional movement of external objects. This fine sensory ability allows for more precise manipulation of tools and small objects (1). Not surprisingly, this area contains abundant Meissner and Merkel–Ranvier corpuscles that are not present in the nail matrix or bed (79).

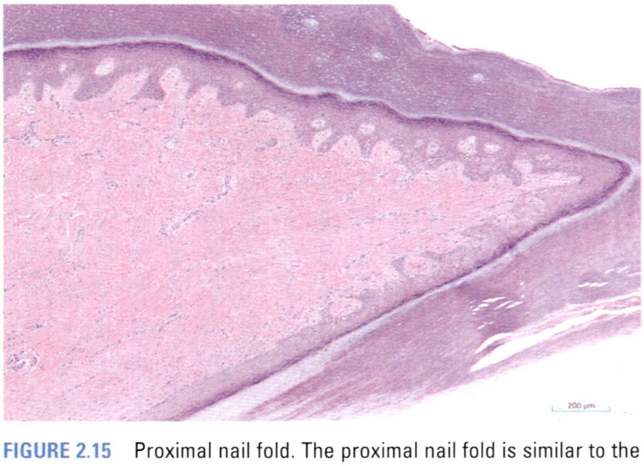

FIGURE 2.15 Proximal nail fold. The proximal nail fold is similar to the epidermis of the proximal digit, though it lacks hair follicles (H&E, ×100).

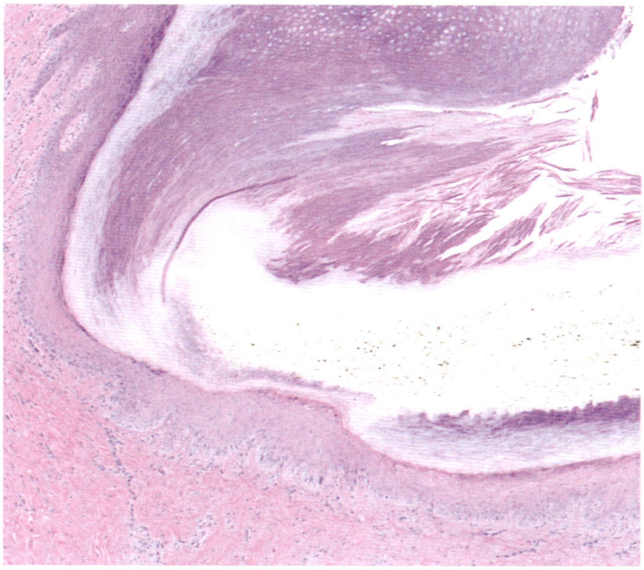

FIGURE 2.16 Transition of the ventral portion of the proximal nail fold into the nail matrix epithelium. The cuticle of the proximal nail fold is evident (H&E, ×50).

Nail Bed

The nail bed is a flat epithelium with small keratinocytes (Fig. 2.17). As expected, the proliferative rate of nail bed keratinocytes is significantly lower than that of the more

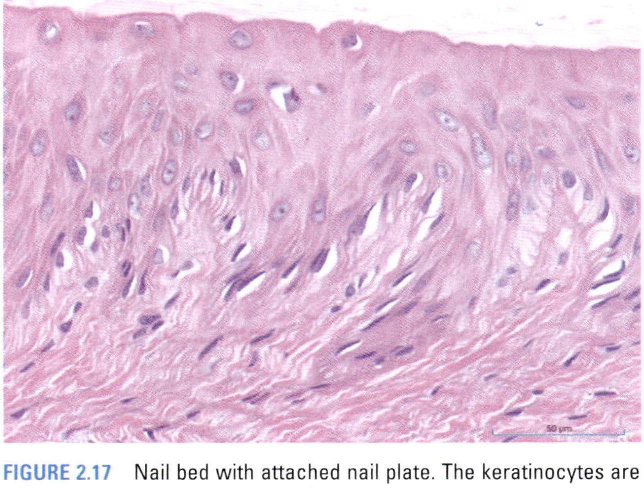

FIGURE 2.17 Nail bed with attached nail plate. The keratinocytes are small and monotonous (H&E, ×400).

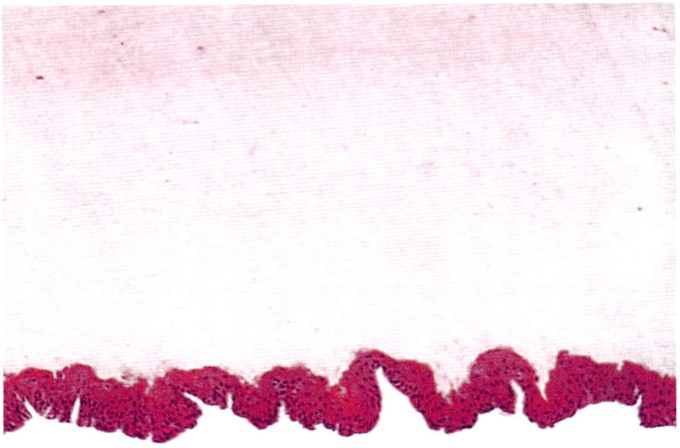

FIGURE 2.18 Junction between nail bed epithelium and nail plate. Note the serrated surface of the nail bed epithelium which contacts the nail plate.

proximal nail plate-producing matrix cells (42). The attachment to the subjacent onychodermis is organized with a series of unique, longitudinal, parallel nail bed ridges placed in a tongue-and-groove fashion (Fig. 2.8). These longitudinal furrows may be visible immediately after avulsion of the nail plate. They can also be appreciated microscopically, where serrated interdigitations of the ventral surface of the nail bed papillae and rete are evident (Fig. 2.18) (80). Fine venules of the papillary onychodermis run within these onychodermal ridges, disruption of which produces splinter hemorrhages.

The nail bed begins proximally where the crescent-shaped lunula ends (Fig. 2.3). Nail bed epithelial cells continually divide to produce the nail bed epithelium along the distal edge of the lunula. Just distal to the lunula, the nascent nail bed epithelium attaches to the deep surface of the nail plate with a thin parakeratotic layer. Human carcinoembryonic antigen (CEA) and CEA-related molecules has been implicated in the adhesion of the nail plate to the nail bed (81). A CEA family antigen, Nonspecific crossreactinve antigen (NCA) (CD with 66c)-like immunoreactivity, is strongly expressed in this transition zone. NCA expression is not identified in the surrounding nail matrix, the hyponychium, or the lateral nail folds. The nail bed epithelium moves distally at the same rate as the nail plate. Though there have been attempts to characterize the mechanism for the concurrent movement of the nail bed epithelium with the nail plate, the molecular details of the movement are not yet understood (70,82).

The onychodermal band is a 0.5 to 1.5 mm transverse band in the distal nail bed that is best visualized upon compression of the distal digit or in low oxygen states when there is acrocyanosis (Fig. 2.19) (58). Some have speculated that these situations facilitate visualization of the onychodermal band because the distal nail bed and proximal nail bed have different vascular supplies (83). The hue of the onychodermal band may be altered in various diseases (78). Most commonly, the proximal nail bed assumes a frosted appearance

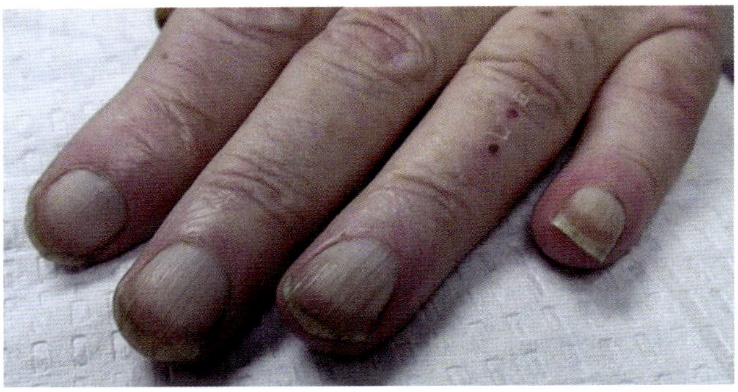

FIGURE 2.19 Nails. Note the prominent onychodermal band in each distal nail unit. (Photograph courtesy of Dr. Martin Zaiac.)

(i.e., leukonychia), and the distal onychodermal band develops a prominent reddish hue (84). This clinical finding has been termed "Terry nails" and develops in the setting of diabetes mellitus, chronic renal insufficiency, and cirrhosis.

Some diseases cause the nail bed abnormalities. Psoriasis, pachyonychia congenita, and the benign neoplasm, onychopapilloma, can cause the nail bed to become acanthotic with a granular layer (85). Chronic lung diseases, lichen planus, and other forms of dermatitis may produce acanthotic yellowing of all the nails, termed "twenty-nail dystrophy." Markedly acanthotic and hyperkeratotic nails, termed "onychogryphosis" or "ram's horn nail" can be due to a variety of congenital and acquired diseases. These nails have a claw-like appearance and often limit the day-to-day use of the fingers (86).

Hyponychium

The hyponychium is the hyperkeratotic zone where the distal nail plate detaches from the nail bed epithelium. The portion of the hyponychium closest to the nail bed has been termed the "nail isthmus." This area contains parakeratotic corneocytes that provide a protective seal for the nail bed, thereby preventing invasion from fungus and

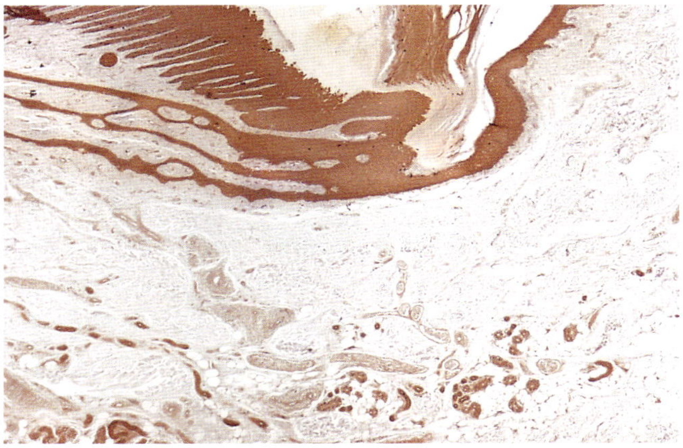

FIGURE 2.20 Distal edge of nail plate. Showing the hyperkeratotic hyponychium subjacent to the free, distal edge of the nail plate (cytokeratin immunostain, ×100).

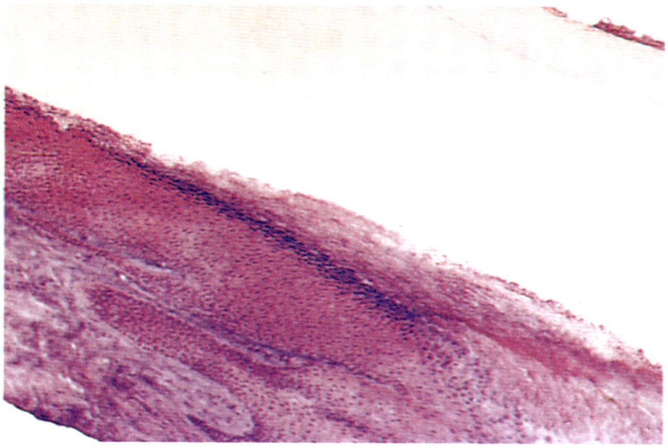

FIGURE 2.21 Hyponychium. The hyponychium characteristically shows marked hypergranulosis (H&E, ×100).

other infectious agents (Fig. 2.20) (87,88). The epithelium immediately distal to detachment is analogous to acral epithelium; it contains eccrine glands, but follicular structures are absent. There is a prominent stratum granulosum with hyperkeratosis, similar to the cuticle of the proximal nail fold (Fig. 2.21). Like the lateral nail folds, the hyponychium contains abundant Meissner and Merkel–Ranvier corpuscles (79) that provide sensation and allow for precise motor movements such as the manipulation of fine tools.

A variety of diseases can alter the hyponychium and subsequent growth of the nail plate. Acral-like epithelium of the hyponychium may grow proximally into the nail bed region when the nail plate becomes detached from the nail bed as a result of trauma, neoplasia, or infection. Termed "pterygium inversum unguis" or "disappearing nail bed," this ingrowth may prevent nail plate reattachment such that it is no longer possible to grow a normal-appearing nail plate (89). The hyponychium is also an entry point for fungus into the nail unit (88). This infection, termed "onychomycosis," produces subungual hyperkeratosis in the hyponychium. Subungual hyperkeratosis can also occur in patients with psoriasis, contact dermatitis, and congenital diseases, such as pachyonychia congenital (78). Subungual material is also sometimes sampled in forensic autopsies since exogenous DNA from an attacker may be identified after the nails are used in self-defense (90).

OTHER TISSUES OF THE NAIL UNIT

Onychodermis

The onychodermis is the specialized dermis of the nail unit and has unique immunohistochemical and histologic features that are not present elsewhere in the skin (91). Although the nail unit does not have a defined subcutis, scattered adipocytes are usually present on histologic sections. The onychodermis subjacent to the nail matrix and bed contains a unique population of fibroblasts that express CD10 and CD13 (i.e., onychofibroblasts) (92,93). The significance of the onychofibroblast is not well understood. However, studies of cultured cells have shown that their presence is required for nail matrix cells to produce hard keratin (94). Interestingly, onychofibroblasts also appear to be important in promoting nail unit regeneration after injury to the distal digit. Onychomatricoma is a benign tumor unique the nail unit that contains CD10+ onychofibroblasts (95).

The basement membrane zone of the onychodermis expresses all the same antigens found elsewhere on the skin, including epidermal-associated antigens, such as the 220- and 180-kDa bullous pemphigoid antigens and a-6 b-4 integrin, and lamina lucida antigens LH39, GB3, and laminin (96). Deeper dermal components are the same as those of the basement membrane elsewhere on the skin; the 285-kDa linear immunoglobulin A (IgA) antigen, extracellular matrix glycoproteins, chondroitin sulfate, type VII collagen and its associated proteins, and antigens for LH24 and LH39 have all been identified (97). The collagen bundles in the onychodermis are dense and show strong collagen-IV immunostaining (Fig. 2.22). They are vertically oriented in the proximal nail bed but slanted at a 45-degree angle in the distal nail bed, thereby providing a strong attachment to the underlying periosteum of the distal phalanx.

Vascular Supply

The nail unit has an arterial blood supply derived from paired digital arteries. The main supply passes into the pulp space of the distal phalanx before reaching the dorsum of the digit (98,99). An accessory supply arises further back on the digit and does not enter the pulp space. The digital arterial system is important for the surgeon to understand, and has three major characteristics: (a) arched anastomotic arteries in the deep dermis; (b) superficial terminal arteries supplying the rete (100); and (c) overall tortuosity of arteries subjacent to the nail apparatus. Systemic oxygen levels are routinely measured from the distal digit arterial blood supply using microelectrodes (101).

The vasculature of the nail bed is unique because it must supply a vascular structure within a limited space between the hard nail plate and bone. The onychodermis contains a generous vascular supply with a high density of thin-walled venules and capillaries as well as thick-walled arteries with longitudinal and circular smooth muscle layers (Fig. 2.23). The venules at the proximal nail fold are particularly tortuous and prominent (Fig. 2.24). It is important to avoid interpreting this normal, high density of vessels as a vascular neoplasm, such as a hemangioma or granulation tissue (i.e., ulcerated pyogenic granuloma). Scanning electron microscopy studies reveal special vascular patterns of microcirculation in the nail unit (102). The vasculature of the eponychium and perionychium follows the direction of nail growth, although there is a wide mesh net of capillaries where the eponychium contacts the nail plate. The vessels of the nail bed are arranged in parallel, longitudinal trabeculae that run throughout its length.

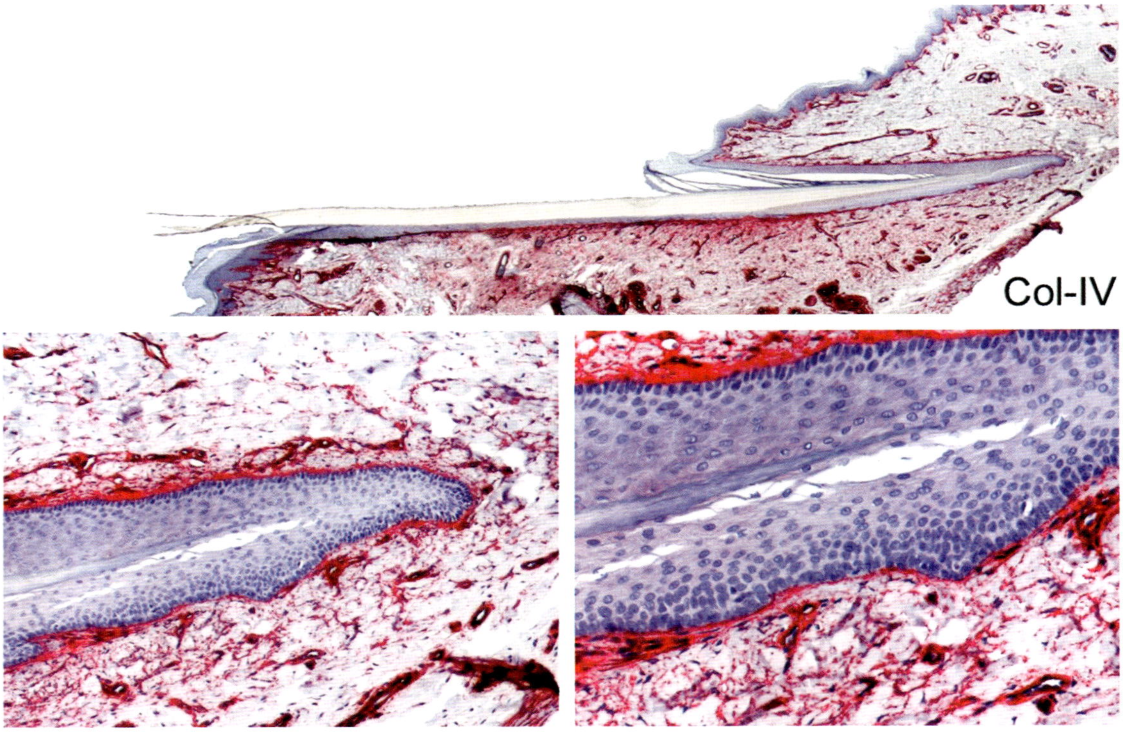

FIGURE 2.22 Collagen IV immunostain. Collagen IV staining is present along the basement membrane of the nail unit, especially in the nail matrix (higher magnification).

The venous drainage of the nail unit occurs through two veins, one on each side of the nail plate in the proximal nail fold. It is essentially the same as the network of the skin but the capillary loops are more horizontal and visible throughout their length. The capillary network of the proximal nail fold is usually examined with dermoscopy, which can visualize abnormalities such as tortuosity of the vessels occurring with collagen vascular diseases (Fig. 2.25) (103).

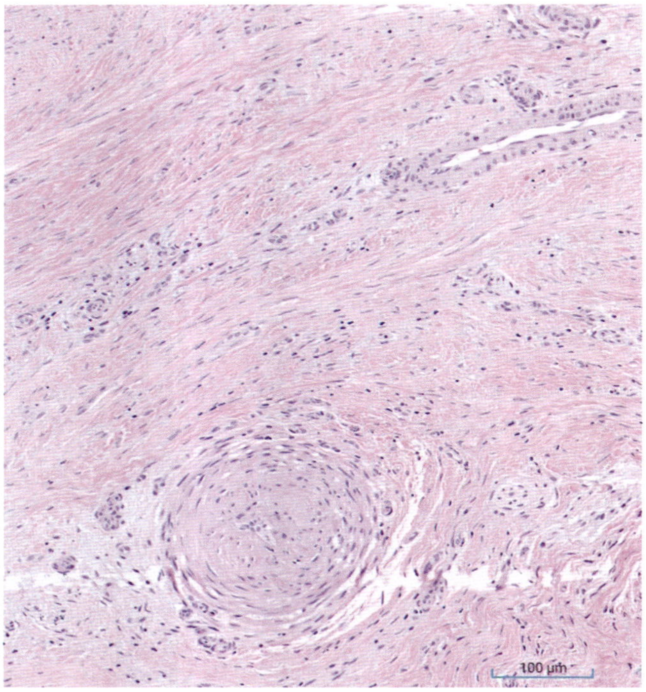

FIGURE 2.23 Onychodermis. The onychodermis contains numerous, variably sized blood vessels. It is important not to over diagnose the normal dermal vasculature as a hemangioma (H&E, ×100).

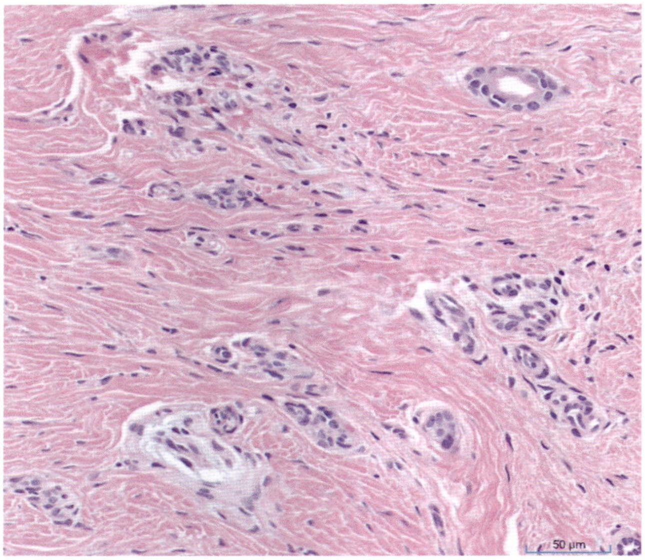

FIGURE 2.24 Onychodermis. Tortuous venules in the onychodermis of the proximal nail fold (H&E, ×200).

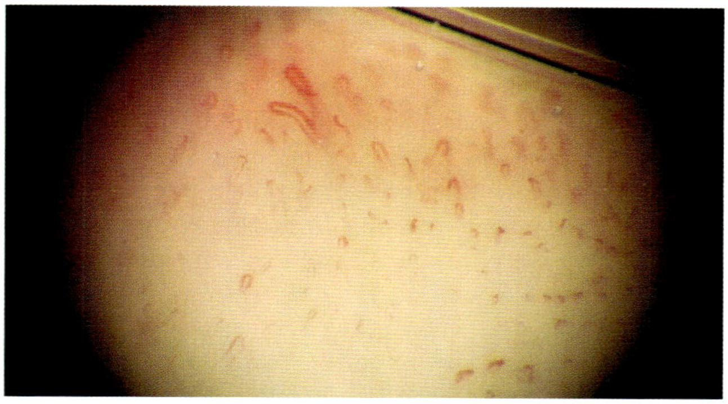

FIGURE 2.25 Dermoscopic image of the proximal nail fold. The vessels are tortuous, which is typical of a collagen vascular disease. (Photo courtesy of Dr. Antonella Tosti.)

Nerves

As previously noted, Vater–Pacini and Meissner corpuscles are present in large numbers within the nail folds, where they enhance sensation and improve fine manipulation of objects. However, these are absent in the nail matrix and bed. Importantly, peripheral nerves travel along the lateral aspects of the digits. These nerves can be treated with local anesthesia prior to nail unit surgery with excellent effect. Of note, in embryonic studies, nerve regrowth is not needed for digit-tip regrowth (104).

Bone

Though the deeper distal phalange is not considered to be part of the nail unit, its close proximity to the nail creates a situation in which the nail unit and bone rely on each other for structure and growth. For example, postamputation regrowth of the nail unit will not occur without the presence of subjacent bone (105). Because of this proximity, tumors originating from bone and cartilage often impinge on the nail unit, producing an altered, dystrophic nail (Fig. 2.26). It is therefore important to rule out the presence of a deeper mass when evaluating new alterations to the nail unit.

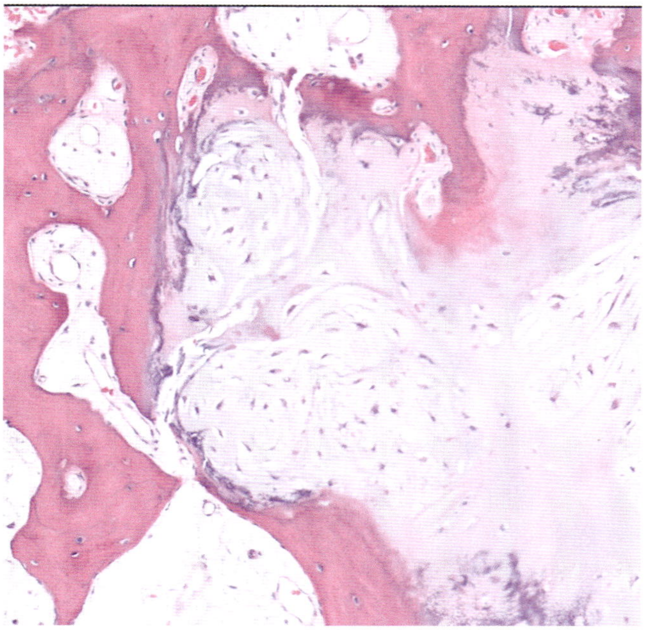

FIGURE 2.26 Enchondroma. An enchondroma of the distal digit features woven bone and cartilage differentiation (H&E, ×400).

SPECIALIZED CELLS OF THE NAIL UNIT

Melanocytes

Melanocytes in the nail matrix produce pigment for the nail plate that protects it from ultraviolet rays (106). The density of the resident melanocytes increases from proximal to distal in the nail matrix (107,108) along the epithelial basalis, and is directly proportional to the Fitzpatrick skin phototype (Fig. 2.27). Darker phototypes have larger and more abundant melanosomes in the melanocytes (109). Caucasians with very light skin and a Fitzpatrick phototype I have nail matrices containing only sparse, poorly developed melanocytes that are usually inapparent on H&E sections (110). The melanocytes are scattered irregularly along the epithelial basalis with occasional small clusters of melanocytes in the suprabasilar region (107,111). This irregular placement may account for the longitudinal streaks seen in the nail plates of individuals with very dark phototypes (106).

Resident and neoplastic melanocytes of the nail unit are dendritic in nature, similar to those seen in the bulb of the hair follicle, on the lips, and in the genital region (Fig. 2.28). The dendritic processes tend to be longer and thicker in individuals with darker phototypes (44).

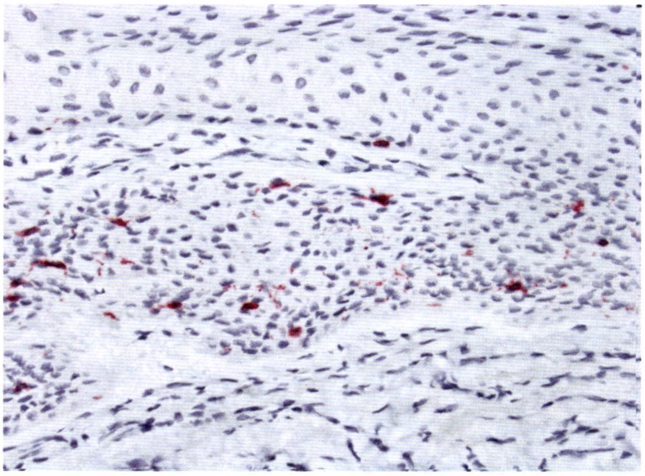

FIGURE 2.27 Melanocytes in the nail unit. Resident melanocytes are scattered along the basalis of the nail matrix epithelium. The density of the melanocytes is directly proportional to the Fitzpatrick skin type (MelanA immunostain, ×200).

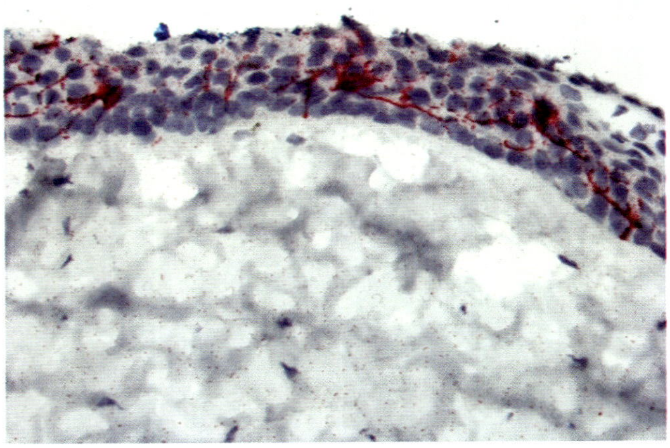

FIGURE 2.28 Dendritic melanocytes in the nail bed epithelium of melanoma in situ. The normal resident melanocytes and most in-situ neoplastic melanocytes of the nail unit are dendritic in nature, similar to those seen in the bulb/matrix of the hair follicle (MelanA immunostain, ×400).

Melanocytes of the nail unit are best visualized using cytoplasmic immunostains such as MelanA/Mart-1 because they stain the dendritic processes of melanocytes (Fig. 2.28). Nuclear markers such as SOX-10 or MITB may fail to detect sparse, irregularly distributed melanocytes when the nuclei are not present in the plane of section examined. Immunostains for S100-protein and HMB-45 will also highlight the resident melanocytes of the nail unit.

Merkel Cells

The concentration and distribution of Merkel cells varies throughout life. In adults, Merkel cells are essentially absent from the nail matrix and bed, but found in high concentrations (50 cells/mm^2) in the proximal nail folds (112). However, Merkel cells are more numerous and widely distributed in the nail unit during embryonic development. They are identifiable by day 60 of gestation but begin to disappear by day 84. The embryonic function of these Merkel cells is not well understood (113). Similar to Merkel cells in the skin, they express cytokeratins 18 and 20.

IMMUNOLOGY OF THE NAIL UNIT

The nail apparatus is constantly exposed to environmental damage. It requires effective immune responses to combat infection, while avoiding the loss of nail production and regeneration by an overly active immune system. Most of the data suggest that the immune activity varies in different regions of the nail unit. Localized suppression of the immune system in the proximal nail matrix helps prevent induction of autoimmune phenomena, while a pro-inflammatory state in the proximal nail fold wards off infection (114).

An overall environment of immune tolerance exists in the proximal nail matrix. Expression of HLA-A, HLA-B, and HLA-C is suppressed in keratinocytes and resident melanocytes, while expression of macrophage migration inhibitory factor and HLA-G is increased; both of these molecules serve to prevent natural killer cells and CD8-positive T-cells from reacting to major histocompatibility complex (MHC) molecules (115). Subpopulations of CD1a-positive, CD4-positive, and CD8-positive immune cells and natural killer cells are decreased in the proximal nail matrix compared with other parts of the nail unit, and resident Langerhans cells show lower expression of class II MHC and CD209 molecules. On the other hand, immunosuppressant molecules such as transforming growth factor-b1, a-melanocyte stimulating hormone, insulin-like growth factor-1, and adrenocorticotropic hormone are increased in this region.

In contrast to the proximal nail matrix, the proximal nail fold shows increased protective immunity against infectious agents. There is a high density of CD4-positive T-cells, both in the epithelium and in the subjacent dermis. Antimicrobial peptides, mainly cathelicidin LL-37, have also been identified in the proximal nail fold. These peptides have activity against relevant nail pathogens without requiring involvement of the cellular immune system (116). Thus, the portion of the nail unit that is most exposed to the environment balances relative immune privilege with increased immune protection against infectious agents.

PROTEIN EXPRESSION IN THE NAIL UNIT

The expression of keratin genes in the nail unit is complex and has been extensively studied in both the developing embryonic nail unit and after birth. The details of keratin expression in this region are complicated, but molecular and immunohistochemical mapping of keratin expression provides the foundation for a deeper understanding of nail unit formation and maintenance (94,117–122).

Keratins comprise a large family of intermediate filament proteins encoded by more than 50 distinct genes (123). Approximately 50% of keratin proteins are hard-keratins, or trichocytes, that form the nail plate and hair shaft. The remainder are soft keratins that are produced by epithelial cells present in the skin, squamous mucosa, and nail unit (118,119,124,125). Both keratin groups have been further divided into type I and type II proteins, based on their size, charge, and amino acid sequences (126). Both keratin type I and II proteins are important to nail unit integrity. Type I keratin genes are grouped in one genomic domain on chromosome 17q21.2 and type II are found on chromosome 12q13.3 (117,123,126,127). Their expression profiles differ significantly (128,129).

Nail plate hardness is due to the presence of globular, sulfur-rich matrix proteins and alpha-fibrillar hard keratins that are low in sulfur but rich in glycine and tyrosine, similar to the follicular hair shaft (130). Importantly, 90% of the total nail plate mass is composed of hard keratins and keratin-associated proteins, and only 10% is from keratins produced by soft keratins from the skin-type epithelial cells (125). Though a large set of keratins contribute to the nail plate, most of the hard keratin comes from a set of proteins named type I Ha keratin (117). Shotgun proteomic analysis of the nail plate has identified 144 proteins with more than 300 constituents of the isopeptide cross-linked proteome. These include a number of keratins, as well as cytoplasmic membrane proteins, junctional proteins, and histones. Crosslinking of proteins with transglutaminases contributes to the hard protein product (127).

The transition from distal nail bed to hyponychium is of particular interest regarding keratin (K) expression. Expression of K1 and K10 is minimal or absent in the matrix and nail bed but is detected just distal to the nail bed in the hyponychium where a stratum granulosum begins to appear (117,120,129). The nail bed and more distal hyponychium can further be distinguished by differences in expression of keratins K5, K6, and K16 (Fig. 2.29) (120). K6 and K16 are expressed in the suprabasilar eponychium of the proximal nail fold, the nail bed, and the apical portion of the nail matrix, but not the ventral portion (129). K6hf expression is exclusive to the nail bed. Variable expression of other proteins exists throughout the nail unit. For example, pancornulin, a component of the stratum corneum of the epidermis elsewhere, is expressed in the proximal and lateral nail folds and in the proximal nail matrix. Sciellin, another component of the stratum, is present in all areas of the nail bed (131). Trichohyalin, in contrast, is only present in the ventral matrix (132).

Keratin provides significant structural support in the entire epidermis. Mutations of different keratin genes can produce widespread cutaneous alterations including nail fragility and hyperkeratosis (133–135). It is not precisely known how mutations in keratins lead to hyperkeratosis, but fragility of the nail bed epithelium may lead to inflammation and a subsequent hyperproliferative state of nail unit epithelium. The type I keratin gene, *K17*, is important in embryogenesis, nail maturation, and maintenance of nail integrity; mutations are responsible for some types of ectodermal dysplasia (136,137).

Further characterization of the nail unit has been aided by immunohistochemical characterization of the differential expression of keratin proteins. As expected, a pan-cytokeratin cocktail, such as AE-1/AE-3 highlights the epithelium of the entire nail unit (Fig. 2.30). On the other hand, an antibody to CK5–8 shows expression limited to the nail matrix and bed (Fig. 2.31). Very little CK15, CK19, and CK20 is identified in the nail matrix. Cytokeratin 15-positive stem cells are found in the proximal nail matrix and the proximal nail fold. For unknown reasons, CK-KL1 is seen in the dorsal nail matrix but not in the ventral matrix (Fig. 2.32). Finally, claudin-1 (CLDN1), a major component of tight junctions that contributes to nail plate hardness, is strongly expressed in the nail matrix (Fig. 2.33).

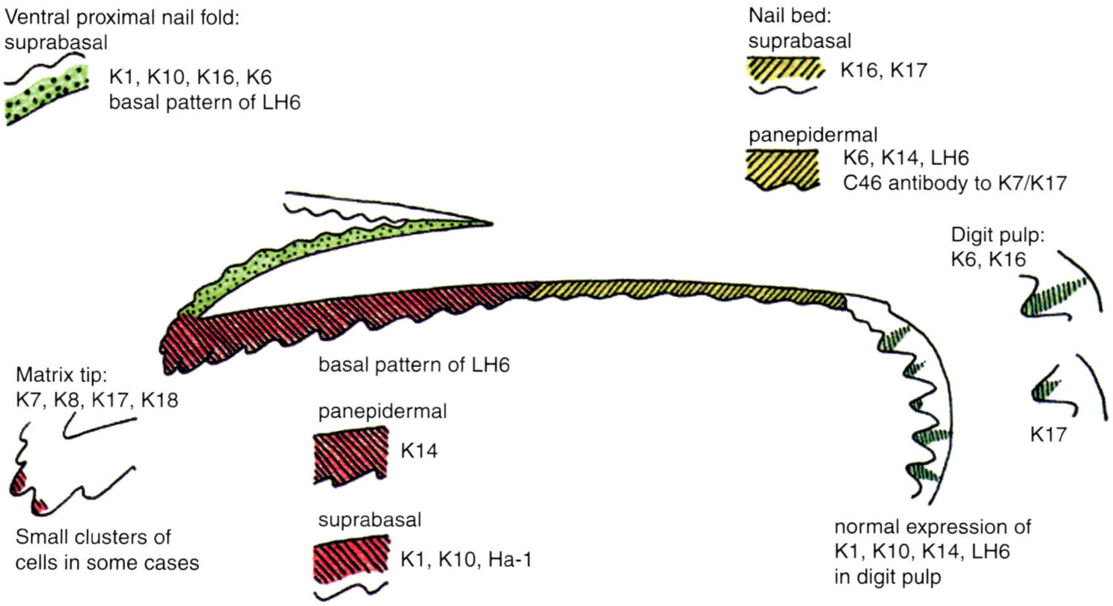

FIGURE 2.29 Keratin expression in the normal nail unit: Markers of regional differentiation. (From De Berker D, Wojnarowska F, Sviland L, et al. Keratin expression in the normal nail unit: markers of regional differentiation. *Br J Dermatol.* 2000;142:89–96.)

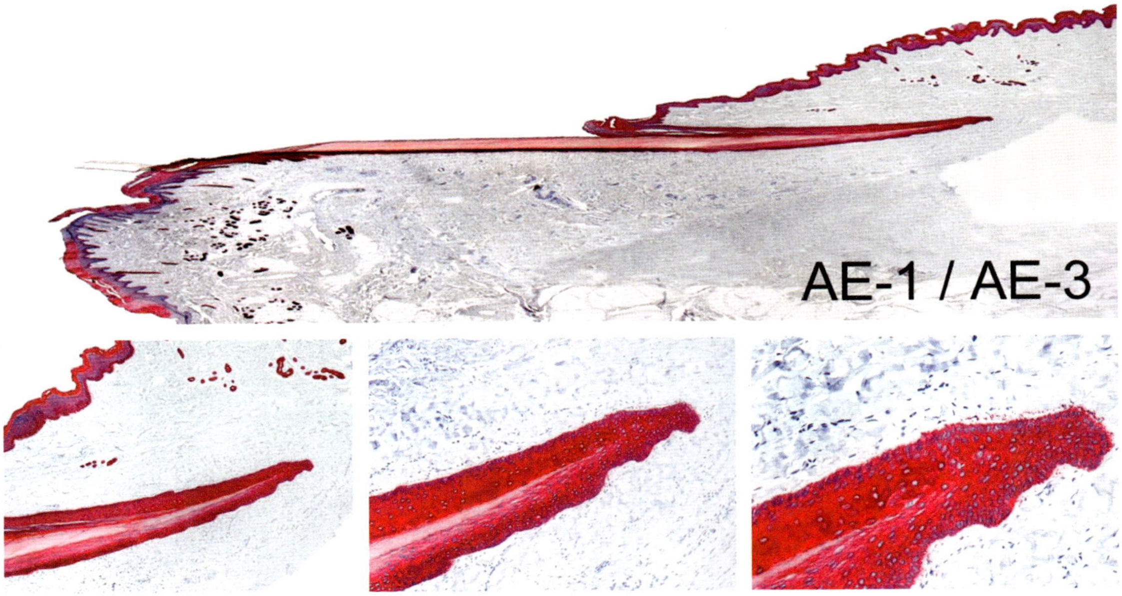

FIGURE 2.30 Keratin immunostain of nail unit. Immunostaining with pan-cytokeratin AE1/AE3 antibody highlights the epithelium of all portions of the nail unit.

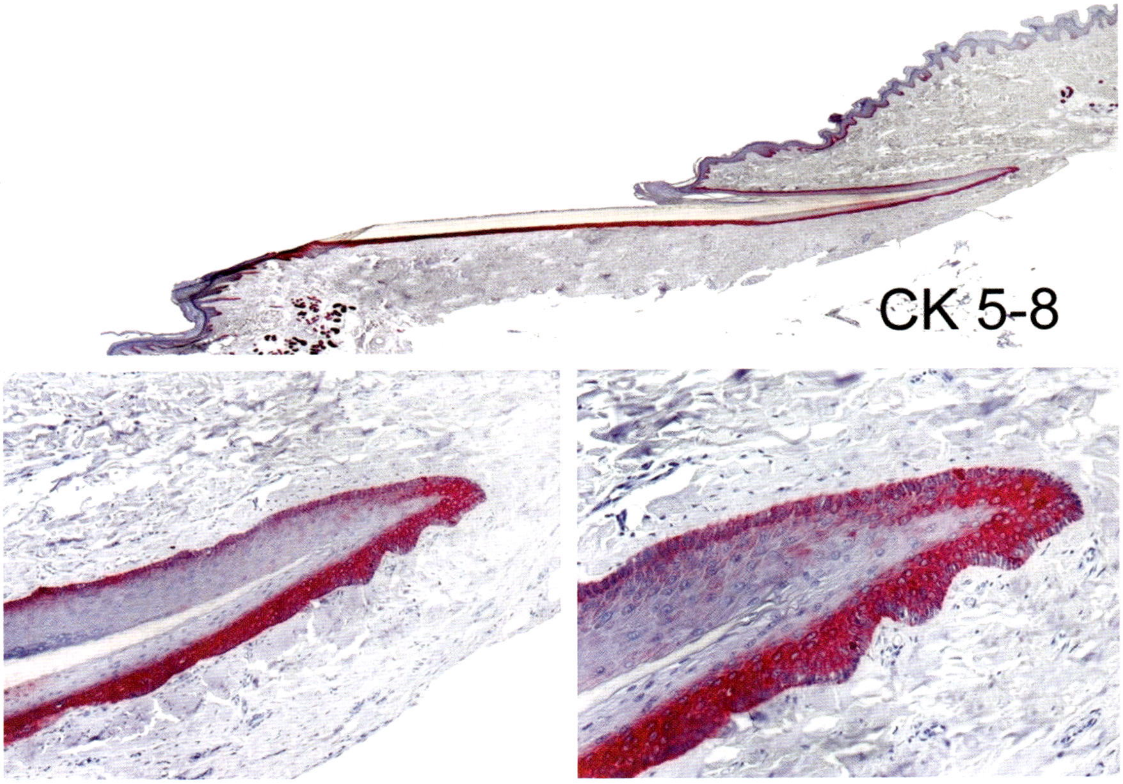

FIGURE 2.31 Keratin immunostain of nail unit. Immunostaining with cytokeratin CK5–8 antibody shows expression mostly in the nail bed and matrix.

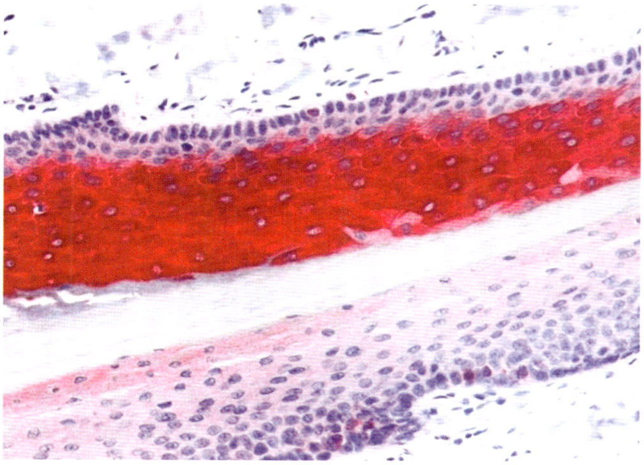

FIGURE 2.32 Keratin immunostain of nail unit. Immunostaining with cytokeratin CK-KL1 antibody shows strong expression in the dorsal nail matrix but none in the ventral nail matrix.

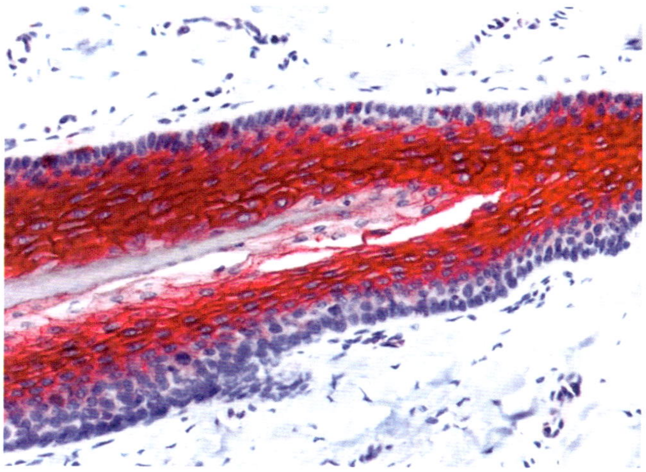

FIGURE 2.33 Immunostaining with claudin-1 (CLDN-1) antibody. CLDN-1, important in cell-cell adhesion, contributes to nail plate hardness. It is expressed in the nail matrix epithelium.

COMPARISONS OF THE NAIL UNIT TO THE HAIR FOLLICLE

Early in the 20th century, the nail unit was described as being similar to the hair follicle (96,137). Although interesting from an evolutionary standpoint, comparisons between hair and nails are of limited utility for pathologists interested in understanding normal histologic structures and distinct disease processes. A brief discussion of the molecular similarities is provided below.

There are several molecular similarities between the nail unit and hair apparatus (138). The nail unit and hair follicle both have hard keratin-producing matrical cells that create a visible keratin structure, namely the nail plate or hair shaft. Keratin expression is significantly homologous between the hair follicle and the nail unit epithelia (80). LH6, an antibody generally specific for the basal cells of complex epithelia (139), is expressed throughout the entire epithelia of both the nail bed and the follicular outer root sheath (121). β-Catenin is present not only in the hard keratin-producing cells of the nail matrix but also in the hard keratin-producing cells of the hair follicle bulb and matrix (140). This protein, which is a component of the (Wnt)/β-catenin signaling pathway, plays an important role in the morphogenesis of hair (141).

A substantial body of evidence suggests that both hair and nails are sites of relative immune privilege (142,143). Both the nail and hair matrix epithelia show downregulation of HLA-A/B/C immunoreactivity. The hair and nail matrices lack MHC class I expression, which serves to prevent attacks by autoreactive CD8-positive T-cells on the melanocyte-associated antigens of melanocytes. In addition, expression of MHC class II and CD209 on CD1a-positive cells is diminished. However, there are some distinct immunologic differences between the nails and hair that suggest a greater extent of immune privilege in the nail unit (114). For example, HLA-G is increased in the nail matrix but not in any part of the hair follicle (115).

NAIL PLATE GROWTH

Stem cells of the nail unit have not been extensively studied. Presumably, the stem cells in the proximal nail matrix are needed for a lifetime of never-ending nail plate production, and some authors have speculated that stem cells in the proximal nail fold maintain the epithelium of the proximal and lateral nail folds. The subject remains of great interest, since regeneration of the distal digit, including the nail unit, would be an important medical advancement (144).

In addition to CK15-positive stem cells in the proximal nail matrix and the proximal nail fold, another stem cell niche exists along the distal edge of the lunula where the nail matrix and bed are in contact. This niche of stem cells is not identifiably histologically, but sustains the ever-growing nail bed epithelium that emerges from this point, attaches to the nail plate, and moves distally with the plate until it is desquamated at the hyponychium. The mechanism for the joint movement of the nail bed and plate is still not well understood. Some have suggested that pressure from the ever-expanding plate drags the subjacent nail bed epithelium along with it (70). Nails grow distally rather than vertically due to a combination of factors. First, the proximal and lateral nail folds guide growth. Second, the matrix is oriented in a proximal-to-distal long axis that promotes growth in the same direction. Third, tight adherence of the nail plate to the nail bed may regulate distal rather than vertical growth.

The rate of nail plate growth varies between 1.9 and 4.4 mm per month (145,146). In general, fingernails grow faster than toenails. A normal fingernail grows out completely

TABLE 2.2 Physiologic and Pathologic Variations That Influence Nail Plate Growth

Physiologic		Pathologic	
Increased	**Decreased**	**Increased**	**Decreased**
Men	Women	Psoriasis	Fever
Daytime	Night	Pityriasis rubra pilaris	Poor nutrition
Summer	Winter	Hyperthyroidism	Hypothyroidism
Pregnancy	First day of life	A-V shunts	Decreased blood supply
Third digit	First and fifth digits	Idiopathic onycholysis in women	Kwashiorkor
Right hand[a]	Left hand[a]	Epidermolytic hyperkeratosis	Beau lines
Youth	Old age	Hyperpituitarism	Denervation/immobilization
Nail biting		Morgagni–Stewart–Morel sd	Acute infection
Avulsion		Brittle nail sd. *Medications:* calcium, vitamin D, benoxaprofen, biotin, cysteine, oral contraceptives, L-dopa, fluconazole, itraconazole, terbinafine, etretinate	Chronic disease, smoking, onychomycosis, yellow nail sd, lichen planus, relapsing polychondritis *Medications:* methotrexate, azathioprine, cyclosporine, lithium, retinoids, sulfonamides, heparin

[a]In a person's dominant right hand.
sd, syndrome.

in approximately 6 months, whereas a normal toenail takes 12 to 18 months to achieve complete growth. Growth is not the same for all fingers or toes, and several physiologic circumstances can cause variations in growth rate (Table 2.2). Nails grow faster in men than women and after complete avulsion (78). Nail growth is faster among individuals younger than 20 years of age (147), as well as chronic nail biters, and those living in warmer climates (148). Conversely, nails grow more slowly during lactation (147). The rate of nail plate growth may also be altered by cutaneous and systemic diseases (Table 2.2) (149–151). Nails grow quicker with psoriasis (149), etretinate treatment (152), and hyperthyroidism (78), whereas growth is slower in patients with onychomycosis (153), local ischemic conditions (145), nutritional deficiency (151), hypothyroidism, and yellow nail syndrome (2). A sudden decrease in nail growth due to acute viral illness results in the development of a transverse, depressed band called a Beau line (154). Some have also suggested that slow nail growth is a predisposing factor to fungal infection, however, the not uncommon presentation of onychomycosis of a single nail does not support this hypothesis (153).

IMAGING MODALITIES OF THE NAIL UNIT

In Vivo Confocal Microscopy of the Nail Unit

In vivo confocal microscopy has shown utility in analyzing the various structures within the complex nail unit (Fig. 2.34). Unfortunately, the limited penetration of confocal microscopy prevents its use in imaging all structures within the nail unit. Confocal imaging within the nail plate has the potential for in vivo diagnosis of onychomycosis and recognizing the prominent vessels of benign neoplasms, such as hemangiomas (155–157).

Dermoscopy

Dermoscopy provides an in vivo, low-power magnification of the nail unit and is mostly used to evaluate pigmented lesions (158,159). With dermoscopy, not only the surface features are appreciated, but subjacent epithelial and onychodermal structures can also be seen with strong lighting. Nail plate analysis can be useful, and dermoscopy after nail plate avulsion can be used to better view the nail matrix, bed epithelium, and the subjacent onychodermis. Dermoscopic analysis of the vasculature of the nail unit is also helpful in evaluating different systemic diseases, particularly collagen vascular diseases (Fig. 2.25).

NAIL UNIT PROCESSING FOR HISTOPATHOLOGIC ANALYSIS

Tissue Processing

Obtaining an adequate specimen is the primary challenge in nail unit pathology. Not only is it difficult to biopsy or excise lesions from the nail unit, but submission of these specimens to the histology laboratory and processing by the laboratory are quite challenging. Pathologists are often reticent to evaluate lesions from the nail unit because the microscopic slides received are often substandard. Tissue segments with nail plate often fail to adhere to the glass slide, and sectioning of the small, delicate specimens from the nail matrix and bed epithelium are not well represented on histologic sections.

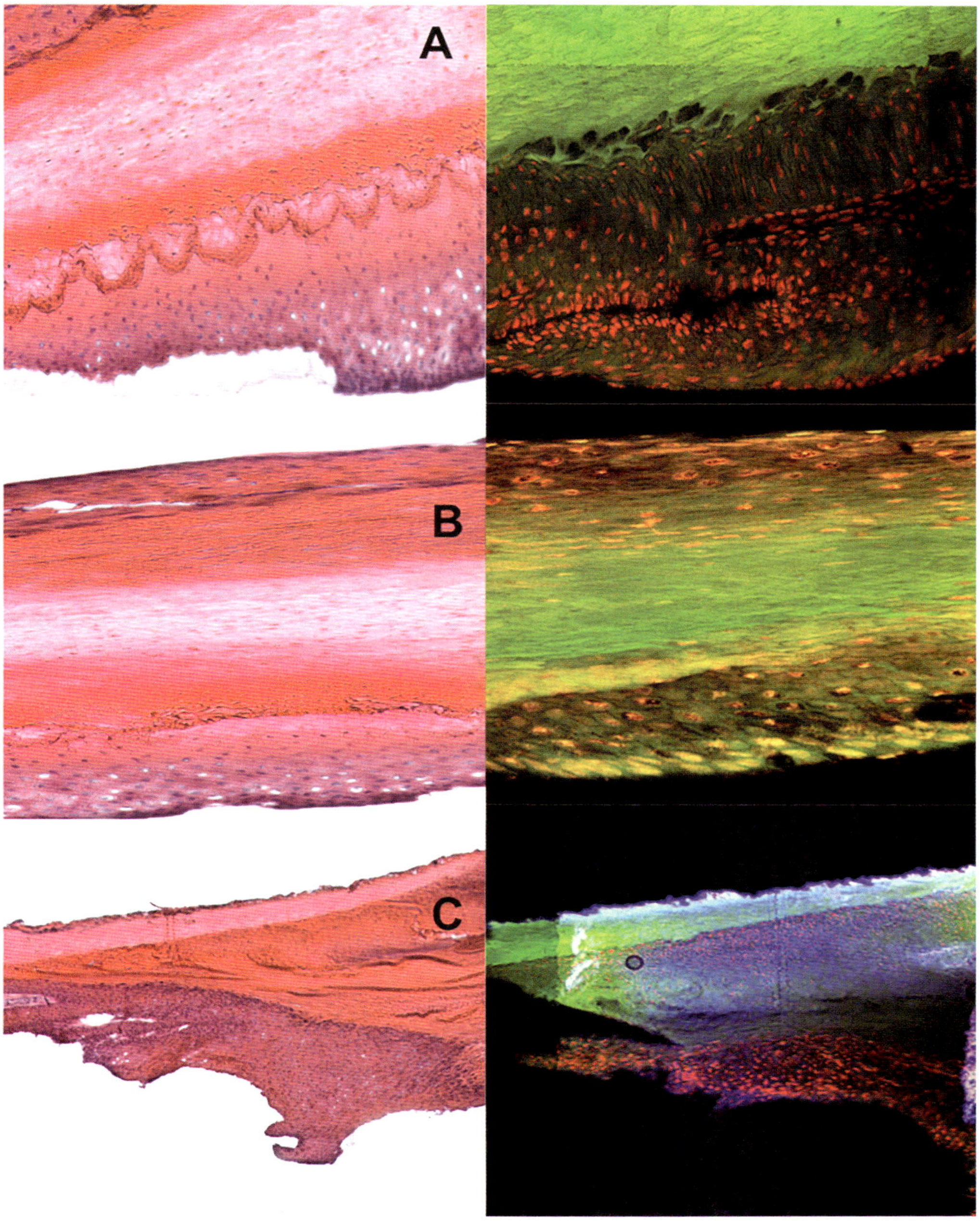

FIGURE 2.34 Confocal microscopy of fragment of the distal (**A**), middle (**B**), and proximal (**C**) nail unit stained with acridine orange and ethidium bromide (**right**). These are compared to traditional H&E staining (**left**). The slight three-dimensional imaging of the confocal microscopy is evident.

This is a particular problem when pigmented lesions are assessed. Orientation is also challenging, especially when samples are obtained from a diseased nail unit that is structurally altered. Inking of the specimen to provide orientation is helpful (160). Fortunately, a few simple techniques, performed not only by the clinician performing the biopsy/excision, but also in the histology laboratory can greatly increase the quality of the histologic sections (161,162).

Submission of a nail matrix/bed biopsy is best performed using a technique borrowed from ophthalmologic pathology (163). After reflection of the nail plate, a tissue segment is excised from the nail matrix and bed. The delicate specimen is then placed on a paper cartoon of a nail unit in a manner that preserves anatomic orientation (Fig. 2.35) (161). The specimen is then submitted in 10% formaldehyde on a paper cartoon inside a tissue cassette with sponges holding it in place. Fixation in a flat manner maintains orientation and simplifies embedding and sectioning.

Upon arrival at the histology laboratory, the fixed flat specimen, still on a paper cartoon, can either be sectioned

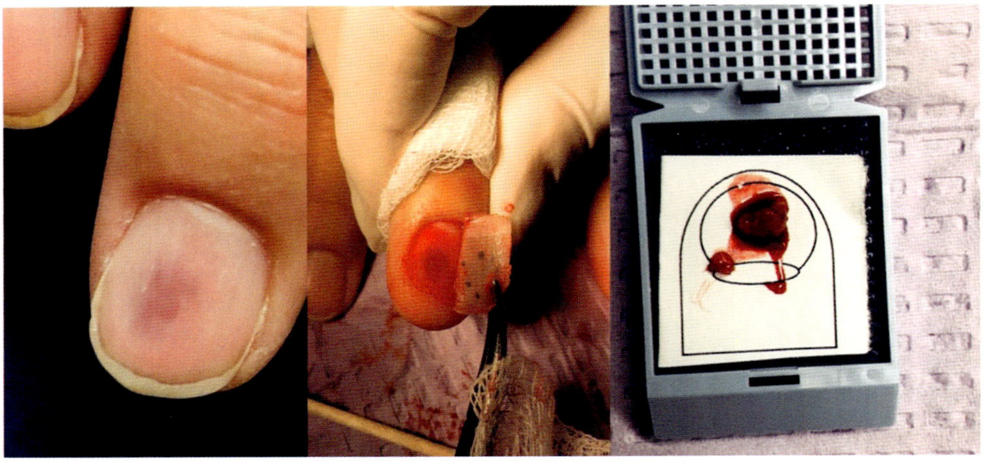

FIGURE 2.35 Surgical photographs showing the lesion of interest before and after reflection of the nail plate. The lesion is then sampled and placed on a paper cartoon of a nail, secured by tissue spongiosis in a cassette. (Photo provided courtesy of Dr. Phoebe Rich.)

immediately, or, if it is quite small, can be sent through tissue processing still on the paper cartoon inside the closed cassette. If orientation is needed, the specimen may be differentially inked by the clinician or upon arrival at the histology laboratory. Initial sectioning of the small nail matrix/bed specimen should include levels, anticipated special stains, and unstained slides to prevent tissue loss when having to reface a tissue block. A proper clinical history can guide the histology laboratory in deciding which special stains might be needed. For instance, a history of a pigmented lesion might include a MelanA immunostain, a Fontana–Masson stain for melanin, and a PAS stain for fungus, since some fungus is pigmented (Tinea nigra). In contrast, a clinical history for possible squamous cell carcinoma might only include level sections, a PAS stain, and unstained slides.

The nail plate is particularly difficult to process and section in the histology laboratory, mostly because it is difficult to adhere to the glass slides. If possible, the nail plate should be submitted in a separate cassette from the soft nail unit tissue so the histologist can prepare sections of the hard and soft tissue from the specimen in an entirely different manner. A wide variety of protocols have been developed to improve adherence, though these protocols have generally not been assessed in a controlled manner. One protocol that works particularly well is detailed here (Table 2.3). Of note, this protocol includes precoating the slides with gelatin. Care should also be taken to move the slide slowly in and out of solutions to further help retain the nail plate fragments on the slide.

Nail plate specimens should be submitted dry in a small, resealable bag, without 10% formaldehyde, so that subungual debris may be included. Subungual debris is particularly important in making a diagnosis of onychomycosis. Unfortunately, the subungual debris will not survive the traditional histologic process, but it is quite useful for PCR molecular analysis or culture. If the specimen is received in formaldehyde, the subungual debris and other small fragments may be retrieved using a cytospin technique (164).

PCR has rapidly replaced traditional tissue culture in identifying and speciating fungal infections of the nail plate (165,166). PCR testing can be performed on the sample without prior processing or it can be performed on nail fragments that have been formalin-fixed, processed, and embedded in paraffin. Some PCR tests now identify a set of mutations that cause terbinafine resistance, a feature which is highly useful in evaluating treatment failures (16). If traditional culture is the only modality available, the microbiology laboratory must use a cycloheximide-free medium so that molds can also be identified.

TABLE 2.3 Procedure for Nail Plate Histopathology Processing (Periodic Acid–Schiff Stain [PAS])

1. Place a small amount of gelatin in water bath.
2. Cut ribbon at 4 μm and float onto gelatin water bath.
3. Using positively charged slide, pick up desired sections for PAS and H&E slides.
4. Place in 65° C oven for 45 min (more time for "difficult" specimens).
5. Deparaffinize slides using xylene or xylene substitutes and hydrate through alcohols (or place PAS slides on programmed deparaffin run and H&E slides on H&E program on stainer).
6. Gently rinse slide in running tap water.
7. Rinse slide in distilled water.
8. Do not digest the slides.
9. Place slide in 1% periodic acid for 10 min.
10. Gently rinse slide quickly in distilled water.
11. Place slide in Schiff solution for 10 min.
12. Gently rinse slides in warm–hot tap water for 5 min.
13. Place slide in light green stain as needed to reach desired background intensity. (approximately 30 sec)
14. Dehydrate slide through three changes of absolute alcohol.
15. Clear slide through three changes of xylene or xylene substitute.
16. Coverslip using permanent mounting media.

Results: Basement membrane, fungi, glycogen and mucin: pink-to-red; other tissue: green

American Master Tech Scientific, Inc. "PAS Kit Procedure."

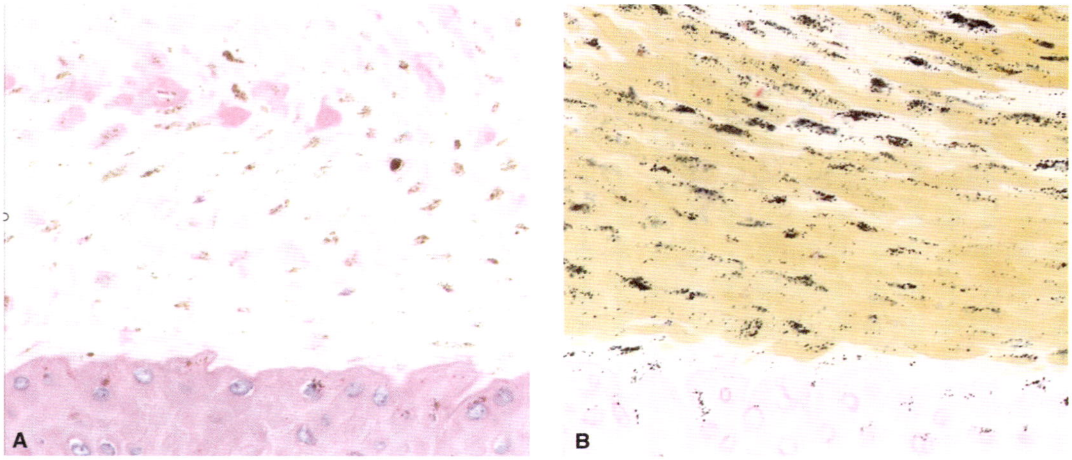

FIGURE 2.36 Melanin pigment in the nail plate. The melanin can be seen on the H&E section (**A**), but it is better seen with a diluted Fontana-Masson stain (**B**) (×400).

ARTIFACTS AND CLUES IN HISTOLOGIC EVALUATION OF THE NAIL UNIT

Analysis of Nail Plate Clippings

Nail plate clipping materials are frequently evaluated because the diagnosis may be made without a biopsy. Melanin and other pigment can generally be seen in routinely stained H&E sections within the nail plate. A Fontana–Masson stain may be used to identify melanin in the nail plate, provided the Fontana–Masson stain is diluted (Fig. 2.36). Failure to dilute the solution results in intense, nonspecific staining that precludes assessment.

Some nail pathologists advocate for nail plate analysis alone when evaluating melanocytic neoplasms of the subjacent nail matrix and bed. This can help triage patients when deciding who needs conservative management and who needs surgical sampling of the lesion. Patterns of nail plate pigmentation and architectural changes may serve as clues to the nature of melanocytic proliferation (167). Grossly evident holes in transverse sections also provide a clue to the diagnosis of a nonmelanocytic neoplasm, such as onychomatricoma (168).

Blood in the Nail Plate

Posttraumatic blood accumulation in the nail plate may simulate a pigmented lesion, even though there is often no history of trauma. Blood within the nail plate does not degrade so the pools of blood will be visible on H&E sections (Fig. 2.37). Importantly, a Perl iron stain does not work well when evaluating nail plate hemorrhage because the iron is not released from the heme structure within the

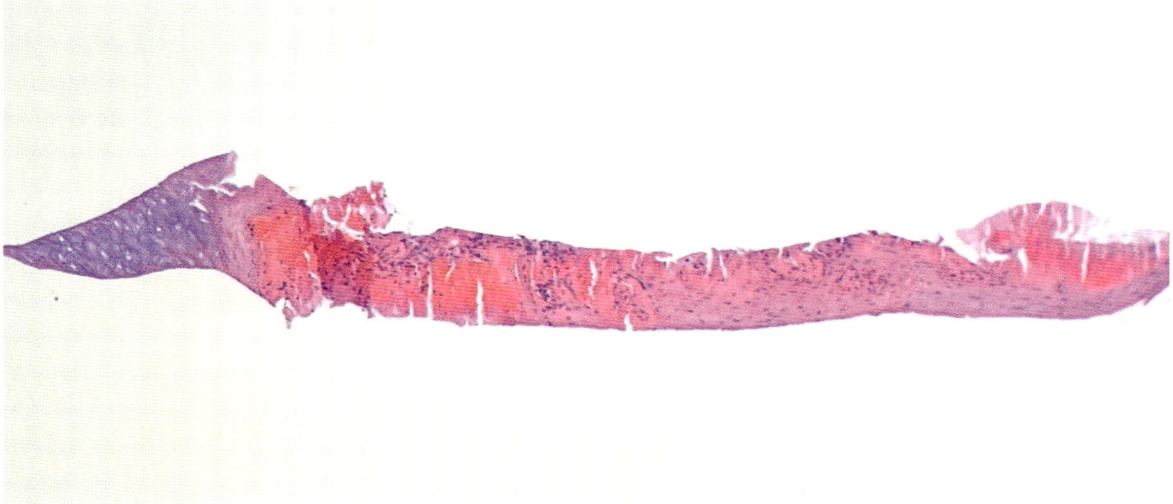

FIGURE 2.37 Blood in the nail plate. Blood is visible within the nail plate on standard H&E sections. A special stain for blood is not necessary (H&E, ×10).

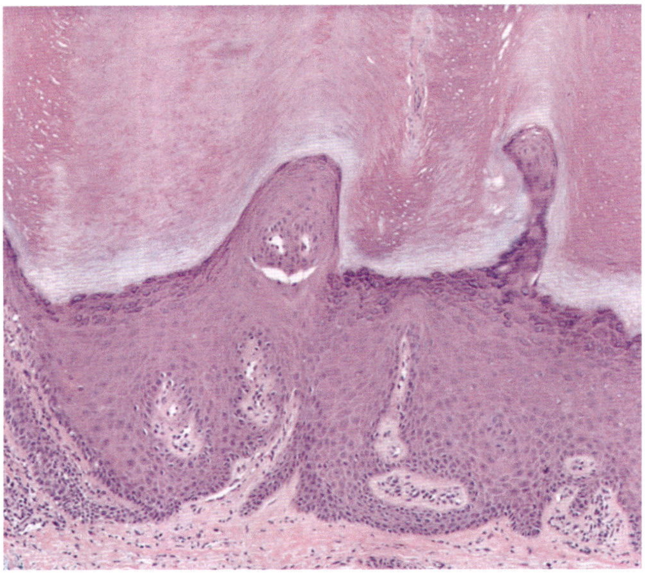

FIGURE 2.38 Verruca of the nail unit. This wart shows verrucous digitations and hypergranulosis (H&E, ×20).

nail. A benzidine stain, which will stain intact blood, is not necessary since blood is always identifiable in H&E sections (169).

Human Papillomavirus (HPV) Infection of the Nail Unit

Most squamous proliferations of the nail unit, whether benign or malignant, are HPV induced. Fortunately, most squamous cell carcinoma of the nail unit is in situ. A biopsy of squamous cell carcinoma may be histologically heterogeneous with benign verruca (wart) and other areas showing malignancy (Fig. 2.38). For this reason, it is important to perform additional biopsies in clinically suspected malignancy when the initial biopsies show only benign verruca. In addition, squamous cell carcinoma of the nail unit has a wide variety of presentations, and it is not uncommon for it to have been treated empirically as an infection. Thus, the pathologist should be on the alert for squamous cell carcinoma (170). Verruca should also be distinguished from epithelial hyperplasia due to dermatitis or external irritation. Reactive hyperproliferative states of the nail unit often cause hypergranulosis of the nail unit epithelium. It is important for the pathologist to identify koilocytic histologic changes before making a diagnosis of a verruca.

Squamous Inclusions in the Onychodermis

Small collections of banal squamous epithelium with no attachment to the nail bed epithelium may be seen in the onychodermis (Fig. 2.39). These nests, which may be cystic, have been variably termed "subungual epidermoid inclusions" or "subungual onycholemmal cysts" (171).

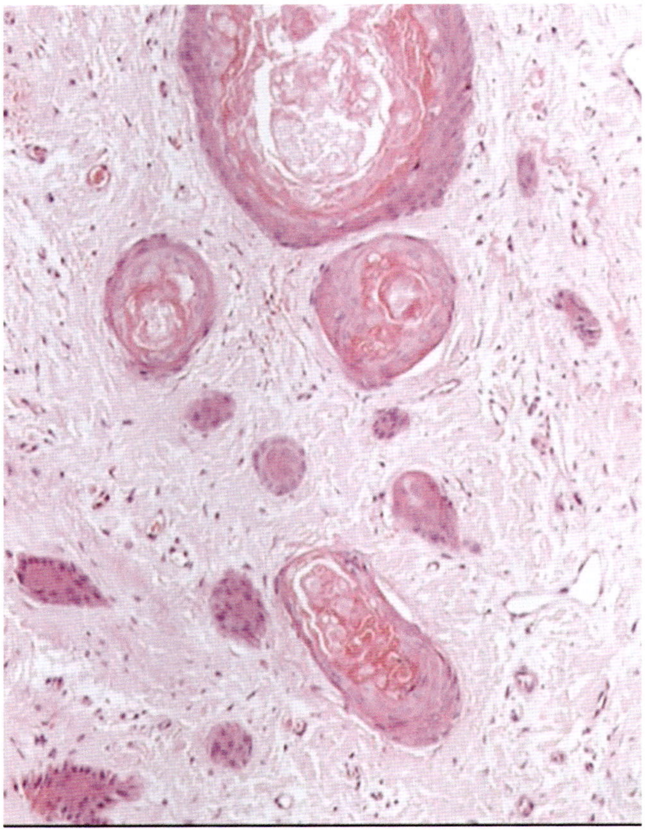

FIGURE 2.39 Subungual onycholemmal cysts. These subungual inclusions are likely the result of trauma and are entirely benign (H&E, ×200).

They are likely the result of trauma and sometimes calcify (172). It is important to know about nail bed inclusions to prevent an overdiagnosis of a benign or malignant epithelial tumor.

Vascularity and Glomus Cells of the Onychodermis

As discussed earlier in the chapter, it is important not to overinterpret the normal vascularity of nail unit as a hemangioma (Fig. 2.23). This is challenging because hemangiomas, especially the lobular capillary hemangioma, are common in the nail unit. Glomus tumors are also common in the nail unit, but, fortunately, these usually present as a painful blanching lesion. Thus, a diagnosis of a glomus tumor is usually clinically suspected. Of note, glomus cells surrounding the normal vasculature of the nail unit can be quite prominent, and an overdiagnosis of a glomangioma should be avoided.

A Clinical Mass with No Histologic Finding

If a mass or other lesion of the nail unit is identified clinically, and proper sampling of the lesion yields no histopathologic findings, a deeper lesion of the underlying bone should be considered (Fig. 2.26). In this scenario, imaging studies, such as ultrasound, CT scanning,

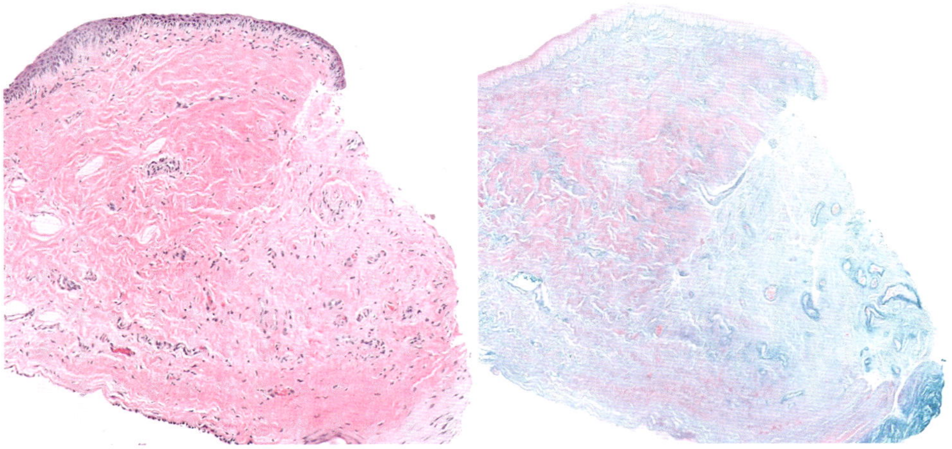

FIGURE 2.40 Digital myxoid/mucous cyst. The interstitial mucin is not evident in the H&E section (left), but is prominently seen with a mucin stain (right) (H&E and Colloidal iron, ×10).

or magnetic resonance imaging (MRI) can be helpful in further evaluation (173,174). If a digital mucous/myxoid cyst is suspected and no mucin is identified, histologic mucin studies can be quite revealing since older lesions are more fibrotic than mucinous (Fig. 2.40). It is important to remember that the digital mucous/myxoid cyst is not a synovial cyst at all, but rather a leak of synovial fluid from the proximal joint.

REFERENCES

1. Birznieks I, Macefield VG, Westling G, Johansson RS. Slowly adapting mechanoreceptors in the borders of the human fingernail encode fingertip forces. *J Neurosci*. 2009; 29(29):9370–9379.
2. Pavlidakey GP, Hashimoto K, Blum D. Yellow nail syndrome. *J Am Acad Dermatol*. 1984;11(3):509–512.
3. Haneke E. *Histopathology of the Nail*. CRC Press; 2017: 281–282.
4. Zander R. Untersuchungen uber den Verhornungsprogress 1. Die Histogenese des Nagels beim menschlichen Fetus. *Arch Anat Entwick*. 1886;(1):273.
5. Kolliker A. Die entwicklung des menschlichen nagels. *Z Wiss Zool*. 1988;(1):1–12.
6. Unna P. Entwicklungsgeschichte und Anatomie. In: von Ziemssen HW, ed. In: *Handbook Der Speciellen Pathologie und Therapie* Vol 14. 1. Nabu Press, 2010.
7. Boas J. Ein Beitrag zur Morphologie der Nagel, Kralien, Hufe and Klauen der Saugetiere. *Mrophol Jahrb*. 1883;9:390.
8. Henle J, ed.*Das Wachstrum de Menschlichen Nagels under Der Pferdehufs*. Dieterich; 1884.
9. Branca A. Notes sur la structure de l'onble. *Ann Dermatol Syphiligr*. 1910;1:353–371.
10. Clark WE, Buxton LHD. Studies in nail growth. *Br J Dermatol*. 1938;50(5):221–235.
11. Lewis BL. Microscopic studies of fetal and mature nail and surrounding soft tissue. *AMA Arch Dermatol Syphilol*. 1954;70(6):732–747.
12. Zaias N. Embryology of the human nail. *Arch Dermatol*. 1963;87(1):37–53.
13. Lee DY, Yang JM, Mun GH, Jang KT, Cho KH. Immunohistochemical study of specialized nail mesenchyme containing onychofibroblasts in transverse sections of the nail unit. *Am J Dermatopathol*. 2011;33(3):266–270.
14. Kim YJ, Jung CJ, Na H, et al. Cyclin D1 and PRAME expression in distinguishing melanoma in situ from benign melanocytic proliferation of the nail unit. *Diagn Pathol*. 2022;17(1):41.
15. Litz CE, Cavagnolo RZ. Polymerase chain reaction in the diagnosis of onychomycosis: a large, single-institute study. *Br J Dermatol*. 2010;163(3):511–514.
16. Kano R, Noguchi H, Harada K, Hiruma M. Rapid molecular detection of terbinafine-resistant dermatophytes. *Méd Mycol J*. 2021;62(2):41–44.
17. Kivell TL, Lemelin P, Richmond BH, Schmitt D. The evolution of the primate hand. In: Barrett E, ed. *Developments in Primatology: Progress and Prospects*. Springer; 2016:17–54.
18. Noro M, Yuguchi H, Sato T, et al. Role of paraxial mesoderm in limb/flank regionalization of the trunk lateral plate. *Dev Dyn*. 2011;240(7):1639–1649.
19. King M, Arnold JS, Shanske A, Morrow BE. T-genes and limb bud development. *Am J Méd Genet Part A*. 2006; 140(13):1407–1413.
20. Ohuchi H, Nakagawa T, Yamamoto A, et al. The mesenchymal factor, FGF10, initiates and maintains the outgrowth of the chick limb bud through interaction with FGF8, an apical ectodermal factor. *Development*. 1997;124(11):2235–2244.
21. Capdevila J, Belmonte JCI. Patterning mechanisms controlling vertebrate limb development. *Annu Rev Cell Dev Biol*. 2001;17(1):87–132.
22. Summerbell D. Reduction of the rate of outgrowth, cell density, and cell division following removal of the apical ectodermal ridge of the chick limb-bud. *J Embryol Exp Morphol*. 1977;40(1):1–21.
23. Zeller R, López-Ríos J, Zuniga A. Vertebrate limb bud development: moving towards integrative analysis of organogenesis. *Nat Rev Genet*. 2009;10(12):845–858.
24. Dudley AT, Ros MA, Tabin CJ. A re-examination of proximodistal patterning during vertebrate limb development. *Nature*. 2002;418(6897):539–544.

25. Sun X, Mariani FV, Martin GR. Functions of FGF signaling from the apical ectodermal ridge in limb development. *Nature*. 2002;418(6897):501–508.
26. Duboule D. Making progress with limb models. *Nature*. 2002;418(6897):492–493.
27. Tickle C, Wolpert L. The progress zone—alive or dead? *Nat Cell Biol*. 2002;4(9):E216–E217.
28. Tabin C, Wolpert L. Rethinking the proximodistal axis of the vertebrate limb in the molecular era. *Genes Dev*. 2007;21(12):1433–1442.
29. Parr BA, McMahon AP. Dorsalizing signal Wnt-7a required for normal polarity of D-V and A-P axes of mouse limb. *Nature*. 1995;374(6520):350–353.
30. Maiolino S, Boyer DM, Rosenberger A. Morphological correlates of the grooming claw in distal phalanges of platyrrhines and other primates: a preliminary study. *Anat Rec: Anat Rec (Hoboken)*. 2011;294(12):1975–1990.
31. Carroll SB. Evo-devo and an expanding evolutionary synthesis: a genetic theory of morphological evolution. *Cell*. 2008;134(1):25–36.
32. Stern DL, Orgogozo V. The loci of evolution: how predictable is genetic evolution? *Evolution*. 2008;62(9):2155–2177.
33. Holbrook KA, Odland GF. The fine structure of developing human epidermis: light, scanning, and transmission electron microscopy of the periderm. *J Invest Dermatol*. 1975;65(1):16–38.
34. Conejo-Mir J, Ambrosiani J, Dorado M. Analisis De La Morfogénesis Ungueal. *Estudio Con Microscopio Electronica De Barrido En El Embrion Humano*. Isdin; 1985:1–8.
35. Caputo R, Dadati E. Preliminary observations about the ultrastructure of the human nail plate treated with thioglycolic acid. *Arch Klin Exp Dermatol*. 1968;231(4):344–354.
36. Forslind B, Thyresson N. On the structure of the normal nail. A scanning electron microscope study. *Arch Dermatol Forsch*. 1975;251(3):199–204.
37. Hashimoto K, Gross BG, Nelson R, Lever WF. The ultrastructure of the skin of human embryos. 3. The formation of the nail in 16–18 weeks old embryos. *J Invest Dermatol*. 1966;47(3):205–217.
38. Hashimoto K. Ultrastructure of the human toenail I. Proximal nail matrix. *J Invest Dermatol*. 1971;56(3):235–246.
39. Hashimoto K. Ultrastructure of the human toenail II. Keratinization and formation of the marginal band. *J Ultrastruct Res*. 1971;36(3–4):391–410.
40. Hashimoto K. Ultrastructure of the human toenail: cell migration, keratinization and formation of the intercellular cement. *Arch Dermatol Forsch*. 1971;240(1):1–22.
41. Jimbow K, Takahashi M, Sato S, Kukita A. Ultrastructural and cytochemical studies of melanogenesis in melanocytes of normal human hair matrix. *J Electron Microsc*. 1971;20(2):87–92.
42. Meyer JC, Grundmann HP. Scanning electron microscopic investigation of the healthy nail and its surrounding tissue. *J Cutan Pathol*. 1984;11(1):74–79.
43. Suchard R. Des modifications des celluies de la matrice et du lit de i'ongle dans quelques cas pathologiques. *Arch Physiol*. 1882;2:445.
44. Holbrook KA. Human epidermal embryogenesis. *Int J Dermatol*. 1979;18(5):329–356.
45. Ogura R, Knox JM, Griffin AC, Kusuhara M. The concentration of sulfhydryl and disulfide in human epidermis, hair and nail. *J Invest Dermatol*. 1962;38(2):69–75.
46. Stern DK, Creasey AA, Quijije J, Lebwohl MG. UV-A and UV-B penetration of normal human cadaveric fingernail plate. *Arch Dermatol*. 2011;147(4):439–441.
47. Garson JC, Baltenneck F, Leroy F, Riekel C, Müller M. Histological structure of human nail as studied by synchrotron X-ray microdiffraction. *Cell Mol Biol (Noisy-le-grand)*. 2000;46(6):1025–1034.
48. Helmdach M, Thielitz A, Röpke EM, Gollnick H. Age and sex variation in lipid composition of human fingernail plates. *Skin Pharmacol Appl Skin Physiol*. 2000;13(2):111–119.
49. Walters KA, Flynn GL, Marvel JR. Physicochemical characterization of the human nail: I. Pressure sealed apparatus for measuring nail plate permeabilities. *J Invest Dermatol*. 1981;76(2):76–79.
50. Farren L, Shayler S, Ennos AR. The fracture properties and mechanical design of human fingernails. *J Exp Biol*. 2004;207(5):735–741.
51. Egawa M, Fukuhara T, Takahashi M, Ozaki Y. Determining water content in human nails with a portable near-infrared spectrometer. *Appl Spectrosc*. 2003;57(4):473–478.
52. Stern DK, Diamantis S, Smith E, et al. Water content and other aspects of brittle versus normal fingernails. *J Am Acad Dermatol*. 2007;57(1):31–36.
53. Krönauer C, Gfesser M, Ring J, Abeck D. Transonychial water loss in healthy and diseased nails. *Acta Derm Venereol*. 2001;81(3):175–177.
54. Nuutinen J, Harvima I, Lahtinen M, Lahtinen T. Water loss through the lip, nail, eyelid skin, scalp skin and axillary skin measured with a closed-chamber evaporation principle. *Br J Dermatol*. 2003;148(4):839–841.
55. Sakai N, Shimawaki S. Strain in the nail at fingertip compression. *Ski Res Technol*. 2007;13(4):449–453.
56. Jarrett A, Spearman RI. The histochemistry of the human nail. *Arch Dermatol*. 1966;94(5):652–657.
57. Pautard FG, Mineralization of keratin and its comparison with the enamel matrix. *Nature*. 1963;199(4893):531–535.
58. Terry RB. The onychodermal band in health and disease. *Lancet*. 1955;265(6856):179–181.
59. Chen E, Pallon J, Forslind B. Distribution of calcium and sulphur in the blue-light-exposed rat retina. *Graefes Arch Clin Exp Ophthalmol*. 1995;233(3):163–167.
60. Bolliger A, Gross R. Non-Keratins of human toenails. *Aust J Exp Biol Méd Sci*. 1953;31(2):127–130.
61. Levitt JI. Creatinine concentration of human fingernail and toenail clippings. Application in determining the duration of renal failure. *Ann Intern Med*. 1966;64(2):312–327.
62. Kopito L, Mahmoodian A, Townley RR, Khaw KT, Shwachman H. Studies in cystic fibrosis: analysis of nail clippings for sodium and potassium. *N Engl J Med*. 1965;272(10):504–509.
63. Johnson M, Shuster S. Determinants of nail thickness and length. *Br J Dermatol*. 1994;130(2):195–198.
64. Martin GM. Copper content of hair and nails of normal individuals and of patients with hepatolenticular degeneration. *Nature*. 1964;202(4935):903–904.
65. Mandal BK, Ogra Y, Anzai K, Suzuki KT. Speciation of arsenic in biological samples. *Toxicol Appl Pharmacol*. 2004;198(3):307–318.
66. Cingolani M, Scavella S, Mencarelli R, Mirtella D, Froldi R, Rodriguez D. Simultaneous detection and quantitation of

morphine, 6-acetylmorphine, and cocaine in toenails: comparison with hair analysis. *J Anal Toxicol*. 2004;28(2):128–131.
67. Drapé JL, Wolfram-Gabel R, Idy-Peretti I, et al. The Lunula: a magnetic resonance imaging approach to the subnail matrix area. *J Invest Dermatol*. 1996;106(5):1081–1085.
68. Pinkus F. The development of the integument. In: Kleibel F, Mali F, eds. *Manual of Human Embryology*. JB Lippincott; 1910.
69. Le Gros Clark WE. The problem of the claw in primates. *Proc Zool Soc Lond*. 1936;106(1):1–24.
70. Zaias N. The movement of the nail bed. *J Invest Dermatol*. 1967;48(4):402–403.
71. Lewin K. The normal fingernail. *Br J Dermatol*. 1965;77: 421–430.
72. Samman PD. The ventral nail. *Arch Dermatol*. 1961; 84(6):1030–1033.
73. Samman P. *The Nail in Disease*. Heinemann Medical Books; 1978.
74. Fernandez-Flores A. Basic concepts in nail pathology. *Am J Dermatopathol*. 2023;45(10):675–693.
75. Berker DD, Angus B. Proliferative compartments in the normal nail unit. *Br J Dermatol*. 1996;135(4):555–559.
76. Cameli N, Picardo M, Tosti A, Perrin C, Pisani A, Ortonne JP. Expression of integrins in human nail matrix. *Br J Dermatol*. 1994;130(5):583–588.
77. Schweitzer TP, Rayan GM. The terminal tendon of the digital extensor mechanism: Part I, anatomic study. *J Hand Surg*. 2004;29(5):898–902.
78. Baran R, de Berker D, Holzberg M, Thomas L. *Baran & Dawber's Diseases of the Nails and their Management*. John Wiley & Sons; 2012.
79. Winkelmann RK. *Nerve Endings in Normal and Pathologic Skin: Contributions to the Anatomy of Sensation*. Charles C Thomas; 1960.
80. Inoue S, Nambu T, Shimomura T. The RAIG family member, GPRC5D, is associated with hard-keratinized structures. *J Invest Dermatol*. 2004;122(3):565–573.
81. Egawa K, Kuroki M, Inoue Y, Ono T. Nail bed keratinocytes express an antigen of the carcinoembryonic antigen family. *Br J Dermatol*. 2000;143(1):79–83.
82. Shi J, Lv Z, Nie M, et al. Human nail stem cells are retained but hypofunctional during aging. *J Mol Histol*. 2018;49(3):303–316.
83. Martin BF, Platts MM. A histological study of the nail region in normal human subjects and in those showing splinter hemorrhages of the nail. *J Anat*. 1959;93:323–330.
84. Raffle EJ. Terry's nails. *Lancet*. 1984;323(8386):1131.
85. Omura EF. Histopathology of the Nail. *Dermatol Clin*. 1985;3(3):531–541.
86. Ko D, Lipner SR. Onychogryphosis: case report and review of the literature. *Skin Appendage Disord*. 2018;4(4):326–330.
87. Perrin C. Peculiar zone of the distal nail unit: the nail isthmus. *Am J Dermatopathol*. 2007;29(1):108–109.
88. Zaias N. Onychomycosis. *Arch Dermatol*. 1972;105(2):263–274.
89. Caputo R, Cappio F, Rigoni C, et al. Pterygium inversum unguis: Report of 19 cases and review of the literature. *Arch Dermatol*. 1993;129(10):1307–1309.
90. Reddy K, Lowenstein EJ. Forensics in dermatology: Part II. *J Am Acad Dermatol*. 2011;64(5):811–824.
91. Lee DY, Park JH, Shin HT, Jang KT, Kwon GY, Shim JS. Onychodermis (specialized nail mesenchyme) containing onychofibroblasts in horizontal sections of the nail unit. *Br J Dermatol*. 2012;166(5):1127–1129.
92. Lee KJ, Kim WS, Lee JH, et al. CD10, a marker for specialized mesenchymal cells (onychofibroblasts) in the nail unit. *J Dermatol Sci*. 2006;42(1):65–67.
93. Park JH, Lee DY, Jang KT, et al. CD13 is a marker for onychofibroblasts within nail matrix onychodermis: comparison of its expression patterns in the nail unit and in the hair follicle. *J Cutan Pathol*. 2017;44(11):909–914.
94. Okazaki M, Yoshimura K, Fujiwara H, Suzuki Y, Harii K. Induction of hard keratin expression in non-nail-matrical keratinocytes by nail-matrical fibroblasts through epithelial-mesenchymal interactions. *Plast Reconstr Surg*. 2003;111(1):286–290.
95. Lee DY. The relation of onychomatricoma to onychodermis in the nail unit. *Ann Dermatol*. 2013;25(3):394–395.
96. Sehgal VN, Aggarwal AK, Srivastava G, Chatterjee K. Nail biology, morphologic changes, and clinical ramifications: part I. *Skinmed*. 2011;9(1):39–46.
97. Sinclair RD, Wojnarowska F, Leigh IM, Dawber RP. The basement membrane zone of the nail. *Br J Dermatol*. 1994; 131(4):499–505.
98. Flint MH. Some observations on the vascular supply of the nail bed and terminal segment of the finger. *Br J Plast Surg*. 1955;8(3):186–195.
99. Smith DO, Oura C, Kimura C, Toshimori K. Artery anatomy and tortuosity in the distal finger. *J Hand Surg Am*. 1991;16(2):297–302.
100. Hale AR, Burch GE. The arteriovenous anastomoses and blood vessels of the human finger. Morphological and functional aspects. *Medicine (Baltimore)*. 1960;39:191–240.
101. Wang W, Winlove CP, Michel CC. Oxygen partial pressure in outer layers of skin of human finger nail folds. *J Physiol*. 2003;549(3):855–863.
102. Sangiorgi S, Manelli A, Congiu T, et al. Microvascularization of the human digit as studied by corrosion casting. *J Anat*. 2004;204(2):123–131.
103. Ross JB. Nail fold capillaroscopy–a useful aid in the diagnosis of collagen vascular diseases. *J Invest Dermatol*. 1966; 47(4):282–285.
104. Reginelli AD, Wang YQ, Sassoon D, Muneoka K. Digit tip regeneration correlates with regions of Msx1 (Hox 7) expression in fetal and newborn mice. *Development*. 1995; 121(4):1065–1076.
105. Zhao W, Neufeld DA. Bone regrowth in young mice stimulated by nail organ. *J Exp Zoöl*. 1995;271(2):155–159.
106. Feibleman GE, Stoll H, Maize JC. Melanomas of the palm, sole, and nailbed: a clinicopathologic study. *Cancer*. 1980;46(11):2492–2504.
107. Higashi N. Melanocytes of nail matrix and nail pigmentation. *Arch Dermatol*. 1968;97(5):570–574.
108. Higashi N, Saito T. Horizontal distribution of the dopa-positive melanocytes in the nail matrix. *J Invest Dermatol*. 1969;53(2):163–165.
109. Baran R, Juhlin L. Photoonycholysis. *Photodermatol Photoimmunol Photomed*. 2002;18(4):202–207.
110. Ruben BS. Pigmented lesions of the nail unit: clinical and histopathologic features. *Semin Cutan Med Surg*. 2010; 29(3):148–158.
111. Scott GA, Haake AR. Keratinocytes regulate melanocyte number in human fetal and neonatal skin equivalents. *J Invest Dermatol*. 1991;97(5):776–781.

112. Lacour JP, Dubois D, Pisani A, Ortonne JP. Anatomical mapping of Merkel cells in normal human adult epidermis. *Br J Dermatol*. 1991;125(6):535–542.
113. Moll I, Moll R. Merkel cells in ontogenesis of human nails. *Arch Dermatol Res*. 1993;285(6):366–371.
114. Ito T, Ito N, Bettermann A, Tokura Y, Takigawa M, Paus R. Collapse and restoration of MHC class-I-dependent immune privilege: Exploiting the human hair follicle as a model. *Am J Pathol*. 2004;164(2):623–634.
115. Paud R, Nicholoff B, Ito T. A 'hairy' privilege. *Trends Immunol*. 2005;26(1):32–40.
116. Dorschner RA, Lopez-Garcia B, Massie J, Kim C, Gallo RL. Innate immune defense of the nail unit by antimicrobial peptides. *J Am Acad Dermatol*. 2004;50(3):343–348.
117. Berker DD, Wojnarowska F, Sviland L, Westgate GE, Dawber RP, Leigh IM. Keratin expression in the normal nail unit: Markers of regional differentiation. *Br J Dermatol*. 2000; 142(1):89–96.
118. Kitahara T, Ogawa H. Cultured nail keratinocytes express hard keratins characteristic of nail and hair in vivo. *Arch Dermatol Res*. 1992;284(4):253–256.
119. Kitahara T, Ogawa H. Coexpression of Keratins characteristic of skin and hair differentiation in nail cells. *J Invest Dermatol*. 1993;100(2):171–175.
120. Perrin C, Langbein L, Schweizer J. Expression of hair keratins in the adult nail unit: An immunohistochemical analysis of the onychogenesis in the proximal nail fold, matrix and nail bed. *Br J Dermatol*. 2004;151(2):362–371.
121. Stark HJ, Breitkreutz D, Limat A, Bowden P, Fusenig NE. Keratins of the human hair follicle: "Hyperproliferative" keratins consistently expressed in outer root sheath cells in vivo and in vitro. *Differentiation*. 1987;35(3):236–248.
122. McLean WHI; Epithelial Genetics Group. Genetic disorders of palm skin and nail. *J Anat*. 2003;202(1):133–141.
123. Hesse M, Magin TM, Weber K. Genes for intermediate filament proteins and the draft sequence of the human genome: Novel keratin genes and a surprisingly high number of pseudogenes related to keratin genes 8 and 18. *J Cell Sci*. 2001;114(14):2569–2575.
124. Heid HW, Moll I, Franke WW. Patterns of expression of trichocytic and epithelial cytokeratins in mammalian tissues II. Concomitant and mutually exclusive synthesis of trichocytic and epithelial cytokeratins in diverse human and bovine tissues (hair follicle, nail bed and matrix, lingual papilla, thymic reticulum). *Differentiation*. 1988;37(3):215–230.
125. Kitahara T, Ogawa H. The extraction and characterization of human nail keratin. *J Dermatol Sci*. 1991;2(6):402–406.
126. Smith TA, Strelkov SV, Burkhard P, Aebi U, Parry DAD. Sequence comparisons of intermediate filament chains: Evidence of a unique functional/structural role for coiled-coil segment 1A and linker L1. *J Struct Biol*. 2002;137(1–2):128–145.
127. Rice RH, Xia Y, Alvarado RJ, Phinney BS. Proteomic analysis of human nail plate. *J Proteome Res*. 2010;9(12):6752–6758.
128. Fuchs E. Keratins and the skin. *Annu Rev Cell Dev Biol*. 1995;11(1):123–154.
129. Perrin C. Expression of follicular sheath keratins in the normal nail with special reference to the morphological analysis of the distal nail unit. *Am J Dermatopathol*. 2007; 29(6):543–550.
130. Gillespie JM, Frenkel MJ. The diversity of keratins. *Comp Biochem Physiol Part B: Comp Biochem*. 1974;47(2):339–346.
131. Baden HP, Kvedar JC. Epithelial cornified envelope precursors are in the hair follicle and nail. *J Invest Dermatol*. 1993;101(S1):72S–74S.
132. O'Keefe EJ, Hamilton EH, Lee SC, Steinert P. Trichohyalin. A structural protein of hair, tongue, nail, and epidermis. *J Invest Dermatol*. 1993;101(1):S65–S71.
133. Irvine AD, Mclean WH. Human keratin diseases: the increasing spectrum of disease and subtlety of the phenotype–genotype correlation. *Br J Dermatol*. 1999;140(5):815–828.
134. Perrin C, Langbein L, Schweizer J, et al. Onychomatricoma in the light of the microanatomy of the normal nail unit. *Am J Dermatopathol*. 2011;33(2):131–139.
135. Wang Z, Wong P, Langbein L, Schweizer J, Coulombe PA. Type II epithelial keratin 6hf (K6hf) is expressed in the companion layer, matrix, and medulla in anagen-stage hair follicles. *J Invest Dermatol*. 2003;121(6):1276–1282.
136. Wong P, Coulombe PA. Loss of keratin 6 (K6) proteins reveals a function for intermediate filaments during wound repair. *J Cell Biol*. 2003;163(2):327–337.
137. Baran R, Dawber RPR, Haneke E. Hair and nail relationship. *Skinmed*. 2005;4(1):18–23.
138. Sellheyer K, Nelson P. The concept of the onychodermis (specialized nail mesenchyme): an embryological assessment and a comparative analysis with the hair follicle. *J Cutan Pathol*. 2013;40(5):463–471.
139. Purkis PE, Steel JB, Mackenzie IC, Nathrath WB, Leigh IM, Lane EB. Antibody markers of basal cells in complex epithelia. *J Cell Sci*. 1990;97(1):39–50.
140. Lane EB, Wilson CA, Hughes BR, Leigh IM. Stem cells in hair follicles. *Ann N Y Acad Sci*. 1991;642(1):197–213.
141. Kim C-R, Shin H-T, Park J-H, et al. Nuclear and cytoplasmic localization of β-catenin in the nail-matrix cells and in an onychomatricoma. *Clin Exp Dermatol*. 2013;38(8):917–920.
142. Chuong CM, Noveen A. phenotypic determination of epithelial appendages: genes, developmental pathways, and evolution. *J Invest Dermatol Symp Proc*. 1999;4(3):307–311.
143. Ito T, Meyer KC, Ito N, Paus R. Immune privilege and the skin. *Curr Dir Autoimmun*. 2008;10:27–52.
144. Takeo M, Chou WC, Sun Q, et al. Wnt activation in nail epithelium couples nail growth to digit regeneration. *Nature*. 2013;499(7457):228–232.
145. Bean WB. Nail growth: 30 years of observation. *Arch Intern Med*. 1974;134(3):497–502.
146. Sibinga MS. Observations on growth of fingernails in health and disease. *Pediatrics*. 1959;24(2):225–233.
147. Orentreich N, Markofsky J, Vogelman JH. The effect of aging on the rate of linear nail growth. *J Invest Dermatol*. 1979;73(1):126–130.
148. Geoghegan B, Roberts DF, Sampford MR. A possible climatic effect on nail growth. *J Appl Physiol*. 1958;13(1):135–138.
149. Landherr G, Braun-Falco O, Hofmann C, Plewig G, Galosi A. [Growth of finger nails in psoriasis patients undergoing PUVA therapy]. *Hautarzt*. 1982;33(4):210–213.
150. Fleckman P. Anatomy and physiology of the nail. *Dermatol Clin*. 1985;3(3):373–381.
151. Geyer AS, Onumah N, Uyttendaele H, Scher RK. Modulation of linear nail growth to treat diseases of the nail. *J Am Acad Dermatol*. 2004;50(2):229–234.
152. Baran R. [Therapeutic assessment and side-effects of the aromatic retinoid on the nail apparatus]. *Ann Dermatol Venereol*. 1982;109(4):367–371.

153. Yu H-J, Kwon H-M, Oh D-H, Kim JS. Is slow nail growth a risk factor for onychomycosis? *Clin Exp Dermatol*. 2004; 29(4):415–418.
154. Alobaida S, Lam JM. Beau lines associated with COVID-19. *CMAJ*. 2020;192(36):E1040.
155. Hongcharu W, Dwyer P, Gonzalez S, Anderson RR. Confirmation of onychomycosis by in vivo confocal microscopy. *J Am Acad Dermatol*. 2000;42(2):214–216.
156. Cinotti E, Fouilloux B, Perrot JL, Labeille B, Douchet C, Cambazard F. Confocal microscopy for healthy and pathological nail. *J Eur Acad Dermatol Venereol*. 2014;28(7):853–858.
157. Kaufman SC, Beuerman RW, Greer DL. Confocal microscopy: a new tool for the study of the nail unit. *J Am Acad Dermatol*. 1995;32(4):668–670.
158. Ronger S, Touzet S, Ligeron C, et al. Dermoscopic examination of nail pigmentation. *Arch Dermatol*. 2002;138(10): 1327–1333.
159. Tosti A, Argenziano G. Dermoscopy allows better management of nail pigmentation. *Arch Dermatol*. 2002;138(10): 1369–1370.
160. George R, Clarke S, Ioffreda M, Billingsley E. Marking of nail matrix biopsies with ink aids in proper specimen orientation for more accurate histologic evaluation. *Dermatol Surg*. 2008;34(12):1705–1706.
161. Reinig E, Rich P, Thompson CT. How to submit a nail specimen. *Dermatol Clin*. 2015;33(2):303–307.
162. Thompson C. Preparing a biopsy specimen. In: Chiacchio ND, Tosti A, et al., eds. *Therapies for Nail Disorders: A Quick Guide to Best Practice*. Appendix B. CRC Press; 2020.
163. Torczynski E. Preparation of ocular specimens for histopathologic examination. *Ophthalmology*. 1981;88(12):1367–1371.
164. Jordan CS, Stokes B, Thompson CT. Subungual debris cytopathology increases sensitivity of fungus detection in onychomycosis. *J Am Acad Dermatol*. 2016;75(1):222–224.
165. Bock M, Maiwald M, Kappe R, Nickel P, Näher H. Polymerase chain reaction-based detection of dermatophyte DNA with a fungus-specific primer system. *Mycoses*. 1994;37(3–4):79–84.
166. Spiliopoulou A, Bartzavali C, Jelastopulu E, Anastassiou ED, Christofidou M. Evaluation of a commercial PCR test for the diagnosis of dermatophyte nail infections. *J Méd Microbiol*. 2015;64(Pt 1):25–31.
167. Rodriguez O, Elenitsas R, Jiang AJ, Abbott J, Rubin AI. A call for nail clipping histopathology to become an essential component of the routine evaluation of melanonychia: Benefitting patients as a triage and surgical planning maneuver. *J Cutan Pathol*. 2023;50(3):279–283.
168. Sanchez M, Hu S, Miteva M, Tosti A. Onychomatricoma has channel-like structures on in vivo reflectance confocal microscopy. *J Eur Acad Dermatol Venereol*. 2014; 28(11):1560–1562.
169. Hafner J, Haenseler E, Ossent P, Burg G, Panizzon RG. Benzidine stain for the histochemical detection of hemoglobin in splinter hemorrhage (subungual hematoma) and black heel. *Am J Dermatopathol*. 1995;17(4):362–367.
170. Lecerf P, Richert B, Theunis A, André J. A retrospective study of squamous cell carcinoma of the nail unit diagnosed in a Belgian general hospital over a 15-year period. *J Am Acad Dermatol*. 2013;69(2):253–261.
171. Fanti PA, Tosti A. Subungual epidermoid inclusions: report of 8 cases. *Dermatology*. 1989;178(4):209–212.
172. Telang GH, Jellinek N. Multiple calcified subungual epidermoid inclusions. *J Am Acad Dermatol*. 2007;56(2):336–339.
173. Jaramillo FA, Mejía DCQ, Ordúz HMM, González Ardila C. Nail unit ultrasound: a complete guide of the nail diseases. *J Ultrasound*. 2017;20(3):181–192.
174. Goettmann S, Drape JL, Idy-Peretti I, et al. Magnetic resonance imaging: a new tool in the diagnosis of tumours of the nail apparatus. *Br J Dermatol*. 1994;130(6):701–710.

SECTION II

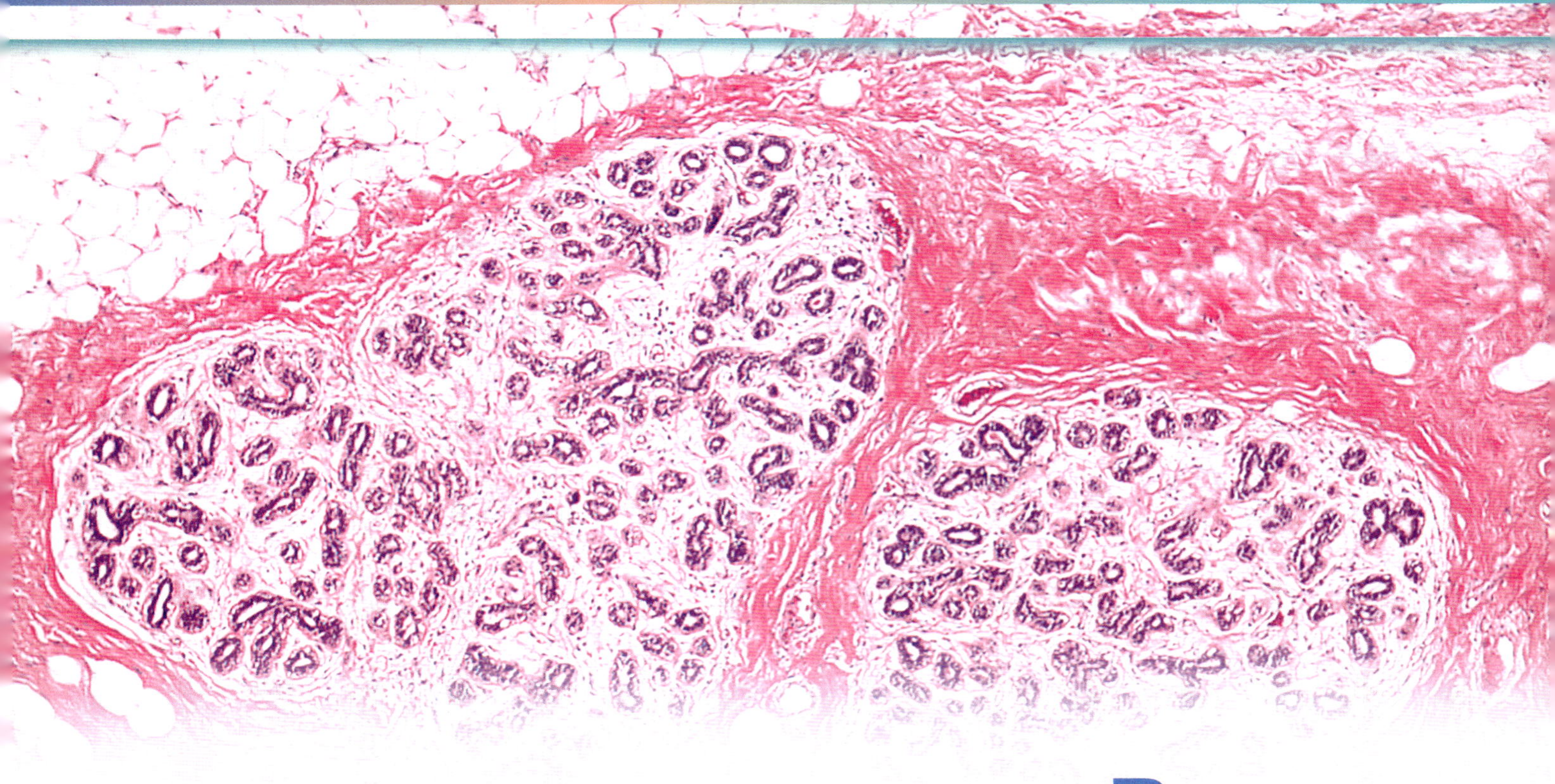

Breast

3

Breast

Stuart J. Schnitt ▪ Laura C. Collins

Remarkable advances in breast imaging have provided a variety of noninvasive means to assist in the evaluation of patients with breast disorders (1–4). Nevertheless, histologic examination of tissue specimens remains the cornerstone for the diagnosis of breast diseases, and an understanding of normal breast histology is essential for accurate evaluation of such specimens. It should be noted, however, that what constitutes "normal" histology in the breast varies according to sex, gender, age, menopausal status, phase of the menstrual cycle, pregnancy, and lactation, among other factors. Therefore, determination of whether a given breast specimen is normal or shows pathologic alterations must take these variables into consideration.

EMBRYOLOGY

Development of the human mammary gland begins during the 5th week of gestation, at which time thickenings of the ectoderm appear on the ventral surface of the fetus. These mammary ridges, also known as milk lines, extend from the axilla to the groin. Except for a small area in the pectoral region, the bulk of these ridges normally regress as the fetus continues to develop. Failed regression of other portions of the milk lines can result in the appearance of ectopic mammary tissue or accessory nipples anywhere along the milk lines in postnatal life; this phenomenon is most commonly encountered in the axilla, inframammary fold, and vulva (5–7).

The earliest stages of breast development are largely independent of sex steroid hormones (8). After the 15th week of gestation, the developing breast exhibits transient sensitivity to testosterone, which acts primarily on the mesenchyme. Under the influence of testosterone, the mesenchyme condenses around an epithelial stalk on the chest wall to form the breast bud, the site of mammary gland development. Solid epithelial columns then develop within the mesenchyme, and these ultimately give rise to the lobes or segments of the mammary gland. Portions of the fetal papillary dermis encase the developing epithelial cords and eventually give rise to the vascularized fibrous connective tissue that surrounds and invests the mammary ducts and lobules. The more collagen-rich reticular dermis extends into the breast to form the suspensory ligaments of Cooper, which attach the breast parenchyma to the skin. Portions of the mesenchyme differentiate into fat within the collagenous stroma between the 20th and 32nd weeks of gestation. During the last 8 weeks of gestation, the epithelial cords canalize and branch, forming lobuloalveolar structures as a result of mesenchymal paracrine effects. A depression in

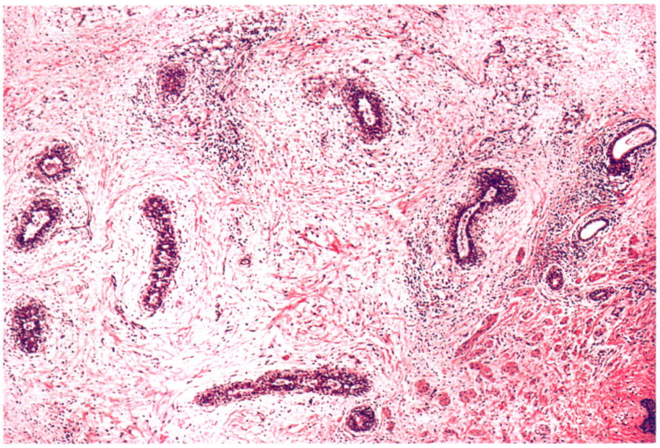

FIGURE 3.1 Breast tissue from an infant showing ducts embedded in a loose connective tissue stroma. The stromal mononuclear cells are hematopoietic elements, indicative of persistent extramedullary hematopoiesis. (Courtesy of Theonia Boyd, MD, Texas Children's Hospital, Houston, TX.)

the epidermis, the mammary pit, forms at the convergence of the lactiferous ducts. The nipple forms by evagination of the mammary pit near the time of birth.

During the last few weeks of gestation, the fetal mammary gland is responsive to maternal and placental steroid hormones, and, as a result, the epithelial cells in the acinar units exhibit secretory activity. At the time of birth, withdrawal of the maternal and placental sex steroids stimulates prolactin secretion, which in turn stimulates colostrum secretion. Both male and female neonates exhibit palpable enlargement of the breast bud. As the serum levels of maternal and placental sex steroid hormones and prolactin decline during the first month of life, secretory activity ends, and the gland regresses and becomes inactive. At this stage, and until puberty, the breast consists primarily of lactiferous ducts that exhibit some branching without evidence of progressive alveolar differentiation, although some rudimentary lobular structures may persist.

Another feature that may be seen in the fetal breast is extramedullary hematopoiesis, and this may persist in the periductal stroma until 4 months of age (9) (Fig. 3.1).

ADOLESCENCE

Adolescent breast development in the female begins with the onset of puberty and the cyclic secretion of estrogen and progesterone. However, a variety of other steroid and peptide hormones are also required for proper mammary gland development (8) (Table 3.1). The ducts elongate, branch, and develop a thickened epithelium due primarily to the influence of estrogen (10) (Fig. 3.2). The process of ductal growth and branching is largely independent of progesterone. There is an increase in the density of periductal connective tissue, also as a result of relative estrogen dominance. Deposition of stromal adipose tissue occurs, and it is this adipose tissue that is largely responsible for the enlargement and protrusion of the breast disk at this time. Cyclical exposure to progesterone following exposure to estrogen during ovulatory cycles promotes lobuloacinar growth, as well as connective tissue growth. Although the majority of breast development occurs during puberty, this process continues into the third decade, and terminal differentiation of the breast is only induced by pregnancy.

The adolescent male breast is composed of fibroadipose tissue and ducts lined by low cuboidal cells.

TABLE 3.1 Major Steroid and Peptide Hormonal Influences on the Breast

Hormone	Effects
Estrogen	Required for ductal growth and branching during adolescence Required for lobuloalveolar growth during pregnancy Required for induction of progesterone receptor Not necessary for maintenance of secretion or lactation
Progesterone	Required for lobuloalveolar differentiation and growth Probable mitogen in normal estrogen-primed breast Not necessary for ductal growth and branching
Testosterone	Stimulates breast mesenchyme during fetal development Causes mesenchymal destruction of mammary epithelium during critical period of testosterone sensitivity
Glucocorticoids	Required for maximal ductal growth Enhances lobuloalveolar growth during pregnancy
Insulin	Enhanced ductal-alveolar growth Enhances protein synthesis in mammary epithelium Required for secretory activity (with glucocorticoids and prolactin)
Prolactin	Stimulates epithelial growth after parturition Required for initiation and maintenance of lactation
Human placental lactogen	Able to substitute for prolactin in epithelial growth and differentiation Stimulates alveolar growth and lactogenesis in second half of pregnancy
Growth hormone	Required for ductal growth and branching during adolescence May contribute to lobuloacinar growth during pregnancy
Thyroid hormone	Increases epithelial response to prolactin May enhance lobuloacinar growth

McCarty KS, Nath M. Breast. In: Sternberg SS, ed. *Histology for Pathologists.* Lippincott-Raven; 1997: 71–82.

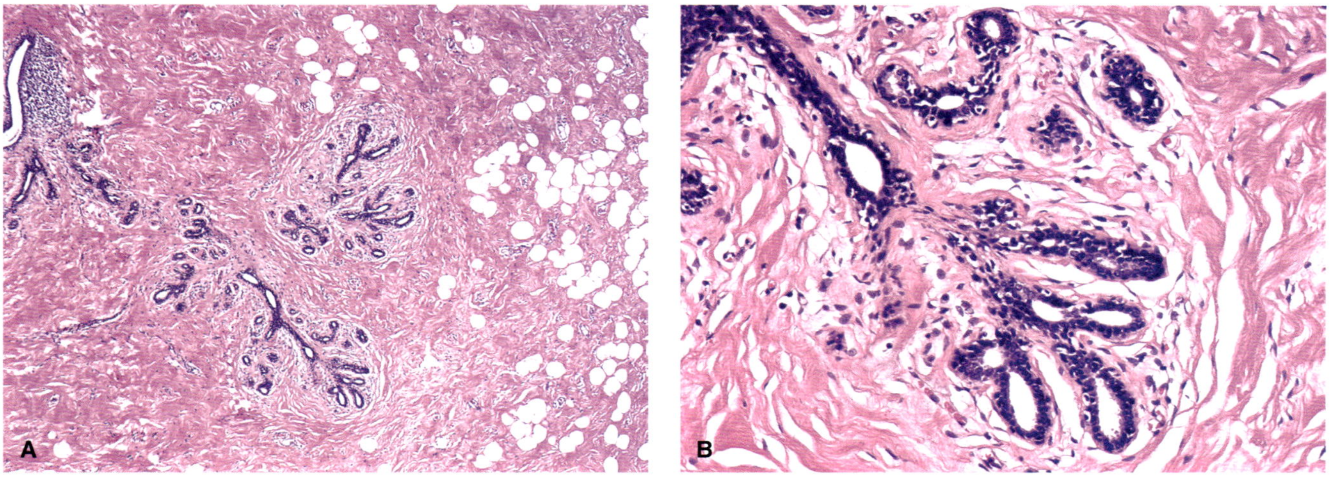

FIGURE 3.2 Adolescent breast tissue composed of branching ducts with rudimentary lobule development (type 1 lobules). The stroma consists of a mixture of fibrous connective tissue and adipose tissue. **A:** Scanning magnification. **B:** High power.

THE ADULT FEMALE BREAST

The size of the breast is greatly influenced by the individual's body habitus since the breast is a major repository for fat; it can range in size from 30 g to more than 1,000 g. The breast lies on the anterior chest wall over the pectoralis major muscle and typically extends from the 2nd to the 6th rib in the vertical axis and from the sternal edge to the mid-axillary line in the horizontal axis. Breast tissue also projects into the axilla as the tail of Spence. The breast extends laterally over the serratus anterior muscle and inferiorly over the external oblique muscle and the superior rectus sheath. The breast lies within a space in the superficial fascia, which is continuous with the cervical fascia superiorly and the superficial abdominal fascia of Cooper inferiorly. The only boundary of the breast that is anatomically well-defined is the deep surface where it abuts the pectoralis fascia. However, despite this macroscopic demarcation, microscopic foci of glandular tissue may extend into and even through the pectoral fascia and may traverse the other anatomic boundaries described above. The clinical significance of this observation is that even total mastectomy does not result in removal of all glandular breast tissue. Bundles of dense fibrous connective tissue, the suspensory ligaments of Cooper, extend from the skin to the pectoral fascia and provide support to the breast.

The adult female breast consists of a series of ducts, ductules, and lobular acinar units embedded within a stroma that is composed of varying amounts of fibrous and adipose tissue. The stroma comprises the major portion of the nonlactating adult breast, and the relative proportions of fibrous tissue and adipose tissue vary with age and among individuals (Fig. 3.3).

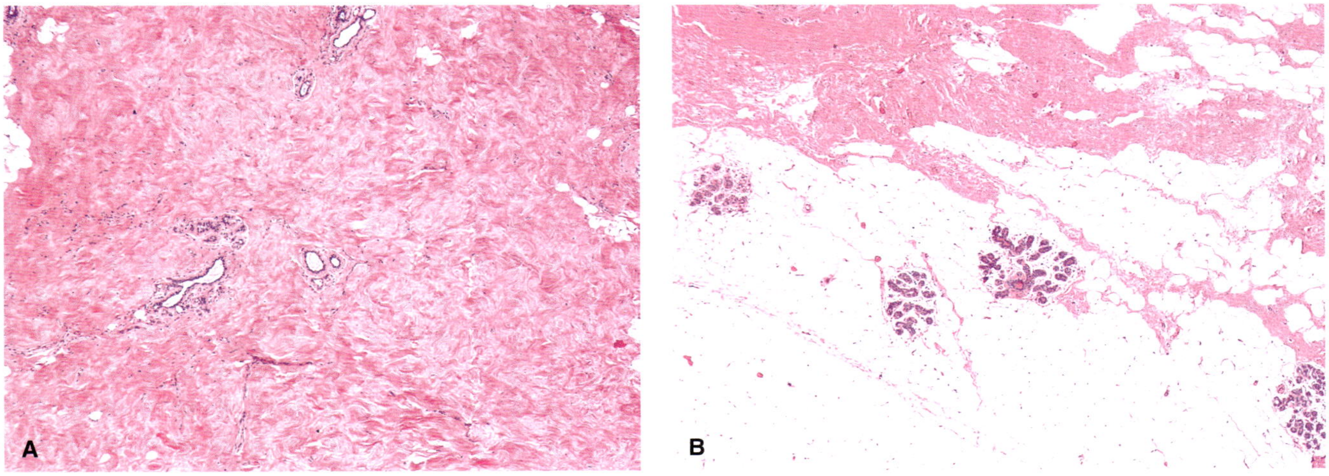

FIGURE 3.3 The stroma is the predominant component of the nonlactating breast and consists of varying amounts of collagen and adipose tissue. **A:** Low-power view of breast with dense, fibrotic stroma. **B:** Low-power view of breast with predominantly fatty stroma.

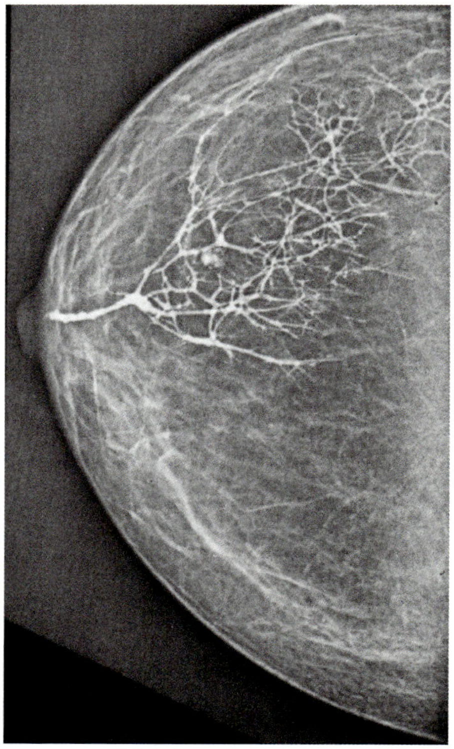

FIGURE 3.4 Ductogram (galactogram). Performed by injecting contrast material into an orifice of a lactiferous duct at the nipple, a ductogram demonstrates the complex ramifications of a single mammary ductal system (also known as a segment or lobe).

The ductal-lobular system of the breast is arranged in the form of segments, or lobes. While these segments can be readily appreciated by injecting the ductal system with dyes or radiologic contrast agents (Fig. 3.4), they are anatomically poorly defined, and no obvious boundaries can be appreciated between these segments during surgery, upon gross inspection of mastectomy specimens, or on histologic examination. In addition, these segments show considerable individual variation with regard to their extent and distribution (11), and the ramifications of individual segments may overlap. The segmental nature of some neoplastic processes in the breast, particularly ductal carcinoma in situ, is now widely appreciated. This recognition, in conjunction with observations in developmental anatomy and morphology, has led to the development of the "sick lobe" hypothesis of breast cancer (12,13). This theory postulates that early breast carcinoma (ductal carcinoma in situ) is a lobar disease, often isolated to a single ductal system (or lobe). Thus, surgical resection of the involved lobe or segment is an important therapeutic goal. Unfortunately, since it is not possible for the surgeon to define intraoperatively the boundaries of the involved segment, performing a "segmentectomy" to remove the entirety of a diseased segment is at this time more of a theoretical concept than a practically attainable goal.

Each segment consists of a branching structure that has been likened to a flowering tree (14) (Fig. 3.5). The lobules represent the flowers, draining into ductules and ducts (twigs and branches); these drain into the collecting ducts (trunk) that open onto the surface of the nipple. Just deep to the nipple, the ducts expand to form lactiferous sinuses, which in turn terminate in cone-shaped ampullae just below the surface of the nipple.

The actual number of segments in the breast and their relationship to each other has long been a matter of debate. Most textbooks indicate that there are 15 to 20 ductal orifices on the nipple surface and suggest that this corresponds to the number of ductal systems, segments, or lobes in the breast (5,6,15,16). In contrast, a number of mammary duct injection studies have suggested that there are only between 5 and 15 discrete breast ductal systems or segments in each breast. The discrepancy between the number of ductal orifices on the nipple and the actual number of breast segments or ductal systems may be explained by the fact that some of the orifices on the nipple represent openings of sebaceous glands or other nonductal tubular structures that do not contribute to the ductal-lobular anatomy of the breast. Another possibility is that some lactiferous ducts bifurcate immediately prior to entering the nipple or end blindly (16,17). The issue of anastomoses between ductal systems is also unresolved. One study indicated that, while ductal systems may lie in close proximity to one another and even intertwine within a particular quadrant, they do not interconnect (16). However, anastomoses between ductal systems have been reported by others (18).

The epithelium throughout the ductal-lobular system is bilayered, consisting of an inner (luminal) epithelial cell layer and an outer (basal) myoepithelial cell layer. The importance of this double cell layer cannot be overemphasized because it is one of the main guides used to distinguish benign from malignant lesions (19). The luminal epithelial cells of the resting breast ducts and lobules are cuboidal to columnar in shape and typically have pale

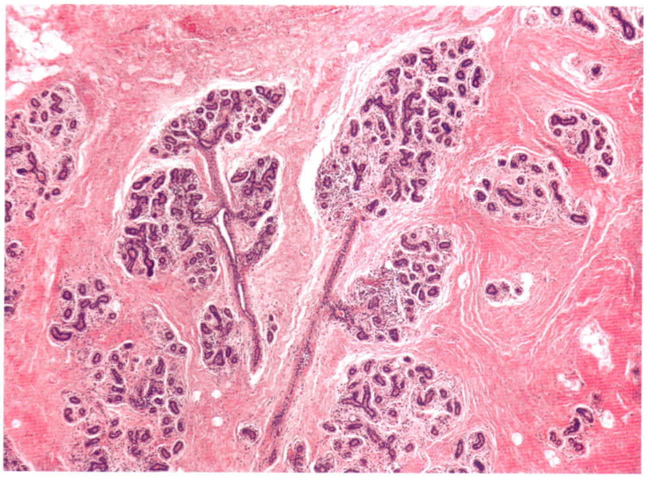

FIGURE 3.5 Microanatomy of normal adult female breast tissue showing extralobular ducts, terminal ducts, and lobules, the latter composed of groups of small glandular structures, the acini.

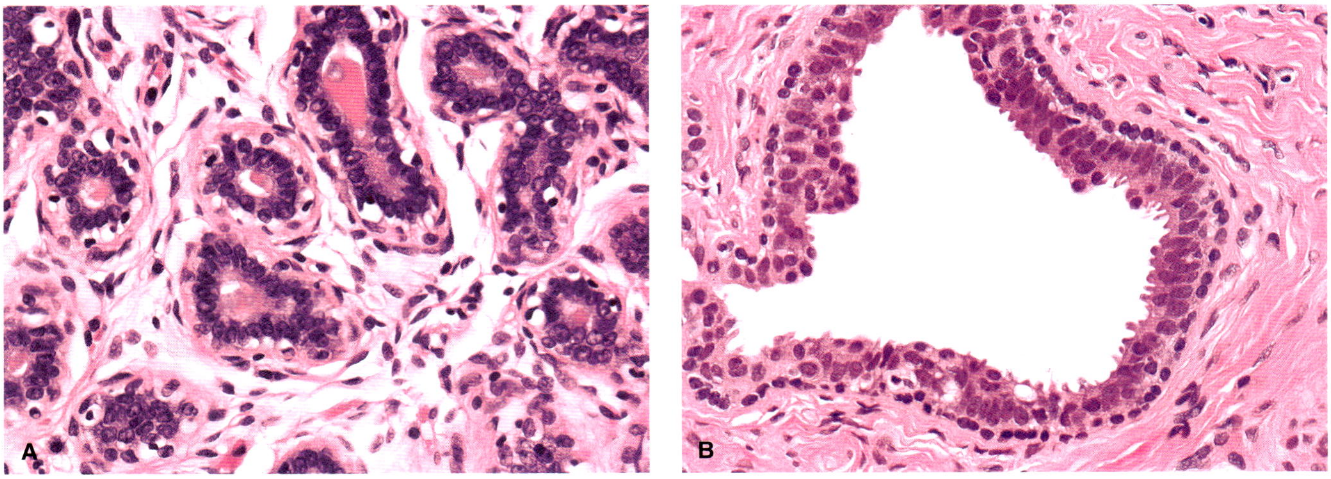

FIGURE 3.6 The mammary ductal-lobular system is lined by a dual cell population, an inner epithelial cell layer and an outer layer of myoepithelial cells. **A:** High-power view of a lobule. The myoepithelial cells surrounding the acinar epithelial cells are variably conspicuous. **B:** High-power view of an extralobular duct, showing distinct epithelial and myoepithelial cell layers.

eosinophilic cytoplasm and relatively uniform oval nuclei. These epithelial cells express a variety of low–molecular-weight cytokeratins, including cytokeratins 7, 8, 18, and 19 (20–24).

The outer (or myoepithelial) cell layer, although always present, is variably distinctive (Fig. 3.6). Myoepithelial cells range in appearance from barely discernible, flattened cells with compressed nuclei to prominent epithelioid cells with abundant clear cytoplasm. In some cases, the myoepithelial cells have a myoid appearance featuring a spindle cell shape and dense, eosinophilic cytoplasm reminiscent of smooth muscle cells (Fig. 3.7). Even when inconspicuous on hematoxylin- and eosin-stained sections, myoepithelial cells can readily be demonstrated using immunohistochemical stains for a variety of markers, including actins, calponin, smooth muscle myosin heavy chain, p63, CD10, and p75, among others (25–30) (Fig. 3.8). However, these markers vary in both sensitivity and specificity for myoepithelium as well as in their expression by location within the terminal duct lobular unit (TDLU). Myoepithelial cells express high–molecular-weight cytokeratins 5/6, 14, and 17 (20–24,31), but the expression of cytokeratin 14 is restricted to myoepithelial cells of the large ducts and terminal ducts; expression is not seen in myoepithelial cells of the intralobular ductules and acini (32).

Normal breast luminal epithelium is composed of cells in various states of differentiation as demonstrated by a panel of biomarkers that includes estrogen receptor (ER), androgen receptor, vitamin D receptor, low– and high–molecular-weight cytokeratins, and Ki-67. Interestingly, each of these cell types appears to correspond to breast cancers with a similar immunophenotype, raising the possibility

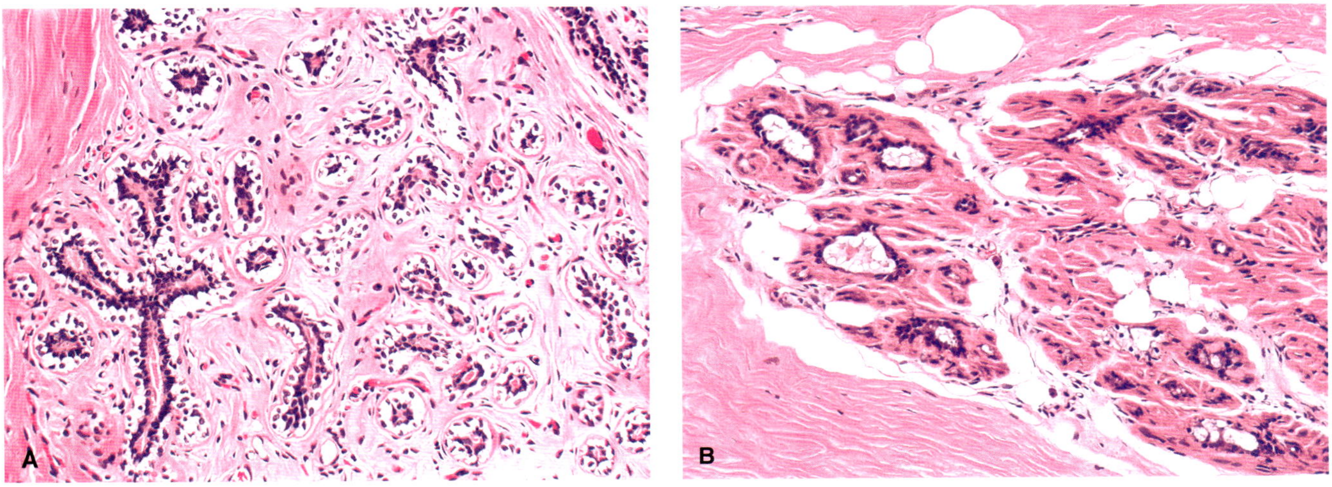

FIGURE 3.7 Myoepithelial cells can vary in their histologic appearance. **A:** Myoepithelial cells in this lobule show prominent cytoplasmic clearing. **B:** In this lobule, the myoepithelial cells show myoid features.

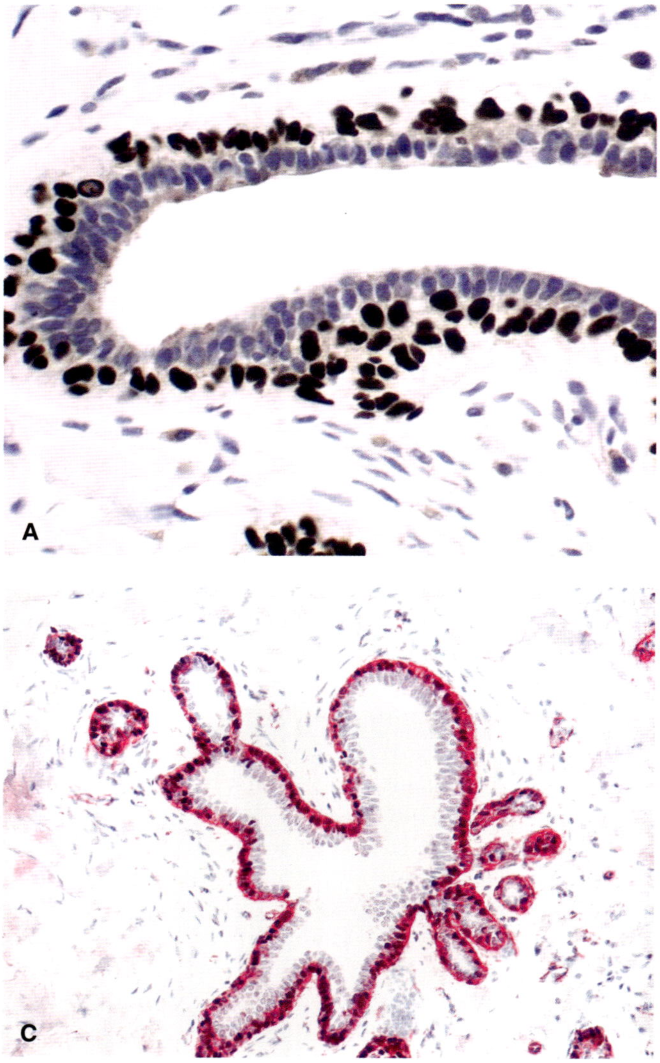

FIGURE 3.8 **A.** Extralobular duct **(A)** and **(B)** lobule immunostained for p63. The myoepithelial cells show strong nuclear reactivity, whereas the epithelial cell nuclei are negative. **C:** Double immunostain for smooth muscle actin (red cytoplasmic staining) and p63 (brown nuclear staining) highlight the myoepithelial cells around this mammary duct. Note the lack of staining of the epithelial cells for both p63 and smooth muscle actin.

that breast tumors have phenotypes or differentiation states that correspond to those of normal cells, akin to those seen in hematologic malignancies (33).

A third cell type dispersed irregularly throughout the ductal-lobular system expresses high–molecular-weight cytokeratins 5 and 14 in the absence of expression of markers of differentiated luminal epithelial cells (e.g., low–molecular-weight cytokeratins) or myoepithelial cells (e.g., smooth muscle myosin heavy chain). These cells have been postulated to represent progenitor cells capable of differentiating into both luminal epithelial cells and myoepithelial cells (34). The relationship of these putative "progenitor" cells to mammary stem cells is an unresolved issue (34,35). However, it is clear that cells with the characteristic features of stem cells (i.e., self-renewal and the ability to differentiate down different cell lineages to form all of the cell types found in the mature tissue) do exist in the breast. Mammary stem cells appear to be important in both breast development (36) and mammary carcinogenesis (35,37,38). Phenotypic features that have been associated with the mammary stem cell population include lack of expression of ER and progesterone receptor (PR), high expression of CD44, low or absent expression of CD24, and expression of aldehyde dehydrogenase 1 (ALDH1) (38,39), although the most accurate combination of markers to reliably identify mammary stem cells is unresolved at this time.

A basal lamina consisting of type IV collagen and laminin surrounds the mammary ducts, ductules, and acini (21,40). This basal lamina is present outside the myoepithelial cell layer and serves to demarcate the breast ductal-lobular system from the surrounding stroma (Fig. 3.9). Beyond the basal lamina, the extralobular ducts exhibit a zone of fibroblasts and capillaries. Elastic tissue is normally present in variable amounts around ducts and is generally more prominent in older than in younger women. Elastic fibers are not typically seen around the terminal ducts or lobular acini.

The lobule, together with its terminal duct forms the TDLU. This represents the structural and functional unit of the breast. During lactation, epithelial cells in both the terminal duct and lobule undergo secretory changes.

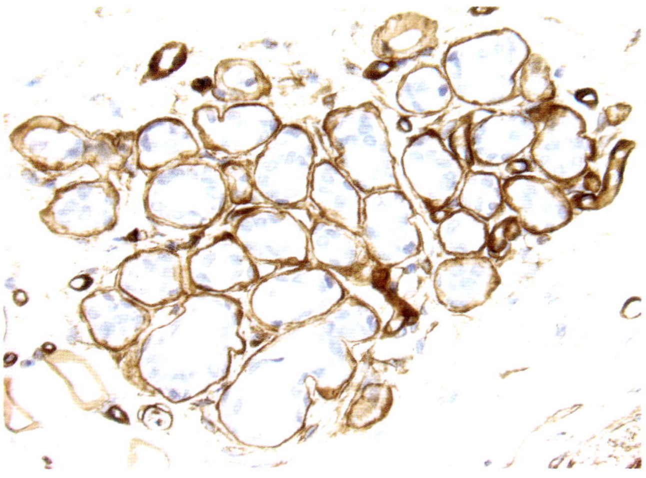

FIGURE 3.9 Immunostain for type IV collagen highlights the basal lamina around the acini of a lobule.

Thus, the terminal ducts are responsible for both secretion and transport of the secretions to the extra-lobular portion of the ductal system (15). Subgross anatomic studies have shown that most lesions originally termed "ductal" (e.g., cysts, ductal epithelial hyperplasia, and ductal carcinoma in situ) actually arise from the TDLU, which "unfolds" with coalescence of the acini to produce larger structures resembling ducts. The majority of pathologic changes in the breast, including in situ and invasive carcinomas, are generally considered to arise from the TDLU (14,41). Indeed, the only common lesion thought to arise from the large- or medium-sized duct rather than from the TDLU is the solitary intraductal papilloma (Fig. 3.10).

The normal lobule consists of a variable number of blind-ending terminal ductules, also called acini, each with its typical double cell layer. The lobular acini are invested by a loose, fibrovascular intralobular stroma with varying numbers of lymphocytes, plasma cells, macrophages, and mast cells. This specialized intralobular stroma is sharply demarcated from the surrounding interlobular stroma, which is less cellular, more dense, and highly collagenized, with variable amounts of adipose tissue (Fig. 3.11). One feature of note that is sometimes encountered in the extralobular stroma is the presence of multinucleated giant cells (42). Their significance is unknown; and, while they may present a disturbing appearance, they should not be mistaken for the malignant cells of an invasive carcinoma (Fig. 3.12).

The size of mammary lobules and number of acini per lobule are extremely variable. Russo et al. have described four lobule types (43–45). Type 1 lobules are the most rudimentary and most prevalent in prepubertal and nulliparous women, comprising 65% to 80% of the lobules in these groups (Fig. 3.2). These lobules are composed primarily of ducts with sprouting alveolar buds. From a practical standpoint, however, it is not possible to reliably distinguish type 1 lobules (i.e., those that have not fully developed) from those in which the number of acini is reduced due to involution. Type 1 lobules gradually evolve to more mature structures (type 2 and type 3 lobules) through the development of additional alveolar buds. The number of alveolar buds per lobule increases from approximately 11 in type 1 lobules to 47 and 80 in type 2 and type 3 lobules, respectively. There are data to suggest that the histologic appearance of normal lobules may influence the risk of subsequent breast cancer. Specifically, women whose breast tissue exhibits predominantly type 1 lobules or lobules that have undergone involution have a reduced risk of subsequent breast cancer compared to women with predominantly type 3 lobules or those that have not undergone involution (46–48). While there is some evidence to suggest the stage of involution is consistent across the breast and may provide an additional assessment of risk, the value of reporting the predominant lobular type in clinical practice remains to be determined (49).

The lobules exhibit morphologic changes during the menstrual cycle, and these are seen in both the epithelial and stromal components (50–53). These changes are summarized in Table 3.2. While the changes that occur during the menstrual cycle are variable among lobules in the same breast, even among immediately adjacent lobules, a

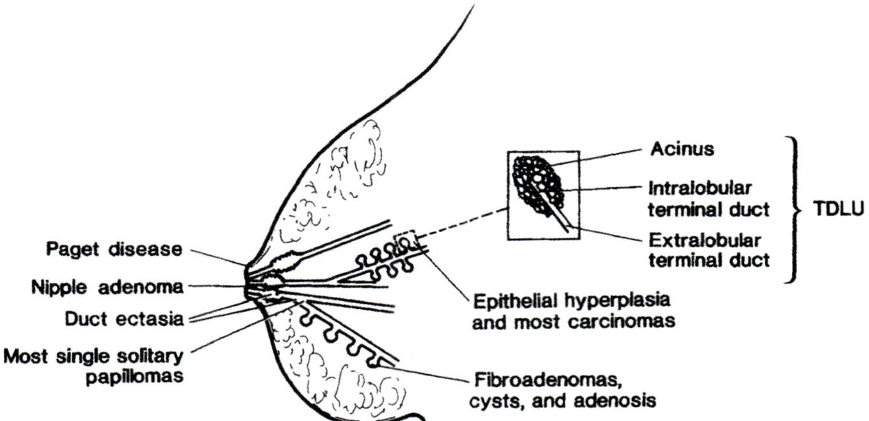

FIGURE 3.10 A schematic representation of the breast, indicating the sites of origin of pathologic lesions. (Reprinted from Schnitt SJ, Millis RR, Hanby AM, Oberman HA. The breast. In: Mills SE, Carter D, Greeson JK, Oberman HA, Reuter VE, Stoler MH, eds. *Sternberg's Diagnostic Surgical Pathology*. 4th ed. Lippincott, Williams & Wilkins; 2004:323–398.)

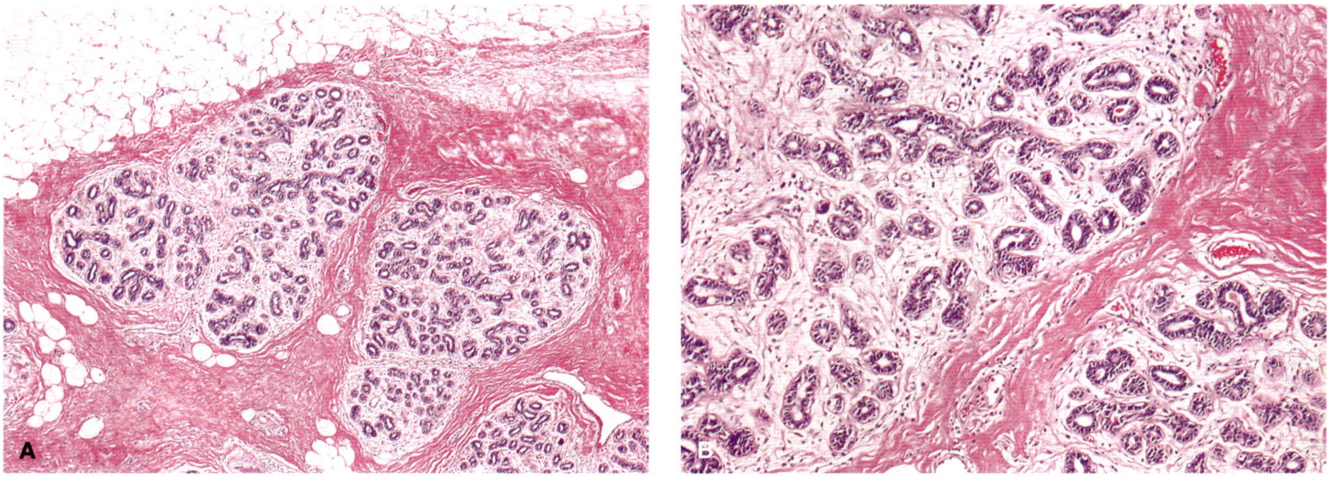

FIGURE 3.11 Intralobular and extralobular stroma. **A:** Low-power view of several lobules that are invested by loose, intralobular stroma. The interlobular stroma is composed primarily of dense collagen with admixed adipose tissue. **B:** Higher-power view contrasts loose intralobular stroma with more collagenized interlobular stroma.

dominant morphologic pattern is typically present in each phase. However, these menstrual cycle–related changes are subtle when compared with the dramatic alterations seen during pregnancy and lactation and when compared with the menstrual cycle–related changes seen in the endometrium. Occasionally, the TDLU epithelial cells show prominent clear cell change in the cytoplasm. This may be seen in both premenopausal and postmenopausal women and appears to be unrelated to pregnancy or exogenous hormone use (54).

The nipple–areola complex is a circular area of skin that exhibits increased pigmentation and contains numerous sensory nerve endings. The nipple is placed centrally and is elevated above the surrounding areola. As noted earlier, the tip of the nipple contains 15 to 20 orifices, though the number of openings may not correlate directly with the number of breast segments. In the nonlactating breast, these duct openings typically possess keratin plugs. The areola surface exhibits numerous small, rounded elevations, the tubercles of Montgomery.

Both the nipple and areola are covered by keratinizing, stratified squamous epithelium, and this extends for a short distance into the terminal portions of the lactiferous ducts. The epidermis of the nipple–areola complex may contain occasional clear cells that are cytologically benign and should not be confused with Paget cells (55,56) (Fig. 3.13). Some of these cells represent clear keratinocytes, whereas others are thought to be derived from epidermally located mammary ductal epithelium (i.e., Toker cells) (56).

The proximal ramifications of the mammary ductal system that are present in the dermis of the nipple typically have a pleated or serrated contour (Fig. 3.14). These ducts

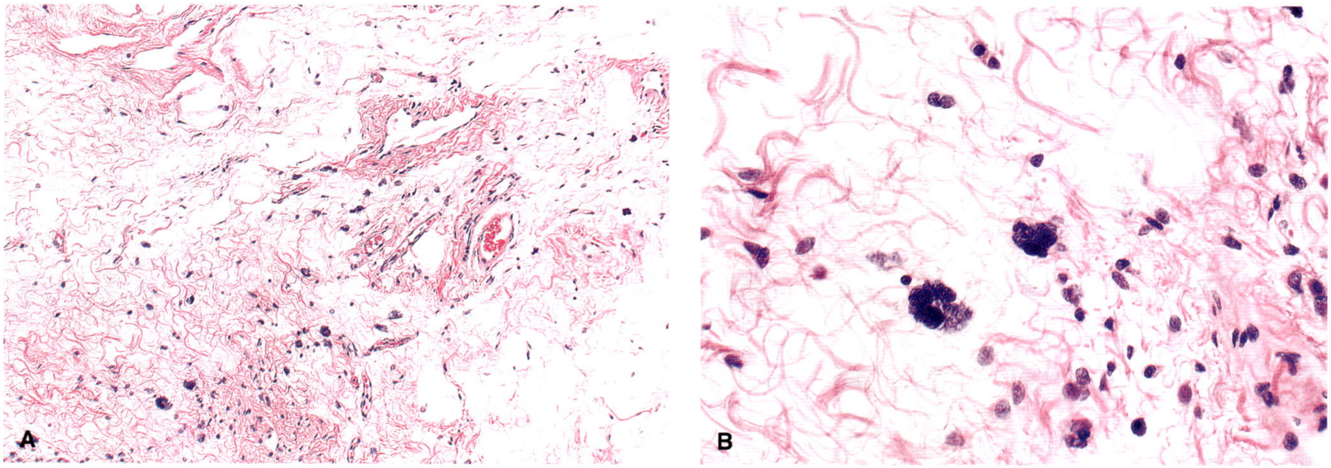

FIGURE 3.12 Multinucleated stromal giant cells. **A:** Low-power view showing multinucleated giant cells scattered in the stroma. **B:** High power illustrates cytologic detail. These cells have a mesenchymal phenotype. Despite their worrisome histologic appearance, they have no known clinical significance.

TABLE 3.2 Histologic Changes in Lobules During the Menstrual Cycle

Menstrual Cycle Phase	Epithelium	Acinar Lumina	Intralobular Stroma
Early follicular	*Cells*: single cell type (small, polygonal cells with pale eosinophilic cells); myoepithelial cells inconspicuous *Orientation*: poor *Secretion*: none *Mitoses/apoptosis*: rare	Largely closed and inapparent	Dense, cellular, with plump fibroblasts
Late follicular	*Cells*: three cell types, including luminal basophilic cells, intermediate pale cells (as seen in early follicular phase), and myoepithelial cells with clear cytoplasm *Orientation*: radial around lumen *Secretion*: none *Mitoses/apoptosis*: rare	Well defined	Less cellular and more collagenized than in early luteal phase
Early luteal	*Cells*: three cell types, including luminal basophilic cells with minimal apical snouting, intermediate pale cells, and myoepithelial cells with prominent cytoplasmic vacuolization and ballooning *Orientation*: radial around lumen *Secretion*: slight *Mitoses/apoptosis*: rare	Open, enlarged compared to follicular phase, with slight secretion	Loose
Late luteal	*Cells*: three cell types, including luminal basophilic cells with prominent apical snouting, intermediate pale cells and myoepithelial cells with prominent cytoplasmic vacuolization *Orientation*: radial around lumen *Secretion*: active apocrine secretion from luminal cells *Mitoses/apoptosis*: frequent (peak of mitotic activity)	Open, with secretion	Loose, edematous, congested blood vessels
Menstrual	*Cells*: two cell types, including luminal basophilic cells with scant cytoplasm and less apical snouting than in late luteal phase, and myoepithelial cells with extensive cytoplasmic vacuolization *Orientation*: radial around lumen *Secretion*: resorbing *Mitoses/apoptosis*: rare	Distended with secretion	Dense, cellular

McCarty KS, Nath M. Breast. In: Sternberg SS, ed. *Histology for Pathologists*. Lippincott-Raven; 1997:71–82.

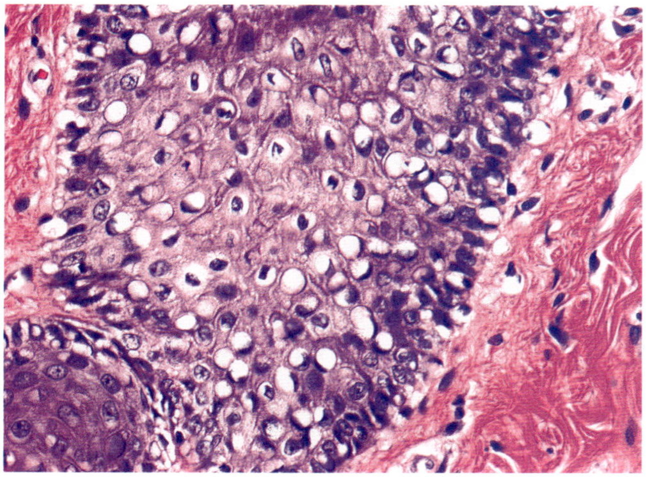

FIGURE 3.13 Clear cells in nipple epidermis. In some cells, the clearing is extreme, with formation of large intracytoplasmic vacuoles. These cells should not be mistaken for the cells of Paget disease.

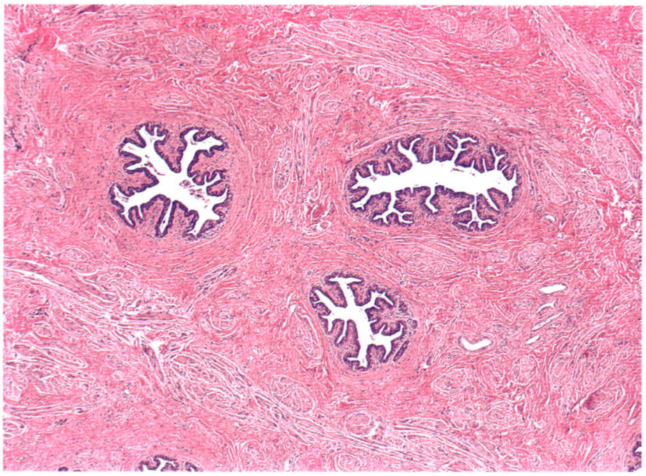

FIGURE 3.14 Cross section through the nipple. The irregular, pleated, or serrated contour of the nipple ducts is evident.

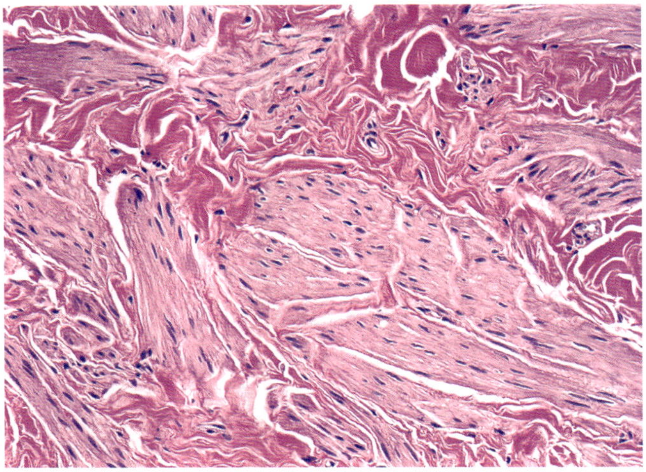

FIGURE 3.15 High-power view of nipple dermis/stroma, demonstrating prominent bundles of smooth muscle fibers.

are surrounded by a stroma rich in circular and longitudinal smooth muscle bundles, collagen, and elastic fibers (Fig. 3.15). Occasionally, lobules may be seen in the nipple (57). Simple mammary ducts are also present throughout the dermis of the areola, even at its periphery, and these may extend to within less than 1 mm of the basal layer of the epidermis (58).

While the nipple–areola complex lacks pilosebaceous units and hairs except at the periphery of the areola, the dermis contains numerous sebaceous glands. Some of these glands open directly onto the surface of the nipple and areola, whereas others drain into a lactiferous duct or share a common ostium with a lactiferous duct. The tubercles of Montgomery represent a unit consisting of a sebaceous apparatus and an associated lactiferous duct (59) (Fig. 3.16). During pregnancy, these tubercles become increasingly prominent. Apocrine sweat glands may also be seen in the dermis of the nipple and areola.

Another finding that may occasionally be encountered within the breast parenchyma is the presence of intramammary lymph nodes (60,61). These lymph nodes may be identified as an incidental finding in breast tissue removed because of another abnormality, or they may be seen as densities on mammograms (62).

PREGNANCY AND LACTATION

It is not until pregnancy that full development of the breast occurs in humans. During pregnancy, epithelial cell proliferation resumes. There is a dramatic increase in the number of lobules, as well as in the number of acinar units within each lobule secondary to epithelial cell proliferation and lobuloalveolar differentiation. These changes occur under the influence of estrogen, progesterone, prolactin, and growth hormone; growth is further enhanced by adrenal glucocorticoids and insulin. Lobular development and expansion occur at the expense of both the intralobular and extralobular stroma. By the end of the first trimester, there is grossly evident breast enlargement, superficial venous dilatation, and increased pigmentation of the areola.

During the second and third trimesters, lobular growth continues, and the acinar units begin to appear monolayered. The myoepithelial cells in the acini are difficult to discern at this time due to the increase in size and volume of the epithelial cells, but they remain clearly evident in the extralobular ducts. The cytoplasm of the epithelial cells becomes vacuolated, and secretion accumulates in the greatly expanded lobules. After parturition, the lactating breast is characterized by distension of the lobular acini as a result of abundant accumulated secretory material and prominent epithelial cell cytoplasmic vacuolization. Many of the epithelial cells have a bulbous or hobnail appearance

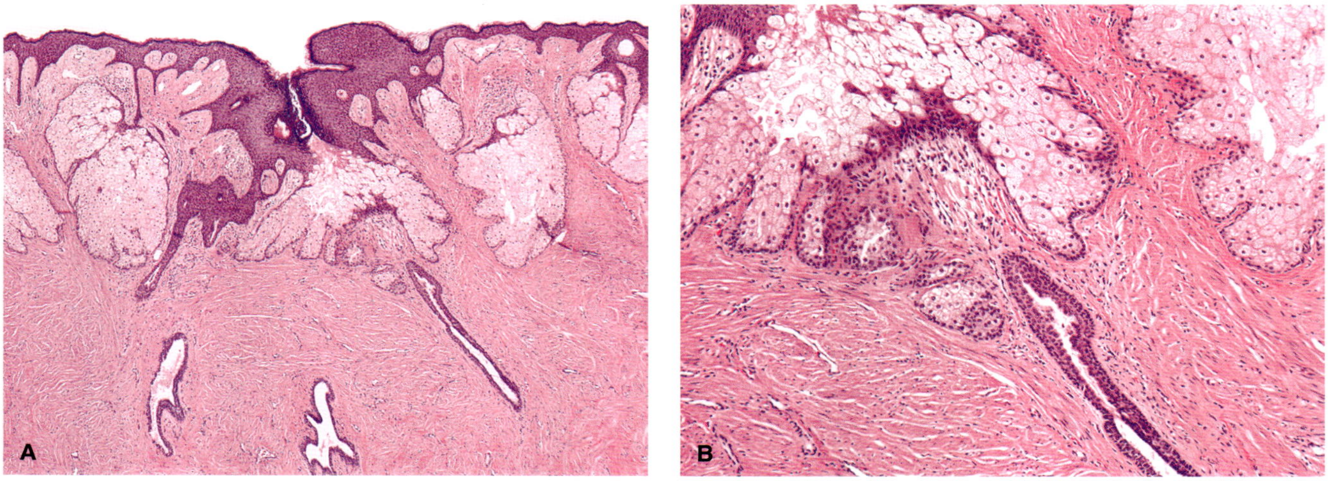

FIGURE 3.16 Montgomery areolar tubercle. **A:** Low-power view. **B:** Higher-power view. These tubercles are units composed of a lactiferous duct and associated sebaceous gland.

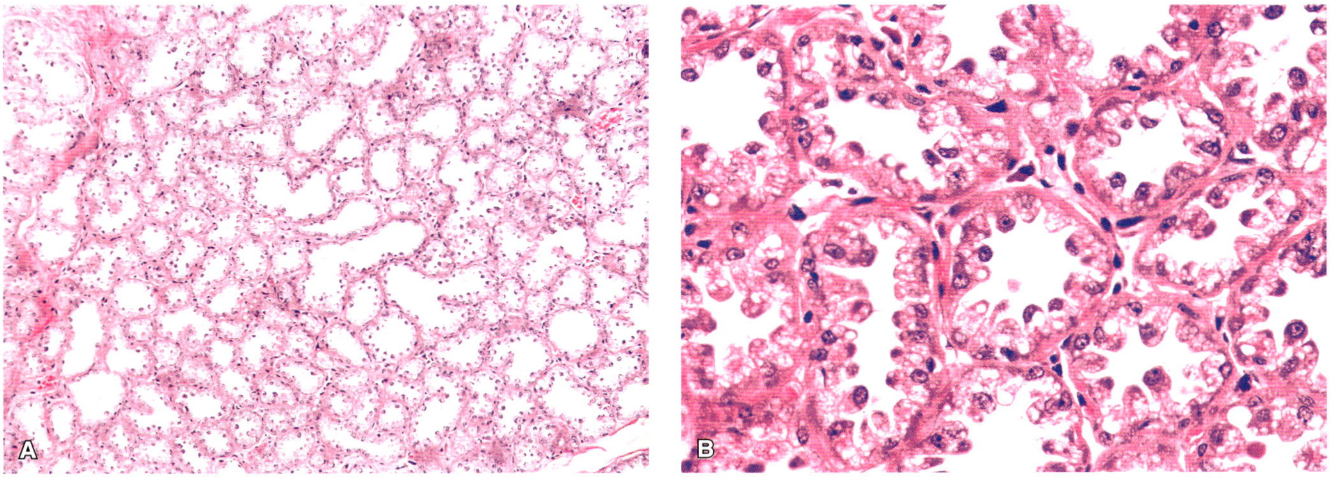

FIGURE 3.17 Lactating breast tissue. **A:** There are numerous acini in this lobule, and these are enlarged and dilated. There is minimal intervening stroma. **B:** Higher-power view illustrates prominent epithelial cell enlargement, cytoplasmic vacuolization, and protrusion of cells into the acinar lumen. Some of the cells have a hobnail appearance. Myoepithelial cells are inconspicuous.

and protrude into the acinar lumina (Fig. 3.17). Myoepithelial cells remain attenuated and inconspicuous. The florid changes seen in pregnancy and lactation can be alarming to the inexperienced observer; areas of infarction, which occasionally occur in the pregnant breast, may compound the problem (63).

When lactation ceases, the lobules involute and return to their normal resting appearance. Involution usually proceeds unevenly and takes several months. Involuting lobules are irregular in contour and are frequently infiltrated by lymphocytes and plasma cells (64,65). Occasionally, an isolated lobule showing secretory changes may be seen in the breasts of women who are not pregnant; this phenomenon may occur in the nulliparous woman as well.

MENOPAUSE

During the postmenopausal period, the reduction of estrogen and progesterone levels lead to involution and atrophy of the mammary TDLUs, reduction in the size and complexity of the acini, and loss of the specialized intralobular stroma (66,67). Ducts may become variably ectatic. The postmenopausal breast is characterized by a marked reduction in glandular tissue and collagenous stroma, often with a concomitant increase in stromal adipose tissue. The end stage of menopausal involution is typified by remnants of the TDLUs, seen as ducts with atrophic acini, surrounded by hyalinized connective tissue or embedded within adipose tissue with little or no surrounding stroma (Fig. 3.18).

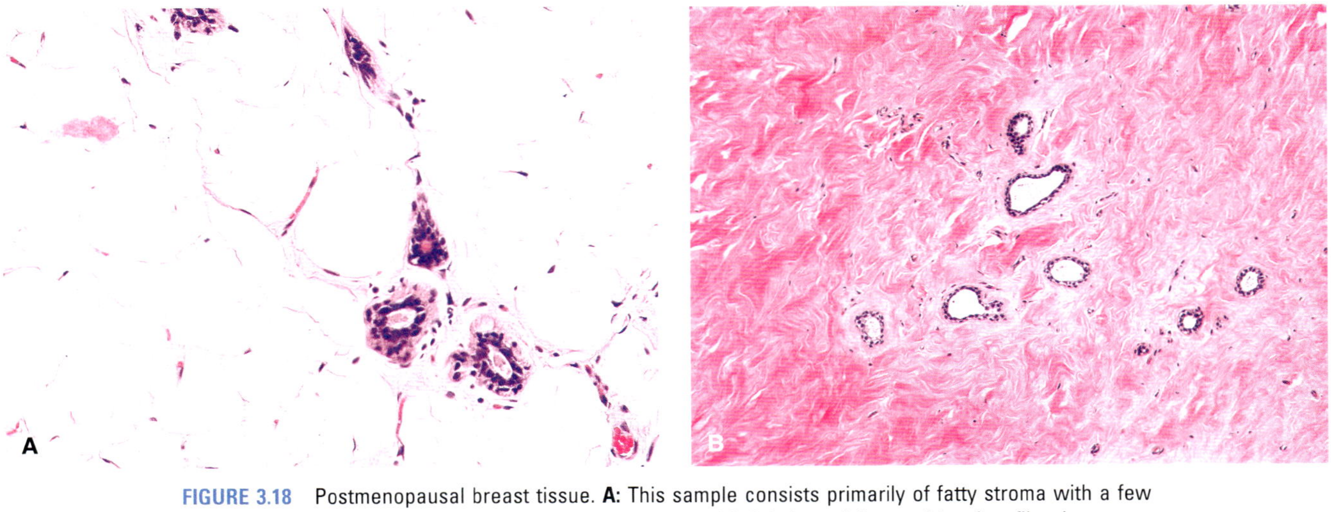

FIGURE 3.18 Postmenopausal breast tissue. **A:** This sample consists primarily of fatty stroma with a few atrophic ductules. **B:** In this specimen, a few residual, atrophic lobular acini are evident in a fibrotic stroma, which has replaced the normal, loose intralobular stroma.

BLOOD SUPPLY

The principal arterial supply to the breast is provided by the internal mammary and lateral thoracic arteries. Perforating branches of the internal mammary artery provide the blood supply to approximately 60% of the breast, mainly the medial and central portions. Approximately 30% of the breast, mainly the upper and outer portions, receives blood from the lateral thoracic artery. Branches of the thoracoacromial, intercostal, subscapular, and thoracodorsal arteries make minor contributions to the mammary blood supply (7).

Venous drainage of the breast, as in other locations, shows considerable individual variation but largely follows the arterial system. There is a superficial venous complex that runs transversely from lateral to medial in the subcutaneous tissue. These vessels then drain into the internal thoracic vein. Deep venous drainage of the breast is via three routes: the perforating branches of the internal thoracic vein, branches of the axillary vein, and tributaries of the intercostal veins, which drain posteriorly into the vertebral veins and the vertebral plexus (5,8).

LYMPHATIC DRAINAGE

Lymphatic drainage of the breast occurs through four routes: cutaneous, axillary, internal thoracic, and posterior intercostal lymphatics. The cutaneous lymphatic drainage system consists of both a superficial plexus of channels that lies within the dermis overlying the breast and a deeper network of lymphatic channels that runs with the mammary ducts in the subareolar area. Most of these cutaneous channels drain to the ipsilateral axilla. Cutaneous lymphatics from the inferior aspect of the breast may drain to the epigastric plexus and ultimately to the lymphatic channels of the liver and intra-abdominal lymph nodes.

There are three lymphatic drainage pathways in the mammary parenchyma. The most important drainage basin for lymphatic flow from the breast is the axilla, and the axillary lymph nodes receive the vast majority of the lymph drained. The internal thoracic route carries less than 10% of the lymphatic flow from the breast and ultimately terminates in the internal mammary lymph nodes (7). Drainage eventually empties into the great veins via the thoracic duct, the lower cervical nodes, or the jugular-subclavian confluence. The third and least important route of mammary lymphatic drainage is the posterior intercostal lymphatics, which drain into the posterior intercostal lymph nodes. An understanding of the lymphatic drainage of the breast is of particular importance in the current era of sentinel lymph node biopsy since this explains the occasional finding of sentinel lymph nodes outside of the axilla (5,7,8).

THE ADULT MALE BREAST

The adult male breast, like the female breast, is composed of glandular epithelial elements embedded in a stroma that is composed of varying amounts of collagen and adipose tissue. However, in contrast to the adult female breast, the epithelial elements of the male breast usually consist of branching ducts without lobule formation.

BREAST TISSUE IN TRANSGENDER INDIVIDUALS

With gender-affirming care becoming increasingly available to individuals in need, pathologists may be required to examine histologically biologically female or male breast tissue influenced by exogenous testosterone or estrogen/progesterone, respectively. In some cases, these agents are used in combination with medications that suppress endogenous hormones. Breast tissue from transgender men reflects the influence of exogenous testosterone in those individuals using gender-affirming hormones. Specifically, lobular atrophy is frequently seen, the degree of which is greater with longer duration of testosterone therapy. Other findings reported in up to 45% of people include stromal fibrosis, cysts, apocrine metaplasia, and fibroadenomatous change, with usual ductal hyperplasia and pseudoangiomatous stromal hyperplasia being less frequently seen (68–70). Atypical lesions are rarely encountered (68,70,71). Breast tissue is less frequently encountered from transwomen, but when specimens are received, they are more likely to show development of lobules along with proliferative and occasionally secretory changes (72).

BIOLOGIC MARKERS, IMMUNOPHENOTYPE, AND MOLECULAR BIOLOGY

Estrogen Receptor and Progesterone Receptor

There are at least two different ERs, ERα and ERβ; ERα has been far more extensively studied. Using immunohistochemistry, ERα expression can be demonstrated in the nuclei of both ductal and lobular epithelial cells, with a higher proportion in lobules than in ducts. However, even in the lobules, only a small proportion of the cells show ERα immunoreactivity. Most often, ERα-positive cells in the lobules are distributed singly,

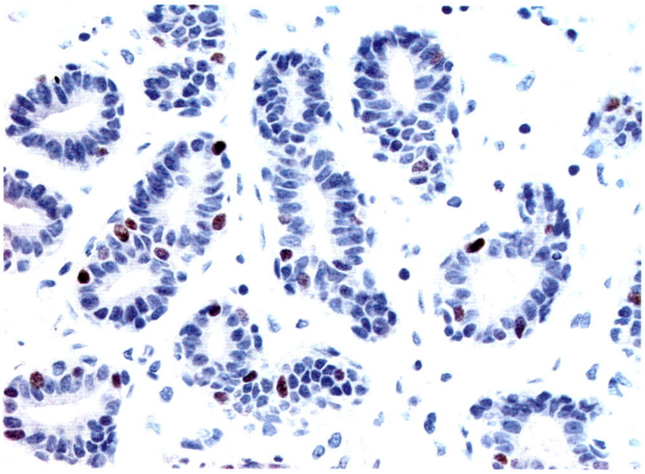

FIGURE 3.19 Immunostain for estrogen receptor-α (ERα) in a normal lobule. A minority of epithelial cells show nuclear staining.

admixed with, and surrounded by, ERα-negative cells (73) (Fig. 3.19). Furthermore, there is considerable heterogeneity in staining for ERα among lobules within the same breast. Of interest, there is generally an inverse relationship between expression of ERα and markers of cell proliferation in breast tissue from premenopausal women. In particular, most ERα-positive cells do not show expression of the proliferation-related antigen Ki-67, and Ki-67–positive cells are typically ERα-negative. The proportion of ERα-positive cells, including proliferating cells, gradually increases with age but remains relatively stable after the menopause. The incidence of lobules showing contiguous patches of ERα-positive cells also increases with age and with involutional changes (73). In premenopausal women, ERα expression varies with the phase of the menstrual cycle, being higher in the follicular phase than the luteal phase (74). Myoepithelial cells do not show ERα immunoreactivity (33).

Studies have indicated that a second form of ER, ERβ, is also expressed in normal breast tissue. Expression of ERβ has been observed not only in epithelial cells of ducts and lobules, but also in myoepithelial cells, endothelial cells, and stromal cells (74–77). The expression of this form of ER does not seem to vary with the phase of the menstrual cycle. It has been speculated that the relative levels of ERβ and ERα may be important in determining the risk of breast cancer development, and that higher levels of ERβ relative to ERα are protective against neoplastic progression in the breast (75). However, additional studies are needed to more clearly elucidate the role of ERβ in normal breast physiology and in breast cancer pathogenesis in order to determine which ERβ isoform provides the greatest specificity in this regard (78).

Expression of PR has not been as extensively studied in normal breast tissue as has ER. Like ERα, PR is expressed in the nuclei of ductal and lobular epithelium. However, in contrast to ERα expression, PR expression does not seem to vary with the menstrual cycle phase (74).

Other Biomarkers and Immunophenotypic Features

Expression of a wide variety of biomarkers has been studied in benign breast tissue (79) and a comprehensive review of these is beyond the scope of this chapter. However, a couple of markers merit brief mention because their expression may cause diagnostic confusion. For unclear reasons, the normal breast epithelium may rarely show HER2 protein overexpression, p53 protein accumulation, or *p53* mutations. Of note, p53 alterations have been identified more frequently in normal epithelium adjacent to breast carcinomas that are negative for ER, PR, and HER2 ("triple negative") and/or in *BRCA1* mutation carriers, raising the possibility that this "p53 signature" may confer a predisposition to p53-associated high-grade carcinomas (80,81).

The antiapoptotic protein bcl-2 is consistently expressed by normal breast epithelial cells (82). S-100 protein is strongly expressed by normal myoepithelial cells and variably expressed by mammary epithelial cells (83). Epithelial cells also show variable expression for casein (84), α-lactalbumin (85), gross cystic disease fluid protein-15 (86), mammaglobin (87), GATA-3, TRPS1 (88), and c-kit (CD117) (89), among other proteins. As noted earlier, keratins 7, 8, 18, and 19 (20–24) are typically expressed by epithelial cells, whereas myoepithelial cells express keratins 5/6, 14, and 17 (20–24,31).

Molecular Markers

The ability to evaluate DNA, RNA, and protein using the modern tools of molecular biology, particularly when guided by techniques such as laser capture microdissection (90), are enhancing our understanding of breast tumorigenesis and may even serve to redefine what constitutes "normal." For example, a number of studies have shown that histologically normal TDLUs can exhibit an abnormal genotype characterized by loss of heterozygosity (91,92), allelic imbalance (93,94) at various chromosomal loci, or altered gene expression profiles (95,96). At this time, however, the significance of these molecular alterations in histologically normal breast tissue remains to be determined. Studies of normal breast tissue using these techniques may also help further define the presence and nature of progenitor cells and their role in breast development and carcinogenesis (23,97–99), as well as patterns of gene and protein expression that distinguish normal from abnormal cells in breast tissue (100–103).

CONCLUSION

The histologic features of the normal breast are dynamic and vary with age and hormonal milieu, among other factors. An understanding of normal breast histology is essential to permit the reliable distinction between physiologic changes and pathologic alterations.

REFERENCES

1. Jochelson M. Breast cancer imaging: the future. *Semin Oncol*. 2001;28(3):221–228.
2. Leung JWT. New modalities in breast imaging: digital mammography, positron emission tomography, and sestamibi scintimammography. *Radiol Clin North Am*. 2002; 40(3):467–482.
3. Koomen M, Pisano ED, Kuzmiak C, Pavic D, McLelland R. Future directions in breast imaging. *J Clin Oncol*. 2005;23(8):1674–1677.
4. Hylton N. Magnetic resonance imaging of the breast: opportunities to improve breast cancer management. *J Clin Oncol*. 2005;23(8):1678–1684.
5. Rosen PP. *Rosen's Breast Pathology*. 3rd ed. Lippincott Williams & Wilkins; 2009.
6. Tavassoli FA. Normal development and anomalies. In: Tavassoli FA, ed. *Pathology of the Breast*. 2nd ed. Appelton and Lange; 1999:1–25.
7. Osborne MP. Breast anatomy and development. In: Harris JR, Lippman ME, Morrow M, Osborne CK, eds. *Diseases of the Breast*. 4th ed. Lippincott Williams and Wilkins; 2010:3–11.
8. McCarty KS, Nath M. Breast. In: Sternberg SS, ed. *Histology for Pathologists*. Lippincott-Raven; 1997:71–82.
9. Anbazhagan R, Bartek J, Monaghan P, Gusterson BA. Growth and development of the human infant breast. *Am J Anat*. 1991;192(4):407–417.
10. Monaghan P, Perusinghe NP, Cowen P, Gusterson BA. Peripubertal human breast development. *Anat Rec*. 1990; 226(4):501–8.
11. Going JJ, Moffat DF. Escaping from Flatland: clinical and biological aspects of human mammary duct anatomy in three dimensions. *J Pathol*. 2004;203(1):538–544.
12. Going JJ, Mohun TJ. Human breast duct anatomy, the 'sick lobe' hypothesis and intraductal approaches to breast cancer. *Breast Cancer Res Treat*. 2006;97(3):285–291.
13. Tot T. The theory of the sick breast lobe and the possible consequences. *Int J Surg Pathol*. 2007;15(4):369–375.
14. Jensen HM. Breast pathology, emphasizing precancerous and cancer-associated lesions. In: Bulbrook RO, Taylor DJ, eds. *Commentaries on Research in Breast Disease. Vol. 2*. Alan R. Liss; 1981:41–86.
15. Page DL, Anderson TJ. *Diagnostic Histopathology of the Breast*. Churchill Livingstone; 1987.
16. Love SM, Barsky SH. Anatomy of the nipple and breast ducts revisited. *Cancer*. 2004;101(9):1947–1957.
17. Rusby JE, Brachtel EF, Michaelson JS, Koerner FC, Smith BL. Breast duct anatomy in the human nipple: three-dimensional patterns and clinical implications. *Breast Cancer Res Treat*. 2007;106(2):171–179.
18. Ohtake T, Kimijima I, Fukushima T, et al. Computer-assisted complete three-dimensional reconstruction of the mammary ductal/lobular systems: implications of ductal anastomoses for breast-conserving surgery. *Cancer*. 2001;91(12):2263–2272.
19. Schnitt SJ, Millis RR, Hanby AM, Oberman HA. The Breast. In: Mills SE, ed. *Diagnostic Surgical Pathology*. 4th ed. Lippincott, Williams & Wilkins; 2004:323–395.
20. Jarasch ED, Nagle RB, Kaufmann M, Maurer C, Böcker WJ. Differential diagnosis of benign epithelial proliferations and carcinomas of the breast using antibodies to cytokeratins. *Hum Pathol*. 1988;19(3):276–289.
21. Bocker W, Bier B, Freytag G, et al. An immunohistochemical study of the breast using antibodies to basal and luminal keratins, alpha-smooth muscle actin, vimentin, collagen IV and laminin. Part I: normal breast and benign proliferative lesions. *Virchows Arch A Pathol Anat Histopathol*. 1992;421(4):315–322.
22. Heatley M, Maxwell P, Whiteside C, Toner P. Cytokeratin intermediate filament expression in benign and malignant breast disease. *J Clin Pathol*. 1995;48(1):26–32.
23. Bocker W, Moll R, Poremba C, et al. Common adult stem cells in the human breast give rise to glandular and myoepithelial cell lineages: a new cell biological concept. *Lab Invest*. 2002;82(6):737–746.
24. Abd El-Rehim DM, Pinder SE, Paish CE, et al. Expression of luminal and basal cytokeratins in human breast carcinoma. *J Pathol*. 2004;203(2):661–671.
25. Yaziji H, Gown AM, Sneige N. Detection of stromal invasion in breast cancer: the myoepithelial markers. *Adv Anat Pathol*. 2000;7(2):100–109.
26. Barbareschi M, Pecciarini L, Cangi MG, et al. p63, a p53 homologue, is a selective nuclear marker of myoepithelial cells of the human breast. *Am J Surg Pathol*. 2001;25(8):1054–1060.
27. Moritani S, Kushima R, Sugihara H, Bamba M, Kobayashi TK, Hattori T. Availability of CD10 immunohistochemistry as a marker of breast myoepithelial cells on paraffin sections. *Mod Pathol*. 2002;15(4):397–405.
28. Yeh IT, Mies C. Application of immunohistochemistry to breast lesions. *Arch Pathol Lab Med*. 2008;132(3):349–358.
29. Bhargava R, Dabbs DJ. Use of immunohistochemistry in diagnosis of breast epithelial lesions. *Adv Anat Pathol*. 2007;14(2):93–107.
30. Popnikolov NK, Cavone SM, Schultz PM, Garcia FU. Diagnostic utility of p75 neurotrophin receptor (p75NTR) as a marker of breast myoepithelial cells. *Mod Pathol*. 2005;18(12):1535–1541.
31. Nielsen TO, Hsu FD, Jensen K, et al. Immunohistochemical and clinical characterization of the basal-like subtype of invasive breast carcinoma. *Clin Cancer Res*. 2004;10(16):5367–5374.
32. Going JJ. *Normal Breast*. In: O'Malley FP, Pinder SE, Goldblum JR, eds. Churchill, Livingstone, Elsevier; 2006:55–65.
33. Santagata S, Thakkar A, Ergonul A, et al. Taxonomy of breast cancer based on normal cell phenotype predicts outcome. *J Clin Invest*. 2014;124(2):859–870.
34. Boecker W, Weigel S, Handel W, Stute P. The normal breast. In: Boecker W, ed. *Preneoplasia of the Breast: A New Conceptual Approach to Proliferative Breast Disease*. Elsevier; 2006:2–27.
35. Cariati M, Purushotham AD. Stem cells and breast cancer. *Histopathology*. 2008;52(1):99–107.
36. Shackleton M, Vaillant F, Simpson KJ, et al. Generation of a functional mammary gland from a single stem cell. *Nature*. 2006;439(7072):84–88.
37. Morimoto K, Kim SJ, Tanei T, et al. Stem cell marker aldehyde dehydrogenase 1-positive breast cancers are characterized by negative estrogen receptor, positive human epidermal growth factor receptor type 2, and high Ki67 expression. *Cancer Sci*. 2009;100(6):1062–1068.
38. Zhou L, Jiang Y, Yan T, et al. The prognostic role of cancer stem cells in breast cancer: a meta-analysis of published literatures. *Breast Cancer Res Treat*. 2010;122(3):795–801.

39. Al-Hajj M, Wicha MS, Benito-Hernandez A, Morrison SJ, Clarke MF. Prospective identification of tumorigenic breast cancer cells. *Proc Natl Acad Sci U S A*. 2003; 100(7):3983–3988.
40. Barsky SH, Siegal GP, Jannotta F, Liotta LA. Loss of basement membrane components by invasive tumors but not by their benign counterparts. *Lab Invest*. 1983;49(2):140–147.
41. Wellings SR, Jensen HM, Marcum RG. An atlas of subgross pathology of the human breast with special reference to possible precancerous lesions. *J Natl Cancer Inst*. 1975;55(2):231–273.
42. Rosen PP. Multinucleated mammary stromal giant cells: a benign lesion that simulates invasive carcinoma. *Cancer*. 1979;44(4):1305–1308.
43. Russo J, Russo IH. Development of the human mammary gland. In: Neville MC, Daniel CW, eds. *The Mammary Gland Development, Regulation, and Function*. Plenum Press; 1987:67–93.
44. Russo J, Rivera R, Russo IH. Influence of age and parity on the development of the human breast. *Breast Cancer Res Treat*. 1992;23(3):211–218.
45. Russo J, Romero AL, Russo IH. Architectural pattern of the normal and cancerous breast under the influence of parity. *Cancer Epidemiol Biomarkers Prev*. 1994;3(3):219–224.
46. Baer HJ, Collins LC, Connolly JL, Colditz GA, Schnitt SJ, Tamimi RM. Lobule type and subsequent breast cancer risk: results from the Nurses' Health Studies. *Cancer*. 2009; 115(7):1404–1411.
47. Milanese TR, Hartmann LC, Sellers TA, et al. Age-related lobular involution and risk of breast cancer. *J Natl Cancer Inst*. 2006;98(22):1600–1607.
48. Radisky DC, Visscher DW, Frank RD, et al. Natural history of age-related lobular involution and impact on breast cancer risk. *Breast Cancer Res Treat*. 2016;155(3):423–430.
49. Vierkant RA, Hartmann LC, Pankratz VS, et al. Lobular involution: localized phenomenon or field effect? *Breast Cancer Res Treat*. 2009;117(1):193–196.
50. Vogel PM, Georgiade NG, Fetter BF, Vogel FS, McCarty KS, Jr. The correlation of histologic changes in the human breast with the menstrual cycle. *Am J Pathol*. 1981;104(1):23–34.
51. Longacre TA, Bartow SA. A correlative morphologic study of human breast and endometrium in the menstrual cycle. *Am J Surg Pathol*. 1986;10(6):382–393.
52. Ramakrishnan R, Khan SA, Badve S. Morphological changes in breast tissue with menstrual cycle. *Mod Pathol*. 2002;15(12):1348–1356.
53. Anderson TJ. Normal breast: myths, realities, and prospects. *Mod Pathol*. 1998;11(2):115–119.
54. Tavassoli FA, Yeh IT. Lactational and clear cell changes of the breast in nonlactating, nonpregnant women. *Am J Clin Pathol*. 1987;87(1):23–29.
55. Toker C. Clear cells of the nipple epidermis. *Cancer*. 1970; 25(3):601–610.
56. Kohler S, Rouse RV, Smoller BR. The differential diagnosis of pagetoid cells in the epidermis. *Mod Pathol*. 1998; 11(1):79–92.
57. Rosen PP, Tench W. Lobules in the nipple. Frequency and significance for breast cancer treatment. *Pathol Annu*. 1985;20(Pt 2):317–322.
58. Schnitt SJ, Goldwyn RM, Slavin SA. Mammary ducts in the areola: implications for patients undergoing reconstructive surgery of the breast. *Plast Reconstr Surg*. 1993;92(7):1290–1293.
59. Smith DM Jr., Peters TG, Donegan WL. Montgomery's areolar tubercle. A light microscopic study. *Arch Pathol Lab Med*. 1982;106(2):60–63.
60. Egan RL, McSweeney MB. Intramammary lymph nodes. *Cancer*. 1983;51(10):1838–1842.
61. Jadusingh IH. Intramammary lymph nodes. *J Clin Pathol*. 1992;45(11):1023–1026.
62. Svane G, Franzen S. Radiologic appearance of nonpalpable intramammary lymph nodes. *Acta Radiol*. 1993;34(6):577–580.
63. Oberman HA. Breast lesions confused with carcinoma. In: McDivitt RW, Oberman HA, Ozello L, Kaufman N, eds. *The Breast*. Williams and Wilkins; 1984:1–3.
64. Battersby S, Anderson TJ. Proliferative and secretory activity in the pregnant and lactating human breast. *Virchows Arch A Pathol Anat Histopathol*. 1988;413(3):189–196.
65. Battersby S, Anderson TJ. Histological changes in breast tissue that characterize recent pregnancy. *Histopathology*. 1989;15(4):415–419.
66. Hutson SW, Cowen PN, Bird CC. Morphometric studies of age related changes in normal human breast and their significance for evolution of mammary cancer. *J Clin Pathol*. 1985;38(3):281–287.
67. Cowan DF, Herbert TA. Involution of the breast in women aged 50–104 years: a histopathologic study of 102 cases. *Surgical Pathology*. 1989;2(4):323–333.
68. Baker GM, Guzman-Arocho YD, Bret-Mounet VC, et al. Testosterone therapy and breast histopathological features in transgender individuals. *Mod Pathol*. 2021;34(1):85–94.
69. Hernandez A, Schwartz CJ, Warfield D, et al. Pathologic evaluation of breast tissue from transmasculine individuals undergoing gender-affirming chest masculinization. *Arch Pathol Lab Med*. 2020;144(7):888–893.
70. Torous VF, Schnitt SJ. Histopathologic findings in breast surgical specimens from patients undergoing female-to-male gender reassignment surgery. *Mod Pathol*. 2019;32(3):346–353.
71. Wolters EA, Rabe KE, Siegel L, Butts J, Klein ME. Histopathologic features of breast tissue from transgender men and their associations with androgen therapy. *Am J Clin Pathol*. 2023;159(1):43–52.
72. Andrews AR, Kakadekar A, Schmidt RL, Murugan P, Greene DN. Histologic findings in surgical pathology specimens from individuals taking feminizing hormone therapy for the purpose of gender transition: a systematic scoping review. *Arch Pathol Lab Med*. 2022;146(2):252–261.
73. Shoker BS, Jarvis C, Sibson DR, Walker C, Sloane JP. Oestrogen receptor expression in the normal and pre-cancerous breast. *J Pathol*. 1999;188(3):237–244.
74. Shaw JA, Udokang K, Mosquera JM, Chauhan H, Jones JL, Walker RA. Oestrogen receptors alpha and beta differ in normal human breast and breast carcinomas. *J Pathol*. 2002;198(4):450–457.
75. Shaaban AM, O'Neill PA, Davies MP, et al. Declining estrogen receptor-beta expression defines malignant progression of human breast neoplasia. *Am J Surg Pathol*. 2003;27(12):1502–1512.
76. Speirs V, Walker RA. New perspectives into the biological and clinical relevance of estrogen receptors in the human breast. *J Pathol*. 2007;211(5):499–506.
77. Younes M, Honma N. Estrogen receptor β. *Arch Pathol Lab Med*. 2011;135(1):63–66.

78. Haldosen LA, Zhao C, Dahlman-Wright K. Estrogen receptor beta in breast cancer. *Mol Cell Endocrinol*. 2014;382(1):665–672.
79. Krishnamurthy S, Sneige N. Molecular and biologic markers of premalignant lesions of human breast. *Adv Anat Pathol*. 2002;9(3):185–197.
80. Wang X, Stolla M, Ring BZ, et al. P53 alteration in morphologically normal/benign breast tissue in patients with triple-negative high-grade breast carcinomas: breast p53 signature? *Hum Pathol*. 2016;55:196–201.
81. Wang X, El-Halaby AA, Zhang H, et al. P53 alteration in morphologically normal/benign breast luminal cells in BRCA carriers with or without history of breast cancer. *Hum Pathol*. 2017;68:22–25.
82. Siziopikou KP, Prioleau JE, Harris JR, Schnitt SJ. bcl-2 expression in the spectrum of preinvasive breast lesions. *Cancer*. 1996;77(3):499–506.
83. Egan MJ, Newman J, Crocker J, Collard M. Immunohistochemical localization of S100 protein in benign and malignant conditions of the breast. *Arch Pathol Lab Med*. 1987;111(1):28–31.
84. Earl HM, McIlhinney RA, Wilson P, Gusterson BA, Coombes RC. Immunohistochemical study of beta- and kappa-casein in the human breast and breast carcinomas, using monoclonal antibodies. *Cancer Res*. 1989;49(21):6070–6076.
85. Bailey AJ, Sloane JP, Trickey BS, Ormerod MG. An immunocytochemical study of alpha-lactalbumin in human breast tissue. *J Pathol*. 1982;137(1):13–23.
86. Mazoujian G, Pinkus GS, Davis S, Haagensen DE Jr. Immunohistochemistry of a gross cystic disease fluid protein (GCDFP-15) of the breast. A marker of apocrine epithelium and breast carcinomas with apocrine features. *Am J Pathol*. 1983;110(2):105–112.
87. Sasaki E, Tsunoda N, Hatanaka Y, Mori N, Iwata H, Yatabe Y. Breast-specific expression of MGB1/mammaglobin: an examination of 480 tumors from various organs and clinicopathological analysis of MGB1-positive breast cancers. *Mod Pathol*. 2007;20(2):208–214.
88. Yoon EC, Wang G, Parkinson B, et al. TRPS1, GATA3, and SOX10 expression in triple-negative breast carcinoma. *Hum Pathol*. 2022;125:97–107.
89. Chui X, Egami H, Yamashita J, et al. Immunohistochemical expression of the c-kit proto-oncogene product in human malignant and non-malignant breast tissues. *Br J Cancer*. 1996;73(10):1233–1236.
90. Simone NL, Paweletz CP, Charboneau L, Petricoin EF III, Liotta LA. Laser capture microdissection: beyond functional genomics to proteomics. *Mol Diagn*. 2000;5(4):301–307.
91. Deng G, Lu Y, Zlotnikov G, Thor AD, Smith HS. Loss of heterozygosity in normal tissue adjacent to breast carcinomas. *Science*. 1996;274(5295):2057–2059.
92. Lakhani SR, Chaggar R, Davies S, et al. Genetic alterations in 'normal' luminal and myoepithelial cells of the breast. *J Pathol*. 1999;189(4):496–503.
93. Larson PS, de las Morenas A, Cupples LA, Huang K, Rosenberg CL. Genetically abnormal clones in histologically normal breast tissue. *Am J Pathol*. 1998;152(6):1591–1598.
94. Larson PS, de las Morenas A, Bennett SR, Cupples LA, Rosenberg CL. Loss of heterozygosity or allele imbalance in histologically normal breast epithelium is distinct from loss of heterozygosity or allele imbalance in co-existing carcinomas. *Am J Pathol*. 2002;161(1):283–290.
95. Graham K, Ge X, de Las Morenas A, Tripathi A, Rosenberg CL. Gene expression profiles of estrogen receptor-positive and estrogen receptor-negative breast cancers are detectable in histologically normal breast epithelium. *Clin Cancer Res*. 2011;17(2):236–246.
96. Danforth DN Jr. Genomic changes in normal breast tissue in women at normal risk or at high risk for breast cancer. *Breast Cancer (Auckl)*. 2016;10:109–146.
97. Dontu G, Al-Hajj M, Abdallah WM, Clarke MF, Wicha MS. Stem cells in normal breast development and breast cancer. *Cell Prolif*. 2003;36(Suppl 1):59–72.
98. Sarrio D, Franklin CK, Mackay A, Reis-Filho JS, Isacke CM. Epithelial and mesenchymal subpopulations within normal basal breast cell lines exhibit distinct stem cell/progenitor properties. *Stem Cells*. 2012;30(2):292–303.
99. Keller PJ, Arendt LM, Skibinski A, et al. Defining the cellular precursors to human breast cancer. *Proc Natl Acad Sci U S A*. 2012;109(8):2772–2777.
100. Sgroi DC, Teng S, Robinson G, LeVangie R, Hudson JR Jr., Elkahloun AG. In vivo gene expression profile analysis of human breast cancer progression. *Cancer Res*. 1999;59(22):5656–5661.
101. Perou CM, Jeffrey SS, van de Rijn M, et al. Distinctive gene expression patterns in human mammary epithelial cells and breast cancers. *Proc Natl Acad Sci U S A*. 1999;96(16):9212–9217.
102. Emmert-Buck MR, Strausberg RL, Krizman DB, et al. Molecular profiling of clinical tissue specimens: feasibility and applications. *Am J Pathol*. 2000;156(4):1109–1115.
103. Espina V, Geho D, Mehta AI, Petricoin EF III, Liotta LA, Rosenblatt KP. Pathology of the future: molecular profiling for targeted therapy. *Cancer Invest*. 2005;23(1):36–46.

SECTION III

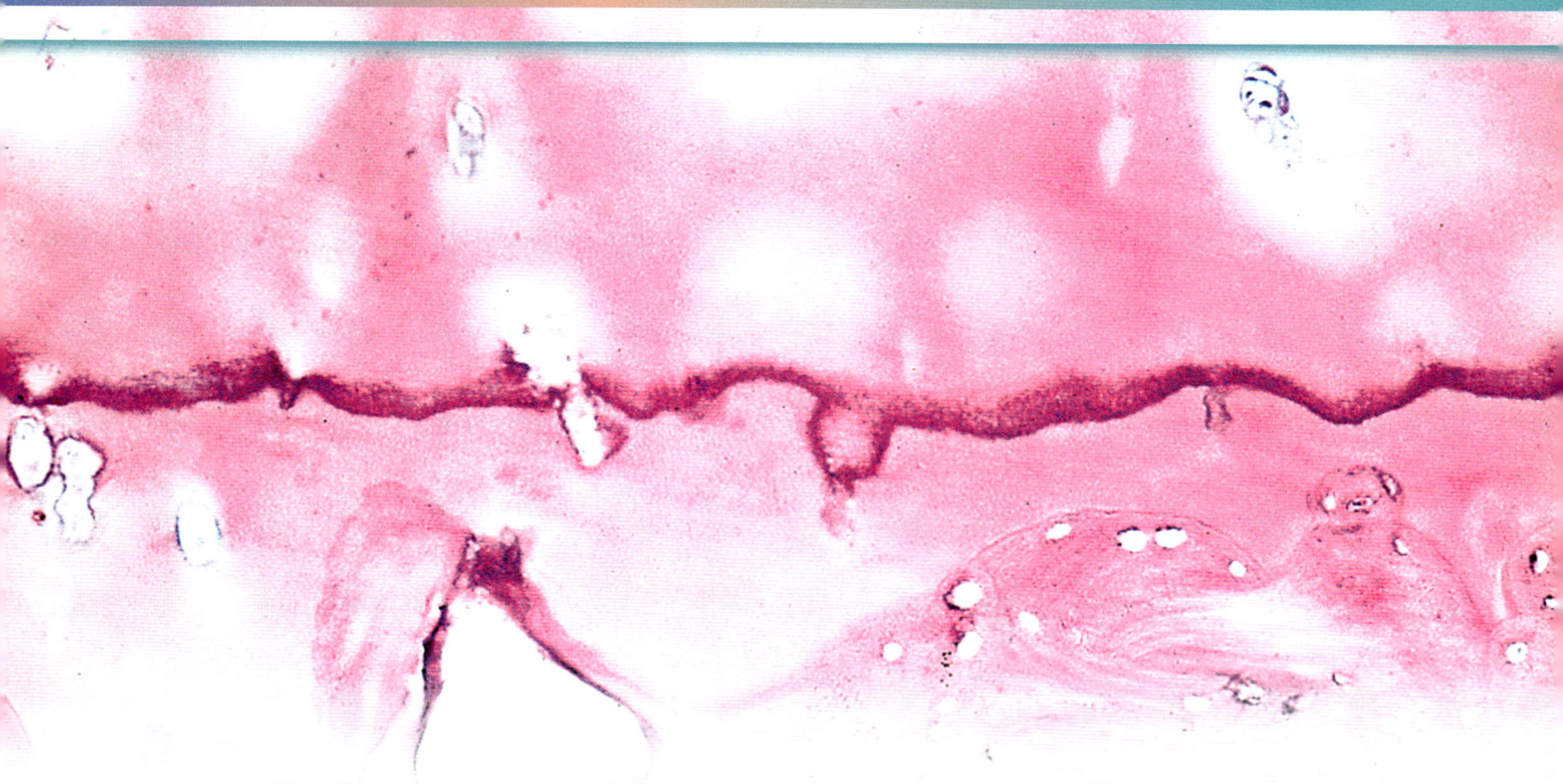

Musculoskeletal System

4

Bone

Darcy A. Kerr ▪ Andrew E. Rosenberg

The skeleton is vital to sustaining life. It is critical to maintaining mineral homeostasis and calcium and phosphorous metabolism, and the structural characteristics of bone are fundamental to locomotion, acting as levers for muscle action and organ protection. In addition, bones form the frameworks of our bodies, determining size and shape, and providing a storehouse and nurturing environment for the hematopoietic elements. Lastly, the cells of bone play an important endocrine role in energy metabolism, the renal excretion of phosphate, and development and maturation of the male phenotype.

The skeleton is composed of approximately 206 bones; the numbers vary slightly depending on the study and age of the individual (1). All bones are continuously remodeling throughout life. The unique biology of the skeletal system is related to its growth and development, ability to change its structure, mass, function, and metabolism in response to biomechanical and systemic demands, and capacity to repair itself when injured.

The term "bone" refers to a specialized type of mesenchymal connective tissue or to a complete organ composed of bone tissue. Individual bones are composed of bone tissue integrated with the cartilage, fibrous tissue, fat, blood vessels, nerves, and hematopoietic elements. The human skeletal system consists of bones and their articulations.

The skeleton can be anatomically divided into the axial skeleton, which includes the skull, vertebral column, ribs, sternum, and hyoid, and the appendicular, or peripheral, skeleton consisting of the upper and lower limbs and the pelvis. The acral skeleton refers to the bones of the hands and feet (Table 4.1).

Bone, whether referring to the organ or a type of connective tissue, is composed of organic elements, namely cells, noncollagenous proteins, and collagen, particularly type 1 collagen. All bone is initially deposited in an unmineralized state and this osteoid is subsequently mineralized with calcium hydroxyapatite. Thus, mineralized bone comprises a unique biphasic blend of organic and inorganic elements. Developing and adult bones have a distinct macroscopic appearance due to their hierarchical structures (Fig. 4.1). Type I collagen is deposited in a woven or lamellar pattern and, together with the cells and noncollagenous proteins, is arranged into cortical or trabecular bone, respectively. The ultimate quality, quantity, and architecture of these components confer important biologic properties to individual bones which are influenced by the mechanical function of, and applied forces to, the bone. The ability of bone tissue to alter its mass and structure in response to its biomechanical environment is termed Wolff law and is unique to the skeletal system.

TABLE 4.1 Location, Type, and Method of Formation for Bones of the Skeleton

Skeletal Region	Bone	Type (Based on Shape and Size)	Method of Formation[a]
Axial	Skull[b]	Flat and irregular	Intramembranous
	Mandible	Irregular	Intramembranous
	Maxilla	Irregular	Intramembranous[c]
	Clavicle	Flat	Intramembranous[c]
	Sternum	Flat	Endochondral
	Vertebral column	Irregular	Endochondral
	Ribs	Flat	Endochondral
	Hyoid	Irregular	Endochondral
Appendicular	Limbs	Tubular—long	Endochondral
	Pelvis	Flat	Endochondral
Acral	Metacarpals	Cuboid	Endochondral
	Carpals	Tubular—short	Endochondral
	Metatarsals	Cuboid	Endochondral
	Tarsals	Tubular—short	Endochondral

[a]Cortices of all bones are formed by intramembranous ossification.

[b]Frontal, parietal, and parts of the occipital bones.

[c]Limited component forms from endochondral ossification.

BONE—THE ORGAN: GROSS AND MICROSCOPIC ANATOMY

Bones are lightweight, and usually cylindrical structures that have a relatively high tensile strength. They are rigid, but not brittle. Tan white and smooth surfaced, they are the hardest and strongest structures of the body, being as strong as cast iron but one-third of the weight because of their unique structure. Bones are reinforced, asymmetric, hollow, and solid structures (Fig. 4.2) designed to provide a maximum strength-to-weight ratio.

Bones vary greatly in size and shape, and these characteristics form their classification. Flat bones are composed of bilaminar plates; other bones are cuboidal, irregular, or tubular. Tubular bones may be long or short and the latter are the most numerous (Table 4.1). Tubular bones are anatomically subdivided along the long axis into the epiphysis, metaphysis, and diaphysis (2). The epiphysis is a slightly flared region that extends from the base of the articular

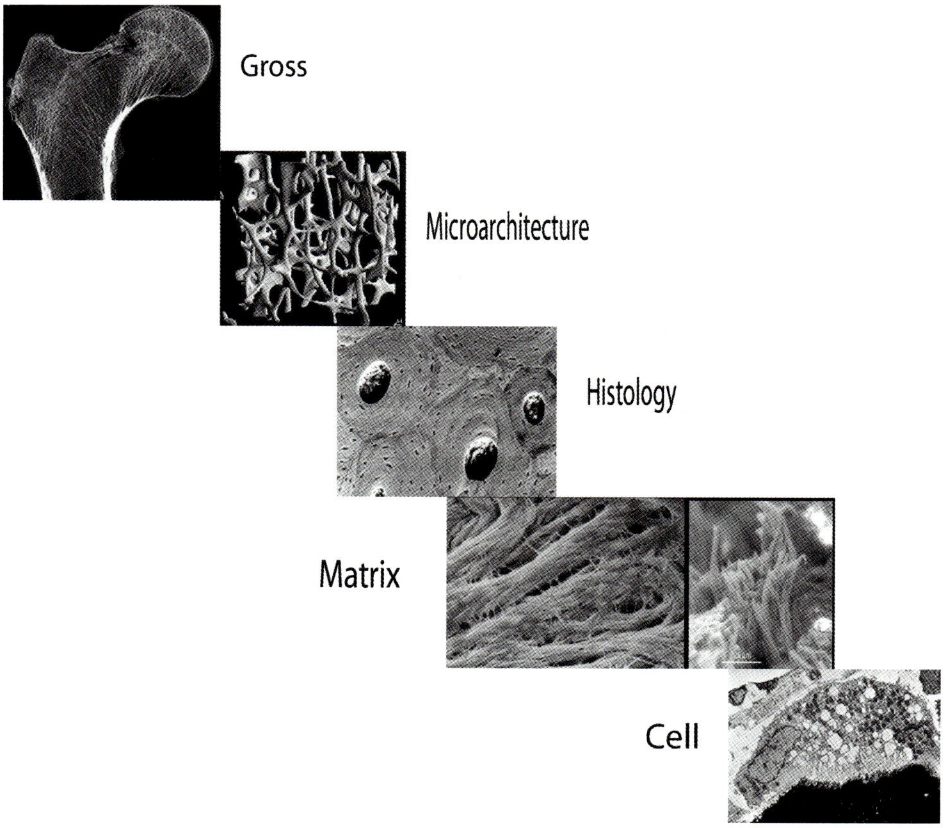

FIGURE 4.1 The hierarchal structure of bone. Gross. A cross-section of the head and upper diaphysis of a macerated femur with the soft tissue removed. The cortex of the metaphysis and diaphysis is of varying thickness depending upon the stress placed upon the bone. The trabeculae of the head follow the lines of stress. Microarchitecture. The cross-linking of the trabeculae demonstrated by micro CT. Histology. A microscopic section of the compact bone of the cortex showing haversian systems and circumferential and interstitial lamellae. Matrix. Bundles of collagen fibers by scanning electron microscopy of two macerated areas of bone in which the mineral has been removed. Cell. An electron micrograph of an osteoclast on the surface of a trabecula. (Supplied by Bauxstein ML, PhD. Department of Orthopaedics, Beth Israel Deaconess Medical Center, Boston, MA.)

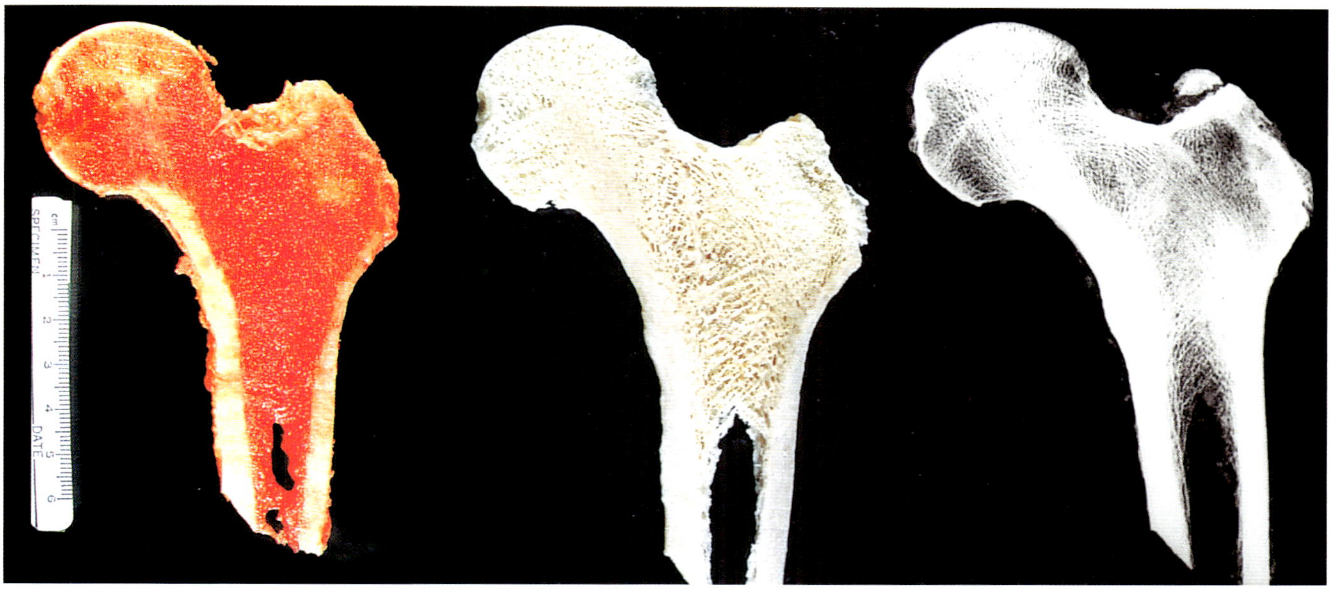

FIGURE 4.2 Gross (*left*) and macerated (*center*) longitudinally cut specimen and the accompanying x-ray (*right*) of a proximal femur, including the head, neck, and upper diaphysis sectioned in the frontal plane. The cortex defines the outer limits of the bone and is thickest along the medial surface (*left*) of the neck and diaphysis, where load bearing is greatest. The medullary cavity is filled with bony trabeculae and red hematopoietic and yellow fatty marrow. The trabeculae are aligned along the lines of stress; this is especially prominent in the medial portion. The horizontal line at the base of the head on the x-ray represents the accrual of bone that occurred during closure of the epiphyseal growth plate.

surface to the point where narrowing of the bone diameter begins. A secondary center of ossification is often formed in the epiphysis during development. The metaphysis extends from the base of the epiphysis to the region where the bone achieves a consistently narrow diameter. The diaphysis, or shaft, extends from the base of one metaphysis to the base of the opposing metaphysis (Fig. 4.3). The diaphysis contains the primary center of ossification in immature or growing bone, the metaphysis contains the cartilaginous growth plate (i.e., physis), and the epiphysis contains a secondary center of ossification. Apophyses are anatomic sites that harbor secondary centers of ossification and growth plates that produce a protuberance to which tendons and muscles attach, such as the greater and lesser trochanters of the femur. The medical and forensic determination of skeletal age and the prediction of ultimate growth are based on the amount and localization of ossification, the degree of growth plate maturation, the formation and size of the secondary ossification centers, and the extent of remodeling.

Despite their differences in size and shape, all bones are of similar composition. They generally have a periosteum attached to the outer surface of the cortex, which is composed of compact cortical bone. Its endosteal surface forms a boundary with the medullary canal which contains variable amounts of cancellous (i.e., trabecular) bone and fatty or hematopoietic marrow (Fig. 4.2), blood vessels, and nerves. The quantity and arrangement of cortical and cancellous bone are directly related to the biomechanical function of any given bone. For example, bones exposed to the largest torsional and load-bearing forces are composed of 80% to 100% cortical bone and 0% to 20% cancellous bone, whereas bones designed to transmit weight-bearing

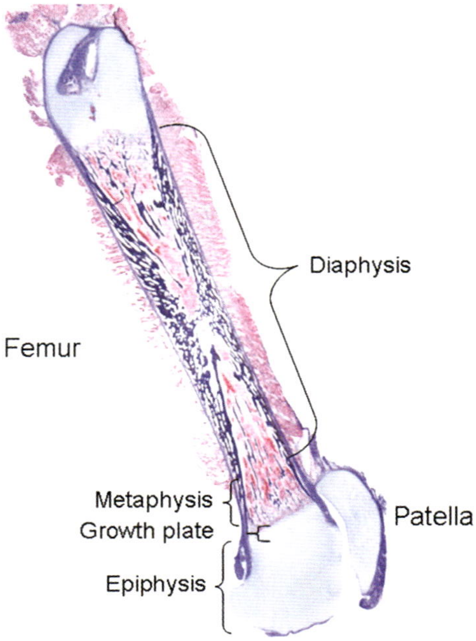

FIGURE 4.3 Longitudinal whole mount of an immature femur and patella prior to the formation of the secondary centers of ossification. The long bone is composed of the proximal and distal epiphyses, metaphyses and growth plates (physes), and the intervening diaphysis.

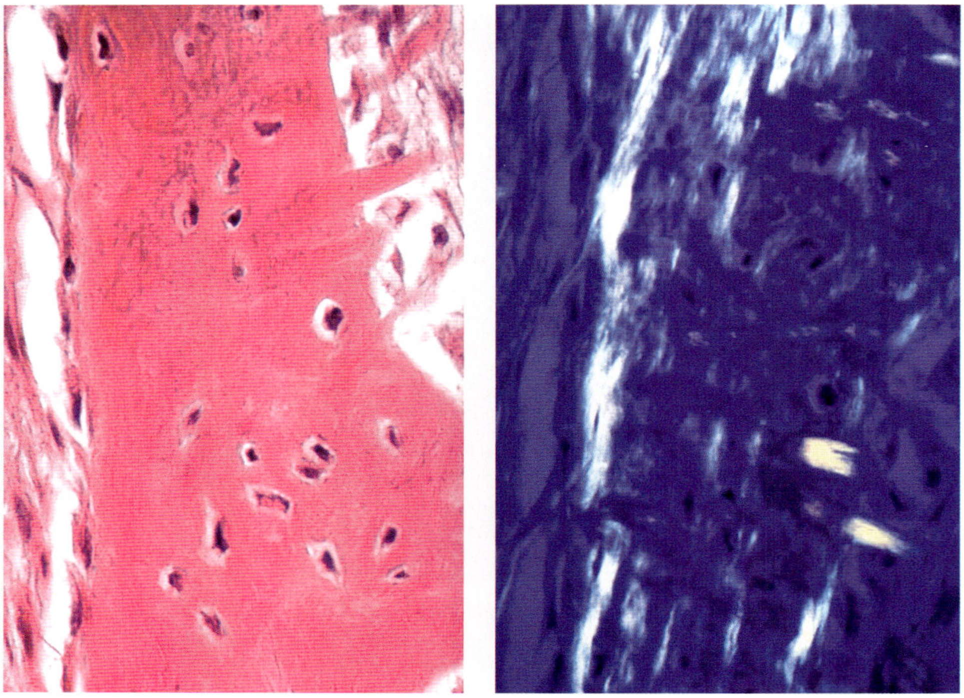

FIGURE 4.4 Woven bone seen on H&E-stained slide (*left*) and with polarized light (*right*). The collagen fibers are oriented in all planes. There are many plump osteocytes, and their long axes appear randomly oriented, following the direction of the neighboring collagen fibers.

forces, such as the vertebral bodies, consist of 80% cancellous bone and 20% cortical bone. The trabeculae of cancellous bone are arranged according to the lines of stress to which they are exposed in their biomechanical environment (Fig. 4.2).

Woven and Lamellar Bone

Bone tissue is categorized into woven and lamellar types based on the organization of its type I collagen fibers and major structural proteins. The collagen fibers of woven bone are arranged in a seemingly haphazard feltwork (Fig. 4.4), while those of lamellar bone are deposited in parallel arrays (Fig. 4.5). The parallel arrays of collagen in lamellar bone are oriented longitudinally or circumferentially around haversian canals forming osteons or haversian systems.

Woven bone is deposited during periods of rapid bone formation. It composes parts of developing cortex and trabeculae during embryogenesis and in the growing infant and the adolescent. It may also be the predominant type of bone in various reactive and neoplastic conditions, including fracture calluses, the involucrum of osteomyelitis, Codman triangle overlying a neoplasm, and the matrix of bone-forming tumors. Woven bone is hypercellular with enlarged osteocytes and lacunae oriented in a haphazard fashion; the long axes of the cells parallel the seemingly random arrangement of neighboring collagen fibers (Fig. 4.4). The mineral content of woven bone is higher than that of lamellar bone. Overall, this structural organization enables woven bone to resist forces equally in all directions, but it is weaker, less rigid, and more flexible than lamellar bone.

The entire mature skeleton is normally composed solely of lamellar bone. In contrast to woven bone, lamellar bone is synthesized at a slower rate and is less cellular. The osteocytes and their lacunae are smaller and their distribution is more organized; the long axes of osteocytes parallel the orderly collagen lamellae (Fig. 4.5). Additionally, the process of mineralization of lamellar bone differs from that of woven bone. It occurs more slowly and the mineral content increases over time as more apatite crystals materialize long after deposition of the organic matrix. Microradiographs of bone sections reveal varying densities with the oldest bone

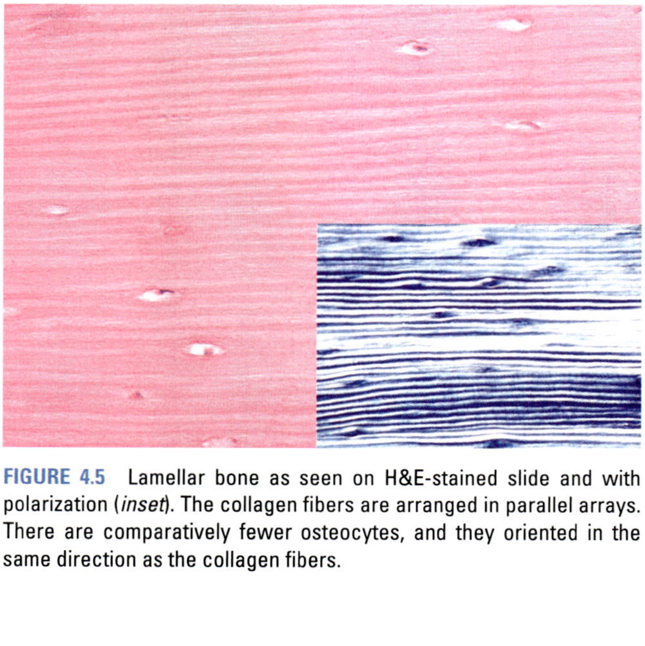

FIGURE 4.5 Lamellar bone as seen on H&E-stained slide and with polarization (*inset*). The collagen fibers are arranged in parallel arrays. There are comparatively fewer osteocytes, and they oriented in the same direction as the collagen fibers.

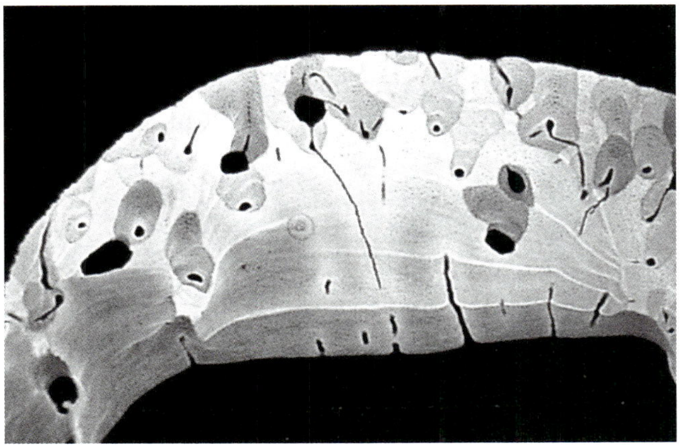

FIGURE 4.6 Microradiograph of the cortex of a 2-month-old female patient. There are various degrees of mineralization, with the radiolucent areas (*dark*) being the most recently deposited and least mineralized. The radiodense portions (*light areas*) represent the oldest areas of formation and the most mineralized.

being most heavily mineralized (Fig. 4.6). Since the mineral and collagen fibers are well organized and intimately bound to one another, lamellar bone has greater rigidity and tensile strength, and less elasticity than woven bone.

Both lamellar and woven bone are made by osteoblasts in discrete units that are deposited in an appositional fashion. Osteoblasts first synthesize and secrete a layer of unmineralized organic matrix (i.e., osteoid) composed of type I collagen admixed with noncollagenous proteins on a preexisting surface consisting of mineralized cartilage, bone, or collagen, and then they regulate osteoid mineralization. Osteoid appears eosinophilic in hematoxylin and eosin (H&E)-stained histologic sections. Discrete units of lamellar bone deposited by cohorts of osteoblasts are separated from one another by cement lines also known as reversal lines. These cement lines are deposited by osteoblasts on the surface of bone previously resorbed by osteoclasts. They appear as thin (1 to 5 μm) and intensely basophilic linear structures H&E-stained sections (Fig. 4.7). Cement lines are composed of mineral-rich collagen and noncollagenous proteins (3). Some investigators have suggested that cement lines represent a residuum of mineralized ground substance that is secreted during the initial reversal phase of new bone formation (4).

Cortical (Compact) Bone

Cortical bone, also known as dense compact bone, is hard and off-white (Figs. 4.2 and 4.8). Its thickness depends on its location and mechanical requirements: it is thickest in areas exposed to large torsional and weight-bearing forces, such as the medial aspect of the mid-diaphysis of

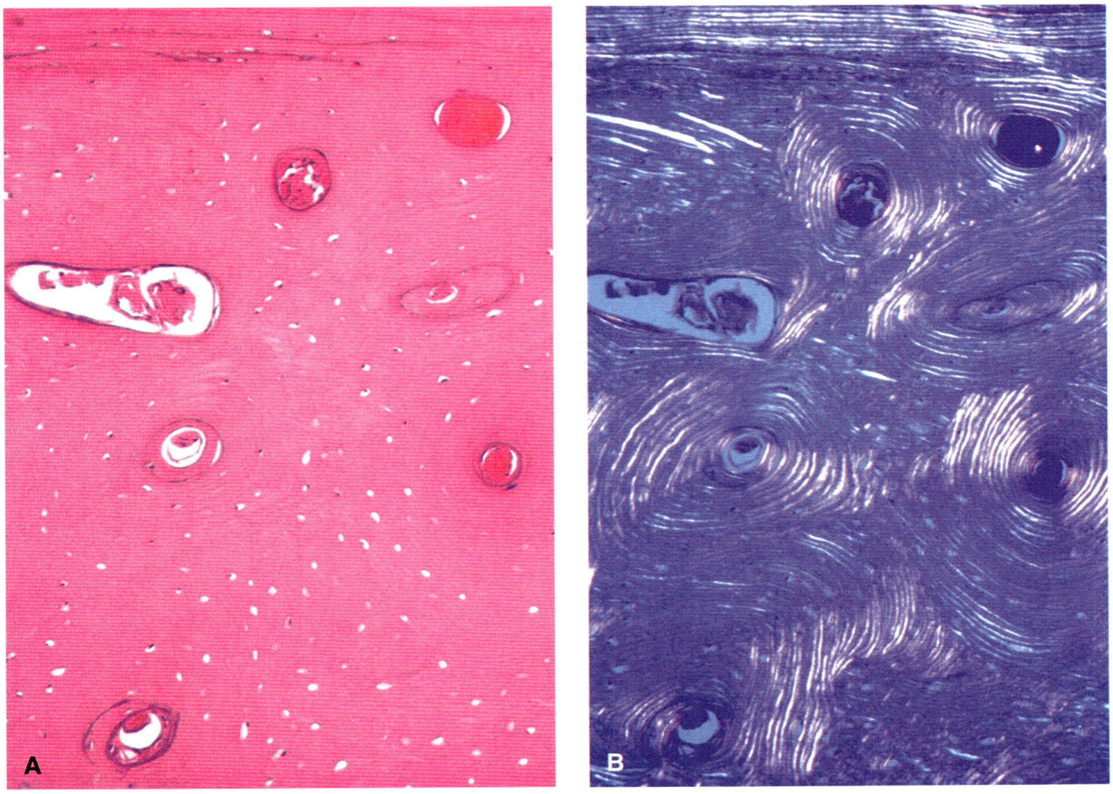

FIGURE 4.7 Cross-sections of cortex with circumferential, concentric, and interstitial lamellae. The circumferential lamellae are beneath the periosteum, the concentric lamellae surround the haversian canals containing blood vessels, and the interstitial lamellae fill the intervening spaces. Scattered intensely basophilic cement lines are present, tethering the units of lamellar bone (**A**, H&E; **B**, polarized light).

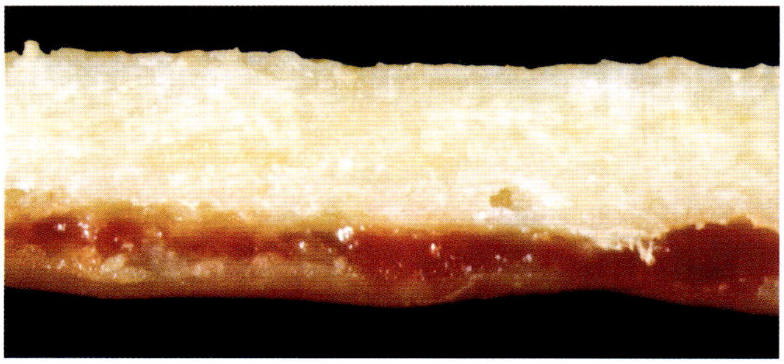

FIGURE 4.8 Photograph of a longitudinal section through a flat bone. The inner cortex is tan white and solid. The round hole within it represents the pathway of the nutrient artery. The *red area* is the central bone marrow.

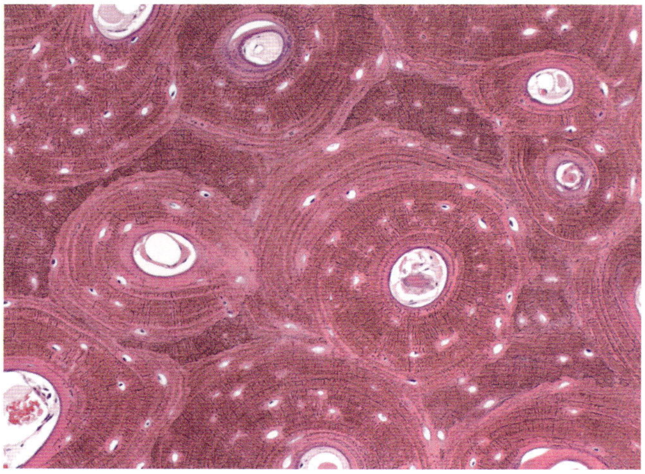

FIGURE 4.10 Cortical bone with Haversian systems surrounded by concentric and interstitial lamellae. The osteocytes have numerous canaliculi.

the femur, and thinnest where the transmission of weight-bearing and torsional forces is least, as seen adjacent to articular surfaces and within vertebral bodies (Fig. 4.2).

Cortical bone is constructed entirely of woven bone during the early stages of growth and development. It is gradually remodeled over time until it is composed entirely of lamellar bone in the mature skeleton. Lamellar bone displays three architectural patterns in adult cortical bone: circumferential, concentric, and interstitial (Fig. 4.7). The circumferential lamellae form outer and inner envelopes to the cortex and consist of several subperiosteal and endosteal layers that are oriented parallel to the long axis of the bone. They are the first cortical lamellae to be deposited and comprise almost the entire cortex in young individuals. As mechanical stresses on the bone increase with age, many of the circumferential lamellae are replaced by the concentric lamellae of the haversian systems, although the lamellae immediately beneath the periosteum and along the endosteum are generally retained (Figs. 4.7 and 4.9 to 4.12).

Haversian systems, or osteons, are created by osteoclastic resorption of the circumferential lamellae that usually begins on the endosteal surface of the cortex, and less frequently on the periosteal surface. The bone resorption proceeds perpendicular or at an angle to the long axis of the bone forming the Volkman canal (Figs. 4.10 and 4.11). The numerous osteoclasts situated in the leading edge of the canal are known as the "cutting cone," and the canal they generate becomes filled with vessels, nerves, mesenchymal cells, and stem cells enmeshed in loose connective tissue. The osteoclastic activity becomes concentrated on one side of the canal within a short distance and, as a consequence, the direction of the newly formed haversian canal becomes aligned with the long axis of the bone. The burrowing osteoclasts elongate the canal, making way for newly formed osteoblasts to deposit lamellae of bone in a

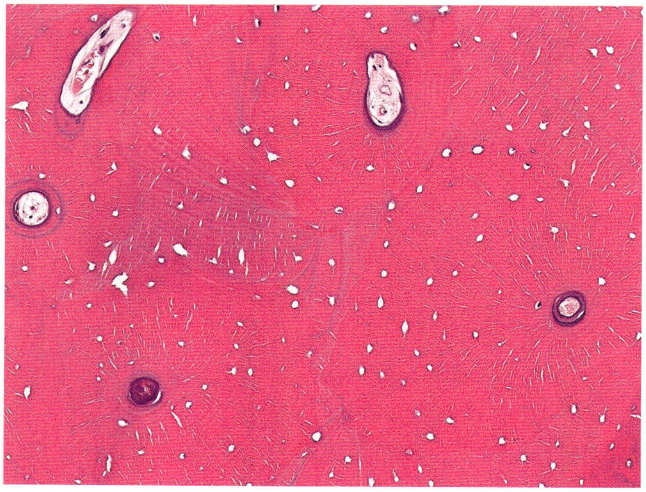

FIGURE 4.9 Multiple adjacent mature haversian systems containing the central canal, blood vessels, and surrounding concentric lamellae. Empty lacunae are seen in areas of the interstitial lamellae.

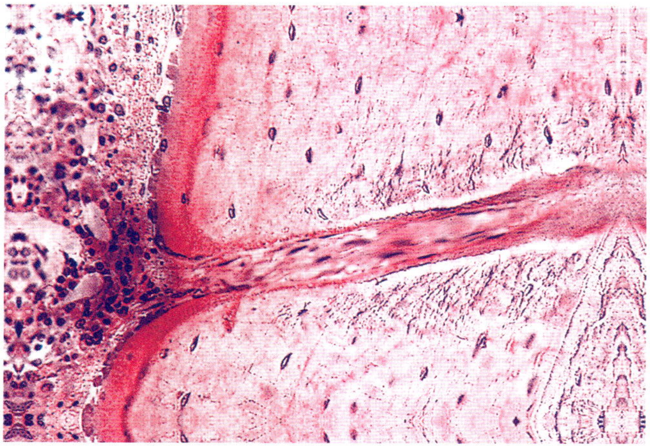

FIGURE 4.11 A forming Volkmann canal coursing through the cortex of the bone. The canal is angled with respect to the bone lamellae and is filled with connective tissue. Osteoid and osteoblasts, indicating new bone formation, are present around the endosteal opening of the canal, but none are seen in the canal. The cortex shows circumferential lamellar bone with regularly placed osteocytes (undemineralized bone section).

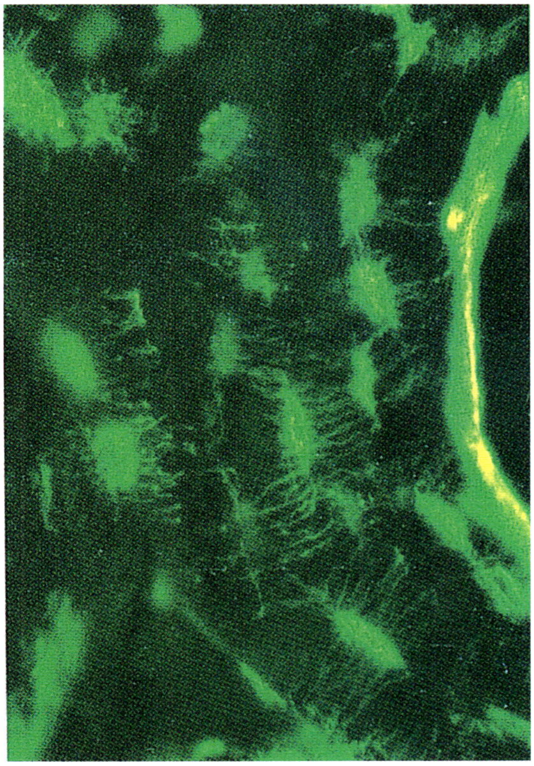

FIGURE 4.12 An undemineralized section of cortical bone, stained in vivo, with tetracycline. A layer of tetracycline appears at the mineralization front of the haversian system where new bone is being formed. The osteocyte lacunae and the canaliculi of their connecting dendritic processes are visible as bright green areas. The lacunae of the adjacent interstitial regions (*left upper and lower corners*) do not connect with the haversian cells surrounding the haversian canal (*dark area right center*) (unstained, fluorescent light).

target-like, or concentric, fashion in their wake. The collagen fibers within a lamella are oriented parallel to each other; however, their pitch is slightly different from those in adjacent lamellae, thereby enhancing the biomechanical strength of the cortex. The accrual of concentric lamellae over time reduces the diameter of the haversian canal, ultimately producing a small region that contains nutritional blood vessels and nerve twigs (Fig. 4.9). Together, these elements define a haversian or osteonal system.

Mature haversian canals are long and cylindrical, range from 25 to 125 μm in diameter (average 50 μm), and are widest nearest the medullary cavity. They form an intricate, branching, spiraling, and interconnecting network that courses throughout the cortex. The number of haversian systems in a particular bone is variable; it is determined by age, the amount of mechanical stress and weight that the bone is subjected to over time, and other biologic and genetic factors (4,5).

Individual haversian systems are relatively self-contained metabolic units because the nutritional support of their cells, especially the surface osteoblasts and osteocytes, depends upon diffusion from their central vessels. Consequently, there is a biologic limit on the maximal number of lamellae contained within any haversian system because osteocyte viability is not sustainable beyond a certain distance from the vessels. Also, the integrated network of osteocytes is generally limited to the osteon within which it develops, as osteocytic cytoplasmic processes usually contact only other osteocytes that dwell in the same system (Fig. 4.12).

Interstitial bone fills the spaces between the haversian systems. The interstitial lamellae represent the remnants of circumferential lamellae of previously formed haversian systems that have become partially destroyed by osteoclastic activity. They are irregular, geometric-shaped units of lamellar bone (Figs. 4.7 and 4.9) that help anneal haversian systems to one another. This arrangement is important in maintaining cortical integrity. The osteocytes confined to the interstitial lamellae may lose their access to nutritional sources and consequently undergo necrosis, leaving behind empty lacunae.

Cement lines denote the physical boundaries of every haversian system and unit of interstitial bone. As the bone ages and is subject to varying forces and remodeling, more haversian systems develop to replace preexisting interstitial lamellae and older haversian systems which subsequently become newly created interstitial lamellae. The endosteum is covered by loose areolar connective tissue that immediately abuts the osteoblasts along the inner surface of the cortex and along the surfaces of the trabeculae.

Cancellous (Trabecular or Spongy) Bone

Cancellous bone is commonly referred to as trabecular or spongy bone. It is tan white, fenestrated, and located within the medullary cavity, appearing as interconnecting plates and struts of trabecular bone (Figs. 4.1 and 4.2). The adult lamellae are oriented parallel to the long axis of a trabecula (Fig. 4.13), whereas developing cancellous bone is composed of significant amounts of woven bone. The woven

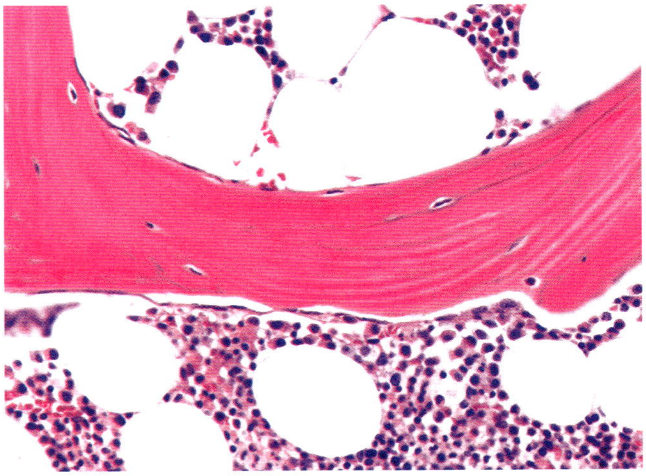

FIGURE 4.13 A mature trabecula composed of lamellar bone. The collagen lamellae are oriented parallel to the long axis of the trabecula.

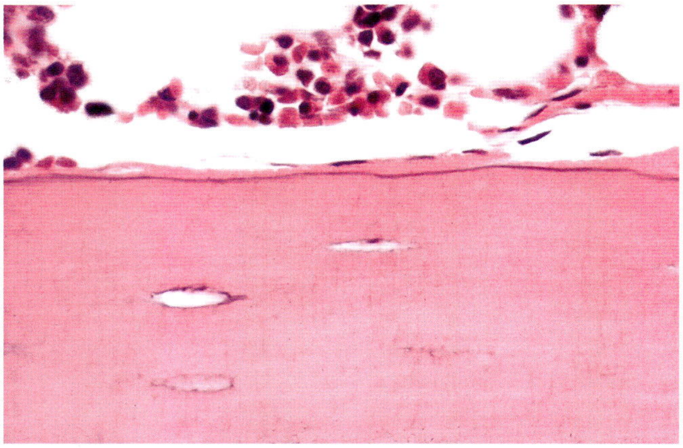

FIGURE 4.14 Quiescent osteoblasts lining a trabecular surface of lamellar bone.

bone has a central core of calcified cartilage when initially formed in infants and children and in reparative bone containing cartilage. This central core of cartilage is termed the primary spongiosa. Enlargement of trabeculae occurs by appositional growth with newly formed bone being deposited on preexisting trabecular surfaces. Mature cancellous bone is lamellar and the trabeculae are oriented according to lines of mechanical stress to provide support and distribute large weight-bearing forces along a variety of different pathways (Fig. 4.2). Consequently, cancellous bone is most abundant in the weight-bearing ends of bones, such as the epiphyses and vertebral bodies, and is present in only small amounts in the mid-diaphysis of tubular bones. Small trabeculae are avascular, while larger ones may contain small haversian-like systems, including concentric lamellae. The surfaces of mature trabeculae are typically lined by quiescent osteoblasts or surface-lining cells and adjacent endosteal connective tissue (Fig. 4.14).

In three dimensions, the trabeculae are usually interconnecting plates (Fig. 4.15) and their global surface area is very large, approaching the square footage of three football fields. This organization facilitates remodeling and allows

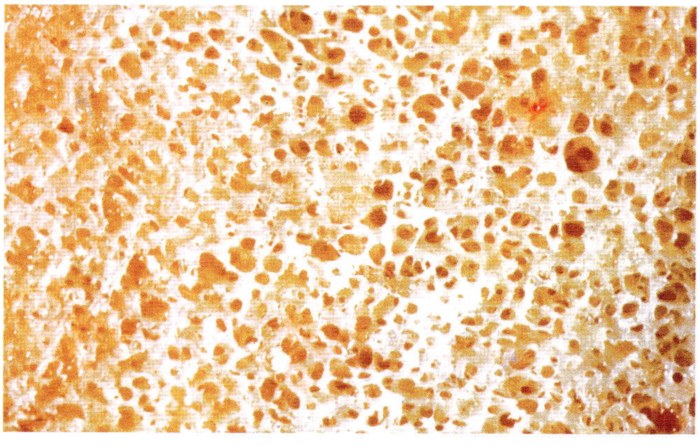

FIGURE 4.15 Gross photograph of a macerated portion of cancellous bone. The trabeculae (*white areas*) are forming interconnecting plates.

the skeleton to rapidly respond to the metabolic demands of the body. The mature trabeculae are heavily mineralized with a thin (1 to 3 μm) layer of osteoid that appears a lighter shade of pink than mineralized bone beneath the relatively inactive flattened osteoblasts.

Periosteum

The periosteum consists of a thin layer of dense white connective tissue that covers the outer surface of all bone cortices. It is attached to the cortex with some looseness in children, but firmly anchored to the bone by collagen fibers in adults. The periosteum is constructed of an outer fibrous layer and an inner cellular (cambium) layer in children. The cambium layer is composed of spindle-shaped fibroblasts, osteoprogenitor cells, and mature polyhedral osteoblasts (Fig. 4.16). In general, the number of osteoprogenitor cells depends on the age of the individual and the amount of bone cell activity in any particular site; these progenitor cells are especially numerous during periods of active bone formation. In contrast, the adult periosteum appears largely as a fibrous layer containing fibroblasts and broad type I collagen fibers that are continuous with those of the joint capsule, tendons, and muscle fascia. The collagen fibers of

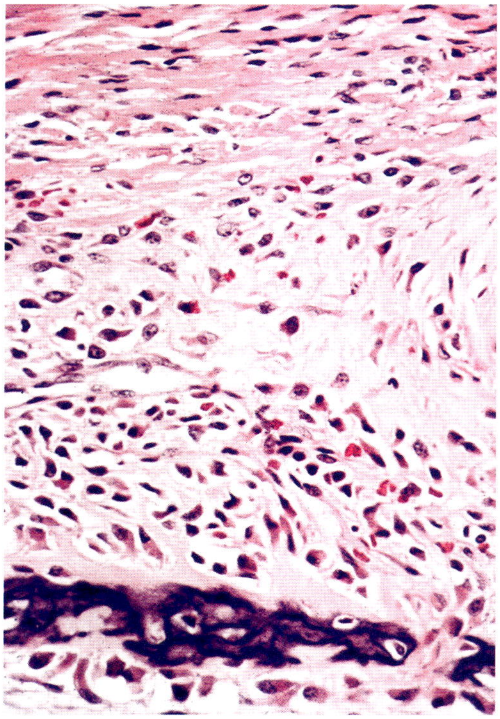

FIGURE 4.16 The outer fibrous layer of a fetal periosteum consists of thick bundles of horizontally oriented collagen fibers. The inner cellular cambium layer contains spindle-shaped osteoprogenitor cells. The maturing osteoblasts steadily become more polyhedral and acquire increasing amounts of amphophilic cytoplasm. A portion of intramembranous woven bone is seen forming the new cortex. The hematoxylin-positive area staining blue is mineralized, while the *pink area* above is unmineralized osteoid.

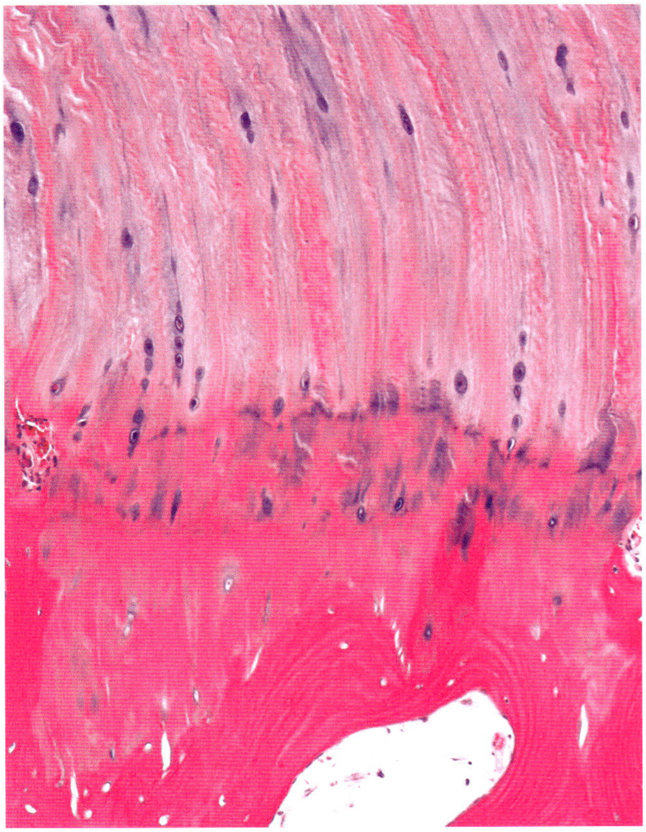

FIGURE 4.17 Dense regular connective tissue inserts into the bone in a tendoligamentous insertion site or enthesis.

the tendoligamentous insertion site pierce the periosteum and become anchored in the bone where they are known as Sharpey fibers (Fig. 4.17).

Vascular Supply and Innervation

Bones are vascular organs and receive their blood supply from three main sources: one to two large nutrient arteries per bone, metaphyseal and epiphyseal vessels, and vessels in the periosteum. Nutrient arteries enter long bones in the diaphysis, traverse the cortex through foramina, and divide into ascending and descending branches within the medullary cavity. Smaller branch arteries, arterioles, capillaries, venules, and veins (Fig. 4.18) course throughout the medullary cavity, nourish the fatty and hematopoietic marrow, and extend into haversian canals where they supply the inner two-thirds of the cortex. Small arteries terminate at the ends of growing bones and give rise to capillary loops at the bases of epiphyseal growth plates. The epiphyseal and metaphyseal vessels access bone through small apertures. They also provide blood flow to regions of the epiphysis and metaphysis in the mature skeleton and to the secondary centers of ossification during active endochondral ossification. Small periosteal vessels supply the outer third of the cortex. The venous drainage system of bone is composed of medullary

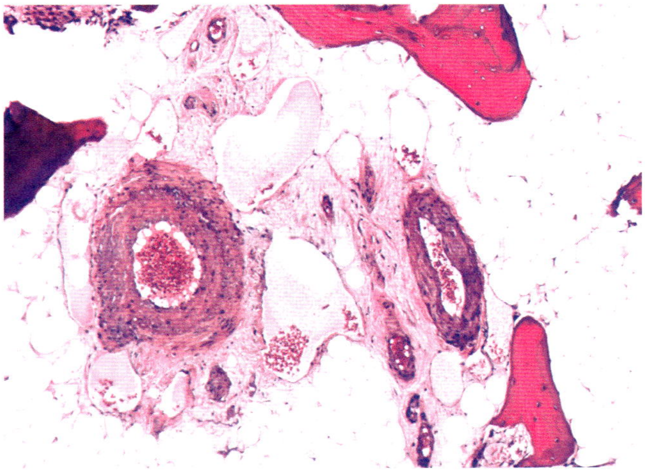

FIGURE 4.18 Small arteries, surrounded by dilated capillaries and nerves, in an area of fatty marrow.

sinusoids that empty into a central venous sinus that merges with nutrient veins. Lymphatic vessels are present within connective tissue adjacent to the periosteum but are absent within normal cortical or cancellous bone (6).

Bones are innervated mainly by nonmyelinated nerves derived from the autonomic nervous system that function to control blood flow. Larger nerve branches are usually associated with arterial vessels (Fig. 4.18), whereas small groups of fibers can be found adjacent to vessels in haversian systems. Nerves supplying the periosteum contain sensory elements and are the source of the sensation of bone pain.

Bone Marrow

Bone marrow is a unique tissue contained within bone cavities. It is composed of an admixture of adipose tissue and hematopoietic elements. Bone marrow is designated "yellow" when composed of fat, and "red" when it contains a predominance of hematopoietic elements. Red marrow is generally composed of 20% to 40% fat admixed with hematopoietic cells. Overall, approximately 95% of human adult marrow consists of fat which accounts for 8% to 10% of body fat (7,8). It can account for 30% of the total body fat volume (8). The ratios and distributions of fat and hematopoietic elements in the skeleton depend on the age of the individual and the location of the bone. Most of the marrow consists of hematopoietic elements at birth, but this decreases to approximately 30% by the age of 25 years with progressive fatty replacement thereafter (7). Hematopoietic marrow is more widely distributed throughout the skeleton in children compared to adults; it tends to be located in proximal portions of the long bones and the axial skeleton rather than the distal appendicular skeleton of adults. Women have about 10% more fat in the marrow than age-matched men and this difference becomes more pronounced at menopause. Systemic conditions that affect energy metabolism also can affect the ratio of hematopoietic elements to fat (9).

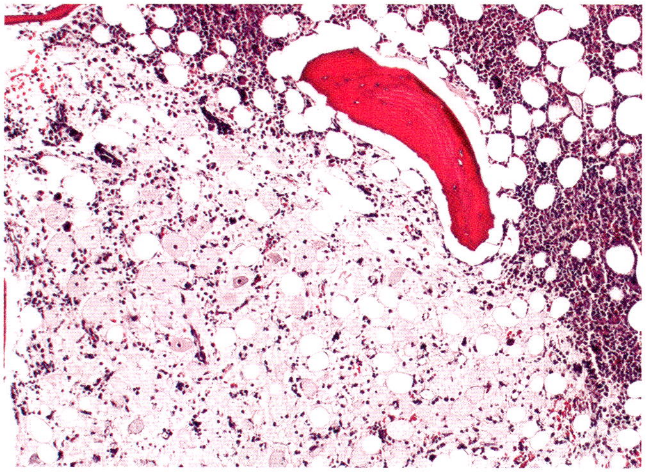

FIGURE 4.19 Normal fat in the marrow is composed of white fat; when brown fat is seen as demonstrated here, it likely represents involvement by hibernoma.

Bone marrow fat is white fat; brown and beige fat are not components of normal marrow fat, although brown fat can manifest as a hibernoma within the bone (Fig. 4.19) (10). Bone marrow fat is not arranged in lobules as in other sites and is distributed as sheets of cells that fill the marrow cavities; adipocytes tend to be smaller than those in subcutaneous and visceral fat and range from 40 to 65 μm in diameter. The white fat of bone marrow also differs from that in other anatomic sites in that it manifests reduced lipid hydrolysis in reaction to energy requirements and diminished insulin responsiveness. Thus, bone marrow fat may not be reduced in the setting of starvation.

Bone marrow adipose tissue plays an important role in the homeostasis of bone and hematopoietic elements. It also influences bone mass, remodeling, immune cell function, and energy metabolism through the secretion of leptin and adiponectin (9). Bone marrow adipose tissue is involved in the development and progression of different diseases including acute myeloid leukemia, multiple myeloma, metastases, osteoporosis, diabetes mellitus, obesity, and anorexia nervosa. Serous atrophy, or gelatinous transformation, of the bone marrow, occurs in a wide variety of conditions including malnutrition, immunosuppression, chronic heart or kidney failure, malignancy, and cytotoxic therapy. In this situation, the marrow has a gelatinous consistency corresponding to an increase in extracellular hyaluronic acid-rich mucopolysaccharides that are not normally present in bone marrow. There is also a decrease in the amount of hematopoietic elements and the number and size of adipocytes (Fig. 4.20) (11–13).

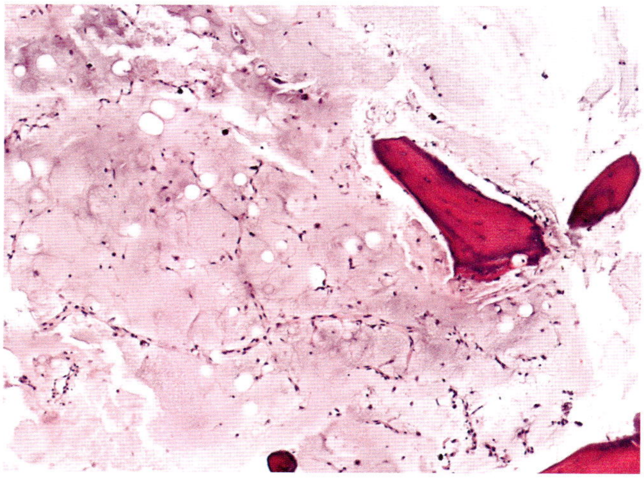

FIGURE 4.20 Serous atrophy or gelatinous transformation of the marrow is characterized by replacement of the marrow space by abundant myxoid extracellular material with a loss of hematopoietic cells and atrophy of fat cells. Scattered thin-walled, branching capillaries and chronic inflammatory cells are present.

ORGANIC AND INORGANIC COMPONENTS OF BONE

The special biphasic amalgamation of organic and inorganic materials found in bone distinguishes it from all other tissues in the body. The organic component consists of proteins and bone cells and the inorganic element is a specialized, calcium-poor form of apatite that resembles hydroxyapatite $[Ca_{10}(PO_4)_6(OH)_2]$ in which the hydroxyl residues are replaced by phosphate and carbonate ions. The integration of the mineral phase with the organic matrix provides bone with hardness, strength, and limited elasticity (14).

Organic Components

Proteins

The organic component accounts for approximately 35% of the wet weight of bone; 90% of this weight is due to the presence of collagen. Collagen is the primary structural protein of bone. It is produced by osteoblasts and the overwhelming majority (90%) is type I (15); type III, V, and fibril-associated collagens with interrupted triple felices (FACIT) collagens are present in trace amounts. FACIT collagens are nonfibrillar collagens that organize and stabilize the extracellular matrix and include collagens IX, XII, XIV, XIX, XX, and XXI (16). Type III collagen may be increased in pathologic conditions. In addition to their contribution to structural support, the numerous, large type I collagen molecules anchor many of the other constituents (17).

The noncollagenous proteins of bone are grouped according to their function as adhesion proteins, calcium-binding proteins, mineralization proteins, enzymes, cytokines, growth factors, and receptors (18,19). These proteins mediate all aspects of bone cell activity and are extremely important to the biologic success of bone as a tissue. Many of these substances are synthesized and secreted by osteoblasts, and others are derived and concentrated from the serum (18). Approximately 25% of these noncollagenous proteins are exogenously derived (16). Osteocalcin is an important osteoblast-produced protein that functions as a hormone involved in the regulation of glucose and insulin metabolism as well as male development and fertility (20). Osteocalcin is made by both osteoblasts and osteoclasts, and its serum quantification is an important clinical marker of bone turnover (17,21).

Osteoprogenitor Cells

Osteoprogenitor cells are derived from tissue-bound mesenchymal stem cells. These mesenchymal stem cells have the capability to form adipocytes, chondrocytes, myocytes, and fibroblasts. The mesenchymal stem cells are located in the perianlage tissue of fetuses, the periosteum, the haversian systems, and Volkmann and medullary canals. Osteoprogenitor cells are primitive committed mesenchymal cells that have the capacity to produce only osteoblasts. The process of osteoblast differentiation and maturation is complex, involving a variety of different elements including transcription factors such as runt domain-containing transcription factor-2 (RUNX2), osterix (OSX), and activating transcription factor-4 (ATF4) as well as signaling pathways such as Wnt and Notch (Table 4.2) (22–27). Osteoprogenitor cells appear as spindle cells and do not have any distinguishing morphologic features so they cannot be identified with certainty in routine histologic sections (Figs. 4.16 and 4.21), except by immunohistochemistry for the relevant transcription factors. Since bone can be formed in skin, soft tissue, muscle, and viscera in both experimental and pathologic conditions, osteoprogenitor cells or inducible stem cells are likely present in these sites as well.

TABLE 4.2 Important Regulators of Osteoblast Differentiation and Function

Transcription Factor	Actions and Regulation
RUNX2	Promotes osteoblast differentiation. Regulation: actions promoted by MAF, TAZ, SATB2, MSX2, BAPX1, RB, GLI2, DLX5 and inhibited by Twist, HAND2, STAT1, Schnurri-3, ZPF521, HOXA2, HES, HEY, and GLI3
OSX	Promotes osteoblast differentiation, required downstream of RUNX2. Regulation: actions promoted NFATC1 and inhibited by p53
ATF4	Promotes mature osteoblast functions. Regulation: actions promoted SATB2 and inhibited by FIAT
Signaling Pathway	**Actions and Regulation**
Wnt	Promotes osteoblast differentiation
Notch	Inhibits osteoblast differentiation
BMP signaling	Stimulates osteoblast differentiation and function
FGF signaling	Promotes preosteoblast proliferation, stimulates osteoblast differentiation and function

RUNX2, runt-related transcription factor 2; MAF, macrophage activating factor; TAZ, transcriptional co-activator with PDZ-binding motif; SATB2, special AT-rich sequence-binding 2; MSX2, Msh homeobox 2; BAPX1, Bagpipe homeobox protein homologue 1; RB, retinoblastoma; GLI2, Gli transcription factor 2; DLX5, distal-less homeobox 5; Twist, Twist-related protein; HAND2, heart and neural crest derivatives expressed 2; STAT1, signal transducer and activator of transcription 1; Schnurri-3, Schnurri-3 zinc finger protein; ZPF521, zinc-finger protein 521; HOXA2, homeobox A2; HES, Hairy and Enhancer of Split; HEY, HES-related with YRPW motif; GLI3, Gli transcription factor 3; OSX, osterix; NFATC1, nuclear factor of activated T cells cytoplasmic 1; p53, tumor protein p53; ATF4, activating transcription factor 4; FIAT, factor inhibiting ATF4-mediated transcription; Wnt, Wingless-related integration site; Notch, Notch type-1 transmembrane protein; BMP, bone morphogenic protein; FGF, fibroblast growth factor.

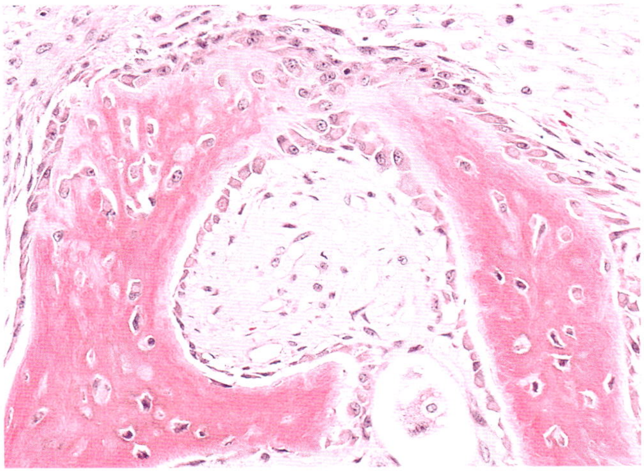

FIGURE 4.21 Metabolically active osteoblasts lining a trabecula of woven bone. Active osteoblasts have abundant amphophilic to basophilic cytoplasm, nuclei polarized away from the bone surface, and perinuclear halos. Some osteoblasts are in various stages of being surrounded by matrix and becoming osteocytes. Spindle cells in the connective tissue adjacent to the trabecula may represent osteoprogenitor cells.

Osteoblasts

Osteoblasts are vital to bone tissue and are responsible for the production, transport, and arrangement of most of the components of the osteoid. They also initiate and regulate matrix mineralization and use autocrine and paracrine mechanisms to control the activity of neighboring osteoblasts, osteocytes, and osteoclasts. Osteoblasts contain alkaline phosphatase, osteopontin, and osteocalcin within their cytoplasm. They also contain receptor activator for nuclear factor κβ (RANK) ligand (RANKL) in their cytoplasm and cell membranes, and they express receptors for parathyroid hormone (PTH), prostaglandins, vitamin D_3, estrogens, and cytokines on their cell membranes (15,17–19,22,28).

Osteoblasts are present on all bone surfaces and their lifespan ranges from months to many years. Their metabolic state is closely related to their morphology; they are spindle-shaped when quiescent and large and polyhedral when rapidly producing bone. Metabolically active osteoblasts vary in size from 10 to 80 μm (average 20 to 30) and have abundant amphophilic to basophilic cytoplasm that is in intimate contact with the bone (Figs. 4.21 and 4.22). Multiple cytoplasmic processes extend from the cells into and through the bone (Fig. 4.22), contacting adjacent osteoblasts and osteocytes via nexus junctions. The nuclei of active osteoblasts are polarized away from the matrix surface and often have a conspicuous nucleolus as well as a prominent perinuclear halo that represents a well-developed Golgi apparatus (Fig. 4.21). Approximately 60% to 80% of osteoblasts undergo

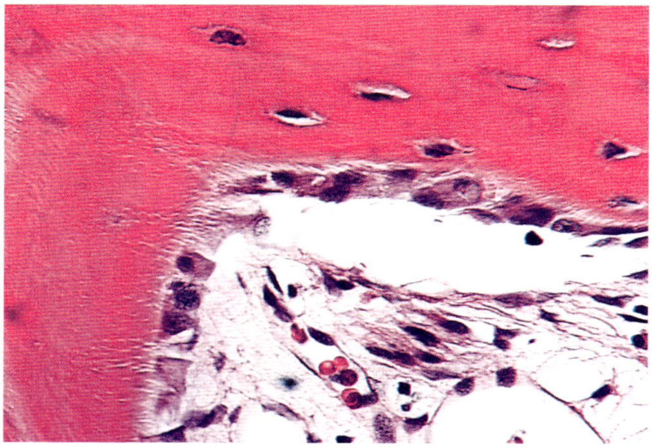

FIGURE 4.22 Metabolically active osteoblasts forming lamellar bone. The thin layer of osteoid cannot be identified in these demineralized sections. At the right of the micrograph, the osteoblasts are becoming inactive, flattening, and being incorporated into the bone as osteocytes. The osteocytes of the lamellar bone are spindle-shaped. The dendrites are identifiable extending from the osteoblast bodies into the bone.

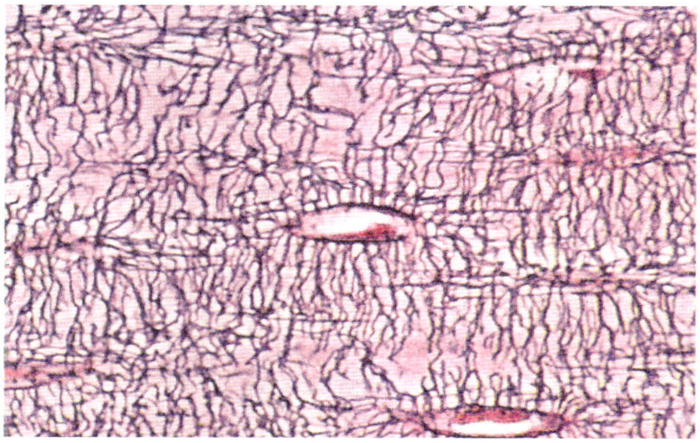

FIGURE 4.23 Osteocytes within lacunar spaces. Numerous cell processes course through the matrix and contact those of neighboring osteocytes.

apoptosis; the remaining osteoblasts either become osteocytes surrounded by matrix or flatten and elongate as their synthetic activity diminishes and ultimately, they remain as a cellular lining of the bone surfaces (29) (Figs. 4.14 and 4.22). The cytoplasm of productive osteoblasts contains extensive, granular endoplasmic reticulum, a large, prominent Golgi apparatus, and numerous mitochondria and lysosomes (30). In contrast, the cytoplasm of inactive osteoblasts resembles that of quiescent fibroblasts (31).

Osteocytes

Osteoblasts enveloped by matrix become osteocytes, and their half-life is estimated to be as long as 25 years (32). Greater than 90% of bone cells are osteocytes in the adult skeleton. In the average adult skeleton, there are approximately 42 billion osteocytes, 9.1 million of which are replenished every day (29,33). The cell body, nucleus, and surrounding scant cytoplasm reside within a lacunar space. The nuclei are comparatively small and are not always visible in every plane of section. Thus, random lacunae appear empty in any given slide of bone tissue. Osteocytes have numerous long and delicate cytoplasmic processes similar to the axonal processes of neurons (Figs. 4.10, 4.12, 4.23, and 4.24). These cell processes traverse the matrix through small canaliculi and provide a very large surface area of contact between the osteocyte, matrix, and extracellular fluid that bathes each cell. It is estimated that the adult skeleton contains 3.7 trillion osteocyte dendritic projections with a total length of 175,000 km and 23 trillion cellular connections (33). Osteocyte cell processes connect to those

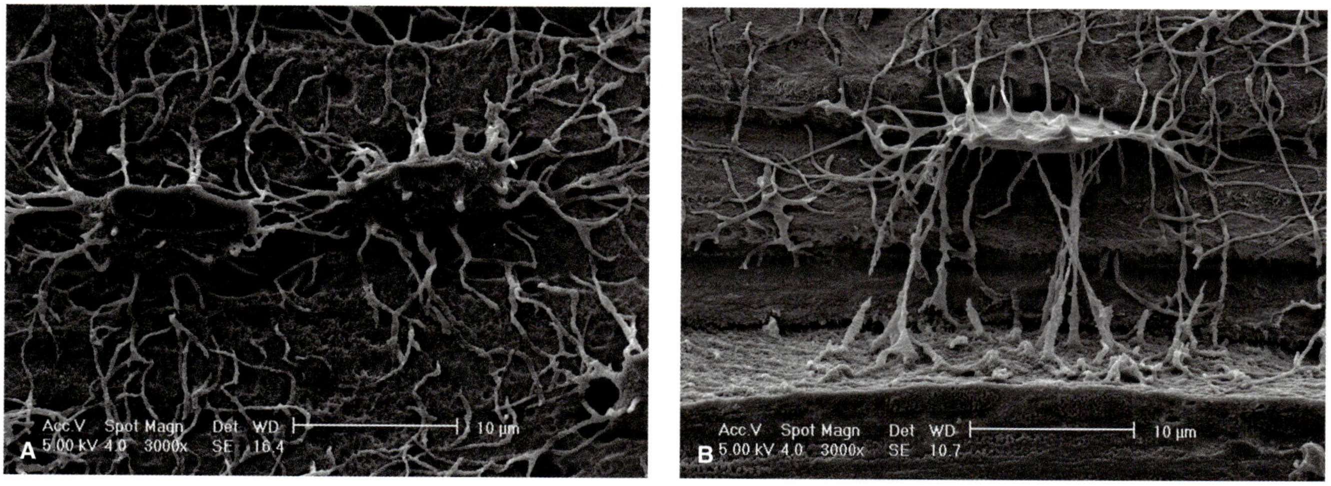

FIGURE 4.24 The osteocyte lacuno-canalicular network demonstrating the contact of cytoplasmic processes of two adjacent osteocytes in the bone (**A**) and an intraosseous osteocyte connecting with an osteoblast on the bone surface (**B**). Osteocyte cell processes (dendrites) pass through the canalicular network. The images are scanning electron micrographs of resin-embedded acid-etched mouse bone samples. (With permission of Bonewald LF. Generation and function of osteocyte dendritic processes. *J Musculoskelet Neuronal Interact.* 2005;5:321–324.)

of neighboring osteocytes and surface osteoblasts via gap junctions. Gap junctions facilitate the transfer of small molecules and biologically generated electrical potentials from cell to cell. In this manner, osteocytes communicate with one another and form a complex and integrated network throughout bone tissue (Fig. 4.24). Osteocytes are thought to function as sensor cells in bone that can mediate the effects of mechanical loading through their extensive communication network.

The number, size, shape, and position of osteocytes vary according to the type of bone they inhabit. In woven bone, they are numerous, large, and plump (Fig. 4.4). Their arrangement appears disorganized because their long axes parallel the direction of the neighboring collagen fibers that often show a random distribution. In lamellar bone, osteocytes are comparatively fewer in number, smaller, more spindle shaped. They appear to be more regularly organized because the cells are oriented in the same direction as the surrounding lamellae (Fig. 4.5). Age is another factor that affects osteocyte quantity as osteocyte apoptosis increases with aging, leading to decreased osteocyte density (29).

The repertoire of biologic activity possessed by osteocytes helps them maintain bone tissue and allows bone to be responsive to the mechanical and metabolic demands of the body. As mechanosensory cells, they translate mechanical forces into biologic activity (32,34). The detection of physical forces stimulates osteocytes to produce and release intercellular messengers that target precursor cells, osteoblasts, and osteoclasts (33,35). These cells, in turn, respond by remodeling the bone regionally and allowing it to change its mass and structure according to demands of the external physical environment. The widespread distribution of osteocytes and their cell processes is fundamental to mineral homeostasis (29). Osteocytes generate and respond to fluxes in ion concentrations and mediate the exchange of calcium and other ions between the bone matrix and extracellular fluid. In certain conditions, they may even be able to rapidly release calcium and phosphorus from the mineralized matrix by a process termed *osteocytic osteolysis*, which manifests histologically as enlarged lacunar spaces (36,37). Osteocytes also produce fibroblast growth factor-23 (FGF-23), which negatively regulates PTH and is a hormone essential in the regulation of serum phosphorus by modulating the reabsorption of phosphorus in the renal tubule (29,38). They also control osteoblastic activity through the secretion of sclerostin, which inhibits osteoblastic bone formation through Wnt/β-catenin pathway, and osteoclast activity through secretion of RANKL and osteoprotegerin (OPG) (29).

Osteoclasts

Osteoclasts are terminally differentiated, multinucleated cells responsible for bone resorption. They are mobile effector cells that have a lifespan of only several weeks. By the time they are recognizable by light microscopy, they are mature, biologically active, and can be found residing within resorption pits termed Howship lacunae that are produced by their digestion of mineralized bone matrix (Fig. 4.25).

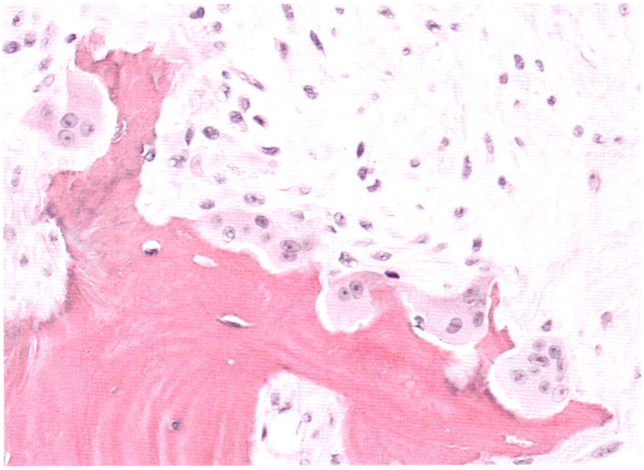

FIGURE 4.25 Multiple osteoclasts associated with resorption pits (Howship lacunae) on a trabecula.

Osteoclasts are 40 to 100 μm in diameter and are polarized with one portion of the cell membrane intimately attached to the bone and the remainder exposed to the extracellular fluid in its microenvironment. The segment of cell membrane that adheres to bone is laden with $\alpha_V\beta_3$ integrins. The integrins bind to specific extracellular bone matrix proteins (e.g., vitronectin, osteopontin, and bone sialoprotein) previously deposited by osteoblasts, and in this manner, the osteoclast can anchor to the bone surface. There is also a network of interconnecting actin filaments within the cytoplasm that extends from the site where the osteoclast cell membrane attaches to bone directly to the nuclei (31). On average, osteoclasts have 4 to 20 nuclei. In normal circumstances, however, the amount is usually not greater than 12. The nuclei and adjacent prominent Golgi apparatus tend to congregate away from the bone-resorbing surface and are surrounded by abundant amphophilic cytoplasm.

The cytoplasm in the vicinity of the resorbing surface is rich in tartrate-resistant acid phosphatase, carbonic anhydrase, and membrane-bound lysosomes (39). The adjacent cell membrane directly apposes the bone-resorbing surface and has numerous finger-like extensions that effectively increase its surface area. These fingerlike extensions have been collectively termed the brush border.

The lysosomes fuse with the brush border and release their contents into the resorption pit, thereby beginning the process of bone digestion. Metabolic activation of osteoclasts is initiated by anchorage, and this process generates a stimulatory signal that is transmitted to the nuclei by the actin network. The nuclei, now activated, orchestrate the complex and transitory cytoplasmic and cell membrane modifications required for bone digestion. Importantly, mineralized bone or cartilage is more efficiently resorbed by osteoclasts than is nonmineralized substrate. Focal, or partial, demineralization of collagen fibers is one of the first steps in matrix resorption and is followed first by catabolism of noncollagenous proteins then the degradation of collagen fibers. Once osteoclast activity ceases and the cell moves to another targeted site, macrophages move into the base of the resorption pit and phagocytize the organic remnants.

Osteoclasts are derived from mononuclear, hematopoietic progenitor cells of the granulocytic-macrophage colony-forming units (GM-CFU) and macrophage colony-forming units (M-CFU) (40,41). The mononuclear preosteoclasts undergo primary fusion to form multinucleated osteoclasts capable of acquiring and shedding nuclei throughout their short lifespans. A variety of cytokines and growth factors are critical to osteoclast development, maturation, and activity. These include interleukin (IL)-1, IL-3, IL-6, IL-11, tumor necrosis factor (TNF), granulocyte-macrophage colony-stimulating factor (GM-CSF), and macrophage colony-stimulating factor (M-CSF) (41). These factors work by either stimulating osteoclast progenitor cells or participating in a paracrine system in which osteoblasts and marrow stromal cells play a central role.

The system regulating osteoclasts is essential to bone metabolism, and its mediators include RANK, RANKL, and OPG (29,42). RANK is a member of the TNF family of receptors expressed mainly on cells of macrophage/monocytic lineage, such as preosteoclasts. When this receptor binds its specific ligand (RANKL) through cell-to-cell contact, a series of signal cascades is activated and osteoclastogenesis is initiated. RANKL is produced by and expressed on the cell membranes of osteoblasts and marrow stromal cells. Its expression is stimulated by apoptotic osteocytes and may be influenced by other osteotrophic factors. Its major role in bone metabolism is stimulation of osteoclast formation, differentiation, activation, and survival. The actions of RANKL can be blocked by another member of the TNF family of receptors, OPG, which is a soluble protein produced by bone, hematopoietic marrow cells, and immune cells. OPG inhibits osteoclastogenesis by acting as a decoy receptor; it binds to RANKL, thereby preventing the interaction of RANK with RANKL (29). The interplay between bone cells and these molecules permits osteoblasts and stromal cells to control osteoclast development, thereby ensuring the tight coupling of bone formation and resorption vital to the success of the skeletal system. This system also provides a mechanism for a wide variety of biologic mediators (hormones, cytokines, and growth factors) to influence the homeostasis of bone tissue.

Inorganic Components

Mineral

Approximately 65% of the bone volume is composed of minerals. The primary inorganic mineral of mature bone is a calcium-deficient variant of hydroxyapatite $[Ca_{10}(PO_4)_6(OH)_2]$, in which the hydroxyl groups have been largely replaced by phosphate and carbonate groups (14,16). The inorganic component of the skeleton represents a major reservoir for calcium and phosphate; it contains more than 99% of the body's calcium and 85% of its phosphorus. Bone crystals also harbor 95% of the body's sodium, 50% of its magnesium, and trace amounts of other essential minerals (43).

It has been suggested that bone contains two distinct types of mineral, one appears as granular, electron-dense bands within the collagen fibrils, and the other manifests as needle or filament-like crystals residing in the interfibrillar spaces. The deposition mechanism and chemical interactions between the collagen matrix and calcium phosphate apatite nanocrystals remain to be clarified (44,45). Two main theories seek to explain the mineralization of collagen: the hole zone theory suggests that initial mineralization begins in holes within the collagen fibrils, while the superhelix theory hypothesizes that mineralization takes place along collagen fibrils arranged in a parallel array (46). However, the use of newer, less invasive tissue sectioning methods reveals that most mineral is not deposited on collagen. Rather, extrafibrillar mineral lamellae of polycrystalline plates are arranged circumferentially around collagen fibrils (47).

Bone mineralization is divided into two stages: primary and secondary mineralization. Primary mineralization is a fast process occurring over the course of several days. It is biologically tightly controlled by osteoblasts that secrete collagen fibrils, noncollagenous proteins, and matrix vesicles. Membrane transporters and enzymes package Ca^{2+} and PO_4^{3-} into matrix vesicles, the small extracellular, plasma membrane-bound vesicles where mineral is first observed. This influx initiates calcium phosphate nucleation followed by crystal growth. Initially amorphous, crystals are then converted into mature crystalline structures of calcium phosphate, principally hydroxyapatite. The mature calcium phosphate crystals elongate radially until they penetrate the membranes of the matrix vesicles and continue growing outward to form calcifying nodules, which are needle-shaped crystals in globular arrangements that still retain some transporters and enzymes. Osteopontin, osteocalcin, and matrix gla protein in the surrounding tissues regulate subsequent growth of the calcifying nodules. The collagen becomes mineralized, beginning at the site of contact with the calcifying nodule. It is estimated that 60% to 70% of the minerals are deposited by primary mineralization. Secondary mineralization is a slower process that occurs over several months and culminates in the full mineral load of the bone. This stepwise process leads to the mosaic-like pattern of bone mineralization seen on microradiographs (Fig. 4.6). The biologic and ultrastructural mechanisms behind secondary mineralization are not known with certainty, but it is hypothesized that osteocytes regulate the transportation of Ca^{2+} and PO_4^{3-}, leading to their physical and chemical deposition (44,46).

Once the crystals are deposited in the bone, they remain there for days to years, only to be dissolved at a future time during bone resorption when the calcium and phosphorous are released into the extracellular fluid and become available for other biologic activities. Initial mineralization of osseous organic matrix takes approximately 2 weeks; therefore, the surfaces of bone are covered by a layer of unmineralized osteoid (Figs. 4.14 and 4.22). The width of this layer is dependent on the relative rate of bone formation. The osteoid seam spans only 1 to 5 μm in thickness in inactive regions where the bone is nearly fully mineralized, while this layer is several times thicker in foci of rapid bone deposition. The actual zone of mineralization can be detected in individuals who receive tetracycline because this antibiotic agent binds to the bone at the mineralization front and can be visualized with fluorescent microscopy (Fig. 4.12).

BONE FORMATION, GROWTH, AND REMODELING

From the moment that skeleton formation begins in the embryo until the stage that adult stature is attained, the bones of the body undergo a marked increase in size, refinement of their shape, and enhancement of their contours. Because bone is a rigid structure, it cannot grow interstitially and only enlarges by the apposition of new bone on its surface. Appositional growth alone is adequate for portions of the skeleton that enlarge slowly during maturation, such as the skull and diameter of long bones. However, it is insufficient for bones that must increase in size at a more rapid rate, such as the length of long and short tubular bones of the extremities, the vertebrae, and the ribs. Cartilage, in contrast, exhibits both appositional and interstitial growth; it increases its volume and enlarges in all dimensions as new cells and freshly synthesized extracellular matrix are added. Consequently, the elongation of tubular bones in embryos and prepubertal children occurs as growing cartilage is replaced by bone; most of the increase in bone length is derived from the cartilage primordium represented in the anlage and growth plate (i.e., physis).

The genetic code for skeletal morphogenesis is encrypted in the homeobox genes. Homeobox genes contain the DNA library of a repository of transcriptional regulators essential for growth and differentiation. The expression of homeobox genes occurs in a specific order and temporal sequence. Homeobox gene activation in the skeletal system results in the generation of localized cellular condensations of primitive mesenchyme at the sites of future bones. The mesenchymal condensations are the earliest precursors of individual bones and are critical to formation of the skeleton. They begin to develop just prior to day 40 of gestation. The origins of the mesenchymal condensations depend upon their anatomic location. Those of the craniofacial skeleton are derived from cells that migrate from the cranial neural crest, whereas those of the axial skeleton and appendicular skeleton are derived from the paraxial mesoderm and lateral plate mesoderm, respectively (48–50). The mesenchymal cells in the condensations begin to alter their genetic expression by the seventh week of gestation and, at this point, assume the morphology of matrix-forming cells. Those cells that mature into chondrocytes form a cartilage anlage of the future bone, which is fundamental to the process of endochondral ossification. Those that develop directly into osteoblasts produce bone *via* intramembranous ossification. Ossification begins around the eighth week of gestation. Mature bone tissue formed from endochondral ossification is morphologically identical to that developing *via* intramembranous ossification.

Endochondral Ossification

The newly formed cartilage anlage is avascular and has the crude shape of the adult bone (Figs. 4.26 and 4.27). The mesenchyme surrounding the anlage forms the perichondrium (Fig. 4.26), which is the precursor to the periosteum that will develop once ossification begins. This process is initiated in each bone is temporally regulated and the sequence is the same in all humans.

Growth of the anlage occurs both interstitially and appositionally as a result of the proliferation of chondrocytes and the accumulation of extracellular matrix (Figs. 4.28 and 4.29). The matrix is composed of proteoglycans and type II collagen with lesser amounts of collagen types IX, X, XI, and XIII (14). This process culminates in three nearly simultaneous events (51). First, the mesenchymal stem cells of the perichondrium produce a layer of osteoblasts around the midportion of the cartilaginous shaft. These osteoblasts deposit a collar of woven mineralized bone on the surface of the anlage, thereby heralding the transformation of the perichondrium into periosteum. The periosteum, osteoblasts, and thin surface layer of bone delineate the middle region of the diaphysis and form the primary center of ossification

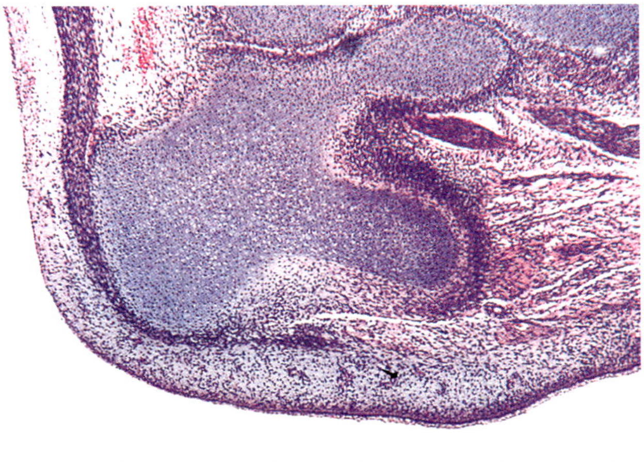

FIGURE 4.26 Photomicrograph of cartilage anlage of the os calcis (calcaneus). The cartilage model is the approximate shape of the adult bone. The attachment site of the Achilles tendon and the tibial–calcaneal joint are present.

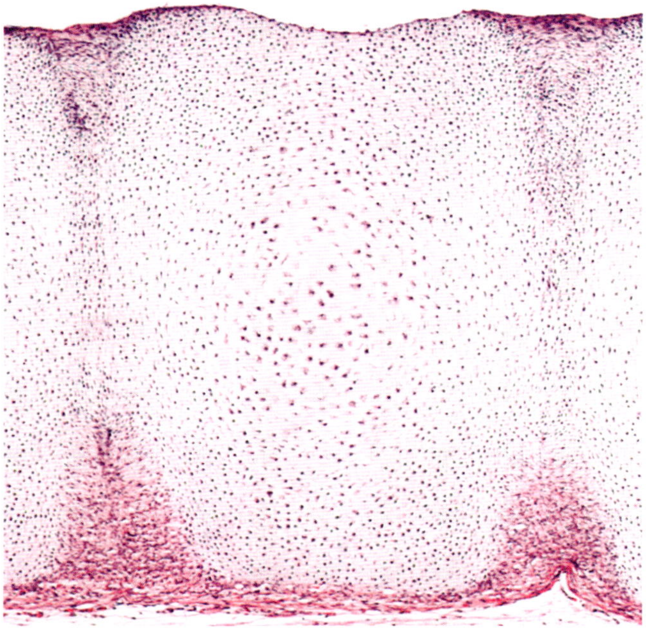

FIGURE 4.27 Sagittal section through vertebral body composed of cartilage anlage. The chondrocytes in the center are undergoing hypertrophy.

(Figs. 4.30 to 4.34). Second, the chondrocytes in the center of the anlage shaft that are encased by the periosteal shell of bone begin to hypertrophy and undergo biologic changes (Figs. 4.29 to 4.33). Chondrocyte enlargement is accompanied by an increase in intracellular glycogen and deposition of type X collagen in the vicinity of the chondrocyte. Soon thereafter, the chondrocytes undergo apoptotic necrosis as the surrounding matrix mineralizes. Mineralization mostly occurs *via* matrix vesicles derived from the chondrocytes, although some of it may occur within the holes of the collagen fibers. Finally, a capillary network originating from periosteal vessels forms. This network penetrates the woven bone of the primary center of ossification with the aid of matrix resorption (Fig. 4.34). The capillaries are the precursor to the future nutrient vessels and are accompanied by pericytes and other primitive mesenchymal cells, including immature osteoprogenitor and osteoclast progenitor cells.

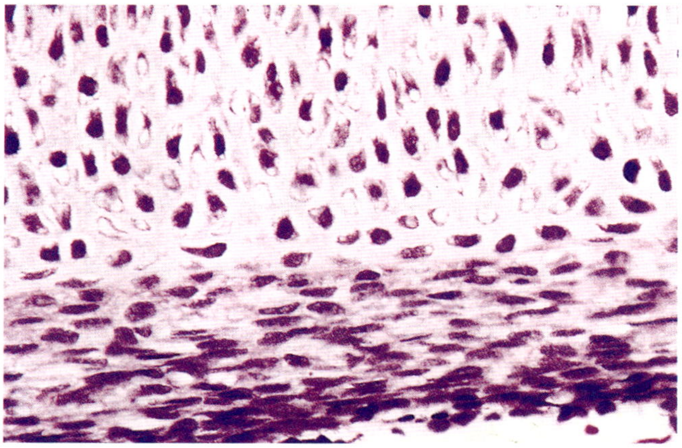

FIGURE 4.28 Cartilage anlage of femur in an embryo. The perichondrium is in intimate contact with the cartilage.

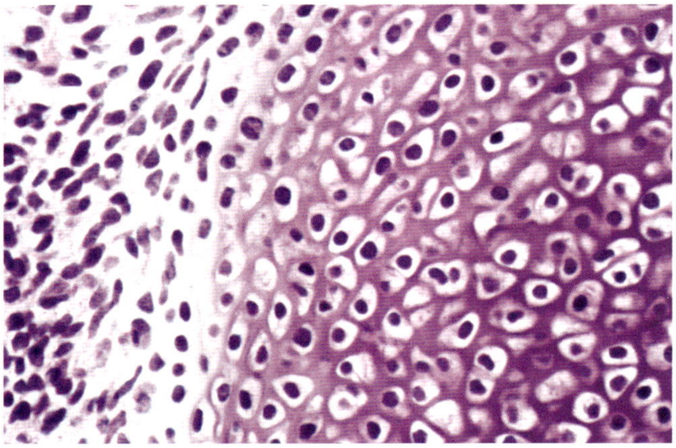

FIGURE 4.29 Cells of mesenchymal condensation surrounding an area in which they have differentiated into hyaline cartilage anlage. The chondrocytes show early hypertrophy.

As the cartilaginous core of the bone undergoes continued resorption, osteoblasts derived from perivascular stem cells deposit layers of osteoid on residual, longitudinally oriented struts of mineralized cartilage. These trabeculae are composed of a central cartilaginous core covered by a layer of woven bone. They are the first trabeculae formed and constitute the primary spongiosa. The spaces that develop as a

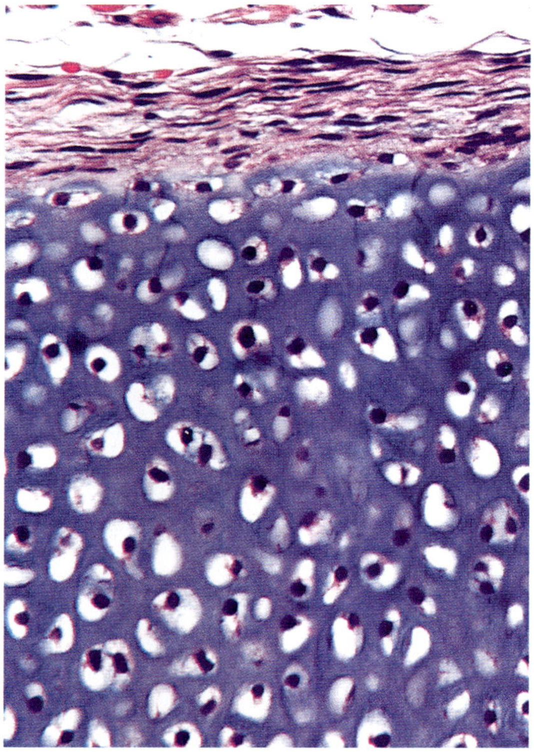

FIGURE 4.30 The perichondrium about the diaphysis of a cartilage anlage containing hypertrophied chondrocytes.

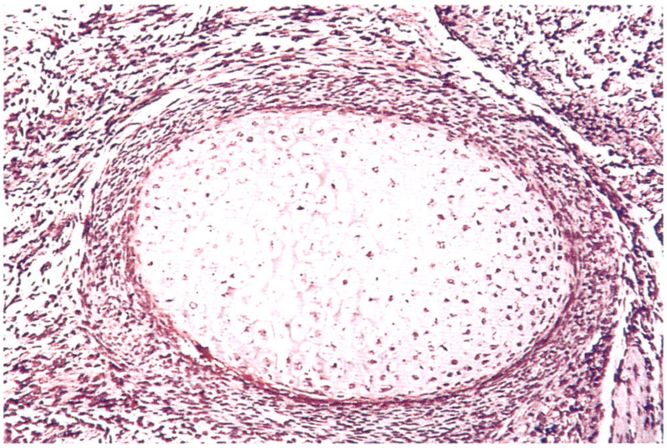

FIGURE 4.31 Cross-section of the diaphysis of an embryonic femur. The thin, collar-like primary center of ossification is between the hypertrophying chondrocytes and the periosteal cells.

consequence of the cartilage resorption ultimately coalesce to form the medullary cavity. The medullary cavity is initially filled with loose connective tissue but is eventually occupied by variable amounts of adipose tissue and hematopoietic elements. This complex process begins within the center of the shaft and progresses toward both ends of the bone as the bone lengthens. When complete with primary trabeculae and an adjacent secondary center of ossification, the entrapped cartilage is recognized as the growth plate, or physis (Figs. 4.35 to 4.38). Longitudinal growth of bones occurs at the physis so that markers placed in either the diaphysis or metaphysis and the epiphysis will separate with bone growth while markers placed only in the diaphysis will remain the same distance apart as the skeleton develops.

The fully developed growth plate is divided into five merging regions that correspond to different stages of chondrocyte maturation (Figs. 4.36 to 4.39). As the chondrocytes pass through the different stages, they do not literally move within the matrix but mature in the position they occupied upon formation. Important regulators of chondrocyte growth and maturation include the *Indian hedgehog* gene (Ihh), PTH, parathyroid hormone–related protein (PTHrP), FGF, RUNX2 and RUNX3, and bone morphogenic proteins (BMPs) (Table 4.3) (52–57). The zones include a reserve zone of resting chondrocytes located nearest the ends of the

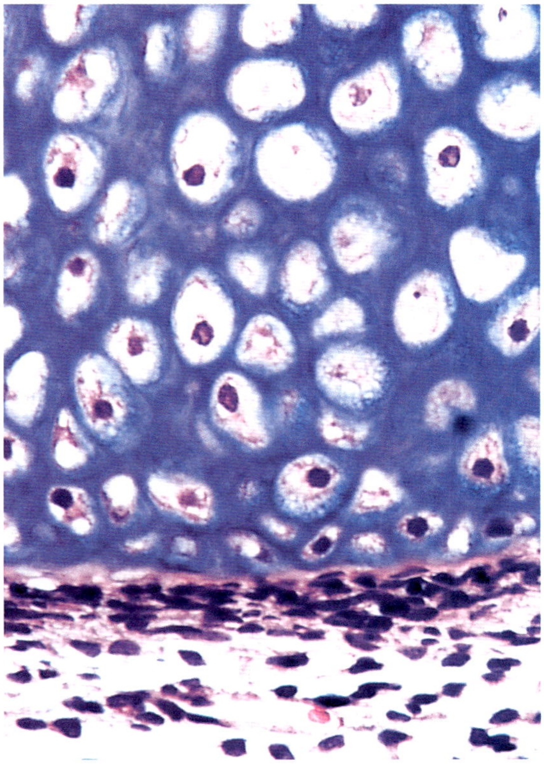

FIGURE 4.33 Primary center of ossification. A thin pink layer of osteoid, containing flattened osteocytes, separates the hypertrophied chondrocytes from the periosteal osteoblasts.

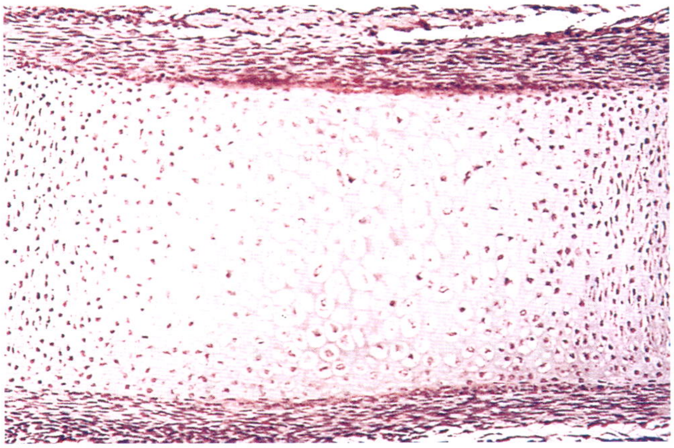

FIGURE 4.32 Longitudinal section of the primary center of ossification in an embryonic femur. The cellular layer of the periosteum is producing osteoblasts, which have formed a layer of pink osteoid. The outer spindle cells of the periosteum are oriented longitudinally along the femoral shaft. The underlying chondrocytes show hypertrophy.

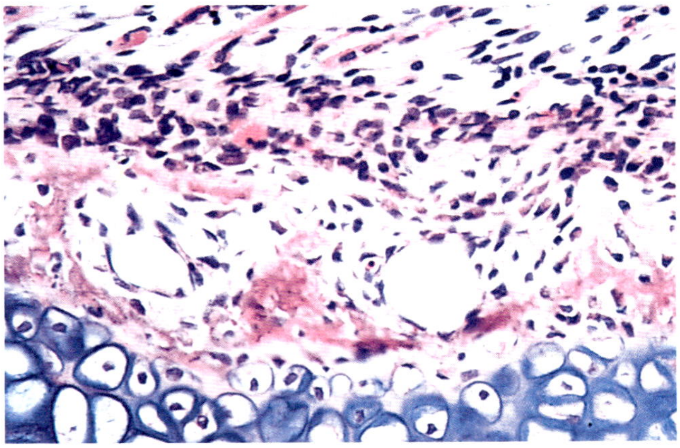

FIGURE 4.34 Primary center of ossification with capillary proliferation indicating the early formation of the nutrient artery. Between these capillaries, new trabecular membranous woven bone is being formed in the cambium layer of the periosteum. The outer fibrous layer of the periosteum is more cellular than in the adult.

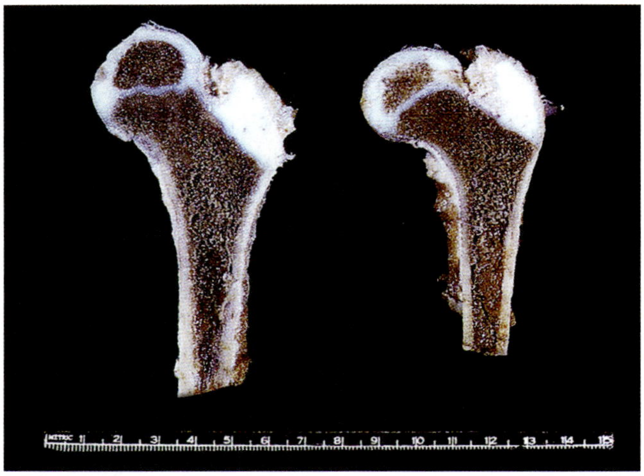

FIGURE 4.35 Femoral heads of a 3.5-year-old male patient. The secondary centers of ossification in the femoral heads are separated from the primary centers by the epiphyseal growth plates (the physes). The secondary center of ossification of the apophysis of the greater trochanter has not as yet formed. The metaphyses and diaphyses resemble their adult shapes.

bone, a region of proliferating chondrocytes arranged in spiral columns, a region of chondrocyte hypertrophy, a region of chondrocyte apoptosis and matrix mineralization, and a region of cartilage resorption by osteoclasts that tunnel into the mineralized matrix.

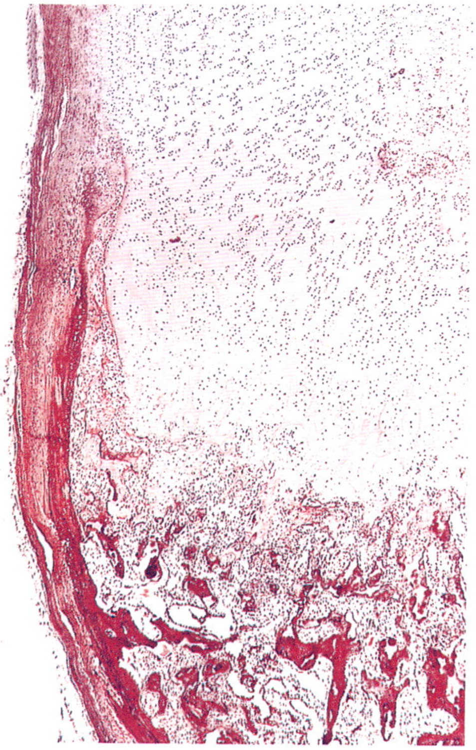

FIGURE 4.36 Epiphyseal growth plate of the costochondral junction from a 2-month-old male patient. The growth plate is surrounded by the ring of Ranvier. No secondary center is present. Primary trabeculae with central cartilaginous cores are seen in the metaphysis and upper diaphysis.

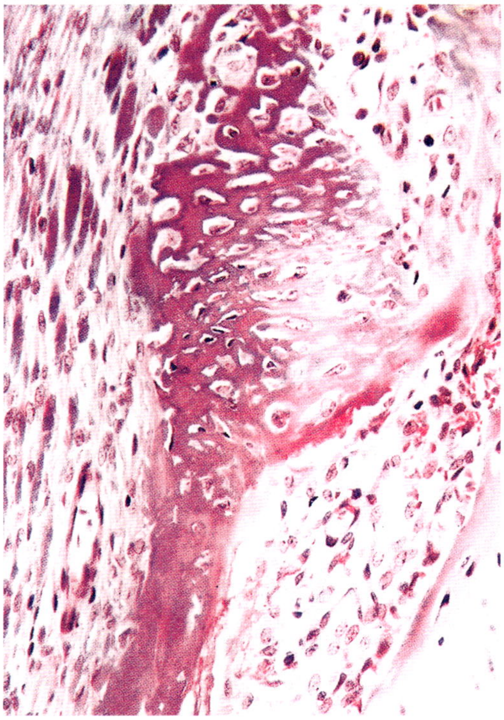

FIGURE 4.37 Ring of Ranvier, composed of bone that forms by the process of intramembranous ossification beneath the periosteum on the left and delineates the peripheral portion of the epiphyseal growth plate near the metaphysis.

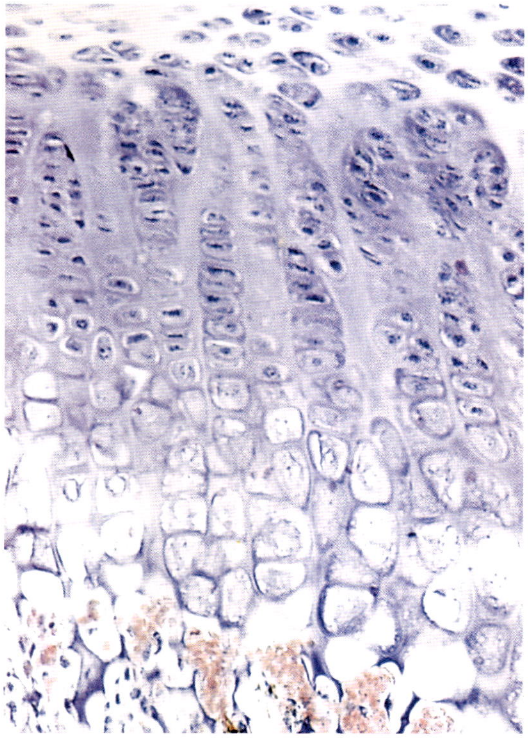

FIGURE 4.38 Maturing epiphyseal growth plate showing the reserve zone (*top*), the proliferating zone, the hypertrophied zone, and zone of mineralization of the cartilage. The primary trabeculae are oriented vertically and are supporting the base of the growth plate.

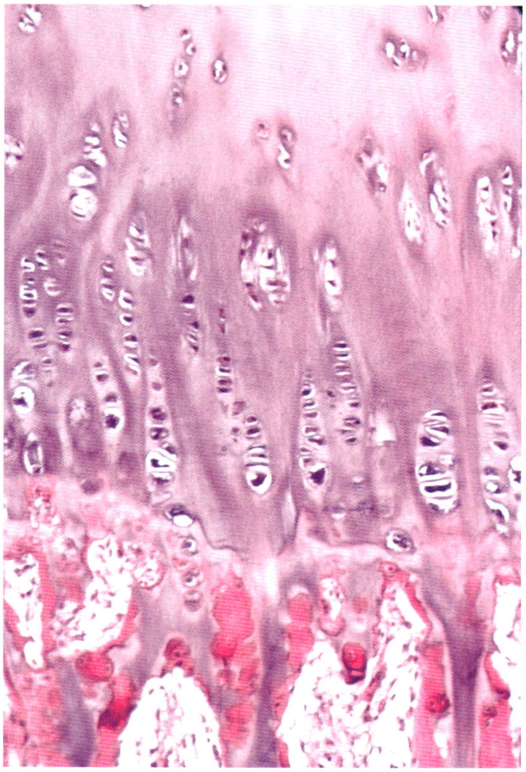

FIGURE 4.39 Junction between the mineralized cartilage columns and the primary trabeculae with the mineralized cartilage cores covered by a layer of woven bone.

The chondrocytes in the reserve zone are relatively small, round, or ovoid, and surrounded by abundant matrix. Those at its base give rise to chondrocytes that form the zone of proliferation. The chondrocytes in the zone of proliferation undergo cell division, become arranged in spiral columns, and elaborate the extracellular matrix. In the zone of hypertrophy, the chondrocytes enlarge with the cytoplasmic volume increasing 10 times resulting in a transverse diameter of up to 30 μm; these cells remodel the surrounding matrix as they increase in size (51). The chondrocytes in the zone of mineralization secrete matrix vesicles that are derived from the cell membranes. These control mineralization of the surrounding extracellular matrix. Three to five longitudinally oriented chondrocytes are associated with a region of calcified matrix that measures 90 to 150 μm in length (58). As the matrix mineralizes, the chondrocytes undergo rapid necrosis in the last row of lacunae before the ossification front. This process is associated with the release of cytokines that attract the ingrowth of endothelial cells and osteoclasts that tunnel into the mineralized matrix. The osteoclasts digest the transverse septa of matrix that separates the chondrocytes from one another and leave behind residual longitudinal struts of mineralized cartilage parallel to the long axis of the bone. The orientation of the struts is determined by the preexisting columnar arrangement of chondrocytes in the proliferative and hypertrophied zones

TABLE 4.3 Important Local and Systemic Regulators of the Growth Plate

Factor	Actions and Regulation
Local Effects	
Ihh	Master regulator of chondrocyte differentiation. Stimulates PTHrP, regulates resting zone to proliferative zone chondrocytes, stimulates proliferation, determines transition from proliferating to hypertrophic chondrocytes. Regulation: CSPG binding influences its distribution throughout growth plate; positively regulated by BMPs, negatively regulated by dEF1.
PTHrP	Maintains proliferative chondrocytes in the proliferative pool. Regulation: positively regulated by ADAMTS-12 and ADAMTS-17.
Cyclin D1	Transitions cells from G1 to S phase.
RUNX2 and RUNX3	Promote chondrocyte differentiation.
TGF-β	Promotes differentiation and matrix synthesis.
C/EBP(β)	Promotes differentiation of chondrocytes from proliferative to hypertrophic.
Panexxin3	Promotes differentiation of chondrocytes from proliferative to hypertrophic.
Capn4	Modulates terminal cell differentiation.
SDF-1	Enhances chondrocyte hypertrophy.
SOX9	Proliferation and differentiation of the fetal growth plate; negatively regulates vascularization, cartilage resorption, and trabecular bone formation.
FGFs	Inhibit proliferation.
Vitamin D3	Organization of growth plate, mineralization, and longitudinal bone growth.
CNP	Promotes chondrogenesis by stimulating expression of cell adhesion molecules and glycosaminoglycans.
Systemic Effects	
Growth hormone (GH)	Longitudinal bone growth. Stimulates production of IGF-1.
Glucocorticoids	Suppress bone growth. Thought to be mediated through disturbance of the GH/IGF-1 axis.
Thyroid hormone	Promotes chondrocyte hypertrophy and terminal differentiation through Wnt/β-catenin signaling, stimulates the IGF-1 pathway.
Estrogens and androgens	Increase skeletal growth in early puberty, fuse growth plate in late puberty.

Ihh, Indian hedgehog; PTHrP, parathyroid hormone-related protein; CSPG, chondroitin sulfate proteoglycan; BMPs, bone morphogenic proteins; dEF1, d-crystallin enhancer factor 1; ADAMTS-12, a disintegrin and metalloproteinase with thrombospondin motifs 12; ADAMTS-17, a disintegrin and metalloproteinase with thrombospondin motifs 17; RUNX, runt-related transcription factor; TGF-β, transforming growth factor beta; C/EBP(β), CCAAT/enhancing binding protein-β; Capn4, Calpain-4; SDF-1, stromal cell-derived factor 1; SOX9, sex-determining region Y-box 9; FGFs, fibroblast growth factors; CNP, C-type natriuretic peptide; IGF-1, insulin-like growth factor-1; Wnt/β-catenin, canonical Wingless-related integration site/beta-catenin signaling pathway.

(Figs. 4.36 and 4.38). This columnar arrangement is thought to be controlled by a combination of physical or mechanical properties of the extracellular matrix and signaling molecules in the resting zone, specifically those involved in the noncanonical frizzled pathway. Primary cilia may also help create polarization gradients and transduce Ihh signaling (52). The

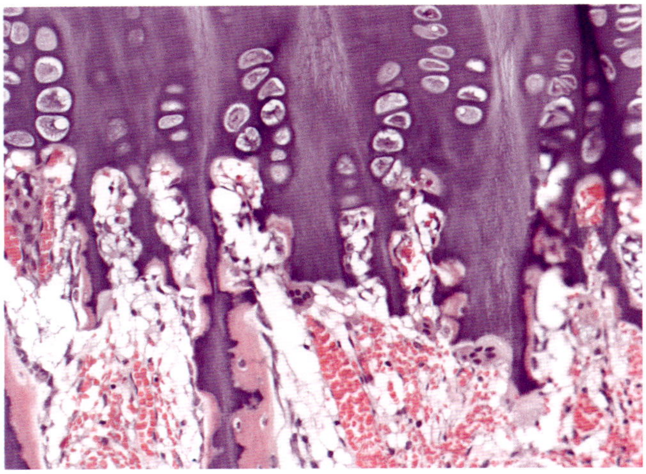

FIGURE 4.40 Base of growth plate showing osteoclastic resorption of residual calcified zone of hypertrophy and primary spongiosa with newly deposited woven bone on surface of strut of mineralized cartilage.

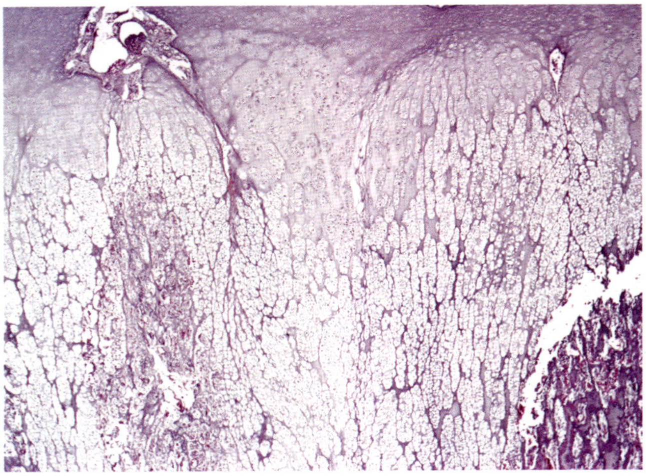

FIGURE 4.41 Mineralization of the cartilage is impaired in rickets, where removal of the cartilage is delayed, and the zone of hypertrophy of the physeal plate becomes massively, irregularly thickened. In this case with co-occurring osteopetrosis, abnormal osteoclasts cannot resorb the primary spongiosa which persists.

mineralized cartilaginous struts act as scaffolding for newly deposited woven bone and form the primary trabeculae (Figs. 4.38 to 4.40).

The rate of resorption of mineralized cartilage at the chondro-osseous junction is balanced by the rate of chondrocyte proliferation in the zone of proliferation in the growing long bone. It is estimated that chondrocytes spend several days in the zones of proliferation and hypertrophy (59). Evidence suggests that in an individual column of chondrocytes, four to seven cells die per day. Studies have found that in the costochondral junction of newborn infants there are 12.6 ± 1.0 chondrocytes per column. Of these, 39.6% ± 6.9% are proliferating chondrocytes, although the number of chondrocytes per column decreases with age (60). The rate of growth differs for each physeal plate and is greatest in the growth plate of the distal femur, followed by that of the proximal tibia. Diseases that impair mineralization of the cartilage result in delayed removal of the cartilage, resulting in massive hypertrophy of the zone of hypertrophy (Fig. 4.41). While most tubular long bones have two epiphyseal growth plates, other bones (e.g., ribs and some of the phalanges, carpals, tarsals, metacarpals, and metatarsals) have only a single physis. The growth plates, located at the diaphyseal–epiphyseal junctions of the bones, are delineated peripherally by a thin circumferential collar of bone that is a continuation of the primary center of ossification. This bony collar is called the ring of Ranvier (61) (Figs. 4.36 and 4.37).

Dramatic changes occur in the cortex at the same time that continued appositional and interstitial growth of epiphyseal and growth plate cartilage result in bone growth. Subperiosteal bone is deposited while bone is resorbed along the endosteum, thereby allowing the cortical thickness to remain proportionally uniform while the medullary cavity enlarges and increases the overall diameter of the bone. The bone that first forms the cortex is woven in nature; but, within the first several years of life, the fabricated bone is lamellar. Variation in the rate of formation and resorption alters the shape of the bone. This process is most pronounced in a region just distal to the base of the growth plate, known as the cut-back zone. The cut-back zone is rich in subperiosteal osteoclasts, which reduce the diameter of the bone to that of the diaphysis and result in "funnelization" of the bone. At the same time, the cortical thickness is maintained or increases by appositional new bone formation on the endosteal surface of the cortex. The diameter of the diaphysis continues to enlarge during growth and development and becomes asymmetric in specific sites. This process not only determines the eventual diameter of the bone but also controls the thickness and contour of the cortex as well as the arrangement of the trabeculae. Conditions altering the balance of bone formation and resorption may cause abnormally thickened or significantly thinned cortices.

Long bones lengthen through a similar process in the epiphysis, which represents the secondary center of ossification (Figs. 4.35 and 4.42). A few long bones also have a similar growth center in the apophyses. The maturation and replacement of the cartilage anlage in a secondary center is identical to that which occurs in the diaphysis, although maturation proceeds from the center and proceeds in a centrifugal fashion toward the periphery. This means that the growing area of the secondary center is initially a sphere. Eventually, the enlarging primary and secondary centers of ossification approach one another, entrapping a cylindrical segment of residual cartilage anlage and delineating the final form of the physeal plate.

Continual growth of the primary and secondary centers of ossification results in merging of their reserve zones. At this time, a plate of bone demarcating the secondary center from the forming growth plate is deposited such that

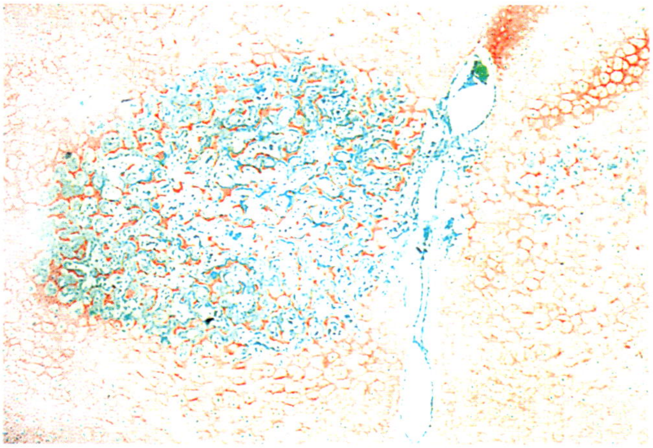

FIGURE 4.42 Photomicrograph of the secondary center of ossification of the femoral head. The nutrient vessels from the ligamentum teres are seen. Osteoid (*green*) is seen on the cartilage trabeculae (*orange*) in the spherical center (Goldner stain).

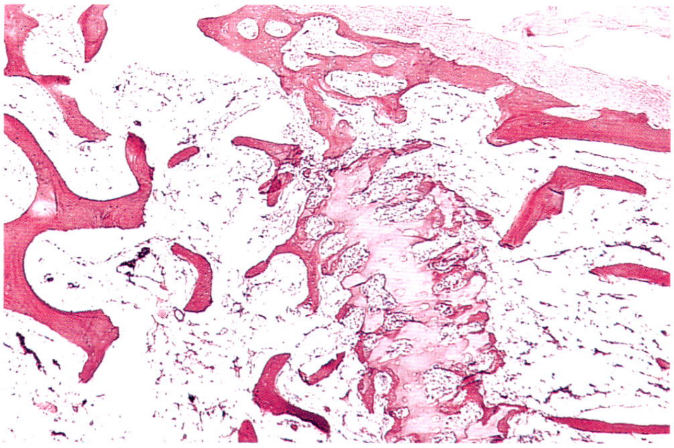

FIGURE 4.43 Closing epiphyseal growth plate in a 16-year-old boy. The periphery of the plate has been bridged by bone connecting the diaphysis metaphysis and the secondary center of ossification. Cellular proliferation in the physis has ceased while the maturation process continues. The secondary center of ossification is at the left of the micrograph above the remnant of the physis.

future centrifugal growth of the epiphysis is hemispheric. The cartilage located at the base of the true articular cartilage is responsible for progressive epiphyseal enlargement, and it has the architectural organization of a physis. Variation in the subarticular growth results in concordance of the shapes of the ends of articulating bones. The epiphysis receives its nutrition primarily from blood vessels within the bone and its adjacent periosteum, while the articular cartilage is nourished by synovial fluid.

The apophyseal cartilage is located on the surface of the bone and represents a secondary-like center of ossification. It is responsible for the development of apophyseal bone of the iliac crests, the greater and lesser trochanters of the femur (Fig. 4.35), and the tibial tuberosities.

Modeling of the newly formed bone begins once endochondral ossification is well underway at the growth plate. The primary spongiosa undergoes complete osteoclastic resorption, and secondary trabeculae composed solely of lamellar bone are deposited. The expanding medullary cavity is largely free of cancellous bone spicules throughout much of the diaphysis. Rather, it is filled with adipose tissue and the hematopoietic marrow. Subperiosteal bone deposition and endosteal resorption of the cortex maintain a proper, tubular shape, and mechanical forces exerted by weight bearing and muscle attachments alter the rates of these processes in specific regions, thereby sculpting the contour of the bone.

Several hormones, including PTH, growth hormone, somatomedins, thyroid hormone, androgens, estrogens, and adrenal cortical hormones, are essential regulators of bone growth (54) (Table 4.3). Low doses of androgens and estrogens at puberty promote cell division in the proliferative zone of the growth plate and the secondary center of ossification. This is accompanied by an increased rate of cartilage maturation, mineralization, osteoclastic removal, and formation of primary trabeculae. The net effect of these changes is the rapid skeletal growth occurring during puberty. As estrogen and androgen levels increase and growth hormone and somatomedin levels fall off, chondrocyte proliferation decreases while maturation and bone formation proceed. This leads to a diminution or thinning of the growth plate, ultimately resulting in endochondral ossification of the cartilage within the growth plate (Fig. 4.43). Fusion of the growth plate occurs as the hypertrophic chondrocytes furthest from the physis undergo cell death. Although this process has been attributed to apoptosis, it morphologically resembles hypoxia-induced necrosis and may not involve classical apoptotic pathways at a molecular level (52,62). After the growth plate is closed, all additional bone growth is appositional. An imprinted gene network is thought to be responsible for growth deceleration (52). Cessation of growth of the secondary centers of ossification occurs in a similar fashion. However, a remnant of mineralized growth cartilage, which is the tidemark cartilage, persists at the base of the articular surface. It is demarcated from the true articular cartilage by a thin undulating layer of densely mineralized matrix known as the tidemark (Fig. 4.44). The biologic potential of the tidemark cartilage persists. Increased circulating hormones, such as growth hormone in the setting of acromegaly, can reactivate the process of endochondral ossification and produce additional growth in the adult. In normal circumstances, however, the vestige of the growth cartilage remains dormant and functions as an anchor of the true articular cartilage to the subchondral bone plate.

Intramembranous Ossification

Intramembranous ossification refers to the process of bone formation in which the tissue occupying the site of the future bone is a fibrous membrane. The membrane is rich in osteoprogenitor cells and, in normal situations, develops from the mesenchymal condensations in the developing embryo,

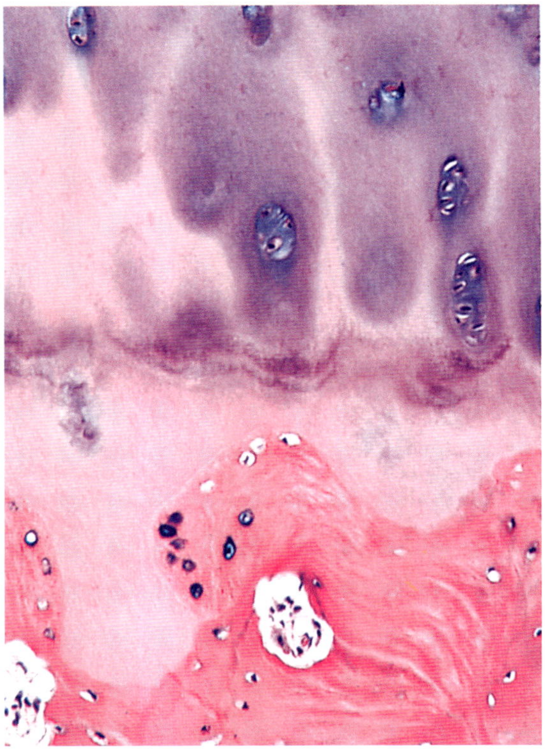

FIGURE 4.44 Section of base of articular cartilage in the region of the tidemark. The tidemark separates the articular nonmineralized cartilage (*above*) from the mineralized cartilage remnant of the physis and the lamellar bone of the subchondral plate (*below*).

the periosteum, or the thin layer of fibrous tissue adjacent to all active bone-forming sites. The osteoprogenitor cells within the membrane produce offspring that differentiate into mature osteoblasts capable of directly depositing the bone matrix (Fig. 4.45). Large portions of the flat bones of the skull form by this process. Also, since the cortices of all bones are largely created by osteoblasts derived from the cambium layer of the periosteum, all bones are formed by intramembranous ossification, at least in some part. Growth of membranous bone occurs only by the apposition of new bone, and the medullary cavities of membranous bones are created and maintained by endosteal osteoclastic activity. Initially, the marrow spaces of these bones are composed of highly vascularized loose connective tissue, which is eventually replaced by adipose and hematopoietic tissues.

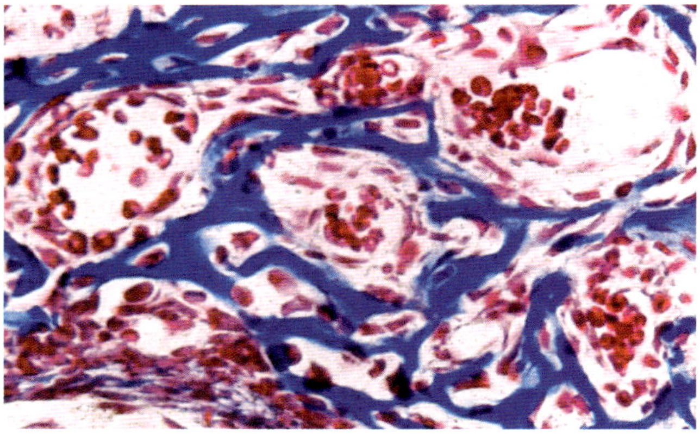

FIGURE 4.45 Intramembranous bone from a fetal skull. The osteoblasts are large along the randomly formed trabeculae. The osteocytes and their lacunae are large, round, and irregularly spaced in the trabeculae (Trichrome stain).

Modeling and Remodeling

The processes of bone formation and resorption are tightly coupled, and their balance determines skeletal mass at any point in time (50). The skeleton undergoes modeling as it grows and enlarges during childhood and young adulthood. Bone formation predominates during this period, whereas bone resorption prevails after the third or fourth decades. Remodeling is defined as the breakdown and renewal of bone and is fundamental to the formation and maintenance of the skeleton. Remodeling is a dynamic process involving the removal and deposition of both cortical and trabecular bone; it continues throughout life to maintain bone mass, skeletal integrity, and skeletal function (63). This process is complex and at least partially controlled by the central nervous system through leptin, serotonin, and other hormones in response to mechanically induced microdamage. It depends on the integrated actions of osteoblasts, osteocytes, and osteoclasts (64). Together these cells form the functional or basic multicellular unit of bone (BMU, or bone remodeling unit of Frost) and, in adults, are responsible for remodeling approximately 10% of the skeleton on an annual basis (Figs. 4.46 and 4.47) (65). This feat is accomplished by approximately one million BMUs that are active at any one time, and which likely first target sites that are experiencing fatigue and microdamage (63). The process may begin on any bony surface and incorporates three phases of cell activity: activation, bone resorption, and bone formation.

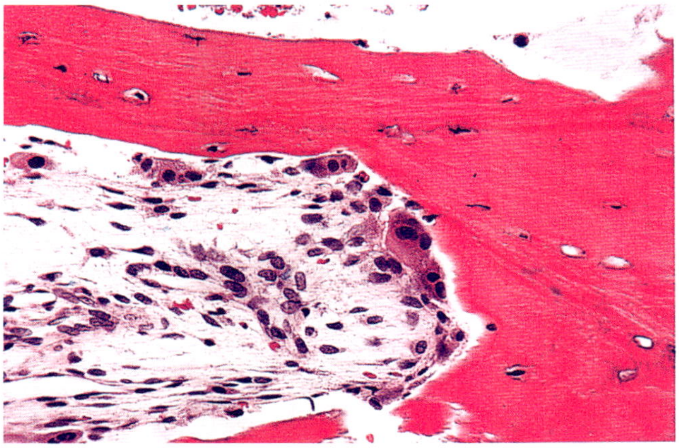

FIGURE 4.46 Basic multicellular unit of bone (bone remodeling unit of Frost). Osteoclasts form the leading edge of the bone resorption ("the cutting cone"), and just behind them are mononuclear macrophages and osteoblasts. The newly created space is filled with a vascular loose connective tissue.

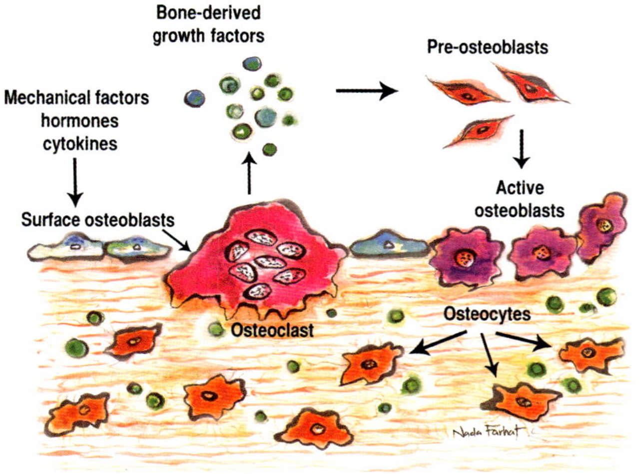

FIGURE 4.47 Illustration of the basic multicellular unit of bone, or bone remodeling unit of Frost. Osteocytes sense mechanical forces through their intricately connected network of dendrites. Mechanical factors, hormones, or cytokines stimulate resting surface osteoblasts to secrete cytokines that activate osteoclasts. Osteoclastic bone resorption through tartrate-resistant acid phosphatase within resorption lacunae releases bone-derived growth factors that stimulate precursor cells to differentiate into active osteoblasts. New osteoblastic bone fills the resorbed lacunae, ultimately coupling bone resorption with bone formation. (Courtesy of Nada Farhat.)

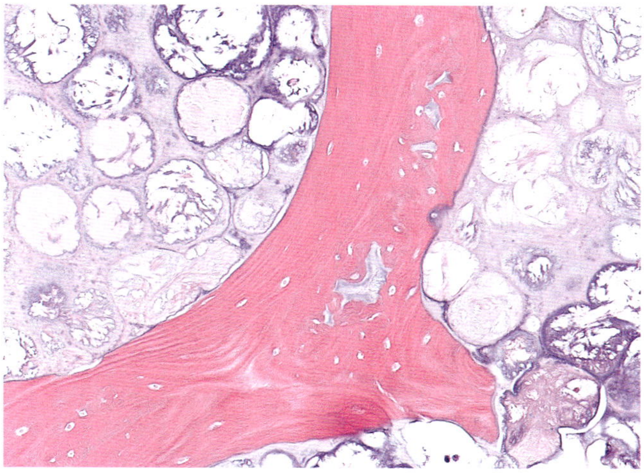

FIGURE 4.48 Osteonecrosis demonstrates a region of lamellar bone containing no viable cells, with adjacent marrow harboring no viable adipocytic or hematopoietic elements. The necrotic marrow is saponified and partially mineralized.

Bone Tissue in Abnormal Conditions

Many pathologic conditions of bone result from abnormalities in bone remodeling. These disorders may be generalized, in the form of a metabolic bone disease, or localized to small regions of the skeleton or individual bones. For instance, the diminished bone mass in postmenopausal osteoporosis, hyperparathyroidism, and hyperthyroidism results from increased osteoclastic bone resorption, which is not adequately compensated for by an appropriate amount of new bone formation. The lytic lesions of early Paget disease or those caused by metastases and myeloma result from a localized increased osteoclastic bone resorption that is disproportionate to the amount of new bone that is deposited. The goal of therapy for this broad spectrum of diseases is to restore bone mass, balance bone formation, and resorption, and protect and maintain structural integrity.

Normal bone tissue can be altered in a variety of different conditions. Bone necrosis secondary to ischemia manifests as groups or regions of lacunae that are empty or contain mummified osteocytes. The surface lining cells and hematopoietic and fatty marrow elements are similarly affected (Fig. 4.48). The reaction to the necrotic tissue can provide information regarding the time course of the process.

Injury to bone, such as fracture, activates the reparative response that can ultimately lead to complete healing so that the changes caused by the initiating event are no longer detectable. Bone tissue is the only tissue that can completely heal itself without development of a scar. This phenomenon is accomplished by activating the processes involved in its ontogeny, namely endochondral and intramembranous ossification. This causes hemorrhage, necrosis of injured tissue, subsequent blood clotting with deposition of a fibrin network, degranulation of platelets with release of cytokines and growth factors, influx of inflammatory cells, stimulation of tissue-fixed mesenchymal stem cells and the production of new cells of different phenotypes and matrix production. The newly formed reparative tissue represents callus and is morphologically similar to bone and cartilage formed during growth and development (Fig. 4.49). Over time, if properly mobilized, callus undergoes remodeling until the ends of the broken bones are completely united and have the appearance of normal bone.

Decrease in bone mass underlies all forms of osteoporosis. Bone mass is diminished by an imbalance in osteoclastic

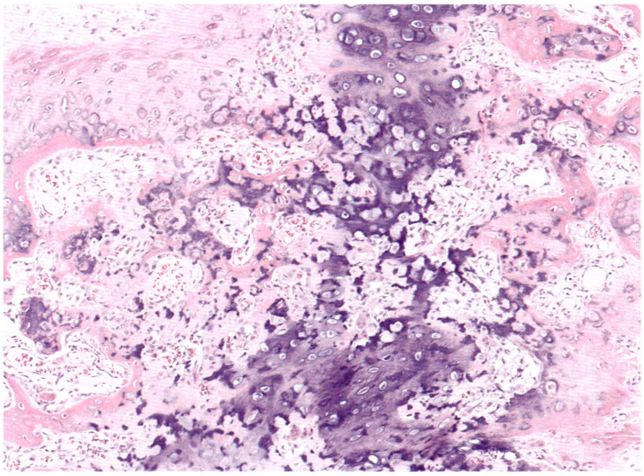

FIGURE 4.49 Fracture callus with repair shows a cartilage scaffolding undergoing endochondral ossification and forming interconnected woven bone trabeculae. Moderately cellular fibrous tissue with neovascularization surrounds the trabeculae.

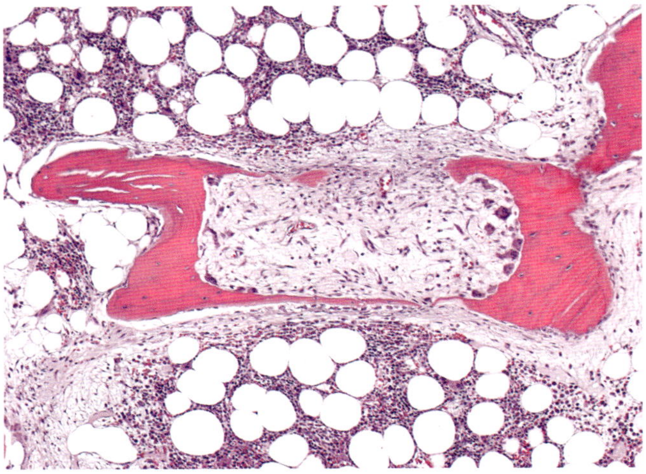

FIGURE 4.50 Dissecting osteitis in hyperparathyroidism is evident as a cone of osteoclasts resorbing the center of a bone trabecula, leaving a central core of fibrous tissue.

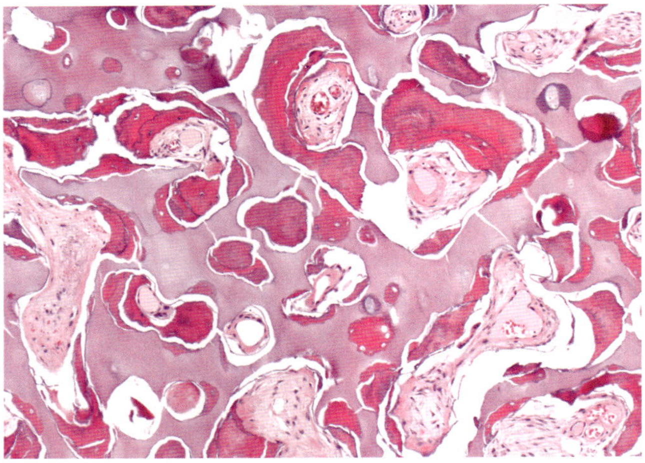

FIGURE 4.52 In osteopetrosis, the primary spongiosa persists in the medullary cavity as it is not effectively resorbed by the abnormal osteoclasts.

and osteoblastic activity. Decreased bone mass manifests as thinned cortices with haversian and Volkman canals that have increased diameters and thinned bony trabeculae that have lost their normal interconnections, such that they appear as round or ovoid islands of bone surrounded by hematopoietic or fatty marrow. In osteogenesis imperfecta, the osteoporotic bone also contains abnormally increased amounts of woven bone and this combination predisposes to fracture.

Osteoid is a normal constituent of bone and its thickness depends on the degree of osteoblastic activity. It is usually less than 20 microns thick and covers less than 20% of trabecular surfaces. As previously mentioned, osteoid appears lighter pink in comparison to mineralized bone in conventional decalcified histologic sections. Increased thickness of osteoid seams and increased surface area that they cover are manifestations of rickets and osteomalacia of any cause. The growth plates are abnormally thickened in patients with rickets because there is diminished calcification of the cartilage matrix and an increased thickness of the zone of hypertrophy in the physeal plate.

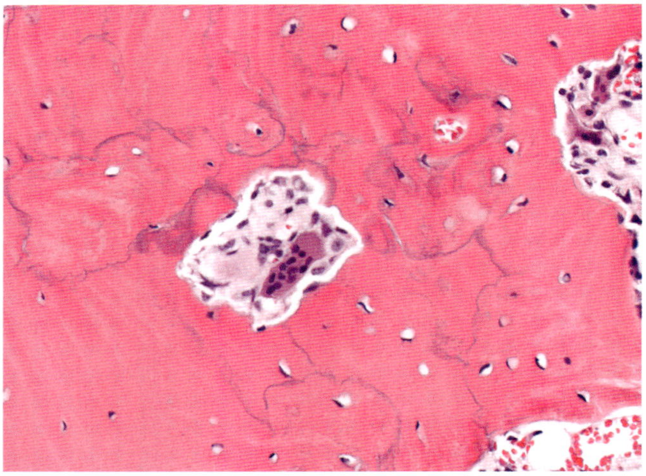

FIGURE 4.51 Classic Paget disease of bone with a prominent mosaic pattern of haphazardly oriented lamellar bone units outlined by prominent cement lines. Osteoclasts are enlarged and increased in number.

Osteoclastic activity is increased in different diseases. Increased numbers of osteoclasts are present states of hyperparathyroidism. They can be found expanding cortical canals (i.e., cortical cutting cones) and boring down the centers of bone trabeculae (i.e., dissecting osteitis) (Fig. 4.50). There is usually osteoblastic activity in the wake of the osteoclastic activity because bone resorption and formation are coupled processes. In this situation, peritrabecular fibrosis may be prominent because this tissue contains activated mesenchymal stem cells that give rise to osteoblasts. In active Paget disease of bone, osteoclasts are increased in number and size, often with a greater number of nuclei and cytoplasmic inclusions. The bone exhibits a classic mosaic pattern formed by haphazardly oriented lamellar units delineated by prominent cement lines that may have cracks or separations along their course. Prominent osteoblastic activity is also frequently present (Fig. 4.51).

Decreased osteoclastic activity can be induced by medications and be a component of osteopetrosis–a group of diseases caused by mutations that affect osteoclasts. In osteopetrosis, the primary spongiosa is not resorbed because the osteoclasts are abnormal. Thus, it persists and fills the medullary cavity, thereby displacing the marrow elements to extramedullary sites (Fig. 4.52). The long bones also have an Erlenmeyer flask deformity because the deficient osteoclasts do not adequately resorb the bone in the metaphyseal cut-back zone.

BONE—HISTOLOGIC ARTIFACTS

Several common artifacts may be seen in histologic sections of bone, and awareness of them can facilitate accurate

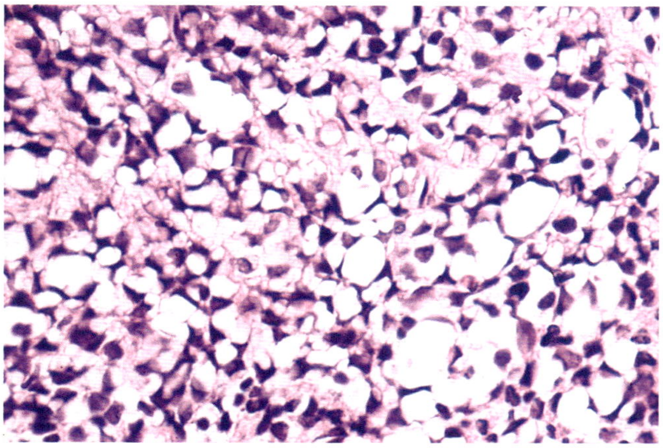

FIGURE 4.53 Bone biopsy specimen that was exposed to a decalcification agent prior to being well-fixed in formalin (in this case, simultaneous exposure). While the cells represent large lymphoid cells of a large cell lymphoma, artifactual vacuolization of the cytoplasm mimics an adenocarcinoma with signet ring cells.

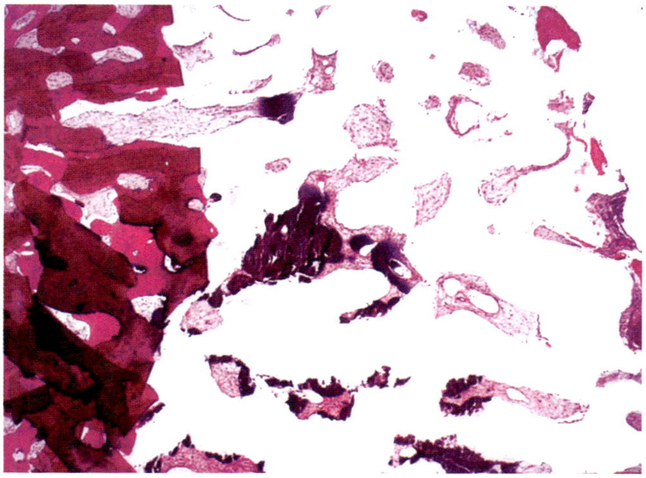

FIGURE 4.55 Section of cancellous bone without decalcification. The bone does not cut well on the microtome, leading to folding and thickening of the tissue section and retraction or separation of the bone from the adjacent tissue. In this case, the separation of the folded bone on the left from the vascular tissue within the marrow space may mimic a vascular neoplasm.

interpretations. The high mineral content of bone requires that it be decalcified during routine histologic processing. Multiple decalcifying agents are available, and they can be categorized as strong acids (e.g., hydrochloric acid), weak acids (e.g., formic acid), chelating agents (e.g., ethylenediaminetetracetic acid [EDTA]), or combinations thereof. If preservation of DNA and RNA is necessary, removal of mineral by EDTA is recommended (66). Specimens must be well fixed in formalin before initiating decalcification. Exposure of tissue to decalcification agents prior to appropriate fixation can lead to poor cellular preservation, such as artifactual cytoplasmic vacuolization (Fig. 4.53). Over-decalcification can lead to very pale stained sections that make it difficult to appreciate the nuclear detail and limit the amount of staining present in normally hematoxylin-rich regions such as those enriched in nucleic acids (Fig. 4.54). Under-decalcification leads to problems cutting bone tissue on the microtome, resulting in thick sections that fold, frequently leading to physical separation of the bone from the surrounding soft tissue causing the tissue to mimic a vascular neoplasm (Fig. 4.55). Bone dust is another common artifact–it is generated during cutting bone in vivo while performing core biopsies, orthopedic saw cuts or reaming, or as a complication of dissecting and sectioning the bone in the pathology laboratory. Bone dust appears as irregular fragments of bone or amorphous pink-purple tissue and can mimic bone necrosis (Fig. 4.56).

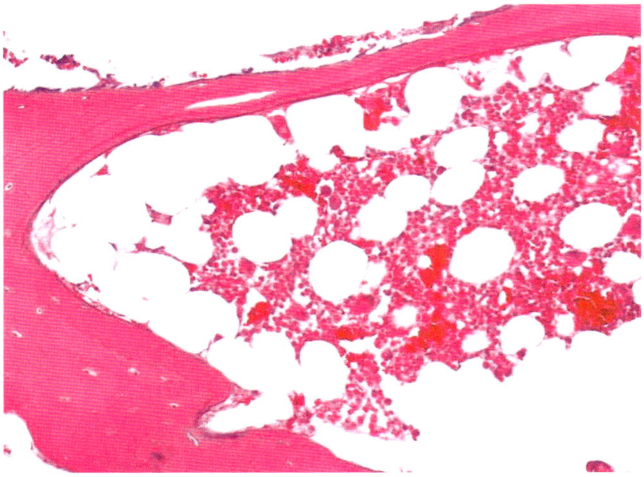

FIGURE 4.54 Trabecular bone and marrow with over-decalcification artifact. The marrow elements have limited hematoxylin staining, leading to pale nuclei and poor contrast between the nucleus and cytoplasm. The specific hematopoietic cell subtypes are more difficult to distinguish from one another.

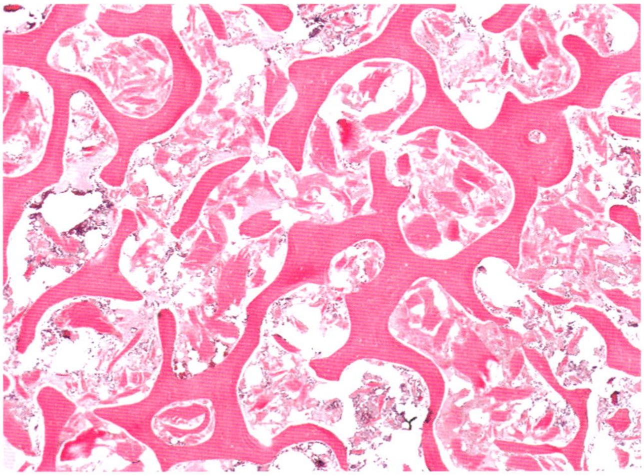

FIGURE 4.56 Bone dust manifests as irregular fragments of cortical bone or amorphous pink-purple tissue and can be mistaken for osteonecrosis or necrotic tumor.

ACKNOWLEDGMENT

Sanford I. Roth contributed to prior editions of this chapter.

REFERENCES

1. Neumann PE, Gest TR. How many bones? Every bone in my body. *Clin Anat*. 2020;33:187–191.
2. Burdan F, Szumilo J, Korobowicz A, et al. Morphology and physiology of the epiphyseal growth plate. *Folia Histochem Cytobiol*. 2009;47:5–16.
3. Milovanovic P, Scheidt AV, Mletzko K, et al. Bone tissue aging affects mineralization of cement lines. *Bone*. 2018;110:187–193.
4. Skedros JG, Holmes JL, Vajda EG, Bloebaum RD. Cement lines of secondary osteons in human bone are not mineral-deficient: New data in a historical perspective. *Anat Rec A Discov Mol Cell Evol Biol*. 2005;286:781–803.
5. Burket J, Gourion-Arsiquaud S, Havill LM, Baker SP, Boskey AL, van der Meulen MCH. Microstructure and nanomechanical properties in osteons relate to tissue and animal age. *J Biomech*. 2011;44:277–284.
6. Edwards JR, Williams K, Kindblom LG, et al. Lymphatics and bone. *Hum Pathol*. 2008;39:49–55.
7. Marinelli Busilacchi E, Morsia E, Poloni A. Bone marrow adipose tissue. *Cells*. 2024;13:724.
8. Pham TT, Ivaska KK, Hannukainen JC, et al. Human bone marrow adipose tissue is a metabolically active and insulin-sensitive distinct fat depot. *J Clin Endocrinol Metab*. 2020;105:2300–2310.
9. Krings A, Rahman S, Huang S, Lu Y, Czernik PJ, Lecka-Czernik B. Bone marrow fat has brown adipose tissue characteristics, which are attenuated with aging and diabetes. *Bone*. 2012;50:546–552.
10. Thorns C, Schardt C, Katenkamp D, Kahler C, Merz H, Feller AC. Hibernoma-like brown fat in the bone marrow: report of a unique case. *Virchows Arch*. 2008;452:343–345.
11. Boutin RD, White LM, Laor T, et al. MRI findings of serous atrophy of bone marrow and associated complications. *Eur Radiol*. 2015;25:2771–2778.
12. Bohm J. Gelatinous transformation of the bone marrow: the spectrum of underlying diseases. *Am J Surg Pathol*. 2000; 24:56–65.
13. Chan N, Ho C, Yip S-F. Cardiac cachexia causing extensive serous degeneration of bone marrow. *Br J Haematol*. 2017;178:660.
14. Gorski JP. Biomineralization of bone: a fresh view of the roles of non-collagenous proteins. *Front Biosci*. 2011;16:2598–2621.
15. Neve A, Corrado A, Cantatore FP. Osteoblast physiology in normal and pathological conditions. *Cell Tissue Res*. 2011; 343:289–302.
16. Clarke B. Normal bone anatomy and physiology. *Clin J Am Soc Nephrol*. 2008;3 Suppl 3:S131–S139.
17. Camozzi V, Vescini F, Luisetto G, Moro L. Bone organic matrix components: Their roles in skeletal physiology. *J Endocrinol Invest*. 2010;33:13–15.
18. Allori AC, Sailon AM, Warren SM. Biological basis of bone formation, remodeling, and repair-part II: extracellular matrix. *Tissue Eng Part B Rev*. 2008;14:275–283.
19. Allori AC, Sailon AM, Warren SM. Biological basis of bone formation, remodeling, and repair-part I: biochemical signaling molecules. *Tissue Eng Part B Rev*. 2008;14:259–273.
20. Oury F, Sumara G, Sumara O, et al. Endocrine regulation of male fertility by the skeleton. *Cell*. 2011;144:796–809.
21. DiGirolamo DJ, Clemens TL, Kousteni S. The skeleton as an endocrine organ. *Nat Rev Rheumatol*. 2012;8:674–683.
22. Chau JF, Leong WF, Li B. Signaling pathways governing osteoblast proliferation, differentiation and function. *Histol Histopathol*. 2009;24:1593–1606.
23. Zhou H, Mak W, Zheng Y, Dunstan CR, Seibel MJ. Osteoblasts directly control lineage commitment of mesenchymal progenitor cells through Wnt signaling. *J Biol Chem*. 2008; 283:1936–1945.
24. Komori T. Regulation of bone development and maintenance by Runx2. *Front Biosci*. 2008;13:898–903.
25. Soltanoff CS, Yang S, Chen W, Li YP. Signaling networks that control the lineage commitment and differentiation of bone cells. *Crit Rev Eukaryot Gene Expr*. 2009;19:1–46.
26. Deng Z-L, Sharff KA, Tang N, et al. Regulation of osteogenic differentiation during skeletal development. *Front Biosci*. 2008;13:2001–2021.
27. Long F. Building strong bones: molecular regulation of the osteoblast lineage. *Nat Rev Mol Cell Biol*. 2011;13:27–38.
28. Burger C, Zhou HW, Wang H, et al. Lateral packing of mineral crystals in bone collagen fibrils. *Biophys J*. 2008;95:1985–1992.
29. Chen H, Senda T, Kubo KY. The osteocyte plays multiple roles in bone remodeling and mineral homeostasis. *Med Mol Morphol*. 2015;48:61–68.
30. Baron R. Anatomy and ultrastructure of bone. In: Flavus MJ, ed. *Primer on the Metabolic Bone Diseases and Disorders of Mineral Metabolism*. Lippincott Williams & Wilkins; 1999:3–10.
31. Marks SC Jr, Popoff SN. Bone cell biology: the regulation of development, structure, and function in the skeleton. *Am J Anat*. 1988;183:1–44.
32. Bonewald LF. The amazing osteocyte. *J Bone Miner Res*. 2011;26:229–238.
33. Buenzli PR, Sims NA. Quantifying the osteocyte network in the human skeleton. *Bone*. 2015;75:144–150.
34. Rochefort GY, Pallu S, Benhamou CL. Osteocyte: The unrecognized side of bone tissue. *Osteoporos Int*. 2010;21:1457–1469.
35. Nakashima T, Hayashi M, Fukunaga T, et al. Evidence for osteocyte regulation of bone homeostasis through RANKL expression. *Nat Med*. 2011;17:1231–1234.
36. Belanger LF. Osteocytic osteolysis. *Calcif Tissue Res*. 1969; 4:1–12.
37. Tsourdi E, Jähn K, Rauner M, Busse B, Bonewald LF. Physiological and pathological osteocytic osteolysis. *J Musculoskelet Neuronal Interact*. 2018;18:292–303.
38. Juppner H. Phosphate and FGF-23. *Kidney Int*. 2011; 79121:S24–S27.
39. Teitelbaum SL. Bone resorption by osteoclasts. *Science*. 2000; 289:1504–1508.
40. Takahashi N, Maeda K, Ishihara A, Uehara S, Kobayashi Y. Regulatory mechanism of osteoclastogenesis by RANKL and Wnt signals. *Front Biosci (Landmark Ed)*. 2011;16:21–30.
41. Edwards JR, Mundy GR. Advances in osteoclast biology: Old findings and new insights from mouse models. *Nat Rev Rheumatol*. 2011;7:235–243.
42. Boyce BF, Yao Z, Xing L. Functions of nuclear factor kappaB in bone. *Ann N Y Acad Sci*. 2010;1192:367–375.

43. Glimcher MJ. The nature of the mineral phase of bone: biological and clinical implications. In: Avioli L KS, ed. *Metabolic Bone Disease and Clinically Related Disorders*. Academic Press; 1998:23–50.
44. Rey C, Combes C. What bridges mineral platelets of bone? *Bonekey Rep*. 2014;3:586.
45. Bonucci E. Bone mineralization. *Front Biosci (Landmark Ed)*. 2012;17:100–128.
46. Hasegawa T. Ultrastructure and biological function of matrix vesicles in bone mineralization. *Histochem Cell Biol*. 2018;149(4):289–304.
47. Schwarcz HP. The ultrastructure of bone as revealed in electron microscopy of ion-milled sections. *Semin Cell Dev Biol*. 2015;46:44–50.
48. Lefebvre V, Bhattaram P. Vertebrate skeletogenesis. *Curr Top Dev Biol*. 2010;90:291–317.
49. Yang Y. Skeletal morphogenesis during embryonic development. *Crit Rev Eukaryot Gene Expr*. 2009;19:197–218.
50. Olsen BR, Reginato AM, Wang W. Bone development. *Annu Rev Cell Dev Biol*. 2000;16:191–220.
51. Mackie EJ, Ahmed YA, Tatarczuch L, Chen K-S, Mirams M. Endochondral ossification: how cartilage is converted into bone in the developing skeleton. *Int J Biochem Cell Biol*. 2008;40:46–62.
52. Lui JC, Nilsson O, Baron J. Recent research on the growth plate: recent insights into the regulation of the growth plate. *J Mol Endocrinol*. 2014;53:T1–T9.
53. Chen H, Ghori-Javed FY, Rashid H, et al. Runx2 regulates endochondral ossification through control of chondrocyte proliferation and differentiation. *J Bone Miner Res*. 2014;29: 2653–2665.
54. Karimian E, Chagin AS, Savendahl L. Genetic regulation of the growth plate. *Front Endocrinol (Lausanne)*. 2012;2:113.
55. Hirai T, Chagin AS, Kobayashi T, Mackem S, Kronenberg HM. Parathyroid hormone/parathyroid hormone-related protein receptor signaling is required for maintenance of the growth plate in postnatal life. *Proc Natl Acad Sci U S A*. 2011;108:191–196.
56. Wuelling M, Vortkamp A. Chondrocyte proliferation and differentiation. *Endocr Dev*. 2011;21:1–11.
57. Marino R. Growth plate biology: New insights. *Curr Opin Endocrinol Diabetes Obes*. 2011;18:9–13.
58. Jerome C, Hoch B. Skeletal system. In: Treuting PM, Dintzis SM, eds. *Comparative Anatomy and Histology: A Mouse and Human Atlas*. Academic Press; 2012:53–70.
59. Farnum CE, Wilsman NJ. Determination of proliferative characteristics of growth plate chondrocytes by labeling with bromodeoxyuridine. *Calcif Tissue Int*. 1993;52:110–119.
60. Gruber HE, Rimoin DL. Quantitative histology of cartilage cell columns in the human costochondral junction: Findings in newborn and pediatric subjects. *Pediatr Res*. 1989;25:202–204.
61. Langenskiold A. Role of the ossification groove of Ranvier in normal and pathologic bone growth: A review. *J Pediatr Orthop*. 1998;18:173–177.
62. Emons J, Chagin AS, Hultenby K, et al. Epiphyseal fusion in the human growth plate does not involve classical apoptosis. *Pediatr Res*. 2009;66:654–659.
63. Eriksen EF. Cellular mechanisms of bone remodeling. *Rev Endocr Metab Disord*. 2010;11:219–227.
64. Karsenty G, Oury F. The central regulation of bone mass, the first link between bone remodeling and energy metabolism. *J Clin Endocrinol Metab*. 2010;95:4795–4801.
65. Pogoda P, Priemel M, Rueger JM, Amling M. Bone remodeling: new aspects of a key process that controls skeletal maintenance and repair. *Osteoporos Int*. 2005;16(Suppl 2):S18–S24.
66. Dimenstein IB. Bone grossing techniques: Helpful hints and procedures. *Ann Diagn Pathol*. 2008;12:191–198.

5

Joints

Fiona Maclean ■ Alison L. Cheah

Bone, cartilage, ligaments, and tendons have a mechanical function, providing movement, stability, and protection. Unlike organs composed of cellular elements with a metabolic function, the connective tissues are largely formed from extracellular material. This matrix is made up of substances that resist the tensile and compressive forces to which they are subjected. The cellular component of the connective tissues provides for their growth and maintenance.

THE NORMAL JOINT

The joint is a functioning unit comprising the ends of contiguous bones together with their soft tissue components, including cartilage, ligaments, and synovium. There are three types of joints. The most common is the diarthrodial joint, which is a mobile unit between two bones. Hyaline cartilage covers the articulating surfaces of the diarthrodial joints, with the exception of the sternoclavicular and temporomandibular joints, which are covered by fibrocartilage. The second type is the amphiarthrodial joint, typified by the intervertebral disc and characterized by limited mobility. The third type is the fibrous synarthroses, such as the skull sutures, which are nonmovable joints and will not be discussed further.

Diarthrodial Joint

In the normal diarthrodial joint, the opposed articular surfaces move painlessly over each other within the required range of motion, with the load distributed across joint tissues. Conversely, clinical joint dysfunction is characterized by instability, loss of motion, maldistribution of load, and pain.

Normal joint function depends on the shape (geometry) of the joint, the mechanical properties of the various tissues that compose the joint and the integrity of the joint, including its neuromuscular control.

The Shape

Perhaps the most obvious feature of any joint is the shape of its articulating surfaces. In general, one surface is convex and the other concave. The convex side usually has the greater surface area. These complementary shapes permit the normal range of motion and provide stability and equal loading. Some joints fit together almost exactly (e.g., the hip and ankle) (1), whereas others (e.g., the knee and finger joints) show greater incongruence (Fig. 5.1).

In some joints, such as the knee, the incongruences of the surfaces are partially compensated by the intra-articular fibrocartilaginous menisci (2). Removal or damage to the menisci results in significant consequences. The same is

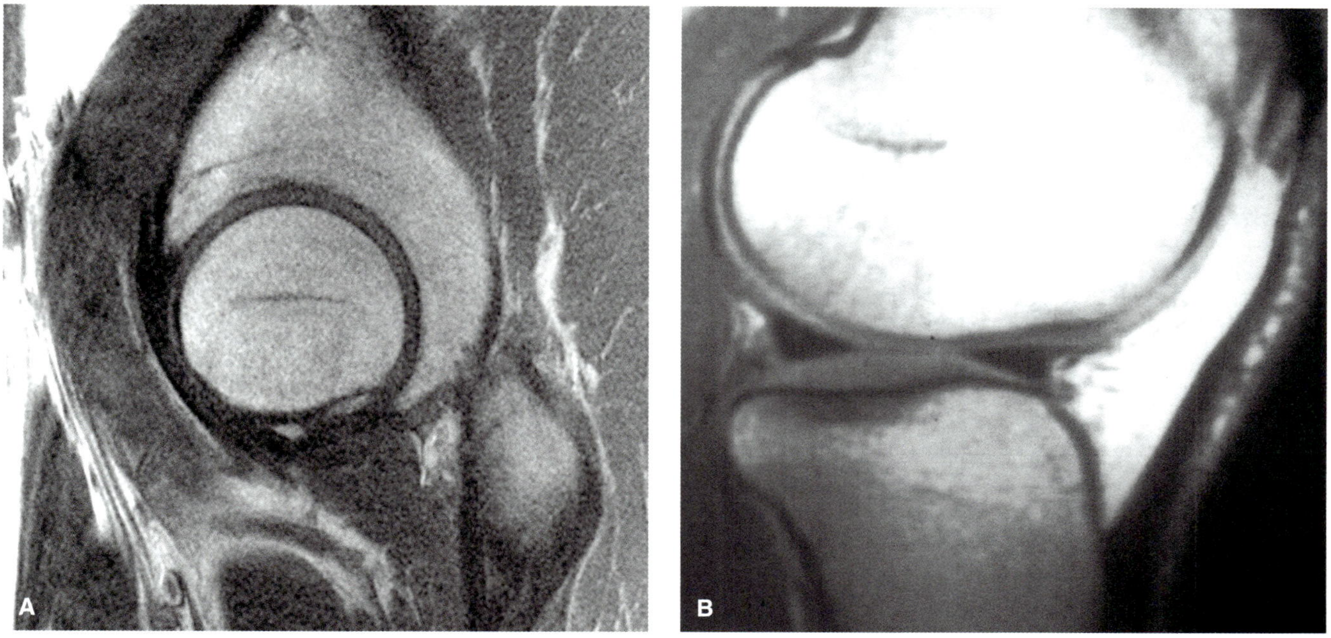

FIGURE 5.1 **A:** Sagittal section through the hip joint on magnetic resonance imaging (MRI) shows a close fit between the acetabulum and femoral head. **B:** Lateral MRI of a normal knee shows the gross incongruity of the articular surfaces. This is partially corrected by the interposed menisci, which act as load-bearing structures.

true of the labrum of the hip joint, which may be regarded as an extension of the articular surface of the acetabulum.

In most joints, the initial contact between the articular surfaces is at the joint periphery. The tissues that make up the articulating surfaces undergo elastic deformation when loaded. As the load increases, the surfaces come into increasing contact with more even distribution of the load (Fig. 5.2).

These mechanical factors, together with the movement of the joint, mix and circulate the synovial fluid. This is essential to the metabolism of the chondrocytes; they derive their oxygen and nutrients from the synovial fluid because the articular cartilage is avascular.

The Mechanical Properties of the Extracellular Matrices

The unique properties of the articulating cartilage allow for ease of movement of the joint while preventing abrasion of the opposing surface. The inherent flexibility and elasticity allow constant adaptation, thus diffusing friction and dissipating shock.

The biomechanical properties of articular cartilage are determined by the extracellular matrices. Connective tissues such as bone, cartilage, and ligaments have a unique composition and structural organization that enables specific mechanical function. Following trauma, disturbances in the structure and/or composition of the extracellular matrix of articular cartilage may result in joint dysfunction. The connective tissue matrices are remodeled by their intrinsic cells (i.e., fibroblasts, osteoblasts, osteoclasts, and chondrocytes) to maintain their physicochemical and mechanical properties.

Collagen fibers, the principal extracellular component of connective tissues, are made up of bundles of fibrils, which are composed of stacked molecules formed

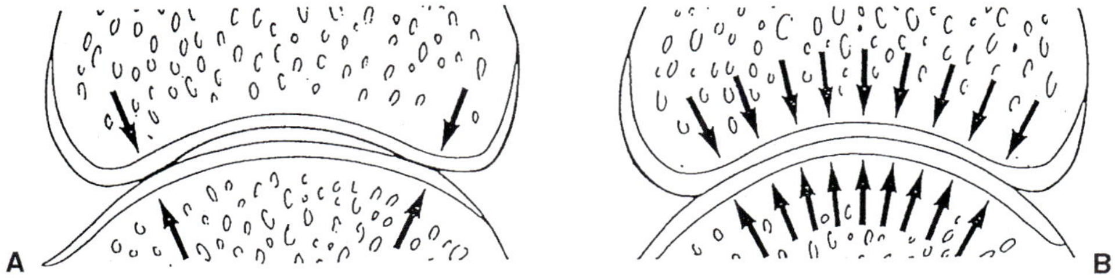

FIGURE 5.2 **A:** Light load. At rest and under light load, limited contact at the joint periphery assures access for synovial fluid to the joint space. **B:** Increased load. With increasing load, deformation of the bone and cartilage allows increased contact of the cartilage surfaces and an equitable distribution of the load. Cyclical loading permits circulation of the synovial fluid in the joint and between the articular surfaces to provide for the metabolic needs of the cartilage.

from polypeptide chains arranged in a helical pattern. The fibrillar collagens provide tensile strength. There are also non–fiber-forming collagens that have varying functions, including acting as binding sites for other matrix components (e.g., type IX collagen) or facilitating calcification (e.g., type X collagen) (3). Type I collagen is the most common form of collagen in tendons, ligaments, and bone. Type II collagen gives articular cartilage its tensile strength and, together with the proteoglycans, is essential for maintaining tissue volume and shape (4).

Connective tissues are also subjected to compression. In bone, the compressive load is resisted by hydroxyapatite. A mucoid filler between the collagen fibers of cartilage provides compressive strength as well as viscoelastic properties. This filler is composed of large, negatively charged macromolecular proteoglycan aggregates.

Proteoglycans are a group of heterogeneous molecules, consisting of protein chains and attached carbohydrates, which have a sticky gel-like quality. The major proteoglycan in cartilage is aggrecan (5), containing a protein core to which carbohydrate side chains (keratan and chondroitin sulfate) are attached. The highly charged proteoglycan molecules attract water and swell considerably. However, within the cartilage, the expansion of the proteoglycans is restricted by the collagen network. When cartilage is loaded, some water is extruded and proteoglycans are further compressed. Removal of the load permits the absorption of water and nutrients into the tissue.

Aggrecan shows age-related changes including decreased size and changes in chemical structure altering the stiffness and water content of the cartilage (6). In addition to aggrecan, the extracellular matrix of cartilage contains many noncollagenous proteins and proteoglycans. The recognition of genetic disorders in which synthesis of the matrix molecules is affected has given insights into their functional role (7).

Prolonged physical activity can cause the total cartilage thickness to decrease by about 5%, although the changes are not uniform throughout the tissue. For example, the superficial zone can experience significant fluid exudation and consolidation in the range of 60%, while the radial zone experiences relatively little fluid flow and consolidation. These changes alter the synthesis and degradation of matrix proteins. Long-term, even minor differences in these cellular processes may affect the micro- and macromorphology of articular cartilage.

Capsular, Pericapsular Tissues and Muscular Control

Through the perception of touch, temperature, pain, and position, there is continuous sensory feedback regarding our movements. Thus, correct joint function is dependent on intact ligaments, muscles and nerves. As recognized by Charcot, a breakdown of neuromuscular coordination can lead to profound arthritis (8).

Amphiarthrodial Joint

The intervertebral disc is a fibrocartilaginous complex that forms the articulation between the vertebral bodies. It can be divided into two components: the outermost fibrous ring (annulus fibrosus) and the innermost gelatinous core (nucleus pulposus). It contributes to spinal mobility and stability, as well as allowing for the transmission of load.

The annulus, when viewed from above, contains fibrous tissue layers arranged in concentric circles. Each layer extends obliquely from vertebral body to vertebral body, with the fibers of each layer running in opposing directions. This arrangement of alternating oblique layers allows for controlled motion that is universal in direction (rotation, flexion, and extension), but restricted in degree (Fig. 5.3).

The fibers of the annulus are attached by Sharpey fibers to the bony endplates of the adjacent vertebral bodies. The

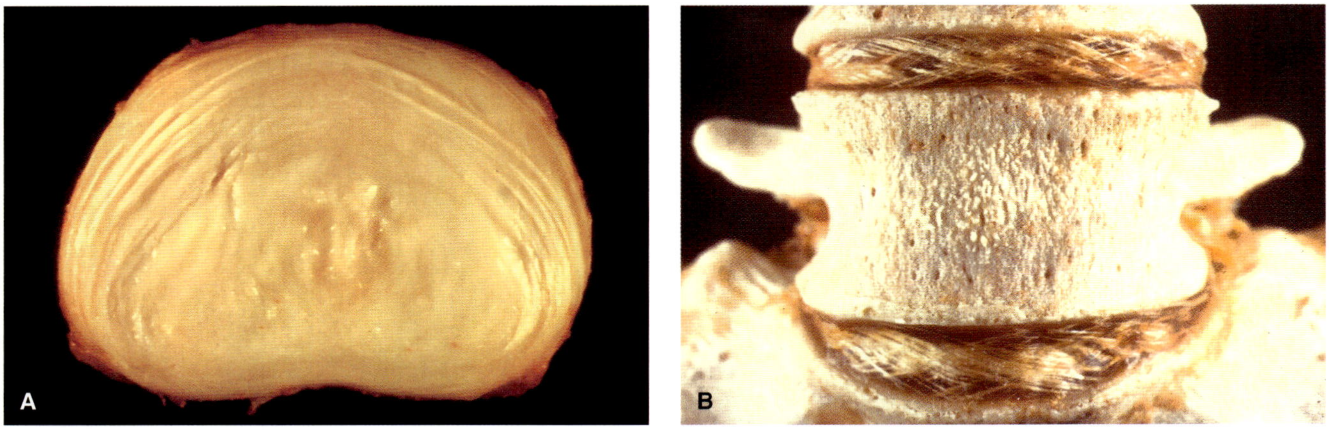

FIGURE 5.3 **A:** An intervertebral disc seen in cross section. Note the layers of circumferential fibers that make up the annulus fibrosus and the well-demarcated bulging central mass of the nucleus pulposus. Note also the decreasing width of the annulus from anterior (*top*) to posterior (*bottom*). **B:** In this desiccated specimen of the lower lumbar spine, the alternating oblique orientation of the collagen fibers in the annulus can be appreciated.

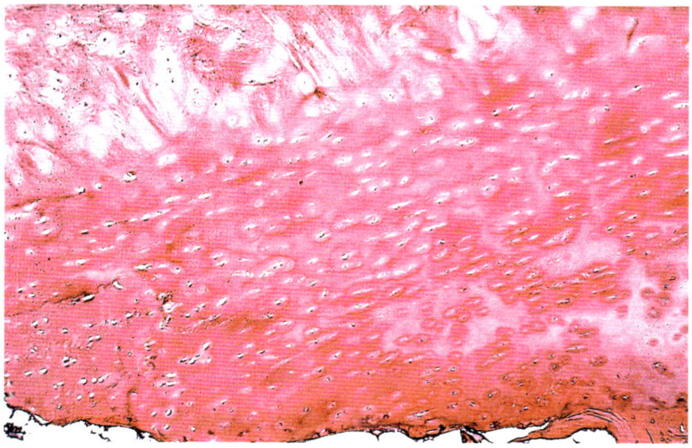

FIGURE 5.4 The nucleus pulposus of the disc (*top*) is separated from the bone (*bottom*) by a dense layer of hyaline cartilage, as demonstrated in this photomicrograph (H&E stain, ×4 objective).

fibers are stronger and more numerous in the anterior and lateral aspects of the disc than in the posterior aspect. As a result, the anterior annulus is almost twice the thickness of the posterior annulus. Due to this variation in thickness, the nucleus pulposus typically occupies an eccentric position within the disc space, closer to the posterior margin. The tissue of the nucleus is separated from that of the bone above and below by a clearly defined layer of hyaline cartilage (Fig. 5.4).

On microscopic examination, the nucleus pulposus shows chondrocytes and stellate and fusiform cells suspended in a loose fibromyxoid matrix rich in proteoglycans. Because no blood vessels are present in most of the adult disc tissue, nutrients must travel by diffusion from the disc margins. A restricted flow of nutrients to the nucleus and inner annulus contributes to disc degeneration. Disc height is not the same in all segments of the spine, with the cervical and thoracic discs being flatter than those of the lumbar region. There is also a variation in disc height from front to back, relative to the curvature of the spine. Age-related dehydration results in disc thinning.

THE NORMAL JOINT TISSUES

Articular Cartilage

Morphology

The articular ends of the bones are covered by hyaline cartilage, which is nerveless, bloodless, firm, and yet pliable. When pressure is applied to hyaline cartilage, it temporarily deforms (9). During development, cartilage is a precursor of the bony skeleton, and it is the means by which the bones increase in length via endochondral ossification at the cartilaginous growth plate (i.e., physis).

In young people, hyaline cartilage is translucent and bluish-white; in older individuals, it is opaque and slightly yellow (Fig. 5.5) (10). This change in the appearance of the articular cartilage is also seen in other connective tissues and is related to a number of factors, including dehydration, increased numbers of collagen cross-linkages, and the possible accumulation of lipofuscin pigment.

On microscopic examination, articular cartilage is characterized by its abundant glassy (hyaline) extracellular matrix with isolated cells located in lacunar spaces (Fig. 5.6). It has four zones: superficial (I), intermediate (II), deep (III), and calcified (IV). In the superficial layer, the cells are flat. In the intermediate zone, the cells form radial groups that follow the pattern of collagen deposition. In the deep zone, the cells are hypertrophied; and in the calcified zone adjacent to the bone, the cells are nonviable (Fig. 5.7).

Within the mineralized bone matrix, the osteocytes are connected to one another by means of cytoplasmic processes. This arrangement is not present within the cartilage. The chondrocytes depend on the diffusion of solutes through the extracellular matrix for their metabolism. The matrix of the deep calcified zone of the articular cartilage forms a barrier to the passage of solutes from the subchondral bone. The articular cartilage is thus dependent on the

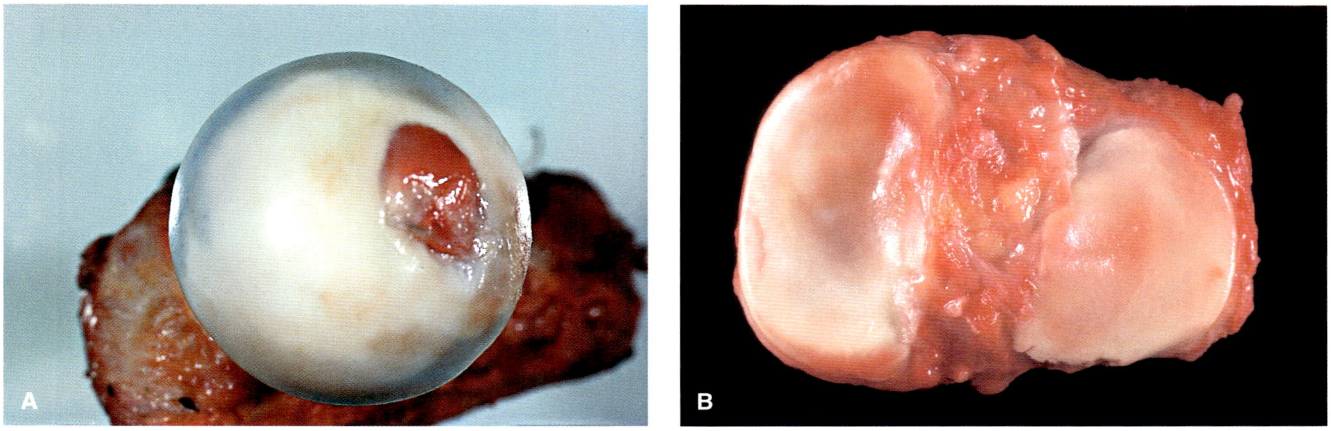

FIGURE 5.5 **A:** A femoral head, resected from a 16-year-old, demonstrates the blue–white translucency of young healthy cartilage. **B:** For comparison, the tibial plateau of a 50-year-old. The cartilage is smooth and healthy in appearance but is more yellowish and opaque than that of the 16-year-old.

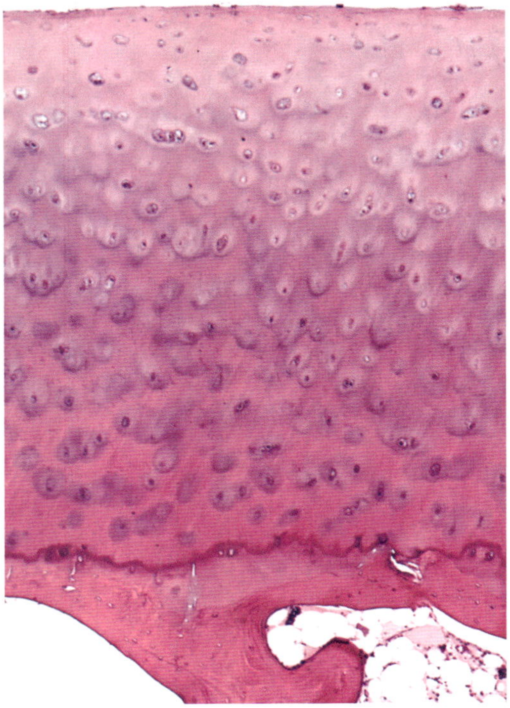

FIGURE 5.6 Normal articular cartilage obtained from the femoral condyle of a middle-aged individual (H&E stain, ×4 objective).

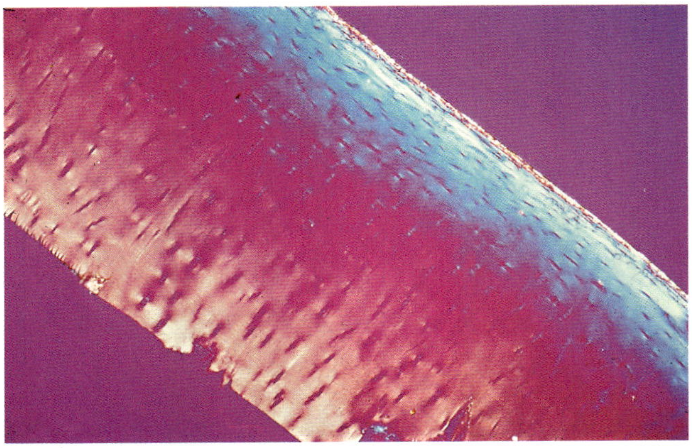

FIGURE 5.8 In this polarized-light photomicrograph, the surface collagen fibers can be visualized as *blue*, and the deeper collagen fibers (which are perpendicular) as *yellow*. Collagen cannot be seen in the intermediate area where the fibers cross (×4 objective).

diffusion of nutrients and metabolites between the synovial fluid and the articular surface (11).

The principal orientation of collagen fibers in articular cartilage is vertical through most of its thickness and horizontal at the surface (Fig. 5.8) (12). The collagen content of cartilage progressively diminishes from superficial to deep. The collagen fibers in the deep layers are vertically aligned, more widely separated, and thicker, forming a web of arch-shaped structures (12). The collagen fibers are continuous with those in the calcified layer. There is discontinuity, however, with the underlying subchondral bone. This network reflects the local stresses and strains in the articular cartilage (13). The principal orientation of the collagen fibers in the menisci of the knee is circumferential, reflecting the tension that occurs during normal loading. Occasional radially disposed fibers are present, acting to resist longitudinal splitting of the menisci resulting from undue compression (Fig. 5.9) (14).

The distribution of proteoglycans in the cartilage matrix is related to the local mechanical requirements and varies from joint to joint. In general, proteoglycan distribution is more even in children than in adults. The surface layers contain much less proteoglycan than the deeper layers. In the deeper layers, the concentration is higher around the chondrocytes (i.e., the pericellular matrix) than between them (i.e., the intercellular matrix) (Fig. 5.10) (15).

The junction between the calcified cartilage and the noncalcified cartilage is marked by a basophilic line known as the tidemark (Fig. 5.11). This basophilic line is not seen in the developing skeleton. In later life, replication of the tidemark is usually evident and in osteoarthritic joints and may be prominent (Fig. 5.12). Mechanical failure in the

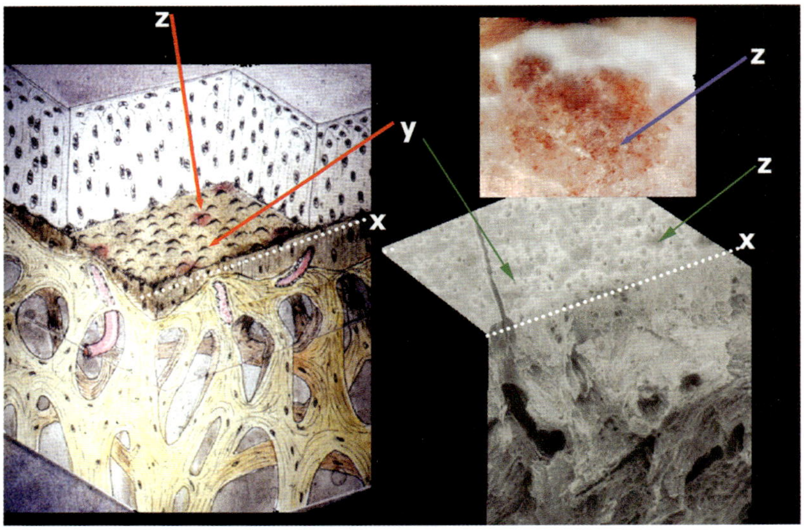

FIGURE 5.7 In this diagram of the articular surface, the organization of the articular surface seen on Figure 5.6 is shown diagrammatically. Noncalcified articular cartilage has been removed to reveal the surface of the tidemark (*x*). The small "volcanic" structures (*y*) represent the location of the cells in the calcification front. The vessels that penetrate the calcification cartilage (*z*) are seen just beneath the calcification front. A dissection (*upper right*) shows the vessels when the cartilage has been pared away, and a scanning electron photomicrograph (*lower right*) shows the section from which the diagram was reconstructed.

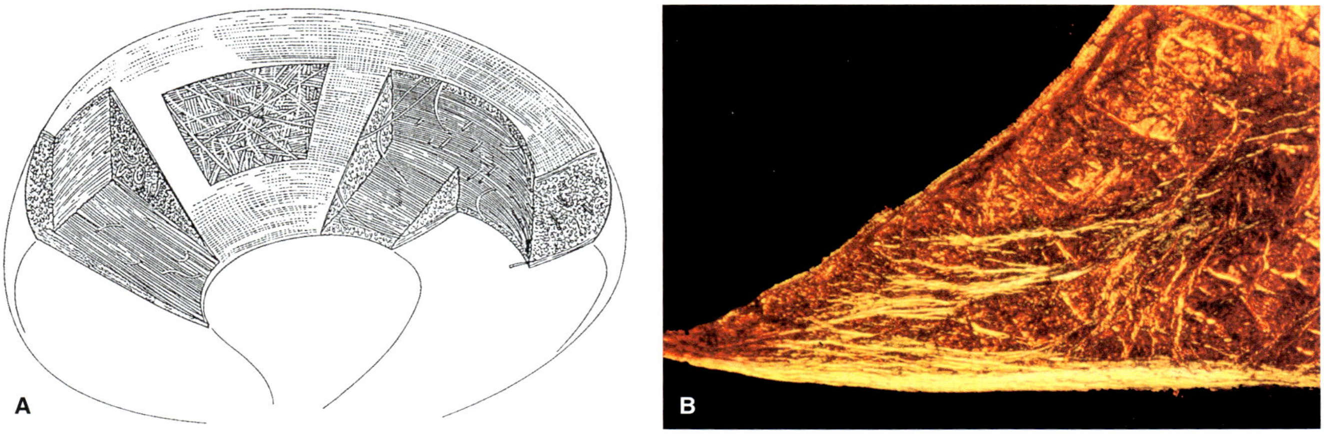

FIGURE 5.9 **A:** Illustration of the distribution of collagen fibers in the meniscus. The majority of the fibers are circumferentially distributed to resist the tension generated in the meniscus when the knee is under compressive load. The radially distributed fibers are most obvious on the tibial surface of the meniscus. **B:** Photomicrograph of a cross section of meniscus seen with polarized light. The tibial surface is the bottom edge where most of the fibers are radially arranged (×1 objective). (Modified from Bullough proteoglycan, Munuera L, Murphy J, et al. The strength of the menisci of the knee as it relates to their fine structure. *J Bone Joint Surg Br.* 1970;52:564–567, with permission.)

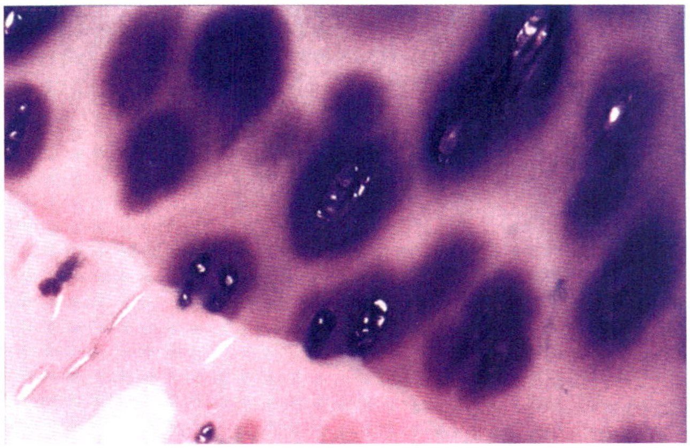

FIGURE 5.10 Portion of cartilage showing intense metachromasia around the chondrocytes in the deep part of the noncalcified cartilage. This represents staining of the proteoglycan. There is much less staining in the interterritorial matrix than around the cell. Even less staining is seen in the calcified cartilage (×25 objective).

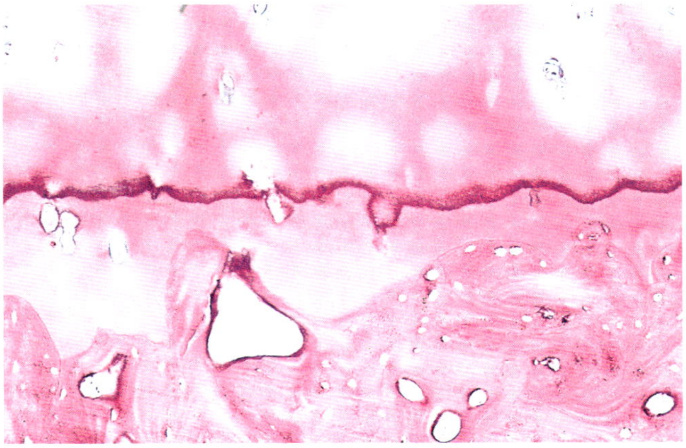

FIGURE 5.11 Deep and calcified layers of the articular cartilage. The deep layer is separated from the calcified layer by a basophilic line referred to as the "tidemark," which represents the mineralizing front (H&E stain, ×4 objective).

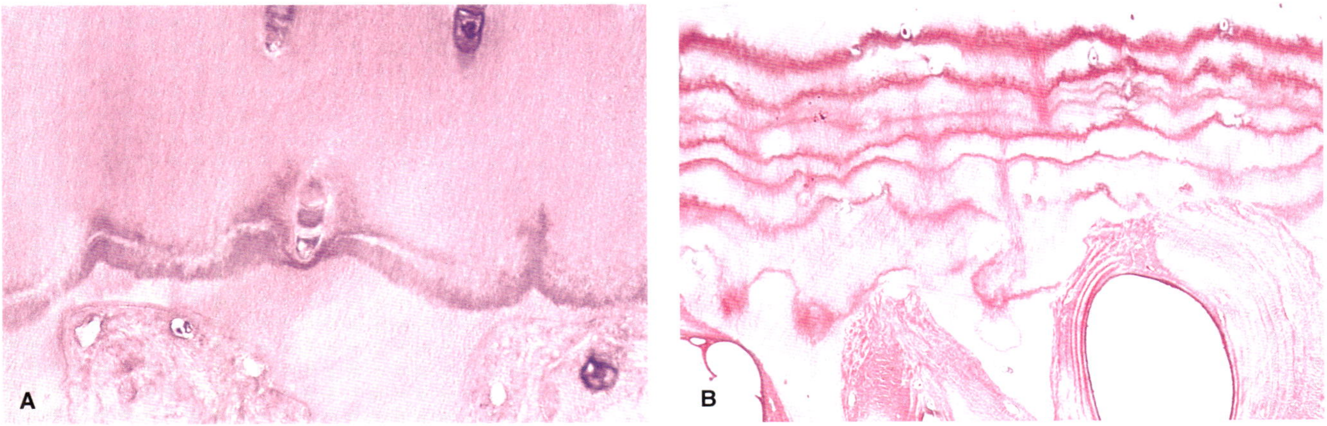

FIGURE 5.12 **A:** Accelerated mineralization with a replicated tidemark. A chondrocyte is seen caught up in the tidemark (H&E stain, ×25 objective). **B:** In most areas of normal cartilage, only one tidemark is observed. However, in an early stage of osteoarthritis, seen here, multiple tidemarks indicating rapid advance of the mineralization front can often be seen (H&E stain, ×10 objective).

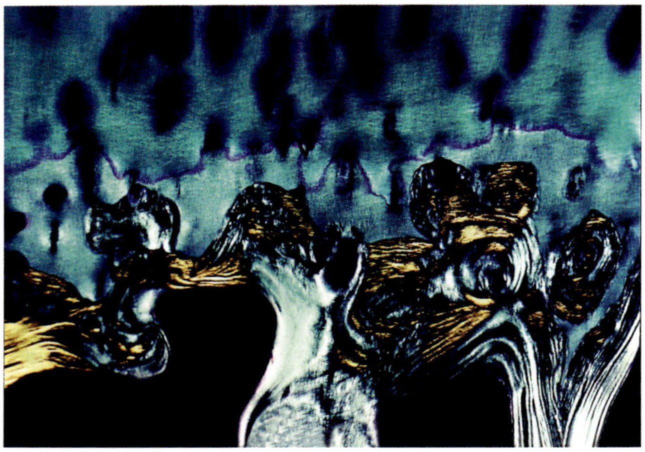

FIGURE 5.13 In this photomicrograph, taken with polarized light, the irregularity of the interface between the subchondral bone and the overlying calcified cartilage is obvious. The functional keying of the bone and cartilage depends on the two tissues having equal rigidity (×4 objective).

deep cartilage rarely gives rise to separation at the bone–cartilage interface. When failure occurs, it is often seen as a horizontal cleft at the tidemark, reflecting the change in the rigidity of the cartilage at this junction. Adult articular cartilage is bordered by the subchondral bone plate, with an irregular interface, somewhat like a jigsaw puzzle (Fig. 5.13). Calcification of this region contributes to its rigid structure.

The insertions of ligaments and tendons into the bone are also calcified, and their adhesion is affected by this arrangement. These structures are dynamic, reflecting growth in skeletally immature individuals and continuous bone turnover in mature individuals.

The chondrocytes embedded in the cartilage matrix are responsible for synthesis and maintenance of the extracellular tissue. At birth, the cartilage shows a relatively homogeneous distribution of chondrocytes, which are small with minimal associated matrix. Changing pressure, shape, and fluid flow in the postnatal period results in the formation of superficial, transitional, and radial zones within the cartilage, each with different histologic appearances (16). The chondrocytes vary in size, shape, and number, both from the superficial to the deep layers and in different anatomic locations. In general, cells at the cartilage surface are more numerous, flatter, smaller, and oriented parallel to the cartilage surface (16). Chondrocytes are more spherical and arranged in columns in the middle zones, reflecting the orientation of collagen fibers present. The pericellular matrix layer is rich in proteoglycans and contains hyaluronic acid with relatively little collagen. A basketlike structure composed of cross-linked type VI collagen provides a protective framework and helps maintain equal hydrostatic pressure.

Cartilage Turnover and Articular Remodeling

Wolff law states that both bone density and bone architecture correlate with the magnitude and direction of the applied load. At the articular end of a bone, the subchondral bone trabeculae are remodeled to accommodate optimal load distribution.

Endochondral ossification is an important mechanism for both bone growth and remodeling. In the epiphyseal growth plate, calcified cartilage is invaded by blood vessels from the metaphyseal bone and is subsequently replaced by bone synthesized by osteoblasts. This process also occurs in adult articular joints. Thus, throughout life, the calcified cartilage is slowly replaced by new subarticular bone (Fig. 5.14).

The tidemark advances into the noncalcified cartilage at a rate in equilibrium with the rate of absorption of the calcified cartilage from the subarticular bone (17). The articular cartilage is not a static tissue; the joint undergoes continuous remodeling, with the extracellular matrix and the chondrocytes replaced throughout life. Apoptosis plays an important role in this process, similar to that of joint development during fetal life (18).

There are regional variations in morphologic, biochemical, and biomechanical components of weight-bearing joints. Variation in cartilage thickness is present in most joints. Stiffness in different areas of the femoral head is related to proteoglycan and water content. Stiffness and cartilage thickness affect the mechanical properties of the calcified cartilage and the underlying cancellous bone (19).

Within the knee joint, the morphology varies depending on whether there is a meniscal covering or not (20). In sites where the meniscus is present, the cartilage is firm and smooth. Where the meniscus is deficient, the cartilage has a rough surface and soft matrix. Even in young adults, cartilage that is not covered by the meniscus shows matrix softening and mild superficial fibrillation (21). These naturally occurring variations are likely related to joint loading. Normally in the knee, load is transmitted through the meniscus and onto the tibial cartilage. The exposed cartilage remains relatively unloaded. Similar areas of disuse atrophy have

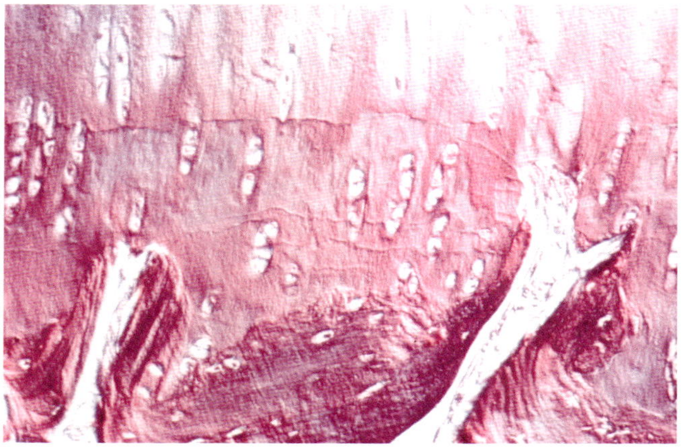

FIGURE 5.14 In this photomicrograph, two vessels can be seen that have extended into the calcified layer of cartilage. Around the circumference of each of these vessels, a thin layer of lamellar bone can be appreciated. By means of continuing endochondral ossification, the articular bone end is continuously modeled (H&E stain, ×10 objective).

been described around the rim of the radial head, in the roof of the acetabulum, and on the perifoveal and inferomedial aspects of the femoral head (17). Changes in the immediate environment of the joint lead to alterations of the cartilage matrix (22). Thus, immobilization or unloading of a joint results in decreased synthesis of glycosaminoglycans. Conversely, exercise appears to increase synthesis (23). Low levels of mechanical stress are associated with enhanced catabolic activity, whereas normal stress is associated with anabolic activity. Under high stress, the chondrocytes are unable to adapt. There is a window of physiologic stress above or below which the chondrocytes cannot maintain an adequate functional matrix.

Fibrocartilage

Apart from hyaline cartilage, two other forms of cartilage are recognized, fibrocartilage and elastic cartilage. The mechanical functions of fibrocartilage and elastic cartilage are very different from those of hyaline cartilage. Hyaline cartilage mainly resists compressive forces, whereas the others function principally to resist tension, with some element of compression.

Fibrocartilage is present in the menisci of the knee, in the annulus fibrosus, at the insertions of ligaments and tendons, and on the inner side of tendons as they angle around pulleys (e.g., at the malleoli). In all these locations, the structures are subjected to compressive and tensile forces. The matrix of fibrocartilage contains a high proportion of type I collagen.

Elastic Cartilage

Elastic cartilage contains a high proportion of elastin in the matrix and is present in the ligamentum flavum (connecting vertebrae), external ear, and epiglottis (Fig. 5.15). Compared to collagen, elastin has much greater elasticity; this is particularly important in the ligaments of the spine, which makes flexion of the vertebral column possible.

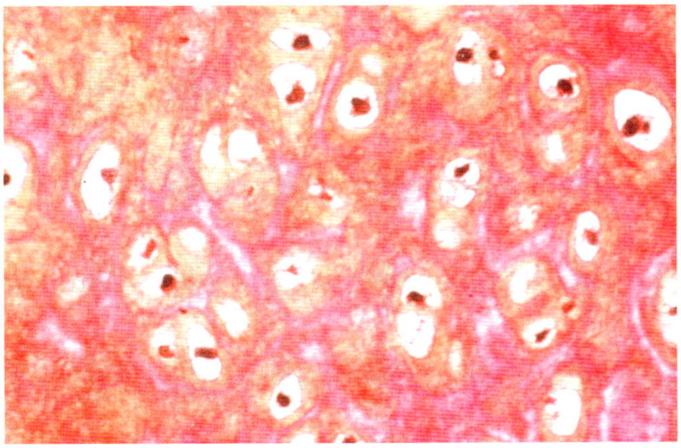

FIGURE 5.15 Although the cells resemble those seen in hyaline cartilage, in this cartilage obtained from the ear the matrix contains many elastic fibers that appear *red* in this section (phloxine and tartrazine stain, ×25 objective).

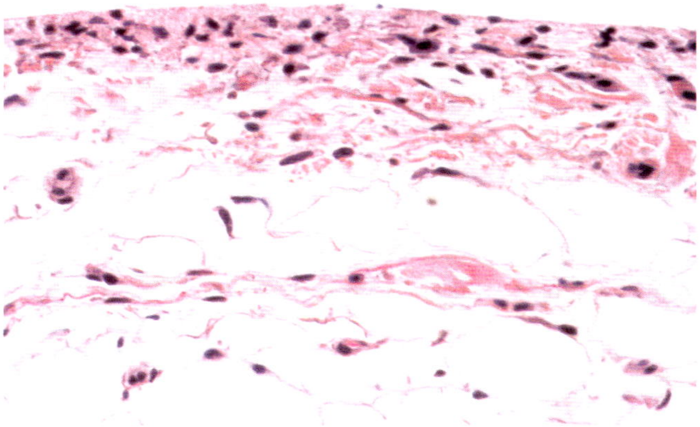

FIGURE 5.16 Normal synovium. The ratio of fat to fibrous tissue varies depending on the joint and the location within the joint (H&E, ×10 objective).

Synovial Membrane

The synovial membrane lines the inner surface of the joint capsule and all other noncartilaginous intra-articular structures. Synovial membranes line the bursae, which permit freedom of movement for adjacent structures and tendon sheaths.

Synovial membrane consists of two components. The first of these is the cellular lining surrounding the joint space. This surface is smooth, moist, and glistening, with a few small villi and fringelike folds. The second component is a supportive backing layer. The surface lining the joint space is normally relatively inconspicuous and consists of a single layer of synoviocytes (Fig. 5.16). The subsynovium comprises fibrovascular tissue with scattered macrophages and mast cells (24).

The synovial membrane has three principal functions: secretion, phagocytosis, and regulation of the movement of metabolites into the synovial fluid (24). These metabolites provide for the metabolic requirements of the joint chondrocytes.

The synovial lining cells consist of two main cell types. Type A synoviocytes (25%) are resident macrophages that serve a phagocytic function within the joint. Type B synoviocytes (75%) are fibroblast-like cells involved in the production of specialized matrix constituents (24,25).

Using immunohistochemistry, type A synoviocytes label with antibodies against macrophages including CD68 and CD163 (26,27). Synoviocytes from normal and diseased synovium additionally labeled with clusterin and podoplanin to varying degrees (28). Strong expression of clusterin may be observed, particularly in type A cells (Fig. 5.17). In arthritic joints, relative expression of clusterin and podoplanin may be altered; for example, increased podoplanin expression and a reduction in clusterin expression has been observed in rheumatoid arthritis (28). Expression of clusterin and podoplanin is additionally present in tenosynovial

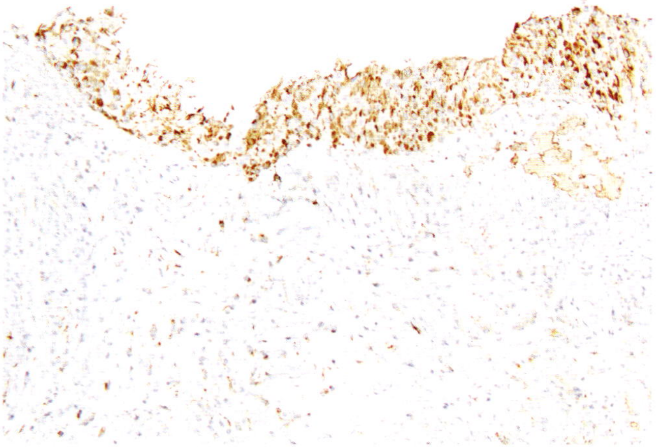

FIGURE 5.17 Synoviocytes show expression of clusterin by immunohistochemistry. Strong expression is observed in type A cells of normal and diseased synovium. (Clusterin immunohistochemical stain, ×10 objective. Image provided courtesy of Dr Jason Hornick, Brigham and Womens' Hospital, Boston, MA.)

giant cell tumor, a neoplasm composed of cells resembling synoviocytes (29).

Ligaments and Tendons

Ligaments, which are structures that join two adjacent bones, are formed mainly of collagen. The arrangement of the collagen bundles within a particular ligament may be complex, reflecting the motion required at that joint. The collagen fibers calcify where they enter the bone, interdigitating with the underlying bone.

Tendons are specialized connective tissues that enable muscles to concentrate or extend their action. The Achilles tendon is an example of a tendon that concentrates the power of several bulky muscles to one area of insertion. Conversely, the long tendons of the hands and feet extend the effect of distant muscles. Adjacent to the insertion point, there is a zone of chondroid metaplasia, with the cells in lacunae-like spaces (30). The tendons enter the bone at an acute angle with resultant shear and compressive forces that induce metaplastic changes at the insertion site (Fig. 5.18).

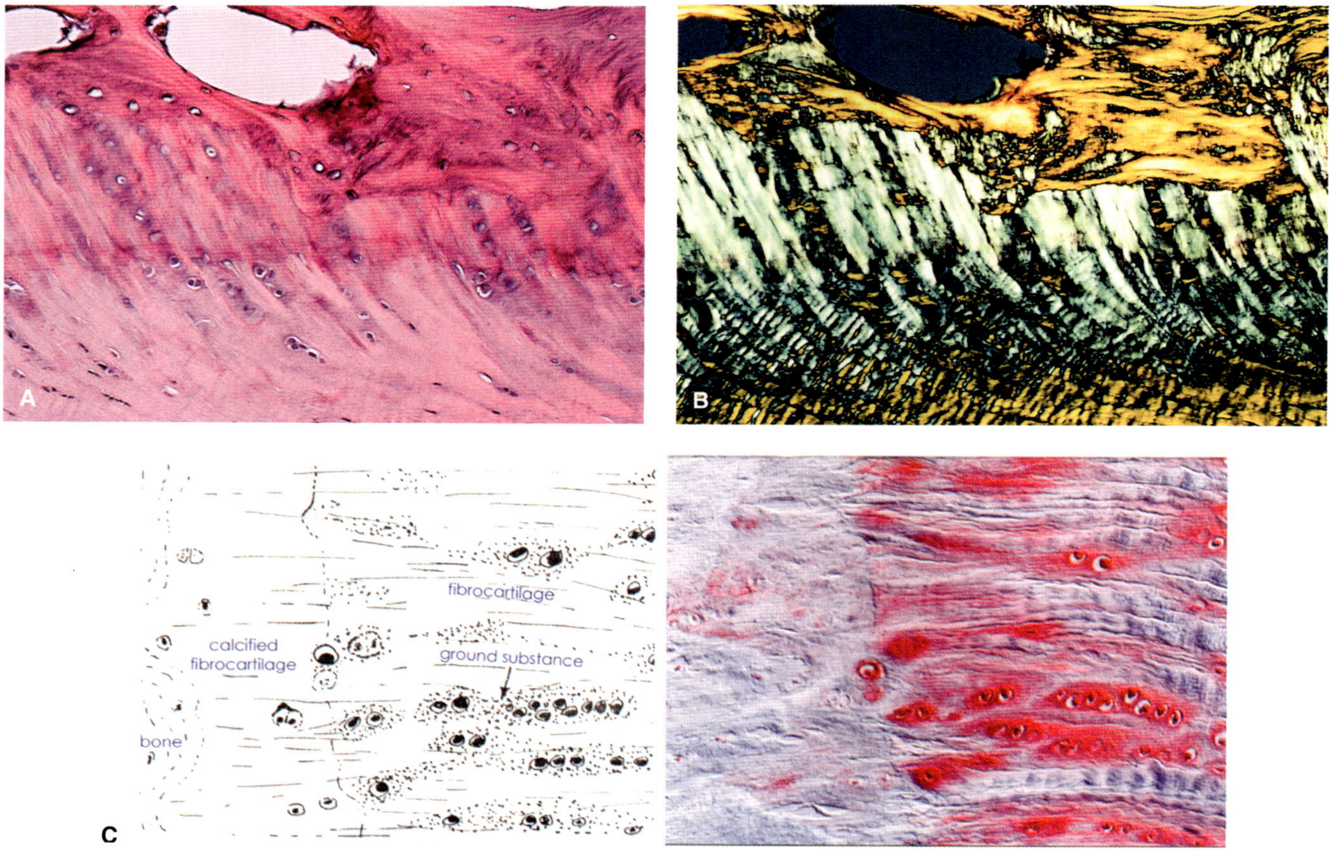

FIGURE 5.18 **A:** Photomicrograph of a ligamentous insertion (H&E stain, ×10 objective). **B:** Polarized light. The portion of the ligament that interfaces with the bone is calcified, and the edge of the calcified portion of the ligament is marked by a basophilic line (*tidemark*) that represents the mineralization front. Note the similarity with the bone–cartilage interface illustrated in Figure 5.14. **C:** A higher-powered view to demonstrate the rounded cells lying in lacunae, which are seen at the insertion site of both ligaments and tendons (fibrocartilaginous metaplasia). The red staining in the matrix indicates the presence of proteoglycans (safranin O stain, ×25 objective).

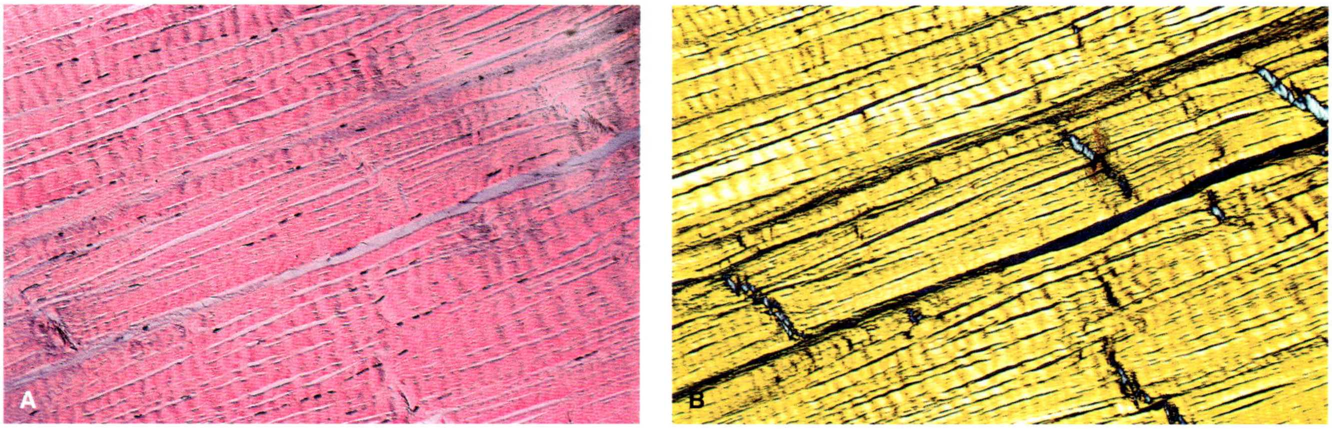

FIGURE 5.19 Photomicrograph of the same field of a tendon that has been photographed in transmitted light (**A**) and in polarized light (**B**). Both images demonstrate the scant and elongated fibroblasts lying between the dense parallel collagen bundles characteristic of tendon (H&E stain, ×4 objective).

Some muscles (e.g., gluteal and paravertebral muscles) have short, fan-shaped insertions without an obvious tendon.

Normal tendons contain relatively sparse tenocytes scattered in a longitudinal pattern between the collagen bundles (Fig. 5.19). There is a slight gradient in the cell population, with the muscular portion of the tendon being more cellular than the distal insertion. Wherever a tendon turns a corner, it is restrained under a pulley and has a surrounding synovial sheath.

The surfaces of the flexor tendons of the hand are covered by a single layer of synovial cells (the endotenon) and a similar layer covers the parietal surface (the epitenon). In the palm of the hand, the tendons are covered by a fine vascular adventitia (paratenon) nourished by vessels from the deep palmar arch (31).

The feeding arteries to the tendons of the hands and feet are long, coiled vessels that can stretch as the tendons move. The sensory nerve supply to tendons is relatively abundant, with free ramifications. The nerve endings provide feedback regarding tendon tension, allowing for proprioceptive impulses.

THE ARTHRITIC JOINT

In much of the Western world, arthritis is the leading cause of disability. Globally, over two decades (1990 to 2020), the prevalence of osteoarthritis has increased by over 130% to 595 million cases (7.6% of the global population), representing a worldwide major public health concern. The prevalence is predicted to further increase by 60% to 100% by 2050 (32). To quote Benjamin Franklin *"in this world nothing can be said to be certain, except death and taxes"* (33), perhaps we should add arthritis to that list.

Clinical arthritis is the consequence of a breakdown in the joint's normal function. It involves loss of capacity for the articulating surfaces to move over one another easily, loss of joint stability, and almost always pain. Loss of motion results from a change in joint shape with resultant incongruities, or a change in the tissue matrices affecting their mechanical properties. Instability may be due to alterations in ligamentous support and neuromuscular control. Pain originates in the bone as a result of maldistribution of load, in the synovium as a result of reactive synovitis, or in the muscle as a consequence of reflex spasm.

Malfunction of a joint results from acute or chronic morbid conditions that produce any of the following (Fig. 5.20):

- Anatomic alterations in the shape of the articulating surfaces (e.g., fracture, increased remodeling, Paget disease, or endocrine disturbances including acromegaly and hyperparathyroidism) (34)
- Loss of structural integrity of the joint (e.g., enzymatic destruction in inflammatory arthritis, traumatic injury which may be acute or low grade and repetitive)
- Alterations in the mechanical properties of the tissue matrices (e.g., brittle collagen as occurs in ochronosis)

Arthritis is broadly classified as degenerative or inflammatory. Most cases of degenerative arthritis fall in the category of osteoarthritis. Inflammatory arthritis is often subclassified based on a constellation of clinical and laboratory findings as seropositive and seronegative arthritis. Seropositive arthritis (e.g., rheumatoid arthritis) is an important immune-mediated condition that can lead to joint destruction. The spondyloarthropathies (e.g., ankylosing spondylosis, reactive arthritis, psoriatic arthritis, and enteropathic arthritis) represent a type of seronegative arthritis along with systemic lupus erythematosus, crystal deposition diseases due to gout and calcium pyrophosphate, and infectious arthritis resulting from bacterial sepsis, granulomatous

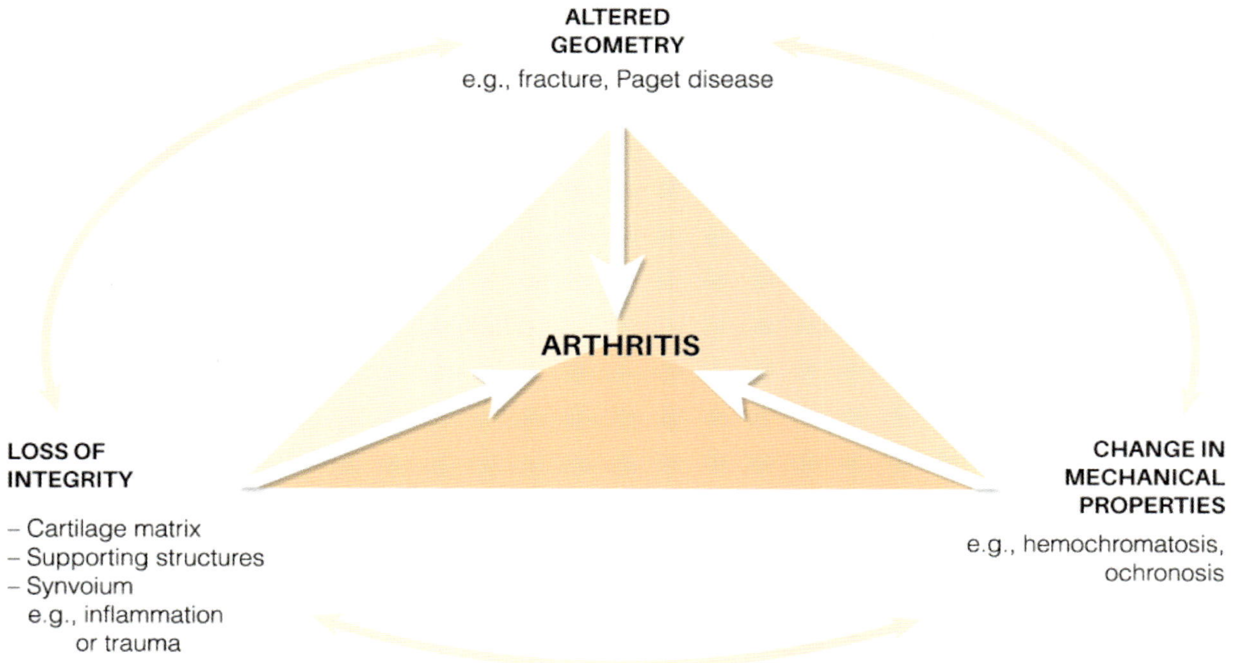

FIGURE 5.20 Causes of arthritis. Changes to shape, constituent materials, and support of joints result in arthritis. Multiple factors are usually present. (Reprinted from Maclean FM. Arthritis and other proliferative joint diseases. *Diagn Histopathol.* 2016;22(10):369–377, with permission.)

mycobacterial disease, or a vector-borne illness such as Lyme disease.

Multiple factors can come into play in an individual, especially when the changes are long-standing. It is inevitable that osteoarthritis will develop when another condition causes joint damage, making assessment of the initial etiology difficult (35). The onset osteoarthritis itself is triggered by a mismatch between the degradation and repair of cartilage, which can have multiple causes, and is promoted by a number of complex interacting pathogenic signaling pathways (36). Pre-existing structural anomalies such as developmental dysplasia of the hip and femoral acetabular impingement often underlie development of osteoarthritis, especially in young individuals (37). Osteoarthritis can be seen secondary to subchondral insufficiency fracture which usually occurs in older individuals subsequent to loading of regions of low bone density. Subchondral insufficiency fracture as a cause of osteoarthritis was long overlooked, possibly due to the fact that it often clinically and radiologically mimics avascular necrosis of bone. It is frequently the cause underlying rapidly destructive osteoarthritis (38). While osteoarthritis is traditionally considered to be a degenerative process, there is a recognized inflammatory component (39). This may play a greater role in certain joints, such as the small joints of the hands, compared to the knee and hip joints in which osteoarthritis is predominately correlated with load (40). Contemporary understanding of the etiology of osteoarthritis includes an appreciation of the complex roles of molecular regulation of the articular cartilage, mechanisms of subchondral bone remodeling, and age-related changes (41). Metabolic syndrome (hypertension, insulin resistance/diabetes, dyslipidemia) has also been identified as a risk factor for developing osteoarthritis (41).

Alteration in Shape

A change in joint shape is characteristic of most forms of arthritis. Inflammatory arthritides often feature tissue loss that results from destruction. On the other hand, although bone and cartilage loss play an important part in the osteoarthritic process, it is the addition of new bone and cartilage in the form of osteophytes that is one of the characteristic features of the disease.

We now recognize that a change in joint shape—either sudden, as with a fracture, or gradual, as in acromegaly or Paget disease—may play an important role in the etiology of arthritis. In other words, although a change in the shape of the joint is an expected result of arthritis, *a change in shape may also be the cause of arthritis.*

TISSUE RESPONSE TO INJURY

Regardless of the etiology, joint injury is characterized by certain basic cellular and tissue responses. There is usually macroscopic and microscopic evidence of both degeneration and repair in the cells and in the extracellular matrix.

These changes may result from direct physical injury, alterations in the cellular synthesis of the matrix, or enzymatic breakdown of the matrix constituents (42). They are most apparent in the surface cartilage.

Injury of vascularized tissue is followed by an acute and then chronic inflammatory response of variable severity. During this process, any injured or devitalized tissue is removed and replaced by granulation tissue, resulting in a fibrous scar. Cartilage does not have this capacity; however, it does have limited regenerative capacity. An inflammatory response is invoked by inevitable involvement of the nearby synovium and bone.

Cartilage

Injury to cartilage is evident in the extracellular matrix, particularly in the collagen component. One of the earliest findings is surface irregularity and roughening (43). Three patterns of surface alteration can be identified: fibrillation (generally age-related), erosion, and cracking (likely trauma-related) (44). The term *fibrillation* is used to describe replacement of the normally smooth, shiny surface by a surface similar to cut velvet. The "pile" of the fibrillated area may be short or shaggy. The junction between the fibrillated area and the adjacent normal-appearing cartilage is distinct.

Fibrillation is not only seen in the context of prior mechanical abrasion in osteoarthritis. The phenomenon may also be related to regional underloading of the cartilage, as it is commonly identified in well-defined areas affecting particular locations in certain joints, even from an early age (45). Fibrillation is more marked in established osteoarthritis, being characterized by deeper clefts and areas of chondrocyte proliferation (Fig. 5.21).

Cartilage *erosion* is characteristic of progressive degenerative changes in the joint. The base of the erosion is initially contoured or smooth, but as further damage occurs, it may eventually be so extensive as to completely denude the joint surface. The exposed subchondral bone, subject to friction and burnishing from the opposing surface, takes on a polished appearance known as *eburnation*. The least common form of cartilage injury is deep *cracking* of the cartilage which results from severe impact loading. These cracks extend vertically into the cartilage and microscopically often have a horizontal component (Fig. 5.22).

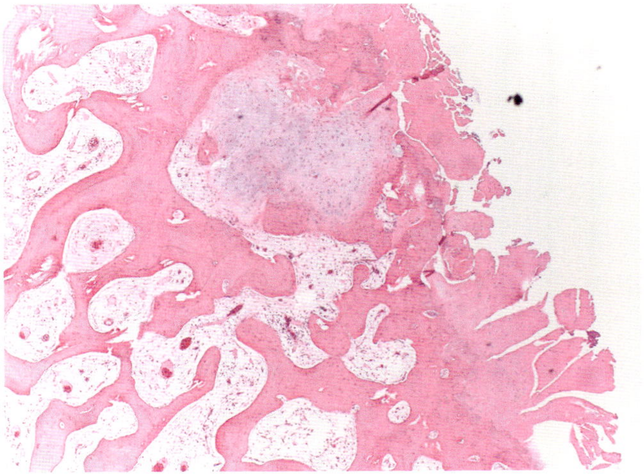

FIGURE 5.21 Fibrillation of the cartilage in osteoarthritis. The clefts are deep and accompanied by cloned proliferation of the chondrocytes. Fragmentation of the cartilage is evident (H&E stain, ×10 objective).

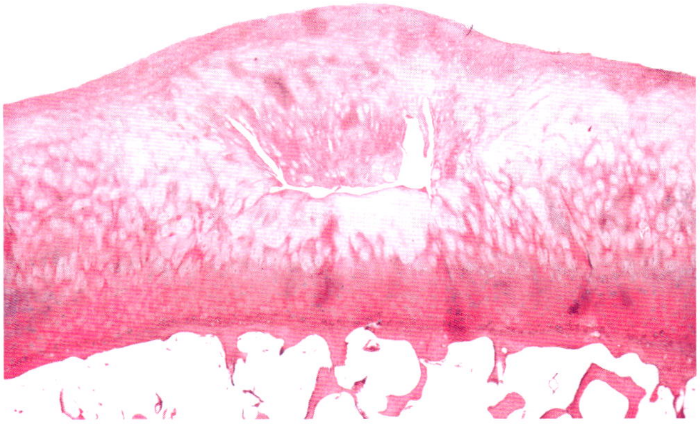

FIGURE 5.22 Deep cracking of the cartilage matrix. The lesion shown is characteristic of a blister-like lesion, which is seen in many cases of chondromalacia patellae (H&E stain, ×4 objective).

Most specimens of osteoarthritis show evidence of intrinsic and extrinsic repair, which may be exuberant, particularly in end-stage disease (46). Intrinsic repair mechanisms in the cartilage are reflected by the histologic response of both cells and matrix. This capacity is, however, limited, often with suboptimal functional and structural outcomes. Regenerating chondrocytes are observed to proliferate in clusters ("cloning") (Fig. 5.23). Chondrocyte proliferation is accompanied initially by increased but altered synthesis of cartilage matrix, as well as cytokines and matrix-degrading enzymes (47). Alterations in the molecular composition and organization of the cartilage matrix contribute to a deterioration in structural integrity of the articular surface. Progressively, there is increased catabolic activity (47). Chondrocytes shift toward terminal differentiation with calcification of the extracellular matrix (47). The tidemark, representing the mineralization front, shifts upward, thereby effectively thinning the cartilage (Fig. 5.12) (47). Vascular ingress of the calcified cartilage from the subchondral bone also plays a role in advancement of the tidemark (48). Chondrocyte necrosis can be identified when ghost outlines of the chondrocytes are present, which is usually a focal phenomenon (Fig. 5.24) but can rarely be widespread. The cartilage of repair has a higher collagen content within its matrix (fibrocartilage). The thick and haphazard collagen network in reparative cartilage can be observed under polarized light (Fig. 5.25) (49).

Extrinsic repair may be initiated from the joint margin or the subchondral bone. Repair initiated at the joint margin consists of a layer of cartilage extending over the joint

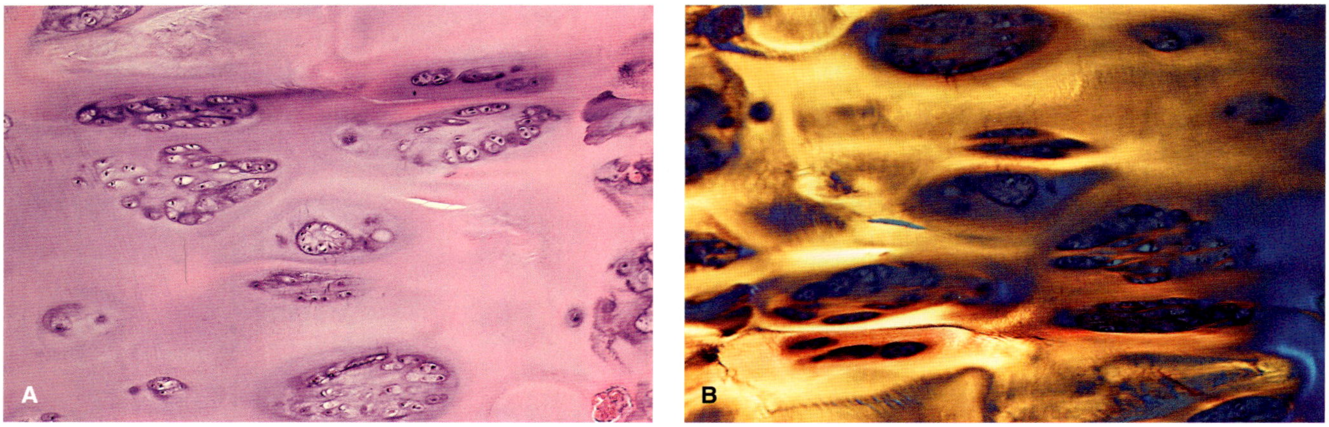

FIGURE 5.23 **A:** Clones of regenerating chondrocytes. Note the basophilia around the clones, which corresponds to increased proteoglycan synthesis by the cells (H&E stain, ×10 objective). **B:** When examined by polarized light, the proliferating clones are visibly displacing the existing collagen matrix.

surface. This cartilage is usually much more cellular and the chondrocytes are evenly distributed (Fig. 5.25). On microscopic examination, this may be easily overlooked. Examination under polarized light will clearly demonstrate the discontinuity between the collagen network of the reparative fibrocartilage and that of the pre-existing hyaline articular cartilage (Fig. 5.26).

Osteoarthritic joints frequently display pits in the bone surface from which protrude small nodules of fibrocartilage that arise in the marrow spaces of the subchondral bone. The fibrocartilage may extend over the previously denuded surface to form a continuous layer of repair tissue.

Bone

Arthritis affects not only the articular cartilage, but also the underlying bone and the structures around the joint. The current paradigm of osteoarthritis suggests that changes in the bone occur in parallel to those occurring in the cartilage (47). There is proliferation of osteoblasts and formation of new bone in the subarticular zone, which occurs both on the surfaces of existing intact trabeculae and around microfractures (Fig. 5.27) (50). In x-rays of arthritic joints, this new bone appears as subchondral sclerosis. Newly formed bone, although thicker, may not be accompanied by equivalent mineralization and adequate stiffness, resulting in biomechanical compromise (47). Changes in the subchondral bone were long thought to be a response to localized overloading from cartilage loss, however, the interaction between the articular cartilage and the subchondral bone is not unidirectional. Subchondral bone deterioration can itself act as a trigger for degeneration of the cartilage (41).

Osteophytes are bony outgrowths that result from a proliferation of periosteal cells at the joint margin, a form of extrinsic repair. These cells differentiate into chondrocytes and the ensuing endochondral ossification enlarges the bony skeleton at the joint periphery. Their localization to sites of joint loading implicates local mechanical factors

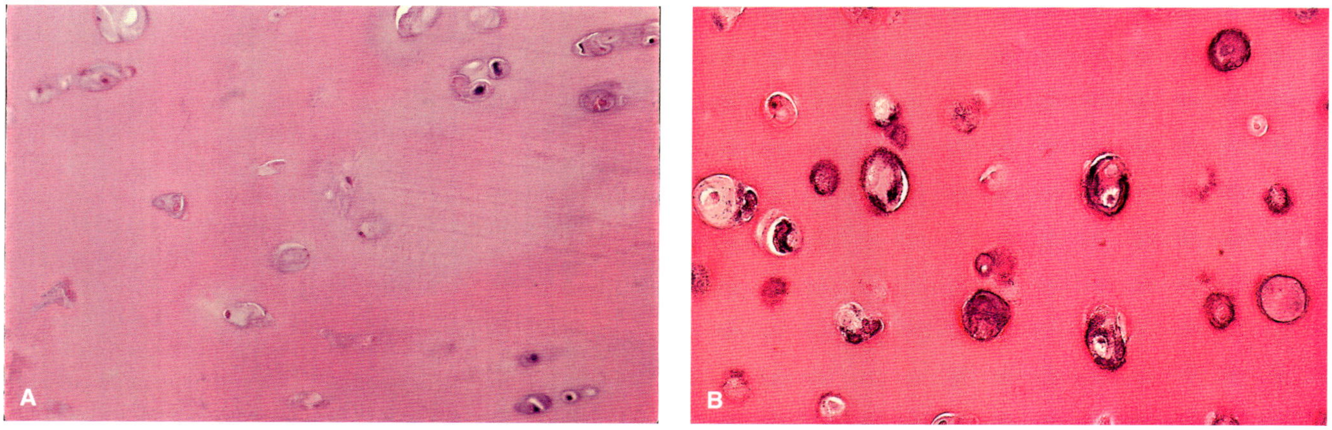

FIGURE 5.24 **A:** Focal chondrocyte necrosis. In cases of degenerative arthritis, focal areas of necrosis (such as seen here) are common. Rarely, the necrosis is extensive. In inflammatory arthritis, chondrocyte necrosis is also common and often associated with an irregular lysis of the matrix around the necrotic cells (H&E stain, ×10 objective). **B:** Focal calcification around necrotic chondrocytes in the deep zone of the cartilage (H&E stain, ×25 objective).

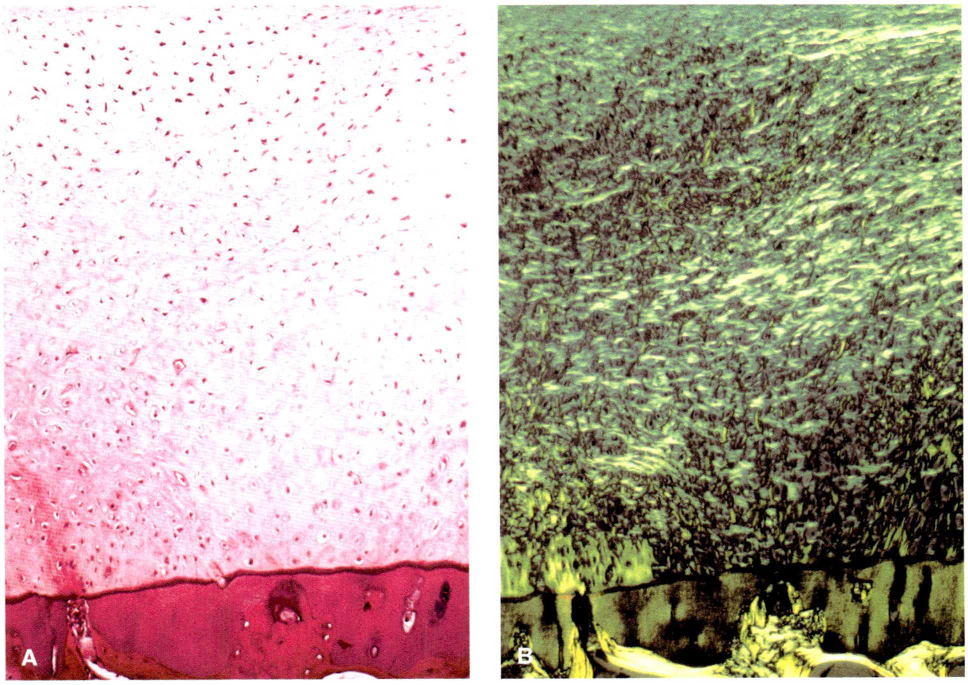

FIGURE 5.25 **A:** A section through the articular surface of an arthritic joint demonstrates extrinsic reparative fibrocartilage, which extends to the tidemark of the original articular hyaline cartilage (H&E stain, ×10 objective). **B:** The same field photographed with polarized light shows the discontinuity of the collagen between the calcified zone and the reparative cartilage.

in their formation, and there is evidence that osteophytes contribute to maintenance of joint stability (48).

Areas of focal pressure necrosis may be identified in the subchondral bone (Fig. 5.28). This should be differentiated from avascular necrosis which itself leads to secondary osteoarthritis. In clinical practice, differentiation may be difficult (51). A clue is whether the involved necrotic bony trabeculae are sclerotic or not, with the former usually seen in the context of osteoarthritis with secondary osteonecrosis.

Subarticular cysts are usually seen only where the overlying cartilage is absent. Such cysts are common in cases of osteoarthritis and are believed to result from transmission of intra-articular pressure through defects in the articular surface into the marrow space of the subchondral bone (52). The cysts increase in size until the pressure within them equals the intra-articular pressure. Cysts may also occur because of focal tissue necrosis (53). Periarticular radiologic "cysts" may be associated with erosion of the marginal

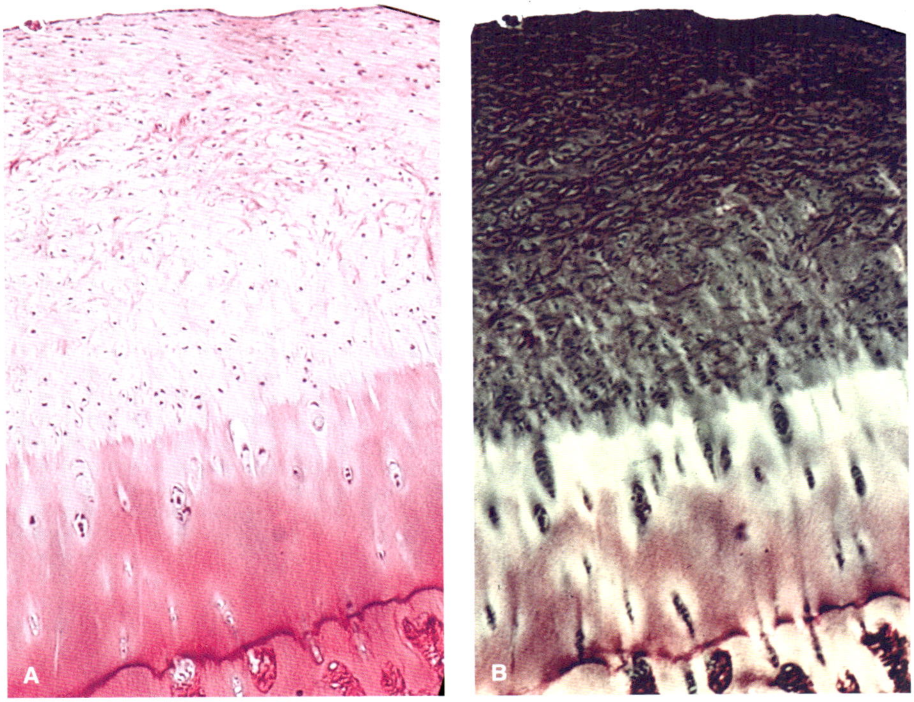

FIGURE 5.26 **A:** Reparative cartilage extending over pre-existing damaged cartilage (H&E stain, ×4 objective). **B:** Same field photographed with polarized light.

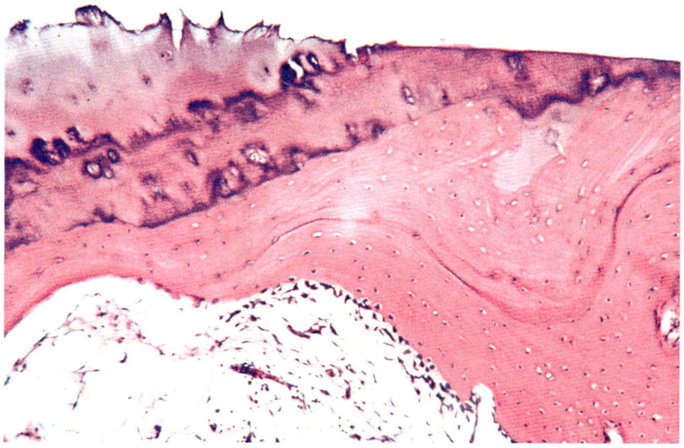

FIGURE 5.27 The edge of an eburnated area of bone in a case of osteoarthritis. There is a very prominent layer of osteoblasts covering the sclerotic bone that underlies the area denuded of cartilage (H&E stain, ×4 objective).

subchondral bone by the diseased synovium in patients with rheumatoid arthritis, and by tophaceous deposits of monosodium urate crystals in patients with gout.

Separated fragments of bone and cartilage from a damaged joint surface may become incorporated into the synovial membrane and digested, or they may remain free as loose bodies in the joint cavity. Under certain circumstances, proliferation of cartilage cells occurs at the surface of these loose bodies and consequently they grow larger (Fig. 5.29). It is possible to visualize periodic extension of this central calcification in the form of concentric rings, which increase in number

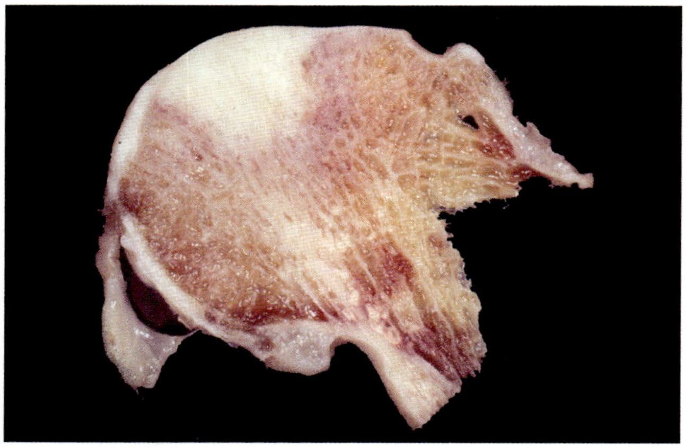

FIGURE 5.28 A section through an osteoarthritic femoral head shows a large wedge-shaped area of necrosis of the superior portion of the head.

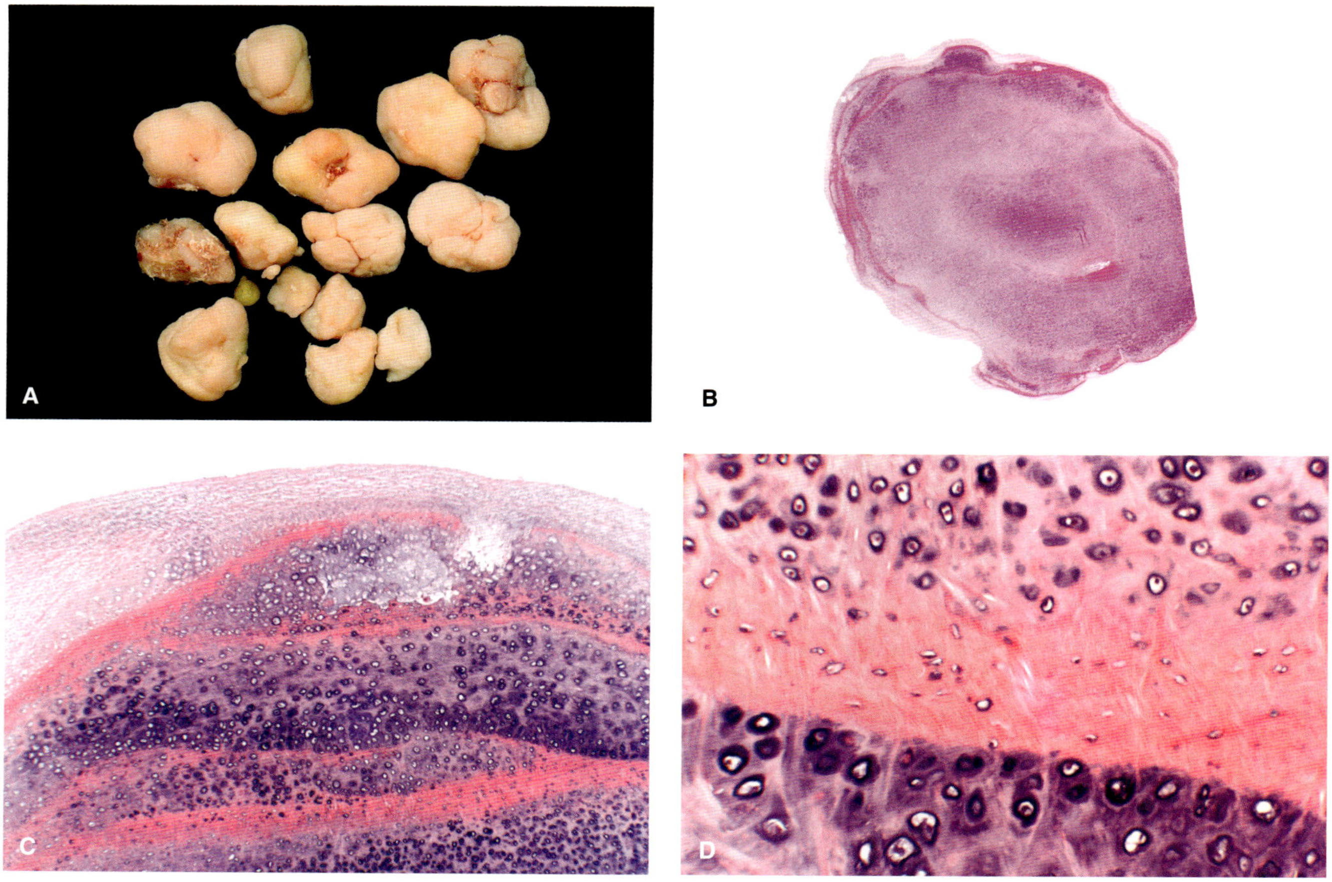

FIGURE 5.29 **A:** Gross photograph of multiple loose bodies in a case of osteoarthritis of the hip joint. **B:** Low-power photograph of a cross section of a loose body showing concentric growth rings (H&E stain, ×1 objective). **C:** Crowded proliferating chondrocytes arranged in a growth ring (H&E stain, ×4 objective). **D:** Benign proliferating chondrocytes (H&E stain, ×25 objective).

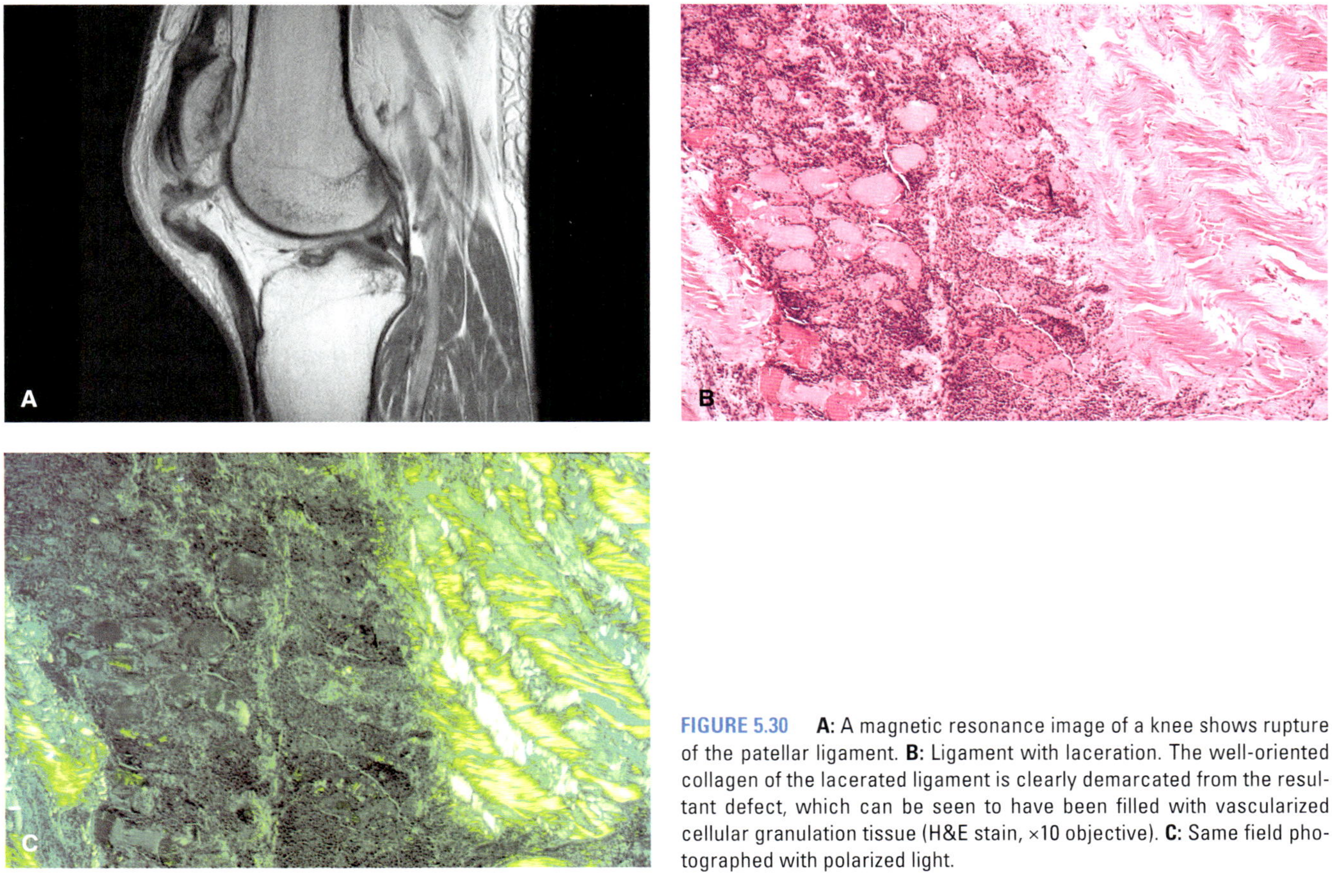

FIGURE 5.30 **A:** A magnetic resonance image of a knee shows rupture of the patellar ligament. **B:** Ligament with laceration. The well-oriented collagen of the lacerated ligament is clearly demarcated from the resultant defect, which can be seen to have been filled with vascularized cellular granulation tissue (H&E stain, ×10 objective). **C:** Same field photographed with polarized light.

as the loose body grows, appearing somewhat akin to the rings of a tree trunk. Sometimes the loose bodies reattach to the synovial membrane, in which case they are invaded by blood vessels.

Although loose bodies can be identified in many forms of arthritis, they are especially prominent in neuropathic (Charcot) joints and in other types of rapidly destructive osteoarthritis. Loose bodies are occasionally so numerous that they must be distinguished from those that occur in primary synovial chondromatosis (54) and are often termed "secondary" synovial chondromatosis.

Ligaments and Tendons

Microscopic evidence of both lacerations and repair by scar tissue is common in the ligamentous and capsular tissue around an arthritic joint. These changes are readily recognized using polarized microscopy, which highlights alterations in collagen organization (Fig. 5.30). Low-grade repetitive trauma results in changes to both ligaments and tendons (tendinosis), including increases in cellularity and myxoid ground substance. When tenocytes proliferate, they may form runs of enlarged and rounded cells (Fig. 5.31). Zones of chondroid metaplasia may be apparent. These changes are maximal in the watershed zone, reflecting abnormal load and compression effect, with the cellular proliferation related to stem cell stimulation.

Synovial Membrane

Injury and breakdown of cartilage and bone result in particulate debris within the joint cavity. This is removed from the synovial fluid by type A cells of the synovial membrane. As a result, the membrane becomes hypertrophic and hyperplastic, with a villiform architecture. The breakdown

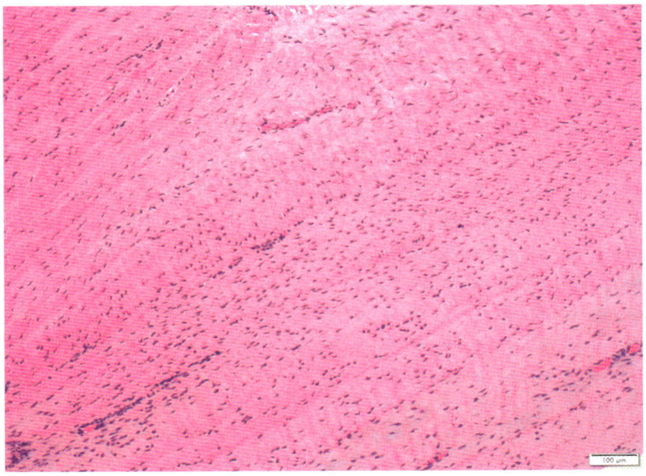

FIGURE 5.31 Linear arrangement of proliferating tenocytes between collagen bundles on a background of increased myxoid ground substance (H&E stain, ×40 objective).

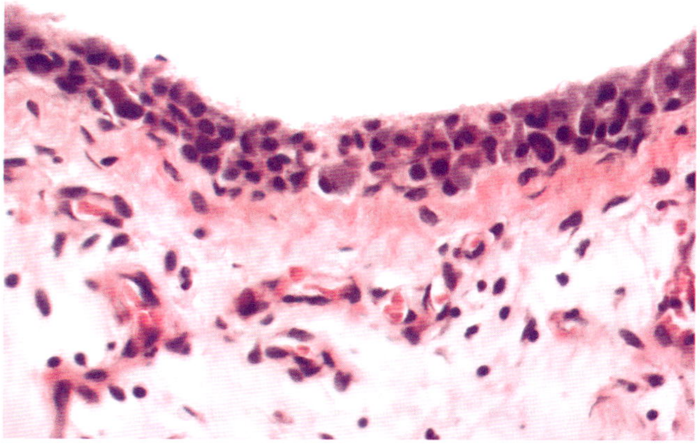

FIGURE 5.32 Synovium removed from the joint of a patient with a moderate degree of osteoarthritis reveals hypertrophy and hyperplasia of synoviocytes. In the subsynovial tissue, there is increased vascularity and a mild chronic inflammatory infiltrate (H&E stain, ×25 objective).

products of the cartilage and bone matrix evoke an inflammatory response (Fig. 5.32). Thus, chronic inflammation can be expected in the synovial membrane of arthritic joints, even when the injury has been purely mechanical. Inflammation is especially prominent in areas of rapid tissue breakdown.

The degree of inflammatory response can be so severe in osteoarthritis that it mimics a true inflammatory arthritis. However, the synovial inflammation of osteoarthritis is the result of cartilage breakdown, whereas the synovial inflammation of rheumatoid arthritis is the cause of cartilage breakdown. Extension of hyperplastic synovium onto the articular surface of the joint (i.e., a pannus) is a common finding in both conditions, likely reflects the effects of local cytokines (Fig. 5.33). However, the extent and aggressiveness of the pannus with associated cartilage damage is generally much less marked in osteoarthritis than in rheumatoid arthritis.

Under normal conditions, the synovial membrane is responsible for the nutrition of articular cartilage. The chronically inflamed and scarred synovial membrane of an arthritic joint functions less effectively than that of a normal joint. The hypertrophied and hyperplastic synovium is also likely to be traumatized as it extends into the joint cavity. Thickening of the joint capsule is a marker of disease in general, rather than being specific for a particular etiology (55). Evidence of bleeding into the joint, with subsequent hemosiderotic synovitis may be present. The orange-brown color of the villous synovium should not be confused with tenosynovial giant cell tumor (formerly known as pigmented villonodular synovitis).

In recent years, there has been an increasing appreciation for the function of stromal cells in disease pathogenesis. There are complex interactions between synovial fibroblasts and immune cells involved in both seronegative and seropositive arthropathies. In rheumatoid arthritis, in particular, synovial fibroblasts participate in joint destruction (56). They function as innate immune cells and interact with other inflammatory cells (T cells, B cells, monocytes, and macrophages), as well as cartilage, endothelial cells, and osteoclasts (56).

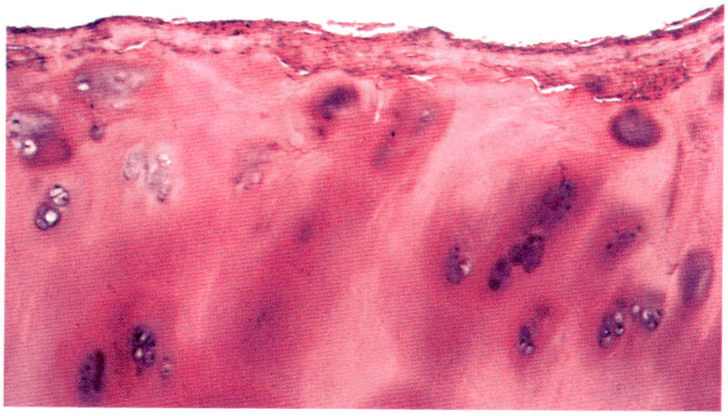

FIGURE 5.33 A portion of the articular surface of a femoral head in a case of osteoarthritis. A fibrous pannus extends over the articular surface (H&E stain, ×10 objective).

Synovial Fluid

Examination of synovial fluid is extremely helpful in the diagnosis of arthritis, both for determining the cause and chronicity of the disease. Whatever the cause of arthritis, the synovial fluid is altered. Normal synovial fluid is viscous, pale yellow, and clear. Even in large joints, the volume is small. In cases of inflammatory arthritis, there is an increased volume of synovial fluid with a high count of inflammatory cells. The amount of hyaluronic acid is markedly diminished, leading to decreased viscosity. In degenerative arthritis, the amount of hyaluronic acid is increased, resulting in an extremely viscous fluid.

Interestingly, in the postarthroplasty setting, there is formation of a pseudocapsule with regeneration of the synovial lining, resulting in the production of synovial fluid facilitating function of the implanted joint. Although this fluid is less viscous than in a native joint, the protein and phospholipid concentrations are maintained (57). Polarized microscopic examination of synovial fluid is the gold standard for detection of a crystal deposition disease.

POTENTIAL PITFALLS AND ARTIFACTS OF NOTE

There are tissue artifacts and potential pitfalls in examining joint tissue of which pathologists should be aware. Elastic fibers, hyalinized or fibrinoid collagen (often seen in the context of low-grade repetitive trauma) and collagen with heat artifact secondary to cautery can show false-positive staining with Congo red (including birefringence). Immunohistochemical stains for amyloid do not show false positivity in these settings (58,59).

Monosodium urate (gout) and calcium pyrophosphate (pseudogout) crystals dissolve in water. For this reason, transporting, fixing, processing, and staining joint tissues in water-based media (including formalin, hematoxylin, and decalcifying agents) can result in loss of the diagnostic crystals. To mitigate this problem, where there is clinical suspicion of crystalline arthropathy, tissue should be submitted to the laboratory either in a fresh state or in 100% alcohol. A smear should be performed. Processing can then proceed in 100% alcohol. The tissue can be cut and examined unstained or stained with eosin-only (60). If no crystals are identified, direct examination of joint fluid may be beneficial as it is a more sensitive technique.

As joint specimens often include calcified material and/or bone, mention should be made of the decalcification in preparation of specimens. Decalcification allows better visualization of calcified structures, including whether the bone is viable, which may be essential in some scenarios. However, rapid decalcification or decalcification without adequate prior fixation, can result in false-negative results on immunohistochemical stains and molecular analyses (61). Consideration can be given to utilizing EDTA-based decalcifying agents for at least part of the tissue, and careful block selection in utilizing these tests (62). Care also needs to be taken to avoid over decalcification, as this results in pallor of osteocyte nuclei, thereby mimicking osteonecrosis.

Other considerations prior to approaching joint specimens include planning the macroscopic/grossing phase. The clinical scenario, the joint involved, and the nature of the specimen will determine the best approach. In general, it is optimal to section the cartilage and bone in such a way that demonstrates the relationship between these structures. In the case of a femoral head, cutting the head in a wedge-shaped arrangement allows for optimal assessment. Ligaments and tendons are best assessed when sectioned longitudinally (63).

ACKNOWLEDGMENT

This chapter is an updated version of previous iterations written by Dr Peter Bullough, whose contributions are gratefully acknowledged.

REFERENCES

1. Hammond BT, Charnley J. The sphericity of the femoral head. *Med Biol Eng*. 1967;5(5):445–453.
2. Bullough PG, Walker PS. The distribution of load through the knee joint and its possible significance to the observed patterns of articular cartilage breakdown. *Bull Hosp Joint Dis*. 1976;37(2):110–123.
3. Mayne, R. Cartilage collagens. What is their function, and are they involved in articular disease? *Arthritis & Rheumatism*. 1989;32:241–246.
4. Hulmes DJS. Collagen diversity, synthesis and assembly. In: Fratzl P, ed. *Collagen: Structure and Mechanics*. Springer; 2008:15–47.
5. Watanabe H, Yamada Y, Kimata K. Roles of aggrecan, a large chondroitin sulfate proteoglycan, in cartilage structure and function. *J Biochem*. 1998;124(4):687–693.
6. Kiani C, Chen L, Wu YJ, Yee AJ, Yang BB. Structure and function of aggrecan. *Cell Res*. 2002;12(1):19–32.
7. Roughley PJ. Articular cartilage and changes in arthritis: noncollagenous proteins and proteoglycans in the extracellular matrix of cartilage. *Arthritis Res*. 2001;3(6):342–347.
8. Smith MM, Ghosh P. Experimental models of osteoarthritis. In: Moskowitz RW, Howell DS, Altman RD, Buckwalter JA, Goldberg VM, eds. *Osteoarthritis: Diagnosis and Medical/Surgical Management*. WB Saunders; 2001:171–199.
9. Kempson GE. The mechanical properties of articular cartilage. In: Sokoloff L, ed. *The Joints and Synovial Fluid*. Academic Press; 1980.
10. Van der Korst JK, Skoloff L, Miller EJ. Senescent pigmentation of cartilage and degenerative joint disease. *Arch Pathol*. 1968;86(1):40–47.
11. Maroudas A. The permeability of articular cartilage. *J Bone Joint Surg Br*. 1968;50(1):166–177.
12. Muir H, Bullough P, Maroudas A. The distribution of collagen in human articular cartilage with some of its physiological implications. *J Bone Joint Surg Br*. 1970;52(3):554–563.
13. Wilson W, van Donkelaar CC, van Rietbergen B, Ito K, Huiskes R. Stresses in the local collagen network of articular cartilage: a poroviscoelastic fibril-reinforced finite element study. *J Biomech*. 2004;37(3):357–366.
14. Bullough PG, Munuera L, Murphy J, Weinstein AM. The strength of the menisci of the knee as it relates to their fine structure. *J Bone Joint Surg Br*. 1970;52(3):564–570.
15. Maroudas A, Evans H, Almeida L. Cartilage of the hip joint. Topographical variation of glycosaminoglycan content in normal and fibrillated tissue. *Ann Rheum Dis*. 1973;32(1):1–9.
16. Decker RS. Articular cartilage and joint development from embryogenesis to adulthood. *Semin Cell Dev Biol*. 2017;62:50–56.
17. Goodfellow JW, Bullough PG. The pattern of ageing of the articular cartilage of the elbow joint. *J Bone Joint Surg Br*. 1967;49(1):175–181.
18. Mori C, Nakamura N, Kimura S, Irie H, Takigawa T, Shiota K. Programmed cell death in the interdigital tissue of the fetal mouse limb is apoptosis with DNA fragmentation. *Anat Rec*. 1995;242(1):103–110.
19. Dar FH, Aspden RM. A finite element model of an idealized diarthrodial joint to investigate the effects of variation in the mechanical properties of the tissues. *Proc Inst Mech Eng H*. 2003;217(5):341–348.
20. Bullough PG, Yawitz PS, Tafra L, Boskey AL. Topographical variations in the morphology and biochemistry of adult canine tibial plateau articular cartilage. *J Orthop Res*. 1985;3(1):1–16.
21. Bennet GA, WH, Bauer W. *Changes in the Knee Joint at Various Ages: With Particular Reference to the Nature and Development of Degenerative Joint Disease*. Commonwealth Fund; 1942.
22. Palmoski MJ, Brandt KD, Effects of static and cyclic compressive loading on articular cartilage plugs in vitro. *Arthritis Rheum*. 1984;27(6):675–681.
23. Palmoski M, Perricone E, Brandt KD. Development and reversal of a proteoglycan aggregation defect in normal canine knee cartilage after immobilization. *Arthritis Rheum*. 1979;22(5):508–517.

24. Henderson B, Pettipher ER. The synovial lining cell: biology and pathobiology. *Semin Arthritis Rheum*. 1985;15(1):1–32.
25. Iwanaga T, Shikichi M, Kitamura H, Yanase H, Nozawa-Inoue K. Morphology and functional roles of synoviocytes in the joint. *Arch Histol Cytol*. 2000;63(1):17–31.
26. Hogg N, Palmer DG, Revell PA. Mononuclear phagocytes of normal and rheumatoid synovial membrane identified by monoclonal antibodies. *Immunology*. 1985;56(4):673–681.
27. Fonseca JE, Edwards JC, Blades S, Goulding NJ. Macrophage subpopulations in rheumatoid synovium: reduced CD163 expression in CD4+ T lymphocyte-rich microenvironments. *Arthritis Rheum*. 2002;46(5):1210–1216.
28. Talmon G, Wake L, Muirhead D. Podoplanin and clusterin are reliable markers of nonneoplastic synovium at various sites. *Int J Surg Pathol*. 2013;21(6):587–590.
29. Boland JM, Folpe AL, Hornick JL, Grogg KL. Clusterin is expressed in normal synoviocytes and in tenosynovial giant cell tumors of localized and diffuse types: diagnostic and histogenetic implications. *Am J Surg Pathol*. 2009;33(8):1225–1229.
30. Cooper RR, Misol S, Stimmel P. Tendon and ligament insertion: a light and electron microscopic study. *J Bone Joint Surg Am*. 1970;52(1):1–170.
31. Lundborg G, Myrhage R. The vascularization and structure of the human digital tendon sheath as related to flexor tendon function: an angiographic and histological study. *Scand J Plast Reconstr Surg*. 1977;11(3):195–203.
32. GBD 2021 Osteoarthritis Collaborators. Global, regional, and national burden of osteoarthritis, 1990-2020 and projections to 2050: a systematic analysis for the Global Burden of Disease Study 2021. *Lancet Rheumatol*. 2023;5(9):e508–e522.
33. Franklin B. *The Private Correspondence of Benjamin Franklin, LL.D, F.R.S., &c. Minister Plenipontentiary from the United States of America at the Court of France, and for the Treaty of Peace and Independence With Great Britain, &c. &c: Comprising a Series of Letters on Miscellaneous, Literary, and Political Subjects Written Between the Years 1753 and 1790, Illustrating the Memoirs of His Public and Private Life, and Developing the Secret History of His Political Transactions and Negociations*. Franklin WT, ed. Printed for Henry Colburn; 1817.
34. Johanson NA. Endocrine arthropathies. *Clin Rheum Dis*. 1985;11(2):297–323.
35. Maclean FM. Arthritis and other proliferative joint diseases. *Diagn Histopathol*. 2016;22(10):369–377.
36. Yao Q, Wu X, Tao C, et al. Osteoarthritis: pathogenic signaling pathways and therapeutic targets. *Signal Transduct Target Ther*. 2023;8(1):56.
37. Ganz R, Leunig M, Leunig-Ganz K, Harris WH. The etiology of osteoarthritis of the hip: an integrated mechanical concept. *Clin Orthop Relat Res*. 2008;466(2):264–272.
38. Yamamoto T, Schneider R, Bullough PG. Insufficiency subchondral fracture of the femoral head. *Am J Surg Pathol*. 2000;24(3):464–468.
39. Rahmati M, Mobasheri A, Mozafari M. Inflammatory mediators in osteoarthritis: a critical review of the state-of-the-art, current prospects, and future challenges. *Bone*. 2016;85:81–90.
40. Long H, Liu Q, Yin H, et al. Prevalence trends of site-specific osteoarthritis from 1990 to 2019: findings from the Global Burden of Disease Study 2019. *Arthritis Rheumatol*. 2022;74(7):1172–1183.
41. Roelofs AJ, De Bari C. Osteoarthritis year in review 2023: biology. *Osteoarthritis Cartilage*. 2024;32(2):148–158.
42. Rieppo J, Töyräs J, Nieminen MT, Kovanen V, Hyttinen MM, Korhonen RK. Structure-function relationships in enzymatically modified articular cartilage. *Cells Tissues Organs*. 2003;175(3):121–132.
43. Collins DH. *The Pathology of Articular and Spinal Diseases*. Edward Arnold & Co; 1949.
44. Heine J. Über die arthritis deformans. *Virchows Archiv für Pathologische Anatomie und Physiologie und für Klinische Medizin*. 1926;260(3):521–663.
45. Barland P, Novikoff AB, Hamerman D. Electron microscopy of the human synovial membrane. *J Cell Biol*. 1962; 14(2):207–220.
46. Macys JR, Bullough PG, Wilson PD. Coxarthrosis: a study of the natural history based on a correlation of clinical, radiographic, and pathologic findings. *Semin Arthritis Rheum*. 1980;10(1):66–80.
47. Lories RJ, Luyten FP. The bone-cartilage unit in osteoarthritis. *Nat Rev Rheumatol*. 2011;7(1):43–49.
48. Goldring MB, Goldring SR. Articular cartilage and subchondral bone in the pathogenesis of osteoarthritis. *Ann N Y Acad Sci*. 2010;1192:230–237.
49. Nakata K, Bullough PG. The injury and repair of human articular cartilage: a morphological study of 192 cases of coxarthrosis. *Nihon Seikeigeka Gakkai Zasshi*. 1986;60(7):763–775.
50. Christensen SB. Osteoarthrosis: changes of bone, cartilage and synovial membrane in relation to bone scintigraphy. *Acta Orthop Scand Suppl*. 1985;56(sup214):1–43.
51. Franchi A, Bullough PG. Secondary avascular necrosis in coxarthrosis: a morphologic study. *J Rheumatol*. 1992;19(8):1263–1268.
52. Landells JW. The bone cysts of osteoarthritis. *J Bone Joint Surg Br*. 1953;35-B(4):643–649.
53. Rhaney K, Lamb DW. The cysts of osteoarthritis of the hip: a radiological and pathological study. *J Bone Joint Surg Br*. 1955;37-B(4):663–675.
54. Villacin AB, Brigham LN, Bullough PG. Primary and secondary synovial chondrometaplasia: histopathologic and clinicoradiologic differences. *Hum Pathol*. 1979;10(4):439–451.
55. Rakhra KS, Bonura AA, Nairn R, Schweitzer ME, Kolanko NM, Beaule PE. Is the hip capsule thicker in diseased hips? *Bone Joint Res*. 2016;5(11):586–593.
56. Ospelt C. Synovial fibroblasts in 2017. *RMD Open*. 2017; 3(2):e000471.
57. Kung M, Markantonis J, Nelson S, Campbell P. The synovial lining and synovial fluid properties after joint arthroplasty. *Lubricants*. 2015;3(2):394–412.
58. Yakupova EI, Bobyleva LG, Vikhlyantsev IM, Bobylev AG. Congo red and amyloids: history and relationship. *Biosci Rep*. 2019;39(1):BSR20181415.
59. Fernandez-Flores A. Positive staining with Congo red in tissues with heat artifact due to cautery. *Rom J Morphol Embryol*. 2009;50(2):203–206.
60. Longacre TA, Greenson JK, Hornick JL, Reuter VE. *Mills and Sternberg's Diagnostic Surgical Pathology*. 7th ed. Lippincott Williams & Wilkins USA; 2022.
61. Susman S, Berindan-Neagoe I, Petrushev B, et al. The role of the pathology department in the preanalytical phase of molecular analyses. *Cancer Manag Res*. 2018;10:745–753.
62. Klein MJ, Memoli VA. Orthopaedic specimen preparation: what pathologists should know and do. *Semin Diagn Pathol*. 2011;28(1):4–12.
63. Dimenstein IB. Bone grossing techniques: helpful hints and procedures. *Ann Diagn Pathol*. 2008;12(3):191–198.

6

Adipose Tissue

Ivy John ■ Karen Fritchie

TYPES OF FAT CELLS

Until recently, it was commonly believed there were only two types of fat cells, namely white adipocytes and brown adipocytes. However, efforts to uncover the functions of fat cells in obesity have discovered a third type of fat cell, or at least a third functional state or class of cell that has been termed "beige" or "brite." These beige adipocytes result from stimuli to white adipocytes and appear in anatomical sites corresponding to white adipose tissue.

WHITE FAT

Prenatal Development

The morphology of developing adipose tissue has been studied in detail. Poissonnet et al. (1) examined serial sections obtained from 805 human fetuses of various ages and determined that adipose tissue primordia cannot be identified by light microscopy prior to the second trimester of pregnancy. In the early phase of adipocyte development (stage I), spindle cells destined to become adipose tissue are loosely arranged in ground substance. Aggregates of these mesenchymal cells begin to condense around proliferating primitive blood vessels after 14 weeks of gestation and represent stage II in adipocyte development (Fig. 6.1). Later, capillaries continue to proliferate and create a rich network around which preadipocytes assume a stellate shape and become organized into a mesenchymal lobule (stage III). These preadipocytes do not contain lipid. Further development leads to accumulation of fine lipid vacuoles in the cytoplasm that define stage IV of development (Fig. 6.2). Continued proliferation of the lobular components results in formation of densely packed aggregates of vacuolated fat cells with a rich capillary vascular network. Condensation of mesenchyme

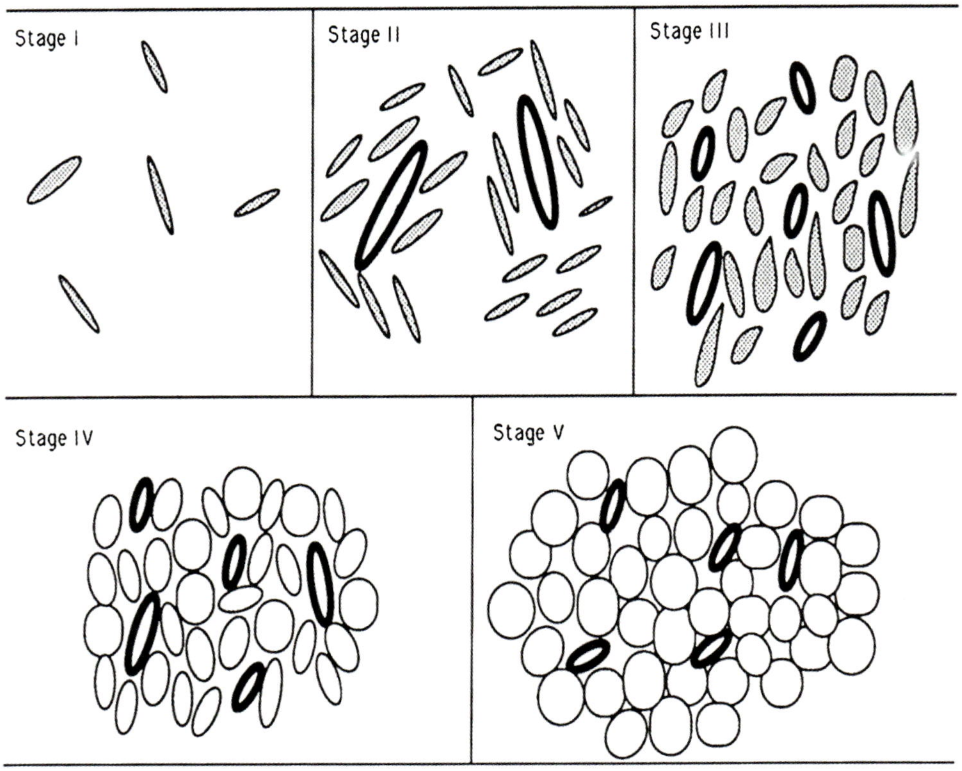

FIGURE 6.1 Developmental stages of adipose tissue. Stage I: Stellate cells (stippled) embedded in amorphous ground substance. Stage II: As angiogenesis begins, mesenchymal cells (stippled) condense around the blood vessels (bold ovals). Stage III: A rich capillary network develops from each vessel, forming a glomerulus-like network around which each lobule forms. The preadipocytes become more stellate. Stage IV: With accumulation of lipid, these adipocytes, with multiple small lipid droplets, become closely packed around the capillaries. Stage V: Further accumulation of lipid with many unilocular cells (clear circles) is evident. The perilobular mesenchyme condenses into interlobular septa at this stage.

at the periphery of the lobule results in formation of fibrous interlobular septa, which heralds the final stage of development (stage V) that concludes in the 24th gestational week. The number of fat lobules remains relatively constant for the next 5 weeks. Continued growth results from proliferating capillaries and an increase in adipocyte size that causes an increase in the size of the fat lobules (Fig. 6.2).

The same sequence of events occurs in adipose tissue throughout the body (2). Primitive white fat lobules first appear in the face, neck, breast, and abdominal wall at 14 weeks of gestation, and they are evident over the back and shoulders by 15 gestational weeks. Development in the upper and lower extremities and anterior chest begins around the 16th week. By the end of the 23rd week, a layer of subcutaneous fat completely covers the extremities.

There is a very close association between adipocyte development and angiogenesis. Fat appears first in well-vascularized regions before differentiation can be identified in the less well-supplied adjacent subcutaneous tissue. There is also an important physiologic significance to this close anatomic relationship. Lipoprotein lipase, the hormone responsible for transfer of triglyceride from circulating lipoproteins to adipose tissue, is synthesized by adipocytes and transferred to the luminal surface of the capillary endothelium (3). Thus, this close spatial relationship provides efficient transfer of enzyme and lipid.

Because of this close developmental association of capillaries and adipocytes, some have proposed that the adipocyte precursor, or preadipocyte, may be derived from endothelial cells (4). Others have proposed that the preadipocyte is derived from a perivascular reticulum cell, perivascular fibroblast-like cell, or undifferentiated mesenchymal cell. The presumptive adipocyte precursor has been characterized ultrastructurally in the newborn rat (5). The preadipocyte is a spindle cell with four to five cytoplasmic extensions along its long axis and abundant rough endoplasmic reticulum (ER). Lipid accumulates first as small droplets adjacent to the nucleus. As more lipid appears, it coalesces into a single large vacuole, and the cell takes on an ovoid then round shape. The amount of rough ER decreases as the cell matures. Although cell shape and abundance of rough ER were taken as supportive evidence for the common origin of the preadipocyte and the fibroblast, which has a similar ultrastructural appearance, these similarities may have been a coincidence. The immature adipocyte needs to synthesize and excrete lipoprotein lipase and, thus, it is expected to contain abundant rough ER. Fibroblasts synthesizing and secreting procollagen have a similar array of organelles.

As the preadipocyte accumulates lipid to become an adipocyte, both multilocular and unilocular adipocytes can be seen. Multilocular adipocytes predominate at first but assume the unilocular appearance characteristic of mature adipocytes with further lipid accumulation. Thus, attempts to differentiate brown from white adipocyte tissue at the light microscopic level, which rely on the presence of multivacuolated cells to characterize and identify brown fat, are not entirely reliable. Reliance on ultrastructural and

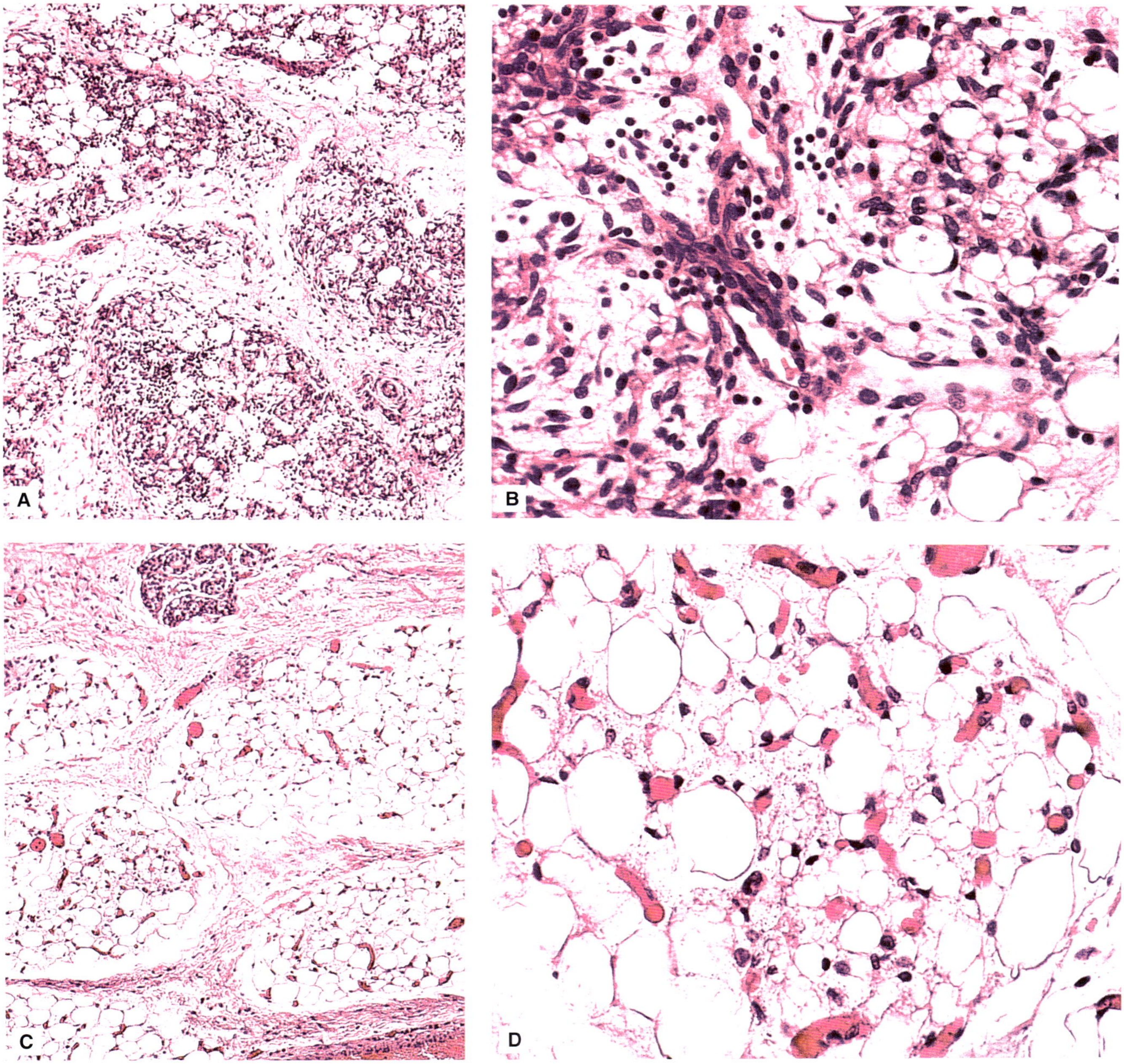

FIGURE 6.2 Fetal fat. **A:** Fat lobules from a 25-week fetus with a myxoid quality and prominent vasculature. **B:** At high power, both univacuolated and multivacuolated cells are noted together with small capillaries. By 37 weeks, the lobules are more developed (**C**) and many of the cells are univacuolated (**D**).

biochemical differences helps distinguish between these different forms of adipose tissue.

Molecular Biology

At least one gene involved in adipocytic differentiation has been identified through work on the adipocytic neoplasm known as myxoid liposarcoma. The translocation t(12;16)(q13;p11) of that tumor disrupts function of *DDIT3* found at 12q13. *DDIT3* is rearranged in nearly all myxoid liposarcomas and its breakpoint has been cloned (6). *DDIT3*, also known as *CHOP* and *GADD153*, encodes a member of the CCAAT/enhancer-binding protein (C/EBP) family and has a DNA-binding domain. Its protein product is a dominant inhibitor of other C/EBP transcription factors known to be important in cell proliferation (7). Members of the C/EBP family are highly expressed in fat and are involved in the differentiation of primitive mesenchymal cells into adipocytes and in the growth arrest of terminally differentiated adipocytes (8). *DDIT3* itself is induced in the differentiation of 3T3-L1 cells to adipocytes.

Postnatal Development

The average-size infant has approximately 5 billion adipocytes at birth, which represents only 16% of the total number of adipocytes in adults (9). Adipose tissue continues to grow in parallel with general growth throughout the first 10 years of life. Fat cells enlarge significantly during the first 6 months of life without much increase in cell number (10). The cell size remains fairly constant until puberty, at which point the number of adipocytes begins to progressively increase. At puberty, there is a substantial increase in adipocyte size and number such that the total number of adipocytes is similar to that of the adult by the end of puberty (10). New adipocytes may continue to form throughout life (11,12). *De novo* adipocyte formation can be triggered by overdistention of existing fat cells and the mass of stored triglycerides (13). Loss of fat cells may also occur following years of strict dietary restriction (11).

Adult white adipose tissue may be a source of mesenchymal stem cells (12) that can differentiate into multiple lineages including adipose cells, chondrocytes, osteoblasts, neuronal cells, endothelial cells, and cardiomyocytes (14).

Gender Differences

The differences in body fat content noted between men and women begin in early childhood, with young girls harboring more fat than boys. Fetal studies, however, have failed to note differences in the pattern of distribution or quantity of fat in prenatal life. Gender differences in the distribution of adipose tissue following puberty are thought to be related to steroid hormone secretion (10). In humans, estrogens and progesterone induce an increase in trochanteric fat that, in women, results in the gynecoid habitus of the lower body. Fat deposits in these regions are reduced by androgens in men. The percentage of body fat also differs in men and women. Males reach a peak in body fat content during early adolescence, whereas women continue to accumulate fat relative to body weight throughout the teen years.

Functions

White adipose tissue is the body's largest energy store, and it possesses the enzymes necessary for the uptake and release of triglycerides. Briefly, triglycerides circulate in the blood in the form of chylomicrons from the intestine and very low-density lipoproteins from the liver (15). Lipoprotein lipase is synthesized by adipocytes and transferred to endothelial cells. It is located on the endothelial cell surface such that it hydrolyzes circulating triglyceride to release free fatty acids. Free fatty acids are then taken up by the fat cells and re-esterified to glycerol phosphate to form triacylglycerol, which is stored within a lipid droplet. The fat is mobilized through the action of hormone-sensitive lipase, which hydrolyzes stored triglycerides. The released free fatty acids may be re-esterified or released to the circulation where they are bound to albumin for transfer to other cells.

Until recently, the main endocrine function of adipose tissue was thought to be conversion of androstenedione to estrone, the major source of estrogen in men and postmenopausal women. We now know, however, that the aromatase action is localized to the stromal cell fraction of adipose tissue rather than the adipocyte (16). Adipose tissue also plays a more dynamic endocrine role than previously appreciated. It expresses several hormones, growth factors, and cytokines, the latter of which may be found in adipocytes, stromal cells, and macrophages localized to adipose tissue. Leptin, a regulator of energy expenditure and appetite; interleukin-6 (IL-6), which may play a role in the metabolic syndrome; and several important regulators of glucose and lipid metabolism, the complement cascade, and the fibrinolytic system are all produced by adipose tissue.

Leptin, the protein product of *OB*, is synthesized exclusively by adipocytes and acts on the hypothalamus to increase energy expenditure and decrease appetite (17). In humans, fasting lowers serum leptin levels and increases appetite (18). Unfortunately, elevated leptin levels do not have the reverse effect, and leptin does not appear to have antiobesity activity in humans. Most obese individuals have elevated serum leptin levels that are proportional to the amount of adipose tissue and, thus, it is postulated that humans are leptin resistant (19). Leptin receptors are present on most tissues, and leptin may play a role outside the adipose tissue to accelerate wound healing, increase vascular tone, and inhibit bone formation (20).

Cytokines are secreted by adipocytes, stromal cells, and resident macrophages. IL-6 is made by adipocytes and macrophages, and approximately 30% of circulating IL-6 is produced in adipose tissue (21). Like leptin, serum IL-6 levels are highly correlated with percent body fat. The IL-6 released from intra-abdominal stores enters the portal circulation and stimulates hepatic triglyceride secretion; this may contribute to the hypertriglyceridemia seen in patients with visceral obesity (20). IL-6 also stimulates hepatic secretion of acute phase reactants, increases platelet number and activity, and increases expression of endothelial adhesion molecules. There are ongoing investigations regarding the role this cytokine plays in the metabolic syndrome and the risk of cardiovascular disease in obesity.

Adipocytes also secrete C3 and adipsin, both of which participate in the alternate complement pathway (22). Plasminogen activator inhibitor-1 (PAI-1) is a potent inhibitor of the fibrinolytic system and favors development of thromboemboli. Insulin induces expression of PAI-1 by adipocytes, and elevated levels are seen in obesity (23).

Mitochondrial dysfunction in adipocytes may be related to the development of obesity and the diabetes mellitus epidemic and appears to be involved in HIV treatment–associated lipodystrophy (24).

Regulation

White adipose tissue contains numerous receptors for hormones, cytokines, catecholamines, and lipoproteins. Catecholamines acting through α-2 receptors inhibit lipolysis, and a predominance of α-2 receptors in gluteal fat of women is thought to impact maintenance of these fat stores despite weight loss (15). Regional differences in lipoprotein lipase levels also occur in women. Gluteal fat in premenopausal women tends to have high lipoprotein lipase levels, and these regions contain larger fat cells. Such regional differences disappear after menopause and are not present in obese men, suggesting that the sex steroids play a role in adipose tissue distribution and activity (25,26). Both androgens and estrogen modulate *OB* expression and control adipose tissue development (27,28).

Insulin stimulates lipogenesis and glucose uptake while inhibiting fat breakdown. Insulin and glucocorticoids stimulate DNA synthesis in cultured human adipocytes and promote conversion of preadipocytes to mature adipocytes. These effects are more pronounced on cells obtained from obese people compared with those who are lean. Estradiol-17β stimulates cultured preadipocytes obtained from men and women. Progesterone acts *in vitro* to stimulate both preadipocyte division and lipoprotein lipase activity (29). This dual role facilitates triglyceride accumulation in women. Fibroblast growth factor 1, secreted by adipose-derived microvascular endothelial cells, stimulates preadipocyte differentiation and accumulation of triglycerides (30).

Tumor necrosis factor alpha (TNF-α) and IL-6 have the opposite effects and are implicated along with leptin in the weight loss and anorexia of chronic wasting illnesses (31,32). TNF-α is expressed in preadipocytes blocks their differentiation to mature adipocytes through CCAAT/enhancer-binding protein alpha (C/EBP-α) (33). It also suppresses lipoprotein lipase and stimulates mobilization of fatty acids. TNF-α also induces the release of IL-6 and leptin from adipose tissue, and the action of these cytokines is closely interrelated.

In addition to adipocyte function, fat cell size and number are regulated. Numerous studies using tritiated thymidine incorporation as a marker of cell division in adipose tissue have identified mitotically active cells within fat. Mature lipid-laden adipocytes are generally considered to be incapable of cell differentiation because of the absence of mitotic figures seen histologically in normal adipose tissue. However, sampling fat from rats injected with tritiated thymidine has shown that the number of labeled cells in subcutaneous fat initially rises due to cell proliferation. The concentration of radioactivity then falls, probably as a result of a dilutional effect resulting from continued cell division (34). Similar studies in which the subcutaneous tissue of rats is separated into stromal and adipose components have shown that the specific radioactivity of the adipocyte fraction does not increase until 2 to 5 days after injection (35). Thus, DNA synthesis likely occurs in non–lipid-laden cells or preadipocytes. As these cells accumulate lipid, labeled cells are detected within the adipocyte fraction.

Gross Aspects

Fatty tissue is typically homogeneous and bright yellow, with a glistening and greasy surface. Adipose tissue is arranged in variably sized, usually small lobules separated by thin fibrous septa. Any variation in color indicates a pathologic process. Fat necrosis and inflamed fat are typically white or pale yellow, and many lipomas are composed of fat lobules that are more pale than normal adipose tissue. Angiolipomas are characteristically reddish or orange/yellow reflecting an increased density of blood vessels. Definite gray or white streaks are present in spindle cell lipomas. Well-differentiated liposarcomas are often composed of large fat lobules and nodules that are pale yellow compared with normal white adipose tissue.

Histology

A mature white fat cell is spherical and measures up to 120 μm in diameter (36). The cytoplasm is compressed at the perimeter of the cell, and only a thin cell membrane is evident in hematoxylin–eosin (H&E)-stained sections. Reticulin and periodic acid–Schiff (PAS) stains highlight the adipocyte basement membrane (Fig. 6.3). The cytoplasm is displaced by a single lipid vacuole, and the cells are fairly uniform in size (Fig. 6.4). The nucleus, although ovoid, is thin and small with finely distributed chromatin; when seen in profile, a central clear vacuole may be seen within the nucleus. This phenomenon has been termed Lochkern, which is German for "hole in nucleus" (Fig. 6.4). Normal subcutaneous fat is finely divided into ill-defined lobules by thin bands of collagen (Fig. 6.5).

Ultrastructure

The ultrastructure of developing adipocytes has previously been discussed. In brief, preadipocytes are spindle-shaped with abundant ER and small spherical mitochondria (36). Lipid accumulates as small perinuclear inclusions that coalesce to form larger lipid droplets. The mitochondria become filamentous and the ER less prominent. Mature adipocytes have nuclei that are flattened against the cytoplasmic membrane by the lipid droplet. There is only a thin, tenuous rim of cytoplasm surrounding it. Pinocytotic vesicles are present in variable numbers but are increased following periods of starvation. Adjacent to the cell membrane are deposits of basement membrane. Capillaries are closely opposed to the adipocyte basement membrane. Nerves may be rarely identified adjacent to white fat cells, but are more commonly seen in intercellular collagenous septa.

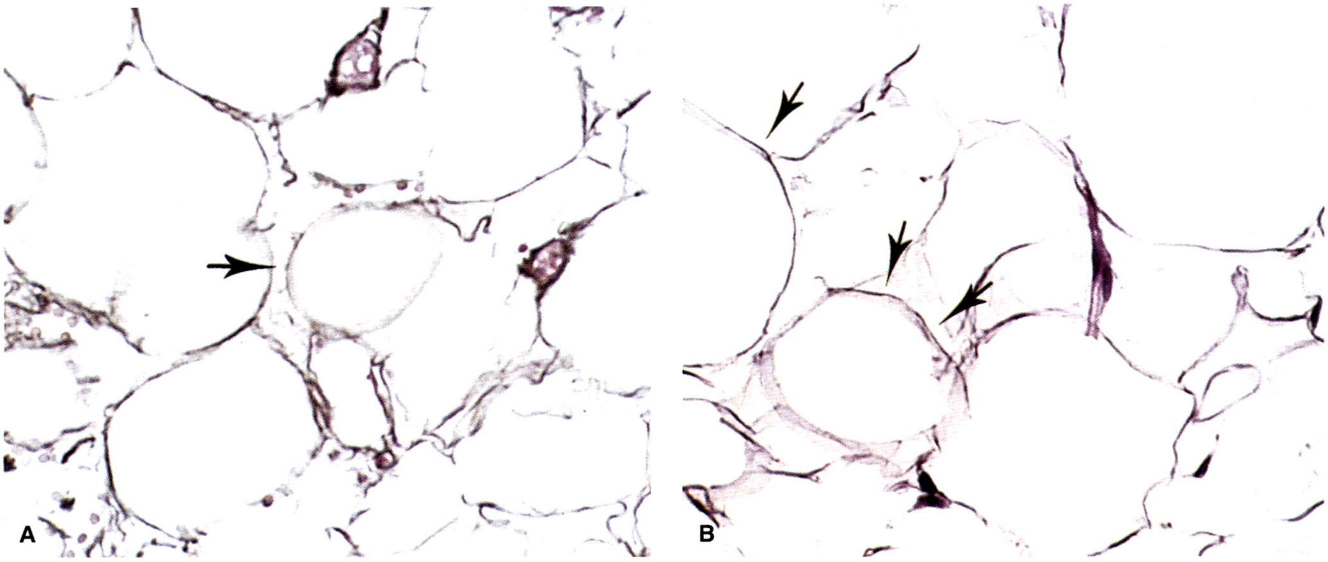

FIGURE 6.3 Normal adult adipocyte. **A:** On a reticulin stain, each adipocyte is outlined by reticulin (*arrow*), which is present outside the cytoplasm. **B:** The same is true on PAS stain, where the basement membrane is highlighted (*arrows*) and encompasses the pale residue of cytoplasm remaining after fixation and embedding.

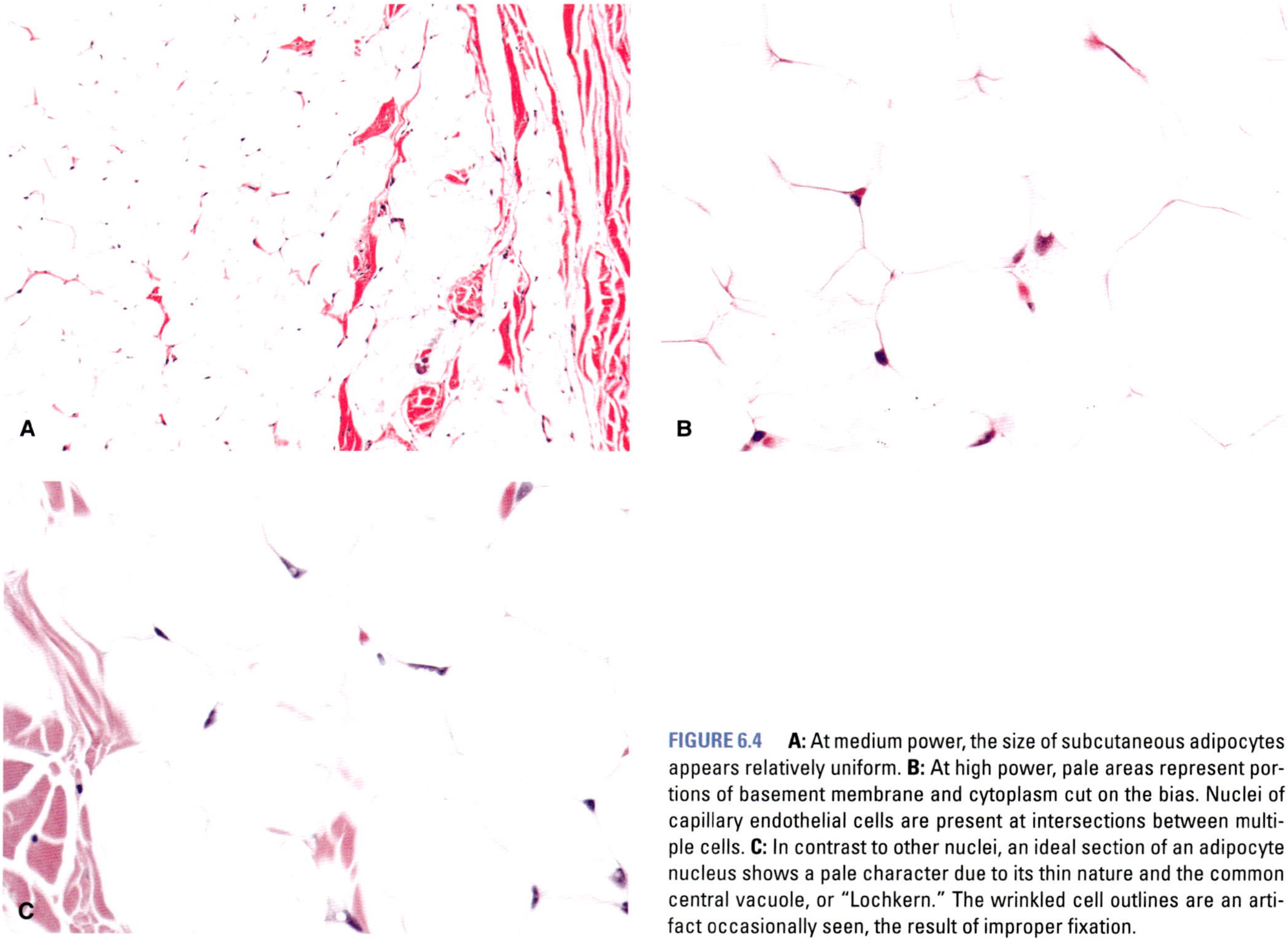

FIGURE 6.4 **A:** At medium power, the size of subcutaneous adipocytes appears relatively uniform. **B:** At high power, pale areas represent portions of basement membrane and cytoplasm cut on the bias. Nuclei of capillary endothelial cells are present at intersections between multiple cells. **C:** In contrast to other nuclei, an ideal section of an adipocyte nucleus shows a pale character due to its thin nature and the common central vacuole, or "Lochkern." The wrinkled cell outlines are an artifact occasionally seen, the result of improper fixation.

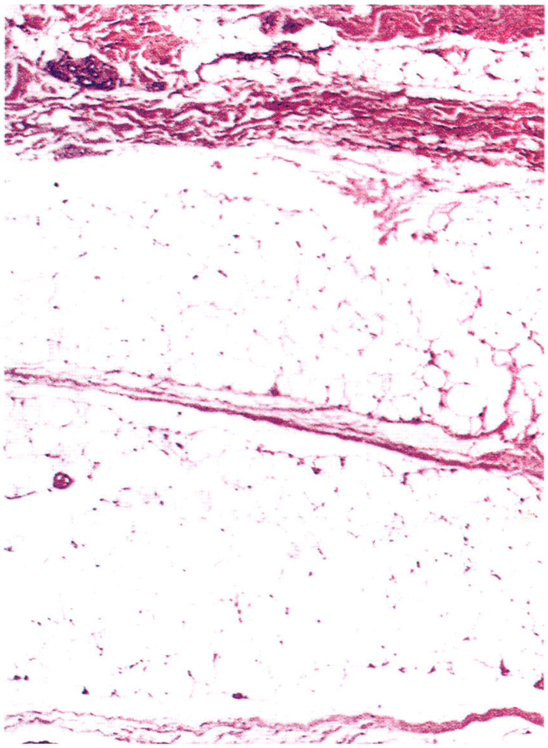

FIGURE 6.5 Adult subcutaneous fat lobule with associated microvasculature; note the thin and delicate fibrous tissue septa.

BROWN FAT

Prenatal Development

The development of brown adipose tissue has been studied in animal models. The brown adipocyte precursors are spindle cells closely related to a network of capillaries (37). As the cells and vessels proliferate, they are organized into lobules by connective tissue septa. As the cells accumulate lipid, they are initially unilocular. However, further lipid accumulation, results in multiple cytoplasmic lipid vacuoles. As in white fat, the close association between developing adipocytes and blood vessels has led some to speculate that adipocytes actually develop from endothelial cells. Although similar ultrastructural features are cited as supportive evidence of theory, more recent investigations have attributed these similarities to a common origin from undifferentiated mesenchymal cells. In fact, ultrastructural and biochemical studies have shown that unique features such as large mitochondria and a unique mitochondrial protein distinguish brown from white fat. Recent studies suggest that brown fat precursors may also be recruited directly from white adipose tissue and skeletal muscle (38).

Necropsy studies have identified lobules of developing brown fat in the human fetus and deposits are well established by the 5th month of gestation and the amount of brown fat increases in proportion to growth throughout life (39). The largest of these are from the posterior cervical, axillary, suprailiac, and perirenal regions. Those in the neck and axillae are closely associated with the major blood vessels of these regions in such a way that they extend along the course of the cervical blood vessels into the root of the neck. The suprailiac collections lie deep in the abdominal muscles and superficial to the peritoneum, investing the anterior abdominal wall to the diaphragm. Intermediate-sized brown fat pads occur in the interscapular paralateral trapezius and deltoid regions. Small collections are evident in the intercostal area. There are no differences in distribution between the sexes or among the races.

Postnatal Development

The presence of brown fat beyond the neonatal period has been debated, although autopsy studies (40), have demonstrated lobules of brown fat throughout life to the eighth decade. Brown fat is most prevalent in young children and, over the next several decades, gradually disappears from most sites. Deposits of brown fat can be seen in the interscapular region, around the neck vessels and muscles, around the structures of the mediastinum, and adjacent to the lung hila in children younger than 10 years. Intra-abdominal and retroperitoneal deposits can also be seen around the kidneys, pancreas, spleen, mesocolon, and omentum, as well as in the anterior abdominal wall in children of this age group. Although brown fat disappears from most areas with increasing age, it may persist around the kidneys, adrenals, and aorta and within the mediastinum and neck throughout adult life.

Function

The main function of brown adipose tissue is heat production. The maximal aerobic capacity per gram of brown adipose tissue is almost 10 times that of skeletal muscle (41). Small quantities of brown fat are capable of raising heat production by over 20% (42). Heat production is closely related to the active sympathetic innervation of brown fat and stimulation by norepinephrine. Release of norepinephrine results in the production of cyclic adenosine monophosphate and lipolysis to release free fatty acids (43). These undergo oxidation within the mitochondria to produce adenosine triphosphate. Brown fat mitochondria contain a unique protein, thermogenin, which uncouples the oxidation of fatty acids from generation of adenosine triphosphate and dissipates heat (44,45).

Brown fat performs several metabolic functions. It is activated by cold temperatures to produce heat in small rodents and hibernating animals resulting in nonshivering thermogenesis. Brown fat functions similarly in neonates, who are unable to alter the external environment in order to maintain body temperature. Brown fat accumulation and activation may also play a role in weight regulation. Overfed rats show a compensatory increase in brown fat activation and metabolic rate, thereby minimizing weight gain (46).

Many types of obesity in laboratory mice and rats are related to defective regulation of brown adipose tissue, including that seen in *ob/ob* mice (47). In contrast, exaggerated leanness may be associated with excessive responsiveness of brown adipose tissue to external factors, such as sympathetic stimulation, although its role in weight regulation, obesity, and thermal regulation in human adults remains controversial (48). Increased amounts of periadrenal brown fat in malnourished people at autopsy suggest a compensatory increase in nonshivering thermogenesis to maintain body temperature in those with diminished subcutaneous fat and cachexia (49).

Regulation

Unlike white fat, brown fat is highly innervated and regulated by sympathetic stimulation. Nerves enter each lobe and branch within the interlobular septa, running along the vessels to terminate on the fat cells (50). Brown fat cells have numerous β-1 and β-2 adrenoreceptors that regulate lipolysis and thermogenesis (43). The α-adrenoreceptors do not act directly in heat production. Rather, norepinephrine likely acts to increase the number and character of brown fat cells. Using continuous infusions of norepinephrine, Mory et al. (51) found that chronic sympathetic stimulation results in increased cellularity, protein content, and mitochondrial density in brown fat. Similarly, patients with pheochromocytomas have functional brown adipose tissue with biochemical features comparable to that of the brown adipose tissue of rodents (52).

Hormones also play a role in brown fat regulation, but it is minor in comparison to the sympathetic system. Thyroid hormone, although active in regulating metabolic rate, has little importance in diet-induced or nonshivering thermogenesis (43). Insulin stimulates glucose intake into brown adipose tissue. Both cortisol and gonadal steroid hormones inhibit thermogenesis, thus promoting energy conservation.

Histology

The term brown fat was applied to this tissue because of its characteristic gross appearance. The abundant vascularity and numerous mitochondria within the cells impart a characteristic reddish-brown color to the tissue. Brown fat is organized into lobules of adipocytes, capillaries, nerves, and connective tissue surrounded by a thin, fibrous capsule (53). The cells are polygonal in shape, with a mixture of multivacuolated and univacuolated cells (Fig. 6.6). The multivacuolated cell, characteristic of brown fat, has a highly granular cytoplasm with numerous lipid inclusions. Its granular appearance is due to the presence of numerous mitochondria necessary for thermogenesis. The nucleus is spherical and often centrally located, although large lipid inclusions may displace it toward the periphery of the cell or, rarely, to the extreme perimeter. Small nucleoli are common. Brown fat cells containing a solitary lipid vacuole are indistinguishable from the mature signet-ring cell–type white adipocytes, even though they are ultrastructurally different. On average, brown fat cells are smaller than white adipocytes, approximately 25 to 40 μm. In animals that hibernate, marked seasonal variation in cell size has been noted. Both exposures to cold and starvation result in lipid depletion, causing reduction in cell size and wrinkling of the cell membrane.

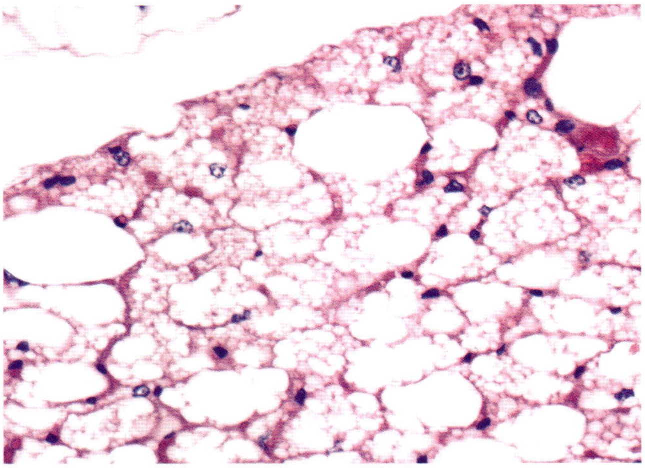

FIGURE 6.6 Normal adult brown fat. Nearly all cells have centrally placed nuclei and multivacuolated cytoplasm. Rare cells (*top left*) are nonvacuolated. An arborizing thin capillary network is noted.

Brown adipose cells are surrounded by a network of collagen fibers that contain numerous minute nerve axons and blood vessels. Nonmyelinated axons terminate on the fat cells, providing an avenue for direct sympathetic regulation. The vascularity is quite prominent with numerous capillaries coursing between the adipocytes. The vascularity of brown fat is substantially greater than that of white fat (53).

Beige Fat

Cells that resemble brown adipocytes may appear after thermogenic stimuli at anatomical sites corresponding to white adipose tissue, in a process called "browning" of white adipose tissue. This third class of fat cells are closer to white adipocytes in terms of cell lineage and are often called "inducible, beige, or brite adipocytes" (54,55). In combating the obesity epidemic, efforts to increase the expenditure of thermogenic energy by brown fat revealed that brown adipocytes were derived from different precursors than white adipocytes. Brown adipocytes originate from a precursor shared with skeletal muscle that expresses Myf5-Cre, while white adipocytes originate from Myf5-negative precursors (55). Beige cells appear in white adipose tissue in response to cold or beta-adrenergic agonists, especially in certain locations like inguinal and retroperitoneal regions, possibly by transdifferentiation. Suffice it to say that these discoveries may lead to better understanding of obesity and perhaps a therapy based on biologic pathways.

Stem Cells

Adipose tissue contains not only mature adult cells but also stem cells. These adipose-derived stem cells are a common source of research and have been shown to generate pericytes, chondrocytes, endothelial cells, and Schwann cells, and have been used for repair of ischemic muscle (56,57).

HISTOCHEMISTRY

Enzyme Histochemistry

Enzyme histochemistry within developing adipocytes is related to the stage of adipocyte differentiation; enzymatic differentiation of adipocytes may even precede morphologic differentiation (58,59). In regions destined to become adipose tissue, undifferentiated morphology is initially present without a capillary bed and without enzymatic capacity. These preadipocytes lack significant lipid or a basal lamina and exhibit developing enzymatic activity, but they cannot release fat due to the absence of active lipase (esterase). In mature lobules, adipocytes contain lipid, a basal lamina, and a well-developed capillary bed; they exhibit high metabolic activity, including nicotinamide adenine dinucleotide (NADH)-tetrazolium reductase, nicotinamide adenine dinucleotide phosphate (NADPH)-tetrazolium reductase, and glucose-6-phosphate dehydrogenase (G6PDH). NADP-dependent malate dehydrogenase activity is acquired only by late-stage adipocytes. This enzymatic differentiation occurs before the cell adopts the fully rounded morphology of a mature adipocyte (58).

Lipoprotein lipase is found at high concentration in fatty tissues. It is involved in the transport of serum triglycerides into adipocytes in the form of fatty acids. However, this enzyme can also be in skeletal muscle (60,61), cardiac muscle, and other tissues (62), where it may be localized to endothelial cells. Concentration in fat is directly related to the serum insulin concentration.

Lipid Histochemistry

Lipids are lost in formaldehyde after prolonged fixation, and thus, cases to be tested using frozen cryostat sections of fixed material should be obtained as soon as possible. Oil red O and Sudan IV stains are the most commonly utilized histochemical stains (58,63,64). Of these, oil red O gives the more intense stain and is more rapid to perform, whereas Sudan black B may stain nonlipid substances nonspecifically. As a rule, neutral fats are detected using these fat stains. However, a differential staining pattern between neutral fats and fatty acid components and phospholipids can be obtained with the Nile blue sulfate stain (65). The latter stains neutral fat pink to red, whereas fatty acids and phospholipids stain bluish. The lipid composition of fatty tissues may also be investigated using new techniques such as the hot-stage polarizing-light microscopic method (66).

The lipid composition of white adipose tissue is 99% triglycerides in the form of neutral fat and less than 1% in the form of phospholipid, cholesterol, and fatty acids (57). Less-differentiated adipocytes, including those found in neoplasms, demonstrate a shift away from neutral fat to phospholipids and cholesterol (65). Unfortunately, lipid stains have little use in the everyday examination of adipocyte lesions. The droplets seen on the stains may represent nonspecific staining, and various other mesenchymal lesions may contain lipid (67).

IMMUNOHISTOCHEMISTRY

There is no commercially available specific immunohistochemical marker for adipose tissue but this is not usually a problem as adipocytes are typically easy to recognize based on routine histologic assessment. However, adipocytes are positive for S100 protein (68,69). S100 protein staining may be altered during lipolysis, suggesting that this protein may act as a carrier for free fatty acids (69). The S100 protein is a highly acidic calcium-binding protein of molecular weight 21,000. It consists of two polypeptide chains (α and β) and may occur as three combinations of dimers: S100a (a, β), S100b (b, β), or S100ao (a, α) (70). When fat cells have been analyzed, they have been shown to contain only S100b (β form), similar to Schwann cells (70,71). In the routine practice of immunohistochemistry, adipose tissue reacts in a variable fashion (Fig. 6.7) (72). Lipomas and liposarcomas are also variably positive with S100 (73,74).

Obesity

Obesity is a complex chronic disease caused by an interplay of genetic predisposition and lifestyle factors (e.g., sedentary behaviors and energy dense diets), leading to neuroendocrine

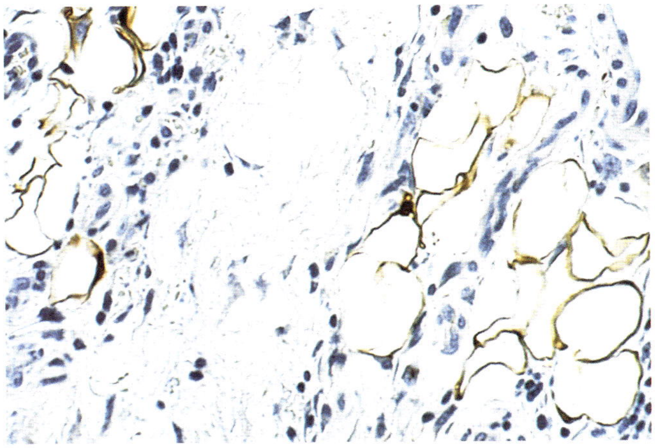

FIGURE 6.7 S100 immunohistochemistry. Reactivity is seen both in the nuclei and in the cytoplasm surrounding the lipid droplets. Such S100-positive results appear to vary considerably from case to case, probably reflecting fixation differences.

adaptations that promote excess adiposity (75). Resultant pathophysiological changes include dysregulation of appetite and satiety signaling (e.g., leptin melanocortin pathways) and adipose tissue dysfunction with chronic inflammation, leading to insulin resistance and metabolic derangements (76). Its prevalence has risen to pandemic levels such that more than one billion people were affected worldwide in 2022. The prevalence of obesity among adults is more than double that of 1990 w and rates have increased even more among pediatric patients. The obesity epidemic has led to a public health crisis with increased rates of type 2 diabetes mellitus, cardiovascular disease, and some malignancies, contributing to increased morbidity and premature mortality (77). Lifestyle modification (e.g., diet, exercise, behavioral therapy) remains the cornerstone of management, typically reducing body mass by 5 to 10 percent which is enough to improve blood pressure and glycemic control (78), but physiological compensations may lead to weight regain, necessitating long term, multimodal treatment (76). Recent advances in pharmacotherapy with semiglutide, a GLP 1 receptor agonist, and tirzapatide, a dual GLP 1/GIP agonist, have demonstrated unprecedented efficacy, inducing approximately 15 to 20 percent weight reduction (79,80). Metabolic (bariatric) surgery is also an effective intervention for severe obesity, yielding approximately 25 to 30 percent sustained weight loss and improvement of comorbid conditions (81). Current policy efforts emphasize comprehensive public health strategies from early life prevention to improved food environments and regulating the food industry to address the root causes of obesity (82).

ADIPOCYTE LESIONS

Terminology

Adipocyte hypertrophy is a recognized phenomenon and is found in obesity. Enlarged fat cells can also be identified in lipomas and liposarcomas, in which case cells appear to have three or four times the normal diameter (>300 μm). Increased numbers of adipocytes can also be seen, especially in obese patients. Mature adipocytes are incapable of regeneration and, thus, new fat cells are added through in situ mesenchymal cell differentiation recruited from primitive perivascular cells. Atrophy of adipocytes may be seen in malnutrition, starvation, or as the effect of chemotherapy. The appearance of mature fat cells as small foci in unusual places is termed metaplasia and is discussed in a later section. Localized new growths of adipocytes are presumably clonal entities and represent neoplasms.

Degeneration

Sclerema adiposum neonatorum is a condition characterized by widespread fat necrosis. The subcutaneous fat is grossly abnormal and contains rubbery plaques due to fat necrosis and degenerative individual fat cells with intracellular needle-shaped crystals (83). This fat crystallization is brown and can be highlighted by polarization. Such crystals may also be identified in up to 30% of stillbirths as a general degeneration following intrauterine demise (83). Neu–Laxova syndrome is another condition that results from a defect in lipid metabolism. It causes a lard-like appearance to the adipose tissue and is lethal.

Atrophy

Fat lobules undergo extensive atrophy during starvation or malnutrition and these changes are particularly noticeable in the subcutaneous region and omentum. Individual fat cells are reduced in size and fat content; those without much lipid appear rounded or epithelioid (84). In the extreme, lobules of these epithelioid cells can simulate tumor nodules histologically (Fig. 6.8). The cytoplasm is variable in amount and is eosinophilic or granular with or without small lipid vacuoles of differing size, depending on the severity of the malnutrition. Some cells have a multivacuolated appearance. Individual cells are separated by a homogeneous eosinophilic or amphophilic substance that may represent an extract of serum or proteoglycan matrix (85). As part of this involution process, lipofuscin is deposited within the shrinking cells. Importantly, each lobule retains its overall oval shape, although markedly reduced in size and considerably separated from other lobules. In extreme cachexia, only streaks of tissue remain.

Nearly identical changes can also be seen in the white adipose tissue of fasted animals. As the cells gradually lose their lipid, the single lipid droplet breaks up into multiple vacuoles. Gradually, all lipid disappears, such that the cells become small and ovoid, sometimes measuring only 15 μm in diameter (86). Pericellular collagen is expanded such that adipocytes appear as clusters of mesenchymal cells in fibrous stroma. Similarly, in cachexia, the fat cells are reduced in volume but not in number (87). Ultrastructurally, multiple pinocytotic vesicles are seen clustered along the entire cell membrane (53). Lipid is not seen within these vesicles, and their significance is unknown.

Chemotherapy causes changes referred to as serous atrophy or gelatinous transformation in the bone marrow (88,89). In this situation, most of the adipocytes are destroyed, leaving scattered adipocytes of varying size. The interstitial compartment is composed of the same eosinophilic myxoid substance described previously. Droplets of lipid are also found and, upon regeneration, may appear as foci of lipogranulomas.

Although the microscopic features of the starvation on human brown fat have not been described, animals maintained on a dextrose–thiamine diet are known to show distinct morphologic changes in brown fat (36). The mitochondria are disrupted, and large, irregular electron-dense inclusions are seen within the mitochondrial matrix. The cristae may assume a mosaic pattern with compartmentalization of the material. These cells revert to normal after 24 hours of a normal diet. Similar changes in white fat mitochondria have not been seen with starvation, suggesting that the active mitochondria of brown fat are particularly labile and sensitive to dietary changes.

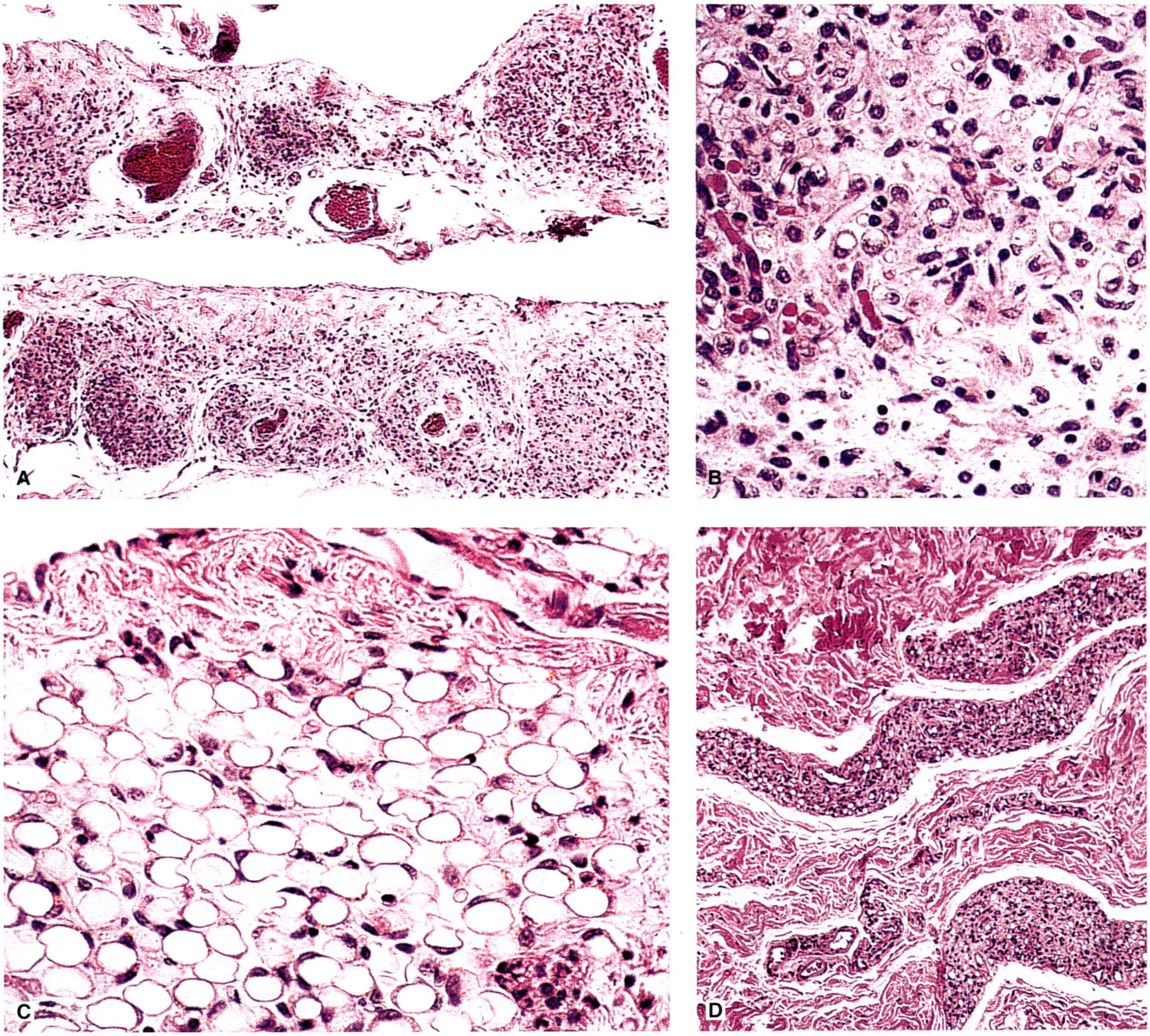

FIGURE 6.8 **A**: The extreme atrophy seen here in the omentum of a patient with anorexia nervosa may mimic tumor deposits. **B**: At high power, shrunken eosinophilic cells are seen with occasional vacuoles and lipofuscin pigment. **C**: In less severe starvation, these omental adipocytes are well recognized, although much smaller than normal size; again, note the presence of pigment. **D**: In the skin, severe cachexia secondary to a cancer resulted in marked involution of the cutaneous fat lobules, which appear only as elongated streaks.

Cellulite

The term "cellulite" is applied to the external skin when it exhibits linear depressed streaks (mattress phenomenon) or frank dimpling. Cellulite is typically found on the thigh and buttocks and is more common in females; it can be divided into incipient cellulite and full-blown cellulite. The former results from an uneven undersurface of the dermal–hypodermal interface, with fibrous tissue surrounding the protruding papillae adiposa; vertical fibrous strands of uneven thickness divide the hypodermal fat (90). In contrast, full-blown cellulite consists of a delicate meshwork of collagen fibers produced by increased hypodermal pressure of fat accumulation and increasing fat volume. Scattered CD34+ fibroblasts are seen in both forms of cellulite. Unlike women, men have a smoother, strand-free dermal interface in the thigh and buttock areas (90).

Ischemia

Ischemic injury to adipocytes elicits minimal alterations that may be difficult to distinguish from normal fat. Ischemia-related are most commonly encountered in the subcutaneous fat of legs removed for atherosclerotic vascular disease. They consist of accentuation of the lobular architecture by thickened fibrous septa that have a myxoid in quality. The

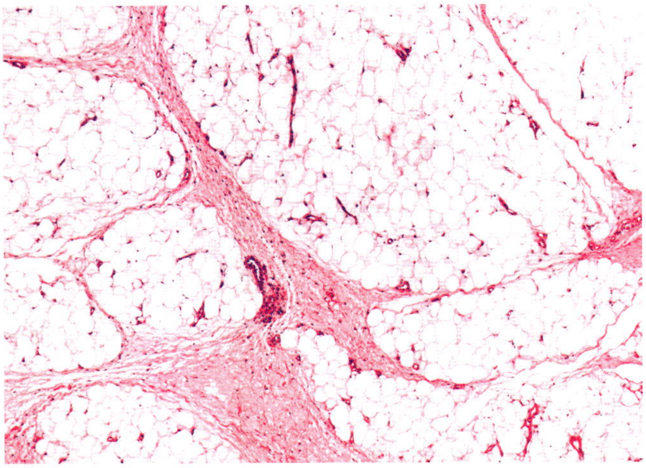

FIGURE 6.9 Accentuated fat lobules in ischemia of the lower extremity. Loose myxoid connective tissue widens the septa between lobules; edema and a mild inflammatory infiltrate are present.

septa are edematous and contain scattered inflammatory cells (Fig. 6.9).

Metaplasia

Adipocytic metaplasia is most frequently encountered in the cardiac valves and it is usually calcified (Fig. 6.10). The emergence of mature adipose tissue seems to parallel the appearance of osteoblasts forming bone within the calcific deposits. Once adipose tissue is present, bone marrow precursors may become resident, presumably from circulating cells, and cause hematopoiesis. Metaplasia may be encountered in any structure that undergoes dystrophic calcification and ossification, including calcified large vessels, laryngeal cartilage, and small, ossified bronchioles.

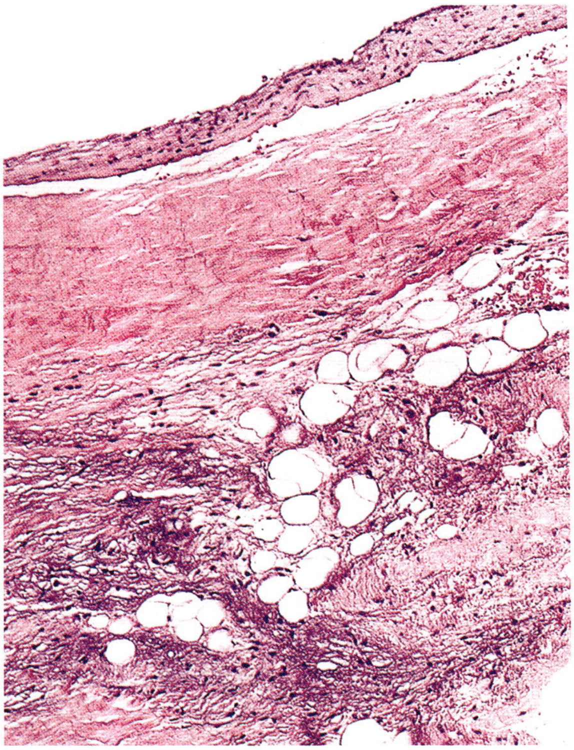

FIGURE 6.10 Fatty metaplasia of cardiac valve. Mature adipocytes are found in a myxoid background but are more commonly seen in association with calcification or ossification.

ADIPOCYTES IN ORGANS

Fatty Infiltration

Fatty infiltration is defined as the presence of mature adipose tissue in sites not normally containing fat. This is a disorder or condition relating to adipocyte cell growth and, therefore, the term fatty degeneration is a misnomer and incorrect. In some situations, such as within extremity muscle groups, the process of fatty infiltration is often related to atrophy of the involved site (91). The association between fatty infiltration and atrophy or involution is also noted in other organs (thymus) (92), bone marrow (93), and kidney (94) and apparently signifies the propensity for adipocytes to fill a void left by atrophic processes (85). Whatever the stimulus may be, the adipocytes probably arise from pluripotent mesenchymal cells adjacent to blood vessels (85). The reversal of this relationship is found in the parathyroid gland, where there is an inverse relationship between parenchymal cells and adipocytes, to the point where no adipocytes are present in complete parathyroid hyperplasia.

Nonatrophic organs can also accumulate fat cells (lipomatosis), and the classic examples are the heart and the pancreas (95). In these locations, no parenchymal damage is discerned (84). In the case of the pancreas, normal parenchymal histology and function are present even though the pancreas may be nearly invisible grossly (95,96). This type of pancreatic lipomatosis is correlated with age and obesity and also occurs in patients with diabetes mellitus (96). Pancreatic atrophy with resultant lipomatosis also exists as a rare condition known as Shwachman syndrome (96). Fatty infiltration of the heart is most often an innocuous condition with no effect on the myocardial fiber or cardiac function (95). However, severe adiposity can rarely result in cardiac rupture (84). Another clinically important lesion, lipomatous hypertrophy of the interatrial septum (97,98), represents a focal enlargement that may cause sudden death, arrhythmias, or congestive failure (99,100). Of course, the occasional appearance of fat in endocardial biopsies in no way indicates cardiac perforation (101).

Isolated fat cells can be found within lymph nodes in childhood, but enlarged nodes with prominent fatty infiltration mainly occur in adults, particularly in obesity (94). Common in the abdomen and retroperitoneum, these fatty lymph nodes can be mistaken for lipomas (102) or a neoplastic process (103). Lipomas rarely occur in the liver (104), but they should not be confused with the hepatic pseudolipoma, which simply represents a bulge on the liver surface resulting from a previously detached appendix epiploica (105). In the mouth, fat is one of the components contributing to macroglossia in certain conditions (106).

Ectopic Adipose Tissue

Ectopic fat may occur in the dermis, where it causes a pedunculated appearance; this has been termed nevus lipomatosis superficialis.

Adipocytic Tumors

Adipocytic tumors are among the most common mesenchymal tumors with lipomas being the most common among adults. These tumors show a wide histologic spectrum, often with variable amounts of adipocytic differentiation, and may present diagnostic challenges because they may closely simulate the features of normal fat. Herein we will focus on lipoma and their atypical/malignant counterpart, atypical lipomatous tumor (ALT)/well-differentiated liposarcoma inasmuch as they can simulate nonneoplastic fat.

Lipoma

Lipoma is defined as a superficial or deep circumscribed and expansile benign neoplasm composed of mature adipose tissue, which is commonly (but need not be) encapsulated. Such a definition emphasizes its well-differentiated and clonal nature and serves to distinguish most lipomas from normal fat.

Clinically, most lipomas are subcutaneous tumors typically in the middle-aged to elderly patients. Males and females are equally affected, and there are no racial differences. Most tumors are located on the trunk or upper extremities; if other sites are encountered, consideration should be given to one of the lipoma subtypes (e.g., forearm for angiolipoma, upper back/posterior neck for spindle cell/pleomorphic lipoma). When one views normal fat histologically, the size of fat cells appears to vary somewhat due to the sectioning plane; however, the variation is relatively small (80 to 120 μm) (Fig. 6.4). There is a tendency for cell size to vary more widely in lipomas. Practically, this means that a medium-power view will often disclose a two- to fivefold size range. Normal fat has a netlike structure of fibrous tissue, wherein such dispersed fibrous bands or septa dissect the adipose tissue randomly. The fibrous tissue is thicker in quality in bodily regions exposed to pressure, such as the hands, feet, and buttocks (84). This netlike fibrous tissue arrangement is recapitulated within lipomas (Fig. 6.12), particularly at the periphery where small lobules are often found. A delicate reticulin network is also present in lipomas, contributed to by the basement membranes of both lipocytes and capillaries; each lipocyte is completely encircled by reticulin in a manner like normal fat cells (Fig. 6.3). Lipomas may show reactive changes or fat necrosis but should lack cytologic atypia (Figs. 6.11 and 6.12). The ultrastructure of lipoma recapitulates that of its normal counterpart (107).

Intramuscular lipomas are composed of mature adipocytes associated with muscle fibers that are widely dispersed throughout the lesion (108) (Fig. 6.13). Like their superficial counterparts, fat necrosis may be present.

Atypical Lipomatous Tumor (ALT)/Well-Differentiated Liposarcoma

ALT/well-differentiated liposarcoma accounts for 40% to 45% of all liposarcomas. The term atypical lipomatous tumor is used for tumors that occur in anatomical locations suitable for complete surgical removal, while the term well-differentiated liposarcoma is applied to tumors originating in deep-seated, central body sites like the retroperitoneum. ALT/well-differentiated liposarcoma presents in middle-aged adults with a peak incidence between the fourth and fifth decades of life. Most of these tumors affect the deep soft tissues of proximal extremities, retroperitoneum, and paratesticular regions. ALT/well-differentiated liposarcoma typically manifests as a painless, deep-seated mass that can gradually grow to a substantial size, especially in the retroperitoneum.

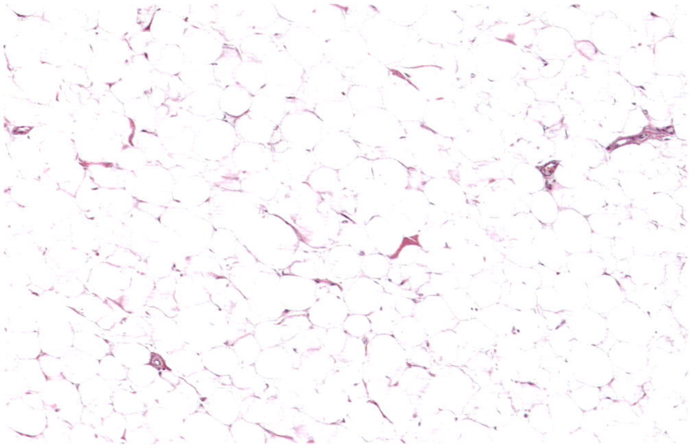

FIGURE 6.11 Lipoma. Lipoma is characterized by uniform proliferation of mature adipocytes with minimal variation in cell size and shape. Note the complete absence of cytologic atypia in the adipocytes.

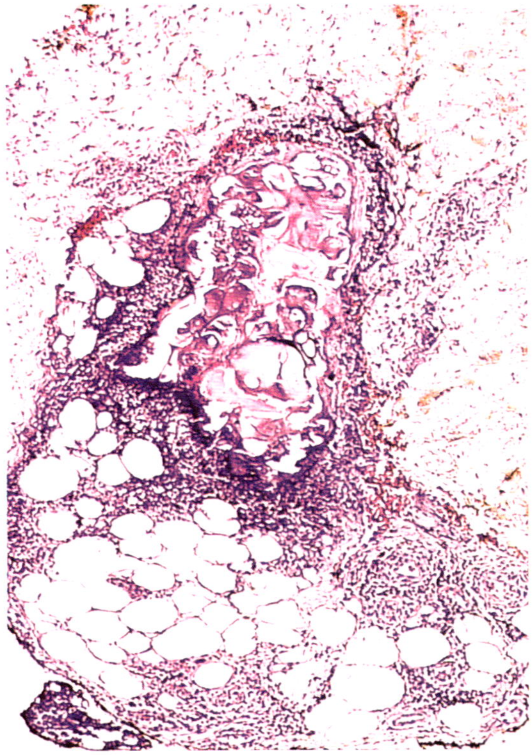

FIGURE 6.12 Lipoma with fat necrosis. Cell size and shape exhibit considerable variation due to fat necrosis. Macrophage nuclei and vacuolated cytoplasm are observed interspersed among mature fat cells.

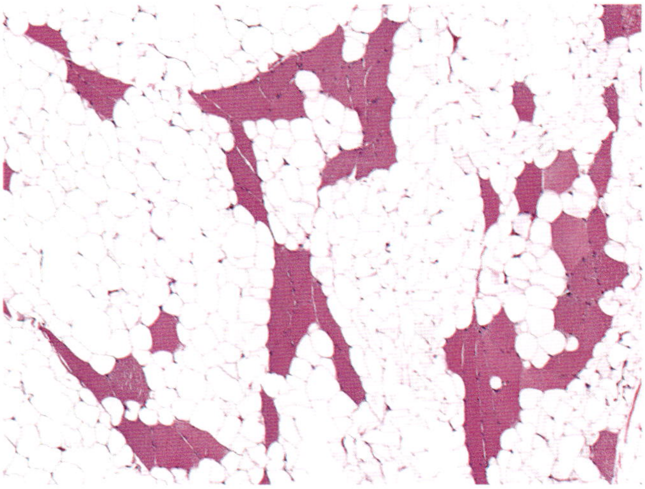

FIGURE 6.13 Intramuscular lipoma. Striated muscle cells are separated by a mature adipocytic proliferation.

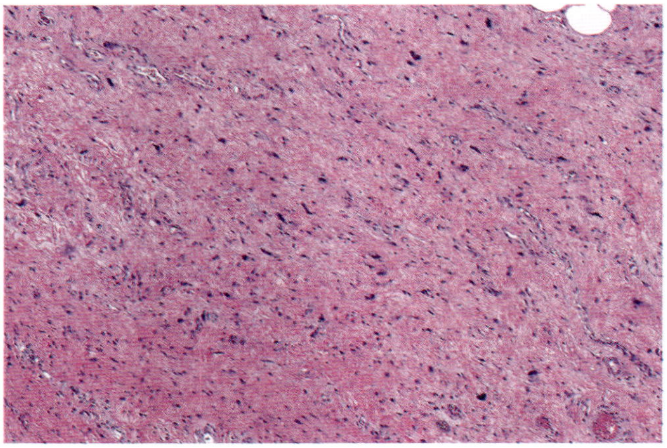

FIGURE 6.15 Atypical lipomatous tumor/well-differentiated liposarcoma, sclerosing variant. This variant is distinguished by the presence of highly atypical stromal cells within a collagenous background.

Imaging often reveals fat density masses with thick septa (>2 mm). ALT/well-differentiated liposarcomas are large, multilobulated lesions that range in color from yellow to ivory. They are classified into three subtypes: lipoma-like, sclerosing, and inflammatory. Of these, lipoma-like tumors are most likely to simulate normal fat. They are characterized by a predominant mature adipocytic proliferation with a variable number of atypical hyperchromatic stromal cell, often located in fibrous septa and blood vessel walls (Fig. 6.14). These atypical cells have smudgy, enlarged, and oftentimes multilobed nuclei, that should be apparent at scanning magnification. Unfortunately, diagnostic cells may be sparse, and careful sampling may be necessary to identify them. Sclerosing forms of ALT/well-differentiated liposarcoma are commonly observed in the groin and retroperitoneum, marked by atypical hyperchromatic stromal cells set in a hyalinized, fibrous, or collagenous stroma (Fig. 6.15). Although adipocytes may be present, they are often rare in this subtype. The inflammatory form features a dense chronic inflammatory infiltrate in a background of otherwise lipoma-like or sclerosing form of ALT/well-differentiated liposarcoma (Fig. 6.16). These tumors are frequently misdiagnosed as hematopoietic neoplasms, idiopathic retroperitoneal fibrosis/sclerosing mesenteritis, or nonneoplastic inflammatory processes due to the prominent inflammatory background. Herniated orbital fat can simulate ALT. This uncommon phenomenon arises from the anterior displacement of intraconal fat caused by the dehiscence of the Tenon capsule. The lesion is characterized by a mixture of mature adipocytes, delicate fibrovascular septae, Lochkern cells, and floret-like giant cells. Fluorescence in situ studies for *MDM2* amplification can help distinguish this entity from ALT. Fat atrophy can induce changes that simulate fatty neoplasia. The reduction in intracellular lipid content leads to fat cells taking on an epithelioid shape with a more prominent nucleus, creating a superficial resemblance to lipoblasts. Useful diagnostic clues include uniformly sized adipocytes, often with lipofuscin pigment in the cytoplasm, and the overall preservation of normal lobules.

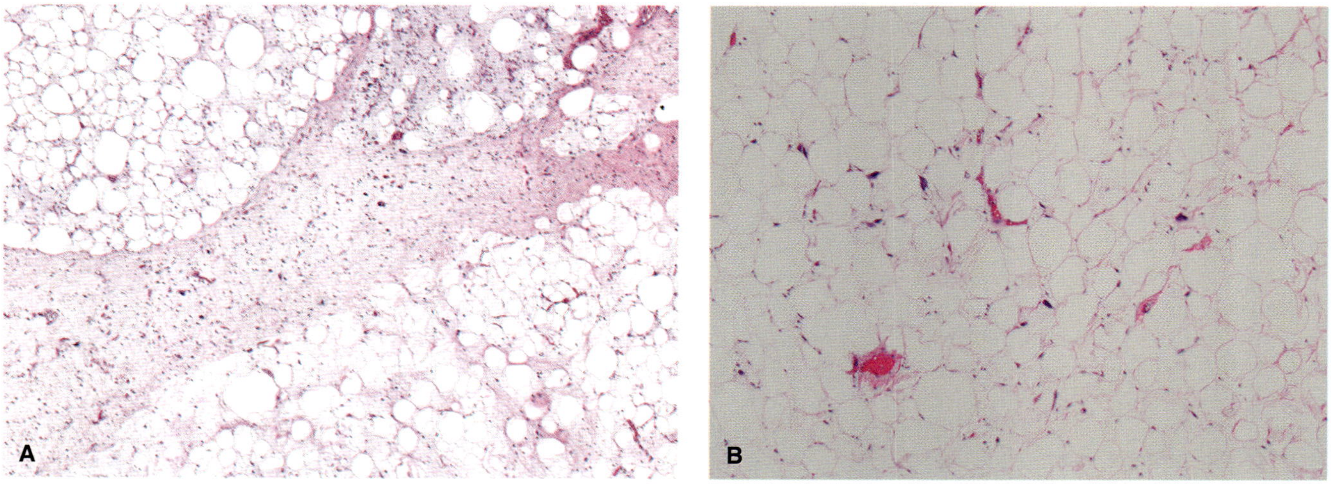

FIGURE 6.14 **A:** Atypical lipomatous tumor/well-differentiated liposarcoma, lipoma-like variant. These tumors frequently exhibit fibrous septa. **B:** Nuclear atypia is observed in both the fat cells and spindle stromal cells.

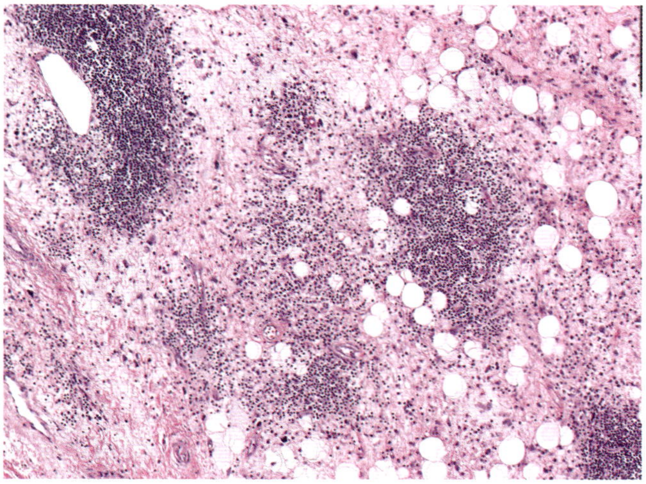

FIGURE 6.16 Atypical lipomatous tumor/well-differentiated liposarcoma, inflammatory variant. When the inflammatory component dominates, recognizing hyperchromatic, atypical stromal cells becomes the most valuable diagnostic clue.

MIMICS OF FAT CELLS

Mature Fat Cells

Pathologists visualize adipocytes as clear cells or "white holes" on routine sections. Therefore, other cells or processes with this white hole appearance may be confused with them. Some lesions are fairly obvious—like the vacuolated lymphadenopathy of lymphangiogram effect. Dilated superficial lymphatic vessels if closely clustered, as they may be in a nasal polyp, remind one of adipocytes at medium power. The submucosal cystic spaces of pneumatosis cystoides intestinalis (109) are composed of gas with a lining of inflammatory cells, macrophages, and giant cells. Cysts very similar in histology are occasionally noted within ovarian teratomas; here, it is probably a reaction to internal rupture. Likewise, small gaseous cysts without any lining in the intestinal mucosa mimic lipocytes in an entity termed pseudolipomatosis (109). Similar clear but artifactual vacuoles in the skin have been called "pseudolipomatosis cutis" (110).

REFERENCES

1. Poissonnet CM, Burdi AR, Bookstein FL. Growth and development of human adipose tissue during early gestation. *Early Hum Dev.* 1983;8:1–11.
2. Poissonnet CM, Burdi AR, Garn SM. The chronology of adipose tissue appearance and distribution in the human fetus. *Early Hum Dev.* 1984;10:1–11.
3. Robinson DS. In: Florkin M, Stotz EH, eds. *Comparative Biochemistry*. Elsevier; 1970; 18: 51–116.
4. Hausman GJ, Campion DR, Martin RJ. Search for the adipocyte precursor cell and factors that promote its differentiation. *J Lipid Res.* 1980;21:657–670.
5. Napolitano L. The differentiation of white adipose cells. An electron microscope study. *J Cell Biol.* 1963;18:663–679.
6. Crozat A, Aman P, Mandahl N, Ron D. Fusion of CHOP to a novel RNA-binding protein in human myxoid liposarcoma. *Nature.* 1993;363:640–644.
7. Ladanyi M. The emerging molecular genetics of sarcoma translocations. *Diagn Mol Pathol.* 1995;4:162–173.
8. Aman P, Ron D, Mandahl N, et al. Rearrangement of the transcription factor gene CHOP in myxoid liposarcomas with t(12;16)(q13;p11). *Genes Chromosomes Cancer.* 1992;5:278–285.
9. Martin RJ, Ramsay T, Hausman GJ. Adipocyte development. *Pediatr Ann.* 1984;13:448–453.
10. Poissonnet CM, LaVelle M, Burdi AR. Growth and development of adipose tissue. *J Pediatr.* 1988;113(1 Pt 1):1–9.
11. Sjostrom L, William-Olsson T. Prospective studies on adipose tissue development in man. *Int J Obes.* 1981;5:597–604.
12. Nakao N, Nakayama T, Yahata T, et al. Adipose tissue derived mesenchymal stem cells facilitate hematopoiesis in vitro and in vivo. *Am J Pathol* 2010;177:547–554.
13. Björntorp P. Adipocyte precursor cells. In: Björntorp P, Cairella M, Howard AN, eds. *Recent Advances in Obesity Research III. Proceedings of the 3rd International Congress on Obesity*. John Libbey; 1981:58–69.
14. Witkowska-Zimny M, Walenko K. Stem cells from adipose tissue. *Cell Mol Biol Lett.* 2011;16:236–257.
15. Kolata G. Why do people get fat? *Science.* 1985;227:1327–1328.
16. Hirsch J, Fried SK, Edens NK, Leibel RL. The fat cell. *Med Clin North Am.* 1989;73:83–96.
17. Cinti S, Frederich RC, Zingaretti MC, De Matteis R, Flier JS, Lowell BB. Immunohistochemical localization of leptin and uncoupling protein in white and brown adipose tissue. *Endocrinology.* 1997;138:797–804.
18. Hukshorn CJ, Saris WHM. Leptin and energy expenditure. *Curr Opin Clin Nutr Metab Care.* 2004;7:629–633.
19. Jequier E. Leptin signaling, adiposity, and energy balance. *Ann N Y Acad Sci.* 2002;967:379–388.
20. Frühbeck G, Gómez-Ambrosi J, Muruzábal FJ, Burrell MA. The adipocyte: a model for integration of endocrine and metabolic signaling in energy metabolism regulation. *Am J Physiol Endocrinol Metab.* 2001;280:E827–E847.
21. Wisse BE. The inflammatory syndrome: the role of adipose tissue cytokines in metabolic disorders linked to obesity. *J Am Soc Nephrol.* 2004;15:2792–2800.
22. Choy LN, Rosen BS, Spiegelman BM. Adipsin and an endogenous pathway of complement from adipose cells. *J Biol Chem.* 1992;267:12736–12741.
23. Birgel M, Gottschling-Zeller H, RöHrig K, Hauner H. Role of cytokines in the regulation of plasminogen activator inhibitor-1 expression and secretion in newly differentiated subcutaneous human adipocytes. *Arterioscler Thromb Vasc Biol.* 2000;20:1682–1687.
24. De Pauw A, Tejerina S, Raes M, Keijer J, Arnould T. Mitochondrial (dys)function in adipocyte (de)differentiation and systemic metabolic alterations. *Am J Pathol.* 2009;175:927–939.
25. Rebuffé-Scrive M, Enk L, Crona N, et al. Fat cell metabolism in different regions in women. Effect of menstrual cycle, pregnancy, and lactation. *J Clin Invest.* 1985;75:1973–1976.
26. Fried SK, Kral JG. Sex differences in regional distribution of fat cell size and lipoprotein lipase activity in morbidly obese patients. *Int J Obes.* 1987;11:129–140.

27. Björntorp P. The regulation of adipose tissue distribution in humans. *Int J Obes Relat Metab Disord*. 1996;20:291–302.
28. Kopelman PG. Effects of obesity on fat topography: metabolic and endocrine determinants. In: Kopelman PG, Stock MJ, eds. *Clinical Obesity*. Blackwell Science; 1998:158–175.
29. Björntorp P. Fat cell distribution and metabolism. *Ann N Y Acad Sci*. 1987;499:66–72.
30. Hutley L, Shurety W, Newell F. et al. Fibroblast growth factor 1: a key regulator of human adipogenesis. *Diabetes*. 2004;53:3097–3106.
31. Hube F, Hauner H. The role of TNF-alpha in human adipose tissue: prevention of weight gain at the expense of insulin resistance? *Horm Metab Res*. 1999;31:626–631.
32. Strassmann G, Fong M, Kenney JS, Jacob CO. Evidence for the involvement of interleukin 6 in experimental cancer cachexia. *J Clin Invest*. 1992;89:1681–1684.
33. Stephens JM, Pekala PH. Transcriptional repression of GLUT4 and C/EBP genes in 3T3-L1 adipocytes by tumor necrosis factor-alpha. *J Biol Chem*. 1991;266:21839–21845.
34. Hellman B, Hellerstrom C. Cell renewal in the white and brown fat of the rat. *Acta Pathol Microbiol Scand*. 1961;51:347–353.
35. Hollenberg CH, Vost A. Regulation of DNA synthesis in fat cells and stromal elements from rat adipose tissue. *J Clin Invest*. 1969;47:2485–2498.
36. Napolitano L. The fine structure of adipose tissues. In: Reynold AE, Cahill GF, eds. *Handbook of Physiology. Section 5: Adipose Tissue*. American Physical Society; 1965:109–123.
37. Nnodim JO. Development of adipose tissue. *Anat Rec*. 1987; 219:331–337.
38. Schulz TJ, Huang TL, Tran TT, et al. Identification of inducible brown adipocyte progenitors residing in skeletal muscle and white fat. *Proc Natl Acad Sci U S A* 2011;108(1): 143–148.
39. Merklin RJ. Growth and distribution of human fetal brown fat. *Anat Rec*. 1974;178:637–646.
40. Heaton JM. The distribution of brown adipose tissue in the human. *J Anat*. 1972;112(Pt 1):35–39.
41. Girardier L. Brown fat: an energy dissipating tissue. In: Girardier L, Stock MJ, eds. *Mammalian Thermogenesis*. Chapman and Hall; 1983:50–98.
42. Rothwell NJ, Stock MJ. Brown adipose tissue. In: Baker PF, ed. *Recent Advances in Physiology*. Vol. 10. Churchill Livingstone; 1984:349–384.
43. Rothwell NJ, Stock MJ. Whither brown fat? *Biosci Rep*. 1986;6:3–18.
44. Bouillaud F, Combes-George M, Ricquier D. Mitochondria of adult human brown adipose tissue contain a 32000-Mr uncoupling protein. *Biosci Rep*. 1983;3:775–780.
45. Cunningham S, Leslie P, Hopwood D, et al, The characterization and energetic potential of brown adipose tissue in man. *Clin Sci (Lond)* 1985;69:343–348.
46. Rothwell NJ, Stock MJ. A role for brown adipose tissue in diet-induced thermogenesis. *Nature* 1979;281:31–35.
47. Himms-Hagen J. Brown adipose tissue thermogenesis: interdisciplinary studies. *FASEB J* 1990;4:2890–2898.
48. Blaza S. Brown adipose tissue in man: a review. *J R Soc Med*. 1983;76:213–216.
49. Santos GC, Araujo MR, Silveira TC, Soares FA. Accumulation of brown adipose tissue and nutritional status: a prospective study of 366 consecutive autopsies. *Arch Pathol Lab Med*. 1992;116:1152–1154.
50. Cottle WH. The innervation of brown adipose tissue. In: Lindberg O, Tissue BA. *Brown Adipose Tissue*. Elsevier; 1970:155–178.
51. Mory G, Bouillaud F, Combes-George M, Ricquier D. Noradrenaline controls the concentration of the protein in brown adipose tissue. *FEBS Lett*. 1984;166:393–396.
52. Ricquier D, Nechad M, Mory G. Ultrastructural and biochemical characterization of human brown adipose tissue in pheochromocytoma. *J Clin Endocrinol Metab*. 1982;54:803–807.
53. Afzelius BA. Brown adipose tissue: its gross anatomy, histology, and cytology. In: Lindberg O, ed. *Brown Adipose Tissue*. Elsevier; 1970:1–31.
54. Giralt M, Villarroya F. White, brown, beige/brite: different adipose cells for different functions? *Endocrinology*. 2013; 154(9):2992–3000.
55. Sanchez-Gurmaches J, Guertin DA. Adipocyte lineages: tracing back the origins of fat. *Biochim Biophys Acta*. 2014; 1842(3):340–351.
56. Zhang J, Du C, Guo W, et al. Adipose tissue-derived pericytes for cartilage tissue engineering. *Curr Stem Cell Res Ther*.2017;12(6):513–521.
57. Almalki SG, Valle YL, Agrawal DK. MMP-2 and MMP-14 silencing inhibits VEGFR2 cleavage and induces the differentiation of porcine adipose-derived mesenchymal stem cells to endothelial cells. *Stem Cells Transl Med*. 2017;6(5):1385–1398.
58. Pearse AGE. *Histochemistry: Theoretical and Applied*. Vol 2. 3rd ed. Williams & Wilkins; 1972.
59. Hausman GJ. Anatomical and enzyme histochemical differentiation of adipose tissue. *Int J Obes*. 1985;9(Suppl 1):1–6.
60. Lithell H, Boberg J, Hellsing K, Lundqvist G, Vessby B. Lipoprotein-lipase activity in human skeletal muscle and adipose tissue in the fasting and the fed states. *Atherosclerosis*. 1978;30:89–94.
61. Lithell H, Hellsing K, Lundqvist G, Malmberg P. Lipoprotein-lipase activity of human skeletal-muscle and adipose tissue after intensive physical exercise. *Acta Physiol Scand*. 1979;105:312–315.
62. Fielding CJ, Havel RJ. *Arch Pathol Lab Med*. 1977;101:225–229.
63. Zugibe FT. *Diagnostic Histochemistry*. CV Mosby; 1970.
64. Hausman GJ. Techniques for studying adipocytes. *Stain Technol*. 1981;56:149–154.
65. Popper H, Knipping G. A histochemical and biochemical study of a liposarcoma with several aspects on the development of fat synthesis. *Pathol Res Pract*. 1981;171:373–380.
66. Waugh DA, Small DM. Methods in laboratory investigation: identification and detection of in situ cellular and regional differences of lipid composition and class in lipid-rich tissue using hot stage polarizing light microscopy. *Lab Invest*. 1984;51:702–714.
67. Spicer SS. *Histochemistry in Pathologic Diagnosis*. Marcel Dekker; 1987.
68. Michetti F, Dell'anna E, Tiberio G, Cocchia D. Immunochemical and immunocytochemical study of S-100 protein in rat adipocytes. *Brain Res*. 1983;262:352–356.
69. Haimoto H, Kato K, Suzuki F, Nagura H. The ultrastructural changes of S-100 protein localization during lipolysis in adipocyte. An immunoelectron-microscopic study. *Am J Pathol*. 1985;121:185–191.
70. Takahashi K, Isobe T, Ohtsuki Y, Akagi T, Sonobe H, Okuyama T. Immunochemical study of the distribution of alpha and beta subunits of S-100 protein in human

neoplasms and normal tissues. *Virchows Arch B Cell Pathol Incl Mol Pathol*. 1984;45:385–396.
71. Nakazato Y, Ishida Y, Takahashi K, Suzuki K. Immunohistochemical distribution of S-100 protein and glial fibrillary acidic protein in normal and neoplastic salivary glands. *Virchows Arch A Pathol Anat Histopathol*. 1985;405:299–310.
72. Kahn HJ, Marks A, Thom H, Baumal R. Role of antibody to S100 protein in diagnostic pathology. *Am J Clin Pathol*. 1983;79:341–347.
73. Nakajima T, Watanabe S, Sato Y, Kameya T, Hirota T, Shimosato Y. An immunoperoxidase study of S-100 protein distribution in normal and neoplastic tissues. *Am J Surg Pathol*. 1982;6:715–727.
74. Hashimoto H, Daimaru Y, Enjoji M. S-100 protein distribution in liposarcoma. An immunoperoxidase study with special reference to the distinction of liposarcoma from myxoid malignant fibrous histiocytoma. *Virchows Arch A Pathol Anat Histopathol*. 1984;405:1–10.
75. Frühbeck G, Busetto L, Dicker D, et al. The ABCD of obesity: An EASO position statement on a diagnostic term with clinical and scientific implications. *Obes Facts*. 2019;12(2):131–136.
76. Heymsfield SB, Wadden TA. Mechanisms, pathophysiology, and management of obesity. *N Engl J Med*. 2017;376(3):254–266.
77. NCD Risk Factor Collaboration. Worldwide trends in underweight and obesity from 1990 to 2022: A pooled analysis of 3,663 population-representative studies with 222 million children, adolescents, and adults. *Lancet*. 2024;403(10379):1378–1396.
78. Wharton S, Lau DCW, Vallis M, et al. Obesity in adults: A clinical practice guideline. *CMAJ*. 2020;192(31):E875–E891.
79. Jastreboff AM, Aronne LJ, Ahmad NN, et al. Tirzepatide once weekly for the treatment of obesity. *N Engl J Med*. 2022;387(3):205–216.
80. Wilding JPH, Batterham RL, Calanna S, et al. Once-weekly semaglutide in adults with overweight or obesity. *N Engl J Med*. 2021;384(11):989–1002.
81. Schauer PR, Bhatt DL, Kirwan JP, et al. Bariatric surgery versus intensive medical therapy for diabetes—5-year outcomes. *N Engl J Med*. 2017;376(7):641–651.
82. Swinburn BA, Kraak VI, Allender S, et al. The global syndemic of obesity, undernutrition, and climate change: The Lancet Commission report. *Lancet*. 2019;393(10173):791–846.
83. Raife T, Landas S. Intracellular crystalline material in visceral adipose tissue: A common autopsy finding [abstract]. *Am J Clin Pathol*. 1990;94:511.
84. Cahill GF. Pathologic anatomy of adipose tissue. In: Renold AE, Cahill GF, eds. *Handbook of Physiology. Section 5: Adipose Tissue*. Waverly Press; 1965.
85. Manthorpe R, Helin G, Kofod B, Lorenzen I. Effect of glucocorticoid on connective tissue of aorta and skin in rabbits. Biochemical studies on collagen, glycosaminoglycans, DNA and RNA. *Acta Endocrinol (Copenh)*. 1974;77:310–324.
86. Napolitano LM. Observations on the fine structure of adipose cells. *Ann N Y Acad Sci*. 1965;131:34–42.
87. Dahlman I, Mejhert N, Linder K, et al. Adipose tissue pathways involved in weight loss of cancer cachexia. *Br J Cancer*. 2010;102:1541–1548.
88. Seaman JP, Kjeldsberg C, Linker A, Gelatinous transformation of the bone marrow. *Hum Pathol*. 1978;9:685–692.
89. Wittels B. Bone marrow biopsy changes following chemotherapy for acute leukemia. *Am J Surg Pathol*. 1980;4:135–142.
90. Piérard GE, Nizet JL, Piérard-Franchimont C. Cellulite: from standing fat herniation to hypodermal stretch marks. *Am J Dermatopathol*. 2000;22:34–37.
91. Adams RD. *Diseases of Muscle. A Study in Pathology*. 3rd ed. Harper & Row; 1975.
92. Rosai J, Levine GD. Tumors of the thymus. In: Firminger HI, ed. *Atlas of Tumor Pathology, 2nd Series, Fascicle 13*. Armed Forces Institute of Pathology; 1976.
93. Rywlin AM. *Histopathology of the Bone Marrow*. Little, Brown & Co; 1976:19.
94. Ackerman LV, Rosai J. *Surgical Pathology*. 5th ed. CV Mosby; 1974:649.
95. Robbins SL, Cotran RS, Kumar V. *The Pathologic Basis of Disease*. 3rd ed. WB Saunders; 1984.
96. Seifert G. Lipomatous atrophy and other forms. In: Kloppel G, Heitz PU, eds. *Pancreatic Pathology*. Churchill Livingstone; 1984.
97. Heyer CM, Kagel T, Lemburg SP, Bauer TT, Nicolas V. Lipomatous hypertrophy of the interatrial septum: a prospective study of incidence, imaging findings, and clinical symptoms. *Chest*. 2003;124(6):2068–2073.
98. O'connor S, Recavarren R, Nichols LC, Parwani AV. Lipomatous hypertrophy of interatrial septum: an overview. *Arch Pathol Lab Med*. 2006;130:397–399.
99. McAllister HA. *Tumors of the Cardiovascular System*. Armed Forces Institute of Pathology; 1978:44–46.
100. Rokey R, Mulvagh SL, Cheirif J, Mattox KL, Johnston DL. Lipomatous encasement and compression of the heart: antemortem diagnosis by cardiac nuclear magnetic resonance imaging and catheterization. *Am Heart J*. 1989;117:952–953.
101. Waller BF. *Pathology of the Heart and Great Vessels*. Churchill Livingstone; 1988.
102. Symmers WSC. The lymphoreticular system. In: Symmers WSC, ed. *Systemic Pathology*. Vol 2. Churchill Livingstone; 1978:647–651.
103. Smith T. Fatty replacement of lymph nodes mimicking lymphoma relapse. *Cancer*. 1986;58:2686–2688.
104. Takayasu K, Shima Y, Muramatsu Y. et al. Imaging characteristics of large lipoma and angiomyolipoma of the liver. Case reports. *Cancer*. 1987;59:916–921.
105. Pounder DJ. Hepatic pseudolipoma. *Pathology* 1983;15:83–84.
106. Shafer WG. *Development Disturbances of Oral and Paraoral Structures. A Textbook of Oral Pathology*. 4th ed. WB Saunders; 1983:24–25.
107. Fu YS, Parker FG, Kaye GI, Lattes R. Ultrastructure of benign and malignant adipose tissue tumors. *Pathol Annu*. 1980;15(Pt 1):67–69.
108. Kindblom LG, Angervall L, Stener B, Wickbom I. Intermuscular and intramuscular lipomas and hibernomas. A clinical, roentgenologic, histologic and prognostic study of 46 cases. *Cancer*. 1974;33:754–762.
109. Petras RE. Nonneoplastic intestinal diseases. In: Mills SE, ed. *Sternberg's Diagnostic Surgical Pathology*. 4th ed. Lippincott Wilkins; 2004:1519–1520.
110. Trotter M, Crawford RI. Pseudolipomatosis cutis: superficial dermal vacuoles resembling fatty infiltration of the skin. *Am J Dermatopathol*. 1998;20:443–447.

7

Skeletal Muscle

Hannes Vogel

Together, the skeletal muscles comprise the largest organ system in the body by weight and volume. Although all muscles share many common attributes, more than 600 muscles differ in size, shape, and function. Not only is the variation in the gross anatomy of these muscles considerable, but histologic characteristics such as fiber size and proportion of fiber types vary with different sites. Since many anatomic pathologists are called upon to interpret microscopic findings in muscle as they relate to disease, a familiarity with the histology of normal muscle is the basis of any accurate assessment. Today, an understanding of muscle histology is dependent on some knowledge of developmental and molecular biology.

EMBRYOLOGY

The prenatal development of skeletal muscle is orchestrated by a host of genes, transcription factors, and microRNAs that is too complex to discuss in detail, although several important ones are mentioned below. Skeletal muscle develops embryologically from somitic mesodermal tissue. The paraxial mesoderm is first apparent on day 17 of embryonic development and is the origin of the somites that are completely formed by day 30. At this time, a series of 42 to 44 pairs of rounded somites can be found adjacent to the notochord. By the 4th week of gestation, the mesodermal somites separate into the dermatomes and segmental myotomes, the latter of which give rise to the muscles of the body wall. Expression of *Hox* genes is important during the early stages of segmentation. This family of homeobox genes acts as transcription factors and is involved in craniocaudal segmentation of the body.

The dorsal division of myotomes, the epimeres, gives rise to the back muscles, whereas the ventral division of hypomeres differentiates into the lateral and ventral muscles of the body wall, including the intercostal muscles, abdominal oblique muscles, and strap muscles of the neck. The muscles of the extremities arise from the limb buds that form from the lateral plate mesoderm, which is also the origin of the bone, tendon, ligaments, and blood vessels. The mesenchyme of the limb buds appears during the 4th week of gestation and is subject to induction by the somites. The muscle tissue derived from somitic mesoderm invades the limb buds in week 5. The primordia of individual muscles can be appreciated at the end of the 8th week.

Although the limb and trunk muscles are derived from somitic mesoderm, cervical and craniobulbar muscles develop from the branchial arches. Differentiation of the limb musculature follows a cephalocaudad and proximal-to-distal progression. In each limb, the somitic mesenchyme subdivides into a dorsal and ventral mass with respect to the skeletal elements. The extensor, abductor, and supinator muscles are derived from the dorsal mass, whereas the flexor, adductor, and pronator muscles originate from the ventral mass.

The myogenic regulatory factors (MRFs) form a select family of transcription factors that determine the fate of the muscle cell lineage. These genes regulate the transformation of mesenchyme into muscle tissue. The MRFs are a group of four muscle-specific proteins including myogenic determining factor (MyoD), myogenic factor 5 (MYF5), myogenin, and myogenic regulatory factor 4 (MRF4). These proteins act at multiple points in development to establish the skeletal muscle phenotype through the regulation of proliferation, irreversible cell cycle arrest of precursor cells, activation of sarcomeric and muscle-specific genes that facilitate differentiation, and assembly of sarcomeres. *Pax-3* is highly expressed in embryonic muscle. This regulatory factor is induced by sonic hedgehog and bone morphogenetic protein (bmp4), and activates MyoD, leading to the formation of mononuclear myoblasts.

The earliest form of mesenchymal cell to become devoted to a myogenic cell lineage is the myoblast, the most immature of muscle cells. These are small, round, mononucleated cells with prominent nucleoli and increased mitotic activity. Myoblasts contain cytoplasm without detectable filaments, but ribosomes can be identified. Myoblasts require *MyoD* and *Myf5* for proliferation. Later in development, noggin, which is encoded by *Nog*, inactivates bmp4, thereby promoting differentiation. *Pax-7* activates *Nfix* to encode nuclear factor 1X, which acts like a switch to turn on fetal genes and repress embryonic genes (1). *Nfix* activates *MCK*, a promoter gene in myogenesis. In addition, *BCK* encodes the B isoform of creatine kinase, which is an early marker for skeletal myogenesis that is expressed at a high level in embryonic neural tissues and developing striated muscle. Myogenin is encoded by *Myf4* and is expressed late in myogenesis under the tight control of *Myf5*. Myogenin is involved in myoblast recognition and adherence mechanisms that rely on adhesion molecules such as M-cadherin, its expression promotes myotube fusion, and it is ultimately important in increasing muscle mass. In turn, MRF4 promotes differentiation into myofibers. Further details of basic muscle embryology may be found elsewhere (2,3).

Masses of proliferating myoblasts represent the source of myotubes, the next step in myogenesis. These myoblasts are recognizable by their spindle shapes and numerous ribosomes, Golgi apparatus, and specialized cytoskeletal proteins including slow (type 1, Myh7) and embryonic (Myh3) myosin heavy chains (MyHC), α-actins (cardiac [Actc1] and skeletal [Acta1]), and desmin, as well as metabolic enzymes such as β-enolase and carbonic anhydrase III (CAIII). They undergo intense proliferation and by 9 weeks of gestation are prepared to fuse with one another to form multinucleated cells called primary myotubes. They range from 8 to 50 μm in diameter, with a large central nucleus in cross section, which are in chains in longitudinal section (Fig. 7.1). The diminutive cytoplasm contains a few myofibrils, glycogen, and perinuclear mitochondria.

Successful cell–fusion requires several pathways including recognition, migration, adhesion, signaling, actin cytoskeletal dynamics, and membrane merger. Although many initial steps of myoblast fusion are similar to those of other fusogenic cell types, the elements and molecular basis of myoblast fusion have not been fully defined. Time-lapse photography and electron microscopy of myoblast fusion have provided important clues about these events. Following adhesion, electron microscopy has demonstrated that alignment occurs through the parallel apposition of the membranes of elongated myoblasts with myotubes or other myoblasts. Unilamellar vesicles are also observed in close apposition to the fusing membranes of muscle cells during development or muscle regeneration. Ultrastructurally, myoblasts are seen to have contact with each other through filopodia. Adjacent myoblasts are often joined by gap junctions. Fusing myoblasts become longitudinally oriented, a process that requires fibronectin. Interestingly, the alignment of myotubes observed in vivo does not occur in the absence of tension or deposition of aligned extracellular matrices in tissue culture. At this stage of myogenesis, groups of primitive myoblasts and myotubes are enclosed by a common basement membrane. Primary myotubes cluster in groups of 4 to 12 with spaces between clusters containing fibroblasts and probably angioblasts. There is usually one larger primary myotube in each cluster.

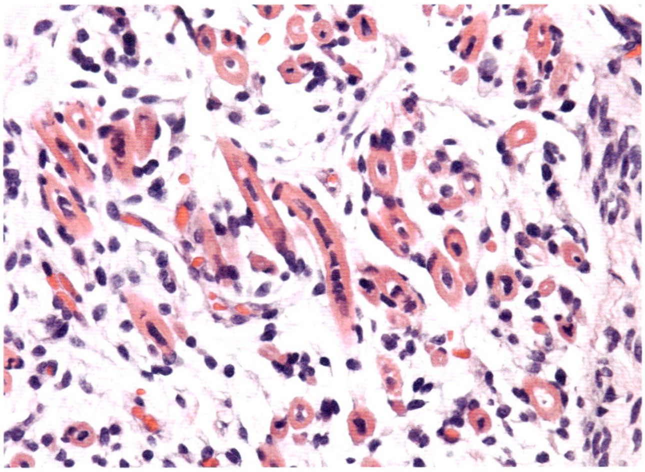

FIGURE 7.1 Myotube stage of muscle development at 14 weeks of gestational age. Myotubes typically have large central nuclei, and longitudinal sections display rows of numerous nuclei.

Extensive cytoskeletal reorganization occurs before and after fusion. Myotubes differ from myoblasts by the presence of multiple nuclei and cytoplasmic filaments. Filaments first form at the peripheral portions of the sarcoplasm and consist of 10-nm fibroblast-like fibrils that disappear during maturation. The myofibrils have a rudimentary form of sarcomeres with A and I bands and Z discs. Immunohistochemical techniques also demonstrate the presence of desmin and vimentin within myotubes. Secondary myotubes continue to arise through the successive waves of fusion of postmitotic primary myotubes. More mature secondary myotubes have a larger-diameter, several hundred nuclei, and prominent myofilaments. Secondary myotubes are initially encased by the same basal lamina of the parent primary myotubes, but they later separate and develop individual basal lamina. These cells also begin to show evidence of contractile activity. Secondary myotubes eventually give rise to muscle fibers. As they approach this stage of development, secondary myotubes cease fusing and develop acetylcholine receptors (AChR) on the cell surface. The receptor proteins are diffusely distributed on the cell surface at first, but they become concentrated in hot spots where motor endplates will develop. Myoblast fusion occurs during development and throughout adulthood as skeletal muscle growth and regeneration require the accumulation of additional nuclei within myofibers. Alterations in myoblast fusion may also contribute to muscle diseases involving loss of muscle mass.

Myoblasts undergo morphologic changes related to normal development and select pathologic conditions including aging, disuse, and exercise (4). A necessary remodeling of muscles occurs through apoptosis during embryogenesis, thereby removing "unwanted" cells or structures to make room for further maturation. The large-diameter primary myotubes are preferentially affected. Apoptosis is accompanied by misshapen nuclei and irregular chromatin condensations along the nuclear envelope, contraction of the cytoskeleton, blebs on the cell membrane, and packaging of cytoplasmic organelles into an apoptotic body. The programmed cell death of both primary and mature myotubes occurs between 10 and 16 weeks of gestation, and involves the TRAIL death ligand signaling pathway (5). TRAIL binds to its upregulated receptor, DR5, thereby initiating the caspase 8 route to apoptosis. At the same time, FLIP, a caspase 8 inhibitor, is reduced. A second pathway involves mitochondrial permeability brought about by proapoptotic members of the Bcl-2 family that is mediated by caspase 9. The Bcl-2/caspase 9 pathway is also important in the sarcopenia of aging during which there is a loss of myocytes. Because it is a multinucleated cell, the muscle fiber does not always undergo apoptosis in the same fashion as other cell types. Damage limited to individual myonuclei seems to be more common than death of the entire cell. Finally, the elevated expression of endogenous caspase inhibitors in muscle seems to confer relatively greater resistance to apoptosis in muscle tissue.

Mature muscle fibers differ from myotubes in that their nuclei are peripheral, and their filaments are organized into sarcomeres. Muscle fibers also develop a sarcotubular system and become innervated over time. Immature muscle fibers often acquire multiple innervation sites, all but one of which eventually disappears. The number of myotubes declines after the 21st week of gestation so that by the time of birth, myotubes are no longer conspicuous. As myotubes become fewer in number during the 5th month of development, the muscle fibers undergo histochemical differentiation in. Between 15 and 20 weeks of gestation, a primitive progenitor of the checkerboard pattern emerges in which all myotubes and myofibers have high ATPase and oxidative enzyme activity. By 20 weeks of gestation, approximately 10% of fibers are larger in diameter, with both high oxidative enzyme activity and reduced ATPase activity. These fibers are basophilic in hematoxylin and eosin (H&E) stains and represent the Wohlfart type B fibers, which are the earliest type 1 fibers to be detected in developing muscle. The remaining 90% of fibers (Wohlfart type A) correspond to type 2 fibers with enhanced ATPase activity. Although type 2A and 2B fibers are not yet visible, a few type 2C fibers that stain darkly in both acid and alkaline ATPase reactions are apparent. These fibers typically express both fast and slow myosin. The more mature checkerboard histochemical pattern, which is stimulated by the innervation of fibers, is almost complete between 26 and 30 weeks of gestation. At birth, the histochemical mosaic begins to resemble that of mature adult muscle; approximately 80% of fibers are clearly identified as type 1 or type 2. The remaining 20% are undifferentiated fibers that have both abundant oxidative enzyme activity and stain darkly in routine ATPase reactions. A few Wohlfart type B fibers remain at birth. Type 2C fibers are not encountered.

POSTNATAL AND DEVELOPMENTAL CHANGES

During the prenatal period and childhood, muscle fibers continue to increase in length until full growth is attained. Muscle fibers lengthen in response to growth of the skeleton by virtue of two fundamental changes in the sarcomeres. Existing sarcomeres lengthen, producing longitudinal fiber growth. This mechanism may account for up to a 25% increase in fiber length and indicates that there is a relative "excess" of sarcomeres that elongates during periods of rapid skeletal growth. Muscle fibers also undergo longitudinal growth with the addition of new sarcomeres and synthesis of contractile proteins. New sarcomeres are added at the end of fibers, usually at the myotendinous junctions. There is also evidence to suggest that new sarcomeres are added within internal segments. Studies suggest a gradual rise in the number of fibers between birth and the end of the fifth decade. In some muscles, the total increase in fibers may reach 80% to 100% of the neonatal level. The mechanism accounting for an increase in the fiber population probably involves a subset of dividing stem cells that undergo fusion to produce new mature fibers.

A major aspect of muscle fiber growth after birth relates to an increase in transverse dimension. There is an almost fivefold increase in muscle fiber diameter between birth and adulthood. For example, the average diameter of mature fibers is between 50 and 60 μm as compared to 7 μm at 20 weeks of gestational age. By birth, fiber diameters have roughly doubled to about 15 μm. Fiber diameters increase at a relatively slow rate until puberty, adding 2 μm each year up to age 5 years, then 3 μm per year between 5 and 9 years. By 10 years of age, type 1 and 2 fibers measure between 38 and 42 μm with little variability between

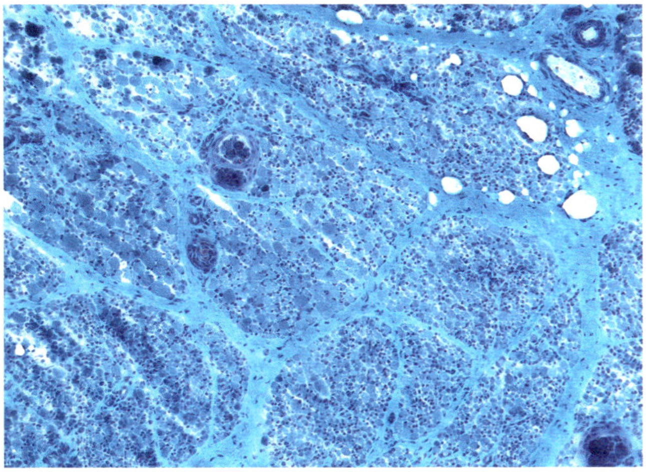

FIGURE 7.2 Infant muscle. A relative increase in perimysial connective tissue is normal (Gomori trichrome).

their average diameters. A burst of growth occurs around the time of puberty at which time fibers gain their adult sizes. Type 1 fibers are typically larger than type 2 fibers in females, while the reverse is noted in males.

A major revision in the histochemical profile of muscle occurs after birth. The checkerboard staining pattern is clearly evident in alkaline ATPase reactions at birth. However, fiber typing may not be distinct in oxidative enzyme reactions. The emergence of type 1, 2A, and 2B fibers in oxidative preparations occurs during infancy. Undifferentiated fibers having both abundant oxidative enzyme and ATPase activity represent approximately 20% of fibers at birth but these gradually differentiate into type 1 and type 2 fibers during the first year of life. The fate of Wohlfart type B fibers, comprising about 1% of myofibers at birth, is unknown. They are not seen in biopsies of children over the age of 12 months.

The connective tissue elements of muscle are prominent at birth, particularly the perimysial components. The perimysium may account for up to 20% of the cross-sectional area of muscle tissue immediately after birth (Fig. 7.2). During early childhood, the perimysium and other connective tissue components rapidly shrink to less than 5% of the cross-sectional area, in part because of the enlargement of the muscle fibers. Blood vessels (especially arteries) appear excessively thickened as a result of the presence of abundant smooth muscle elements in the immediate postnatal period. The noncontractile, supporting connective tissue contains abundant collagen and scattered fibroblasts, as well as scattered foci of hematopoiesis that persist after birth. These foci are more likely to be seen in the distal muscles of the extremities and disappear within 1 month after birth.

ANATOMY

There are over 600 voluntary muscles in the human body, most of which are paired. They comprise 25% of the total body weight at birth and 40% to 50% of the total weight in adults. A greater muscle mass is encountered in males than in females. The smallest muscle in the body, the stapedius, measures only 2 mm in length, whereas the sartorius and other large muscles of the extremities measure up to 2 ft in length (61 cm). Skeletal muscles are composed of varying numbers of muscle fibers that range up to approximately 1,000,000 in the gastrocnemius. These muscles are connected at both ends to tendons or the epimysium.

Because the fibers work in conjunction with each other, they are aligned in the same direction. A few skeletal muscles are modeled after the lumbricals, small intrinsic muscles of the hands and feet, in which all the fibers are arranged in a fusiform structure that tapers at either end at the site of tendinous insertion. The more familiar unit is a parallelogram composed of muscle fibers that insert at both ends on a flat tendon composed of dense collagen. Parallel muscles contain fascicles that run parallel to the longitudinal axis of the muscle. Oblique muscles are pennate or featherlike. They are associated with a tendon that runs within the muscle or on its surface, and the muscle fibers insert obliquely on the tendon. However, some are bipennate, much like a feather with a central shaft from which a series of barbs radiate on either side (e.g., peroneus longus). Other muscles are simple pennate with only one set of parallel muscle fibers attached obliquely on a shaft-like tendon (e.g., extensor digitorum longus). Muscles are designated as complex pennate when the muscle consists of multiple parallelograms attaching to several tendons in the muscle mass. Not all skeletal muscles precisely fit into these models; some may be triangular (e.g., pectoralis minor) or spiral (e.g., forearm supinators). Although most muscles are attached to bony skeletal structures, some voluntary muscles of the larynx and esophagus do not have attachments to bone.

The blood supply to skeletal muscles is of paramount importance to their proper function. Skeletal muscle requires 20% of cardiac output for basal metabolic needs, although this number increases drastically with vigorous exercise that requires use of large muscle groups, such as running and swimming. Not only is an intact supply of oxygen and substrates of importance, but also the efficient removal of metabolites such as lactic acid and carbon dioxide. It is known that the arterial supply to muscles varies somewhat with the individual. In general, the skeletal muscles are subserved by several vessels rather than a single artery, which renders them rather resistant to ischemia from an embolus or disease of a single vessel. The vascular supply to skeletal muscle falls into one of the following five categories:

1. The blood supply is derived from a single nutrient artery that divides in a longitudinal fashion within the muscle itself. The medial and lateral head of the gastrocnemius is an example of such a system and represents a risk for greater damage as the consequence of arterial occlusion.

2. The muscle is supplied by several separate arteries entering the muscle along its length. Anastomoses are formed within the muscle between the territories of each artery. This pattern is typical of the soleus.
3. The blood supply arises from a single main artery that enters the belly of the muscle and subsequently forms a radiating pattern of collaterals, as in the biceps brachii.
4. A pattern of anastomosing arcades is derived from a series of penetrating arteries. This vascular pattern is considered to be the most efficient form of vascularization and is seen in the tibialis anterior.
5. A less efficient form of the anastomosing arcade pattern is the rectangular pattern of anastomoses formed by a series of penetrating arteries. This so-called quadrilateral pattern is seen in the extensor hallucis longus muscle.

Once a main artery enters the muscle substance, it branches into a number of primary intramuscular arteries that ramify in the epimysium and perimysium. The primary arteries, with a diameter that ranges from 80 to 360 μm, give rise to numerous secondary arterioles that run parallel to the direction of the muscle fibers. The secondary arterioles often connect to primary arteries, forming artery-to-artery anastomoses. The secondary arterioles range from 50 to 100 μm in diameter and have a thin adventitia composed of fibroblasts and collagen. The smooth muscle coat is much thinner than that of the primary arteries, usually having only two to three layers of cells. The internal elastica is prominent and continuous. The secondary arterioles branch to form terminal arterioles, which measure 15 to 50 μm in diameter. Their smooth muscle coats are usually only one layer of cells thick. The internal elastica becomes discontinuous and is lost in smaller vessels. The distal portions of the terminal arterioles have precapillary sphincters, which are formed from the smooth muscle cells of the media. These sphincters are found in blood vessels with an inner diameter of less than 15 μm. Footlike processes between the smooth muscle cells and the endothelium may be seen in the region of the sphincters.

As in other tissues, the arterioles end in an elaborate system of capillaries. Relatively few capillaries are open at rest but there is a considerable increase in the number of open capillaries during muscle activity. A marked difference in capillary density is observed in different muscles, as well as in trained versus untrained subjects. Studies of capillary density reveal that the average single muscle fiber is surrounded by 1.7 capillaries. Capillary density may also be expressed as the number of capillaries per fiber, which averages 0.7 in cross sections.

The density of capillaries also reflects oxygen consumption within muscle. Therefore, increased numbers of capillaries are evident where larger numbers of type 1 fibers are present. This phenomenon is less evident in humans than in animals with muscles composed chiefly or totally of one fiber type. For example, the cat soleus muscle is composed almost entirely of type 1 fibers and the density of capillaries is 1,600/mm^2. In contrast, the cat gastrocnemius muscle contains far fewer type 1 fibers and has a capillary density of 600/mm^2. The capillaries within skeletal muscle travel primarily in a longitudinal direction, although they are frequently linked by short transverse branches.

Capillaries are composed of endothelial cells surrounded by a basement lamina. Occasional pericytes are encountered outside the basement membrane. Endothelial cells typically contain numerous pinocytotic vesicles. Where endothelial cells are joined, they lack tight junctions. Hence, the capillary endothelium of muscles is freely permeable to tracers such as horseradish peroxidase. The capillary pericytes are essentially smooth muscle cells that contain large numbers of filaments, and they are innervated by small-diameter unmyelinated nerve fibers. The basement membrane lies between the endothelium and pericytes and measures 20 to 30 nm, although some thickening and reduplication of the basal lamina occur in older patients.

The nerve supply to individual skeletal muscles usually enters the surface of the muscle at the belly and is accompanied by one or more major penetrating arteries. Myelinated and unmyelinated axons are present within the main nerve trunk. Contributions to the nerve are made from myelinated efferent motor fibers that innervate the muscle fibers and unmyelinated autonomic efferent fibers, as well as somatic afferent sensory fibers from muscle spindles, Golgi tendon organs, and ac corpuscles. At least 50% of the fibers are sensory in function. The motor fibers that innervate the myofibers demonstrate a bimodal size distribution. The large-diameter α-fibers innervate fast motor units, whereas the β-fibers are distributed to slow motor units and some intrafusal fibers of the muscle spindle. The γ-fibers have very small diameters and supply the remainder of the muscle spindle fibers. The large motor fibers are relatively uniform in diameter, measuring between 10 and 15 μm. The small motor fibers vary from 2 to 7 μm in diameter.

As the distal motor axon approaches the muscle fiber, it is transformed into the terminal axon, which represents the proximal portion of the neuromuscular junction, or motor endplate. The neuromuscular junction, measuring about 50 μm in diameter, is composed of the presynaptic portion or terminal axon and the postsynaptic portion, which is formed by a unique region in the muscle fiber. The presynaptic and postsynaptic domains are separated by a specialized, 50-nm wide intercellular space, the synaptic cleft. The myelinated motor nerve terminates at the presynaptic region as an unmyelinated axonal segment that is enveloped by the teloglia, the distal projections of Schwann cells. The terminal axon and teloglia are covered by a layer of endoneurium, the sheath of Henle, which becomes continuous with the endomysium of the muscle fiber in the area of the motor endplate. Numerous synaptic vesicles, each 45 to 50 nm in diameter, are found in the terminal axon. The vesicles are most plentiful around thickened zones of increased electron density at the presynaptic membrane. Studies utilizing freeze–fracture electron microscopy have demonstrated that parallel pairs of double rows of intramembranous particles, each measuring 10 nm in diameter, are located at these electron-dense zones. The particles are considered to represent voltage-sensitive calcium channels known as active zones.

At the postsynaptic region of the muscle fiber, the cell surface is elevated to form the hillock of Doyère, or sole plate. Within the sole plate, the sarcoplasm is granular, and a cluster of sarcolemmal nuclei is often seen. Nuclei in this location are plump and vesicular. The terminal axon ramifies in the sole plate as a series of branches called telodendria; these indent the surface of the fiber, producing gutters or troughs. The surface of the fiber at the motor endplate is undulating and redundant, creating the complex of postjunctional folds that can be demonstrated by supravital staining as the subneural apparatus of Couteaux (Fig. 7.3). The spaces between the folds denote the secondary synaptic clefts. As a result of the formation of these clefts, the surface area of the postsynaptic membrane is increased to approximately 10 times the surface area of the presynaptic portion. The postsynaptic membrane of the folds is thicker and more densely stained at the crests than in the depths of the clefts. By electron microscopy, the juxtaneural membrane at the crests of the folds contains irregularly spaced densities measuring 11 to 14 nm in diameter. In freeze–fracture preparations, the crests of the inner membrane are studded with rows of particles that are similar in size to these densities (about 10 nm). These large intramembranous particles are considered to represent the AChR, a pentameric 275-kDa glycoprotein.

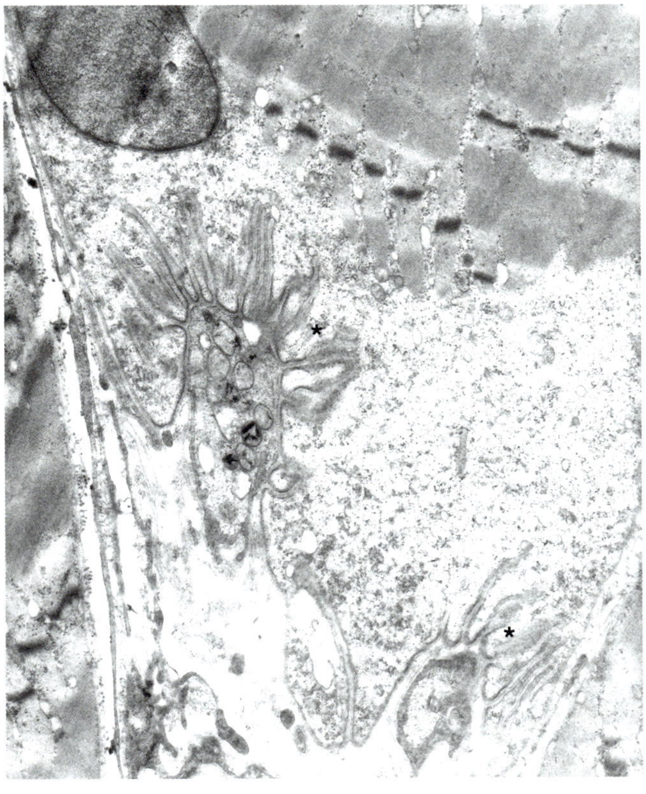

FIGURE 7.3 Electron micrograph of motor endplate. Ultrastructurally, the motor endplate consists of a terminal axon and a postsynaptic region formed by a specialized portion of the muscle fiber. The surface of the fiber is undulating, representing the postjunctional folds (*asterisks*).

LIGHT MICROSCOPY

Familiarity with the normal structure of skeletal muscle provides a useful background for the pathologist in the evaluation of muscle biopsies. Other sources offering a more comprehensive discussion of the light microscopy, histochemistry, and electron microscopy of normal muscle than is possible here are found in the literature (6–9). The muscle fiber is a multinucleated, syncytial-like unit, shaped like a long, narrow cylinder that varies in length depending on the muscle. For example, a single cell may be as small as 1 mm in length in the stapedius or several centimeters long in the sartorius. The average length of skeletal muscle cells in humans is about 3 cm. The normal adult myocyte is not perfectly round; it is polygonal, producing a multifaceted profile in cross section. The nuclei are usually located subsarcolemmally and number four to six per cell when sectioned transversely. There are approximately 30 nuclei for each millimeter of fiber length. The sarcolemmal nuclei are slender and flat, with an orientation that is parallel to the long axis of the fiber. These nuclei measure 5 to 12 μm in length and 1 to 3 μm in width. The nucleoli are small and may not be visible in many fibers. Random fibers may have internally placed nuclei but these should not account for more than 3% of visible nuclei. Increased numbers of internal nuclei may suggest a condition involving increased turnover of myofibers. The sarcoplasm is light pink and textured in H&E-stained cross sections (Fig. 7.4A). In most cases, the mitochondria can be identified, especially in type 1 fibers, as tiny red granules within the sarcoplasm, normally more numerous closer to subsarcolemmal regions than in the center of the fiber (Fig. 7.4B). Notable accumulations of mitochondria may be seen in subsarcolemmal regions in normal muscle, and while a reliable marker of mitochondrial proliferation, the clinical significance is unknown or doubtful. The cross-striations are best appreciated in longitudinal sections and may be accentuated by increasing the refractile index by lowering the microscope condenser (Fig. 7.5). The inability to detect cross-striations by H&E in formalin-fixed paraffin-embedded tissue or sarcotubular details in cryosections should raise the possibility of myofiber necrosis. Cross-striations are best demonstrated in periodic acid–Schiff (PAS) and PTAH stains or in resin-embedded material where alternating dark and light bands are evident (Fig. 7.6). Sections stained with Gomori trichrome are composed of fibers and connective tissue that stain green whereas nuclei are blue-black.

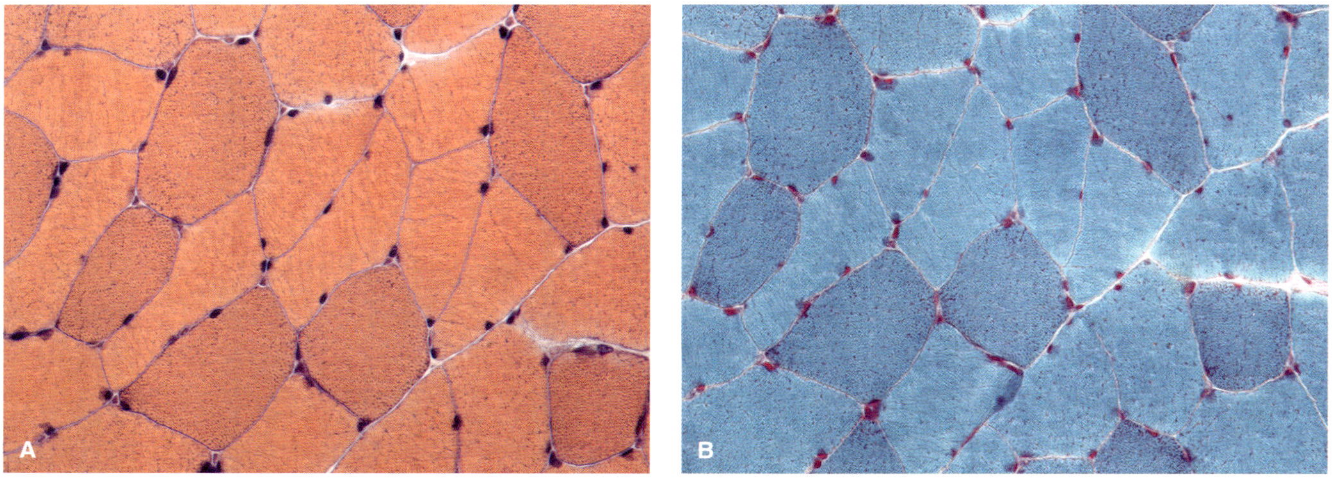

FIGURE 7.4 Cross section of frozen section of muscle. **A:** The sarcoplasm is textured and the sarcolemmal nuclei are peripheral in location (H&E). **B:** Gomori trichrome stain. Mitochondria appear as red granular areas, especially prominent in the subsarcolemmal regions of the fiber.

Red muscle contains high numbers of mitochondria and lipid with a higher capillary density because it depends on aerobic respiration and is designed for the fatigue resistance required for postural function or sustained activity. The color of red muscles is due to relatively greater myoglobin content than white muscles. White muscle fibers contain fewer mitochondria and abundant glycogen, rendering them better suited to anaerobic glycolytic respiration and sudden and intermittent contraction. In vertebrates, particularly in birds, "red" muscle (e.g., soleus) can easily be distinguished from "white" (e.g., pectoralis) muscles upon external inspection, since an entire muscle in such species may be composed of either red or white fibers. Human muscles, on the contrary, contain both fiber types, which typically assume a mixed mosaic arrangement reminiscent of a checkerboard. Depending on anatomic location and function, the proportion of type 1 and type 2 fibers varies; a typical human muscle contains approximately twice as many type 2 fibers (60% to 65%) as type 1 fibers (35% to 40%). This may be conceptualized in the "average" muscle as roughly one-third portions of type 1, 2A, and 2B fibers resulting in the two-thirds majority of type 2 fibers overall.

The demonstration of the histochemical properties of the muscle fibers comprising a biopsy, which is known as fiber typing, is accomplished by applying histochemical techniques (Table 7.1). Fiber typing is best appreciated using enzyme histochemical reactions performed on frozen sections, although an H&E or trichrome stain of high quality in well-preserved muscle will allow for the discernment of darker type 1 fibers versus paler type 2 fibers (Fig. 7.4).

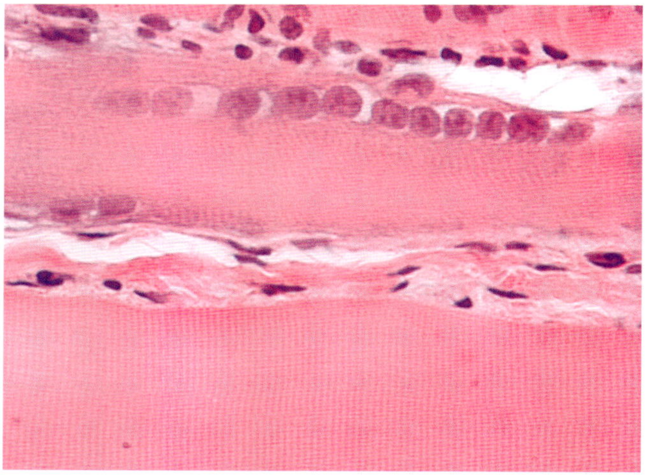

FIGURE 7.5 Longitudinal section of formalin-fixed paraffin-embedded muscle demonstrated cross-striation, a valuable finding to all pathologists in recognizing skeletal muscle in diverse situations. Note the more basophilic fiber above with large vesicular nuclei within the sarcoplasm and not in the usual subsarcolemmal position, representing features of muscle fiber regeneration, usually associated with myopathic, not neuropathic, conditions (H&E).

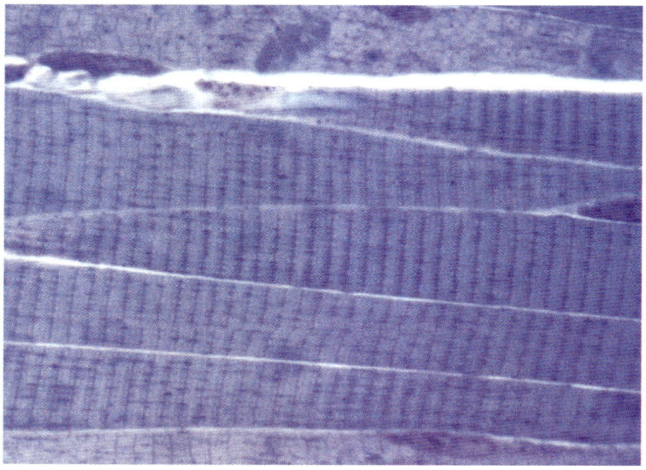

FIGURE 7.6 Resin section. Sarcomere pattern is shown in longitudinal section (toluidine blue).

TABLE 7.1 Fiber Typing

Stain/Reaction	Type 1 Fibers	Type 2 Fibers
ATPase, pH 9.4	Light	Dark
NADH-TR	Dark	Light
PAS/phosphorylase	Light	Dark
Oil red O	Dark	Light

A number of histochemical procedures may be employed for the detection of fiber types, but the most traditional method for this purpose is the myosin ATPase reaction. By changing the pH during the procedure, a spectrum of staining reactions can be produced. In the standard or alkaline ATPase reaction, which is conducted at a pH of 9.4, two fiber types are seen. Type 2 fibers are dark in staining intensity, whereas type 1 fibers are pale (Fig. 7.7A). Fibers of intermediate staining intensity are not observed in the alkaline incubation. If the pH of the incubating solution is brought into the acidic range at pH 4.3 (i.e., reverse ATPase reaction), the reverse staining pattern develops; type 1 and 2 fibers are dark and light, respectively (Fig. 7.7B). Higher pH reveals two populations of type 2 fibers: type 2A fibers are virtually unstained and type 2B fibers are intermediately stained at pH 4.6, whereas type 1 fibers are much darker (Fig. 7.7C). These staining patterns reflect the excellent fatigue resistance of type 1 fibers versus the greater fatigability of type 2B fibers and the intermediate properties of type 2A fibers. Immunohistochemistry for fast and slow myosin is an established adjunct or replacement for ATPase enzyme histochemistry (Fig. 7.7D). It is also possible to subdivide type 2 fibers into types 2A and 2B using myosin antibodies (10). Immunohistochemistry for fiber typing

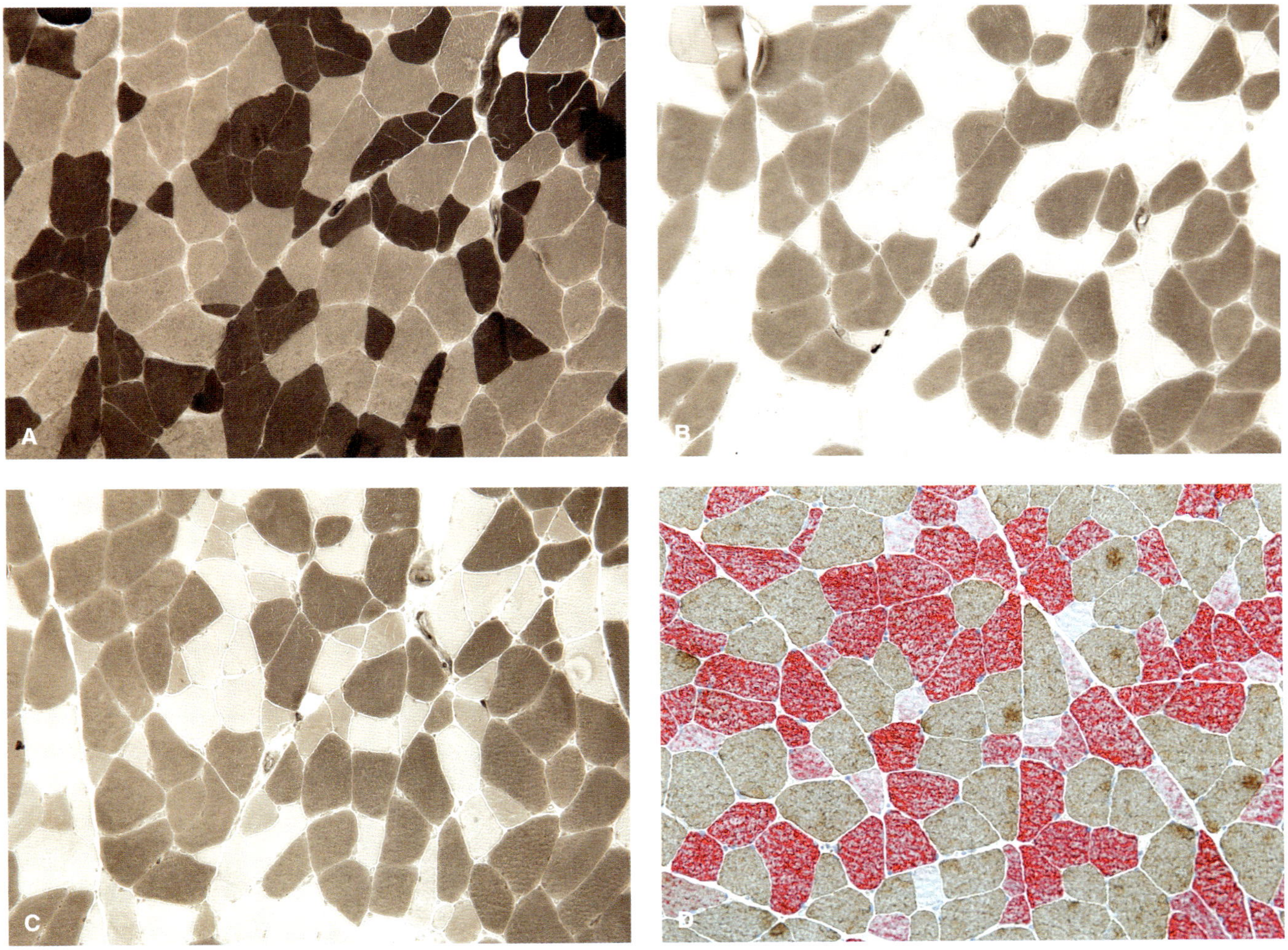

FIGURE 7.7 Skeletal muscle, fiber type determination. **A:** In the myosin ATPase reaction at pH 9.4, type 1 fibers are light and type 2 fibers are dark. **B:** At pH 4.3, the reverse staining pattern is evident. **C:** At pH 4.6, 2A and 2B fibers are distinguishable, with intermediate staining of 2B fibers, which are selectively atrophic in this case of steroid myopathy. **D:** Use of dual antimyosin antibody immunohistochemistry may also be used, showing brown staining of type 1 fibers expressing slow myosin, red-stained type 2A expressing fast myosin, and pale-staining type 2B fibers.

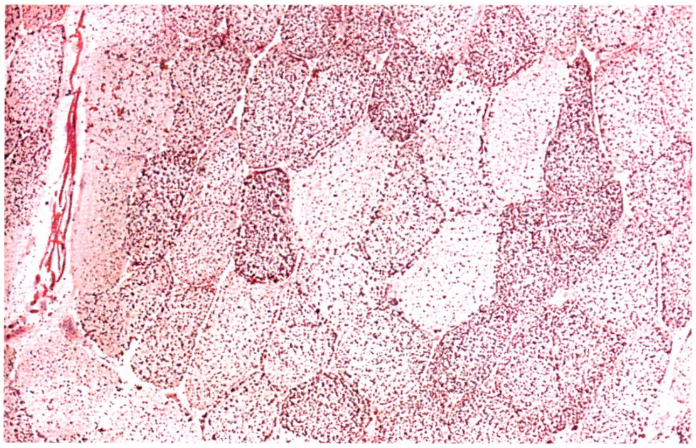

FIGURE 7.8 Lipid content of fibers as demonstrated with the oil red O stain. Type 1 oxidative fibers have a denser lipid concentration.

may be also applicable to fixed tissues, unlike the requirement for cryosections with the ATPase technique.

The nicotinamide adenine dinucleotide tetrazolium reductase (NADH-TR) and other oxidative enzyme reactions show a bimodal pattern of staining intensity, with the darker fibers containing more mitochondria corresponding to type 1 specificity. Oxidative enzyme reactions may further subdivide type 2 fibers into two categories, although the difference is subtle and not as reliable as the result of myosin ATPase enzyme histochemistry at pH 4.6. Although all muscle fibers contain glycogen and the companion enzyme phosphorylase, they are more abundant in type 2 (glycolytic) fibers. The PAS stain, a method of detecting glycogen, and the histochemical reaction for phosphorylase can be used as a means of fiber typing, although these methods are primarily used to investigate possible cases of enzyme deficiency (e.g., type 5 glycogen storage disease, McArdle disease). In fact, almost all stains and enzyme histochemical preparations in frozen muscle pathology may show differences between type 1 and 2 fiber staining intensities, although these techniques are not entirely reliable. Type 1 fibers are rich in neutral lipid, which can be visualized in oil red O and other fat stains (Fig. 7.8).

Striated muscles are partitioned into fascicles, each of which is invested by a connective tissue sheath known as the perimysium. The intramuscular nerves, primary arteries, secondary and terminal arterioles, and veins travel within this sheath throughout the muscle. Intramuscular nerve bundles or twigs are especially numerous at the innervation zone in the belly of the muscle (Fig. 7.9). Up to 10 myelinated nerve fibers may be present in an individual twig, which is surrounded by a thin mantle of perineurial connective tissue. The myelinated nerve fibers are perhaps best demonstrated in Gomori trichrome–stained sections, in which the bright red–colored myelin sheaths surround the unstained axons. Tangential sections of nerve twigs may be mistaken for areas of focal fibrosis or abnormal vascular structures. In addition, the examination of intramuscular nerves should not be considered a reliable means to diagnose diseases of peripheral nerves, which is more appropriately accomplished through a dedicated nerve biopsy. The perimysium is a framework that lends stability to the fascicles, in part by its attachment to the epimysium. The epimysium forms the septa that sequester groups of fascicles, as well as the fascia that encircles the entire muscle and merges with the dense collagenous connective tissue of the tendons.

The perimysium within each fascicle gives way to a normally unobtrusive network, the endomysium. Each muscle fiber may appear to be partly or completely invested by endomysium, a mesenchymal matrix composed of collagen, elastic, and reticulin fibers that support the preterminal arterioles and capillary blood supply to the fascicles. The interface is enlarged at the site where the tendon junction interdigitates with the cell membrane, thereby transferring tension into shear stress. Two transmembrane proteins, the dystrophin–glycoprotein complex and $\alpha 7\beta 1$ integrin, are especially abundant at the myotendinous junction. Presumably, the dystrophin–glycoprotein complex maintains the integrity of the sarcolemma, whereas $\alpha 7\beta 1$ integrin, a receptor for laminin-2, plays a role in the organization of the basement membrane at the myotendinous junction. The muscle fibers become more variable at the interface between muscle and either fascia or tendon, often appearing small in size with increased numbers of internal nuclei. The fibers are separated by dense collagenous trabeculae as they attach to the tendon or fascia (Fig. 7.10). These regions should not be interpreted to signify pathologic alterations. The muscle biopsy should be obtained from the belly of the muscle, avoiding the tendinous insertions or subfascial muscle. However, sampling of fascia may be important when an inflammatory condition is suspected clinically.

Several specialized structures are found within the connective tissue–supporting framework. Muscle spindles, first described in the 19th century, were once considered to be a pathologic finding but are now known to be mechanoreceptors that sense the length and tension of skeletal

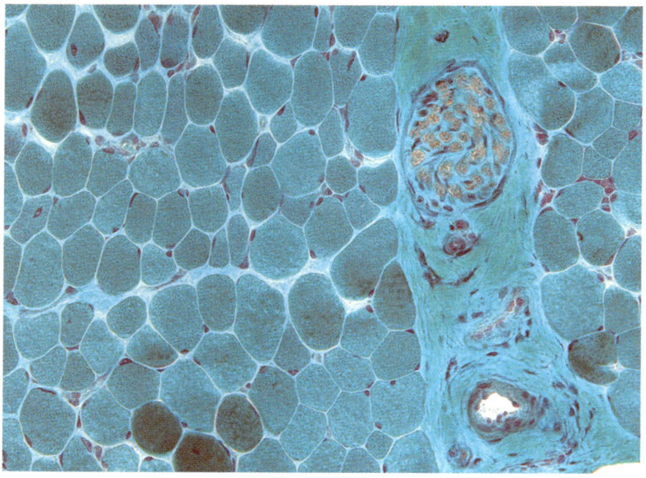

FIGURE 7.9 Intramuscular nerves. Nerve twigs contain axons surrounded by red-staining myelin sheaths (Gomori trichrome).

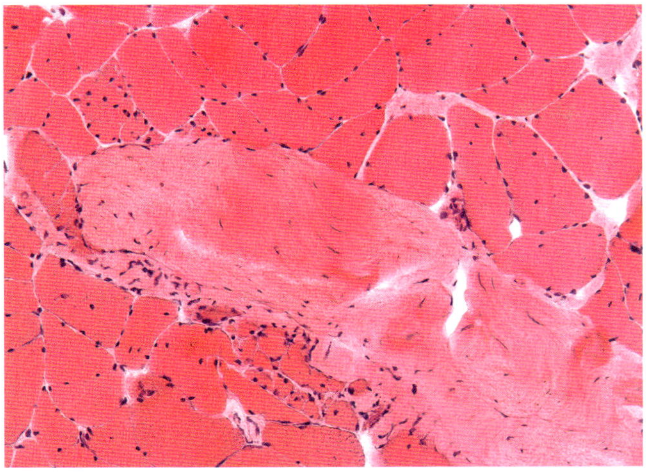

FIGURE 7.10 Subfascial muscle. At the interface, the muscle fibers normally vary in size, have internally placed nuclei, and increased endomysial connective tissue around muscle fibers (H&E).

muscle, governing integrated muscle activity. Although they are encountered in virtually all muscles, they are more frequently detected in smaller muscles devoted to finely coordinated activities, such as those of the hand. They are more numerous in distal than in girdle muscles. Quantitative studies have shown that 70 to 100 muscle spindles may be located in an individual muscle. Muscle spindles tend to lie in the deeper portions of the muscle, particularly in the muscle belly. They are often found where type 1 fibers are more plentiful. As the name implies, muscle spindles are fusiform in shape with a swollen center and tapering ends. They measure 3 to 4 mm in length and 200 μm in diameter. A thin fibrous capsule represents the outer boundaries of the muscle spindle. The capsule is an extension of the perimysium, where spindles are usually located. In certain muscles, such as those of the eye, face, and mouth, the capsule merges with the perimysium and is somewhat indistinct. The capsule is composed of 10 to 15 layers of flattened pavement cells that are specialized fibroblasts. These pavement cells are tightly adherent and separated only by thin layers of delicate collagen fibrils. The pavement cells are epithelial-like, in that each is surrounded by a basement membrane. As one proceeds from the equatorial region of the spindle toward the poles, the number of layers of pavement cells progressively diminishes.

There are 3 to 15 intrafusal fibers within the capsule of the typical muscle spindle (Fig. 7.11). The number of intrafusal fibers is less in small muscles than in larger axial muscles. Two distinct populations of intrafusal fibers are found, both of which are smaller in diameter than the extrafusal fibers. There are usually one to three bag fibers that measure about 20 μm in diameter. The chain fibers number two to seven per spindle and have a diameter of 10 μm or less. The bag fibers are longer, sometimes extending beyond the polar ends of the capsule, and measure 4 to 8 mm in length. The chain fibers are shorter, measuring 2 to 4 mm. The bag fibers are recognized in the equatorial region of the spindle by the presence of large aggregates of nuclei. The nuclei remain internal or central in the bag fibers but are far less numerous distant from the equatorial region. The smaller chain fibers are distinguished by a row of central nuclei that extends along the length of the fiber. There are two types of histologically distinct bag fibers. Bag 1 fibers have considerable oxidative enzyme activity and are pale in ATPase reactions. Bag 2 fibers also have high oxidative enzyme activity and show intermediate staining in ATPase reactions. Chain fibers stain darkly in ATPase reactions despite high oxidative enzyme activity, and are considered by many to represent type 2 fibers.

The innervation of muscle spindles, which is both motor and sensory, is complex and will only be summarized here. The intrafusal efferent fibers are derived from branches of β- and γ-efferent axons. The β-axons terminate primarily on nuclear bag fibers. The γ-fibers supply both nuclear bag and chain intrafusal fibers. It is not uncommon for intrafusal fibers to have polyhedral innervation. Two types of sensory innervation are seen in the muscle spindle. The larger-diameter group 1A afferent fibers emanate from the equatorial region. They originate as the annulospiral endings, a series of neural coils and spirals that attach to the nuclear bag and chain fibers. Smaller-diameter group 2 afferent fibers come from the paraequatorial regions of the spindle and are associated mainly with the flower-spray endings of Ruffini. The majority of these endings project from the nuclear chain fibers. The secondary, or flower-spray, endings consist of a branching network that enwraps the intrafusal fiber between its polar and equatorial regions.

The Golgi tendon organ is an encapsulated sensory nerve terminal located at the junction of muscle with tendon or aponeurosis. The location of these structures allows them to sense changes in muscle tension. They have an inhibitory function in the event of strong muscle contraction. These

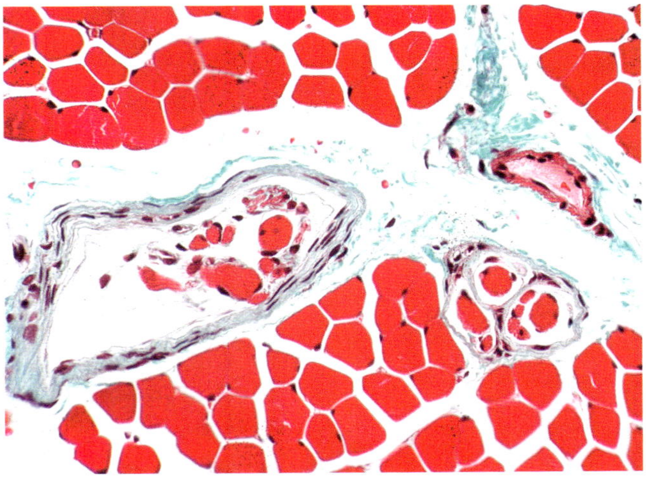

FIGURE 7.11 Muscle spindle. A fibrous capsule encloses a nerve twig and several intrafusal fibers, which are normally smaller than the extrafusal fibers (Gomori trichrome).

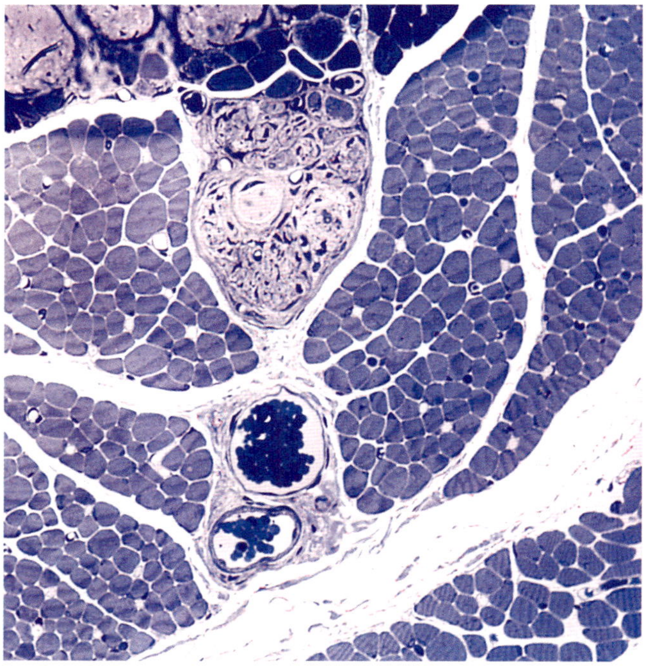

FIGURE 7.12 Golgi tendon organ. Fascicles of collagen surrounded by several nerve bundles (resin section, toluidine blue).

fusiform structures measure about 1.5 mm in length and 120 μm in diameter. They consist of one or more fascicles of collagen fibrils that are attached to tendon or aponeurosis and enveloped by a multilamellar capsule (Fig. 7.12). Each structure is connected to 20 to 30 muscle fibers. The Golgi tendon organ is innervated by a myelinated 1B afferent axon, measuring 7 to 15 μm in diameter. The afferent nerve typically divides and arborizes around the individual collagen bundles.

Pacinian corpuscles are distributed widely in the subcutaneous tissues of the body, although they may also be encountered within the muscular fascial planes and adjacent to tendons or aponeuroses. There is a rodlike nerve terminal in the center of the Pacinian corpuscle which is innervated by fast-conducting group 1 or 2 afferent axons. The central axon is surrounded by a capsule composed of concentric layers of cells (Fig. 7.13). The elongated cells forming the capsule are surrounded by basal lamina and separated by fine collagen fibrils. Pacinian corpuscles are receptor organs that are sensitive to vibration.

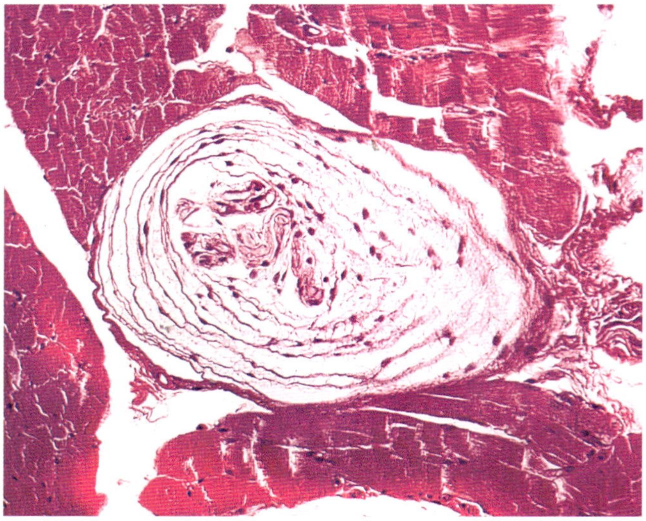

FIGURE 7.13 Pacinian corpuscle. A central nerve terminal is surrounded by a capsule composed of concentric layers of cells (H&E).

ULTRASTRUCTURE

The ultrastructural examination of skeletal muscle is conventionally performed on sections oriented longitudinally, wherein deviations from the orderly striated architecture are more easily detected than in cross sections. The sarcoplasm of each muscle fiber is divided into multiple parallel subunits, the myofibrils, which are minute, cylindroid contractile structures measuring approximately 1 μm in diameter. Myofibrils are segmented into a series of identical sarcomeres that are equal in length whether the muscle is contracted or at rest. The myofibrils are aligned in register with the sarcomeres of surrounding myofibrils. The unique periodicity of the fiber fine structure is a function of the regimentation of this contractile system. The rectangular banding pattern within each sarcomere is produced by the arrangement of filaments (Fig. 7.14). The Z band, which forms the lateral boundaries of the sarcomere, is an electron-dense bar-shaped structure oriented perpendicular to the long axis of the myofibril. The distance between consecutive Z bands represents the sarcomere length, an average of 2.5 to 3.0 μm. The I bands are the most electron-lucent portions of the sarcomere and stand in dramatic contrast to the dark Z bands that bisect them. The I bands are shorter than the moderately dense A bands located at the centers of the sarcomeres. Within each sarcomere are stacks of parallel filaments that are of two types. The thicker filaments

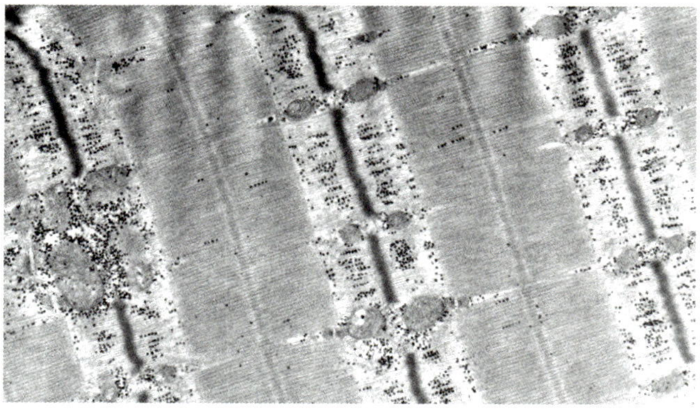

FIGURE 7.14 Ultrastructurally, the several myofibrils can be seen in register in a longitudinal section of the contractile apparatus. Each is composed of a series of sarcomeres that contain A, I, and Z band regions.

measure 15 nm in diameter and are principally composed of myosin. The thinner filaments, containing chiefly actin, are 8 nm in diameter. The thin filaments are attached to the Z band and extend across the I band, where only thin filaments are found. They penetrate the A band in which alternating thick and thin filaments are visualized. Thick filaments are restricted to the A band region of the sarcomere and determine its length.

The sarcoplasmic organelles are more concentrated around the sarcolemmal nuclei and between the myofibrils. Most of the mitochondria are ovoid or elliptoid in cross section, which belies their more elongated or tubular shapes. They are most easily recognized adjacent to the Z bands where their long axes are parallel to those of the myofibrils. Both mitochondria and lipid vacuoles are more conspicuous in oxidative fibers. Mitochondria are not normally longer than a sarcomere in length. However, mitochondrial enlargement should not be equated with bona fide mitochondrial disease. Recent studies employing high-resolution, three-dimensional scanning electron microscopy have demonstrated connectivity between muscle cell mitochondria forming a reticulum connecting complex IV/COX-active subsarcolemmal paravascular mitochondria and complex V/ATP synthase–rich intramyofibrillar mitochondria.

Glycogen is composed of granules with a diameter of 15 to 30 nm, and is more abundant in glycolytic fibers, particularly in the I band region of the sarcomere. Intermyofibrillar glycogen is more abundant in newborn skeletal muscle but is relatively sparse in more mature muscle. Glycogen content varies with diet and is also dependent on the amount of exertion sustained by the muscle. The sarcoplasmic reticulum and the transverse tubules together comprise the sarcotubular complex. Sarcotubular complexes are more abundant in type 2 fibers and function in the excitation–contraction coupling process. The sarcoplasmic reticulum is analogous to the endoplasmic reticulum of other cells and represents an elaborate system of tubules that, by branching in all directions, surrounds the myofibrils. In contrast to the sarcoplasmic reticulum, which has no communication with the extracellular space, the transverse tubules arise as invaginations from the cell membrane. They are observed at regular intervals along the length of the fiber, particularly at the junction of the A and I bands. The transverse tubules encircle the myofibrils and are disposed in a predominantly transverse direction. Branches of the sarcotubular complex join together as triads at the A–I band junctions. Here, pairs of terminal cisterns derived from the sarcoplasmic reticulum are positioned on either side of a central transverse tubule. In this location, the sarcoplasmic reticulum tubules appear as hollow, membrane-bound profiles, whereas the transverse tubules are somewhat more electron dense.

In a normal muscle fiber, the nuclei (myonuclei) occur beneath the sarcolemma, although they may occasionally occur internally. Satellite cells represent approximately 10% of the myonuclei seen in cross sections of muscle and comprise a separate population of myoblastic stem cells that are a source of nuclei during muscle growth, particularly hypertrophy. Satellite cells also have the capacity to synthesize new muscle after myocyte injury. There is a decline in the number of satellite cells as a result of the aging process so that they constitute only 2% to 3% of myofiber nuclei in older individuals. They appear as small, mononuclear, fusiform cells situated beneath the basement membranes of neighboring muscle fibers. They cannot be reliably distinguished from the muscle fiber nuclei under the light microscope but are immunopositive for neural cell adhesion molecule (NCAM). Satellite cells are not randomly distributed along the length of the muscle fiber; they are more numerous in the sole plates of the neuromuscular junction and the polar regions of the muscle spindles. The nuclei of satellite cells differ somewhat from the nuclei of muscle fibers. Satellite cells lie beneath the external lamina of the muscle fiber and are separated from the fiber by their own plasma membrane and a slender gap of 50 nm or less. They are more elongated, their nuclear chromatin is peripherally dense, and nucleoli are lacking. The satellite cell nuclei are usually asymmetrical within the cytoplasm, which contains only a few filaments without evidence of sarcomere formation. The sarcoplasm also contains free ribosomes, microtubules, and centrioles, which may be associated with cilia. Where the cell membranes of the satellite cell and muscle fiber are opposed, numerous pinocytotic vesicles are seen.

SPECIAL TECHNIQUES

Perhaps more than any other tissue, skeletal muscle in humans has been studied using a wide variety of specialized techniques, in part because human muscle biopsies are frequently collected in such a way as to make both fresh, unfixed tissue and material for special studies available. In addition to routine histochemical methods that are focused primarily on the identification of fiber types, a number of other histochemical procedures have been developed on the basis of several aspects of muscle metabolism. Among these are histochemical techniques to identify various enzymes involved in glycogen metabolism and glycolysis. Familiar examples are histochemical stains for phosphorylase and phosphofructokinase.

Enzyme histochemical procedures have been developed to study mitochondrial function. Cytochrome oxidase (COX)–deficient fibers are a reliable indicator of mitochondrial dysfunction and increase in number as a function of age. COX-deficient fibers are most reliably detected through the use of combined enzyme histochemistry for COX and succinate dehydrogenase (SDH), whereby normal fibers show dual staining and COX-deficient fibers appear in varying shades of blue (Fig. 7.15). COX-deficient fibers are increased in a variety of diseases, not only primary mitochondrial myopathies, but also in some inflammatory myopathies such as inclusion body myositis and dermatomyositis. In determining the possible pathologic significance of the number of COX-deficient fibers, the following quantification can be considered within

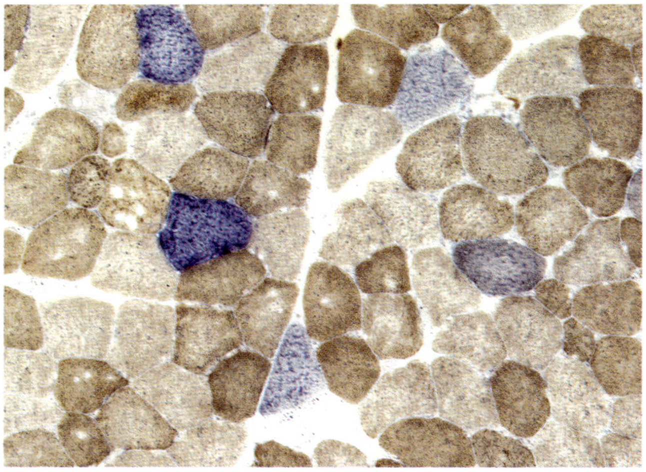

FIGURE 7.15 COX-deficient fibers are detected in a highly sensitive method by performing both the COX and SDH enzyme histochemical reactions on the same section. Fibers lacking the brown stain signifying COX activity will stain preferentially blue, marking SDH activity.

normal limits for the listed age ranges within a 100× magnified microscopic field: 0.3 (30 to 39 years), 0.8 (40 to 49 years), 2.0 (50 to 59 years), 2.5 (60 to 69 years), 4.2 (70 to 79 years), and 6.5 (80 to 89 years). When a primary mitochondriopathy is suspected on this basis, respiratory chain complex enzyme activities may be assayed using the frozen muscle, or the mitochondrial and nuclear DNA relevant to mitochondrial function may be analyzed for pathologic point mutations and deletions.

Other enzyme histochemical techniques offer particular applications in diagnosing certain myopathies or neurogenic diseases of muscle. The NADH-TR preparation highlights the cytoarchitecture of myofibers by staining the sarcotubular system and is well-suited to reveal a host of abnormalities such as target fibers in reinnervated muscle, central or multicores in those respective diseases, and a variety of nonspecific but abnormal morphologies represented by moth-eaten, lobulated, and trabecular fibers (Fig. 7.16). The nonspecific esterase preparation

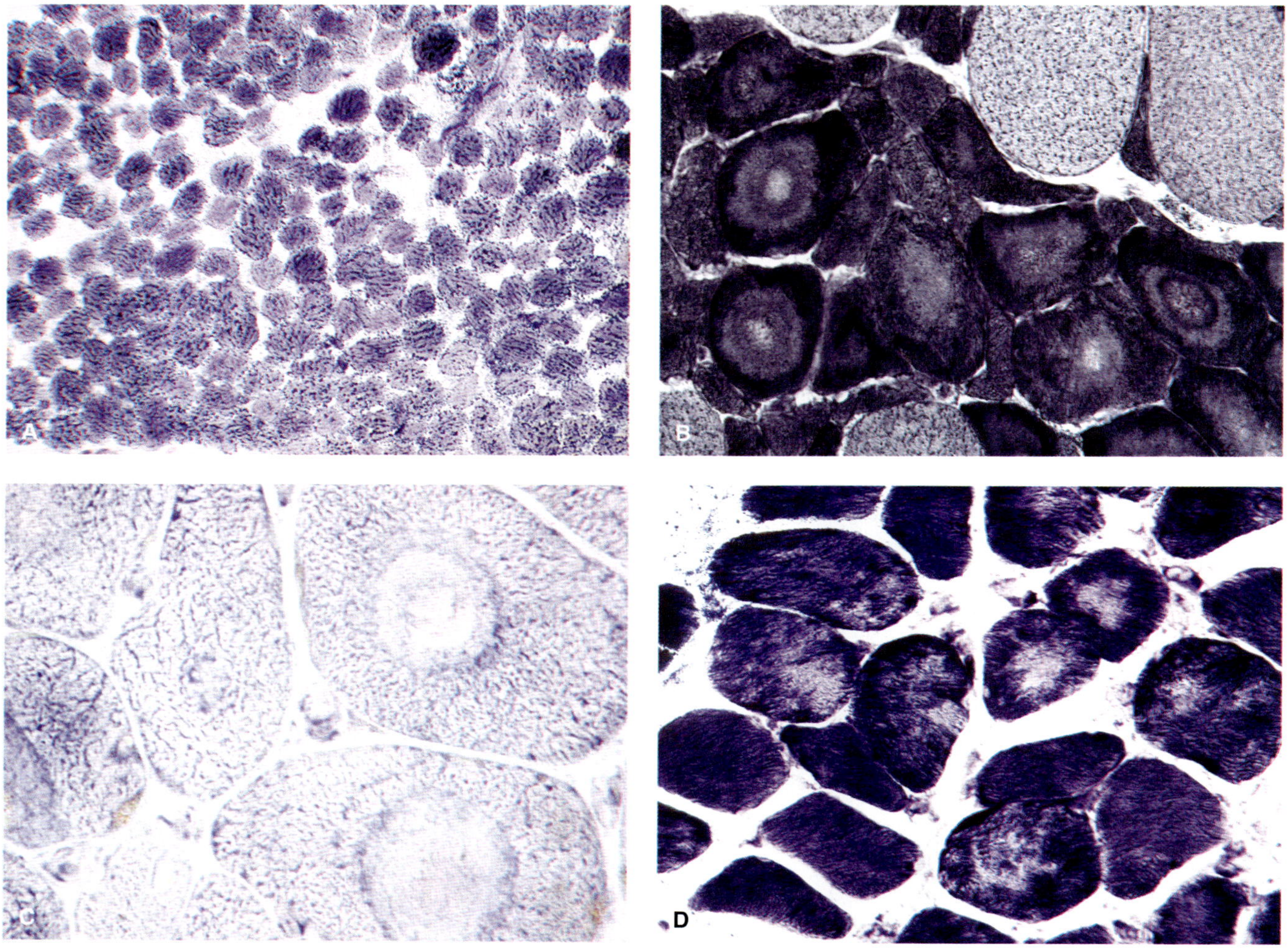

FIGURE 7.16 Aspects of NADH-TR enzyme histochemistry. **A:** Newborn muscle. Indistinct fiber typing is evident in oxidative enzyme reactions. **B:** Target fibers, with an inner unstained zone surrounded by a rim of increased enzyme activity. **C:** Central core myopathy. **D:** Moth-eaten fibers are a nonspecific finding in both myopathic and neuropathic conditions.

identifies denervated fibers in a highly sensitive manner as appearing overly dark, angular, and atrophic. The neuromuscular junction is normally identified by esterase enzyme histochemistry. It may show subtle abnormalities in diseases affecting structure such as myasthenia gravis. The esterase preparation also carries the ancillary advantage of highlighting infiltrative inflammatory cells by virtue of the esterase present in their cytoplasm. The alkaline phosphatase reaction identifies endomysial capillaries as well as regenerating fibers. The acid phosphatase reaction is characteristic of fibers with increased lysosomal activity; its utility in recognizing acid maltase deficiency/type 2 glycogen storage disease is a classic example. The acid phosphatase reaction in perimysial connective tissue has been advocated as a means of recognizing immune myopathies with perimysial pathology. Finally, in the workup of human disease, enzyme histochemical analysis of muscle tissue can be supplemented by biochemical analysis, specifically when histochemical techniques are unavailable.

Immunohistochemistry is an indispensable technique in pathology that has found a niche in the study of muscle. As already described, it may serve as an adjunct or replacement to traditional myosin ATPase preparations for fiber type analysis or in cases with vague fiber type differentiation by the ATPase technique. Fibers undergoing regeneration can be detected by immunohistochemical methods: regenerating fibers contain fetal myosins and react strongly with antibodies to vimentin and desmin. Denervated fibers may show nestin immunopositivity.

Among the dystrophic myopathies, visualization of the pathogenic proteins or their loss has become a mainstream tool in myopathology. Antibodies to the C- and N-termini and rod domain of dystrophin may permit the diagnosis of Duchenne and Becker dystrophies in the approximately 30% of cases in which blood-derived DNA is not diagnostic. Several congenital and limb girdle muscular dystrophies are also diagnosable by immunohistochemistry, usually revealing partial or complete losses of the relevant protein (Fig. 7.17, Table 7.2).

The nerve supply to muscle, including the intramuscular nerve twigs and motor endplates, cannot be adequately studied in routine samples. The anatomic location of nerve endings and endplates is variable depending on the muscle selected. They may be restricted to a narrow band across the muscle, or they may be more widely distributed throughout the muscle tissue. Some investigators prefer to biopsy shorter muscles, maximizing the chance of finding

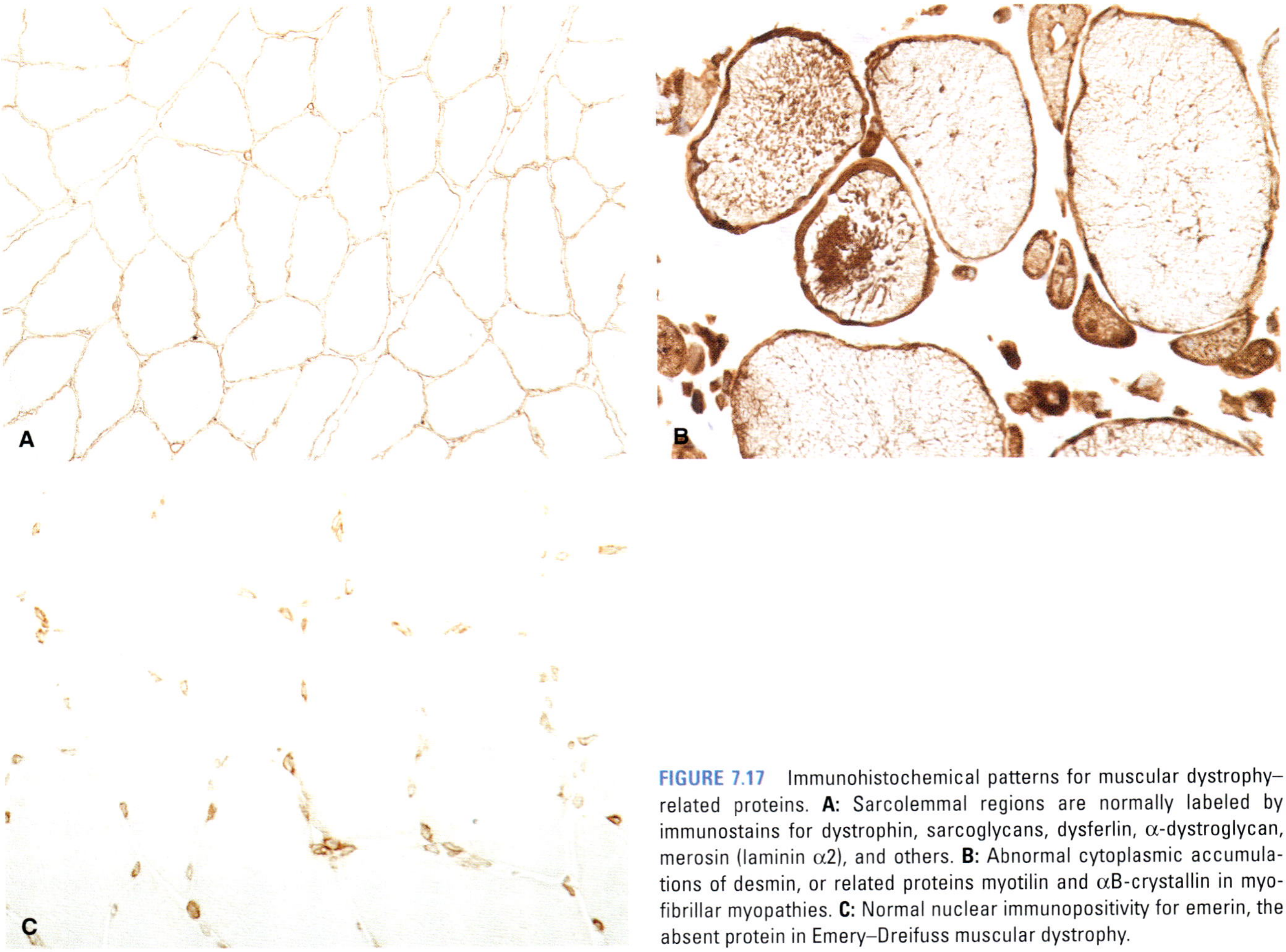

FIGURE 7.17 Immunohistochemical patterns for muscular dystrophy–related proteins. **A:** Sarcolemmal regions are normally labeled by immunostains for dystrophin, sarcoglycans, dysferlin, α-dystroglycan, merosin (laminin α2), and others. **B:** Abnormal cytoplasmic accumulations of desmin, or related proteins myotilin and αB-crystallin in myofibrillar myopathies. **C:** Normal nuclear immunopositivity for emerin, the absent protein in Emery–Dreifuss muscular dystrophy.

TABLE 7.2 Immunohistochemical Identification of Proteins in Selected Skeletal Muscle Diseases

Proteins	Muscle Disease
α-Actinin, actin, tropomyosin	Nemaline myopathies
α-Dystroglycan	Congenital muscular dystrophies of the muscle–eye–brain disorders
Caveolin	LGMD 1C; hyperCKemia; rippling muscle disease
Desmin; myotilin; α-actin; αB-crystallin	Myofibrillar AKA desmin myopathies
Dysferlin	LGMD 2B
Dystrophin, utrophin	Duchenne and Becker muscular dystrophy Internalized in some dystrophies, i.e., myofibrillar myopathies
Emerin	Emery–Dreifuss MD
MHC-1; membrane attack complex of C5b-9 (MAC)	Inflammatory myopathies
Merosin	Congenital muscular dystrophy
Sarcoglycans	LGMD 2C–2F
SMI-31; TDP-43; LC3; αB-crystallin	Inclusion body myositis

LGMD, limb girdle muscular dystrophy.

the intramuscular nerves. The external intercostal muscle has been used for this reason. Many limb muscles have a single band of terminal motor innervation that corresponds to the motor point. The motor point can be identified with the use of an electrical stimulator. After the administration of local anesthesia and incision of the skin, the muscle is stimulated using a metallic electrode before any tissue is removed. The nerve endings can be located at sites where a single fascicle, rather than the whole muscle, contracts after stimulation with a very weak current. Once the innervation zone is established electrically, the biopsy is removed. Different portions of the muscle innervation can be subsequently evaluated using a variety of techniques.

Vital staining with methylene blue has been used to demonstrate the intramuscular nerve twigs as well as the endplates. This technique requires that the muscle be injected with a methylene blue solution before the muscle sample is actually taken. An undesirable complication of this technique is muscle pain, which many patients experience during the injection of the dye. In order to preserve the staining of the nerve endings, the biopsy must be oxygenated for 1 hour. This technique is obviously complicated and not recommended for most laboratories. A simpler but less elegant technique for the demonstration of nerve twigs is the staining of muscle with silver methods such as Bodian stain. The postjunctional portion of the endplate can be stained histochemically for acetylcholinesterase activity. The reaction product is not restricted to the postjunctional membrane, and consequently, this is a relatively crude method of studying endplates.

More precise methods of studying endplates involve the use of α-bungarotoxin and freeze–fracture electron microscopy. α-Bungarotoxin is derived from cobra venom and binds specifically with the AChR. Immunoperoxidase techniques using α-bungarotoxin allow direct ultrastructural visualization of the postjunctional region of the motor endplate. With the use of freeze–fracture preparations, both the active zones of the presynaptic membrane and the AChRs of the postsynaptic membrane can be studied in greater detail. Freeze–fracture microscopy may be a useful ancillary diagnostic tool in certain disorders of the neuromuscular junction.

Finally, morphometric analysis of muscle tissue is indicated in the event that normal or abnormal findings, such as variations in fiber diameters, are minimal and subtle. Morphometry has been performed manually in the past, but more recently it has been possible to conduct morphometric analysis using computer-based image analysis (11).

GENDER, TRAINING, AND AGING

Some of the earliest studies addressing differences between males and females with regard to muscle fiber size and composition were conducted by Brooke et al. (12). In a seminal study of the biceps muscle in six patients, they established certain principles that remain generally true concerning gender differences in skeletal muscle. Individual muscle fibers are, on average, larger in males than in females. Type 2 fibers are usually larger than type 1 fibers in males, whereas type 1 fibers tend to be of equal or greater diameter in females. However, some of these differences are dependent on the muscles sampled, as examination of the vastus lateralis indicates no significant difference in diameter between type 1 and type 2 fibers in males. Similarly, males have a much higher percentage of type 2 fibers in the biceps muscle compared with women, but males and females have similar proportions of type 1 and type 2 fibers, in the vastus lateralis. Androgens likely play a role in the size of muscle fibers in some muscles, since it is known that testosterone supplements produce muscle fiber hypertrophy.

The effect of exercise and training on skeletal muscle has been examined over the years. The results of many of these studies are conflicting, but certain general principles have emerged. It is clear that exercise and training of any type increases muscle fiber diameter. Most authorities agree that power training such as weightlifting results in remarkable hypertrophy of type 2 fibers and less, if any, enlargement of type 1 fibers. Anaerobic activities promote hypertrophy of type 2 fibers and slow age-related atrophy of type 2 fibers. In long-distance runners, for whom aerobic, fatigue-resistant metabolism is more important, type

1 fibers tend to be larger and are accompanied by greater capillary density. Elite long-distance runners may have a greater proportion of type 1 fibers, whereby runners have genetically determined fiber type compositions and little, if any, conversion of fiber types takes place during training. Furthermore, improvement in muscle function through training may be more a reflection of changes in metabolism, including mitochondrial function that is not manifested in fiber sizes and type distribution. Animal studies have shed minimal light on these questions, in part because animal muscle responds differently to exercise and training than does human muscle. In fact, animal experiments have more often clouded the issues of exercise and fiber composition instead of resolving the controversy.

There is a functional and structural decline in skeletal muscle beginning in the sixth decade and accelerating after the age of 70 years. By the age of 75 years, there is a 30% to 50% decline in muscle strength, the causes of which are complex. Part of the answer lies in the reduction of fiber diameters from approximately 65 microns in the 30- to 50-year age group to around 43 microns in an 81- to 89-year age group for type 1 fibers and from 70 microns to 40 microns for type 2 fibers. The reduction in fiber diameter may be due, in part, to an alteration in growth factors, including myostatin, a member of the transforming growth factor family. Myostatin is a negative regulator of muscle, a mediator of catabolic pathways in muscle cells.

Because of the alterations in the composition of their connective tissues, associated with decreased elasticity and flexibility, and because many older patients have joint disease of varying severity, the elderly become less active with a corresponding reduction in muscle volume and contractile strength. Some experts view this condition as a form of disuse. Their conclusions are supported by the fact that aging individuals, like young patients who do not use their muscles (e.g., as a result of immobilization in a cast), have selective atrophy of type 2 fibers. The effect of poor nutrition in the elderly is probably a factor, as it is known that cachexia is accompanied by atrophy of type 2 fibers.

A second problem in the elderly population is an insidious degeneration of motor units, specifically to the anterior horn cells in the spinal cord. Due to degenerative spine disease, there is also injury to nerve roots, with subsequent radiculopathy. The integrity of the muscle fiber is closely related to the maintenance of its nerve supply. Any sustained interruption of trophic influences from the motor neuron or nerve will culminate in atrophy of the denervated muscle fiber. Acute denervation of muscle results in randomly distributed atrophic fibers with a preference for type 2 fiber involvement. Atrophic fibers are characteristically angular or ensate (sword-like) when sectioned transversely. They appear flattened and bipolar, with tapering ends and may lie in tandem. Selective atrophy of type 2 fibers may be the only pathologic abnormality in acute denervation, so that the proper diagnosis requires corroborative clinical information to exclude other causes of selective type 2 myofiber atrophy such as disuse, advanced age in females, polymyalgia rheumatica, and many others. The proportion of atrophic type 1 and type 2 fibers tends to equalize in progressive denervation.

The esterase enzyme histochemical preparation is very useful under these circumstances because denervated fibers are overly dark whereas atrophic fibers in other conditions are not (Fig. 7.18A). Atrophic fibers are also excessively dark in oxidative enzyme reactions, but such staining applies to fiber atrophy of almost any cause. Small dark fibers in oxidative preparations are probably explained by the fact that mitochondria are relatively spared or may even proliferate in the atrophic process and occupy a proportionately greater volume of sarcoplasm.

Prima facie evidence of advanced denervation is a progression from random fiber atrophy to grouped atrophy in which multiple collections of small, angular atrophic fibers are present in the biopsy sample (Fig. 7.18B). As a consequence of chronic denervation and of reinnervation, the normal checkerboard staining profile observed in histoenzymatic reactions is lost. In an effort to reestablish the nerve supply to denervated muscle fibers, intact intramuscular nerves undergo collateral sprouting, and new synapses are formed with atrophic fibers. As motor units enlarge, reinnervated fibers occupying a large area are converted to one histochemical type. The phenomenon of type grouping (Fig. 7.18C) is explained by the fact that all muscle fibers within a single motor unit are of the same type and the motor neuron, through the trophic influences of its axon and collaterals, governs the histochemical properties of its fibers. The plasticity of muscle fibers allows conversion from one histochemical type to the other when there is renervation by a motor neuron of the opposite type.

Type grouping and target fibers are pathognomonic of renervation. Despite their unique specificity, *bona fide* target fibers are present in less than 25% of cases of neuropathic disease. Although targets and central cores are similar morphologically, they differ in three ways. Both can occur singly within a fiber but the target is larger in diameter. The target is limited in length, only extending across a few sarcomeres, whereas the central core may run the entire length of the fiber. Most important is the three-zone architecture of the target fiber (Fig. 7.16B). The central zone is ultrastructurally indistinguishable from the unstructured core, and is surrounded by an intermediate zone that forms an intensely stained rim in oxidative enzyme reactions. By definition, the intermediate zone, difficult to identify in most other stains, is absent from a core. It is a zone of transition between the central zone of severe sarcoplasmic disruption and the third zone, which represents the normal portion of the muscle fiber. Targetoid fibers lack the intermediate zone of increased oxidative enzyme activity and are morphologically identical to core fibers. The term *core* is conventionally used in cases of congenital central core disease (Fig. 7.16C), and the term *targetoid* is applied to cores that are found in any other condition. Targetoid fibers are more commonly

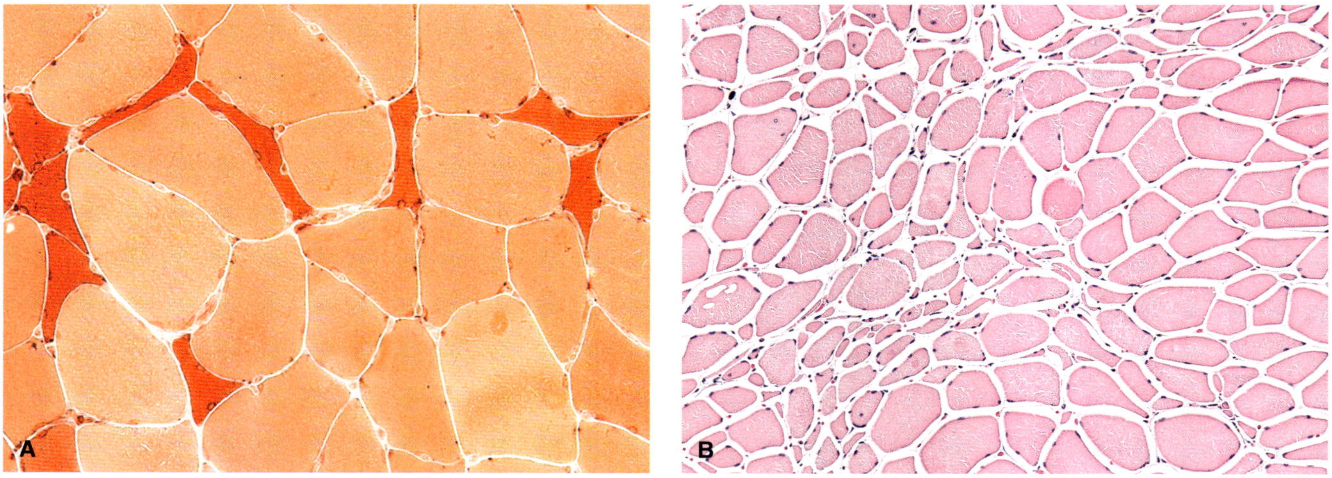

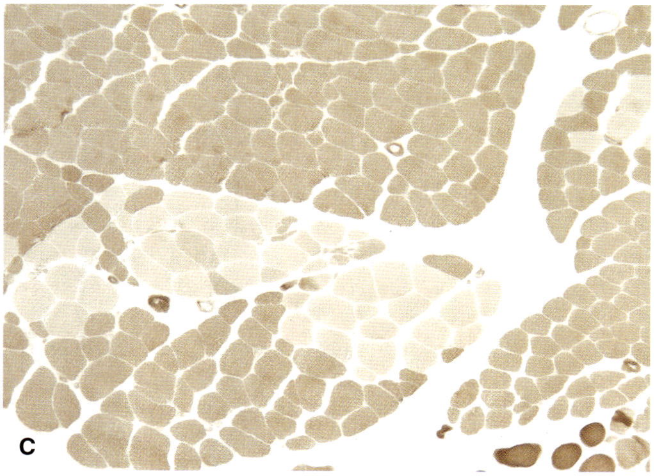

FIGURE 7.18 Neuropathic abnormalities. **A:** Neurogenic atrophy. Angular atrophic fibers overly dark in the esterase enzyme histochemical preparation. **B:** Grouped atrophy in a confirmed case of amyotrophic lateral sclerosis. Note the compensatory hypertrophy of some fibers and the notable degree of myofiber shrinkage with formalin fixation (H&E, formalin-fixed paraffin-embedded). **C:** Chronic denervation with reinnervation producing fiber type grouping and alteration of the normal checkerboard staining profile, with groups, defined as a fiber completely surrounded by its own fiber type, of both types (ATPase @ pH 9.4).

encountered in neurogenic atrophy than any other condition and are more frequently seen than target fibers.

ARTIFACTS

The most common artifacts are related to unsuspected or inadvertent injury to the muscle specimen in vivo (precollection phase), collection and processing errors at the time of removal, and improper tissue holding, freezing, and sectioning. Needle tracts, such as those produced during electromyography (EMG) studies, may be easily recognized when they are linear. More often, needle tracts are cut tangentially so that the pathologist may be misled by a histologic picture of myopathy exemplified by fiber necrosis, regeneration, inflammation, and interstitial fibrosis (Fig. 7.19A). This kind of precollection artifact is generally traceable to poor communication between the physician requesting the biopsy and the individual performing the procedure, who is unaware of the previous intramuscular injections. Since routine immunizations are usually administered in the thighs of infants, it is not unusual for a diagnostic muscle biopsy in this age group to include a prior injection site. Aggregates of macrophages containing adjuvant material will be observed microscopically in such instances (Fig. 7.19B).

In an uncommon but very misleading collection error, large numbers of neutrophils are occasionally observed within the intramuscular blood vessels. These cells may be marginated or even penetrate the vascular walls to enter the perimysium or endomysium, simulating an acute vasculitis. In the absence of other pathologic changes within the specimen, the presence of neutrophils usually means that the muscle has been crushed or squeezed by an inexperienced hand during the biopsy procedure.

Vacuolar artifact results from improper freezing technique and can be minimized by using proper techniques that permit rapid freezing and appropriate storage to prevent thawing. Mild vacuolar artifacts may be tolerable, but large vacuoles that disrupt the sarcoplasm are especially troublesome (Fig. 7.20A). Larger vacuoles can interfere with accurate biopsy interpretation by distorting the pathologic changes in the sample or by simulating the picture of vacuolar myopathy, such as glycogen or lipid storage disease.

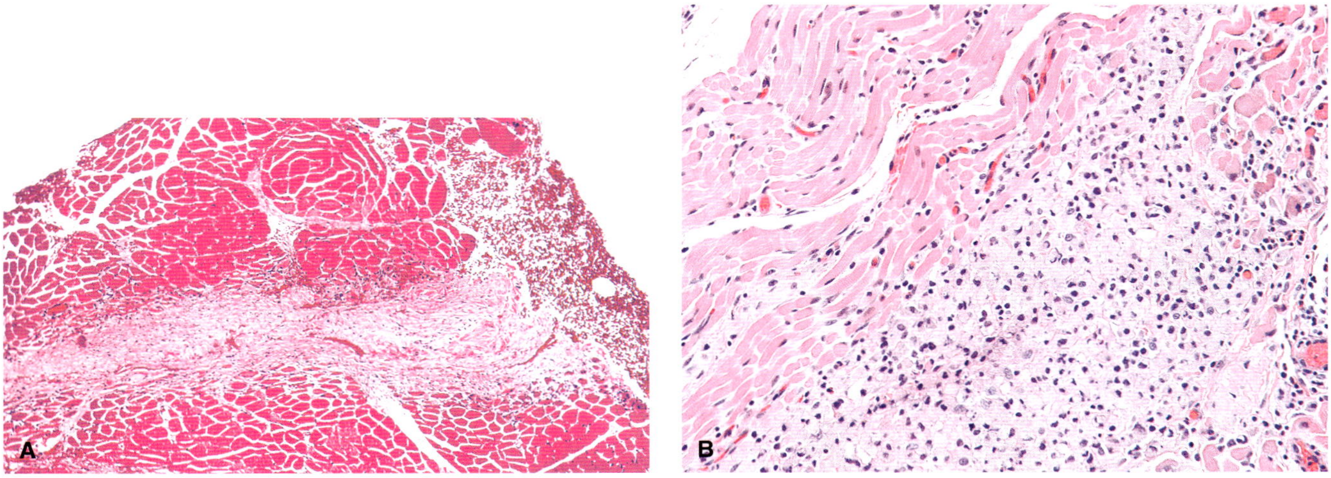

FIGURE 7.19 **A:** Needle tract. Area of injury contains necrotic fibers and a small focus of lymphocytic inflammation (H&E). **B:** Thigh immunization injection site showing aggregates of macrophages containing adjuvant material.

Spaces representing freeze artifact can be lessened by thawing and snap freezing the specimen using the method described below (Fig. 7.20B). Also, use of powdered gloves in specimen handling can leave behind tiny talc granules that are confirmed by viewing the section under polarized illumination.

Hyaline fibers are evident in specimens damaged by contraction artifact. These fibers are abnormally increased in diameter and rounded in configuration. Contraction artifact is best observed in longitudinal sections where dark hypercontracted regions are separated by pale, ghostlike zones of myofibrillar disruption (Fig. 7.21A). Their sarcoplasm in both paraffin and frozen sections is smudged, glassy, and more deeply stained than in normal fibers. Contraction artifact is particularly undesirable when either immunohistochemical or electron microscopic studies are needed, even if the artifact is subtle and cannot be appreciated at the light microscopic level. The detection of ultrastructural abnormalities, which is dependent on the normal alignment of the myofibrils and myofilaments, is compromised by distortion of sarcomeric structures. An undue number of these fibers may be seen in muscle biopsies taken from infants in which the smallness of the biopsy results in excessive mechanical irritation and contraction artifact. True hypercontracted, hyaline fibers in a carefully removed muscle biopsy are a common feature of Duchenne muscular dystrophy or any condition with diminished sarcolemmal integrity (Fig. 7.21B). Hyaline fiber formation may also be seen as a harbinger of in vivo necrosis of a myofiber, aside from the aforementioned iatrogenic effects described above.

Electrocautery as a means of hemostasis is a notorious cause of contraction artifact and should be strictly avoided

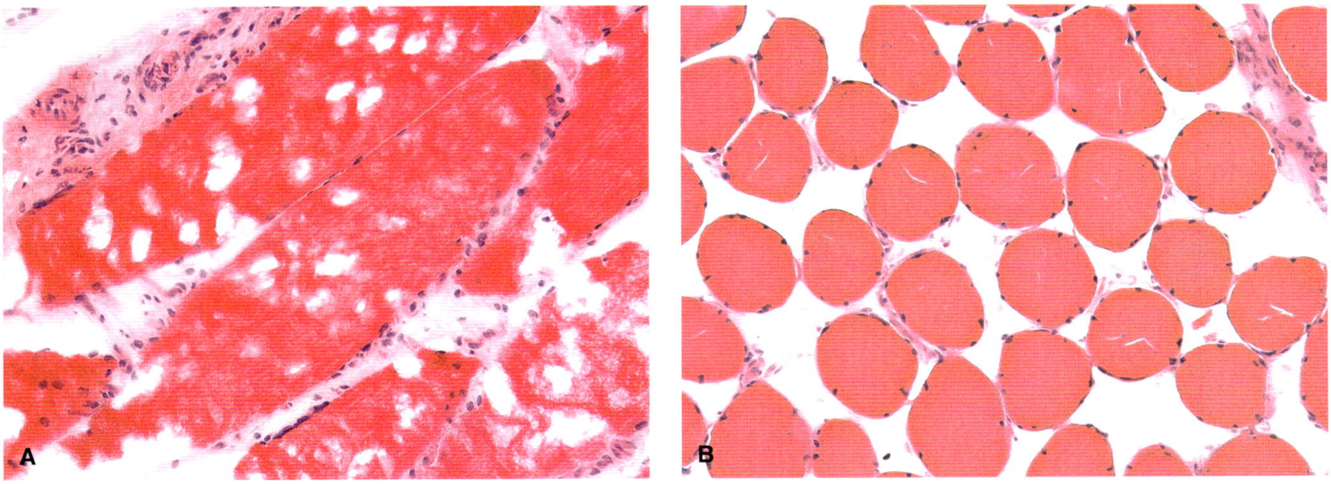

FIGURE 7.20 **A:** Vacuolar artifact. Improper freezing has caused numerous clear holes to form within the fibers (H&E). **B:** Thawing and refreezing of improperly frozen specimens may result in diminished freeze artifact but renders the fibers rounded with empty spaces between fibers, and slit-like spaces representing collapsed vacuoles.

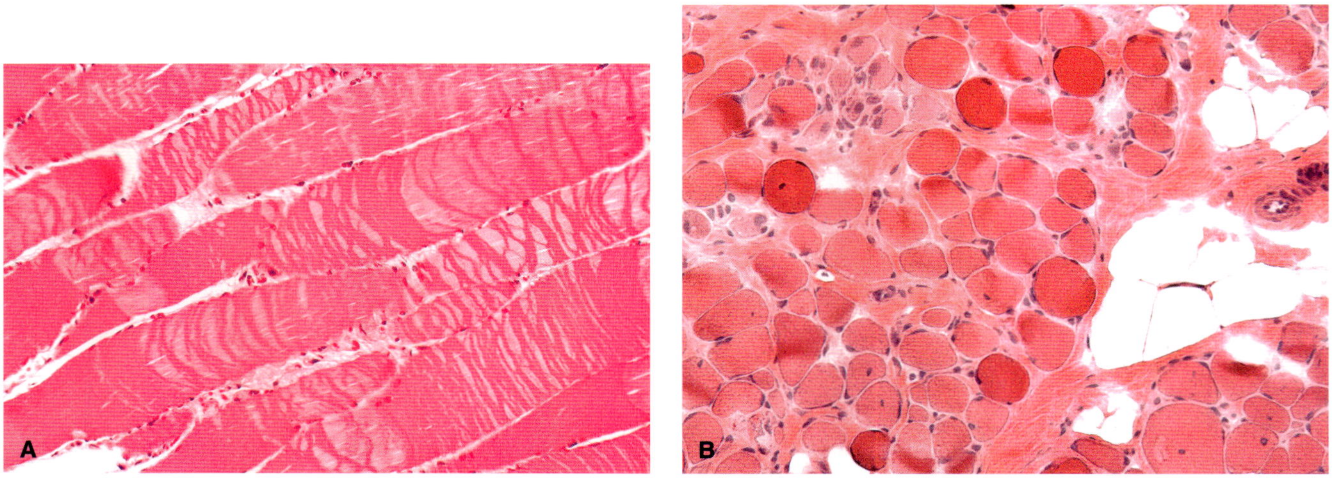

FIGURE 7.21 **A:** Contraction artifact. Dark contraction bands and lucent zones of fiber disruption are seen in longitudinally oriented fibers (H&E). **B:** Hyaline, contracted fibers. Several fibers are enlarged, rounded, with darkly stained sarcoplasm, seen in this example of Duchenne muscular dystrophy (H&E).

until the biopsy is completely excised. Isometric clamping during collection is advocated by some as a means to avoid contraction artifact; however, many surgeons are unfamiliar with their use and consequently they may produce more artifact than was meant to avoid.

Excessively pale histochemical reactions can result from the degradation of enzyme systems in the sarcoplasm. Artifacts are distinguished from legitimate abnormal staining if all histochemical reactions in the biopsy are pale. Laboratories that accept transported consultation specimens should instruct the originating institution to keep the specimen cool in a sealed specimen container, on a very slightly moistened gauze or nonstick specimen pad, and ensure transport as quickly as possible. If a specimen is inadvertently exposed to melting ice or excessive amounts of saline, the excess fluid infiltrates the endomysium and perimysium simulating edema and promotes ice crystal artifact. Moreover, this hypotonic fluid produces a rat-bite effect in some fibers, the result of which is a peripheral irregular defect in the fiber, looking like a bite has been taken out of the sarcoplasm.

Other artifacts are due to sectioning and staining errors. Of unknown origin is nuclear vacuolization, which seems to occur in frozen sections rather than paraffin sections (Fig. 7.22). Scattered nuclei have peripherally marginated chromatin. The center of the nuclei is clear or light pink in H&E stains. The color, greater frequency, and size distinguish this artifact from most pathologic intranuclear inclusions. A disfiguring and distracting artifact occurs in frozen sections stained with Gomori trichrome. The normal green staining of the tissue is distorted by irregular, red-stained areas that interfere with biopsy interpretation (Fig. 7.23). This seems to be

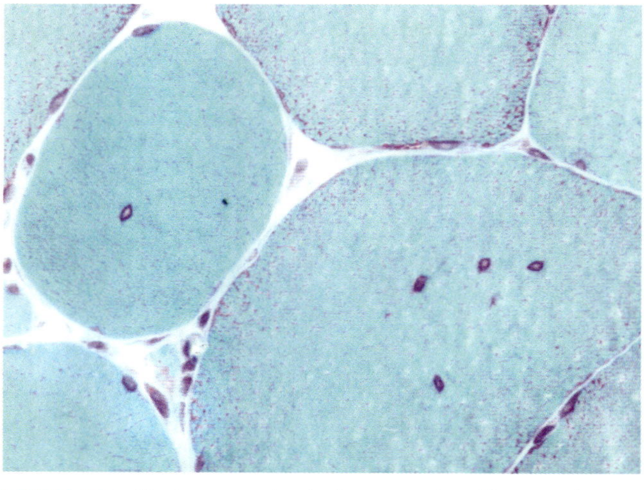

FIGURE 7.22 Nuclear vacuoles in frozen section. The vacuoles are small and round with a thick collar of chromatin as opposed to true intranuclear inclusions (H&E).

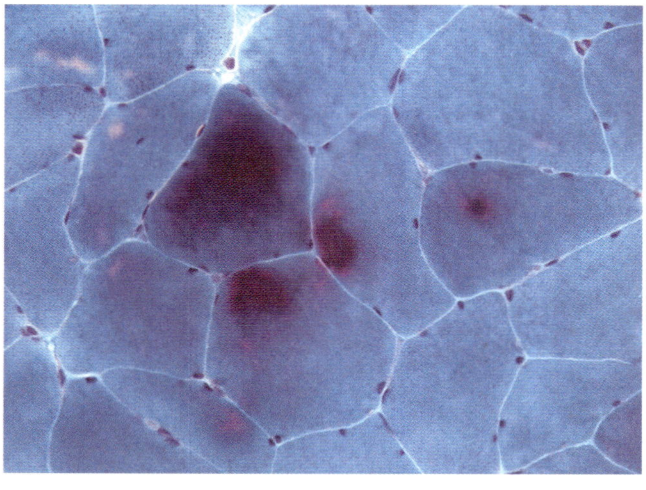

FIGURE 7.23 Of uncertain origin is an abnormal artifact in frozen section stained with Gomori trichrome. The internal detail of the sarcoplasm is distorted by a red discoloration of the normal green-staining properties of the fibers.

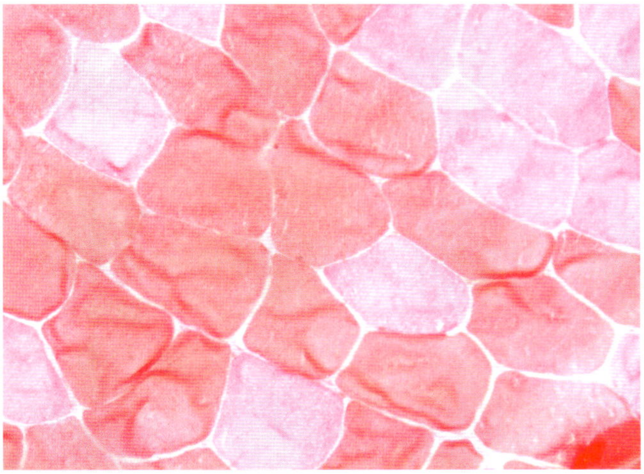

FIGURE 7.24 Piecrust artifact. Linear dark-staining creases caused by wrinkling or folding of the sarcoplasm (ATPase @ pH 9.4).

more common when specimens are shipped in ice and the ice melts, leaving the exposed fresh tissue floating in water. Dark staining of the sarcoplasm in random fibers is often due to variations in section thickness. Fibers adjacent to the connective tissue of the perimysium are especially susceptible to this artifact. Inconsistencies of section thickness may be recognized when linear, band-like regions of intense staining are visible within muscle fibers. Another sectioning artifact involves wrinkling or folding of the fiber that is especially evident in ATPase reactions (Fig. 7.24). This phenomenon is sometimes called the piecrust artifact and it has been attributed to coverslip lifting or to improper transfer from the cryostat.

DIFFERENTIAL DIAGNOSIS

Several findings in skeletal muscle biopsies are normal minor variations that may be mistaken for pathologic change, or are entirely nonspecific. These include internal nuclei, ring fibers, increased endomysial connective tissue, inflammation, variations in fiber diameter, lipofuscin, cytoplasmic bodies, tubular aggregates, and ragged red fibers.

One of the most common pathologic abnormalities in muscle biopsies is nuclear internalization (Fig. 7.25). Quantitative analyses have demonstrated that the nuclei are peripherally located in 97% to 99% of normal muscle fibers. Internal nuclei may be seen in 5% to 10% of fibers in a variety of conditions, particularly those associated with mild atrophy. Nuclear internalization has no specific diagnostic significance and can represent a reaction to virtually any type of injury, including neuropathic conditions. The diagnosis of myotonic dystrophy should be strongly considered if the vast majority of fibers contain internal nuclei.

One must exercise caution in interpreting the significance of ring fibers ("ringbinden"). The ring is formed by a bundle of peripheral myofibrils that are circumferentially oriented such that they encircle the internal portion of sarcoplasm, which is normal in structure and orientation (Fig. 7.26). The ring is especially well visualized in stains performed on cross sections that demonstrate the relationships between the striations of the transversely oriented peripheral myofibrils and the inner sarcoplasmic contents. Rings are also seen to advantage in PTAH stains, resin sections, and under phase-contrast microscopy. The pathologically oriented myofibrils are generally normal in structure except for hypercontraction of the sarcomeres. In properly processed, uncontracted muscle biopsies, ring fibers may be a pathologic criterion for all types of muscular dystrophy, particularly myotonic disorders. However, they can also signify regeneration whereby reorganization of the fiber architecture involves failure to align myofibrils correctly.

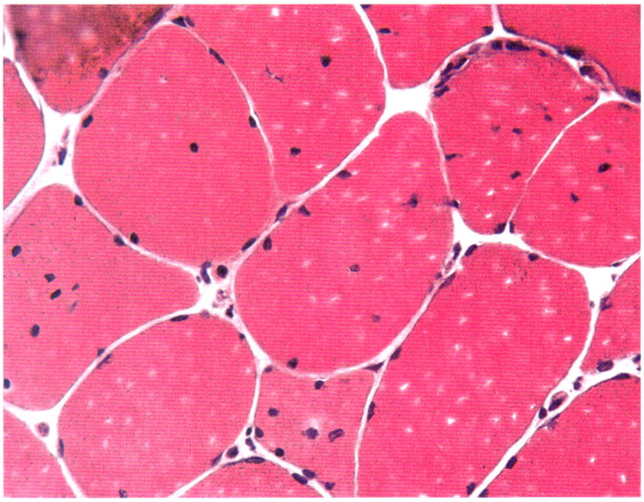

FIGURE 7.25 Nuclear internalization. Several fibers contain internal pyknotic nuclei, a common nonspecific finding if seen in less than 3% of fibers. Greater numbers may be seen in a wide variety of neuropathic and myopathic conditions (H&E).

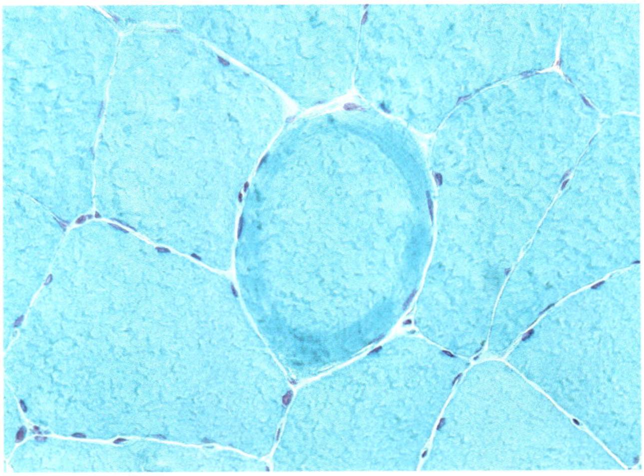

FIGURE 7.26 Ring fibers ("ringbinden"). Bundles of myofibrils are circumferentially oriented, forming rings that encircle transversely sectioned fibers (H&E).

Excessive quantities of endomysial connective tissue usually reflect reactive fibrosis accompanying neuromuscular disease. However, abundant connective tissue is normally present at the interface between muscle and tendons or fascia and should not be regarded as reactive fibrosis. Although endomysial connective tissue is not prominent in the biopsies of infants, the perimysial connective tissue far exceeds the amount present in older children and adults.

Interstitial and perivascular inflammatory cells almost always reflect clinical disease, most frequently immunologically mediated myopathies such as polymyositis or dermatomyositis. Moreover, some dystrophic myopathies are well known to display prominent inflammatory infiltrates, including facioscapulohumeral dystrophy, merosin deficiency, and the dysferlinopathies. However, small foci of hematopoiesis can be present and do not represent pathologic inflammatory infiltrates. Muscles subjected to trauma such as EMG needles often harbor foci of inflammation for months following the diagnostic study and are not clinically significant.

One of the most demanding challenges to the diagnostic pathologist is the muscle biopsy characterized by a variation in fiber diameters. Infrequently biopsied muscles such as paraspinal or abdominal wall muscles may be proposed as an alternative biopsy site when spinal surgery or gastrostomy placement is planned, in order to spare the patient a separate incision. However, these muscles normally display conspicuous variation in fiber diameters that would be considered abnormal if seen in biopsy samples obtained from the proximal muscles of the extremities. Smaller muscles, especially those devoted to finely coordinated activities, have smaller-fiber diameters than large, bulky muscles. In evaluating fiber size, it may be necessary to measure fiber diameters. Morphometric analysis of the muscle biopsy is imperative when the changes in fiber diameters are minimal and subtle. Average fiber diameters should not vary more than 12% between type 1 and 2 fibers. While the diameter of fibers is determined by several factors, it is important to realize that there is up to 30% diminution of fiber diameters in formalin-fixed paraffin-embedded tissue as compared with cryosections.

In order to obtain statically significant morphometric data, the lesser diameter of each muscle fiber should be determined, based on a minimum number of 200 fibers in the sample. The atrophic or hypertrophic process may be selective, affecting only one fiber type, or it may be nonselective. Selective atrophy of type 1 fibers is most commonly encountered in myotonic dystrophy, but conspicuous type 1 fiber hypotrophy along with type 2 hypertrophy is a definitional feature of congenital fiber type disproportion. Type 2 fiber atrophy is a common finding in acute denervation, disuse, upper motor neuron deficit, protein malnutrition, chronic corticosteroid administration, Cushing disease, myasthenia gravis, primary hyperparathyroidism, some rheumatologic disorders, aging females, and paraneoplasia (Fig. 7.27). Hypertrophy of type 1 fibers is relatively specific for infantile spinal muscular atrophy. The pattern of atrophy is important in distinguishing between normal and abnormal. Randomly distributed small or large fibers may be normal, depending on other factors discussed above. Grouped atrophy, where five or more small angular fibers cluster together, is essentially diagnostic of chronic neurogenic disease (Fig. 7.18B). Perifascicular atrophy is typical of dermatomyositis and other immune myopathies with perimysial pathology.

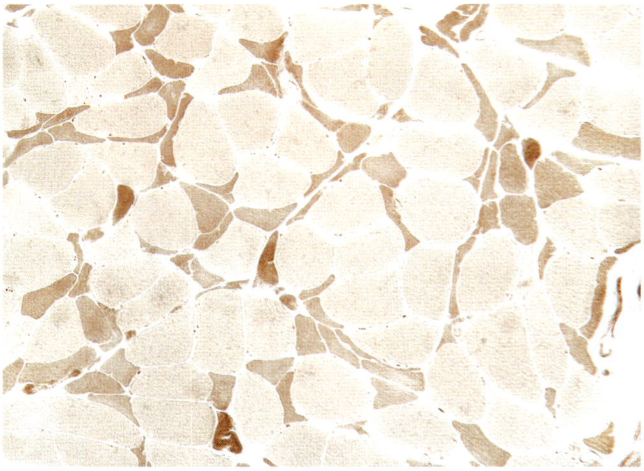

FIGURE 7.27 Type 2 fiber atrophy of an extreme degree in a case of polymyalgia rheumatica (ATPase @ pH 9.4).

Lipofuscin is a common finding in paranuclear and subsarcolemmal locations in muscle fibers, and likely reflects intracellular degradation of waste products. It is a nonspecific finding associated with aging, but also seen in vitamin E deficiency and a wide variety of myopathic conditions. Cytoplasmic bodies are easily recognizable sarcoplasmic masses composed of a central core with a pale periphery and are best visualized with the Gomori trichrome stain (Fig. 7.28). They are well documented to be associated and contiguous with

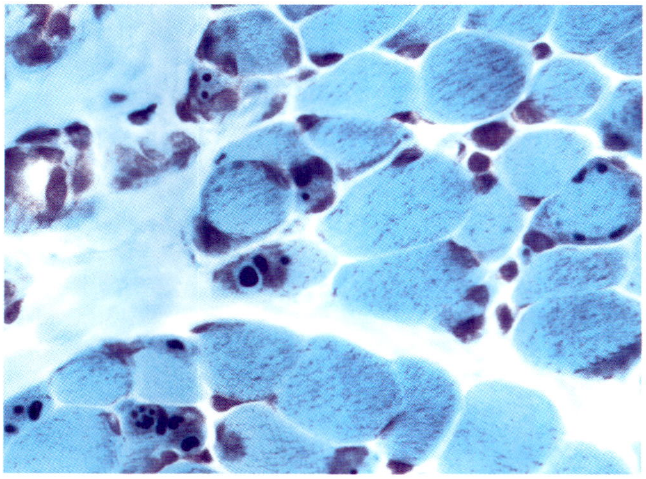

FIGURE 7.28 Cytoplasmic bodies are a highly recognizable but nonspecific find, showing red color in the Gomori trichrome stain with the characteristic haloes.

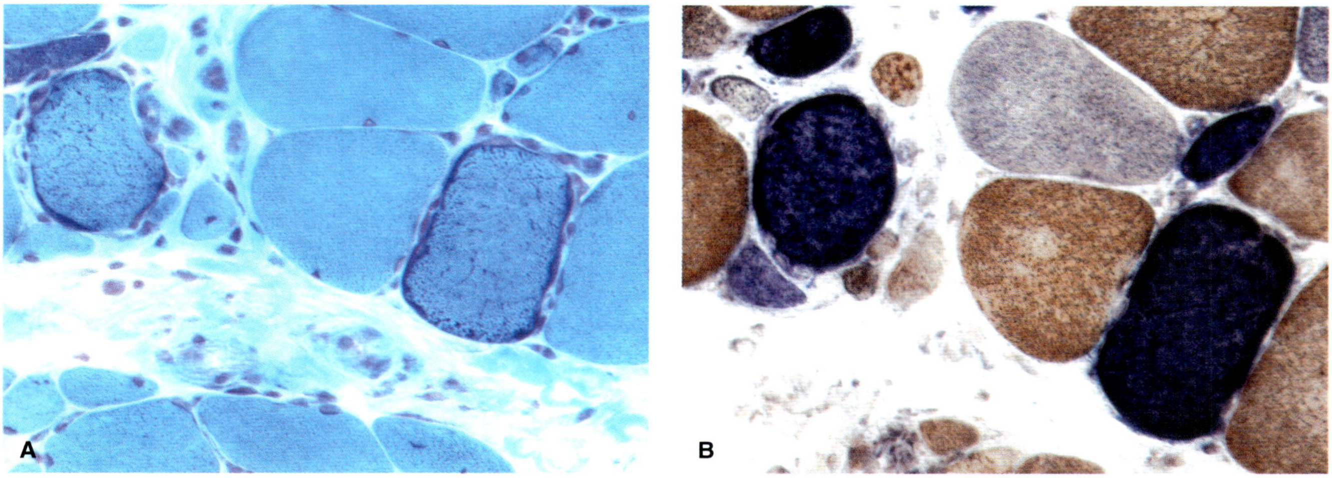

FIGURE 7.29 **A:** Ragged red fibers. Ragged red fibers are recognized in Gomori trichrome stain as having an irregular sarcolemmal surface with collections of red-staining material. **B:** The adjacent section shows the same fiber as a ragged blue fiber by the COX-SDH enzyme histochemical preparation.

varying degrees of Z-band streaming. They may be seen in a wide variety of unrelated disorders, including denervation atrophy, inflammatory myopathies, myotonic and other dystrophies, periodic paralysis, mitochondrial myopathies, and others. When seen as the predominant finding in some myopathies, the term cytoplasmic body myopathy has been used.

Ragged red fibers can be observed in the aging process. These fibers are recognized in Gomori trichrome stains performed on frozen sections because they exhibit an irregular surface and collections of red-staining subsarcolemmal material (Fig. 7.29A). The ragged red areas represent foci of increased, often abnormal mitochondria. An SDH enzyme histochemical preparation of a ragged red fiber yields the same evidence for subsarcolemmal mitochondrial proliferation, known as a "ragged blue fiber" (Fig. 7.29B). Ragged red or blue fibers in younger individuals are hallmarks of the mitochondrial myopathies, which are characterized by mitochondrial dysfunction and mutations of mitochondrial genes. It is now known that mitochondrial damage occurs in the aging cell, including skeletal muscle, as a result of increased sporadic mitochondrial DNA deletions. Ragged red fibers are considered to be a reflection of this phenomenon, which may be associated with late-onset mitochondrial disease in some cases.

SPECIMEN HANDLING

Muscle biopsies should be performed by physicians with expertise in biopsy technique and a sincere interest in obtaining the best possible specimen. The physician who has direct responsibility for the patient's care needs to be sure that the biopsy comes from an appropriate muscle so that it is representative of the disease process. The disease process is widespread in some conditions such as metabolic diseases, and virtually any muscle is suitable for biopsy. However, some disorders are characterized by symptoms limited to a specific muscle or muscle group and, in this situation, sampling of the affected area is more likely to yield meaningful results. The tissue sample should be obtained from a region in which the disease process remains active rather than quiescent. Ultrasound or MRI guidance can be useful in choosing a biopsy site, especially in inflammatory myopathies or dystrophic myopathies in which there may be drastic differences between involved muscles. The pathologic findings are likely to be those of end-stage disease and defy conclusive pathologic classification, particularly when marked weakness or wasting is present. Muscles subjected to previous traumatic injury, such as needle tracts incurred during EMG or intramuscular injections of medications, and muscles altered by an unrelated disease process should not be biopsied.

The special handling of the muscle biopsy is of paramount importance. Ideally, muscle biopsies are performed on weekdays when histotechnical support is available for immediate processing. Submission of the specimen "after hours" may be clinically unavoidable. When a vasculitis is suspected and rapid treatment is contemplated, specimens can be processed for formalin fixation and paraffin embedding on an urgent basis.

Ideally, two separate specimens from the same site are desirable. The major drawback to needle biopsy is the limited size of the sample, although basic histologic determinations may be made in the hands of skilled histotechnologists with such small biopsies. Genetic testing may be done when an adequate amount of material is obtained (approximately 30 to 50 mg); however, enzyme activity assays predictably require amounts of muscle tissue (at least 150 mg) only achievable with open biopsies.

The primary fixative for paraffin embedding of muscle biopsies is 10% formalin, buffered to a pH of 7.4 in a 0.1 M phosphate buffer. Strips of muscle 1 mm in width are dissected from the edges of the sample and postfixed in phosphate-buffered 2% glutaraldehyde for electron microscopy. A second unfixed specimen measuring 1 × 0.5 × 0.5 cm is obtained for the preparation of frozen sections. The portion designated for cryotomy should be carefully assessed for orientation to yield cross sections. A dissecting microscope is sometimes useful toward this goal. The ideal method of rapid freezing is to cool isopentane in liquid nitrogen to below −100 °C, then immerse the muscle tissue this solution. Many laboratories place the cylinder of muscle into gum tragacanth on a cork bed, a semisolid substance that solidifies as an embedding base upon freezing. Whatever technique is employed, the freezing technique must be carried out rapidly. Freezing the tissue in a cryostat in a fashion similar to most specimens submitted for frozen section diagnosis from the operating room is contraindicated because of the propensity to cause freeze artifact. When a muscle biopsy has been suboptimally frozen, it is likely to demonstrate unacceptable freeze artifact; thawing to room temperature with subsequent refreezing can result in significant resolution of freeze artifact. The unavoidable result is that the muscle fibers will assume an abnormal rounded profile with an artificial separation of the fibers by spaces not found in nature. The frozen sample should be oriented so that cross sections of muscle are cut. It is highly desirable that the histotechnologists obtain sequential and consecutive facing cryosections so that individual fibers can be tracked through the diverse stains performed. Serial frozen sections are stained with H&E, Gomori trichrome, and by the enzyme histochemical reactions ATPase (at pH 9.4, 4.3, and 4.6), NADH-TR, nonspecific esterase, and combined COX-SDH. Other stains such as PAS for glycogen, myophosphorylase, alkaline and acid phosphatase, and lipid stains are performed when indicated. Frozen tissue may also be used for biochemical analysis, for immunohistochemical preparations, and for immunofluorescence microscopy. Inasmuch as frozen tissue may be needed for future additional studies, muscle biopsies can be sealed in airtight plastic capsules or bags to prevent dessication and freezing artifact while stored in an ultralow freezer at −70 °C.

REFERENCES

1. Messina G, Biressi S, Monteverde S, et al. Nfix regulates fetal-specific transcription in developing skeletal muscle. *Cell*. 2010;140(4):554–566.
2. Gilbert SF, Barresi MJF. *Development Biology*. 11th ed. Sinauer Associates; 2016.
3. Sadler TW. *Langman's Medical Embryology*. 13th ed. Lippincott Williams & Wilkins; 2016.
4. Adhihetty PJ, Hood DA. Mechanisms of apoptosis in skeletal muscle. *Basic Appl Myol*. 2003;13(4):171–179.
5. O'Flaherty J, Mei Y, Freer M, Weyman CM. Signaling through the TRAIL receptor DR5/FADD pathway plays a role in the apoptosis associated with skeletal myoblast differentiation. *Apoptosis*. 2006;11(12):2103–2113.
6. Heffner RR Jr, Moore SA, Balos LL. Muscle biopsy in neuromuscular diseases. In: Mills SE, ed. *Sternberg's Diagnostic Surgical Pathology*. Vol. 1. 6th ed. Lippincott Williams & Wilkins; 2015:113–147.
7. Dubowitz V, Sewry CA, Oldfors A. *Muscle Biopsy. A Practical Approach*. Saunders Elsevier; 2013.
8. Banker BQ, Engel AG. Basic reactions of muscle. In: Engel AG, Franzini-Armstrong C, eds. *Myology: Basic and Clinical*. 3rd ed. McGraw-Hill; 2004:691–747.
9. Curtis E, Sewry C. Electron microscopy in skeletal muscle pathology. In: Stirling JW, Curry A, Eyden B, eds. *Diagnostic Electron Microscopy: A Practical Guide to Interpretation and Technique*. 1st ed. John Wiley & Sons, Ltd; 2013.
10. Raheem O, Huovinen S, Suominen T, Haapasalo H, Udd B. Novel myosin heavy chain immunohistochemical double staining developed for the routine diagnostic separation of I, IIA and IIX fibers. *Acta Neuropathol*. 2010;119(4):495–500.
11. Pertl C, Eblenkamp M, Pertl A, et al. A new web-based method for automated analysis of muscle histology. *BMC Musculoskelet Disord*. 2013;14:26.
12. Brooke MH, Kaiser KK. Muscle fiber types: how many and what kind? *Arch Neurol*. 1970;23(4):369–379.

8

Blood Vessels

Carolyn H. Glass

GROSS AND LIGHT MICROSCOPIC FEATURES

The normal structures of the human vascular and lymphatic systems are designed to accommodate their functional roles within the circulatory system. Arteries are phenotypically different from veins and lymphatic vessels, given their primary function of delivering oxygenated blood to the rest of the body. The structures of vessels change progressively throughout life, and this is particularly true for the aorta, elastic and muscular arteries, and larger veins (Table 8.1). Over time, these and other age-related changes lead to decreased distensibility of both arteries and veins, although arteries tend to be more severely affected. Aging arteries are especially affected by common diseases such as atherosclerosis, hypertension, and diabetes mellitus (Table 8.2). Surgical pathologists must be fully aware of the nature and extent of these alterations, as well as their variation from site to site.

Aorta

The length and the breadth of the aorta increase progressively throughout life, although there are some variations in the rates of these changes between men and women and from decade to decade. Aortic enlargement produces the characteristic aortic unfolding seen in chest radiographs and can result in valvular incompetence if the aortic valve annulus is involved.

The aorta is made up of three layers: (1) the adventitia which houses the vaso vasorum, peripheral nerves, lymphoid and adipose tissue, (2) the media, which is mostly composed of elastic and collagen fibers, smooth muscle cells, and a mucopolysaccharide-rich ground substance, and (3) the endothelium-lined intima. The medial layers of the aorta and carotid, innominate, and proximal axillary arteries contain a high density of elastic fibers that allow their expansion and resilience to high pressures. Parallel lamellar units of elastin envelop smooth muscle cells, ground substance, and collagen (Fig. 8.1). There are approximately 40 lamellar units at birth and at least 50 in adult life, each measuring roughly 11 μm in thickness. Interconnecting bands of collagen and elastin fibers provide strength, while their lamellar arrangement distributes stress evenly across the wall, smoothing the cyclical pressure waves of cardiac contraction (1).

The changes associated with vascular aging include progressive thickening of the aortic wall due to accumulation

TABLE 8.1 Age-Related Changes in Blood Vessels

	Major Macroscopic and Histologic Features
Aorta	Progressive and linear increase in diameter with age Eccentric or diffuse fibrous intimal thickening Fragmentation of elastic lamellae with widening of interlamellar spaces Focal amyloid deposits Thickening of walls of vasa vasorum
Muscular arteries	Progressive dilatation and tortuosity Caliber of vessels usually less in females, especially coronary arteries Intimal fibrosis, sometimes suggesting reduplication of the internal elastic lamella Focal fragmentation and calcification of internal elastic lamella Increased fibrosis and hyalinization of media No significant inflammation in atheroma-free segments
Arterioles	Intimal thickening, usually as concentric layers of fibroelastic tissue Hyalinization of media
Capillaries	Basement membrane thickening, approximately twofold increase in thickness from puberty to old age
Venules and veins	Larger veins show intimal fibrosis and hypertrophy of both circular and longitudinal bundles

of smooth muscle cells and ground substance with thinning and fragmentation of elastic fibers as well as fibrosis and proteoglycan deposition. These changes in the structure of the extracellular matrix likely result from upregulation of genes in smooth muscle cells that mediate matrix metalloproteinase production (2). Apoptosis can be demonstrated in a number of cell types within atheromatous plaques (3). While it is unlikely to be a key factor in the aging of the arterial wall, experimental evidence suggests that chronic apoptosis of vascular smooth muscle cells accelerates atherosclerosis while promoting calcification and medial degeneration (4,5). Age-related changes are the result of decades of "wear and tear." In the long term, they account for mural weakening that leads to aortic dilatation, especially in patients with systemic hypertension. In this respect, the effects of hypertension on the wall of arteries can be seen as accelerated aging. Vascular calcification is a common complication and, although it is most frequent in atheromatous segments, it may occur in areas where the intima is virtually devoid of plaques (6). Aortic and coronary arterial calcifications are especially common in chronic kidney disease and may, in part, be related to hyperphosphatemia (7). In its most pronounced form, layers of calcifications replace large parts of the media of arteries, a condition termed Monckeberg sclerosis (Fig. 8.2). Small amounts of amyloid can also be detected in aortic atheromatous lesions of middle-aged and elderly subjects and may be derived from serum amyloid A or other apolipoproteins (8,9).

TABLE 8.2 Manifestations of Common Diseases in Arteries and Arterioles

Condition	Major Histologic Features
Normal adults	Minimal intimal thickening may be eccentric or diffuse Intact internal elastic lamella, or occasional small breaks only No significant inflammation
Atherosclerosis	Eccentric fibrous intimal thickening, intimal and medial foam cells with lipid deposition Neovascularization with intimal and medial hemorrhage Dystrophic calcification Adventitial aggregates of plasma cells, lymphocytes, and macrophages Intimal and medial aggregates of T-lymphocytes, especially at shoulders of lesion Rupture or erosion of fibrous cap with associated thrombus formation
Systematic hypertension	Concentric fibrous intimal thickening and medial hypertrophy, especially in arterioles Fibrinoid necrosis in accelerated (malignant) phase Aneurysmal dilatation of intracerebral arterioles and capillaries Increased atherosclerosis
Diabetes mellitus	Hyaline change in arterioles Capillary microaneurysms with basement membrane thickening Loss of pericytes; retinal neovascularization Increased atherosclerosis in arteries
Active arteritis	Acute or chronic inflammatory cell inflammation of adventitia and media. Mural edema, reactive intimal thickening, and endothelial necrosis Fibrinoid necrosis of wall, occasionally aneurysmal dilatation
Healed arteritis	Bizarre patterns of disordered fibrous intimal thickening Medial scarring with patchy aggregates of chronic inflammatory cells Abnormally prominent medial blood vessels

The spectrum of diseases usually encountered in aortic surgical specimens is limited to a few diseases that often show overlapping histologic features. These include cystic medial degeneration and cystic medial necrosis, aortic dissections, aortic aneurysms with atherosclerotic changes, and aortitis. The most common adult surgical specimens submitted for evaluation include aortic repairs due to aortic dissection or aneurysmal disease. Pediatric cardiac surgeons may also submit aortic samples from coarctation repairs, other narrowed segments of the aorta, or the ductus arteriosus (arterial duct).

Cystic Medial Degeneration

Cystic medial degeneration is characterized by fragmentation of the elastin fibers of the media and formation of mucoid pools (Fig. 8.3). However, it is important for surgical pathologists to recognize that there is wide variation in the extent of degeneration, particularly in the ascending aorta.

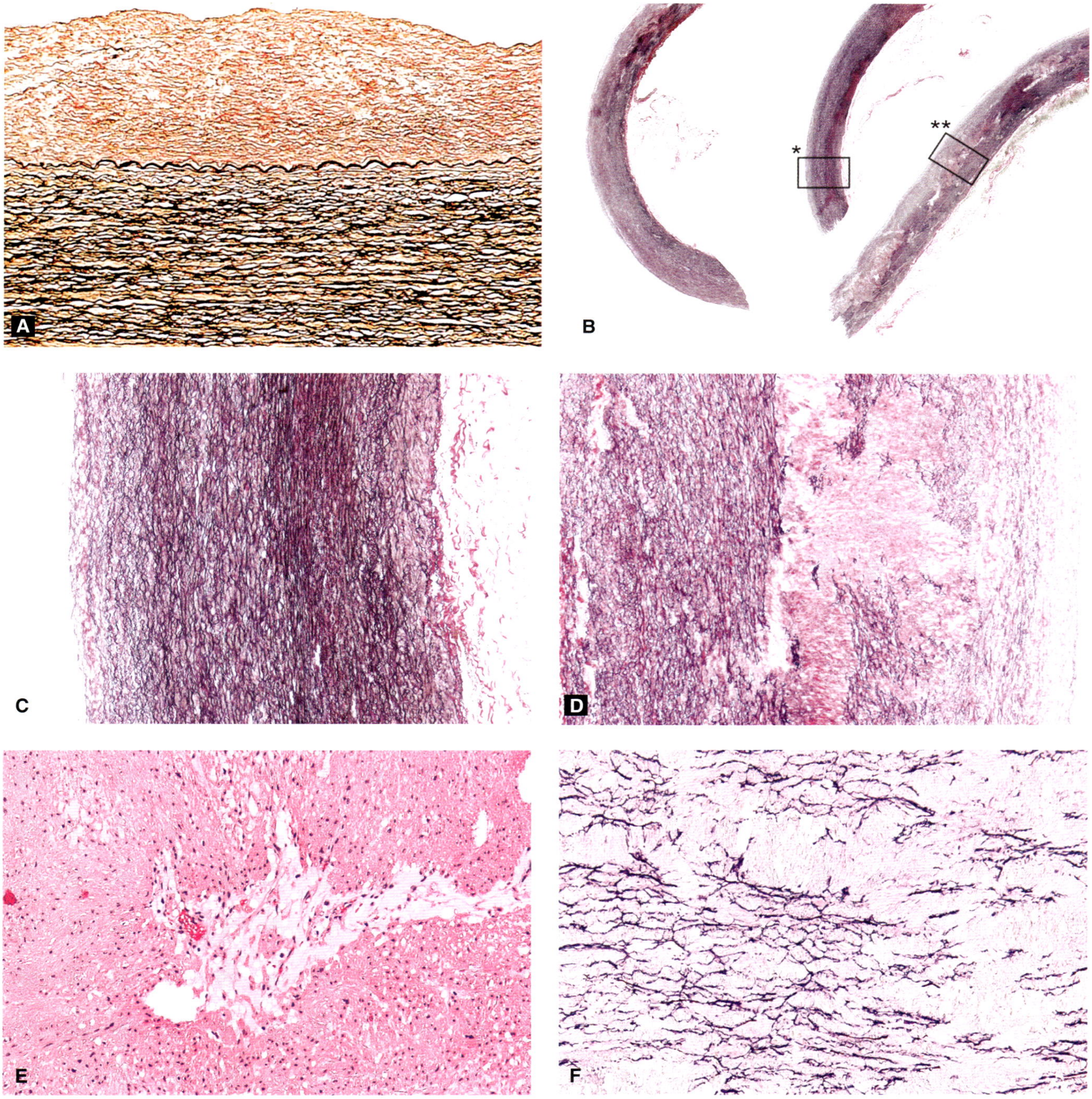

FIGURE 8.1 **A:** Inner half of the aortic wall of a 62-year-old male patient. There is a moderate degree of fibrous intimal thickening, which has no immediate clinical relevance but may predispose to atherosclerosis. There is only slight fragmentation of the elastic lamellae; the overall appearance is well within normal limits for a patient of this age (elastic van Gieson). **B:** Overview of three full-thickness cross sections through the dilated thoracic aorta of a 74-year-old man. There is marked variation in the degree of degenerative change of the media. The boxed area in the central section (enlarged in **C**) has an almost normal appearance. In contrast, the boxed area on the right (enlarged in **D**) shows marked elastic degeneration. **E:** The typical appearance of cystic medial degeneration in an H&E-stained section. Note the prominent pool of mucoid material. **F:** This shows a corresponding section to **E** but is stained for elastic tissue. There is extensive loss of the normal elastic framework.

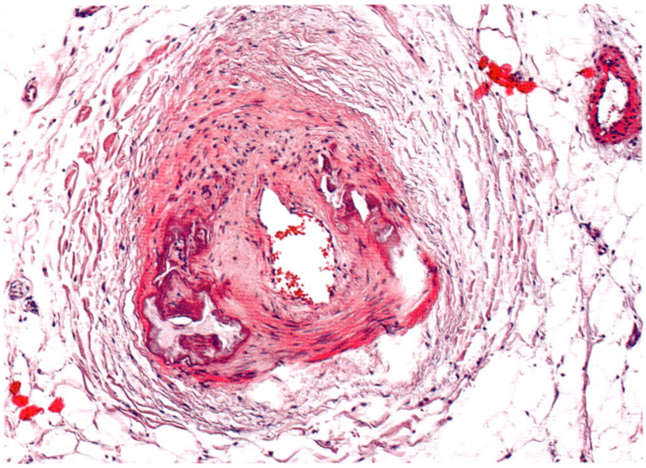

FIGURE 8.2 Arterial calcification. This muscular artery was dissected from a lower limb amputation specimen from a patient with diabetes mellitus. Note the prominent shell-like calcification of the media. This is termed Monckeberg sclerosis (hematoxylin and eosin).

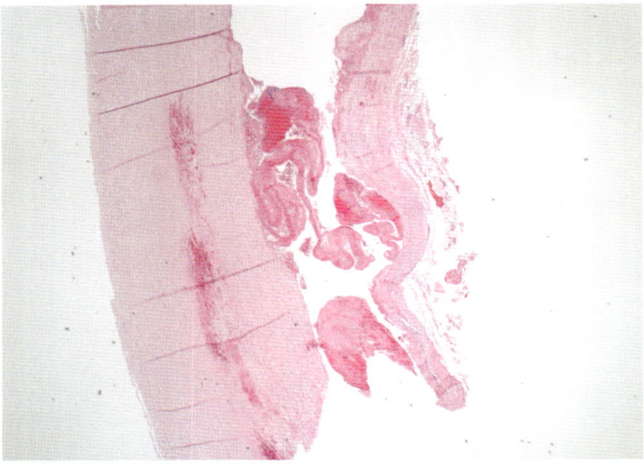

FIGURE 8.4 Acute aortic dissection of the thoracic aorta. Fibrin is identified within the false lumen between the intimal and medial layers (e.g., dissection of the blood vessel).

Some areas have few, if any, stainable nuclei, reflecting smooth muscle cell death. More recently, areas of smooth muscle cell apoptosis and disorganized proliferation, fibrosis, and angiogenesis have been described, suggesting that medial degeneration is a process of degenerative injury and repair (3,10). However, Schlatmann and Becker (11) showed that the histologic alterations of medial degeneration showed a striking correlation with age and may represent the normal aging process of elastic arteries. Similar features are seen in hypertensive patients who have an altered hemodynamic profile and individuals with a history of cocaine abuse (12). These findings suggest that medial degeneration may be related to an imbalance between the mechanical forces imposed on the aortic wall during systole and the capacity of the aortic wall to resist these forces. It is interesting that patients with genetic disorders of connective tissue, such as Marfan syndrome and some types of Ehlers–Danlos syndrome, also have similar aortic changes due to abnormalities in fibrillin and collagens, respectively. Fibrillin is a glycoprotein that is closely associated with elastin fibers and may act as a scaffold upon which elastin fibers are laid down. There is also growing evidence for abnormal TGFβ signaling in Marfan disease and other inherited aortopathies (13–15).

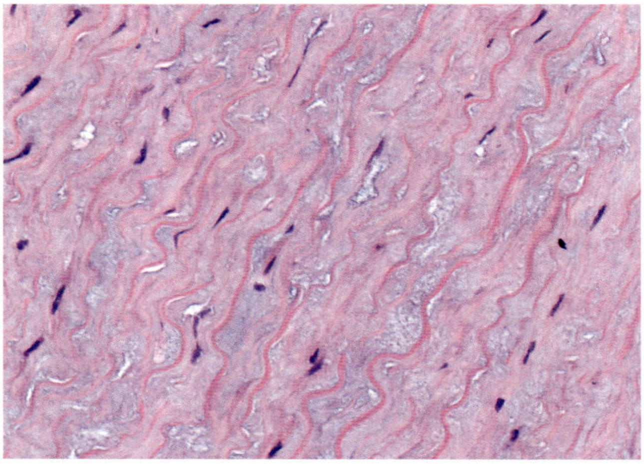

FIGURE 8.3 A histologic feature of cystic medial necrosis includes accumulation of mucoid material within the smooth muscle medial layer or the aorta.

Aortic Dissection

An aortic dissection results from a tear in the inner layer of the aorta that expands when highly pressurized blood creates a false lumen between the endothelial and medial layers. Fibrin deposition can be identified along the medial layer of the dissection (Fig. 8.4). Cardiac tamponade can occur when the dissection extends through the adventitial layer, and sudden cardiac death may result from retrograde dissection that involves the coronary arteries. Aortic dissections are classified as type A when they involve the ascending aorta or type B when they involve the descending aorta (16).

Aortic dissections usually occur in older men with uncontrolled high blood pressure. Postmortem specimens also indicate that severe calcific atherosclerotic plaques involving the aortic arch and great vessels may predispose to tears in the intimal flap. Turner syndrome, Marfan syndrome, Ehlers–Danlos syndrome, Loeys–Dietz syndrome, and other genetic disorders are also associated with an increased the risk of developing aortic dissections (17).

Aortic Aneurysm

Abdominal aortic aneurysms have been historically considered purely atheromatous in origin, but this is an oversimplification. Although most examples show substantial atherosclerotic changes, there seems to be a genetic predisposition to aneurysm development. At least 17 genetic disorders have been associated with aortic aneurysms in younger patients (18,19) and genetic studies of abdominal

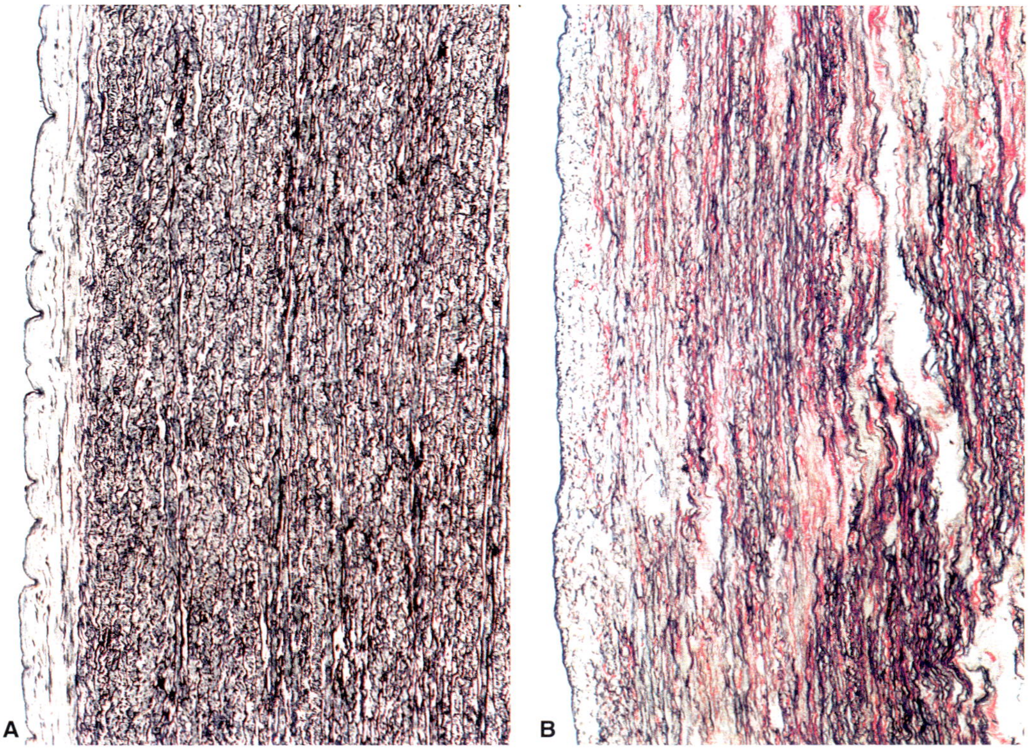

FIGURE 8.5 **A:** The normal appearance of the aortic media of a 48-year-old male. There are many parallel lamellae of elastic tissue. There is no significant intimal thickening. **B:** The aortic wall of a 31-year-old male with Marfan syndrome. The medial elastic tissue is extensively fragmented, and there is fibrosis and loose mucopolysaccharide-rich areas. Such extensive changes would be unusual even in an elderly patient (elastic van Gieson).

aortic aneurysms have provided compelling evidence for an inherited basis of disease. Approximately 75% of cases show an autosomal dominant pattern of inheritance, and both males and females are affected. To date, several potential susceptibility genes have been described, but causative genetic mutations have not (14).

Aortic aneurysms show a variety of histologic abnormalities. Elastic fragmentation and associated medial necrosis are the most common histologic findings in both ascending and thoracic aortic aneurysms (Fig. 8.5). Presumably, shear forces cause alterations in the secretion pattern of smooth muscle cells or promote their death by apoptosis. Vascular smooth muscle cell apoptosis and regeneration, p53 accumulation, and *bax* upregulation have been demonstrated in areas of cystic medial necrosis (20). Whether the atherosclerosis is the primary cause or a secondary complication, the inflammation and medial scarring that accompany it further damage a wall already weakened by normal aging or by specific genetically determined alterations in its matrix. Aggregates of lymphocytes and plasma cells are often present in the adventitia of atheromatous segments of the aorta and coronary arteries. These chronic adventitial infiltrates must not be interpreted as evidence of aortitis when dealing with samples from the ascending aorta during repair of dissecting aneurysms or aortic reconstructions for root dilatation.

Aortitis

Giant cell aortitis and Takayasu aortitis are the two most common types of aortitis encountered in the United States. Giant cell arteritis predominantly affected adults 50 to 80 years of age and is the most common systemic vasculitis in the Western hemisphere. It is a granulomatous vasculitis associated with autoimmune disease, most often involving the temporal artery and ascending aorta. Untreated giant cell aortitis may cause fatal dissections with up to 80% of patients (21). Active aortitis features an inflammatory infiltrate composed of neutrophils, lymphocytes, macrophages, and giant cells within the medial layer (Fig. 8.6). Acute on chronic aortitis may show a "parenthetical" pattern with inflammatory cells surrounding an area of healed fibrosis which. Importantly, one should not exclude the possibility of giant cell aortitis when giant cells are not detected, as not all active cases contain giant cells. Active aortitis should prompt a notification to the rheumatology service for further workup. Of note, small collections of lymphocytes, macrophages, and giant cells are occasionally seen in the media of aortic specimens that show extensive atherosclerotic changes (22).

Takayasu arteritis also shows a predilection for the aorta, primarily affecting girls and young women of Asian descent. It is most likely an autoimmune condition with reported elevated antiaortic and antiendothelial autoantibodies,

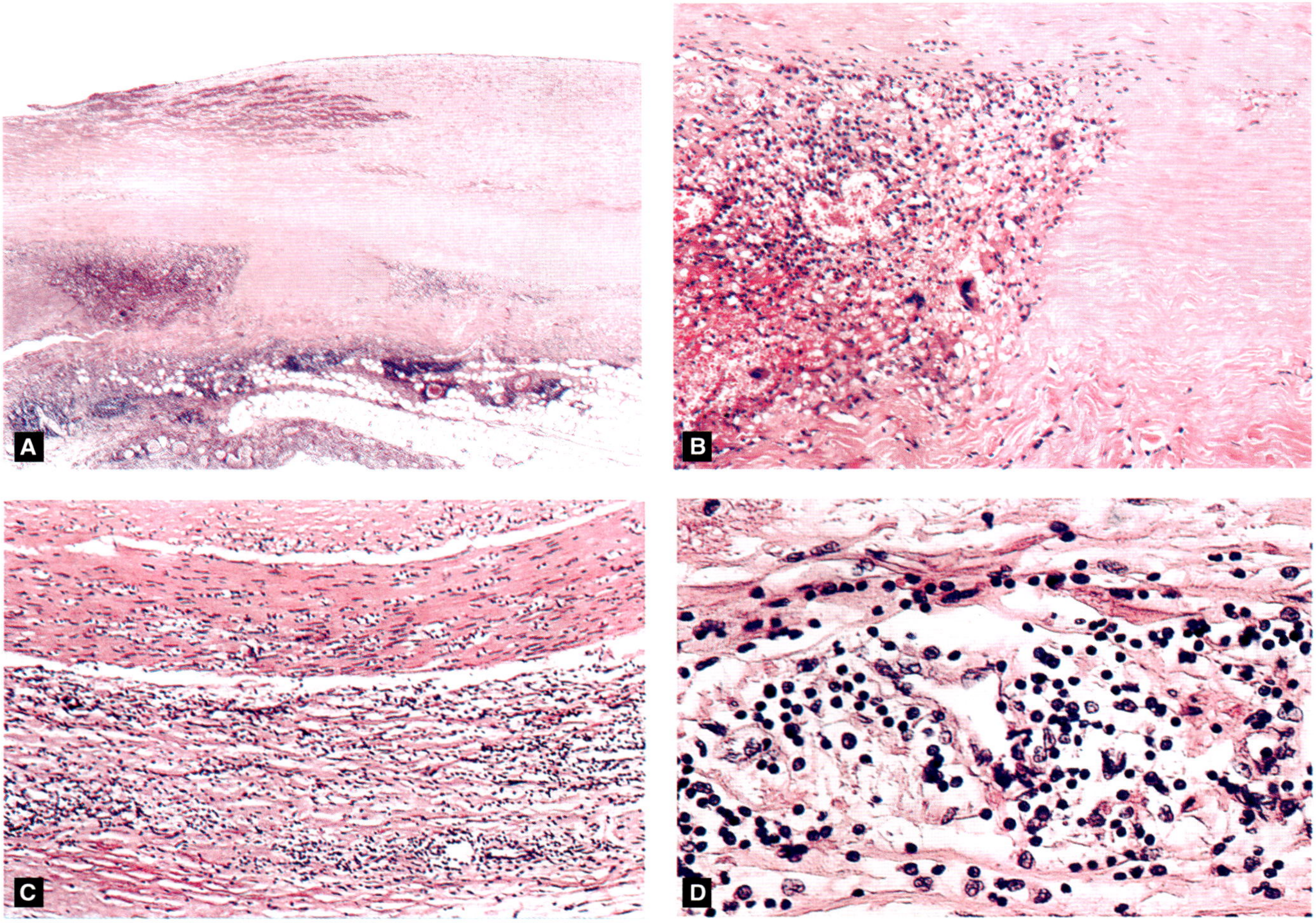

FIGURE 8.6 **A:** Low-power view of ascending aorta from a 58-year-old female with chronic aortitis and periaortitis. There is marked thickening of the aortic wall and multifocal presence of inflammatory infiltrates in the media of the artery. Also note the nodular lymphocytic infiltrates in the periadventitial tissues. They are not unique for chronic aortitis, but can also be seen in severely atherosclerotic arteries. **B:** Detail of inflammatory infiltrate showing presence of multiple giant cells (giant cell aortitis). **C, D:** Adventitial chronic inflammatory infiltrates in the wall of an atheromatous coronary artery. A few inflammatory cells have infiltrated into the media. The magnified view on the right confirms that most of the inflammatory cells are lymphocytes or plasma cells (hematoxylin and eosin).

as well as circulating immune complexes (23). Histology shows a granulomatous mixed inflammatory infiltrate, including giant cells, necrosis and elastic lamina disruption (24). Associated chronic complications may include arteriosclerosis and hypertension.

Findings in Patients with Congenital Aortic Disorders

Cardiac surgeons have several techniques for repairing aortic coarctations and may submit samples of aorta, the narrowed aortic segment, the subclavian artery, or the ductus arteriosus (arterial duct) for histologic identification. The aorta around the coarctation may show reactive intimal thickening, even in neonates, but the underlying elastic structure is usually well preserved. The coarctation itself can have a variety of appearances. In long-standing cases, there may be dense intimal and medial fibrosis. In neonates, the intima may have a distinctly irregular pattern of fibroelastic intimal thickening, resembling some forms of arterial dysplasia (Fig. 8.7). The structure of the arterial duct changes progressively during intrauterine growth and in the postnatal period (25,26) and can be influenced by prostaglandin treatment. Unlike the aorta and the proximal subclavian artery, which are elastic vessels, the arterial duct has a muscular media and a defined internal elastic lamella. From 35 weeks of gestation there is progressive fragmentation of the internal elastic lamina. Small intimal cushions contribute to functional closure of the duct at birth. The internal elastic lamella is preserved in patients with postnatal persistent ductus arteriosus (27).

Peripheral Arteries

It is only in children and young adults that muscular arteries conform to the classical descriptions of textbooks. The

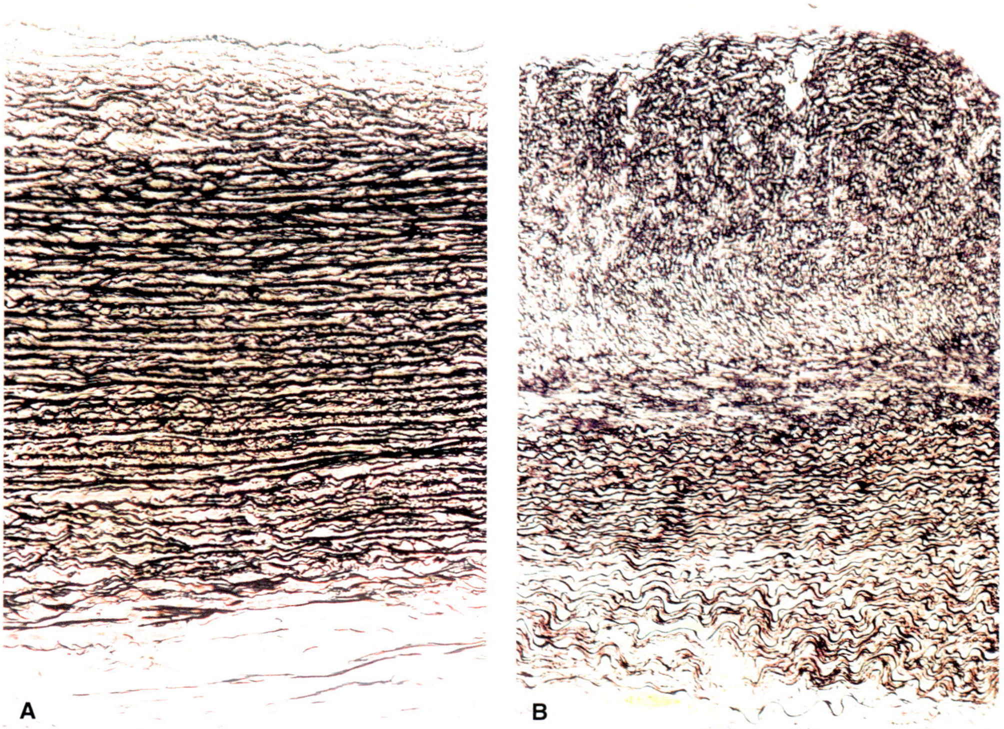

FIGURE 8.7 Coarctation of the aorta. **A:** The aortic wall distal to a coarctation in a 3-month-old child. There is slight intimal edema only. **B:** The coarctation itself; note the irregular arrangement of the intimal fibroelastic tissue (elastic van Gieson).

intima of arteries is defined as the region from and including the endothelium to the luminal margin of the media (28). At birth, the intima is a virtual space with the endothelium closely opposed to the internal elastic lamella. This layer thickens slowly with age in an eccentric fashion at branching points or bifurcations, or diffusely. Intimal thickening occurs preferentially at sites of altered blood flow or mechanical stress, suggesting that it is an adaptive response to injury. Vascular smooth muscle cells derived from the underlying media and extracellular matrix proteins accumulate in the thickened intima. Nutrients reach the media of elastic or muscular arteries by direct diffusion through the intima or via small branches of the vasa vasorum that reenter the media from the adventitial aspect. Vasa are best seen in biopsy samples taken from the ascending aorta during root repairs and sometimes have remarkably thick muscular walls (Fig. 8.8). The transition from elastic to muscular arteries is abrupt and is usually at the point of a major arterial orifice.

Progressive intimal fibrosis affects nearly all arteries (Fig. 8.9), but in surgical pathology material it is especially noticeable in the spleen, myometrium, and thyroid (Fig. 8.10). As in the aorta, fragmentation of the elastic tissue, usually the internal elastic lamella, is common and is of no specific significance. Some aging arteries show reduplication of the internal lamella, producing a pattern of concentric intimal thickening. Small calcifications can be identified in otherwise normal vessels, usually just to the medial aspect of the internal elastic lamella. These aging changes, often loosely termed arteriosclerosis, have been studied most extensively in the coronary arteries, where they tend to be less substantial in women compared with age-matched men.

Approximately 75% of the mass of the media is composed of smooth muscle cells. These cells are arranged in a

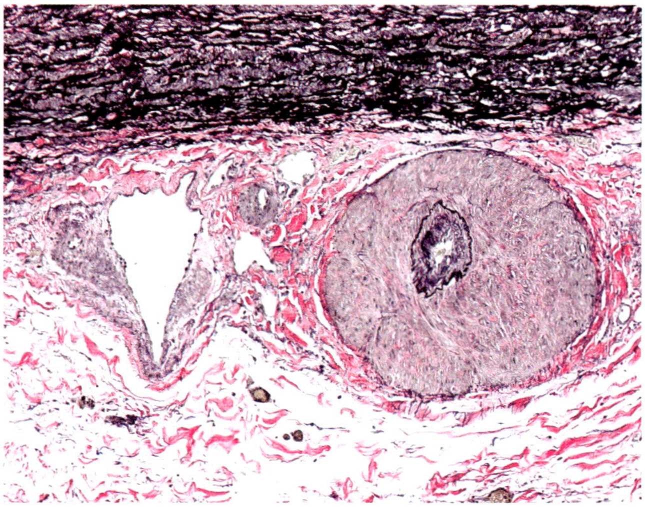

FIGURE 8.8 The aortic adventitia. The thick-walled vessel is a vasa vasorum. The thin-walled vessel (**left**) is a small vein.

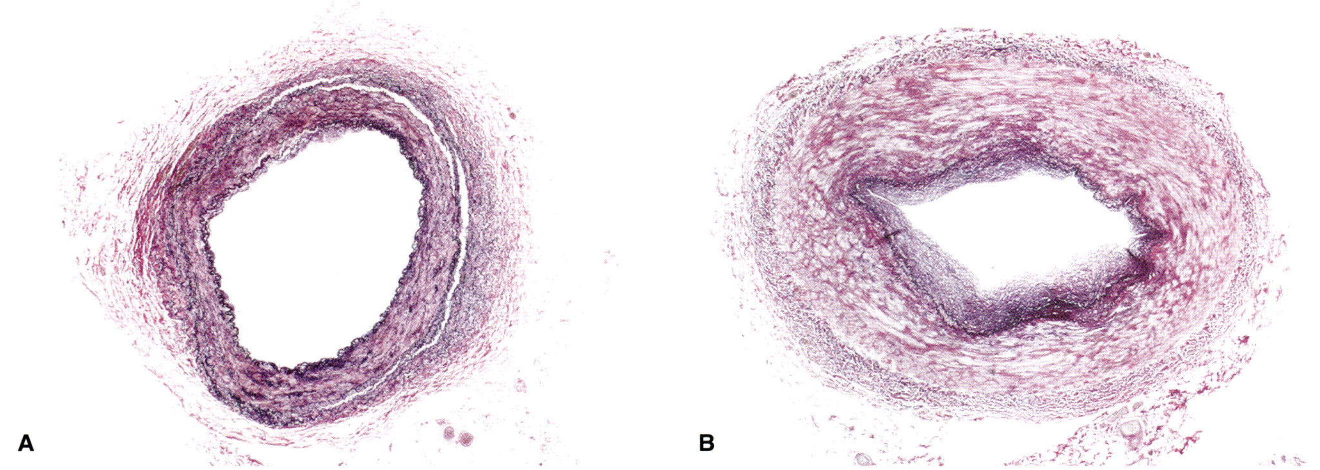

FIGURE 8.9 Transverse sections of the mammary artery (**A**) and the radial artery (**B**) illustrate the marked differences in artery architecture from site to site. These arteries were retrieved from the same 58-year-old male patient and were used as coronary artery bypass grafts. These sections were made from redundant segments (elastic van Gieson stains).

spiral or circumferential pattern around the wall (29). As in the intima, collagen and elastin fibers are normally present in small amounts and they increase throughout life. Arteries dilate and become more tortuous with increasing age, although their caliber is less in women than age-matched men, particularly with respect to the coronary arteries. This could contribute to the poorer results recorded in women after both coronary artery surgery and coronary interventions (30,31). If arterial dilatation is pronounced and irregular, spontaneous thrombosis may result.

Arteries in chronically inflamed tissues and within tumors often show pronounced fibrous intimal thickening, sometimes termed endarteritis obliterans. In the early stages of this process, the fibrous tissue contains abundant basophilic ground substance. Although inflammatory or tumor cells often closely surround the adventitia, they do not usually penetrate far into the muscular wall. These changes must be carefully distinguished from those of systemic vasculitis. In general terms, vasculitis tends to affect vessels of a specific size, cause necrosis of vessel walls with associated hemorrhage, and leads to tissue infarction. In healed vasculitis, there is irregular fibrosis of the muscular wall (Fig. 8.11).

Atherosclerosis and Acute Arterial Thrombosis

Pathologists are frequently called upon to evaluate coronary arteries for atherosclerotic changes or acute coronary artery thrombosis in autopsy specimens or ischemic heart transplant explants. A modified form of the American Heart Association classification of atheromatous lesions is now

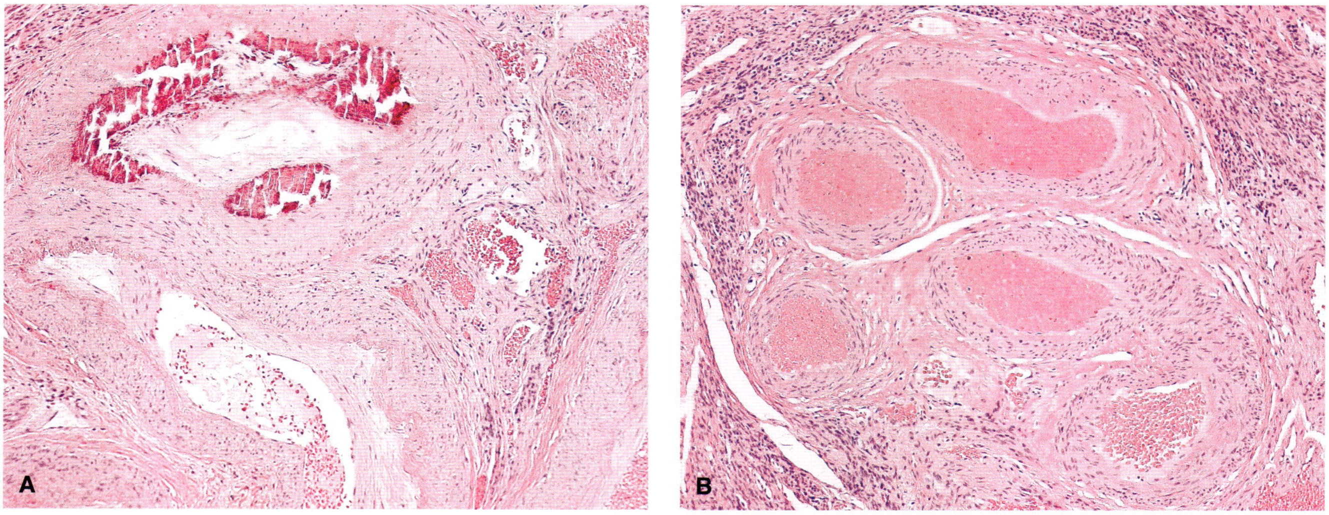

FIGURE 8.10 **A, B**: Age-related changes in arteries. These thick-walled vessels were close to the serosa of the myometrium in a 52-year-old female. Note the prominent calcification in **A** and the increased tortuosity in **B**.

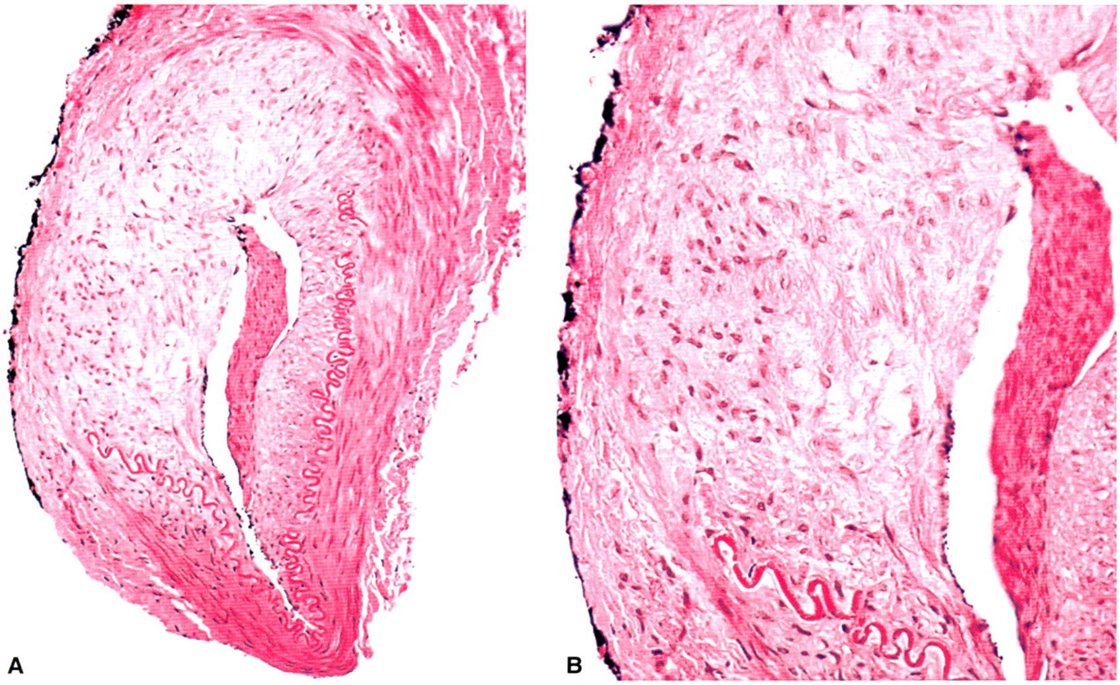

FIGURE 8.11 Healed temporal arteritis. This patient had been receiving steroid therapy for 2 weeks when this biopsy was performed. **A:** The low-power view shows irregular thickening of the wall and a loss of about 50% of the internal elastic lamella. **B:** This higher-power view shows fresh fibrous scarring of the media. Changes such as these are not part of the normal aging process.

widely used in studies of both coronary and carotid arteries (Table 8.3) (28). There is much interest in the concept of the vulnerable plaque, which is a plaque at high risk for hemorrhage or thrombosis. Several postmortem studies evaluating coronary arteries from patients with myocardial infarction or sudden cardiac death have shown that vulnerable lesions have specific features such as a large lipid core, a thin fibrous cap, and marked inflammatory activity in the plaque tissue (28). Radiofrequency intravascular ultrasonography can identify and measure many structural features during coronary angiography. A prospective study of nearly 700 patients found that the lesions responsible for a second coronary event were usually thin fibrous cap atheromas (32). The pathology of recently symptomatic carotid plaques is similar to that of culprit coronary plaques, with strong correlations between macrophage instability and plaque instability (33). In atheromatous arteries, there is often marked neovascular proliferation. Hemorrhage from these vessels contributes to the growth of lesions and their lipid content (34).

TABLE 8.3 Modified American Heart Association Classification of Atheromatous Lesions

Lesion	Histologic Features
Pathologic intimal thickening	Smooth muscle cell proliferation, intimal fibrosis, extracellular lipid but no lipid core or necrosis
Fibrous cap atheroma	Well-formed lipid core with thick fibrous cap, free of inflammatory cells (>80 μm coronary artery >200 μm carotid artery)
Thin fibrous cap atheroma	Thin cap of inflamed fibrous tissue with underlying lipid core
Ruptured plaque	Luminal thrombus communicating with lipid core via a ruptured fibrous cap
Eroded plaque	Luminal thrombus with endothelial ulceration; lipid core may be absent or small and does not communicate with lumen
Calcified lesions	Heavily calcified plaques with or without thrombus or lipid core

This is a simplified version of the classification of Virmani R, Kolodgie FD, Burke AP, et al. Lessons from sudden cardiac death: a comprehensive morphological classification for atherosclerotic lesions. *Arterioscler Thromb Vasc Biol.* 2000;20:1262–1275.

Infection and Acute Arterial Thrombosis

Infection with SARS-CoV-2 and other viruses is known to increase risk of thrombosis. Respiratory infections are associated with a three- to sixfold increased risk of arterial thrombosis and subsequent myocardial infarction and/or ischemic stroke, and a two- to threefold increased risk of deep vein thrombosis (35). Myocardial infarction and stroke can occur after acute infection or vaccination, and risk is highest during the first month of infection (36). Ackermann et al. reported that acute respiratory distress syndrome due to H1N1 influenza showed more severe endothelial injury associated with intracellular virus and disrupted cell membranes by electron microscopy, widespread pulmonary vessel thrombosis, and a ninefold increase in alveolar

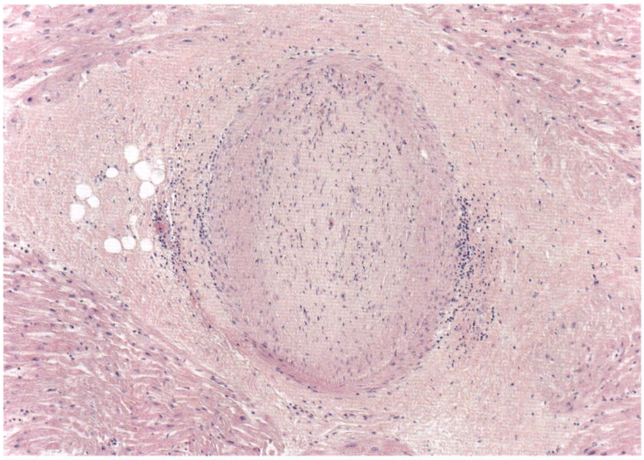

FIGURE 8.12 Cardiac allograft vasculopathy is characterized by concentric fibromuscular intimal and medial thickening with luminal stenosis, subendothelial lymphocytic accumulation, and perivascular fibrosis.

capillary microthrombi when compared with age-matched, uninfected controls (36). Bois et al. demonstrated nonocclusive fibrin microthrombi in 80% of autopsy patients with COVID-19 compared with 33% of influenza and normal controls (37).

Cardiac Allograft Vasculopathy in Transplant Patients

Cardiac allograft vasculopathy is a long-term complication following heart transplantation. Host allorecognition of donor human leukocyte antigen (HLA) on coronary endothelium triggers T-cell activation, altered cytokine expression, endothelial activation, and downstream effects such as vascular smooth muscle proliferation, extracellular matrix deposition, and intimal or medial thickening and vasculopathy (38,39). The formation of anti-HLA or antiendothelial antibodies increases risk of cardiac allograft vasculopathy and mortality (40). Coronary angiography demonstrates diffuse, longitudinal and concentric luminal narrowing of primary or branch vessels (41) and intravascular ultrasound shows increased maximal intimal thickness of the coronary arteries, commonly involving the left anterior descending coronary artery (42). In explant hearts, there is concentric fibromuscular intimal and medial thickening with luminal stenosis, subendothelial lymphocytic accumulation, and perivascular fibrosis (Fig. 8.12).

Arterioles

As arteries approach their distal target organs, they ramify into smaller and smaller branches to bring oxygen and other nutrients to the target tissues. There are no specific histologic features that accurately distinguish small arteries from larger arterioles. The small arteries of the coronary circulation are said to have a diameter of greater than 500 μm, whereas prearterioles have a diameter of 500 to 100 μm and true arterioles have a diameter of less than 100 μm. Prearterioles are largely responsible for autoregulation of coronary blood flow while arterioles react to metabolic changes in the circulation (30,31). These are important physiologic differences but in biopsy material, there is so much variation in the contours of these small vessels that accurate distinction is often impossible and probably unnecessary. Larger arterioles ("prearterioles") have an obvious media and an adventitial layer of connective tissue. In the smallest arterioles, an internal elastic lamella may not be identified. The smooth muscle cells are arranged circumferentially, each cell winding around the wall several times. This is the structural basis of the precapillary sphincter. Small arterioles have a very thin adventitia but are richly supplied by sympathetic nerve fibers.

Hyalinization is a common lesion of arterioles and small arteries. It increases with age and in chronic conditions such as hypertension and diabetes mellitus. The glassy uniform appearance of the vascular wall reflects accumulation of plasma proteins and small amounts of lipids. As in arteries, reduplication of elastic tissue and intimal fibrosis are common changes in the aged arterioles. In severe longstanding benign hypertension and in the malignant phase, the arteriolar lumen can be substantially narrowed by concentric layers of fibrous tissue and smooth muscle cells.

Fibrinoid necrosis of the arteriolar media is the hallmark of malignant hypertension and some forms of acute vasculitis (Fig. 8.13). It must always be regarded as pathologic. In the earliest changes of diabetic microangiopathy, arterioles and capillaries often show prominent basement membrane thickening, which can be readily identified in renal and peripheral nerve biopsies (43). Although there is physiologic evidence of small-vessel disease in the heart and the peripheral vasculature, characteristic histologic changes of diabetic microangiopathy are seldom seen in these sites (44). Changes related to diabetes mellitus include increased type IV collagen and laminin in the basement membrane with diminished amounts of proteoglycans. Albumin and

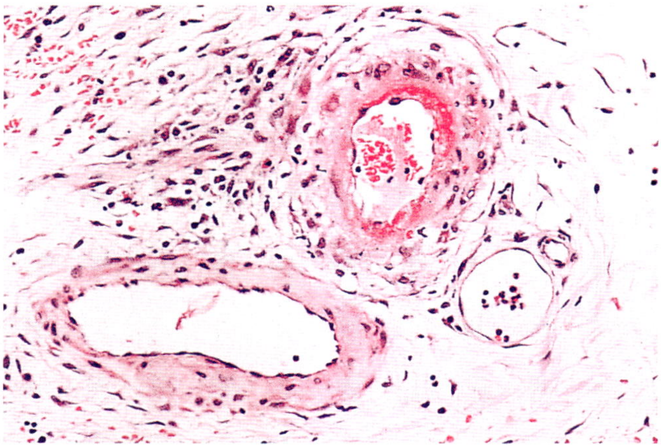

FIGURE 8.13 Florid fibrinoid necrosis in a small intestinal vessel of a patient with systemic lupus erythematosus. Fibrinoid necrosis is not a feature of normal aging or uncomplicated hypertension. It always should be regarded as pathologic.

immunoglobulins accumulate in these abnormal basement membranes, binding to glycosylated protein residues and contributing to the overall eosinophilic appearance.

Capillaries, Sinusoids, Venules, and Lymphatic Vessels

Capillaries, sinusoids, and small (postcapillary) venules are collectively termed exchange vessels and, along with arterioles, comprise the microvascular bed that forms the structural basis of the microcirculation. In contrast to arterioles, capillaries are not invested with a muscular media or an elastic lamella. Rather, they contain a complete, single layer of endothelial cells that rests on a basement membrane, the thickness of which varies from site to site. Basement membrane thickness increases with age, almost doubling in muscle capillaries from patients 10 to 70 years of age. Capillaries have no adventitial fibrous tissue support but pericytes are present in and around the basement membrane. It is difficult to identify pericytes in routine sections, but they are easily seen by electron microscopy and stain with actin antibodies, especially the α1 isoform (vascular smooth muscle actin). Pericytes provide structural support; and, because they contain several forms of myosin, can also regulate blood flow. It is likely that pericytes help maintain the vascular basement membrane and are capable of phagocytosis. It is generally accepted that the turnover of pericytes is increased in the capillaries of diabetic patients, and this may contribute to the development of small-vessel disease. Pericytes may also have a role in the control of peripheral insulin activity (45).

The capillary endothelium has circular fenestrations that act as pores through the full thickness of the endothelial cell. Fenestrations are especially prominent in renal glomerular endothelial cells and are found in the capillaries of the intestinal mucosa, skin, and endocrine glands. The liver, spleen, pituitary gland, adrenal glands, and bone marrow contain sinusoids, rather than capillaries. These vascular structures have diameters ranging up to 30 to 40 μm and are generally more distended than capillaries. They have prominent fenestrations, and there are significant gaps between endothelial cells. The basement membrane is attenuated or entirely lacking. This structure allows for a rapid interchange between blood and the surrounding tissue. In contrast, fenestrations are poorly developed or absent in brain, muscle, lung, and connective tissue.

The transition from venous capillary to muscular venule and small collecting vein is characterized by the gradual acquisition of a muscular media. The paracortical or high endothelial venules of lymph nodes have an important role in T-lymphocyte recirculation (46). The endothelial cells of postcapillary venules have a prominent cuboidal or columnal appearance, usually with an ovoid nucleus and a single central nucleolus. They stain specifically with the HECA-452 antibody, as discussed in detail subsequently.

At the light microscopic level, small lymphatic vessels closely resemble capillaries, although lymphatic channels generally have a larger diameter and a less regular cross-sectional profile. They begin as dilated channels with closed ends and anastomose freely. Small lymphatic spaces have an incomplete basal lamina and lack both pericytes and tight junctions. They are permeable to large proteins, cell debris, and microorganisms. Lymphatic channels have numerous valves and are often slightly distended at these sites, producing a slightly beaded appearance. Large lymphatic vessels, such as the thoracic duct, have a thin muscular media with no clear division into circular or longitudinal coats, as well as a fibrous adventitia. A longitudinal muscular layer is present in both the right lymphatic and thoracic ducts. Although lymphatic vessels are present in most tissues, they are rarely found in the epidermis, nails, cornea, articular cartilage, central nervous system, or bone marrow. As detailed and illustrated later, lymphatic vessels show immunopositivity for D2-40, LYVE-1, and VEGFR-3.

Small-Vessel Vasculitis

Small-vessel vasculitides affect the capillaries and venules, and are often mediated by antibody deposition that results in blood flow restriction. The two main types of small-vessel vasculitis are antineutrophil cytoplasmic antibody (ANCA)-associated vasculitides (e.g., microscopic polyangiitis, granulomatous polyangiitis, and eosinophilic granulomatosis with polyangiitis) and immune complex–mediated vasculitides (e.g., antiglomerular basement membrane, cryoglobulinemic, and IgA vasculitis) (47,48).

Antibody-Mediated Rejection in Cardiac Transplant Patients

Two types of rejection may occur in heart transplant recipients: acute cellular rejection and antibody-mediated rejection. The latter is associated with accelerated graft failure and is immune mediated with a vasculocentric distribution involving interstitial capillaries of the myocardium. Antibody-mediated rejection was first described in 1987 by Herskowitz et al. who identified a subset of heart transplant patients with "arteriolar vasculitis" and poor outcomes (49). Hammond et al. subsequently identified antibody deposition and complement activation (50) to be driving etiologic factors underlying antibody-mediated rejection. This condition develops when recipient antibodies are directed against donor HLA on the endothelial layer of the allograft, thereby inducing fixation and activation of the complement cascade. Antibodies are typically directed against donor HLA class I or class II molecules and termed donor-specific antibodies. Donor-specific antibodies are preformed and present prior to transplantation or arise de novo after transplantation. The endomyocardial biopsy remains the gold standard for the diagnosis of pathologic antibody mediated rejection. The 2004 International Society for Heart and Lung Transplantation (ISHLT) revision of the cardiac antibody mediated rejection criteria details the two most important histologic criteria as (1) capillary injury with endothelial swelling and (2) intravascular macrophage accumulation.

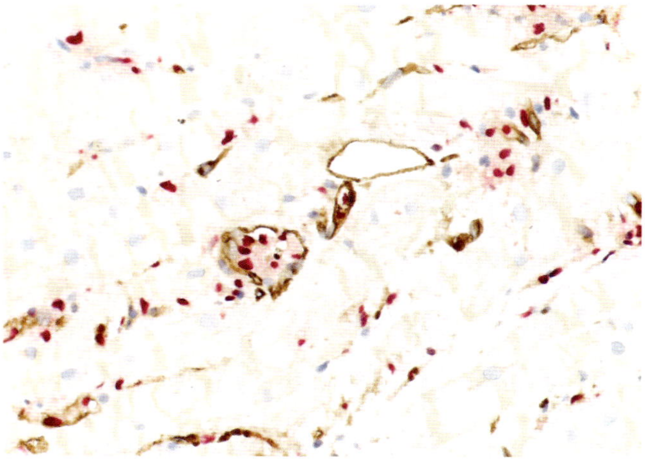

FIGURE 8.14 Antibody-mediated rejection in heart transplant patients is associated with an increased risk of cardiac allograft vasculopathy and poor survival. IHC double stains such as the Pu.1 (histiocyte marker—*pink*) within an endothelial lined capillary (*brown*) precisely highlights histiocytes within capillary lumens have been developed for clinical diagnosis.

Ancillary studies to confirm these findings include immunofluorescence to detect immunoglobulin deposition (IgG, IgM, and/or IgA) with complement (C1q, C4d, and/or C3d) on frozen tissue sections, as well as CD68 staining of intracapillary macrophages, and/or C4d staining of capillaries by immunohistochemistry (IHC). Dual stains that precisely highlight macrophages within capillary lumens (51) are widely used in clinical practice (Fig. 8.14). Antibody-mediated rejection has also been associated with an increased risk of cardiac allograft vasculopathy and poor survival.

Veins

The primary function of venous system is to return blood to the heart *via* the vena cavae, although it also acts as a capacitance reservoir for the circulatory system, especially in the splanchnic system. In contrast to arteries of a similar diameter, veins have much thinner walls with a circular layer of muscle. The saphenous, iliac, brachiocephalic, portal and renal veins, and the vena cavae contain a longitudinal muscular layer but these can be difficult to distinguish in surgical histology sections (Fig. 8.15). Placental, dural, and retinal veins and the veins of erectile tissue contain very little muscle. In general, the veins of the lower limbs have thicker walls than those of the arms and abdomen. Most veins have valves to prevent the reflux of blood. Larger veins are supported by a well-developed fibrous adventitia. Some veins are tethered to surrounding connective tissue fascia.

The increasing use of the saphenous vein as an arterial conduit has led to a greater understanding of the normal structure of larger veins and the changes that occur in them as a result of aging (52). The connective tissue adventitia is often well developed. Saphenous veins in middle-aged and elderly patients show intimal fibrosis and longitudinal and circular muscle hypertrophy with a substantial increase in medial connective tissue (Fig. 8.15) (53). These changes must be distinguished from the form of atherosclerosis that develops in vein bypass grafts.

Pulmonary Arteries and Veins

Although the basic histologic structure of pulmonary vessels resembles that of their systemic counterparts, there are differences that reflect the much lower pressure of the pulmonary circuit. The lumina of major pulmonary arteries are widely dilated despite a relatively thin wall thickness and the intima is hardly discernible. In an adult, the pulmonary arterial media is composed of only 10 to 15 parallel elastic lamellae, whereas, even in a young child, 40 aortic lamellae can be identified. The thickness of the pulmonary trunk is 40% to 80% of the size of the aorta (Fig. 8.16).

Pulmonary embolism is a serious medical condition that occurs when a blood clot is usually formed in the lower extremities and travels to a segment of the pulmonary artery. A saddle embolus is a clot lodged in the main pulmonary artery or at the bifurcation. Lobar, segmental, and subsegmental emboli are located in the pulmonary arterial branches corresponding to the specific anatomic lung locations and can result in serious hemodynamic compromise and lung injury. The risk of pulmonary embolus increases with both primary and secondary hypercoagulable states. Primary hypercoagulable states include inherited clotting disorders such as factor V Leiden, prothrombin gene mutation, protein C and S deficiency and antithrombin III deficiency. Secondary hypercoagulable states result from acquired factors such as immobilization, cancer, oral contraceptives, obesity, pregnancy, major trauma or surgery, inflammatory or autoimmune disorders. Early in the pandemic, approximately 20% of patients with severe COVID-19 have coagulation abnormalities, and almost all critically ill patients had major coagulation disorders. Risk factors associated with hypercoagulability among patients with SARS-CoV-2 infection included (1) decreased circulating protein C and antithrombin, (2) increased plasminogen activator inhibitor-1, (3) instrumentation causing endothelial injury, (4) severe gastrointestinal symptoms with dehydration, and (5) recent or current pregnancy. A recent meta-analysis of 27 studies that included 3,342 COVID-19 patients reported pooled incidence rates of pulmonary embolus and deep venous thrombosis of 16.5% and 14.8%, respectively (54).

Pulmonary artery branches contain more elastic tissue than systemic arteries of similar size; pulmonary arteries with a diameter of 1 mm are elastic vessels, with gradual transitions to vessels that contain more smooth muscle. Smaller pulmonary arteries and arterioles have a thin layer of smooth muscle cells enveloped by both an internal and an external layer of elastic tissue (55).

Arterioles give rise to a rich network of alveolar capillaries. Pericytes are not easily identified, and the endothelium

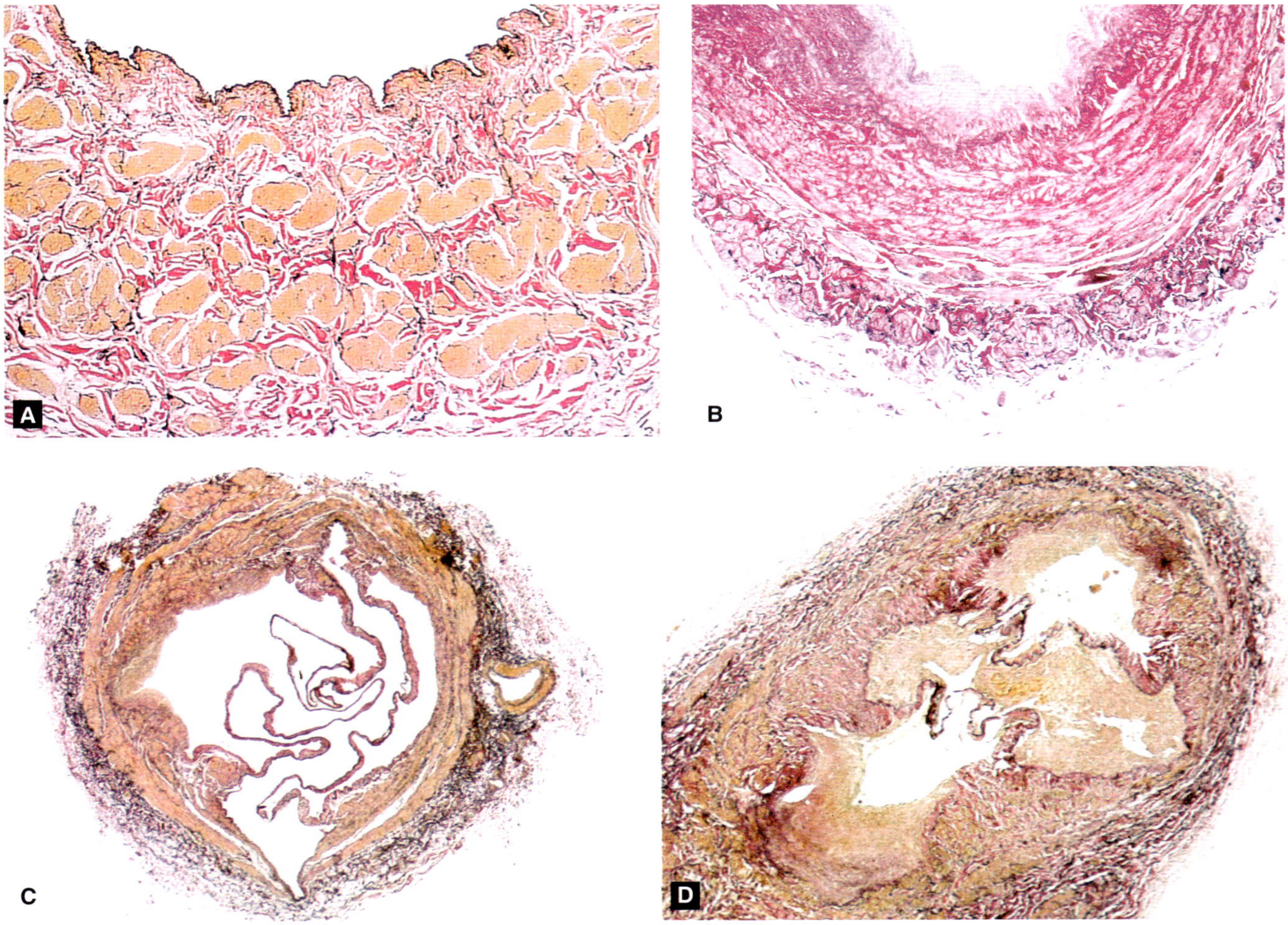

FIGURE 8.15 **A:** Renal vein from a 58-year-old female, close to the junction with the inferior vena cava. There is no significant intimal thickening, and a thin internal elastic lamella can be identified. Note the thin layer of subendothelial collagen. The muscular wall is composed of coarse fascicles, which are not clearly arranged into circular and longitudinal layers. **B:** Detail of the wall of the proximal saphenous vein. In contrast to the renal vein the media is composed of distinct layers of circular and longitudinal smooth muscles. **C:** Cross section of normal saphenous vein. This vein is commonly used by cardiovascular surgeons for arterial bypass grafting procedures. There is a regular pattern of alternating muscle layers with some fibrous tissue and scant elastin. Note the valvular structures in the lumen of the vein. **D:** Vein affected by phlebosclerosis, showing cushions of intimal thickening fibrosis of the media and marked fibrotic thickening of the valves (elastic van Gieson).

and alveolar epithelium appear to share a basement membrane in many areas. The walls of pulmonary veins are less structured than their systemic counterparts. The media is composed of a rather haphazardly arranged but roughly circular layer of connective tissue and smooth muscle cells. No distinct and continuous elastic lamellae are present, and valves are absent (Fig. 8.17). It can be very difficult to distinguish between small pulmonary arteries and veins. The distinction can only be made with certainty when the vessel in question is clearly seen in continuity with a larger artery or vein (Fig. 8.17A).

It can be difficult to distinguish the early vascular changes of pulmonary hypertension from those of normal aging (Table 8.4). The initial changes in both conditions include intimal fibrosis and medial muscular hypertrophy, and each of these features is most prominent in muscular arteries and larger arterioles (56). In fact, the absence of significant changes in the larger arteries may be misleading. Long-standing pulmonary hypertension leads to complex changes in muscular arteries that include florid intimal thickening, medial hypertrophy, and prominent dilatation of small branches of parent vessels (Fig. 8.18). In the most extreme examples, angiomatoid malformations may develop, and occasionally there is fibrinoid necrosis of the vessel wall. Mooi and Grünberg have summarized and extensively illustrated the current WHO classification of pulmonary hypertensive diseases (55). Lung biopsy is no longer used for the assessment of pulmonary hypertension in children with congenital heart disease nor in adults with primary pulmonary hypertension. However, surgical

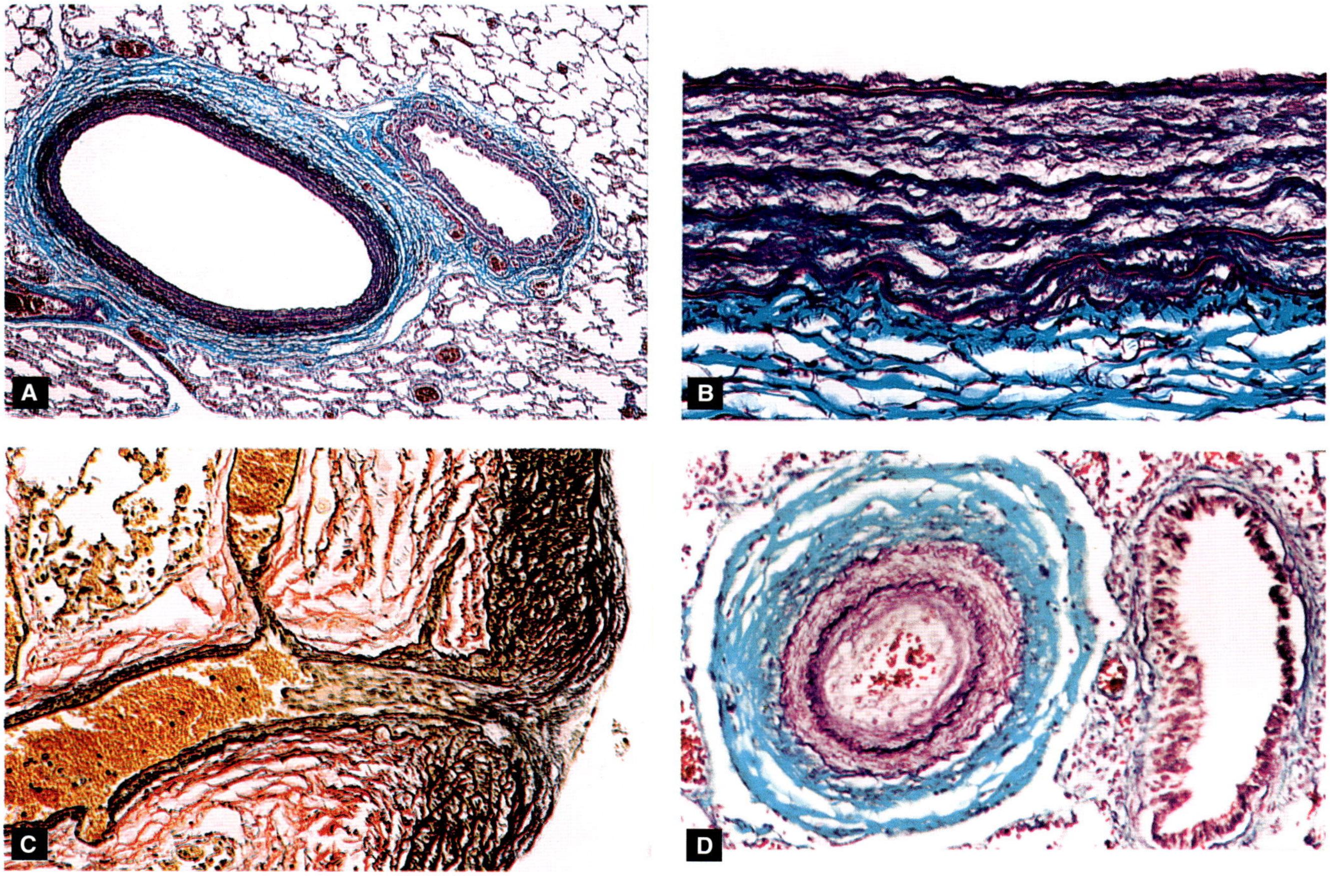

FIGURE 8.16 **A:** Elastic pulmonary artery from a 1-year-old child. The lung was inflated via the main pulmonary artery, which therefore appears much larger than the corresponding bronchus. **B:** A magnified view of the elastic wall (Gömöri trichrome). **C:** The transition from elastic to muscular pulmonary arteries in a 73-year-old male. Note the larger number of elastic lamellae. There is slight fibrous intimal thickening only (elastic van Gieson). **D:** A small pulmonary artery from a patient with long-standing pulmonary hypertension and chronic obstructive airways disease. There is hypertrophy of the muscular wall and pronounced fibrous intimal thickening (Gomori trichrome).

TABLE 8.4 Histologic Features of Pulmonary Vessels

Vessel	Normal	Age-Related Changes	Pulmonary Hypertension
Elastic arteries (>500 μm)	Widely patent lumen, media of 10–15 parallel lamellae of elastic tissue	Slight intimal fibrosis; increased medial thickness due to occasional atheromatous plaques	Atherosclerosis and dilation of main pulmonary arteries; medial thickening due to hypertrophy of admixed muscular elements
Muscular arteries or arterioles	Thin muscular wall often with distinct internal and external elastic laminae	Increased muscular media, eccentric intimal fibrosis, especially in vessels less than 300 μm in diameter	Complex changes include florid intimal thickening, medial hypertrophy, dilation of small branches, angiomatoid (plexiform) lesions, and fibrinoid necrosis
Veins	Thin media of irregularly arranged fibrous tissue and muscle without distinct elastic lamella	The media may appear hyalinized	Intimal fibrosis, medial muscular hypertrophy—occasionally sufficient to mimic appearance of arteries

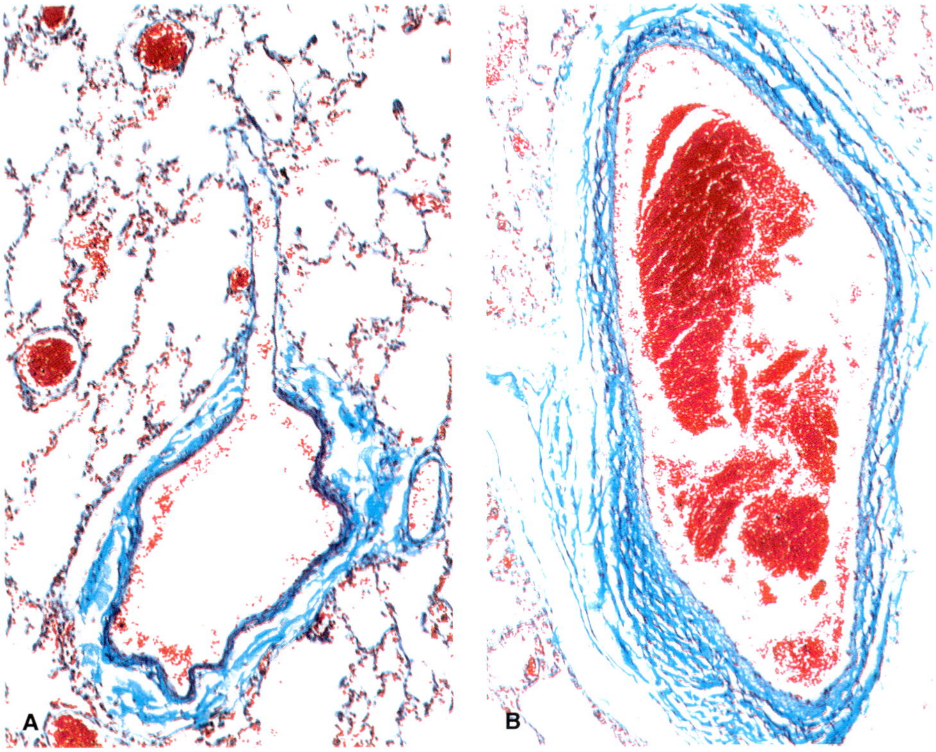

FIGURE 8.17 Normal pulmonary veins. **A:** A pulmonary venule draining into a small vein. Very little muscle is present in the wall. **B:** A large pulmonary vein close to the hilum of the lung (Gomori trichrome).

pathologists must make a careful assessment of the pulmonary arteries and veins in lung biopsy specimens and be able to describe and grade these alterations accurately.

Age-related changes in pulmonary vessels are seldom described in detail (57). Long-standing cardiac failure results in intimal fibrosis, medial hypertrophy, and hyalinization of the pulmonary veins. Marked medial hypertrophy may confer an arterialized appearance to pulmonary veins, and they may develop an incomplete internal and/or external elastic lamina. In this situation, multiple levels stained for elastin and trichrome can be helpful because the elastic lamellae are seldom complete in these abnormal veins, and there is often more medial fibrosis than in corresponding pulmonary arteries. Extensions of atrial myocardium (myocardial sleeves) are common findings in pulmonary veins and the vena cavae close to their junction with the heart (Fig. 8.19). Abnormal electrical activity associated with these sleeves may cause ectopic beats and atrial arrhythmias, especially when the sleeves are fibrotic or infiltrated with amyloid (58).

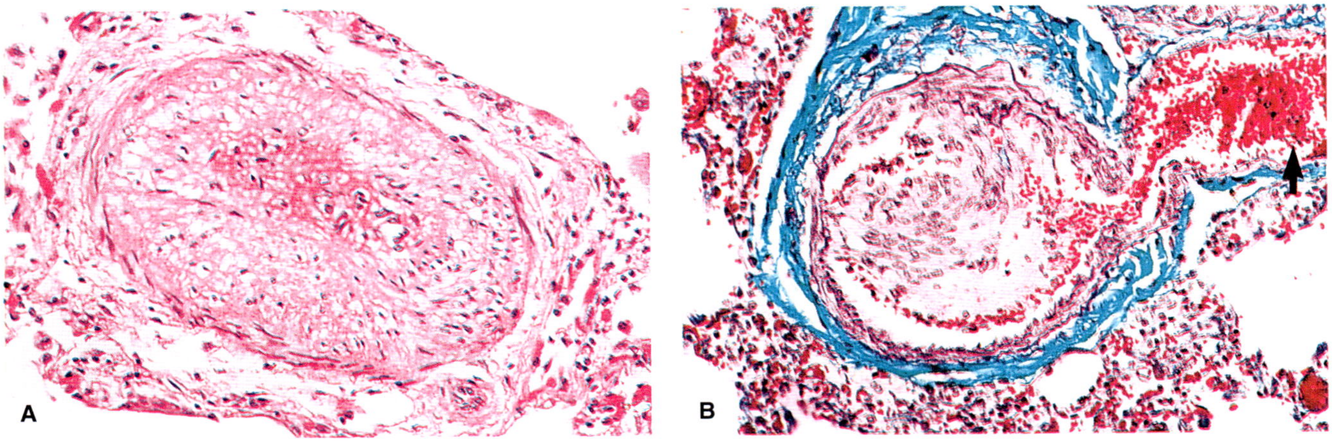

FIGURE 8.18 Advanced pulmonary hypertensive changes. **A:** There is marked hypertrophy of the medial muscle in a small pulmonary artery. **B:** An early plexiform lesion with nearby dilated thin-walled branches (*arrow*) (Gomori trichrome).

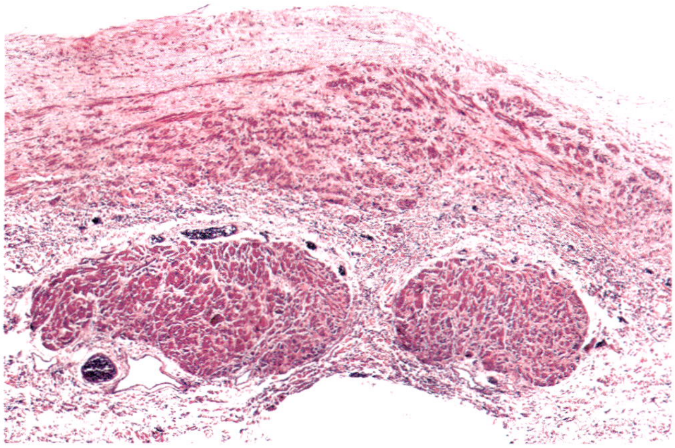

FIGURE 8.19 Myocardial sleeves in pulmonary vein. Section through a pulmonary vein close to the left atrial wall. Rounded extensions of myocardial tissue are present in the vein wall. These myocardial sleeves are important sites for generation of atrial arrhythmias, especially in the elderly when the sleeves develop fibrosis or amyloid deposits.

Anastomoses, Angiodysplasias, and Vascular Malformations

Anastomoses commonly develop between arteries and veins in the systemic circulation. They vary in size from about 200 to 800 μm and in some sites, such as the nail bed, have a complex structure. Those that occur in the anal canal have been termed "anal glomeruli." In this area, as well as the uterus and bladder, venous plexuses may be composed of prominent conglomerates of venous channels that may be difficult to distinguish from angiodysplasias or vascular malformations. The nasal mucosa, clitoris, and deep parts of the glans penis contain abundant erectile vascular tissue characterized by interconnected vessels accompanied by bundles of smooth muscle cells that may be mistaken for angiomas or a malformation (Fig. 8.20). There are also anastomoses between pulmonary and bronchial veins and between the portal and systemic circulations. Portal-systemic shunts lead to increased venous pressure in the portal system and varices that can develop in the esophageal or rectal submucosa, as well as in periumbilical or diaphragmatic region. Peripheral glomus tumors almost certainly

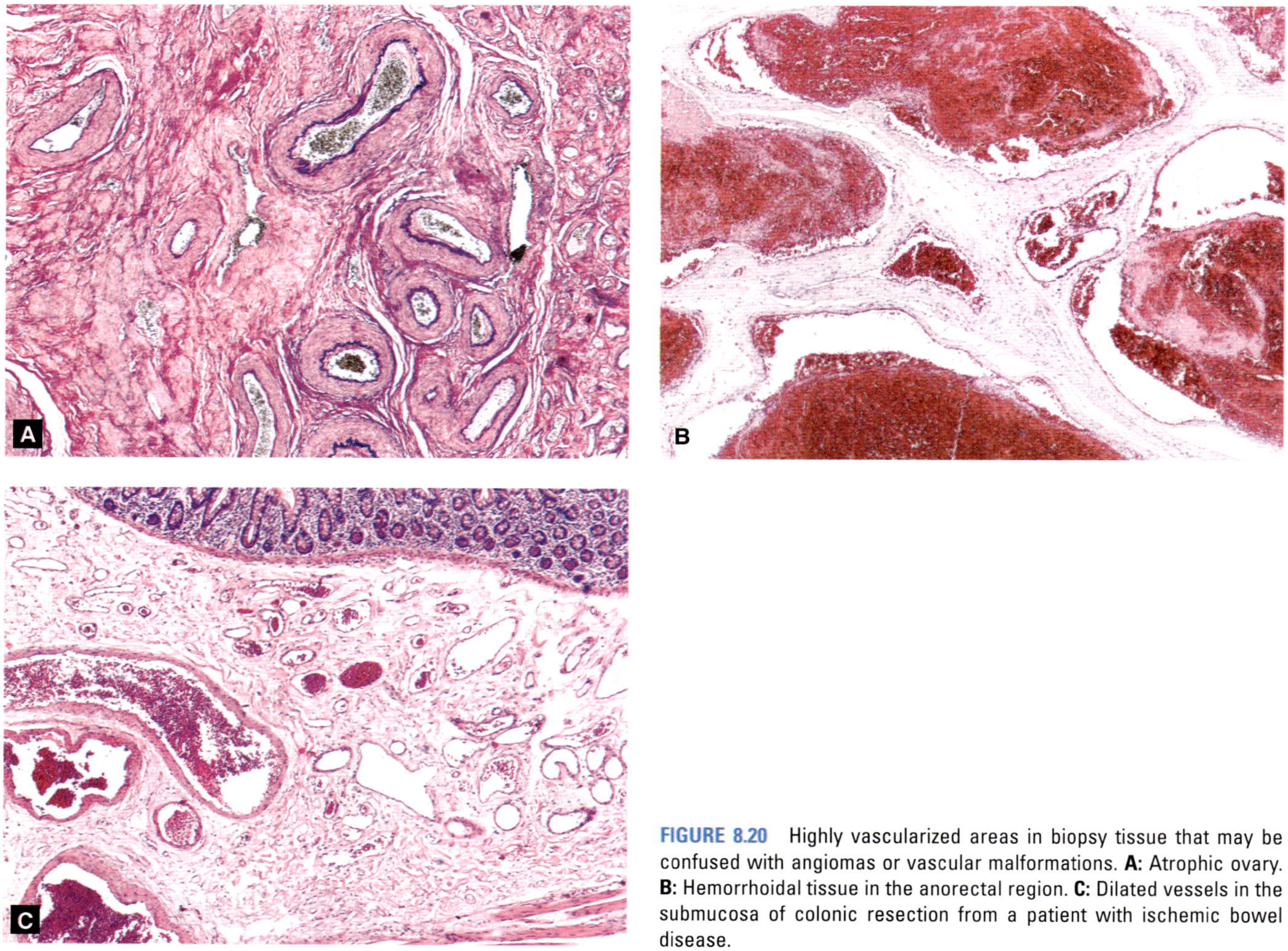

FIGURE 8.20 Highly vascularized areas in biopsy tissue that may be confused with angiomas or vascular malformations. **A:** Atrophic ovary. **B:** Hemorrhoidal tissue in the anorectal region. **C:** Dilated vessels in the submucosa of colonic resection from a patient with ischemic bowel disease.

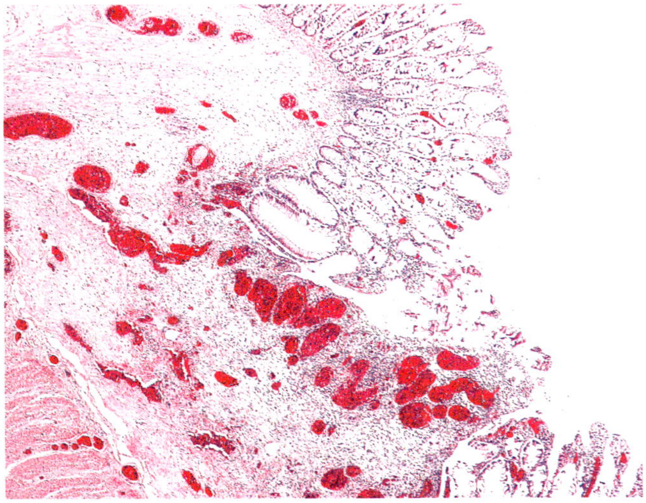

FIGURE 8.21 Angiodysplasia of the colon. Note the many dilated thin-walled blood vessels in the submucosa. Although these vessels are distended, their basic structure is unaltered.

arise from supporting cells that surround the normal but rather complex anastomosing channels between digital arterioles and venules. Glomus cells do not express endothelial markers but, because they stain with smooth muscle actin and vimentin, may be related to vascular smooth muscle (59).

Surgical pathologists must be familiar with the normal vascular patterns at various sites in order to avoid misinterpreting normal structures as arteriovenous malformations or angiodysplasias. Malformations and angiodysplasias must only be diagnosed when the vessel walls are abnormal. For example, aging changes and atherosclerosis seldom involve the smaller leptomeningeal arteries. Thus, the presence of eccentric fibrous intimal thickening or disruption of the elastic lamellae in vessels from this region would support a diagnosis of a malformation. Similarly, colonic angiodysplasias are usually located on the antimesenteric border of the colon, often close to the ileocecal valve (60). They represent dilatations of pre-existing capillary rings and veins due to intermittent outflow obstruction secondary to increased muscular pressure during peristalsis (Fig. 8.21). Submucosal arteries of the large intestine may show pronounced age-related tortuosity, and this must not be interpreted as an abnormality. Some cases of massive gastrointestinal hemorrhage result from abnormally large submucosal arteries that arise directly from omental vessels rather than the submucosal plexus. These abnormal vessels have been termed caliber-persistent arteries or Dieulafoy lesions (61).

Vascular Surgery

The changes commonly seen in vessels after surgical procedures and interventions are summarized in Table 8.5.

TABLE 8.5 Pathologic Changes After Vascular Surgery

Procedure	Spectrum of Histologic Change
Endarterectomy	*Acutely:* Surface platelet and fibrin deposits on inner face of surgical dissection, occasionally progressing to occlusive thrombosis *Chronically:* Variable degrees of fibrous intimal hyperplasia, occasionally progressing to restenosis. "False" aneurysm formation
Vein bypass grafting	*Acutely:* Thrombosis, dissection at anastomosis site *Chronically:* Dilatation with fibrous intimal thickening and medial muscular hypertrophy (to be contrasted with preimplant state). Occasionally, marked intimal fibrosis with lipid deposition and hemorrhage, leading to occlusion ("vein atherosclerosis")
Internal mammary and radial artery grafting	*Acutely:* Thrombosis at anastomosis sites *Chronically:* Occasional grafts become fibrosed. Graft atherosclerosis uncommon
Angioplasty and stenting	*Acutely:* Acute inflammation, dissection, and thrombosis *Chronically:* Restenosis due to reactive fibrous intimal thickening, now reduced by drug-eluting stents
Prosthetic vessels	*Acutely:* Thrombotic occlusion *Chronically:* Extensive macrophage and giant cell infiltration of fabric wall Formation of fibrin-rich pseudointima, occasionally progressing to partial or complete occlusion. Graft failure and thrombosis

Endarterectomy

Patency can be restored to a partially occluded artery by drawing out a proportion of the atherosclerotic intima. The procedure is usually applied to the carotid bifurcation but occasionally to iliac, femoral, or even coronary arteries. Meta-analysis has confirmed that carotid endarterectomy is highly beneficial in symptomatic patients with 70% + atheromatous stenosis (62). Ideally, the surgeon should establish a plane between the innermost media and the intima, and the atheromatous material should be removed in its entirety. At its bifurcation the carotid artery has an elastic wall, and the material removed will include layers of elastic tissue, atheromatous debris, and thrombus. Stroke is the most important immediate complication of the procedure and is usually the result of acute thrombosis or thromboembolism. Longer-term complications are recurrent thrombosis, aneurysmal dilatation, and restenosis due to fibrous intimal proliferation (63).

Bypass Grafts

Care must be taken to distinguish the pathologic changes seen in autologous saphenous vein bypass grafts from those associated with normal aging. When subjected to arterial pressure, many vein grafts dilate and most show some fibrous intimal thickening and medial muscular hypertrophy. In time, many develop pronounced fibrous intimal thickening with areas of lipid deposition, intramural hemorrhage, and thrombosis. These appearances closely mimic atherosclerosis and are an important cause of graft failure. In one postmortem study in which saphenous vein conduits were sampled throughout their length, more than 75% narrowing was demonstrated in 11% to 26% of the segments examined (64). Intensive lipid-lowering therapy can attenuate graft atherosclerosis. External synthetic sheaths are sometimes placed around vein grafts and may prevent vein wall thickening. Careful surgical harvesting of vein grafts may minimize damage to the endothelial layer (65,66). Grafts can sometimes be dilated by angioplasty, but redo coronary bypass procedures are now a significant part of the work of all cardiac surgery departments.

Coronary artery stenoses are routinely bypassed with the left or the right internal mammary artery. The origin of the artery from the subclavian artery is preserved, and it is then dissected away from the chest wall. There is usually a surrounding cuff of soft tissue, but some surgeons dissect this away, producing a so-called "skeletalized" graft (67). Long-term patency rates are superior to those of saphenous vein grafts because an artery is used and the caliber of the normal internal mammary artery is similar to that of distal coronary arteries. Pre-existing occlusive disease is present in fewer than 5% of patients, and only occasional grafts develop atheromatous obstructions. In its proximal portion, the internal mammary is an elastic artery, but the media is muscular from about the level of the 4th rib. Segments of the radial artery are also used as free grafts. Like saphenous vein grafts, they are anastomosed proximally to the aortic root and distally to the coronary arteries. The radial artery is muscular and is invariably free of significant atheroma (Fig. 8.12).

Angioplasty

Percutaneous coronary angioplasty (PTCA) with stent emplacement is now the treatment of choice for many proximal coronary stenoses and is increasingly used as a primary intervention to open thrombosed coronary arteries after myocardial infarction (68). The mortality rate in most centers is now less than 1%, and over 90% of procedures are initially successful. The design of stents and the antiproliferative agents that they release undergo continual modification and development (69).

In order to dilate the vessel, the heavily fibrous and focally calcified atheromatous plaque must be cracked open. Only when this has occurred can the deeper intima and underlying media be distended by the inflated balloon and held open by the expandable metallic stent. Early histologic studies of patients dying soon after angioplasty demonstrated a characteristic pattern of radial tears or splits, sometimes with dissections extending into the underlying media. Stents minimize the complications of these changes. Stent thrombosis occurs in about 1% of patients within a year of stent implantation. Antiproliferative agents released from drug-eluting stents may inhibit endothelialization of the stent struts (69). Restenosis is the result of fibrous intimal proliferation, thrombus formation, and an overall reduction in the size of the vessel lumen. These changes can now be visualized by either intravascular ultrasound or optical coherence tomography that creates a virtual histology image of the arterial wall (70). If death occurs soon after the procedure, the stent can be carefully extracted from the opened artery which is then processed in the usual way. After late closure, stents can be cut with an electric diamond saw and then embedded in hard plastic (Fig. 8.22). A technique for dissolving metallic stents has also been developed (71).

Prosthetic Vessels

Vascular surgeons use various techniques to anastomose grafts to native vessels. Grafts made of Dacron are successful in large-caliber bypass procedures such as aortoiliac grafting. Expanded polytetrafluorethylene (ePTFE) is preferred for below-knee grafts and in the construction of arteriovenous fistulae for chronic hemodialysis, although long-term patency rates are poor. At present, prosthetic grafts are not used to bypass coronary stenosis. The many factors that contribute to chronic graft failure include inherent thrombogenicity secondary to lack of an endothelial lining and compliance mismatch between the graft and the native vessel (72).

Prosthetic grafts develop a pseudointima composed of fibrin and enmeshed leukocytes, often with a partial endothelial lining. They often contain a giant cell reaction around the woven fibers of the graft with a moderate degree of adventitial fibrosis that binds the prosthesis to the surrounding tissues and reduces its elasticity. Long-term complications include thrombosis, particularly at flexures or surgical anastomoses, infection, and deterioration of the fibers of the graft. Modifications under study in industry include coating grafts with anticoagulants such as heparin, thrombomodulin, or hirudin, seeding grafts with endothelial cells or incorporating nitric oxide into grafts (72).

ELECTRON MICROSCOPY

Ultrastructural studies have made enormous contributions to our understanding of vascular biology. However, even surgical pathologists with a specific interest in vascular pathology have only limited experience and expertise in electron microscopy. Some of the most important ultrastructural features of vessels are summarized in Table 8.6.

Endothelial Cells

The entire vascular system is lined internally by a single layer of rather spindle-shaped endothelial cells. Small finger-like microvilli, 200 to 400 nm long, may be seen on their surfaces. A thin network of membrane-bound proteoglycans and glycoproteins, the glycocalyx, coats the luminal

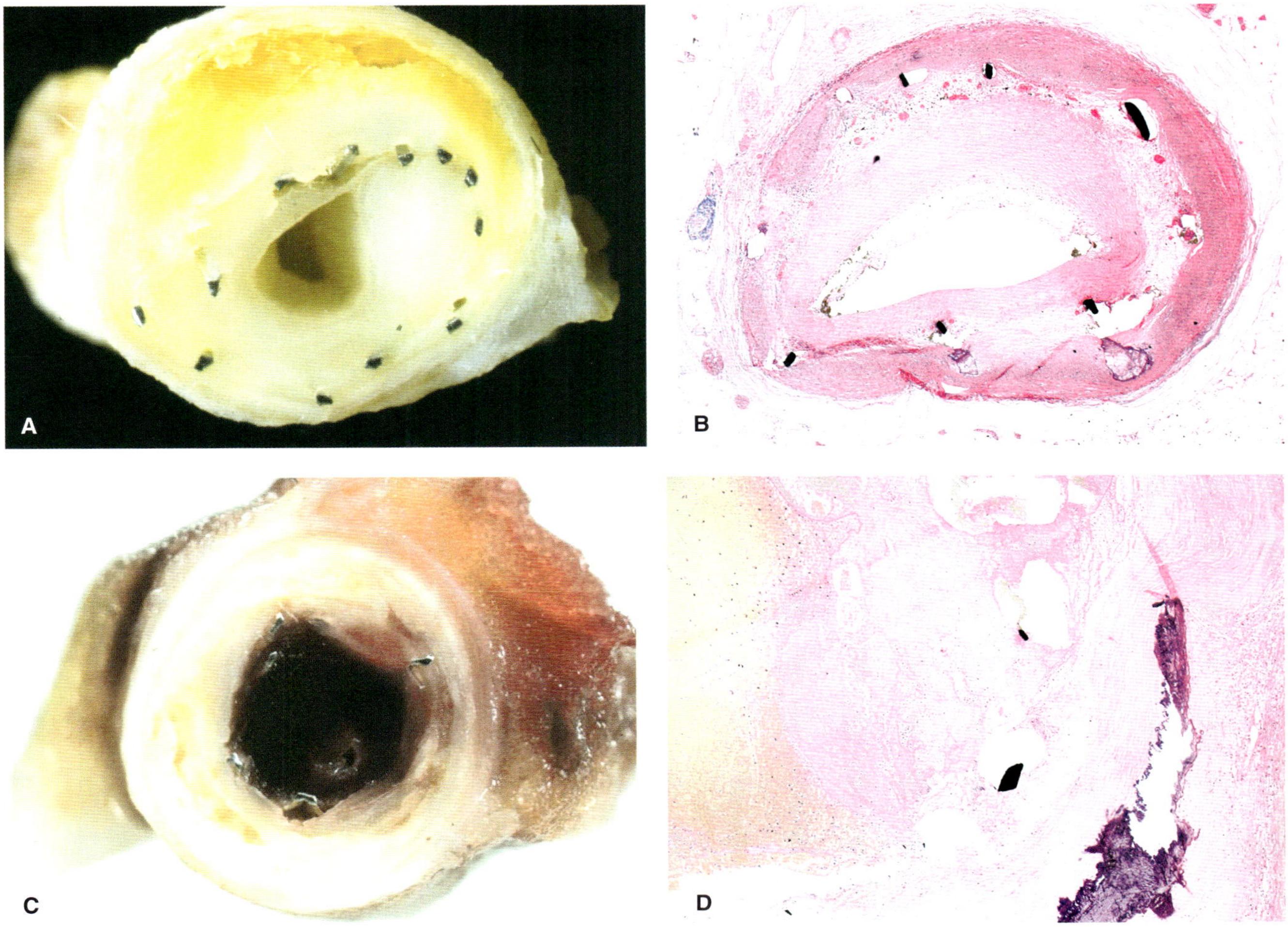

FIGURE 8.22 **A:** Macroscopic cross section of a stented coronary artery. Note the thick layer of concentric neointimal tissue that caused restenosis. The stent was cut with a diamond saw. This leaves the metal struts more or less in place, although some distortion is inevitable. **B:** Histology of the same artery after plastic embedding shows the stent struts at the interface of fibrocellular intimal tissue and pre-existent calcified atherosclerotic plaque. **C:** Macroscopic view of cross section through coronary artery with stent affected by thrombotic occlusion. **D:** Detail of the histology of the same segment after plastic embedding, showing black stent strut and empty spaces (where the struts had been in situ) adjacent to fresh thrombus.

TABLE 8.6 Ultrastructural Features of Vascular Tissues

Endothelial cells are joined by tight, adherans, or gap junctions.
Transendothelial channels characterize fenestrated endothelium, as in hepatic sinusoids, glomeruli, and endocrine organs.
Cytoplasmic inclusions of endothelium include lysosomes, plasmalemmal vesicles, and Weibel–Palade bodies (see Fig. 8.19).
Capillary endothelium is surrounded by basement membrane in which pericytes are embedded. There is very little basement membrane around lymphatic vessels. There are direct appositions between processes of pericytes and endothelium through gaps in the basement membrane.
Smooth muscle cells are invested in basement membrane and are linked by communicating (gap) junctions. Elastin and collagen fibers may be closely opposed to the surfaces of smooth muscle cells.
No significant media in small arterioles, capillaries, or lymphatic vessels.
The adventitia is composed of collagen and some elastin fibers. It has a well-developed structure in large veins but is very thin in some arteries.

surface of the endothelium. It varies in thickness from 0.5 μm in small capillaries to 4.5 μm in the carotid artery. The glycocalyx plays important roles with respect to vascular permeability, blood cell–vessel wall interactions, and shear stress sensing. The glycocalyx modulates inflammatory reactions by binding cytokine molecules and superoxide dismutase and can influence coagulation by binding proteins such as antithrombin III, thrombomodulin, and inhibitors of tissue factor. Glycocalyx proteins are responsible for transmission of shear stress signals into specific processes such as nitric oxide production (i.e., mechanotransduction). There is growing evidence that the glycocalyx is attenuated or absent in pathologic conditions such as atherosclerosis, diabetes mellitus, ischemia–reperfusion injury, and smoking (73). Although endothelial cells contain relatively sparse endoplasmic reticulum, a small number of free ribosomes, and an inconspicuous Golgi apparatus, they produce a variety of molecules that are important in blood coagulation and regulation of vascular tone.

Junctional complexes between endothelial cells are tight, adherens, or gap junctions (74). Tight junctions play a barrier function and help maintain cell polarity. A variety of molecules, notably those of the claudin family, create the barrier and regulate electrical resistance between cells (75). Loss of this barrier function may be important in disorders such as diabetic retinopathy (76). Adherens junctions regulate permeability to white cells and soluble molecules, and have a role in contact inhibition (77). Gap junctions are assembled from proteins known as connexins and form channels between adjacent cells (78). Alterations in gap junction proteins have been documented in human heart disease, including arrhythmias and cardiomyopathies (79).

Inclusions of Endothelial Cells

Lysosomes are readily identified in most endothelial cells and are involved in intracytoplasmic digestion of foreign debris and products of metabolism. In many areas of the vascular system, membrane-bound vesicles measuring up to 80 to 90 nm can be identified (Fig. 8.23). They are most prominent on the abluminal surface of the endothelial cell. They were originally known as plasmalemmal vesicles but are now termed caveolae. Their functions include sequestration and concentration of small molecules, maintaining endothelial barrier function, regulating nitric oxide synthesis, and cholesterol metabolism (80). Weibel–Palade bodies are elongated secretory organelles specific to endothelial cells. Von Willebrand factor is stored inside Weibel–Palade bodies as tubules, but forms long strings that recruit platelets to sites of endothelial injury upon release (81).

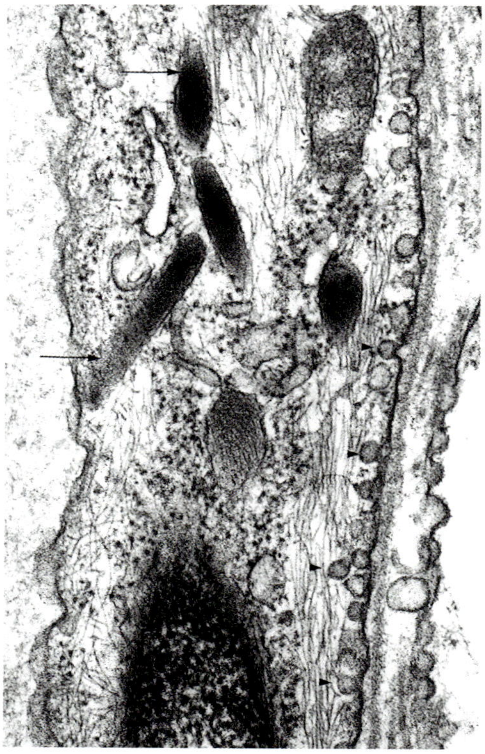

FIGURE 8.23 Transmission electron micrograph of an endothelial cell from a small subcutaneous capillary. Plasmalemmal vesicles (caveolae) are present on the abluminal surface (*arrowheads*). There are conspicuous Weibel–Palade bodies (*arrows*). Only part of the endothelial cell nucleus is included (**bottom**) (original magnification ×15,000).

Media

The aorta media consists of homogeneous parallel elastic lamellae that alternate with layers containing smooth muscle cells and a variety of extracellular components. Smooth muscle predominates in muscular arteries. The power of contraction of smooth muscle is as great as skeletal muscle and can be maintained for longer periods. The structure of smooth muscle cells is maintained by vimentin and desmin intermediate filaments, and the contractile forces are generated by actin and myosin filaments. Smooth muscle cells are arranged in parallel longitudinal bundles with the wide part of one cell opposed to the tapering part of another (Fig. 8.24). Each smooth muscle cell is invested with a basal lamina that merges with fine collagen and elastin fibers.

Individual smooth muscle cells are often linked by communicating (gap) junctions, but tight junctions are not generally seen. There are gap junctions between the smooth muscle cells and the overlying endothelium in the microcirculation, as well as in some larger arteries and arterioles (82). These myoendothelial junctions could have an important role in relaying physiologic or pharmacologic stimuli between the blood vessel lumen and the media.

Adventitia and Supporting Cells

The adventitial layer consists almost entirely of collagen and elastic fibers. The thickness of this layer varies with the size of the vessel, and it may be continuous with the surrounding connective tissue. It is particularly well developed in some medium-sized veins, but may be as thin as 80 μm in some cerebral arteries. A layer of elastic tissue, the external elastic lamella, is present at the junction of the media and adventitia. The pericytes that are present in and among the basement membrane of capillaries and small venules superficially resemble fibroblasts. The ultrastructural appearance of their cytoplasmic filaments suggests that they are contractile, and this is further evidence that they are of mesenchymal origin.

Lymphatic Vessels and Veins

The smallest lymphatic vessels have wider lumina than blood capillaries and a discontinuous basement membrane. A variety of anchoring filaments bind the lymphatic endothelium to the surrounding collagenous tissues, perhaps providing the sort of support normally produced by

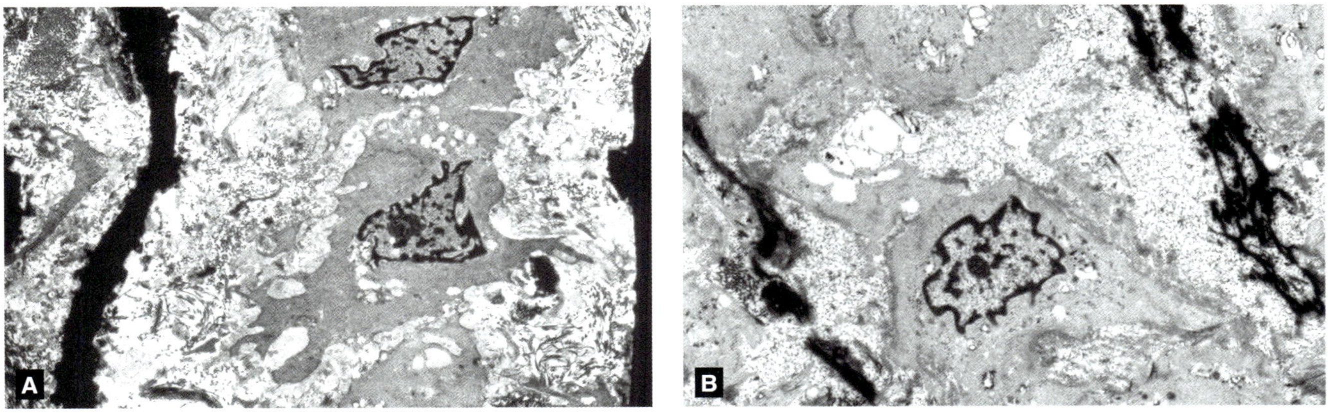

FIGURE 8.24 Transmission electron microcopy of aortic media. **A:** Normal aorta. There are musculoelastic lamellar units with regular thick elastic lamellae with numerous small extensions. Smooth muscle is present in the interlamellar space. **B:** Marfan syndrome. There is distinct thinning and fragmentation of both elastic lamellae. There is an increase in fibrillary collagen and mucoid deposits in the matrix of the interlamellar space.

basement membrane and enmeshed pericytes in capillaries. The ultrastructural appearances of venous capillaries, venules, and small veins mirror those seen at the light microscopic level.

ANTIGEN EXPRESSION OF NORMAL VASCULAR TISSUE

Endothelium

There are differences in the size, shape, orientation, and antigen expression between arterial and venous endothelium (83). Several genes have been identified that are preferentially expressed in either arterial or venous endothelium. An increasing panel of monoclonal antibodies is available to investigate aspects of this heterogeneity, some of which may have a role in pathology. The most widely used antibodies are directed against factor VIII, CD31, and CD34 (Fig. 8.25). Since these antigens are present in all types of endothelial cells, they are considered to be pan-endothelial markers. The lectin Ulex europaeus agglutinin 1 binds to some α–L-fructose–containing glycocompounds and virtually all human endothelia. The staining pattern is sometimes more intense than that of factor VIII antibodies, especially in immature vessels. However, all endothelial markers cross-react to some extent with other cell components. For example, CD31 reacts strongly with platelets, macrophages, and lymphocytes in areas of hemorrhage or thrombosis. Factor VIII antibodies can produce distinct diffuse extracellular staining patterns due to leakiness of the endothelium in atherosclerotic plaques and at sites of

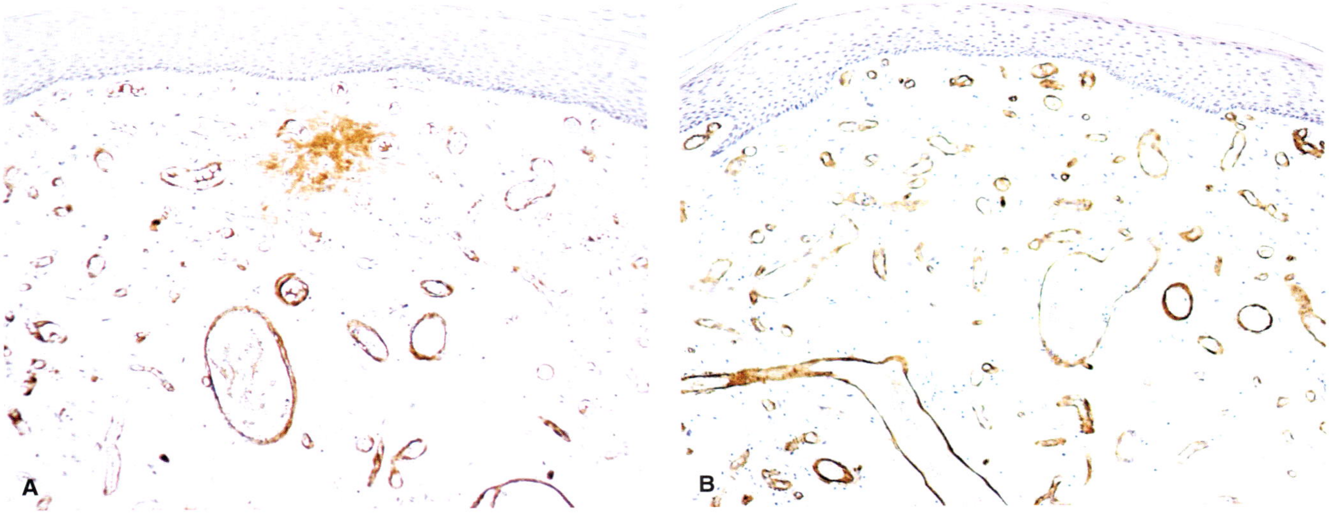

FIGURE 8.25 Staining of small vessels in a hemangioma with antibodies to factor VIII (**A**) and CD34 (**B**). As in these illustrations, the staining with CD34 is usually sharper than with factor VIII. Some nonspecific extravascular staining is often seen with factor VIII but has no significance.

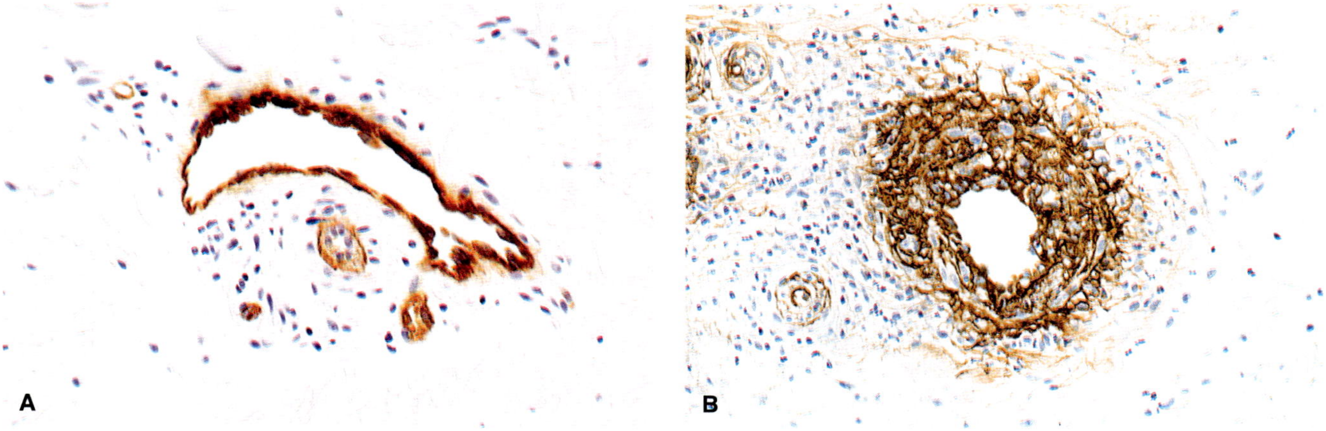

FIGURE 8.26 Staining patterns with factor VIII antibodies. **A:** Microvessel of the superior dermal vascular plexus showing sharp endothelial immunostaining. **B:** Microvessel of the superior dermal vascular plexus in case of cutaneous small vessel vasculitis. There is prominent perivascular staining due to leakage of factor VIII from damaged vessels.

vasculitis (Fig. 8.26). These antibodies can be helpful in confirming that tumor deposits are in vascular or lymphatic channels, rather than in artifactual tissue spaces (Fig. 8.27). ERG is a relatively new marker that may be used to detect nuclear staining of endothelial cells.

Antibodies that recognize proteins involved in the early steps in angiogenesis include antiendoglin (CD105) and anti-VEGF. CD105 is an interesting epitope, since it expresses on activated endothelial cells, especially under conditions of hypoxia. Increased expression is noted on the microvessels of tumors such as glioma and breast carcinoma and also nonneoplastic situations such as the microvessels that develop in the hypoxic environment deep in long-standing atherosclerotic plaques. In carcinomas of the breast, a relationship between microvascular CD105 expression and invasive potential of the tumor has been observed, which appears to correlate with poor outcomes (84) (Fig. 8.28). In different sites in the vascular system, the endothelium may show marked heterogeneity in morphology, gene expression patterns, and related differences in functional status. Antibodies to glucose transporter protein 1 (GLUT-1 antibodies) react with the endothelium of cerebral capillaries, the placental vasculature, and one specific type of angioma—the juvenile capillary angioma (Fig. 8.29) (85). Another site-specific antibody is anti-HECA 452, which reacts specifically with the plump endothelial cells of high endothelial venules in lymphoid tissue and postcapillary venules in chronically inflamed tissue (Fig. 8.30) (86). Endothelial cells undergo profound functional alterations (endothelial activation) in inflamed tissues and in atheromatous lesions. These include upregulation of cell surface adhesion molecules such as ICAM-1, VCAM-1, and PECAM (CD31)

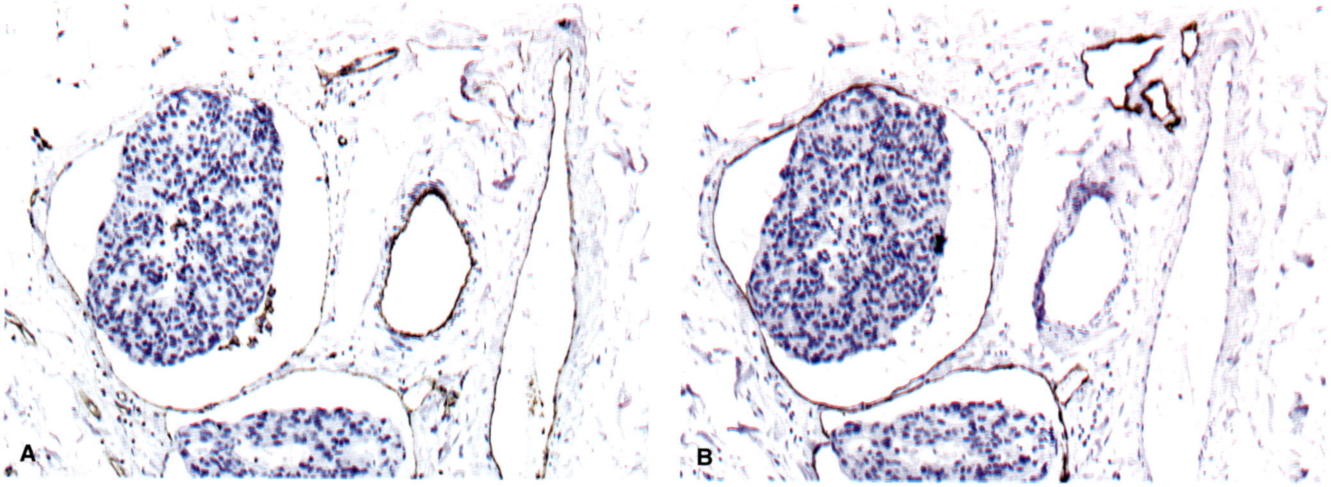

FIGURE 8.27 Ductal carcinoma of the breast with angioinvasion. CD31 immunostaining **(A)** highlights several empty vascular channels. Note that the spaces containing the tumor deposits do stain positively with CD31. However D2-40 immunostaining **(B)**, which is specific for lymphatic endothelium, confirms that the tumor deposits are within lymphatic vessels rather than tissue spaces.

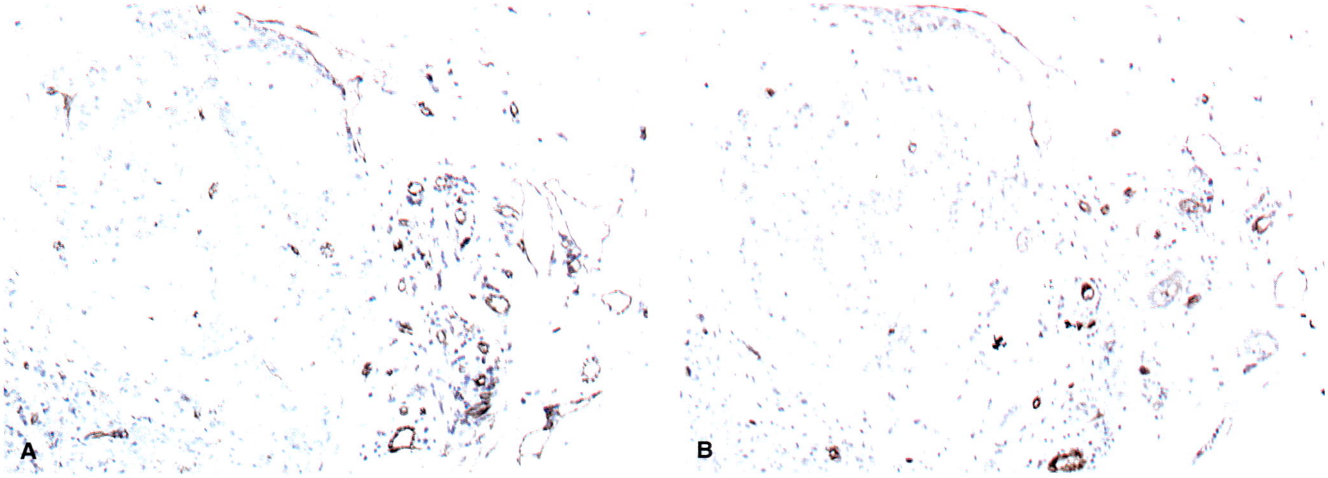

FIGURE 8.28 Microvasculature in breast carcinoma. CD31 staining (**A**) demonstrates a large number of microvessels. There is expression of endoglin (anti-CD105) (**B**) in only a subfraction of the microvessels.

or de novo expression of leukocyte adhesion molecules. D2-40, LYVE-1, and VEGFR3 stain lymphatic endothelium specifically, and D2-40 staining is commonly seen in examples of Kaposi sarcoma (Fig. 8.31) (87).

Smooth Muscle

Biochemical and immunohistologic studies have demonstrated that vascular smooth muscle has a distinctive component of contractile and intermediate filament proteins (88). Desmin and γ-smooth muscle actin predominate in most smooth muscle cells, whereas α-smooth muscle actin and vimentin are more common in vascular smooth muscle cells. Antibodies directed against smooth muscle actin (SMA-1) are excellent markers of medial muscle; SMA-1 recognizes the full spectrum of proliferating (or synthetic) and mature (or contractile) smooth muscle phenotypes. As SMA-1 reacts with pericytes, it clearly outlines capillaries in reactive microvascular proliferations and in pyogenic granulomas and juvenile angiomas during their growth phase. All benign vascular proliferations, including glomus tumors, stain strongly with SMA-1 antibodies (Fig. 8.32). SMA-1 immunostaining can be helpful in the detection of early organization of thromboembolic material, especially in situations where age determination of the thrombus is required such as investigations of sudden (coronary) cardiac death, pulmonary embolism, and in medicolegal autopsies.

Other Useful Antibodies for Diagnostic Vascular Pathology

Immunohistochemical studies of the inflammatory infiltrates in atheromatous lesions (Fig. 8.33) have contributed greatly to our understanding of the pathogenesis of

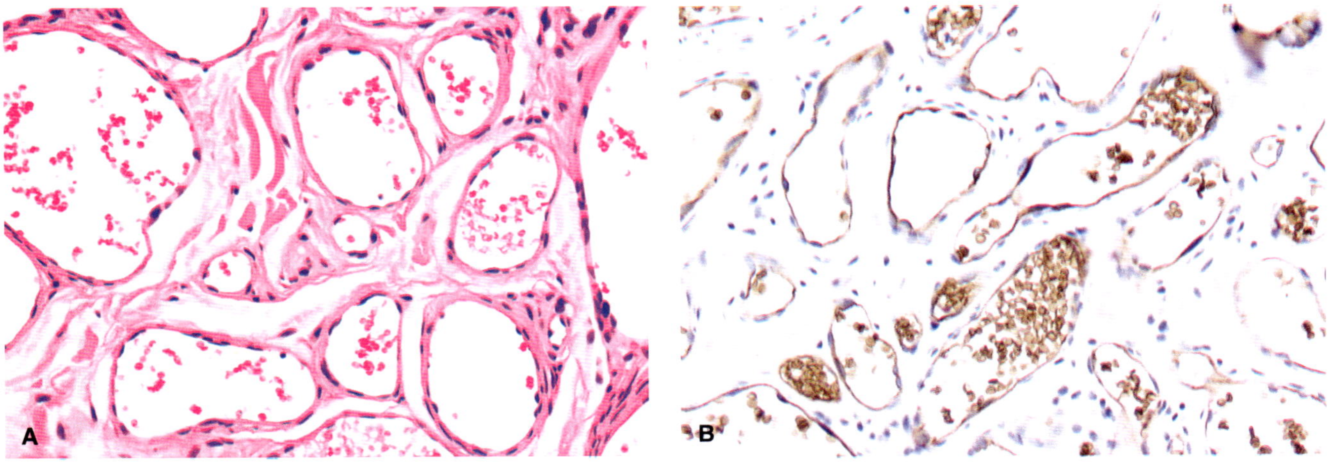

FIGURE 8.29 Site-specific staining of vascular endothelium. The vascular endothelium in this juvenile capillary hemangioma from a 3-year-old male (**A**) is specifically stained with the GLUT-1 antibody (**B**). This antibody also stains the endothelium of cerebral capillaries and the placenta. In contrast, the more commonly used endothelial antibodies such as factor VIII, CD31, and CD34 stain most types of normal and neoplastic endothelia.

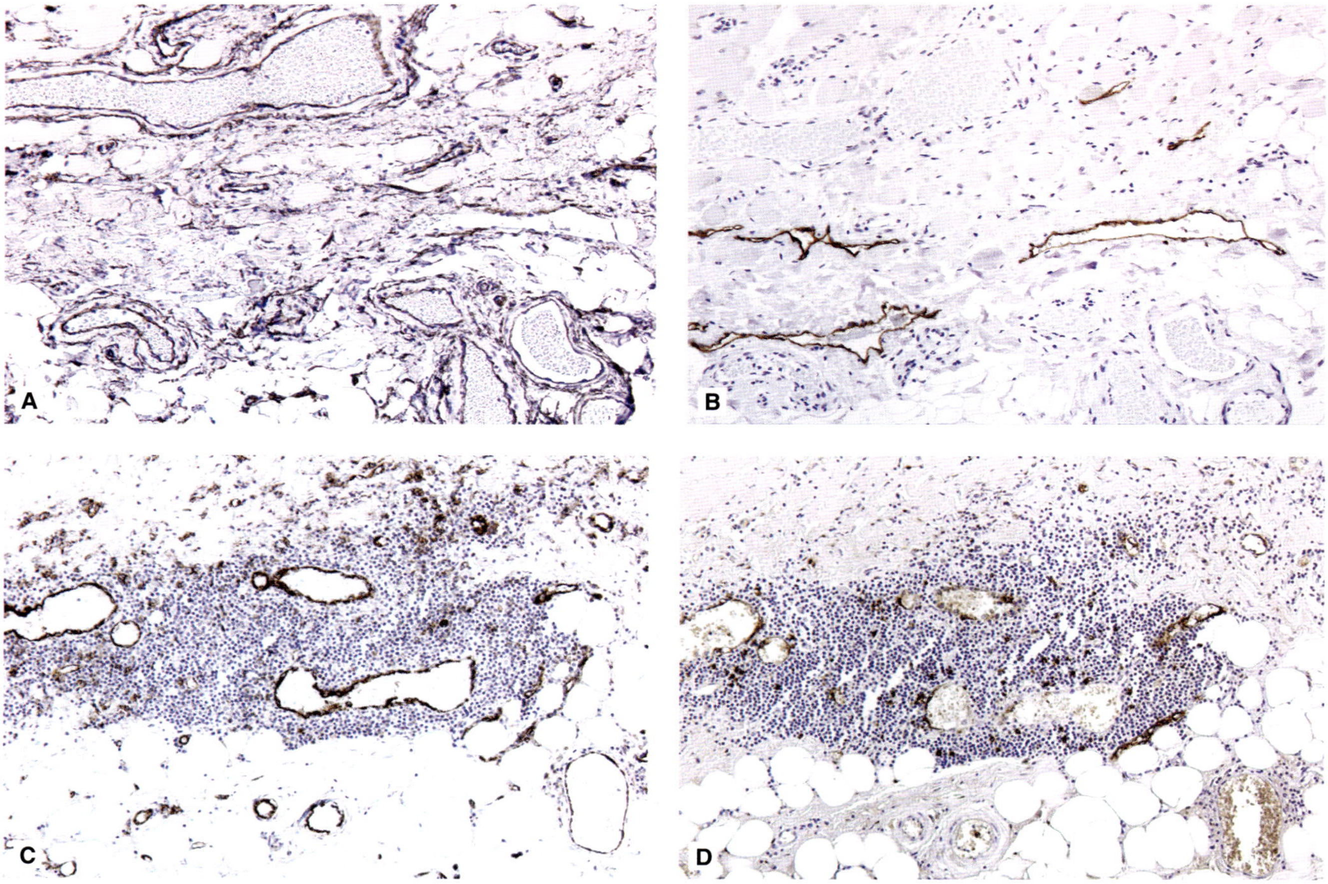

FIGURE 8.30 Immunohistochemical staining of vessels. **A** and **B** show a mixture of vessels from the subcutaneous tissues of a 68-year-old female from close to a leg ulcer: CD31 antibody staining identifies many vascular spaces (**A**), and a similar section is stained with the antibody D2-40, which recognizes lymphatic endothelium only (**B**). **C** and **D** are from a nodular inflammatory infiltrate in the aortic adventitia adjacent to a large atheromatous plaque. **C** has been stained with CD31, which recognizes most vessels, and **D** was stained with HECA-452, which recognizes high endothelial venules.

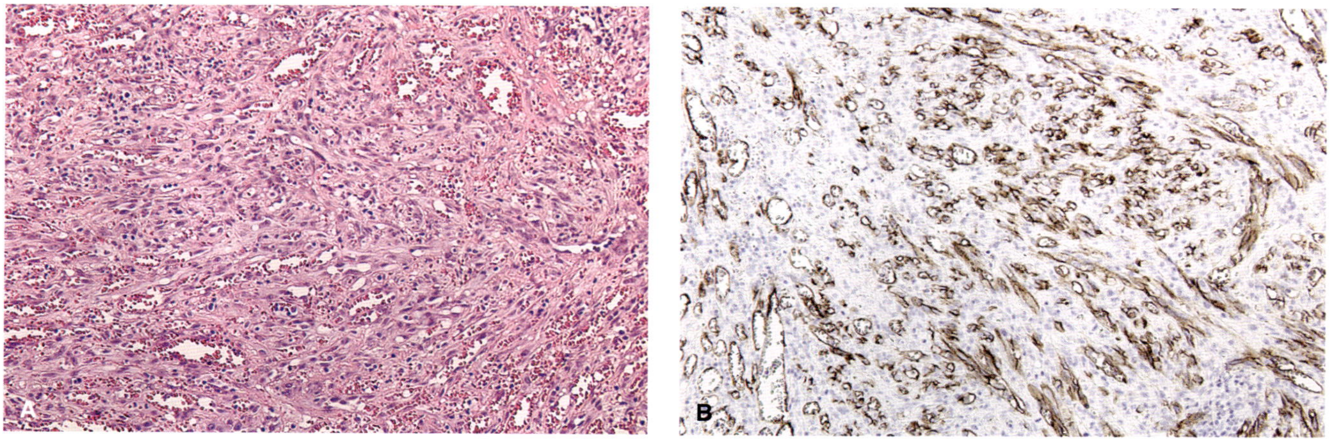

FIGURE 8.31 Kaposi sarcoma. **A** has been stained with H&E. **B** was immunostained with the D2-40 antibody, a specific marker of the lymphatic endothelium. Note the strong positive staining; LYVE-1 is another antibody that specifically stains lymphatic endothelium.

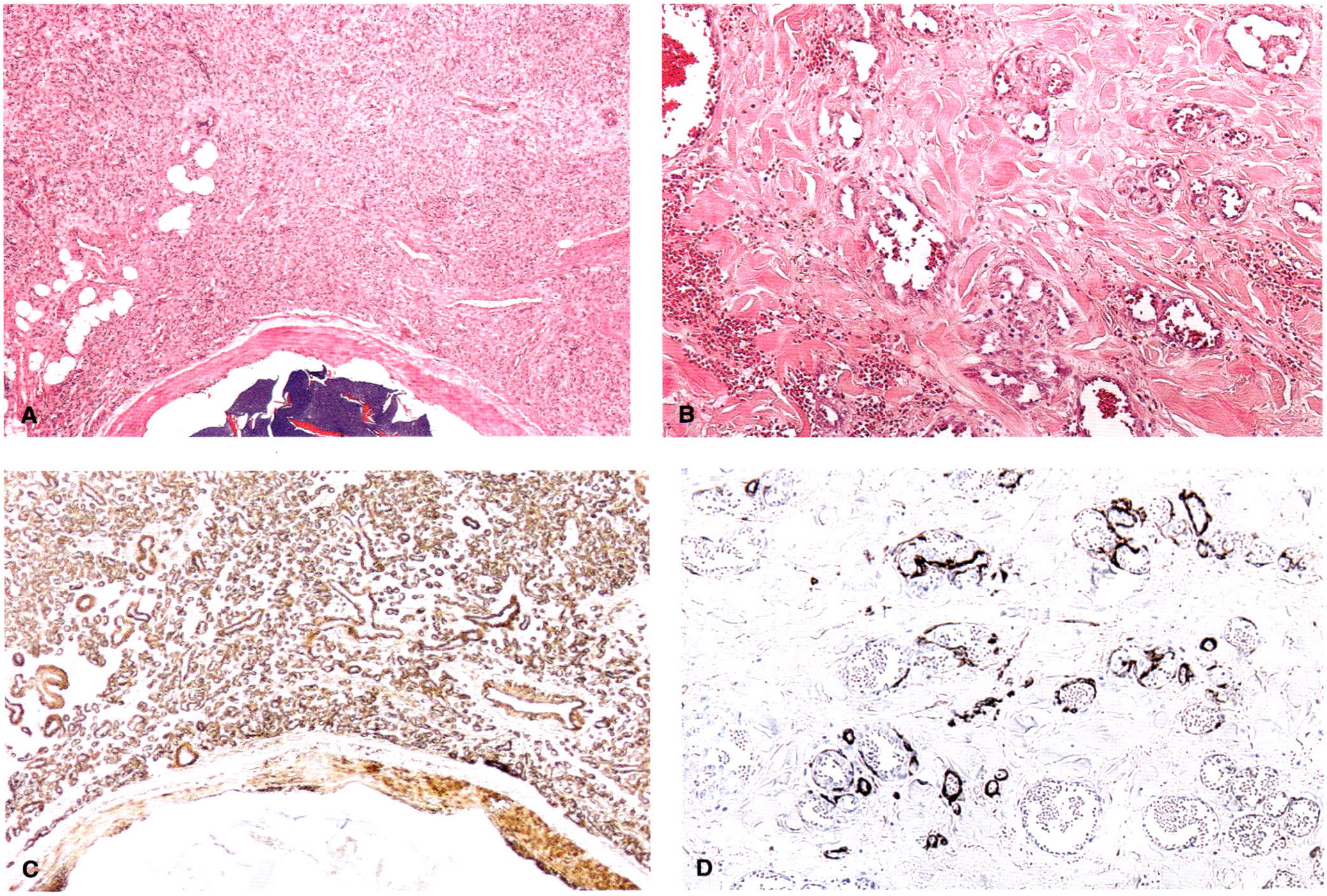

FIGURE 8.32 Patterns of staining with smooth muscle actin antibody. **A** and **C** are from a benign vascular proliferation. Note the intense staining of the walls of these small vessels. **B** and **D** are from an angiosarcoma. Only small amounts of actin are present in the walls of the malignant blood vessels (**D**).

atherosclerosis (89) but are not of proven value in everyday surgical pathology. T-lymphocyte markers, such as CD3 and CD4, may be of use in the diagnosis of vasculitis, especially temporal arteritis with minimal inflammatory activity (90). Transplant-related arteriosclerosis features a high relative proportion of CD8-positive T-lymphocytes that also express granzyme B. Anti-factor VIII and antifibrinogen antibodies demonstrate vascular leakiness and tissue damage, while CD61 stains platelet aggregates in microvessels (91). Glycophorin A is a specific marker of erythrocytes and their

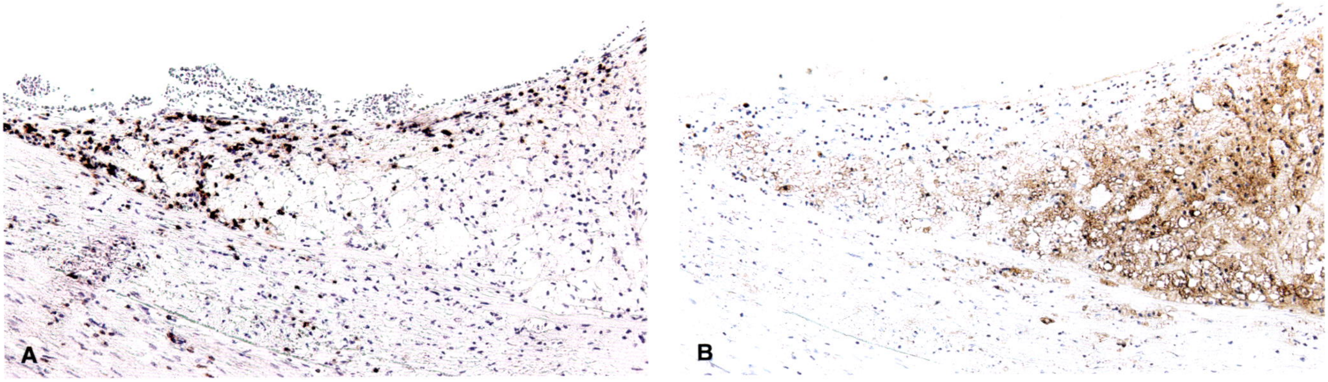

FIGURE 8.33 Immunohistochemical staining in atherosclerosis. **A:** CD3-positive lymphocytes are present at the edge of a lesion. **B:** Macrophages react for CD68.

precursors in the bone marrow. The epitopes are preserved in tissues for long periods and, thus, the antibody aids detection of old hemorrhage (92,93).

ACKNOWLEDGMENTS

This chapter is an updated version of previous iterations written by Patrick J. Gallagher and Allard C. van der Wal, whose contributions are gratefully acknowledged.

REFERENCES

1. Laurent S, Boutouyrie P, Lacolley P. Structural and genetic basis of arterial stiffness. *Hypertension*. 2005;45:1050–1055.
2. Jacob MP. Extracellular matrix remodeling and matrix metalloproteinases in the vascular wall during aging and in pathological conditions. *Biomed Pharmacother*. 2003;57:195–202.
3. Kavurma MM, Bhindi R, Lowe HC, Chesterman C, Khachigian LM. Vessel wall apoptosis and atherosclerotic plaque instability. *J Thromb Haemost*. 2005;3:465–472.
4. Kentish JC. Smooth muscle and the cardiovascular and lymphatic systems. In: Stranding S, ed. *Gray's Anatomy: The Anatomical Basis of Clinical Practice*. 40th ed. Churchill Livingstone Elsevier; 2008:127–144.
5. Clarke MC, Littlewood TD, Figg N, et al. Chronic apoptosis of vascular smooth muscle cells accelerates atherosclerosis and promotes calcification and medial degeneration. *Circ Res*. 2008;192:1529–1538.
6. Roijers RB, Debernardi N, Cleutjens JPM, Schurgers LJ, Mutsaers PH, van der Vusse GJ. Microcalcifications in early intimal lesions of atherosclerotic human coronary arteries. *Am J Pathol*. 2011;178:2879–2887.
7. Jono S, Shioi A, Ikari Y, Nishizawa Y. Vascular calcification in chronic kidney disease. *J Bone Miner Metab*. 2006;24:176–181.
8. Maier W, Altwegg LA, Corti R, et al. Inflammatory markers at the site of ruptured plaque in acute myocardial infarction: locally increased interleukin-6 and serum amyloid A but decreased C-reactive protein. *Circulation*. 2005;111:1355–1361.
9. Mucchiano GI, Häggqvist B, Sletten K, Westermark P. Apolipoprotein A-1-derived amyloid in atherosclerotic plaques of the human aorta. *J Pathol*. 2001;193:270–275.
10. Boddaert J, Mallat Z, Fornes P, et al. Age and gender effects on apoptosis in the human coronary arterial wall. *Mech Ageing Dev*. 2005;126:678–684.
11. Schlatmann TJ, Becker AE. Histologic changes in the normal aging aorta: implications for dissecting aortic aneurysm. *Am J Cardiol*. 1977;39:13–20.
12. Hsue PY, Salinas CL, Bolger AF, Benowitz NL, Waters DD. Acute aortic dissection related to crack cocaine. *Circulation*. 2002;105:1592–1595.
13. Gelb BD. Marfan's syndrome and related disorders—more tightly connected than we thought. *N Engl J Med*. 2006;355:841–844.
14. Loeys BL, Schwarze U, Holm T, et al. Aneurysm syndromes caused by mutations in the TGF-beta receptor. *N Engl J Med*. 2006;355:788–798.
15. Judge DP, Dietz HC. Marfan's syndrome. *Lancet*. 2005;366:1965–1976.
16. Perez ZG, Zafar MA, Velasco JJ, et al. Aortic size at the time of type A and type B dissections. *Ann Thorac Surg*. 2023;116(2):262–268.
17. Bossone E, Eagle KA. Epidemiology and management of aortic disease: aortic aneurysms and acute aortic syndromes. *Nat Rev Cardiol*. 2021;18(5):331–348.
18. Hinterseher I, Tromp G, Kuivaniemi H. Genes and abdominal aortic aneurysm. *Ann Vasc Surg*. 2011;25:388–412.
19. Kuivaniemi H, Shibamura H, Arthur C, et al. Familial aortic aneurysms: collection of 233 multiplex families. *J Vasc Surg*. 2003;37:340–345.
20. Ihling C, Szombathy T, Nampoothiri K, et al. Cystic medial degeneration of the aorta is associated with p53 accumulation, Bax upregulation, apoptotic cell death, and cell proliferation. *Heart*. 1999;82:286–293.
21. Lee A, Luk A, Phillips KRB, Lim KD, David TE, Butany J. Giant cell aortitis: a difficult diagnosis assessing risk for the development of aneurysms and dissections. *Cardiovasc Pathol*. 2011;20(4):247–253.
22. Ryder HF, Tafe LJ, Burns CM. Fatal aortic dissection due to fulminant variety of isolated aortitis. *J Clin Rheumatol*. 2009;15:295–299.
23. Legendre P, Régent A, Thiebault M, Mouthon L. Anti-endothelial cell antibodies in vasculitis: a systematic review. *Autoimmun Rev*. 2017;16(2):146–153.
24. Li Cavoli G, Mulè G, Vallone MG, Caputo F. Takayasu's disease effects on the kidneys: current perspectives. *Int J Nephrol Renovasc Dis*. 2018;11:225–233.
25. Szyszka-Mroz J, Wozniak W. A histological study of human ductus arteriosus during the last embryonic week. *Folia Morphol (Warsz)*. 2003;62:365–367.
26. Anderson RH, Becker AE, Robertson WB. The arterial duct. In: Symmers WS, ed. *The Cardiovascular System Part A*. Churchill Livingstone; 1993:1193, 197–202.
27. Jain D, Dietz HC, Oswald GL, Maleszewski JJ, Halushka MK. Causes and histopathology of ascending aortic disease in children and young adults. *Cardiovasc Pathol*. 2011;20:15–25.
28. Stary HC, Blankenhorn DH, Chandler AB, et al. A definition of the intima of human arteries and of its atherosclerosis-prone regions. A report from the Committee on Vascular Lesions of the Council on Arteriosclerosis, American Heart Association. *Circulation*. 1992;85:391–405.
29. Lakatta EG, Levy D. Arterial and cardiac aging: major shareholders in cardiovascular disease enterprises: Part I: aging arteries: a 'set up' for vascular disease. *Circulation*. 2003;107:139–146.
30. van der Wal AC. Coronary artery pathology. *Heart*. 2007;93:1484–1489.
31. Camici PG, Crea F. Coronary microvascular dysfunction. *N Engl J Med*. 2007;356:830–840.
32. Kim C, Redberg RF, Pavlic T, Eagle KA. A systematic review of gender differences in mortality after coronary artery bypass graft surgery and percutaneous interventions. *Clin Cardiol*. 2007;30:491–495.
33. Antoniadis AP, Chatzizisis YS, Giannoglou GD. Pathogenetic mechanisms of coronary ectasia. *Int J Cardiol*. 2008;130:335–343.
34. Kolodgie FD, Gold HK, Burke AP, et al. Intraplaque hemorrhage and progression of coronary atheroma. *N Engl J Med*. 2003;349:2316–2325.

35. Smeeth L, Thomas SL, Hall AJ, Hubbard R, Farrington P, Vallance P. Risk of myocardial infarction and stroke after acute infection or vaccination. *N Engl J Med.* 2004;351(25): 2611–2618.
36. Ackermann M, Verleden SE, Kuehnel M, et al. Pulmonary vascular endothelialitis, thrombosis, and angiogenesis in Covid-19. *N Engl J Med.* 2020;383(2):120–128.
37. Bois MC, Boire NA, Layman AJ, et al. COVID-19-associated nonocclusive fibrin microthrombi in the heart. *Circulation.* 2021;143(3):230–243.
38. Hiemann NE, Wellnhofer E, Knosalla C, et al. Prognostic impact of microvasculopathy on survival after heart transplantation: evidence from 9713 endomyocardial biopsies. *Circulation.* 2007;116(11):1274–1282.
39. Bishawi M, Bowles D, Pla MM, et al. PD-1 and PD-L1 expression in cardiac transplantation. *Cardiovasc Pathol.* 2021; 54:107331.
40. Kfoury AG, Stehlik J, Renlund DG, et al. Impact of repetitive episodes of antibody-mediated or cellular rejection on cardiovascular mortality in cardiac transplant recipients: defining rejection patterns. *J Heart Lung Transplant.* 2006; 25(11):1277–1282.
41. Mehra MR, Crespo-Leiro MG, Dipchand A, et al. International Society for Heart and Lung Transplantation working formulation of a standardized nomenclature for cardiac allograft vasculopathy-2010. *J Heart Lung Transplant.* 2010;29(7): 717–727.
42. Kobashigawa JA, Tobis JM, Starling RC, et al. Multicenter intravascular ultrasound validation study among heart transplant recipients: outcomes after five years. *J Am Coll Cardiol.* 2005;45(9):1532–1537.
43. Hammes H-P. Pericytes and the pathogenesis of diabetic retinopathy. *Horm Metab Res.* 2005;37(Suppl 1):39–43.
44. Ashgar O, Al-Sunni A, Khavandi K, et al. Diabetic cardiomyopathy. *Clin Sci (Lond).* 2009;116:741–760.
45. Richards OC, Raines SM, Attie AD. The role of blood vessels, endothelial cells and vascular pericytes in insulin secretion and peripheral insulin action. *Endocr Rev.* 2010;31:343–363.
46. Hayasaka H, Taniguchi K, Fukai S, Miyasaka M. Neogenesis and development of the high endothelial venules that mediate lymphocyte trafficking. *Cancer Sci.* 2010;101:2302–2308.
47. Kamesh L, Harper L, Savage COS. ANCA-positive vasculitis. *J Am Soc Nephrol.* 2002;13(7):1953–1960.
48. Aymonnier K, Amsler J, Lamprecht P, Salama A, Witko-Sarsat V. The neutrophil: a key resourceful agent in immune-mediated vasculitis. *Immunol Rev.* 2023;314(1):326–356.
49. Herskowitz A, Soule LM, Ueda K, et al. Arteriolar vasculitis on endomyocardial biopsy: a histologic predictor of poor outcome in cyclosporine-treated heart transplant recipients. *J Heart Transplant.* 1987;6(3):127–136.
50. Hammond EH, Yowell RL, Nunoda S, et al. Vascular (humoral) rejection in heart transplantation: pathologic observations and clinical implications. *J Heart Transplant.* 1989;8(6):430–443.
51. Glass C, Butt YM, Gokaslan ST, Torrealba JR. CD68/CD31 immunohistochemistry double stain demonstrates increased accuracy in diagnosing pathologic antibody-mediated rejection in cardiac transplant patients. *Am J Transplant.* 2019;19(11): 3149–3154.
52. Owens CD. Adaptive changes in autologous vein grafts for arterial reconstruction: clinical implications. *J Vasc Surg.* 2010; 51:736–746.
53. Langes K, Hort W. Intimal fibrosis (phlebosclerosis) in the saphenous vein of the lower limb: a quantitative analysis. *Virchows Arch A Pathol Anat Histopathol.* 1992;421:127–131.
54. Suh YJ, Hong H, Ohana M, et al. Pulmonary embolism and deep vein thrombosis in COVID-19: a systematic review and meta-analysis. *Radiology.* 2021;298(2):E70–E80.
55. Mooi WJ, Grünberg K. Histopathology of pulmonary hypertensive diseases. *Curr Diagn Pathol.* 2006;12:429–440.
56. Patchefsky AS. Nonneoplastic pulmonary disease. In: Mills SE, ed. *Sternberg's Diagnostic Surgical Pathology.* 5th ed. Vol 1. Lippincott Williams & Wilkins; 2009:1035–1039.
57. Warnock ML, Kunzmann A. Changes with age in muscular pulmonary arteries. *Arch Pathol Lab Med.* 1977;101:175–179.
58. Steiner I, Hájková P, Kvasnicka J, Kholová I. Myocardial sleeves of pulmonary vein and atrial fibrillation: a postmortem histopathological study on 100 subjects. *Virchows Arch.* 2006;449:88–95.
59. Gombos Z, Zhang PJ. Glomus tumor. *Arch Pathol Lab Med.* 2008;132:1448–1452.
60. Warkentin TE, Moore JC, Anand SS, Lonn EM, Morgan DG. Gastrointestinal bleeding, angiodysplasia, cardiovascular disease, and acquired von Willebrand syndrome. *Transfus Med Rev.* 2003;17:272–286.
61. Baxter M, Aly EH. Dieulafoy's lesion: current trends in diagnosis and management. *Ann R Coll Surg Engl.* 2010;92:548–554.
62. Rerkasam K, Rothwell PM. Systematic review of the operative risks of carotid endarterectomy for recently symptomatic stenosis in relation to the timing of surgery. *Stroke.* 2009;40: e564–e572.
63. Riles TS, Rockman CB. Cerebrovascular disease. In: Townsend CM, Beauchamp RD, Evers BM, Mattox KL, eds. *Sabiston Textbook of Surgery.* 18th ed. WB Saunders; 2008: 1895–1899.
64. Kalan JM, Roberts WC. Morphologic findings in saphenous veins used as coronary arterial bypass conduits for longer than 1 year: Necropsy analysis of 53 patients, 123 saphenous veins, and 1865 five-millimetre segments of veins. *Am Heart J.* 1990;119:1164–1184.
65. Jeremy JY, Gadsdon P, Shukla N, et al. On the biology of saphenous vein grafts fitted with external synthetic sheaths and stents. *Biomaterials.* 2007;28:895–908.
66. Parang P, Arora R. Coronary vein graft disease: pathogenesis and prevention. *Can J Cardiol.* 2009;25:e57–e62.
67. Ali E, Saso S, Ahmed K, Athanasiou T. When harvested for coronary artery bypass surgery, does a skeletonized or pedicled radial artery improve conduit patency? *Interact Cardiovasc Thorac Surg.* 2010;10:289–292.
68. D'Souza SP, Mamas MA, Fraser DG, Fath-Ordoubadi F. Routine early coronary angioplasty versus ischaemia-guided angioplasty after thrombolysis in acute ST-elevation myocardial infarction: a meta-analysis. *Eur Heart J.* 2011;32:972–982.
69. Popma JJ, Bhatt DL. Percutaneous coronary and valvular intervention. In: Bonow RO, Mann DL, Zipes DP, Libby P, eds. *Braunwald's Heart Disease.* 9th ed. Elsevier Saunders; 2011:1270–1300.
70. Garcia-Garcia HM, Gonzalo N, Regar E, Serruys PW. Virtual histology and optical coherence tomography: from research to a broad clinical application. *Heart.* 2009;95:1362–1374.
71. Bradshaw SH, Kennedy L, Dexter DF, Veinot JP. A practical method to rapidly dissolve metallic stents. *Cardiovasc Pathol.* 2009;18:127–133.

72. Kapadia MR, Popowich DA, Kibbe MR. Modified prosthetic vascular conduits. *Circulation.* 2008;117:1873–1882.
73. Reitsma S, Slaaf DW, Vink H, van Zandvoort MAMJ, oude Egbrink MGA. The endothelial glycocalyx: composition, functions, and visualization. *Pflugers Arch.* 2007;454:345–359.
74. Bazzoni G, Dejana E. Endothelial cell-to-cell junctions: molecular organization and role in vascular homeostasis. *Physiol Rev.* 2004;84:869–901.
75. Van Itallie CM, Anderson JM. The molecular physiology of tight junction pores. *Physiology (Bethesda).* 2004;19:331–338.
76. Hsueh WA, Quinones MJ. Role of endothelial dysfunction in insulin resistance. *Am J Cardiol.* 2003;92:10J–17J.
77. Baum B, Georgiou M. Dynamics of adherens junctions in epithelial establishment, maintenance, and remodeling. *J Cell Biol.* 2011;192:907–917.
78. Maeda S, Tsukihara T. Structure of the gap junction channel and its implications for its biological functions. *Cell Mol Life Sci.* 2011;68:1115–1129.
79. Hesketh GG, Van Eyk JE, Tomaselli GF. Mechanisms of gap junction traffic in health and disease. *J Cardiovasc Pharm.* 2009;54:263–272.
80. Chidlow JH Jr, Sessa WC. Caveolae, caveolins, and cavins: complex control of cellular signaling and inflammation. *Cardiovasc Res.* 2010;86:219–225.
81. Valentijn KM, Sadler JE, Valentijn JA, Voorberg J, Eikenboom J. Functional architecture of Weibel–Palade bodies. *Blood.* 2011;117:5033–5043.
82. Giepmans BNG. Gap junctions and connexin-interacting proteins. *Cardiovasc Res.* 2004;62:233–245.
83. Aird WC. Phenotypic heterogeneity of the endothelium: II. Representative vascular beds. *Circ Res.* 2007;100:174–190.
84. Kumar S, Ghellal A, Li C, et al. Breast carcinoma: vascular density determined using CD105 antibody correlates with tumor prognosis. *Cancer Res.* 1999;59:856–861.
85. North PE, Waner M, Mizeracki A, Mihm MC Jr. GLUT1: a newly discovered immunohistochemical marker for juvenile hemangiomas. *Hum Pathol.* 2000;31:11–22.
86. Jackson DG. Biology of the lymphatic marker LYVE-1 and applications in research into lymphatic trafficking and lymphangiogenesis. *APMIS.* 2004;112:526–538.
87. Kahn HJ, Bailey D, Marks A. Monoclonal antibody D2–40, a new marker of lymphatic endothelium, reacts with Kaposi's sarcoma and a subset of angiosarcomas. *Mod Pathol.* 2002; 15:434–440.
88. Desmouliere A, Chaponnier C, Gabbiani G. Tissue repair, contraction, and the myofibroblast. *Wound Repair Regen.* 2005; 13:7–12.
89. Hansson GK. Inflammation, atherosclerosis and coronary artery disease. *N Engl J Med.* 2005;352:1685–1695.
90. Weyand CM, Goronzy JJ. Medium- and large-vessel vasculitis. *N Engl J Med.* 2003;349:160–169.
91. Meijer-Jorna LB, Mekkes JR, van der Wal AC. Platelet involvement in cutaneous small vessel vasculitis. *J Cutan Pathol.* 2002;29:176–180.
92. Arbustini E, Morbini P, D'Armini AM, et al. Plaque composition in plexogenic and thromboembolic pulmonary hypertension: the critical role of thrombotic material in pultaceous core formation. *Heart.* 2002;88:177–182.
93. Virmani R, Kolodgie FD, Burke AP, et al. Atherosclerotic plaque progression and vulnerability to rupture: angiogenesis as a source of intraplaque hemorrhage. *Arterioscler Thromb Vasc Biol.* 2005;25:2054–2061.

SECTION IV

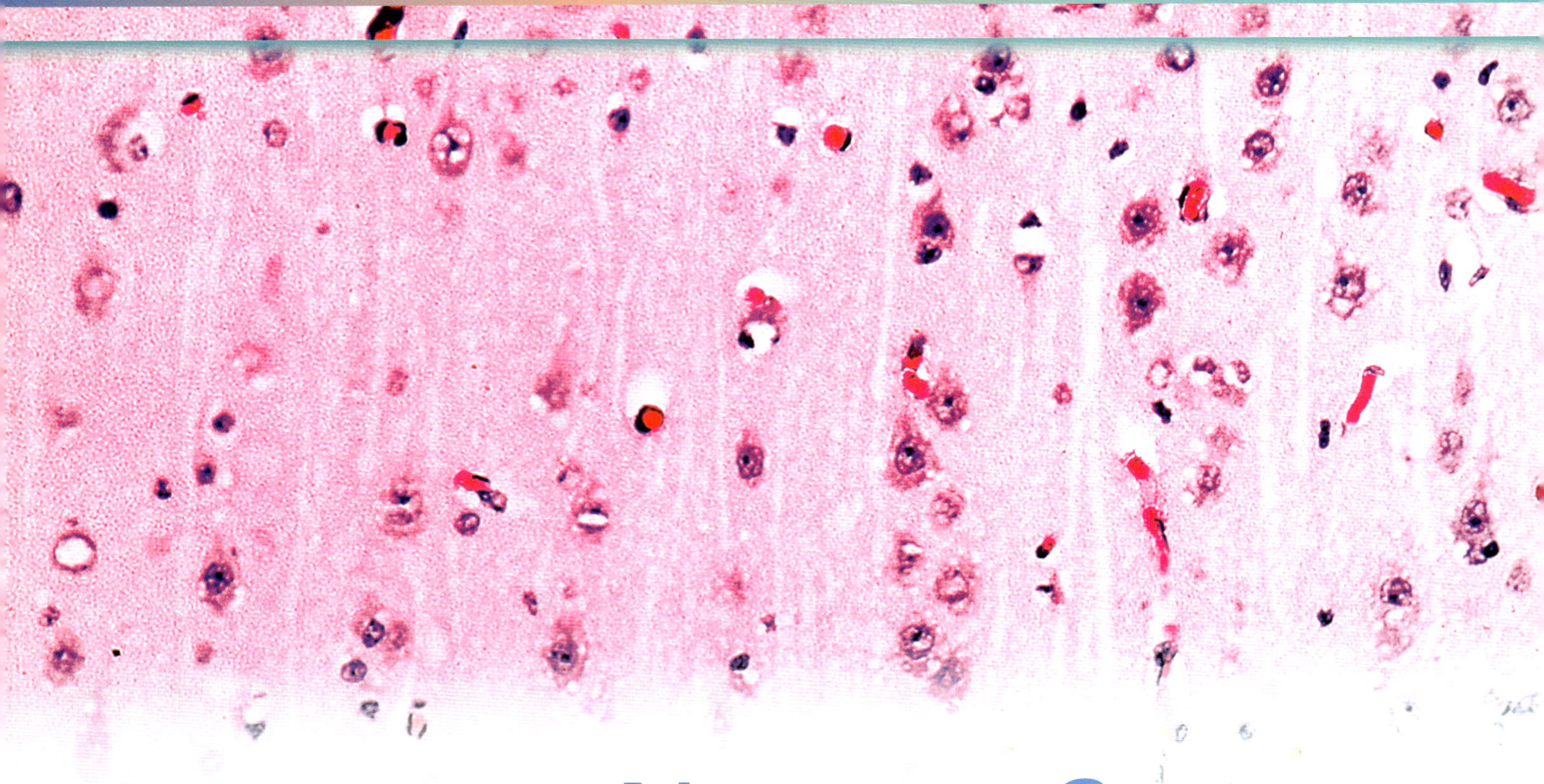

Nervous System

9

Central Nervous System

Gregory N. Fuller ▪ Leomar Y. Ballester ▪ Maria A. Gubbiotti

INTRODUCTION

The central nervous system (CNS) is unparalleled among natural systems in terms of structural and functional complexity. As a consequence of its intricate regional architecture, heterogeneous cellular constituents, and an associated extensive and somewhat arcane lexicon, the nervous system is often viewed as a formidably Byzantine realm by many non-neuropathologists; and yet, a working familiarity with the normal morphology of this complex organ must precede competent evaluation of the many disease states that afflict it. To this end, this chapter will present the salient features of regional neuroanatomy followed by a description of the essentials of microscopic anatomy of the CNS, with special emphasis on those aspects that constitute potential diagnostic pitfalls, including normal anatomic variations, alterations associated with advancing age, reactive changes, and common artifacts. Although the focus of the book is on histology, we have included gross images of the brain to highlight key points of surface and gross anatomy that provide context to the microscopic descriptions.

REGIONAL NEUROANATOMY

We have limited this discussion of regional neuroanatomy to those principles of structural organization that are of practical value to the diagnostician, emphasizing the rudiments of neuroembryology by which the basic organization of the nervous system is best understood. Further details about topographical neuroanatomy can be found in the list of suggested readings provided at the end of the chapter.

Organization of the Spinal Cord and Brain Stem

Embryologically, the nascent CNS begins as a hollow tube formed by the invagination of the neural plate ectoderm. This primitive cylinder is subdivided functionally into a dorsal sensory ("alar") plate and a ventral motor ("basal") plate. The two

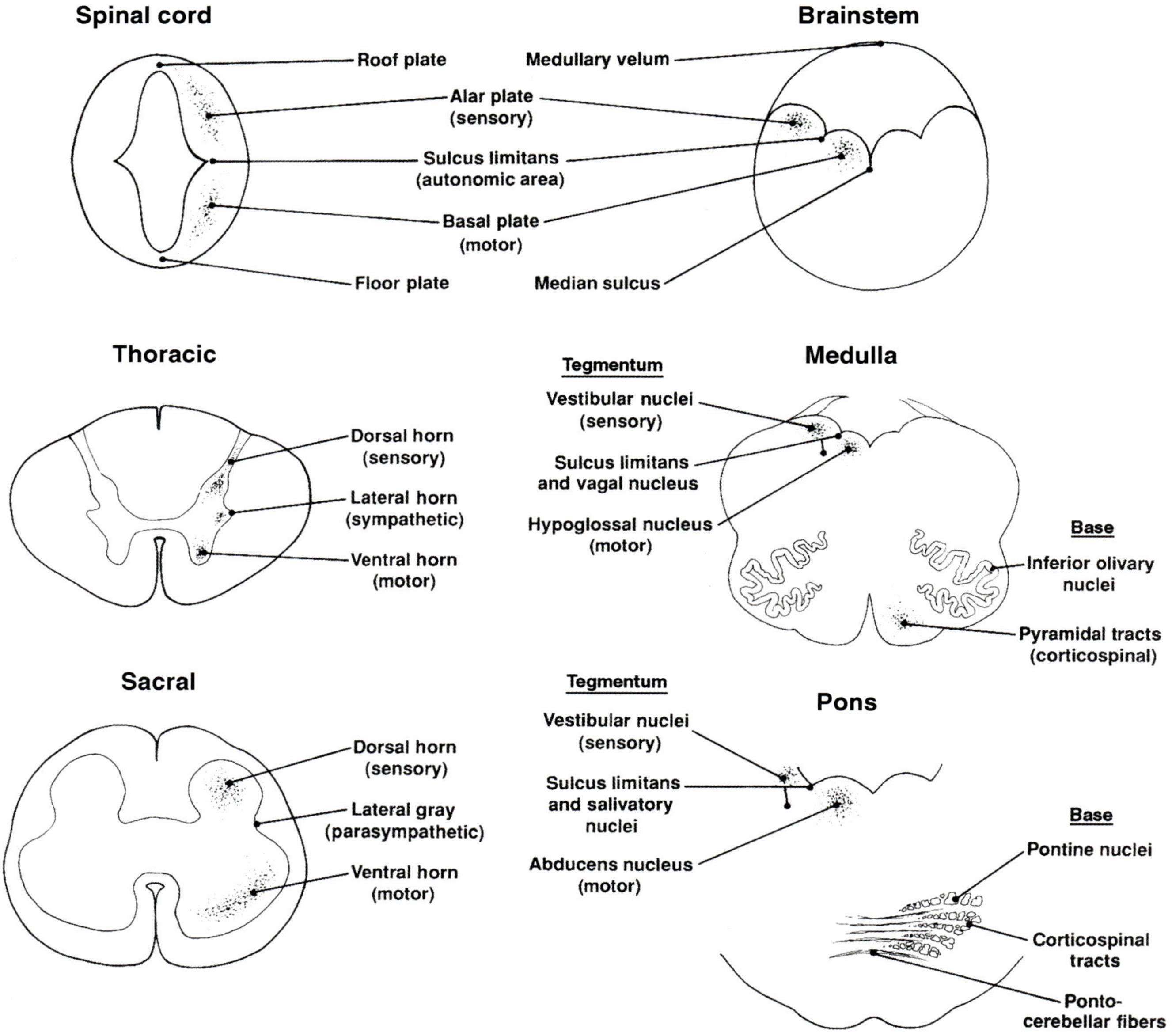

FIGURE 9.1 Structural organization of the spinal cord and brain stem. (See text for discussion.)

are separated by a lateral groove, termed the "sulcus limitans," along which develops the efferent autonomic system (Fig. 9.1). This primitive organizational pattern is retained, essentially unaltered, in the mature spinal cord. The central gray matter consists of: (a) dorsal horns that receive sensory input from the dorsal roots, (b) ventral horns that contain motor neurons whose axons are conducted to the somatic periphery by the ventral roots, and (c) the lateral autonomic gray matter. Spinal autonomic neurons are confined to thoracic (sympathetic) and sacral (parasympathetic) levels, forming the intermediolateral cell columns. The axons of these preganglionic neurons exit the spinal cord through the ventral roots, ultimately to synapse on postganglionic neurons in the peripheral autonomic ganglia. The sympathetic intermediolateral cell column produces a third horn of gray matter in the thoracic cord, termed "the lateral horn" (Figs. 9.1 and 9.2). The parasympathetic intermediolateral cell column occupies a similar lateral position in the sacral cord (at the S-2, S-3, and S-4 levels), but does not form a distinct horn.

Spinal Cord

The anatomy of the spinal cord varies according to the level (Fig. 9.2). Two enlargements of the ventral horns, one in the cervical region (Fig. 9.2A) and another in the lumbosacral region (Fig. 9.2C), provide motor innervation for the upper and lower extremities, respectively. In contrast, the ventral horns of the thoracic cord provide innervation for the more limited axial musculature of the trunk and are, accordingly, much smaller (Fig. 9.2B). As mentioned earlier, the lateral horns of the gray matter are a unique feature of the thoracic

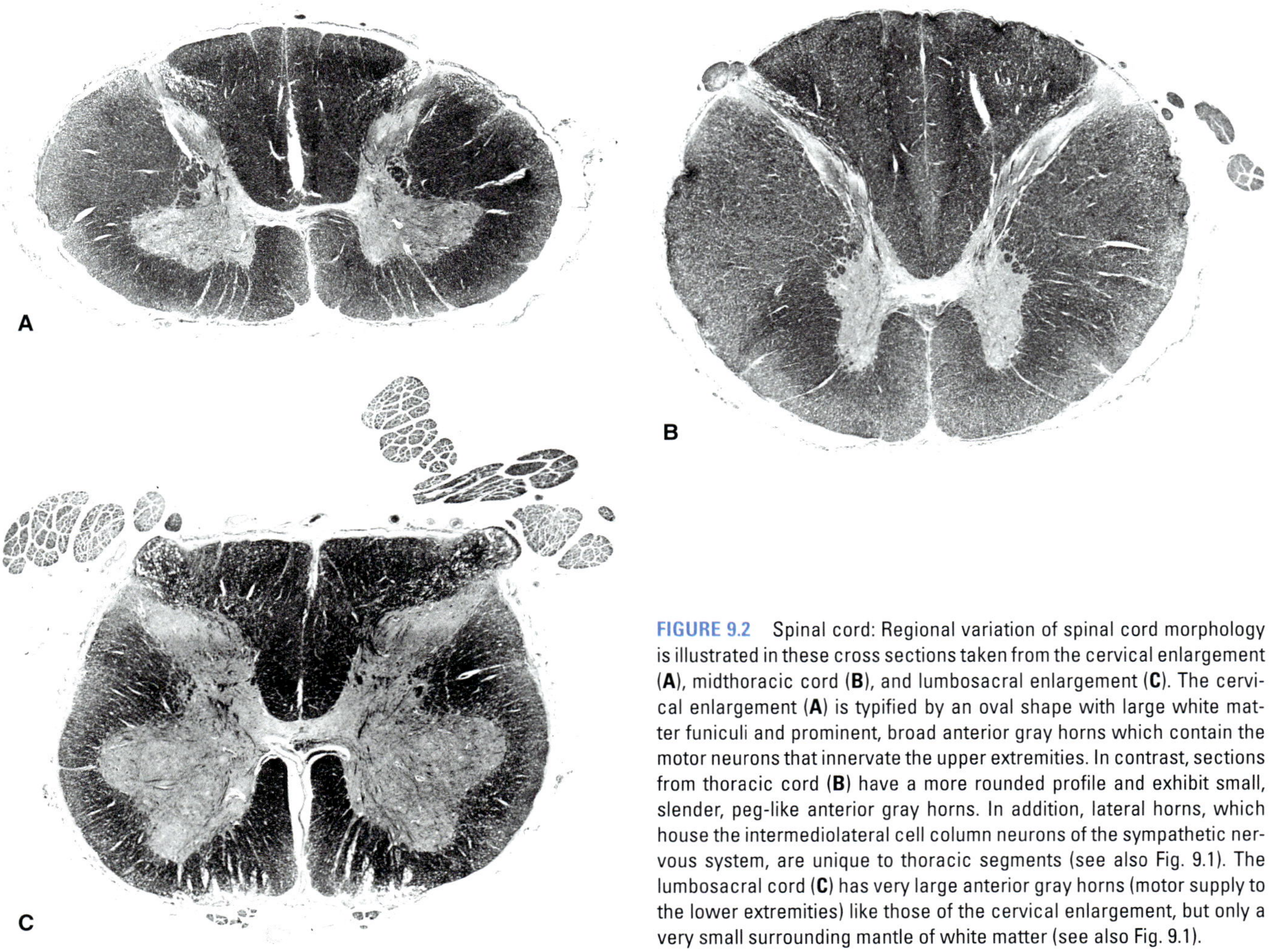

FIGURE 9.2 Spinal cord: Regional variation of spinal cord morphology is illustrated in these cross sections taken from the cervical enlargement (**A**), midthoracic cord (**B**), and lumbosacral enlargement (**C**). The cervical enlargement (**A**) is typified by an oval shape with large white matter funiculi and prominent, broad anterior gray horns which contain the motor neurons that innervate the upper extremities. In contrast, sections from thoracic cord (**B**) have a more rounded profile and exhibit small, slender, peg-like anterior gray horns. In addition, lateral horns, which house the intermediolateral cell column neurons of the sympathetic nervous system, are unique to thoracic segments (see also Fig. 9.1). The lumbosacral cord (**C**) has very large anterior gray horns (motor supply to the lower extremities) like those of the cervical enlargement, but only a very small surrounding mantle of white matter (see also Fig. 9.1).

cord. The thickness of the surrounding white matter fiber bundles, termed "funiculi," also varies with the cord level. It is greatest in the cervical cord where the thickness reflects the summated accrual of ascending fiber tracts that have successively entered at lower levels, as well as the maximum content of descending tracts that are en route to lower levels. The cord is thinnest in the lumbosacral region. The terminus of the spinal cord, the filum terminale, is composed primarily of meningeal connective tissue and is discussed separately with the pia-arachnoid (see Fig. 9.66).

Brain Stem

The brain stem (Fig. 9.3) is innately more complex than the spinal cord, but its organization is readily understood when viewed as a slightly modified version of the basic structure of the spinal cord. The stem is also a neural tube, but one that has been stretched dorsally and splayed laterally so that the ventrally located embryonic motor plate is medial and the dorsal sensory plate is lateral (Fig. 9.1). Therefore, within the brain stem, the cranial nerve motor nuclei are located medially, the sensory nuclei laterally, and the autonomic nuclei are intermediate in position.

The brain stem can be further subcategorized in cross section into tectum, tegmentum, and base (Figs. 9.4 to 9.6). The tectum is the roof of the ventricular system, as exemplified by the superior and inferior colliculi (corpora quadrigemina) of the midbrain and the superior medullary vela of the pons and the medulla. The tegmentum forms the floor of the cerebral aqueduct and fourth ventricle and is divisible into the medial motor and lateral sensory areas discussed previously (Fig. 9.1). The base is located subjacent to the tegmentum and is the most ventral portion of the stem. It is composed principally of the so-called "long tracts," that is, the descending motor pathways and ascending sensory pathways that link the spinal cord with higher neural centers. The combination of long tract signs with dysfunction of specific cranial nerves allows for the precise anatomic localization of brain stem lesions by clinical examination.

Cerebellum

Embryologically, the cerebellum arises as a dorsal outgrowth of the fetal brain stem and remains connected to it in the adult by the three pairs of cerebellar peduncles:

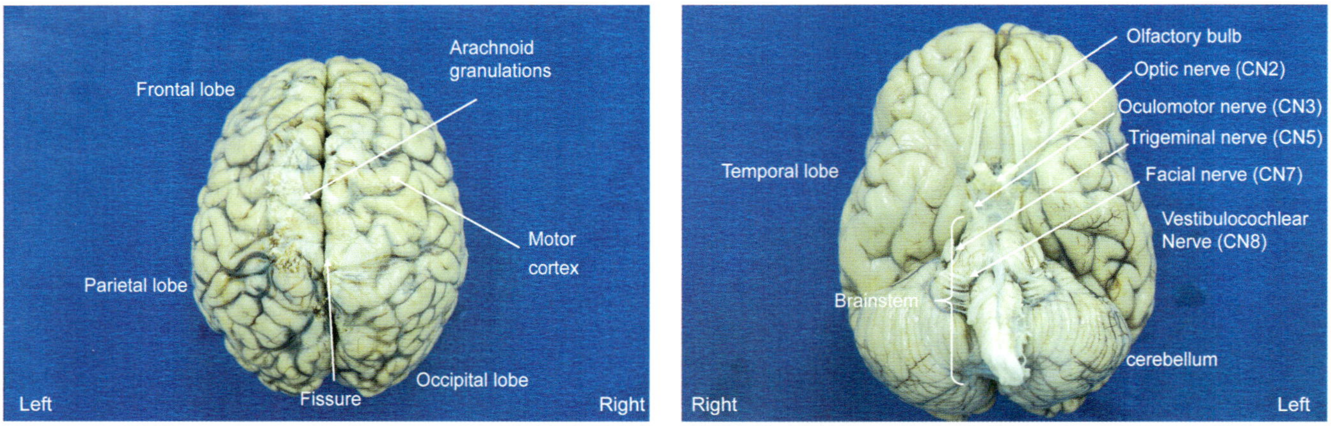

FIGURE 9.3 Surface anatomy of the brain: The brain is covered by meninges which include the dura (not shown here, see Fig. 9.60), the arachnoid with its granulations (see Fig. 9.61), and in closest proximity to the parenchyma, the pia (see Fig. 9.61). The interhemispheric fissure separates the left and right hemispheres. Anterior to the central sulcus is the prefrontal gyrus which harbors the motor cortex. The primary visual cortex (see Fig. 9.12) is located in the medial aspect of the occipital lobes. The olfactory bulb and tract (see Figs. 9.51 and 9.52) course along the inferior aspect of the frontal lobes. Some fibers from the left and right optic nerves (see Fig. 9.65) decussate in the chiasm before making synaptic connections with the lipofuscin-rich neurons of the lateral geniculate nucleus (see Fig. 9.10). The cerebellum consists of multiple folia (see Fig. 9.7), which leads to its undulated surface appearance.

the superior brachium conjunctivum, the middle brachium pontis, and inferior restiform body that join with the midbrain, pons, and medulla, respectively. The cerebellum is composed of three structural and functional compartments: cortex, medulla, and deep nuclei (Fig. 9.7). The cortex displays three distinct laminae: an outer hypocellular molecular layer, an intermediate single-cell thick Purkinje cell layer, and a deep hypercellular granular cell layer (Fig. 9.8). Before 1 year of age, the cerebellar cortex is conspicuous for remnants of a fourth layer of small neurons, the fetal external granular cell layer, which is located immediately subjacent to pia (Fig. 9.8). The external granular cells are gradually depleted during the first year of life as they descend the processes of Bergmann glia to reach their final positions in the internal granular cell layer. Embedded within the white matter of the cerebellar medulla are four

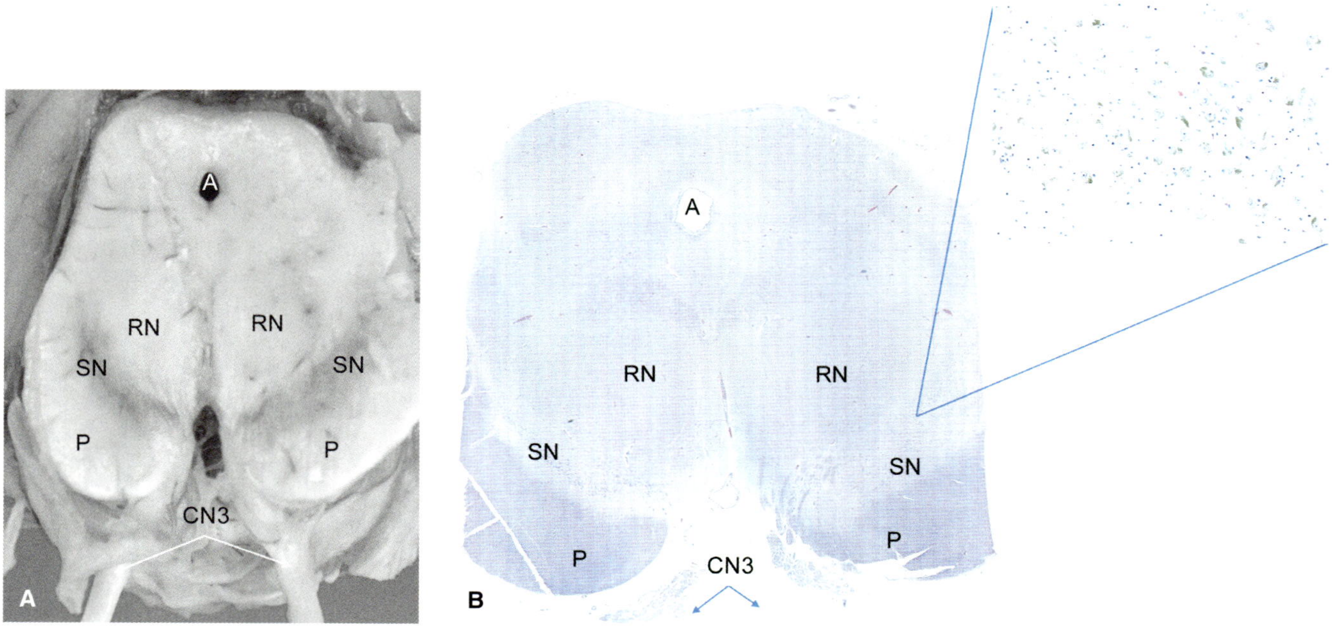

FIGURE 9.4 Midbrain: In cross section of the midbrain (**A**), the aqueduct (*A*), red nucleus (*RN*), substantia nigra (*SN*) and peduncles (*P*) are easily identified. The substantia nigra contains pigmented catecholaminergic neurons (*B*) (see Fig. 9.23).

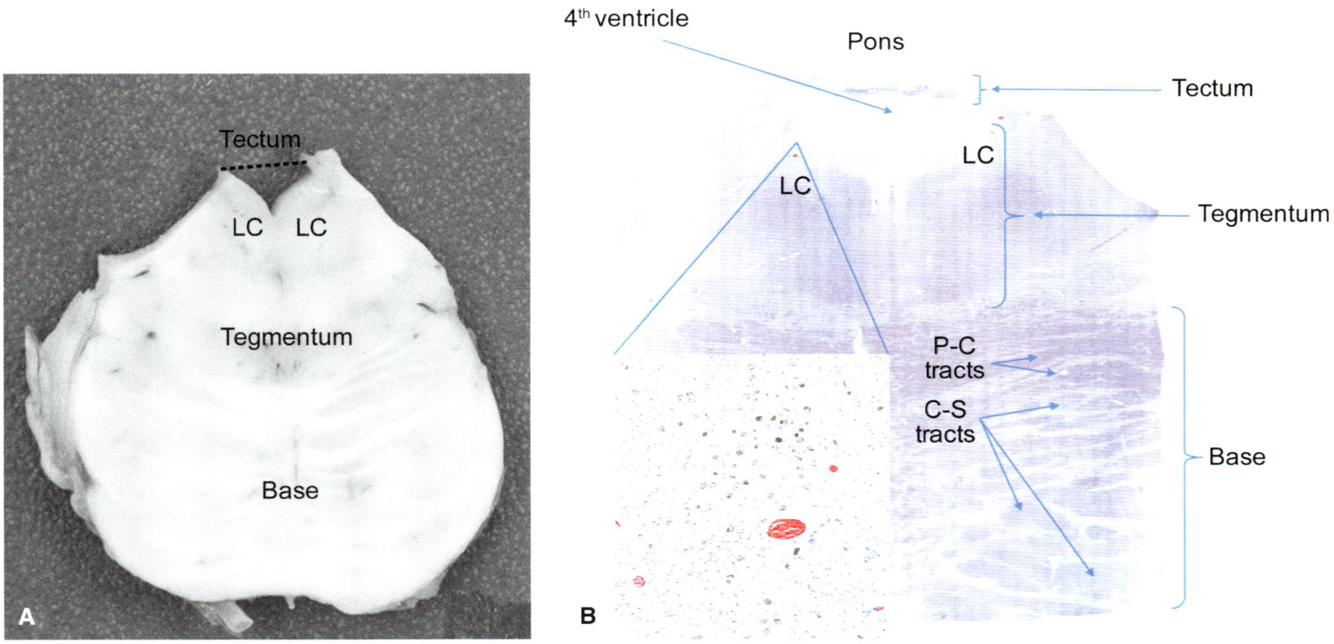

FIGURE 9.5 Pons: In cross section of the pons (**A**), the aqueduct has given rise to the fourth ventricle with its roof (tectum) and floor (tegmentum). (**B**) Two nuclei (locus ceruleus, *LC*) with pigmented catecholaminergic neurons (see Fig. 9.23) are evident on cross section of the pons. The base of the pons is formed by pontocerebellar tracts (*P-C*) and cortico-spinal tracts (*C-S*).

pairs of nuclei. From medial to lateral, these are the fastigial, globose, emboliform, and dentate nuclei. The dentate is by far the largest and is usually the only deep nucleus seen on routine sections (Fig. 9.7). Its serpiginous profile is strikingly similar to that of the inferior olivary nucleus of the medulla oblongata (Fig. 9.1), which is a major source of afferent fibers to the cerebellum.

The Purkinje cell dendritic arbor extends into the molecular layer like a hand with outstretched fingers (1). Its broad, flat palm and radiating fingers are oriented

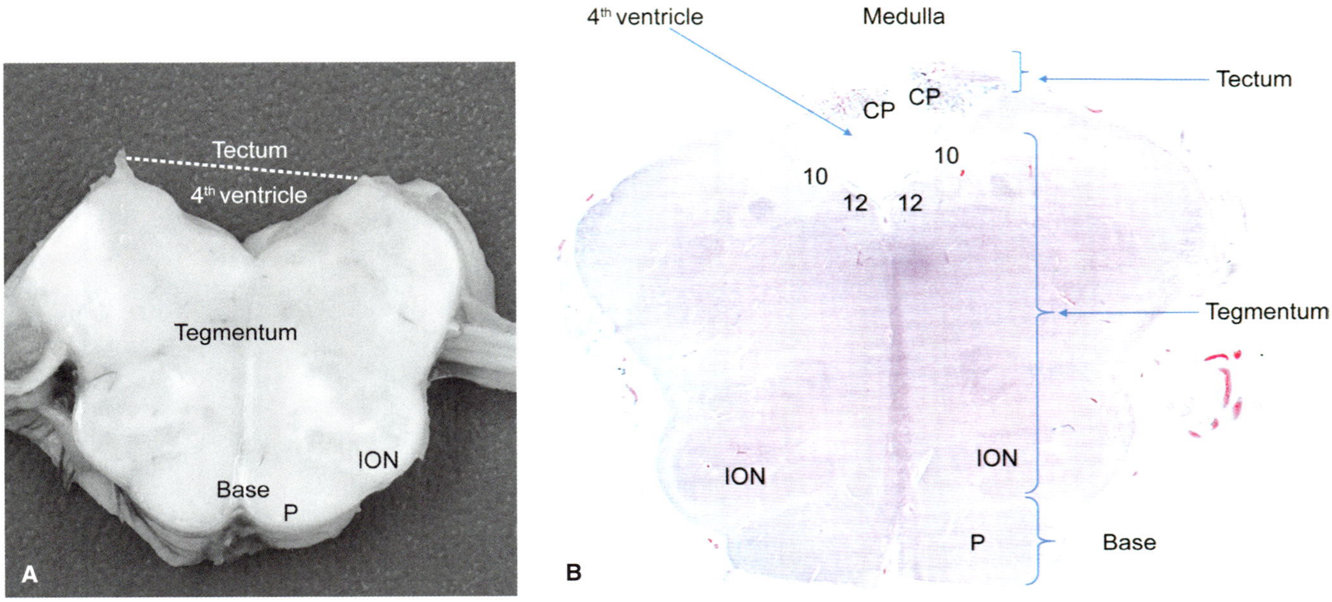

FIGURE 9.6 Medulla: In cross section of the medulla, the fourth ventricle continues, with its roof (tectum) and floor (tegmentum). Tufts of choroid plexus (see Figs. 9.53 and 9.54) can be appreciated here. The nuclei for two cranial nerves (Vagus #10 and Hypoglossal #12) can be identified microscopically in the periventricular region. Although not pigmented on gross examination, the dorsal nucleus of the vagus nerve contains pigmented neurons that can be appreciated microscopically. The convoluted inferior olivary nucleus (*ION*) is located above the peduncles (*P*).

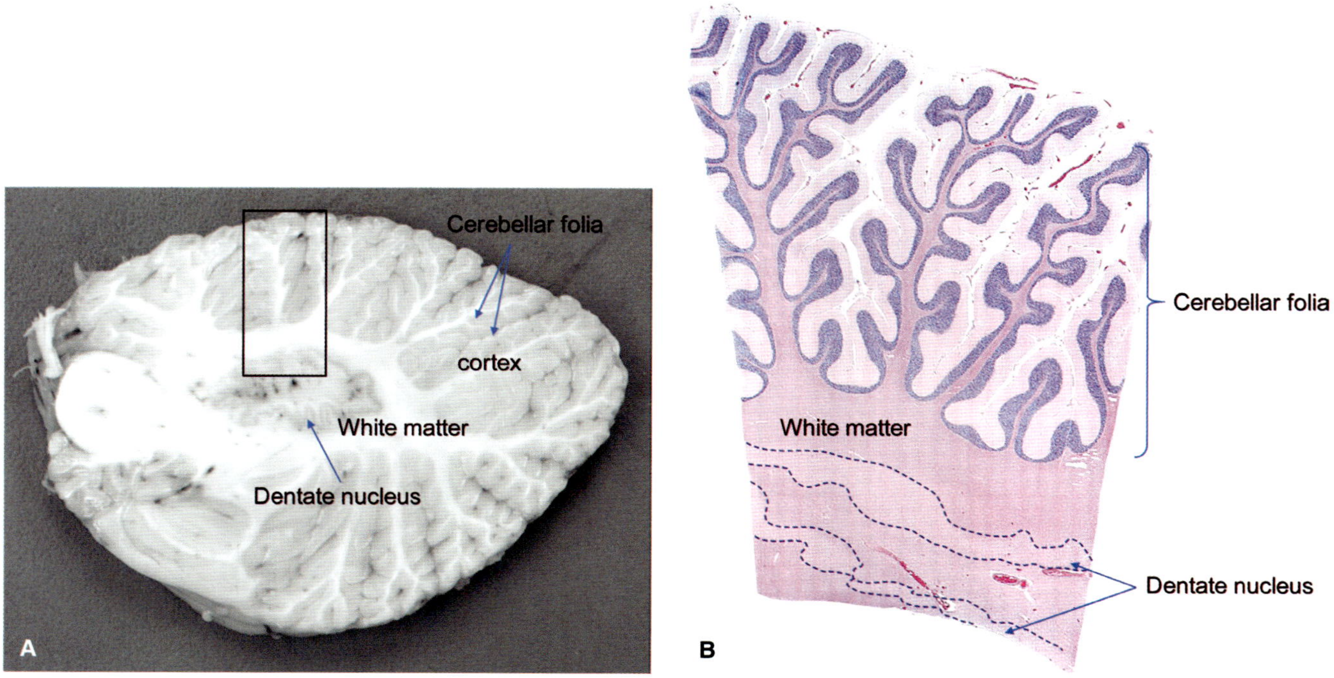

FIGURE 9.7 Cerebellum: In cross section of the cerebellum, the dentate nucleus is easily identified with the naked eye (**A**). The complex architecture of the cerebellar folia is apparent in this low magnification view of a hematoxylin and eosin-stained slide (**B**). The cell bodies of the neurons that form the dentate nucleus have a convoluted arrangement that mimics that of the ION (see Fig. 9.6).

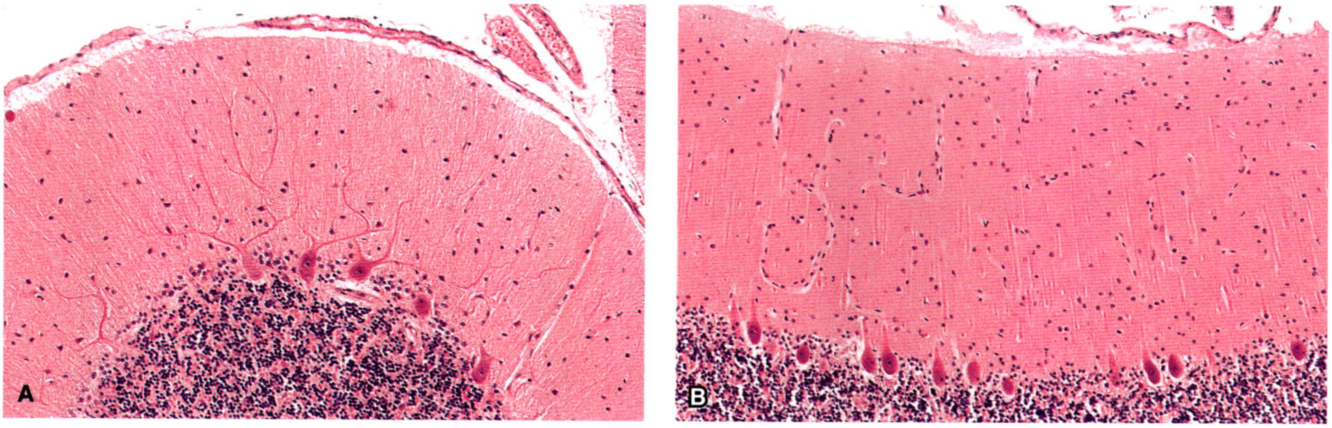

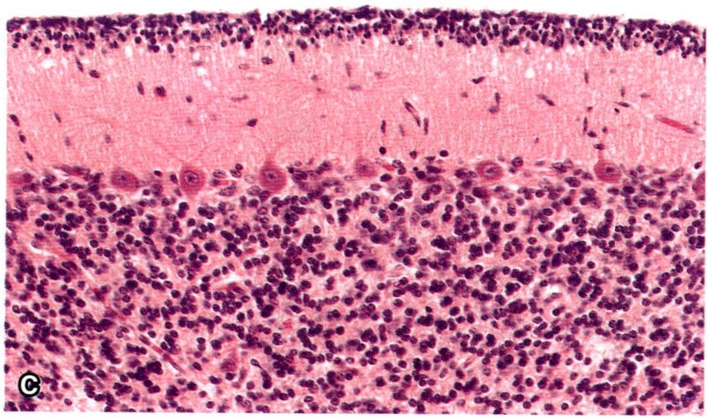

FIGURE 9.8 Cerebellar cortex: The adult cerebellar cortex is composed of three layers—an outer hypocellular molecular layer, a middle Purkinje cell layer, and an inner densely populated granular cell layer. Whereas the Purkinje cells are prototypically neuronal in appearance, the small cells of the granular layer are hardly recognizable as neurons by traditional histologic criteria (see Fig. 9.7). A cross section of a cerebellar folium (**A**) shows the typical broadly branching Purkinje cell dendritic arbor. However, sections taken parallel to the folia (**B**) reveal the streamlined "on edge" appearance of the arbor, which should not be interpreted as pathologic pruning. The fetal cerebellum (**C**) has an additional cortical lamina, the external granular layer, applied to the surface of the cortex. This pool of cells populates the internal granular cell layer during development and is, thereby, depleted by the end of the first year of postnatal life.

perpendicular to the long axis of the cerebellar convolutions, or folia. Thus, routine folia cross sections show the typical, elaborate dendritic branching pattern, whereas longitudinal sections present a dramatically different "on edge" view of the arbor (Fig. 9.8). This should not be mistaken for pathologic dendritic tree "pruning" seen in some disease states.

Diencephalon

The diencephalon is interposed between the brain stem and the cerebrum. Four major divisions are recognized: epithalamus, including the pineal gland and habenula, thalamus, subthalamus, and hypothalamus. The medial and lateral geniculate nuclei of the thalamus are sometimes considered together as the metathalamus. The strategic location of the thalamus is related to its major role in processing and relaying information passing between the cerebral cortex, brain stem, and spinal cord. With the exception of olfaction, all sensory data are processed by specific thalamic nuclei before distribution to the primary sensory cortices.

Certain portions of the diencephalon immediately subjacent to the third ventricle are characteristically involved in Wernicke encephalopathy, particularly the large dorsomedial nuclei of the thalamus and the mammillary bodies of the hypothalamus. The lesions at these sites are postulated to account for the memory disturbance that accompanies this disorder.

Cerebrum

The CNS is much more complicated supratentorially. It remains a hollow structure, but one that is no longer easy to consider as a tube of foldings and regional overgrowths. In light of this complexity, it is appropriate to review only those areas that are of particular diagnostic relevance.

Basal Ganglia

The term basal ganglia refers to the deep gray matter masses of the telencephalon and encompasses the caudate nucleus, putamen, globus pallidus, and amygdala (Fig. 9.9). The term ganglion was formerly used interchangeably with nucleus,

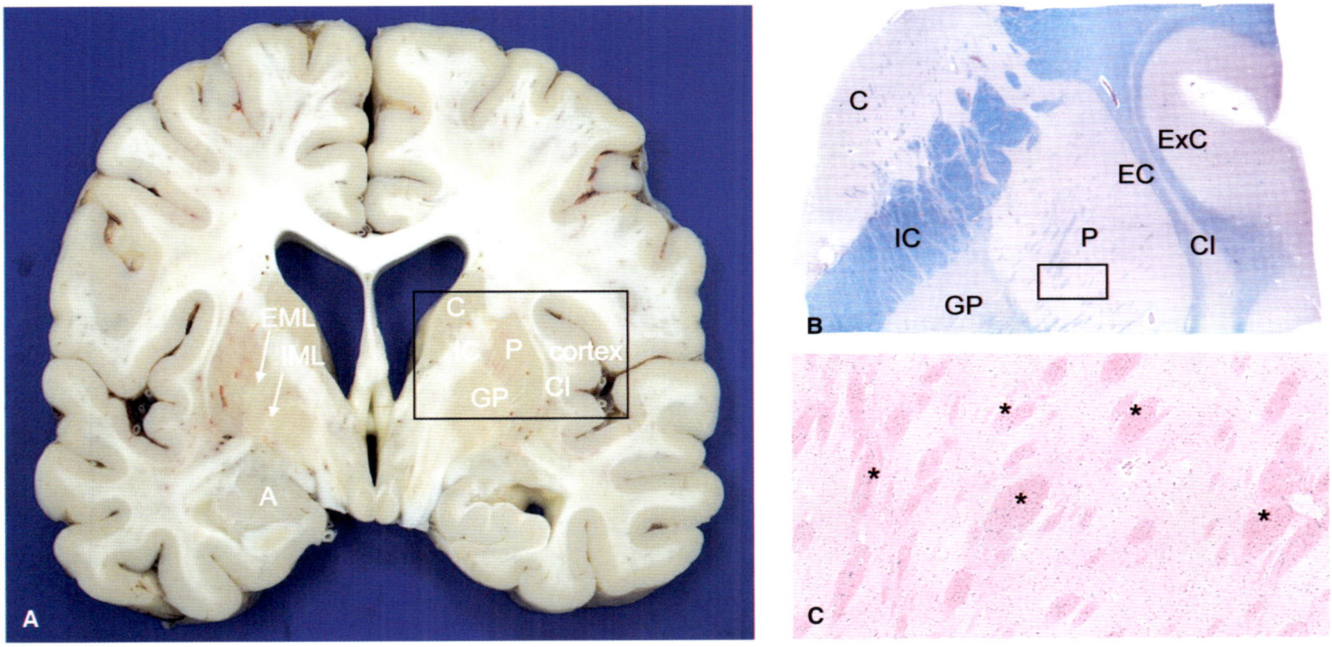

FIGURE 9.9 Basal ganglia: The gray matter of the telencephalon is broadly divisible into two components—the superficial cortical gray matter mantle and the deep gray nuclei. The latter are known as the basal ganglia, and consist of the caudate nucleus (*C*), putamen (*P*), globus pallidus (*GP*), and amygdala (**A**). The lenticular nucleus is composed of the medially situated, diffusely myelinated globus pallidus and the laterally placed putamen (**B**), whose myelinated fibers are grouped into slender fascicles known as the pencil bundles of Wilson (*asterisks*) (**C**). The internal and external medullary laminae (*IML* and *EML*, respectively) divide the globous pallidus into lateral and medial segments. The internal capsule (*IC*) separates the lenticular nucleus from the caudate and the thalamus. Lateral to the putamen there are the external capsule (*EC*), the claustrum (*CI*), and the extreme capsule (*ExC*). Gray matter bridges occasionally span the capsule to connect the caudate and putamen, a reflection of the close functional relationship between these two nuclei. The lenticular nucleus receives its blood supply from several lenticulostriate arteries, which are direct branches of the middle cerebral artery, and is the most common site of intracerebral hypertensive hemorrhage and lacunar infarction. The large lenticulostriate artery coursing through the lateral putamen was known in former times as Charcot artery, or, more colorfully, as the artery of internal hemorrhage. The lenticulostriate vessels are often surrounded by dilated perivascular spaces that should not be mistaken for lacunar infarcts.

and ganglion cell was synonymous with neuron to earlier neuroanatomists. With the exception of the basal ganglia, the current definition of a ganglion is now generally restricted to mean a collection of neuronal cell bodies located outside the CNS, namely, the sensory and autonomic ganglia of the peripheral nervous system. Reference to CNS neurons as ganglion cells is still occasionally encountered, and this historical sense of the term is reflected in the names of such neoplastic entities as ganglioglioma, ganglioneuroma, and ganglion cell tumor.

The amygdala, or archistriatum (Fig. 9.9), is located in the mesial temporal lobe immediately rostral to the hippocampus and is functionally related to the limbic system (2). The remaining nuclei of the basal ganglia play an integral role in the modulation of motor function, and probably participate in other higher neural systems as well. The caudate nucleus has a long tapering tail that intimately follows the curvature of the lateral ventricle (Fig. 9.9). It is morphologically and functionally related to the putamen, and these two nuclei are collectively referred to as the neostriatum, or simply striatum. For descriptive purposes, the putamen and the medially situated globus pallidus (i.e., paleostriatum or pallidum) are collectively termed the lentiform (i.e., lenticular) nucleus. The putamen and pallidum are separated from one another by the external medullary lamina of the pallidum, whereas the pallidum is itself divided into medial and lateral segments by the internal medullary lamina (Fig. 9.9). The globus pallidus is so named because of its pale appearance in the fresh state compared to the putamen. This contrast is attributable to the dense meshwork of myelinated fibers in the pallidum. In contrast, myelinated axons in the putamen are grouped into slender fascicles termed the pencil bundles of Wilson that project medially to the pallidum and to the substantia nigra (Fig. 9.9C). The histologic appearance of the lentiform nucleus is distinctive and permits unambiguous identification of even very limited amounts of tissue from this site.

The basal ganglia are prominently involved in a variety of pathologic processes, including "kernicterus" in the neonate and lacunar infarction in adults. Carbon monoxide poisoning classically produces selective necrosis of the inner segment of the pallidum. A frequent incidental finding of no diagnostic significance is micronodular mineralization of small blood vessels of the lentiform nucleus, which is typically most prominent in the globus pallidus. Histologically, similar micronodular mineralization is also commonly seen in the hippocampus.

Hippocampal Formation

The hippocampal formation comprises the subiculum, Ammon horn, and dentate gyrus (Fig. 9.10). The subiculum forms the inferior base of the hippocampal formation,

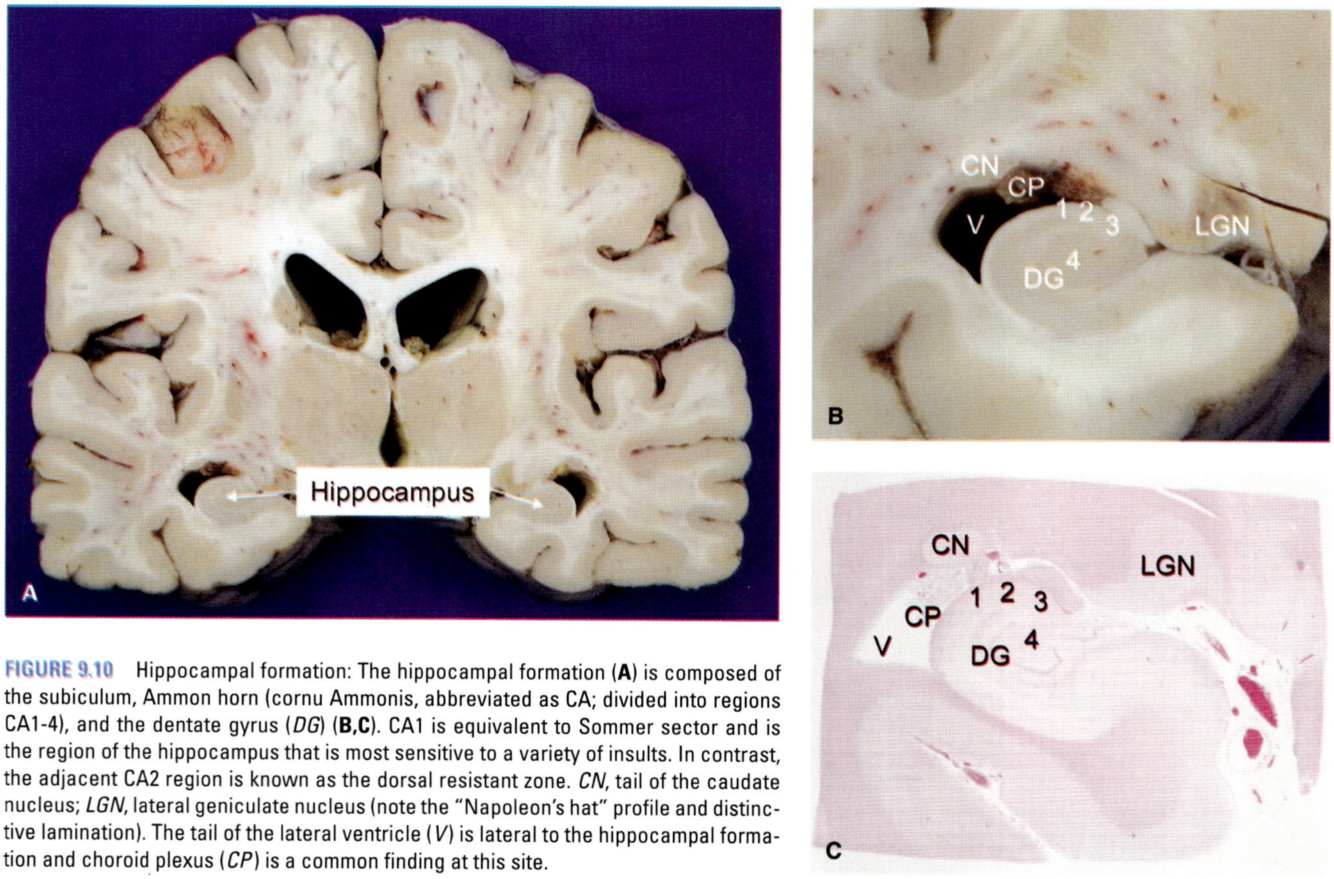

FIGURE 9.10 Hippocampal formation: The hippocampal formation (**A**) is composed of the subiculum, Ammon horn (cornu Ammonis, abbreviated as CA; divided into regions CA1-4), and the dentate gyrus (*DG*) (**B,C**). CA1 is equivalent to Sommer sector and is the region of the hippocampus that is most sensitive to a variety of insults. In contrast, the adjacent CA2 region is known as the dorsal resistant zone. *CN*, tail of the caudate nucleus; *LGN*, lateral geniculate nucleus (note the "Napoleon's hat" profile and distinctive lamination). The tail of the lateral ventricle (*V*) is lateral to the hippocampal formation and choroid plexus (*CP*) is a common finding at this site.

joining the parahippocampal gyrus with Ammon horn. Ammon horn, routinely abbreviated as CA to denote cornu Ammonis, is divided into four regions, on the basis of cytologic architecture and synaptic connectivity. CA1 arches superiorly, forming the medial floor of the temporal horn of the lateral ventricle along with CA2. The dorsally situated CA2 is usually recognizable by the greater compactness of the pyramidal cell layer, as compared to CA1. CA3 forms a descending medial arch that terminates in the hilus of the dentate gyrus. The final segment of Ammon horn, CA4, lies within the hilus of the dentate gyrus and is often referred to as the end plate. CA1 is the zone that is most sensitive to seizures, ischemia, and Alzheimer disease. In contrast, the adjacent CA2 segment is known as the dorsal resistant zone, in recognition of its relative sparing compared to the other three sectors. The exquisite sensitivity of CA1 to injury, with sparing of the adjacent CA2, is routinely observed as mesial sclerosis of the hippocampus, which is seen in many temporal lobes resected for intractable epilepsy. The classic histologic description of the patterns of neuronal loss in Ammon horn was based on observations made on the brains of such epileptic patients by Wilhelm Sommer in 1880; E. Brotz coined the term Sommer sector to denote CA1 in 1920.

There are several notable features that are frequently encountered incidentally in the examination of routine hippocampal sections and can be mistakenly interpreted as evidence of disease. One is an asymptomatic micronodular mineralization comparable to that seen in the pallidum. In the hippocampal formation, it is most commonly seen just outside the apex of the dentate gyrus (Fig. 9.11). A second common finding is a residual hippocampal fissure, which produces a rarefied lamina or cystic cleft that can be mistaken for a healed infarct. In addition, pyramidal neurons of Ammon horn are often dark and shrunken in autopsy material and care must be taken not to overinterpret such changes as evidence of antemortem ischemia (see Fig. 9.24).

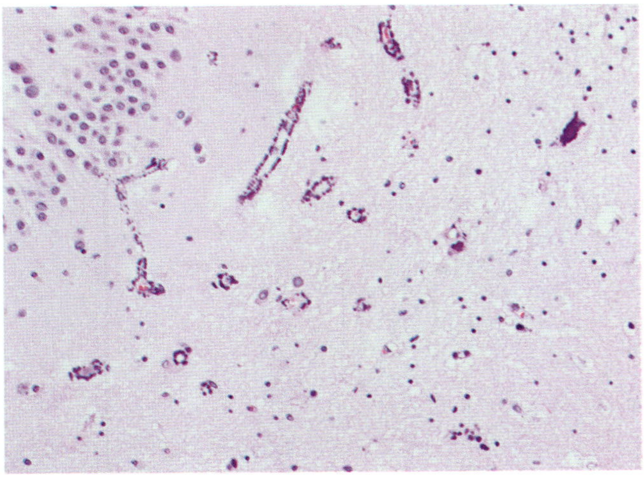

FIGURE 9.11 Micronodular mineralization: Micronodular mineralization is a common incidental finding, which is also seen in the globus pallidus.

Cerebral Cortex

From antiquity, neuroscientists have sought to divide the cortical mantle into discrete, functionally significant units. Early efforts yielded fanciful maps akin to those of phrenology and physiognomy. More recently, the application of light microscopy and special staining techniques for cell bodies (Nissl stains), dendritic arbors and unmyelinated axons (Golgi stains), and the myelin sheaths of myelinated axons (myelin stains such as the Weil, Weigert, and Luxol fast blue methods) have permitted a more scientific approach, although the details are beyond the scope of this chapter. In brief, the parcellation of the cortex is based on regional variation in the relative number, composition, and distribution of cortical neurons and their processes (cytoarchitectonics and myeloarchitectonic). Many neuroanatomists have divided the cortex into a number of regions varying from 20 to more than 200, depending on the particular morphologic criteria and degree of subtlety employed. Currently, the most popular cortical map is that devised by Korbinian Brodmann in 1909. Brodmann map and the classical nomenclature for the gross anatomy of the cerebral sulci and gyri are the two systems most commonly used at present for reference purposes in the neuroanatomical and clinical literature.

Within the context of even the simplest cortical map, it is generally not possible to assign a given histologic section of cortex to a precise anatomical locus without prior knowledge of the section's provenance. However, two cortical areas do exhibit distinctive features: Primary motor cortex and primary visual cortex (Fig. 9.12). The motor cortex, located on the precentral gyrus of the frontal lobe, is distinguished by the presence of the giant pyramidal cells of Betz (3). Pyramidal cells generally range from 10 to 50 μm in soma height from base to origin of the apical dendrite. By comparison, Betz cells may exceed 100 μm in soma height. The primary visual cortex, located on the banks of the calcarine fissure of the medial occipital lobe, is remarkable for the presence of a prominent external band of Baillarger, termed the line (or stria) of Gennari. This myelinated stratum located in lamina IV is usually visible to the naked eye and permits exact delineation of the primary visual cortex (i.e., Brodmann area 17) from the adjacent visual association cortex (i.e., Brodmann area 18).

CELLULAR CONSTITUENTS OF THE CENTRAL NERVOUS SYSTEM

Gray Matter and White Matter

By volume, most of the CNS is composed of gray matter and white matter (Fig. 9.13). Specialized types of CNS tissues include the choroid plexus, the pineal gland, the circumventricular organs, and the infundibulum and neurohypophysis. The hallmark of gray matter (Fig. 9.13B) is the presence of neuronal cell bodies embedded within a finely textured eosinophilic background termed neuropil (Fig. 9.13C,D).

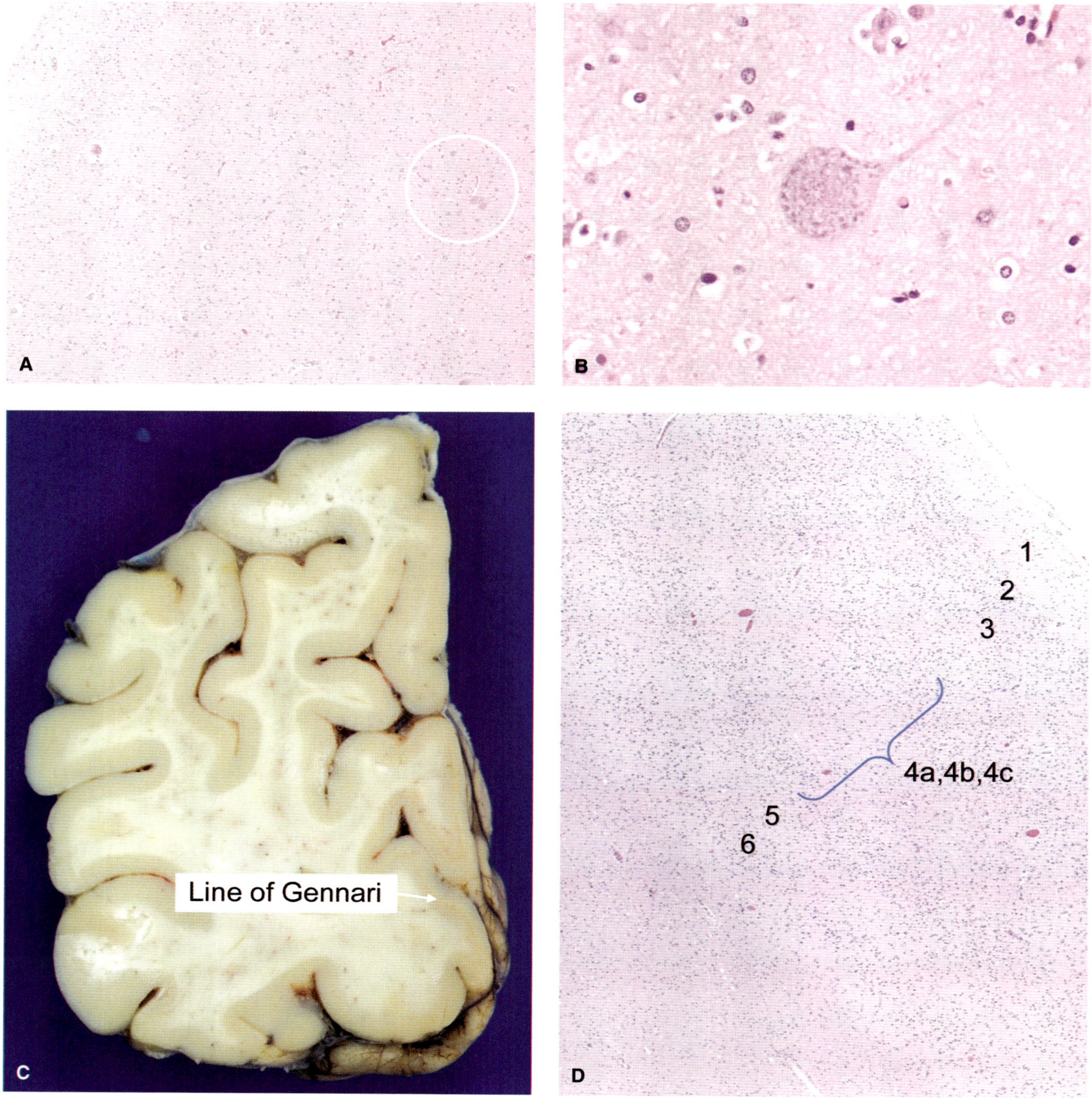

FIGURE 9.12 Primary motor cortex and primary visual cortex: The motor cortex is distinguished by the presence of the giant pyramidal cells of Betz (**A,B**). Primary visual cortex can be identified with the naked eye by the presence of the "external band of Baillarger" a thin white band running through the gray matter in the medial aspect of the occipital lobe (**C**). Microscopically the six layers of cortical neurons can be identified (**D**).

Neuropil is an interwoven meshwork of neuronal and glial cell processes. The individual neurites that compose the neuropil are not generally distinguishable in routine hematoxylin and eosin (H&E)-stained sections (Fig. 9.13C) but are resolvable at the ultrastructural level (Fig. 9.13D). White matter, in contrast, is composed primarily of myelinated axons and the supporting cells, oligodendroglia, that produce and maintain the myelin sheaths (Fig. 9.13E).

Neurons

Normal Microscopic Anatomy

The prototypical neuron is exemplified by the large multipolar Betz cells of the motor cortex, the alpha motor neurons of the ventral horn of the spinal cord, and the Purkinje cells of the cerebellum. These neurons are characterized by large perikarya (i.e., cell bodies) with abundant Nissl substance

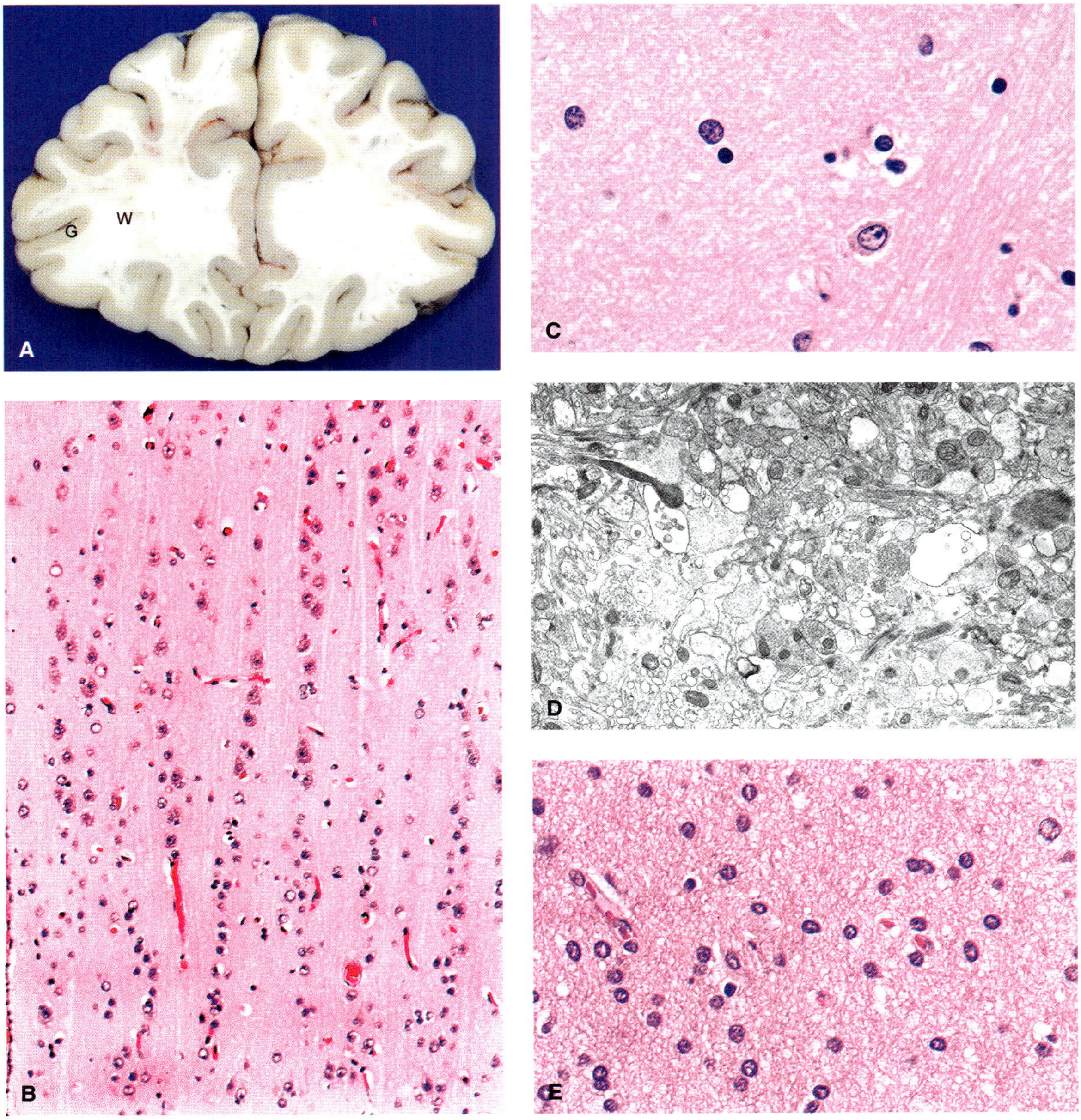

FIGURE 9.13 Gray matter and white matter: Gray (*G*) and white (*W*) matter can be distinguished with the naked eye on cross section of the brain (**A**). Gray matter contains abundant neuropil surrounding large neurons and smaller astrocytes and oligodendroglia (**B**). Neuropil is the term used for the fine amorphous eosinophilic background matrix of the CNS that fills the space between the cell bodies of the various cellular constituents as seen on H&E stains (**C**). Ultrastructural examination shows the neuropil to be composed of myriad intimately intermingling processes of the cellular constituents (**D**). White matter, in contrast, is composed primarily of oligodendroglia and the axons that they myelinate, and displays a much more uniform, homogeneous appearance (**E**).

that represents rough endoplasmic reticulum, robust dendritic arborizations, and large nuclei with prominent single nucleoli (Fig. 9.14). Such large multipolar forms, however, represent only one type of neuron; the diapason of neuronal morphologies is exceedingly broad. This is readily apparent by comparison of alpha motor neurons with granular cell neurons (Fig. 9.14A,D). These two neuronal populations typify the classical dichotomous subdivision of CNS neurons into large, extroverted projection neurons with long axons (Golgi type I neurons) and small introverts that function regionally with restricted connections (Golgi type II neurons). Between these two poles is a full spectrum of neuronal sizes and shapes, with an equally impressive variety of dendritic arbor configurations. The details of the latter are generally appreciable only with special stains for neuronal processes. The cell processes of neurons are separated into two categories: Axons (Fig. 9.14C), of which each neuron only has one, and dendrites (Fig. 9.14A), which are often multiple.

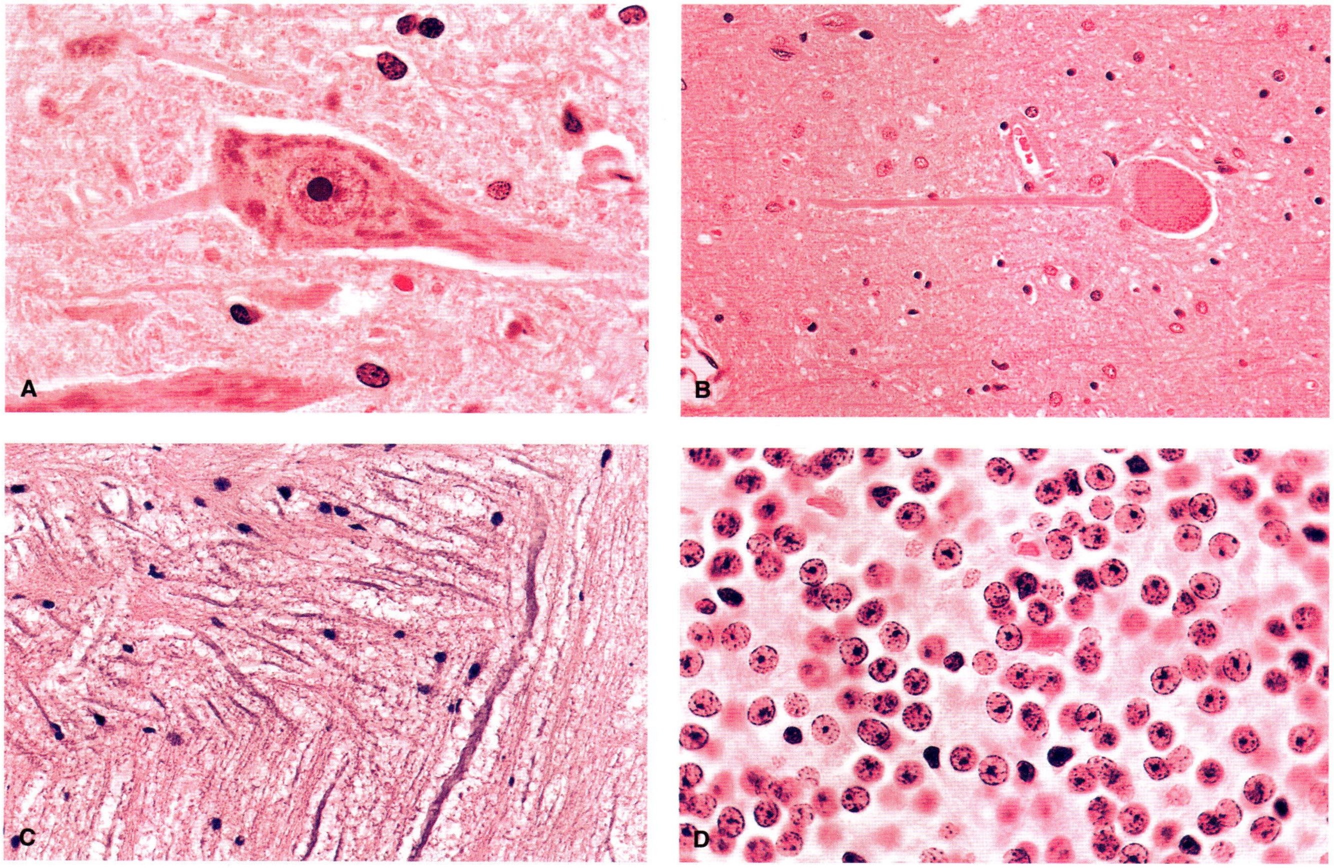

FIGURE 9.14 Neurons: Classical neuronal features, as illustrated by a motor neuron from the ventral horn of the spinal cord, include a large cell body (soma, perikaryon) with abundant cytoplasmic Nissl substance (rough endoplasmic reticulum, the "tigroid substance" of early microscopists), cytoplasmic processes, and a large nucleus with a single prominent nucleolus (**A**). The large process extending to the right is clearly recognizable as a dendrite by its content of Nissl substance, whereas in this fortuitous section, the smaller process extending to the left is identified as the neuron's axon by its lack of Nissl substance. Axons are further distinguished from dendrites by their nontapering profile (**B**). The nontapering profile of axons is easily recognized in white matter (**C**). The extremes of neuron size and shape are readily apparent from a comparison of a large motor neuron (**A**) with the small granular cell neurons of the cerebellar cortex (**D**) that approximately the same size as a motor neuron's nucleolus!

With respect to morphologic variants, one unique population of CNS neurons merits brief mention. The mesencephalic nucleus of the trigeminal nerve, which is concerned chiefly with the mediation of jaw proprioception, is composed of first-order sensory neurons that possess only a single process emanating from the cell body (Fig. 9.15A). This nucleus constitutes the only intraparenchymal example of this class of neurons; all other primary sensory neuronal perikarya are gathered outside the CNS in the spinal and cranial nerve ganglia (Fig. 9.15B–D).

Immunohistochemistry

Antibodies have been raised against a wide variety of neuronal proteins that are being isolated and characterized at an ever-increasing rate. Most of these markers are confined to use for research purposes but several have found utility in the diagnostic laboratory. One of the earliest such markers, neuron-specific enolase (NSE), has proven notoriously unreliable as a marker of neuronal differentiation. Its use in evaluating neoplasms is not recommended. Antibodies directed against epitopes on the constituent proteins of neurofilaments, which are major cytoskeletal elements of the neuronal perikaryon and cytoplasmic processes, have been used extensively in both experimental and clinical studies (Fig. 9.16). One of the most useful and widely employed neuronal markers is synaptophysin. Synaptophysin is an integral membrane protein of synaptic vesicles. In the normal nervous system, antisynaptophysin antibodies yield a diffuse, finely granular pattern throughout the gray matter neuropil (Fig. 9.17A). In addition, punctate granular decoration is seen along the cell bodies and proximal dendrites of several types of large, projection class neurons, including the Purkinje cells of the cerebellum, alpha motor neurons of the spinal cord, extraocular motor neurons

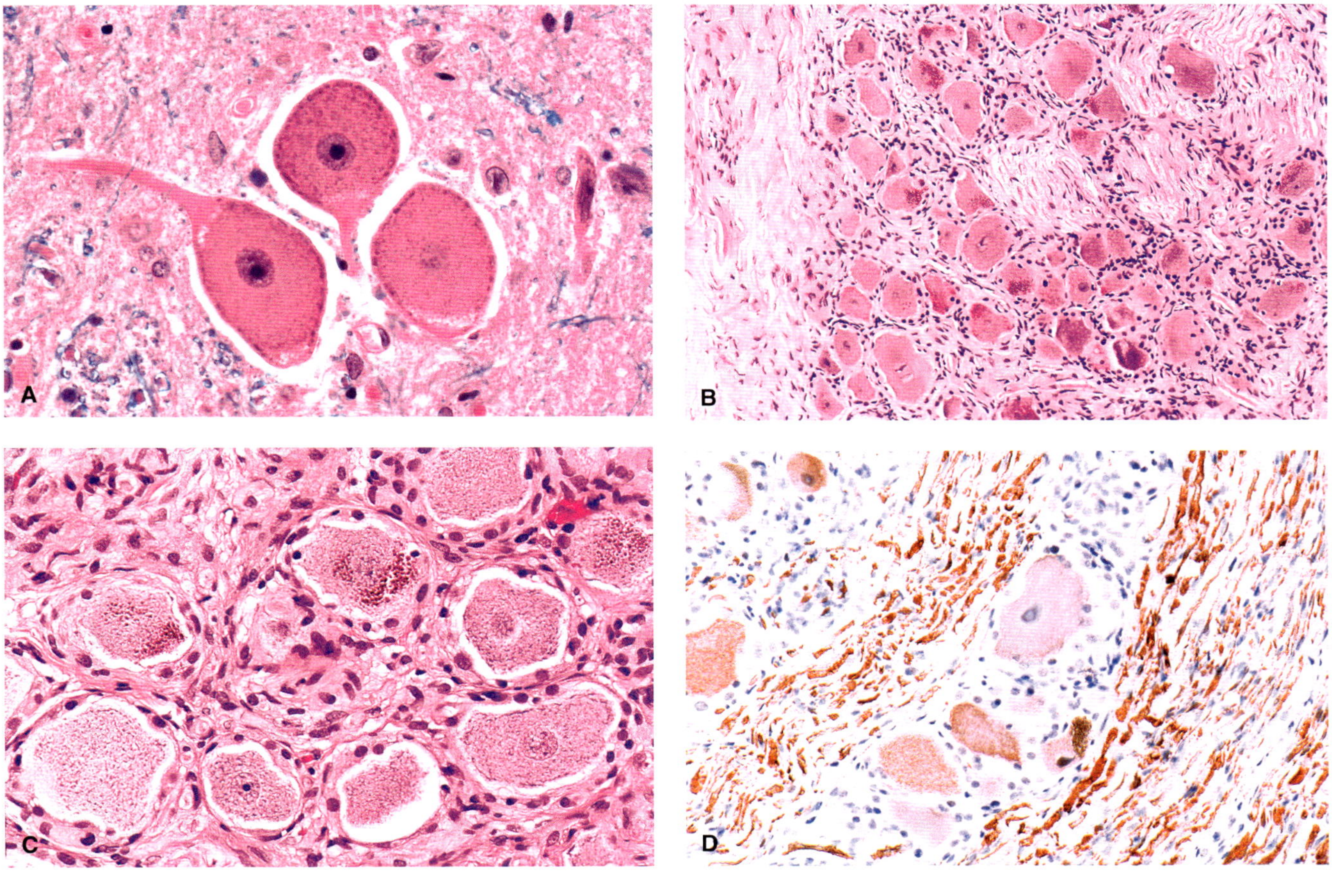

FIGURE 9.15 Unipolar neurons: Another neuronal morphologic variant is the unipolar (pseudounipolar) neuron. These large neurons possess only a single cell process, an axon, with an absence of dendrites. Unipolar neurons are primary sensory neurons and the only example of this class of neuron within the CNS is the mesencephalic nucleus of the trigeminal nerve in the upper pons and midbrain lateral to the periaqueductal gray matter (**A**). All other unipolar neurons are located in the peripheral nervous system ganglia (**B**). The dorsal root ganglia of the spinal cord provide a good example, with large unipolar neurons surrounded by satellite cells (**B**). Ganglionic neurons typically display cytoplasmic pigment (**C**) and their cell bodies and axons are strongly positive for phosphorylated neurofilament proteins (**D**).

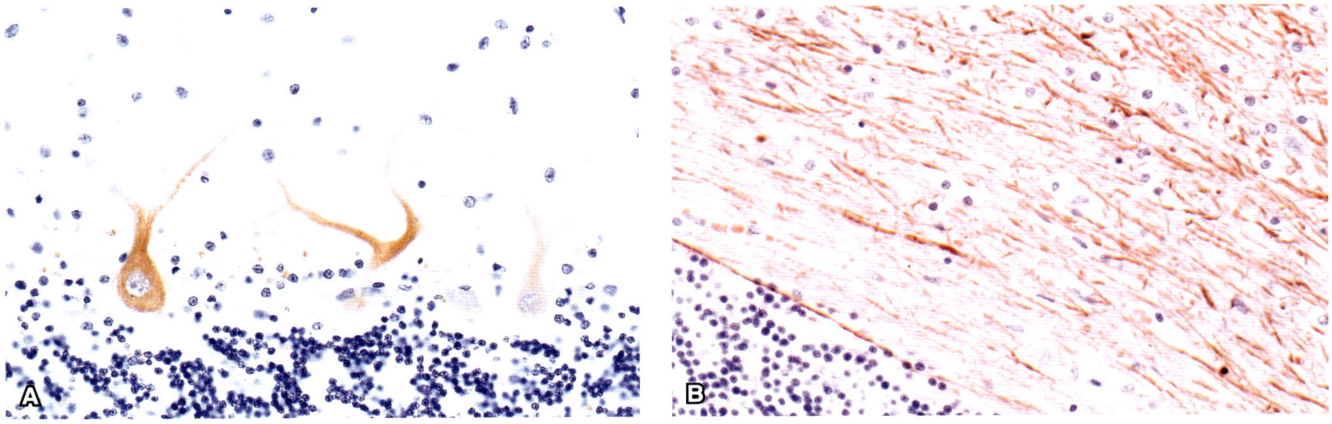

FIGURE 9.16 Neurofilament proteins (NFPs): Antibodies directed against NFP epitopes illuminate the cytoskeleton of neurons and their processes. As illustrated in the cerebellar cortex, specific antibodies directed against either nonphosphorylated (**A**) or phosphorylated (**B**) NFPs differentially identify cell bodies and dendrites or axons, respectively.

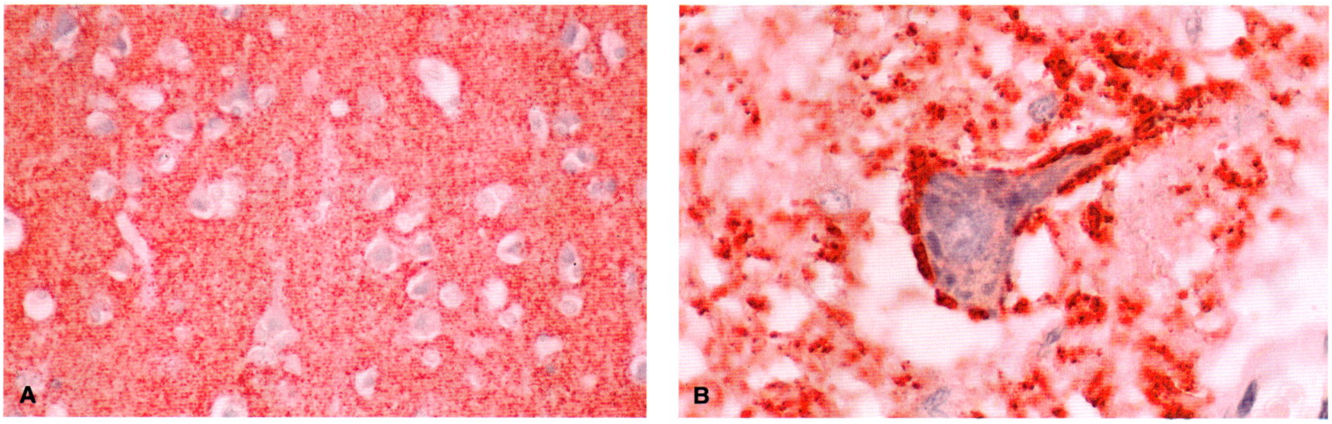

FIGURE 9.17 Synaptophysin is one of the most useful and widely employed markers of neuronal differentiation. The neuropil of gray matter, which is rich in synaptic contacts, shows a diffuse, finely granular pattern (**A**). Several specific types of large projection class neurons show prominent punctate decoration of the cell body and proximal dendrites, as illustrated here by a motor neuron in the hypoglossal nucleus of the medulla (**B**). Other groups of large neurons exhibiting this pattern of synaptophysin immunopositivity include Purkinje cells of the cerebellum, motor neurons of the ventral horn of the spinal cord, and Betz cells of the precentral gyrus in the cerebral cortex.

of the brain stem, and Betz cells of the precentral gyrus (Fig. 9.17B). Another marker that is useful in the identification of most neurons is the antibody neuronal nuclei (NeuN) that highlight postmitotic neurons (Fig. 9.18) (4).

Interestingly, while NeuN stains the granular cells of the cerebellum, its expression is not seen in larger Purkinje cells. This disparity in expression may relate to differences in the development of these two types of neurons. Purkinje cell precursors undergo a final mitosis within the ventricular region after which they subsequently migrate outwards before reaching their final resting place. In contrast, granular cell neurons originate early during development in the proliferating ventricular zone and migrate outward to form an external granule layer where they continue to divide throughout the neonatal period, after which they migrate to the mature granular cell layer. Notable NeuN-negative neuronal populations include mitral cells of the olfactory bulbs and retinal photoreceptor cells.

Age-Related Neuronal Inclusions

A variety of inclusions, largely intracytoplasmic, appear with increasing frequency upon aging. By far, the most common is lipofuscin, which is yellow-to-pale brown and unaltered by most histologic procedures, including the H&E method (Fig. 9.19) (5). Its autofluorescence and partial avidity for the acid-fast stain can be used to visualize differentially this "wear and tear" pigment, although little functional significance is generally assigned to lipochrome accumulation in normal aging. Lipofuscin may accumulate to such an extent in larger neurons that it displaces organelles and creates an

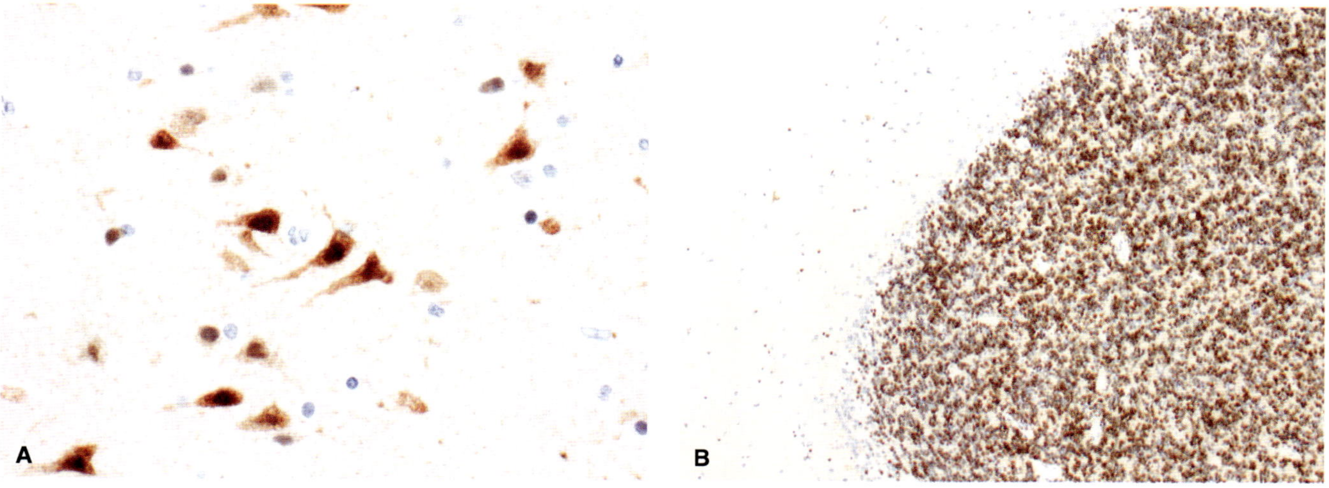

FIGURE 9.18 Neuronal nuclei: NeuN is a very useful marker for identifying neurons. The nuclear pattern of expression of this protein makes it easy to interpret (**A**). Although the neurons in the internal granular cell layer of the cerebellum are strongly positive, Purkinje cell neurons do not express this marker (**B**).

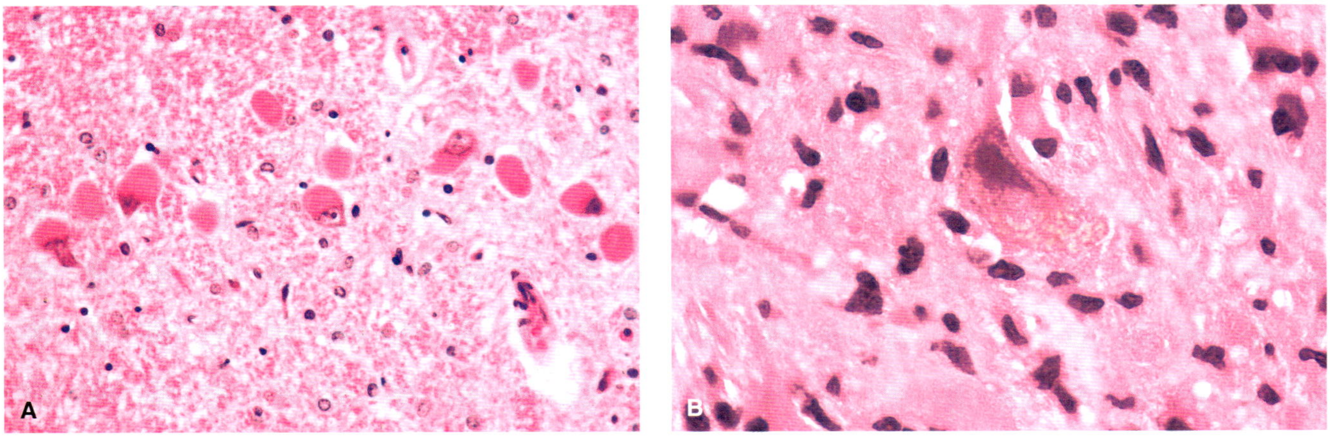

FIGURE 9.19 Lipofuscin: Prominent accumulation of lipofuscin pigment in large neurons with increasing age, as seen for example in the lateral geniculate nucleus (**A**), can result in peripheral displacement of the nucleus and Nissl substance, mimicking central chromatolysis. As a practical application of normal neurohistology, the presence of lipofuscin can sometimes aid in the identification of neurons in intraoperative frozen sections (**B**). As illustrated here, a neuron, identified by its lipofuscin content, is surrounded by tumor cells, thereby supporting a diagnosis of infiltrating glioma.

appearance similar to the cell swelling of central chromatolysis described below (see Fig. 9.26). The lateral geniculate body provides an example of a densely populated nucleus filled with neurons that display prominent accumulation of lipofuscin; it is often discernible by its distinctly mahogany hue compared to adjacent cortex (see Fig. 9.10B). Lipofuscin accumulation is not simply a function of cell size; some classes of large neurons such as cerebellar Purkinje cells are immune to significant accrual.

Functionally, more significant neuronal inclusions that may be seen in asymptomatic individuals are those associated with Alzheimer disease, namely neurofibrillary tangles, neuritic plaques, granulovacuolar degeneration, and Hirano bodies. These illustrate the often ill-defined distinction between health and disease, because these changes can be seen in the elderly, albeit in limited numbers, in the absence of antemortem disturbances of mentation. Neurofibrillary tangles may be found in asymptomatic individuals in occasional neurons of the subiculum or Ammon horn. Although silver stains greatly aid visualization and quantitation, these structures may be identified on routinely stained H&E sections if the observer is familiar with their appearance. They appear as slightly basophilic wisps of faintly fibrillar material in pyramidal cells. These flame-shaped tangles extend out into cell processes, most notably the apical dendrite, and are often more prominent on one side of the nucleus (Fig. 9.20A). This morphology reflects the fact that tangles generally conform to the shape of the cell body. For example, the pigmented neurons of the locus coeruleus are multipolar and lack the dominant apical dendrite of pyramidal neurons. Thus, their tangles are globular in shape.

The senile plaque is also a manifestation of cell injury, but one that, like slight atherosclerosis, is not an unexpected finding in the brains of asymptomatic adults. The plaque is usually seen in its primitive form as a somewhat ill-defined, roughly circular region of abnormal argyrophilic neurites that is not visualized in the H&E-stained section. As the plaques mature, they become visible in the latter preparation, particularly when a central core of eosinophilic amyloid appears (Fig. 9.20B). The latter can be more readily seen by Congo red, periodic acid–Schiff (PAS) staining or staining for β-amyloid. As noted earlier, both neurofibrillary tangles and neuritic plaques are more easily identified and quantitated with special techniques such as immunofluorescence or silver stains (Fig. 9.20C). Immunostains for Tau protein and β-amyloid (Fig. 9.20D) are also helpful in the identification of tangles and plaques, respectively. A finding that can also be encountered in the brains of patients with amyloid plaques is the accumulation of amyloid in the wall of meningeal and cortical blood vessels (Fig. 9.20E).

Two additional intraneuronal inclusions that are seen in Alzheimer disease, but only rarely in nondemented individuals, are granulovacuolar degeneration and Hirano bodies (Fig. 9.21). As the name implies, the inclusion of granulovacuolar degeneration consists of a dark, basophilic granule inside a small, clear vacuole. Clusters of these cytoplasmic inclusions may be present within a single neuron (Fig. 9.21A). Depending on the plane of section, Hirano bodies appear in H&E-stained sections as brightly eosinophilic ovoid, elliptical, or elongated rod-like refractile inclusions that are located either in very close apposition to a neuronal perikaryon (Fig. 9.21B), or within the neuropil. Ultrastructural examination supports localization in neuronal cell bodies and processes, and immunohistochemical studies reveal the presence of actin and actin-associated proteins. Unlike neurofibrillary tangles and neuritic plaques, both granulovacuolar degeneration and Hirano bodies exhibit a very limited neuroanatomic distribution and are, in fact, virtually confined to the hippocampal formation. Encountering more than one or two cells with these alterations should

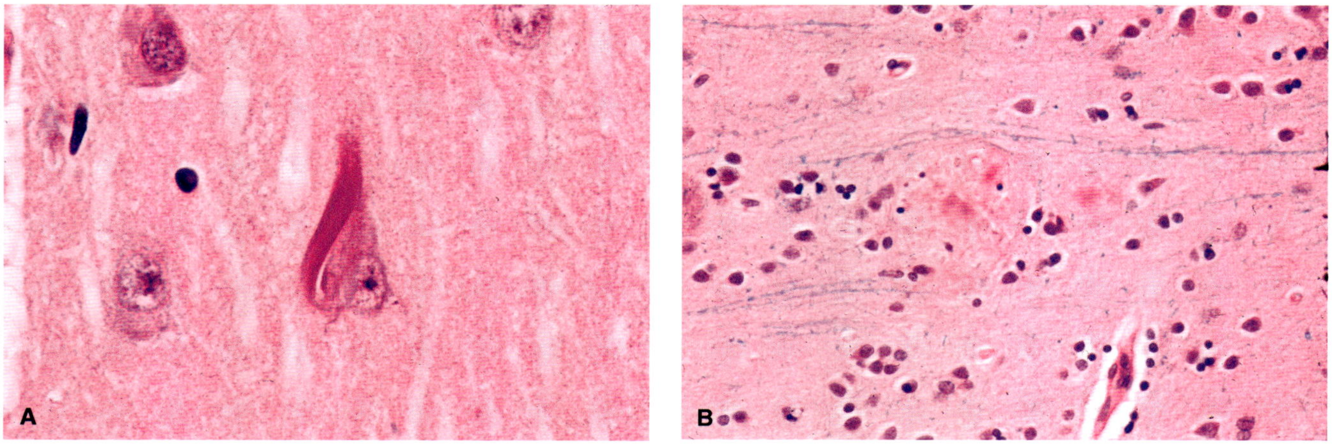

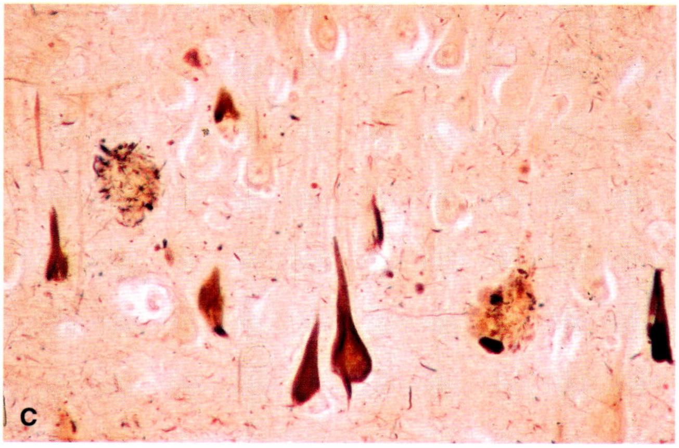

FIGURE 9.20 Neurofibrillary tangles and neuritic plaques: Neurofibrillary tangles (**A**) may be seen sporadically in the hippocampal formation of aging brains and have a fibrillary texture in H&E-stained sections. Mature senile (neuritic) plaques (**B**) deform the smooth texture of the neuropil and appear in H&E-stained sections as spherical, somewhat granular foci with a central eosinophilic core that is composed of amyloid. Note that adjacent myelinated axons are focally displaced as they pass by the plaque. In earlier stages, the plaques are less well defined and are not identifiable in H&E-stained sections. Although both neurofibrillary tangles and large mature neuritic plaques can be seen in H&E-stained sections, use of special techniques, such as silver stains (**C**), greatly facilitates visualization and quantitation.

raise the issue of Alzheimer disease and prompt a search for other attendant histologic features.

A particularly striking cytoplasmic inclusion occasionally encountered in routine sections of the hypoglossal nuclei of the medulla (less often in the ventral horn motor nuclei of the spinal cord) is the hyaline (colloid) inclusion (Fig. 9.22). These inclusions consist of ectatic cisternae of endoplasmic reticulum and are rarely seen in the first few decades of life, but appear with increasing frequency thereafter. They are occasionally mistaken for viral inclusions.

Catecholaminergic neurons throughout the brain stem gradually accumulate neuromelanin as a by-product of

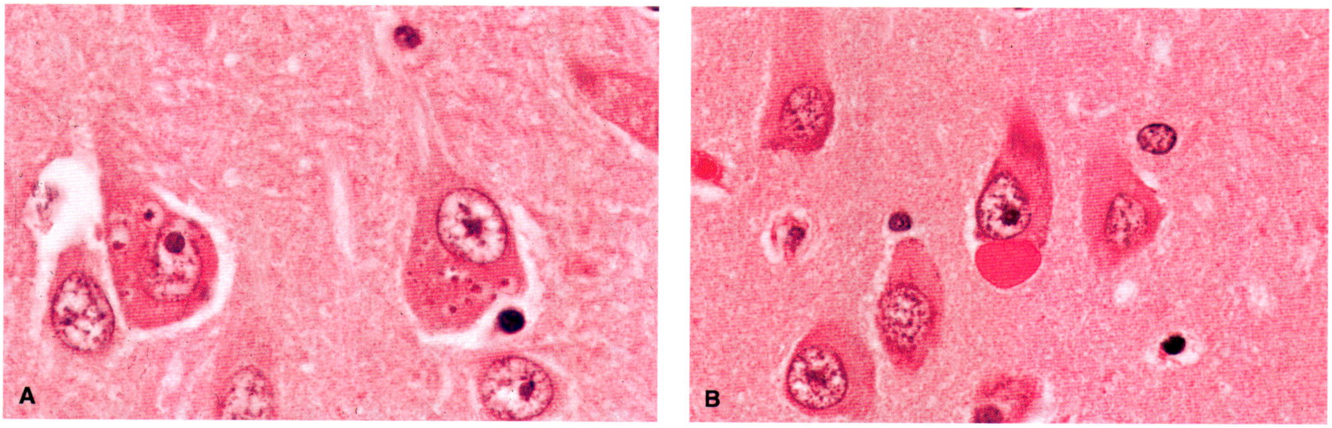

FIGURE 9.21 Granulovacuolar degeneration (of Simchowitz) (**A**) and Hirano bodies (**B**): Like neurofibrillary tangles, both of these intracytoplasmic inclusions can occasionally be seen in the hippocampal formation of normal older individuals. However, although these alterations are not pathognomonic for dementing illness, an appreciable number of affected cells should prompt a search for the evidence of Alzheimer disease in the form of a thorough examination for senile (neuritic) plaques and neurofibrillary tangles.

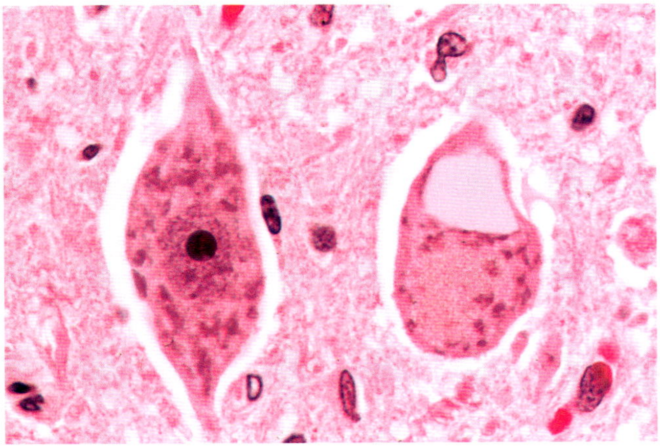

FIGURE 9.22 Hyaline (colloid) inclusion: These eye-catching inclusions, as seen in the neuron on the right, may be observed sporadically throughout the neuraxis but are most commonly encountered in the large motor neurons of the hypoglossal nuclei in the medulla (as in this micrograph). Less frequently, they may be seen in the motor neurons of the ventral horn of the spinal cord. Electron microscopic examination reveals ectatic cisternae of endoplasmic reticulum.

neurotransmitter synthesis. The largest and most densely populated of these nuclei is the substantia nigra, which contains dopaminergic neurons. The locus ceruleus ("blue spot"), which is also seen by the unaided eye, is a collection of noradrenergic neurons in the rostral pontine tegmentum. It is of practical importance that the Lewy bodies of Parkinson disease can be found in both of these neuroanatomic locales. Of the smaller and more diffusely distributed pigmented neurons, those in the vicinity of the dorsal motor nucleus of the vagus nerve in the medulla oblongata are most commonly encountered during routine histologic examination. Microscopically, neuromelanin appears as coarse, dark brown granules (Fig. 9.23A), and should not be confused with melanocytic melanin. The latter is also present in the CNS but is confined to leptomeningeal melanocytes as discussed below (see Fig. 9.67).

Several eosinophilic inclusions may be seen in pigmented brain stem neurons. The most striking of these are the commonly encountered Marinesco bodies (Fig. 9.23B) (6). These bright red, hyaline-appearing structures are located within the nucleus, often adjacent to and about the same size as a nucleolus. As such, they are alternatively designated as an alternative designation is paranucleolar bodies. Multiple Marinesco bodies may occur within a single nucleus and, in some cases, a large percentage of pigmented neurons exhibit these eye-catching inclusions. In such cases, they may raise concern about a viral infection, but are not pathologic and have yet to be correlated with any significant process except advancing age. Two types of eosinophilic inclusions may be encountered in the cytoplasm of pigmented neurons. Clusters of diminutive acidophilic granules are occasionally noted (Fig. 9.23C) but have no pathologic significance. Lewy bodies are much larger, notably displace the cytoplasmic neuromelanin from which they are separated by a small clear halo and are associated with Parkinson disease (Fig. 9.23D).

Autolysis and Basic Neuronal Reactions to Injury

As captured in their normal state by perfusion fixation or rapid immersion fixation, neurons are generally rotund with lightly eosinophilic cytoplasm that is stippled with basophilic Nissl substance in the case of the larger neurons. The surrounding glia are inconspicuous, and few clear vacuoles are seen. However, virtually all autopsy and surgical specimens suffer some degree of autolysis that alters this ideal appearance. Autolysis may cause some changes that simulate a disease state; neurons may be rendered somewhat contracted and basophilic with condensed nuclei. Simultaneously, the processes of glia that surround neurons and blood vessels imbibe water to produce clear vacuoles (see Fig. 9.37). The neuronal response to injury overlaps in some cases with these autolytic changes, and it may not be possible to distinguish agonal hypotensive injury from autolysis in autopsy specimens. In the former setting, the neuronal contraction is pronounced and the perineuronal and perivascular spaces are exceptionally prominent.

There are, however, three neuronal changes that provide unequivocal evidence of antemortem injury. One is the red neuron of ischemic damage, the second is central chromatolysis, and the third is ferruginization. The red neuron is characterized by a shrunken cell body and intense cytoplasmic eosinophilia with complete loss of Nissl basophilia (Fig. 9.24). The nucleus is dark and usually lacks a distinguishable nucleolus but may be pale and demonstrate early karyolysis. At times, it has a somewhat fragmented look suggesting karyorrhexis, although clearly defined karyorrhexis is rare. In surgical specimens, crush artifact due to tissue handling also results in dark, shrunken neuronal perikarya, although these cells lack the distinctive cytoplasmic eosinophilia of ischemia (Fig. 9.25).

Central chromatolysis, the second unequivocally abnormal finding, consists of a loss of central basophilic staining of the cell body with peripheral margination of the Nissl substance (Fig. 9.26). It is seen in poliomyelitis as well as a number of other pathologic states. There are numerous mimickers that lie in wait for the unwary. For example, some normal neuronal populations, such as the supraoptic and paraventricular nuclei of the hypothalamus and the dorsal nucleus of Clarke of the thoracic spinal cord, display Nissl substance that is preferentially distributed peripherally in the soma. Other neurons, such as those of the mesencephalic nucleus of the trigeminal nerve discussed previously (see Fig. 9.15), have large, exquisitely rounded somas that mimic chromatolysis. The giant pyramidal cells of Betz in the motor cortex are so large in comparison to surrounding neurons that on they may give an initial impression of chromatolytic swelling. Finally, as discussed earlier, one must be careful not to mistake the accumulation of various substances that displace the Nissl substance peripherally, such as lipofuscin, for central chromatolysis (see Fig. 9.19).

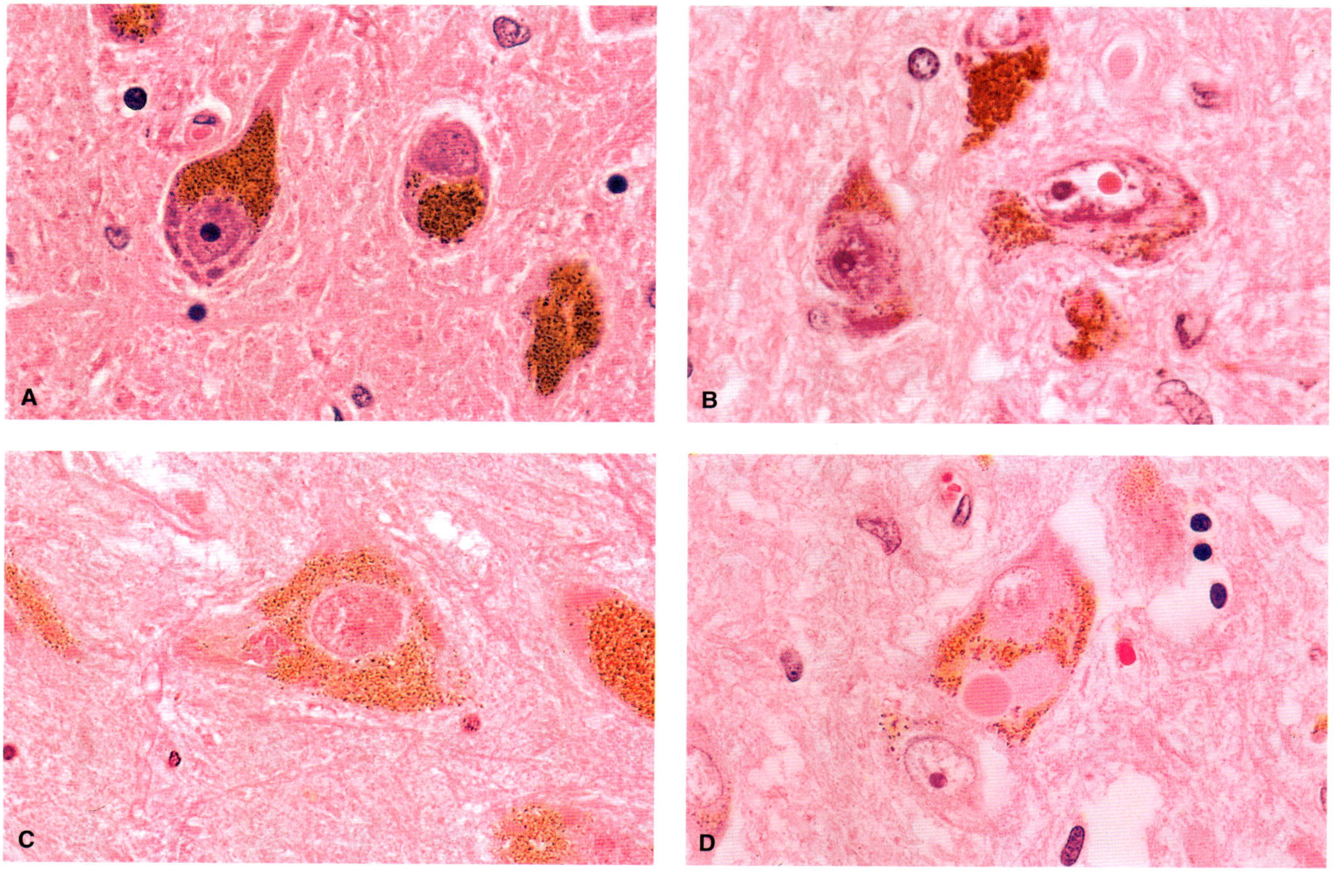

FIGURE 9.23 Pigmented neurons of the brain stem: Neuromelanin is a coarse, dark brown cytoplasmic pigment (**A**) that is formed as a by-product of catecholamine synthesis and is frequently encountered microscopically in scattered catecholaminergic neurons distributed widely throughout the brain stem. Two large populations are visible grossly: The substantia nigra ("black substance") of the midbrain and the locus ceruleus ("blue spot") of the pons. Marinesco bodies (**B**) are eosinophilic, spheroidal, paranucleolar bodies that are often observed in the nuclei of pigmented neurons, especially those of the substantia nigra. The number of Marinesco bodies increases with advancing age and can be quite striking in some individuals. They should not be mistaken for intranuclear viral inclusions. Clusters of minute intracytoplasmic eosinophilic granules (**C**), seen in this micrograph to the left of the nucleus, may occasionally catch the eye of an obsessive observer. They have no known pathologic significance and are much smaller than Lewy bodies (**D**), which are the characteristic intracytoplasmic inclusions of Parkinson disease.

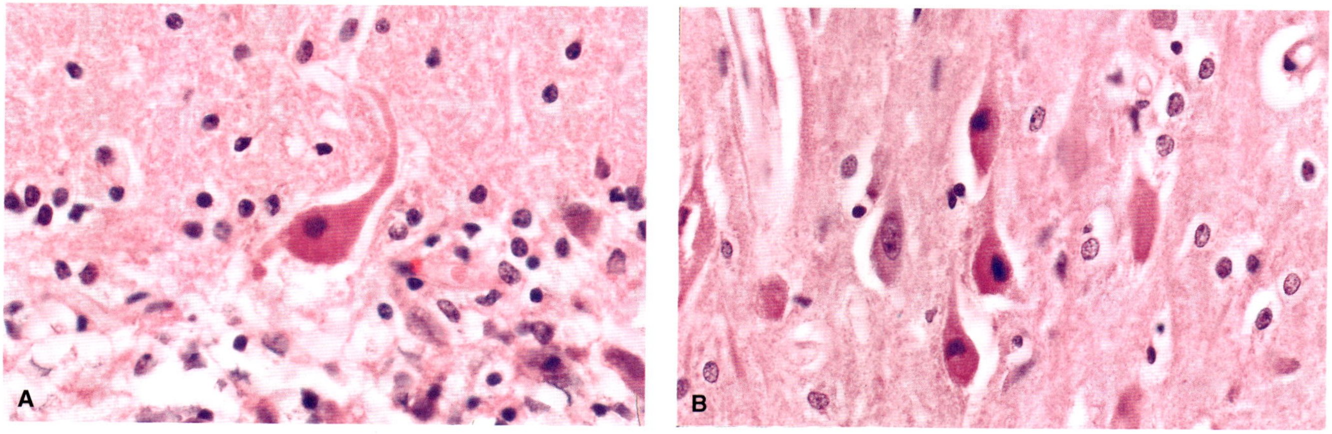

FIGURE 9.24 Ischemic injury: The sine qua non of ischemic damage to the nervous system is the so-called red neuron. As illustrated here by a Purkinje cell of the cerebellum (**A**) and pyramidal neurons of the hippocampal CA1 region (**B**), the soma (cell body) is shrunken, the cytoplasm is intensely eosinophilic, and the nucleus is pyknotic with no discernible nucleolus. It is largely the pronounced eosinophilia that distinguishes this cellular alteration from autolytic neuronal condensation, in which the cytoplasm is dark and basophilic (see Fig. 9.25).

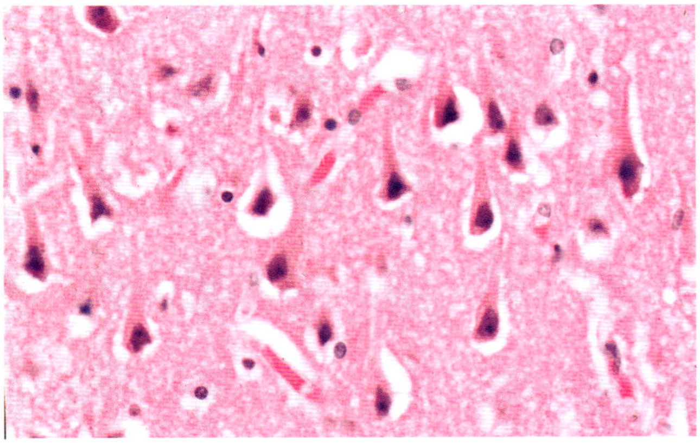

FIGURE 9.25 Neuronal contraction as a tissue-handling artifact: In contrast to ischemic insult (Fig. 9.24), in "crush artifact" the cytoplasm is dark and basophilic rather than brightly eosinophilic.

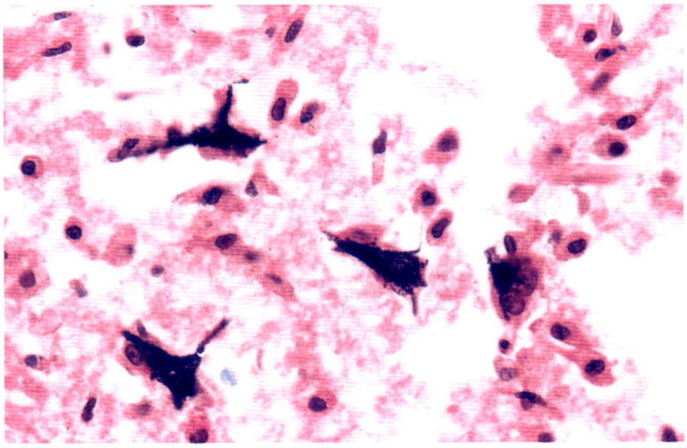

FIGURE 9.27 Mineralized (ferruginized) neurons: These encrusted relics resembling petrified tree trunks are most commonly encountered around the margins of old infarcts.

A striking finding sometimes encountered near old infarcts is the presence of ferruginized or fossilized neurons, in which both perikarya and axons are encrusted by blue-staining minerals (Fig. 9.27). This arresting phenomenon is not limited to the vicinal tissue of old infarcts, although this is the most common context, nor is it confined to the adult nervous system, because prenatal insults may result in similar findings. Clusters of axons thus affected can be mistaken for fungal hyphae (Fig. 9.28).

A common reaction of axons to injury seen in a wide variety of pathologic states is the formation of localized dilatations known as axonal spheroids or axon "retraction balls" (Fig. 9.29A). These structures can be highlighted with immunohistochemical stains against the amyloid precursor protein (APP) and ubiquitin. Ultrastructural examination shows greatly distended axis cylinders filled with bundles of neurofilaments and cellular organelles. A regional variant of this process may be observed in the granular cell layer of the cerebellum where focal dilatations of Purkinje cell axons are termed torpedoes. These structures are seen in a number of cerebellar degenerative diseases, as well as in normal aging. The most common site in the CNS where scattered axonal spheroids are routinely encountered as an incidental finding in aged individuals is in the rostral fasciculus gracilis of the medulla (Fig. 9.29B). Spheroids in this location are often mineralized.

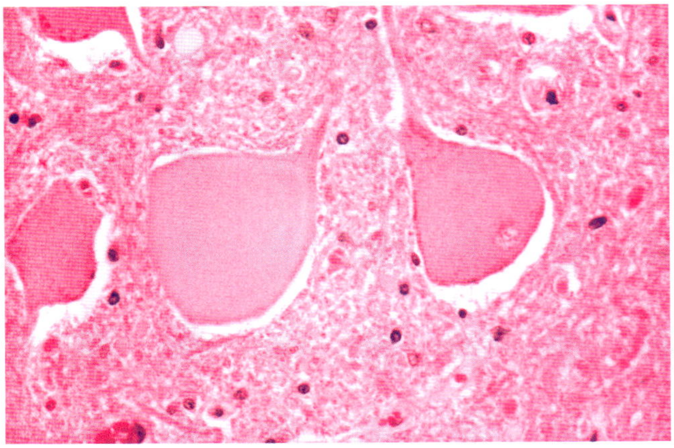

FIGURE 9.26 Central chromatolysis: Cytoplasmic hyalinization and swelling with peripheral displacement of the nucleus and lipofuscin can be a response to either intrinsic neuronal disease (such as in poliomyelitis or other viral infections) or to interruption of the axon in close proximity to the cell body. In the latter setting, the term axonal reaction is applied.

Astrocytes

Normal Microscopic Anatomy

Like neurons, astrocytes are also heterogeneous. The cells of one class conform to the classic star shape and occur as either the fibrillary or protoplasmic form. Fibrillary astrocytes populate the white matter, whereas the latter inhabit the gray matter. Other important subtypes of astrocytes include the "pilocytic" astrocytes of the periventricular region, cerebellum, spinal cord, and the Bergmann astrocytes, which are distributed in a narrow lamina between the cell bodies of Purkinje neurons in the cerebellar cortex (Fig. 9.30). In gray matter,

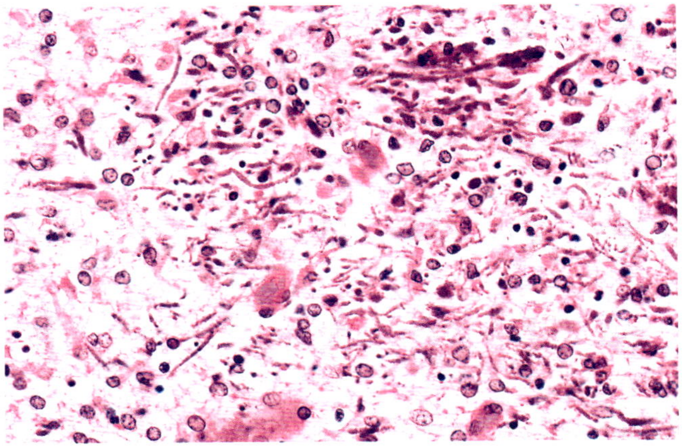

FIGURE 9.28 Mineralized axons: Clusters of mineralized axons have a superficial resemblance to fungal hyphae.

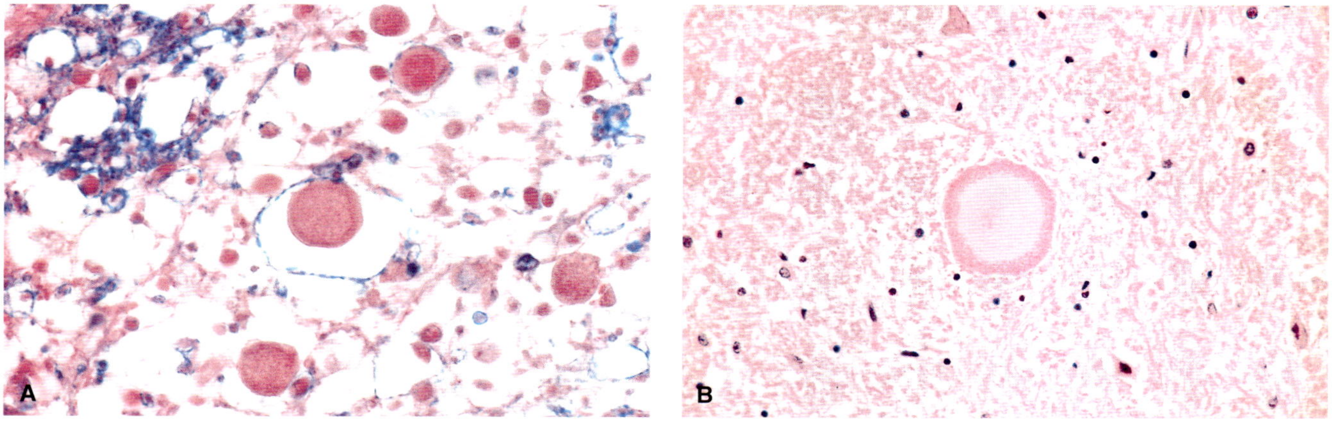

FIGURE 9.29 Axonal spheroids: Focal dilatations known as spheroids (**A**) are a common axonal reaction to injury that is seen in a wide array of pathologic conditions including radiation damage and post-traumatic diffuse axonal injury. Axonal spheroids are also frequently encountered incidentally in older individuals in the dorsal medulla oblongata (rostral fasciculus gracilis near nucleus gracilis) where, as in this example, they are often mineralized (**B**).

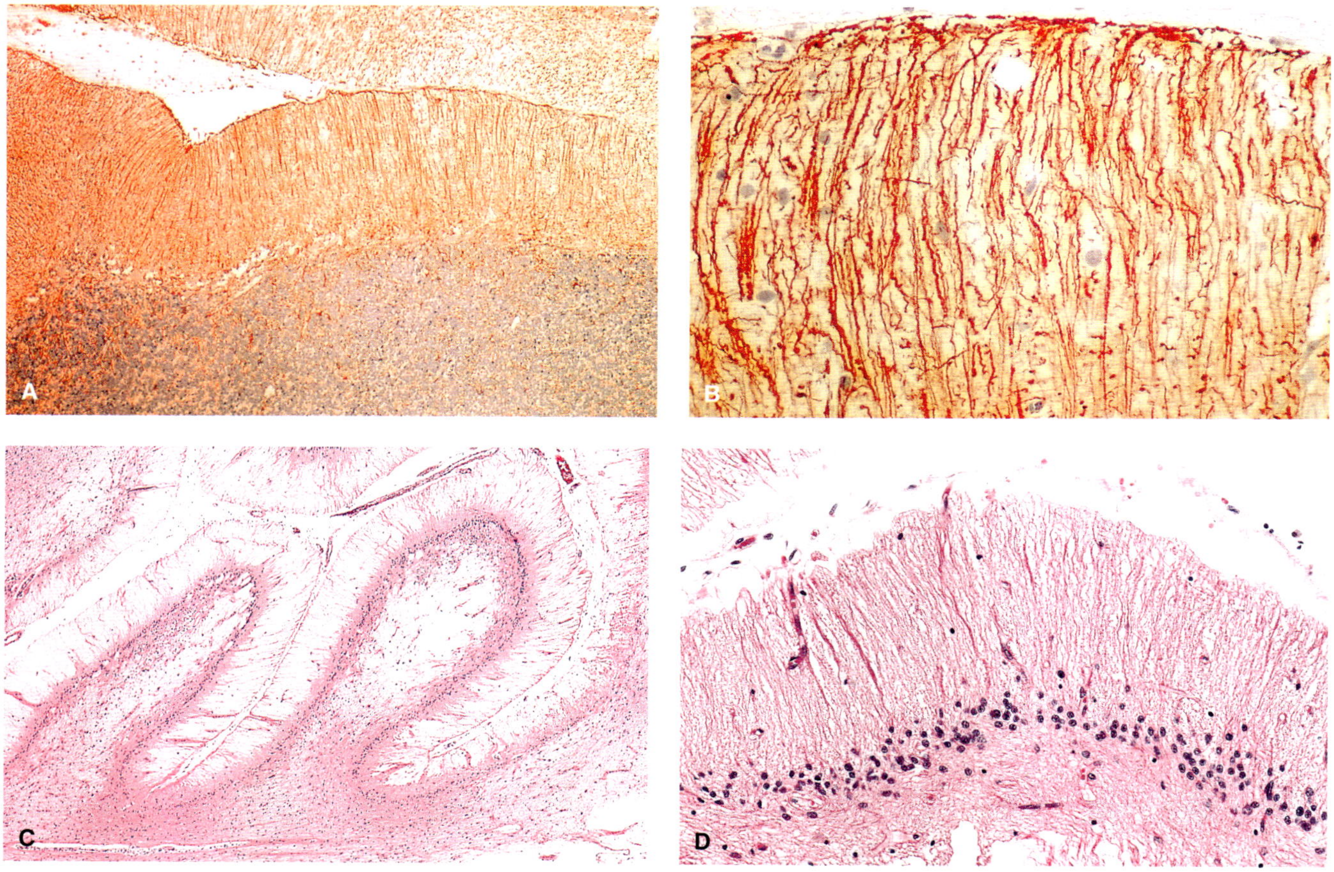

FIGURE 9.30 Bergmann glia: These astrocytes illustrate the fact that specialization is not confined to neurons. Bergmann astrocytes have cell bodies distributed in a narrow lamina of the cerebellar cortex coextensive with that of the Purkinje cells. Each cell sends an elongated process through the molecular layer to the subpial surface. These processes are not usually well seen in healthy cerebellum with routine H&E staining (see Fig. 9.8), but can be exquisitely visualized with immunohistochemistry for GFAP (**A,B**). An equally striking unmasking of this elegant architecture is often seen without the use of specialized staining techniques in areas of cerebellar cortex adjacent to healed infarcts, in which the degree of ischemia was sufficient to kill the indigenous neuronal populations but spared the more resistant Bergmann glia (**C,D**). Like other astrocytes throughout the CNS, Bergmann glia respond to ischemic insult by proliferating, resulting in an increased thickness of the cell body lamina referred to as "Bergmann gliosis" (**D**).

nuclei of protoplasmic astrocytes cannot generally be distinguished from those of small neurons because the cytoplasm of both, blends imperceptibly into the surrounding neuropil and is not normally discernible as a discrete entity. In white matter, it is usually difficult in H&E-stained sections to distinguish fibrillary astrocytes from the much more numerous oligodendroglia. The nuclei of oligodendrocytes are smaller and more hyperchromatic, but usually these two cell types do not fall into two clearly defined groups. In sections stained for myelin, a very small amount of eosinophilic cytoplasm may occasionally, but not invariably, be seen surrounding normal astrocytic nuclei. This helps distinguish this cell from the oligodendrocyte whose cytoplasm, other than the myelin sheath, is not usually apparent by conventional light microscopy (see Fig. 9.38). Astrocytic cytoplasm becomes much more prominent when astrocytes respond to CNS injury, culminating in the abundant glassy cytoplasm of the gemistocyte (Fig. 9.31). To appreciate the distinctive morphology of the star cell, one must visualize its radiating processes. These thread-like extensions reach out to define a sphere of influence that is many times greater in extent than one would have suspected by looking at an H&E-stained section alone. Historically, this tinctorial feat was achieved through the technically capricious metallic impregnations, but now is accomplished with considerably greater ease and predictability by the immunohistochemical localization of glial fibrillary acidic protein (GFAP) (Fig. 9.31). In the case of the fibrillary astrocyte, processes branch infrequently, whereas those of the protoplasmic astrocyte are more numerous and divide more frequently. They are often less well stained with GFAP than the fibrillary types. Neither type of resting astrocyte is as apparent immunohistochemically as are reactive astrocytes.

The polar forms of astrocytes include the pilocytic and Bergmann types. The pilocytic astrocyte is not conspicuous in its native state but becomes so when responding as gliosis and forming Rosenthal fibers. The latter are hyaline, often

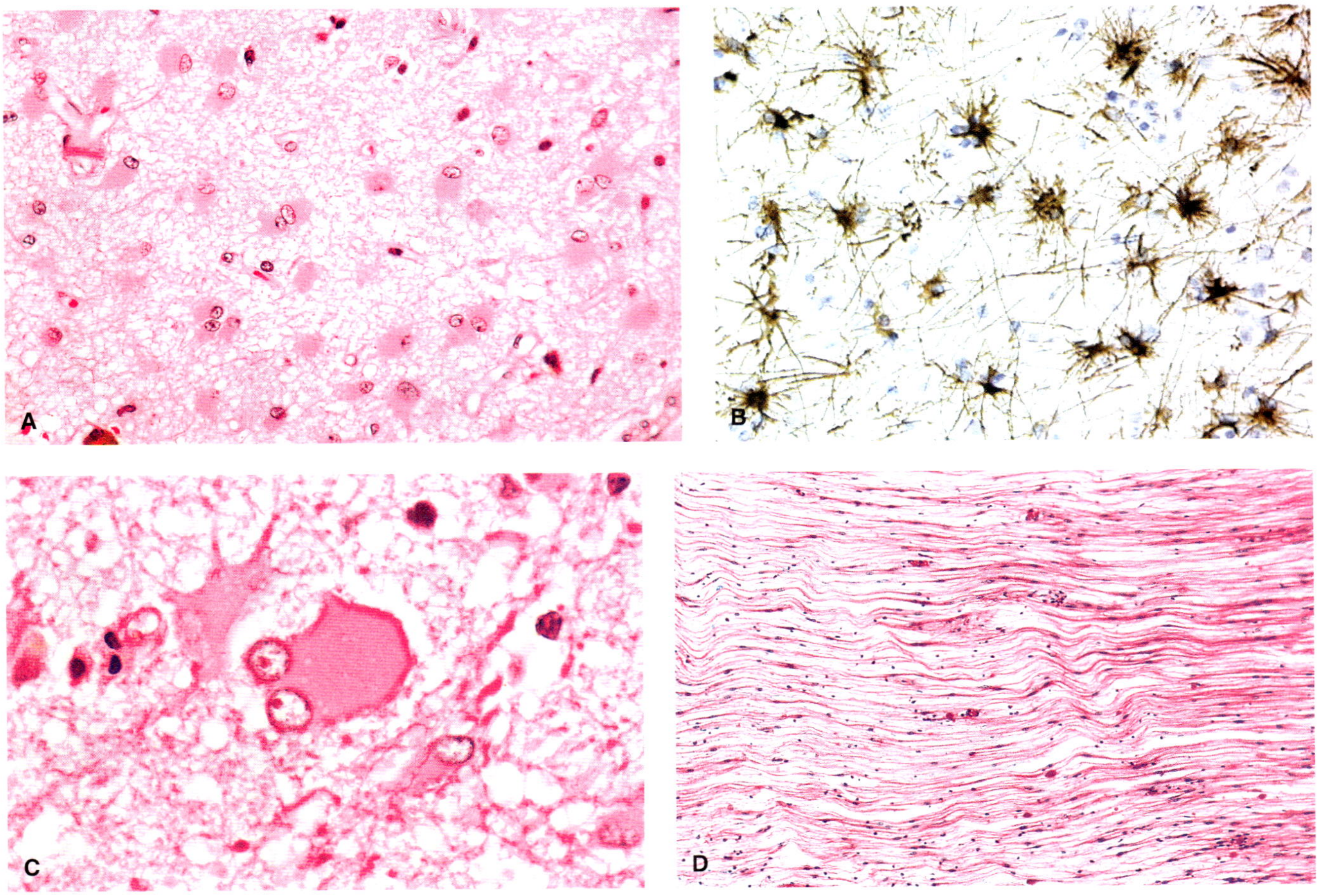

FIGURE 9.31 Reactive astrocytosis: The plainly visible cytoplasm and processes of the astrocytes seen in panel (**A**) is proof of an insult to the nervous system. Under normal conditions, only bare nuclei are usually seen. The extensive, radiating cytoplasmic processes for which the astrocyte received its name are most readily appreciated when reactive astrocytes are immunostained for GFAP (**B**). Reactive astrocytes have been descriptively classified according to the amount and configuration of visible cytoplasm, and include the aptly named gemistocytic "laden" or "stuffed" cell (**C**), and pilocytic "hair cell" (**D**) types. Reactive gemistocytes are typical of the acute astrocytic reaction to CNS damage whereas dense fibrillary gliosis is commonly seen in longstanding lesions such as healed infarcts.

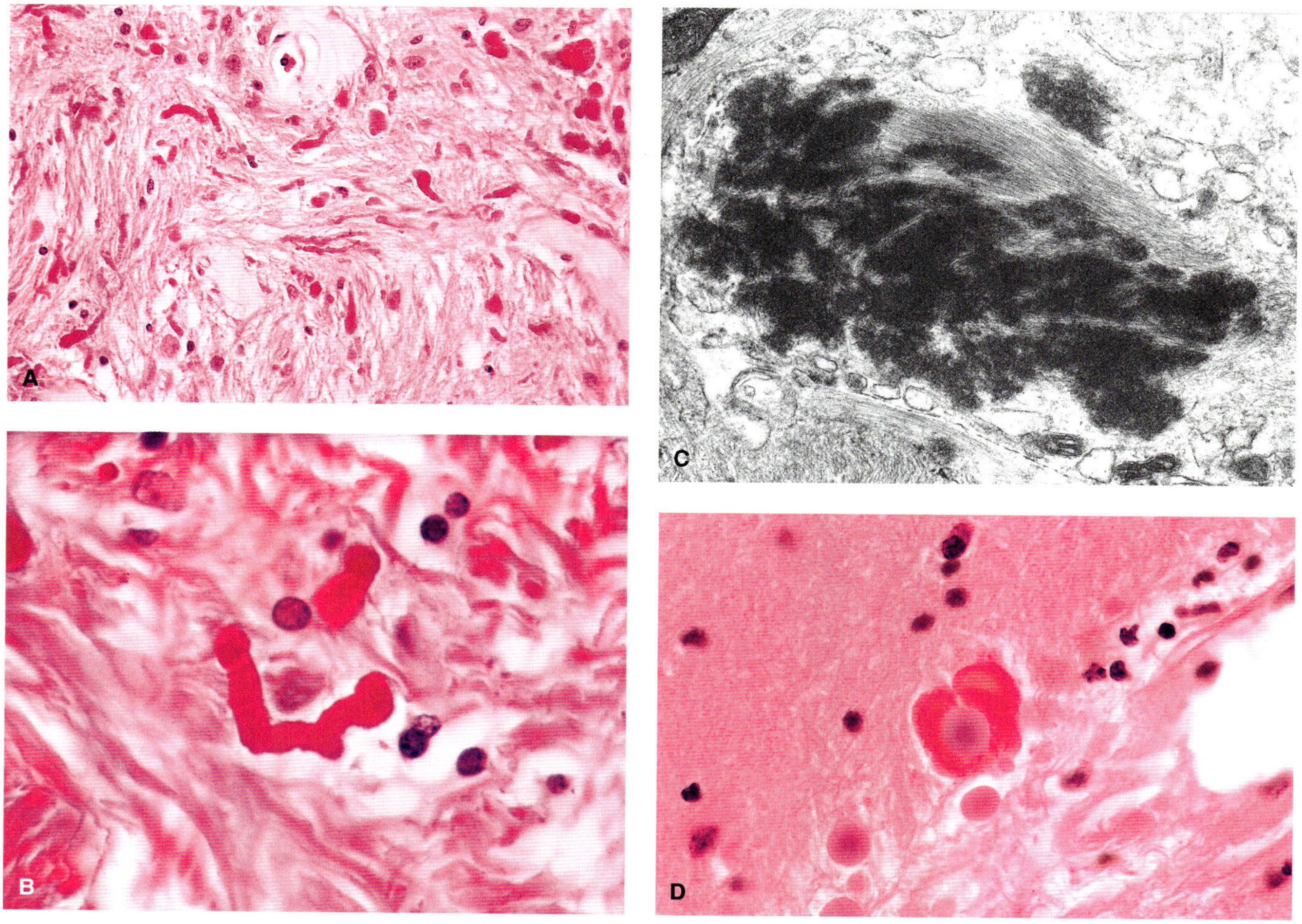

FIGURE 9.32 Rosenthal fibers: Chronic reactive fibrillary astrogliosis is often accompanied by Rosenthal fiber formation (**A**). Rosenthal fibers are brightly eosinophilic, lumpy, elongated structures (**B**) that by ultrastructural examination, appear as electron-dense amorphous masses surrounded by and merging with dense bundles of glial filaments (**C**). Occasionally, the two most common intracytoplasmic inclusions of astrocytes, corpora amylacea, and Rosenthal fibers, may be seen together in the same astrocytic process (**D**).

corkscrew-shaped, eosinophilic structures that are wedged within one of the bipolar processes of the cell (Fig. 9.32). These structures are occasionally seen in the normal hypothalamus or pineal gland but become much more prominent in gliosis around craniopharyngiomas, pineal cysts, cerebellar hemangioblastomas, and chronic lesions of the spinal cord.

The Bergmann astrocytes are confined to a one-to-two-cell thick lamina. Their polar processes extend to the pial surface of the cerebellum and are only faintly seen with difficulty in standard sections. Yet, they are well visualized with immunohistochemistry for GFAP and at the margins of old cerebellar infarcts (Fig. 9.30). The Bergmann glia provide an excellent illustration of astrocytic specialization. Their processes are a form of scaffold and serve as a reminder of the cooperative interplay between astrocytes and neurons during embryologic development. At that time, the small neurons of the external granular cell layer spiral down the Bergmann processes to reach their final destination in the internal granular cell layer.

Age-Related Inclusions in Astrocytes: Corpora Amylacea

The ubiquitous corpora amylacea are, by far, the most salient astrocytic inclusions encountered in routine sections (7). These faintly laminated, slightly basophilic polyglucosan bodies accumulate with age and are observed in greatest numbers where astrocytic foot processes are most numerous, particularly around blood vessels and beneath the pia (Fig. 9.33). The olfactory tracts of adults are also typically rich in corpora amylacea (see Fig. 9.55). The similarity between corpora amylacea and fungal yeast forms such as cryptococcus is a source of potential diagnostic error since both are strongly positive for methenamine silver, alcian blue, and PAS (Fig. 9.33). In some individuals, corpora amylacea are strikingly numerous although no pathologic significance has yet been attributed to this abundance.

Astrocytic Reactions to Injury

Astrocytes respond rapidly and dramatically to CNS injury. This response typically consists of two components:

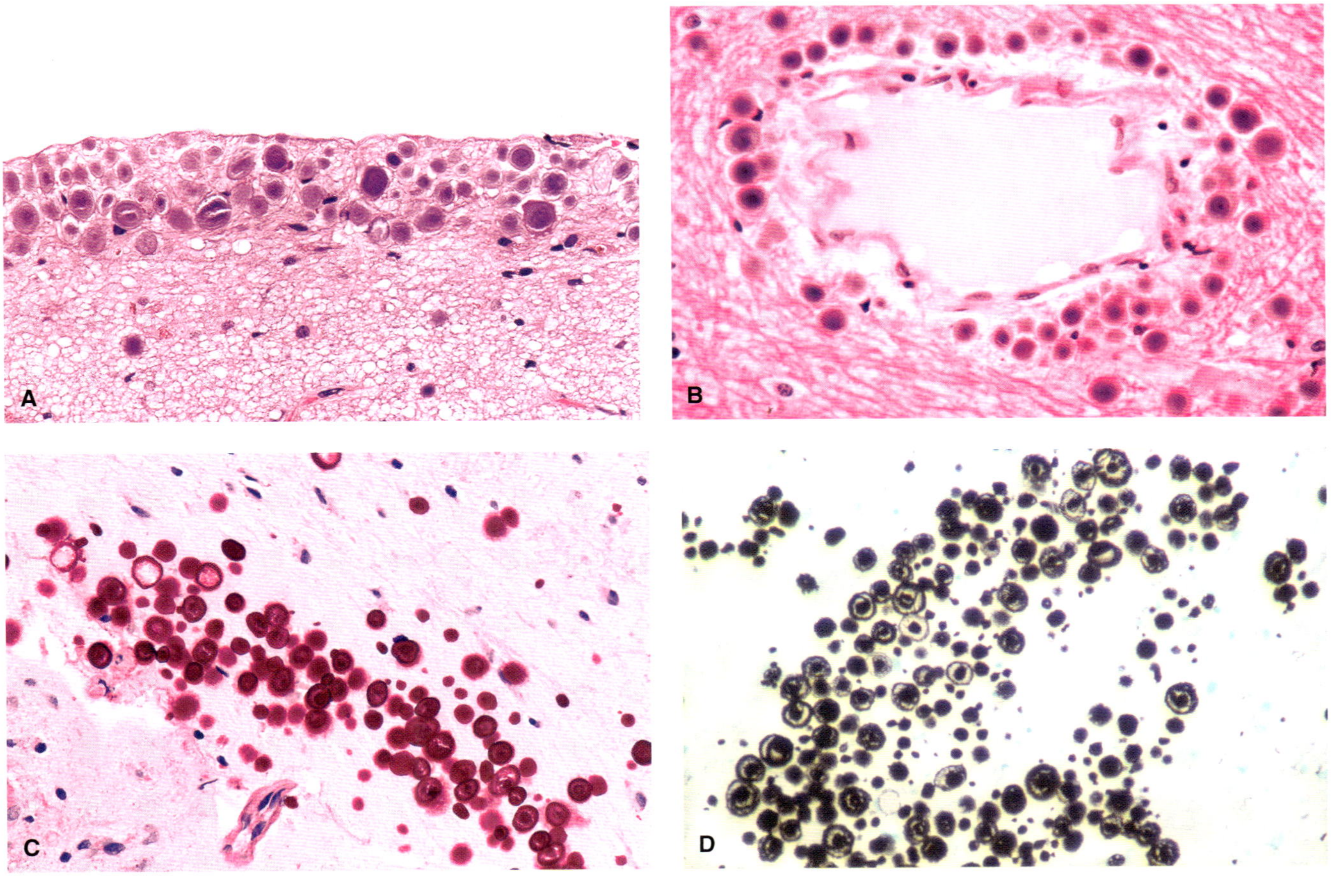

FIGURE 9.33 Corpora amylacea: These basophilic, lamellated polyglucosan bodies accumulate in astrocytic processes with age, most prominently in subpial (**A**) and perivascular (**B**) locations. Corpora amylacea can resemble fungal yeast forms and show strong positivity for fungal stains such as PAS-fungus (**C**) and Gomori methenamine silver (GMS) (**D**).

hypertrophy and hyperplasia. The initial hypertrophic response, an increase in cell size and cytoplasmic prominence, occurs rapidly following CNS insult. Conspicuous cytoplasm is generally indicative of reactive gliosis and constitutes prima facie evidence of CNS injury. Reactive astrocytes display a broad range of cytoplasmic quantity, from just barely perceptible to robustly embonpoint (Fig. 9.31). The latter cells are known as gemistocytes, or stuffed cells. Gliosis may, of course, also present as an increase in the number and density of astrocytic nuclei without attendant cytoplasmic prominence. This chronic type of gliosis is frequently subtle and often requires special stains for confirmation and quantitation.

The end result of acute reactive astrogliosis, such as that accompanying cerebral infarction, is frequently a dense fibrillary gliosis (Fig. 9.31). The often-invoked analogy between the astrocyte and fibroblasts of the peripheral tissues is quite apt in that astrocytes are ubiquitously distributed cells with mitotic capability that responds with alacrity to a wide range of deleterious stimuli. A distinctive cytoplasmic inclusion seen in fibrillary astrogliosis is the Rosenthal fiber (Fig. 9.32). These strikingly eosinophilic, elongated, anfractuous structures are observed in a wide variety of reactive states that share in common significant chronicity. Rosenthal fibers are also characteristic of several specific nosologic entities, including Alexander disease and juvenile pilocytic astrocytoma. It should be stressed, however, that Rosenthal fibers may be strikingly abundant in the chronically compressed glial stroma surrounding a large number of nonneoplastic conditions including syringomyelia, pineal cysts, and slowly expanding nonglial tumors such as craniopharyngiomas.

There are several specialized forms of reactive astrogliosis that deserve brief description. Reactive astrocytes with multiple small nuclei ("micronuclei") are termed Creutzfeldt astrocytes, and may be seen in a number of reactive or neoplastic states but are especially typical of demyelinating processes (Fig. 9.34). A specific type of astrocytic reaction is seen in a variety of hepatic diseases that produce hyperammonemia. The reaction consists of nuclear changes exclusively: swelling with contortion of the nuclear membrane, chromatin clearing, and development of one or two prominent nucleoli (Fig. 9.35). In sharp contrast to all of the other types of reactive astrocytes, these Alzheimer type II astrocytes fail to exhibit prominent cytoplasm by

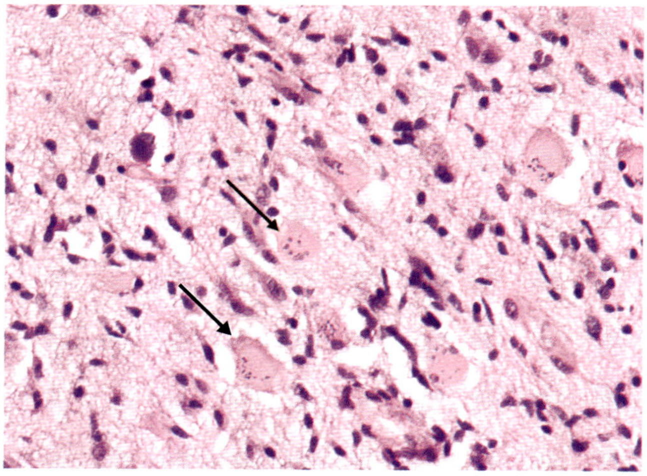

FIGURE 9.34 Granular mitoses (*white arrow*) and Creutzfeldt astrocytes (*black arrow*). These distinctive reactive astrocytes can be seen in a variety of pathologic conditions including tumors (as shown in this figure) but are particularly characteristic of demyelinating diseases.

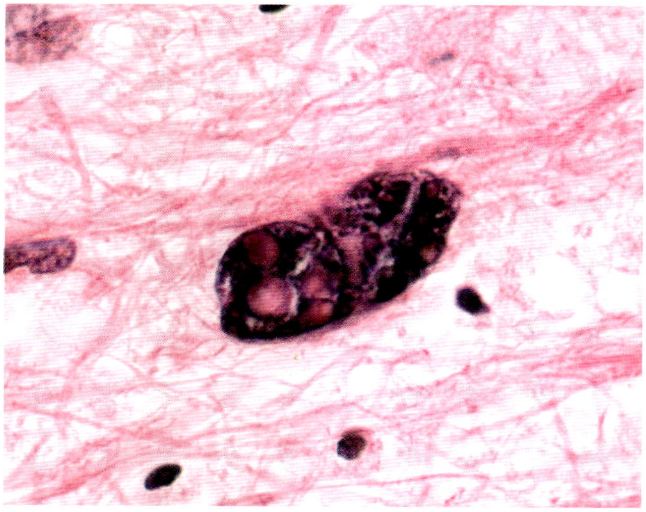

FIGURE 9.36 Bizarre reactive astrocytes of PML: Atypical-appearing reactive astrocytes are sometimes the most striking finding in a PML biopsy and can be mistaken for neoplasia by the unprepared.

routine H&E microscopy. Alzheimer type II astrocytes may be seen throughout the neuraxis but are particularly prominent in certain locations, most notably the globus pallidus. Alzheimer type I astrocytes differ from type II astrocytes in displaying abundant eosinophilic cytoplasm (Fig. 9.35) and are only seen with frequency in Wilson disease. As the eponyms imply, both types of reactive astrocytic morphologies were described by Alois Alzheimer and have no relationship to the dementing disease of the same ilk.

Among astrocytic reactions to injury, none is more striking than that observed in progressive multifocal leukoencephalopathy (PML). Not infrequently, the most eye-catching histologic aspect of PML is the alarming degree of nuclear hyperchromatism and pleomorphism exhibited by scattered astrocytes (Fig. 9.36). Perivascular clearing is a routinely observed artifact of autolysis (Fig. 9.37). By electron microscopy, these clear spaces are revealed to be greatly dilated astrocytic perivascular foot processes. This phenomenon of water imbibition by astrocytes is seen both as an autolytic change in virtually all autopsy specimens and, when extreme, as a marker of antemortem hypoxic/ischemic injury.

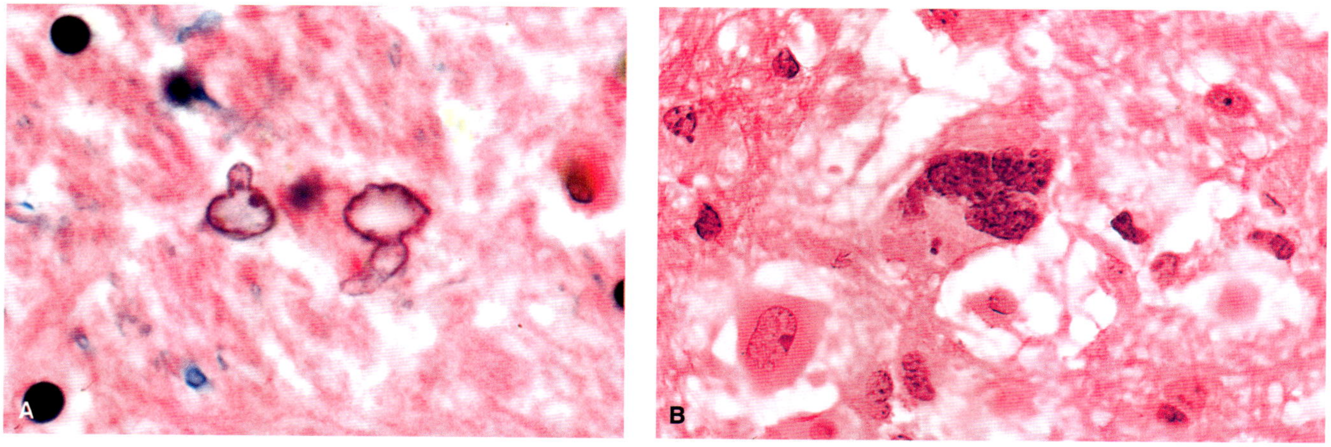

FIGURE 9.35 Alzheimer astrocytes in hyperammonemia: Two types of reactive astrocytic morphologies, termed Alzheimer type I and Alzheimer type II astrocytes, are associated with hyperammonemic conditions. They were described by Alois Alzheimer and bear his name but have nothing to do with the dementing disease that was also a subject of the famous neurologist's investigations. By far, the most frequently encountered are Alzheimer type II astrocytes (**A**). Typical features include an enlarged pale nucleus with an irregular contour and one or more small nucleoli. In marked contrast to other types of reactive astrocytes, visible cytoplasm is lacking. Alzheimer type II astrocytes are commonly seen in a wide variety of diseases that result in increased blood ammonia. In contrast, Alzheimer type I astrocytes (**B**) have large, irregularly lobulated or multiple nuclei and clearly discernible eosinophilic cytoplasm. These cells are not seen in most hyperammonemic diseases, with the exception of hepatolenticular degeneration (Wilson disease).

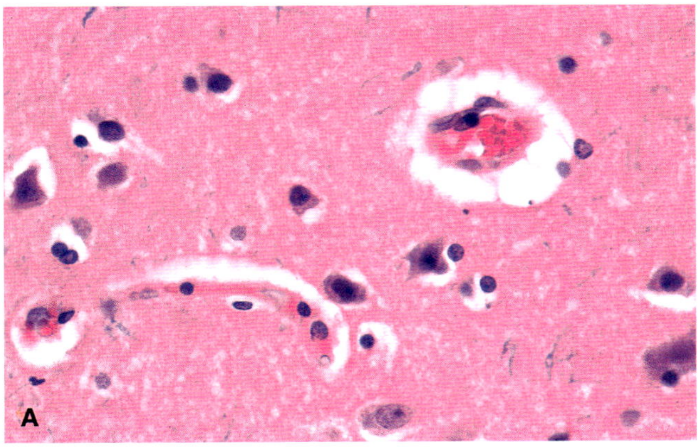

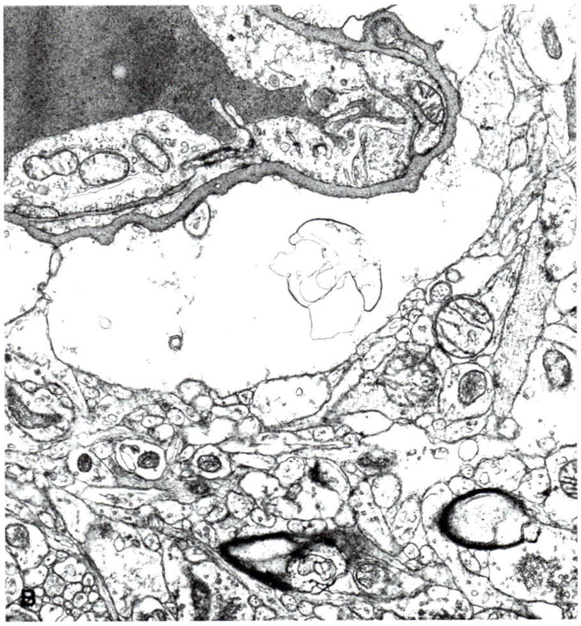

FIGURE 9.37 Perivascular astrocytic foot process swelling in autolysis: This common artifact of routine tissue processing is observed by light microscopy as apparent perivascular clearing of the neuropil (**A**). As seen by electron microscopy, the clearing is due to dilated astrocyte foot processes (**B**).

Oligodendroglia

The oligodendroglia are small cells that are active in the formation and maintenance of myelin and in the, as yet poorly understood capacity of attending to neuronal cell bodies, which has been termed satellitosis. The obligatory orientation of oligodendrocytes to fiber pathways in white matter is occasionally made apparent by a fortuitous plane of section wherein the fascicular distribution of these cells is seen (Fig. 9.38). Oligodendrocytes are encountered as two-to-three small, dark nuclei that are pressed against the cell membranes of larger neurons in gray matter (Fig. 9.39). These normal satellite oligodendroglia must be distinguished from infiltrating neoplastic cells in surgical specimens obtained from infiltrating gliomas. Both astrocytomas and oligodendrogliomas may also exhibit satellitosis, but it is most prominent in the latter neoplasm. In general, the nuclei of the neoplastic satellites are larger, more pleomorphic, and more coarsely constructed than the normal orbiting cortical oligodendrocytes.

Identification of normal oligodendroglia in both gray and white matters is greatly facilitated by the presence of a perinuclear halo that imparts a fried egg appearance. This appearance results from swelling and vacuolation of the cytoplasm and is analogous to the perivascular swelling and vacuolation of astrocytic foot processes (Fig. 9.40A). The perinuclear halo is a well-known, distinctive, and diagnostically useful feature of oligodendrogliomas (Fig. 9.40B). Oligodendroglia and their neoplastic counterparts exhibit strong immunopositivity for S100 protein (Fig. 9.32C) and the transcription factor Olig2 (8). However, Olig2 is not specific for oligodendroglia or oligodendroglioma. Both oligodendroglioma and astrocytoma diffusely express Olig2.

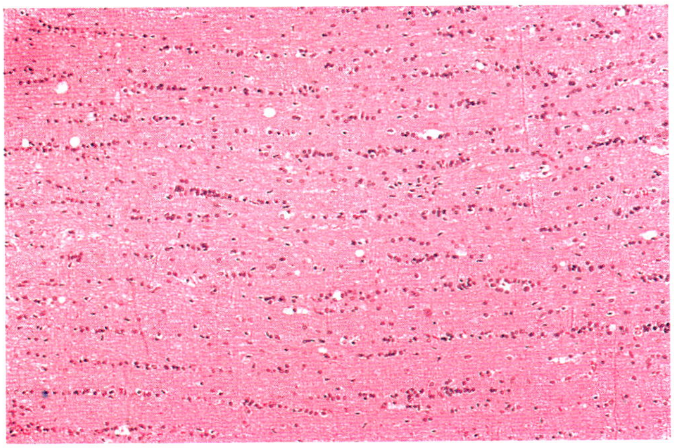

FIGURE 9.38 Oligodendroglia: As seen here in a white matter tract (the corpus callosum) cut in longitudinal section, these glia may be identified, even at low power, as rows of nuclei queuing up between fascicles of myelinated axons.

Ependyma

This cuboidal-to-columnar epithelium provides a lining for the CNS ventricular system (Fig. 9.41) and specializes focally as a covering for the choroid plexus (see Fig. 9.53). The ciliated nature of the ependyma is readily appreciable in the child but is generally less so thereafter. Tanycytes are specialized constituents of the ependyma with elongated abluminal processes that reach the subependymal vasculature. Thus, these cells provide a physical link between the ventricular, vascular, and intraparenchymal compartments of the CNS. Tanycytes are most numerous in the modified ependyma covering many of the circumventricular organs. These cells are best visualized through the use of the Golgi stain.

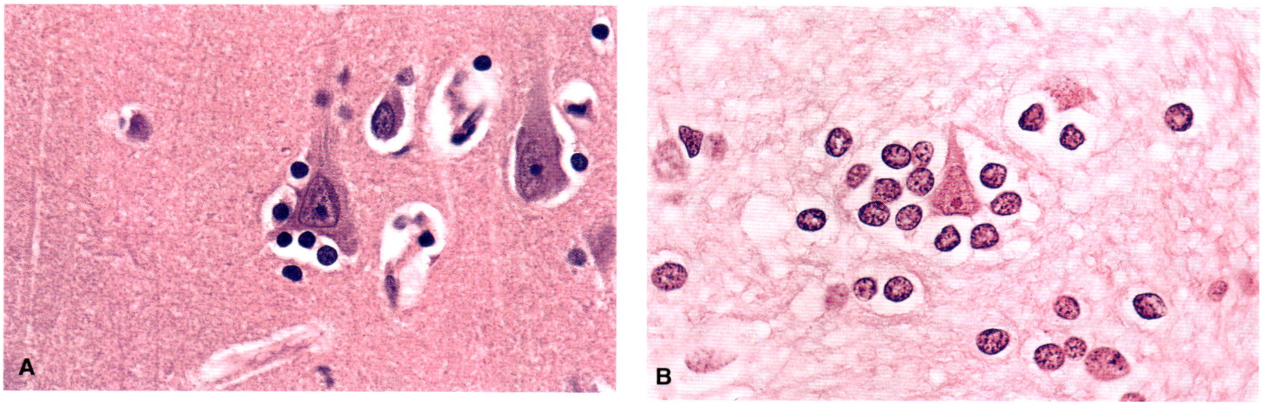

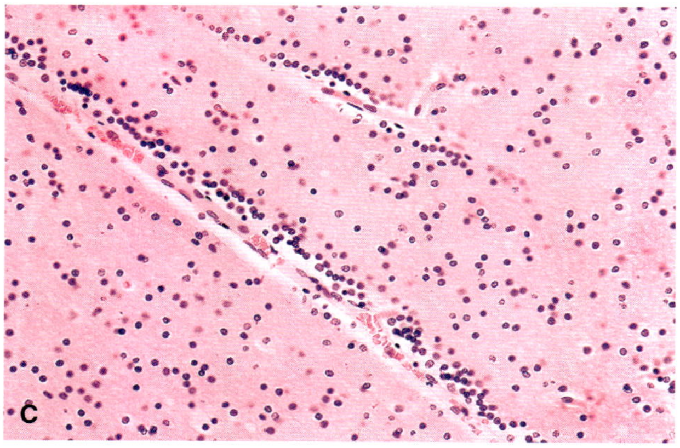

FIGURE 9.39 Perineuronal satellitosis: Normal perineuronal glia (**A**) consist primarily of oligodendroglial satellite cells, together with occasional astrocytes and microglia. This affinity of normal oligodendroglia for neuronal perikarya is often retained by their neoplastic counterparts, oligodendrogliomas, in the form of "neoplastic satellitosis" (**B**). Nonneoplastic oligodendroglial hyperplasia (**C**) can also be seen, as for example in some cases of longstanding epilepsy.

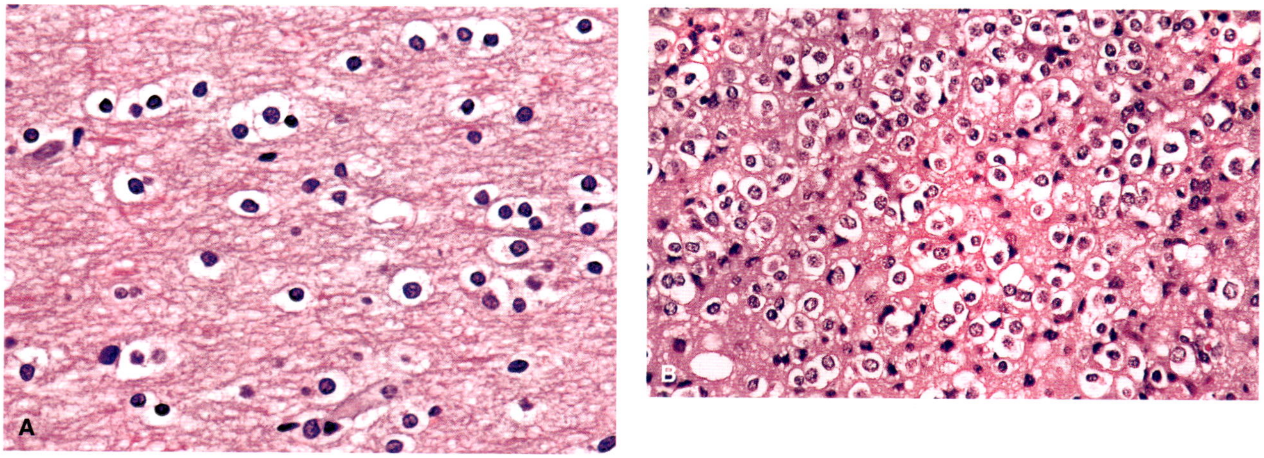

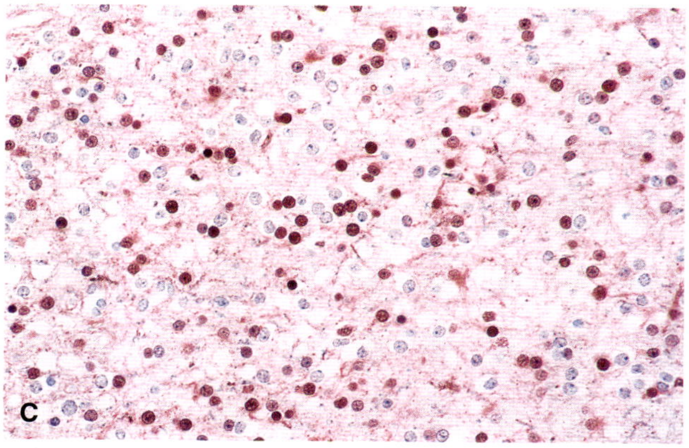

FIGURE 9.40 Oligodendroglia: In many specimens, oligodendroglia exhibit characteristic perinuclear halos (**A**). This "fried egg" appearance is an artifact of hypoxia/ischemia and delayed fixation, and is a useful diagnostic feature that is also exhibited by oligodendrogliomas (**B**). Oligodendroglia, both normal and neoplastic, typically show strong nuclear immunopositivity for S100 protein (**C**).

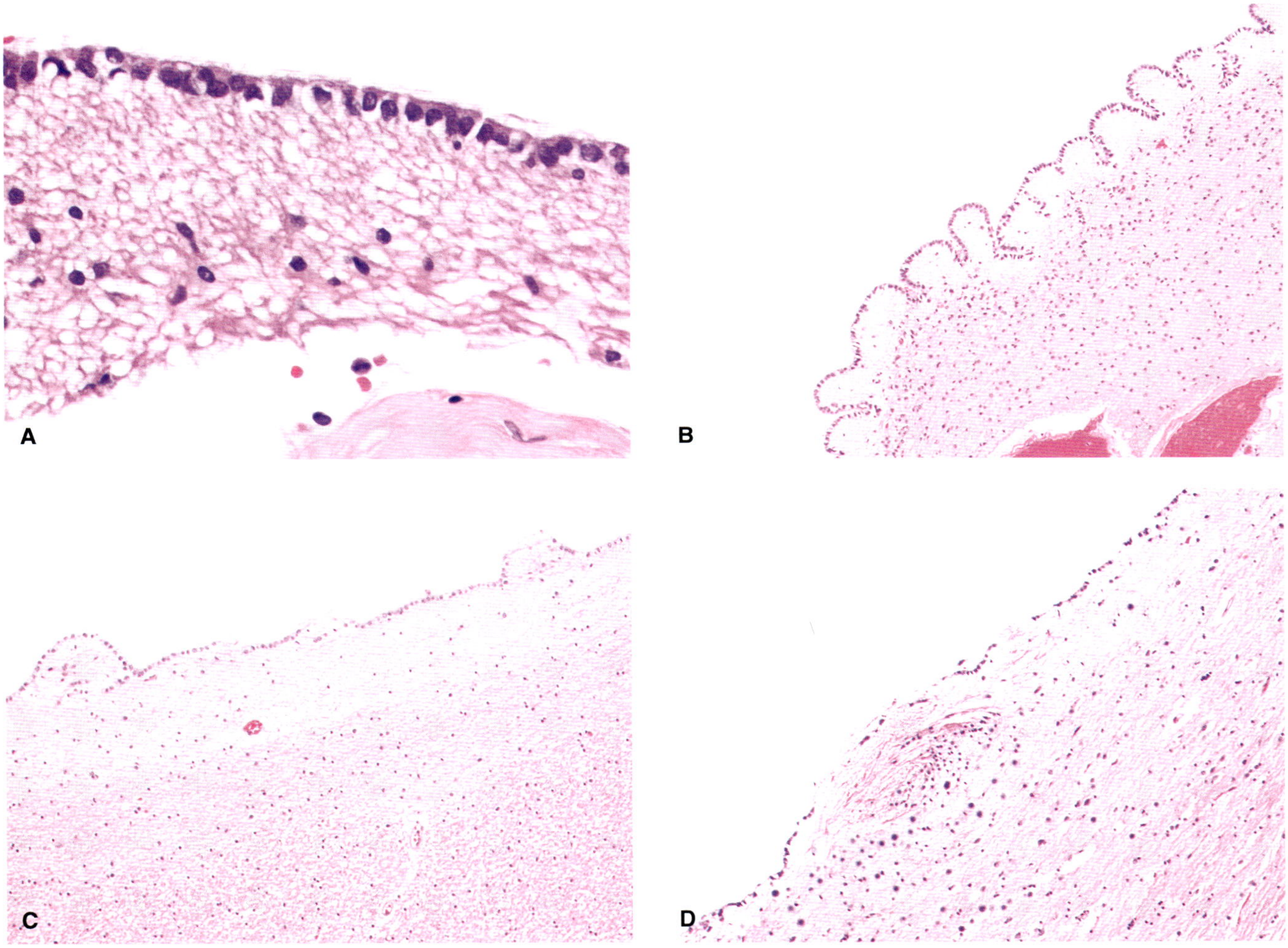

FIGURE 9.41 Ependyma and the subependymal plate: The lining of the ventricular system varies from a robust ciliated-columnar epithelium (**A**) to nearly squamous flattened cuboidal. The relative abundance of cilia and the height of the ependyma vary with anatomic location and both decrease with age. The hypocellular fibrillary zone located immediately subjacent to the ependyma is known as the subependymal plate and contains scattered glia as single cells and in small clusters. Glia of the subependymal plate respond to ependymal injury with a proliferative response termed granular ependymitis (**C,D**). Granular ependymitis, the combination of focal denudation of the ependyma, coupled with an exophytic fusiform proliferation of the subependymal glia constitutes granular ependymitis (**C,D**). Despite the implication of an inflammatory etiology inherent in the name, this common alteration can result from many diverse insults, ranging from hydrocephalus to viral infections. Normal undulations of the ependyma (**B**), termed *plicae*, should not be confused with granular ependymitis. Subependymomas originate from the glia of the ependyma and subependymal plate.

The closely apposed ependymal surfaces of the tips of the lateral ventricle horns frequently fuse during development, resulting in cords of ependymal cell nests and rosettes. This is especially typical of the distal portions of the posterior horns in the occipital lobes (Fig. 9.42). The white matter in such areas appears pale and can simulate the rarefaction seen after ischemic insult. Detached ependymal rosettes may be encountered subjacent to the ventricular lining at any location throughout the neuraxis (Fig. 9.42).

The ependyma-lined central canal of the spinal cord is patent in the child (Fig. 9.43) but generally becomes obliterated around the time of puberty, unless obstructive hydrocephalus is present. In the latter case, the canal may remain patent and even become dilated (hydromyelia). In the normal adult, however, the spinal ependymal cells have completed their role as a generative epithelium and remain as scattered clumps and rosettes (Fig. 9.43B). Occasional sections of adult spinal cord may exhibit a focally patent central canal.

The primary ependymal response to injury is loss. The resultant focal denudation of the ventricular wall is often accompanied by a proliferation of local cells, the subependymal glia. This nonspecific reaction, termed granular

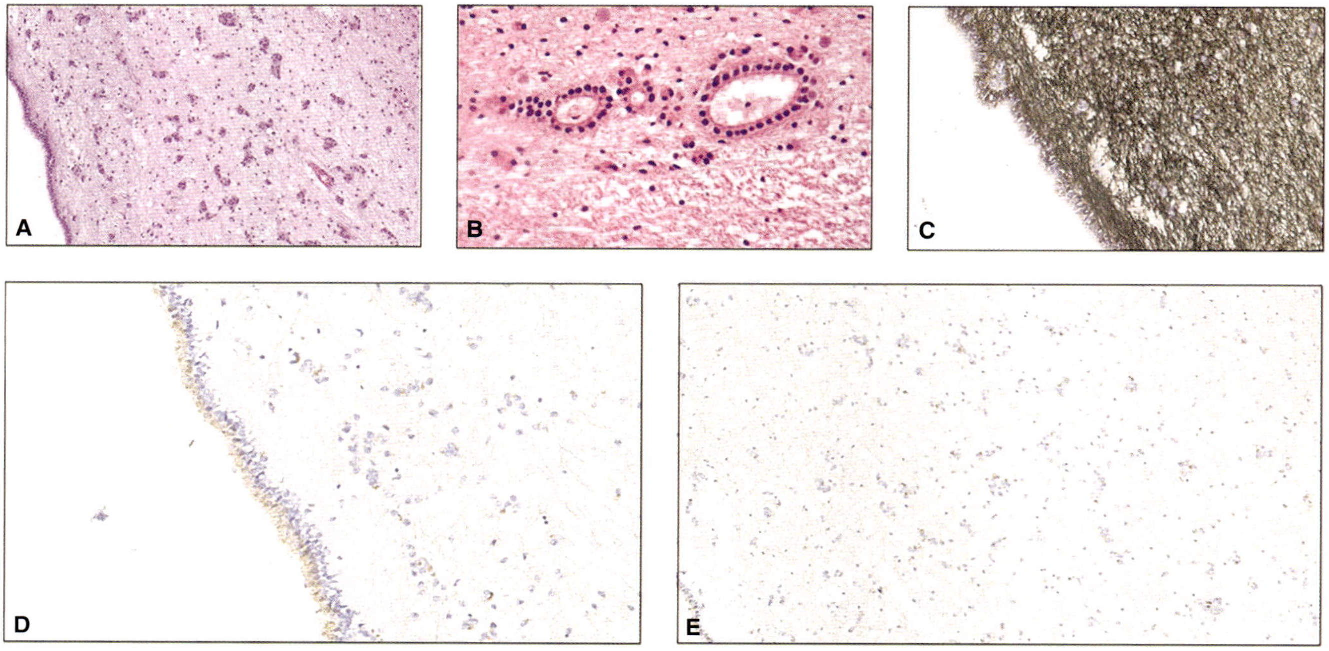

FIGURE 9.42 Ependymal rosettes: Clusters of ependymal rosettes may be found subjacent to the ependymal lining of the ventricular system throughout the neuraxis **(A,B)**. They are particularly common in areas where opposed ventricular surfaces fuse during development, such as at the tips of the lateral ventricle horns, especially the occipital horns (as illustrated here), and at the lateral angles of the fourth ventricle. These normal rosette clusters are occasionally sampled in surgical specimens and should not be misinterpreted as evidence of disease. The ependymal cells demonstrate basal staining for GFAP **(C)** with luminal staining for EMA **(D)**. EMA shows dot-like staining in the ependymal rosettes **(E)**.

ependymitis (Fig. 9.41C,D), is the potential product of a broad range of disparate etiologies, from viral infection to hydrocephalus. It is seen frequently at autopsy as a very focal, limited response and has, in this setting, little diagnostic significance. The normal ependyma is commonly thrown into folds, termed plicae, in many parts of the ventricular system (Fig. 9.41B). These normal undulations should not be confused with granular ependymitis.

Microglia and the Monocyte—Macrophage System

Normal Microscopic Anatomy

The small, dark, elongated nuclei of microglia are ubiquitous in the normal brain. They are so small and inconspicuous in H&E-stained sections, however, that they are rarely noticed (Fig. 9.44A). They must be distinguished

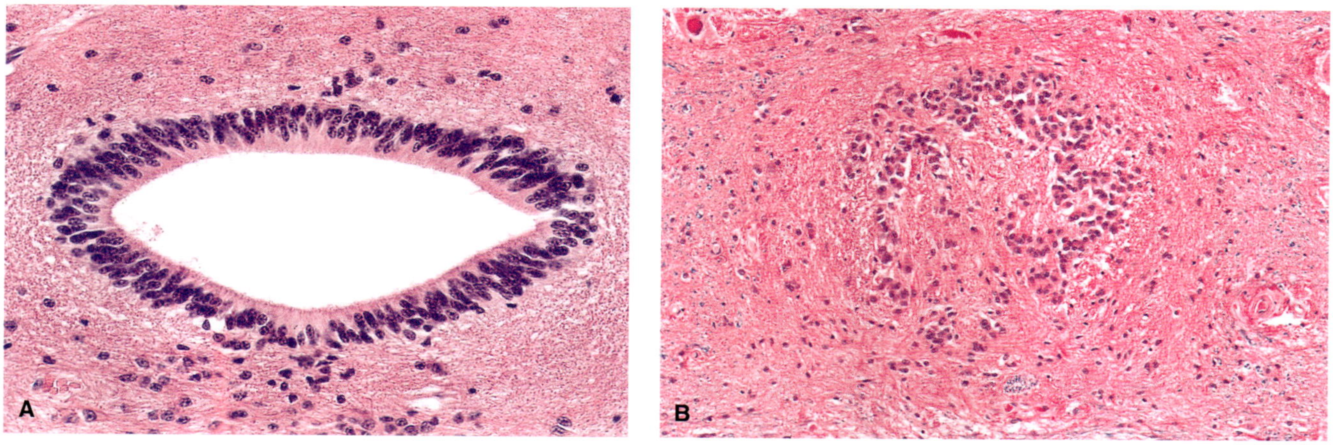

FIGURE 9.43 Central canal of the spinal cord: In the child (**A**), the central canal is widely patent and exhibits the ciliated columnar ependymal lining expected in a young individual. In contrast, the central canal of adults is typically obliterated over much of its length, with only residual small nests and occasional rosettes of ependymal cells (**B**).

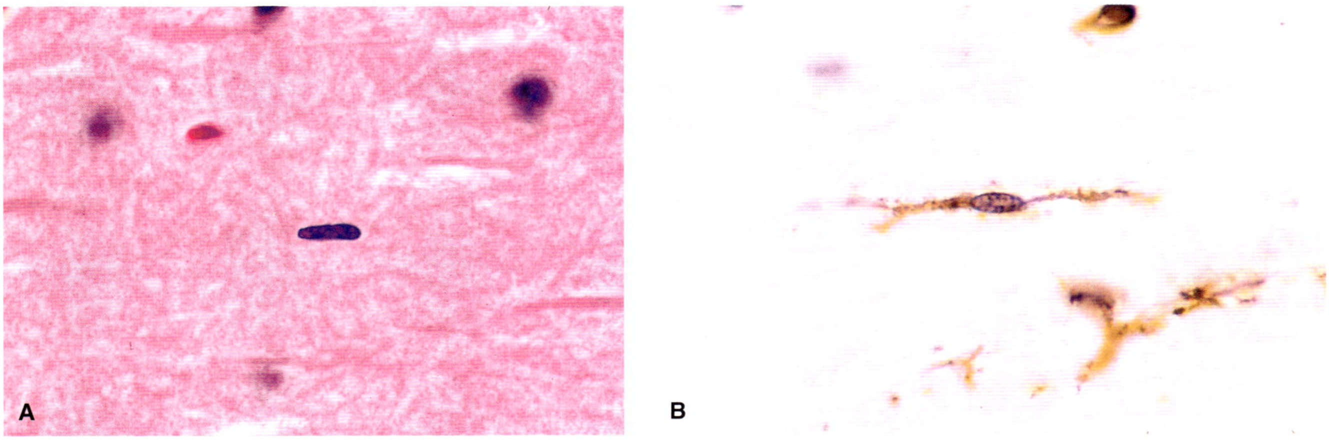

FIGURE 9.44 Microglia: These normally inconspicuous residents of the CNS parenchyma are identifiable by their classical rod-shaped nuclei on routine H&E staining (**A**). Dendritic processes, often bipolar, are vividly demonstrated by lectin staining (**B**).

from the commonly encountered tangential or en face sections of endothelial cells, which possess similarly elongated, albeit somewhat larger and plumper, nuclei. Special staining techniques such as the classic silver carbonate method, the more predictable lectin histochemistry (Fig. 9.44B), and immunohistochemical markers such as HAM-56, Iba1, and CD68 uncloak the dendritic processes of microglia and permit unambiguous visualization. The staining patterns depicted by these antibodies correspond to the localization of each respective antigen. For example, CD68 is a marker of lysosomes, and is inherently the least specific stain for microglia. Ramified microglia do not show significant CD68 staining because the lysosomes in this microglial subset are found in close proximity to the nucleus. In contrast, Iba1 is localized to the cytoplasm, including the thin protrusions of ramified microglia, making it a more specific marker and useful for studying activated microglia. Iba1 is expressed more strongly in ramified microglia within grey matter than in white matter, and some ramified microglia do not show expression of Iba1 at all (9). Additional study is needed to parse the specific differences in expression of these microglial markers. By these techniques, microglia are seen to be strikingly pervasive throughout the CNS parenchyma.

Response to Injury

In contrast to the relative passivity and anonymity of microglia in healthy nervous tissue, their activity is by no

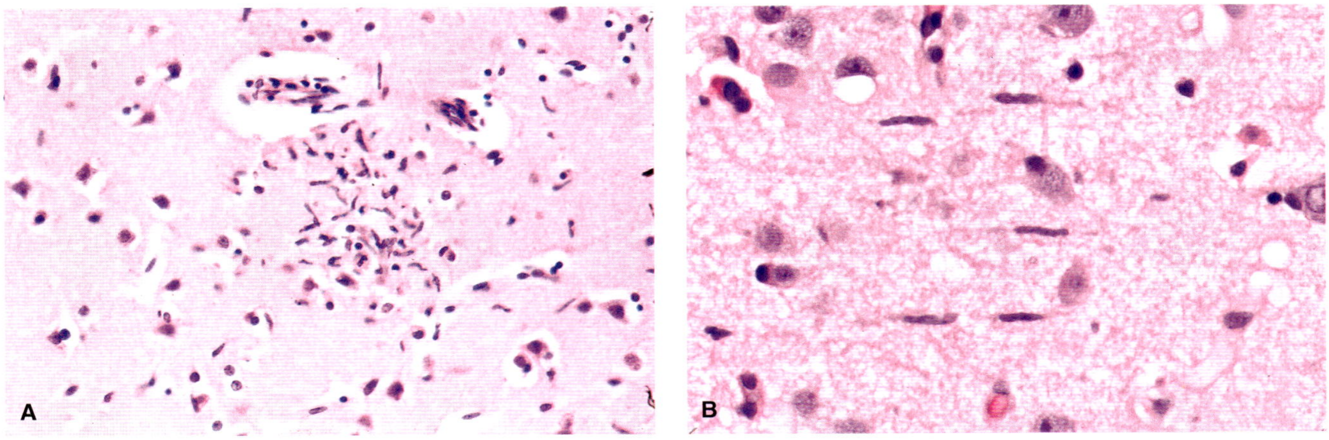

FIGURE 9.45 Microglia: Reactive microglia may assume several forms. Microglial nodules (**A**) are focal hypercellular collections of microglia together with reactive astrocytes that commonly form as a response to viral and rickettsial infections. In diffuse microgliosis (**B**), seen in a variety of conditions, including ischemia, the characteristic elongated rod-shaped nuclei can be identified on H&E-stained sections, but the true extent of their presence is more accurately visualized through immunohistochemistry, as seen here with HAM-56 (**C**). (*continued*)

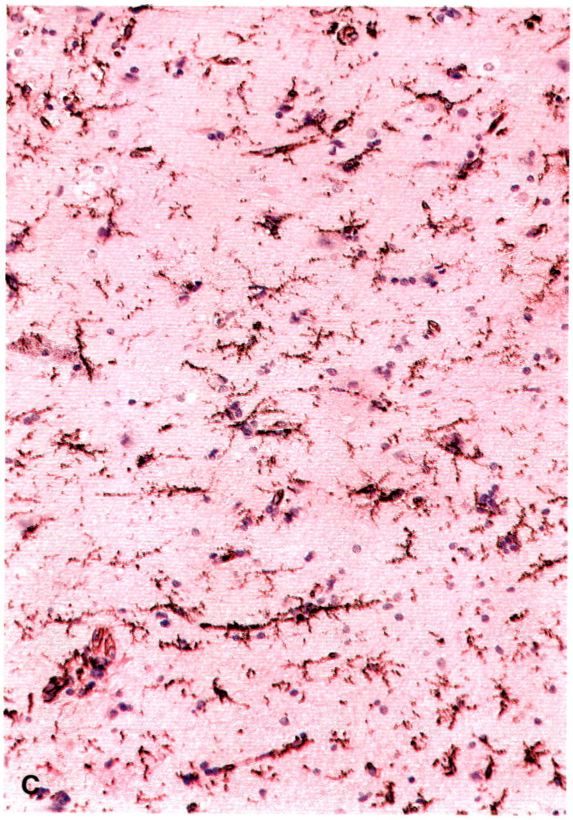

FIGURE 9.45 *(Continued)*

means subtle when called to action by parenchymal injury. Two variants are seen: microglial nodules and diffuse microgliosis. Microglial nodules, or microglial stars, are frequent concomitants of viral or rickettsial infection, and they consist of both astrocytes and microglia (Fig. 9.45A). The microglia have an elongated shape and are known as rod cells. A form of glial nodule is also seen around degenerating neurons, as in amyotrophic lateral sclerosis. Diffuse microgliosis is equally distinctive. In this context, the rod-shaped nuclei of microglia may be present in such numbers as to be easily identified (Fig. 9.45B), although the full extent of microgliosis is often best appreciated through immunohistochemistry for markers such as CD68, CD163, or HAM-56 (Fig. 9.45C) (10).

Destruction of nervous tissue, by whatever mechanism, generally elicits a macrophage response that serves to clear nonviable debris (Fig. 9.46A,B). Both the activation of autochthonous tissue microglia and the diapedesis of blood monocytes are sources for these scavengers. The weight of evidence suggests that recruitment of blood monocytes plays a predominant role in large lesions such as infarcts, but that the supply of indigenous cells is sufficient for lesser insults.

Macrophages are proliferative cells (Fig. 9.47). Mitotic figures will, therefore, usually be present in disease processes that elicit a macrophage response, such as infarction and demyelination. They should not be interpreted as suggestive of a neoplastic process. Macrophages often contribute substantially to the cellularity of tissue samples and, depending on preservation and fixation conditions, their identity may not always be obvious. For example, in some specimens, clearing of the macrophage cytoplasm lends an appearance similar to that of oligodendroglial cells to the extent that, together with the attendant hypercellularity, an infiltrating glioma might be suspected. In such instances a number of antibodies, such as KP-1, Ham-56, Iba1, or CD163, can be used to identify the macrophage component (Fig. 9.46C,D).

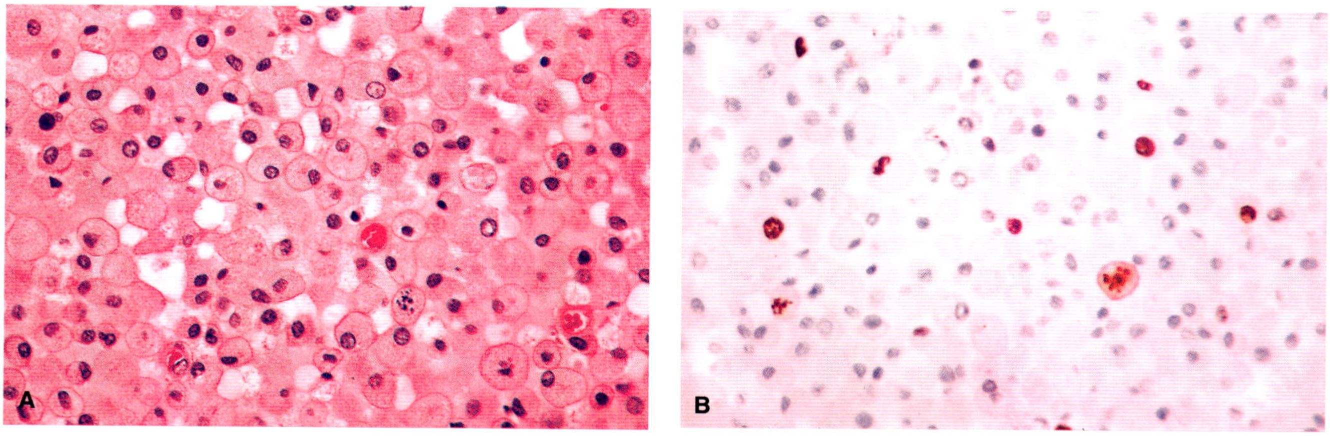

FIGURE 9.46 Macrophages: Discrete cell boundaries and vesicular cytoplasm serve to distinguish macrophages from other cellular constituents of the nervous system (**A**). Cognoscenti of the literature will be familiar with a number of colorful appellations given these cells in former times, including *Gitter cells (lattice cells)* and *compound granular corpuscles.* Macrophages are mitotically active cells and populations responding to CNS injury are readily labeled with proliferation markers such as the monoclonal antibody MIB-1 (**B**). Mitotic figures are, thus, to be expected in tissue samples from a wide range of nonneoplastic conditions that elicit a macrophage response, including infarcts and demyelinative diseases.

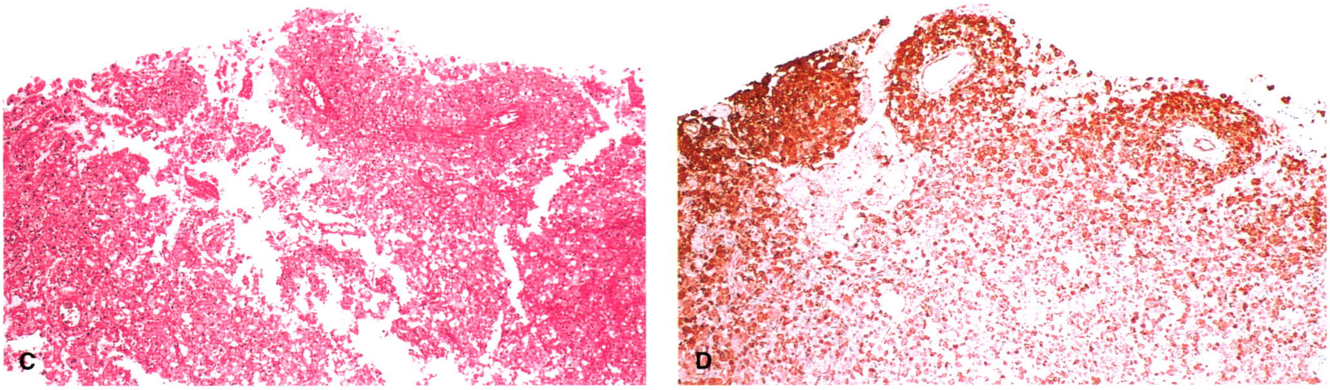

FIGURE 9.46 (*Continued*) In hypercellular biopsies (**C**), macrophages can be separated from other cellular constituents by a number of commercially available antibodies such as HAM-56, CD68, and CD163 (**D**).

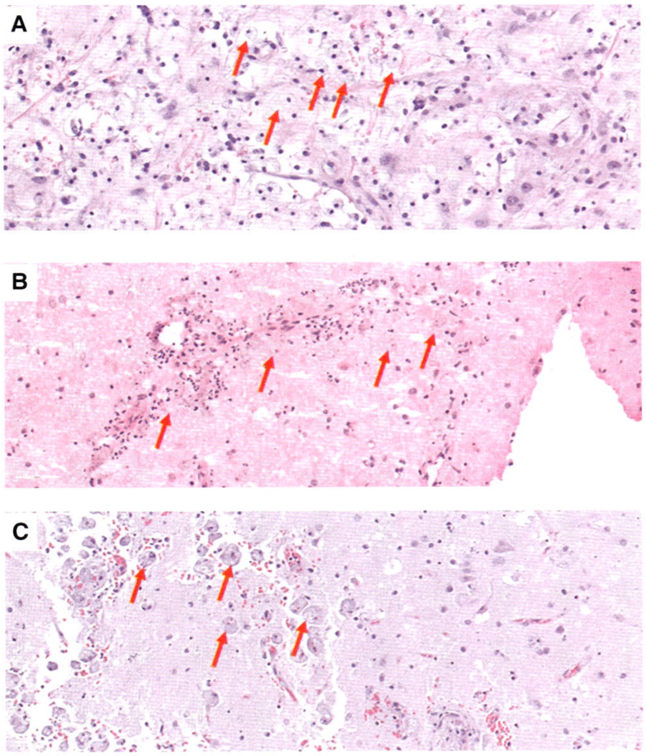

FIGURE 9.47 H&E-stained formalin-fixed paraffin-embedded (FFPE) tissue sections of brain tissue adjacent to an infarct, with numerous macrophages (*red arrows*) **(A)**. Notice that the macrophages are spherical, with pale or clear cytoplasm, a small round nucleus, and a small nucleus-to-cytoplasm ratio. **(B)** H&E-stained frozen section showing amoebae (*red arrows*) in a perivascular distribution in a patient with amoebic encephalitis **(C)**. H&E-stained FFPE sections of brain parenchyma showing the amoeba organisms. Note that, similar to macrophages, the amoebae have pale or clear cytoplasm, a small round nucleus, and a small nucleus-to-cytoplasm ratio. Some amoebae in **(C)** display intracytoplasmic red blood cells (*red arrows*), which can help in their identification.

SPECIALIZED ORGANS OF THE CENTRAL NERVOUS SYSTEM

Pineal Gland

The pineal body (epiphysis or conarium) presents a singular histologic appearance among CNS tissues with a prominently lobulated architecture (Figs. 9.48 and 9.49) (11). This glandular appearance might be mistaken for carcinoma by the unwary, and it can be difficult to distinguish the normal pineal gland from a well-differentiated pineocytoma in small surgical specimens.

Generally present in the pineal gland after puberty are corpora arenacea (i.e., acervuli cerebri or brain sand). These mineralized concretions accrue with age and confer the radiologic hyperdensity that, before the era of computerized tomography and magnetic resonance imaging (MRI), made the normal midline position of the pineal gland a useful radiologic landmark (Fig. 9.48). The increase in corpora arenacea with senescence is accompanied by gradual gliosis and cystic change, with attendant effacement of the lush glandular appearance of the pineal gland seen in the earlier decades of life. The ubiquitous incidental pineal cysts typically have densely gliotic walls with scattered Rosenthal fiber formation (Fig. 9.49C). The investing leptomeninges of the pineal gland contain arachnoid cell nests that occasionally give rise to meningiomas of the pineal region (Fig. 9.48). A rare, but documented finding, includes the presence of melanin pigmentation, likely originating from leptomeningeal melanocytes (Fig. 9.50). While generally an incidentally found curiosity, they may harbor clinical relevance as some posit these cells as a possible source of primary pineal melanoma. Pinealocytes express strong immunopositivity for the neuronal marker synaptophysin (Fig. 9.51A). This useful phenotypic marker is retained by most pineal parenchymal neoplasms. In addition to pineocytes, the pineal gland also contains an indigenous population of astrocytes whose distribution is revealed by immunostaining for GFAP (Fig. 9.51B).

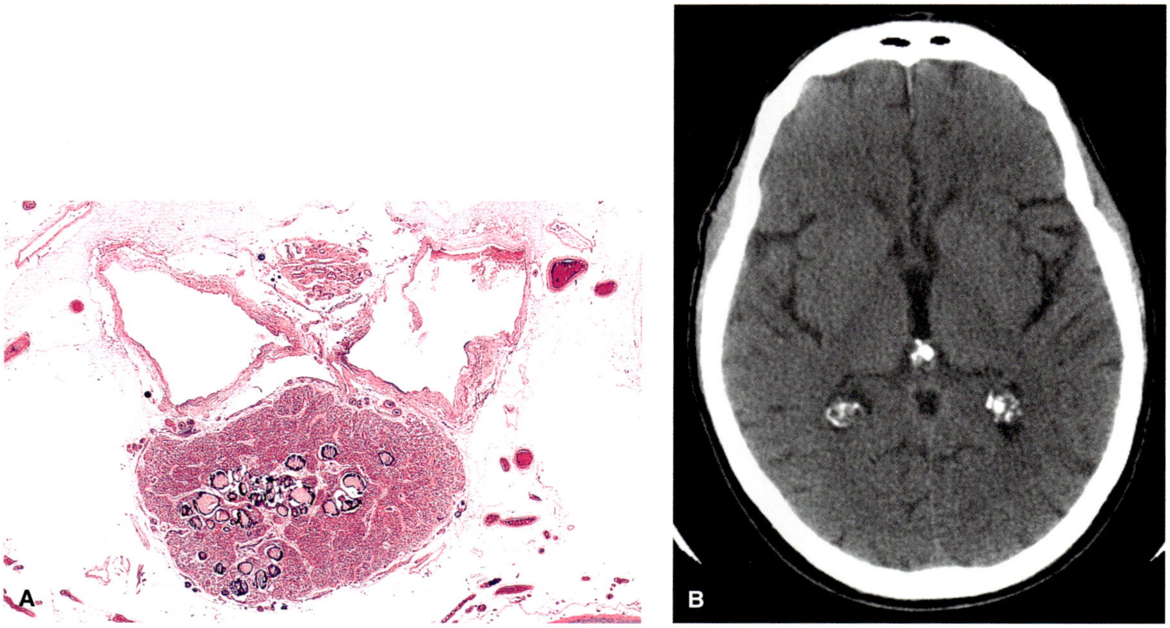

FIGURE 9.48 Pineal gland (epiphysis): Whole mount of a cross section of the pineal gland and its environs in situ (**A**) reveals the typical mineralized concretions variously referred to as corpora arenacea ("sand bodies"), acervuli cerebri ("little heaps"), or simply brain sand. Superior to the pineal gland are the paired internal cerebral veins, and between them is the suprapineal recess of the third ventricle which is lined with ependyma and often contains a tuft of choroid plexus. The loose connective tissue (redundant leptomeninges), in which all of these structures are located, is called velum interpositum. The calcification of the pineal gland increases with age and was, thereby, quite useful as a radiographic midline marker prior to the advent of contemporary high-resolution neuroimaging modalities (**B**). Also seen in this radiograph of a normal adult brain are prominently calcified tufts of choroid plexus (glomera choroidea; see Fig. 9.56) in the atria of both lateral ventricles.

Median Eminence and Infundibulum

The median eminence, infundibulum, and neurohypophysis display a unique constellation of morphologic features that reflect their specialized neuroendocrine functions (12). The background stroma is highly spindled (Fig. 9.52A) and contains nodular microvascular tangles termed gomitoli (Fig. 9.52B), spherical granular bodies called Herring bodies (Fig. 9.52C, Table 9.1), which are storage sites for oxytocin and vasopressin, and scattered cells bearing lipofuscin-like brown pigment (Fig. 9.52D). The constellation of features comprising a highly spindled background, vascular tangles, and granular bodies gives this region of the CNS more than a passing resemblance to pilocytic astrocytoma. An additional incidental finding, particularly in tissue sections of the infundibulum, is the presence of small clusters of granular cells, termed granular cell

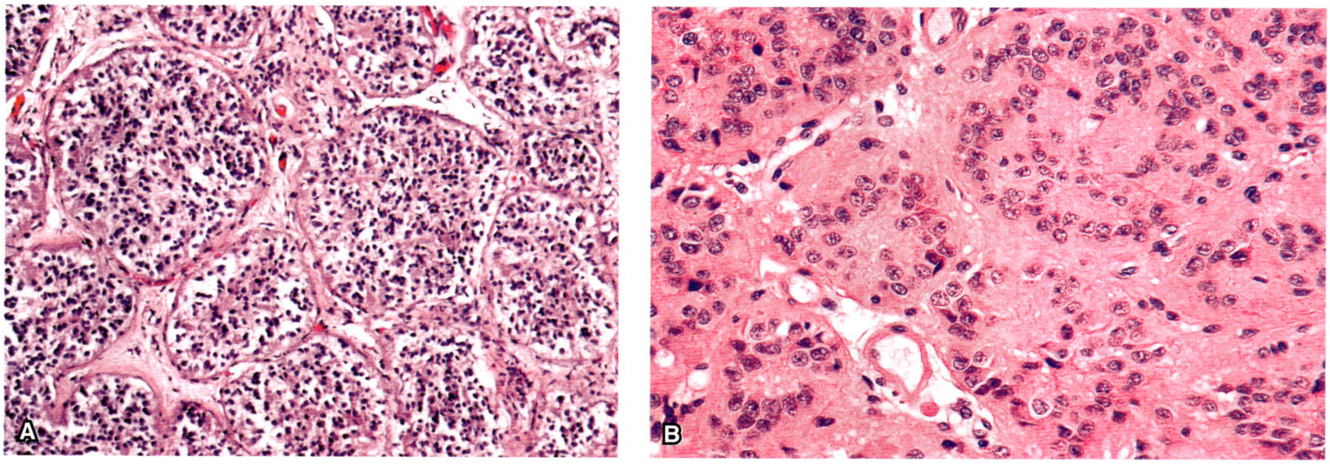

FIGURE 9.49 Pineal histology: The pineal gland has a richly glandular architecture that is unlike any other region of the CNS. Salient features include a prominent lobular organization with connective tissue septa (**A**) and pineocytic rosettes (**B**).

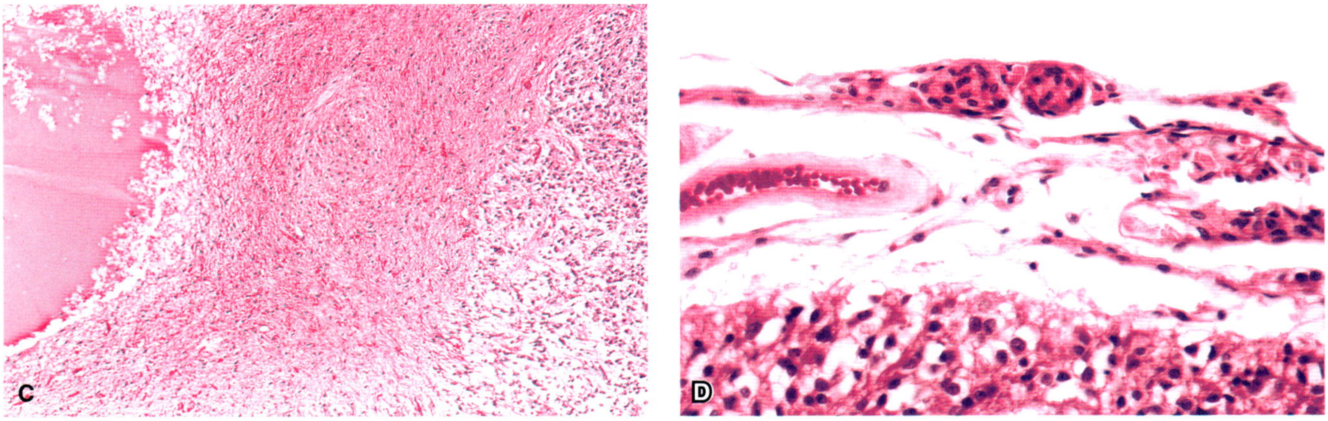

FIGURE 9.49 (*Continued*) The latter impart a distinctly neuroendocrine character. Two additional histologic features of note are the ubiquitous incidental pineal cysts, whose walls (**C**) typically exhibit astrogliosis with scattered Rosenthal fibers, and arachnoid cell nests of the investing velum interpositum (**D**) that occasionally provide a source for meningiomas arising in this region.

tumorlets (Fig. 9.52E). The cells from the neurohypophysis express the thyroid transcription factor 1 (TTF1) (13). This protein is not expressed by cells of the adenohypophysis, therefore, this is a useful tool for identifying tumors that originate from the neurohypophysis; such as pituicytoma, spindle cell oncocytoma, and granular cell tumor. Above (i.e., superior to) the pituitary stalk and infundibulum is the median eminence, which separates these structures from the lumen of the third ventricle (Fig. 9.53). Here you can find the hypothalamic nuclei (ventromedial and lateral nuclei highlighted in Fig. 9.53A–D). TTF1-positive cells can also be encountered in this region including within the ependymal lining (Fig. 9.53E). Ependymal cells in other locations do not express TTF1 (Fig. 9.53F).

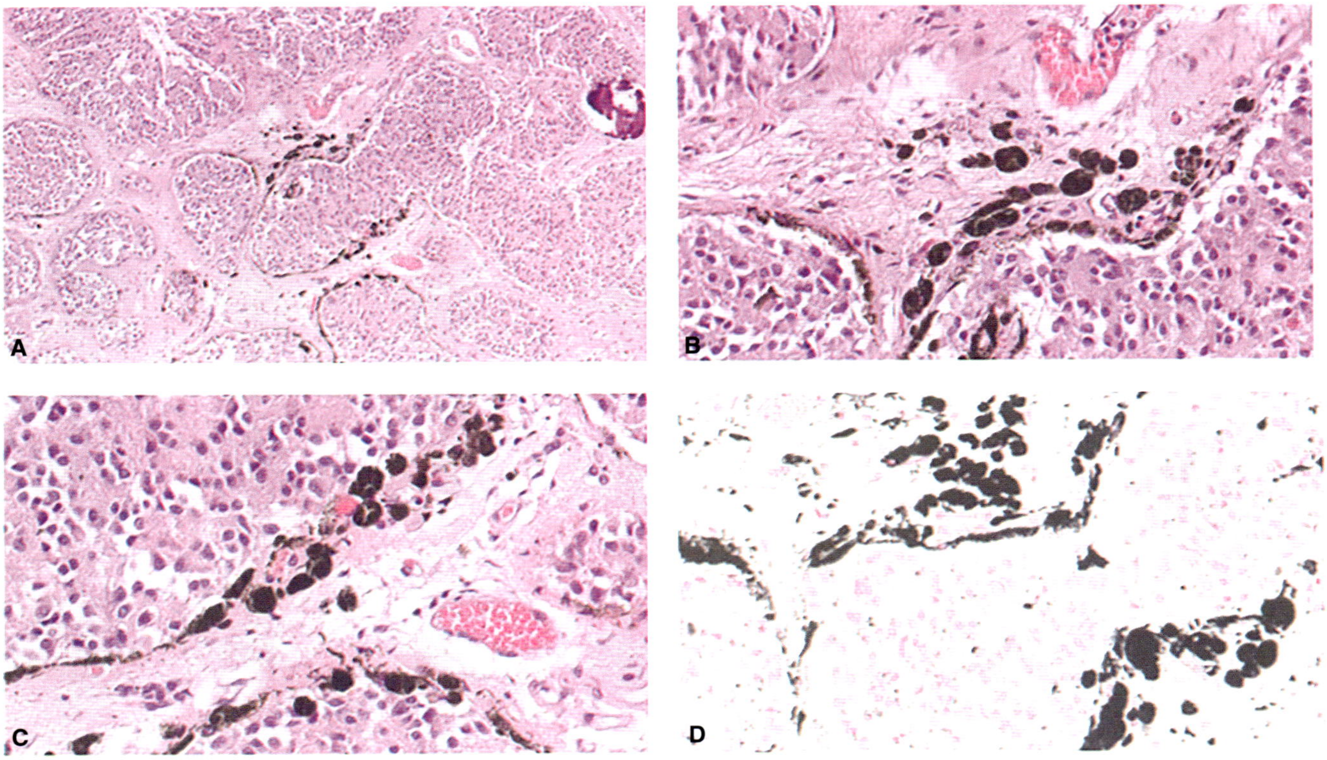

FIGURE 9.50 Incidental finding within the pineal gland demonstrating the presence of melanin, with a likely embryologic origin from leptomeningeal melanocytes. Such cells could be the possible origin of primary pineal melanoma. **A**: Pigmentation with thin linear profiles at the periphery of the pineal lobules. **B,C**: High power images of the large globular melanocytic perikarya. **D**: Fontana stain confirming the identity of melanin.

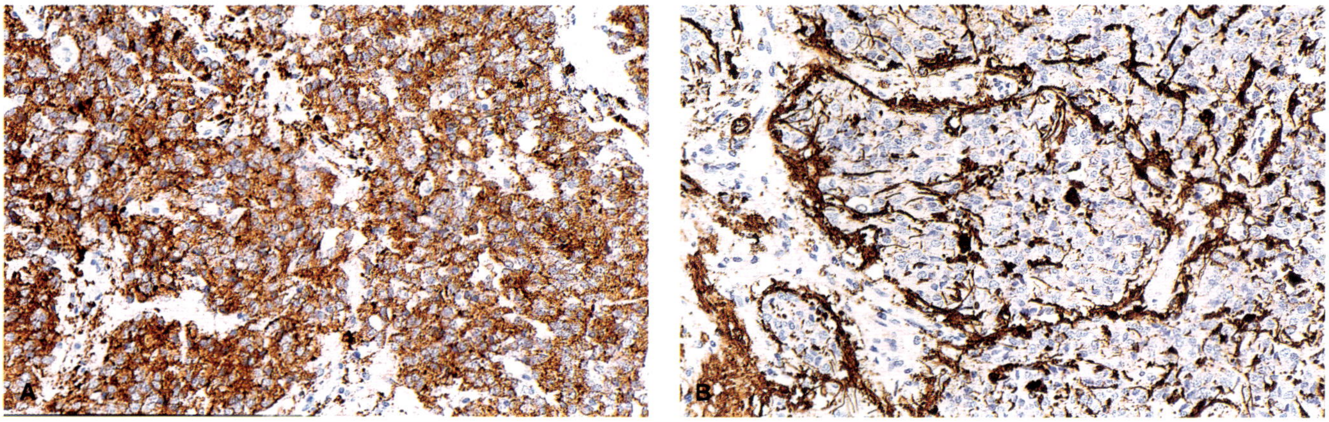

FIGURE 9.51 Pineal immunohistochemistry: Pineocytes (and their neoplastic progeny, the pineal parenchymal tumors) are strongly immunopositive for the neuronal marker synaptophysin (**A**). As expected, the indigenous populations of pineal astrocytes are well visualized with antibodies directed against GFAP (**B**).

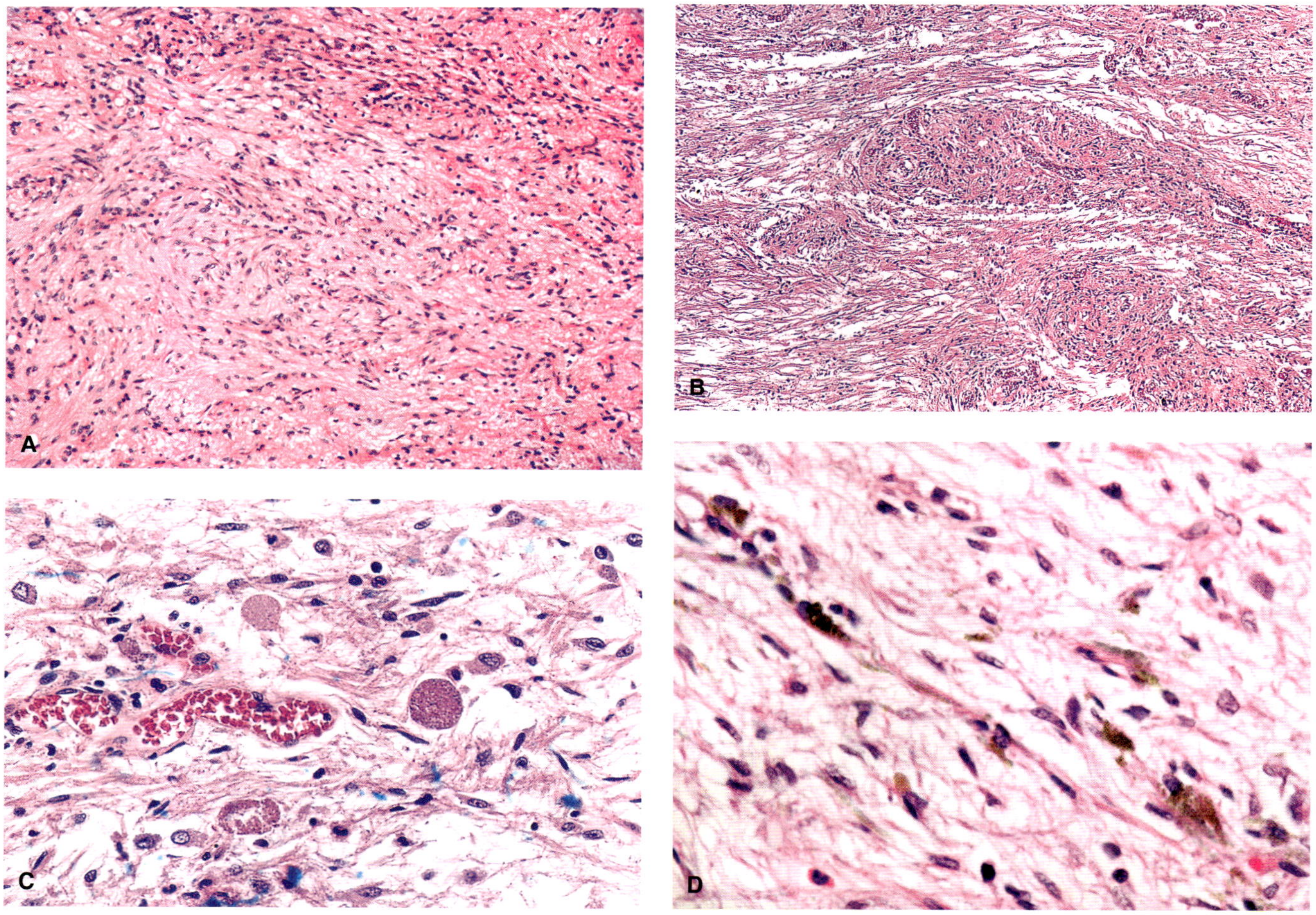

FIGURE 9.52 Median eminence, infundibulum, and neurohypophysis: This unique region of the CNS exhibits three distinguishing histologic features: A highly spindled stroma composed of pituicytes (**A**), prominent capillary tangles of the hypothalamo-hypophyseal portal system called gomitoli (**B**), and spherical, eosinophilic axonal specializations for the storage of oxytocin and vasopressin called Herring bodies (**C**). This constellation of highly spindled stroma, vascular tangles, and granular bodies bears a resemblance to pilocytic astrocytoma. The presence of lipofuscin-like pigment is also very characteristic (**D**).

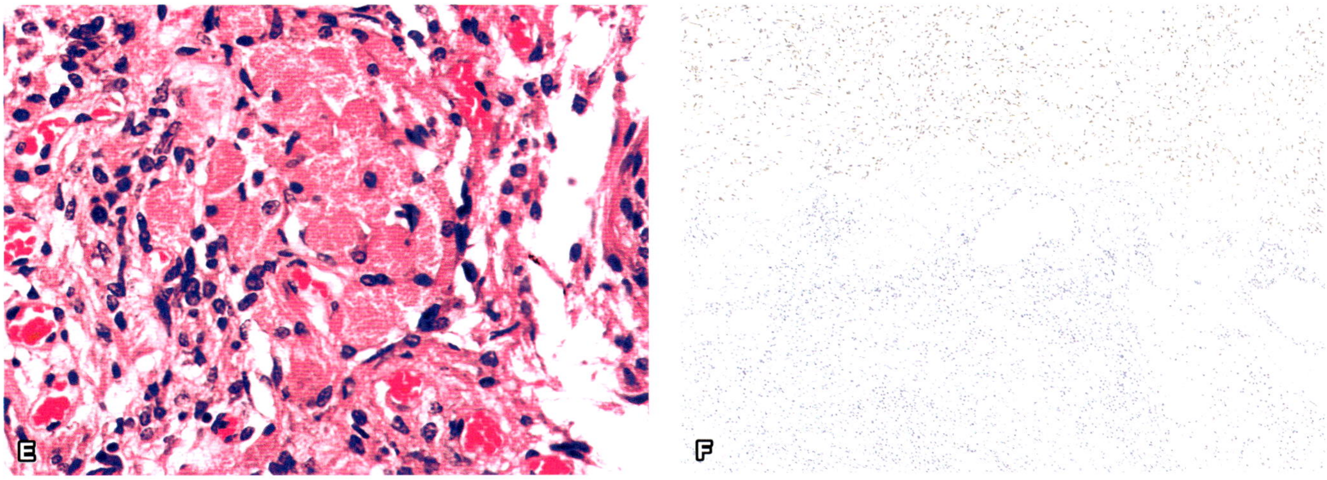

FIGURE 9.52 (*Continued*) Small clusters of granular cells ("granular cell tumorlet") may be seen, particularly in the infundibulum (**E**), as an incidental finding, but occasionally reach a sufficient size to produce compression of the infundibulum and subsequent clinical presentation with mildly elevated serum prolactin ("stalk effect"). Cells from the neurohypophysis express thyroid transcription factor 1 (TTF1), a useful marker for the identification of neoplasms that originate from this site (**F**). In contrast, cells from the adenohypophysis do not express TTF1.

Olfactory Bulbs and Tracts

The intracranial components of the olfactory apparatus include the olfactory bulbs and tracts, which have a very distinctive histologic appearance (14). Familiarity with these structures is useful, not only for the neuropathologist, but also for the general surgical pathologist who may encounter them in resections performed as part of the surgical treatment for regionally invasive entities of the nasal and paranasal sinuses. In such situations, the surgical pathologist may be called upon to render an intraoperative frozen-section assessment of tissue resected superior to the cribriform plate. The ability to recognize the normal histologic features of olfactory bulb tissue is, thus, of more than pedantic importance. The olfactory bulb has a laminar

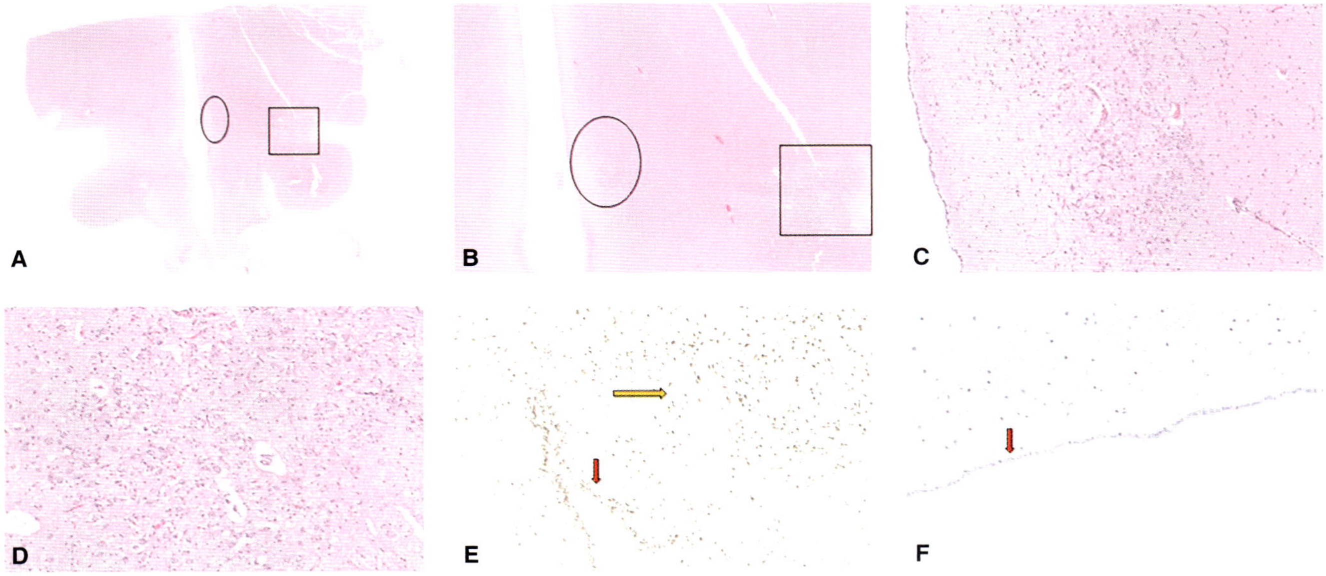

FIGURE 9.53 **A:** Third ventricle with median eminence/portion of neurohypophysis and hypothalamic nuclei, including the ventromedial nuclei (*circle*) and lateral nuclei (*square*). **B:** Higher-power view highlighting the ventromedial nuclei (*circle*) and lateral nuclei (*square*). **C:** High-power view of ventromedial nuclei of hypothalamus. **D:** High-power view of lateral nuclei of hypothalamus. **E:** TTF1 immunostain highlighting cells of the median eminence/superior third ventricle ependymal cells (*red arrow* highlights ependymal cells; *yellow arrow* highlights associated TTF1-positive cells of the infundibulum). **F:** In contrast, lateral ventricle ependymal cells are negative for TTF1 expression (*red arrow*).

TABLE 9.1 Granular Bodies in the CNS: Three Etiologic Classes

Normal
Herring bodies of the neurohypophysis
Reactive
Granular bodies adjacent to vascular malformations and conditions with iron deposition
Neoplastic
Pilocytic astrocytoma
Pleomorphic xanthoastrocytoma
Ganglioglioma

organization (Fig. 9.54). The outer layer is composed of spindled bundles of entering olfactory nerve fascicles intermixed with distinctive spherical, anuclear areas (termed "glomeruli") that constitute specialized zones of synaptic contact between olfactory nerve collaterals and the dendrites of intrinsic olfactory bulb neurons. Mitral cells are large neurons, so named for a resemblance of the perikaryon shape to a bishop's mitre. Their cell bodies are located in a lamina deep to that of the glomeruli (Fig. 9.54). The deepest layer consists of a thick lamina of granular cell neurons that are comparable in size to those of the cerebellum and dentate gyrus. The olfactory tracts extend posteriorly from the olfactory bulbs. They are triangular in cross section and, in adults, are notable for their profuse numbers of corpora amylacea (Fig. 9.55).

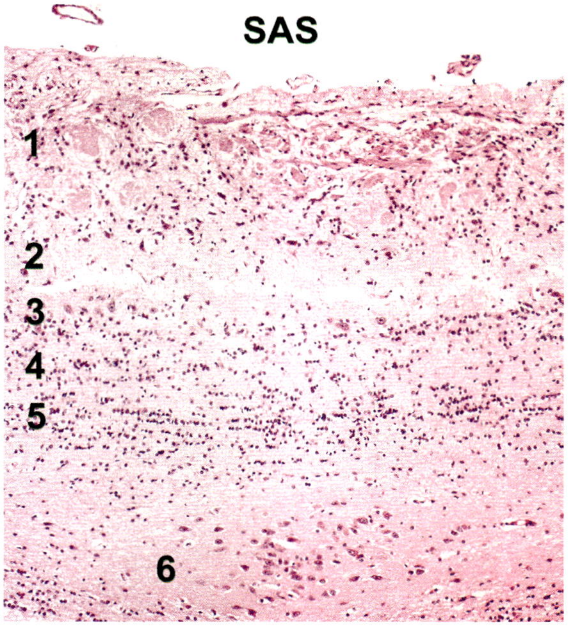

FIGURE 9.54 Olfactory bulb: The olfactory bulbs (see Fig. 9.3) have a very distinctive laminar organization. The most superficial layer, termed the glomerular layer (*1*) is covered by the leptomeninges (pia-arachnoid) and the subarachnoid space (*SAS*). The glomerular layer displays a unique architecture, with spindled olfactory nerve fascicles intermixing with spherical hypocellular synaptic zones called glomeruli (*1*). Deeper layers include the external plexiform (*2*), mitral cell (*3*), internal plexiform (*4*), and granular cell layer (*5*). Deeper still is the anterior olfactory nucleus (*6*). A working familiarity with olfactory bulb histology is essential for the surgical pathologist as this structure is frequently seen in the frozen-section laboratory during resection of superior nasal cavity tumors that may invade the cribriform plate and overlying olfactory bulbs.

Choroid Plexus

The choroid plexus is a specialized organ of the CNS that is responsible for the production of cerebrospinal fluid (15). It is found in the body, atrium, and temporal horns of the lateral ventricles, in the interventricular foramina of Monro, in the roof of the third ventricle, and in the roof and lateral recesses of the fourth ventricle. The frontal and occipital horns of the lateral ventricles and the aqueduct of Sylvius are devoid of choroid plexus. The plexus is most obvious in the atria of the lateral ventricles (Fig. 9.56), where prominent bilateral tufts, the glomera choroidea, are formed. Cystic xanthomatous change is a common incidental finding

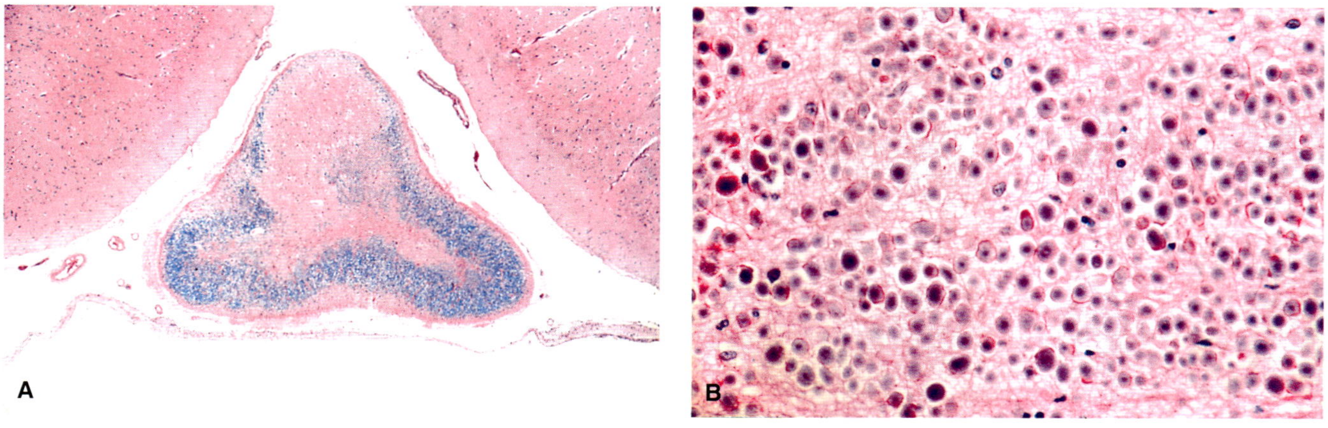

FIGURE 9.55 Olfactory tracts: The olfactory tracts contain myelinated fiber bundles and are roughly triangular in cross section (**A**). A distinctive feature of the tracts frequently observed in adults is their remarkable content of corpora amylacea (**B**).

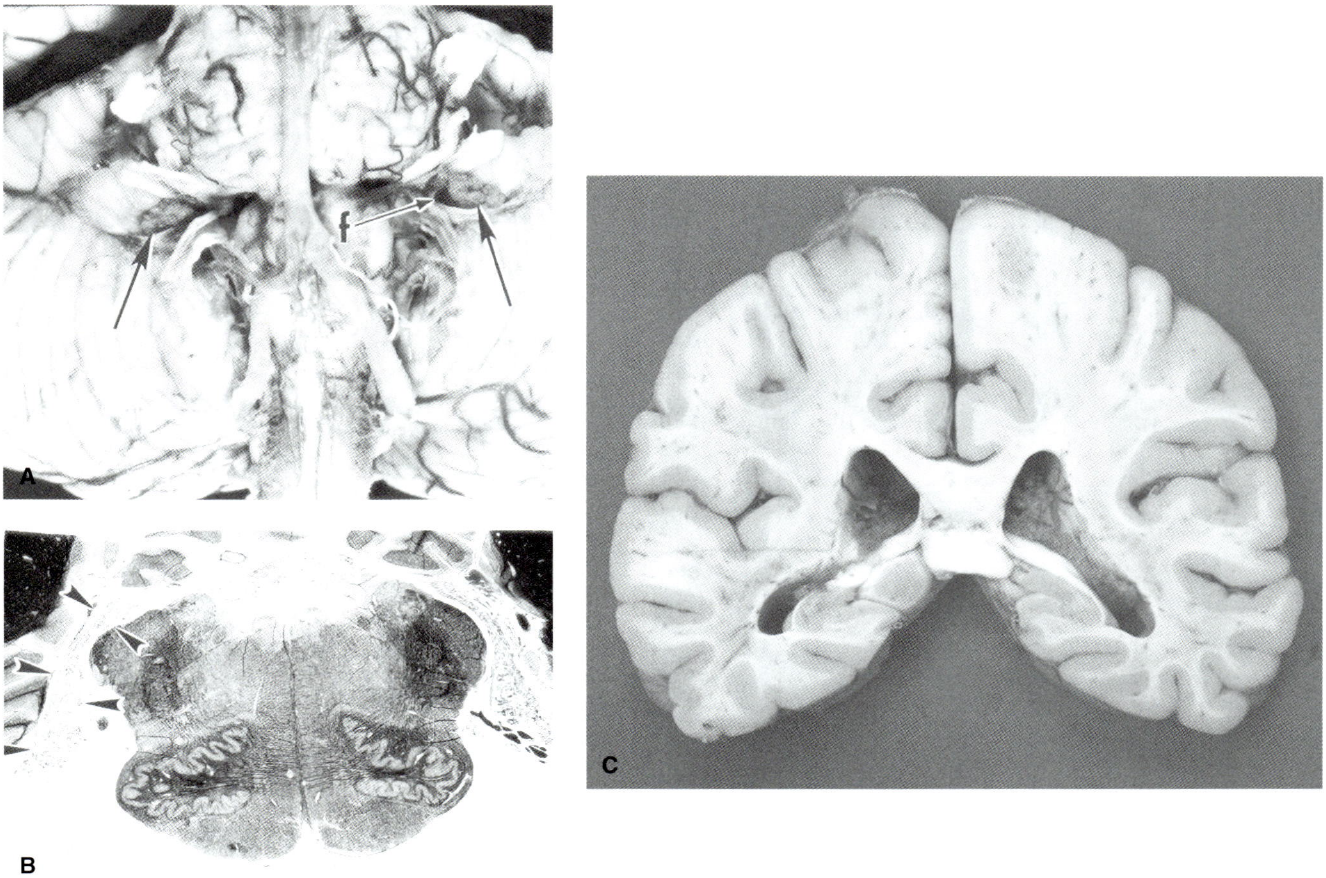

B

FIGURE 9.56 Choroid plexus: Choroid plexus produces the cerebrospinal fluid and is found in the lateral ventricles, foramen of Monro, roof of the third ventricle, and fourth ventricle. **A**: Small tufts of choroid plexus are normally visible on the basal surface of the brain stem in the cerebellopontine angle (*arrows*), and indicate the location of the lateral foramina of Luschka (*f*) from which they protrude. The dusty discoloration of the inferior medulla is due to the presence of leptomeningeal melanocytes (see Fig. 9.67). **B**: The ependyma-lined sleeve of the lateral recess of the fourth ventricle (*arrowheads*), together with the protruding tuft of choroid plexus, is referred to in the older literature as the "flower basket of Bochdalek" or "cornucopia." Pieces of the ependymal cuff are often seen adherent to the lateral aspect of the medulla in autopsy brain stem sections and should be recognized as a normal finding. The largest tufts of choroid plexus are called the glomera choroidea and are located in the atria of the lateral ventricles (**C**).

in these botryoid structures. The plexus is also a normal resident in the subarachnoid space of the cerebellopontine angle cisterns, which it reaches from the lateral recesses of the fourth ventricle by protruding through the foramina of Luschka (Fig. 9.56). The paired foramina of Luschka open laterally into the ventral basilar cerebellopontine angle cisterns and are to be distinguished from the single foramen of Magendie that opens in the dorsal midline into the cisterna magna.

Choroid plexus consists of invaginated fronds of vascular leptomeninges covered by an ependyma that is modified to become a highly secretory epithelium (Fig. 9.57). The cells are larger and more cobblestoned than those of the adjacent ependyma (Fig. 9.41). In addition to collagen and blood vessels, small nests of meningothelial cells are common normal habitués of the choroid plexus; whorls of these cells frequently give rise to psammoma bodies (Fig. 9.57). These cells also explain the occurrence of intraventricular meningiomas. Nonspecific deposition of mineral salts also occurs commonly throughout the connective tissue core with increasing age and accounts for most of the plexuses' radiodensity. An additional aging change is cytoplasmic vacuolization of the ependyma-derived lining cells.

Circumventricular Organs

The circumventricular organs comprise a diverse group of specialized CNS centers that share two anatomical features: a periventricular location and vasculature that lacks the characteristic blood–brain barrier properties found

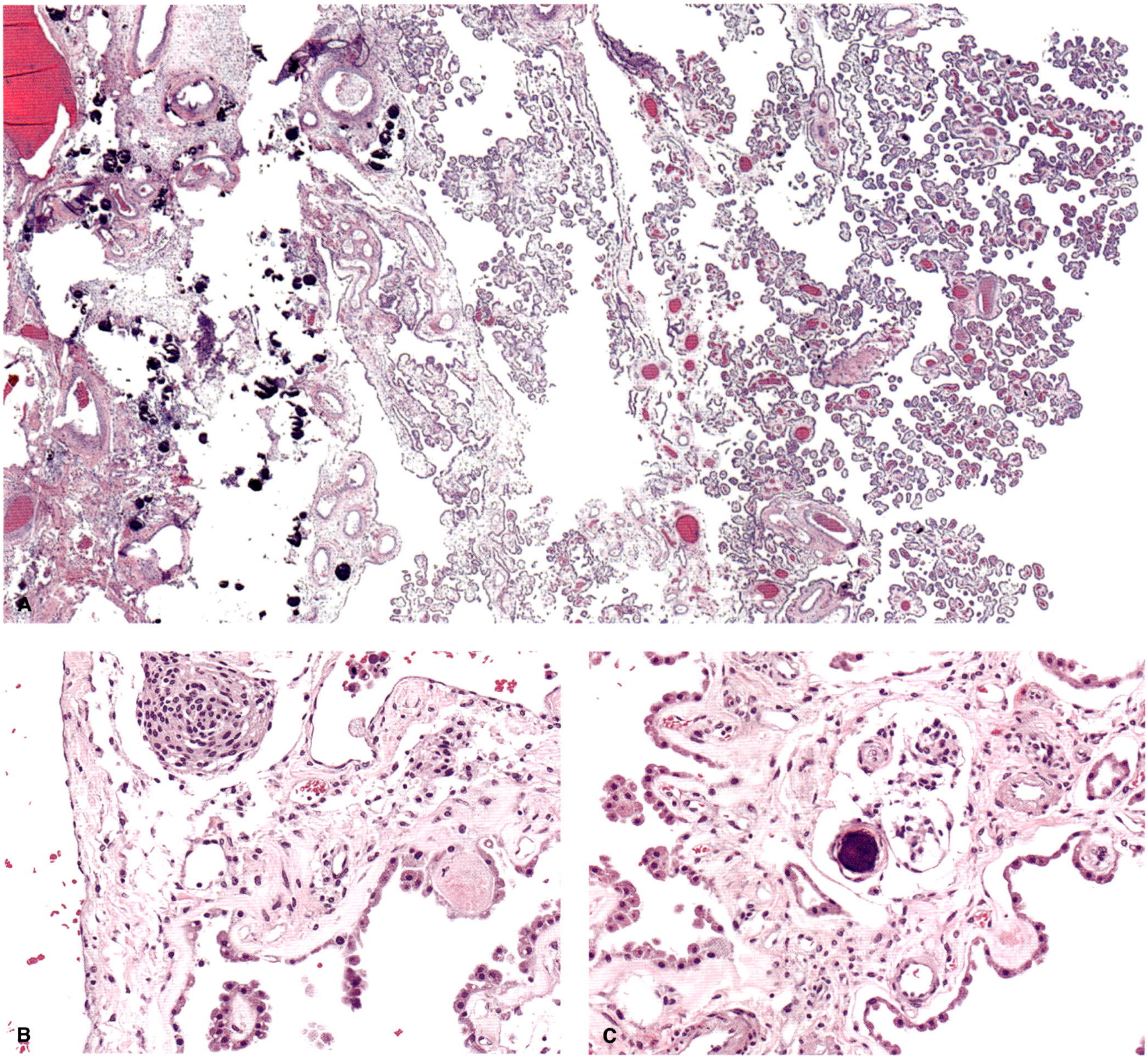

FIGURE 9.57 Choroid plexus: Histologically, choroid plexus is seen to be covered by simple cuboidal epithelium (modified ependyma) (**A**). In adults, each choroid epithelial cell bears a single prominent paranuclear cytoplasmic vacuole. An additional aging change seen in normal choroid plexus is calcification, which occurs in two forms: Nonspecific deposition of calcium salts in the collagenous stroma (**A**, left side of image) and as psammoma bodies (**C**). The latter arise from meningothelial cell nests (**B**) that are normally present in the choroid plexus as a result of the stroma's embryologic derivation from the pia-arachnoid meninges.

throughout the rest of the brain and the spinal cord (16). There are six circumventricular organs: the pineal gland, subfornical organ, organum vasculosum of the lamina terminalis, area postrema, subcommissural organ, and the median eminence–infundibulum–neurohypophysis (Figs. 9.58 and 9.59). Of these, all except the subcommissural organ are fully developed in the adult human (Fig. 9.59). The subcommissural organ, which is located on the ventral surface of the posterior commissure just caudal to the pineal gland, is a very prominent circumventricular organ in most vertebrates (Fig. 9.59B), but it regresses near the end of gestation in humans such that only vestigial remnants may be present (Fig. 9.59C). It is thought that papillary tumors of the pineal region may originate from these remnants, which may explain the keratin expression seen in such neoplasms.

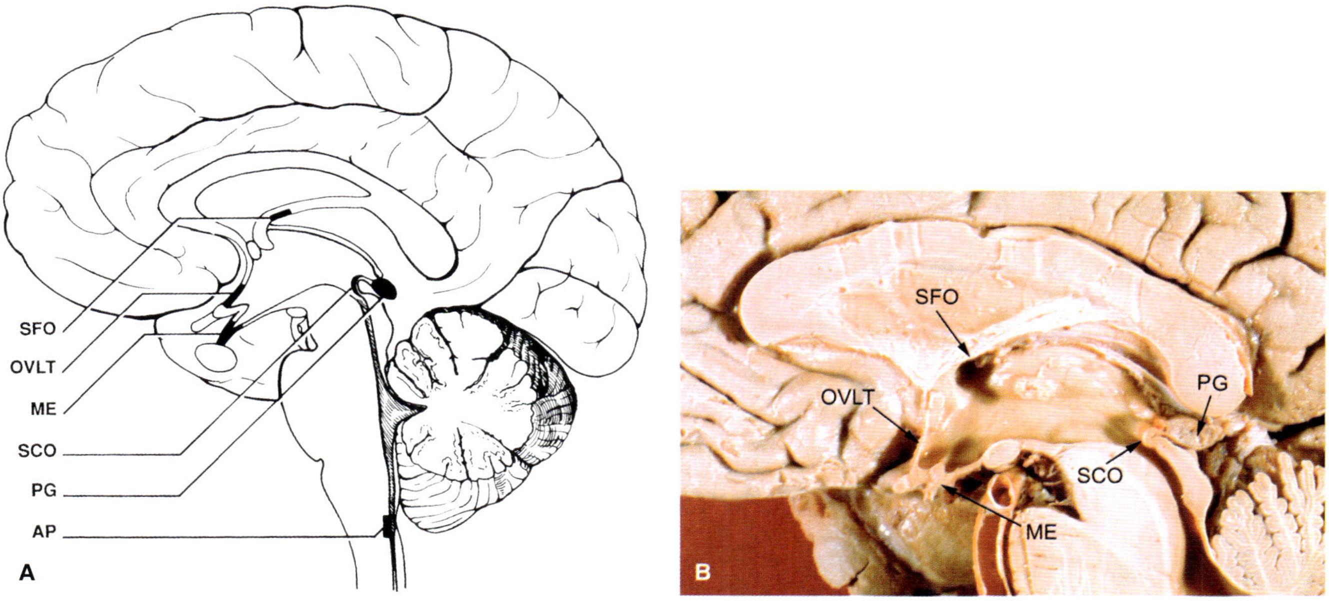

FIGURE 9.58 Circumventricular organs: As illustrated in the diagram (**A**) and mid-sagittally sectioned autopsy brain (**B**), the circumventricular organs share a midline or paramidline position, proximity to the ventricular system, and lack of the usual blood–brain barrier. The subcommissural organ is present in the developing fetus but is vestigial in the adult. *SFO*, subfornical organ; *OVLT*, organum vasculosum of the lamina terminalis; *ME*, median eminence and infundibulum; *SCO*, subcommissural organ; *PG*, pineal gland; *AP*, area postrema.

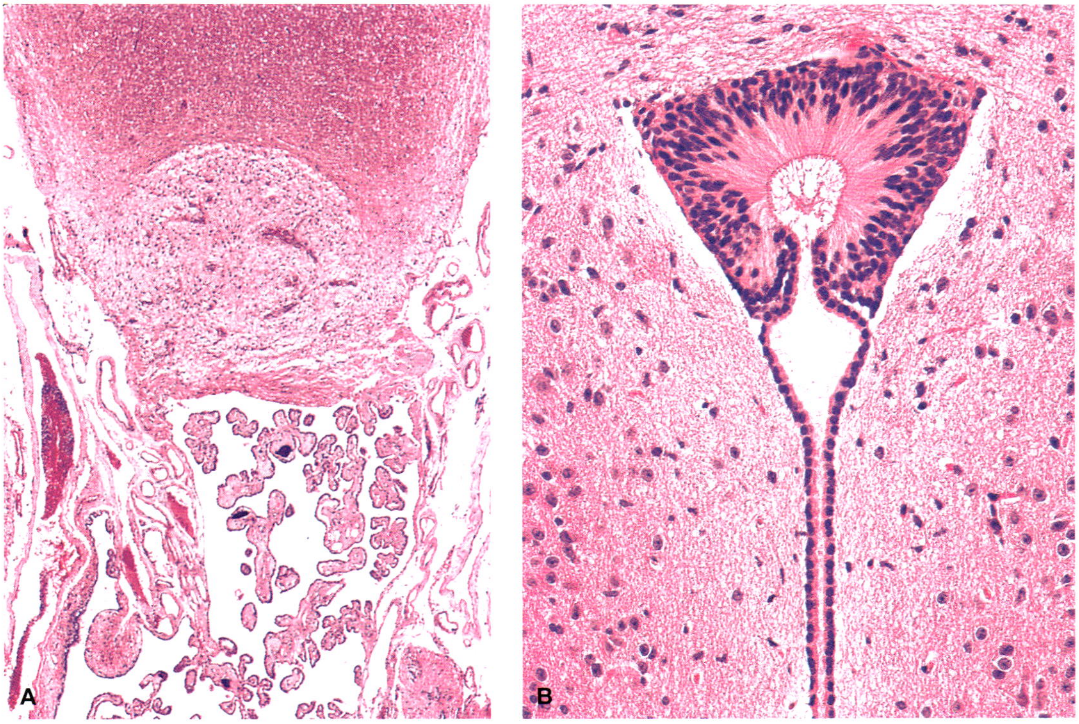

FIGURE 9.59 Circumventricular organs: Histologically, all of the circumventricular organs except for the subcommissural organ are very similar, with a loose neuropil that is highly vascular and lacks a blood–brain barrier as illustrated by the subfornical organ (**A**). The subcommissural organ is located in the region of the pineal gland just beneath the posterior commissure in the posterior dorsal third ventricle and is highly developed in most mammalian species, as illustrated by the mouse (**B**). (*continued*)

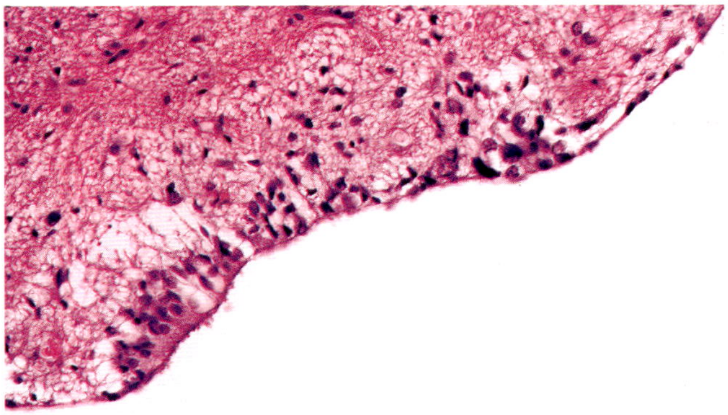

C

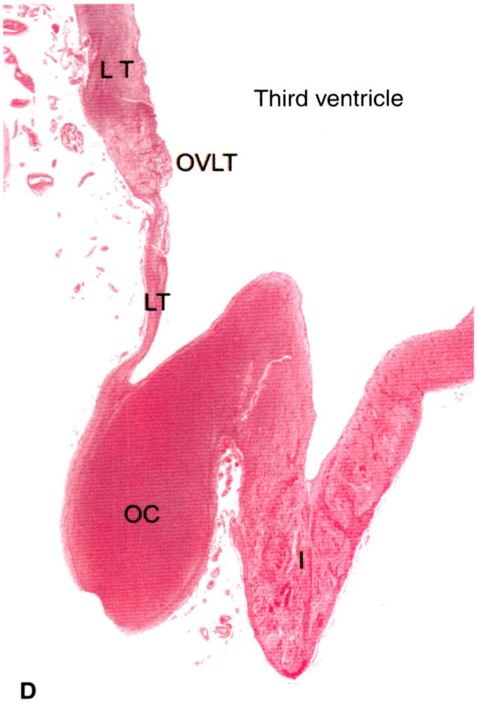

D

FIGURE 9.59 (*Continued*) In humans, the subcommissural organ is vestigial but remnants are occasionally encountered (**C**). The organum vasculosum of the lamina terminalis (*OVLT*) and the median eminence–infundibulum–neurohypophysis are two additional circumventricular organs that are in contact with the third ventricle (**D**). *I*, infundibulum; *LT*, lamina terminalis; *OC*, optic chiasm.

INTRADURAL ELEMENTS OF THE PERIPHERAL NERVOUS SYSTEM

The major intradural representatives of the peripheral nervous system are the cranial and spinal nerves and small autonomic fibers in the adventitia of blood vessels. In all of the cranial nerves except cranial nerve eight, the transition from central to peripheral nervous system occurs within 2 mm of the pial surface. In the eighth cranial nerve, the CNS extends out along the nerve for a centimeter or so to the level of the internal auditory meatus. At this point, the transition occurs between the medial CNS segment and the lateral peripheral segment that emerges from the apparatus for hearing and balance (Fig. 9.60). The myelin of the CNS is formed by oligodendrocytes, whereas that of the peripheral nervous system is formed by Schwann cells. Peripheral nerve is noted for its content of interstitial collagen and the elongated nuclei of Schwann cells. Two additional intrathecal components of the peripheral nervous system may pique interest on fortuitous encounters. The first is the microneuroma, which is usually found in the parenchyma of the spinal cord or, more rarely, the medulla (Fig. 9.61). These structures consist of a Gordian knot of

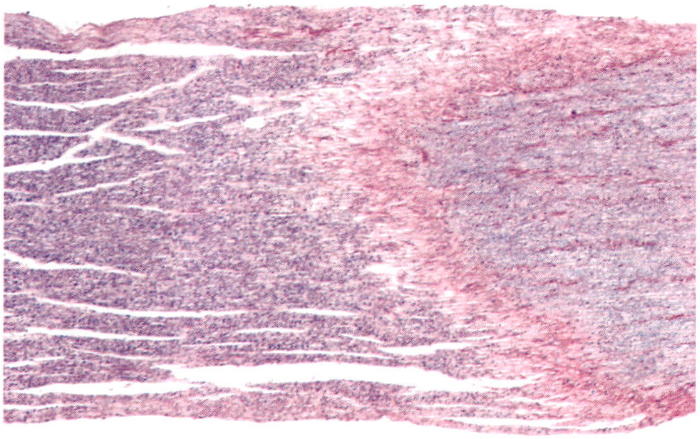

FIGURE 9.60 Transition zone from central to peripheral nervous system myelin: for cranial nerve eight (vestibulocochlear) this transition occurs in the vicinity of the internal acoustic meatus. This transition zone is also known as the "Obersteiner–Redlich zone (ORZ)."

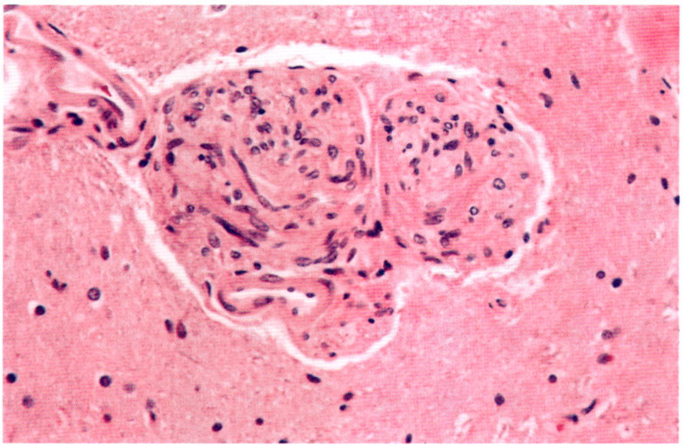

FIGURE 9.61 Microneuroma: These tangled balls of unmyelinated axons are most often encountered in the spinal cord, less often in the medulla, as an incidental finding in an otherwise unremarkable specimen.

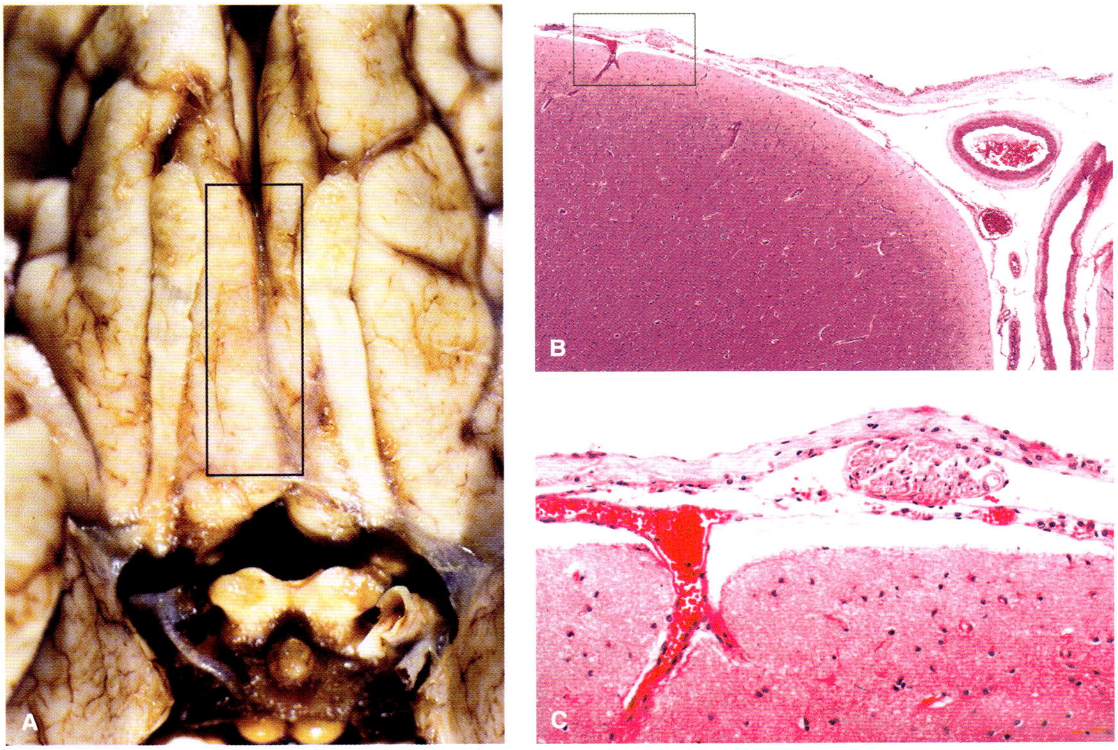

FIGURE 9.62 Cranial nerve zero: Cranial nerve zero (CN0), also known as the terminal nerve or nervus terminalis, is present in humans as a plexus of small peripheral nerve fascicles found in the subarachnoid space that covers the gyri recti that lie between the olfactory bulbs and tracts (**A**). Tissue sections taken through the gyri recti that include the overlying leptomeninges (**B**) will often include a terminal nerve fascicle cut in cross section (**C**). The small peripheral nerve fascicles of cranial nerve zero are one potential source of subfrontal schwannomas.

unmyelinated axons that may arise secondary to traumatic injury of peripheral nerve roots. Presumably, the regenerating axons follow the penetrating spinal or medullary arteries into the CNS parenchyma along the Virchow–Robin spaces until the tapering perivascular spaces ultimately block further advance of the regenerating axons and result in the neuroma. An additional component of the peripheral nervous system that occasionally arouses interest is the unmyelinated terminal nerve that is variously termed nervus terminalis, cranial nerve zero, and cranial nerve T. This nerve courses in the subarachnoid space covering the gyri recti of the orbital surface of the frontal lobes (Fig. 9.62). Although usually composed of multiple small anastomosing fascicles, it occurs as a single trunk in some specimens and can be quite striking. Rarely, intrafascicular ganglion cells may be observed. The anatomic location of the cranial nerve zero fascicles position it as a potential source of origin of subfrontal schwannoma.

MENINGES

Dura Mater (Pachymeninx)

The dura is composed of two tightly annealed layers of fibrous connective tissues (Fig. 9.63) (17). The outer layer functions as the periosteum of the cranium, whereas the inner meningeal layer is joined to the arachnoid membrane by weak intercellular junctions and focally forms the four dural reduplications that compartmentalize the cranial cavity: the falx cerebri, falx cerebelli, tentorium cerebelli, and diaphragma sellae. The two layers of the dura separate to accommodate the dural venous sinuses (Fig. 9.63B); the inner meningeal layer is pierced by draining veins and by arachnoid villi. The latter conduct cerebrospinal fluid back into the venous circulation and are obvious over the superior parasagittal convexities of the cerebral hemispheres where they project into the superior sagittal sinus (Fig. 9.63). They are present in all other major venous sinuses, and are observed along the posterior margin of the cerebellar hemispheres in relation to the sinus confluens and the transverse venous sinuses. Small villi are also present intraspinally. The epithelial properties of arachnoid granulations are reflected ultrastructurally in elongated, interdigitating cell processes bonded together with desmosomes and immunohistochemically, by positivity for epithelial membrane antigen (EMA) (Fig. 9.63). These features are also characteristic of meningiomas, as is the more specific marker, somatostatin receptor 2 (SSTR2) (18).

With age, the deposition of collagen enlarges the arachnoid villi that are then referred to as pacchionian bodies (Fig. 9.63). Such large granulations frequently press through the overlying roof of the superior sagittal sinus and its lateral lacunae to produce small pits or depressions in the inner table of the calvarium. These are known as the foveolae granulares or pacchionian foveolae. Portions of the dura, particularly the falx cerebri and parasagittal dura associated with the superior sagittal sinus, often calcify with age. Calcification may also be seen in association with chronic renal failure. Focal ossification is sometimes encountered as an incidental finding.

Pia-Arachnoid (Leptomeninges)

The arachnoid forms a continuous sheet immediately subjacent to the dura. The dura and arachnoid exist in vivo as a physically continuous tissue, with sparse but unequivocal intercellular junctions linking these two historically discrete membranes. The storied subdural space has, thus, taken its rightful place in the pantheon of neuromythology, alongside the syncytial theory of the neural net. It has been proposed that the term spatium subdurale be eliminated from the standardized nomenclature.

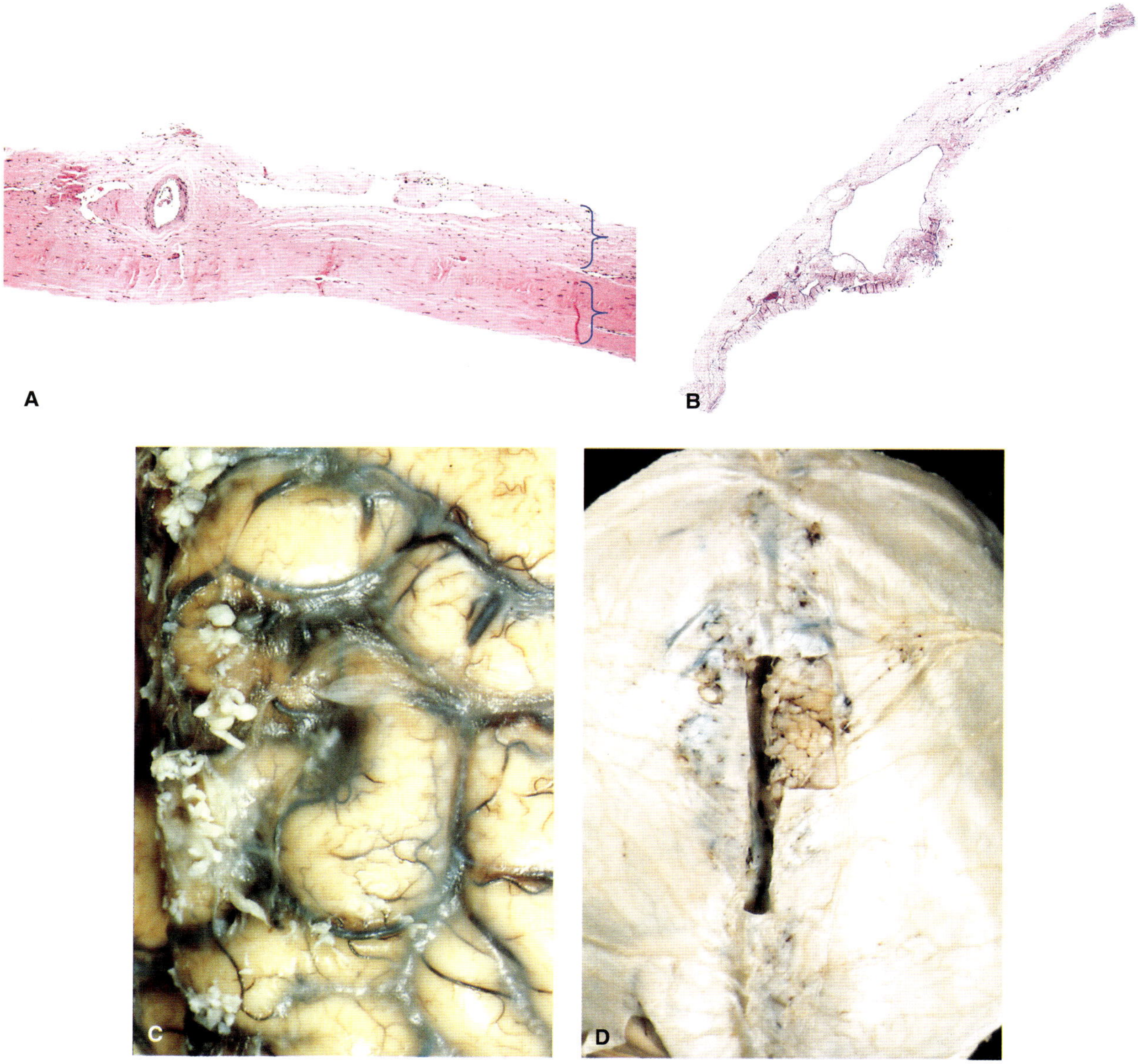

FIGURE 9.63 The dura is formed by two annealed layers of fibrous connective tissue (**A**). The two layers separate to accommodate the dural venous sinuses (**B**). Arachnoid granulations (villi) (**C**) are specialized structures of the arachnoid membrane serve to return cerebrospinal fluid from the subarachnoid space to the venous circulation and are accordingly found in relation to all major dural venous sinuses. The villi are most prominent in the superior sagittal (**D**) and transverse sinuses. With advancing age, they undergo collagenous hypertrophy, as seen in these micrographs, and may then be referred to as pacchionian bodies. The enlarged villi remodel the overlying bone of the inner table of the calvarium to produce small pits termed pacchionian foveolae, or foveolae granulares.

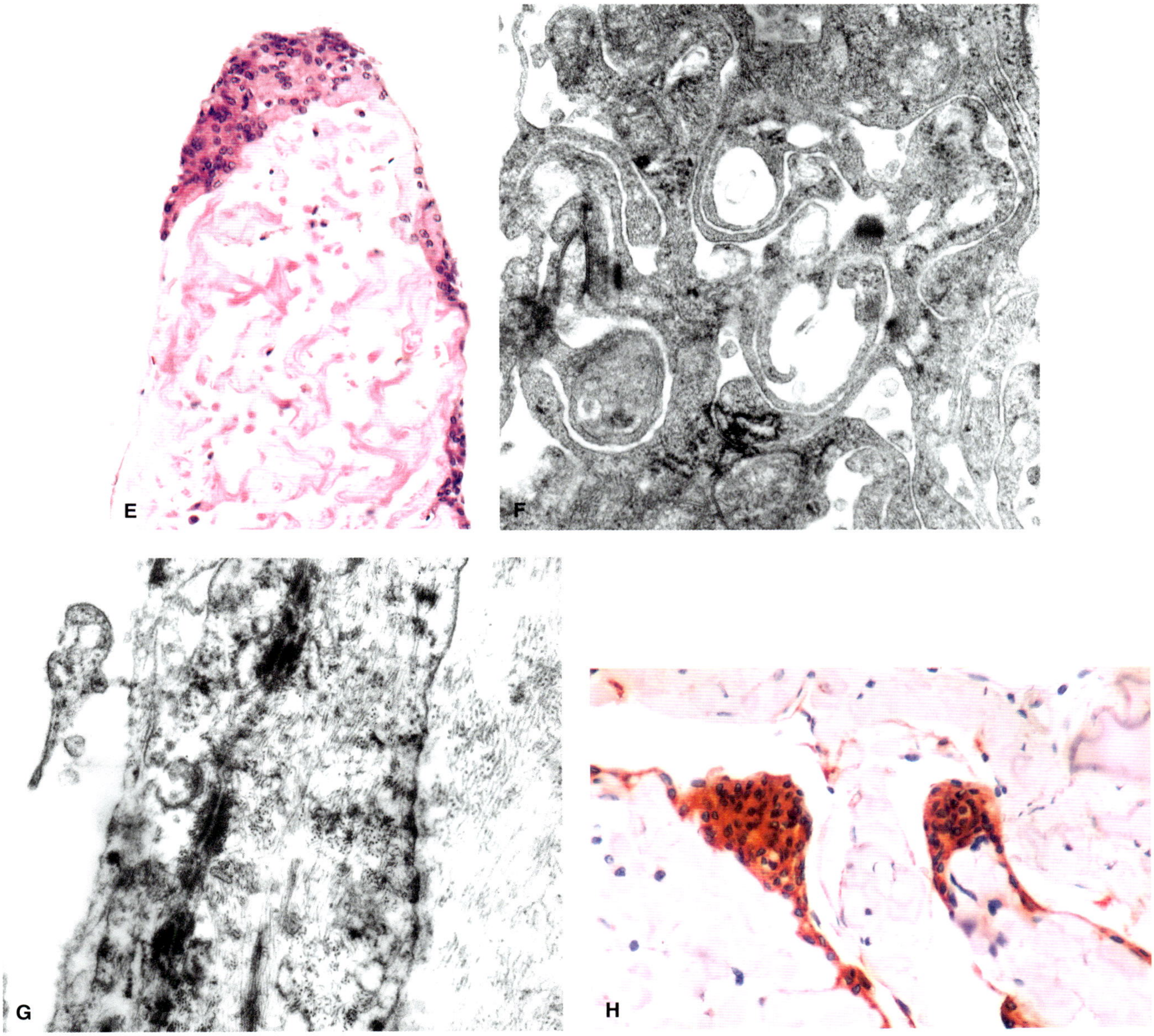

FIGURE 9.63 (*Continued*) Nests of meningothelial cells may be seen anywhere along the arachnoid membrane, but are especially prominent in the apical regions of arachnoid granulations where they are termed arachnoid cap cells (**E**), and in the arachnoid covering the orbitofrontal cortex. Normal meningothelial cells are innately inclined to form whorls and psammoma bodies, two features that are often retained by their neoplastic counterparts, meningiomas. The meningothelial cells of the arachnoid membrane, including the cap cells, serve an epithelial function. Accordingly, they possess elongated, intertwined cell processes (**F**) that are tightly spot welded together by numerous desmosomes (**G**) and exhibit strong immunopositivity for epithelial membrane antigen (EMA) (**H**). Like the tendency to form whorls and psammoma bodies, these epithelial phenotypic traits are retained by neoplastically transformed meningothelial cells and serve as useful diagnostic features of the vast majority of meningiomas which otherwise exhibit a very broad range of light microscopic morphologies.

Nevertheless, there is no disputing the fact that the interface between dura and arachnoid constitutes the weak link or path of least resistance for pathologic processes that tend to disrupt the meninges. It seems unlikely that such venerable terms as subdural hematoma will be cashiered.

The pia mater and arachnoid are often considered as a single delicate covering of the brain and spinal cord and together comprise the pia-arachnoid or leptomeninges. The arachnoid is connected to the pia by delicate strands, the arachnoid trabeculae (Fig. 9.64A). In the young, the arachnoid is crystal clear, but with age it becomes gradually thickened. The extent of this change varies considerably. In some cases, it is severe enough to raise concern about a pathologic process—meningitis and meningeal carcinomatosis being

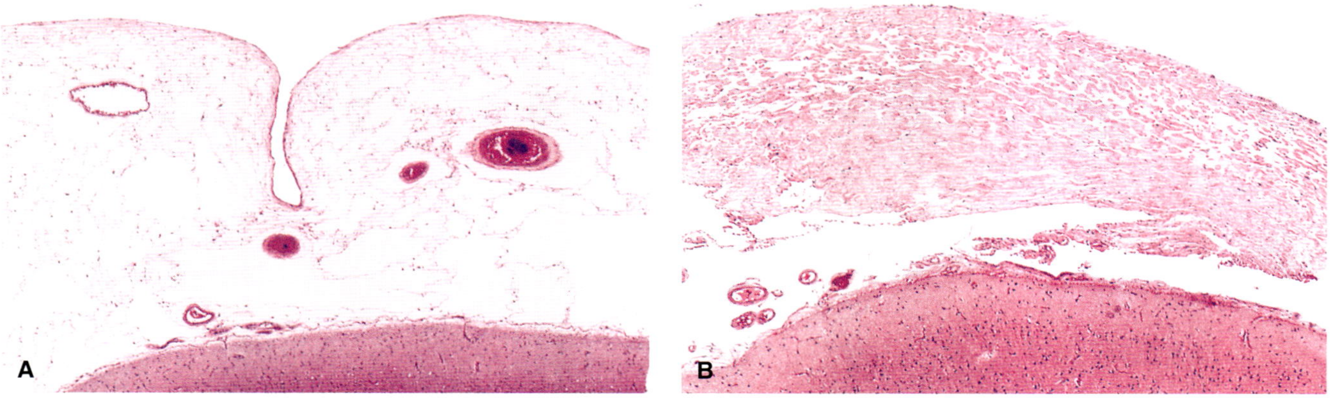

FIGURE 9.64 Subarachnoid space: The subarachnoid space (**A**) is delimited by the arachnoid membrane externally and by the pia mater internally. Delicate arachnoid trabeculae course between these two membranes. In adults, gradual collagen deposition in the subarachnoid space (**B**) results in grossly appreciable "clouding" of the leptomeninges. This aging fibrosis appears grossly as diffuse opacification with focal plaques and small punctate nodules. It is characteristically most prominent along the dorsal cerebral convexities adjacent to the superior sagittal sinus.

the two usual suspects. This normal age-related arachnoid thickening is typically most pronounced over the dorsal parasagittal cerebral convexities. Microscopically, it results from the deposition of dense bundles of collagen (Fig. 9.61B) analogous to the collagenous hypertrophy of arachnoid villi that occurs prominently in the same vicinity. Nests of arachnoid cells (also called meningothelial cells) may be seen throughout the arachnoid membrane (Fig. 9.63) but are concentrated over the arachnoid villi. These distinctive elements become more obvious and more clustered with advancing age and, in the adult, often form whorls with centrally placed psammoma bodies, thereby resembling the nests of a meningioma. As mentioned previously, small nests of arachnoid cells are also present in the vascular connective tissue core of the choroid plexus (see Fig. 9.57). Both normal and neoplastic meningothelial cells are immunoreactive for EMA—an understandable property considering the epithelial phenotype of the desmosome-containing meningothelial cell (Fig. 9.63).

The dorsal leptomeninges of the thoracic and lumbosacral spinal cord occasionally contain white wafer-like plaques (Fig. 9.65), a finding that is often termed arachnoiditis ossificans. These brittle lesions are usually roentgenographically and histologically devoid of bone or mineral, and most often consist of laminated and hyalinized fibrous tissues. True arachnoiditis ossificans generally occur in the context of prior symptomatic inflammation or trauma to the leptomeninges. Hyaline plaques, in contrast, are typically discovered as an incidental finding at autopsy in the absence of any relevant clinical history.

Like the dura, the pia is traditionally divided into two layers: The epipia, which covers the surface of the CNS parenchyma and surrounds the vasculature, and the intima pia, which extends into the CNS parenchyma as the posterior median and intermediate septa of the spinal cord. Three specialized structures of the epipia are recognized: the denticulate ligaments on either side of the spinal cord, the linea splendens adjacent to the anterior spinal artery, and the filum terminale. All three structures are composed primarily of dense bundles of collagen.

Of note, attention has recently been turned to a novel fourth layer of the meninges: the subarachnoid lymphatic-like membrane. This layer is immunohistochemically distinct from the dura, arachnoid, and pia, and is likely related to another relatively new concept coined the glymphatic system. The glymphatic system is a clearance pathway involving the exchange of solutes in cerebrospinal fluid, primarily during sleep, via the aquaporin-4 water channels in astrocyte foot processes. In vivo imaging has demonstrated a Prox1-positive monolayer of cells beneath the dura that divides the subarachnoid space into outer and inner components. Functional studies have shown

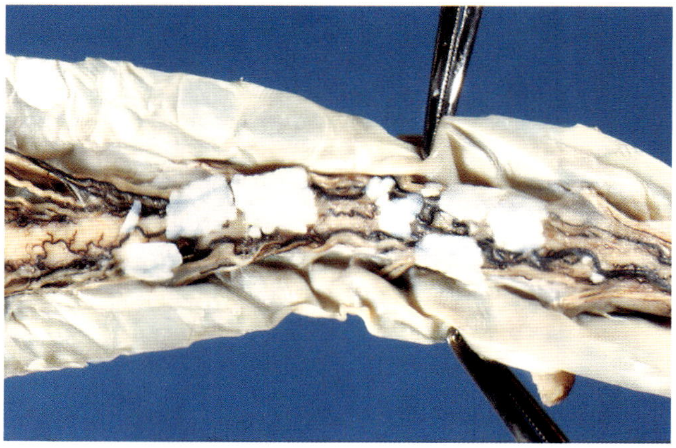

FIGURE 9.65 Hyaline plaques of the spinal leptomeninges: These plaques are common incidental findings at autopsy and occur most frequently in the dorsal spinal arachnoid, although they may occasionally be seen in the cerebral leptomeninges as well.

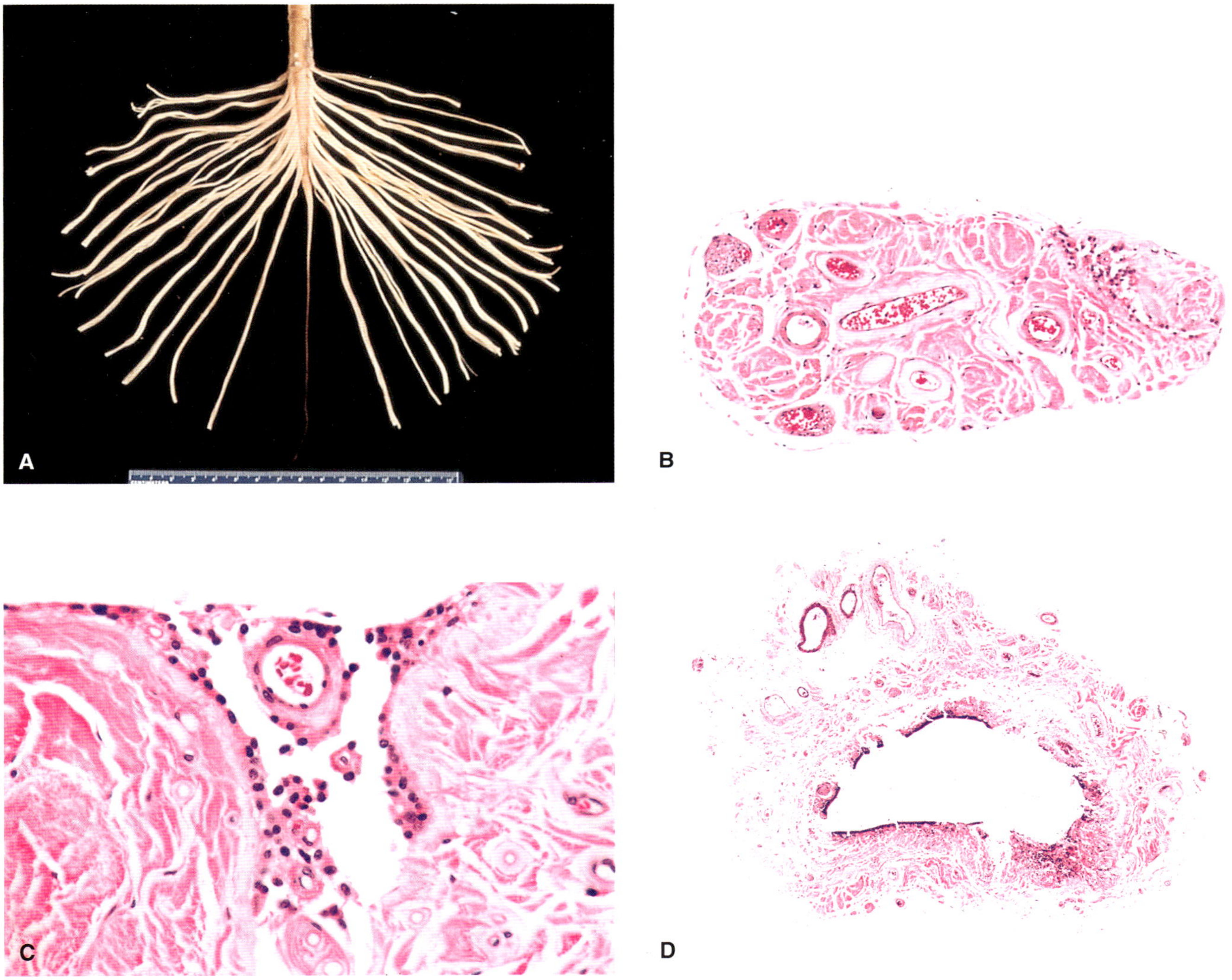

FIGURE 9.66 Filum terminale: The filum terminale is the terminus of the spinal cord and extends downward from the conus medullaris surrounded by the nerve roots of the cauda equina (**A**). As seen in cross section (**B**), the filum is composed primarily of dense collagenous tissue and contains blood vessels, small peripheral nerve fascicles, and, of significant clinical importance, a small, often eccentrically located, ependymal remnant of the central canal (*upper right*). The latter structure, shown at higher magnification in (**C**), is the origin of myxopapillary ependymoma. A remnant of the embryonic terminal ventricle of Krause (ventriculus terminalis), which consists of a focal dilatation of the central canal located in the region of the junction of the conus medullaris with the filum, may be encountered in sections from this vicinity (**D**).

that solutes <3 kDa are able to pass from one compartment to the other.

The filum terminale forms the terminus of the spinal cord and warrants additional brief description. It is composed largely of leptomeningeal collagen but also contains small blood vessels, occasional small nerve fascicles, and may harbor focal collections of adipocytes in a minority of normal individuals. Most importantly, however, it is an ependymal remnant of the central canal (Fig. 9.66). This structure is the source of origin for a unique neoplasm of the conus medullaris and filum terminale: the myxopapillary ependymoma.

Leptomeningeal Melanocytes

True melanocytes like those found in the skin are normal cellular constituents of the meninges (19). They are most concentrated in the leptomeninges of the ventral aspect of the upper cervical spinal cord and medulla oblongata (Fig. 9.67). In individuals with an abundant melanocytic presence, the distribution territory extends upward through the pontine cistern and mesencephalic interpeduncular fossa, lateral to the inferior cerebellar hemispheres and mesial aspects of the temporal lobes, and as far rostrally as the gyri recti of the orbitofrontal cortex. It is not unusual for melanocytes to follow the investing leptomeninges of

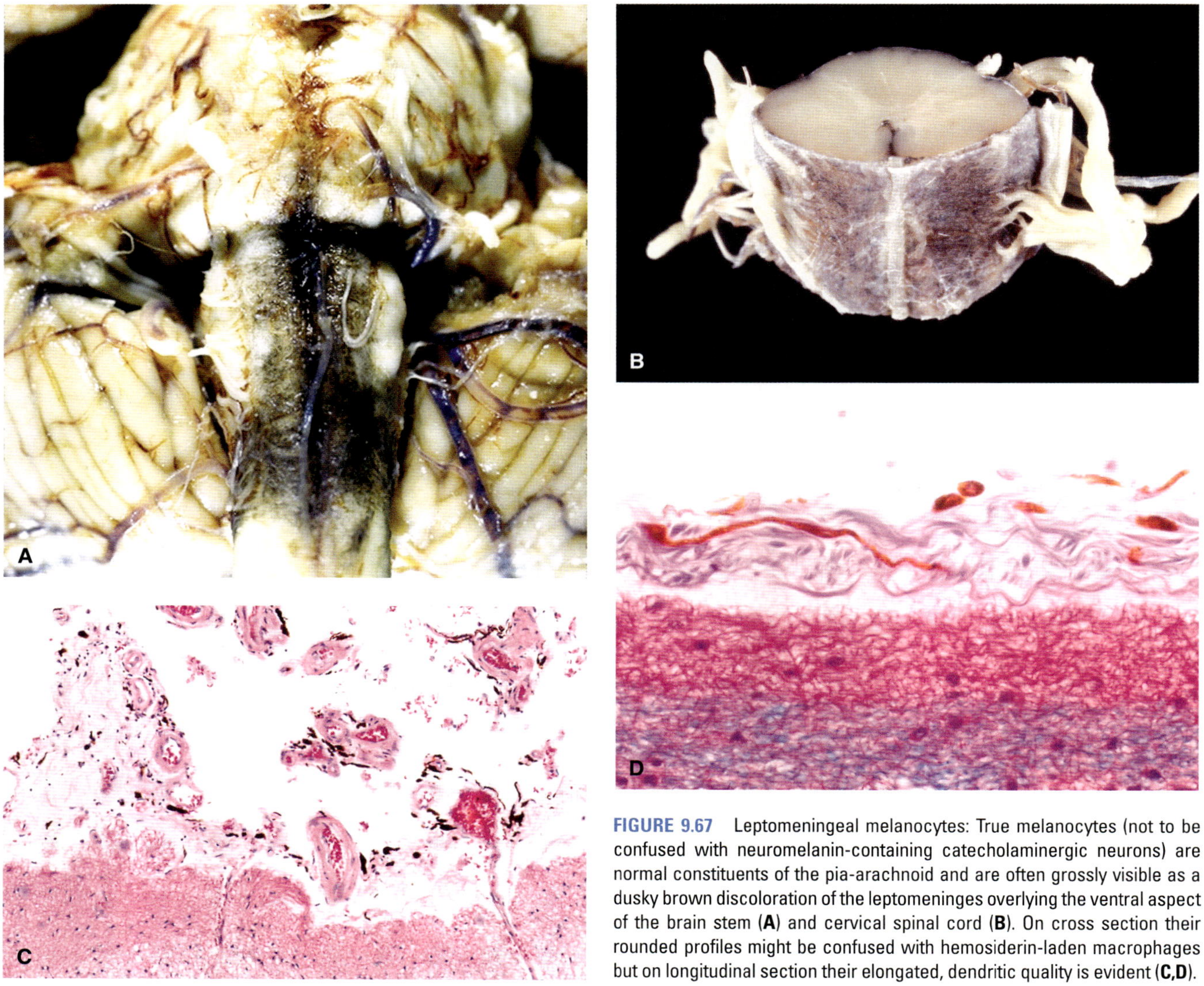

FIGURE 9.67 Leptomeningeal melanocytes: True melanocytes (not to be confused with neuromelanin-containing catecholaminergic neurons) are normal constituents of the pia-arachnoid and are often grossly visible as a dusky brown discoloration of the leptomeninges overlying the ventral aspect of the brain stem (**A**) and cervical spinal cord (**B**). On cross section their rounded profiles might be confused with hemosiderin-laden macrophages but on longitudinal section their elongated, dendritic quality is evident (**C,D**).

the perivascular Virchow–Robin spaces around large penetrating arteries for short distances into the CNS parenchyma. Intrinsic melanocytes of the leptomeninges may be involved in a spectrum of proliferative conditions ranging from benign melanocytoma to primary CNS melanoma, with all of these entities being exceptionally rare. In contrast, the normal presence of melanocytes in the leptomeninges must always be borne in mind when examining surgical biopsies from CNS sites known to harbor these distinctive elements; one must avoid misinterpreting them as evidence of a melanocytic neoplasm or as hemosiderin-laden macrophages (Table 9.2). With regard to the latter, the long dendritic processes of the melanocytes are generally quite distinctive (Fig. 9.67D).

Optic Nerve

The optic nerves, optic tracts, and optic chiasm are direct extensions of the CNS and not peripheral nerves. Myelin of

TABLE 9.2 "Brown Pigment" in the Central Nervous System

Normal
Lipofuscin in large neurons
Lipofuscin-like pigment in the infundibulum and neurohypophysis
Neuromelanin (catecholaminergic neurons)
Melanocytic melanin (leptomeningeal melanocytes)
Melanin in fetal pineal gland
Abnormal
Hemosiderin from hemorrhage
Melanin of metastatic melanoma
Melanin of pigmented primary CNS tumors
Lipochrome-like pigment in many "melanotic" CNS tumors
Pigmented fungi
Malaria pigment
Artifact
"Formalin pigment"

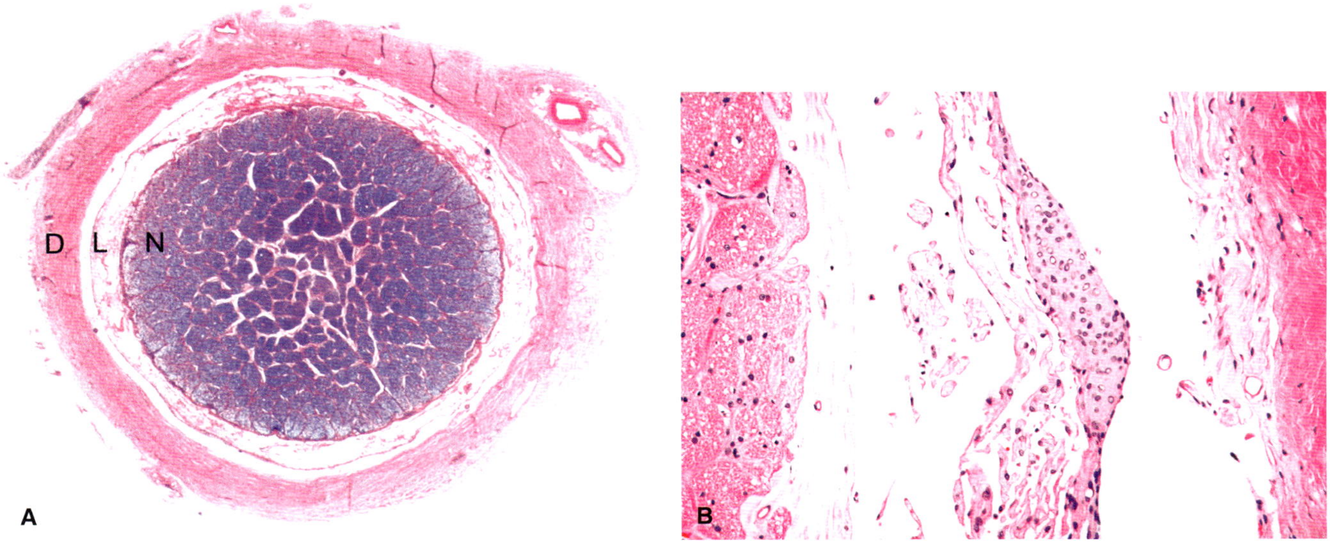

FIGURE 9.68 Optic nerve: A whole-mount cross section (**A**) reveals that the optic nerve (*N*) is surrounded by leptomeninges (*L*), which includes the pia mater and arachnoid together with the enclosed cerebrospinal fluid-containing subarachnoid space. The leptomeninges and subarachnoid space are in turn covered by the densely fibrous dura mater (*D*), which in this location, is often referred to as the optic nerve sheath. At higher magnification (**B**), leptomeningeal arachnoid cell clusters are clearly seen. The presence of arachnoid cell nests around the optic nerve must be borne in mind when examining intraoperative frozen tissue sections from this neuroanatomic vicinity.

the optic nerves is of the central type and is produced by oligodendroglia, not Schwann cells. Thus, the optic nerves are susceptible to diseases of CNS white matter, such as multiple sclerosis. Being extensions of the CNS, the optic nerves are surrounded by the three meninges, pia mater, arachnoid, and dura mater, with the enclosed subarachnoid space (Fig. 9.68). The presence of an arachnoid layer surrounding the optic nerve explains the occurrence of optic sheath meningiomas.

FETAL BRAIN

The two most distinctive histologic features of fetal brain compared to adult brain are active neurogenesis and paucity of myelin. The former is observed as a prominent, dense aggregation of neuroblasts and immature neurons in the periventricular and subpial zones. A similarly transient layer of migrating neurons in the fetal and infant cerebellum (the external granular layer) has been discussed. These generative laminae begin involuting during the latter part of gestation; remnants are present during the first year of postnatal life (see Fig. 9.3C).

ARTIFACTS

A variety of gross and macroscopic artifacts may complicate evaluation of tissue specimens from the CNS. Many of these are seen frequently in surgical neuropathology practice, while a few are limited primarily to autopsy neuropathology (Table 9.3). Artifacts can be broadly separated into those that hinder diagnostic evaluation versus those that mimic histopathologic lesions. Among the former are artifacts of the crush–burn–freeze–suck–soak group (Fig. 9.69). The cavitational ultrasonic surgical aspirator (CUSA) is widely employed by neurosurgeons for the safe removal of diseased CNS tissue, particularly soft tumors, and a trap can be used to collect the aspirated tissue and saline irrigation solution for the submission for histologic evaluation. Although microscopic examination of CUSA material can be very informative, the pathologist must be aware of the artifacts that frequently accompany such specimens, including artificial distortion and smearing, and the introduction of extraneous material. CUSA artifact is one cause of pseudonecrosis in CNS tissue samples (Table 9.4). Among artifacts that mimic lesions, the most common are perinuclear halo artifact, collapsed leptomeningeal vessels, and bone dust (Fig. 9.70).

Iatrogenically introduced foreign material is also encountered with regularity by pathologists who examine CNS specimens and warrants brief mention (Table 9.5, Fig. 9.71). A variety of foreign agents are used to control bleeding during surgery and may be introduced preoperatively by the interventional radiologist for embolization of vascular lesions or intraoperatively by the surgeon to control hemorrhage during and after surgery. All of these agents periodically appear in tissue sections. Since they are designed to be resorbable and can therefore be left in place, the morphologic appearance will vary depending on the time interval from placement at the initial surgery and subsequent

TABLE 9.3 Artifacts

Postal Service Artifacts
Crush (glass slides)
Paraffin pox (paraffin blocks melted into bubble-wrap during shipping in hot weather)
Cautery Artifact
Tissue
Blood vessels
Freeze Artifact
Crush Artifact
Embedding sponge
Surgical or histology forceps
Cavitational Ultrasonic Surgical Aspiration (CUSA, Cavitron) Artifacts
Pseudonecrosis
Admixed exogenous constituents
Cranial bone fragments ("bone dust")
Hemostatic agents
Pseudomineralization
Bone dust
Laminar pseudomineralization
Delayed Fixation Artifact
Perinuclear halos—oligodendroglial (useful)
Perinuclear halos—neurons (mimics oligodendroglioma)
Pseudohypercellularity on smear preparations (uneven thickness mimics glioma)
Air-drying Artifact on Touch/Smear/Drag Preps
Collapsed Leptomeningeal Vessels (Mimics Vascular Malformation)
Formalin Pigment
Extraneous Contaminants
Iatrogenic foreign material
Tissue fragments ("floaters")
Microtome water bath fungal mold
Airborne plant pollen spores
Autopsy Artifacts
Cerebellar conglutination
Mechanical herniation ("toothpaste") artifact of spinal cord
Macroscopic gas-forming bacteria vacuolation ("Swiss cheese brain")

TABLE 9.4 Pseudonecrosis Etiologies

Artifact
Cavitron ultrasonic surgical aspirator (CUSA)/saline solution artifact
Hematoxylin absence artifact
Normal Regional Histology
Cerebellar cortex on smear preparation (hypocellular eosinophilic molecular layer mimics necrosis)
Iatrogenic
Degenerating microfibrillar collagen (Avitene) textiloma (gossypiboma)

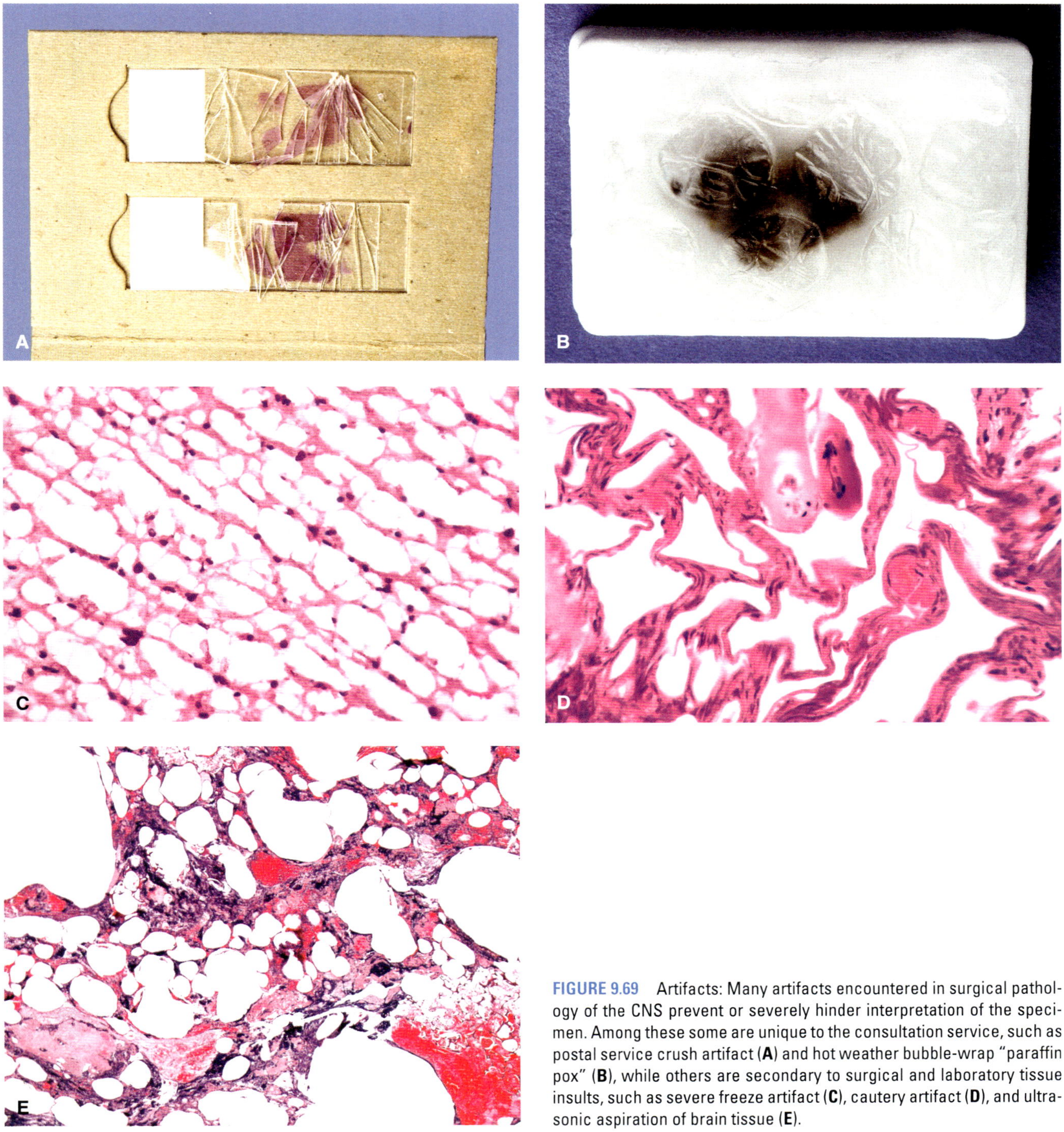

FIGURE 9.69 Artifacts: Many artifacts encountered in surgical pathology of the CNS prevent or severely hinder interpretation of the specimen. Among these some are unique to the consultation service, such as postal service crush artifact (**A**) and hot weather bubble-wrap "paraffin pox" (**B**), while others are secondary to surgical and laboratory tissue insults, such as severe freeze artifact (**C**), cautery artifact (**D**), and ultrasonic aspiration of brain tissue (**E**).

resection during a second surgery (e.g., resection of recurrent tumor). Resorbable hemostatic agents elicit a chronic inflammatory reaction of variable intensity, which occasionally may be severe enough to create mass effect and clinical symptoms.

There are a few artifacts with which the pathologist who examines postmortem CNS specimens should be familiar (Fig. 9.72), the most common being autolysis of the cerebellar granular cell layer (i.e., cerebellar conglutination), mechanical distortion of the spinal cord produced by forceps pressure during removal, and the production of cystic cavities of varying size in the brain by the postmortem proliferation of gas-forming bacteria, which has been termed Swiss cheese brain.

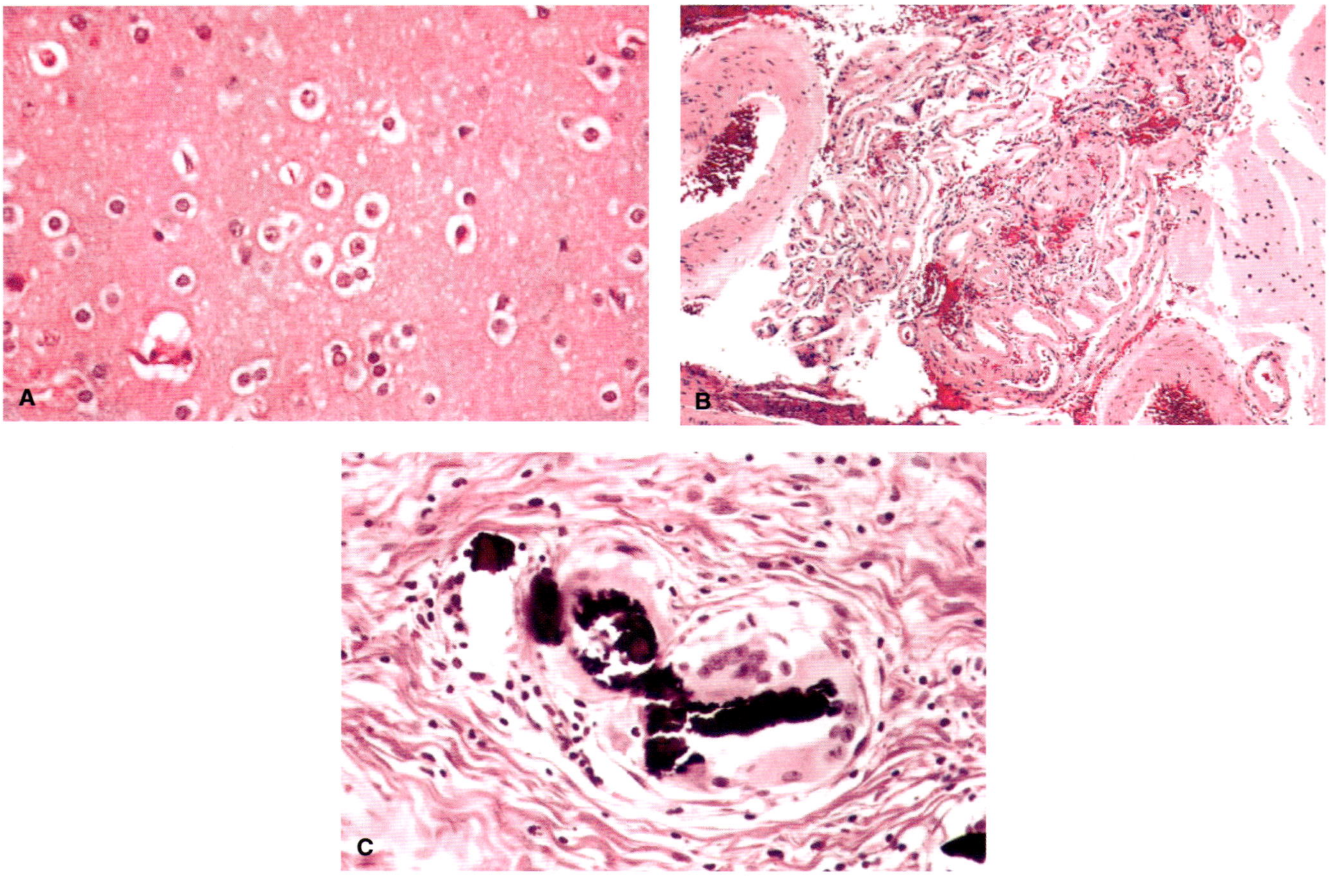

FIGURE 9.70 Artifacts: Several types of artifacts may not be recognized as artifactual in nature to the unaware and so may be particularly misleading. For example, one of the most characteristic morphologic features of normal oligodendrocytes and their derivative tumors, oligodendrogliomas, in formalin-fixed paraffin-embedded tissue sections is the presence of perinuclear halos. However, depending on fixation conditions and other factors, prominent halos may sometimes be seen around other cell types, including neurons (**A**); care must be exercised in such situations to avoid misdiagnosis. Another example of misleading artifact is the tangle of normal blood vessels that results from collapsed vascular leptomeninges (**B**). The result can mimic vascular malformation. Before rendering a diagnosis of vascular abnormality in such circumstances, the adjacent brain or spinal cord tissue should be examined for the evidence of associated features, such as gliosis, hemosiderin deposition, and granular bodies. Finally, also under the category of misleading artifact is "bone dust," which consists of microscopic fragments of cranial bone derived from the surgeon's drill that become intermixed with the tissue sample and can mimic calcification or ossification. In repeat operations, such bone dust fragments left in situ at the previous operation can be seen accompanied by a foreign body-type giant cell reaction (**C**).

TABLE 9.5 Iatrogenically Introduced Foreign Material

Preoperative Embolic Agents Gelatin foam (Gelfoam) Polyvinyl alcohol particles Acrylic microspheres (Embospheres)
Intraoperative Hemostatic Agents
Resorbable Hemostatic Agents Gelatin foam (Gelfoam) Oxidized cellulose (Surgicel, Oxycel) Microfibrillar bovine collagen (Avitene)
Nonresorbable Hemostatic Agents "Retained" cotton ball/pledgets
Surgically Introduced Therapeutic Materials Gliadel wafers (chemotherapy) Muslin (cotton) fabric (aneurysm wrapping)

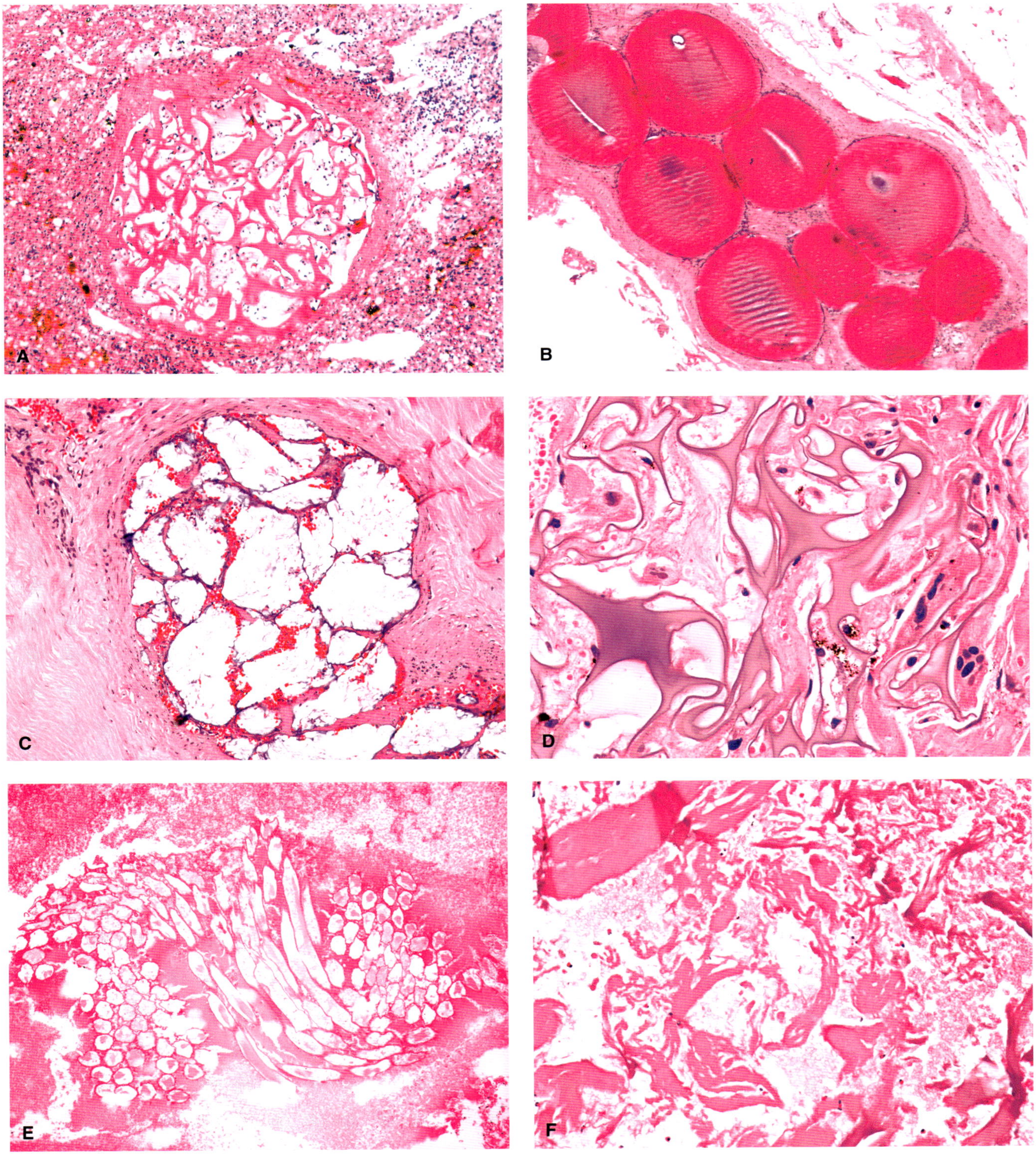

FIGURE 9.71 Artifacts: An additional category of artifacts seen in surgical neuropathology consists of foreign material placed by the interventional radiologist or neurosurgeon and subsequently encountered by the pathologist upon tissue resection. The most common examples of this are embolic and hemostatic agents. Embolic materials are introduced by catheter prior to surgery for highly vascular lesions to reduce intraoperative bleeding; the most common are gelatin foam (**A**), acrylic resin spheres (**B**), and polyvinyl alcohol particles (**C**). Hemostatic agents, in contrast, are placed in the surgical site to stop bleeding during the operation and often left in place after closing to prevent postoperative bleeding. The most commonly employed agents are gelatin foam (**D**), oxidized cellulose (**E**), and microfibrillar bovine collagen (**F**).

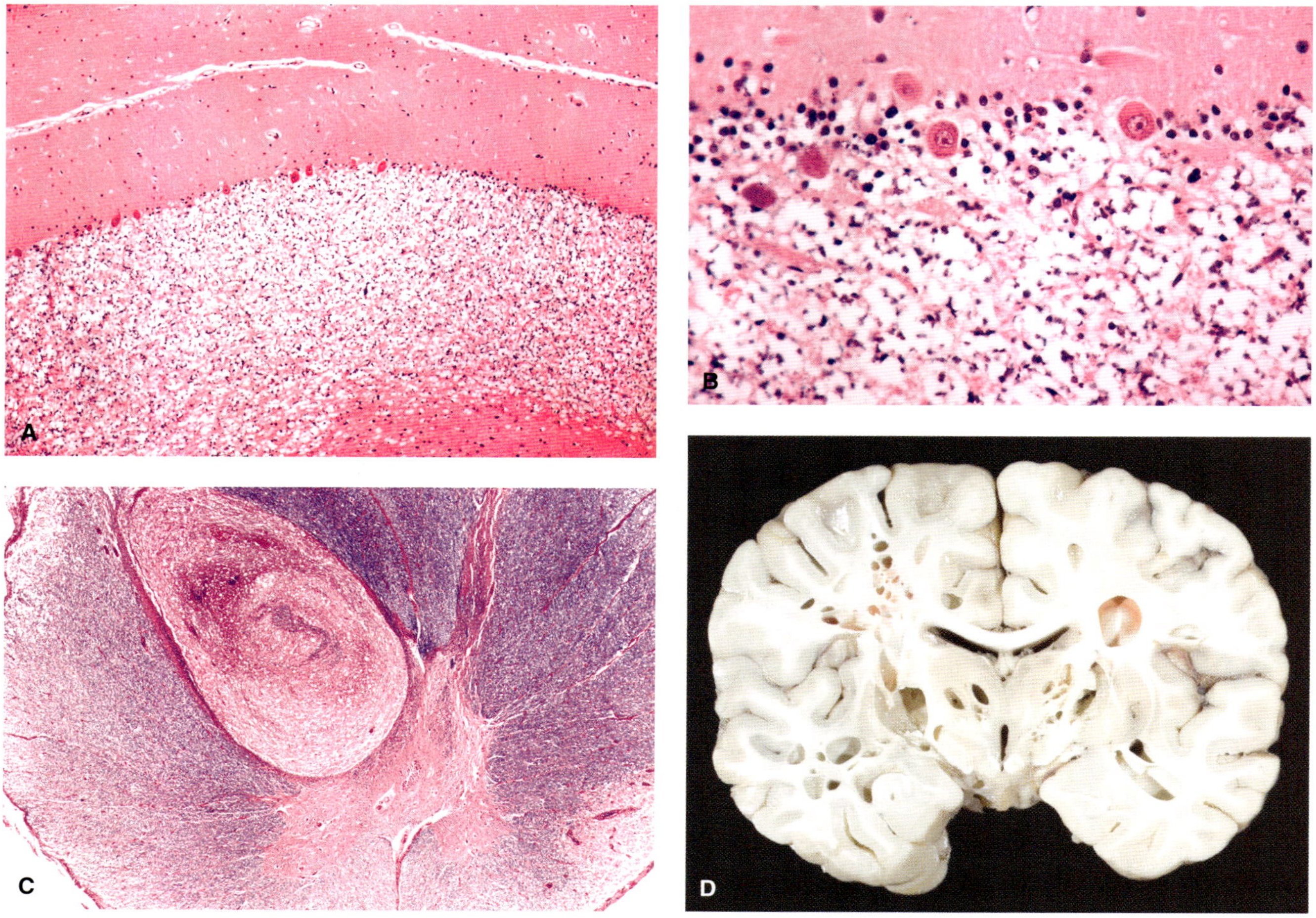

FIGURE 9.72 Artifacts: There are a number of gross and histologic artifacts that are usually encountered only in autopsy specimens of the CNS. The most common of these are cerebellar conglutination (**A,B**), also known as "etat glace," which consists of autolytic dissolution of the granular cell layer of the cerebellum; "toothpaste" or "squeeze" artifact of the spinal cord (**C**), which results from focal crushing of the cord by forceps during removal at autopsy and resultant internal herniation of the central gray matter, mimicking malformation or heterotopia; and "Swiss cheese brain" (**D**), which is a striking macroscopic vacuolization of the brain resulting from postmortem proliferation of gas-forming bacteria.

SUGGESTED READINGS

Neuropathology Textbooks

Burger PC. *Smears and Frozen Sections in Surgical Neuropathology*. PB Medical Publishing; 2009.

Burger PC, Scheithauer BW, Vogel FS. *Surgical Pathology of the Nervous System and its Coverings*. 4th ed. Churchill Livingstone; 2002.

Perry A, Brat DB, eds. *Practical Surgical Neuropathology: A Diagnostic Approach*. 2nd ed. Elsevier; 2018.

Smith C, Perry A, Kovacs, G, Jacques T. *Greenfield's Neuropathology*. 10th ed. CRC Press; 2024.

Neuropathology Review Books

Gray F, De Girolami U, Poirer J. *Escourolle and Poirer's Manual of Basic Neuropathology*. 6th ed. Butterwoth-Heinemann; 2018.

Osborn, AG, Linscott, LL, Salzman, KL. *Osborn's Brain*. 3rd ed. Elsevier; 2023.

Neuropathology Atlases

Ellison D, Love S, Chimelli L, et al. *Neuropathology*. 3rd ed. Mosby; 2013.

Hirano A. *Color Atlas of Pathology of the Nervous System*. 2nd ed. Igaku-Shoin; 1988.

Okazaki H, Scheithauer BW. *Atlas of Neuropathology*. Gower Medical; 1988.

Schochet SS, Nelson J. *Atlas of Clinical Neuropathology*. Appleton & Lange; 1989.

Weller RO. *Color Atlas of Neuropathology*. Oxford University Press; 1984.

Veterinary Neuropathology

Summers BA, Cummings JF, de Lahunta A, eds. *Veterinary Neuropathology*. Mosby; 1995.

REFERENCES

1. Paul MS, Limaiem F. Histology, purkinje cells. In: *StatPearls*. 2024.
2. Ding SL, Royall JJ, Sunkin SM, et al. Comprehensive cellular-resolution atlas of the adult human brain. *J Comp Neurol*. 2016;524:3127–3481.
3. Nolan M, Scott C, Hof PR, Ansorge O. Betz cells of the primary motor cortex. *J Comp Neurol*. 2024;532:e25567.
4. Mullen RJ, Buck CR, Smith AM. NeuN, a neuronal specific nuclear protein in vertebrates. *Development*. 1992; 116:201–211.
5. Goyal VK. Lipofuscin pigment accumulation in human brain during aging. *Exp Gerontol*. 1982;17:481–487.
6. Yuen P, Baxter DW. The morphology of Marinesco bodies (paranucleolar corpuscles) in the melanin-pigmented nuclei of the brain-stem. *J Neurol Neurosurg Psychiatry*. 1963;26: 178–183.
7. Molnar J. Corpora amylacea in the central nervous system. *Nature*. 1951;168:39.
8. Marie Y, Sanson M, Mokhtari K, et al. OLIG2 as a specific marker of oligodendroglial tumour cells. *Lancet*. 2001; 358:298–300.
9. Ahmed Z, Shaw G, Sharma VP, Yang C, McGowan E, Dickson DW. Actin-binding proteins coronin-1a and IBA-1 are effective microglial markers for immunohistochemistry. *J Histochem Cytochem*. 2007;55:687–700.
10. Hogger P, Dreier J, Droste A, Buck F, Sorg C. Identification of the integral membrane protein RM3/1 on human monocytes as a glucocorticoid-inducible member of the scavenger receptor cysteine-rich family (CD163). *J Immunol*. 1998;161:1883–1890.
11. Cooper ER. The human pineal gland and pineal cysts. *J Anat*. 1932;67:28–46.
12. Lechan RM, Toni R. Functional anatomy of the hypothalamus and pituitary. In: Feingold KR, Anawalt B, Blackman MR, et al., eds. *Endotext*. MDText.com, Inc.; 2000.
13. Lazzaro D, Price M, de Felice M, Di Lauro R. The transcription factor TTF-1 is expressed at the onset of thyroid and lung morphogenesis and in restricted regions of the foetal brain. *Development*. 1991;113:1093–1104.
14. Levine C, Marcillo A. Origin and endpoint of the olfactory nerve fibers: as described by Santiago Ramón y Cajal. *Anat Rec (Hoboken)*. 2008;291:741–750.
15. Mortazavi MM, Griessenauer CJ, Adeeb N, et al. The choroid plexus: A comprehensive review of its history, anatomy, function, histology, embryology, and surgical considerations. *Childs Nerv Syst*. 2014;30:205–214.
16. Kiecker C. The origins of the circumventricular organs. *J Anat*. 2018;232:540–553.
17. Greenberg RW, Lane EL, Cinnamon J, Farmer P, Hyman, RA. The cranial meninges: Anatomic considerations. *Semin Ultrasound CT MR*. 1994;15:454–465.
18. Reubi JC, Schaer JC, Waser B, Mengod G. Expression and localization of somatostatin receptor SSTR1, SSTR2, and SSTR3 messenger RNAs in primary human tumors using in situ hybridization. *Cancer Res*. 1994;54:3455–3459.
19. Goldgeier MH, Klein LE, Klein-Angerer S, Moellmann G, Nordlund JJ. The distribution of melanocytes in the leptomeninges of the human brain. *J Invest Dermatol*. 1984;82: 235–238.

10

Pituitary and Sellar Region

Benjamin B. Cho ■ M. Beatriz S. Lopes

EMBRYOLOGY

To fully appreciate the anatomy of the pituitary (hypophysis), an understanding of its embryogenesis is essential. The gland consists of an anterior lobe (adenohypophysis), a posterior lobe (neurohypophysis), and an intermediate zone (Fig. 10.1). The development of each differs significantly.

The adenohypophysis has its origin in a thickening of oral ectoderm (1,2). During the 3rd week of gestation, this thickened plate invaginates in a cephalad direction to form the Rathke pouch, which retains its connection to the stomodeum via a narrow stalk. In the 6th week, the stalk becomes so attenuated that the pouch loses its stomodeal attachment as it comes into contact with the infundibulum. Cellular proliferation in the anterior wall of Rathke pouch gives rise to the pars distalis, the principal and largest portion of the anterior lobe. In addition, a tongue-like extension of the pars distalis, the pars tuberalis, grows upward to partially surround the anterior surface of the infundibulum. The posterior portion of Rathke pouch gives rise to a thin segment of pituitary gland, the pars intermedia or intermediate lobe. In this zone, microcystic remnants of Rathke pouch containing colloid-like material are commonly seen (Fig. 10.1). Gross cystic dilatation of such remnants is common but infrequently produces clinically significant intermediate lobe or Rathke cleft cysts.

A remnant of the pharyngohypophyseal stalk is demonstrable in fetuses and occasionally encountered in adults; it comprises the pharyngeal pituitary (3). Located in the midline, beneath the mucoperiosteum of the nasopharynx, the pharyngeal pituitary extends from the posterior border of the vomer along the sphenoid bone. Although the full spectrum of anterior pituitary hormone–producing cells may be demonstrated in pharyngeal pituitaries, they are rarely the seat of medical or surgical disease (3).

The neurohypophysis develops from a neuroectodermal bud first noticeable in the floor of the diencephalon at 4 weeks of gestation (4). Two weeks later, the protrusion grows ventrally to abut the posterior portion of Rathke pouch. This specialized portion of the nervous system comprises the magnocellular nuclei, their axons within the median eminence and infundibular stalk, and their terminations in the pars nervosa (posterior lobe). Oxytocin and vasopressin, as well as their carrier proteins the neurophysins, are detectable in the supraoptic nucleus and paraventricular nucleus at 19 weeks and in the posterior lobe at 23 weeks (5).

As early as 7 to 8 weeks of gestation, the portal system begins to develop. By 12 weeks, both the median eminence and the anterior lobe are vascularized but the circulation of the hypothalamic–pituitary portal system is not completed until 18 to 20 weeks gestational age (5).

By 12 weeks, a number of cartilaginous plates have fused to form the cartilaginous neurocranium (1). The body of the sphenoid bone and the sella turcica results from fusion of hypophyseal cartilaginous plates located on either side of the developing pituitary. The sella is well formed by 7 weeks and matures through a process of enchondral ossification.

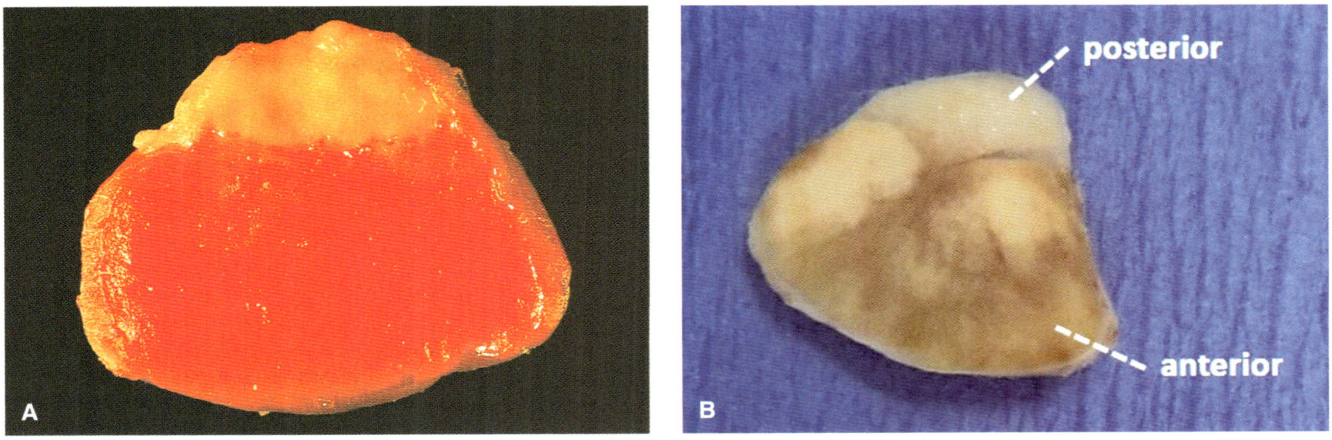

FIGURE 10.1 **A:** Normal unfixed adult pituitary gland cut in the horizontal plane. The posterior lobe is located at the top of the field. A few intermediate lobe cysts are present. The deep red color of the anterior lobe is a reflection of its extensive vascularity. **B:** Normal formalin-fixed adult pituitary gland highlighting the posterior and anterior lobes.

Rare developmental malformations of the pituitary gland, including ectopic pituitary gland in the suprasellar region and pituitary dystopia, have been reported (6,7).

The understanding of pituitary development has expanded as a result of the identification of molecular mechanisms that may specify cell determination and differentiation. Recent advances in pituitary development in mammals have suggested that pituitary organogenesis is controlled by a combination of sequential exogenous and endogenous signals (8–11). These signals induce the expression of interacting transcriptional regulators in temporal and spatial patterns. Pituitary cell types appear to emerge from a common pluripotent stem cell under the response of lineage-specific transcription factors and locally produced growth factors critical for determination and differentiation of specific cell types. Briefly, pituitary organogenesis and early differentiation require multiple transcription factors, one of which is the Prophet of PIT1 (PROP1). Pituitary-specific POU-class homeodomain transcription factor 1 (PIT1) is a transcription factor that regulates the functional differentiation of the somatotrophs, lactotrophs, and thyrotrophs. High levels of estrogen receptors in cells that contain PIT1 favor prolactin (PRL) expression, whereas thyrotroph embryonic factor (TEF) induces thyroid-stimulating hormone (TSH) expression. Familial and sporadic mutations affecting PROP1 expression result in combined growth hormone (GH), PRL, TSH, and gonadotropin deficiency (12). The basic helix–loop–helix transcription factor NeuroD1/beta2 and the T-box family member TBX19 (TPIT) appear to synergistically play a role in the functional differentiation for pituitary proopiomelanocortin (POMC) cell lineage and corticotroph cells. PROP1 induces the pituitary development of PIT1-specific lineages as well as gonadotrophs. The nuclear receptor steroidogenic factor 1 (SF1) and GATA2 are key factors for further differentiation of gonadotroph cells (13). These transcription factors have been localized in human embryonic and adult pituitary gland and have served as diagnostic tools for characterization of pituitary adenomas in surgical pathology practice (Table 10.1).

TABLE 10.1 Adenohypophysial Cell Lineage Transcription Factors

Lineage	Main Transcription Factors and Other Cofactors	Pituitary Cell
PIT1 lineage	PIT1	Somatotroph cell
	PIT1, ERα	Lactotroph cell
	PIT1, GATA2	Thyrotroph cell
Corticotroph lineage	TPIT	Corticotroph cell
Gonadotroph lineage	SF1, ERα, GATA2	Gonadotroph cell

PIT1, pituitary-specific POU-class homeodomain transcription factor 1; ERα, estrogen receptor α; GATA2, member of the GATA family of zinc-finger transcriptional regulatory proteins; TPIT, T-box family member TBX19; SF1, steroidogenic factor 1.

The time period of recognition of the various human pituitary hormone–producing cells during embryonal development have been well determined by immunohistochemical stains. Corticotrophs are the first cells to differentiate in the human fetal pituitary (at around 5 weeks of gestational age). Somatotrophs appear around 8 to 9 weeks, followed by thyrotrophs and gonadotrophs at 12 to 15 weeks. Lactotrophs, although seen in small numbers as early as 12 weeks, are only fully recognizable at 23 weeks (5,14,15).

GROSS ANATOMY

Bony Sella

The pituitary gland is centrally situated at the base of the brain, where it lies safely nestled in the sella turcica, a saddle-shaped concavity within the sphenoid bone (Figs. 10.2 to 10.4). It is attached to the hypothalamus by both the

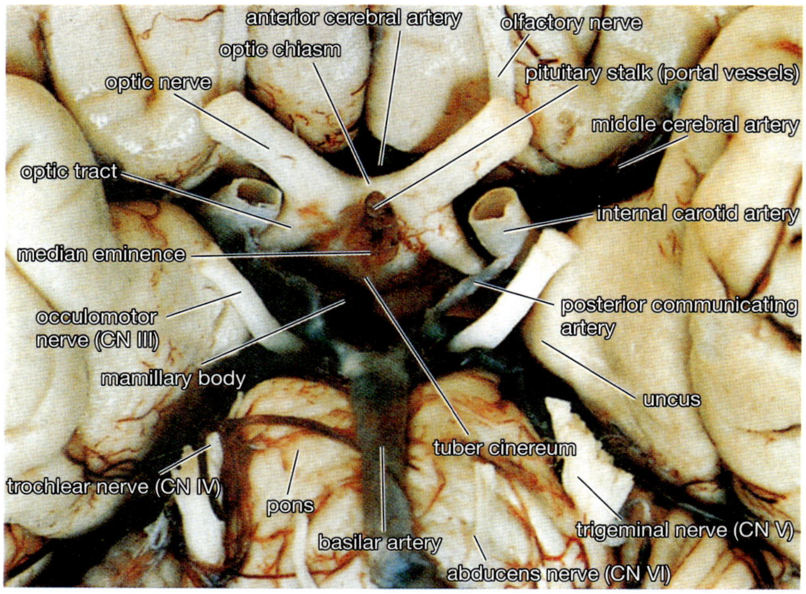

FIGURE 10.2 Ventral surface of normal brain showing the pituitary stalk and surrounding structures. The pituitary gland has been removed. The proximity of the optic chiasm to the pituitary is the basis for the visual field deficits accompanying suprasellar extension of pituitary adenomas. Visible along the posterior aspect of the stalk are tributaries of the portal system.

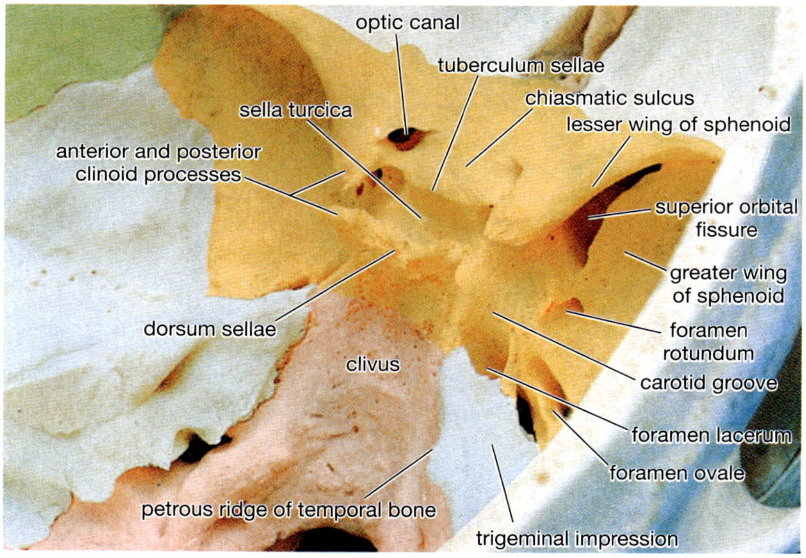

FIGURE 10.3 Oblique view of the normal skull base; the various skull bones are indicated in color. The sella turcica is centrally located with several foramina nearby. The foramen spinosum is not visible in this view. Yellow, sphenoid; pink, occipital; light blue, temporal; green, parietal; white, frontal.

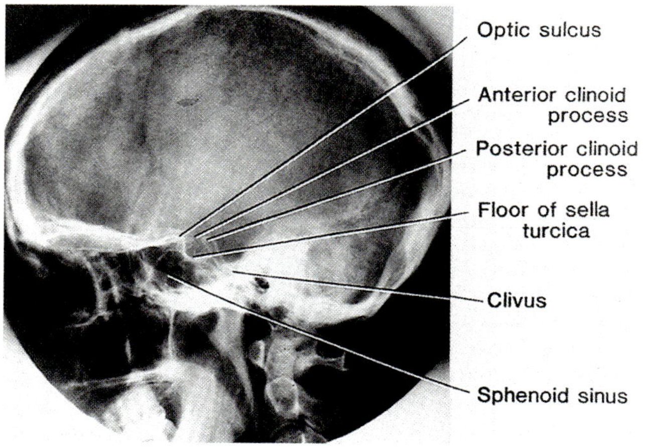

FIGURE 10.4 This normal lateral skull radiograph shows the central location of the sella turcica and surrounding bony anatomy.

pituitary stalk and a tenuous vascular network (Figs. 10.5 to 10.7). By virtue of its location, the pituitary gland has many important anatomic relations. Anterior to the sella, the sphenoid bone forms a midline slope, the tuberculum sella, as well as a transverse indentation, the chiasmal sulcus, so named for the overlying optic chiasm (Figs. 10.3 and 10.4). The optic canals, which transmit the optic nerves, lie anterolateral to the sulcus, whereas the optic tracts are posterolateral. In view of the pituitary's proximity to the optic apparatus, pituitary lesions that extend superiorly may cause significant visual field deficits (Figs. 10.2, 10.5 to 10.11). Specifically, compression of decussating fibers in the chiasm produces bitemporal hemianopsia, whereas compromise of an optic tract leads to homonymous hemianopsia. Suprasellar extension may cause hypothalamic dysfunction and hydrocephalus.

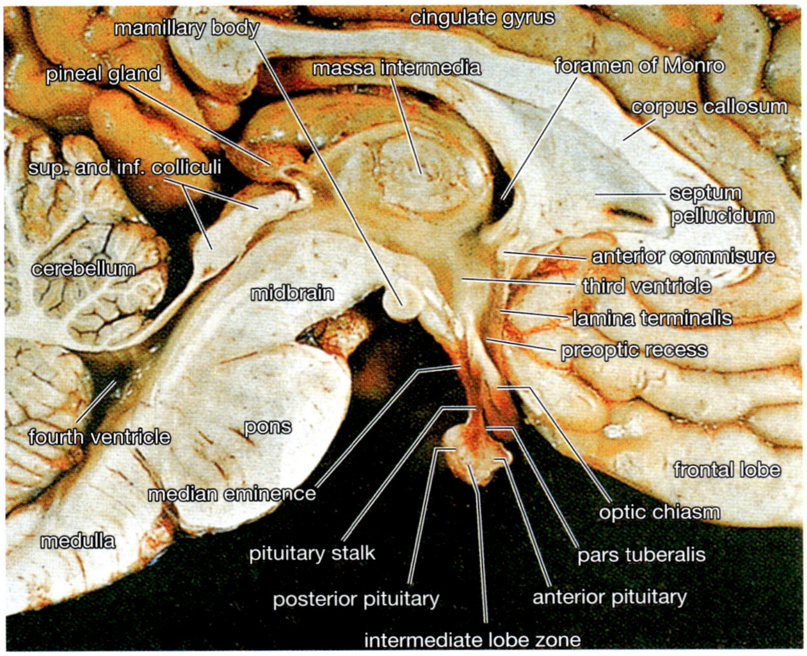

FIGURE 10.5 Midline sagittal section through the brain at the level of the pituitary stalk and pituitary gland, showing the pituitary gland with surrounding structures including hypothalamus, third ventricle, and optic chiasm.

The floor of the sella forms a portion of the roof of the sphenoid air sinus, a fortuitous relationship that permits ready surgical access (Figs. 10.4, 10.6 to 10.12) (16). Indeed, the transsphenoidal approach to the pituitary initially involves mobilization of the nasal septal cartilage followed by resection of a portion of the ethmoid plate. Sublabial or endonasal access to the sphenoid sinus and the sellar region are commonly used for surgical resection of pituitary lesions. Upon breaking through the anterior sphenoid wall, the sella may be seen bulging into the roof of the sinus. A septated sphenoid sinus may affect the surgeon's orientation at surgery (Figs. 10.8 and 10.9). The pituitary is then exposed by traversing the bony sellar floor and incising the dural investment around the gland.

The sloping anterior sellar wall terminates in posterolateral projections, the anterior clinoid processes (Figs. 10.3 and 10.4). Posterior to the sella, the sphenoid bone continues as the dorsum sellae, anterolateral portions of which form the posterior clinoid processes (Figs. 10.3 and 10.4). Posterior to the dorsum sellae is situated the downward-sloping clivus, notorious as the site of predilection of chordomas (Figs. 10.3, 10.4, 10.6, and 10.7). A number of neurovascular foramina are situated in the sellar region; by name and contents from anterior to posterior, they include

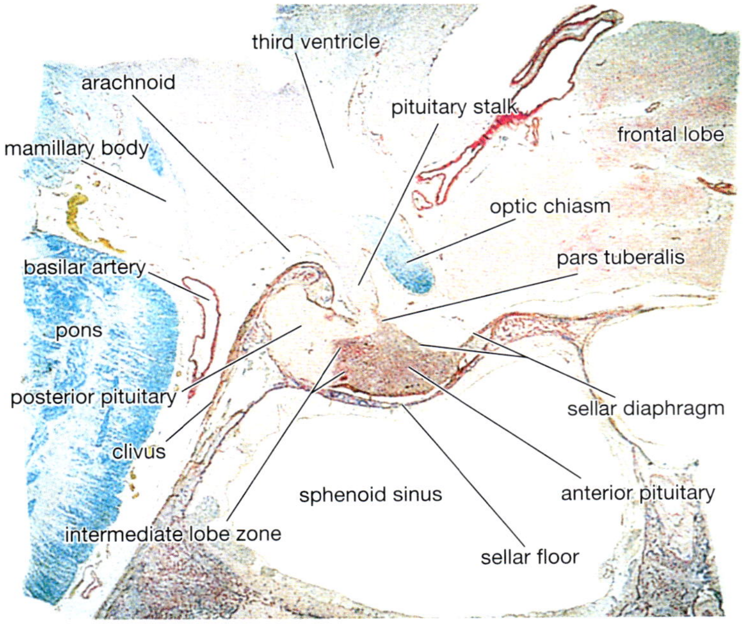

FIGURE 10.6 Sagittal whole-mount section of normal pituitary gland and surrounding structures. The anterior (*at right*) and posterior (*at left*) lobes are clearly delineated. The pars tuberalis is the thin tongue-shaped portion of the anterior lobe that extends for a short distance up the stalk. This diagram illustrates the proximity of the optic chiasm to the pituitary. Superior extension of a pituitary tumor may compress the optic chiasm with resultant visual field deficits, whereas downward extension may fill the sphenoid sinus (Luxol-Fast Blue–PAS).

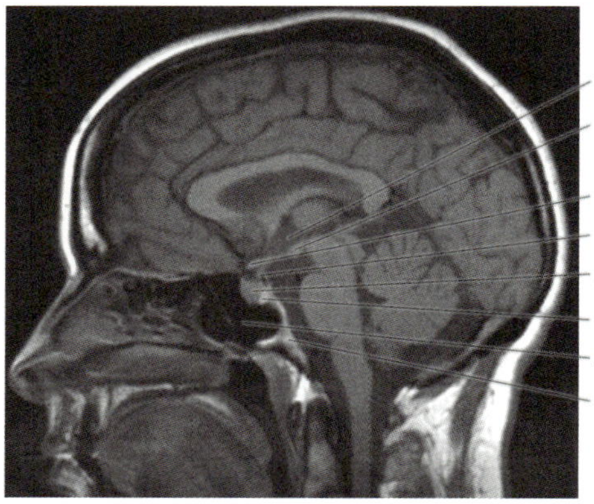

FIGURE 10.7 Magnetic resonance imaging (MRI): Sagittal view of the brain at the level of the pituitary stalk and gland. The clarity of the pituitary gland, stalk, hypothalamus, and optic chiasm is remarkable, making MRI an excellent imaging modality for the assessment of pituitary lesions. One advantage of MRI over computed tomography (CT) is the absence of bony artifact with MRI.

the foramen rotundum that permits passage of the maxillary nerves, ovale traversed by the mandibular nerves, spinosum that contains the middle meningeal arteries, and lacerum that contains the internal carotid arteries (Fig. 10.3).

Meninges

The physical relationship of the meninges to the pituitary and sella is unusual in that the pituitary lacks a leptomeningeal investment. Periosteal dura lines the sella turcica, whereas the dura proper covers the lateral aspects of the cavernous sinuses and forms the sellar diaphragm. The diaphragm is usually thin at the center and thick at its periphery; it possesses a variably sized central aperture through which the pituitary stalk passes (17). Leptomeninges do encircle the stalk; however, below the level of the sellar diaphragm, they reflect back upon themselves to form a circumferential channel, the infradiaphragmatic hypophyseal cistern. This arrangement explains the higher incidence of development of meningiomas in the suprasellar surface of the diaphragm rather than in the intrasellar compartment.

In some individuals, the leptomeninges exhibit an important anatomic variation. It consists of extension or herniation of the arachnoid through an exceedingly large diaphragmatic opening. A deficiency of the diaphragma sellae is assumed to be a prerequisite for the formation of an empty sella. In one study, the incidence of an intrasellar arachnoidocele was found to exceed 20% (17). In such cases, transsphenoidal surgery may result in cerebrospinal fluid (CSF) leak because of inadvertent violation of the subarachnoid space. With prolonged exertion of even normal CSF pressure, enlargement of such arachnoidoceles may produce sellar enlargement and pituitary compression, the gland being reduced to a thin crescent on the posterior sellar floor (Fig. 10.13). The so-called empty sella may be present in as many as 5.5% of autopsies (18,19) and shows a distinct predilection for obese, multiparous females

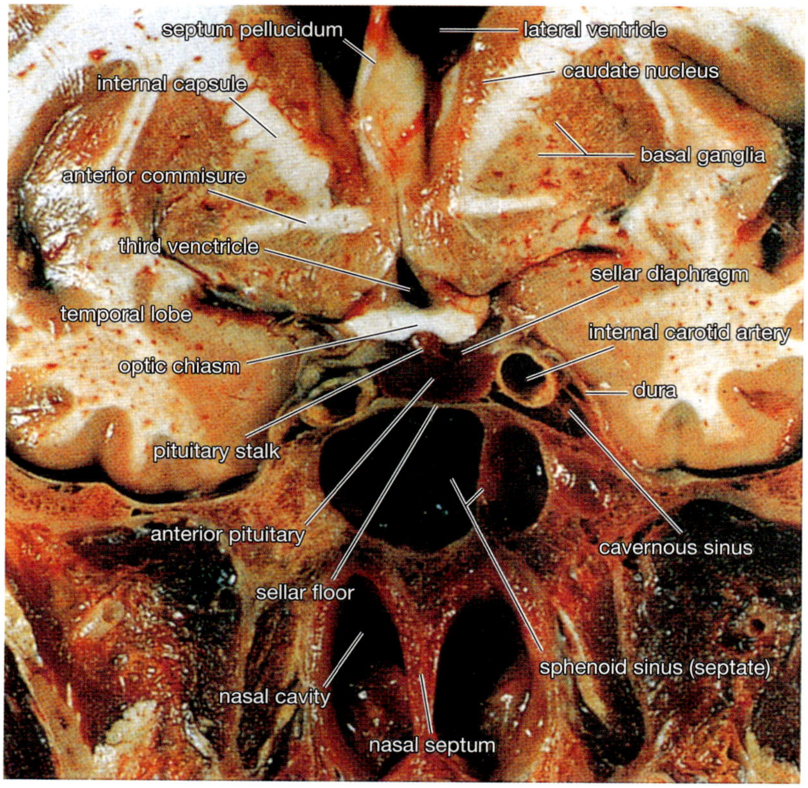

FIGURE 10.8 Coronal section of the head at the level of the pituitary stalk and gland. This photograph clearly illustrates the intimate relationships between the cavernous sinuses, the sphenoid sinus, and the pituitary gland. Invasive adenomas may extend laterally into one or both cavernous sinuses or inferiorly into the sphenoid sinus. Note the proximity of the optic chiasm to the pituitary.

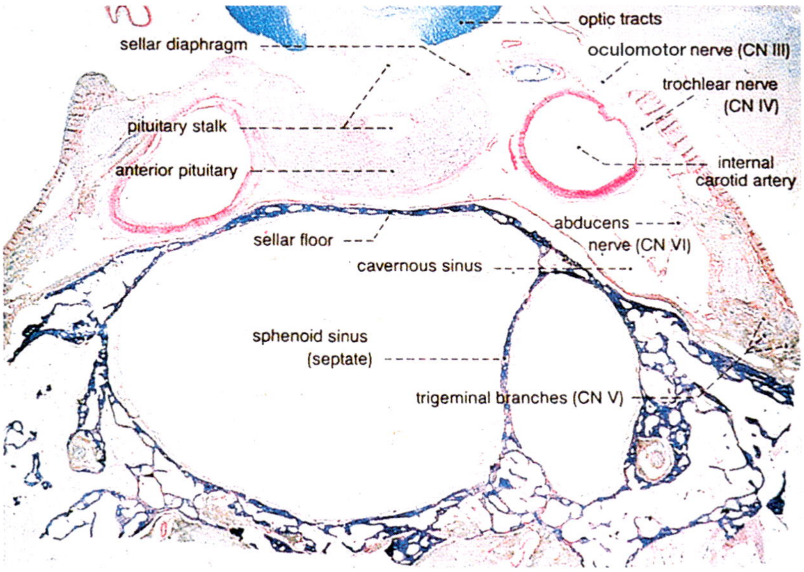

FIGURE 10.9 Coronal whole-mount view of the normal pituitary gland and surrounding structures. Note the location of cranial nerves III, IV, VI, and branches of cranial nerve V within the cavernous sinuses, a relationship explaining the occurrence of cranial nerve palsy in association with invasive pituitary adenomas. This section also illustrates the proximity of the internal carotid arteries to the pituitary gland (Luxol–Fast Blue–PAS).

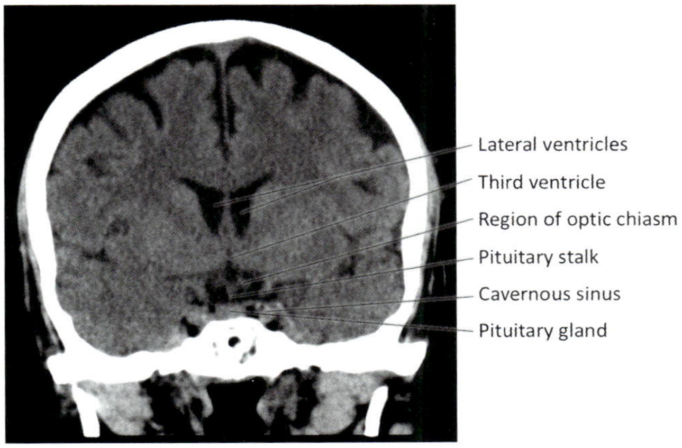

FIGURE 10.10 A computed tomographic (CT) coronal view of the normal skull and brain at the level of the pituitary stalk and gland. The CT scan is a good imaging modality for assessing the pituitary gland; however, radiologists often encounter problems with bony artifact. For this reason, MRI is superior.

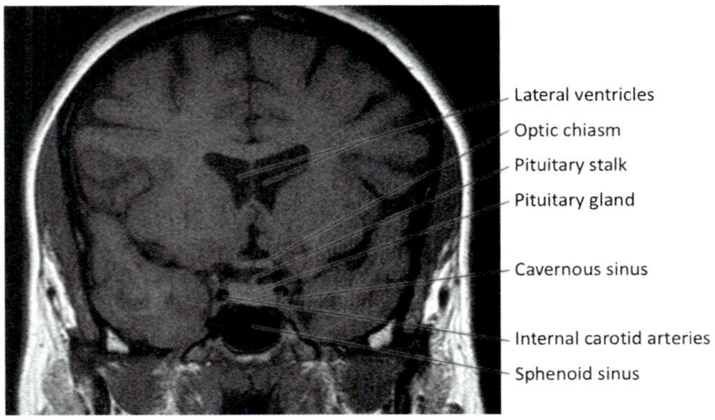

FIGURE 10.11 Coronal MRI of the skull and brain at the level of the pituitary stalk and gland.

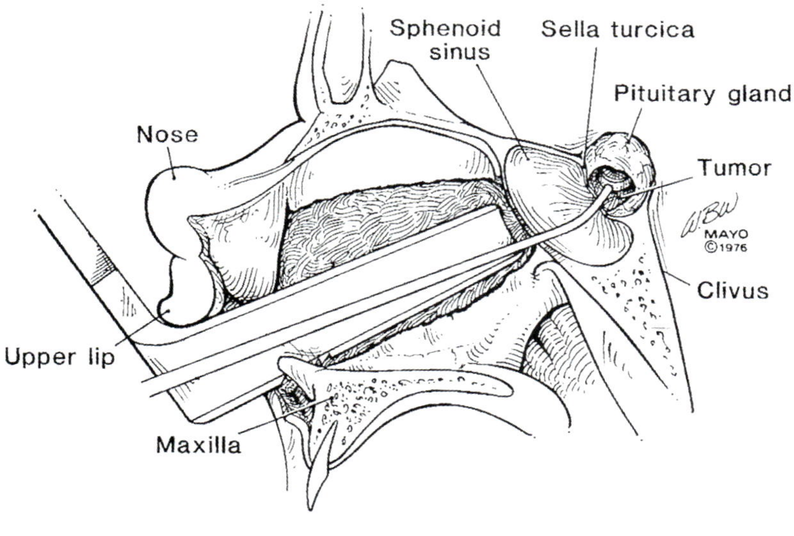

FIGURE 10.12 Transsphenoidal approach to the pituitary gland. After mobilizing the nasal septal cartilage and resecting a portion of ethmoid plate, a sublabial incision is made, and a sphenoid speculum is placed. Next, the floor of the sphenoid sinus and the floor of the sella turcica are traversed. Finally, the dural investment of the pituitary gland is incised, and the gland is exposed. This diagram illustrates a curette in place for removal of a pituitary adenoma.

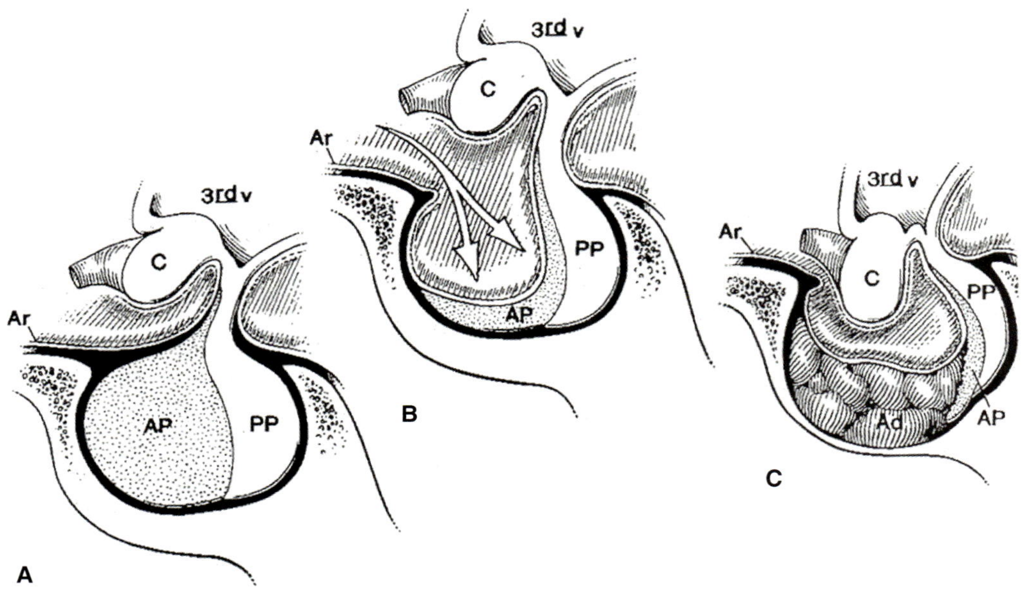

FIGURE 10.13 Illustration of normal as well as variants of the empty sella. **A (left):** The normal pituitary–sellar relationships. The leptomeninges cover the stalk and sellar diaphragm but do not extend into the sella. **B (middle):** In primary empty sella syndrome, an excessively large diaphragmatic orifice permits herniation of leptomeninges into the sella. Prolonged CSF pressure compresses the gland against the sellar floor. **C (right):** Secondary empty sella may result from infarction of a pituitary adenoma, infarction of the pituitary gland, and surgical or radioablation of the gland. (Ar, arachnoid; AP, anterior pituitary; PP, posterior pituitary; C, optic chiasm; 3rd v, third ventricle.)

(Fig. 10.14). Compression of the gland and traction deformation of the pituitary stalk may cause mild to moderate hypopituitarism and hyperprolactinemia, respectively.

Vasculature

Vascular structures of major surgical significance abound in the sellar region.

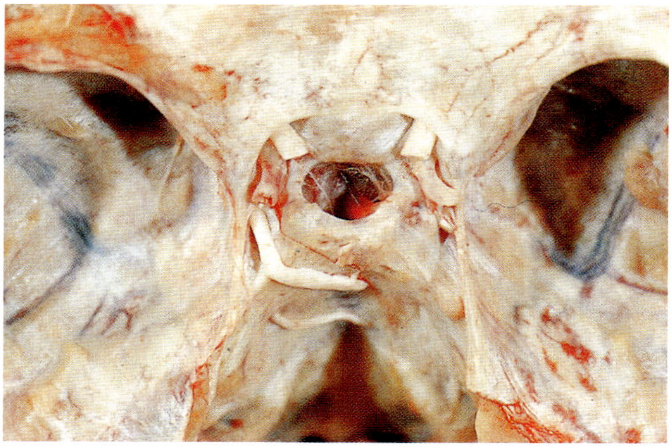

FIGURE 10.14 Superior view of the skull base, demonstrating an incidentally encountered primary empty sella. Normally in this view the upper surface of the pituitary gland would be visible through the diaphragmatic aperture, but here the sella appears empty. Only rarely is primary empty sella syndrome symptomatic. This specimen is from a 57-year-old obese diabetic woman.

The Cavernous Sinuses

The paired cavernous sinuses are situated on either side of the sella and, in part, lie lateral and superior to the sphenoid sinuses (Figs. 10.8 to 10.11). Each cavernous sinus is partially invested by dura of the middle fossa, as well as by thin bony walls of the sphenoid sinus. Venous drainage to the sinuses comes from a number of sources, including the eye from the superior ophthalmic vein, the brain from inferior and middle cerebral veins, and sphenoparietal sinus. Communication between right and left cavernous sinuses takes place through intercavernous sinuses bordering the anterior and posterior aspects of the sella (17). The complex thus forms a venous ring around the sella and its contents. Additional intercavernous sinuses are located along the ventral surface of the pituitary. The cavernous sinuses proper contain, in addition to their content of venous sinuses, a number of vital neurovascular structures (17). These include the cavernous segments of the internal carotid arteries and segments of cranial nerves III (oculomotor), IV (trochlear), V (trigeminal), and VI (abducens) (Figs. 10.8 and 10.9). Delicate areolar tissue fills the interstices between venous channels, arteries, and nerves. The location of the horizontal portions of the internal carotid arteries within the cavernous sinuses varies, not only from person to person but also from left to right. As a result, the carotids may lie immediately adjacent to the sella, in which case they may create a surgical risk (Figs. 10.8 and 10.9). Several branches of the internal carotid artery arise within the cavernous sinus, including the meningohypophysial trunk, which is the largest intracavernous branch, the artery of the inferior cavernous sinus, and McConnell capsular arteries (17).

Given their location, the cavernous sinuses may be directly involved by pituitary tumors. For example, extension of an invasive adenoma into the cavernous sinuses may produce neuropathies of cranial nerves III through VI (Fig. 10.9), including ptosis, facial pain, or diplopia.

The Arterial Supply

The principal arterial supply of the pituitary originates in two branches of the internal carotids: the superior and inferior hypophyseal arteries (Fig. 10.15) (20,21). A single superior hypophyseal artery springs from each carotid shortly after its entry into the cranial cavity and promptly divides into posterior and anterior branches, each of which anastomoses with the corresponding branch from the opposite side to form an arterial ring around the upper pituitary stalk. The anterior branches give rise to trabecular or loral arteries, which descend on the upper surface of the anterior lobe, course toward the pituitary stalk, and terminate in long-stalk arteries along the pars tuberalis. In their brief course along the anterior lobe, trabecular arteries each give rise to a small artery of the fibrous core (20). The posterior and anterior branches of the superior hypophyseal arteries are also the source of short-stalk arteries, which penetrate the superior aspect of the pituitary stalk to run upward or downward within it.

In contrast to the superior hypophyseal arteries, the inferior branches originate from the meningohypophyseal

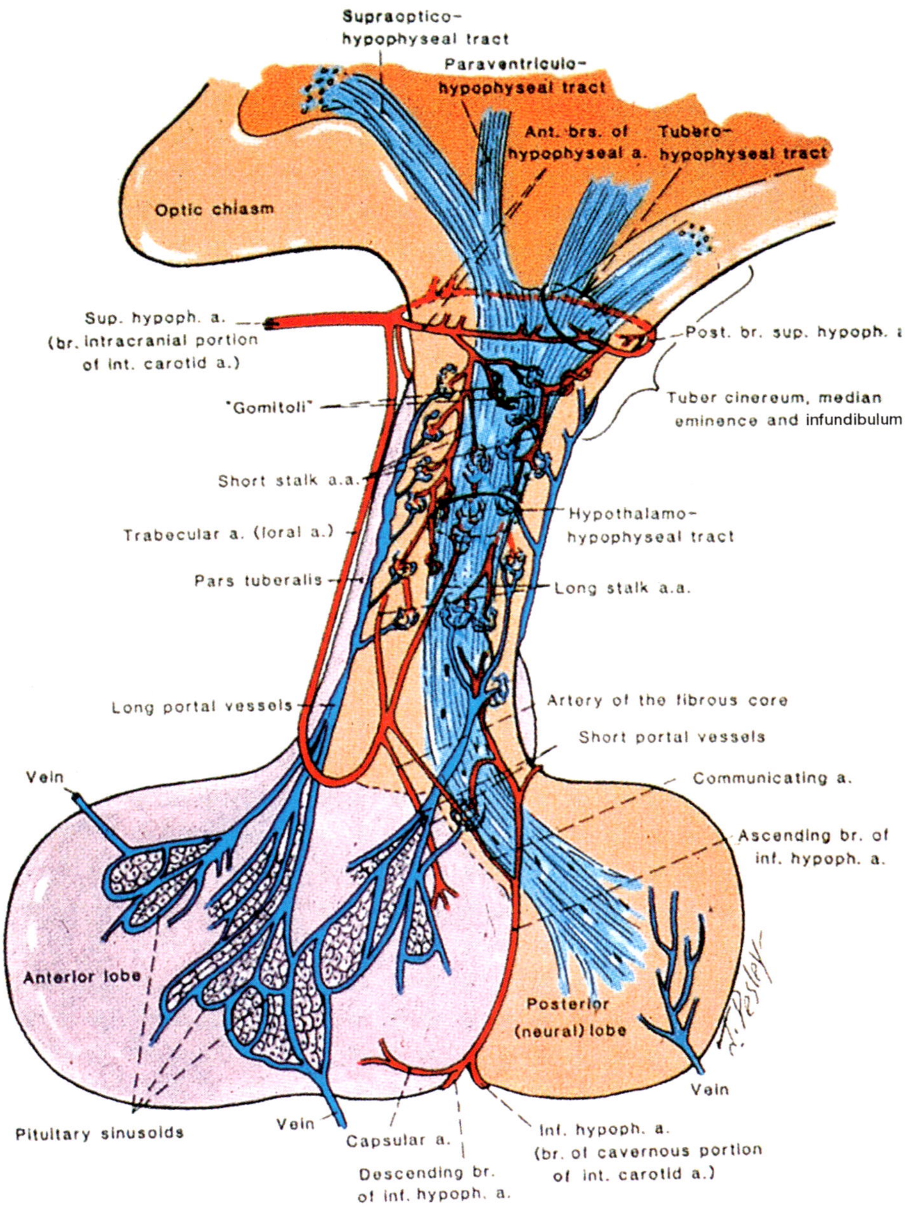

FIGURE 10.15 Diagrammatic representation of the vasculature of the pituitary gland. The superior and inferior hypophyseal arteries and branches of the internal carotid arteries comprise the major blood supply of the gland. Small terminal branches of the superior and inferior hypophyseal arteries give rise to tangled capillary loops termed gomitoli, which drain into portal vessels. The latter traverse the length of the stalk and terminate as a capillary bed in the anterior lobe. The anterior pituitary thus receives the majority of its blood supply not from arteries but from the portal system. The portal system forms a vital link between the hypothalamus and the pituitary gland.

trunk within the cavernous sinuses. The meningohypophyseal trunk gives rise to several vessels, one of which is the inferior hypophyseal artery. They contact the inferolateral portions of the gland and bifurcate into medial and lateral branches that anastomose with their opposite counterparts to form an arterial circle about the posterior lobe. Thus, branches of the inferior hypophyseal arteries supply primarily the posterior lobe and lower portion of the stalk, contributing only small capsular branches to the periphery of the anterior lobe (22). Although many of the arterial branches in the pituitary stalk and infundibulum form arterioles and capillaries, some give rise to unique vascular complexes termed gomitoli (Fig. 10.15). These "balls of thread" consist of a central artery surrounded by a glomeruloid tangle of capillaries. The transition from central arteries to the capillaries is via short specialized arterioles endowed with thick smooth muscle sphincters that serve to regulate blood flow. The mixture of periarteriolar capillaries drains into an extensive pampiniform network, the portal system, which envelopes the stalk (Figs. 10.2 and 10.15).

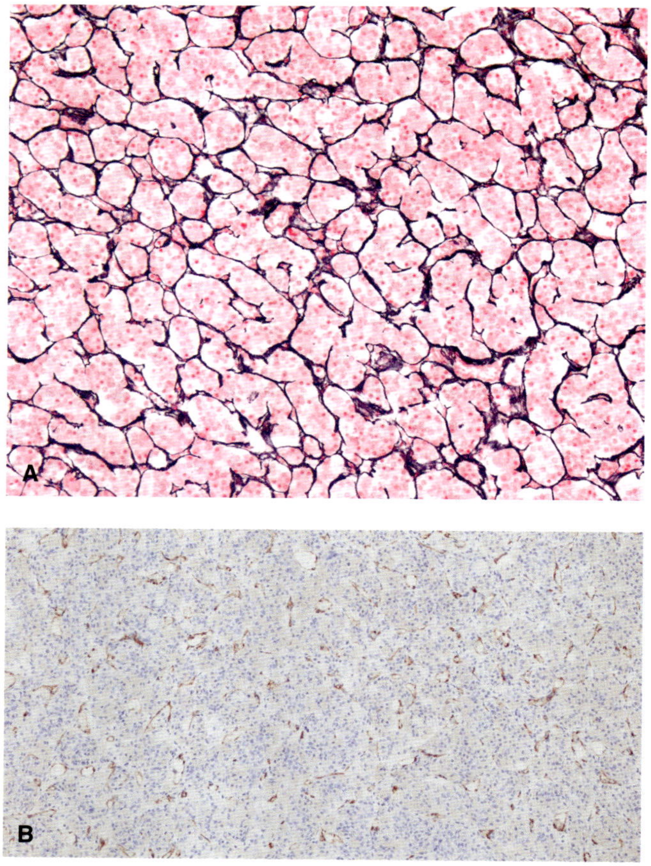

FIGURE 10.16 Adenohypophysis. **A:** The intricate capillary and connective tissue network outlined in reticulin stain. The reticulin stain is invaluable in the evaluation of pituitary adenomas, which are largely devoid of reticulin, whereas the surrounding normal gland retains it (Wilder reticulin, original magnification ×40). **B:** The capillary endothelium of the anterior lobe capillary network stains strongly for CD31 (immunostain, original magnification ×100).

The Hypophyseal Portal System

The hypophyseal portal system, the critical link between hypothalamus and pituitary, takes its origin from the capillary plexus of the median eminence and stalk, which itself is derived from terminal ramifications of the superior and inferior hypophyseal arteries (21). The capillary plexus in the median eminence and superior stalk, the site of uptake of hypophysiotropic (hypothalamic) factors, drains into the long portal vessels that course along the surface of the stalk to supply the majority (90%) of the anterior lobe, whereas the smaller capillary plexus in the lower stalk gives rise to the short portal vessels that descend into its central portion, including that bordering the posterior lobe (22). Distally, the portal system communicates with a delicate capillary network in the anterior lobe, which carries hypophysiotropic factors into the pituitary and conveys anterior lobe hormones to the general circulation (Fig. 10.16). Therefore, aside from a minor direct arterial supply via capsular branches of the inferior hypophyseal arteries, the majority of the anterior lobe's circulation is venous, originating from the portal vessels (20–22). In contrast to that of the anterior lobe, the blood supply of the posterior lobe is direct and arterial, a characteristic that explains the predilection of metastatic carcinomas for the neural lobe.

Venous outflow of the pituitary is via collecting vessels that drain into the subhypophyseal sinus, cavernous sinus, and superior circular sinus (21).

PHYSIOLOGY AND HISTOLOGY

Hypothalamus

The pituitary is known to be under significant hypothalamic control. In fact, the pituitary and the hypothalamus form a complex neurohormonal circuit, a vital element for maintenance of a normal endocrine status. Weighing approximately 5 g and forming the lower walls and floor of the third ventricle, the hypothalamus lies above the pituitary to which it is connected by the pituitary stalk (Figs. 10.5 to 10.7, 10.16, and 10.17). As the name suggests, the hypothalamus lies inferior to the thalamus. Although at first glance it appears poorly demarcated, the hypothalamus is bordered anteriorly by the anterior commissure, optic chiasm, and lamina terminalis; posteriorly and superiorly by the midbrain tegmentum and mammillary bodies; dorsally by the hypothalamic sulcus; and laterally by the subthalamic nuclei (23,24). The region consists of several ill-defined but functionally related neuronal nuclei (Figs. 10.17 and 10.18). Afferent and efferent connections bring these nuclei into contact with nearby as well as remote portions of the central nervous system, including other diencephalic structures, the cerebrum, brainstem, and spinal cord. Two major and distinct hypothalamohypophyseal secretory systems are recognized. The first is composed of the SON and PVN and their projections to

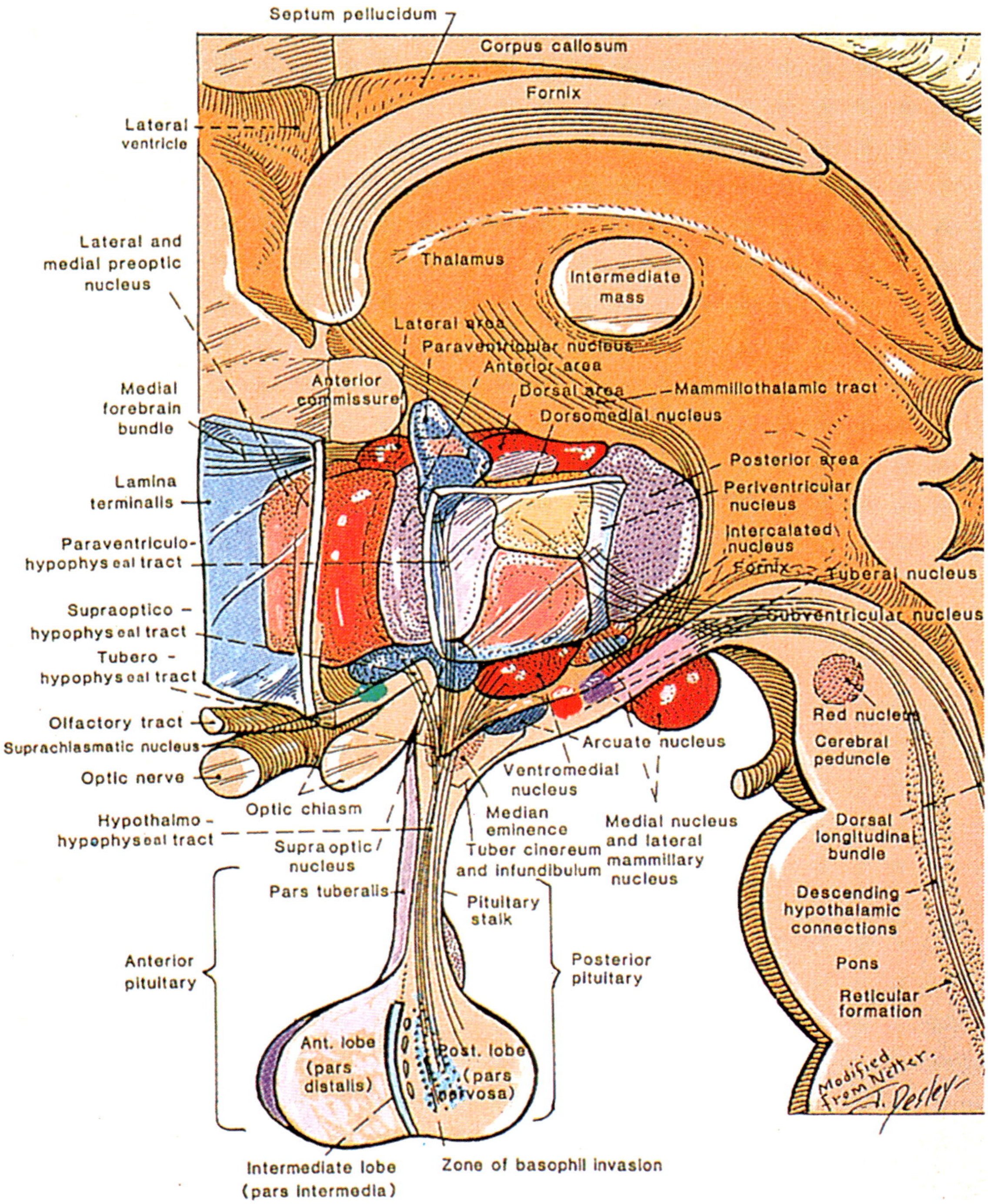

FIGURE 10.17 Diagrammatic representation of the hypothalamic nuclei, showing the supraopticohypophyseal and paraventriculohypophyseal tracts, as well as the tuberohypophysial tract. The former carry vasopressin and oxytocin along axons to the posterior lobe, whereas the latter carries hypothalamic releasing and inhibiting hormones to the median eminence, where they enter the portal system for transport to the anterior lobe.

the neurohypophysis (i.e., supraopticohypophyseal and paraventriculohypophyseal tracts, respectively); the other is composed mainly of nuclei of the tuberal region, the funnel-shaped floor of the third ventricle, and their processes terminating in the median eminence (i.e., tuberohypophyseal tract) (Fig. 10.17).

The supraoptico nucleus is located superior to the optic tracts, whereas the wedge-shaped PVN lies ventromedial to the fornix and abut the walls of the third ventricle (Fig. 10.18). Due to their predominant composition of large neurons measuring up to 25 μm, these are termed magnocellular nuclei (Fig. 10.19). Each contains vasopressin- and

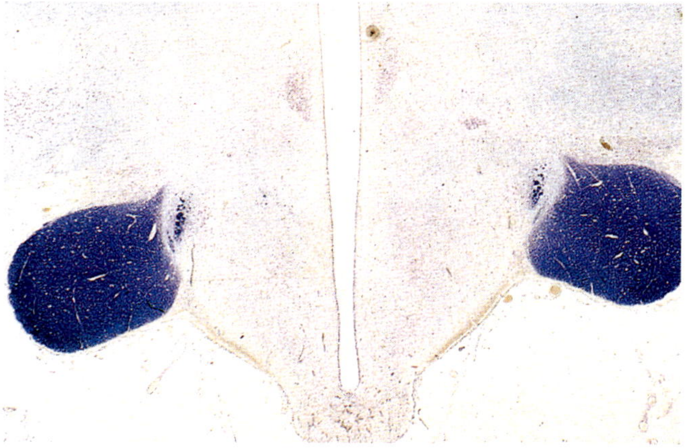

FIGURE 10.18 Coronal whole-mount section through the hypothalamus and third ventricle. The PVN is visible as darkly staining areas beneath the ependyma of the third ventricle (*upper field*), whereas the SON lies above the heavily myelinated optic tracts. The arcuate nucleus lies inferior to the base of the third ventricle (*cresyl violet*).

oxytocin-producing neurons, but only one hormone is produced by a given neuron. Their long axons form the supraopticohypophyseal and paraventriculohypophyseal tracts, which carry vasopressin and oxytocin (the so-called neurohypophyseal hormones), as well as their respective carrier proteins, the neurophysins, to the posterior lobe of the pituitary gland.

Oxytocin and vasopressin, both nonapeptides, are synthesized mainly in the perikarya of magnocellular neurons in the supraoptico nucleus and paraventricular nucleus; parvocellular neurons in the paraventricular nucleus also synthesize vasopressin. The two hormones differ by only two amino acids. Both are synthesized as a 20-kDa precursor peptide that is rapidly cleaved into the active hormone and its neurophysin carrier protein (neurophysin I and II), packaged into secretory granules in the Golgi as a hormone–neurophysin complex, and secreted from nerve endings that terminate primarily in the neurohypophysis (25). Large intra-axonal accumulations of these hormones, named Herring bodies, are often visible by light microscopy as round, granular structures that appear eosinophilic on hematoxylin and eosin (H&E) sections (Fig. 10.20). In transit from the hypothalamus to the posterior lobe, prohormones undergo extensive processing and cleavage to form the final products, vasopressin and oxytocin. Although these hormones differ by only two amino acids, oxytocin exhibits virtually no antidiuretic activity, and vasopressin has negligible oxytocic effect. Oxytocin mediates the "milk let-down reflex" by stimulating contraction of myoepithelial cells surrounding terminal mammary lobules. In addition, it serves a role in parturition, binding, and facilitating contraction in the final stages of parturition. The major physiologic role of vasopressin, also called antidiuretic hormone (ADH), is the formation of hypertonic urine. Acting via cyclic adenosine 3′,5′-monophosphate (AMP),

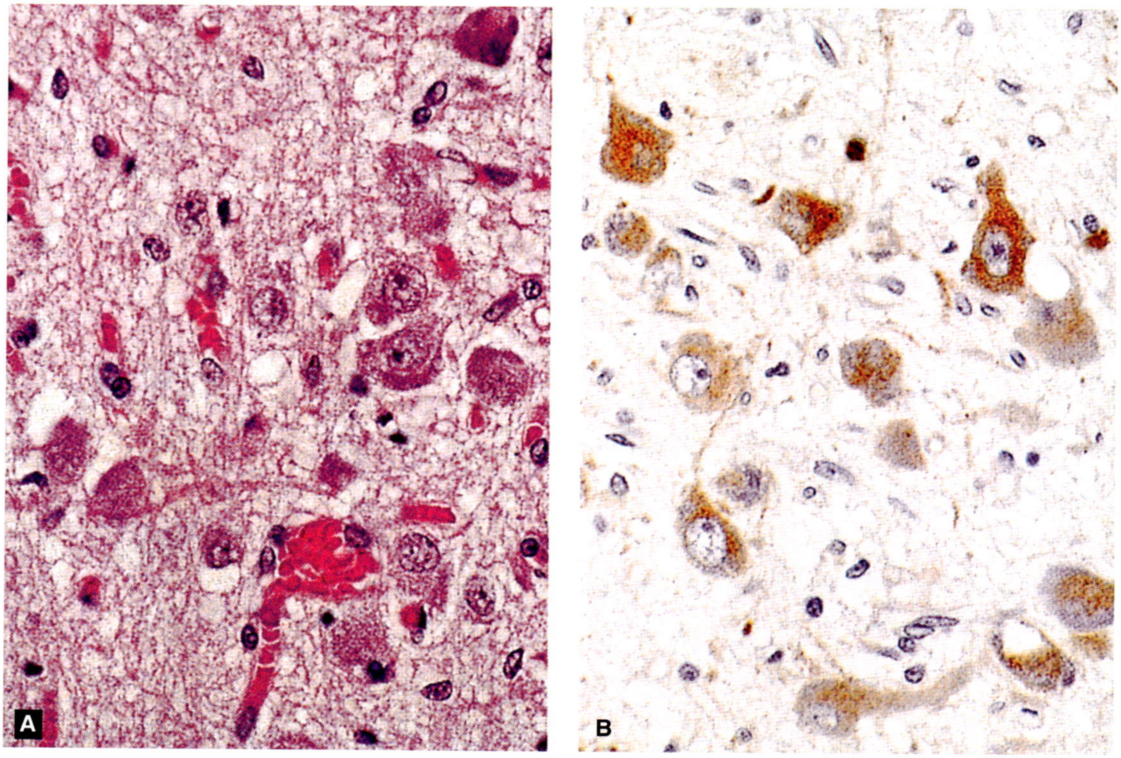

FIGURE 10.19 Paraventricular nucleus of the hypothalamus. **A:** Its high degree of vascularity is a characteristic of magnocellular nuclei (H&E, original magnification ×100). **B:** The nerve cell bodies stain positively for vasopressin (immunostain, original magnification ×100).

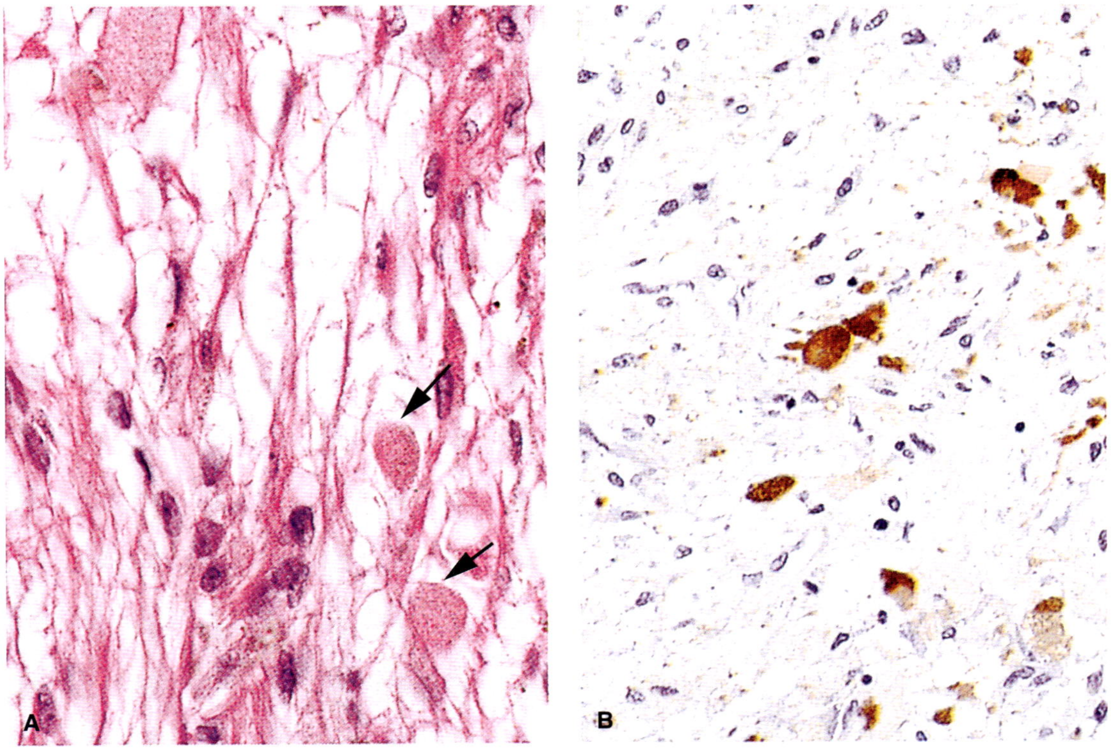

FIGURE 10.20 Pituitary stalk. Axonal swellings termed Herring bodies characterize its axons. **A:** They are distinguished by their ovoid shape and granular character on H&E (*upper left* and *lower right; arrows*). Herring bodies represent intra-axonal accumulations of oxytocin- and vasopressin-containing granules en route to the posterior lobe (H&E, original magnification ×100). **B:** Herring bodies stain for vasopressin (immunostain, original magnification ×100).

vasopressin increases the water permeability of renal collecting ducts, allowing the hypotonic intraductal fluid to equilibrate with the hypertonic fluid in the medullary interstitium. The results are concentrated urine and conservation of body water. Genetic mutations in either the signal peptide or VP-neurophysin give rise to central diabetes insipidus. Likewise, damage to the neurohypophysis from head trauma, surgery, inflammatory processes, or neoplasms may destroy vasopressin-producing neurons and cause diabetes insipidus.

The second component of the hypothalamohypophyseal system is the tuberoinfundibular tract. Its fibers originate in a number of hypothalamic nuclei lying within the walls of the inferior third ventricle and tuberal region (Figs. 10.17 and 10.18) (23). Products of these nuclei, targeted for the anterior pituitary, consist of releasing and inhibiting hormones. Unlike the magnocellular neurons of the supraoptical nucleus and paraventricular nucleus, these are small neurons and for this reason are termed parvicellular neurons (Fig. 10.21). Their processes project to the median eminence, a highly vascular zone located in the posterior proximal portion of the pituitary stalk. Here, the hypothalamic hormones are released into the first portion of the portal system for transport to the anterior lobe.

The hypothalamic hormones include five peptide hormones: corticotropin-releasing hormone (CRH), growth hormone–releasing hormone (GHRH), gonadotropin-releasing hormone (GnRH) or luteinizing hormone-releasing hormone (LHRH), thyrotropin-releasing hormone (TRH), and somatostatin (SST) or somatotropin release-inhibiting factor (SRIF). The sites of synthesis of the various hypothalamic hormones, their characteristics, and target cells are summarized in Tables 10.2 and 10.3. In addition to these hormones, several bioactive

TABLE 10.2 Pituitary Hormones and Hypothalamic Stimulatory and Inhibitory Hormones

	Hypothalamic Hormone	
Pituitary Hormone	**Stimulatory**	**Inhibitory**
GH	GHRH	Somatostatin
PRL	TRH VIP	Dopamine
ACTH	CRH AVP	?
FSH/LH	GnRH	?
TSH	TRH	Somatostatin

GH, growth hormone; GHRH, growth hormone–releasing hormone; PRL, prolactin; TRH, thyrotropin-releasing hormone; VIP, vasoactive intestinal peptide; ACTH, adrenocorticotropic hormone; CRH, corticotropin-releasing hormone; AVP, arginine vasopressin; FSH, follicle-stimulating hormone; GnRH, gonadotropin-releasing hormone; LH, luteinizing hormone; TSH, thyroid-stimulating hormone.

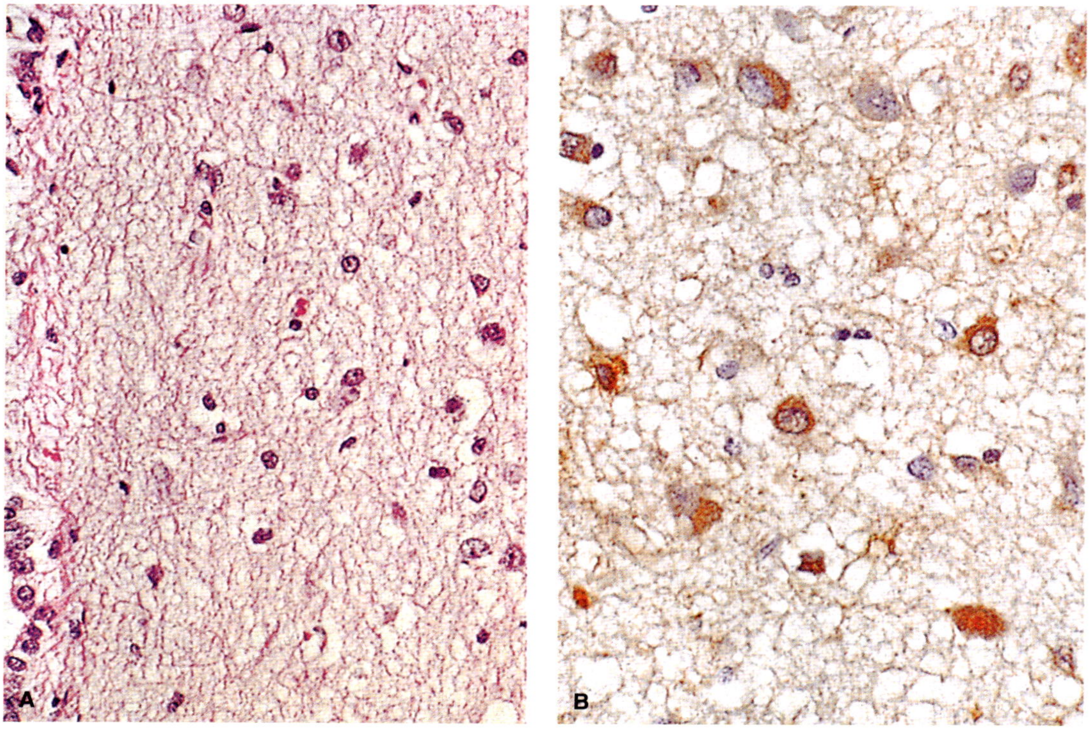

FIGURE 10.21 Periventricular nucleus. **A:** Lying beneath the ependyma of the third ventricle (*bottom left*), this ill-defined nucleus is composed of small nerve cell bodies (H&E, original magnification ×63). **B:** Its constituent neurons stain for CRH, a tropic hormone that exerts its effect on corticotrophs in the anterior lobe (immunostain, original magnification ×100).

substances produced in the hypothalamus participate in the regulation of anterior pituitary hormone secretion. The most significant one is dopamine, which inhibits the secretion of PRL by pituitary lactotrophs (25).

Adenohypophysis

The pituitary gland is a tan to brown, bean-shaped structure varying in weight from 500 to 700 mg (Fig. 10.1). An average-sized gland of 600 mg measures about 13 × 10 × 6 mm. In general, the weight of the female pituitary gland is greater than that of the male (26). Among women, the gland is smaller in nulliparas than in multiparas. In pregnancy the gland enlarges significantly (up to 30%) primarily as a result of lactotroph cell hyperplasia (26). The anterior lobe comprises 80% of the pituitary gland and includes the pars distalis, intermedia, and tuberalis. The body and stalk

TABLE 10.3 Major Hypothalamic–Pituitary Axis Hormones

Hormone	Hypothalamic Sites	Pituitary Hormone	Target Hormone
GHRH	Arcuate nucleus	GH	IGF-1
TRH	Widely distributed in CNS with concentration in ventromedial, dorsal, and paraventricular nuclei, particularly on left	TSH	T4, T3
GnRH	Widespread distribution with concentration in the arcuate, ventromedial, dorsal, and paraventricular nuclei	FSH LH	Estrogen Progesterone/Estrogen (Female) Testosterone (Male)
CRH	Periventricular, medial paraventricular nuclei; colocalized with arginine vasopressin	ACTH	Cortisol DHEA
Somatostatin	Periventricular, paraventricular (parvocellular neurons), arcuate nuclei	GH	N/A
DA	Arcuate nucleus	PRL	N/A

GHRH, growth hormone–releasing hormone; GH, growth hormone; IGF-1, insulin-like growth factor 1; TRH, thyrotropin-releasing hormone; TSH, thyroid-stimulating hormone, T4, thyroxine; T3, triiodothyronine; GnRH, gonadotropin-releasing hormone; FSH, follicle-stimulating hormone; LH, luteinizing hormone; CRH, corticotropin-releasing hormone; ACTH, adrenocorticotropic hormone; DHEA, dehydroepiandrosterone; DA, dopamine; PRL, prolactin.

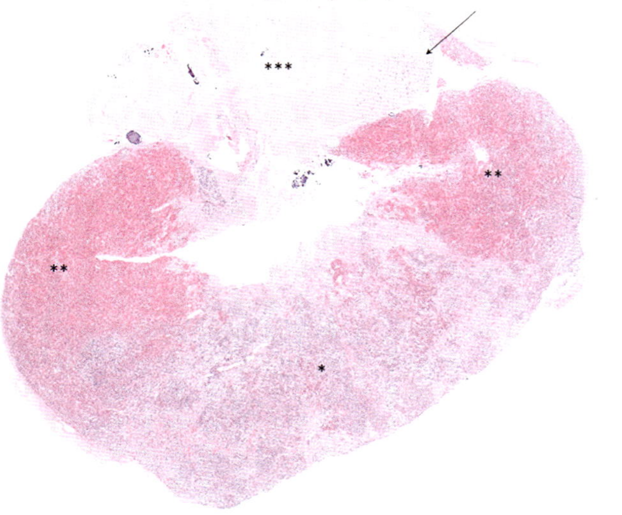

FIGURE 10.22 H&E-stained low-power view of the pituitary sectioned horizontally demonstrates the regional distribution of cellularity in the anterior lobe pars distalis: the basophil-predominant central mucoid wedge (*) is flanked by two lateral wings (**) composed of predominantly acidophilic cells. Note focal basophil invasion (*arrow*) of the posterior lobe (***) (original magnification ×10).

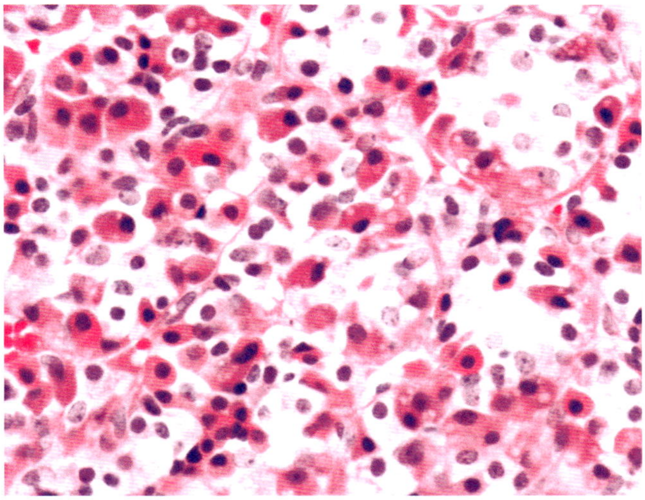

FIGURE 10.23 Anterior lobe. An H&E-stained section shows chromophobic, acidophilic, and basophilic cells. Acidophils are most numerous in the lateral wings, whereas basophils are found in greatest number in the central or mucoid wedge. The different cell types are arranged in acinar formations that can be highlighted by reticulin stain (see Fig. 10.16) (original magnification ×200).

of the gland are surrounded by a delicate capsule derived from the meninges (17). Staining characteristics roughly divide the pars distalis into a central mucoid wedge and two lateral wings, zones best visualized in coronal and horizontal sections (Fig. 10.22).

By light microscopy, the cells of the anterior lobe show variation, not only in size and shape, but also in their histochemical staining characteristics (Figs. 10.23 and 10.24). They are arranged in nests, cords, and small acini bounded by an interlacing capillary network that is best seen on reticulin stain (Fig. 10.16). This architectural pattern, altered in hyperplasia and conspicuously absent in adenomas, is of considerable diagnostic significance (see *Differential Diagnosis*).

The pars intermedia, poorly developed in humans, consists in large part of epithelium-lined spaces containing periodic acid–Schiff (PAS)-positive colloid; the constituent cells are ciliated, goblet, and a few neuroendocrine cells, the latter of which may show variable immunoreactivity for pituitary hormones including adrenocorticotropic hormone (ACTH), luteinizing hormone (LH), and follicle-stimulating hormone (FSH) (Fig. 10.25) (27). These Rathke cleft remnants, the vast majority of which are microscopic in size, are present in about 25% of pituitaries of autopsies (28). Cysts of the Rathke cleft are usually incidental postmortem findings; symptomatic cysts are uncommon, representing about 5% to 9% of surgical lesions of the sellar region (29).

In H&E-stained sections, three principal cell types are identified in the normal anterior lobe: acidophils, basophils, and chromophobes (Figs. 10.23 and 10.24). The designations reflect their staining affinities for acidic and basic dyes, with the chromophobic cells lacking affinity for either. These reactivities form the basis of an outdated classification of pituitary adenomas that offers little in specifying their hormone content or endocrine function. On the other hand, advances in immunohistochemistry permit morphologic and functional correlation (Fig. 10.26). Numerous cells in the central or mucoid wedge are basophilic and stain strongly via the PAS method. Such cells produce ACTH and TSH. In contrast, most cells in the lateral wings are acidophilic and are engaged either in GH or, less frequently, in PRL production. The essential morphologic features of the five principal cell types and the biochemical

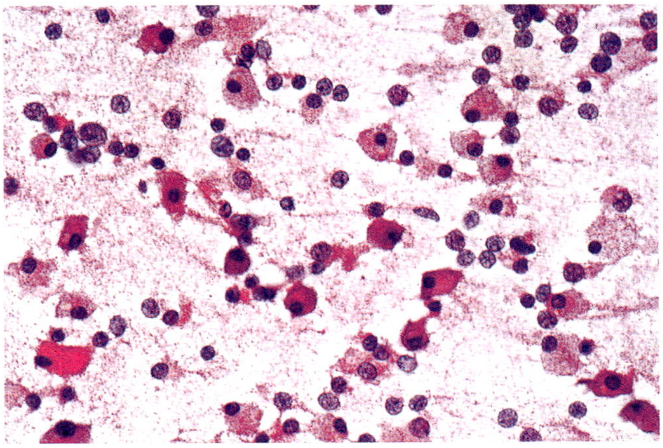

FIGURE 10.24 H&E-stained cytologic smear of the normal anterior lobe, demonstrating acidophils, chromophobes, and basophils. Delicate nuclei, inconspicuous nucleoli, and variable cytoplasmic staining characterize normal cells and permit their distinction from adenoma cells (original magnification ×100).

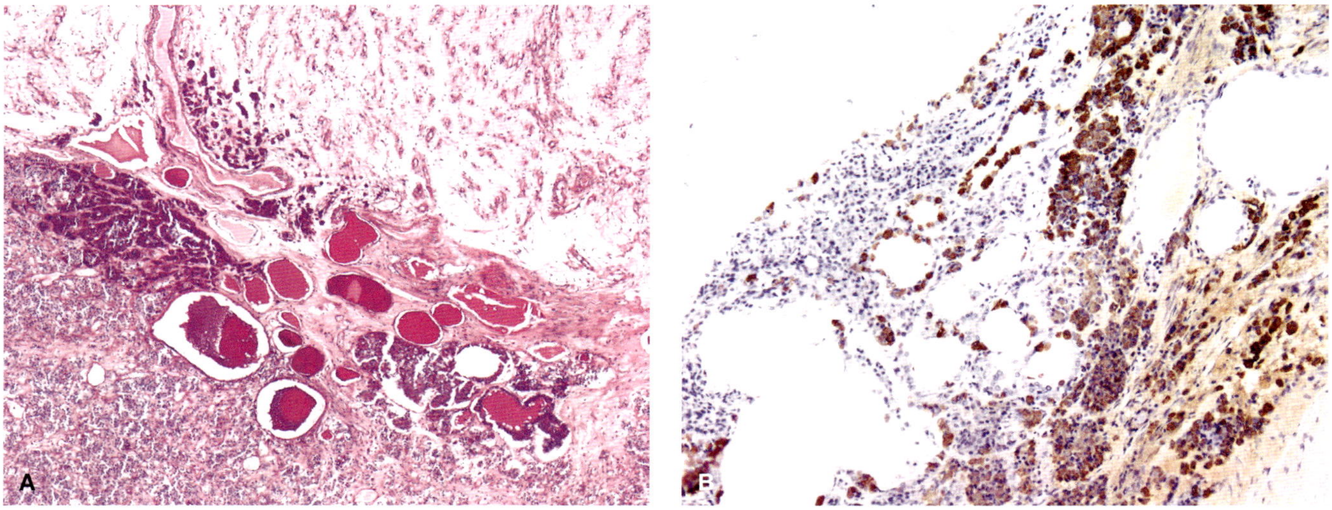

FIGURE 10.25 Intermediate lobe remnant. **A:** Intermediate lobe cysts are lined by a single layer of cuboidal to columnar epithelium that may be nonciliated, ciliated, mucin-producing cells, or neuroendocrine cells. The cyst contains an eosinophilic colloid (PAS, original magnification ×40). **B:** Many of the cells lining the cysts are neuroendocrine cells showing variable reactivity for pituitary hormones, in particular ACTH (ACTH immunostain, original magnification ×100).

characteristics of the six hormones of the anterior pituitary are summarized in Table 10.4.

Somatotroph Cells

Somatotrophs, or GH cells, occur in greatest density in the lateral wings and comprise approximately 50% of all adenohypophysial cells. A minority of somatotroph cells is scattered throughout the median portion of the gland. They are medium-sized, ovoid cells with round, centrally located nuclei, relatively prominent nucleoli, and abundant acidophilic granules (Fig. 10.27). Immunohistochemical stains for GH show strong and diffuse cytoplasm staining, consistent with the numerous secretory granules present at the ultrastructural level, and PIT1 is expressed in somatotroph nuclei (Fig. 10.27). Somatotroph cells are rather stable during life, and their number, morphology, and immunoreactivity are unchanged by age.

TABLE 10.4 The Normal Anterior Pituitary: Morphologic and Functional Features of Secretory Cells

Cell Type	Product	Location	Percentage of Cells (%)	Histochemical Staining[a]	Immunoperoxidase Staining
Somatotroph[b]	GH; 21,000 dalton polypeptide	Lateral wings	50	Acidophilic; PAS (–)	GH, PIT1
Lactotroph[b]	PRL; 23,500 dalton polypeptide	Resting cells: generalized Secreting cells: posterolateral wings	15–20	Acidophilic; PAS (–)	PRL, PIT1, ERα
Corticotroph	ACTH; 4,507 dalton polypeptide	Mucoid wedge	15–20	Basophilic; PAS and lead hematoxylin (+)	ACTH, TPIT
Gonadotroph	FSH and LH; 35,100 and 28,260 dalton glycoprotein	Generalized	10	Basophilic; PAS, lead hematoxylin, aldehyde fuchsin, and aldehyde thionine (+)	β-FSH, β-LH,[c] α-subunit, SF1, GATA2/3
Thyrotroph	TSH; 28,000 dalton glycoprotein	Anterior mucoid wedge	<5	Same as for gonadotroph	TSH, α-subunit, PIT1, GATA2/3

[a]The tinctorial characteristics of normal pituitary cells depend on adequate cytoplasmic granular storage. If sparsely granulated, cells that are otherwise functional may appear nonreactive or "chromophobic."

[b]Rare acidophilic stem cells (presumed somatotroph and lactotroph precursor cells producing both GH and PRL, i.e., mammosomatotroph cells) are present in the normal pituitary.

[c]Many gonadotrophs are capable of producing both FSH and LH, although immunohistochemical and ultrastructural evidence suggests that some gonadotrophs may produce only one hormone.

GH, growth hormone; PRL, prolactin; ACTH, adrenocorticotropic hormone; FSH and LH, follicle-stimulating hormone and luteinizing hormone; TSH, thyroid-stimulating hormone; PIT1, pituitary-specific POU-class homeodomain transcription factor 1; ERα, estrogen receptor α; TPIT, T-box family member TBX19; SF1, steroidogenic factor 1; GATA2/3, members of the GATA family of zinc-finger transcriptional regulatory proteins.

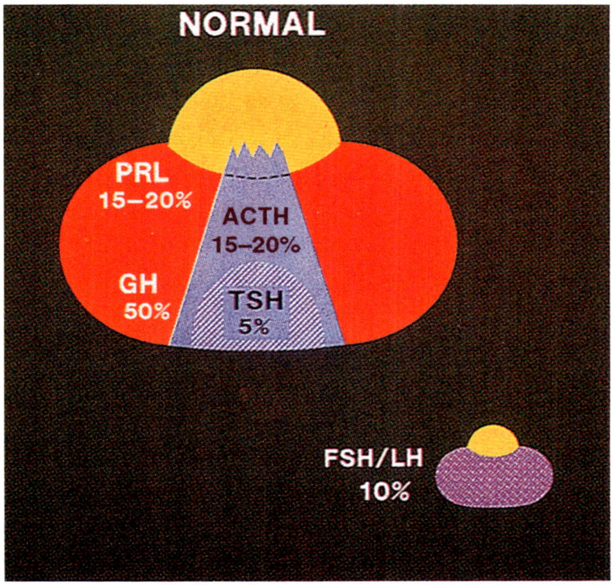

FIGURE 10.26 Preferential localization and relative frequency of functional anterior lobe cell types in the normal pituitary gland. Note that gonadotrophs (LH/FSH cells), represented in the small gland (*lower right*), are distributed diffusely and show no preferential localization.

The development of somatotrophs and GH transcription is determined by the expression of the nuclear transcription factor PIT1. Five distinct genes on chromosome 17q22 encode GH and related proteins. Two alternatively spliced peptides are produced by the pituitary GH gene (*hGH-N*), a 22-kDa GH (191 amino acids) and a less abundant 20-kDa GH molecule, both with similar biologic activity (30). During pregnancy, placental syncytiotrophoblast cells express a GH variant gene (*hGH-V*), which plays a significant role in human fetal growth and development (31).

The product of GH cells has extensive effects by direct action of GH and actions through mediators of hepatic origin called insulin-like growth factor 1 (IGF-1) (30). GH functions as the major promoter of growth but also in several other metabolic pathways, including glucose, insulin, and fatty acid metabolism. The pulsatile secretion of GH from the anterior lobe is under the control of the two hypothalamic regulatory hormones: GHRH and somatostatin. GHRH controls GH synthesis by regulating transcription of GH mRNA via control of the intracellular cyclic AMP pathway. Somatostatin appears to determine the timing and amplitude of GH pulses but has no effect on GH synthesis. IGF-1 causes negative feedback inhibition of GH release by acting at the hypothalamus and the pituitary.

Hypersecretion of GH in children and adolescents causes gigantism, whereas its overproduction in adults leads to acromegaly.

Lactotroph Cells

Lactotrophs, or PRL cells, comprise approximately 20% of anterior pituitary cells and are concentrated in the posterior portions of the lateral wings. Histologically, they appear either acidophilic (densely granulated) or chromophobic (sparsely granulated). Chromophobic cells are more numerous, possess elongate processes, and, despite an abundance of endoplasmic reticulum and well-developed Golgi complexes, contain relatively few cytoplasmic granules (Fig. 10.28). Densely granulated lactotrophs are thought to represent a storage phase, whereas sparsely granulated cells are engaged in active secretion. A characteristic pattern of PRL staining of lactotrophs by immunohistochemistry, the so-called "Golgi pattern," is the presence of paranuclear staining corresponding to PRL in the Golgi apparatus (Fig. 10.28). Lactotrophs also show nuclear expression of PIT1 and ERα by immunohistochemistry.

With the exception of pregnancy and lactation, there is no significant difference in PRL cell number between men and women. The doubling in volume of the pituitary during pregnancy is because of striking hyperplasia as well as hypertrophy of chromophobic lactotrophs, termed pregnancy cells. They persist until shortly after delivery or the termination of lactation (26). PRL cell hyperplasia also may accompany estrogen administration and hypothyroidism (30). The existence of mammosomatotrophs, cells engaged in PRL and GH production and possessing unique ultrastructural features, attests to the existence of a histogenetic relationship between PRL and GH cells (32).

PRL secretion is unique among the anterior pituitary hormones in that it is under tonic hypothalamic inhibition by dopamine produced by tuberoinfundibular dopamine neurons. Dopamine type 2 (D2) receptors in lactotrophs mediate PRL inhibition. Several PRL-releasing factors participate in PRL secretion, including TRH and vasoactive intestinal peptide (VIP). Glucocorticoids, thyroid hormone, and a fragment of GnRH weakly suppress PRL secretion. PRL synthesis is also regulated by effects of estrogen on PRL gene expression.

Disruption of the hypothalamus or the hypothalamic–hypophyseal stalk may impede dopamine delivery to the anterior lobe, causing hyperprolactinemia, a phenomenon termed "stalk effect." This regulatory pathway accounts for the spontaneous PRL hypersecretion often secondary to a space-occupying sellar or parasellar mass (e.g., pituitary macroadenoma, Rathke cleft cyst, craniopharyngioma, meningioma) that compresses the pituitary stalk.

PRL acts through specific PRL receptors in multiple tissues, including breast, liver, ovary, testis, and prostate. The main site of PRL action is in the breast, where it induces and maintains lactation stimulating the formation of casein, lactalbumin, lipids, and carbohydrates, all essential components of breast milk. During pregnancy, high levels of estrogen, progesterone, placental lactogen, and PRL induce acinar development and promote milk formation. As previously noted, milk secretion is under the control of oxytocin, a potent stimulator of myoepithelial cell contraction in breast tissue. PRL also decreases reproductive function and suppresses sexual drive by suppressing hypothalamic GnRH and pituitary gonadotropin secretion and impairing gonadal steroidogenesis (30).

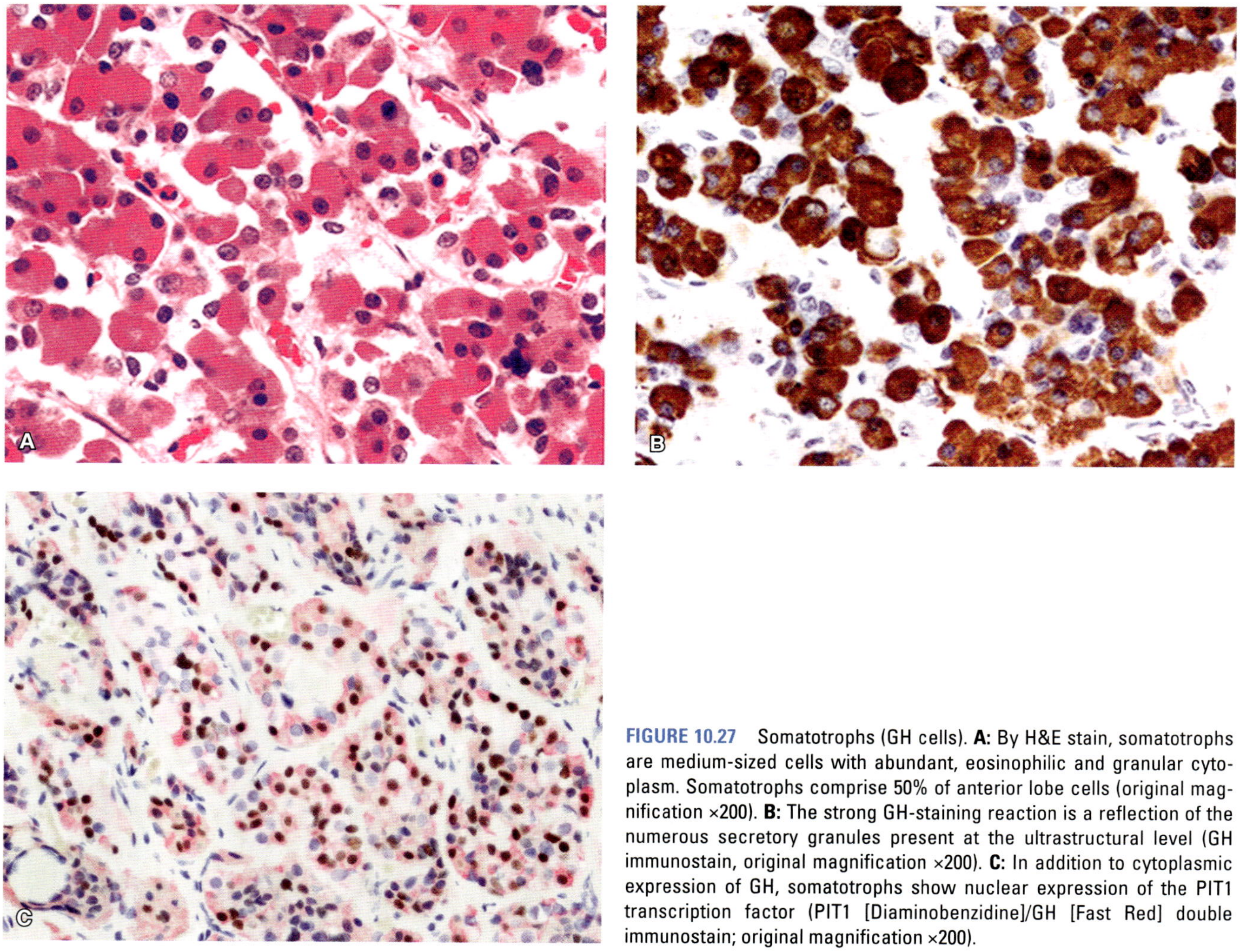

FIGURE 10.27 Somatotrophs (GH cells). **A:** By H&E stain, somatotrophs are medium-sized cells with abundant, eosinophilic and granular cytoplasm. Somatotrophs comprise 50% of anterior lobe cells (original magnification ×200). **B:** The strong GH-staining reaction is a reflection of the numerous secretory granules present at the ultrastructural level (GH immunostain, original magnification ×200). **C:** In addition to cytoplasmic expression of GH, somatotrophs show nuclear expression of the PIT1 transcription factor (PIT1 [Diaminobenzidine]/GH [Fast Red] double immunostain; original magnification ×200).

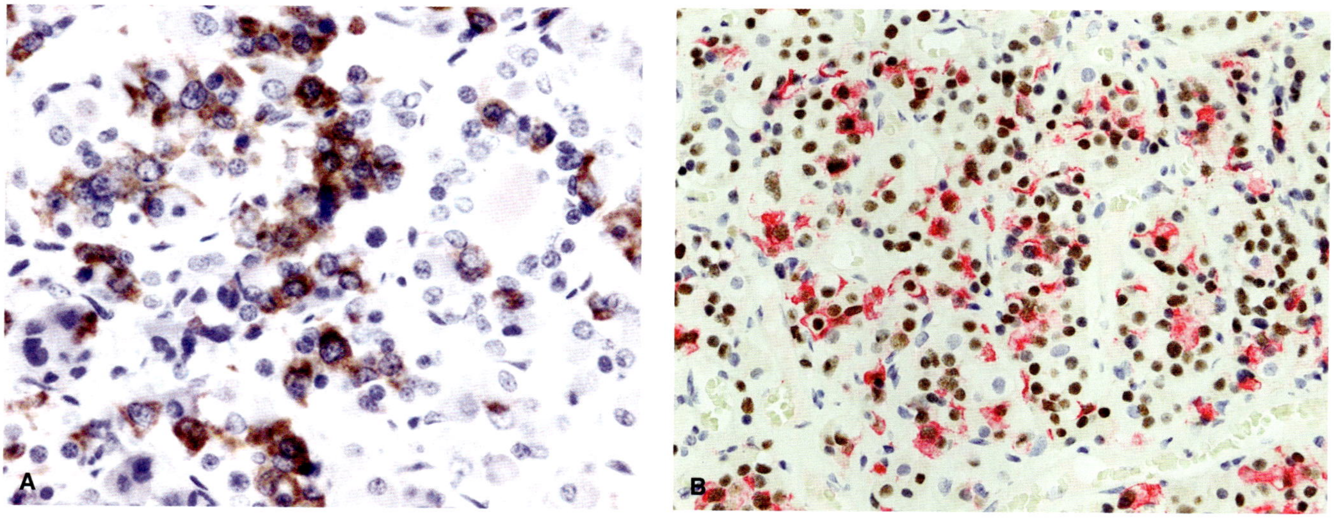

FIGURE 10.28 Lactotrophs (PRL cells). **A:** Lactotrophs comprise 15% to 20% of anterior lobe cells. The majority are sparsely granulated, angular cells with processes that may wrap around adjacent cells. Many lactotrophs in this field show paranuclear staining, a pattern corresponding to PRL in the Golgi apparatus (PRL immunostain, original magnification ×100). **B:** In addition to cytoplasmic expression of PRL, lactotrophs show nuclear expression of the PIT1 transcription factor [PIT1 (Diaminobenzidine)/PRL (Fast Red) double immunostain; original magnification ×200].

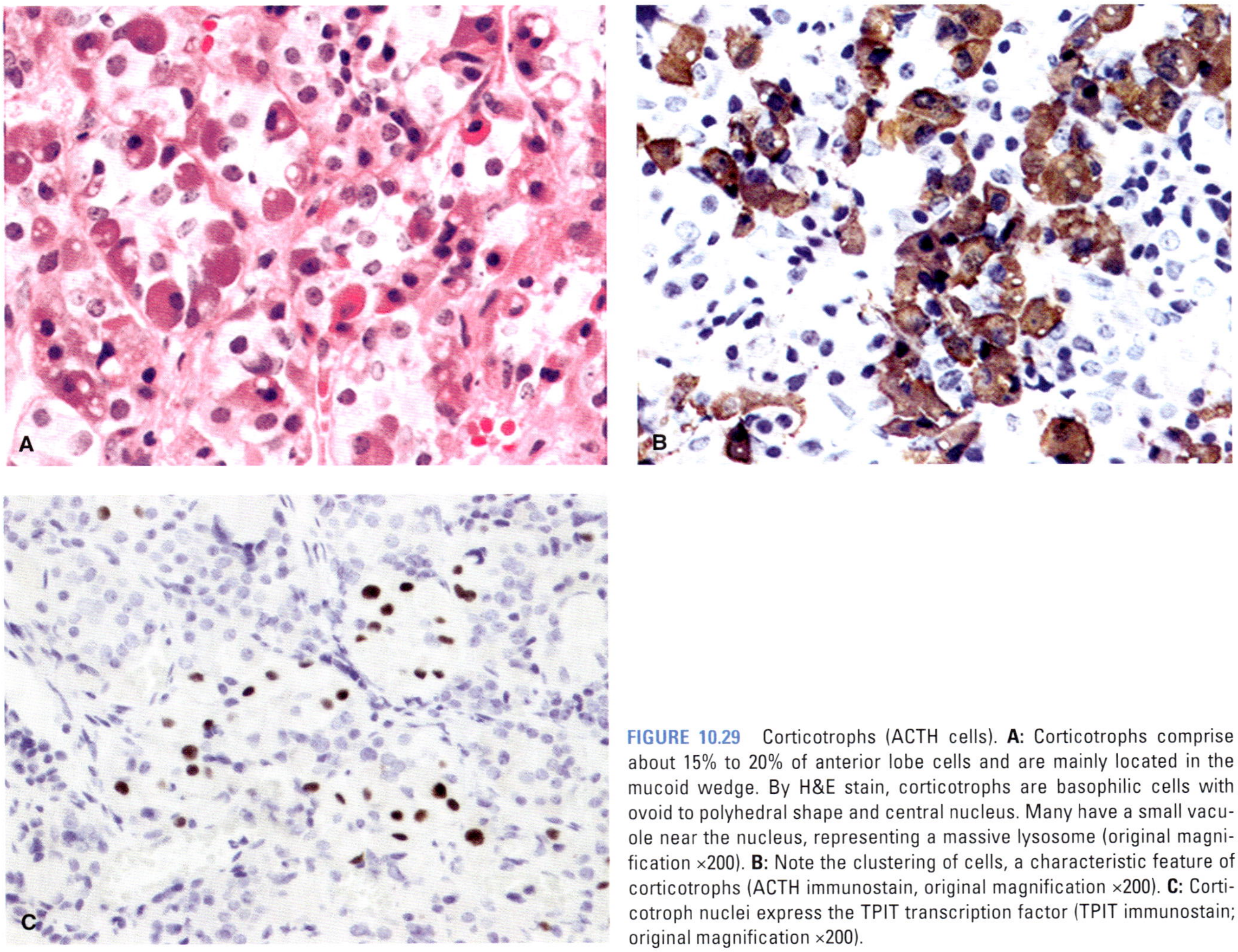

FIGURE 10.29 Corticotrophs (ACTH cells). **A:** Corticotrophs comprise about 15% to 20% of anterior lobe cells and are mainly located in the mucoid wedge. By H&E stain, corticotrophs are basophilic cells with ovoid to polyhedral shape and central nucleus. Many have a small vacuole near the nucleus, representing a massive lysosome (original magnification ×200). **B:** Note the clustering of cells, a characteristic feature of corticotrophs (ACTH immunostain, original magnification ×200). **C:** Corticotroph nuclei express the TPIT transcription factor (TPIT immunostain; original magnification ×200).

In patients with hyperprolactinemia, besides galactorrhea, decreased libido and reduced fertility are common clinical symptoms.

Corticotroph Cells

Corticotrophs, or ACTH cells, comprise 15% to 20% of adenohypophysial cells and are most numerous in the mid and posterior portions of the mucoid wedge (Figs. 10.22 and 10.26). Histologically, corticotroph cells are polygonal, medium- to large-sized, and basophilic (Fig. 10.29). A paranuclear vacuole that corresponds to one or several lysosomal structures at the ultrastructural level is typically seen in the cytoplasm (Fig. 10.30) (27,33). Corticotroph cells are strongly PAS positive because of a carbohydrate moiety contained in POMC, the precursor molecule of ACTH. In immunostained preparations, corticotrophs show nuclear TPIT expression and strong and diffuse cytoplasmic ACTH staining (Fig. 10.29), consistent with the numerous secretory granules present at the ultrastructural level. Corticotrophs may also contain other POMC derivatives, including β-lipotropin (β-LPH), melanocyte-stimulating hormone (MSH), endorphin, and encephalin, which are

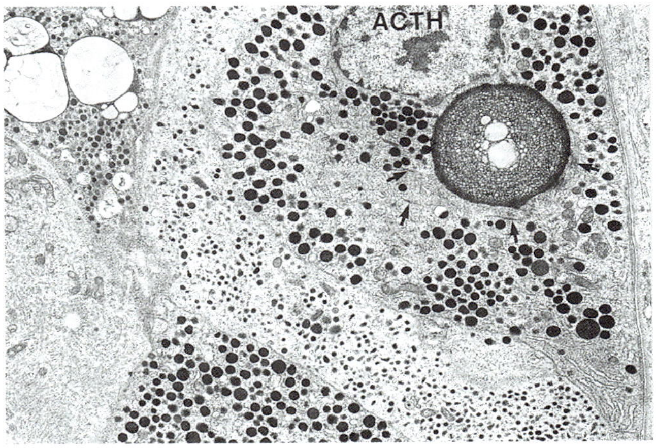

FIGURE 10.30 Corticotroph (ACTH cell). The cell contains the typical large lysosome and spherical to slightly pleomorphic and variably electron-dense secretory granules measuring 150 to 450 nm. Bundles of intermediate filaments (*arrows*) are a regular feature of corticotrophs. The adjacent cells with small granules are likely TSH cells (original magnification ×8,250).

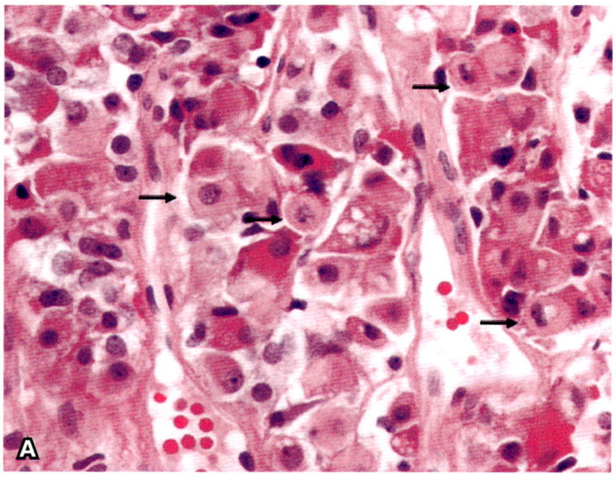

FIGURE 10.31 Crooke hyaline change. **A:** Crooke cells (*arrows*) are characterized by a conspicuous eosinophilic perinuclear ring consisting of cytokeratin (H&E, original magnification ×100). **B:** ACTH stains show central displacement of the nucleus and organelles by filament accumulation (ACTH immunostain, original magnification ×100). **C:** Staining for cytokeratin shows strong reactivity of the Crooke cells (CK immunostain, original magnification ×100).

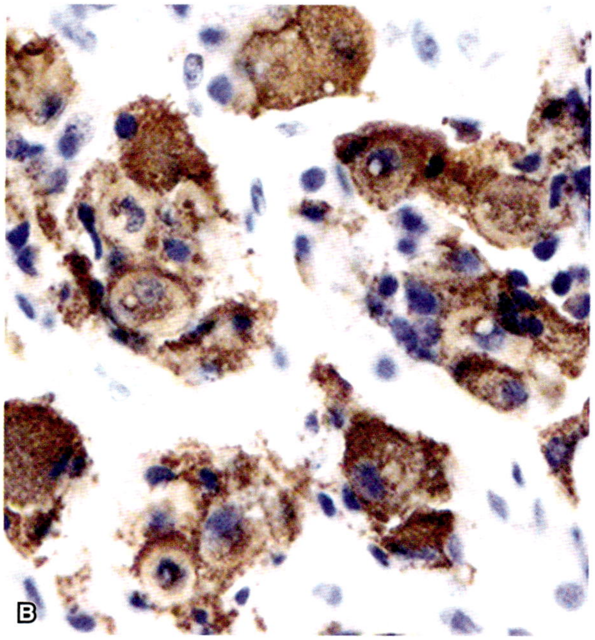

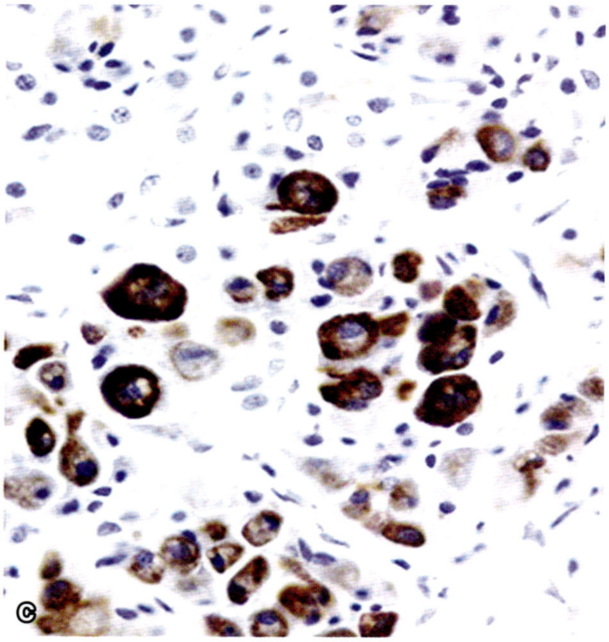

not of significance in the daily practice of surgical pathology. Perinuclear bundles of cytokeratin filaments are also a characteristic feature of ACTH cells. Under conditions of glucocorticoid excess, either exogenous or endogenous, corticotrophs accumulate cytokeratin as a manifestation of Crooke hyaline change (Figs. 10.31 and 10.32).

ACTH, the principal product of corticotrophs, is part of a larger precursor molecule, the POMC, which is enzymatically cleaved in the anterior pituitary into β-LPH and ACTH. In the intermediate lobe, ACTH is cleaved into MSH and corticotropin-like intermediate lobe protein (CLIP) (30). ACTH stimulates the adrenal cortex to secrete glucocorticoids, mineralocorticoids, and androgens. It plays a critical role in both the transport of amino acids and glucose into muscle, as well as the stimulation of insulin release from the pancreas. Pituitary ACTH secretion is regulated by hypothalamic CRH, arginine vasopressin (AVP), and proinflammatory cytokines, including interleukin-6 (IL-6) as well as leukemia inhibitory factor.

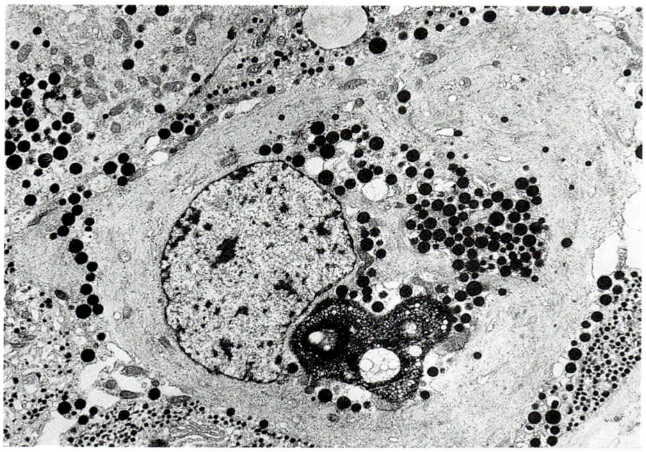

FIGURE 10.32 Crooke cell. This electron micrograph of the pituitary adjacent to a corticotroph cell adenoma displays massive accumulation of cytokeratin filament. Secretory granules are displaced to the perinuclear zone or to the periphery of the cytoplasm. Note the large lysosome in the lower portion of the field (original magnification ×6,720).

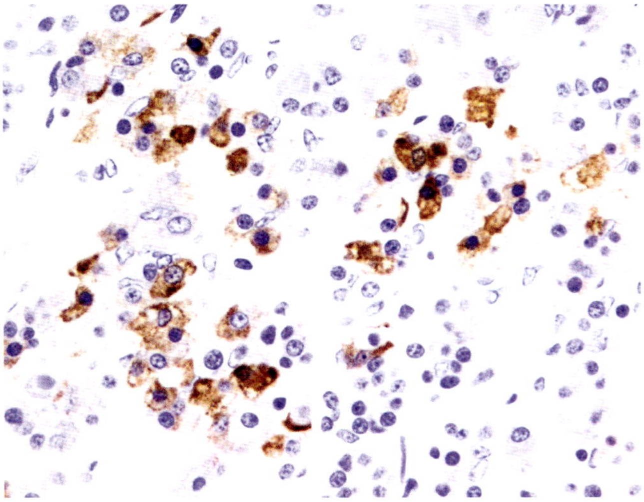

FIGURE 10.33 Thyrotrophs (TSH cells). Thyrotrophs are medium-sized angulated cells with some demonstrating elongate processes. They comprise only about 5% of anterior lobe secretory cells and show strong TSH immunoreactivity (TSH immunostain, original magnification ×100). Like somatotrophs and lactotrophs, thyrotrophs show nuclear expression of the PIT1 transcription factor (not pictured).

Excess secretion of ACTH, such as occurs in Cushing disease, leads to stereotypic abnormalities such as truncal obesity, hypertension, diabetes mellitus, amenorrhea, hirsutism, muscle atrophy, striae, impaired wound healing, and mental status changes. Hyperpigmentation also may occur in this setting and is because of the effects of MSH.

Thyrotroph Cells

Thyrotrophs, or TSH cells, are located primarily in the anterior part of the mucoid wedge and comprise only 5% of adenohypophysial cells (34). They are medium-sized and angular or elongate (Fig. 10.33). Like corticotrophs, normal TSH cells are basophilic and are PAS positive.

TSH, like the other two glycoprotein hormones of the pituitary (LH and FSH), consists of two noncovalently bound subunits, alpha (α) and beta (β); the α-subunit is common to all three glycoprotein hormones; the β-subunit is specific for each hormone and confers biologic specificity. Immunohistochemical staining of thyrotrophs reflects this composition, as thyrotrophs show cytoplasmic expression of α-subunit and β-TSH along with nuclear expression of PIT1 and GATA3.

TSH binds to thyroid cells, inducing RNA and protein synthesis and thereby the production of thyroglobulin and thyroid hormones. Secretion of TSH is regulated by both hypothalamic hormones and circulating thyroid hormones. TRH stimulates TSH release, while thyroid hormones, dopamine, somatostatin, and glucocorticoids inhibit TSH secretion.

In the setting of primary hypothyroidism, thyrotrophs undergo hypertrophy and hyperplasia. When excessive, the response may produce sufficient hypertrophy of the pituitary gland to mimic adenoma. Pituitary adenomas that elaborate TSH are rare; most occur in the setting of hypothyroidism, although a minority results in hyperthyroidism (27,30).

Gonadotroph Cells

Gonadotrophs, or LH and FSH cells, comprise 10% of the adenohypophysis, show a strong affinity for both basic and PAS stains, and generally are evenly distributed throughout the anterior lobe. Immunohistochemical and ultrastructural studies have shown that LH and FSH may be produced in isolation or by the same cell (Fig. 10.34) (35). Like TSH,

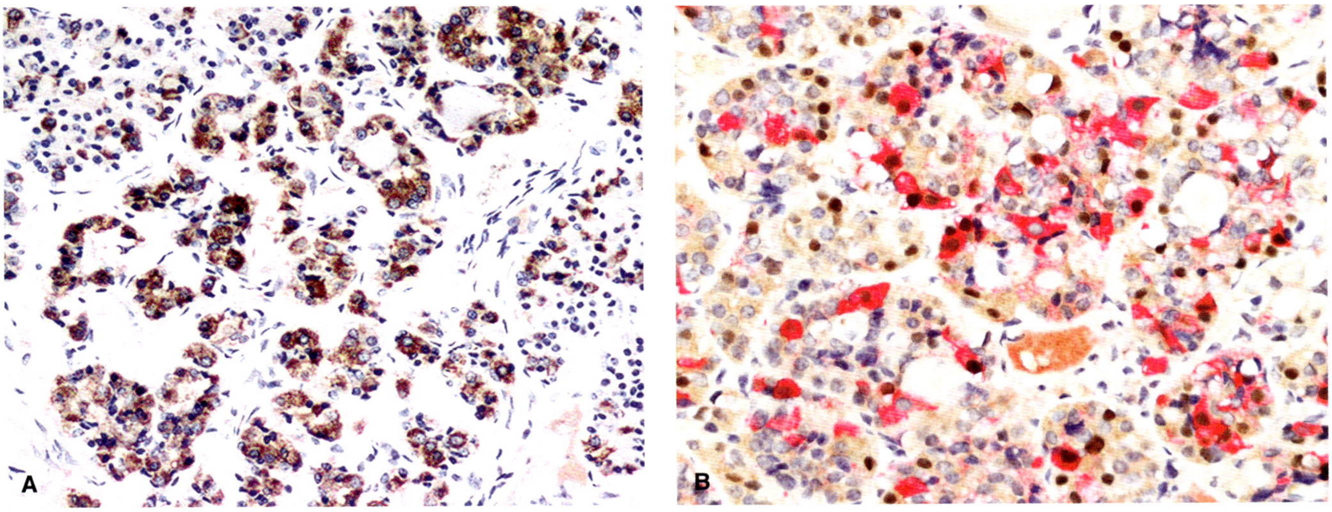

FIGURE 10.34 Gonadotrophs (FSH/LH cells). **A:** Gonadotrophs manufacture both LH and FSH. The paucity of strongly staining cells in this field reflects the fact that only 10% of anterior lobe secretory cells are gonadotrophs (β-FSH immunostain, original magnification ×100). **B:** Gonadotrophs show nuclear expression of the SF1 transcription factor with a subset demonstrating cytoplasmic expression of β-FSH [SF1 (Diaminobenzidine)/β-FSH (Fast Red) double immunostain; original magnification ×200).

LH and FSH are glycoprotein hormones consisting of α- and β-subunits. In the daily practice of surgical pathology, monoclonal antibodies against the specific β-subunits of LH and FSH are universally applied. In addition, antibodies against the α-subunit are useful for determination of abnormal production and/or secretion of this subunit hormone. Gonadotrophs also express the nuclear transcription factors SF1 and GATA3.

Both LH and FSH play distinct but essential roles in the reproductive physiology of men and women. In women, LH is required for ovulation and follicular luteinization. In men, it stimulates interstitial Leydig cells to produce testosterone. FSH promotes follicular development in women, whereas in men it induces Sertoli cells to produce an androgen-binding protein.

Pars Tuberalis

The pars tuberalis, an upward extension of the anterior lobe along the pituitary stalk, is composed of normal acini of pituitary cells scattered among surface portal vessels. These cells consist of mainly gonadotrophs intermixed with a few corticotrophs and thyrotrophs, histologically resembling those of the pars distalis (anterior lobe) (36). They often show immunoreactivity for ACTH, β-FSH, β-LH, and α-subunit. Although in functional terms they may or may not differ from similar cells in the anterior lobe, these cells do show a distinct tendency to undergo squamous metaplasia (see *Variation in Normal Morphology of the Pituitary Gland*).

Folliculostellate Cells

The folliculostellate cell is the sixth cellular element of the adenohypophysis, a specialized sustentacular-like cell that appears to have multiple functions related to phagocytosis, secretion of growth factors, and intercellular communication (37,38). They comprise less than 5% of the anterior lobe cells, are scattered about the anterior lobe, and contribute to the formation of anterior lobe follicles and cysts of the intermediate lobe (38). They are readily identified by their immunoreactivity for S100 protein (Fig. 10.35) and can also be identified by their expression of glial fibrillary acidic protein (GFAP) and vimentin. They participate in regulating several activities of anterior pituitary including scavenger activity by engulfing degenerated cells and paracrine regulation of endocrine cells through the production of cytokines (such as IL-6 and leukemia inhibitory factor) and various growth factors (including basic fibroblast growth factor and vascular endothelial growth factor) (38).

Age-Related Changes of the Adenohypophysis

The cytology of the anterior pituitary varies with age. For instance, the late fetal or term pituitary gland shows PRL cell hyperplasia, a reflection of high maternal estrogen levels. Also, when compared with the adult pituitary, the prepubertal gland shows gonadotroph cells to be poorly developed.

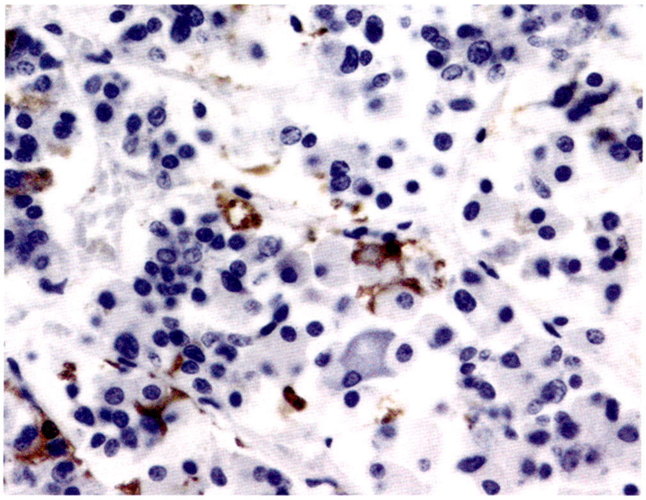

FIGURE 10.35 Anterior lobe, folliculostellate cells. Folliculostellate cells comprise less than 5% of anterior lobe cells and are scattered throughout the pituitary, including the intermediate lobe zone. Folliculostellate cells stain for S100 protein, GFAP, and vimentin (S100 immunostain, original magnification ×100).

The gland weight remains stable throughout adult life, decreasing only slightly in the elderly associated with interstitial and perivascular fibrosis (see later) (39,40). Pregnancy is the period in which the adenohypophysis undergoes major changes resulting in a doubling of its weight because of a gradual increase in large chromophobic PRL cells ("pregnancy cells") (Fig. 10.36) (26,41). The increase in PRL-producing cells during pregnancy results not only from proliferation of PRL cells but also from recruitment of mammosomatotroph cells to enhance PRL production (42). Lactotroph hyperplasia gradually disappears within months after the end of a pregnancy. The process is often incomplete; hence, the pituitaries of multiparas are larger than those of women who were never pregnant. Pregnancy also results in a significant decrease in gonadotropin immunoreactivity, a reflection of the production of gonadotropic hormones by the placenta.

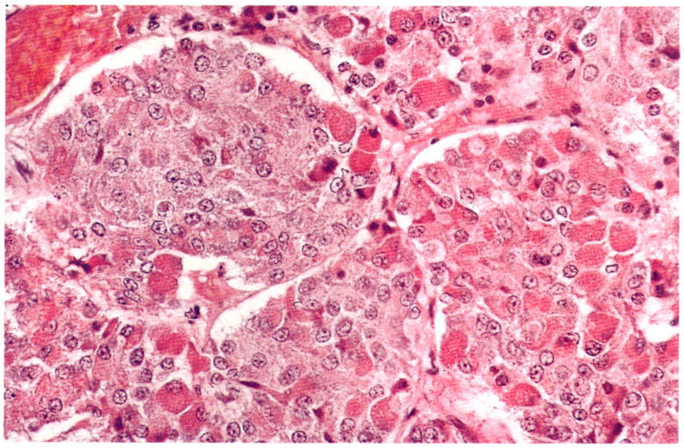

FIGURE 10.36 The pituitary in pregnancy features abundant pale chromophobic PRL cells (pregnancy cells) (H&E, original magnification ×100).

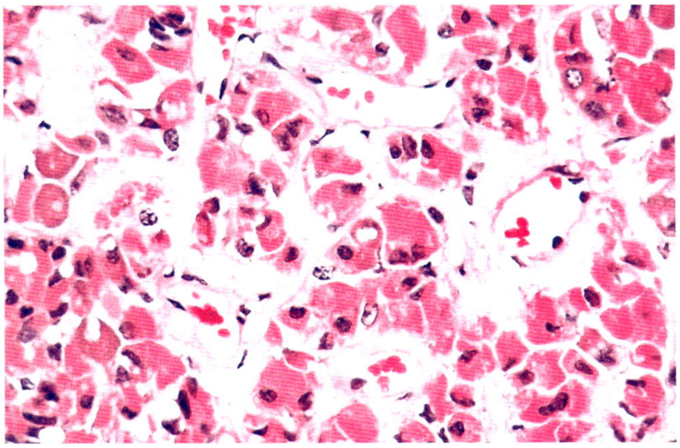

FIGURE 10.37 Perivascular fibrosis is a common feature of the aging pituitary (H&E, original magnification ×100).

Fibrosis is the most frequent age-related change (39,40). It is generally perivascular in distribution (Fig. 10.37) but is on occasion patchy, suggesting a remote microinfarct. The effects of aging on the cellular contents of several pituitary hormones have been studied. Specifically, GH and PRL cells have been shown to undergo no significant decrease in number, granularity, distribution, or immunoreactivity with increasing age (43,44). Both ACTH and TSH cells also appear to be unaffected by age, but no data are available regarding the effects of senescence on FSH and LH cells.

Neurohypophysis

As a functional unit, the neurohypophysis consists of the infundibulum, pituitary stalk, and posterior lobe. The posterior lobe, a ventral extension of the central nervous system, is the site of release of the hypothalamic hormones oxytocin and vasopressin. Its cellular elements consist of: (1) unmyelinated axons originating from the supraoptic nucleus and paraventricular nucleus and, to a lesser extent, from cholinergic neurons of the hypothalamus; (2) an extensive vascular network; and (3) specialized glial cells termed pituicytes (Fig. 10.38).

Pituicytes, the most numerous cells of the neurohypophysis, are elongated, uni- or bipolar cells that display the prolongation of the cytoplasm into one or more

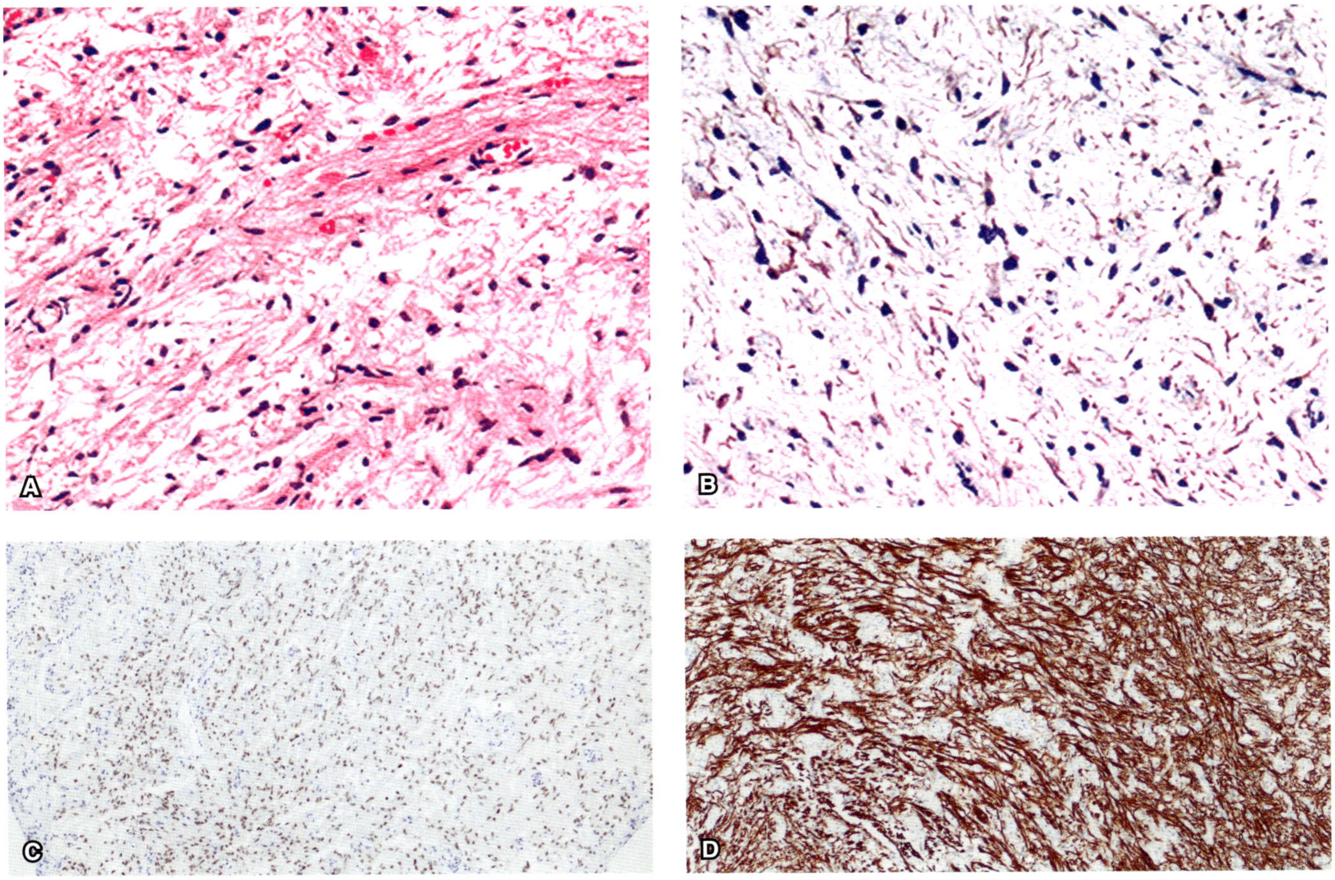

FIGURE 10.38 Posterior lobe. **A:** Pituicytes have elongated nuclei dispersed within the neuropil of the neurohypophysis (H&E, original magnification ×100). **B:** These cells stain positively for glial fibrillary acid protein (GFAP) (immunostain, original magnification ×100). **C:** Pituicytes are embryologically derived from the diencephalon and show nuclear expression of thyroid transcription factor 1 (TTF-1) (immunostain, original magnification ×100). **D:** Axons coursing through the posterior lobe express neuronal markers such as neurofilament protein (NFP) (immunostain, original magnification ×100).

processes. Pituicytes appear to exist in five principal forms: major, dark, ependymal, oncocytic, and granular (45). Their morphologic diversity is thought to be a reflection of their, still not completely understood, physiologic roles (45). Pituicytes are positive for GFAP (Fig. 10.38) as well as S100 protein and vimentin. Similar to glial cells of other areas of the central nervous system, pituicytes expand processes to adjacent connective tissue or to a blood vessel wall. Pituicytes are embryologically derived from the floor of the diencephalon, an area under the influence of thyroid-specific enhancer-binding protein (T/EBP, also known as NKX2-1 or TTF-1) during normal development (46). TTF-1 is expressed in normal fetal and adult pituicytes as well as in neoplastic pituicytes (Fig. 10.38) (47). The axons of the posterior lobe are histologically identified using silver stains and/or immunostains for neuronal markers (Fig. 10.38). Focal axonal dilatations, known as Herring bodies, represent intra-axonal accumulations of posterior lobe hormones (Fig. 10.20). The most important function of the neurohypophysis is the transfer of hormonal substances from neurosecretory granules to the intravascular space. The complex anatomy of the neuronal, vascular, and perivascular compartments forms the basis for this elaborate process. Beginning at the neuronal side, neurohormonal factors appear to be released into minute channels that traverse the outermost, or abluminal, basement membrane of vessels to communicate with the perivascular space. They then traverse the inner, or luminal, basement membrane and endothelium in order to gain access to the vascular space (48).

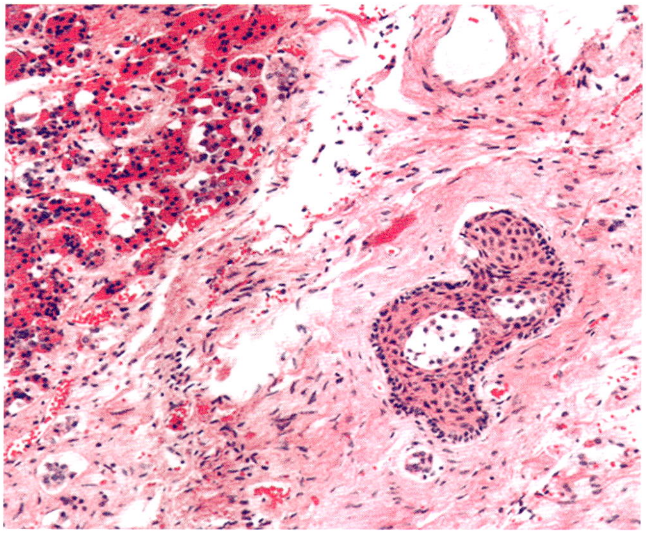

FIGURE 10.39 Anterior pituitary, squamous metaplasia. Squamous metaplasia of secretory cells is a common feature of the pars tuberalis.

Variation in Normal Morphology of the Pituitary Gland

A number of normal histologic variations in the pituitary gland may mimic clinically significant lesions. Examples

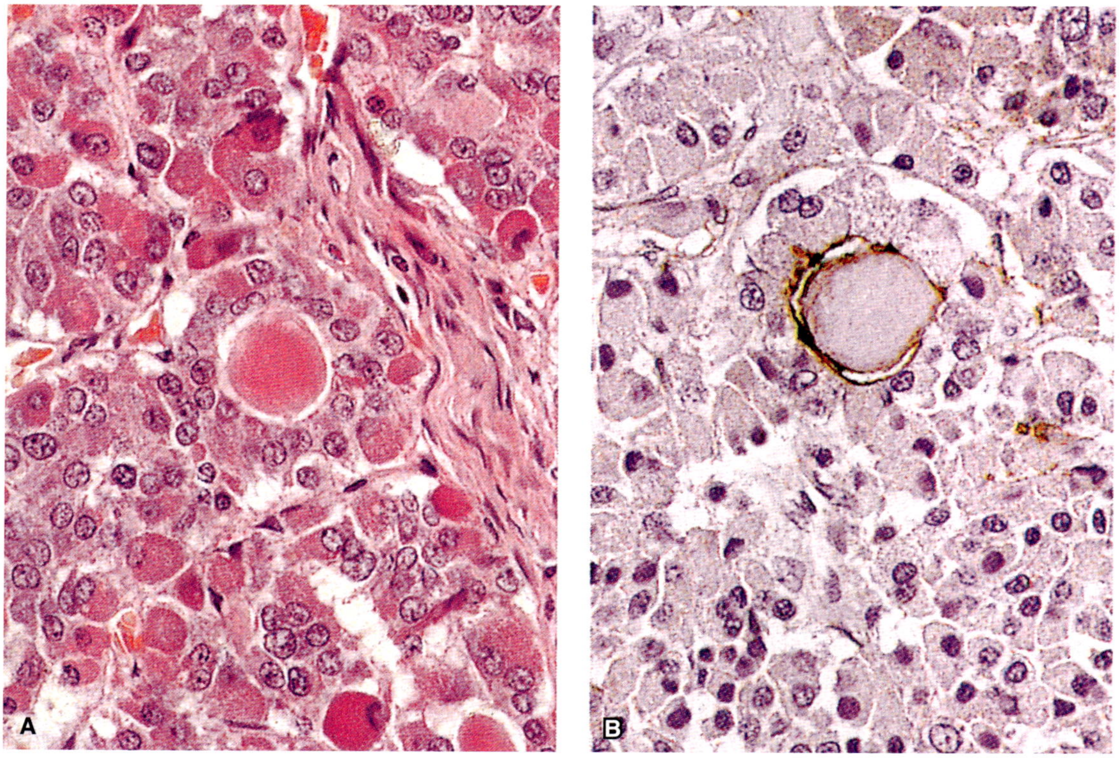

FIGURE 10.40 Anterior pituitary, follicle formation. Follicles, some containing a small quantity of colloid-like material, are commonly found in the anterior lobe (H&E, original magnification ×100).

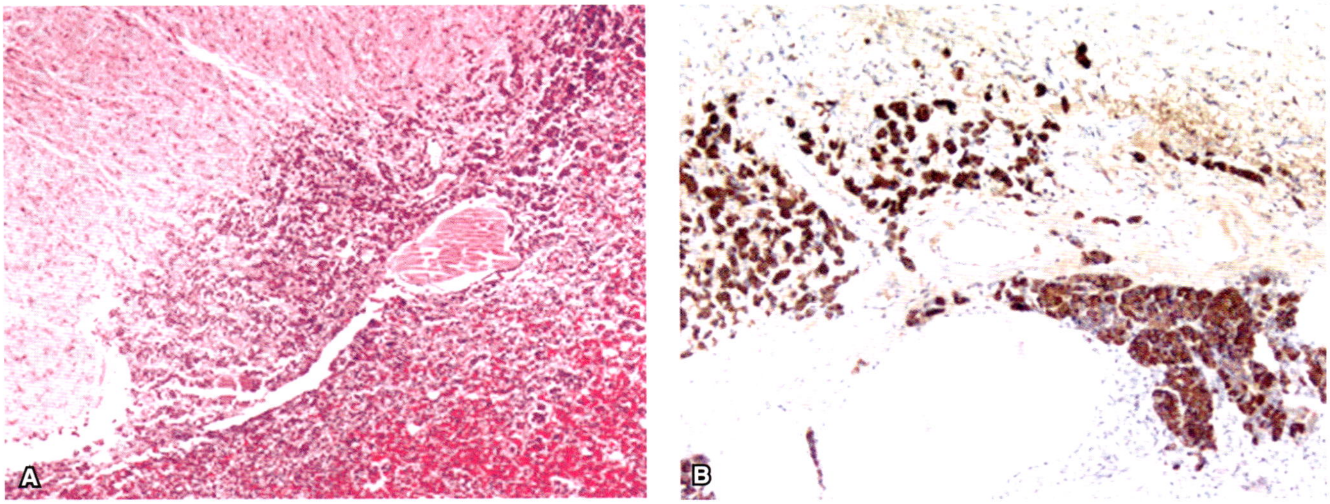

FIGURE 10.41 Basophil invasion. **A:** This subpopulation of corticotroph cells appears to infiltrate the substance of the posterior lobe (H&E, original magnification ×4). **B:** The invading cells are strongly immunoreactive for ACTH and other POMC derivative hormones (ACTH immunostain, original magnification ×40).

include squamous cell nests in the pars tuberalis, follicles in the anterior lobe, basophil invasion of the posterior lobe, salivary gland rests, and granular cell clusters and tumorlets of the stalk and neurohypophysis.

Squamous cell nests show a definite predilection for the pars tuberalis (Fig. 10.39); they have been found in up to 24% of autopsy cases, occur more commonly in elderly patients, and show no sex predilection (36,49). They arise through a process of metaplastic transformation from adenohypophyseal cells, as evidenced by simultaneous expression of keratin and pituitary hormones, most often β-FSH, β-LH, or ACTH (36). Because squamous metaplasia also may accompany foci of ischemic infarction in the anterior lobe, it appears to be an inherent property of pituitary secretory cells.

Follicles are not an uncommon feature of the normal anterior pituitary. Their functional constituent cells, termed

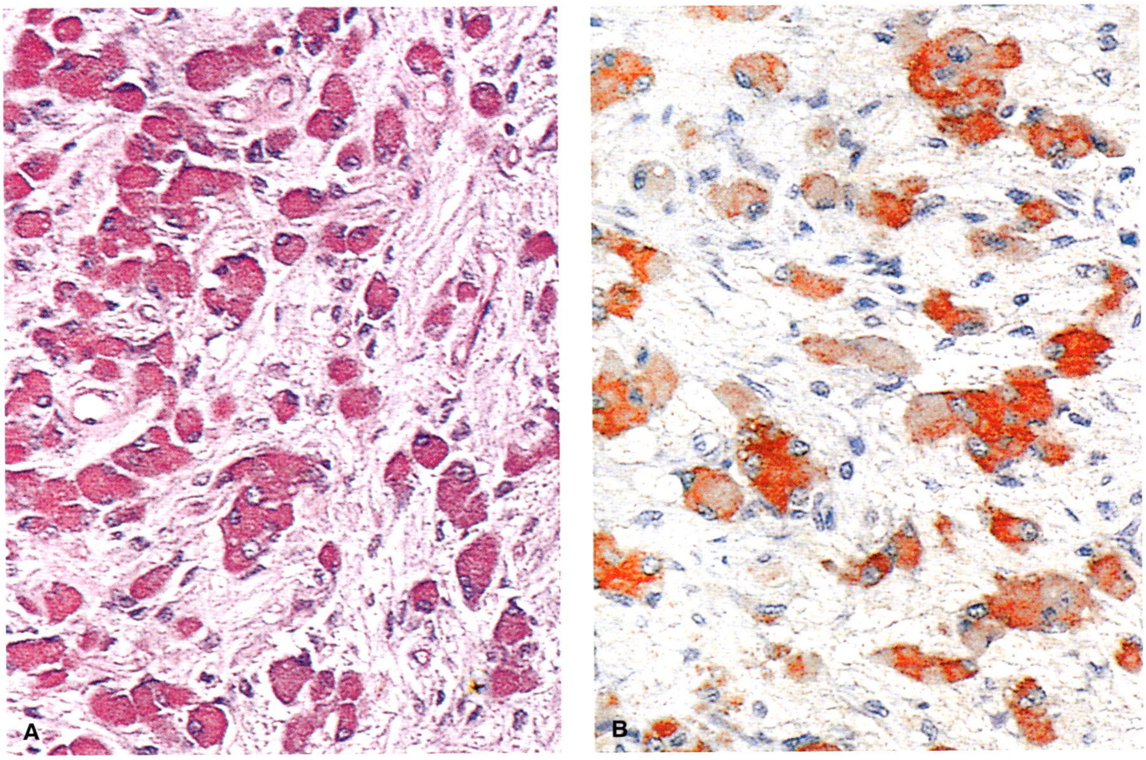

FIGURE 10.42 Basophil invasion. **A:** PAS staining (original magnification ×63). **B:** ACTH immunostain (original magnification ×63).

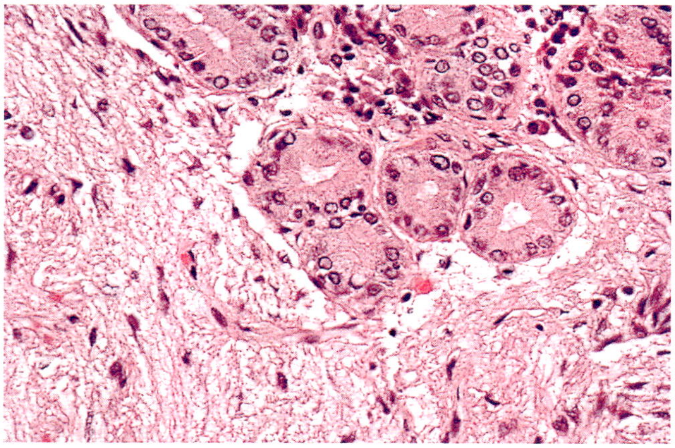

FIGURE 10.43 Salivary gland rest in the posterior lobe. The glands are composed of a single layer of cuboidal to columnar epithelium with basally oriented nuclei and granular PAS-positive cytoplasm. Salivary gland rests are encountered both on the surface of or within the posterior lobe, where it abuts the intermediate lobe zone (H&E, original magnification ×63).

follicular cells, appear to be derived in large part from various secretory cells (Fig. 10.40). Ultrastructurally, they are often poorly granulated or agranular and linked by apical junctional complexes. Within follicular lumina, one often finds cellular debris. The stimulus for follicle formation is, therefore, thought to be damage or rupture of anterior lobe secretory cells (50).

Basophil invasion consists of corticotropic basophils extending from the pars intermedia into the neurohypophysis (Figs. 10.41 and 10.42). Basophil invasion is more common in men and the elderly and may at first glance mimic an adenoma. Similar to ordinary corticotrophs, these basophilic cells are immunoreactive for ACTH and other POMC derivatives; however, they contain few cytokeratin filaments and are less susceptible to Crooke hyalinization in response to hypercortisolism (27).

Salivary gland rests appear as tubular glands upon the surface or in the substance of the neurohypophysis, often just posterior to the pars intermedia (Fig. 10.43). They are composed of a single layer of cuboidal to columnar epithelium with basally oriented nuclei and finely granular, strongly PAS-positive cytoplasm. Salivary gland rests are often oncocytic.

Granular cell nests or tumorlets, most located in the stalk or posterior lobe, are found in about 6% of autopsy pituitaries and are more common among the elderly (51). Varying from scattered cells to compact tumor-like nodules, they are composed of plump cells with granular acidophilic and strongly PAS-positive cytoplasm and relatively small nuclei (Fig. 10.44). Only rarely do granular cells form clinically significant tumors (52). The origin of granular cell tumorlets and tumors of the neurohypophysis is still not

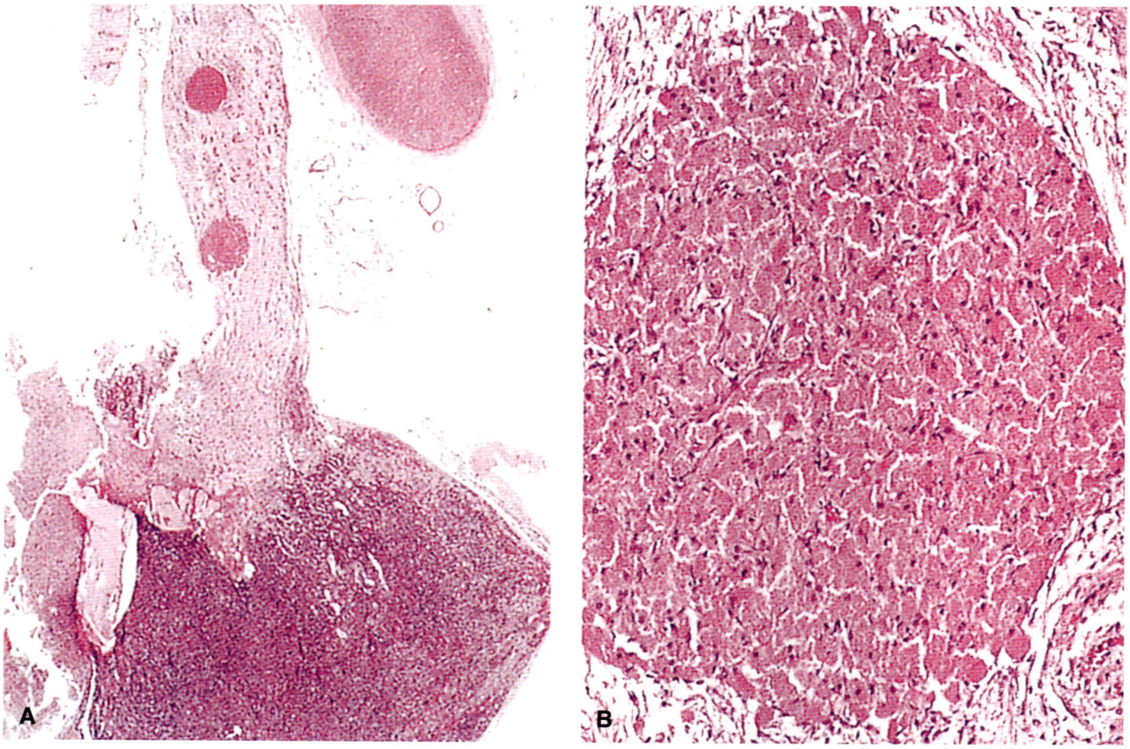

FIGURE 10.44 Pituitary stalk, granular cell tumorlets. **A:** Low-power view of the stalk shows two tumorlets. The optic chiasm is at the upper portion of the field; the anterior lobe is at the lower portion of the field (H&E, original magnification ×20). **B:** High-power view of a granular cell tumorlet. Such nodules are composed of pituicytes, modified glial cells, with abundant lysosome-rich eosinophilic cytoplasm. Tumorlets, as well as individual granular cells, are of no clinical significance (H&E, original magnification ×100).

completely understood, but TTF-1 expression by granular cell nests and tumorlets (as well as by pituicytomas and spindle cell oncocytomas) supports a common derivation from a pituicyte precursor (53).

DIFFERENTIAL DIAGNOSIS

The principal consideration in differential diagnosis of pituitary lesions is the distinction of normal pituitary tissue from a pituitary neuroendocrine tumor, that is, adenoma. The most conspicuous architectural feature of the adenohypophysis is the arrangement of its cells in acini that, depending on orientation of section, vary from round to oval or somewhat elongate. The acini are surrounded by a delicate reticulin- or PAS-positive capillary network (Fig. 10.16). In contrast, pituitary adenomas lack this uniform acinar architecture, showing only scant reticulin that is limited to scattered vessels (Fig. 10.45). Although most normal pituitary acini are heterogeneous in their cellular content, thus permitting the distinction of normal from adenomatous tissue on H&E sections alone, some parts of the pituitary contain largely a single cell type and appear fairly monomorphous. For instance, eosinophilic GH cells are present in large numbers in the lateral wings. On the other hand, occasional adenomas composed of mixed cell populations superficially resemble normal adenohypophysis. As a result, the distinction of normal from adenomatous tissue may be more easily achieved by reticulin staining than by immunohistochemistry alone.

Small biopsies of the intermediate lobe may include the so-called basophilic invasion normally seen in the posterior pituitary (Figs. 10.41 and 10.42). Since these basophilic cells are arranged diffusely or in clusters and lack the typical acinar formation seen in the anterior pituitary, they may be mistaken for corticotroph microadenomas.

A limited biopsy of the pituitary may occasionally include intermediate zone cysts, which are normal derivatives of Rathke cleft. Correlation with radiologic and operative data usually obviates confusion with Rathke cleft cyst. Clinically significant cysts are readily evident on neuroimaging and are identified as sizable cysts by the experienced surgeon.

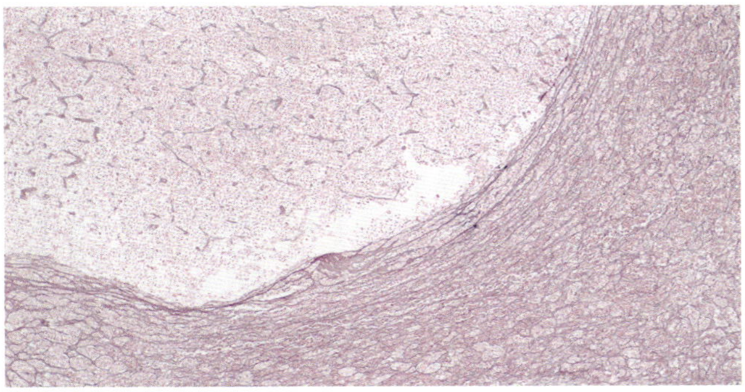

FIGURE 10.45 Disruption of the reticulin network in a pituitary adenoma. Compare the acinar pattern of the normal adenohypophysis (*bottom right*) with the adenoma (*upper left*) (Wilder reticulin, original magnification ×100).

As previously noted above, adenohypophyseal cells may undergo squamous metaplasia, particularly in the pars tuberalis. This location is only occasionally sampled in surgical specimens. Scant in extent, intimately associated with adenohypophyseal cells, and cytologically benign, they are unlikely to be confused with either cysts (epidermoid or dermoid cysts) or with neoplasms (craniopharyngiomas). As papillary craniopharyngiomas are characterized by *BRAF* V600E mutations, it is worth noting that normal adenohypophysis (corticotrophs in particular) may cross-react with the VE1 antibody and should not be interpreted as evidence of *BRAF* V600E mutation without additional testing (54). Similar nonspecific cross-reactivity can be seen in the cilia of Rathke remnant epithelial cells (55).

In very small numbers, cytologic benign lymphocytes are seen in the intermediate zone of the normal pituitary in 10% of autopsied subjects (56). Unassociated with endocrine disease, such cells are readily distinguished from the far more widespread and dense infiltrates of lymphocytic hypophysitis or abscess.

A limited biopsy of the neurohypophysis can readily be mistaken for glioma in that the vast majority of its nucleated cells are specialized astrocytes (pituicytes). Unlike pituicytomas, the tumor most closely mimicked by posterior pituitary tissue, the neurohypophysis contains large numbers of axons that may be highlighted by immunostains for neurofilament protein (Fig. 10.38) that are absent in pituicytomas.

REFERENCES

1. Moore KL. *The Developing Human: Clinically Oriented Embryology*. WB Saunders; 1988:170–205.
2. Falin LI. The development of human hypophysis and differentiation of cells of its anterior lobe during embryonic life. *Acta Anat (Basel)*. 1961;44:188–205.
3. Hori A, Schmidt D, Rickels E. Pharyngeal pituitary: development, malformation, and tumorigenesis. *Acta Neuropathol*. 1999;98(3):262–272.
4. Solov'ev GS, Bogdanov AV, Panteleev SM, Yanin VL. Embryonic morphogenesis of the human pituitary. *Neurosci Behav Physiol*. 2008;38(8):829–833.
5. Asa SL, Kovacs K. Functional morphology of the human fetal pituitary. *Pathol Annu*. 1984;19(Pt 1):275–315.
6. Hori A. Suprasellar peri-infundibular ectopic adenohypophysis in fetal and adult brains. *J Neurosurg*. 1985;63(1):113–115.
7. Lennox B, Russell DS. Dystopia of the neurohypophysis; two cases. *J Pathol Bacteriol*. 1951;63(3):485–490.
8. Scully KM, Rosenfeld MG. Pituitary development: Regulatory codes in mammalian organogenesis. *Science*. 2002;295(5563):2231–2235.
9. Zhu X, Rosenfeld MG. Transcriptional control of precursor proliferation in the early phases of pituitary development. *Curr Opin Genet Dev*. 2004;14(5):567–574.

10. Asa SL, Ezzat S. Molecular determinants of pituitary cytodifferentiation. *Pituitary.* 1999;1(3–4):159–168.
11. Lamolet B, Pulichino AM, Lamonerie T, et al. A pituitary cell-restricted T box factor, Tpit, activates POMC transcription in cooperation with Pitx homeoproteins. *Cell.* 2001;104(6):849–859.
12. Mody S, Brown MR, Parks JS. The spectrum of hypopituitarism caused by PROP1 mutations. *Best Pract Res Clin Endocrinol Metab.* 2002;16(3):421–431.
13. Dasen JS, O'Connell SM, Flynn SE, et al. Reciprocal interactions of Pit1 and GATA2 mediate signaling gradient-induced determination of pituitary cell types. *Cell.* 1999;97(5):587–598.
14. Dubois PM, Begeot M, Dubois MP, Herbert DC. Immunocytological localization of LH, FSH, TSH and their subunits in the pituitary of normal and anencephalic human fetuses. *Cell Tissue Res.* 1978;191(2):249–265.
15. Begeot M, Dubois MP, Dubois PM. Growth hormone and ACTH in the pituitary of normal and anencephalic human fetuses: Immunocytochemical evidence for hypothalamic influences during development. *Neuroendocrinology.* 1977;24(3–4):208–220.
16. Dunn IF, Laws ER. Microsurgical approaches for transsphenoidal surgery. In: Laws ER, Lanzino G, eds. *Transsphenoidal Surgery*. 1st ed. Elsevier; 2010:120–127.
17. Bergland RM, Ray BS, Torack RM. Anatomical variations in the pituitary gland and adjacent structures in 225 human autopsy cases. *J Neurosurg.* 1968;28(2):93–99.
18. Berke JP, Buxton LF, Kokmen E. The 'empty' sella. *Neurology.* 1975;25(12):1137–1143.
19. Kaufman B, Chamberlin WB Jr. The ubiquitous "empty" sella turcica. *Acta Radiol Diagn (Stockh).* 1972;13(1):413–425.
20. Stanfield JP. The blood supply of the human pituitary gland. *J Anat.* 1960;94(Pt 2):257–273.
21. Xuereb GP, Prichard MM, Daniel PM. The arterial supply and venous drainage of the human hypophysis cerebri. *Q J Exp Physiol Cogn Med Sci.* 1954;39(3):199–217.
22. Gorczyca W, Hardy J. Arterial supply of the human anterior pituitary gland. *Neurosurgery.* 1987;20(3):369–378.
23. Scheithauer BW. The hypothalamus and neurohypophysis. In: Kovacs K, Asa SL, eds. *Functional Endocrine Pathology*. Blackwell; 1991:170–224.
24. Waxman SG. Diencephalon. In: Waxman SG, ed. *Clinical Neuroanatomy*. 28th ed. McGraw-Hill; 2017.
25. Molitch ME, Schimmer BP. Introduction to endocrinology: The hypothalamic-pituitary axis. In: Brunton LL, Hilal-Dandan R, Knollmann BC, eds. *Goodman & Gilman's: The Pharmacological Basis of Therapeutics*. 13th ed. McGraw-Hill; 2017.
26. Scheithauer BW, Sano T, Kovacs KT, Young WF Jr, Ryan N, Randall RV. The pituitary gland in pregnancy: A clinicopathologic and immunohistochemical study of 69 cases. *Mayo Clin Proc.* 1990;65(4):461–474.
27. Kovacs K, Horvath E. *Tumors of the Pituitary Gland*. Armed Forces Institute of Pathology; 1986.
28. Teramoto A, Hirakawa K, Sanno N, Osamura Y. Incidental pituitary lesions in 1,000 unselected autopsy specimens. *Radiology.* 1994;193(1):161–164.
29. Zada G, Lin N, Ojerholm E, Ramkissoon S, Laws ER. Craniopharyngioma and other cystic epithelial lesions of the sellar region: A review of clinical, imaging, and histopathological relationships. *Neurosurg Focus.* 2010;28(4):E4.
30. Melmed S, Jameson LJ. Anterior pituitary: Physiology of pituitary hormones. In: Kasper D, Fauci A, Hauser S, et al, eds. *Harrison's Principles of Internal Medicine*. 19th ed. McGraw-Hill; 2014.
31. Handwerger S, Freemark M. The roles of placental growth hormone and placental lactogen in the regulation of human fetal growth and development. *J Pediatr Endocrinol Metab.* 2000;13(4):343–356.
32. Mulchahey JJ, Jaffe RB. Detection of a potential progenitor cell in the human fetal pituitary that secretes both growth hormone and prolactin. *J Clin Endocrinol Metab.* 1988;66(1):24–32.
33. Horvath E, Ilse G, Kovacs K. Enigmatic bodies in human corticotroph cells. *Acta Anat (Basel).* 1977;98(4):427–433.
34. Phifer RF, Spicer SS. Immunohistochemical and histologic demonstration of thyrotropic cells of the human adenohypophysis. *J Clin Endocrinol Metab.* 1973;36(6):1210–1221.
35. Phifer RF, Midgley AR, Spicer SS. Immunohistologic and histologic evidence that follicle-stimulating hormone and luteinizing hormone are present in the same cell type in the human pars distalis. *J Clin Endocrinol Metab.* 1973;36(1):125–141.
36. Asa SL, Kovacs K, Bilbao JM. The pars tuberalis of the human pituitary. A histologic, immunohistochemical, ultrastructural and immunoelectron microscopic analysis. *Virchows Arch A Pathol Anat Histopathol.* 1983;399(1):49–59.
37. Marin F, Stefaneanu L, Kovacs K. Folliculo-stellate cells of the pituitary. *Endocr Pathol.* 1991;2(4):180–192.
38. Devnath S, Inoue K. An insight to pituitary folliculo-stellate cells. *J Neuroendocrinol.* 2008;20(6):687–691.
39. Shanklin WM. Age changes in the histology of the human pituitary. *Acta Anat (Basel).* 1953;19(3):290–304.
40. Sano T, Kovacs KT, Scheithauer BW, Young WF Jr. Aging and the human pituitary gland. *Mayo Clin Proc.* 1993;68(10):971–977.
41. Stefaneanu L, Kovacs K, Lloyd RV, et al. Pituitary lactotrophs and somatotrophs in pregnancy: A correlative in situ hybridization and immunocytochemical study. *Virchows Arch B Cell Pathol Incl Mol Pathol.* 1992;62(5):291–296.
42. Frawley LS, Boockfor FR. Mammosomatotropes: Presence and functions in normal and neoplastic pituitary tissue. *Endocr Rev.* 1991;12(4):337–355.
43. Calderon L, Ryan N, Kovacs K. Human pituitary growth hormone cells in old age. *Gerontology.* 1978;24(6):441–447.
44. Kovacs K, Ryan N, Horvath E, Penz G, Ezrin C. Prolactin cells of the human pituitary gland in old age. *J Gerontol.* 1977;32(5):534–540.
45. Takei Y, Seyama S, Pearl GS, Tindall GT. Ultrastructural study of the human neurohypophysis. II. Cellular elements of neural parenchyma, the pituicytes. *Cell Tissue Res.* 1980;205(2):273–287.
46. Kimura S, Hara Y, Pineau T, et al. The T/ebp null mouse: Thyroid-specific enhancer-binding protein is essential for the organogenesis of the thyroid, lung, ventral forebrain, and pituitary. *Genes Dev.* 1996;10(1):60–69.
47. Lee EB, Tihan T, Scheithauer BW, Zhang PJ, Gonatas NK. Thyroid transcription factor 1 expression in sellar tumors: A histogenetic marker? *J Neuropathol Exp Neurol.* 2009;68(5):482–488.
48. Seyama S, Pearl GS, Takei Y. Ultrastructural study of the human neurohypophysis. III. Vascular and perivascular structures. *Cell Tissue Res.* 1980;206(2):291–302.

49. Luse SA, Kernohan JW. Squamous-cell nests of the pituitary gland. *Cancer.* 1955;8(3):623–628.
50. Horvath E, Kovacs K, Penz G, Ezrin C. Origin, possible function and fate of "follicular cells" in the anterior lobe of the human pituitary. *Am J Pathol.* 1974;77(2):199–212.
51. Luse SA, Kernohan JW. Granular-cell tumors of the stalk and posterior lobe of the pituitary gland. *Cancer.* 1955;8(3):616–622.
52. Lopes MBS, Mete O, Roncaroli F, et al. Pituicytoma, granular cell tumor of the sellar region, and spindle cell oncocytoma. In: Louis DN, Oghaki H, Wiestler OD, et al, eds. *WHO Classification of Tumours Editorial Board. Central Nervous System.* 5th ed. International Agency for Research on Cancer; 2021: 401–405.
53. Mete O, Lopes MB, Asa SL. Spindle cell oncocytomas and granular cell tumors of the pituitary are variants of pituicytoma. *Am J Surg Pathol.* 2013;37(11):1694–1699.
54. Mordes DA, Lynch K, Campbell S, et al. VE1 antibody immunoreactivity in normal anterior pituitary and adrenal cortex without detectable BRAF V600E mutations. *Am J Clin Pathol.* 2014;141(6):811–815.
55. Jones RT, Abedalthagafi MS, Brahmandam M, et al. Cross-reactivity of the BRAF VE1 antibody with epitopes in axonemal dyneins leads to staining of cilia. *Mod Pathol.* 2015;28(4):596–606.
56. Shanklin WM. Lymphocytes and lymphoid tissue in the human pituitary. *Anat Rec.* 1951;111(2):177–191.

11 Peripheral Nervous System

Peter Pytel

Peripheral nerves are often simplistically described as cables that connect the brain to all parts of the body or, in an analogy allowing for slightly more complexity, wires with surrounding insulation (Fig. 11.1). Often, we also describe the basic histology in similarly simplistic terms by referring to a nerve as a structure comprised of parallel arranged axons and associated Schwann cells. The long list of single-gene defects linked to distinct peripheral neuropathies and the increasingly complex immunobiology underlying inflammatory neuropathies remind us that nerves are delicate structures that require the well-orchestrated interaction between many different components to achieve successful signal transduction. This chapter will therefore aim to create an appreciation for the hidden complexities behind the deceivingly simple structure of peripheral nerves.

Small peripheral nerve branches are found in the background of most tissues and often pathologists treat them as part of normal anatomy. On some occasions a nerve is specifically sampled (a) as peripheral nerve biopsy to further investigate causes of a peripheral neuropathy, (b) as nerve sample to evaluate a patient in preparation of a possible peripheral nerve repair after injury, (c) as peripheral nerve margin of a tumor resection. Beyond these targeted samples, peripheral nerves may be examined to assess the presence of perineurial invasion.

Nerves are typically organized into fascicles (Fig. 11.2). Each fascicle is surrounded by a dedicated cellular barrier, the perineurium, that defines the endoneurial space and separates it from the epineurium. The following discussion will outline the development of the peripheral nervous system and discuss the components of peripheral nerves.

DEVELOPMENT OF THE PERIPHERAL NERVOUS SYSTEM

Development of the peripheral nervous system requires three main processes: the initial lineage determination and migration of cells that later give rise to its components, mechanisms to guide the outgrowth of axonal processes, and development of myelination or associated Schwann cells. Some of the processes that occur during initial development also play a role during response to injury and repair.

The Main Cells

The neural plate represents the first anatomical evidence of nervous system differentiation. It develops as a thickened specialized area in the middorsal ectoderm during the late

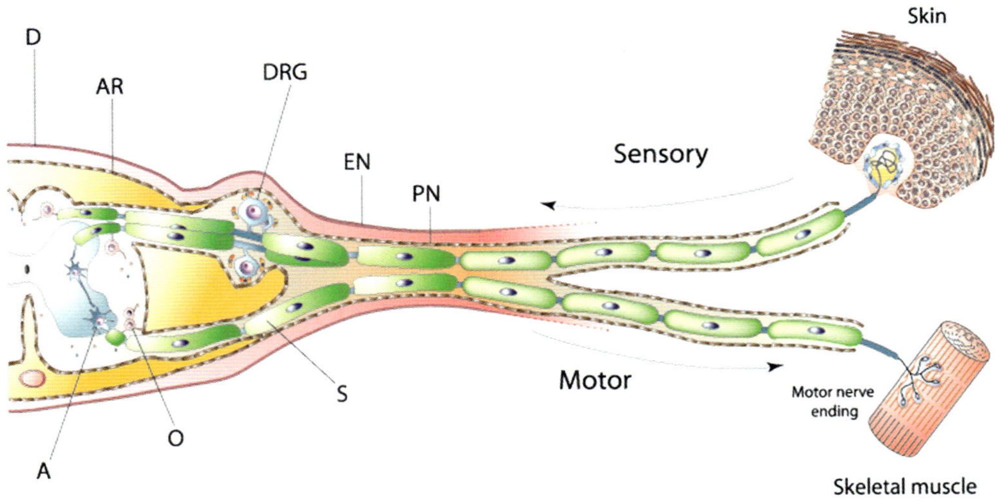

Sergio Piña-Oviedo

FIGURE 11.1 Anatomy of spinal nerve roots. Motor axons arising from the anterior horn cell (*A*) are initially myelinated by oligodendrocytes (*O*) and then pass into the anterior root to be myelinated by Schwann cells (*S*). Sensory nerve axons pass into the dorsal root ganglion (*DRG*) and the central extension of the sensory neuron passes via the dorsal root into the spinal cord. Arachnoid (*AR*) appears to be continuous with the perineurium of the peripheral nerve (*PN*). Dura (*D*) extends from the spinal cord to coat the roots within the intervertebral foramen and is continuous with the epineurium (*EN*).

gastrula stage of the developing embryo. This zone later becomes depressed along the axial midline to form a neural groove that folds inward to form the neural tube (1). Before fusion is completed, groups of cells become detached from the lateral folds of the neural plate to form the neural crests. Anteriorly, neural crests are located at the level of the presumptive diencephalon and extend backward along the whole neural tube (2).

In the peripheral nervous system, the neural crest is the source of neurons, Schwann cells, and satellite cells in the autonomic and sensory ganglia. The ectodermal placodes may also give rise to ganglion cells in the cranial region (1). Migrating multipotent neural crest cells and their subsequent development are determined and progressively limited by the inductive effect of neuregulins and their ErbB receptors, environmental factors, and relations with other cell types (1,3,4). The transcription factors Oct-6, Krox20, and SOX10 are initially expressed in the earliest migrating neural crest cells and appear to be intimately involved in the development of Schwann

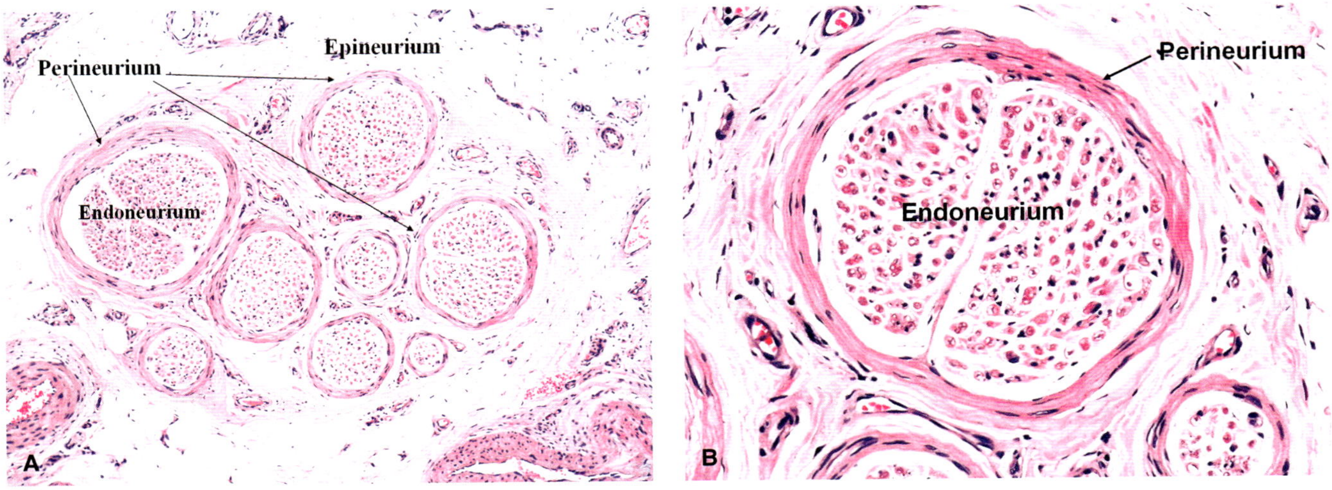

FIGURE 11.2 Peripheral nerve sheaths and compartments. **A:** A low-power view of a transverse section of a normal sural nerve. The nerve fascicles with roughly circular outlines are surrounded by perineurium and embedded in the connective tissue of the epineurium. Epineurial blood vessels (**A**) are also cut in cross section and there is adherent adipose tissue. The 1-μm sections are stained with hematoxylin and eosin (×16). **B:** The endoneurial compartment containing myelinated and nonmyelinated nerve fibers and their accompanying Schwann cells is surrounded by perineurium. Paraffin section stained with hematoxylin and eosin (×45).

cells (3–5). Melanoblasts also express SOX10 and, thus, it has been suggested that Schwann cell precursors associated with nerve endings could be the source of melanocytes (6). This hypothesis could explain the association between alterations in skin pigmentation and neurologic disorders as observed in patients with neurofibromatosis type 1 (NF1) while also accounting for the presence of melanin in some peripheral nerve sheath tumors (6).

The dorsal root ganglia represent the posterior sensory root ganglia. Developing neuroblasts in this region extend neurites both centrally into the neural tube and toward the periphery. Motor neurons in the anterior lateral parts of the neural tube extend their neurites toward the periphery as they develop. Schwann cells become associated with the peripheral nerves and eventually form myelin around many of the axons (3,7). The proximal portions of the anterior horn cell axons and the central axons of the sensory ganglion cells are myelinated within the neural tube by oligodendrocytes (Fig. 11.1).

Growth of Axons

It is fascinating to consider how neuronal processes grow over long distances and arrive at specific terminal regions. Genetic determinants, growth factors, and the extracellular matrix appear to play important roles in the appropriate guidance of neuronal processes (3,7). Ramón y Cajal first proposed the concept of neurotrophic substances in 1909 to explain the directionality and specificity of axonal growth in the developing nervous system. It was not until the 1960s that nerve growth factor (NGF) was discovered by Levi-Montalcini, Cohen, and Hamburger as a target-derived neurotrophic factor that supports the survival and differentiation of sensory and autonomic ganglia in the peripheral nervous system (8). NGF is a protein composed of three subunits—alpha (α), beta (β), and gamma (γ). The γ-subunit of this complex acts as a serine protease and cleaves the N-terminus of the β-subunit, thereby activating the protein into functional NGF (8,9). Other substances that participate in axon growth are members of the NGF family and include brain-derived neurotrophic factor (BDNF) semaphorin-3A, neuropilin-1, ephrin, and neurotrophins 3 (NT-3), 4/5 (NT-4/5), and 6 (NT-6) (8–10). The tips of growing axons possess multiple surface receptors for soluble and bound molecules that provide information for the growth course of an axon (4,10). NGF interacts with its receptor on the axonal surface and promotes motility of the growing tip of the axon by interaction with its cytoskeleton (10,11).

Schwann cells produce NGF and possess NGF receptors on their surface membranes, but expression of these receptors diminishes markedly as peripheral nerves mature. As NGF binds to Schwann cell receptors and becomes concentrated on the surface of the primitive Schwann cell, it provides a chemotactic stimulus for growing axons (7,10). Failed trophic interactions between the target organ and its innervation may result in nerve dysfunction (11). Indeed, some human neuropathies have been attributed to deficiency of neurotrophic factors, thereby providing a rational basis for the clinical use of neurotrophic agents in select peripheral neuropathies (9,11).

The extracellular matrix also plays an important role in axonal growth and guidance. The tip of the growing axon has receptors for adhesion to extracellular substances such as collagen, fibronectin, laminin, and entactin; binding of extracellular components to these receptors promotes elongation of axons and stimulates cytoskeletal protein synthesis, thereby promoting movement and axon growth (11,12). Some of these extracellular components are found within or near basement membranes surrounding Schwann cells (12).

Schwann Cells and Myelination

Schwann cells move freely between and around developing peripheral nerve axons, forming primitive sheaths around the neurites and growing in parallel with them. Contact with axons stimulates Schwann cell division *in vitro* (13). *In vivo* Schwann cell multiplication virtually ceases in the normal adult animal, but mitotic activity is induced by peripheral nerve damage. It is thought that exposure of the axon to the Schwann cell following demyelination or during axonal regeneration promotes Schwann cell division, whereas the relationship between Schwann cells and axons in the normal nerve induces some sort of contact inhibition in the Schwann cells (13). If axon regeneration does not occur following axon damage, Schwann cells gradually decrease in number, suggesting that their growth and survival depend on contact with axons (7,13). Experimental evidence also suggests that continued axon regeneration depends on the presence of Schwann cells (7,14).

Fascicles containing large axon bundles surrounded by Schwann cell processes are identifiable in the sural nerve by the 9th week of gestation (15,16). Schwann cells extend several long, flattened processes that wrap around large clusters of fine axons between weeks 10 and 15. At this stage, two to four Schwann cells are located within a common basement membrane and form Schwann cell families (16).

Myelination of peripheral nerves in humans commences between the 12th and 18th weeks of gestation (16,17). An intricate signaling network between axon and Schwann cells influences axonal diameter and the developing myelin as outlined here. The initiation of myelination depends on the diameter of the axon and its association with Schwann cells. The axon itself determines whether it is myelinate or not based on the amount of type III Neuregulin 1 exposed on its membrane (17). By the time that axons span 1.0 and 3.2 μm in diameter, they are in a 1:1 relationship with Schwann cells and have either formed mesaxons or membrane spirals with compact myelin sheaths of 3 to 15 layers (3,16,17). Schwann cells around myelinated fibers and around unmyelinated fibers are able to produce myelin and certain transcription factors, such as SOX10, Krox20, and Oct-6/Scip, are known to be involved in the myelination program (3,4,7). Schwann cells in developing and regenerating peripheral nerves also express high levels of the neurotrophin receptor p75NTR (18). Certain glycoproteins, such as myelin-associated glycoproteins (MAGs), are believed to participate in establishing specific Schwann cell–axon interactions in the developing peripheral nervous system (19). It also appears that Schwann cells may influence the caliber of axons since the axonal diameter may be decreased markedly in some hereditary demyelinating neuropathies in which there is a genetic defect in Schwann cells (13,14,17). Myelinating Schwann cells control the number and phosphorylation state of neurofilaments in the axon, leading to enlargement of the axon itself. Conversely, absence of myelin results in fewer neurofilaments, reduced phosphorylation levels, and smaller axonal diameters (16). MAGs act as a myelin signal that modulates the caliber of myelinated axons (20). Maintenance of an axon therefore appears to depend not only on influences from the neuron cell body but also on interactions of the axon with the accompanying Schwann cells (11,13,20).

Some 70% of axons within a mixed sensory nerve such as the sural nerve are very small and will become segregated into groups of 8 to 15 axons lying in longitudinal grooves within one Schwann cell; these will form the unmyelinated (Remak) fibers within the peripheral nerve. Thus, all axons in the peripheral nervous system are invaginated into the surfaces of Schwann cells, but myelin sheaths only form around the larger axons, which represent only a small proportion of peripheral nerve fibers (16).

ANATOMY OF PERIPHERAL NERVES

An understanding of the anatomy of peripheral nerves is essential for the interpretation of clinical signs and symptoms of patients with peripheral neuropathies, analysis of peripheral nerve specimens, and planning an autopsy to investigate a patient with a peripheral neuropathy (16,21). Most major nerves, such as the sciatic and median nerves, contain motor, sensory, and autonomic nerve fibers. Motor axons end peripherally at muscle endplates and many of the sensory nerves are associated with peripheral sensory endings. The cell bodies of sensory nerves lie outside the central nervous system in the dorsal root ganglia or in the cranial nerve ganglia (16). Each ganglion contains numerous, almost spherical neurons and their surrounding satellite cells.

Dorsal root ganglion cells were first described by von Kölliker in 1844. They are examples of pseudounipolar cells, which means that a single, highly coiled axon arises from each perikaryon and shows a T- or Y-shaped bifurcation at a node of Ranvier some distance from the neuron, resulting in formation of central and peripheral axons. Thus, the initial segment of axon gives the impression that the cell is a unipolar neuron, but it actually has two axons (Fig. 11.1). The central axon passes into the spinal cord either to synapse in the posterior sensory horn of gray matter or to pass directly into the dorsal columns. Peripheral axons pass into the peripheral nerves (16,21). Satellite cells are derived from the neural crest and have an origin similar to that of Schwann cells (22). They have been referred to by various names including amphicyte, capsular cells, perisomatic gliocyte, and perineuronal satellite Schwann cells (16,22).

Sir Charles Bell first demonstrated that motor function lay in the anterior roots, while François Magendie showed that the sensory function lay in the posterior roots. This anterior-motor, posterior-sensory organization is known as the Bell–Magendie law (16,21). Motor nerves are derived from anterior horn cells in the spinal cord or from defined nuclei in the brain stem. The initial segment of the axon lies within the central nervous system and is ensheathed by myelin formed by oligodendrocytes (Fig. 11.1). As the axons pass out of the brain stem or spinal cord, they become myelinated by Schwann cells. Anterior spinal roots join the posterior roots as they pass through the intervertebral foramina to form spinal nerves. Cranial nerves leave the skull through a number of different foramina. The junction between oligodendrocytes and the Schwann sheath of the cranial nerves is known as Obersteiner–Redlich zone or the glial/Schwann junction and has some clinical significance (23). The pulsatile compression of the Obersteiner–Redlich zone by a vessel in some exit foramina may be responsible for the clinical symptoms of trigeminal and glossopharyngeal neuralgia, hemifacial spasm, torticollis spasmodicus or even symptoms of essential hypertension when a vascular cross compression of the left vagus nerve occurs (23).

Autonomic nerves are either parasympathetic or sympathetic in nature. Preganglionic parasympathetic fibers pass out of the brain stem in the III, VII, IX, and X cranial nerves and from the sacral cord in the second and third sacral nerves. Postganglionic parasympathetic neurons are situated near or within the structures being innervated. Sympathetic preganglionic fibers arise from neurons in the intermediolateral cell columns of gray matter in the thoracic spinal cord and pass out in thoracic anterior roots (16). These preganglionic fibers are myelinated and reach the sympathetic trunk through the corresponding anterior spinal roots. In the paravertebral sympathetic trunk, the preganglionic fibers synapse with the sympathetic ganglion cells. The postganglionic fibers of the parasympathetic and sympathetic nervous system innervate viscera, blood vessels, and smooth muscle of the eye and skin (16,21).

HISTOLOGY, IMMUNOCYTOCHEMISTRY, AND ULTRASTRUCTURE OF PERIPHERAL NERVES

The careful examination of normal larger, named nerves typically allows the identification of distinct glistening, white fascicles that are bound together by connective tissue. Grossly appreciable loss of the myelin that imparts the white color can result in gray discoloration. This is probably best appreciated in the color change of atrophic anterior nerve roots in autopsies of patients with amyotrophic lateral sclerosis (ALS). Microscopically, transverse sections of a peripheral nerve (Fig. 11.2) reveal a distinct, organoid arrangement with the endoneurial compartments that contain axons as well as Schwann cells and that are surrounded by perineurium to form the individual fascicles (16,21).

Epineurium

The epineurium consists of moderately dense connective tissue binding nerve fascicles together. It merges with the adipose tissue that surrounds peripheral nerves (Fig. 11.2A), particularly in the subcutaneous tissue. In addition to fibroblasts, the epineurium contains mast cells. Although mostly composed of collagen, there are true elastic fibers composed of elaunin and oxytalan fibers in the epineurium that impart some elastic recoil to a fresh, unfixed nerve specimen (21,24). The amount of epineurial tissue varies and is more abundant in nerves adjacent to joints. As nerve branches become smaller to consist of only one fascicle, epineurium is no longer present. In larger nerves, one or more arteries, veins, and lymphatics run in parallel to the nerve fascicles within the epineurium and are termed the *vasa nervorum* (16,21,25). Inflammation and occlusion of such arteries is an important cause of nerve damage in vasculitic diseases (25). Overgrowth of epineurial adipose tissue produces a localized area of tissue distortion termed the lipofibromatous hamartoma. This condition is most commonly seen in the median nerve and can be associated with soft tissue overgrowth resulting in macrodactyly or megadactyly (26). In some nerves, such as in the sciatic nerve, there is an outermost common layer of connective tissue known as the paraneurium (mesoneurium). This layer of loose connective tissue surrounds nerves and may play a role in compression or entrapment neuropathies (27).

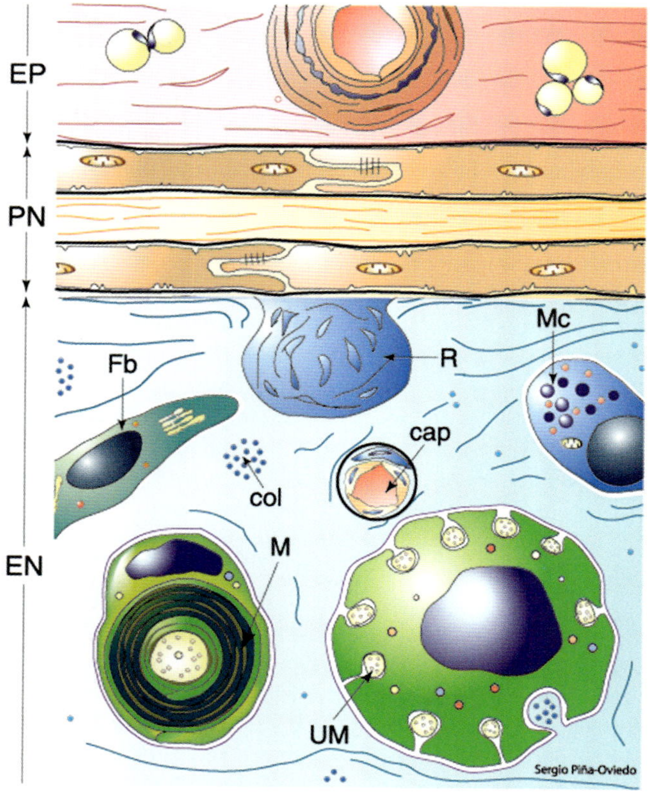

FIGURE 11.3 Diagram to show the major elements of peripheral nerve compartments. The epineurium (*EP*) contains collagen, blood vessels, and some adherent adipose tissues. The flattened cells of the perineurium (*PN*) are joined by tight junctions and form flattened layers separated by collagen fibers. Renaut bodies (*R*) project into the endoneurium (*EN*). Schwann cells forming lamellated myelin (*M*) (drawn uncompacted in this diagram) surround the larger axons. Multiple unmyelinated axons (*UM*) are invaginated into the surface of Schwann cells. Other elements include fibroblasts (*Fb*), mast cells (*Mc*), capillaries (*cap*), and collagen (*col*).

Perineurium

Originally described by Henle in the 19th century, the perineurium has been known by a variety of different terms such as laminated sheath of Ranvier, mesothelium, perilemma, neurothelium, perineurothelium, and perineurial epithelium (21,28).

The intercellular signaling molecule secreted by Schwann cells, Desert Hedgehog (Dhh), was discovered as a result of the pioneering work of Nüsslein-Volhard and Wieschaus. This molecule is important to formation of the perineurium (29). Dhh –/– mice only have thin, disorganized perineurium that does not adequately function as a normal blood nerve barrier (30).

The perineurium consists of concentric layers of flattened cells separated by layers of collagen (Figs. 11.2 to 11.4). The number of cell layers varies from nerve to nerve and depends on the size of the nerve fascicle. For example, there are 8 to 12 layers of perineurial cells in the sural nerve but that number decreases progressively such that a single layer of perineurial cells surrounds fine distal nerve branches (21,28). Perineurial cells eventually fuse to form the outer core of the terminal sensory endings in Pacinian corpuscles and muscle spindles (21,31). The perineurial cells of motor nerves form an open funnel as the nerve ends at the motor endplate. Paraganglia of the vagus nerve may lie just underneath the perineurium (28,32).

Perineurial cells contain thin sheets of cytoplasm with small amounts of endoplasmic reticulum, prominent vimentin filaments, and numerous pinocytotic vesicles that open onto the external and internal surfaces of the cell. There are fibrillary patches of actin bundles adjacent to the plasma membrane and it has been suggested that the perineurium has contractile properties (21). Basement membrane is usually seen on both sides of each perineurial lamina (27,28). Numerous cell junctions, including well-formed tight junctions (i.e., *zonulae occludentes*), are present between adjacent perineurial cells and appear to

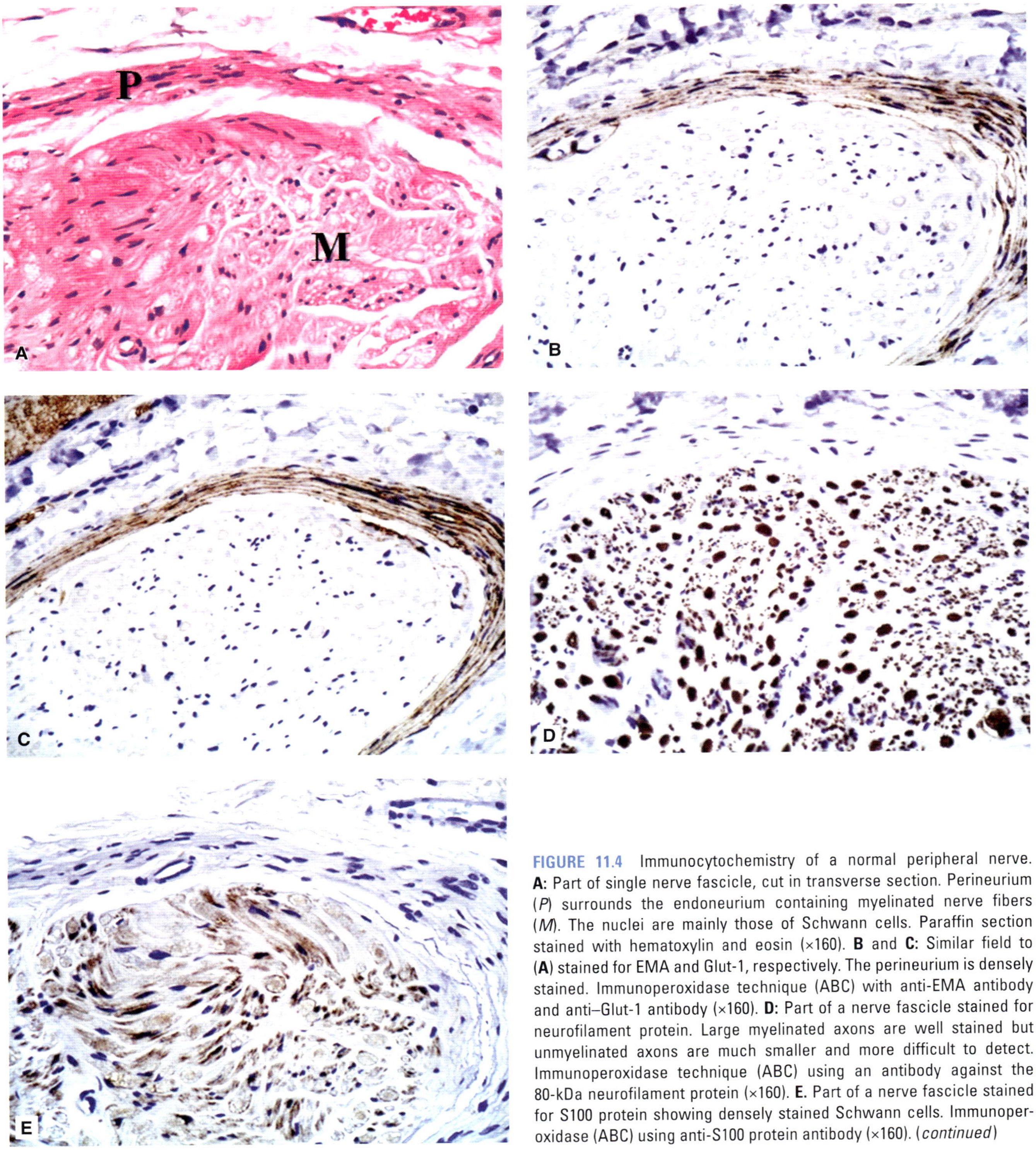

FIGURE 11.4 Immunocytochemistry of a normal peripheral nerve. **A:** Part of single nerve fascicle, cut in transverse section. Perineurium (*P*) surrounds the endoneurium containing myelinated nerve fibers (*M*). The nuclei are mainly those of Schwann cells. Paraffin section stained with hematoxylin and eosin (×160). **B** and **C:** Similar field to (**A**) stained for EMA and Glut-1, respectively. The perineurium is densely stained. Immunoperoxidase technique (ABC) with anti-EMA antibody and anti–Glut-1 antibody (×160). **D:** Part of a nerve fascicle stained for neurofilament protein. Large myelinated axons are well stained but unmyelinated axons are much smaller and more difficult to detect. Immunoperoxidase technique (ABC) using an antibody against the 80-kDa neurofilament protein (×160). **E.** Part of a nerve fascicle stained for S100 protein showing densely stained Schwann cells. Immunoperoxidase (ABC) using anti-S100 protein antibody (×160). (*continued*)

be critical for the formation of the blood–nerve barrier (27,28). Claudins are integral membrane proteins that play a major role in tight junctions and are present in normal and neoplastic perineurium. Claudin-1 expression is largely limited to perineurial cells of peripheral nerves but is also present in paranodal regions and in the outer mesaxon along internodes (28,33). When tracer substances such as ferritin and horseradish peroxidase are injected into the blood, they do not enter peripheral nerves because there are tight junctions in endoneurial capillaries as well as the inner layers of the perineurium. This blood–nerve barrier is analogous to the blood–brain barrier and represents a complex

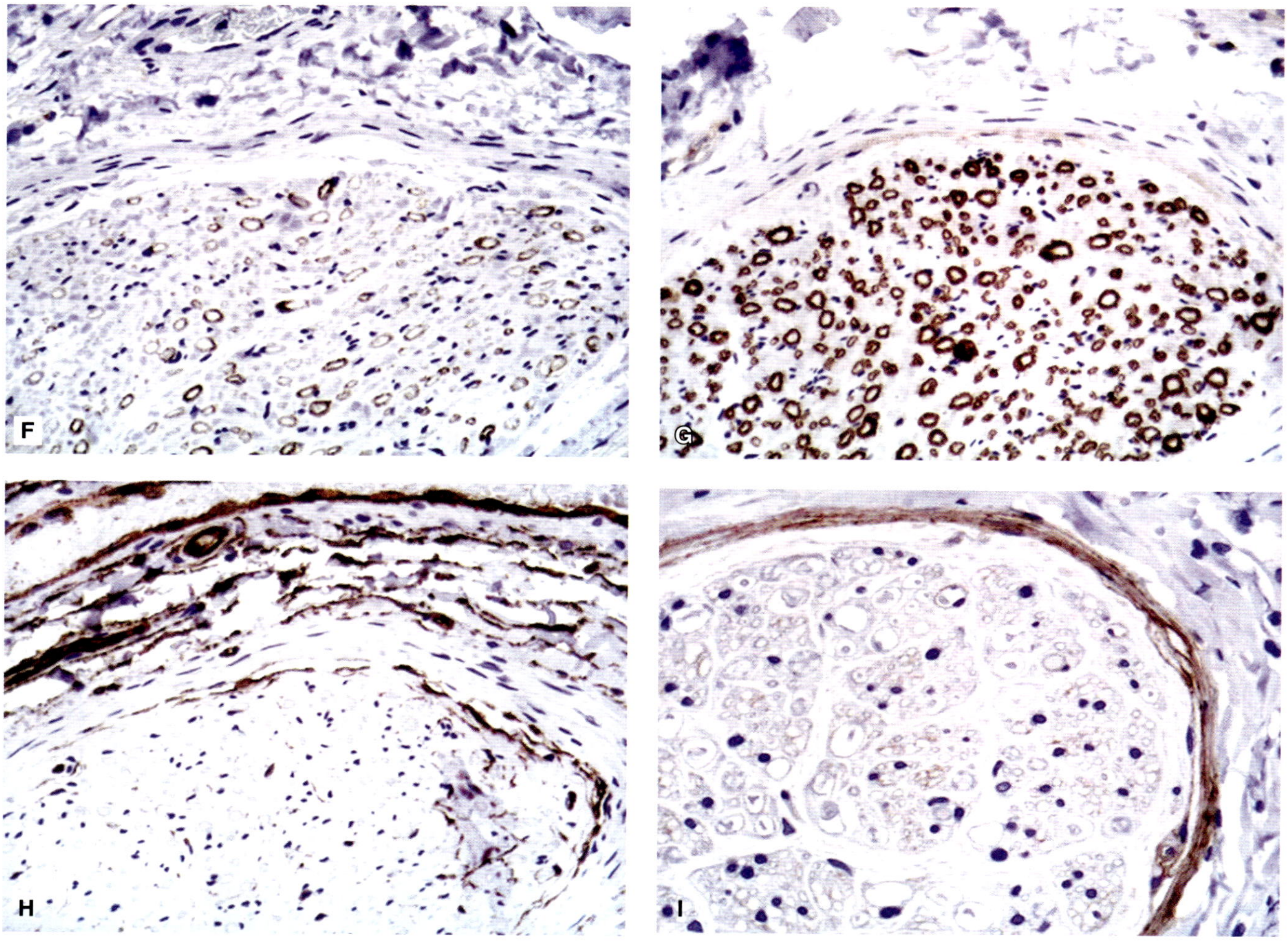

FIGURE 11.4 (*Continued*) **F** and **G:** Part of a nerve fascicle stained for E-cadherin and CD57 (Leu7), respectively, showing stained Schwann cells. Immunoperoxidase (ABC) using anti–E-cadherin and anti-CD57 antibodies (×160). **H:** Part of a nerve showing few endoneurial and numerous epineurial CD34-positive cells. These cells are clearly distinct from Schwann cells that comprise the bulk of the cells in the nerve. Immunoperoxidase (ABC) using anti-CD34 (QBend10) antibody (×160). **I:** Part of a nerve showing collagen IV staining the basement membrane of the perineurial cells. Immunoperoxidase (ABC) using anticollagen IV antibody (×160).

functional system involving endoneurial cells, endothelial cells, perineurial cells, pericytes, Schwann cells, and basement membranes. The blood–nerve barrier is present soon after birth and prevents the entry of drugs and other substances into nerves that may otherwise interfere with or block conduction (27,28,34). No such blood–nerve barrier exists in the dorsal root ganglia or in autonomic ganglia; these sites in the peripheral nervous system are vulnerable to mercury and other toxins (35).

The epineurial sheath of the nerve is continuous with the dura mater at the junction of spinal nerves and spinal nerve roots (Fig. 11.1), and the perineurium blends with the pia-arachnoid (28). There are some morphologic similarities between the perineurium and arachnoid cells, although arachnoid cells are not usually surrounded by basement membrane. Perineurial cells and pia-arachnoid cells are immunopositive for epithelial membrane antigen (EMA) (Fig. 11.4B) and vimentin but are negative for S100 protein, SOX10, and CD57 (28,36). Perineurial cells also express insulin-dependent glucose transporter protein I (Glut-1) (Fig. 11.4C) and claudin-1 (28,33,37). Immunohistochemistry has demonstrated that perineurial cells proliferate in traumatic neuroma, Morton neuroma, neurofibroma, solitary circumscribed neuroma, and perineurioma (28,38).

Some cancer cells can break through the perineurial sheath to grow along the perineurial space; perineurial invasion has been correlated with decreased survival times among cancer patients (39). Immunohistochemistry for Glut-1, EMA, and claudin-1 may be used to rapidly and accurately assess the presence of perineurial invasion (28,40). However, there are some situations in which benign proliferating ductules may be found within the perineurium and endoneurium;

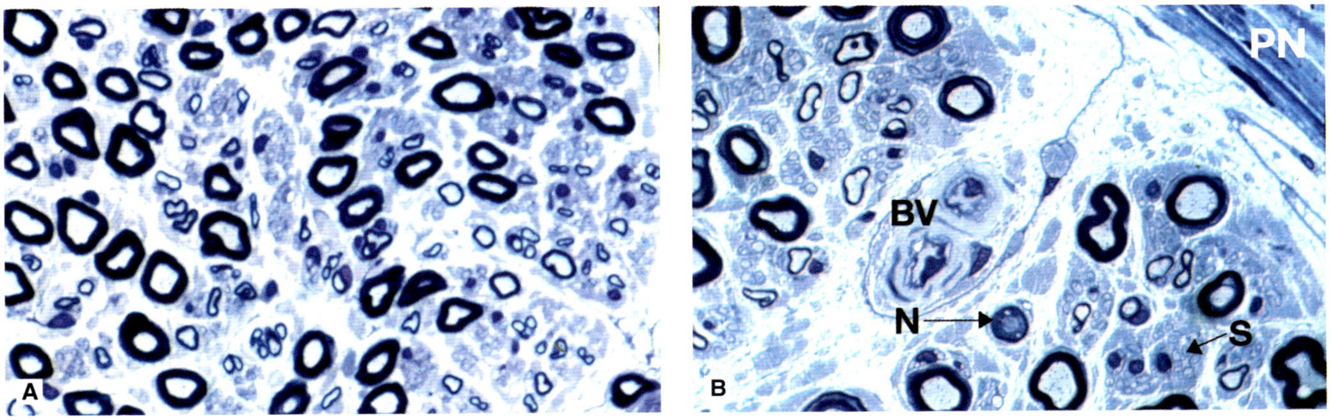

FIGURE 11.5 High-power histology of human sural nerve in transverse section. **A:** Large- and small-diameter myelinated fibers are seen. In a normal nerve, myelinated fibers are separated from each other but in this nerve, there are also small numbers of regenerating clusters (see also Fig. 11.12B). The 1-µm resin section is stained with toluidine blue (×160). **B:** Part of a sural nerve fascicle cut in transverse section. Perineurium is at the top right (*PN*). Both large and small myelinated fibers vary in cross-sectional outline. Splits within the sheath are Schmidt–Lanterman incisures. Endoneurial blood vessel (*BV*). Section through a fiber near the node of Ranvier (*N*). Unmyelinated axons are seen as unstained circles within Schwann cells (*S*). The 1-µm resin section (×310).

this finding has been termed *vasitis nodosa* (41). Nerve involvement has also been reported in fibrocystic disease of the breast, normal and hyperplastic prostatic tissue, and the normal pancreas (41).

Endoneurium

The endoneurium is the compartment that contains axons and their surrounding Schwann cells, collagen fibers, fibroblasts, capillaries, macrophages, and a few mast cells (Figs. 11.3 to 11.5).

In cross sections of peripheral nerves, some 90% of the nuclei belong to Schwann cells, 5% to fibroblasts, and 5% to other cells. The different cells in the endoneurial compartment, axons, myelin, and matrix will be discussed in the following sections.

Endoneurial Fibroblast-Like Cells, Macrophages, and Mast Cells

There are CD34-positive bipolar cells with delicate dendritic processes within the endoneurium and peripheral nerve sheath tumors (Fig. 11.4H) (42,43). These endoneurial fibroblast-like cells are frequently located near blood vessels and under the perineurium where they are arranged parallel to perineurial cells; they may function as phagocytes under certain conditions (42,43). Fibroblasts are ultrastructurally identical to fibroblasts elsewhere in the body.

An intrinsic population of immunocompetent and potentially phagocytic endoneurial macrophages has been described within the human endoneurium. This population shares several lineage-related and functional markers with macrophages and may represent the peripheral counterpart of microglial cells (i.e., del Rio-Hortega cells) of the central nervous system (44). Mast cells are normal constituents of the endoneurium and are seen in sensory ganglia as well as the epineurial sheath of peripheral nerves. There is an increase in the number of mast cells in some pathologic conditions such as axonal, or Wallerian, degeneration, and in some peripheral nerve sheath tumors (45). A characteristically high number of mast cells are seen in neurofibromas, and they are present in the Antoni B areas of schwannomas (46). Mast cells may influence growth of neurofibromas because some of their mediators may also act as growth factors. Apparently, the inciting factor for mast cell migration into nerve sheath tumors is Kit ligand that is hypersecreted by NF1 deficient Schwann cell popluations (46). Mast cell stabilizers potentially reduce the proliferation of neurofibromas and diminish clinical symptoms of itching that often accompany them (46). Nerve injury causes breakdown of the blood–nerve barrier as endoneurial vessels become permeable to fluid and protein. The increase in permeability may be related to the release of biogenic amines from mast cells within the endoneurium. Proteases released from mast cells have a high myelinolytic activity and may play a role in certain demyelinating diseases (46,47).

Endoneurial Extracellular Matrix

Collagen within the endoneurial compartment is highly organized and forms two distinct sheaths around myelinated and unmyelinated nerve fibers and their Schwann cells. The outer endoneurial sheath of Key and Retzius is composed of longitudinally oriented large-diameter collagen fibers; the inner endoneurial sheath of Plenk and Laidlaw is composed of fine collagen fibers oriented obliquely or circumferentially to the nerve fibers. The term neurilemma has been applied

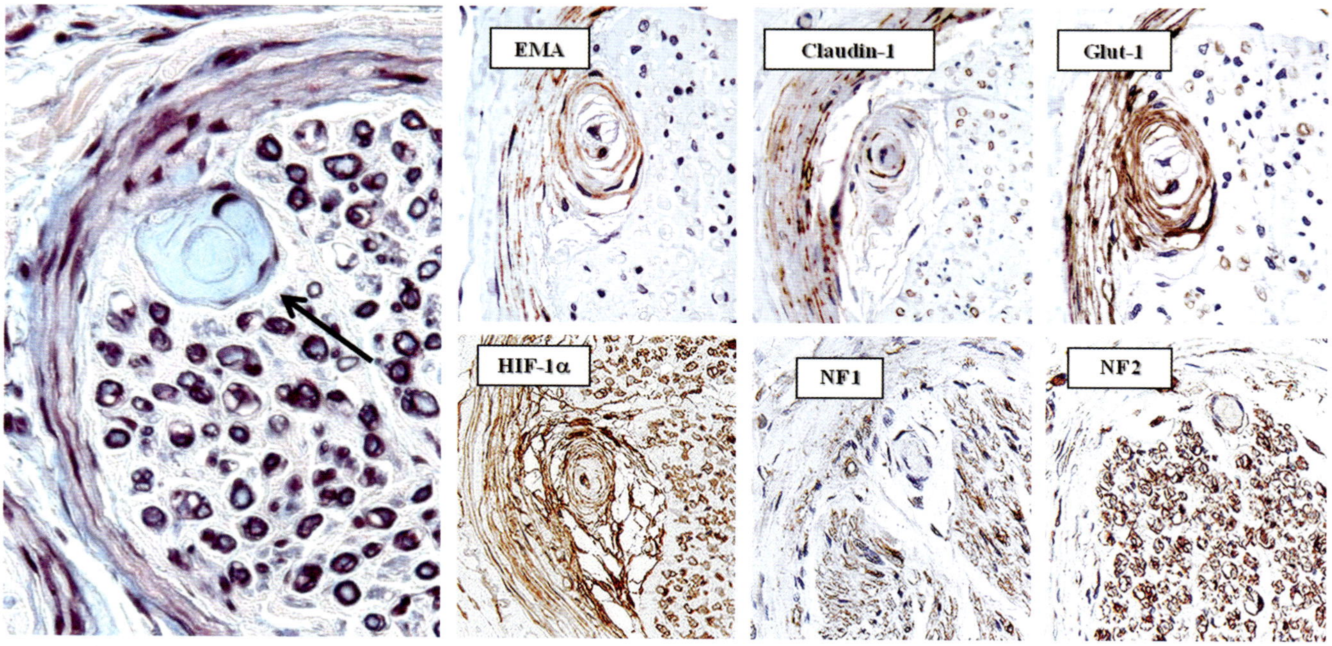

FIGURE 11.6 Nerve fascicles showing Renaut bodies (RB) (*arrow*). **Left panel:** Russell–Movat pentachrome, showing the Renaut corpuscles in blue. Classical perineurial markers such as EMA, claudin-1, and Glut-1 are expressed in Renaut bodies. In addition, HIF-1α, a transcription factor that regulates Glut-1 is also expressed in Renaut bodies and the perineurium. Expression of NF2 is also detected in Renaut bodies and scattered perineurial cells. In contrast, NF1 is negative in RB (see Ref. (51)).

to the combined sheath formed by the basement membrane of the Schwann cell and the adjacent inner endoneurial sheath of collagen fibers (16,21). The historic term neurilemmoma is therefore inappropriate when used to describe tumors of Schwann cell origin. The longitudinal orientation of collagen fibers in the outer endoneurial sheath, together with the Schwann cell basement membrane tubes, may play an important role in guiding axons as they regenerate following peripheral nerve damage (48).

Renaut bodies (Figs. 11.3 and 11.6) are not infrequently seen in the endoneurium of human peripheral nerves. Described in the 19th century by the French physician J. L. Renaut, they are cylindrical hyaline structures attached to the inner aspect of the perineurium (28,49). Composed of randomly oriented collagen fibers, spidery fibroblasts, and perineurial cells, Renaut bodies stain positively with Alcian blue due to the presence of acid glycosaminoglycans. The rest of the endoneurium also contains Alcian blue–positive mucoproteins (28). Renaut bodies produce extracellular matrix highly enriched in elastic fiber components and express vimentin, collagen IV, EMA, Glut-1, and claudin-1; interestingly the adjacent perineurium shows identical pattern of immunoreactivity (50). They may extend for some distance along the nerve and end in a blunt and abrupt fashion (50). Their precise function is not known, but Renaut himself thought that they may act as protective cushions within the nerve. They increase in number in compressive neuropathies, trauma, and in a number of other neuropathies including hypothyroid neuropathy (49).

Axons

Most peripheral nerves contain a mixture of myelinated and unmyelinated nerve fibers. As the axons are oriented longitudinally along the nerve, quantitative estimates of the number of fibers in the nerve and their diameters are only adequately assessed in exact transverse sections. Staining techniques that can be used to identify nerve fibers and other components within peripheral nerves are summarized in Table 11.1. Longitudinal sections of peripheral nerve are less valuable than transverse sections, but teased nerve fibers are very valuable for detecting segmental demyelination and remyelination and for assessing past axonal degeneration and regeneration (51).

There are approximately 8,000 myelinated axons or nerve fibers per mm^2 and some 30,000 unmyelinated axons per mm^2 in a transverse section of a human sural nerve (16,52). The axons within peripheral nerves are categorized as class A, class B, and class C fibers according to their diameter, function, and the speed at which they conduct nerve impulses. In general, axonal diameter correlates with the extend of myelination and conduction speed. Class A axons are myelinated and further subdivided into six groups covering three size ranges. The largest are 10 to 20 μm diameter myelinated axons that conduct at 50 to 100 m/sec; myelinated axons of 5 to 15 μm in diameter conduct at 20 to 90 m/sec, and those myelinated axons of 1 to 7 μm diameter conduct at 12 to 30 m/sec. Class B axons are myelinated preganglionic autonomic fibers spanning approximately 3 μm in diameter that conduct at 3 to 15 m/sec. Unmyelinated

TABLE 11.1 **Histologic Techniques for Peripheral Nerves**

Technique	Application
A. General	
1. Hematoxylin and eosin (H&E)	Detection of inflammation, myelin, and axons (Figs. 11.2B and 11.4A)
2. Hematoxylin–van Gieson	Collagen stains red; myelin black
3. Reticulin stains	Basement membrane around each Schwann cell in normal (e.g., Gordon–Sweet) nerve and schwannomas
4. Masson trichrome	Fibrinoid necrosis in vasculitis
5. Alcian blue	Glycosaminoglycans stain blue
6. Toluidine blue	(a) Mast cells in paraffin section, (b) general stain for 1-µm resin section, (c) metachromatic stain for sulfatide lipid.
B. Stains for myelin	
1. Luxol fast blue	Myelin stains blue
2. Loyez	Myelin stains black
3. Osmium	Myelin stains black
4. Periodic acid–Schiff (PAS)	Myelin stains bright pink (good for detecting small number of nerve fibers in muscle biopsies)
5. Polarized light (frozen section)	Normal myelin, birefringent; degenerating myelin isotropic (nonbirefringent)
6. Marchi method	Degenerating myelin stains black (due to the presence of cholesterol esters); normal myelin is unstained
7. Oil red O	Degenerating myelin stains bright red; normal myelin pink
C. Stains for axons	
1. Palmgren's of Bodian's (silver stains)	Axons stains black
D. 0.5- to 1-µm sections	
1. Toluidine blue	Myelin stains black; axons unstained; Schwann cells and other cells, blue; collagen, blue (Figs. 11.5, 11.12, and 11.14A,B)
2. Toluidine blue and carbol fuschin	Myelin stains black; axons unstained; cell and collagen, pink/blue (Fig. 11.14C)
E. Electron microscopy	(Figs. 11.8, 11.10, and 11.16)
F. Teased fibers	
G. Immunohistochemistry	
1. S100 protein	Schwann cells (Fig. 11.4E); schwannomas, neurofibromas
2. CD57	Schwann cells (Fig. 11.4G); schwannomas, neurofibromas
3. CD56	Neurons, axons, Schwann cells; schwannomas
4. Calretinin	Schwann cells and some schwannomas
5. CD146 (Mel-CAM)	Schwann cells and some schwannomas
6. SOX10	Schwann cells, melanocytes, schwannomas, neurofibromas, and melanomas
7. GFAP (glial fibrillary acid protein)	Some schwannoma cells; possibly unmyelinated
8. PGP 9.5	Neurons, axons, Schwann cells and some schwannomas
9. Myelin basic protein	Myelin
10. Neurofilaments	Axons (Fig. 11.10C)
11. Epithelial membrane antigen/Glut-1/ claudin-1	Perineurium (Fig. 11.4B,C); perineurioma
12. CD34	-CD34-positive endoneurial fibroblasts (Fig. 11.4H)
13. CD68/CD163	Endoneurial macrophages
14. GAP 43	Regenerating axons

class C axons are 0.2 to 1.5 µm in diameter and conduct impulses at 0.3 to 1.6 m/sec; these include postganglionic autonomic and some afferent sensory axons, including pain fibers (16,21).

› *ULTRASTRUCTURE* Although myelinated nerve axons can be demonstrated in paraffin sections (Fig. 11.4), they are best visualized by light microscopy in 0.5 to 1 µm thick toluidine blue–stained resin sections (Fig. 11.5). Most axons larger than 3 µm in diameter are myelinated. Although along much of the length of a myelinated nerve fiber has a circular outline in cross section, there is considerable variation in shape within the normal nerve, especially in the perinuclear regions and near the node of Ranvier (Fig. 11.7).

The axon itself is limited by a smooth plasma membrane, the axolemma, that is separated from the encompassing Schwann cell by a 10 to 14 nm gap, the periaxonal space of Klebs (Fig. 11.8). The axonal cytoplasm, or axoplasm, contains mitochondria, cisternae of smooth endoplasmic reticulum, occasional ribosomes and glycogen granules, peroxisomes, and neurotransmitter-containing vesicles. The most prominent component of the axoplasm and a key functional element is the cytoskeleton. The cytoskeleton is composed of three main elements: microfilaments, neurofilaments, and microtubules. Microfilaments span 5 to 7 µm in diameter and are composed of chains of actin; they comprise approximately 10% of the total axonal protein. Microfilaments are virtually confined to the cortical

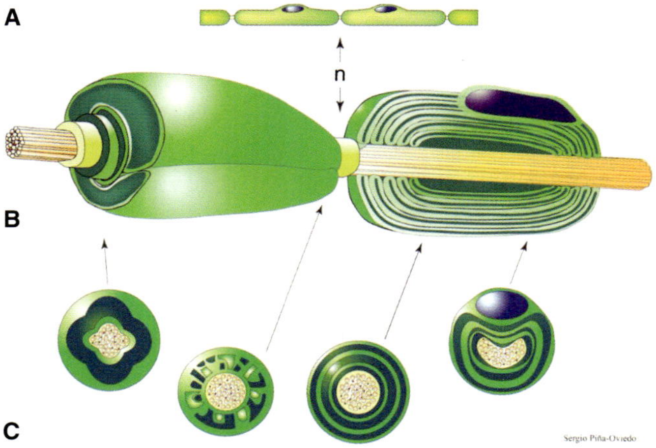

FIGURE 11.7 Diagram to show the relationships between (**A**) teased fibers, (**B**) nerve fibers in longitudinal section, and (**C**) nerve fibers in transverse section. **A:** In teased fibers, nodes of Ranvier (*n*) are separated by internodal portions of the Schwann cell and myelin sheath. The Schwann cell nucleus is roughly in the center of the internode. **B:** Longitudinal section through the node of Ranvier shows how the myelin sheath terminates as a series of end loops. The axon narrows as it passes through the node of Ranvier. **C:** Transverse sections of peripheral nerve as seen in electron micrographs and 1-µm resin sections are here related to the different portions of the internode and the node of Ranvier. From left to right, the paranodal region shows crenation of both axon and myelin sheath in larger fibers. At the node of Ranvier, the axon is small and coated by radially arranged Schwann cell processes and myelin end loops. Throughout most of the internode the myelinated fiber is circular. In the region of the nucleus, the axon and the myelin sheath may be ovoid rather than circular in outline.

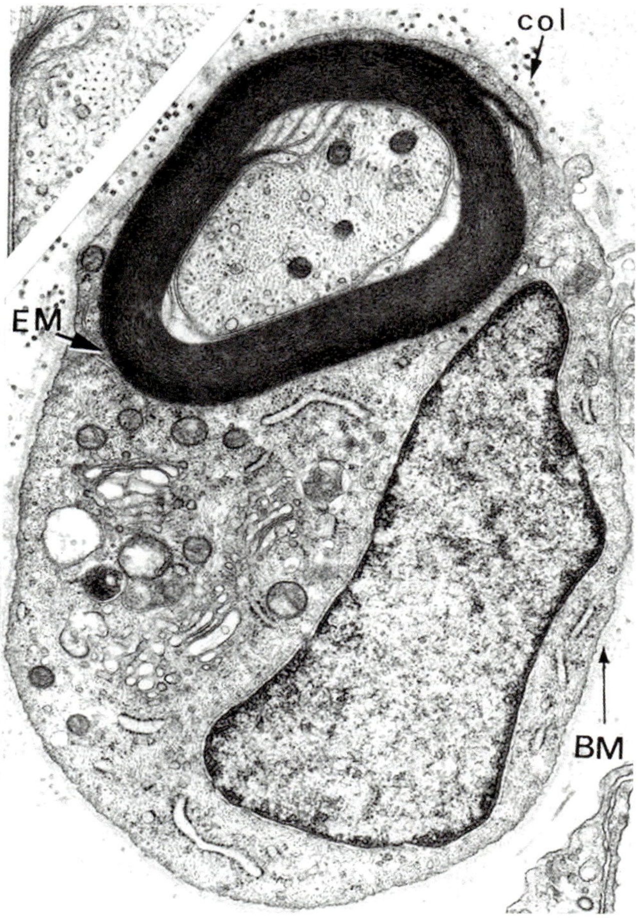

FIGURE 11.8 Transverse section of a myelinated nerve fiber in the perinuclear region. The axon contains mitochondria, small vesicles, and numerous neurofilaments and neurotubules cut in cross section (inset, *top left*). A distinct periaxonal space separates the axon from its encompassing Schwann cell. Myelin is compacted except at the external mesaxon (*EM*) and internally around the internal mesaxon near the axon itself. Part of a Schmidt–Lanterman incisure is seen on the inside of the myelin sheath. Abundant rough- and smooth-surfaced endoplasmic reticulum is seen in the perinuclear cytoplasm of the Schwann cell. A basement membrane (*BM*) surrounds the Schwann cell plasma membrane and endoneurial collagen fibers are seen cut in cross section (*col*). Electron micrograph (×18,400); inset (×40,000).

zone of the axoplasm immediately beneath the axolemma (16). Neurofilaments (Figs. 11.4C and 11.8) are 8 to 10 µm intermediate filaments. They constitute a major filamentous component of larger axons (53). They were described originally by Ramón y Cajal and Bielschowsky as argentophilic neurofibrillae.

Neurofilaments tend to appear in multiple whorled bundles in the perikaryon with no clear orientation to other elements of the cell, although they appear in longitudinal, mostly parallel orientation within the axons (53). Small arm-like filaments are seen by electron microscopy and project from the surfaces of the neurofilaments to form an irregular polygonal lattice. Neurofilaments are composed of protein triplets that are chemically and immunochemically distinct (53). Three major subunits are recognized and classified according to their molecular weights of 68,000 (NF-L), 150,000 (NF-M), and 200,000 (NF-H) Daltons. The neurofilaments within axons are phosphorylated and immunocytochemically distinct from the nonphosphorylated filaments within the perikaryon. Immunocytochemistry for neurofilament protein (Fig. 11.4D) is often valuable for detecting large- or medium-sized axons in normal nerves, traumatic lesions, tumors involving peripheral nerves, and occasionally for detecting axonal processes in neuronal tumors (53). The third cytoskeletal component in the axoplasm is microtubules, or neurotubules (54). Microtubules are cylindrical (Fig. 11.8), unbranched, longitudinally oriented, hollow tubules. They are roughly 25 nm in diameter and composed of globular subunits of tubulin ranging from 4 to 5 nm in diameter. Periodic radial projections of high–molecular-weight proteins arise from the surfaces of the neurotubules and represent one component of the microtubule-associated proteins (MAPs). These arm-like projections bind neurofilaments and actin filaments, creating the neurotubule–neurofilament–actin filament lattices. These three-dimensional lattices form an ordered structure in the axoplasm and play important roles in axonal transport and maintaining the shape of the axon (54). Microtubules also direct the transport of vesicular organelles between the cell body and the axon, thereby determining the composition of the axon to some degree (54).

› *Axoplasmic Flow* Scott proposed that neuronal cell bodies secreted growing substances in order to maintain the function of the axon. He suggested that such substances pass down the axon cytoplasm to the axon terminals. This suggestion was endorsed by Ramón y Cajal when he observed how regeneration occurs from the proximal stump of a damaged axon as long as continuity with the cell body is maintained (16,55). More definitive evidence of axonal transport was later provided by experimental studies using autoradiography and other techniques. Not only can tritiated leucine and other substances be traced by autoradiography as they are transported along axons from the cell body, but the transport of organelles within the axon can also be directly observed using dark-field microscopy or Nomarsky optics. Axoplasmic flow, the movement of different materials along the axoplasm, occurs both away from (i.e., anterograde) and toward (i.e., retrograde) the cell body (55). Both anterograde and retrograde flow occur at the same rate and by the same mechanisms.

Anterograde axoplasmic transport occurs at two velocities—fast and slow. Most organelles and large molecular weight substances within the axon are conveyed by fast axoplasmic transport, up to 400 mm/day. If a ligature is placed around a nerve, transported material accumulates proximal to the ligature as well as distal to it, reflecting interference with anterograde and retrograde transport. The filamentous lattice component of microtubules, neurofilaments, and actin filaments is responsible for fast axoplasmic flow, and these three elements probably act as rails along which the various transported organelles and substances move. Fast axoplasmic transport is dependent on oxidative energy mechanisms and adenosine triphosphate (ATP). It also depends on calcium and magnesium ions and is blocked by calcium channel-blocking agents. Several substances block axoplasmic flow by interfering with these mechanisms. Trifluoperazine, for example, blocks calmodulin (calcium-activating protein) and, thus, also blocks axoplasmic flow. Microtubules are depolymerized by cold and by colchicine. Vincristine and vinblastine are known to bind tubulin and prevent the normal assembly of microtubules in the axons (54,55).

Retrograde axoplasmic transport may convey information and organelles back to the cell body. During development, NGF is taken up by nerve terminals and transported in a retrograde fashion to the cell body, where it may play a role in neuronal maturation (55). It has been suggested that the transport of such growth factors may also influence the metabolism of mature neurons and the absence of such signals from the distal part of the neuron when the axon is severed may trigger chromatolysis (55). Retrograde transport is also a pathway by which some toxins (e.g., tetanus neurotoxin) and metals (e.g., lead, cadmium, and mercury) may bypass the blood–brain barrier and accumulate in neurons (56). Neurotropic viruses (e.g., herpes, rabies, and polio) are transported to the central nervous system by retrograde transport (56). Defects in axonal transport have now been implicated in various disease processes including at least some forms of ALS, diabetic neuropathy, and inherited neuropathies (57–60).

Slow axoplasmic transport at 1 to 3 mm/day concerns the distal movement of cytoskeletal elements such as neurofilaments, microtubules, and actin. It is a one-way process. Neurofilaments are broken down by calcium-activated proteases at the distal end of the axon and microtubules are depolymerized distally (61). Various toxins such as hexocarbons and their derivatives may interfere with slow axoplasmic transport so that neurofilaments accumulate and form large swellings within the axon (35,62). It is thought that neurofilaments within an axon maintain the bulk and the shape of large axons; neurofilaments are less numerous in small axons.

› *The Periaxonal Space of Klebs* As the Schwann cell enwraps the axon, it leaves a 12 to 14 nm space between the Schwann cell membrane and the axolemma (Fig. 11.8) termed the periaxonal space of Klebs (3,63). This space is in continuity with the extracellular space at the node of Ranvier through a narrow helical channel at the site where the terminal cytoplasmic processes of the Schwann cell approach the axolemma (Fig. 11.7) (16). The maintenance of the periaxonal space of Klebs is mediated by an intrinsic 100-kDa MAG) in the periaxonal membrane of the Schwann cell (20,63,64). This protein has a heavily glycosylated domain with sialic acid and sulfate residues on the external surface of the plasma membrane extending into the periaxonal space. In fact, nearly half of the peptide of MAG lies in the periaxonal space (20).

Schwann Cells

Theodore Schwann identified a vague sheath of cells within nerve fibers that have subsequently borne his name as Schwann cells. Schwann cells are derived from the neural crest and migrate with growing axons into the developing peripheral nerves (3,65). They produce NGF in development and during regeneration. As the nerves grow, Schwann cells divide axons into groups and eventually establish 1:1 relationships with the larger fibers that they will ultimately myelinate (16,65). Immature proliferating Schwann cells have a relatively large volume of cytoplasm compared with mature Schwann cells. The Schwann cell cytoplasm is rich in mitochondria, polyribosomes, Golgi cisterns, and rough endoplasmic reticulum (Fig. 11.8). The cytoskeleton within the cells includes vimentin intermediate filaments and is particularly obvious during the active proliferative and migrating phases of development and regeneration (66).

Schwann cells in a normal adult peripheral nerve are associated with both myelinated fibers and unmyelinated (Remak) fibers. The Schwann cytoplasm is divided into two compartments in myelinated fibers: around the nucleus and on the outside of the myelin sheath, and a thin rim of cytoplasm inside the myelin sheath and around the internal mesaxon (Fig. 11.8). Schwann cells within a nerve can be ultrastructurally identified by their relationship with myelinated or unmyelinated fibers. In damaged peripheral nerves, however, Schwann cells can be identified most easily by the presence of an investing basement membrane (Fig. 11.8). Other cells within the endoneurium, such as fibroblasts, do not have a basement membrane and, although macrophages may invade the basement membrane tubes, they have a distinct ruffled border that distinguishes them from Schwann cells. Perineurial cells may be found in the endoneurial compartment, particularly in damaged nerves. They possess a basement membrane but also have tight junctions that are not a feature of Schwann cells (3,65). Schwann cells normally accumulate lipofuscin with increasing age, and they may have lamellated structures in the paranuclear cytoplasm in the form of Pi (π) granules of Reich. Such granules are composed of wide-spaced lamellated structures and amorphous osmiophilic material; they are rich in acid phosphatase and stain metachromatically with toluidine blue in frozen sections (67). Other inclusions such as the corpuscles of Elzholz are seen in Schwann cell cytoplasm. These bodies are spherical, 0.5 to 2.0 μm in diameter, and stain intensely with the Marchi method. Few π granules remain in Schwann cells following nerve damage in which there has been extensive Schwann cell mitotic activity and proliferation (67).

The investing basal lamina of Schwann cells is composed of laminins 1, 8, 10, fibronectin, and entactin/nidogen. They also synthesize and secrete heparan sulfate, N-syndecan, glypican, perlecan, collagen types I, III, IV, V, and XVIII, β-1 and β-4 integrin, and BM-40 (3,16,65). Schwann cells can be identified in paraffin sections by a close investment by reticulin staining as well as basement membrane highlighted by immunohistochemical stains for laminin and collagen IV. There is a rich reticulin network investing each cell, not only in the normal peripheral nerve but also in Schwann cell tumors. Cytoplasmic and nuclear staining for S100 protein can be identified by immunocytochemistry (Fig. 11.4E); this acidic protein is 100% soluble in ammonium sulfate at neutral pH and represents a calcium-binding EF-hand type molecule. It has no known function. Although it is present in Schwann cells, it is not found in fibroblasts or perineurial cells (68). Schwann cells are also immunolabeled with SOX10, E-cadherin (Fig. 11.4F), CD56, and CD57 (Fig. 11.4G), whereas perineurial cells are negative for these markers. Calretinin, the 29-kD, calcium-binding protein that also belongs to the family of EF-hand proteins, is expressed in Schwann cells and in up to 94% of schwannomas (69). Normal and neoplastic Schwann cells also express vimentin, PGP 9.5, and CD271, a low-affinity NGF receptor or p75 NTR—neurotrophin receptor.

Schwann cells are occasionally labeled by anti-GFAP antibodies (36). GFAP immunoreactivity in the peripheral nervous system has been demonstrated in enteric ganglia, and in Schwann cells in the sciatic, splenic, and vagus nerves (36,70). Several distinct subsets of Schwann cells or related cells exist. There are those associated with the perisynaptic region, as well as amphicytes that are the perineuronal satellite cells of the dorsal root ganglia and the autonomic ganglia (65). Schwann cells also participate in the formation, function, and maintenance of neuromuscular junctions and Meissner corpuscles (3,30,65). These "terminal Schwann cells" may be identified by their expression of Herp-protein, which is not present in nonterminal myelinating Schwann cells (71). An interesting and peculiar intermediate glial cell type known as the olfactory ensheathing cell is associated with neuronal processes of the olfactory bulb; these cells share astrocytic and Schwann cell phenotypes in that they promote axonal regeneration (72).

Myelin

Myelin sheaths appear as slightly basophilic rings in H&E-stained transverse paraffin sections of nerve (Fig. 11.4). They can be more prominently stained by Luxol fast blue or by hematoxylin stains such as Loyez (Table 11.1). Myelin is well depicted by Sudan black staining in frozen section slides and it can be identified due to its birefringence in polarized light, a technique that is particularly suitable for identifying myelin in enzymatic histochemical preparations. Myelin is formed by the fusion of Schwann cell membranes and, by electron microscopy, it is seen as a regularly repeating lamellated structure with a 12- to 18-nm periodicity (73,74). External and internal mesaxons can be traced from the cell surface on the outer and inner aspects of the sheath (Fig. 11.8). The myelin membrane is divided into two structurally and biochemically distinct domains: the compact and the noncompact myelin, each of which is characterized by a unique set of proteins. Compact myelin contains P0, PMP-22, P2 protein, and MBP, whereas noncompact myelin contains MAG, Cx32, α6β4 integrin, and E-cadherin (3,17). As the external aspects of the Schwann cell membranes fuse to produce compact myelin, an interrupted interperiod, or intermediate line forms in the myelin. The more densely stained period line (i.e., major dense line) is formed by fusion of the cytoplasmic aspects of the cell membrane. A narrow cleft can be resolved between the components of the interperiod line. In myelinating Schwann cells, noncompact myelin is present in paranodal loops, Schmidt–Lanterman incisures, nodal microvilli, and inner and outer edges of the myelin (3,17). The cytoplasm that connects the Schwann cell perikaryon to remote parts of the myelin internode is retained in noncompact myelin (3). Several types of cell junctions, including tight, gap, and adherens junctions are seen between the myelin lamellae. These are known as autotypic or reflexive junctions, denoting complexes between membranes of the same cell (3,17,75).

The biochemical makeup of myelin is approximately 75% lipid and 25% protein. The major lipids are cholesterol, sphingomyelin, and galactolipids; these molecules are present in higher proportions than they are in other cell membranes. The arrangement of lipids accounts for the liquid crystalline anisotropic nature of normal myelin sheaths that are birefringent in polarized light. Esterification of the cholesterol in degenerating myelin produces the isotropic, nonbirefringent lipid droplets that can be detected by Sudan dyes, oil red O, and the Marchi technique (Table 11.1). As myelin degenerates and the cholesterol becomes esterified, the ultrastructural lamellated pattern is lost and replaced by the amorphous osmiophilic globules seen in electron micrographs. More than half the protein in myelin is a transmembrane 28- to 30-kDa glycoprotein P0 (72,76); other proteins include P1 and P2. P0 mediates hemophilic adhesive interactions between Schwann cell plasma membranes; it is a key structural constituent of both the major dense line and interperiod line of compact myelin, and it is involved in myelin compaction (73,76). Numerous mutations in P0 have been described in a variety of demyelinating diseases (see below) (76).

Although the lipid composition of myelin in the peripheral nervous system is very similar to that of the central nervous system, the protein components are markedly different (77). Myelin of the central nervous system has no P0 protein but contains proteolipid that is soluble in organic solvents. It also contains an 18-kDa basic protein that is probably homologous with the P1 protein of peripheral nerve myelin. These biochemical differences may account for differences in the structure between peripheral and central nervous system myelin (77,78).

Myelin sheaths are essential for the normal functioning of the peripheral nervous system and in those hereditary neuropathies in which myelination is defective, severe disability and retardation of development are seen (79,80). Acting as a biologic electrical insulator, myelin allows discontinuous and very rapid conduction of a wave of depolarization along the nerve fiber. Myelination is an evolutionary adaptation that allows increased conduction velocities without excessive increases in axon diameter (17,81).

Myelination in the peripheral nervous system occurs well in advance of that of the central nervous system. Although there is little myelin in human cerebral hemispheres at birth, myelin sheaths have already started to form around peripheral nerves at this time. Myelination is initiated by contact between Schwann cells and future myelinated axons (3,16,19). The Schwann cell rotates around the axons and may form 50 or more spirals, resulting in formation of the myelin sheath.

As the Schwann cell differentiates and produces a basement membrane, it acquires polarity *via* interaction between its cytoskeleton and some basement membrane components (3,21). The Schwann cell then begins to extend processes around individual axons, generating traction to pull the whole cell round and creating a spiral composed of many lamellae (21). The importance of the basement membrane as a prerequisite to the formation of myelin is emphasized by the lack of myelination when the basement membrane is deficient (82). MAG also plays an important role in myelination (76,83); it is present in the membranes of Schwann cells around myelinated fibers but not in those cells associated with unmyelinated fibers. MAG probably functions through its interaction with the Schwann cell cytoskeleton, and this facilitates process lengthening and rotation during myelination (82). Periaxin is a 47-kDa protein constituent of the dystroglycan–dystrophin-related protein 2 complex that links the Schwann cell cytoskeleton to the extracellular matrix; it is located in the periaxonal region of Schwann cell plasma membranes and possibly interacts with MAG during myelination (84). Mutations in the periaxin gene result in the autosomal recessive demyelinating Charcot–Marie–Tooth (CMT4F) and Déjerine–Sottas diseases (80,85). As myelination proceeds, cytoplasm is expressed from the spiral of Schwann cell processes and membranes to form the 12- to 18-nm lamellated structure of myelin.

The length of an embryonic Schwann cell is 30 to 60 μm and it becomes associated with the length of axon in the developing nerve. As the nerve lengthens with growth of the body and limbs, so does the Schwann cell such that the length of the Schwann cell or internodal distance (Fig. 11.7) in myelinated fibers reaches some 190 μm at 18 weeks of gestation and 475 μm at birth. Normal Schwann cells in adults may extend for up to 1 mm in length along the myelinated fibers. Schwann cells associated with unmyelinated fibers lengthen to reach approximately 250 μm in the adult sural nerve. Following damage to a peripheral nerve, Schwann cells revert to their embryonic lengths and thus give short internodes in regenerating and remyelinating nerve fibers (see Figs. 11.12 and 11.14) (16).

› Schmidt–Lanterman Clefts or Incisures (S–L I)

Once viewed as artifacts, the clefts or incisures described by Schmidt and Lanterman (Fig. 11.9) are now known to be fixed components of the myelin sheath (16,21). Each Schmidt–Lanterman incisure consists of a continuous spiral of Schwann cell cytoplasm that runs from the outer (nuclear) to the inner (paraxonal) Schwann cell compartment in an oblique fashion at about 9 degrees to the long axis of the sheath. The funnel-shaped cleft splits the cytoplasmic membranes at the major dense line and forms a route for the passage of substances from the outer cytoplasmic layer through the myelin sheath to the inner cytoplasm. Near the external surface of the cleft, stacks of desmosome-like structures and Cx32-rich gap junctions are sometimes seen, possibly maintaining the integrity of the spiral (16). The cell junction proteins, claudin-5, MUPP1, E-cadherin, and neurofascin, have been selectively detected at the Schmidt–Lanterman incisure (16,21). Cytoplasm in the clefts contains membrane-bound dense bodies, lysosomes, an occasional mitochondria, intermediate filaments, and a single microtubule (Fig. 11.9) that runs circumferentially around the fiber. This microtubule may be associated with transport and stabilization of the cytoplasmic spiral (16). The number of Schmidt–Lanterman incisures correlates with the diameter of the axon; the larger the fiber, the more clefts in the myelin sheath per Schwann cell. The presence of these clefts suggests that they are an

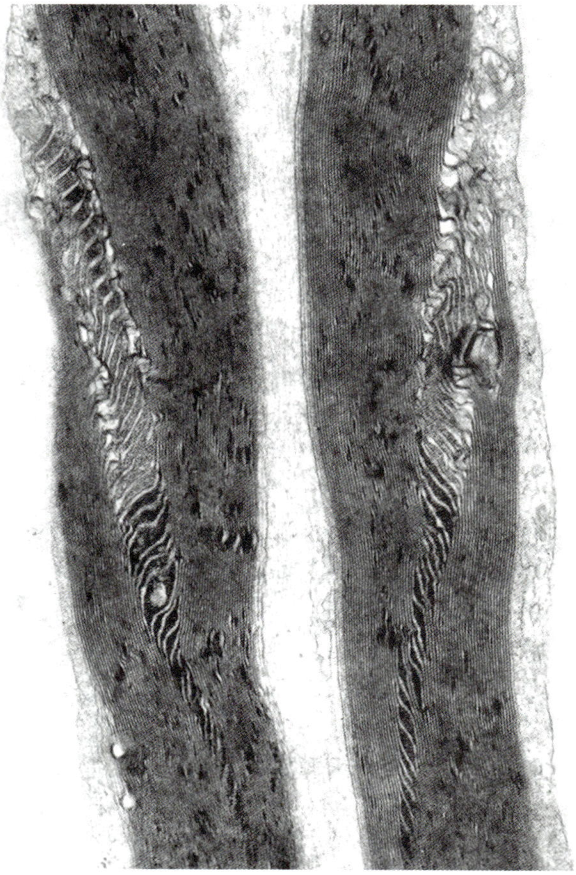

FIGURE 11.9 Longitudinal section of peripheral nerve: a Schmidt–Lanterman incisure. Blebs of cytoplasm are seen running through the myelin sheath. Densities in the cytoplasm (*top left*) suggest some form of junction between the spiral turns of the incisure. The axon is cut tangentially. (Electron micrograph (×30,000). (Reprinted from Weller RO, Cervos Navarro J. *Pathology of Peripheral Nerves: A Practical Approach*. Butterworth; 1977 with permission.)

important functional part of the sheath. It also seems obvious that they are pathways of communication between the inner and outer Schwann cell cytoplasm, but their full significance remains to be elucidated.

Ramón y Cajal described longitudinal cytoplasmic channels, the longitudinal bands of Cajal, which are similar to Schmidt–Lanterman incisure and lie beneath the surface of the Schwann cell plasma membrane. They are known to be separated from each other by the appositions formed by the periaxin–dystrophin-related protein 2 (Drp2) dystroglycan complex (17,86). Microtubule-based transport in these bands permits the Schwann cells to lengthen in response to axonal growth. Interestingly, Schwann cells lacking longitudinal bands of Cajal are unable to keep pace with axon growth, suggesting a role in regulating the ability of Schwann cells to increase in length (17,86).

› ***Nodes of Ranvier*** With the introduction of techniques whereby individually separated myelinated nerve fibers could be stained black with osmium tetroxide, a new view of nerve fibers was obtained. In his publication of 1876, Louis A. Ranvier described and illustrated the constrictions or "étranglements annulaires," that are now known as the nodes of Ranvier (21,87). The functions of the node at that time were not known but Ranvier did suggest that the constrictions may prevent displacement or flow of the semiliquid myelin along the nerve fibers. He also suggested that the gap in the myelin sheath at the node of Ranvier might allow diffusion of nutrients into the axon (16).

In teased fibers stained with osmium tetroxide or viewed in polarized light, the nodal gap is readily visible, as is the bulbous swelling of the fiber on either side of the node of Ranvier (Fig. 11.15). The distance between each node along a myelinated fiber (Fig. 11.7) is approximately proportional to the thickness of the myelin sheath. Internodal segments between the nodes of Ranvier vary from 200 to 1500 μm in length in normal adults; the Schwann cell nucleus is usually sited around the middle of the internode.

Histologic study of 1 μm transverse resin sections of nerve and electron microscopic observations reveal a complex structure at the node of Ranvier and in the paranodal regions. As the axon approaches a node of Ranvier, it may become cruciform in cross section, especially in large fibers (Fig. 11.7). Deep furrows develop in the surrounding myelin sheath, and those furrows are filled with cytoplasm rich in mitochondria. As the axon passes through the node, it is reduced to one-third or one-sixth of its internodal diameter.

Three important zones that define the node of Ranvier are the node, the paranode, and the juxtaparanode (Fig. 11.10). The myelin sheaths that surround the internode end in myelin loops that are anchored to the membrane of the axon, the axolemma, through a protein complex that includes neurofascin 155 (NF155) anchored in the membranes of Schwann cells, as well as Caspr1 and CNTN1 that are anchored in the axolemma (88). Ultrastructurally, this zone is visualized as dense transverse bands. Microvillus-like processes extend from the Schwann cell cytoplasm to cover the nodal region. The basement membranes of the two Schwann cells on each side of the node are continuous and cover the nodal region. The central nodal region lacks any covering by Schwann cells and is characterized by a high density of Nav1.6 sodium channels. Neurofascin 186 (NF186) is anchored in the microvillous processes of the Schwann cells and, along with gliomedin, helps to ensure the proper localization of sodium channels. The juxtaparanodal region is characterized by a high density of Kv 1.2 potassium channels that are anchored through Caspr2 anchored in the axolemma and Tag2 anchored in the Schwann cell membrane. The paranode is thought to aid in maintaining the distinct localization of potassium and sodium channels at and around the node. The localization of these ion channels is critical for the saltatory conduction along myelinated axons (88).

The abundance of mitochondria in the paranodal cytoplasm is an indication of the high-energy requirements of

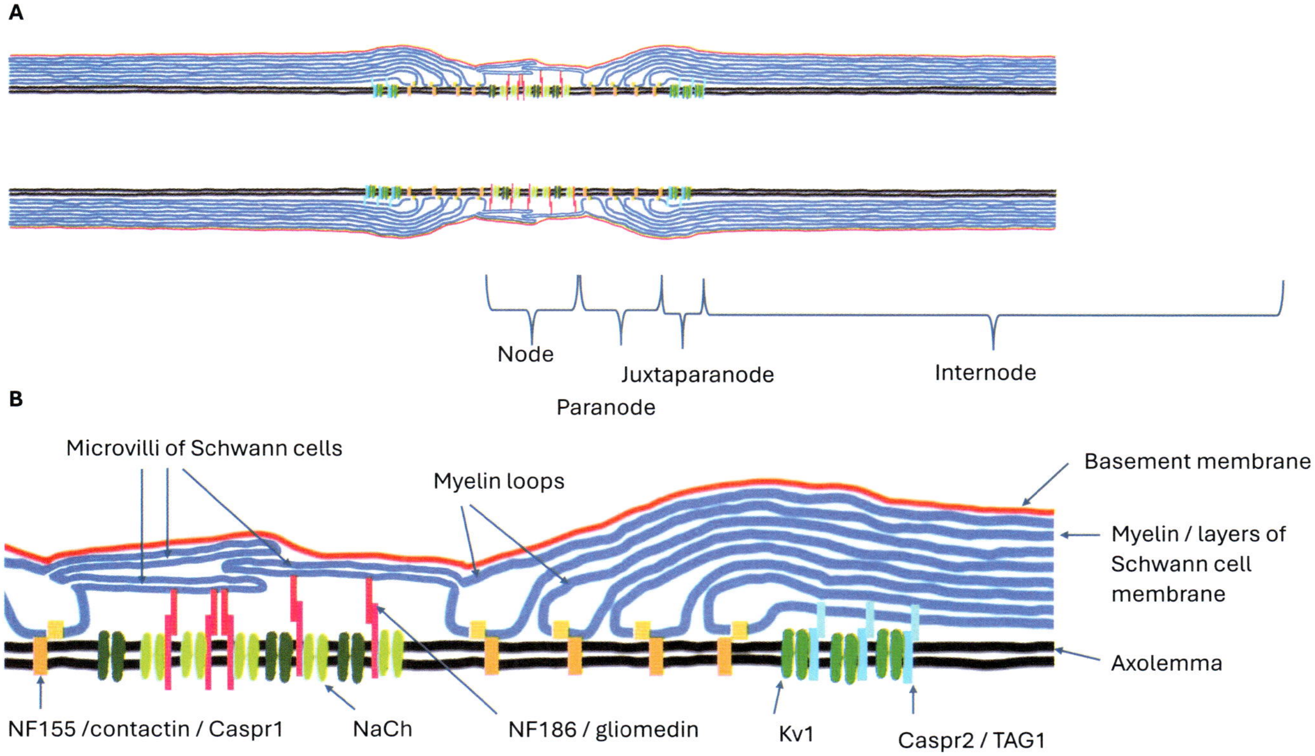

FIGURE 11.10 **A:** The main segments along a node of Ranvier include (a) the node, (b) the paranode, (c) the juxtaparanode, and beyond that (d) the internode. **B:** The node is devoid of any myelin but is covered by overlapping microvillous processes that extend over the nodal area from the adjacent Schwann cells and the basement membrane that is continuous between the two Schwann cells. The axolemma, the axonal membrane, in the nodal segment is characterized by a high density of sodium channels (light grain). These are localized through this region by a protein complex that includes NF186 (anchored in the sarcolemma) and gliomedin (anchored in the microvillous processes of the Schwann cells). The paranodal region is characterized by the termination of the Schwann cell myelin in myelin loops. These attach to the axolemma through a protein complex that includes contactin 1 and Caspr1 (anchored in the axolemma) and NF155 (anchored in the Schwann cell membrane of the myelin loops). These junctional complexes can be visualized ultrastructurally as dense bands and they are thought to aid in maintaining the distinct localization of ion channels around toward the nodal and juxtaparanodal region. The juxtaparanodal region, lastly, is characterized by a high density of potassium channels that are linked to a protein complex that includes Caspr2 (anchored in the axolemma) and Tag1 (anchored in the Schwann cell membrane of the myelin).

the node. There is also an electron-dense polyanionic-rich material around the Schwann cell processes that constitutes the extracellular matrix of the node. This gap substance creates a ring-like structure (i.e., ring of Nemiloff) and may provide an ion pool necessary for nodal function. It has been demonstrated that the gap substance contains glycosaminoglycans with cation-binding substances (16, 86).

› ***REMAK CELLS*** The specialized Schwann cells associated with unmyelinated axons are called Remak cells. Unmyelinated fibers make up approximately 80% of the axons in most peripheral nerves and can be detected as unstained structures by light microscopy in toluidine blue–stained 0.5 μm transverse resin sections of peripheral nerve (Fig. 11.5) (16,21,65). However, at 1 to 3 μm diameter, they are almost at the limit of resolution and are only seen in good-quality sections. Such fibers can be stained by Palmgren or Bodian silver techniques but are poorly visualized in immunocytochemical preparations using antineurofilament antibodies (Fig. 11.4D).

The structure of unmyelinated fibers and their quantitation are most adequately studied by transmission electron microscopy (Fig. 11.11). They are more numerous than myelinated fibers in mixed peripheral nerves by a factor of 3 or 4:1 (25,89) and were first recognized in 1838 by the Polish physician Robert Remak as “fimbriae organicae.” The Schwann cells associated with unmyelinated axons are referred to as Remak cells (7). Schwann cells have the potential to differentiate into either a myelinating or nonmyelinating ensheathing cells, depending upon the signals received from the axons that they contact (7,17). A minimum caliber of 1 μm is required before an axon can be myelinated (17). Certain molecules such as L1, NCAM, integrins, and neurofascin, are known to be expressed on unmyelinated axons and are downregulated during axonal myelination (7). Schwann cells around myelinated and

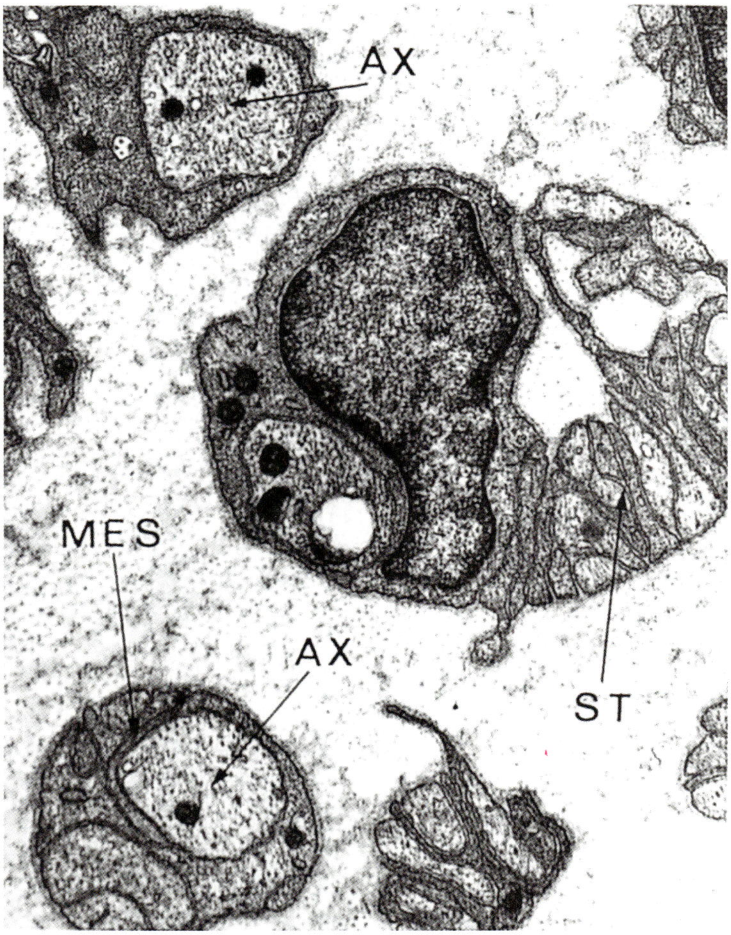

FIGURE 11.11 Unmyelinated axons (1.3 μm in diameter) cut in transverse section. The axons (*AX*) are surrounded by Schwann cells. Mesaxons (*MES*). Stacks of Schwann cell processes (*ST*) are commonly seen in adult nerves. Electron micrograph (×13,000).

unmyelinated axons may be regarded as originating from the same cell type but developing morphologic, biochemical, and physiologic differences. The main populations of axons surrounded by nonmyelinating Schwann cells are the small, nociceptive C-type axons, the postganglionic sympathetic axons, and some of the preganglionic sympathetic and parasympathetic fibers (3). Other axons are unmyelinated for some of their length, specifically at regions proximal to neuromuscular junctions, at the most distal segments of sensory and autonomic neurons, and specialized sensory endings of the pacinian and Meissner corpuscles (3).

The cytoplasm of nonmyelinating Schwann cells contains a Golgi apparatus, rough endoplasmic reticulum, mitochondria, microtubules, and microfilaments and may exhibit centrioles near the nucleus. Pi (π) granules, however, are not present. The nuclei of these cells are ellipsoid with one or more prominent nucleoli. A continuous basement membrane surrounds each cell (3). Schwann cells associated with unmyelinated fibers express different phenotypic characteristics from Schwann cells around myelinated axons. Although both types of Schwann cell contain immunocytochemically detectable vimentin intermediate filaments, S100 proteins, and similar basement membrane components, Schwann cells associated with unmyelinated axons are more likely to express GFAP (90). Such cells also lack MAG that is apparently necessary for segregation and myelination of axons. *Mycobacterium leprae* colonize nonmyelinating Schwann cells by attaching to laminin-2 and its receptor α-dystroglycan, while myelin-forming (this was correct in the submitted word document) Schwann cells seem to be relatively free from infection by *M. leprae*. There is often a strong cell-mediated immune response in leprosy with extensive inflammation and peripheral nerve damage that causes paralysis and loss of sensation, frequently leading to unintentional mutilation of hands and feet (91). An important function of nonmyelinating Schwann cells is as a reservoir of potentially mitotic cells (3). They retain the capacity to undergo mitosis in mature nerves and are prompted to divide by local myelin damage. Therefore, Remak cells may be the source of new cells during nerve repair (3). Unmyelinated Schwann cells of the nerve terminal at the neuromuscular junction actively modulate synapse formation, respond to nerve conduction and neurotransmitter signaling, and play a role in the repair of the neuromuscular junction (63).

Electron microscopy of transverse sections of normal peripheral nerve shows how numerous unmyelinated axons are associated with a single Schwann cell. Short mesaxons extend from the surface of the cell (Fig. 11.11) and the Schwann cell is separated from the axon plasma membrane by a space 10 to 15 nm wide that is analogous to the periaxonal space of Klebs seen around myelinated fibers. Although many axons may be gathered close to the cell body in the perinuclear region of the Schwann cell away from the nuclear region, single axons become more widely separated and are enclosed by thin Schwann cell processes (89) (Figs. 11.3 and 11.11). Each Schwann cell associated with unmyelinated axons in the sural nerve is between 200 and 500 μm in length. As axons pass from one Schwann cell to another, they are surrounded by flattened irregular, finger-like processes that interlock and become telescoped into the adjacent Schwann cell. The surface of the axon is always in contact with the Schwann cell. In young children, only a single thin layer of Schwann cytoplasm surrounds each axon away from the nuclear region; but in adult nerves, the picture is more complex with several Schwann cell processes stacked together and associated with each unmyelinated axon (16).

Pockets of collagen bundles are frequently invaginated into the surface of Schwann cells associated with unmyelinated fibers (Fig. 11.3), particularly in aging nerves and those with loss of unmyelinated fibers. The pockets of collagen fibers are separated from the surface of the Schwann cell by a layer of basement membrane. The significance of this phenomenon is not fully known.

Interesting immunologic properties have been ascribed to Schwann cells. Numerous studies have shown that Schwann cells display a large repertoire of properties ranging from the participation in antigen presentation to secretion of pro- and anti-inflammatory cytokines, chemokines, and neurotrophic factors (92,93). Schwann cells express Ia determinants on their membranes and are able to present foreign antigens to specific synergic T cells (93). When Schwann cells are exposed to inflammatory cytokines, they have the capacity to induce selective damage to T cells and regulate the immune response in the peripheral nervous system (93). A role for Schwann cells has been suggested in myasthenia gravis (94).

Schwann cells also express complement receptor CR1 (CD35) and CD59, a 19- to 25-kDa glycoprotein that binds to complement proteins C8 and C9 in the assembling cytolytic membrane attack complex (95). This may indicate that regulation of complement activation by these proteins is important in neural host defense mechanisms and may be implicated in the complement-mediated damage occurring in inflammatory demyelinating disease such as Guillain–Barré syndrome (96).

Unmyelinated Schwann cells ensheathing sympathetic fibers within the bone marrow play an important role in hematopoietic stem cell regulation (97). Schwann cells express bone marrow niche factor genes and the TGFβ activator molecule that converts the inactive form of TGFβ present in the stem cells into the active form. This action may downregulate lipid raft clustering, essential for hematopoietic stem cell activation (63). Schwann cells also maintain the hibernation state of the stem cells (63,98).

Blood Supply of Peripheral Nerves

Vasa nervorum supplying peripheral nerves are derived from a series of branches from associated regional arteries. Branches from those arteries enter the epineurium (Figs. 11.2 and 11.3) to form an anastomosing plexus. Vessels penetrate the perineurium obliquely and enter the endoneurium as capillaries surrounded by pericytes (Fig. 11.5). Tight junctions between the endothelial cells of the endoneurial capillaries constitute the blood–nerve barrier (21,99).

Complete infarction of peripheral nerves is very uncommon, probably due to the rich anastomotic connections of epineurial arteries. However, inflammation and thrombotic occlusion of epineurial arteries is seen in vasculitides (52) and occlusion by emboli occurs in patients with atherosclerotic peripheral vascular disease. Both these disorders result in ischemic damage to peripheral nerves with axonal degeneration and consequent peripheral neuropathy (51).

CORRELATION OF NORMAL HISTOLOGY WITH THE PATHOLOGY OF PERIPHERAL NERVES

Handling and Preparation of Peripheral Nerve Biopsy and Autopsy Specimens

The sural nerve is the nerve that is most commonly biopsied to assess for peripheral neuropathies (100,101). Sampling the sural nerve, a sensory nerve, will not leave the patient with a motor deficit. However, it is important to remember that findings in this nerve may not be fully representative of changes in patients with a motor neuropathy; examination of small branches of motor nerves within a muscle biopsy may be more fruitful (51,101). A wider range of motor and sensory nerves may be sampled at autopsy depending on the clinical picture. Whether taken at biopsy or autopsy, peripheral nerves are very easily damaged. The myelin sheaths are semiliquid and may be crushed by indelicate handling (Fig. 11.12). The specimen should only be gripped at one end and then gently dissected free before laying it, very gently stretched, on a piece of dry card and placing it in fixative or in liquid nitrogen for snap freezing. Fresh, frozen nerve should be used for enzyme and lipid histochemical studies whereas formalin-fixed nerve can be embedded in paraffin for the application of routine stains and immunocytochemistry (Table 11.1). Although formalin-fixed material can be used for the preparation of 0.5 to 1 μm resin-embedded sections and for electron microscopy, ideally the tissue should be fixed in glutaraldehyde and postfixed in osmium for ultrastructural studies. Teased fibers can be prepared from either glutaraldehyde- or formalin-fixed material (51,101).

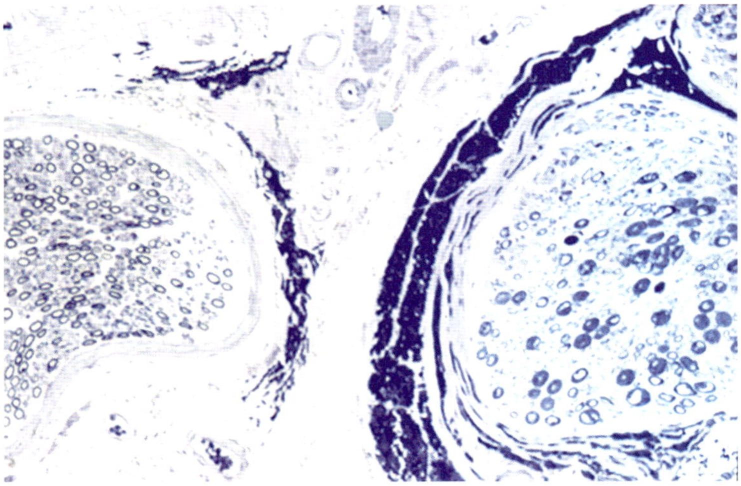

FIGURE 11.12 Histologic artifact in a peripheral nerve. In this transverse resin section, the fascicle to the left of the picture is well preserved. However, there is extensive recent hemorrhage (*center*) that occurred during the biopsy procedure; the myelinated axons in the nerve fascicle are squeezed and distorted (*right*). The 1-μm resin section is stained with toluidine blue (×40).

The method of preparation really depends on the information sought. Frozen sections are ideal for detecting abnormal lipids, such as sulfatide in metachromatic leukodystrophy, and for detecting cholesterol ester droplets of degenerating myelin by staining for Sudan red or oil red O. Increased lysosomal enzyme activity can be detected in frozen sections stained histochemically for acid phosphatase and may help detect disorders characterized by axonal degeneral or segmental demyelinization, as well as Krabbe leukodystrophy (101). Brief formalin or glutaraldehyde fixation can be used in some cases for electron microscopic enzyme histochemistry. Frozen sections can also be used for immunofluorescence for the detection of immunoglobulin binding to myelin sheaths in paraproteinemia. Transverse frozen sections of nerve are ideal for these purposes, although they are often more difficult to prepare than longitudinal sections.

A variety of methods are used for preparing and examining fixed specimens of peripheral nerve, and each method reveals different information (100,101). Transverse sections of the nerve generally provide the most information, although longitudinal sections facilitate identification of regenerating axons by immunocytochemistry (Fig. 11.4F). Paraffin-embedded sections can be stained with a variety of histologic stains and for immunocytochemistry to reveal nerve components (Table 11.1). Blood vessels and inflammatory exudates are ideally studied in paraffin sections, but quantitation of nerve fibers, the detection of axon degeneration and regeneration, and the assessment of segmental demyelination and remyelination are more satisfactory in 0.5 to 1 μm toluidine blue–stained resin sections or by electron microscopy. The presence of amyloid in the endoneurium or giant axons in some hereditary neuropathies and in some toxic neuropathies can be detected both in paraffin and in resin-embedded sections. Teased preparations are most useful for detecting segmental demyelination and remyelination and for assessing whether axonal degeneration and regeneration have occurred within the nerve in the past through the detection of short internodes.

Peripheral Neuropathies

Peripheral neuropathies are arguably the most common neurologic diseases with an overall prevalence of 1 to 12% and a prevalence in the older population up to 30% (102). In most cases, the diagnosis is established based on clinical features, nerve conduction studies, and laboratory testing such as serologic studies, vitamin B12 levels, and metabolic analyses. Nerve biopsies are reserved for situations of a rapidly progressive neuropathy or a setting in which confirmation of a clinically suspected inflammatory disorder or amyloidosis is attempted. The pathologic diagnosis of peripheral neuropathy usually requires close clinicopathologic correlation and knowledge of the electrophysiologic data such as nerve conduction velocities and electromyography (85,88). Although there are a number of specific histopathologic features that aid the diagnosis of peripheral neuropathy (e.g., amyloid, the presence of lepra bacilli, abnormal lipids such as sulfatide within the nerve, giant axons, and vasculitis) (16), for the most part, assessment of peripheral nerve pathology depends on detection and quantitation of general pathologic features and good clinicopathologic correlation (51,85,101,103).

Introduction

Peripheral neuropathies are often classified as either demyelinating or axonal. This distinction is usually made based on results of nerve conduction studies that identify slowed conduction speeds in demyelinating diseases and decreased signal amplitude in axonal neuropathies. Histopathologic features can also reveal features that support a diagnosis of a demyelinating or an axonal disease process. This bimodal classification often falls short because some diseases result in mixed features that don't as neatly fall into these two categories and, more importantly, peripheral neuropathies that result from very specific changes to distinct structural domains of myelinated axons also exist. The latter is often characterized by antibodies such as those targeting proteins specifically expressed at the node of Ranvier itself (nodopathy), the paranodal region (paranodopathy), or the juxtaparanode (juxtaparanodopathy). Some authors classify these disorders based on the target of the causative antibody (88). This section will still start with the description of the more traditional concepts of axonal and demyelinating neuropathy before expanding on these latter diseases.

Axonal Degeneration and Regeneration

There is little to no regenerative capacity in the central nervous system after axonal injury (104). Such neuronal destruction is seen in poliomyelitis, ALS, spinal muscular atrophy, and infarction of the spinal cord. Dorsal root ganglion cells may be lost in *Varicella zoster* viral infection or in a variety of hereditary sensory neuropathies. There is, however, the potential of repair after damage to peripheral nerves. The axons of peripheral nerves typically arise from the lower motor neurons in the spinal cord or the dorsal root ganglia and may reach 1 m or more in length. These axons may be hundreds of thousands of times longer than the diameter of the cell body from which they arise. The function of these long axons is dependent on the already outlined intricate mechanisms of axonal transport and is susceptible to physical, chemical, and biologic injury. If an axon in a peripheral nerve is injured, its distal end degenerates but subsequent regeneration may still occur (Fig. 11.13). The success of the regeneration depends on the distance between the site of damage and the nerve end organ as well as the amount of scarring or other obstruction laid in the path of the regenerating axons.

Axonal degeneration was first described by Waller in 1850 and the eponym Wallerian degeneration is still used to describe this process. Much of the fundamental work on nerve degeneration and regeneration, however, was performed by Ramón y Cajal in the early part of the 20th century (105). In broad terms, the stereotypic response of Wallerian degeneration results in beading, fragmentation, and disintegration of the distal axon and goes along with activation of Schwann cells and the immune system. NMNAT-NAD$^+$-SARM1 (sterile alpha and TIR motif-containing 1 protein)

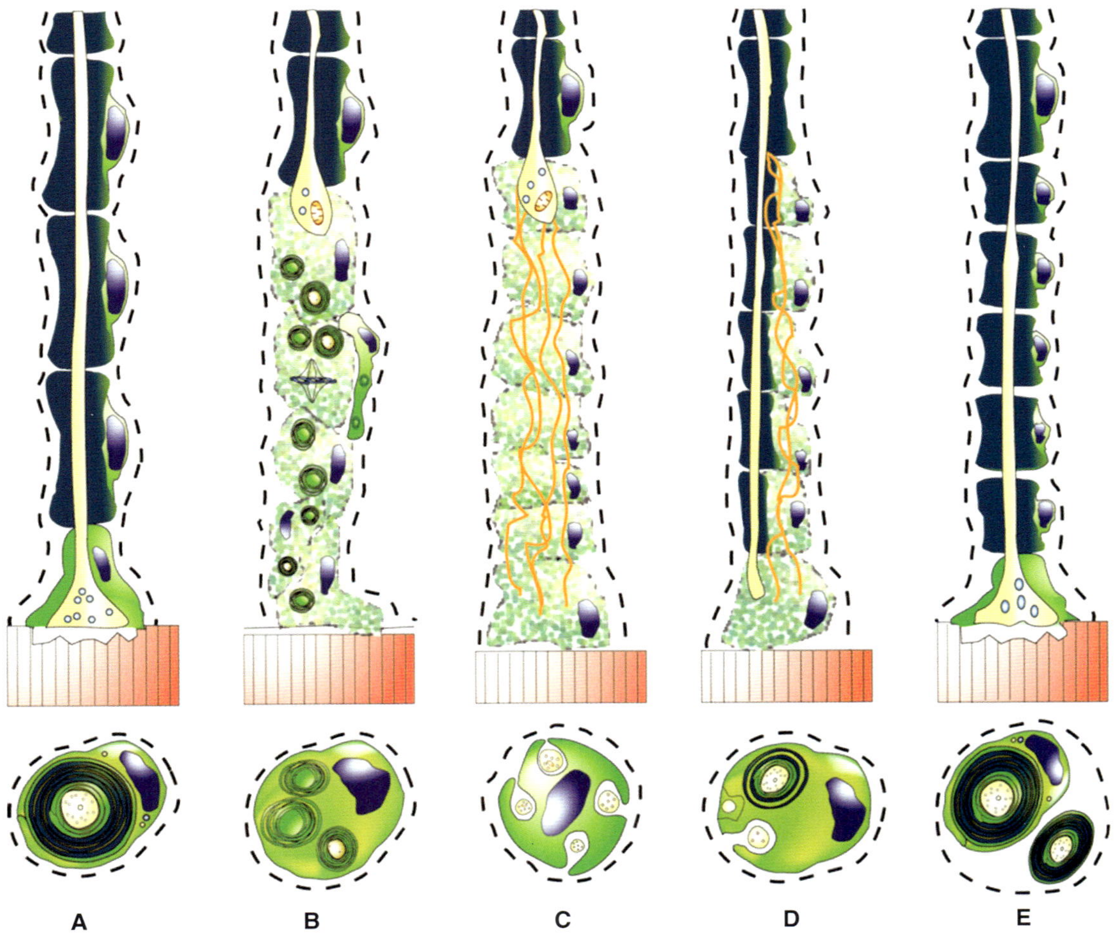

FIGURE 11.13 Diagram summarizing the events occurring during axonal degeneration and regeneration. **A:** Normal nerve. **B:** By 7 days after axonal damage, Schwann cells containing axon and myelin debris have divided to form bands of Büngner. **C:** Axon sprouts grow from the swollen end-bulb of the proximal axon. **D:** An axon becomes myelinated. **E:** Connection with the end organ is reestablished; regenerated internodes are short. (Reprinted from Weller RO, Cervos Navarro J. *Pathology of Peripheral Nerves: A Practical Approach*. Butterworth; 1977 with permission.)

signaling plays a crucial role in regulating the speed at which the process of Wallerian degeneration progresses after axonal transection (106,107). NAD^+ is a co-enzyme and abundant metabolite that is crucial for key cellular processes. Its rapid depletion in the distal axonal segment after axonal transection is a key driver of the degeneration that occurs in the distal axonal segment. Increased NMMNAT activity leads to increased NAD^+ production and helps to maintain NAD^+ levels after axonal transection. The activated form of SARM1 results in the degradation of NAD^+.

Autophagy also appears to play a key role in Wallerian degeneration and may be largely independent of the NMNAT-NAD^+-SARM1 signaling. Injury leads to induction of axonal autophagy while inhibition of autophagy can reduce axonal degeneration after mechanical or toxic injury. Disruption of axonal transport and the movement of autophagosomes can contribute to axonal degeneration after injury while the overexpression of vacuolar protein sorting 4 (Vps4) offers neuroprotection (108).

The different processes that contribute to Wallerian degeneration occur on a typical timeline (109). Activation of the MAP kinase system can occur within minutes after axonal transection, the degradation of NMNAT, activation of SARM1, and NAD^+ depletion after about 2 hours. Axonal autophagy starts within 2 to 4 hours. The breakdown of neurofilament and the disintegration of axons starts within 8 hours to 2 days. The first morphologic changes become obvious after about 24 hours with retraction of myelin from the nodes of Ranvier. By 48 hours, myelin and axonal changes become more obvious as the axon disintegrates, and myelin sheaths become disrupted and form globules (Fig. 11.14) in which axon fragments are enclosed (110). Myelin debris is initially birefringent and has a lamellated ultrastructure, as does normal myelin. As proteins break down and lysosomal enzymes become active around the myelin debris, cholesterol within the myelin is esterified to cholesterol esters and lipid debris loses its birefringence in polarized light; the subsequent amorphous globular lipid stains strongly with Sudan dyes and oil red O (110,111). The Marchi technique also differentiates between normal myelin (unstained) and degenerating myelin (stained black).

During the 2nd week after nerve injury, much of the myelin debris is removed from the distal part of the nerve and regenerative features become more prominent. Axon fragments and myelin debris are broken down by both Schwann cells and macrophages (110,112). Schwann cells directly attract macrophages by secretion of different proteins, probably regulated by autocrine circuits involving the neuropoietic cytokines IL-6 and leukemia inhibitory factor (112,113). Macrophages promote nerve repair through the elaboration of Schwann cell mitogens and may also affect neurons and axonal growth directly through the release of neurotrophins (112,114).

The peak proliferation of Schwann cells is between 3 and 15 days after nerve damage (115). As Schwann cells proliferate, they form columns termed the bands of Büngner that are surrounded by basement membrane (Fig. 11.12). Regenerating axons grow along the bands of Büngner; failed regeneration causes the bands to shrink and Schwann cells may disappear, becoming replaced by fibrous tissue (116).

Several easily detectable histologic changes occur during axonal regeneration. The neuron cell bodies in the anterior horns of the spinal cord or in the dorsal root ganglia show chromatolysis during the first 3 weeks after axonal injury. The nerve cell perikaryon swells by nearly 20% and the nucleus becomes eccentric. Nissl substance is dispersed so that the cytoplasm becomes pale and there is a marked increase in polyribosomal ribonucleic acid (RNA) with upregulation of genes encoding growth-associated protein 43 (GAP 43), cytoskeleton protein 23, and β-tubulin. There is also an increase in peptides such as galanin and

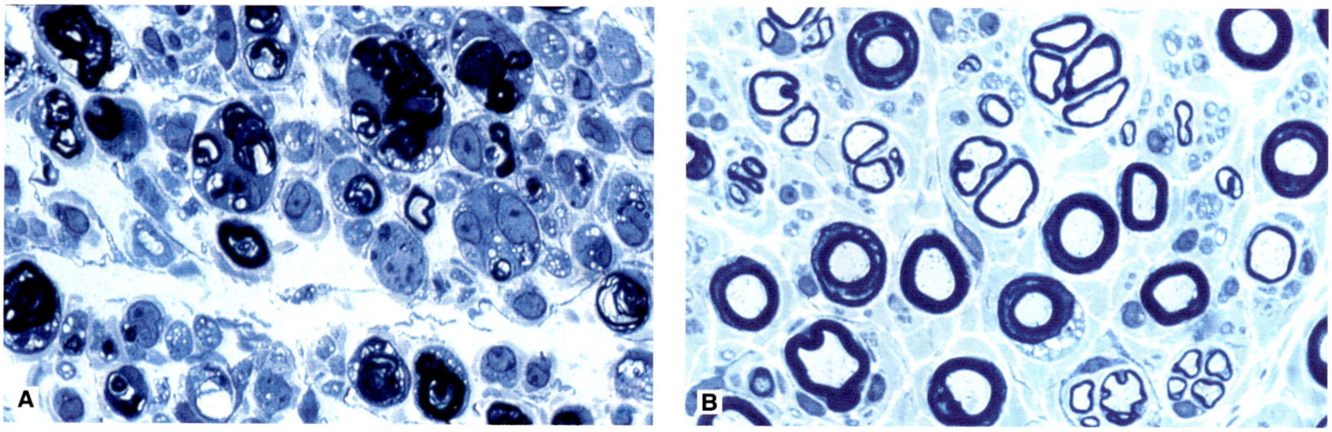

FIGURE 11.14 Axonal degeneration and regeneration in transverse sections of peripheral nerve. **A:** Axonal degeneration 4 days after nerve section. Few axons are visible and myelin is forming globules in Schwann cells and macrophages. The 1-μm section is stained with toluidine blue (×310). **B:** Axonal regeneration in a human nerve biopsy. Normal myelinated fibers are interspersed with clusters of closely associated thinly myelinated regenerating fibers (*top and bottom right*). The 1-μm section is stained with toluidine blue (×310).

vasointestinal polypeptide (VIP), reflecting the metabolic events involved in axon regeneration (117,118).

Nerve injury triggers the change of Schwann cells to a cell phenotype specialized to promote repair. The conversion to repair-mode Schwann cells involves dedifferentiation and reprogramming that consist of downregulation of myelin genes combined with upregulation of trophic factors and an increase of diverse cytokines as part of the innate immune response. This repair program is controlled by mechanisms involving the transcription factor c-Jun, which is rapidly upregulated in Schwann cells after injury. In the absence of c-Jun, damage results in the formation of a dysfunctional repair cell, neuronal death, and failure of functional recovery (119).

Starting around the 4th day after injury, multiple nerve sprouts, or neurites, extend from the axonal growth cone and grow distally at a rate of 1 to 2.5 mm/day. As the neurites enter the bands of Büngner, they invaginate into the surface of the Schwann cell and become myelinated. Unmyelinated fibers regenerate in a similar way, but they are smaller and no myelin sheaths form around them. Regenerating neurites can be detected in the classical way by silver staining; but, in cross sections of peripheral nerve, they are best demonstrated in 0.5- to 1-μm resin sections or by electron microscopy. Characteristically, regenerating axons form clusters encircled by a single basement membrane. These clusters (Fig. 11.14) are recognized by the close association of small, thinly myelinated axons within the nerve; myelinated nerve fibers in a normal nerve are well separated from each other by endoneurial collagen (Fig. 11.5).

Axonal growth and regeneration are stimulated by NGF synthesized by Schwann cells, fibroblasts, and macrophages, and transported back along the axon by retrograde axoplasmic transport to stimulate nerve cell protein synthesis (3,8,9,54). Connective tissue elements may also play a role in guiding regenerating axons (120). Neurite outgrowth–promoting factors on cell surfaces or in the extracellular matrix promote extension of the axon by providing an appropriate "adhesiveness" in the substrate (3,120). NGF, brain-derived growth factor (BDGF), glial cell line–derived neurotrophic factor (GDNF), neurotrophin (NT-3), and neurite outgrowth–promoting factor released by Schwann cells, are essential for axonal growth after injury (3,118,121).

The success of regeneration, with axons reaching effective end organs, may be influenced by several factors. If the injury is far proximal from the end organ, few regenerating axons may make effective reconnections (119). The presence of scar tissue or discontinuity of anatomical pathways may also inhibit regeneration. If regeneration to the distal stump of a nerve is blocked by scar tissue, axons may grow outside the original course of the nerve and even back alongside the proximal stump. This results in the formation of traumatic neuromas. Accumulation of sodium channels within the tips of injured axons that may be responsible, in part, for the ectopic axonal excitability and the resulting abnormal sensory phenomena that frequently complicate peripheral nerve injury. Macrophages may also participate in the genesis of chronic pain after the neuroma has formed by creating demyelinating axonal regions susceptible to external stimuli, releasing substances that influence axonal regeneration, and direct action on the denuded remodeling membranes.

Axon degeneration, often with regeneration, is a feature of numerous peripheral neuropathies Most toxic neuropathies (35,122) result in chronic axonal degeneration at the extreme distal ends of sensory and motor nerves. Diseases associated axonal degeneration does not occur in a monophasic manner through a simple mechanical injury as that described in Wallerian degeneration. Defective interplay between NMNAT2 and SARM1 as well as autophagy promotes other neuropathies caused by chemotherapies or inherited neuropathies such as certain forms of Charcot–Marie–Tooth disease (123–126).

Segmental Demyelination and Remyelination

When demyelination occurs in peripheral nerves, it has a segmental distribution with each segment representing the degeneration one internode that is established around one axon by a single Schwann cell (Figs. 11.7 and 11.15). Such segments can be contiguous and thus demyelination may occur over long lengths of the nerve or in short sporadic segments (16). The axon remains intact except in severe demyelinating neuropathies in which secondary axonal degeneration occurs. Remyelination can occur with restoration of nerve function. Demyelination results from direct interference with Schwann cell metabolism or the lysosomal actions of Schwann cells (112,115). Other demyelinating peripheral neuropathies result from inherited defects or autoimmune conditions. In all of these settings segmental demyelination followed by remyelination (96) (Figs. 11.15 and 11.16).

The first stages of segmental demyelination occur at the node of Ranvier, where the nodal gap becomes widened (85). This results in severe slowing of conduction of nerve impulses across the demyelinated segment. Preserved axons remain invaginated within Schwann cells as the myelin sheaths are broken down (Figs. 11.16 and 11.17). Schwann cells proliferate and within a few days start to remyelinate the demyelinated axons such that remylenation may be well advanced within 2 weeks of the demylenating episode.

Segmental demyelination can be detected in teased nerve fibers, first by the presence of widening of the gap of the node of Ranvier and then by the presence of axons devoid of myelin sheaths. Intercalated thin myelin sheaths along the axon are seen as remyelination proceeds (Fig. 11.15). Naked axons can be recognized in transverse section and remyelinating fibers are detected by the presence of inappropriately thin myelin sheaths by electron microscopy (Figs. 11.13 to 11.15). Segmental demyelination is a feature of a number of peripheral neuropathies due to mild vascular damage due to rheumatoid arthritis (85,127), diabetes, Guillain–Barré syndrome, occasional toxic neuropathies, and metabolic neuropathies (Fig. 11.15) (16,85, 122). Throughout the range of neuropathies, however,

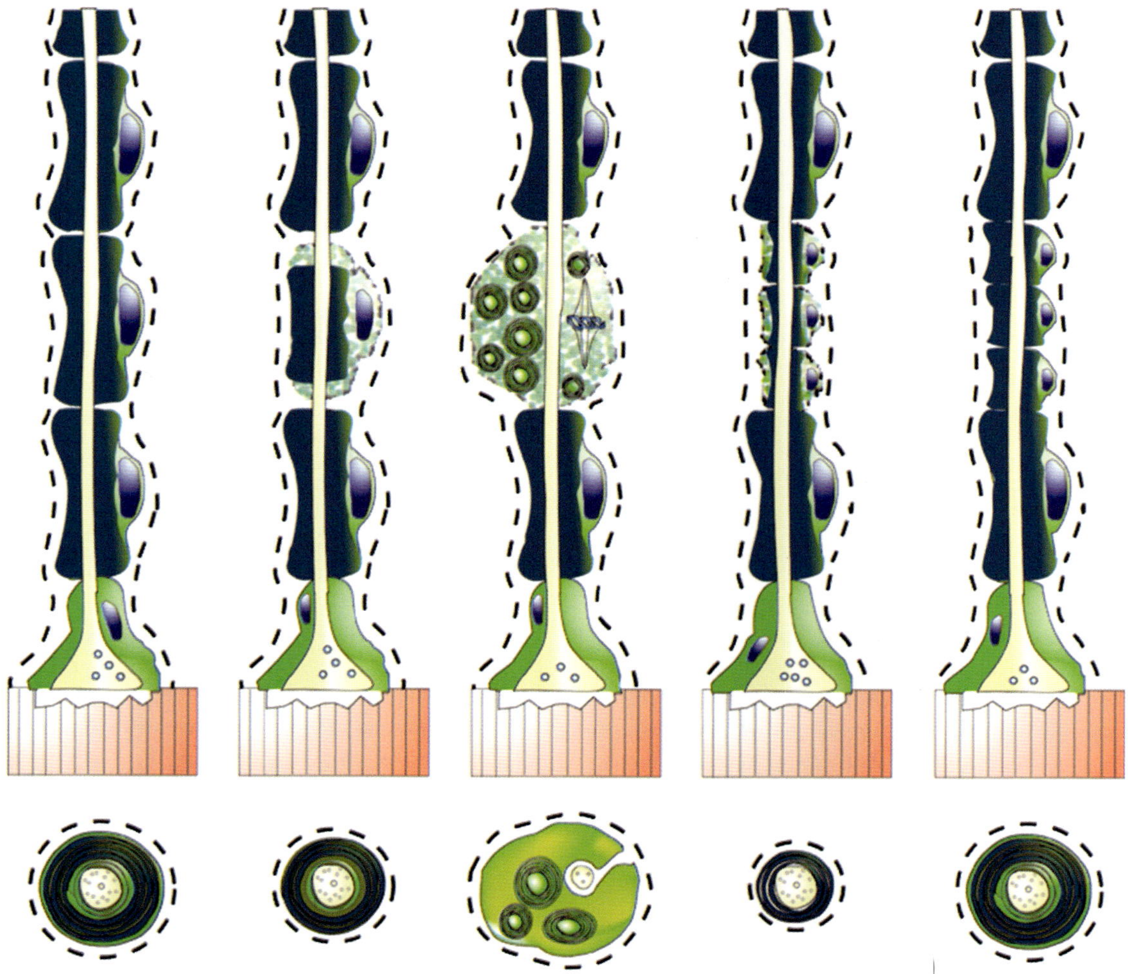

FIGURE 11.15 Summary of events occurring in primary segmental demyelination and remyelination. **A:** Normal nerve. **B:** Early segmental demyelination; retraction of paranodal myelin with widening of the nodal gap. **C:** Destruction of myelin sheath and Schwann cell mitosis. **D** and **E:** Remyelination; intercalated short segments. (Reprinted from Weller RO, Cervos Navarro J. *Pathology of Peripheral Nerves: A Practical Approach*. Butterworth; 1977, with permission.)

segmental demyelination is less common than axonal degeneration.

Recurrent segmental demyelination is a feature of a number of chronic hereditary neuropathies, particularly Charcot–Marie–Tooth disease, Déjerine–Sottas disease (hereditary motor and sensory neuropathy types I and III), and Refsum disease (80). In such diseases, repeated segmental demyelination is responsible for a florid proliferation of Schwann cells and formation of onion-bulb formations that represent whorled arrangements of Schwann cells around a single central axon (Figs. 11.17 and 11.18) (79). These onion-bulb formations reflect chronic and recurrent demyelination and remyelination. Chronic inflammatory demyelinating polyneuropathy can also lead to the formation of prominent onion-bulb type structures. So far, at least 30 causative genes have been identified for primary heritable demyelinating neuropathies. Panel-based genetic studies have now largely replaced the use of peripheral nerve biopsies in those patients in whom an inherited peripheral neuropathy is considered (79,80).

Nodopathies

Guillain–Barré syndrome, or acute demyelinating polyradiculoneuropathy, and chronic inflammatory demyelinating polyradiculoneuropathy are often listed as prototypical examples of demyelinating diseases. These are autoimmune demyelinating neuropathies that go along with T-lymphocytes and macrophage infiltration (128). Complement activation and formation of membrane attack complex on the outward surface of myelin membranes cause myelin destruction, resulting in decreased conduction velocities, prolonged distal motor latencies, and partial

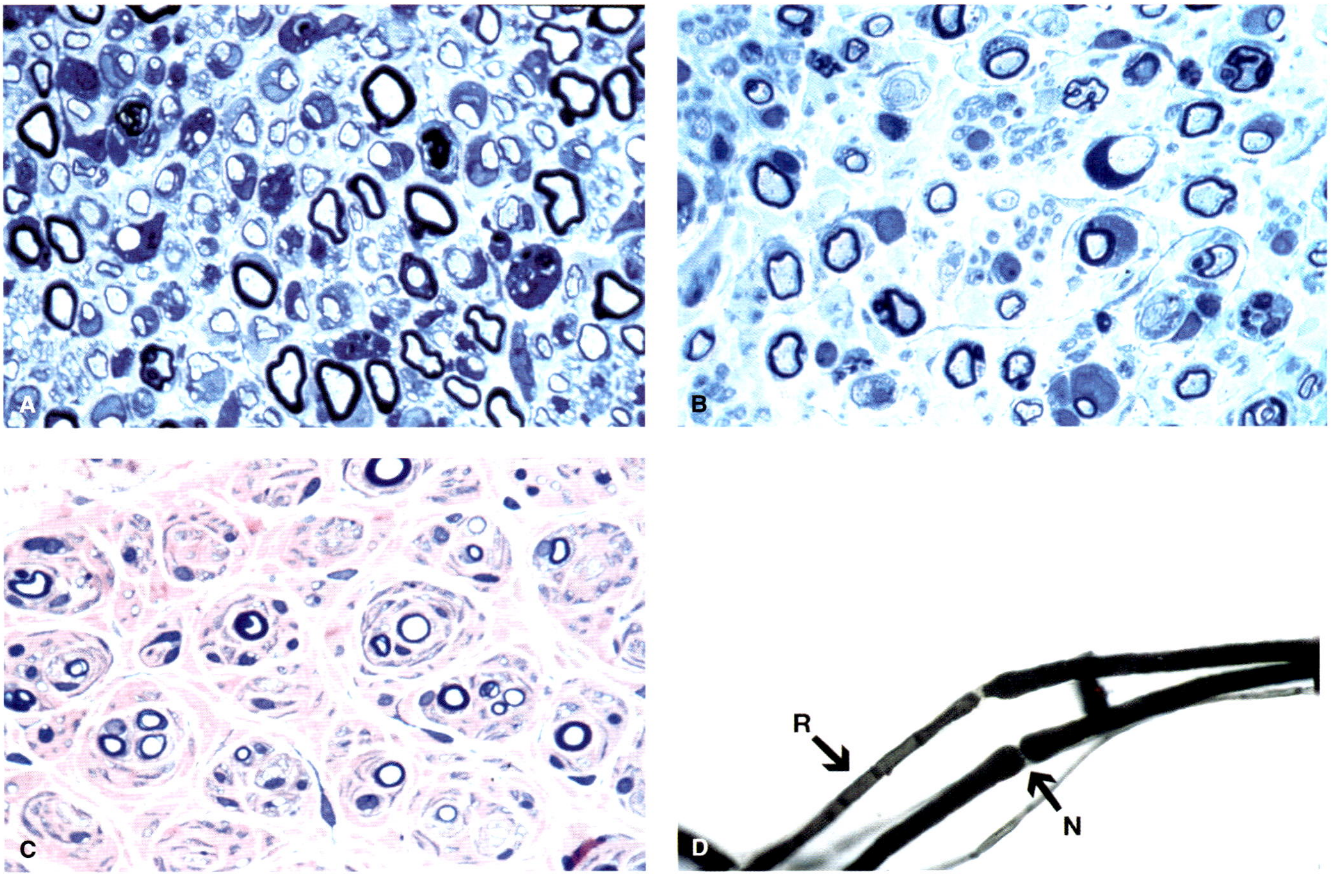

FIGURE 11.16 Segmental demyelination and remyelination. **A:** Transverse section of peripheral nerve showing early remyelination in an experimental animal. There are normal, large myelinated fibers with axons 8 to 10 μm in diameter and axons 3 to 5 μm in diameter, which have thin myelin sheaths and are remyelinating. Myelin debris is seen in Schwann cells and macrophages. The 1-μm resin section is stained with toluidine blue (×310). **B:** Nerve biopsy from a child with metachromatic leukodystrophy (sulfatide lipidosis). Large axons with either thin myelin sheaths (remyelination) or with no myelin sheath at all (demyelinated) (*right of center*) are seen in the biopsy. Unmyelinated fibers (*center*) are unaffected. The 1-μm resin section is stained with toluidine blue (×310). **C:** Hypertrophic neuropathy (Charcot–Marie–Tooth disease—HSMN type I). Demyelinated and remyelinated axons are seen at the centers of onion-bulb whorls formed by Schwann cell processes. There is abundant endoneurial collagen (*pink*). The 1-μm resin section is stained with toluidine blue and carbol fuchsin (×240). **D:** Teased fibers. Normal fiber (*lower*) with a normal node of Ranvier (*N*). The fiber above has a thin, remyelinating segment (*R*) on one side of the node and a normal segment on the other side. Osmium-stained teased fibers (×200). (Reprinted from Weller RO. *Colour Atlas of Neuropathology*. Oxford University Press and Harvey Miller; 1984 with permission.)

motor conduction. Some immune-mediated demyelinating polyradiculoneuropathies are associated with antibodies that specifically target regions around the node of Ranvier (Fig. 11.10) (88,128).

GM1 gangliosides are enriched at the nodal axolemma and at the Schwann cell microvilli at the node of Ranvier (88,128). Anti-GM1 antibodies are a common finding in nodopathies classified as acute motor axonal neuropathy or multifocal motor neuropathy. The former is associated with IgG antibodies to GM1 or GD1a that deposit at the node of Ranvier, whereas the latter results from anti-GM1 IgM antibodies. Antibody deposition results in binding of complement and formation of membrane attack complex, thereby causing loss of nodal sodium channels and disruption of paranodal myelin.

Some patients have been shown to have IgG antibodies targeting NF186, a key component of the protein complex that helps with the clustering of sodium channels at the node of Ranvier (88,128). These can either be complement-activating IgG3 antibodies or IgG4 antibodies that interfere with normal function of the targeted protein and disrupt normal nerve conduction. Ultrastructural studies show loss of microvilli and overgrowth of the processes of the two adjoining Schwann cells that can lead to complete

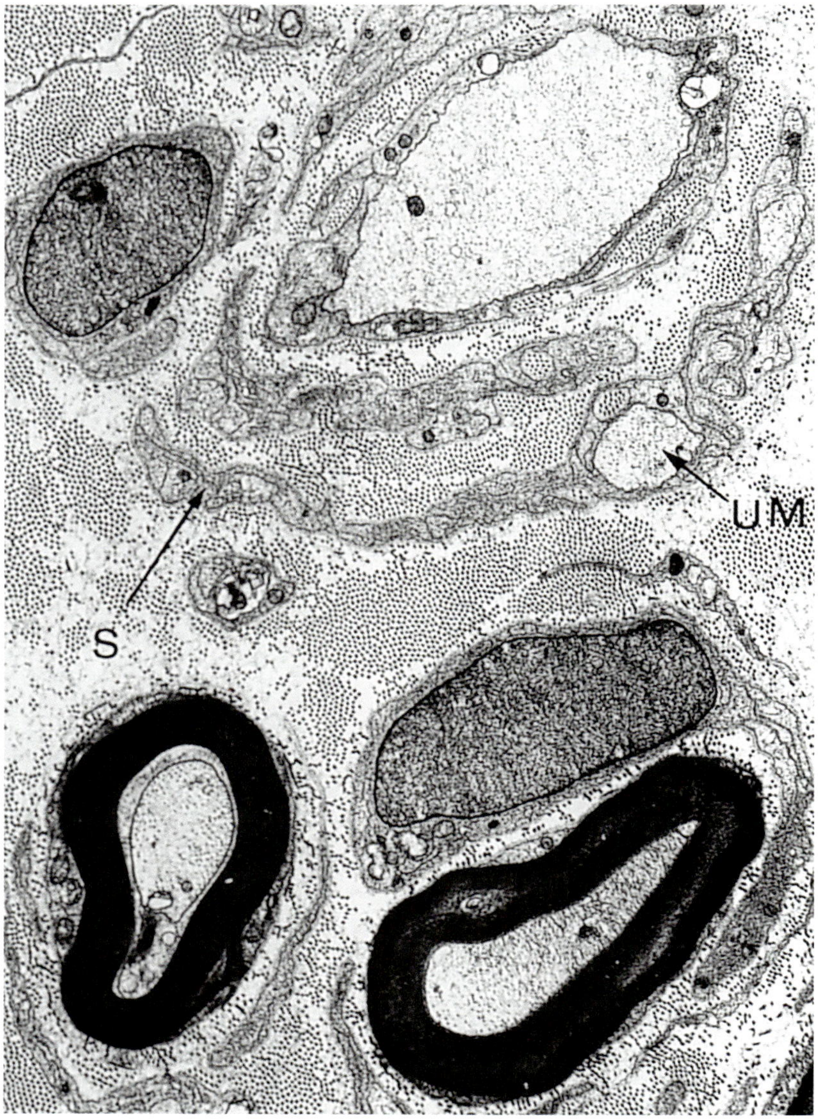

FIGURE 11.17 Demyelination. A large-diameter axon (*top*) is demyelinated and devoid of a myelin sheath. A small onion-bulb whorl has formed around this axon with the encirclement by Schwann cell processes (*S*), one of which contains an unmyelinated fiber (*UM*). Thickly myelinated fibers are seen at the bottom of the picture. Electron micrograph (×6,600).

occlusion of the nodal gap (129). Rare genetic nodopathies related to germline variants in NF186 or gliomedin have recently been described.

Antibodies to NF155 expressed on the paranodal Schwann cell loops, contactin 1, and contactin-associated protein 1 (Caspr1) expressed on the axolemmal can result in a clinical picture mimicking the historical description of chronic inflammatory demyelinating polyradiculoneuropathy. These proteins form a complex that represents an important part of the paranodal transverse bands. Antibodies targeting NF155 lead to disruption of paranodal structures that can be seen ultrastructurally as loss of the transverse bands, irregular myelin loops, and widening of the nodes (129). Recent descriptions of rare cases carrying germline variants in *CNTN1* encoding contactin and CNTNAP1 encoding Caspr1 have also been described (130).

Another distinct form of peripheral neuropathy results from IgM anti-MAG antibodies. These may preferentially affect noncompacted myelin at the outer myelin loops at the node of Ranvier as well as at the Schmidt Lantermann Incisures. These antibody deposits may lead to detachment of myelin loops and complement activation. Ultrastructural studies show widening of the space between myelin layers. A suggestion has been made to regard this disease process as demyelinating para-internodopathy (88).

Localized Reactive Lesions

An understanding of the structure and staining reactions of normal peripheral nerve is essential for unraveling the complexities of traumatic lesions of nerve. Identification of cell types, recognition of patterns of organization, and the detection of normal elements within a traumatic lesion allow a more confident diagnosis and description to be formulated. Several of these lesions are classified as types of "neuromas." This term should always be used in conjunction with a distinct descriptor though.

Traumatic Neuroma. Traumatic neuromas (amputation neuromas) may develop as painful swellings at the distal ends of amputated limbs or at sites of nerve damage without amputation (131,132). They develop when the regenerating, outgrowing axons cannot find their distal target after a transecting nerve injury. The elongating, unguided axons attract Schwann cells and perineurial cells resulting in a nodular or multinodular proliferation that includes axons, Schwann cells, and variable numbers of perineurial cells in a random admixture that lacks a typical organoid arrangement. Schwann cells typically predominate in these lesions that may mimic Schwann cell neoplasms histologically. Special stains can confirm a dense population of axons that would not be found in a true Schwann cell neoplasm. The lineage of the Schwann cells can be confirmed by stains for SOX10 and S100 and the admixture of variably positive perineurial cells by staining for EMA, Glut-1, as well as claudin-1. A trichrome stain can also sometimes aid in confirming the random admixture of collagen between axons and Schwann cells in a pattern different from what is seen in normal nerves.

Morton Neuroma. In 1835 F. Civinini from Pisa, and later in 1876 T. G. Morton, from Pennsylvania, described a neuroma that involves the plantar interdigital nerves, almost always between the third and fourth toes (Civinini–Morton metatarsalgia), and consists of small painful swellings on the nerves (131–133). Histologically, there are fibrosis and edema of the endoneurium and perineurium and the accumulation of mucosubstances similar to endoneurial glycosaminoglycans. The detection of axons and Schwann cells by immunohistochemistry is often a useful adjunct to the diagnosis of this lesion.

Nerve Sheath Ganglion. A pseudocyst or nerve sheath ganglion, containing mucinous material that stains with Alcian blue, may form on a peripheral nerve, generally at a site of repeated trauma. Although the fibrous capsule of the cyst and its mucinous contents may dominate the picture, damaged nerve components can usually be detected adjacent to the cyst (132). This lesion finds a more detailed description in the radiology literature than the pathology literature (132).

Localized Hypertrophic Neuropathy. Localized hypertrophic neuropathy is a lesion that typically presents as a localized enlargement of a nerve on the extremities. True localized hypertrophic neuropathy results from extensive, onion-bulb–like proliferation of Schwann cells around axonal processes. The lesional Schwann cells exhibit the expected expression of S100 and SOX10. This staining profile clearly sets this lesion apart from intraneural perineurioma, a neoplastic condition that it has often been confused in the literature. Intraneural perineuriomas are, however, characterized by onion-bulb–like arrangements of perineurial cells and often show clonal alterations of TRAF7 or NF2 (134). One recent report describes a case of localized hypertrophic neuropathy in the context of a mosaic RASopathy KRAS alteration suggesting the possibility of localized hypertrophic neuropathy occurring in the context of distinct mutational changes (135).

ACKNOWLEDGMENTS

The author gratefully acknowledges the work of the authors of the prior edition of this chapter, Dr. Carlos Ortiz-Hidalgo and Dr. Roy O. Weller.

REFERENCES

1. Dupin E, Le Douarin NM. The neural crest, a multifaceted structure of the vertebrates. *Birth Defects Res C Embryo Today*. 2014;102(3):187–209.
2. Petersen J, Adameyko I. Nerve-associated neural crest: peripheral glial cells generate multiple fates in the body. *Curr Opin Genet Dev*. 2017;45:10–14.
3. Kidd GJ, Ohno N, Trapp BD. Biology of Schwann cells. *Handb Clin Neurol*. 2013;115:55–79.
4. Castelnovo LF, Bonalume V, Melfi S, Ballabio M, Colleoni D, Magnaghi V. Schwann cell development, maturation and regeneration: a focus on classic and emerging intracellular signaling pathways. *Neural Regen Res*. 2017;12(7):1013–1023.
5. Karamchandani JR, Nielsen TO, van de Rijn M, West RB. Sox10 and S100 in the diagnosis of soft-tissue neoplasms. *Appl Immunohistochem Mol Morphol*. 2012;20(5):445–450.
6. Graham A. Melanocyte production: dark side of the Schwann cell. *Curr Biol*. 2009;19(24):R116–R117.
7. Monk KR, Feltri ML, Taveggia C. New insights on Schwann cell development. *Glia*. 2015;63(8):1376–1393.
8. Skaper SD. Nerve growth factor: a neuroimmune crosstalk mediator for all seasons. *Immunology*. 2017;151(1):1–15.
9. Önger ME, Delibas¸ B, Türkmen AP, Erener E, Altunkaynak BZ, Kaplan S. The role of growth factors in nerve regeneration. *Drug Discov Ther*. 2017;10(6):285–291.
10. Giger RJ, Hollis ER III, Tuszynski MH. Guidance molecules in axon regeneration. *Cold Spring Harb Perspect Biol*. 2010;2(7):a001867.
11. Madduri S, Gander B. Schwann cell delivery of neurotrophic factors for peripheral nerve regeneration. *J Peripher Nerv Syst*. 2010;15(2):93–103.

12. Masu M. Proteoglycans and axon guidance: a new relationship between old partners. *J Neurochem*. 2016;139(Suppl 2): 58–75.
13. Taveggia C. Schwann cells-axon interaction in myelination. *Curr Opin Neurobiol*. 2016;39:24–29.
14. Quintes S, Goebbels S, Saher G, Schwab MH, Nave KA. Neuron-glia signaling and the protection of axon function by Schwann cells. *J Peripher Nerv Syst*. 2010;15(1):10–16.
15. Eid EM, Hegazy AMS. Anatomical variations of the human sural nerve and its role in clinical and surgical procedures. *Clin Anat*. 2011;24(2):237–245.
16. Thomas PK, Ochoa J. Microscopic anatomy of peripheral nerve fibers. In: Dyck PJ, Thomas PK, Lambert EH, et al., eds. *Peripheral Neuropathy*. Vol I. 2nd ed. Saunders; 1984:34–96.
17. Herbert AL, Monk KR. Advances in myelinating glial cell development. *Curr Opin Neurobiol*. 2017;42:53–60.
18. Xiao J, Kilpatrick TJ, Murray SS. The role of neurotrophins in the regulation of myelin development. *Neurosignals*. 2009;17(4):265–276.
19. Nave KA. Myelination and the trophic support of long axons. *Nat Rev Neurosci*. 2010;11(4):275–283.
20. Han H, Myllykoski M, Ruskamo S, Wang C, Kursula P. Myelin-specific proteins: a structurally diverse group of membrane-interacting molecules. *Biofactors*. 2013;39(3): 233–241.
21. King R. Microscopic anatomy: normal structure. *Handb Clin Neurol*. 2013;115:7–27.
22. Streit A. The cranial sensory nervous system: specification of sensory progenitors and placodes. In: *Stem Book [Internet]*. Harvard Stem Cell Institute; 2008:1–20.
23. Alfieri A, Fleischhammer J, Strauss C, Peschke E. The central myelin-peripheral myelin transitional zone of the nervus intermedius and its implications for microsurgery in the cerebellopontine angle. *Clin Anat*. 2012;25(7):882–888.
24. Tassler PL, Dellon AL, Canoun C. Identification of elastic fibers in the peripheral nerve. *J Hand Surg Br*. 1994;19(1): 48–54.
25. Cavanagh JG. Pathology of peripheral nerve diseases. In: Weller RO, ed. *Systemic Pathology: Nervous System Muscle and Eyes*. Vol 4. 3rd ed. Churchill Livingstone; 1990: 544–578.
26. Shekhani HN, Hanna T, Johnson JO. Lipofibromatous hamartoma of the median nerve: a case report. *J Radiol Case Rep*. 2016;10(11):1–7.
27. Estebe J-P, Atchabahian A. The nerve: a fragile balance between physiology and pathophysiology. *Eur J Anaesthesiol*. 2017;34(3):118–126.
28. Piña-Oviedo S, Ortiz-Hidalgo C. The normal and neoplastic perineurium: a review. *Adv Anat Pathol*. 2008;15(3): 147–164.
29. Küspert M, Weider M, Müller J, Hermans-Borgmeyer I, Meijer D, Wegner M. Desert hedgehog links transcription factor Sox10 to perineurial development. *J Neurosci*. 2012; 32(16):5472–5480.
30. Parmantier E, Lynn B, Lawson D, et al. Schwann cell-derived Desert hedgehog controls the development of peripheral nerve sheaths. *Neuron*. 1999;23(4):713–724.
31. Zimmerman A, Bai1 L, Ginty DD. The gentle touch receptors of mammalian skin. *Science*. 2014;346(6212): 950–954.
32. Goehler LE, Relton JK, Dripps D, et al. Vagal paraganglia bind biotinylated interleukin-1 receptor antagonist: a possible mechanism for immune-to-brain communication. *Brain Res Bull*. 1997;43(3):357–364.
33. Folpe AL, Billings SD, McKenney JK, Walsh SV, Nusrat A, Weiss SW. Expression of claudin-1, a recently described tight junction-associated protein, distinguishes soft tissue perineurioma from potential mimics. *Am J Surg Pathol*. 2002; 26(12):1620–1626.
34. Stubbs EB Jr. Targeting the blood-nerve barrier for the management of immune-mediated peripheral neuropathies. *Exp Neurol*. 2020;331:113385.
35. Katona I, Weis J. Diseases of the peripheral nerves. *Handb Clin Neurol*. 2017;145:453–474.
36. MacKeever PE. Immunohistochemistry of the nervous system. In: Dabbs DJ, ed. *Diagnostic Immunohistochemistry*. 4th ed. Elsevier; 2014:762–828.
37. Piña AR, Martínez MM, de Almeida OP. Glut-1, best immunohistochemical marker for perineurial cells. *Head Neck Pathol*. 2015;9(1):104–106.
38. Macarenco RS, Ellinger F, Oliveira AM. Perineurioma: a distinctive and under recognized peripheral nerve sheath neoplasm. *Arch Pathol Lab Med*. 2007;131(4):625–636.
39. Tsai CY, Yeh CJ, Chao YK, Chang HK, Tseng CK, Liu YH. Perineural invasion through the sheath in posttherapy esophagectomy specimens predicts poor survival in patients with esophageal squamous cell carcinoma. *Eur J Surg Oncol*. 2017;43(10):1970–1976.
40. Fogt F, Capodieci P, Loda M. Assessment of perineural invasion by Glut-1 immunohistochemistry. *Appl Immunohistochem*. 1995;3:194–197.
41. Ronaghy A, Yaar R, Goldberg LJ, Mahalingam M, Bhawan J. Perineurial involvement: what does it mean? *Am J Dermatopathol*. 2010;32(5):469–476.
42. Richard L, Topilko P, Magy L, et al. Endoneurial fibroblast-like cells. *J Neuropathol Exp Neurol*. 2012;71(11): 938–947.
43. Hirose T, Tani T, Shimada T, Ishizawa K, Shimada S, Sano T. Immunohistochemical demonstration of EMA/Glut1-positive perineurial cells and CD34-positive fibroblastic cells in peripheral nerve sheath tumors. *Mod Pathol*. 2003;16(4):293–298.
44. Müller M, Leonhard C, Krauthausen M, Wacker K, Kiefer R. On the longevity of resident endoneurial macrophages in the peripheral nervous system: a study of physiological macrophage turnover in bone marrow chimeric mice. *J Peripher Nerv Syst*. 2010;15(4):357–365.
45. Chen S, Burgin S, McDaniel A, et al. Nf1-/- Schwann cell-conditioned medium modulates mast cell degranulation by c-Kit-mediated hyperactivation of phosphatidylinositol 3-kinase. *Am J Pathol*. 2010;177(6):3125–3132.
46. Staser K, Yang FC, Clapp DW. Mast cells and the neurofibroma microenvironment. *Blood*. 2010;116(2):157–164.
47. Esposito B, De Santis A, Monteforte R, Baccari GC. Mast cells in Wallerian degeneration: morphologic and ultrastructural changes. *J Comp Neurol*. 2002;445(3):199–210.
48. Harris GM, Madigan NN, Lancaster KZ, et al. Nerve guidance by a decellularized fibroblast extracellular matrix. *Matrix Biol*. 2017;60–61:176–189.
49. Kazamel M, Boes CJ. Renaut corpuscles or peripheral nerve infarcts? A historical overview. *Can J Neurol Sci*. 2017;44(2): 184–189.

50. Piña-Oviedo S, Del Valle L, Baquera-Heredia J, Ortiz-Hidalgo C. Imunohistochemical characterization of Renaut bodies in superficial digital nerves: further evidence supporting their perineurial cell origin. *J Peripher Nerv Syst*. 2009;14(1): 22–26.
51. King R, Ginsberg L. The nerve biopsy: indications, technical aspects, and contribution. *Handb Clin Neurol*. 2013; 115:155–170.
52. Lindemuth R, Ernzerhof C, Schimrigk K. Comparative morphometry of myelinated nerve fibers in the normal and pathologically altered human sural and tibial nerve. *Clin Neuropathol*. 2002;21(1):29–34.
53. Yuan A, Rao MV, Veeranna, Nixon RA. Neurofilaments and neurofilament proteins in health and disease. *Cold Spring Harb Perspect Biol*. 2017;9(4):a018309.
54. Fainikar A, Bass PW. Critical roles for microtubules in axonal development and disease. *Results Probl Cell Differ*. 2009;48:47–64.
55. Tasdemir-Yilmaz OE, Segal RA. There and back again: Coordinated transcription, translation and transport in axonal survival and regeneration. *Curr Opin Neurobiol*. 2016;39:62–68.
56. Salinas S, Schlavo G, Kremer EJ. A hitchhiker's guide to the nervous system: the complex journey of viruses and toxins. *Nat Rev Microbiol*. 2010;8(9):645–655.
57. Tosolini AP, Sleigh JN, Surana S, Rhymes ER, Cahalan SD, Schiavo G. BDNF-dependent modulation of axonal transport is selectively impaired in ALS. *Acta Neuropathol Commun*. 2022;10(1):121.
58. Fumagalli L, Young FL, Boeynaems S, et al. *C9orf72*-derived arginine-containing dipeptide repeats associate with axonal transport machinery and impede microtubule-based motility. *Sci Adv*. 2021;7(15):eabg3013.
59. Van de Vondel L, De Winter J, Timmerman V, Baets J. Overarching pathomechanisms in inherited peripheral neuropathies, spastic paraplegias, and cerebellar ataxias. *Trends Neurosci*. 2024;47(3):227–238.
60. Peters A. The effects of normal aging on myelin and nerve fibers: a review. *J Neurocytol*. 2002;31(8–9):581–593.
61. Terada S. Where does slow axonal transport go? *Neurosci Res*. 2003;47(4):367–372.
62. Cavanagh JB. Toxic and deficiency disorders. In Weller RO ed. *Systemic Pathology: Nervous System, Muscle and Eye*. Vol. 3. 3rd ed. Churchill Livingstone; 1990:244–308.
63. Katalymov LL, Glukhova NV. Some characteristics of periaxonal space in myelinated nerve fibers. *Dokl Biol Sci*. 2003;388:9–11.
64. Nave K-A, Trapp BD. Axon-glial signaling and the glial support of axon function. *Annu Rev Neurosci*. 2008;31: 535–561.
65. Armati PJ, Mathey EK. An update on Schwann cell biology—immunomodulation, neural regulation and other surprises. *J Neurol Sci*. 2013;333(1–2):68–72.
66. Woodhoo A, Sommers L. Development of the Schwann cell lineage: from the neural crest to the myelinated nerve. *Glia*. 2008;56(14):1481–1490.
67. Weller RO, Herzog I. Schwann cell lysosomes in hypertrophic neuropathy and in normal human nerves. *Brain*. 1970;93(2):347–356.
68. González-Martínez T, Pérez-Piñera P, Díaz-Esnal B, Vega JA. S-100 proteins in the human peripheral nervous system. *Microsc Res Tech*. 2003;60(6):633–638.
69. Fine SW, McCline SA, Li M. Immunohistochemical staining for calretinin is useful for differentiating schwannomas from neurofibromas. *Am J Clin Pathol*. 2004;122(4):552–559.
70. Koirala S, Reddy LV, Ko CP. Roles of glia cell in the formation, function and maintenance of the neuromuscular junction. *J Neurocytol*. 2002;32(5–8):987–1002.
71. Oda R, Yaoi T, Okajima S, Kobashi H, Kubo T, Fushiki S. A novel marker for terminal Schwann cells, homocysteine-responsive ER-resident protein, as isolated by a single cell PCR-differential display. *Biochem Biophys Res Commun*. 2003;308(4):872–877.
72. Wewetzer K, Verdú E, Agelov DN, Navarro X. Olfactory ensheathing glia and Schwann cells: two of a kind? *Cell Tissue Res*. 2002;309(3):337–345.
73. Snaidero N, Simons M. Myelination at a glance. *J Cell Sci*. 2014;127(Pt 14):2999–3004.
74. Salzer JL. Schwann cell myelination. *Cold Spring Harb Perspect Biol*. 2015;7(8):a020529.
75. Pereira JA, Lebrun-Julien F, Suter U. Molecular mechanisms regulating myelination in the peripheral nervous system. *Trends Neurosci*. 2012;35(2):123–134.
76. Nave KA, Werner HB. Myelination of the nervous system: mechanisms and functions. *Annu Rev Cell Dev Biol*. 2014; 30:503–533.
77. Garbay B, Heape AM, Sargueil F, Cassagne C. Myelin synthesis in the peripheral nervous system. *Prog Neurobiol*. 2000;61(3):267–304.
78. Simons M, Trotter J. Wrapping it up: the cell biology of myelination. *Curr Opin Neurobiol*. 2007;17(5):533–540.
79. Stojkovic T. Hereditary neuropathies: an update. *Rev Neurol (Paris)*. 2016;172(12):775–778.
80. Weis J, Senderek J. Introduction to the hereditary neuropathies. In: Vallat JM, Weis J, eds. *Peripheral Nerve Disorders. Pathology and Genetics*. ISNP Wiley Blackwell; 2014: 59–61.
81. Werner HB. On the evolution of myelin. *Brain Res*. 2016; 1641(Pt A):1–3.
82. Podratz JL, Rodriguez E, Windebank AJ. Role of the extracellular matrix in myelination of peripheral nerve. *Glia*. 2001;35(1):35–40.
83. Quarles RH. Myelin-associated glycoprotein (MAG): past, present and beyond. *J Neurochem*. 2007;100(6): 1431–1448.
84. Yang Y, Shi Y. L-periaxin interacts with S-periaxin through its PDZ domain. *Neurosci Lett*. 2015;609:23–29.
85. Schroeder JM, Weis J. Basic pathology of the peripheral nervous system. In: Vallat JM, Weis J, eds. *Peripheral Nerve Disorders. Pathology and Genetics*. ISNP Wiley Blackwell; 2014:38–58.
86. Sherman DL, Wu LM, Grove M, Gillespie CS, Brophy PJ. Drp2 and periaxin form Cajal bands with dystroglycan but have distinct roles in Schwann cell growth. *J Neurosci*. 2012;32(27):9419–9428.
87. Rasband MN, Peles E. The nodes of Ranvier: molecular assembly and maintenance. *Cold Spring Harb Perspect Biol*. 2015;8(3):a020495.
88. Uncini A, Mathis S, Vallat JM. New classification of autoimmune neuropathies based on target antigens and involved domains of myelinated fibres. *J Neurol Neurosurg Psychiatry*. 2022;93(1):57–67.

89. Weller RO, Cervos Navarro J. *Pathology of Peripheral Nerves: A Practical Approach*. Butterworth; 1977.
90. Yang Z, Wang KK. Glial fibrillary acidic protein: from intermediate filament assembly and gliosis to neurobiomarker. *Trends Neurosci*. 2015;38(6):364–374.
91. Scollard DM, Truman RW, Ebenezer GJ. Mechanisms of nerve injury in leprosy. *Clin Dermatol*. 2015;33(1):46–54.
92. Kieseier BC, Hu W, Hurtung H-P. Schwann cells as immunomodulatory cells. In: Armati PK, ed. *The Biology of Schwan cells: Development, Differentiation and Immunomodulation*. Cambridge University Press; 2007:118–125.
93. Bonetti B, Valdo P, Ossi G, et al. T-cell cytotoxicity of human Schwann cells: TNFalpha promotes fasL-mediated apoptosis and IFN gamma perforin-mediated lysis. *Glia*. 2003;43(2):141–418.
94. Petrov KA, Girard E, Nikitashina AD, et al. Schwann cells sense and control acetylcholine spillover at the neuromuscular junction by α7 nicotinic receptors and butyrylcholinesterase. *J Neurosci*. 2014;34(36):11870–11183.
95. Vedeler C, Ulvestad E, Borge L, et al. Expression of CD59 in normal human nervous tissue. *Immunology*. 1994;82(4):542–547.
96. Wijdicks EF, Klein CJ. Guillain–Barré syndrome. *Mayo Clin Proc*. 2017;92(3):467–479.
97. Maestroni GJM. Adrenergic modulation of hematopoiesis. *J Neuroimmune Pharmacol*. 2020;15(1):82–92.
98. Yamazaki S, Ema H, Karlsson G, et al. Nonmyelinating Schwann cells maintain hematopoietic stem cell hibernation in the bone marrow niche. *Cell*. 2011;147(5):1146–1158.
99. Ahmed AM, Weller RO. The blood-nerve barrier and reconstitution of the perineurium following nerve grafting. *Neuropathol Appl Neurobiol*. 1979;5(6):469–483.
100. Chkheidze R, Pytel P. What every neuropathologist needs to know: peripheral nerve biopsy. *J Neuropathol Exp Neurol*. 2020;79(4):355–364.
101. Brandner S. The pathological diagnosis of nerve biopsies. A practical approach. *Diagn Histopathol*. 2016;22(9):333–344.
102. Lehmann HC, Wunderlich G, Fink GR, Sommer C. Diagnosis of peripheral neuropathy. *Neurol Res Pract*. 2020;2:20.
103. Bouche P. Clinical assessment and classification of peripheral nerve diseases. In: Vallat JM, Weis J, eds. *Peripheral Nerve Disorders. Pathology and Genetics*. ISNP Wiley Blackwell; 2014:1–11.
104. Geuna S, Raimondo S, Ronchi G, et al. Chapter 3: histology of the peripheral nerve and changes occurring during nerve regeneration. *Int Rev Neurobiol*. 2009;87:27–46.
105. Segura LMG. Ramón y Cajal y la neurociencia del siglo XXI. *Jano Extra*. 2005;1:16–22.
106. Mack TG, Reiner M, Beirowski B, et al. Wallerian degeneration of injured axons and synapses is delayed by a Ube4b/Nmnat chimeric gene. *Nat Neurosci*. 2001;4(12):1199–1206.
107. Osterloh JM, Yang J, Rooney TM, et al. dSarm/Sarm1 is required for activation of an injury-induced axon death pathway. *Science*. 2012;337(6093):481–484.
108. Wang H, Wang X, Zhang K, et al. Rapid depletion of ESCRT protein Vps4 underlies injury-induced autophagic impediment and Wallerian degeneration. *Sci Adv*. 2019;5(2):eaav4971.
109. Zhang K, Jiang M, Fang Y. The drama of Wallerian degeneration: the cast, crew, and script. *Annu Rev Genet*. 2021;55:93–113.
110. Conforti L, Gilley J, Coleman MP. Wallerian degeneration: an emerging axon death pathway linking injury and disease. *Nat Rev Neurosci*. 2014;15(6):394–409.
111. Gámez-Sánchez JA, Carty L, Iruarrizaga-Lejarreta M, et al. Schwann cell autophagy, myelinophagy, initiates myelin clearance from injured nerves. *J Cell Biol*. 2015;210(1):153–168.
112. Thumm M, Simons M. Myelinophagy: Schwann cells dine in. *J Cell Biol*. 2015;210(1):9–10.
113. Burnett MG, Zager EL. Pathophysiology of peripheral nerve injury: a brief review. *Neurosurg Focus*. 2004;16(5):E1.
114. Carr MJ, Johnston AP. Schwann cells as drivers of tissue repair and regeneration. *Curr Opin Neurobiol*. 2017;47:52–57.
115. Blesch A, Tuszynski MH. Nucleus hears axon's pain. *Nat Med*. 2004;10(3):236–237.
116. Wong KM, Babetto E, Beirowski B. Axon degeneration: make the Schwann cell great again. *Neural Regen Res*. 2017;12(4):518–524.
117. Panayotis N, Karpova A, Kreutz MR, Fainzilber M. Macromolecular transport in synapse to nucleus communication. *Trends Neurosci*. 2015;38(2):108–116.
118. Holahan MR. A shift from a pivotal to supporting role for the growth-associated protein (GAP-43) in the coordination of axonal structural and functional plasticity. *Front Cell Neurosci*. 2017;11:266.
119. Jessen KR, Mirsky R. The repair Schwann cell and its function in regenerating nerves. *J Physiol*. 2016;594(13):3521–3531.
120. Chernousov MA, Yu WM, Chen ZL, Carey DJ, Strickland S. Regulation of Schwann cell function by the extracellular matrix. *Glia*. 2008;56(14):1498–1507.
121. Dyck PJ, Dyck PJB, Giannini C, et al. Peripheral nerves. In: Graham DI, Lantos PL, eds. *Greenfield's Neuropathology*. Arnold; 2002:551–675.
122. Brandner S. Toxic neuropathies. In: Vallat JM, Weis J, eds. *Peripheral Nerve Disorders. Pathology and Genetics*. ISNP Wiley Blackwell; 2014:238–246.
123. Tarasiuk O, Molteni L, Malacrida A, Nicolini G. The role of NMNAT2/SARM1 in neuropathy development. *Biology (Basel)*. 2024;13(1):61.
124. Sato-Yamada Y, Strickland A, Sasaki Y, Bloom J, DiAntonio A, Milbrandt J. A SARM1-mitochondrial feedback loop drives neuropathogenesis in a Charcot-Marie-Tooth disease type 2A rat model. *J Clin Invest*. 2022;132(23):e161566.
125. Li Y, Pazyra-Murphy MF, Avizonis D, et al. Sarm1 activation produces cADPR to increase intra-axonal Ca^{++} and promote axon degeneration in PIPN. *J Cell Biol*. 2022;221(2):e202106080.
126. Bosanac T, Hughes RO, Engber T, et al. Pharmacological SARM1 inhibition protects axon structure and function in paclitaxel-induced peripheral neuropathy. *Brain*. 2021;144(10):3226–3238.
127. Weller RO, Bruckner FE, Chamberlain MA. Rheumatoid neuropathy: a histological and electrophysiological study. *J Neurol Neurosurg Psychiatry*. 1970;33(5):592–604.
128. Vallat JM, Mathis S. Pathology explains various mechanisms of auto-immune inflammatory peripheral neuropathies. *Brain Pathol*. 2024;34(2):e13184.

129. Vallat JM, Magy L, Corcia P, Boulesteix J-M, Uncini A, Mathis S. Ultrastructural lesions of nodo-paranodopathies in peripheral neuropathies. *J Neuropathol Exp Neurol*. 2020; 79(3):247–255.
130. Eshed-Eisenbach Y, Brophy PJ, Peles E. Nodes of Ranvier in health and disease. *J Peripher Nerv Syst*. 2023;28(Suppl 3): S3–S11.
131. Ortiz Hidalgo C, Weller RO. Traumatic and compressive lesions of peripheral nerves. In: Vallat JM, Weis J, eds. *Peripheral Nerve Disorders. Pathology and Genetics*. ISNP Wiley Blackwell; 2014:244–308.
132. Abreu E, Aubert S, Wavreille G, Gheno R, Canella C, Cotten A. Peripheral tumor and tumor-like neurogenic lesions. *Eur J Radiol*. 2013;82(1):38–50.
133. Stecco C, Fantoni I, Macchi V, et al. The role of fasciae in Civinini–Morton's syndrome. *J Anat*. 2015;227(5):654–664.
134. Klein CJ, Wu Y, Jentoft ME, Mer G, et al. Genomic analysis reveals frequent TRAF7 mutations in intraneural perineuriomas. *Ann Neurol*. 2017;81(2):316–321.
135. Vizcaino MA, Belzberg A, Ahlawat S, et al. Localized hypertrophic neuropathy as a neoplastic manifestation of KRAS-mediated rasopathy. *J Neuropathol Exp Neurol*. 2020;79(6):647–651.

SECTION V

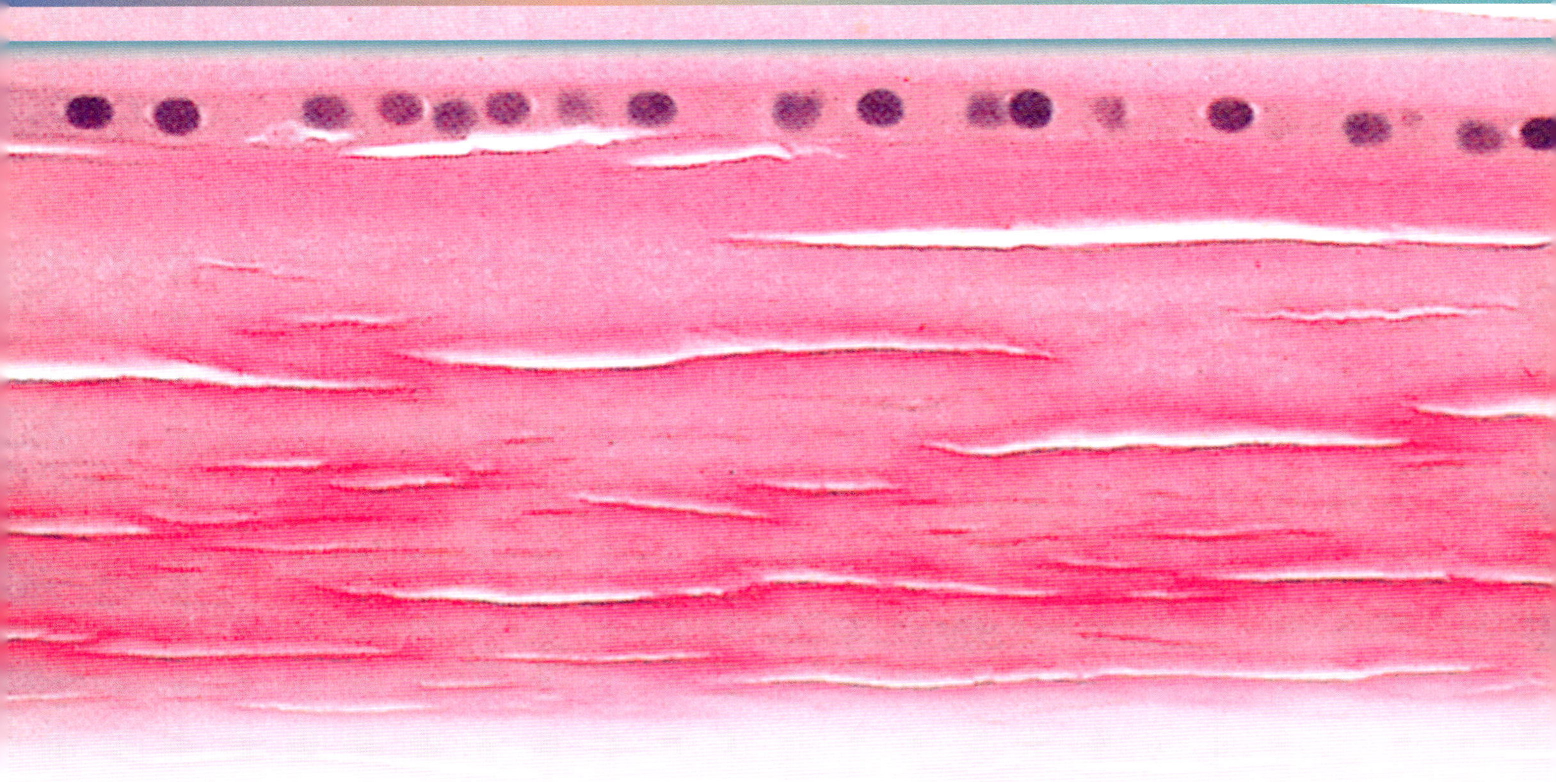

Head and Neck

12

Eye and Ocular Adnexa

Alan D. Proia ▪ Thomas J. Cummings

The eye and surrounding tissues are subject to a wide variety of primary ocular and systemic diseases. Understanding ocular anatomy will enable the general surgical pathologist to appreciate morphologic abnormalities and facilitate the diagnosis of many pathologic conditions affecting those structures. Similar to other specimens received by the surgical pathologist, a discussion with the ophthalmologist submitting the tissue may be necessary for its proper handling, sectioning, and processing to prevent artifactually masking the diagnostic histologic features, such as the common age-related arcus lipoides.

This chapter presents an overview of the normal histology of the eye and ocular adnexa. Several excellent texts are available for more detailed information on ocular anatomy, development, and age-related changes (1–8).

The eye is roughly spherical, and external measurements are routinely obtained in three dimensions. In utero, the eye's anterior–posterior (axial) length is approximately 7 mm at 15 weeks of gestation, 10 mm at 20 weeks, and 15 mm at 30 weeks (9,10). At birth, the anterior–posterior diameter of the eye is approximately 17 to 18 mm (11), while by age 3, the axial length is about 22 mm (9). In adults, the anterior–posterior plane of the eye measures approximately 24 mm, whereas the vertical and horizontal dimensions are about 23 to 23.5 mm, though the eyes of men tend to be slightly larger than those of women (1). The globe's equator is midway between the eye's anterior and posterior poles.

Several external landmarks allow the pathologist to orient the globe and to determine whether an eye is from the right or the left side (Fig. 12.1). By establishing the nasal (medial) and temporal (lateral) sides of the globe and the superior surface of the eye, one can easily deduce the side of the eye. The six extraocular muscles (four rectus and two oblique muscles), arising in the posterior orbit and running forward to insert upon the sclera, are used for identifying the eye's side. The rectus muscles arise from a fibrous ring at the apex of the orbit, the annulus of Zinn, and are enveloped by a fascial membrane that creates a cone-shaped structure posterior to the globe. The levator palpebrae superioris also arises at the orbital apex and extends anteriorly to the eyelids. Of the extraocular muscles, only the inferior oblique has a muscular insertion upon the sclera; the other muscles have tendinous insertions. The surgeon usually removes the extraocular muscle insertions during enucleation of the globe, though a short stump of the inferior oblique muscle is often present. The extraocular

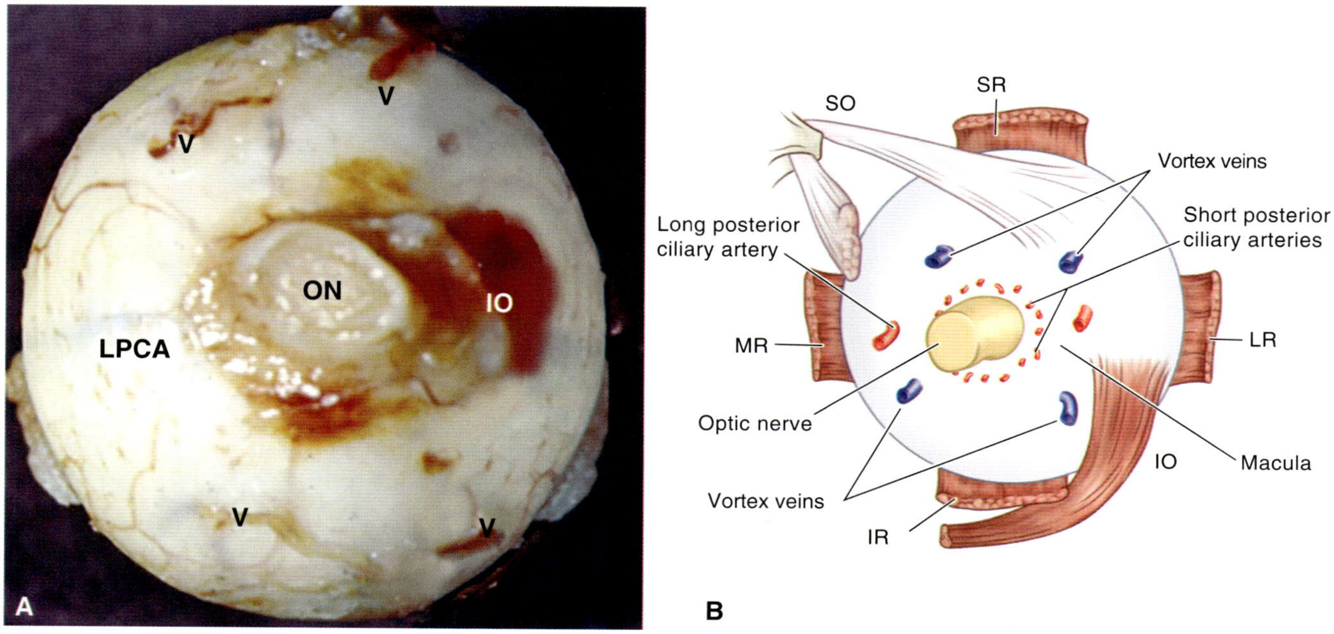

FIGURE 12.1 **A:** This posterior view of a right eye shows the transected optic nerve (*ON*) with its dura mater, the medial long posterior ciliary artery (*LPCA*), four vortex veins (*V*), and the insertion of the inferior oblique muscle (*IO*). **B:** This drawing depicts the right eye as seen from behind. Several external landmarks are useful in determining the orientation of the globe. The optic nerve is located approximately 1 mm inferior to and 3 mm nasal to the posterior pole of the eye. The long posterior ciliary arteries are located in the horizontal plane, and four vortex veins exit the sclera posteriorly. The superior oblique muscle (*SO*) inserts on the top of the globe, whereas the inferior oblique muscle (*IO*) inserts temporally and its fibers run posteriorly and nasally. The rectus muscles insert medially (*MR*), laterally (*LR*), inferiorly (*IR*), and superiorly (*SR*). Short posterior ciliary arteries form a ring surrounding the optic nerve. (Reproduced with permission from Freddo TF, Chaum E. *Anatomy of the Eye and Orbit. The Clinical Essentials.* Wolters Kluwer; 2018.)

muscles are frequently present on eyes obtained postmortem. The superior and inferior oblique muscles are most useful in orientating the globe. The tendinous insertion of the superior oblique muscle behind the superior rectus muscle insertion indicates the top of the eye. The inferior oblique muscle inserts on the sclera temporally in the horizontal meridian, and its fibers run inferiorly toward the back of the orbit. The optic nerve is also helpful in assessing orientation because it exits the globe slightly nasally to the posterior pole of the eye. Adjacent to the optic nerve, the prominent long posterior ciliary arteries course through the superficial sclera in opposite directions in a horizontal plane. Anteriorly, the dimensions of the cornea may be helpful in topographic orientation. The cornea is usually elliptical in the adult, with its horizontal diameter slightly greater than its vertical breadth. In young children, this difference is less apparent.

The eye is traditionally described as having three tissue layers that surround the vitreous, the lens, and the spaces of the anterior and posterior chambers (Fig. 12.2). The outermost part of the eye is composed of the transparent cornea and the opaque sclera. The ocular middle layer, termed the "uvea" or "uveal tract," comprises the iris, the ciliary body, and the choroid. The innermost retina is in direct contact with the vitreous body.

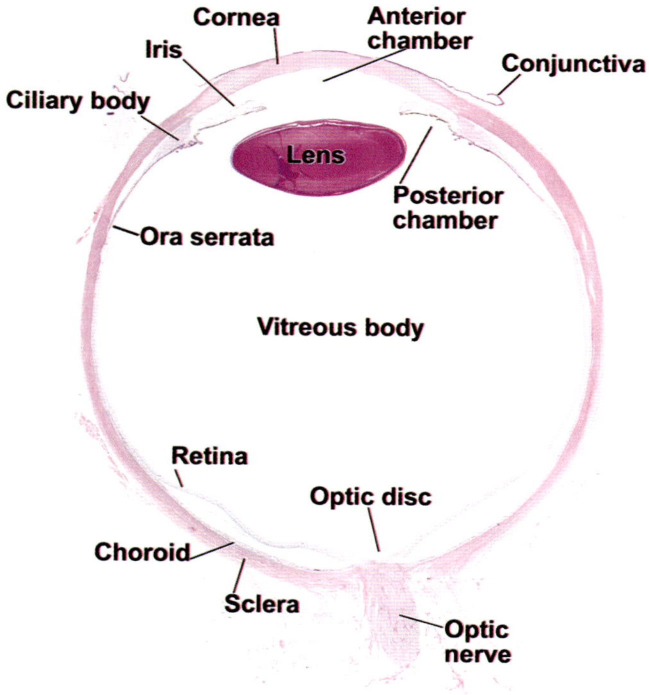

FIGURE 12.2 This photomicrograph of a human eye histologic section illustrates the major tissue layers of the eye, the lens, the vitreous body, and the spaces of the anterior and posterior chambers. The posterior retina exhibits artifactual detachment due to formalin fixation.

CORNEA

The transparent cornea occupies one-sixth of the surface of the globe and refracts the entering light (12,13). Although individual variation is common, the cornea measures approximately 11.7 mm in the horizontal plane and 10.6 mm in the vertical plane (1). Centrally, the cornea is about 0.5 mm thick, but peripherally it thickens to about 0.67 mm (1). Histologically, the cornea consists of six distinct layers: (a) the epithelium, (b) the basal lamina of the epithelium, (c) the Bowman layer, (d) the stroma, (e) the Descemet membrane, and (f) the endothelium (Fig. 12.3).

The corneal epithelium, composed of stratified nonkeratinized squamous cells, is about five to seven cell layers thick. The basilar epithelial cells are polygonal in shape, and they acquire a more flattened appearance as they become displaced to the corneal surface during differentiation. In the normal cornea, even the most superficial epithelial cells retain their nuclei (Fig. 12.4). Mitotic figures are uncommon in the epithelium but are occasionally observed in the basal cells. Some Langerhans cells are present within the corneal epithelium and are more abundant peripherally (14,15). Langerhans cells are not ordinarily recognizable in routinely stained tissue sections, but they can be visualized by immunohistochemical methods using antibodies to Langerin/CD207, CD1a, or S100 protein. Apoptosis is uncommonly seen in the epithelium of the normal cornea, but it may be apparent in corneas with bullous keratopathy (corneal edema).

The corneal epithelium rests upon a basal lamina, which is difficult to see in hematoxylin and eosin (H&E)–stained tissue sections. Staining with the periodic acid–Schiff (PAS) reaction makes this layer apparent (Fig. 12.4). In certain pathologic conditions, most commonly bullous keratopathy, the epithelial basal lamina assumes an intraepithelial location.

The Bowman layer is an acellular structure located just posterior to the epithelial basal lamina (Fig. 12.4). It is approximately 8 to 14 μm thick. Transmission electron microscopy shows that the Bowman layer is not a true basement membrane but is composed of randomly oriented delicate collagen fibers. The anterior face of the Bowman layer ends distinctly at its junction with the epithelial basal lamina. Posteriorly, the Bowman layer merges inconspicuously with the underlying corneal stroma. Unmyelinated sensory nerves reach the epithelium from the stroma after crossing the Bowman layer. However, nerve processes are challenging to detect in the cornea in standard tissue sections but are visible using antibodies to S100 protein (16).

The stroma accounts for approximately 90% of the cornea's thickness. It is composed of 200 to 250 layers (lamellae) of collagen fibers embedded in a proteoglycan-rich extracellular matrix (6). The stroma contains keratan sulfate proteoglycans (lumican, keratocan, mimecan) and a galactosaminoglycan-rich proteoglycan (decorin). Transmission electron microscopy has disclosed that the corneal collagen fibers are regularly spaced and of uniform diameter, contributing to the cornea's transparency. Surrounded by the stromal collagen lamellae are the corneal fibroblasts (keratocytes). The lamellae of the superficial one-third

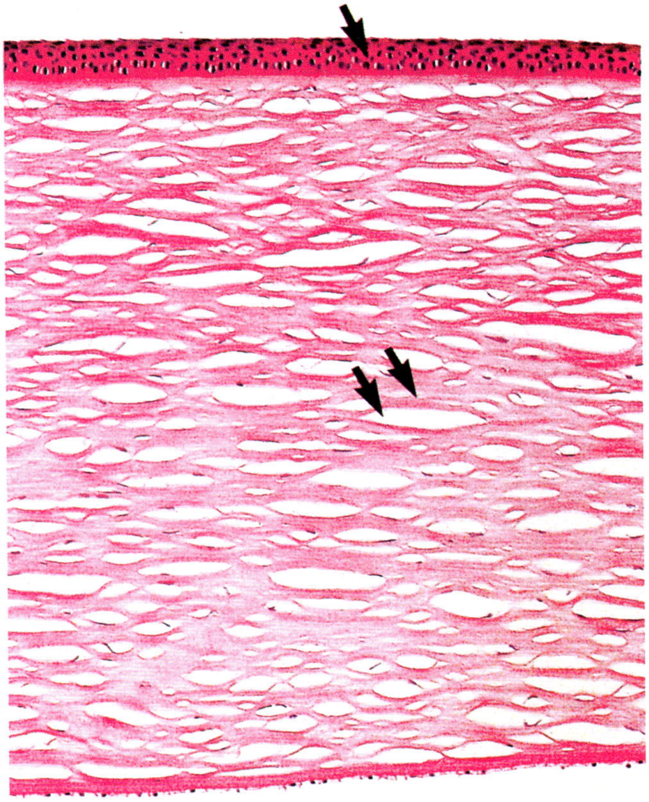

FIGURE 12.3 The stratified squamous epithelium of the cornea (*arrow*) overlies the basal lamina and the Bowman layer. The clefts within the collagenous stroma (*double arrows*) represent artifacts of tissue processing. No blood vessels or lymphatics are normally present within the cornea. The Descemet membrane and the corneal endothelium are located just posterior to the stroma (H&E).

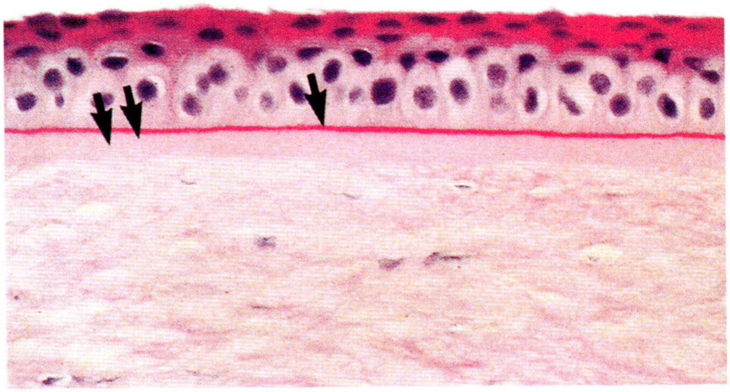

FIGURE 12.4 The corneal epithelium rests upon a thin basal lamina (*arrow*), which is prominent in this section following periodic acid–Schiff staining. The acellular band directly underneath the basal lamina is the Bowman layer (*double arrows*) (periodic acid–Schiff).

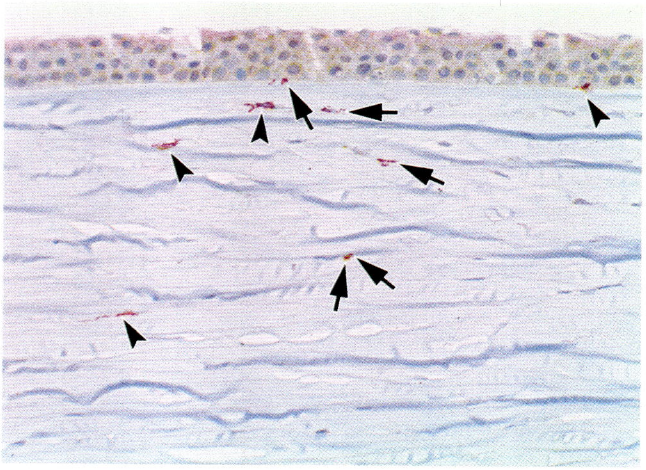

FIGURE 12.5 Variable numbers of CD68+ (Discovery Yellow [Ventana]) and CD163+ (Discovery Purple [Ventana]) macrophages are present beneath the corneal epithelium and in the stroma of normal corneas, with CD163+ cells (*arrows*) predominating over those expressing CD68 (*double arrows*) or coexpressing CD68 and CD163 (*arrowheads*).

of the stroma are arranged less orderly than those in the deeper two-thirds (17), and the anterior stroma is sometimes less eosinophilic than the posterior stroma (6). Other cell types are seldom identified in tissue sections of the normal corneal stroma using H&E stain, but the cornea is endowed with a sparse resident population of macrophages (18) demonstrable using antibodies to CD68 and CD163 (Fig. 12.5). The normal cornea lacks blood vessels, and its nutrition is obtained from an arterial plexus at the junction of the cornea and sclera and direct contact with the aqueous of the anterior chamber. Clefts are almost invariably present between the collagen lamellae in tissue sections of routinely processed formalin-fixed corneas. Interpreted as lymphatic channels by early histologists, these clefts are tissue processing artifacts. Lymphatic vessels are not present in the normal cornea.

The Descemet membrane, a true basal lamina elaborated by the underlying corneal endothelial cells, begins to form during fetal life. At birth, it is approximately 3 to 4 μm thick (Fig. 12.6). Basal laminar material is continuously added to the posterior part of the Descemet membrane throughout life so that by adulthood this structure attains a thickness of approximately 10 to 12 μm (Fig. 12.6). The fetal and postnatal regions of the Descemet membrane differ ultrastructurally, leading to the fetal zone, or anterior banded layer, and the postnatal zone, or posterior nonbanded layer. This difference is occasionally discernible by light microscopy, with the fetal region having a slightly basophilic tinge compared to the postnatal region (Fig. 12.6B).

The junction of the stroma and the Descemet membrane has assumed importance with the advent of corneal graft surgery during which the Descemet membrane is stripped from the host cornea and replaced by either a lamella of posterior corneal stroma and the Descemet membrane/corneal endothelium (i.e., the Descemet stripping automated endothelial keratoplasty) (19), the Descemet membrane/corneal endothelium without stroma (i.e., the Descemet membrane endothelial keratoplasty) (20), or the Descemet membrane with the pre-Descemet layer (i.e., the pre-Descemet endothelial keratoplasty) (21). The interface between the stroma and Descemet membrane has no distinctive features by routine light microscopy except for a lack of keratocytes. However, transmission electron microscopy reveals a zone of interfacial matrix containing aggregates of homogeneous amorphous material and irregularly arranged collagen fibrils traversing between collagenous lamellae of the stroma and the fetal zone of the Descemet membrane (17,22). This

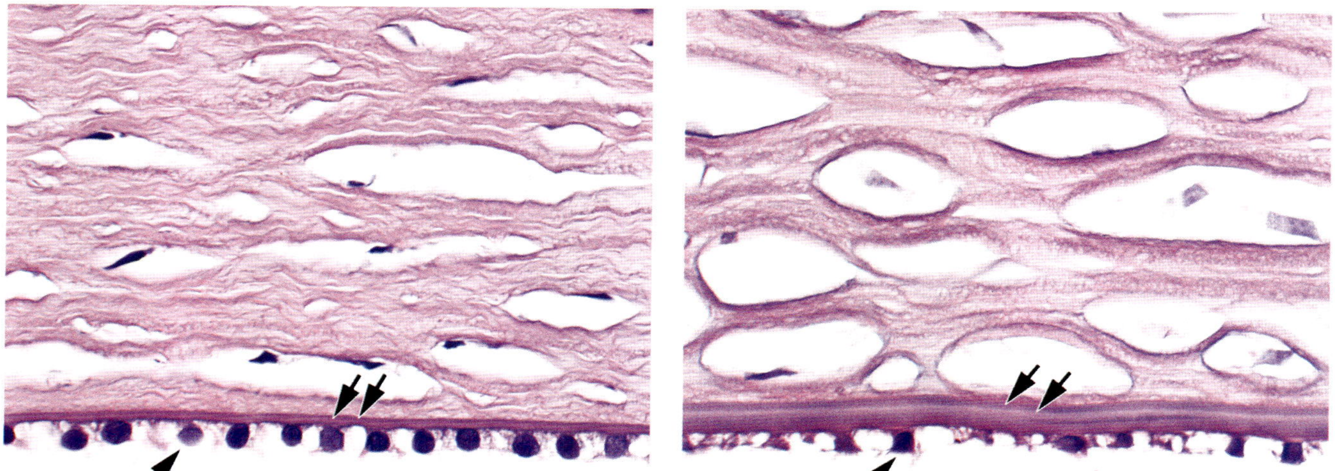

A B

FIGURE 12.6 A thin monolayer of corneal endothelial cells (*arrow*) covers the posterior surface of the Descemet membrane (*double arrows*). These cells are in direct contact with the aqueous of the anterior chamber. The number of corneal endothelial cells decreases, and the Descemet membrane thickness increases with advancing age. The cornea on the left (**A**) is from a 30.5-week-old girl, while that on the right (**B**) is from a 78-year-old woman (H&E).

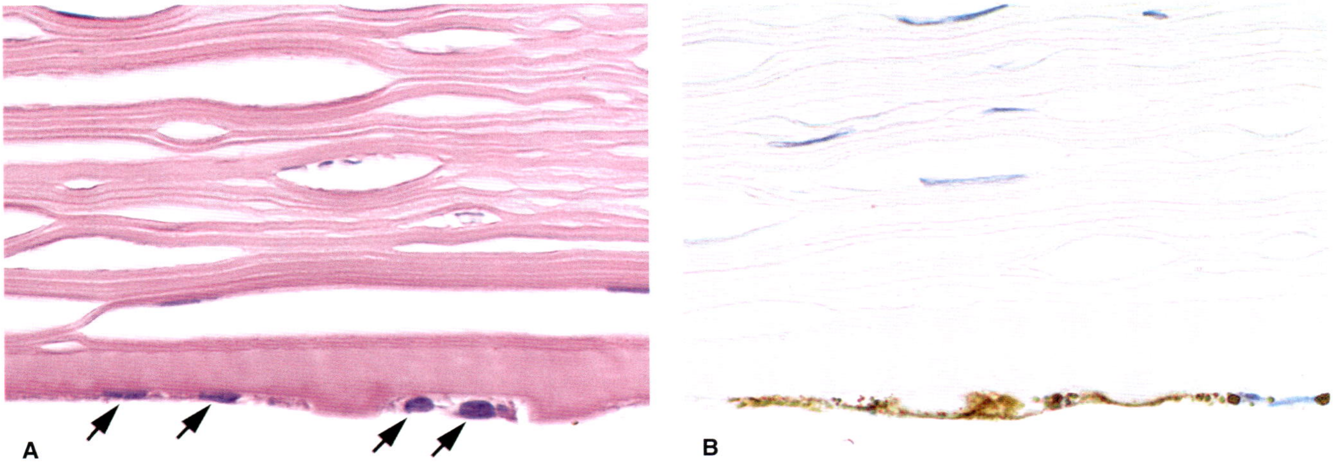

FIGURE 12.7 Flattened, metaplastic corneal endothelial cells (*arrows*) from corneas of eyes with bullous keratopathy or other conditions requiring corneal transplantation may exhibit cytokeratin expression. Normal corneal endothelial cells do not express cytokeratins. This penetrating keratoplasty graft corneal button is from a woman with Fuchs' corneal dystrophy (**A:** H&E; **B:** pan-cytokeratin antibody cocktail, diaminobenzidine chromogen).

interfacial matrix represents a cleavage zone during stripping of the Descemet membrane (22). Dua et al. found that cleavage occurred most commonly between the last row of stromal keratocytes and an acellular layer of 5 to 8 lamellae of longitudinally, horizontally, and obliquely oriented collagen bundles that was termed "the Dua layer" (23).

The corneal endothelium (Fig. 12.6) is directly exposed to the aqueous in the anterior chamber. Although this cell layer does not line blood vessels or lymphatic spaces, the term "endothelium" is firmly entrenched in the literature. These cells function as an osmotic pump to regulate a necessary state of stromal dehydration that preserves corneal clarity. Endothelial decompensation results in corneal edema and diminished optical transparency. The corneal endothelium is S100 protein-positive by immunohistochemistry, supporting a neural crest origin (24). Endothelial cells react with the monoclonal antibody 2B4.14.1 that recognizes the renal Tamm–Horsfall glycoprotein (THGP) antigen, raising the possibility that the corneal endothelium expresses a molecule with homeostatic properties similar to that ascribed to THGP (25). The endothelial cells of the cornea normally form a single flattened layer and virtually never regenerate by mitosis in human eyes (13,26), though they can regenerate in organ culture (27,28). The number of corneal endothelial cells decreases with advancing age (29,30), and it is essential to consider the patient's age when assessing whether or not the number of endothelial cells is normal (30) (Fig. 12.6). Under pathologic conditions such as epithelial ingrowth/downgrowth (31) and posterior polymorphous corneal dystrophy (32), cytokeratin-containing squamous cells replace the endothelium and form a layer more than one cell thick. It is essential to observe a multilayered cytokeratin-positive layer before rendering a diagnosis of epithelial ingrowth/downgrowth or posterior polymorphous dystrophy since a monolayer of flattened, cytokeratin-positive metaplastic endothelial cells may be found covering the posterior surface of the Descemet membrane or a retrocorneal fibrous membrane in corneas from eyes with bullous keratopathy or other conditions precipitating corneal transplantation (Fig. 12.7) (31).

After the second decade of life, age-related focal excrescences termed Hassall–Henle warts commonly form on the peripheral part of the Descemet membrane (Fig. 12.8). Virtually identical focal thickenings (i.e., corneal guttae) occur on the central part of the Descemet membrane under pathologic circumstances, most notably in Fuchs' corneal dystrophy. The presence of excrescences on the Descemet membrane in tissue sections of corneal buttons removed at penetrating keratoplasty (full-thickness corneal transplant) is always abnormal. Hassall–Henle warts are too peripheral in location to be present in a surgically excised corneal button.

Corneal epithelium and endothelium are prone to be "rubbed off" during the sectioning of corneal buttons, and it is essential to consider this artifact versus the actual loss of corneal epithelium and endothelium.

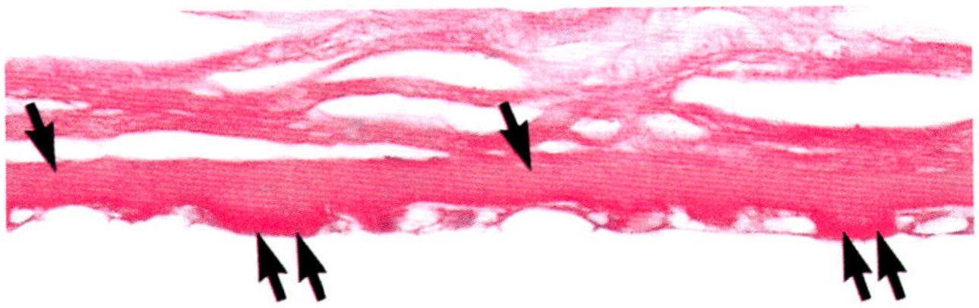

FIGURE 12.8 The Descemet membrane (*single arrows*) is located immediately posterior to the corneal stroma. Excrescences on the peripheral portion of the Descemet membrane (Hassall–Henle warts) (*double arrows*) represent an aging change (H&E).

SCLERA

The sclera, which accounts for approximately five-sixths of the eye's surface area, begins at the periphery of the cornea and extends posteriorly to the optic nerve. The sclera is relatively rigid, protecting the eye from trauma and helping to maintain intraocular pressure. Anteriorly, the sclera is visible underneath the transparent conjunctiva and is normally white in adults. The sclera varies in thickness depending on location. It is 0.8 mm thick near its junction with the cornea and is thinnest at the insertions of the four rectus muscles approximately 5 to 8 mm posterior to the corneoscleral junction, where it measures only 0.3 mm in thickness. From this point posteriorly, the sclera gradually thickens and attains its maximal width of about 1.0 mm adjacent to the optic nerve.

The sclera has three components: the episclera, stroma, and lamina fusca. The episclera is the most superficial part and is located between Tenon capsule, the fibrovascular connective tissue structure that envelops the globe, and the underlying scleral stroma with which it merges. The Tenon capsule (i.e., fascia bulbi) essentially represents a thin envelope of connective and elastic tissue and, though it is very well defined anatomically (33–36), is often challenging to recognize histologically due to its thinness and the way it merges with conjunctival connective tissue anteriorly (Fig. 12.9), the episclera diffusely, and the orbital fibroadipose tissue posteriorly (Figs. 12.10 and 12.11) (37–40). Smooth muscle is present in the superomedial, superolateral, inferomedial, and inferolateral anterior Tenon capsule (37) and the musculofibroelastic pulleys in the posterior Tenon capsule around the insertions of the rectus and inferior oblique muscles (38–40). The episclera is composed of loosely arranged collagen fibers and fibroblasts embedded in an extracellular matrix. Occasional melanocytes and mononuclear leukocytes are also present. Anteriorly, the episclera is richly vascularized. The largest component of the sclera is its stroma, which consists of fibrous bands of collagen, occasional elastic fibers, and scattered fibroblasts (Fig. 12.12). The corneal and scleral stroma are microscopically similar, but when viewed by transmission electron microscopy, the individual collagen fibers within the sclera vary in diameter and are randomly arranged, in contrast to the orderly packed corneal collagen fibers.

Although the scleral stroma is relatively avascular, blood vessels, accompanying nerves, and scattered melanocytes, are present in perforating emissarial canals (Fig. 12.13).

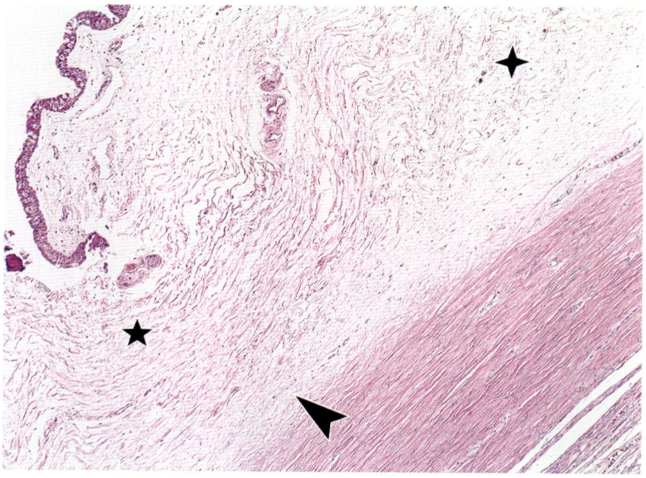

FIGURE 12.9 The Tenon capsule, which is not present in surgically enucleated eyes, is thickest anteriorly and is composed of loose fibrous connective tissue (*five-pointed star*) that merges with the looser connective tissue of the conjunctival substantia propria (*four-pointed star*) approximately 1.5 to 2 mm (35) posterior to the corneoscleral limbus. The Tenon capsule is attached to the adjacent episclera (*arrowhead*), which is sometimes, but not always, slightly less eosinophilic than the capsule (H&E).

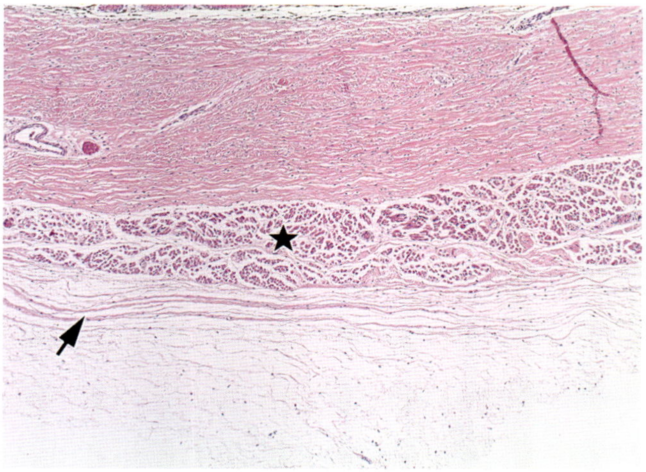

FIGURE 12.10 The Tenon capsule (*arrow*) merges with orbital fibroadipose tissue posteriorly, as seen in this infant's eye. The Tenon capsule is easiest to identify where it envelopes the extraocular muscles (*five-pointed star*) at their insertion sites into the sclera (*arrow*; H&E).

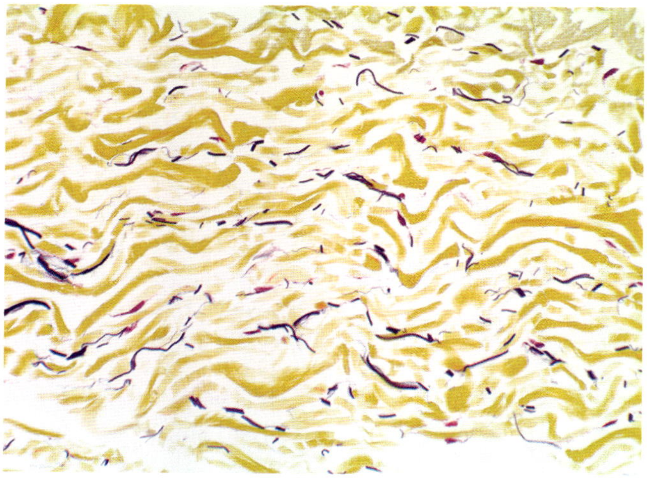

FIGURE 12.11 The Tenon capsule contains collagen and elastic fibers (Movat pentachrome; collagen is *yellow*, and elastic fibers are *deep purple*).

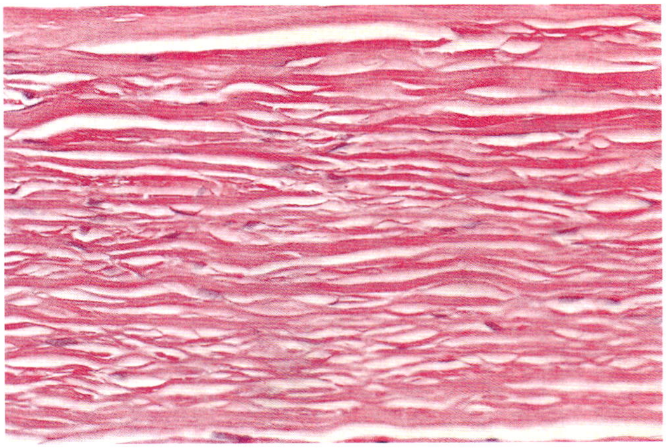

FIGURE 12.12 The scleral stroma is composed predominantly of a haphazard array of collagen bundles that vary in diameter. Scattered fibroblasts occur between the collagen bundles (H&E).

The anterior ciliary arteries perforate the sclera near the insertion of the rectus muscles. Venous channels draining the iris, ciliary body, and choroid (i.e., vortex veins) exit the sclera several millimeters posterior to the equator of the eye. The posterior ciliary arteries pass through the sclera near the optic nerve. In some individuals, a nerve in an emissarial canal near the corneoscleral junction may be prominent and attain a thickness of 1 to 2 mm diameter. The nodular and often pigmented clinical appearance of this "nerve loop of Axenfeld" may mimic a neuroma, cyst, pigmented neoplasm, or foreign body (41–43).

The innermost layer of the sclera, the lamina fusca, contains loose collagen fibers, fibroblasts, and scattered melanocytes. It represents a region of transition between the sclera and the underlying choroid. The sclera is weakly attached to the choroid by thin fibers of collagen.

With increasing age, several histologic changes occur in the sclera. Calcium may deposit diffusely between the individual collagen fibers throughout the entire scleral stroma. Localized abnormalities, known as senile scleral plaques, may occur just anterior to the insertion of the horizontal rectus muscles. These lesions are characterized by decreased stromal cellularity, abnormal collagen, and, in advanced cases, calcification (44).

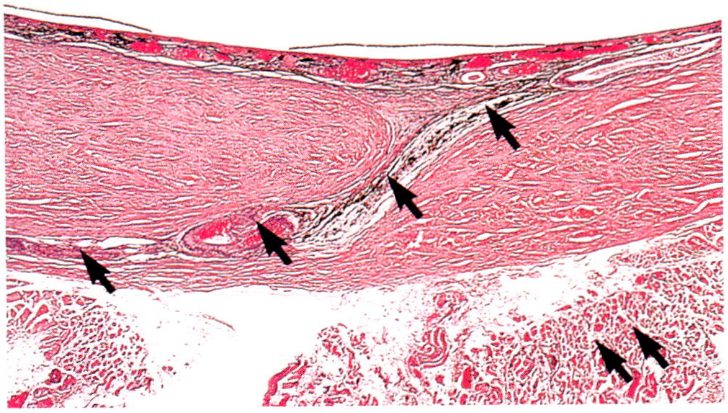

FIGURE 12.13 This figure illustrates a blood vessel penetrating the sclera and extending to the prominently vascularized choroid through an emissarial canal (*single arrows*). Pigmented melanocytes are also present. The fibers of the inferior oblique muscle are present at its insertion site upon the outer sclera (*double arrows*) (H&E).

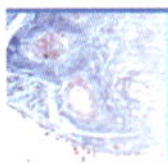

CORNEOSCLERAL LIMBUS

The corneoscleral junction, or limbus, is not a distinct anatomic site but is a significant clinical landmark. Many surgical procedures on the anterior part of the eye are accomplished after access *via* an incision in the limbal area. For this discussion, the trabecular meshwork and the Schlemm canal will be considered part of the corneoscleral limbus.

The limbus is approximately 1.5 to 2.0 mm wide, and separate layers of the cornea merge with components of the sclera or conjunctiva in this area (Fig. 12.14). The squamous epithelium of the cornea extends centrifugally beyond the limbus until it meets the epithelium of the bulbar conjunctiva. At the limbus, the Bowman layer of the cornea blends into the subepithelial tissues of the conjunctiva, and the corneal and scleral stroma become continuous. The Descemet membrane abruptly terminates in the limbal region, giving rise to the clinically significant landmark known as the Schwalbe ring. In about 15% of eyes, a prominent area of thickening is identified histologically at this site (Fig. 12.15) (6). Immediately adjacent to the Schwalbe ring is the most anterior aspect of the trabecular meshwork. The trabecular meshwork and the Schlemm canal constitute the apparatus mostly responsible for removing the aqueous from the eye (Fig. 12.16). Aqueous drainage occurs in the angle between the anterior surface of the iris and the sclera. Histologically, the meshwork appears as a collection of finely branching and delicately pigmented connective tissue bands. The cells lining the trabecular meshwork are continuous with the corneal endothelium. Posteriorly, the trabecular meshwork extends to a roughly triangular projection of scleral connective tissue, known as the scleral spur.

Located slightly anterior and superficial to the trabecular meshwork is the Schlemm canal, an endothelial-lined venous channel that completely encircles the limbus. Since the Schlemm canal sometimes gives off smaller branches, two lumina are occasionally seen on histologic sections of the anterior chamber angle. Although the trabecular meshwork and the Schlemm canal appear to be in intimate contact in tissue sections, they are separated by a thin layer of connective tissue and separate endothelial linings. The aqueous percolates among the delicate beams of the trabecular meshwork before becoming transported to the Schlemm canal. Ultrastructural examination of this region discloses giant cytoplasmic vacuoles in the endothelial lining of the Schlemm canal adjacent to the trabecular meshwork. Some think these vacuoles contain fluid in the

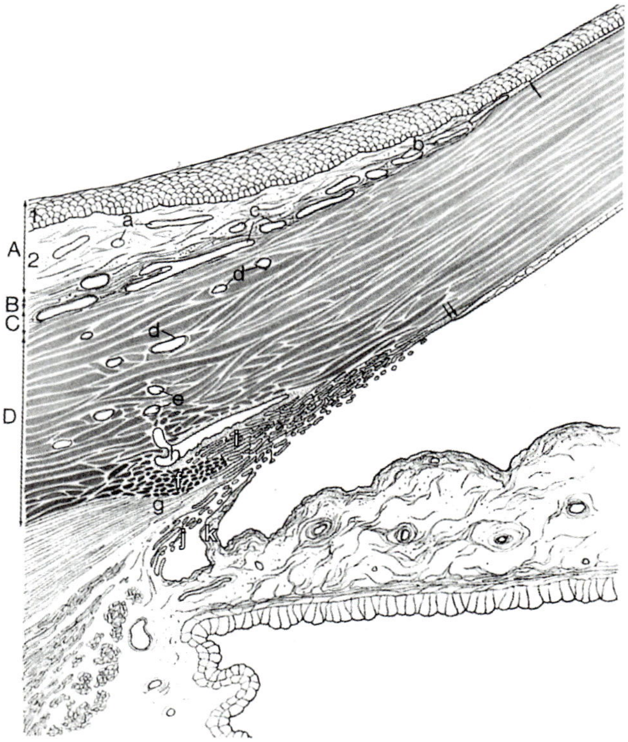

FIGURE 12.14 The corneoscleral limbus represents the junction of the peripheral cornea with the anterior sclera and is not a distinct anatomic site. Clinically, the limbus is an important landmark. The conjunctiva of the limbus (*A*) is composed of the epithelium (*1*) and stroma (*2*). The thin connective tissue layer of the Tenon capsule (*B*) overlies the episclera (*C*). The corneal and scleral stroma merge gradually in the area marked "*D*." Vessels of the conjunctival stroma (*a, b*), episclera (*c*), and limbal plexus (*d, e*) are illustrated. The projection of collagen fibers known as the scleral spur (*f*) merges with the smooth muscle fibers of the ciliary body (*g*). The Schlemm canal (*h*) and the trabecular meshwork (*i, j*) are responsible for the removal of the aqueous from the eye. Occasionally, processes from the iris (*k*) insert upon the trabecular meshwork. The Bowman layer (*arrow*) and Descemet membrane (*double arrows*) both terminate in the area of the limbus. (Reproduced with permission from Hogan MJ, Alvarado JA, Weddell JE. *Histology of the Human Eye*. WB Saunders; 1972, with permission from Elsevier.)

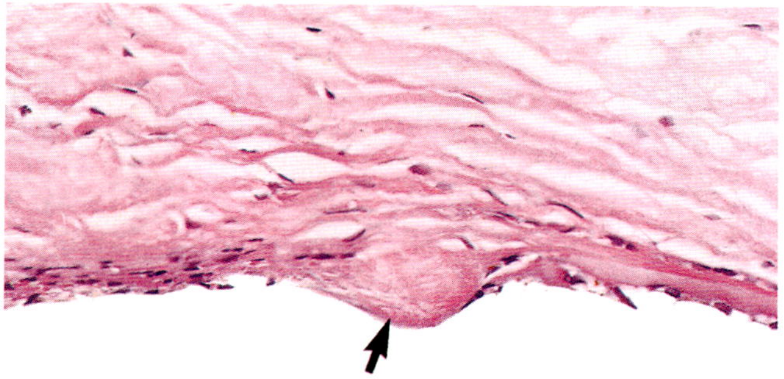

FIGURE 12.15 The Schwalbe ring is a significant clinical landmark in the limbal area and represents the peripheral termination of the Descemet membrane. Prominent Schwalbe rings (*arrow*) are identified histologically in about 15% of eyes (H&E).

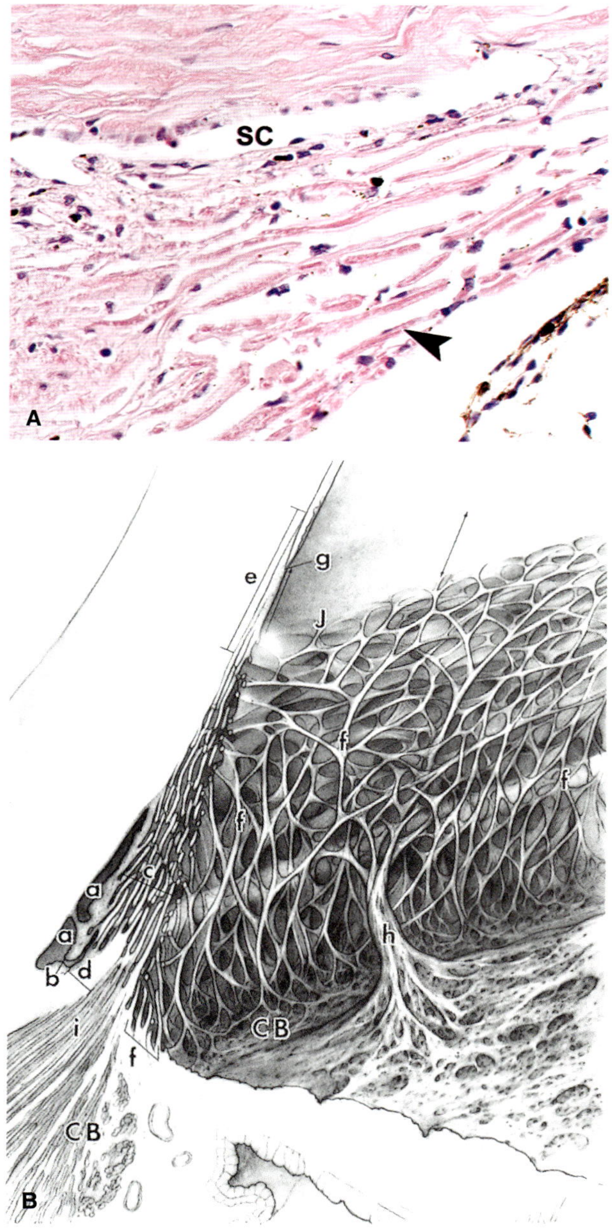

FIGURE 12.16 **A:** Located at the angle of the anterior chamber are the Schlemm canal (*SC*) and the trabecular meshwork (*arrowhead*). The Schlemm canal is an endothelial channel that drains the aqueous from the eye. The aqueous reaches the Schlemm canal after percolating through the connective tissue strands of the trabecular meshwork (H&E). **B:** This drawing depicts structures within and near the angle of the anterior chamber. In this illustration, the Schlemm canal (*a*) has two channels, one in communication with a small collecting channel (*b*). The collecting channel is intimately associated with the limbal part of the trabecular meshwork (*c*). The scleral spur (*d*) is closely associated with the trabecular meshwork. The Descemet membrane terminates peripherally in the area denoted "*e*" and "*g*." Some components (*f*) of the trabecular meshwork arise at the ciliary body (*CB*). Isolated strands of meshwork merge with a nearby process (*h*) from the anterior surface of the iris. A ciliary body muscle (*i*) attaches to the trabecular meshwork. The corneal endothelium merges with endothelial cells of the meshwork (*j*). (Reproduced with permission from Hogan MJ, Alvarado JA, Weddell JE. *Histology of the Human Eye*. WB Saunders; 1971, with permission from Elsevier.)

process of being transported from the trabecular meshwork into the lumen of the Schlemm canal (45). Once in the Schlemm canal, the aqueous drains into the episcleral venous plexus through numerous small collector channels. Prolonged obstruction to the outflow of the aqueous results in increased intraocular pressure and glaucoma.

CONJUNCTIVA, CARUNCLE, AND PLICA SEMILUNARIS

The conjunctiva is a thin continuous mucous membrane lining the inner surface of the eyelids and much of the anterior surface of the eye (46). In addition to its protective function, the conjunctiva allows the eyelids to move smoothly over the globe. The conjunctival epithelium is composed of two to five layers of cells and rests upon a basal lamina. The epithelium of the eyelid conjunctiva is composed of stratified columnar cells, while those of the conjunctiva covering the eye are typically cuboidal. Within the conjunctival epithelium are goblet cells that secrete mucoid material that becomes incorporated into the tear film (Fig. 12.17). Melanocytes are present in the basal epithelial layers and, like melanocytes in the skin, they transfer melanosomes into the adjacent epithelial cells. Pigmented epithelial cells are numerous in dark-skinned individuals (Fig. 12.18), giving rise to clinically apparent complexion-associated conjunctival melanosis. Conjunctival epithelial cells are immunoreactive using antibodies to AE1, AE3, cytokeratin (CK) 7, CK14 (basal cells), CK19, and sometimes CK17 and CK18 (basal cells) (47). The epithelium is not immunoreactive to CK8, CK10, or CK20 antibodies (47). The conjunctival stroma is composed of loose, fibrovascular subepithelial connective tissue that contains nerve cells, melanocytes, and accessory lacrimal glands. Lymphoid follicles with germinal centers reside in the conjunctiva (Fig. 12.19), particularly in areas where the conjunctiva lining the inner surface of the eyelid merges with the portion covering the eyeball at the superior and inferior fornices; scattered lymphocytes are not unusual within the conjunctiva and their presence does

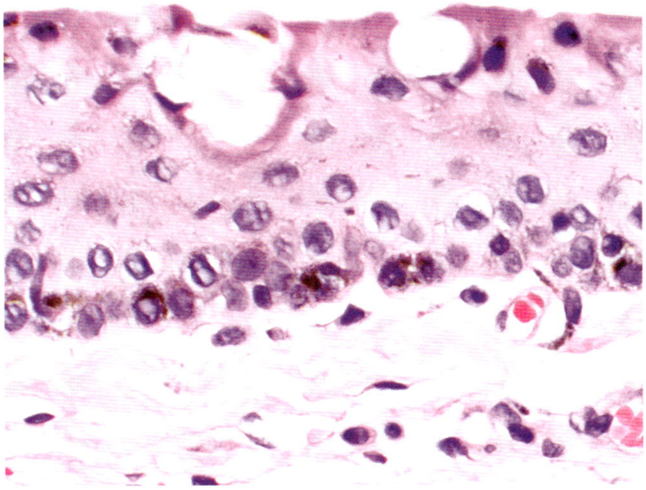

FIGURE 12.18 In dark-skinned individuals, the basal layers of the conjunctival epithelium contain intracellular melanin granules (H&E).

not indicate chronic conjunctivitis unless both plasma cells and significant numbers of lymphocytes are present. Three distinct areas of the conjunctiva are recognized (Fig. 12.20): the palpebral (eyelid) conjunctiva, the bulbar conjunctiva, and the conjunctiva lining the fornices.

The morphologic attributes of the conjunctiva vary in different parts of this tissue. Although goblet cells exist throughout the conjunctival epithelium, the numbers vary widely and the location must be considered when deciding whether the number of goblet cells is normal or abnormal. The goblet cell count is highest in the lower eyelid and lowest at the limbus (48). The conjunctival stroma is thickest in the fornices and bulbar areas and thinnest in the palpebral conjunctiva and at the corneoscleral limbus, where small conjunctival papillae, the palisades of Vogt, are evident (49,50) and serve as a site for putative limbal stem cells (51). The palpebral conjunctiva is firmly attached to the inner surface of the eyelids, but the bulbar conjunctiva is loosely adherent to the underlying sclera by thin connective tissue strands.

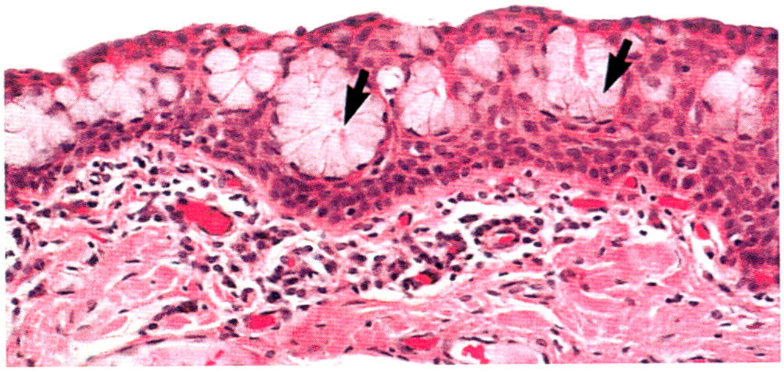

FIGURE 12.17 Goblet cells (*arrows*) are prominent in this section of conjunctival epithelium. Scattered mononuclear cells are often present in apparently healthy individuals in the underlying conjunctival stroma (H&E).

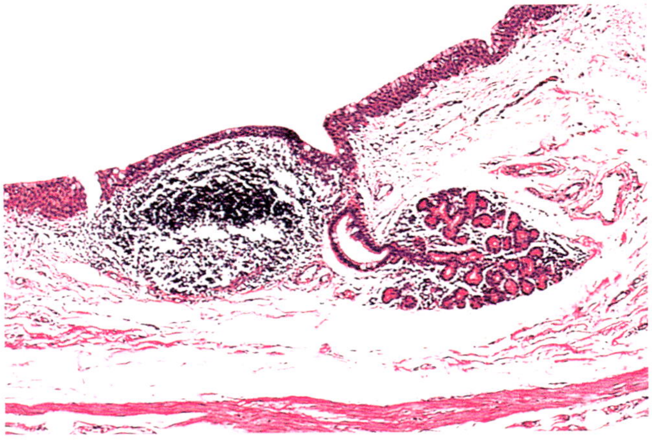

FIGURE 12.19 A small lymphoid follicle and an island of accessory lacrimal tissue are present in the stroma of the palpebral conjunctiva (H&E).

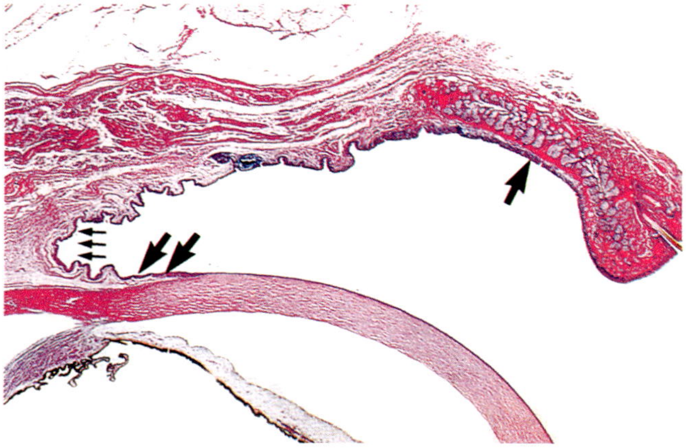

FIGURE 12.20 The conjunctiva can be divided into three parts. The palpebral conjunctiva (*arrow*) lines the posterior surface of the eyelid. The bulbar conjunctiva (*double arrows*) extends from the limbus over the anterior sclera. The bulbar and palpebral conjunctivae converge upon the conjunctiva of the superior and inferior fornices (*triple arrows*) (H&E).

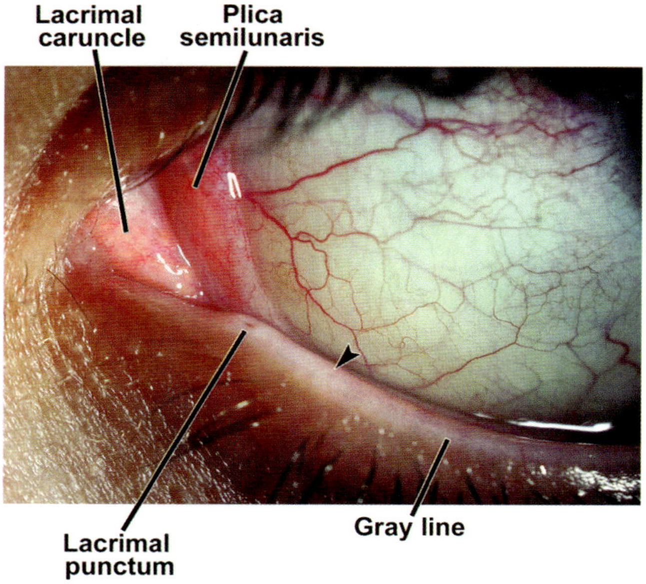

FIGURE 12.21 The caruncle and the plica semilunaris (semilunar fold) are specialized portions of the conjunctiva located in the eye's medial interpalpebral angle. Before tears enter the lacrimal drainage apparatus through the lacrimal punctum, they accumulate at the medial canthus (lacrimal lake). The demarcation between the conjunctival and cutaneous portions of the eyelid is discernible clinically at the so-called "gray line." The secretions of the meibomian glands (tarsal glands) reach the surface of the eyelids at small orifices, one of which is marked with an *arrowhead*. (Courtesy of Paola Torres, COT.)

The palpebral conjunctiva, which lines the posterior surface of the eyelids, extends from the fornices to the mucocutaneous junction at the eyelid margins, where the epithelium of the conjunctiva merges abruptly with the epidermis of the anterior surface of the eyelids. The palpebral conjunctiva contains several infoldings of epithelium termed the crypts of Henle (52). Islands of accessory lacrimal glands, morphologically identical to the main tear-producing gland within the orbit, are found within the palpebral conjunctiva. The subconjunctival tissue of the upper fornix may contain over 40 such glands, but fewer than 10 accessory lacrimal glands are present in the lower fornix (glands of Krause). The upper eyelids have approximately two to five accessory lacrimal glands (glands of Wolfring) located at the superior aspect of the tarsus. The bulbar conjunctiva begins at the limbus, at which point the corneal epithelium gradually becomes replaced by conjunctival epithelium, which continues over the sclera to the superior and inferior fornices. The conjunctiva is thrown into small folds in the fornices before becoming the palpebral conjunctiva.

Both the caruncle and the plica semilunaris (semilunar fold) represent specialized segments of the conjunctiva (Fig. 12.21). The caruncle is the nodular mass of fleshy tissue located in the medial interpalpebral angle of the eye. Its surface is covered by stratified nonkeratinized squamous and goblet-cell–rich conjunctival epithelium. The subepithelial stroma of the caruncle contains hair follicles with associated sebaceous glands, smooth muscle, adipose connective tissue, and occasionally, accessory lacrimal glands and sweat glands (Fig. 12.22) (53,54). The plica semilunaris, an arc-shaped fold of conjunctiva located immediately lateral to the caruncle, is thought to be a vestigial remnant of the nictitating membrane of lower species. The histologic features of the plica semilunaris are similar to those in other areas of the conjunctiva, except that the epithelium contains abundant goblet cells and, occasionally, smooth muscle fibers within the stroma (8). The semilunar fold may also feature nodules of cartilage, though the frequency varies between races and ethnic groups (55).

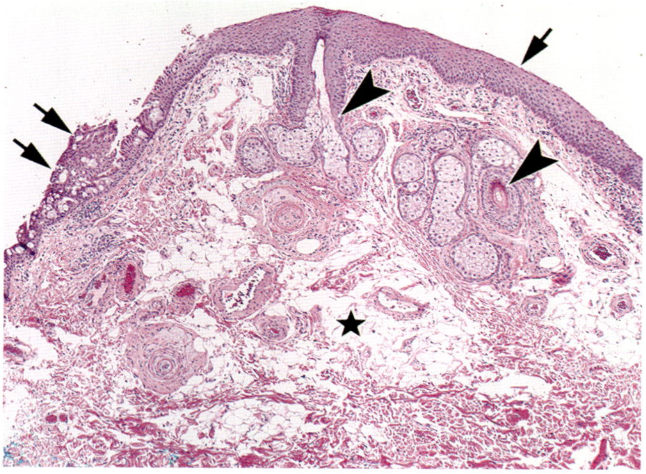

FIGURE 12.22 Stratified nonkeratinized squamous (*arrow*) and goblet-cell–rich conjunctival epithelium (*double arrows*) cover the caruncle's surface. The subepithelial stroma of the caruncle contains hair follicles with associated sebaceous glands (*arrowheads*), smooth muscle, adipose connective tissue (*five-pointed star*), and occasionally, accessory lacrimal glands and sweat glands, though not all of these structures may be apparent in routine sections from individual cases (H&E).

THE UVEAL TRACT

The uveal tract is located between the outer scleral covering and the inner retina. It begins anteriorly as the iris, extends to the ciliary body, and posteriorly to the choroid. The term *uvea* is derived from the Latin word *uva* (grape) because this portion of the eye was thought to somewhat resemble the dark color of a grape after the sclera and cornea are stripped from the globe.

The Iris

The iris is a thin diaphragm of tissue with a central opening, the pupil, which functions to regulate the amount of light reaching the retina (Fig. 12.23). Muscles within the iris dilate or constrict the pupil in response to sympathetic or parasympathetic nerve impulses. The diameter of the iris is approximately 21 mm, whereas the pupil's diameter ranges from 1 to 8 mm. The iris is thinnest at its point of attachment with the ciliary body peripherally, the iris root. Normally, the iris rests gently upon the crystalline lens and bulges slightly forward. Structurally and developmentally, the iris consists of two main parts: the stroma and the posterior epithelial lining.

Numerous ridges and depressions may be identified in tissue sections of the anterior iris stroma. These correspond to the contraction folds and furrows seen on clinical examination. The anterior surface of the iris lacks a cellular lining. The anterior border layer is a stromal modification composed of a discontinuous layer of fibroblasts and melanocytes (6,56). The stroma contains melanocytes, nerve cells, blood vessels, and smooth muscle in a loose connective tissue background. The color of the iris is unrelated to the number of melanocytes but reflects the size and possibly the number of melanin granules within the melanocytes (Fig. 12.24) (57–60). In addition to melanocytes,

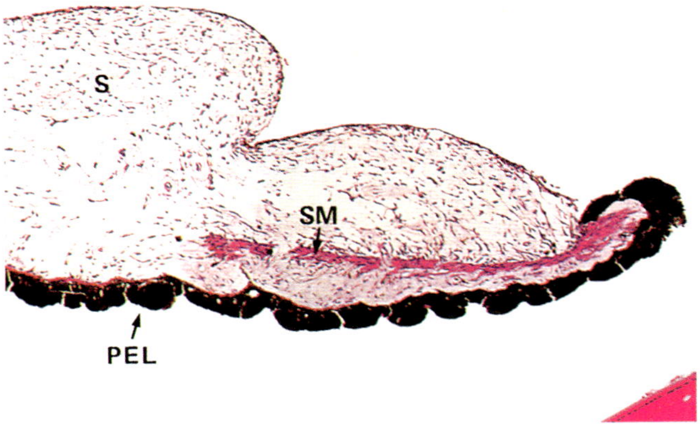

FIGURE 12.23 The iris is composed of stroma (*S*) and a posterior epithelial lining (*PEL*). The sphincter muscle (*SM*) of the iris is evident within the stroma. The pigmented posterior epithelial lining normally extends around the lip of the pupil anteriorly for a short distance (H&E).

melanosome-containing macrophages ("clump cells") are also scattered within the iris stroma, particularly in the vicinity of the sphincter muscle and the iris root (61). A thick collar of collagen fibers normally surrounds the blood vessels within the iris stroma and is especially prominent in adults. To the inexperienced observer, these normal vessels may appear to have arteriolosclerosis. In the pathologic process of iris neovascularization, thin-walled blood vessels, which lack a collagenous coat, cover the anterior surface of the iris.

The sphincter muscle (sphincter pupillae), a bundle of circularly arranged smooth muscle innervated by parasympathetic nerves, acts to constrict the pupil. It is located within the posterior stroma of the pupillary zone and is nearly 1 mm wide. Radially oriented smooth muscle fibers with scattered cytoplasmic melanosomes are also located within the posterior stroma of the iris (dilator pupillae). Innervated by sympathetic nerves, this muscle is active in pupil dilatation.

Posteriorly, the iris has two separate but closely apposed epithelial layers derived from the neuroectoderm (6). The cells of the anterior epithelial layer, which are in direct contact with the posterior aspect of the stroma, are continuous with smooth muscle fibers of the dilator pupillae. The posterior iris pigment epithelial layer is in direct contact with the aqueous of the posterior chamber. The cytoplasm of both epithelial layers contains numerous melanosomes (approximately 1 μm in diameter), larger than those of the iris stroma (diameter of about 0.5 μm). The number of melanosomes in the iris epithelial layers does not vary significantly between lightly and darkly pigmented individuals. In persons with ocular and oculocutaneous albinism, melanin synthesis is reduced or absent in the pigmented epithelia and the stromal melanocytes (62). The pigmented epithelia of the iris normally extend around the lip of the pupil, anteriorly, for a short distance. In certain pathologic conditions, fibrovascular tissue on the anterior surface of the iris everts the pupillary margin and pulls the pigmented epithelia onto the anterior surface of the iris. This displaced pigmented epithelium may be apparent clinically and is known as ectropion uveae.

The Ciliary Body

The middle segment of the uveal tract, the ciliary body, is located between the iris and the choroid. It is situated interior to the anterior sclera and is composed of two ring-shaped components: the pars plicata and the pars plana (Fig. 12.25). The anterior most aspect of the ciliary body, the pars plicata, begins at the scleral spur and contains approximately 70 sagittally oriented folds (approximately 2 mm long and 0.8 mm high). Continuous with these folds is the flat pars plana, which is approximately 4 mm wide and merges posteriorly with the serrated anterior border of the retina (ora serrata). Both portions of the ciliary body consist of epithelium, stroma, and smooth muscle.

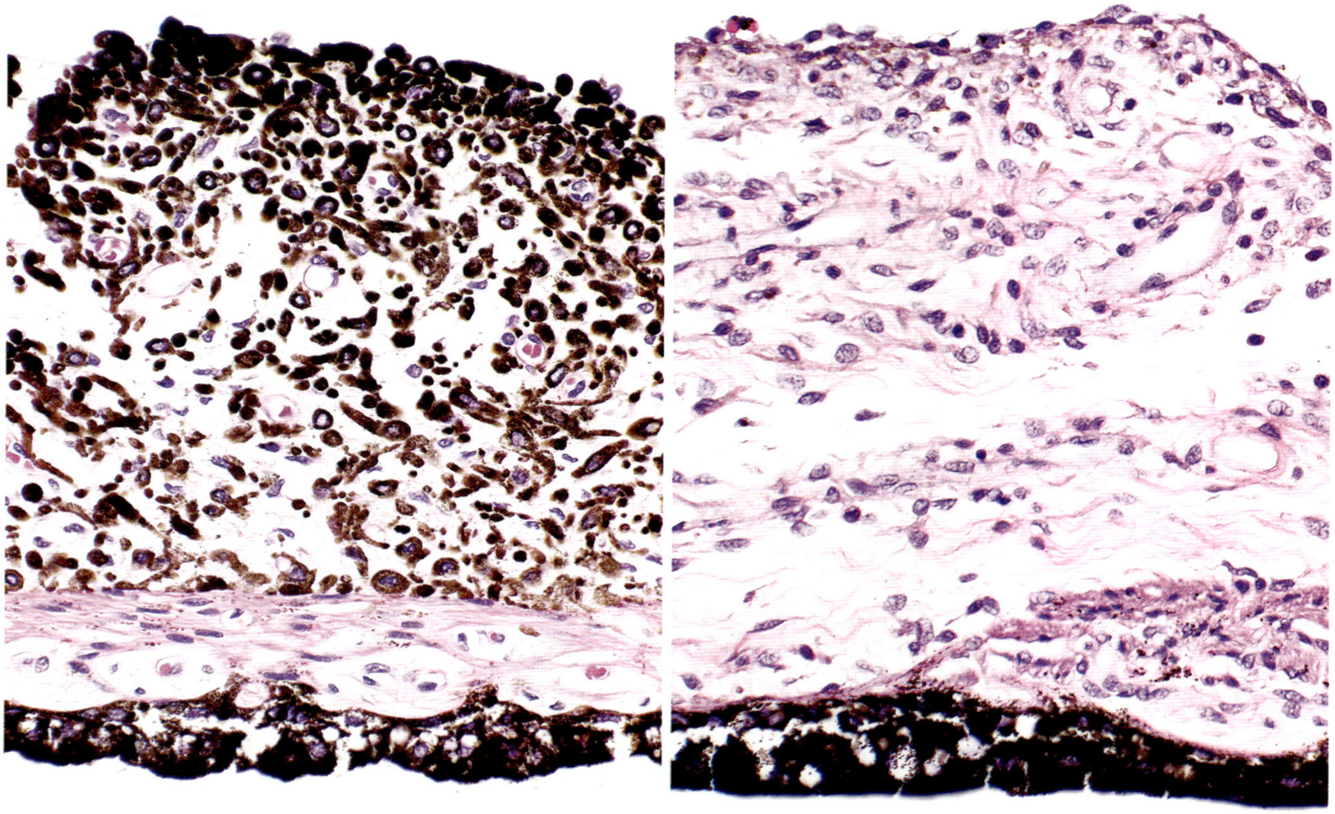

FIGURE 12.24 The color of the iris is unrelated to the number of melanocytes but reflects the size and possibly the number of melanin granules within the melanocytes (brown iris [**left**] and blue iris [**right**]). The amount of pigment in the posterior epithelial lining is similar in irises of different colors. In contrast to the posterior surface of the iris, the anterior iris lacks a cellular lining but has a mixture of fibroblasts and melanocytes (H&E).

FIGURE 12.25 The lens and the ciliary body are viewed from behind in this photograph. The ciliary body has two components: the pars plicata and the pars plana. The pars plicata contains about 70 sagittally oriented folds or ciliary processes (*arrow*). The pars plicata gradually merges with the flat pars plana (*arrowhead*).

The ciliary epithelium has two distinct layers that share a similar developmental derivation from neuroectoderm (Fig. 12.26). The inner epithelial layer is virtually nonpigmented and is contiguous with the aqueous of the posterior chamber. At the ora serrata, the neurosensory retina converges into the nonpigmented ciliary epithelial monolayer, which extends anteriorly until it abuts the posterior epithelial layer of the iris. In contrast, the outer ciliary epithelial layer is pigmented and unites with the retinal pigment epithelium at the ora serrata. The pigmented epithelium of the ciliary body overlies a PAS-positive basal lamina that is closely adherent to the adjacent stroma. The basal lamina of the pigmented epithelium can become conspicuously thickened in diabetes mellitus. Acellular fibers, known as zonules (Fig. 12.26), attach the crests of the nonpigmented ciliary epithelium in the pars plicata to the crystalline lens capsule.

The stroma of the ciliary body is composed of fibroblasts, blood vessels, nerve cells, and melanocytes, and is most abundant in the ciliary processes of the pars plicata and least abundant in the valleys between these processes and in the pars plana. During infancy, the stroma is

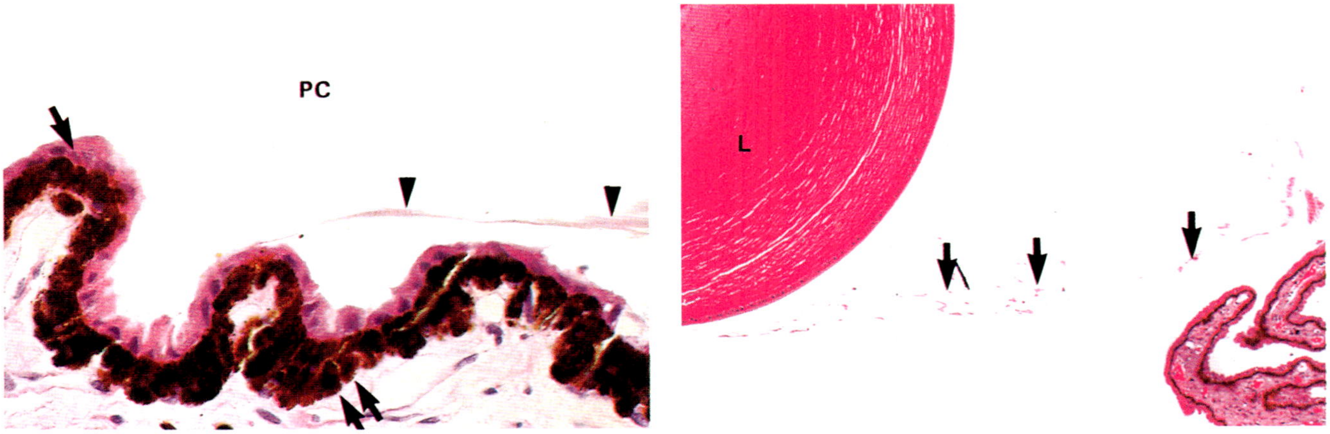

FIGURE 12.26 **Left:** The epithelium of the ciliary body has two distinct layers. The inner nonpigmented layer (*arrow*) is in direct contact with the aqueous of the posterior chamber (*PC*). The outer pigmented epithelial layer (*double arrows*) is adjacent to the underlying stroma. Acellular eosinophilic fibers (zonules) attach to the crests of the nonpigmented epithelium of the pars plicata (*arrowheads*). Zonules do not originate in the valleys between the ciliary processes (H&E). **Right:** Zonular fibers (*arrows*) span between the pars plicata of the ciliary body (on the *right*) and the lens (*L*) and hold the lens in place (H&E).

sparse (Fig. 12.27, *left*) but expands until adulthood. With advanced age, the ciliary body stroma becomes hyalinized (Fig. 12.27, *right*) and may calcify.

The smooth muscle of the ciliary body (Fig. 12.28) forms three distinct bundles. The outermost muscle runs in a longitudinal or meridional direction, whereas the middle layer contains radially oriented fascicles of cells, and the innermost muscle cells are aligned in a circular fashion. In routinely processed globes, histologic differentiation of these three muscular layers is difficult. Muscles of the ciliary body attach in large part to the scleral spur. The ciliary muscle assists in accommodation. As it contracts, the ciliary body extends forward, reducing pressure on the zonules and enabling the lens to become less concave, thereby increasing its refractive power.

Choroid

The richly vascularized choroid (Fig. 12.29) extends from the ciliary body to the optic nerve. Its inner aspect is firmly adherent to the retinal pigment epithelium. The outer surface of the choroid is loosely attached to the overlying sclera.

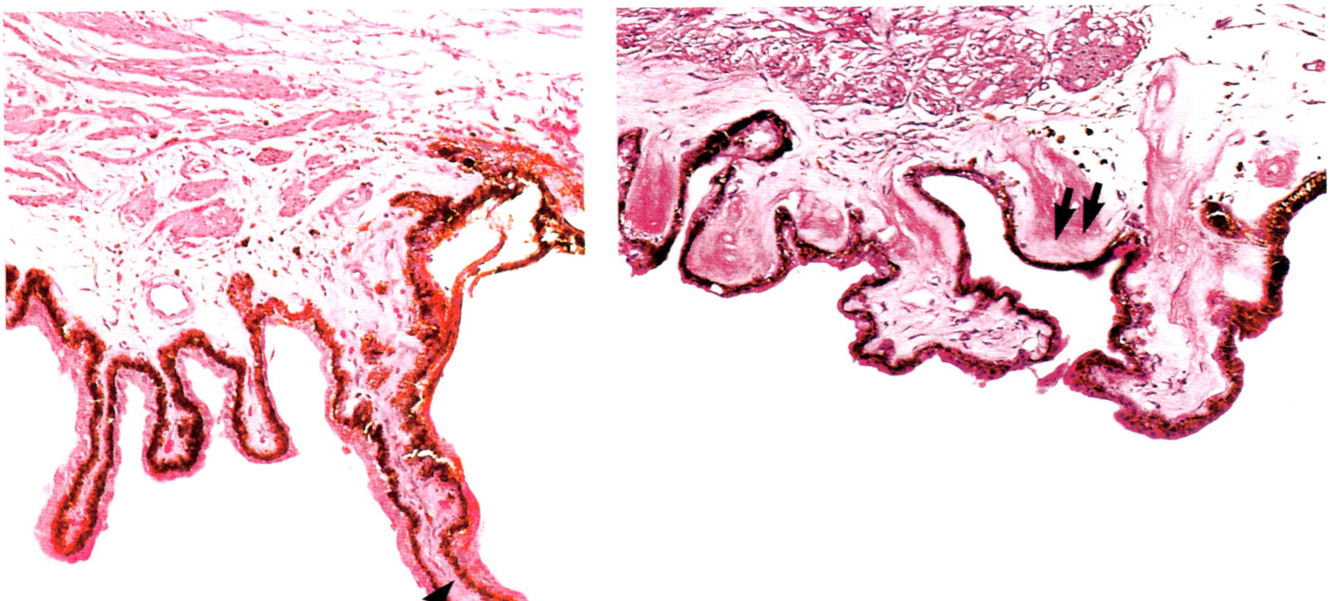

FIGURE 12.27 The pars plicata of the ciliary body changes with age. In infancy (*left*), the stroma of the ciliary processes is sparse (*arrow*). The stroma continues to expand until adulthood (*right*), and, with advancing age, the ciliary process stroma becomes hyalinized (*double arrows*) (*left and right*, H&E).

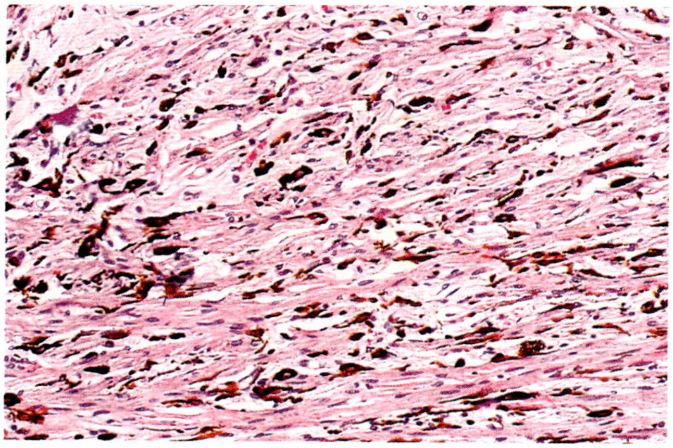

FIGURE 12.28 Smooth muscle constitutes a large portion of the ciliary body. Pigmented melanocytes are often present between the smooth muscle bundles and are more evident in dark-skinned individuals (H&E).

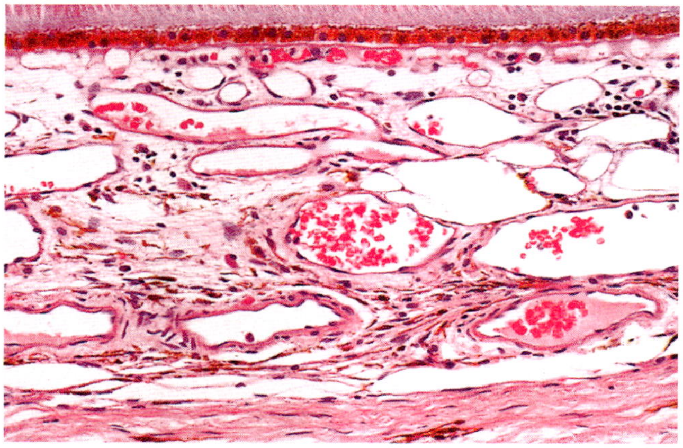

FIGURE 12.29 This photomicrograph illustrates the well-vascularized choroid. At the bottom of the figure, the choroid abuts the sclera. The single layer of retinal pigment epithelium is present at the top of the figure (H&E).

The Bruch membrane delineates the choroid from the overlying retinal pigment epithelium and is approximately 2 to 4 μm thick. Although the Bruch membrane appears as a thin eosinophilic layer in tissue sections, ultrastructural analysis has disclosed it to be composed of five distinct layers: the basal lamina of the overlying retinal pigment epithelium, a collagenous layer, an elastic fiber-rich component, another collagenous portion, and the basal lamina of the endothelial cells of the underlying capillary network (choriocapillaris). Located in the innermost choroidal stroma adjacent to the Bruch membrane, the choriocapillaris connects with arterial and venous channels from vessels in the outer choroidal stroma. Its function is to nourish the outer retinal layers. With age, the Bruch membrane thickens and commonly acquires focal excrescences known as drusen (Fig. 12.30, *left*). Both the drusen and the Bruch membrane may calcify. Another common, if not universal, aging change is the development of a layer of blood vessels between the Bruch membrane and

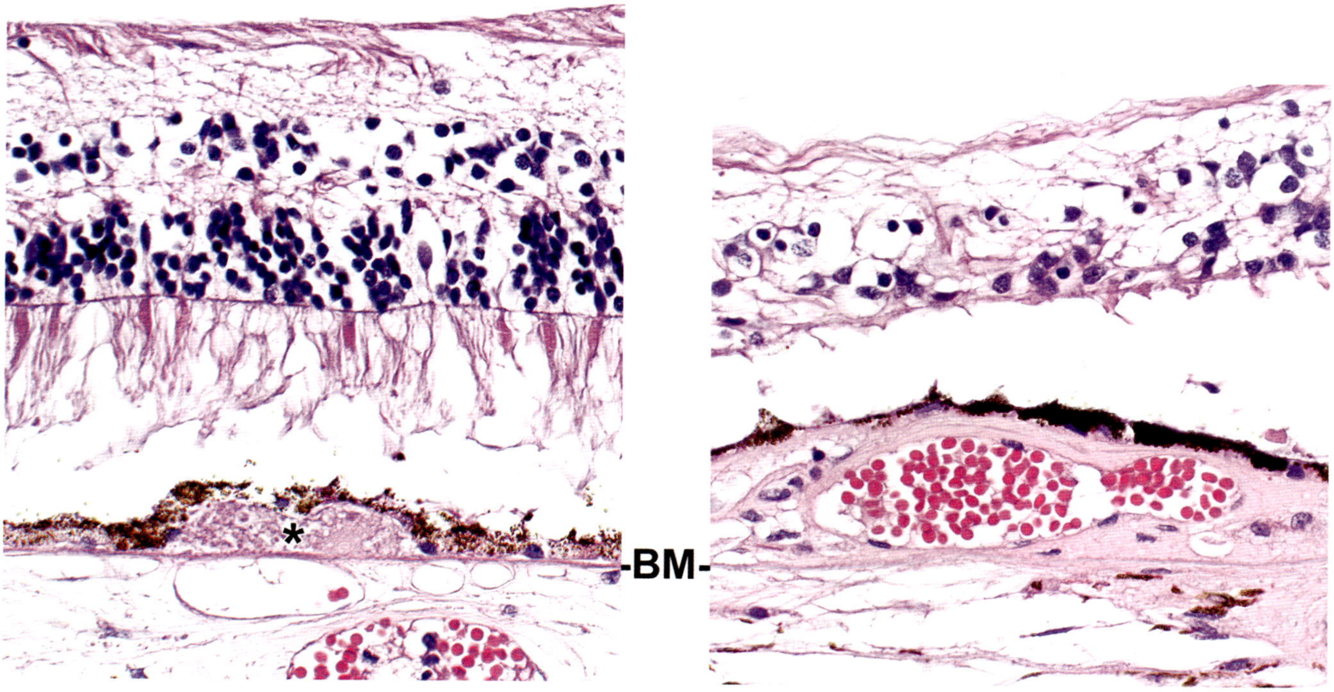

FIGURE 12.30 **Left:** An excrescence (*asterisk*) on the Bruch membrane (*BM*) bulges into the overlying retina. Such excrescences (drusen) are common as one ages and occasionally calcify (H&E). **Right:** Another very common age-related change in the eye is the development of a layer of blood vessels between the Bruch membrane and the retinal pigment epithelium (termed "choroidal neovascularization") in the peripheral choroid just posterior to the ora serrata (H&E).

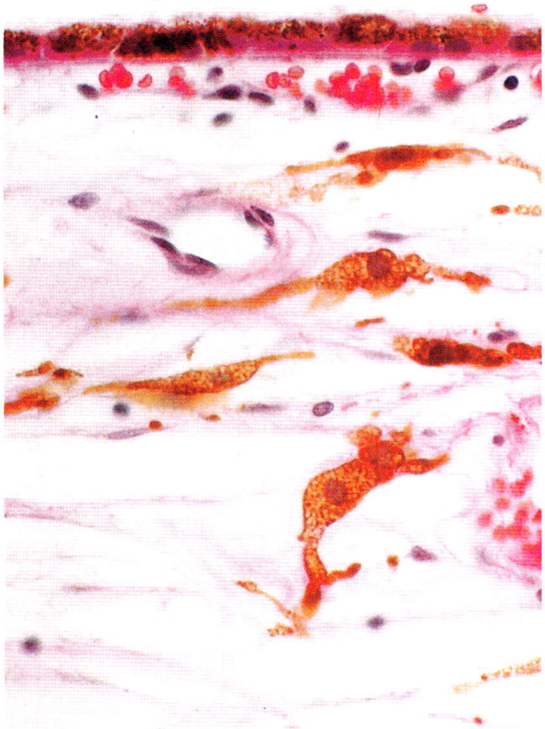

FIGURE 12.31 Numerous pigmented melanocytes are located within the choroidal stroma (H&E).

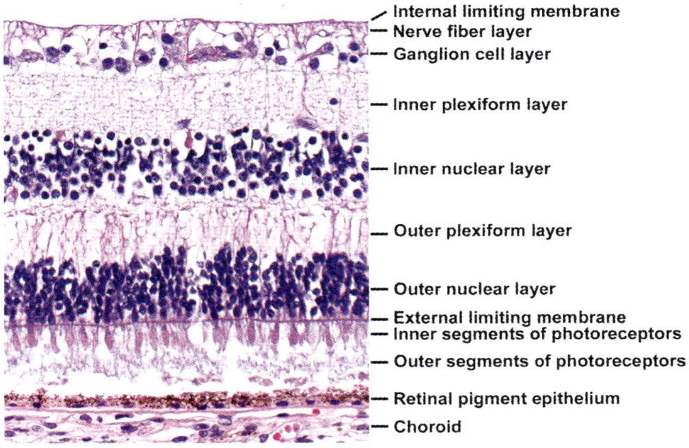

FIGURE 12.32 The cellular components of the retina are organized in well-defined layers. The choroid is directly below the retinal pigment epithelium in this figure. Specialized extensions of the photoreceptors, known as the outer and inner segments, are located immediately adjacent to the retinal pigment epithelium; many outer and inner segments are artifactually separated in this postmortem eye. Cell bodies of the photoreceptors are present in the outer nuclear layer; synapses between the bipolar cells, horizontal cells, and the photoreceptors occur in the outer plexiform layer; the inner nuclear layer embraces nuclei of the amacrine, bipolar, horizontal, and Müller cells; the inner plexiform layer contains axons and dendrites of amacrine, bipolar, and ganglion cells; ganglion cell bodies are located in the ganglion cell layer; the nerve fiber layer contains ganglion cell axons; and the internal limiting membrane separates the neurosensory retina from the vitreous.

the retinal pigment epithelium in the peripheral choroid just posterior to the ora serrata (Fig. 12.30, *right*) (63,64). This layer of neovascularization is usually more prominent temporally than nasally in horizontal sections through the eye (65).

The choroidal stroma is thinnest anteriorly, near the ciliary body, where it is approximately 0.1 mm thick. Posteriorly, the choroidal stroma thickens to nearly 0.22 mm at the optic nerve. The tenuous connection between the choroidal stroma and the sclera is responsible for the pathologic and artifactual separations often seen between these two layers in histologic sections. The stroma contains abundant pigmented melanocytes (Fig. 12.31), which are more numerous in individuals with dark choroidal pigmentation and may contain larger and more numerous pigment granules (66). Collagen fibers, some smooth muscle, neurons of the autonomic nervous system, mast cells (67), lymphocytes and monocytes/macrophages (18,68), antigen-presenting cells (69), and a prominent vascular system are also present. Large- and medium-sized arteries (branches of the posterior ciliary arteries) and veins (vortex veins) are situated in the outermost choroid.

RETINA

The cellular components of the retina include the photoreceptors (rods and cones), a variety of different neurons (ganglion, bipolar, horizontal, and amacrine cells), and neuroglial cells (Müller cells and astrocytes). Many of these types of cells can only be detected with specific staining techniques. The constituents of the retina are stratified into distinct layers (Fig. 12.32). The rods and cones comprise the outermost part of the neurosensory retina and are closely apposed to the retinal pigment epithelium. The anterior boundary of the retina has a serrated edge (ora serrata; Fig. 12.33), which is approximately 0.1 mm thick. Here, the retina converges into a single layer of nonpigmented epithelium, which continues anteriorly to merge with the nonpigmented

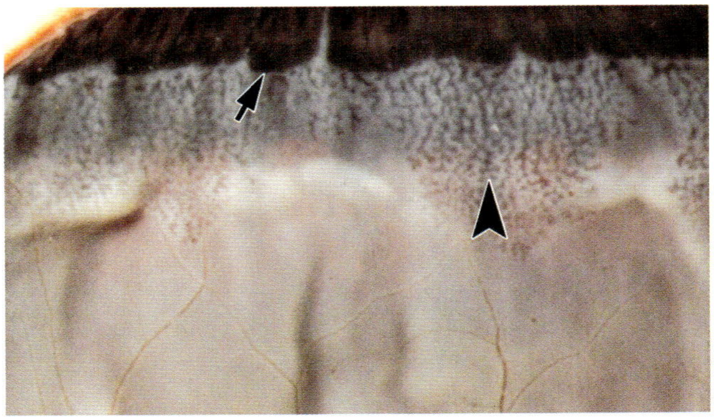

FIGURE 12.33 The ora serrata marks the anterior boundary of the retina. Peripheral cystoid degeneration is an almost invariable finding in the retina of all human eyes after the age of 20. Macroscopically, the peripheral retina immediately behind the ora serrata (*arrow*) has a vacuolated appearance (*arrowhead*).

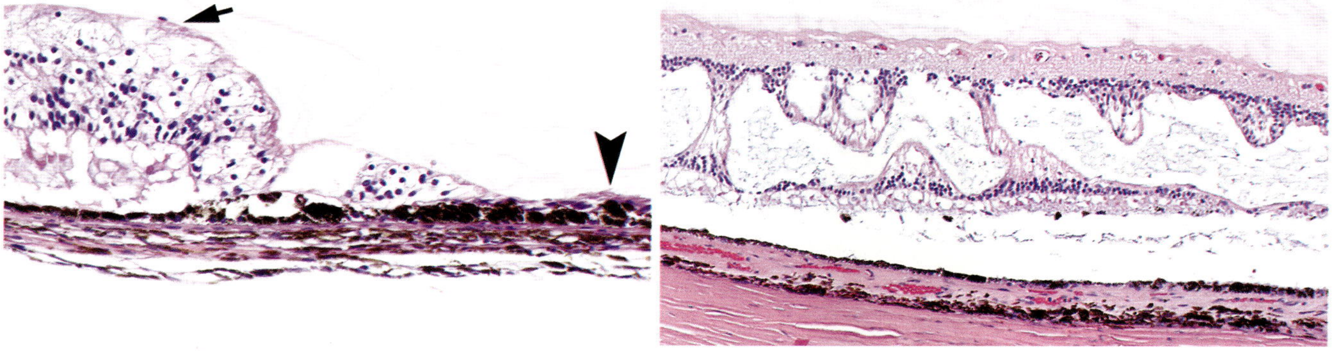

FIGURE 12.34 **Left:** At the ora serrata, the multilayered retina (*arrow*) converges with the single layer of nonpigmented epithelium of the ciliary body (*arrowhead*). The retina exhibits ischemic atrophy, a common age-related change in the peripheral retina (H&E). **Right:** Numerous cyst-like spaces within the retina characterize peripheral cystoid degeneration microscopically (H&E).

epithelium of the ciliary body (Fig. 12.34, *left*). Cysts develop in the peripheral retina (peripheral cystoid degeneration) in virtually everyone over age 20 (Figs. 12.33 and 12.34, *right*) (70,71). Ischemic degeneration of the peripheral retina is also extremely common in the eyes of elderly individuals (Fig. 12.33, *left*) (72,73). Posteriorly, the retina extends to the optic nerve, where it is approximately 0.5 to 0.6 mm thick. The neurosensory retina is in direct contact with the vitreous and lies interior to the retinal pigment epithelium, which defines the outermost border of the retina.

The retinal pigment epithelium is a monolayer of cells. These epithelial cells contain numerous intracytoplasmic melanosomes; cellular processes envelop part of the overlying rods and cones, as shown by transmission electron microscopy. The phagocytic function of the retinal pigment epithelium assists in the turnover of the photoreceptor elements. Undigested products of phagolysosomes culminate in the progressively increasing number of lipofuscin granules that accumulate within the retinal pigment epithelium with time. In the eyes of older individuals, patches of basal linear deposit that are finely granular and eosinophilic in H&E-stained sections and stain for collagen using the picro-Mallory or other histochemical methods accumulate between the retinal pigment epithelial cells and Bruch membrane (74). To be considered a normal aging change and not indicative of age-related macular degeneration, the patches of basal liner deposit should not extend beyond eight contiguous retinal pigment epithelial cells (74). Another very common age-related change involving the retinal pigment epithelium is peripapillary chorioretinal atrophy, characterized by retinal pigment epithelial basal lamina that is narrowed, duplicated, or thickened by basal laminar deposit, thickened inner Bruch membrane, atrophic and degenerating retinal pigment epithelium, and photoreceptor degeneration at the edges of the optic nerve (Fig. 12.35) (75).

Some photoreceptors are cylindrical in appearance (rods), whereas others are conical and somewhat longer and thicker (cones). Internal to the photoreceptors is the outer plexiform layer, formed from cell processes of the horizontal and bipolar cells and axonal extensions of the rods and cones. The inner nuclear layer embraces the nuclei of several cell types (the bipolar, Müller, horizontal, and amacrine cells). Constituents of the inner plexiform layer include bipolar and amacrine cell axons and dendrites of the ganglion cells. Near the vitreal aspect of the retina is the ganglion cell layer, composed predominantly of ganglion cell bodies. The axons of these large neurons make up the nerve fiber layer; these processes are usually unmyelinated, but bundles of some nerve fibers are occasionally myelinated as an incidental developmental anomaly (76). In older individuals, basophilic PAS-positive intracellular rounded bodies (corpora amylacea), indistinguishable from similar structures in the brain, often accumulate in the nerve fiber layer of the

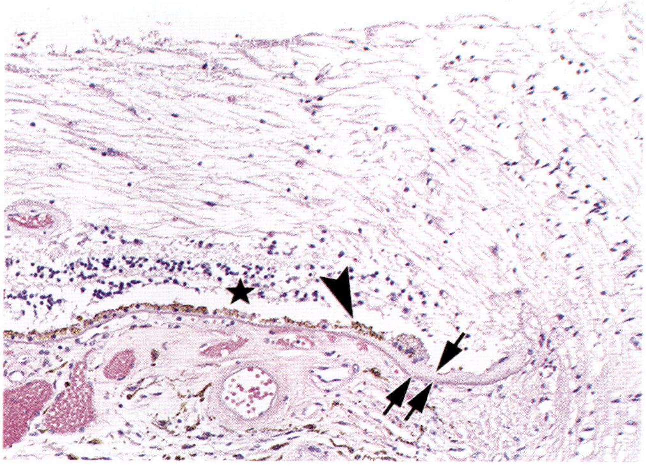

FIGURE 12.35 Peripapillary chorioretinal atrophy is a frequent age-related change characterized by retinal pigment epithelial basal lamina that is narrowed, duplicated, or thickened by basal laminar deposit (*arrow*), thickened inner Bruch membrane (*double arrows*), atrophic and degenerating retinal pigment epithelium (*arrowhead*), and photoreceptor degeneration (*five-pointed star*) at the edges of the optic nerve (H&E).

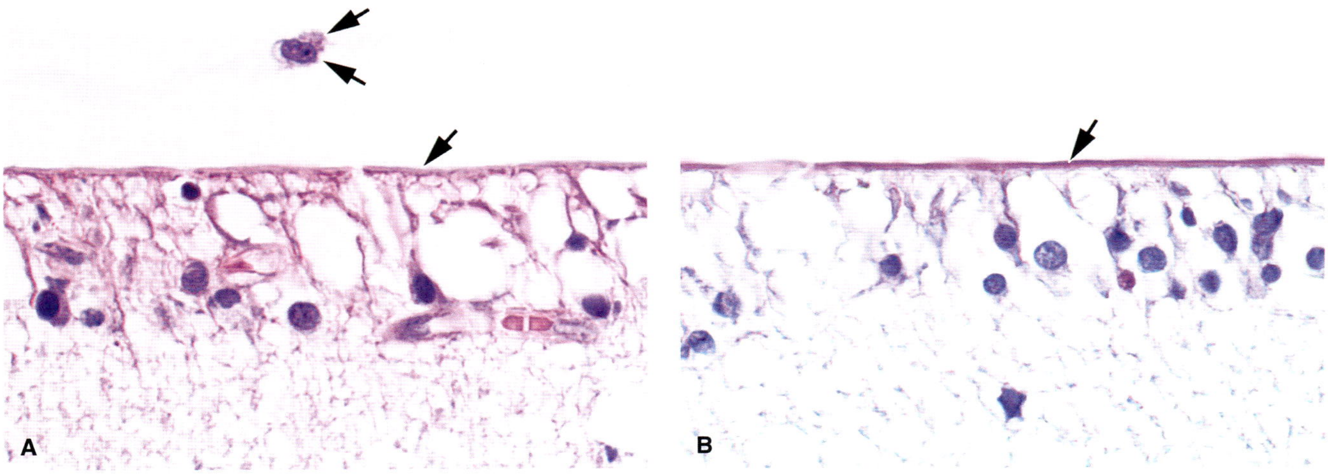

FIGURE 12.36 The retina's internal limiting membrane separates the neurosensory retina from the vitreous. In H&E-stained sections (*A*), the internal limiting membrane (*arrow*) is thin, acellular, and eosinophilic and is highlighted using periodic acid–Schiff stain (*B*). A hyalocyte (vitreous macrophage) is above the internal limiting membrane in *A* (*double arrows*).

retina near the optic disc. By light microscopy, two acellular zones can be distinguished within the retina: the external and internal limiting membranes. The so-called "external limiting membrane" is located between the photoreceptors and the outer nuclear layer. The membrane represents firm junctions between Müller cells and adjacent photoreceptors (zonula adherens). The basal lamina of the Müller cells accounts for the hyaline structure seen on light microscopy and is known as the "internal limiting membrane." The internal limiting membrane is a thin, acellular, eosinophilic membrane in H&E-stained sections (Fig. 12.36, *left*) and is highlighted using PAS (Fig. 12.36, *right*) (77,78). Simple epiretinal membranes, composed of a delicate layer of glial cells, are a common microscopic finding in enucleated eyes and often produce no symptoms or macroscopically detectable changes (79,80). The glial cells are immunoreactive using antibodies to glial fibrillary acidic protein (GFAP) (81) and connective tissue growth factor (82).

Similar to the neuroglial tissue of the brain, the neuronal cells of the retina show strong immunopositivity to synaptophysin (Fig. 12.37) and NeuN (*Neu*ronal *N*uclei) (Fig. 12.38) (83). The intensity of ganglion cell staining for synaptophysin is markedly influenced by the antibody and methods employed, and the ganglion cell staining is often much less than that of the inner and outer plexiform layers. Neurofilament protein highlights the axons of the nerve fiber

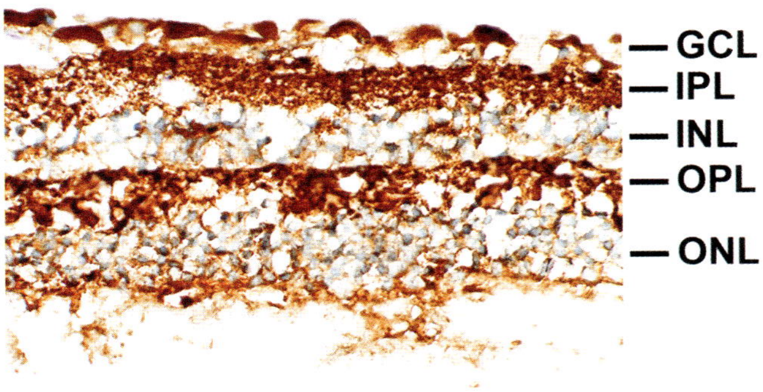

FIGURE 12.37 Synaptophysin immunopositivity is present within the ganglion cell layer (*GCL*), inner (*INL*) and outer (*ONL*) nuclear layers, and inner (*IPL*) and outer (*OPL*) plexiform layers.

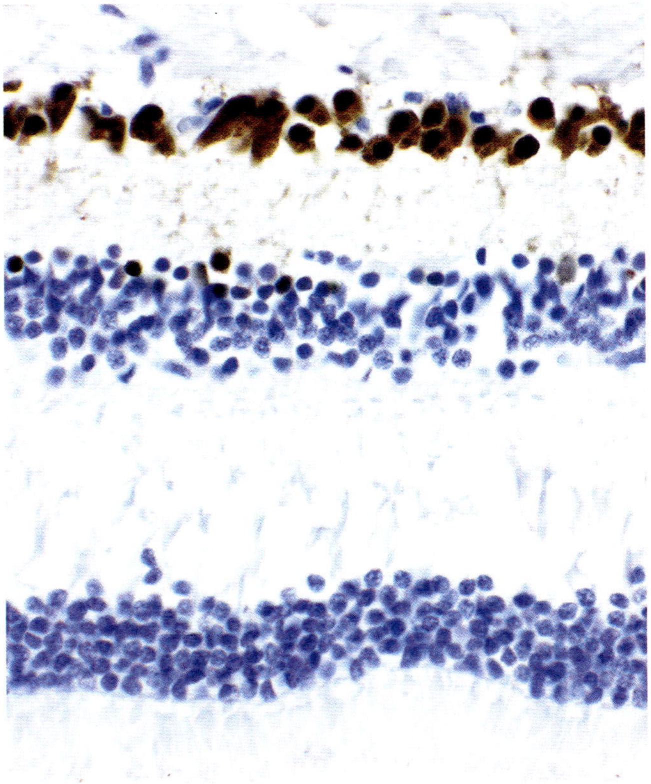

FIGURE 12.38 Reactivity with the immunohistochemical marker NeuN is restricted to neurons of the ganglion cell layer and a few cells in the inner nuclear layer.

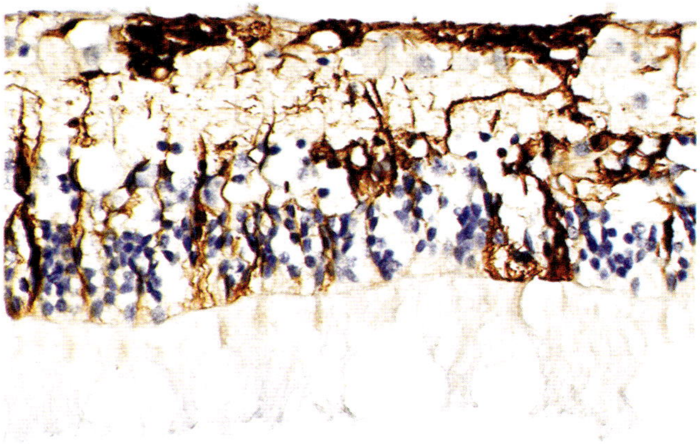

FIGURE 12.39 Antibodies to glial fibrillary acidic protein (GFAP) highlight the retinal glia and their processes.

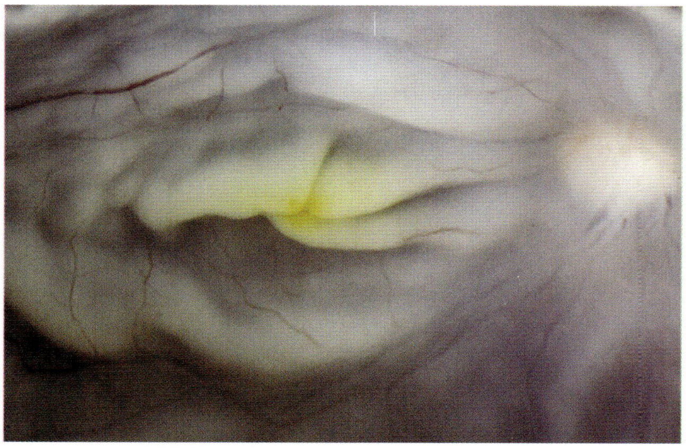

FIGURE 12.41 The macula in this eye from a 76-year-old man shows the macula lutea (*yellow spot*), where the bipolar and ganglion cells contain the pigment xanthophyll. The optic nerve is to the right of the macula.

layer as they continue posteriorly to enter the optic nerve. Glial cells and their processes react with GFAP (Fig. 12.39).

Light passes through the entire neurosensory retina before it is converted by the photoreceptor cells into electric impulses. The impulses are eventually transmitted to the visual cortex in the occipital lobe of the brain through a complex series of intercellular connections.

The retina varies in structure in different sites (Fig. 12.40). A yellow specialized portion of the retina is located in the posterior pole of the eye (in an area slightly temporal to the optic disc). This is the macula lutea (yellow spot), where the bipolar and ganglion cells contain the pigment xanthophyll (Fig. 12.41). The ganglion cells are several layers thick in the retina's macular region. The center of the macula contains a slightly depressed area (the fovea centralis) measuring almost 1.5 mm in diameter; it is responsible for most visual acuity. The walls of the fovea centralis are known as

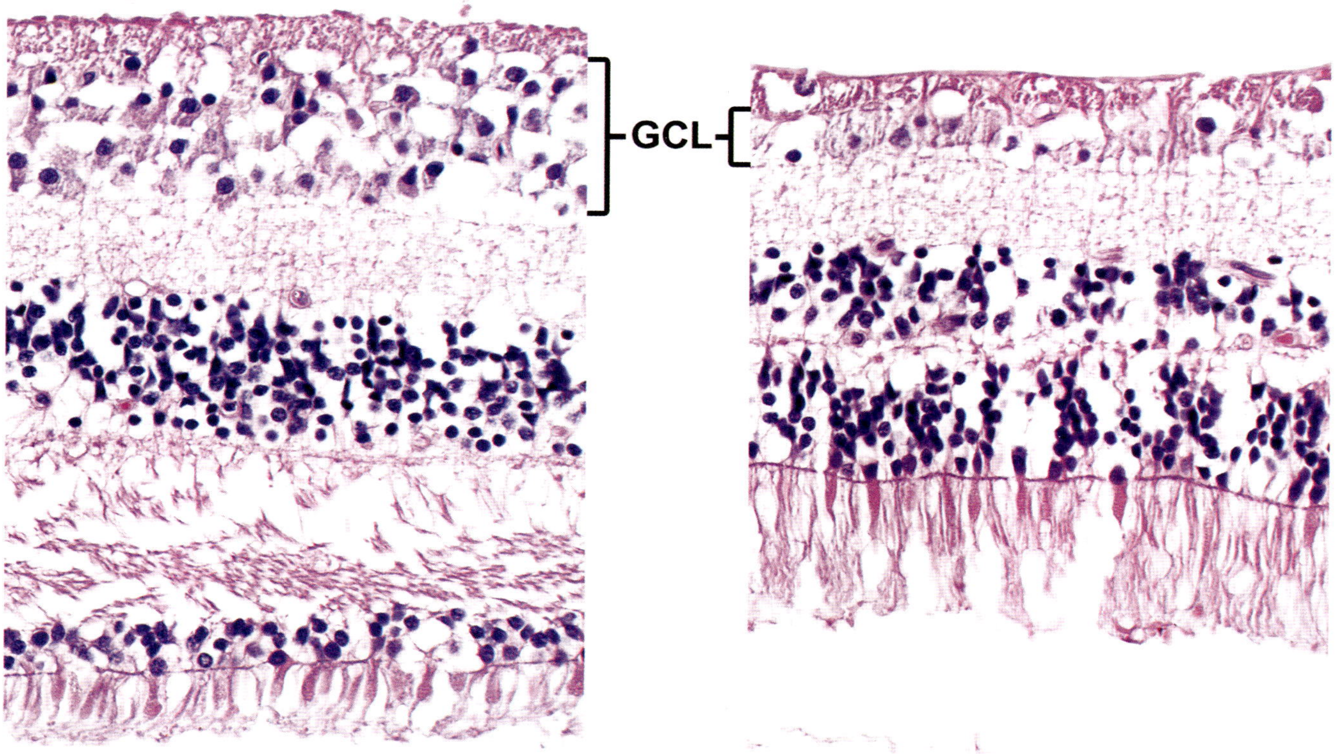

FIGURE 12.40 The retina has regional histologic variations. In the macular region (**left**), the neurosensory retina is thick, and ganglion cells (*GCL*) are multilayered. In areas outside of the macula (**right**), the neurosensory retina thins and ganglion cells (*GCL*) form a single layer and then become sparse as one approaches the ora serrata (**left** and **right**, H&E).

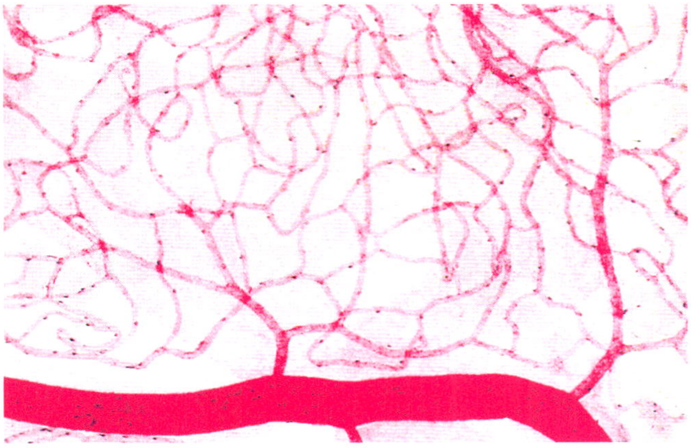

FIGURE 12.42 Following trypsin digestion, this flat preparation of a normal retina discloses retinal capillaries adjacent to a retinal arteriole (H&E).

the clivus, and the precise center is designated the foveola. Blood vessels are absent in the foveola, which measures approximately 0.4 mm in diameter. The inner layers of the retina are displaced peripherally in the foveola so that only the photoreceptors, the outer nuclear layer, and the outer plexiform layer are present. Cones are located within the foveola, but rods are absent.

The microvasculature of the normal retina is composed of branches of the central retinal artery and tributaries of the central retinal vein. It contains arterioles, venules, and intervening capillaries (Fig. 12.42). In capillaries from normal individuals, endothelial cells and pericytes are present in a ratio of approximately 1:1. The retinal microvasculature is affected in hypertension, diabetes mellitus, and other conditions. Capillary microaneurysms and the loss of capillary pericytes are characteristics of diabetic retinopathy. These are best visualized in flat preparations of the retina after trypsin digestion of the retinal cells (84).

Artifacts of the Retina

It is necessary to distinguish a true detachment of the neurosensory retina from the retinal pigment epithelium from an artifactitious retinal detachment in the same location. True detachments of the neurosensory retina are characterized by the presence of blood or eosinophilic proteinaceous fluid in the space between the two retinal layers (Fig. 12.43), rounded edges at the site of the retinal break, photoreceptor elements of one fold of retina adjacent to the internal limiting membrane of another fold (i.e., Zimmerman sign), presence of cyst-like spaces in the detached retina, and absence of photoreceptor outer segments, although the last change may not be present in cases of very acute detachment. Artifacts that simulate retinal detachments typically lack subretinal fluid rich in eosinophilic protein or blood, have squared-off edges at the site of the break with intact photoreceptor outer segments, and fragments of pigment epithelium cell debris are adherent to the photoreceptor outer segments (Fig. 12.44) (85).

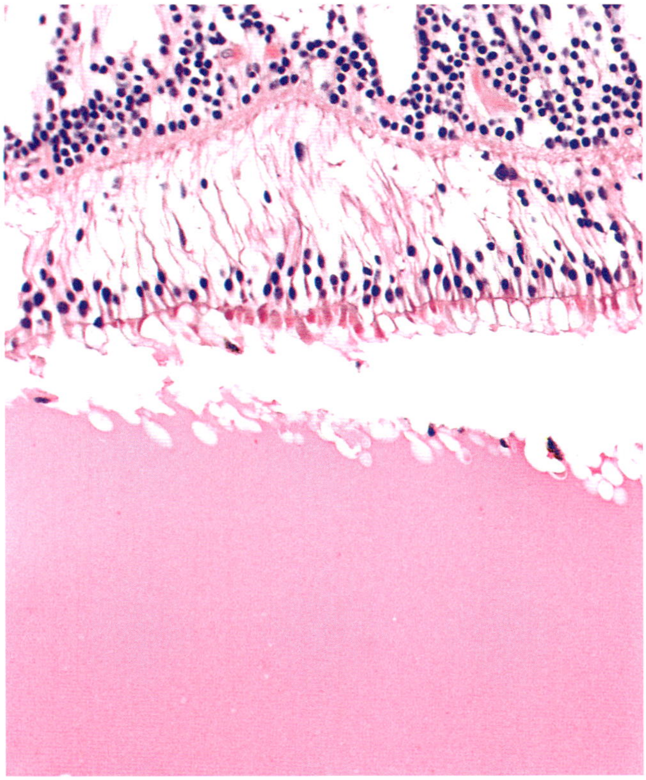

FIGURE 12.43 A feature of a true retinal detachment is the presence of eosinophilic proteinaceous fluid within the subretinal space (H&E).

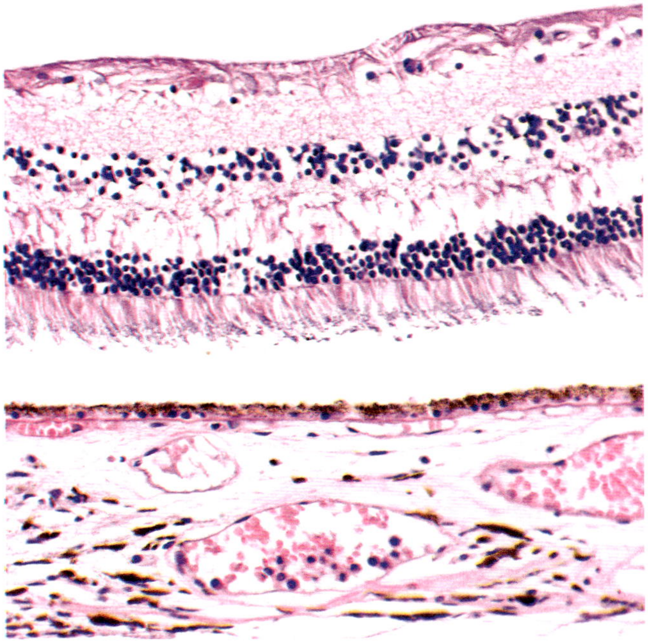

FIGURE 12.44 Artifactual retinal detachments feature retinal pigment epithelium granules within the tips of the photoreceptors and the absence of subretinal eosinophilic fluid (H&E).

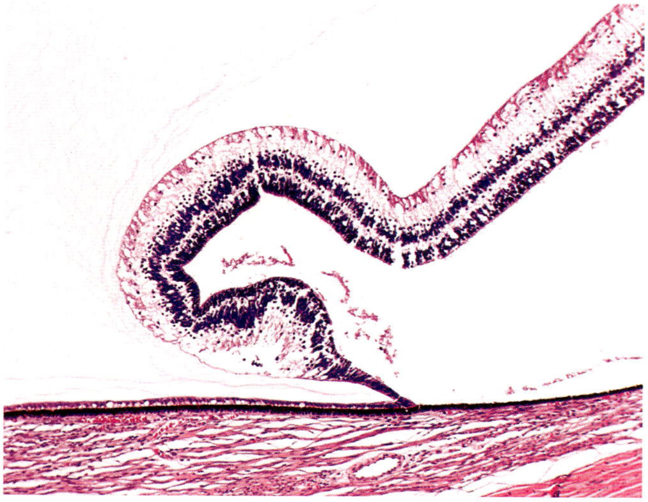

FIGURE 12.45 The Lange fold is a postmortem artifact usually seen in infant eyes. At the ora serrata, the peripheral retina typically has a bowed or concave appearance anteriorly. The absence of subretinal fluid distinguishes this from a true retinal detachment (H&E).

The neurosensory retina of neonates and children folds inwardly upon itself (i.e., the Lange fold) at the ora serrata when eyes are subjected to a fixative such as formalin (Fig. 12.45) (86,87). This artifact of fixation is not observed in the living eye or in unfixed enucleated eyes that have been sectioned to observe the peripheral retina. The Lange fold is thought to result from traction on the peripheral retina by a shortening of the vitreous base and posterior lens zonules caused by tissue fixation. The Lange fold is not observed after the age of 20 years, presumably because the peripheral retina becomes firmly bound to the subjacent retinal pigment epithelium at that point. The convexity of this artifact of fixation is directed anteriorly and axially in neonates, but the fold is initiated some distance from the ora serrata in older infants and children, apparently because of a propensity for peripheral retinal adhesions to the subjacent retinal pigment epithelium with increasing age. Subretinal fluid is not present between the layers of the neurosensory retina in Lange folds (87).

The pathologist should be aware of two other retinal artifacts. Fetal eyes are prone to retinal autolysis and edema resulting in retinal folds and fragmentation (88–90). Lipid crystals may result from prolonged formalin fixation and appear as white opacities on the retinal surface (Fig. 12.46) (91). Lipid crystals are common after several years of formalin fixation, but they are histologically detectable only using frozen sections where they stain strongly using the Baker chromate-hematein procedure or weakly with Sudan IV (91). The lipid crystals are found in all retina layers except the photoreceptor layer and are most abundant in the nerve fiber layer.

THE OPTIC NERVE

More than one million axons from the retinal nerve fiber layer converge at the optic nerve head, which accounts for the physiologic blind spot in the normal visual field and represents the beginning of the optic nerve. The central retinal artery and vein traverse the optic nerve and are surrounded by glial tissue within a slight depression at the origin of the nerve (Fig. 12.47). The axons extend approximately 1 mm from the optic nerve head to a sieve-like partition of connective tissue in the sclera termed the lamina cribrosa, through which nerve fibers pass on their way to the brain (92,93). Over one thousand nerve fiber bundles surrounded by astrocytes, oligodendroglia, and collagenous septa (Fig. 12.48) can be identified in cross sections of the optic nerve. Like the brain, the optic nerve is surrounded by pia mater, arachnoid, and dura. Small focal meningothelial proliferations occasionally form within the leptomeninges

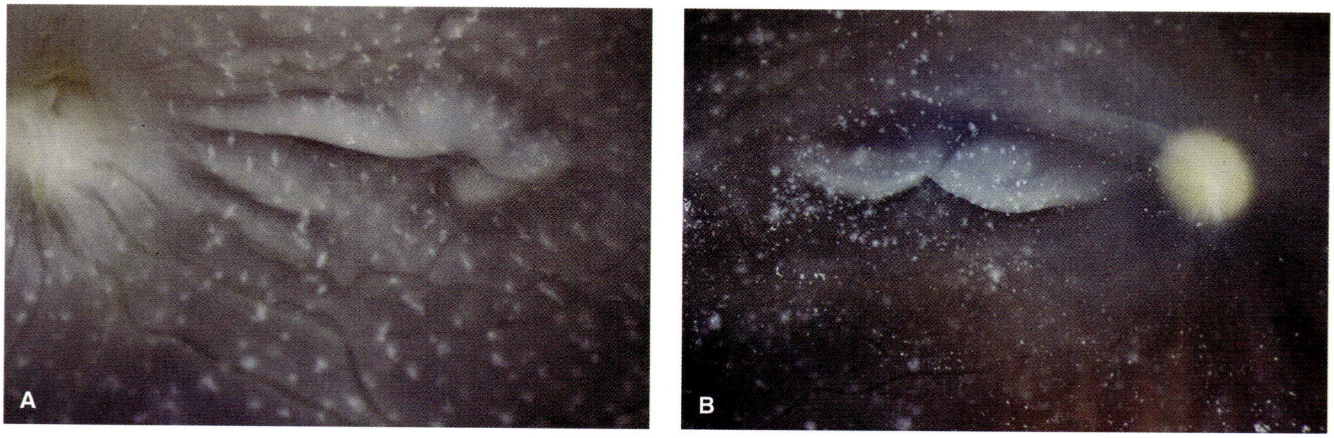

FIGURE 12.46 Lipid crystals resulting from prolonged formalin fixation appear as white opacities on the retina's surface. These photographs, prepared using a dissecting microscope, show lipid crystal opacities on the surface of the macula and surrounding retina in the right eye of a 56-year-old woman (*left*) and the left eye of a 63-year-old man (*right*).

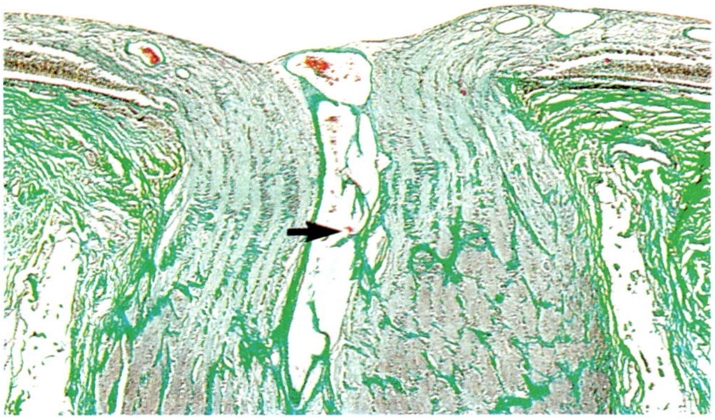

FIGURE 12.47 The optic nerve penetrates the sclera near the posterior pole of the eye. This histologic section contains the central retinal artery (*arrow*) in the central part of the optic nerve. The central retinal artery and the central retinal vein traverse the optic nerve until they exit the nerve about 8 to 15 mm posterior to the eyeball (Masson trichrome).

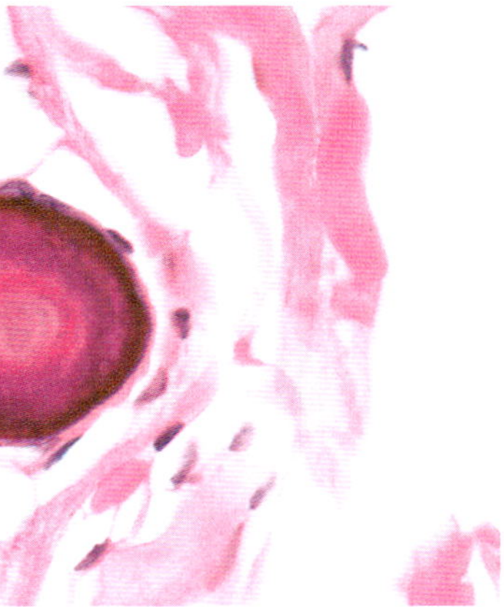

FIGURE 12.49 Laminated psammoma bodies, such as this one, are often closely associated with the meningothelial cells of the optic nerve (H&E).

surrounding the optic nerve, and these may give rise to some orbital meningiomas. Psammoma bodies sometimes occur in the arachnoid mater (Fig. 12.49). Pigmented melanocytes are sometimes encountered within the leptomeninges and the optic nerve head.

After leaving the globe, each optic nerve continues posteriorly through the orbit to its respective optic foramen, and then to the optic chiasm before terminating in the lateral geniculate bodies.

At the level of the lamina cribrosa, the axons within the optic nerve become myelinated by concentric membranous processes of the oligodendroglia. The rather abrupt transition between myelinated and nonmyelinated nerve fibers is readily appreciated in tissue sections stained with Luxol fast blue or other dyes with an affinity for myelin (Fig. 12.50). Myelination of the optic nerve proceeds from the brain toward the eye, is evident by 32-week gestation, and progresses rapidly following a full-term gestation through 2 years of age and less rapidly thereafter (94). As the axons acquire myelin coats, the diameter of the optic nerve doubles to nearly 3 mm. The central retinal artery and vein are located within the central core of the optic nerve, adjacent to the globe. Both of these vascular channels exit the nerve some 8 to 15 mm posterior to the lamina cribrosa; the channels are not evident within tissue sections of the optic nerve closer to the brain. The orbital portion of the optic nerve extends some 25 mm from the lamina cribrosa to the optic foramen at the apex of the orbit.

If the optic nerve becomes compressed during enucleation of the globe, some optic nerve tissue may extrude into

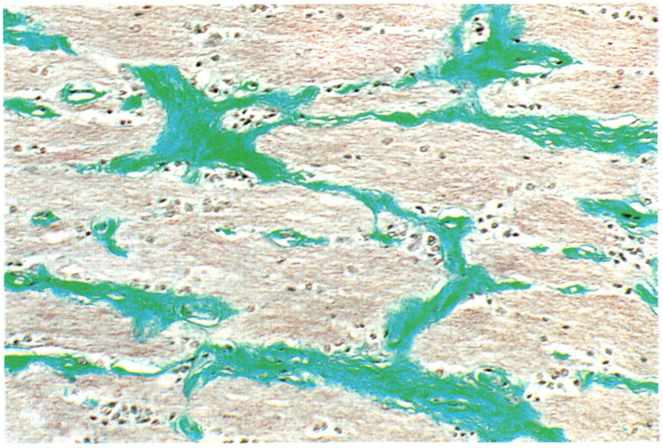

FIGURE 12.48 Nerve fiber bundles within the optic nerve are surrounded by thin collagenous septa (Masson trichrome).

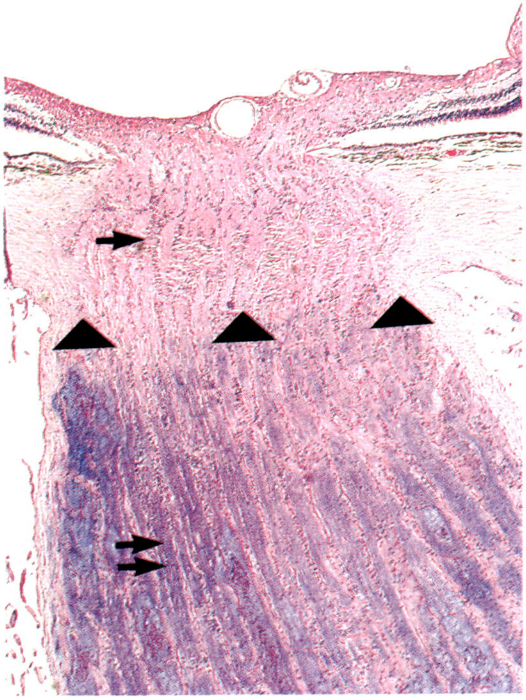

FIGURE 12.50 The abrupt transition between nonmyelinated (*arrow*) and myelinated (*double arrows*) nerve fibers of the normal optic nerve at the level of the lamina cribrosa (*arrowheads*) is dramatically illustrated in this tissue section stained with a dye having an affinity for myelin (Luxol fast blue).

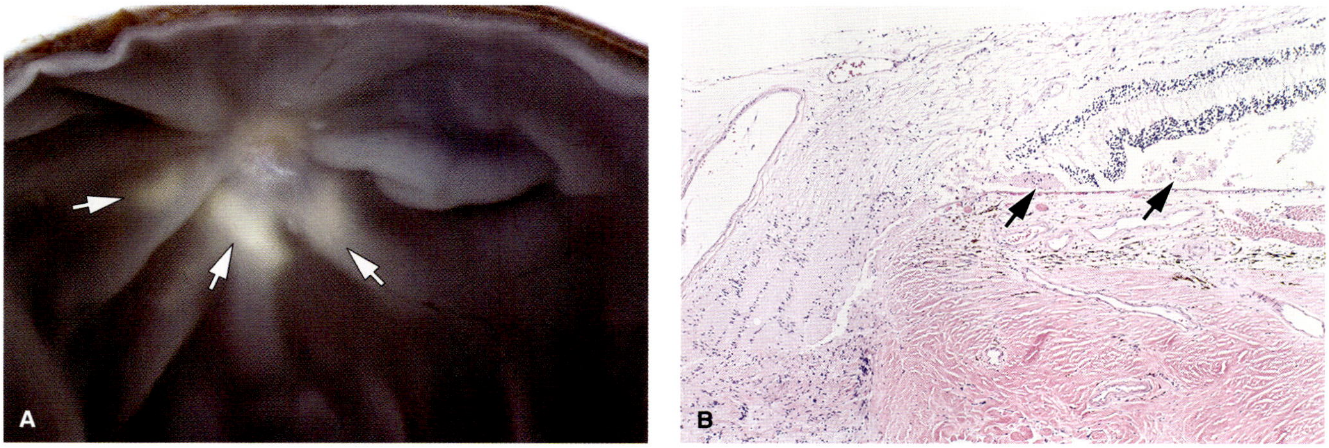

FIGURE 12.51 **A:** Subretinal myelin artifact appears as cream-colored nodules (*arrows*) beneath the neurosensory retina below the edge of the optic disc in the left eye of an 80-year-old woman (*left*). **B:** The myelin (*arrows*) is insinuated between the outer neurosensory retina and Bruch membrane (*right*, H&E). The retinal pigment epithelium is absent secondary to peripapillary chorioretinal atrophy.

the eye and become dislodged into the lumen of blood vessels near the optic discs, between the neurosensory retina and the retinal pigment epithelium (Fig. 12.51), and even into the vitreous (Fig. 12.52) (91,95,96). Neural tissue within the optic nerve may become displaced, in a manner comparable to the "toothpaste" artifact of the spinal cord that follows a traumatic insinuation of white matter into the gray matter. This artifact should not be mistaken for ectopic intraocular nervous tissue, tumors, giant drusen, vitreous worms, or subretinal exudates (96). With age, corpora amylacea, similar to those in the retina and brain, may become evident in the optic nerve.

The pathologist may occasionally encounter swollen axons within and immediately posterior to the lamina cribrosa that resemble ganglion cells (Fig. 12.53). With H&E staining, the swollen axons exhibit an eosinophilic and light purple fibrillar tincture distinct from the neuropil and fibrovascular septa of the optic nerve, the connective tissue of the lamina cribrosa, and corpora amylacea. They are often triangular-shaped with indentations and emanating processes. Though ganglion-like, they lack the characteristic histological features of ganglion cells elsewhere, such as nuclei, nucleoli, Nissl substance, and intracytoplasmic lipofuscin pigment. They are immunoreactive to neurofilament protein (Fig. 12.53) and synaptophysin antibodies and are immunonegative to NeuN, GFAP, and ß-amyloid precursor protein antibodies. The swollen axons occur in glaucoma-free, normotensive eyes (97) and eyes with glaucoma (98), and likely result from constriction of axoplasmic flow (97) that may progress to hydropic degeneration in glaucomatous eyes (98).

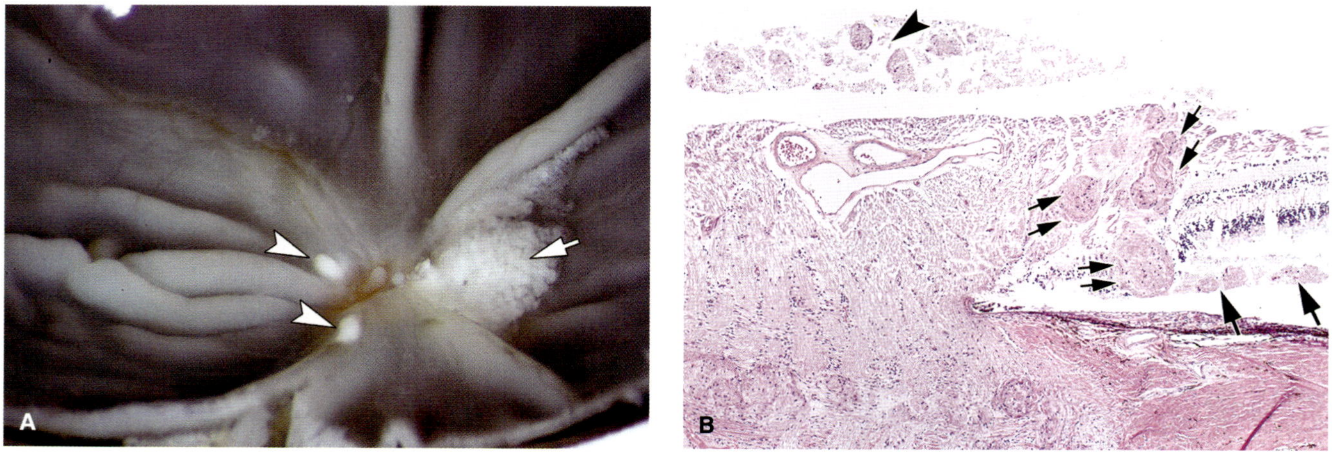

FIGURE 12.52 **A:** Intravitreal myelin artifact forms a mulberry-like mass on the retinal surface at the edge of the optic disc (*arrow*), while intraretinal/subretinal myelin artifact (*arrowheads*) manifests as two opaque smaller white nodules on the opposite side of the optic disc (*left*) in the left eye from a 28-year-old woman. **B:** Nodules of myelin are beneath (*arrows*), rupturing through the neurosensory retina (*double arrows*), and overlying the optic disc (*arrowhead*). The retinal pigment epithelium was artifactually displaced during sectioning (*right*; H&E).

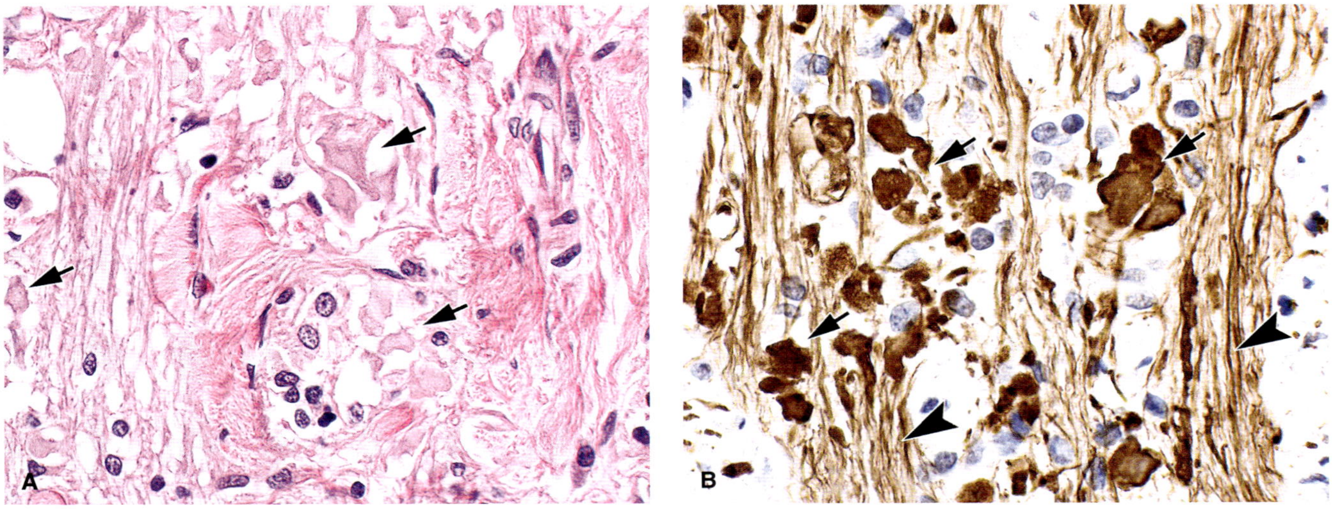

FIGURE 12.53 **A:** This section through the lamina cribrosa shows numerous swollen axons (*arrows*) with an eosinophilic and light purple fibrillar tincture distinct from the neuropil and fibrovascular septa of the optic nerve, the connective tissue of the lamina cribrosa, and corpora amylacea (*left*, H&E). **B:** The swollen (*arrows*) and normal (*arrowheads*) axons stain intensely using antibodies to neurofilament protein, which also highlights their irregular shapes with indentations and emanating processes (*right*).

THE CRYSTALLINE LENS

The biconvex ocular lens (Fig. 12.54) is located directly behind the pupil and in front of the anterior face of the vitreous. In the adult, it measures approximately 10 mm in diameter and 4 to 5 mm in width. The lens is held in place by zonules that connect it to the pars plicata of the ciliary body. The lens is encircled by a collagen- and carbohydrate-rich capsule, which serves as the attachment site of the zonules. The capsule over the anterior surface of the lens thickens with time. At 2 to 5 years of age, the anterior capsule varies from 8 μm wide at the pole to 15 μm wide at its maximum, and the capsular thickness increases to 14 μm at the pole and 21 μm at its maximum by 35 years of age (Fig. 12.55).

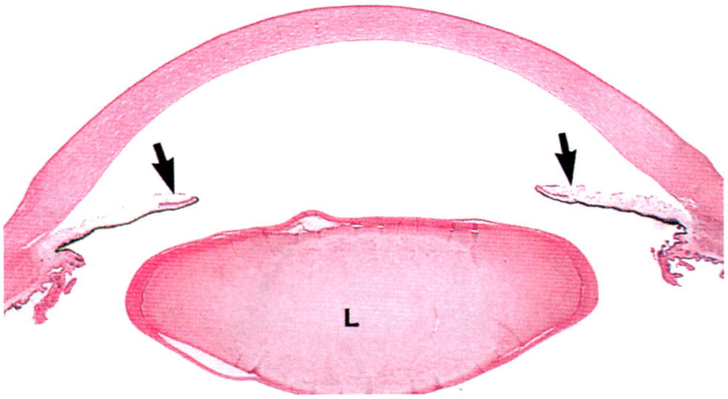

FIGURE 12.54 The crystalline lens (*L*) is situated just posterior to the pupil and iris (*arrows*) (H&E).

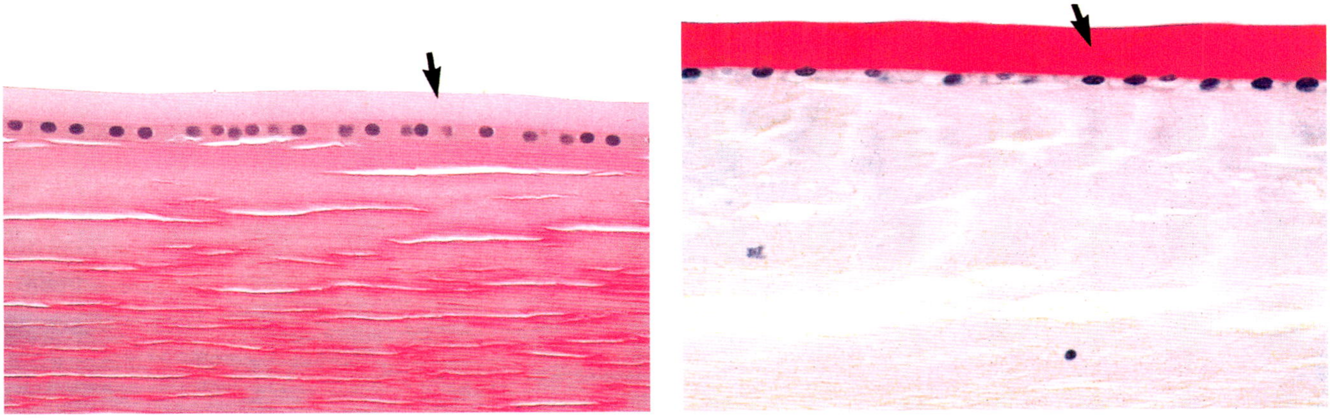

FIGURE 12.55 The anterior lens capsule (*arrow*) appears as an eosinophilic acellular band overlying a single layer of epithelial cells in hematoxylin and eosin–stained preparations (**left**). The lens capsule is rich in carbohydrates and reacts intensely with the periodic acid–Schiff stain (*arrow*) (**right**) (**left**, H&E; **right**, periodic acid–Schiff).

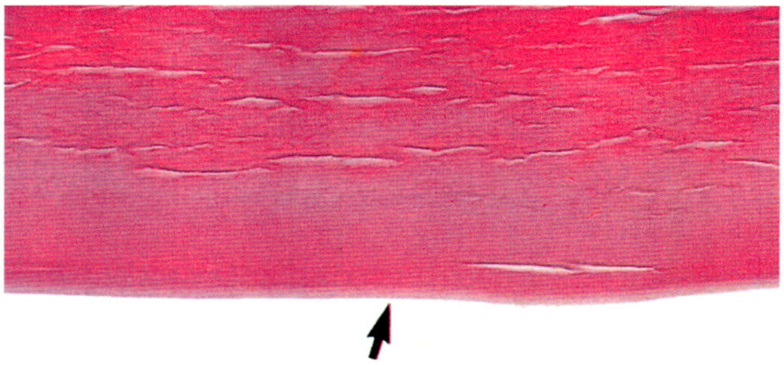

FIGURE 12.56 Posteriorly, the lens capsule (*arrow*) is thinner than anteriorly, and epithelial cells are absent (H&E).

The posterior lens capsule is 2 μm at the pole and 18 μm maximally near the posterior periphery at 2 to 5 years of age, and it reaches its maximum thickness of 4 to 23 μm at about 35 years of age then diminishes to 2 to 9 μm by the age of 71 (Fig. 12.56) (3,6). Directly interior to the anterior lens capsule is a single layer of cuboidal epithelium that extends to the level of the lens equator. These cells do not normally exist posterior to this point. Proliferating epithelial cells elongate at the lens equator and become displaced toward the center of the lens, or lens nucleus, where they are retained for life. This process continues throughout life, and the long, slender cells are designated lens fibers. The fibers retain their nuclei in the peripheral part of the lens near the equator, but the nuclei disintegrate as the fibers become displaced toward the center of the lens such that the center of the lens lacks nuclei. In some cataractous lenses, such as the cataract of rubella, the fibers within the center of the lens retain their nuclei.

The normally transparent lens commonly opacifies with age. Discrete globules of degenerate lens fibers may form and are frequently accompanied by an extension of epithelial cells posterior to the equator. The high density of the lens fibers makes it difficult to obtain histologic sections of the lens that are free from artifacts.

Infant eyes can demonstrate an artifact of fixation resulting in an umbilicated, dimpled, or concave configuration of the posterior surface of the lens (Fig. 12.57) (99,100).

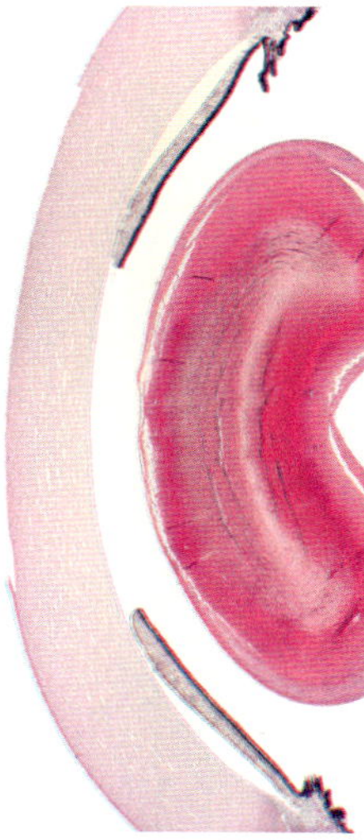

FIGURE 12.57 This infant's eye demonstrates an artifact of fixation resulting in a posterior concave or umbilicated appearance to the crystalline lens. This figure also has artifactual absence of most of the corneal epithelium, a very common finding in postmortem eyes (H&E).

INTRAOCULAR COMPARTMENTS

The eye accommodates two major fluid-containing intraocular compartments. One is filled with aqueous humor, the other with vitreous humor. The aqueous compartment is divided into an anterior and a posterior chamber (Fig. 12.58). The anterior chamber is delineated in front by the cornea, peripherally by the drainage angle of the eye, and posteriorly by the pupil and the iris. The small posterior chamber is located between the pigmented epithelia of the iris, the ciliary body, the anterior face of the vitreous, and the lens. The aqueous humor, a watery solution that does not normally stain with routine histologic techniques, is produced by the ciliary body and flows forward through the aperture of the pupil to the anterior chamber where it leaves the eye through the trabecular meshwork and the Schlemm canal. The anterior chamber contains approximately 0.25 mL of aqueous humor; the posterior chamber has a volume of only approximately 0.06 mL. Normal human aqueous humor has a density slightly greater than water, and like plasma, it contains protein, ascorbic acid, electrolytes, and

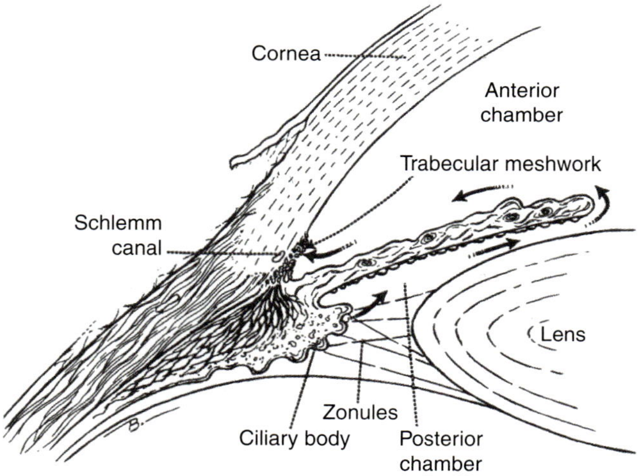

FIGURE 12.58 The anterior chamber is defined by the cornea, the anterior surface of the iris, and the pupil. The boundaries of the much smaller posterior chamber include the posterior surface of the iris, the ciliary body, and the anterior face of the vitreous. The aqueous is produced by the ciliary body and circulates from the posterior chamber through the pupil into the anterior chamber. The aqueous drains from the eye through the trabecular meshwork and the Schlemm canal. (Reproduced with permission from Klintworth GK, Landers MB 3rd. *The Eye. Structure and Function in Disease Monograph Series*. Williams & Wilkins; 1976.)

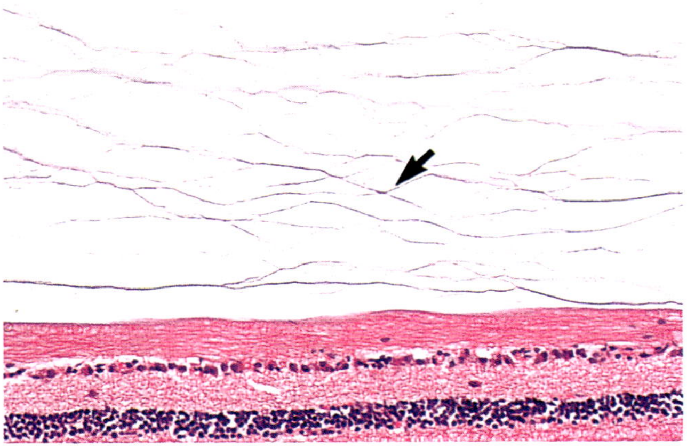

FIGURE 12.59 The vitreous (*arrow*) appears as an amorphous material, usually strands, in standard tissue sections (H&E).

glucose. Aqueous humor has relatively low-protein and high-ascorbic acid concentrations compared to plasma.

The vitreous humor extends from the neurosensory retina to the lens and contains a gel-like material composed of water, protein, hyaluronic acid, and a small population of cells, designated hyalocytes, that are rarely noted in standard tissue sections. The hyalocytes are tissue macrophages that are thought to synthesize collagen and hyaluronic acid. The gelatinous consistency of the vitreous is due to a framework of numerous, randomly oriented collagen fibrils. The concentration of glucose and ascorbic acid is much lower than in the aqueous humor, whereas the concentration of soluble protein is similar (101,102). The vitreous humor is attached securely to the retina at the ora serrata and near the optic disc. Occasionally, vitreous humor may be identified as an amorphous acellular material in H&E-stained sections (Fig. 12.59).

THE EYELIDS

The eyelids (Fig. 12.60) can be divided into cutaneous and conjunctival portions (103). The cutaneous segment of the eyelid is composed of a stratified squamous epidermis overlying a loosely arranged dermis, beneath which is the muscular tissue. The eyelids contain several types of skin appendages. Sebaceous glands termed glands of Zeis

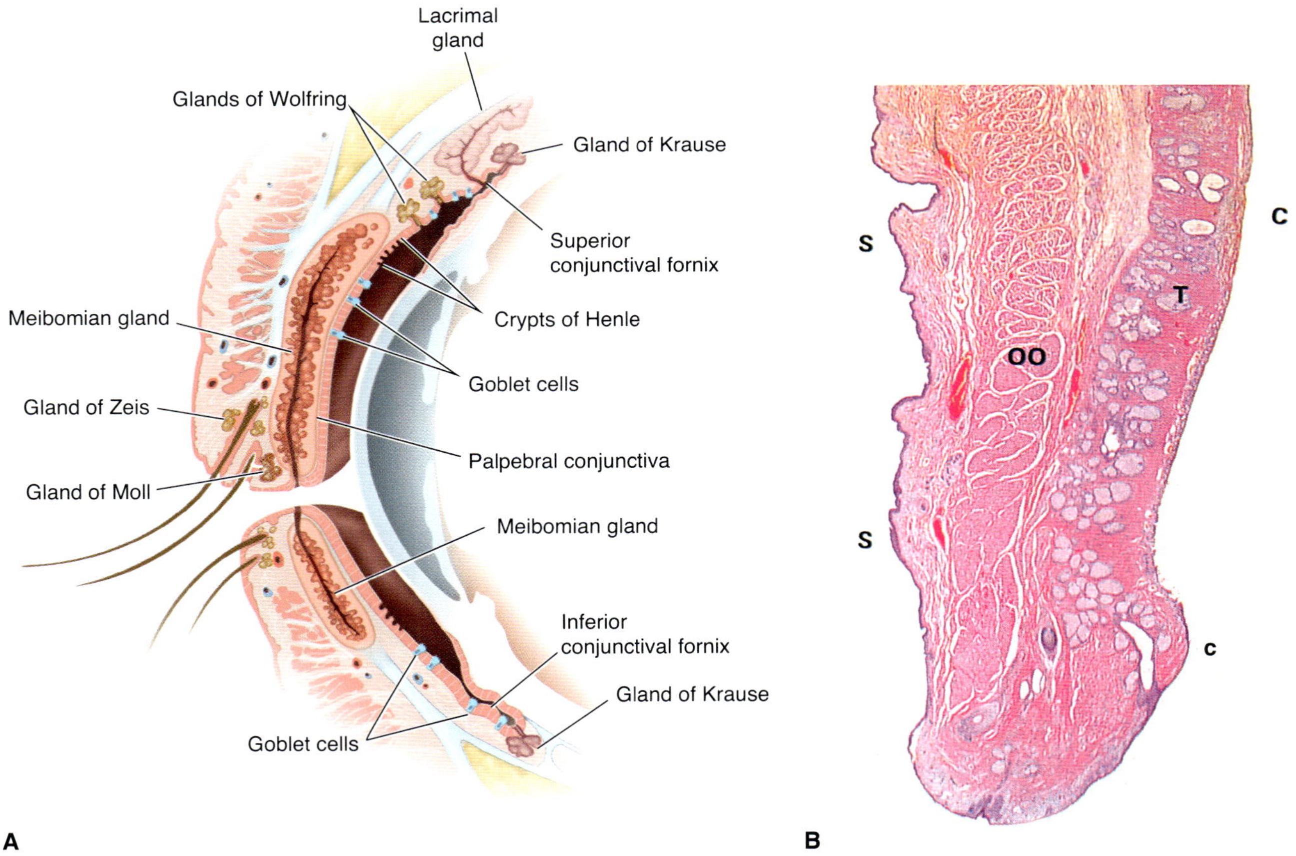

FIGURE 12.60 **A:** Components of the eyelid include skin, cutaneous appendages (glands of Zeiss and Moll), muscle, connective tissue, meibomian glands, conjunctiva, and accessory lacrimal gland tissue (glands of Krause and Wolfring). **B:** The skin surface (*S*), the orbicularis oculi muscle (*OO*), the tarsus (*T*), and the conjunctiva (*c*) are evident in this histologic section of an eyelid (H&E). (**A:** Reproduced with permission from Freddo TF, Chaum E. *Anatomy of the Eye and Orbit. The Clinical Essentials*. Wolters Kluwer; 2018.)

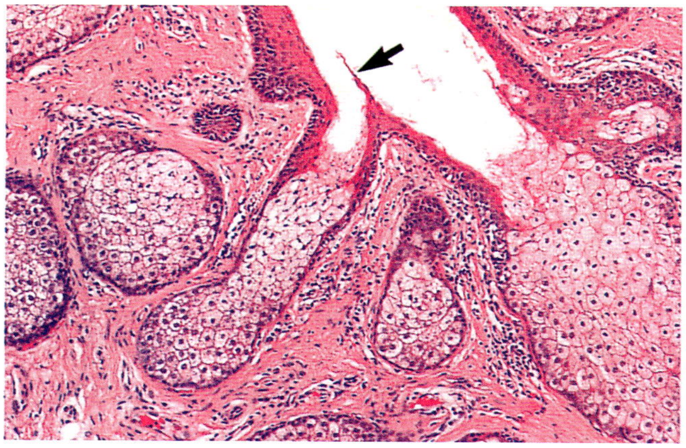

FIGURE 12.61 Modified sebaceous glands, the meibomian glands, deposit secretions into ducts opening onto the eyelids. A valve (*arrow*) is evident in this meibomian gland duct (H&E).

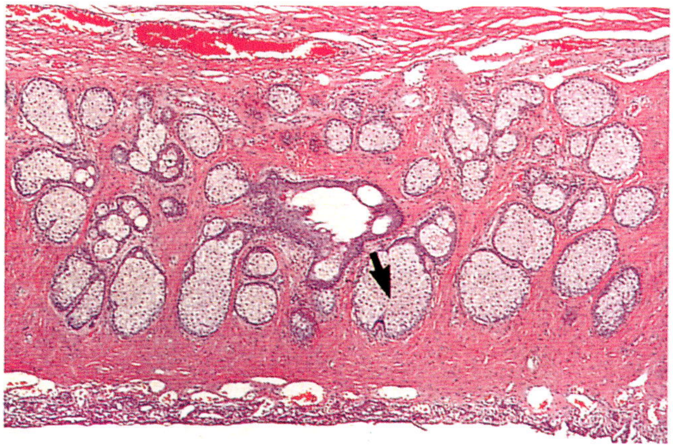

FIGURE 12.63 The tarsus, composed of dense fibrous tissue, contains the meibomian glands (*arrow*). The palpebral conjunctiva is immediately beneath the tarsus at the bottom of this figure (H&E).

deposit their secretions together with decomposed whole cells into hair follicles of the eyelashes, while other sebaceous glands, termed meibomian glands, empty into ducts that that open into the lid margins (Fig. 12.61). The secretions of apocrine glands represent the pinched-off luminal aspect of the lining acinar cells; these glands also open into the follicles of the eyelashes (glands of Moll) (Fig. 12.62). In addition, the dermis of the eyelid contains eccrine sweat glands that discharge secretions directly onto the skin via a convoluted duct. The subcutaneous portion of the upper and lower eyelids contains concentrically arranged skeletal muscle fibers (orbicularis oculi), but very little adipose connective tissue. The striated muscle of the palpebral portion of the levator palpebrae superioris is also present in the upper eyelid; it terminates in a dense fibrocollagenous aponeurosis. Small bundles of smooth muscle fibers, the Müller muscle, are located within the upper and lower eyelids and aid in raising the upper eyelids, lowering the lower lids, and keeping the eyelids open.

The junction between the cutaneous and conjunctival parts of the eyelid is demarcated clinically by a sulcus (the gray line [see Fig. 12.21]) located between the orifices of the meibomian glands and the eyelashes. The conjunctival portion, or tarsus, of the eyelid is made up of dense connective tissue containing the meibomian glands and the palpebral conjunctiva (Fig. 12.63). Meibomian acinar epithelial cells express CK5/6, CK14, and epithelial membrane antigen (EMA), but are not immunoreactive to CK7, Cam5.2, AE1/AE3, and carcinoembryonic antigen antibodies (104). Meibomian duct epithelial cells express CK5/6 and CK14, and they exhibit inconsistent immunostaining for CK17, AE1/AE3, and EMA (104). The tarsus, located immediately posterior to the muscles of the eyelid, accounts for most of the rigidity of the eyelids and is covered posteriorly by conjunctival epithelium and a thin subepithelial stroma. As described earlier, accessory lacrimal glands are present in the palpebral conjunctiva.

The presence of more prominent subcutaneous, suborbicularis, and pretarsal fat pad in the upper eyelid distinguishes an Asian eyelid from a Caucasian eyelid (105).

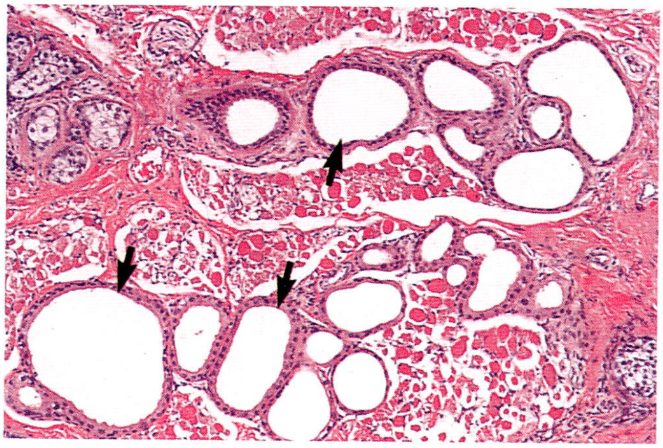

FIGURE 12.62 Apocrine glands (glands of Moll) (*arrows*) occur in the eyelid and open into the follicles of the eyelashes (H&E).

THE ORBIT

Bones of the skull, face, and nose define the posterior and peripheral borders of the orbit. At the anterior orbital margin, the periosteum of the orbital bones gives rise to the orbital septum, which is a dense connective tissue sheet that extends forward to insert into the eyelids (Fig. 12.60). Tissue posterior to this septum is considered to be within the orbit. The adult orbit is pyramidal and has a volume of approximately 25 cm^3, with the globe occupying about 7.2 to 8.2 cm^3 of the volume (4). The dimensions of the orbit are quite variable, but the anterior entrance is on average 3.5 cm in height and 4.0 cm in width (3). Orbital depth averages about 4.9 cm with a range from 4.1 to 5.7 cm (106). Several bony canals allow for transmission of blood vessels and nerves into and out of the orbit, posteriorly.

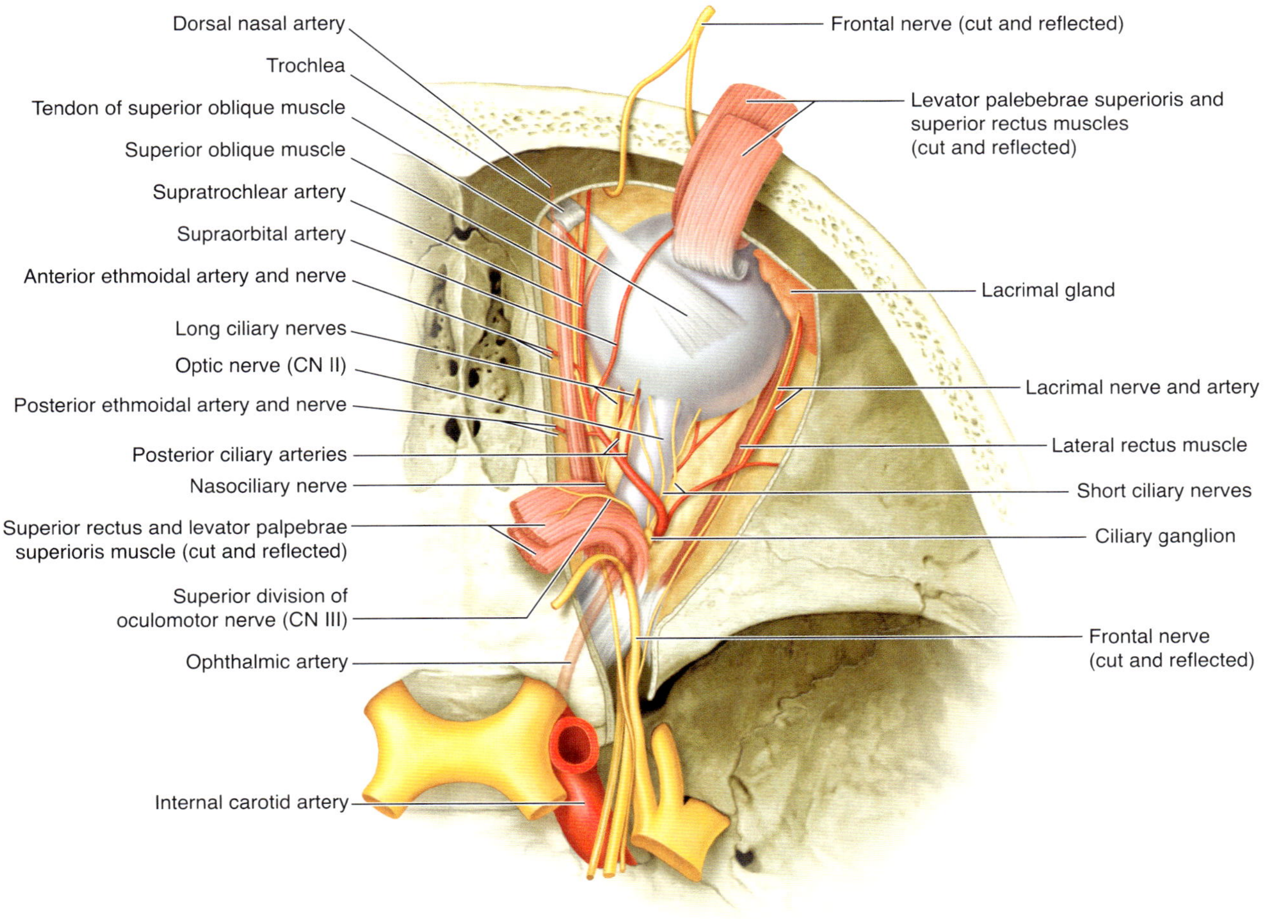

FIGURE 12.64 The bony cavity of the orbit contains the eyeball and its fibrous covering (the Tenon capsule), the cartilaginous trochlea, the lacrimal gland, the extraocular muscles, blood vessels, nerves, and fibrovascular connective tissue. The trochlea and the lacrimal gland are located within the superonasal and superotemporal aspects of the orbit, respectively. Some of the extraocular muscles originate from a ring of fibrous tissue in the posterior orbit known as the annulus of Zinn. (Reproduced with permission from Freddo TF, Chaum E. *Anatomy of the Eye and Orbit. The Clinical Essentials*. Wolters Kluwer; 2018.)

The contents of the orbit are organized in a complex three-dimensional arrangement (Fig. 12.64). Aside from the eye, the optic nerve and its meningeal coverings, the Tenon capsule, the extraocular muscles, the lacrimal gland, blood vessels, and a delicate framework of fibroadipose connective tissue constitute the major components of the orbit.

The only epithelial structure normally present in the orbit is the lacrimal gland (Fig. 12.65). Closely apposed to the globe and situated in the superolateral aspect of the orbit, this gland is traditionally divided into two parts: a larger orbital lobe and a smaller palpebral lobe. Approximately a dozen ducts from the lacrimal gland open into the superior conjunctival fornices and transmit their secretions into the tear film. The lacrimal gland is not encapsulated, and thin fibrovascular connective tissue septa divide the tissue into lobules composed of acini lined by columnar epithelial cells. Occasionally, some lobules extend posteriorly behind the globe. Most cells are serous and contain

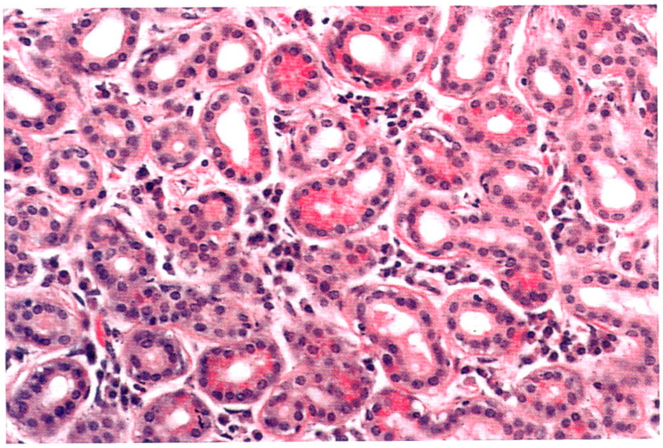

FIGURE 12.65 Columnar epithelial cells line the lacrimal gland acini. Scattered lymphocytes and plasma cells are normally present in the gland. Flattened myoepithelial cells are evident around some of the acini, such as the one near the center of the photomicrograph (H&E).

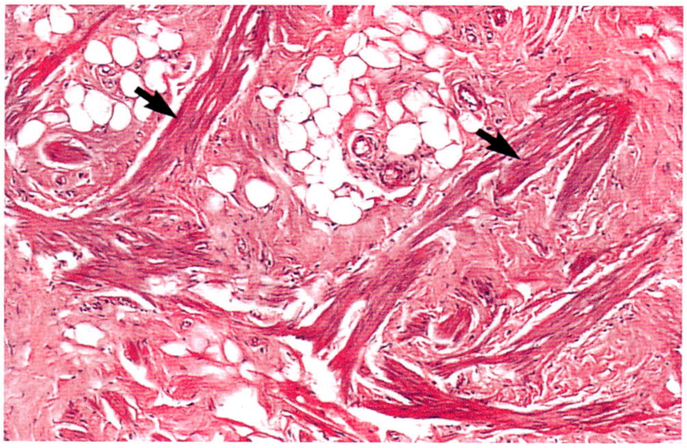

FIGURE 12.66 The fibrous septa of the orbit, which are critical for normal support and motion of the globe, may contain smooth muscle bundles (*arrows*) (H&E).

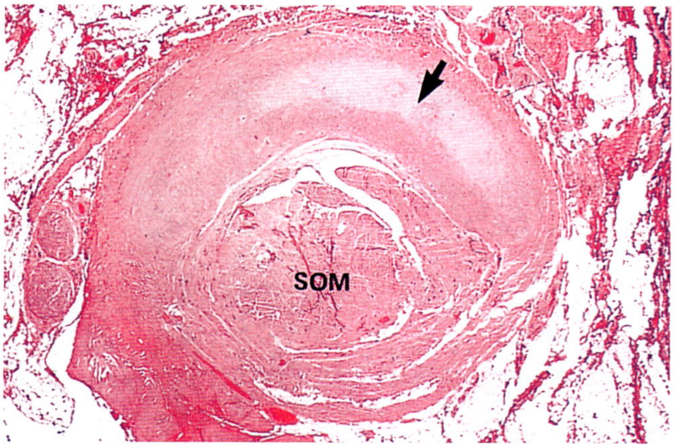

FIGURE 12.67 The arc-shaped trochlea (*arrow*), the only cartilaginous structure of the normal orbit, envelops the skeletal muscle fibers of the superior oblique muscle (*SOM*) (H&E).

scattered intracytoplasmic fat droplets and many PAS-positive secretory granules. Myoepithelial cells, inconspicuous in routine H&E-stained sections, surround the acini and intercalated ducts (107). Mucinous cells similar to those of salivary glands are not usually present in the acini but may be identified in the ducts. Occasional lymphocytes and plasma cells are commonly present between the acini of the lacrimal gland. With increasing age, the orbital lobes of lacrimal glands in women exhibit increasing diffuse fibrosis, diffuse atrophy, and periductal fibrosis, while men have increasing periductal fibrosis of their palpebral lobes (107).

The orbit contains the oculomotor, trochlear, and abducens nerves. These cranial nerves innervate the extrinsic muscles of the eye and branches of the ophthalmic division of

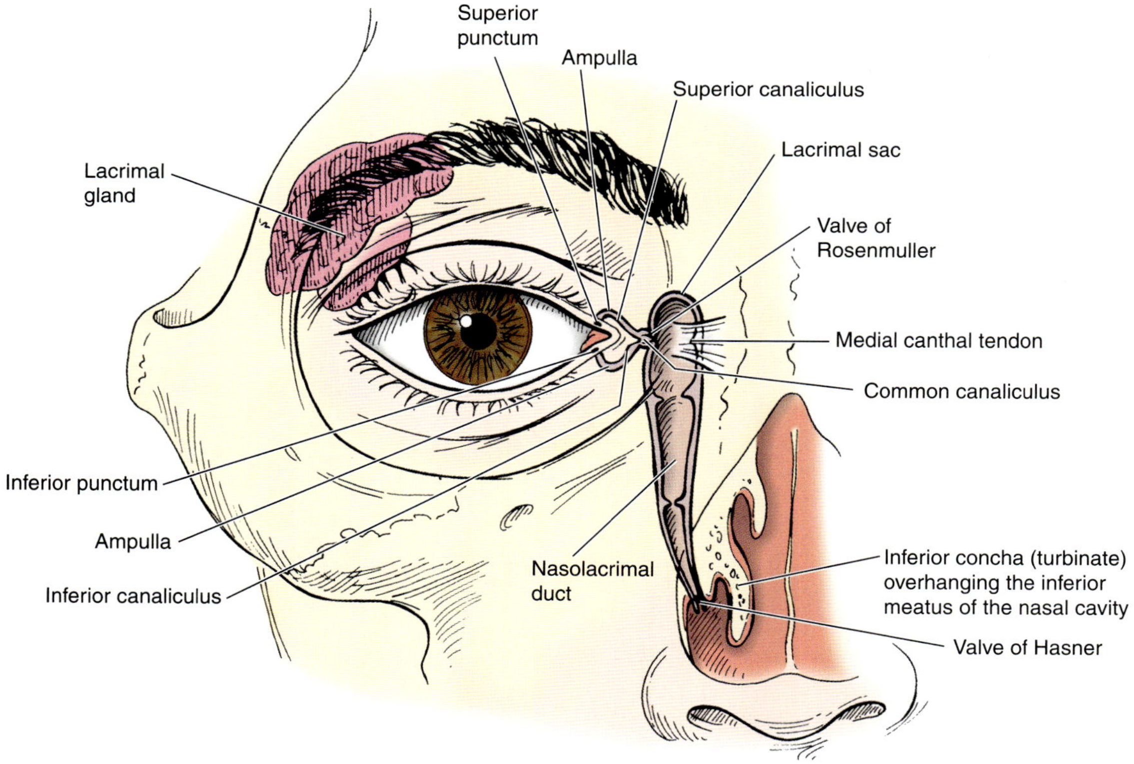

FIGURE 12.68 The lacrimal gland and drainage apparatus are illustrated here. The lacrimal gland is located in the superotemporal orbit and contributes secretions to the tear film. Tears enter the canaliculi through the puncta and drain through the nasolacrimal sac and duct, eventually reaching the inferior meatus within the nose. (Reproduced with permission from Freddo TF, Chaum E. *Anatomy of the Eye and Orbit. The Clinical Essentials*. Wolters Kluwer; 2018.)

the trigeminal nerve, as well as the parasympathetic and sympathetic nerves that innervate the cornea, conjunctiva, and the muscles of the ciliary body and iris. Neurons of the ciliary ganglion, located near the optic nerve close to the orbital apex and which measures approximately 2 mm in diameter, receive parasympathetic and sympathetic nerve fibers.

Other constituents of the orbit include smooth muscle (Fig. 12.66) (4,36) and the arc-shaped structure (trochlea), through which the tendon of the superior oblique muscle passes before insertion upon the eyeball (Fig. 12.67). The trochlea is the only cartilaginous structure normally present in the orbit. It arises from the superior nasal aspect of the frontal bone. Smooth muscle forms the orbital muscle of Müller (4) and is present within some of the fibrous septa that are crucial for normal support and motion of the globe (4,36). Smooth muscle is also present in the Tenon capsule anteriorly (37) and in the posteriorly around the insertions of the rectus and inferior oblique muscles (38–40).

Lymphatic channels do not exist in the orbit according to traditional teaching, but they have now been identified using immunohistochemical staining in the extraocular muscles, lacrimal gland, and optic nerve sheath (108,109). These may possibly arise within the orbital adipose tissue as a result of inflammation (108). The orbit normally lacks lymphoid tissue but contains scattered lymphocytes. These cells presumably give rise to the monoclonal and polyclonal lymphoid proliferations that frequently develop within the orbit, creating diagnostic and prognostic difficulty for the pathologist (110,111).

LACRIMAL DRAINAGE APPARATUS

The lacrimal drainage apparatus (Fig. 12.68), composed of the puncta, canaliculi, lacrimal sac, and nasolacrimal duct,

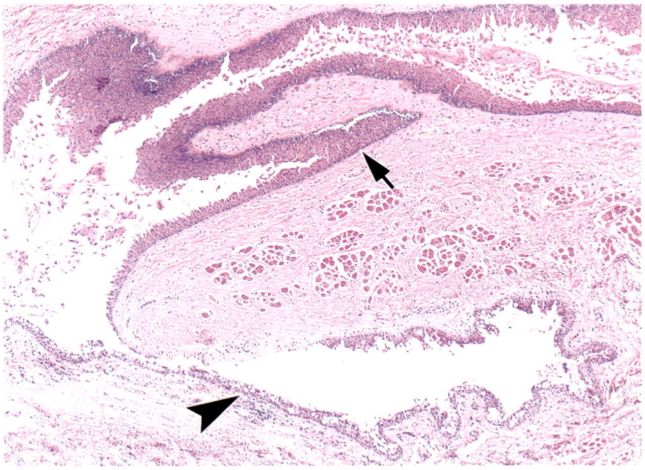

FIGURE 12.69 The nonkeratinizing stratified squamous epithelium of the lacrimal canaliculi (*arrow*) transitions to a double-layered epithelium containing goblet cells in the lacrimal sac (*arrowhead*). There is postmortem desquamation of surface epithelial cells in the canaliculus (H&E).

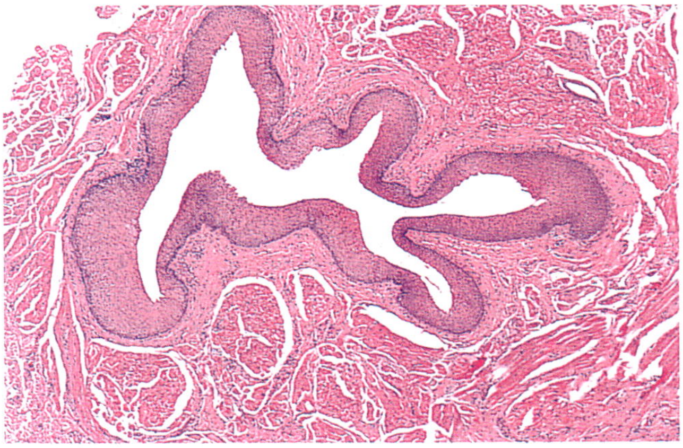

FIGURE 12.70 The lacrimal canaliculi are lined by nonkeratinizing stratified squamous epithelium and are surrounded by fibrous tissue (H&E).

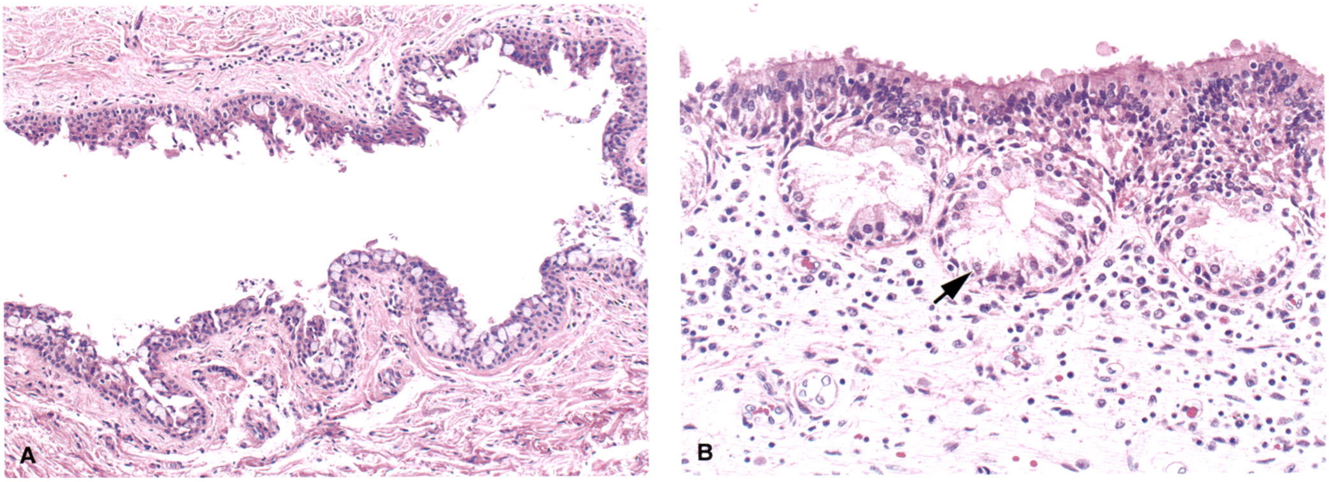

FIGURE 12.71 **A:** The epithelium of the lacrimal sac is double-layered with a basal-cell layer and an overlying columnar layer containing mucus-secreting goblet cells. The epithelium may appear multilayered due to oblique sectioning and the thickness of the histological sections (H&E). **B:** The goblet cells may be solitary or form mucous glands (*arrow*) (H&E).

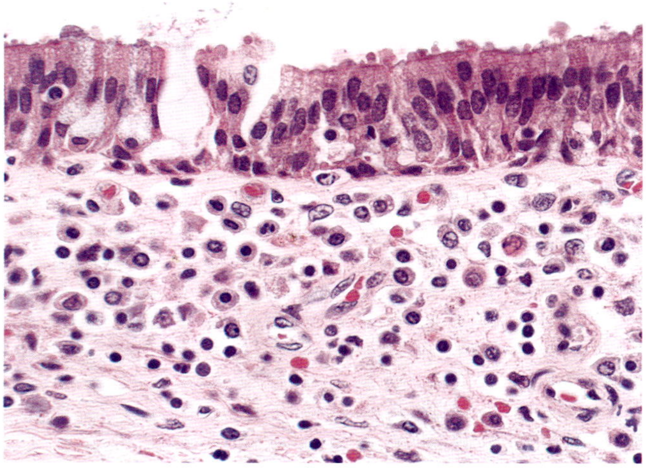

FIGURE 12.72 The lacrimal duct epithelium resembles the lacrimal sac epithelium with a bilayered epithelium having mucus-secreting goblet cells in the inner columnar layer (H&E).

collects the tears and drains them to the nose (112–115). Tear fluid drains toward the medial canthus and then passes through openings in the medial aspect of each eyelid, known as the lacrimal puncta. The puncta drain into the lacrimal canaliculi, tubular structures approximately 0.5 mm in diameter. Initially, the canaliculi are oriented vertically, but within 2 mm of their origin, bend at right angles to become almost horizontal within the eyelids. The distal portions of the canaliculi exit the upper and lower eyelids. They merge to form the lacrimal sac, encased by bones located in the inferomedial wall of the orbit. A duct (the nasolacrimal duct) that is nearly 1 cm long drains the lacrimal sac into the inferior nasal meatus of the nose. The epithelium lining the lacrimal drainage apparatus varies in different regions (Fig. 12.69) (114,115). In the canaliculi, it is a nonkeratinizing stratified squamous epithelium (Fig. 12.70). The lacrimal sac and nasolacrimal duct are lined by a double-layered epithelium having a basal-cell layer and an overlying columnar layer with surface microvilli and mucus-secreting goblet cells (Figs. 12.71 and 12.72). The goblet cells may be solitary or form mucous glands, more commonly in the lacrimal sac. The lacrimal sac fundus, but not the nasolacrimal duct, may have small serous glands resembling sweat glands, with an excretory duct penetrating the lining epithelium and emptying into the sac lumen. Beneath the epithelium is loose connective tissue with elastic fibers, lymphocytes, sometimes lymphoid follicles, and a venous plexus that connects to the cavernous body of the nasal inferior turbinate (114,115).

ACKNOWLEDGMENTS

The authors acknowledge the contributions of Gordon K. Klintworth and Mark W. Scroggs to previous editions of this chapter. Mr. Steven Conlon provided invaluable assistance with the figures.

REFERENCES

1. Bron AJ, Tripathi RC, Tripathi BJ. *Wolff's Anatomy of the Eye and Orbit*. 8th ed. Chapman & Hall Medical; 1997.
2. Cavallotti CAP, Cerulli L, eds. *Age-Related Changes in the Human Eye. Aging Medicine*. Humana Press; 2008.
3. Duke-Elder S, Wybar KC. *The Anatomy of the Visual System, Volume II, System of Ophthalmology*. The C.V. Mosby Company; 1961.
4. Dutton JJ. *Atlas of Clinical and Surgical Orbital Anatomy*. 2nd ed. Elsevier Saunders; 2011.
5. Freddo TF, Chaum E. *Anatomy of the Eye and Orbit: The Clinical Essentials*. Wolters Kluwer; 2018.
6. Hogan MJ, Alvarado JA, Weddell JE. *Histology of the Human Eye: An Atlas and Textbook*. W.B. Saunders; 1971.
7. Jakobiec FA. *Ocular Anatomy, Embryology, and Teratology*. Harper & Row; 1982.
8. Fine BS, Yanoff M. *Ocular Histology: A Text and Atlas*. 2nd ed. Harper & Row; 1979.
9. Fledelius HC, Christensen AC. Reappraisal of the human ocular growth curve in fetal life, infancy, and early childhood. *Br J Ophthalmol*. 1996;80(10):918–921.
10. Harayama K, Amemiya T, Nishimura H. Development of the eyeball during fetal life. *J Pediatr Ophthalmol Strabismus*. 1981;18(4):37–40.
11. Sorsby A, Sheridan M. The eye at birth: measurement of the principal diameters in forty-eight cadavers. *J Anat*. 1960; 94(Pt 2):192–197.
12. DelMonte DW, Kim T. Anatomy and physiology of the cornea. *J Cataract Refract Surg*. 2011;37(3):588–598.
13. Nishida T, Saika S, Morishige N. Cornea and sclera: anatomy and physiology. In: Mannis MJ, Holland EJ, eds. *Cornea*. 5th ed. Elsevier; 2022:1–222.
14. Zhivov A, Stave J, Vollmar B, Guthoff R. In vivo confocal microscopic evaluation of Langerhans cell density and distribution in the normal human corneal epithelium. *Graefes Arch Clin Exp Ophthalmol*. 2005;243:1056–1061.
15. Gillette TE, Chandler JW, Greiner JV. Langerhans cells of the ocular surface. *Ophthalmology*. 1982;89(6):700–711.
16. Wisely CE, Rafailov L, Cypen S, Proia AD, Boehlke CS, Leyngold IM. Clinical and morphologic outcomes of minimally invasive direct corneal neurotization. *Ophthalmic Plast Reconstr Surg*. 2020;36(5):451–457.
17. McTigue JW. The human cornea: a light and electron microscopic study of the normal cornea and its alterations in various dystrophies. *Trans Am Ophthalmol Soc*. 1967;65:591–660.
18. Chinnery HR, McMenamin PG, Dando SJ. Macrophage physiology in the eye. *Pflugers Arch*. 2017;469:501–515.
19. Price MO, Gorovoy M, Price FW Jr, Benetz BA, Menegay HJ, Lass JH. Descemet's stripping automated endothelial keratoplasty: three-year graft and endothelial cell survival compared with penetrating keratoplasty. *Ophthalmology*. 2013;120(2):246–251.
20. Guerra FP, Anshu A, Price MO, Price FW. Endothelial keratoplasty: fellow eyes comparison of Descemet stripping automated endothelial keratoplasty and Descemet membrane endothelial keratoplasty. *Cornea*. 2011;30(12):1382–1386.
21. Dua H, Said D. Pre-Descemets endothelial keratoplasty: the PDEK clamp for successful PDEK. *Eye*. 2017;31(7): 1106–1110.

22. Schlötzer-Schrehardt U, Bachmann BO, Laaser K, Cursiefen C, Kruse FE. Characterization of the cleavage plane in Descemet's membrane endothelial keratoplasty. *Ophthalmology*. 2011;118(10):1950–1957.
23. Dua HS, Faraj LA, Said DG, Gray T, Lowe J. Human corneal anatomy redefined: a novel pre-Descemet's layer (Dua's layer). *Ophthalmology*. 2013;120(9):1778–1785.
24. Shamsuddin AKM, Nirankari VS, Purnell DM, Chang SH. Is the corneal posterior cell layer truly endothelial? *Ophthalmology*. 1986;93(10):1298–1303.
25. Howell DN, Burchette J, Paolini JF, Geier SS, Fuller JA, Sanfilippo F. Characterization of a novel human corneal endothelial antigen. *Invest Ophthalmol Vis Sci*. 1991; 32(9):2473–2482.
26. Van den Bogerd B, Dhubhghaill SN, Koppen C, Tassignon M-J, Zakaria N. A review of the evidence for in vivo corneal endothelial regeneration. *Surv Ophthalmol*. 2018;63(2): 149–165.
27. Treffers WF. Human corneal endothelial wound repair: in vitro and in vivo. *Ophthalmology*. 1982;89(6):605–613.
28. Okumura N, Koizumi N. Regeneration of the corneal endothelium. *Curr Eye Res*. 2020;45(3):303–312.
29. Abib FC, Barreto J Jr. Behavior of corneal endothelial density over a lifetime. *J Cataract Refract Surg*. 2001;27(10): 1574–1578.
30. Williams KK, Noe RL, Grossniklaus HE, Drews-Botsch C, Edelhauser HF. Correlation of histologic corneal endothelial cell counts with specular microscopic cell density. *Arch Ophthalmol*. 1992;110(8):1146–1149.
31. Jakobiec FA, Bhat P. Retrocorneal membranes: a comparative immunohistochemical analysis of keratocytic, endothelial, and epithelial origins. *Am J Ophthalmol*. 2010;150(2): 230–242.e2.
32. Ross JR, Foulks GN, Sanfilippo FP, Howell DN. Immunohistochemical analysis of the pathogenesis of posterior polymorphous dystrophy. *Arch Ophthalmol*. 1995;113(3):340–345.
33. Lockwood CB. The anatomy of the muscles, ligaments, and fasciae of the orbit, including an account of the capsule of Tenon, the check ligaments of the recti, and the suspensory ligaments of the eye. *J Anat Physiol*. 1885;20(Pt 1):i2–i25.
34. Whitnall SE. *The Anatomy of the Human Orbit and Accessory Organs of Vision*. Henry Frowde and Hodder & Stoughton; 1921.
35. Scobee RG. The fascia of the orbit: its anatomy and clinical significance. *Am J Ophthalmol*. 1948;31(12):1539–1553.
36. Koornneef L. New insights in the human orbital connective tissue. Result of a new anatomical approach. *Arch Ophthalmol*. 1977;95(7):1269–1273.
37. Kakizaki H, Takahashi Y, Nakano T, et al. Anatomy of Tenons capsule. *Clin Experiment Ophthalmol*. 2012;40(6):611–616.
38. Demer JL, Miller JM, Poukens V, Vinters HV, Glasgow BJ. Evidence for fibromuscular pulleys of the recti extraocular muscles. *Invest Ophthalmol Vis Sci*. 1995;36(6):1125–1136.
39. Demer JL, Poukens V, Miller JM, Micevych P. Innervation of extraocular pulley smooth muscle in monkeys and humans. *Invest Ophthalmol Vis Sci*. 1997;38(9):1774–1785.
40. Demer JL, Oh SY, Clark RA, Poukens V. Evidence for a pulley of the inferior oblique muscle. *Invest Ophthalmol Vis Sci*. 2003;44(9):3856–3865.
41. Reese AB. Intrascleral nerve loops. *Arch Ophthalmol*. 1931; 6(5):698–703.
42. Spencer WH. Sclera. In: Spencer WH, ed. *Ophthalmic Pathology: An Atlas and Textbook*. 4th ed. W.B. Saunders Company; 1996:334–371.
43. Crandall AS, Yanoff M, Schaffer DB. Intrascleral nerve loop mistakenly identified as a foreign body. *Arch Ophthalmol*. 1977;95(3):497–498.
44. Scroggs MW, Klintworth GK. Senile scleral plaques: a histopathologic study using energy-dispersive x-ray microanalysis. *Hum Pathol*. 1991;22(6):557–562.
45. Tripathi RC. Aqueous outflow pathway in normal and glaucomatous eyes. *Br J Ophthalmol*. 1972;56(3):157–174.
46. Hou JH. The conjunctiva: anatomy and physiology. In: Mannis MJ, Holland EJ, eds. *Cornea*. 5th ed. Elsevier; 2022:23–30.
47. Jakobiec FA, Mendoza PR, Colby KA. Clinicopathologic and immunohistochemical studies of conjunctival large cell acanthoma, epidermoid dysplasia, and squamous papilloma. *Am J Ophthalmol*. 2013;156(4):830–846.
48. Kessing SV. Mucus gland system of the conjunctiva. A quantitative normal anatomical study. *Acta Ophthalmol (Copenh)*. 1968;95:1–133.
49. Goldberg MF, Bron AJ. Limbal palisades of Vogt. *Trans Am Ophthalmol Soc*. 1982;80:155–171.
50. Townsend WM. The limbal palisades of Vogt. *Trans Am Ophthalmol Soc*. 1991;89:721–756.
51. Dua HS, Shanmuganathan V, Powell-Richards A, Tighe P, Joseph A. Limbal epithelial crypts: a novel anatomical structure and a putative limbal stem cell niche. *Br J Ophthalmol*. 2005;89(5):529–532.
52. Greiner JV, Henriquez AS, Covington HI, Weidman TA, Allansmith MR. Goblet cells of the human conjunctiva. *Arch Ophthalmol*. 1981;99(12):2190–2197.
53. Kathuria SS, Howarth D, Hurwitz JJ, Oestreicher J. An anatomic and histologic study of the caruncle. *Ophthalmic Plast Reconstr Surg*. 1999;15(6):407–411.
54. Jakobiec FA, Lam H, Bhat P, Pineda R. Non-syndromic supernumerary caruncles causing ocular irritation after cataract surgery: a critical review of caruncular dysgeneses. *Am J Ophthalmol*. 2010;149(3):398–404. e1–e2.
55. Duke-Elder SS. *Normal and Abnormal Development, Part 2: Congenital Deformities. Volume III System of Ophthalmology*. The C.V. Mosby Company; 1963.
56. Freddo TF. Ultrastructure of the iris. *Microsc Res Tech*. 1996;33(5):369–389.
57. Eagle R Jr. Iris pigmentation and pigmented lesions: an ultrastructural study. *Trans Am Ophthalmol Soc*. 1988;86:581–687.
58. Wilkerson CL, Syed NA, Fisher MR, Robinson NL, Albert DM. Melanocytes and iris color: light microscopic findings. *Arch Ophthalmol*. 1996;114(4):437–442.
59. Imesch PD, Bindley CD, Khademian Z, et al. Melanocytes and iris color: electron microscopic findings. *Arch Ophthalmol*. 1996;114(4):443–447.
60. Imesch PD, Wallow IH, Albert DM. The color of the human eye: a review of morphologic correlates and of some conditions that affect iridial pigmentation. *Surv Ophthalmol*. 1997;41:S117–S123.
61. Wobmann PR, Fine BS. The clump cells of Koganei: a light and electron microscopic study. *Am J Ophthalmol*. 1972; 73(1):90–101.
62. Neveu MM, Padhy SK, Ramamurthy S, et al. Ophthalmological manifestations of oculocutaneous and ocular albinism: current perspectives. *Clin Ophthalmol*. 2022;16:1569–1587.

63. Ring HG, Fujino T. Observations on the anatomy and pathology of the choroidal vasculature. *Arch Ophthalmol*. 1967;78(4):431–444.
64. Friedman E, Smith TR, Kuwabara T. Senile choroidal vascular patterns and drusen. *Arch Ophthalmol*. 1963;69(2):220–230.
65. Sarks S. New vessel formation beneath the retinal pigment epithelium in senile eyes. *Br J Ophthalmol*. 1973;57(12):951–965.
66. Harbour J, Brantley M, Hollingsworth H, Gordon M. Association between choroidal pigmentation and posterior uveal melanoma in a white population. *Br J Ophthalmol*. 2004;88(1):39–43.
67. May CA. Mast cell heterogeneity in the human uvea. *Histochem Cell Biol*. 1999;112(5):381–386.
68. Ezzat M, Hann C, Vuk-Pavlovic S, Pulido JS. Immune cells in the human choroid. *Br J Ophthalmol*. 2008;92(7):976–980.
69. Chang JH, McCluskey P, Wakefield D. Expression of toll-like receptor 4 and its associated lipopolysaccharide receptor complex by resident antigen-presenting cells in the human uvea. *Invest Ophthalmol Vis Sci*. 2004;45(6):1871–1878.
70. Straatsma BR, Foos RY. Typical and reticular degenerative retinoschisis. *Am J Ophthalmol*. 1973;75(4):551–575.
71. O'Malley PF, Allen RA. Peripheral cystoid degeneration of the retina: incidence and distribution in 1,000 autopsy eyes. *Arch Ophthalmol*. 1967;77(6):769–776.
72. Wolter JR, Wilson WW. Degeneration of the peripheral retina: report of an instructive case. *Am J Ophthalmol*. 1959;47(2):153–166.
73. O'Malley P, Allen RA, Straatsma BR, O'Malley CC. Paving-stone degeneration of the retina. *Arch Ophthalmol*. 1965;73(2):169–182.
74. Sarks S. Ageing and degeneration in the macular region: a clinico-pathological study. *Br J Ophthalmol*. 1976;60(5):324–341.
75. Curcio CA, Saunders PL, Younger PW, Malek G. Peripapillary chorioretinal atrophy: Bruch's membrane changes and photoreceptor loss. *Ophthalmology*. 2000;107(2):334–343.
76. Ramkumar HL, Verma R, Ferreyra HA, Robbins SL. Myelinated retinal nerve fiber layer (RNFL): a comprehensive review. *Int Ophthalmol Clin*. 2018;58(4):147–156.
77. Foos R. Vitreoretinal juncture; epiretinal membranes and vitreous. *Invest Ophthalmol Vis Sci*. 1977;16(5):416–422.
78. Zimmerman LE, Straatsma BR. Anatomic relationships of the retina to the vitreous body and to the pigment epithelium. In: Schepens CL, ed. *Importance of the Vitreous Body in Retina Surgery with Special Emphasis on Reoperations*. The C.V. Mosby Company; 1960:15–28.
79. Smith TR. Pathologic findings after retina surgery. In: Schepens CL, ed. *Importance of the Vitreous Body in Retina Surgery with Special Emphasis on Reoperations*. The C.V. Mosby Company; 1960:61–93.
80. Foos RY. Vitreoretinal juncture—simple epiretinal membranes. *Albrecht Von Graefes Arch Klin Exp Ophthalmol*. 1974;189(4):231–250.
81. Snead D, James S, Snead MP. Pathological changes in the vitreoretinal junction 1: epiretinal membrane formation. *Eye*. 2008;22(10):1310–1317.
82. Arnon E, Brecheisen M, Uhles S, Westenskow PD, Lad EM, Proia AD. Histopathological characterization of connective tissue growth factor (CTGF) expression in human eyes with age-related macular degeneration and diabetic retinopathy. *Invest Ophthalmol Vis Sci*. 2023;64(8):2749–2749.
83. Mullen RJ, Buck CR, Smith AM. NeuN, a neuronal specific nuclear protein in vertebrates. *Development*. 1992;116(1):201–211.
84. Kuwabara T, Cogan DG. Studies of retinal vascular patterns: Part I. Normal architecture. *Arch Ophthalmol*. 1960;64(6):904–911.
85. Folberg R. The eye. In: Spencer WH, ed. *Ophthalmic Pathology: An Atlas and Textbook*. 4th ed. W.B. Saunders Company; 1996:1–37.
86. Kalina RE. A histopathologic postmortem and clinical study of peripheral retinal folds in infant eyes. *Am J Ophthalmol*. 1971;71(2):446–448.
87. Gartner S, Henkind P. Lange's folds: a meaningful ocular artifact. *Ophthalmology*. 1981;88(12):1307–1310.
88. Herwig MC, Müller AM, Holz FG, Loeffler KU. Morphologic analysis of artifacts in human fetal eyes confounding histopathologic investigations. *Invest Ophthalmol Vis Sci*. 2011;52(5):2712–2718.
89. Verdijk R, Herwig-Carl M. *Fetal and Neonatal Eye Pathology*. Springer; 2020.
90. Dubucs C, Chassaing N, Khung-Savatovsky S, Aziza J, Courtade-Saïdi M. Artifacts and main lesions in fetal ocular histology. *Morphologie*. 2023;107(357):207–218.
91. Cogan DG, Kuwabara T. Some common artifacts in the retina. *J Histochem Cytochem*. 1958;6(4):290–293.
92. Anderson DR. Ultrastructure of human and monkey lamina cribrosa and optic nerve head. *Arch Ophthalmol*. 1969;82(6):800–814.
93. Radius RL, Gonzales M. Anatomy of the lamina cribrosa in human eyes. *Arch Ophthalmol*. 1981;99(12):2159–2162.
94. Magoon EH, Robb RM. Development of myelin in human optic nerve and tract: a light and electron microscopic study. *Arch Ophthalmol*. 1981;99(4):655–659.
95. Cogan DG, Kuwabara T. Myelin artifacts. *Am J Ophthalmol*. 1967;64(3):622/104–626/108.
96. Zimmerman LE, Fine BS. Myelin artifacts in the optic disc and retina. *Arch Ophthalmol*. 1965;74(3):394–398.
97. Hollander H, Makarov F, Stefani F, Stone J. Evidence of constriction of optic nerve axons at the lamina cribrosa in the normotensive eye in humans and other mammals. *Ophthalmic Res*. 1995;27(5):296–309.
98. Knox DL, Eagle RC, Green WR. Optic nerve hydropic axonal degeneration and blocked retrograde axoplasmic transport: histopathologic features in human high-pressure secondary glaucoma. *Arch Ophthalmol*. 2007;125(3):347–353.
99. Eagle RCJ. *Eye Pathology: An Atlas and Text*. 3rd ed. Wolters Kluwer; 2017.
100. Herwig MC, Müller AM, Klarmann-Schulz U, Holz FG, Loeffler KU. Lens artifacts in human fetal eyes—the challenge of interpreting the histomorphology of human fetal lenses. *Graefes Arch Clin Exp Ophthalmol*. 2014;252(1):155–162.
101. Bito LZ. The physiology and pathophysiology of intraocular fluids. *Exp Eye Res*. 1977;25:273–289.
102. Reddy DVN, Kinsey VE. Composition of the vitreous humor in relation to that of plasma and aqueous humors. *Arch Ophthalmol*. 1960;63(4):715–720.
103. Dutton JJ, Tawfik HA, Proia AD. *Comprehensive Textbook of Eyelid Disorders and Diseases*. Wolters Kluwer; 2023.

104. Jakobiec FA, Mehta M, Iwamoto M, Hatton MP, Thakker M, Fay A. Intratarsal keratinous cysts of the Meibomian gland: distinctive clinicopathologic and immunohistochemical features in 6 cases. *Am J Ophthalmol*. 2010;149(1):82–94.
105. Jeong S, Lemke BN, Dortzbach RK, Park YG, Kang HK. The Asian upper eyelid: an anatomical study with comparison to the Caucasian eyelid. *Arch Ophthalmol*. 1999;117(7):907–912.
106. Nitek S, Wysocki J, Reymond J, Piasecki K. Correlations between selected parameters of the human skull and orbit. *Med Sci Monit*. 2009;15(12):BR370–BR377.
107. Obata H. Anatomy and histopathology of the human lacrimal gland. *Cornea*. 2006;25:S82–S89.
108. Dickinson A, Gausas R. Orbital lymphatics: do they exist? *Eye*. 2006;20(10):1145–1148.
109. Nakao S, Hafezi-Moghadam A, Ishibashi T. Lymphatics and lymphangiogenesis in the eye. *J Ophthalmol*. 2012;2012(1): 783163.
110. Andrew NH, Coupland SE, Pirbhai A, Selva D. Lymphoid hyperplasia of the orbit and ocular adnexa: a clinical pathologic review. *Surv Ophthalmol*. 2016;61(6):778–790.
111. Mulay K, Honavar SG. An update on ocular adnexal lymphoma. *Semin Diagn Pathol*. 2016;33(3):164–172.
112. Kominami, Rieko, Yasutaka S, Taniguchi Y, Shinohara H. Anatomy and histology of the lacrimal fluid drainage system. *Okajimas Folia Anat Jpn*. 2000;77(5):155–160.
113. Takahashi Y, Kakizaki H, Nakano T, Asamoto K, Ichinose A, Iwaki M. Anatomy of the vertical lacrimal canaliculus and lacrimal punctum: a macroscopic study. *Ophthalmic Plast Reconstr Surg*. 2011;27(5):384–386.
114. Paulsen F. *The Human Nasolacrimal Ducts*. Springer-Verlag; 2003.
115. Paulsen F, Thale A, Kohla G, et al. Functional anatomy of human lacrimal duct epithelium. *Anat Embryol (Berl)*. 1998;198:1–12.

13

The Ear and Temporal Bone

Bruce M. Wenig

The ear can be considered as three distinct regions, or compartments, to include the external ear, the middle ear and temporal bone, and the inner ear (Fig. 13.1). The external ear consists of the auricle (pinna), external auditory canal (meatus), and the tympanic membrane at the medial end of the auditory canal. The middle ear cavity includes the ossicles, the Eustachian tube that connects the middle ear space to the nasopharynx, and the expansion of the middle ear cavity in the form of air cells in the temporal bone. The inner ear is embedded in the petrous portion of the temporal bone. It consists of a membranous (otic) labyrinth that lies within a dense bone referred to as the otic capsule that

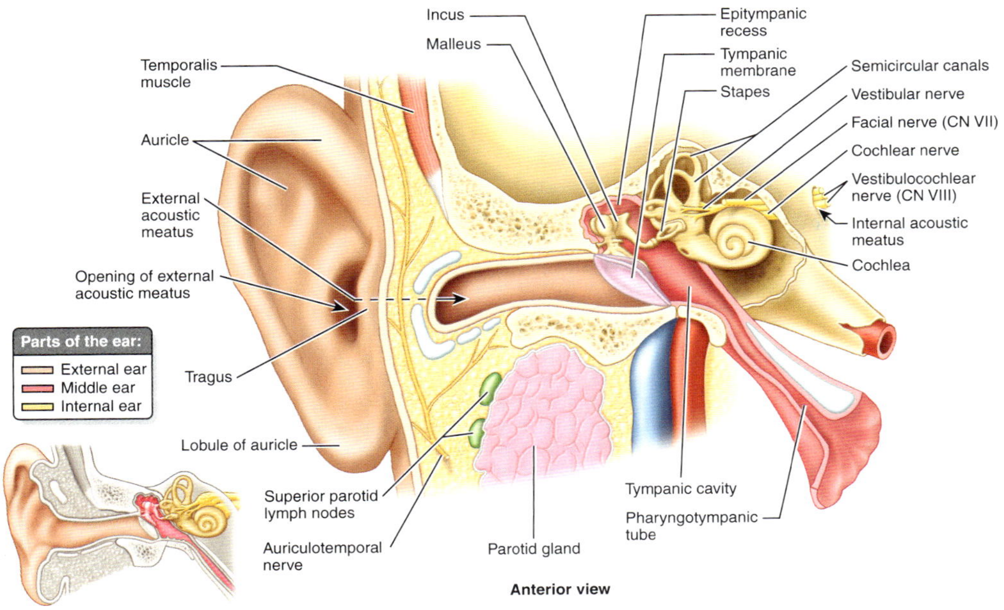

FIGURE 13.1 Parts of ear. A coronal section of the ear, with accompanying orientation figure, demonstrates that the ear has three parts: external, middle, and internal. The external ear consists of the auricle and external acoustic meatus. The middle ear is an air space in which the auditory ossicles are located. The internal ear contains the membranous labyrinth; its chief divisions are the cochlear labyrinth and the vestibular labyrinth. (Reprinted with permission from Moore KL, Dalley AF, Agur AMR. *Clinically Oriented* Anatomy. 9th ed. Wolters Kluwer Heath; 2023. Figure 8.111 on Page 984.)

is excavated to form the osseous (periotic) labyrinth (1). The external and middle ears form the sound-conducting apparatus for the auditory part of the inner ear. The inner ear is the sense organ for hearing and for balance.

EXTERNAL EAR

Embryology

The external ear develops from the first branchial groove. The external auricle (pinna) forms from the fusion of the auricular hillocks (tubercles), a group of mesenchymal tissue swellings from the first and second branchial arches, that lie around the external portion of the first branchial groove, a remnant of which ultimately becomes the external auditory canal (2). The tympanic membrane forms from the first branchial groove, as well as the first and second branchial pouches (2). The ectoderm of the first branchial groove gives rise to the epithelium on the external side of the tympanic membrane, while the endoderm from the first branchial pouch gives rise to the epithelium on the internal side. The mesoderm of the first and second branchial pouches gives rise to the connective tissue lying between the external and internal epithelia (Fig. 13.2) (2).

ANATOMY

The anatomy of the external ear is seen in Figure 13.3. The outer portion of the external ear includes the auricle or pinna leading into the external auditory canal. The skeleton of the auricle consists of a single plate of elastic cartilage that conforms to the shape of the ear and is continuous with that of the external auditory canal. The lobule is the only part of the auricle that is devoid of skeletal support. The auricle is attached to the bony skull by three ligaments, including anterior, superior, and posterior (1). The anterior ligament attaches the helix and the tragus to the zygomatic

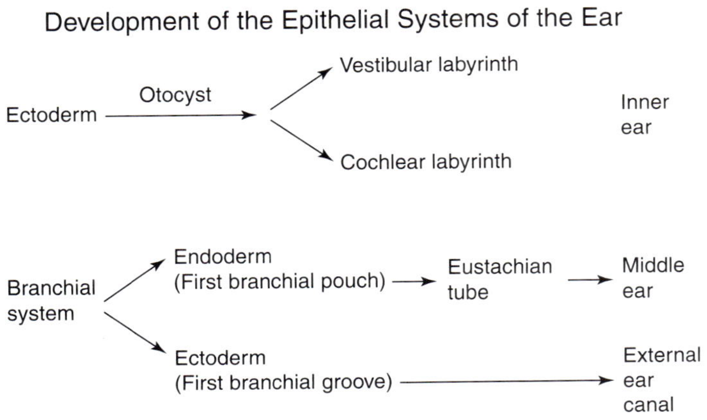

FIGURE 13.2 Diagrammatic representation of the embryology of the epithelia of the inner, middle, and external ears.

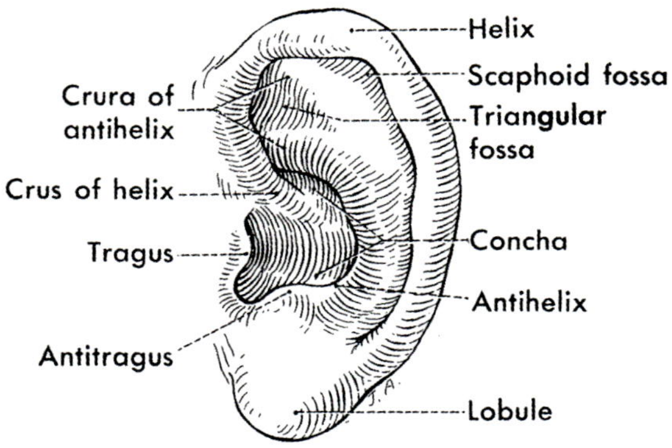

FIGURE 13.3 Anatomy of the external ear.

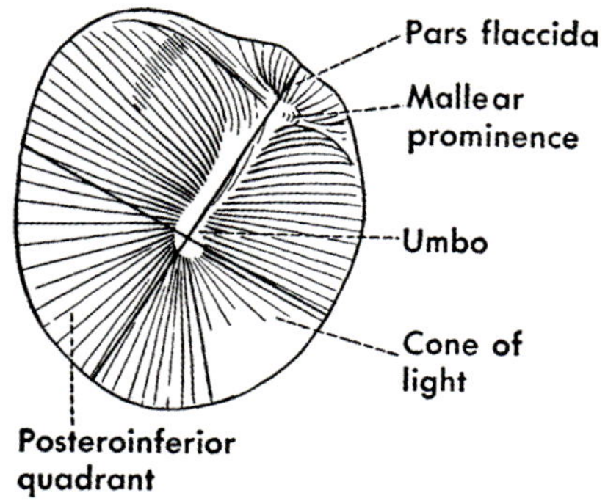

FIGURE 13.4 External (lateral) view of the tympanic membrane.

process. The superior ligament attaches the spine of the helix to the superior margin of the bony external meatus. The posterior ligament attaches the medial surface (eminence) of the concha to the mastoid process. The auricle is anchored through its continuity with the cartilage of the meatus and through the skin and extrinsic muscles. The extrinsic muscles of the ear include the anterior, superior, and posterior auricular muscles. These muscles are usually functionless but may be subject to voluntary control as in ear "wiggling." There are also small intrinsic muscles in connection with the cartilage of the external ear, but they are of no apparent importance. The extrinsic and intrinsic muscles of the ear are innervated by the facial nerve.

The external auditory canal or meatus extends from the concha to its medial limit on the external aspect of the tympanic membrane. The lateral portion of its wall consists of cartilage and connective tissue while the medial portion of its wall consists of bone (1). The cartilaginous part of the external auditory canal constitutes slightly less than half its total length. Inconstant fissures of Santorini occur in the cartilage; these fissures may transmit infection from the canal to the parotid gland and the superficial mastoid regions, or vice versa. The bony part of the canal is formed by both the tympanic and petrous parts of the temporal bone. The anterior, inferior, and lower posterior parts of the bony wall are formed by the C-shaped part of the temporal bone that develops from the annulus tympanicus in the fetus. Of note, the annulus is incomplete in the posterosuperior part of the wall; this part of the wall is formed by the squamous and petrous parts of the temporal bone in adults. There are close relationships between (1) the anterior and inferior walls of the cartilaginous canal and the parotid gland, (2) the anterior wall of the bony canal and the mandibular condyle, (3) the posterior wall and the mastoid air cells, and (4) the medial portion of the superior wall and the epitympanic recess.

The tympanic membrane (ear drum) is situated obliquely at the end of the external auditory canal sloping medially both from above downward and from behind forward. The tympanic membrane is a fibrous sheet interposed between the external auditory canal and the middle ear cavity (Fig. 13.4). The connective tissue interposed between these two layers consists of radiating fibers attached to the manubrium of the malleus that are reinforced peripherally by circular fibers. The latter are thickened at the margin of the tympanic membrane to form a fibrocartilaginous ring (annulus fibrocartilaginous) attaching the tympanic membrane to the tympanic sulcus of the temporal bone. There is a limited area in the upper portion of the tympanic membrane where the connective tissue fibers are lacking; this area is referred to as the pars flaccida or Shrapnell membrane. The tympanic portion of the temporal bone is also deficient in this area; this gap is referred to as the tympanic incisure or the notch of Rivinus, and refers to the site where tympanic membrane attaches to the temporal bone. The remainder of the tympanic membrane contains intact connective tissue fibers and is referred to as the pars tensa.

The outer aspect of the tympanic membrane is concave. The center of the concavity is referred to as the umbo and represents the strong point of attachment of the manubrium of the malleus to the tympanic membrane. The lateral process of the malleus is attached at the anterosuperior portion of the tympanic membrane. From this point of attachment, the anterior and posterior mallear folds pass to the cartilaginous annulus and separate the pars flaccida from the pars tensa. In otoscopic examinations of the tympanic membrane, the bright area of light reflection present downward and forward from the umbo is referred to as the "cone of light."

HISTOLOGY

The auricle is essentially a cutaneous structure composed of keratinizing, stratified squamous epithelium, and adnexal structures that include hair follicles, sebaceous glands, and eccrine sweat glands (Fig. 13.5). In addition to the hair follicles and sebaceous glands, the outer third of the external auditory canal contains modified apocrine glands called ceruminal glands that replace the eccrine glands of the auricular dermis (Fig. 13.6). Ceruminal glands produce cerumen and are arranged in clusters. The glands are composed of cuboidal cells with eosinophilic cytoplasm often containing a granular, golden-yellow pigment. These cells have secretory droplets along their luminal border. Peripheral to the secretory cells are flattened myoepithelial cells. The ducts of the ceruminal glands terminate in the hair

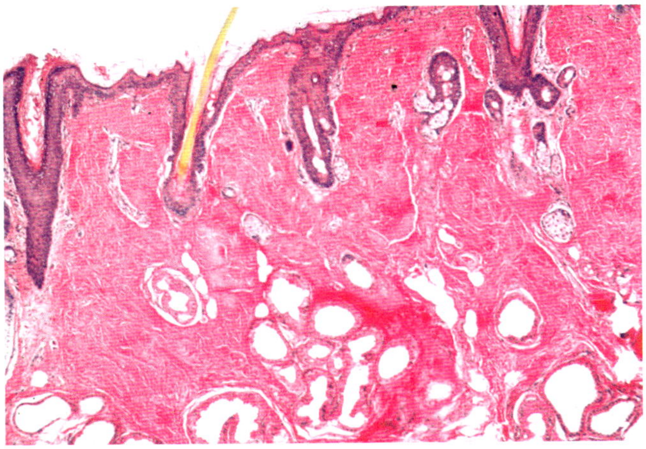

FIGURE 13.5 The auricle is a cutaneous structure composed of keratinizing stratified squamous epithelium with associated cutaneous adnexal structures that include hair follicles, sebaceous glands, and eccrine sweat glands.

follicle or on the skin. The ducts of the ceruminal glands lack apocrine and myoepithelial cells. Ceruminal glands and other adnexal structures are absent in the inner portion of the external auditory canal. The subcutaneous tissue is composed of fibroconnective tissue, fat, and elastic-type fibrocartilage, which gives the auricle its structural support (Fig. 13.7). The earlobe is devoid of cartilage and is replaced by a pad of adipose tissue. The perichondrium is composed of loose vascular connective tissue.

Similar to the auricle, the external auditory canal is lined by keratinizing squamous epithelium that extends to include the external aspect of the tympanic membrane. The tympanic membrane has a central bilaminated zone including lateral radially arranged and medial circularly arranged collagenous fibers (Fig. 13.8). The inner two-thirds of the external auditory canal contain bone rather than cartilage. Due to the absence of adnexal structures, there is relatively close apposition of the epithelium to the subjacent bone.

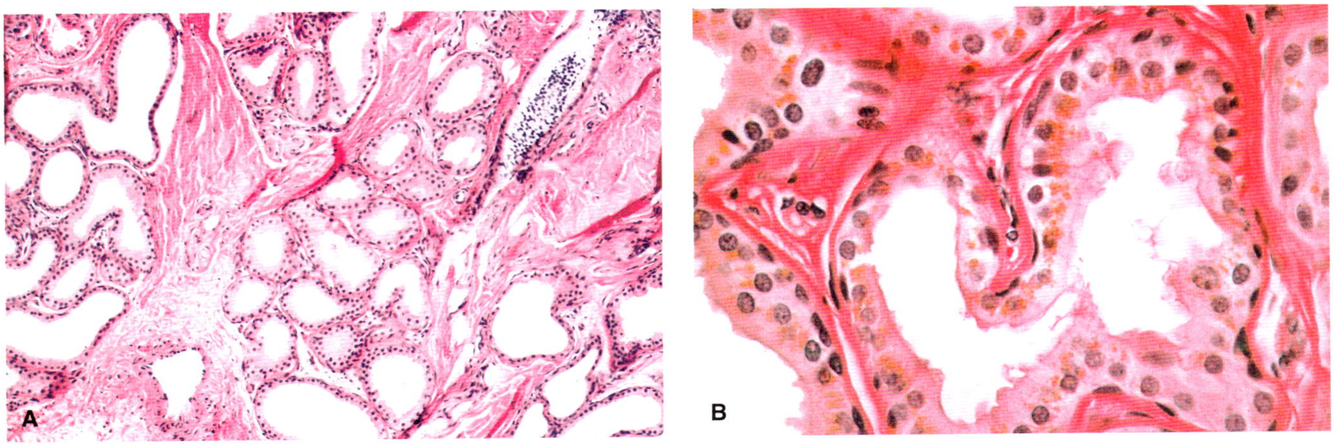

FIGURE 13.6 In addition to the hair follicles and sebaceous glands, the outer third of the external auditory canal is noteworthy due to the presence of modified apocrine glands called ceruminal glands that replace the eccrine glands seen in the auricular dermis. **A:** Ceruminal glands are submucosal in location and are arranged in clusters or lobules. **B:** They are composed of two cell layers, including the inner or secretory cells containing intracytoplasmic cerumen appearing as granular, golden-yellow pigmentation, and flattened-appearing myoepithelial cells located peripheral to the secretory cells. Focally, the secretory cells show holocrine (decapitation) type secretion.

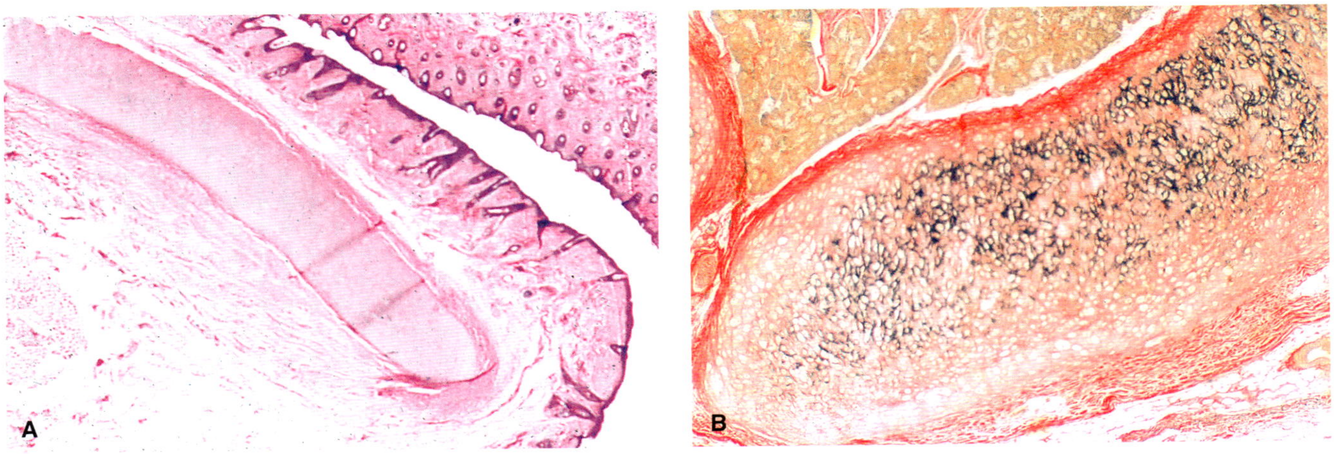

FIGURE 13.7 **A:** The cartilage of the external ear is elastic. **B:** Elastic stains show the abundant elastic fibers (black staining) in the auricular cartilage.

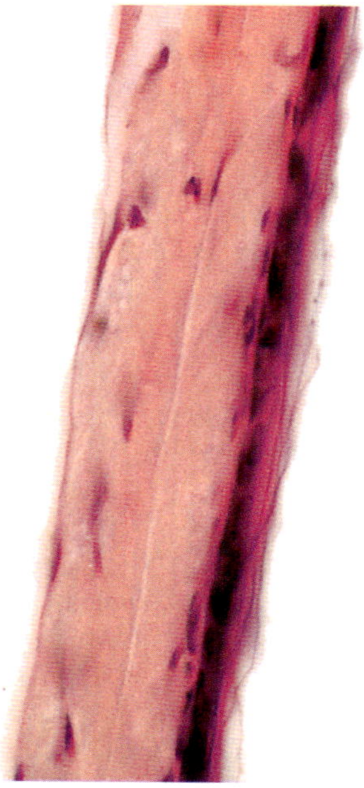

FIGURE 13.8 Section of pars tensa of tympanic membrane. The following layers may be distinguished from right to left: stratified squamous epithelium, lamina propria, radial arrangement of collagenous fibers, circular arrangement of collagenous fibers (i.e., *at right angles to former layer*), lamina propria, and middle ear epithelium.

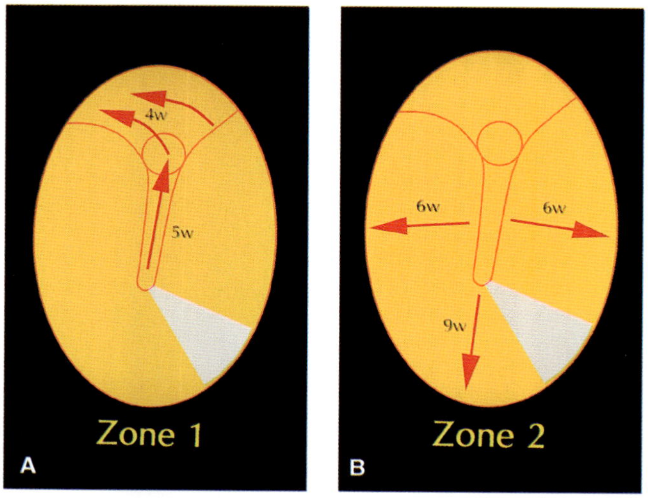

FIGURE 13.9 Summary of pathways of migration on tympanic membrane as determined by serial photography of dye markings. The tympanic membrane and adjacent deep external canal epithelium are depicted as being viewed on face. Two discrete pathways are present: **A:** Passing upward along a tongue of epithelium over the handle of the malleus to join epithelium moving in a posterosuperior direction over the pars flaccida region (*zone 1*). **B:** A radial pathway moving centrifugally from the pars flaccida and handle of malleus regions to the periphery (*zone 2*). The times given for each region are the weeks required for the dye to be completely cleared from that region.

AUDITORY EPITHELIAL MIGRATION

Auditory epithelial migration represents the mechanism by which keratin is removed from the tympanic membrane. Without such a self-cleaning process, the keratin squames normally produced by the stratified squamous epithelium of the tympanic membrane would continuously build up and interfere with the conduction of sound via the tympanic membrane. The entire epithelium including keratin moves from the tympanic membrane onto the deep external auditory canal. From the deep external auditory canal, the epithelium moves laterally to the junction of the deep (osseous) canal and the cartilaginous canal where it is desquamated (3–5).

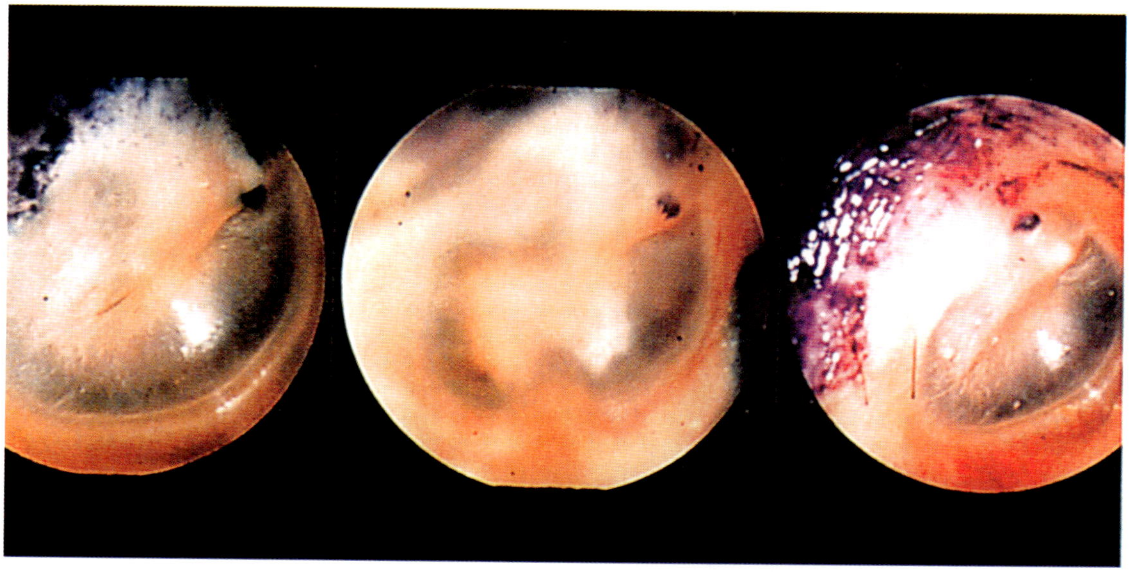

FIGURE 13.10 Pathway of auditory epithelial migration as shown by movement of blue dye daubed on tympanic membrane. Dye is seen in the first photograph on the day at which it was daubed, just anterior and inferior to the lateral process of the malleus. In the next photograph, taken 9 days later, it has moved posteriorly and superiorly to lie over that structure. Thirteen days later, in the third photograph, it has crossed the pars flaccida region, moving in the same direction toward the external canal.

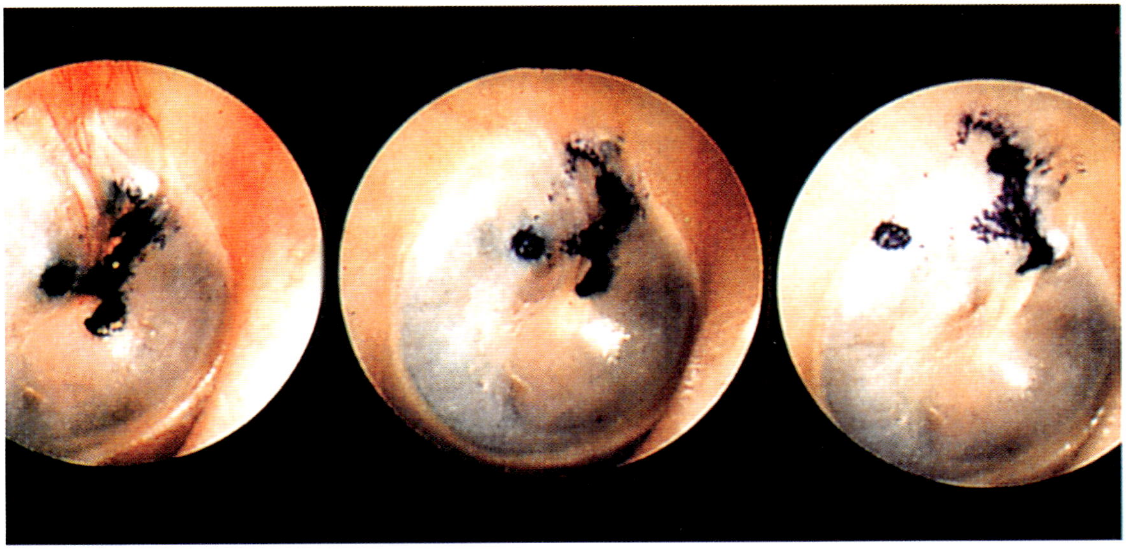

FIGURE 13.11 In this daubed tympanic membrane, an irregular array of dye is seen on the handle of the malleus region on the sixth day after its deposition, in the first photograph. By the fifteenth day, in the second photograph, a round dot that was just posterior to the handle of the malleus has separated and is commencing to travel backward, the main mass of dye moving discretely upward along the handle of the malleus. This process has advanced on the 27th day in the third photograph, the posterior dye having reached the back edge of the tympanic membrane and the large mass being now situated across the pars flaccida at an angle that has now changed to a posterosuperior one.

Auditory epithelial migration occurs in two separate and discrete pathways (Fig. 13.9) (6). In one pathway, the epithelium first moves upward over the handle of the malleus then posterosuperiorly across the pars flaccida and laterally over the deep canal (Fig. 13.10). The other pathway is radially moving centrifugally away from the handle of the malleus and the pars flaccida to the periphery of the tympanic membrane and then to the deep canal (Fig. 13.11). Michaels and Soucek have extensively evaluated the process of auditory epithelial migration (3–5) and correlated the pathways with development of the tympanic membrane epithelia and deep external canal in the embryo and fetus (6). The process of auditory epithelial migration may play a role in cholesteatoma development, although there is no definitive evidence linking auditory epithelial migration to the pathogenesis of cholesteatoma.

MIDDLE EAR

Embryology

The middle ear space develops from invagination of the first branchial pouch (pharyngotympanic tube) from the primitive pharynx. The Eustachian tube and the tympanic cavity develop from the endoderm of the first branchial pouch; the malleus and the stapes are derived from the mesoderm of the first branchial arch (Meckel cartilage) while the incus develops from the mesoderm of the second branchial arch (Reichert cartilage) (Fig. 13.2) (2).

Anatomy

The middle ear of the tympanic cavity lies within the temporal bone. It is located between the tympanic membrane and the squamous portions of the temporal bone laterally and the petrous portion of the temporal bone surrounding the inner ear medially. The anatomic limits of the tympanic cavity include: (a) lateral or internal aspect made up by the tympanic membrane and squamous portion of the temporal bone, (b) medial aspect bordered by the petrous portion of the temporal bone, (c) superior (roof) delimited by the tegmen tympani, a thin plate of bone which separates the middle ear space from the cranial cavity, (d) inferior (floor) aspect bordered by a thin plate of bone separating the tympanic cavity from the superior bulb of the internal jugular vein, (e) anterior aspect delimited by a thin plate of bone separating the tympanic cavity from the carotid canal housing the internal carotid artery, and (f) posterior aspect delimited by the petrous portion of the temporal bone containing the mastoid antrum and mastoid air cells (1,7,8). The tympanic cavity communicates anteriorly with the nasopharynx by way of the Eustachian (auditory or pharyngotympanic) tube and it communicates posteriorly with the mastoid air cells by way of the aditus and mastoid antrum. The contents of the tympanic cavity include the ossicles (malleus, incus, and stapes), the ligaments of the ossicles, the tendons of the ossicular muscles, the Eustachian tube, the tympanic cavity proper, the epitympanic recess, the mastoid cavity, and the chorda tympani of the facial (VII) nerve. The middle ear and external ear function as conduits for sound conduction for the auditory part of the internal ear.

Lateral Wall

The tympanic cavity extends above the level of the tympanic membrane as the epitympanic recess (attic). The head of the malleus and the body of the short process of the incus lie in this area. The epitympanic recess projects laterally above the external acoustic meatus; it is this portion of the tympanic cavity that has a part of the squamous portion of the temporal bone as its lateral wall.

Roof

The roof of the tympanic cavity is the tegmen tympani, a thin plate of bone separating the middle ear cavity from the cranial cavity. In children, the unossified petrosquamous suture of the tegmen tympani may allow the direct passage of infection from the middle ear to the meninges of the middle cranial fossa (1). In adults, especially in the setting of a long-standing history of chronic otitis media, compromise of the tegmen tympani, an already thin plate of bone may result in acquired encephalocele, in which glial-type tissue is present within the middle ear cavity. In addition, in adults, veins from the middle ear perforate the petrosquamous suture to end in the petrosquamous sinus and the superior petrosal sinus and may potentially transmit infection directly to the cranial venous sinuses (1).

Floor

The floor of the tympanic cavity is usually a thin plate of bone separating the cavity from the internal jugular vein. In the presence of a large superior bulb of the internal jugular vein, it may bulge into the middle ear and present dehiscences (9).

Posterior Wall

The posterior wall of the tympanic cavity opens through the narrow aditus ad antrum in the wider mastoid (tympanic) antrum. Below the aditus is a relatively thin bone separating the tympanic cavity from the antrum and it is from this posterior wall that the pyramidal eminence projects with an aperture at its apex from which the tendon of the stapedius muscle is transmitted. Above and behind the pyramidal eminence, the facial nerve curves downward to change its course from horizontal to vertical. The chorda tympani, arising from the facial nerve, then enters the tympanic cavity through the canaliculus of the chorda in the posterior wall.

Anterior Wall

The lower part of the anterior wall is part of the petrous apex. This area consists of a thin plate of bone, which may be incomplete or contain air cells, separating the cavity from the carotid canal in which the internal carotid artery is located. The upper part of the anterior wall is deficient since the canal containing the tensor tympani muscle opens in this location and immediately below this area is the tympanic orifice of the auditory tube.

Medial Wall

The medial wall of the tympanic cavity is the petrous portion of the temporal bone that surrounds the internal ear and separates the middle ear cavity and the inner ear cavity. Several important markings are found on its surface. These include the broad prominence produced by the anterior end of the lateral semicircular canal and the prominence of the facial (fallopian) canal produced by the horizontal portion of the facial nerve in its course between the inner and middle ears. The cochleariform process transmits the tendon of the tensor tympani muscle. Its apex is the landmark for the position of the turn (geniculum or external genu) between the anterolaterally and posteriorly directed horizontal portions of the facial nerve. Immediately below the facial canal is the *fossula fenestrae vestibuli* also referred to as the stapes niche, which contains the oval window, closed by the base of the stapes. Below the oval window is the promontory formed by the basal turn of the cochlea. The tympanic nerve plexus lies on the promontory. Below the back part of the promontory, the *cochlear fossula,* or round window niche, leads to the round window or fenestrae cochlea. Behind the promontory is a depression referred to as the sinus tympani. This site may harbor infections or transmit them to the ampullary end of the posterior canal and posterior end of the lateral canal (1).

MIDDLE EAR OSSICLES AND MUSCLES

The middle ear ossicles include the malleus, the incus, and the stapes (Fig. 13.12). The malleus is composed of a head, upper and lower manubria (handle), lateral process, and anterior process. The malleus is closely attached to the tympanic membrane by its manubrium (handle) and its lateral process, while its head projects above the epitympanic recess to articulate with the body of the incus. The anterior (long) process (processus gracilis) of the malleus extends obliquely downward from the neck toward the tympanosquamous fissure. The anterior process may reach the tympanosquamous fissure in infants, but the distal part is composed of connective tissue forming the anterior ligament of the malleus in adults (1). The malleus is also attached to the tympanic wall by superior and lateral mallear ligaments. The lateral ligament attaches the neck of the malleus to the margin of the tympanic notch.

The parts of the incus include its body, long and short processes. The body of the incus is fitted against the head of the malleus and lies in the epitympanic recess. The short process (crus) rests in a depression referred to as the fossa of the incus that is situated in the posterior wall of the tympanic cavity below the aditus ad antrum. The long process (crus) of the incus descends parallel and slightly posteromedial to the manubrium, of the malleus, but at its lower end turns medial to articulate with the stapes (1). The incus is held in place by a posterior ligament that attaches to its short process, and by a superior ligament to attach to the body. A knob-like expansion of the long crus of the incus occurs at the incudomalleal joint and is referred to as the lenticular process.

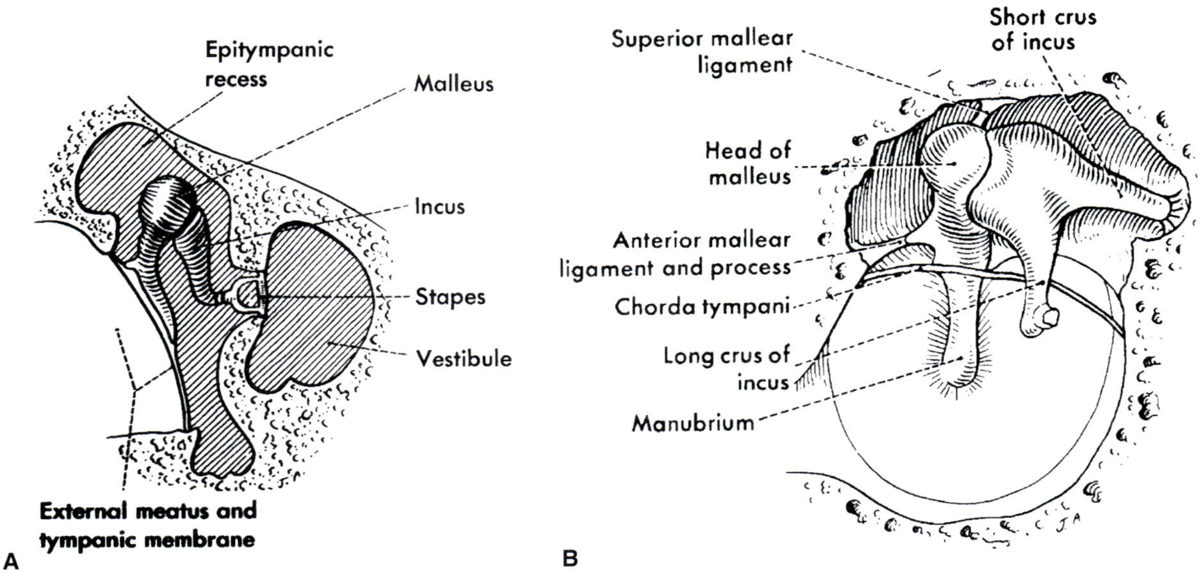

FIGURE 13.12 Diagrammatic depiction of the anatomy of the middle ear ossicles: **(A)** Frontal aspect (rotated through approximately 90 degrees) and **(B)** medial aspect.

The stapes is formed by its two crura, a head that lies at the junction of the crura and a footplate that lies on the oval window (Fig. 13.13). The head of the stapes articulates with the incus. From its articulation with the incus, the stapes passes almost horizontally to the oval (vestibular) window. The footplate of the stapes is attached to the oval window by the annular ligament. The latter, a ring of elastic fibers, allows movement of the stapes but seals any potential space between its footplate and the edges of the oval window.

The incudomalleolar and incudostapedial joints are diarthrodial synovial joints (see below). In addition to their ligaments, the stapes and the manubrium of the malleus have muscles attached to them. The stapedius muscle diminishes the excursion of the base of the stapes by its reflex contraction. Important functions ascribed to the stapedius muscle are to protect the inner ear from excessive sound and to improve discrimination for higher frequencies in speech. The stapedius muscle is innervated by a branch of the facial nerve.

FIGURE 13.13 Intact resected stapes showing from left to right its head, two crura and footplate.

The tensor tympani muscle draws the manubrium medially thereby tightening the tympanic membrane. The tensor tympani muscle primarily protects against excessive noise, but also functions in conjunction with the tensor veli palatini muscle to respond to swallowing and electric stimulation from the tongue. The action of these two muscles pump air from the tympanic cavity into the Eustachian tube forcing air into the nasopharynx and helping to open the isthmus (10). The tensor tympani muscle is innervated by a branch of the mandibular nerve.

EUSTACHIAN (AUDITORY) TUBE

The Eustachian tube extends from its tympanic ostium high on the anterior wall of the tympanic cavity to a nasopharyngeal ostium situated posterior to the inferior nasal concha (1). The tube is slightly S shaped and runs downward, medially and anteriorly to the nasopharynx because the tympanic ostium is approximately 2 to 2.5 cm higher than the nasopharyngeal end in adults. The length of the tube in adults varies from 31 to 38 mm (11). In infants, the tube is shorter, relatively wider, and more horizontal in its course and, therefore, an easier pathway for infections ascending from the nasopharynx to the tympanic cavity.

The tube can be divided into an osseous portion and a cartilaginous portion. The osseous portion is a bony canal that accounts for the lateral or tympanic third of the tube.

The anteromedial two-thirds have a cartilaginous and connective tissue wall and are referred to as the cartilaginous portion of the tube. The cartilaginous and osseous tubes meet at an obtuse angle.

HISTOLOGY

Tympanic Cavity Proper

The epithelial lining of the tympanic cavity is a single layer of respiratory epithelium of flattened to cuboidal epithelium (Fig. 13.14). Under normal conditions, there are no glandular elements within the middle ear (see below under Select Abnormalities and Pathology). Further, stratified squamous epithelium is not present in the tympanic cavity under normal conditions nor does squamous metaplasia occur in the middle ear (5). Ciliated pseudostratified columnar epithelium may be found in limited patches among the flattened or the cuboidal epithelium.

Eustachian Tube

The lining of the Eustachian (auditory) tube is a low ciliated epithelium throughout much of its length except as it approaches its nasopharyngeal end where it becomes ciliated pseudostratified columnar epithelium containing goblet cells. The cartilage of the nasopharyngeal portion of the Eustachian tube is of hyaline type and is surfaced by mucosa containing seromucinous glands (Fig. 13.15). Lymphoid tissue may be prominent, particularly in children, and is referred to as Gerlach tubal tonsil (Fig. 13.16). Reactive hyperplasia of this lymphoid tissue may occlude the Eustachian tube and provide a desirable milieu for otitis media, especially in children. The mucosa of the osseous portion of the Eustachian tube is separated from the carotid canal by a plate of bone measuring 1 mm in thickness (5). Dehiscence of the carotid canal is fairly frequent (12). Squamous carcinoma of the middle ear or Eustachian tube, a rare occurrence, may easily penetrate this area and gain access to the carotid artery with the potential for widespread dissemination (13).

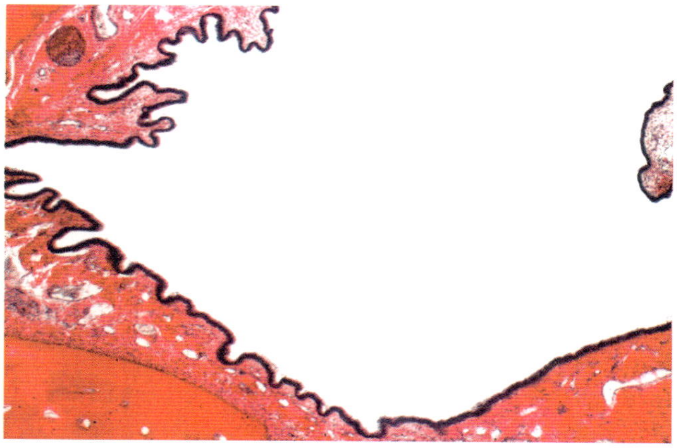

FIGURE 13.14 The epithelial lining of the tympanic cavity is a single layer of epithelium (cuboidal to respiratory). Under normal conditions glands are not identified within the tympanic cavity.

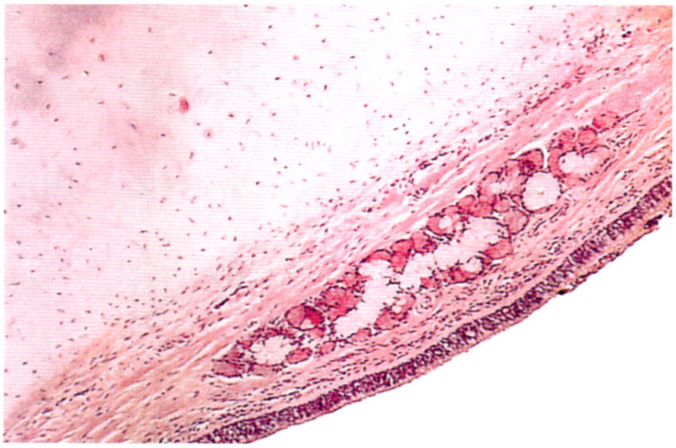

FIGURE 13.15 Cartilaginous portion of Eustachian tube with seromucinous glands.

Mastoid Air Cells

The mastoid air cells represent a network of intercommunicating spaces that emanate from the tympanic cavity (5). Each air cell is lined by flattened to cuboidal epithelium which rests on periosteum that covers a thin frame of lamellar bone (Fig. 13.17).

Pneumatization of the Temporal Bone

The rudimentary mastoid bone of the newborn contains a single air space, the antrum, surrounded by diploic bone

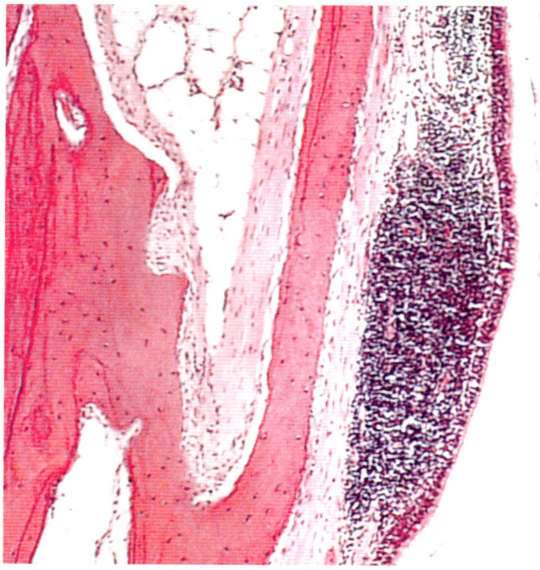

FIGURE 13.16 Mucosa of Eustachian tube. The lining is of ciliated columnar epithelium. In the lamina propria beneath, there are numerous lymphocytes, which are probably the result of inflammation.

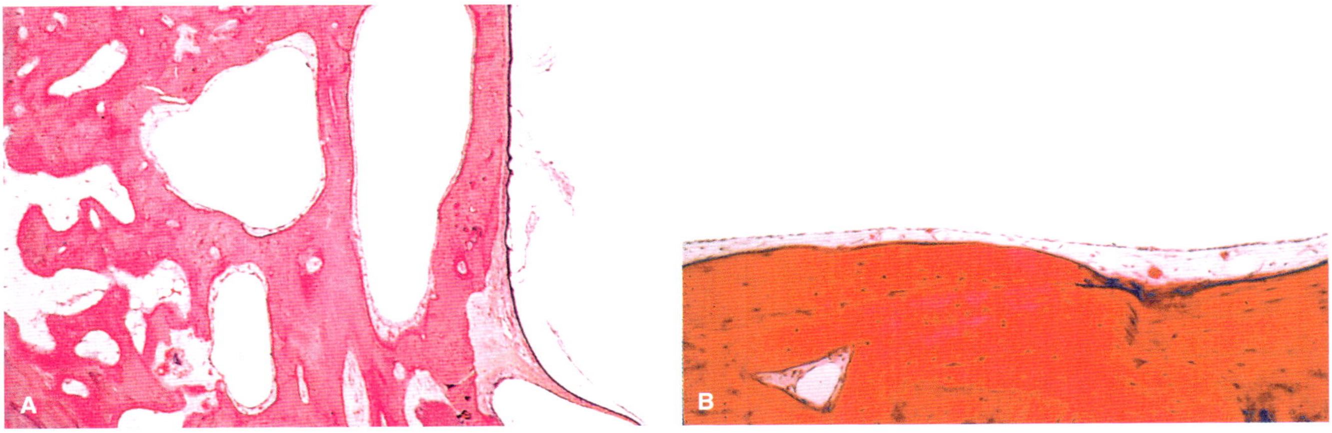

FIGURE 13.17 **A:** Mastoid air cells (*center*), tympanic membrane (*lower right*), and squamous epithelium of the osseous portion of the external canal (*right*). Note the thin covering of skin over the external ear canal and the proximity of bone to it. **B:** Higher magnification shows the very thin epithelium of the mastoid air cells resting on periosteum covering lamellar bone.

containing hematopoietic elements (14). As the mastoid process develops, the marrow spaces hollow out. The mesenchymal component occupying the space is resorbed and the developing air-containing cells become lined by the advancing endodermal epithelium. The mastoid process is constantly pneumatized in adults, although not in infants. The cells grow out from the antrum, as well as from each other forming complex interlocking chains of thin-walled cavities opening into each other. The antrum always contains air cells; the mastoid process is usually one of several types including pneumatized (containing air cells), diploic (containing marrow), mixed (containing air cells and marrow) or sclerotic. Approximately 80% of mastoid is well pneumatized by 3 to 4 years of age, although normal pneumatization fails to occur in approximately 20% of the individuals (1,15).

Middle Ear Ossicles

The middle ear ossicles develop from cartilage with a single center of ossification for bone; there is no epiphyseal ossification. The persistence of cartilage in each of the ossicles and the bifurcation of the stapes to form the crura with the obturator foramen between them distinguish the middle ear ossicles from other long bones (Fig. 13.18) (5).

The head of the stapes is formed of endochondral bone with a cartilaginous cap at the incudostapedial joint. The crura of the stapes are formed of periosteal bone only. From the middle ear aspect of the stapes footplate to its vestibular surface, the histologic findings include the flattened to cuboidal epithelium of the tympanic cavity, a thin layer of bone, cartilage, and a single flattened (perilymphatic) epithelial cell layer (Fig. 13.19).

The malleus and the incus, similar to long bones, have an outer covering of periosteal bone layer and an inner core of endochondral bone with well-formed haversian systems. The manubrium (handle) of the malleus is predominantly covered by retained cartilage rather than periosteal bone. The entire inner core of the manubrium, as well as the rest of the malleus is composed of endochondral bone. The anterior process is formed in the membrane early in fetal

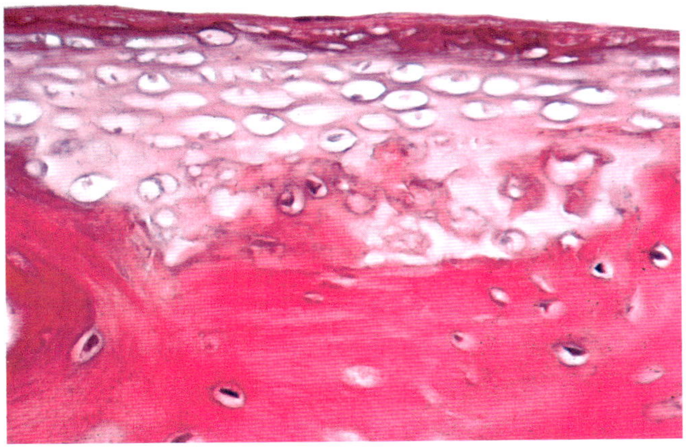

FIGURE 13.18 Stapes footplate showing persistence of cartilage.

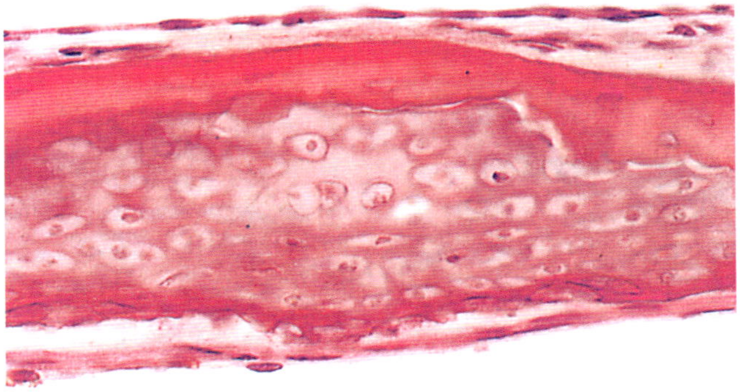

FIGURE 13.19 Stapes footplate. Beneath the cubical epithelium of the middle ear above, there is a thin layer of bone. Below this, the footplate consists of cartilage and there is a basal flattened layer of cells comprising the lining of the vestibule.

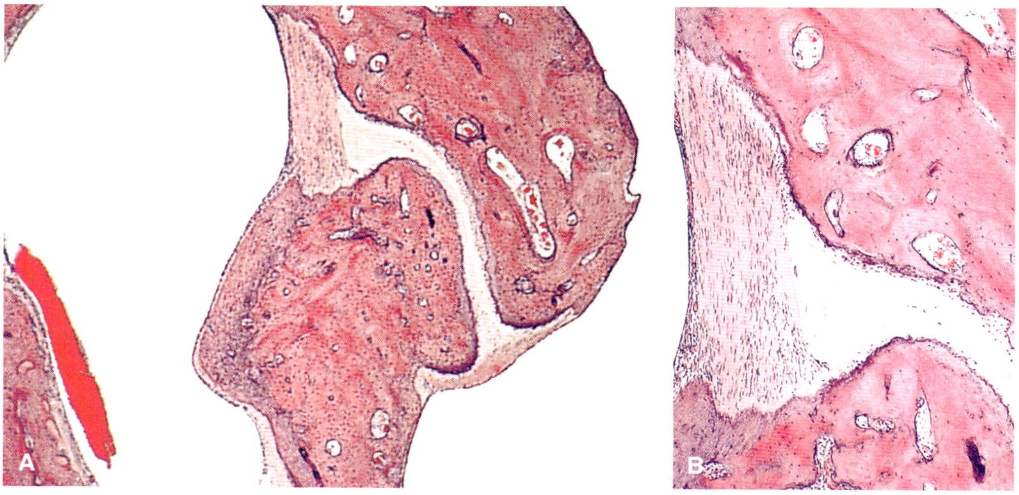

FIGURE 13.20 **A:** Incudomalleal joint. Note the joint capsule at each end of the joint. The joint space is occupied by the fibrocartilage of the articular disc. **B:** Higher power, note one end of joint capsule and articular disc.

life and merges with the malleus after its formation (5). At its superior aspect, the manubrium is separated from the tympanic membrane by a ligament covered by the middle ear epithelium. The short process of the incus shows a tip of the unossified cartilage.

Middle Ear Joints

Both the incudomalleal and incudostapedial joints are diarthrodial. Middle ear epithelium is present on the outer surface of the joint capsule and synovial membrane is present on its inner surface. The joint capsule is composed of fibrous tissue with a high elastic fiber content (5). The articular disc representing the space in between the articular ends is comprised predominantly of fibrocartilage (Fig. 13.20). The articular processes of both the malleus and the incus are covered by cartilage.

The annular ligament binds the cartilaginous edge of the stapes footplate to the cartilaginous rim of the vestibular window (stapediovestibular joint) (Fig. 13.21) and is composed of fibrous tissue with elastic fibers being prominent near the ligament surfaces (16). Cartilage also covers the articular surfaces of the stapediovestibular joints.

The fissula ante fenestram is the canal linking the middle ear with the vestibule, lies in the bone just anterior to the stapediovestibular joint and develops as a slit filled with fibrous tissue often with associated cartilage (Fig. 13.22).

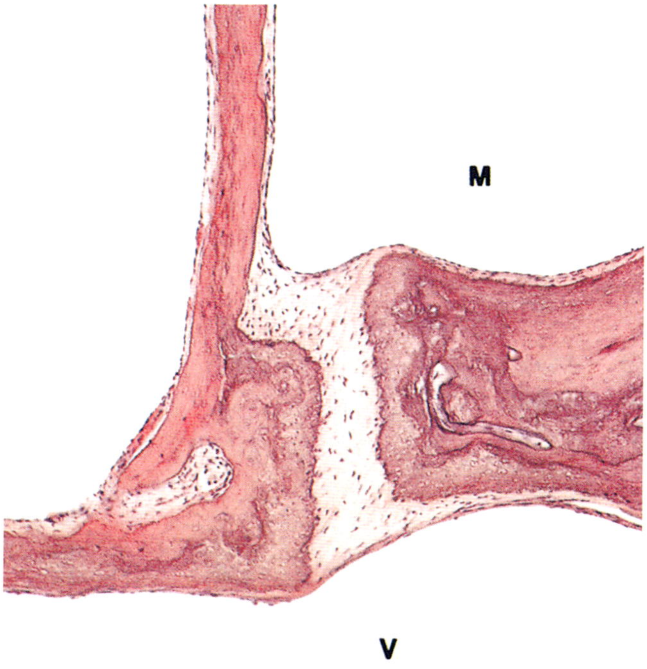

FIGURE 13.21 Stapediovestibular joint, part of footplate of stapes, adjacent bony labyrinthine wall, and crus of stapes. The footplate shows a lamina of cartilage on its vestibular surface, which is continuous with the cartilage of the stapediovestibular joint. *M*, middle ear cavity; *V*, cavity of vestibule.

Middle Ear Muscles

The muscles of the middle ear including the tensor tympani and stapedius muscles are composed of a central tendon formed by elastic tissue with muscle fibers radiating from

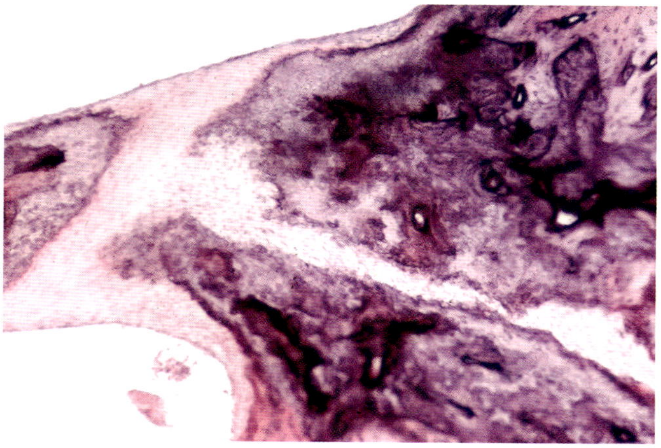

FIGURE 13.22 Fissula ante fenestram is the canal within the bone linking the middle ear with the vestibule and develops as a slit filled with fibrous tissue.

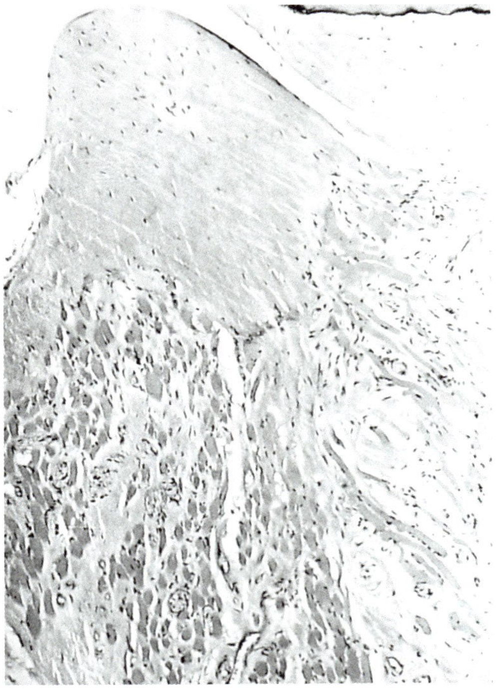

FIGURE 13.23 Stapedius muscle and tendon. The skeletal muscle fibers and fibrous bands between them radiate to a tendon.

it (Fig. 13.23). This configuration has been described as feather shaped. The tensor tympani muscle has a prominent mature adipose tissue component (Fig. 13.24) which is believed to function as insulation for the cochlea against electric effects from its contraction (5).

INNER EAR

Embryology

The first division of the ear to develop is the inner ear which appears toward the end of the first month of gestation (2,17).

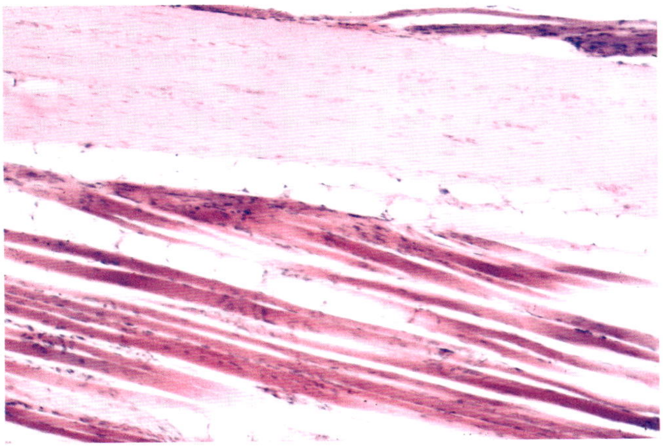

FIGURE 13.24 Tensor tympani muscle showing the presence of a mature adipose tissue component.

The membranous labyrinth, including the utricle, saccule, semicircular ducts, and cochlear duct arises from the otic vesicle (otocyst). The otic vesicle forms from the invagination of the surface ectoderm, located on either side of the neural plate, into the mesenchyme (Fig. 13.2). This invagination eventually loses its connection with the surface ectoderm. The bony labyrinth, including the vestibule, semicircular canals and cochlea arises from the mesenchyme around the otic vesicle (2,17,18).

The first division of the ear to develop is the inner ear, which appears toward the end of the first month of gestation (2,17). The membranous labyrinth, including the utricle, saccule, three semicircular ducts, cochlear duct, and endolymphatic sac arises from the placodal thickening of the ectoderm to become a closed otic vesicle (otocyst). The otic vesicle forms from the invagination of the surface ectoderm, located on either side of the neural plate, into the mesenchyme (Fig. 13.2). This invagination eventually loses its connection with the surface ectoderm. The membranous labyrinth, which is essentially tubular and saccular in turn, is filled with endolymphatic fluid. The early development of the membranous labyrinth takes place in mesenchyme and subsequently in the cartilage destined to form the petrous portion of the temporal bone (1). The membranous labyrinth lies in cavities excavated from this mesenchyme or cartilage. The space lying between the inner surface of the bony wall and the outer surface of the membranous labyrinth is the perilymphatic space, which develops by fusion of mesenchymal spaces to form larger ones surrounding the membranous portion. The bony labyrinth, including the vestibule, semicircular canals, and cochlea arises from the mesenchyme around the otic vesicle (2,17,18).

ANATOMY

The internal (inner) ear, or labyrinth, is embedded within the petrous portion of the temporal bone and comprises the medial portion of the temporal bone adjacent to the cranial cavity. The inner ear contains the membranous labyrinth, which is surrounded by the osseous labyrinth (Fig. 13.25). The membranous labyrinth contains the cochlea, which is the organ of hearing and the vestibular system, which maintains equilibrium.

Osseous Labyrinth (Otic Capsule)

The osseous labyrinth consists of the vestibular and the cochlear capsule. The central portion of the osseous labyrinth cavity is the vestibule, a large ovoid perilymphatic space approximately 4 mm in diameter. It contains both the saccule and the utricle of the membranous labyrinth. The floor of the bony vestibule contains several structures. There is an elliptical recess for the anterior end of the utricle, as well as a spherical recess for the saccule; the latter is

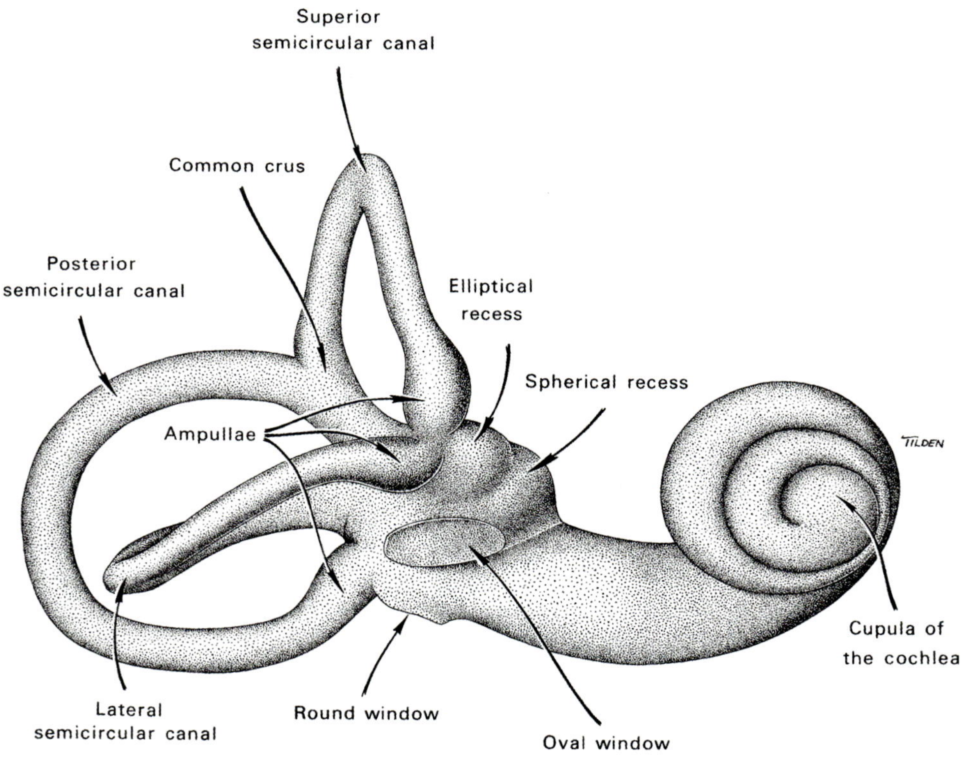

FIGURE 13.25 Schematic depicting the osseous labyrinth.

located anterior and lateral to the elliptical recess. The base of the stapes is located in the oval window on the lateral wall of the vestibule. Through the stapes, the perilymph of the vestibule receives vibrations from the tympanic membrane and the ossicular chain set up by sound waves reaching the tympanic cavity. Along the medial wall and floor of the vestibule, where it abuts the lateral end of the internal acoustic meatus, are small openings for the entrance of the nerve branches to the vestibular portion of the ear (1).

The bony cochlea, a part of the otic capsule, is a hollowed spiral with two- to three-fourths turns diminishing from a relatively broad base to a pointed cupula or apex. It is so named due to its resemblance to a snail shell. The base of the cochlea lies against the anteromedial surface of the vestibule and next to the anterior surface of the lateral (blind) end of the internal auditory canal. A central core of bone called the modiolus runs forward from the cochlea but does not reach the cupula. It is around this central core that the spiral channels of the cochlea (perilymphatic and endolymphatic) are arranged. A layer of bone arranged in a spiral fashion unites the modiolus to the peripheral wall of the bony cochlea and separates successive spiral cavities from each other (1). The modiolus is hollow to accommodate the cochlear nerve. The base of the modiolus lies against the lateral end of the internal auditory canal to where the cochlear nerve runs.

The vestibular aqueduct extends through the otic capsule from the vestibule to the posterior cranial fossa transmitting the endolymphatic duct. The terminal end of the vestibular aqueduct is the endolymphatic sac, a dilated area that ends blindly outside the dura (1). The cochlear duct opens at one end into the lower end of the scala tympani and at the other end into the subarachnoid cavity (1). The issue as to whether the cochlear duct represents an open channel between the subarachnoid space and the perilymphatic space at the lower end of the scala tympani remains controversial (1). A possible role ascribed to the cochlear duct is to serve as part of the pressure-adjusting mechanism of the perilymph in conjunction with the round window (19–21).

Membranous Labyrinth

The membranous (otic) labyrinth is the spiral-appearing structure that resembles the shell of a snail. Its principle components are the cochlear duct, the utricle, the saccule, the ductus reuniens, the semicircular canals with their ampullae, and the endolymphatic sac and duct.

Cochlear Duct

The membranous cochlea or cochlear duct is a cone-shaped, spirally oriented membranous tube between the osseous spiral lamina and the outer osseous wall of the cochlea to which it is attached (7). The cochlear duct, also referred to as the scala media, lies between the scala vestibuli and the scala tympani (Fig. 13.26). These three compartments are fluid-filled. The cochlear duct and membranous labyrinth contain endolymph, whereas the scala vestibuli and the scala tympani contain perilymph. The cerebrospinal fluid communicates directly with the perilymphatic space through the cochlear aqueduct (Fig. 13.27). The cochlear

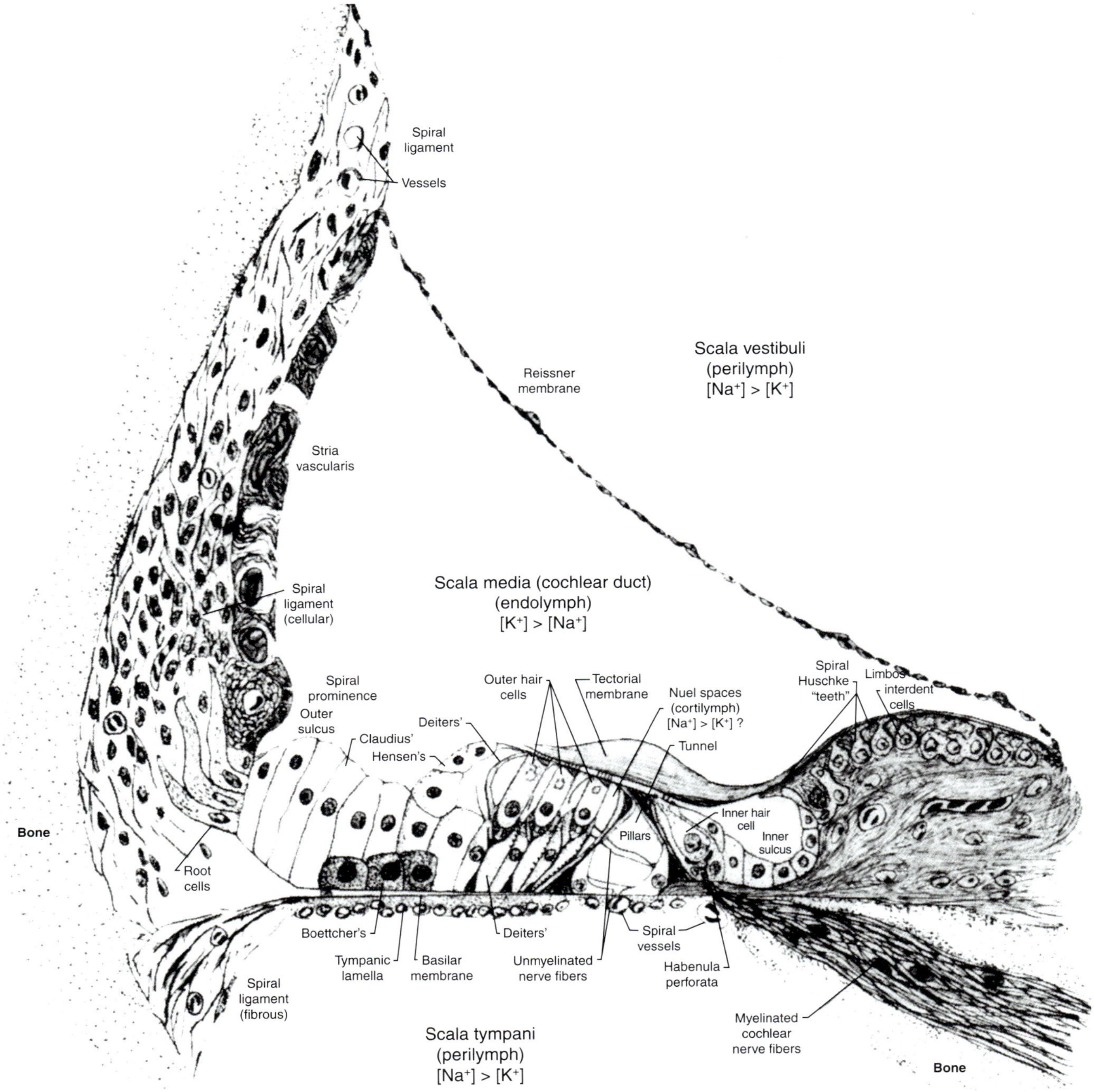

FIGURE 13.26 Schematic illustration of the membranous labyrinth, the latter containing the cochlea (organ of hearing) and the vestibular (system of balance) systems, showing the relationship between the endolymph containing scala media (cochlear duct) and the perilymph containing scala vestibuli and scala tympani. (From Brown JL. Sensory systems. In: Brobeck JR, ed. *Physiological Basis of Medical Practice*. 10th ed. Williams & Wilkins; 1979:8.1–8.135.)

duct contains the sensory (end) organ of hearing known as the spiral organ of Corti. The organ of Corti rests on the basilar membrane, which separates the cochlear duct from the scala tympani (Fig. 13.26). Together, the organ of Corti and the basilar membrane form the spiral membrane, which is the floor or tympanic wall of the cochlear duct. The spiral ligament is a thickened modified portion of periosteum on the bony cochlea. It forms the outer curved wall of the cochlea duct and adjacent parts of the scalae. The scala tympani lies below the basilar membrane, while the scala vestibuli lies above the cochlear duct and is separated from it by Reissner membrane. Reissner membrane forms the roof of the cochlear duct. The scala tympani and scala vestibuli communicate with each other only at the apex known

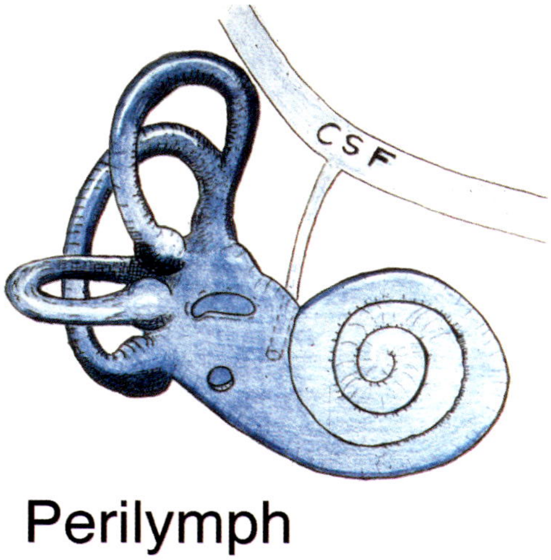

FIGURE 13.27 Schematic depicting the direct communication between the perilymphatic space and cerebrospinal fluid (CSF) through the cochlear aqueduct (perilymphatic duct).

as the helicotrema. The scala vestibuli winds toward the apex of the cochlea at the helicotrema becoming the scala tympani, which, in turn, coils back toward the round window (Fig. 13.28). The scala vestibuli and the scala tympani communicate with the middle ear via the oval window and round window, respectively. The scala tympani ends blindly at the round window membrane but the scala vestibuli opens up at that level into the perilymphatic space of the vestibulum. The cochlear duct connects with the vestibular system via ductus reuniens located at the saccule. In this way, the three semicircular canals that comprise the vestibular system are filled with endolymph.

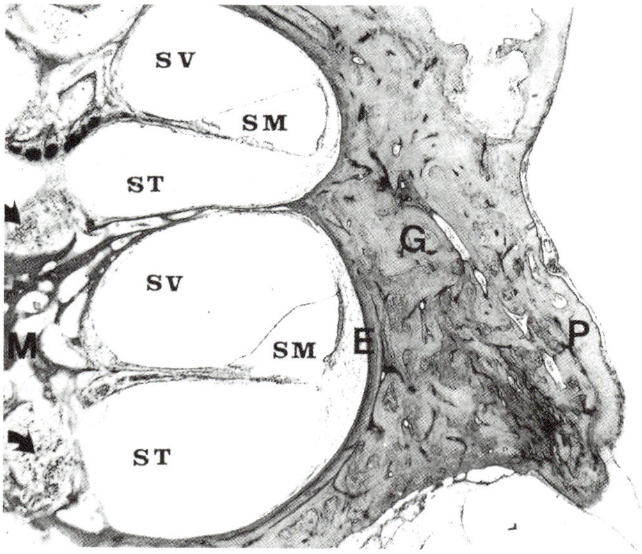

FIGURE 13.28 Cochlea, bony cochlea, and modiolus. *Arrows*, spiral ganglion cells of basal and middle coils in modiolus; *E*, endosteal layer of bone; *G*, endochondral layer containing globuli interossei; *P*, periosteal layer; *SM*, scala media; *ST*, scala tympani; *SV*, scala vestibuli.

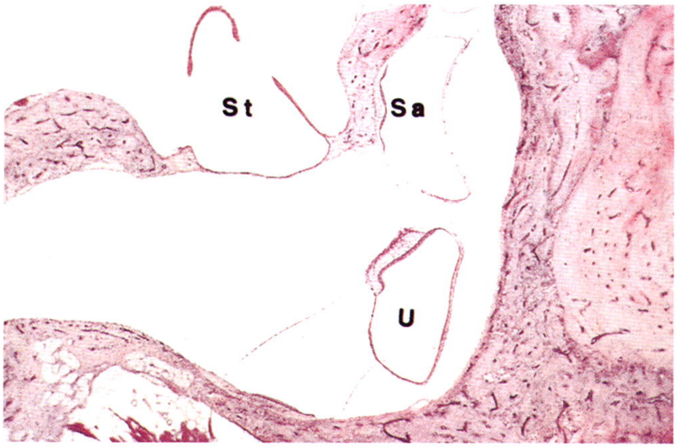

FIGURE 13.29 Ventricle of cat showing utricle (*U*) and saccule (*Sa*). *St*, stapes.

Utricle

The utricle is an oval-shaped portion of the membranous labyrinth that lies superior to the saccule in the medial wall of the vestibule (Fig. 13.29). It is larger than the semicircular ducts and receives both ends of each semicircular duct (total of five) since the anterior and posterior ducts share a common opening. The macule of the utricle, located on the inferior surface of the utricle (utricular recess), is a sensory end organ. The utriculosaccular duct usually arises from the utricle and communicates with the endolymphatic duct and also connects the utricle to the saccule. As previously indicated, the three semicircular canals communicate with the utricle via openings formed by the union of the nondilated or nonampullary ends of the superior and posterior canals termed the common crus.

Saccule

The saccule is located anteromedial to the upper (anterior) end of the utricle (Fig. 13.29) and tends to be more round than the utricle. The saccule and the utricle are continuous via the utriculosaccular duct and with the cochlear duct by the ductus reuniens (i.e., canalis reuniens of Hensen) (1). The macule of the saccule contains the sensory nerve endings of this portion of the inner ear and is an oval thickening on the lateral wall.

Semicircular Canals

The semicircular ducts include the anterior or superior duct, the posterior duct and the lateral ducts. The end of each semicircular duct is expanded to form the ampulla. The anterior duct is directed anterolaterally, the posterior duct is directed posterolaterally, and the lateral ducts form a laterally directed angle of approximately 90 degrees between themselves. The bony or osseous canals follow a similar direction. The three semicircular canals communicate with the utricle via openings formed by the union of the nondilated or nonampullary ends of the superior and posterior canals termed the common crus. From the common crus, the anterior duct curves upward while the posterior duct curves backward and then downward. The other or (membranous) ampullary ends of the semicircular canals contain the sensory endings of the ducts. At the ampullary ends, the anterior and posterior ducts empty into the utricle. The lateral semicircular duct lies in an approximate horizontal plane; both of its ends also connect to the utricle, with the anterior end being the ampulla.

Endolymphatic Duct and Sac

The endolymphatic duct traverses the medial portion of the petrous pyramid in its own bony canal, the vestibular aqueduct (7). The endolymphatic duct can be divided into segments. The first segment is the dilated portion referred to as the sinus. It represents the common channel into which the utricular and saccular ducts open. The next segment of the duct is the narrow isthmus. After the isthmus, the duct widens again to become the endolymphatic sac. Most of the endolymphatic sac is located within the funnel-shaped cranial aperture of the cochlear duct lying within a duplication of the posterior fossa dura partially covered medially by a thin bony shelf referred to as the operculum. The endolymphatic sac ends in a terminal dilatation, or fovea, of the sac. Two portions of the endolymphatic sac are recognized: a proximal rugose portion with an irregular lumen caused by numerous folds of the epithelial lining and a distal portion with a smooth epithelial lining. Both the endolymphatic sac and the duct show irregularly placed thin papillary outgrowths from their epithelial base into the lumen. These may sometimes be identified on radiologic imaging in the living patient. The membranous vestibular system contains the receptor organs for sense of motion and position.

INNER EAR INNERVATION

The neural structures of the inner ear, including the VIII cranial nerve (vestibulocochlear) and the VII cranial nerve (facial) enter the inner ear through the internal auditory canal. The VIII cranial nerve is variably referred to as the acoustic, auditory, or vestibulocochlear nerve. This nerve functionally consists of vestibular and cochlear divisions. In the internal auditory canal, these two parts are closely associated but at the lateral end of the canal the nerve trunk divides into three parts including two vestibular and one cochlear.

The vestibular nerve arises from the bipolar cells of the superior and inferior divisions of the (afferent) vestibular or Scarpa ganglion located at the lateral end of the internal auditory canal. Peripherally, the vestibular nerve divides into two main divisions, the superior and inferior divisions. The superior part of the ganglion gives nerves to the ampullae of the lateral and anterior (superior) canals and to the saccular and utricle maculae. The inferior part of the ganglion gives rise to a posterior ampullary nerve and a nerve to the saccule. The inferior part also gives rise to a branch to the cochlear division.

The cells of origin for the cochlear nerve form the spiral ganglion, which represents the first of the four neurons between the auditory end organ and the auditory cortex. The spiral ganglion is located in coils of the modiolus at the base of attachment of the osseous spiral lamina (Fig. 13.28). The osseous spiral lamina is a thin trabecula of bone termed the habenulae perforate. It surrounds afferent nerve fibers that run from the organ of Corti to the acoustic nerve, as well as efferent fibers that arise from the olivocochlear system of Rasmussen and extend to the outer hair cells. The latter make their exit from the brain with the vestibular part of the nerve and join the cochlear branch of the VIII cranial nerve via the vestibulocochlear communicating branch or nerve of Oort through the modiolus (1). Before reaching the modiolus, the nerve fibers are unmyelinated but become myelinated upon reaching the cochlear modiolus. The central fibers or axons of these bipolar neurons unite to form nerve bundles and pass from the cochlear modiolus through nerve channels in the osseous spiral foraminous tract into the internal auditory canal where they form the cochlear nerve.

The vestibulocochlear nerve is usually connected to the facial nerve within the internal auditory canal. Together, the three nerves enter the posterior cranial fossa, traverse the cerebellopontine angle, and enter the brain stem at the lower posterolateral aspect of the pons. The central auditory pathways consist of three additional neurons that form numerous connections with nuclei throughout the central nervous system as part of a complex auditory reflex system that ultimately converges on the auditory cortex in the anterior transverse gyrus of the superior temporal lobe (7,8).

The facial nerve enters the temporal bone through the internal auditory meatus within the petrous portion of the temporal bone in company with (lying above) the VIII cranial nerve and the internal auditory artery. The facial nerve then passes Bill bar, which represents a pointed bony projection separating the facial nerve from the superior division of the vestibular nerve. The facial nerve pierces the arachnoid and dura at the outer end of the canal where it enters the facial canal (i.e., aqueduct of Fallopius). This canal continues for a short distance such that the facial nerve comes to lie just above the cochlea where it bears the geniculate ganglion. The greater petrosal nerve comes off the geniculate ganglion and passes anteriorly and medially to enter the middle cranial fossa. Immediately beyond the geniculate ganglion, the facial nerve turns sharply at the external genu (i.e., geniculum) laterally and posteriorly. As it runs backward through the lateral wall of the vestibule/medial wall of the tympanic cavity, the facial nerve inclines downward and laterally at the bony prominence of the facial canal. This bulge is a normal finding that may be large enough to cover the oval window and base of the stapes. The facial nerve then curves broadly downward to run almost vertically through the mastoid process to the stylomastoid foramen. At this point, the facial canal ends and the nerve exits the skull. Shortly before leaving the stylomastoid foramen, the facial nerve gives off the chorda tympani, which is composed of sensory and preganglionic motor fibers. Slightly above the stylomastoid foramen, the chorda tympani leaves the facial trunk and takes a recurrent course upward and forward in its canaliculus ("iter chordae posterius") to enter the tympanic cavity through its posterior wall. It passes between the malleus and incus within the tympanic cavity and leaves the tympanic membrane through a canal in the pterygotympanic fissure ("iter chordae anterius"). The chorda tympani then joins the lingual nerve to be distributed to the anterior two-thirds of the tongue (taste buds) and to the submandibular ganglion through which postganglionic fibers reach the submandibular and sublingual salivary glands.

HISTOLOGY

Osseous Labyrinth

The cavity of osseous labyrinth (otic capsule) surrounds and replicates the outline of the membranous labyrinth lying within it. The osseous labyrinth is extremely dense and includes three layers: an outer periosteal layer, an inner layer abutting the membranous labyrinth lined by a thin layer of endosteum, and a middle layer that contains calcified cartilaginous matrix referred to as globuli interossei or globuli ossei (Fig. 13.30) (5). The density of the osseous labyrinth is necessary to insulate and safeguard the delicate vibrations of the fluids contained within it and is necessary in maintaining the integrity and functions of hearing and balance (5). As noted by Michaels (5), the bone of the adult osseous labyrinth is neither fully lamellar nor entirely woven; the osseous labyrinth lacks the normal developmental process of removal and replacement of calcified cartilaginous matrix and primitive bone (5).

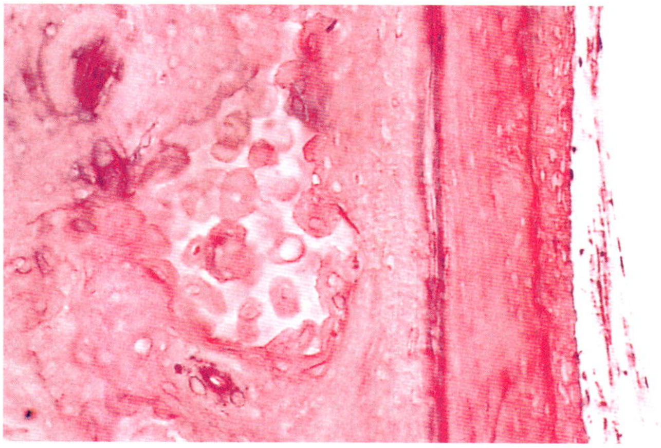

FIGURE 13.30 Globuli ossei (*left*) and endosteum (*right*) of cochlea.

Membranous Labyrinth

The membranous labyrinth consists of epithelium-lined channels surrounded by connective tissue. Its three basic divisions include the semicircular canals, the utricle and saccule, and the cochlear duct. These structures are histologically similar, containing a specialized thickened epithelium surrounded by and attached to a fibrogelatinous membrane. The specialized epithelium consists of supporting cells and neuroepithelium or hair cells. The neuroepithelia have hair-like processes that project from the free edge of the cells.

Cochlea

The organ of Corti consists of neurotransmitting hair cells that rest on the basilar membrane and is arranged in a spiral like the duct itself (Fig. 13.31). The organ of Corti consists of supporting cells and hair cells. The supporting, or pillar, cells are mostly phalangeal cells (cells of Deiters) that are arranged in two groups: a single row of cells on the inner aspect and three to five rows of cells on the outer aspect. The number of cells present on the outer aspect of the organ of Corti varies depending on their location within the cochlea; there are more rows of cells toward the apex and fewer rows of cells toward the base. The inner row of phalangeal cells is associated with a single layer of hair cells; the outer rows of phalangeal cells alternate with rows of hair cells. The phalangeal cells are so-named due to the shape of the stiff processes that project from the cell membranes and contribute to the reticular membrane that covers the free surface of the organ (1). The hair cells have numerous

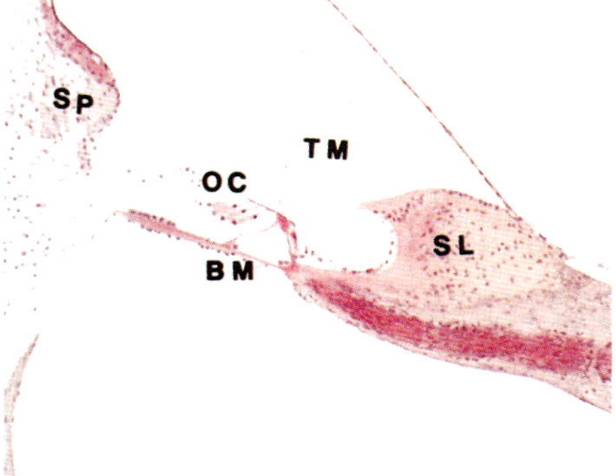

FIGURE 13.31 Scala media of cat. *BM*, basilar membrane; *OC*, organ of Corti; *RM*, Reissner membrane; *SL*, spiral limbus; *SP*, spiral prominence; *SV*, stria vascularis; *TM*, tectorial membrane.

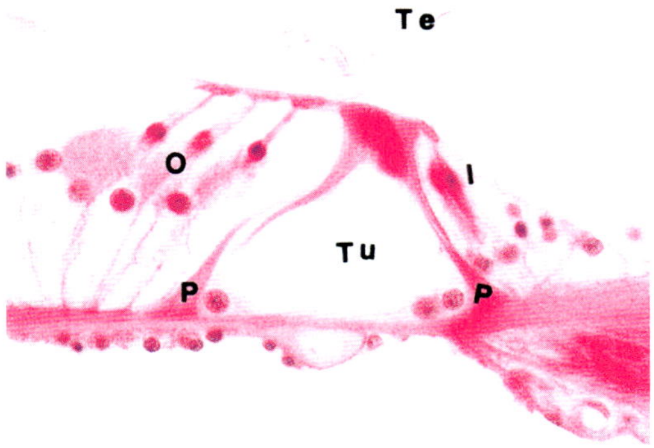

FIGURE 13.32 Higher power of organ of Corti from Figure 13.31. *I*, inner hair cells; *O*, outer hair cells; *P*, pillar cells (walls of tunnel); *Te*, tectorial membrane; *Tu*, tunnel of Corti.

(40 to 100 per cell) "hairs" projecting from the reticular (cuticular) surface. The outer hair cells are most sensitive and are believed to be responsible for the cochlear microphonics. They are short and wedge-shaped and are located between the apices of the phalangeal cells in order to reach the basilar membrane (1). The inner hair cells are long, less susceptible to damage than the outer hair cells and are believed to be less sensitive to sound. Intercellular spaces among the cells of the organ of Corti are apparently filled with intercellular substance. The largest of these spaces is the canal of Corti. It runs the entire length of the organ of Corti between inner and outer rows of phalangeal and hair cells and (Fig. 13.32). The tunnel of Corti is bounded by special supporting cells, the inner and outer pillars (Corti rods) (Fig. 13.32). The tunnel and pillars together form the Corti arch.

The basilar membrane is a fibrous tissue that supports the organ of Corti and is attached to the tectorial membrane. The basilar membrane contains fibers that pass from the bony spiral lamina to the spiral crest of the spiral ligament. The basilar membrane increases in size from the base to the apex of the cochlea and is felt to have resonator action with deformation of the membrane by sound beginning at its lower end traveling toward the helicotrema. The tectorial membrane is a gelatinous structure with numerous fine fibers. Like the basilar membrane, the tectorial membrane increases in size from the base to the apex of the cochlea, and is believed to have vibratory effect on the hair cells.

Together, the organ of Corti and the basilar membrane form the spiral membrane, which is the floor or tympanic wall of the cochlear duct. The spiral ligament is a thickened, modified portion of periosteum of the osseous cochlea, which forms the outer or curved wall of the cochlear duct and adjacent parts of the scalae. The vestibular, or Reissner membrane, is thin and consists of two layers of cells: an inner cell layer of ectodermal origin that contains epithelial-like cell clusters and an outer mesoderm-derived layer consisting of large, flat and elongated cells (5). This membrane forms the roof of the cochlear duct. The vestibular membrane bulges toward the scala vestibuli in Ménière disease. The outer (vertical) wall of the cochlear duct contains the stria vascularis, which is supplied by 30 to 35 small arteries originating from the modiolar region of the scala vestibuli and pass outward to the lateral wall of the osseous labyrinth (Fig. 13.31) (22). The stria vascularis is the likely source of the endolymph (23). The tissue spaces of the spiral ligament serve as a site of absorption (24). The stria vascularis is altered in ototoxic conditions as it may occur secondary to use of cisplatin, diuretic agents, and other drugs as discussed subsequently (5).

Semicircular Canals, Utricle, and Saccule

The end of each semicircular duct is expanded to form the ampulla. The sensory endings in the ampullae of the ducts are the cristae. Each crista consists of thickened epithelium; above each crista rests a gelatinous formation of viscous protein polysaccharide called the cupola. The hairs of the neuroepithelial hair cells project into the base of the cupola. As a result of the gelatinous nature of the cupola, it may be bent by the pressure of the endolymph, which apparently stimulates the hair cells and, therefore, the nerve endings of the cristae.

The utricle and the saccule represent the two main membranous structures of the vestibule. They are lined by a sensory epithelium known as the macula (Fig. 13.33). The

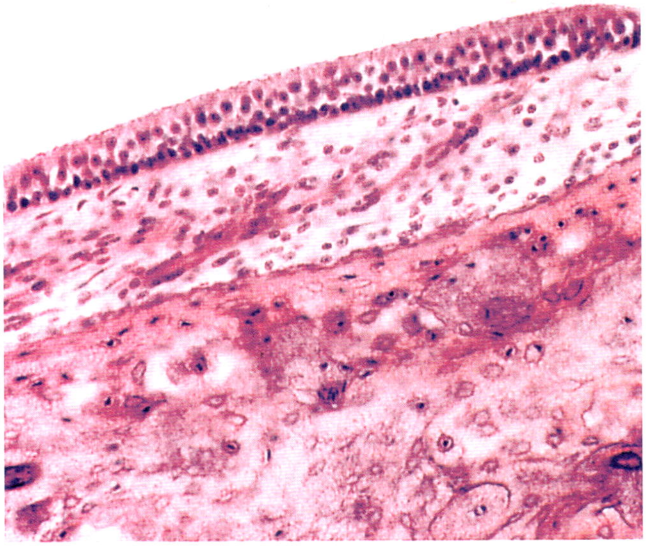

FIGURE 13.33 Higher power of part of Figure 13.29, showing macule of saccule.

FIGURE 13.34 Schematic representation of the ultrastructure of vestibular hair cells showing the principal features of type I (**A**) and type II (**B**) hair cells and their supporting cells. (Courtesy of the estate of Dr. George T. Nager.)

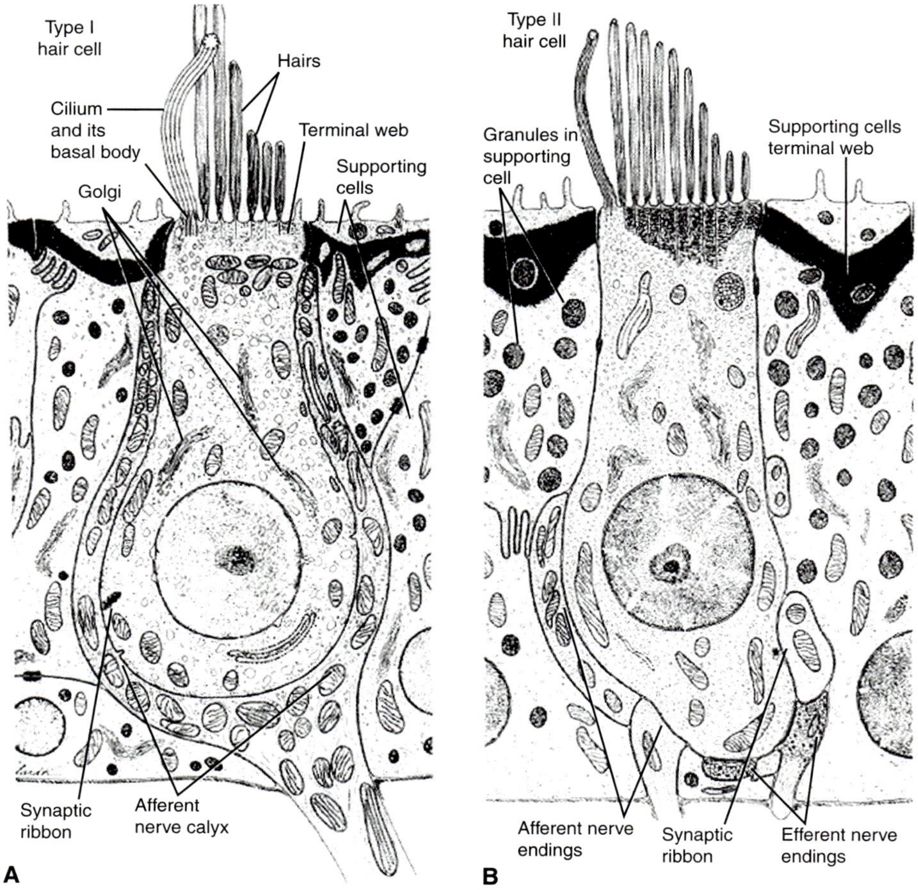

maculae are identical to one another in structure and are similar to the cristae of the semicircular canals. By transmission electron microscopy, these sensory cells are of two types: the type 1 cell is flask-shaped with a swollen basal portion and the type 2 cell is cylindrical. Type 1 cells are attached to fibers of the sensory nerves by a wide chalice-like terminal and the terminal of type 2 cells is connected by button-like attachments of the nerve (Fig. 13.34) (5). The sensory epithelium consists of hair cells which, in turn, have stiff, immotile projections embedded in the gelatinous otolithic membrane. The otolithic membrane also contains embedded crystalline bodies (i.e., otoliths) that contain calcium carbonate and a protein suspended in a jelly-like polysaccharide. It is only in the presence of otoliths that the maculae differ from the other sensory areas of the ear.

Nerves and Paraganglia

Most cranial and spinal nerves have glia extending only a fraction of a millimeter beyond their external origins (25,26). The optic nerve contains neuroglia throughout its length and, thus, is a tract of brain rather than a true nerve. The vestibulocochlear nerve (cranial nerve VIII) typically has glia extending from 6 to 8 mm along its course (Fig. 13.35). This distribution of glia helps to explain the more frequent occurrence of glial tumors on this nerve as compared to the other cranial nerves (25). The vestibular and cochlear divisions of the nerve are fused near the entrance to the internal auditory meatus. At this point, there is an abrupt transition of the coverings of the nerve fibers from the pale staining oligodendroglia proximally to the darker staining Schwann cells distally (Fig. 13.36). This glial–Schwann sheath junction is termed the Obersteiner–Redlich line. Vestibular schwannomas (also referred to as acoustic neuromas) may arise anywhere between this junction and the cribrosa area at the fundus of the canal (27).

Paraganglia similar in structure to the carotid body are normally present in the ear and may give rise to jugulotympanic paragangliomas. Most the paraganglia are found in relation to the jugular bulb, and a minority are found under the mucosa of the medial side of the middle ear promontory (Fig. 13.37).

Endolymphatic Sac and Duct

The lining epithelium of the endolymphatic duct is low cubical (Fig. 13.38) and the epithelium of the endolymphatic sac is taller with a papillary appearance (Fig. 13.39). An aggressive neoplasm, termed endolymphatic sac tumor, is presumed to originate from the endolymphatic sac

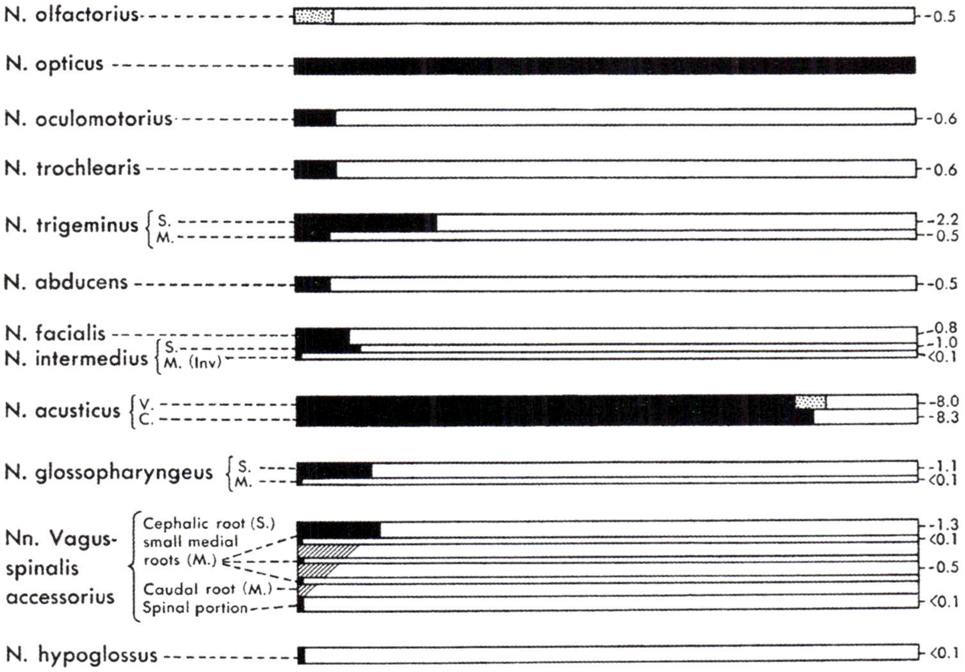

FIGURE 13.35 Schematic illustration of the cranial and spinal nerves showing that most of the nerves have glia extending for only a fraction of a millimeter beyond their external origins with the exception of the optic nerve which is really a tract of brain given the presence of neuroglia throughout its length and the VIII (vestibulocochlear) nerve. The VIII cranial nerve typically has glia extending from 6 to 8 mm along its course. (From Hollinshead WH. The cranium. In: Hollinshead WH, ed. Anatomy for Surgeons. 3rd ed. Philadelphia, PA: Harper and Row; 1982:27.)

epithelium (28). This tumor, initially considered to represent a low-grade malignancy (i.e., adenocarcinoma), is locally destructive but does not metastasize. It is characterized by low-cuboidal epithelium or columnar cells with a papillary or glandular architecture (28,29). The neoplastic cells in the endolymphatic sac papillary tumor are immunoreactive for cytokeratins, cytokeratin 7, EMA, carbonic anhydrase IX (membranous), PAX8 (nuclear) and GLUT1 with limited staining for S100 protein, GFAP, and vimentin. The tumor cells are negative for renal cell carcinoma marker (RCC), CD10, thyroglobulin, thyroid transcription

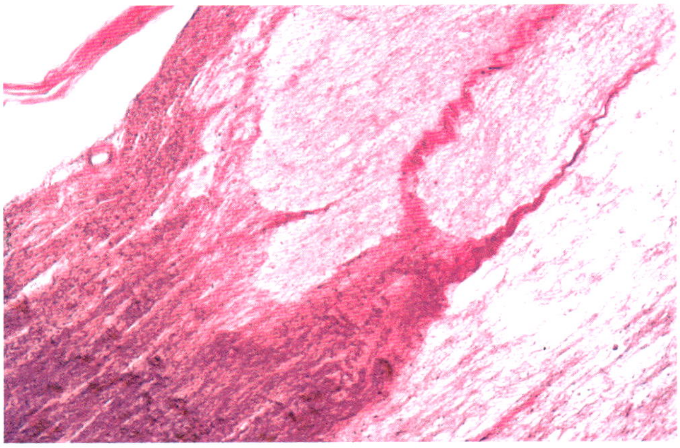

FIGURE 13.36 The vestibular and cochlear divisions of the vestibulocochlear nerve are fused near the entrance to the internal auditory meatus. At this location is the glial–Schwann sheath junction also referred to as the Obersteiner–Redlich line where the nerve changes in appearance from pale staining proximally to dark staining distally owing to the abrupt transition of the coverings of the nerve fibers from the pale staining oligodendroglia to the darker staining Schwann cells.

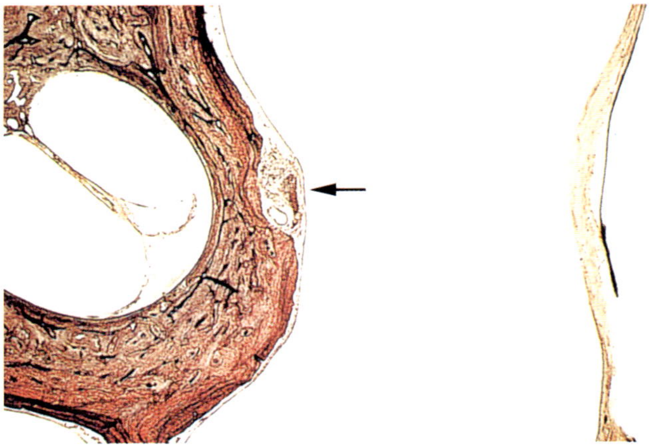

FIGURE 13.37 Normal tympanic paraganglion under mucosa of medial side of middle ear over promontory. The tympanic membrane is on the right. Gomori reticulin stain.

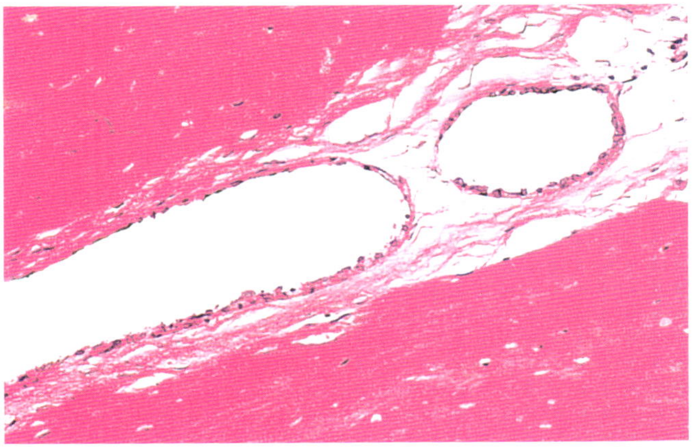

FIGURE 13.38 Endolymphatic duct within the vestibular aqueduct. The duct is lined by low cubical epithelium.

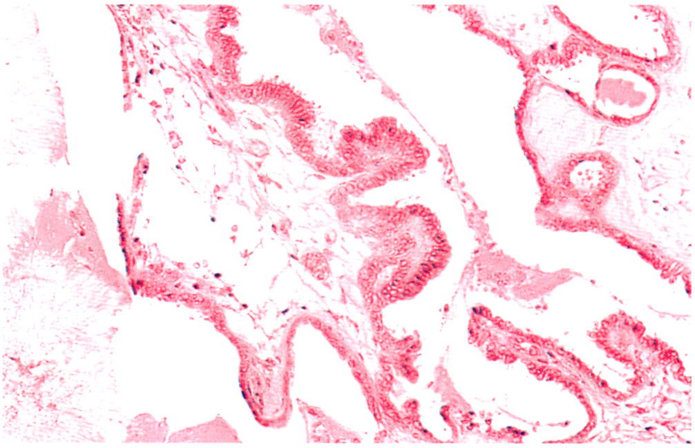

FIGURE 13.39 Endolymphatic sac, which is lined by tall columnar epithelium arranged on papillae.

factor 1 (TTF1), PSA, AMACR, synaptophysin, GATA3, and p63 (30). Patients with endolymphatic sac tumors often describe symptoms similar to those occurring in Ménière disease. This tumor is associated with von Hippel–Lindau disease, an autosomal dominant–inherited disease, caused by constitutional pathogenic variants in the tumor suppressor gene *VHL* (3p25) (31–33). When inherited, a second mutation affecting the wild-type allele results in a lack of functional VHL protein and risk for tumor development (34). Sporadic endolymphatic sac tumors also occur (35,36).

COMPOSITION AND CIRCULATION OF THE PERILYMPH AND ENDOLYMPH

Perilymph has a similar chemical composition as cerebrospinal fluid with low potassium and high sodium concentrations. The similarities of perilymph and cerebrospinal fluid support the concept that perilymph is derived, at least in part, from cerebrospinal fluid. The opening of the cochlear aqueduct (perilymphatic duct) into both the subarachnoid and perilymphatic spaces (1) presumably results in flow of cerebrospinal fluid into the labyrinth.

The perilymphatic spaces of each osseous semicircular canal are continuous on both ends with the perilymphatic space of the vestibule. The vestibule, in turn, is continuous with the scala vestibuli, which is continuous with the scala tympani at the helicotrema. All perilymphatic spaces open wide into each other. Due to areas of discontinuity or deficiency in the compact bone of the petrous portion of the temporal bone, communications may exist between the perilymphatic space and other cavities. Such areas of potential communication include the middle and inner ear via the round and oval windows. In addition, the vestibular and cochlear aqueducts and the foramina for the nerves and blood vessels of the inner ear serve as potential channels between the inner ear and the cranial cavity.

Endolymph is a physiologic intracellular-like fluid that fills the labyrinth and serves crucial sensory functions. Vibrational waves transmitted following displacement of this fluid from outside stimuli convey information about sound, position, and balance to central sensory neural structures. The endolymph is biochemically specialized and distinct from other central nervous system fluids (37). Endolymph contains more than 30 times as much potassium as does perilymph or cerebrospinal fluid with only one-tenth as much sodium (37). Endolymph also has a low-protein content; its protein is entirely globulin instead of an admixture of globulin and albumin (1). Its viscosity is similar to that of the eye vitreous due to its high mucopolysaccharide content. The electrolyte concentration of the endolymph is critical for normal functioning of the sensory organs. It is generally believed that the main source of endolymph is the stria vascularis, as well as the epithelia of the ampullae of the semicircular ducts and the maculae of the utricle and saccule. Human endolymphatic sacs may have endocrine/paracrine capacity through expression of peptides with potent natriuretic activity (38). Such peptides include uroguanylin and brain natriuretic peptide, but also peptides regulating vascular tone, including adrenomedullin 2. Further, both neurophysin and oxytocin (OXT) were reported to be significantly expressed (38). The endolymphatic sac may influence the hypothalamo–pituitary–adrenal axis and regulate vasopressin receptors and aquaporin-2 channels in the inner ear via OXT expression. The endolymphatic sac may also influence systemic and/or intracranial blood pressure through direct and indirect action on the renal vascular system (39). Given its crucial role in the sensation and transmission of balance, hearing, and position information, disorders of endolymph have profoundly disabling symptoms with hearing loss and vertigo attacks that can be difficult to diagnose and treat effectively (37).

Endolymph circulates through the cochlear duct (scala media) downward to the base of the cochlea, then through the ductus reuniens into the saccule, finally arriving in the endolymphatic sac and duct where it is reabsorbed. The cochlear duct communicates with the vestibular endolymph-containing sacs through two canals so that the endolymphatic system is, like the perilymphatic system, a continuous one.

Conduction of sound occurs via air and bone. The pinna and external auditory canal conduct sound waves in air to the tympanic membrane. Conduction of sound by air is less efficient when compared with the ossicular route. The ossicular chain, including the malleus, the incus, and the stapes, enhances the sound energy transmission by conveying vibrations from the tympanic membrane to the footplate of the stapes lying on the oval window of the vestibule which in turn is in contact with the perilymph. From the vestibular perilymph, vibrations derived from sound waves pass directly to the perilymphatic spaces of the cochlea, first via the scala vestibuli (upper compartment) ascending from the oval window and then to the scala tympani (lower compartment) descending to the round window. The walls of the endolymph-containing cochlear duct lie between the perilymph-containing scala vestibuli and scala tympani and receive waves of vibrations from the perilymph. The vibration waves are carried through the endolymph and affect the sensory cells of the organ of Corti located in the cochlear duct. These signals are passed to the cochlear nerve and subsequently transmitted to the cerebral cortex.

SELECT ABNORMALITIES AND PATHOLOGY

External Ear

Abnormalities of the first or second branchial arches can result in alterations in the structure and function of the external ear. The first and second branchial arch syndromes include otologic and nonotologic abnormalities. Otologic abnormalities include malformed or absent external ears, an atretic external auditory canal, and impaired hearing. The nonotologic abnormalities include asymmetric facies, alterations of the temporomandibular joint, neuromuscular deficiencies, and associated anomalies of the cardiovascular, renal, and central nervous systems. Goldenhar syndrome, also known as oculoauriculovertebral dysplasia, is a first and second branchial arch syndrome characterized by ear tags, preauricular pits and fissures, epidermoids, lipodermoids, and vertebral column abnormalities (38). Ear abnormalities can also be seen in association with Down syndrome (Fig. 13.40). Cryptotia is a rare anomaly in which the superior portion of the auricle is buried in the scalp (Fig. 13.41). Microtia represents gross hypoplasia of the pinna with a blind or absent external auditory canal. Microtia is typically bilateral although the degree of hypoplasia may differ on the two sides. Accessory tragi (i.e., supernumerary ears, accessory auricle, and polyotia) appear at birth as skin-covered nodules or papules and are phenotypically heterogeneous. They may be solitary or multiple, unilateral or bilateral, sessile or pedunculated, and soft or cartilaginous. They are usually located on the skin surface anterior to the auricle. Accessory tragi recapitulate the normal external auricle and include skin, cutaneous adnexal structures, and a central core of cartilage. They are thought to be related to second branchial

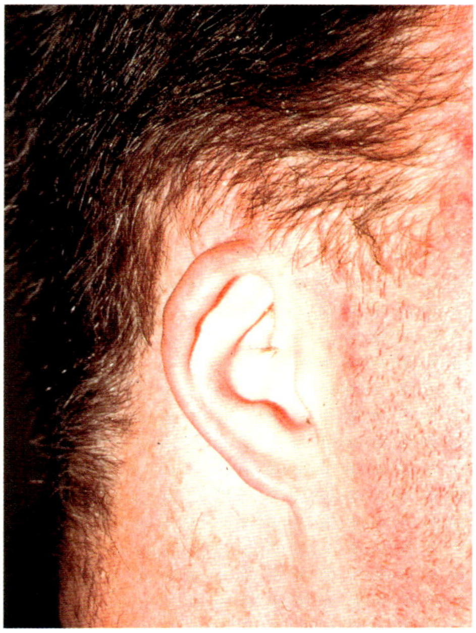

FIGURE 13.40 Individual with Down syndrome; in comparison to normal patients with Down syndrome have external ears that are smaller, low set and have an incompletely developed superior helix.

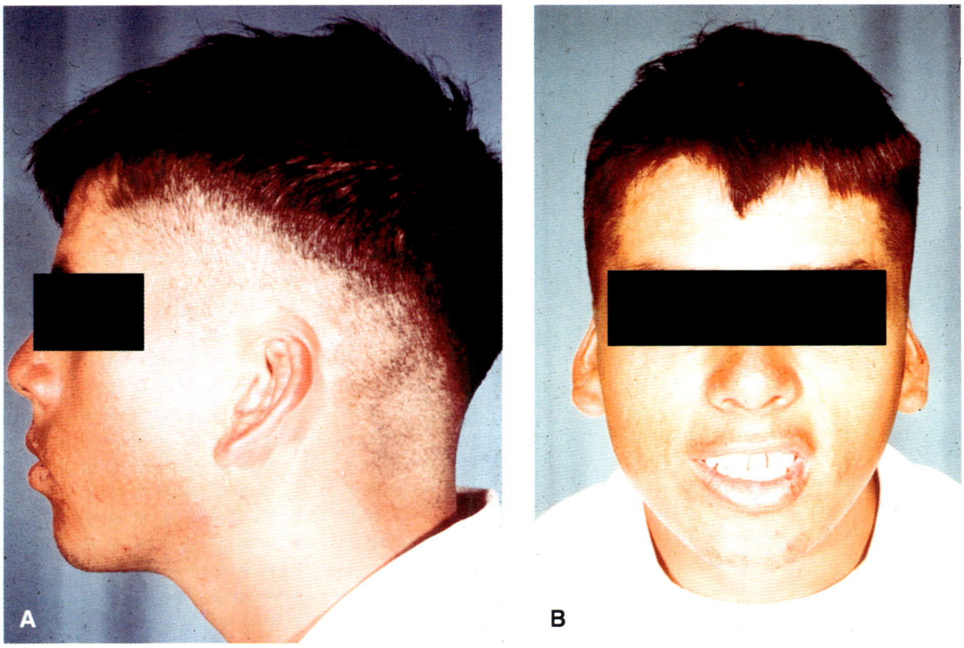

FIGURE 13.41 **A, B:** Individual with cryptotia, a rare anomaly in which the superior portion of the auricle is buried in the scalp.

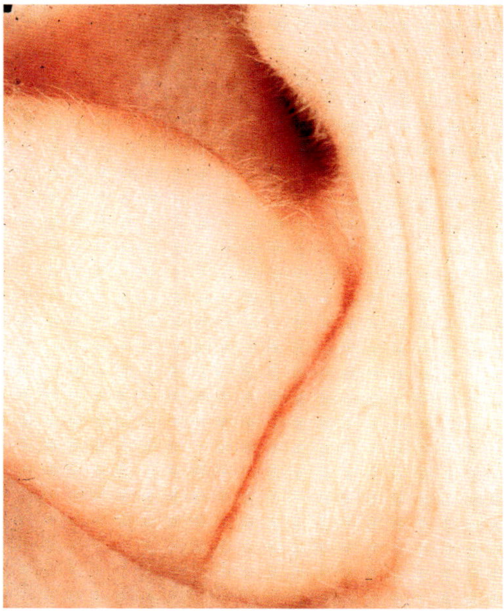

FIGURE 13.42 Individual with a grade 3 earlobe crease represented by a deep crease along 100% of the earlobe. This person had similar earlobe crease on the opposite ear. Bilateral grades 2 and 3 creases are associated with a significantly higher risk of death from atherosclerosis and myocardial infarction.

arch anomalies, and may occur independent of other congenital anomalies or in association with cleft palate or lip, mandibular hypoplasia, or Goldenhar syndrome (oculoauriculovertebral dysplasia) and other abnormalities (40).

Diagonal earlobe creases (i.e., Frank sign) in adults have been associated with coronary artery disease (41). In this situation, the crease runs diagonally backward and downward across the lateral surface of the earlobe from the external meatus (Fig. 13.42). The extent and depth of the crease can be assessed with a three-tiered grading system: grade 1 is least obvious, grade 2 includes a superficial crease across 100% of the earlobe or a deep crease across 50% of the earlobe, and grade 3 features a deep crease along 100% of the earlobe. Bilateral grades 2 and 3 creases are associated with a significantly higher risk of death from atherosclerosis and myocardial infarction (42).

MIDDLE EAR

Otitis Media

Otitis media is either an acute or a chronic infectious disease of the middle ear space that usually occurs in childhood. The most common microorganisms implicated in causing disease of the middle ear are *Streptococcus pneumoniae* and *Hemophilus influenzae* (43). Otoscopic examination reveals a hyperemic, opaque, bulging tympanic membrane with limited mobility; purulent otorrhea may be present. Bilateral involvement is not uncommon. The middle ear infection most likely results from bacterial or viral pharyngitis that spreads via the Eustachian tube.

Acute and/or chronic otitis media is usually managed medically, although some patients undergo procedures that yield tissue histopathologic examination. Pathologic alterations of chronic otitis media include variable numbers of lymphocytes, histiocytes, plasma cells, and eosinophils. Multinucleated giant cells and foamy histiocytes may be present and acute inflammatory cells are frequently present. Other findings include fibrosis, granulation tissue, tympanosclerosis, cholesterol granulomata, and reactive bone formation. The middle ear ossicles may be immobilized by scar tissue or show partial or complete destruction. Perforation of the tympanic membrane pars tensa results in squamous epithelial ingrowth that can lead to the development of cholesteatoma. Glandular metaplasia of the middle ear epithelium can develop in patients with chronic otitis media and simulate a gland-forming neoplasm (Fig. 13.43). Metaplastic glands are unevenly distributed in the tissue specimens,

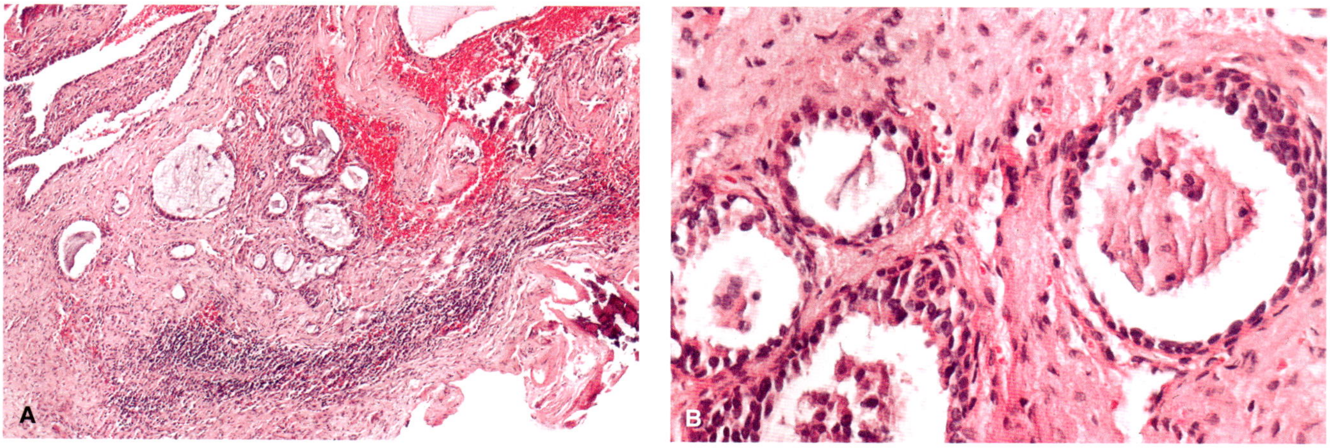

FIGURE 13.43 Under normal conditions glands are not identified in the middle ear space. However, glandular metaplasia can be found in the setting of otitis media. **A:** Otitis media showing chronic inflammation, fibrosis, glandular metaplasia, and foci of calcifications (*lower right*); residual normal cuboidal epithelium of the middle ear is seen in the upper left. **B:** Higher magnification showing glandular metaplasia is the setting of chronic otitis media.

are variably shaped, and are separated by abundant stromal tissue. The glands are lined by a columnar to cuboidal epithelium with or without cilia or goblet cell metaplasia. Thin (serous) or thick (mucoid) secretions may be present in their lumina. The identification of cilia is confirmatory of middle ear glandular metaplasia and is not found in association with middle ear adenomas (43). Further, the haphazard arrangement of the glands in the background of changes of chronic otitis media should allow one to distinguish metaplastic glands from neoplastic glands.

Tympanosclerosis represents dystrophic mineralization (calcification or ossification) of the tympanic membrane or middle ear. It is usually due to recurrent episodes of otitis media in the pediatric population, occurring in 3% to 33% of patients with otitis media (44). Tympanosclerosis can also be seen in children following myringotomy and tube insertion. In this setting, the tympanosclerotic foci may or may not be permanent. Tympanosclerosis of older patients represents irreversible accumulation of mineralized material and is associated with conductive hearing loss (45,46). Tympanosclerotic foci may be localized or diffuse and appear as white nodules or plaques (Fig. 13.44). Dense clumps of calcified or ossified material or debris can be seen within the stromal tissues or in the middle (connective tissue) aspect of the tympanic membrane (Fig. 13.45). Tympanosclerosis may cause scarring and ossicular fixation.

Cholesterol granuloma is a histologic designation describing a foreign body–type granulomatous response to cholesterol crystals derived from ruptured erythrocytes and breakdown of the lipid layer in their cell membranes. Cholesterol granulomata arise in the middle ear and mastoid and can result from any condition characterized by hemorrhage combined with incomplete drainage and ventilation of the middle ear space (47).

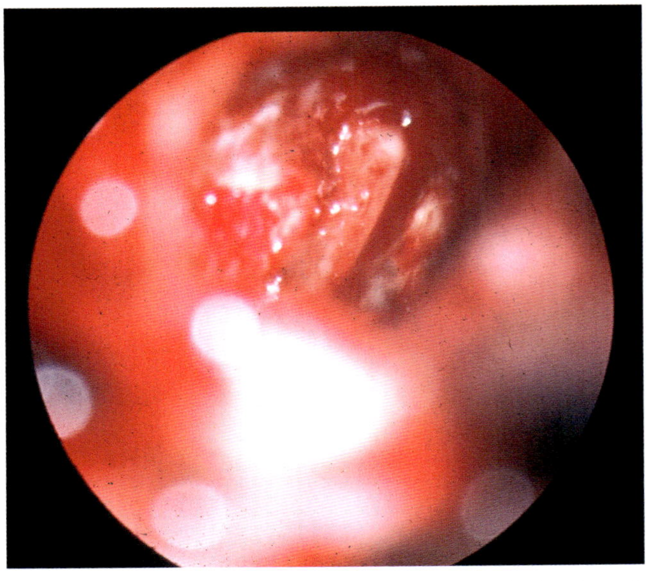

FIGURE 13.44 Tympanic membrane in tympanosclerosis showing calcified plaque on the tympanic membrane.

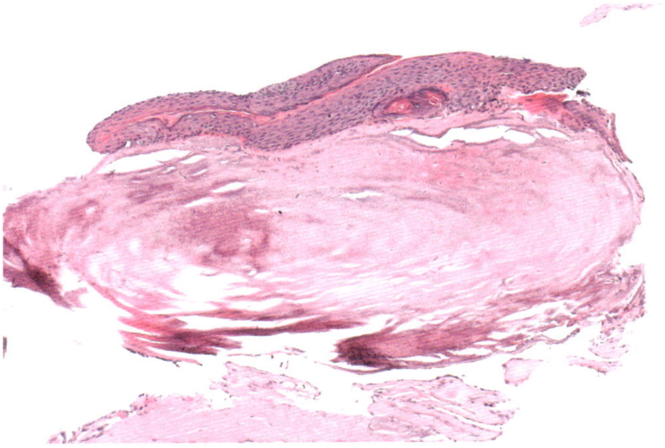

FIGURE 13.45 Tympanosclerosis. The tympanic membrane is thickened and calcified and covered on its external (external auditory canal) aspect by keratinizing squamous epithelium (*top*) and internal (tympanic cavity) aspect by cuboidal epithelium (*bottom right*).

Cholesterol granuloma of the middle ear may present as idiopathic hemotympanum; patients may also complain of hearing loss and tinnitus. Most cholesterol granulomata in the middle ear and temporal bone have an indolent biologic behavior and cause no significant bone resorption (48). However, cholesterol granulomata of the petrous apex may produce a large tumor-like mass with erosion and/or destruction of adjacent structures that mimics a neoplasm (49). Depending on the direction of expansion, apical cholesterol granulomata may invade the cochlea, cerebellopontine angle, jugular foramen, cranial nerves V–XI, brainstem, and cerebellum, and cause life-threatening symptoms (50). Involvement of the petrous apex is more likely to be associated with sensorineural hearing loss; additional signs and symptoms may include headaches, cranial nerve deficits (e.g., facial paralysis, others), and bone erosion with involvement of the posterior or middle cranial fossa (49–51). Apical cholesterol granulomata appear as smooth or irregular cysts with smooth margins and evidence of bone remodeling on axial computed tomography.

Cholesterol granulomata are composed of irregularly shaped clear spaces surrounded by macrophages and multinucleated giant cells (Fig. 13.46). The clear spaces often have sharp, needle-like or jagged edges that represent areas in which the cholesterol dissolved during tissue processing. Cholesterol granulomata are not related to cholesteatomas but may occur in association with cholesteatomas that rupture and elicit an inflammatory reaction.

Cholesteatoma (Keratoma)

Cholesteatoma is a pseudoneoplastic lesion akin to an epidermal inclusion cyst. It is composed of stratified squamous epithelium that forms a sac-like accumulation of keratin within the middle ear space. These lesions can cause local tissue destruction and even appear to be invasive, but they are not considered to be neoplasms. Initially, cholesteatomas may remain clinically silent until extensive invasion of

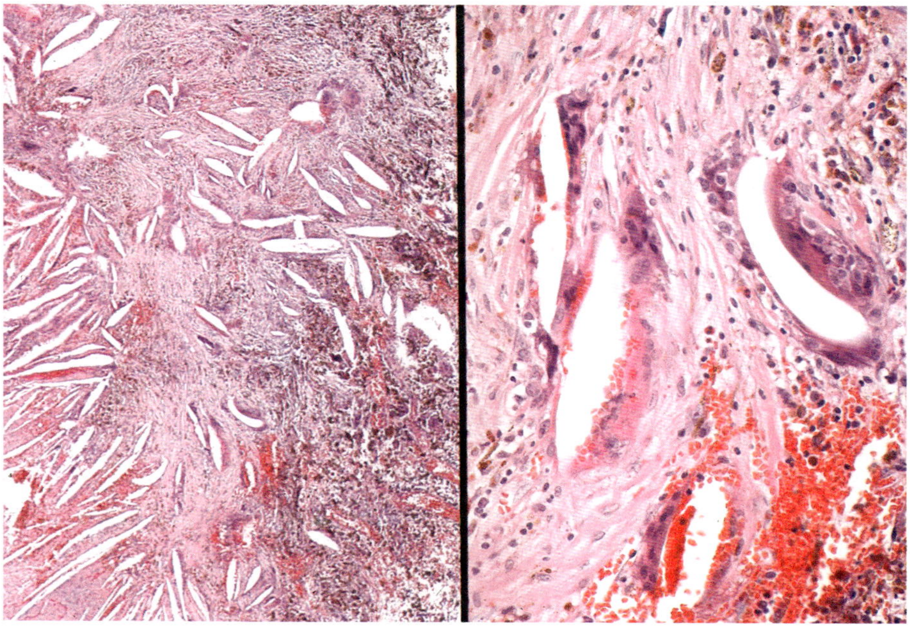

FIGURE 13.46 Cholesterol granuloma appears as empty, irregularly shaped clefts or spaces surrounded by histiocytes and multinucleated giant cells. Fresh hemorrhage and hemosiderin pigment are readily apparent.

the middle ear space and mastoid occurs. Symptoms include hearing loss, malodorous discharge, and pain and may be associated with a polyp arising in the attic of the middle ear or perforation of the tympanic membrane. Otoscopic examination may reveal the presence of white debris within the middle ear, which is considered diagnostic. The term cholesteatoma is a misnomer in that this type of lesion not a neoplasm nor does it contain cholesterol. Cholesteatomas of the inner and middle ear take three forms: acquired cholesteatoma, congenital cholesteatoma, and cholesteatoma of the petrous apex. Depending on the site of origin in the tympanic membrane, they may be subdivided into pars flaccida (Shrapnell membrane) and pars tensa cholesteatomas.

Acquired Cholesteatoma

Acquired cholesteatoma is the most common type of cholesteatoma. It tends to be more common in men than in women and occur in older children and young adults. Acquired cholesteatoma develops when the epidermis of the external ear canal is introduced into the middle ear. This can occur in several ways: (1) perforation of the tympanic membrane, (2) ingrowth of stratified squamous epithelium into the middle ear following localized retraction of the tympanic membrane, (3) entrapment of squamous epithelium following surgery and/or trauma, and (4) squamous metaplasia of the middle ear mucosa (52). A decrease in middle ear pressure can induce retraction of the tympanic membrane in the pars flaccida, pars tensa, or both (52). Retraction pockets are felt to represent the precursors for the development of cholesteatoma (Fig. 13.47) (53,54). Dysfunction of the Eustachian tube leading to recurrent otitis media could also be a causative factor (49). The triggers for cholesteatoma onset are diverse, and may involve tympanic membrane trauma, tympanic membrane disease, and/or tympanic cavity mucosa disease. These situations may lead to alterations in cell migration and squamous hyperplasia that contribute to the development of retraction pockets and cholesteatoma (55,56). The upper posterior part of the middle ear space is the most common site of acquired cholesteatomas (52).

Congenital Cholesteatoma

Congenital cholesteatomas are found in infants and young children and may cause hearing loss (57). Congenital

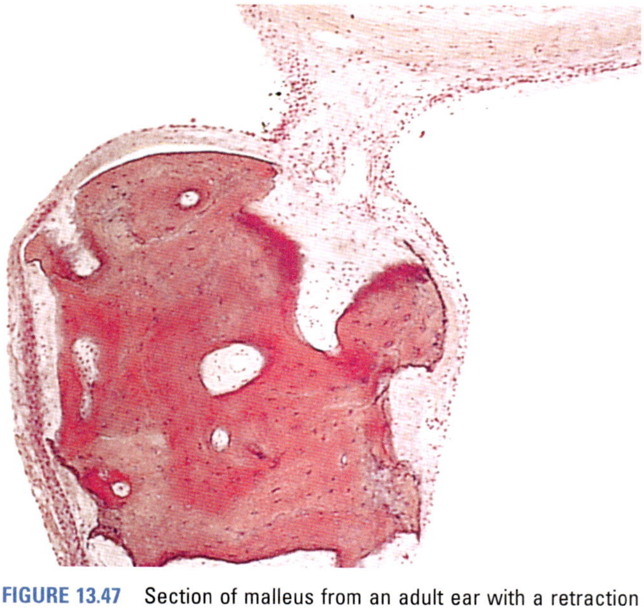

FIGURE 13.47 Section of malleus from an adult ear with a retraction pocket of the tympanic membrane at autopsy. There is a thin layer of stratified squamous epithelium between the bone and middle ear epithelium. This was found on serial section to be an ingrowth of the stratified squamous epithelium from the outer epithelial covering of the retraction pocket.

cholesteatoma is a cholesteatoma of the middle ear that exists in the presence of an intact tympanic membrane, presumably in the absence of chronic otitis media. Small colonies of epidermoid cells (i.e., epidermoid formations) are found on the lateral anterior superior surface of the middle ear in temporal bones after 15 weeks of gestation (56). These epidermoid colonies normally disappear during the first year of life. If, however, the epidermoid cells do not disappear and continue to grow, they will become a congenital cholesteatoma, which has also been termed an epidermoid cyst (58). Most congenital cholesteatomas are found in the anterosuperior part of the middle ear. Early lesions produce no symptoms and are discovered by otoscopic examination. Larger lesions can produce signs and symptoms similar to those of acquired cholesteatoma. Patients with congenital cholesteatoma rarely have Eustachian tube dysfunction and, thus, they usually have reasonable preoperative hearing with few postoperative complications (59). Surgical removal is curative.

Cholesteatoma of the Petrous Apex

Cholesteatoma of the petrous apex is an epidermoid cyst and bears no relation to cholesteatoma of the middle ear. It is likely of congenital origin, but no cell rests have been discovered to explain their origin. Symptoms usually relate to involvement of the VII and VIII cranial nerves in the cerebellopontine angle (58). When located at the petrous apex, these lesions can damage nearby structures causing hearing loss (60). Due to its rarity, diagnostic evaluation and treatment can vary. The primary goal of petrous apex cholesteatoma resection is gross total removal with tympanoplasty and canal-wall up or canal-wall down tympanomastoidectomy (60).

Pathology

Acquired and congenital cholesteatomas are histologically similar. The typical lesion is a cystic, white-to-pearly appearing mass of varying size that contains creamy or waxy granular material. They are composed of stratified keratinizing squamous epithelium supported by subepithelial fibroconnective tissue and contain keratin debris; granulation tissue may be present (Fig. 13.48). The essential diagnostic feature is the keratinizing squamous epithelium; the presence of keratin debris alone is not diagnostic of a cholesteatoma. The keratinizing squamous epithelium is cytologically bland and shows cellular maturation. Cholesteatomas do not metastasize but they are locally invasive and destructive. The destructive properties of cholesteatomas result from a mass effect with pressure erosion of surrounding structures, and the production of collagenase, which has osteodestructive capabilities due to its influence on bone resorption. Collagenase is produced by both the squamous and fibrous tissue components of the cholesteatoma. Erosion may occur in the lateral wall of the attic, the middle ear ossicles, the tegmental bone over the attic and antrum, or the mastoid cortex depending on the location and size of the cholesteatoma (52). Less frequent progression includes erosion of the lateral sinus and jugular bulb, the vestibular and cochlear capsules, the fallopian canal, the dura of the middle and posterior cranial fossa, the semicircular canals, and the facial nerve (49). Sequela of such erosions may include semicircular canal fistulas, exposed tympanic facial nerve, or brain herniation through the tegmen.

Dornelles et al. (61) attempted to evaluate the quantification of angiogenesis and matrix metalloproteinases (MMPs) as markers of aggressiveness in cholesteatoma (56). They compared expression of CD31, MMP2, and MMP9 in pediatric and adult patients and found that pediatric cholesteatomas were more inflammatory and associated with a greater degree of MMP expression, suggesting they had a more aggressive phenotype than similar lesions occurring in adults (61). Park et al. (62) evaluated differential expression of p63 and survivin in human middle ear cholesteatoma epithelium. They found that p63 protein was diffusely expressed in acquired cholesteatoma with variable

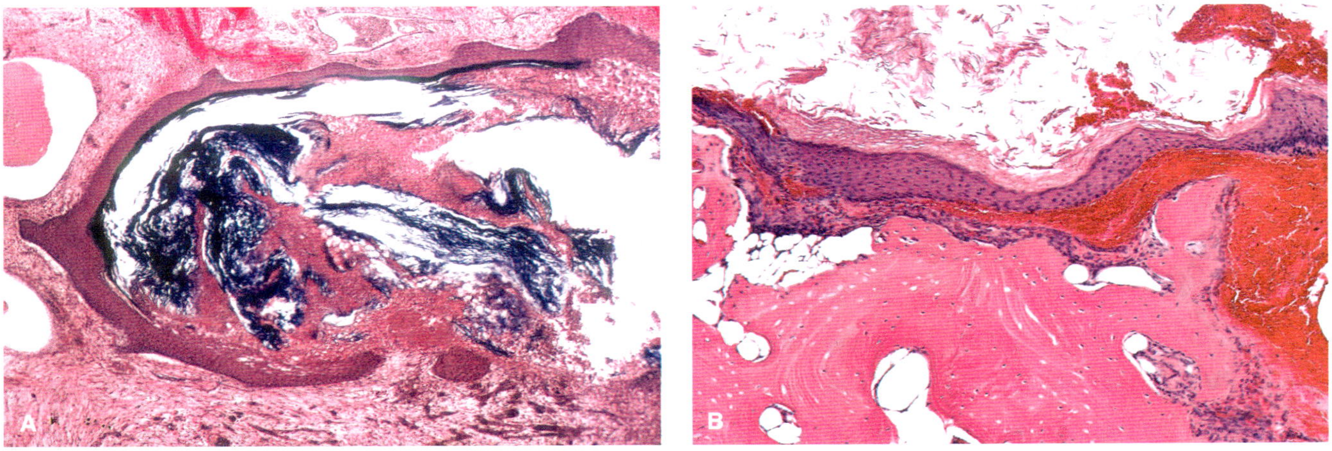

FIGURE 13.48 Cholesteatoma of the middle ear. **A:** The histologic diagnosis of cholesteatoma is based on the presence of finding keratinizing squamous epithelium within the middle ear space. **B:** Osseous involvement.

basal cell p63 staining in congenital lesions. Survivin staining was more common in acquired cholesteatomas, suggesting putative roles for p63 and survivin in the development of acquired cholesteatomas. Genetic alterations including upregulation of EGFR, TGF-alpha and metalloproteinases as well as downregulation of tumor suppressor genes and altered expression of proto-oncogenes have been reported in cholesteatomas (63,64) but the mechanisms underlying their development remain unclear (65). Of note, cholesteatomas do not transform into squamous cell carcinomas. Data from ploidy analysis suggest that cholesteatomas lack molecular evidence of genetic instability (66).

Otosclerosis

Otosclerosis is a disorder of the bony labyrinth and stapedial footplate that exclusively occurs in humans. Otosclerosis means hardening of the ear and is derived from Greek (ous, ear; skleros, hard; osis, condition); osseous ankylosis (Greek: ankoulon, to stiffen); chronic metaplastic osteitis; progressive otospongiosis. Otosclerosis primarily causes conductive hearing loss beginning in the second and third decades of life. Hearing loss is slowly progressive and its extent directly correlates with the degree of stapedial footplate fixation. It is not uncommon for patients with otosclerosis to also have vestibular disturbances (67,68). Otosclerosis usually involves both ears, although unilateral disease can occur up to 15% of cases (69). Surgical management of the conductive hearing loss caused by stapes fixation (stapedectomy) is the treatment of choice with replacement of the fixed stapes by a prosthesis. The resected bone may include the entire stapes including the footplate or only the superstructure that includes the head and crura without the footplate.

Although many theories regarding the etiology of otosclerosis appear in the literature, the etiology of otosclerosis is unclear. Hereditary factors are often cited among the causes of otosclerosis with half of all cases occurring in families with more than one affected member. Schrauwen et al. (70) used a genome-wide analysis to identify genetic factors involved in otosclerosis and found two highly associated single-nucleotide polymorphisms (SNPs) that replicated in two additional independent populations. They then genotyped 79 tag SNPs to fine map the two genomic regions defined by the associated SNPs. The region with the strongest association signal was on chromosome 7q22.1, spanning intron 1 to intron 4 of reelin (RELN), a gene known for its role in neuronal migration. Expression of RELN was confirmed in the inner ear and in stapes footplate specimens. The authors provided evidence that implicates RELN in the pathogenesis of otosclerosis. The same authors completed a replication study that included 1141 samples from four additional populations from Europe (71). Several SNPs in this region replicated in these populations separately. While the power to detect significant association in each population is small, six of the seven SNPs replicated and showed an effect in the same direction as in the previous populations. The authors also confirmed the presence of allelic heterogeneity in this region, further implicating RELN in the pathogenesis of otosclerosis.

Recently, Hodge et al. (72) investigated the role and distribution of various molecular markers using immunohistochemistry and immunofluorescence to further elucidate the pathogenesis of otosclerosis. These authors reported that Nidogen-1, BSP, and collagen-IX were expressed in the otospongiotic regions, and to lesser extent, in the otosclerotic regions. β2-laminin and ubiquitin were uniformly expressed in both otospongiotic and otosclerotic regions. TGF β-1, though present in the otosclerosis bones, was absent in the normal hearing and sensorineural hearing loss controls. Hodge et al. (72) proposed that the activity and function of TGF-1 may play a key role in the development and pathogenesis of otosclerosis.

Environmental factors, particularly fluoride and measles infection, are potentially linked to the development of otosclerosis. Measles virus RNA has been detected in stapes footplate samples (73) and molecular detection of measles virus has been reported in primary cell cultures of otosclerotic tissue (74). Michaels and Soucek (75) found that the main plaque of otosclerosis is a histologic replica of the external layer of the otic capsule and seems to arise from similar cells in the periosteum that follow a defined course into the footplate of the stapes, the basal coil of the cochlea, and the saccule. This process most often begins from the adjacent temporal bone anterior to the oval window, eventually involving the footplate of the stapes and moving across the annulus fibrosus or the stapediovestibular joint (Fig. 13.49). Stapedial involvement causes fixation of the stapes with inability to transmit sound waves. Otosclerotic changes can be seen in the resected stapedial footplate as well as the adjacent temporal bone (5). On the basis of its invasive growth into preexisting normal structures with replacement of these structures, Michaels and Soucek (76) introduced the notion that otosclerosis represents an invasive (low-grade) osseous neoplasm.

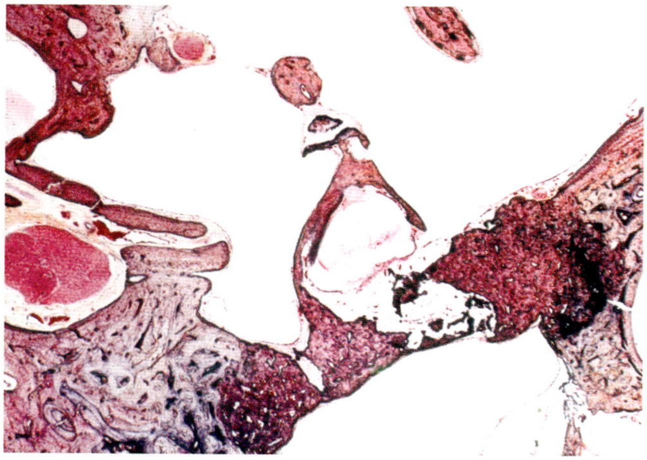

FIGURE 13.49 Otosclerosis of the temporal bone (anterior to the oval window) involving the footplate of the stapes moving across the annulus fibrosus or stapediovestibular joint. Stapedial involvement causes fixation of the stapes with the inability to transmit sound waves resulting in conductive hearing loss.

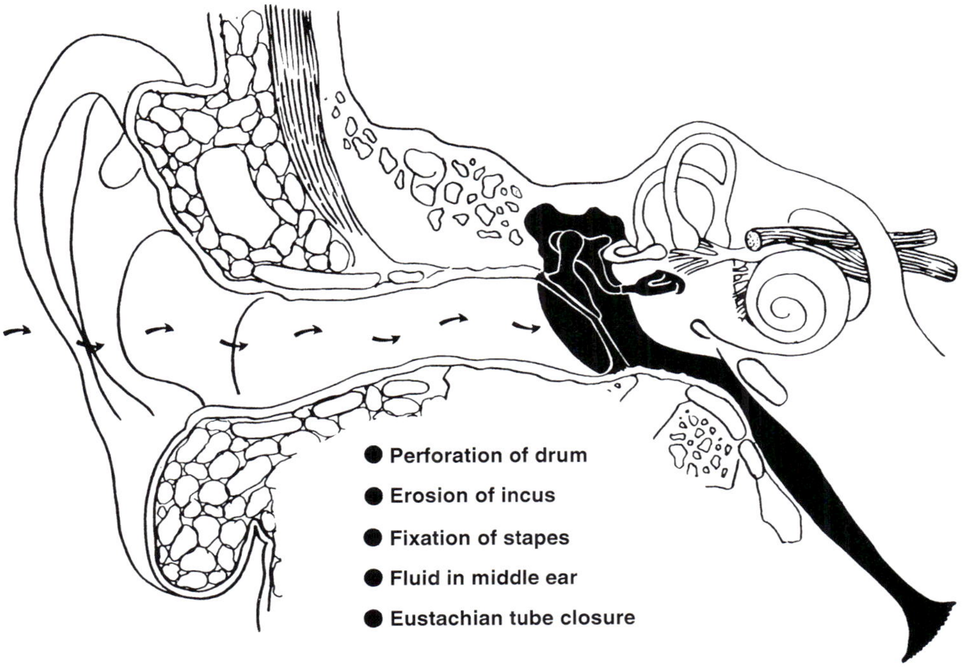

FIGURE 13.50 Causes of conductive hearing loss.

INNER EAR

Presbycusis and Other Hearing Loss

Hearing loss may include conductive hearing loss and sensorineural hearing loss. There are many causes for conductive and sensorineural hearing loss, respectively (Figs. 13.50 and 13.51). Hearing loss that occurs with increasing age is referred to as presbycusis. There is still controversy as to the underlying electrophysiologic and histopathologic alterations associated with the development of presbycusis. Degenerative alterations within various microanatomic structures of the cochlea have been invoked as the cause of presbycusis (77–80). Studies of aged ears (79,80) have demonstrated scanty to absent outer hair cells of the third row (Fig. 13.52) with complete loss of both inner and outer

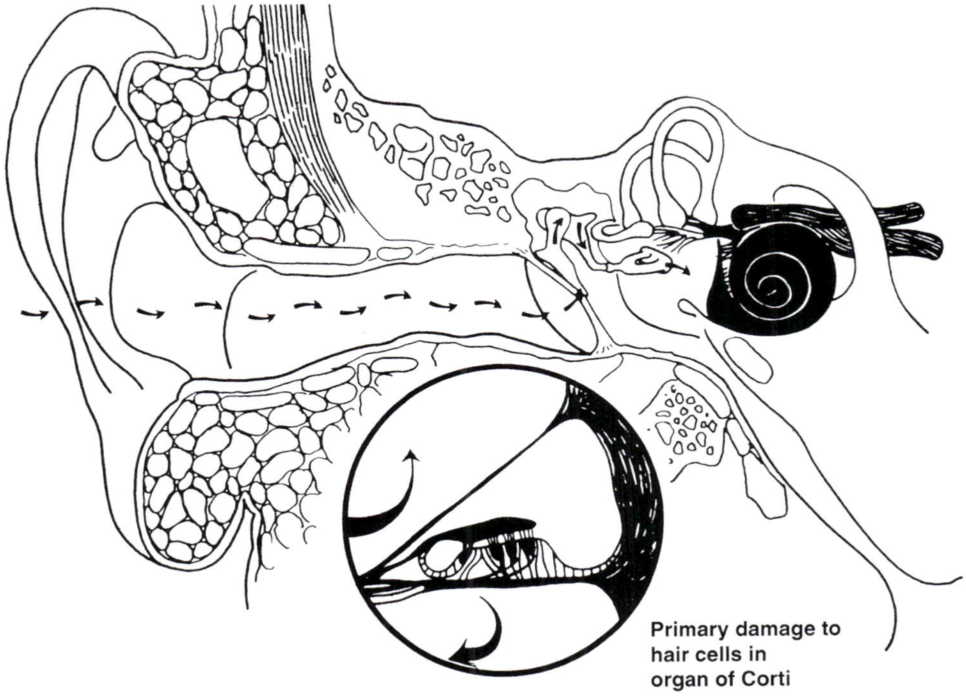

FIGURE 13.51 Cause of sensorineural hearing loss.

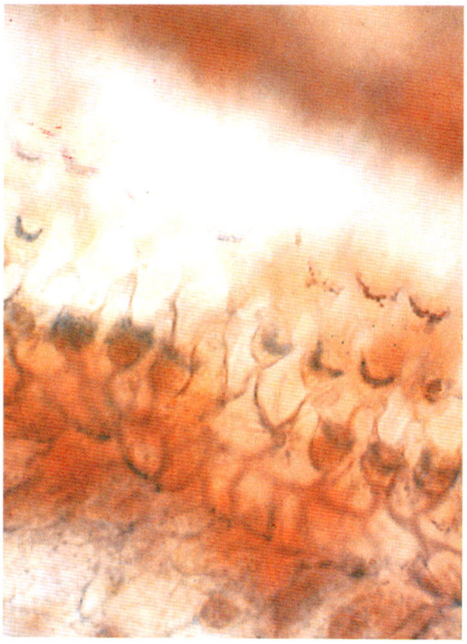

FIGURE 13.52 Surface preparation of outer hair cells from basal coil of cochlea of an elderly man. There are many gaps among the hair cells of the first two rows. Osmic acid, alcian blue, and phloxine eosin (oil immersion).

hair cells of all rows at the extreme lower end of the basal coil. In contrast, the inner hair cells sustained minimal loss, the first row of outer hair cells had greater loss and the second row more loss but not to the extent seen in the third row of outer hair cells or at the extreme lower end of the basal coil (79,80). Enormously lengthened and thickened stereocilia and giant stereocilia have also been identified in aged ears and may contribute to the development of presbycusis. Degenerative alterations within various microanatomic structures of the membranous labyrinth resulting in sensorineural hearing loss may occur secondary to infectious disease (Fig. 13.53), metabolic abnormalities (Fig. 13.54), trauma (Fig. 13.55), sensory presbycusis (Fig. 13.56), and

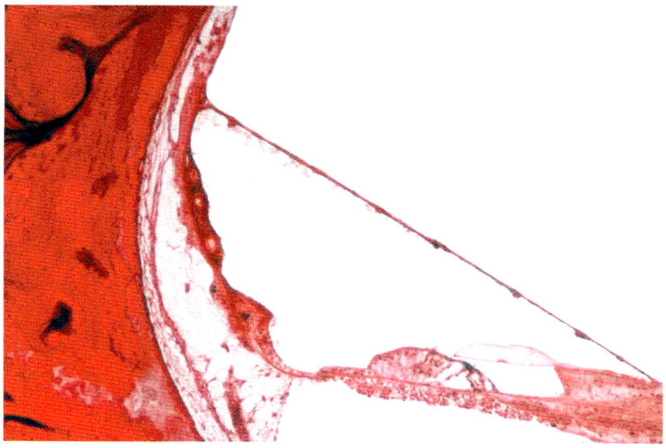

FIGURE 13.54 Metabolic-related hearing loss in a patient with diabetes characterized by the presence of hyalinized vasculature structures in the stria vascularis.

secondary to use of certain medications including cisplatin, diuretic agents, and other drugs (Figs. 13.57 and 13.58).

Ménière Disease

Ménière disease, also known as endolymphatic hydrops, is an idiopathic disorder of the inner ear associated with a symptom complex of spontaneous, episodic attacks of vertigo, sensorineural hearing loss, tinnitus, and a sensation of aural fullness. Ménière disease is characterized by a set of symptoms including fluctuating sensorineural hearing loss, episodes of vertigo, tinnitus, and pressure sensation in the ear. The onset of vertigo is frequently sudden reaching maximum intensity within a few minutes lasting for an hour or more and either subsiding completely or continuing as a sensation of unsteadiness for hours to days.

The pathogenesis of Ménière disease is multifactorial (81). Genetic studies have contributed to the understanding of the genetic basis of vestibular disorders. Exome sequencing has identified three single nucleotide variants in *PRKCB*, *DPT*, and *SEMA3D* linked with familial Ménière

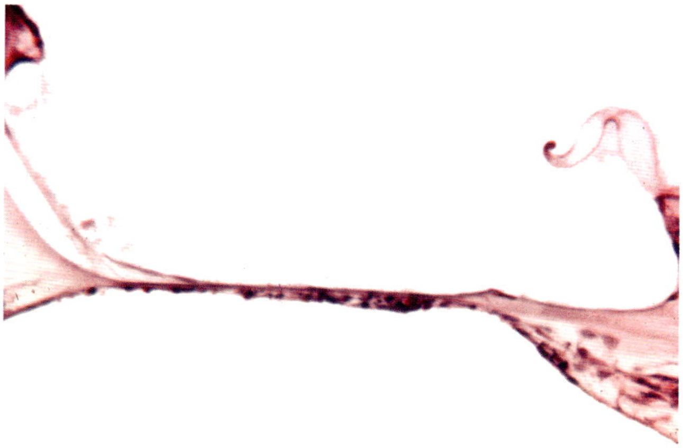

FIGURE 13.53 Viral labyrinthitis (end-stage) related hearing loss with total degeneration of the organ of Corti.

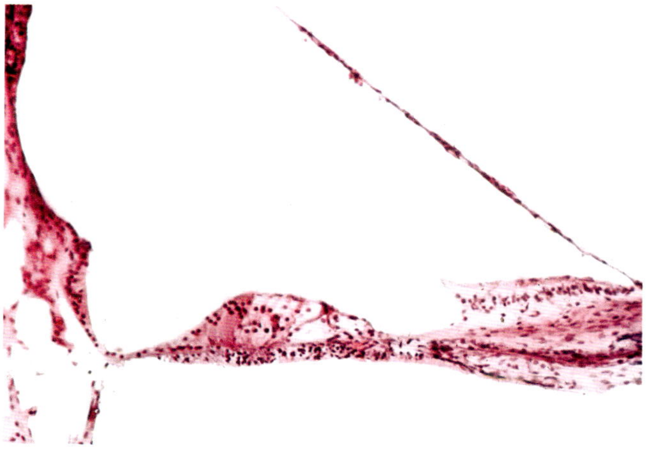

FIGURE 13.55 Posttraumatic hearing loss with focal avulsion of the organ of Corti.

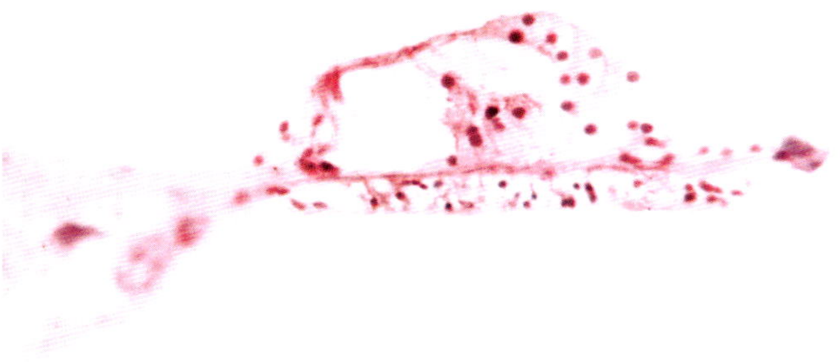

FIGURE 13.56 Sensory presbycusis characterized by the loss of hair cells in the organ of Corti.

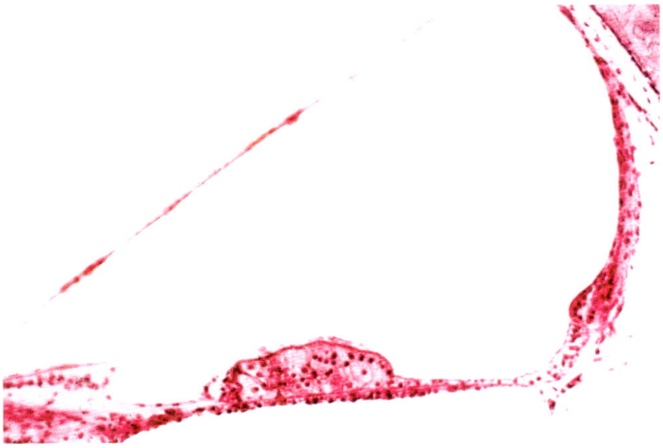

FIGURE 13.57 Strial presbycusis secondary to ototoxic effect caused by cisplatin therapy with strial atrophy.

disease further supporting the genetic background for most episodic or progressive vestibular-related syndromes (82). Ménière disease is characterized by distortion of the membranous labyrinth as a consequence of the overaccumulation of endolymph (i.e., endolymphatic hydrops) and at the expense of the perilymphatic space (83,84). Endolymph normally circulates in a radial and a longitudinal fashion, but inadequate absorption of endolymph by the endolymphatic sac leads to accumulation in Ménière disease (84). Serum biomarkers may prove useful in identifying other risk factors (81). Distinct microRNA expression profiles in the perilymph and serum suggest the possible presence of novel Ménière disease biomarkers (85).

Endolymphatic hydrops primarily involves the cochlear duct and the saccule in its early stages ultimately involves the entire endolymphatic system. Alterations of the membranous labyrinth include dilatation, outpouching, rupture, and collapse (Fig. 13.59). Fistulae may occur. Severe cytoarchitectural and atrophic changes may occur in the sense organs with loss of neurons in the cochlea.

Michaels and Soucek (86) described a thin highly vascular layer of bone, the vestibular arch, surrounding most of the intravestibular endolymphatic duct. In the normal ear, this contains osteoblasts that may help regulate the potassium composition of the endolymph. These authors

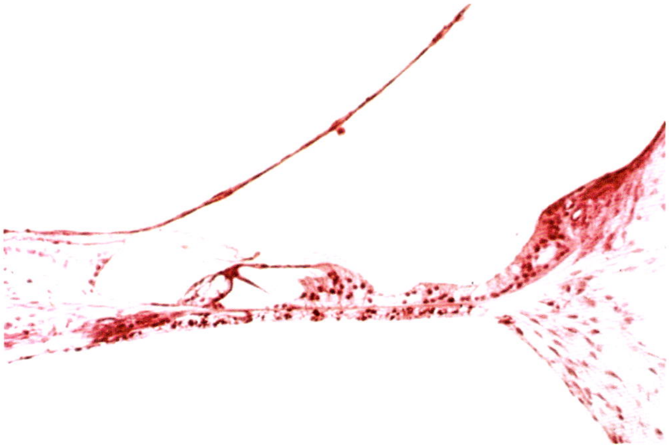

FIGURE 13.58 Kanamycin ototoxic effects on the organ of Corti with loss of hair cells.

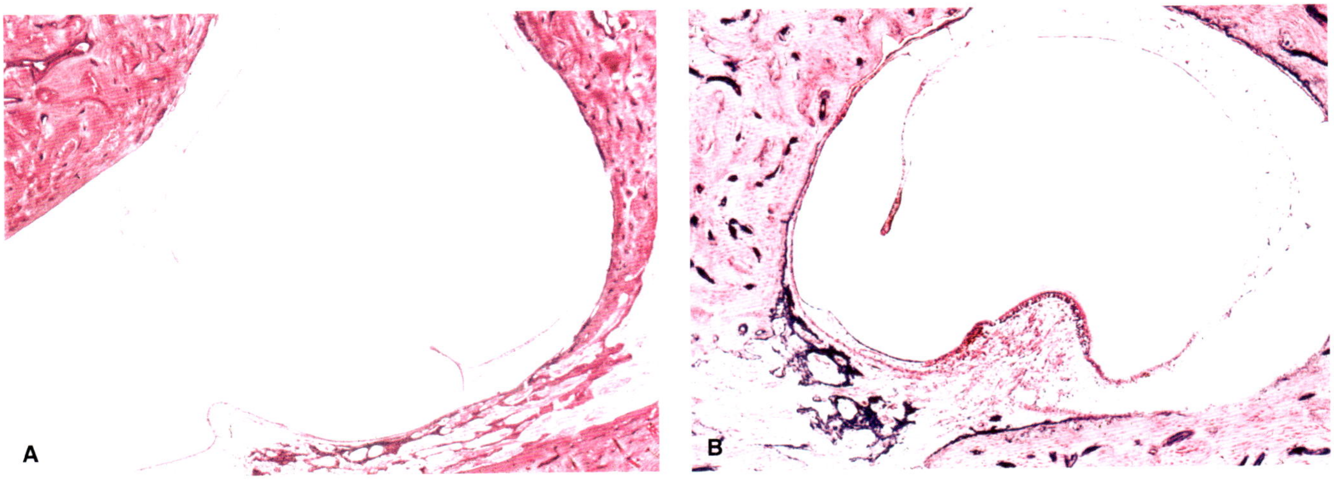

FIGURE 13.59 Ménière disease: **A:** Dilatation. **B:** Rupture of the membranous labyrinth.

found widespread apoptosis and death of vascular and other structures in the arch of patients Ménière disease, possibly explaining the presence of hyperkalemia in the nearby endolymph (86–88).

TEMPORAL BONE DISSECTION

For a detailed discussion of proper postmortem removal, sectioning, and processing for microscopic evaluation of the temporal bone, the reader is referred to other texts (5,89).

REFERENCES

1. Hollinshead WH. The ear. In: Hollinshead WH, ed. *Anatomy for Surgeons*. 3rd ed. Harper and Row; 1982:159–221.
2. Moore KL, ed. The ear. In: *The Developing Human: Clinically Oriented Embryology*. 4th ed. W.B. Saunders; 1988:412–420.
3. Michaels L, Soucek S. Development of the stratified squamous epithelium of the tympanic membrane and external canal: the origin of auditory epithelial migration. *Am J Anat*. 1989;184:334–344.
4. Michaels L, Soucek S. Stratified squamous epithelium in relation to the tympanic membrane: its development and kinetics. *Int J Pediatr Otorhinolaryngol*. 1991;22:135–149.
5. Michaels L. The ear. In: Sternberg SS, ed. *Histology for Pathologists*. 2nd ed. Lippincott-Raven; 1997:337–366.
6. Michaels L, Soucek S. Auditory epithelial migration on the human tympanic membrane. II. The existence of two discrete migratory pathways and their embryologic correlates. *Am J Anat*. 1990;189:189–200.
7. Nager GT. Anatomy of the membranous cochlea and vestibular labyrinth. In: Nager GT, ed. *Pathology of the Ear and Temporal Bone*. Williams and Wilkins; 1993:3–48.
8. Schuknecht HF, ed. Anatomy. In: *Pathology of the Ear*. 2nd ed. Lea & Febiger; 1993:31–74.
9. Maybaum JL, Goldman JL. Primary jugular bulb thrombosis. A study of twenty cases. *Arch Otolaryngol*. 1933;17:70.
10. Kamerer DB. Electromyographic correlation of tensor tympani and tensor veli palatini muscles in man. *Laryngoscope*. 1978;88:651–662.
11. Graves GO, Edwards LF. The Eustachian tube: a review of its descriptive, microscopic, topographic and clinical anatomy. *Arch Otolaryngol*. 1944;39:359.
12. Moreano EH, Paparella MM, Zelterman D, Goycoolea MV. Prevalence of carotid canal dehiscence in the human middle ear: a report of 1000 temporal bones. *Laryngoscope*. 1994;104:612–618.
13. Michaels L, Wells M. Squamous cell carcinoma of the middle ear. *Clin Otolaryngol Allied Sci*. 1980;5:235–248.
14. Nager GT. Pneumatization of the temporal bone. In: Nager GT, ed. *Pathology of the Ear and Temporal Bone*. Williams and Wilkins; 1993:53–62.
15. Tremble GE. Pneumatization of the temporal bone. *Arch Otolaryngol*. 1934;19:172–182.
16. Davies DV. A note on the articulations of the auditory ossicles and related structures. *J Laryngol Otol*. 1948;62:533–536.
17. Dayal VS, Farkashidy J, Kokshanian A. Embryology of the ear. *Can J Otolaryngol*. 1973;2:136–142.
18. Lysakowski A, McCrea RA, Tomlinson RD. Anatomy of vestibular end organs and neural pathways. In: Cummings CW, Frederickson JM, Harker LA, Krause CJ, Richardson MA, Schuller DE, eds. *Otolaryngology: Head Neck Surgery*. 3rd ed. Mosby; 1998:2561–2583.
19. Kobrak H. Influence of the middle ear on labyrinthine pressure. *Arch Otolaryngol*. 1935;21:547.
20. Lindsey JR, Schuknecht HF, Neff WD, Kimura RS. Obliteration of the endolymphatic sac and the cochlear aqueduct. *Ann Otol Rhinol Laryngol*. 1952;61:697–716.
21. Tonndorf J, Tabor JR. Closure of the cochlear windows: its effect upon air- and bone-conduction. *Ann Rhinol Laryngol*. 1962;71:5–29.
22. Belemer JJ. The vessels of the stria vascularis: with special reference to their functions. *Arch Otolaryngol*. 1936;23:93–97.
23. Guild SR. The circulation of the endolymph. *Am J Anat*. 1927;39:57–81.
24. Altmann F, Waltner JG. New investigations on the physiology of the labyrinthine fluids. *Laryngoscope*. 1950;60:727–739.
25. Tarlov IM. Structure of the nerve root. II. Differentiation of sensory from motor roots; observations on identification of function in roots of mixed cranial nerves. *Arch Neurol Psychiatr*. 1937;37:1938.
26. Hollinshead WH. The cranium. In: Hollinshead WH, ed. *Anatomy for Surgeons*. 3rd ed. Harper and Row; 1982:26–27.
27. Hyams VJ, Batsakis JG, Michaels L. Acoustic neuroma. In: Hartmann WH, Sobin LH, eds. *Tumors of the Upper Respiratory Tract and Ear. Atlas of Tumor Pathology, Fascicle 25, Second Series*. Armed Forces Institute of Pathology; 1988;323–326.
28. Heffner DK. Low-grade adenocarcinoma of probable endolymphatic sac origin. A clinicopathologic study of 20 cases. *Cancer*. 1989;64:2292–2302.
29. Wenig BM, Heffner DK. Endolymphatic sac tumors: fact or fiction? *Adv Anat Pathol*. 1996;3:378–387.
30. Thompson LDR, Magliocca KR, Stelow E, Wenig BM, Bishop JA. CAIX and PAX8 are commonly immunoreactive in endolymphatic sac tumors: differential with renal cell carcinoma in von-Hippel-Lindau patients. *Modern Pathol*. 2018;98(suppl):487–488.
31. Megerian CA, McKenna MJ, Nuss RC, et al. Endolymphatic sac tumors: histopathologic confirmation, clinical characterization, and implication in von Hippel–Lindau disease. *Laryngoscope*. 1995;105:801–808.
32. Sgambati MT, Stolle C, Choyke PL, et al. Mosaicism in von Hippel–Lindau disease: lessons from kindreds with germline mutations identified in offspring with mosaic parents. *Am J Hum Genet*. 2000;66:84–91.
33. Findeis-Hosey JJ, McMahon KQ, Findeis SK. Von Hippel-Lindau disease. *J Pediatr Genet*. 2016;5:116–123.
34. Maher ER, Neumann HP, Richard S. von Hippel-Lindau disease: a clinical and scientific review. *Eur J Hum Genet*. 2011; 19:617–623.
35. Schnack DT, Kiss K, Hansen S, Miyazaki H, Bech B, Caye Thomasen P. Sporadic endolymphatic sac tumor-A very rare cause of hearing loss, tinnitus, and dizziness. *J Int Adv Otol*. 2017;13:289–291.
36. Jegannathan D, Kathirvelu G, Mahalingam A. Three sporadic cases of endolymphatic sac tumor. *Neurol India*. 2016; 64:1336–1339.

37. Casale J, Agarwal A. Anatomy, head and neck, ear endolymph. [Updated 2023 February 9]. In: *StatPearls [Internet]*. StatPearls Publishing; 2024.
38. Møller MN, Kirkeby S, Vikeså J, Nielsen FC, Cayé-Thomasen P. The human endolymphatic sac expresses natriuretic peptides. *Laryngoscope*. 2017;127:E201–E208.
39. Smith CA, Lowry OH, Wu M-L. The electrolytes of the labyrinthine fluids. *Laryngoscope*. 1954;64:141–153.
40. Schuknecht HF, ed. Developmental defects. In: *Pathology of the Ear*. 2nd ed. Lea & Febiger; 1993:115–189.
41. Frank ST. Aural sign of coronary heart disease. *N Engl J Med*. 1973;289:327–328.
42. Patel V, Champ C, Andrews PS, Gostelow BE, Gunasekara NP, Davidson AR. Diagonal earlobe creases and atheromatous disease: a postmortem study. *J R Coll Physicians Lond*. 1992;26:274–277.
43. Wenig BM. Otitis media. In: Wenig BM, ed. *Atlas of Head and Neck Pathology*, 3rd ed. Saunders Elsevier: 2016:1108–1113.
44. Bhaya MHJ, Scachern PA, Morizono T, Paparella MM. Pathogenesis of tympanosclerosis. *Otolaryngol Head Neck Surg*. 1993; 109:413–420.
45. Pace A, Iannella G, Riminucci M, Corsi A, Magliulo G. Tympanomastoid cholesterol granuloma: case report and review of the literature. *Clin Med Insights Case Rep*. 2020;13:1179547620958728.
46. Gibb AG, Pang YT. Current considerations in the etiology and diagnosis of tympanosclerosis. *Eur Arch Otorhinolaryngol*. 1994;251:439–451.
47. Nager GT, Vanderveen TS. Cholesterol granuloma involving the temporal bone. *Ann Otol Rhinol Laryngol*. 1976;85:204–209.
48. Nager GT. Cholesterol granulomas. In: Nager GT, ed. *Pathology of the Ear and Temporal Bone*. Williams & Wilkins; 1994:914–939.
49. Olcott C, Strasnick B. A blue middle ear mass: cholesterol granuloma mimicking a glomus tumor and endolymphatic sac tumor. *Am J Otolaryngol*. 2017;38:100–102.
50. Thedinger BA, Nadol JB Jr, Montgomery WW, Thedinger BS, Greenberg JJ. Radiographic diagnosis, surgical treatment, and long-term follow-up of cholesterol granulomas of the petrous apex. *Laryngoscope*. 1989;99:896–907.
51. Nager GT. Cholesteatomas of the middle ear. In: Nager GT, ed. *Pathology of the Ear and Temporal Bone*. Williams & Wilkins; 1994:298–350.
52. Michaels L. The biology of cholesteatoma. *Otolaryngol Clin North Am*. 1989;22:869–881.
53. Wells M, Michaels L. Role of retraction pockets in cholesteatoma formation. *Clin Otolaryngol Allied Sci*. 1983;8:39–45.
54. Schuknecht HF. Cholesteatoma. In: Schuknecht HF, ed. *Pathology of the Ear*. 2nd ed. Lea & Febiger; 1993:204–206.
55. Louw L. Acquired cholesteatoma pathogenesis: stepwise explanations. *J Laryngol Otol*. 2010;124:587–593.
56. Michaels L. Origin of congenital cholesteatoma from a normally occurring epidermoid rest in the developing middle ear. *Int J Pediatr Otorhinolaryngol*. 1988;15:51–65.
57. Wei B, Zhou P, Zheng Y, Zhao Y, Li T, Zheng Y. Congenital cholesteatoma clinical and surgical management. *Int J Pediatr Otorhinolaryngol*. 2023;164:111401.
58. de Souza CD, Sperling NM, da Costa SS, Yoon TH, Abdel Hamid M, de Souza RA. Congenital cholesteatomas of the cerebellopontine angle. *Am J Otol*. 1989;10:358–363.
59. Bennett M, Warren F, Jackson GC, Kaylie D. Congenital cholesteatoma: theories, facts and 53 patients. *Otolaryngol Clin North Am*. 2006;39:1081–1094.
60. Reardon T, Turnow M, Elston S, et al. Surgical management of petrous apex cholesteatomas in the pediatric population: a systematic review. *Surg Neurol Int*. 2022;13:494.
61. Dornelles Cde C, da Costa SS, Meurer L, Rosito LP, da Silva AR, Alves SL. Comparison of acquired cholesteatoma between pediatric and adult patients. *Eur Arch Otorhinolaryngol*. 2009;266:1553–1561.
62. Park HR, Min SK, Min K, Jun SY, Seo J, Kim HJ. Increased expression of p63 and surviving in cholesteatomas. *Acta Otolaryngol*. 2009;129:268–272.
63. Kuo CL, Shiao AS, Yung M, et al. Updates and knowledge gaps in cholesteatoma research. *Biomed Res Int*. 2015; 2015:854024.
64. Kuo CL. Etiopathogenesis of acquired cholesteatoma: prominent theories and recent advances in biomolecular research. *Laryngoscope*. 2015;125:234–240.
65. Sanderson A. Cholesteatoma. In: El-Naggar AK, Chan JKC, Grandis JR, Takata T, Slootweg PJ, eds. *WHO Classification of Head and Neck Tumours*. 4th ed. IARC. 2017:269–270.
66. Desloge RB, Carew JF, Finstad CL, et al. DNA analysis of human cholesteatomas. *Am J Otol*. 1997;18:155–159.
67. Cody DTR, Baker HL Jr. Otosclerosis: vestibular symptoms and sensorineural hearing loss. *Ann Otol Rhinol Laryngol*. 1978;87:778–796.
68. Morales-Garcia C. Cochleo-vestibular involvement in otosclerosis. *Acta Otolaryngol*. 1972;73:484–492.
69. Schuknecht HF, ed. Otosclerosis. In: *Pathology of the Ear*. 2nd ed. Lea & Febiger; 1993:365–379.
70. Schrauwen I, Ealy M, Huentelman MJ, et al. A genome-wide analysis identified variants in the RELN gene associated with otosclerosis. *Am J Hum Genet*. 2009;84:328–338.
71. Schrauwen I, Ealy M, Fransen E, et al. Genetic variants in RELN gene are associated with otosclerosis in multiple European populations. *Hum Genet*. 2010;127:155–162.
72. Hodge S, Lopez IA, Cronkite A, et al. Dynamic molecular markers of otosclerosis in the human cochlea. *Ann Otol Rhinol Laryngol*. 2024;133(4):390–399.
73. Karosi T, Kónya J, Szabó LZ, Sziklai I. Measles virus prevalence in otosclerotic stapes footplate samples. *Otol Neurotol*. 2004;25:451–456.
74. Gantumur T, Niedermeyer HP, Neubert WJ, Arnold W. Molecular detection of measles virus in primary cell cultures of otosclerotic tissue. *Acta Otolaryngol*. 2006;126:811–816.
75. Michaels L, Soucek S. Origin and growth of otosclerosis. *Acta Otolaryngol*. 2011;131:460–468.
76. Michaels L, Soucek S. Atypical mature bone in the otosclerotic otic capsule as the differentiated zone of an invasive osseous neoplasm. *Acta Otolaryngol*. 2014;134:118–123.
77. Schuknecht HF, ed. Disorders of growth, metabolism, and aging. In: *Pathology of the Ear*. 2nd ed. Lea & Febiger; 1993.
78. Keithley EM. Pathology and mechanisms of cochlear aging. *J Neurosci Res*. 2020;98:1674–1684.
79. Soucek S, Michaels L, Frohlich A. Pathological changes in the organ of Corti in presbycusis as revealed in microslicing and staining. *Acta Otolaryngol Suppl*. 1987;436:93–102.
80. Soucek S, Michaels L, Frohlich A. Evidence of hair cell degeneration as the primary lesion in hearing loss of the elderly. *J Otolaryngol*. 1986;15:175–183.
81. Hoskin JL. Ménière's disease: new guidelines, subtypes, imaging, and more. *Curr Opin Neurol*. 2022;35:90–97.

82. Roman-Naranjo P, Gallego-Martinez A, Lopez Escamez JA. Genetics of vestibular syndromes. *Curr Opin Neurol*. 2018; 31:105–110.
83. Paparella MM. The cause (multifactorial inheritance) and pathogenesis (endolymphatic malabsorption) of Ménière's disease and its symptoms (mechanical and chemical). *Acta Otolaryngol*. 1985;99:445–451.
84. Klis SFL, Buijs J, Smoorenburg GF. Quantification of the relationship between electrophysiologic and morphologic changes in experimental endolymphatic hydrops. *Ann Otol Rhinol Laryngol*. 1990;99:566–570.
85. Shew M., Wichova H, St Peter M, Warnecke A, Staecker H. Distinct microrNA profiles in the perilymph and serum of patients with Ménière's disease. *Front Neurol*. 2021;12:646928.
86. Michaels L, Soucek S, Linthicum F. The intravestibular source of the vestibular aqueduct: its structure and pathology in Ménière's disease. *Acta Otolaryngol*. 2009;129:592–601.
87. Michaels L, Soucek S. The intravestibular source of the vestibular aqueduct. III: Osseous pathology of Ménière's disease, clarified by a developmental study of the intraskeletal channels of the otic capsule. *Acta Otolaryngol*. 2010;130: 793–798.
88. Michaels L, Soucek S, Linthicum F. The intravestibular source of the vestibular aqueduct. II: its structure and function clarified by a developmental study of the intraskeletal channels of the otic capsule. *Acta Otolaryngol*. 2010;130:420–428.
89. Schuknecht HF, ed. Histological method. In: *Pathology of the Ear*. 2nd ed. Lea & Febiger; 1993.

14

Mouth, Nose, and Paranasal Sinuses

Liron Pantanowitz ■ William Faquin

EMBRYOLOGY AND PRENATAL CHANGES

The discussion of the development of this highly specialized part of the head and neck is restricted to structures of importance to the surgical pathologist. For details, the reader is referred to other sources (1,2).

The oral region develops from an ectodermal depression, the stomodeum. The deep oral cavity is formed by the forward growth of structures around the margins of the stomodeum, giving rise to superficial parts of the face and jaws, as well as the walls of the oral cavity. The stomodeal prominence is surrounded bilaterally by the maxillary and mandibular processes and rostrally by the unpaired frontal prominence. The upper lip, maxilla, and nose are derived from structures surrounding the stomodeum. The caudal boundary of the oral cavity is formed by the paired mandibular processes, which, during the second year of life, fuse in the midline to form the mandible. The paired maxillary processes likewise meet in the midline, crowding the nasal elevation to ultimately form the maxilla and, by fusion in the midline, the palate. The contours of the face change with the rapid growth of the nose and jaws (2). The nose is formed on either side of the frontonasal elevation by an invagination of ectoderm into the mesoderm to form two nasal pits that gradually converge toward the midline, where they merge with each other. The underlying mesenchyme develops into bone, cartilage, and skeletal muscle.

At the end of the second month of fetal life, the formation of the bony structures begins; the maxilla is one of the first bones to calcify. Simultaneously, the nasal pits become progressively deeper and extend downward toward the oral cavity. Later, elevations appear on the lateral walls of the right and left nasal cavities that will become the conchal portions of the nasal turbinates. The nasal cavities communicate with chambers in the adjacent bones known as paranasal sinuses. Named for the bones in which they lie, they comprise the frontal, maxillary, sphenoidal, and ethmoidal sinuses. The paranasal sinuses are first identified around the fourth month of fetal life, but most of their expansion occurs after birth, and they attain full size many years later. The mucosa lining the nasal cavities invaginates into the surrounding bone, thereby lining the expanding sinus. While the palate has been taking shape from the roof of the mouth, the tongue has been forming in the floor. The posterior part of the tongue behind the sulcus terminalis is derived from the midventral areas of branchial arches II, III, and IV (3,4).

The tonsils first develop as endodermal epithelial buds that arise from the lining of the primitive oronasal cavity and grow into the subjacent mesenchyme to eventually give rise to the tonsillar crypts. Crypt formation is simple in the lingual tonsil, but more complex in the palatine tonsils. Lymphoid tissue begins to accumulate and surround the crypts when secondary budding of the crypts takes place. This development occurs in close association with mucous glands, which explains the close anatomic proximity of such glands to the tonsils.

Tooth development is of considerable importance to an understanding of the pathogenesis of odontogenic tumors

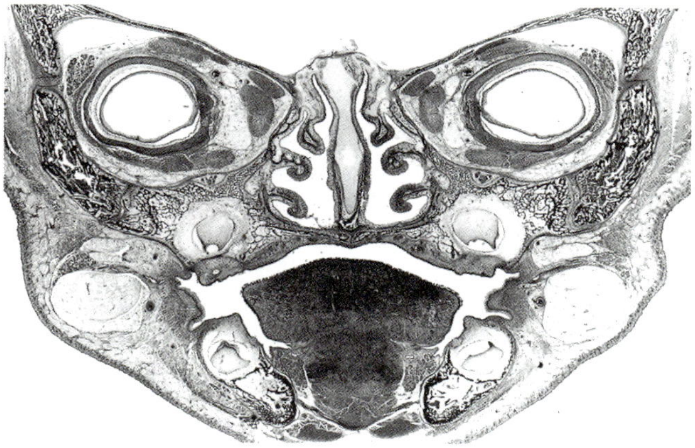

FIGURE 14.1 Coronal section of the head of a human fetus about 30 weeks of age (284-mm crown-rump length). The bell-shaped enamel organs are present in each quadrant. The tongue is relatively large. The paranasal sinuses are not yet discernible at this stage of fetal development.

and cysts. Odontogenesis is a highly coordinated and complex process that relies upon several genes, growth factors, structural proteins (e.g., amelogenin, tuftelin, predentin, cementum, enamelin), and extracellular matrix molecules being expressed in temporal- and space-specific patterns (5). The teeth begin to develop inside the gums of the upper and lower jaws (Fig. 14.1). Such regulatory interactions occur during the early stages of morphogenesis, particularly when the dental epithelium induces the condensation of mesenchymal cells around the epithelial bud. Teeth pass through three stages of development: growth, mineralization, and eruption. The growth period is further subdivided into the bud, cap, and bell stages.

The oral epithelium initially thickens and begins to grow into the subjacent mesenchyme around the entire arc of each jaw. The free margin of this epithelial band gives rise to two invaginating processes. During the bud stage, the outer process (i.e., vestibular lamina) will form the vestibule that demarcates the cheeks and lips. From the inner horseshoe-shaped process (i.e., dental lamina), tooth buds arise at the site of each future tooth. Thus, the primordia for the temporary deciduous teeth are formed. Shortly afterward, the primordia of the succedaneous teeth develop in the same way. The permanent tooth germs lie in a hollow of the alveolar sockets on the lingual side of the deciduous teeth. The developing enamel organ of each tooth takes the shape of a goblet with the dental lamina as its stem. As the dental lamina disintegrates, the inner lining cells of the enamel organ differentiate to become columnar epithelial cells called ameloblasts that constitute the inner enamel epithelium. The outer layer of cells flattens into a layer of closely packed outer enamel epithelial cells. Loosely arranged epithelium of the stellate reticulum lies between the ameloblasts and the outer enamel epithelium. Mesenchymal cells inside the goblet-shaped enamel organ proliferate to form a dense aggregate termed the dental papilla during the cap stage (6).

The dental papilla ultimately forms the dentin, cementum, and pulp. The dentin is the internal layer of the tooth, the cementum is the bony tissue covering the root of the tooth, and the pulp is the soft inner part of the tooth. More peripherally, the condensing mesenchymal cells extend around the enamel organ as the dental follicle which contains stem cells that can be used in periodontal and bone regeneration work (7). The cells of the dental follicle eventually produce alveolar bone and collagen fibers of the periodontium. In the final bell stage of growth, the epithelium of the cap forms the enamel. The outer and inner enamel epithelium meet at their apical ends during this stage, and they proliferate to form Hertwig epithelial root sheath. This structure initiates differentiation of the outermost cells of the papilla to become arranged in a row of single columnar cells to form odontoblasts (Figs. 14.2 and 14.3). Nerves and blood vessels in the dental papilla begin to form the primitive dental pulp. The dental papilla grows toward the gum, crowding in on the enamel organ, which by then has lost its connection with the oral epithelium. Nonmineralized predentin is produced by the odontoblasts against the inner surface of the enamel organ during dentinogenesis. As the odontoblasts produce predentin, their cell bodies recede toward the center of the tooth, so that each odontoblast leaves behind a thin process (i.e., Tomes fiber) that occupies a dentinal tubule. The organic matrix of predentin eventually mineralizes to become dentin, which is arranged in tubules running from the pulp chamber toward the periphery (Fig. 14.4B,C). Meanwhile the enamel cap of the tooth is formed by the ameloblasts (8). The formation of dentin and enamel begins at the tip of the crown and progresses toward the root of the tooth (9). As the developing

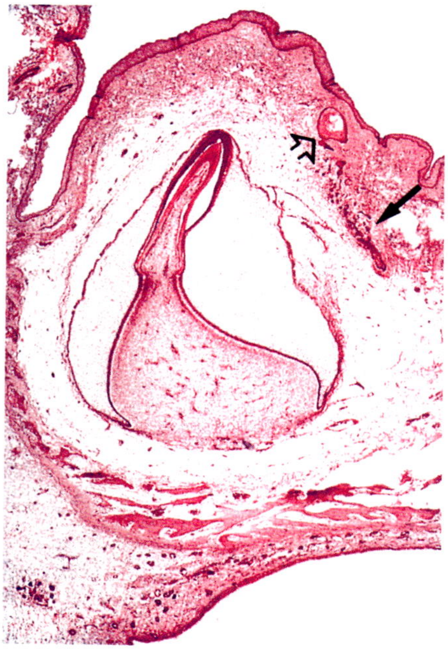

FIGURE 14.2 Enamel organ of deciduous tooth. Formation of enamel and dentin has begun at the crown area of the tooth. (*Arrow*, remnants of dental lamina; *arrowhead*, a small epithelial cyst [rest of Serres].)

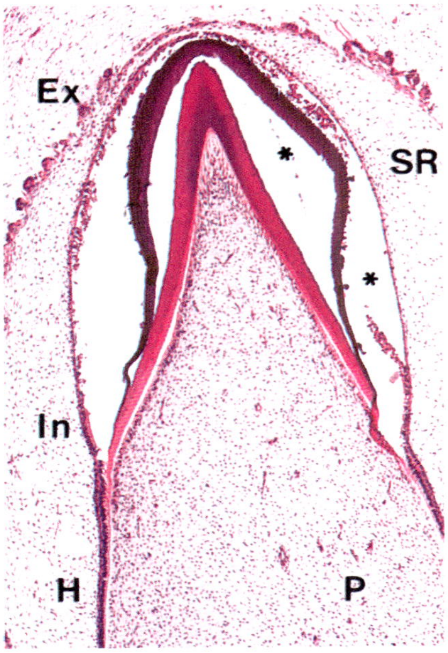

FIGURE 14.3 Enamel organ. Amelogenesis and dentinogenesis progress from crown to root. Early enamel and dentin appear as black bands, widest at the crown area, and become thinner toward the root. (*Ex*, external enamel epithelium; *In*, inner enamel epithelium; *SR*, stellate reticulum; *H*, Hertwig epithelial root sheath; *P*, dental papilla; *, artifact [separation].)

root increases in length, the previously formed crown moves closer to the surface of the gum. Even when the crown of the tooth begins to erupt, the root is still incomplete and continues growing until the crown has completely emerged (10,11) (Figs. 14.4A and 14.5). The enamel is made by differentiated ameloblasts that produce long, thin enamel prisms, or rods; these rods become calcified and are surrounded by a thin organic matrix. Enamel is avascular and acellular, and it has been described as being among the hardest substances in the body. Microscopically, the appearance of enamel is reminiscent of fish scales (Fig. 14.5B). Enamel production is completed when the crown is mineralized, and its final size is attained (12). The flattened ameloblasts and remaining cells of the enamel organ form a cuticle on the surface of the enamel at this point, but this membrane is subsequently shed (13). Patients with cystic fibrosis may have increased developmental defects of enamel (14).

After the root has attained its full length and definitive position in the jaw, bone-like cementum is deposited on it. Cementum is produced by the mesenchymal cells adjacent to the root. These cells become differentiated into a cementoblast layer that resembles the osteogenic (cambial) layer of the periosteum (15,16). Cementum is similar to bone, but it lacks haversian canals, nerves, and blood vessels (Fig. 14.6B) (4). Cementum may contain cementocytes or be completely acellular. Fibers from the rest of the dental sac form the periodontal ligament, which firmly attaches the tooth in the bony alveolar socket (17).

As the jaws approach their adult size, the latent primordia of the permanent teeth follow the same developmental

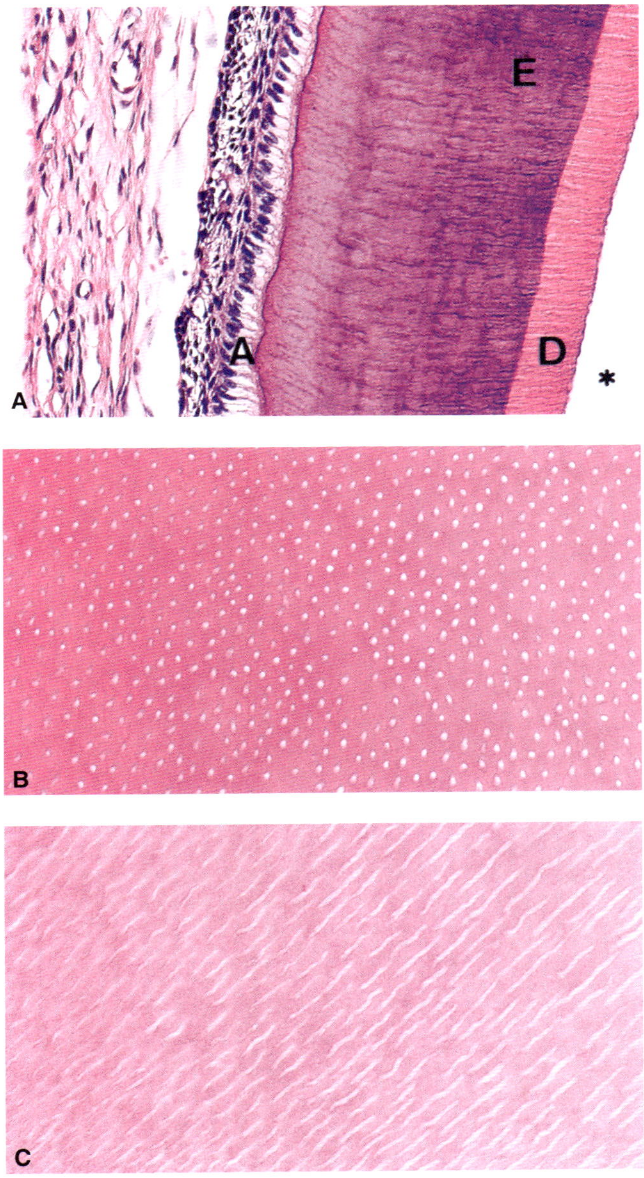

FIGURE 14.4 **A:** Developing tooth, sagittal section. Ameloblasts (*A*) line the surface of the enamel matrix (*E*) which is partly mineralized. Dentin (*D*) is not mineralized (predentin) and has been separated by an artifact (*) from the pulp (missing). High magnification view of the tubular structure of dentin seen in (**B**) cross-section and (**C**) sagittally.

process as did the deciduous teeth (18) (Fig. 14.6A). When a developing permanent tooth increases in size, the root of the corresponding deciduous tooth is partly resorbed by osteoclastic activity (Fig. 14.7). Thus, anchorage of the deciduous tooth is weakened and the tooth is shed, permitting the underlying permanent tooth to erupt.

Innumerable minor salivary glands, many of which are the seromucinous type, are distributed within the subepithelial stroma throughout the upper aerodigestive tract and oral cavity (6,19,20). An additional set of predominantly mucinous minor salivary glands are found in the nasopharynx between the pharyngeal wall and the Rosenmüller fossa.

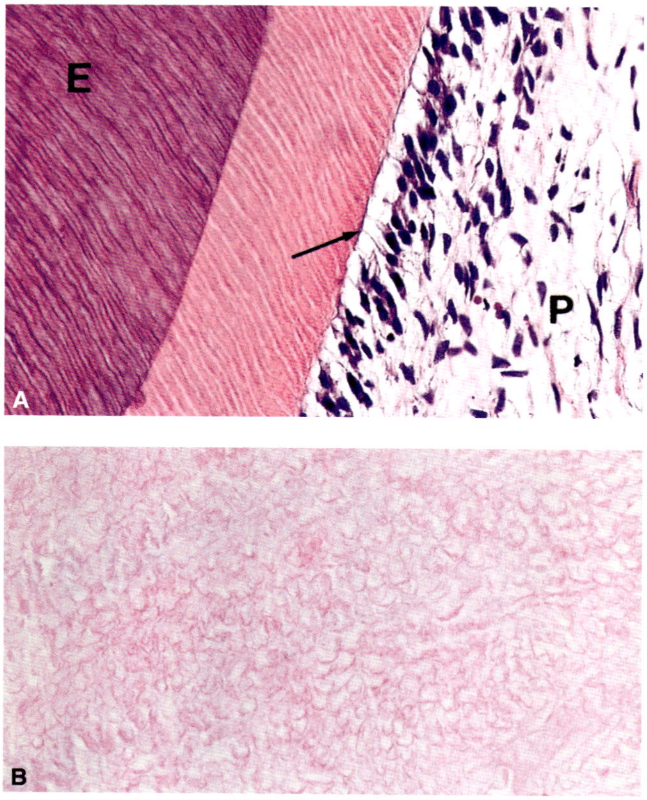

FIGURE 14.5 **A:** Developing tooth. Layer of polarized odontoblasts with Tomes fibers (*arrow*) extending into tubules of predentin. Fibroblasts of pulp (*P*) are loosely arranged. (*E*, enamel matrix.) **B:** High-magnification view showing the fish scale pattern of enamel which is acellular and avascular.

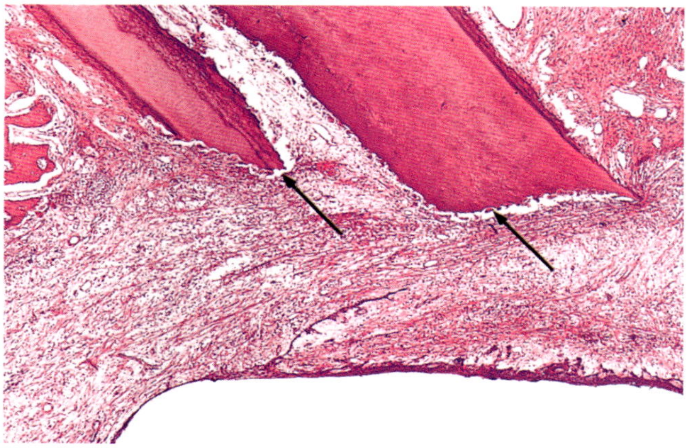

FIGURE 14.7 Roots of deciduous molar tooth before shedding. Resorption of dentin is indicated by numerous Howship lacunae and osteoclasts (*arrows*). The thin band (*bottom*) is the reduced internal enamel epithelium that covered the crown of the underlying permanent tooth; the enamel dissolved with decalcification.

These glands are called the tubarial salivary glands to reflect their location near the torus tubarius (21). The minor salivary glands in the oral cavity arise from the oral ectoderm and develop following a pattern of epithelial–mesenchymal interactions between the outgrowth of an ectodermal bud from the lining of the stomodeum and the underlying mesenchyme. This development begins between the 6th and 8th weeks of gestation. The proliferation, differentiation, and morphogenesis of these glands depend on intrinsic and extrinsic factors. The intrinsic factors reflect a programmed pattern of cell-specific gene expression while extrinsic factors include cell–cell and cell–matrix interactions, as well as a variety of growth factors, including fibroblast growth factor, transforming growth factor-beta, and sonic hedgehog (22).

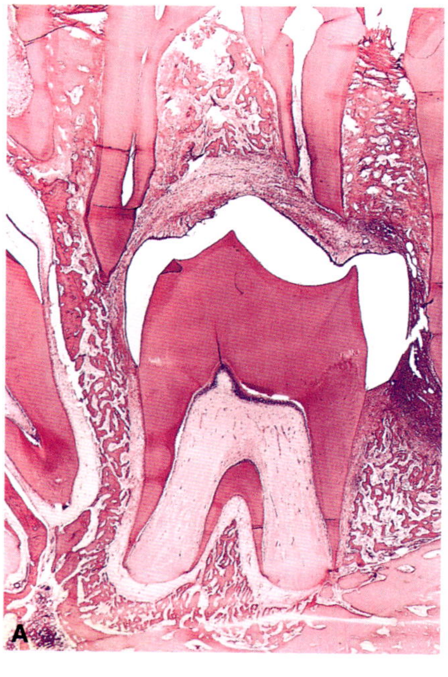

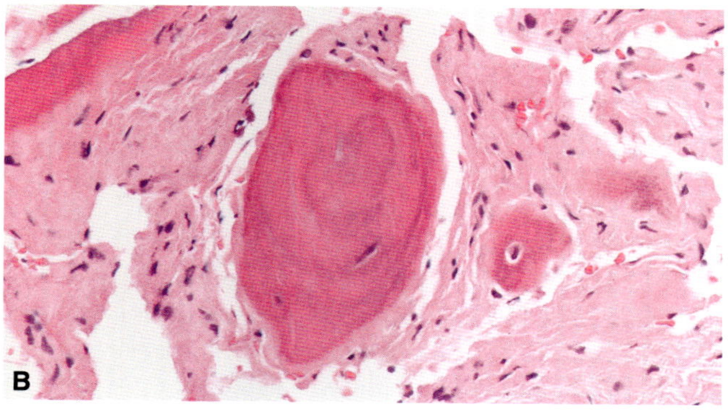

FIGURE 14.6 **A:** Deciduous and unerupted permanent teeth in a 10-year-old child. Perpendicular section through the roots of two deciduous teeth with underlying permanent molar tooth. Empty space was left by enamel that was dissolved by decalcification. Note the relationship between teeth and bone, respectively. **B:** Microscopically, cementum, as shown in this example of cementoosseous dysplasia, is similar to compact bone.

GROSS ANATOMY

In this chapter we will consider only those features that are important for the surgical pathologist.

Jawbone

The mandible and maxilla form the basic framework of the mouth. The mandible is the largest of the jawbones and has a horseshoe configuration. Its horizontal part forms the body, which is continuous with the vertical parts of the two sides, the rami. The body has a thick cortex, and the compact shell contains plates of cancellous bone arranged along the trajectories. The upper part of the body is termed the alveolar process; it is hollowed into sockets that carry eight teeth on each side. The mental foramina are openings on the anterior surface of the mandible that serve as a conduit for nerves and blood vessels. Each mandibular ramus is a nearly vertical, flattened, oblong plate of bone surmounted by two processes. The posterior articular process ends in the condyle, articulating with the articular disk of the temporomandibular joint. The anterior, or coronoid, process serves for the insertion of the temporal muscle.

The right and left maxillae jointly form the upper jaw and comprise the boundaries of four cavities: the roof of the mouth, the floor and lateral wall of the nose, the floor of the maxillary sinus, and the floor of the orbit. The alveolar processes of the maxillae together form the alveolar arch.

Nose

The external nose and nasal septum are partly composed of hyaline cartilage and bone. The two nares of the external nose are separated from each other by the median vertical soft tissue columella. The columella is attached posteriorly to the nasal septum, which forms the medial wall of two approximately symmetrical chambers, the right and left nasal cavities. The nasal cavity extends from the nares anteriorly to the choanae posteriorly. The nasal cavity widens to form the hair-bearing vestibule behind the nares. The limen nasi is a ridge on the lateral nasal wall that separates the vestibule from the rest of the nasal cavity proper. Each nasal fossa contains an olfactory region that occupies the superior portion of the nasal cavity. The rest of the nasal fossa consists of the respiratory region. The superior, middle, and inferior turbinates are located on the lateral walls of each nasal fossa. Rarely, turbinates may be bifid or trifurcate, and there may even be an anomalous accessory middle turbinate. A fourth supreme turbinate may be present at the uppermost portion of the lateral nasal wall. The scroll-shaped turbinates hang over the corresponding funnel-shaped nasal passages, or meatuses, into which the paranasal sinuses open. The concha bullosa is a normal anatomic variant present in approximately 50% of individuals. It has been termed the pneumatized turbinate and may be associated with nasal deviation and sinusitis. The nasal mucosa is in continuity with the mucosa of these sinuses through their corresponding openings. The nasal mucous membrane is the most vascular and thickest over the turbinates and nasal septum.

Paranasal Sinuses

The air-filled paranasal sinuses are located in the bones around the nasal cavity (Fig. 14.8). The maxillary and frontal sinuses open into the middle meatus of the nose, while the sphenoid sinus opens into the sphenoethmoidal recess above the superior turbinate. There are numerous ethmoidal sinuses that form small communicating cavities comprising the ethmoid air cells (i.e., ethmoid labyrinth). The ethmoid air cells have thin bony walls and can be divided into three groups based on location: the anterior and middle ethmoidal sinuses open into the middle meatus, and the posterior ethmoidal sinus opens into the superior meatus of the nose. The most posterior sphenoethmoidal cell is known as the Onodi air cell. Its importance is related to the fact that it can be damaged during endoscopy. The frontal sinus is located within the frontal bone above the orbits and drains into the nasal cavity middle meatus via the frontonasal duct (23). The paranasal sinuses vary in size, shape, and location; one or more of them may be underdeveloped or even absent.

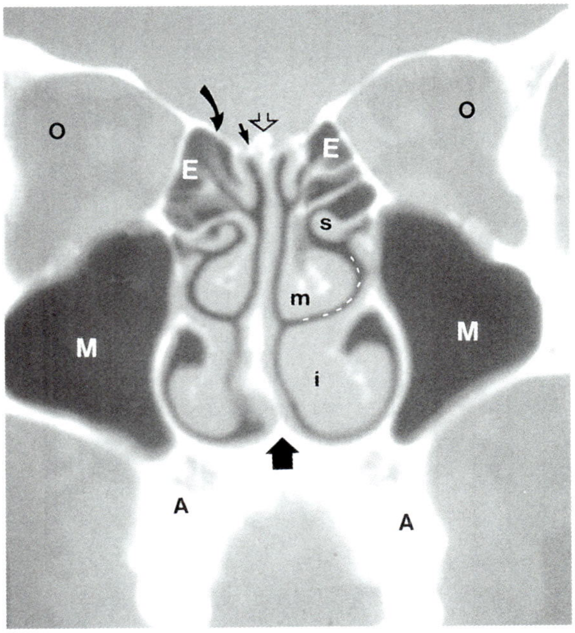

FIGURE 14.8 Coronal computed tomography (CT) image through the midportion of the paranasal sinuses, just behind the globes. The *large arrow* indicates the bottom of the nasal septum, the *open arrow* indicates the crista galli, *small arrow* the cribiform plate, and the *curved arrow* shows the fovea ethmoidalis (roof of the ethmoid sinus). The *dotted line* highlights the middle meatus. (*M*, maxillary sinus; *A*, alveolar ridge; *O*, orbit; *E*, ethmoid sinus; *i*, inferior turbinate; *m*, middle turbinate; *s*, superior turbinate.)

Blood Vessels

The nasal cavity has an extraordinarily rich blood supply derived from branches of the internal and external carotid arteries. The anterior and posterior ethmoidal branches of the ophthalmic artery supply the frontal and ethmoidal sinuses, as well as the roof of the nasal cavity. The sphenopalatine branch of the maxillary artery supplies the mucosa of the turbinates, meatuses, and the nasal septum. The mucosa of the maxillary sinus is supplied by branches of the maxillary artery and the sphenoidal sinus by the pharyngeal artery. The branches of all these vessels form a plexiform network in and below the mucous membrane. Kiesselbach plexus is a vascular area on the anterior portion of the nasal septum formed by the anastomoses of several arteries that can be a source of epistaxis. The veins of the nasal mucosa are well developed, particularly in the inferior turbinate and in the posterior part of the nasal fossa where they form a cavernous plexus. The veins of the lower jaw drain via the inferior central vein into the pterygoid plexus. Blood from the upper jaw and facial structures drains in two directions: the more anterior areas drain into the anterior facial vein and the more posterior parts, including the teeth, soft palate, fauces, and pharynx, drain into the pterygoid plexus. These anatomic relationships are particularly important because infections and tumors of the face, mouth, nose, and paranasal sinuses can reach the intracranial cavernous sinus either via the emissary vein of Vesalius or by way of veins communicating with the inferior or superior ophthalmic veins. The latter veins drain the structures of the anterior face, such as lips, cheek, and external nose, as well as the mucosa of the frontal sinuses, ethmoidal cells, and upper lateral nasal walls. The cavernous sinus also receives venous blood from the mucosa of the sphenoidal sinus.

Nerves

The regions discussed here have a rich and complex innervation including branches of the olfactory (i.e., cranial nerve I) and trigeminal (i.e., cranial nerve V) nerves, the description of which is beyond the scope of this chapter. However, many small and intermediate-sized nerve branches lie in close proximity to the mucosa and submucosa of the upper aerodigestive tract and are, therefore, prone to early invasion by carcinoma.

Lymph Nodes

Lymph nodes of the head and neck are abundant and can be divided into 10 groups: the occipital, mastoid, parotid, facial, sublingual, submaxillary, submental, retropharyngeal, anterior cervical, and lateral cervical nodes. Only those lymph nodes related to the regions covered in this chapter are discussed here. The location of an enlarged lymph node involved by metastatic carcinoma can be a very important clue about the site of origin of the carcinoma based upon knowledge of the lymphatic drainage to particular lymph node groups.

There are three principal parotid lymph node groups: superficial (e.g., suprafascial or preauricular) nodes; subfascial (i.e., extraglandular) lymph nodes contained in the parotid sheath; and deep intraglandular lymph nodes. The parotid lymph nodes drain the parotid gland, external ear, frontotemporal facial region, eyelids, upper lip, root of the nose, floor of the nasal cavity, soft palate, and buccal mucosa. Their efferents pass into the superior deep cervical nodes.

The small superficial group of facial lymph nodes includes, from rostral to caudal, the infraorbital (i.e., nasolabial) lymph nodes situated in the nasolabial fold, the buccal lymph nodes placed external to the buccinator muscle and its fascia, and the mandibular lymph nodes located external to the mandible. The facial lymph nodes receive afferent lymphatic vessels from the eyelids, nose, cheek, upper lip, and subjacent lymph nodes, and they empty into the submaxillary lymph nodes. There are also deep facial lymph nodes situated deep to the ramus of the mandible near the maxillary artery. Heterotopic buccal lymph nodes may rarely be encountered immediately beneath the buccal mucosa near the orifice of Stensen duct, giving the false clinical impression of a neoplasm (24,25). The sublingual (or lingual) lymph nodes are intercalated along the course of the collecting lymph trunks of the tongue. The submaxillary (submandibular) lymph nodes, situated in the space lodging the submaxillary salivary gland, are located around or within the fascial sheath of this gland. They receive lymph vessels from the chin, lips, cheeks, nose (including the anterior nasal cavity), gums, teeth, floor of the mouth, hard palate, tongue, and other nearby lymph nodes. The submental lymph nodes receive lymphatic vessels from similar regions. Their afferent vessels connect partly with the submaxillary and internal jugular lymph nodes. The retropharyngeal lymph nodes, which lie between the posterior wall of the pharynx and prevertebral fascia, may project anteriorly onto the soft palate and, therefore, may be mistaken for palatine tonsils. The more lateral retropharyngeal lymph nodes atrophy with age. These lymph nodes drain the nasal cavities, palate, middle ear, nasopharynx, and oropharynx. They send efferent vessels to the internal jugular chain of lymph nodes.

Lymphatics

For a detailed description of the lymphatic system draining the head and neck, the reader is referred to the compendium by Tobias, which is a translation of the original work by Rouvière (26). Lymphatic vessels arising from the upper lip terminate in the parotid, submental, and submaxillary lymph nodes. Central lymphatic vessels from the lower lip drain into the submental lymph node, while those originating more laterally empty into the submaxillary lymph nodes. Lymphatic vessels arising from the mucosa of the cheeks traverse the buccinator muscle to eventually terminate in the submaxillary lymph nodes. Their course may be interrupted by the buccinator lymph nodes. Cutaneous lymphatic vessels of the cheek end in the submaxillary, submental, and parotid lymph nodes. The lymphatic network of the tongue is divided into a superficial and deep (muscular) set. They drain to the lingual and submaxillary lymph nodes but end mainly in the deep cervical lymph nodes. The lymph node at the bifurcation of the common carotid artery is considered to be the principal lymph node of the tongue (27). Lymphatic vessels in the region of the tongue may cross the midline to reach lymph nodes of the opposite side. The gingiva contains a similar superficial and deep anastomosing lymphatic network that drains into the sublingual, submental, submaxillary, internal jugular, and occasionally the retropharyngeal lymph nodes. Lymphatic vessels that exit from the dental pulp of the teeth are in direct communication with those of the gingiva while those from the floor of the mouth are continuous with those of the tongue and gums. Afferent lymphatic vessels from the floor of the mouth drain into the sublingual, submental, and deep cervical lymph nodes, while those from the posterolateral region terminate in the submaxillary and deep cervical lymph nodes. The draining lymphatic vessels of the palate, which are continuous with those of the gums and palatine tonsils, reach the submaxillary, retropharyngeal, and deep lymph nodes of the neck.

The cutaneous lymphatic vessels of the nose terminate in the submaxillary lymph nodes. Lymph from the nasal vestibule goes to the parotid and submaxillary lymph nodes. Lymphatic vessels originating from the anterior nasal cavities drain to the submental lymph nodes, while those from the posterior cavities pass to the retropharyngeal and deep superior cervical lymph nodes. The lymphatic vessels from the olfactory region do not communicate with those of the respiratory region (27). Lymphatic vessels arising from the olfactory region communicate to the subarachnoid space of the brain via small canaliculi passing through the foramina of the cribriform plate along with the olfactory nerve filaments. The lower aspect of the inferior turbinate drains to the internal jugular lymph nodes. The upper portion of the inferior turbinate, along with lymphatic vessels from the middle turbinate, drains into the retropharyngeal and internal jugular lymph nodes. The superior turbinates drain to the retropharyngeal and deep cervical lymph nodes. Lymphatic vessels from the frontal and maxillary sinuses, along with the anterior and medial group of ethmoidal sinuses, drain to the submaxillary lymph nodes. The posterior ethmoidal group and sphenoidal sinuses drain lymph into the retropharyngeal lymph nodes.

Tonsils

Nonencapsulated or partially encapsulated lymphoid tissue present in the oropharynx is normally organized into tonsils composed of epithelium-covered lymphoid structures. When used alone, the most prominent tonsillar tissue comprises the palatine tonsil. Waldeyer ring represents the circular collection of submucosal lymphoid tissues near the superior portion of the upper aerodigestive tract (28). Waldeyer ring is comprised of the palatine, pharyngeal, tubal, and lingual tonsils, as well as the lateral pharyngeal lymphoid bands and intervening isolated lymphoid follicles termed pharyngeal granulations (28,29). The pharyngeal bands are located on the posterolateral wall of the oropharynx just behind the posterior tonsillar pillar. The ovoid palatine tonsils are situated laterally in the oropharynx within the triangular tonsillar fossa, which is bounded by the palatoglossal arch anteriorly and the palatopharyngeal arch posteriorly. Their tonsillar crypts often contain desquamated epithelium, debris, and microorganisms, particularly *Actinomyces* that may appear as white spots on the tonsillar surface. These aggregates of organisms may calcify and, thus, have been referred to as tonsilloliths. The palatine tonsils are the only partially encapsulated tonsils. The capsule is formed by compressed connective tissue at the interface between the tonsil and the pharynx and separates the tonsils from the underlying musculature of the pharyngeal wall. The tonsils are largest in early childhood but begin to gradually atrophy after the age of 4 years such that they are increasingly fibrotic after puberty. Bilateral enlargement of the tonsils may be seen with obesity, obstructive sleep apnea, Down syndrome, and secondary to diseases affecting the tonsils. Unilateral enlargement of the tonsil is usually because of a pathologic condition including carcinoma which is often associated with infection by high-risk HPV (30). The pharyngeal tonsil of Luschka is more commonly known as the adenoid and appears as a single pyramidal aggregate of lymphoid tissue located superiorly in the midline of the nasopharyngeal wall. Unlike the other tonsils, the adenoid does not contain typical crypts, but rather numerous surface folds extending from the tonsillar base anterolaterally. The surface of the pharyngeal tonsil also forms a median recess known as the pharyngeal bursa. The tubal tonsil (i.e., Eustachian tonsil or Gerlach tonsil) is that small portion of the pharyngeal tonsil that is located in the nasopharynx behind the pharyngeal opening of the Eustachian tube. The lingual tonsil is situated on the dorsum of the tongue posteriorly between the sulcus terminalis and the valleculae. In most individuals, the median glossoepiglottic ligament divides the lingual tonsil into bilateral lobes. Benign hairy polyps (i.e., naso-oropharyngeal choristoma) may be encountered at birth on the tonsil or tonsillar pillars (31).

Additional tonsillar structures may also be found in the normal human oropharynx. These include the oral tonsils, which are structurally similar to the palatine tonsils (32). Oral tonsils occur chiefly in the palate, floor of the mouth, and on the ventral surface of the tongue. They span only 1 to 3 mm in diameter and are firm, circumscribed, mobile lymphoid aggregates present under intact oral mucosa. Oral tonsils contain a single central crypt. It has been proposed by some authors that intraoral lymphoepithelial cysts originate from occluded crypts of the oral tonsils (33,34). Reactive tonsillar tissue has also been noted in the region of the pyriform sinus, palate, and lateral surface of the tongue (35,36).

MICROSCOPY

Mouth

Lips and Vermilion Border

The lips at the superior portion of the digestive tract are partly covered by hair-bearing skin with sweat glands and sebaceous glands along with a rich supply of sensory nerves. The inner surfaces of the lips are covered by the oral mucosa and form part of the wall of the oral cavity. Between the external integument and the oral mucosa are the orbicularis oris muscle, the labial vessels, nerves, and adipose tissue with numerous minor salivary glands. The latter are easily accessible for biopsies to diagnose Sjögren syndrome. The junction between the skin and oral mucosa is known as the vermilion border, where the keratinized squamous epithelium of the skin transitions to the mucous membrane of the oral cavity. The squamous epithelium of the vermilion border is thin, and the tall connective tissue papillae are close to the surface. The blood in the prominent capillary network is visible through the thin epithelium, accounting for the redness of the lips. Many venous tributaries extend up to the nose and down to the chin (37). The transition zone has no hairs. In adults, ectopic sebaceous glands are commonly observed in the vermilion border, at the corners of the mouth, or in the buccal mucosa; these are termed Fordyce spots or Fox–Fordyce granules. These sebaceous glands increase with age, such that 70% to 80% of elderly persons have them (Fig. 14.9) (38,39). Like the skin, the vermilion border is exposed to physical forces and chemical agents. For this reason, actinic keratosis and solar elastosis can be seen on the vermilion border. The squamous epithelium of the transitional zone imperceptibly merges with the stratified squamous epithelium of the oral mucosa.

Oral Mucosa and Submucosa

The oral mucosa consists of an epithelial layer and an underlying layer of lamina propria (Fig. 14.10). The mucosa of the oral cavity shows regional modifications in structure and cytokeratin expression that correspond to functional requirements. The stratified squamous epithelium of the oral mucosa has three functional types: the lining mucosa, masticatory mucosa, and specialized mucosa (40). Most of the oral mucosa is lined by nonkeratinized squamous epithelium, representing the lining of mucosa. The palate, gingiva, and dorsum of the tongue are exposed to the forces of mastication and are covered by keratinized epithelium of the masticatory mucosa type. The mucosa of the palate is orthokeratinized, whereas the epithelium of the gingiva is often parakeratinized. The masticatory mucosa has longer rete ridges that reflect the pressure associated with mastication (41). The epithelium throughout the oral cavity is worn off by mastication and speaking; hence, exfoliated epithelial cells are normal constituents of the saliva and are frequently encountered by the pathologist in sputum or as "contaminants" in bronchoscopic specimens. Squamous cells of the buccal mucosa are a convenient source for the microscopic demonstration of the sex chromosome (i.e., Barr body) (Fig. 14.11). The shed epithelial

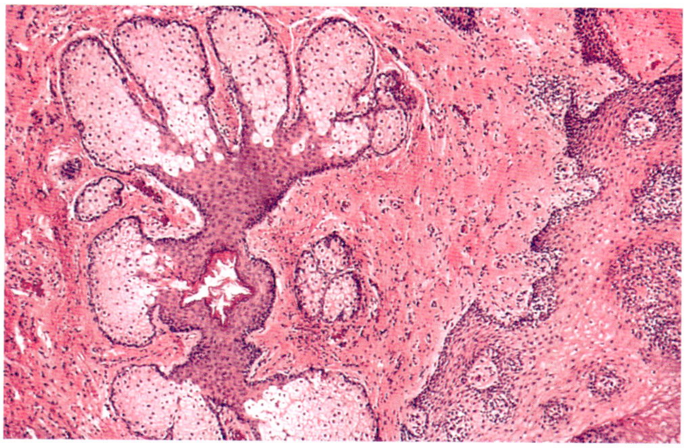

FIGURE 14.9 Ectopic sebaceous glands in the vermilion border (Fox–Fordyce granule).

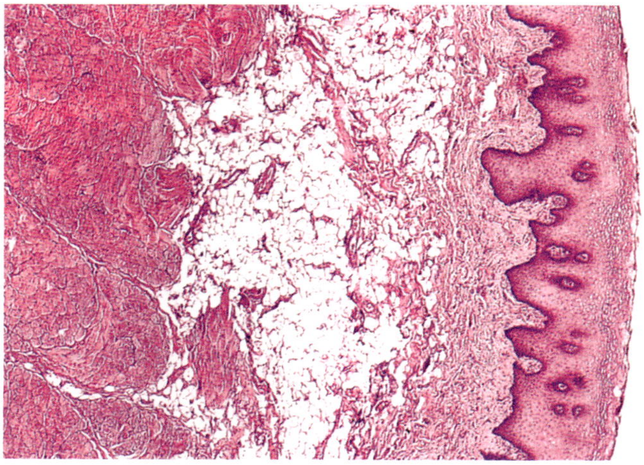

FIGURE 14.10 Inner aspect of cheek, cross section. *Left* to *right:* cross-striated muscle, adipose tissue, lamina propria, buccal mucosa.

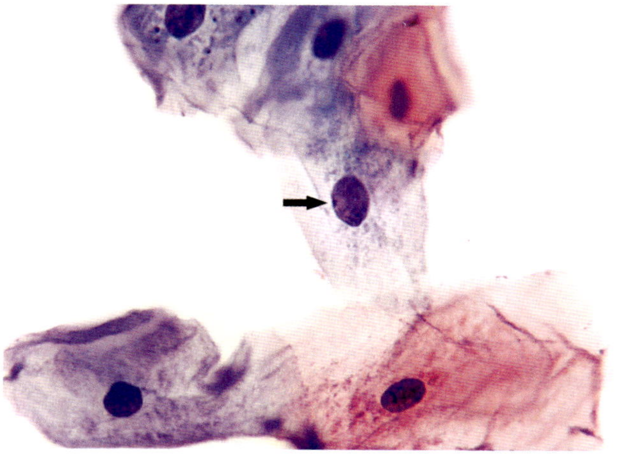

FIGURE 14.11 Cytologic image of a scraping of the buccal mucosa. Intermediate squamous cell with sex chromatin body (Barr body) (*arrow*) lying against the inner nuclear membrane (Papanicolaou stain).

cells are replaced by the basal cells, which divide and migrate to the surface before they themselves are eventually worn off. The renewal of the oral mucosa takes about 12 days (42). As the name implies, the squamous epithelium of the mucosa is kept moist and glistening by mucus secreted by the numerous minor and paired major salivary glands. This thin film of mucus covers and protects all intraoral structures, including the teeth, which are bathed by saliva. Saliva rinses away bacteria, provides buffering agents (e.g., phosphate and bicarbonate) that neutralize acids created by bacteria inhabiting dental plaques, contains antibacterial agents, and minerals required for tooth remineralization. Not surprisingly, conditions that decrease saliva production or cause dry mouth (i.e., xerostomia) promote tooth decay.

The interface of the epithelium and lamina propria is delineated by the basal lamina, or basement membrane which is a key structure breached when invasive squamous cell carcinoma is present. The basal lamina can be difficult to discern in hematoxylin and eosin (H&E) stained sections, but reticulin and other special stains demonstrate it well (Fig. 14.12). The basal lamina is secreted by the epithelial cells and serves supportive and filtering functions. It also regulates differentiation, migration, and polarity of the epithelial cells. The basal lamina is composed of type IV collagen, heparan sulfate, and two glycoproteins, laminin and entactin; the latter interact with other components of the extracellular matrix. A single layer of basal cells rests on the basal lamina. These basal cells continuously divide, and the new cells push the overlying ones toward the surface. During this process of differentiation, the small cuboidal basal cells become polyhedral and larger, forming the stratum spinosum. These cells contain abundant intracytoplasmic tonofilaments that attach to desmosomes, connecting the squamous epithelial cells with each other. The cells gradually become flat in the superficial layers. The nonkeratinized squamous epithelium lacks a stratum

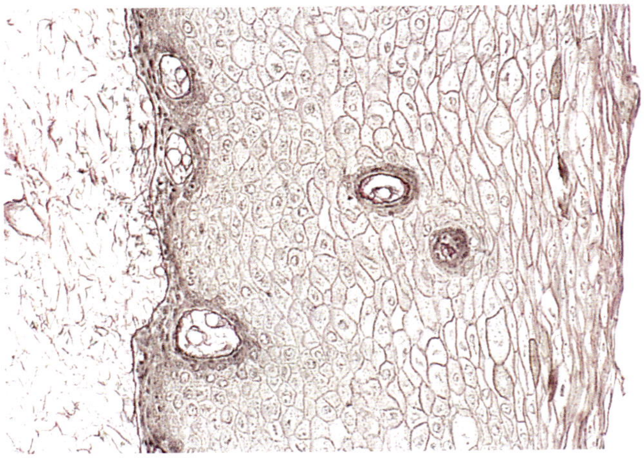

FIGURE 14.12 Buccal mucosa. Reticulin stain delineates the cell membrane of nonkeratinized squamous epithelial cells and the delicate basement membrane. Papillae of the lamina propria contain blood vessels.

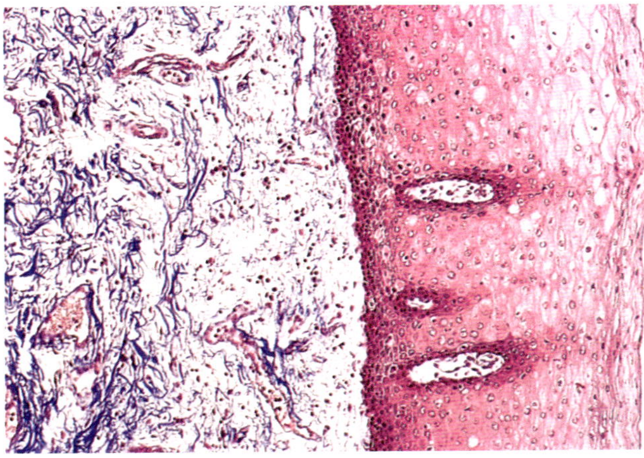

FIGURE 14.13 Buccal mucosa showing maturation of squamous epithelium: there is a row of small basal cells, larger cells of stratum spinosum, and parallel arranged flat surface cells. No keratinization is seen. Lamina propria shows delicate strands of connective tissue, blood vessels, and a few lymphocytes (Mallory trichrome).

granulosum and stratum corneum. The surface cells may retain their nuclei, and their cytoplasm does not contain keratin filaments (Fig. 14.13) (43). The cells of keratinizing epithelium form a stratum granulosum, which is a prominent layer three to five cells thick. The cells of this layer contain numerous intracytoplasmic keratohyalin granules that stain with hematoxylin. As the process of keratinization advances, the nucleus and cytoplasmic organelles become disrupted and disappear while the cell becomes filled with intracellular protein keratin. Thus, the stratum corneum, is formed at the surface.

All oral epithelia show expression of cytokeratins 5 and 14 in the basal cells of stratifying epithelium. The oral mucosa exhibits striking site-specific differences in cytokeratin synthesis that are usually apparent in the fetus by 23 weeks gestational age (Table 14.1) (44,45). The differences in distribution of these cytoskeletal proteins reflect the relationship between morphology and function of these epithelia (46). The gingiva shows more complex cytokeratin expression, similar to the epidermis. For example, gingival epithelia are immunoreactive for CK1 and CK10 in keeping with epithelial properties of toughness and rigidity, as well as CK4 and CK13 which promote flexibility and elasticity. In contrast, the lining mucosa resembles the stratified nonkeratinizing squamous epithelium of the esophagus and shows a paucity of cytokeratins. Malignant transformation is often associated with alterations in the cytokeratin pattern.

Normal oral mucosal epithelial cells, express the ABO blood group antigens (47). In general, the blood group antigen expression on epithelial cells follows the general phenotype of the host individual as determined by routine serologic methods. The loss of these antigens on malignant epithelial cells may be a valuable marker for primary carcinoma. All epithelial cell layers of the enamel organ, however, are devoid of the blood group antigens.

TABLE 14.1 Cytokeratin Expression Profiles of Different Oral Epithelia, as Determined by Immunohistochemical and Electrophoretic Studies

Cytokeratin Type	CK1	CK4	CK5	CK6	CK7	CK8	CK10	CK13	CK14	CK16	CK18	CK19
Epidermis (for comparison)	++	/	++	+	/	–	++	/	++	+	/	/
Oral lining mucosa	/	++	++	+	/	–	–	++	+	–	–	+
Oral gingival epithelium	++	+	++	+	–	–	++	+	++	++	–	+
Epithelium of oral sulcus	–	++	++	/	–	+	–	++	++	++	+	++
Junctional epithelium	–	–	++	/	–	+	–	++	++	+	+	++
Masticatory mucosa (e.g., hard palate)	++	/	++	/	/	/	/	/	++	++	/	/
Epithelia of enamel organ	/	–	++	/	+	+	–	–	/	/	–	++
Rests of Malassez	/	–	++	/	–	–	–	–	+	–	–	++

+, weak or occasional expression; ++, strong expression; –, expression absent; /, no data.

Modified from Mackenzie IC, Rittman G, Gao Z, et al. Patterns of cytokeratin expression in human gingival epithelia. *J Periodontal Res* 1991;26:468–478.

The lamina propria is a delicate layer of connective tissue situated beneath the squamous epithelium. It contains few elastic and collagenous fibers and is rich in blood vessels, lymphatic vessels, and nerves. The nerves belong to the sensory branches of the trigeminal nerve. The lamina propria also contains scattered lymphocytes, which are often found migrating through the epithelium. Consequently, there are normally a few lymphocytes in the saliva.

The submucosa under the lining mucosa is composed of loosely arranged connective tissue that contains larger blood vessels, lymphatic vessels, nerves, adipose tissue, and numerous minor salivary glands. Lymphatic vessels in the oral submucosa are more numerous and of larger caliber than those in the lamina propria. This lymphatic network increases progressively from normal tissue to precancerous lesions and tumor tissue. There is no submucosa in areas where the mucosa is in close proximity to the hard palate. The fibers of the lamina propria directly attach to bone in these areas in these areas; the mucosa, lamina propria, and periosteum are joined together as one membrane referred to as a mucoperiosteum.

Palate and Uvula

The roof of the oral cavity is formed by the palate; the anterior two-thirds consist of the hard palate, and the posterior one-third is comprised of the soft palate, which is a component of the oropharynx. The palate separates the oral and nasal cavities. It is bounded by the alveolar arches and gums anteriorly and laterally, and is continuous with the soft palate posteriorly. The hard palate is covered by masticatory mucosa, which has a series of palatal rugae that run across the midline but do not cross it. These ridges are readily visible. They are supported by dense connective tissue fibers that pass directly from the papillary layer of the lamina propria into the underlying bone. The submucosa contains adipose tissue in the anterolateral regions of the hard palate, whereas the posterolateral regions contain pure mucous glands (Fig. 14.14). These minor salivary glands are termed the palatine glands.

The soft palate is mobile without bony support. It is suspended from the posterior border of the hard palate like a curtain. Its oral surface is covered by squamous mucosa whereas the nasal surface is continuous with the floor of the nasal cavity and mostly lined by ciliated, pseudostratified columnar epithelium with goblet cells. The soft palate contains striated muscle fibers, blood vessels, and nerves (Fig. 14.15). The striated muscle of the palate has a different cytoskeletal structure compared with the limb muscles,

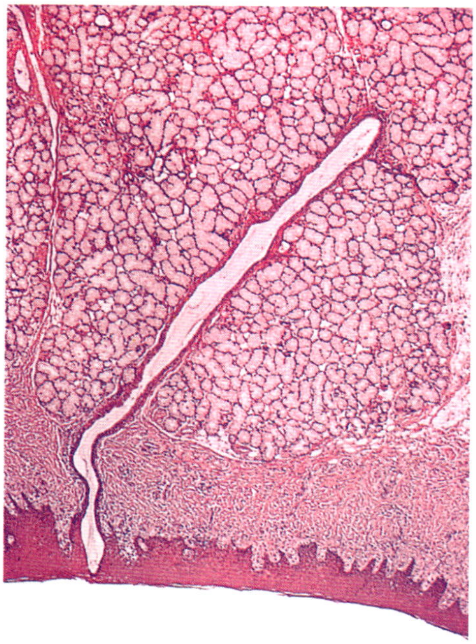

FIGURE 14.14 Hard palate with pure mucous minor salivary glands and duct. Note dense connective tissue of lamina propria (H&E).

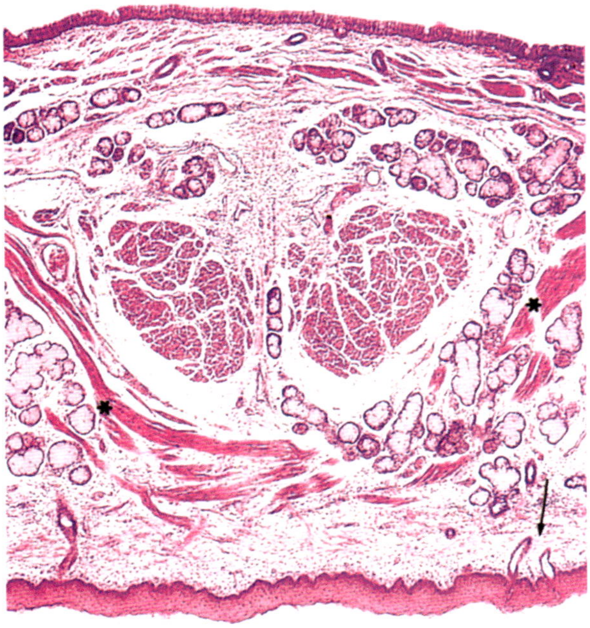

FIGURE 14.15 Soft palate sectioned in the coronal plane. Nasal respiratory mucosa (*top*), oral squamous mucosa (*bottom*). Fascicles of pharyngopalatine muscle are to the right and left of the midline. Fibers of levator veli palatini muscle (*) are obliquely descending. Ducts of minor salivary glands are near the oral mucosa (*arrow*).

and may lack or show only minimal desmin positivity (48). Larger mucous glands underlie the oral epithelium of the soft palate, whereas smaller groups of mixed glands are present on the nasal surface under the respiratory epithelium. The uvula is a small conical process of soft tissue that hands from the posterior border of the soft palate and, in some cases, may be bifid (49). The uvula is microscopically similar to the soft palate (Fig. 14.16A,B). It contains predominantly mucous glands and muscle fibers that become progressively sparse from the proximal to the distal end. Mucous glands in this area have been shown to express prostate-specific membrane antigen (50). The musculus uvulae muscle inserts into the actual mucosa of the uvula. Mast cells, usually located around blood vessels, are a frequent finding (51).

Floor of the Mouth

The mucous membrane of the floor of the mouth is thin and loosely attached to the underlying structures. Its rete ridges are short and the submucosa contains some adipose tissue with numerous sublingual mucous glands. Extruded mucin around these minor salivary glands or their ducts (i.e., ranula) may present intraorally or plunge downward into the soft tissues of the neck. The former are termed simple ranulas whereas the latter are classified as deep ranulas.

Tonsils

The tonsils are organized aggregates of lymphoid tissue covered by mucosa. The close proximity of lymphocytes to the surface epithelium facilitates the direct internal transport of foreign material from the exterior. Epithelium-lined crypts and folds further aid in trapping foreign material. The mucosa lining the palatine and lingual tonsils consists of stratified squamous epithelium (Fig. 14.17A,B), whereas the mucosa overlying the pharyngeal tonsil is a pseudostratified ciliated respiratory-type epithelium that contains occasional goblet cells. The epithelium lining the crypts or folds represents a specialized epithelial extension of the regional surface epithelium. The palatine tonsil contains 10 to 30 crypts that may extend to the deep juxtacapsular region. In the lingual tonsil (Fig. 14.17A) and pharyngeal tonsil, the

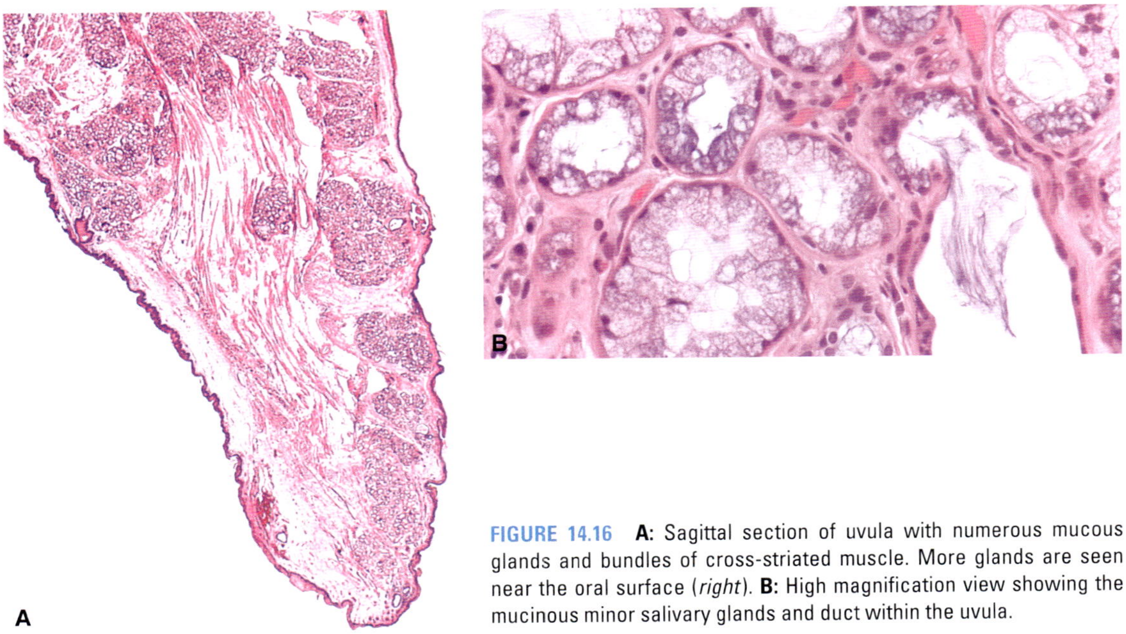

FIGURE 14.16 **A:** Sagittal section of uvula with numerous mucous glands and bundles of cross-striated muscle. More glands are seen near the oral surface (*right*). **B:** High magnification view showing the mucinous minor salivary glands and duct within the uvula.

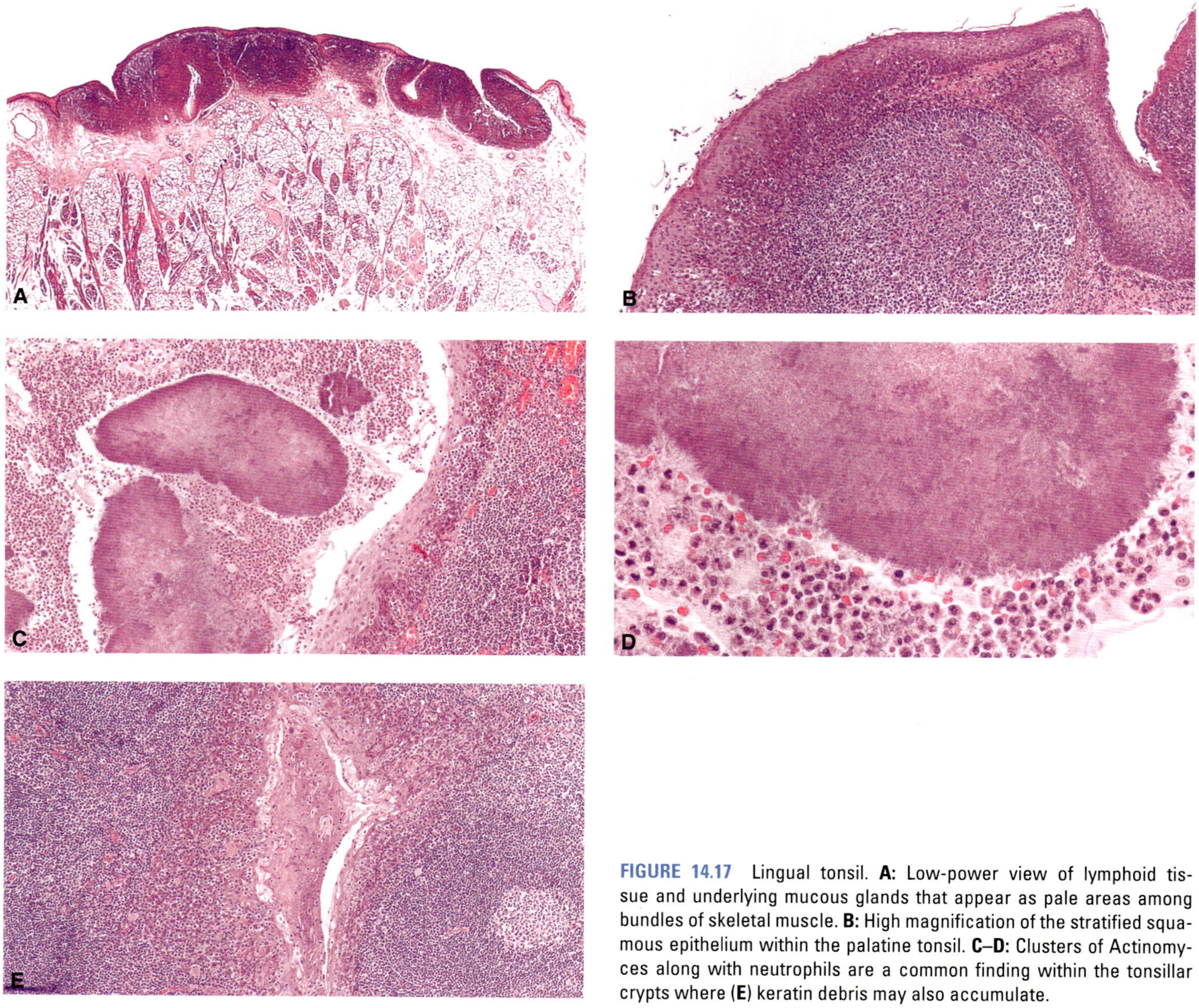

FIGURE 14.17 Lingual tonsil. **A:** Low-power view of lymphoid tissue and underlying mucous glands that appear as pale areas among bundles of skeletal muscle. **B:** High magnification of the stratified squamous epithelium within the palatine tonsil. **C–D:** Clusters of Actinomyces along with neutrophils are a common finding within the tonsillar crypts where **(E)** keratin debris may also accumulate.

lining epithelium forms only shallow folds 0.5- to 1.0-cm deep. Sulfur granules comprised of *Actinomyces* and *Actinomyces*-like oral flora are a frequent finding within tonsillar crypts (Fig. 14.17C–E) along with neutrophils and keratin debris. As in lymph nodes, the lymphoid component may contain lymphoid follicles, some with active germinal centers (Fig. 14.19B). Intraepithelial lymphocytes within the crypt-lining epithelium are commonly observed (Figs. 14.18 and 14.19). Sometimes the epithelium is so heavily infiltrated by lymphocytes that it is scarcely distinguishable. This specialized lymphoepithelium of the tonsillar crypts lacks a basement membrane layer and is the source of oropharyngeal HPV-associated squamous cell carcinomas (Fig. 14.18B). Intraepithelial lymphocyte trafficking primarily overlies lymphoid follicles (Figs. 14.20 and 14.21), resembling Peyer patches of the small intestinal mucosa. Intraepithelial lymphocytes of the surface mucosa are predominantly T cells (CD3+, CD5+, CD7+, and CD8+), whereas those present within the crypt epithelium include both T cells and B cells (52). For unknown reasons, lymphoid and dendritic cells within tonsillar tissue may express low levels of estrogen receptor (ER) (53). The epithelium overlying lymphoid nodules also contains M cells that exhibit numerous surface microvilli and microfolds similar to those of the enteric mucosa. At the ultrastructural level, intraepithelial lymphocytes have been shown to be located within intracytoplasmic compartments that communicate with each other to form an intraepithelial network of channels (54); as noted, this specialized lymphoepithelium lacks a basement membrane (30). Langerhans cells found in the crypt epithelium may contain intracytoplasmic Birbeck granules (55).

The tonsils are normally present in close association with minor salivary glands (Fig. 14.22A,B). Frequently, the excretory ducts of these mucous glands empty into the tonsillar crypts. The minor salivary glands (i.e., Weber glands)

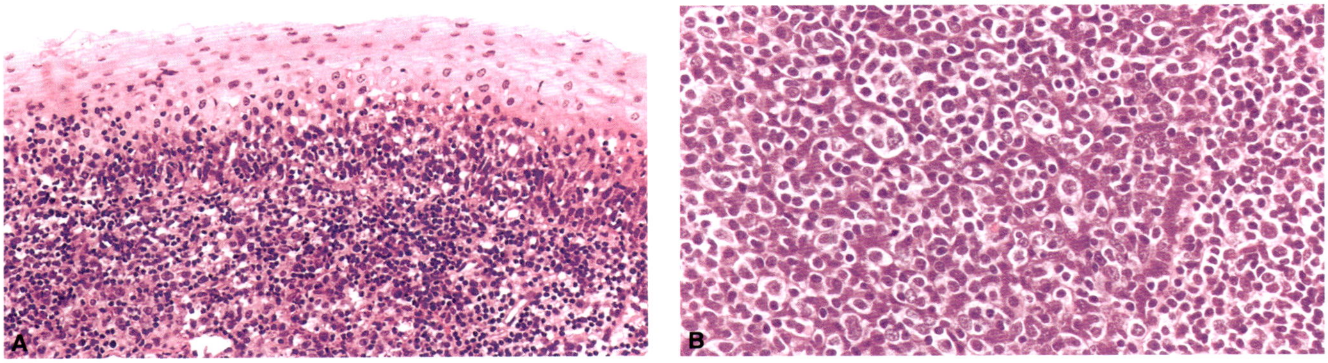

FIGURE 14.18 Lingual tonsil. **A:** The mucosa (nonkeratinized stratified squamous epithelium) is infiltrated with numerous lymphocytes that obscure the basement membrane. **B:** The specialized lymphoepithelial crypt lining of the palatine tonsil lacks a defined basement membrane.

adjacent to the palatine tonsils represent a putative reservoir of pathogenic bacteria and, therefore, should be removed along with the tonsil during a tonsillectomy (56). Small foci of hyaline cartilage and/or bone may be present close to the fibrous capsule of the palatine tonsil, likely representing an embryologic remnant of Reichert cartilage that originates from the second branchial arch (57).

Minor Salivary Glands

Numerous small salivary glands are scattered throughout the submucosa of the oral cavity, sinonasal cavities, and adjacent to the palatine and pharyngeal tonsils. These glands are unencapsulated and are named by their location. They can be classified as mucous, serous, and mixed sero-mucinous types (Table 14.2). These glands empty onto the mucosal surface through numerous small excretory ducts. The secretory activity of these glands appears to be continuous, although they can respond to specific local chemical or physical stimuli.

Structurally, the minor salivary glands are compound tubular or tubuloacinar glands. Mixed acini contain mucous cells that are nearest to the excretory duct, while the serous cells are located at the cul-de-sac of the acini, appearing as crescents called the demilunes of Giannuzzi (Figs. 14.23, 14.24, and 14.46). The mucous acini are more tubular and larger than those of the serous type and contain flattened nuclei at the cell bases. The mucous cells have a pale-appearing

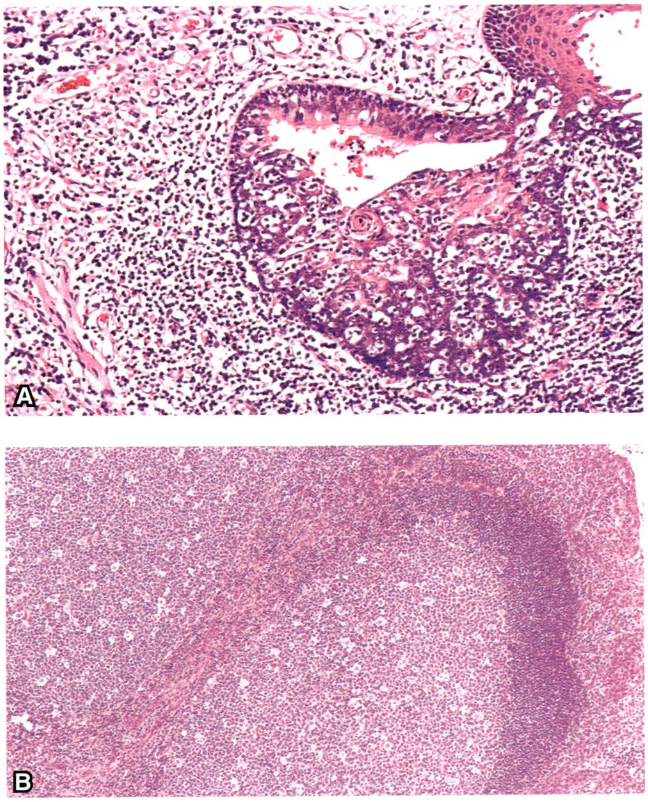

FIGURE 14.19 Lingual tonsil. **A:** Squamous epithelium of crypt is disrupted by lymphocytes that have migrated into it. **B:** Organized lymphoid follicle within the stroma of the palatine tonsil.

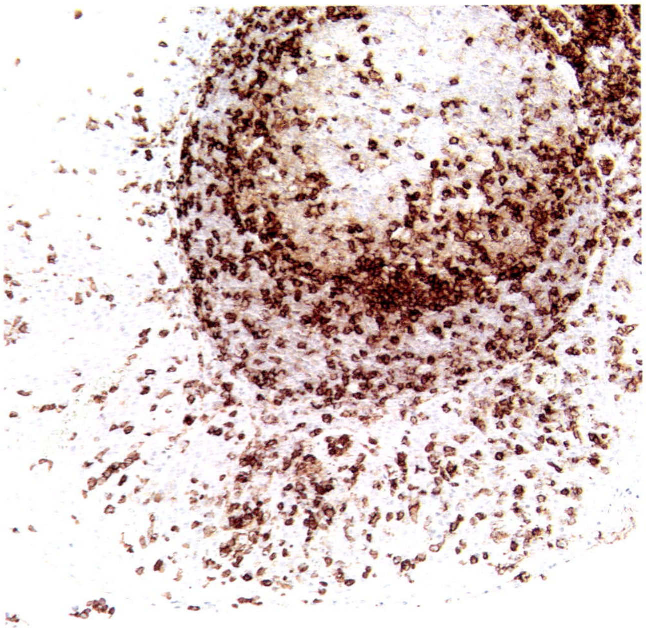

FIGURE 14.20 Tonsil lymphoepithelium overlying a secondary lymphoid follicle with a germinal center. The CD3 (T-cell marker) immunostain highlights abundant T cells trafficking through the epithelium.

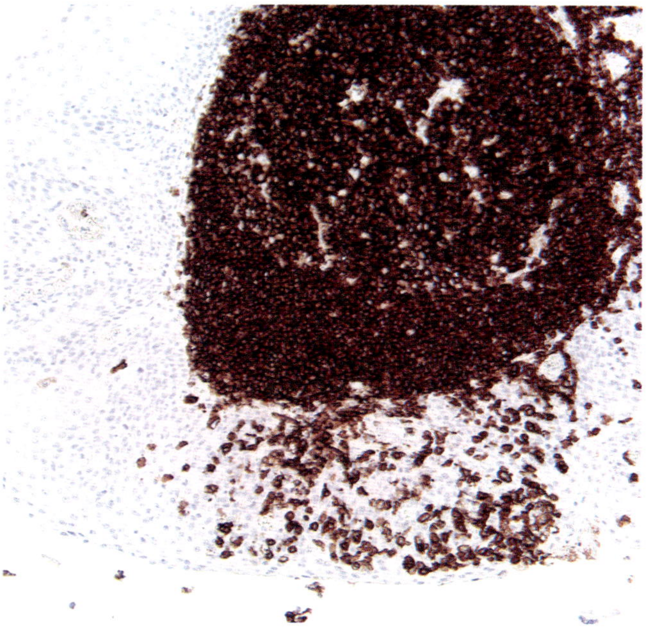

FIGURE 14.21 Tonsil showing CD20 (B-cell marker) immunoreactive lymphocytes located within a lymphoid follicle. B cells focally pass through the lymphoepithelium.

TABLE 14.2 Minor Salivary Glands of the Mouth, Nose, and Paranasal Sinuses

Name	Location	Type of Acini
Labial (superior and inferior)	Lips	Mixed (predominantly mucous)
Buccal	Cheek	Mixed (predominantly mucous)
Glossopalatine	Anterior faucial pillar Glossopalatine fold	Pure mucous
Palatine	Hard palate Soft palate Uvula	Pure mucous (mixed in uvula) Mixed (predominantly mucous)
Palatine tonsil (Weber glands)	Subjacent to tonsil capsule	Pure mucous
Sublingual	Floor of mouth	Mixed (predominantly mucous)
Lingual (glands of Blandin and Nuhn)	Anterior tongue	Mixed (predominantly mucous)
Tongue (Ebner glands)	Circumvallate papillae	Pure serous
Lingual tonsil	Base of tongue	Pure mucous
Nasal and paranasal	Nose and sinuses	Mixed

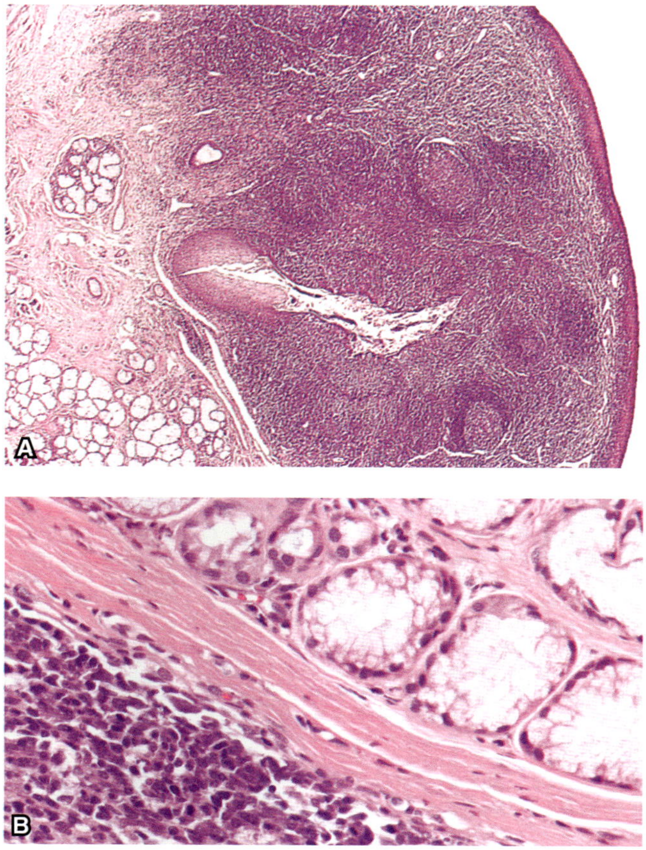

FIGURE 14.22 Lingual tonsil. **A:** Lymphoid tissue with follicles under mucosa and around its infolding (crypt). Note mucous glands and ducts. **B:** Minor salivary glands closely associated with the palatine tonsil.

cytoplasm in H&E-stained sections (Fig. 14.14) and will also stain with Alcian blue, periodic acid–Schiff (PAS), or mucicarmine stains (58,59). The smaller serous cells have more rounded basal nuclei with eosinophilic cytoplasm that contains zymogen granules. Normal acinar and ductal cells

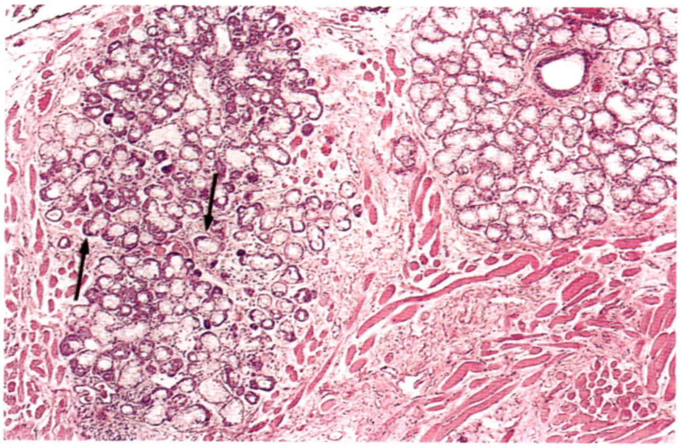

FIGURE 14.23 Minor salivary glands in the uvula are surrounded by cross-striated muscle. The gland with the distinct duct is of the pure mucous type. The other gland is seromucinous, with the serous cells forming darker-staining crescents (*arrows*).

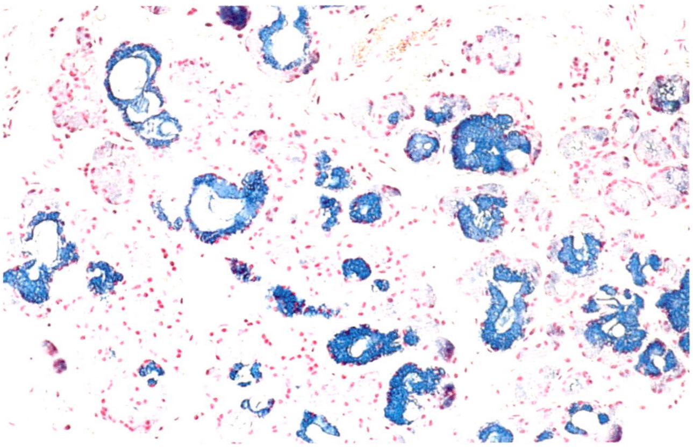

FIGURE 14.24 Mixed minor salivary gland of the uvula. Acid mucopolysaccharides in mucous cells are stained turquoise (Alcian blue at pH 2.6).

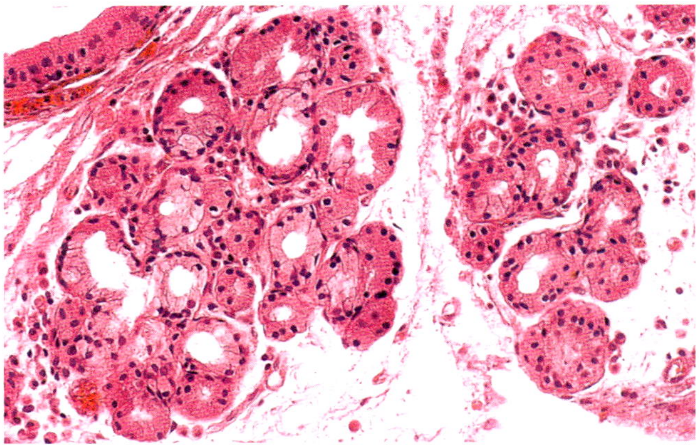

FIGURE 14.25 Oncocytes in minor salivary gland appear as cuboidal swollen cells with a finely granular oxyphilic cytoplasm.

express androgen receptors (60). In women, however, estradiol is positive in ductal epithelial cells, but testosterone and dihydrotestosterone are negative (61). This sex difference of hormonal receptors in minor salivary gland epithelia is thought to play a role in Sjögren disease and explain its strong predilection for females. Normal salivary gland tissue is not immunoreactive for pleomorphic adenoma gene-1 (PLAG1), a zinc finger transcription factor gene consistently rearranged and overexpressed in pleomorphic adenomas of the salivary glands. Normal minor salivary glands also do not appear to undergo appreciable proliferation and apoptosis. By comparison, the ductal epithelial and acinar cells of salivary glands from Sjögren disease patients do exhibit increased apoptosis (62).

Contractile myoepithelial cells are wrapped around the acinus and assist by squeezing the secretion from the acinar cells into the excretory ducts (63). The myoepithelial cells of the minor salivary glands are variably immunoreactive with antibodies to cytokeratins (e.g., CK5, CK14, and CK17), smooth muscle markers (e.g., smooth muscle actin [SMA], h-caldesmon, calponin), p63 and, rarely, S-100 protein (64–66). While the protein p63 is present within myoepithelial cells in all stages of salivary gland morphogenesis, calponin is only detected with increasing salivary gland maturation. Evidence for a neuroectodermal phenotype and immunoexpression of S-100, GFAP, and NSE has not been observed in myoepithelial cells of salivary glands (65,66). It should be noted that melanocytes may be found in $<2\%$ of minor salivary glands (67).

It is important to note that the minor salivary glands, wherever they occur in the mouth, nose, or paranasal sinuses, can become involved by the same pathologic processes as the major salivary glands. This is especially true for neoplasms, which arise from various cellular components of the minor salivary glands. Excess adipose tissue in association with these glands may represent a sialolipoma (68).

Squamous metaplasia can occur in the excretory ducts and acini of minor salivary glands and, in extreme cases of regeneration, can mimic squamous cell carcinoma. A good example of this is seen in necrotizing sialometaplasia or in irradiated salivary glands which is most commonly observed in the palate. Oncocytic cells may be found in minor salivary glands in varying numbers and distribution patterns (Fig. 14.25). Oncocytes are large, cuboidal, or columnar epithelial cells with finely granular eosinophilic cytoplasm due to increased mitochondrial content. They have been identified in many exocrine or endocrine glands (69). Mitochondria may occupy up to 60% of the cell cytoplasm and can be visualized after 48 hours of staining with phosphotungstic acid–hematoxylin, histochemical demonstration of mitochondrial enzyme activity in frozen tissue sections, immunohistochemistry for antimitochondrial antibodies, or by electron microscopy. Oncocytes are benign cells that occur with increasing frequency in the minor salivary glands throughout the upper aerodigestive tract in older persons. They develop because of metaplasia of ductal and acinar epithelium. However, the etiology and functional significance of oncocytic metaplasia is not clear. In the minor salivary glands, oncocytes can be seen in nodular or diffuse oncocytosis as well as oncocytoma (70,71). Oncocytic features can be seen in a variety of salivary gland cancers, especially mucoepidermoid carcinomas (72,73).

Cheeks

The skin of the cheek is part of the facial skin. The intraoral surface of the cheeks is covered by stratified squamous mucosa supported by lamina propria. The submucosa consists of some fat cells and mixed-type salivary glands embedded in fibrovascular connective tissue. The mucosa and submucosa are bound to the underlying buccal musculature by connective tissue fibers (Fig. 14.13).

Juxtaoral Organ of Chievitz

Deep in the wall of the cheeks overlying the angle of the mandible sits an anatomically well-defined small fusiform neuroepithelial structure, the juxtaoral organ of Chievitz, which is normally present in the buccotemporal space (74,75). It measures 0.7 to 1.7 cm in length and 0.1 to 0.2 cm

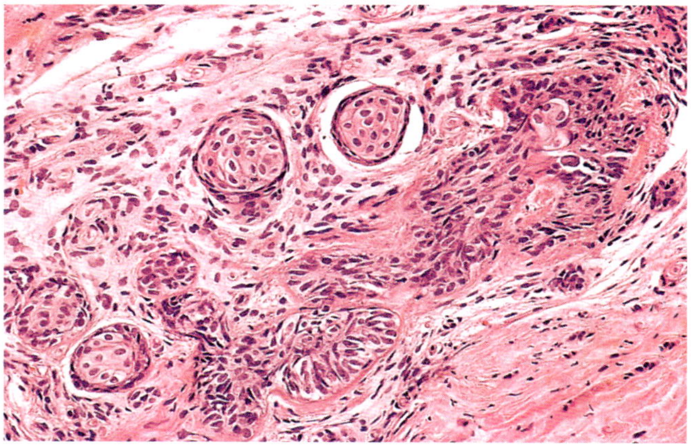

FIGURE 14.26 Juxtaoral organ of Chievitz. Small nests of nonkeratinized squamous epithelial cells are delineated by basement membrane. The elongated cells outside the epithelial islands are fibroblasts and Schwann cells. Nerve is seen below and to the right.

in diameter in adults and persists throughout life. It is multilobulated, has a dense fibrous capsule, and consists of round or elongate nests of nonkeratinizing squamous-like epithelial cells embedded in an organized connective tissue stroma rich in small nerves and sensory receptors derived from two to four branches of the buccal nerve (76). The connective tissue envelope is divided into a thin inner stratum fibrosum internum, middle stratum nervosum, and outer stratum fibrosum externum. These nests of epithelial cells are clustered in histologic sections, but serial sectioning shows them to be small sprouts and folds in continuity with a mass of epithelium. Thus, cross sections through different portions of this structure can show considerable variation in number and shape of epithelial sprouts (Fig. 14.26). The larger nests of epithelium are composed of cells with a clear PAS-positive cytoplasm and round or oval nuclei, forming "light centers." Intercellular bridges can be seen toward the center of the cell nests. The cells in the smaller nests can show a whorl-like or concentric arrangement. Occasionally, a gland-like lumen or a follicle filled with colloid-like mucin-negative material is encountered. Melanin pigmentation has been reported in the juxtaoral organ (77). Its central epithelial cells are immunoreactive for cytokeratins, whereas the outer basaloid cells are usually keratin-negative. The cell nests are also positive for vimentin, epithelial membrane antigen (EMA), CK10, CK14, CK19, GLUT-1, p63, KL-1, EGFR, and NGF but are negative for S-100 protein, glial fibrillary acidic protein (GFAP), neuron-specific enolase (NSE), synaptophysin, and chromogranin (78–80).

The juxtaoral organ has considerable importance for the surgical pathologist because the presence of squamous epithelial nests intimately admixed with numerous small nerves may be misinterpreted as perineural invasion by squamous cell carcinoma or mucoepidermoid carcinoma (81,82). The juxtaoral organ may be hyperplastic and enlarged (83,84), but carcinomas originating from it have not been reported.

The function of the juxtaoral organ is unknown. Since its first description in a 10-week-old human embryo (85), the juxtaoral organ has been widely believed to be a rudimentary structure, representing an abortive salivary gland anlage. More recently, the possibility of a neurosecretory and receptor function has been raised (86). The juxtaoral organ contains pacinian corpuscles (87), supporting its mechanosensory function (88). Immunoreactivity of the epithelial cells and nerve fibers for the ion channel marker Piezo2 supports its putative role as a mechanoreceptor (88). For a thorough review of the literature, the reader is referred to the small monograph by Zenker (76) or review paper by Pantanowitz and Balogh (86). Finally, we would like to alert the reader to the fact that similar benign epithelial islands may reside within peripheral nerves in the maxilla (89) and mandible (90).

Tongue

Situated on the floor of the mouth, the tongue is an organized functional structure of cross-striated muscle invested by a mucous membrane. Some of its muscles are extrinsic in that they have their origins outside the tongue, whereas others are intrinsic and contained entirely within it. The bundles of cross-striated muscle are embedded in connective tissue with some adipocytes arranged three dimensionally. The tongue is well-supplied with interanastomosing blood vessels. Age-related changes include lingual varicosities, glossitis, and atrophy of the taste papillae. The tongue has a rich supply of myelinated and nonmyelinated nerves containing motor, sensory, and vegetative nerve fibers, and some of these are associated with ganglion cells. The ventral surface of the tongue is covered by smooth squamous mucosa with short, blunt rete ridges (Fig. 14.27). Its submucosa merges with the connective tissue that intersects with the ventral muscle bundles of the tongue. The lingual frenulum anchors the tongue in the mouth. It is covered by squamous mucosa and contains a submucosal layer of

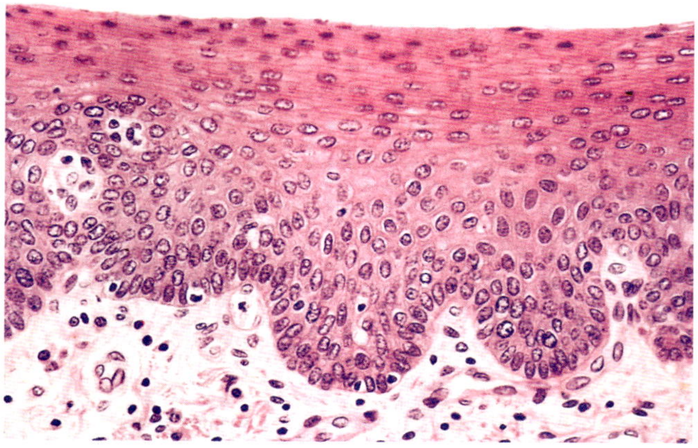

FIGURE 14.27 Ventral surface of tongue. The stratified nonkeratinized squamous epithelium of the lining mucosa has a slightly wavy interface with the lamina propria. A few scattered lymphocytes are seen in the lamina propria and in the epithelium.

fascia of variable thickness (91). Layers of the fascia may envelop the sublingual glands and vessels. The genioglossus muscle is also inserted into the fascia. Malocclusion can result in a long frenulum. Tongue-tie (ankyloglossia) occurs when the frenulum is too short and tight (92).

The dorsal surface of the tongue is divided into anterior and posterior parts by a V-shaped shallow groove, the sulcus terminalis. The anterior two-thirds of the dorsum of the tongue is lined by specialized mucosa bound by connective tissue fibers to the underlying skeletal muscle. This specialized mucosa is modified keratinized squamous epithelium covered with small papilla that appear as grossly visible projections. The pathologist has to be aware of these papillae so as not to mistake them for papillary epithelial hyperplasia, papillomas, or oral hairy leukoplakia. Papillae can be filiform, fungiform, foliate, or circumvallate. Fungiform, foliate, and circumvallate papillae can contain taste buds, in which case they are termed gustatory papillae (93). Most papillae are filiform and appear as conical projections of keratinized epithelium (Fig. 14.28). The fungiform papillae are rounded elevations above the surface of the tongue that are scattered among the filiform papillae. The fungiform papillae are surfaced by nonkeratinized mucosa; they appear as small red nodules because the thin epithelium does not mask the underlying vascular connective tissue (Fig. 14.29). Fungiform papillae can mimic denture-induced fibrous hyperplasia or small traumatic fibromas and may be pigmented in healthy individuals, especially in African Americans. Pigmentation is attributed to increased melanocytes, melanin, and melanophages. Pigmented fungiform papillae may also be associated with systemic conditions (e.g., iron deficiency, Peutz–Jeghers syndrome, Addison disease) (94). The foliate papillae are located posteriorly along the sides of the tongue while the circumvallate papillae are found at the junction of the anterior two-thirds and the posterior one-third of the tongue. These are the largest papillae, measuring 0.1 to 0.2 cm in diameter, and they are arranged in a V-shape immediately anterior to the sulcus terminalis. The circumvallate papillae number 6 to 12. A small, ring-shaped furrow surrounds each circumvallate papilla and separates it from a circular, palisade-like mucosal elevation termed the vallum that represents the outer border of the circumvallate papilla (Fig. 14.30). Occasionally, there may be a round or rhomboid-shaped midline patch of smooth mucosa on the dorsal surface of the tongue that lacks both papillae and taste buds (called median rhomboid glossitis). This benign disorder is thought to arise from an embryologic defect.

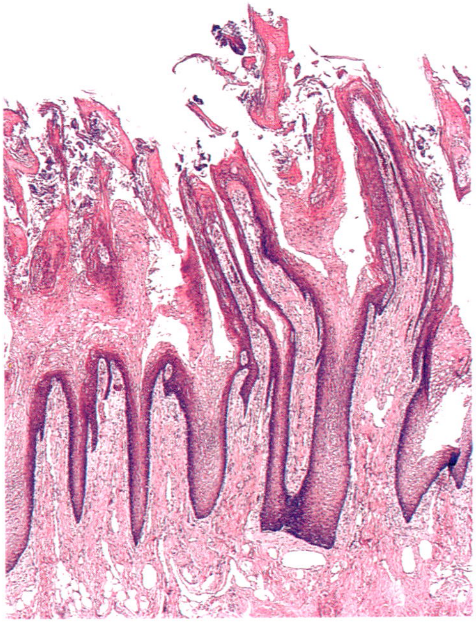

FIGURE 14.28 Dorsal surface of tongue. Filiform papillae have a connective tissue core beset with secondary papillae with pointed ends. The superficial squamous cells are keratinized. Note the slender, pointed rete ridges.

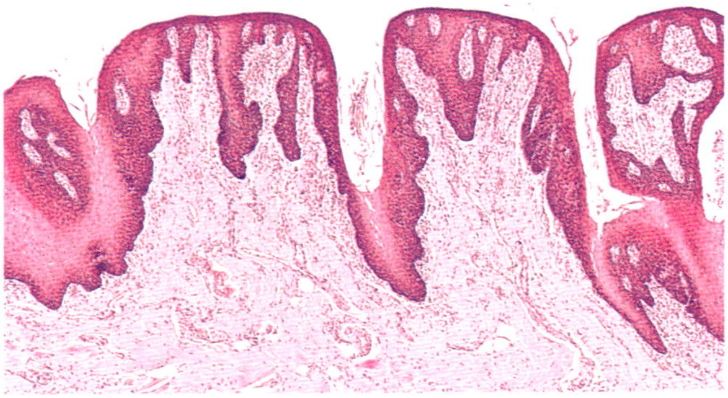

FIGURE 14.29 Fungiform papillae. Slightly rounded, elevated structures with a larger connective tissue core. Smaller intervening connective tissue papillae project into the base of the surface epithelium.

Taste buds are present in large numbers on the sides of the circumvallate papillae and in lesser numbers on the fungiform and foliate papillae, as well as elsewhere on the dorsal and lateral aspects of the tongue. These small structures stain lighter than the surrounding epithelium. Like other simple epithelia, taste buds express low–molecular-weight

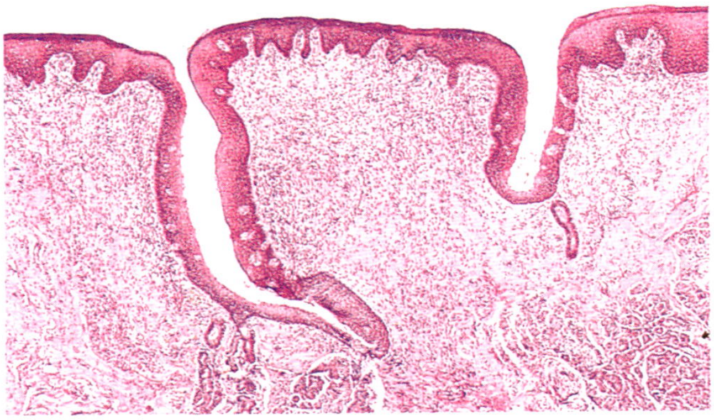

FIGURE 14.30 Circumvallate papilla. Numerous taste buds are on the lateral walls of the papilla and on the epithelium facing the papilla within the furrow. Ducts of serous glands open into the furrow surrounding the circumvallate papilla.

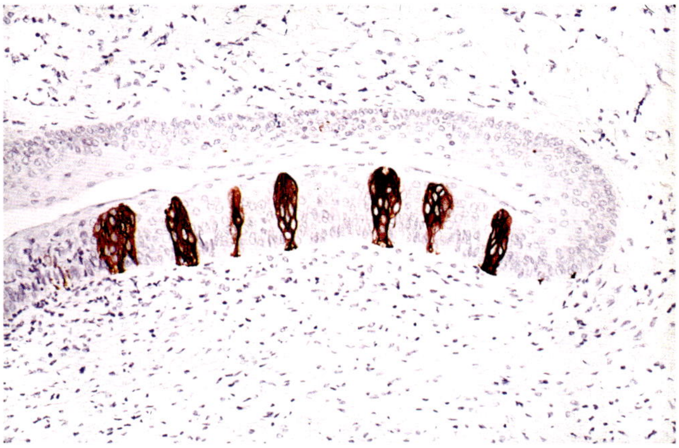

FIGURE 14.31 Taste buds showing immunoreactivity for CAM 5.2.

keratins such as CK18, and are immunoreactive with the antibody CAM5.2 (Fig. 14.31). Numerous small serous glands (i.e., Ebner glands) are located under the circumvallate papillae. The ducts of these glands empty their secretions into the small furrows around each papilla; this serous secretion flushes out the furrows, thus facilitating the perception of new tastes. The taste buds are barrel-shaped intramucosal sensory receptors that occupy the full thickness of the mucosa and communicate with the surface through a small opening, the gustatory pore (Fig. 14.32A,B). The taste buds are composed of three types of cells: gustatory, or taste cells, supporting, or sustentacular, cells; and basal cells. The taste cells are crescent shaped, have lightly staining cytoplasm, and possess numerous fine microvilli that protrude through the taste pore as gustatory hairs. The basal ends of the taste cells are in intimate contact with many fine nerve terminals that leave through the basement membrane and become myelinated outside the taste bud. The supporting cells are likewise crescentic, extending between the basement membrane and the surface; they form a shell for the taste bud and are also scattered between the gustatory cells. The supporting cells have a dark cytoplasm and possess microvilli that also protrude through the taste pore. The small basal cells are situated between the bases of the other cells and give rise to the other cell types of the taste bud (95,96). The taste buds of the anterior two-thirds of the tongue are innervated by sensory nerve branches of the facial nerve and those of the posterior third of the tongue are innervated by the glossopharyngeal nerve.

One may normally encounter biphasic neuroepithelial structures in the posterolateral region of the tongue that are similar to the juxtaoral organ of Chievitz. The term "subgemmal nerve plexus" or "subgemmal neurogenous plaque" has been used interchangeably in the literature for this structure (97–99). Subgemmal neurogenous plaques can range from 0.6 to 3.6 mm in diameter (100). They are more commonly detected in adults in the 5th to 7th decades of life and show a slight predilection for females (101). They appear as clusters of squamous epithelial cells with bland cytomorphology in close association with the circumscribed subepithelial nerve plexus (Fig. 14.33) as well as ganglion cells and taste buds. The overlying epithelium may have a pseudoepitheliomatous appearance due to hyperplasia. A subgemmal neurogenous plaque can mimic a nerve sheath tumor or squamous cell carcinoma. In many cases, they may be associated with lymphoid tissue with reactive germinal centers (102). Immunohistochemical evaluation shows that the epithelium is positive for cytokeratin and CAM5.2 while the nerve fibers show positivity for S-100 protein, GFAP, CD56, and synaptophysin (99,103). The origin and function of these neuroepithelial structures and their relationship to the nerve plexus of taste buds have yet to be completely

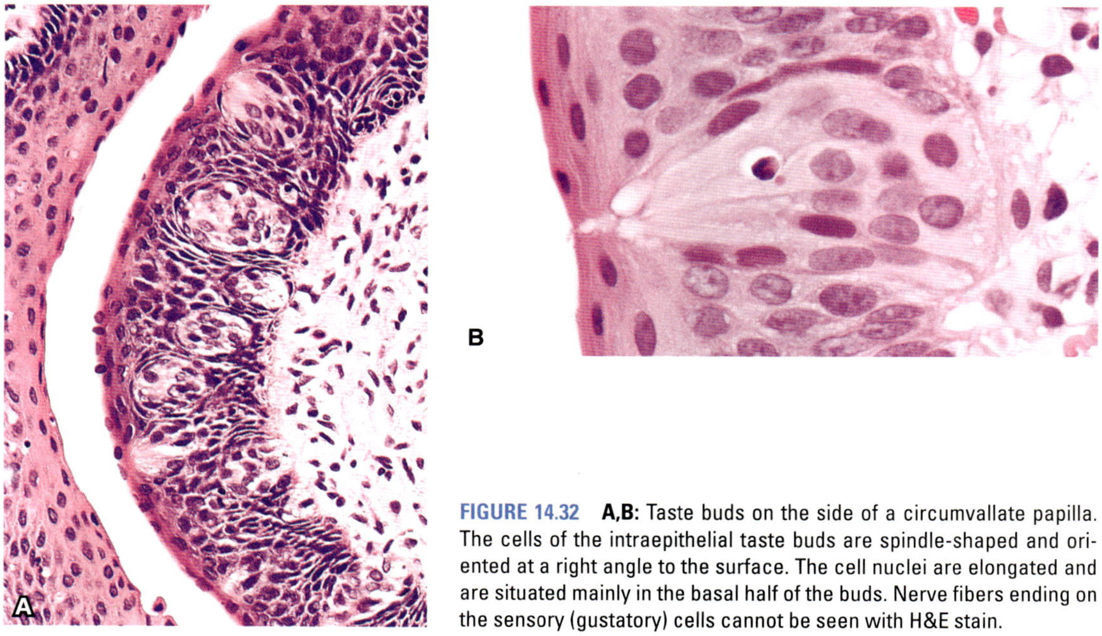

FIGURE 14.32 **A,B:** Taste buds on the side of a circumvallate papilla. The cells of the intraepithelial taste buds are spindle-shaped and oriented at a right angle to the surface. The cell nuclei are elongated and are situated mainly in the basal half of the buds. Nerve fibers ending on the sensory (gustatory) cells cannot be seen with H&E stain.

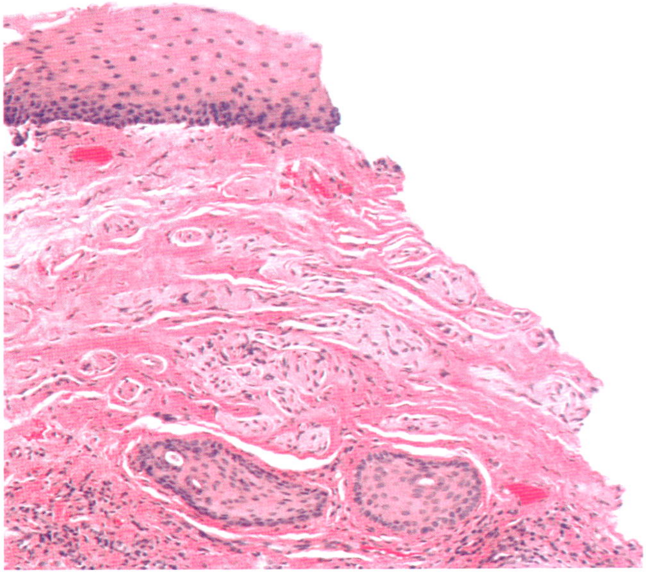

FIGURE 14.33 Superficial tongue biopsy showing a subgemmal neurogenous plaque comprised of superficial nerves in close proximity to bland squamous cell nests.

elucidated. The intimate association of epithelial remnants with peripheral nerves is a potential diagnostic pitfall for pathologists as they may be mistaken for a neural lesion or squamous cell carcinoma with perineural invasion.

The apex of the V-shaped sulcus terminalis projects backward and is marked by a small pit, the foramen cecum, which is an embryologic remnant indicating the upper end of the thyroglossal duct. Correspondingly, ectopic thyroid tissue can occur at the base of the tongue or anywhere along the tract of the thyroglossal duct caudally (Fig. 14.34A,B). Ectopic thyroid resembles normal thyroid tissue, but the surgical pathologist should be aware that ectopic thyroid glands can be present within the muscle, thereby mimicking carcinoma. In contrast to carcinoma, however, the cells of benign thyroidal rests contain nuclei that are small, dark, round, and uniform; they are the nuclear features of papillary thyroid carcinoma (Fig. 14.34B). They lack the nuclear features of papillary thyroid carcinoma.

Gingiva

The gingiva (gum) is that portion of the oral mucosa surrounding the neck of the teeth like a collar. It is attached to both the teeth and alveolar process. Gingival width is wider in the posterior regions of the mouth (104). Masticatory mucosa covers the gum. It is composed of parakeratinized or keratinized stratified squamous epithelium and lamina propria that lie directly over the bone without a submucosal layer (105). Rather, the connective tissue of the lamina propria contains collagenous fibers that bind the epithelium tightly to the alveolar periosteum and bone. The gingival epithelium interdigitates with the underlying connective tissue, forming long, interconnected rete ridges that are separated by connective tissue plates and papillae (Figs. 14.35 and 14.36). The

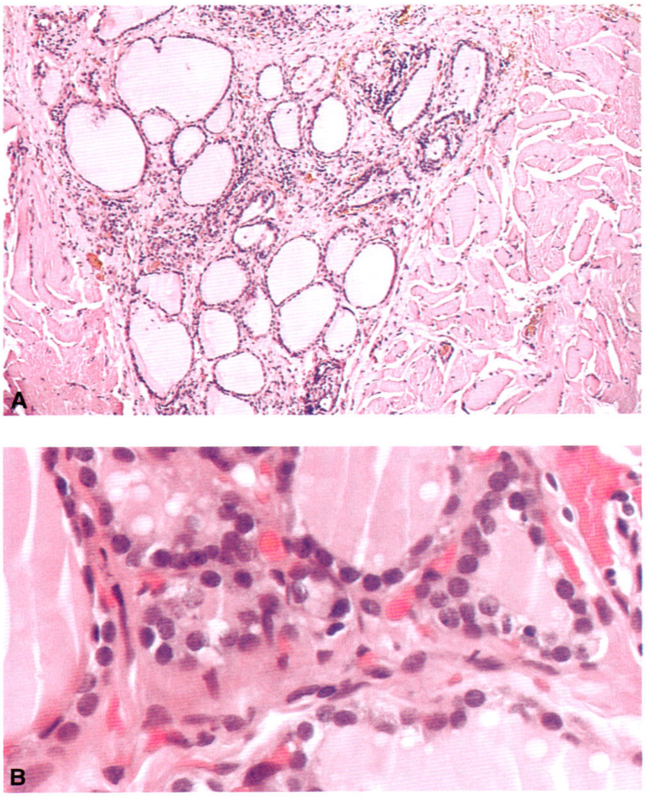

FIGURE 14.34 Lingual thyroid. **A:** The glandular parenchyma is embedded in skeletal muscle. The thyroid tissue is not encapsulated, and care should be taken to avoid mistaking it for invasive growth. **B:** High magnification of lingual thyroid demonstrating the small, dark round nuclei of benign thyroid follicles.

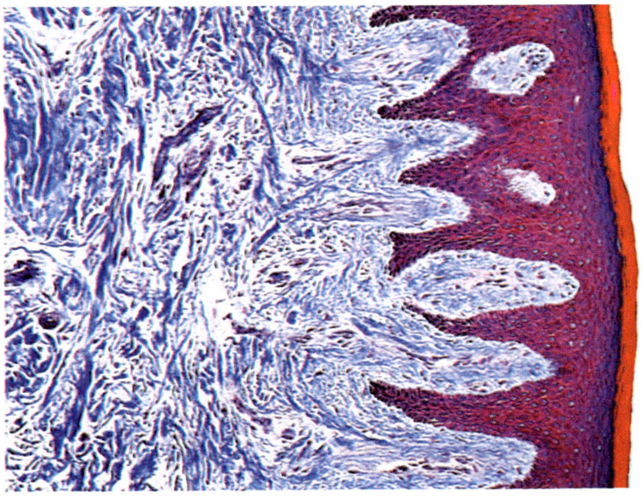

FIGURE 14.35 Gingiva (gum). The masticatory mucosa has tall rete ridges. A dense network of collagen fibers (*blue*) tightly anchors the epithelium to the underlying bone (not shown); the keratin layer (*orange band*) on the surface of the epithelium imparts further strength to it (Mallory trichrome).

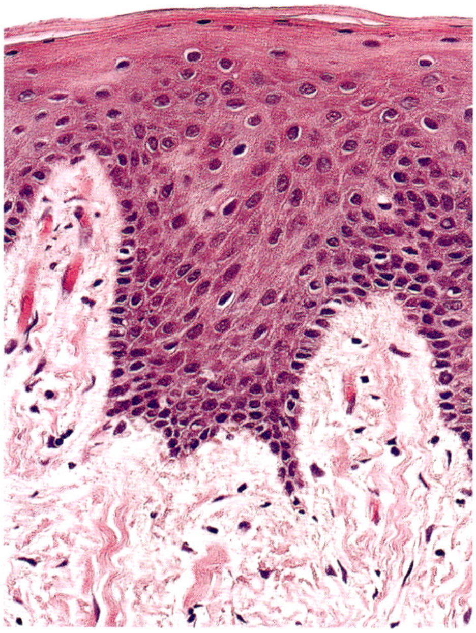

FIGURE 14.36 Gingiva. Tall rete ridges and dense lamina propria with blood vessels in papillae.

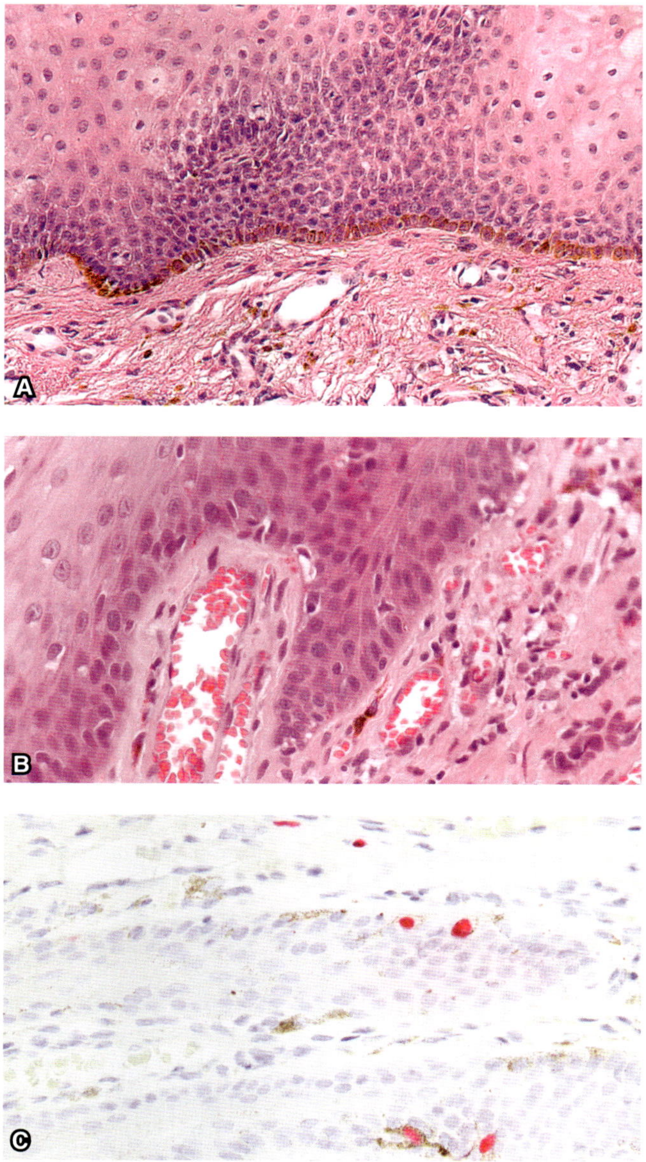

FIGURE 14.37 Mucosal melanosis of the lip. **A:** Numerous melanocytes above the basal lamina appear as a brown ribbon. **B:** Melanotic macule of oral mucosa with pigment incontinence and (**C**) immunostain for SOX-10 highlighting melanocytes.

gingival epithelium is divided anatomically into oral gingival, oral sulcular, and junctional epithelia. The cytokeratin expression profile of gingival epithelium corresponds to this anatomical division (Table 14.1). Oral gingival epithelium that extends onto the oral surface of the gum best fits the general pattern of masticatory mucosa. The short portion of gingiva apposed to the tooth (sulcular gingiva) differs from the rest of the gingival epithelium in that it is thinner, lacks the characteristic rete ridges, and is nonkeratinizing (106,107).

The gingiva is highly vascular; its vessels originate in the periodontium and extend into the lamina propria, forming well-organized capillary loops. Biopsies of gingiva can be performed to evaluate for the presence of amyloidosis and other systemic diseases.

Intraepithelial Nonkeratinocytes

Four different nonepithelial cell types normally occur in the oral mucosa: melanocytes, Merkel cells, Langerhans cells, and lymphocytes.

Melanocytes are found mainly in a basal location in the oral mucosa; they are more common in people with darker complexions (108,109). Small areas of melanin pigmentation, mostly <10 mm in diameter, may occur. These are most common on the gingiva but are also seen in the lips, palate, and buccal mucosa and are called mucosal melanosis or melanotic macules (110) (Fig. 14.37A–C). Melanin pigment formed by melanocytes in the basal layer is transferred to adjacent epithelial cells. Such changes can sometimes occur after inflammatory reactions, in smokers, patients with Peutz–Jeghers syndrome or Addison disease, or secondary to drugs and human immunodeficiency virus (HIV) infection. Oral melanin incontinence may also occur when melanin pigment accumulates within macrophages in the lamina propria (Fig. 14.37B). Melanocytic hyperplasia (i.e., lentigo) and pigmented nevi may also occur in the oral mucosa but less commonly than on the skin (111–114). Lentigo shows increased melanin pigmentation in the basal cell layer without an increase in the number of melanocytes. Intramucosal nevi are the most common nevi. Other types of nevi, such as compound nevi, junctional nevi, blue nevi, and combined forms can be observed on the vermilion border, cheek mucosa, gingiva, palate, and tongue. Understandably, primary malignant melanomas can arise anywhere from the oral mucosa (115).

Merkel cells also occur in the basal layer of the oral epithelium, either individually or in clusters (116). Clusters (named Merkel complexes) are preferentially located in masticatory mucosa in close contact with the tongue and with sensory nerves, which supports the notion that they have a mechanoreceptor function. Merkel cells are activated by mechanical stimulation (117). They tend to be conspicuous on the mandibular lingual gingiva where they are believed to play a role in the proprioceptive positioning of the tongue. Increased numbers of Merkel cells can be seen in edentulous individuals (118). On routine H&E-stained sections, their cytoplasm appears lighter than that of the surrounding basal cells. These neuroendocrine cells are morphologically and functionally identical to those in the skin. They are immunoreactive with antibodies to S-100 protein, chromogranin, synaptophysin, NSE, CK20, and villin (119,120). Apart from marking Merkel cells in the oral mucosa, CK20 shows specificity for lingual taste buds. Their ultrastructure is characterized by many intracytoplasmic dense-core, membrane-bound granules, 80 to 100 μm in diameter (Fig. 14.38). Merkel cells of the oral region differ from those found in the skin because they express S-100 protein, and at the ultrastructural level also contain a nuclear rodlet. There is a higher prevalence of oral Merkel cells in edentulous subjects, and they may also increase as a consequence of tissue injury.

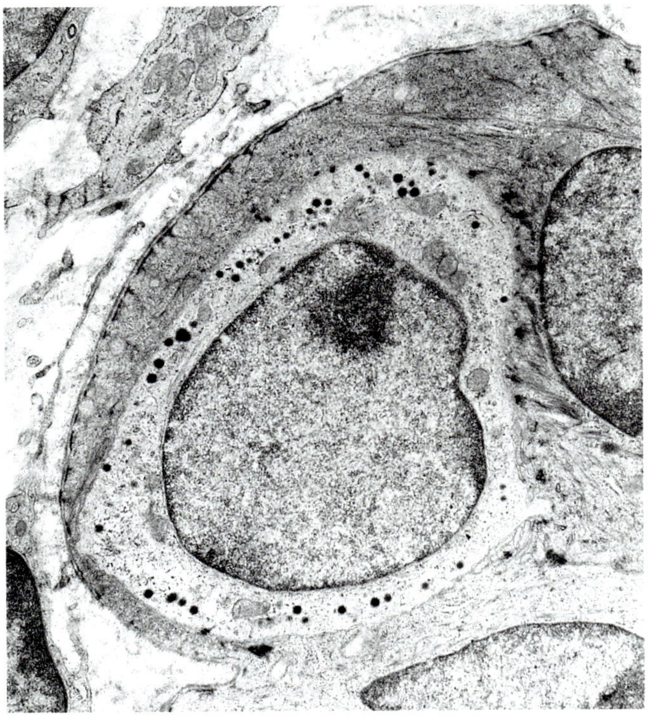

FIGURE 14.38 Merkel cell surrounded by keratinocytes in the basal layer. Transmission electron micrograph shows the lighter-staining cytoplasm with characteristic and subplasmalemmal, membrane-bound, dense-core granules. The nucleus has no fibrous lamina, unlike melanocytes. The adjacent dermis contains fibroblasts. (Original magnification ×11,500.)

Langerhans cells, mostly located suprabasally, are microscopically similar to melanocytes and cannot be distinguished with certainty in routine H&E-stained sections. In the gingiva, the Langerhans cells are structurally and functionally similar to those in the skin (121,122). Their cytoplasm appears clear, and their small indented nuclei stain heavily with hematoxylin. Their slender, spider-like dendritic processes, spread among the cells of the stratum spinosum and can be seen well with the immunohistochemical stain for S-100 protein (Fig. 14.39A,B), as well as CD1a, langerin, the MHC class II molecules (e.g., HLA-DR), and various adhesion molecules (123,124). Electron microscopy shows characteristic Birbeck granules in their cytoplasm and a lack of tonofilaments and desmosomes. The frequency of oral mucosal Langerhans cells varies inversely with the degree of keratinization. They are quantitatively lowest in the floor of the mouth and do not occur around the basal portion of the taste buds or in epithelium lining periodontal pockets. The number of Langerhans cells is a significant prognostic factor for head and neck squamous cell carcinoma (125), and has been associated with other head and neck cancers (126).

Langerhans cells play an important role in the immune response, being involved in the processing and presentation of antigens to subjacent lymphocytes. An increased number of Langerhans cells is seen in biopsy samples of the oral mucosa in patients with oral lichen planus, dental caries, and in tumor epithelium of invasive oral squamous cell carcinoma. They are also associated with tobacco and alcohol consumption.

Teeth and Supporting Structures

The deciduous and permanent teeth have a similar microscopic appearance. The teeth are set in bony sockets on the alveolar processes of the maxillae and the mandible. The part of the tooth that lies within the socket is called the root; there may be multiple roots. The tip of each root is called the apex. The alveolar processes are covered by the gum, and the crowns of the teeth project above the gums. The roots of the teeth are held securely in their sockets by bundles of collagen fibers called the periodontal ligaments, or periodontal membrane (16). The periodontium includes the tissues investing and supporting the teeth: the cementum, periodontal membrane, alveolar bone, and gingiva (Figs. 14.40 and 14.41). The center of each tooth has a pulp chamber, or pulp cavity, that is filled with dental pulp containing loosely arranged fibroblasts, nerves, blood vessels, and lymphatic vessels (127) (Fig. 14.42). The pulp chamber narrows toward the root and becomes the root canal. The vessels and postganglionic sympathetic and sensory nerve fibers enter and leave the root canal through a small opening, the apical foramen. The pulp chamber of the growing tooth is lined by a single continuous layer of odontoblasts, which are tall columnar cells with oval nuclei. An elongated cell process, also known as Tomes fiber, reaches from each odontoblast into the extracellular matrix

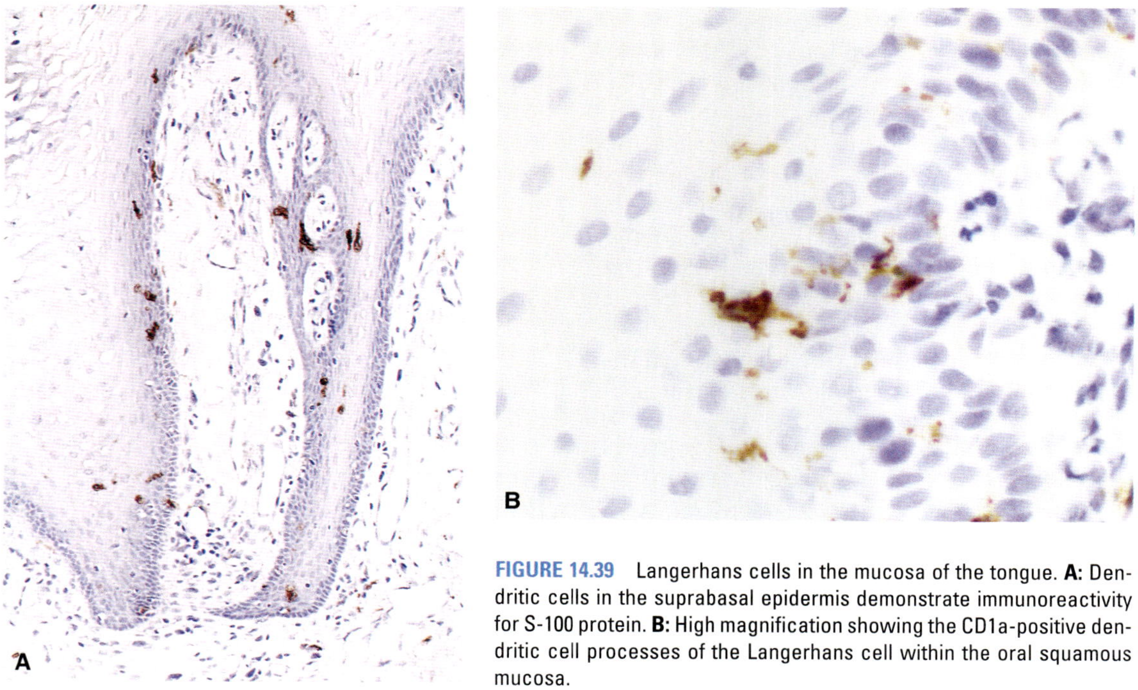

FIGURE 14.39 Langerhans cells in the mucosa of the tongue. **A:** Dendritic cells in the suprabasal epidermis demonstrate immunoreactivity for S-100 protein. **B:** High magnification showing the CD1a-positive dendritic cell processes of the Langerhans cell within the oral squamous mucosa.

secreted by them (Fig. 14.5). The matrix around the odontoblastic process eventually mineralizes, so that Tomes fiber comes to lie within a dentinal tubule (128,129). Besides the odontoblastic process, the tubules also contain 200-μm long unmyelinated nerve fibers, which account for the well-known sensitivity of dentin. Dentin is arranged in the shape of tubules running from the pulp chamber toward the periphery. The dentinal tubules and the meshwork of collagen between them are embedded in hydroxyapatite crystals. On a weight basis, 80% of dentin consists of inorganic calcium salts and 20% of organic material. It is harder than bone but softer than enamel. Dentin makes up most of the wall of the tooth. In the mature tooth, many odontoblasts become inactive, but some continue producing

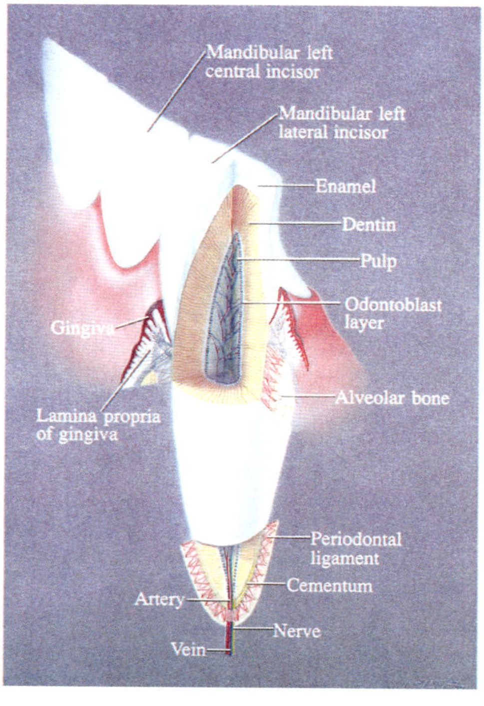

FIGURE 14.40 Schematic drawing depicts the periodontium including the gingiva, alveolar bone, periodontal ligament, and cementum.

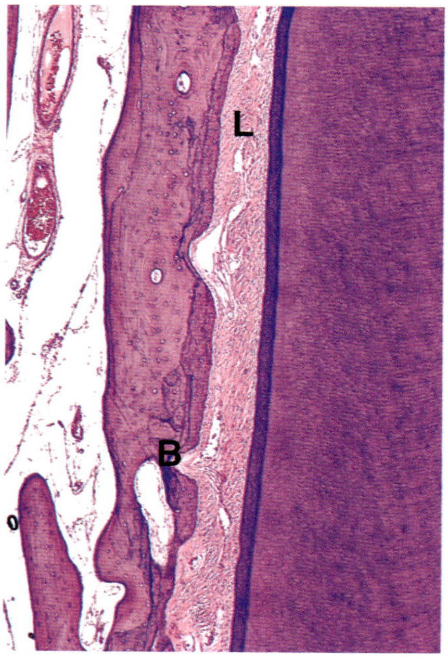

FIGURE 14.41 Root of tooth and supporting structures, perpendicular section. Dentin is covered by a thin layer of cementum (appearing as a blue band). The periodontal ligament (*L*) holds the tooth in the bony socket of the alveolar process (*B*).

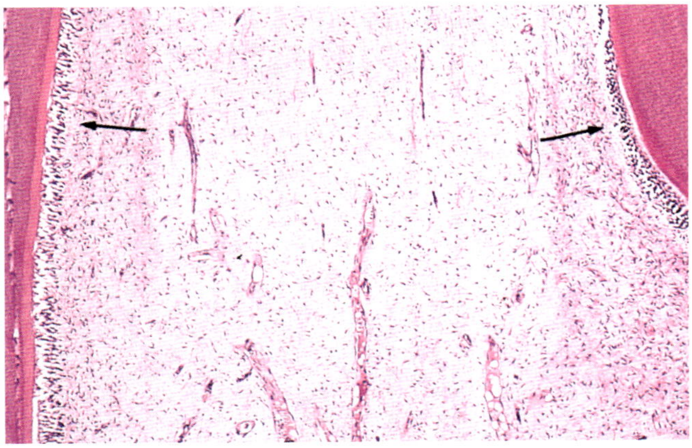

FIGURE 14.42 Pulp of developing tooth. Odontoblasts (*arrows*) line dentin. The pulp consists mostly of loosely arranged fibroblasts. Note the delicate wall of blood vessels.

predentin at a reduced rate throughout the life of the tooth. In response to physiologic or pathologic stimuli, odontoblasts can upregulate their protein synthetic activity.

As the dental pulp ages, the number of fibroblasts decreases and, concomitantly, the number and size of collagen fibers increases. Pulp stones, or denticles, are commonly observed in the dental pulp of aging individuals. True denticles contain dentinal tubules within a mineralized matrix and are surrounded by odontoblasts. False denticles are composed of a mineralized matrix arranged in concentric lamellae. Most denticles are asymptomatic.

Enamel covers the crown of the tooth. It is the hardest material found in the body and consists of 99.5% apatite crystals (130). Mature enamel is made up of long thin rods that dissolve during decalcification and are, therefore, not seen on conventional histologic sections. Enamel proteins such as amelogenin, enamelin, and the basement membrane-type proteoglycan perlecan are immunolocalized in particular odontogenic tumors such as ameloblastomas (131). Perlecan is found mainly in the intercellular spaces of the enamel organ, as well as in the dental papilla/pulp and dental follicle. The dentin–enamel junction lies at the former interface between the inner enamel epithelium and dental mesenchyme. Coronal dentin is covered with enamel, and radicular dentin is covered with cementum. Thus, cementum covers the root of the tooth. A slight indentation (cervical line) encircles the tooth and marks the junction of the crown with the root. The cementum joins the enamel at this junction (cementoenamel junction). Cementum is similar in structure and composition to bone but has fewer cells, called cementocytes, that occupy lacunae (15,16) (Fig. 14.6B). It is composed of 55% organic material and 45% inorganic material. Cementum is attached by the periodontal ligament to the surrounding bone (Fig. 14.41). In older persons, many cementocytes die, and only the surface layer appears viable. Another aging phenomenon is cementicles, which are small round calcified bodies on or in the cementum and in the periodontal ligament. Cementicles are of no clinical significance.

From the pathologist's point of view, it is noteworthy that examination of teeth involved in neoplastic growth in most cases shows tumor involving the periodontal ligament and the alveolar bone, destroying these structures. The roots of the teeth may remain intact or may undergo resorption. However, the dental pulp is rarely invaded by neoplasm.

Odontogenic epithelium changes over time. Before tooth eruption, it consists of the external enamel epithelium, inner enamel epithelium, stellate reticulum, Hertwig epithelial root sheath, and dental lamina. After complete tooth formation, it comprises the epithelial rests of Malassez. Most odontogenic epithelia express cytokeratins 5, 8, and 19 (132). In addition, the external enamel epithelium and stellate reticulum express cytokeratins 7, 13, 14, 17, and 18; the inner enamel epithelium expresses cytokeratins 14 and 18; the dental lamina expresses cytokeratins 7, 13, and 14; and epithelial rests of Malassez stain for cytokeratins 14 and 19. The marker keratin 17 (K17) distinguishes junctional epithelium from epithelial rests of Malassez after tooth eruption (133). During odontogenesis, CK14 expression in ameloblasts is gradually replaced by CK19. Odontogenic epithelium is the source of odontogenic cysts and some neoplasms, which explains why CK19 is present in almost all epithelial cells of odontogenic cysts.

Pathologic Correlates of the Rests of Serres and Rests of Malassez

Developmental remnants of the dental lamina, called the rests of Serres, commonly occur under the gum as small nests of squamous epithelium (15,22) (Figs. 14.2 and 14.43). These epithelial islands may proliferate and undergo cystic degeneration, which leads to the formation of gingival cysts. Similarly, epithelial remnants of Hertwig root sheath, called the rests of Malassez, are universally present in the periodontal membrane and, exceptionally, even in the bone of the alveolar ridge. These normally quiescent epithelial cell remnants have the potential to differentiate into ameloblasts (134). When dental caries cause a bacterial infection and necrosis of the pulp, the infection usually spreads

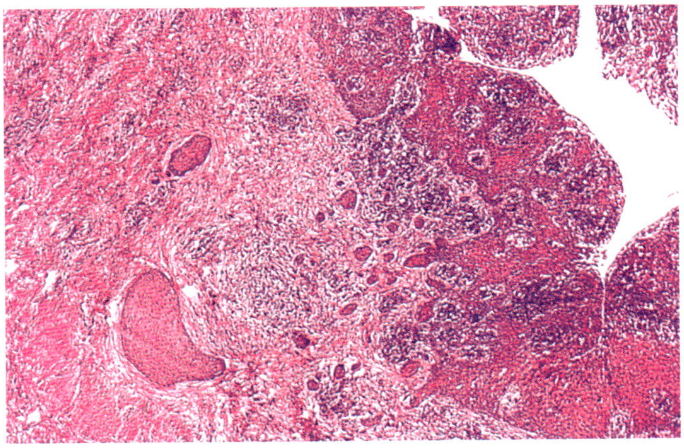

FIGURE 14.43 Rests of Malassez in periodontium near the apical granuloma.

toward the apical foramen, and a periapical granuloma may develop. As a result of the inflammatory stimulus, the nearby epithelium of the rests of Malassez begins to proliferate and forms an epithelial lining in these apical granulomas (Fig. 14.43). It is also assumed that the rests of Serres and the rests of Malassez are potential sources of ameloblastomas and various odontogenic cysts including radicular cysts and dentigerous cysts.

Nose and Paranasal Sinuses

External Nose and Nasal Vestibule

The external nose is covered by skin that is rich in sebaceous glands, sweat glands, and small hairs. The anterior skin of the nares and widened nasal vestibule, lined by skin that is continuous with the integument of the nose, similarly contains many hair follicles and sebaceous and sweat glands. The squamous epithelium of the vestibule merges with the respiratory mucosa, which covers the respiratory portion of the nasal cavity and all of the paranasal sinuses.

Nasal Mucosa

At the level of the limen nasi, the lining of the nasal cavity gradually changes from squamous epithelium to nonciliated cuboidal or columnar epithelium. Farther into the nasal cavity, this becomes continuous with the pseudostratified ciliated respiratory-type columnar epithelium. The mucosa of the respiratory portion is continuous with that of the ostia and contiguous paranasal sinuses. The mucosal lining contains three main cell types: basal (reserve) cells, goblet cells, and ciliated cells (Fig. 14.44A,B). Basal cells, confined to the vicinity of the basal lamina, divide to produce new daughter cells that differentiate to become mucous or ciliated cells. Interestingly, it has been observed that not just the surface cells have cilia, but so do some of the basal cells (135). The mucous cells rest on the basement membrane with a slender stem, and the cytoplasm of their apical portion contains varying amounts of mucus. As these cells fill up with mucus, they resemble a goblet with a foot and a stem. The cells discharge their contents to form a thin sheet of mucus that covers the surface of the respiratory epithelium. When the mucosa is irritated, the number of goblet cells increases. Another source of this mucous coat is the secretion from the seromucinous glands in the lamina propria.

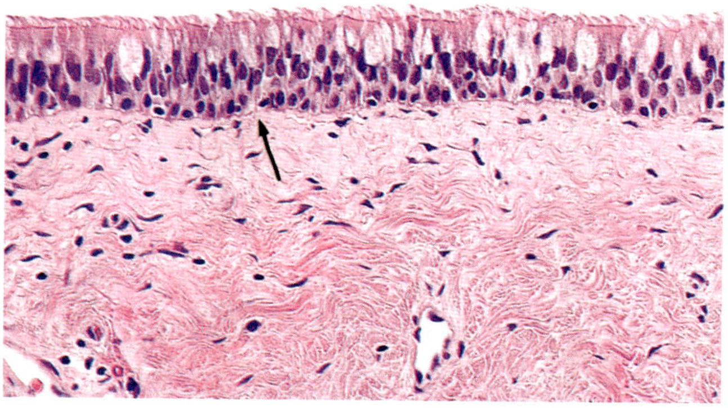

FIGURE 14.44 Pseudostratified respiratory mucosa of the nose with predominantly ciliated cells. Goblet cells have clear cytoplasm. The basal cells (*arrow*) are lying on thin basal lamina.

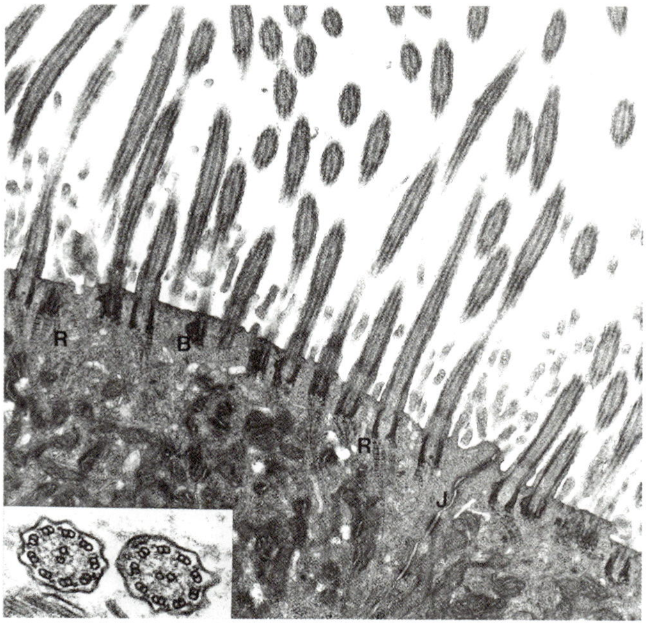

FIGURE 14.45 Transmission electron micrograph of nasal mucosa. Vertical section shows ciliated epithelial cells extending to the surface. The shafts of the slender long cilia have peripheral and central microtubules that appear as darker linear structures extending from the kinetosome (basal body) (*B*). Cross-banded filamentous rootlets (*R*) are associated with kinetosomes. Note junctional complexes (*J*). *Inset*: Cross section of the shaft of two cilia. Note the symmetrically arranged nine peripheral doublets and the central pair of single microtubules. (Original magnification ×25,000; *inset*, original magnification ×87,500.)

Most cells of the respiratory epithelium bear small finger-like cell processes, the cilia, which project into the lumen. Each cell has over 200 cilia, each about 5- to 7-μm long. Ultrastructurally, on cross section, the ciliary shafts show a highly characteristic circular arrangement of nine pairs of microtubules (i.e., doublets) arranged symmetrically around two central microtubules (Fig. 14.45). Longitudinal sections of the cilia show that the ciliary shaft ends in a basal body termed the kinetosome, from which it is derived (136). The nasal mucosa is a convenient site to biopsy when ultrastructural abnormalities of cilia are suspected, such as in immotile cilia syndrome (137,138). The function of the ciliated cells in the nose is to constantly move the protective mucous by the coordinated sweeping motion of the cilia toward the pharynx. In order to perform this function, the cilia beat at 10 to 20 cycles/second. They have an effective whip-like stroke forward and a recovery stroke backward. The propulsive forward phase is much faster and more vigorous than the recovery phase (139). Ciliary activity is optimal at the normal nasal temperature of about 30° C. However, cilia are hardy and persist under unfavorable conditions, including extreme cold and heat. They will also beat normally and forcefully in a pus-filled cavity. When injured or destroyed by an acute infection, they regenerate rapidly. They do not, however, tolerate excessive drying. Accordingly, they are dependent always upon a coating of moisture for their activity and preservation. The mucus covering the nasal mucosa and paranasal sinuses forms a conveyor of the bacteria and other foreign matter. The nasal mucociliary layer is the first line of defense against bacterial invasion in this location.

Melanocytes are also present in the normal mucosa of the upper airways. They can be seen in the respiratory epithelium and glands of the nasal cavity. Melanocytes are commonly encountered in the lamina propria of the septum and turbinates, particularly in dark-skinned adults (140). This explains why primary malignant melanomas may arise in the nose and paranasal sinuses (141).

Intraepithelial CD8+ T-lymphocytes are diffusely scattered throughout the nasal mucosa (52). These T cells coexist with a population of dendritic cells that lacks CD1a expression. The nasal mucosa and subepithelial tissue contain virtually no B cells. This finding may explain why most primary nasal lymphomas are of CD8+ T-cell derivation.

The lamina propria (i.e., tunica propria) lies beneath the mucous membrane and contains numerous small mucous and serous glands that discharge their secretions through lobular ducts onto the surface (Fig. 14.46A–C). These minor salivary glands are embedded in vascular fibroconnective tissue rich in elastic fibers. This connective tissue is firmly attached to the perichondrium and periosteum of the cartilages and bones that form the nasal cavity (142–144). Nasal seromucinous hamartoma (i.e., microglandular adenosis of the nose) is a rare tumor characterized by a disordered arrangement of small serous tubules, ducts and glands that lack a myoepithelial layer (45). They may resemble inflammatory nasal polyps, respiratory epithelial adenomatoid hamartomas, or low-grade sinonasal adenocarcinomas.

The turbinates are somewhat curved structures supported by an osseous axis enveloped by relatively thick mucosa (Figs. 14.1 and 14.47). Their convex surfaces protrude toward the nasal cavity. Located beneath their mucosa is a tunica propria that attaches the mucosa to underlying structures. The tunica propria is of variable thickness, being thickest over the nasal septum and medial aspects of the inferior and middle turbinates that are most exposed to inhaled and exhaled air. The epithelium in these areas contains many goblet cells and the basement membrane is prominent (146). These areas also have abundant blood vessels and clusters of mixed seromucinous glands with 6 to 10 glands/mm^2 (Fig. 14.48). The glands vary from simple straight tubules lined with goblet cells to tubuloalveolar glands. The chief ducts of the latter open onto the mucosal surface by minute orifices. The glands tend to be at a level between the mucosa and the underlying bone.

The turbinates and the lower part of the septum are rich in venous sinuses, which are of variable size and shape, forming a dense network of large veins that resemble erectile tissue (Fig. 14.49A,B). These blood vessels are of irregular shape, have muscular walls, and can rapidly dilate and constrict, thereby permitting fast adjustments of mucosal

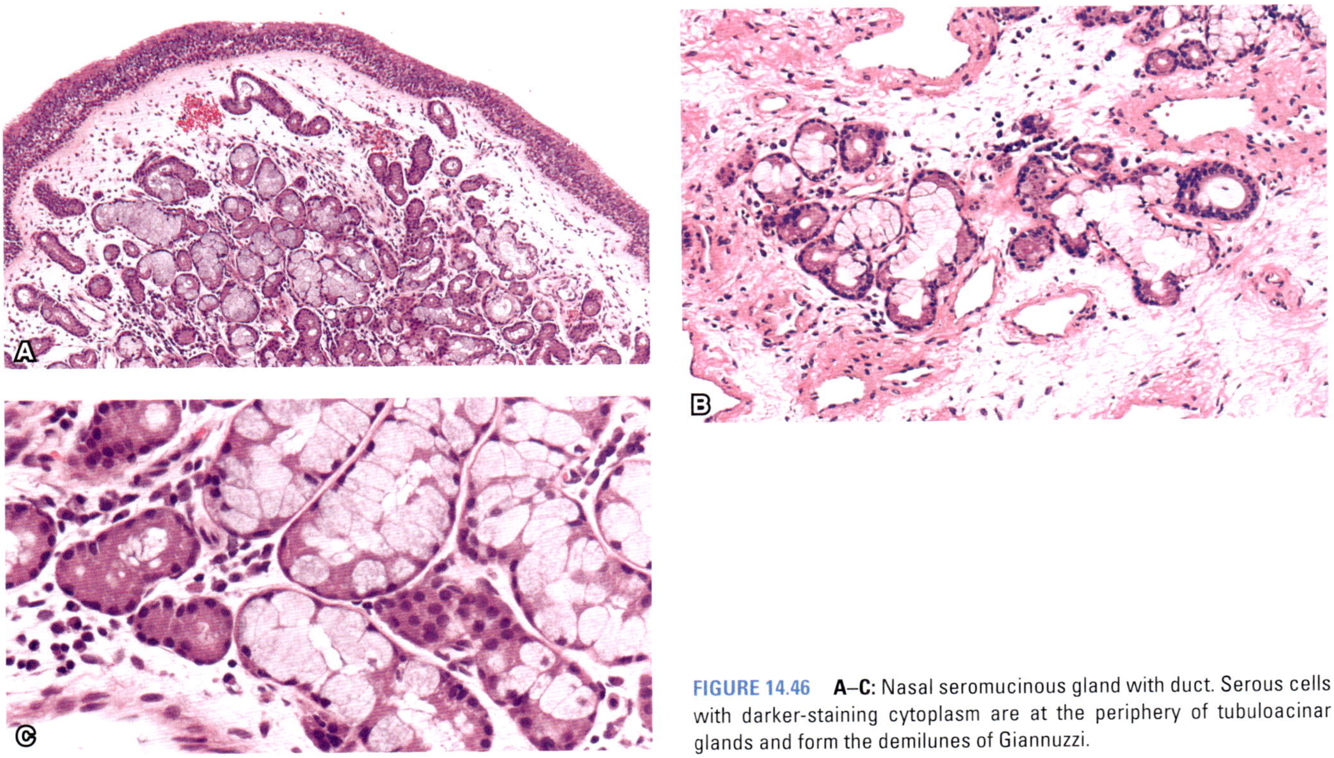

FIGURE 14.46 **A–C:** Nasal seromucinous gland with duct. Serous cells with darker-staining cytoplasm are at the periphery of tubuloacinar glands and form the demilunes of Giannuzzi.

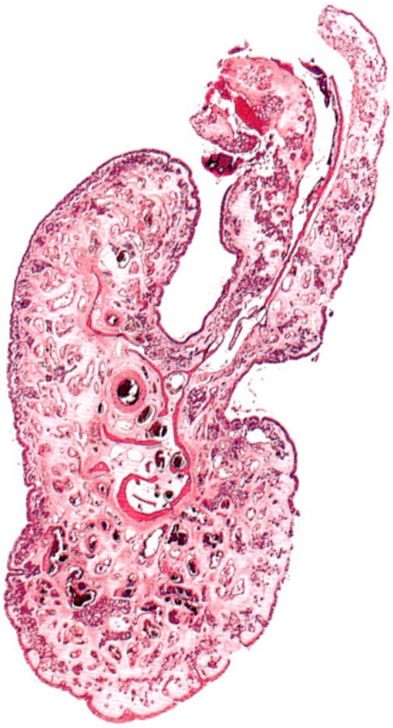

FIGURE 14.47 Middle turbinate, coronal section. Bone in the axis of the turbinate appears as a delicate, curled structure. The covering mucous membrane and tunica propria are rich in blood vessels and mucous glands, particularly on the convexity and inferior border of the turbinate, areas that are most exposed to the airstream in the nasal cavity. (Low-power view.)

temperature and secretion to climatic changes. It is important for the surgical pathologist to know about the normal rich vascular anatomy of the turbinates in order to avoid mistaking them for a hemangioma, angiofibroma, or angioleiomyoma. The stroma of the nasal mucosa also contains

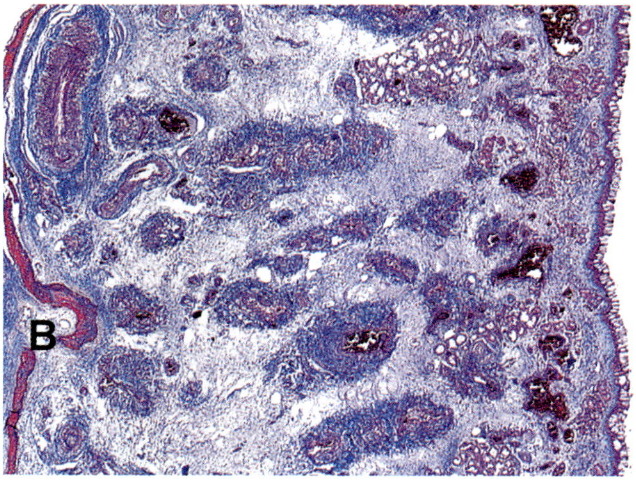

FIGURE 14.48 Turbinate, coronal section. Mucosal blood vessels are surrounded by a thick sheath of connective tissue (*blue*); note the large artery near the bone (*B*) (Mallory trichrome).

lymphatics, small nerves, and a sprinkling of lymphocytes and plasma cells but no lymphoid aggregates. A few mast cells and eosinophils are normally also present.

The osseous portion of the turbinates consists of thin, interconnecting laminae of lamellar bone, forming a continuous shell that is interconnected with bone trabeculae. The interosseous spaces contain numerous large veins, arteries, and some nerves. In contrast to the submucosal veins, the intraosseous veins have a rather large round or oval lumen on cross section and have proportionately thin walls. Occasional adipocytes, but no hematopoietic marrow, are seen

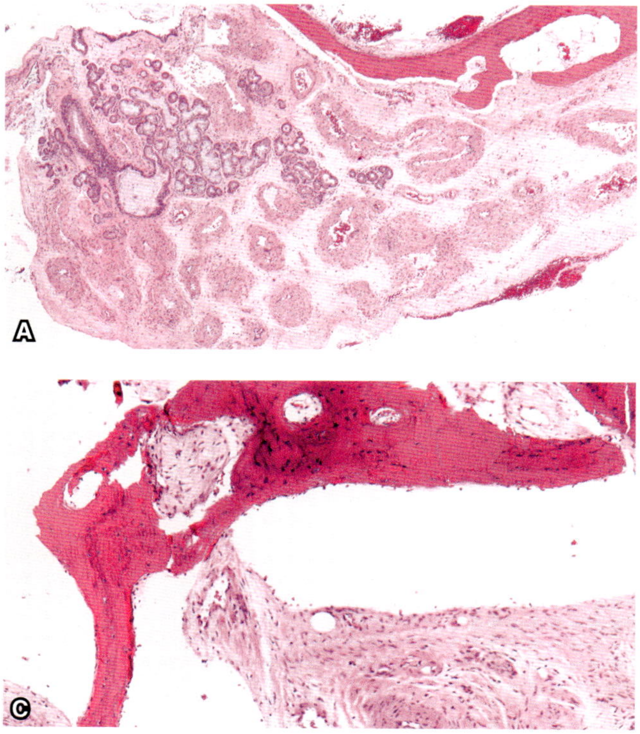

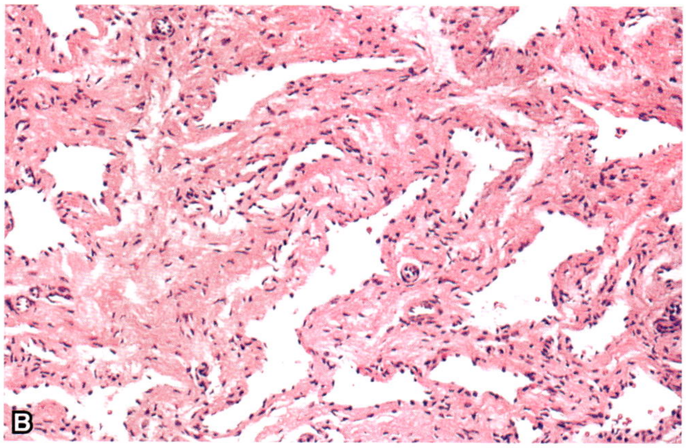

FIGURE 14.49 **A–B:** Inferior turbinate. Numerous larger blood vessels of variable size and shape are closely packed and form a sponge-like vascular system resembling erectile tissue. The endothelial cells are evenly distributed. Capillaries are lacking. **C:** Pneumatized pocket within the middle turbinate forming the concha bullosa.

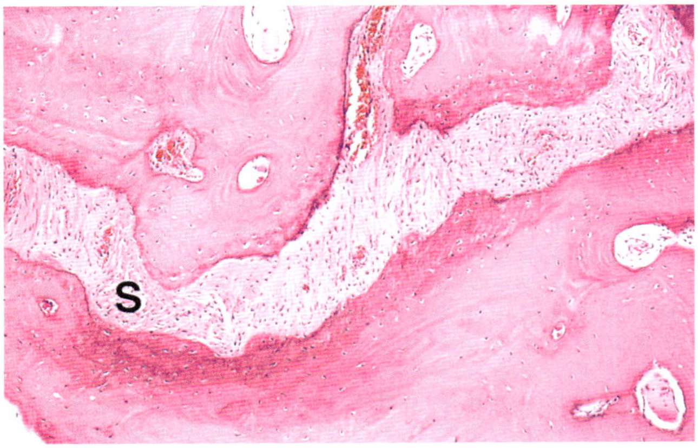

FIGURE 14.50 Suture (*S*) in the nasal septum. Parallel edges of the vomer (*top*) and maxilla are connected by parallel, densely arranged collagen fibers that are anchored in the bones.

in turbinated bone. The presence of prominent hematopoiesis in the facial and nasal bones or the paranasal sinuses is abnormal and may be observed in conditions such as thalassemia major. In many individuals, the bony portion of the middle turbinate becomes pneumatized (Fig. 14.49C), an anatomic variation known as concha bullosa. It has been reported in over 40% of individuals who are usually asymptomatic, but in some patients, it can result in nasal obstruction and sinusitis (147).

The nasal septum consists of a large cartilaginous plate and four small osseous plates, all of which firmly unite with sutures (Fig. 14.50). The nasal mucosa is closely apposed to the underlying structures of the nasal septum. The periosteum and perichondrium of the nasal septum attach so closely to the overlying submucosa as to constitute one membrane called the mucoperiosteum. A common site of nosebleed is Little area, or Kiesselbach area, on the anterior part of the cartilaginous nasal septum above the intermaxillary line (148). The submucosa of this area is richly supplied with thin-walled dilated blood vessels (Fig. 14.51). Although rarely seen in surgical specimens, Little area should not be mistaken for a pyogenic granuloma, which frequently occurs in this location.

Vomeronasal Organ of Jacobson

The vestigial remains of a paired embryonic structure, the vomeronasal organ of Jacobson, are situated under the mucosa of the lower anterior side of the nasal septum covering an area of 0.2 to 0.6 cm. It consists of a small tubular sac lined by columnar epithelium with microvilli in adults, but it has no sensory cells with cilia and lacks other well-differentiated olfactory structures (Fig. 14.52). The duct-like organ is surrounded by numerous exocrine glands (149). The organ is best developed in the 20th week of embryonic life, after which regressive changes occur as it becomes rudimentary. It has no known function in humans and is of no pathologic significance (150). In many vertebrates, Jacobson organ is highly developed, particularly in animals of keen olfactory sensibility (151–153). In humans, this vestigial organ does not have a sensory function, and there is no significant impact if it gets damaged during nasal septal surgery (154).

Olfactory Mucosa

The roof of the nasal cavity and contiguous portions of the nasal septum and superior turbinate form the olfactory region (155). Here, the ciliated columnar epithelium of the nasal mucosa is modified by liberally scattered cells of the sensory organ of smell. The nasal anatomy in detailed relation to olfactory function is dealt with by Hummel and Welge-Lüssen in their recent update (156). The olfactory epithelium consists of three types of cells: olfactory nerve cells, supporting, or sustentacular, cells, and basal cells that lie on the basal lamina (157,158) (Fig. 14.53). The olfactory nerve cells are heterogeneous throughout the olfactory epithelium (159). The olfactory nerve cells are spindle shaped and have a spherical nucleus. The dendrites of these bipolar cells extend to the surface of the pseudostratified olfactory

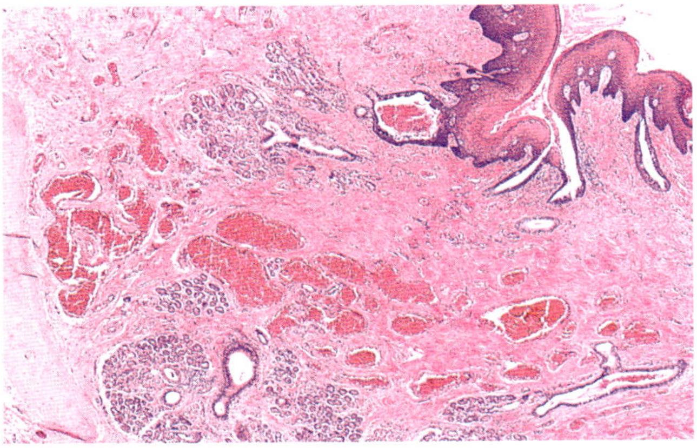

FIGURE 14.51 Little area. Teleangiectatic, thin-walled blood vessels are clustered under the epithelium in the anterior portion of the nasal septum. This area is above the level of mucous glands.

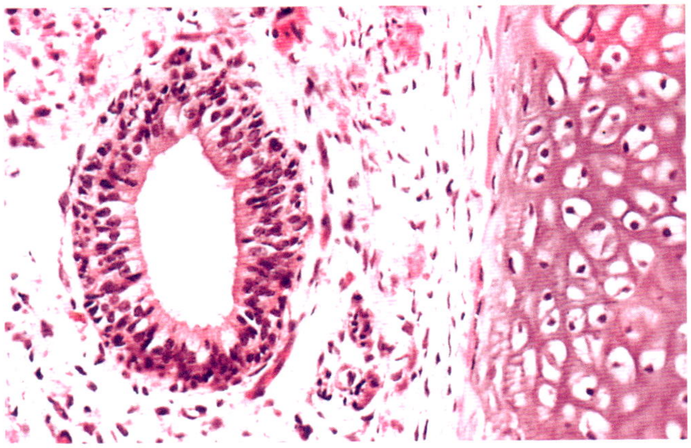

FIGURE 14.52 Vomeronasal organ of Jacobson in a 22-week-old embryo (180-mm crown-rump length), including a cross section of the tubular structure adjacent to vomeronasal cartilage. In humans the columnar epithelium has microvilli but no sensory epithelium.

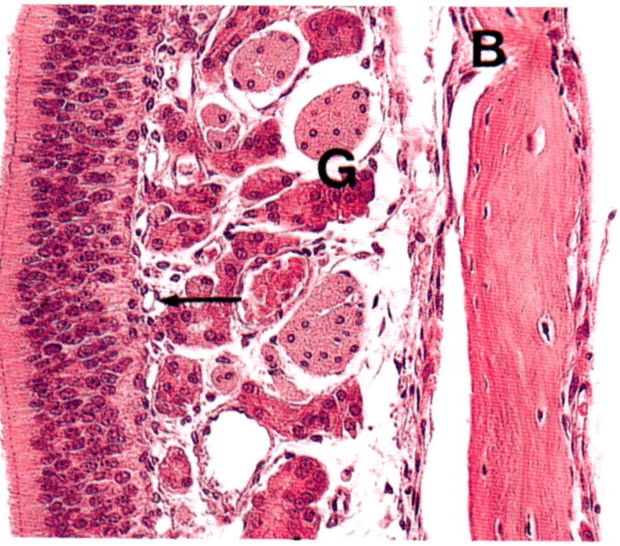

FIGURE 14.53 Olfactory mucosa. The population of olfactory nerve cells and supporting cells forms a pseudostratified columnar epithelium with distinct microvilli. Basal cells lie on basal lamina (*arrow*). Bowman glands (*G*) with excretory ducts and nerves are between the epithelium and bone of the nasal septum (*B*).

epithelium and send out a tuft of fine processes known as olfactory cilia (hairs). The cilia function as receptors for the detection of odorants. They are 2-μm long and lie along the surface of the mucosa embedded in mucus. The deep processes of these bipolar cells form axons that find their way through the basal lamina and join neighboring processes to become bundles of unmyelinated olfactory nerve fibers (122). These fibers collect to form myelinated nerves, which then pass through the cribriform plate of the ethmoid bone to end on the mitral cells in the olfactory bulb. The supporting cells are tall, cylindrical cells. They can contain lipofuscin in elderly individuals, imparting the yellow hue characteristic of the olfactory mucous membrane. The free surface of these cells possesses many slender microvilli that protrude into the covering mucus. The basal cells are small and conical, lying with their bases on the basement membrane. They likely represent stem cells that can give rise to new supporting and sensory cells. The mucosa is supported by lamina propria composed of loose connective tissue that houses the olfactory glands of Bowman (160). The secretions of these tubuloalveolar glands are carried to the mucosal surface by narrow ducts.

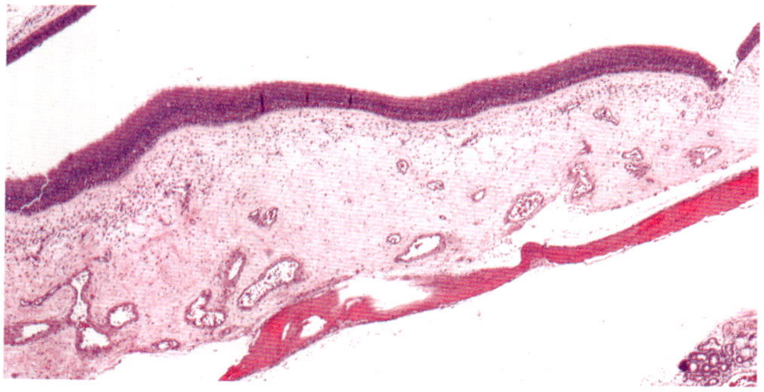

FIGURE 14.54 Sinus (Schneiderian) mucosa which is thinner and less vascular than the nasal mucosa.

Paranasal Sinuses

The sinuses are lined with a mucous membrane that is continuous with the nasal mucosa and similar to that of the nasal cavity. However, the schneiderian epithelium of the paranasal sinuses is thinner than that of the nasal cavity and its and lamina propria is both thinner and less vascular (Fig. 14.54). Prolonged irritation of the mucosa can result in an increase in the number of goblet cells, such that they

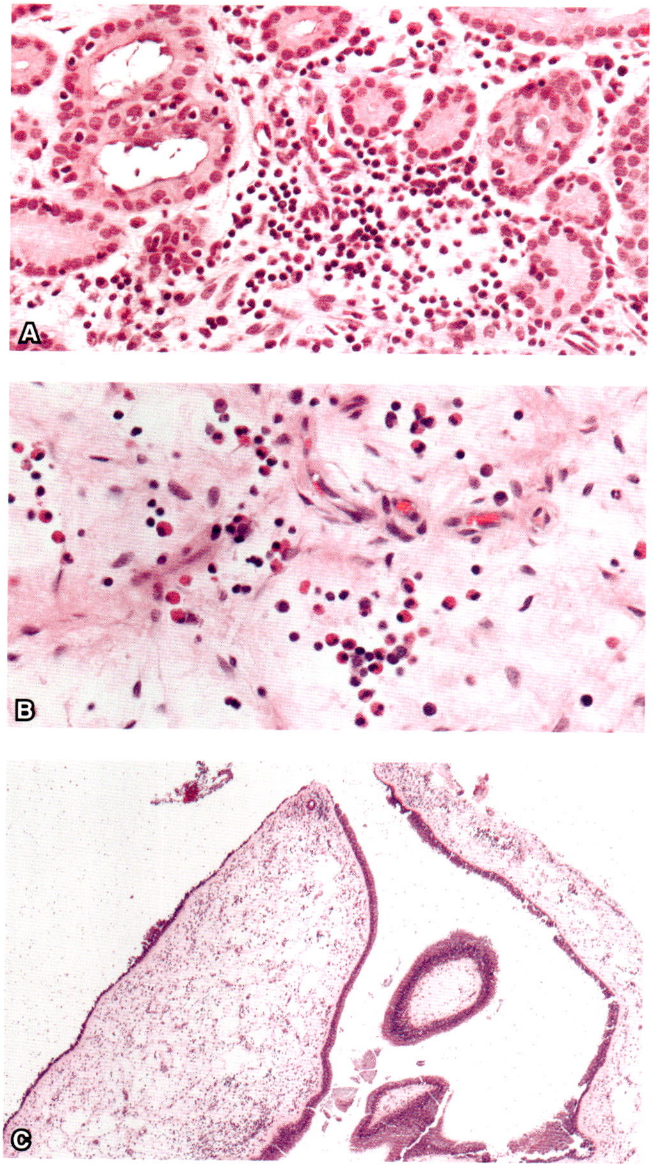

FIGURE 14.55 **A:** Sinusitis consisting of mixed lymphocytes and plasma cells within the stroma of the sinus mucosa. **B:** A characteristic feature of allergic sinusitis is the presence of eosinophils, and (**C**) some cases may show polypoid features.

predominate or completely replace the ciliated cells. Seromucous glands present in the sinuses are less numerous than those of the nasal mucosa and they are largely concentrated at the ostium of the maxillary sinus. The mucus formed in the sinuses is moved by the action of the cilia through the apertures to the nasal cavities. Longstanding inflammation of the sinuses results is a common problem more than 10% of the population suffers from varying degrees of chronic rhinosinusitis (161). The inflammatory infiltrate consists of a mixture of plasma cells and lymphocytes. Eosinophils and allergic mucin may be present in cases of allergic rhinosinusitis (Fig. 14.55A–C). A subset of patients with chronic rhinosinusitis develop sinonasal polyps (161).

REFERENCES

1. Moore KL, Persaud TVN. *The Developing Human: Clinically Oriented Embryology*. 5th ed. WB Saunders; 1993.
2. Sperber GH. *Craniofacial Embryology*. 4th ed. Butterworth-Heinemann; 1989.
3. Wen SL. Human enamel structure studied by high resolution electron microscopy. *Electron Microsc Rev.* 1989;2(1):1–16.
4. Saito MM, Onuma K, Yamakoshi Y. Cementum is key to periodontal tissue regeneration: a review on apatite microstructures for creation of novel cementum-based dental implants. *Genesis.* 2023;61(3-4):e23514.
5. Thesleff I, Vaahtokari A, Kettunen P, et al. Epithelial-mesenchymal signaling during tooth development. *Connect Tissue Res.* 1995;32:9–15.
6. Tucker AS. Salivary gland development. *Semin Cell Dev Biol.* 2007;18(2):237–244.
7. Honda MJ, Imaizumi M, Tsuchiya S, Morsczeck C. Dental follicle stem cells and tissue engineering. *J Oral Sci.* 2010; 52:541–552.
8. Listgarten MA. Phase-contrast and electron microscopic study of the junction between reduced enamel epithelium and enamel in unerupted human teeth. *Arch Oral Biol.* 1966;11:999–1016.
9. Schour I, Massler M. Studies in tooth development: the growth pattern of human teeth. *J Am Dent Assoc.* 1940; 27:1918–1931.
10. Gorski JP, Marks SC Jr. Current concepts of the biology of tooth eruption. *Crit Rev Oral Biol Med.* 1992;3:185–206.
11. Marks SC Jr, Gorski JP, Cahill DR, et al. Tooth eruption—a synthesis of experimental observations. In: Davidovitch Z, ed. *The Biological Mechanisms of Tooth Eruption and Root Resorption*. EBSCO Media; 1988:161–169.
12. Orban B, Sicher H, Weinmann JP. Amelogenesis (a critique and a new concept). *J Am Coll Dent.* 1943;10:13–22.
13. Schroeder HE. Development and structure of the tissues of the tooth. In: *Schroeder HE. Oral Structural Biology*. Thieme Medical; 1986:4–184.
14. O'Leary F, Coffey N, Burke FM, et al. The prevalence of developmental defects of enamel in a cohort of adults with cystic fibrosis—A cross sectional study. *J Dentistry.* 2024;144:104893.
15. Bhaskar SN, Orban BJ, eds. *Orban's Oral Histology and Embryology*. 11th ed. CV Mosby Year Book; 1990.
16. Held AJ. Cementogenesis and the normal and pathologic structure of cementum. *Oral Surg Oral Med Oral Pathol.* 1951;4:53–67.
17. Smukler H, Dreyer CJ. Principal fibres of the periodontium. *J Periodont Res.* 1969;4:19–25.
18. Logan WHG, Kronfeld R. Development of the human jaws and surrounding structures from birth to the age of fifteen years. *J Am Dent Assoc.* 1933;20:379–427.
19. Chibly AM, Aure MH, Patel VN, Hoffman MP. Salivary gland function, development, and regeneration. *Physiol Rev.* 2022;102(3):1495–1552.
20. Miletich I. Introduction to salivary glands: structure, function and embryonic development. *Front Oral Biol.* 2010;14:1–20.
21. Valstar MH, de Bakker BS, Steenbakkers R, et al. The tubarial salivary glands: a potential new organ at risk for radiotherapy. *Radiother Oncol.* 2021;154:292–298.
22. Schroeder H. *Oral Structure Biology: Embryology, Structure and Function of Normal Hard and Soft Tissues of the Oral Cavity and the Temporomandibular Joints*. Thieme; 1991.
23. Vaid S, Vaid N. Normal anatomy and anatomic variants of the paranasal sinuses on computed tomography. *Neuroimaging Clin N Am.* 2015;25:527–548.
24. Gorlin RJ. Heterotopic lymphoid tissue; a diagnostic problem. Oral Surg Oral *Med Oral Pathol.* 1957;10:87–89.
25. Sadeghi EM, Ashrafi MH. Heterotopic lymph node of the buccal mucosa simulating a tumor: a clinicopathological appraisal. *ASDC J Dent Child.* 1982;49:304–306.
26. Tobias MJ. *Anatomy of the Human Lymphatic System*. Edwards Brothers, Inc.; 1938.
27. Richtsmeier WJ, Shikhani AH. The physiology and immunology of the pharyngeal lymphoid tissue. *Otolaryngol Clin North Am.* 1987;20:219–228.
28. Goeringer GC, Vidić B. The embryogenesis and anatomy of Waldeyer's ring. *Otolaryngol Clin North Am.* 1987;20: 207–217.
29. Dolen WK, Spofford B, Selner JC. The hidden tonsils of Waldeyer's ring. *Ann Allergy.* 1990;65:244–248.
30. Lewis JS, Jr., Beadle B, Bishop JA, et al. Human papillomavirus testing in head and neck carcinomas: guideline from the college of American pathologists. *Arch Pathol Lab Med.* 2018;142(5):559–597.
31. Simmonds JC, Jabbour J, Vaughn JA, Paulson VA, Poe DS, Rahbar R. Hairy polyps: a new case presentation and a pathogenetic hypothesis. *Laryngoscope.* 2019;129:2398–2402.
32. Knapp MJ. Oral tonsils: location, distribution, and histology. *Oral Surg Oral Med Oral Pathol.* 1970;29:155–161.
33. Knapp MJ. Pathology of oral tonsils. *Oral Surg Oral Med Oral Pathol.* 1970;29:295–304.
34. Buchner A, Hansen LS. Lymphoepithelial cysts of the oral cavity. A clinicopathologic study of thirty-eight cases. *Oral Surg Oral Med Oral Pathol.* 1980;50:441–449.
35. Napier SS, Newlands C. Benign lymphoid hyperplasia of the palate: report of two cases and immunohistochemical profile. *J Oral Pathol Med.* 1990;19:221–225.
36. Simpson HE. Lymphoid hyperplasia in foliate papillitis. *J Oral Surg Anesth Hosp Dent Serv.* 1964;22:209–214.
37. Moorefield AK, Rose-Reneau Z, Wright B, Surek CC. Venous tributaries of the lip: implications for lip filler injection. *Plast Reconstr Surg.* 2023;152(2):257e–263e.
38. Miles AEW. Sebaceous glands in the lip and cheek mucosa of man. *Br Dent J.* 1958;105:235–248.

39. Sewerin I. The sebaceous glands in the vermilion border of the lips and in the oral mucosa of man. *Acta Odontol Scand.* 1975;33(Suppl 68):13–226.
40. Meyer J, Squier CA, Gerson SJ, eds. *The Structure and Function of the Oral Mucosa*. Pergamon Press; 1984.
41. Wu T, Xiong X, Zhang W, et al. Morphogenesis of rete ridges in human oral mucosa: a pioneering morphological and immunohistochemical study. *Cells Tissues Organs.* 2013;197:239–248.
42. Skougaard MR. Cell renewal, with special reference to the gingival epithelium. *Adv Oral Biol.* 1970;4:261–288.
43. Squier CA, Finkelstein MW. Oral mucosa. In: Ten Cate AR, ed. *Oral Histology, Development, Structure and Function*. CV Mosby; 1989:341–382.
44. Mackenzie IC, Rittman G, Gao Z, Leigh I, Lane EB. Patterns of cytokeratin expression in human gingival epithelia. *J Periodontal Res.* 1991;26:468–478.
45. Sawaf MH, Ouhayoun JP, Forest N. Cytokeratin profiles in oral epithelial: a review and a new classification. *J Biol Buccale.* 1991;19:187–198.
46. Ouhayoun JP, Gosselin F, Forest N, et al. Cytokeratin patterns of human oral epithelia: differences in cytokeratin synthesis in gingival epithelium and the adjacent alveolar mucosa. *Differentiation.* 1985;30:123–129.
47. Mandel U. Carbohydrates in oral epithelia and secretions: Variation with cellular differentiation. *APMIS Suppl.* 1992;27:119–129.
48. Shah F, Berggren D, Holmlund T, Levring Jäghagen E, Stål P. Unique expression of cytoskeletal proteins in human soft palate muscles. *J Anat.* 2016;228:487–494.
49. Prasad P, Khalil E, Desa VB, et al. Bifid uvula-an enigma. *J Pharm Bioallied Sci.* 2023;15(Suppl 1):S806–S809.
50. Glantschnig L, Dierks A, Wienand G, et al. Intense PSMA expression of the uvula detected by [68Ga]Ga-PSMA-PET/CT. *Eur J Nucl Med Mol Imaging*. 2024;51(5):1488–1489.
51. Olofsson K, Mattsson C, Hammarström ML, Hellström S. Structure of the human uvula. *Acta Otolaryngol.* 1999; 119:712–717.
52. Graeme-Cook F, Bhan AK, Harris NL. Immunohistochemical characterization of intraepithelial and subepithelial mononuclear cells of the upper airways. *Am J Pathol.* 1993;143:1416–1422.
53. Chang CP, Hang JF, Chen CJ, et al. Tonsil tissue control is ideal for monitoring estrogen receptor immunohistochemical staining. *J Chin Med Assoc.* 2023;86(12):1053–1059.
54. Winther B, Innes DJ. The human adenoid. A morphologic study. *Arch Otolaryngol Head Neck Surg.* 1994;120:144–149.
55. Jovic M, Avramović V, Vlahović P, Savić V, Veličkov A, Petrović V. Ultrastructure of the human palatine tonsil and its functional significance. *Rom J Morphol Embryol.* 2015;56(2):371–377.
56. Vanhapiha N, Sanmark E, Blomgren K, Wikstén J. Minor salivary gland infection as origin of peritonsillitis—novel theory and preliminary results. *Acta Otolaryngol.* 2022; 142(2):182–186.
57. Bhargava D, Raman R, Khalfan Al Abri R, et al. Heterotopia of the tonsil. *J Laryngol Otol.* 1996;110:611–612.
58. Eversole LR. The histochemistry of mucosubstances in human minor salivary glands. *Arch Oral Biol.* 1972;17:1225–1239.
59. Munger BL. Histochemical studies on seromucous- and mucous-secreting cells of human salivary glands. *Am J Anat.* 1964;115:411–429.
60. Laine M, Bläuer M, Ylikomi T, et al. Immunohistochemical demonstration of androgen receptors in human salivary glands. *Arch Oral Biol.* 1993;38:299–302.
61. Kumagami H, Onitsuka T. Estradiol and testosterone in minor salivary glands of Sjögren's syndrome. *Auris Nasus Larynx.* 1993;20:137–143.
62. Herrera-Esparza R, Bollain-y-Goytia J, Ruvalcaba C, Ruvalcaba M, Pacheco-Tovar D, Avalos-Díaz El. Apoptosis and cell proliferation: the paradox of salivary glands in Sjögren's disease. *Acta Reumatol Port.* 2008;33:299–303.
63. Tandler B, Denning CR, Mandel ID, Kutscher AH. Ultrastructure of human labial salivary glands. 3. Myoepithelium and ducts. *J Morphol.* 1970;130:227–245.
64. Ogawa Y. Immunocytochemistry of myoepithelial cells in the salivary glands. *Prog Histochem Cytochem.* 2003;38:343–426.
65. Grandi D, Campanini N, Becchi G, Lazzaretti M. On the myoepithelium of human salivary glands. An immunocytochemical study. *Eur J Morphol.* 2000;38:249–255.
66. Ianez RF, Buim ME, Coutinho-Camillo CM, Schultz R, Soares FA, Lourenço SV. Human salivary gland morphogenesis: myoepithelial cell maturation assessed by immunohistochemical markers. *Histopathology.* 2010;57:410–417.
67. Takeda Y. Existence and distribution of melanocytes and HMB-45-positive cells in the human minor salivary glands. *Pathol Int.* 2000;50:15–19.
68. Nonaka CF, Pereira KM, de Andrade Santos PP, de Almeida Freitas R, da Costa Miguel MC. Sialolipoma of minor salivary glands. *Ann Diagn Pathol.* 2011;15:6–11.
69. Hamperl H. Über das Vorkommen von Onkocyten in verschiedenen Organen und ihren Geschwülsten: (Mundspeicheldrüsen, Bauchspeicheldrüse, Epithelkörperchen, Hypophyse, Schilddrüse, Eileiter). *Virchows Arch A.* 1936;298:327–375.
70. Chang A, Harawi SJ. Oncocytes, oncocytosis, and oncocytic tumors. *Pathol Annu.* 1992;27:263–304.
71. Balogh K Jr, Roth SI. Histochemical and electron microscopic studies of eosinophilic granular cells (oncocytes) in tumors of the parotid gland. *Lab Invest.* 1965;14:310–320.
72. Goode RK, Corio RL. Oncocytic adenocarcinoma of salivary glands. *Oral Surg Oral Med Oral Pathol.* 1988;65:61–66.
73. Sripodok P, Poomsawat S, Juengsomjit R, Kosanwat T. Intraoral oncocytic mucoepidermoid carcinoma—a rare case report and review of the literature. *Ann Maxillofac Surg.* 2021;11(2):313–316.
74. Tschen JA, Fechner RE. The juxtaoral organ of Chievitz. *Am J Surg Pathol.* 1979;3:147–150.
75. Pantanowitz L, Tschen JA. Organ of Chievitz. *Ear Nose Throat J.* 2004;83:230.
76. Zenker W. *Juxtaoral Organ (Chievitz' Organ). Morphology and Clinical Aspects*. Urban & Schwarzenberg; 1982.
77. Ide F, Mishima K, Saito I. Melanin pigmentation in the juxtaoral organ of Chievitz. *Pathol Int.* 2003;53:262–263.
78. Pantanowitz L, Tschen JA, Balogh K. The juxtaoral organ of Chievitz. *Int J Surg Pathol.* 2003;11:37.
79. Pantanowisz L. Immunophenotype of the juxtaoral organ. *Int J Oral Maxillofac Surg.* 2004;33:113.
80. Constantini A, Mihu CM, Bosca AB, et al. Short histological kaleidoscope—recent findings in histology. Part III. *Rom J Morphol Embryol.* 2023;64(2):115–133.
81. Lutman GB. Epithelial nests in intraoral sensory nerve endings simulating perineural invasion in patients with oral carcinoma. *Am J Clin Pathol.* 1974;61:275–284.

82. Mikó T, Molnár P. The juxtaoral organ—a pitfall for pathologists. *J Pathol.* 1981;133:17–23.
83. Soucy P, Cimone G, Carpenter B. An unusual intraoral mass in a child: the organ of Chievitz. *J Pediatr Surg.* 1990;25:1200.
84. Leibl W, Pflüger H, Kerjaschki D. A case of nodular hyperplasia of the juxtaoral organ in man. *Virchows Arch a Pathol Anat Histol.* 1976;371:389–391.
85. Chievitz JH. Beiträge zur Entwicklungsgeschichte der Speicheldrüsen. *Arch Anat Physiol.* 1885;9:401–436.
86. Pantanowitz L, Balogh K. Significance of the juxtaoral organ (of Chievitz). *Head Neck.* 2003;25:400–405.
87. Ide F, Mishima K, Saito I. Pacinian corpuscle in the juxtaoral organ of Chievitz. *J Oral Pathol Med.* 2004;33:443–444.
88. Suárez-Quintanilla J, Martín-Cruces J, Martín-Biedma B, et al. The juxta-oral organ of Chievitz (organum yuxtaorale) updated: embryology, anatomy, function and pathology. *Ann Anat.* 2020;232:151582.
89. Eversole LR, Leider AS. Maxillary intraosseous neuroepithelial structures resembling those seen in the organ of Chievitz. *Oral Surg Oral Med Oral Pathol.* 1978;46:555–558.
90. Jensen JL, Wuerker RB, Correll RW, Erickson JO. Epithelial islands associated with mandibular nerves. Report of two cases in the walls of mandibular cysts. *Oral Surg Oral Med Oral Pathol.* 1979;48:226–230.
91. Mills N, Geddes DT, Amirapu S, Mirjalili SA. Understanding the lingual frenulum: histological structure, tissue composition, and implications for tongue tie surgery. *Int J Otolarnygol.* 2020;2020:1820978.
92. Biswas P, Agarwal N, Saha P, et al. Lingual frenulum length: a prospecting link to craniofacial morphology in adults. *J Orthod Sci.* 2023;12:54.
93. Jung HS, Akita K, Kim JY. Spacing patterns on tongue surface-gustatory papilla. *Int J Dev Biol.* 2004;48(2–3):157–161.
94. Tao J, Hashemi DA, Wang JV, Geronemus R, Shelton R. Treatment of pigmented fungiform papillae of the tongue with 694 nm Q-switched ruby laser. *Lasers Surg Med.* 2024; 56(4):334–336.
95. Kruger L, Mantyh PW. Gustatory and related chemosensory systems. In: Björklund A, Hökfelt T, Swanson LW, eds. Handbook of Chemical Neuroanatomy. Vol. 7. *Integrated Systems of the CNS Part II.* Elsevier; 1989:323–411.
96. Oakley B. Neuronal–epithelial interactions in mammalian gustatory epithelium. In: Bock GR, ed. *Regeneration of Vertebrate Sensory Receptor Cells.* Wiley; 1991:277–287.
97. McDaniel RK. Subepithelial nerve plexus (with ganglion cells) associated with taste buds. *Oral Surg Oral Med Oral Pathol Oral Radiol Endod.* 1999;87:605–609.
98. Triantafyllou A, Coulter P. Structural organization of subgemmal neurogenous plaques in foliate papillae of tongue. *Hum Pathol.* 2004;35:991–999.
99. Val-Bernal JF, Garijo MF, Fontanil N. Subgemmal neurogenous plaques of the tongue: a systematic autopsy study. *Rom J Morphol Embryol.* 2022;63(3):545–553.
100. Meirelles DP, Felix FA, de Sousa SF, de Aguiar MCF. Subgemmal neurogenous plaque: a series of six cases and a literature overview. *Oral Maxillofac Surg.* 2024;28(2):975–983.
101. Palazzolo MJ, Fowler CB, Magliocca KR, et al. Neuroepithelial structures associated with the subepithelial nerve plexus of taste buds: a fortuitous finding resembling the juxtaoral organ of Chievitz. *Oral Surg Oral Med Oral Pathol Oral Radiol.* 2014;117:497–501.
102. Pellicioli AC, Fonseca FP, Silva RN, et al. Histomorphometric characterization of subgemmal neurogenous plaques. *Oral Surg Oral Med Oral Pathol Oral Radiol.* 2017;123:477–481.
103. Soares CD, de Souza LL, Laffitte CM, et al. Symptomatic subgemmal neurogenous plaque in patients with COVID-19: Is there an association? *J Oral Pathol Med.* 2023;52(6):514–520.
104. Lambrou P, Kalfas S, Arhakis A. Gingival thickness and gingival width in children: a cross-sectional study utilizing ultrasonography. *Eur Arch Paediatr Dent.* 2024;25(2):217–225.
105. Schroeder HE, Listgarten MA. The gingival tissues: the architecture of periodontal protection. *Periodontol.* 1997;13: 91–120.
106. Ainamo J, Loe H. Anatomical characteristics of gingiva. A clinical and microscopic study of the free and attached gingiva. *J Periodontol.* 1966;37:5–13.
107. Melcher AH, Bowen WH, eds. *Biology of the Periodontium.* Academic Press; 1969.
108. Squier CA, Waterhouse JP. The ultrastructure of the melanocyte in human gingival epithelium. *Arch Oral Biol.* 1967;12:119–129.
109. Schroeder HE. Melanin containing organelles in cells of the human gingiva. *J Periodontal Res.* 1969;4:1–18.
110. Kaugars GE, Heise AP, Riley WT, Abbey LM, Svirsky JA. Oral melanotic macules. A review of 353 cases. *Oral Surg Oral Med Oral Pathol.* 1993;76:59–61.
111. Buchner A, Merrell PW, Hansen LS, Leider AS. Melanocytic hyperplasia of the oral mucosa. *Oral Surg Oral Med Oral Pathol.* 1991;71:58–62.
112. Buchner A, Ledier AS, Merrell PW, Carpenter WM. Melanocytic nevi of oral mucosa: a clinicopathologic study of 130 cases from northern California. *J Oral Pathol Med.* 1990; 19:197–201.
113. Buchner A, Hansen LS. Pigmented nevi of the oral mucosa: a clinicopathologic study of 36 new cases and review of 155 cases from the literature. Part I: a clinicopathologic study of 36 new cases. *Oral Surg Oral Med Oral Pathol.* 1987;63:566–572.
114. Buchner A, Hansen LS. Pigmented nevi of the oral mucosa: A clinicopathologic study of 36 new cases and review of 155 cases from the literature. Part II: analysis of 191 cases. *Oral Surg Oral Med Oral Pathol.* 1987;63:676–682.
115. Trodahl JN, Sprague WG. Benign and malignant melanocytic lesions of the oral mucosa. An analysis of 135 cases. *Cancer.* 1970;25:812–823.
116. Hashimoto K. Fine structure of Merkel cell in human oral mucosa. *J Invest Dermatol.* 1972;58:381–387.
117. Tong CK, Moayedi Y, Lumpkin EA. Merkel cells and keratinocytes in oral mucosa are activated by mechanical stimulation. *Physiol Rep.* 2024;12(2):e15826.
118. Kingsmill VJ, Berkovitz BK, Barrett AW. An immunohistochemical analysis of human Merkel cell density in gingival epithelium from dentate and edentulous subjects. *Arch Oral Biol.* 2005;50(10):883–887.
119. Mahomed F. Neuroendocrine cells and associated malignancies of the oral mucosa: a review. *J Oral Pathol Med.* 2010;39:121–127.
120. Righi A, Betts CM, Marchetti C, et al. Merkel cells in the oral mucosa. *Int J Surg Pathol.* 2006;14(3):206–211.
121. Waterhouse JP, Squier CA. The Langerhans cell in human gingival epithelium. *Arch Oral Biol.* 1967;12:341–348.

122. Chou JM, Daniels TE. Langerhans cells expressing HLA-DQ, HLA-DR and T6 antigens in normal oral mucosa and lichen planus. *J Oral Pathol Med.* 1989;18:573–576.
123. Barrett AW, Cruchley AT, Williams DM. Oral mucosal Langerhans' cells. *Crit Rev Oral Biol Med.* 1996;7:36–58.
124. Anjana R, Joseph L, Suresh R. Immunohistochemical localization of CD1a and S100 in gingival tissues of healthy and chronic periodontitis subjects. *Oral Dis.* 2012; 18(8):778–785.
125. Kindt N, Descamps G, Seminerio I, et al. Langerhans cell number is a strong and independent prognostic factor for head and neck squamous cell carcinomas. *Oral Oncol.* 2016; 62:1–10.
126. Pusztaszeri MP, Sadow PM, Faquin WC. Association of CD1a-positive dendritic cells with papillary thyroid carcinoma in thyroid fine-needle aspirations: a cytologic and immunocytochemical evaluation. *Cancer Cytopathol.* 2013; 121(4):206–213.
127. Baume LJ. The biology of pulp and dentine. A historic terminologic-taxonomic, histologic-biochemical, embryonic and clinical survey. *Monogr Oral Sci.* 1980;8:1–220.
128. Holland GR. The odontoblast process: form and function. *J Dent Res.* 1985;64:499–514.
129. Thomas HF. The dentin-predentin complex and its permeability: anatomical review. *J Dent Res.* 1985;64:607–612.
130. Nylen UM, Termine JD. Tooth enamel III. Its development, structure, and composition. *J Dent Res.* 1979;58:675–1031.
131. Ida-Yonemochi H, Ikarashi T, Nagata M, et al. The basement membrane-type heparan sulfate proteoglycan (perlecan) in ameloblastomas: its intercellular localization in stellate reticulum-like foci and biosynthesis by tumor cells in culture. *Virchows Arch.* 2002;441:165–173.
132. Domingues MG, Jaeger MM, Araújo VC, Araújo NS. Expression of cytokeratins in human enamel organ. *Eur J Oral Sci.* 2000;108:43–47.
133. Li S, Li S, Cao M. Developmental relationship between junctional epithelium and epithelial rests of Malassez. *Int J Dev Biol.* 2024;68(1):39–45.
134. Shinmura Y, Tsuchiya S, Hata K, Honda MJ. Quiescent epithelial cell rests of Malassez can differentiate into ameloblast-like cells. *J Cell Physiol.* 2008;217(3):728–38.
135. Joiner AM, Green WW, McIntyre JC, Allen BL, Schwob JE, Martens JR. Primary cilia on horizontal basal cells regulate regeneration of the olfactory epithelium. *J Neurosci.* 2015;35:13761–13772.
136. Fawcett DW, Porter KR. A study of the fine structure of ciliated epithelium. *J Morphol.* 1954;94:221–281.
137. Afzelius BA. The immotile-cilia syndrome and other ciliary diseases. *Int Rev Exp Pathol.* 1979;19:1–43.
138. Howell JT, Schochet SS Jr, Goldman AS. Ultrastructural defects of respiratory tract cilia associated with chronic infections. *Arch Pathol Lab Med.* 1980;104:52–55.
139. Sleigh MA, Blake JR, Liron N. The propulsion of mucus by cilia. *Am Rev Respir Dis.* 1988;137:726–741.
140. Zak FG, Lawson W. The presence of melanocytes in the nasal cavity. *Ann Otol Rhinol Laryngol.* 1974;83:515–519.
141. Cove H. Melanosis, melanocytic hyperplasia, and primary malignant melanoma of the nasal cavity. *Cancer.* 1979;44: 1424–1433.
142. Rhys-Evans PH. Anatomy of the nose and paranasal sinuses. In: Kerr AG, Groves J, Scott-Brown WG, eds. *Scott-Brown's Otolaryngology*. 5th ed. Vol I. Butterworth; 1987:138–161.
143. Drake-Lee AB. Physiology of the nose and paranasal sinus. In: Kerr AG, Groves J, Scott-Brown WG, eds. *Scott-Brown's Otolaryngology*. 5th ed. Vol I. Butterworth; 1987:162–182.
144. Ballenger JJ. The clinical anatomy and physiology of the nose and accessory sinuses. In: Ballenger JJ, ed. *Diseases of the Nose, Throat, Ear, Head and Neck*. 14th ed. Lea & Febiger; 1991:3–22.
145. Ambrosini-Spaltro A, Morandi L, Spagnolo DV, et al. Nasal seromucinous hamartoma (microglandular adenosis of the nose): a morphological and molecular study of five cases. *Virchows Arch.* 2010;457:727–734.
146. Trotter CM, Hall GH, Salter DM, et al. Histology of the mucous membrane of the human inferior nasal concha. *Clin Anat.* 1990;3:307–316.
147. Kar M, Altıntaş M. The incidence of concha bullosa: a retrospective radiologic study. *Eur Arch Otorhinolaryngol.* 2023;280(2):731–735.
148. MacArthur FJ, McGarry GW. The arterial supply of the nasal cavity. *Eur Arch Otorhinolaryngol.* 2017;274:809–815.
149. Jahnke V, Merker HJ. Electron microscopic and functional aspects of the human vomeronasal organ. *Am J Rhinol.* 2000; 14:63–67.
150. Zuckerkandl E. Das Jacobsonsche organ. *Erg Anat Entwicklungsgesch.* 1910;18:801–843.
151. Pearlman SJ. Jacobson's organ (Organon vomero-nasale Jacobsoni): Its anatomy, gross, microscopic and comparative, with some observations as well on its function. *Ann Otol Rhinol Laryngol.* 1934;43:739–768.
152. Negus VE. The organ of Jacobson. *J Anat.* 1956;90:515–519.
153. Seifert K. Licht- und elektronenmikroskopische Untersuchungen am Jacobsonschen Organ (Organon vomeronasale) der Katze. *Arch Klin Exp Ohr Nas Kehlk Heilk.* 1971;200:223–251.
154. Bruintjes TD, Bleys RLAW. The clinical significance of the human vomeronasal organ. *Surg Radiol Anat.* 2023; 45(4):457–460.
155. Naessen R. The identification and topographical localization of the olfactory epithelium in man and other mammals. *Acta Otolaryngol.* 1970;70:51–57.
156. Hummel T, Welge-Lüssen A. *Taste and Smell. An update.* Karger; 2006.
157. Schneider RA. The sense of smell in man: its physiologic basis. *N Engl J Med.* 1967;277:299–303.
158. Palay SL. The general architecture of sensory neuroepithelia. In: Bock GR, ed. *Regeneration of Vertebrate Sensory Receptor Cells*. Wiley; 1991:3–24.
159. Omura K, Han B, Nishijima H, et al. Heterogeneous distribution of mature olfactory sensory neurons in human olfactory epithelium. *Int Forum Allergy Rhinol.* 2022;12(3):266–277.
160. Seifert K. Licht- und elektronenmikroskopische Untersuchungen der Bowman-Drüsen in der Riechschleimhaut makrosmatischer Säuger. *Arch Klin Exp Ohren Nasen Kehlkopfheilkd.* 1971;200:252–274.
161. Simpson T, Talati V, Baird AM, et al. Histopathology of allergic fungal rhinosinusitis versus chronic rhinosinusitis with nasal polyps. *Laryngoscope*. 2024;134(6):2617–2621.

15

Larynx and Pharynx

Lisa M. Rooper

LARYNX

Definition and Boundaries

The larynx is a structurally and functionally complex organ composed of multiple connective tissue and epithelial elements that act in concert to play an essential role in airflow, speech, and swallowing. It is defined superiorly by the epiglottis and aryepiglottic folds and inferiorly by the inferior rim of the cricoid cartilage. Its anterior aspect consists of the lingual surface of the epiglottis, lamina of the thyroid cartilage, arch of the cricoid cartilage, and thyrohyoid and cricothyroid membranes (Fig. 15.1), while the posterior lamina of the cricoid cartilage and arytenoid cartilages form its posterior surface (Fig. 15.2). Although it closely abuts the larynx, the pyriform sinus is included in the hypopharynx and is not part of the larynx.

Embryology

The superior and inferior aspects of the larynx have separate developmental origins. The supraglottic larynx originates from the third and fourth pharyngeal arches, similar to the oral cavity and oropharynx. The glottic and subglottic larynx arise from the sixth pharyngeal arch, which also gives rise to the trachea and lungs. Notably, the separate origins of the superior and inferior aspects of the larynx lead to largely independent lymphatic circulations of these structures and help dictate the spread of laryngeal carcinoma (1). The respiratory apparatus first appears at the third week of gestation with emergence of the larynx during the fourth and fifth weeks. The lumen of the larynx is temporarily occluded by epithelium but undergoes recanalization during the ninth and tenth weeks to form the vocal cords. Due to the complexity of laryngeal development, at least 30 different congenital malformations are recognized (2).

Gross and Functional Anatomy

The larynx is composed of mucosa and submucosa overlying an elastic cone, cartilages, and intrinsic and extrinsic muscles. It is anatomically divided in three major mucosal compartments that are essential to differentiate for cancer staging. These compartments include the supraglottis, glottis, and subglottis. The supraglottic larynx spans from the tip of the epiglottis to the superior border of the true vocal cord (3) and includes the aryepiglottic folds (i.e., arytenoepiglottic folds), false vocal cords, and ventricles. The glottis includes the true vocal cords and anterior commissure (4,5). The subglottis is the mucosal area spanning from the lower border of the true vocal cords to the first tracheal cartilage (6).

The elastic cone of the larynx provides structural support to the true vocal cords. The vocal ligament is an area of thickened elastic tissue on the free edge of the true cord that is visible as a white band beneath the mucosa (Fig. 15.3). Notably, the false vocal cords are loose folds of mucosa that lack this defining elastic cone.

The larynx is supported by a core framework of cartilages (Fig. 15.4). The major structural cartilages of the larynx include the cricoid, thyroid, and paired arytenoid cartilages, all of which are of hyaline type and tend to be

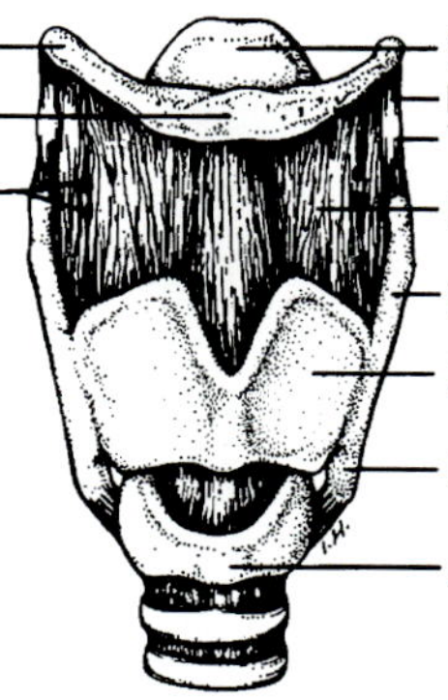

FIGURE 15.1 Anterior view of an unopened larynx shows the lamina of the thyroid cartilage, arch of the cricoid cartilage, and thyrohyoid membrane as the major structures that define the anterior external surface of the larynx. (Reprinted with permission from Mills SE, Fechner RE. Pathology of the larynx. *Atlas of Head and Neck Pathology Series.* American Society of Clinical Pathologists Press; 1985.)

resistant to the spread of carcinoma. Both the thyroid and cricoid cartilages undergo calcification starting as early as the second decade of life such that they demonstrate extensive ossification and harbor hematopoietic marrow elements in older individuals. In contrast, the epiglottis is composed of elastic cartilage with numerous fenestrations that may facilitate the spread of invasive carcinoma. Several dense ligaments anchor the cricoid and thyroid cartilages together, thereby limiting their movement. However, the arytenoid cartilages more freely articulate with the cricoid process; their positions determine the tension of the vocal ligament which inserts onto the vocal process and facilitates vocal cord movement. The arytenoid cartilages articulate at tiny diarthrodial cricothyroid and cricoarytenoid joints lined by flattened synovial cells. As such, they are susceptible to conditions such as gout and rheumatoid arthritis.

The laryngeal muscles include two key groups: extrinsic muscles that originate from extra-laryngeal structures and insert onto laryngeal structures, and intrinsic muscles that both originate from and insert onto laryngeal structures. Extrinsic muscles include the omohyoid, sternohyoid, sterothyroid, and thyrohyoid muscles, all of which act on the larynx during swallowing. The intrinsic muscles include the cricothyroid, posterior cricoarytenoid, lateral cricoarytenoid, and thyroarytenoid muscles. The vocalis muscle is also a member of this group. It includes small strands of muscle that insert along the length of the vocal ligaments and may be regarded as a component of the thyroarytenoid muscle. The intrinsic muscles act on laryngeal cartilages and move the vocal cord during phonation.

Although not technically part of the larynx, the pre-epiglottic space is an adjacent area that is important for the spread of laryngeal carcinoma. This triangular space is filled with fat and loose connective tissue. It is defined posteriorly by the epiglottis, anteriorly by the thyroid cartilage and thyrohyoid membrane, and superiorly by the hyoepiglottic ligament.

Microscopic Anatomy

The true vocal cords are lined by squamous epithelium at birth, while the remainder of the larynx is lined by ciliated respiratory type epithelium (7) (Fig. 15.5). Before 6 months of age, however, other areas of the larynx also begin developing squamous epithelium. Squamous epithelium first appears on the false cords but does not uniformly replace ciliated respiratory mucosa (8,9). By adulthood, the lingual surface of the epiglottis is covered by stratified squamous epithelium, while the laryngeal surface of the epiglottis is lined by squamous epithelium in its upper portion and respiratory-type epithelium in its inferior portion. Patches of squamous epithelium can be seen in both the supraglottic and subglottic regions in approximately 50% of nonsmoking adults, while the ciliated respiratory epithelium in the larynx can be entirely replaced by squamous epithelium in smokers.

The stratified squamous epithelium of the larynx varies from 5 to 25 cells in thickness (Fig. 15.6). The basal layer is composed of small cells with scant cytoplasm and ovoid nuclei

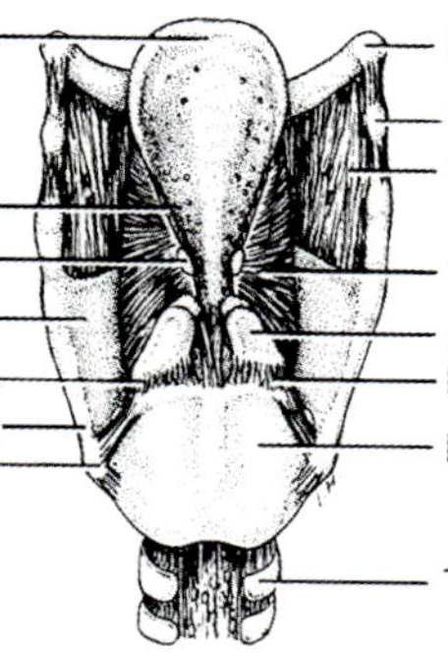

FIGURE 15.2 Posterior view of an unopened larynx highlights the lamina of the cricoid cartilage and position of the arytenoid cartilages as the major structures that define the posterior external surface of the larynx. (Reprinted with permission from Mills SE, Fechner RE. Pathology of the larynx. *Atlas of Head and Neck Pathology Series.* American Society of Clinical Pathologists Press; 1985.)

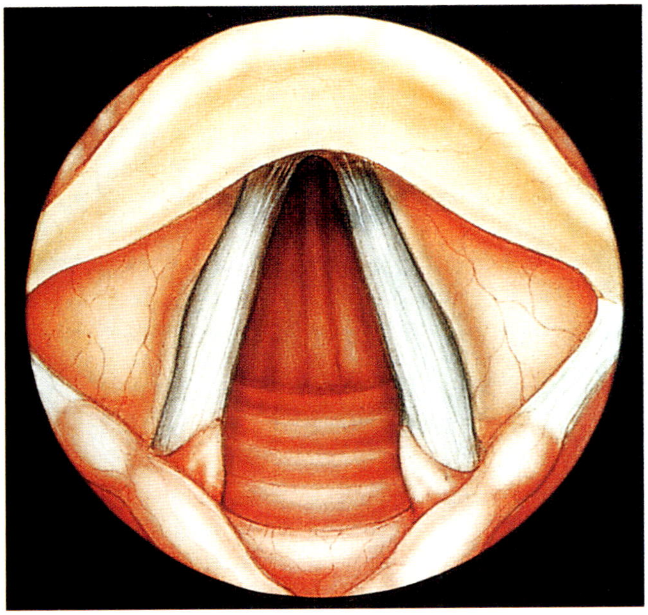

FIGURE 15.3 The vocal ligament of the larynx is visible through the mucosa of the true cord as a gray-to-white zone when viewed endoscopically from above.

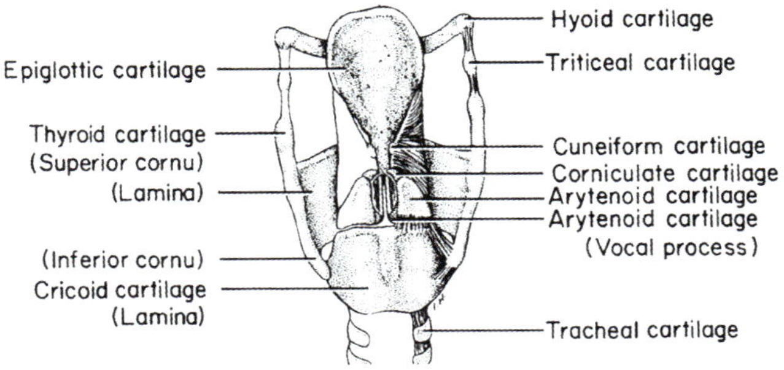

FIGURE 15.4 The major cartilages of the larynx are better seen in the absence of the associated soft tissues. (Reprinted with permission from Mills SE, Fechner RE. Pathology of the larynx. *Atlas of Head and Neck Pathology Series*. American Society of Clinical Pathologists Press; 1985.)

that are oriented perpendicularly to the surface. Mitotic figures are normally confined to this layer or the adjacent parabasal layer. Dendritic melanocytes may be present in the basal layer, especially in people with more pigmented skin, although the frequency of this phenomenon is unclear (10,11). Rare malignant melanomas of the larynx presumably arise from these melanocytic foci. The broadest layer of the epithelium is referred to as the prickle cell, or malpighian layer. Cells in this layer have abundant eosinophilic cytoplasm and prominent

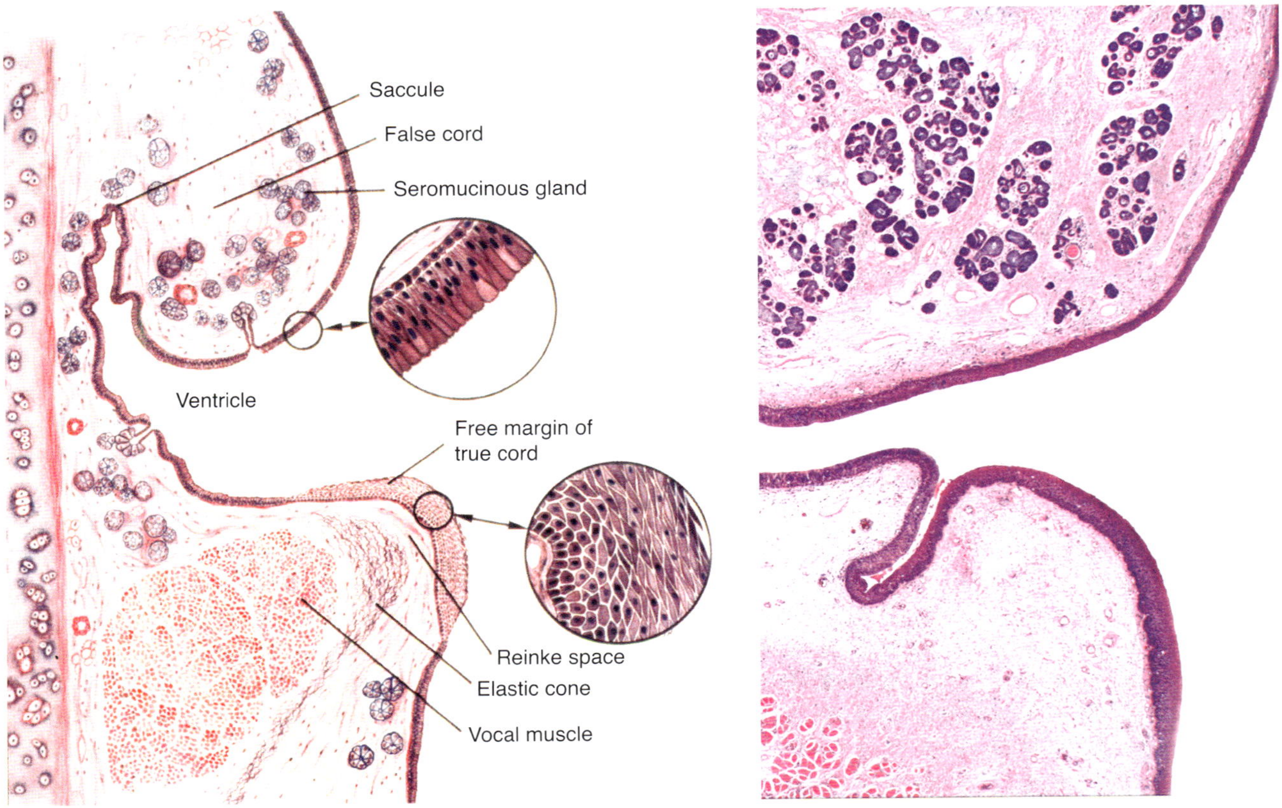

FIGURE 15.5 Drawings of the normal microscopic anatomy of the larynx at low magnification (*left*) and higher magnification (*right*) illustrate that seromucinous glands are prominent in the false cord, which is lined by ciliated columnar epithelium, while the vocalis muscle, elastic cone, and Reinke space are all present in the true cord. (Left image reprinted with permission from Mills SE, Fechner RE. Pathology of the larynx. *Atlas of Head and Neck Pathology Series*. American Society of Clinical Pathologists Press; 1985.)

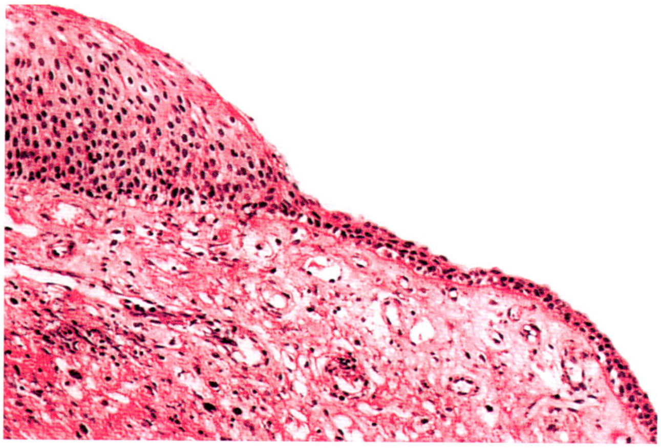

FIGURE 15.6 The squamous epithelium of the larynx can vary from approximately 5 to over 25 cells in thickness, even within the same larynx.

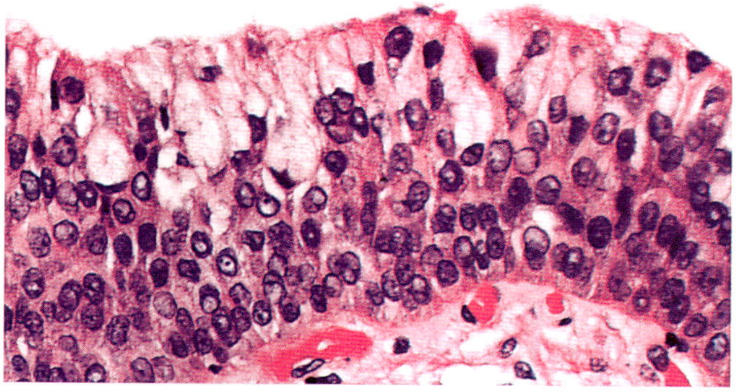

FIGURE 15.8 Goblet cells and columnar mucinous cells may be present in variable numbers within the nonsquamous epithelium of the larynx.

thin cytoplasmic strands connecting desmosomes of adjacent cells that reflect cell shrinkage during fixation. As squamous cells mature through the prickle cell layer, their nuclei enlarge, become oriented parallel to the surface, develop nucleoli with somewhat vesicular chromatin, and subsequently shrink to become pyknotic. The superficial layer is composed of flattened squamous cells with small, condensed nuclei. Although the larynx lacks surface keratinization, exposure to cigarette smoke or other irritants may lead to development of parakeratosis. Laryngeal squamous epithelium may rarely demonstrate orthokeratosis with keratohyaline granules, thereby resembling the skin. This finding is sometimes referred to as epidermoid metaplasia and likely represents a well-differentiated type of dysplasia or predysplastic proliferation (12).

The ciliated epithelium of the larynx is a pseudostratified respiratory epithelium. The innermost basal or reserve cell layer is composed of a single layer of small round cells with scant cytoplasm. The second row of cells is composed of ciliated columnar cells and varies from one to several cell layers in thickness (Fig. 15.7). The nuclei of these cells have a pseudostratified appearance due to the fact that they are not uniformly located adjacent to the basement membrane. There is also variation in the appearance and frequency of mucus-secreting cells; both goblet cells and nongoblet mucinous cells with ill-defined mucin vacuoles can be seen at the surface or in the middle portion of the epithelium (Fig. 15.8).

The junction between squamous and ciliated epithelium is consistently seen inferior and superior to the true vocal cords, although comingling of these cell types can be encountered throughout the larynx in zones of squamous metaplasia. The transitional zone between these epithelial types can vary from a few cells to an area spanning 1 to 2 mm in breadth. This transitional area essentially represents

FIGURE 15.7 The ciliated columnar epithelium of the larynx may be only a few cells in thickness (*left*), or it may form a considerably thicker layer (*right*). (Reprinted with permission from Mills SE, Fechner RE. Pathology of the larynx. *Atlas of Head and Neck Pathology Series.* American Society of Clinical Pathologists Press; 1985.)

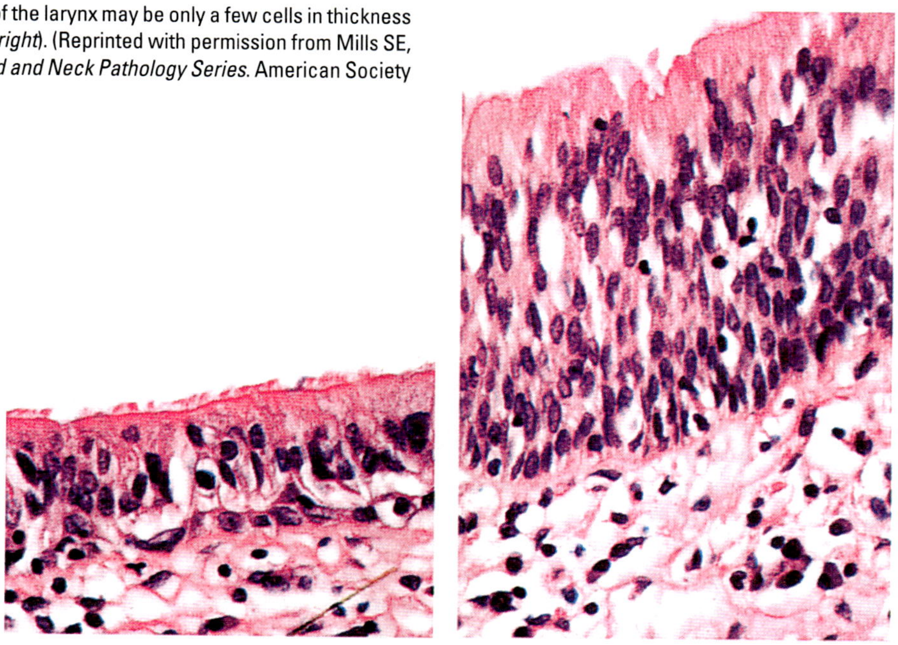

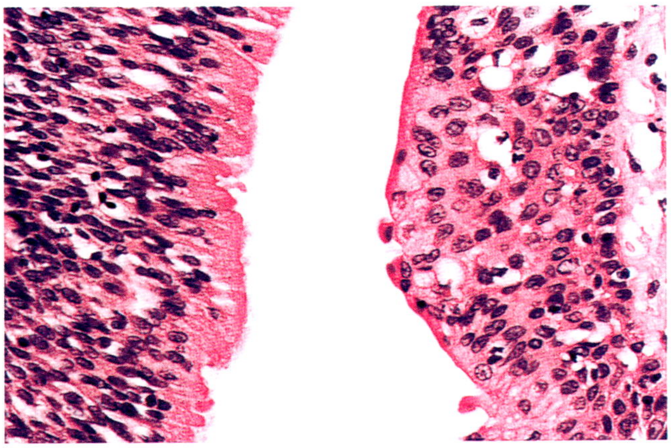

FIGURE 15.9 Ciliated columnar epithelium lines the false cord on the left, while a transitional zone composed of immature squamous metaplasia is seen on the true cord on the right; the disorganized appearance of this transitional zone should not be mistaken for dysplasia. (Reprinted with permission from Mills SE, Fechner RE. Pathology of the larynx. *Atlas of Head and Neck Pathology Series*. American Society of Clinical Pathologists Press; 1985.)

immature squamous metaplasia, wherein columnar cells are gradually replaced by small, basaloid cells that progressively enlarge until they are recognizable as mature squamous epithelium (Fig. 15.9). The epithelium of the transition zone is somewhat disorganized and contains basaloid cells that extend throughout its full thickness. Although these cells have uniform nuclei without pleomorphism, mitotic figures may be seen above the basal layer. These transitional zones can be confused with squamous dysplasia, particularly in frozen sections or limited samples.

Laryngeal epithelium is susceptible to infection by low-risk human papillomavirus types, most commonly including types 6 and 11. In fact, these HPV types have been detected in up to 25% of microscopically normal larynx specimens (13). Low-risk HPV is implicated in the development of squamous papillomas but not laryngeal squamous cell carcinoma. On the other hand, high-risk HPV subtypes including 16 and 18 can be associated with laryngeal squamous cell carcinoma, although it is extremely uncommon. High-risk HPV subtypes are recognized in less than 5% of laryngeal squamous cell carcinomas and are not associated with improved prognosis (14). They have not been demonstrated in normal laryngeal mucosa (15).

The lamina propria of the true vocal cord (i.e., Reinke space) is the connective tissue that lies between the vocal ligament and squamous epithelium (Fig. 15.10). The Reinke space lacks lymphatic vessels and contains only a few capillaries and scattered seromucinous glands. The paucity of draining vessels in this area helps limit the spread of carcinomas of the true vocal cords and, thus, many of these tumors may be curable via conservative surgery or radiation. However, poor draining also contributes to edema of true vocal cords in patients with upper respiratory tract infections, frequently leading to hoarseness or dysphonia. When

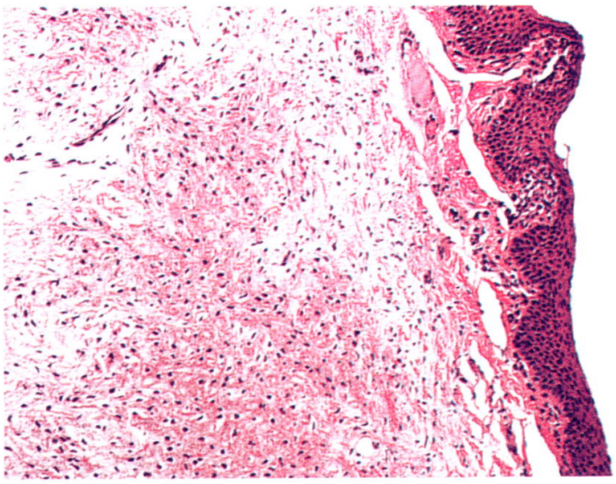

FIGURE 15.10 The true vocal cord has a narrow, sparsely vascular zone known as the Reinke space that lies between the squamous epithelium and the underlying vocal ligament.

the edema is pronounced, it can result in the development of vocal cord nodules and polyps. In contrast, the lamina propria of the anterior commissure contains more abundant capillaries and lymphatic vessels.

Seromucinous glands are present throughout most of the larynx but are most abundant in the false cords (Fig. 15.11) and just below the anterior commissure. They are generally absent beneath the squamous epithelium of the free edge of the vocal cord, although glands rarely may be present in the underlying vocalis muscle (Fig. 15.12). The fenestrations of the epiglottic cartilage can contain abundant seromucinous glands. As in other mucosal sites, laryngeal seromucinous glands are composed of both serous and mucinous acini surrounded by myoepithelial cells. They communicate with the surface epithelium via ducts that can be lined by either squamous or columnar epithelium and may be ciliated (16) (Fig. 15.13). Seromucinous glands of the larynx frequently undergo oncocytic metaplasia, presumably as a result of aging. Oncocytes are not seen in the seromucinous glands of people younger than 18 years of age, but are detected in 80% of glands from people more

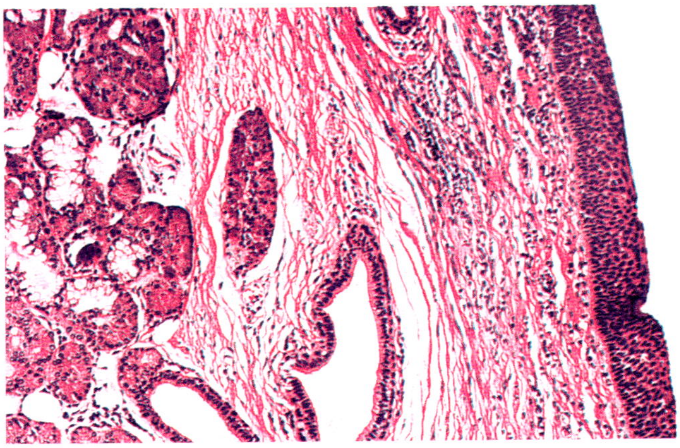

FIGURE 15.11 Seromucinous glands and their ducts are most prominent in the false cord.

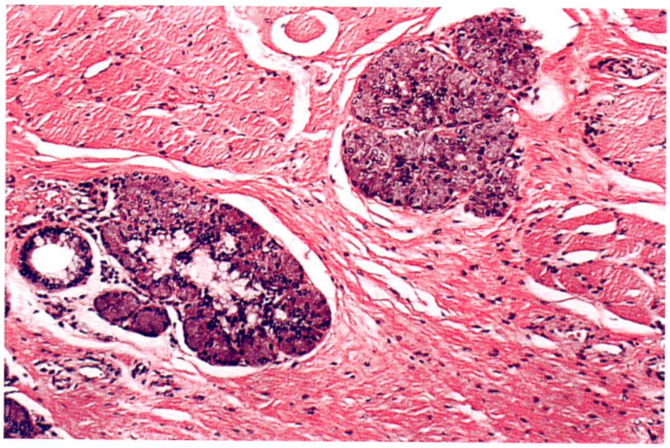

FIGURE 15.12 Seromucinous glands are occasionally located deep within the vocalis muscle.

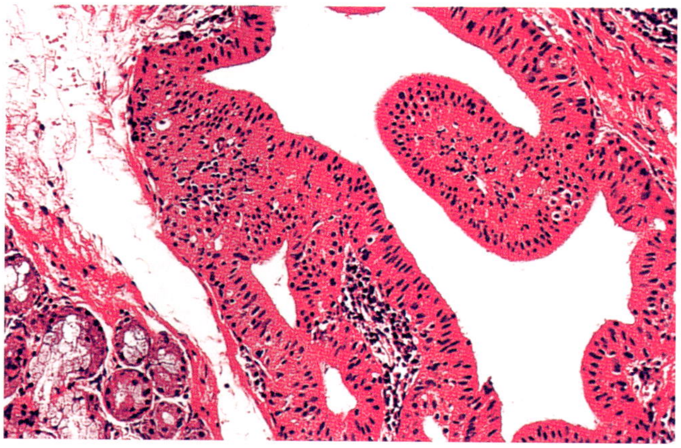

FIGURE 15.14 Oncocytic transformation of seromucinous epithelium can result in cystic structures with variable papillary architecture, sometimes described as oncocytic papillary cystadenoma.

than 50 years of age (17,18). Oncocytic metaplasia may come to clinical attention if associated cystic growth creates symptoms. Excised glands sometimes have papillary fronds and can be described as oncocytic papillary cystadenoma (Fig. 15.14).

Altered seromucinous glands of the larynx may mimic carcinoma in various ways. Necrotizing sialometaplasia can occur in the larynx when infarction or trauma results in squamous metaplasia of seromucinous glands (19) (Fig. 15.15). The metaplastic squamous nests may display mild-to-moderate nuclear atypia and mitotic activity that raise concern for squamous cell carcinoma or mucoepidermoid carcinoma, particularly when seen in frozen section specimens. Retained acinar architecture and transitions between squamous nests and preserved glands provide important clues to a benign diagnosis of sialometaplasia (Fig. 15.16). In addition, ducts leading to the seromucinous glands can be involved by squamous metaplasia with basal cell hyperplasia (Fig. 15.17) or dysplasia (Fig. 15.18) that mimic invasive squamous cell carcinoma. Retention of normal duct architecture including columnar cells in part of the lumen can help make this distinction. Notably, seromucinous glands of the larynx can give rise to the full spectrum of salivary gland tumors.

Two mechanisms can give rise to cartilaginous rests in the laryngeal submucosa. First, the vocal process of the arytenoid cartilage can occasionally be encountered in biopsies. It appears as a sharply circumscribed nodule of mature elastic cartilage (Fig. 15.19). Second, chondroid metaplasia can occur in the middle to posterior portions of the true vocal cord, as reported in 1% to 2% of larynges at autopsy (20,21). These metaplastic foci are not well-circumscribed and are surrounded by connective tissue rich in acid mucopolysaccharides (Fig. 15.20). Both the vocal process and chondroid metaplasia lack the multinodular pattern of cartilaginous neoplasms.

Microscopic islands of normal thyroid tissue may be seen within the fibrous capsule of the larynx and trachea external to the cricothyroid membrane or adjacent muscles (Fig. 15.21). Rarely, thyroid follicles may also be identified interior to the laryngeal cartilage, usually at the junction of the cricoid cartilage and first tracheal ring (22). This thyroid

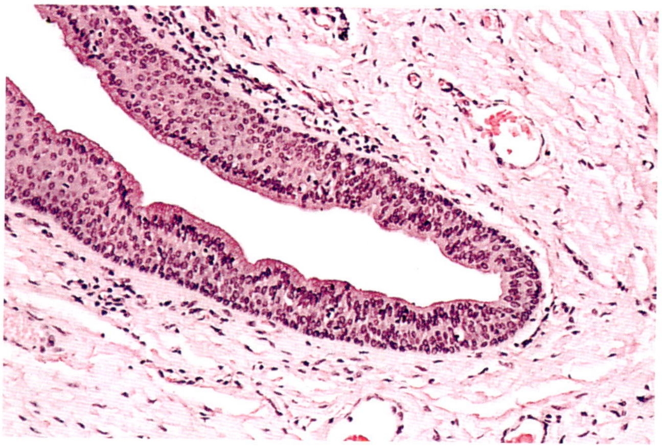

FIGURE 15.13 Ducts from seromucinous glands may be lined by squamous cells, ciliated columnar epithelium, or a mixture of the two.

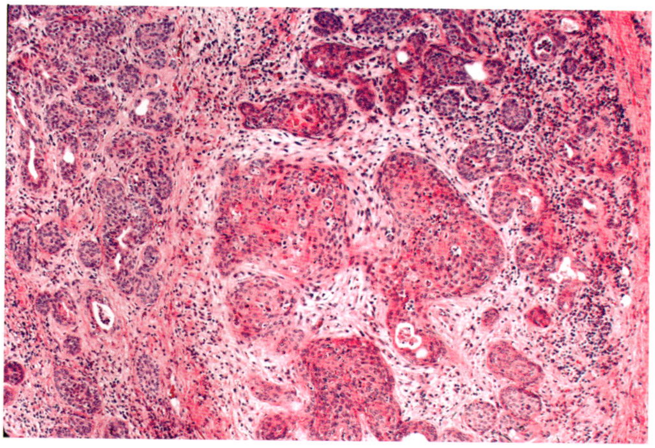

FIGURE 15.15 Necrotizing sialometaplasia shows replacement of the seromucinous lobules by aggregates of squamous cells with associated inflammation.

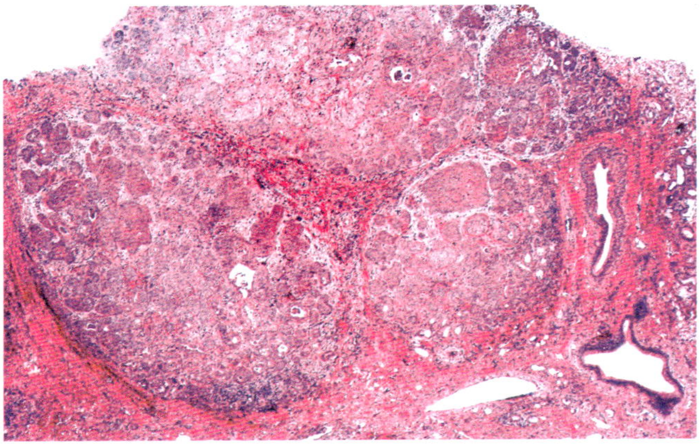

FIGURE 15.16 At low power, preservation of the preexisting lobular architecture of the seromucinous glands can help confirm a diagnosis of necrotizing sialometaplasia.

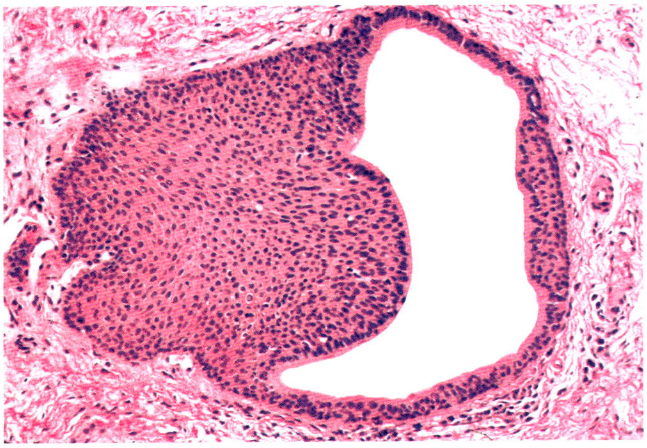

FIGURE 15.17 This seromucinous gland duct is involved by squamous metaplasia.

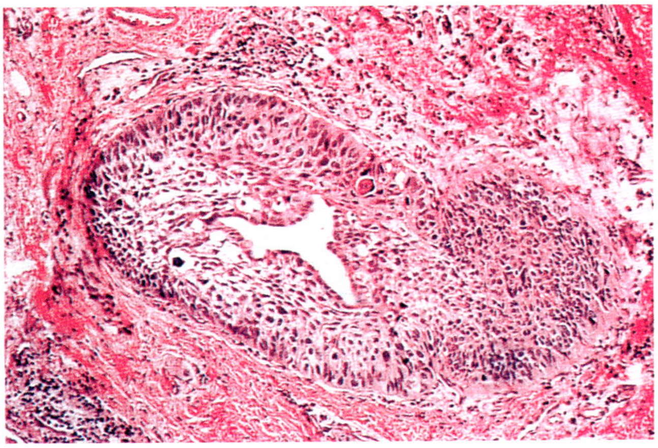

FIGURE 15.18 When surface dysplasia extends into seromucinous gland ducts, they should not be misinterpreted as invasive carcinoma.

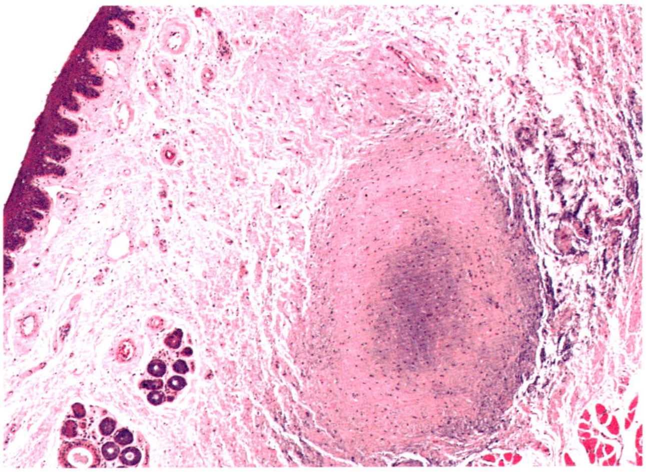

FIGURE 15.19 The vocal process of the arytenoid cartilage may appear in a laryngeal biopsy as a sharply circumscribed nodule of elastic-type cartilage that should not be mistaken for a cartilaginous neoplasm.

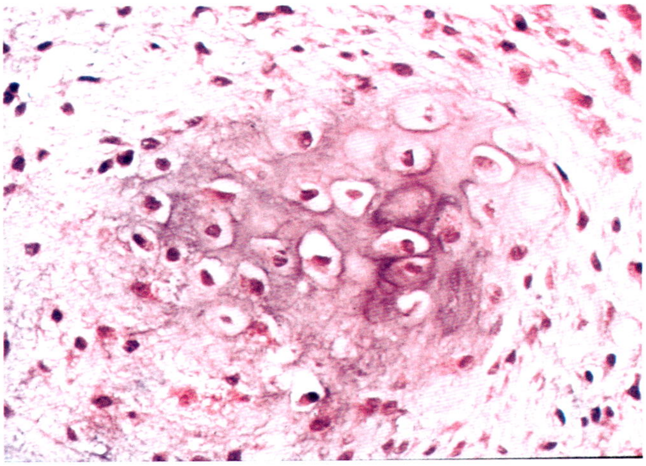

FIGURE 15.20 Chondroid metaplasia of the larynx has ill-defined margins. (Reprinted with permission from Mills SE, Fechner RE. Pathology of the larynx. *Atlas of Head and Neck Pathology Series.* American Society of Clinical Pathologists Press; 1985.)

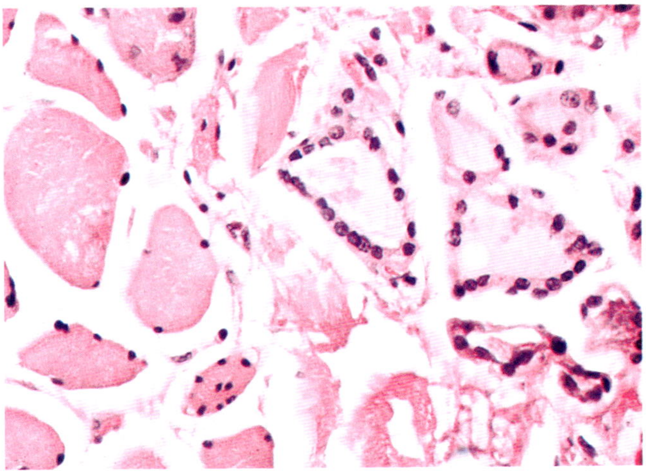

FIGURE 15.21 Small nests of normal thyroid follicles may be found in the soft tissue or skeletal muscle immediately external to the laryngeal cartilages.

tissue is not continuous with the main thyroid gland, likely losing its connection during embryologic development (23). The cells of these rests are small and may have an atrophic appearance with minimal cytoplasm and no cytologic atypia. Awareness of this phenomenon will help differentiate these foci from involvement by thyroid carcinoma.

Neural, Vascular, and Lymphatic Components

The intrinsic muscles of the larynx are innervated by branches of the vagus nerve with the superior laryngeal branch supplying the cricothyroid muscle and the terminal portion of the recurrent laryngeal nerve (i.e., inferior laryngeal nerve) innervating the other muscles. The branches of the superior and recurrent laryngeal nerves form anastomoses within the interarytenoid muscle and, less commonly, the pyriform sinus. Branches from the superior laryngeal nerve, also known as the communicating nerve, pass through the cricothyroid muscle to partially innervate the vocalis muscle (24,25).

The laryngeal nerve is associated with at least two pairs of paraganglia in the normal larynx. The superior paraganglia localize to the upper anterior third of the false cords adjacent to the margin of the thyroid cartilage and the internal branch of the superior laryngeal nerve (26,27). The inferior paraganglia can be found between the thyroid and cricoid cartilages or just below the cricoid cartilage in close association with the inferior laryngeal nerve (26,27). These paraganglia are minute neuroendocrine structures ranging from 0.1 to 0.4 mm that likely serve as chemoreceptors and may give rise to paragangliomas of the larynx. They are composed of nests of cells with amphophilic to clear cytoplasm and round nuclei with speckled chromatin (Fig. 15.22). Their small size should help distinguish paraganglia from laryngeal neuroendocrine tumors.

The superior laryngeal artery, which is derived from the superior thyroid artery, provides the blood supply to the superior aspect of the larynx. The inferior laryngeal artery, a small branch of the inferior thyroid artery, supplies the lower portions of the larynx. There may be anastomoses between the superior and inferior laryngeal arteries. Both laryngeal arteries are accompanied by similarly named veins, with the superior laryngeal vein joining the superior thyroid vein and the inferior laryngeal vein joining the inferior thyroid vein. Numerous anastomoses exist between the left and right inferior laryngeal veins across the front of the trachea, sometimes leading to contralateral venous return.

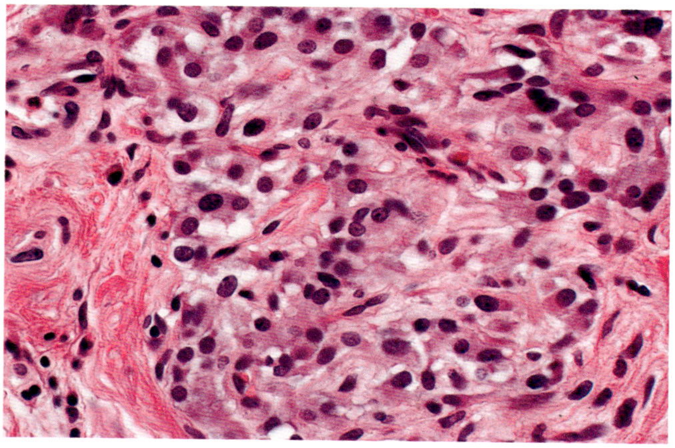

FIGURE 15.22 A nest of normal paraganglia cells from the lateral supraglottic larynx should not be confused with the insular growth pattern of a neuroendocrine tumor.

The lymphatic vessels of the larynx drain along with the vasculature (1). Supraglottic lymphatics are prominent and drain superiorly, typically terminating in the anterior jugular chain. Subglottic lymphatic vessels drain inferiorly, terminating in the midline pretracheal lymph nodes or, rarely, the lower cervical lymph nodes. Some laryngeal lymphatic vessels end in very small lymph nodes associated with the thyrohyoid membrane, cricotracheal ligament, or superior trachea although they ultimately drain into deep cervical lymph nodes. As noted above, the glottis has scant lymphatic drainage.

PHARYNX

Definition and Boundaries

The pharynx is divided into three functionally and structurally separate areas: the nasopharynx, oropharynx, and hypopharynx (Fig. 15.23). All of these areas play an important role in airflow but the oropharynx and hypopharynx are also essential to swallowing and the oropharynx contributes to speech. The palatine tonsil and lingual tonsil in the oropharynx and adenoids in the nasopharynx form a wreath of lymphoid tissue that is referred to as Waldeyer ring; these structures function in antigen sampling and stimulation of immune responses.

The nasopharynx is the portion of the pharynx that lies above the soft palate. Its anterior walls are perforated by posterior nasal choanae, the lateral walls harbor the openings of the Eustachian tubes, the posterior wall abuts the

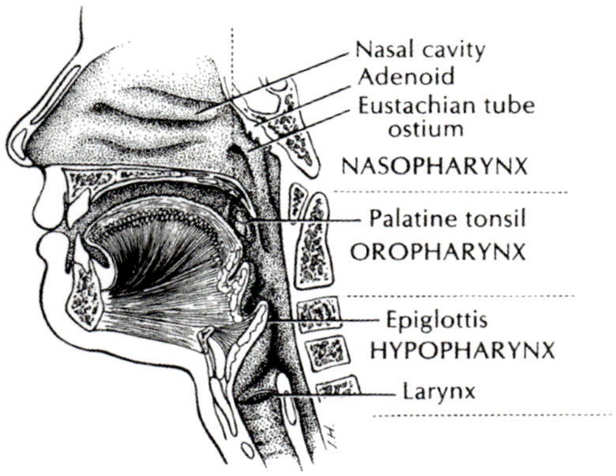

FIGURE 15.23 This sagittal section delineates the boundaries of the nasopharynx, oropharynx, and hypopharynx.

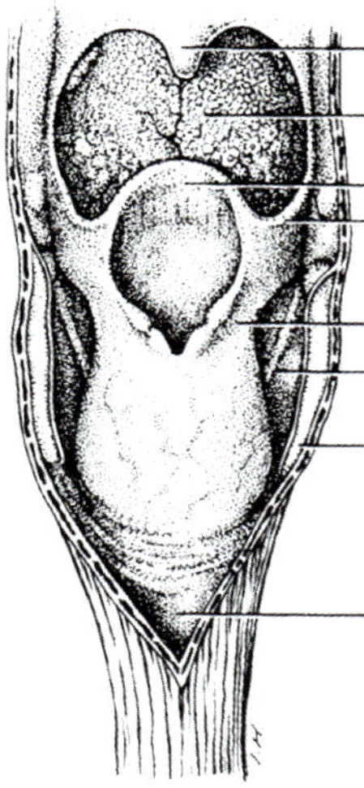

FIGURE 15.24 The pyriform sinuses are a part of the hypopharynx that surround the larynx laterally.

ventral surfaces of the first and second cervical vertebrae, and the roof is formed by the skull base and underside of the sphenoid sinus. The oropharynx is immediately inferior to the nasopharynx and extends from the soft palate to the tip of the epiglottis. It is bounded anteriorly by the faucial opening from the mouth and the posterior third of the tongue, laterally by the pharyngeal constrictor muscles, and posteriorly by the second and third cervical vertebrae. The hypopharynx extends below the tip of the epiglottis to the opening of the esophagus. The pyriform sinuses, also known as pyriform recesses, represent the lateral extensions of the hypopharynx that partially surround the larynx lateral to the aryepiglottic folds (Fig. 15.24).

Embryology

The pharynx derives from the third and fourth pharyngeal arches, which are formed from the endoderm of the foregut, although it is in direct continuity with the ectoderm at its cephalic end. During embryonic development, the stomodeum and primitive pharyngeal portion of the foregut are separated by the buccopharyngeal membrane, which is lined externally by ectoderm and internally by endoderm. This membrane ruptures at the third week of gestation, establishing continuity between the oral cavity and oropharynx (28). The primitive nasal cavity forms at the superior aspect of the buccopharyngeal membrane and enlarges during the fifth through seventh weeks of development, to form the bucconasal membrane. This membrane subsequently ruptures, establishing the connection between the nasal cavity and nasopharynx (28). The endodermal origin of the nasopharynx and ectodermal origin of the sinonasal tract may explain some of the different types of tumors that affect these sites. The secondary palate develops from the eighth through tenth gestational weeks, establishing the basic pharyngeal structures. From this point, the pharynx enlarges as the palate and tongue move downward (28).

Gross and Functional Anatomy

The roof of the nasopharynx is composed of mucosa overlying the basal portions of the sphenoid and occipital bones, whereas its lateral and posterior walls are composed of the superior constrictor muscles and pharyngobasilar fascia. The soft palate forms the floor of the anterior portion of the nasopharynx; it moves posteriorly and superiorly during speech and keeps food and water out of the nasal region during swallowing. The nasopharynx contains three structures frequently encountered by pathologists. The adenoids, or pharyngeal tonsils, are mass-like aggregates of lymphoid tissue present in the roof of the nasopharynx. They are prominent in children but become atrophic in adults. The fossa of Rosenmuller, also known as the pharyngeal recess, is a depression in the posterolateral nasopharynx. The Eustachian tube openings are just anterior to the fossa of Rosenmuller in the lateral nasopharyngeal wall and are surrounded superiorly and posteriorly by the torus tubarius, an area of mucosa-covered cartilage.

The oropharynx is bounded anteriorly by the opening of the mouth, also known as the fauces, and the posterior third of the tongue. This area of the tongue is fixed and termed the base of tongue. The lateral walls of the faucial opening are composed of the two tonsillar pillars. The anterior tonsillar pillar, or palatoglossal arch, extends forward from the soft palate to the tongue, while the posterior tonsillar pillar, or palatopharyngeal arch, extends laterally from the soft palate along the pharyngeal wall. These pillars contain similarly named palatoglossal and palatopharyngeal muscles. The tonsillar fossa lies between the tonsillar pillars and is home to the palatine tonsil. The palatine tonsil can have tremendous size variation due to fluctuations in the amount of reactive lymphoid tissue. The underlying pharyngobasilar fascia forms the tonsillar capsule, which forms a plane of cleavage that facilitates surgical removal. The base of tongue is inferior to the fauces. It contains abundant submucosal lymphoid tissue that forms the lingual tonsil. Although they are located at the anterior opening to the oropharynx, both the base of tongue and palatine tonsils are considered to be part of the oropharynx and not the oral cavity-a distinction which is key for cancer staging.

The parapharyngeal space, also known as the lateral pharyngeal space, is located deep to the tonsil and lateral to the pharynx. It is an inverted pyramid, with the skull base forming the superior base of the pyramid and the attachment of the cervical fascia to the hyoid bone forming the inferior apex. It is bounded medially by the superior constrictor muscle of the pharynx and laterally by the pterygoid lamina, inner surface of the mandibular ramus, and deep lobe of the parotid gland (29). It normally contains the internal carotid artery, internal jugular vein, cranial nerves IX to XII, the cervical sympathetic chain, vagal and carotid bodies, and multiple lymph nodes (29). Any of these structures within the parapharyngeal space may produce mass-forming lesions, including nerve sheath tumors and paragangliomas. It is also common for salivary gland tumors from the deep lobe of the parotid gland to extend into the

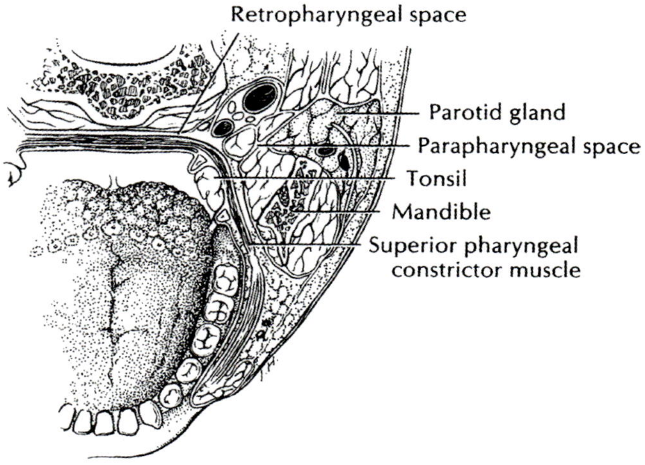

FIGURE 15.25 The parapharyngeal space lies deep to the tonsil and contains several vital structures. It is in continuity, posteriorly, with the retropharyngeal space.

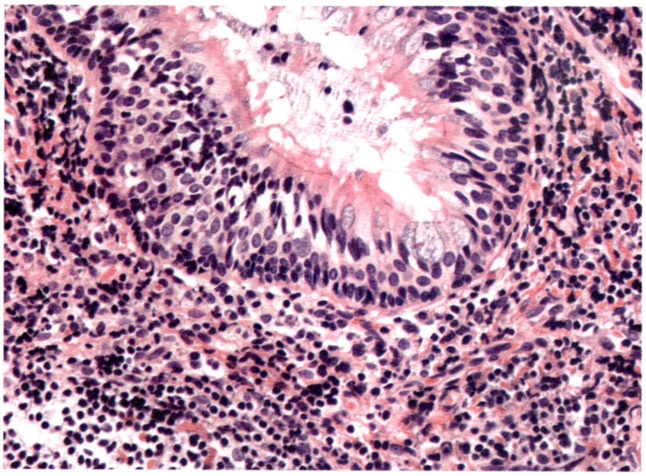

FIGURE 15.26 The adenoids are composed of epithelial crypts lined by ciliated respiratory-type mucosa unlike the exclusively squamous-lined crypts of the palatine tonsils.

parapharyngeal space. Tonsillar abscesses may involve and spread rapidly through the parapharyngeal space. Posteriorly, the parapharyngeal space is in direct continuity with an area of loose connective tissue behind the pharyngeal wall and anterior to the prevertebral fascial (29). This area is known as the retropharyngeal space (Fig. 15.25).

The hypopharynx is a cone-shaped region with a wide superior opening to the oropharynx and a narrow inferior opening to the esophagus. Pathologically, it can be divided into three regions: the pyriform sinuses, postcricoid region, and posterior pharyngeal wall. The most important of these structures are the pyriform sinuses, which represent elongated gutters that extend laterally on both sides of the larynx to the opening of the esophagus. The pyriform sinus is bounded medially by a thin layer of muscles derived from the aryepiglottic fold and esophageal opening and laterally by the thyroid cartilage. The common carotid artery runs posterior and lateral to the pyriform sinus. Notably, tumors arising in the hypopharynx frequently extend into the larynx. It is important to distinguish site of origin for these tumors since hypopharyngeal squamous cell carcinomas have a worse prognosis than those arising from adjacent structures.

Microscopic Anatomy

The nasopharynx is lined by a mix of stratified squamous epithelium and respiratory-type columnar epithelium. The squamous epithelium is most commonly seen in the lower portion of the anterior and posterior nasopharyngeal walls and the anterior half of the lateral nasopharyngeal walls, while respiratory epithelium is most common at the posterior nasal choanae and the roof of the nasopharynx. Similar to the larynx, the junction between the squamous and respiratory epithelium usually includes some amount of transitional mucosa that contains basaloid cells with minimal cytoplasm. These areas are particularly prominent at the junction of the nasopharynx and oropharynx and may pose a pitfall for squamous dysplasia when encountered in frozen sections.

The adenoid tissue of the nasopharynx forms a characteristic lymphoepithelial structure that plays an important role in the mucosal immune system (30). It is characterized by a dense submucosal lymphoid infiltrate with frequent follicles and germinal center formation. The adenoid crypts are composed of rounded invaginations of ciliated respiratory mucosa which should not be mistaken for a neoplastic glandular process (Fig. 15.26). Even outside the adenoids, lymphoid follicles can be present in the submucosa throughout the nasopharynx (Fig. 15.27). These follicles are particularly prominent in the Eustachian tube opening, sometimes known as the Gerlach tonsil, but may also be seen in the lateral and posterior nasopharyngeal walls and in the nasopharyngeal surface of the soft palate. This submucosal lymphoid infiltrate is a normal component of nasopharyngeal biopsies and should not be overinterpreted as a pathologic inflammatory process. Although the vast majority of nasopharyngeal carcinomas in both endemic and nonendemic regions are driven by Epstein–Barr virus (EBV),

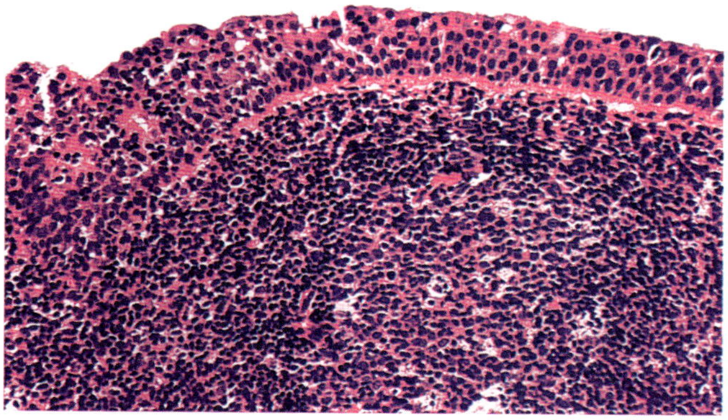

FIGURE 15.27 Submucosal lymphoid aggregates are present normally throughout the nasopharynx and should not be overly interpreted as severe chronic inflammation.

the normal nasopharyngeal epithelium does not appear uniquely sensitive to EBV infection and carcinogenesis likely results from a complex interplay of genetic alterations in host cells and stroma (31).

Both the oropharynx and hypopharynx are lined by stratified squamous epithelium similar to that seen in the squamous portions of the larynx and nasopharynx. This mucosa is consistently nonkeratinizing, with parakeratin or orthokeratin developing only in the setting of chronic irritation.

The palatine and lingual tonsils in the oropharynx also represent lymphoepithelial structures that are part of mucosal immune system (30). Their dense lymphoid components contain numerous follicles with frequent germinal center formation, separated by interfollicular areas. Unlike lymph nodes, there are no sinuses, trabeculae, or capsules. While the surface of the tonsil is lined by stratified squamous epithelium identical to other areas of the pharynx and larynx, the tonsillar crypt epithelium is composed of basaloid cells with scant cytoplasm and uniform nuclei with vesicular chromatin. These cells are arranged in a reticulated pattern with an irregular border and frequent extension of the lymphoid cells into the epithelium (Fig. 15.28). This feature can result in the appearance of isolated irregular nests of basaloid cells deep within the tonsil that raises concern for carcinoma (Fig. 15.29). Recognition of the contiguity of this epithelium with normal crypts at low power and the absence of increased cytologic atypia or expansile growth can help confirm that such nests are benign. Crypts frequently contain keratinaceous and inflammatory debris. Aggregates of *Actinomyces* species, commonly known as sulfur granules, are also seen regularly in crypt lumina and do not indicate pathogenic infection.

Like the transformation zone of the uterine cervix, the tonsillar crypt epithelium is particularly sensitive to infection with high-risk HPV. Indeed, HPV-associated oropharyngeal squamous cell carcinomas almost entirely arise from the crypt epithelium of the lingual and palatine tonsils (32). These tumors also recapitulate the tonsillar crypt epithelium

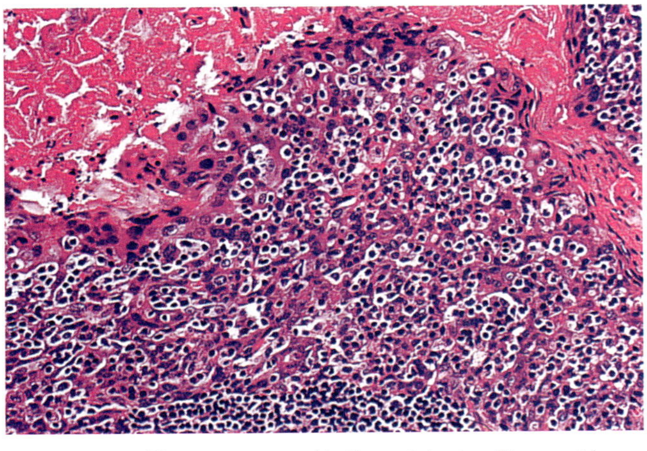

FIGURE 15.28 The squamous epithelium of the tonsillar crypt has a reticulated pattern with an irregular junction with lymphoid tissue and squamous cells.

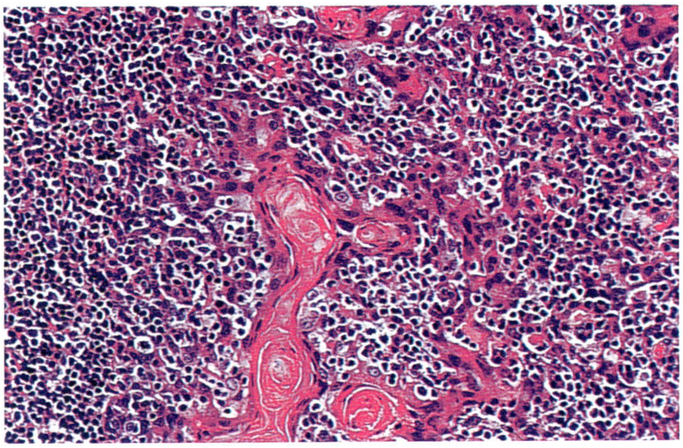

FIGURE 15.29 Irregular nests of normal tonsillar crypt epithelium are frequently present deep within the tonsil and should not be misinterpreted as carcinoma.

in many ways, with basaloid cytology and minimal keratinization. Some tumors have a lymphoepithelial appearance with reticulated architecture that closely mimics the reticulated architecture of the crypts. Normal crypt epithelium can show patchy p16 expression but does not show the block-like staining in greater than 70% of cells required for diagnosis of HPV-associated squamous cell carcinoma (33,34) (Fig. 15.30).

The basement membrane is discontinuous in the areas of the tonsillar crypts, thereby affording easy access of intraepithelial carcinoma to lymphatic vessels despite the absence of overtly invasive growth (35). This finding accounts for the common phenomenon of HPV-associated squamous cell carcinomas that present with large cervical lymph node metastases and microscopic or undetectable primary tumors. As an important consequence of this unique anatomic feature, no in situ HPV-associated dysplasia or carcinoma in situ is recognized in the oropharynx (34).

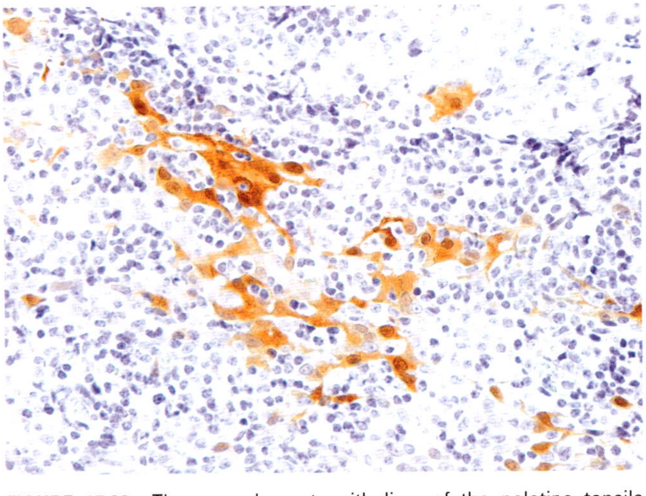

FIGURE 15.30 The normal crypt epithelium of the palatine tonsils can show patchy positivity for p16, but does not display the block-like expression in greater than 70% of cells required for diagnosis of HPV-associated squamous cell carcinoma.

Numerous submucosal seromucinous glands are present throughout the nasopharynx, oropharynx, and hypopharynx, most of which have predominantly mucinous acini. These are particularly numerous in nasopharynx in the torus tubarius adjacent to the Eustachian tube opening. As in the larynx, oncocytic metaplasia can affect the glandular and ductal epithelium, becoming increasingly frequent with advancing age (36,37). Rarely, oncocytic metaplasia in the nasopharynx can also undergo melanocytic change, sometimes termed melanotic oncocytic hyperplasia (38). Seromucinous glands of this location can also give rise to salivary gland neoplasms. Islands of metaplastic cartilage and bone are uncommonly identified adjacent to the tonsils, being present in less than 5% of all resected tonsils (39). This finding likely represents a secondary reactive change.

There are several developmental remnants that can occur in the nasopharynx. The Rathke pouch is an intracranial invagination of epithelium at the anterior portion of the pituitary gland. Remnants of this structure, sometimes called the pharyngeal pituitary, are present in the roof of the nasopharynx in 95% to 100% of people, most commonly at the vomerosphenoidal articulation (40–42). These consist of nests of cytologically bland small round cells located deep in the mucosa or underlying periosteum that range in size from 0.2 to 6 mm (Fig. 15.31). Such remnants may be the source of ectopic pituitary adenomas that arise in the nasopharynx, although most pituitary adenomas that involve the nasopharynx do so via invasion from the pituitary fossa (43,44). The pharyngeal bursa is an embryonic structure situated posterior to Rathke pouch. Remnants of this bursa are found in 3% of adults (45). These remnants can come to clinical attention with formation of cysts that are separated from the adenoid by a fibrous membrane, clinically known as Tornwaldt cysts (46) (Fig. 15.32). The cranial end of the embryonic notochord is also closely associated with the roof of the developing nasopharynx (47). Rarely, notochordal remnants may also be present in the submucosa of the nasopharynx (48). While these notochordal remnants may give rise to primary nasopharyngeal chordomas, most chordomas involving the nasopharynx represent extension from skull base tumors (47).

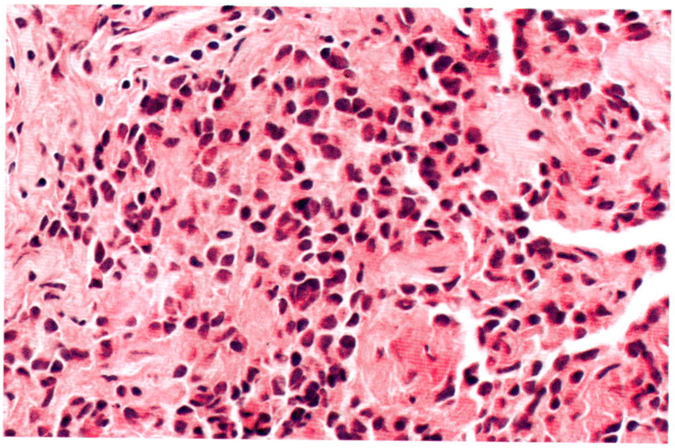

FIGURE 15.31 Nests of ectopic pituitary cells present in the superior portion of the nasopharynx.

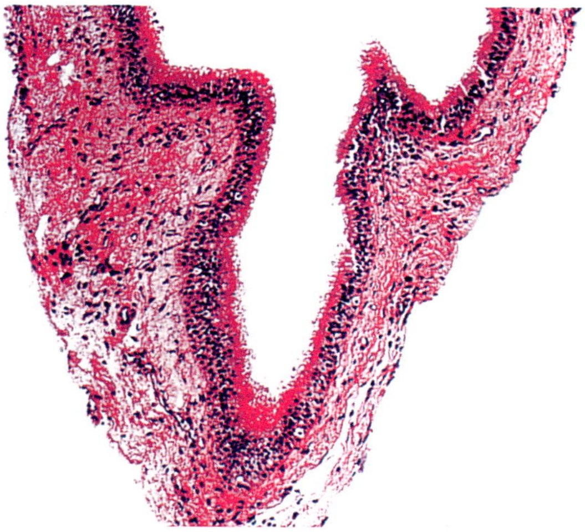

FIGURE 15.32 Nasopharyngeal cysts, also known as Tornwaldt cysts, are rimmed by fibrous tissue and lined by ciliated columnar epithelium.

Neural, Vascular, and Lymphatic Components

The pharyngeal plexus is formed by the union of the pharyngeal branches of the vagus and glossopharyngeal nerves and provides almost all of the nerves supplying the muscle of the pharynx. However, the external laryngeal nerve, a separate branch of the vagus nerve that primarily supplies the larynx, may provide a portion of innervation of the inferior constrictor muscle.

The ascending pharyngeal artery runs upward along the posterior lateral pharyngeal wall and provides the blood supply to the superior portion of the pharynx (Fig. 15.33). Branches from the superior and inferior thyroid arteries provide the blood supply to the inferior portion of the pharynx. The veins draining the pharynx merge posteriorly to form the pharyngeal plexus, which drains into the pterygoid plexus and superior and inferior thyroid veins.

Lymphatic vessels from the nasopharyngeal roof and posterior wall join in the midline, pass through the pharyngeal fascia, and split to the right or left retropharyngeal

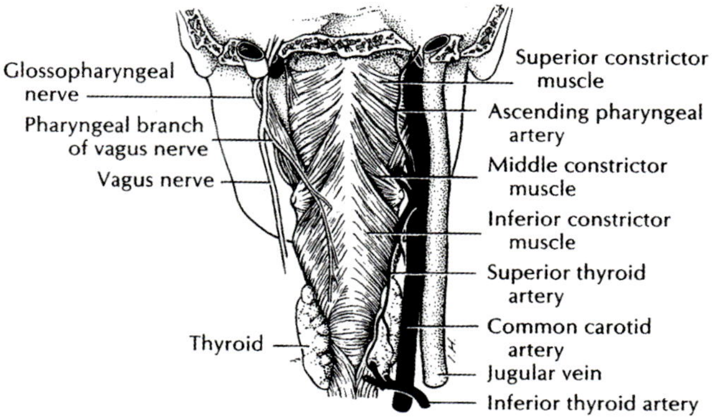

FIGURE 15.33 Posterior view of the pharynx with emphasis on vascular and neural components.

lymph nodes with a small number terminating in the highest lymph nodes of the internal jugular and spinal chains. The soft palate lymphatic vessel drainage converges in a group of lymph nodes below the anterior belly of the digastric muscle in front of the jugular chain. Lymphatic vessels from the tonsil drain through the lateral wall of the pharynx and terminate in subdigastric lymph nodes anterior to the jugular chain. The hypopharynx also contains numerous lymphatic vessels that converge at an orifice in the thyrohyoid membrane and ramify into several trunks that drain into the internal jugular chain lymph nodes.

REFERENCES

1. Bocca E, Pignataro O, Mosciaro O. Supraglottic surgery of the larynx. *Ann Otol Rhinol Laryngol.* 1968;77(6):1005–1026.
2. Hartnick CJ, Cotton RT. Congenital laryngeal anomalies. Laryngeal atresia, stenosis, webs, and clefts. *Otolaryngol Clin North Am.* 2000;33(6):1293–1308.
3. Stell PM, Gudrun R, Watt J. Morphology of the human larynx. III. The supraglottis. *Clin Otolaryngol Allied Sci.* 1981;6(6):389–393.
4. Andrea M, Guerrier Y. The anterior commissure of the larynx. *Clin Otolaryngol Allied Sci.* 1981;6(4):259–264.
5. Stell PM, Gregory I, Watt J. Morphometry of the epithelial lining of the human larynx. I. The glottis. *Clin Otolaryngol Allied Sci.* 1978;3(1):13–20.
6. Stell PM, Gregory I, Watt J. Morphology of the human larynx. II. The subglottis. *Clin Otolaryngol Allied Sci.* 1980;5(6):389–395.
7. Hopp ES. The development of the epithelium of the larynx. *Laryngoscope.* 1955;65(7):475–499.
8. Scott GB. A quantitative study of microscopical changes in the epithelium and subepithelial tissue of the laryngeal folds, sinus and saccule. *Clin Otolaryngol Allied Sci.* 1976;1(3):257–264.
9. Tucker J, Vidic B, Tucker GF Jr, Stead J. Survey of the development of laryngeal epithelium. *Ann Otol Rhinol Laryngol.* 1976;85(5 Suppl 30 Pt 2):1–16.
10. Busuttil A. Dendritic pigmented cells within human laryngeal mucosa. *Arch Otolaryngol.* 1976;102(1):43–44.
11. Goldman JL, Lawson W, Zak FG, Roffman JD. The presence of melanocytes in the human larynx. *Laryngoscope.* 1972;82(5):824–835.
12. Singhi AD, Arnold CA, Lam-Himlin DM, et al. Targeted next-generation sequencing supports epidermoid metaplasia of the esophagus as a precursor to esophageal squamous neoplasia. *Mod Pathol.* 2017;30(11):1613–1621.
13. Nunez DA, Astley SM, Lewis FA, Wells M. Human papilloma viruses: a study of their prevalence in the normal larynx. *J Laryngol Otol.* 1994;108(4):319–320.
14. Fakhry C, Westra WH, Wang SJ, et al. The prognostic role of sex, race, and human papillomavirus in oropharyngeal and nonoropharyngeal head and neck squamous cell cancer. *Cancer.* 2017;123(9):1566–1575.
15. Morshed K, Polz-Dacewicz M, Szymanski M, Polz D. Short-fragment PCR assay for highly sensitive broad-spectrum detection of human papillomaviruses in laryngeal squamous cell carcinoma and normal mucosa: clinico-pathological evaluation. *Eur Arch Otorhinolaryngol.* 2008;265(Suppl 1):S89–S96.
16. Nassar VH, Bridger GP. Topography of the laryngeal mucous glands. *Arch Otolaryngol.* 1971;94(6):490–498.
17. Gallagher JC, Puzon BQ. Oncocytic lesions of the larynx. *Ann Otol Rhinol Laryngol.* 1969;78(2):307–318.
18. Lundgren J, Olofsson J, Hellquist H. Oncocytic lesions of the larynx. *Acta Otolaryngol.* 1982;94(3-4):335–344.
19. Wenig BM. Necrotizing sialometaplasia of the larynx. A report of two cases and a review of the literature. *Am J Clin Pathol.* 1995;103(5):609–613.
20. Hill MJ, Taylor CL, Scott GB. Chondromatous metaplasia in the human larynx. *Histopathology.* 1980;4(2):205–214.
21. Iyer PV, Rajagopalan PV. Cartilaginous metaplasia of soft tissues in the larynx. Case report and literature review. *Arch Otolaryngol.* 1981;107(9):573–575.
22. Bone RC, Biller HF, Irwin TM. Intralaryngotracheal thyroid. *Ann Otol Rhinol Laryngol.* 1972;81(3):424–428.
23. Richardson GM, Assor D. Thyroid tissue within the larynx. Case report. *Laryngoscope.* 1971;81(1):120–125.
24. Sanders I, Wu BL, Mu L, Li Y, Biller HF. The innervation of the human larynx. *Arch Otolaryngol Head Neck Surg.* 1993;119(9):934–939.
25. Wu BL, Sanders I, Mu L, Biller HF. The human communicating nerve. An extension of the external superior laryngeal nerve that innervates the vocal cord. *Arch Otolaryngol Head Neck Surg.* 1994;120(12):1321–1328.
26. Kleinsasser O. [The inferior laryngeal glomus. A nonchromaffin paraganglion, unknown so far, of the structure of the so-called carotid gland in human larynx.] *Arch Ohren Nasen Kehlkopfheilkd.* 1964;184:214–224.
27. Lawson W, Zak FG. The glomus bodies ("paraganglia") of the human larynx. *Laryngoscope.* 1974;84(1):98–111.
28. Sumida S, Masuda Y, Watanabe S, Nishizaki K, Słipka J. Development of the pharynx in normal and malformed fetuses. *Acta Otolaryngol Suppl.* 1994;517:21–26.
29. Batsakis JG, Sneige N. Parapharyngeal and retropharyngeal space diseases. *Ann Otol Rhinol Laryngol.* 1989;98(4 Pt 1):320–321.
30. Arambula A, Brown JR, Neff L. Anatomy and physiology of the palatine tonsils, adenoids, and lingual tonsils. *World J Otorhinolaryngol Head Neck Surg.* 2021;7(3):155–160.
31. Tsao SW, Yip YL, Tsang CM, et al. Etiological factors of nasopharyngeal carcinoma. *Oral Oncol.* 2014;50(5):330–338.
32. Gelwan E, Malm IJ, Khararjian A, Fakhry C, Bishop JA, Westra WH. Nonuniform distribution of high-risk human papillomavirus in squamous cell carcinomas of the oropharynx: rethinking the anatomic boundaries of oral and oropharyngeal carcinoma from an oncologic HPV perspective. *Am J Surg Pathol.* 2017;41(12):1722–1728.
33. Begum S, Cao D, Gillison M, Zahurak M, Westra WH. Tissue distribution of human papillomavirus 16 DNA integration in patients with tonsillar carcinoma. *Clin Cancer Res.* 2005;11(16):5694–5699.
34. Lewis JS Jr, Beadle B, Bishop JA, et al. Human papillomavirus testing in head and neck carcinomas: guideline from the College of American Pathologists. *Arch Pathol Lab Med.* 2018;142(5):559–597.
35. Westra WH. The pathology of HPV-related head and neck cancer: implications for the diagnostic pathologist. *Semin Diagn Pathol.* 2015;32(1):42–53.
36. Benke TT, Zitsch RP 3rd, Nashelsky MB. Bilateral oncocytic cysts of the nasopharynx. *Otolaryngol Head Neck Surg.* 1995;112(2):321–324.

37. Morin GV, Shank EC, Burgess LP, Heffner DK. Oncocytic metaplasia of the pharynx. *Otolaryngol Head Neck Surg.* 1991;105(1):86–91.
38. Li JJX, Ng JKM, Chan ABW. Clinicopathological features of melanotic and non-melanotic oncocytic lesions of the nasopharynx. *Pathology.* 2019;51(6):600–604.
39. Ciris IM, Erkilinc G, Bozkurt KK, Karahan N, Yasan H, Sivrice ME. Cartilaginous choristomas in tonsillectomy specimen: A prospective analysis. *Int J Pediatr Otorhinolaryngol.* 2019;122:191–195.
40. Boyd JD. Observations on the human pharyngeal hypophysis. *J Endocrinol.* 1956;14(1):66–77.
41. McGrath P. Extrasellar adenohypophyseal tissue in the female. *Australas Radiol.* 1970;14(3):241–247.
42. Melchionna RH, Moore RA. The pharyngeal pituitary gland. *Am J Pathol.* 1938;14(6):763–772.1.
43. Kikuchi K, Kowada M, Sasaki J, Sageshima M. Large pituitary adenoma of the sphenoid sinus and the nasopharynx: report of a case with ultrastructural evaluations. *Surg Neurol.* 1994;42(4):330–334.
44. Langford L, Batsakis JG. Pituitary gland involvement of the sinonasal tract. *Ann Otol Rhinol Laryngol.* 1995;104(2):167–169.
45. Hollender AR. The nasopharynx; a study of 140 autopsy specimens. *Laryngoscope.* 1946;56:282–304.
46.. Miyahara H, Matsunaga T. Tornwaldt's disease. *Acta Otolaryngol Suppl.* 1994;517:36–39.
47. Binkhorst CD, Schierbeek P, Petten GJ. Neoplasms of the notochord; report of a case of basilar chordoma with nasal and bilateral orbital involvement. *Acta Otolaryngol.* 1957;47(1):10–20.
48. Bonneville JF, Belloir A, Mawazini H, et al. Calcified remnants of the notochord in the roof of the nasopharynx. *Radiology.* 1980;137(2):373–377.

16

Major Salivary Glands

Elizabeth M. Kurian ▪ Justin A. Bishop ▪ Doreen N. Palsgrove

INTRODUCTION

Salivary glands are part of the exocrine system of glands, secreting substances into ducts that empty onto the surface of target tissues. Their primary function is to produce and secrete saliva. Saliva within the oral cavity facilitates speech, preserves oral hygiene, allows taste, mastication, and moistens the oral cavity/aerodigestive tract (1). There are three pairs of major salivary glands: parotid glands, submandibular (or submaxillary) glands, and sublingual glands. They are connected symmetrically to the oral cavity, where they empty their secretion only under specific stimuli. The quality of the saliva produced by the major glands is variable and depends on both the stimuli (e.g., gustatory, olfactory, mechanical, and/or neurohormonal influences) and the predominant participating gland (2). Most of the stimulated saliva is produced by the parotid gland, while baseline/resting state saliva is produced by the submandibular and sublingual glands (3,4). The parotid gland is composed of serous acini, so it secretes predominantly watery saliva, whereas the sublingual and submandibular glands secrete viscous saliva: the sublingual gland secretes a greater amount of mucoid than serous fluid and the submandibular gland produces more serous than mucoid saliva (5).

A small amount of saliva is continuously produced by the minor salivary glands throughout the oral cavity, pharynx, and upper airways (1,3,6). Seventy percent of salivary mucins, a large amount of immunoglobulins (mainly IgA), salivary acid phosphatase, and lysozymes are produced by minor salivary glands (6). These glands are usually submucosal, composed primarily of mucous acini with loose connective tissue, and lack a true capsule (6,7). They are characterized by a complex ductal system like that of the major glands but they have shorter tracts. Of note, there are specific minor glands at the base of the circumvallate papilla of the tongue called Von Ebner glands, which are uniquely composed of serous acini (6). Other notable agglomerated salivary tissue occurs in the nasopharynx and around the Eustachian tube, recently referred to as tubarial glands. The predominance of mucous acinar cells and notable absence of striated and intercalated ducts makes them most comparable to the minor salivary glands of the palate, but some suggest reclassification as a separate organ (8–12).

Saliva is multifunctional. The saliva produced by major salivary glands (approximately 90% of the daily 1.5-L production) plays an important role in preparing food for digestion by maintaining ionic composition, lubrication, and controlling oral bacterial flora (3). Secretions also include a variety of nucleic acids, peptides, proteins, ions, metabolites, and microorganisms (13–15). Viruses (e.g., mumps, enteric viruses, hepatitis C and D viruses, severe acute respiratory syndrome coronavirus 2, human immunodeficiency virus, human cytomegalovirus, rabies) may also directly infect and replicate in salivary gland tissue (16–22). Hence, the overall physiologic state of an individual can be noninvasively evaluated by assessing the composition of salivary secretions (2,3,23–25).

EMBRYOLOGIC AND POSTNATAL DEVELOPMENTAL CHANGES

Branching Morphogenesis

Salivary gland primordia are formed by interactions among stem cells and progenitor cells. As the primordia grow, they ramify into a bush-like system surrounded by mesenchymal tissue. The mesenchyme and the basal lamina produced by salivary gland epithelium, play an important role in the timely branching of ducts, cytologic differentiation within the extracellular matrix milieu of growth factors, and lobular organization of the gland (6,26). Animal models have provided much-needed insights into this developmental process, known as branching morphogenesis, by which arborized structures with a high surface-to-volume ratio are formed in vertebrate organs (e.g., salivary gland, lung, and kidney). The mouse salivary gland, specifically the submandibular gland, is the most frequently used model that has contributed most of our knowledge regarding the general process of branching morphogenesis (27–29). However, differences in progenitor cell types and mechanisms regulating cell maturation may exist in the parotid and sublingual glands. Major salivary glands develop at the back of the mouth from epithelial buds that form behind the molar tooth germs (30).

Branching morphogenesis begins as a thickening of oral epithelium at the site of gland development during pre-bud stage, followed by ingrowth of the epithelium into underlying mesenchymal tissue (26). During the bud stage, the epithelial rudiment continues to grow, forming an elongated cord that terminates in a globular mass of cells referred to as the primary end-bud. Differential growth from the tip of the end-bud leads to the initial branching of the rudiment. Additional branching buds with stalks form at the distal end and early lumen formation begins via programmed cell death apoptosis during the pseudoglandular stage (6). Increased degradation of the basement membrane is mediated by production of hyaluronidase in the surrounding mesenchyme and occurs at the distal ends of growing branches. These changes are accompanied by a concomitant shift in the pattern of glycosaminoglycan synthesized by branching ducts. During early glandular development, hyaluronic acid and chondroitin sulfate are the predominant glycosaminoglycans produced. However, the pattern of glycosaminoglycan synthesis shifts with decreasing amounts of hyaluronic acid and chondroitin sulfate and increasing amounts of heparan sulfate as branching continues and becomes more pronounced (31). Several studies have shown differential expression of fibronectin, laminin, gamma 2, and TIMP-3 in developing buds (32), while early secretion of glycoproteins MUC1 and MUC4, cell polarization, and the presence of electrolytic fluid regulate luminal expansion (6). Duct and bud growth continue throughout the canalicular stage until the final branched duct, lumina, and acini with surrounding myoepithelial cells are formed during terminal bud stage (6). Extracellular matrix, myoepithelial cells, myofibroblasts, immune cells, endothelial cells, and nerve fibers surround acinar cells. Development continues in all salivary glands, with fully functional glands at birth (33).

Alterations in pathways involved in branching morphogenesis can result in a variety of genetic disorders involving salivary glands. For example, mutations in the signaling components of the ectodysplasin pathway lead to hypohidrotic ectodermal dysplasia (HED) which is characterized by missing and malformed teeth, sparse hair, and reduced sweating and salivary flow (34,35). The participation of the fibroblastic growth factor receptor (FGFR) signaling pathway is also essential to salivary gland development (29). Germline mutations in *FGF10* may result in aplasia of lacrimal and salivary glands (ALSG), an autosomal dominant disorder characterized by atresia, hypoplasia, or aplasia of the lacrimal and salivary glands (36). Mutations in *FGFR2* result in lacrimo-auriculo-dento-digital (LADD) syndrome, which is a multiple congenital anomaly disorder mainly affecting lacrimal and salivary glands and their ducts, ears, teeth, and distal limb segments (37).

Parotid Gland

The parotid gland is the first of the three major glands to appear during embryogenesis and is seen by the 6th week of gestation (1,6). It derives from the ectoderm as an epithelial bud from the primitive floor of mouth epithelium at the angle between the maxillary process and the mandibular arch (26). In the 7th week, the gland transitions to

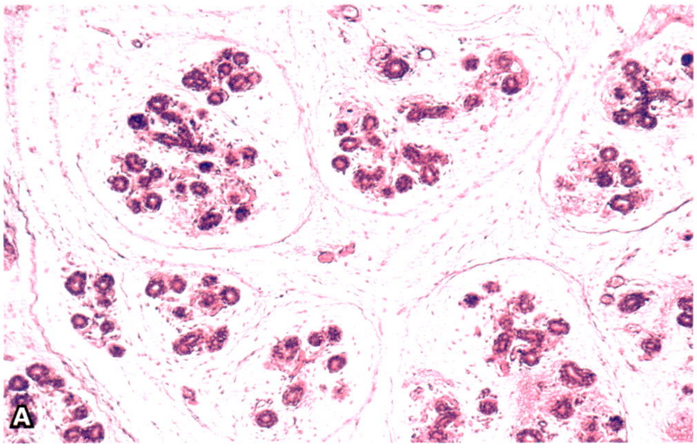

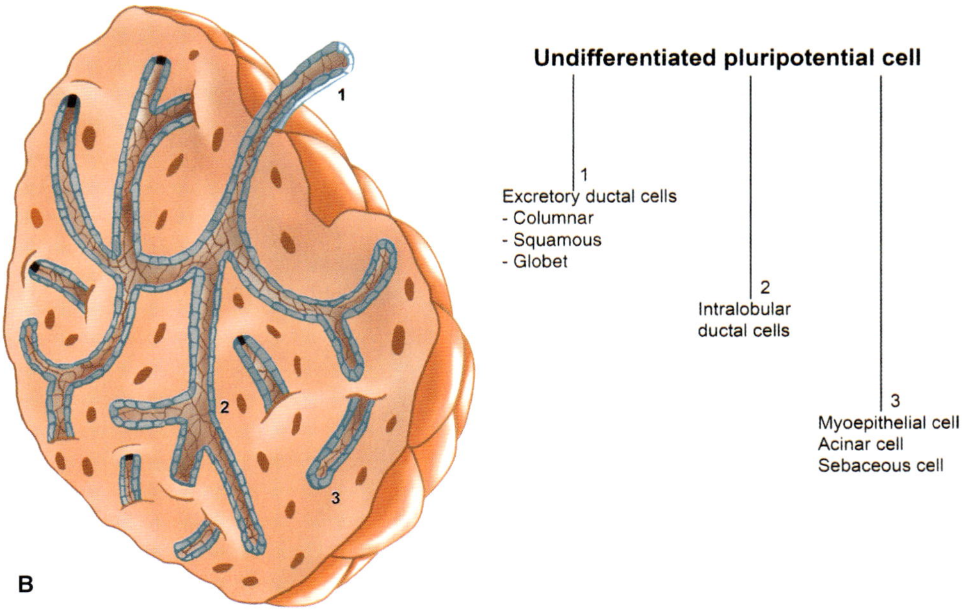

FIGURE 16.1 Development and histogenesis of the epithelial constituents of the salivary gland. **A:** Fetal parotid gland at 4 months; a lobular architecture is present. **B:** Schematic representation of the gland showing excretory ducts: (*1*) secretory ducts; (*2*) intralobular ducts; (*3*) and acini; the cellular lines of differentiation.

the preauricular region (26). The facial nerve is created in the 10th week and arbitrarily divides the gland into superficial and deep lobes (26). By the 12th week, the gland has attained its general pattern of organization. The epithelial structures are arranged in lobules limited by a capsule of loose connective tissue (Fig. 16.1). The mesenchyme is then colonized by numerous lymphocytes that later develop into intraglandular and extraglandular lymph nodes (26). The epithelial bud undergoes rounds of branching morphogenesis with terminal differentiation of end buds such that secretory lobules and acini are apparent by week 14 (33,38,39). By 24 weeks, canalization occurs with cell differentiation of the excretory ducts and progressive transformation of ciliated cells to columnar, squamous, and goblet cells (40). Intralobular duct and acinar differentiation, including myoepithelial cell formation, begins around the 8th month and myoepithelial cell differentiation by the 19th-to-24th week period. Myoepithelial cells are arranged in the basal portions of the acini and intercalated ducts where they appear as clear cells by electron microscopy but they become flattened and show cytoplasmic prolongations between the 25th and 32th gestational weeks (41). Myoepithelial markers, including smooth muscle actin (SMA), calponin, S-100 protein, and p63 protein, may be detected by immunohistochemical methods at the earliest stages of development (42). Secretion in the parotid gland may begin as early as 18 weeks (43), but several studies in rodents suggest that full maturation is completed only after birth (26).

The final location of the parotid gland is behind the maxillary branch of the inferior facial nerve, below and in front of the external ear. It is enclosed within a fibroadipose capsule in a depression termed the parotid space that is delimited by the masseter muscle anteriorly, the zygomatic arch superiorly, the tragus posteriorly, the mastoid process posteroinferiorly, the anterior border of the sternocleidomastoid muscle inferiorly, the posterior belly of the digastric and stylohyoid muscles medially, and skin and superficial musculoaponeurotic system fascia laterally (44). The parotid gland contains the retromandibular vein, the external carotid artery, and the facial nerve. The retromandibular vein marks the arbitrary division of the deep and superficial lobes and approximates the depth of the facial nerve on imaging studies.

The adult parotid gland is the largest of the three major salivary glands and weighs between 14 and 28 g (26). The facial nerve arbitrarily divides the gland into superficial and deep lobes (38). The main portion, or superficial lobe, is flattened along the lateral aspect with a quadrilateral shape while the deep lobe is wedge-shaped. The deep lobe is positioned between the mastoid tip, bony and cartilaginous external auditory canal, the ramus of the mandible, the posterior border of the medial pterygoid muscles, and the prestyloid region of the parapharyngeal space (45). Part of the deep lobe lies in the parapharyngeal space. The anterior extension of the parotid gland along the course of the parotid duct is designated the facial process and maintains continuity with the main gland (46).

A normal parotid gland cannot be palpated. Since most of the gland is superficial, it may become palpable with pathologic processes at the angle of the mandible. The superficial lobe is also where most salivary tumors develop, which permits conservative surgical treatment of many parotid gland neoplasms. Tumors arising in the deep lobe may present with asymptomatic preauricular, posterior cheek swelling or as swelling of the lateral pharyngeal wall, tonsil, or palate (47).

The surgical anatomic area where the parotid gland is located is referred to as the parotid region. It is important to keep in mind the anatomical relation between the facial nerve, the gland, and the subcutaneous planes. Although the adult parotid gland is covered by superficial musculoaponeurosis and skin, there is no natural plane between the parotid gland and the overlying skin (48,49); therefore, surgical exposure requires separation in the subcutaneous plane just beneath hair follicles (39). There are also notable differences in the extracranial course of the facial nerve between children and adults, largely because of the relative absence of the mastoid process in young children. This results in the facial nerve lying more superficially just deep to subcutaneous tissue. As the mastoid air cell system increases in size and volume postnatally, the facial nerve subsequently takes a more medial and protected course behind the mastoid tip and tragal cartilage (39). A deep lamina invests the medial portion of the gland to cover the adjacent styloid muscles and forms the stylomandibular ligament, a fusion of the superficial and deep layers of the deep cervical fascia to the parotid gland. The stylomandibular ligament separates the parotid gland tail from the posterior submandibular space (44).

Accessory parotid tissue is a nodule separated from the main gland with its own blood supply and independent drainage. Previously known as Socia parotidis (44), it is considered a normal anatomic variant found on the masseter muscle and/or associated with the parotid duct, with or without aplasia of the main ipsilateral parotid gland (38,50–52). The incidence of this finding in the general population is 20% to 56% with no differences in laterality or distribution by sex (26,53).

The parotid gland is composed of numerous ducto-acinar units connected through excretory ducts to the main parotid duct (i.e., Stensen duct) located in the anterior portion of the gland. The parotid duct crosses over the masseter muscle and, at the edge of the muscle, turns inward ~90 degrees to the buccinator muscle/fat pad before opening into the oral vestibule at the level of the second maxillary molar tooth via a small papilla (33,44,50). The papilla is a small in-folding of mucosal tissue at the duct opening and limits retrograde air or secretions into the duct (44). The buccal branch of the facial nerve is intimately associated with Stensen duct.

Although the parotid gland is the first major salivary gland to develop, it is the last major salivary gland to be encapsulated by fascia, which allows the inclusion of lymph nodes (intraparotid and extraglandular) and lymphatic channels. The submandibular and sublingual glands, on the other hand, undergo encapsulation prior to the development of their regional lymphatic systems (44). Superficial parotid lymph nodes receive drainage from the auricle, preauricular skin, eyelids, and lacrimal glands, while deep intraparotid lymph nodes receive drainage from the parotid gland, soft palate, nasopharynx, middle ear, and external auditory meatus (39,44). Lastly, the temporal scalp, upper face, and anterior pinna drain into the preauricular lymph nodes in the superficial fascia (39). Tumors from any of these locations can metastasize to their respective draining intraparotid or periparotid lymph nodes.

Blood to the parotid gland is supplied the superficial temporal artery and posterior auricular artery, both branches of the external carotid artery (44). Deoxygenated blood is drained from the parotid gland via the maxillary and superficial temporal veins, which form the retromandibular vein and empties into the external jugular vein (44).

Innervation is derived from sympathetic and parasympathetic nerve fibers. Sympathetic stimulation increases the secretion of proteins and is supplied by postganglionic sympathetic fibers from the superior cervical ganglion of the sympathetic chain traveling with the external carotid artery to the parotid gland (54). Parasympathetic innervation increases the serous (watery and ion-rich) salivary secretion. The glossopharyngeal nerve (cranial nerve IX) carries preganglionic parasympathetic fibers to the otic ganglion from which postganglionic parasympathetic fibers travel to the parotid gland via the auriculotemporal nerve, which is tributary of the mandibular division of the trigeminal nerve (cranial nerve V) (33,55). Communicating rami, or bypass fibers, of the auriculotemporal nerve join the facial nerve within the parotid gland and provide sensory fibers for the zygomatic, buccal, and mandibular divisions of the facial nerve to the skin (56). If perineural tumor involves this point of the facial nerve, it may be a bridge to involvement of the trigeminal nerve (44,56). The great auricular nerve is a cutaneous sensory branch of the cervical spine plexus (arising from C2 to C3) that supplies the skin overlying the parotid gland, the parotid fascia, mastoid process, and auricle (44). It emerges from the mid-portion of the posterior border of the sternocleidomastoid muscle, which is known as the "punctum nervosum" or Erb point (44,57). Any distension of the parotid fascia (either inflammation, tumor, etc.) results in pain transmission to the cervical spinal cord (44). As an aside, the accessory nerve is located 1 cm superior to Erb point and is easily damaged when level V lymph nodes are biopsied (57). This site is a common site of injury to a cranial nerve and iatrogenic major nerve injury overall (57).

Submandibular Gland

The submandibular (submaxillary) gland begins developing at the end of the 6th week in the groove between the lower jaw and tongue (26). Lineage tracing studies have recently shown that all major salivary glands, including the submandibular gland, are derived from ectoderm, which would explain submandibular gland involvement in HED (30,58,59). Striated and intercalated ducts are completed between 13 and 16 weeks, while acinar development, composed of mixed mucous and serous acinar cells, is completed by 20 to 24 weeks (33). The submandibular gland extends backward beneath the jaw as it develops, wrapped in the fibrous superficial layer of the deep cervical fascia (finely encapsulated) within the submandibular triangle. The submandibular triangle is bounded by the inferior border of the mandibular body and the anterior and posterior bellies of the digastric muscle (3). The submandibular gland is U-shaped with the anterior portion of the gland forming an uncinate process as it wraps around the posterior edge of the mylohyoid muscle laterally and medially, while the deep aspect of the gland rests on the hyoglossus muscle (60,61). Superficially, the gland is covered by platysma muscle followed by subcutaneous fat and skin. The marginal mandibular nerve (branch of cranial nerve VII) passes just superficial to the fascia of the submandibular gland and deep to the plane of the platysma (62).

The submandibular glands are the second largest salivary glands, weighing 7 to 15 g (6,26). The acini are mixed type with predominantly serous cells and some mucous cells. In contrast to the parotid gland, the intercalated ducts are shorter, the striated ducts are more arborized, and there is no lymph node formation inside the gland (6).

The submandibular duct (i.e., Wharton duct) is the main excretory duct of the submandibular gland and is approximately 5 cm in length. It empties into the oral cavity via a narrow opening in the summit of a papilla termed the sublingual caruncle, which is located at the base of the frenulum linguae on each side (26,33). Saliva within the submandibular gland maintains an alkaline pH, a condition conducive to calcium salt formation. Congenital failure of submandibular duct canalization is a rare developmental anomaly that can lead to saliva retention posterior to an imperforate orifice in the oral cavity and a bluish cyst-like swelling in the mouth floor (63,64). Surgical marsupialization with or without ductoplasty is typically indicated to avoid sequelae associated with an imperforate orifice, such as glandular atrophy, sialadenitis, airway compromise, and feeding difficulties (65).

The lymph of the submandibular gland is carried to lymph nodes located along the bony, deep, and cutaneous surfaces of the gland within the submandibular triangle (level IB) (33,66). Level IB nodes also receive drainage from the submental lymph nodes (level IA), the lower anterior nasal cavity, the hard and soft palate, the maxillary and mandibular alveolar ridges, the floor of mouth, the cheek, the upper and lower lips, and most of the anterior tongue (67,68). Nodes in level IB are, therefore, at risk of developing metastases from a variety of anatomic sites. In contrast to the parotid gland, however, metastatic tumor cannot spread via lymphatics directly to the submandibular gland itself because there are no intraglandular lymph nodes (69).

Blood is primarily supplied to the submandibular gland via facial artery and submental artery and partly by the lingual and external jugular arteries (70). Venous drainage primarily occurs through the anterior facial vein, venae comitantes of the facial artery, and hilum vein close to the submandibular duct (70).

Parasympathetic innervation of the submandibular gland is not only responsible for the secretion of saliva, but it also plays a main role in the development and regeneration of the gland (71,72). Parasympathetic fibers carried by the facial nerve (i.e., cranial nerve VII) join the lingual nerve, which is a branch of the marginal mandibular division of the trigeminal nerve, and synapse at the submandibular ganglion (54). Short postsynaptic parasympathetic fibers then leave the ganglion to innervate the submandibular gland (6). Sympathetic innervation regulates secretions through vasoconstriction and is derived from postganglionic sympathetic fibers arising from the superior cervical ganglion of the sympathetic chain traveling with the external carotid artery to the submandibular gland (54).

Sublingual Gland

The sublingual gland is the last and smallest of the three major salivary glands to develop between 7 and 8 weeks of gestation (6). The head of the gland is the lesser sublingual gland and lies superficially in the anterior floor of the mouth, while the greater sublingual gland is formed by the tail and runs posteriorly to contact the submandibular gland (73).

Primordia are located immediately lateral to the submandibular glands for the greater sublingual glands and in the linguogingival sulcus for the lesser sublingual glands. The epithelial buds grow downward from the floor of mouth, between the lower jaw and the tongue, for duct and acinar development (66). The epithelial buds grow downward from the groove between the lower jaw and the tongue and, similar to the parotid and submandibular glands, are ectoderm-derived (30). Their final positions are anterior and superior to the submandibular glands (26,66).

The sublingual gland is the only unencapsulated major salivary gland, weighing approximately 3 g (6,26,66,74). It sits within a V-shaped sublingual space and lacks a fascial lining. The sublingual space is delimited by the mucosa of the floor of mouth superiorly, mylohyoid muscle inferiorly, medial surface of the mandible laterally, and genioglossus and geniohyoid muscles medially. The submandibular duct, deep portion of the submandibular gland, lingual nerve, and lingual artery and vein also lie within the sublingual space (73). The major submandibular gland is positioned medial to the sublingual fossa (6,74) with 8 to 30 surrounding minor sublingual glands (66). The greater sublingual gland, when present, drains via the major sublingual duct, termed Bartholin duct, which either joins the submandibular duct or opens independently at its own orifice adjacent to the submandibular duct (73,74). Numerous small ducts, the ducts of Rivinus, drain each of the discrete minor sublingual gland units by joining the submandibular duct or directly emptying onto the mucosa of the floor of mouth along the crest of the sublingual folds (73,75).

Ranulas form by extravasation of saliva after trauma to the sublingual gland or ruptured main duct. The ranula may be intraoral and confined to the floor of mouth as a mucus retention cyst or extravasation pseudocyst or cervical cyst without intraoral involvement. Cervical, or plunging, ranulas typically present as a submandibular mass with herniation through or behind the mylohyoid muscle; they may or may not extend into the parapharyngeal and retropharyngeal spaces (76,77).

Sublingual acini are mixed type with predominantly mucous acinar cells and some serous acinar cells. Unlike the other major salivary glands, intercalated and striated ducts are very short to seemingly nonexistent (66,78).

Blood comes from the sublingual and submental arteries and drains via tributaries of the external jugular system (26,66). Lymphatic drainage of the sublingual glands is to the submandibular lymph nodes and innervation is like that of the submandibular gland (66).

SECRETORY UNIT

The salivary glands are exocrine glands with numerous secretory units composed of acinar cells (serous or mucous), ductal cells, myoepithelial cells, and basal cells (Fig. 16.2). While acini are supported by myoepithelial cells, parts of the ductal system may be invested by myoepithelial or basal cells (79,80).

The lobular architecture of the glands is defined by septa of interconnected connective tissue carrying vascular and neural branches, as well as excretory ducts.

Acini

Serous acinar cells are polarized, pyramid-shaped cells with small, round, basally located nuclei and cytoplasm packed with basophilic zymogen granules (Fig. 16.3). Their cellular contents are periodic acid–Schiff (PAS) positive, resistant to diastase digestion, and unreactive with mucicarmine and Alcian blue. Granules vary in number depending on the phase of the secretory cycle (81). The primary enzyme present in zymogen granules is amylase, which splits starch into smaller water-soluble carbohydrates (82). Zymogen granules also contain agglutinin, proline-rich proteins, and histatins. Other enzymes such as nonspecific antibacterial lysozyme, lactoferrin, trypsin and chymotrypsin-like proteases, lysyl endopeptidase, and histidine peptidase have also been detected (83,84). Serous acinar cells are arranged in spherical structures with narrow central lumens, rarely visible by light microscopy, through which secretions drain into intercalated ducts. Ultrastructurally, acinar cells contain abundant endoplasmic reticulum, Golgi vesicles, mitochondria,

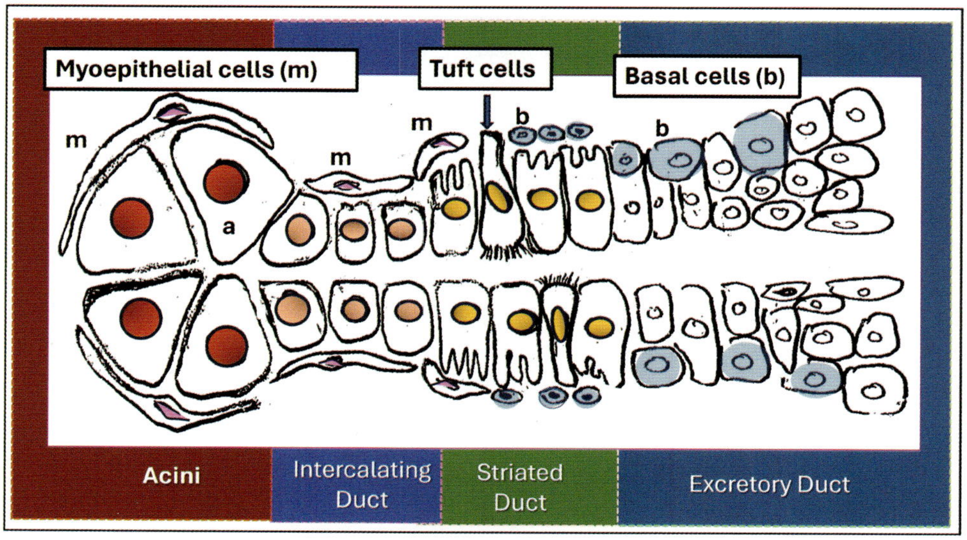

FIGURE 16.2 Illustration of salivary gland secretory unit.

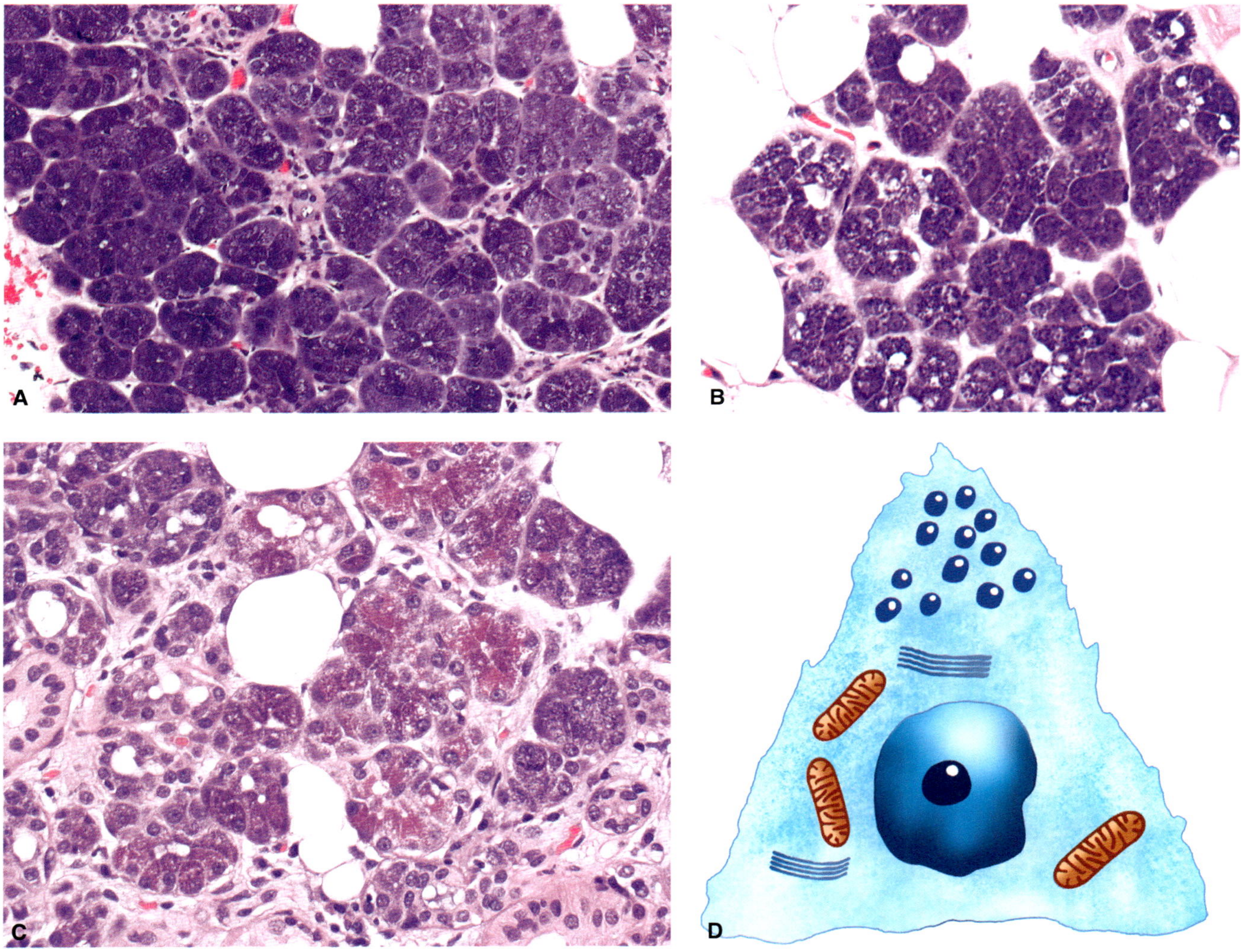

FIGURE 16.3 **A–C:** Serous acini with variable granule size and density. **D:** Schematic representation of an acinar cell.

and secretory granules. Variably sized secretory granules occupy most of the apical portion of the cytoplasmic compartment. The cells are connected by junctional complexes at the apical pole, while desmosomes attach myoepithelial cells along the basal side. Numerous microvilli protrude into the acinar lumen (81).

Mucous acinar cells have basally located nuclei and voluminous cytoplasm filled with a clear mucous substance (Fig. 16.4A). They contain variable amounts of acid sialomucins that stain blue with alcian blue and pink with mucicarmine (Fig. 16.4B) and neutral sialomucins that are magenta with PAS (Fig. 16.4C). The characteristics of these sialomucins also differ between submandibular and sublingual glands (85). Mucous acini are negative for SOX10 and S-100 protein but positive for NKX3.1 (Fig. 16.4D).

Mixed acini are found in the submandibular and sublingual glands. Mixed secretory units often exhibit a predominance of mucous cells in direct contact with the intercalated ducts with an outer, crescent-shaped border of serous cells termed a demilune (Fig. 16.5).

In diabetic patients, even with well-controlled glycemic levels, there are morphologic changes that may manifest symptomatically (86). These include increased acinar and secretory granule size, decreased number of granules, and a reduction of mitochondrial size. High-resolution electron microscopy demonstrates increased protrusions along the luminal membranes that correspond to secretory activity, as well as a significantly increased density of microbuds that represent recycling membranes, and a reduction in the density of microvilli (86). The parotid gland is less affected, suggesting the submandibular gland is more vulnerable to metabolic changes (86,87).

Ducts

The salivary gland ductal system modifies the water and electrolyte concentration of saliva and transports it from the gland to the oral cavity. The first two segments, the intercalated duct and striated duct components, are located within the lobules. Secretions drain from acini into intercalated

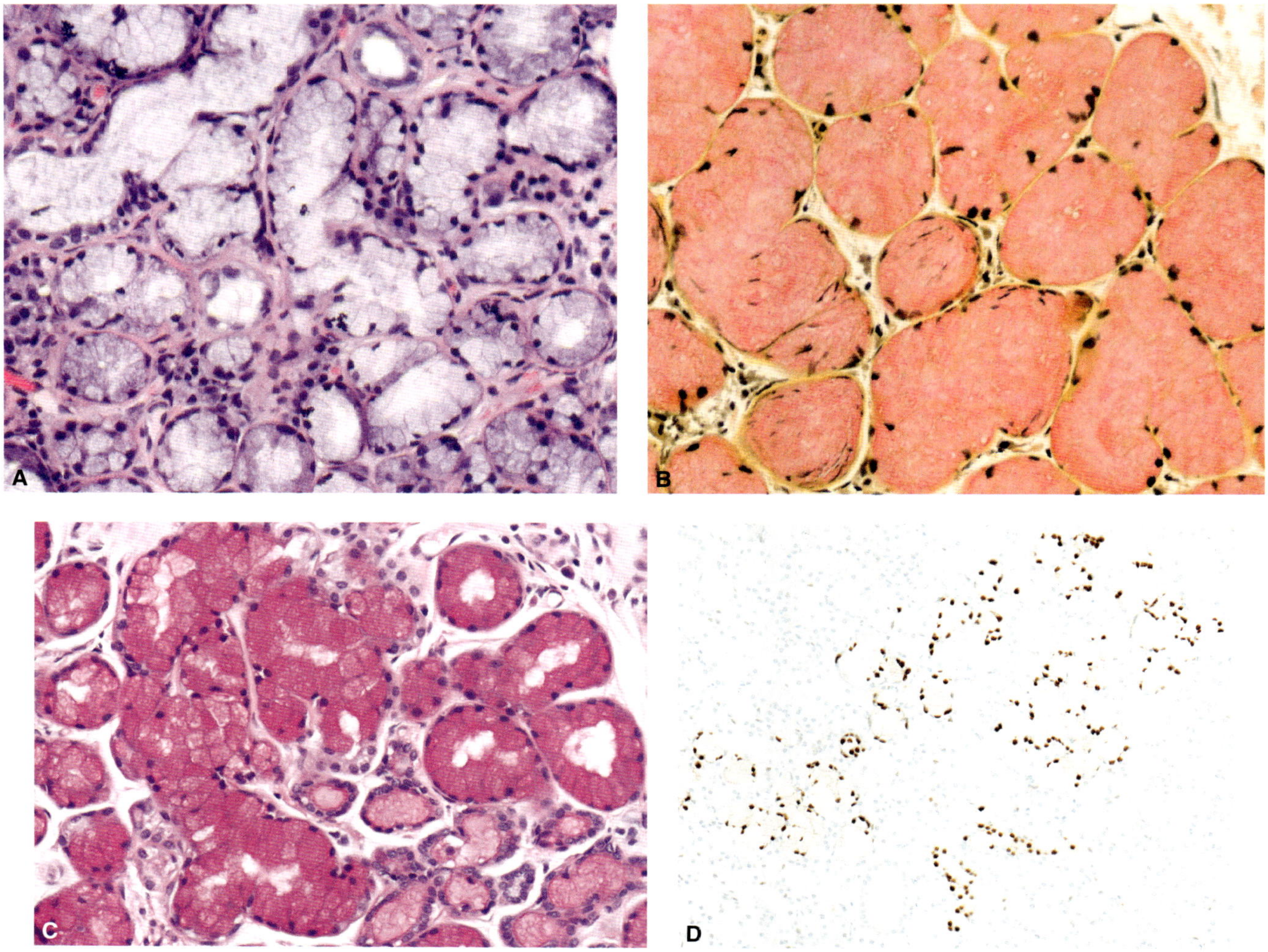

FIGURE 16.4 **A:** Mucous-type acini of the sublingual gland. **B:** Mucous acini, mucicarmine positive, highlighting acidic glycoproteins within mucous cells. **C:** Mucous acini, PAS positive (magenta), highlighting the neutral glycoproteins of the mucous cells. Note: PAS can also be positive in serous type acini. **D:** Mucous acini of the submandibular gland are positive for NKX3.1, while serous acini are negative.

ducts, followed by striated ducts, and then the interlobular excretory ducts located in the septal connective tissue between the lobules.

Intercalated ducts lie directly in contact with the acini (Fig. 16.6). They are lined by a single layer of cuboidal epithelial cells with small amounts of amphophilic to eosinophilic cytoplasm and small, round nuclei. Interestingly, recent structural and functional data indicate these cells have secretory function similar to native acinar cells (88). The cuboidal epithelium is supported by an irregular layer of myoepithelial cells (89). The lengths of the intercalated ducts vary between the three major glands: they are longest in the parotid gland (Fig. 16.7A), short in the submandibular gland (Fig. 16.7B), and hardly visible in the sublingual gland (Fig. 16.7C). Intercalated ducts join the striated ducts at various angles.

Striated ducts are lined by simple columnar epithelial cells with eosinophilic cytoplasm and round, centrally located nuclei (Fig. 16.8). Invaginations of the plasma membrane along the basal surface of the cells result in parallel vertical striations that can be seen with routine light microscopy. Ultrastructurally, the cells demonstrate numerous mitochondria and the basal folds form a complex interdigitation with the folds of adjacent cells. Scattered bottle-shaped, microvilli-projecting, chemosensory tuft cells are also present between the columnar epithelial cells in the submandibular gland (90). Striated ducts are supported by both myoepithelial and basal cells (91).

Most interlobular excretory ducts are lined by pseudostratified, columnar epithelium with occasional goblet-type mucous cells and tuft cells, lined by an abluminal basal cell layer (Fig. 16.9) (92). However, simple cuboidal, simple columnar, pseudostratified, stratified cuboidal, or stratified columnar epithelium may be present at any level of the duct (93). Proximal excretory ducts, for example, are situated within an interlobular septum and positioned at the edge of

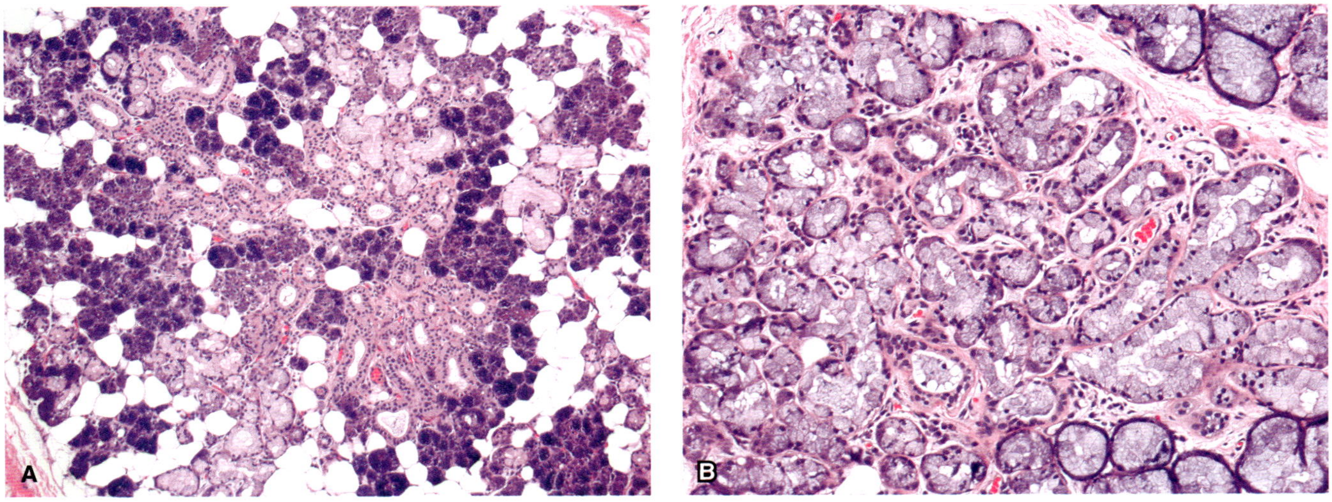

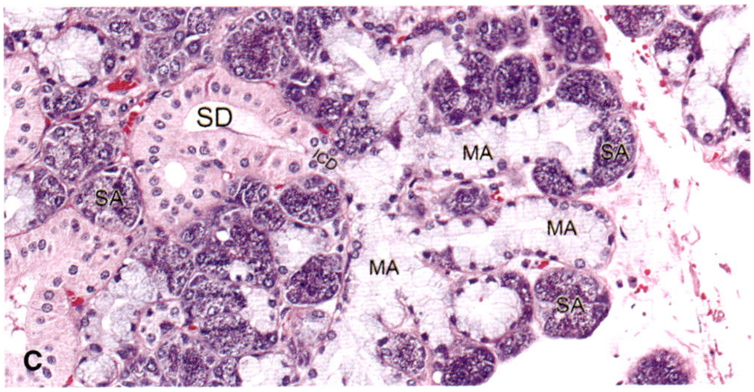

FIGURE 16.5 **A:** Mixed acini of the submandibular gland (serous > mucous). **B:** Mixed acini of the sublingual gland (mucous > serous). **C:** In mixed acini, serous acinar cells (SA) may aggregate at the periphery of the acini forming crescents or half-moons (demilunes) around mucous acinar cells (MA), which are in direct contact with the intercalated duct system. *SD*, striated duct; *ICD*, intercalated duct.

a lobule. They may exhibit characteristics of striated ducts, including basal striations. Excretory ducts become progressively larger before joining the principal duct.

Tuft cells are bottle-shaped, microvilli-projecting, cholinergic chemosensory cells. They are singly interspersed between the principal ductal cells of striated and excretory ducts (Fig. 16.2) and presumably modulate the expression of neuronal and inflammatory gene pathways (90,92,94). Ultrastructurally, they have a narrow apex that protrudes into the lumen with apical microvilli and a basally located nucleus with narrow projections toward the basal lamina (95). Although found in all major salivary glands, the ratio of tuft cells to principal cells is highest in the submandibular gland and lowest in the sublingual gland (92).

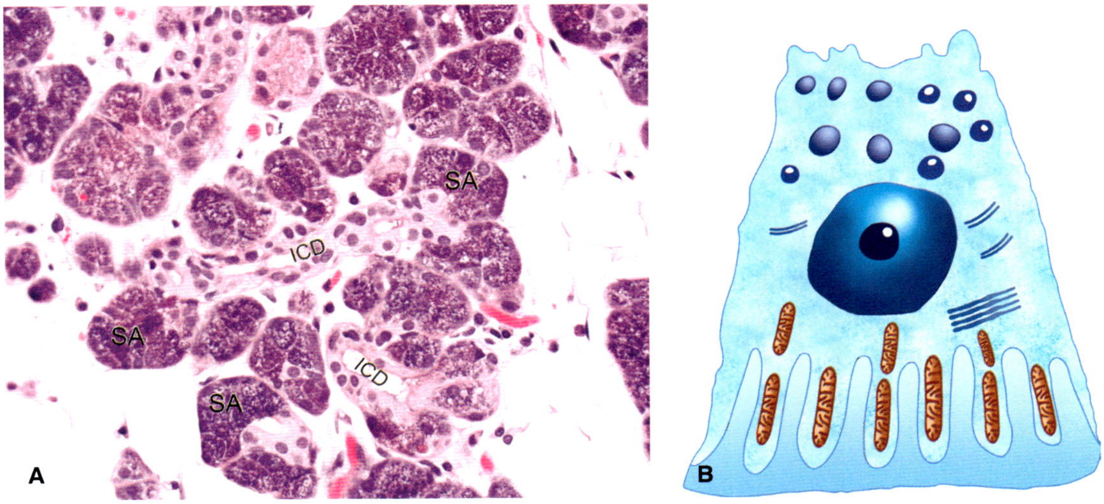

FIGURE 16.6 **A:** Intercalated ducts lie in direct contact with acinar cells. **B:** Schematic representation of a columnar cell of the striated duct. *ICD*, intercalated duct; *SA*, serous acini.

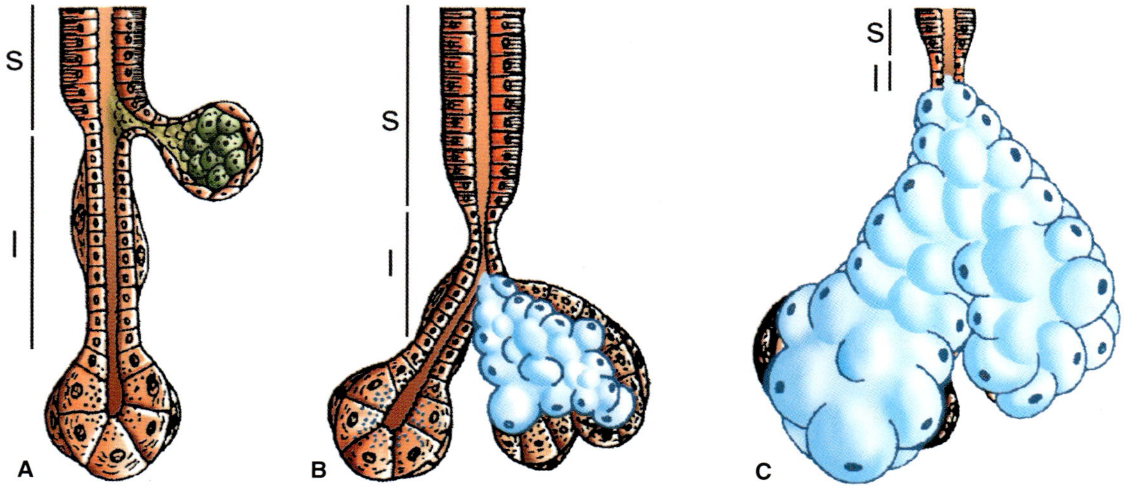

FIGURE 16.7 The morphology of the major salivary glands is characterized by three types of secretory units. In the parotid (**A**) the intercalated duct (I) is longer than in the submandibular (**B**) and sublingual glands (**C**). In contrast, the striated duct (S) is longer in the submandibular gland. Sebaceous glands are more frequent in the parotid gland; in the sublingual gland the intralobular ducts are inconspicuous.

The principal duct or main excretory duct is lined by stratified, squamous epithelium near the opening in the mucous membrane of the oral cavity (Fig. 16.10). Small basal cells are distributed continuously along the basement membrane.

Sebaceous Glands

Sebaceous glands or small collections of sebaceous cells are sometimes encountered in surgical pathology specimens, usually in the wall of a duct, and represent differentiation of the salivary duct unit. They are most commonly found in the parotid gland (10% to 42%), occasionally encountered in the submandibular gland (5% to 6%), and rarely seen in the sublingual gland or intraparotid lymph nodes (96–98). They have been described both as isolated cells in the intercalated or striated ducts and as larger aggregates demarcated by a basal membrane in the interlobular ducts (97,99). The histologic features include a central group of cells with irregular pyknotic nuclei and abundant vacuolated cytoplasm. The nuclei are larger than those of cells in the parotid gland acini or intercalated duct cells, with small single to few double nucleoli (99). In larger foci, the cell size and cytoplasmic volume increase toward the center or in proximity to the duct (Fig. 16.11) (99). The cytoplasm is lipid-rich and therefore can be stained with Sudan III and IV, oil red O (26). Sebaceous cells are also positive for androgen receptor and adipophilin but negative for mucicarmine. Upon development to a certain size, the holocrine secretion is emptied into the duct containing saliva (26).

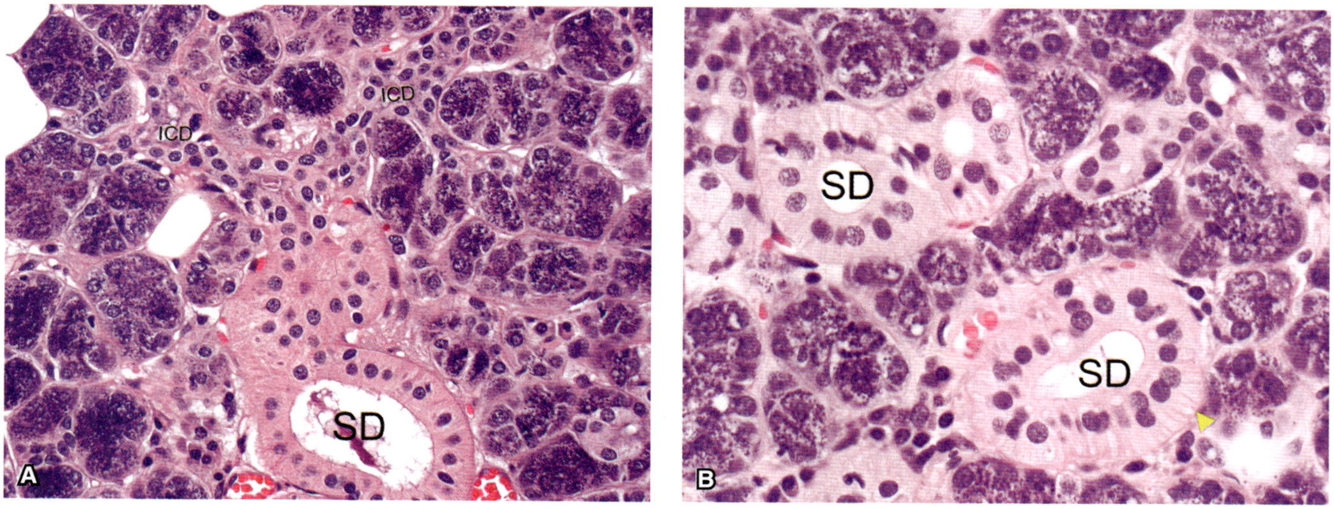

FIGURE 16.8 **A:** Striated ducts (SD) are lined with a columnar ductal epithelium and drain secretions from the intercalated duct (ICD) system. **B:** Striations (*yellow arrowhead*), created by invaginations of the cell membrane, can be seen perpendicular to the basal lamina.

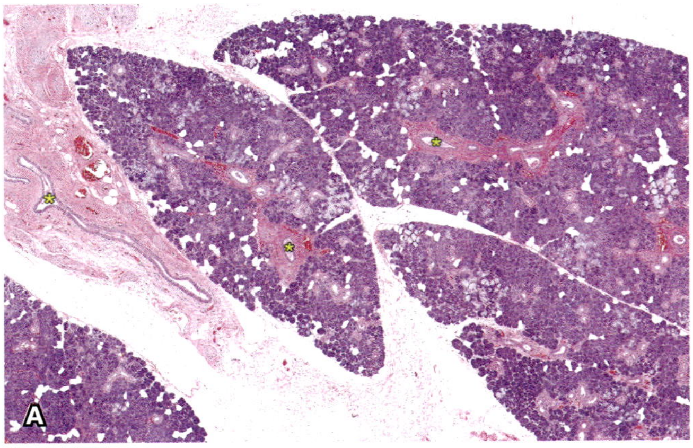

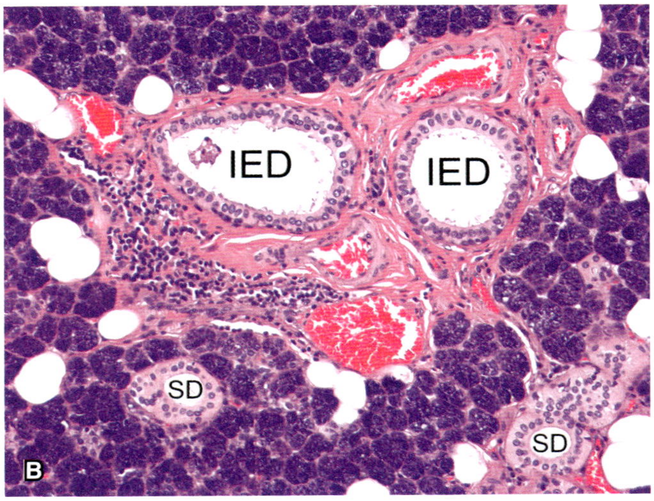

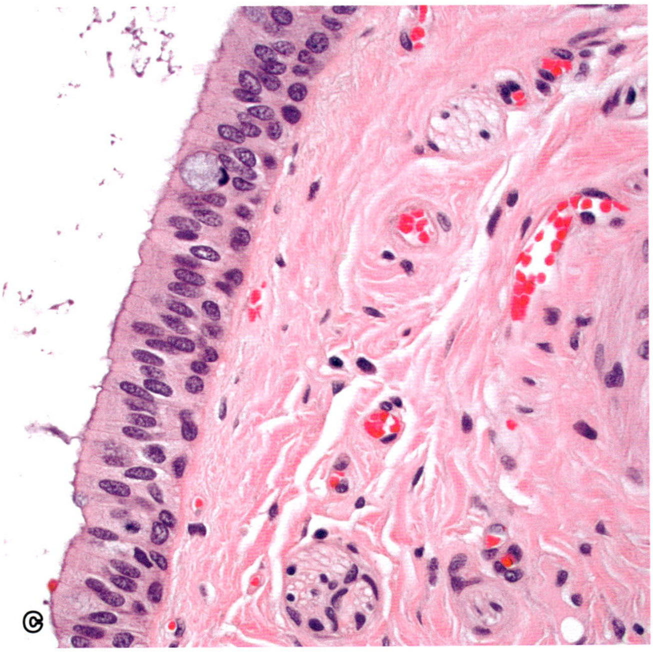

FIGURE 16.9 **A:** Interlobular excretory ducts (*yellow asterisks*) of the submandibular gland located within septal connective tissue adjacent to vessels and nerves. **B:** Interlobular excretory ducts (IED) of the parotid gland compared to intralobular striated ducts (SD). **C:** Interlobular excretory duct with goblet-type mucous cell, high power.

Sebaceous neoplasms of salivary gland origin are rare and include sebaceous adenoma, sebaceous lymphadenoma, sebaceous carcinoma, and sebaceous lymphadenocarcinoma. Sebaceous differentiation may also occur in salivary gland neoplasms with ductal components such as sialolipoma, pleomorphic adenoma, epithelial-myoepithelial carcinoma, mucoepidermoid carcinoma, and adenoid cystic carcinoma (96).

Chronic sialadenitis can show interlobular and intralobular ductal foci of sebaceous differentiation, likely representing a metaplastic process in response to the inflammatory background (97).

Myoepithelial Cells

Myoepithelial cells lie between the luminal epithelial cells and the basal lamina of acini, intercalated ducts, and portions of the striated ducts. The cells are flat and stellate-shaped with long, interwoven cytoplasmic processes that extend over the epithelial surface (91). In major salivary glands, myoepithelial cells cover ducto-acinar units including portions of striated ducts, whereas, in the minor salivary glands, they invest only acini and intercalated ducts. Excretory ducts are devoid of myoepithelial cells (100).

Structurally myoepithelial cells have features of both epithelial cells and smooth muscle with surface pinocytotic vesicles, ectoplasmic bundles of 6-nm microfilaments with randomly dispersed fusiform densities, and perinuclear tonofilaments joined by mature desmosomes and partially encompassed by basement membrane (100–102). Cytoplasmic filaments are located along the basal aspect of the cell and consist of actin, tropomyosin, and myosin myofilaments arranged parallel arrays gathered in "dense bodies"

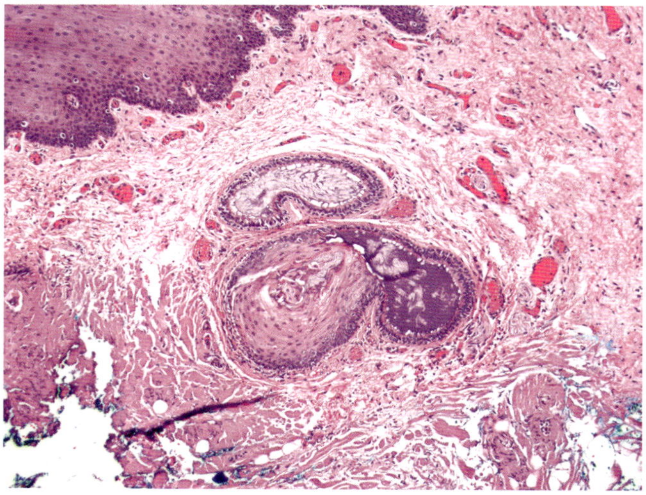

FIGURE 16.10 The main excretory duct (principal duct) is lined by stratified squamous epithelium near the oral surface where it empties.

and anchored to the basal plasma membrane in attachment plaques and focal densities, similar to that of smooth muscle (100). Contraction of myoepithelial cells facilitates mucous cell secretion and drainage. Myoepithelial cells are typically positive for cytokeratins (AE1/AE3, CAM5.2, CK14), squamous markers (CK5/6, p63, p40), and muscle markers (e.g., α-SMA, muscle-specific actin, smooth muscle myosin heavy chains, calponin, and/or h-caldesmon), negative for vimentin, rarely positive for glial fibrillary acidic protein (GFAP), variably reactivity for S-100 protein, and frequently positive for SOX10 (Fig. 16.12) (42,103–105).

Modified neoplastic myoepithelial cells can transform/differentiate into various morphologic cell types (e.g., chondroid, squamous, oncocytic, spindled) and secrete substantial amounts of basement membrane constituents (e.g., laminin, type IV collagen) and nonbasement membrane matrix (e.g., chondroitin sulfate proteoglycan, fibronectin, type I/II collagen, elastin) (Fig. 16.13) (89,100,101). Myoepithelial markers

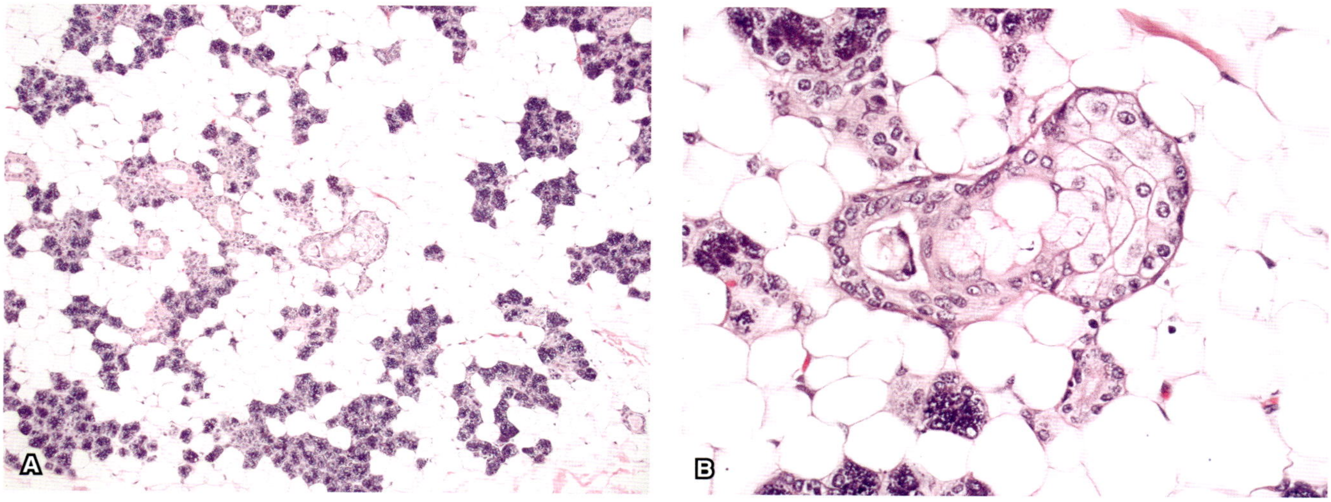

FIGURE 16.11 Intraparotid sebaceous elements, low power (**A**) and high power (**B**).

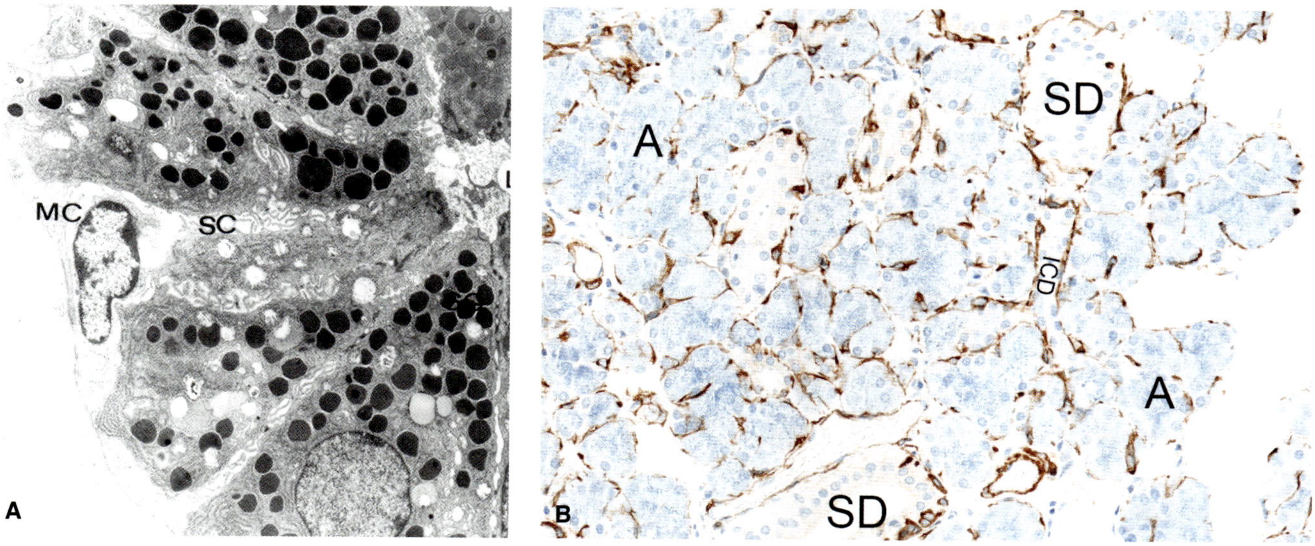

FIGURE 16.12 **A:** Ultrastructure of myoepithelial cell (MC) between epithelial secretory cells and secretory capillaries (SC) (×6,000).

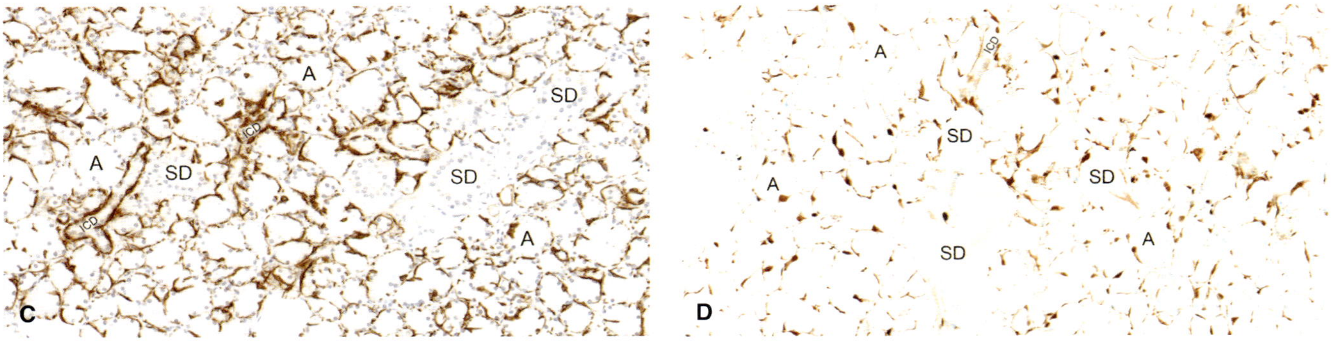

FIGURE 16.12 (*Continued*) **B–D:** Myoepithelial cells are positive for smooth muscle actin (**B**), calponin (**C**), and S-100 protein (**D**) around acini and intercalated ducts (ICD). Striated ducts (SD) may exhibit both myoepithelial and basal cells. *A*, acini.

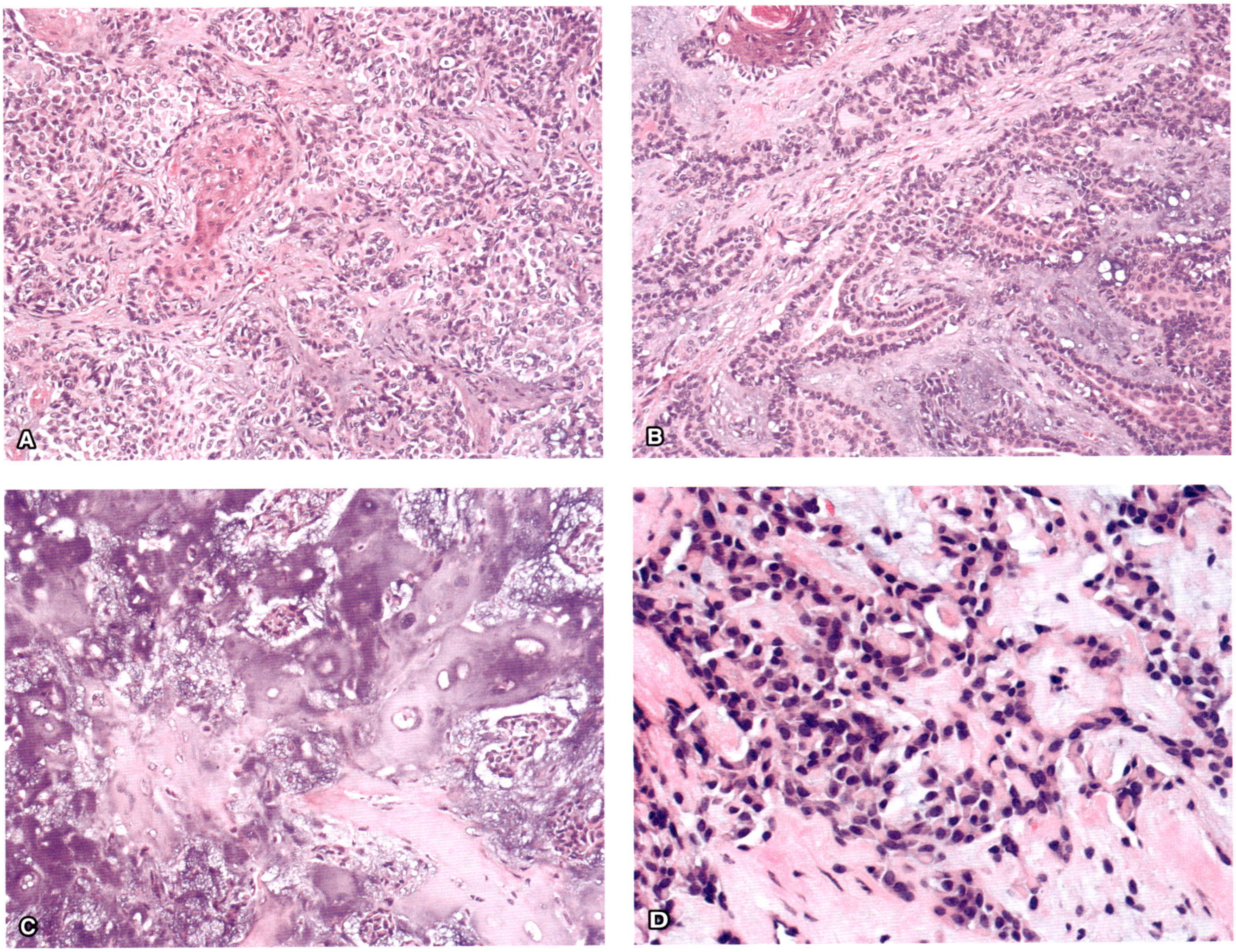

FIGURE 16.13 Modified neoplastic myoepithelial cells of a pleomorphic adenoma demonstrating squamous differentiation and plasmacytoid morphology (**A**), abluminal arrangement around ductal structures (**B**), and chondromyxoid matrix production (**C, D**).

TABLE 16.1 Salivary Gland Tumors with Myoepithelial or Basal Cell-Like Components

	Myoepithelial Cells	Basal Cells	Neither
Benign	• Salivary gland myoepithelioma • Pleomorphic adenoma • Basal cell adenoma • Intercalated duct adenoma • Sclerosing polycystic adenoma • Salivary gland anlage tumor	• Warthin tumor • Lymphadenoma • Striated duct adenoma • Sialadenoma papilliferum • Ductal papilloma • Oncocytoma • Cystadenoma of the salivary gland • Sebaceous lymphadenoma	• Canalicular adenoma
Malignant	• Myoepithelial carcinoma • Epithelial–myoepithelial carcinoma • Adenoid cystic carcinoma • Microsecretory adenocarcinoma • Basal cell adenocarcinoma • Sclerosing microcystic adenocarcinoma • Intraductal carcinoma • Sialoblastoma	• Sebaceous adenocarcinoma or lymphadenocarcinoma	• Polymorphous adenocarcinoma • Acinic cell carcinoma • Salivary duct carcinoma • Mucoepidermoid carcinoma • Secretory carcinoma • Hyalinizing clear cell carcinoma

may be variably expressed in salivary gland neoplasms, even those without a recognized myoepithelial cell component (Table 16.1). GFAP expression, for example, is frequently prominent in pleomorphic adenomas, focal in canalicular adenomas, and negative in adenoid cystic carcinoma (106,107).

Basal Cells

Basal cells are the abluminal cells of striated and excretory ducts (91). Similar to myoepithelial cells, basal cells stain with intermediate filament CK14, p63, and p40 (Fig. 16.14). However, unlike myoepithelial cells, basal cells are negative for S-100 protein, SOX10, GFAP, SMA, and calponin (Table 16.2) (105). Ultrastructurally, basal cells contain tonofilament bundles, parallel arrays of microfilaments with short, elongated densities, scattered intermediate filaments, micropinocytotic vesicles, as well as desmosomes joining cell processes to the base of neighboring columnar cells (91). Irregularly shaped cellular processes also extend into the underlying collagenized stroma. Unlike myoepithelial cells, most basal cells lack myofibrils.

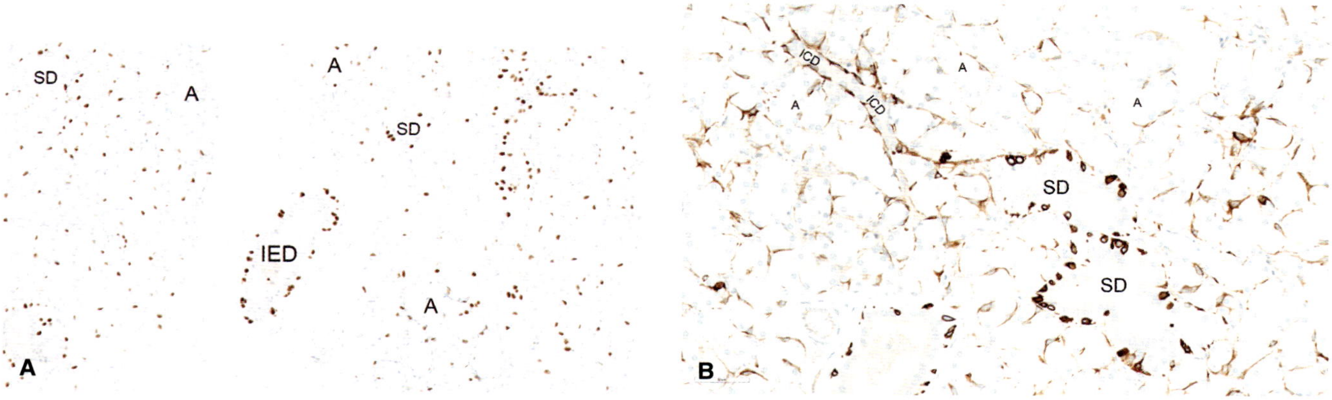

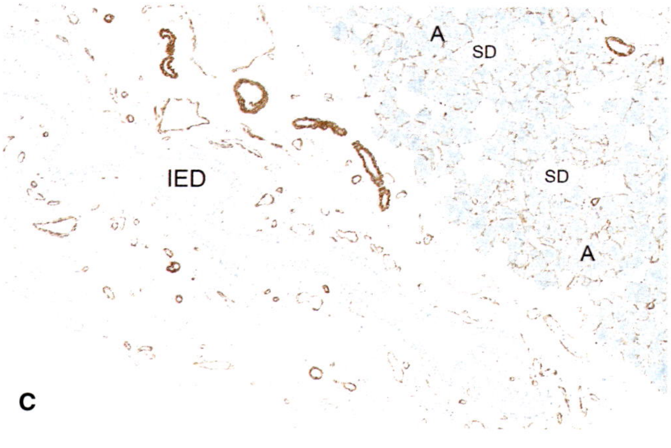

FIGURE 16.14 **A:** p40 highlights nuclei of basal cells around interlobular excretory ducts (IED) and striated ducts (SD) as well as myoepithelial cells around intercalated ducts (ICD) and acini (**A**). **B:** CK5/6 shows the same pattern of expression. **C:** Basal cells are negative for smooth muscle actin.

TABLE 16.2 Immunoprofile of Non-Neoplastic Myoepithelial and Basal Cells

	Myoepithelial Cells	Basal Cells
EMA	–	–
CAM5.2	Weak +	Weak +
CK5/6	+	+
CK7	Weak +	Weak +
CK20	–	–
CK903 (34 beta E12)	+	+
p63/p40	+	+
SMA, calponin, SMMH	+	–
S-100 protein	+	–
SOX10	+	–
GFAP	Rare subset +	–

STROMA

Supporting the secretory units of the adult salivary gland is connective tissue (fibroblasts, myofibroblasts, pericytes, adipocytes, and telocytes) with blood vessels and nerves and organ-specific extracellular matrix. The extracellular matrix is a dynamic and complex microenvironment that provides physical and biochemical signaling to regulate cell behavior and maintain organ stability (108,109). Multiple molecules, including components of the extracellular matrix (laminin, glycosaminoglycans, hyaluronic acid, tenascin, fibronectin, type IV collagen), cell adhesion receptors, proteases, and growth factors, mediate these cell interactions.

Most of what we know about the extracellular matrix is gleaned from studies focused on salivary gland regeneration, biogenesis, and remodeling in the setting of disease or injury (108). Collagens, which stain blue with Masson trichrome, are the principal proteins of the extracellular matrix. Laminin and type IV collagen are present close to the glandular epithelium at all stages of development. Tenascin-C is a multimodular extracellular matrix protein with several molecular forms that can bind with fibronectin, collagen, fibrillin-2, periostin, proteoglycans, and pathogens. Interestingly, stromal tenascin-C is consistently found in tissue remodeling regardless of etiology (110). In normal tissue, tenascin-C is involved with cell proliferation and migration; therefore, it is found in periductal tissue and along vessels and nerves (111,112). In the neoplastic setting, such as pleomorphic adenoma, the proximity of tenascin-C to myxoid/chondroid areas suggests interplay with myoepithelial cells for this type of matrix differentiation (112).

Telocytes are within the parotid gland stroma and function to mediate and integrate neural and/or vascular functions. This is achieved by interactions with other telocytes or other cell types in nerves and vessels through direct stromal synapses or microvesicles and exosomes (113). Telocytes are specifically found in a periacinar distribution and bordering ductal segments of striated to interlobular ducts (113). Ultrastructural telocytes are characterized by long, thin, and moniliform prolongations termed telopodes (113). Telopods are defined by the presence of alternating thin fibrillar-like segments (i.e., podomers) and dilated, cisternae-like regions (i.e., podoms), rich in caveolae, endoplasmic reticulum and mitochondria (113). Based on this definition, electron microscopy is required for identifying telocytes. Immunohistochemically, these cells are positive for CD117 and vimentin (113).

LYMPHOID TISSUE

The parotid gland is unique in that it is seeded with lymphocytes during embryogenesis at about the 3rd month of gestation that eventually develop into intraparotid (intraparenchymal) and periparotid (capsular and extraglandular) lymph nodes as well as ill-defined intraparenchymal lymphoid nodules that lack the complete organization of a lymph node. Intraparenchymal lymph nodes are mostly located in the superficial lobe and frequently contain intranodal salivary ducts and acini (Fig. 16.15). It is the intranodal salivary tissue that develops into primary intranodal salivary gland tumors such as Warthin tumors. The presence of extrafollicular reticulum cell scan be used to distinguish lymph node tissue from tertiary lymphoid aggregates; these cells are positive for low–molecular-weight cytokeratins (114–116).

Lymph nodes and lymphoid nodules are not present within the submandibular or sublingual glands (69).

Chronic sialadenitis is characterized by lymphocytic aggregates and infiltrates and parenchymal atrophy with or without fibrosis. The most common cause is secretory inactivity which leads to the accumulation of sialoliths, resulting in focal obstructive atrophy. It may also be induced by radiation, radioiodine therapy, medications, infection, and autoimmune disorders such as Sjögren syndrome and IgG4-related disease or granulomatous diseases (117). When lymphoepithelial lesions are present in the setting of chronic sialadenitis it is referred to as lymphoepithelial sialadenitis. Lymphoepithelial sialadenitis is often associated with Sjögren syndrome, an autoimmune disorder that carries an increased risk of lymphoma, particularly extranodal marginal zone lymphoma of mucosa-associated lymphoid tissue type (118). However, B-cell clonality may also occur in the absence of histologic evidence of progression to extranodal marginal zone lymphoma (118).

Lymphoepithelial lesions may involve intranodal salivary tissue, sometimes undergoing cystic degeneration. HIV-associated salivary gland disease is also associated with

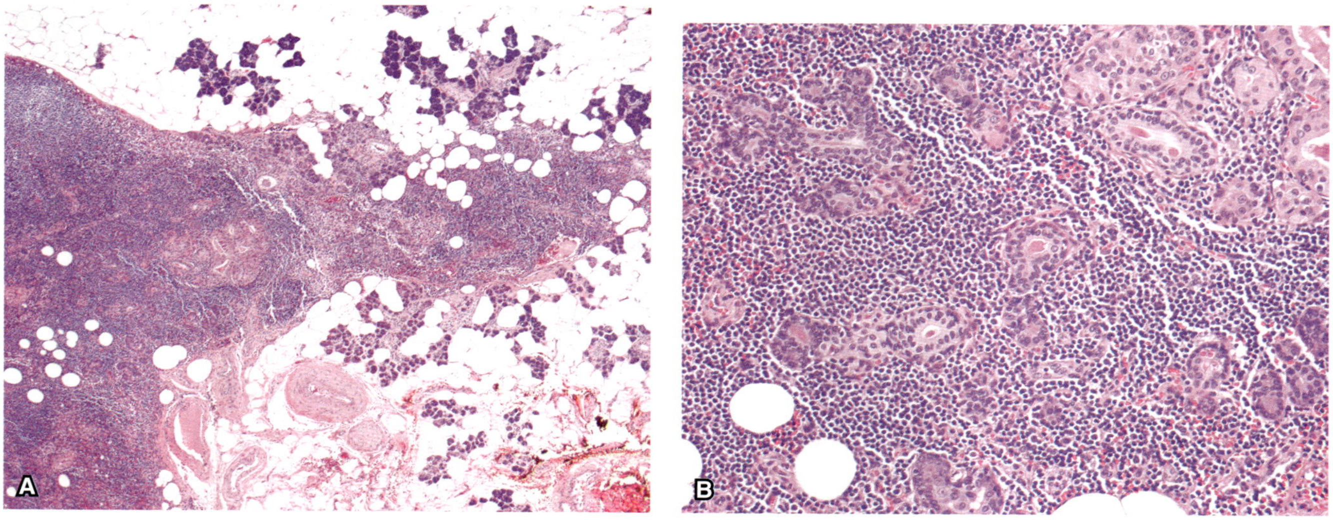

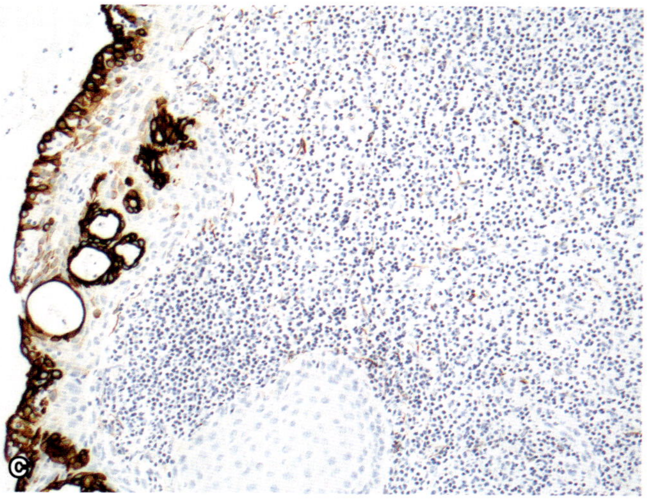

FIGURE 16.15 Intraparotid lymph node with acinar and ductal elements, low power (**A**) and high power (**B**). CAM5.2 immunostain highlights dendritic-like extrafollicular reticulum cells with delicate cytoplasmic processes in a case of mucoepidermoid carcinoma arising in an intraparotid lymph node (**C**).

cystic lymphoepithelial lesions involving salivary gland tissue and intraparotid lymph nodes. In the setting of HIV, cystic lesions are typically lined by squamous epithelium, bilateral and multifocal, and associated with diffuse cervical lymphadenopathy (119).

HETEROTOPIC SALIVARY TISSUE AND ITS SIGNIFICANCE

Heterotopic or ectopic salivary tissue is termed a choristoma. Although heterotopia mainly occurs in the head and neck (Fig. 16.16), it may rarely occur in other regions such as the gastrointestinal tract and prostate (120,121). Sites of predilection include the neck along the sternocleidomastoid muscle (122), capsules of the thyroid and parathyroid glands (123), middle ear (124), within mandibular bone, and cervical lymph nodes (Figs. 16.17 and 16.18). Heterotopias consist of aggregates of seromucous acini and ducts of salivary type that may develop the same

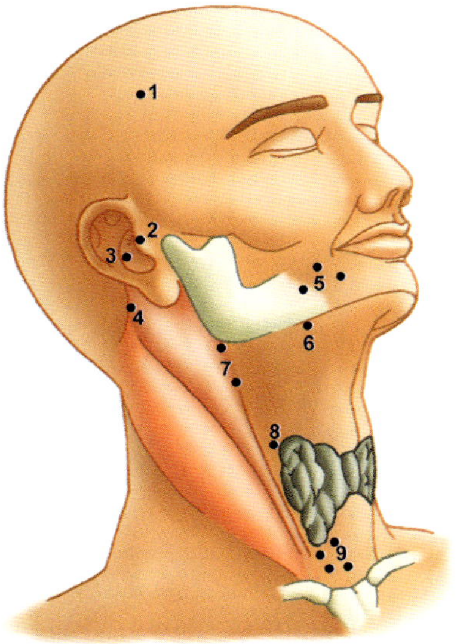

FIGURE 16.16 Extranodal salivary heterotopias. Pituitary gland (*1*), middle ear (*2*), external auditory canal (*3*), cerebellopontine angle (*4*), mandible (*5*), oropharynx (*6*), cervical superior (*7*), thyroid capsule (*8*), and lower anterolateral neck (*9*).

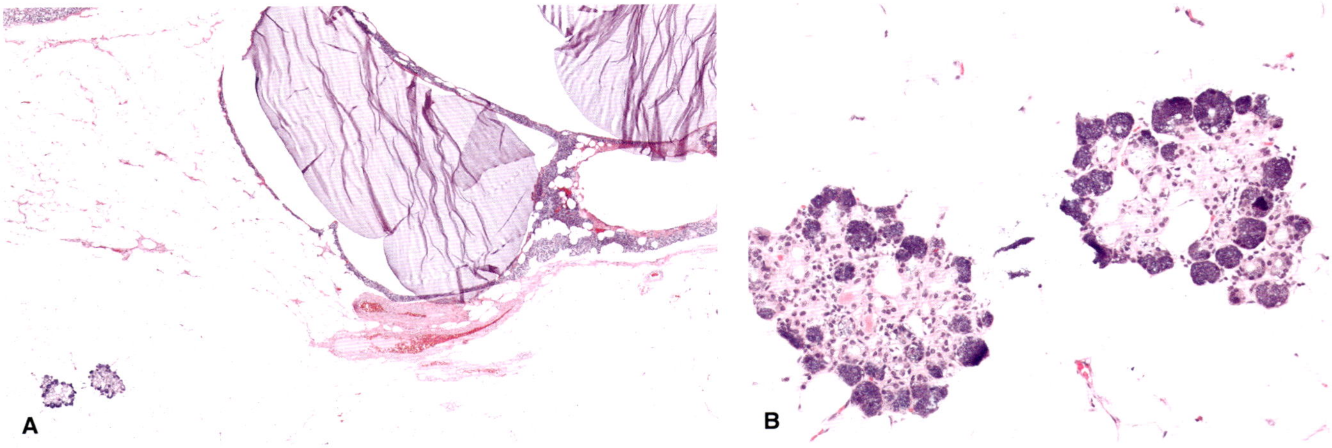

FIGURE 16.17 Parathyroid salivary gland heterotopia, low power (**A**) and high power (**B**).

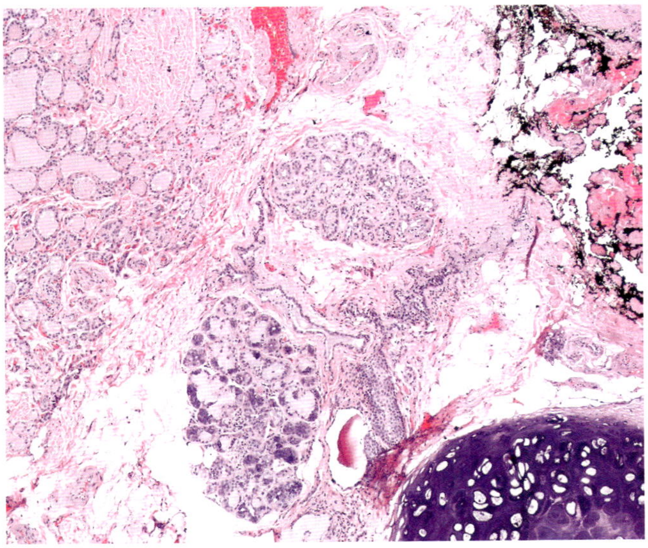

FIGURE 16.18 Perithyroidal salivary gland heterotopia, low power.

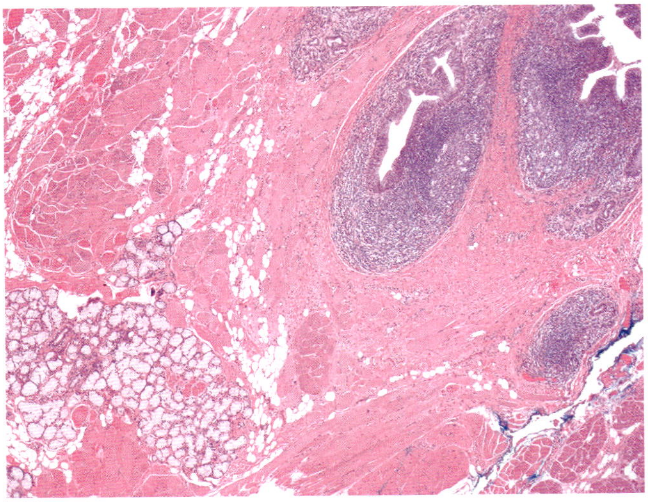

FIGURE 16.19 Salivary tissue associated with thyroglossal duct cyst, low power.

pathologic changes as usual salivary gland tissue including oncocytic hyperplasia and neoplasia. Although the pathogenesis remains unclear, association with embryologic remnants (e.g., branchial apparatus or Rathke pouch) can occur (Fig. 16.19) (125,126). Very rarely, an anomalous connection between any of the salivary glands or their ducts and the cutaneous surface may occur (127,128). Such congenital salivary gland fistulas are referred to as ectopic accessory parotid system and usually open on the facial skin near the angle of the mouth or, less commonly, the retroauricular region, oral mucosa, or neck (129). The fistulae are typically unilateral and present with clear salivary discharge from a cutaneous pit that increases with eating.

CHANGES ASSOCIATED WITH AGING

Salivary hypofunction can occur as a result of aging due to natural changes of the salivary glands. Aging does not affect the number of ducts; however, it is associated with decreased acinar cell secretion and increased fat and fibrovascular tissue in both the parotid and submandibular glands (4).

In minor glands, hypofunction may be due to age-related decreases in both stimulation and lower responsiveness to stimulation due to decreased numbers of receptors (4). Xu et al. state "with age, there is a reduction in the number of olfactory and taste receptors, diminished neuronal saliva stimulation (less transmitters acting on the receptors) and a decrease in the blood perfusion at the gland level" (4). These factors lead to decreased salivary flow rate (in comparison to young patients), resulting in changes in concentration of ions/inorganic compounds and alterations in mucin content. Since mucins form the immobile pellicle and the mobile salivary film, decreases in these components result in inferior quality and decreased quantity of saliva. Lactoferrin and peroxidase, which are needed for bacterial defense, have reduced activity as well. Poor mucosal salivary protection also results in increased risk of oral disease in the elderly (4).

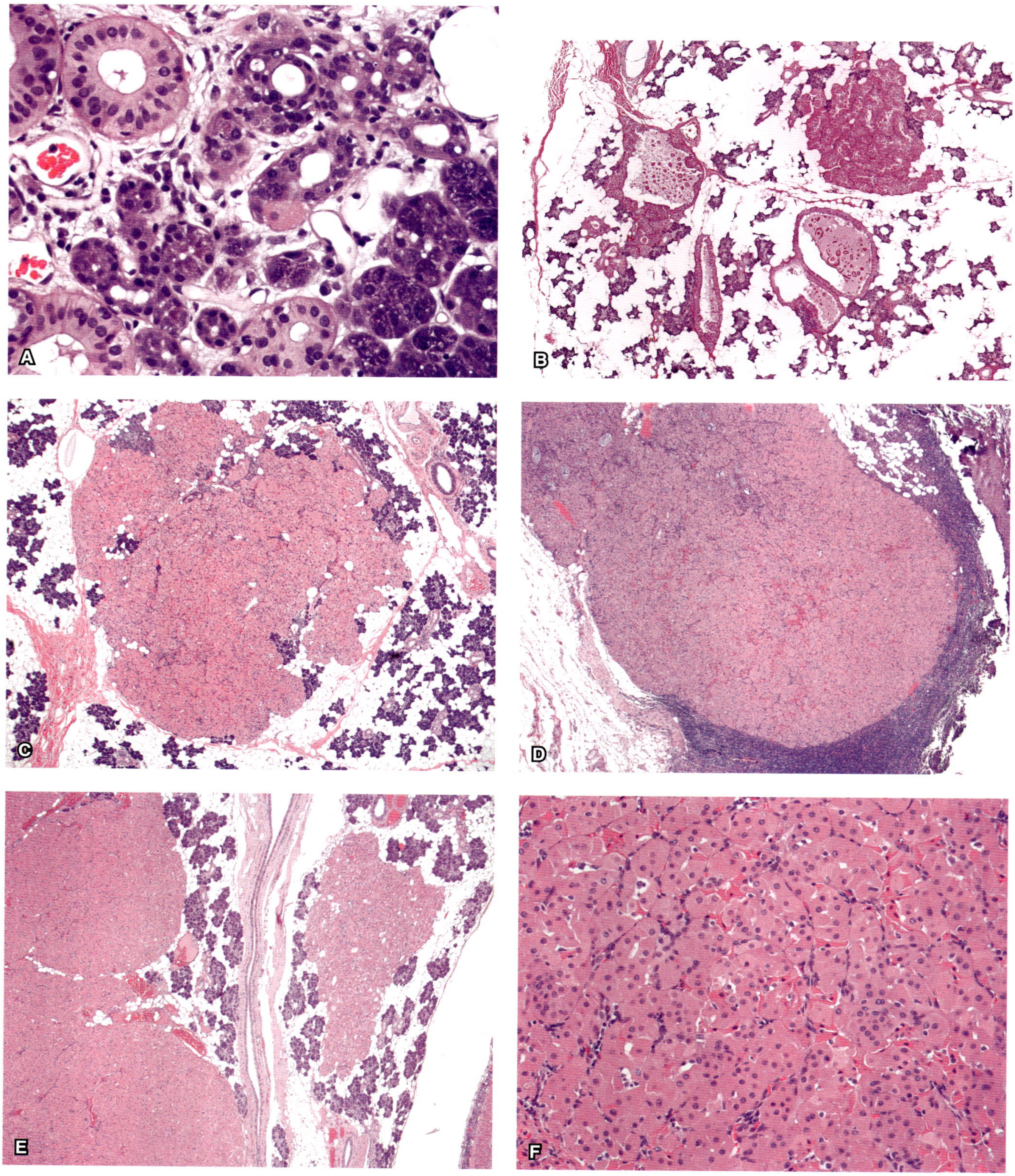

FIGURE 16.20 **A:** Single-cell oncocytic metaplasia involving a ducto-acinar unit. **B:** Multifocal oncocytic metaplasia with cystic change in a background of fatty infiltration. **C:** Microscopic nodule of oncocytic metaplasia. **D:** Nodular oncocytic hyperplasia involving an intraparotid lymph node. **E:** Multinodular oncocytic hyperplasia, dominant and microscopic nodules. **F:** Oncocytic hyperplasia, high power. Note the abundant granulated cytoplasm.

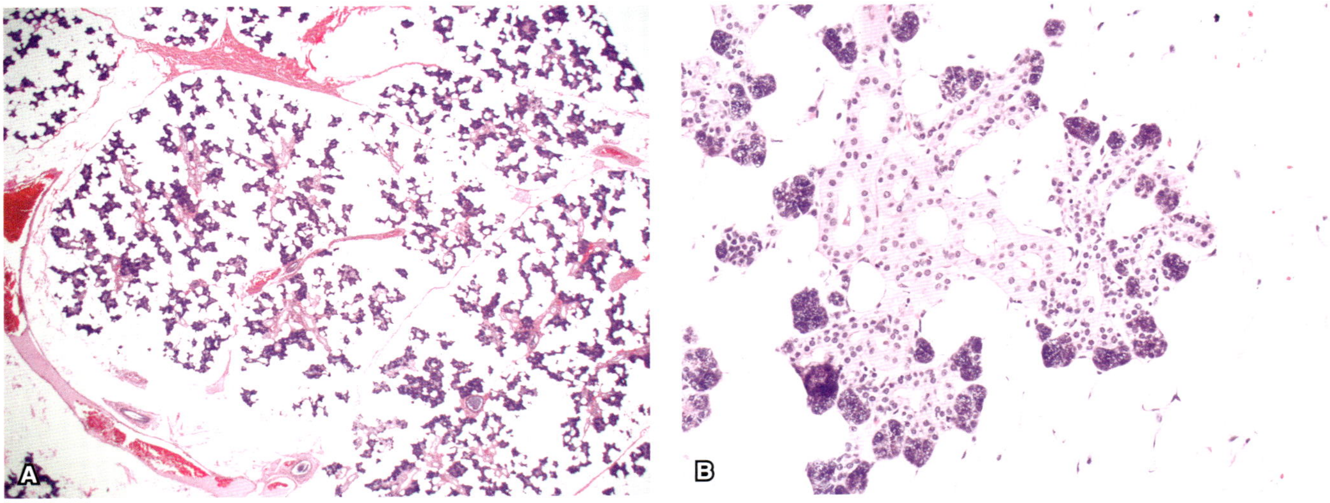

FIGURE 16.21 Fatty infiltration of the parotid, low power (**A**) and high power (**B**).

Taste receptors respond to food components dissolved within saliva. Changes in inorganic ion concentrations, a decrease in the number of taste buds, and a lag in the turnover of taste receptor cells (4) all occur with advanced and, thus, likely account for a true decline in taste in older patients.

Based on animal studies, it seems that mesenchymal stem cells from older individuals do not have the same performance ability as progenitor cells from younger individuals due to epigenetic modifications (109). Age-related loss in number and function of mesenchymal stem cells likely contributes to decreased taste sensation among older adults. Decreased mesenchymal stem cell function may also be induced by pharmacotherapy, including corticosteroids and antidepressants (109).

Oncocytes

Oncocytic change is a histologic reflection of increased cytoplasmic mitochondria and appears as abundant eosinophilic cytoplasm with fine granules and centrally located round nuclei with uniformly conspicuous nucleoli (130). Aging is thought to cause a functional exhaustion of mitochondrial enzymes resulting in compensatory hyperplasia of mitochondria, resulting in oncocytic change (131,132). Oncocytic change can occur within acini and ducts as single cells, small clusters, variably sized nodules, or diffusely involving all ducts and acini without formation of discrete nodules or tumors (Fig. 16.20) (133,134).

Fatty Infiltration

Adipose tissue deposition increases with age and is gender dependent, being greater in men than in women by approximately 10% (n = 100, even gender distribution) (135). The gradual increase in fibroadipose connective tissue in salivary glands is referred to as fatty infiltration (Fig. 16.21). Diffuse fatty infiltration, or lipomatosis, results in diffuse swelling of the gland without forming a discrete tumor mass. A variable degree of increased interstitial fat may also be seen in the parotid glands of patients with underlying disease conditions including alcoholic cirrhosis, obesity, and hyperlipidemia (135–138). Lipomatosis limited to one lobule of the gland also is known as lobular lipomatous atrophy, and may mimic a lipoma or sialolipoma in limited biopsy material. Distinction may be made by the frequent association with underlying obstructive and inflammatory changes in the former.

REACTIVE CHANGES

Metaplasia

Squamous metaplasia may be present in large salivary ducts in the setting of chronic inflammation, particularly when associated with calculi (Fig. 16.22). Acinar-intercalated duct complexes also demonstrate a propensity for squamous metaplasia following ischemic and radiation injury (139). Necrotizing sialometaplasia is a benign, self-limiting process whereby ischemic lobular necrosis of salivary gland tissue results in squamous metaplasia of acini and intercalated ducts, marked inflammation, and granulation tissue with or without pools of mucin (Fig. 16.23). These lesions can clinically and histologically simulate squamous cell carcinoma or mucoepidermoid carcinoma (140). Distinguishing features of a benign process include the presence of uniform squamous cell nests arranged in a lobular pattern with residual duct lamina in one or more of the squamous nests (141).

Mucous metaplasia occurs in the setting of obstructive and postradiation sialadenitis and is often found in interlobular excretory ducts and less frequently in intralobular ducts. Proliferation of goblet-type mucous cells may be

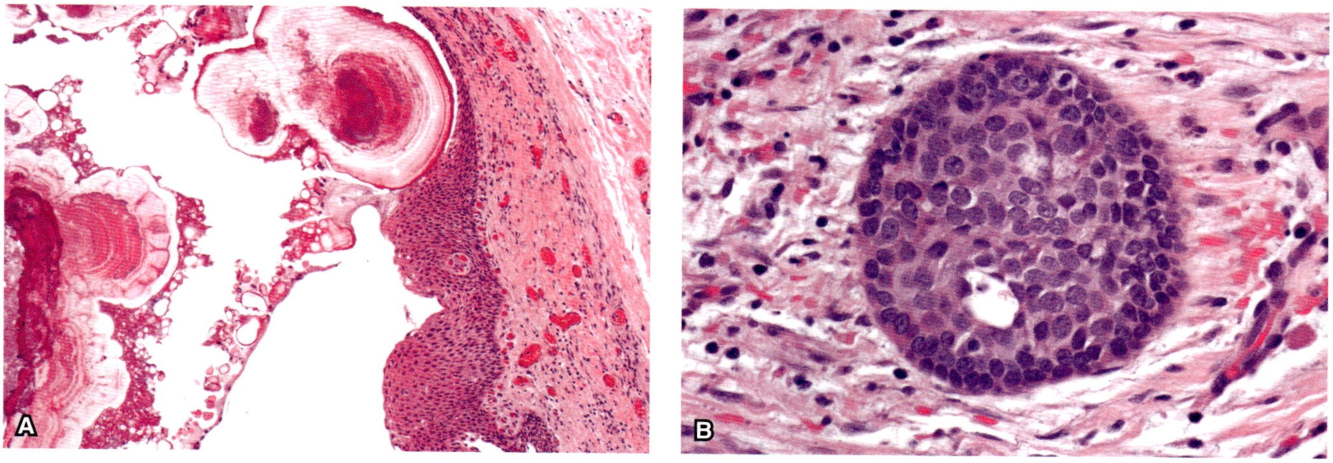

FIGURE 16.22 Squamous metaplasia associated with sialolithiasis, low power (**A**) and high power (**B**).

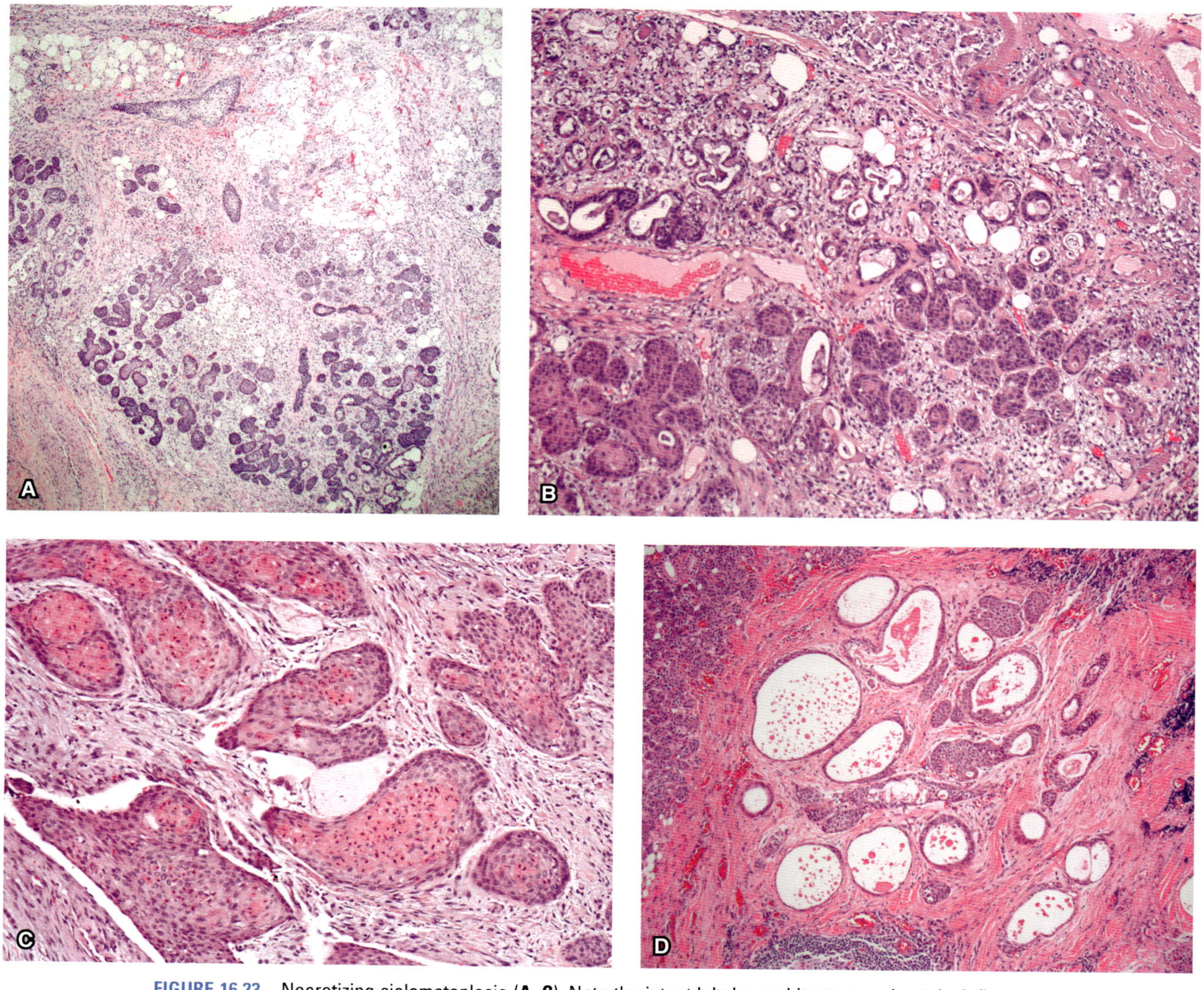

FIGURE 16.23 Necrotizing sialometaplasia (**A–C**). Note the intact lobular architecture and patchy inflammatory infiltrate (**A, B**). Low-grade mucoepidermoid carcinoma (**D**) mimicking necrotizing sialometaplasia.

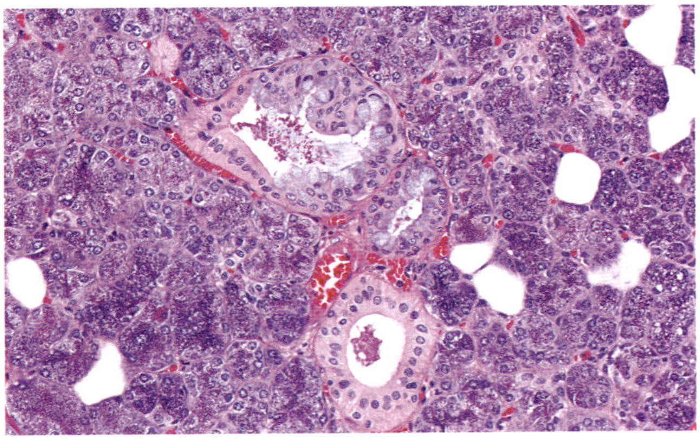

FIGURE 16.24 Mucous metaplasia of intralobular ducts.

accompanied by ciliated cells mimicking respiratory epithelium (Fig. 16.24).

Hyperplasia

Sialadenosis is a diffuse, noninflammatory, non-neoplastic swelling of the salivary glands that feature uniform hypertrophy and hyperplasia of the serous acinar parenchyma. It has been associated with altered nutritional states such as starvation, malnutrition, bulimia, anorexia nervosa, endocrinopathies, and metabolic dysregulation due to diabetes mellitus, chronic alcohol consumption, and other causes (142). In most cases, enlarged acinar cells are packed with accumulated secretory granules (Fig. 16.25). These changes may be accompanied by loss and thinning of the myofilament component of myoepithelial cells, suggesting that myoepithelial cells lack the ability to resist an increase in acinar pressure in the setting of sialadenosis (143). Sialadenosis eventually gives way to fatty infiltration as the process becomes more chronic.

Proliferations of intercalated ducts are seen in a variety of clinicopathologic conditions including chronic sialadenitis, atrophy, and postradiation, but they may also be seen in association with neoplasms (144,145). Hyperplastic proliferations do not form discrete masses but rather blend imperceptibly with the acinic and mucous cells of the surrounding salivary gland parenchyma. The ducts are closely packed with uniform cuboidal to columnar epithelium and abluminal myoepithelial cells, intermingled serous acinar

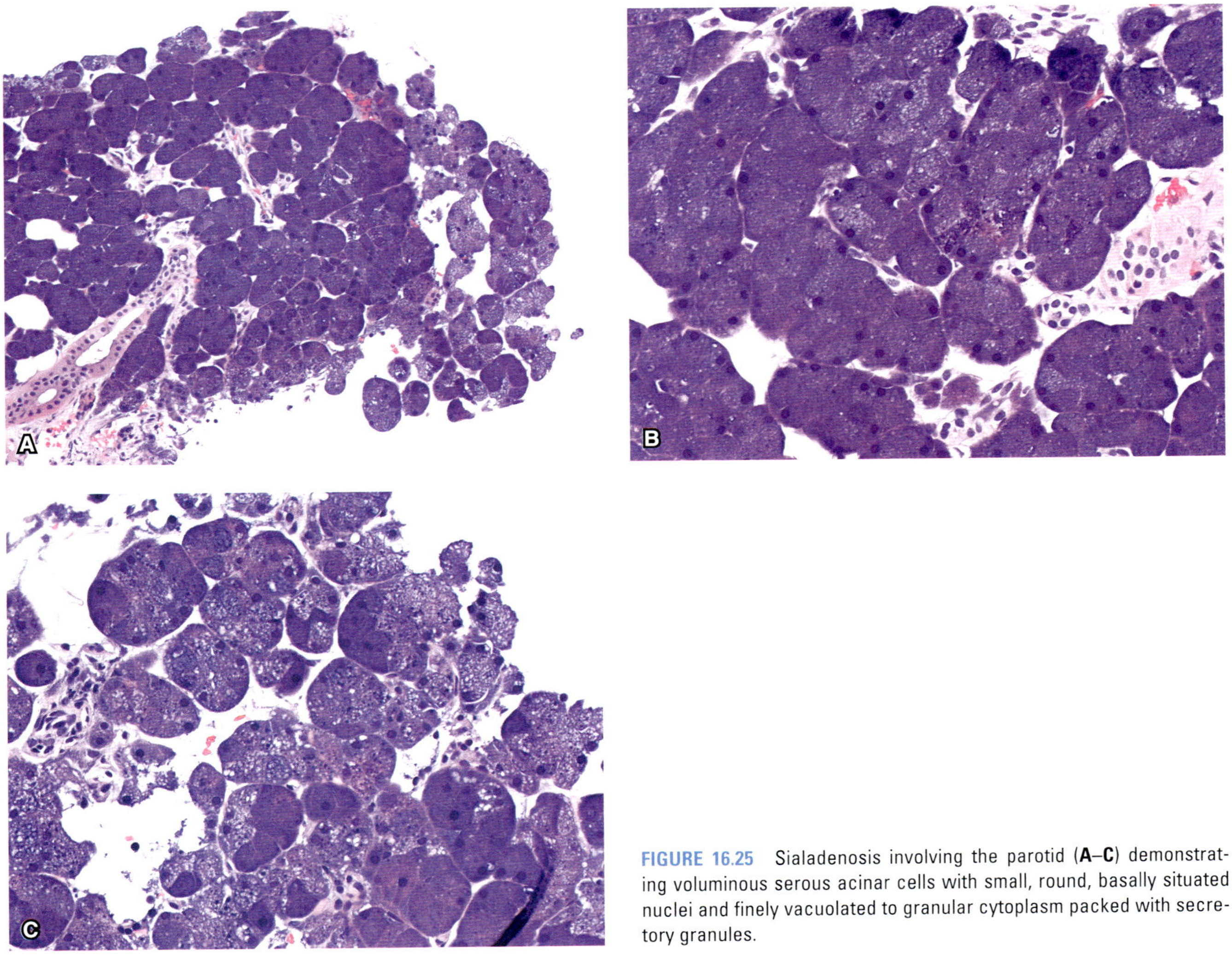

FIGURE 16.25 Sialadenosis involving the parotid (**A–C**) demonstrating voluminous serous acinar cells with small, round, basally situated nuclei and finely vacuolated to granular cytoplasm packed with secretory granules.

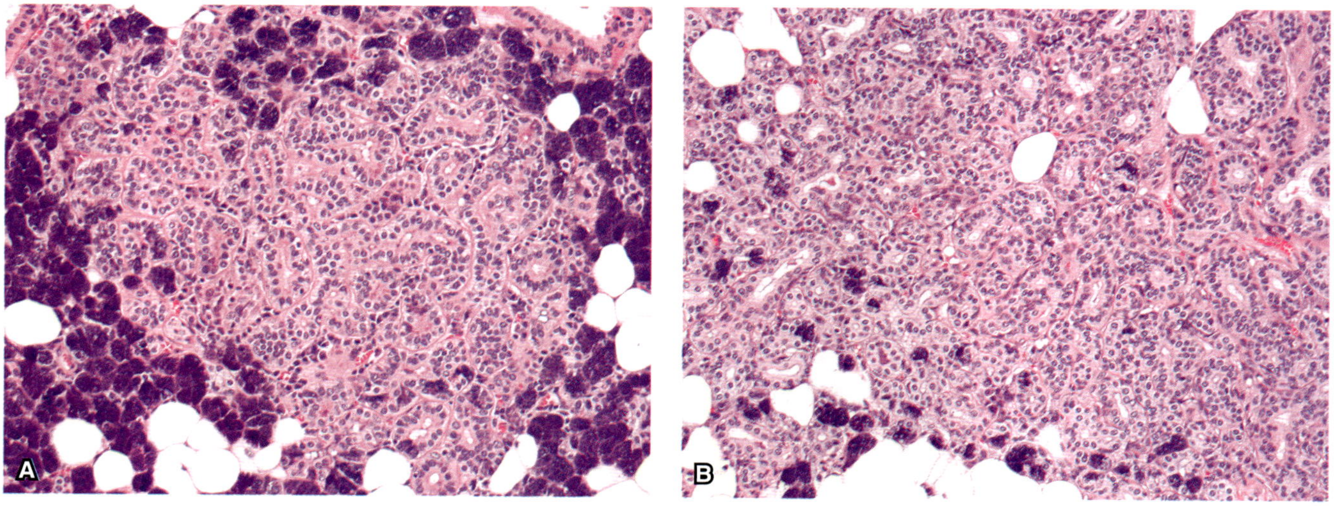

FIGURE 16.26 Intercalated duct hyperplasia (**A, B**) of the parotid mixed with serous acinar cells.

cells, and minimal intervening stromal tissue (Fig. 16.26). It has been suggested that some salivary gland neoplasms such as epithelial–myoepithelial carcinomas and basal cell adenomas may be derived from hyperplasia of the intercalated ducts (145–148).

Oncocytic hyperplasia is a non-neoplastic proliferation of oncocytes that can occur as variably sized nodules, occasionally showing a vaguely organoid pattern, peripheral fibrosis, and expansile growth in clinically detectable masses (Fig. 16.20). These hyperplastic nodules are typically unencapsulated, multifocal, bilateral, and most commonly involve the parotid. Rarely, it may present in the setting of Birt–Hogg–Dubé syndrome, an autosomal dominant condition caused by germline variants in the folliculin gene (*FLCN*) (149). Detection of an alteration involving *FLCN* should suggest the possibility of Birt–Hogg–Dubé syndrome, rather than be considered evidence of a neoplasm.

Atrophy

Atrophy of one or more lobules of salivary tissue is a common finding in surgical specimens of tumoral salivary glands (Fig. 16.27). This atrophy is typically caused by partial or total obstruction of an excretory duct. Accordingly, secretory units distal to the obstructed duct are dilated, and the acinar lumina become visible. The secretory cells lose their granules and have a similar appearance to that of the intercalated duct. An inflammatory component may be present. The cells gradually disappear, and the parenchyma is replaced by adipose tissue and collagen fibers.

Diffuse atrophy of the submandibular gland frequently occurs following radiotherapy. Acinar cells disappear while residual ducts become dilated and exhibit nuclear hyperchromasia and pleomorphism, loss of cell polarity, and prominent myoepithelial cells (150,151).

Regeneration

Salivary gland parenchyma has the capacity for regeneration a few weeks after partial resection. In studies with ligation of the main salivary gland excretory duct and subsequent restoration of the duct, the acinar component recuperates with glandular acinar normalization by 7 to 14 days (152). In general, regenerating tissue follows a lobular embryonic pattern with solid buds and branching columns of undifferentiated cells that eventually form excretory units (153). Regeneration after atrophy due to duct obstruction is completed from residual ducts that differentiate to form acinar cells (154,155).

Artifacts

In mixed acini, serous acinar cells can appear like outer crescent-shaped or half-moon formations (i.e., demilunes) around mucous cells in formalin-fixed tissue. Although some authors have suggested these findings represent artifactual structures produced by immersion methods of fixation, data from other studies seemingly contradict these conclusions (156,157).

Artifactual changes in salivary gland surgical specimens are commonly seen in the setting of electrocautery and post fine needle aspiration (FNA). Electrocautery can induce oncocytic changes in serous acinar cells, leading to a misdiagnosis of an oncocytic neoplasm. Unlike true oncocytes, acinar cells with oncocytoid change do not have centrally placed nuclei (Fig. 16.28). Rather, they exhibit uniform, round nuclei in the basal half of the cell, similar to their unaffected acinar counterparts.

FNA and/or core needle biopsy can also pose challenges in subsequent resection specimens (158). Post-FNA infarction and hemorrhage are the most common tissue effects followed by vascular proliferation, fibrosis, and metaplastic changes. FNA-associated ischemic type necrosis is thought to be secondary to disrupted blood flow either in the tumor

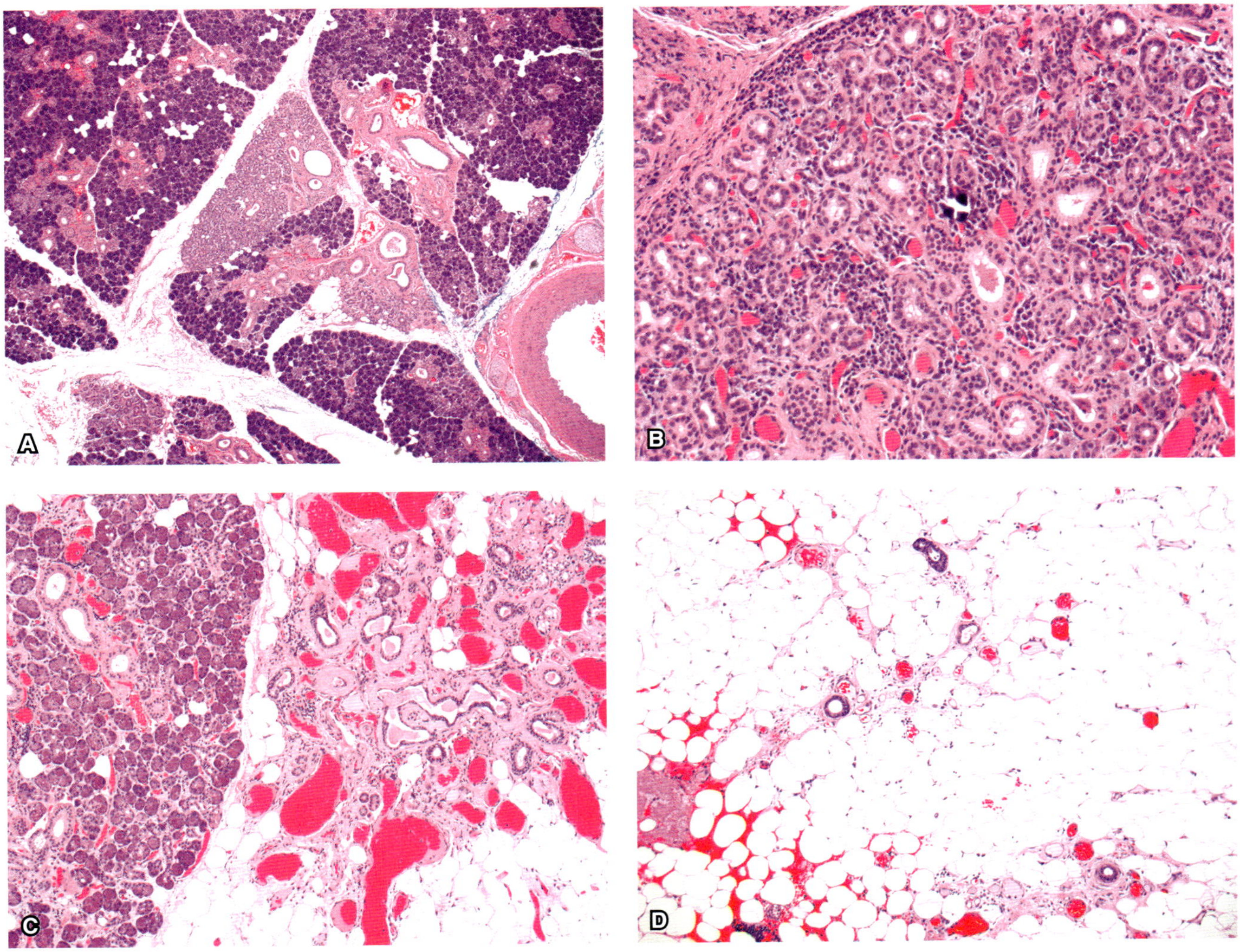

FIGURE 16.27 Lobular atrophy of the salivary gland, low power (**A**) and high power (**B**) demonstrating acinar cell shrinkage, chronic inflammation, and microscopic sialolithiasis. Advanced atrophy with marked acinar loss, dilated ducts, and increased stromal fibroadipose tissue (**C, D**).

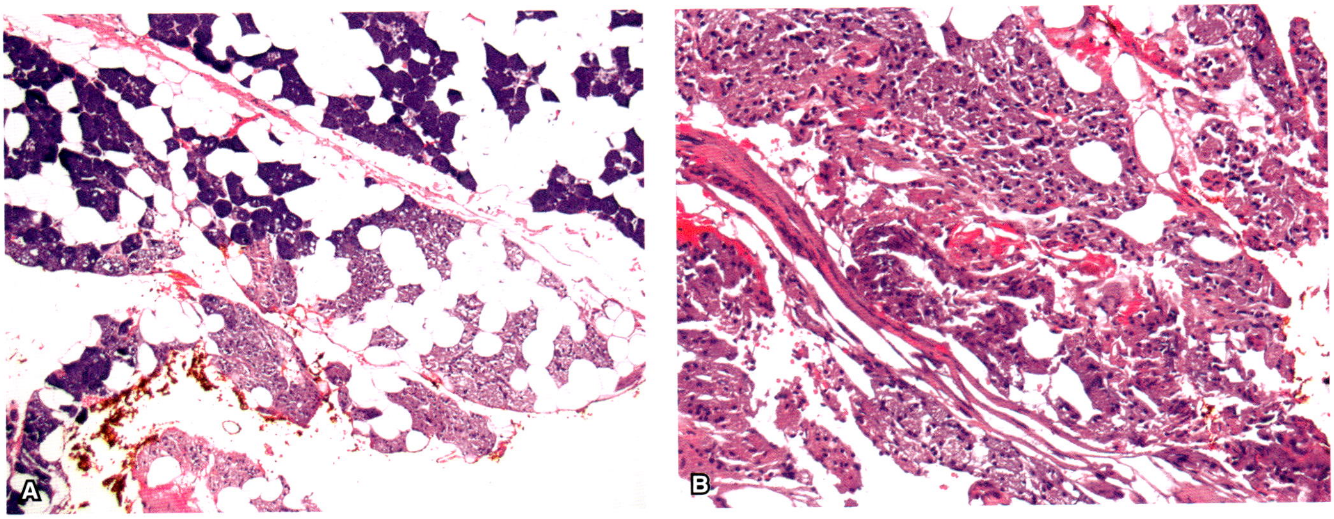

FIGURE 16.28 Cautery artifact mimicking oncocytic metaplasia, low power (**A**) and high power (**B**).

or in the non-neoplastic gland. Tumors rich in mitochondria may be more susceptible. Ischemic necrosis can usually be distinguished from true tumor necrosis by the absence of single-cell necrosis in the form of cell shrinkage with condensation of nuclear chromatin and cytoplasm and apoptotic bodies. A rim of histiocytes and granulation tissue may also be present. Squamous metaplasia and necrotizing sialometaplasia can occur and be confused with squamous cell carcinoma or mucoepidermoid carcinoma; however, they typically lack cytologic atypia and maintain the lobular architecture of the acinar units. Tumor cells can also theoretically seed along a core needle biopsy track, mimicking an invasive process (159). However, the risk of tumor cell dissemination along the biopsy track is extremely low (160).

Radiation

Ionizing radiation therapy is a common modality for treating head and neck malignancies. Salivary gland radiosensitivity is paradoxical considering the mature cell type with low cell proliferation and cell turnover in excess of 60 days (152). Damage results in increased proliferation then reverts to homeostasis (152). Radiation changes are comparable between the parotid gland and submandibular gland, but specific regions of the glands may be more sensitive than others (161). Early radiation changes include damage to the plasma membrane of acinar cells impairing water secretion; while late changes include lack of functional acinar cells secondary to a lack of progenitor cells for regeneration/field damage, inhibition of acinar self-duplication, and/or a damaged microenvironment (3,161). Postradiation damage features replacement of glands with extracellular matrix deposition and fibrosis (3). Serous acini of the parotid and submandibular gland are the most at risk for necrosis without regeneration, while mucous acini are relatively resistant and myoepithelial cells are largely unaffected. The main symptom of radiation-related salivary gland injury is hyposalivation with mucositis. Hyposalivation is due to DNA damage, death of acinar cells, disruption of calcium-mediated signaling, and increased creation of reactive oxygen species (3). Up to 90% of patients with radiation-related salivary gland injury have permanent loss of function (3).

CORRELATIVE NORMAL AND NEOPLASTIC HISTOLOGY

The classification of salivary gland neoplasms is an ever-evolving process. Current diagnostic categories are based on tumor cell differentiation and histomorphologic similarities to normal ducts and acini. For example, low-grade acinic cell carcinoma recapitulates normal serous acinar cells with variable amounts of zymogen-like secretory granules, while biphasic salivary gland neoplasms show both ductal and myoepithelial/basal cell differentiation (Table 16.1). Monophasic tumors are characterized by a predominance of epithelial or myoepithelial cells.

Histologic and immunophenotypic similarities between normal and neoplastic salivary gland components do not imply that a particular tumor arises from the structures it mimics. There is remarkable plasticity of terminally differentiated ductal and myoepithelial cells to replenish lost secretory acinar components (109,152). For example, lineage tracing studies suggest that basal and myoepithelial cells function as multipotent progenitors with the ability to give rise to all epithelial cell lineages during development and salivary gland maintenance (162). Data from recent studies have also shown that acinar cells can undergo acinar-to-ductal metaplasia in response to ligation-induced injury (152,163). These findings suggest a dynamic process of differentiation akin to branching morphogenesis whereby cells can assume a facultative, multipotent, stem cell-like state that is influenced by the biochemical milieu of their niche native microenvironment to modify form and function (164). However, further studies are needed to better understand the molecular mechanisms that dictate lineage commitment and differentiation programs in neoplastic and non-neoplastic settings.

IMMUNOHISTOCHEMISTRY AND SPECIAL STAINS

Our understanding of salivary gland cytodifferentiation is supported by identifying expression of different intermediate filaments, enzymes, sialomucins, and proteins using special stains and/or immunohistochemical methods.

The expression of low–molecular-weight cytokeratins varies according to cell type in normal salivary glands. CK7 expression is stronger in luminal ductal cells and less intense in basal, myoepithelial, and acinar cells (Fig. 16.29A). CK18 expression is stronger in both serous and mucous cells relative to ductal epithelium, whereas CK8 and CK19 react very strongly in all parts of the duct system with weak to no expression within acinar cells. Acini and ductal cells of intercalated, striated, and interlobular ducts are also positive for epithelial membrane antigen (165). All cellular elements of normal salivary glands are negative for CK20. Ductal cells of striated and excretory ducts are also positive for high–molecular-weight cytokeratin CK903 (Fig. 16.29B, Table 16.3).

High–molecular-weight keratins (e.g., CK5/6, 34 beta E12, CK14, CK17, CK19) stain basal and myoepithelial cells (Figs. 16.14B and 16.29B). Both myoepithelial and basal cells also express p63 and p40 (Fig. 16.14A); however, only myoepithelial cells express SMA, calponin, caldesmon, and myosin heavy chain. Of note, SMA immunoexpression can vary depending on the stage of myoepithelial cell differentiation (166). Rarely, a subset of myoepithelial cells may also stain with GFAP (Table 16.2) (167,168).

SRY-related HMG-box 10 (SOX10) protein, a transcription factor often used to support neural, melanocytic, and

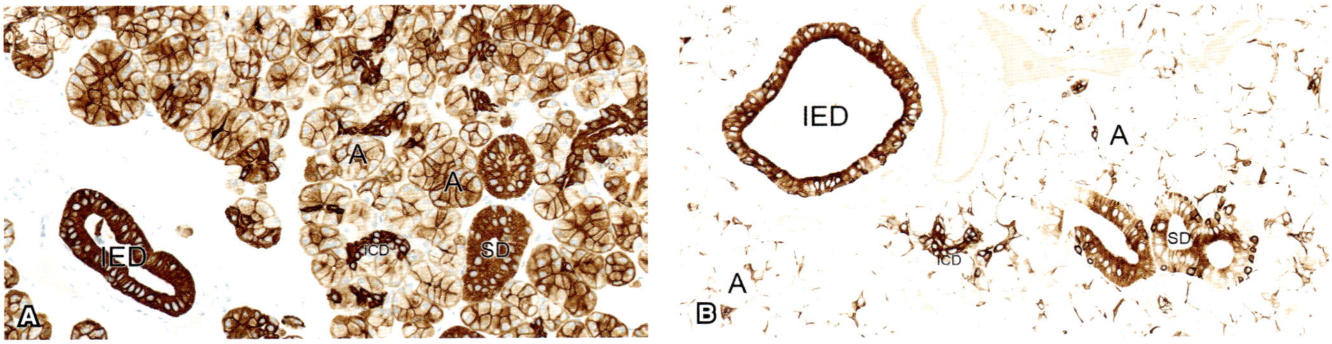

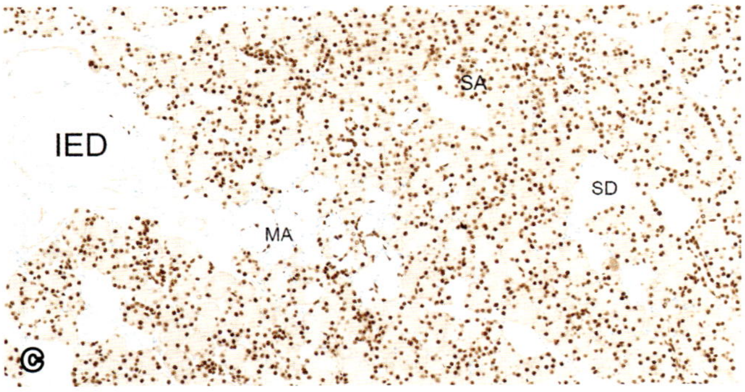

FIGURE 16.29 CK7 (**A**), CK903 (34 beta E12) (**B**), and SOX10 (**C**) expression in normal submandibular gland tissue. *IED*, interlobular excretory duct; *SD*, striated duct; *ICD*, intercalated duct; *A*, acini; *SA*, serous acini; *MA*, mucous acini.

breast origin, is expressed in the nuclei of serous acinar cells, myoepithelial cells, and ductal epithelium of intercalated ducts (Fig. 16.29C) (105,169–171). S-100 protein is similarly expressed in myoepithelial cells but not basal cells with variable expression within intercalated ductal epithelium (Fig. 16.12D) (172).

Special stains can also be used to highlight acidic glycoproteins and neutral glycoproteins. Serous acini are negative for mucicarmine and positive with PAS while mucous acini are positive for alcian blue and mucicarmine with only faint PAS staining (173). Serous acinar cells are positive for amylase and lactoferrin, while intercalated ducts express lysozyme, striated ducts express alpha-amylase, and myoepithelial cells express alkaline phosphatase and ATPase (83,84,145). Sebaceous lipid-rich cytoplasm can be stained with Sudan III and IV, oil red O, and osmic acid (26).

GATA binding protein 3 (GATA-3), a transcription factor that plays a role in luminal epithelial differentiation in the adult mammary gland, estrogen receptor β, and progesterone receptor can show weak nuclear staining of both acini and ducts of normal salivary gland tissue (174–176). Rare, scattered cells may also be positive for mammaglobin (177).

TABLE 16.3 Immunoprofile of Non-Neoplastic Acinar and Ductal Cells

	Mucous Acinar Cells	Serous Acinar Cells	Intercalated Ductal Cells	Striated Ductal Cells	Excretory Ductal Cells
EMA	+	+	+	+	+
CAM5.2	+	+	+	+	+
CK5/6	–	–	–	–	–
CK7	+	+	+	+	+
CK20	–	–	–	–	–
CK903 (34 beta E12)	–	–	+	+	+
p63/p40	–	–	–	–	–
SMA, calponin, SMMH	–	–	–	–	–
S-100 protein	–	–	Weak +	–	–
SOX10	–	+	+	–	–

SMA, smooth muscle actin; SMMH, smooth muscle myosin heavy chain; EMA, epithelial membrane antigen.

Anoctamin-1 (DOG1), a calcium-activated chloride channel, shows diffuse apical membranous staining of normal acini and variable luminal staining of intercalated ducts (178). Serous acini are also positive for gross cystic disease fluid protein-15 (GCDFP-15), somatostatin receptor 2 (SSTR2), and polyclonal carcioembryonic antigen (CEA), but they are negative for monoclonal CEA. Nuclei of mucous acinar cells are positive for NKX3.1, an androgen-regulated, homeobox-containing transcription factor also expressed in prostate epithelial cells and normal testis (Fig. 16.4D) (179–181). Salivary gland tissue is also positive for prostate-specific membrane antigen in acinar cells and prostate-specific antigen in serous acinar cells and striated ducts (182). In fact, prostate-specific membrane antigen (PSMA) PET/CT consistently shows tracer uptake in healthy salivary, seromucous, and lacrimal glands and has been used to detect and stage salivary gland malignancies (183,184).

SPECIMEN HANDLING

The parotid gland is the most common location of major salivary gland tumors. The choice of surgical resection is largely dependent on the location of the tumor within the gland and its relationship to the branches of the facial

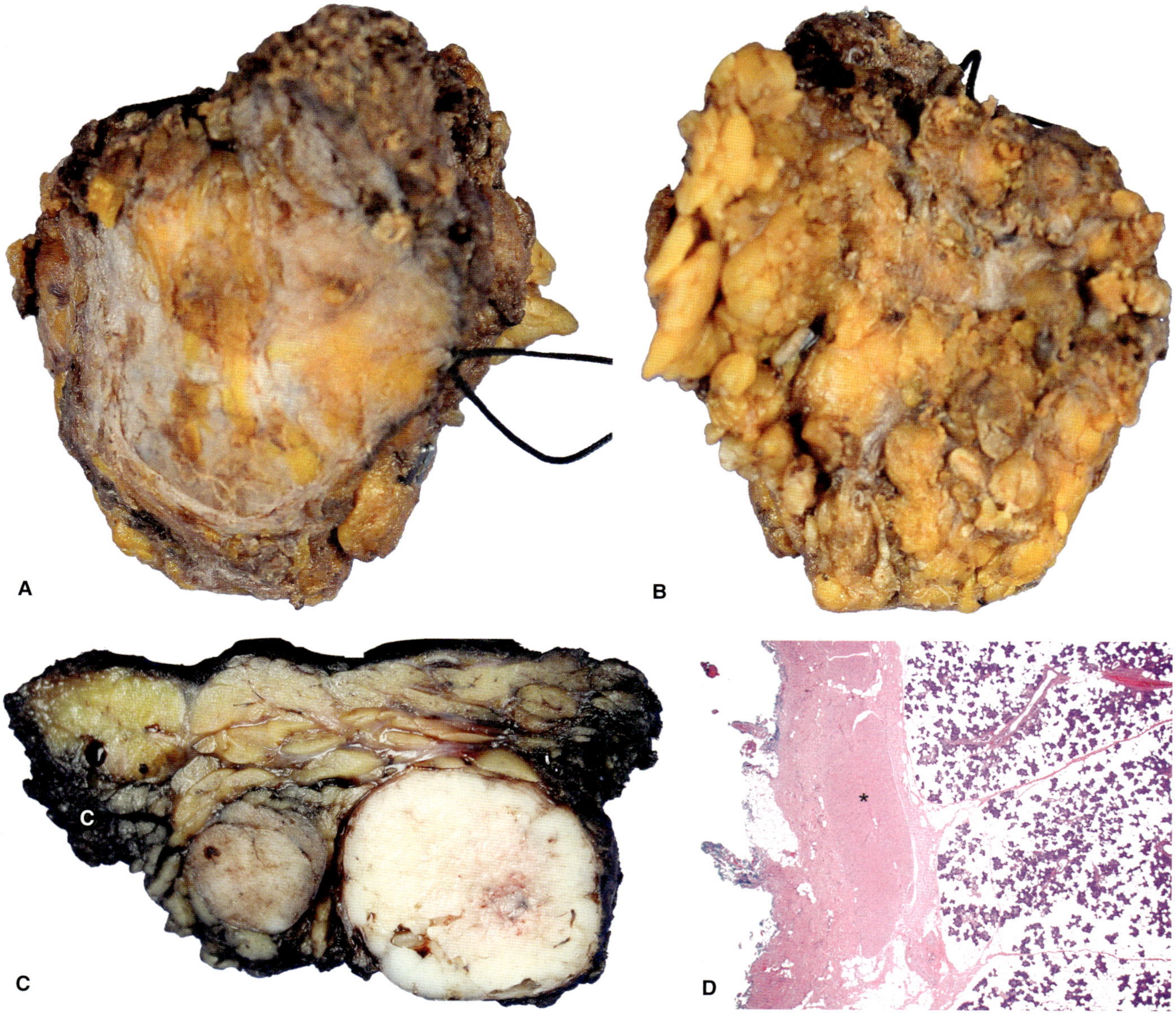

FIGURE 16.30 Skin-sparing, superficial parotidectomy specimen: lateral/superficial aspect (**A**), medial/deep aspect (**B**), and cut surface (**C**). The lateral/superficial aspect of the parotid is covered by parotid fascia which separates the gland from the subcutaneous soft tissue of the cheek. On cut section (**C**), the lateral/superficial aspect can be identified as the longest and smoothest surface of the specimen. Microscopically (**D**), the lateral/superficial is characterized by fibrous connective tissue and platysma muscle (*asterisk*).

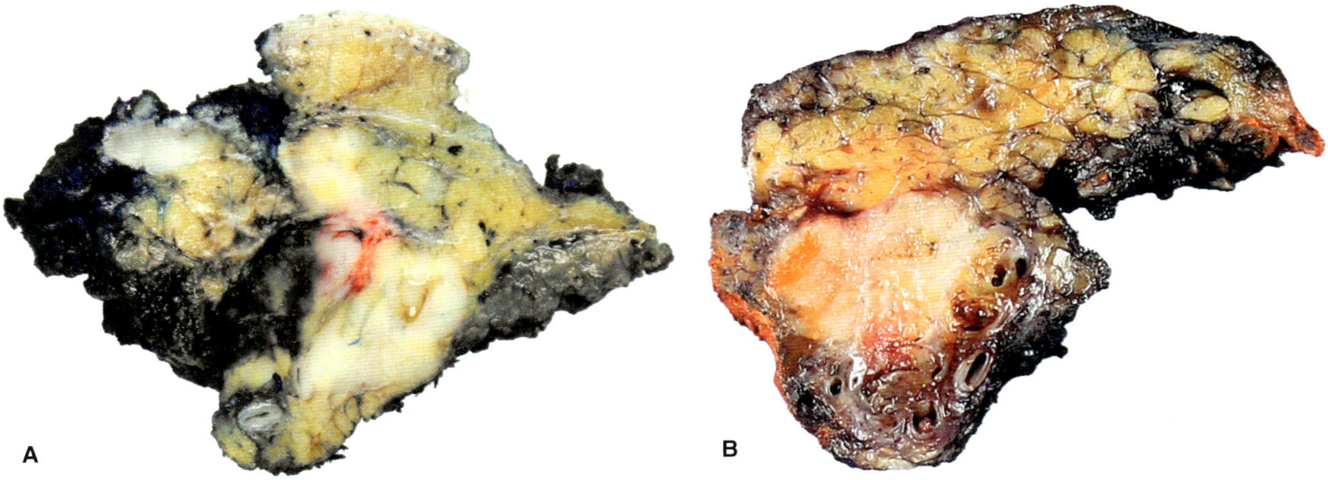

FIGURE 16.31 Gross sections of total parotidectomy specimens with (**A**) and without (**B**) overlying skin. The lateral/superficial aspect is obvious when there is overlying skin present (**A**); the parotid fascia can then be seen as a thin white plane of tissue overlying the gland and separating it from the subcutaneous fat. Without overlying skin (**B**), the lateral/superficial aspect can be identified as the longest and smoothest surface of the wedge-shaped specimen.

nerve. When in the superficial lobe, partial parotidectomy is the most common procedure of choice for diagnosis and treatment. Although the goal of the partial parotidectomy is to leave the facial nerve and its branches intact, complications such as temporary and permanent damage to the facial nerve may still occur.

When overlying skin is not resected with the parotid gland, a skin incision is carried down through the subcutaneous tissue and just deep to the superficial muscular aponeurotic system, which is continuous with the platysma inferiorly and the temporoparietal fascia superiorly, but above the parotid fascia. This creates a relatively smooth lateral/superficial surface that can be used to orient the specimen for inking and margin assessment (Figs. 16.30 and 16.31).

Total parotidectomy that includes both superficial and deep parotid lobes and preserves the facial nerve may be performed for recurrent or multifocal benign tumors. *En bloc* facial nerve sacrifice is otherwise indicated for malignant tumors that have compromised the facial nerve.

All specimens should be serially sectioned perpendicular to the long axis to evaluate periparotid tissue infiltration. Intraparotid and periparotid lymph nodes should also be identified. When radical parotidectomy includes selective neck dissection, the lymph nodes should be separated according to the different anatomical regions when possible.

Submandibular gland tumors are typically resected with level 1B selective neck dissection. Contrary to protocols for lymph node staging, resections for primary neoplasms of the submandibular gland should be inked for margin assessment.

REFERENCES

1. Porcheri C, Mitsiadis TA. Physiology, pathology and regeneration of salivary glands. *Cells*. 2019;8:976.
2. Proctor GB. The physiology of salivary secretion. *Periodontol 2000*. 2016;70:11–25.
3. Chibly AM, Aure MH, Patel VN, et al. Salivary gland function, development, and regeneration. *Physiol Rev*. 2022; 102:1495–1552.
4. Xu F, Laguna L, Sarkar A. Aging-related changes in quantity and quality of saliva: where do we stand in our understanding? *J Texture Stud*. 2019;50:27–35.
5. Becerra L, Soares RV, Bruno LS, et al. Patterns of secretion of mucins and non-mucin glycoproteins in human submandibular/sublingual secretion. *Arch Oral Biol*. 2003;48: 147–154.
6. de Paula F, Teshima THN, Hsieh R, et al. Overview of human salivary glands: highlights of morphology and developing processes. *Anat Rec (Hoboken)*. 2017;300:1180–1188.
7. Hand AR, Pathmanathan D, Field RB. Morphological features of the minor salivary glands. *Arch Oral Biol*. 1999;44 Suppl 1:S3–S10.
8. Valstar MH, de Bakker BS, Steenbakkers RJHM, et al. The tubarial salivary glands: a potential new organ at risk for radiotherapy. *Radiother Oncol*. 2021;154:292–298.
9. Banik S, Gaikwad MR, Bag ND. Tubarial salivary gland—the new member of nasopharynx. *J Oral Maxillofac Surg*. 2023;81:663–664.
10. Mudry A, Jackler RK. Are "tubarial salivary glands" a previously unknown structure? *Radiother Oncol*. 2021;154:314–315.
11. Iwanaga J, Ibaragi S, Nakano K, Takeshita Y, Tubbs RS. No convincing evidence for the presence of tubarial salivary glands: a letter to the editor regarding "The tubarial salivary glands: a potential new organ at risk for radiotherapy". *Radiother Oncol*. 2021;154:321–322.
12. Pringle S, Bikker FJ, Vogel W, et al. Immunohistological profiling confirms salivary gland-like nature of the tubarial glands and suggests closest resemblance to the palatal salivary glands. *Radiother Oncol*. 2023;187:109845.
13. Diesch T, Filippi C, Fritschi N, et al. Cytokines in saliva as biomarkers of oral and systemic oncological or infectious diseases: a systematic review. *Cytokine*. 2021;143:155506.

14. Szydlarska D, Grzesiuk W, Kondracka A, Bartoszewicz Z, Bar-Andziak E. Measuring salivary androgens as a useful tool in the diagnosis of polycystic ovary syndrome. *Endokrynol Pol.* 2012;63:183–190.
15. Voultsios A, Kennaway DJ, Dawson D. Salivary melatonin as a circadian phase marker: validation and comparison to plasma melatonin. *J Biol Rhythms.* 1997;12:457–466.
16. Rubin S, Eckhaus M, Rennick LJ, et al. Molecular biology, pathogenesis and pathology of mumps virus. *J Pathol.* 2015; 235:242–252.
17. Ghosh S, Kumar M, Santiana M, et al. Enteric viruses replicate in salivary glands and infect through saliva. *Nature.* 2022;607:345–350.
18. Atyeo N, Maldonado JO, Warner BM, Chiorono JA. Salivary glands and viral pathogenesis. *J Dent Res.* 2024;103: 227–234.
19. Beauregard M, Casey GA. Demonstration of rabies antigen in salivary glands of rabies suspected animals. *Can J Comp Med.* 1969;33:55–58.
20. Kloover JS, Hillebrands JL, de Wit G, et al. Rat cytomegalovirus replication in the salivary glands is exclusively confined to striated duct cells. *Virchows Arch.* 2000;437:413–421.
21. Matuck BF, Dolhnikoff M, Duarte-Neto AN, et al. Salivary glands are a target for SARS-CoV-2: a source for saliva contamination. *J Pathol.* 2021;254:239–243.
22. Labouyrie E, Merlio JP, Beylot-Barry M, et al. Human immunodeficiency virus type 1 replication within cystic lymphoepithelial lesion of the salivary gland. *Am J Clin Pathol.* 1993;100:41–46.
23. Shiiki N, Tokuyama S, Sato C, et al. Association between saliva PSA and serum PSA in conditions with prostate adenocarcinoma. *Biomarkers.* 2011;16:498–503.
24. Song M, Bai H, Zhang P, Zhhhou X, Ying B. Promising applications of human-derived saliva biomarker testing in clinical diagnostics. *Int J Oral Sci.* 2023;15:2.
25. Almeslet A, Alnamlah S, Alanzan L, Aldriwesh R, AlWehaiby S. Role of salivary biomarkers in cystic fibrosis: a systematic review. *Biomed Res Int.* 2022;2022:5818840.
26. Martinez-Madrigal F, Micheau C. Histology of the major salivary glands. *Am J Surg Pathol.* 1989;13:879–899.
27. Larsen M, Yamada KM, Musselmann K. Systems analysis of salivary gland development and disease. *Wiley Interdiscip Rev Syst Biol Med.* 2010;2:670–682.
28. Hsu JC, Yamada KM. Salivary gland branching morphogenesis–recent progress and future opportunities. *Int J Oral Sci.* 2010;2:117–126.
29. Mattingly A, Finley JK, Knox SM. Salivary gland development and disease. *Wiley Interdiscip Rev Dev Biol.* 2015;4:573–590.
30. Rothova M, Thompson H, Lickert H, Tucker AS. Lineage tracing of the endoderm during oral development. *Dev Dyn.* 2012;241:1183–1191.
31. Cutler LS. The role of extracellular matrix in the morphogenesis and differentiation of salivary glands. *Adv Dent Res.* 1990;4:27–33.
32. Sakai T. Epithelial branching morphogenesis of salivary gland: exploration of new functional regulators. *J Med Invest.* 2009;56 Suppl:234–238.
33. Holmberg KV, Hoffman MP. Anatomy, biogenesis and regeneration of salivary glands. *Monogr Oral Sci.* 2014;24:1–13.
34. Haara O, Fujimori S, Schmidt-Ullrich R, Hartmann C, Thesleff I, Mikkola MJ. Ectodysplasin and Wnt pathways are required for salivary gland branching morphogenesis. *Development.* 2011;138:2681–2691.
35. Melnick M, Phair RD, Lapidot SA, Jaskoll T. Salivary gland branching morphogenesis: a quantitative systems analysis of the Eda/Edar/NFkappaB paradigm. *BMC Dev Biol.* 2009;9:32.
36. Entesarian M, Dahlqvist J, Shashi V, et al. FGF10 missense mutations in aplasia of lacrimal and salivary glands (ALSG). *Eur J Hum Genet.* 2007;15:379–382.
37. Rohmann E, Brunner HG, Kayserili H, et al. Mutations in different components of FGF signaling in LADD syndrome. *Nat Genet.* 2006;38:414–417.
38. Chalard F, Hermann AL, Elmaleh-Berges M, Ducou le Pointe H. Imaging of parotid anomalies in infants and children. *Insights Imaging.* 2022;13:27.
39. Carlson GW. The salivary glands. Embryology, anatomy, and surgical applications. *Surg Clin North Am.* 2000;80:261–273, xii.
40. Wells KL, Patel N. Lumen formation in salivary gland development. *Front Oral Biol.* 2010;14:78–89.
41. Azuma M, Sato M. Morphogenesis of normal human salivary gland cells in vitro. *Histol Histopathol.* 1994;9:781–790.
42. Ianez RF, Buim ME, Coutinho-Camillo CM, Schultz R, Soares FA, Lourenco SV. Human salivary gland morphogenesis: myoepithelial cell maturation assessed by immunohistochemical markers. *Histopathology.* 2010;57:410–417.
43. Kumagai M, Sato I. Immunolocalization of fibronectin and collagen types I and III in human fetal parotid and submandibular glands. *Cells Tissues Organs.* 2003;173:184–190.
44. McGeary R, Rhyner PA, Desai AB, Bhatt AA. Navigating the parotid glands: anatomy, imaging work-up and next steps. *Clin Neuroradiol.* 2022;32:615–623.
45. Kochhar A, Larian B, Azizzadeh B. Facial nerve and parotid gland anatomy. *Otolaryngol Clin North Am.* 2016;49: 273–284.
46. Ahn D, Yeo CK, Han SY, Kim JK. The accessory parotid gland and facial process of the parotid gland on computed tomography. *PLoS One.* 2017;12:e0184633.
47. Aasen MH, Hutz MJ, Yuhan BT, Britt CJ. Deep lobe parotid tumors: a systematic review and meta-analysis. *Otolaryngol Head Neck Surg.* 2022;166:60–67.
48. Gasser RF. The early development of the parotid gland around the facial nerve and its branches in man. *Anat Rec.* 1970;167:63–77.
49. Gray H, Williams PL. *Gray's Anatomy.* Lea & Febiger; 1973.
50. Spuntarelli G, Santecchia L, Urbani U, Zama M. Minor salivary gland neoplasm in children. *J Craniofac Surg.* 2013; 24:664–667.
51. Frommer J. The human accessory parotid gland: its incidence, nature, and significance. *Oral Surg Oral Med Oral Pathol.* 1977;43:671–676.
52. Capaccio P, Luca N, Sigismund PE, Pignataro L. Recurrent inflammation of accessory parotid tissue associated with unilateral parotid gland aplasia: diagnostic and therapeutic implications. *Eur Arch Otorhinolaryngol.* 2012;269:1551–1554.
53. Toh H, Kodama J, Fukuda J, et al. Incidence and histology of human accessory parotid glands. *Anat Rec.* 1993;236: 586–590.
54. Ferreira JN, Hoffman MP. Interactions between developing nerves and salivary glands. *Organogenesis.* 2013;9:199–205.
55. Steel SJ, Robertson CE. First bite syndrome: what neurologists need to know. *Curr Pain Headache Rep.* 2021;25:31.

56. Schmalfuss IM, Tart RP, Mukherji S, Mancuso AA. Perineural tumor spread along the auriculotemporal nerve. *AJNR Am J Neuroradiol*. 2002;23:303–311.
57. Smith AM. Observations: gross anatomy revisited. *J Grad Med Educ*. 2015;7:495.
58. Singh P, Warnakulasuriya S. Aplasia of submandibular salivary glands associated with ectodermal dysplasia. *J Oral Pathol Med*. 2004;33:634–636.
59. Nordgarden H, Johannessen S, Storhaug K, Jensen JL. Salivary gland involvement in hypohidrotic ectodermal dysplasia. *Oral Dis*. 1998;4:152–154.
60. Schrank TP, Mikhaylov Y, Zanation AM. Surgical excision of the submandibular gland. *Oper Tech Otolaryngol Head Neck Surg*. 2018;29:162–167.
61. Hoffman HT, Koch M, Witt RL, et al. Proposal for standardized ultrasound analysis of the salivary glands: part 1 submandibular gland. *Laryngoscope Investig Otolaryngol*. 2024;9:e1224.
62. Beahm DD, Peleaz L, Nuss DW, et al. Surgical approaches to the submandibular gland: a review of literature. *Int J Surg*. 2009;7:503–509.
63. Mandel L, Alfi D. Diagnostic imaging for submandibular duct atresia: literature review and case report. *J Oral Maxillofac Surg*. 2012;70:2819–2822.
64. Prosdocimo ML, Barreto Nogueira AP, Aparecida de Albuquerque Cavalcante M, et al. Congenital dilatation of the submandibular duct. *Int J Pediatr Otorhinolaryngol*. 2018;113:16–18.
65. Pownell PH, Brown OE, Pransky SM, et al. Congenital abnormalities of the submandibular duct. *Int J Pediatr Otorhinolaryngol*. 1992;24:161–169.
66. Grewal JS, Jamal Z, Ryan J. Anatomy, head and neck, submandibular gland. *StatPearls*; 2024.
67. Gregoire V, Ang K, Budach W, et al. Delineation of the neck node levels for head and neck tumors: a 2013 update. DAHANCA, EORTC, HKNPCSG, NCIC CTG, NCRI, RTOG, TROG consensus guidelines. *Radiother Oncol*. 2014;110:172–181.
68. Abou-Foul AK. Surgical anatomy of the lymphatic drainage of the salivary glands: a systematic review. *J Laryngol Otol*. 2020:1–7.
69. Fives C, Feeley L, Sadadcharam M, O'Leary G, Sheahan P. Incidence of intraglandular lymph nodes within submandibular gland, and involvement by floor of mouth cancer. *Eur Arch Otorhinolaryngol*. 2017;274:461–466.
70. Li L, Gao X-L, Song Y-Z, et al. Anatomy of arteries and veins of submandibular glands. *Chin Med J (Engl)*. 2007;120:1179–1182.
71. Carpenter GH, Khosravani N, Ekstrom J, Osailan SM, Paterson KP, Proctor GB. Altered plasticity of the parasympathetic innervation in the recovering rat submandibular gland following extensive atrophy. *Exp Physiol*. 2009;94:213–219.
72. Nedvetsky PI, Emmerson E, Finley JK, et al. Parasympathetic innervation regulates tubulogenesis in the developing salivary gland. *Dev Cell*. 2014;30:449–462.
73. Bradley PJ. Sublingual gland. *Oper Tech Otolaryngol Head Neck Surg*. 2018;29:P168–176.
74. Ab Rahim NAC, Liew YT, Ghauth S, Narayanan P, Abu Bakar Z. A single institution cadaveric study on anatomical variation of the sublingual gland duct. *Indian J Otolaryngol Head Neck Surg*. 2023;75:347–351.
75. Zhang L, Xu H, Cai ZG, et al. Clinical and anatomic study on the ducts of the submandibular and sublingual glands. *J Oral Maxillofac Surg*. 2010;68:606–610.
76. Zhang B, Yang Z, Zhang RM, et al. Are the patients with anatomic variation of the sublingual/Wharton's duct system predisposed to ranula formation? *Int J Pediatr Otorhinolaryngol*. 2016;83:69–73.
77. Zhi K, Gao L, Ren W. What is new in management of pediatric ranula? *Curr Opin Otolaryngol Head Neck Surg*. 2014;22:525–529.
78. Tandler B, Gresik EW, Nagato T, Philips CJ. Secretion by striated ducts of mammalian major salivary glands: review from an ultrastructural, functional, and evolutionary perspective. *Anat Rec*. 2001;264:121–145.
79. Takakura IKS, Yashuhara R, Tanaka J, Ide F, Mishima K. Assessment of MEF2C as a novel myoepithelial marker using normal salivary gland and pleomorphic adenoma: an immunohistochemical study. *J Oral Maxillofac Surg Med Pathol*. 2022;34:523–530.
80. Hardy G, Kramer B. The myoepithelium of human major salivary glands revisited. *SADJ*. 1998;53:371–375.
81. Takano K, Malamud D, Bennick A, Oppenheim F, Hand AR. Localization of salivary proteins in granules of human parotid and submandibular acinar cells. *Crit Rev Oral Biol Med*. 1993;4:399–405.
82. Gorr S-U, Venkatesh SG, Darling DS. Parotid secretory granules: crossroads of secretory pathways and protein storage. *J Dent Res*. 2005;84:500–509.
83. Caselitz J, Jaup T, Seifert G. Lactoferrin and lysozyme in carcinomas of the parotid gland. A comparative immunocytochemical study with the occurrence in normal and inflamed tissue. *Virchows Arch A Pathol Anat Histol*. 1981;394:61–73.
84. Reitamo S, Konttinen YT, Segerberg-Konttinen M. Distribution of lactoferrin in human salivary glands. *Histochemistry*. 1980;66:285–291.
85. Eversole LR. The mucoprotein histochemistry of human mucous acinar cell containing salivary glands: submandibular and sublingual glands. *Arch Oral Biol*. 1972;17:43–53.
86. Lilliu MA, Solinas P, Cossu M, et al. Diabetes causes morphological changes in human submandibular gland: a morphometric study. *J Oral Pathol Med*. 2015;44:291–295.
87. Lilliu MA, Loy F, Cossu M, Solinas P, Isola R, Isola M. morphometric study of diabetes related alterations in human parotid gland and comparison with submandibular gland. *Anat Rec (Hoboken)*. 2015;298:1911–1918.
88. Wahl AM, Takano T, Su S, et al. Structural and functional analysis of salivary intercalated duct cells reveals a secretory phenotype. *J Physiol*. 2023;601:4539–4556.
89. Dardick I, Rippstein P, Skimming L, Boivin M, Parks WR, Dairkee SH. Immunohistochemistry and ultrastructure of myoepithelium and modified myoepithelium of the ducts of human major salivary glands: histogenetic implications for salivary gland tumors. *Oral Surg Oral Med Oral Pathol*. 1987;64:703–715.
90. Tavares Dos Santos H, Nam K, Maslow FM, et al. Tuft cells are present in submandibular glands across species. *J Histochem Cytochem*. 2022;70:659–667.
91. Riva A, Serra GP, Proto E, Faa G, Puxeddu R, Riva FT. The myoepithelial and basal cells of ducts of human major salivary glands: a SEM study. *Arch Histol Cytol*. 1992;55 Suppl:115–124.

92. Sato A, Miyoshi S. Tuft cells in the main excretory duct epithelia of the three major rat salivary glands. *Eur J Morphol.* 1996;34:225–228.
93. Tandler B, Pinkstaff CA, Phillips CJ. Interlobular excretory ducts of mammalian salivary glands: structural and histochemical review. *Anat Rec A Discov Mol Cell Evol Biol.* 2006;288:498–526.
94. Kotas ME, O'Leary CE, Locksley RM. Tuft Cells: context- and tissue-specific programming for a conserved cell lineage. *Annu Rev Pathol.* 2023;18:311–335.
95. Sato A, Miyoshi S. Ultrastructure of the main excretory duct epithelia of the rat parotid and submandibular glands with a review of the literature. *Anat Rec.* 1988;220:239–251.
96. Gnepp DR. My journey into the world of salivary gland sebaceous neoplasms. *Head Neck Pathol.* 2012;6:101–110.
97. Batsakis JG, el-Naggar AK. Sebaceous lesions of salivary glands and oral cavity. *Ann Otol Rhinol Laryngol.* 1990;99:416–418.
98. Linhartova A. [Sebaceous cells in repeat biopsies of normal salivary glands and in sebaceous carcinoma of the parotid gland]. *Cesk Patol.* 1982;18:21–25.
99. Meza-Chavez L. Sebaceous glands in normal and neoplastic parotid glands; possible significance of sebaceous glands in respect to the origin of tumors of the salivary glands. *Am J Pathol.* 1949;25:627–645.
100. Shah AAK, Mulla AF, Mayank M. Pathophysiology of myoepithelial cells in salivary glands. *J Oral Maxillofac Pathol.* 2016;20:480–490.
101. Erlandson RA, Cardon-Cardo C, Higgins PJ. Histogenesis of benign pleomorphic adenoma (mixed tumor) of the major salivary glands. An ultrastructural and immunohistochemical study. *Am J Surg Pathol.* 1984;8:803–820.
102. Dardick I. A role for electron microscopy in salivary gland neoplasms. *Ultrastruct Pathol.* 1985;9:151–161.
103. Prasad AR, Savera AT, Gown AM, Zarbo RJ. The myoepithelial immunophenotype in 135 benign and malignant salivary gland tumors other than pleomorphic adenoma. *Arch Pathol Lab Med.* 1999;123:801–806.
104. Nakazato Y, Ishida Y, Takahashi K, Suzuki K. Immunohistochemical distribution of S-100 protein and glial fibrillary acidic protein in normal and neoplastic salivary glands. *Virchows Arch A Pathol Anat Histopathol.* 1985;405:299–310.
105. Ohtomo R, Mori T, Shibata S, et al. SOX10 is a novel marker of acinus and intercalated duct differentiation in salivary gland tumors: a clue to the histogenesis for tumor diagnosis. *Mod Pathol.* 2013;26:1041–1050.
106. Curran AE, Allen CM, Beck FM, Damm DD, Murrah VA. Distinctive pattern of glial fibrillary acidic protein immunoreactivity useful in distinguishing fragmented pleomorphic adenoma, canalicular adenoma and polymorphous low grade adenocarcinoma of minor salivary glands. *Head Neck Pathol.* 2007;1:27–32.
107. Gorski Z, Purgina B, Wasserman JK. Glial fibrillary acidic protein expression helps distinguish pleomorphic adenoma from histologic mimics. *Head Neck Pathol.* 2022;16:695–702.
108. Marinkovic M, Tran ON, Wang H, et al. Extracellular matrix turnover in salivary gland disorders and regenerative therapies: obstacles and opportunities. *J Oral Biol Craniofac Res.* 2023;13:693–703.
109. Hart DA. One of the primary functions of tissue-resident pluripotent pericytes cells may be to regulate normal organ growth and maturation: implications for attempts to repair tissues later in life. *Int J Mol Sci.* 2022;23:5496.
110. Sakakura T. Serendipity; close encounter of tenascin C. *Front Immunol.* 2020;11:620182.
111. Abedsaeidi M, Hojjati F, Tavassoli A, Sahebkar A. Biology of tenascin C and its role in physiology and pathology. *Curr Med Chem.* 2024;31:2706–2731.
112. Sunardhi-Widyaputra S, Van Damme B. Immunohistochemical expression of tenascin in normal human salivary glands and in pleomorphic adenomas. *Pathol Res Pract.* 1993;189:138–143.
113. Nicolescu MI, Bucur A, Dinca O, Rusu MC, Popescu LM. Telocytes in parotid glands. *Anat Rec (Hoboken).* 2012;295:378–385.
114. Kurian EM, Miller R, McLean-Holden AL, Oliai BR, Bishop JA. Low molecular weight cytokeratin immunostaining for extrafollicular reticulum cells is an effective means of separating salivary gland tumor-associated lymphoid proliferation from true lymph node involvement. *Head Neck Pathol.* 2020;14:593–597.
115. Xu X, Roberts SA, Pasha TL, Zhang PJ. Undesirable cytokeratin immunoreactivity of native nonepithelial cells in sentinel lymph nodes from patients with breast carcinoma. *Arch Pathol Lab Med.* 2000;124:1310–1313.
116. Franke WW, Moll R. Cytoskeletal components of lymphoid organs. I. Synthesis of cytokeratins 8 and 18 and desmin in subpopulations of extrafollicular reticulum cells of human lymph nodes, tonsils, and spleen. *Differentiation.* 1987;36:145–163.
117. Harrison JD. Causes, natural history, and incidence of salivary stones and obstructions. *Otolaryngol Clin North Am.* 2009;42:927–947.
118. Routsias JG, Goules JD, Charalampakis G, Tzima S, Papageorgiou A, Voulgarelis M. Malignant lymphoma in primary Sjogren's syndrome: an update on the pathogenesis and treatment. *Semin Arthritis Rheum.* 2013;43:178–186.
119. Shanti RM, Aziz SR. HIV-associated salivary gland disease. *Oral Maxillofac Surg Clin North Am.* 2009;21:339–343.
120. Abdul Karim L, Kwon DH, Ozdemirli M. Salivary gland heterotopia in the gastroesophageal junction: a case series and review of the literature. *Case Rep Gastrointest Med.* 2018;2018:6078581.
121. Dikman SH, Toker C. Seromucinous gland ectopia within the prostatic stroma. *J Urol.* 1973;109:852–854.
122. Haemel A, Gnepp DR, Carlsten J, Robinson-Bostom L. Heterotopic salivary gland tissue in the neck. *J Am Acad Dermatol.* 2008;58:251–256.
123. Edwards PC, Bhuiya T, Kahn LB, Fantasia JE. Salivary heterotopia of the parathyroid gland: a report of two cases and review of the literature. *Oral Surg Oral Med Oral Pathol Oral Radiol Endod.* 2005;99:590–593.
124. Purnell PR, Interval E, Williams HJ, Cassis A. Middle ear choristoma presenting as cholesteatoma with conductive hearing loss. *J Surg Case Rep.* 2019;2019:rjz129.
125. Schochet SS Jr, McCormick WF, Halmi NS. Salivary gland rests in the human pituitary. Light and electron microscopical study. *Arch Pathol.* 1974;98:193–200.
126. Goodman RS, Daly JF, Valensi Q. Heterotopic salivary tissue and branchial cleft sinus. *Laryngoscope.* 1981;91:260–264.
127. Yamasaki H, Tashiro H, Watanabe T. Congenital parotid gland fistula. *Int J Oral Maxillofac Surg.* 1986;15:492–494.

128. Pattanshetti S, Menon P, Solanki S, Saxena A. Congenital parotid fistula: surgical management in two infants and review of literature. *J Indian Assoc Pediatr Surg*. 2021;26:48–50.
129. Dutta M. The ectopic accessory parotid system with congenital cheek fistula: an overview and current update. *Laryngoscope*. 2017;127:1351–1360.
130. Resta L, Piscitelli D, Fiore MG, Sanguedolce F. Role of electron microscopy in diagnosis of parotid tumours. *Acta Otorhinolaryngol Ital*. 2005;25:150–152.
131. Dardick I, Birek C, Lingen MW, Rowe PE. Differentiation and the cytomorphology of salivary gland tumors with specific reference to oncocytic metaplasia. *Oral Surg Oral Med Oral Pathol Oral Radiol Endod*. 1999;88:691–701.
132. Li N, Ye Y, Wu Y, et al. Alterations in histology of the aging salivary gland and correlation with the glandular inflammatory microenvironment. *iScience*. 2023;26:106571.
133. Mandel L, Carrao V. Bilateral parotid diffuse hyperplastic oncocytosis: case report. *J Oral Maxillofac Surg*. 2005;63:560–562.
134. Azevedo LR, Damante JH, Lara VS, Lauris JRP. Age-related changes in human sublingual glands: a post mortem study. *Arch Oral Biol*. 2005;50:565–574.
135. Chang H-C, Juan C-J, Chiu H-C, et al. Effects of gender, age, and body mass index on fat contents and apparent diffusion coefficients in healthy parotid glands: an MRI evaluation. *Eur Radiol*. 2014;24:2069–2076.
136. Katona K, Elekes E, Farkas N, et al. Image analysis of fatty infiltration in labial salivary gland biopsies: extent and its correlation to age, obesity and diabetes. *J Oral Pathol Med*. 2017;46:537–542.
137. Garcia DS, Bussoloti Filho I. Fat deposition of parotid glands. *Braz J Otorhinolaryngol*. 2013;79:173–176.
138. Cunha JLS, de Sousa SF, Mota CP, et al. Sialolipomas of minor salivary glands: a multi-institutional study and literature review. *J Oral Pathol Med*. 2021;50:210–219.
139. Dardick I, Jeans MT, Sinnott NM, et al. Salivary gland components involved in the formation of squamous metaplasia. *Am J Pathol*. 1985;119:33–43.
140. Zhurakivska K, Maiorano E, Nocini R, et al. Necrotizing sialometaplasia can hide the presence of salivary gland tumors: a case series. *Oral Dis*. 2019;25:1084–1090.
141. Gnepp DR. Warthin tumor exhibiting sebaceous differentiation and necrotizing sialometaplasia. *Virchows Arch A Pathol Anat Histol*. 1981;391:267–273.
142. Davis AB, Hoffman HT. Management options for sialadenosis. *Otolaryngol Clin North Am*. 2021;54:605–611.
143. Ihrler S, Rath C, Zengel P, et al. Pathogenesis of sialadenosis: possible role of functionally deficient myoepithelial cells. *Oral Surg Oral Med Oral Pathol Oral Radiol Endod*. 2010;110:218–223.
144. Yu GY, Donath K. Adenomatous ductal proliferation of the salivary gland. *Oral Surg Oral Med Oral Pathol Oral Radiol Endod*. 2001;91:215–221.
145. Weinreb I, Seethala RR, Hunt JL, Chetty R, Dardick I, Perez-Ordonez B. Intercalated duct lesions of salivary gland: a morphologic spectrum from hyperplasia to adenoma. *Am J Surg Pathol*. 2009;33:1322–1329.
146. Naunheim MR, Lin HW, Faquin WC, Lin DT. Intercalated duct lesion of the parotid. *Head Neck Pathol*. 2012;6:373–376.
147. Chetty R. Intercalated duct hyperplasia: possible relationship to epithelial-myoepithelial carcinoma and hybrid tumours of salivary gland. *Histopathology*. 2000;37:260–263.
148. Di Palma S. Epithelial-myoepithelial carcinoma with coexisting multifocal intercalated duct hyperplasia of the parotid gland. *Histopathology*. 1994;25:494–496.
149. Lindor NM, Kasperbauer J, Lewis JE, Pittelkow M. Birt-Hogg-Dube syndrome presenting as multiple oncocytic parotid tumors. *Hered Cancer Clin Pract*. 2012;10:13.
150. Walker NI, Gobe GC. Cell death and cell proliferation during atrophy of the rat parotid gland induced by duct obstruction. *J Pathol*. 1987;153:333–344.
151. Coppes RP, Stokman MA. Stem cells and the repair of radiation-induced salivary gland damage. *Oral Dis*. 2011;17:143–153.
152. Rocchi C, Barazzuol L, Coppes RP. The evolving definition of salivary gland stem cells. *NPJ Regen Med*. 2021;6:4.
153. Carpenter GH, Cotroneo E. Salivary gland regeneration. *Front Oral Biol*. 2010;14:107–128.
154. Takahashi S, Schoch E, Walker NI. Origin of acinar cell regeneration after atrophy of the rat parotid induced by duct obstruction. *Int J Exp Pathol*. 1998;79:293–301.
155. Alvares EP, Sesso A. Cell proliferation, differentiation and transformation in the rat submandibular gland during early postnatal growth. A quantitative and morphological study. *Arch Histol Jpn*. 1975;38:177–208.
156. Yamashina S, Tamaki H, Katsumata O. The serous demilune of rat sublingual gland is an artificial structure produced by conventional fixation. *Arch Histol Cytol*. 1999;62:347–354.
157. Tandler B. Are demilunes in mixed salivary glands real or fixation artifacts? A critique. *J Cytol Histol*. 2014;5:218.
158. Mukunyadzi P, Bardales RH, Palmer HE, Stanley MW. Tissue effects of salivary gland fine-needle aspiration. Does this procedure preclude accurate histologic diagnosis? *Am J Clin Pathol*. 2000;114:741–745.
159. Douville NJ, Bradford CR. Comparison of ultrasound-guided core biopsy versus fine-needle aspiration biopsy in the evaluation of salivary gland lesions. *Head Neck*. 2013;35:1657–1661.
160. Shah KS, Ethunandan M. Tumour seeding after fine-needle aspiration and core biopsy of the head and neck–a systematic review. *Br J Oral Maxillofac Surg*. 2016;54:260–265.
161. Konings AW, Coppes RP, Vissink A. On the mechanism of salivary gland radiosensitivity. *Int J Radiat Oncol Biol Phys*. 2005;62:1187–1194.
162. Song EC, Min S, Oyelakin A, et al. Genetic and scRNA-seq analysis reveals distinct cell populations that contribute to salivary gland development and maintenance. *Sci Rep*. 2018;8:14043.
163. Ninche N, Kwak M, Ghazizadeh S. Diverse epithelial cell populations contribute to the regeneration of secretory units in injured salivary glands. *Development*. 2020;147:dev192807.
164. Athwal HK, Murphy G III, Tibbs E, et al. Sox10 regulates plasticity of epithelial progenitors toward secretory units of exocrine glands. *Stem Cell Reports*. 2019;12:366–380.
165. Tatemoto Y, Kumasa S, Watanabe Y, Mori M. Epithelial membrane antigen as a marker of human salivary gland acinar and ductal cell function. *Acta Histochem*. 1987;82:219–226.
166. Raman S, Sherlin HJ. Utility of smooth muscle actin and CD117 as reliable markers in the diagnosis of salivary gland neoplasms. *J Oral Maxillofac Pathol*. 2019;23:218–223.
167. Stead RH, Qizilbash AH, Kontozoglou T, Daya AD, Riddell RH. An immunohistochemical study of pleomorphic adenomas of the salivary gland: glial fibrillary acidic protein-like

immunoreactivity identifies a major myoepithelial component. *Hum Pathol.* 1988;19:32–40.
168. Katori Y, Hayashi S, Takanashi Y, et al. Heterogeneity of glandular cells in the human salivary glands: an immunohistochemical study using elderly adult and fetal specimens. *Anat Cell Biol.* 2013;46:101–112.
169. Zhu S, Schuerch C, Hunt J. Review and updates of immunohistochemistry in selected salivary gland and head and neck tumors. *Arch Pathol Lab Med.* 2015;139:55–66.
170. Hsieh MS, Lee YH, Chang YL. SOX10-positive salivary gland tumors: a growing list, including mammary analogue secretory carcinoma of the salivary gland, sialoblastoma, low-grade salivary duct carcinoma, basal cell adenoma/adenocarcinoma, and a subgroup of mucoepidermoid carcinoma. *Hum Pathol.* 2016;56:134–142.
171. Lee JH, Kang HJ, Yoo CW, et al. PLAG1, SOX10, and Myb expression in benign and malignant salivary gland neoplasms. *J Pathol Transl Med.* 2019;53:23–30.
172. Zarbo RJ, Regezi JA, Batsakis JG. S-100 protein in salivary gland tumors: an immunohistochemical study of 129 cases. *Head Neck Surg.* 1986;8:268–275.
173. Redman RS. Histologic changes in the salivary glands following radiation therapy. In: Cha S, ed. *Salivary Gland Development and Regeneration.* Springer International Publishing; 2017:75–93.
174. Kouros-Mehr H, Slorach EM, Sternlicht MD, Werb Z. GATA-3 maintains the differentiation of the luminal cell fate in the mammary gland. *Cell.* 2006;127:1041–1055.
175. Valimaa H, Savolainen S, Soukka T, et al. Estrogen receptor-beta is the predominant estrogen receptor subtype in human oral epithelium and salivary glands. *J Endocrinol.* 2004;180:55–62.
176. Shick PC, Riordan GP, Foss RD. Estrogen and progesterone receptors in salivary gland adenoid cystic carcinoma. *Oral Surg Oral Med Oral Pathol Oral Radiol Endod.* 1995;80:440–444.
177. Gorbokon N, Timm P, Dum D, et al. Mammaglobin-A expression is highly specific for tumors derived from the breast, the female genital tract, and the salivary gland. *Diagnostics (Basel).* 2023;13:1202.
178. Chenevert J, Duvvuri U, Chiosea S, et al. DOG1: a novel marker of salivary acinar and intercalated duct differentiation. *Mod Pathol.* 2012;25:919–929.
179. Sumitomo S, Kumasa S, Mitani H, Mori M. Comparison of CEA distribution in lesions and tumors of salivary glands as determined with monoclonal and polyclonal antibodies. *Virchows Arch B Cell Pathol Incl Mol Pathol.* 1987;53:133–139.
180. Patel S, Wald AI, Bastaki JM, Chiosea SI, Singhi AD, Seethala RR. NKX3.1 expression and molecular characterization of secretory myoepithelial carcinoma (SMCA): advancing the case for a salivary mucous acinar phenotype. *Head Neck Pathol.* 2023;17:467–478.
181. Meirovitz A, Shouchane-Blum K, Maly A, et al. The potential of somatostatin receptor 2 as a novel therapeutic target in salivary gland malignant tumors. *J Cancer Res Clin Oncol.* 2021;147:1335–1340.
182. Isola M, Maxia C, Murtas D, et al. Prostate-specific antigen: an unfamiliar protein in the human salivary glands. *J Anat.* 2024;244:873–881.
183. Klein Nulent TJW, Valstar MH, de Keizer B, et al. Physiologic distribution of PSMA-ligand in salivary glands and seromucous glands of the head and neck on PET/CT. *Oral Surg Oral Med Oral Pathol Oral Radiol.* 2018;125:478–486.
184. Civan C, Kasper S, Berliner C, et al. PSMA-directed imaging and therapy of salivary gland tumors: a single-center retrospective study. *J Nucl Med.* 2023;64:372–378.

SECTION VI

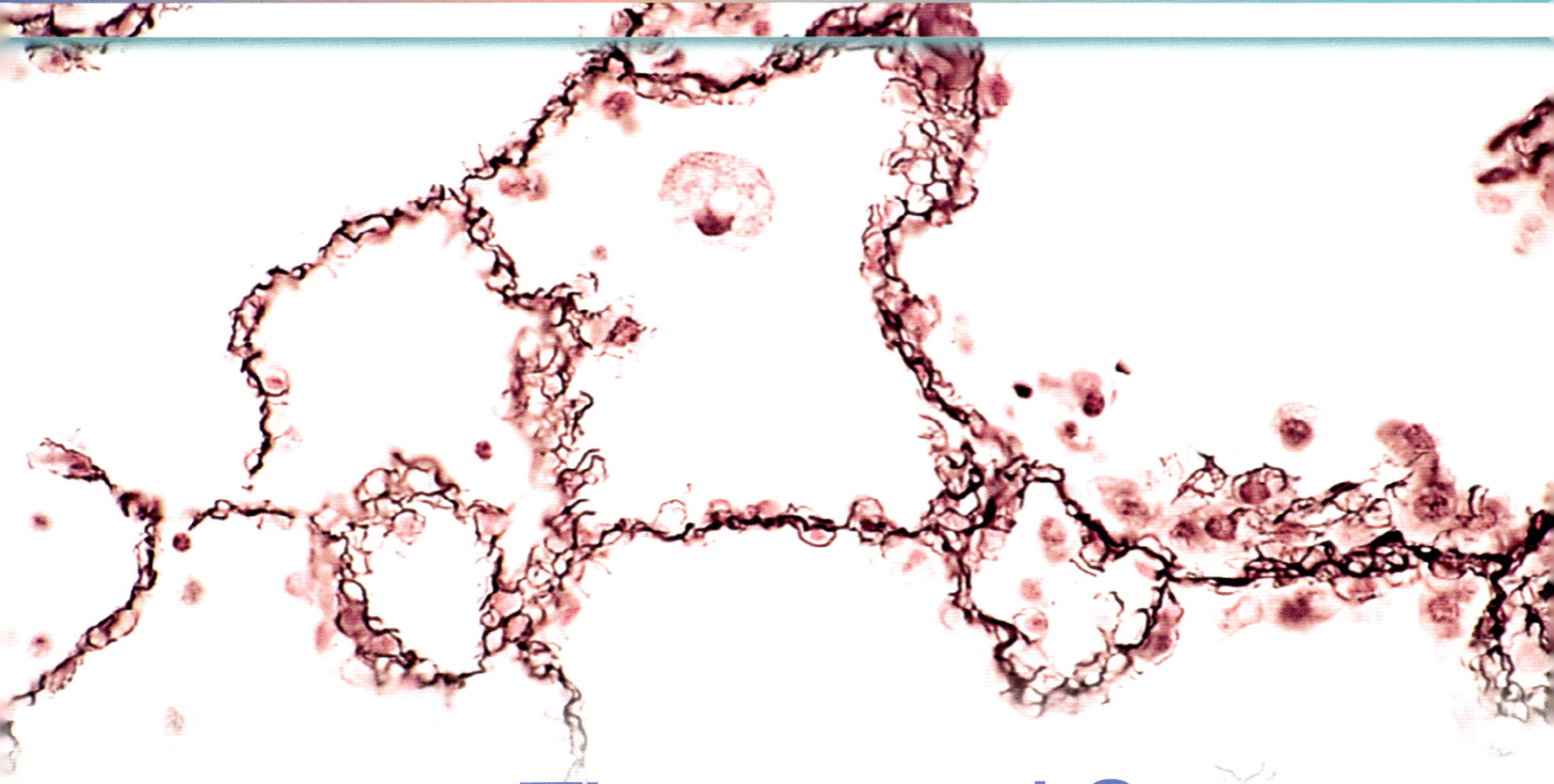

Thorax and Serous Membranes

17

Lungs

Yasmeen M. Butt ■ Henry D. Tazelaar

NORMAL STRUCTURE AND HISTOLOGY

The following review is based on multiple, including some classic, references (1–19).

General

The lungs are paired intrathoracic organs that are divided into lobes. The right lobe consists of right upper, right middle, and right lower lobes while the left is composed of the left upper and left lower lobes. The lingula is a rudimentary appendage arising from the left upper lobe and is analogous to the right middle lobe. The lobes are further divided into bronchopulmonary segments (Table 17.1).

The segmental anatomy of the lung is important for radiologists, interventional pulmonologists, and pathologists in defining the location of lesions. The lobes are divided by fissures, each with its own pleural investments, although these fissures are incomplete or poorly developed in some cases. The segments are not separated by fissures and do not normally have separate pleural investments, although they are recognizable on the basis of their supplying bronchi termed the segmental bronchi.

The primordial lungs arise as a ventral bud off the foregut that extends caudally into the primitive thoracic mesenchyme. The bud elongates and bifurcates to form the right and left main bronchi. Bronchial cartilage, smooth muscle, and other connective tissues are derived from the surrounding mesenchyme. The right and left main bronchial buds elongate into their respective thoracic cavities, branching repeatedly until the bronchial tree is complete. The phases of airway and lung parenchymal development are summarized in Table 17.2.

A complex series of epithelial–mesenchymal reactions occur as the lung progresses through successive phases of development. These stages are modified by physiologic mechanical forces and humoral factors. They are regulated by homeobox genes, nuclear transcription factors, hormones, and other soluble mediators such as growth factors, chemokines, and cytokines (17,18).

TABLE 17.1 Bronchopulmonary Segments[a]

Right upper lobe	Left upper lobe
1. Apical	1, 2. Apical posterior
2. Posterior	3. Anterior
3. Anterior	Lingula
Right middle lobe	4. Superior
4. Lateral	5. Inferior
5. Medial	Left lower lobe
Right lower lobe	6. Superior
6. Superior	7. Anterior-medial basal
7. Medial basal	8. Lateral basal
8. Anterior basal	9. Posterior basal
9. Lateral basal	
10. Posterior basal	

[a]Modified from Kuhn C III. Ultrastructure and cellular function in the distal lung. In: Thurlbeck WM, Abell MR, eds. *The Lung*. Williams & Wilkins; 1978.

Airways

The airways serve as conduits for air traveling to and from the alveoli. They remove inhaled foreign material *via* the mucociliary escalator and play import roles in immune surveillance, air-moisturizing, and air-warming.

The airways arise by unequal dichotomous branching of the bronchial buds. The normal lung contains approximately 20 generations (range: 10–30 divisions) extending from the trachea to the respiratory bronchioles. The diameter of an airway is approximately the same as its accompanying artery in histologic sections (19). Together, these structures form a bronchovascular bundle. Disparities in size of either the airway or artery suggest a pathologic condition. Airways are divided into "large" and "small" types from a radiographic and clinical standpoint. "Small" airways are those with an internal diameter of less than 2 mm (20). On the other hand, airways are histologically classified based on their composition, as follows:

- Bronchi are cartilaginous airways and are usually more than 2 mm in diameter (Fig. 17.1). They are conducting airways and the cartilage plates in their walls help prevent complete collapse during bronchial constriction and expiration. Submucosal seromucinous glands are a characteristic landmark of bronchi. The external wall also contains a circular layer of smooth muscle cells. The cartilage plates may calcify or ossify with aging (Fig. 17.2).
- Bronchioles are membranous airways that are usually less than 2 mm in diameter. They lack cartilage and submucosal glands and normally have very few goblet cells (Fig. 17.3).
- Nonrespiratory bronchioles are defined as bronchioles located proximal to respiratory bronchioles.
- Terminal bronchioles are nonrespiratory bronchioles located just proximal to respiratory bronchioles.

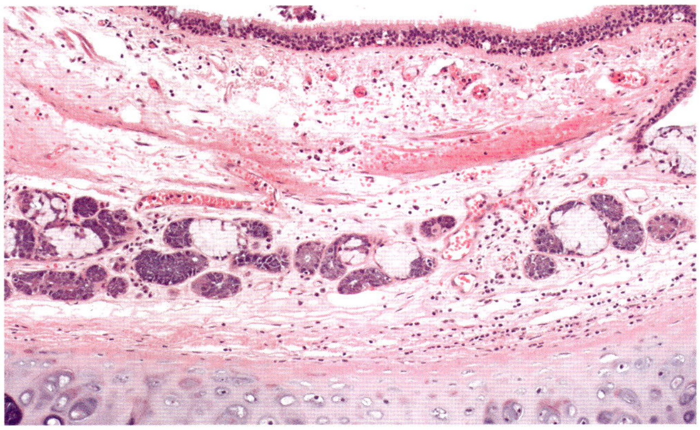

FIGURE 17.1 Bronchus. Scattered goblet cells can be appreciated in the mucosa with clusters of bronchial submucosal glands. Beneath the epithelial basement membrane, there is a vascularized layer of connective tissue with wisps of smooth muscle above the submucosal glands. Cartilage can be appreciated at the bottom of the image.

TABLE 17.2 Phases of Lung Development

Phase	Gestation (approx. weeks)	Major Events
Embryonic	3½–6	Development of major airways and pleura
Pseudoglandular	6–16	Development of airways to terminal bronchioles and parts of prospective lung parenchyma including the beginning of the development of the acinus
Canalicular	16–28	Completion of the branching morphogenesis. Further development of the acinus and its vascularization. Earliest alveolocapillary barrier formation. Beginning of surfactant production
Saccular	28–36	Subdivision of saccules by secondary crests
Alveolar	36 wks to term (and up to 21 yrs of age)	Acquisition of alveoli

Modified from Schitny JC. Development of the lung. *Cell Tissue Res.* 2017;367:427–444.

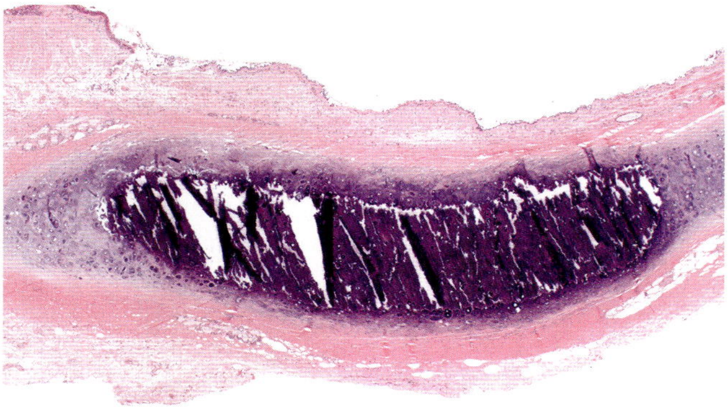

FIGURE 17.2 Calcification of cartilage in bronchi can happen with aging. Ossification (not pictured) may also occur.

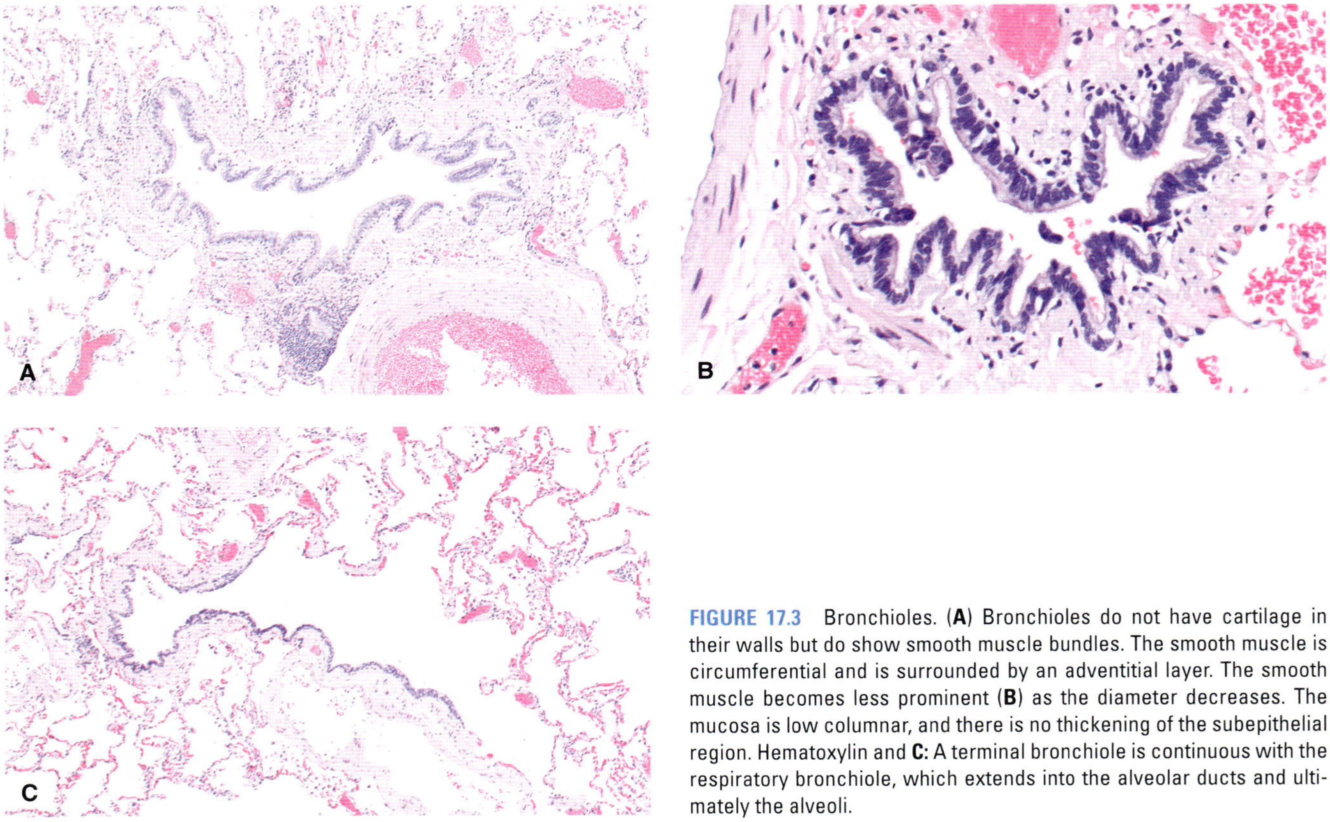

FIGURE 17.3 Bronchioles. (**A**) Bronchioles do not have cartilage in their walls but do show smooth muscle bundles. The smooth muscle is circumferential and is surrounded by an adventitial layer. The smooth muscle becomes less prominent (**B**) as the diameter decreases. The mucosa is low columnar, and there is no thickening of the subepithelial region. Hematoxylin and **C:** A terminal bronchiole is continuous with the respiratory bronchiole, which extends into the alveolar ducts and ultimately the alveoli.

- Respiratory bronchioles are airways that have alveoli budding from their walls. They respire and serve as conducting airways.

The surface epithelium of large bronchi rests on a basement membrane below which there is an elastin-rich layer of connective tissue. Together these elements comprise the bronchial mucosa. Beneath the bronchial mucosa lies the submucosa, in which submucosal glands, cartilage, nerves, ganglia, and branches of the bronchial arteries may be found. There is no clear histologic boundary between the mucosa and submucosa; however, the basement membrane has been used as an arbitrary boundary. Deep to the submucosa lies a peribronchial sheath composed of loose connective tissue, which is continuous with that of the accompanying pulmonary artery. The bronchial epithelium is a pseudostratified columnar epithelium composed primarily of ciliated columnar cells with interspersed mucous-containing goblet cells. Lesser numbers of neuroendocrine cells, basal cells, brush cells, and migratory inflammatory cells are also normally present. The height of the pseudostratified columnar epithelium decreases with progressive branching of the bronchial tree.

The walls of bronchioles are normally much thinner than those of the bronchi. The surface epithelium rests on a basement membrane, which overlies a thin layer of loose elastin-rich connective tissue. This is surrounded by a muscularis composed of a layer of smooth muscle cells. The muscularis is, in turn, invested by a peribronchiolar connective tissue sheath which is continuous with that of the adjacent artery.

Direct communications between nonrespiratory bronchioles and alveoli have been identified and termed Lambert canals (21). These are thought to promote collateral ventilation in the peripheral lung. Lambert canals are rarely visible in histologic sections but may be conspicuous in scarred airways when surrounded by the metaplastic bronchiolar epithelium best described as peribronchiolar metaplasia (Fig. 17.4). Peribronchiolar metaplasia is a relatively common incidental finding in a variety of lung diseases and can be seen in association with aspiration, infection, smoking and other chronic inhalation diseases, and hypersensitivity pneumonitis, among other diseases. It is typically indicative of chronic injury to small airways (22–24).

Cell types of the airway epithelium include basal cells, neuroendocrine cells, ciliated cells, club cells, goblet cells, intermediate cells, and brush cells. Ultrastructural abnormalities in the ciliated cells of the respiratory tract are associated with primary ciliary dyskinesia as well as a handful of other pathologic conditions. Goblet cells and ciliated cells decrease in number as the terminal bronchioles are approached and the mucosa becomes less columnar and more cuboidal in appearance. These changes are accompanied by a concomitant increase in club cells. Club, or Clara, cells (25) have secretory functions and produce a surfactant-like protein that probably stabilizes surfactant. They also serve as progenitor

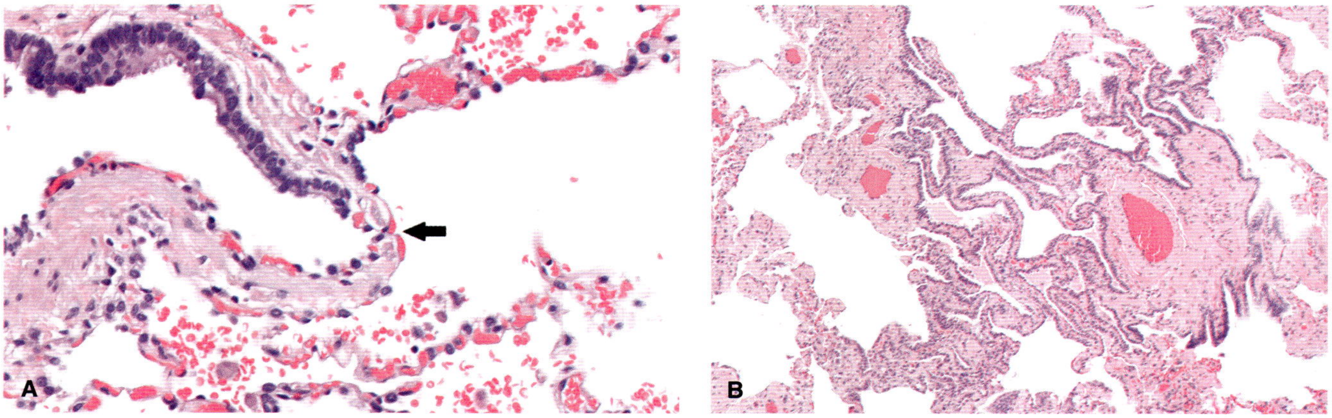

FIGURE 17.4 Lambert canals. **A:** These canals represent communications between nonrespiratory bronchioles and adjacent alveoli and are only rarely observed in histologic sections (*arrow*). **B:** It is these canals that are thought to be the origin of peribronchiolar metaplasia, which is seen as a repair phenomenon involving peribronchiolar alveoli after bronchiolar injury.

cells for reconstituting the epithelium after bronchiolar injury. Club cell differentiation in tumors of the lung may be appreciated by the presence of apical PAS-positive, diastase-resistant granules that are also visible as dense granules ultrastructurally. Neuroendocrine cells of the lung were previously referred to as Kulchitsky cells. They contain dense core granules and are part of the diffuse neuroendocrine system. Aggregates of neuroepithelial cells, termed neuroepithelial bodies, are often found at airway bifurcations. The presence of goblet cells in bronchioles (i.e., goblet cell metaplasia) usually signifies chronic airway injury and may be seen in patients with bronchiectasis, asthma, and a smoking history. The cell types in the airways and parenchyma are summarized in Table 17.3 and illustrated in Figures 17.5 and 17.6.

TABLE 17.3 Major Cell Types of the Lower Respiratory Tract[a]

Cell Type	Features	Function(s)	Location	Histochemical and/or Immunohistochemical Staining
Ciliated	Columnar, cuboidal, ciliated bronchial lining cells; each cell has approximately 250 cilia at the apical surface, and each cilium is approximately 6 μm long	Proximal transport of mucous stream (mucociliary escalator)	Bronchi and bronchioles	Epithelial markers,[b] tubulin, BRAF V600E reacts with cilia
Goblet	Columnar mucus-secreting cells; contain mucous glycoprotein, which discharges apically	Contribute to airway mucous mucin content	Bronchi (more numerous proximally); small numbers in bronchioles	Epithelial markers,[b] histochemical mucin stains, MUC5AC
Basal	Short cells with relatively little cytoplasm; oriented along the basement membrane; do not reach the luminal surface of the epithelium	Precursor cell of ciliated and goblet cells	Bronchi; rare in bronchioles	Epithelial markers[b], p40, p63
Neuroendocrine (Kulchitsky or K cells)	Basal-oriented cells with numerous dense-core (neurosecretory) granules; single or in groups (neuroepithelial bodies), the latter near sites of airway bifurcation	Specific functions not known; considered part of the diffuse neuroendocrine system. Believed to be involved in oxygen sensing, smooth muscle tonus, and immune responses	Bronchi; rare in bronchioles	Chromogranin-A, synaptophysin, CD56, INSM1
Brush	Found infrequently at all levels of the airways; some have termed these type III pneumocytes; they are named for a brush border of microvilli approximately 2 μm in length	Thought to be involved in fluid absorption or chemoreceptor function	All airways	Ultrastructurally identified

TABLE 17.3 Major Cell Types of the Lower Respiratory Tract[a] (*Continued*)

Cell Type	Features	Function(s)	Location	Histochemical and/or Immunohistochemical Staining
Serous	Identical to serous cells in the minor salivary gland tissues	Produce secretion of lower viscosity than that from mucous cells	Primarily bronchioles	Lysozyme
Neuroendocrine bodies	Clusters of 4–10 neuroendocrine cells adjacent to the subepithelial basement membrane	Unknown; hypotheses include chemoreceptor, tactile receptor, vasoconstrictive functions	Bronchi, bronchioles, and alveoli	Chromogranin-A, synaptophysin, CD56, calcitonin, INSM1
Oncocytic	Eosinophilic mitochondria-rich cells in submucosal gland ducts	Ion secretory functions	Submucosal glands	Epithelial markers[b]
Squamous	Stratified squamous epithelium is an abnormal metaplastic replacement of normal pseudostratified respiratory epithelium	Protective, reparative	Bronchi, bronchioles, and occasionally alveoli	Epithelial markers,[b] desmoglein-3
Club	Cuboidal, nonciliated, nonmucous bronchiolar cells; protuberant apical cytoplasm with large, ovoid electron-dense granules; comprise the majority of nonciliated distal bronchiolar cells	Secretory functions of proteins of the extracellular lining fluid in distal airways, regulating its content; progenitor for other bronchiolar cells; role in protection of surfactant; metabolism of toxic substances; regulate the pulmonary immune system	Predominantly in terminal and respiratory bronchioles	CC10, CC16, diastase-resistant PAS-positive apical granules
Type I alveolar pneumocyte	Large, flat, squamous alveolar lining cells; cover some 93% of alveolar surface area; incapable of division	Provide a thin air–blood interface for gas transfer	Alveoli	Epithelial markers,[b] caveolin, and aquaporin
Type II alveolar pneumocyte	Columnar alveolar lining cells; microvillous surface; synthesize and secrete surfactant (lamellar ultrastructural inclusions); capable of division	Maintain alveolar stability; progenitor for type I pneumocytes	Alveoli	Epithelial markers,[b] surfactant protein C, TTF-1, Napsin-A
Minor salivary tissue: serous, mucous, ductal cells	Submucosal minor salivary glands identical to other sites with serous and mucinous acinar cells that secrete into the ducts, which empty at the mucosal surface	Secretion and contribution to airway mucous stream	Bronchial submucosa	Epithelial markers,[b] histochemical stains for mucin, Alcian blue/PAS
Smooth muscle	Bundled smooth muscle surrounds the conducting airways to the level of the alveolar ducts	Contraction of the airway	Peripheral in the airway and external to the cartilage in bronchi	Muscle-specific actin, smooth muscle actin, desmin, vimentin
Other cells[c]				

[a]Table is modified from Castranova V, Rabovsky J, Tucker JH, Miles PR. The alveolar type II epithelial cell: a multifunctional pneumocyte. *Toxicol Appl Pharmacol.* 1988;93:472–483; Colby TV, Koss MN, Travis WD. Tumors of the lower respiratory tract. In: Rosai J, ed. *Atlas of Tumor Pathology.* 3rd series, Fascicle 13. Armed Forces Institute of Pathology; 1995:465–471; Corrin B. *Pathology of the Lungs.* Churchill Livingstone; 2000.; Kasper M, Reimann T, Hempel U, et al. Loss of caveolin expression in type I pneumocyte as an indicator of subcellular alterations during lung fibrogenesis. *Histochem Cell Biol.* 1998;109:41–48; Kreda SM, Gynn MC, Fenstermacher DA, et al. Expression and localization of epithelial aquaporins in the adult human lung. *Am J Respir Cell Mol Biol.* 2001;24:224–234; Lou YP, Takeyama K, Grattan KM, et al. Platelet-activating factor induces goblet cell hyperplasia and mucin gene expression in airways. *Am J Respir Crit Care Med.* 1998;157(pt 1):1927–1934; Rogers AV, Dewar A, Corrin B, et al. Identification of serous-like cells in the surface epithelium of human bronchioles. *Eur Respir J.* 1995;6:498–504; Ryerse JS, Hoffmann JW, Mahmoud S, et al. Immunolocalization of CC10 in Clara cells in mouse and human lung. *Histochem Cell Biol.* 2001;115:325–332; Branchfield K, Nantie L, Verheyden JM, et al. Pulmonary neuroendocrine cells function as airway sensors to control lung immune response. *Science.* 2016;351(6274):707–710; Rokicki W, Rokicki M, Wojtacha J, et al. The role and importance of club cells (Clara cells) in the pathogenesis of some respiratory diseases. *Kardiochir Torakochirurgia Pol.* 2016;13(1):26–30.

[b]For example, pancytokeratin (AE1/AE3, OSCAR), CAM5.2, epithelial membrane antigen, cytokeratin 5/6, others.

[c]Endothelial cells and pericytes; interstitial fibrocytes, fibroblasts, and myofibroblasts; macrophages; lymphoid cells, including Langerhans cells; mast cells; mesothelial pleural lining; cartilage and bone; smooth muscle; peripheral nerves; and myoepithelial cells.

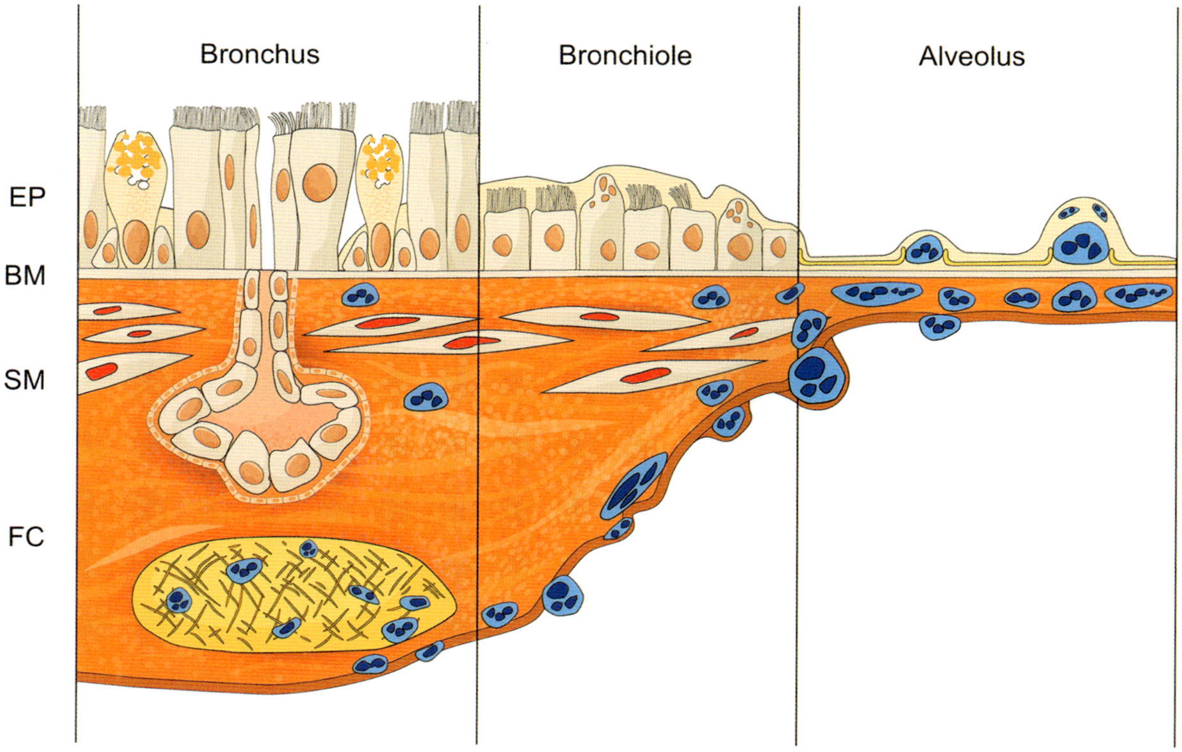

FIGURE 17.5 Respiratory tract epithelia. There is a progression of pseudostratified ciliated columnar epithelium in the large airways, to a more cuboidal epithelium in the small airways, to type I pneumocytes in the alveoli. The epithelium in the large airways is designed for maintaining and moving the mucous stream, whereas the flat pneumocytes, type I, lining the airspaces facilitate gas transfer. *EP*, epithelium; *BM*, basement membrane; *SM*, smooth muscle; *FC*, fibroconnective tissue. (Weibel ER, Taylor CR. Functional design of the human lung for gas exchange. In: Fishman AP, ed. *Pulmonary Diseases and Disorders*. Vol. 1. 3rd ed. McGraw-Hill; 1988:21–61.)

The layers of smooth muscle cells in the airways play an important functional role in regulating airflow in the lungs. Smooth muscle cell layers are arranged in a complex spiral pattern around the airways that becomes progressively less prominent in the distal conducting airways. The airway receives its nutrition from the bronchial arteries. Bundles of smooth muscle cells can be seen in the airways at the level of the alveolar ducts where they are interrupted by alveolar openings and associated with increased lamellae of elastic fibers. They may appear as isolated round or oblong aggregates of smooth muscle cells in cross sections (Fig. 17.7).

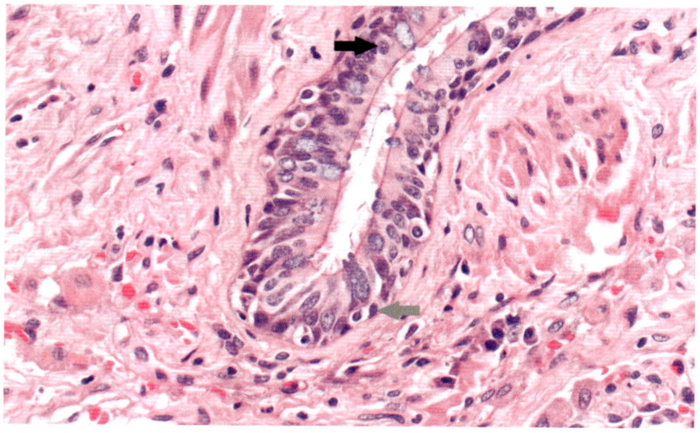

FIGURE 17.6 Bronchial epithelium. Normal bronchial epithelium is pseudostratified and columnar, with numerous ciliated cells and scattered basophilic and flocculent-appearing goblet cells. Club cells (*black arrow*) are interspersed throughout and there are smaller inconspicuous basal cells above the basement membrane (*gray arrow*).

Submucosal salivary-type glands containing both serous and mucous cells are found in the larger bronchi and

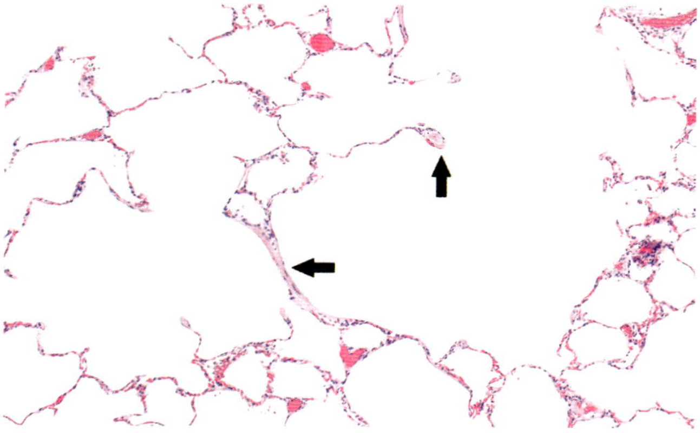

FIGURE 17.7 Bundles of smooth muscle can be seen at the level of alveolar ducts and may appear as isolated round or oblong aggregates on cross section (*arrows*).

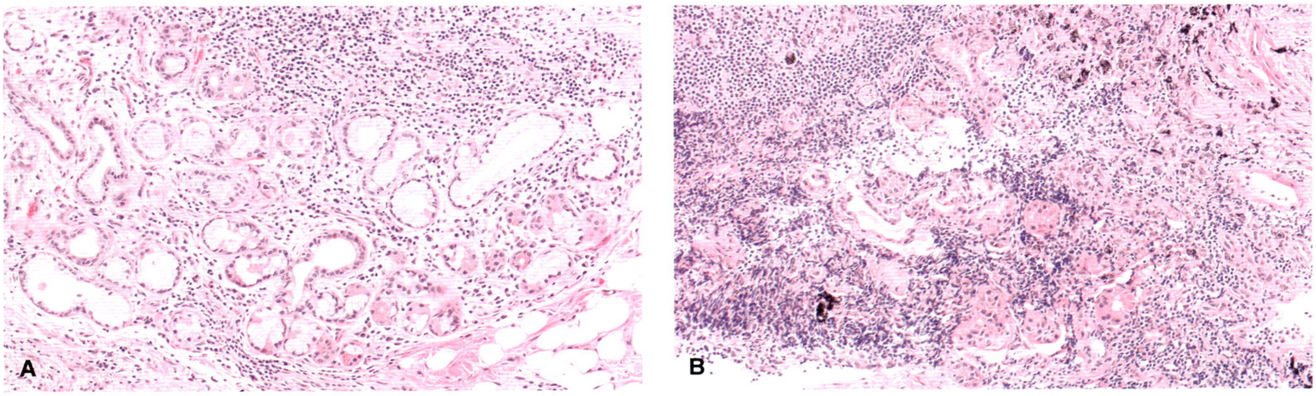

FIGURE 17.8 Bronchial submucosal glands. **A:** The bronchial submucosal glands are normally located in the submucosa above the bronchial cartilage and contain mixed seromucous glands with a duct leading to the bronchial mucosa (also Fig. 17.1). **B:** It is not uncommon for these glands to be present within bronchial lymph nodes and their presence should not be mistaken for adenocarcinoma.

adjacent lymph nodes (Fig. 17.8A,B). Oncocytic metaplasia can be seen in the glands of older adults. Large airway walls contain ganglia, nerves, and bronchial arteries.

Lobule and Acinus

The lungs are divided into lobes, segments, and lobules. The smallest of these subdivisions is the lobule, an aggregate of pulmonary acini bounded by the connective tissue of the interlobular septa (Fig. 17.9). The lobules span 1 to 2 cm in diameter and are made visible from both the pleura and cut surface of the parenchyma by their septal demarcations. Lobules are accentuated in fibroinflammatory conditions in which the septa become thickened by collagen and/or cellular infiltrates. Pulmonary lobules are identifiable with high-resolution computed tomography (CT) scanning of the lung (12). Although some authors subclassify lobules, this practice is discouraged because the distinction between secondary and primary lobules is inconsistent and of limited clinical significance.

The functional unit of the lung is the acinus where gas transfer takes place (Fig. 17.10). The precise definition of the acinus is variable. Some define it as the lung tissue supplied by a single terminal bronchiole and, thus, suggest that each pulmonary lobule comprises some 3 to 10 acini (12,15). Others define the acinus as a respiratory bronchiole and its supplied alveolar ducts and sacs (10,13). According to this definition, each lobule comprises some 20 to 30 acini that are 1 to 2 mm in diameter.

The epithelial cells lining alveoli can be classified as types I and II pneumocytes (Table 17.3). Type I pneumocytes are flat and responsible for gas exchange. Type II pneumocytes are cuboidal are present sparsely in the alveolus and produce surfactant. They are progenitors of type I

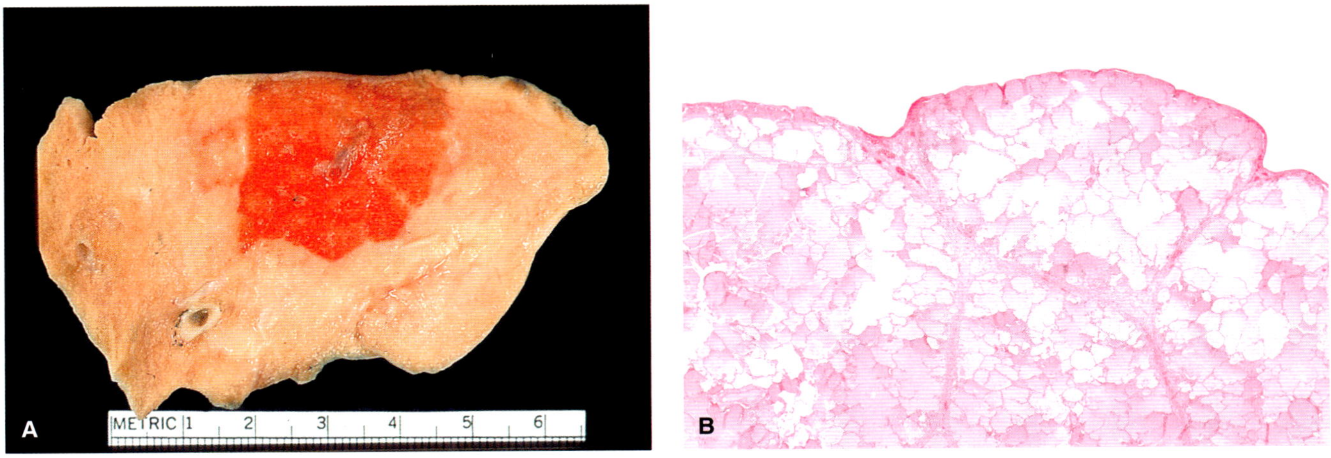

FIGURE 17.9 Pulmonary lobule. **A:** This cut section of normal lung tissue shows focal hemorrhage highlighting a pulmonary lobule. The hemorrhage stops abruptly at the interlobular septa and, centrally, a bronchovascular bundle can be appreciated. The lobule is approximately 2 cm in diameter. **B:** Interlobular septa, which serve as the boundaries of pulmonary lobules, are often inconspicuous. This case has extensive pulmonary edema, allowing for easy visualization of the interlobular septa.

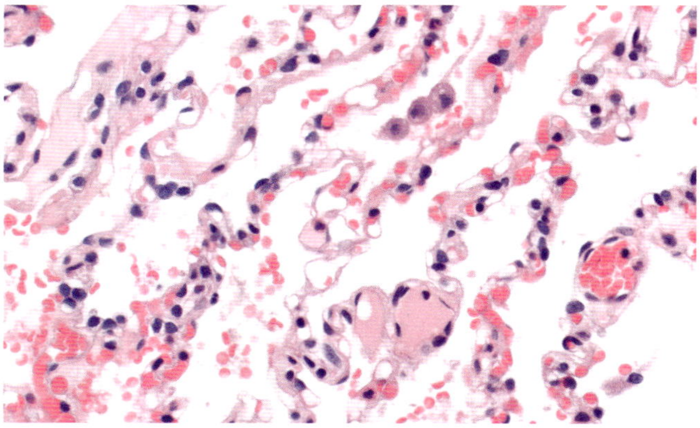

FIGURE 17.10 Distal lung parenchyma. The acinus is the functional unit of the lung where gas transfer takes place. Individual capillaries can be seen within the alveolar septa with inconspicuous overlying type II pneumocytes.

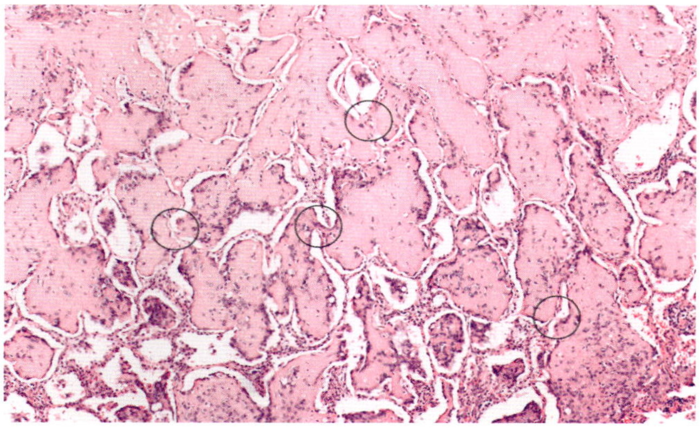

FIGURE 17.12 Pores of Kohn. The pores of Kohn (*circles*) can be challenging to distinguish and are best appreciated in process that traverse them such as organizing pneumonia or an epithelioid hemangioendothelioma (pictured).

cells and proliferate after injury to restore alveolar epithelial integrity. Type II cell hyperplasia represents a nonspecific marker of alveolar injury and repair (Fig. 17.11). Hyperplastic pneumocytes can appear very reactive/atypical and can be confused with neoplastic processes. A key feature of hyperplastic pneumocytes is a preserved N:C ratio despite the presence of an enlarged nucleus with a prominent nucleolus. Macrophages are a normal finding in the lung, scattered over the surfaces of the alveoli and percolating into the interstitium; a number of subpopulations of pulmonary macrophages are definable, based on their anatomic locations and their role in host defense and mucous clearance functions (26). Pulmonary macrophages are greatly increased in cigarette smokers. Increased Langerhans cells are also found in the bronchiolar epithelium of smokers, but their accurate recognition requires immunohistochemical stains such as S100 protein, CD1a, and langerin.

The epithelial and capillary basement membranes in the alveolar septum are irregularly fused. Gas transfer takes place across the alveolar–capillary membrane, which includes the attenuated cytoplasm of the type I cell, the endothelial cell cytoplasm, and their fused basement membranes. Pores of Kohn represent direct communications between adjacent alveoli through a "pore" in the alveolar wall. They are thought to be involved in collateral ventilation. Pores of Kohn are rarely visible with the light microscope, but can sometimes be appreciated when areas of organizing pneumonia span adjacent alveoli (Fig. 17.12).

The lung is invested with a rich framework of connective tissue coursing throughout the interstitium. It is well developed and easily visible along bronchovascular sheaths and in the septa that delimit lobules. This framework is continuous from the hilum to the pleura and encompasses the interstitial compartment of the lung down to the level of the alveolar wall and perivascular areas. Collagen, elastic fibers, mesenchymal cells, and a few inflammatory cells can be identified in the alveolar walls. This alveolar interstitial space is normally inconspicuous by light microscopy in adults, although some interstitial widening and increased cellularity are normal histologic findings in young children.

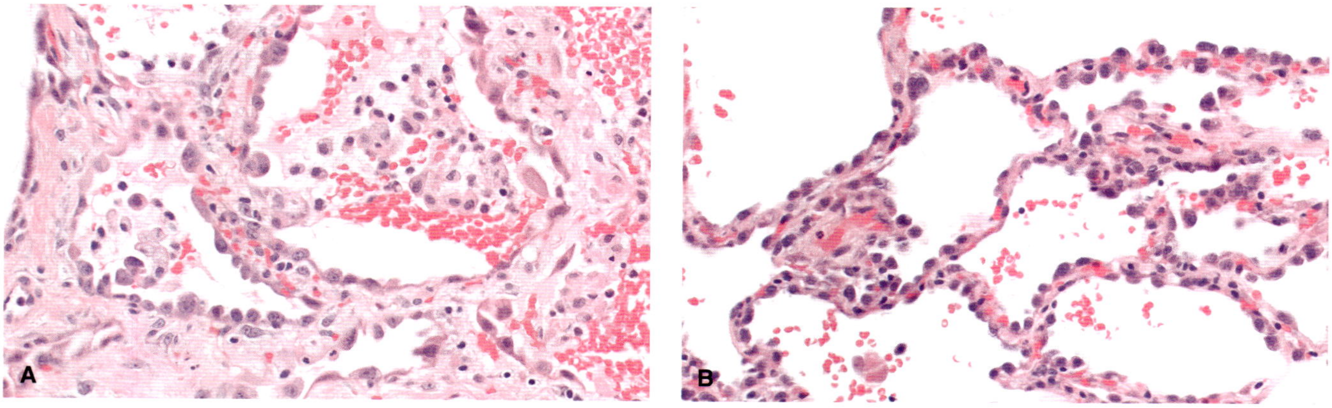

FIGURE 17.11 Reactive alveolar cell hyperplasia. **A:** A reactive proliferation of type II pneumocytes (alveolar cell hyperplasia) is common after injury. This case illustrates how this appears as a single row of cells protruding from the alveolar surface. **B:** Significant cytologic atypia is lacking. The interstitium underlying this process may show edema, inflammatory cells, and some fibrinous exudate consistent with recent lung injury.

Given the concentration of nuclei of the various cell types comprising the alveolar wall, it may be difficult to separate the cell types and decide when the degree of cellularity is pathologic. Markers such as CD45 may be beneficial, as is comparing various foci in the biopsy since interstitial infiltrates are rarely uniform in their distribution.

Vasculature

The lungs have two independent vascular systems: the arteries and veins of the bronchial, or systemic system and those of the pulmonary system. The alveolar microvasculature is designed for oxygenation of venous blood derived from the pulmonary arteries. Large elastic pulmonary arteries in infants are similar to the aorta in structure but the elastic fiber lamellae become more irregular, fragmented, and less compact in adulthood. Elastic tissue remains relatively prominent in the pulmonary arterial tree to approximately the point where bronchi become bronchioles. At this juncture, the pulmonary arteries transition to primarily muscular arteries. The muscular pulmonary arteries and arterioles have both internal and external elastic membranes (Fig. 17.13A) while the pulmonary veins have only a lamellated single (outer) elastica (Fig. 17.13B). Small intra-acinar pulmonary veins merge into larger veins in the interlobular septa. The bronchial arteries in the walls of bronchi are part of the systemic circulation and have pressures similar to systemic arterial pressures. The pulmonary circulation is a low-pressure system with a normal mean pressure of approximately 10 mm Hg, although it is somewhat higher at higher elevations above sea level.

It may be difficult to distinguish small pulmonary arteries from venules, especially since a single elastic lamina forms as arteries diminish in caliber. In pathologic states, the veins may develop muscular hypertrophy and increased mural thickness, making the distinction of arteries from veins even more difficult. The location of the vessel, particularly if it is in a septum or accompanied by an airway, is extremely helpful, and sometimes the only way to confidently distinguish pulmonary veins from small pulmonary arteries. The pulmonary arteries may appear histologically normal in patients with mild pulmonary hypertension and, thus, pressure measurements taken at catheterization are more reliable than histologic subtleties such as myointimal thickening, in determining the degree of pulmonary hypertension. Only plexiform and dilatation lesions connote unequivocal pulmonary hypertension. The presence of some degree of diffuse interstitial fibrosis or localized pulmonary scars is commonly associated with mural thickening of pulmonary vessels and should not be interpreted as evidence of pulmonary hypertension.

Lymphatic Vessels and Lymphoid Tissue

The lung is invested with a rich supply of lymphatic vessels and lymphoid tissue. Lymphatic drainage proceeds toward the hilum of the lung. Lymph fluid from the lower lobes tends to drain to mediastinal lymph nodes, with lymph fluid from the remaining portions of the lung draining to tracheobronchial lymph nodes. On the left side, the lymph fluid drains into the thoracic duct; and, on the right side, it drains into the right bronchomediastinal trunk. Both of these ultimately drain into their respective subclavian veins. Lymphatic channels are found along bronchovascular structures, interlobular septae, and within the pleura, while pulmonary veins can be identified in the alveolar septae, interlobular septae, and pleura. Valves may be apparent in some sections. Lymphatic vessels do not extend into alveolar walls so interstitial water is "pumped" out by lymphatic channels present either centrally in the lobule or at the lobule periphery. The lymphatic vessels are inconspicuous except in pathologic states, such as pulmonary edema or lymphangitic carcinoma. Lymphoid infiltrates, sarcoidosis, and some pneumoconioses tend to be distributed along the lymphatic routes and, thus, these infiltrates are referred to as perilymphatic/lymphangitic radiologically. However, the lymphatic vessels themselves may not be prominent in these conditions.

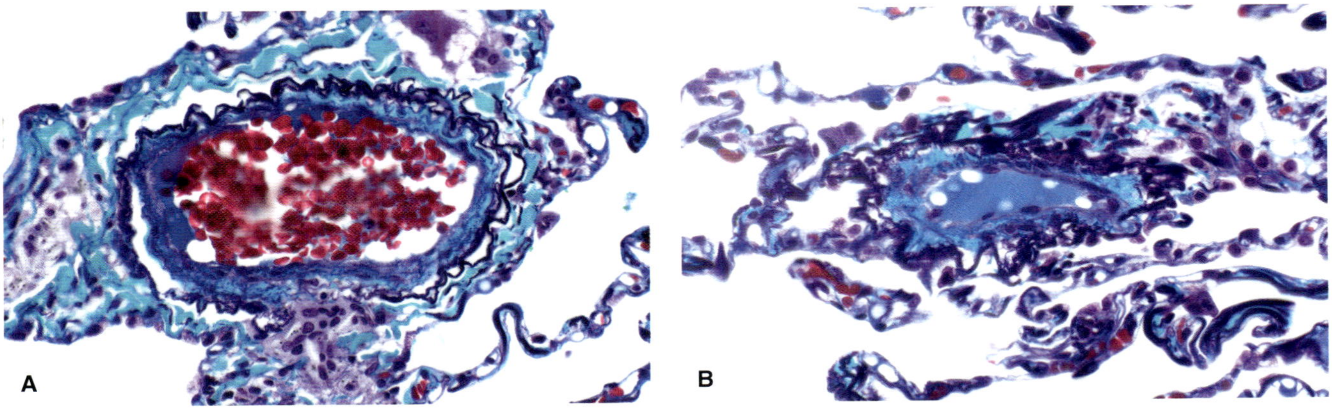

FIGURE 17.13 Pulmonary vasculature (elastic-trichrome stains). Normal pulmonary arteries contain two elastic laminae (**A**), whereas veins have a single elastic lamina (**B**). The location of a vein within a septum (**B**) is also very helpful in identifying it as a vein.

Lymphoid tissue may be seen as small collections of lymphocytes along the lymphatic routes, especially branch points of the bronchovascular bundles; lymphoid tissue is generally absent or inconspicuous except in pathologic states. Lymphoid tissue along the airways is part of the diffuse mucosa-associated lymphoid tissue (MALT) and in the lung is referred to as bronchus-associated lymphoid tissue (BALT). Submucosal lymphoid tissue in the intermediate and small airways is associated with flattening and attenuation of the overlying respiratory mucosa (lymphoepithelium), which has increased HLA class II antigens. At these sites, B-lymphocyte emperipolesis is common and is thought to reflect active antigen processing by the BALT. BALT displays the same reactions observed in the lymphoid tissue at other sites; for example, reactive hyperplasia and immunoblastic proliferation, which can be confused with lymphoid malignancies.

The lymphoid tissue of BALT has four immunophenotypically identifiable compartments, including B-cell–rich follicles, B-cell follicular mantle and marginal zones, and T-cell–rich interfollicular regions. A follicular dendritic cell network is present. Polyclonal plasma cells are identified in the perifollicular tissue. The follicles tend to be polarized into a darker side and a paler side, with the latter oriented toward the epithelial surface; BALT also shares features of MALT as seen at other sites. In normal adult lungs, BALT is rarely present, and its presence correlates with some form of chronic antigenic stimulation (27–29). BALT is not present at birth but is found normally with increasing age (29), probably as a result of exposure to environmental antigens (i.e., inducible BALT). After the individual has been exposed to most of the common antigens, the inducible BALT regresses and dendritic cells in the airways assume the role of antigen uptake and presentation. Its reappearance in adults follows chronic antigen stimuli, such as chronic infection, allergens, response to tumors, or in autoimmunity (30). Hyperplasia of BALT is frequently accompanied by lymphoid hyperplasia at other sites.

BALT refers only to lymphoid tissue along the airways (31) and not to the lymphoid tissue that may be seen in the pleura and septa as part of the diffuse lymphoid system in the lung. Diffuse lymphoid tissue is normally present at the bifurcations of lymphatic vessels; such aggregates often consist of small round lymphocytes, Langerhans cells, and pigmented macrophages with cytoplasmic anthracosilicotic material.

Intrapulmonary peribronchial lymph nodes are a normal finding, but peripheral intraparenchymal lymph nodes are less common. However, intraparenchymal lymph nodes are increasingly recognized in smokers and others with high dust exposure using current imaging techniques (32). Intrapulmonary lymph nodes are usually septal or subpleural in location. Anthracosis and small amounts of silica and silicates in these nodes are common and nonspecific.

Pleura

The visceral pleura is composed of connective tissue, elastic tissue, and an outer mesothelial layer (Fig. 17.14). Two elastic tissue laminae may be seen but these are not sharply defined. In disease, particularly pleural fibrosis and adhesions, several layers may be apparent and the elastic tissues are increased. Elastic tissue stains are useful in assessing whether a given pathologic process, such as carcinoma, has transgressed the visceral pleura (33), but interpretation is difficult when the pleura is fibrotic and the elastic tissue is increased. Pleural involvement has been shown to be a prognostically important feature of non–small-cell lung cancer and, in fact, this finding is used to upstage carcinomas independent of tumor size (33,34). Lymphatic vessels that are continuous with those in the interlobular septa are also identifiable in the pleura. When the pleura is fibrotic or pleural adhesions are present, vessels may be thick and sclerotic or even pseudoangiomatous in appearance. Fatty metaplasia is often striking in the foci of pleural and

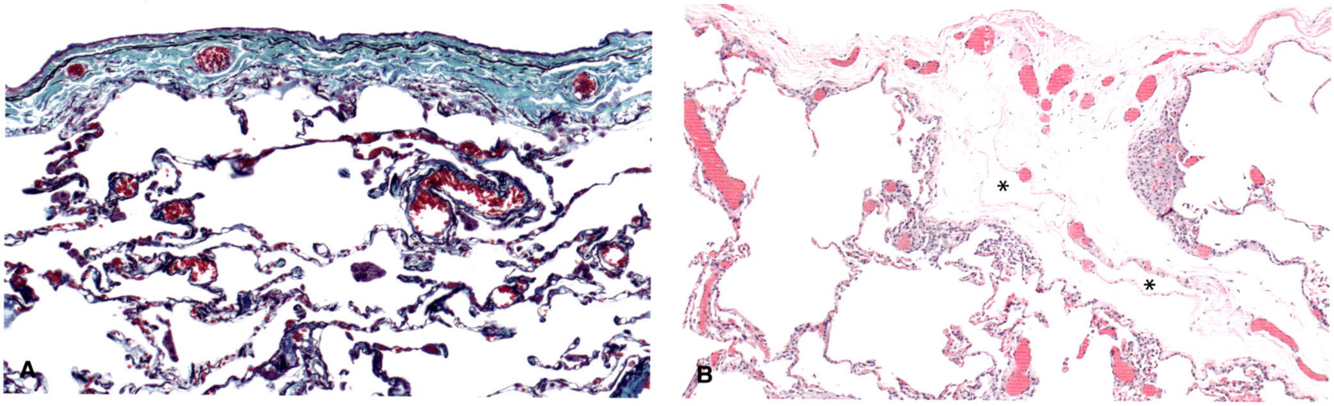

FIGURE 17.14 Pleural elastic tissue (elastic-trichrome stain). The pleura contains an elastic tissue membrane, which may appear as a single thin layer of elastic tissue or as two layers. Typically, the outermost layer is thicker (**A**) or as a richer network with occasional elastic fibers distributed through the thickness of the visceral pleura (**B**). Normal lymphatic spaces (*asterisks*) can be appreciated at the juncture between an interlobular septum and the visceral pleura.

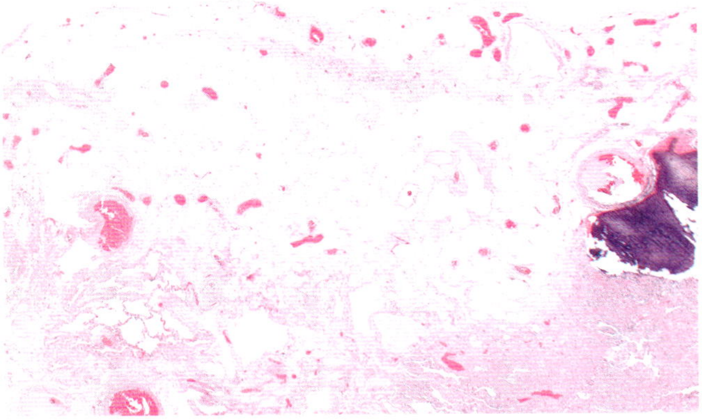

FIGURE 17.15 Fatty metaplasia is often present in areas of pleura with associated scarring. Prominent vascular changes, as seen here, can also be present.

subpleural scarring (Fig. 17.15). The visceral pleura usually remains viable over pulmonary infarcts because it has a separate vascular supply. Submesothelial fibroblasts may undergo mesothelial metaplasia (and become cytokeratin-positive) in inflammatory conditions of the pleura.

SPECIAL STAINS AND THE EVALUATION OF LUNG HISTOLOGY

While most diagnostic work in lung histopathology can be performed with routine hematoxylin and eosin (H&E) staining, there are a number of special stains that aid interpretation and may highlight findings. Elastic tissue staining is helpful in evaluating the pulmonary vasculature, airways, and pleura. Smaller arteries and veins often can be distinguished from each other with elastic tissue staining, and these stains may highlight damage to the intima and media of vessels that are inapparent on H&E staining. Elastic tissue stains are also useful in highlighting abnormalities of bronchioles and may confirm the bronchiolar or alveolar duct origin of small scars in cases of complete obliteration of the small airways. Elastic tissue staining is helpful in highlighting the pleural elastica and its relationship to tumors, as well as distinguishing adenocarcinoma in situ with its solely lepidic growth pattern from areas of invasion characteristic of acinar pattern carcinomas. Lastly, elastic stains can aid for the assessment of vascular invasion.

Trichrome stains highlight normal and pathologic findings in the lung, but they do not provide quite as much information as elastic tissue stains.

Immunostains for epithelial markers may be beneficial in atelectatic or fibrotic lung when it is difficult to appreciate the architectural features on H&E staining. Comparing these stains with stains for endothelial cells (such as CD31 or CD34) may help in assessing lung architecture and structural relationships. The use of CD31 can be confusing when alveolar macrophages are abundant since alveolar macrophages frequently show prominent staining. The lymphoid tissue in the lung is assessed with the panoply of lymphoid markers used at other sites. Both CD3 and CD20 facilitate assessment of the proportion and distribution of T and B cells, respectively. In general, most inflammatory conditions have a preponderance of T cells as the diffuse component of the infiltrates, with scattered B-cell–rich follicles. The presence of a dense diffuse population of CD20-positive B cells in lung parenchyma is very worrisome for lymphoma and should prompt additional immunohistochemical and molecular studies S100 protein, CD1a, and langerin are useful in identifying Langerhans cells, although the latter stains are much more specific.

Normal mesothelial cells stain with calretinin (nuclear and cytoplasmic), cytokeratin 5/6, D2-40, and WT-1 (nuclear reactivity), especially when mesothelial reactive changes or hyperplasia are present. In some cases, separation of mesothelial hyperplasia from mesothelioma can be very challenging. Mesothelial cells can assume a myofibroblastic phenotype and express smooth muscle actin, desmin, and other myogenic-type proteins when they are activated. FISH studies for *CDKN2A* (p16 protein) gene deletion and BAP1 and MTAP nuclear expression loss by immunohistochemistry and other testing might be of help in assessing the possibility of mesothelioma (35).

PATTERN RECOGNITION BASED ON NORMAL ANATOMIC LANDMARKS

It is useful to define pathologic conditions in the lung in relation to normal anatomic landmarks (Fig. 17.16). This can usually be done with diffuse diseases and often with localized processes. This correlation is extremely useful when assessing for possible diffuse lung diseases, especially in conjunction with gross findings, as well as with high-resolution CT scans (36). Histologic patterns and their corresponding high-resolution CT patterns can be recognized and are shown in Table 17.4. Certain lung diseases are associated with specific patterns. For example, sarcoidosis has a lymphangitic distribution while chronic hypersensitivity pneumonitis is usually bronchiolocentric. Idiopathic usual interstitial pneumonia demonstrates subpleural and paraseptal distribution with more or less involvement of the center of the lobule.

It is apparent that there is close correlation (but not a 1:1 relationship) between histology and high-resolution CT. It can be extremely beneficial to discuss cases' pathologic findings with an experience thoracic radiologist especially when there are indeterminate or confusing pathologic findings. The radiologist may suggest diseases that are not apparent in biopsy material due to limited size of the biopsy or sampling issues. Since 2002, it has been extensively demonstrated that the classification of difficult interstitial lung

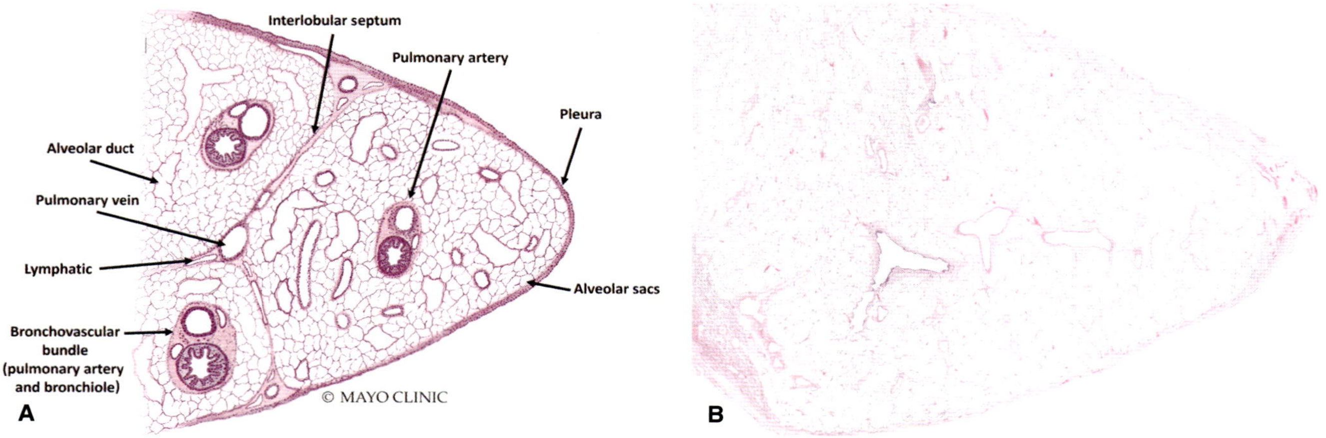

FIGURE 17.16 Wedge lung biopsy. **A:** This stylized diagram is used to depict anatomic landmarks. Structures depicted in (**A**) can be appreciated in an actual wedge biopsy specimen (**B**).

diseases is better achieved based on a multidisciplinary discussion, rather than on histologic grounds alone (37).

SITE-RELATED CHANGES COMMONLY SEEN IN SURGICAL PATHOLOGY MATERIAL

Site-specific changes that may be primary lesions or incidental findings in surgical material are shown in Table 17.5.

TABLE 17.4 Corresponding Histologic and Radiologic Patterns[a]

Histologic	Radiologic (High-Resolution Computed Tomography)
Broncho/bronchiolocentric	Centrilobular Bronchovascular Nodular
Angiocentric	Bronchovascular (arterial) Interlobular septal (venous)
Pleural/subpleural	Pleural/subpleural
Lymphatic	Bronchovascular Interlobular septal Pleural
Peripheral acinar	Subpleural peripheral distribution (paraseptal)
Septal	Septal
Random nodular	Random nodular
Parenchymal consolidation	Consolidation, ground glass
Diffuse interstitial	Diffuse interstitial, ground glass, reticular, septal lines
Mixed and unclassified	Mixed/unclassifiable

[a]Modified from Colby TV, Swensen SJ. Anatomic distribution and histopathologic pattern in diffuse lung disease: Correlation with HRCT. *J Thorac Imaging.* 1996;11:1–26.

Lung Parenchyma

Biopsies from lobar tips were often obtained in the era of open lung biopsy because of their accessibility. However, samples from these areas, particularly the lingula or right middle lobe, may show incidental inflammatory and fibrotic changes (38), including interstitial fibrosis, epithelial metaplasia, and even focal honeycombing, none of which are necessarily representative of a diffuse process. Myointimal proliferation is common in the arteries and veins of these biopsies, and the airspaces may contain aggregates of macrophages and neutrophils. Incidental changes in lobar tips are usually obvious as such since the more proximal lung tissue is either not affected or significantly less affected, although these changes may become a problem when dealing with small wedge biopsies, particularly those that are 2 cm or less in greatest dimension. Thus, surgeons should be encouraged to obtain biopsy samples that are at least 3 cm deep into the underlying parenchyma, or a biopsy spanning at least 5 cm in greatest dimension. The possibility of middle lobe syndrome (which may affect the lingula, the right middle lobe, or both) (39,40) should be considered for persistent infiltrates in these sites. Currently, the majority of surgical lung biopsies are performed by video thoracoscopy. The video-assisted thoracoscopic surgery (VATS) biopsy is performed with the lung deflated and nearly all sites are accessible.

Apical caps (Fig. 17.17) were once thought to be the result of healed tuberculosis, but they are common in patients who have never had tuberculosis and are now

TABLE 17.5 Site-Specific Changes in Lung Tissue

- Inflammatory changes at lobar tips, especially lingula and right middle lobe
- Apical caps: Pleural and subpleural fibroelastosis in the apex and the upper lobes and superior segments of the lower lobes
- Upper lobe centriacinar emphysema
- Visceral pleural blebs

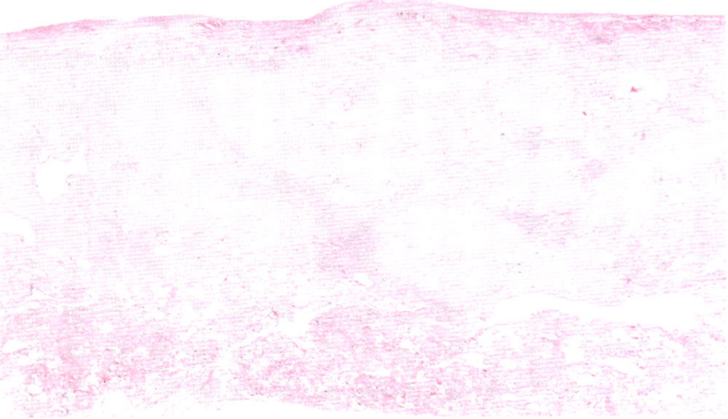

FIGURE 17.17 Apical cap. Apical caps occur most frequently in the apices. They represent regions of pleural and subpleural thickening by predominantly elastotic fibrous tissue characterized by eosinophilic to violet-appearing elastic fibers.

believed to represent an ischemic alteration related to the intrinsic physiologic underperfusion of these regions (41–44). Apical caps occur most frequently in the apex of the upper lobes, but they are also encountered in the upper regions of the lower lobes (44). They appear as regions of fibroelastosis in the pleura and subpleural lung parenchyma; the background structure of collapsed alveolar walls and interposed eosinophilic collagen can often be discerned in the tissue and enhanced with elastin stains. Ossification and nests of metaplastic pneumocytes may be present. The changes are sufficiently distinctive that apical caps can often be suspected histologically, even when one does not know the site of origin. Apical caps share some histologic similarities with, pleuroparenchymal fibroelastosis (PPFE). As opposed to PPFE, however, apical caps are often asymptomatic and localized. Larger apical caps are sometimes biopsied or resected in older individuals to exclude a carcinoma or a visceral pleural tumor. In the era of high-resolution CT scans and imaging surveillance of smokers, smaller apical caps are now sampled with increasing frequency. They can be PET-avid secondary to nearby infection and show remarkable cytologic atypia of pneumocytes that should not be confused with invasive carcinoma.

Centriacinar, or centrilobular, emphysema is a pathologic abnormality that is more common and more severe in the upper lobes (45). It is found predominantly in cigarette smokers and is commonly seen in lobes resected for bronchogenic carcinoma. Emphysematous changes are frequently accompanied by some degree of fibrosis, particularly when bullous change is present. The fibrosis appears as strands of dense, hypocellular, brightly eosinophilic collagenous septa traversing the emphysematous spaces. Anthracosis is also a common finding in smokers and urban dwellers. Focal pleural and subpleural fibrosis (Fig. 17.18) is extremely common, especially as an incidental microscopic finding in lobar resections from smokers (46). Abnormal airspaces formed by irregular fibrous septa accompanied by pigmented macrophages in alveolar spaces, fibroblast foci, honeycomb change, and smooth muscle hyperplasia are known features of smoking-related interstitial fibrosis (SRIF). Usually, this is localized, but it can be diffuse and actually represent a form of interstitial lung disease (see below).

Subpleural bullous change (Fig. 17.19) and intrapleural blebs can be the only pathologic change found in patients with recurrent pneumothoraces who do not have diffuse lung disease (47). Both are typically evident in the upper lobes. Bullae are emphysematous sharply demarcated air-containing spaces 1 cm or greater in diameter with lung parenchyma lined by TTF-1–positive pneumocytes (48,49). Blebs are intrapleural and lined by bland fibrous tissue, macrophages, and occasional giant cells. Rupture of these spaces introduces air into the pleural space, inciting a mesothelial proliferation with numerous macrophages, giant cells, and eosinophils. This observation has been referred to as eosinophilic pleuritis and may accompany any condition that is associated with pneumothorax (50). Collections of interstitial air in the lung tissue, with or without

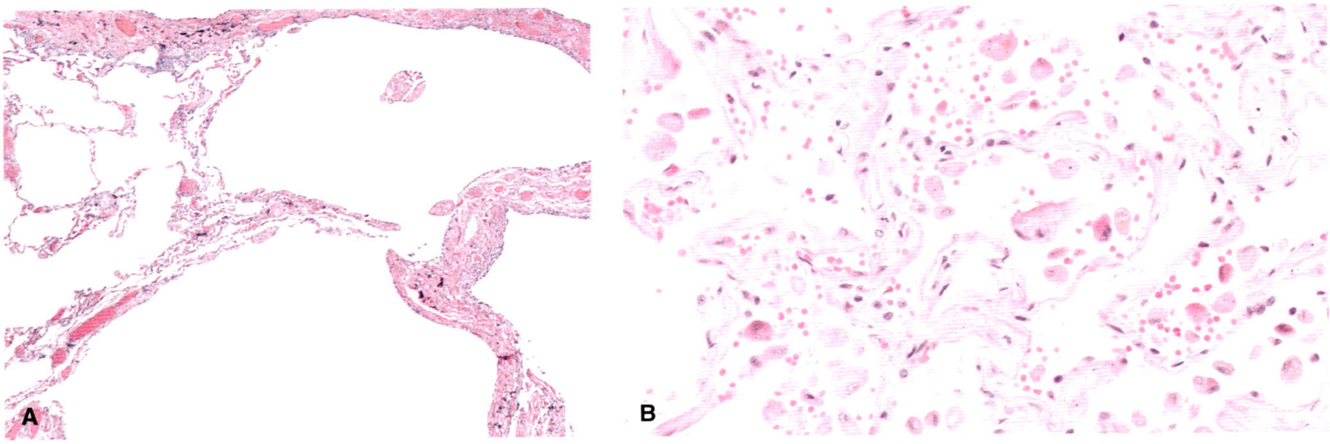

FIGURE 17.18 Subpleural emphysematous change in smoking. **A:** There is simplification of airspaces. **B:** Some of the alveolar walls show mild fibrosis with hyaline-appearing collagen (smoking-related interstitial fibrosis). Scattered pigmented alveolar macrophages (also referred to as smoker's macrophages) are also noted.

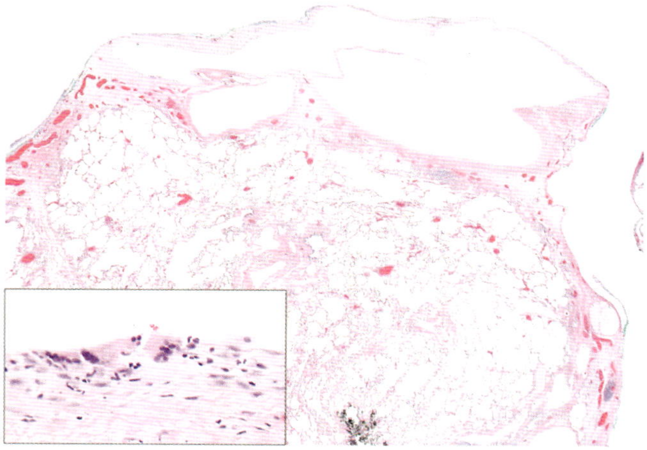

FIGURE 17.19 Pneumothorax. An intrapleural bleb may predispose to pneumothorax. These result from interstitial air, and are lined by connective tissue. They can sometimes have an associated giant cell reaction. (also Fig. 17.18). Inset: Pneumothorax is often associated with a pleural reaction, which has been labeled eosinophilic pleuritis because of the association of eosinophil infiltrate and mesothelial and macrophage reaction on the pleural surface.

giant cell and/or eosinophilic reaction, may also be present (i.e., pulmonary interstitial emphysema). Focal subpleural scarring can usually be distinguished from chronic fibrosing interstitial pneumonias by its restriction to the subpleural region, lack of active fibroblastic proliferation, and the relative abrupt transition to normal alveolar walls. Clinical history and radiographic findings may also be supportive. When a diffuse lung disease such as pulmonary Langerhans cell histiocytosis or lymphangioleiomyomatosis has resulted in pneumothorax, both the primary process and the secondary eosinophilic pleuritis may be present.

Pleura

Adhesions between the pleura and chest wall are composed of fibrovascular tissue with pockets of bland mesothelial cells. The adhesions identified in surgical specimens and even autopsy tissue often have no obvious cause and represent an incidental finding. Dense, hyalinized fibrotic pleural (visceral or parietal) plaques (51–53) are the consequence of a variety of prior inflammatory events. Their presence is often an incidental finding, although one should consider the possibility of prior asbestos exposure if the lesions are bilateral and there is no obvious inflammatory cause (54). The fibrous pleural plaques of asbestos exposure are frequently calcified and are typically radiographically visible.

ARTIFACTS SEEN IN LUNG BIOPSY AND RESECTION MATERIAL

Artifacts related to lung biopsy or prior procedures are described in Table 17.6.

TABLE 17.6 Artifacts Seen in Lung Biopsies and Resections

- Changes related to prior instrumentation including bronchoscopy, bronchoalveolar lavage, needle aspiration
- Airway changes due to ventilatory assistance
- Apparent edema from freezing of cryobiopsies
- Compression/squish
- Pseudolipoid change ("bubble artifact")
- Telescoping of vessels and airways
- Airway epithelium sloughing/detachment
- Hemorrhage/inflammatory changes
- Material from surgical gloves (e.g., talc and starch)
- Inflation-induced alveolar distension resembling emphysema and patchy compression in underinflated zones
- Sponge artifact
- Lesions of aggregated mesothelial cells and macrophages (LAMM)

Knowledge of the clinical course of events prior to lung biopsy usually allows the pathologist to avoid misinterpreting changes related to prior instrumentation. Previous bronchial biopsies may cause hemorrhage, airway inflammation, ulceration, granulation tissue reaction, and fibrous scarring. Strips of epithelium may be dislodged by mechanical trauma and embedded in inspissated mucus; residual basal cells may be all that is left adherent to the basement membrane. Bronchoalveolar lavage can produce vacuolation of alveolar pneumocytes and macrophages. Previous needle biopsies may induce necrosis and hemorrhage in the parenchyma, followed by organization and reactive epithelial atypia (Fig. 17.20). Patients who have received positive-pressure ventilation (Fig. 17.18) may have disproportionate bronchiolar and alveolar duct distension, especially when high inspiratory pressures are used in the acute respiratory distress syndrome (ARDS). Some associated acute inflammatory exudate in the lumen is common, often with relative absence of associated inflammation in the airway wall or surrounding alveolar spaces. Occasionally, one can find exogenous thrombotic foreign material in bronchial arteries when patients have been treated with bronchial artery embolization for persistent hemoptysis.

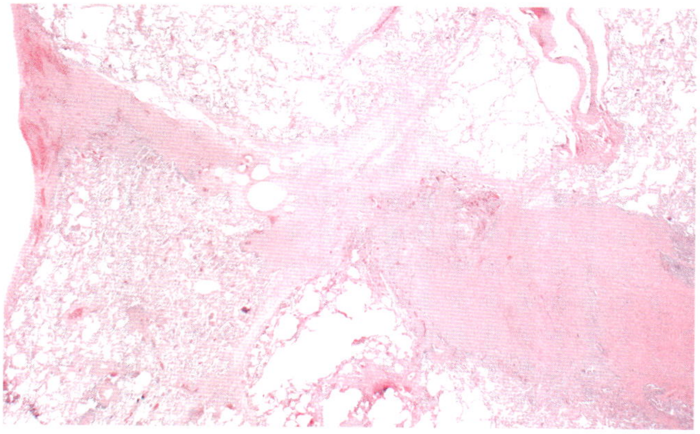

FIGURE 17.20 Needle tract to a peripheral adenocarcinoma. The biopsy had been performed a month earlier, and the needle tract shows evidence of organization.

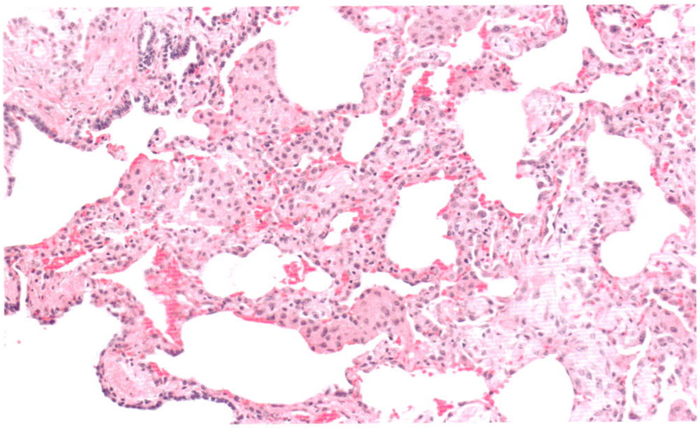

FIGURE 17.21 Bubble artifact. This fragment from a wedge resection shows artifactual compression and cutout appearing spaces consistent with bubble artifact.

Compression of the lung tissue, particularly in forceps transbronchial biopsies, may produce rounded spaces in the alveoli that resemble the fat vacuoles of exogenous lipoid pneumonia (Fig. 17.21). This artifactual change, often called "bubble artifact" (55,56), can be recognized because there are no associated macrophages or giant cells with distending intracytoplasmic lipid vacuoles; all the vacuoles are extracellular. There is also very little fibrosis, which would be an expected feature of chronic exogenous lipoid pneumonia.

Compression of airways produces crinkling and telescoping of the epithelium. Sometimes entire strips of mucosa are displaced into the bronchiolar or alveolar spaces. Similarly, procedural telescoping of the endothelium can be seen in small vessels, mimicking the appearance of intimal thickening or organizing/recanalized thrombus (Fig. 17.22).

Compression-induced nuclear smearing can occur in any cellular tumor or reactive lymphoid tissue of the bronchial mucosa and/or mediastinal lymph nodes. The resulting changes may simulate small cell carcinoma. Recognition of the diagnostic pitfall, examining multiple levels, assessing the periphery of the tissue where the cells are more likely to be intact, and enlisting the aid of concomitant cytology specimens and immunohistochemistry allows resolution of most cases.

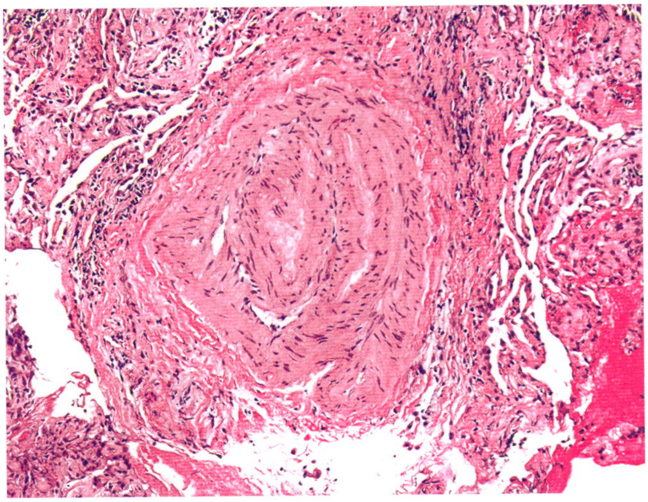

FIGURE 17.22 Telescoping of vessels. This vessel has the artifactual appearance of showing intimal thickening or thrombosis. This common artifact occurs secondary to intussusception of the wall of the vessel into the lumen.

Procedural compression may be encountered in all types of lung samples since lung tissue is soft and readily compressible (55,56). It is particularly common in VATS wedge biopsies because this procedure is typically performed with the lung collapsed and samples may become further compressed during retrieval through a small hole in the chest wall (57). The use of forceps by the surgeon or pathology assistant can also compress the parenchyma of small biopsies. Artifactual compression may be misinterpreted as interstitial pneumonia or interstitial fibrosis because the apposition of alveolar walls produces apparent scarring and hypercellularity (Fig. 17.23). In a compressed lung, the vessels and septa may appear to have more collagen than normal and to be thickened because they are contracted and shortened. With experience, this change can be recognized using routine H&E sections; connective tissue stains show an absence of scarring and a normal background of supporting fibrous tissue along vessels and in septa. Careful assessment of the nuclei in the atelectatic lung shows that most are endothelial or epithelial in origin, rather than inflammatory. Leukocyte common antigen (CD45) immunostaining and immunohistochemical stains for other lymphoid markers may be useful in this setting. It is also unusual in surgical biopsies (and larger specimens) for the entire specimen to be uniformly atelectatic; and, therefore, a low-power survey of the entire pattern generally helps one appreciate compressed areas merging with a more normal lung. In fact, when significant fibrosis is present (usually associated with foci of honeycombing), this artifact is rarely a problem because the more rigid fibrotic lung tissue tends to retain its inflated configuration. Injecting the stapled lung with formalin helps alleviate this artifact, but introduces artifacts of its own, so we do not perform this procedure. Agitating

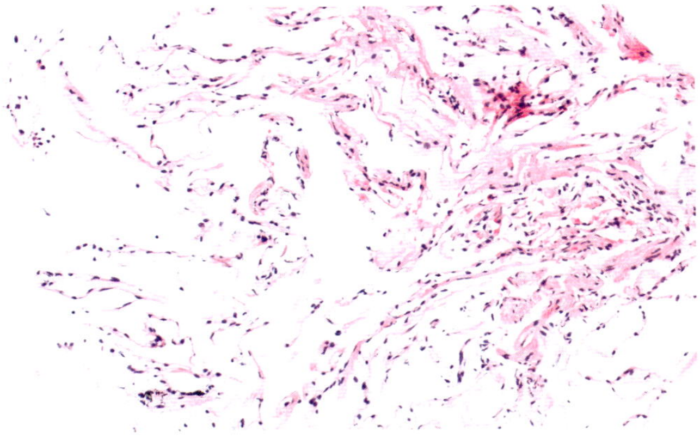

FIGURE 17.23 Compression/squish artifact. Recognizable alveolar walls can be traced into the region of compression on the right side of the image.

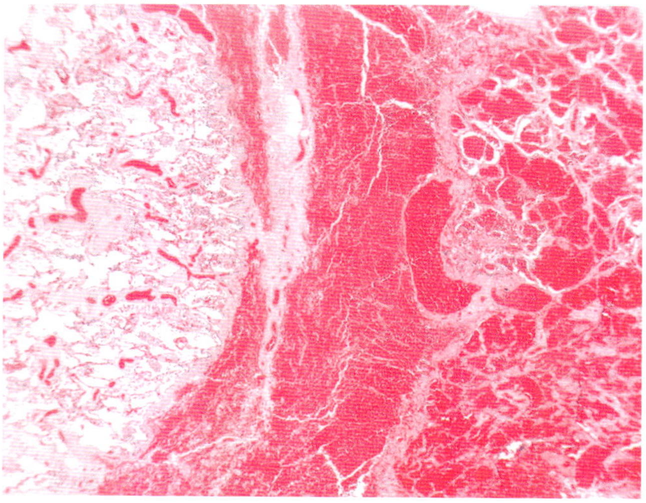

FIGURE 17.24 Traumatic hemorrhage. The right side of the field shows extensive hemorrhage into alveolar spaces, whereas the tissue on the left shows a complete absence of hemorrhage. There is also dissection of red blood cells into the interlobular septal region (*center*). Such a well-demarcated focus of hemorrhage and dissection of red blood cells into the connective tissue would be very unusual in a diffuse alveolar hemorrhage syndrome. The findings are typical of traumatic/procedural hemorrhage.

the tissue in a formalin container after slicing a specimen can expand the lung significantly.

Fresh intra-alveolar blood due to the trauma of surgery is extremely common in biopsy material and should not be overinterpreted as a pathologic finding (Fig. 17.24). Trauma of the procedure is the most common cause of fresh blood in alveolar spaces. One can approach this problem from three points of view: the statistical, histologic, and clinical. Statistically, the vast majority of cases of fresh alveolar blood are traumatic since pathologic alveolar hemorrhage is relatively uncommon. Thus, in any given case, acute blood is unlikely to be significant. When pathologic hemorrhage is present, there is usually (but not always) an associated fibrinous exudate and there may be hyaline membranes, reactive type II cells, microfoci of organizing pneumonia, and evidence of prior hemorrhage manifested by hemosiderin-filled macrophages in the interstitium or airspaces. This last finding is not a reliable histologic criterion in patients who have been smokers and those with venous obstruction or chronic passive congestion. Lastly, correlation with the clinical history is essential. Suspected clinical diagnosis of hemoptysis, alveolar hemorrhage, or vasculitis should prompt careful interpretation of the morphologic finding of intra-alveolar blood. The clinician can usually confirm whether an alveolar hemorrhage syndrome or alveolar hemorrhage because of some other cause is in the realm of possibility in any given patient.

The most common cause of macrophages staining positively with iron stains is respiratory bronchiolitis in smokers. The hemosiderin in smoker's macrophages is finely granular, in contrast to the coarse, dark blue staining in chronic hemorrhage (Fig. 17.25). Both types of macrophages will stain positively for a Prussian blue iron special stain. Pigmented macrophages can also be seen in lungs of those who smoke marijuana.

Prolonged surgical manipulation of the lung, and even multiple transbronchial biopsies, can lead to margination of neutrophils in capillaries (especially those in the pleura), which can mimic capillaritis (57). Capillaritis is usually associated with other features of diffuse alveolar hemorrhage, including blood in the airspaces, fibrin deposition, and the presence of hemosiderin-laden macrophages. Instrument clamps applied to the biopsy specimen prior to removal can result in lymphatic obstruction, dilatation, and septal edema.

Lung pathologists are divided regarding the issue of inflation fixation of biopsy specimens. This can be easily accomplished with a syringe filled with formalin and a fine gauge needle. Careful inflation of a lung biopsy specimen (58) may be helpful diagnostically and is aesthetically pleasing since the lung architecture is more easily appreciated. A heavy hand can create overdistension of the alveoli and an emphysematous appearance. If this occurs, clinical correlation may be required to assess whether emphysema is actually a clinical

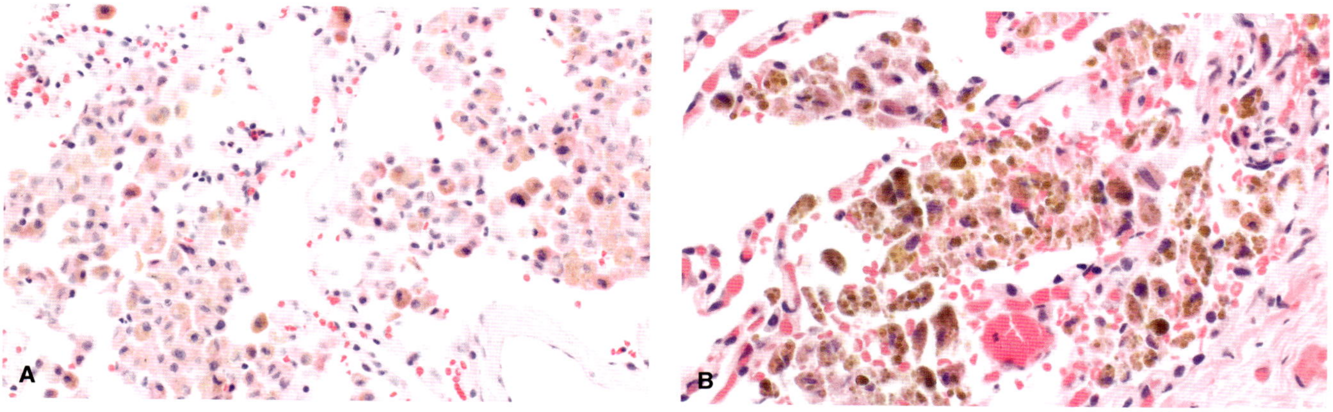

FIGURE 17.25 Respiratory bronchiolitis. **A:** An accumulation of pigmented macrophages in alveolar spaces. The pigment is fine and powdery appearing with a golden-brown color and occasional anthracotic flecks. **B:** In contrast, hemosiderin-laden macrophages have a darker brown color and coarser granules. Both types of macrophages will stain with Prussian blue, which cannot be used to tell them apart.

consideration. Patchy uninflated portions of lung tissue are common in biopsies that have been nonuniformly inflated. One problem that may be encountered in inflated biopsies is that cells and fluid may be "washed out" of the airspaces. This is especially true of smoker's bronchiolitis, which may be quite subtle in inflated specimens. In addition, lymphatic vessels and other spaces may appear artifactually prominent.

Plastic sponges are sometimes put in processing cassettes by pathology staff to ensure that small specimens are not lost during processing. Sponge artifact, with triangular "holes" in the tissue, is a well-known artifact in any tissue that is processed this way, particularly if it is placed on the sponges prior to complete formalin fixation (56).

As mentioned earlier, VATS lung biopsies have largely replaced traditional open lung biopsy as the diagnostic procedure for obtaining wedge biopsies of lung tissue for histologic evaluation. Although there are minor disadvantages in comparison to traditional open lung biopsy, diagnostic accuracy does not appear to be compromised (57,59–62). Allowing for the fact that bimanual palpation is not as feasible with VATS biopsies and that the tissue is forcibly pulled through a small hole in the chest wall, it is remarkable that sampling error and specimen artifacts are only a minor problem. The specimen size approaches that achieved with traditional open lung biopsy, and the minor degrees of hemorrhage, compression, and neutrophil margination represent artifactual changes that do not compromise diagnostic accuracy. Moreover, VATS lung biopsies provide excellent sampling of the peripheral/distal lung parenchyma, which are often not amenable to a transbronchial biopsy approach. This becomes very important for the diagnosis of pathologies like usual interstitial pneumonia.

The use of transbronchial biopsy with a cryoprobe, referred to as cryobiopsy, has emerged as an alternative to forceps and open lung biopsies for the sampling of the peripheral lung parenchyma. In this technique, the flexible cryoprobe is inserted *via* a bronchoscopic approach, rapidly freezes an area of peripheral lung parenchyma which is then extracted as a cryobiopsy. Specimens procured using this technique are larger than those obtained using forceps and demonstrate less crush and compression artifact, resulting in the better appreciation of the histology and the morphologic features. Cryobiopsies have been reported to be of use for the diagnosis of diffuse parenchymal lung disease, with a 70% to 80% diagnostic yield (62,63), particularly when combined with multidisciplinary discussion. Proteinaceous intra-alveolar exudate may be an artifact and should not necessarily be construed as alveolar edema.

TABLE 17.7 Incidental Findings in Lung Tissue

Effects of smoking: emphysema, chronic lymphocytic bronchitis, respiratory (smoker's) bronchiolitis with or without fibrosis, dust macules
Changes of asthma
Ossification and marrow formation in bronchial cartilages, bronchial submucosal fatty infiltration, and elastotic changes
Parenchymal nodules Carcinoid tumorlets Meningothelial-like nodules Atypical adenomatous hyperplasia Healed granulomatous disease, infarcts, etc. Hamartomas Focal scars Occasional (anthraco-) silicotic nodules (in the absence of clinical/radiologic pneumoconiosis) Metaplastic bone Intrapulmonary lymph nodes Small carcinomas (and other tumors)
Intracellular/intra-alveolar/interstitial structures Macrophages Corpora amylacea Blue bodies Schaumann bodies Asteroid bodies Calcium oxalate crystals Mallory hyaline-like material in type II pneumocytes Ferruginous bodies Anthracotic pigment/birefringent material (including silica and silicates) Metaplastic bone
Epithelioid and/or cholesterol granulomas, giant cells, lipogranulomas
Intravascular/vascular Megakaryocytes Thrombi (mimic emboli) Bone marrow emboli Calcification and iron encrustation of the elastic tissue (mineralizing elastosis) Foreign material (but usually pathologic)
Interstitial air in the lung parenchyma
Hilar/peribronchial lymph nodes Sinus and paracortical histiocytes (may simulate granulomas) Occasional silicotic or anthracosilicotic nodules Anthracosis/birefringent material Hamazaki–Wesenberg bodies
Pleura
Adhesions Pleural plaques Nodular mesothelial hyperplasia Lesions of aggregated mesothelial cells and macrophages (LAMM) Fatty metaplasia

INCIDENTAL FINDINGS IN LUNG BIOPSY AND RESECTION TISSUE

Lung Parenchyma

There are a number of incidental findings in the lung tissue that may not be related to the primary process that prompted the biopsy or resection (Table 17.7). Sometimes these "incidental" findings are of clinical and pathologic significance, and other times they are of no significance. Correlation of the individual finding(s) with the clinical and radiologic presentation helps determine their significance.

Findings related to smoking and emphysema are extremely common, especially in lungs resected for carcinoma (46).

They may be divided into three broad groups: large airway changes, small airway lesions, and abnormalities of the alveolar parenchyma. Lesions of the airways are the substrates for the clinical diagnosis of "chronic bronchitis." In the large airways, one sees goblet cell hyperplasia, squamous metaplasia, basement membrane thickening, hypertrophy and hyperplasia of bronchial glands with dilated ducts and mucostasis, and, often, a mild submucosal chronic inflammatory infiltrate (45,48). The changes in the small airways may be quite dramatic and may mimic or even produce an interstitial lung disease (46,64–66). Smoking-induced respiratory bronchiolitis (Fig. 17.23) includes goblet cell metaplasia, a mild inflammatory infiltrate in the airway walls, metaplasia of type II cells in the surrounding alveoli, mild peribronchiolar fibrosis, and prominent accumulations of pigmented macrophages in the adjacent airspace and the bronchiolar lumen (67). The pigmented macrophages contain phagocytosed debris from inhaled cigarette smoke and have prominent secondary lysosomes in their cytoplasm. This results in a dirty granular tan or brown appearance to the cytoplasm. These cells contain PAS-positive lysosomes and finely granular Prussian blue–positive (hemosiderin) material, as well as tiny, irregular flecks of brown-black material.

Smoking-related changes in the pulmonary parenchyma usually manifest as centriacinar emphysema with airspace enlargement and loss of alveolar walls (45,48). Histologic quantification of emphysema on biopsy material is difficult, although one can often determine whether it is present and subjectively quantify it. Bullous emphysema with abnormal airspaces >1 cm in diameter is more common in the upper lobes and is usually associated with some fibrosis in the septa of the bullae and adjacent alveolar walls. The fibrosis is relatively acellular, noninflammatory, eosinophilic, and poorly vascularized. Metaplasia or ulceration of the epithelium and interstitial dissection of air occur in bullae and may induce a distinctive giant cell reaction analogous to that seen in persistent interstitial emphysema. SRIF (46), is a distinctive type of fibrosing interstitial lung disease seen in smokers characterized by the presence of diffuse or patchy pauciinflammatory alveolar septal widening by eosinophilic hyalinizing collagenous fibrosis. It normally occurs in close relationship with areas of emphysema and respiratory bronchiolitis, and it can also be associated with areas of smooth muscle hyperplasia, honeycomb change, and fibroblastic foci. While these changes are usually incidental, they can rarely reflect diffuse lung disease (68).

Asthmatics are predisposed to a number of lung conditions that may lead to lung biopsy, although the asthmatic changes themselves may not be the dominant lesion. These airway changes include goblet cell hyperplasia/metaplasia in the airway epithelium, thickening of the basement membrane and submembranous region, smooth muscle hypertrophy and hyperplasia, lymphoid hyperplasia, fibrosis, and a variable infiltrate of eosinophils, mast cells, lymphocytes, and neutrophils (45,48,69). Mucostasis with Curschmann spirals may be an accompanying feature. When mucostasis is extensive and is associated with sloughing of epithelial fragments into the airways (i.e., Creola bodies), one should suspect that the asthma itself is the main lesion. Such an appearance is typical of status asthmaticus. Other features of asthma include distal atelectasis with a few airspace macrophages and eosinophils, and poorly formed nonnecrotic granulomata (69,70). Patients with quiescent asthma or past history of asthma may have airways that are entirely normal or show only minor inflammatory or fibrotic changes.

Metaplastic bone, including bone marrow and calcification, is an age-related change that is occasionally seen in bronchial cartilages (71). Metaplastic bone may also be seen in regions of scarring, particularly in apical caps, secondary to aspiration and heart failure, or unassociated with any pathologic changes. In some older individuals with chronic bronchitis, the bronchial submucosa may have a gray elastotic appearance that reflects calcification of elastic fibers in the submucosa and basal lamina.

Carcinoid tumorlets (72–75), also referred to simply as tumorlets, and minute pulmonary meningothelial-like nodules (76,77) are nodular proliferations that may be mistaken for other types of lesions or even metastases, especially in frozen sections. Tumorlets (Fig. 17.26) represent well-circumscribed proliferations of neuroendocrine cells that usually occur around and within the walls of small airways, particularly in scarred or bronchiectatic airways, and in a small number of patients with unexplained airflow destruction as part of the diffuse idiopathic neuroendocrine hyperplasia syndrome. These foci lack mitotic figures and necrosis and share morphologic features with spindle cell carcinoid tumors.

Tumorlets are often multiple and some may become large enough to be recognized radiographically, especially when they occur in areas of bronchiectasis. Exactly where one draws the line between a carcinoid tumorlet and a carcinoid tumor, particularly in cases with multiple lesions, is a bit arbitrary, although a cutoff point of 0.5 cm diameter or larger for carcinoid tumor is used (40,75). Importantly, this measurement cutoff does not apply to tumorlets that may extend for >5 mm along the length of an airway, while maintaining a <5 mm width or thickness. Diffuse idiopathic pulmonary neuroendocrine cell hyperplasia is closely related to tumorlets, and the two often coexist. According to the World Health Organization (WHO), diffuse idiopathic pulmonary neuroendocrine cell hyperplasia is defined as a generalized proliferation of pulmonary neuroendocrine cells that may be confined to the mucosa of airways with or without luminal protrusion, locally invade to form tumorlets, or develop into carcinoid tumors (75). When this proliferation expands and invades the bronchiolar submucosa, the tumorlet designation is appropriate. When diffuse idiopathic pulmonary neuroendocrine cell hyperplasia is encountered, usually multiple airways are involved, and the patients typically have evidence of airflow obstruction. Similarly, the presence of multiple tumorlets or carcinoid

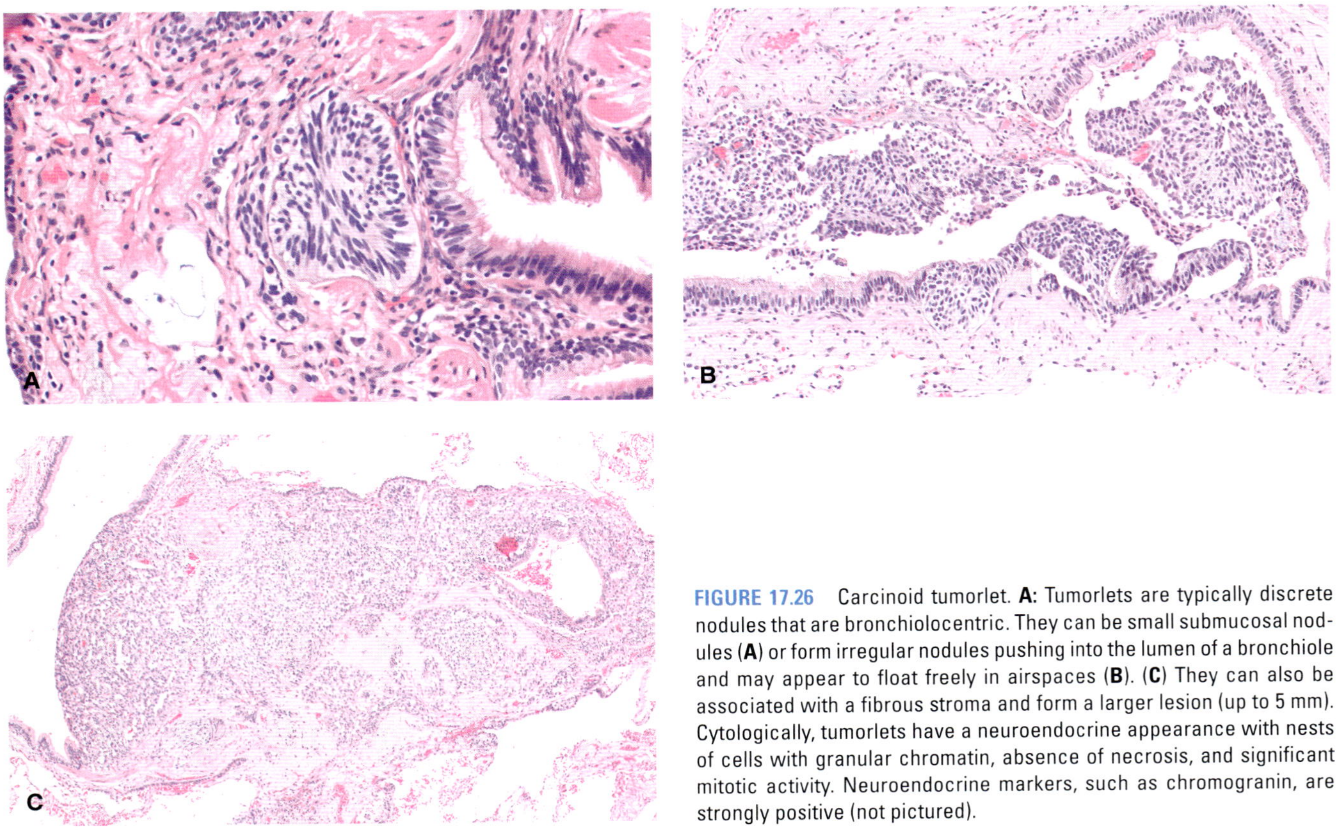

FIGURE 17.26 Carcinoid tumorlet. **A:** Tumorlets are typically discrete nodules that are bronchiolocentric. They can be small submucosal nodules (**A**) or form irregular nodules pushing into the lumen of a bronchiole and may appear to float freely in airspaces (**B**). (**C**) They can also be associated with a fibrous stroma and form a larger lesion (up to 5 mm). Cytologically, tumorlets have a neuroendocrine appearance with nests of cells with granular chromatin, absence of necrosis, and significant mitotic activity. Neuroendocrine markers, such as chromogranin, are strongly positive (not pictured).

tumors should prompt one to search for evidence of diffuse idiopathic pulmonary neuroendocrine cell hyperplasia, and consider the possibility of clinical evidence of airflow obstruction (78).

Meningothelial-like nodules (Fig. 17.27) were originally thought to represent an intrapulmonary proliferation of perivenular chemoreceptor cells (75–77). However, the accumulated evidence suggests that the cellular constituents are more closely related to meningothelial cells (75–77,79). Meningothelial-like nodules are almost always positive for vimentin and approximately one-third stain for EMA; they are negative with cytokeratin and synaptophysin by immunohistochemistry. Loss of heterozygosity is relatively common in multiple meningothelial-like nodules and affects more loci compared to cases with solitary meningothelial-like nodules (79). These observations suggest that isolated meningothelial-like nodules are probably reactive, while multiple meningothelial-like nodules

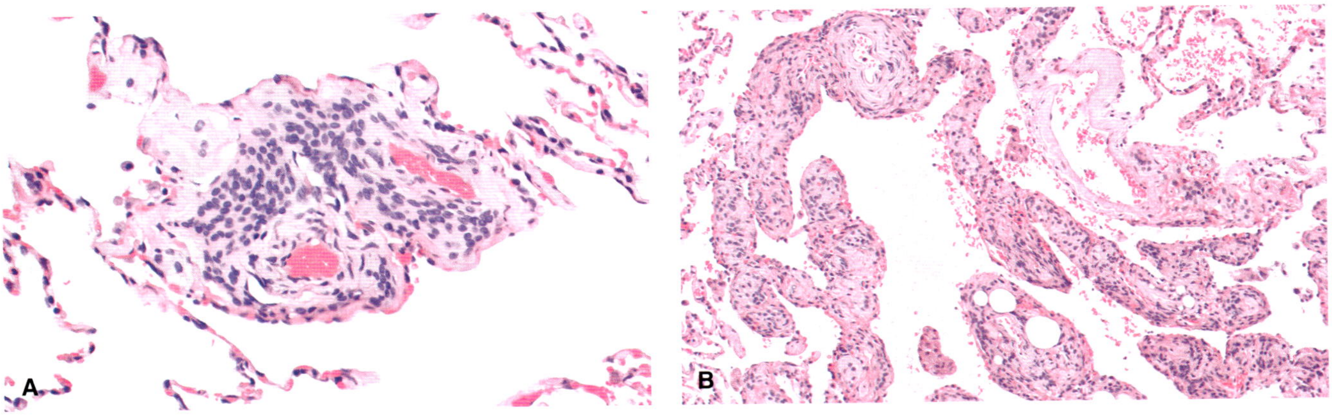

FIGURE 17.27 Minute pulmonary meningothelial-like nodule. **A:** These small parenchymal nodules are sometimes associated with pulmonary veins and are associated with syncytial-appearing cells in a collagenous stroma. Cytologically, they bear a distinct resemblance to meningothelial cells. **B:** They can also appear to track along the alveolar septa.

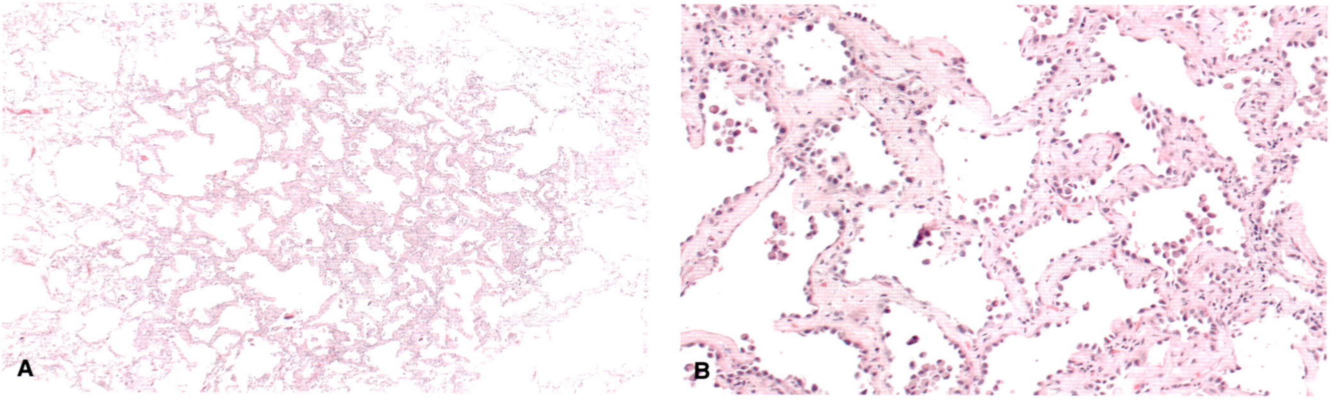

FIGURE 17.28 Atypical adenomatous hyperplasia. **A:** A 4- to 5-mm diameter lesion is apparent as slight alveolar septal thickening and a proliferation of knob-shaped cells along alveolar walls. **B:** Cytologically, the cells lining the alveoli have mild-to-moderate atypia, but lack the crowding and increased atypia associated with adenocarcinoma in situ.

might represent a transition from a reactive to a neoplastic proliferation. Minute pulmonary meningothelial-like nodules are composed of interstitial clusters of fusiform cells with pale eosinophilic cytoplasm, forming small stellate nodules within the alveolated parenchyma that are rarely grossly appreciable. Meningothelial-like nodules express progesterone receptors similar to meningothelial cells and they are distinctively negative for vascular, melanoma, and smooth muscle markers. They are largely negative for synaptophysin and chromogranin but can show some CD56 reactivity. The distinction between meningothelial-like nodules and tumorlets can be a challenge. In such cases, the characteristic location within the alveolar septae favors a diagnosis of a meningothelial-like nodule. Meningothelial-like nodules are incidental, generally asymptomatic lesions that have been reported to occur in 4.9% of autopsies and up to 9.5% on cancer resections. They can be seen in up to 47% of extensively sampled resection specimens. Recent use of high-resolution CT scans has increased clinical awareness of these lesions and increased the likelihood that they will be resected. They can present as multiple bilateral micronodules, a phenomenon referred to as diffuse pulmonary meningotheliomatosis.

Atypical adenomatous hyperplasia is a proliferative epithelial process occurring as small nodular lesions in the lung parenchyma (Fig. 17.28). These were first recognized in resection specimens for carcinoma and named bronchioloalveolar cell adenomas (80). Atypical adenomatous hyperplasia has been reported to be present in approximately 10% of resection specimens obtained for carcinoma (80). They are most easily recognized grossly in cases that are inflated with the Bouin fixative. The WHO defines atypical adenomatous hyperplasia as "a small (≤0.5 cm), localized proliferation of mild to moderately atypical type II pneumocytes and/or club cells lining alveolar walls and sometimes respiratory bronchioles" (75). These lesions are considered to represent preinvasive neoplasms that can develop into some nonmucinous adenocarcinomas. The evidence supporting this association is epidemiologic, morphologic, morphometric, cytofluorometric, and genetic. Atypical adenomatous hyperplasia generally represents an incidental histologic finding although it can occasionally be identified on imaging and is found in 2% to 4% of routine autopsies of non–cancer-bearing patients and up to 35% of lobectomies for adenocarcinoma of the lung. Atypical adenomatous hyperplasia occurs close to respiratory bronchioles, in areas lacking significant inflammation or fibrosis, and demonstrates no more than moderate atypia. The presence of cytoplasmic snouts suggests a club cell origin. Inconspicuous pseudopapillae can be seen and should not be confused with invasive carcinoma. In general, we advocate a conservative approach to these lesions, since (a) there are many causes of type II cell proliferation, and many of them are reactive; and (b) while some studies support the concept that atypical adenomatous hyperplasia is a direct precursor of lung adenocarcinoma, there are no unequivocal histologic/pathologic studies documenting progression of atypical adenomatous hyperplasia to adenocarcinoma. Moreover, there are no data regarding the risk of transformation of atypical adenomatous hyperplasia to adenocarcinoma. Atypical adenomatous hyperplasia should not be confused with in situ adenocarcinoma, which is usually larger, with a more monotonous proliferative cell population and with a greater degree of cytologic atypia.

Focal scars, healed granulomatous disease, and organized infarcts are among other incidental nodular lesions occasionally encountered in the lung. Early infarcts have a wedge shape with hemorrhagic necrosis, while the overlying pleura is viable with a fibrinous pleuritis with a peripheral rim of granulation tissue. Older infarcts are often rounded and have a rim of fibrous tissue; they can even be mistaken for healed granulomata. Squamous metaplasia is common in the airspaces adjacent to organizing infarcts and around bronchioles. It may be sufficiently exuberant to be mistaken for a neoplastic process.

A single anthracosilicotic nodule is an occasional finding that may be accepted as incidental and insignificant if

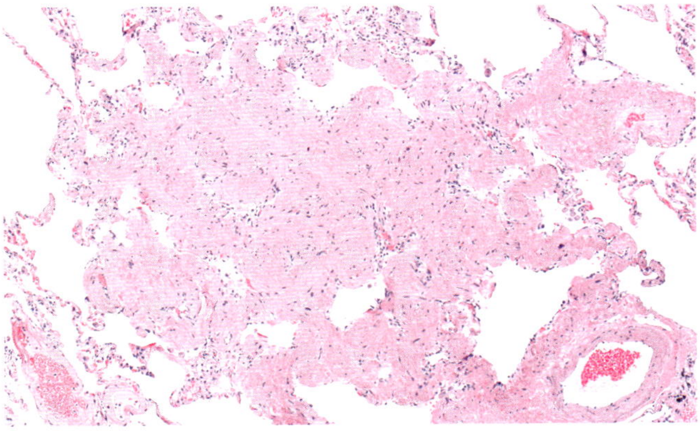

FIGURE 17.29 Incidental parenchymal scar. These are typically subpleural and appear to center on alveolar ducts and commonly have fascicles of normal-appearing smooth muscle.

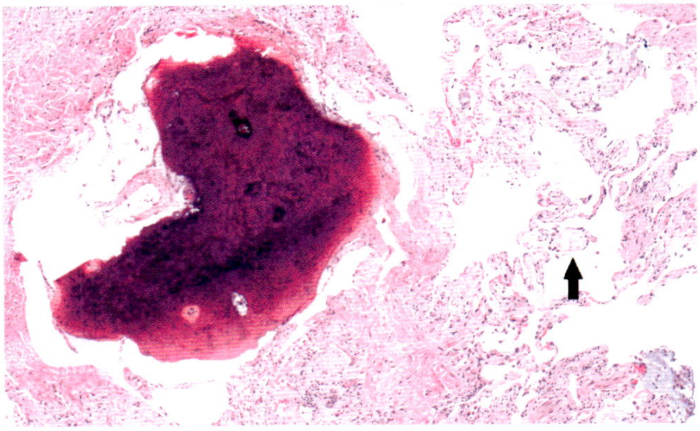

FIGURE 17.30 Ossification. An incidental focus of pulmonary ossification is noted. While not specific, bone formation is often seen in association with aspiration (*arrow*—degenerated aspirated food material).

there is no clinical or radiologic evidence of pneumoconiosis. When anthracosilicotic nodules are multiple, the occupational history and the possibility of pneumoconiosis should be explored.

Although extensive parenchymal scarring is usually a pathologic process, focal scars a few millimeters in diameter are common incidental findings in biopsy material. One distinctive form of scar that is frequently observed (and may be identifiable as 2- to 3-mm centrilobular nodules in CT scans) consists of scars centered on alveolar ducts in the periphery of the lung (Fig. 17.29). These round or somewhat stellate lesions are particularly common among ex-smokers and contain abundant fascicles of smooth muscle, thereby simulating lymphangioleiomyomatosis.

There are a number of histologic changes that accompany pulmonary scarring, regardless of cause. These include vascular intimal and medial thickening, sometimes to the point of luminal occlusion, smooth muscle and myofibroblastic hyperplasia in the interstitium, metaplasia and hyperplasia of type II cells or bronchiolar-type epithelium, accumulations of intra-alveolar macrophages and mucostasis, carcinoid tumorlets, dystrophic calcification or ossification, microscopic paracicatricial airspace enlargement, and metaplastic adipose tissue in the pleural and peribronchial regions. The proliferation of type II cells associated with scars may be confused with adenocarcinoma, although carcinomas show uniform cellular crowding with abrupt transitions to normal alveolar walls, significant cytologic atypia, and a lack of ciliated cells. Although atypical adenomatous hyperplasia tends to have relatively little interstitial widening and fibrosis, it enters into the differential diagnosis in this situation. Of note, both proliferations would be thyroid transcription factor-1 (TTF-1) and Napsin-A positive. The differential diagnosis also includes micronodular pneumocyte hyperplasia, usually seen associated with tuberous sclerosis and sometimes lymphangioleiomyomatosis, although this is a much rarer entity to encounter (83).

In rare instances, mature bone is observed in normal alveoli; it may represent the residue of an organized airspace exudate due to aspiration, chronic passive congestion of mitral stenosis, or an ancient organized pneumonia. Extensive dystrophic ossification occasionally accompanies lesions with diffuse pulmonary fibrosis (Fig. 17.30) (81). Focal dystrophic ossification is a common finding in lung scar, regardless of cause. When ossification is extensive the differential diagnosis includes dendriform ossification (82).

Intrapulmonary lymphoid tissue (Fig. 17.31) is not uncommonly encountered in wedge biopsies for pulmonary nodules and it is sometimes removed for suspicion of neoplasm. Lymphoid tissues in the lung vary from loosely organized aggregates of lymphocytes to fully developed lymph nodes. The nodes characteristically occur along interlobular septae.

A number of intra-alveolar and intracellular structures are seen in the lung. Small numbers of intra-alveolar macrophages are a normal finding; an increase in their numbers is a nonspecific finding and a common reaction in smokers (65). Focal desquamative interstitial pneumonia-like reactions are seen in many pathologic conditions, especially in smokers and in conditions characterized by fibrosis and architectural distortion (46,84) for the presence of hemosiderin in alveolar macrophages should prompt consideration for causes of primary or secondary alveolar hemorrhage.

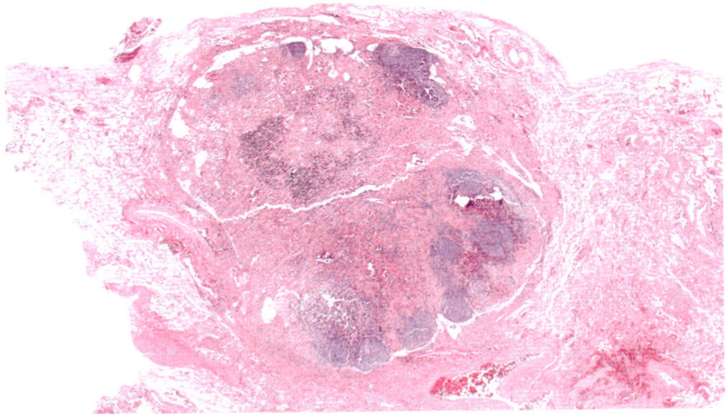

FIGURE 17.31 Intrapulmonary lymph node. There is a relatively large lymph node in the pulmonary parenchyma.

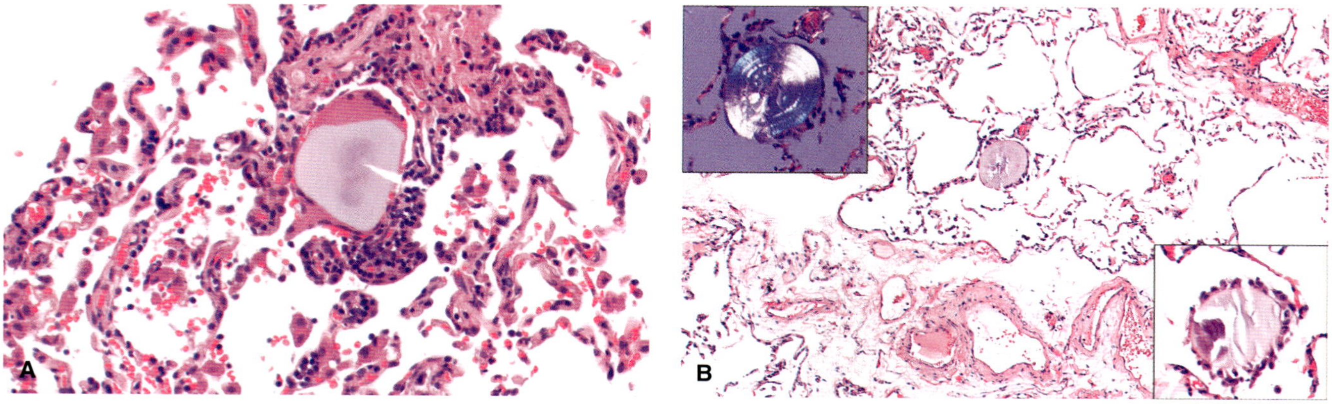

FIGURE 17.32 Pulmonary corpora amylacea. **A:** These show a central bluish nidus and a giant cell or macrophage reaction around them. They may have radiating proteinaceous arrays and show cracking in histologic sections. **B:** They are often weakly polarizable (*inset, upper left*) and appear somewhat degenerated (*inset, lower right*), leading to a misdiagnosis as aspirated material. Careful examination will typically reveal characteristic lamellations.

Corpora amylacea (85) are eosinophilic, rounded, slightly lamellated proteinaceous bodies (Fig. 17.32) that stain positively with PAS and faintly with Congo red stains. They are more common in the lungs of older individuals. Sometimes there is a blue-gray, calcified, or polarizable crystalline particulate body in their centers and a macrophage or giant cell response around them. The exact nature and cause of corpora are unclear, but they are of no clinical significance and should not be confused with aspirated exogenous material or food particles.

Blue bodies (Fig. 17.33) are intra-alveolar, lamellated, basophilic, calcified structures found in airspaces associated with alveolar macrophages and giant cells (86). They can be seen in a number of diffuse lung diseases related to accumulation of macrophages and are of no diagnostic significance. They are thought to be related to macrophage catabolism and are composed primarily of calcium carbonate.

Schaumann bodies (Fig. 17.34) are lamellated, calcified bodies seen in giant cells and associated with granulomas (86–88). They are of no diagnostic significance and may be found in granulomata from diverse causes. Schaumann bodies are endogenously derived and may be mistaken for exogenous material since they may be partially birefringent due to the concomitant presence of oxalate crystals.

Asteroid bodies (Fig. 17.35) represent a starlike array of crystallized intracellular proteins seen in giant cells of many granulomatous conditions. Other than being aesthetically pleasing, they are nonspecific.

Calcium oxalate crystals are lucent, birefringent, platelike crystals that are often found in giant cells and may be mistaken for exogenous material (Fig. 17.36) (87). Accumulation of oxalate crystals around aspergillomas and in patients with forms of invasive aspergillosis is also common.

Material similar to the Mallory hyaline (Fig. 17.37) may be found in the reactive type II cells of a number of interstitial diseases. It is distinctive, but nonspecific, and may be immunopositive for keratin and ubiquitin (89).

Ferruginous bodies are indicative of significant inhalational exposure to ferruginated material, but their presence

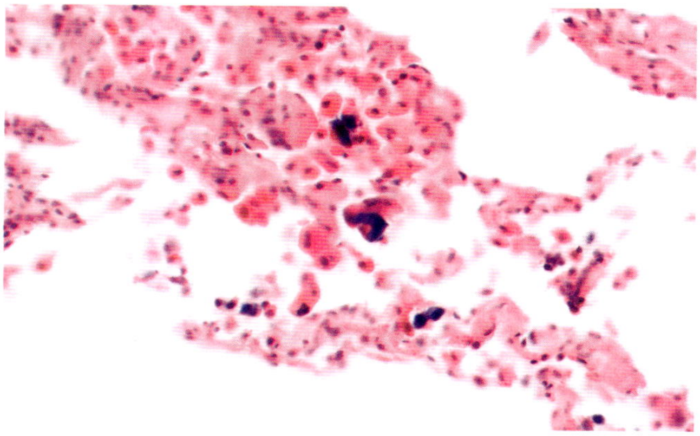

FIGURE 17.33 Blue bodies. Blue bodies are gray or blue, intra-alveolar, calcified, lamellated structures often associated with clusters of macrophages, including giant cells.

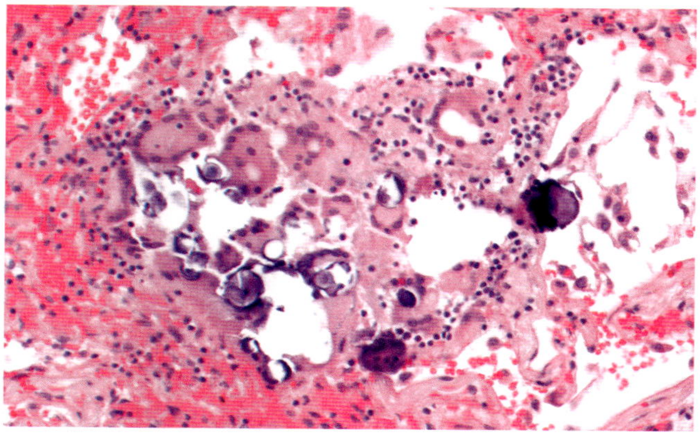

FIGURE 17.34 Schaumann body. There is a bluish calcified structure associated with the surrounding granulomatous inflammation. A lamellated appearance similar to a psammoma body is focally apparent.

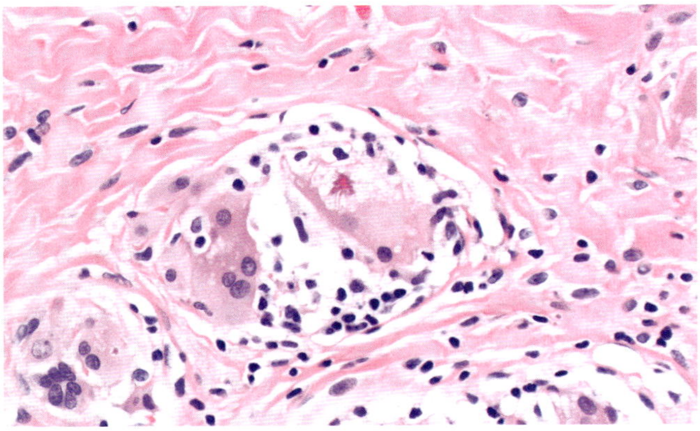

FIGURE 17.35 Asteroid body. An asteroid body in a giant cell is illustrated from a case of sarcoidosis.

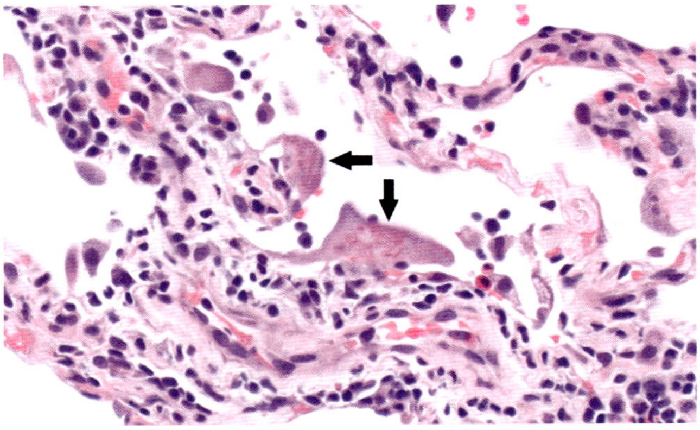

FIGURE 17.37 Hyaline material resembling Mallory hyaline (*arrows*) may be seen in reactive type II pneumocytes in a number of acute and chronic conditions. This is also known as Kuhn hyaline.

does not necessarily correspond to clinically significant lung disease. Studies with electron-probe analysis have helped elucidate the variety of materials that may become iron-encrusted, of which asbestos fibers comprise only a small portion (90).

One may find short, needlelike, birefringent material (usually silica or silicates) in association with anthracotic pigment in virtually any urban adult and the amount may be quite impressive (Fig. 17.38). These materials tend to be more common along lymphatic routes in the lung or in the regional lymph nodes. This material should be distinguished from formalin pigment, as well as from surgical glove talc or starch; the latter is limited to handled surfaces and often demonstrates a Maltese-cross configuration with polarization. Silicates are more brightly birefringent and larger than silica. The diagnosis of silicosis (and silicatosis) is not based solely on the presence of birefringent material but requires clinicopathologic correlations, appropriate imaging findings, and parenchymal silicotic nodules or masses of histiocytes along lymphatic routes. Precise characterization of any material identified requires special techniques such as electron-probe analysis.

An occasional nonnecrotizing epithelioid granuloma may be found in the lungs of patients who have no evidence of granulomatous disease; they are analogous to the occasional granuloma seen at many sites in the body. Granulomatous inflammation can be seen in parenchyma adjacent to mass lesions in approximately 8% of resection specimens (91). Likewise, cholesterol granulomata or single giant cells containing cholesterol clefts may also be an occasional incidental finding (Fig. 17.39). In some instances, the presence of cholesterol granulomata has been linked to prior alveolar hemorrhage, mucostasis, or pulmonary hypertension (92), but usually no significance can be ascribed to them. Such lesions can also be seen in association with hypersensitivity pneumonitis. Lipogranulomata are an occasional nonspecific finding said to be more common in diabetics (93).

An interesting finding is the presence of scattered megakaryocytes in alveolar walls (Fig. 17.40), predominantly within alveolar capillaries. Large numbers can be seen, particularly during sepsis. They are of no diagnostic significance but should not be overinterpreted as malignant or virally infected cells. Along with the bone marrow and spleen, the lung acts as a major reservoir for megakaryocytes (94).

Bone marrow emboli, common in autopsy material, may also be seen in biopsy specimens (Fig. 17.41). Rarely

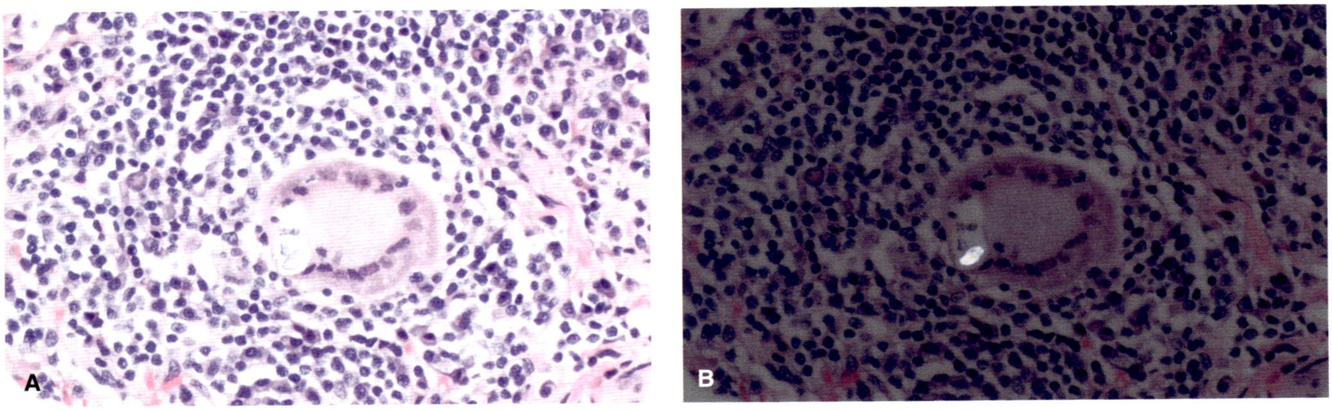

FIGURE 17.36 Calcium oxalate crystals. **A:** Calcium oxalate crystals are commonly associated with Schaumann bodies and granulomatous inflammation and show bright birefringence, as noted in this partially polarized photomicrograph (**B**). The association with giant cells is characteristic.

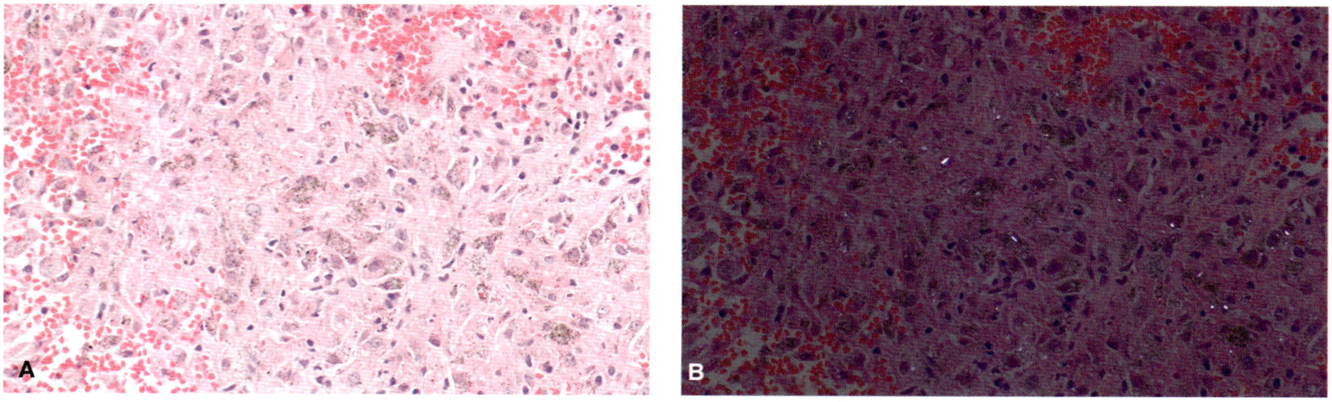

FIGURE 17.38 **A:** Silica and silicates in the normal lung. **B:** Birefringent silica and silicate particles are a common nonspecific finding in and around the anthracotic pigment that is so commonly seen in urban adults and smokers.

correlated with any clinically significant process, they may be a consequence of bony trauma, excision of ribs, or secondary to strenuous cardiopulmonary resuscitation. In some cases, however, bony trauma cannot be implicated and the marrow emboli are an unexplained incidental finding.

Recent intravascular thrombi are relatively common in any severe acute inflammatory lung disease and likely form in situ (Fig. 17.42). They should not be considered as evidence of pulmonary emboli without corroborating clinical information.

Calcification and iron encrustation of the pulmonary elastic tissue may occur in patients with metabolic abnormalities or chronic pulmonary congestion such as occurs as in association with congestive heart failure. In this situation, calcifications may be associated with a giant cell reaction (Fig. 17.43). This phenomenon was once termed endogenous pneumoconiosis but has since been described as mineralizing elastosis, as the former is biologically inaccurate and potentially misleading (95). The presence of mineralizing elastosis should prompt one to search for causes of the pulmonary hemorrhage.

Intravascular foreign material is usually birefringent and may elicit a giant cell reaction. It is most commonly detected in patients with a history of intravenous drug abuse, although fragments of foreign material are occasionally seen in patients without a history of drug abuse. In this situation, the findings may be related to intravenous lines during hospitalization or surgery.

The finding of interstitial air, either localized or diffuse, is well known to pediatric lung pathologists and may be an incidental finding in adults who have been on ventilators. Interstitial air can occur as an incidental finding in a region of subpleural fibrosis but may also be a significant pathologic finding in its own right (Fig. 17.44). The abnormal air-filled spaces resemble honeycomb lung with the exception that they lack the expected bronchiolar metaplastic epithelial lining and the airspaces either show a complete lack of a cellular lining or show a macrophage-rich response either with, or without, a concomitant giant cell reaction. Interstitial air presenting in this way may be encountered in patients with interstitial lung disease, in patients with a history of pneumothorax, and in and around bullae and blebs. It has also been described in patients with COVID-related pneumonia (96) Interstitial air may be completely missed, being interpreted as tissue tearing.

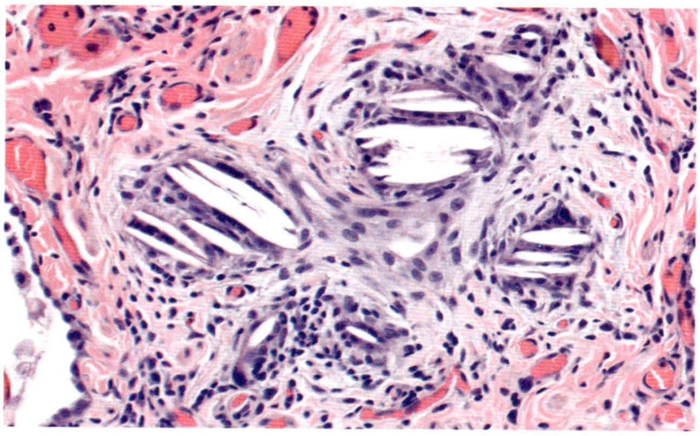

FIGURE 17.39 Cholesterol granulomas. Granulomas and clusters of giant cells containing cholesterol clefts are a frequent nonspecific finding in interstitial lung disease.

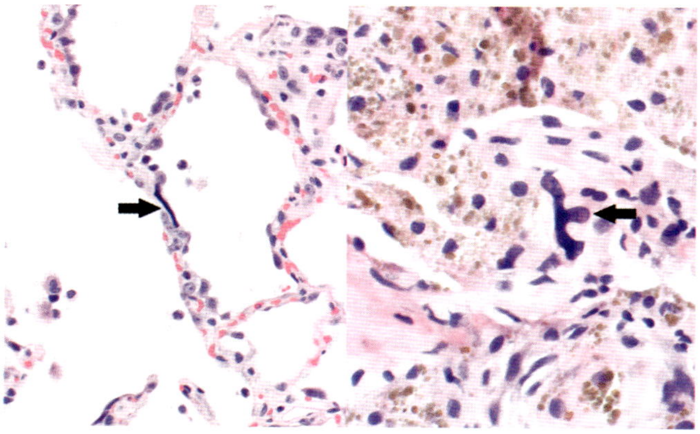

FIGURE 17.40 Megakaryocytes. Megakaryocytes are a common finding in normal lung, often present within alveolar capillaries (*arrows*).

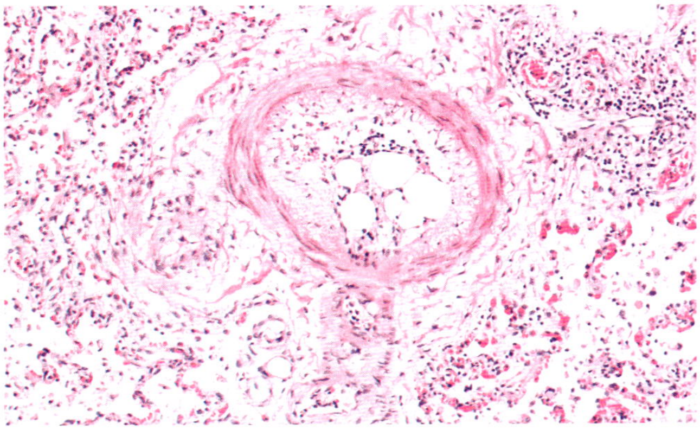

FIGURE 17.41 Bone marrow embolus. An incidental bone marrow embolus (*right center*) identified in a pulmonary artery in a biopsy from a patient with lymphangioleiomyomatosis (*left*).

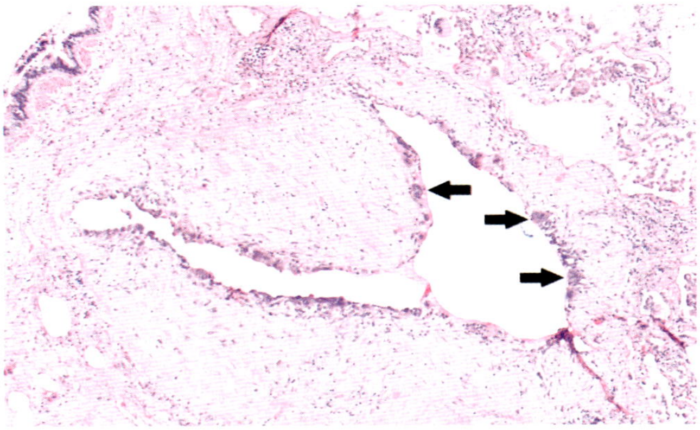

FIGURE 17.44 Interstitial air. When air dissects into the lung interstitium and becomes persistent (typically during positive end-expiratory pressure ventilation) peculiarly shaped airspaces may be seen within interlobular septa in the peripheral lung. These artificial spaces represent interstitial air and become lined by giant cells (*arrows*). In children on ventilator support, this process is referred to as persistent interstitial pulmonary emphysema (PIPE).

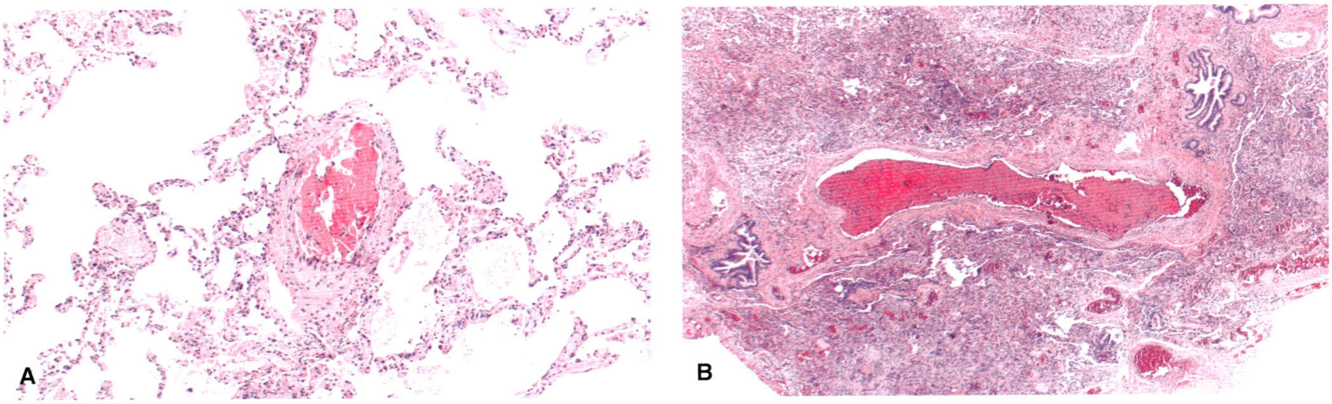

FIGURE 17.42 Thromboemboli in acute lung injury. Small (**A**) or even somewhat large (**B**) fibrin thrombi are commonly present in biopsies showing extensive acute lung injury. Organizing diffuse alveolar damage can be appreciated around the artery.

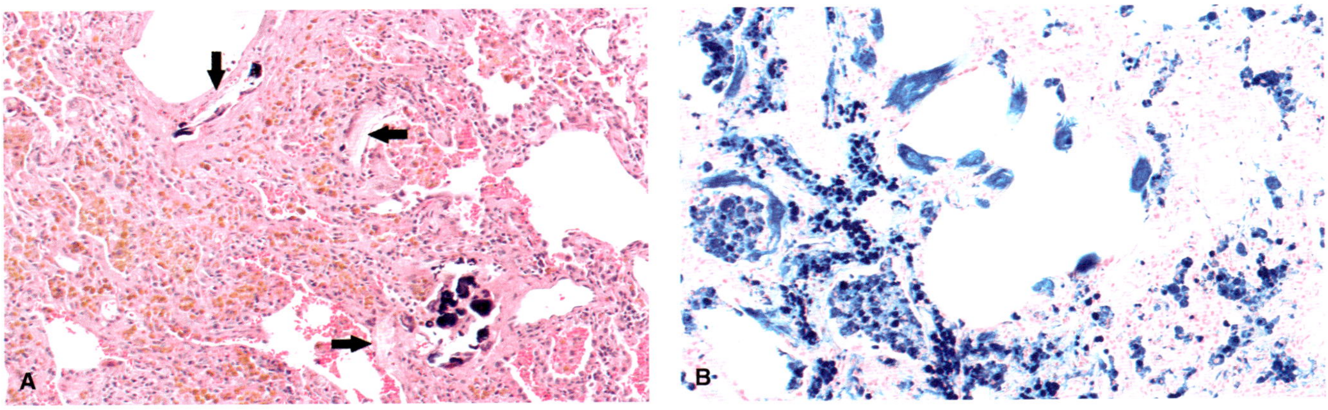

FIGURE 17.43 Iron deposition on elastic tissue or mineralizing elastosis. **A:** Chronic hemorrhage, may result in encrustation of interstitial and vascular elastic fibers by hemosiderin that has a gray to brown semi-translucent appearance (*arrows*). A giant cell reaction is a frequent accompanying finding. **B:** A Prussian blue special stain highlights the diffuse iron deposition.

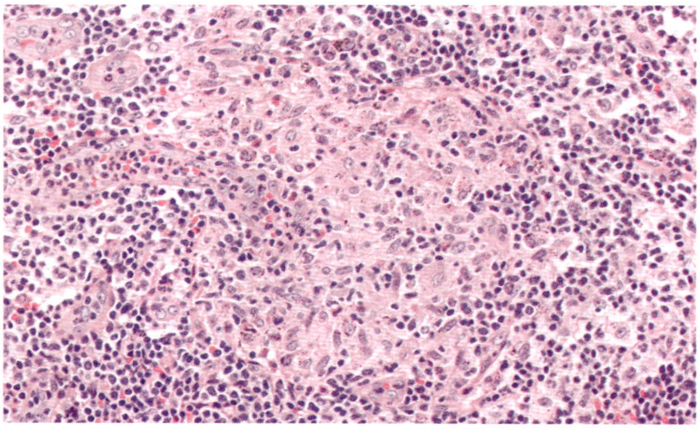

FIGURE 17.45 Histiocyte clusters in hilar nodes. Normal hilar and mediastinal lymph nodes commonly have clusters of histiocytes with variable amounts of dust particles/anthracotic pigment in them. Sometimes they may appear somewhat granulomatous and may be difficult to distinguish from a true granulomatous reaction.

Lymph Nodes

Hilar and peribronchial lymph nodes frequently exhibit a number of characteristic changes. Clusters of dust-filled macrophages in the sinuses and paracortical areas may resemble small granulomata (Fig. 17.45). Silicotic or anthracosilicotic nodules, old/healed infectious granulomata, and sinus histiocytosis are also common. When sarcoidosis is a consideration, one may have difficulty distinguishing the normal histiocytosis of hilar nodes from the granulomata of sarcoidosis. In general, this problem can be approached by maintaining a high threshold for the diagnosis of a granuloma, requiring well-formed, rounded granulomatous masses, accumulations of intragranulomatous fibrinoid or hyalinized material, giant cells, and careful clinicopathologic correlation.

Although multiple intraparenchymal anthracosilicotic nodules should make one suspect the possibility of pneumoconiosis, anthracosilicotic nodules in hilar and intraparenchymal nodes are quite common in the absence of pneumoconiosis. The nodules are composed of concentric whorled and layered hyalinized collagen and are usually surrounded by a nonpalisaded rim of dust-filled macrophages that contain birefringent material when examined with polarized light. Central degenerative changes may be evident. Anthracosilicotic nodules should be distinguished from old/healed granulomatous disease. The presence of palisaded macrophages and giant cells is more characteristic of old granulomata. Of note, anthracosilicotic nodules can be secondarily infected, particularly with mycobacterial organisms (i.e., silicotuberculosis). The presence of any necrotic granulomatous inflammation in an anthracosilicotic nodule is an indication to perform histochemical stains for microorganisms.

Hamazaki–Wesenberg bodies (97) are yellow-brown intracellular or extracellular structures that measure approximately 5 μm in length. They are associated with sinus histiocytes in lymph nodes, especially hilar nodes (Fig. 17.46). The cause of these bodies is unknown, but they resemble lipofuscin. Positivity with methenamine silver and PAS stains may lead to their confusion with yeast forms, but their H&E appearance, lack of associated necrosis or inflammation, and positive reaction with the Fontana–Masson stain should facilitate their recognition and distinction from fungi.

Pleura

Changes in the pleura in wedge biopsies may be incidental or part of the underlying pathologic process, and their significance needs to be assessed on an individual case basis.

Two different but similar lesions that are related to the pleura include nodular mesothelial hyperplasia and lesions of aggregated mesothelial cells and macrophages. Both are benign and of no clinical significance, but the latter can cause considerable confusion for pathologists, often raising concern for a neoplasm. Both are comprised of a mixed population of macrophages and mesothelial cells. The former, however, are in situ lesions, attached to the pleural surface, have a distinct vascular or capillary network, and are often associated with chronic inflammatory cells. They may be related to prior pleural inflammation. Lesions of aggregated

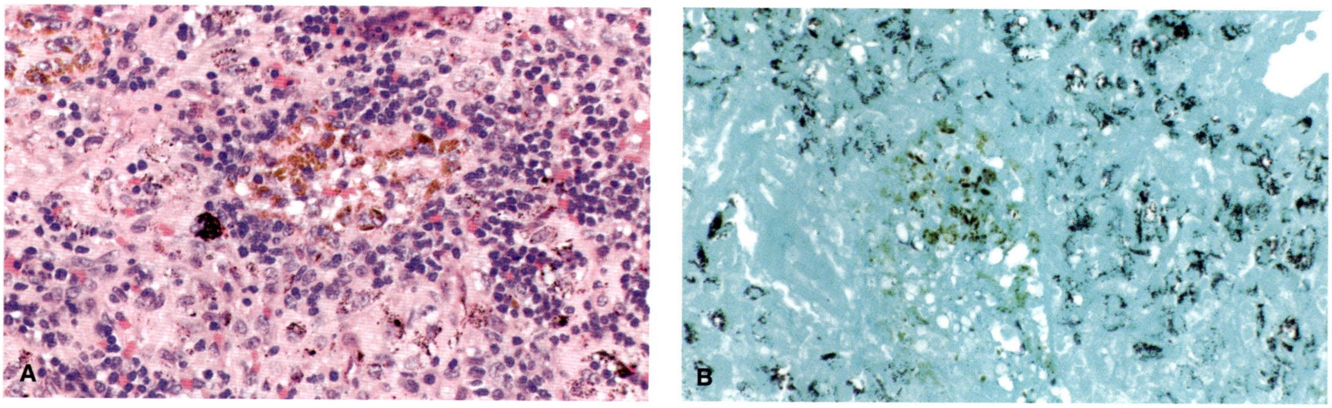

FIGURE 17.46 Hamazaki–Wesenberg bodies. **A:** Hamazaki–Wesenberg bodies represent yellow-brown oval structures associated with histiocytes, occasionally encountered in hilar and mediastinal lymph nodes. **B:** Their positive staining with silver stains sometimes leads to confusion with fungi.

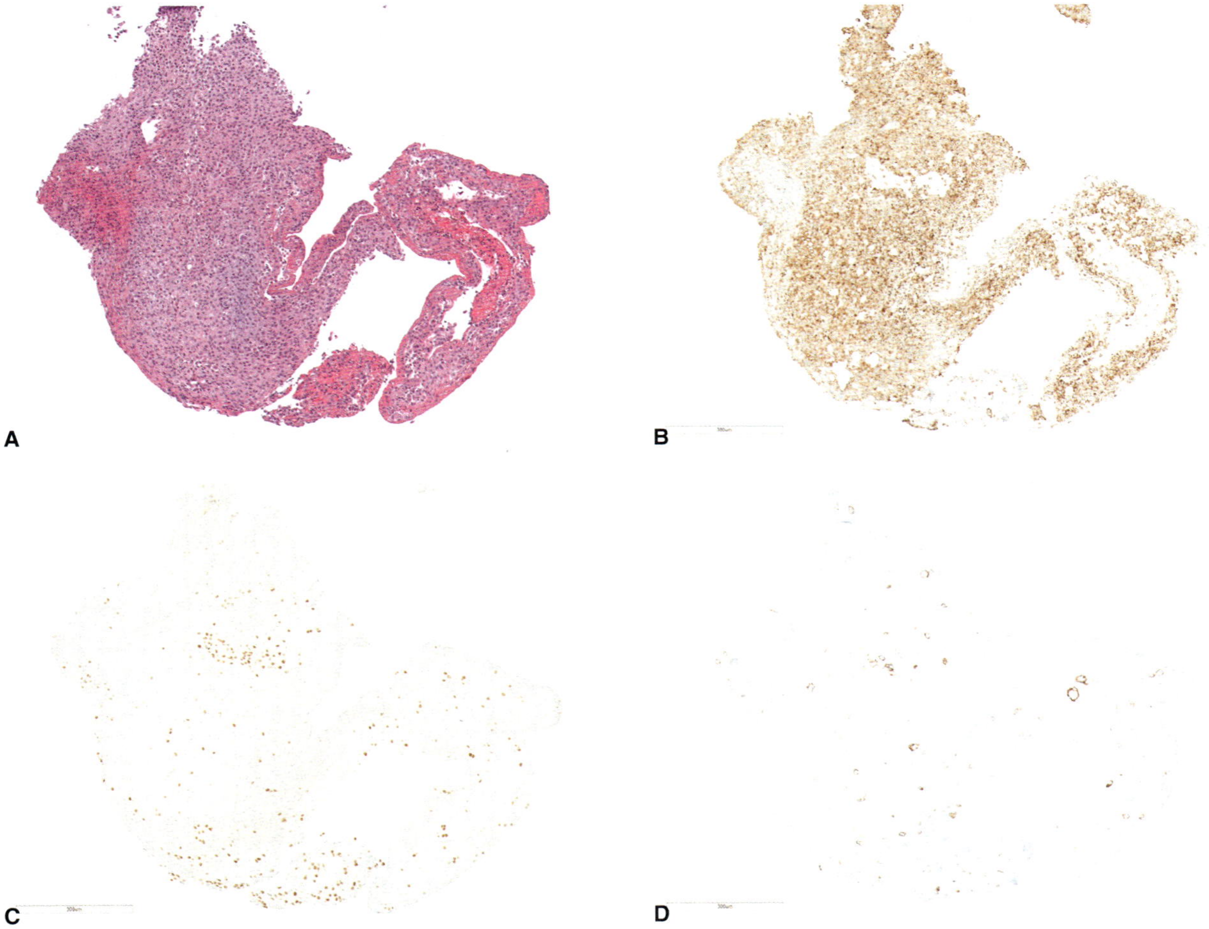

FIGURE 17.47 Lesion of aggregated mesothelial cells and macrophages (LAMM). **A:** An apparent mass-forming lesion demonstrates a mixed cell population of macrophages and mesothelial cells highlighted by CD68 for macrophages (**B**) and WT-1 highlighting mesothelial cells (**C**). CD34 (**D**) is negative for blood vessels, highlighting the lack of vascularized growth.

mesothelial cells and macrophages are iatrogenically produced artifacts of instrumentation, either needle, transbronchial or surgical biopsies. They result from the agglomeration of macrophages and mesothelial cells. Rarely they present as mass lesions and may be submitted by the surgeon as a mass or apparent lymph node (Fig. 17.47), or they can show up on slides from these cases, where they can cause concern for a neoplasm. These aggregates of macrophages and mesothelial cells have no vasculature, as they are not in situ lesions (11).

INCIDENTAL FINDINGS IN TRANSBRONCHIAL BIOPSIES

The major artifacts or incidental findings in transbronchial biopsies that lead to misinterpretation include compression/squish that mimics interstitial pneumonia, bubble artifact simulating lipoid/aspiration pneumonia, and portions of the pleura that are misinterpreted as neoplastic or suspicious for being neoplastic (55,56). Such portions of pleura may even include some pleural fat, a particularly common finding in fibrosing interstitial pneumonias. The finding of pleural tissue in transbronchial biopsies (Fig. 17.48) is not uncommon but also is

FIGURE 17.48 Visceral pleura in transbronchial biopsies. Transbronchial biopsies directed toward the periphery of the lung may sample portions of the visceral pleura. These may lead to misdiagnosis, particularly if there is an associated reactive pleuritis.

TABLE 17.8 Effects of Aging Seen in Lung Biopsies

- Tracheobronchial cartilage ossification
- Submucosal fatty metaplasia
- Oncocytic metaplasia and/or hyperplasia in bronchial glands
- Elastosis of submucosal connective tissue
- Emphysema and small cysts
- Pulmonary arterial and venous intimal thickening and hyalinization of arterioles
- Alveolar enlargement (rarely appreciated in biopsy material)
- Medial calcification in bronchial arteries
- Amyloidosis (ATTR)
- Anthracosis (urban dwellers)

not well recognized among most pathologists. Strips of reactive mesothelial cells in such specimens may be confused with carcinoma. The use of calretinin and/or WT-1 stains could aid in the proper identification of mesothelial origin in these cases.

As in wedge biopsies, the most common cause of red blood cells with some fibrin deposition in airspaces in transbronchial biopsies is trauma related to the procedure of the biopsy rather than an alveolar hemorrhage syndrome.

EFFECTS OF AGING

Some of the effects of aging on the lungs are shown in Table 17.8. Calcification and ossification of cartilages in the large airways may be seen. Intimal thickening with atherosclerosis can be seen in pulmonary arteries and veins (Fig. 17.49), and apical pulmonary arteries are affected more often (2). Intimal thickening of veins is often hyaline and sclerotic in character, and should be distinguished from the more cellular intimal and muscular proliferation seen in pulmonary veno-occlusive disease and chronic passive congestion. Mural hyalinization of small arterioles is also an aging change. This should not be confused with septal amyloidosis, which tends to appear more homogenous than collagen. Congo red stain could also be of help. True amyloid deposits may be an incidental finding and increase with age (98,99). However, amyloid should be typed, preferably by mass spectrometry when it is identified. Depending on type additional evaluation of the patient may be necessary to exclude a lymphoproliferative disorder.

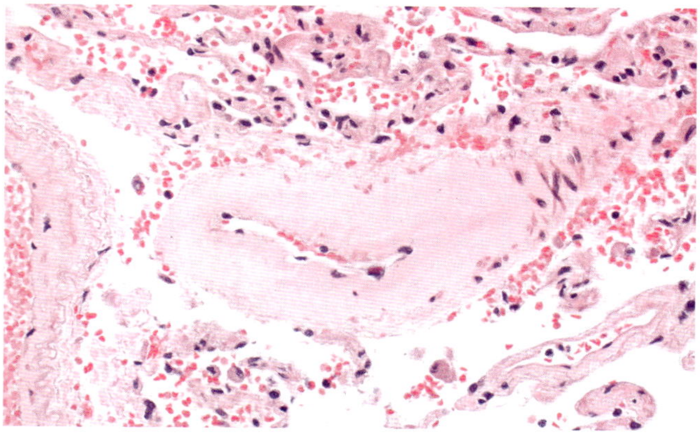

FIGURE 17.49 Aging change in vessels. In elderly individuals, a hyaline thickening of the intima may be encountered as an incidental finding. Clinical correlation is suggested, since this feature may also be encountered in the setting of pulmonary hypertension.

Some degree of alveolar enlargement accompanies aging (100), which can either manifest as small cysts or emphysema. Longitudinal elastic tissue fibers are a normal finding immediately beneath the surface epithelium in the large airways. These may take on an elastotic appearance in the lungs of older individuals.

TABLE 17.9 Situations in Which a Lung Biopsy May Appear Normal

- Sampling error
- Pulmonary vascular disease
- Small airway (bronchiolar) disease, subtotal or complete obliterative bronchiolitis
- Pulmonary edema and early diffuse alveolar damage
- Emboli, including fat emboli, silicone emboli
- A very subtle interstitial infiltrate
- Intravascular lymphoma
- Cardiac disease with secondary pulmonary abnormalities

THE BIOPSY THAT LOOKS NORMAL AT FIRST GLANCE

When a lung biopsy for presumed diffuse lung disease initially appears normal, a number of possibilities should be considered (Table 17.9), especially pulmonary edema (Fig. 17.9).

Although one can argue that interstitial infiltrates that are so subtle as to be overlooked are probably not clinically significant, their recognition is necessary. This situation can arise, when biopsy material is obtained from the less severely affected portions of lungs in patients with interstitial pneumonias, in cases in which the inflammatory infiltrate has been suppressed by steroids or immunosuppressive therapy, and in conditions characterized by patchy inflammation. The presence of inflammatory cells in the perivenular regions and alveolar wall with reactive type II cells is usually indicative of an interstitial pneumonia.

IMMUNOHISTOCHEMISTRY

Immunohistochemistry is applied frequently to the study of primary and metastatic neoplasms in the lung. Knowledge of the expected immunoreactivity of normal lung and pleura is helpful in these situations and such knowledge can be put to good use in the study of certain nonneoplastic

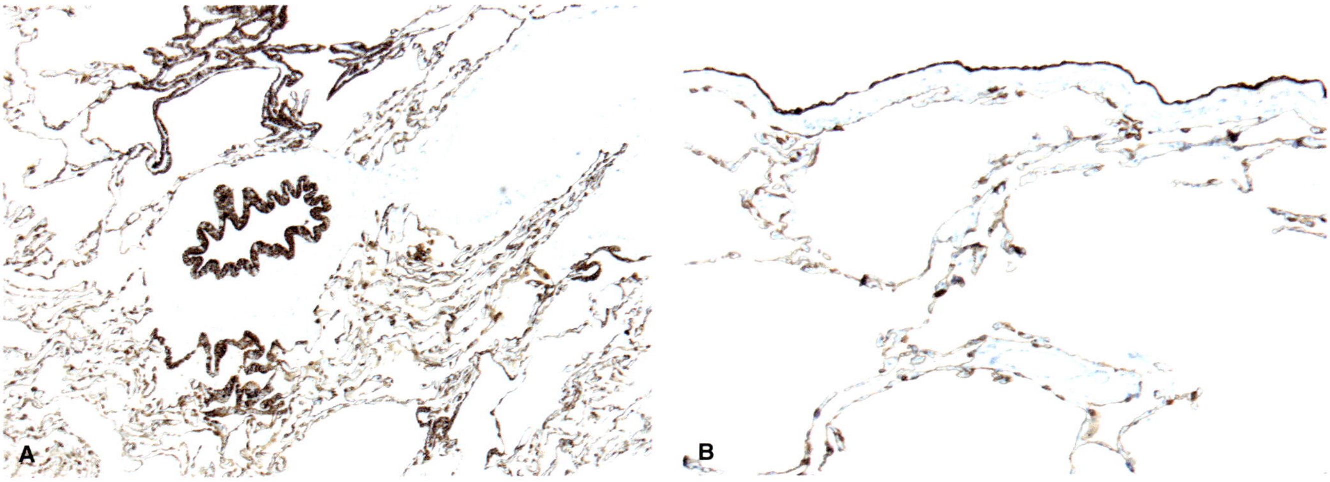

FIGURE 17.50 Pancytokeratin (CAM5.2). Cytokeratin immunoreactivity in normal peripheral lung varies somewhat by the molecular weight of the target cytokeratin peptide against which the antibody is directed (1,2). Broad-spectrum antibodies to cytokeratin stain all lung epithelial and pleural mesothelial cells, similar to the epithelial membrane antigen (EMA). **A:** Note the intense immunoreactivity of CAM5.2 in bronchiolar epithelium and the surrounding alveolar lining cells. **B:** Normal mesothelial cells can be better and more consistently visualized with broad-spectrum antibodies as compared to more targeted cocktails of specific cytokeratin polypeptides (e.g., CK5/6) used for characterizing mesothelioma (101,102).

disorders when lung histology is obscured by a disease process. The immunostaining pattern of normal lung cells is summarized in Table 17.3.

Some examples of lung tissue stained with several commonly employed antibodies are presented in Figures 17.50 to 17.52 (101–106). The most frequent immunostains used in pulmonary pathology are summarized in Table 17.10. When applying immunostains for the diagnosis of lung malignancies, it is important to recognize staining of the uninvolved underlying benign lung architecture. For example, one should be aware of TTF-1/Napsin-A reactivity of entrapped pneumocytes in otherwise solid tumors. Likewise, lepidic spreading tumors may show TTF-1/Napsin-A reactivity of the underlying benign alveolated parenchyma. This reactivity should not be construed as evidence of adenocarcinoma. Also, the lung contains large amounts of airspace macrophages, that can show nonspecific staining, particularly with Napsin-A antibodies. Of note, p63 is not a specific marker of squamous cell carcinoma. The WHO recommends the use of p40 as the most specific marker of squamous differentiation (75). Importantly, the lung is a common site of metastases. It is imperative to recognize that certain immunostains, otherwise considered to be relatively specific of an organ of origin can be expressed in primary lung malignancies. For example, GCDFP15, estrogen and progesterone receptor, and GATA3 immunostains can be reactive in some lung non–small-cell carcinomas (107–110), and CDX2 may be expressed in primary mucin-producing adenocarcinomas of the lung (111). Likewise, TTF-1 is expressed in many thyroid neoplasms with the exception of anaplastic thyroid carcinomas, a small percentage of ovarian serous carcinomas, endometrial and endocervical adenocarcinomas, and colonic adenocarcinomas. Napsin-A can be expressed in papillary renal cell carcinomas and in some clear cell carcinomas of the gynecologic tract (106). Lastly, failure of a tumor to react with antibodies to TTF-1 and Napsin-A does not exclude lung adenocarcinoma. In such cases, the use of histochemical mucin stains for mucin to demonstrate glandular differentiation may be helpful.

TABLE 17.10 Immunostains Commonly Used for the Diagnosis of Lung Primary Neoplasms

Neoplasm	Immunoperoxidase Stains
Adenocarcinoma	Keratins TTF-1 (WHO recommended) Napsin A
Squamous Carcinoma	P40 (WHO recommended, most specific) P63 CK5/6
Small Cell Carcinoma	Keratins (most often dot-like) TTF-1 Ki67 (usually >80% proliferative rate) Synaptophysin (may be negative) Chromogranin (may be negative) INSM-1 (may be negative) RB1 loss Pou2F3 (in low keratin, low neuroendocrine tumors)
Carcinoid Tumors	CAM 5.2 Chromogranin Synaptophysin INSM1 Ki67 (lower rate than small cell, which is helpful in small crushed biopsies)
SMARCA4 Deficient Thoracic Tumor	BRG-1 loss

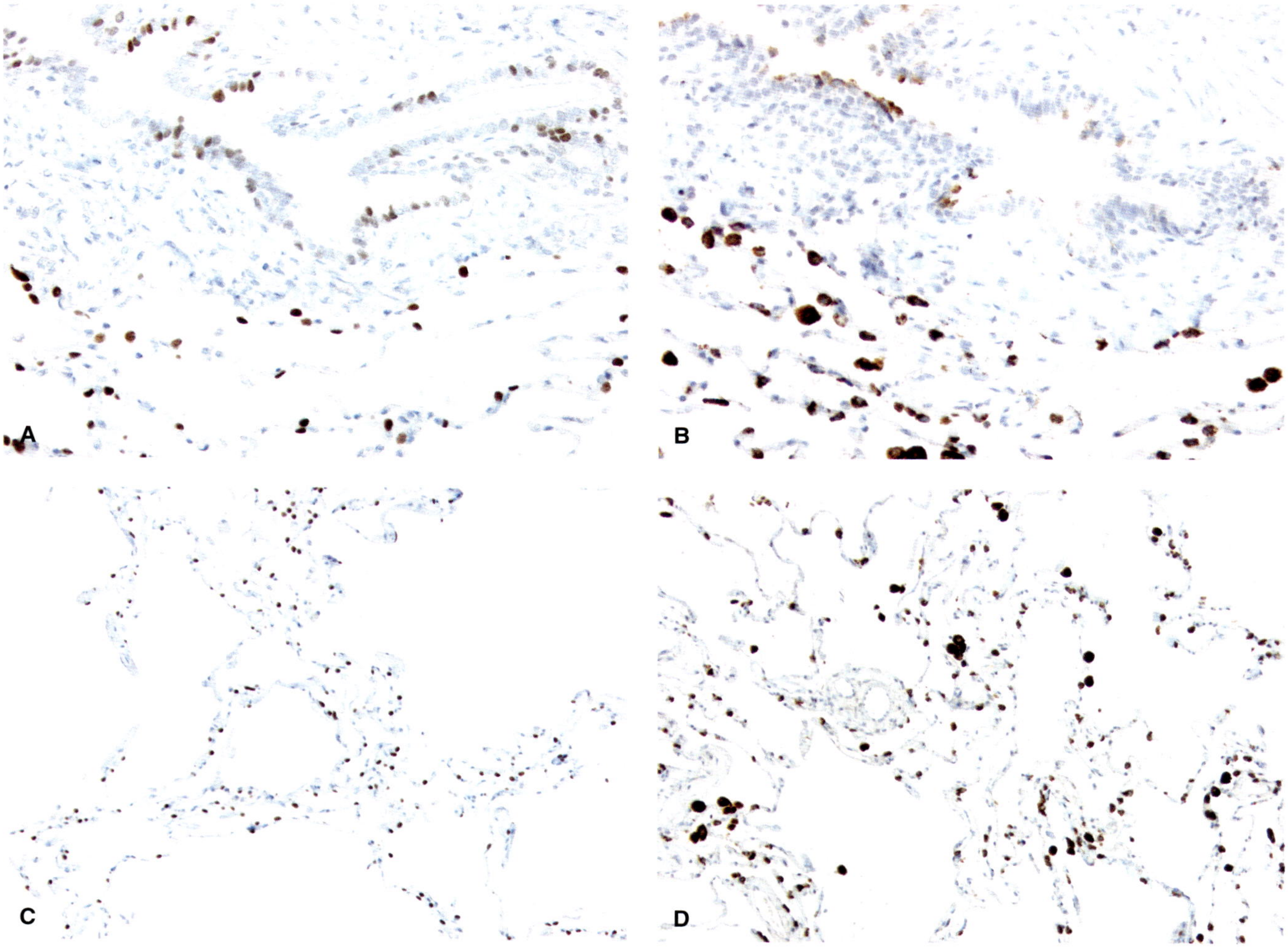

FIGURE 17.51 TTF-1 and Napsin-A. These two antibody targets are useful in narrowing potential origins for carcinomas in lung or present in the setting of the metastatic disease of unknown origin (3,4). Both stain lung epithelial cells; thyroid transcription factor-1 (TTF-1) with a nuclear localization, the aspartic proteinase Napsin-A with a cytoplasmic localization. TTF-1 in a normal adult lung is highly restricted to lung epithelium of the airways and alveoli. Antibodies directed against Napsin-A, however, also decorate lung macrophages to variable extent and produce more abundant staining in peripheral lung. (**A, C, TTF-1; B, D, Napsin-A**) (103,104).

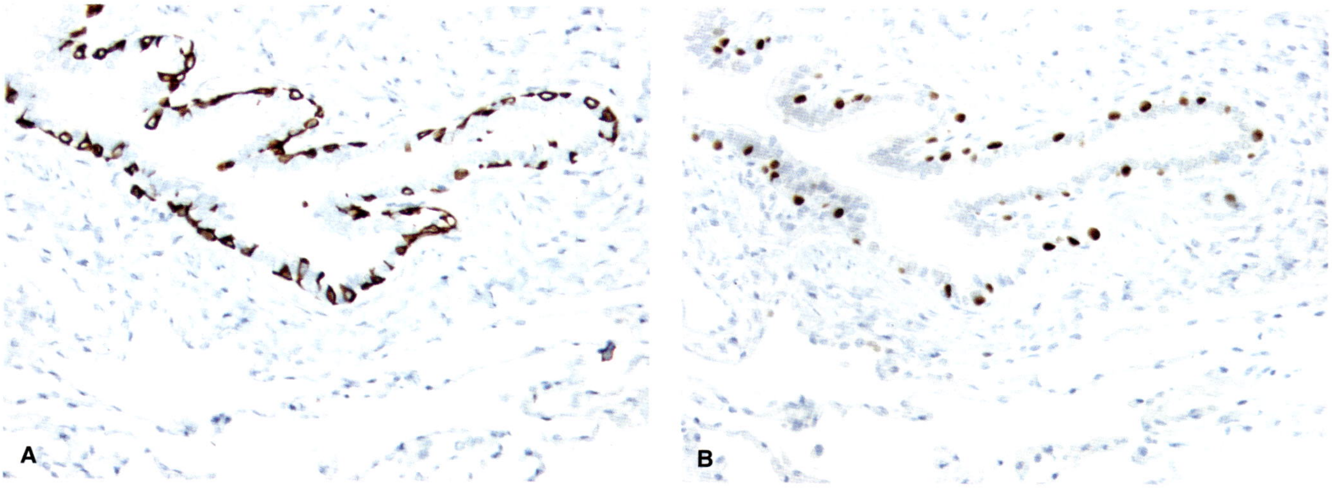

FIGURE 17.52 Cytokeratin 5/6 and p63. These antibodies are useful markers for basal-type epithelial and myoepithelial cells in many organs (5). In the lung tissue, both show immunoreactivity in peripheral airway epithelium but not alveolar epithelial cells (105). (**A, Cytokeratin 5/6; B, p63**)

REFERENCES

1. Nagaishi C, Nagasawa N. *Functional Anatomy and Histology of the Lung*. University Park Press; 1972.
2. Wagenvoort CA, Wagenvoort N. *Pathology of Pulmonary Hypertension*. John Wiley; 1977.
3. Kuhn C 3rd. Ultrastructure and cellular function in the distal lung. In: Thurlbeck WM, Abell MR, eds. *The Lung*. Williams & Wilkins; 1978.
4. Scadding JG, Cumming G, eds. *Scientific Foundations of Respiratory Medicine*. WB Saunders; 1981.
5. Gail DB, Lenfant CJ. Cells of the lung: biology and clinical implications. *Am Rev Respir Dis.* 1983;127:366–387.
6. Bienenstock J, Befus AD. Gut-and-bronchus-associated lymphoid tissue. *Am J Anat.* 1984;170:437–445.
7. Langston C, Kida K, Reed M, Thurlbeck WM. Human lung growth in late gestation and in the neonate. *Am Rev Respir Dis.* 1984;129:607–613.
8. Schitny JC. Development of the lung. *Cell Tissue Res.* 2017;367:427–444.
9. Fawcett DW, Jensh RP. *Bloom and Fawcett: Concise Histology*. 2nd ed. Hodder Education Publishers; 2002.
10. Coalson JJ. The adult lung: structure and function. In: Saldana MJ, ed. *Pathology of Pulmonary Disease*. JB Lippincott; 1994:3–14.
11. Butt YB, Tazelaar HD. The unremarkable lung. In: *Atlas of Pulmonary Pathology: A Pattern Based Approach*. Wolters Kluwer; 2022:1–13.
12. Kuhn C III. Normal anatomy and histology. In: Thurlbeck WM, Churg AM, eds. *Pathology of the Lung*. 2nd ed. Thieme Medical Publishers; 1995:1–36.
13. Ochs M. Functional design of the human lung for gas exchange. In: Grippi MA, Antin-Ozerkis DE, Dela Cruz CS, Kotloff RM, Kotton C, Pack AI, eds. *Fishman's Pulmonary Diseases and Disorders.* 6th ed. McGraw-Hill Education; 2023.
14. Albertine KH, Ramirez MI, Morty RE. Anatomy. In: Courtney Broaddus V, Ernst JD, King TE Jr, eds., et al. *Murray and Nadel's Textbook of Respiratory Medicine*. 7th ed. Elsevier; 2022:3–23.
15. Corrin B. *Pathology of the Lungs*. Churchill Livingstone; 2000.
16. Smith ML, Wick MR. Lung anatomy. In: Smith ML, Leslie KO, Wick MR, eds. *Practical Pulmonary Pathology*. Elsevier; 2022:1–19.
17. Chinoy MR. Lung growth and development. *Front Biosci.* 2003;8:392–415.
18. Meyer KB, Wilbrey-Clark A, Nawijn M, Teichmann SA. The human lung cell atlas: a transformational resource for cells of the respiratory system. In: Nikolic MZ, Hogan BLM, eds. *Lung Stem Cells in Development, Health and Disease*. European Respiratory Society; 2021.
19. Yaegashi H, Takahashi T. The airway dimension in ordinary human lung. A standardized morphometry of lung sections. *Arch Pathol Lab Med.* 1994;118:969–974.
20. Churg A, Muller NL. *Atlas of Interstitial Lung Disease Pathology*. 2nd ed. Wolters Kluwer; 2019.
21. Lambert MW. Accessory bronchiolealveolar communications. *J Pathol Bacteriol.* 1955;70:311–314.
22. Yousem SA, Dacic S. Idiopathic bronchiolocentric interstitial pneumonia. *Mod Pathol.* 2002;15:1148–1153.
23. Churg A, Meyers J, Suarez T, et al. Airway-centered interstitial fibrosis: a distinct form of aggressive diffuse lung disease. *Am J Surg Pathol.* 2004;28:62–68.
24. Fukuoka J, Franks TJ, Colby TV, et al. Peribronchiolar metaplasia: a common histologic lesion in diffuse lung disease and a rare cause of interstitial lung disease; clinicopathologic features of 15 cases. *Am J Surg Pathol.* 2005;29(7):948–954.
25. Winkelmann A, Noack T. The Clara cell: a "third Reich eponym." *Eur Respir J.* 2010;36(4):722–727.
26. Lehnert BE. Pulmonary and thoracic macrophage subpopulations and clearance of particles from the lung. *Environ Health Perspect.* 1992;97:17–46.
27. Gould SJ, Isaacson PG. Bronchus-associated lymphoid tissue (BALT) in human fetal and infant lung. *J Pathol.* 1993;169:229–234.
28. Richmond J, Pritchard GE, Ashcroft T, Avery A, Corris PA, Walters EH. Bronchus associated lymphoid tissue (BALT) in human lung: its distribution in smokers and non-smokers. *Thorax.* 1993;48:1130–1134.
29. Tschernig T, Kleemann WJ, Pabst R. Bronchus-associated lymphoid tissue (BALT) in the lungs of children who had died from sudden infant death syndrome and other causes. *Thorax.* 1995;50:658–660.
30. Randall TD. Bronchus-associated lymphoid tissue (BALT) structure and function. *Adv Immunol.* 2010;107:187–241.
31. Bienenstock J. Bronchus-associated lymphoid tissue. *Int Arch Allergy Appl Immunol.* 1985;76(suppl 1):62–69.
32. Kradin RL, Spirn PW, Mark EJ. Intrapulmonary lymph nodes. Clinical, radiologic, and pathologic features. *Chest.* 1985;87:662–667.
33. Gallagher B, Urbanski SJ. The significance of pleural elastica invasion by lung carcinomas. *Hum Pathol.* 1990;21:512–517.
34. Amin MB, Edge SB, Greene FL, et al, eds. *AJCC Cancer Staging Manual*. 8th ed. Springer; 2017.
35. Churg A. New developments in mesothelial pathology. *Histopathology.* 2024;84:136–152.
36. Colby TV, Swensen SJ. Anatomic distribution and histopathologic pattern in diffuse lung disease: correlation with HRCT. *J Thorac Imaging.* 1996;11:1–26.
37. Travis WD, Cotabel U, Hansell DM, et al; ATS/ERS Committee on Idiopathic Interstitial Pneumonias. An official American Thoracic Society/European Respiratory Society statement: update of the International Multidisciplinary Classification of the Idiopathic Interstitial Pneumonias. *Am J Respir Crit Care Med.* 2013;188(6):733–748.
38. Newman SL, Michael RP, Wang NS. Lingular lung biopsy: is it representative? *Am Rev Respir Dis.* 1985;132:1084–1086.
39. Albo RJ, Grimes OF. The middle lobe syndrome: a clinical study. *Dis Chest.* 1966;50:509–518.
40. Kwon KY, Myers JL, Swensen SJ, Colby TV. Middle lobe syndrome: a clinicopathological study of 21 patients. *Hum Pathol.* 1995;26:302–307.
41. Renner RR, Markarian B, Pernice NJ, Heitzman ER. The apical cap. *Radiology.* 1974;110:569–573.
42. McCloud TC, Isler RJ, Novelline RA, Putman CE, Simeone J, Stark P. The apical cap. *AJR Am J Roentgenol.* 1981;137:299–306.
43. Lagstein A. Pulmonary apical cap-what's old is new again. *Arch Pathol Lab Med.* 2015;139(10):1258–1262.
44. Yousem SA. Pulmonary apical cap: a distinctive but poorly recognized lesion in pulmonary surgical pathology. *Am J Surg Pathol.* 2001;25:679–683.

45. Thurlbeck WM, Wright JL. *Thurlbeck's Chronic Airflow Obstruction*. 2nd ed. BC Decker; 1999.
46. Katzenstein AL, Mukhopadhyay S, Zanardi C, Dexter E. Clinically occult interstitial fibrosis in smokers: classification and significance of a surprisingly common finding in lobectomy specimens. *Hum Pathol*. 2010;41(3):316–325.
47. Schneider F, Murali R, Veraldi KL, Tazelaar HD, Leslie KO. Approach to lung biopsies from patients with pneumothorax. *Arch Pathol Lab Med*. 2014;138:257–265.
48. Thurlbeck WM. *Chronic Airflow Obstruction in Lung Disease*. WB Saunders; 1976.
49. Raoof S, Bondalapati P, Vydyula R, et al. Cystic lung disease: algorithmic approach. *Chest*. 2016;150:945–965.
50. Askin FB, McCann BG, Kuhn C. Reactive eosinophilic pleuritis: A lesion to be distinguished from pulmonary eosinophilic granuloma. *Arch Pathol Lab Med*. 1977;101:187–191.
51. Churg A. Asbestos fibers and pleural plaques in a general autopsy population. *Am J Pathol*. 1982;109:88–96.
52. Meurman L. Asbestos bodies and pleural plaques in a Finnish series of autopsy cases. *Acta Pathol Microbiol Scand*. 1966;(Suppl 181):1–107.
53. Roberts GH. The pathology of parietal pleural plaques. *J Clin Pathol*. 1971;24:348–353.
54. Hillerdal G. Pleural plaques and risk for bronchial carcinoma and mesothelioma. A prospective study. *Chest*. 1994; 105:144–150.
55. Katzenstein ALA. *Katzenstein and Askin's Surgical Pathology of Non-neoplastic Lung Disease*. 3rd ed. WB Saunders; 1997.
56. Kendall DM, Gal AA. Interpretation of tissue artifacts in transbronchial lung biopsy specimens. *Ann Diagn Pathol*. 2003;7:20–24.
57. Kadokura M, Colby TV, Myers JL, et al. Pathologic comparison of video-assisted thoracic surgical lung biopsy with traditional open lung biopsy. *J Thorac Cardiovasc Surg*. 1995; 109:494–498.
58. Churg A. An inflation procedure for open lung biopsies. *Am J Surg Pathol*. 1983;7:69–71.
59. Bensard DD, McIntyre RC Jr, Waring BJ, Simon JS. Comparison of video thoracoscopic lung biopsy to open lung biopsy in the diagnosis of interstitial lung disease. *Chest*. 1993;103:765–770.
60. Ferson PF, Landreneau RJ, Dowling RD, et al. Comparison of open versus thoracoscopic lung biopsy for diffuse infiltrative pulmonary disease. *J Thorac Cardiovasc Surg*. 1993;106:194–199.
61. Carnochan FM, Walker WS, Cameron EW. Efficacy of video-assisted thoracoscopic lung biopsy: a historical comparison with open lung biopsy. *Thorax*. 1994;49:361–363.
62. Doberer D, Trejo Bittar HE, Wenzel SE. Should lung biopsies be performed in patients with severe asthma? *Eur Respir Rev*. 2015;24(137):525–539.
63. Ravaglia C, Bonifazi M, Wells AU, et al. Safety and diagnostic yield of transbronchial lung cryobiopsy in diffuse parenchymal lung diseases: a comparative study versus video-assisted thoracoscopic lung biopsy and a systematic review of the literature. *Respiration*. 2016;91:215–227.
64. Kheir F, Becerra PBU, Bissell B, et al. Transbronchial lung cryobiopsy in patients with interstitial lung disease: a systemic review. *Ann Am Thorac Soc*. 2022;19:1193–1202.
65. Niewoehner DE, Kleinerman J, Rice DB. Pathologic changes in the peripheral airways of young cigarette smokers. *N Engl J Med*. 1974;291:755–758.
66. Myers JL, Veal CF Jr, Shin MS, Katzenstein AL. Respiratory bronchiolitis causing interstitial lung disease. A clinicopathologic study of six cases. *Am Rev Respir Dis*. 1987;135:880–884.
67. Yousem SA. Respiratory bronchiolitis-associated interstitial lung disease with fibrosis is a lesion distinct from fibrotic nonspecific interstitial pneumonia: a proposal. *Mod Pathol*. 2006;19(11):1474–1479.
68. Vehar SJ, Yadev R, Mukhopadhyay S, Nathani A, Tolle LB. Smoking-related interstitial fibrosis in patients presenting with diffuse parenchymal lung disease. *Am J Clin Pathol*. 2023;159:146–157.
69. Trejo Bittar HE, Yousem SA, Wenzel SE. Pathobiology of severe asthma. *Annu Rev Pathol*. 2015;10:511–545.
70. Trejo Bittar HE, Doberer D, Mehrad M, et al. Histologic findings of severe/therapy-resistant asthma from video-assisted thoracoscopic surgery biopsies. *Am J Surg Pathol*. 2017;41(2):182–188.
71. Ashley DJ. Bony metaplasia in trachea and bronchi. *J Pathol*. 1970;102:186–188.
72. Churg A, Warnock ML. Pulmonary tumorlet. A form of peripheral carcinoid. *Cancer*. 1976;37:1469–1477.
73. Gould VE, Linnoila RI, Memoli VA, Warren WH. Neuroendocrine components of the bronchopulmonary tract: hyperplasias, dysplasias, and neoplasms. *Lab Invest*. 1983;49:519–537.
74. Ranchod M. The histogenesis and development of pulmonary tumorlets. *Cancer*. 1977;39:1135–1145.
75. *World Health Organization Classification of Tumours of the Lung, Pleura, Thymus and Heart*. 5th ed. IARC Press; 2021.
76. Kuhn C 3rd, Askin FB. The fine structure of so-called minute pulmonary chemodectomas. *Hum Pathol*. 1975;6:681–691.
77. Gaffey MJ, Mills SE, Askin FB. Minute pulmonary meningothelial-like nodules. A clinicopathologic study of so-called minute pulmonary chemodectoma. *Am J Surg Pathol*. 1988;12:167–175.
78. Miller RR, Muller NL. Neuroendocrine cell hyperplasia and obliterative bronchiolitis in patients with peripheral carcinoid tumors. *Am J Surg Pathol*. 1995;19:653–658.
79. Mukhopadhyay S, El-Zammar OA, Katzenstein A-L. Pulmonary meningothelial-like nodules: new insights into a common but poorly understood entity. *Am J Surg Pathol*. 2009;33:487–495.
80. Miller RR. Bronchioloalveolar cell adenomas. *Am J Surg Pathol*. 1990;14:904–912.
81. Egashira R, Jacob J, Kokosi MA, et al. Diffuse pulmonary ossification in fibrosing interstitial lung diseases: prevalence and associations. *Radiology*. 2017;285:255–263.
82. Bisceglia M, Chiaramonte A, Panniello G, Tucci A, Orcioni GF, Colby TV. Selected case from the Arkadi M. Rywlin international pathology slide series: diffuse dendriform pulmonary ossification: report of 2 cases with review of the literature. *Adv Anat Pathol*. 2015;22:59–68.
83. Muir TE, Leslie KO, Popper H, et al. Micronodular pneumocyte hyperplasia. *Am J Surg Pathol*. 1998;22:465–472.
84. Bedrossian CW, Kuhn C 3rd, Luna MA, Conklin RH, Byrd RB, Kaplan PD. Desquamative interstitial pneumonia-like reaction accompanying pulmonary lesions. *Chest*. 1977; 72:166–169.
85. Hollander DH, Hutchins GM. Central spherules in pulmonary corpora amylacea. *Arch Pathol Lab Med*. 1978;102:629–630.
86. Koss MN, Johnson FB, Hochholzer L. Pulmonary blue bodies. *Hum Pathol*. 1981;12:258–266.

87. Visscher D, Churg A, Katzenstein AL. Significance of crystalline inclusions in lung granulomas. *Mod Pathol.* 1988;1:415–419.
88. Schaumann J. On the nature of certain peculiar corpuscles present in the tissue of lymphogranulomatosis benigna. *Acta Med Scand.* 1941;106:239–253.
89. Warnock ML, Press M, Churg A. Further observations on cytoplasmic hyaline in the lung. *Hum Pathol.* 1980;11:59–65.
90. Churg A, Warnock ML. Asbestos and other ferruginous bodies: their formation and clinical significance. *Am J Pathol.* 1981;102:447–456.
91. Hung Y, Hunninghake G, Putman RK, et al. Non-neoplastic pulmonary parenchymal findings in patients undergoing lung resection for mass lesions. *Mod Pathol.* 2016;29:472A.
92. Glancy DL, Frazier PD, Roberts WC. Pulmonary parenchymal cholesterol-ester granulomas in patients with pulmonary hypertension. *Am J Med.* 1968;45:198–210.
93. Reinila A. Perivascular xanthogranulomatosis in the lungs of diabetic patients. *Arch Pathol Lab Med.* 1976;100:542–543.
94. Borges I, Sena I, Azevedo P. Lung as a niche for hematopoietic progenitors. *Stem Cell Rev.* 2017;13(5):567–574.
95. Walford RL, Kaplan L. Pulmonary fibrosis and giant-cell reaction with altered elastic tissue: endogenous pneumoconiosis. *AMA Arch Pathol.* 1957;63:75–90.
96. Roden AC, Boland JM, Johnson TF, et al. Late complications of COVID19. A morphologic, imaging and droplet digital polymerase chain reaction study of lung tissue. *Arch Pathol Lab Med.* 2022;146:791–804.
97. Ro JY, Luna MA, Mackay B, Ramos O. Yellow-brown (Hamazaki-Wesenberg) bodies mimicking fungal yeasts. *Arch Pathol Lab Med.* 1987;111:555–559.
98. Riehani A, Soubani AO. The spectrum of pulmonary amyloidosis. *Respir Med.* 2023;218:107407.
99. Khoor A, Colby TV. Amyloidosis of the lung. *Arch Pathol Lab Med.* 2017;141(2):247–254.
100. Gillooly M, Lamb D. Airspace size in lungs of lifelong non-smokers: effect of age and sex. *Thorax.* 1993;48:39–43.
101. Blobel GA, Moll R, Franke WW, Vogt-Moykopf I. Cytokeratins in normal lung and lung carcinomas. I. Adenocarcinomas, squamous cell carcinomas and cultured cell lines. *Virchows Arch B Cell Pathol Incl Mol Pathol.* 1984;45(4):407–429.
102. Moll R, Divo M, Langbein L. The human keratins: biology and pathology. *Histochem Cell Biol.* 2008;129(6):705–733.
103. Chuman Y, Bergman A, Ueno T, et al. Napsin A, a member of the aspartic protease family, is abundantly expressed in normal lung and kidney tissue and is expressed in lung adenocarcinomas. *FEBS Lett.* 1999;462(1–2):129–134.
104. Nakamura N, Miyagi E, Murata S, Kawaoi A, Katoh R. Expression of thyroid transcription factor-1 in normal and neoplastic lung tissues. *Mod Pathol.* 2002;15(10):1058–1067.
105. Saad RS, Liu YL, Silverman JF. Distribution of basal/myoepithelial markers in benign and malignant bronchioloalveolar proliferations of the lung. *Appl Immunohistochem Mol Morphol.* 2010;18(3):219–225.
106. Woo JS, Reddy OL, Koo M, Xiong Y, Li F, Xu H. Application of immunohistochemistry in the diagnosis of pulmonary and pleural neoplasms. *Arch Pathol Lab Med.* 2017;141(9):1195–1213.
107. Wang LJ, Greaves WO, Sabo E. GCDFP-15 positive and TTF-1 negative primary lung neoplasms: a tissue microarray study of 381 primary lung tumors. *Appl Immunohistochem Mol Morphol.* 2009;17(6):505–511.
108. Striebel JM, Dacic S, Yousem SA. Gross cystic disease fluid protein-(GCDFP-15): expression in primary lung adenocarcinoma. *Am J Surg Pathol.* 2008;32(3):426–432.
109. Su JM, Hsu HK, Chang H, et al. Expression of estrogen and progesterone receptors in non-small-cell lung cancer: immunohistochemical study. *Anticancer Res.* 1996;16(6B):3803–3806.
110. Berg KB, Churg A. GATA3 immunohistochemistry for distinguishing sarcomatoid and desmoplastic mesothelioma from sarcomatoid carcinoma of the lung. *Am J Surg Pathol.* 2017;41(9):1221–1225.
111. Cowan ML, Li QK, Illei PB. CDX-2 expression in primary lung adenocarcinoma. *Appl Immunohistochem Mol Morphol.* 2016;24(1):16–19.

18

Thymus

David Suster ■ Saul Suster

The thymus is a prototypical lymphoepithelial organ composed of intimately admixed epithelial and lymphoid elements that act in concert to perform their assigned roles. In addition, the thymus harbors other cellular constituents, such as a variety of mesenchyme-derived elements, scattered neuroendocrine cells, and presumably germ cells, all of which may take part in the development of neoplastic and nonneoplastic processes of this organ. Although much progress has been made in the immunohistochemical and molecular characterization of the cellular components of the thymus, the diagnosis of thymic lesions remains largely dependent on the light microscopic interpretation of the findings by the pathologist.

EMBRYOLOGY

The thymus is derived from the third and, to a lesser extent, the fourth pharyngeal pouches, which contain elements derived from all the three germinal layers. During the 6th week of gestation, the endodermal lining of the ventral wing of the third pharyngeal pouch forms a pronounced saccule, which subsequently detaches from the pharyngeal wall, giving rise to the thymic primordia (1,2). At approximately the same time, the cervical sinus, which is an ectodermal structure that results from the fusion of the second, third, and fourth branchial clefts, attaches to the thymic primordia, investing it with a layer of ectodermal cells (1,3). As development continues, the thymic primordia migrate in a caudal and a medial direction along with the lower parathyroid glands. During the 8th week, these primordia enlarge toward their lower ends, forming two epithelial bars that fuse along the midline to occupy their definitive position within the anterosuperior mediastinum. The tail portion of the organ becomes thin and elongated during this descent and breaks up into small fragments that usually disappear, although they may persist in the soft tissues of the neck, often in intimate connection with the lower parathyroid gland and sometimes embedded within the thyroid gland (1,4).

Once migration is complete, the thymic endoderm-derived epithelial cells develop into a reticular network of cells. The surrounding mesenchymal elements form a capsule around it; and, as a result of ingrowth of the capsule, trabeculae form that divide the organ into numerous lobules. By the 10th week, small lymphoid cells originating in the fetal liver and bone marrow populate the thymus, and the organ differentiates into a cortex and a medulla (5). Small tubular structures composed of epithelial cells (sometimes referred to as medullary duct epithelium) also make their appearance at this stage and will later give rise to Hassall corpuscles (6). The thymus progressively enlarges until puberty and from then on and thereafter begins to involute, although persisting in an atrophic state into old age.

DEVELOPMENTAL ABNORMALITIES

Disturbances in the embryologic development of the thymus may give rise to a series of congenital anomalies. One of the anomalies most frequently encountered is the presence of parathyroid gland tissue within the thymus (Fig. 18.1) (7). Such ectopically located tissue is most frequently encountered within the thymic capsule or in close proximity to it (8). This abnormality is easily explained by the close developmental relationship that exists between the two organs, as described in the section on embryology. The thymus itself also may be found in ectopic locations; this usually results from failure of the organ to migrate to its final destination during embryonic development. Undescended thymuses are most often located in the lateral neck, in close association with or even buried within the parathyroid or thyroid glands (9). The thymic rests tend to undergo cystic changes (9). They also may give rise to thymomas, ectopic examples of which have been described in submandibular (10), paratracheal (11,12), and intratracheal locations (13), as well as within the thyroid gland (14,15). A morphologically distinctive lesion of the thymus, now termed "branchioma," displaying combined features of neoplasia and hamartoma occurring in the lower neck has also been recognized (16). Ectopic nodules composed of thymic tissue have been reported in several other locations, including the base of the skull (17) and the pulmonary hilus at the root of the bronchus (18). Given that thymomas can also occur within lung parenchyma and in the pleura away from the mediastinum, it is logical to assume that thymic rests may also be present in such locations. The lungs and pleura, however, develop much earlier than the thymus during embryogenesis, making the misplacement of thymic tissue during embryologic migration unlikely as an explanation for the development of these tumors in pulmonary and pleural locations. Some authors have postulated origin from uncommitted pluripotential stem cells for such tumors (19). Ectopic thymomas have also been identified in the middle and posterior mediastinum; although rare, it has been postulated that such tumors may have had a direct connection to the thymic gland or its surrounding adipose tissue (20).

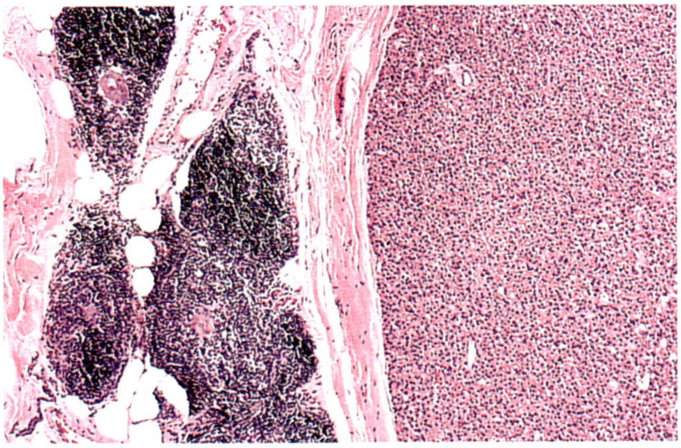

FIGURE 18.1 Ectopic parathyroid tissue adjacent to the normal thymus.

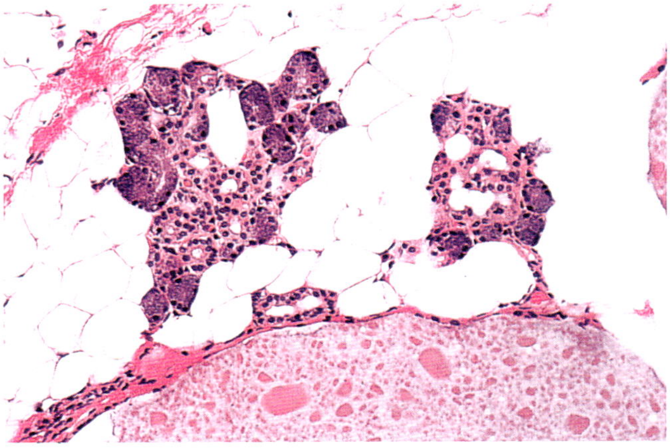

FIGURE 18.2 Mature-appearing salivary gland acini are seen adjacent to cystically dilated Hassall corpuscle.

Ectopic sebaceous glands have been reported in the thymus (21) and are felt to be related to the contribution of the ectodermally derived cervical sinus to the developing thymus. Mature-appearing salivary gland tissue also may be present in the thymus (Fig. 18.2) and has been reported as a component of an intrathoracic cyst that contained normal thymus and parathyroid tissue within its walls (22). It was postulated that this finding could be related to a developmental malformation whereby salivary gland anlage had been incorporated into the uppermost portion of the third pharyngeal pouch during embryogenesis.

Morphologic abnormalities of the thymus have been observed in association with combined immunodeficiency syndromes and T-cell defects. In this situation, the thymus has an embryonal appearance with lobules of small spindle epithelial cells that lack small lymphocytes and Hassall corpuscles (23). Such morphologic alterations have been termed thymic dysplasia and are believed to represent disturbances of normal development. In conditions such as reticular dysgenesis, Swiss-type hypogammaglobulinemia, thymic alymphoplasia, and ataxia–telangiectasia, the immune deficit is accompanied by aplasia or hypoplasia of the thymus (24–26). DiGeorge syndrome is believed to result from an arrest in the development of the third and possibly fourth branchial arches; it features a vestigial but normal thymus associated with absence or hypoplasia of the parathyroid glands (27). Nezelof syndrome (28) is similar in that a vestigial thymus is present but the parathyroid glands are normal. The failure of development is thought to involve only that part of the branchial endoderm that will differentiate into thymic epithelium. Histologic changes in the thymus similar to those of thymic dysplasia of primary immunodeficiency disease have been observed in infants as a consequence of graft versus host disease following blood transfusions (29). A case of congenital aplasia of the parathyroid glands and thymus in the newborn has also been described (30). The term *thymic dysplasia* in the setting of immunodeficiency conditions may be misleading because it is unrelated to neoplasia.

APOPTOSIS

Apoptosis is the name that has been given to the process of programmed cell death. Most thymic lymphocytes die in situ by such a process. During T-cell ontogeny in the thymus, the T cells (T lymphocytes) may undergo a process of positive or negative selection. The accumulated evidence suggests that apoptosis plays a major role in the process of negative selection and that the majority of cortical thymocytes die by this mechanism. Autoreactive thymocytes or harmful cells in the thymus with injured DNA or alterations of their metabolism are also thought to be eliminated by the process of apoptosis at a specific stage of their differentiation (31,32). Therefore, a disturbance of the apoptotic process within the thymus can be responsible for the appearance of autoreactive cells in the circulation which may give rise to the development of autoimmune disease. In addition, failed apoptosis may play a role in carcinogenesis. On the other hand, massive induction of apoptosis within the thymus can lead to immunodeficiency due to a decrease in the number of lymphocytes. Thus, thymocyte apoptosis plays an integral role in the pathophysiology of the immune system.

The exact biochemical mechanism involved in thymocyte apoptosis has not yet been completely elucidated; however, various complex physiologic and nonphysiologic mechanisms play a role (33). Several groups have noted that messenger RNA (mRNA) levels for various proteins are increased early in thymocytes after treatment with glucocorticoids or radiation, two agents known to induce massive apoptosis of cortical thymocytes (34,35). Several genes show increased expression in cells undergoing apoptosis. Among them, three proto-oncogene products have been shown to act as important regulators of the apoptotic process in mammals: c-myc, bcl-2, and p53. Bcl-2 mRNA is present in the surviving mature thymocytes of the medulla and also in most immature ($CD4^+/CD8^+$) thymocytes, although the majority of cortical thymocytes that ultimately undergo apoptosis display no bcl-2, suggesting that this oncogene may be involved in the preservation of T cells (36). Alterations of *TP53* also play a role in thymocyte apoptosis. Malfunction of *TP53* may promote carcinogenesis by permitting mutant cells to duplicate their DNA before it is repaired. In mice, p53-deficient thymocytes have been shown to display drastic resistance to the apoptotic effects of radiation (37). The latter observation suggests that the integrity of *TP53*, the protein product of which arrests cell proliferation, may be necessary for the normal apoptotic process in the thymus.

Another pathway of thymocyte apoptosis involves the T-cell receptor (TCR)/CD3 complex. Rearrangement of TCR genes during thymocyte development leads to expression of unique, clonally expressed TCRs. Potentially autoreactive thymocytes bearing TCRs with high avidity for self undergo negative selection through TCR-mediated, activation-induced apoptosis, whereas thymocytes with TCRs of low self-avidity are positively selected for survival (38,39). Activation of the TCR/CD3 complex leads to the preferential elimination of immature ($CD4^+/CD8^+$) thymocytes (40). These observations suggest that TCR-mediated signals are involved in apoptosis, and this phenomenon is strongly related to the mechanism of negative selection, although the precise mechanisms of apoptosis have not yet been elucidated (41,42).

ANATOMY

The fully mature human thymus is an encapsulated midline structure predominantly located in the anterosuperior mediastinum. It is composed of two lobules joined in the midline by loose connective tissue and thymic parenchyma. The base of the organ lies on the pericardium and great vessels. The upper poles of each lobe extend into the lower neck and are closely applied to the trachea. The lower poles extend down over the pericardium for a variable distance, generally up to the level of the fourth costal cartilage. The anterior border of the gland is made up by the cervical fascia, strap muscles of the neck, sternum, costal cartilages, and intercostal muscles, and the lateral borders are covered by reflections of the parietal pleura.

The size and weight of the gland vary considerably depending on the age of the person, although wide variations among individuals in the same age group have also been observed (43). The mean weight at birth is approximately 20 g. The organ exhibits continuous growth in size until puberty when it reaches a mean weight of 35 to 50 g; thereafter, it undergoes atrophy and progressive fatty replacement of the parenchyma.

The blood supply of the organ is derived from the internal mammary, superior and inferior thyroid arteries, and, to a lesser degree, the pericardiophrenic arteries. The arterial branches course along fibrous septa near the corticomedullary junction, where they branch into the cortex and the medulla. The capillaries descend from the outer cortex toward the medulla to form postcapillary venules, and they exit the thymus through the septa as interlobular veins. The venous system drains into the left brachiocephalic, internal thoracic, and inferior thyroid veins. There are no true intraparenchymal afferent lymphatic vessels in the thymus. Lymphatic vessels arise in the interstitium of the lobular septa and merge to form larger vessels that course alongside the arteries in the septa. The innervation of the organ is derived from branches of the vagus nerve and the cervical sympathetic nerves.

HISTOLOGY

The lobule is the basic structural unit of the thymus. Each lobule is composed of two morphologically distinctive areas, the cortex and the medulla, both of which are composed of varying proportions of epithelial cells and thymic lymphocytes (i.e., thymocytes) (Fig. 18.3). The sparse epithelial cells present are overshadowed by the numerous, closely packed small lymphocytes in the cortex. The medulla, in contrast, contains a larger number of epithelial cells and fewer lymphocytes. The cortex and the medulla together correspond to the thymic epithelial compartment, which is the site of T-cell maturation. Another important anatomic compartment of the thymus is the perivascular space. The perivascular space in the mature infant thymus represents a virtual space containing thymic blood vessels; it is contained within the capsule but outside of the thymic epithelial network (44). The perivascular space becomes more prominent with aging and in thymoma, and is eventually replaced with fat and lymphocytes in the involuted thymus of the adult (45). In addition to the epithelial and perivascular space compartments, there is also a stromal compartment that harbors a variety of other cell types. A thin fibrous capsule is generally present surrounding the entire gland. The interplay of these various cellular elements and compartments contributes to define the various organotypical features of the normal thymus.

One of the problems involved in defining the "normal" thymus is that the histologic appearance of the organ changes over time, depending on its stage of maturation or involution. Some data suggest that the thymus gland does not actually decrease in size or alter its shape during the process of involution (45). While the overall size and shape of the gland are retained throughout adulthood into old age, the various cellular constituents are replaced: the epithelial compartment is gradually replaced by mature adipose tissue until only a few epithelial elements remain. Residual epithelial elements consist of small, oval- to spindle-shaped cells with hyperchromatic nuclei and scant cytoplasm, rather than large, round cells with vesicular nuclei and abundant cytoplasm typically seen in the thymus of a young person (46). Other involutional changes include the formation of cystic structures, usually resulting from dilatation of Hassall corpuscles, and small, abortive glandular or epithelial rosette-like structures within lymphopoietic islands.

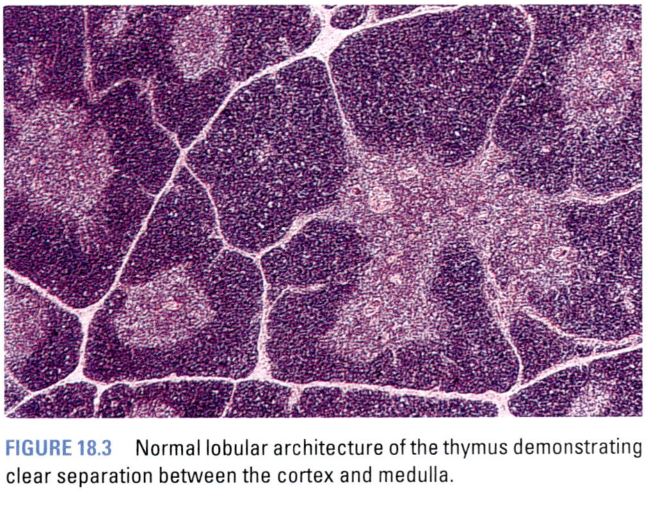

FIGURE 18.3 Normal lobular architecture of the thymus demonstrating clear separation between the cortex and medulla.

The most distinctive organotypical features of the mature thymus of childhood and adolescence include a fibrous capsule, a lobular architecture with sharp separation between the cortex and the medulla, a dual cell population of large, round thymic epithelial cells and immature T cells, and perivascular spaces. The organotypical features of the "normal" involuted thymus of the adult, on the other hand, include the presence of small, spindle-shaped thymic epithelial cells, abundant cystic and rosette-like epithelial structures, and the paucity of immature T cells. These features represent a continuum that will manifest in various proportions depending on the age, functional status of the organ, various physiologic conditions, and disease states affecting the thymus.

Epithelial Cells

The epithelial cells of the thymus have traditionally been divided into cortical and medullary; some of the latter are arranged in round keratinized structures known as Hassall corpuscles. The active thymus of infancy features mostly plump epithelial cells with round or oval nuclei. Their cytoplasm is endowed with numerous prolongations that join with those of adjacent cells to form a veritable network or reticulum. This feature has led to the previous designation of "reticuloepithelial cells" and is more pronounced in the cortex. These various epithelial elements play an active role in promoting T-cell maturation in the thymus, either through action of their humoral substances or through direct contact with thymocytes (47).

A subset of thymic epithelial cells localized to the cortex has been designated as nurse cells. These epithelial cells contain abundant cytoplasm within which numerous engulfed mature T cells are present. A ring-like immunostaining pattern has been observed in these cells with antibodies against epithelial cells (48). These nurse cells presumably provide a specialized microenvironment for T-cell maturation, differentiation, and selection in the thymus (48,49).

The medullary thymic epithelial cells are involved in the regulation of autoimmunity; they screen marrow-derived progenitor lymphoid cells for autoreactivity. Dendritic

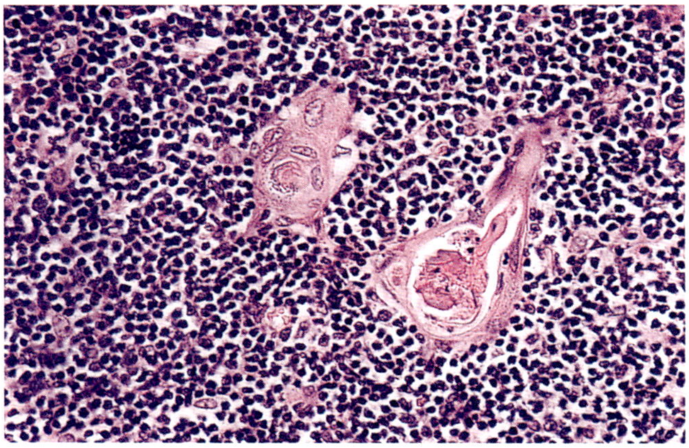

FIGURE 18.4 Normal Hassall corpuscles showing characteristic concentric arrangement of keratinizing epithelial cells.

cells in the medulla delete potentially auto-aggressive cells directed against major histocompatibility complex (MHC) antigens, and the medullary thymic epithelial cells screen against tissue-restricted antigens under the control of an autoimmune regulator transcription protein (AIRE) (50). AIRE is normally expressed in most medullary thymic epithelial cells but, with the exception of B1 thymoma, is absent in the majority of thymomas. Thus, its absence has been suggested to represent a marker for immature phenotype in thymomas; AIRE-positive B1 thymomas most closely resemble the organotypical normal thymus, with good corticomedullary definition and presence of Hassall corpuscles (51).

Another distinctive type of epithelial cell recently identified in the thymus is the "tuft cell." Tuft cells represent a unique subset of medullary thymic epithelial cells involved in type 2 immunity. They play a role in establishing an IL4-enriched medullary microenvironment by promoting the development and intrathymic function of type 2 invariant natural killer cells (NKT2) (52).

Hassall Corpuscles

Hassall corpuscles constitute the most readily identifiable feature of the thymus at the light microscopic level. They are restricted to the medulla and are characterized by a concentric pattern of keratinization, the keratin formed being of high–molecular-weight (epidermal) type (Fig. 18.4). These structures may show morphologic variability as a result of inflammation-induced reactive changes. This includes cystic degeneration with accumulation of cellular debris, dystrophic calcification, and infiltration by lymphocytes, foamy macrophages, and eosinophils (Fig. 18.5). The thymic lesions known as Dubois microabscesses associated with congenital syphilis (53) represent an exaggeration of the cystic changes in Hassall corpuscles. We believe that

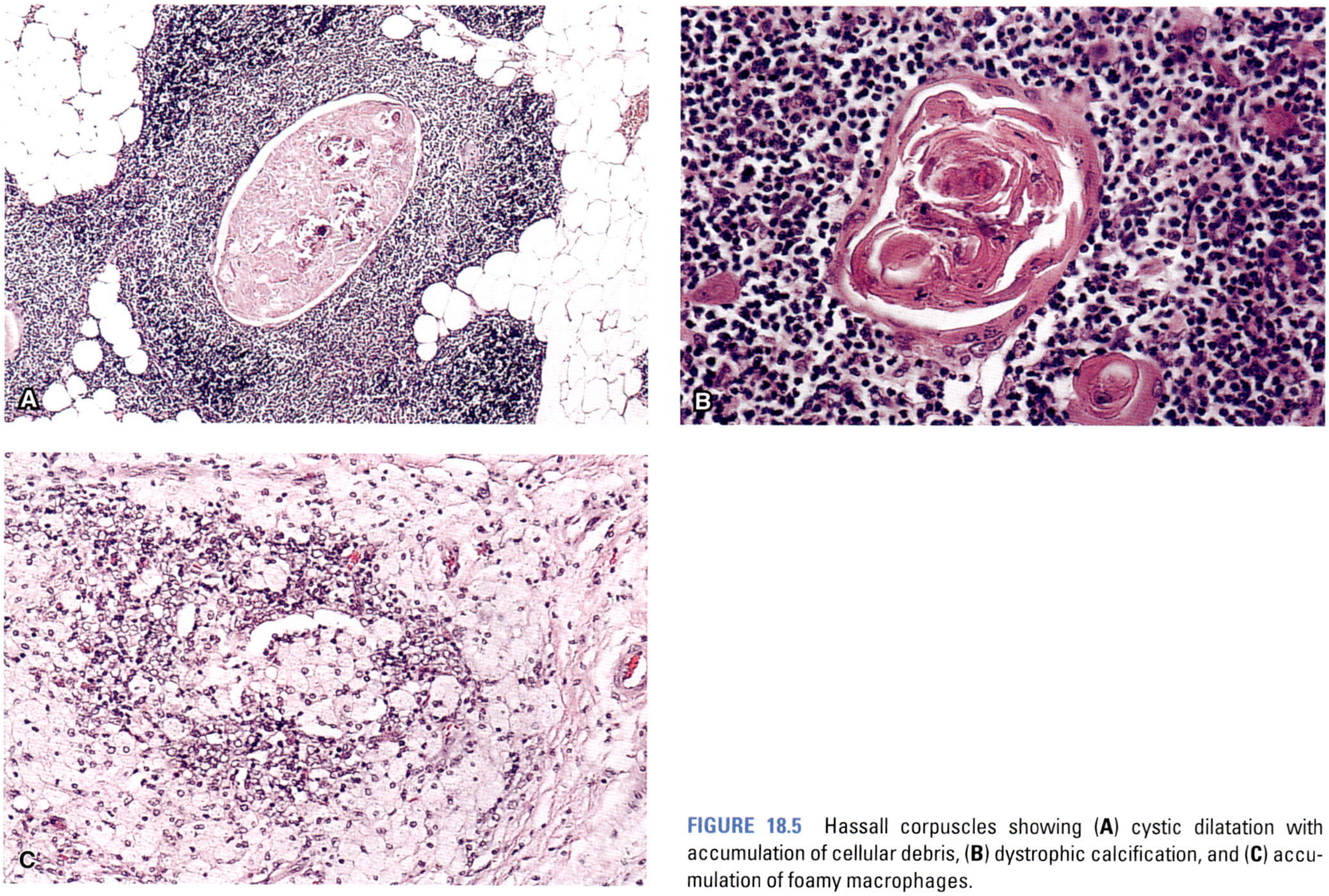

FIGURE 18.5 Hassall corpuscles showing (**A**) cystic dilatation with accumulation of cellular debris, (**B**) dystrophic calcification, and (**C**) accumulation of foamy macrophages.

most multilocular thymic cysts are not congenital abnormalities but, rather, represent cystic enlargement of Hassall corpuscles secondary to inflammatory changes (54). An additional feature that can be seen in relation to cystic Hassall corpuscles is the presence of glandular elements, as manifested by the appearance of ciliated or goblet cell–type columnar epithelium and secretion of sulfated acid mucopolysaccharides in the lumen (55). It is likely that these glandular changes are related to the embryologic origin of Hassall corpuscles (6).

Thymic Lymphocytes (Thymocytes)

The predominant cell population of the thymic cortex is made up of lymphocytes that may be large, medium, or small. Large, mitotically active lymphoblasts comprise about 15% of the lymphoid cells and are found predominantly in the subcapsular portion of the outer cortex (56). These lymphocytes have round or oval nuclei, one or two prominent nucleoli, and relatively abundant, basophilic cytoplasm. A gradient of smaller, less mitotically active cells occurs from the outer cortex to the corticomedullary junction and, to a lesser degree, into the medulla of the normal thymus (Fig. 18.6). In the capsular region and deep cortex, most of the thymic lymphocytes are short-lived and die in situ (57). This results in lympholysis and active phagocytosis, features that impart these areas with a prominent "starry sky" appearance. They can become particularly prominent in stress-related thymic involution but can also be seen in neoplastic conditions such as thymoma and lymphoblastic lymphoma (Fig. 18.7).

Other Cell Types

In addition to epithelial cells and lymphocytes of T-cell lineage, the thymus contains an array of additional cell types. B cells aggregate in lymphoid follicles or may be scattered as individual cells. Lymphoid follicles with active germinal centers are found in otherwise normal thymuses, especially in children and adolescents (Fig. 18.8) (57,58). Germinal centers in the thymus arise within perivascular spaces that are clearly separated from the thymic parenchyma by a basal lamina (59). This observation has led to the proposal that the thymus may be divided into two major functional compartments: (a) the cortex and the medulla that constitute the true thymic parenchyma and (b) the extraparenchymal perivascular spaces (60). Germinal centers in the thymus can thus be explained as being derived from preexisting perivascular B cells in the extraparenchymal compartment. They are prominent in patients with myasthenia gravis and other immune-mediated diseases (9), but their presence in an otherwise normal thymus should not necessarily be taken as an indicator of an underlying immune disorder. The incidence of germinal centers in the thymuses of normal individuals without septicemia or autoimmune disease ranges from 2.1% to 40% (58,59). This wide variation may be related to a sampling factor or to the age of the patients; germinal centers would be expected to be less frequent in the older age groups. Also, stress has been shown to be a

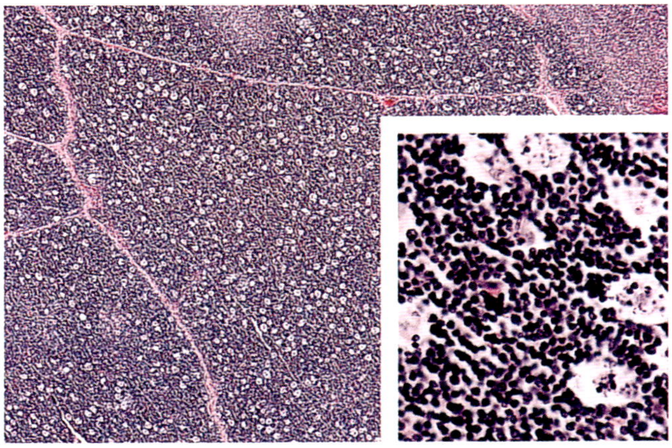

FIGURE 18.7 Prominent starry sky appearance in the deep cortex of a patient with accidental thymic involution. Note the abundance of tingible body macrophages (*inset*).

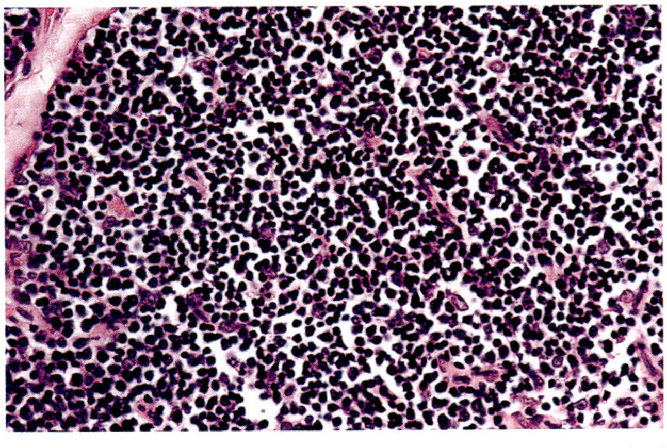

FIGURE 18.6 Normal thymic cortex. There are numerous cortical thymocytes, most of which have small nuclei with densely packed chromatin.

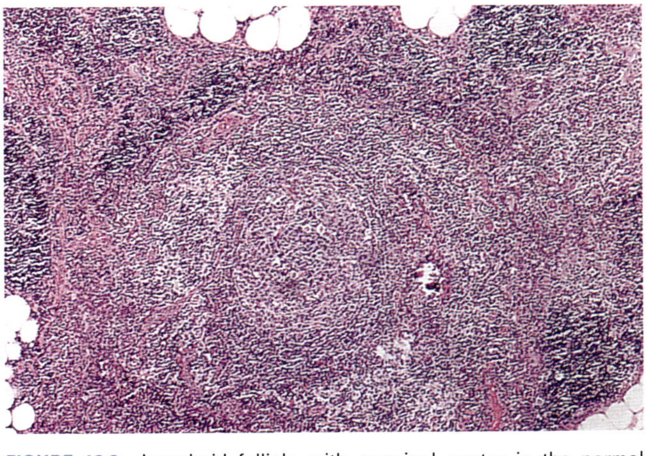

FIGURE 18.8 Lymphoid follicle with germinal center in the normal thymus.

factor responsible for the decrease in the number of these structures (61).

Isolated B cells are found in both fetal and normal adult thymuses. They tend to be distributed along the septa and in close proximity to small vessels at the corticomedullary junction and in the medulla (62). Intrathymic B cells are significantly increased in patients with myasthenia gravis, a finding that some investigators consider to be a more specific change for this disorder than the presence of germinal centers (62,63). Some intramedullary B cells show a tendency to cluster around Hassall corpuscles in fetuses, newborn babies, children, and adults; these lymphocytes show evidence of activation and bear a distinctive immunophenotype (64). Unlike B cells of germinal centers and surrounding mantle zones, these medullary B cells are negative for CD21 and do not express surface or cytoplasmic immunoglobulins. These cells share immunophenotypic features with parafollicular marginal zone B lymphocytes in lymphoid follicles and are believed to be part of the mucosa-associated lymphoid tissue (MALT). It has been proposed that a significant proportion of non-Hodgkin B-cell lymphomas located in the anterior mediastinum arises from this intrathymic B-cell population. In addition, a newly described type of MALT lymphoma composed of monocytoid B cells has been identified arising from the thymus (65); this finding raises the possibility that extranodal parafollicular B cells may represent another cellular constituent of the normal thymus (66).

Several other types of hematolymphoid cells are present in small but constant numbers in the normal thymus (67–69). Macrophages are mainly located in the cortex, show phagocytic activity, are markedly α-naphthyl acetate esterase-positive, acid phosphatase-positive, HLA-DR negative, and are antigenically indistinguishable from macrophages of other organs (47,69). Interdigitating dendritic cells are mainly located in the medulla; they are HLA-DR positive, have little lysosomal enzyme activity, and show S100 protein reactivity (70). Both cell types are thought to be involved in lymphocyte and epithelial cell interaction.

Langerhans cells can be found in the thymic medulla; both interdigitating dendritic cells and Langerhans cells are increased in the thymuses of patients with myasthenia gravis (71). The latter cells provide the anatomic substrate for the development of Langerhans cell histiocytosis within the thymus (72). Eosinophils are usually present in the thymus of children, sometimes in large numbers (73). They appear in fetal life and persist until puberty, after which they become infrequent. They are found mainly in the connective tissue septa or within the medulla and occasionally may be present within Hassall corpuscles. Mast cells are also normally present in human thymuses. They are usually found within and scattered parallel to the connective tissue septa, often in a perivascular location. Increased numbers of mast cells in the thymus have been observed in patients with severe combined immunodeficiency and thymic alymphoplasia (74), but the significance of this finding is not understood. Plasma cells are rare in the normal thymus; they are usually located in the connective tissue septa or, more rarely, in the medulla (75). Plasma cells may be numerous in the involuting thymus and are increased in numbers in patients with myasthenia gravis (76).

Neuroendocrine cells are a minor but constant component of the normal thymus (77). Peptide- and amine-producing neuroendocrine cells have been identified in the thymus glands of reptiles and birds (78–80) and, to a lesser degree, in mammalian thymuses. Some of these cells may be embryologically and functionally analogous to C cells of thyroid (81). Their presence has been offered as the probable substrate for the development of neuroendocrine tumors and other neuroendocrine neoplasms in this organ, including calcitonin-positive medullary carcinomas (82,83).

Myoid cells can be found in the thymic medulla. They are common in reptilian and avian thymuses but can also be found in human thymuses, particularly in infants. They have microscopic, immunohistochemical, and ultrastructural features of striated muscle cells (84,85). Their histogenesis remains a subject of debate with some studies pointing toward a derivation from the neural crest (86) and others showing the existence of shared epitopes with thymic epithelial cells (87). Thymus myoid cells are reportedly increased in patients with myasthenia gravis (62,88) and those with true thymic hyperplasia (89), suggesting a role in immunoregulatory mechanisms. Thymic neoplasms thought to harbor the neoplastic counterpart of myoid cells also have been described (90).

Germ cells are normally present in the thymus. Scattered germ cells presumably reach the thymus during ontogenesis and some of them persist into adult life. However, direct evidence of their presence in the normal thymus gland has never been demonstrated, and the only presumptive evidence of their occurrence is the fact that nearly all mediastinal germ cell tumors arise within the thymus (9). Rosai et al. (91) proposed that thymic germ cells may develop into somatic cells, thus making their detection and identification difficult by conventional means. They further hypothesized that thymic myoid cells may be of germ cell origin, thereby providing indirect evidence for the existence of the latter in this organ (91).

Connective tissue elements of the thymus include vessels, fibrous tissue, nerves, and fat. The vasculature of the thymus arises from arteries that enter the organ via fibrous trabeculae from the capsule. In contrast to other major lymphoid organs, the thymus lacks a hilum. The thymic vessels are ensheathed by a layer of thymic epithelial cells. This anatomic configuration and the failure of the thymus to produce antibodies against circulating antigens led to the concept of the blood–thymus barrier (92). In an elaborate study using electron-opaque tracers of different molecular dimensions (93), Raviola and Karnovsky demonstrated that, although some bloodborne macromolecules do penetrate the thymus, their distribution is limited to the medulla, suggesting the existence of a blood–thymus barrier operating at the level of the cortex. More recent studies have challenged the concept of the blood–thymus barrier by showing that the thymic cortex is permeable to immunoglobulin molecules present in the extravascular compartment (94).

The presence of a distinct anatomic compartment bound by the vessel wall and the sheath of epithelial cells has been postulated. It has further been suggested that it is through these perivascular spaces that the mature T cells exit the thymus to colonize peripheral lymphoid organs once their maturation is complete. The spaces are inconspicuous in the normal thymus; they may appear dilated in atrophic and involuting organs, and acquire their greatest prominence in thymomas (9).

ULTRASTRUCTURE

Elongated cytoplasmic processes can be appreciated in the epithelial cells of the thymic cortex, which are also covered by basal lamina material (95–97). Medullary epithelial cells are more densely packed and have blunted cytoplasmic projections. A consistent feature of medullary epithelial cells is the greater frequency of desmosomes and the presence of dense tonofilaments that often insert into desmosomes. This feature is particularly obvious in Hassall corpuscles. Cortical and medullary epithelial cells may display a range of ultrastructural appearances depending on the electron density of their cytoplasm, including pale cells with electron-lucent cytoplasm, intermediate cells with variable electron density, and dark cells with electron-dense cytoplasm. The medulla also contains undifferentiated epithelial cells with sparse cytoplasm (98).

Thymic myoid cells display ultrastructural features of skeletal muscle, including myofilaments with dense patches (Fig. 18.9). Some investigators have identified cells displaying both tonofilaments and myofilaments by electron microscopy, and in some instances, myoid cells have been observed to display desmosomal connections with epithelial cells (76).

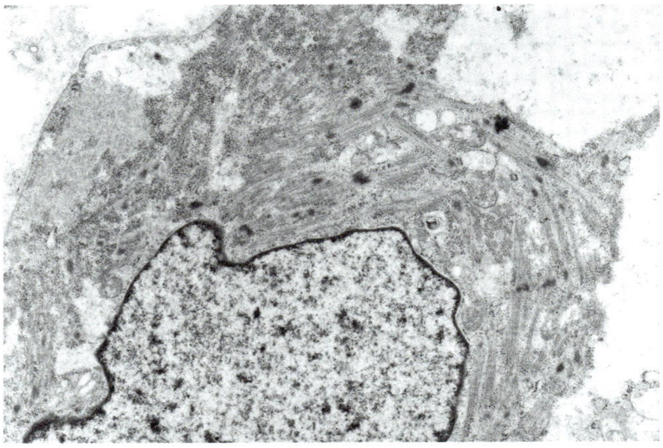

FIGURE 18.9 Ultrastructural appearance of myoid cell in the thymus. Note the bundles of actin and myosin myofilaments with clearly discernible Z lines in the cytoplasm.

IMMUNOHISTOCHEMISTRY

Thymic Epithelial Cells

Thymic epithelial cells may express a variety of differentiation antigens. At least four antigenically distinctive types of epithelial cells are recognized in the normal thymus: inner cortical, subcapsular cortical, medullary, and the cells of Hassall corpuscles (Table 18.1) (99–102). Inner cortical epithelium exhibits keratin positivity and strongly reacts with TE-3, a murine monoclonal antibody raised against human thymic stroma (103). Subcapsular cortical epithelial cells and medullary epithelial cells react strongly with TE-4 monoclonal antibody. Both cell types also label with A2B5, a monoclonal antibody directed against a complex neuronal ganglioside found on the surfaces of neurons and neuroendocrine cells (104).

Anti-p19 is another marker of subcapsular cortical epithelium and medullary epithelium in normal thymus. This antibody defines the structural core protein of the human T-cell lymphoma virus and is likely acquired during normal thymic ontogeny. In the normal human thymus, the antigen defined by anti-p19 parallels the reactivity of A2B5 in epithelial cells (105). Interestingly, some data suggest that expression of the p19 antigen is lost with malignant transformation of thymomas (106).

Subcapsular cortical, inner cortical, and medullary epithelial cells have been reported to express class I and class II major histocompatibility antigens (101,107). However, studies using double immunolabeling have shown that HLA-DR expression is absent in medullary epithelial cells and that the previously reported positivity may reflect diffusion of the stain from surrounding interdigitating dendritic cells (71). Epithelial cells in the normal thymus and thymomas have

TABLE 18.1 Summary of Main Antigenic Determinants Found in Epithelial Cells of the Normal Thymus

Cells	Keratin	TE-3	TE-4	A2B5	Anti-p19	Antithymosin $\alpha 1$	Antithymopoietin	HLA/Ia
Inner cortical epithelium	+	+	–	–	–	–	–	+
Subcapsular cortical epithelium	+	–	+	+	+	+	+	+
Medullary epithelium	+	–	+	+	+	+	–	–
Hassall corpuscles	++	–	–	–	–	–	–	–

also been reported to express epithelium-associated glycoprotein H (tissue blood group O antigen), peanut agglutinin receptor antigen (PNA-r), and Sophora japonica agglutinin receptor antigen (SJA-r), which are detectable by lectin binding (108). Other studies have demonstrated expression of the epidermal growth factor receptor and transforming growth factor-alpha in subcapsular, cortical, and medullary epithelial cells, suggesting that this substance plays a role in the growth and differentiation of these cells (109). Subcortical epithelial cells lining the boundaries between the thymic parenchyma and its surrounding fibrous tissue also demonstrate positivity for Leu-7, a differentiation antigen found in human null/killer cells and neuroendocrine cells (110). The cells of Hassall corpuscles show more keratin positivity than other thymic epithelial cells and react particularly strongly with the high–molecular-weight keratin AE2 antibody, which is considered a marker of terminal epithelial maturation (111). Conversely, Hassall corpuscles are unreactive for the other antigens mentioned above.

There are no reliable markers that distinguish thymic epithelial cells from other types of epithelial cells. Thymic epithelial neoplasms, as a group, are immunoreactive with cytokeratin antibodies and react with a wide variety of keratin subsets. The most common keratins expressed in thymoma are CK8/CK18 low–molecular-weight cytokeratins (CAM5.2), CK19, and CK5/6 (112). Broad-spectrum keratin and keratin cocktails such as AE1/AE3 are also helpful for labeling the epithelial cells in these tumors. Another marker, p63, strongly labels thymic epithelial cells (113). Virtually all thymic epithelial cells in all types of thymomas show strong nuclear expression for this marker. Thus, p63 positivity may serve as an aid for the diagnosis of thymoma, although this marker is not specific and may also be seen in squamous cell carcinomas, urothelial and myoepithelial neoplasms, and other tumors. More specific markers of squamous differentiation (e.g., p40 and desmoglein-3) are expressed in normal thymic tissue, thymomas, and thymic carcinoma (114). Their role in diagnosis parallels that of p63 and is, however, quite limited. Other markers that may also react to a more limited extent with epithelial cells in thymomas include EMA, MUC1, bcl-2, and CD57.

Thymic Lymphocytes

The normal lymphoid population of the thymus shows marked immunophenotypic heterogeneity, reflective of their functional status. The very term *thymocyte*, which was originally introduced to designate all thymic lymphocytes, has acquired a more specific immunologic meaning and is now restricted to immature thymic lymphocytes of T-cell lineage, exclusive of pre–T-cell lymphocytes. Thymocytes can be divided into three types in accordance with their stages of maturation (21,115–117). Marrow-derived precursors are first committed to a T-lymphocyte lineage in the cortex, where they first become CD4$^-$/CD8$^-$, then CD4$^+$/CD8$^-$, and later turn into CD4$^+$/CD8$^+$ cells. The double-positive cells then undergo TCR recombination followed by positive selection of double-positive lymphocytes through interactions with MHC ligands on cortical epithelial cells. Signals arising from these interactions cause the lineage to split into CD4$^+$ and CD8$^+$ mature thymocytes. The positively selected cells then migrate into the medulla where they undergo negative selection by medullary epithelial cells and dendritic cells and are either deleted or converted into suppressive thymic regulatory cells involved in the enforcement of self-tolerance (6).

The use of lymphoid markers is of limited value for the identification or subclassification of thymic epithelial neoplasms (118–123). However, identification of immature T lymphocytes admixed with the neoplastic epithelial cells in the appropriate context may be helpful for supporting the diagnosis of thymoma. The best markers in this setting are CD1a, CD3, and nuclear expression of TdT. Another marker, CD99, strongly labels immature T lymphoblasts in thymoma (124,125).

MOLECULAR BIOLOGY

Molecular studies have shown rearrangement and expression of T-cell antigen receptors (TCRs) in thymic lymphocytes. Expression of the TCR represents a critical step in the development of thymic T cells (126,127). The TCR molecules are heterodimeric proteins that are structurally analogous to immunoglobulin molecules. Rearrangement and expression of the TCRs are needed for the T cell to recognize antigens in association with self-MHC antigens. There are two major types of TCRs: one contains α- and β-polypeptide chains and the other contains γ- and δ-chains. The α-β–TCR is expressed in nearly all T cells, whereas the γ-δ–TCR is expressed in approximately 2% of T cells. Immature thymic lymphocytes (CD4$^-$/CD8$^-$) contain mRNA for the β-chain of the TCR. Mature thymic lymphocytes (CD4$^+$/CD8$^+$), on the other hand, contain mRNA for both β- and α-chains (128).

As immature thymocytes begin to undergo TCR gene rearrangements, they first acquire cytoplasmic and then cell surface expression of CD3, CD4, and CD8. The CD4$^+$/CD8$^+$ lymphocytes are located within the thymic cortex; they coexpress CD1a and show strong positivity for the Ki-67 nuclear proliferation antigen (45). As these double-positive thymocytes complete their TCR gene rearrangements and undergo positive and negative selection, they progressively lose expression of CD1a, Ki-67, and either CD4 or CD8. The resulting "single-positive" thymocytes are either CD4$^+$/CD8$^-$ or CD4$^-$/CD8$^+$ and migrate into the thymic medulla where they complete maturation before being released into the circulation.

The genetic mechanisms that promote lineage commitment in the thymus are not well understood. Recent studies have suggested that there are at least 25 genes involved in the segregation of different T-cell lineages. Commitment to the CD4 lineage is controlled by the upregulation of the genes associated with increased survival followed by expression of genes that regulate nucleosome remodeling and TCR signaling. Commitment to the CD8 lineage is influenced by upregulation of genes that regulate lymphocyte homing followed by suppression of genes that inhibit apoptosis (129).

FUNCTION

The thymus plays a central role in cell-mediated immunity. In early embryonic development, prothymocytes enter the thymus from the bone marrow and migrate to the outer cortex where they undergo a process of maturation. Mature thymocytes move from the outer cortex to the medulla and then migrate into the peripheral circulation, where they function as mature T cells. During their sojourn in the thymus, T cells learn to distinguish self from nonself and acquire the ability to recognize antigens bound to cell surface molecules encoded by the major histocompatibility complex. Circulating $CD4^+$ helper T cells and $CD8^+$ suppressor T cells derived from the thymus play a variety of roles in cell-mediated immunity. These include induction of cytotoxicity, delayed-type hypersensitivity reactions, and transplant rejection.

A controversial aspect of thymic function is related to production of thymic hormones, which may play a role in the induction of differentiation of early T-cell precursors in the thymus. Four types of thymic hormones have been identified: thymopoietin, thymosin, thymulin (formerly known as facteur thymique serique), and thymic humoral factor (128,130,131). Thymopoietin and thymulin are produced by thymic epithelium, whereas thymosins comprise a family of peptides synthesized in many organs. Production of thymic hormones by thymic epithelial cells is a controversial topic. Immunohistochemical studies using polyclonal and monoclonal antibodies reportedly show the presence of thymulin, thymosin α1, and thymopoietin in the cytoplasm of murine and human thymic epithelial cells (131–134). In some studies, hormone localization seems restricted to A2B5+ cells in subcapsular cortical and medullary epithelial cells (135). This finding has led some to suggest that these subtypes represent the functional (secretory or endocrine) portion of the thymus, as contrasted with the nonsecretory epithelium of the inner cortical region and of Hassall corpuscles. In a murine system, thymic hormones also have been described in a subtype of medullary epithelial cells characterized ultrastructurally by numerous cytoplasmic membrane–bound vacuoles containing amorphous material (136). Hirokawa et al. (137) used rabbit antisera against synthetic thymosin α1 and bovine thymosin β3 in 45 cases of human thymomas and normal thymus of newborns; they found reactivity of thymoma cells and normal thymic epithelial cells in 80% and 89% of cases, respectively. The thymosin-containing cells were predominantly localized to the medulla and the subcapsular cortex, and the reaction was most intense in lymphocyte-rich thymomas. Neither thymosin α1 nor thymosin β3 could be detected in other epithelial malignancies of stomach, lung, or liver. However, sufficiently reliable specific antibodies have not yet become available for routine use in diagnostic pathology. In general, thymic hormones have as their main role to induce in situ T-cell differentiation by direct, cell-to-cell, receptor-based interactions as well as by in situ paracrine information in the thymic microenvironment. Thymic hormones have also been used as immunostimulants for enhancing the humoral immunity in cancer patients. Thymosin fraction 5, thymic humoral factor γ2, thymosin α1, and thymopentin have all been utilized in clinical immunotherapeutic protocols as an adjunct to treatment of various cancers (138).

Thymic epithelial cells contain a wide variety of neuropeptides such as oxytocin, vasopressin, β-endorphin, somatostatin, and other anterior pituitary hormones (139–141). They also produce various cytokines and growth factors, including interleukin (IL)-1, IL-6, and granulocyte–monocyte colony-stimulating factor (142,143). The biologic effects of these cytokines are determined by the expression of cytokine receptors on the thymic cell surface. Thymic cytokines may act by autocrine or paracrine mechanisms. Examples of paracrine thymic cytokines are IL-7, which induces thymocyte growth and differentiation, and interferon (IFN)-γ, which induces thymic epithelial cell activation. An example of an autocrine factor is IL-2, for which the producers and targets are thymocytes. The ability of thymocytes to produce cytokines and express cytokine receptors is gradually reduced as they mature from precursor cells to cortical thymocytes. After the completion of the selection process of maturation, the ability of thymocytes to produce cytokines and respond to their action is restored (144). The role and function of cytokines in the thymus are different from those of cytokines in the peripheral compartments of the immune system. Those of the thymus are involved with migration and development of thymocytes and autoregulation of thymocyte cell numbers, while peripheral cytokines are mainly involved in the regulation of inducible inflammatory processes.

AGE-RELATED AND OTHER TROPHIC CHANGES

Thymic Involution

The thymus undergoes a slow physiologic process of age-related involution and atrophy (145–147). This process starts at puberty, at which time the organ reaches its maximum absolute weight. This process is accompanied by gradual changes in thymocyte populations relative to different rates of involution of the cortical and the medullary epithelium. Early changes consist primarily of a decrease in the number of cortical thymocytes with relative sparing of the epithelial elements (148). In the more advanced stages, the parenchyma of the thymus reverts to a more primitive appearance and is replaced by islands of epithelial cells depleted of lymphocytes, with partly cystic, closely aggregated Hassall corpuscles and abundant intervening adipose tissue (149). Although thymic involution can proceed to a point where no thymic tissue can be appreciated grossly, microscopic thymic remnants are almost always present. The best way to locate them is to examine the pre-epicardial fat in a serial fashion. Studies have suggested that the process of involution in the thymus may be quantitative rather than qualitative. Thymopoiesis continues to occur in the thymus of adult humans late into life and thymic activity and function appear to be well maintained into old age (150–152).

A type of change unrelated to senescence is accidental or stress involution. This condition results from the dramatic response of the thymus gland to episodes of severe stress, in which the sudden release of corticosteroids from the adrenal cortex leads to rapid depletion of thymic cortical lymphocytes (153). Microscopically, there is prominent karyorrhexis of lymphocytes with active phagocytosis by macrophages, which creates a prominent starry sky appearance characteristically confined to the cortex. If the stimulus persists, a loss of corticomedullary distinction ensues, with accentuation of the epithelial elements, cystic dilatation of Hassall corpuscles, and emergence of elongated, epithelium-lined cystic spaces that recapitulate the early stages of Hassall corpuscle formation. Further loss of thymocytes results in collapse of the lobular architecture and fibrosis. The thymus is thus transformed into a mass of adipose tissue containing scattered islands of parenchyma with a few lymphocytes.

Acute thymic involution in infancy and childhood significantly correlates with the duration of acute illness. Morphologic parameters such as abundant macrophages in the cortex, increase of interlobular fibrous tissue, and lymphoid depletion of the cortex may enable the pathologist to estimate the duration of acute disease before death (154). A precocious type of thymic involution manifested by epithelial injury has also been observed in pediatric and adult patients with acquired immunodeficiency syndrome (AIDS) (155,156). Some authors have suggested these changes indicate that the thymus is a primary target organ in human immunodeficiency virus infection (157), whereas others have considered these changes as an expression of stress involution (156,158).

TABLE 18.2 Weight and Volume of Normal Human Thymuses

N	Age (Years)	Weight (g)[a]	Volume (cm^3)[a]
6	0–1	27.3 ± 16.4	26.8 ± 16.1
4	1–4	28.0 ± 19.3	27.9 ± 10.4
7	5–9	22.1 ± 9.2	21.5 ± 8.8
5	10–14	21.5 ± 6.1	21.1 ± 6.4
9	15–19	20.2 ± 10.3	19.3 ± 10.1
18	20–24	21.6 ± 9.5	23.0 ± 10.6
9	25–29	23.1 ± 11.8	23.7 ± 11.9
5	30–34	25.5 ± 9.9	27.6 ± 11.2
17	35–44	21.9 ± 9.2	22.2 ± 10.5
14	45–54	24.8 ± 12.8	26.5 ± 12.4
15	55–64	21.3 ± 9.5	23.5 ± 10.4
17	65–84	23.8 ± 16.1	25.6 ± 17.0
5	85–90	18.2 ± 5.4	20.4 ± 6.8
5	91–107	12.4 ± 6.9	13.4 ± 7.2
136	Total	22.8 ± 12.5	23.4 ± 11.9

[a]Values are means ± SD.

Reprinted with permission from Steinmann GG, Klaus B, Müller-Hermelink HK. The involution of the ageing human thymic epithelium is independent of puberty. A morphometric study. *Scand J Immunol.* 1985;22(5):563–575. Table 1.

Thymic Hyperplasia

True thymic hyperplasia is defined as an enlargement of the thymus gland beyond that considered as the upper limit of normal for any given age group (159). In the past, a diagnosis of true thymic hyperplasia was based on reference to standard weight charts of normal thymus glands for comparison. Hammar (160–163) made the first extensive study on the normal weights of the thymus in fresh autopsy specimens in 1906. This work was updated in subsequent studies, the most comprehensive of which was that of Steinman (147), who examined the weight and volume of human thymuses in 136 healthy individuals (Table 18.2). He concluded that determination of the volume of the gland, as measured by the displacement of a physiologic saline solution, was more reliable than weighing the gland and constitutes the optimal parameter for this type of evaluation. In practice, volume determination may not be easy to accomplish in the setting of standard processing of specimens in surgical pathology laboratories. Because of the difficulties involved in defining this process, much confusion has been generated by the use of the term "thymic hyperplasia." The most commonly observed type of thymic hyperplasia in adults is lymphoid follicular hyperplasia, which has been commonly associated with myasthenia gravis and other autoimmune disorders. Thymic hyperplasia unassociated with lymphoid follicular hyperplasia is more commonly seen in children and adolescents and may be mistaken clinically for thymoma. Such cases used to be designated "true thymic hyperplasia" to distinguish them from lymphoid follicular hyperplasia (69).

Recent studies have demonstrated that the thymus can show expansion of the reticular network of thymic cortical epithelial cells and increase in the numbers of immature thymocytes in some situations. Because the growth is at the expense of the native thymic parenchyma rather than B cells in the extraparenchymal compartment, this type of growth has been termed "intraparenchymal thymic hyperplasia"

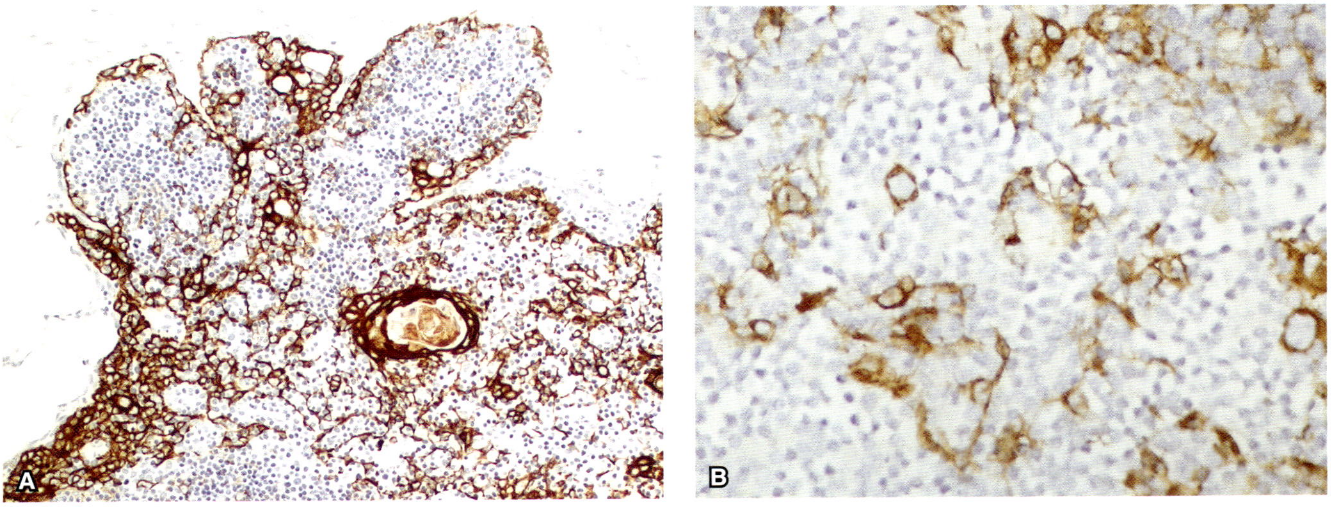

FIGURE 18.10 **A:** Immunoperoxidase stain for cytokeratin in patient with thymic parenchymal hyperplasia shows a continuous branching network of thymic epithelial cells concentrated along the periphery of the lobule and gradually penetrating into the medulla. **B:** Contrast with distribution of the proliferating thymic epithelial cells in type B1 thymoma in which the keratin+ epithelial cells are randomly scattered among the lymphocytes rather than forming a syncytium.

(164). This type of hyperplasia is characterized by the maintenance of the normal corticomedullary architecture with a preserved cortical-to-medullary ratio and increased size of the lobules. Immunohistochemistry for keratin, p40, or p63 antibodies may be helpful in distinguishing hyperplasia from thymoma; these markers will demonstrate a net-like expansion of cortical thymic epithelial cells in parenchymal hyperplasia, rather than isolated scattered epithelial cells as seen in lymphocyte-rich thymomas (Fig. 18.10). Because of difficulties associated with determining the upper limit of the normal weight of the thymus, it is recommended that a diagnosis of parenchymal thymic hyperplasia be rendered only when patients have clinically symptomatic or radiologically abnormal enlargement of the gland as well as the above histologic features.

Thymic parenchymal hyperplasia is a recognized complication of chemotherapy for Hodgkin disease in children (165,166) and germ cell tumors in adults (167,168), in which case it has been attributed to an immunologic rebound phenomenon. Similar thymic enlargement has been observed in children recovering from thermal burns (169) and after cessation of administration of corticosteroids in infants (170). Marked thymic enlargement also occurs in the setting of thymic reconstitution after chemotherapy (171), after the institution of antiretroviral therapy in HIV-infected patients (172), and in patients with acute respiratory distress syndrome due to COVID-19.

Parenchymal ("true") thymic hyperplasia must be distinguished from lymphoid hyperplasia. In the latter condition, the term *hyperplasia* refers to an increased number of lymphoid follicles in the medullary region of the gland that results from migration of mature T cells and B cells into the perivascular space adjacent to but outside of the thymic epithelial meshwork. The problem in establishing the presence of lymphoid hyperplasia vis-à-vis the occurrence of lymphoid follicles in the normal thymus has already been discussed. Lymphoid hyperplasia of the thymus has been associated with myasthenia gravis as well as several other immune-mediated disorders, including systemic lupus erythematosus, rheumatoid arthritis, scleroderma, allergic vasculitis, and thyrotoxicosis.

ARTIFACTS AND OTHER POTENTIAL PITFALLS IN DIFFERENTIAL DIAGNOSIS

Microscopic changes related to involution may be a source of considerable confusion in the interpretation of thymic biopsies. Such changes are primarily related to the distribution, architectural arrangement, and cytologic appearance of the epithelial cells (Fig. 18.11). Some cells may acquire a spindle, atrophic appearance, whereas others arrange themselves in rosette-like formations devoid of central lumina (Fig. 18.12). Given that these changes are found with some frequency in thymic dysplasia and thymoma suggests they represent regressive and functionally inactive states of the epithelial cells (9). Involuting thymic epithelial cells may appear as thin, elongated, and sometimes serpiginous strands of oval to spindle thymic epithelial cells admixed with occasional small lymphocytes that course through the mediastinal fat (Fig. 18.13A). The epithelial cells can be nicely highlighted by p63 or p40 immunostaining (Fig. 18.13B). These strands of epithelial cells can sometimes be traced back to solid nests of involuting thymic remnants. Microscopic clusters containing oval to spindle epithelial cells with dense chromatin and scant eosinophilic cytoplasm admixed with a few small lymphocytes can also

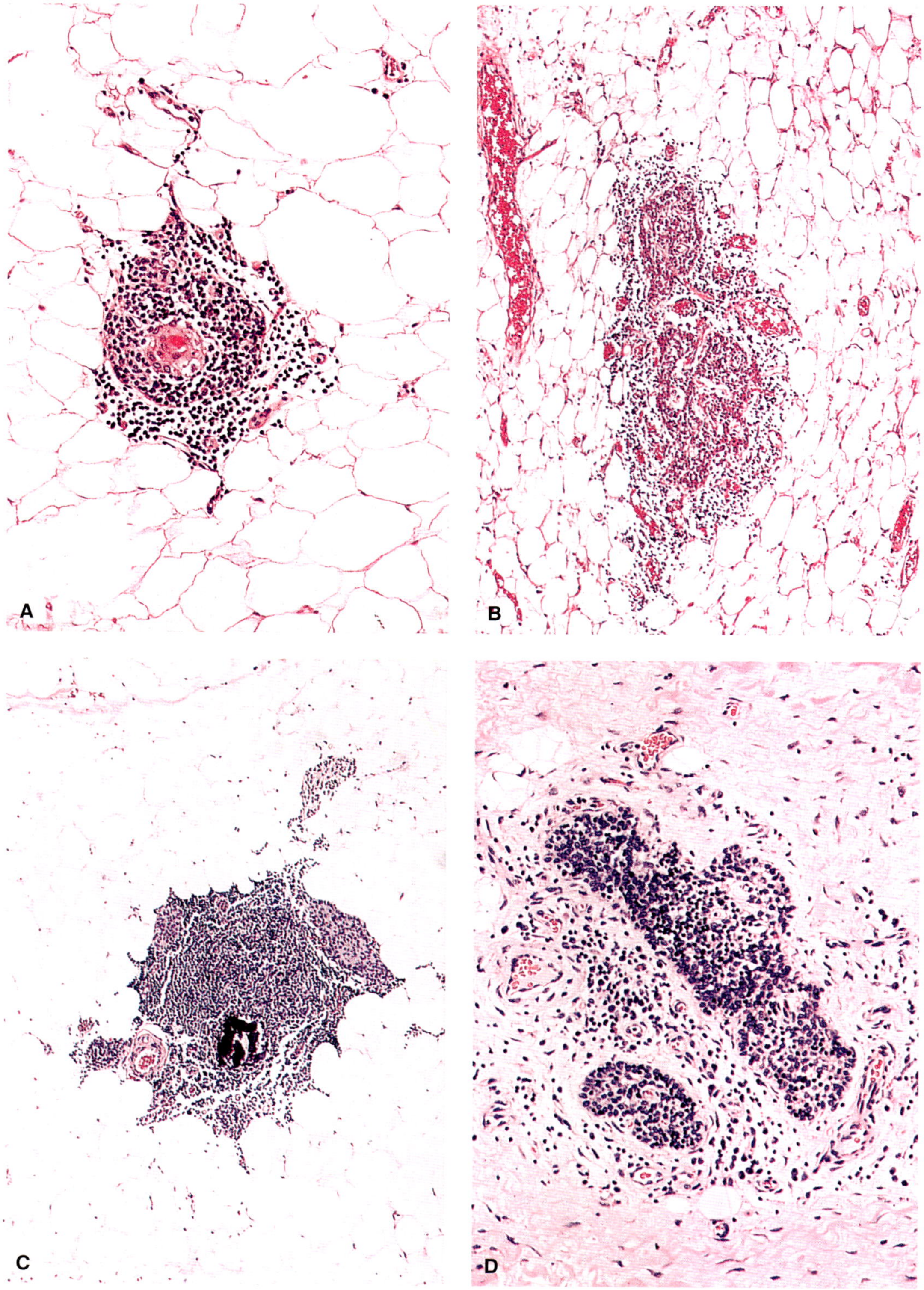

FIGURE 18.11 Epithelial remnants in involuting thymus. **A:** Abortive Hassall corpuscle surrounded by small lymphocytes and scattered epithelial cells. **B:** Anastomosing strands of epithelial cells surrounded by small lymphocytes embedded within the pre-epicardial fat. **C:** Residual thymic island with predominance of lymphocytes, small solid epithelial cell clusters at the periphery, and calcified Hassall corpuscle. **D:** Anastomosing strands of epithelial cells admixed with small lymphocytes embedded within a collagenized stroma.

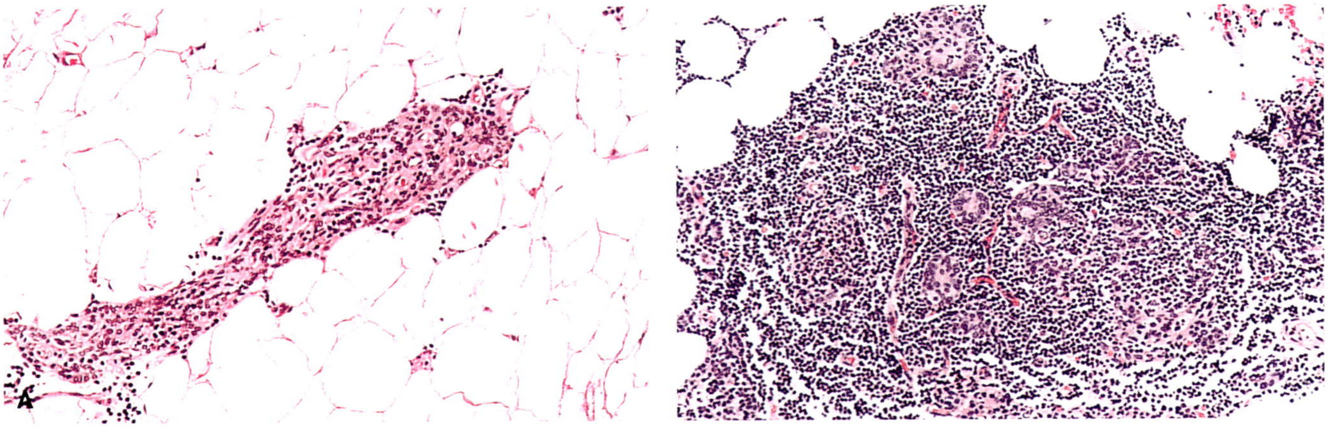

FIGURE 18.12 **A:** Thymic remnant showing elongated configuration with prominent spindling of the cells. **B:** Involuting thymus with epithelial rosettes.

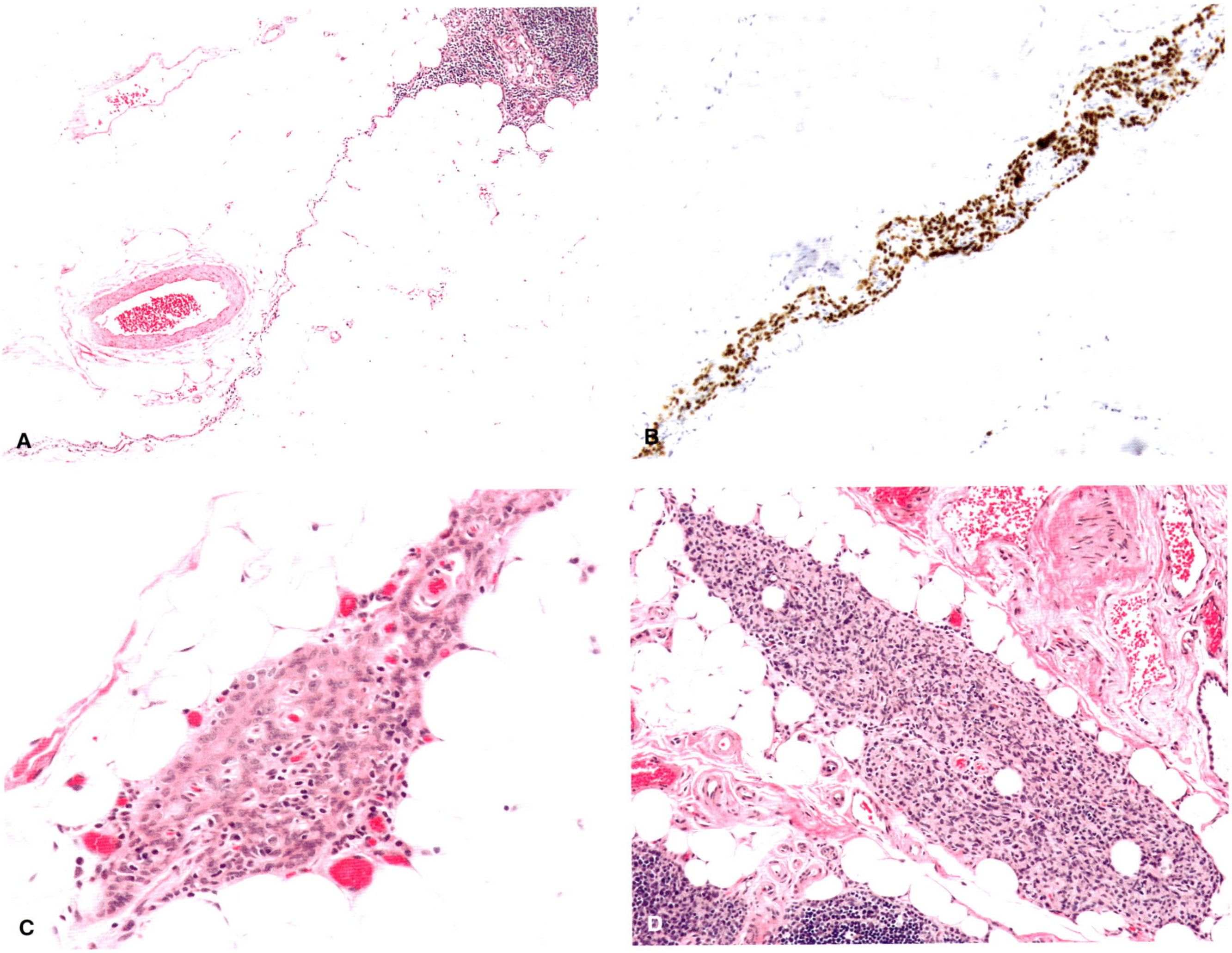

FIGURE 18.13 **A:** A slender, thin, serpiginous strand of atrophic thymic epithelium admixed with scant lymphocytes is seen coursing through the fat and originating from a small island of involuting thymus (*top right*). **B:** Immunohistochemical stain for p63 at higher magnification shows strong nuclear positivity in the thymic epithelial cells. **C:** Microscopic island of involuting thymic epithelium is present in the fat which is composed of oval to spindle epithelial cells with eosinophilic cytoplasm and a sprinkling of small lymphocytes. **D:** Another microscopic island of involuting thymic epithelium is shown composed of atrophic and tightly packed spindle cells admixed with scattered small lymphocytes.

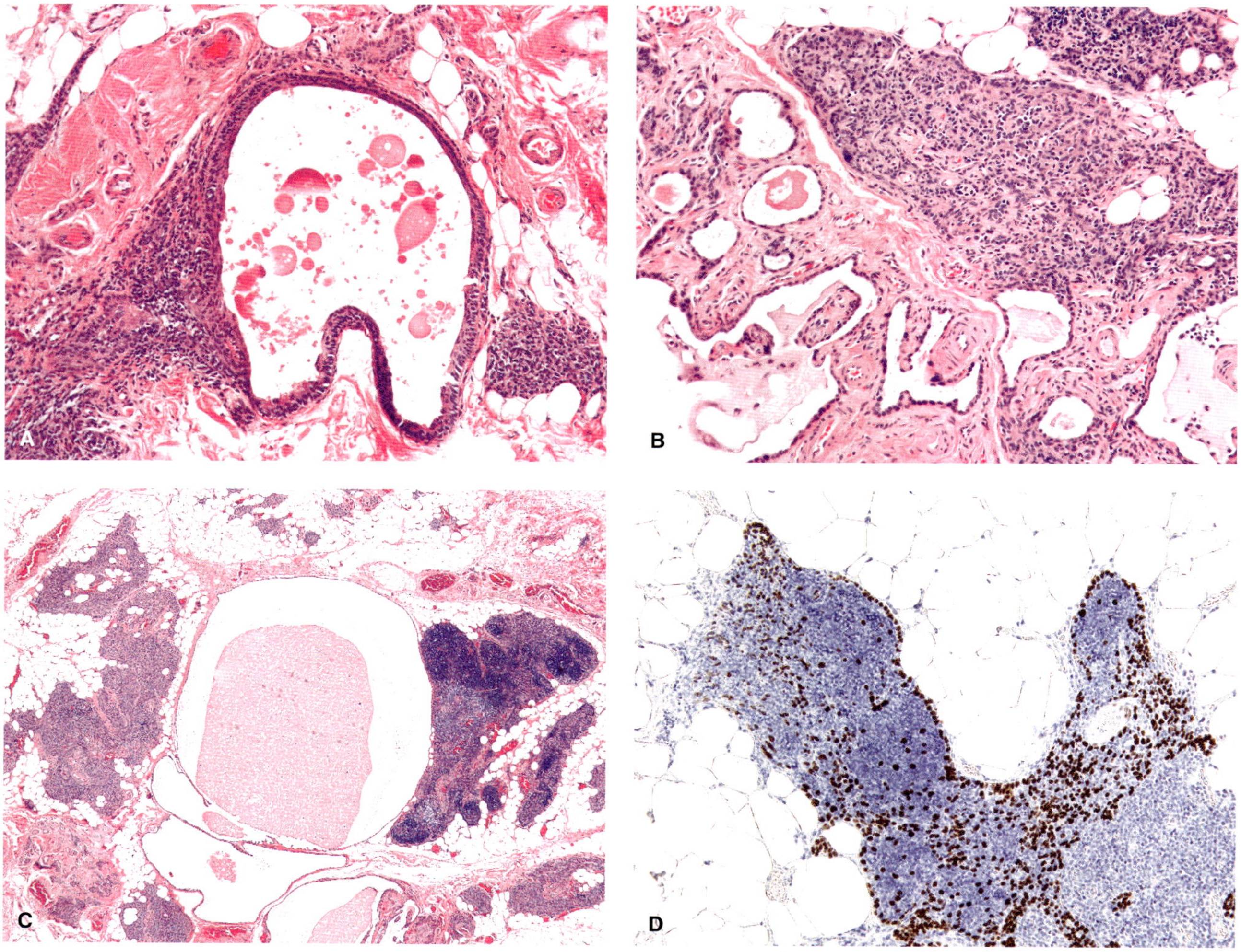

FIGURE 18.14 **A:** Focus of involuting thymic epithelium showing cystic dilatation. The cells lining the cyst range from flat- to cuboidal- to columnar-ciliated epithelium. **B:** Another focus of involuting thymic epithelium shows cystically dilated spaces lined by a layer of cuboidal epithelial cells simulating glandular structures. **C:** Large, cystically dilated spaces are seen flanked by islands of involuting thymic epithelium. The island on the left shows the typical appearance of involuted thymus in adults; the island on the right simulates a lymph node. **D:** Immunostaining for p63 in the solid nodule resembling a lymph node highlights numerous scattered involuting thymic epithelial cells with strong nuclear positivity.

be seen, and may raise the possibility of metastatic malignancy (Fig. 18.13C,D). Similar clusters can be observed in association with cystic and glandular structures (Fig. 18.14A,B). Some of the cystic spaces are adjacent to dense lymphoid collections that resemble lymph nodes; immunostaining for p63 will highlight the thymic epithelial cells scattered throughout these structures and identify them as involuting thymic elements (Fig. 18.14C,D). Tissues showing the above features are a frequent finding in the fat submitted for central lymph node dissection in patients undergoing surgery for thyroid carcinoma.

Sometimes thymic remnants are almost exclusively formed of epithelial elements arranged in well-defined round nests that may simulate neuroendocrine growths (Fig. 18.15) (173). A particularly distinctive appearance is that of thin, elongated strands of thymic epithelium

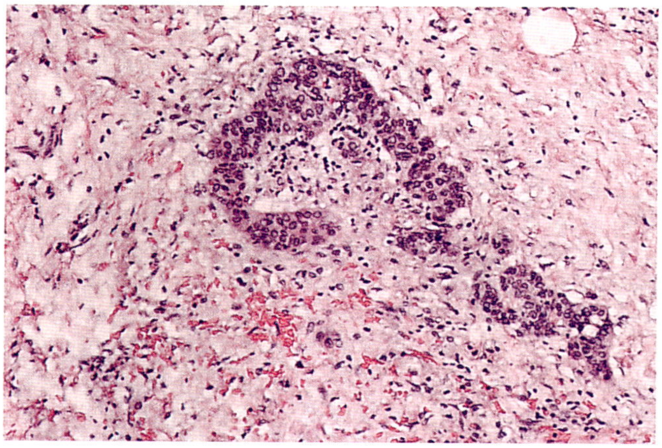

FIGURE 18.15 Strands of residual thymic epithelium arranged in small nests resembling neuroendocrine growths.

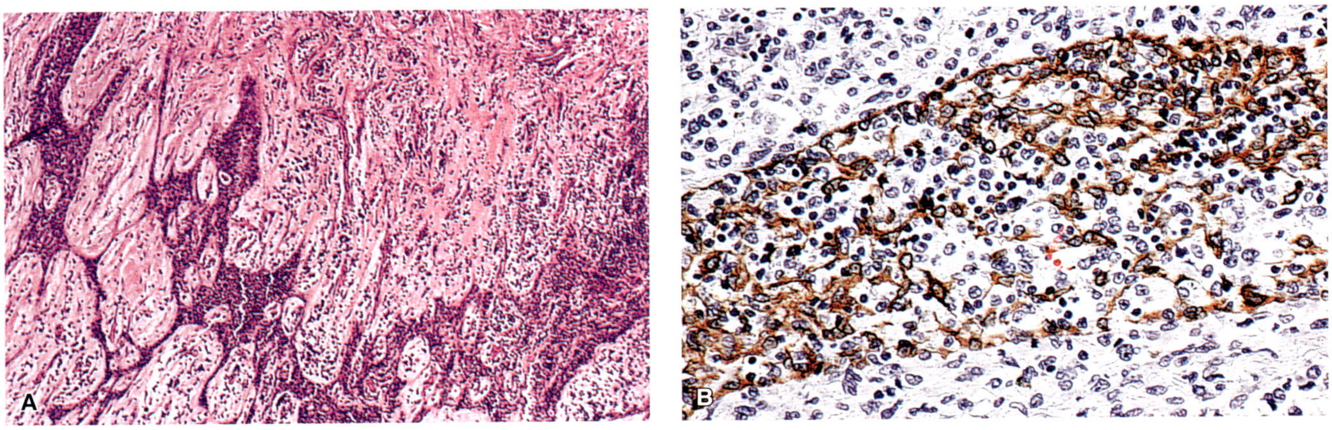

FIGURE 18.16 **A:** Wall of thymic cyst showing branching strands of thymic epithelial cells surrounded by a fibrous stroma. **B:** Entrapped thymic epithelial elements within anterior mediastinal malignant lymphoma. The strong keratin positivity seen in these cells may lead to an erroneous diagnosis of thymic carcinoma.

composed of a single or double cell layer surrounded by, or circumscribing, dense connective tissue in a fibroepitheliomatous fashion. These thin, elongated epithelial strands may be flanked by small lymphocytes and occasionally exhibit an antler-like branching configuration (Fig. 18.16A). They may be seen by themselves or at the periphery of thymic cysts, thymomas, thymic lymphomas, and other thymic neoplasms. In the case of thymic lymphomas and seminomas, these strands can be present not only around but also within the tumor, surrounded and infiltrated by the neoplastic elements. Their presence, whether detected at the hematoxylin and eosin (H&E) level or by immunohistochemistry, can be interpreted erroneously as evidence supporting an epithelial nature for the lesion (Fig. 18.16B). As a matter of fact, these structures constitute the single most important cause of misdiagnosis in cases of large cell lymphoma of the thymus. Conversely, some thymic remnants are made up almost exclusively of lymphocytes, thus simulating lymph nodes. A clue to their real nature should be sought at the very periphery, in which epithelial cells can sometimes be identified encircling the nests, perhaps representing the residual coat of subcapsular cortical cells of the normal organ (Fig. 18.17).

Certain patterns of tissue response to injury in the thymus may also constitute a major source of confusion in the interpretation of biopsies of this organ (174). A common form of response to injury, particularly in cases associated with inflammation, is cystic degeneration of thymic epithelium. The cystic degeneration is thought to be an acquired process, which in its fullest expression will lead to the formation of a multilocular thymic cyst (54). The main histologic features of such cysts include the creation of large cavities lined by squamous, columnar, or cuboidal epithelium, often in continuity with remnants of normal thymic epithelium within the cyst walls; severe acute and chronic inflammation accompanied by fibrovascular proliferation, necrosis, hemorrhage, cholesterol granulomas; and reactive lymphoid hyperplasia with prominent germinal centers. In some instances, the cyst lining may show a moderate degree of cytologic atypia with features of pseudoepitheliomatous hyperplasia that can be easily misinterpreted for malignancy

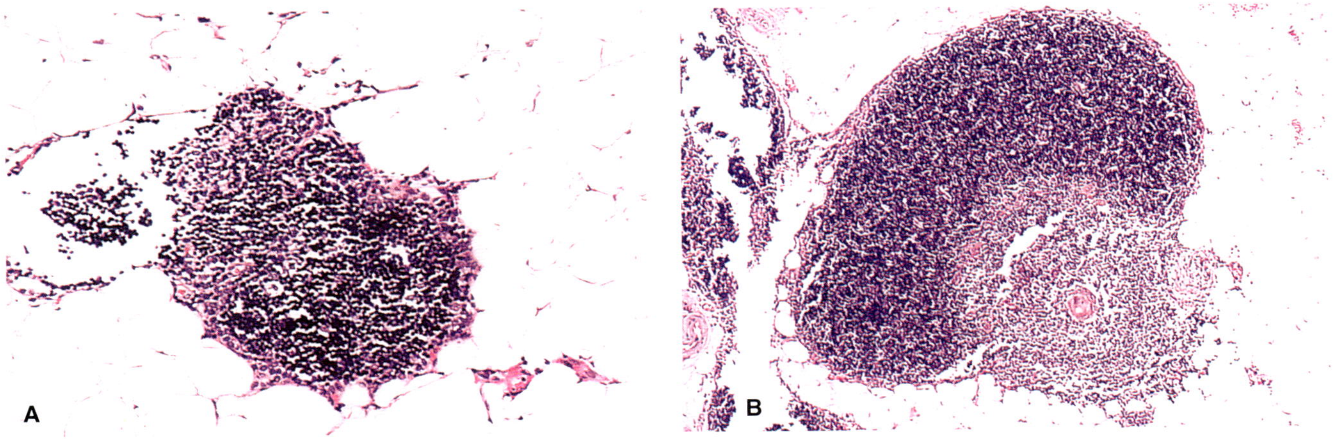

FIGURE 18.17 **A:** Thymic remnant composed predominantly of small lymphocytes simulating a lymphoid nodule. Note the single row of flattened epithelial cells at the periphery. **B:** Thymic remnant composed of cortical and medullary portion, the latter containing a small Hassall corpuscle (*bottom half*).

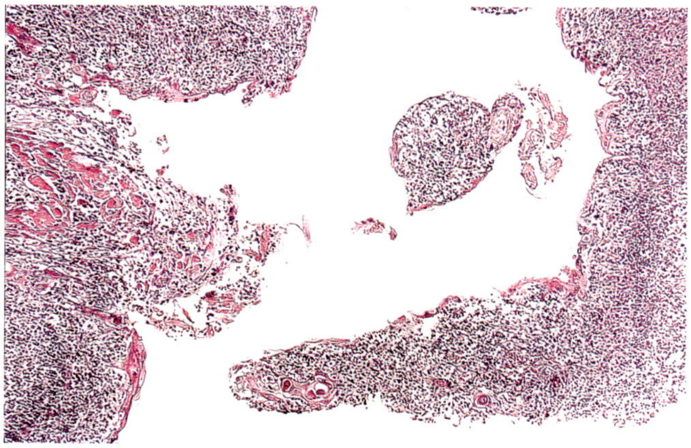

FIGURE 18.18 Cystic dilatation of Hassall corpuscles is seen in an acquired multilocular thymic cyst. Note the cyst lining is in continuity with Hassall corpuscles, and the walls of the cysts contain abundant inflammatory infiltrate.

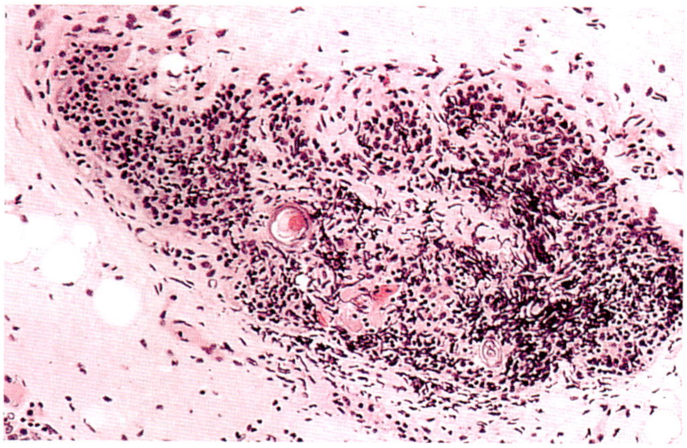

FIGURE 18.19 Extensive crush artifact is seen in this thymic remnant obtained through mediastinoscopic biopsy.

(175). In most cases, the cystic structures are closely associated with Hassall corpuscles, many of which show dilatation and may be found to be in continuity with the lining of the cystic cavities (Fig. 18.18). This type of reaction may be the result of an exaggerated response of medullary duct epithelium-derived structures of the thymus to an underlying inflammatory process (54). However, it is important to point out that seemingly identical changes may take place in uninvolved thymic parenchyma in cases of Hodgkin disease, mediastinal seminoma, and thymoma, to the extent that the neoplastic elements may be overshadowed by the cystic/inflammatory process (176,177). Other primary thymic neoplasms that may also be associated with prominent cystic changes include basaloid carcinoma and mucoepidermoid carcinoma (178,179). Careful search and extensive sampling must be undertaken in cystic mediastinal lesions for proper identification of the diagnostic neoplastic areas.

Another form of tissue response to injury of the thymus that may introduce difficulties for diagnosis is stromal fibrosis. Fibrous overgrowth of the stroma may result from a variety of nonneoplastic conditions such as ionizing radiation or fungal infection (180,181), as well as idiopathic sclerosing mediastinitis (182,183). Several malignant conditions can be accompanied by prominent fibrous changes of the stroma, including primary diffuse large cell lymphoma of the mediastinum, Hodgkin disease, and thymic seminoma (174). In many such instances, the gland may show extensive sclerosis with entrapment of a few scattered foci harboring the diagnostic atypical cells. Such cases may prove literally impossible to diagnose in small mediastinoscopic biopsies and require extensive sampling of the mass to identify the diagnostic areas. The surgeon must be informed of the need for obtaining additional tissue for diagnosis at the time of frozen section examination.

Another potential pitfall for diagnosis is sampling of the normal thymic cortex which is highly cellular, contains immature cells, and shows increased mitotic activity. Limited material obtained via mediastinoscopically obtained biopsies can show features that simulate malignant lymphoma, particularly of the lymphoblastic type. The paucity or absence of cells with convoluted nuclei and the identification of regularly scattered epithelial cells throughout the lymphoid population should point toward the correct interpretation. Finally, an additional source of difficulty for diagnosis lies in the presence of biopsy-induced artifacts, one of the most common being the crush artifact that leads to nuclear elongation and smearing reminiscent of changes seen in small cell carcinoma (Fig. 18.19).

REFERENCES

1. Norris EH. The morphogenesis and histogenesis of the thymus gland in man: In which the origin of the Hassall's corpuscles of the human thymus is discovered. *Contrib Embryol Carnegie Inst.* 1938;27:193–207.
2. Weller GL Jr. Development of the thyroid, parathyroid and thymus glands in man. *Contrib Embryol Carnegie Inst.* 1933;24:93–139.
3. Cordier AC, Haumont SM. Development of thymus, parathyroids and ultimo-branchial bodies in NMRI and nude mice. *Am J Anat.* 1980;157:227–263.
4. Gilmour JR. The embryology of parathyroid glands: the thymus and certain associated rudiments. *J Pathol Bacteriol.* 1937;45:507–522.
5. Jotereau FV, Houssaint E, Le Douarin NM. Lymphoid stem cell homing to the early thymic primordium of the avian embryo. *Eur J Immunol.* 1980;10:620–627.
6. Yayon N, Kedlian VR, Boehme L, et al. A spatial human thymus cell atlas mapped to a continuous tissue axis. *Nature.* 2024;635:708–718.
7. Gilmour JR. Some developmental abnormalities of the thymus and parathyroids. *J Pathol Bacteriol.* 1941;52:213–218.
8. Nathaniels EK, Nathaniels AM, Wang CA. Mediastinal parathyroid tumors: a clinical and pathological study of 84 cases. *Ann Surg.* 1970;171:165–170.
9. Rosai J, Levine GD. Tumors of the thymus. In: *Atlas of Tumor Pathology*. 2nd series, fascicle 13. Armed Forces Institute of Pathology; 1976.

10. Domaniewski J, Ukleja Z, Rejmanowski T. Immunological and clinical problems in ectopic thymoma. *Otolaringol Pol.* 1975;29:579–585.
11. Martin JM, Rundhawa G, Temple WJ. Cervical thymoma. *Arch Pathol Lab Med.* 1986;110:345–357.
12. Yamashita H, Murakami N, Noguchi S, et al. Cervical thymoma and incidence of cervical thymus. *Acta Pathol Jpn.* 1983;33:189–194.
13. Wadon A. Thymoma intratracheale. *Zentralbl Allg Pathol Anat.* 1934;60:308–312.
14. Harach HR, Saravia Day E, Fransilla KO. Thyroid spindle-cell tumor with mucous cysts. An intrathyroid thymoma? *Am J Surg Pathol.* 1985;9:525–530.
15. Miyauchi A, Kuma K, Matsuzuka F, et al. Intrathyroidal epithelial thymoma: an entity distinct from squamous cell carcinoma of the thyroid. *World J Surg.* 1985;9:128–135.
16. Bradova M, Thompson LDR, Hyrcza M, et al. Branchioma: immunohistochemical and molecular genetic study of 23 cases highlighting frequent loss of retinoblastoma 1 immunoexpression. *Virchows Arch.* 2024;484:103–117.
17. Gagens EW. Malformation of the auditory apparatus in the newborn associated with ectopic thymus. *Arch Otolaryngol.* 1932;15:671–680.
18. Castleman B. Tumors of the thymus gland. In: *Atlas of Tumor Pathology*. 1st series, fascicle 19. Armed Forces Institute of Pathology; 1955.
19. Marchevski AM. Lung tumors derived from ectopic tissues. *Semin Diagn Pathol.* 1995;12:172–184.
20. Weissferdt A, Moran CA. The spectrum of ectopic thymomas. *Virchows Arch.* 2016;469:245–254.
21. Wolff M, Rosai J, Wright DH. Sebaceous glands within the thymus: report of three cases. *Hum Pathol.* 1984;15:341–343.
22. Breckler IA, Johnston DG. Choristoma of the thymus. *AMA J Dis Child.* 1956;92:175–178.
23. Landing B, Yutuc I, Swanson V. Clinicopathologic correlation in immunologic deficiency diseases of children, with emphasis on thymic histologic patterns. In: *Proceedings of the International Symposium on Immunodeficiency*. Tokyo University Press; 1976:3–33.
24. Blackburn WR, Gordon DS. The thymic remnant in thymic alymphoplasia. Light and electron microscopic studies. *Arch Pathol.* 1967;84:363–375.
25. Hoyer JR, Cooper MD, Gabrielsen AE, Good RA. Lymphopenic forms of congenital immunologic deficiency diseases. *Medicine (Baltimore).* 1968;47:201–226.
26. Peterson RD, Kelly WD, Good RA. Ataxia-telangiectasia. Its association with a defective thymus, immunological-deficiency disease, and malignancy. *Lancet.* 1964;1:1189–1193.
27. Cooper MD, Petersen RD, Good RA. A new concept of the cellular basis of immunity. *J Pediatr.* 1965;67:907–908.
28. Nezelof C, Jammet ML, Lortholary P, Labrune B, Lamy M. L'hypoplasie hereditaire du thymus: Sa place et sa responsabilite dans une observation d'aplasie lymphocytaire, normoplasmocytaire et normoglobulinemique du nourrisson. *Arch Fr Pediatr.* 1964;21:897–920.
29. Seemayer TA, Bolande RP. Thymus involution mimicking thymic dysplasia: a consequence of transfusion-induced graft versus host disease in a premature infant. *Arch Pathol Lab Med.* 1980;104:141–144.
30. Huber J, Cholnoky P, Zoethout HE. Congenital aplasia of parathyroid glands and thymus. *Arch Dis Child.* 1967; 42:190–192.
31. Fowlkes BJ, Pardoll DM. Molecular and cellular events of T cell development. *Adv Immunol.* 1989;44:207–264.
32. MacDonald HR, Lees RK. Programmed death of autoreactive thymocytes. *Nature.* 1990;343:642–644.
33. Kizaki H, Tadakuma T. Thymocyte apoptosis. *Microbiol Immunol.* 1993;37:917–925.
34. Colbert RA, Young DA. Glucocorticoid-induced messenger ribonucleic acids in rat thymic lymphocytes: rapid primary effects specific for glucocorticoids. *Endocrinology (Baltimore).* 1986;119:2598–2605.
35. Domashenko AD, Nazarova LF, Umansky SR. Comparison of the spectra of proteins synthesized in mouse thymocytes after irradiation or hydrocortisone treatment. *Int J Radiat Biol.* 1990;57:315–329.
36. Korsmeyer SJ. Bcl-2: a repressor of lymphocyte death. *Immunol Today.* 1992;13:285–288.
37. Lowe SW, Schmitt EM, Smith SW, Osborne BA, Jacks T. p53 is required for radiation-induced apoptosis in mouse thymocytes. *Nature.* 1993;362:847–849.
38. Smith CA, Williams GT, Kinsgton R, Jenkinson EJ, Owen JJ. Antibodies to CD3/T-cell receptor complex induce death by apoptosis in immature T cells in thymic cultures. *Nature.* 1989;337:181–184.
39. von Boehmer H. Positive selection of lymphocytes. *Cell.* 1994;76:219–228.
40. Shi YF, Bissonnette RP, Parfrey N, Szalay M, Kubo RT, Green DR. In vivo administration of monoclonal antibodies to the CD3 T cell receptor complex induces cell death (apoptosis) in immature thymocytes. *J Immunol.* 1991;146:3340–3346.
41. Blackman M, Kappler J, Marrack P. The role of the T cell receptor in positive and negative selection of developing T cells. *Science.* 1990;248:1335–1341.
42. Mountz JD, Zhou T, Wu J, Wang W, Su X, Cheng J. Regulation of apoptosis in immune cells. *J Clin Immunol.* 1995;15:1–16.
43. Hammar JA. Die Menschen thymus in Gesundheit und krankheit. *Z Mikrosk Anat Forsch.* 1926;6(suppl):107–208.
44. Flores KG, Li J, Sempowski GD, Haynes BF, Hale LP. Analysis of the human thymic perivascular space during aging. *J Clin Invest.* 1999;104:1031–1039.
45. Hale LP. Histologic and molecular assessment of human thymus. *Ann Diagn Pathol.* 2004;8:50–60.
46. Suster S, Moran CA. Thymoma, atypical thymoma, and thymic carcinoma. A novel conceptual approach to the classification of thymic epithelial neoplasms. *Am J Clin Pathol.* 1999;111:826–833.
47. Lobach DF, Haynes BF. Ontogeny of the human thymus during fetal development. *J Clin Immunol.* 1987;7:81–97.
48. Dipasquale B, Tridente G. Immunohistochemical characterization of nurse cells in normal human thymus. *Histochemistry.* 1991;96:499–503.
49. von Gaudecker B. Functional histology of the human thymus. *Anat Embryol (Berl).* 1991;183:1–15.
50. Peterson P, Org T, Rebane A. Transcriptional regulation by AIRE: molecular mechanisms of central tolerance. *Nat Rev Immunol.* 2008;8:948–957.
51. Marx A, Hohenberger P, Hoffmann H et al. The autoimmune regulator AIRE in thymoma biology: *autoimmunity and beyond*. *J Thorac Oncol.* 2010;5:S266–S272.

52. Miller CN, Proekt I, von Moltke J, et al. Thymic tuft cells promote IL4-enriched medulla and shape thymocyte development. *Nature*. 2018;559:627–631.
53. Rippert H. Die entwicklungsstörung der thymusdrüse bei kongenitaler Lues. *Frankfurt Z Pathol*. 1912;11:209–218.
54. Suster S, Rosai J. Multilocular thymic cyst: an acquired reactive process. Study of 18 cases. *Am J Surg Pathol*. 1991;15:388–398.
55. Henry K. Mucin secretion and striated muscle in the human thymus. *Lancet*. 1966;1:183–185.
56. Cantor H, Weissman I. Development and function of subpopulations of thymocytes and T lymphocytes. *Prog Allergy*. 1976;20:1–64.
57. Everett NB, Tyler RW. Lymphopoiesis in the thymus and other tissues: functional implications. *Int Rev Cytol*. 1967;22:205–237.
58. Middleton G. The incidence of follicular structures in the human thymus at autopsy. *Aust J Exp Biol Med Sci*. 1967;45:189–199.
59. Vetters JM, Barclay RS. The incidence of germinal centres in thymus glands of patients with congenital heart disease. *J Clin Pathol*. 1973;26:583–591.
60. Levine GD, Rosai J. Light and electron microscopy of the human fetal thymus. In: Johannessen JV, ed. *Electron Microscopy in Human Medicine*. Vol. 5. McGraw-Hill; 1980.
61. Goldstein G, Mackay IR. The thymus in systemic lupus erythematosus: a quantitative histopathological analysis and comparison with stress involution. *Br Med J*. 1967;2:475–478.
62. Palestro G, Tridente G, Micca FB, Novero D, Valente G, Godio L. Immunohistochemical and enzyme histochemical contributions to the problem concerning the role of the thymus in the pathogenesis of myasthenia gravis. *Virchows Arch B Cell Pathol Incl Mol Pathol*. 1983;44:173–186.
63. Shirai T, Miyata M, Nakase A, Itoh T. Lymphocyte subpopulation in neoplastic and non-neoplastic thymus and in blood of patients with myasthenia gravis. *Clin Exp Immunol*. 1976;26:118–123.
64. Isaacson PG, Norton AJ, Addis BJ. The human thymus contains a novel population of B lymphocytes. *Lancet*. 1987;2: 1488–1491.
65. Isaacson PG, Chan JK, Tang C, Addis BJ. Low-grade B-cell lymphoma of mucosa-associated lymphoid tissue arising in the thymus. A thymic lymphoma mimicking myoepithelial sialadenitis. *Am J Surg Pathol*. 1990;14:342–351.
66. Cardoso De Almeida P, Harris NH, Bhan AK. Characterization of immature sinus histiocytes (monocytoid cells) in reactive lymph nodes by use of monoclonal antibodies. *Hum Pathol*. 1984;15:330–335.
67. Duijvestijn AM, Schutte R, Köhler YG, Korn C, Hoefsmit EC. Characterization of the population of phagocytic cells in thymic cell suspensions. A morphological and cytochemical study. *Cell Tissue Res*. 1983;231:313–323.
68. Kaiserling E, Stein H, Muller-Hermelink HK. Interdigitating reticulum cells in the human thymus. *Cell Tissue Res*. 1974;155:47–55.
69. Ruco LP, Rosati S, Monardo F, Pescarmona E, Rendina EA, Baroni CD. Macrophages and interdigitating reticulum cells in normal thymus and in thymoma: an immunohistochemical study. *Histopathology*. 1989;14:37–45.
70. Lauriola L, Michetti F, Stolfi VM, Tallini G, Cocchia D. Detection by S-100 immunolabelling of interdigitating reticulum cells in human thymomas. *Virchows Arch B Cell Pathol Incl Mol Pathol*. 1984;45:187–195.
71. Bofill M, Janossy G, Willcox N, Chilosi M, Trejdosiewicz LK, Newsom-Davis J. Microenvironments in the normal thymus and the thymus in myasthenia gravis. *Am J Pathol*. 1985;119:462–473.
72. Siegal GP, Dehner LP, Rosai J. Histiocytosis X (Langerhans' cell granulomatosis) of the thymus. A clinicopathologic study of four childhood cases. *Am J Surg Pathol*. 1985;9:117–124.
73. Bhathal PS, Campbell PE. Eosinophil leucocytes in the child's thymus. *Australas Ann Med*. 1965;14:210–213.
74. Wise WS, Still WJ, Joshi VV. Severe combined immunodeficiency with thymic mast cell hyperplasia. *Arch Pathol Lab Med*. 1976;100:283–286.
75. Goldstein G. Plasma cells in the human thymus. *Aust J Exp Biol Med Sci*. 1966;44:695–699.
76. Henry K. The human thymus in disease with particular emphasis on thymitis and thymoma. In: Kendall MD, ed. *The Thymus Gland*. Academic Press; 1981:85–111.
77. Moll UM, Lane BL, Robert F, Geenen V, Legros JJ. The neuroendocrine thymus. Abundant occurrence of oxytocin-, vasopressin-, and neurophysin-like peptides in epithelial cells. *Histochemistry*. 1988;89:385–390.
78. Ciaccio C. Contributo all'istochimica delle cellule cromaffini. II. Cellule cromaffini del timo di gallum domesticus. *Bull Soc Ital Biol Sper*. 1942;17:619–620.
79. Håkanson R, Larsson LI, Sundler F. Peptide and amine producing endocrine-like cells in the chicken thymus. A chemical, histochemical and electron microscopic study. *Histochemistry*. 1974;39:25–34.
80. Vialli M, Casati C. Sulla presenza di cellule enterocromaffini nel timo dei rettili. *Riv Istochim Norm Patol*. 1958;4:343.
81. Vialli M. Elementi del sistema delle cellule enterocromaffinie cellule C nel timo. *Ann Histochem*. 1973;18:3–7.
82. Rosai J, Higa E. Mediastinal endocrine neoplasm, of probable thymic origin, related to carcinoid tumor. Clinicopathologic study of 8 cases. *Cancer*. 1972;29:1061–1074.
83. Wick MR, Rosai J. Neuroendocrine neoplasms of the thymus. *Pathol Res Pract*. 1988;183:188–199.
84. DrenckhahnD,vonGaudeckerB,Müller-HermelinkHK,Unsicker K, Gröschel-Stewart U. Myosin and actin containing cells in the human postnatal thymus. Ultrastructural and immunohistochemical findings in normal thymus and in myasthenia gravis. *Virchows Arch B Cell Pathol Incl Mol Pathol*. 1979;32:33–45.
85. Hayward AR. Myoid cells in the human fetal thymus. *J Pathol*. 1972;106:45–48.
86. Nakamura H, Lièvre CA-L. Neural crest and thymic myoid cells. *Curr Top Dev Biol*. 1986;20:111–115.
87. Dardenne M, Savino W, Bach JF. Thymomatous epithelial cells and skeletal muscle share a common epitope defined by a monoclonal antibody. *Am J Pathol*. 1987;126:194–198.
88. Van de Velde RL, Friedman NB. Thymic myoid cells and myasthenia gravis. *Am J Pathol*. 1970;59:347–368.
89. Judd RL, Welch SL. Myoid cell differentiation in true thymic hyperplasia and lymphoid hyperplasia. *Arch Pathol Lab Med*. 1988;112:1140–1144.
90. Murakami S, Shamoto M, Miura K, Takeuchi J. A thymic tumor with massive proliferation of myoid cells. *Acta Pathol Jpn*. 1984;34:1375–1383.
91. Rosai J, Parkash V, Reuter VE. On the origin of mediastinal germ cell tumors in men. *Int J Surg Pathol*. 1995;2:73–78.

92. Marshall AH, White RG. The immunological reactivity of the thymus. *Br J Exp Pathol.* 1961;42:379–385.
93. Raviola E, Karnovsky MJ. Evidence for a blood–thymus barrier using electron-opaque tracers. *J Exp Med.* 1972; 136:466–498.
94. Stet RJ, Wagenaar-Hilbers JP, Nieuwenhuis P. Thymus localization of monoclonal antibodies circumventing the blood-thymus barrier. *Scand J Immunol.* 1987;25:441–446.
95. Bearman RM, Levine GD, Bensch KG. The ultrastructure of the normal human thymus: a study of 36 cases. *Anat Rec.* 1978;190:755–781.
96. Hirokawa K. Electron microscopic observation of the human thymus of the fetus and the newborn. *Acta Pathol Jpn.* 1969;19:1–13.
97. Pinkel D. Ultrastructure of the human fetal thymus. *Am J Dis Child.* 1968;115:222–238.
98. van de Wijngaert FP, Kendall MD, Schuurman HJ, Rademakers LH, Kater L. Heterogeneity of epithelial cells in human thymus. An ultrastructural study. *Cell Tissue Res.* 1984;237:227–237.
99. de Maagd RA, MacKenzie WA, Schuurman HJ, et al. The human thymus microenvironment: heterogeneity detected by monoclonal anti-epithelial cell antibodies. *Immunology.* 1985;54:745–754.
100. Haynes BF. The human thymic microenvironment. *Adv Immunol.* 1984;36:87–142.
101. Janossy G, Thomas JA, Bollum FJ, et al. The human thymic microenvironment. An immunohistologic study. *J Immunol.* 1980;125:202–212.
102. Van Ewijk W. Immunohistology of lymphoid and non-lymphoid cells in the thymus in relation to T lymphocyte differentiation. *Am J Anat.* 1984;170:311–330.
103. McFarland EJ, Scearce RM, Haynes BF. The human thymic microenvironment: cortical thymic epithelium is an antigenically distinct region of the thymic microenvironment. *J Immunol.* 1984;133:1241–1249.
104. Eisenbarth GS, Shimizu K, Bowring MA, Wells S. Expression of receptors for tetanus toxin and monoclonal antibody A2B5 by pancreatic islet cells. *Proc Natl Acad Sci U S A.* 1982;79:5066–5070.
105. Haynes BF, Robert-Guroff M, Metzgar RS, et al. Monoclonal antibodies against human T cell leukemia virus p19 defines a human thymic epithelial antigen acquired during ontogeny. *J Exp Med.* 1983;157:907–920.
106. Savino W, Berrih S, Dardenne M. Thymic epithelial antigen, acquired during ontogeny and defined by the anti-p19 monoclonal antibody, is lost in thymomas. *Lab Invest.* 1984;51:292–296.
107. Bhan AK, Reinherz EL, Poppema S, McCluskey RT, Schlossman SF. Location of T cell and major histocompatibility complex antigens in the human thymus. *J Exp Med.* 1980;152:771–782.
108. Wiley EL, Nosal JM, Freeman RG. Immunohistochemical demonstration of H antigen, peanut agglutinin receptor, and Sophora japonica receptor expression in infant thymuses and thymic neoplasias. *Am J Clin Pathol.* 1990;93:44–48.
109. Le PT, Lazorick S, Whichard LP, Haynes BF, Singer KH. Regulation of cytokine production in the human thymus: epidermal growth factor and transforming growth factor alpha regulate mRNA levels of interleukin 1 alpha (IL-1 alpha), IL-1 beta, and IL-6 in human thymic epithelial cells at a post-transcriptional level. *J Exp Med.* 1991;174:1147–1157.
110. Chan WC, Zaatari GS, Tabei S, Bibb M, Brynes RK. Thymoma: an immunohistochemical study. *Am J Clin Pathol.* 1984;82:160–166.
111. Lobach DF, Scearce RM, Haynes BF. The human thymic microenvironment. Phenotypic characterization of Hassall's bodies with the use of monoclonal antibodies. *J Immunol.* 1985;134:250–257.
112. Chu PG, Weiss LM. Expression of cytokeratin 5/6 in epithelial neoplasms: an immunohistochemical study of 509 cases. *Mod Pathol.* 2002;15:6–10.
113. Dotto J, Pelosi G, Rosai J. Expression of p63 in thymomas and normal thymus. *Am J Clin Pathol.* 2007;127:415–420.
114. Walts AE, Hiroshima K, Marchevski AM. Desmoglein 3 and p40 immunoreactivity in neoplastic and nonneoplastic thymus: a potential adjunct to help resolve selected diagnostic and staging problems. *Ann Diagn Pathol.* 2015;19:216–220.
115. Hsu SM, Jaffe ES. Phenotypic expression of T lymphocytes in thymus and peripheral lymphoid tissues. *Am J Pathol.* 1985;121:69–78.
116. Janossy G, Bofill M, Trejdosiewicz LK, et al. Cellular differentiation of lymphoid subpopulations and their microenvironments in the human thymus. In: Muller-Hermelink HK, ed. *The Human Thymus: Histophysiology and Pathology.* Springer-Verlag; 1986:89–125.
117. Tidman N, Janossy C, Bodger M, Granger S, Kung PC, Goldstein G. Delineation of human thymocyte differentiation pathways utilizing double-staining techniques with monoclonal antibodies. *Clin Exp Immunol.* 1981;45:457–467.
118. Chilosi M, Iannucci AM, Pizzolo G, Menestrina F, Fiore-Donati L, Janossy G. Immunohistochemical analysis of thymoma. Evidence of medullary origin of epithelial cells. *Am J Surg Pathol.* 1984;8:309–318.
119. Mokhtar N, Hsu SM, Lad RP, Haynes BF, Jaffe ES. Thymoma: lymphoid and epithelial components mirror the phenotype of normal thymus. *Hum Pathol.* 1984;15:378–384.
120. Sato Y, Watanabe S, Mukai K, et al. An immunohistochemical study of thymic epithelial tumors. II. Lymphoid component. *Am J Surg Pathol.* 1986;10:862–870.
121. Lauriola L, Maggiano N, Marino M, Carbone A, Piantelli M, Musiani P. Human thymoma: immunologic characteristics of the lymphocytic component. *Cancer.* 1981;48:1992–1995.
122. van der Kwast TH, van Vliet E, Cristen E, van Ewijk W, van der Heul RO. An immunohistologic study of the epithelial and lymphoid components of six thymomas. *Hum Pathol.* 1985;16:1001–1008.
123. Eimoto T, Teshima K, Shirakusa T, et al. Heterogeneity of epithelial cells and reactive components in thymomas: an ultrastructural and immunohistochemical study. *Ultrastruct Pathol.* 1986;10:157–173.
124. Robertson PB, Neiman RS, Worapongoaiboon S, John K, Orazi A. 013 (CD99) positivity in hematologic proliferations correlated with TdT positivity. *Hum Pathol.* 1997;10:277–282.
125. Katzin WE, Fishleder AJ, Linden MD, Tubbs RR. Immunoglobulin and T-cell receptor genes in thymomas: genotypic evidence supporting the nonneoplastic nature of the lymphocytic component. *Hum Pathol.* 1988;19:323–328.
126. Swerdlow SH, Angermeier PA, Hartman AL. Intrathymic ontogeny of the T cell receptor associated CD3 (T3) antigen. *Lab Invest.* 1988;58:421–427.
127. Nikolic-Zugic J. Phenotypic and functional stages in the intrathymic development of alpha beta T cells. *Immunol Today.* 1991;12:65–70.

128. Bach JF, Dardenne M, Pleau JM, Rosa J. Biochemical characteristics of a serum thymic factor. *Nature.* 1976;266: 55–56.
129. McCarty N, Shinohara ML, Lu L, Cantor H. Detailed analysis of gene expression during development of T cell lineages in the thymus. *Proc Natl Acad Sci U S A.* 2004;101:9339–9344.
130. Goldstein AL, Low TLK, McAdoo M, et al. Thymosin alpha1: isolation and sequential analysis of an immunologically active thymic polypeptide. *Proc Natl Acad Sci U S A.* 1977;74:725–729.
131. Goldstein G. The isolation of thymopoietin (thymin). *Ann NY Acad Sci.* 1975;249:177–185.
132. Fabien N, Auger C, Monier JC. Immunolocalization of thymosin alpha 1, thymopoietin and thymulin in mouse thymic epithelial cells at different stages of culture: a light and electron microscopic study. *Immunology.* 1988;63:721–727.
133. Jambon B, Montague P, Bene MC, Brayer MP, Faure G, Duheille J. Immunohistologic localization of "facteur thymique serique" (FTS) in human thymic epithelium. *J Immunol.* 1981;127:2055–2059.
134. Savino W, Dardenne M. Thymic hormone-containing cells. VI. Immunohistologic evidence for the simultaneous presence of thymulin, thymopoietin and thymosin alpha 1 in normal and pathological human thymuses. *Eur J Immunol.* 1984;14:987–991.
135. Haynes BF, Warren RW, Buckley RH, et al. Demonstration of abnormalities in expression of thymic epithelial surface antigens in severe cellular immunodeficiency diseases. *J Immunol.* 1983;130:1182–1188.
136. Clark SL Jr. The thymus in mice of strain 129/J, studied with the electron microscope. *Am J Anat.* 1963;112:1–33.
137. Hirokawa K, Utsuyama M, Moriizumi E, Hashimoto T, Masaoka A, Goldstein AL. Immunohistochemical studies in human thymomas. Localization of thymosin and various cell markers. *Virchows Arch B Cell Pathol Incl Mol Pathol.* 1988;55:371–380.
138. Bodey B, Bodey B Jr, Siegel SE, Kaiser HE. Review of thymic hormones in cancer diagnosis and treatment. *Int J Immunopharmacol.* 2000;22:261–273.
139. Geenen V, Robert F, Defresne MP, Boniver J, Legros JJ, Franchimont P. Neuroendocrinology of the thymus. *Horm Res.* 1989;31:81–84.
140. Jevremovic M, Terzic M, Kartaljevic G, Popovic V, Rosic B, Filipovic S. The determination of immunoreactive beta-endorphin concentration in the human fetal and neonatal thymus. *Horm Metab Res.* 1991;23:623–624.
141. Batanero E, de Leeuw FE, Jansen GH, van Wichen DF, Huber J, Schuurman HJ. The neural and neuroendocrine components of the human thymus. II. Hormone immunoreactivity. *Brain Behav Immun.* 1992;6:249–264.
142. Le PT, Lazorick S, Whichard LP, et al. Human thymic epithelial cells produce IL-6, granulocyte-monocyte-CSF, and leukemia inhibitory factor. *J Immunol.* 1990;145:3310–3315.
143. Wainberg MA, Numazaki K, Destephano L, Wong I, Goldman H. Infection of human thymic epithelial cells by human cytomegalovirus and other viruses: effect on secretion of interleukin 1–like activity. *Clin Exp Immunol.* 1988;72:415–421.
144. Yarilin AA, Belyakov IM. Cytokines in the thymus: production and biological effects. *Curr Med Chem.* 2004;11:447–464.
145. Hirokawa K. Age-related changes of thymus. Morphological and functional aspects. *Acta Pathol Jpn.* 1978;28:843–857.
146. Simpson JG, Gray ES, Beck JS. Age involution in the normal adult thymus. *Clin Exp Immunol.* 1975;19:261–265.
147. Steinman GG. Changes in the human thymus during aging. In: Muller-Hermelink HK, ed. *The Human Thymus. Histophysiology and Pathology*. Springer-Verlag; 1986:43–88.
148. Steinmann GG, Klaus B, Muller-Hermelink HK. The involution of the aging human thymic epithelium is independent of puberty. A morphometric study. *Scand J Immunol.* 1985;22:563–575.
149. Smith SM, Ossa-Gomez LJ. A quantitative histologic comparison of the thymus in 100 healthy and diseased adults. *Am J Clin Pathol.* 1981;76:657–665.
150. Jamieson BD, Douek DC, Killian S, et al. Generation of functional thymocytes in the human adult. *Immunity.* 1999; 10:569–575.
151. Bertho JM, Demarquay C, Moulian N, Van Der Meeren A, Berrih-Aknin S, Gourmelon P. Phenotypic and immunohistological analyses of the human adult thymus: evidence for an active thymus during adult life. *Cell Immunol.* 1997;179:30–40.
152. Shanker A. Is thymus redundant after adulthood? *Immunol Lett.* 2004;15:79–86.
153. Selye H. Thymus and adrenals in the response of the organism to injuries and intoxications. *Br J Exp Pathol.* 1936; 17:234–248.
154. van Baarlen J, Schuurman HJ, Huber J. Acute thymus involution in infancy and childhood: a reliable marker for duration of acute illness. *Hum Pathol.* 1988;19:1155–1160.
155. Joshi VV, Oleske JM, Saad S, et al. Thymus biopsy in children with acquired immunodeficiency syndrome. *Arch Pathol Lab Med.* 1986;110:837–842.
156. Seemayer TA, Laroche AC, Russo P, et al. Precocious thymic involution manifest by epithelial injury in the acquired immune deficiency syndrome. *Hum Pathol.* 1984;15:469–474.
157. Grody WW, Fligiel S, Naeim F. Thymus involution in the acquired immunodeficiency syndrome. *Am J Clin Pathol.* 1985;84:85–95.
158. Schuurman HJ, Krone WJ, Broekhuizen R, et al. The thymus in acquired immune deficiency syndrome. Comparison with other types of immunodeficiency diseases, and presence of components of human immunodeficiency virus type 1. *Am J Pathol.* 1989;134:1329–1338.
159. Kendall MD, Johnson HR, Singh J. The weight of the human thymus gland at necropsy. *J Anat.* 1980;131(pt 3):485–497.
160. Lack EE. Thymic hyperplasia with massive enlargement: report of two cases with review of diagnostic criteria. *J Thorac Cardiovasc Surg.* 1981;81:741–746.
161. Katz SM, Chatten J, Bishop HC, Rosenblum H. Massive thymic enlargement. Report of a case of gross thymic hyperplasia in a child. *Am J Clin Pathol.* 1977;68:786–790.
162. Judd RL. Massive thymic hyperplasia with myoid cell differentiation. *Hum Pathol.* 1987;18:1180–1183.
163. Hammar JA. Uber gewicht, involution und persistenz der thymus im postfotalleben der menschen. *Arch Anat Physiol Anat Abt.* 1906;(suppl):91–182.
164. Suster D, Ronen N, Pierce DC, Suster S. Thymic parenchyma hyperplasia. *Mod Pathol.* 2023;36(8):100207.
165. Durkin W, Durant J. Benign mass lesions after therapy for Hodgkin's disease. *Arch Intern Med.* 1979;139:333–336.
166. Shin MS, Ho KJ. Diffuse thymic hyperplasia following chemotherapy for nodular sclerosing Hodgkin's disease. An immunologic rebound phenomenon? *Cancer.* 1983;51:30–33.

167. Carmosino L, DiBenedetto A, Feffer S. Thymic hyperplasia following successful chemotherapy. A report of two cases and review of the literature. *Cancer.* 1985;56:1526–1528.
168. Due W, Dieckmann KP, Stein H. Thymic hyperplasia following chemotherapy of a testicular germ cell tumor. Immunohistological evidence for a simple rebound phenomenon. *Cancer.* 1989;63:446–449.
169. Gelfand DW, Goldman AS, Law AJ, et al. Thymic hyperplasia in children recovering from thermal burns. *J Trauma.* 1972;12:813–817.
170. Caffey J, Silbey R. Regrowth and overgrowth of the thymus after atrophy induced by the oral administration of adrenocorticosteroids to human infants. *Pediatrics.* 1960;26:762–770.
171. Mackall CL, Fleisher TA, Brown MR, et al. Age, thymopoiesis, and $CD4^+$ T-lymphocyte regeneration after intensive chemotherapy. *N Engl J Med.* 1995;332:143–149.
172. Markert ML, Alvarez-McLeod AP, Sempowski GD, et al. Thymopoiesis in HIV-infected adults after highly active antiretroviral therapy. *AIDS Res Hum Retroviruses.* 2001;17:1635–1643.
173. Croxatto OC. Cordones epiteliales con aspecto endocrino observado en restos timicos del adulto. *Medicina (B Aires).* 1972;32:203–208.
174. Suster S, Moran CA. Malignant thymic neoplasms that may mimic benign conditions. *Semin Diagn Pathol.* 1995;12:98–104.
175. Suster S, Barbuto D, Carlson G, Rosai J. Multilocular thymic cysts with pseudoepitheliomatous hyperplasia. *Hum Pathol.* 1991;22:455–460.
176. Suster S, Rosai J. Cystic thymomas. A clinicopathologic study of ten cases. *Cancer.* 1992;69:92–97.
177. Moran CA, Suster S. Mediastinal seminomas with prominent cystic changes. A clinicopathologic study of 10 cases. *Am J Surg Pathol.* 1995;19:1047–1053.
178. Suster S, Rosai J. Thymic carcinoma. A clinicopathologic study of 60 cases. *Cancer.* 1991;67:1025–1032.
179. Moran CA, Suster S. Mucoepidermoid carcinomas of the thymus. A clinicopathologic study of six cases. *Am J Surg Pathol.* 1995;19:826–834.
180. Penn CR, Hope-Stone HF. The role of radiotherapy in the management of malignant thymoma. *Br J Surg.* 1972;59:533–539.
181. Goodwin RA, Nickell JA, Prez RMD. Mediastinal fibrosis complicating healed primary histoplasmosis and tuberculosis. *Medicine (Baltimore).* 1972;51:227–246.
182. Light AM. Idiopathic fibrosis of the mediastinum: a discussion of three cases and review of the literature. *J Clin Pathol.* 1978;31:78–88.
183. Sobrinho-Simoes MA, Vaz Saleiro JV, Wagenvoort CA. Mediastinal and hilar fibrosis. *Histopathology.* 1981;5:53–60.

19

Heart

Gerald J. Berry

With the introduction of the cardiopulmonary bypass technique into clinical practice over 60 years ago, pathologists found themselves examining cardiac specimens in the surgical pathology laboratory for the first time. For many years, the specimens consisted of excised valves, pericardial specimens, ventricular aneurysm excisions, and interventricular specimens. It was not until cardiac transplantation began in the late 1960s that surgical pathologists began handling entire heart specimens. Currently, more than 140,000 heart and combined heart–lung transplant procedures in adults have been recorded in the registries of the International Society of Heart and Lung Transplantation (1,2). Thoracic transplantation in the pediatric age group is now commonly performed, and close to 15,000 heart and combined heart–lung cases have been recorded (3,4). The explanted hearts from infants and children often have complicated congenital lesions that have undergone corrective surgical interventions. Surgical pathologists need to have expertise in the morphologic evaluation of end-stage heart disease of all types. Since the early 1970s, endomyocardial biopsies are routinely performed for the assessment of allograft rejection, infiltrative disorders, and inflammatory conditions.

Today, many institutions have ventricular assist devices (VADs) and total artificial heart programs, so ventricular apical specimens are routinely submitted. Like other biopsy specimens, these fall under the auspices of the surgical pathologist. Further, the introduction of complex electrophysiologic mapping techniques and imaging modalities such as transesophageal and 3D echocardiography, intravascular ultrasound, multidetector-row computed tomography (MDCT) imaging, coronary CT angiography (CCTA), and magnetic resonance imaging (MRI) have renewed interest in structural–functional correlations (5,6). Pathologists assume a critical role in the clinical evaluation and application of these technologies.

The purpose of this chapter is to review normal cardiac anatomy and histology. The histology of the great vessels is described elsewhere in this book. For the most part, adult histology is described, but where important differences exist, the histology in infants and children is also addressed. Furthermore, the purpose is not intended to be encyclopedic. Rather, we highlight those areas that are of practical importance to the surgical pathologist to enable the distinction between normal changes and subtle cardiac pathology. As each of the major anatomic divisions of the heart is described, relevant aspects of aging, gender changes, and/or applied anatomy are mentioned. We have also included recommendations for handling endomyocardial biopsy specimens and common artifactual alterations in this discussion.

HEART WEIGHTS

The weight of the normal adult heart is generally achieved by the end of the second decade of life. Heart weight in infants and children is related to age and body size, and tables are readily available (7,8). A variety of approaches to estimating the weight of the adult heart are used in clinical practice. These differ in complexity and accuracy; for example, the practical guidelines for unfixed adult heart weights published by Hudson in 1965 (9) propose that the male heart weighs 0.45% of the body weight (325 ± 75 g; average 300 g) and the female heart weighs 0.40% of the body weight (275 ± 75 g; average 250 g). Heart weight varies with age, sex, and body height and weight. Using this type of approach, the heart weight in young athletic adults may approach or exceed the upper limits of normal by up to 25%, whereas in the elderly it may approach or be slightly below the lower limits of normal.

Kitzman et al. (10) proposed a more comprehensive approach that predicts the normal heart weight of a formalin-fixed specimen based on body weight or height. In their study, body height as a predictor proved less accurate than body weight. The impact of formalin fixation on the weight of the heart has been addressed with conflicting findings. Some authors have found that fixation may increase heart weight by up to 5% after fixation, although considerable individual variation exists and some hearts actually decrease in weight (11). Descriptive tables of normal ventricular wall thickness, chamber size, valvular orifice measurements, and techniques for specimen handling and dissection are available in textbooks and postmortem dissection manuals (8,12–14).

PRENATAL FETAL CIRCULATION

It is beyond the scope of this chapter to discuss cardiac embryology. Developmental molecular biology has provided many new insights into cardiogenesis and the developmental genetics of congenital heart defects, some of which have radically altered previously held tenets (15–21). The human heart begins to beat spontaneously after the 2nd week of gestation and is fully formed by the 8th to 9th week. Well-oxygenated blood leaves the placenta *via* the umbilical vein, a portion of which passes through the hepatic sinusoids, and the remainder bypasses the liver by way of the ductus venosus to enter the inferior vena cava. Caval blood enters the right atrium together with an admixture of deoxygenated blood from the lower body. Approximately one-third of this blood is diverted through the interatrial septum (ostium secundum) into the left atrium, where it mixes with the small amount of deoxygenated blood returning from the lungs. The blood then passes into the left ventricle and exits into the ascending aorta to supply the coronary arteries, brain, and upper limbs. Approximately 50% of the blood is returned to the placenta for reoxygenation via the umbilical arteries. The remainder of the blood supplies the lower half of the body. Blood returning to the heart from the upper body reaches the right atrium through the superior vena cava. It mixes with the residual two-thirds of the blood from the IVC and enters the right ventricle. A small portion is distributed to the lungs, but the majority is diverted across the ductus arteriosus to the descending aorta. Blood is returned to the placenta with only a small amount reaching the lower body of the fetus.

POSTNATAL CIRCULATION

A series of anatomic and physiologic changes occur at birth, beginning with the infant's first respiratory efforts. Circulation of fetal blood through the placenta ceases and the lungs expand. As alveoli become aerated, the pulmonary vascular bed dilates causing an increase in pulmonary blood flow and abrupt decline in pulmonary vascular resistance. Active remodeling of the pulmonary vasculature with reduction of the muscular medial layer of arterioles and small arteries begins at this point and continues over a period of a few weeks. Increased blood return to the left atrium by way of the pulmonary veins results in slightly higher left atrial pressures compared with the right atrium. This causes the flap covering the foramen ovale to seal, creating an indentation called the fossa ovale. By the end of the 1st month after birth, the significant differences in ventricular hemodynamic load result in thickening of the left ventricular wall and thinning of the right ventricular wall. The ductus arteriosus, located between the left pulmonary artery and the aortic arch, functionally closes within 15 hours of birth and the umbilical veins begin to constrict.

PERICARDIUM

The pericardium surrounds the heart and forms a dual-layered, enclosed sac composed of the outer fibrous and inner serous layers. The fibrous parietal pericardium envelops the heart and is reflected off the ascending aorta, the pulmonary arterial trunk, the terminal 2 to 4 cm of the superior vena cava, the distal segment of the inferior vena cava, and the pulmonary veins. The pericardium serves a protective function as a physical barrier between the heart and diseases of the lung and mediastinum, has an immunologic role, and provides essential mechanical and geo-mechanical functions (22,23). The fibrous pericardium surrounding the heart normally remains unattached to the serous (visceral) pericardium except at the pericardial reflections. The parietal pericardium is composed predominantly of collagenous fibrous tissue, the fibrosa layer, intermixed with occasional elastic fibers, lymphatic vessels, and other small vessels (Fig. 19.1). Its inelastic tendencies account for the development of cardiac tamponade when it is acutely stretched with more than 200 mL of fluid. The parietal pericardium can contain variable amounts of adipose tissue, particularly toward the apex of the heart and is lined by a thin layer of mesothelial cells on its inner surface. The mesothelial cell layer is normally less than 1 mm in thickness, although it varies in thickness over different regions of the heart and great vessels. Both CT and MRI provide detailed resolution of the parietal pericardium and are routinely used to

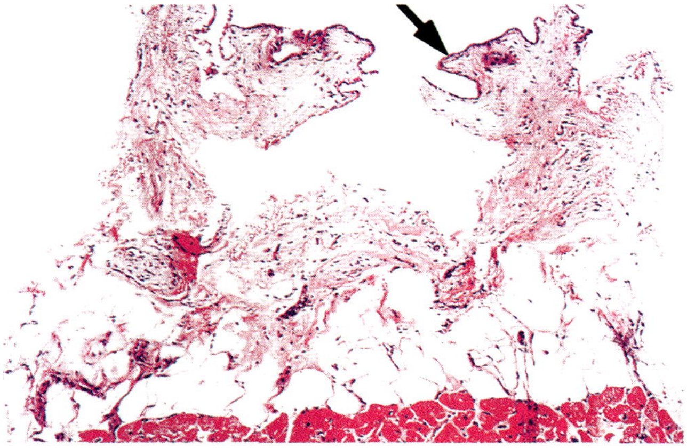

FIGURE 19.2 Section showing the epicardium lined by mesothelial cells (*arrow*) and covering the subepicardial adipose tissue and myocardium (H&E).

assess for developmental, posttraumatic, inflammatory, and neoplastic conditions (24,25). Heterotopic tissues within this layer can include thyroid, thymic, and hepatic elements. Neoplastic and hyperplastic alterations have been described, including intracardiac thyroid in the right ventricular outflow tract (26).

The serous or visceral pericardium is also called the epicardium of the heart. It is a single layer of mesothelium that envelops the heart and is in continuity with the fibrous pericardium at the pericardial reflections at the great vessels (Fig. 19.2). This delicate membrane contains a fibroconnective submesothelial layer and variable amounts of adipose tissue within which are embedded the coronary arteries and veins, lymphatic vessels, nerves, fibroblasts, and macrophages (Fig. 19.3). The normal epicardium often contains small aggregates of lymphocytes that are present from birth.

Between the two mesothelial layers of the parietal and visceral pericardium is a potential space containing up to 50 mL of straw-colored fluid that allows the mesothelium-lined surfaces to glide over one another. The histologic

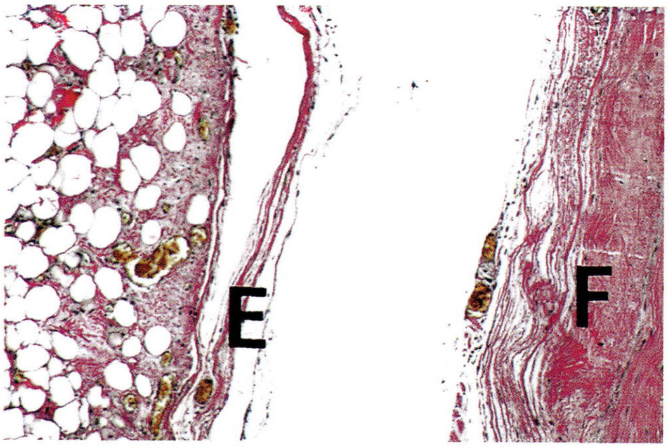

FIGURE 19.1 Section showing fibrous pericardium (*F*) separated from the thinner epicardium (*E*) (elastic van Gieson).

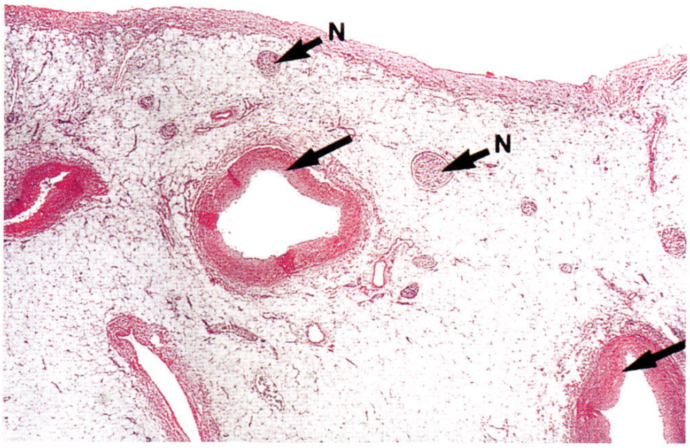

FIGURE 19.3 Section of the epicardium showing the relationship of coronary vessels (*arrows*) and nerves (*N*) to the epicardium (H&E).

composition of the two layers of the pericardium does not vary with age and is similar in infants, children, and adults.

Applied Anatomy

Awareness of the pericardial anatomy and its normal variants is important for cardiac surgeons and radiologists. Two sinuses, the transverse and oblique sinuses, and various recesses are found at the points of reflection of the parietal and the visceral pericardium near the great vessels and atria. The transverse sinus provides access to the left atrium and the mitral valve (27). Surgical pathologists may receive pericardiectomy specimens in cases of constrictive pericarditis, recurrent effusions, tamponade, or neoplasia. Nonspecific fibrosis and chronic inflammation are observed in constrictive pericarditis, but granulomatous infection and neoplastic deposits should be excluded. The pericardial surfaces of explanted hearts are often thickened by previous disease or surgical interventions, in which case the epicardium and parietal pericardium may be fused. Occasional nodular collections of epicardial fat may simulate the gross features of metastases. Congenital cysts of the pericardium are rare. Heterotopic thyroid and thymic tissue should be distinguished from cardiac involvement by contiguous spread or metastasis. Obese subjects may have excessive epicardial deposits of adipose tissue, in which case measurement of ventricular thickness should be carefully performed. The epicardial fat may undergo serous atrophy in conditions of cachexia, resulting in small gelatinous tissue tags (14). Data from a number of studies have shown a relationship between epicardial fat deposits and the severity of atrial fibrillation, metabolic syndrome, and coronary artery disease (28,29).

CARDIAC SKELETON

The cardiac skeleton at the base of the heart is the central supporting structure to which most of the fibers of the myocardium are attached and into which the atrioventricular and aortic valves are anchored. A separate mass of fibrous tissue termed the conus ligament joins the pulmonary ring to the aortic ring. It also serves to separate the atrial and ventricular chambers. The conus ligament is composed of the right and left fibrous trigones, the membranous part of the interventricular septum, and the fibrous annuli of the atrioventricular foramina (30). The fibrous skeleton is composed of layers of dense collagenous connective tissue admixed with small numbers of elastic fibers and, on occasion, small aggregates of adipose tissue. The fibrous tissue of the mitral and aortic valve rings is more substantial than that of the right-sided valves, and the right and left trigones create the direct mitral–aortic continuity (31). The right trigone and membranous septum form the central fibrous body within which the atrioventricular bundle is embedded. Bone and cartilage are not normally found in the fibrous skeleton of human hearts, although they have been described in some animal species.

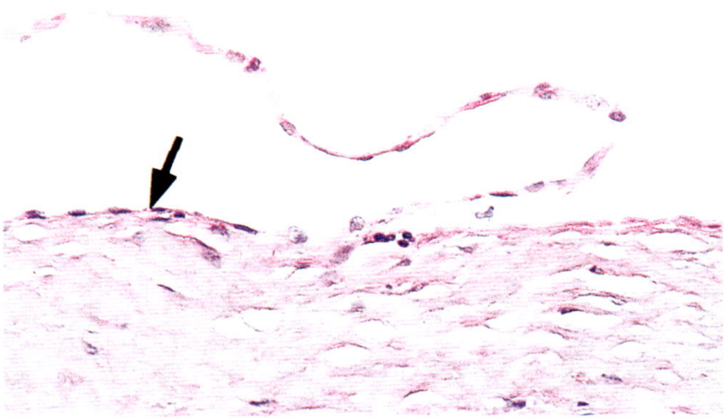

FIGURE 19.4 Section showing a single layer of endothelial cells covering the myocardium (*arrow*) (H&E).

Applied Anatomy

Portions of the cardiac skeleton may be seen in explanted valves. The close proximity of the fibrous skeleton to the conduction system and valvular leaflets is an important consideration for surgeons. Embryologic variations and surgical alterations can significantly affect the functional outcomes of all these components.

INTERNAL STRUCTURE OF THE HEART WALL

The wall of the heart in all chambers consists of three main layers: (a) the endocardium, (b) the intermediate or central muscular portion or myocardium, and (c) the external portion or epicardium. Most investigators believe that the endocardium is homologous with the tunica intima of blood vessels, the myocardium with the smooth muscle media, and the external epicardium with the adventitial layer. The endocardium consists of a single layer of endothelial cells (Figs. 19.4 and 19.5) with a

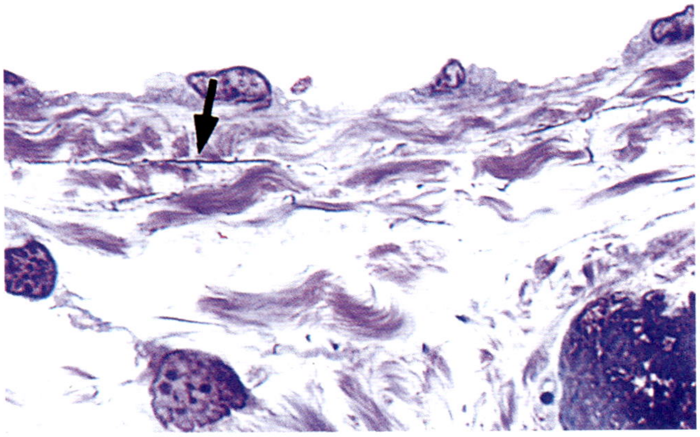

FIGURE 19.5 A 1-µm thick section (plastic embedded) of the endothelium covering the fenestrated elastic fibers (*arrow*) and collagen fibers of the subendocardium (toluidine blue).

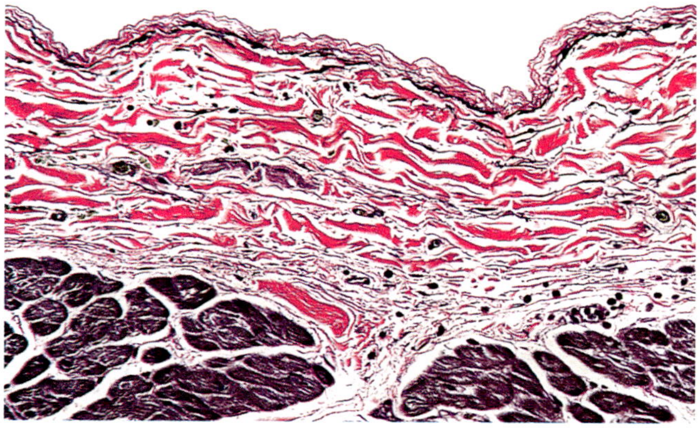

FIGURE 19.6 Section of the endocardium showing the distribution of the elastic fibers (*black*) and the collagen bundles (*red*), as well as vessels and nerves in the subendocardial layer.

subendothelial portion containing a loose elastic framework and collagen bundles, as well as nerves and delicate blood vessels (Fig. 19.6). The subendothelial layer in the atria is situated between the endothelial layer and the prominent elastic layer. A more detailed discussion of the myoarchitecture of the heart, its relationship to its supporting fibrous matrix, and the controversies surrounding the configuration of myocytes has been published (32). This has importance in ventricular remodeling during systolic heart failure, as dilatation results in alteration of the normal helical fiber orientation. For example, the oblique fibers transition to a transverse alignment contributing to diminished ejection fraction (33).

While the chambers of the heart are composed of these three layers, there can be variation in the composition and thickness of one or more components. A rudimentary layer of smooth muscle is often found in the endocardium of both the atria and the ventricles (Fig. 19.7).

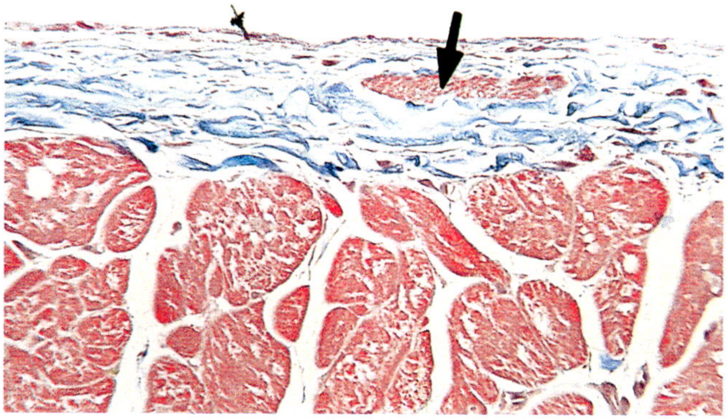

FIGURE 19.7 Section of the endocardium showing smooth muscle bundle (*arrow*) (Masson trichrome).

The endocardium of the atria is thicker than that of the ventricles. Small bundles of smooth muscle cells are often found but are of questionable functional significance. The myocardium is composed of bundles of myocytes separated by fibrous bands or fibrovascular structures (Fig. 19.8). The individual myocytes form a syncytium with end-to-end junctions called intercalated discs, and sometimes have side-to-side junctions. Individual myocytes contain a central ovoid nucleus with a clear zone at the poles (Fig. 19.9). The normal cardiac myocyte contains small amounts of lipofuscin granules in lysosomes; these increase with age and are found in large quantities in a variety of acquired myocardial disorders. The myocytes are filled with contractile myofibrils (actin and myosin). There is a striking difference between pediatric and adult myocardial biopsy specimens with respect to the cellularity at scanning magnification. The smaller amounts of sarcoplasm (the lower sarcoplasm to nuclear ratio) in the myocytes of infants and children produces a more "cellular" appearance (Fig. 19.10). Atrial myocytes contain atrial dense-core bodies that are demonstrable by electron microscopy (Fig. 19.11).

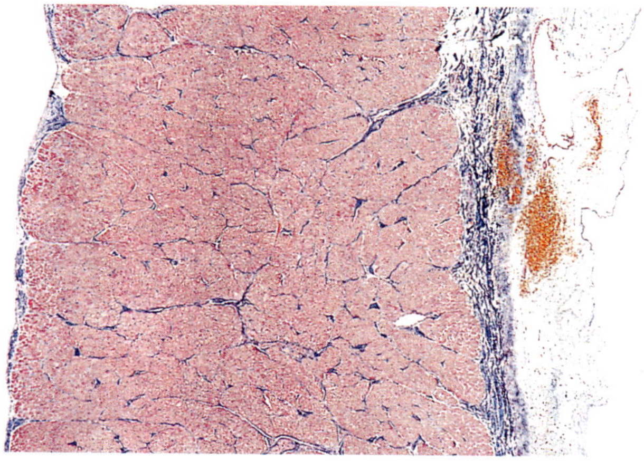

FIGURE 19.8 Section through the right ventricular wall showing the distribution of collagen from the endocardium (*left*) through the myocardium and to the epicardium (*right*) in the normal heart (Masson trichrome).

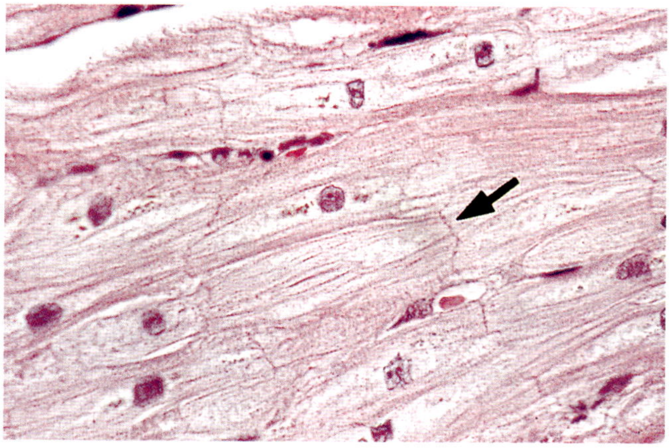

FIGURE 19.9 Section of myocardium showing ovoid nuclei and intercalated discs (*arrow*) (H&E).

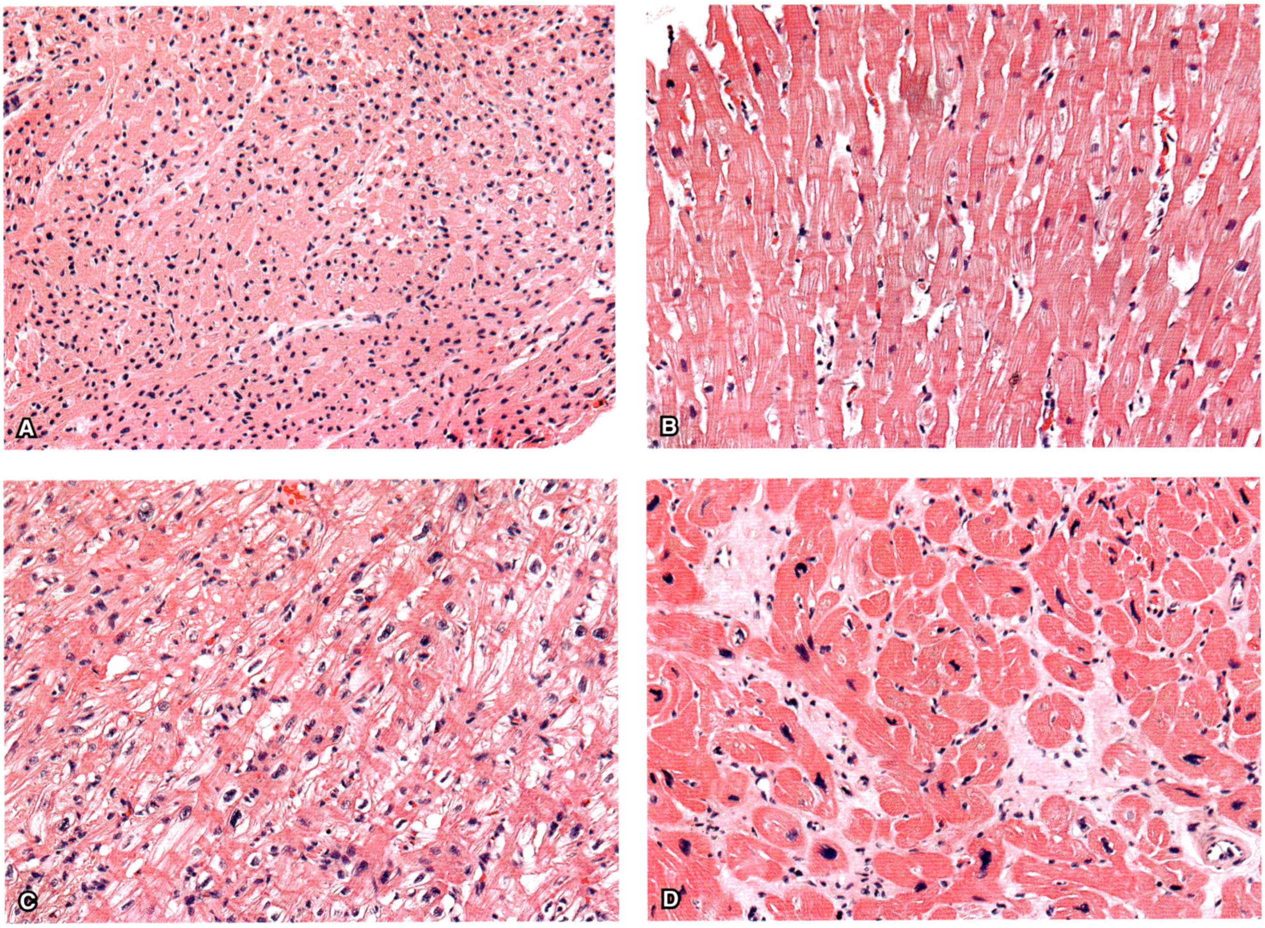

FIGURE 19.10 Pediatric and adult myocardium. Normal pediatric myocardium (**A**) compared to normal adult myocardium (**B**) at the same magnification. **C:** Right ventricular biopsy from an infant with idiopathic dilated cardiomyopathy (IDCM) compared to adult biopsy with IDCM (**D**). Note: The same magnification is utilized for all four panels.

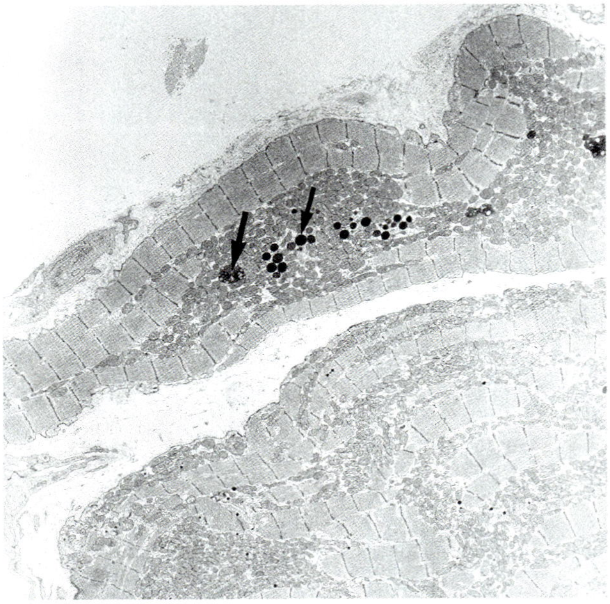

FIGURE 19.11 Electron micrograph of atrial dense-core granules (*small arrow*) that can be differentiated from the lipofuscin granules (*large arrow*).

INTERATRIAL SEPTUM

The interatrial septum separates the right and left atrial chambers. This separation is normally complete, although the foramen ovale may retain a small communication at its floor or along its superior border where the septum has not fused (i.e., patent foramen ovale). This is reported in up to one-third of adults and is usually without clinical complications (34,35). The foramen ovale is closed by the septum to form the circular, dime-sized oval fossa or fossa ovalis. This represents the true interatrial septum, with the remainder of the anatomic structure comprising caval or pulmonary venous infoldings and the muscular and membranous components of the atrioventricular region (31). The membrane covering the oval fossa is a paper-thin, translucent layer early in life that becomes progressively fibrotic and opaque with age, ultimately measuring up to 2 mm in thickness. The histology of the septum displays variable amounts of atrial muscle, fibrous tissue, mature adipose tissue, and, on occasion, brown fat

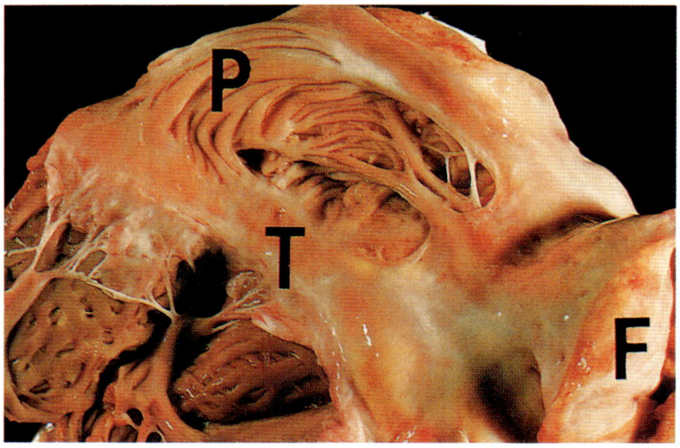

FIGURE 19.12 The right atrium showing the pectinate muscles (*P*), the crista terminalis (*T*), and the interatrial septal fat (*F*).

(Fig. 19.12). The thickness of the atrial septum varies considerably.

Applied Anatomy

Lipomatous hypertrophy of the interatrial septum is characterized by accumulations of adipose tissue within the U-shaped lip, or limbus, of the oval fossa. It may produce bulging of the interatrial septum. In some patients, the patent foramen ovale can cause interatrial shunting and paradoxical embolism.

RIGHT ATRIUM

Venous return to the heart occurs via the inferior and superior vena cavae into the right atrium. The right atrium forms the right lateral cardiac border and is located right and anterior to the left atrium. The right atrial appendage protrudes anteromedially from the right atrium and overlaps the aortic root (31). The atrium is composed of the auricular appendage, a smooth-walled venous sinus that contains the openings of the cavae and coronary sinus, the septal component, and the portion near the opening of the tricuspid valve (36,37). The terminal crest is a prominent muscular ridge extending between the right sides of the openings of both cavae that underlies the sulcus terminalis. The broad-based triangular appendage defines the morphologic right atrium. The interior surface of this appendage is trabeculated due to the muscular bands of the pectinate muscles. The portion of the right atrium lateral to the terminal crest is smooth walled and is derived embryologically from the sinus venosus (Fig. 19.12). The right atrial wall measures 2 mm in thickness.

The orifice of the superior vena cava is valveless while the opening of the inferior vena cava has an inconstant, rudimentary valve called the Eustachian valve. It forms a crescentic fold over the anterior border of the IVC orifice and serves to direct blood toward the foramen ovale and into the left atrium during fetal life. The opening of the coronary sinus has a rudimentary crescent-shaped flap of tissue called the thebesian valve; these valves vary in size and may be fenestrated or absent. The Chiari network usually appears as a lacelike veil of tissue and represents the remnant of the right sinus venosus valve; it is occasionally found in normal hearts (Fig. 19.13). A fascinating review of the historical figures that have produced eponyms in cardiac anatomy is noteworthy (38). The right atrial chamber is covered by endocardium. Elastic fibers in the subendocardium pass into a typical fenestrated elastic membrane that contains blood vessels, nerves, and branches of the conducting system. The wall is so thin in the spaces between the muscle bundles of the atria, that the connective tissue of the endocardium blends with the epicardium. The epicardial surface of the atrium is rich in nerves and ganglia (Fig. 19.14).

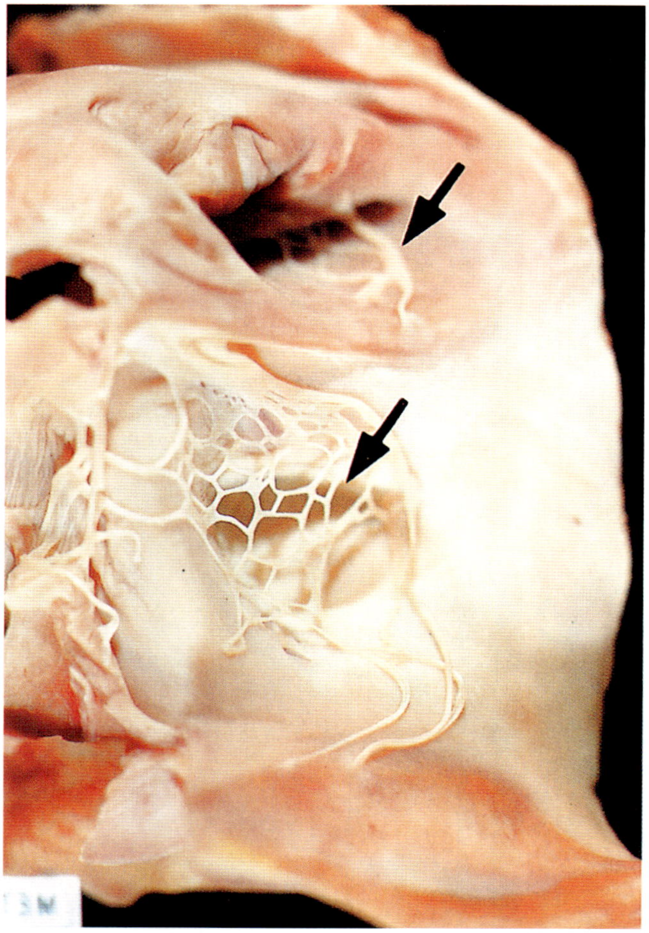

FIGURE 19.13 Picture of Chiari network (*arrows*), or lacelike pattern of the remnants of the thebesian valve, in the IVC opening into the right atrium.

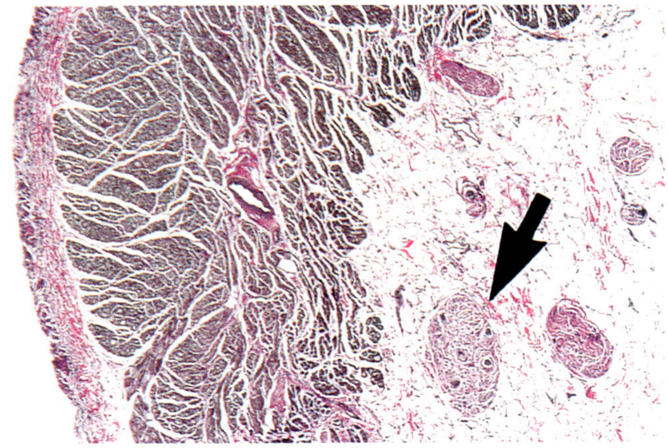

FIGURE 19.14 Section through the right atrium showing the nerves and ganglion cells (*arrow*) in the overlying subepicardial adipose tissue (elastic van Gieson).

LEFT ATRIUM

Like the right atrium, this chamber consists of an appendage, a venous component containing the ostia of the four pulmonary veins, a septal component, and a region near the mitral valve orifice. On average, the wall of the left atrium is 3 mm thick. The septal surface is smooth. The four pulmonary veins open into the left atrium on the posterior wall. The atrial appendage has a narrow, angulated tubular shape resembling a hockey stick, although there can be some variation in appearance. It is lined by pectinate muscles and lacks a terminal crest at its interface with the venous portion of the atrium (39). The endocardial layer is thicker and more opaque than that of the right atrium. This is due, in part, to the higher pressures of the pulmonary veins emptying into the atrium. Sometimes a patch of thickened rough endocardium is seen on the posterior wall above the mitral valve as a result of mitral regurgitation. This is called MacCallum patch but is not usually seen in the normal heart. The atrial septum is smooth on the left side with the exception of a central shallow depression corresponding to the oval fossa. The endocardium of the left atrium is thicker as a result of increased collagen layers, particularly near the openings of the pulmonary veins. The two left pulmonary veins enter on the posterolateral aspect, and the right veins enter on the posteromedial side; there are no true valves at the venous–atrial junction. Small collections of atrial muscle can be found within the walls of the pulmonary veins at this junction and may act as physiologic valves.

The composition of the left atrium is similar to that of the right atrium with two main differences (40). First, the myocytes of the left atrium are arranged in a more irregular fashion than those of the right atrium and second, sleeves of myocardium are in continuity between the roof of the left atrium and the distal pulmonary veins at their interface. These pulmonary venous sleeves are sites of development of atrial fibrillation and targets for ablation techniques. The left atrium shares a number of important anatomic relationships with adjacent structures including the pulmonary artery superiorly, the left phrenic nerve, the left superior pulmonary vein posteriorly, and the mitral valve inferiorly (41).

RIGHT VENTRICLE

Recently, the right ventricle, both in healthy and diseased states, has become the subject of intensive investigation, particularly with respect to techniques for morphologic and functional assessment and targeted therapeutic interventions (42,43). The morphologic right ventricle lies anterior to the other heart chambers. It has an inflow portion (i.e., sinus), an apical trabecular component, and an outflow portion, or conus, that has also been termed the infundibulum (44). The outflow tract is separated anatomically from the inflow tract by a muscular arch called the crista supraventricularis. The endocardial surface is coarsely trabeculated, particularly in the apical region.

The trabeculations form the inner two-thirds of the ventricular wall. The septal surface is deeply trabeculated with coarse trabeculae carneae and a thick muscular column called the moderator band (Fig. 19.15). This moderator band is found in most hearts and connects the distal portion of the septum to the free wall. It terminates in the region of the anterior papillary muscle (45). Myocyte disarray is a common finding within trabecular muscles and should not be confused with hypertrophic cardiomyopathy. The right bundle branch travels through the muscular ventricular septum and courses down to end in the moderator muscle. The papillary muscles of the right ventricle are relatively constant, with an anterior

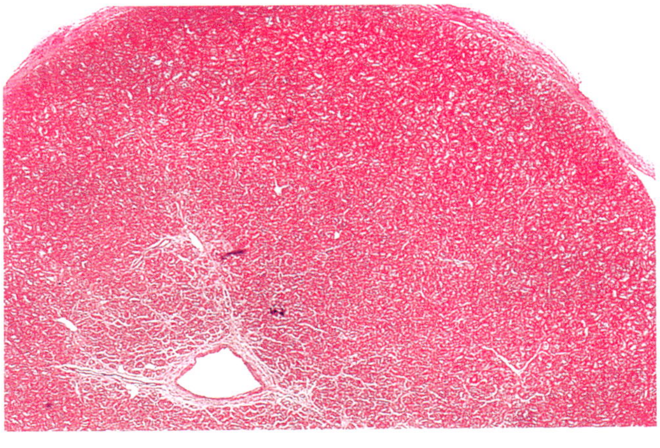

FIGURE 19.15 A section of trabeculae carneae from the normal right ventricle.

papillary muscle located on the anterior wall near its junction with the septum and a small posterior papillary muscle arising under the crista supraventricularis at the inferior border of the right ventricular outflow tract. In addition, there is an inconstant group of posterior papillary muscles that can arise from the diaphragmatic wall of the right ventricle.

The right ventricle contains a thin endocardial layer; the thickness of myocardial wall in normal human adults is approximately one-third of the thickness of the left ventricle, measuring up to 5 mm. The endocardium is similar to that of the other chambers, although it displays more variability from region to region with the thickest areas found in the septum. The subendocardial space includes a fenestrated elastic membrane and, on occasion, bundles of smooth muscle, particularly in the interventricular septum (Fig. 19.8). The interventricular septum also contains blood vessels, nerves, as well as the left bundle branch of the conduction system. The ventricular free wall has numerous vascular channels consisting of intertrabecular channels that lead into myocardial sinusoids and thebesian veins. Myocardial sinusoids are also found within the trabeculae. Arterioluminal vessels lead directly from the systemic coronary circulation into the capillary beds, and empty into the myocardial sinusoids. The myocardium is richly supplied with small vascular channels that form an intramural circulation (46). There is an extensive web of capillaries that course among the cardiac muscle fibers; these are fed by branches of the coronary arteries and are drained, in part, by the coronary veins. They are also directed to the intramyocardial sinusoids and then into the lumen of the heart. Richly anastomosing networks of irregular channels, the myocardial sinusoids, and a capillary bed are present deep within the myocardial musculature. These sinusoids receive vessels from the coronary arteries and the capillaries and communicate with coronary veins. The connections between the coronary arteries and the cardiac chambers are called arterioluminal vessels (Fig. 19.16). Variable amounts of adipose tissue can be found within the myocardium, particularly in the outer half of the free wall. When extensive, it is called fatty infiltration of the right ventricle and represents a metaplastic change. It should not be confused with arrhythmogenic cardiomyopathy.

Applied Anatomy

The importance of recognizing the presence of myocyte disarray and adipose tissue as normal findings within the trabeculated musculature of the right ventricle of the normal adult heart has been discussed. The distinction of "physiologic adipose tissue" from arrhythmogenic cardiomyopathy is problematic and rests upon a combination of clinical, molecular, and morphologic criteria (47–49). Currently, CT and MRI are being evaluated to aid in this distinction (50). The apical trabecular region is also the site of pacemaker wire placement and endomyocardial biopsy sampling.

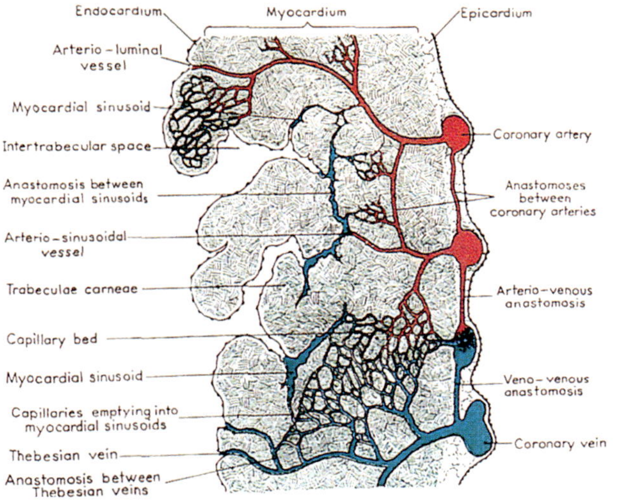

FIGURE 19.16 Diagrammatic representation of the various intramural vascular channels. (From Barry A, Patten B. The structure of the adult heart. In: Gould SE, ed. *Pathology of the Heart and Blood Vessels*. Charles C Thomas; 1968:104–105; Courtesy of Charles C Thomas Publisher, Ltd., Springfield Illinois.)

LEFT VENTRICLE

The left ventricle receives blood from the left atrium during ventricular diastole and ejects blood into the systemic arterial circulation across the aortic valve during ventricular systole. The left ventricle is somewhat bullet- or triangular-shaped, with the blunt tip directed anteriorly and inferiorly to the left (51). Like the right ventricle, it is composed of inflow, septal or apical, and outflow components. It lacks the moderator band and septomarginal trabeculations of the right ventricle (36). The left ventricular chamber is surrounded by a muscular wall measuring up to 15 mm in thickness; the papillary muscles are not included in the measurement. The medial wall of the left ventricle is the interventricular septum, which is shared with the right ventricle. The septum is roughly triangular in shape with the base of the triangle at the level of the aortic cusps. It is entirely muscular except for the small membranous septum located just below the right coronary and posterior cusps. The upper one-third of the septum is the outflow tract and is lined by smooth endocardium. The inferior two-thirds of the septum and the remaining ventricular walls are composed of the crisscrossing trabeculae carneae, which are thinner and less prominent than the trabeculae in the right ventricle. The free wall of the left ventricle is the portion that is exclusive of the septum.

The left ventricle is histologically similar to the right ventricle, although its endocardium is slightly thicker as a result of higher hemodynamic pressures. The small arterioles adjacent to the endocardium have slightly thicker walls than those of the right ventricle, likely on account of the higher pressure in the left ventricle. The myocardium of the left ventricle is arranged in such a way that it appears to spiral inward from the superficial layers. The superficial layers run at right angles to the layers deeper within the wall. The layers are intimately interdigitated to prevent dissection into lamina structures. The attachment of the muscular layers is from the fibrous skeleton at the base of the heart. The deeper muscle layers of the interventricular septum are composed of the deep bulbospiral muscle in the center of the septum that originate from the septal portion of the atrioventricular annulus. The fascicles of the deep bulbospiral and sinospiral muscles interdigitate within the muscular interventricular septum. The spiral muscle pulls the base of the ventricle toward the apex during systolic contraction. The blood supply is similar to that of the right ventricle. In cut sections, the left ventricle displays three layers of myocardium that represent planes of orientation rather than distinct subdivisions of myocardium (52).

Applied Anatomy

Portions of the left ventricular subaortic outflow tract may be submitted from myomectomy procedures in patients with asymmetric types of hypertrophic cardiomyopathy. Portions of the left ventricular free wall may be removed during aneurysm resection or VAD placement. The intermediate bulbospiral muscle of the interventricular septum is primarily involved in idiopathic hypertrophic subaortic stenosis. Since it lies deep within the septum, the disarray associated with idiopathic hypertrophic subaortic stenosis may not always be seen in the superficial myomectomy specimen. The mechanisms of myocyte replacement, senescence, and apoptosis in the normal aging heart and the role of human cardiac stem cells are currently under investigation (53).

CARDIAC VALVES

Semilunar Valves

The semilunar valves include the pulmonic and aortic valves. The normal circumference of the aortic valve is 5.7 to 7.9 cm for women and 6.0 to 8.5 cm for men. The normal pulmonic valve circumference is 5.7 to 7.4 cm for women and 9.2 to 9.9 cm for men (54). Interestingly, when adjusted for body surface area, the valves in women are slightly larger than those in men (10). Each semilunar valve consists of three semicircular cusps (55), each of which is attached by its semicircular border, or annulus, to the aortic or pulmonic ring. The three points of lateral attachment of adjacent cusps are the commissures (Fig. 19.17). The lines of cusp apposition are not at the free margin but are angulated lines extending from well below the point of attachment in the commissure to just below the midpoint of the free edge. These lines (i.e., linea alba) and the central nodules (i.e., noduli arantii) can be seen in the aortic valve (55) and are often the site of attachment of Lambl excrescences. These landmarks are less obvious in the pulmonic valve because of the lower right-sided pressures. The lunulae are thin, delicate areas of the cusp between the linea alba and the free edge.

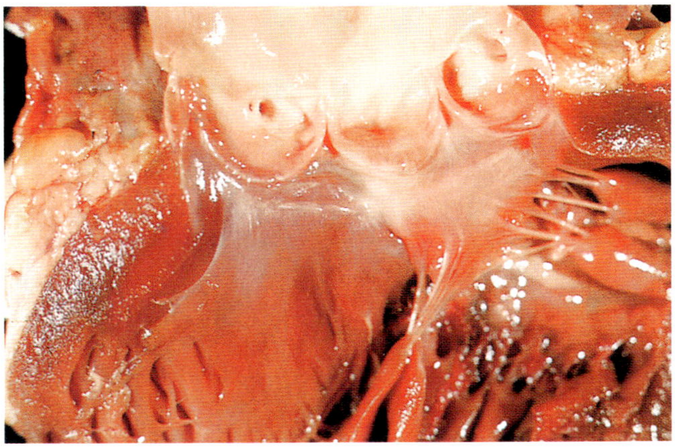

FIGURE 19.17 Aortic valve from a normal heart showing the cusps and the coronary ostia.

The semilunar aortic and pulmonic valves are similar in configuration, except that the aortic cusps are slightly thicker and contain coronary ostia in the aortic root. They are situated at the outflow tract summits of their corresponding ventricles. The pulmonic valve is anterior, superior, and slightly to the left of the aortic valve in the normal heart. The cusps, which are often slightly unequal in width, circle the inside of the pulmonary artery trunk or aortic root. The vessel wall bulges outward behind each cusp to create a pouch-like dilatation known as the sinus of Valsalva. The portion of the cusp adjacent to the rim is thin and may normally contain small perforations. The noduli arantii meet in the center and contribute to the support of the leaflets. Since the plane of the aortic valve is oblique such that the right posterior side is lower than the left anterior side, the origin of the left coronary artery is slightly superior to that of the right coronary artery. The coronary artery ostia are located in the upper one-third of their respective sinuses and measure from 3 to 4 mm in diameter; the left ostium is slightly wider than the right ostium. The right coronary artery passes anteriorly and to the left. In some hearts, there is a separate ostium in the right coronary sinus for the conus artery, sometimes called the third coronary artery (51).

The molecular mechanisms of valvular development during embryogenesis have been clarified in recent studies (56). The histology of the semilunar valves is that of a well-defined multilayered structure. Three distinct layers are recognizable: the fibrosa, the spongiosa, and the ventricularis. The fibrosa is a layer of dense collagen that constitutes the major structural component of the cusp and extends to its

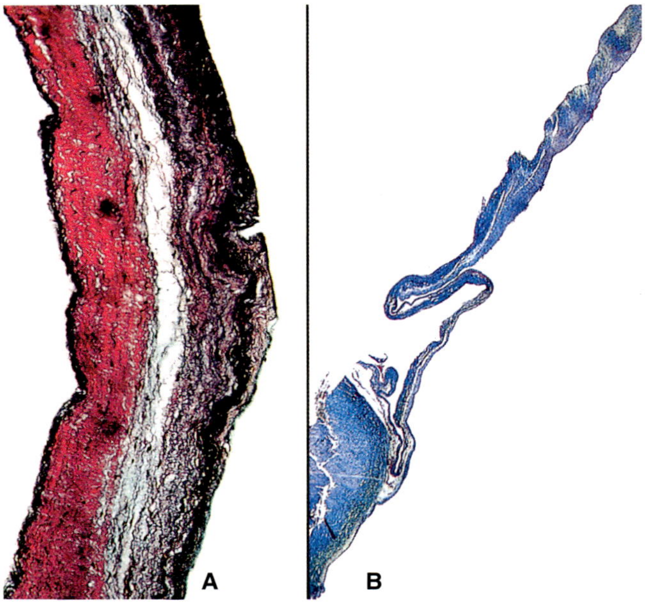

FIGURE 19.18 High- (**A**) and low-power (**B**) views of a section of the aortic cusp showing the three distinct layers as described in the text.

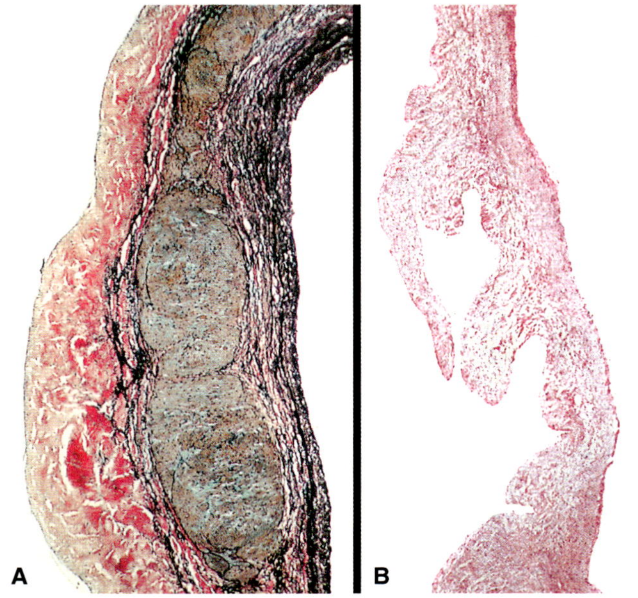

FIGURE 19.19 High- (**A**) and low-power (**B**) views of a section of the mitral leaflet showing the three distinct layers with muscle in the central portion near the base of the valve.

free edge (52). The densely packed collagen bundles blend into the collagen of the valvular ring in the region of the commissures. Some fibroblasts are present in this layer, as are some very fine elastic fibers. The spongiosa is adjacent to the fibrosa and occupies a central position in the thickness of the cusp (Fig. 19.18). It is best developed in the basal third of the cusp. It does not extend to the free edge, which is composed only of the fibrosa and ventricularis layers. The spongiosa is composed of large amounts of proteoglycans and glycosaminoglycans, loosely arranged collagen fibrils, scattered fibroblasts, and mesenchymal cells; it serves as a "shock absorber" for the valve (57). The ventricularis is closest to the ventricular surface. It is adjacent to the spongiosa and is in direct contact with the endothelial layer of the cusp inflow surface. It is distinguished from the fibrosa by its abundance of elastic fibers. This feature is often helpful in orienting and identifying the layers of excised aortic valvular cusps. The surface lining of the cusps consists of a single layer of endothelial cells. A superficial accumulation of elastic fibers called the atrialis is present at the base of the cusp; some authors consider it to represent a fourth layer in the semilunar valve apparatus. The surface topography of the aortic cusp varies according to its state of stress. The bundles are wavy and the inflow surface is smooth in the stressed state, whereas it is rougher in the relaxed state (57).

Atrioventricular Valves

The atrioventricular valves include the mitral valve and the tricuspid valve. The normal circumference of the tricuspid valve is 10 to 11.1 cm in women and 11.2 to 11.8 cm in men. The normal circumference of the mitral valve is 8.2 to 9.1 cm in women and 9.2 to 9.9 cm in men (54). The valvular apparatus is made up of the annulus, commissures, leaflets, chordae tendineae, and papillary muscles. The annulus is composed of a ring of circumferentially oriented collagen and elastic fibers with extensions into the ventricle and the atrium. Currently, a number of different descriptors are used in clinical practice to distinguish the two leaflets of the mitral valve (58). The smaller but broader leaflet is designated as the anterior or aortic leaflet and comprises a third of the circumference. It is arbitrarily divided into three segments (A1, A2, A3). The posterior, or mural, leaflet is likewise separated into three segments, beginning near the anterolateral commissure (P1) and ending near the posteromedial commissure (P3). The central segment (P2) can vary considerably in size (59).

The atrioventricular valves have four histologic layers: atrialis/auricularis, spongiosa, fibrosa, and ventricularis (Fig. 19.19). As in the semilunar valves, all layers contain valve interstitial cells that are essential for extracellular matrix remodeling and maintaining structural integrity (60). There is an abundance of collagen fibers and the different types include type I (74%), type III (24%), and type V (2%) (54). The collagen bundles of the annulus spread down into the majority of the cusps of the mitral valve (except at the free edge of the leaflet) and are known as the fibrosa. They continue into the chordae tendineae and finally spread out into a network that covers the tip of the papillary muscles (Fig. 19.20). The ventricularis is adjacent to the fibrosa layer on the ventricular side of the valve. It contains many elastic and collagen fibers and is covered by endothelium. Some of the elastic fibers extend into the chordae tendineae, but in general, this layer is incomplete and does not extend to the free edge of the leaflet. The spongiosa is situated on the atrial side of the fibrosa layer and contains a rich matrix of proteoglycans and glycosaminoglycans with a few elastic fibers,

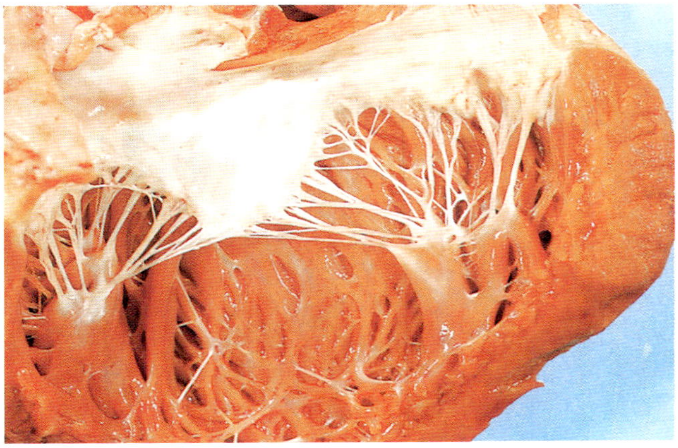

FIGURE 19.20 A normal mitral valve with chordae tendineae inserting into the papillary muscles.

collagen fibrils, and connective tissue cells, including interstitial fibroblasts (57,59). This layer and the fibrosa extend throughout the entire length of the leaflet. The spongiosa in the anterior and posterior mitral leaflets can contain cardiac myocytes that represent a direct extension of the left atrial myocardium but are not in continuity (Fig. 19.19). This layer extends into the middle third of the anterior mitral leaflet, whereas it is limited to the proximal third of the posterior leaflet. Neural elements and lymphatic vessels can be found in the leaflets. The atrial aspect of the spongiosa (i.e., the layer closest to the atrium) is covered by the atrialis or auricularis, which in turn has a continuous endothelial lining. The auricularis contains collagen, elastic fibers, and smooth muscle cells. It is prominent near the annulus but thins out in the distal third of the leaflet such that the most distal aspect of the valves is composed only of spongiosa and fibrosa. The endothelial cells on the atrial aspect of the atrioventricular valves are plump with irregular nuclei compared with the flatter endothelial cells on the ventricular aspect.

The architecture of the tricuspid valve apparatus and the layered arrangements of its leaflets are similar to those in the mitral valve; however, the individual layers are thinner in the tricuspid valve. Cardiac muscle bundles insert fairly low into the base of the tricuspid leaflets but do not extend into the leaflet substance. In the posterior and septal leaflets, the auricularis is thicker and contains more abundant smooth muscle cells.

Chordae Tendineae

The chordae tendineae in the normal state are thin fibrous cords that emanate in a fan-like manner from the broad leaflets of the atrioventricular valves and insert into the papillary muscles (Fig. 19.20). There can be variation in chordae dimensions, especially between the primary and secondary cords. The central cores of the chordae are composed of longitudinally oriented collagen fibrils. The core is surrounded more peripherally by loosely arranged collagen fibrils and elastic fibers and is embedded in a proteoglycan-rich matrix (Fig. 19.21) (57,59). Some chordae contain a small central

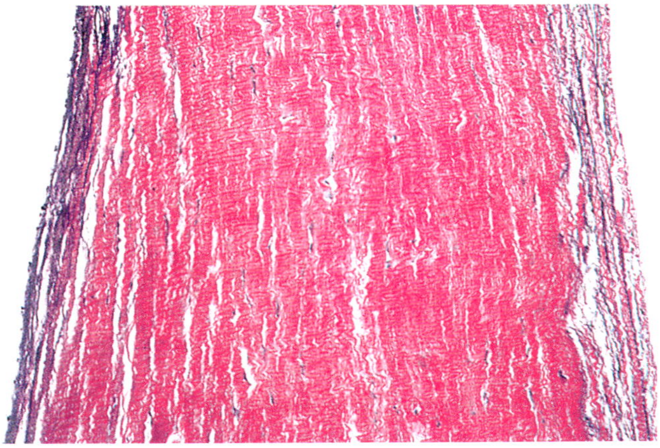

FIGURE 19.21 Longitudinal section of chordae tendineae showing the relationships of elastic fibers on the surface and collagen in the center (elastic van Gieson).

core of muscle known as chordae muscularis (Fig. 19.22), although others contain blood vessels and collagen in variable amounts and appear fleshier in color (Fig. 19.23). The endothelial cells on the chordae resemble the flattened ovoid nuclei of the ventricular surface of the leaflets.

Applied Anatomy of Intracardiac Valves

Semilunar and atrioventricular valves are frequently encountered in the surgical pathology laboratory. Indications for valvular replacement or repair include a variety of congenital, infectious, inflammatory, degenerative, and paraneoplastic diseases. In many cases, the chordae and portions of the papillary muscle may be attached. In the setting of myocardial infarction, infection, or valvular prolapse, ruptured papillary muscles may become surgical specimens. Lambl excrescences and fibrous nodules are papillary projections along the lines of closure or free edge of the valve, respectively.

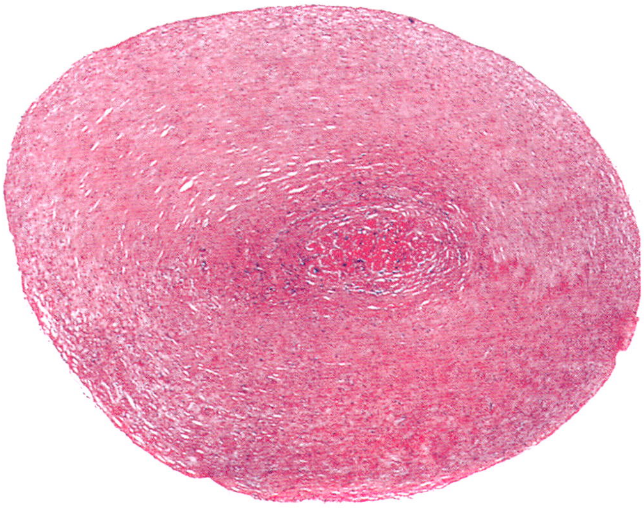

FIGURE 19.22 Transverse section of chordae tendineae showing central core of muscle (H&E).

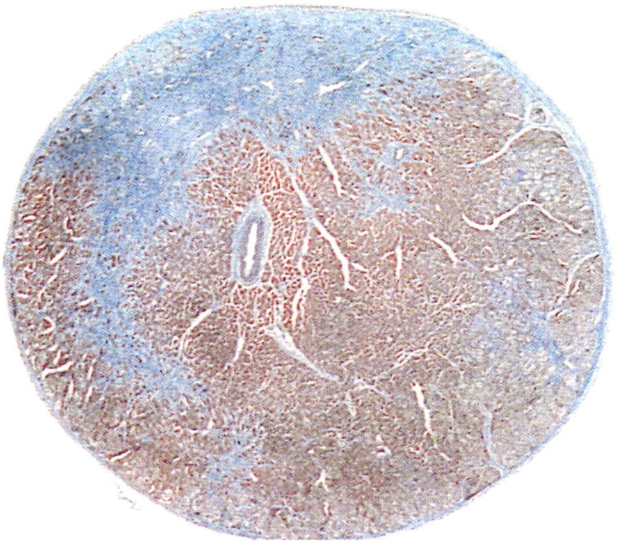

FIGURE 19.23 Transverse section of a muscular chorda showing fibrous tissue, muscle, and small vessel within the chorda (Masson trichrome).

Aging Changes of Intracardiac Valves

All of the cardiac valves become thicker, opaque, and less pliable with age, particularly the left-sided valves. Increased collagen content may account for the loss of plasticity, and calcifications or lipid accumulations may be present; these changes have been collectively termed aortic valve sclerosis. Small windows or fenestrations may be found above the lines of closure, but these do not affect valvular function. The anterior leaflet of the mitral valve often shows yellow atheromatous alteration or atheromatosis due to lipid deposition, while the posterior leaflet may become more opaque. Myxomatous degeneration is more common in the mitral valve leaflets of the elderly. The mitral valve annulus may become calcified with age. Age-related changes in the pulmonic valve can include opacification of the leaflets, central nodular thickening along the edge of cusps (i.e., nodules of Morgagni), Lambl excrescences, and fenestrations. Myxomatous degeneration can occur in the tricuspid valve. Leaflets that contact pacemaker wires often become thickened and nodular. The circumferences of all valves increase with age (10).

PAPILLARY MUSCLES

There are two papillary muscles of the right ventricle, the anterior and conal muscles, and they are relatively constant. There is also a group of inconstant posterior papillary muscles on the inferior wall of the right ventricle. In the left ventricle, there are two constant papillary muscles: the anterolateral and posteromedial forms. The papillary muscles receive the chordae. They are variable in shape and width and on occasion may have multiple heads (Fig. 19.20).

The papillary muscles are surfaced by a fibrous cap, into which the chordae insert (Fig. 19.24). This is a normal constituent and should not be interpreted as abnormal fibrosis. The small arteries and arterioles in the papillary muscle are notable for their wall thickness and irregularity in comparison to other intracardiac small vessels (Fig. 19.25). The myocardium and the endocardial coverings are similar to their counterparts described elsewhere. In marked ventricular

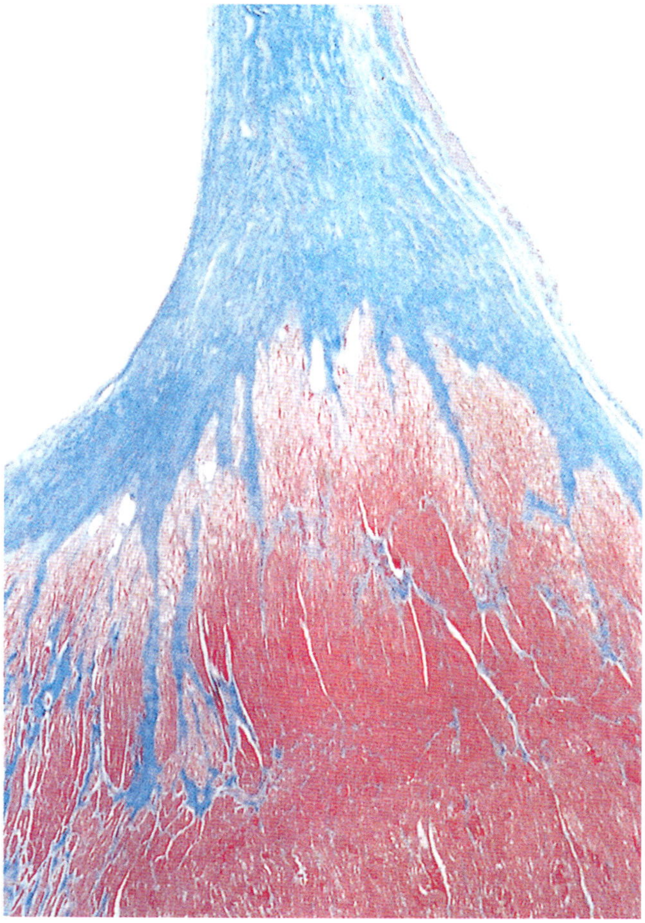

FIGURE 19.24 Longitudinal section of papillary muscle showing the fibrous cap of the insertion of the chordae tendineae (Masson trichrome).

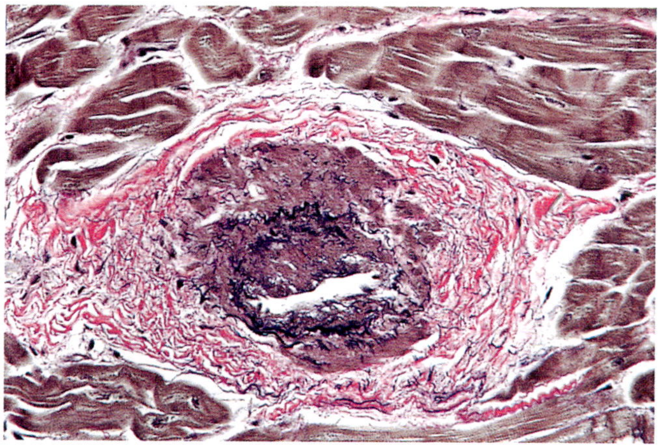

FIGURE 19.25 Section of abnormally thickened arteriole within a papillary muscle (elastic van Gieson).

dilatation, the papillary muscles may become thinned and flattened.

CONDUCTION SYSTEM

Myocardial fibers are delineated along two functional pathways in humans: (a) contractile fibers and (b) myocardial fibers specialized for the initiation and propagation of an impulse for contraction. The conduction system is recognized to be myogenic in origin, with nerves playing only a subsidiary controlling function.

For most pathologists, examining the conduction system is often regarded as a daunting exercise. This is due in large part to the fact that cases requiring detailed morphologic analysis are infrequent and often limited to specific requests from clinicians. In addition to detailing the microscopic features of the sinoatrial and atrioventricular nodes, we will present our practical approach to the dissection of these structures. We recommend careful attention to key anatomic landmarks to ensure successful retrieval of these structures.

Sinoatrial Node

The sinoatrial node has the highest intrinsic rate and is recognized as the primary pacemaker of the heart. This node is situated within the terminal groove at the junction of the superior vena cava and the lateral border of the right atrium (Fig. 19.26). Its position is constant and is marked by the apex of the crest of the atrial appendage. It is ovoid, or cigar-shaped, in most hearts, but cases of horseshoe-shaped sinoatrial nodes that extend into the interatrial groove are reported. Removal of a rectangular block of tissue that includes the distal superior vena cava and atrium on either side of the terminal groove is preferred. Serial sectioning in a longitudinal plane parallel to the terminal groove at 2-mm intervals is recommended, and all the tissue slices can be accommodated in two tissue cassettes. Grossly identifying the sinoatrial nodal artery is also helpful in procuring the node (Fig. 19.26). This is a branch of the right coronary artery that is found in slightly more than half of the general population. Microscopically, the node is arranged around a central artery adjacent to the epicardial adipose tissue. The node is composed of dense connective tissue within which the small muscle fibers are embedded. The muscle fibers contain sparse myofibrils, the striations are not prominent, and the whole mass has a pseudosyncytial appearance. Connective tissue stains such as Masson trichrome and elastic van Gieson stains highlight the abundant collagen and elastic fibers, respectively (Fig. 19.26). Abundant nerve fibers run into the node. The exact pathway(s) carrying the electrical impulse from the sinoatrial node to the atrioventricular node remains controversial. Some investigators think that several specialized bundles of conducting system cells (e.g., anterior, middle, and posterior internodal tracts) conduct the impulse around the atrium. Others argue that the arrangement of myocardial fibers within the atrium and interatrial septum serves to propagate the impulse.

Atrioventricular Node

In our experience, a heart opened along the lines of flow provides optimum exposure for atrioventricular node dissection. From a right-sided approach, the interventricular septum is oriented with the tip of the apex pointing downward. The important landmarks include the oval fossa, ostium of the coronary sinus, and tricuspid valve annulus and leaflets. The atrioventricular node lies within the subendocardial tissues on the right side of the interatrial septum just anterior to the opening of the coronary sinus, posterior to the membranous interventricular septum (i.e., tendon of Todaro), and above the tricuspid valve annulus within the triangle of Koch (Fig. 19.27) (61). A rectangular block of tissue, beginning with a vertical incision adjacent to the ostium of the coronary sinus and extending 1 to 2 cm below the annulus, is removed. After careful trimming of valvular structures, the block will contain components of the tricuspid valve (septal leaflet) and mitral and aortic valves (Fig. 19.27). Serial sectioning at 2- to 3-mm intervals and sequential placement in tissue cassettes generally yields a total of 8 to 10 cassettes.

In histologic sections, the atrioventricular node is flattened against the central fibrous body and is composed of a network of muscle fibers with the superficial zone being arranged in a parallel manner. These specialized fibers retain their intercalated discs and striations but are characterized by their pale eosinophilic appearance (Fig. 19.27). A small atrioventricular nodal artery is often identified adjacent to the atrioventricular node. At the anterior end of the atrioventricular node, the muscle fibers become arranged in parallel lines to form the main bundle of His, or penetrating atrioventricular bundle. To reach the ventricle, the atrioventricular bundle pierces the central fibrous body and runs forward on the upper margin of the muscular ventricular septum. This penetrating portion of the main bundle is surrounded by dense connective tissue and is closely related to the aortic and mitral valve rings (Fig. 19.27). Connective tissue stains aid the localization of the nodal tissue. The fibers of the main bundle are arranged in parallel. The penetrating atrioventricular bundle terminates as the left and right bundle branches. The left fascicle runs downward over the endocardial surface of the interventricular septum to the base of the anterior papillary muscle, and the right fascicle ends in the moderator band of the right ventricle. Direct connection of both bundle branches to a complex ramifying system of subendocardial conduction fibers can be demonstrated in mammalian hearts. Light microscopy shows the fibers in the bundle of His and the conduction bundles to be small and contain few myofibrils (Fig. 19.28).

Aging Changes in the Human Conduction System

With advancing age, the sinoatrial node displays progressive increase in fibrous and adipose tissues and reduction

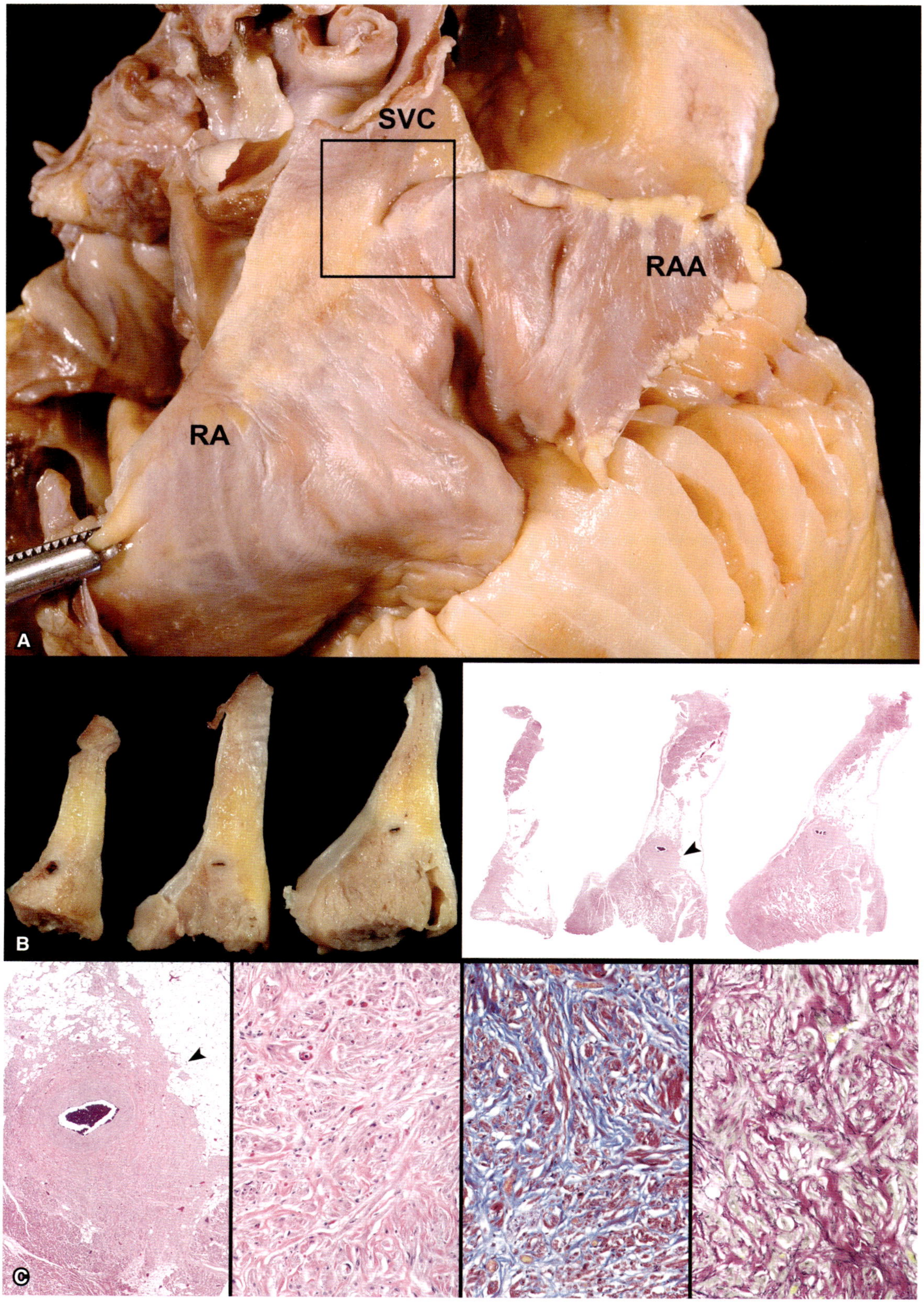

FIGURE 19.26 The sinoatrial (SA) node. **A:** The location of the SA node within the terminal groove at the junction of the superior vena cava (*SVC*) and crest of the atrial appendage (*box*). *RA*, right atrium; *RAA*, right atrial appendage. **B:** Macroscopic and low-power magnification of serial sections of nodal tissue. Note the nodal artery that is adjacent to the SA node (*arrowhead*). **C:** High-power magnification of the SA node showing the specialized fibers embedded within collagen and elastic tissue (H&E, trichrome, and elastic van Gieson stains).

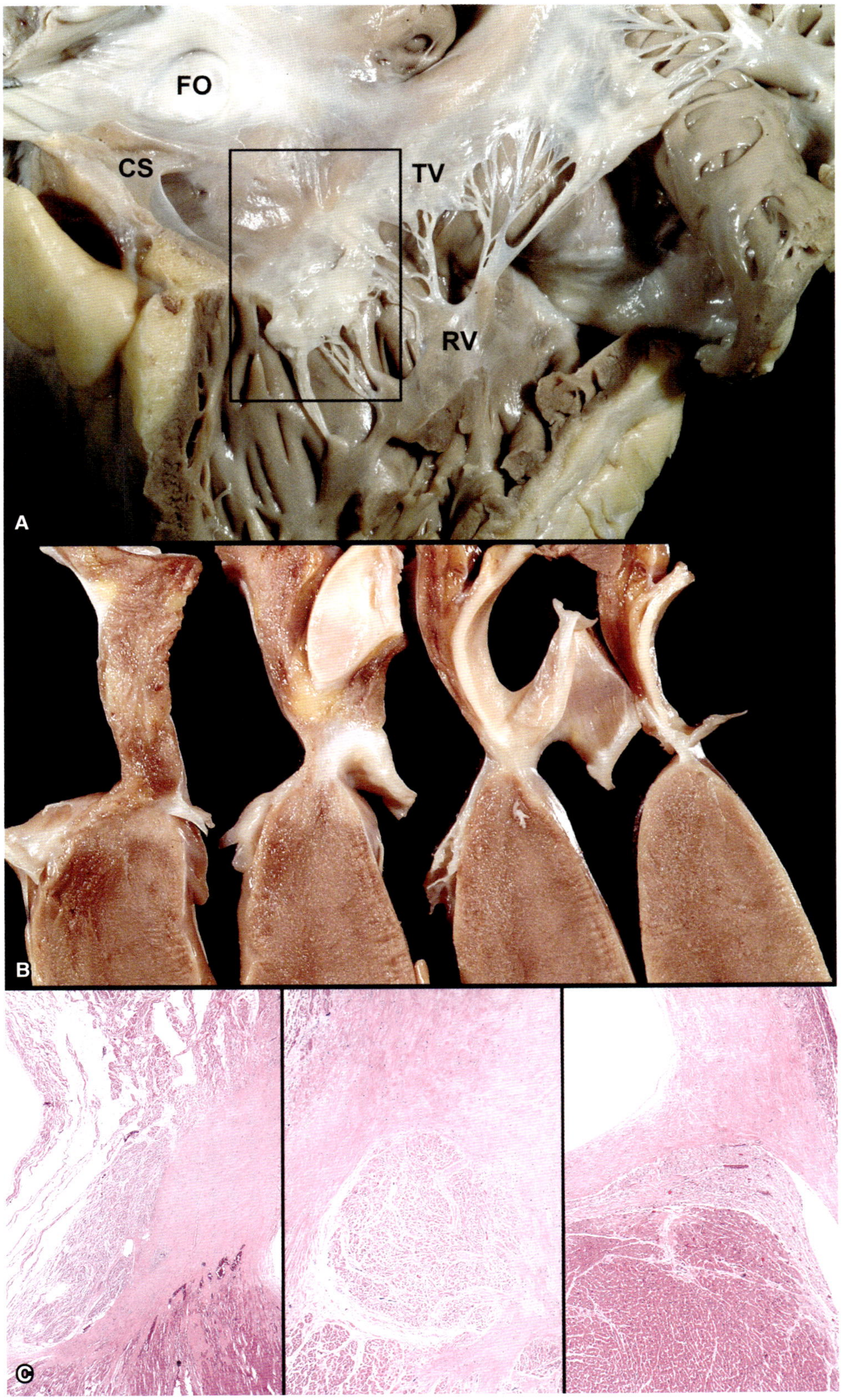

FIGURE 19.27 The atrioventricular (AV) nodal apparatus. **A:** The location of the AV node (*box*) viewed from the right ventricle (*RV*). Important landmarks include the coronary sinus (*CS*), tricuspid valve annulus and leaflet (*TV*), and fossa ovale (*FO*). **B:** Serial sections of the rectangular block of tissue show the relationship of the nodal tissue to the tricuspid valve, mitral valve, atrioventricular valve, and fibrous skeleton of the heart. **C:** The AV node (*left*), AV penetrating bundle (*middle*), and AV bundle with fascicle (*right*) are shown.

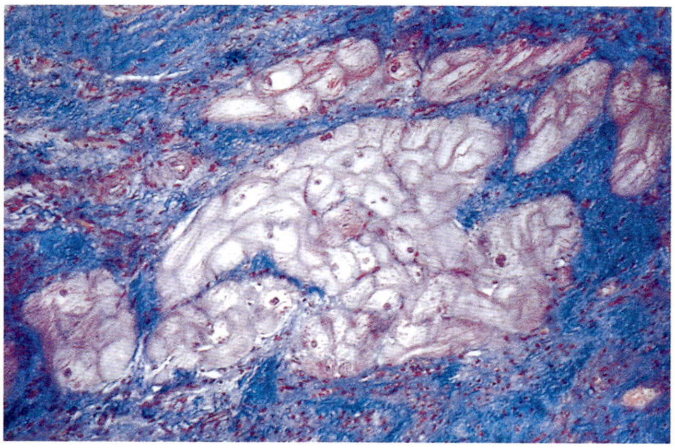

FIGURE 19.28 Section showing the pale cells of the mammalian conducting system. These cells contain glycogen and only sparse myofibrils (Masson trichrome).

of constituent nodal cells, while the atrioventricular node remains relatively unchanged, typically showing only minimal increases in adipose tissue or elastic fibers. Fibrous tissue also increases in the upper portion of the interventricular septum. These changes are associated with a loss of conduction fibers in the region of the left bundle branch. Up to 50% of the left bundle origin may be lost in people over 60 years of age (62).

CARDIAC INNERVATION

The nerve supply of the heart is autonomic, including both the sympathetic and parasympathetic supply via both efferent and afferent fibers. Large nerves can be seen in the epicardium and adjacent to the coronary blood vessels. Small nerves within the myocardium are hard to identify unless special stains are used. Myocardial nerves are best viewed using electron microscopic examination, by which the autonomic nerves can be distinguished. Cardiac parasympathetic ganglia can be found over the surface of the atria and in the atrioventricular groove (Fig. 19.14).

Autonomic Nerves

Axonal varicosities occur at irregular intervals along autonomic fibers, and their morphology is considered useful in determining whether the nerve is adrenergic or cholinergic (57). The varicosities contain accumulations of agranular vesicles and a few mitochondria in cholinergic nerves, whereas the varicosities contain vesicles rich in electron-dense cores in adrenergic nerves. Each of these cores is separated from the limiting membrane of the vesicle by an electron-lucent zone (Fig. 19.29). Presumptive sensory nerve terminals have large diameters and contain numerous mitochondria. They are located in perivascular regions and are surrounded by Schwann cells. An individual Schwann

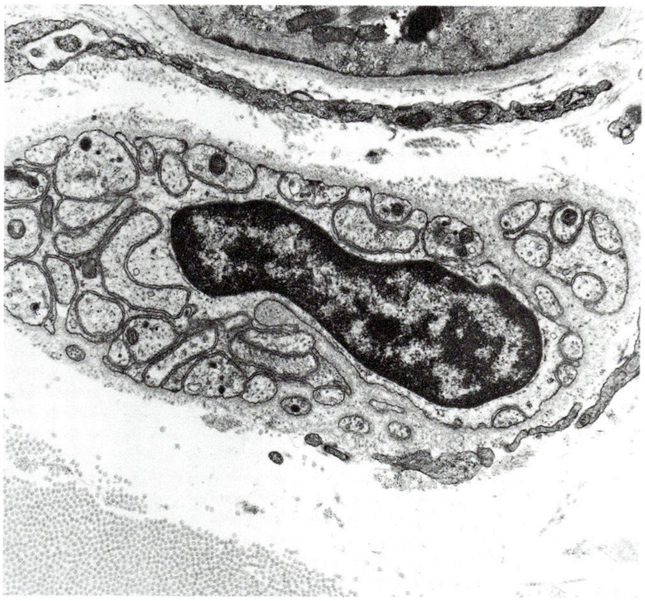

FIGURE 19.29 Electron micrograph of sympathetic nerve showing dense-core granules in the myocardium (original magnification ×22,500).

cell may enclose adrenergic and cholinergic axons together with sensory axons (57). Autonomic ganglia are found in the subepicardial tissue of the atria and atrial appendages and at the root of the great vessels, along the interatrial and atrioventricular grooves in the atrial septum, and in the vicinity of the sinoatrial and atrioventricular nodes. Large nerves can be seen in the subepicardial layer adjacent to the epicardial coronary arteries.

LYMPHATIC VESSELS

There are two networks of lymphatic vessels in the heart: (a) in the endocardium and (b) in the epicardium. The route of drainage of the endocardial network is through channels in the myocardium into the epicardial lymphatics. The epicardial meshwork of channels contains many valves and drains toward the atrioventricular sulcus by means of several longitudinal channels that mostly run parallel to the coronary veins in the anterior and posterior longitudinal sulci of the ventricles (57). Lymphatic vessels leave the pericardial cavity to empty into one of the pulmonary hilar lymph nodes and join the lymphatic drainage system of the mediastinum or into the thoracic duct. Lymphatic channels are also found in the myocardial valves and lie within the grooves of the coronary blood vessels.

The lymph capillaries and larger lymphatic vessels accompany blood vessels in the myocardial interstitium. The walls of the myocardial lymphatic vessels consist of extremely thin endothelial cells, the nuclei of which bulge into the lumen (Fig. 19.30). In contrast to endothelial cells of blood capillaries, those of the lymphatic capillaries do

FIGURE 19.30 Electron micrograph showing an intramyocardial lymphatic with thin walls and no basement membrane (original magnification ×20,000).

not have a well-defined external basal lamina. The larger lymphatic vessels are confined to the outer third of the myocardial wall and occasionally contain valves. These flap-like structures contain a core of collagen embedded in microfibrils and are covered by endothelia.

SMALL INTRAMURAL CORONARY ARTERIES

The structure of the intramural coronary arteries and the larger coronary arteries is similar and consists of the endothelium (*i.e.*, intima), spirally aligned smooth muscle (i.e., media), and adventitia (Fig. 19.31). Muscular arteries have a thick adventitial layer of collagen and some elastic fibers. The thickness of the adventitia is often similar to the medial layer and should not be misinterpreted as perivascular fibrosis. The smallest muscular arteries contain three or four layers of smooth muscle cells, while larger arteries have up to 40 layers. Arterioles are <100 microns

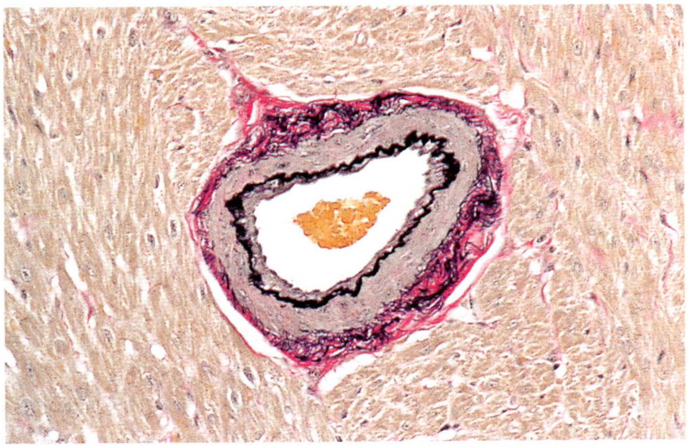

FIGURE 19.31 Transverse section of intramyocardial arteriole (elastic van Gieson).

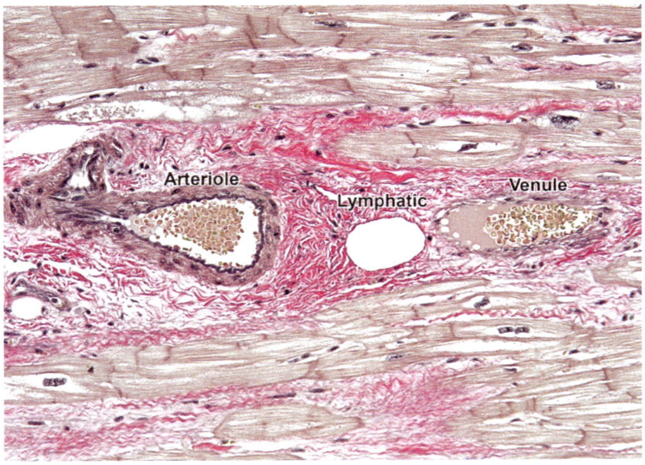

FIGURE 19.32 Section showing the relationship between arterioles, lymphatics, and venules (elastic van Gieson).

in diameter and have flattened, elongated endothelial cells that do not protrude into the lumen (Fig. 19.32). Their internal elastic lamina is discontinuous. Metarterioles are also known as precapillary sphincters. The endothelial cells in metarterioles have numerous surface projections that bulge into the lumen (57). Although the medial smooth muscle cells form a single discontinuous layer, it gradually disappear as capillaries begin. Capillaries are distinguished by the fact that their walls are composed of only a single layer of endothelial cells. They do not contain smooth muscle cells but may be closely associated with pericytes (57). Capillary endothelial cells may contain microvilli and cytoplasmic processes, or filopodia. The myocardium harbors a rich network of capillaries (Fig. 19.33). These branches undergo anastomoses and eventually become thin-walled venules measuring up to 100 μm in diameter.

VEINS AND VENULES

Venules are lined by thin, flat endothelial cells and characteristically contain a large amount of connective tissue in the vicinity of their external surface; they contain collagen fibrils that approach the endothelial layer and are anchored on its outer surface (Fig. 19.32) (57). Venules gradually increase in size to become veins. Veins have larger lumens but thinner walls than their arterial counterparts.

Veins are composed of three layers: intima, media, and adventitia. The intima is thin, lacks smooth muscle cells, and has a poorly defined internal elastic lamina. The media is also thin and contains few smooth muscle cells and elastic fibers. The adventitia is thick with abundant collagen and elastic fibers. Cardiac veins drain blood into either the coronary sinus or directly into the chambers through the thebesian veins.

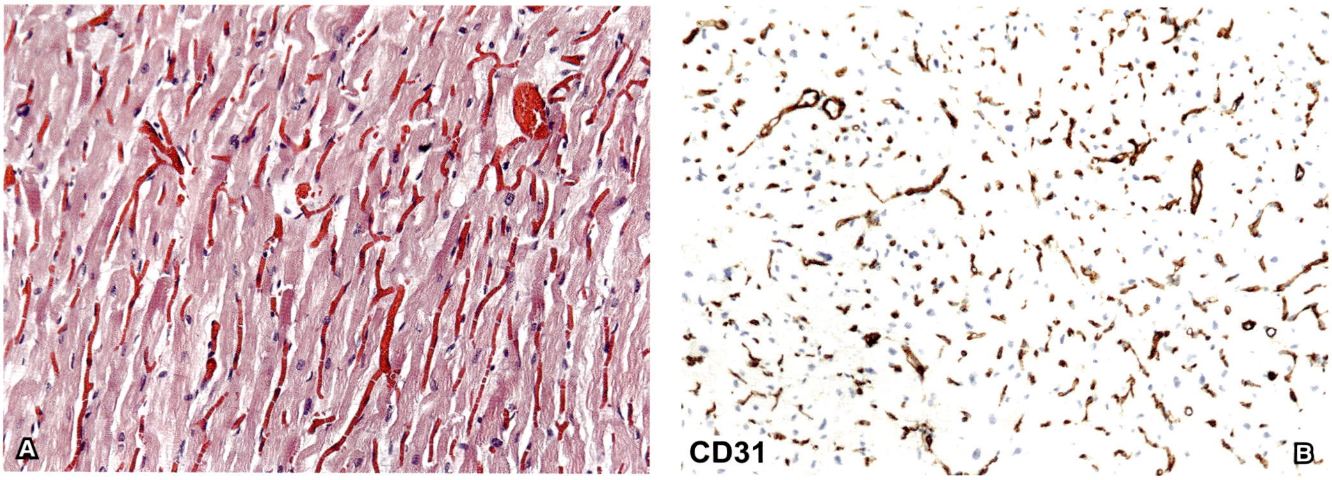

FIGURE 19.33 Rich capillary network of capillaries within the myocardium. **A:** Engorged capillaries showing interdigitations (H&E). **B:** *CD31* immunohistochemical stain highlights endothelial cells.

THE ENDOMYOCARDIAL BIOPSY

The transvenous endomyocardial biopsy is currently utilized for the diagnosis of allograft rejection and a variety of inflammatory, metabolic, and neoplastic conditions that affect the heart. Introduced in the early 1960s, the bioptome and technique have undergone modifications that now permit clinicians the opportunity to obtain cardiac tissue in a safe outpatient setting (63). The right internal jugular vein or femoral vein approaches are commonly used. Complications are uncommon and include local issues (e.g., hematomas and nerve injury) and cardiac problems such as arrhythmias, tricuspid valve apparatus damage, and ventricular perforation.

Tissue Handling and Processing

Proper tissue procurement and handling are critical for optimal diagnostic evaluation (64). Biopsy specimens should be gently extracted from the bioptome with a needle tip to limit crush artifactual distortion. The clinical indications for the biopsy determine, in large part, the method of tissue handling. For example, the tissue should immediately be placed in a standard fixative such as 10% neutral buffered formalin for standard light microscopy. To demonstrate the type of amyloid fibril in cardiac amyloidosis by immunofluorescence (e.g., AL, AA, or transthyretin), one or two pieces should be received in saline or Zeus medium and then snap frozen in a plastic BEEM capsule containing an embedding medium. The diagnosis of chronic anthracycline cardiotoxicity requires that at least 3 to 5 biopsy pieces be obtained, all of which are submitted in fixative for transmission electron microscopy (e.g., 2.5% glutaraldehyde with 2% paraformaldehyde in 0.1M sodium cacodylate buffer, pH 7.2).

For routine diagnostic evaluation, overnight processing and paraffin embedding are sufficient. For emergent cases, a 60- to 90-minute rapid ("ultra") processing cycle is available, and microscopic slides can be prepared within 3 to 4 hours. All the biopsy pieces should be embedded in the same block. We recommend that a minimum of three slides be prepared, with each sectioned at 4 to 5 μm thickness from various depths within the paraffin block. Multiple tissue "ribbons," are placed on each slide.

We routinely stain with hematoxylin and eosin (H&E) and use stains such as Masson trichrome to confirm the presence of myocyte damage or fibrosis, Congo red stain for amyloid fibrils, and the Prussian blue stain for iron deposition. Immunohistochemical, immunofluorescent, and molecular studies are utilized for specific indications. Paraffin-section immunohistochemistry is used to evaluate for the presence of infectious myocarditis (e.g., cytomegalovirus or toxoplasmosis) and antibody-mediated rejection (e.g., intravascular collections of $CD68^{+}$/CD163/PU.1 macrophages and complement deposition with C4d on the microvasculature). Posttransplant lymphoproliferative disorders can be evaluated with a variety of methods to assess for B-cell clonality, light chain restriction, Epstein–Barr virus-encoded RNAs (EBER), and anomalous coexpression of B-cell and T-cell antigens.

Biopsy Limitations and Tissue Artifacts

Sampling error in the diagnosis of rejection, myocarditis, and infection remains a major consideration in the clinical management of patients and the evaluation of new noninvasive diagnostic modalities. In general, the false-negative rate is low, particularly when four or more pieces of tissue are submitted. The issue of how many lymphocytes are normally found in the myocardium has been addressed in a number of studies. The mean number of lymphocytes reported is normally fewer than 5 per high-power field in an endomyocardial biopsy (65). Tazelaar and Billingham reported foci of mononuclear inflammatory cells in 9.3% of cases in which biopsy samples were obtained from donor hearts just before

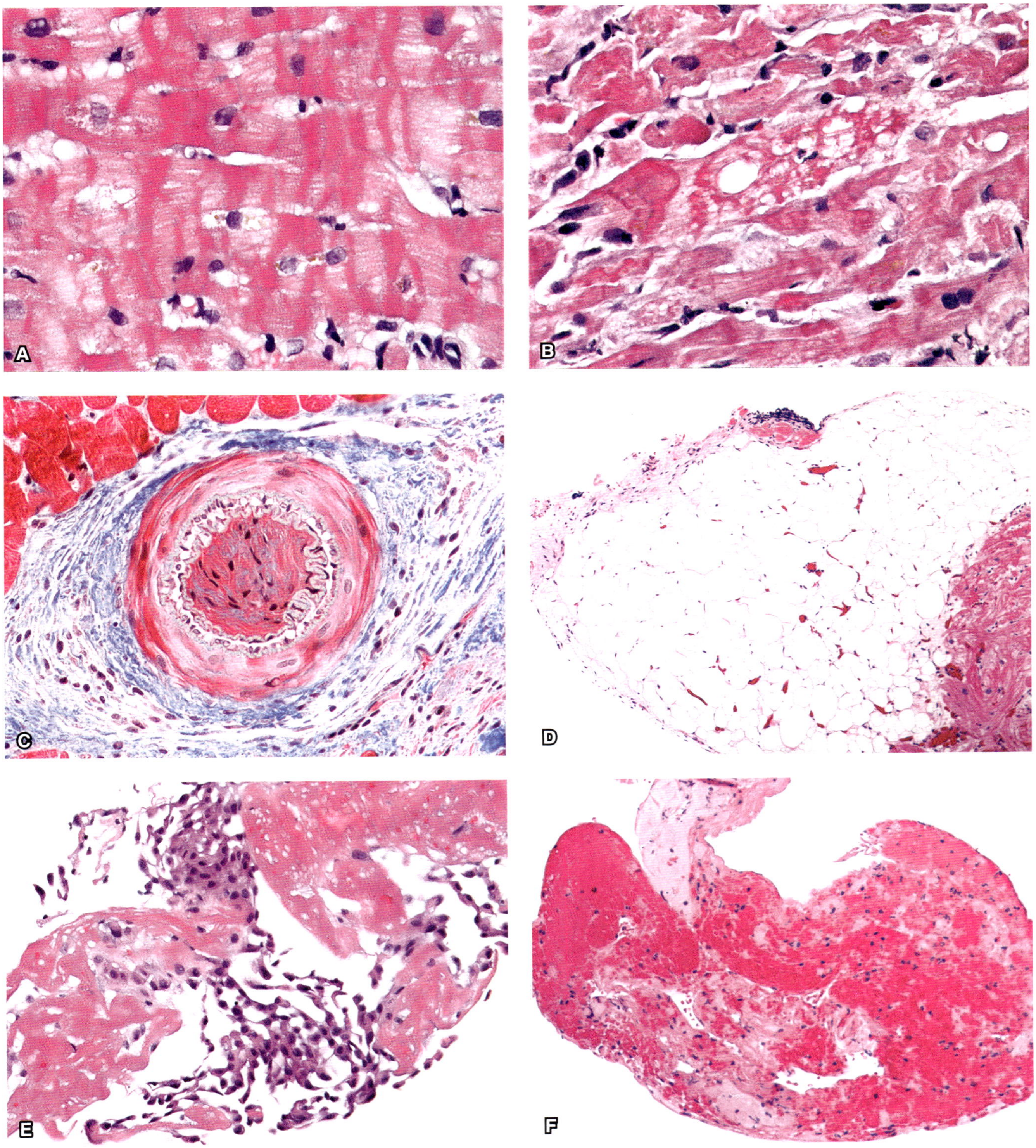

FIGURE 19.34 Artifacts of endomyocardial biopsy specimens. **A:** Contraction band artifact (H&E); note the normal appearance of myocyte nuclei. **B:** Contraction band necrosis with hyperchromatic pyknotic nuclei and eosinophilic cytoplasm. The changes are contrasted with the common contraction band artifact in **A**. **C:** Telescoping of intramyocardial artery is highlighted by a trichrome stain. **D:** Intramyocardial adipose tissue. The presence of fat does not imply epicardial localization or perforation. **E:** Mesothelial cells admixed with fibrin indicative of ventricular perforation. **F:** Thrombus without attached myocardial tissue. (*continued*)

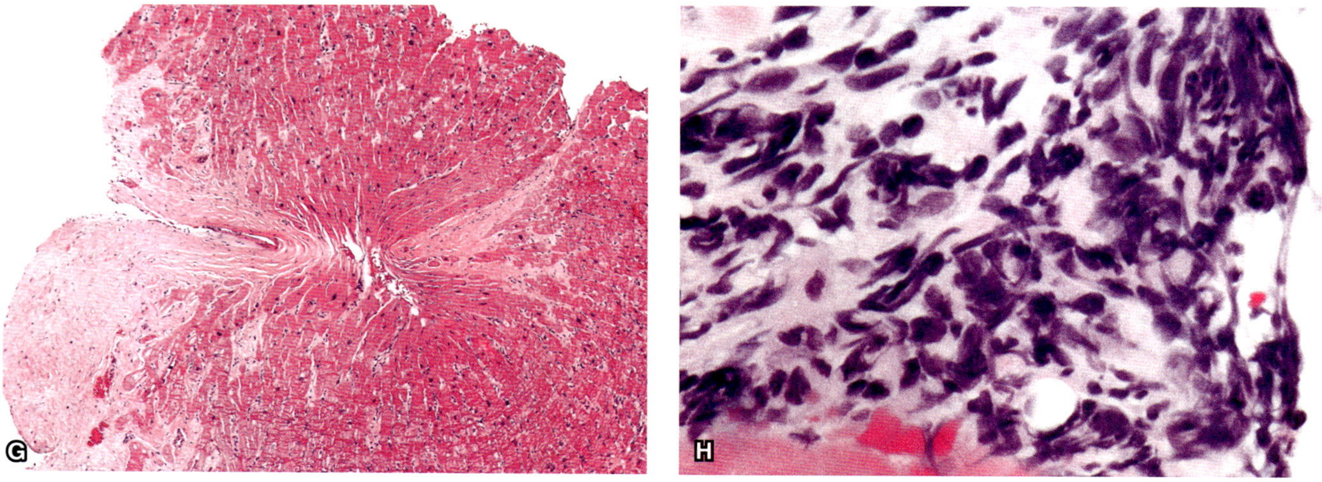

FIGURE 19.34 (*Continued*) **G:** Bioptome-induced "Victorian waistband" artifact. **H:** Crush artifactual distortion of cells.

transplant. These foci ranged from 6 to at least 50 cells in number (66). In an autopsy study of young men who died from acute traumatic injury, focal collections of at least 100 mononuclear cells were found in 5% of cases. That study predated the Dallas criteria and the term "focal myocarditis" was utilized (67). These studies support the concept that small clusters of mononuclear cells, including lymphocytes and macrophages, are normal within the myocardium and should not be indiscriminately classified as myocarditis.

A variety of artifacts occur in endomyocardial biopsy specimens that may mimic pathologic lesions. The surgical pathologist should be aware of these patterns to avoid a misdiagnosis that could lead to unnecessary therapeutic interventions. These have been reviewed in detail in a recent publication and only selected topics will be briefly reviewed (68). The most common biopsy artifact is the presence of contraction bands in myocytes (Fig. 19.34A). They are identical to the linear bands observed in acute ischemic necrosis and catecholamine ("pressor") effect. These changes are induced by the biopsy procedure itself and can be diminished by using fixatives stored at room temperature. In ischemic injury, the nuclei of surrounding myocytes are usually pyknotic (Fig. 19.34B), while in artifactually induced contraction bands, the nuclei remain normal in appearance.

Another frequent artifact is intussusception, or "telescoping," of small arteries that has been confused with luminal occlusion by thrombus and transplant-related arteriosclerosis. Connective tissue stains such as Masson trichrome or elastic van Gieson can be used to highlight the internal elastic membranes of both vessel segments (Fig. 19.34C). Intramyocardial accumulations of mature adipose tissue can simulate the epicardial tissue, especially if associated with vessels of relatively large caliber (Fig. 19.34D). Both can be found in the right ventricular apical region, and adipose tissue is found not uncommonly in women and elderly patients. This should not be confused with arrhythmogenic cardiomyopathy; clinical–pathologic correlation is essential for this purpose.

Ventricular perforation is identified by the presence of mesothelial cells (Fig. 19.34E). Accumulations of fresh platelet- and fibrin-rich thrombus may be identified along the endocardial surface of biopsy fragments (Fig. 19.34F). These form as a result of the repetitive placement of the bioptome along the endocardial surface and do not indicate chronic mural thrombi. A number of patterns of bioptome-induced tissue distortion or crush artifact can be observed in biopsy samples. The "hour-glass" or "Victorian waistband" effect is caused by central constriction of the tissue by the bioptome mechanism (Fig. 19.34G). A more problematic artifact is the smearing of cytoplasmic and nuclear components of cells that yield strands of basophilic material (Fig. 19.34H). In this setting, it may not be possible to distinguish between lymphocytes, endothelial cells, macrophages, and myocytes that might be present; we do not attempt to evaluate these foci for allograft rejection or myocarditis. In some cases, procurement of additional leveled H&E-stained sections can provide less distorted foci in the deeper aspects of the biopsy sample. In our experience, immunohistochemical stains have not been consistent or helpful in this situation.

SUMMARY

Because of the unique structural–functional nature of cardiac disease, the surgical pathologist should have a working knowledge of anatomy, histology, and physiology. Moreover, the alterations produced by the endomyocardial biopsy and the bioptome require familiarity with the myriad of tissue artifacts. With a practical understanding of these points, the evaluation of specimens ranging from endomyocardial samples to explanted hearts will enhance the diagnostic information provided to clinicians and patients.

ACKNOWLEDGMENTS

This Chapter was originally authored by Dr. Margaret Billingham and has been revised by the two of us in previous editions. Dr. Billingham died on July 14, 2009. She was Professor of Pathology, emerita and the former Director of Cardiac Pathology at Stanford University. Dr. Billingham made innumerable contributions to the fields of cardiovascular pathology and cardiovascular medicine, including cardiac transplant pathology, myocarditis, anthracycline cardiotoxicity and cardiomyopathies. She was a founding member of the Society of Cardiovascular Pathology and was awarded its Distinguished Achievement Award in 2001. She was a member of the Society of Heart and Lung Transplantation and served as its president in 1990. She published over 500 peer-reviewed articles, abstracts, and book chapters and directed numerous teaching workshops on the endomyocardial biopsy in transplant and nontransplant pathology. Dr. Billingham is warmly remembered for the impact that her teaching, friendship, and mentorship had on scores of residents, fellows and colleagues around the world. It is an honor to update her work for this edition.

REFERENCES

1. Hsich E, Singh TP, Cherikh WS, et al; International Society for Heart and Lung Transplantation. The International Thoracic Organ Transplant Registry of the International Society for Heart and Lung Transplantation: Thirty-Ninth Adult Heart Transplantation Report-2022; focus on transplant for restrictive heart disease. *J Heart Lung Transplant*. 2022; 41(10):1366–1375.
2. Chambers DC, Perch M, Zuckermann A, et al; International Society for Heart and Lung Transplantation. The International Thoracic Organ Transplant Registry of the International Society for Heart and Lung Transplantation: thirty-eighth adult lung transplantation report –2021; focus on recipient characteristics. *J Heart Lung Transplant*. 2021;40(10):1060–1072.
3. Singh TP, Cherikh WS, Hsich E, et al; International Society for Heart and Lung Transplantation. The International Thoracic Organ Transplant Registry of the International Society for Heart and Lung Transplantation: Twenty-Fourth Pediatric Heart Transplantation Report –2021; focus on recipient characteristics. *J Heart Lung Transplant*. 2021;40(10):1050–1059.
4. Hayes D Jr, Harhay MO, Cherikh WS, et al; International Society for Heart and Lung Transplantation. The International Thoracic Organ Transplant Registry of the International Society for Heart and Lung Transplantation: Twenty-Fourth Pediatric Lung Transplantation Report –2021; focus on recipient characteristics. *J Heart Lung Transplant*. 2021;40(10): 1023–1034.
5. Clemente A, Seitun S, Mantini C, et al. Cardiac CT angiography: normal and pathological anatomical features-a narrative review. *Cardiovasc Diagn Ther*. 2020;10(6):1918–1945.
6. Lindner JR. Contrast echocardiography: current status and future directions. *Heart*. 2021;107(1):18–24.
7. Scholz DG, Kitzman DW, Hagen PT, Ilstrup DM, Edwards WD. Age-related changes in normal human hearts during the first 10 decades of life. Part I (growth): a quantitative anatomic study of 200 specimens from subjects from birth to 19 years old. *Mayo Clin Proc*. 1988;63:126–136.
8. Connolly AJ, Finkbeiner WE, Ursell PC, Davis RL. *Autopsy Pathology: A Manual and Atlas*. 3rd ed. Elsevier; 2015.
9. Hudson R. Structure and function of the heart. In: Hudson R, ed. *Cardiovascular Pathology*. Vol. 1. Edward Arnold; 1965: 12–23.
10. Kitzman DW, Scholz DG, Hagen PT, Ilstrup DM, Edwards WD. Age-related changes in normal human hearts during the first 10 decades of life. Part II (maturity): a quantitative anatomic study of 765 specimens from subjects 20 to 99 years old. *Mayo Clin Proc*. 1988;63:137–146.
11. Hutchins GM, Anaya OA. Measurements of cardiac size, chamber volumes and valve orifices at autopsy. *Johns Hopkins Med J*. 1973;133:96–106.
12. Sheaff MT, Hopster DJ. Organ dissection—cardiovascular system. In: Sheaff MT, Hopster DJ, eds. *Post Mortem Technique Handbook*. Springer; 2001:87–116.
13. Maleszewski JJ, Lai CK, Nair V, Veinot JP. Anatomic considerations and examination of cardiovascular specimens (excluding devices). In: Buja LM, Butany J, eds. *Cardiovascular Pathology*. 5th ed. Elsevier; 2022:27–84.
14. Edwards WD. Applied anatomy of the heart. In: Brandenberg RO, Fuster V, Guiliani ER, McGoon ER, eds. *Cardiology: Fundamentals and Practice*. Year Book Medical; 1987:47–112.
15. Cook AC, Yates RW, Anderson RH. Normal and abnormal fetal cardiac anatomy. *Prenat Diagn*. 2004;24:1032–1048.
16. Zhao Y, Deng W, Wang Z, et al. Genetics of congenital heart disease. *Clin Chim Acta*. 2024;552:117683.
17. Chaudhry B, Ramsbottom S, Henderson DJ. Genetics of cardiovascular development. *Prog Mol Biol Transl Sci*. 2014;124: 19–41.
18. Baban A, Lodato V, Parlapiano G, Drago F. Genetics in congenital heart diseases: unraveling the link between cardiac morphogenesis, heart muscle disease, and electrical disorders. *Heart Fail Clin*. 2022;18(1):139–153.
19. MacGrogan D, Münch J, de la Pompa JL. Notch and interacting signalling pathways in cardiac development, disease, and regeneration. *Nat Rev Cardiol*. 2018;15(11):685–704.
20. Kelly RG, Buckingham ME, Moorman AF. Heart fields and cardiac morphogenesis. *Cold Spring Harb Perspect Med*. 2014;4(10): a015750.
21. Morton SU, Quiat D, Seidman JG, Seidman CE. Genomic frontiers in congenital heart disease. *Nat Rev Cardiol*. 2022;19(1): 26–42.
22. Rodriguez ER, Tan CD. Structure and anatomy of the human pericardium. *Prog Cardiovasc Dis*. 2017;59:327–340.
23. Fatehi Hassanabad A, Zarzycki A, Deniset JF, Fedak PW. An overview of human pericardial space and pericardial fluid. *Cardiovasc Pathol*. 2021;53:107346.
24. Kligerman S. Imaging of pericardial disease. *Radiol Clin North Am*. 2019;57(1):179–199.
25. Wang TKM, Ayoub C, Chetrit M, et al. Cardiac magnetic resonance imaging techniques and applications for pericardial diseases. *Circ Cardiovasc Imaging*. 2022;15(7):e014283.
26. Burke A, Tavora FR, Maleszewski JJ, Frazier AA, Tumors of the heart and great vessels. In: *Atlas of Tumor Pathology*. 4th series, fascicle 16. Armed Forces Institute of Pathology; 2015:254–257.

27. Wilcox BR, Cook AC, Anderson RH. *Surgical Anatomy of the Heart*. Cambridge University Press; 2004.
28. Le Jemtel TH, Samson R, Ayinapudi K, Singh T, Oparil S. Epicardial adipose tissue and cardiovascular disease. *Curr Hypertens Rep*. 2019;21(5):36.
29. Iacobellis G. Epicardial adipose tissue in contemporary cardiology. *Nat Rev Cardiol*. 2022;19(9):593–606.
30. Anderson RH, Becker AE. *The Heart: Structure in Health and Disease*. Gower Medical Publishing; 1992.
31. Malouf JF, Edwards WD, Tajik AJ, et al. Functional anatomy of the heart. In: Fuster V, Alexander RW, O'Rouke RA, eds. *Hurst's the Heart*. 11th ed. McGraw-Hill; 2004:45–86.
32. Anderson RH, Smerup M, Sanchez-Quintana D, Loukas M, Lunkenheimer PP. Three-dimensional arrangement of the myocytes in the ventricular walls. *Clin Anat*. 2009;22:64–76.
33. Buckberg G, Hoffman JI, Mahajan A, Saleh S, Coghlan C. Cardiac mechanics revisited: the relationship of cardiac architecture to ventricular function. *Circulation*. 2008;118: 2571–2587.
34. Sweeney LJ, Rosenquist GC. The normal anatomy of the atrial septum in the human heart. *Am Heart J*. 1979;98:194–199.
35. Homma S, Messé SR, Rundek T, et al. Patent foramen ovale. *Nat Rev Dis Primers*. 2016;2:15086.
36. Sheppard M, Davies MJ. *Practical Cardiovascular Pathology*. Arnold Publishers; 1998.
37. Corradi D, Maestri R, Macchi E, Callegari S. The atria: from morphology to function. *J Cardiovasc Electrophysiol*. 2011;22: 223–235.
38. Conti AA. Calling the heart by name: distinguished eponyms in the history of cardiac anatomy. *Heart Surg Forum*. 2011; 14:e183–e187.
39. DeSimone CV, Prakriti BG, Tri J, Syed F, Sm AN, Asirvatham SJ. A review of the relevant embryology, pathohistology, and anatomy of the left atrial appendage for the invasive cardiac electrophysiologist. *J Atr Fibrillation*. 2015;8(2):81–87.
40. Anderson RH, Cook AC. The structure and components of the atrial chambers. *Europace*. 2007;9:vi3–vi9.
41. Evora PRB, Menardi AC, Celotto AC, Albuquerque AAS, Chagas HMA, Rodrigues AJ. The left atrial appendage revised. *Braz J Cardiovasc Surg*. 2017;32(6):517–522.
42. Banerjee D, Haddad F, Zamanian RT, Nagendran J. Right ventricular failure: a novel era of targeted therapy. *Curr Heart Fail Rep*. 2010;7:202–211.
43. Walker LA, Buttrick PM. The right ventricle: biologic insights and response to disease: updated. *Curr Cardiol Rev*. 2013;9:73–81.
44. Crucean A, Spicer DE, Tretter JT, Mohun TJ, Anderson RH. Revisiting the anatomy of the right ventricle in the light of knowledge of its development. *J Anat*. 2024;244(2):297–311.
45. Davies MJ. Introduction to normal cardiac anatomy. In: Davies MJ, Mann JM, eds. *The Cardiovascular System. Part B: Acquired Diseases of the Heart*. Churchill Livingstone; 1995:1–6.
46. Barry A, Patten B. The structure of the adult heart. In: Gould SE, ed. *Pathology of the Heart and Blood Vessels*. Charles C Thomas; 1968:104–105.
47. Gandjbakhch E, Redheuil A, Pousset F, Charron P, Frank R. Clinical diagnosis, imaging, and genetics of arrhythmogenic right ventricular cardiomyopathy/dysplasia: JACC state-of-the-art review. *J Am Coll Cardiol*. 2018;72(7):784–804.
48. Corrado D, Basso C, Judge DP. Arrhythmogenic cardiomyopathy. *Circ Res*. 2017;121(7):784–802.
49. Corrado D, Zorzi A, Cipriani A, et al. Evolving diagnostic criteria for arrhythmogenic cardiomyopathy. *J Am Heart Assoc*. 2021;10(18):e021987.
50. Kübler J, Burgstahler C, Brendel JM, et al. Cardiac MRI findings to differentiate athlete's heart from hypertrophic (HCM), arrhythmogenic right ventricular (ARVC) and dilated (DCM) cardiomyopathy. *Int J Cardiovasc Imaging*. 2021;37(8):2501–2515.
51. James TN, Sherf L, Schlant RC, et al. Anatomy of the heart. In: Hurst JW, Logue RB, Rackley CE, et al., eds. *The Heart*. 5th ed. McGraw-Hill; 1982:22–74.
52. Ho SY. Anatomy and myoarchitecture of the left ventricular wall in normal and in disease. *Eur J Echocardiogr*. 2009; 10:iii3–iii7.
53. Kajstura J, Gurusamy N, Ogórek B, et al. Myocyte turnover in the aging human heart. *Circ Res*. 2010;107:1374–1386.
54. Silver MM, Freedom RM. Gross examination and structure of the heart. In: Silver MD, ed. *Cardiovascular Pathology*. Vol. 1. 2nd ed. Churchill Livingstone; 1991:1–42.
55. Davies MJ, Pomerance A, Lamb D. Techniques in examination and anatomy of the heart. In: Pomerance A, Davies MJ, eds. *Pathology of the Heart*. Blackwell Scientific; 1975:1–48.
56. Henderson DJ, Eley L, Turner JE, Chaudhry B. Development of the human arterial valves: understanding bicuspid aortic valve. *Front Cardiovasc Med*. 2022;8:802930.
57. Ferrans VJ, Rodríguez ER. Ultrastructure of the normal heart. In: Silver MD, ed. *Cardiovascular Pathology*. 2nd ed. Churchill Livingstone; 1991:43–101.
58. Carpentier A, Branchini B, Cour JC, et al. Congenital malformations of the mitral valve. Pathology and surgical treatment. *J Thorac Cardiovasc Surg*. 1976;72:854–866.
59. McCarthy KP, Ring L, Rana BS. Anatomy of the mitral valve: understanding the mitral valve complex in mitral regurgitation. *Eur J Echocardiogr*. 2010;11:i3–i9.
60. Liu AC, Joag VR, Gotlieb AI. The emerging role of valve interstitial cell phenotypes in regulating heart valve pathobiology. *Am J Pathol*. 2007;171:1407–1418.
61. Edwards WD. Cardiovascular system. In: Ludwig J, ed. *Handbook of Autopsy Practice*. 3rd ed. Humana Press; 2002:21–44.
62. Davies MJ, Anderson RH. The pathology of the conduction system. In: Pomerance A, Davies MJ, eds. *The Pathology of the Heart*. Blackwell Scientific; 1975:367–412.
63. Baughman KL. History and current techniques of endomyocardial biopsy. In: Baumgartner WA, Reitz B, Kasper E, Theodore J, eds. *Heart and Lung Transplantation*. 2nd ed. WB Saunders; 2002:267–281.
64. Berry GJ, Billingham ME. The pathology of human cardiac transplantation. In: Baumgartner WA, Reitz B, Kasper E, Theodore J, eds. *Heart and Lung Transplantation*. WB Saunders; 2002:286–306.
65. Edwards WD, Holmes DR Jr, Reeder GS. Diagnosis of active lymphocytic myocarditis by endomyocardial biopsy: quantitative criteria for light microscopy. *Mayo Clin Proc*. 1982; 57:419–425.
66. Tazelaar HD, Billingham ME. Myocardial lymphocytes: Fact, fancy or myocarditis? *Am J Cardiovasc Pathol*. 1986;1:47–50.
67. Stevens PJ, Ground KEU. Occurrence and significance of myocarditis in trauma. *Aerosp Med*. 1970;41:776–780.
68. Hauck AJ, Edwards WD. Histopathologic examination of tissues obtained by endomyocardial biopsy. In: Fowles RE, ed. *Cardiac Biopsy*. Futura Publishing; 1992:95–153.

20

Serous Membranes

Julia R. Naso ■ Andrew Churg

ANATOMY

The serous membranes are derived from the mesoderm and line the pleural, pericardial, and peritoneal cavities with mesothelial cells, which normally appear as a simple or cuboidal epithelium (Fig. 20.1). The parietal and visceral mesothelia are separated by a thin layer of fluid and supported by a fibrous submesothelial layer, which becomes continuous with the outer layer of invested organs and the chest or abdominal walls. The serous membranes show functional differentiation according to their location and are capable of a great variety of reactive changes, which may mask or mimic neoplastic changes.

Pleura

The pleura is a continuous membrane that covers the chest wall and the lungs. The visceral pleura coats the entire pulmonary surface, including the major and minor fissures that divide the lungs into lobes, whereas the parietal pleura extends over the ribs, sternum, and supporting structures and is reflected over the mediastinal structures on both right and left. In the posterior mediastinum, the two layers of parietal pleura are separated by a thin band of fibrovascular connective tissue. Superiorly, the cervical pleura is reflected into the retroclavicular area over the apex of the lung and is coated by a thickened layer of fibrous tissue and skeletal muscle. Inferiorly, the diaphragmatic pleura represents its caudal extent. Anteriorly, the pleura is reflected over part of the pericardium. The posterior visceral pleura becomes continuous with the diaphragmatic pleura over the pulmonary ligament.

Pericardium

The heart and great vessels lie in the pericardium, which are lined by a continuous layer of mesothelium. The visceral (epicardial) side overlies the myocardium, and the parietal (pericardial) layer rests on a layer of dense fibrous tissue containing branches of the internal mammary and musculophrenic vessels, descending aorta, and branches of the vagus, phrenic, and sympathetic nerves. The thoracic surface of the pericardium is coated with parietal pleura.

Peritoneum

The peritoneum is a nearly continuous membrane lining the potential space between the intra-abdominal viscera and

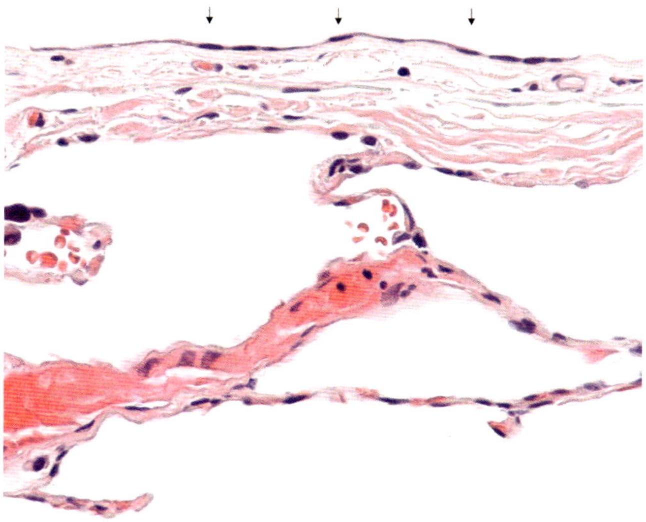

FIGURE 20.1 Normal visceral pleural, with a single layer of flat mesothelial cells on the pleural surface (*arrows*).

the abdominal wall. In females, it is normally interrupted by the lumina of the fallopian tubes. The peritoneum is more complex than either the pleura or the pericardium. The parietal layer covers the abdominal wall, diaphragm, anterior surfaces of the retroperitoneal viscera, and the pelvis. The visceral peritoneum invests the intestines and other intra-abdominal viscera. The elongated structures in which the parietal and visceral layers come together are the mesentery, which contains blood vessels, lymphatics, lymph nodes, and nerves.

The greater omentum is a double sheet consisting of four layers of mesothelium between which there are numerous blood vessels and adipose tissue; lymphatic vessels and lymph nodes are less prominent than those in the mesentery. The peritoneal cavity is grossly divided into the greater sac over the intestines, the retrogastric lesser sac, the right and left retrocolic areas, and the pelvis. Outpouchings of peritoneum are often seen as pathology specimens. Inguinal hernia sacs, for example, are pouches of parietal peritoneum invested with variable amounts of fibrous and skeletal muscle that have been pushed through the abdominal musculature into the inguinal canal. Umbilical or ventral hernias are also outpouchings of peritoneum, but the specimens received by pathologists after surgical repair are usually composed of the preperitoneal fibroadipose tissue pushed ahead of the parietal peritoneum rather than the mesothelial sac itself.

The scrotum acquires a lining of parietal mesothelium, the processus vaginalis, into which the testes descend during the 7th month of gestation. A mesothelial layer forms the surface of the tunica vaginalis. Distention of this mesothelial sac on the tunica vaginalis results in a hydrocele. Congenital hydroceles communicate with the peritoneal cavity, whereas acquired hydroceles do not. The sac of an inguinal hernia communicates with the peritoneal cavity but not with the mesothelium-lined space of the scrotum. Both the hernia and hydrocele sacs are capable of a wide range of reactive changes.

FUNCTIONAL ANATOMY

The serous membranes serve as a selective barrier for fluid and cells. Control of fluid volumes appears to be at the capillary level because fluid is freely diffusible through the visceral mesothelium and is collected in parietal lymphatics via stomata in the parietal mesothelium. Fluid turnover in the pleural cavity is estimated at 0.7 mL/hr (Fig. 20.2). Protein contents are maintained at relatively low level (1.0 to 1.5 g/dL), potentially at the level of mesothelial microvilli. Small molecules spanning less than 4 nm in diameter diffuse through the intercellular spaces and junctions between mesothelial cells. Larger molecules that span up to 50 nm in diameter are transferred across the mesothelium by pinocytotic uptake and transcellular transport. Whole cells may enter the pleural space via the stomata. Notably, loss of control of mechanisms of fluid volume and protein content regulation can result in effusions.

Pleura

Sahn (1) and Pistolesi et al. (2) described the functional anatomy of the pleura, which is a continuous membrane

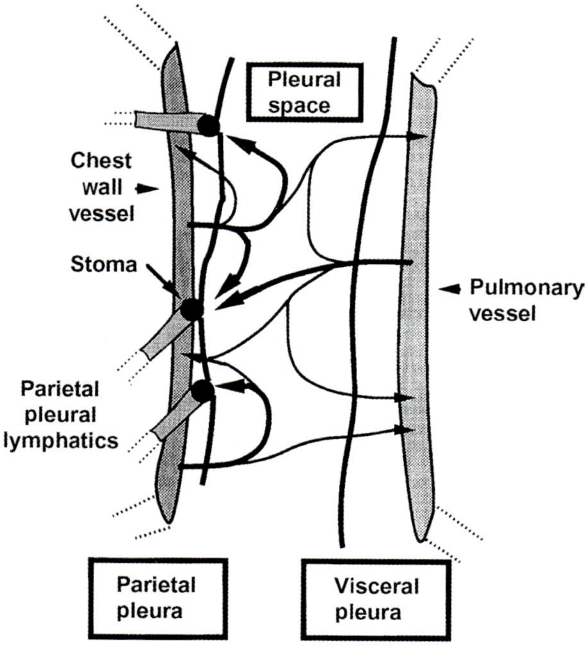

FIGURE 20.2 Model of the dynamics of pleural fluid formation. A transudate from capillaries in visceral and parietal pleurae is partly reabsorbed by those capillaries and the rest diffuses into the pleural space, where it is resorbed via stomata into parietal pleural lymphatics.

surrounding a space that normally contains approximately 10 mL of clear colorless fluid; this fluid is rich in glycoproteins and acts as a lubricant to minimize friction between visceral and parietal pleurae. The pleural surface is lined by a single layer of mesothelial cells anchored to a basement membrane that lies on layers of collagen and elastic tissues containing vascular and lymphatic vessels. Mesothelial cells express the secretory component of IgA, which is otherwise limited to surfaces with direct environmental contact (3). Although the visceral and parietal pleurae are opposing parts of the same continuous membrane, there are major functional differences between them.

The visceral pleura is innervated by branches of the vagus nerves and sympathetic nerve trunks. It has an arterial blood supply from the bronchial arteries and a venous return that passes first into the pulmonary veins and then into the left atrium with the exception of some hilar regions that are drained by bronchial veins into the right atrium (4). The lymphatic vessels that pass through the visceral pleura represent the superficial layer of pulmonary lymphatic system, which are intersegmental, especially in the lower lobes (5). An external elastic lamina supports the mesothelial cells and an internal layer invests the vessels to become continuous with the pulmonary interstitium. It is important for pathologists to determine the integrity of the visceral pleural elastin when assessing for pleural invasion by primary lung carcinomas and this factor affects pathologic tumor staging (6).

The parietal pleura is anatomically, histologically, and functionally different from the visceral pleura. The parietal pleura is innervated by branches of the intercostal nerves, which are responsible for the pain associated with pleurisy. The arterial and venous blood supplies to the parietal pleura are derived from the intercostal vessels. Although the mesothelial cells that lie on the surface of the parietal pleura are cytologically similar to those that form the continuous membrane over the visceral pleura, this surface is interrupted by stomata that range from 2 to 12 μm in diameter. Li (7) described the stomata on the human diaphragmatic pleura as usually penetrating deep through connective tissue with apparent communication between the pleural cavity and the underlying lymphatic lacunae. In some areas, stomata are covered with mesothelial cells that have great microvilli; these microvilli are longer than those in other mesothelial cells and contain a denser network of filaments. The underlying lymphatic vessels drain directly into intercostal channels and then into the mediastinum (8–12) where the mesothelium covers collections of macrophages called Kampmeier foci (13). Boutin et al. (14) showed the concentration of asbestos fibers in these areas, which are sometimes termed "black spots" when they contain a concentration of carbon in individuals who have inhaled coal dust. Miserocchi et al. (15) described asbestos-fiber accumulation in "black spots" corresponding to the stomata.

Peritoneum

Stomata have also been described on the peritoneum of the abdominal wall, omentum, mesentery, ovaries, and pelvis, as well as on the underside of the diaphragm (16,17). Milky spots on the peritoneum, particularly the omentum, have been described as glomerular-like capillary networks of blood vessels that enable fluid exchange between the peritoneal cavity, bloodstream, and surrounding omental tissue. They lie directly beneath the stomata and are associated with macrophages, T and B lymphocytes, and plasma cells. They are said to be a preferred site for implantation of cancer cells in patients with peritoneal carcinomatosis (18).

MORPHOLOGY

Mesothelial Cells

The mesothelial cells lining pleural and peritoneal spaces are 16 to 40 μm in diameter with rounded nuclei and a relatively large amount of cytoplasm. In histologic sections taken at right angles to the serosal surface, normal mesothelial cells are very thin, whereas *en face* sections show regular sheets with evenly spaced cells. Defoliated sheets of normal mesothelial cells may also be evident in cytologic preparations of peritoneal washes taken during a laparotomy (Fig. 20.3).

Effusion cytology preparations typically contain normal mesothelial cells. They have abundant cytoplasm with centrally placed round or elliptical nuclei (Fig. 20.4). The cell borders may appear fuzzy reflecting the presence of microvilli, and clear intercellular windows may be seen between clumps of mesothelial cells (19).

Mesothelial cells undergo variably pronounced proliferative and hyperplastic changes in a variety of reactive processes. They generally retain their abundant cytoplasm, but the cell borders are sometimes less sharply defined. The nuclei are larger, both absolutely and relatively, and the

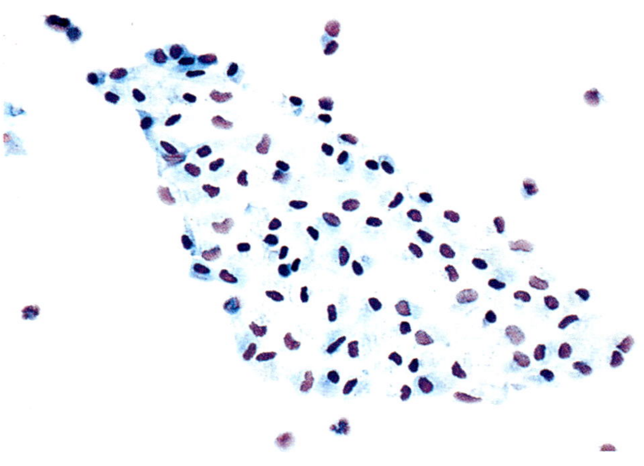

FIGURE 20.3 In this peritoneal wash specimen, a sheet of normal mesothelial cells has been detached.

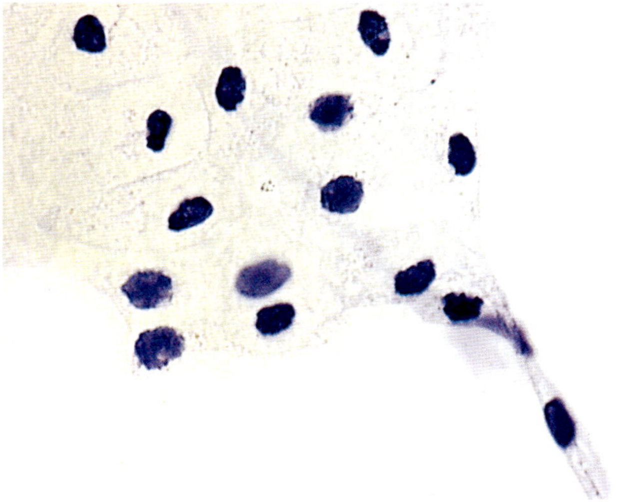

FIGURE 20.4 At higher magnification, a sheet of relatively normal mesothelial cells with abundant, clear cytoplasm and crisply defined cell borders. The centrally placed nuclei are small and have a homogeneous chromatin.

chromatin pattern is more hyperchromatic. Nucleoli are often present and prominent (Figs. 20.5 to 20.8) and multinucleation may occur (19). Normal mitotic figures are commonly seen. The cytoplasm may acquire large vacuoles as the cells degenerate and imbibe fluid (Figs. 20.9 to 20.13).

Submesothelial Layer

Much of the submesothelial layer is composed of types I and III collagen, fibronectin, proteoglycans, elastin, and other extracellular proteins. Basement membrane proteins, including type IV collagen and laminin, are found at the mesothelial cell–stromal interface. Normally, the submesothelial layer contains few cells, but during reactive processes, the submesothelial layer may become much more prominent as keratin-positive spindled mesothelial cells, myofibroblasts, inflammatory cells, and capillaries proliferate there. The term "visceral fasciae" has been used to

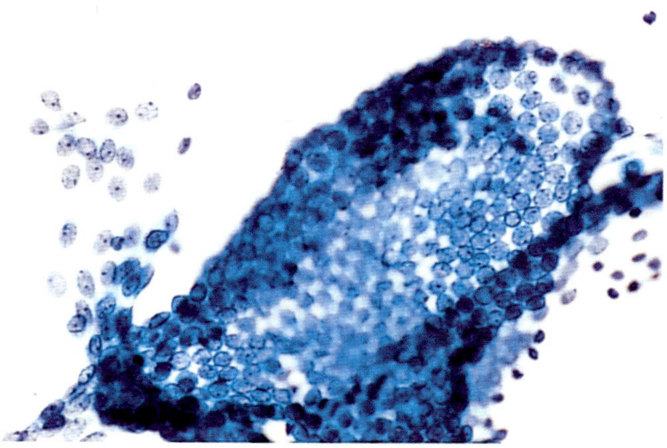

FIGURE 20.5 This detached fragment of reactive mesothelium shows an intact mesothelial layer with cells in two phases of the reactive process shown in Figures 20.6 and 20.7.

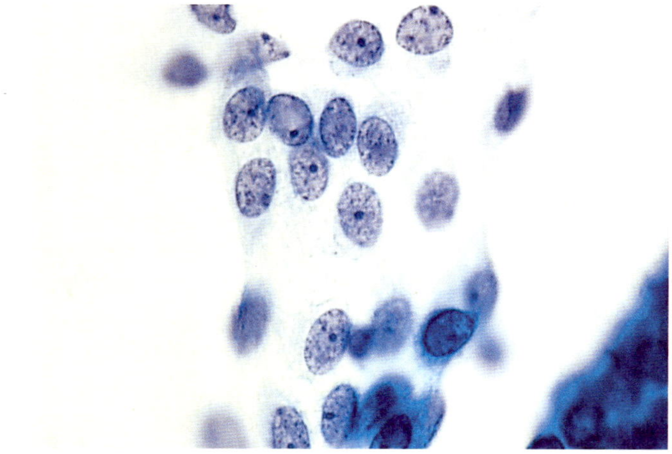

FIGURE 20.6 The reactive mesothelial cells from the left side of Figure 20.5 have abundant cytoplasm, and the nuclei are larger with a more vesicular chromatin pattern. Nucleoli are present but not prominent.

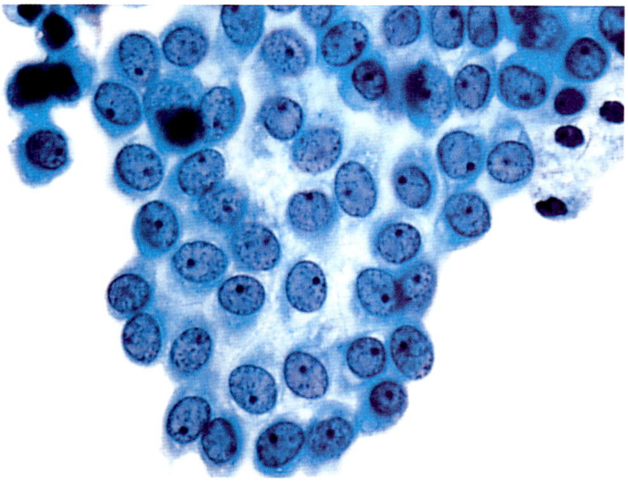

FIGURE 20.7 The more reactive mesothelial cells from the right side of Figure 20.6 have less cytoplasm, larger nuclei with a more vesicular chromatin pattern, and more prominent nucleoli.

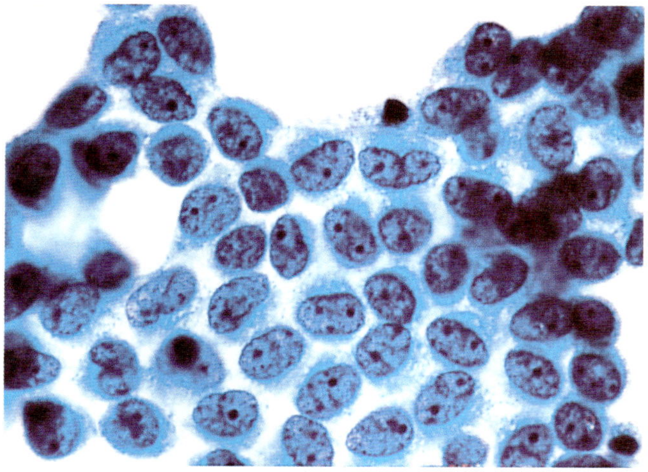

FIGURE 20.8 In this sheet of reactive mesothelial cells, the cytoplasm is smaller and the nuclei are relatively larger and have a more irregular chromatin pattern.

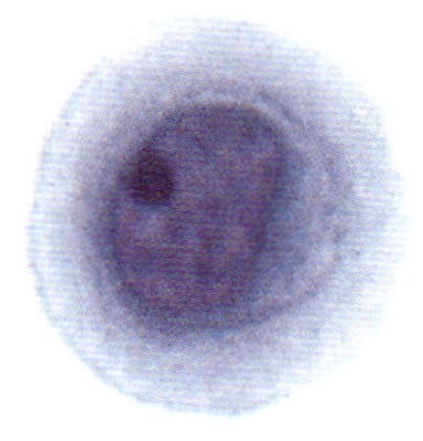

FIGURE 20.9 This individual reactive mesothelial cell has a limited amount of cytoplasm and a relatively large nucleus with a nucleolus. The cell border is highly irregular and fuzzy, consistent with the presence of the numerous elongated microvilli, which are evident on electron microscopy (see Fig. 20.15). The cytoplasm is divided into an outer less dense layer and an inner denser layer, which, ultrastructurally, corresponds to the presence of intermediate filaments with the characteristics of keratin (see Fig. 20.16).

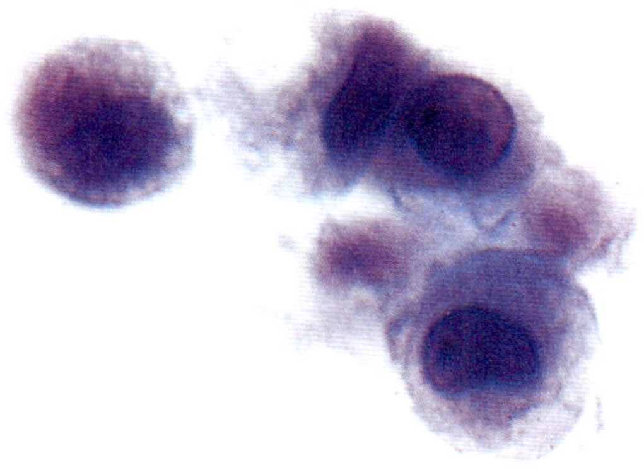

FIGURE 20.10 These reactive mesothelial cells are loosely joined together. The uppermost cell has a vacuole in the cytoplasm, which could be either a vesicle or an intracytoplasmic lumen.

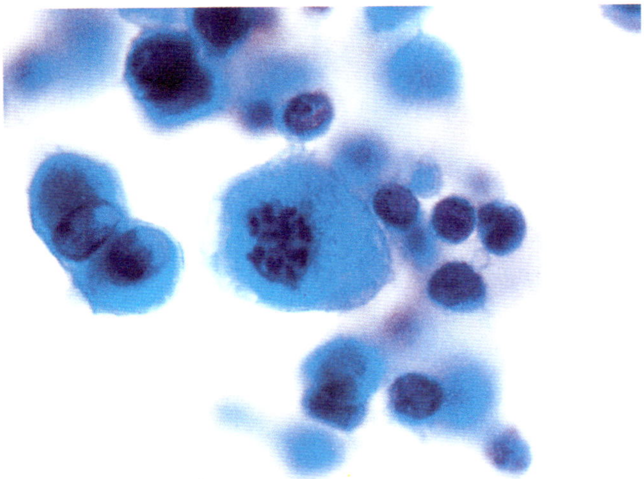

FIGURE 20.11 Normal mitotic figures may be seen in the proliferating cells of reactive mesothelium.

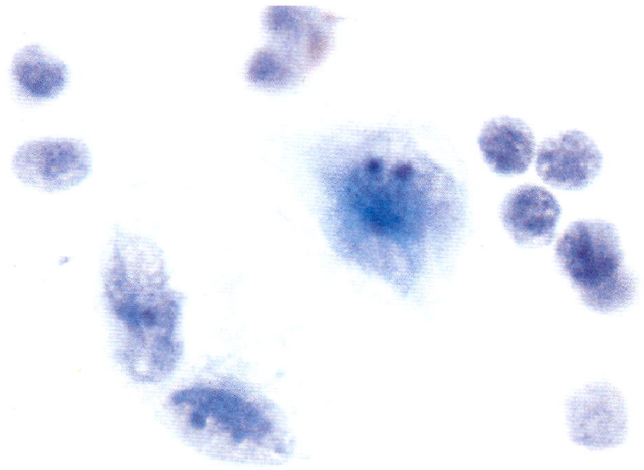

FIGURE 20.12 Reactive mesothelial cells may degenerate and swell. These three cells have abundant multivacuolated cytoplasm and large nuclei with prominent nucleoli.

describe the connective tissue lying immediately beneath the mesothelium and surrounding the viscera. It varies in thickness from approximately 134 μm around the lung to 792 μm around the heart, and 987 μm around the abdominal viscera.

The submesothelial mesenchymal cell population serves as the anchoring substratum for the mesothelium. Both mesothelial and submesothelial cells contribute to the extracellular proteins that comprise the matrix. Collagens and laminin make account for approximately 3% of the total protein synthesized by mesothelial cells.

It is not clear whether submesothelial cells serve as a source of mesothelial cell renewal in normal development or in conditions of rapid mesothelial cell turnover. Data from early ultrastructural and kinetics studies utilizing thymidine incorporation suggested that stromal cells contribute to the repopulation of denuded mesothelium (20,21). This notion is consistent with the observation that submesothelial cells, when stimulated to proliferate, synthesize keratin, and assume a more epithelioid morphology. However, later

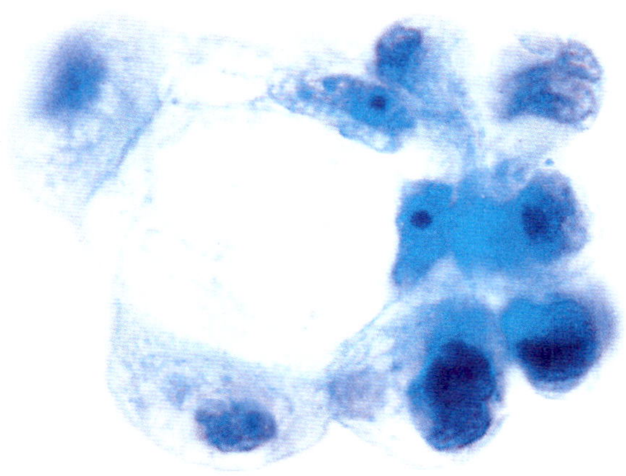

FIGURE 20.13 When markedly reactive mesothelial cells form irregular groups and combine with degenerating forms, they may mimic the appearance of mucin-producing adenocarcinoma.

studies have demonstrated that healing of injured serosa is usually accomplished by multiplication and migration of surface mesothelial cells at the edges of the wounded area (22).

HISTOCHEMISTRY

Mesothelial cells secrete hyaluronan which can sometimes be visualized as mucicarmine-positive intracytoplasmic vacuoles, but is often completely removed during specimen processing resulting in the presence of optically clear vacuoles (23,24). Cytoplasmic vacuoles in mesothelioma are expected to be negative for periodic acid-Schiff stain with diastase digestion, unlike the mucin-containing vacuoles in adenocarcinoma (24,25).

The main constituents of the submesothelial tissue are elastic fibers, staining with Verhoeff van Gieson; collagen, staining with Masson trichrome; and glycosylated proteins, including glycosaminoglycans, staining with Alcian blue.

ULTRASTRUCTURE

The most characteristic morphologic feature of mesothelial cells is the presence of numerous long and slender microvilli (Figs. 20.14 and 20.15), measuring up to 3 μm in length and 0.1 μm in diameter. They are more numerous in caudal portions of the parietal pleura and in the visceral pleura.

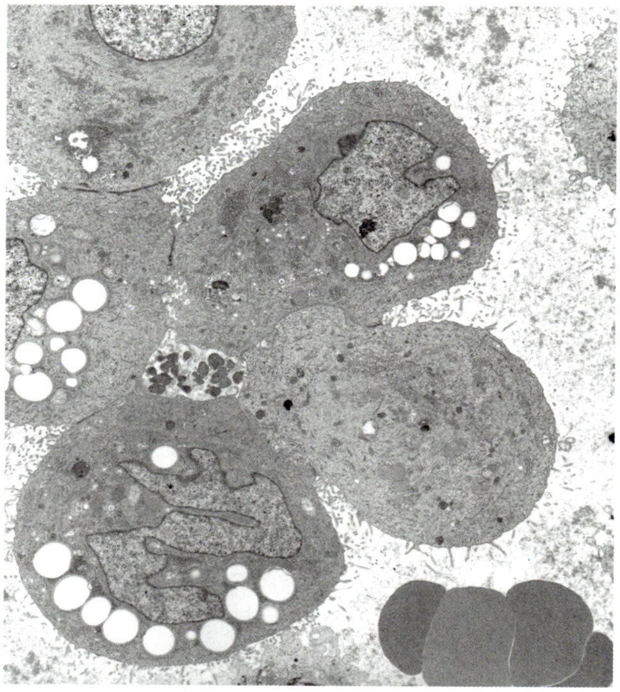

FIGURE 20.15 Ultrastructure of a cluster of detached mesothelial cells within a pleural effusion. Cytoplasmic lipid droplets impart a vacuolated appearance to some cells. Note the numerous long microvilli, which impart the "fuzzy" appearance to these cells at the light microscopic level.

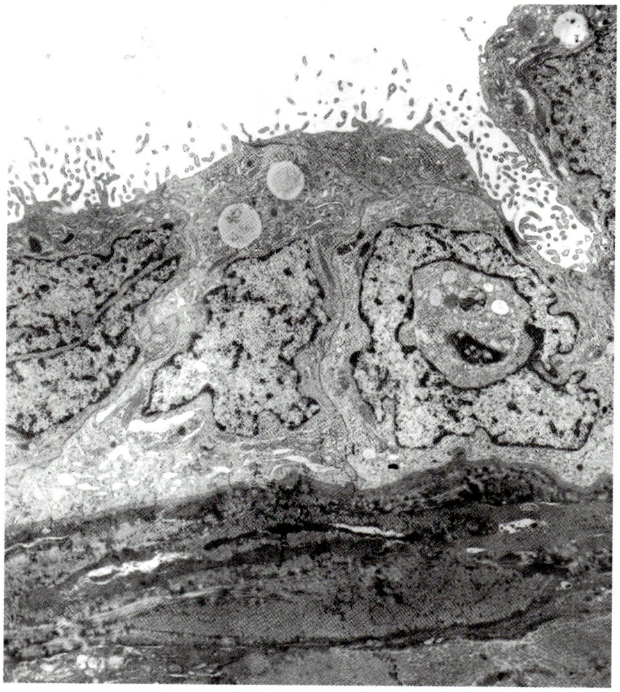

FIGURE 20.14 Mesothelial cells with their elongated microvilli, cover the surface of the serosa. The subjacent stroma is composed of collagen and fibroblasts.

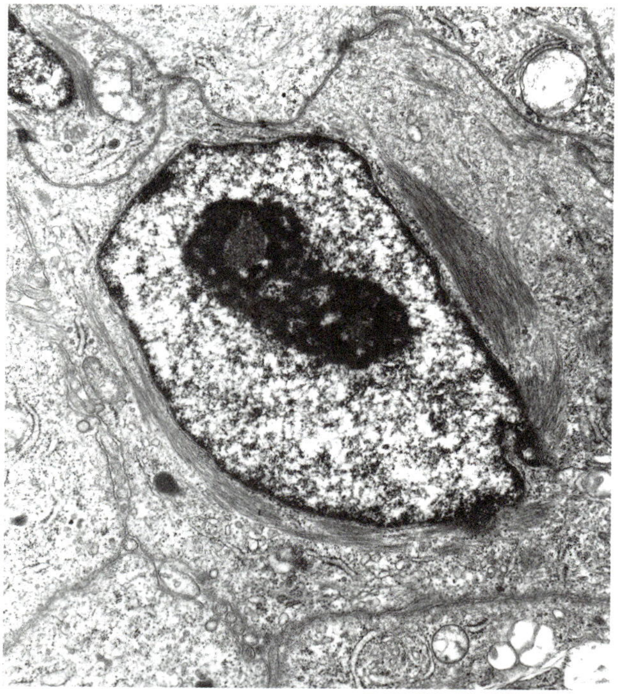

FIGURE 20.16 Ultrastructure of a mesothelial cell. Intermediate filaments are arranged in a perinuclear distribution. This distribution accounts for the appearance of a so-called "two-tone" cytoplasm by light microscopy.

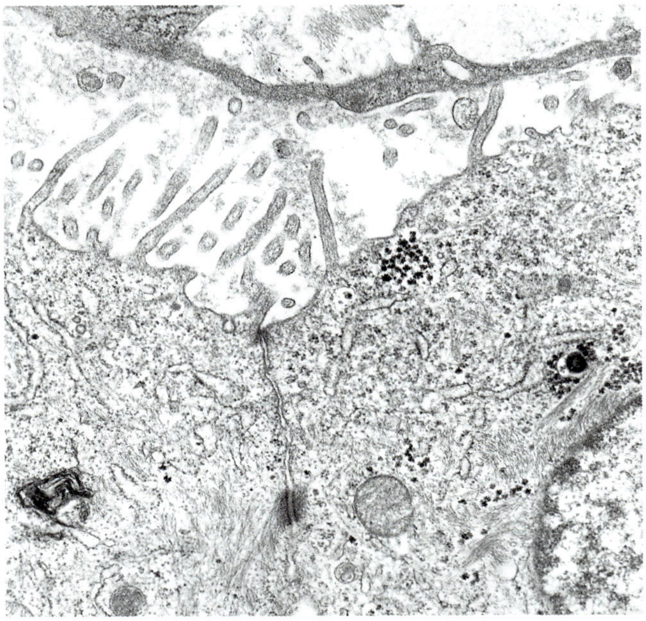

FIGURE 20.17 High magnification of the luminal aspect of two mesothelial cells. Note the small tight junction, subjacent desmosome, and the cytoskeletal filaments within the microvilli.

Tight junctions, gap junctions, and desmosomes may all be seen. Intermediate filaments are often prominent and tend to be arranged in a perinuclear, circumferential distribution, although they do not aggregate into bundles (Figs. 20.16 and 20.17).

REACTIVE MESOTHELIUM AND ITS DIFFERENTIAL DIAGNOSIS

Reactive mesothelial cells have a very wide range of appearances. The simplest form is a single layer of epithelioid cells that assume a cuboidal shape, but columnar forms and very large forms also occur. Reactive epithelioid mesothelial cells may also aggregate in heaped-up groups on the serosal surface (Figs. 20.18 to 20.21).

The capacity for reactive mesothelial cells to produce morphologic patterns mimicking neoplasia is well known and may be a source of diagnostic confusion with malignancy. Reactive processes can be diffused or localized, and show epithelioid or spindled morphology. Reactive mesothelial cell populations are often quite cellular with high mitotic rates and considerable cytologic atypia. Moreover, reactive mesothelial cells may form papillary groups, or become entrapped within fibrosis in a manner that mimics invasion (25,26). Consequently, mesothelial proliferations can mimic a variety of epithelioid and sarcomatoid malignancies. Clinical presentation and follow-up information can aid distinction between mesothelioma and these entities, but morphology and immunohistochemistry also play key roles.

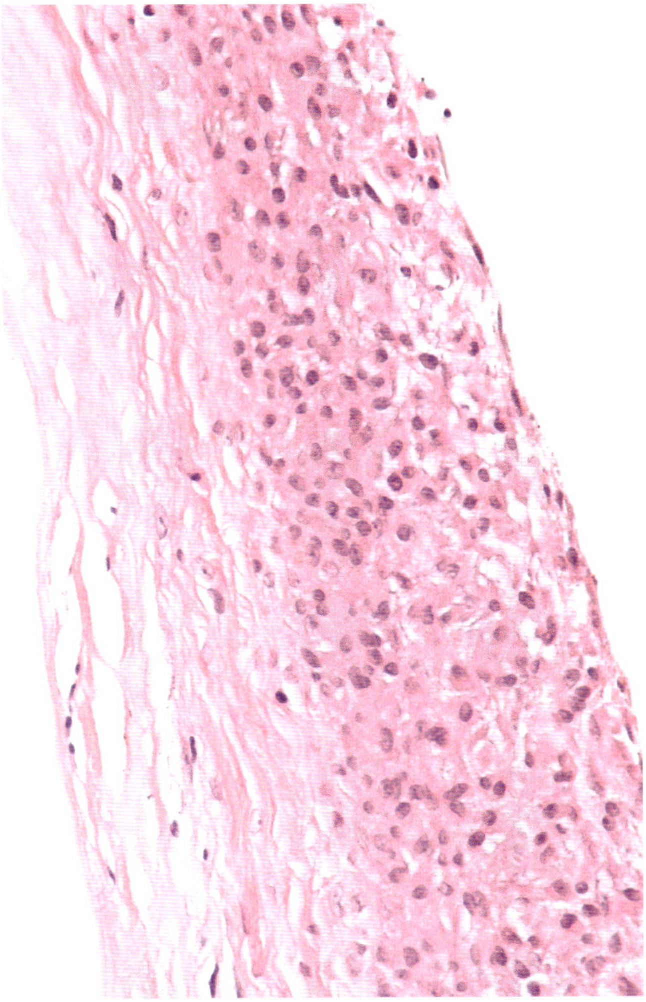

FIGURE 20.18 Reactive mesothelial cells forming a stratified lining along the visceral pleura, from a case with pleural blebs.

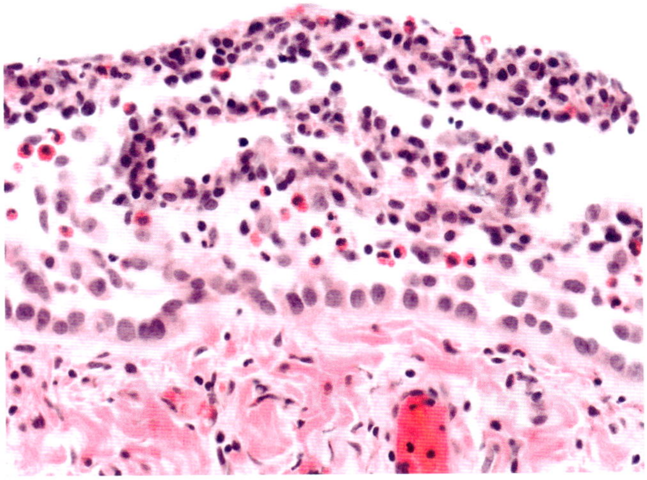

FIGURE 20.19 Reactive mesothelial cells with an overlying membrane of eosinophils and chronic inflammatory cells from a patient with a spontaneous pneumothorax.

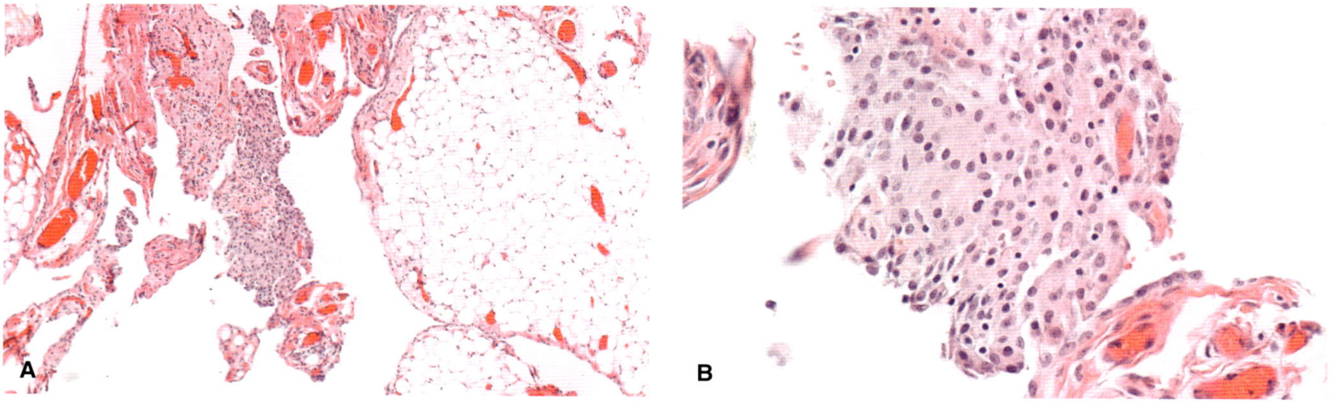

FIGURE 20.20 Reactive mesothelial cells in omental tissue, resected as part of staging for a patient with endometrial carcinoma (**A:** 100×; **B:** 400×).

Morphologic Distinction Between Reactive Mesothelial Proliferations and Malignancy

True stromal invasion, such as invasion into adipose tissue or lung parenchyma, allows definitive diagnosis of malignancy (25,26), but is not always captured in the sampled tissue. Complex papillae, cellular stratification, severe cytologic atypia, expansile nodules, and the absence of inflammatory infiltrates all support a diagnosis of mesothelioma. Other features are more characteristic of reactive mesothelial hyperplasia. Nonneoplastic mesothelial proliferations typically show zonation with areas of highest cellularity near the mesothelial surface and decreased cellularity with increased fibrosis away from the surface. On the other hand, malignant proliferations maintain high cellularity deep in the stroma or show reverse zonation (Table 20.1).

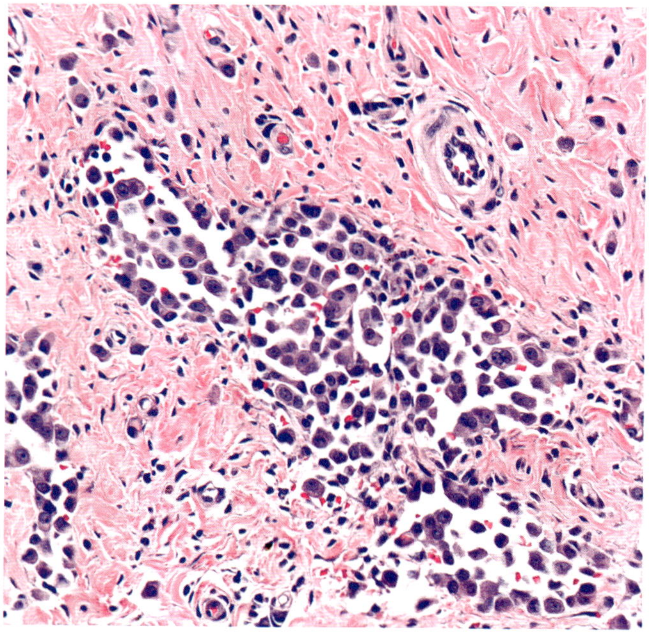

FIGURE 20.21 Hernia sac with reactive mesothelial cells.

For cases with bland spindled mesothelial proliferations, the differential diagnosis includes a fibrous pleurisy in the setting of organizing pleural effusions (Fig. 20.22) and desmoplastic mesothelioma. Both disorders may contain regions that show increased cellularity in a predominantly fibrous background. Storiform growth, bland necrosis without significant inflammation or cellular debris, disorganized and/or uneven growth, and abrupt changes in cellularity are more common in mesothelioma, whereas abundant perpendicularly oriented vessels and restriction of hypercellularity to a surface zone are more suggestive of a reactive process (25,26). Definitive fat invasion remains diagnostic of malignancy in this morphologic context but must be distinguished from “fake fat,” a term used to refer to horizontally aligned artifactual spaces that may be seen in densely fibrotic pleural tissue (Fig. 20.23) (27).

Psammoma bodies (Fig. 20.24) are laminated calcified structures that most likely arise through concentric calcification of dead cells; they are nonspecific and have been observed in inflammatory processes accompanied by mesothelial hyperplasia, as well as in mesothelioma and epithelial neoplasia. Decidual changes should also not be mistaken for metastatic carcinoma or malignant mesothelioma. Although

TABLE 20.1 Morphologic Features of Reactive Mesothelium versus Mesothelioma

Feature	Reaction	Mesothelioma
Proliferation	Zonal	Diffuse/irregular
Capillaries	“Sunburst” perpendicular to the surface	Random
Stromal expansion	Flat	Nodular
Bland necrosis	None	Often
Spread	Confined to pleura	Invades normal stroma

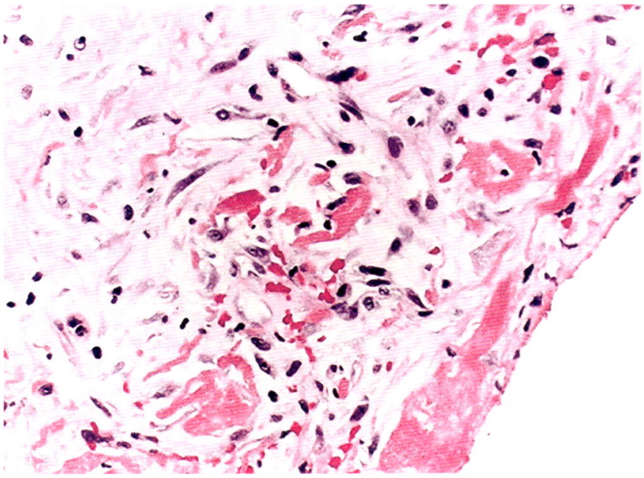

FIGURE 20.22 Organizing pleuritis. There is zonation (more cellular under the effusion and less cellular/more fibrotic away from the effusion) and capillaries perpendicular to the pleural surface.

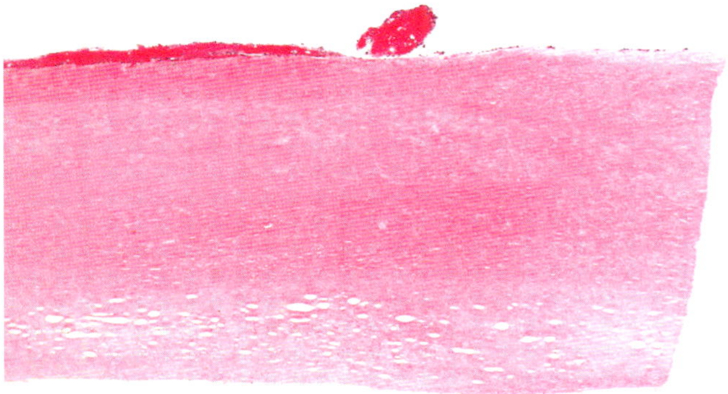

FIGURE 20.23 "Fake fat" refers to horizontally aligned artifactual spaces that may be seen in densely fibrotic pleural tissue and should not be confused with fat invasion by a malignancy.

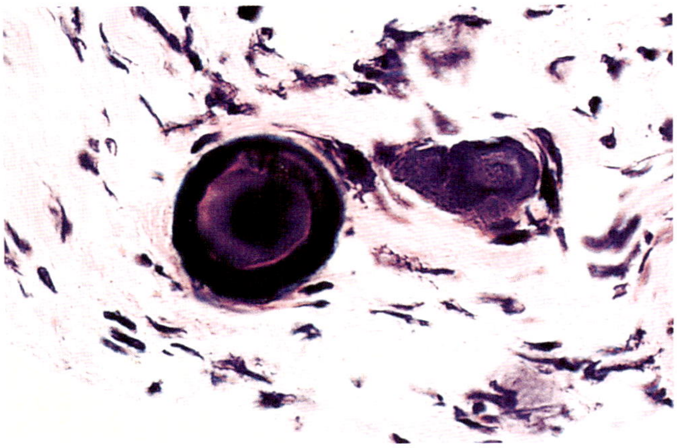

FIGURE 20.24 Psammoma bodies in the left pelvic peritoneum of a 58-year-old female.

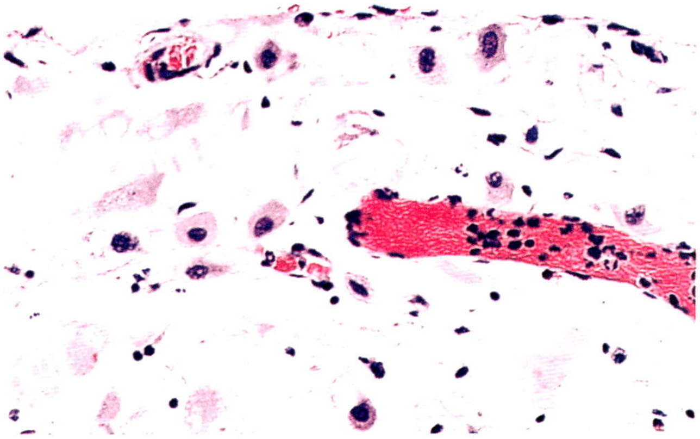

FIGURE 20.25 Decidual change in the pelvis during pregnancy is seen in the subserosal tissue. Loosely cohesive cells with abundant eosinophilic cytoplasm are present.

usually encountered in the submesothelial layer of the pelvic peritoneal surfaces, decidual changes may be seen in the serosal surfaces of the gastrointestinal tract, liver, spleen, diaphragm, and within lymph nodes (Fig. 20.25). Finally, benign mesothelial cells can be detected in mediastinal and abdominal lymph nodes (Figs. 20.26 and 20.27) (28–30) and should not be misinterpreted as malignant solely due to their presence within a lymph node.

Immunohistochemical Features of Mesothelial Lineage

Mesothelial cells show strong diffuse immunoreactivity with pan-keratin antibodies including AE1/AE3, OSCAR, CK8/18, and Cam5.2 (Fig. 20.28). Immunohistochemistry for CK7 may be positive but CK20 is typically negative (31). Reactive mesothelial cells commonly show immunoreactivity for desmin (32–34), and SMA immunoreactivity is common among spindled mesothelial cells (32,35). Sarcomatoid mesotheliomas also commonly show SMA positivity (36), and a minority are positive for desmin (32–34); this

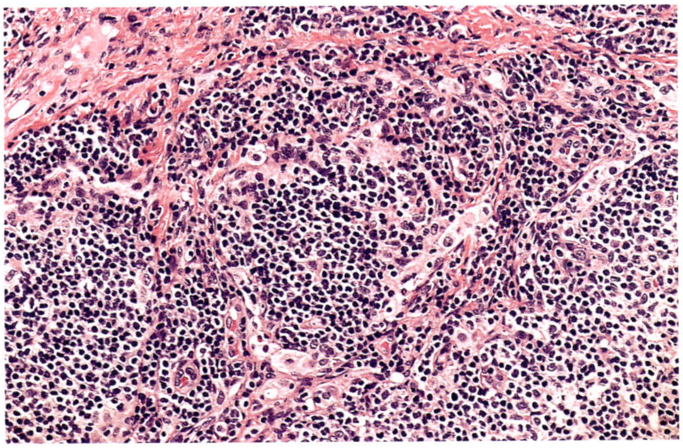

FIGURE 20.26 Internal mammary lymph node with large epithelioid cells in the sinuses found during coronary artery bypass graft surgery in a 61-year-old man. No pleural lesion was present.

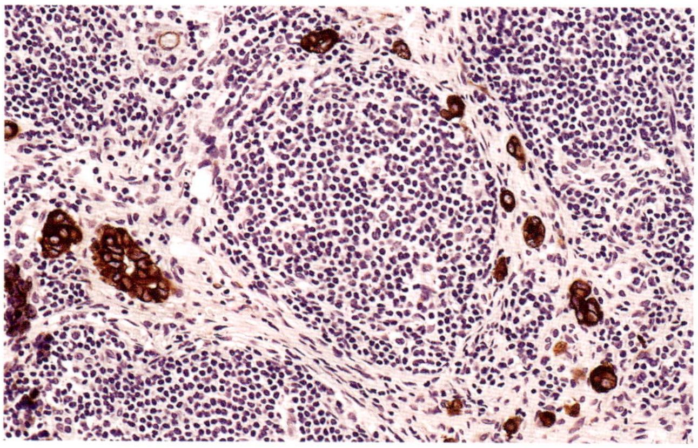

FIGURE 20.27 Immunohistochemical stain for keratin (AE1/AE3) demonstrates the epithelioid cells singly and in groups. No tumor was found and the cells were considered as mesothelial inclusions, an uncommon but well-described phenomenon in mediastinal lymph nodes.

staining pattern should not be interpreted to reflect smooth muscle differentiation and does not have adequate specificity to be useful in the distinction of benign and malignant proliferations.

Immunohistochemical markers that are typically positive in mesothelial cells and negative in most adenocarcinomas include nuclear and cytoplasmic calretinin staining, nuclear WT1 staining, membranous staining for D2-40 (i.e., podoplanin), and cytoplasmic expression of CK5/6 (Table 20.2 and Figs. 20.29 to 20.31). The specificity of these stains is less than 100%, with "false-positive" staining seen in variable proportions among carcinomas that may enter the differential diagnosis for epithelioid mesothelioma. For instance, pulmonary adenocarcinomas show calretinin staining in 5% to 10% of cases, CK5/6 staining in 2% to 20% of cases, and D2-40 staining in up to 15% of cases (25,37). Almost all pulmonary squamous cell carcinomas show CK5/6 staining, and both calretinin and D2-40 staining are quite common, being present in 23% to 40% and 15% to 50% of cases, respectively (25,37–39). WT1,

TABLE 20.2 Immunohistochemical Markers Useful in the Distinction of Mesothelium From Adenocarcinoma

Predominantly Positive in Mesothelial Proliferations	Predominantly Positive in Adenocarcinoma
HEG1 (membrane staining)	Claudin-4 (membrane staining)
Calretinin (nuclear and cytoplasmic staining)	Ber-EP4 (membrane staining)
CK5/6 (cytoplasmic staining)	MOC-31 (membrane staining)
WT-1 (nuclear staining)	
D2-40 (membrane staining)	

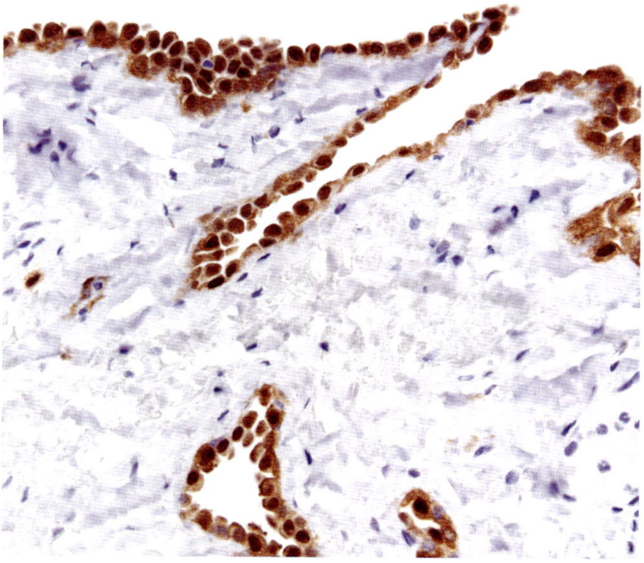

FIGURE 20.29 Calretinin immunohistochemical staining of both nucleus and cytoplasm in benign reactive mesothelial cells.

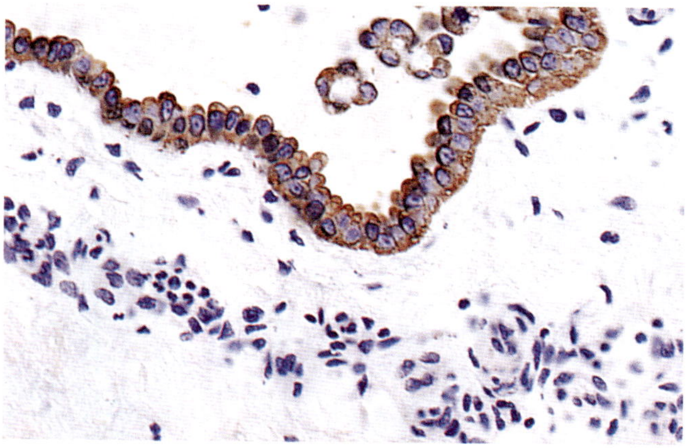

FIGURE 20.28 Keratin expression in reactive mesothelium and detached mesothelial cells, stained with a cocktail of monoclonal antibodies (AE1/AE3).

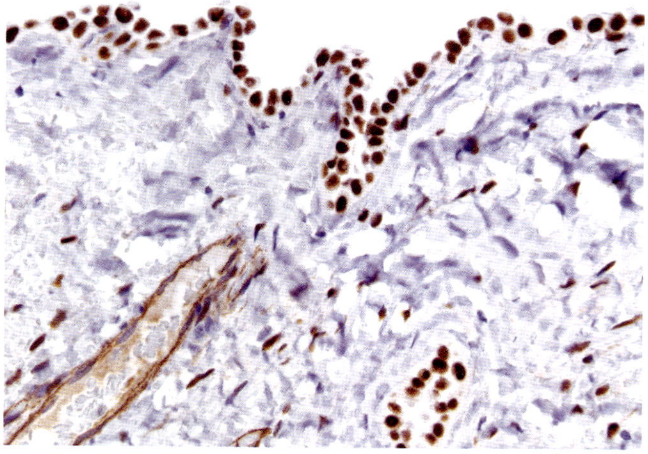

FIGURE 20.30 WT-1 immunoreactivity in benign reactive mesothelium is nuclear (original magnification 40×).

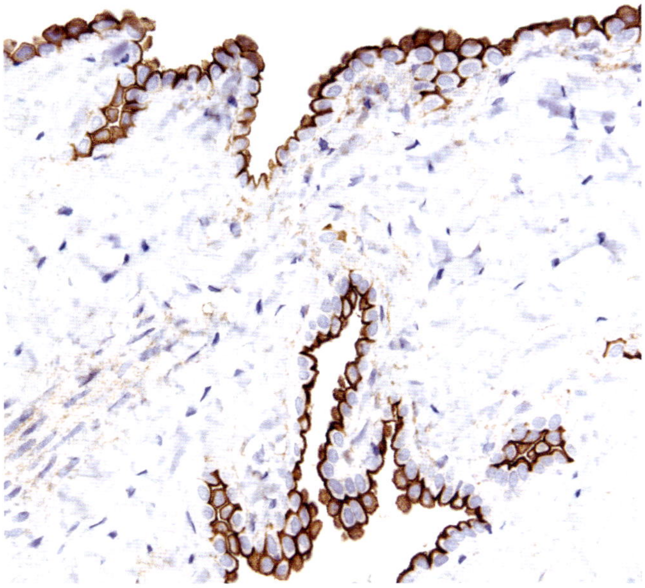

FIGURE 20.31 D2-40 immunoreactivity in benign reactive mesothelium is predominantly membranous.

although highly specific when excluding non–small-cell lung carcinoma, is positive in most serous carcinomas (25,40). D2-40 stains a considerable proportion of ovarian serous carcinomas (41) and calretinin is frequently positive in sex cord stromal tumors and adrenocortical carcinoma (37).

Membranous staining with the HEG1 SKM9-2 antibody (Fig. 20.32A) shows high specificity for mesothelial proliferations versus non–small-cell lung carcinoma (99.7%) and similar sensitivity for epithelioid mesothelioma (91%) when compared to the aforementioned markers (42,43). This antibody clone recognizes glycopeptide modifications on HEG1 protein, a mucin-like membrane protein (44,45). However, HEG1 is less useful for distinguishing peritoneal lesions, as it stains an approximately 17% to 67% of serous carcinomas (42,46).

The above markers stain a high proportion of both benign reactive mesothelial proliferations and epithelioid mesotheliomas, but only a minority of sarcomatoid mesotheliomas; their sensitivities for sarcomatoid mesothelioma range up to 30% to 40% in most studies (25,47–50), HEG1 is reportedly positive in 63% of cases (43). Keratin immunostaining is at least focally retained in most (~95%) sarcomatoid mesothelioma (25,47) and can be of use for highlighting fat invasion. Osteosarcomatous,

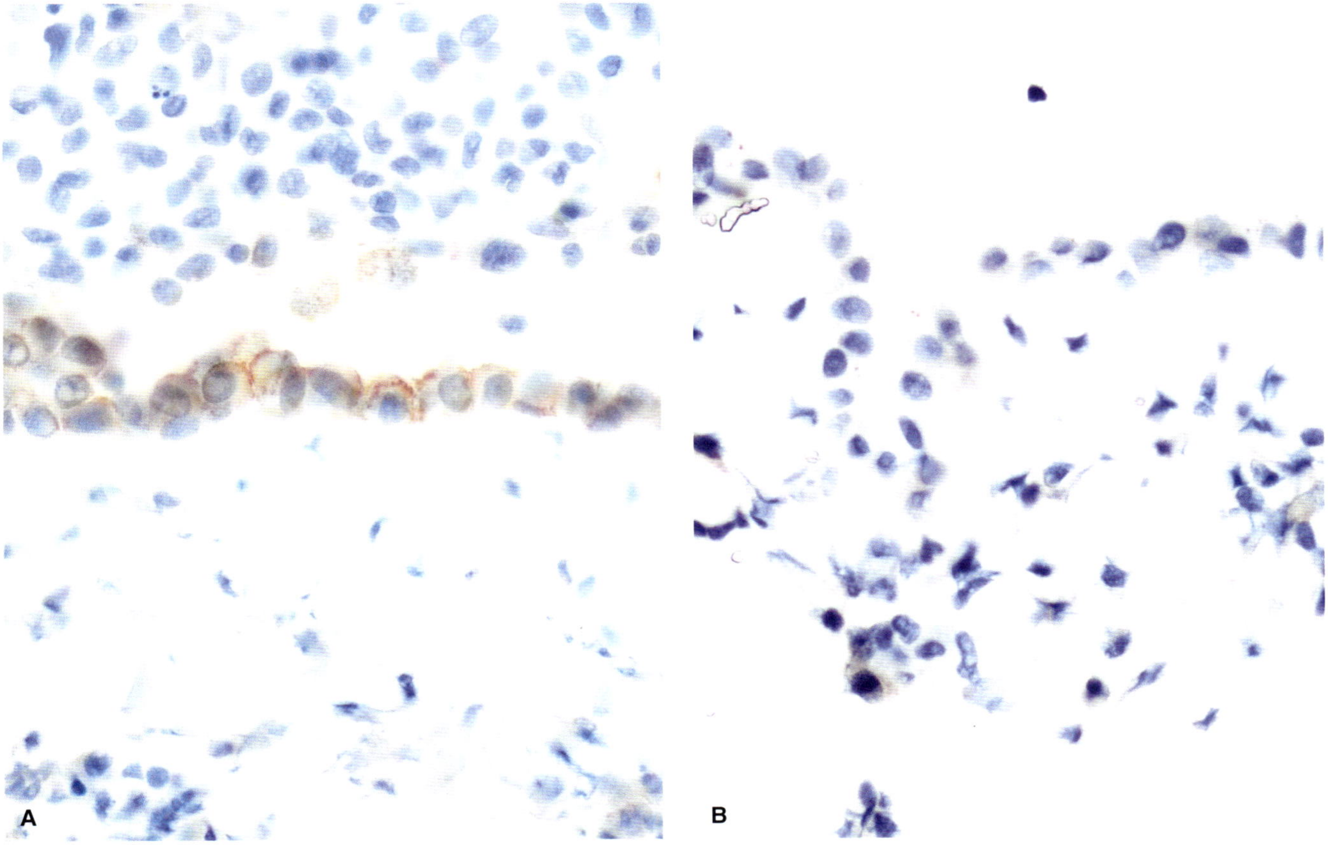

FIGURE 20.32 (**A**) HEG1 and (**B**) claudin-4 staining in reactive mesothelial cells from a case of spontaneous pneumothorax.

chondrosarcomatous, and rhabdomyosarcomatous differentiation can occur in mesothelioma, but the majority of mesotheliomas with such heterologous elements still show keratin immunoreactivity (51).

Also of use are markers expected to be negative in cells of mesothelial lineage and positive in those of epithelial lineage. Among markers commonly used for this purpose are claudin-4 (Fig. 20.32B), MOC-31, and Ber-EP4. Claudin-4 has shown higher specificity for carcinoma versus epithelioid mesothelioma (99.9% vs. 65–68%) and similar sensitivity for non–small-cell lung carcinoma (93%) compared with MOC31 and Ber-EP4 (43,52). The sensitivity of claudin-4 for ovarian carcinomas may be slightly lower, although literature is conflicting: one study reported 72% sensitivity across 212 ovarian serous carcinomas (53), whereas a meta-analysis including 174 serous carcinomas reported 99% sensitivity (43). Notably, claudin-4, MOC-31, and Ber-EP4 staining are present in approximately one-third of sarcomatoid carcinomas (52), such that a negative result in a malignant spindle cell proliferation does not exclude the possibility of carcinoma.

Additional markers have previously been proposed for distinguishing mesothelial and epithelial lineage, including the 'mesothelial' markers HBME-1, thrombomodulin, mesothelin, N-cadherin, and EMA, and the "carcinoma:" markers CEA, Leu-M1, B72.3 and BG-8 (54), although these markers tend to have more limited sensitivity and specificity than those previously mentioned. The use of combinations of two or more mesothelial and two or more epithelial markers was previously recommended (25,54), but more recently the use of two stains, HEG1 and claudin-4, has been proposed to provide sufficient diagnostic accuracy in most cases (43).

Immunohistochemical Distinction of Benign and Malignant Mesothelial Proliferations

One of the earliest immunostains useful for distinguishing benign and malignant mesothelial proliferations is BAP1 (Fig. 20.33A), which has now been assessed for this purpose in numerous studies. Data from a large meta-analysis showed that loss of nuclear BAP1 is always an indicator of malignancy in mesothelial cells (55,56). Other immunohistochemical or FISH-based markers that are now recognized as diagnostic of malignancy in this context are MTAP immunohistochemistry, *CDKN2A* (which encodes p16^{INK4A} protein) fluorescent in situ hybridization (Fig. 20.33B), and Merlin (the protein product of *NF2*) immunohistochemistry (Table 20.3) (56–58). The *MTAP* gene is located within 200 Kb of the *CDKN2A* locus and is commonly codeleted with *CDKN2A*. Consequently, loss of MTAP cytoplasmic immunohistochemical staining is seen in most, but not all, cases with *CDKN2A* homozygous deletion (59–61).

TABLE 20.3 Differentiation of Reactive From Neoplastic Mesothelium by Immunohistochemistry and FISH

Assay	Reactive	Mesothelioma
BAP-1 (IHC; nuclear staining)	Present	Absent
CDKN2A (FISH)	Present	Homozygous deletion
MTAP (IHC; cytoplasmic staining)	Present	Absent
Merlin (IHC; cytoplasmic/ membrane staining)	Present	Absent
NF2 (FISH)	Present	Hemizygous deletion

CDKN2A, *NF2*, and *BAP1* are all tumor suppressor genes. Their loss of activity is thought to drive mesothelioma pathogenesis, explaining the near-perfect specificity of these markers for malignancy. However, they also serve as tumor suppressors in other malignancies, necessitating confirmation of mesothelial lineage by other means.

Only a subset of mesotheliomas has loss of each of the above tumor suppressors, such that sensitivities of these markers for malignancy are limited. BAP1 has a sensitivity of ~70%

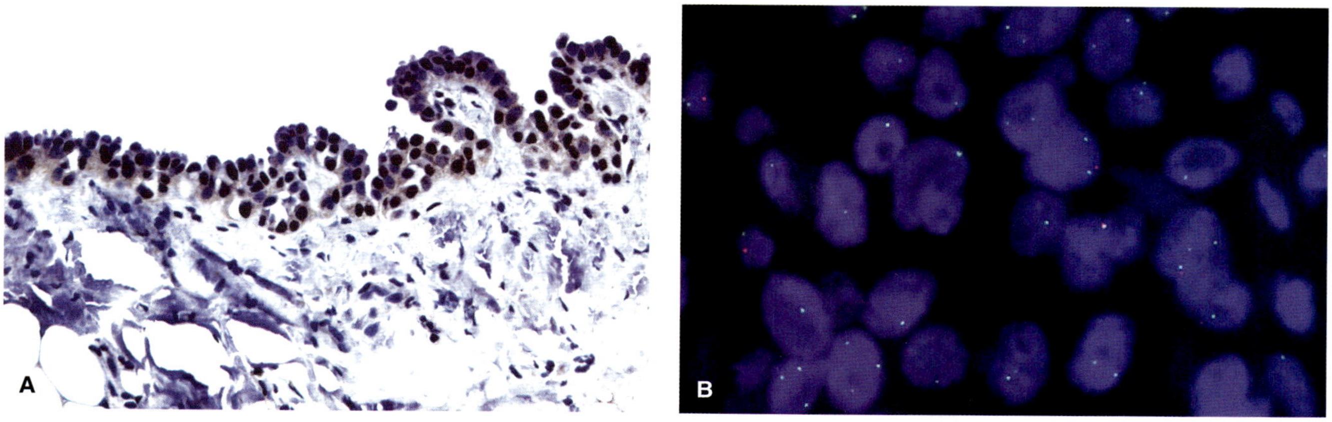

FIGURE 20.33 **A:** Immunoperoxidase stain for BAP-1 stains the nuclei of normal mesothelial cells in this section of pleura. Immunostaining is lost with BAP-1 mutation in mesothelioma. **B:** Fluorescence in situ hybridization (FISH) analysis with *CDKN2A* (*red signal*) and CEN-9 (*green signal*) dual color probe set demonstrates homozygous deletion of the *CDKN2A* gene (0 *red signals*, 1–2 *green signals*). Homozygous deletion of the *CDKN2A* gene is useful to distinguish malignant mesothelioma from reactive mesothelial hyperplasia.

for epithelioid mesothelioma and less than ~20% for sarcomatoid mesothelioma (56). *CDKN2A* FISH has a sensitivity of 60% to 70% for epithelioid pleura mesothelioma, <50% for peritoneal mesothelioma, and over 80% for pleural sarcomatoid mesothelioma (56). Merlin immunohistochemistry has a sensitivity of 35% to 41% for epithelioid pleural mesothelioma and 39% to 67% for sarcomatoid pleural mesothelioma (57,58). When used in combination, BAP1, MTAP, and Merlin immunostains have a sensitivity of 90% for pleural mesothelioma (57). Thus, a panel of those markers can identify the majority of malignant cases, but retention of these markers does not entirely exclude the possibility of malignancy.

Promising new immunohistochemical markers for the distinction of benign and malignant mesothelial proliferations include p53 (57,62), EZH2 (63,64), 5-HMC (65), PD-L1 (66,67), and CD47 (68), but additional studies are needed to further define the performance of these markers.

MESOTHELIOMA IN SITU

Mesothelioma in situ represents an important mimic of reactive atypia in mesothelial cells, but the idea that it is a morphologically recognizable entity has a long history of controversy. In 1992 Whitaker et al (69) described what they believed was mesothelioma in situ, but all of their cases also had invasive mesothelioma so the authors could not exclude the possibility that what they were seeing was simply surface growth of an underlying mesothelioma. A further problem was that reactive mesothelial cells can show a wide variety of patterns, as discussed above. Thus, for many years, the recommendation to pathologists was that they not diagnose mesothelioma in situ.

This situation changed when molecularly based immunochemical markers of mesothelial malignancy were described. Churg et al. (70) reported the first modern series mesothelioma in situ cases in 2020 using BAP1 and MTAP immunohistochemistry. Diagnostic requirements included (1) no radiologic and/or direct visual evidence of pleural or peritoneal tumor, (2) no invasive tumor appeared within one year to avoid inclusion of inadequately sampled invasive mesothelioma, and (3) loss of BAP1 in surface mesothelial cells. Defined as such, mesothelioma in situ consisted mostly of a single layer of cuboidal, minimally atypical, mesothelial cells (Fig. 20.34). Follow-up revealed that 7 of the 10 patients ultimately developed invasive mesothelioma and an 8th patient was thought to develop an invasive mesothelioma based on clinical grounds (median time to invasion: 60 months). One patient still had not developed invasive disease at 120 months, and one did not have follow-up data available.

One of the 10 cases had papillary areas that resembled well-differentiated papillary mesothelial tumor and subsequent reports have shown that mesothelioma in situ can exactly mimic this lesion on routine morphology (71–73). Well-differentiated papillary mesothelial tumors have broad fibrovascular, myxoid, or hyalinized papillae with a single layer of lining mesothelial cells, and lack invasive growth (74,75). The phenomenon of mesothelioma in situ mimicking well-differentiated papillary mesothelial tumor appears to be more common in the peritoneal compared to the pleural cavity. Well-differentiated papillary mesothelial tumor morphology is uncommon in the pleural cavity, and when it does occur it has a high likelihood of being mesothelioma in situ (72,73).

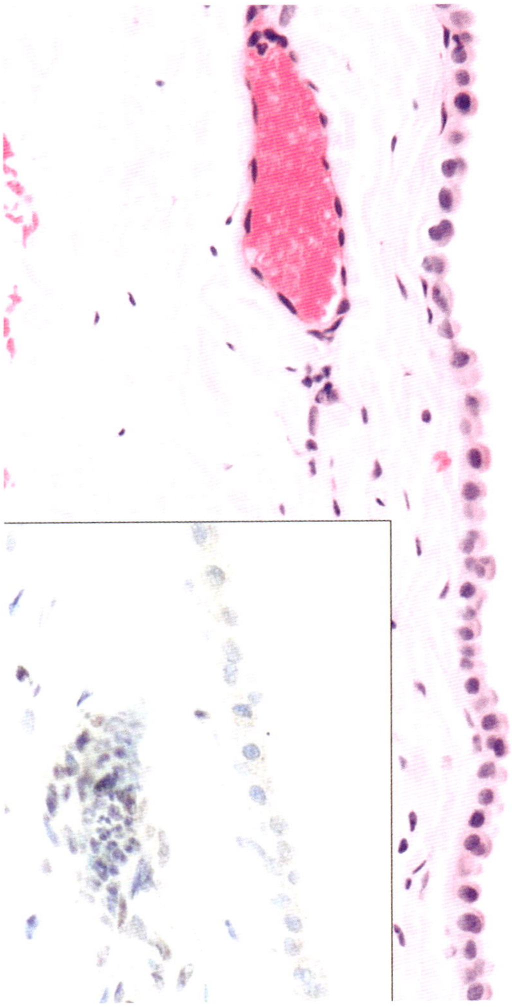

FIGURE 20.34 Typical picture of mesothelioma in situ that appears as a single layer of cuboidal, minimally atypical cells. BAP1 is lost (*inset*).

There are now roughly 40 cases of mesothelioma in situ in the literature (76). Most data suggest a relatively slow progression to invasive disease, although the transition to invasive mesothelioma is much faster in cases that have lost both BAP1 and MTAP by immunochemistry or *CDKN2A* by FISH, compared to those that have only lost BAP1 (77). Curiously, pleural mesothelioma in situ mimicking well-differentiated papillary mesothelial tumor with associated microinvasion at the time of diagnosis still had a remarkably indolent course (72), unlike the usual aggressive course of invasive mesothelioma. A small number of mesothelioma in

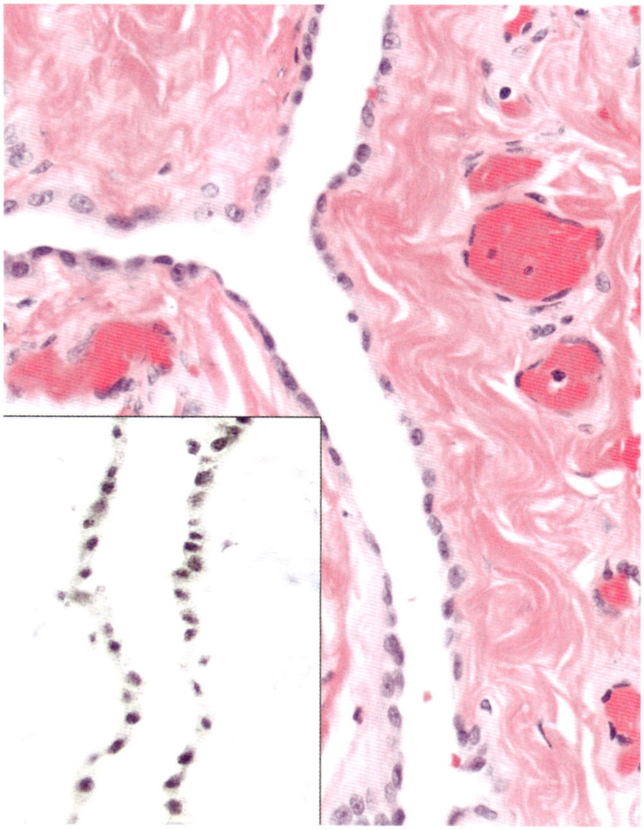

FIGURE 20.35 Reactive mesothelial cells from a case of spontaneous pneumothorax. Morphologically this picture mimics mesothelioma in situ, but BAP (*inset*) is not lost.

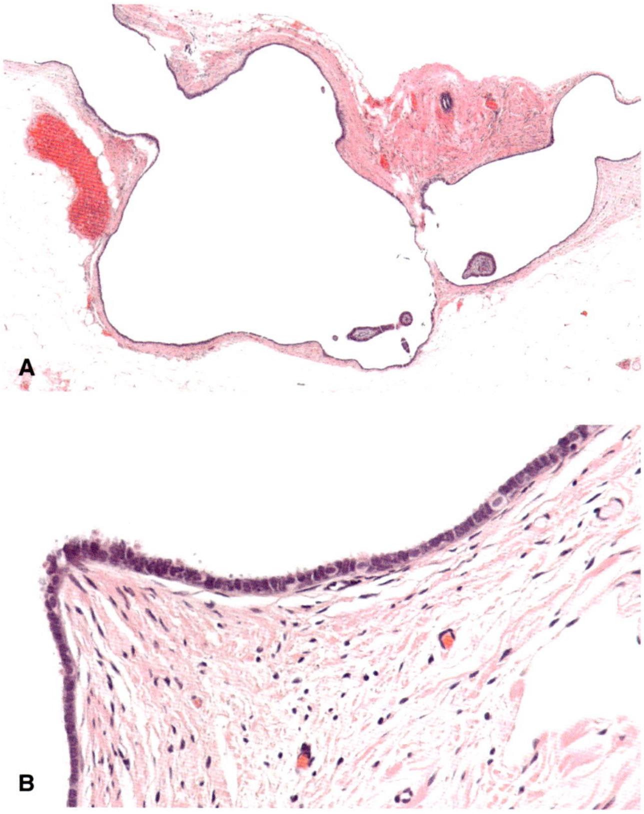

FIGURE 20.36 Endosalpingiosis identified in this omentum has (**A**) cystic structures with papillary invaginations, with (**B**) a nonstratified epithelial lining composed of ciliated cells, secretory and intercalated (peg) cells, similar to fallopian tube epithelium.

situ cases have been sequenced, with the usual result that only the *BAP1* gene shows abnormalities (76). This finding accords with the idea that *BAP1* loss/mutations occur very early in the development of invasive mesothelioma.

The problem for the diagnostic pathologist is that benign mesothelial reactions can exactly mimic flat mesothelioma in situ (Fig. 20.35). A helpful hint is that patients with mesothelioma in situ often have repeated effusions/ascites without any discernible etiology, and by definition do not have evidence of tumor on imaging or visual inspection. In such settings, and in all cases with well-differentiated papillary mesothelial tumor morphology, a BAP1 stain is recommended.

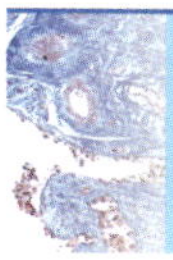

ENDOSALPINGIOSIS AND ENDOMETRIOSIS INVOLVING SEROSAL MEMBRANES

Endosalpingiosis refers to glandular spaces lined by epithelium similar to fallopian tube epithelium, with three cell types: ciliated, secretory, and intercalated (Fig. 20.36) (78). On occasion, psammoma bodies are present. Periglandular stroma containing chronic inflammatory cells is separated from the epithelium by a PAS-positive basement membrane. Endosalpingiosis may be differentiated from endometriosis by the lack of endometrial stroma or evidence of stromal hemorrhage associated with endometriosis (79–81).

This condition is seen exclusively in women and has been reported in 12.5% of omenta removed during surgery in females. A large proportion of these women have coexisting benign disease of the fallopian tube. The origin of the glandular inclusions is debated but is most likely either related to the influence of müllerian development on the peritoneal mesothelium (coelomic lining) or is a sequela of disease within the tube resulting in extratubal growth of displaced tubal epithelium. Although definitive evidence of neoplasia arising in endosalpingiosis has not been documented, considerable difficulty may be encountered when differentiating extraovarian tumor implants removed in the setting of common epithelial ovarian tumors from endosalpingiosis with cellular atypia. Evaluation of the severity of epithelial atypia, mitotic activity, the presence of ciliated cells, and the presence of invasive characteristics may aid in establishing a diagnosis of malignancy in this setting.

Endometriosis is defined by the presence of glands lined by endometrial-type epithelium surrounded by endometrial stroma, outside the uterine endometrial mucosa and myometrium (82). The condition occurs most frequently in women of childbearing age. It may occur in a variety of body

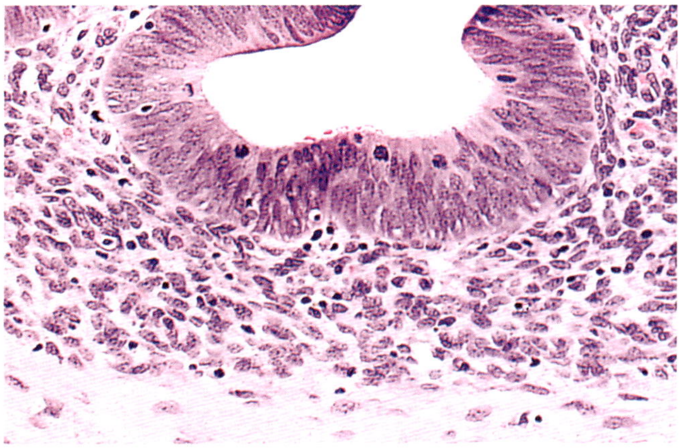

FIGURE 20.37 Endometriosis involving the peritoneum with extension into the soft tissue of the anterior abdominal wall of a 23-year-old woman. Endometrioid glands and stroma are present.

sites, ranging from the pelvic peritoneum to distant organs such as lungs, kidney, and skin, but the most frequent site is the peritoneal lining of the pelvic organs (Fig. 20.37). Although the histogenesis of endometriosis remains unclear, two general theories have been proposed: (1) ectopic growth of endometrial elements displaced from endometrial tissue through local means, such as "retrograde menstruation" of endometrial tissue into the pelvis through the uterine tubes or via vascular routes to distant organs (83,84) and (2) metaplastic change of the pelvic peritoneum along müllerian lines of differentiation (85,86). Although the mesothelium normally lacks sex steroid receptors, reactive mesothelium adjacent to endometriosis expresses focal immunoreactivity for estrogen and progesterone receptors (87).

Endometriosis may appear as brown–maroon foci on the peritoneal surfaces and may be accompanied by fibrosis or adhesions. Microscopically, endometrial stroma surrounding endometrial epithelium is present. Response to hormonal influences is often seen and may be synchronous with intrauterine endometrium. Metaplasia occurs in both epithelial and stromal elements, similar to metaplasias encountered in the endometrium of the uterus. The presence of hemosiderin-laden macrophages and fibrosis may be the only evidence that endometriosis had once been present.

REFERENCES

1. Sahn SA. State of the art. The pleura. *Am Rev Respir Dis.* 1988; 138:184–234.
2. Pistolesi M, Miniati M, Giuntini C. Pleural liquid and solute exchange. *Am Rev Respir Dis.* 1989;140:825–847.
3. Ernst CS, Brooks JJ. Immunoperoxidase localization of secretory component in reactive mesothelium and mesotheliomas. *J Histochem Cytochem.* 1981;29:1102–1104.
4. Albertine KH, Wiener-Kronish JP, Roos PJ, Staub NC. Structure, blood supply, and lymphatic vessels of the sheep's visceral pleura. *Am J Anat.* 1982;165:277–294.
5. Fourdrain A, Lafitte S, Iquille J, et al. Lymphatic drainage of lung segments in the visceral pleura: a cadaveric study. *Surg Radiol Anat.* 2018;40:15–19.
6. Gallagher B, Urbanski SJ. The significance of pleural elastica invasion by lung carcinomas. *Hum Pathol.* 1990;21:512–517.
7. Li J. Ultrastructural study on the pleural stomata in human. *Funct Dev Morphol.* 1993;3:277–280.
8. Leak LV. Gross and ultrastructural morphologic features of the diaphragm. *Am Rev Respir Dis.* 1979;119:3–21.
9. Wang NS. The preformed stomas connecting the pleural cavity and the lymphatics in the parietal pleura. *Am Rev Respir Dis.* 1975;111:12–20.
10. Wang NS. Anatomy and physiology of the pleural space. *Clin Chest Med.* 1985;6:3–16.
11. Li J, Zhao Z, Zhou J, Yu S. A study of the three-dimensional organization of the human diaphragmatic lymphatic lacunae and lymphatic drainage units. *Ann Anat.* 1996;178:537–544.
12. Lee KF, Olak J. Anatomy and physiology of the pleural space. *Chest Surg Clin N Am.* 1994;4:391–403.
13. Kampmeier O. Concerning certain mesothelial thickenings and vascular plexuses of the mediastinal pleura associated with histiocyte and fat-cell production, in the human newborn. *Anat Rec.* 1928;39:201–208.
14. Boutin C, Dumortier P, Rey F, Viallat JR, De Vuyst P. Black spots concentrate oncogenic asbestos fibers in the parietal pleura. Thoracoscopic and mineralogic study. *Am J Respir Crit Care Med.* 1996;153:444–449.
15. Miserocchi G, Sancini G, Mantegazza F, Chiappino G. Translocation pathways for inhaled asbestos fibers. *Environ Health.* 2008;7:4.
16. Wassilev W, Wedel T, Michailova K, Kühnel W. A scanning electron microscopy study of peritoneal stomata in different peritoneal regions. *Ann Anat.* 1998;180:137–143.
17. Li JC, Yu SM. Study on the ultrastructure of the peritoneal stomata in humans. *Acta Anat (Basel).* 1991;141:26–30.
18. Liu J, Geng X, Li Y. Milky spots: omental functional units and hotbeds for peritoneal cancer metastasis. *Tumour Biol.* 2016;37:5715–5726.
19. Sheaff M, Singh N. Serous fluids. In: *Cytopathology*. Springer-Verlag; 2013:101–141.
20. Raftery AT. Regeneration of parietal and visceral peritoneum in the immature animal: a light and electron microscopical study. *Br J Surg.* 1973;60:969–975.
21. Raftery AT. Regeneration of parietal and visceral peritoneum: an electron microscopical study. *J Anat.* 1973;115:375–392.
22. Whitaker D, Papadimitriou J. Mesothelial healing: morphological and kinetic investigations. *J Pathol.* 1985;145:159–175.
23. MacDougall DB, Wang SE, Zidar BL. Mucin-positive epithelial mesothelioma. *Arch Pathol Lab Med.* 1992;116:874–880.
24. Hammar SP, Bockus DE, Remington FL, Rohrbach KA. Mucin-positive epithelial mesotheliomas: a histochemical, immunohistochemical, and ultrastructural comparison with mucin-producing pulmonary adenocarcinomas. *Ultrastruct Pathol.* 1996;20:293–325.
25. Husain AN, Colby TV, Ordóñez NG, et al. Guidelines for pathologic diagnosis of malignant mesothelioma 2017 update of the consensus statement from the International Mesothelioma Interest Group. *Arch Pathol Lab Med.* 2018;142:89–108.
26. Churg A, Galateau-Salle F. The separation of benign and malignant mesothelial proliferations. *Arch Pathol Lab Med.* 2012;136:1217–1226.

27. Churg A, Cagle P, Colby TV, et al; US-Canadian Mesothelioma Reference Panel. The fake fat phenomenon in organizing pleuritis: a source of confusion with desmoplastic malignant mesotheliomas. *Am J Surg Pathol.* 2011;35:1823–1829.
28. Pizzi M, Valentini E, Galligioni A, et al. Benign mesothelial cells in lymph nodes and lymphatic spaces associated with ascites. *Histol Histopathol.* 2016;31:747–750.
29. Brooks JS, LiVolsi VA, Pietra GG. Mesothelial cell inclusions in mediastinal lymph nodes mimicking metastatic carcinoma. *Am J Clin Pathol.* 1990;93:741–748.
30. Parkash V, Vidwans M, Carter D. Benign mesothelial cells in mediastinal lymph nodes. *Am J Surg Pathol.* 1999;23:1264–1269.
31. Moll R, Löwe A, Laufer J, Franke WW. Cytokeratin 20 in human carcinomas. A new histodiagnostic marker detected by monoclonal antibodies. *Am J Pathol.* 1992;140:427–447.
32. Churg A, Sheffield BS, Galateau-Salle F. New markers for separating benign from malignant mesothelial proliferations: are we there yet? *Arch Pathol Lab Med.* 2016;140:318–321.
33. King J, Thatcher N, Pickering C, Hasleton P. Sensitivity and specificity of immunohistochemical antibodies used to distinguish between benign and malignant pleural disease: a systematic review of published reports. *Histopathology.* 2006;49:561–568.
34. Minato H, Kurose N, Fukushima M, et al. Comparative immunohistochemical analysis of IMP3, GLUT1, EMA, CD146, and desmin for distinguishing malignant mesothelioma from reactive mesothelial cells. *Am J Clin Pathol.* 2014;141:85–93.
35. Pitt MA, Haboubi NY. Serosal reaction in chronic gastric ulcers: an immunohistochemical and ultrastructural study. *J Clin Pathol.* 1995;48:226–228.
36. Kung IT, Thallas V, Spencer EJ, Wilson SM. Expression of muscle actins in diffuse mesotheliomas. *Hum Pathol.* 1995;26:565–570.
37. Ordóñez NG. Value of calretinin immunostaining in diagnostic pathology: a review and update. *Appl Immunohistochem Mol Morphol.* 2014;22:401–415.
38. Chu PG, Weiss LM. Expression of cytokeratin 5/6 in epithelial neoplasms: an immunohistochemical study of 509 cases. *Mod Pathol.* 2002;15:6–10.
39. Ordóñez NG. The diagnostic utility of immunohistochemistry in distinguishing between epithelioid mesotheliomas and squamous carcinomas of the lung: a comparative study. *Mod Pathol.* 2006;19:417–428.
40. Hecht JL, Lee BH, Pinkus JL, Pinkus GS. The value of Wilms tumor susceptibility gene 1 in cytologic preparations as a marker for malignant mesothelioma. *Cancer.* 2002;96:105–109.
41. Chu AY, Litzky LA, Pasha TL, Acs G, Zhang PJ. Utility of D2-40, a novel mesothelial marker, in the diagnosis of malignant mesothelioma. *Mod Pathol.* 2005;18:105–110.
42. Naso JR, Tsuji S, Churg A. HEG1 is a highly specific and sensitive marker of epithelioid malignant mesothelioma. *Am J Surg Pathol.* 2020;44:1143–1148.
43. Churg A, Naso JR. Hypothesis: HEG1 and claudin-4 staining will allow a diagnosis of epithelioid and biphasic mesothelioma versus non-small-cell lung carcinoma with only two stains in most cases. *Histopathology.* 2023;82:385–392.
44. Tsuji S, Washimi K, Kageyama T, et al. HEG1 is a novel mucin-like membrane protein that serves as a diagnostic and therapeutic target for malignant mesothelioma. *Sci Rep.* 2017;7:45768.
45. Matsuura R, Kaji H, Tomioka A, et al. Identification of mesothelioma-specific sialylated epitope recognized with monoclonal antibody SKM9-2 in a mucin-like membrane protein HEG1. *Sci Rep.* 2018;8:14251.
46. Hiroshima K, Wu D, Koh E, et al. Membranous HEG1 expression is a useful marker in the differential diagnosis of epithelioid and biphasic malignant mesothelioma versus carcinomas. *Pathol Int.* 2021;71:604–613.
47. Klebe S, Brownlee NA, Mahar A, et al. Sarcomatoid mesothelioma: a clinical-pathologic correlation of 326 cases. *Mod Pathol.* 2010;23:470–479.
48. Hinterberger M, Reineke T, Storz M, Weder W, Vogt P, Moch H. D2-40 and calretinin – a tissue microarray analysis of 341 malignant mesotheliomas with emphasis on sarcomatoid differentiation. *Mod Pathol.* 2007;20:248–255.
49. Attanoos RL, Dojcinov SD, Webb R, Gibbs AR. Anti-mesothelial markers in sarcomatoid mesothelioma and other spindle cell neoplasms. *Histopathology.* 2000;37:224–231.
50. Naso JR, Cheung S, Ionescu DN, Churg A. Utility of SOX6 and DAB2 for the diagnosis of malignant mesothelioma. *Am J Surg Pathol.* 2021;45:1245–1251.
51. Klebe S, Mahar A, Henderson DW, Roggli VL. Malignant mesothelioma with heterologous elements: clinicopathological correlation of 27 cases and literature review. *Mod Pathol.* 2008;21:1084–1094.
52. Naso JR, Churg A. Claudin-4 shows superior specificity for mesothelioma vs non-small-cell lung carcinoma compared with MOC-31 and Ber-EP4. *Hum Pathol.* 2020;100:10–14.
53. Boylan KLM, Misemer B, De Rycke MS, et al. Claudin 4 is differentially expressed between ovarian cancer subtypes and plays a role in spheroid formation. *Int J Mol Sci.* 2011;12:1334–1358.
54. Ordóñez NG. The immunohistochemical diagnosis of mesothelioma: a comparative study of epithelioid mesothelioma and lung adenocarcinoma. *Am J Surg Pathol.* 2003;27:1031–1051.
55. Wang L-M, Shi Z-W, Wang J-L, et al. Diagnostic accuracy of BRCA1-associated protein 1 in malignant mesothelioma: a meta-analysis. *Oncotarget.* 2017;8:68863–68872.
56. Churg A, Naso JR. The separation of benign and malignant mesothelial proliferations: new markers and how to use them. *Am J Surg Pathol.* 2020;44:e100–e112.
57. Chapel DB, Hornick JL, Barlow J, Bueno R, Sholl LM. Clinical and molecular validation of BAP1, MTAP, P53, and Merlin immunohistochemistry in diagnosis of pleural mesothelioma. *Mod Pathol.* 2022;35:1383–1397.
58. Martin SD, Cheung S, Churg A. Immunohistochemical demonstration of Merlin/NF2 loss in mesothelioma. *Mod Pathol.* 2023;36:100036.
59. Chapel DB, Schulte JJ, Berg K, et al. MTAP immunohistochemistry is an accurate and reproducible surrogate for CDKN2A fluorescence in situ hybridization in diagnosis of malignant pleural mesothelioma. *Mod Pathol.* 2020;33:245–254.
60. Hida T, Hamasaki M, Matsumoto S, et al. Immunohistochemical detection of MTAP and BAP1 protein loss for mesothelioma diagnosis: comparison with 9p21 FISH and BAP1 immunohistochemistry. *Lung Cancer.* 2017;104:98–105.
61. Berg KB, Dacic S, Miller C, Cheung S, Churg A. Utility of methylthioadenosine phosphorylase compared with BAP1 immunohistochemistry, and CDKN2A and NF2 fluorescence in situ hybridization in separating reactive mesothelial

proliferations from epithelioid malignant mesotheliomas. *Arch Pathol Lab Med.* 2018;142:1549–1553.

62. Naso JR, Tessier-Cloutier B, Senz J, Huntsman DG, Churg A. Significance of p53 immunostaining in mesothelial proliferations and correlation with TP53 mutation status. *Mod Pathol.* 2022;35:77–81.
63. Yoshimura M, Kinoshita Y, Hamasaki M, et al. Highly expressed EZH2 in combination with BAP1 and MTAP loss, as detected by immunohistochemistry, is useful for differentiating malignant pleural mesothelioma from reactive mesothelial hyperplasia. *Lung Cancer.* 2019;130:187–193.
64. Shinozaki-Ushiku A, Ushiku T, Morita S, Anraku M, Nakajima J, Fukayama M. Diagnostic utility of BAP1 and EZH2 expression in malignant mesothelioma. *Histopathology.* 2017;70:722–733.
65. Chapel DB, Husain AN, Krausz T. Immunohistochemical evaluation of nuclear 5-hydroxymethylcytosine (5-hmC) accurately distinguishes malignant pleural mesothelioma from benign mesothelial proliferations. *Mod Pathol.* 2019;32:376–386.
66. Derakhshan F, Ionescu D, Cheung S, Churg A. Use of Programmed Death Ligand-1 (PD-L1) staining to separate sarcomatoid malignant mesotheliomas from benign mesothelial reactions. *Arch Pathol Lab Med.* 2020;144:185–188.
67. Chung YS, Kim M, Cha YJ, Kim KA, Shim HS. Expression of V-set immunoregulatory receptor in malignant mesothelioma. *Mod Pathol.* 2020;33:263–270.
68. Schürch CM, Forster S, Brühl F, Yang SH, Felley-Bosco E, Hewer E. The "don't eat me" signal CD47 is a novel diagnostic biomarker and potential therapeutic target for diffuse malignant mesothelioma. *Oncoimmunology.* 2017;7:e1373235.
69. Whitaker D, Henderson DW, Shilkin KB. The concept of mesothelioma in situ: implications for diagnosis and histogenesis. *Semin Diagn Pathol.* 1992;9:151–161.
70. Churg A, Galateau-Salle F, Roden AC, et al. Malignant mesothelioma in situ: morphologic features and clinical outcome. *Mod Pathol.* 2020;33:297–302.
71. Lee HE, Molina JR, Sukov WR, Roden AC, Yi ES. BAP1 loss is unusual in well-differentiated papillary mesothelioma and may predict development of malignant mesothelioma. *Hum Pathol.* 2018;79:168–176.
72. Galateau-Salle F, Hamilton T, MacNeill A, et al. Mesothelioma in situ mimicking well-differentiated papillary mesothelial tumor. *Am J Surg Pathol.* 2023;47:611–617.
73. Churg A, Galateau-Salle F. Well differentiated papillary mesothelial tumor: a new name and new problems. *Mod Pathol.* 2022;35:1327–1333.
74. Nicholson A, Znaor A, Dacic S. Tumours of the pleura and pericardium. In: *WHO Classification of Tumours Editorial Board Thoracic Tumours*. Vol 5. 5th ed. International Agency for Research on Cancer; 2021.
75. Longacre T, Malpica A, Kim K-R. Tumours of the peritoneum. In: *WHO Classification of Tumours Editorial Board Female Genital Tumours*. Vol 4. 5th ed. International Agency for Research on Cancer; 2020.
76. Symes E, Tjota M, Cody B, et al. Mesothelioma in situ of the peritoneum: report of three cases and review of the literature. *Histopathology.* 2024;84(3):492–506.
77. Nishikubo M, Jimbo N, Tanaka Y, Tachihara M, Itoh T, Maniwa Y. Sarcomatoid mesothelioma originating from mesothelioma in situ: are methylthioadenosine phosphorylase loss and CDKN2A homozygous deletion poor prognostic factors for preinvasive mesothelioma? *Virchows Arch.* 2022; 481:307–312.
78. Zinsser KR, Wheeler JE. Endosalpingiosis in the omentum: a study of autopsy and surgical material. *Am J Surg Pathol.* 1982; 6:109–117.
79. Hsu YK, Parmley TH, Rosenshein NB, Bhagavan BS, Woodruff JD. Neoplastic and non-neoplastic mesothelial proliferations in pelvic lymph nodes. *Obstet Gynecol.* 1980;55:83–88.
80. Horn LC, Bilek K. Frequency and histogenesis of pelvic retroperitoneal lymph node inclusions of the female genital tract. An immunohistochemical study of 34 cases. *Pathol Res Pract.* 1995;191:991–996.
81. Schnurr RC, Delgado G, Chun B. Benign glandular inclusions in para-aortic lymph nodes in women undergoing lymphadenectomies. *Am J Obstet Gynecol.* 1978;130:813–816.
82. Clement P. Endometriosis, lesions of the secondary müllerian system, and pelvic mesothelial proliferations. In: *Blaustein's Pathology of the Female Genital Tract*. 3rd ed. Springer-Verlag; 1987:517–559.
83. Simpson JA. The pathogenesis of postsalpingectomy endometriosis in laparotomy scars. *Am J Obstet Gynecol.* 1945; 50:597–620.
84. Sampson JA. Metastatic or embolic endometriosis, due to the menstrual dissemination of endometrial tissue into the venous circulation. *Am J Pathol.* 1927;3:93–110.43.
85. Ferguson BR, Bennington JL, Haber SL. Histochemistry of mucosubstances and histology of mixed müllerian pelvic lymph node glandular inclusions. Evidence for histogenesis by müllerian metaplasia of coelomic epithelium. *Obstet Gynecol.* 1969;33:617–625.
86. Lauchlan SC. The secondary Müllerian system. *Obstet Gynecol Surv.* 1972;27:133–146.
87. Nakayama K, Masuzawa H, Li SF, et al. Immunohistochemical analysis of the peritoneum adjacent to endometriotic lesions using antibodies for Ber-EP4 antigen, estrogen receptors, and progesterone receptors: implication of peritoneal metaplasia in the pathogenesis of endometriosis. *Int J Gynecol Pathol.* 1994;13:348–358.

SECTION VII

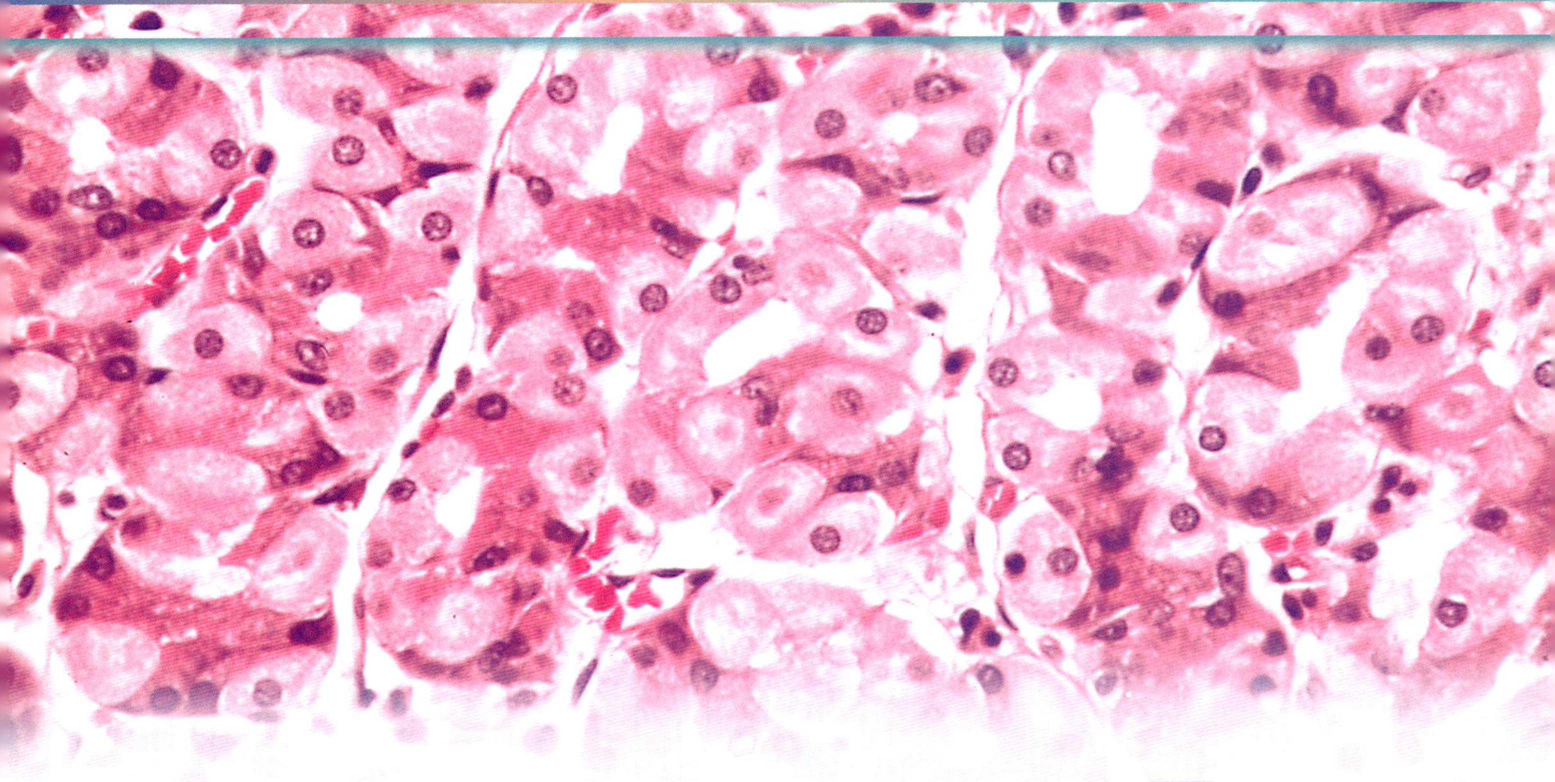

Alimentary Tract

21

Esophagus

Elizaveta Chernetsova ■ Shawn Winer ■ Robert Riddell

ANATOMICAL/MACROSCOPIC FEATURES

Embryology

In the early stages of development, the notochord induces formation of the foregut from the endoderm (1). Septa arise from the lateral walls of the foregut at 21 days gestation, fuse, and divide the foregut into the esophagus and trachea. This process of septation begins at the carina and extends cephalad, being completed by 5 to 6 weeks of gestation (Fig. 21.1) (2). Failure of this process can result in esophageal atresia with or without tracheoesophageal fistulae, tracheal atresia or stenosis, foregut duplications, bronchogenic cysts, bronchopulmonary sequestrations, bronchial atresias, and congenital cystic lesions of the lung (3–6).

Separation of the esophagus from the tracheal tube is regulated by diverse signaling between the epithelial cells and surrounding mesenchyme and is highly coordinated by transcriptional factors and signaling pathways (7,8). The dorsal foregut endoderm expresses Sox2 and gives rise to the esophagus, while the ventral foregut endoderm expresses Nkx2.1 and forms the trachea. Reciprocal inhibition between the two transcription factors drives the process.

The esophagus is initially lined by a thin layer of stratified columnar epithelium that proliferates to nearly occlude the lumen (9). New vacuoles appear between these luminal cells and coalesce to form a single esophageal lumen with a superficial layer of ciliated epithelial cells (Fig. 21.2) (9). Ciliated cells appear in the middle third of the esophagus as early as 8 weeks of gestation. These extend cephalad and

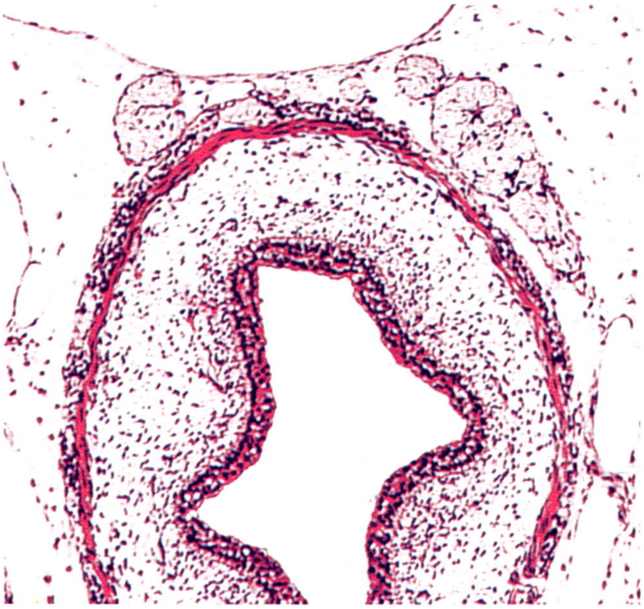

FIGURE 21.1 Fetal esophagus (late first trimester). Transverse section overview of the esophagus demonstrating inner mucosal layer, middle submucosal layer, and thin outer muscle layer. Note the vagus nerves lying over the esophagus.

caudally to almost cover the entire stratified columnar epithelium such that the esophagus is entirely surfaced by a single layer of columnar cells by approximately 10 weeks (9). Cardiac-type glands form as a result of the downward growth of these columnar cells into the lamina propria with subsequent proliferation and differentiation at approximately 4 months gestational age.

Stratified squamous epithelium initially appears in the middle one-third of the esophagus at 5 months gestational age. It extends cephalad and caudally and replaces the ciliated epithelium under the influence of transcriptional factors and signaling pathways (Fig. 21.3) (10). p63, a member of the p53 family, is one transcription factor involved in the conversion of columnar into stratified squamous epithelium

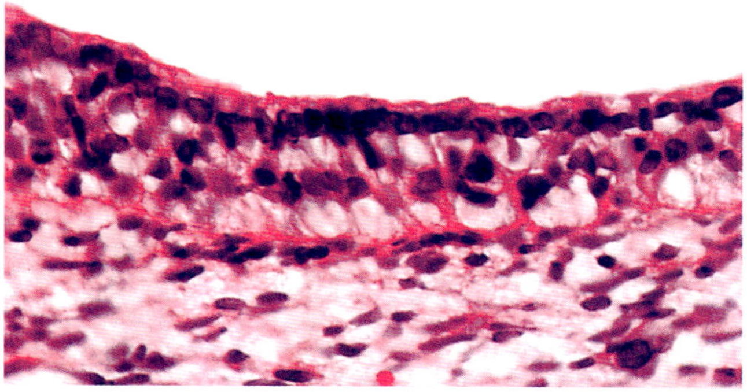

FIGURE 21.2 Fetal esophagus (late first trimester). The epithelial layer is composed of stratified columnar epithelium. Note the lack of muscularis mucosae.

in the esophagus (11). Sox2 is expressed during this process and represses Nkx2.1. It is required for the stratification and lineage differentiation of the esophageal epithelial cells (12). Manipulation of this transcriptional network can be used to direct pluripotent stem cells into esophageal mucosal organoids, which may provide a new modality to study and treat esophageal diseases (13).

The upper esophagus is the last area to be replaced by squamous epithelium; and, if this process is not completed by the time of birth, there may be persistence of ciliated cells in the upper esophagus (9,10). These residual cells are usually short-lived, being replaced by squamous epithelium within 2 to 3 days postpartum (10,14). However, they may persist into adult life or differentiate into gastric glands (15). The submucosal glands develop after the appearance of the squamous epithelium and are likely derived from this squamous epithelial layer (10,14). Interestingly, the mucosa of the fetal upper esophagus lacks a muscularis mucosae (Fig. 21.2).

Development of the gastrointestinal neuromuscular system begins at 4 weeks of gestation, with neural crest cells entering the foregut and migrating rostro-caudally. The myenteric plexus develops first, followed by formation of the submucosal plexus 2 to 3 weeks later. The circular muscle layer develops at 6 weeks of gestation, followed by the development of the longitudinal layer by 9 weeks. The muscularis propria consists entirely of smooth muscles initially but striated muscle cells gradually develop in the upper esophagus such that the normal ratio and arrangement of smooth and skeletal muscle is established by 5 months of gestation (10).

Multiple genes, transcription factors, and signal pathways are involved in the regulation of esophageal muscular development (2,16). Specifically, the cell surface receptor Cdo is required for setting up the striated-smooth muscle boundary and the Wnt signaling receptor Fz4 is required for proper striated muscle differentiation and development; a number of transcription factors (e.g., Myf5, Foxp1, Foxp2, and Pax7) also play roles in this process (2,7,17). ICCs appear at 9 weeks of gestation and become closely associated with the myenteric plexus (18). By week 14, the fetal gut has a mature appearance (18).

Topography and Relations

The esophagus begins in the neck at the level of the cricoid cartilage, passes through the thorax within the posterior mediastinum, and traverses the diaphragm, which forms part of the lower esophageal sphincter to its junction with the stomach. The overall length varies with trunk length but averages approximately 8.3 ± 2.4 cm in adults. Endoscopic distances are measured from the incisor teeth; the junction of the esophagus and stomach is approximately 40 cm from the incisors in the average adult male but may vary from 38 to 43 cm. Esophageal length also correlates with height in children (19).

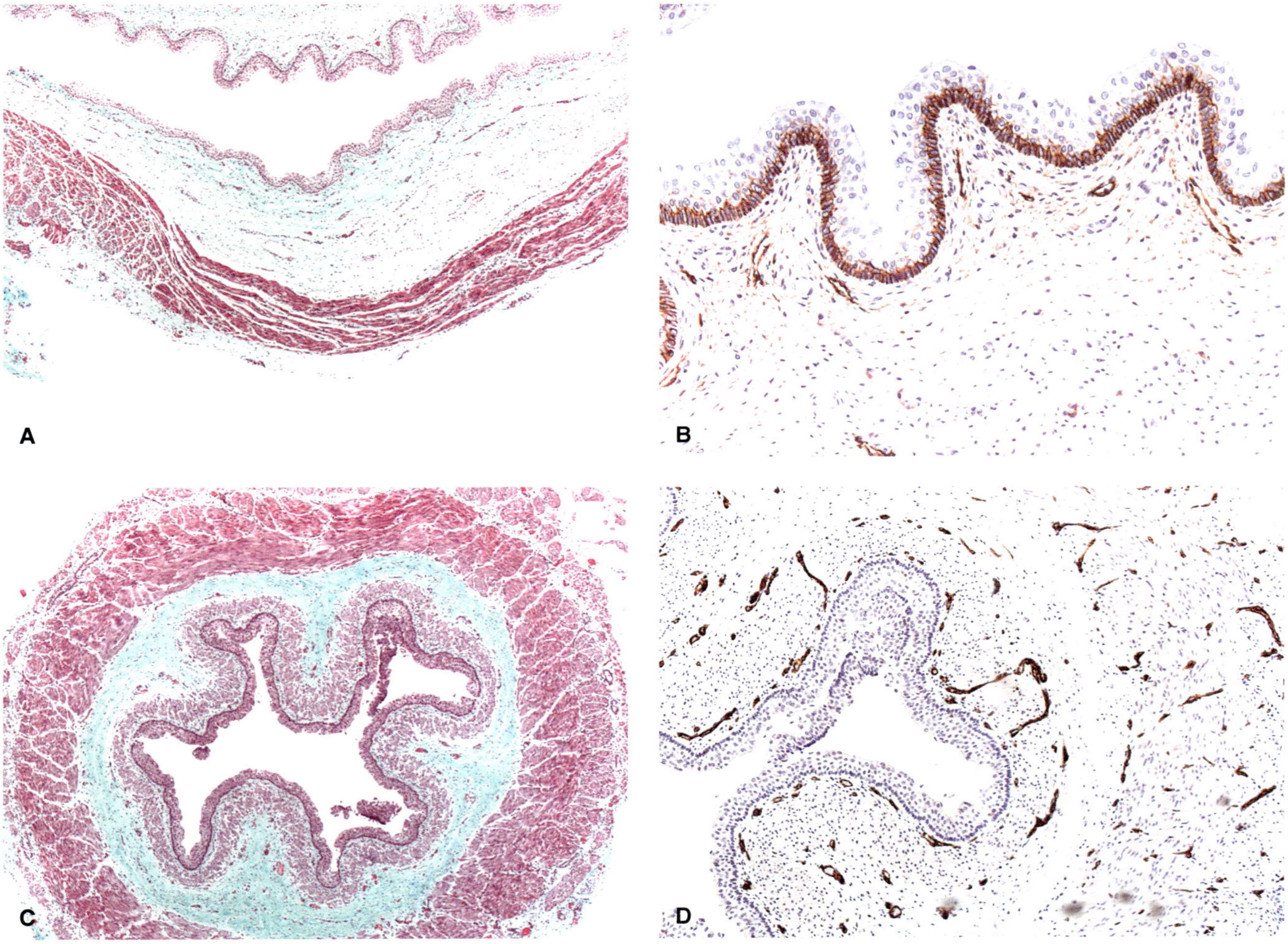

FIGURE 21.3 Fetal esophagus (third trimester). **A:** Upper esophagus at 33 weeks. The epithelial layer at this stage consists of stratified squamous mucosa but note the lack of a muscularis mucosae (trichrome stain). **B:** Third-trimester mucosa showing mature nonkeratinized squamous mucosa, here illustrated with D2-40 that highlights both the basal layer and the relative paucity of subepithelial lymphatics. A mild condensation of stromal cells immediately beneath the epithelium shows the location of the future muscularis mucosae. **C:** Transverse section of midesophagus (same esophagus as "A") showing a well-formed muscularis mucosae. Note also the individual smooth muscle cells of developing muscularis mucosae (trichrome stain). **D:** Muscularis mucosae of midesophagus showing the anastomosing plexus of capillaries on both sides of the muscularis mucosae (CD31).

Points of Constrictions

The normal esophagus is externally compressed at several points: the cricoid origin of the esophagus, the left side at the level of the aortic arch, at the crossing of the left main bronchus and left atrium, and at the lower esophageal sphincter where the esophagus passes through the diaphragm (Fig. 21.4). These constrictions may become clinically significant if food or pills become lodged at these sites of luminal narrowing and cause contact mucosal injury. The most common sites of injury are at the levels of the aortic arch and left atrium, especially in patients with left atrial enlargement (20–22). Dysphagia aortica is a rare cause of dysphagia resulting from the extrinsic compression of the esophagus by a thoracic aortic aneurysm or tortuous thoracic aorta (23). Kommerell diverticulum refers to the aneurysmal dilatation of the descending aorta at the origin of an aberrant subclavian artery. Approximately 5% of patients with Kommerell diverticulum present with dysphagia due to the esophageal compression, chest pain, or recurrent pneumonia (24). Knowledge of the relationships between the esophagus and other anatomic structures is important because they may be directly affected by esophageal diseases such as carcinoma or diverticula. Disease of adjacent structures may also cause local compression of the esophagus.

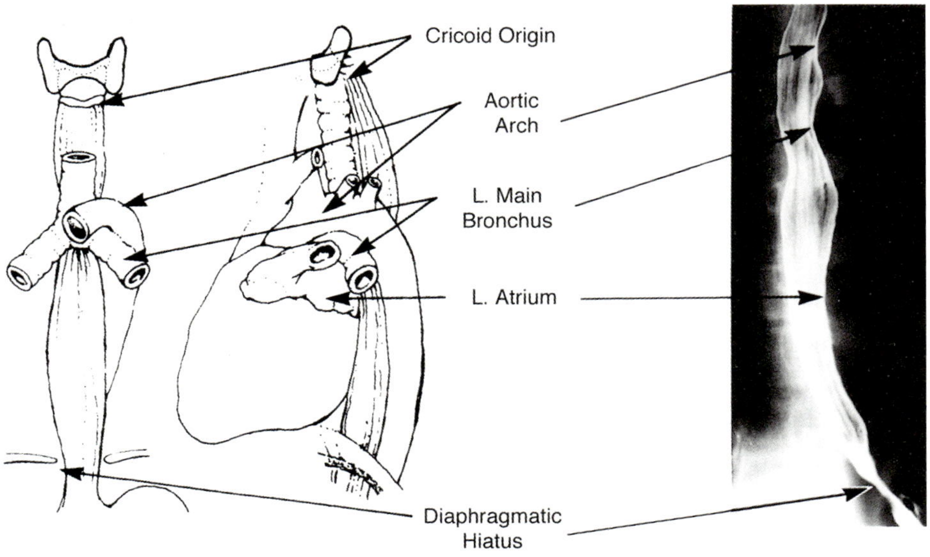

FIGURE 21.4 Relationship of the esophagus with normal esophageal constrictions. Barium swallow of the normal esophagus (**right**) demonstrates narrowing of the lumen at the sites of constriction.

Anatomical Division of the Esophagus

The International Classification of Diseases recognized three anatomic compartments traversed by the esophagus: cervical, thoracic, and abdominal (Table 21.1). The esophagus is divided into upper, middle, and lower sections, forming the basis of the Union for International Cancer Control (UICC)/American Joint Committee on Cancer (AJCC) and College of American Pathologists (CAP) schemas for location of tumors (Fig. 21.5) (25).

The cervical portion of the esophagus is located in the anterior neck along with the trachea adjacent to the common carotid artery and lobes of the thyroid gland; the recurrent laryngeal nerves ascend between it and the trachea and the thoracic duct is located to the left.

The thoracic segment of the esophagus lies posterior to the trachea and slightly left of its bifurcation; midesophageal diverticula can develop in this area as a result of traction on the esophagus secondary to inflamed mediastinal lymph nodes (26). The esophagus courses posterior to the left atrium and is flanked by azygos veins on both sides. The right and left vagus nerves run laterally, giving off branches that form plexuses along the posterior and anterior esophageal surfaces and reuniting to form the anterior and posterior vagal trunks that course to the stomach.

The abdominal esophagus is nestled in the esophageal groove on the posterior surface of the left hepatic lobe. It measures only 1.25 cm in length and is surfaced by peritoneum anteriorly on its left aspect. The esophagus enters the abdomen by passing through the esophageal hiatus, which is formed by muscles of the diaphragm and contains the phrenoesophageal ligament. The muscular sling encircling the esophagus is usually formed entirely from the right diaphragmatic crus (27), although anatomic variations do occur. The phrenoesophageal ligament arises from the fascia of the abdominal diaphragm and divides into an ascending and a descending leaf. The former passes up through the hiatus to insert 2 to 3 cm above the hiatus, whereas the descending leaf variably inserts at or

TABLE 21.1 **Anatomic Compartments of the Esophagus. Regions of the Esophagus, Their Boundaries, and Approximate Distances From the Incisors**

Anatomic Name	Esophageal Name	Anatomic Boundaries	Typical Esophagectomy (Variation ++)
Cervical	Upper	Hypopharynx to sternal notch	15–<20 cm
Thoracic	Upper	Sternal notch to azygos vein	20–<25 cm
	Middle	Lower border of azygos vein to inferior pulmonary vein	<25–<30 cm
	Lower	Lower border of inferior pulmonary vein to esophagogastric junction	<30–<40 cm
Abdominal	Lower	Esophagogastric junction to 5 cm below esophagogastric junction	<40–45 cm
	Esophagogastric junction/cardia	Esophagogastric junction to 5 cm below esophagogastric junction	<40–45 cm

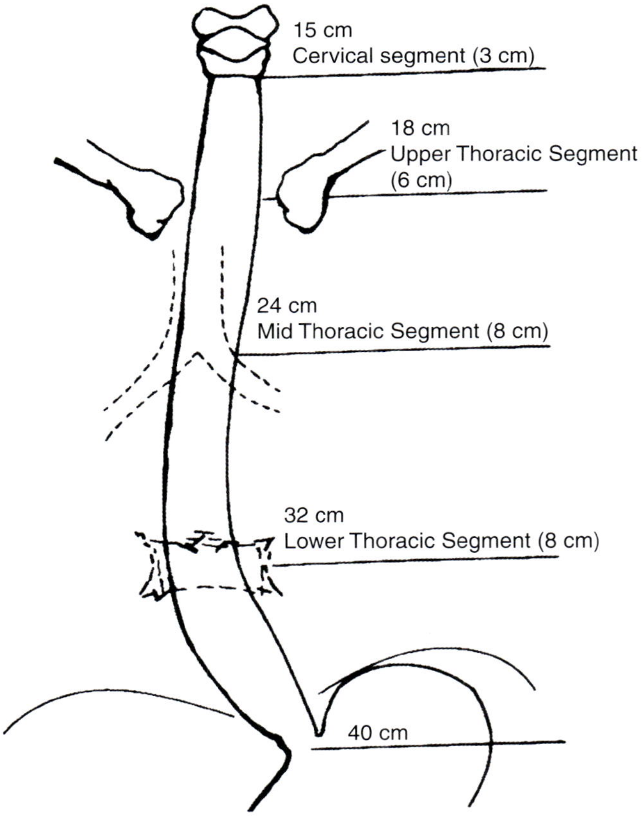

FIGURE 21.5 Esophageal segments with approximate lengths and distances from the incisors.

below the gastroesophageal junction or even into the gastric fundus (28). The gastroesophageal junction is smooth on the right side, whereas the gastroesophageal junction is sharply angled on the left, forming the incisura or angle of His.

The proposed functions of the phrenoesophageal ligament include maintaining the pressure differential between the thorax and the abdomen, fixing the gastroesophageal junction within the abdomen during episodes of increased intra-abdominal pressure, and contributing to the competence of the lower esophageal sphincter (29).

Arterial Supply

The cervical portion of the esophagus is supplied by branches of the inferior thyroid artery with contributions from the intercostal arteries. Branches of the bronchial arteries, intercostal arteries, and aorta supply the thoracic segment, whereas the abdominal segment is supplied by branches of the left gastric and inferior phrenic arteries (29,30). Branches from these arteries run within the muscularis propria, giving rise to branches that course within the submucosa. Anastomoses are extensive, explaining the rarity of esophageal infarction (30,31).

Venous Drainage

The venous return from the upper two-thirds of the esophagus drains into the inferior thyroidal vein and azygos system, eventually reaching the superior vena cava. The lower esophageal segment drains into the systemic circulation through branches of the azygos vein and left inferior phrenic vein. The lower esophageal segment also drains into the portal system from branches of the left gastric vein and through the short gastric veins that empty into the splenic vein (29,30).

The anatomy of the lower esophageal venous system consists of four layers (32). Radially arranged intramucosal channels drain into the superficial venous plexus located in the upper submucosa. This superficial venous plexus consists of three to five main trunks located in the lower submucosa that communicate with the deep intrinsic veins. Perforating veins connect this layer with the layer of veins located on the esophageal adventitia (33).

The portal and caval systems communicate through the esophageal and gastric submucosal veins. Portal hypertension results in dilatation of all the venous channels to form varices. Major hemorrhage occurs as a result of variceal rupture in the deep intrinsic veins, whereas minor hemorrhages result from variceal rupture in the superficial venous plexus or even intramucosal channels (32).

Submucosal and Lamina Propria Vessels at the Gastroesophageal Junction

The submucosal vessels of the distal esophagus are connected to the gastric submucosal vessels by a series of longitudinal (vertical) vessels at the first gastric folds. These vessels of the lamina propria span 2 to 4 cm in length and can often be seen through the squamous or columnar mucosa (34,35). Their lower visible limit also marks the original site of the gastroesophageal junction.

Lymphatic Drainage

A rich network of lymphatic vessels in the lamina propria and submucosa connects with lymphatic vessels in the muscularis propria and adventitial layers, although results of several studies suggest that some lymphatic vessels proximally to drain into recurrent laryngeal and supraclavicular lymph nodes and distally to the paracardial and celiac lymph nodes, thereby bypassing the network in the muscularis propria/adventitia and regional lymph nodes (36). Thus, frozen sections of margins should always include the submucosa to improve the likelihood of detected carcinoma when it is present.

Lymphatic vessels run parallel to the epithelium in the lamina propria while those in the muscularis propria are longitudinally oriented (37).

The cervical esophagus drains into the internal jugular and upper tracheal lymph node groups. The thoracic esophagus drains into the superior, the middle, and the lower mediastinal lymph node groups, while the abdominal segment drains into superior gastric, celiac axis, common hepatic artery, and splenic artery lymph nodes (29). Direct drainage into the thoracic duct has also been documented in autopsy studies (38). Drainage pathways may be highly variable among individuals, and extensive communication between lymphatic vessels results in a varied metastatic pattern among patients with esophageal carcinoma (39,40).

Innervation: Nerves and Interstitial Cells of Cajal (ICCs)

The esophagus receives both parasympathetic and sympathetic nerves from the vagus nerve. It contains afferent and efferent fibers that innervate glands, blood vessels, and muscles. Parasympathetic nerves predominate throughout the esophagus; there are a few sympathetic fibers originating in the cervical and paravertebral chains that run with vascular structures (41).

The esophagus contains two neuronal plexuses: Meissner plexus in the submucosa and Auerbach plexus between the circular and longitudinal layers of the muscularis propria. These structures are not as well developed in the upper esophagus as they are distally and are less developed in the esophagus compared with the remaining gastrointestinal tract (29,41).

The esophageal plexuses receive input from postganglionic sympathetic fibers, preganglionic and postganglionic parasympathetic fibers, and other intrinsic ganglion cells (41). These cells stimulate secretory activity and supply both the muscularis propria and muscularis mucosae. ICCs have been identified in the distal one-third of the esophagus in close association with smooth muscle cells, as well as in the middle one-third, where they are associated with both smooth muscle cells and striated muscle fibers (42). They are found within the Auerbach plexus and in the intramuscular and intermuscular layers. These cells are immunoreactive for CD117 and DOG1. Degeneration of the Auerbach plexus and ICCs can be seen in patients with achalasia (23,25). Gastrointestinal stromal tumors recapitulate the morphologic and immunohistochemical features of the ICCs. However, most mesenchymal tumors of the esophagus are leiomyomas rather than gastrointestinal stromal tumors (43).

Vasoactive intestinal peptide and nitric oxide are regulatory peptides that can be identified within parasympathetic nerve fibers and around smooth muscle bundles, whereas tyrosine hydroxylase is a marker of sympathetic nerves. Some nerves are immunoreactive for substance P, enkephalin, and neuropeptide Y (44,45). Nerve fibers containing vasoactive intestinal peptide and neuropeptide Y are the most abundant types present in the esophagus, and the pattern of innervation by these peptide-containing neurons differs from those of the stomach and small intestine (46). Cholecystokinin (CCK) receptors are found in both the mucosa and nerves of the gastric cardia (47).

HISTOLOGY

The wall of the esophagus consists of four layers: (1) mucosa (including the muscularis mucosae), (2) submucosa, (3) muscularis propria, and (4) adventitia.

Mucosa

The esophageal mucosa is surfaced by nonkeratinizing, stratified squamous epithelium (Fig. 21.6). The squamous mucosa transitions to glandular mucosa in the distal esophagus. The latter may consist of cardiac-type foveolar epithelium, cardio-oxyntic mucosa, or oxyntic mucosa. Metaplasias, heterotopias, and multilayered epithelium can also be seen in glandular mucosa.

Squamous Epithelium

› **Squamous Cells** The squamous epithelium can be divided into basal, prickle (i.e., stratum spinosum), and superficial cell layers. The basal layer occupies approximately 5% to 15% of the epithelial thickness. It is normally one to three cells thick, but approximately 60% of individuals without objective or subjective evidence of gastroesophageal reflux may have basal cell hyperplasia occupying more than 15% of the mucosal thickness in the distal esophagus (48). The boundary between the upper extent of the basal zone and the prickle cell layers has been arbitrarily defined as the level where the nuclei

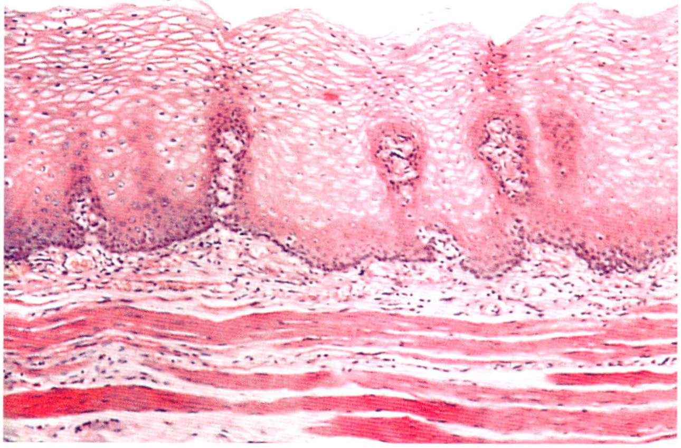

FIGURE 21.6 Midesophagus. The esophageal mucosa consists of a surface epithelial layer, middle lamina propria, and lower muscularis mucosae, which consists of longitudinally oriented smooth muscle bundles.

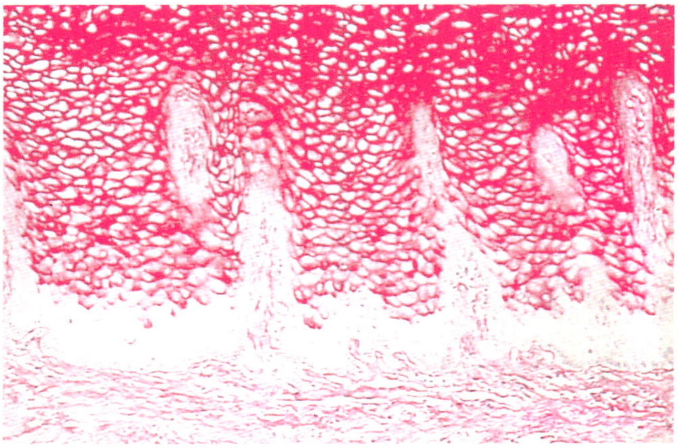

FIGURE 21.7 Midesophagus. The basal cell layer of the esophageal epithelium shows lack of glycogen, allowing for ready distinction from the overlying glycogen-rich cells (PAS).

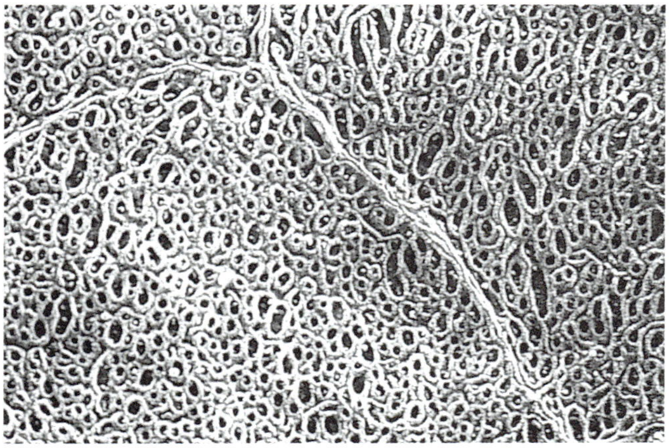

FIGURE 21.9 Scanning electron micrograph of the surface esophageal mucosa in which intercellular junctions are readily appreciated.

are separated by a distance equal to their diameters (49). Periodic acid–Schiff (PAS) stain may be used to demonstrate the upper extent of the glycogen-poor basal cells (Fig. 21.7). The prickle cell layer is so named because of the shrinkage artifact that exposes the desmosomal junctions between cells (40). The cells in the superficial layer maintain their nuclei, and although they do not normally keratinize, they may occasionally demonstrate keratohyalin granules. Keratohyaline granules (Fig. 21.8) consist mainly of keratin and other proteins that contribute to cornification or keratinization, expanding in size and leading to the conversion of keratin tonofilaments into a homogenous keratin matrix (50). The prickle and superficial cell layers consist of glycogen-rich cells that become progressively flatter toward the surface. The electron microscopic appearance of the epithelial layer demonstrates similarities to nonkeratinizing squamous epithelium elsewhere (Fig. 21.9).

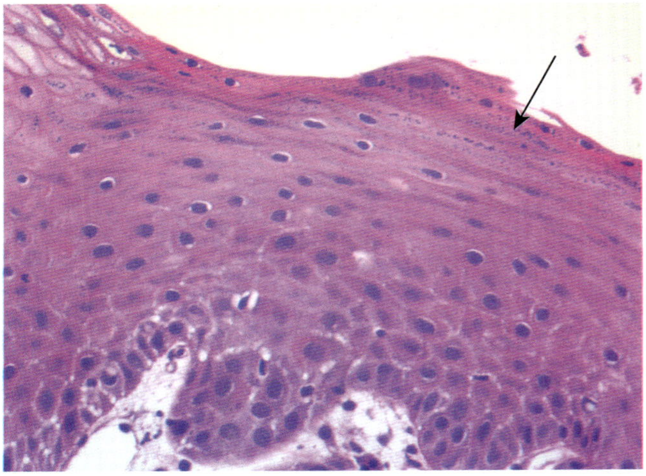

FIGURE 21.8 Keratohyaline granules. Squamous mucosa with keratohyaline granules (*arrow*).

A variety of cell types exist within the squamous mucosa, as described below.

› **MELANOCYTES** Melanocytes are visible in the basal regions of the epithelium in 3% to 8% of esophagi (51–53). Melanocytes contain melanin pigments within melanosomes. They lack desmosomes and tonofilaments but possess long dendritic processes that extend between the keratinocytes, often passing through several layers of cells (54). Melanocytes in the basal layer can coalesce to form visible pigmented patches (i.e., melanosis esophagi) (55) (Fig. 21.10). Esophageal melanosis should be distinguished from *pseudomelanosis*, which is caused by accumulation of exogenous pigmented deposits including various dyes and anthracosis, lipofuscin deposition in macrophages, and iron (55,56).

› **ENDOCRINE CELLS** Some cell populations originally described to be endocrine-based on their argyrophilia are most likely melanocytes. However, endocrine cells of the esophagus stain with endocrine markers similar to those of other sites (57,58). Endocrine cells are most numerous in the distal esophagus (59). Chromogranin staining can also be seen in LgR stem cells in the papillary basal layer (60). Some authors have hypothesized that endocrine cells of the esophagus represent Merkel cells, but there is no solid evidence that Merkel cells exist in the esophagus. In fact, CK20 staining of basal keratinocytes can be seen in esophageal biopsy samples, yet these cells are negative for synaptophysin and chromogranin (57,61) (Fig. 21.11).

› **CYTOKERATIN 20-POSITIVE CELLS** Keratinocytes that express CK20 are not identified in the neonatal esophagus, but are rarely seen singly and in clusters in adults (57), particularly in the midesophagus (62). They possess keratin tonofilaments and occasional desmosomes linking them to adjacent cells. They also contain characteristic

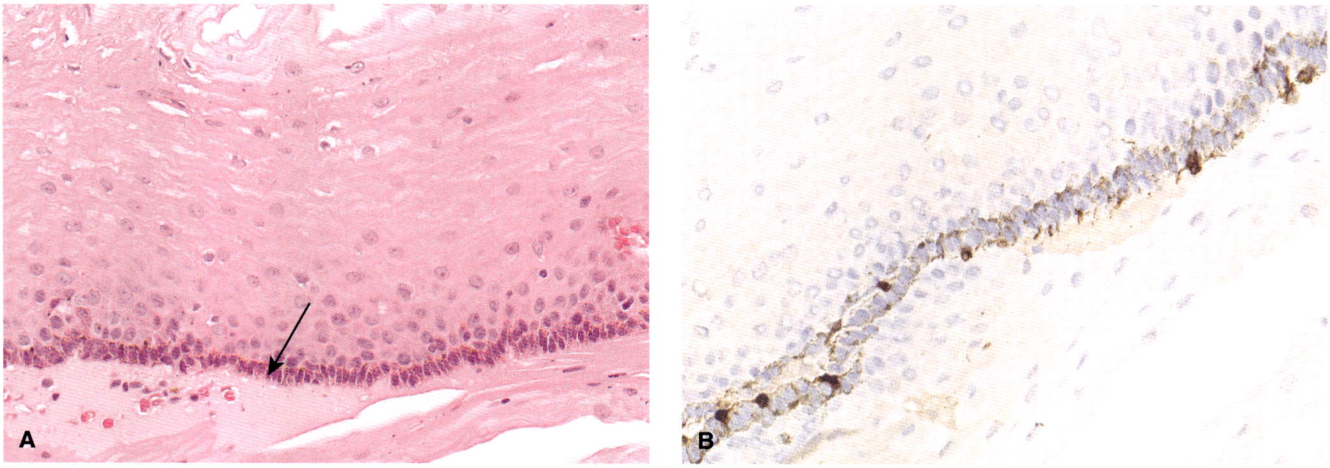

FIGURE 21.10 Melanosis esophagi. Melanosis esophagi. **A:** Hyperpigmentation of the basal layer (*arrow*). **B:** MART-1 immunostain highlights melanocytes.

small, membrane-bound cytoplasmic vesicles and are frequently situated adjacent to a nerve fiber. These vesicles may liberate a transmitter substance across the synapse-like junction between the CK20^{+} cell and the nerve fiber, thereby triggering an impulse. This arrangement suggests a possible role of these cells as mechanical sensory cells that detect food as it passes (54). Approximately one-third of esophageal small cell carcinomas express CK20, raising the

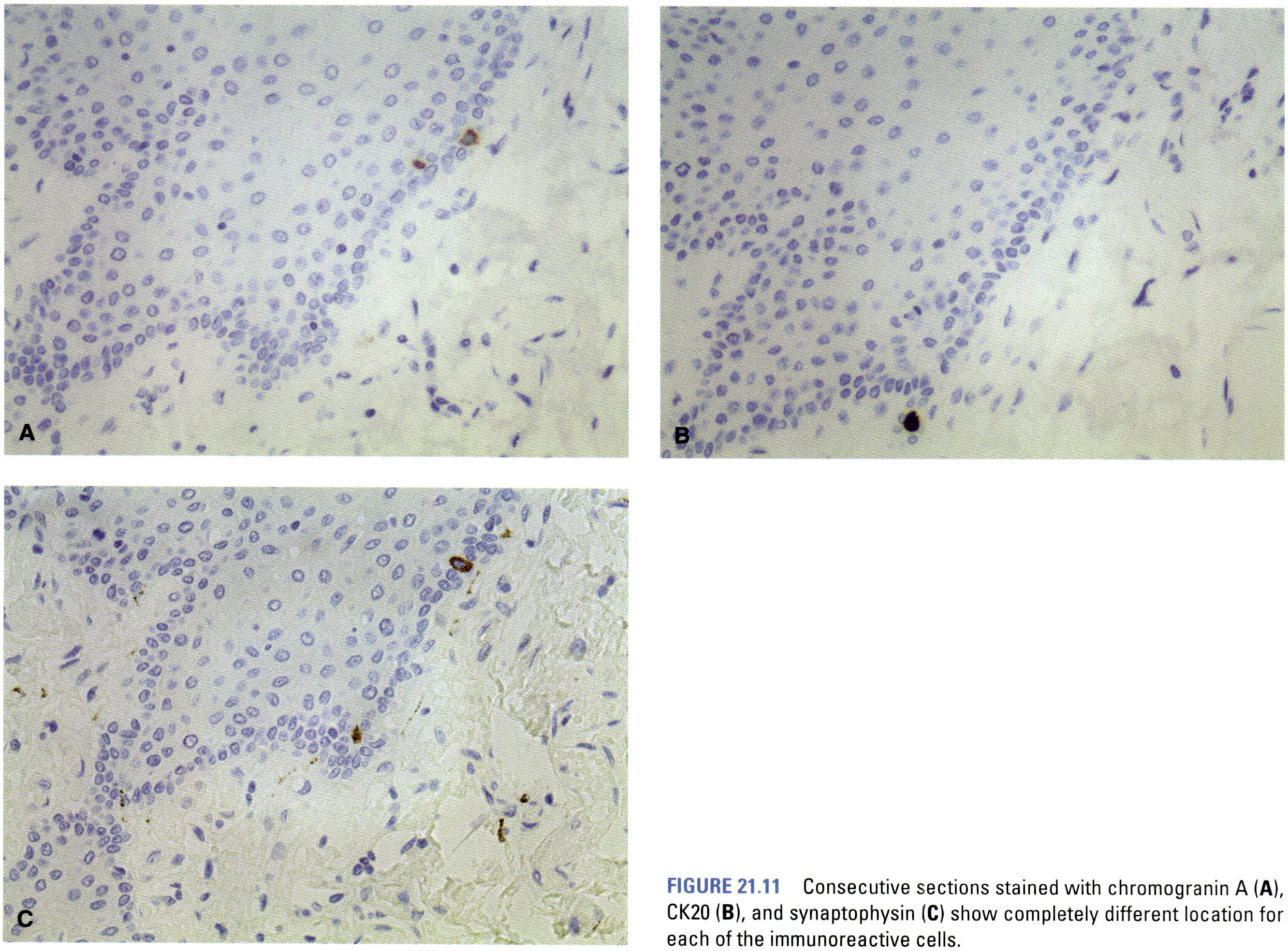

FIGURE 21.11 Consecutive sections stained with chromogranin A (**A**), CK20 (**B**), and synaptophysin (**C**) show completely different location for each of the immunoreactive cells.

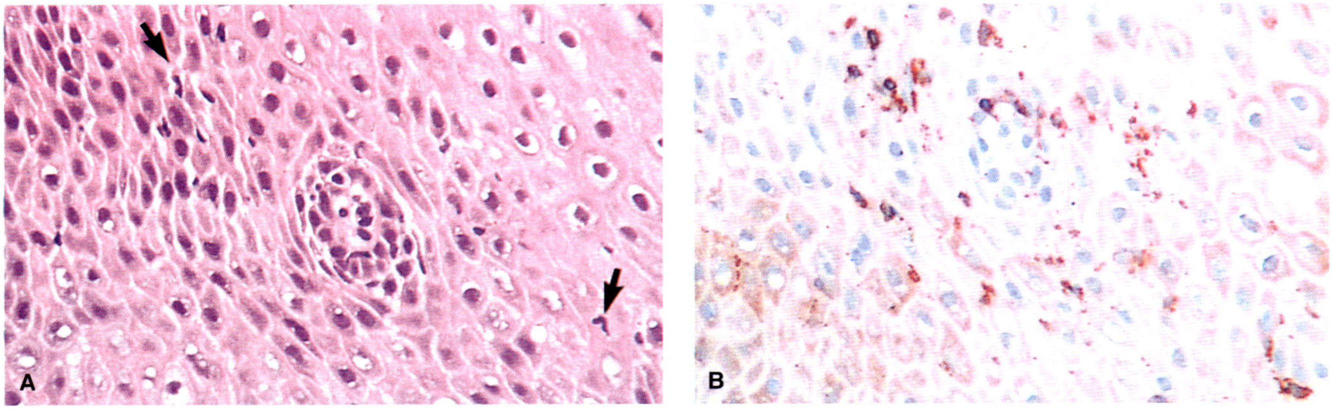

FIGURE 21.12 **A:** Numerous lymphocytes within the esophageal epithelium, some of which have a "squiggle" appearance (*arrows*). In addition, the intercellular spaces are dilated causing very prominent prickles, while some of these have formed small "bubbles" primarily at the junction between epithelial cells (best seen in the upper right quadrant). Care needs to be taken not to include perinuclear vacuolization/cytoplasmic retraction or paranuclear vacuoles as dilated intercellular spaces. **B:** Intraepithelial lymphocytes are of T-cell origin, as demonstrated using T-cell markers, and are primarily suppressor ($CD3^+$, $CD8^+$) cells.

possibility that this tumor type recapitulates the phenotype of such cells in the esophagus (20,57,63).

› *INTRAEPITHELIAL LYMPHOCYTES* Intraepithelial lymphocytes and antigen-presenting cells are normally present within the esophageal squamous epithelium, while the presence of neutrophils, eosinophils, and mast cells is considered to be pathologic (64,65). Plasma cells are not found within the squamous epithelium of the esophagus. It is virtually impossible to distinguish intraepithelial lymphocytes from antigen-presenting cells in routine sections, and thus, they are best collectively termed intraepithelial mononuclear cells. Intraepithelial mononuclear cells reside in the intercellular spaces between keratinocytes. The intercellular spaces also allow passage of lymphokines, cytokines, chemokines, and numerous other molecules directionally from the papillae to the lumen or lumen to the papillae. Intraepithelial mononuclear cells are normally present in a peripapillary distribution, or they are found in a suprabasal location (66,67). As they interdigitate between the epithelial cells, their nuclei become convoluted, leading some to refer to them as "squiggle cells" (Fig. 21.12).

› *ANTIGEN-PRESENTING CELLS* Langerhans cells are the best-recognized population of antigen-presenting cells and are characterized by the presence of racket-shaped Birbeck granules by electron microscopy. They are derived from monocytes and are able to divide as differentiated cells in situ with a turnover time of 8 days. They are located throughout the epithelium but are more numerous in the suprabasilar epithelium of the distal esophagus (68,69) (Fig. 21.13). It is possible that the distribution of Langerhans cells is related to clinical or subclinical gastroesophageal reflux. Langerhans cells can migrate from the epithelium to regional lymph nodes (54). Langerhans cells are immunopositive for S100 protein, CD68, CDla, and langerin (CD207); they are negative for MHC2 in their inactive state.

Conventional dendritic cells can also be found in the esophageal mucosa. These antigen-presenting cells are mostly immunoreactive for MHC2 but negative for CD1a. These cells are derived from the bone marrow and, unlike Langerhans cells, do not divide in the esophagus. Conventional

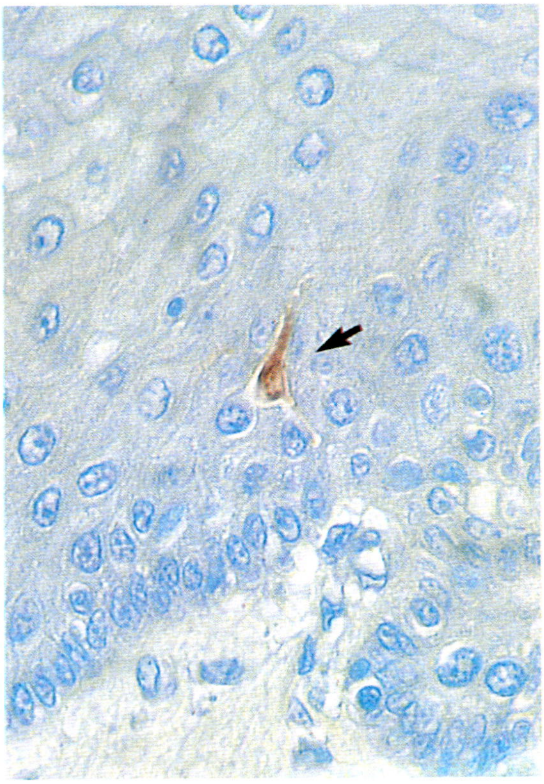

FIGURE 21.13 Langerhans cell (*arrow*) in a suprabasal position (S100).

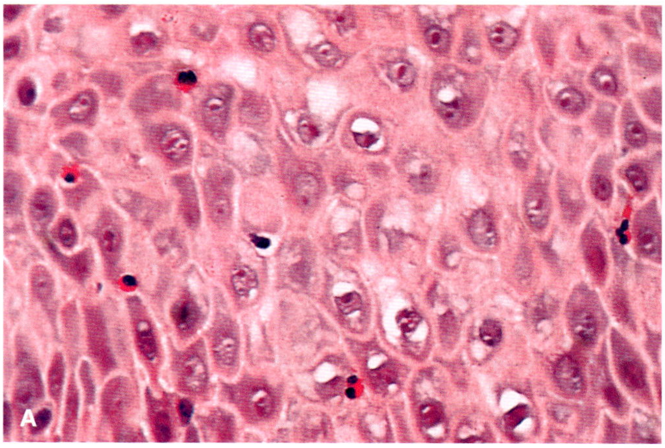

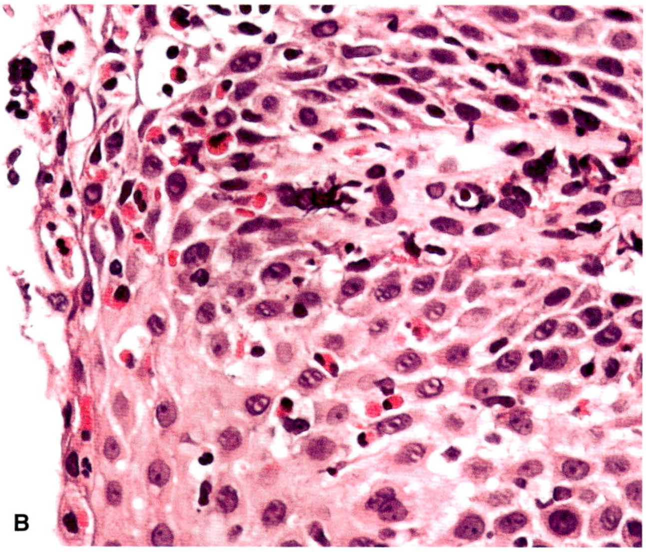

FIGURE 21.14 Eosinophilic esophagitis. **A:** Reflux esophagitis. Intraepithelial eosinophils are present between the epithelial cells. **B:** Allergic esophagitis (feline esophagus). There is marked basal cell hyperplasia, spongiosis with dilated intercellular spaces, and numerous eosinophils.

dendritic cells are found throughout the epithelium and can presumably migrate to local lymph nodes (70,71).

› ***Intraepithelial Eosinophils and Neutrophils*** Intraepithelial eosinophils and neutrophils are abnormal constituents of the epithelium throughout its entire length, although rare intraepithelial eosinophils may be seen in the distal 3 cm of approximately one-third of adult patients and are considered to represent a normal finding if they are unaccompanied by other features of mucosal injury (49). Intraepithelial eosinophils are commonly seen in esophageal biopsy samples from patients with gastroesophageal reflux disease, although eosinophilic esophagitis can show similar histologic features (Fig. 21.14). Intraepithelial neutrophils are rare unless the patient has erosive disease (71,72). Intraepithelial neutrophils can be found in association with erosions or ulcers of any cause and, thus, are not specific for a diagnosis of gastroesophageal reflux disease.

› ***Multinucleated Cells*** Multinucleated intraepithelial histiocytes can be seen in biopsy samples from patients with active gastroesophageal reflux in the absence of viral infections (73,74). Multinucleated epithelial cells may be seen in patients with esophagitis due to a variety of causes and probably represent a regenerative response to injury (74).

Intercellular Spaces and Basement Membrane

All intraepithelial inflammatory cells migrate through the intercellular spaces. Nutrients must also pass to the epithelial cells through these spaces while antigens pass from the lumen to antigen-presenting cells, where they are processed and recruit inflammatory cells to sites of injury.

The epithelium is supported by basement membrane that spans 50 to 80 nm in thickness and anchors the epithelium to the lamina propria (75). The basement membrane is composed of a variety of proteins including type IV collagen. Destruction of the basement membrane by malignant cells originating in the epithelium is the first step of cancer invasion and thus defines an invasive carcinoma, distinguishing it from high-grade dysplasia. The basement membrane also contains antigenic targets of autoimmune esophageal dermatologic diseases such as bullous pemphigoid and epidermolysis bullosa (76).

Cell Kinetics and Stem Cells in Squamous Mucosa

Although one might expect proliferation to begin in the most basal epithelial cell layer, the proliferative activity of the interpapillary basal cells is very low, and these cells are almost entirely Ki-67 negative (Fig. 21.15). Increased epithelial proliferation due to esophagitis results in increased proliferative activity and basal cell hyperplasia (77).

Epithelial cell turnover in the esophagus is slower than that of the small bowel (78). One human study using intravenous 5-iodo-2′-deoxyuridine to detect label-retaining cells found the turnover time of the esophageal squamous mucosa to be 11 days, although labeled cells were still present in the basal layer 67 days after injection. Stem cells of the squamous mucosa express LgR and frequently show chromogranin A staining, although they rarely express Ki-67 (60). Data from experimental models suggest that cell division results in one basal cell remaining attached to the basement membrane, while the other differentiates. Donor bone marrow-derived stem cells home to the esophagus and differentiate into esophageal stem cells in patients who have undergone bone marrow transplantation (79). As such,

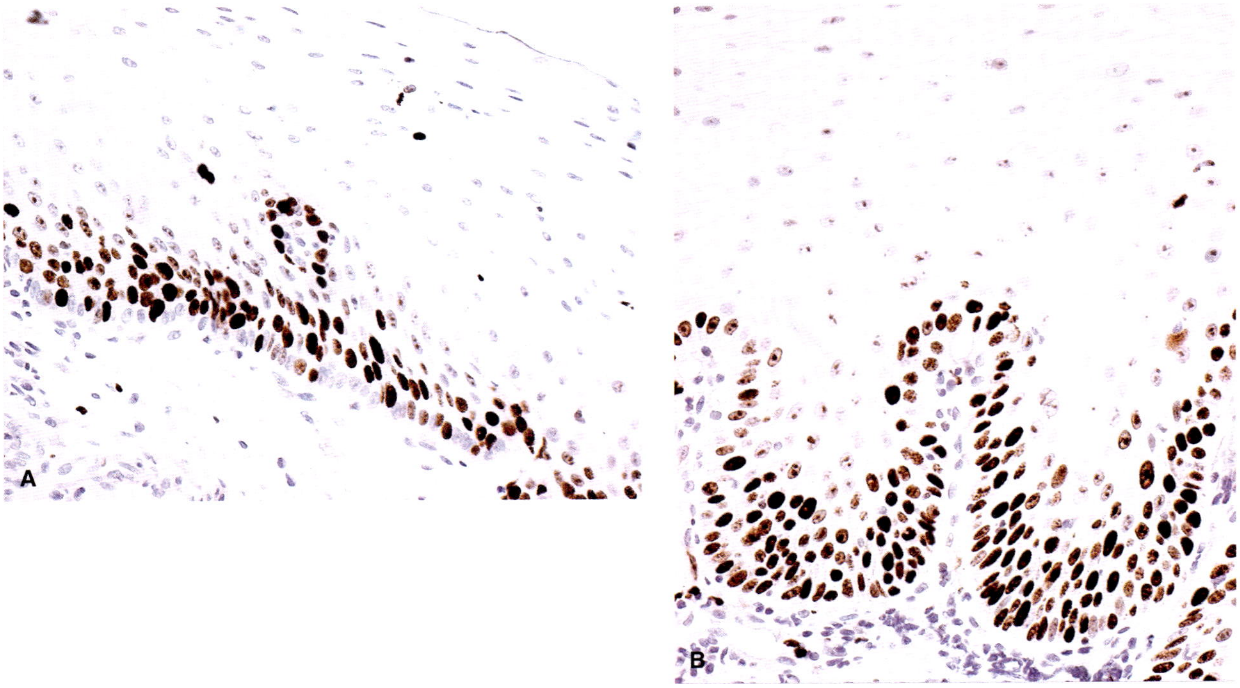

FIGURE 21.15 Basal layer of esophagus immunostained with MIB-1. **A:** The basal cell layer is relatively thin and unstimulated, consisting of only about three cell layers. Many of the cells in the basal layer are completely unstained, therefore, representing likely stem cells, while the proliferating cells with black nuclei are in the cell layer immediately above. **B:** In this biopsy, the basal layer is much thicker and, therefore, more proliferative, and there are fewer noncycling stem cells in the basal layer.

basally situated stem cells possess a different immunophenotype from the more differentiated cells, as discussed subsequently.

Immunohistochemistry of the Normal Esophageal Epithelium

The normal esophageal squamous epithelium is positive for immunomarkers of mature squamous differentiation (e.g., CK4, CK5, CK13), as well as CK17, CK19, and involucrin (80–82). It is negative for CK7 (83,84), CK18 (85,86), and predominantly for CK20 (87–89). Esophageal basal cells express SOX2, as well as keratins 5 and 14. In the normal esophagus, p63 is expressed in the basal and suprabasal layers of the squamous epithelium and in basal cells that line the mucosal gland ducts (90). This protein is related to p53 and binds DNA to help regulate the differentiation and proliferation of epithelial progenitor cells.

Basal cells express CK13; this marker of mature epithelium is expressed at high levels in both the papillary basal layer and epibasal layers, but it is absent from the keratinocytes of the interpapillary basal layer. Expression of CK14 and CK15 is patchy in the interpapillary basal layer in contrast to the papillary basal layer and epibasal layer, both of which show high levels of expression (85). CK4 is detectable in the papillary region from the second epibasal layer upward (89,91). Therefore, interpapillary basal layer appears to contain the least differentiated cells (91) and is home to the stem cell compartment (60). Suprabasal integrin expression is a consistent finding at the tips of the esophageal papillae, suggesting that the papillary basal layer cells migrate to this site (92). This concept is illustrated in Figure 21.16.

Multilayered epithelium is a mixed epithelium composed of immature squamoid cells and acid mucin-containing columnar cells that resemble goblet cells (Fig. 21.17). This composite mucosa shows a pattern of mucin production and cytokeratin expression similar to that of Barrett esophagus. The glandular cells contain neutral mucin, sialomucin, and sulfomucin. The squamoid cells express CK13 and CK19, whereas the columnar cells are positive for CK7, CK8/18, CK19, and CK20 (88,93). Multilayered epithelium does not express CK14 but shows evidence of intestinal-type differentiation with staining for villin in 64% of cases (88). This immunophenotype is similar to that of the ductal epithelium of the mucosal glands, which shows staining for CK19 but not for CK14, a mature squamous epithelium marker. The mucosal gland epithelium also stains for the columnar cell markers CK7, CK8/18, and variably for CK20, CK13, whereas the duct is negative for these markers (88,90).

The esophageal submucosal gland duct epithelium stains for CK7, CK8/18, and CK19. However, this type of epithelium does not stain for CK13, with the exception of the intermediate cell layers and a few dispersed basal

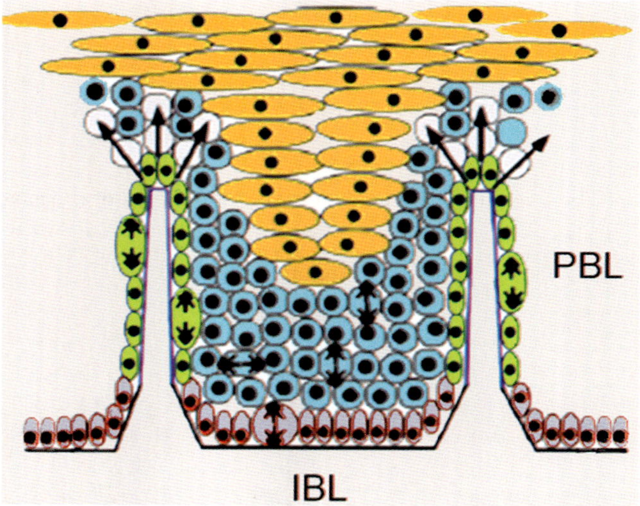

FIGURE 21.16 A model of the cellular organization in the esophageal epithelium. The interpapillary basal layer (*IBL*) cells (blue-gray) constitute the epithelial stem cell compartment. IBL cells proliferate infrequently and asymmetrically. Proliferating cells reside in the epibasal (suprabasal) layers (blue). Papillary basal cells (*PBL*) (green) are proliferative and intermediate in behavior between IBL and epibasal cells (see text). Differentiated squamous cells are shown in orange. (Modified with permission from Seery JP, Watt FM. Asymmetric stem-cell divisions define the architecture of human oesophageal epithelium. *Curr Biol.* 2000;10:1447–1450, with permission from Elsevier.)

cells of stratified excretory (88,89). This epithelium usually stains for CK14 and can express CK20 in some cases (88). CK4 is negative in submucosal glands, except in the intermediate cell layers of the stratified epithelium lining the upper part of the excretory duct at the connection with the surface epithelium, similar to CK6 expression (89).

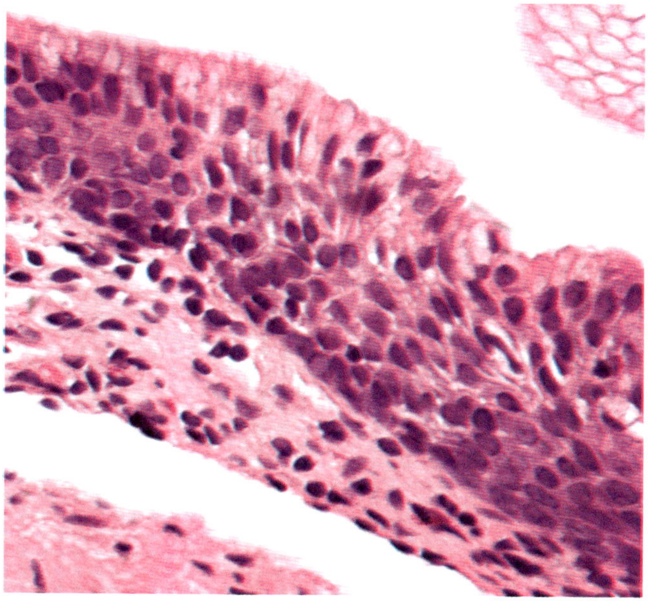

FIGURE 21.17 Multilayered epithelium, in which there is apical mucin production at the luminal surface (*top*), but the remainder of the epithelium appears squamous with intercellular bridges.

Dispersed CK15-positive cells can be found in both cell layers of the excretory duct (89).

Normal cardiac-type epithelium expresses both CK7 and CK20 in a patchy distribution (94). Notably, CDX2 is not expressed in the cardiac-type mucosa of the proximal stomach or in the squamous mucosa of the esophagus (95). However, it is expressed in columnar cell metaplasia of the esophagus, both with and without goblet cells (96–99). CDX2 is downregulated in the progression to esophageal adenocarcinoma, suggesting that CDX2 may function as a tumor suppressor gene (100). Similarly, TTF-1 is not expressed in normal esophagus or gastric mucosae, but may be an important player in the development of Barrett esophagus; it is also known to regulate the expression of CLDN18 (101).

Esophageal Microbiome

The normal esophagus was once thought to be mostly sterile but is now known to harbor bacterial flora. The use of high throughput gene sequencing on mucosal biopsies has led to a better characterization of the esophageal microbiome, which is diverse and unique.

The healthy esophagus is inhabited predominately by Gram-positive bacteria similar to those found in the oropharynx. These bacteria are represented mostly by six phyla with a high relative abundance of *Firmicutes* and *Streptococci* (102). Gram-negative bacteria such as *Veillonella, Prevotella, Haemophilus*, and are also present, although less abundant (103). The microbiome is thought to participate in esophageal homeostasis through promotion of barrier function, interactions with the immune system, and protection against pathogens (104). Occasionally, they can be visualized between and surrounding superficial squamous cells in routinely stained sections. The microbiome composition changes in response to environmental stimuli and other factors, including dietary fiber content, age, endoluminal pH, and use of antibiotics and proton pump inhibitors.

Dysbiosis, or disruption of the normal balance of microbes, contributes to the development and progression of esophageal disease. For example, the composition of bacteria shifts away from Gram-positive bacteria to Gram-negative anaerobes and microaerophiles in patients with gastroesophageal reflux disease and Barrett esophagus (105). There is also a shift away from Gram-positive bacteria to Gram-negative bacteria with a significant increase in *Haemophilus* among patients with eosinophilic esophagitis. Changes in the microbiome are also seen in patients with both squamous cell carcinoma and adenocarcinoma of the esophagus (103,106).

While the precise mechanisms of carcinogenesis are not well characterized, it has been suggested that the lipopolysaccharide present in Gram-negative bacteria is central to this process. Lipopolysaccharide can activate resident immune cells through toll-like receptors, resulting in elaboration of the NF-κB (nuclear factor kappa B) pathway, which leads to increases in cytokines that promote inflammation, tissue damage, and development of Barrett esophagus (107). Lipopolysaccharide also enhances release of inducible nitric oxide synthase (iNOS), which promotes relaxation of the lower esophageal sphincter tone, resulting in increased acid reflux and further tissue damage (107). The resulting increase in chronic inflammation and associated tissue damage ultimately predisposes to the development of adenocarcinoma. Further understanding of how alterations in the microbiome influence the progression of esophageal disease through immune-dependent and independent mechanisms may lead to novel and improved diagnostic, prognostic, preventative, and therapeutic strategies.

The Gastroesophageal Junction and Squamocolumnar Junction

The squamocolumnar junction generally represents the mucosal component of the gastroesophageal junction, although there are some situations in which the squamocolumnar junction is located proximal to the gastroesophageal junction. Glandular metaplasia of the distal esophagus can result in proximal migration of the squamocolumnar junction, either in a circumferential fashion or with irregular tongues and islands of velvety glandular mucosa above the level of the gastroesophageal junction. Proximal displacement of the squamocolumnar junction by cardiac-type glandular mucosa with or without goblet cells (i.e., Barrett esophagus); the length of metaplastic mucosa is proportionate to risk of esophageal adenocarcinoma.

The GEJ can be defined physiologically, anatomically, microscopically, or endoscopically. The muscular gastroesophageal junction is most accurately defined by manometric studies in which the distal segment of the recorded high-pressure zone of the lower esophageal sphincter defines the junction (108). However, severe gastroesophageal reflux disease or Barrett esophagus can result in decreased pressures, such that localization cannot be determined by these means. Anatomical landmarks include the peritoneal reflection from the stomach onto the diaphragm, the incisura cardialis or angle of His, which is defined as the acute angle between the lesser curvature of the stomach and the esophagus (29). The upper margin of the diaphragmatic indentation can also be used as a landmark, although it can move in the presence of a hiatal hernia (29).

The two most practically relevant definitions of the gastroesophageal junction are based on endoscopy (109). The first is used in "the West" and the second in Asia, and especially Japan and neighboring countries.

1. Gastroenterologists in Europe and North America define the gastroesophageal junction as the upper limit of the proximal gastric folds (110). This anatomic landmark can be minimally, but not practically, affected by respiration, gut motor activity, and the degree of distension of the esophagus and stomach. The proximal limit of the gastric folds closely approximates the muscular gastroesophageal junction and may provide a fixed anatomic landmark for the muscular gastroesophageal junction (111).
2. Gastroenterologists in Japan and parts of Asia define the gastroesophageal junction as the distal extent of the esophageal palisade vessels (33,34). This definition is less specific because these vessels can be present above, at, or below the squamocolumnar junction in normal individuals.

› ***The Squamocolumnar Junction*** The mucosal squamocolumnar junction (i.e., Z-line or ora serrata) is an irregular line of contrast with the paler white squamous mucosa above and red glandular mucosa distally (Fig. 21.18) (112). The squamocolumnar junction coincides with the gastroesophageal junction in normal subjects, and a squamocolumnar junction proximal to the gastroesophageal junction is also considered to be within normal limits, provided the glandular mucosa extends no more than 1 cm above the gastroesophageal junction. The squamocolumnar junction can be accompanied by a ring-like narrowing of the lumen, which has been termed Schatzki ring. Schatzki rings are superiorly surfaced by squamous mucosa and covered by glandular mucosa inferiorly.

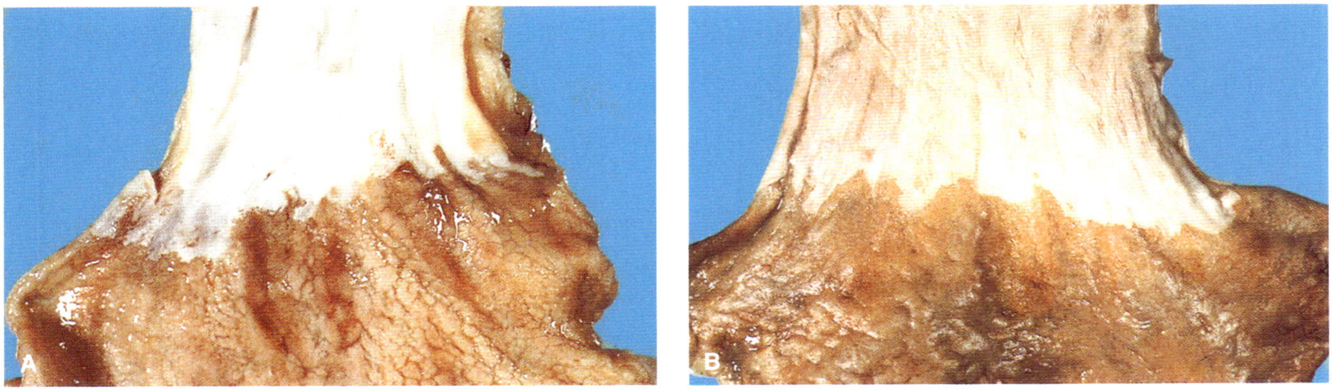

FIGURE 21.18 Gastroesophageal region. **A** and **B:** Formalin-fixed specimen demonstrates the variation of the normal squamocolumnar junction (Z line).

› ***Distal Glandular Esophagus*** The glandular mucosa distal to the Z-line is surfaced by foveolar-type columnar epithelial cells in addition to small aggregates of simple mucinous glands (i.e., cardiac-type mucosa), oxyntic glands (i.e., oxyntic mucosa), or a combination of these types of glands (i.e., cardio-oxyntic mucosa). Cardiac-type and cardio-oxyntic mucosae can show intestinal metaplasia, usually in the form of scattered goblet cells on a background of nongoblet columnar epithelial cells (i.e., incomplete intestinal metaplasia). Goblet cells are less frequently accompanied by absorptive enteric cells and/or Paneth cells (i.e., complete intestinal metaplasia). The squamocolumnar junction may also contain multilayered epithelium consisting of immature squamoid and columnar cells as previously described. Aggregates of pancreatic-type acini are frequently encountered. These have been variably termed "pancreatic heterotopias" and "pancreatic metaplasia," although the former label is more likely to be correct. Cardiac-type mucosa in this location is frequently inflamed due to gastroesophageal reflux, although abundant plasma cell-rich inflammation can signify the presence of *Helicobacter pylori* infection in the proximal stomach.

Most available data suggest that cardiac-type mucosa in the distal esophagus and proximal stomach is acquired; a direct transition from squamous to oxyntic mucosa is found primarily in fetuses. The length of cardia mucosa is variable, ranging up to approximately 1 cm. Its length increases with age and severity of gastroesophageal reflux (113–116). Data from both autopsy and endoscopy-based studies of pediatric patients have shown that 65% of patients do not have cardiac-type mucosa distal to the squamocolumnar junction. When present, it is combined with oxyntic or oxynto-cardiac mucosa and is of variable extent (97,98). Endocrine cells that express chromogranin A and/or serotonin are frequently detected in cardiac-type mucosa (117) but gastrin-producing cells are only found in cases with intestinal metaplasia.

Multilayered epithelium is a distinctive type of epithelium histologically characterized by multiple layers of immature squamoid cells with an overlying layer of columnar epithelial cells with immunohistochemical qualities similar to those of goblet cells in Barrett esophagus (105) (Fig. 21.17). Multilayered epithelium is associated with gastroesophageal reflux and may represent a precursor to intestinal metaplasia of Barrett esophagus (106,107). Some authors have suggested that multilayered epithelium represents an early phase of columnar cell metaplasia in Barrett esophagus, although others believe that it represents simple metaplasia or persistent fetal epithelium (118,119). Multilayered epithelium may display cilia at the luminal surface.

Lamina Propria

The lamina propria of the esophagus is the non-epithelial portion of the mucosa above the muscularis mucosae. It consists of loose connective tissue and contains vascular structures and mucus-secreting cardiac-type glands with accompanying ducts. The presence of scattered inflammatory cells, including lymphocytes, histiocytes, plasma cells, and mast cells, is considered a normal finding and does not correlate with clinical symptoms of gastroesophageal reflux (49). Lymphocytes are predominantly CD4-positive, although CD8-positive cells are also present (120). Most lamina propria plasma cells secrete immunoglobulin (Ig) A. Smaller populations of IgG- and IgM-producing plasma cells are also present (120).

Esophageal papillae are infoldings of the lamina propria that stabilize the epithelium and support capillaries, venules, and lymphatic vessels, thereby facilitating exchange of molecules and inflammatory cells traveling in both directions.

Cardiac-type glands are diffusely scattered in the lamina propria through all levels of the esophagus, predominating in the distal and proximal regions (10,121). While they have been variably considered as heterotopias (14,121), normal constituents, or embryologic remnants, there is little doubt that they perform a lubricating function and are physiologically necessary to facilitate bolus passage. They resemble the cardiac-type glands of the stomach and are composed of neutral mucin-secreting cells (Fig. 21.19). They resemble pyloric glands, even to the extent that they are MUC-6 immunoreactive, and may contain variable numbers of both parietal and chief cells in the distal esophagus. Paneth-like lysozyme-producing granules can be seen in these glands at any level of the esophagus. Their ducts are lined by simple mucin-producing cells in contrast to the submucosal glands, the ducts of which are frequently lined by squamous cells. These duct-lining cells may also extend onto the surface for a variable distance, even producing small islands of simple mucus-producing glands with an abrupt junction at the surface of squamous mucosa.

Normal esophageal mucosa lacks structured lymphoid follicles. However, chronic injury may lead to formation of lymphoid follicles (122,123). Lymphoid follicles are most commonly associated with chronic gastroesophageal reflux disease and eosinophilic esophagitis (123). They may impart a nodular appearance to the mucosa that is endoscopically evident (122).

Muscularis Mucosae (Lamina Muscularis Mucosae)

The muscularis mucosae is composed of longitudinally oriented bundles of smooth muscle cells (108) rather than having both a circular and longitudinal arrangement as is seen in the rest of the tubular gut. The muscularis mucosae begins at the cricoid cartilage of the pharynx and becomes thicker distally (37,124). The muscularis mucosae is underdeveloped in the *cervical* esophagus, merely appearing as thin strands of smooth muscle cells in some areas. Therefore, cancer cells occurring in the cervical esophagus

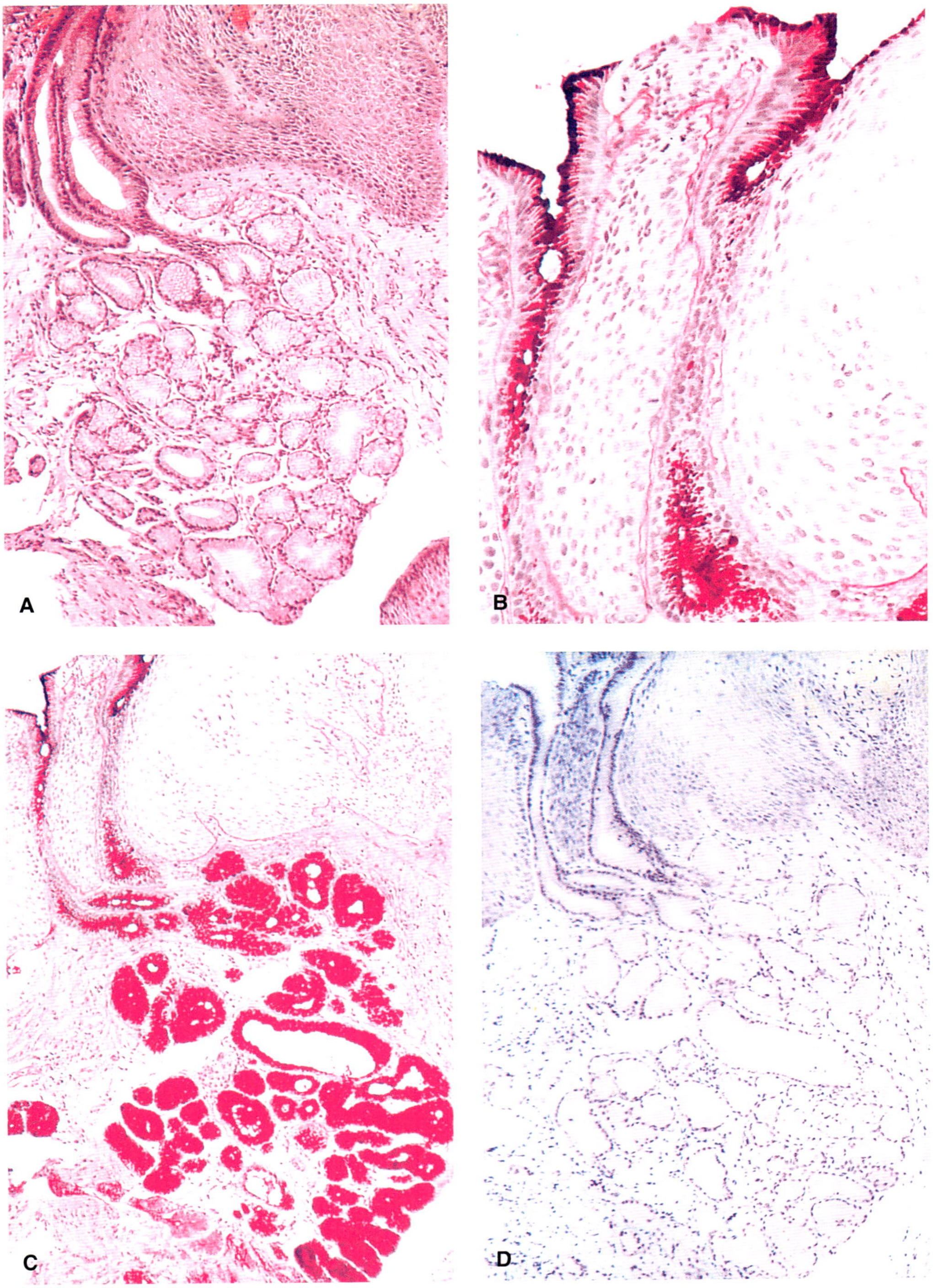

FIGURE 21.19 Midesophagus. **A:** Esophageal cardiac-type glands are located within the lamina propria. The ducts are lined by gastric foveolar–like cells. **B:** The duct-lining cells may extend over the stratified squamous epithelium for variable distances (PAS-D). **C,D:** The glands stained PAS-D positive and Alcian blue at pH 2.5 negative, characteristic of neutral mucins.

can easily infiltrate the submucosa (124). The muscularis mucosae of the thoracic esophagus is well-developed and consists of several cell bundles forming a plate-like structure, the thickness of which varies among different individuals (37,124).

The muscularis mucosae is thickest at the gastroesophageal junction and it may appear as duplicated layers in patients with a history of Barrett esophagus or prior esophageal ulcers, thereby simulating the muscularis propria in biopsy material (Fig. 21.20).

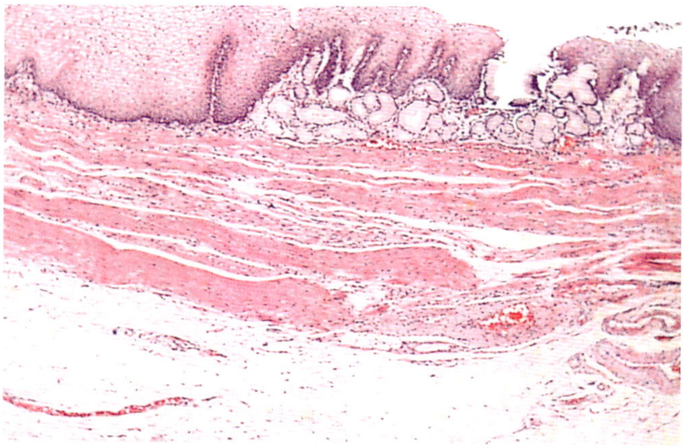

FIGURE 21.20 Distal esophagus. The mucosal layer at the gastroesophageal junction is characterized by muscularis mucosae that are thicker than the muscularis mucosae of the more proximal esophagus (compare with Fig. 21.6). Note the esophageal cardiac-type gland situated above the muscularis mucosae.

Submucosa

The submucosa consists of loose connective tissue containing numerous vessels, adipose tissue, nerve fibers, Meissner plexus, lymphatic vessels, and submucosal glands (Fig. 21.21). Submucosal glands are rarely encountered in mucosal biopsies as it is rare for these samples to even include the muscularis mucosae. However, the submucosal glands are encountered in endoscopic mucosal resections and submucosal dissections. The ducts of the glands may be seen in the lamina propria of mucosal biopsies, particularly samples from the lower esophagus (125,126).

The submucosal glands resemble the minor salivary glands of the oropharynx and are scattered throughout the entire esophagus; they are more concentrated in the upper and lower regions (10). Submucosal glands produce mucous to lubricate the squamous mucosa, facilitating swallowing and maintaining the integrity of the squamous mucosa. They secrete biologically active peptides, including trefoil factor family-3, epidermal growth factor, transforming growth factor-alpha, and prostaglandin E2 (127). Submucosal glands may also protect the mucosa by elaborating a variety of defensive cell products: neutral and sialated mucins prevent viruses and bacteria from entering the mucosa, lysozymes are bactericidal, and pepsinogen is activated to pepsin and contributes to proteolysis (3).

Submucosal glands and their ducts are arranged in rows parallel to the long axis of the esophagus. They consist of acini invested by a peripheral rim of myoepithelial cells. The acini contain mostly mucous cells with smaller numbers of serous and oncocytic cells. Mucous cells generally predominate in lobules at all levels in the esophagus. They contain sulphomucins (Fig. 21.22) that impart a faintly basophilic cytoplasmic hue, and they show strong staining for Alcian blue. Serous cells contain deeply basophilic, granular cytoplasm and peripherally arranged, small, round nuclei; they may be absent from some glands. Oncocytic cells are cuboidal with abundant, densely eosinophilic cytoplasm and uniform, round nuclei with conspicuous nucleoli (3).

Aggregates of two to five lobules drain into a common duct that penetrates the muscularis mucosae and squamous epithelium before extending to the surface. These ducts contain two cell layers: an inner layer of short columnar epithelial cells is supported by an outer layer of small cuboidal cells. As the ducts extend onto the luminal surface, the flattened cuboidal epithelium gradually transitions to a stratified squamous epithelium subjacent to short columnar cells that line the duct lumen (3). Microscopic periductal aggregates of chronic inflammatory cells and duct dilatation are not uncommon findings in the normal esophagus (128). Notably, the presence of submucosal glands is indicative of an esophageal origin as these glands are not present in the stomach. This observation can be helpful in determining whether a biopsy sample consisting of cardiac-type mucosa was obtained from a metaplastic area in the distal esophagus or the proximal gastric mucosa.

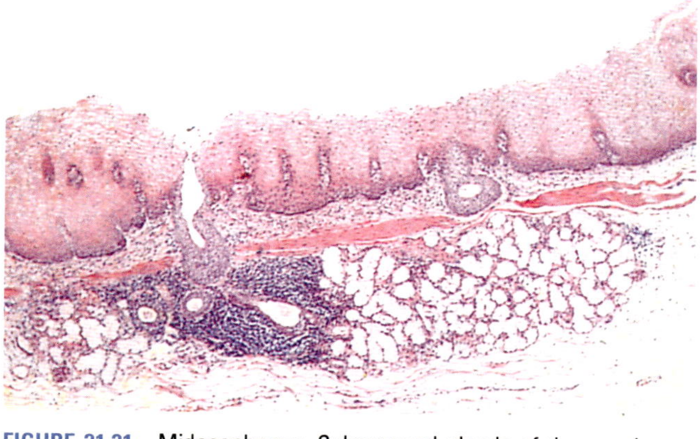

FIGURE 21.21 Midesophagus. Submucosal glands of the esophagus are located within the submucosa just beneath the muscularis mucosae. Periductal and periglandular chronic inflammation can be a normal finding. Note that in contrast to the columnar-lined glands of the lamina propria, the ducts of the submucosal glands are lined by squamous epithelium.

Muscularis Propria

The muscularis propria of the esophagus consists of outer longitudinal and inner circular layers. The esophageal orifice is bounded superiorly by the cricopharyngeal and inferior pharyngeal constrictor muscles, both of which contribute muscle fibers to the esophageal musculature (29). Horizontal fibers from both these muscles form the upper esophageal sphincter, which is a localized zone of increased pressure measuring 2 to 4.5 cm in length (27,108,129). Together, these muscle groups act in concert to initiate and control swallowing. The upper quarter to one-third of the proximal muscularis propria is composed of striated muscles (29), but only a short length (approximately 5%) of the proximal muscularis propria is composed entirely

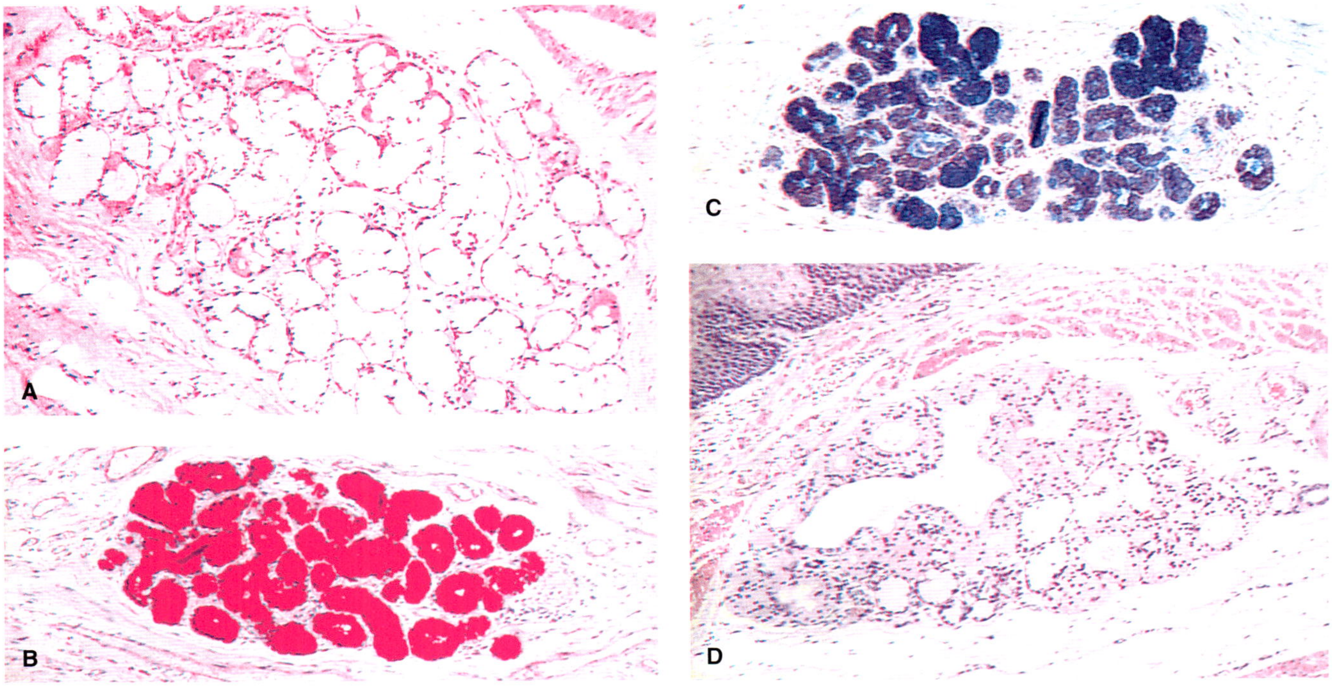

FIGURE 21.22 Midesophagus. **A:** The submucosal glands are composed predominantly of mucus-secreting cells with a variable serous component. **B,C:** The submucosal glands stained positive with PAS-D and Alcian blue at pH 2.5, a characteristic of acid mucins. **D:** Submucosal glands may demonstrate oncocytic metaplasia.

of striated muscle (130). The muscularis propria is composed of smooth and striated muscle distal to this region. The amount of smooth muscle in the wall progressively increases such that smooth muscle predominates throughout much of the wall, and more than 50% of the distal muscularis propria is composed solely of smooth muscle (Fig. 21.23) (130).

The longitudinal layer of the muscularis propria is composed of two bands of smooth muscle cells where it originates at the cricoid cartilage. The muscles sweep dorsally and incompletely interdigitate, leaving a bare V-shaped area (i.e., area of Laimer) exposing the underlying circular layer at the posterior wall of the pharyngoesophageal junction, specifically below the cricopharyngeal muscle. This area represents an area of potential weakness and is prone to development of posterior pulsion diverticula (i.e., Zenker diverticulum). In the upper third, the external longitudinal layer consists of striated muscle cells but their numbers diminish such that they finally disappear around the junction of the upper and middle thirds of the esophagus (29). The longitudinal layer is continuous with the outer longitudinal layer of the stomach.

In contrast to the rest of the gastrointestinal tract, the circular layer of the esophagus is slightly thinner than the longitudinal layer (29). The circular layer continues over the stomach, dividing in the region of the cardia to form the middle circular and inner oblique muscle layers. Fascicles of smooth muscle cells in the inner oblique layer pass in a sling-like manner at the incisura and cross at right angles with the horizontally oriented fascicles of the middle layer, thereby forming a muscular ring to which a possible sphincter function has been ascribed (29). The circular muscle layer of the distal 2 to 6 cm of the esophagus thickens to approximately twice that of the immediately adjacent muscularis mucosae (131).

Some diseases preferentially affect one of the muscle layers. For example, scleroderma results in collagen deposition and atrophy predominantly involving the circular layer. Achalasia results in a loss of ganglion cells and associated neurons of the myenteric plexus with concomitant hypertrophy of the circular layer (10).

Lower Esophageal Sphincter

The lower esophageal sphincter is manometrically defined as a 2- to 4-cm zone of increased luminal pressure, the distal end of which defines the muscular component of the gastroesophageal junction (132). The sphincter maintains an average pressure of 20 mm Hg (range: 10 to 26) (27). The lower esophageal sphincter keeps the lumen closed during rest and relaxes during swallowing to allow food to pass through (133–135). However, physiologic reflux of gastric contents into the esophagus still occurs, even among patients with a competent lower esophageal sphincter.

Adventitia/Serosa

The esophagus lies in a bed of fat and neurovascular tissue supported by collagen and elastic fibers collectively termed the adventitia (76). The adventitia forms a fascial sheath to support the esophagus in the upper mediastinum

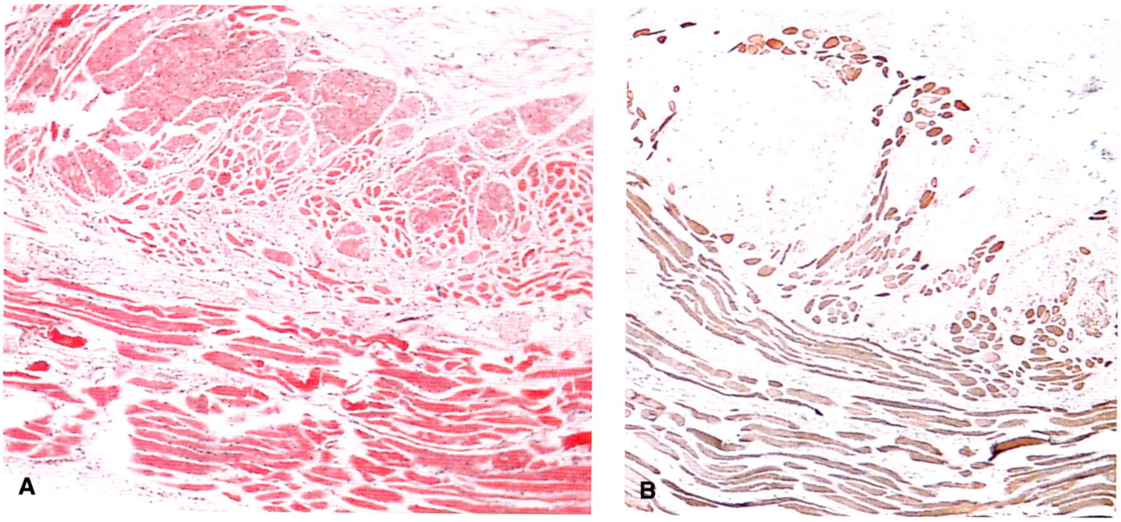

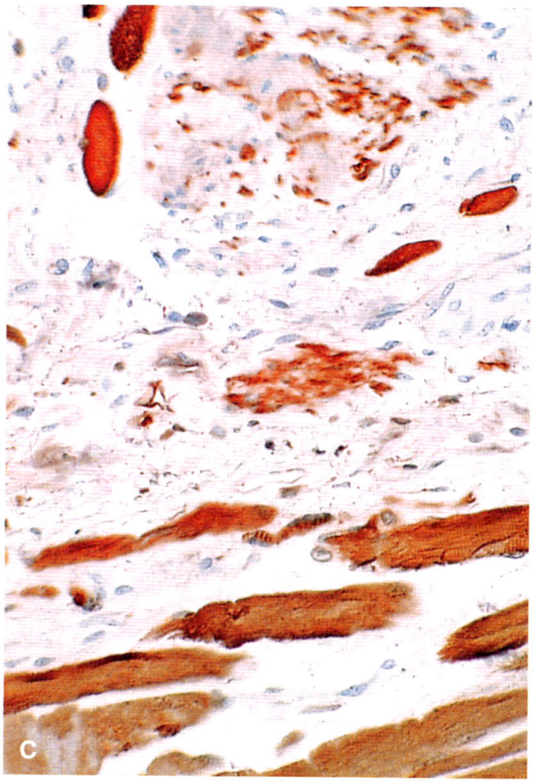

FIGURE 21.23 Proximal esophagus. **A:** The muscularis propria demonstrates a mixture of smooth and striated muscle bundles. **B:** This mix of smooth (weakly stained) and striated (strongly stained) muscles is demonstrated using a myoglobin antibody. **C:** Detailed photomicrographs of myoglobin stained section demonstrate peripherally located nuclei typical of striated muscle.

and thorax (29). However, it is surfaced by serosa after passing through the diaphragm. The lack of a serosa above the diaphragm allows esophageal tumors to spread more easily once they penetrate the muscularis propria (136).

DEVELOPMENTAL DEFECTS

Developmental defects of the esophagus can be attributed to errors in morphogenesis and may occur in isolation or as part of a syndrome.

Heterotopias

Gastric-type mucosa occurs in the upper one-third of the esophagus in 2% to 8% of patients and is designated as the "*inlet patch*" (Fig. 21.24) (10,14,121,137–139). The inlet patch typically has a deep pink, velvety appearance and presents either as a single patch or, less commonly, as multiple patches of gastric mucosa situated just below the upper esophageal sphincter or in the postcricoid region of the esophagus (140). Inlet patches of pediatric patients may consist of simple ciliated columnar mucosa (Fig. 21.25) (141).

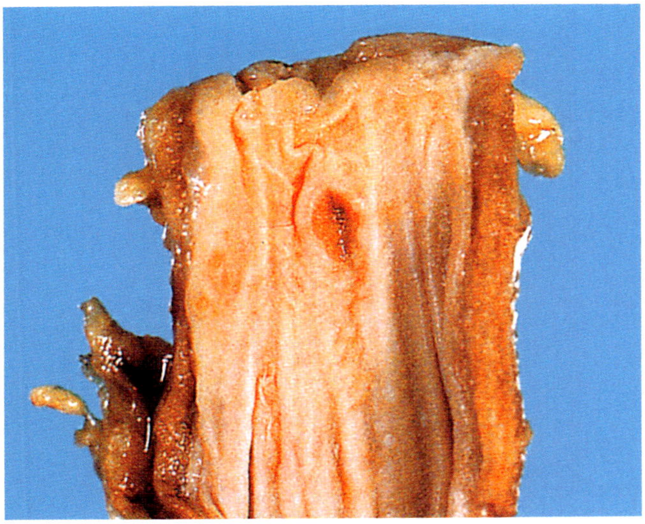

FIGURE 21.24 Proximal esophagus. Gastric body heterotopia situated slightly distal from the esophageal origin.

Pancreatic heterotopias are common in glandular mucosa near the gastroesophageal junction (Fig. 21.26) (142). Macroscopically visible rests of pancreatic tissue can occur in the esophagus and appear as an asymptomatic submucosal mass (143). The heterotopic pancreatic tissue contains a variable combination of ducts, acini, and/or islets (144). Heterotopic hepatic tissue is rarely encountered in the esophagus (145). Sebaceous glands are found in up to 0.05% of asymptomatic subjects and show a predilection for the upper esophagus (Fig. 21.27) (146). Thyroid tissue can also be found in the esophagus (147).

Esophageal Atresia, Congenital Esophageal Stenosis, and Congenital Short Esophagus

Esophageal atresia is the most common significant esophageal malformation, with an incidence of approximately 1 in 3,500 live births. This anomaly is caused by a failure of the primitive gut to recanalize in week 8 (148). Likewise, congenital esophageal stenosis results from incomplete esophageal recanalization during the 8th week of human embryologic development (148). Congenital esophageal stenosis is more frequent in the distal third of the esophagus and appears either as a web or long segment with fibromuscular stenosis and a thread-like lumen. This anomaly may result from incomplete separation of the respiratory tract and may contain cartilage or respiratory glands. It occurs once in every 25,000 live births (149) and is often associated with other developmental abnormalities such as atresia, tracheoesophageal fistula, esophageal duplication, congenital heart disease, trisomy 21, anorectal malformation, duodenal atresia, tracheomalacia, and esophageal hiatal hernia (150,151).

Congenital short esophagus occurs when the esophagus fails to reach its definitive length in the embryonic period. This diagnosis requires the presence of supporting evidence that the stomach developed in the thorax before birth (152,153).

Esophageal Duplication (ED)

The notochord induces the formation of the neural tube, gastrointestinal tract, and other organ systems. A split notochord can result in ED cysts or completely duplicated

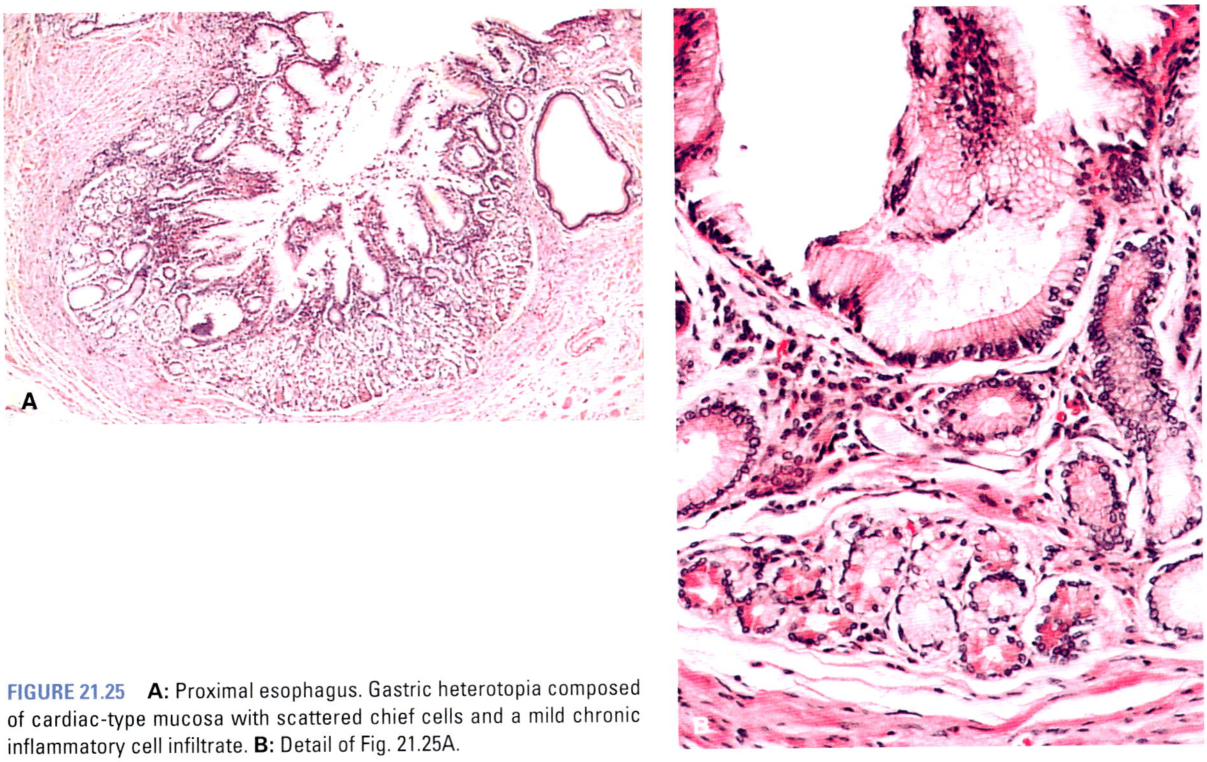

FIGURE 21.25 **A:** Proximal esophagus. Gastric heterotopia composed of cardiac-type mucosa with scattered chief cells and a mild chronic inflammatory cell infiltrate. **B:** Detail of Fig. 21.25A.

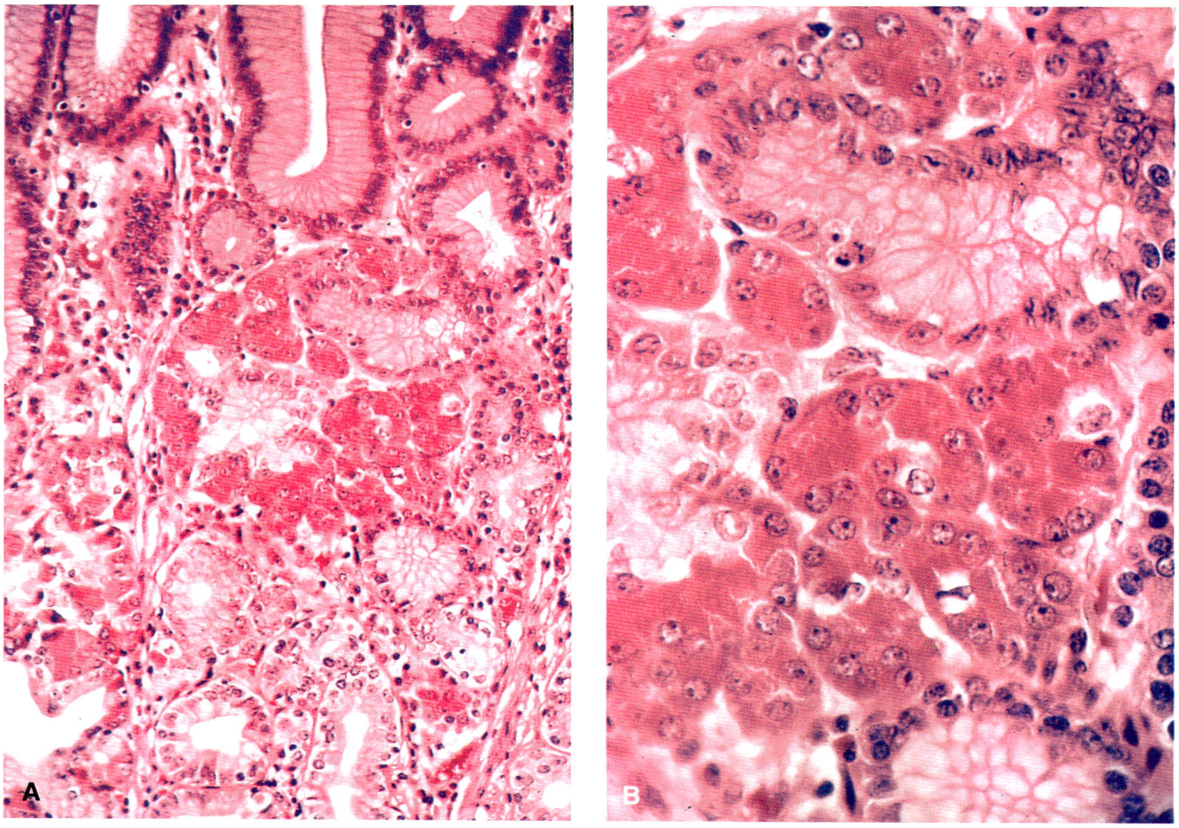

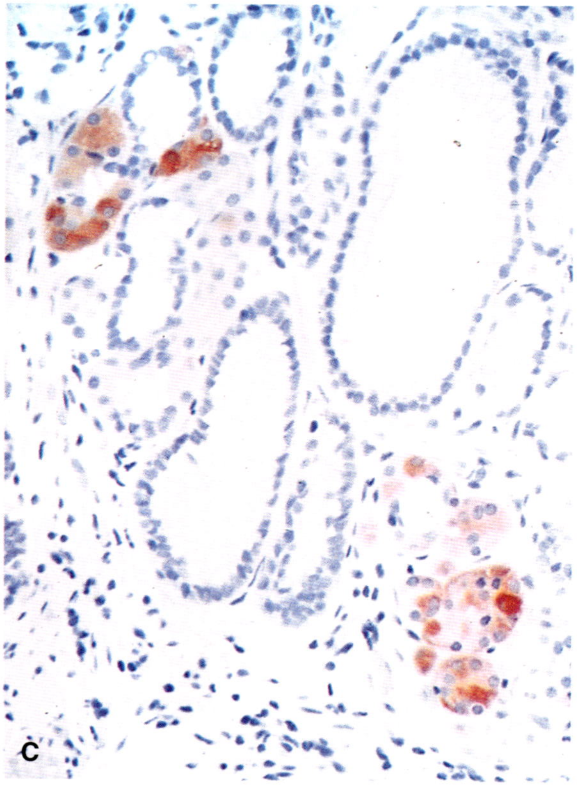

FIGURE 21.26 Pancreatic metaplasia in a biopsy from the cardiac side of the Z line. **A, B:** Glands are an admixture of mucus glands and eosinophilic granular cells superficially resembling a cross between gastric chief cells and Paneth cells. **C:** Immunohistochemical reactivity is found in pancreatic exocrine hormones, amylase in this case.

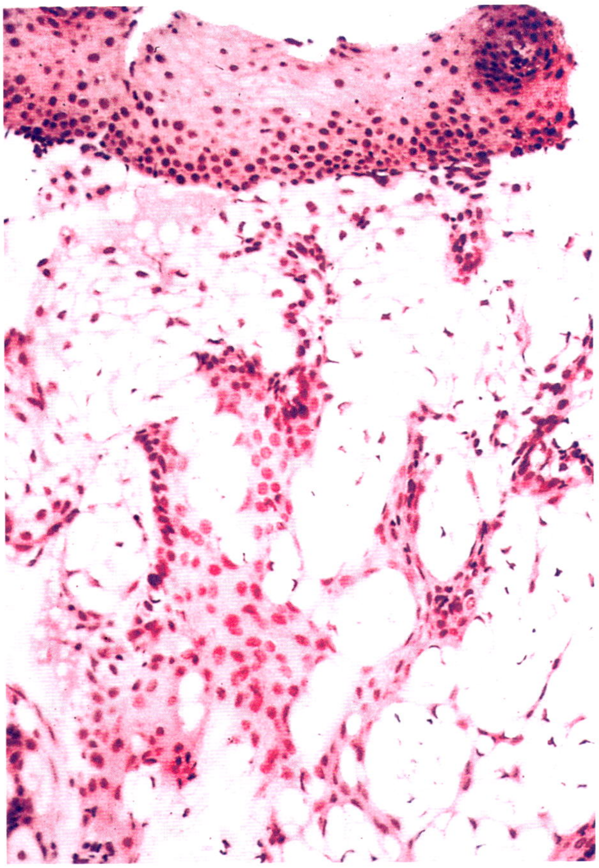

FIGURE 21.27 Sebaceous glands that, in this case, formed a nodule that was examined via biopsy.

esophageal segments of variable length (1,154,155). Any patient presenting with segmental or cystic should undergo radiologic evaluation to assess for the presence of axial skeletal defects because these anomalies can be associated with defects affecting other organ systems. EDs can be cystic, tubular, or diverticular. All three types feature (a) a well-developed coat of smooth muscle, (b) an epithelial lining representing some portion of the alimentary tract, and (c) an attachment to the esophagus (156). The epithelium is generally squamous in nature, but may also contain respiratory, gastric, or pancreatic tissue elements (157,158). ED cysts are usually found close to the wall (159) and may communicate with the esophageal lumen (160).

Esophageal Rings and Webs

Congenital esophageal rings and esophageal webs are thought to result from incomplete vacuolization of the esophageal columnar epithelium during early embryonic life. Most are found in the distal esophagus (161).

Schatzki ring is found in 6% to 14% of subjects undergoing an upper gastrointestinal series and is located at the squamocolumnar junction. Typically, it is associated with the proximal margin of a hiatal hernia. It consists of two layers, mucosa and submucosa, having squamous epithelium on its upper surface and columnar epithelium on its lower surface (Fig. 21.28) (162).

The congenital esophageal web is a thin, eccentric, transverse membrane covered by normal squamous epithelium (161). They are usually asymptomatic but may be associated

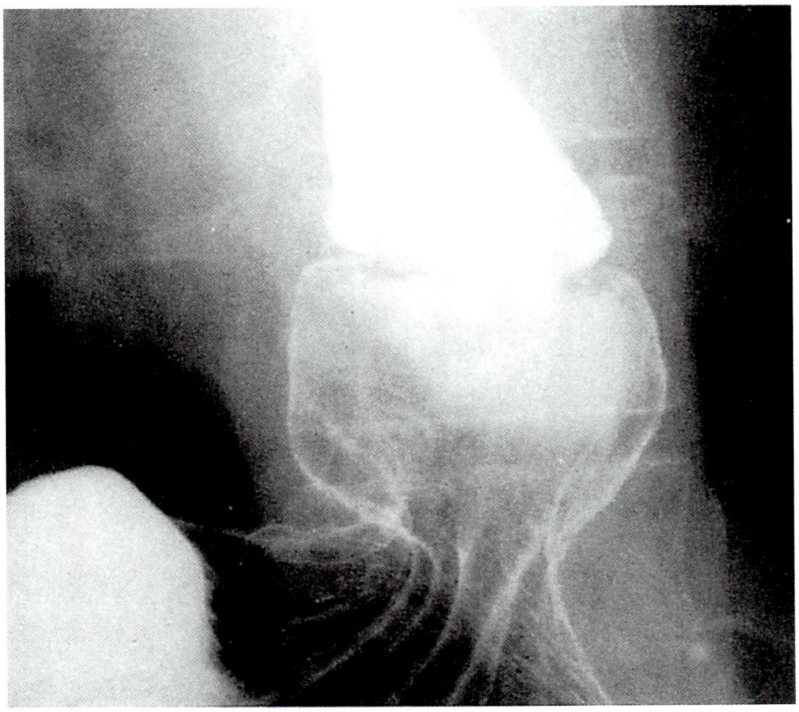

FIGURE 21.28 Lower esophageal mucosal ring (Schatzki ring). The mucosal ring is outlined by the column of barium.

with intermittent dysphagia, sometimes becoming progressive or associated with attacks of sudden dysphagia (27).

Esophageal Diverticula

Zenker diverticulum represents an acquired structural abnormality at the level of the cricopharyngeus muscle (163). As previously described, the outer longitudinal muscle layer is incomplete in this area and is prone to posterior pulsion pseudodiverticula if the cricopharyngeal muscle fails to relax quickly enough upon swallowing. Zenker diverticula contain squamous mucosa but lack a muscularis propria and muscularis mucosae (Fig. 21.29).

Epiphrenic diverticula are pulsion-type pseudodiverticula that lack a muscularis propria and show a predilection for the distal esophagus above the lower esophageal sphincter (164). They develop secondary to esophageal motility disorders (165).

Rokitansky diverticula occur at the level of the thoracic esophagus, particularly near mediastinal lymph nodes. They are formed by cicatricial contraction due to extraesophageal fibrosis and inflammation that tethers the esophageal wall outward. Thus, they involve the entire wall of the esophagus and represent true diverticulum. They most commonly occur in patients with pulmonary tuberculosis (166).

DIAGNOSTIC CONSIDERATIONS

The esophagus is generally sampled to evaluate patients who present with dysphagia or odynophagia for the possibility of esophagitis. A comprehensive discussion of the various forms of esophagitis is beyond the scope of this text. However, a few entities that may simulate normal findings are discussed below.

Glycogenic Acanthosis

Glycogenic acanthosis appears as white nodules or small plaques, primarily in the distal one-third of the esophagus. These lesions usually span less than 1 cm and, if extensive, may coalesce into larger plaques. Glycogenic acanthosis reflects hyperplasia of the squamous cells in the prickle layer that contain abundant glycogen (Fig. 21.30). It can be mistaken for a normal finding because the features can

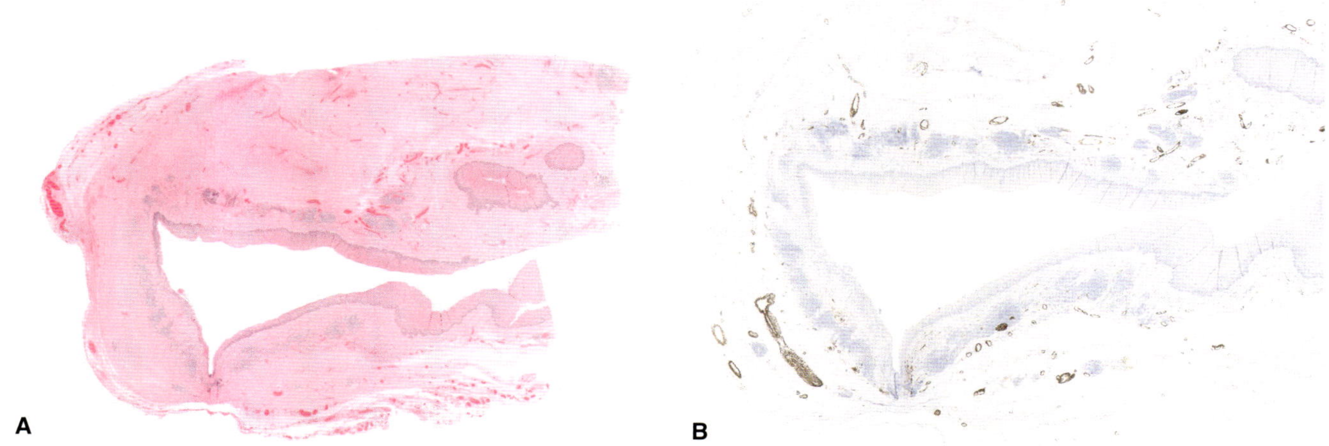

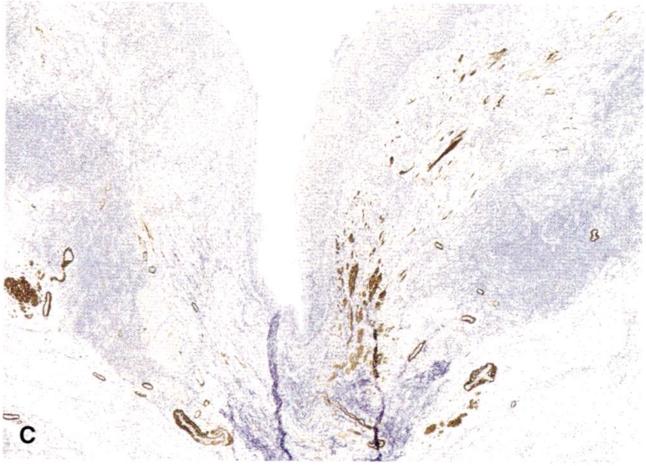

FIGURE 21.29 Zenker diverticulum. **A:** H&E section shows an outpouching of the esophageal wall. **B:** Caldesmon immunostain highlights the smooth muscle bundles within the walls of blood vessels and shows markedly attenuated and nearly absent muscularis mucosae and absent muscularis propria within the wall of a Zenker diverticulum. **C:** Detail of Fig. 21.29B.

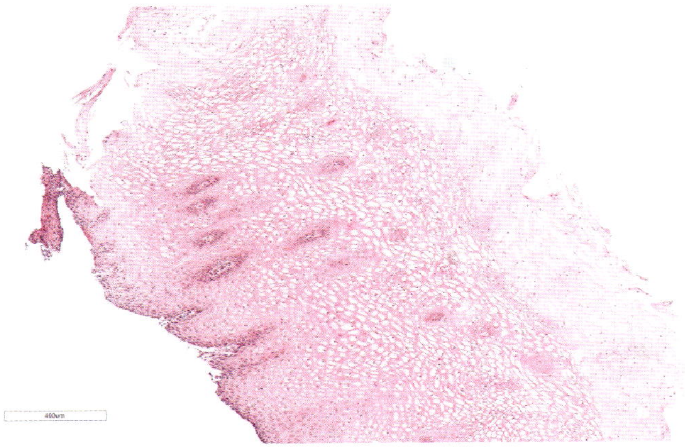

FIGURE 21.30 Glycogenic acanthosis. The superficial layers of the squamous epithelium show clear cytoplasm due to accumulation of glycogen.

be subtle. On the other hand, glycogenic acanthosis may simulate the endoscopic features of candidiasis or leukoplakia. Glycogenic acanthosis should be considered a variant of normal with as yet no defined relationship to infection or malignancy. However, multifocal lesions may be associated with Cowden syndrome (167).

Diffuse Esophageal Hyperkeratosis (DEH)

DEH is a rare condition associated with gastroesophageal reflux disease (168,169). It features thick, white esophageal plaques that reflect a prominent layer of orthokeratotic anucleate squamous cells unaccompanied by hyperplasia of the squamous epithelium. These histologic characteristics distinguish DEH from premalignant conditions that feature parakeratosis (168).

Epidermoid Metaplasia (aka Esophageal Leukoplakia)

Epidermoid metaplasia of the esophagus is an uncommon condition that may be associated with squamous dysplasia or carcinoma in some cases (170,171). It is most commonly described in middle-aged adults with a history of alcohol or tobacco abuse who present with dysphagia or gastroesophageal reflux disease (170). Histologic features of epidermoid esophageal metaplasia included hyperorthokeratosis and a prominent granular cell layer (Fig. 21.31).

Sloughing Esophagitis (Esophagitis Dissecans Superficialis)

Sloughing esophagitis is frequently unrecognized or misinterpreted to represent a normal finding (172). In this condition, squamous mucosa separates, resulting in casts of desquamated material that slough into the lumen. The epithelium is characteristically surfaced by a two-toned layer of parakeratosis; the superficial layers are necrotic, whereas the deeper layers contain viable nuclei. The separation may occur between the squamous epithelium and parakeratotic layer, or between the viable and non-viable parakeratosis (Fig. 21.32). Patients tend to be elderly with debilitating illnesses who take multiple medications, particularly those that cause dry mouth (173).

FIGURE 21.31 Epidermoid metaplasia. The squamous epithelium shows the presence of granular cell layer (as normally seen in skin but not in mucosa) and dense layer of hyperkeratosis.

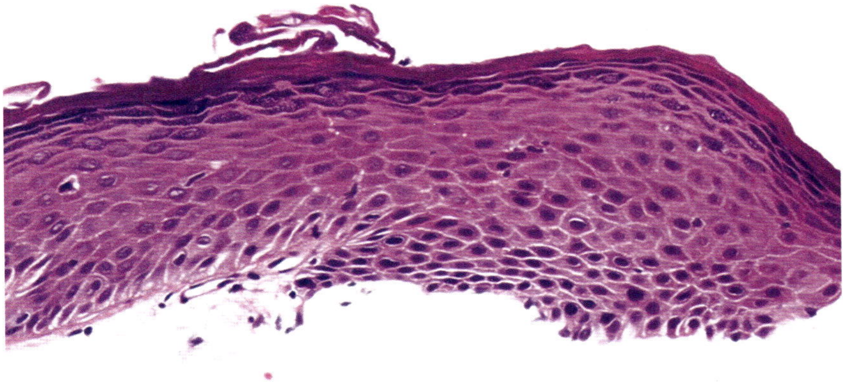

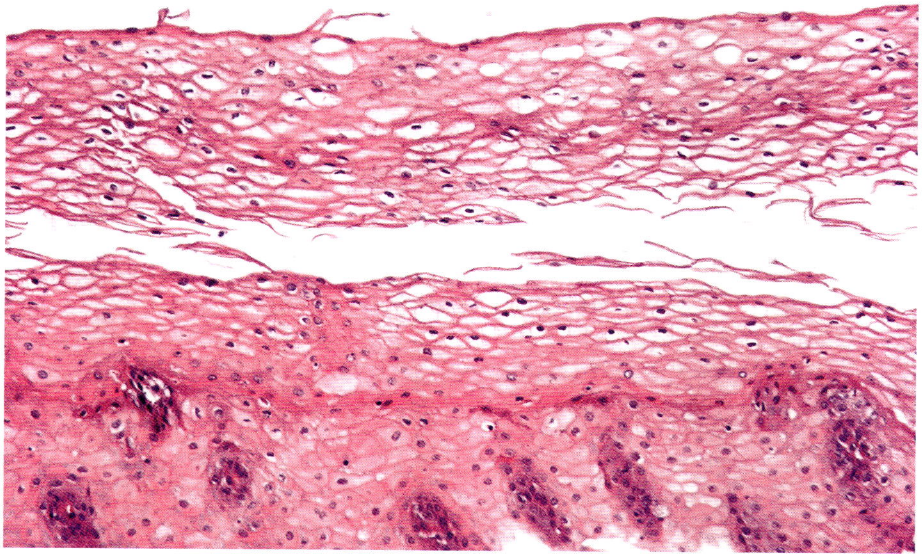

FIGURE 21.32 Sloughing esophagitis. Notice the split in the epithelium, which may or may not be accompanied by vacuolization and edema. Sometimes, just the superficial part is included in biopsies, so its appearance is well worth recognizing. (Courtesy of Dr M Vieth, Bayreuth.)

Inflammatory-Type (Hyperplastic) Polyps

Inflammatory polyps most commonly occur in males as an exaggerated response to injury in a setting of gastroesophageal reflux disease, pill-induced esophagitis, infections, and surgical anastomoses (174). Histologic features depend on the anatomical site. Those above the gastroesophageal junction show polypoid and inflamed lamina propria covered by hyperplastic and irregular tongues of squamous epithelium (174). Inflammatory polyps at the gastroesophageal junction may feature hyperplastic foveolar epithelium, squamous epithelium, or both overlying the inflamed and edematous lamina propria (175). Erosions, ulcerations, and acutely inflamed granulation tissue are common.

Squamous Papilloma

Squamous papilloma appears as a whitish-pink, wart-like exophytic growth with a prevalence of 0.01% to 0.45% (176,177). Most lesions are solitary and asymptomatic (176). Histologically, they are characterized by proliferation of non-dysplastic squamous epithelium surrounding fibrovascular cores of the lamina propria (Fig. 21.33). Most squamous papillomas occur in the lower esophagus and

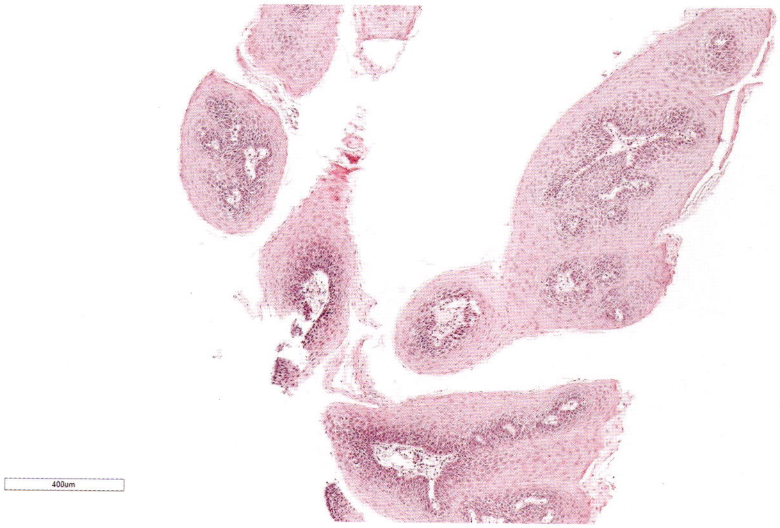

FIGURE 21.33 Squamous cell papilloma. H&E image shows benign squamous proliferation centered around fibrovascular papillary cores.

are associated with gastroesophageal reflux disease. Thus, they are more appropriately classified as inflammatory-type polyps. Those that occur in association with human papillomavirus are extremely rare and usually occur in association with squamous papillomatosis of the respiratory tree.

GASTROESOPHAGEAL REFLUX DISEASE

Problems encountered in gastroesophageal reflux disease are related to squamous epithelial changes that are maximal distally and result in reactive changes (basal cell hyperplasia and increased length of papillae), which can proceed to erosions, ulcers, strictures, and Barrett esophagus. Gastroesophageal reflux disease often elicits mucosal inflammation characterized by increased intraepithelial lymphocytes, eosinophils, and variable numbers of neutrophils.

Histologic Features of Gastroesophageal Reflux Disease

Reactive Squamous Hyperplasia

Reflux-associated squamous hyperplasia displays basal cell hyperplasia that usually occupies greater than 15% of the total epithelial thickness and extension of the papillae into the upper one-third of the epithelium (Fig. 21.34). The maximal extent of papillary length normally varies from 50% to 75% (178).

Rule of thirds. Practically, it is easy to use the "rule of thirds" when examining biopsies, in which papillae should not extend into the upper one-third, and basal cells should be present more than halfway up the basal one-third of the epithelial thickness. Reflux-type changes have been described in the distal 3 cm of the esophagus in approximately 60% of patients without objective or subjective evidence of acid reflux (179).

Mucosal Edema

Mucosal edema, or dilated intercellular spaces, can be demonstrated at both the electron and the light microscopic level. Keratinocytes are separated from each other accumulation of extracellular fluid within the epithelium such that the intercellular bridges between them are readily apparent (Fig. 21.12).

Pseudoepitheliomatous Hyperplasia

Pseudoepitheliomatous hyperplasia occurs during the healing phase of an ulcer. It features mucosal hyperplasia with elongated and expanded papillae. The peripapillary epithelium features basal cell hyperplasia with increased mitotic activity and mild cytologic atypia that may raise concern for a neoplasm. In contrast to the deep aspect of dysplasias and well-differentiated (i.e., verrucous) carcinomas, however, the peripapillary epithelium is arranged as regularly spaced, long and thin projections that taper to a point at their deepest extent in the lamina propria.

Inflammatory Changes on the Distal Side of the Z-Line

The glandular mucosa of the squamocolumnar junction may be chronically inflamed and show variable neutrophilic inflammation. This finding has been termed "chronic carditis" in the literature and generally occurs in the absence of Helicobacter infection. Inflamed cardiac-type mucosa at the gastroesophageal junction may be a more sensitive marker for gastroesophageal reflux disease than inflammatory changes in the squamous mucosa (180).

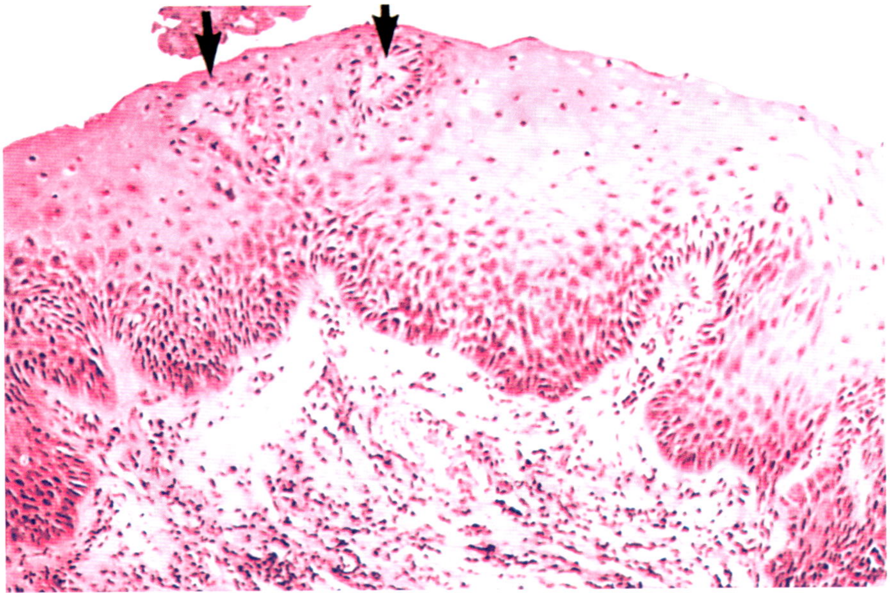

FIGURE 21.34 Reflux esophagitis. Basal cell hyperplasia and lengthening of the papillae are present. The papillae have extended almost to the surface of the mucosa (*arrows*).

LYMPHOCYTIC ESOPHAGITIS

Lymphocytic esophagitis is an inflammatory pattern characterized by an overt increase in intraepithelial lymphocytes. It was originally defined by the presence of >20 intraepithelial lymphocytes/HPF with rare or absent intraepithelial granulocytes, but we now know that this number overlaps with expected number of lymphocytes seen in asymptomatic patients with normal endoscopic examinations (Fig. 21.35A,B) (181). We also know that esophageal lymphocytosis is a common manifestation of gastroesophageal reflux disease, eosinophilic esophagitis, graft-versus-host disease, pediatric Crohn disease, candidiasis, and achalasia or other motility disorders. In these situations, the diagnosis should reflect the underlying disease process, rather than "lymphocytic esophagitis," as the latter does not help the clinician manage the patient. On the other hand, esophageal lymphocytosis accompanied by edema, dyskeratosis, or other features of mucosal injury is more frequently associated with medications (e.g., checkpoint inhibitors, gold, thiazides, and antimalarials), celiac disease, lichen planus, and other immune-mediated disorders (182). It is more common among women who present

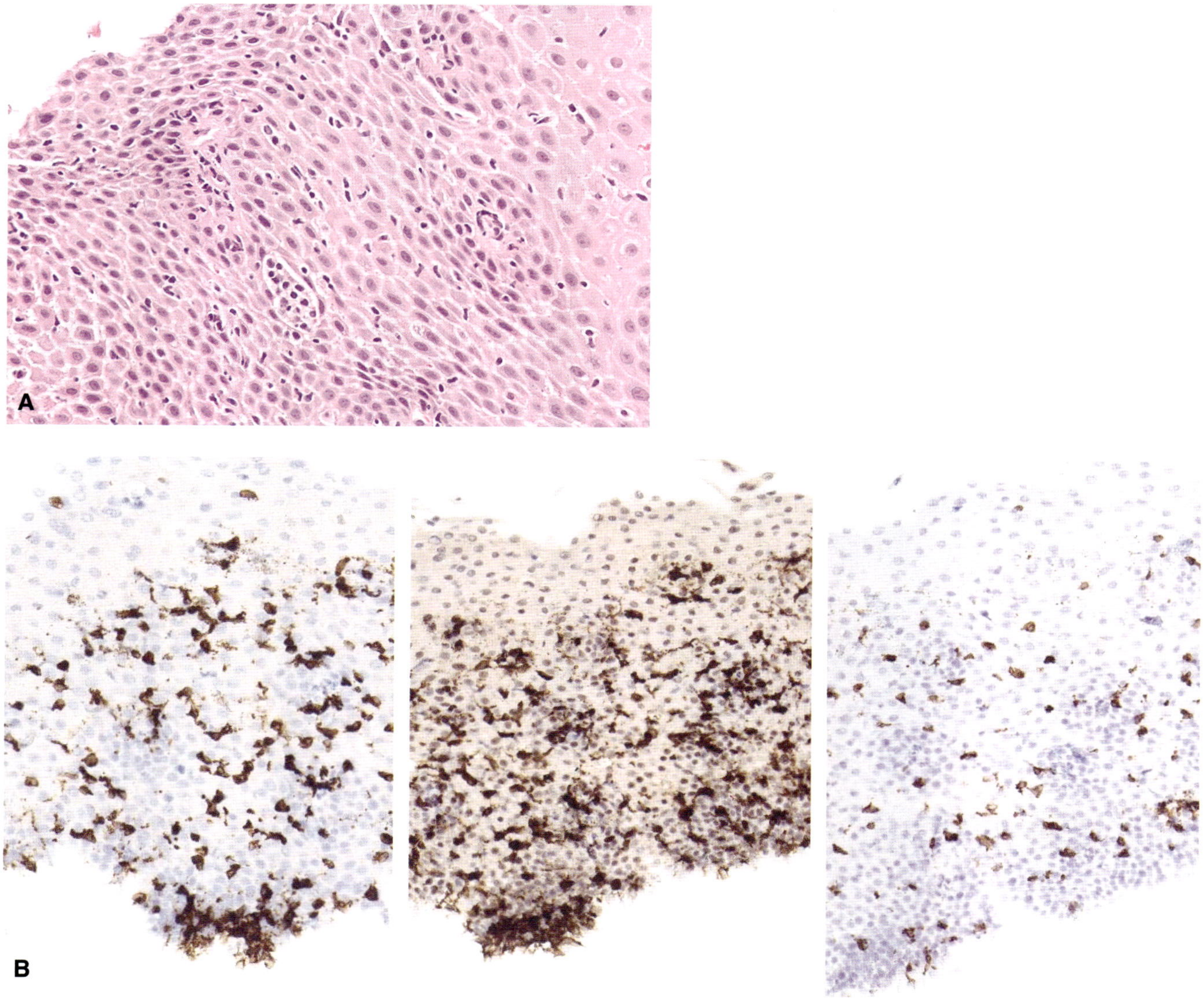

FIGURE 21.35 Lymphocytic esophagitis. **A**: H&E shows numerous intraepithelial lymphocytes centered around the papillae and associated with reactive changes (e.g., spongiosis). **B**: Lymphocytic esophagitis. CD3 (**left**), CD4 (**center**), and CD8 (**right**) immunostains. Normally CD3+/CD8+ cells predominate, but in this example, CD4 cells predominate, and there has to be a CD3-CD4+ subpopulation to account for the number of CD4+ cells. This pattern tends to be seen in motility disorders and Candida infection.

with dysphagia, and frequently affects the esophagus in a diffuse fashion with rings and furrows that simulate eosinophilic esophagitis.

BARRETT ESOPHAGUS AND CARCINOMAS OF THE ESOPHAGUS

The diagnosis of Barrett esophagus requires endoscopic demonstration of a shift in the Z-line ≥1 cm proximal to the gastroesophageal junction (Fig. 21.36) in combination with detection of intestinal metaplasia on biopsy (Fig. 21.37) (183,184). Although some patients have glandular metaplasia of the distal esophagus without goblet cells, they do not have an appreciable risk of developing esophageal adenocarcinoma and thus, are not considered for surveillance endoscopy.

Histologic Appearances of Barrett Esophagus

The goblet cells of Barrett esophagus may occur on a background of foveolar epithelial cells (i.e., incomplete intestinal metaplasia) or, less frequently, in association with fully intestinalized epithelium containing variable numbers of absorptive cells, Paneth cells, and neuroendocrine cells (i.e., complete intestinal metaplasia). They can grow under the normal squamous epithelium in both treatment-naïve patients and those treated with radiofrequency ablation or photodynamic therapy. Duplication of the muscularis mucosae occurs as a result of recurrent ulcers and is accompanied by an increase in the number of blood vessels and lymphatic channels as well as resident inflammatory cells.

Goblet cells should be distinguished from distended foveolar cells (i.e., pseudogoblet cells) that develop in response to injury. Pseudogoblet cells contain clear to pink, homogeneous cytoplasmic mucin and tend to be clustered, whereas true goblet cells are usually singly dispersed among foveolar cells.

Adenocarcinoma of the Esophagus-Depth of Invasion

Endoscopic Mucosal Resections and Submucosal Dissections

Endoscopic resection therapy has revolutionized the treatment of early epithelial neoplasia as these procedures can be curative with low morbidity. Precise histologic assessment of depth of invasion in these excisional specimens is important because it predicts the likelihood of lymphovascular invasion and regional lymph node metastasis (185).

The duplicated muscularis mucosae of Barrett esophagus complicates assessment for the depth of invasion. Four levels of invasion within the duplicated muscularis mucosae are described: lamina propria only (m1); newly formed superficial muscularis mucosae (m2); lamina propria between the superficial and deep muscularis mucosae

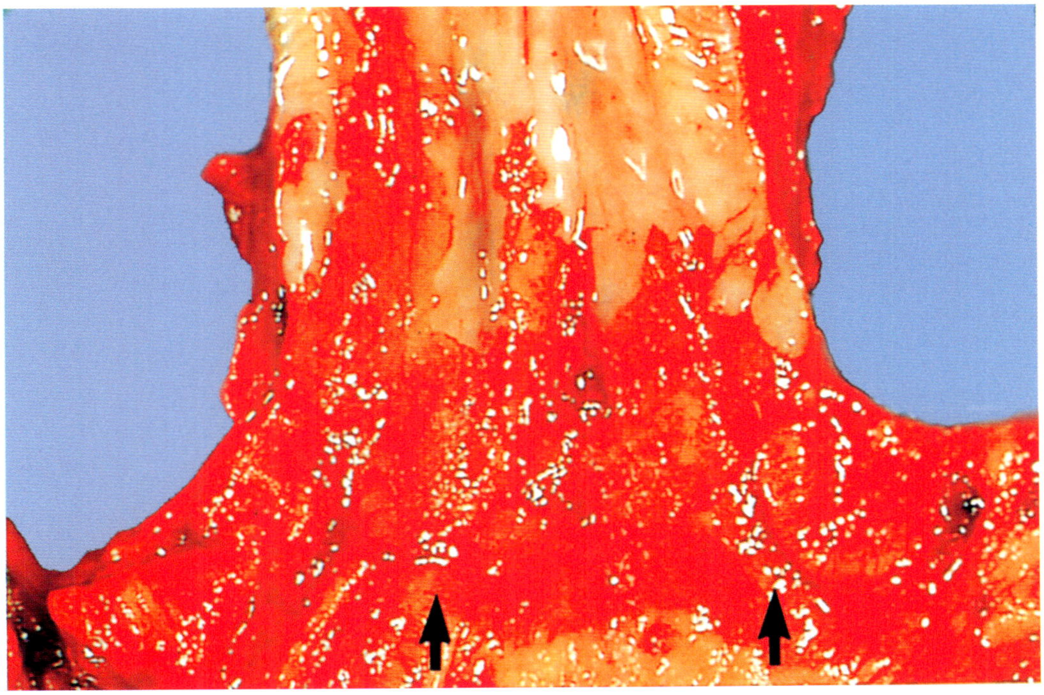

FIGURE 21.36 Gastroesophageal region. Barrett esophagus, demonstrating proximal extension of columnar-lined mucosa well into the tubular esophagus. This columnar-lined mucosa extends more than 2 cm from the proximal gastric folds (*arrows*).

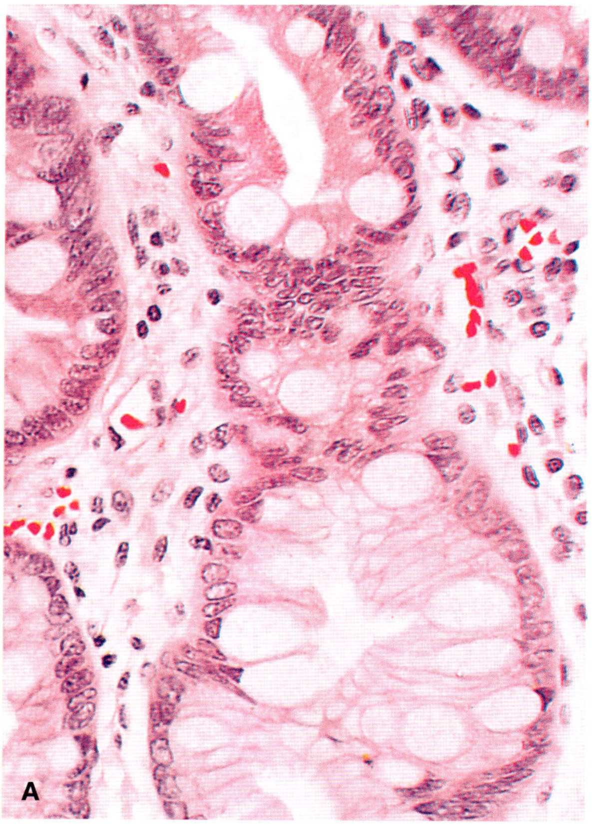

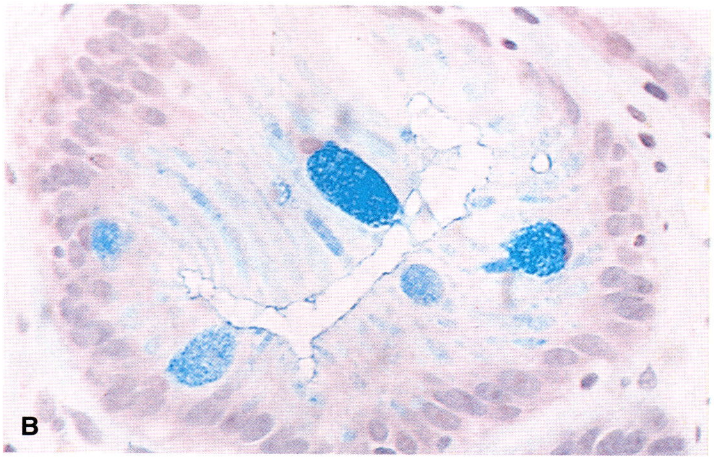

FIGURE 21.37 Barrett esophagus. **A**: Intestinal metaplasia is recognized by the presence of goblet cells. Incomplete intestinal metaplasia (lower half of gland) is characterized by goblet cells associated with gastric foveolar-like columnar cells, whereas complete intestinal metaplasia (upper half of the gland) is characterized by goblet cells associated with small intestinal, absorptive-like columnar cells. **B**: Goblet cells in intestinal metaplasia stain positive with Alcian blue at pH 2.5. It is not recommended to do this either routinely or for diagnostic purposes.

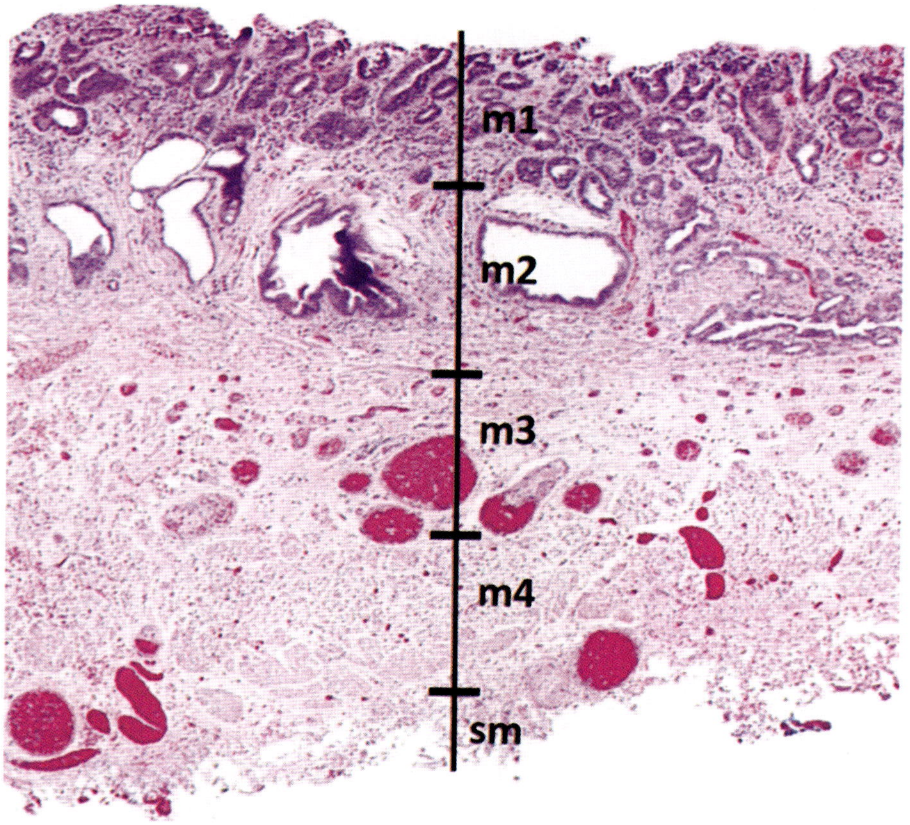

FIGURE 21.38 pT1a (M2) intramucosal adenocarcinoma with subdivision of depth of invasion according to the Vieth and Stolte system. M1, into the lamina propria. M2, into the first/superficial/duplicated layer of the muscularis mucosae, into which the carcinoma is infiltrating. M3, into the space between the duplicated muscularis mucosa and the original (deep) muscularis mucosae at the bottom of the illustration (M4). Submucosa (SM), when completely present (i.e., muscularis propria present), can be divided into thirds Sm1, Sm2, and Sm3 (not shown). Seen here is an ESD sample from the distal esophagus of a well-differentiated intestinal type intramucosal adenocarcinoma invading into the superficial layer of the muscularis mucosa (M2). Strands of smooth muscle can be seen throughout the M2 denoted area.

(m3); and invasion into the outer original deeper muscularis mucosae (m4) (185) (Fig. 21.38).

The risk of lymph node metastases is related to the depth of invasion with reported incidence rates of 0.8% to 4.3% for m1 to m4 and 13.3% to 46% for submucosal invasion (185,186). The depth of submucosal invasion should be measured from the deepest layer of the muscularis mucosae to the deepest point of invasion. Depth of the submucosal invasion is assessed by dividing the submucosa into sm1 (≤500 μm), sm2 (500 to 1,000 μm), and sm3 (>1,000 μm) (187,188). Patients without invasion of the submucosa are invariably followed with surveillance, while those with deep submucosal invasion (i.e., ≥500 mm), lymphovascular invasion, or high-grade features are best treated with esophagectomy unless the patient is a poor surgical candidate (189).

Squamous Cell Carcinoma-Depth of Invasion

The depth of invasion for early squamous cell carcinomas is subdivided based on whether the tumor is limited to the mucosa or submucosa. Mucosal invasion is classified with a three-tiered system: m1 is high-grade dysplasia/carcinoma in situ, m2 is defined as invasion into the lamina propria, and m3 is defined as invasion into the muscularis mucosae. Submucosal invasion is classified as sm1, sm2, and sm3 depending on whether the tumor is limited to the upper one-third of the submucosa, the middle third, or the lower one-third of the submucosa (Fig. 21.39) (190). Tumors confined to the upper two-thirds of the submucosa that are completely resected are considered cured, provided that poorly differentiated components and lymphovascular invasion are lacking (190). Similar to adenocarcinomas, the rate of lymph node metastases increases with the depth of

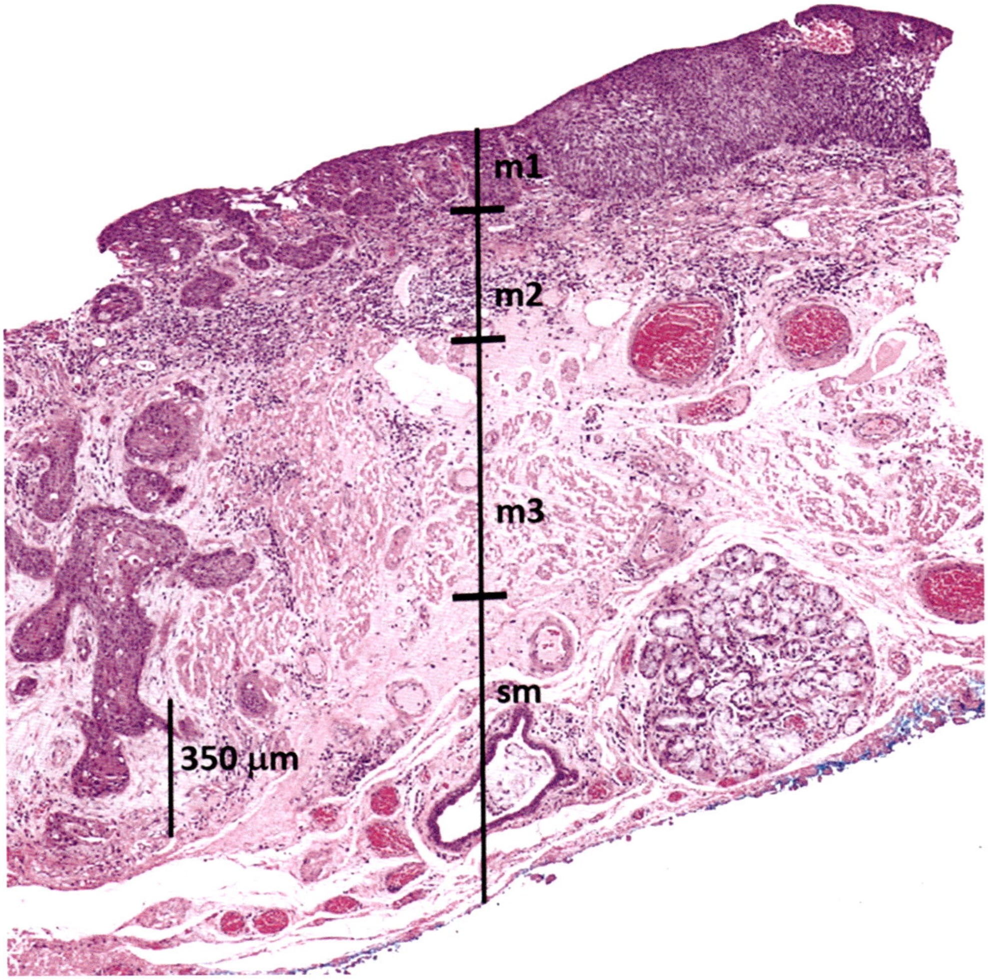

FIGURE 21.39 pT1b invasive squamous cell carcinoma with AJCC 8th edition subdivision of the mucosa. M1, limited to the epithelium/dysplasia only. M2, into the lamina propria. M3, into the muscularis mucosae. The submucosa (SM) can be divided into thirds Sm1, Sm2, and Sm3 (in an esophagectomy specimen where the entire submucosa is present) (not shown). Seen here is an ESD sample from the esophagus of a moderately differentiated invasive squamous cell carcinoma that invades into the submucosa. The depth of submucosal invasion is about 350 μm in this sample (shown with the smaller vertical bar), the measurement taken from the bottom of the muscularis mucosae. The deep resection margin is inked with blue ink.

invasion, ranging up to 18% for m3, 50% for sm1, and 54% to 69% for sm2 to sm3 (185,190,191).

REFERENCES

1. Strobel CT, Byrne WJ, Ament ME, Euler AR. Correlation of esophageal lengths in children with height: application to the Tuttle test without prior esophageal manometry. *J Pediatr.* 1979;94(1):81–84.
2. Abid S, Mumtaz K, Jafri W, et al. Pill-induced esophageal injury: endoscopic features and clinical outcomes. *Endoscopy.* 2005;37(8):740–744.
3. Gulsen MT, Buyukberber NM, Karaca M, Kadayifci A. Cyproterone acetate and ethinylestradiol-induced pill oesophagitis: a case report. *Int J Clin Pract Suppl.* 2005;(147): 79–81.
4. McCullough RW, Afzal ZA, Saifuddin TN, Alba LM, Khan AH. Pill-induced esophagitis complicated by multiple esophageal septa. *Gastrointest Endosc.* 2004;59(1):150–152.
5. Tsai SJ, Lin CC, Chang CW, et al. Benign esophageal lesions: endoscopic and pathologic features. *World J Gastroenterol.* 2015;21(4):1091–1098.
6. Erhiawarie F, Chioma OE, Rahim O, et al. Right subclavian artery with Kommerell's diverticulum: a rare cause of dysphagia. *Cureus.* 2023;15(3):e35812.
7. Esophagus. In: Greene FL, Page DL, Fleming ID, et al., eds. *AJCC Cancer Staging Manual.* Springer New York; 2002: 91–98.
8. Marshall JB, Singh R, Demmy TL, Bickel JT, Everett ED. Mediastinal histoplasmosis presenting with esophageal involvement and dysphagia: case study. *Dysphagia.* 1995;10(1):53–58.
9. Feldman M, Friedman LS, Brandt LJ. Pathophysiology, diagnosis and management. In: Feldman M, Friedman LS, Sleisinger MH, eds. *Sleisenger and Fordtran's Gastrointestinal and Liver Disease.* 7th ed. WB Saunders; 2002:549–671.
10. Geboes K, Geboes KP, Maleux G. Vascular anatomy of the gastrointestinal tract. *Best Pract Res Clin Gastroenterol.* 2001;15(1):1–14.
11. Bazira PJ. Anatomy of the oesophagus. *Surgery (Oxford).* 2023; 41(11):691–697.
12. Kitano S, Terblanche J, Kahn D, Bornman PC. Venous anatomy of the lower oesophagus in portal hypertension: practical implications. *Br J Surg.* 1986;73(7):525–531.
13. Vianna A, Hayes PC, Moscoso G, et al. Normal venous circulation of the gastroesophageal junction. A route to understanding varices. *Gastroenterology.* 1987;93(4):876–889.
14. Choi DW, Oh SN, Baek SJ, et al. Endoscopically observed lower esophageal capillary patterns. *Korean J Intern Med.* 2002;17(4):245–248.
15. Hoshihara Y, Kogure T, Yamamoto T, Hashimoto M, Hoteya O. [Endoscopic diagnosis of Barrett's esophagus]. *Nihon Rinsho.* 2005;63(8):1394–1398.
16. Tachimori Y. Pattern of lymph node metastases of squamous cell esophageal cancer based on the anatomical lymphatic drainage system: efficacy of lymph node dissection according to tumor location. *J Thorac Dis.* 2017;9(Suppl 8): S724–S730.
17. Shibata T, Takita K, Inomata M. Observation of the cytoarchitecture of the human esophageal mucosa with special attention to the lamina muscularis mucosae and the distribution of lymphatic vessels. *Esophagus.* 2019;16(1):44–51.
18. Kuge K, Murakami G, Mizobuchi S, Hata Y, Aikou T, Sasaguri S. Submucosal territory of the direct lymphatic drainage system to the thoracic duct in the human esophagus. *J Thorac Cardiovasc Surg.* 2003;125(6):1343–1349.
19. Saito H, Sato T, Miyazaki M. Extramural lymphatic drainage from the thoracic esophagus based on minute cadaveric dissections: fundamentals for the sentinel node navigation surgery for the thoracic esophageal cancers. *Surg Radiol Anat.* 2007;29(7):531–542.
20. Rice TW, Bronner MP. The esophageal wall. *Thorac Surg Clin.* 2011;21(2):299–305, x.
21. Hirano-Kawamoto A, Honkura Y, Kobayashi Y, Murakami G, Abe S-I, Katori Y. Enteric neurons of the esophagus: an immunohistochemical study using donated elderly cadavers. *Surg Radiol Anat.* 2017;39(5):477–484.
22. Faussone-Pellegrini MS, Cortesini C. Ultrastructure of striated muscle fibers in the middle third of the human esophagus. *Histol Histopathol.* 1986;1(2):119–128.
23. Gockel I, Bohl JR, Eckardt VF, Junginger T. Reduction of interstitial cells of Cajal (ICC) associated with neuronal nitric oxide synthase (n-NOS) in patients with achalasia. *Am J Gastroenterol.* 2008;103(4):856–864.
24. Gockel I, Bohl JR, Doostkam S, Eckardt VF, Junginger T. Spectrum of histopathologic findings in patients with achalasia reflects different etiologies. *J Gastroenterol Hepatol.* 2006;21(4):727–733.
25. Miettinen M, Sarlomo-Rikala M, Sobin LH, Lasota J. Esophageal stromal tumors: a clinicopathologic, immunohistochemical, and molecular genetic study of 17 cases and comparison with esophageal leiomyomas and leiomyosarcomas. *Am J Surg Pathol.* 2000;24(2):211–222.
26. Aggestrup S, Uddman R, Jensen SL, et al. Regulatory peptides in the lower esophageal sphincter of man. *Regul Pept.* 1985;10(2–3):167–178.
27. Aggestrup S, Uddman R, Sundler F, et al. Lack of vasoactive intestinal polypeptide nerves in esophageal achalasia. *Gastroenterology.* 1983;84(5 Pt 1):924–927.
28. Wattchow DA, Furness JB, Costa M. Distribution and coexistence of peptides in nerve fibers of the external muscle of the human gastrointestinal tract. *Gastroenterology.* 1988; 95(1):32–41.
29. Mantyh CR, Pappas TN, Vigna SR. Localization of cholecystokinin A and cholecystokinin B/gastrin receptors in the canine upper gastrointestinal tract. *Gastroenterology.* 1994; 107(4):1019–1030.
30. Weinstein WM, Bogoch ER, Bowes KL. The normal human esophageal mucosa: a histological reappraisal. *Gastroenterology.* 1975;68(1):40–44.
31. Groben PA, Siegal GP, Shub MD, Ulshen MH, Askin FB. Gastroesophageal reflux and esophagitis in infants and children. *Perspect Pediatr Pathol.* 1987;11:124–151.
32. Freeman SC, Sonthalia S. *Histology, Keratohyalin Granules. StatPearls.* StatPearls Publishing LLC; 2024.
33. Ohashi K, Kato Y, Kanno J, Kasuga T. Melanocytes and melanosis of the oesophagus in Japanese subjects–analysis of factors effecting their increase. *Virchows Arch A Pathol Anat Histopathol.* 1990;417(2):137–143.
34. Lam KY, Law S, Chan GS. Esophageal blue nevus: an isolated endoscopic finding. *Head Neck.* 2001;23(6):506–509.

35. De La Pava S, Nigogosyan G, Pickren JW, Cabrera A. Melanosis of the esophagus. *Cancer.* 1963;16:48–50.
36. Squier CA, Kremer MJ. Biology of oral mucosa and esophagus. *J Natl Cancer Inst Monogr.* 2001(29):7–15.
37. Chang F, Deere H. Esophageal melanocytosis morphologic features and review of the literature. *Arch Pathol Lab Med.* 2006;130(4):552–557.
38. Destek S, Gul VO, Ahioglu S, Erbil Y. A rare disease of the digestive tract: esophageal melanosis. *Gastroenterology Res.* 2016;9(2-3):56–60.
39. Harmse JL, Carey FA, Baird AR, et al. Merkel cells in the human oesophagus. *J Pathol.* 1999;189(2):176–179.
40. Tateishi R, Taniguchi K, Horai T, Iwanaga T, Taniguchi H. Argyrophil cell carcinoma (apudoma) of the esophagus. A histopathologic entity. *Virchows Arch A Pathol Anat Histol.* 1976;371(4):283–294.
41. Kaduk B, Barth H. [The localisation of endocrine cells in the distal esophagus (author's transl)]. *Virchows Arch A Pathol Anat Histol.* 1978;377(4):311–328.
42. Pan Q, Nicholson AM, Barr H, et al. Identification of lineage-uncommitted, long-lived, label-retaining cells in healthy human esophagus and stomach, and in metaplastic esophagus. *Gastroenterology.* 2013;144(4):761–770.
43. Saint Martin MC, Chejfec G. Barrett esophagus-associated small cell carcinoma. *Arch Pathol Lab Med.* 1999;123(11):1123.
44. Tripathi M, Streutker CJ, Marginean EC. Relevance of histology in the diagnosis of reflux esophagitis. *Ann N Y Acad Sci.* 2018;1434(1):94–101.
45. Zand Irani M, Talley NJ, Ronkainen J, et al. Neutrophils, eosinophils, and intraepithelial lymphocytes in the squamous esophagus in subjects with and without gastroesophageal reflux symptoms. *Hum Pathol.* 2021;115:112–122.
46. Kirsch R, Bokhary R, Marcon MA, Cutz E. Activated mucosal mast cells differentiate eosinophilic (allergic) esophagitis from gastroesophageal reflux disease. *J Pediatr Gastroenterol Nutr.* 2007;44(1):20–26.
47. Mulder DJ, Pooni A, Mak N, Hurlbut DJ, Basta S, Justinich CJ. Antigen presentation and MHC class II expression by human esophageal epithelial cells: role in eosinophilic esophagitis. *Am J Pathol.* 2011;178(2):744–753.
48. Carmack SW, Lash RH, Gulizia JM, Genta RM. Lymphocytic disorders of the gastrointestinal tract: a review for the practicing pathologist. *Adv Anat Pathol.* 2009;16(5):290–306.
49. Vieth M, Haringsma J, Delarive J, et al. Red streaks in the oesophagus in patients with reflux disease: is there a histomorphological correlate? *Scand J Gastroenterol.* 2001;36(11):1123–1127.
50. Nagai K, Noguchi T, Fujiwara S, Kawahara K, Shimada T. Distribution of Langerhans cells in the human esophagus, as revealed by immunohistochemistry. *Acta Histochem Cytoc.* 2005;38(2):115–119.
51. Zavala WD, De Simone DS, Sacerdote FL, Cavicchia JC. Variation in Langerhans cell number and morphology between the upper and lower regions of the human esophageal epithelium. *Anat Rec.* 2002;268(4):360–364.
52. Geboes K, De Wolf-Peeters C, Rutgeerts P, Janssens J, Vantrappen G, Desmet V. Lymphocytes and Langerhans cells in the human oesophageal epithelium. *Virchows Arch A Pathol Anat Histopathol.* 1983;401(1):45–55.
53. Amon L, Hatscher L, Heger L, Dudziak D, Lehmann CHK. Harnessing the complete repertoire of conventional dendritic cell functions for cancer immunotherapy. *Pharmaceutics.* 2020;12(7):663.
54. Somja J, Demoulin S, Roncarati P, et al. Dendritic cells in Barrett's esophagus carcinogenesis: an inadequate microenvironment for antitumor immunity? *Am J Pathol.* 2013;182(6):2168–2179.
55. Cucchiara S, D'Armiento F, Alfieri E, et al. Intraepithelial cells with irregular nuclear contours as a marker of esophagitis in children with gastroesophageal reflux disease. *Dig Dis Sci.* 1995;40(11):2305–2311.
56. Vieth M, Peitz U, Labenz J, et al. What parameters are relevant for the histological diagnosis of gastroesophageal reflux disease without Barrett's mucosa? *Dig Dis.* 2004;22(2):196–201.
57. Mastracci L, Bruzzone M, Pacella E, et al. The contribution of intraepithelial inflammatory cells to the histological diagnosis of microscopic esophagitis. *Esophagus.* 2015;13(1):80–87.
58. Mastracci L, Grillo F, Parente P, et al. Gastro-esophageal reflux disease and Barrett's esophagus: an overview with an histologic diagnostic approach. *Pathologica.* 2020;112(3):117–127.
59. Noffsinger AE. Update on esophagitis: controversial and underdiagnosed causes. *Arch Pathol Lab Med.* 2009;133(7):1087–1095.
60. Davis-Yadley AH, Keshishian J, Champeaux A, Richter JE. Multinucleated histiocytes in esophageal squamous mucosa secondary to gastroesophageal reflux disease in a patient with esophageal stricture. *Dis Esophagus.* 2016;29(8):1159–1161.
61. Singh SP, Odze RD. Multinucleated epithelial giant cell changes in esophagitis: a clinicopathologic study of 14 cases. *Am J Surg Pathol.* 1998;22(1):93–99.
62. Hashimoto T, Noguchi T, Nagai K, Uchida Y, Shimada T. The organization of the communication routes between the epithelium and lamina propria mucosae in the human esophagus. *Arch Histol Cytol.* 2002;65(4):323–335.
63. Livstone EM, Sheahan DG, Behar J. Studies of esophageal epithelial cell proliferation in patients with reflux esophagitis. *Gastroenterology.* 1977;73(6):1315–1319.
64. Karam SM. Lineage commitment and maturation of epithelial cells in the gut. *Front Biosci.* 1999;4:D286–D298.
65. Burns ER, Scheving LE, Fawcett DF, Gibbs WM, Galatzan RE. Circadian influence on the frequency of labeled mitoses method in the stratified squamous epithelium of the mouse esophagus and tongue. *Anat Rec.* 1976;184(3):265–273.
66. Eastwood GL. Gastrointestinal epithelial renewal. *Gastroenterology.* 1977;72(5 Pt 1):962–975.
67. Epperly MW, Guo H, Shen H, et al. Bone marrow origin of cells with capacity for homing and differentiation to esophageal squamous epithelium. *Radiat Res.* 2004;162(3):233–240.
68. Singh A, Kapur S, Chattopadhyay I, et al. Cytokeratin immunoexpression in esophageal squamous cell carcinoma of high-risk population in Northeast India. *Appl Immunohistochem Mol Morphol.* 2009;17(5):419–424.
69. Lam KY, Loke SL, Shen XC, Ma LT. Cytokeratin expression in non-neoplastic oesophageal epithelium and squamous cell carcinoma of the oesophagus. *Virchows Archiv.* 1995;426(4):345–349.
70. Leedham SJ, Preston SL, McDonald SA, et al. Individual crypt genetic heterogeneity and the origin of metaplastic glandular epithelium in human Barrett's oesophagus. *Gut.* 2008;57(8):1041–1048.
71. Viaene AI, Baert JH. Expression of cytokeratin mRNAs in normal human esophageal epithelium. *Anat Rec.* 1995;241(1):88–98.

72. Glickman JN, Chen YY, Wang HH, Antonioli DA, Odze RD. Phenotypic characteristics of a distinctive multilayered epithelium suggests that it is a precursor in the development of Barrett's esophagus. *Am J Surg Pathol.* 2001;25(5):569–578.
73. Seery JP. Stem cells of the oesophageal epithelium. *J Cell Sci.* 2002;115(Pt 9):1783–1789.
74. Seery JP, Watt FM. Asymmetric stem-cell divisions define the architecture of human oesophageal epithelium. *Curr Biol.* 2000;10(22):1447–1450.
75. Boch JA, Shields HM, Antonioli DA, Zwas F, Sawhney RA, Trier JS. Distribution of cytokeratin markers in Barrett's specialized columnar epithelium. *Gastroenterology.* 1997; 112(3):760–765.
76. Jovanovic I, Tzardi M, Mouzas IA, et al. Changing pattern of cytokeratin 7 and 20 expression from normal epithelium to intestinal metaplasia of the gastric mucosa and gastroesophageal junction. *Histol Histopathol.* 2002;17(2):445–454.
77. Huo X, Zhang HY, Zhang XI, et al. Acid and bile salt-induced CDX2 expression differs in esophageal squamous cells from patients with and without Barrett's esophagus. *Gastroenterology.* 2010;139(1):194–203.e1.
78. Saller J, Al Diffalha S, Neill K, et al. CDX-2 expression in esophageal biopsies without goblet cell intestinal metaplasia may be predictive of Barrett's esophagus. *Dig Dis Sci.* 2020;65(7):1992–1998.
79. Johnson DR, Abdelbaqui M, Tahmasbi M, et al. CDX2 protein expression compared to alcian blue staining in the evaluation of esophageal intestinal metaplasia. *World J Gastroenterol.* 2015;21(9):2770–2776.
80. Liu T, Zhang X, So CK, et al. Regulation of Cdx2 expression by promoter methylation, and effects of Cdx2 transfection on morphology and gene expression of human esophageal epithelial cells. *Carcinogenesis.* 2007;28(2):488–496.
81. Niimi T, Nagashima K, Ward JM, et al. Claudin-18, a novel downstream target gene for the T/EBP/NKX2.1 homeodomain transcription factor, encodes lung- and stomach-specific isoforms through alternative splicing. *Mol Cell Biol.* 2001;21(21):7380–7390.
82. Pei Z, Bini EJ, Yang L, Zhou M, Francois F, Blaser MJ. Bacterial biota in the human distal esophagus. *Proc Natl Acad Sci U S A.* 2004;101(12):4250–4255.
83. Park CH, Lee SK. Exploring esophageal microbiomes in esophageal diseases: a systematic review. *J Neurogastroenterol Motil.* 2020;26(2):171–179.
84. Kaymak T, Hruz P, Niess JH. Immune system and microbiome in the esophagus: implications for understanding inflammatory diseases. *Febs J.* 2022;289(16):4758–4772.
85. Yang L, Lu X, Nossa CW, Francois F, Peek RM, Pei Z. Inflammation and intestinal metaplasia of the distal esophagus are associated with alterations in the microbiome. *Gastroenterology.* 2009;137(2):588–597.
86. May M, Abrams JA. Emerging insights into the esophageal microbiome. *Curr Treat Options Gastroenterol.* 2018;16(1):72–85.
87. Zou Q, Feng L, Cai X, Qian Y, Xu L. Esophageal microflora in esophageal diseases. *Front Cell Infect Microbiol.* 2023;13:1145791.
88. Goyal RK. Columnar cell-lined (Barrett's) esophagus. A historical perspective. In: Spechler SJ, ed. *Barrett's Esophagus.* Elsevier; 1985:1–17.
89. Spechler SJ, Sharma P, Souza RF, Inadomi JM, Shaheen NJ. American Gastroenterological Association medical position statement on the management of Barrett's esophagus. *Gastroenterology.* 2011;140(3):1084–1091.
90. Sharma P, Dent J, Armstrong D, et al. The development and validation of an endoscopic grading system for Barrett's esophagus: the Prague C & M criteria. *Gastroenterology.* 2006;131(5):1392–1399.
91. McClave SA, Boyce HW Jr, Gottfried MR. Early diagnosis of columnar-lined esophagus: a new endoscopic diagnostic criterion. *Gastrointest Endosc.* 1987;33(6):413–416.
92. Mastracci L, Piol N, Molinaro L, et al. Interobserver reproducibility in pathologist interpretation of columnar-lined esophagus. *Virchows Arch.* 2016;468(2):159–167.
93. Kim A, Park WY, Shin N, et al. Cardiac mucosa at the gastroesophageal junction: an Eastern perspective. *World J Gastroenterol.* 2015;21(30):9126–9133.
94. DeMeester SR, DeMeester TR. Columnar mucosa and intestinal metaplasia of the esophagus: fifty years of controversy. *Ann Surg.* 2000;231(3):303–321.
95. Naini BV, Souza RF, Odze RD. Barrett's esophagus: a comprehensive and contemporary review for pathologists. *Am J Surg Pathol.* 2016;40(5):e45–e66.
96. Takubo K, Aida J, Vieth M, et al. Cardiac mucosa in neonates. *Pathol Res Pract.* 2023;246:154498.
97. Chandrasoma P. Histopathology of the gastroesophageal junction: a study on 36 operation specimens. *Am J Surg Pathol.* 2003;27(2):277–278.
98. Zhou H, Greco MA, Daum F, Kahn E. Origin of cardiac mucosa: ontogenic consideration. *Pediatr Dev Pathol.* 2001; 4(4):358–363.
99. De Hertogh G, Van Eyken P, Ectors N, Tack J, Geboes K. On the existence and location of cardiac mucosa: an autopsy study in embryos, fetuses, and infants. *Gut.* 2003;52(6):791–796.
100. Odze R, Spechler SJ, Podgaetz E, Nguyen A, Konda V, Souza RF. Histologic study of the esophagogastric junction of organ donors reveals novel glandular structures in normal esophageal and gastric mucosae. *Clin Transl Gastroenterol.* 2021;12(5):e00346.
101. Huang Q. Controversies of cardiac glands in the proximal stomach: a critical review. *J Gastroenterol Hepatol.* 2011; 26(3):450–455.
102. Voutilainen M, Juhola M, Pitkänen R, Färkkilä M, Sipponen P. Immunohistochemical study of neuroendocrine cells at the gastric cardia mucosa. *J Clin Pathol.* 2002;55(10):767–769.
103. Sawhney RA, Shields HM, Allan CH, Boch JA, Trier JS, Antonioli DA. Morphological characterization of the squamocolumnar junction of the esophagus in patients with and without Barrett's epithelium. *Dig Dis Sci.* 1996;41(6): 1088–1098.
104. Shields HM, Zwas F, Antonioli DA, Doos WG, Kim S, Spechler SJ. Detection by scanning electron microscopy of a distinctive esophageal surface cell at the junction of squamous and Barrett's epithelium. *Dig Dis Sci.* 1993;38(1):97–108.
105. Brien TP, Farraye FA, Odze RD. Gastric dysplasia-like epithelial atypia associated with chemoradiotherapy for esophageal cancer: a clinicopathologic and immunohistochemical study of 15 cases. *Mod Pathol.* 2001;14(5):389–396.
106. Wieczorek TJ, Wang HH, Antonioli DA, Glickman JN, Odze RD. Pathologic features of reflux and Helicobacter pylori-associated carditis: a comparative study. *Am J Surg Pathol.* 2003;27(7):960–968.

107. Shields HM, Rosenberg SJ, Zwas FR, Ransil BJ, Lembo AJ, Odze R. Prospective evaluation of multilayered epithelium in Barrett's esophagus. *Am J Gastroenterol.* 2001;96(12): 3268–3273.
108. Takubo K, Vieth M, Aryal G, et al. Islands of squamous epithelium and their surrounding mucosa in columnar-lined esophagus: a pathognomonic feature of Barrett's esophagus? *Hum Pathol.* 2005;36(3):269–274.
109. Takubo K, Honma N, Arai T. Multilayered epithelium in Barrett's esophagus. *Am J Surg Pathol.* 2001;25(11): 1460–1461.
110. Seefeld U, Krejs GJ, Siebenmann RE, Blum AL. Esophageal histology in gastroesophageal reflux. Morphometric findings in suction biopsies. *Am J Dig Dis.* 1977;22(11): 956–964.
111. Tang P, McKinley MJ, Sporrer M, Kahn E. Inlet patch: prevalence, histologic type, and association with esophagitis, Barrett esophagus, and antritis. *Arch Pathol Lab Med.* 2004; 128(4):444–447.
112. Enterline H, Thompson J. *Pathology of the Esophagus.* Springer-Verlag; 1984.
113. Rector LC, Connerley ML. Aberrant mucosa in the esophagus in infants and in children. *Arch Pathol Lab Med.* 1941;31:285–294.
114. Mahjoub FE, Seighaly F, Mirahmad M, Asili P. Previously unreported esophageal lymphoid polyp in a child: a case report with histopathology. *Fetal Pediatr Pathol.* 2023; 42(3):506–511.
115. Mahjoub FE, Fallahi GH, Niknejad N. Vascular Changes in Eosinophilic Esophagitis (EOE), report of an unusual case. *Fetal Pediatr Pathol.* 2020;39(6):533–538.
116. Nagai K, Noguchi T, Hashimoto T, Uchida Y, Shimada T. The organization of the lamina muscularis mucosae in the human esophagus. *Arch Histol Cytol.* 2003;66(3):281–288.
117. Muhletaler CA, Lams PM, Johnson AC. Occurrence of oesophageal intramural pseudodiverticulosis in patients with pre-existing benign oesophageal stricture. *Br J Radiol.* 1980;53(628):299–303.
118. Wolf C, Timmer R, Breumelhof R, Seldenrijk CA, Smout AJ. Prolonged measurement of lower oesophageal sphincter function in patients with intestinal metaplasia at the oesophagogastric junction. *Gut.* 2001;49(3):354–358.
119. Liebermann-Meffert D, Allgöwer M, Schmid P, Blum AL. Muscular equivalent of the lower esophageal sphincter. *Gastroenterology.* 1979;76(1):31–38.
120. Boyce H. Chapter 3. Esophagus: anatomy and structural anomalies. In: Yamada T, Alpers DH, Kaplowitz N, Laine L, Owyang C, Powell DW, eds. *Textbook of Gastroenterology.* 4th ed. Lippincott Williams & Wilkins; 2003:1148–1165.
121. Vaage SKO. Congenital duplications of the alimentary tract with special regard to their embryogenesis. A follow-up study in surgically corrected cases. *Prog Pediatr Surg.* 1974;7:103–123.
122. Zhang Y, Jiang M, Kim E, et al. Development and stem cells of the esophagus. *Semin Cell Dev Biol.* 2017;66:25–35.
123. Fausett S, Klingensmith J. Compartmentalization of the foregut tube: developmental origins of the trachea and esophagus. *Wiley Interdiscip Rev Dev Biol.* 2012;1:184–202.
124. Freedom RM, Yoo SJ, Goo HW, Mikailian H, Anderson RH. The bronchopulmonary foregut malformation complex. *Cardiol Young.* 2006;16(3):229–251.
125. Jacobs IJ, Ku WY, Que J. Genetic and cellular mechanisms regulating anterior foregut and esophageal development. *Dev Biol.* 2012;369(1):54–64.
126. Perin S, McCann CJ, Borrelli O, De Coppi P, Thapar N. Update on foregut molecular embryology and role of regenerative medicine therapies. *Front Pediatr.* 2017;5:91.
127. Johns BA. Developmental changes in the oesophageal epithelium in man. *J Anat.* 1952;86(4):431–442.
128. Berardi RS, Devaiah KA. Barrett's esophagus. *Surg Gynecol Obstet.* 1983;156(4):521–538.
129. Borrelli O, Hassall E, D'Armiento F, et al. Inflammation of the gastric cardia in children with symptoms of acid peptic disease. *J Pediatr.* 2003;143(4):520–524.
130. Chandrasoma P, Makarewicz K, Wickramasinghe K, Ma Y, Demeester T. A proposal for a new validated histological definition of the gastroesophageal junction. *Hum Pathol.* 2006;37(1):40–47.
131. Daniely Y, Liao G, Dixon D, et al. Critical role of p63 in the development of a normal esophageal and tracheobronchial epithelium. *Am J Physiol Cell Physiol.* 2004;287(1): C171–C181.
132. Que J, Okubo T, Goldenring JR, et al. Multiple dose-dependent roles for Sox2 in the patterning and differentiation of anterior foregut endoderm. *Development.* 2007;134(13):2521–2531.
133. Trisno SL, Philo KED, McCracken KW, et al. Esophageal organoids from human pluripotent stem cells delineate Sox2 functions during esophageal specification. *Cell Stem Cell.* 2018;23(4):501–515.e7.
134. Takubo K, Vieth M, Honma N, et al. Ciliated surface in the esophagogastric junction zone: a precursor of Barrett's mucosa or ciliated pseudostratified metaplasia? *Am J Surg Pathol.* 2005;29(2):211–217.
135. Que J. The initial establishment and epithelial morphogenesis of the esophagus: a new model of tracheal-esophageal separation and transition of simple columnar into stratified squamous epithelium in the developing esophagus. *Wiley Interdiscip Rev Dev Biol.* 2015;4(4):419–430.
136. Chihara D, Romer AI, Bentzinger CF, Rudnicki MA, Krauss RS. PAX7 is required for patterning the esophageal musculature. *Skelet Muscle.* 2015;5:39.
137. Ciocalteu A, Popa P, Ionescu M, Gheonea DI. Issues and controversies in esophageal inlet patch. *World J Gastroenterol.* 2019;25(30):4061–4073.
138. Jacobs E, Dehou MF. Heterotopic gastric mucosa in the upper esophagus: a prospective study of 33 cases and review of literature. *Endoscopy.* 1997;29(8):710–715.
139. Borhan-Manesh F, Farnum JB. Incidence of heterotopic gastric mucosa in the upper oesophagus. *Gut.* 1991;32(9): 968–972.
140. von Rahden BHA, Stein HJ, Becker K, Liebermann-Meffert D, Siewert JR. Heterotopic gastric mucosa of the esophagus: literature-review and proposal of a clinicopathologic classification. *Am J Gastroenterol.* 2004;99(3):543–551.
141. Gutierrez O, Akamatsu T, Cardona H, Graham DY, El-Zimaity HMT. Helicobacter pylori and hetertopic gastric mucosa in the upper esophagus (the inlet patch). *Am J Gastroenterol.* 2003;98(6):1266–1270.
142. Polkowski W, van Lanschot JJ, ten Kate FJ, et al. Intestinal and pancreatic metaplasia at the esophagogastric junction in patients without Barrett's esophagus. *Am J Gastroenterol.* 2000;95(3):617–625.

143. Kung JW, Brown A, Kruskal JB, Goldsmith JD, Pedrosa I. Heterotopic pancreas: typical and atypical imaging findings. *Clin Radiol.* 2010;65(5):403–407.
144. Gaspar Fuentes A, Campos Tarrech JM, Fernández Burgui JL, et al. [Pancreatic ectopias]. *Rev Esp Enferm Apar Dig.* 1973;39(3):255–268.
145. Zhang ZL, Wang JL, Guo CL, et al. Ectopic liver tissue in the esophagus: a case report. *Medicine (Baltimore).* 2019; 98(17):e15260.
146. Park A, Lee JH, Park A, et al. Prevalence rate and clinical characteristics of esophageal ectopic sebaceous glands in asymptomatic health screen examinees. *Dis Esophagus.* 2017;30(1):1–5.
147. Postlethwait RW, Detmer DE. Ectopic thyroid nodule in the esophagus. *Ann Thorac Surg.* 1975;19(1):98–100.
148. El-Gohary Y, Gittes GK, Tovar JA. Congenital anomalies of the esophagus. *Semin Pediatr Surg.* 2010;19(3):186–193.
149. Katzka DA, Levine MS, Ginsberg GG, et al. Congenital esophageal stenosis in adults. *Am J Gastroenterol.* 2000; 95(1):32–36.
150. Murphy SG, Yazbeck S, Russo P. Isolated congenital esophageal stenosis. *J Pediatr Surg.* 1995;30(8):1238–1241.
151. Romeo E, Foschia F, de Angelis P, et al. Endoscopic management of congenital esophageal stenosis. *J Pediatr Surg.* 2011;46(5):838–841.
152. Woodward M, Crabbe DC, Kiely EM, Spitz L. Congenital intrathoracic stomach. *Pediatr Surg Int.* 1996;11(2–3):116–118.
153. Alzaghal AA, Salman R, Berjawi G, Haddad M, Naffaa L. The short esophagus: review of a neglected entity. *Clin Imaging.* 2018;50:43–50.
154. Le Roux BT. Intrathoracic duplication of the foregut. *Thorax.* 1962;17(4):357–362.
155. Tarnay TJ, Chang CH, Nugent RG, Warden HE. Esophageal duplication (foregut cyst) with spinal malformation. *J Thorac Cardiovasc Surg.* 1970;59(2):293–298.
156. Azzie G, Beasley S. Diagnosis and treatment of foregut duplications. *Semin Pediatr Surg.* 2003;12(1):46–54.
157. Holcomb GW III, Gheissari A, O'Neill JA Jr, Shorter NA, Bishop HC. Surgical management of alimentary tract duplications. *Ann Surg.* 1989;209(2):167–174.
158. Bajpai M, Mathur M. Duplications of the alimentary tract: clues to the missing links. *J Pediatr Surg.* 1994;29(10): 1361–1365.
159. Hoem D, Odegaard S, Morild I, Viste A. Dysphagia caused by an embryonic triple epithelial duplication cyst. *Eur J Surg.* 2001;167(7):546–547.
160. Trappey AF III, Hirose S. Esophageal duplication and congenital esophageal stenosis. *Semin Pediatr Surg.* 2017; 26(2):78–86.
161. Tobin RW. Esophageal rings, webs, and diverticula. *J Clin Gastroenterol.* 1998;27(4):285–295.
162. Varadarajulu S, Noone T. Symptomatic lower esophageal muscular ring: response to botox. *Dig Dis Sci.* 2003;48(11):2132–2134.
163. van Overbeek JJM. Pathogenesis and methods of treatment of Zenker's diverticulum. *Ann Otol Rhinol Laryngol.* 2003;112(7):583–593.
164. Sato H, Takeuchi M, Hashimoto S, et al. Esophageal diverticulum: new perspectives in the era of minimally invasive endoscopic treatment. *World J Gastroenterol.* 2019;25(12):1457–1464.
165. Soares R, Herbella FA, Prachand VN, Ferguson MK, Patti MG. Epiphrenic diverticulum of the esophagus. From pathophysiology to treatment. *J Gastrointest Surg.* 2010; 14(12):2009–2015.
166. Constantin A, Constantinoiu S, Achim F, Socea B, Costea DO, Predescu D. Esophageal diverticula: from diagnosis to therapeutic management-narrative review. *J Thorac Dis.* 2023;15(2):759–779.
167. Pilarski R, Burt R, Kohlman W, Pho L, Shannon KM, Swisher E. Cowden syndrome and the PTEN hamartoma tumor syndrome: systematic review and revised diagnostic criteria. *J Natl Cancer Inst.* 2013;105(21):1607–1616.
168. Choudhry V, Williams J, Choudhry U, Kaul V. Diffuse esophageal hyperkeratosis: a rare case presentation. *Cureus.* 2023;15(5):e38757.
169. Sánchez-Delgado L, Fuentes-Valenzuela E, Burgueño-Gómez B, et al. Esophageal hyperkeratosis as an uncommon manifestation of gastroesophageal reflux disease. *Rev Esp Enferm Dig.* 2022;114(8):508.
170. Singhi AD, Arnold CA, Crowder CD, Lam-Himlin DM, Voltaggio L, Montgomery EA. Esophageal leukoplakia or epidermoid metaplasia: a clinicopathological study of 18 patients. *Mod Pathol.* 2014;27(1):38–43.
171. Singhi AD, Arnold CA, Lam-Himlin DM, et al. Targeted next-generation sequencing supports epidermoid metaplasia of the esophagus as a precursor to esophageal squamous neoplasia. *Mod Pathol.* 2017;30(11):1613–1621.
172. Takubo K. *Pathology of the Esophagus: An Atlas and Textbook.* Springer Verlag; 2007.
173. Purdy JK, Appelman HD, McKenna BJ. Sloughing esophagitis is associated with chronic debilitation and medications that injure the esophageal mucosa. *Mod Pathol.* 2012;25(5):767–775.
174. Abraham SC, Singh VK, Yardley JH, Wu TT. Hyperplastic polyps of the esophagus and esophagogastric junction: histologic and clinicopathologic findings. *Am J Surg Pathol.* 2001; 25(9):1180–1187.
175. Melton SD, Genta RM. Gastric cardiac polyps: a clinicopathologic study of 330 cases. *Am J Surg Pathol.* 2010;34(12): 1792–1797.
176. Mosca S, Manes G, Monaco R, Bellomo PF, Bottino V, Balzano A. Squamous papilloma of the esophagus: long-term follow up. *J Gastroenterol Hepatol.* 2001;16(8): 857–861.
177. Szántó I, Szentirmay Z, Banai J, et al. Squamous papilloma of the esophagus. Clinical and pathological observations based on 172 papillomas in 155 patients. *Orv Hetil.* 2005;146(12):547–552.
178. Mastracci L, Grillo F, Parente P, et al. Gastro-esophageal reflux disease and Barrett's esophagus: an overview with an histologic diagnostic approach. *Pathologica.* 2020;112(3): 117–127.
179. Weinstein WM, Bogoch ER, Bowes KL. The normal human esophageal mucosa: a histological reappraisal. *Gastroenterology.* 1975;68(1):40–44.
180. Pinto D, Plieschnegger W, Schneider NI, et al. Carditis: a relevant marker of gastroesophageal reflux disease. Data from a prospective central European multicenter study on histological and endoscopic diagnosis of esophagitis (histoGERD trial). *Dis Esophagus.* 2019;32(1).
181. Pittman ME. Lymphocytic esophagitis: current understanding and controversy. *Am J Surg Pathol.* 2022;46(1): e55–e63.

182. Zaver HB, Ghoz H, Malviya BJ, et al. Lymphocytic esophagitis: assessing risk factors and clinical outcomes. *Dig Dis Sci*. 2021;66(11):3976–3984.
183. Shaheen NJ, Falk GW, Iyer PG, Gerson LB. ACG clinical guideline: diagnosis and management of Barrett's esophagus. *Am J Gastroenterol*. 2016;111(1):30–50; quiz 1.
184. Montgomery EA, Canto MI, Srivastava A. Evaluation and reporting of biopsies from the columnar-lined esophagus and gastro-esophageal junction (GEJ). *Ann Diagn Pathol*. 2019;39:111–117.
185. Vieth M, Stolte M. Pathology of early upper GI cancers. *Best Pract Res Clin Gastroenterol*. 2005;19(6):857–869.
186. Leers JM, DeMeester SR, Oezcelik A, et al. The prevalence of lymph node metastases in patients with T1 esophageal adenocarcinoma a retrospective review of esophagectomy specimens. *Ann Surg*. 2011;253(2):271–278.
187. Kumarasinghe MP, Bourke MJ, Brown I, et al. Pathological assessment of endoscopic resections of the gastrointestinal tract: a comprehensive clinicopathologic review. *Mod Pathol*. 2020;33(6):986–1006.
188. Fotis D, Doukas M, Wijnhoven BP, et al. Submucosal invasion and risk of lymph node invasion in early Barrett's cancer: potential impact of different classification systems on patient management. *United European Gastroenterol J*. 2015;3(6):505–513.
189. Pimentel-Nunes P, Libanio D, Bastiaansen BAJ, et al. Endoscopic submucosal dissection for superficial gastrointestinal lesions: European Society of Gastrointestinal Endoscopy (ESGE) Guideline – Update 2022. *Endoscopy*. 2022;54(6):591–622.
190. Al-Haddad MA, Elhanafi SE, Forbes N, et al. American Society for Gastrointestinal Endoscopy guideline on endoscopic submucosal dissection for the management of early esophageal and gastric cancers: methodology and review of evidence. *Gastrointest Endosc*. 2023;98(3):285–305.e38.
191. Bollschweiler E, Baldus SE, Schröder W, et al. High rate of lymph-node metastasis in submucosal esophageal squamous-cell carcinomas and adenocarcinomas. *Endoscopy*. 2006;38(2):149–156.

22

Stomach

Kevin Waters ▪ Matthew Leong ▪ Elizabeth A. Montgomery

EMBRYOLOGY AND POSTNATAL DEVELOPMENT

The stomach develops as a dilated foregut structure distal to the esophagus when the embryo is less than 1 cm in length. The stomach is first attached to the back of the abdomen by the dorsal mesogastrium and to the primitive diaphragm (i.e., septum transversum) by the ventral mesogastrium. The dorsal mesogastrium of the growing stomach forms the greater omentum and the ventral mesogastrium develops into the lesser omentum.

The stomach is an endodermal structure. Early columnar cell differentiation of the mucosa occurs when the embryo is approximately 80 mm in length. Enzyme and acid production are underway by the 4th month of fetal life and are well established at birth. The newborn stomach displays features similar to those of the adult stomach.

GROSS MORPHOLOGIC FEATURES

The stomach is flattened and curved and situated in the left upper quadrant of the abdomen. It is joined by the esophagus a few centimeters below the diaphragmatic opening. It merges distally with the duodenum, just right of the midline. Its size varies with its contents; it readily distends substantially with ingestion of a large food bolus.

Historically, the stomach has been anatomically divided into four regions that adjoin one another: cardia, fundus, corpus (or body), and antrum (Fig. 22.1) (1). The superior medial edge forms the lesser curvature and the inferior lateral edge is termed the greater curvature. The gastroesophageal junction is the point where the tubular esophagus joins the stomach. It is normally found 40 cm distal to the incisor teeth, although this distance can vary somewhat with the height of the patient. The flat squamous esophageal mucosa transitions to the rugal folds of the stomach at the gastroesophageal junction in healthy patients, although the two may be separated by a 1 to 3 cm segment of cardiac-type mucosa in adults. Some authors consider this region to represent the gastric cardia and most proximal portion of the stomach, while a growing body of literature suggests that it represents metaplasia and dilatation of the distal esophagus due to physiologic, or pathologic, gastroesophageal reflux. In other words, the very existence of a gastric cardia as a normal anatomic structure is the subject of debate. The fundus is defined as the gastric segment that protrudes above the gastroesophageal junction below the

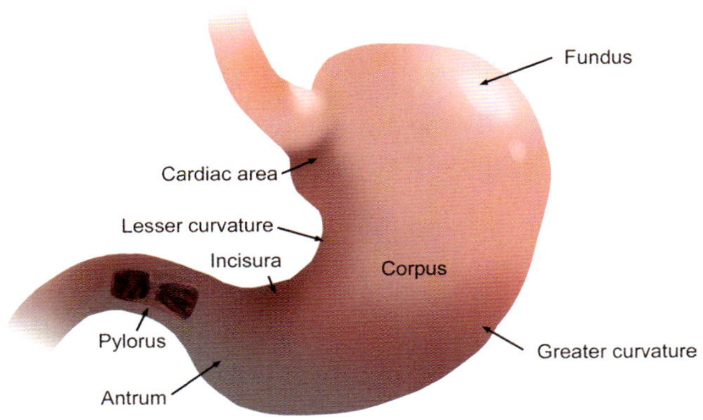

FIGURE 22.1 Stomach with delineated gross anatomical zones.

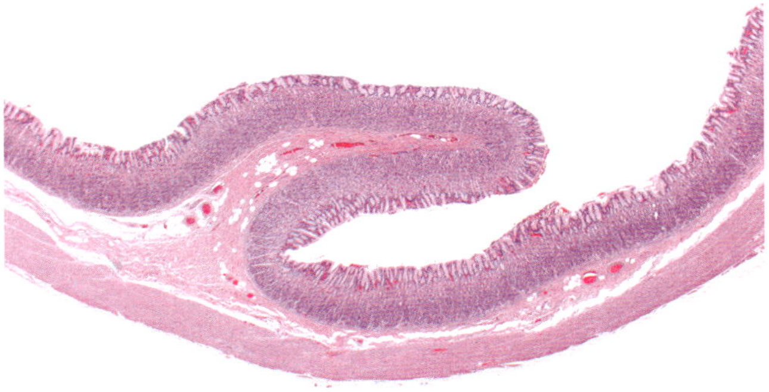

FIGURE 22.3 Low-power magnification of normal stomach with oxyntic mucosa. *Areae gastricae* are the grooves in the mucosa that can be seen at low power.

left hemidiaphragm. The antrum comprises the distal third of the stomach proximal to the pyloric sphincter; the narrow segment of antrum proximal to the pyloric sphincter is the pylorus, or pyloric channel. The remainder of the stomach is the corpus (body) (Fig. 22.2). The fundus and corpus are lined by the same type of oxyntic mucosa and, thus, they are not histologically distinguishable. As a result, some authors use the term fundus and/or body to denote both regions. Although this approach is practical for describing mucosal diseases, it is suboptimal when describing gross anatomy and the location of disease within the proximal regions of the stomach. The histologic junction between the antrum and the corpus is a gradual transition, rather than a sharp demarcation, and roughly corresponds to a notch on the lesser curvature termed the incisura, (2). The normal gastric mucosa is thrown into thick rugae that disappear when the organ is distended. Since the corpus and fundus are the areas that distend most to accommodate food, the folds in these zones are the most prominent when the stomach is empty. The antrum features a flatter mucosa.

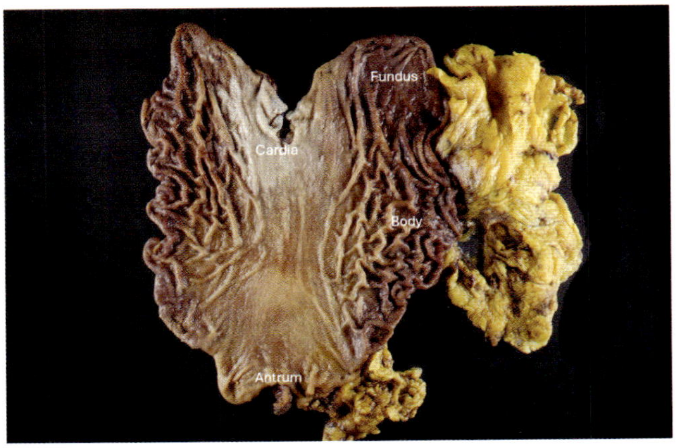

FIGURE 22.2 Stomach mucosal zones. The small focus of the esophagus at the superior aspect of the specimen leads distally into the cardiac mucosa. The fundus and body have oxyntic mucosa and prominent rugal folds. The pyloric mucosa has a triangular shape located at the distal end of the stomach that leads into the duodenum.

The gastric wall has four layers: a mucosa consisting of epithelium, lamina propria, and muscularis mucosae, as well as the submucosa, muscularis propria, and subserosa. The layers are structurally similar to those in the rest of the tubular gastrointestinal tract, although the mucosa differs in the various parts of the stomach. When magnified using an endoscope or dissecting lens, the surface of the mucosa is dissected by slim shallow grooves termed *areae gastricae* (3) that do not disappear in the distended stomach. They are best seen when the mucosa is viewed *en face* using a magnifying glass or dissecting microscope. *Areae gastricae* can be recognized in histologic sections, particularly from gastrectomy specimens, as subtle depressions on an otherwise smooth surface (Fig. 22.3).

Blood Supply

Blood reaches the stomach *via* five arteries, namely (1) the left gastric artery, which arises directly from the celiac axis and supplies the proximal stomach, (2) the right gastric artery supplying the lesser curvature, (3) the right gastroepiploic artery, which supplies the greater curvature arise from the hepatic artery, (4) the left gastroepiploic, and (5) the short gastric arteries. The left gastroepiploic and short gastric arteries emanate from the splenic artery, further supplying the greater curvature. Branches from these arteries anastomose in the subserosal layer, the muscularis propria, and the submucosa, resulting in a rich and redundant vascular supply and the rare occurrence of gastric infarction. The mucosal arteries are derived from the submucosal plexus. These are end arteries supplying specific areas of the mucosa (4).

Nerve Supply

The celiac plexus nerves that adjoin the gastric and gastroepiploic arteries provide sympathetic innervation to the stomach. Additional branches derive from the left and right phrenic nerves. The vagus nerve supplies parasympathetic innervation through anterior and posterior trunks near the esophagogastric junction. After it enters the abdomen, the anterior vagus nerve gives rise to a hepatic branch, and the posterior vagus nerve gives rise to a celiac branch. For this reason, truncal vagotomy above these branches results in denervation of the entire gastrointestinal tract while vagotomy below these nerves results in gastric denervation alone. Selective vagotomy (i.e., gastric corpus denervation) is performed by sectioning lateral branches as the two main gastric nerves pass along the lesser curvature, thereby preserving the antral supply. Nerve plexuses are concentrated in the submucosal Meissner plexus and the myenteric Auerbach plexus that lies between the circular and longitudinal layers of the muscularis propria.

Lymphatic Vessels

Although early authors believed that the entire mucosal thickness was served by a rich lymphovascular supply, we now know that lymphatic channels are limited to the deep lamina propria just superficial to the muscularis mucosae (5,6). From there, efferent vessels traverse the muscularis mucosae and communicate with larger submucosal lymphatic channels. As a result, intramucosal gastric adenocarcinomas occasionally metastasize to regional lymph nodes.

The gastric lymphatic trunks accompany the routes of the main arteries and veins. There are four areas of drainage with unique groups of lymph nodes. The largest comprises the lower end of the esophagus and most of the lesser curvature, accompanying the left gastric artery to the left gastric lymph nodes. The lesser curvature of the pylorus drains to the right gastric and hepatic lymph nodes. The proximal portion of the greater curvature is drained by pancreaticosplenic lymph nodes in the splenic hilum, and the distal greater curvature by the right gastroepiploic lymph nodes in the greater omentum and pyloric nodes near the pancreatic head. Efferents from all groups eventually reach celiac lymph nodes.

It is worthwhile for pathologists to familiarize themselves with the location and nomenclature of the groups of lymph nodes. Japanese colleagues created a system that recognizes the following stations: (1) perigastric along the greater curvature, (2) perigastric along the lesser curvature, right and left paracardial (cardioesophageal), (3) suprapyloric, (4) infrapyloric, (5) left gastric artery, (6) celiac artery, (7) common hepatic artery, (8) hepatoduodenal (portal), (9) splenic artery, and (10) splenic hilum (7). Lymph nodes at these locations are considered regional lymph nodes and inform the N category of the TNM system, whereas more distant involved nodes are noted in the M category. The TNM system currently bases the N assignment on the number of involved regional lymph nodes without concern for their precise locations in the gastric drainage (1).

GENERAL HISTOLOGIC FEATURES

The gastric mucosa has a similar overall pattern throughout the various regions of the stomach. The surface layer contains foveolae or pits that represent invaginations of the surface mucus-producing epithelium, and a deeper layer formed by coiled glands that empty into the bases of the foveolae. The glands below the surface differ in structure and function in the various zones of the stomach that roughly correspond to the gross anatomic components of the stomach.

The gastroesophageal junction is defined as the proximal extent of the rugal folds. It generally corresponds to an abrupt transition from nonkeratinizing squamous epithelium to oxyntic mucosa in children and young adults. However, the squamocolumnar junction is often located proximal to the tips of the rugal folds (i.e., gastroesophageal junction) and it may be irregular rather than forming a circumferential Z line. Proximal displacement and irregularity of the squamocolumnar junction is more common among older adults and individuals with chronic gastroesophageal reflux disease. Columnar epithelium in flat mucosa proximal to the rugal folds is considered to be within the tubular esophagus and, thus, regarded as metaplastic in nature within the tubular esophagus.

The antrum has a triangular shape. It extends 5 to 7 cm proximally from the pylorus along the lesser curvature but only 3 to 4 cm proximally along the greater curvature. There are subtle differences between the mucosae of the pyloric and antral regions, although many observers use these terms interchangeably. Unfortunately, the anatomic incisura has no clear relationship to the proximal extent of antral-type mucosa.

The oxyntic mucosa of the corpus and fundus is specialized for secretion of acid and pepsin. The transition between antral and oxyntic mucosae is gradual, with a 1 to 2 cm wide, variably long zone displaying hybrid features of antral and oxyntic mucosa. There is also a transitional zone in the pyloric channel where pyloric mucosa merges with that of the proximal duodenum. This gastroduodenal mucosa normally contains scattered goblet cells that are typically accompanied by enteric epithelium; these areas should be interpreted as representative of the normal distal stomach and should not be classified as intestinal metaplasia of the gastric mucosa.

Surface Epithelium

The gastric mucosa is surfaced by tall, columnar, mucus-secreting cells with intervening foveolae lined by a similar epithelium (Fig. 22.4). The surface and foveolar lining cells are essentially identical throughout the stomach. Gastric glands beneath the surface empty into the bases of the foveolae. Between the foveolae and the glands is the lamina propria. In the cardiac and pyloric zones, the foveolae are wider than in other areas, imparting a slightly villous appearance.

The surface cells are tall and columnar with basally situated nuclei and superficial cytoplasm distended by mucus. The nuclei show evenly distributed chromatin and smooth nuclear membranes with inconspicuous single nucleoli.

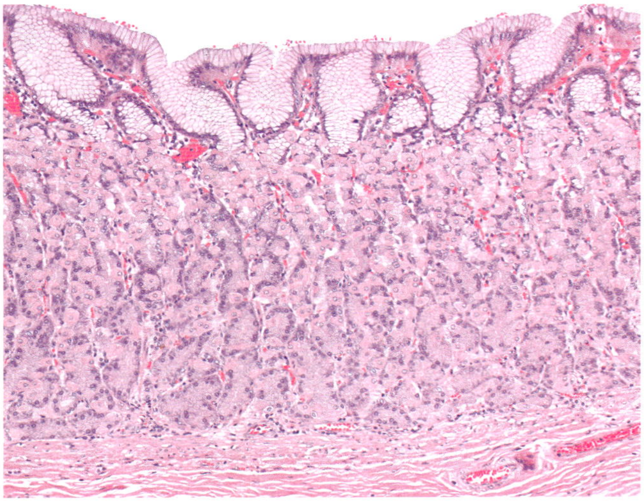

FIGURE 22.4 High-power magnification of normal oxyntic mucosa. In all types of gastric mucosa, each cell has a mucus globule in the surface foveolar epithelium. In the oxyntic mucosa, the pink cells concentrated at superficial aspect of the glands are parietal cells while the amphophilic (blue-gray) zymogenic Chief cells are concentrated at the base of the glands.

The appearance of the mucus depends on the precise preparation of hematoxylin and eosin (H&E)–stained section. The mucus appears as a single vacuole when using alcoholic eosin, whereas aqueous eosin produces multiple smaller vacuoles and a more heavily eosinophilic appearance. Foveolar mucus is neutral mucin that is positive with periodic acid–Schiff (PAS) stain, but negative with Alcian blue (at pH 2.5 and lower) stain (8).

Cardiac and Antral Mucosae

The foveolae account for about half of the mucosal thickness in both cardiac-type and antral mucosae (Fig. 22.5). Both cardiac-type and antral mucosae contain mucin-producing glands that occupy the deep mucosa and are loosely packed in abundant intervening lamina propria. These

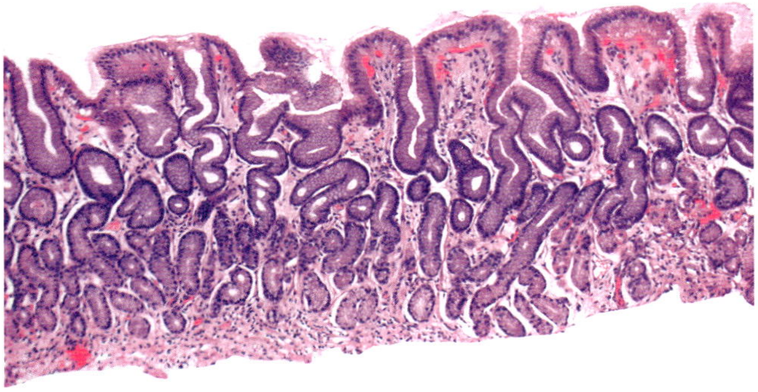

FIGURE 22.5 Gastric pyloric mucosa. In antral mucosa, the surface foveolar epithelium extends down the about half the thickness of the mucosa and can appear slightly villous. The pyloric glands with mucus cells are loosely packed in the lower half of the mucosa. Rare parietal cells can be seen (pink cells on the right sight).

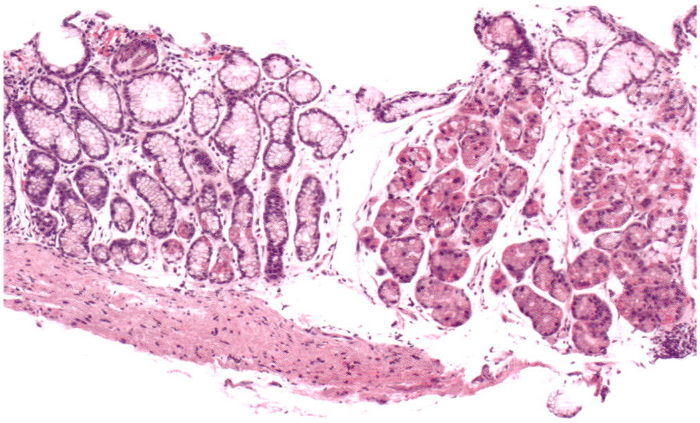

FIGURE 22.6 Transition between antral and oxyntic mucosae. The left side of photomicrograph shows a hybrid-type mucosa with loosely packed glands with scattered pink (eosinophilic) parietal cells and deep antral glands. The right side shows oxyntic mucosa with tightly packed glands with numerous parietal cells at the superficial aspect of the glands and chief cells at the base.

glands are mostly small, although cystic glands are not infrequently in cardiac-type mucosa. The cells comprising the mucus glands have ill-defined borders and bubbly cytoplasm that differs from the cytoplasm of the foveolar and surface epithelia but resembles that of cells in duodenal Brunner glands. Although pyloric-type glands of the antrum secrete only neutral mucin, the secretions of cardiac-type glands are composed predominantly of neutral mucin with small amounts of sialomucin, the type abundantly found in the small intestine (8). Scattered parietal cells can be found either singly or in small groups and become more numerous where the mucosa gradually merges with oxyntic mucosa of the body or fundus (Fig. 22.6). However, zymogenic (chief) cells are seldom detected outside of the oxyntic mucosa and junctional area.

The length of the cardiac mucosa and even its existence as a normal structure have been the subject of debate. When Chandrasoma et al. (9) studied the gastroesophageal region in adult autopsies, less than half (about 45%) had zones of pure cardiac mucosa. Other investigators have reported similar results and noted that the length of the cardia correlates with increased patient age and the presence of reflux-type symptoms (10). In contrast, Kilgore et al. (11) and Zhou et al. (12) examined pediatric autopsy material and found pure cardiac mucosa in every case, measuring 1.0 to 4.0 mm in length (average 1.8 mm). These findings suggest that pure cardiac mucosa and cardio-oxyntic mucosa are normal structures (Fig. 22.7). Despite these observations, the term cardia is now generalized to describe a loosely defined anatomical zone lying distal to the squamocolumnar junction that may contain cardio-oxyntic and pure oxyntic mucosa.

Cardiac-type mucosa can be damaged by gastroesophageal reflux or infection with *Helicobacter pylori*. The

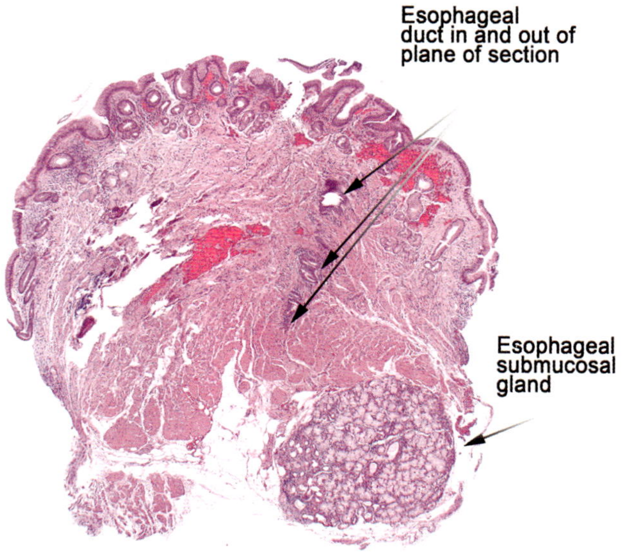

FIGURE 22.7 Low-power view of gastroesophageal junctional mucosa. Whether cardiac mucosa is normal or metaplastic is not always clear. While there is columnar cardiac-type mucosa at the surface, there are esophageal ducts and submucosal glands in the underlying tissue at the right, which indicates that the tissue at the right is from the tubular esophagus and that the surface cardiac-type glands are metaplastic.

alterations include inflammation-related nuclear atypia, intestinal metaplasia, and the presence of mucosa showing hybrid features between squamous and cardiac-type mucosa (13) that has been termed multilayered epithelium (Fig. 22.8). This type of epithelium consists of immature squamous epithelial cells intimately admixed with acid mucin–containing columnar cells that show immunohistochemical evidence of intestinal differentiation but no well-formed goblet cells. Multilayered epithelium is associated with gastroesophageal reflux disease and thought to represent a precursor to Barrett esophagus (14).

Oxyntic Mucosa

The oxyntic mucosa is surfaced by a slim layer of foveolae that accounts for less than one-quarter of the mucosal thickness. The glands below the surface are tightly packed specialized oxyntic glands that contain parietal cells, chief cells, and endocrine cells. Three portions can be described: base, neck, and isthmus. Zymogenic chief cells that secrete pepsinogen predominate in the base. These cuboidal cells feature basally situated nuclei with one or more small nucleoli, as well as pale blue–gray cytoplasm rich in rough endoplasmic reticulum containing ribosomal ribonucleic acid (RNA). The isthmic portion of the glands is most superficial and contains plentiful parietal cells that produce acid and intrinsic factors. These cells are roughly triangular in shape with centrally placed nuclei and evenly distributed chromatin. Their cytoplasm is brightly eosinophilic and somewhat granular, and they contain abundant microcanaliculi that consist entirely of protein. The neck portion of the fundic glands features chief and parietal cells, as well as admixed mucous neck cells (Fig. 22.9). Mucous neck cells can be highlighted with PAS stains and appear similar to the mucus-secreting cells of cardiac and pyloric glands. They produce both neutral and acidic mucin, especially sialomucin, which stains positively with PAS/diastase and Alcian blue at pH 2.5 (15). Mucous neck cells are also found in the isthmic portion of the glands, and occasional parietal cells can appear in the basal portion of the glands. Mucous neck cells are also constituents of the antral mucosa.

The mucous neck cells of the glands are believed to represent the stem cell compartment throughout the stomach, migrating luminally to differentiate into parietal cells, foveolar and surface epithelia, or downward to repopulate chief cells and endocrine cells (16). Although the gastric surface epithelium is replenished every 4 to 8 days, parietal and chief cells are believed to turn over much more slowly, possibly over the course of 1 to 3 years.

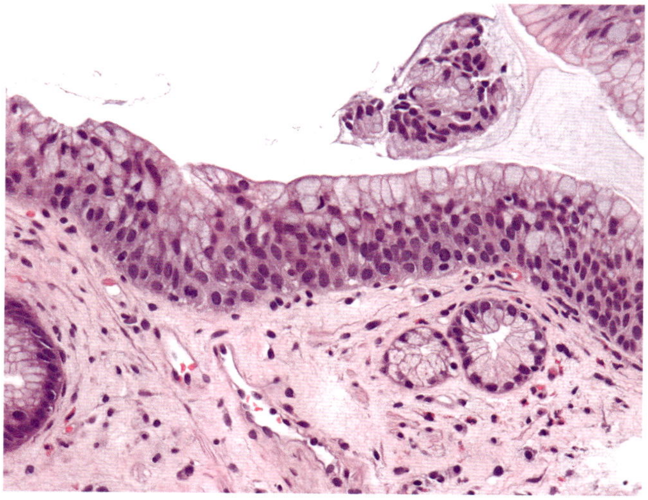

FIGURE 22.8 Multilayered epithelium. Multilayered epithelium shows a hybrid of mucus-producing columnar-type cells at the surface with underlying squamous epithelium.

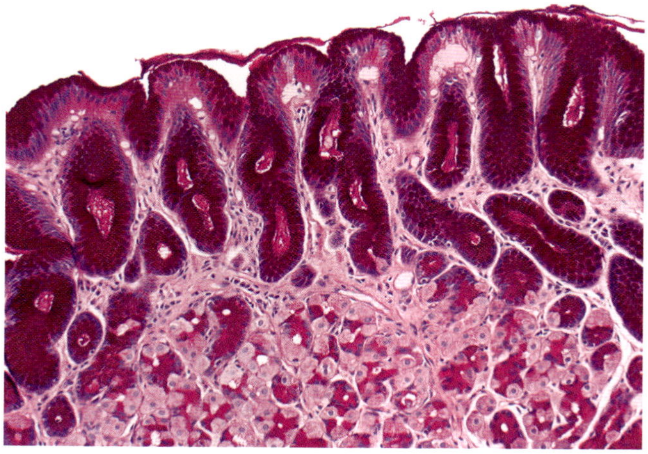

FIGURE 22.9 PAS stain of oxyntic mucosa. The neutral mucin of the surface foveolar epithelium shows strong magenta staining with PAS. In the glands, the PAS stain shows paler staining of the mucous neck cells.

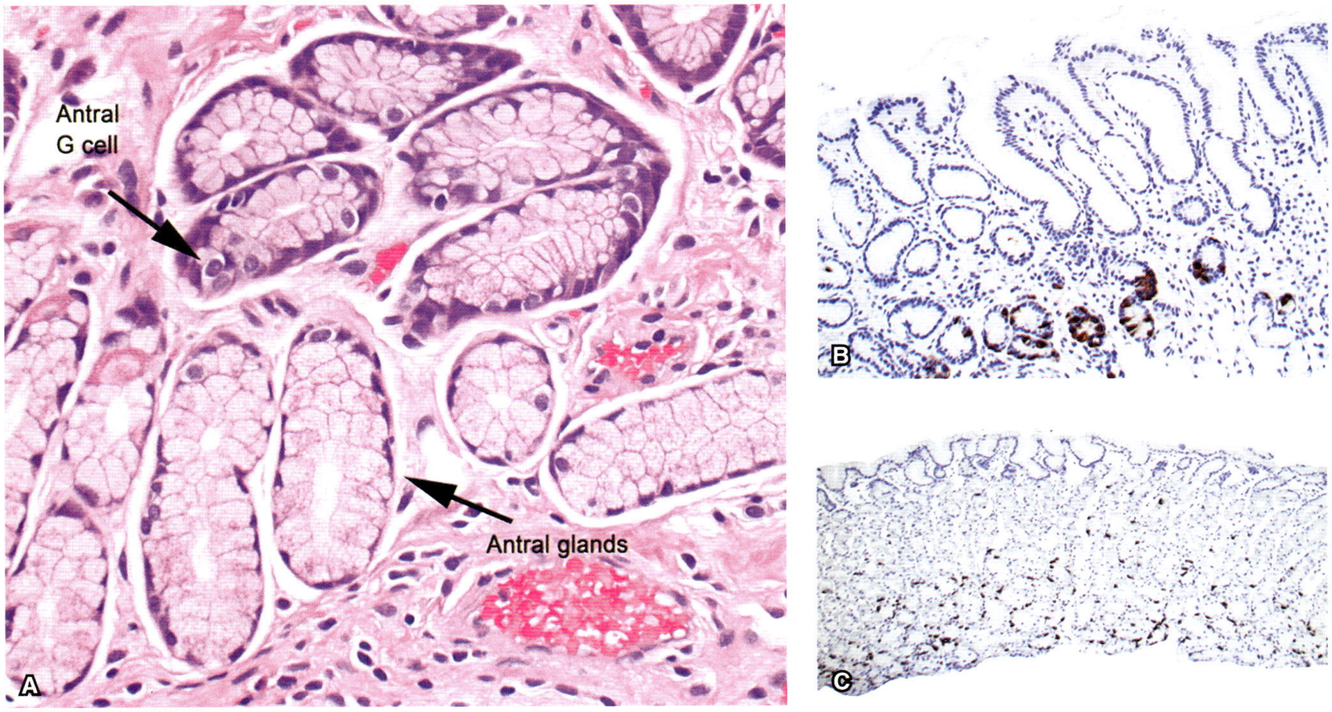

FIGURE 22.10 **A:** The high-power photomicrograph of antral glands shows scattered G cells. They have a "fried egg" appearance with clear cytoplasm. **B:** An immunohistochemical stain for gastrin highlights G cells in pyloric mucosa. **C:** A chromogranin immunohistochemical stain in oxyntic mucosa shows scattered neuroendocrine cells that are mostly comprised of ECL cells.

Endocrine Cells

The stomach contains many types of hormone-producing endocrine cells (Fig. 22.10). Hormones released by these cells manifest their effects by entering the bloodstream or through paracrine actions on nearby cells. Nearly half of the endocrine cells of the antrum are gastrin-producing G cells, 30% are serotonin-producing enterochromaffin (EC) cells, and 15% are somatostatin-producing D cells. On the other hand, most endocrine cells of the body and fundus are enterochromaffin-like (ECL) cells that secrete histamine. The oxyntic mucosa also contains small numbers of EC cells as well as X cells, the function of which is unknown. Endocrine cells are most abundant in the antral mucosa where they number 20 to 50 cells per gland. Endocrine cells of the antrum are found mostly in the mucous neck region. They have regular rounded nuclei and clear cytoplasm that results from a fixation artifact. Assessment for G-cell hyperplasia is challenging in biopsy material due to the wide variation normally seen in healthy patients. Endocrine cells of the body and fundus are present in the oxyntic glands and are concentrated at the base of the mucosa. They can be difficult to detect without immunohistochemical stains. Their numbers vary but most glands contain fewer than 10 endocrine cells but they may number up to 20 cells per gland.

The EC cells and some of the ECL cells have argentaffin or argyrophilic granules. The argentaffin granules can be stained by Fontana, Masson, or diazo techniques, whereas cells with argyrophilic granules are highlighted using the Grimelius technique (17). In modern clinical practice, however, silver stains have been replaced by more sensitive immunohistochemical staining techniques (e.g., synaptophysin, chromogranin, and insulinoma-associated protein-1) (18). Individual hormones such as gastrin and somatostatin can also be readily demonstrated with use of specific antibodies. Gastrin stains are especially helpful when distinguishing antral mucosa from atrophic oxyntic mucosa, as the former contains numerous G cells whereas the latter lacks G cells. In addition to their presence in epithelial cells, some hormones can be detected in neurons and nerve endings present in the gastric wall and mucosa. It is generally believed that vasoactive intestinal peptide (VIP) is predominant in neural tissue and that catecholamines, bombesin, substance P, enkephalins, and possibly gastrin may also be found in these locations.

Endocrine cell hyperplasia can occur due to the trophic effects of increased gastrin secretion, as may occur in the setting of pernicious anemia-type autoimmune gastritis, Zollinger–Ellison syndrome, and long-term acid suppressive therapy. G-cell hyperplasia typically occurs in a linear fashion, whereas ECL-cell hyperplasia appears as linear or nodular aggregates of glands in the deep mucosa (19).

Lamina Propria

Similar to the rest of the columnar intestinal tract, all gastric epithelial cells rest on a basement membrane. The stromal component subjacent to the basement membrane is called the lamina propria. The lamina propria is a well-developed network of reticulin, collagen, and elastic fibers that provides structural support for the gastric epithelium. The lamina propria is most abundant in the antrum and in the superficial aspect of the mucosa where the glands are not as tightly packed. The lamina propria contains numerous cell types including fibroblasts, macrophages, plasma cells, lymphocytes, and occasional eosinophils. Scattered neutrophils and mast cells are also present in normal lamina propria. The lamina propria contains capillaries, arterioles, and nonmyelinated nerve fibers. A few bundles of smooth muscle cells may also extend superficially from the muscularis mucosae into the lamina propria. This is seen more often in the distal antrum.

Mucosal plasma cells mostly secrete immunoglobulin A (IgA). The lymphocytes in the lamina propria are mostly of B-cell lineage, whereas intraepithelial lymphocytes are usually T cells. Formalin fixation artifact can cause the appearance of a halo around these cells.

Small numbers of primary lymphoid follicles are a normal component of the stomach, particularly in oxyntic mucosa of older adults. These appear as round clusters of lymphocytes unassociated with infiltration of adjacent oxyntic glands (20). Secondary lymphoid follicles containing germinal centers are always pathologic and found exclusively in the setting of gastritis. They are most commonly related to *H. pylori* infection but can be seen in other forms of chronic gastritis as well.

Submucosa

The submucosa is the layer between the muscularis mucosae of the deep mucosa and the muscularis propria. This layer helps form the gastric rugal folds. Ganglion cells, nerves, arteries, and veins can all be seen in the loose connective tissue of the submucosa along with numerous elastic fibers. The ganglion cells and nerves are a part of the submucosal autonomic nerve plexus of Meissner.

Muscular Components

The main muscle mass of the stomach can be referred to as the muscularis externa. In North America, the term muscularis propria is preferred. This is because the term *muscularis externa* is ambiguous, as it is not always known whether it refers to the entire muscle thickness or only the outer layer.

The muscularis propria has three identifiable layers: innermost oblique, inner circular, and outer longitudinal. The outer longitudinal smooth muscle cell layer is continuous within the muscularis propria of the esophagus at the level of the gastroesophageal junction. The inner circular layer is aggregated into a palpable thick sphincter at the pylorus. This sphincter is clearly separated from the circular layer of the duodenal muscularis propria by a connective tissue septum. The innermost oblique muscular layer is incomplete; it is present interior to the circular layer and is most obvious in the gastric cardia. However, the presence of a circular sphincter at the cardia is a subject of debate (21). Histologic examination is inconclusive, and although radiologic techniques show swallowed food halted at this level, this could also reflect external compression from the adjacent diaphragmatic crura.

The muscularis mucosae consist of inner circular and outer longitudinal smooth muscle cell layers as well as a minor component of elastic fibers. Slender strands of

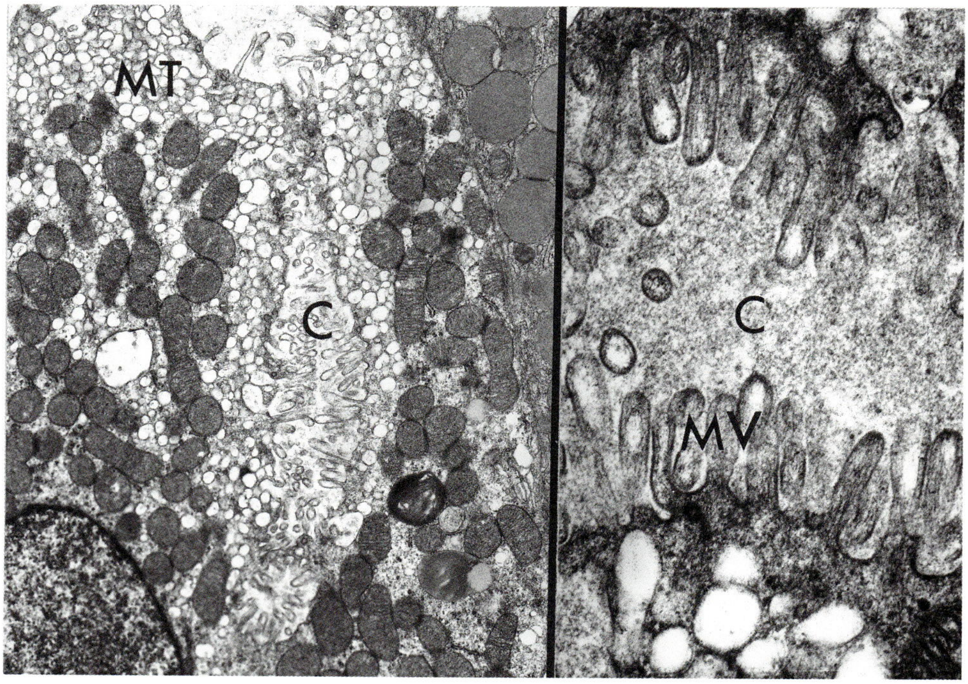

FIGURE 22.11 Ultrastructural appearances of the parietal cell canaliculus (*C*). Note the fingerlike microvilli (*MV*) and the microtubular invaginations (*MT*) (Original magnifications: *left*, ×9,000; *right*, ×41,000). (This figure is courtesy of Dr. David A. Owen.)

smooth muscle cells can extend into the lamina propria, where they terminate at the basement membrane. This is most pronounced in the antrum.

ULTRASTRUCTURE

Foveolar epithelial cells are characterized by multiple-rounded, electron-dense mucous vacuoles in the superficial cytoplasm and stubby microvilli projecting from the luminal surface. The basal cytoplasm contains moderate amounts of rough endoplasmic reticulum and some mitochondria. Adjacent epithelial cells are connected by tight junctions (zonula occludens) at their luminal aspects and by adherence junctions along the rest of the cell-to-cell interfaces. These tight junctions are thought to play an integral role in sustaining mucosal integrity and the gastric mucosal barrier.

Parietal cells are ultrastructurally unique (Fig. 22.11) (22). In the unstimulated state, the cytoplasm contains an apical crescent-shaped canaliculus lined by stubby microvilli. Elongated membrane invaginations, or microtubules, are present between the microvilli. Upon stimulation, the microtubules are replaced by a dense meshwork of intracellular canaliculi (23). The canalicular system is considered essential for hydrochloric acid production. Hydrogen ions are actively transported across the canalicular membrane. The majority of the rest of the parietal cell cytoplasm contains mitochondria to supply the required energy for this process. The abundant mitochondria account for the eosinophilic cytoplasm of parietal cells.

The zymogenic chief cells have similar ultrastructure to protein-secreting exocrine cells elsewhere in the body. The cells contain rough-surfaced vesicles in the superficial cytoplasm and ample rough endoplasmic reticulum.

GASTRIC FUNCTION

The stomach initiates the digestive process by pooling and mixing food. It secretes acid, pepsin, and electrolytes. This process is partially controlled by the vagus nerve and partially by the gastrin levels produced by G cells in the antrum. Gastrin release is initiated by distension of the antrum or direct stimulation from ingested food, particularly amino acids and peptides. Hydrochloric acid is produced by the active transport of hydrogen ions across the cell membranes of the parietal cells. The highly acidic environment of the stomach destroys most ingested microorganisms.

Gastric mucus is secreted in two forms: a soluble form produced by the gastric glands and an insoluble form produced by the surface and foveolar lining cells. Biochemically, the mucus is a complex glycoprotein consisting of a protein core with branched carbohydrate side chains. Histochemically, gastric mucin is almost entirely neutral, although the mucous neck cells secrete small amounts of sulfomucin and sialomucin (15). By immunohistochemistry, the normal stomach contains mucin core proteins MUC5, MUC1, and MUC6 (24). MUC5 is present in foveolar epithelium and mucous neck cells. MUC1 is essentially epithelial membrane antigen; it is present in the foveolar epithelium, as well as chief and parietal cells. MUC6 is present in the antral glands and mucous neck cells. The exact physiologic roles of gastric mucin subtypes remain undetermined, although soluble mucin likely facilitates lubrication. Mucin forms a surface coating with alternating layers of MUC5 and MUC6 (25). This creates a barrier that, along with bicarbonate secreted by the superficial epithelial cells, prevents back diffusion of acid and gastric autodigestion. The actual structural barrier is formed by the continuous layer of luminal mucosal cells and the tight junctions between them. This process is likely modulated by prostaglandins, which promote mucosal blood flow.

SPECIAL TECHNIQUES AND PROCEDURES

A simple H&E stain is sufficient for evaluation of most gastric samples, but there are instances for which special techniques and procedures can have utility. Special stains can be useful to highlight the mucin composition in gastric samples. The PAS/Alcian blue stain can be used to distinguish gastric and intestinal-type mucins, highlighting intestinal metaplasia and distinguishing it from its mimics. Neutral mucins of the stomach stain magenta while acidic mucins of goblet cells stain light blue, and combinations of both mucin types stain purple. PAS stains without Alcian blue are not particularly useful as they do not distinguish between mucin types. Likewise, mucicarmine stains do not distinguish between mucin types and can be negative in some acidic mucins. A combined high-iron diamine and Alcian blue stain can distinguish sialomucins from sulfomucins, as the former stain light blue and the latter are colonic-type mucins that stain black. These stains are infrequently used because the clinical value of mucin subtyping has not been convincingly shown.

Usually, a well-prepared H&E stain is sufficient to distinguish chief and parietal cells. Special stains, such as a Maxwell stain (26), can help with this distinction, but are not of clinical use in daily practice. Parietal cells can also be highlighted and quantified with the use of an antibody for human milk fat globulin (27).

AGE CHANGES

Reduced gastric acid production occurs with aging and is generally due to a reduction of oxyntic gland mass with expansion of pyloric-type glands. The expansion of the pyloric-type glands in the proximal stomach accompanying

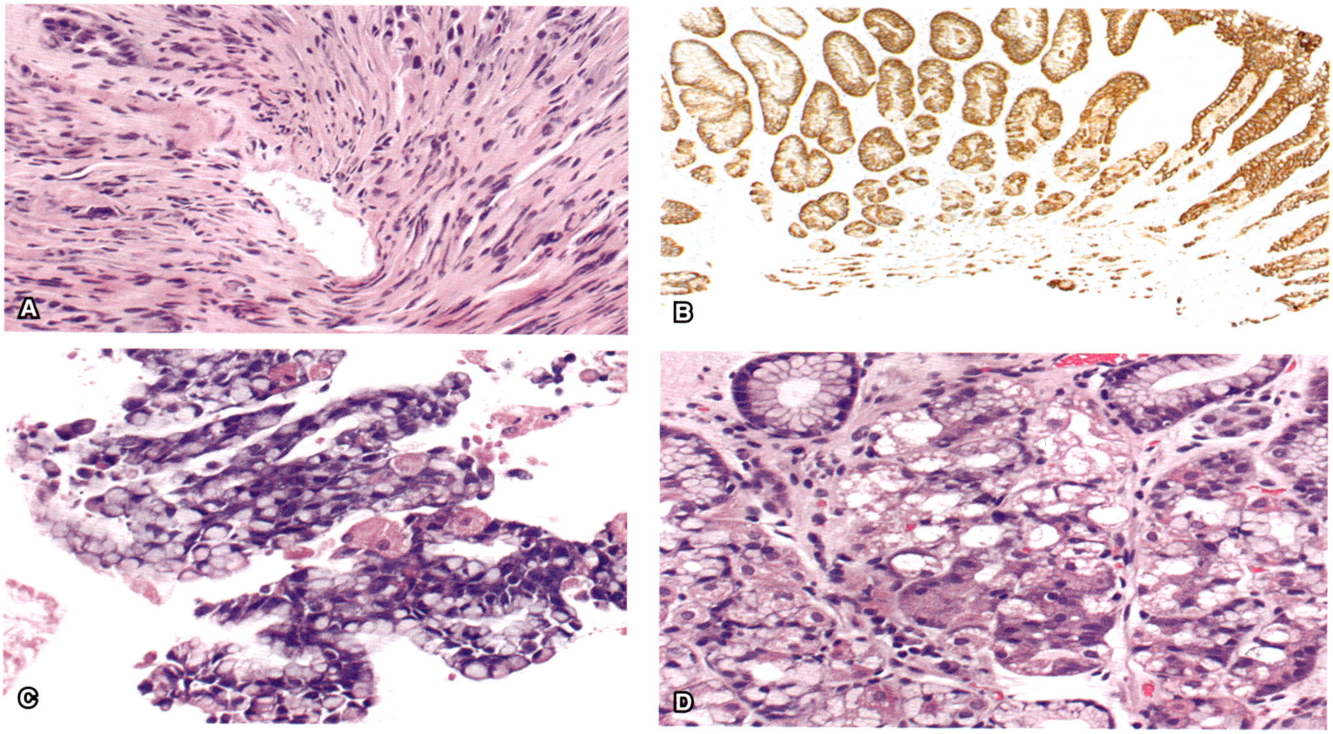

FIGURE 22.12 Biopsy artifacts: **A:** Crushing/stretching artifact at the base of the mucosa can result in overestimation of inflammation and concern for infiltrating single cells. **B:** Pancytokeratin immunohistochemical stain in crushed/stretched tissue can highlight single epithelial cells that have been artifactually separated from the rest of the glands. Note that these cells do not appear enlarged and caution should be taken to not overinterpret these cells. **C:** Sloughed/crushed mucous neck cells can have a pseudosignet ring cell appearance. A diagnosis of signet ring cell adenocarcinoma should be reserved for when invasion is observed in intact tissue. **D:** Vacuolation of parietal cells in oxyntic mucosa may represent a degenerative change.

mucosal atrophy is most likely due to chronic gastritis, which is very common and often clinically silent, rather than aging *per se* (28).

ARTIFACTS

A multitude of artifacts can alter the histology of gastric samples (Fig. 22.12). These artifacts mostly occur due to rough handling either at the time of endoscopic collection or by forceps at the grossing bench. Crush artifact compresses the lamina propria and this crushed appearance can cause the lamina propria to be misinterpreted as a significant inflammatory infiltrate. The glands can also be crushed in a fashion that leads to telescoping of the foveolar lining cells. Stretching gastric mucosa causes increased space between the glands, imparting a false impression of edema. Rough handling can cause procedure-related lamina propria hemorrhage that needs to be distinguished from true hemorrhagic gastritis. While this can be challenging in small biopsies, usually the microscopic appearance of hemorrhagic gastritis is characteristic (superficial epithelial damage with erosions) in acute hemorrhagic gastritis. Sometimes the mucous neck cells in crushed mucosa can extrude and mimic signet ring cell adenocarcinoma (29).

DIFFERENTIAL DIAGNOSIS

It is appropriate to discuss a few patterns of gastritis as they can contain features and cell types that are also seen in normal mucosa, but these patterns differ from those of normal mucosa in the distribution and quantities of the inflammatory cells and in the composition of the epithelium. *H. pylori* gastritis is an infection that is often encountered in routine practice. While scattered plasma cells can be seen in the superficial lamina propria, the early stages of *H. pylori* infection are characterized by a marked expansion of the superficial lamina propria by numerous plasma cells well above the quantity seen in the normal stomach (Fig. 22.13), imparting a blue appearance at scanning magnification. Higher-power examination of the H&E slide may reveal patchy acute inflammation of the gastric glands. Abundant organisms can often be seen on the mucosal surface using routine H&E-stained sections (Fig. 22.14) and ancillary techniques are not indicated when this is the case (30). However, organisms may be sparse in patients with partially treated infections or extensive intestinal metaplasia and identification can be enhanced by using immunohistochemical stains; histochemical stains such as Giemsa, Diff-Quik®, methylene blue, and Warthin Starry do not substantially improve

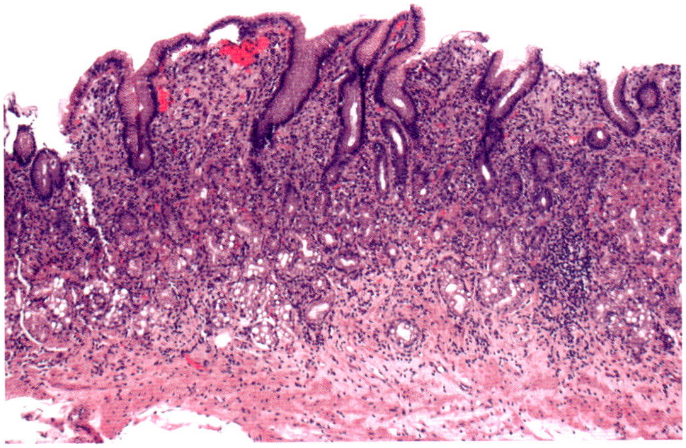

FIGURE 22.13 Low-power magnification of gastric mucosa with *Helicobacter pylori* gastritis. The thick band of plasma cells separating the surface epithelium from the glands seen at low power is strong clue for *H. pylori* gastritis.

organism detection over H&E when only a few organisms are present (Fig. 22.15A) (30). As infection progresses, inflammation can cause atrophy of gastric glands and replacement with intestinal metaplasia. Occasionally, *H. pylori* antigen (Fig. 22.15B) can be detected in the lymphoid follicles of successfully treated patients despite an absence of remaining viable organisms (31).

An especially difficult challenge for pathologists is developing a threshold for distinguishing chronic gastritis

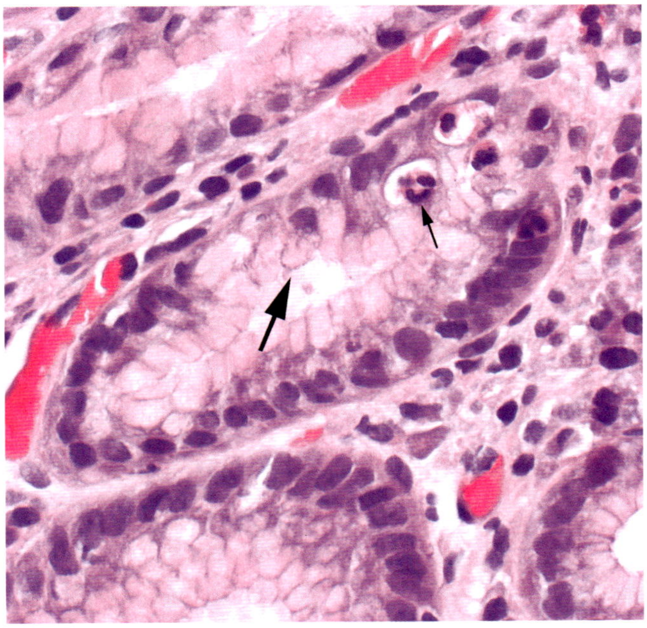

FIGURE 22.14 High-power photomicrograph of *Helicobacter pylori* gastritis. Comma-shaped organisms (*large arrow*) and active inflammation (*small arrow*) are present in the gland.

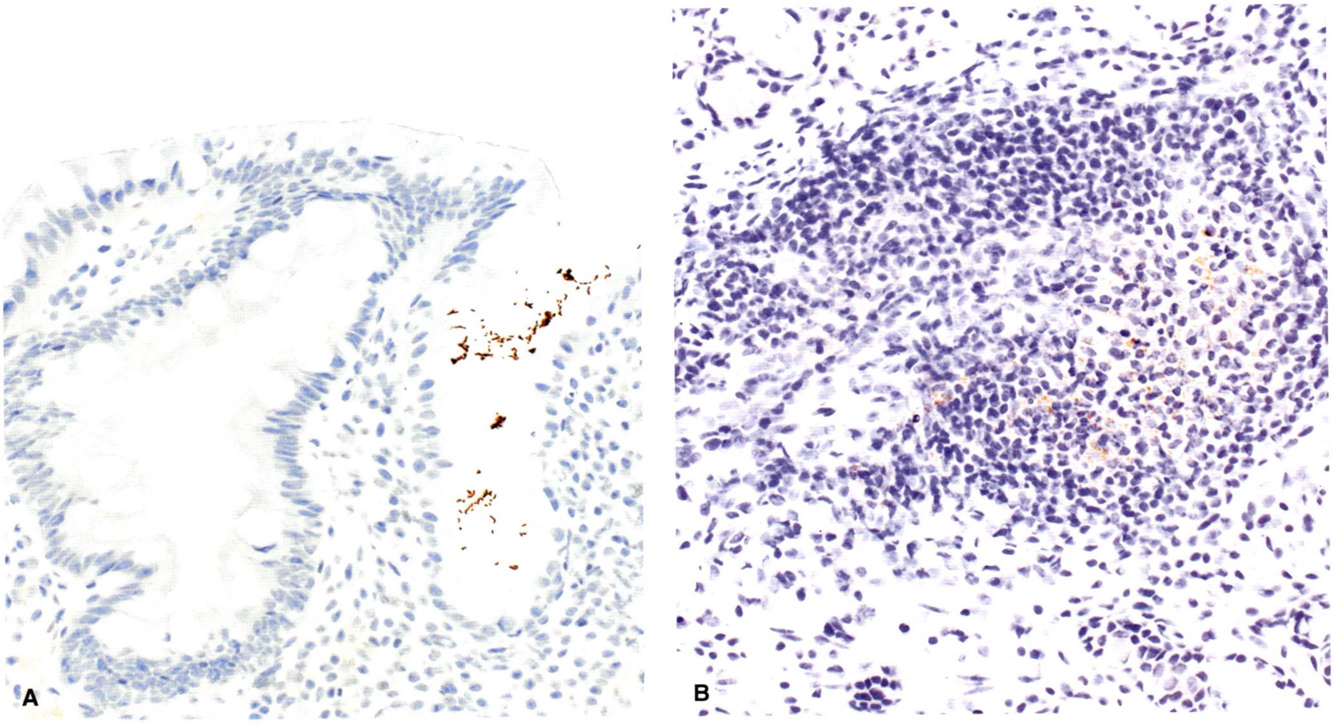

FIGURE 22.15 **A:** An immunohistochemical stain for *Helicobacter pylori* highlights organisms at the surface gastric foveolar epithelium. Note the absence of organisms in the area with intestinal metaplasia on the left. **B:** An immunohistochemical stain for *H. pylori* shows staining of *H. pylori* antigen in a lymphoid follicle in a treated case of *H. pylori* gastritis without residual organisms.

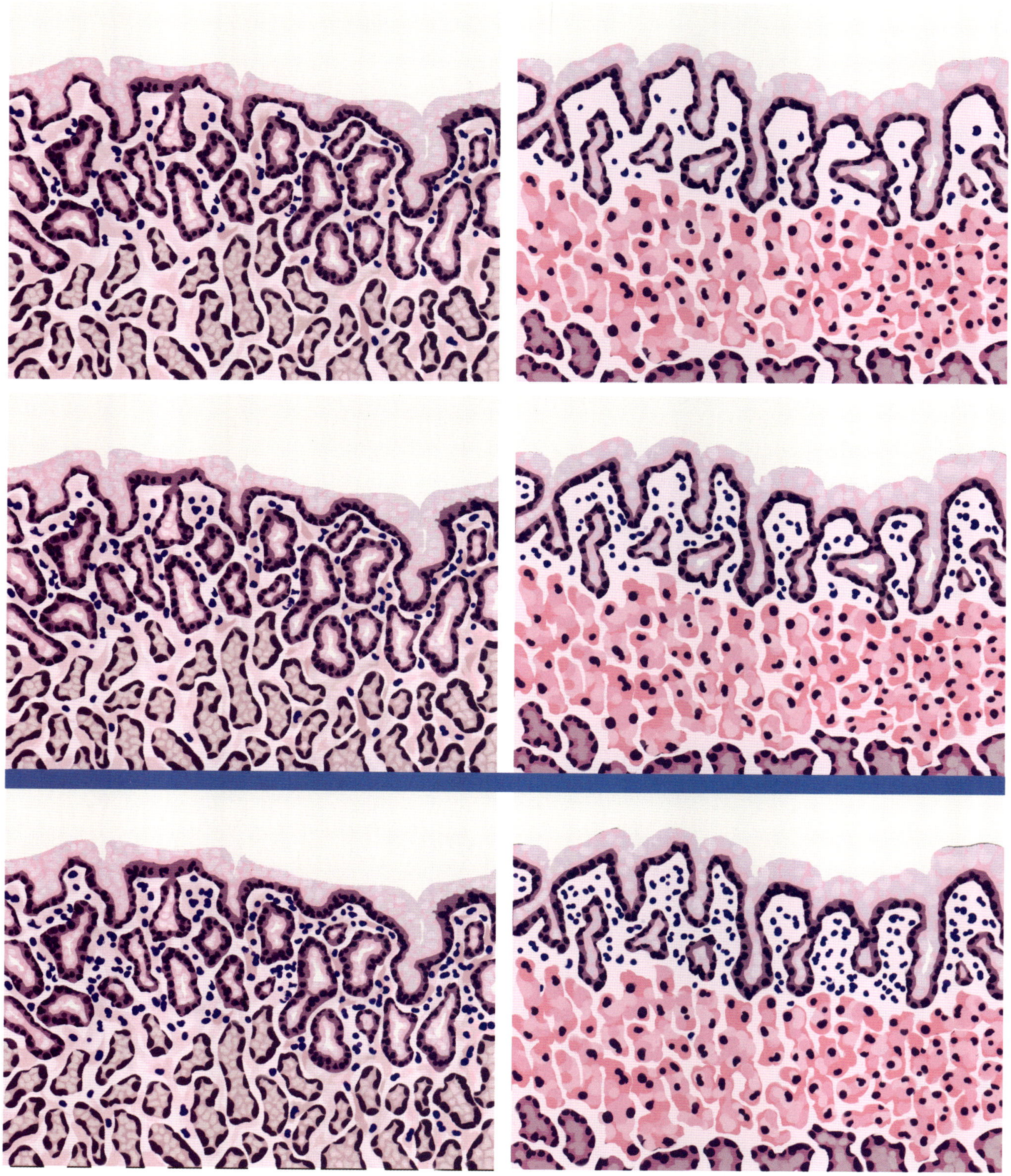

FIGURE 22.16 Visual analog scale of increasing inflammation in antral (**left column**) and oxyntic (**right column**) mucosa. The *blue line* denotes the cutoff for ordering a special stain for *Helicobacter pylori* when organisms cannot be detected on an H&E stain. (Figure adapted from Boylan et al. (32)).

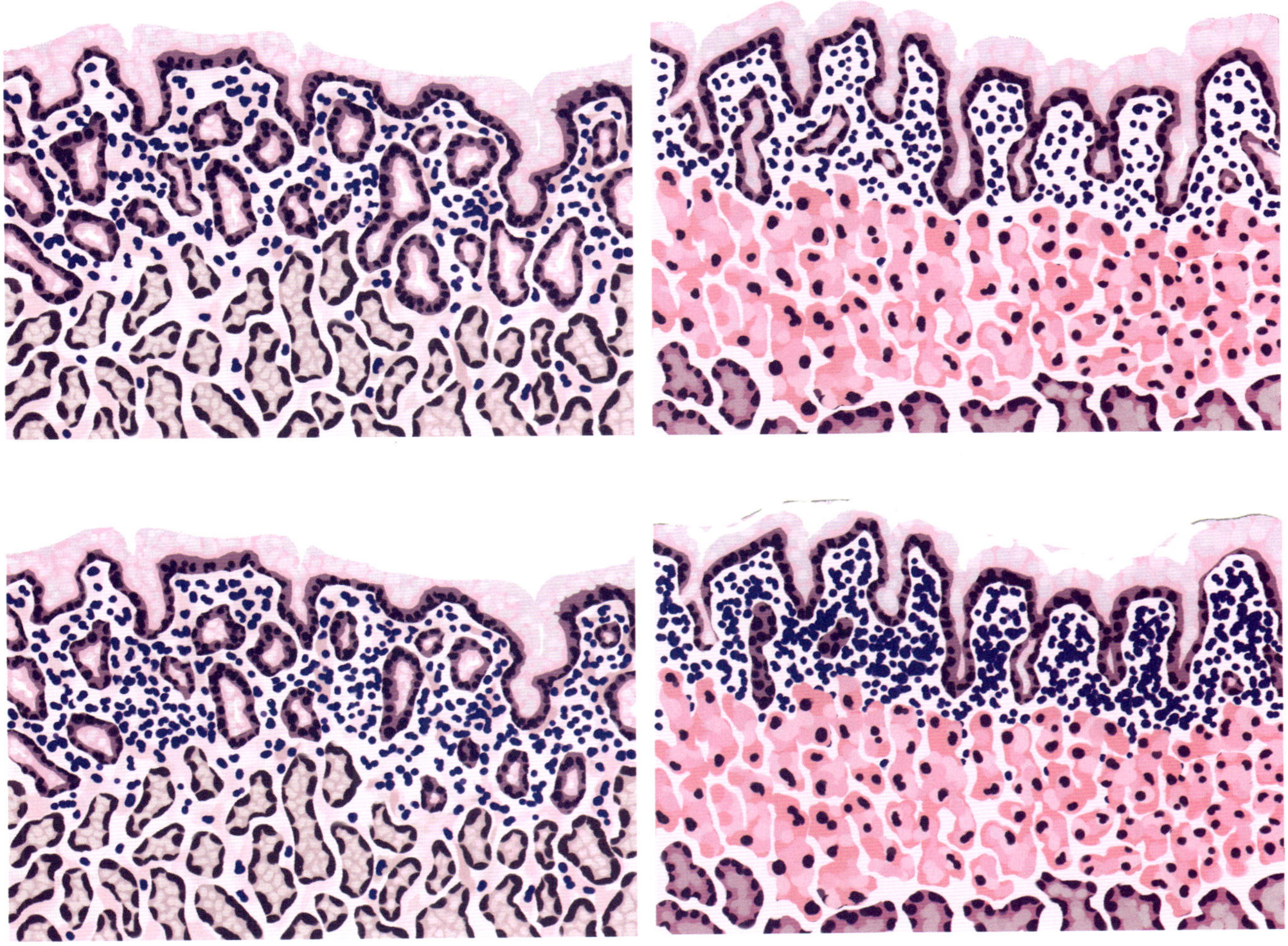

FIGURE 22.16 (*Continued*)

from normal samples because scattered chronic inflammatory cells are present in the superficial lamina propria of normal stomachs. An analog scale was created to suggest a cut point for lamina propria cellularity in order to determine when to obtain immunohistochemical stains for *H. pylori* (Fig. 22.16) (32). Evaluation of the epithelium can also aid in this situation as reactive changes (mucin loss, increased nuclear size, and prominent nucleoli) can be seen in early mild chronic gastritis. Increased mitotic figures in the basal glands indicate increased cell turnover in mucosa regenerating from injury. These features are not specific to a cause of injury and can be seen in any form of gastritis including reactive/chemical gastropathy.

Atrophy of the mucosal glands can be accompanied by increased inflammatory cells in longstanding chronic gastritis. Atrophy is visualized as space between a diminished number of glands. It can be difficult to confidently diagnose mild degrees of atrophy when there is significant artifact, or the precise anatomic location of the biopsy is not known. In these instances, a reticulin stain can confirm atrophy by demonstrating coarse condensation of fibers in the lamina propria.

Chemical, or reactive, gastropathy occurs when there is increased cell turnover at the gastric foveolar surface. This pattern of injury is due to mucosal irritation by acid or bile and represents the final common pathway of a variety of insults that lead to diminished mucosal barrier function. Common etiologies include medications (nonsteroidal anti-inflammatory drugs, doxycycline, and oral iron supplements), alcohol, and slow blood flow through the mucosa in the context of portal hypertension. The condition is classified as a gastropathy rather than gastritis because it is characterized by epithelial injury in the absence of inflammation. In fact, chemical gastropathy cases generally lack infiltrating inflammatory cells unless there is erosion/ulceration. Biopsy features include cytoplasmic depletion of foveolar epithelium with a corkscrew appearance, decreased cytoplasmic mucin, and variable degrees of nuclear atypia (Fig. 22.17) that may raise concern for a neoplasm. Clues to the nonneoplastic nature of this process include the lack of an abrupt transition

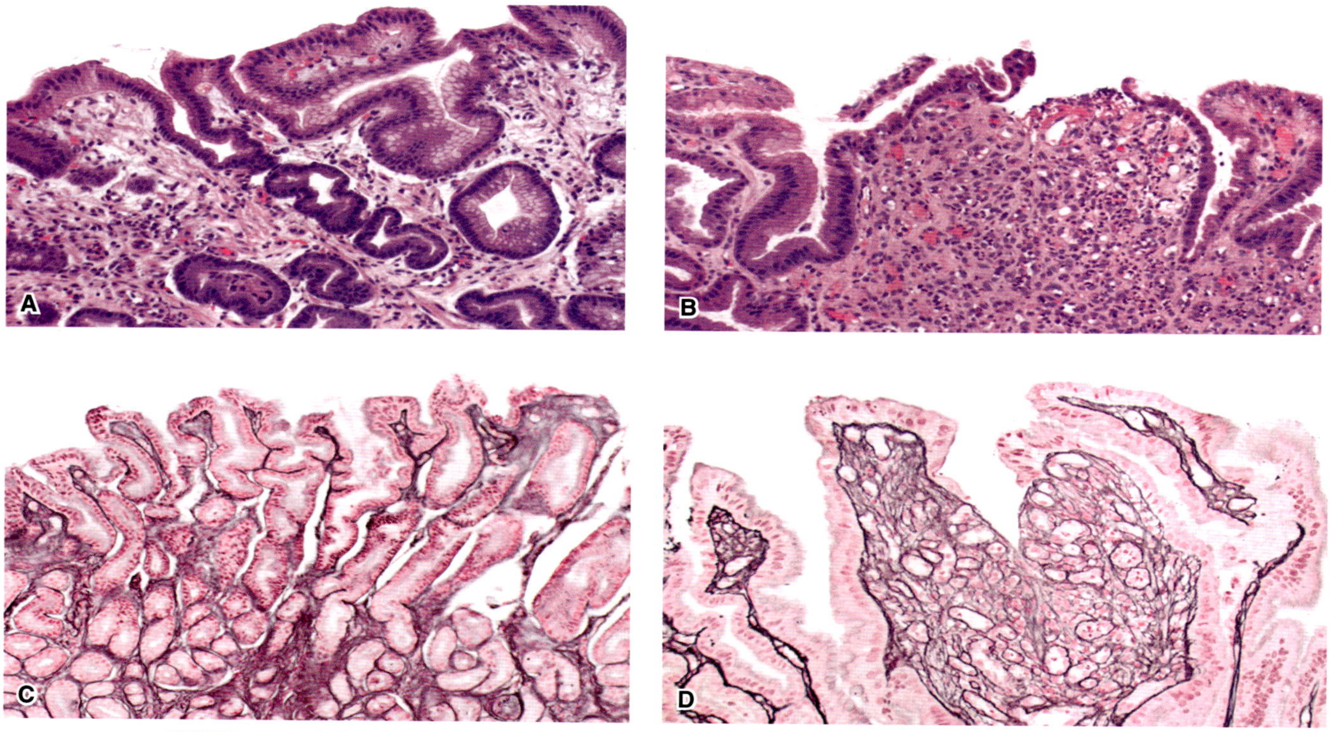

FIGURE 22.17 **A:** Chemical/reactive gastropathy showing twisted "corkscrew" surface foveolar epithelium and extension of strands of the muscularis mucosae into the superficial lamina propria. Note the relative paucity of inflammatory cells compared to other types of gastritis. **B:** Erosion in chemical/reactive gastropathy with increased capillaries and fibrin at the surface. **C:** Reticulin special stain in chemical/reactive gastropathy shows dense fibers in the spaces between the glands. **D:** A reticulin special stain highlights dense fibers in the area of erosion.

between the atypical foci and background mucosa, retained cell polarity, and even distribution of small apical mucin vacuoles at the luminal surface (Fig. 22.18) (33).

Metaplasia

Intestinal metaplasia and pyloric (pseudopyloric) metaplasia are the two most frequently encountered types of metaplasia in the stomach; pancreatic metaplasia is also common in the setting of autoimmune gastritis (34). In pyloric metaplasia, neutral mucus-secreting glands replace the specialized acid- and enzyme-secreting cells of oxyntic glands. This metaplasia begins at the junction of antral and oxyntic mucosae near the incisura and can extend proximally into body and fundus. As this change occurs, the amount of pyloric mucosa expands while the quantity of oxyntic mucosa diminishes (26). As the pyloric metaplasia resembles normal mucosa of the antrum, this change may be difficult to recognize without precise knowledge of the anatomic location of the sample. Though the metaplastic fundic glands lose zymogenic and parietal cells, they retain pepsinogen I activity. Though not of clinical value, immunohistochemical methods can demonstrate retained pepsinogen activity to identify areas of pyloric metaplasia (35).

Intestinal metaplasia affects cells of the surface and deep glandular epithelia resulting in mimicry of either the small or large intestines; intestinal metaplasia may be complete (type I) or incomplete (type II) (Fig. 22.19) (35). Complete intestinal metaplasia resembles the small intestine; fully developed goblet cells are accompanied by enterocytes with a brush border. Advanced cases show alterations in the contour of the mucosa with the development of villi

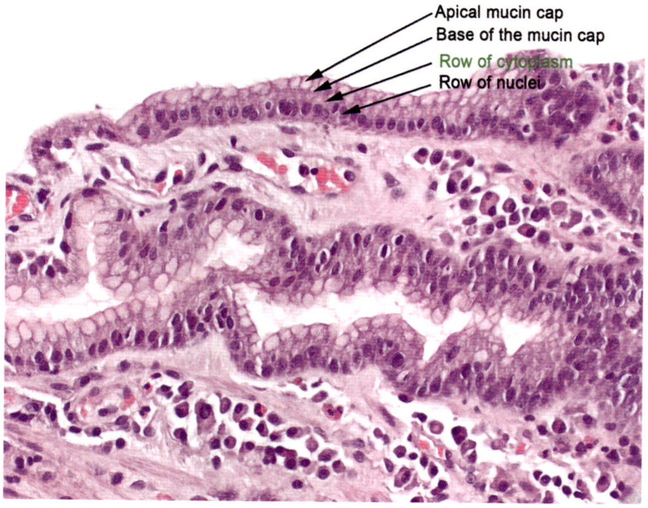

FIGURE 22.18 Cell-to-cell polarity is maintained in chemical/reactive gastropathy with the four lines: Line 1 the apical mucin cap; Line 2 the base of the mucin cap; Line 3 the row of cytoplasm; and Line 4 the row of nuclei.

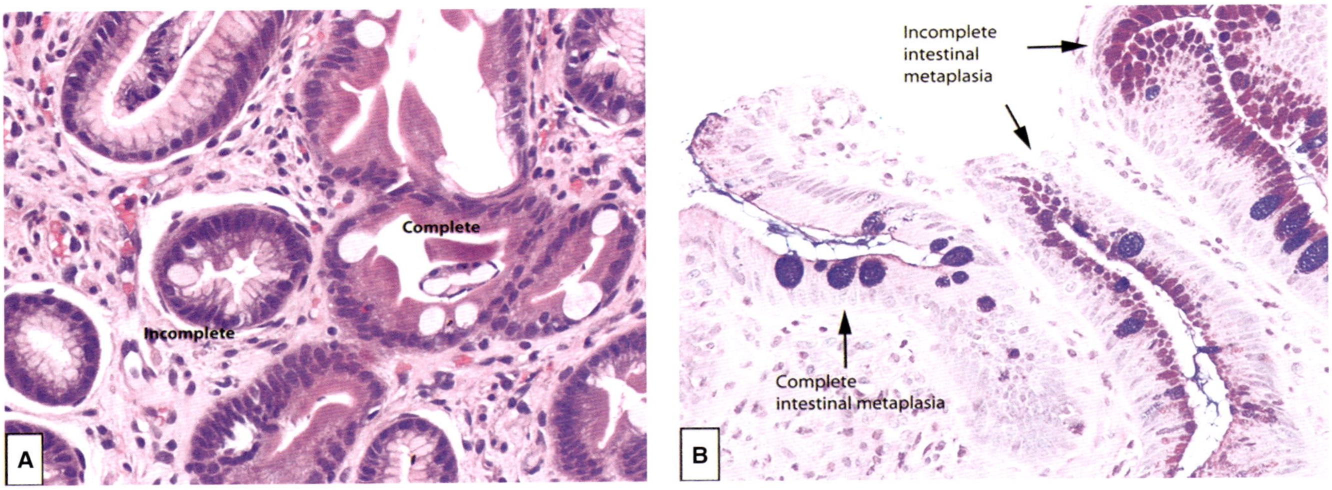

FIGURE 22.19 **A:** The area of complete intestinal metaplasia shows only intestinal-type cells while the area of incomplete IM shows a mixture of intestinal and gastric surface mucin. **B:** A PAS-Alcian blue stain is useful to highlight IM. The area of complete IM shows only the blue acidic intestinal mucin (goblet cells) while the area of incomplete IM shows a mixture of blue goblet cells and magenta neutral gastric mucin.

and crypts. Paneth cells can be seen in the bases of the glands. Incomplete intestinal metaplasia lacks associated absorptive cells. The epithelium consists of a mixture of intestinal-type goblet cells and columnar mucus-secreting cells resembling normal gastric epithelium. However, the background epithelium contains a combination of mucins not normally seen in the stomach as described below.

Histochemical changes characterizing the mucus carbohydrate composition in the various types of intestinal metaplasia are complicated (15,36). The mucus secreted by the normal stomach is neutral type, recognized histochemically as PAS positive and Alcian blue negative (stains magenta). In complete intestinal metaplasia, the enterocyte cytoplasm, apart from the brush border, is mucin-negative, but the goblet cells secrete either sialomucin (a PAS-positive acid mucin that is Alcian blue positive at pH 2.5, and negative at pH 0.5) or sulfomucin (a strongly acidic colonic type mucin that is weakly PAS positive and Alcian blue positive at pH 2.5 and pH 0.5). In incomplete small intestinal metaplasia, sialomucin is present in the nongoblet columnar cells, and in incomplete large bowel metaplasia (also called type III metaplasia) (34), the nongoblet columnar cells contain sulfomucin (Fig. 22.20). Positive staining of sulfomucin with high-iron diamine can help distinguish it from sialomucin (37). The details of these methods are well described in standard textbooks of histochemistry and mirror those described in the 1960s (38).

Mucin core protein expression is altered in the setting of intestinal metaplasia. Normal gastric surface foveolar epithelium expresses MUC5 and lesser amounts of MUC1. By contrast, complete gastric intestinal metaplasia no longer expresses MUC5 and MUC1 and the goblet cells instead express MUC2 (intestinal mucin). Incomplete intestinal metaplasia shows a mixture of mucin core proteins. Mucin core protein expression is also different in the intestinal metaplasia of Barrett esophagus and of the gastric cardia, but the differences are inconsistent and adding immunohistochemistry is not clinically useful (22).

Varying amounts of gastric intestinal metaplasia are often encountered by practicing pathologists in routine gastric biopsies. Though intestinal metaplasia is not a rare histologic finding, it is a reaction to chronic gastric injury and is not a component of normal histology. The types of gastric intestinal metaplasia are often mixed within a biopsy set and are not usually seen as one homogeneous type except in autoimmune gastritis, in which case essentially all the intestinal metaplasia is of the complete type unless the patient has also had infection with *H. pylori* (34).

Rarely seen alterations include subnuclear vacuolation that indents the nucleus (39) and ciliated metaplasia in pyloric mucus glands (40). As it does not approximate

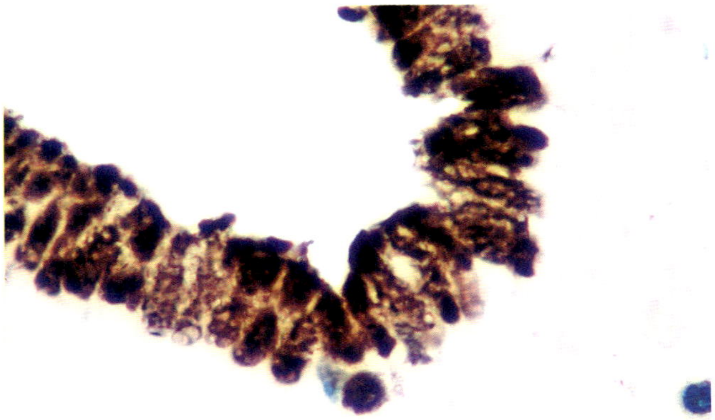

FIGURE 22.20 Incomplete IM of the large bowel. The gland contains cells with cytoplasmic sulfomucin (high-iron diamine and Alcian blue). (This figure is courtesy of Dr. David A. Owen from a previous version of this chapter.)

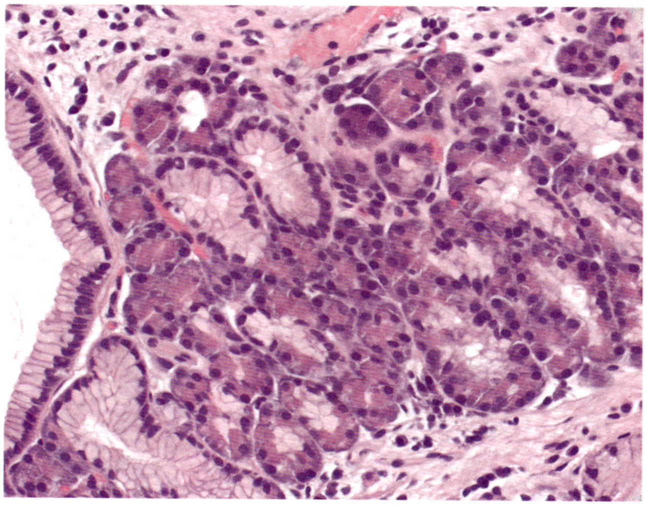

FIGURE 22.21 High-power magnification of pancreatic metaplasia/heterotopia. The acinar cells are morphologically identical to those seen in the pancreas and also produce a lipase and trypsinogen. Pancreatic metaplasia is most frequently encountered in the setting of autoimmune metaplastic atrophic gastritis in oxyntic mucosa and in reflux esophagitis in cardiac mucosa.

a cell type from another location, subnuclear vacuolation is not considered a type of metaplasia and is thought to likely be a degenerative change. Ultrastructurally, the vacuoles consist of a membrane-lined space derived either from endoplasmic reticulum or Golgi and probably contain nonglycoconjugated mucus core protein (41). In the presence of superficial intestinal metaplasia, ciliated cells can be seen at the base of antral glands (40). The mechanism of this change and whether it has any clinical significance is not known.

As above, pancreatic acinar metaplasia (42) can be seen in up to 1.2% of gastric biopsy specimens or 13% of gastrectomy specimens and is particularly common in persons with autoimmune gastritis and immunologic loss of parietal cells (34). The foci of metaplasia resemble normal pancreatic acinar cells and also produce lipase and trypsinogen (Fig. 22.21). Amylase is expressed in 75% of instances. Acinar cells can also be present in patchy nests and lobules in cardiac and oxyntic mucosae where they are probably heterotopic, rather than metaplastic in nature. Islets of Langerhans are absent from pancreatic metaplasias and heterotopias in the gastric mucosa.

SPECIMEN HANDLING

Stomach mucosa is fragile and should be handled delicately. Ideally, the sample should be oriented on filter paper or gelfoam, but this is seldom done. Formalin is most commonly used to fix gastric biopsies, but other fixatives perform well. The biopsies are cut in ribbons and typically two to six levels are included on each slide.

To achieve optimal sections, it is recommended that opened gastrectomy specimens be pinned flat on a corkboard or wax platform before being submerged in formalin for overnight fixation. While fresh sections can be taken, they curl, leading to distortion/loss of orientation on the slide.

REFERENCES

1. Amin M, Edge S, Greene F, et al. *AJCC Cancer Staging Manual*. 8th ed. Springer International Publishing; 2017.
2. Owen DA. Normal histology of the stomach. *Am J Surg Pathol*. 1986;10(1):48–61.
3. Mackintosh CE, Kreel L. Anatomy and radiology of the areae gastricae. *Gut*. 1977;18(11):855–864.
4. Piasecki C. Blood supply to the human gastroduodenal mucosa with special reference to the ulcer-bearing areas. *J Anat*. 1974;118(Pt 2):295–335.
5. Lehnert T, Erlandson RA, Decosse JJ. Lymph and blood capillaries of the human gastric mucosa. A morphologic basis for metastasis in early gastric carcinoma. *Gastroenterology*. 1985;89(5):939–950.
6. Listrom MB, Fenoglio-Preiser CM. Lymphatic distribution of the stomach in normal, inflammatory, hyperplastic, and neoplastic tissue. *Gastroenterology*. 1987;93(3):506–514.
7. Schmidt B, Yoon SS. D1 versus D2 lymphadenectomy for gastric cancer. *J Surg Oncol*. 2013;107(3):259–264.
8. Filipe MI. Mucins in the human gastrointestinal epithelium: a review. *Invest Cell Pathol*. 1979;2(3):195–216.
9. Chandrasoma PT, Der R, Ma Y, Dalton P, Taira M. Histology of the gastroesophageal junction: an autopsy study. *Am J Surg Pathol*. 2000;24(3):402–409.
10. Sarbia M, Donner A, Gabbert HE. Histopathology of the gastroesophageal junction: a study on 36 operation specimens. *Am J Surg Pathol*. 2002;26(9):1207–1212.
11. Kilgore SP, Ormsby AH, Gramlich TL, et al. The gastric cardia: fact or fiction? *Am J Gastroenterol*. 2000;95(4):921–924.
12. Zhou H, Greco MA, Daum F, Kahn E. Origin of cardiac mucosa: ontogenic consideration. *Pediatr Dev Pathol*. 2001;4(4):358–363.
13. Glickman JN, Chen YY, Wang HH, Antonioli DA, Odze RD. Phenotypic characteristics of a distinctive multilayered epithelium suggests that it is a precursor in the development of Barrett's esophagus. *Am J Surg Pathol*. 2001;25(5):569–578.
14. Glickman JN, Spechler SJ, Souza RF, Lunsford T, Lee E, Odze RD. Multilayered epithelium in mucosal biopsy specimens from the gastroesophageal junction region is a histologic marker of gastroesophageal reflux disease. *Am J Surg Pathol*. 2009;33(6):818–825.
15. Goldman H, Ming SC. Mucins in normal and neoplastic gastrointestinal epithelium. Histochemical distribution. *Arch Pathol*. 1968;85(6):580–586.
16. Matsuyama M, Suzuki H. Differentiation of immature mucous cells into parietal, argyrophil, and chief cells in stomach grafts. *Science*. 1970;169(3943):385–387.
17. Grimelius L. A silver nitrate stain for alpha-2 cells in human pancreatic islets. *Acta Soc Med Ups*. 1968;73(5–6):243–270.
18. Rindi G, Buffa R, Sessa F, Tortora O, Solcia E. Chromogranin A, B and C immunoreactivities of mammalian endocrine cells. Distribution, distinction from costored

hormones/prohormones and relationship with the argyrophil component of secretory granules. *Histochemistry*. 1986;85(1):19–28.

19. Solcia E, Fiocca R, Villani L, Luinetti O, Capella C. Hyperplastic, dysplastic, and neoplastic enterochromaffin-like-cell proliferations of the gastric mucosa. Classification and histogenesis. *Am J Surg Pathol*. 1995;19(Suppl 1):S1–S7.
20. Genta RM, Hamner HW, Graham DY. Gastric lymphoid follicles in Helicobacter pylori infection: frequency, distribution, and response to triple therapy. *Hum Pathol*. 1993;24(6):577–583.
21. Bowden RE, el-Ramli HA. The anatomy of the oesophageal hiatus. *Br J Surg*. 1967;54(12):983–989.
22. Rubin W, Ross LL, Sleisenger MH, Jefries GH. The normal human gastric epithelia. A fine structural study. *Lab Invest*. 1968;19(6):598–626.
23. Forte JG, Forte TM, Black JA, Okamoto C, Wolosin JM. Correlation of parietal cell structure and function. *J Clin Gastroenterol*. 1983;5(Suppl 1):17–27.
24. Glickman JN, Shahsafaei A, Odze RD. Mucin core peptide expression can help differentiate Barrett's esophagus from intestinal metaplasia of the stomach. *Am J Surg Pathol*. 2003;27(10):1357–1365.
25. Ho SB, Takamura K, Anway R, Shekels LL, Toribara NW, Ota H. The adherent gastric mucous layer is composed of alternating layers of MUC5AC and MUC6 mucin proteins. *Dig Dis Sci*. 2004;49(10):1598–1606.
26. Maxwell A. The Alcian dyes applied to gastric mucosa. *Stain Technol*. 1963;38:286–287.
27. Walker MM, Smolka A, Waller JM, Evans DJ. Identification of parietal cells in gastric body mucosa with HMFG-2 monoclonal antibody. *J Clin Pathol*. 1995;48(9):832–834.
28. Kekki M, Samloff IM, Ihamaki T, Varis K, Siurala M. Age- and sex-related behaviour of gastric acid secretion at the population level. *Scand J Gastroenterol*. 1982;17(6):737–743.
29. Arnason T, Lauwers GY. Extruded highly proliferative benign mucous neck cells: a peculiar histologic mimic of poorly cohesive gastric carcinoma. *Int J Surg Pathol*. 2014;22(7):623–628.
30. Batts KP, Ketover S, Kakar S, et al. Appropriate use of special stains for identifying Helicobacter pylori: recommendations from the Rodger C. Haggitt Gastrointestinal Pathology Society. *Am J Surg Pathol*. 2013;37(11):e12–e22.
31. Nelson ND, Tondon R, Fortuna D, Westerhoff M, Swanson PE, Furth E. Helicobacter pylori antigen but not the organism is occasionally present within germinal centers: implications for patient management and biology. *Am J Surg Pathol*. 2020;44(11):1528–1234.
32. Boylan KE, Patrey S, McMullen PD, et al. Objective visual analog scale for biopsy diagnosis of Helicobacter pylori infection in clinical practice. *Am J Surg Pathol*. 2021;45(5):672–679.
33. Waters KM, Salimian KJ, Assarzadegan N, et al. Cell polarity (the 'four lines') distinguishes gastric dysplasia from epithelial changes in reactive gastropathy. *Histopathology*. 2021;78(3):453–458.
34. Poveda JC, Chahar S, Garcia-Buitrago MT, Montgomery EA, McDonald OG. The morphologic spectrum of gastric type 1 enterochromaffin-like cell neuroendocrine tumors. *Mod Pathol*. 2023;36(4):100098.
35. Dixon MF, Genta RM, Yardley JH, Correa P. Classification and grading of gastritis. The updated Sydney System. International Workshop on the Histopathology of Gastritis, Houston 1994. *Am J Surg Pathol*. 1996;20(10):1161–1181.
36. Jass JR, Filipe MI. The mucin profiles of normal gastric mucosa, intestinal metaplasia and its variants and gastric carcinoma. *Histochem J*. 1981;13(6):931–939.
37. Filipe MI, Potet F, Bogomoletz WV, et al. Incomplete sulphomucin-secreting intestinal metaplasia for gastric cancer. Preliminary data from a prospective study from three centres. *Gut*. 1985;26(12):1319–1326.
38. Spicer SS. Diamine methods for differentiating mucosubstances histochemically. *J Histochem Cytochem*. 1965;13:211–234.
39. Rubio CA, Slezak P. Foveolar cell vacuolization in operated stomachs. *Am J Surg Pathol*. 1988;12(10):773–776.
40. Rubio C, Hayashi T, Stemmermann G. Ciliated gastric cells: a study of their phenotypic characteristics. *Mod Pathol*. 1990; 3(6):720–723.
41. Thompson IW, Day DW, Wright NA. Subnuclear vacuolated mucous cells: a novel abnormality of simple mucin-secreting cells of non-specialized gastric mucosa and Brunner's glands. *Histopathology*. 1987;11(10):1067–1081.
42. Doglioni C, Laurino L, Dei Tos AP, et al. Pancreatic (acinar) metaplasia of the gastric mucosa. Histology, ultrastructure, immunocytochemistry, and clinicopathologic correlations of 101 cases. *Am J Surg Pathol*. 1993;17(11):1134–1143.

23

Small Intestine

Nicole C. Panarelli

GROSS ANATOMY AND SURGICAL PERSPECTIVE

The small intestine is a coiled, tubular segment of intestine that extends from the gastric pylorus to the ileocecal valve. It spans approximately 7 m in adults and is divided into three segments, namely the duodenum, jejunum, and ileum. The duodenum comprises the proximal 20 to 25 cm, beginning at the pylorus at the level of the first lumbar vertebra and ending at the duodenojejunal flexure. It forms a C shape around the head of the pancreas and is subdivided into the first, second, third, and fourth segments. The first segment spans approximately 5 cm and includes the duodenal bulb. It is the most superior segment of the duodenum, lies within the peritoneum, and is connected to the undersurface of the liver by the hepatoduodenal ligament. The second portion is approximately 10 cm long, descends retroperitoneally, and contains the major papilla into which the main pancreatic duct and common bile duct drain at the ampulla of Vater, as well as the minor papilla into which the accessory pancreatic duct drains (1). The third portion travels horizontally for approximately 8 cm in the retroperitoneum, and the fourth portion ascends anteriorly for 2 cm at the level of the second lumbar vertebra, just left of midline (2).

The origin of the jejunum is marked by the ligament of Treitz, a strip of fibromuscular tissue that anchors the terminal duodenum and the duodenojejunal flexure to the posterior abdominal wall (3). Distal to the ligament of Treitz, the remainder of the small bowel is arbitrarily subdivided into the jejunum and ileum. The jejunum accounts for the proximal two-fifths of intestine and lies within the upper abdominal cavity, while the distal three-fifths comprise the ileum, which terminates at the ileocecal junction within the right iliac fossa in the lower abdominal cavity and pelvis (2,4). No anatomic landmark divides the jejunum from ileum, but several gross features help surgeons identify these segments. The proximal jejunum has a thicker wall and is about twice the diameter of the distal ileum. Jejunal segments have prominent circular mucosal folds (i.e., *plicae circulares* or *valvulae conniventes*) that can be palpated intraoperatively (1,4). The jejunal mesentery is thin and translucent, whereas that of the ileum is thicker and more opaque.

The small intestine derives its blood supply from the celiac and superior mesenteric trunks (5). The duodenum is supplied by branches and interanastomosing arcades of both trunks, and its blood supply is intimately associated with that of the pancreatic head. The jejunum and ileum receive their blood from more distal branches of the superior mesenteric artery (5). The lymphatic and venous drainage systems follow the arterial supply and flow into regional lymphatic vessels and lymph nodes or the portal venous system, respectively. Capillaries and lymphatic lacteals ascend in the villi. Lacteals of adjacent villi fuse to form a wider sinus that eventually drains into the lymphatic network of the submucosa (6).

Sympathetic neural input to the small bowel is carried by the celiac and superior mesenteric plexuses, whereas the parasympathetic supply is derived from distal branches of the vagus nerve; both of these closely follow the arterial paths into the bowel wall (7).

PHYSIOLOGY

The small intestine has several functional roles, the most important of which is the breakdown and absorption of ingested nutrients. Salivary gland secretions, gastric secretions, and pancreatic enzymes break down ingested carbohydrates and proteins to produce more appropriately sized molecules for further digestion in the small intestine. The brush border created by the numerous apical microvilli on absorptive epithelial cells offers an array of aminopeptidases, disaccharidases, and oligosaccharidases which act as key enzymes in completing the process of peptide and carbohydrate hydrolysis (8–10). The resulting monosaccharides, free amino acids, dipeptides, and tripeptides are subsequently absorbed across the epithelial layer, and most pass into the portal venous system for storage or systemic distribution (11).

Fat breakdown is initiated by lingual lipase produced by glands in the tongue and further promoted by gastric lipase. Fat digestion is mostly catalyzed by pancreatic lipases in the lumen of the small intestine, which acts on emulsified droplets of dietary fat admixed with bile salts. The products of digestion, including free fatty acids and monoglycerides, then diffuse across the lipid bilayer membrane of enterocytes (12). Most undergo intracytoplasmic resynthesis to form triglycerides which are eventually combined with cholesterol, phospholipids, and apoproteins to form chylomicrons. Mature chylomicrons exit the Golgi complex for exocytosis across the basolateral enterocyte membrane and subsequent entry into regional lymphatic vessels (11,13).

Water and electrolytes, vitamins, minerals, and various drugs also are absorbed at points along the small bowel mucosa (14). Therefore, the structural integrity of this viscus is critical to the maintenance of nutritional status, as well as in appropriate drug handling. There are also important regional differences in absorption along the length of the bowel. Iron is absorbed proximally whereas vitamin B_{12} (cobalamin) is preferentially absorbed in the ileum. Deficiencies in these substances should prompt investigation for celiac sprue and Crohn disease, respectively.

The small intestine also functions to propel and segmentally mix both newly accepted gastric contents and the residual material left after initial digestive efforts. Intestinal motility is mediated by complex interactions between interstitial cells of Cajal, which are mesenchymal pacemaker cells in the small intestine, smooth muscle cells, and the enteric nervous system (15). With feeding, a distended bowel segment initiates peristalsis and creates forward pulsive motions, whereas a series of slow yet recurring contractions clears the enteric lumen of any residual debris during fasting or between meals. Motilin is an important hormone that is believed to modulate the generation of these migratory motor complexes (16). Other endocrine influences, as well as the autonomic and central nervous systems, play a modulatory role in these intrinsic activities.

The small bowel plays a crucial function in mucosal immunity. The mucosa/gut-associated lymphoid tissues, which are discussed in detail later in this chapter, are important in local defense against microorganisms and generating the initial immunologic responses to pathogens (17).

HISTOLOGY

Although regional histologic differences exist within the small intestine, the general microscopic structure is similar throughout its length. The wall of the small bowel is divided into mucosa, submucosa, muscularis propria (externa), subserosa, and serosa.

Mucosa

Mucosal Architecture and Design

The gross and microscopic architecture of the small intestine maximizes surface area for nutrient absorption (18). The plicae circulares are nearly circumferential folds that course perpendicularly to the longitudinal axis of the bowel (Fig. 23.1) (18,19). These folds are surfaced by mucosa and contain supporting submucosal tissue. In addition to enhancing the surface area, they regulate flow of intraluminal contents, thus increasing the time of contact with absorptive surfaces.

The mucosa comprises epithelium, lamina propria, and muscularis mucosae. A continuous monolayer of epithelium lines the alternating villi and crypts. Villi form leaflike projections over the entire luminal surface and enhance the surface area (Figs. 23.2 and 23.3) (18,20,21). Each villus contains a core of lamina propria, a centrally located, blind-ended lymphatic channel (lacteal), an arteriovenous capillary network, and resident inflammatory cells (Fig. 23.4) (22,23).

The crypts of Lieberkühn are contiguous with and lie between the villi, descending from their bases to the muscularis mucosae (Fig. 23.3). The ratio of villous length to crypt

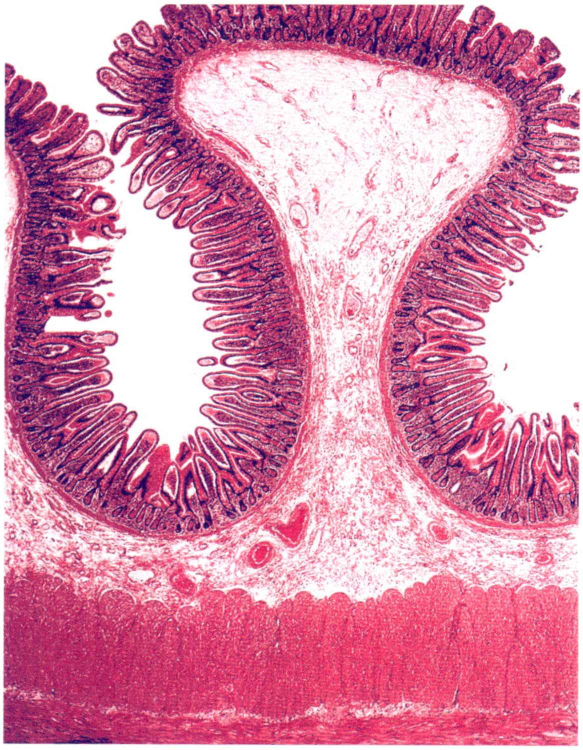

FIGURE 23.1 A single permanent circular fold is composed of a submucosal core and mucosal surface. The absorptive surface area is further augmented by intraluminal mucosal projections (villi).

length in normal small bowel varies from approximately 3:1 in the duodenum to 5:1 in the proximal ileum (Fig. 23.3) (18). The villi become wider and flatter in the terminal ileum, near the transition to the cecum (24). The muscularis mucosae, a thin layer of interrupted smooth muscle cell bundles runs longitudinally and forms the lower limit of the mucosa.

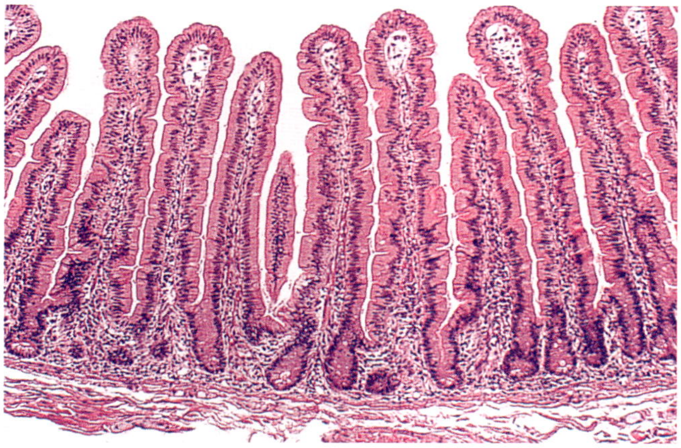

FIGURE 23.3 Normal jejunal villi. These villi are long and slender mucosal projections with a core of lamina propria covered by a luminal epithelial layer. A single row of intestinal glands (crypts) is found at the base of the mucosa. These crypts lie between adjacent villi and are surrounded by the same lamina propria that form the villous cores.

Mucosal Components and Their Composition

› *Epithelium* The distribution and relative density of epithelial cell types differ in the villi and crypts. Crypts represent the regenerative compartment of the epithelium and contain both stem cells and mature epithelial cells. Between 1 and 12 mitotic figures may be seen in crypts and are concentrated at their bases (25,26). Stem cells give rise to the four mature epithelial cell types of the mucosa, namely absorptive cells, goblet cells, endocrine cells, and Paneth cells. Endocrine cells are not derived from the neural crest, nor do they show any evidence of neural differentiation and, thus, their classification as neuroendocrine cells is not appropriate in the tubular gut. Differentiation and maturation occur over the course of 4 to 6 days as the cells

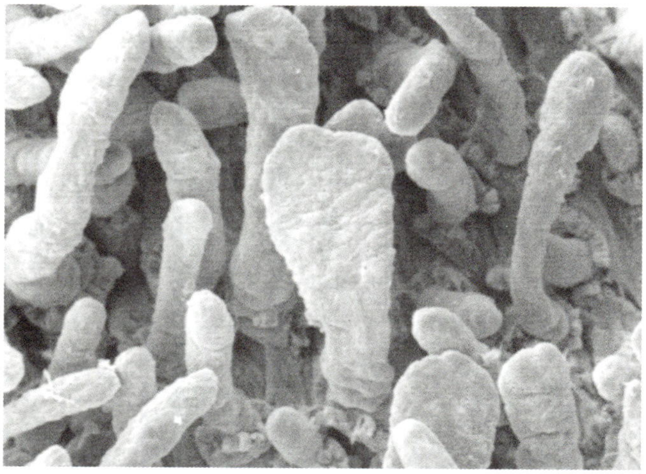

FIGURE 23.2 Scanning electron micrograph of small intestinal mucosa discloses the fingerlike and leaflike appearance of villi. Fingerlike villi predominate in the more distal segments of small bowel (jejunum and ileum), whereas leaflike villi are more common in the duodenum. Mixed populations, as in this micrograph, are considered normal.

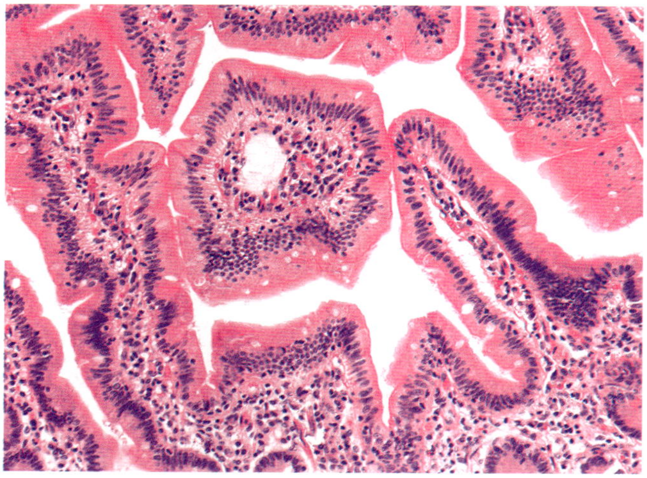

FIGURE 23.4 The duodenal villous surface is covered by a single layer of tall columnar epithelial cells. The underlying lamina propria core contains lymphoid and plasma cells and a connective tissue framework, including a lymphatic vessel (lacteal) and a subepithelial capillary network.

migrate from the regenerative compartments of the crypts to the villous tips, where they are subsequently sloughed into the lumen (27–31). Complex molecular pathways, especially Wnt signaling, play key roles in this proliferation and differentiation (32).

Enteroendocrine cells comprise <1% of the total cell population of the lower gastrointestinal tract; they are most abundant in duodenum, becoming progressively less frequent along the length of the small intestine and colon, then increasing in numbers again in the rectum. These cells produce over 30 different hormone products and, as such, comprise the largest endocrine "organ" in humans. Enteroendocrine cells are typically located in the lower third of crypts and are singly dispersed among other epithelial cell types (33). Ultrastructurally, they are identified by the presence of secretory vesicles, not neurosecretory granules, and apical cytoplasmic processes with microvilli that extend toward the luminal surface. Individual cells containing cholecystokinin, secretin, gastric inhibitory polypeptide, and motilin populate more proximal segments of the small bowel, whereas enteroglucagon-, substance P–, and neurotensin-storing cells are seen with greater frequency in the ileum. Serotonin- and somatostatin-containing cells are present throughout the gastrointestinal tract (34) (Fig. 23.5). All enteroendocrine cells label with immunohistochemical markers of endocrine differentiation (e.g., chromogranin, synaptophysin, INSM1) (Fig. 23.6) (33–36). Three major enteroendocrine cell types populate the small intestine. Enterochromaffin cells are the most abundant; their main secretory product is serotonin which functions in activating excitatory cholinergic neurons and increased smooth muscle contractility. Enterochromaffin cells are pyramidal in shape with their apices oriented toward crypt lumina and nuclei that are usually located in the superficial cytoplasm rather than polarized against the basement membrane. They contain numerous red granules in their cytoplasm; these are smaller and darker than the granules of Paneth cells that, in contrast, are polarized with the nucleus at the basement membrane and the granules in the luminally oriented cytosol. D cells secrete somatostatin, the major inhibitory hormone of the small intestine. They are spindle-shaped with one long apical process and a shorter basal extension. L cells are described as "flask-shaped." They secrete the peptide, tyrosine-tyrosine (YY), which functions to reduce appetite. Other products of L cells include proglucagon-derived peptides, including GLP1 and GLP2 which control glucose homeostasis (37).

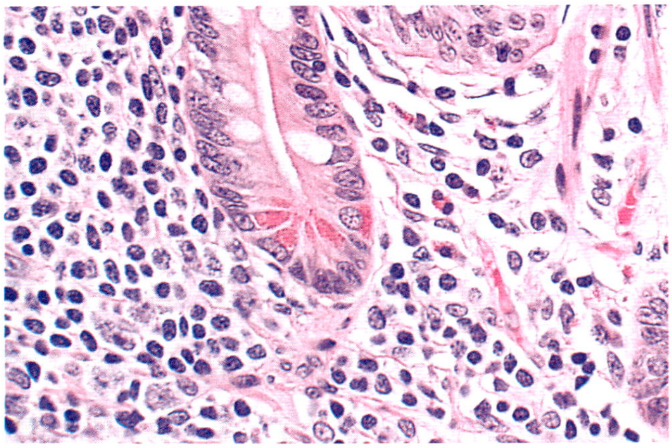

FIGURE 23.5 A single crypt surrounded by normal cellular lamina propria with abundant migratory cells. Both absorptive columnar and goblet cells are seen lining the crypt. In addition, an endocrine cell (infranuclear eosinophilic granules) and several Paneth cells (supranuclear granules) are clearly evident.

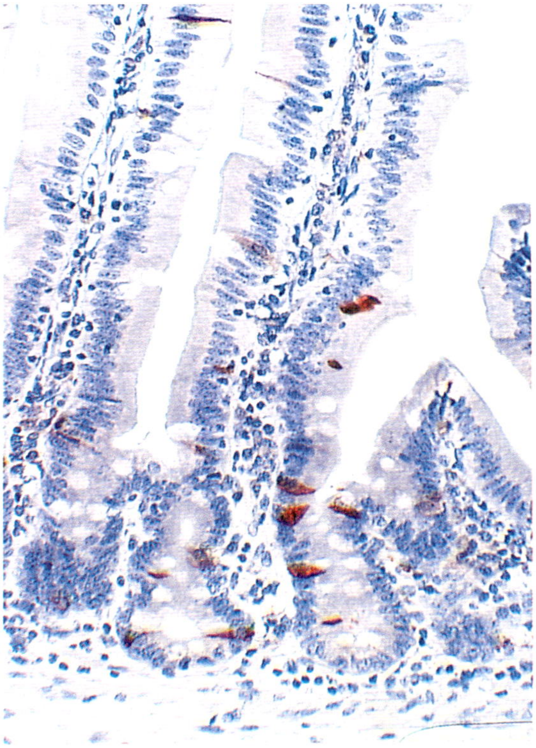

FIGURE 23.6 Immunostain for chromogranin shows numerous endocrine cells within the crypts and several scattered along the villous surface.

Paneth cells are limited to the crypts and are densest at their bases (38). Paneth cells are tall and columnar with basally located nuclei. Their cytoplasm contains supranuclear, intensely eosinophilic granules (Fig. 23.5). Their round nuclei often contain a prominent nucleolus. These cells contain lysozyme, defensins, and immunoglobulins and are capable of phagocytosis. Paneth cell alpha-defensin HD-5 is active against bacteria and can be localized to the cell with immunohistochemistry and in situ hybridization (Fig. 23.7) (39,40). Their contents and location suggest that they may function as a protector of the stem cell and help regulate intestinal microbes (39,40).

Absorptive cells are present in crypts and villi and comprise the most prominent epithelial cell type of the villi. They are tall columnar cells with small, dark, basally oriented ovoid nuclei and eosinophilic cytoplasm (Fig. 23.8).

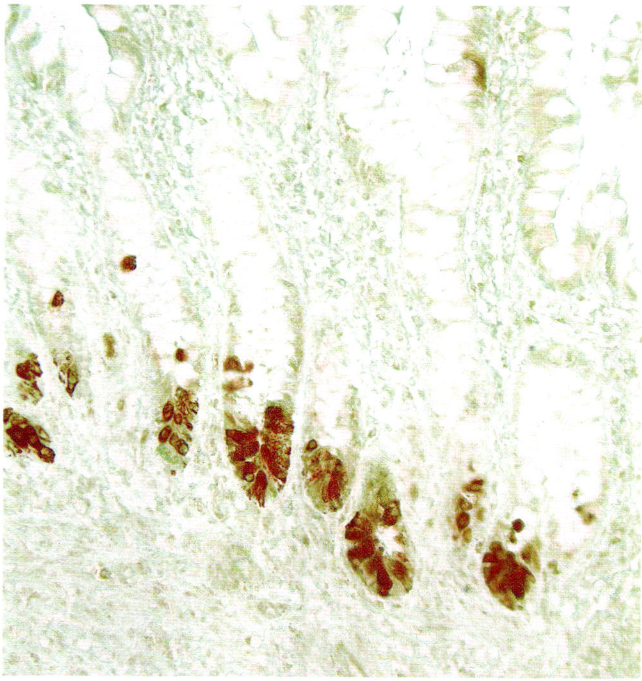

FIGURE 23.7 Immunostain for defensin HD5 highlights Paneth cells at the base of the crypts.

The apical surface contains a densely eosinophilic brush border composed of microvilli and glycocalyx that stains positively with periodic acid–Schiff (PAS) and CD10 (Figs. 23.9). Microvilli, which are best seen on ultrastructural examination, are evenly spaced surface projections that also augment the mucosal surface area of the small intestine (Fig. 23.10) (25). Multiple filamentous structures

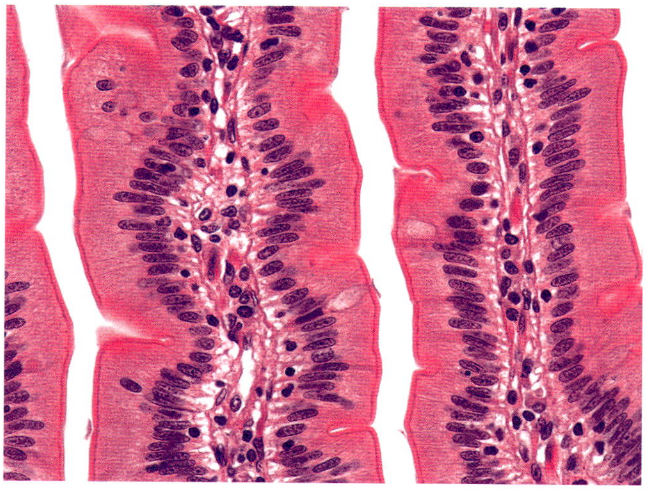

FIGURE 23.8 High-magnification view of jejunal villi disclosing the general features of villous morphology. Both columnar absorptive cells and goblet cells (with apical clear vacuole) cover the villous surfaces; each cell type has a basally situated oval-to-round nucleus. Microvilli (brush border) are seen extending from the columnar absorptive cell surface. Note the intraepithelial lymphocytes scattered among and between the epithelial cells.

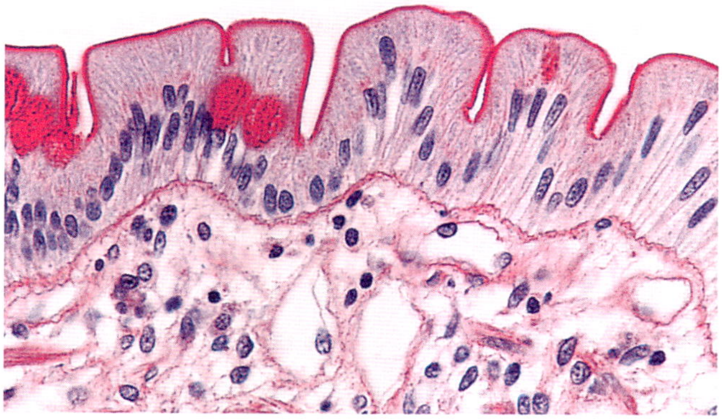

FIGURE 23.9 Periodic acid–Schiff (PAS) stain highlights the microvillus membrane–glycocalyx complex along the apical surface of the absorptive cells. The thin subepithelial basement membrane that separates the lamina propria from the epithelial compartment also stains with PAS but to a lesser degree. The neutral subgroup of mucins contained within the goblet cells are PAS positive as well.

emanating from and contiguous with their surface comprise the glycocalyx (22). The microvillus membrane–glycocalyx complex houses important enzymes—peptidases and disaccharidases—that function in terminal digestive processes. This layer also acts as a physical barrier to microorganisms and other foreign matter (27). The small intestine has a single gel-like mucus layer composed predominantly

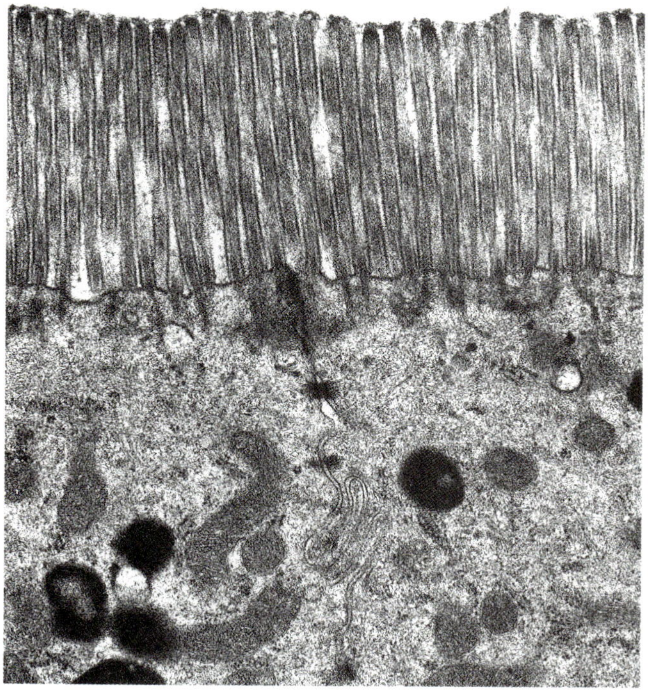

FIGURE 23.10 Transmission electron micrograph of microvilli emanating from the absorptive columnar cell surfaces. The glycocalyx component is the filamentous layer overlying the microvilli, but most of this has been artifactually removed during processing.

of MUC2 (41,42). MUC2 and other components forming the glycocalyx are continually synthesized within absorptive cells and transported to the surface to replace the preexisting coat (22,43). Antibacterial proteins produced by Paneth cells are also contained in the mucus layer.

Goblet cells are interspersed among absorptive cells; each contains a single luminally oriented, discrete blue mucin vacuole and a small, round bland nucleus near the basement membrane (Fig. 23.8). They secrete neutral and acid mucins to provide a moist viscid environment for passage of luminal contents (25). Further examination of goblet cell function has revealed a possible gate-keeping role with antigen presentation via endocytosis at the rim of the goblet cell theca, resulting in controlled secretion and an immunomodulatory response (42). In a combined Alcian blue/PAS stain, the droplets usually appear blue-purple (Fig. 23.11). The acid mucins of the small intestine are primarily sialomucins, in contrast to the colonic goblet cells, which contain predominantly acid sulfomucins (44,45). The number of goblet cells increases with distal progression along the small bowel constituting 4%, 6%, and 12% in the epithelial compartments of the duodenum, jejunum, and ileum, respectively (25,41).

› ***Immunostaining Patterns of the Epithelium*** Native small intestinal epithelial cells show uniform membranous staining for cytokeratin (CK) 7 and are negative

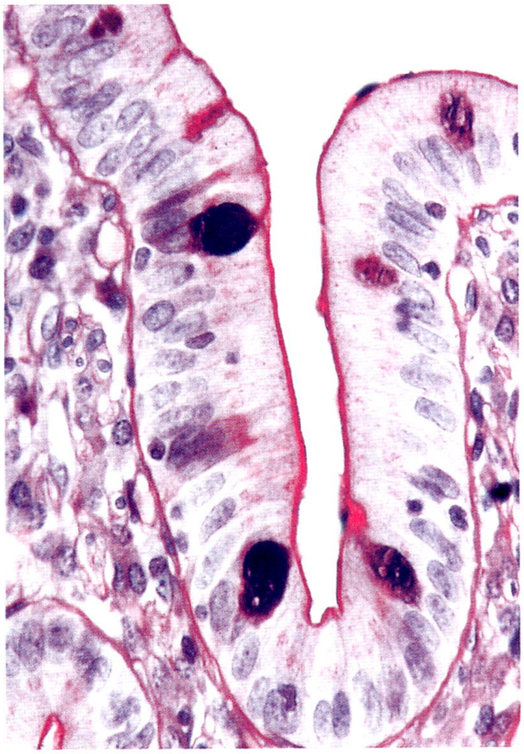

FIGURE 23.11 Alcian blue/PAS combination stain showing characteristic blue-purple apical mucin droplet of intestinal goblet cells. The heterogeneous composition (neutral and acid mucins) allows both stains to be incorporated into the droplet, imparting this distinctive color.

for CK20. They also show nuclear expression of the transcription factor, CDX2, but lack SATB2 expression. On the other hand, approximately half of small intestinal adenocarcinomas gain expression of CK20 and show weak or patchy SATB2 staining. Nearly all retain CK7 expression and more than half are positive for CDX2 (46,47).

Carcinoembryonic antigen (CEA) and CD10 are present on the apical surfaces of cells covering the villi and lining the crypts. These proteins localize to the glycocalyx surface component (48,49). The mucin droplets of goblet cells mark intensely with polyclonal anti-CEA whereas, no intracytoplasmic immunostaining for CEA is evident in the normal small bowel (50). CEA and CD10 immunohistochemistry can be used to demonstrate pathologic conditions such as the microvillus inclusion disease (51,52).

Human leukocyte antigen (HLA)-DR–like antigens are irregularly distributed on the apices of small intestinal columnar cells (50). Immunostains for anti–HLA-DR show reactivity with decreasing intensity from the villous surfaces to the crypt bases. Immune-related cells such as lymphocytes and macrophages along with the walls of capillaries in the lamina propria also show immunoreactivity for HLA-DR (50).

› ***Lamina Propria*** The lamina propria functions both structurally and immunologically. It rests upon the muscularis mucosae, surrounds the crypts, and extends upward to support the villi. A slender eosinophilic, PAS-positive band of basement membrane separates the lamina propria from the epithelial cell layer (Figs. 23.4 and 23.10). The basement membrane comprises basal lamina and a deeper network of collagenous/reticular fibers and ground substance (22). A mesenchymal cell layer, the pericryptal fibroblastic sheath, lies immediately beneath the epithelial cells and basement membrane (51). Interweaving collagen bundles and other connective tissue fibers, fibroblasts, mature fibrocytes, and smooth muscle cells comprise the framework of the lamina propria, whereas capillaries, lymphatics, and nerves course through this layer on their various routes to and from all portions of the bowel wall.

› ***Immune Cell Populations*** The intestinal tract is presented with an antigenic load of commensal flora, potential pathogens, and allergens on a continuous basis. Its immune cell population represents the largest compartment of the human immune system. The intestinal immune system or gut-associated lymphoid tissue system (GALT) comprises Peyer patches (Fig. 23.12), lymphoid follicles and mesenteric lymph nodes (the induction arm), and lamina propria inflammatory cells (the effector arm) (52).

Peyer patches and isolated lymphoid follicles are most prominent in the ileum. Peyer patches are distributed mostly along the antimesenteric border of the small intestine, with the greatest density in the terminal ileum (53). Larger Peyer patches are grossly and endoscopically visible. The lymphoid tissue in Peyer patches can become

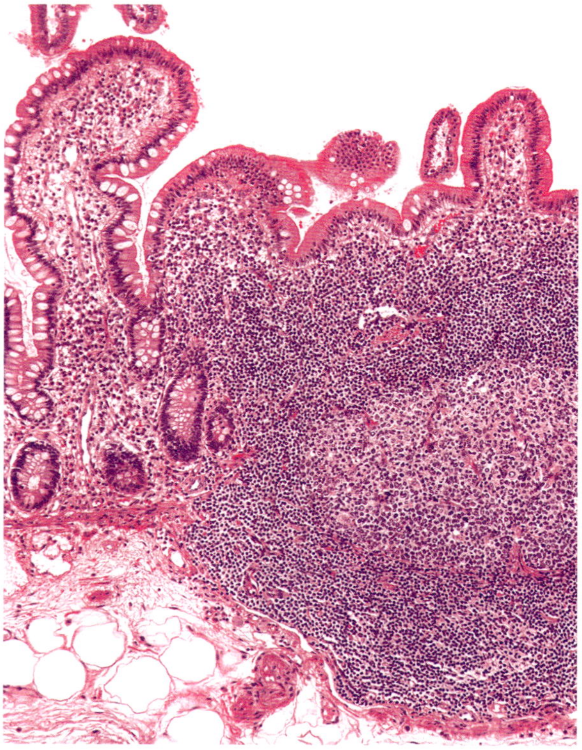

FIGURE 23.12 Lamina propria lymphoid aggregates of the Peyer patch. This organized lymphoid tissue typically extends into the underlying submucosa. The four components of the Peyer patch are seen and include a lymphoid follicle with prominent germinal center, an overlying flattened follicle-associated epithelium, an intervening pale-staining subepithelial dome region, and the T-cell–rich interfollicular zone.

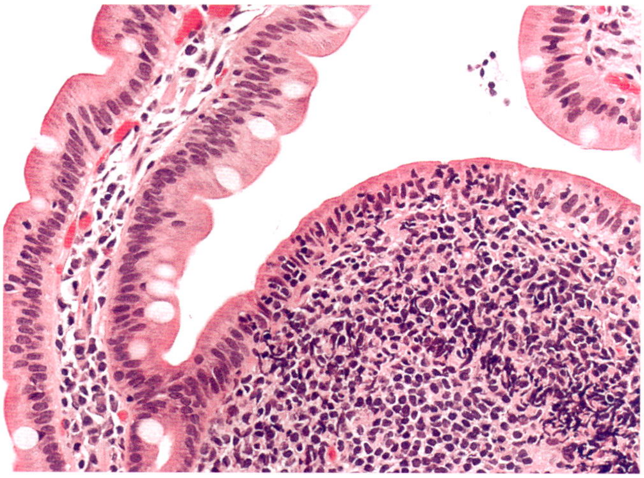

FIGURE 23.13 High magnification of surface epithelium above a lymphoid nodule within the Peyer patch. The polymorphous germinal center (*below*) is surrounded by monotonous, small, round lymphocytes that comprise the nodule's mantle zone. Above this lies the subepithelial dome region with lymphocytes, plasma cells, and macrophages. The follicle-associated epithelium characteristically has few, if any, goblet cells; ultrastructurally and phenotypically, most of these epithelial cells would be identified as M cells.

hyperplastic and this florid lymphoid hyperplasia has been linked to more than one-third of childhood cases of idiopathic intussusception in the region of the ileocecal valve (54). Marked lymphoid hyperplasia in this area can result from infection with enteric viruses, including adenovirus.

Peyer patches occupy the mucosa and a variable portion of the submucosa. The villi over these lymphoid aggregates are often poorly formed or absent. Peyer patches have four distinct compartments including the follicle, the subepithelial dome, the interfollicular zone, and the follicle-associated epithelium (53). Lymphoid follicles within Peyer patches can vary from as few as five to as many as several hundreds. Like their counterpart in lymph nodes, the follicles within Peyer patches contain a predominance of B cells together with follicular dendritic cells, and macrophages. Luminal antigens are sampled by follicle-associated epithelium, specialized epithelial cells overlying Peyer patches, and dendritic cells (Fig. 23.13). The follicle-associated epithelium layer includes specialized microfold cells which have a unique pocket at the basolateral membrane for interaction with B and T cells (52).

Most follicles contain a germinal center with CD10-positive and BCL-2–negative B cells that undergo immunoglobulin class-switching due to the influences of CD4-positive T_h cells, dendritic cells macrophages, TGF-β, and IL-10, undergo resulting in the production of IgA-expressing B cells (55,56). This B-cell population then migrates from Peyer patches to mesenteric lymph nodes and ultimately home to the lamina propria under the influences of CCL25, CXCL12, and CCL28 (53).

The germinal center is surrounded by a mantle zone of small IgD-positive, IgM-positive B cells. The subepithelial dome is situated between the lymphoid follicle and the overlying follicle-associated surface epithelium, and contains a heterogeneous population of B cells, macrophages, dendritic cells, and plasma cells (53). The interfollicular zone, a T-cell–rich zone absent in isolated lymphoid follicles, is a site of interaction between antigen-loaded dendritic cells and T cells; T-cell priming occurs in this area (53).

Plasma cells are the most abundant cellular constituents of the lamina propria and most contain cytoplasmic IgA (Fig. 23.14) (52). Intraluminal secretion of IgA requires active participation by intestinal surface epithelial cells. Polymeric IgA first binds to receptors on the epithelial basolateral membrane where it undergoes receptor-mediated endocytosis. Following directed transport across the cytoplasm, luminal secretion occurs following fusion of the vesicles with the luminal membrane. T lymphocytes localize to the small intestine as a result of selective expression of the integrin, a4b7 and chemokine, CCR9. Both CD4-positive and CD8-positive T cells are present in the lamina propria. The lamina propria also contains B lymphocytes and eosinophils, as well as a smaller number of macrophages, dendritic cells, and mast cells (52).

Eosinophils are generated in the bone marrow and primarily home to mucosal sites, including the lamina propria of

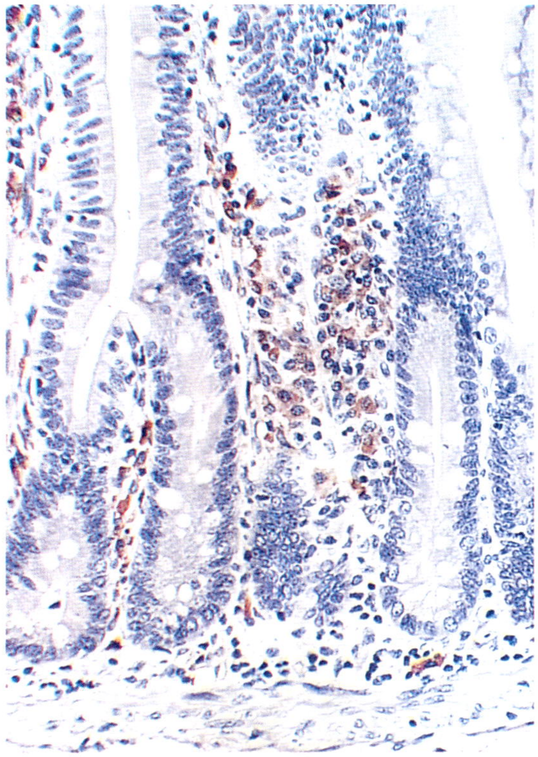

FIGURE 23.14 Numerous IgA-containing plasma cells (red-brown cytoplasmic staining) within the lamina propria of normal small intestine, immunostained for α-heavy chain. Note that, in contrast to T cells (see Fig. 23.9), the plasma cells localize to the lamina propria and are not normally found in the epithelium.

the gastrointestinal tract (57). Eosinophils are an essential component of the innate immune system and are thought to play an important role in the homeostatic maintenance of the intestinal barrier against luminal bacteria. Their density varies across populations, intestinal sites, and with seasonal variations in allergen exposure (58). Their granules contain major basic protein, eosinophil cationic protein, and eosinophil peroxidase, all of which have bactericidal activity (59). Mucosal eosinophils are increased in various disease conditions including primary eosinophilic gastrointestinal disorders as well as secondary eosinophilic disorders such as food allergies and parasitic infection (57,60–62).

Macrophages and dendritic cells are less conspicuous but may be highlighted with CD68 immunostains (macrophages) (63) and CD11b, CD11c, or CD209 (dendritic cells) (64,65). In addition to the phagocytic and bactericidal activities of macrophages, they present antigens to T cells (66). In the small intestine, macrophages are mostly located in the lamina propria just below the surface epithelium, often concentrated near the tips of the villi (67). In contrast to blood monocytes, intestinal macrophages appear to be maintained in a state of only partial activation and fail to produce the levels of proinflammatory cytokines and reactive oxygen and nitrogen intermediates characteristic of conventional macrophages and monocytes. This state of refractoriness may be critical for mucosal homeostasis, in which macrophages function in the uptake and killing of microbes without initiating a potentially injurious inflammatory cascade. A dramatic expression of the phagocytic capacity of macrophages in the small intestine can be seen in cases of disseminated *Mycobacterium avium–intracellulare* complex infection and Whipple disease, both of which are characterized by expansion of the lamina propria with engorged macrophages (68,69).

Mast cells are also relatively inconspicuous in routine H&E-stained sections. They are round with centrally located hyperchromatic nuclei and faintly granular cytoplasm. Immunostains for CD117 and mast cell tryptase highlight mast cells (70). The range of normal mucosal mast cell counts varies widely. Mast cell counts in biopsy samples from the duodenum and terminal ileum can range up to 40 to 50 per high-magnification field (71). Mast cells have surface receptors for IgE and participate in hypersensitivity reactions. Cytoplasmic granules of mast cells contain histamine and serotonin, which together with other factors from eosinophils can lead to both nerve stimulation and smooth muscle contraction. Activated mast cells also release various cytokines, leukotrienes, and prostaglandins. Given their potent armament, activation of gastrointestinal mast cells can have local effects including abdominal pain, diarrhea, and vomiting, as well as distant effects including cutaneous hypersensitivity reactions or asthma (72,73). A potential role for mast cells in the pathogenesis of irritable bowel syndrome may be related to their proximity to enteric nerve fibers and potential to stimulate smooth muscle contraction (74).

Intraepithelial lymphocytes are scattered among and lie between individual epithelial cells, usually just above the basement membrane. Their distribution follows a "decrescendo" pattern with more lymphocytes present in crypt epithelium and bases of villi and fewer at the villous tips. Normally, up to 20 to 25 lymphocytes are present per 100 epithelial cells in the duodenum. Intraepithelial lymphocytosis heralds a variety of disease states (e.g., gluten-sensitive enteropathy, tropical sprue, medication-related injury) in the small intestine and should be assessed in well-oriented villi. The "villous tip" counting method is often employed for practical purposes; there should be no more than 12 lymphocytes infiltrating the 20 epithelial cells lining the tips of well-oriented villi, although some authors suggest that an upper limit of 6 lymphocytes/20 enterocytes is more sensitive for pathologic lymphocytosis (75,76) (Figs. 23.8 and 23.15). Intraepithelial lymphocytes are CD3-positive T cells (Fig. 23.12), and most express CD8 (66,77,78). The majority bear ab T-cell receptors. Approximately 5% to 30% bear γδ T-cell receptors and are either CD4-negative/CD8-negative or CD8-positive (79). CD4-negative/CD8-negative intraepithelial lymphocytes are more common in the ileal mucosa (40) where intraepithelial lymphocyte counts are far less numerous (less than 5 per 100 enterocytes) (79). γδ T cells act as mediators of host-microbial homeostasis by stimulating goblet cell function and regulating mucin expression and glycosylation (80).

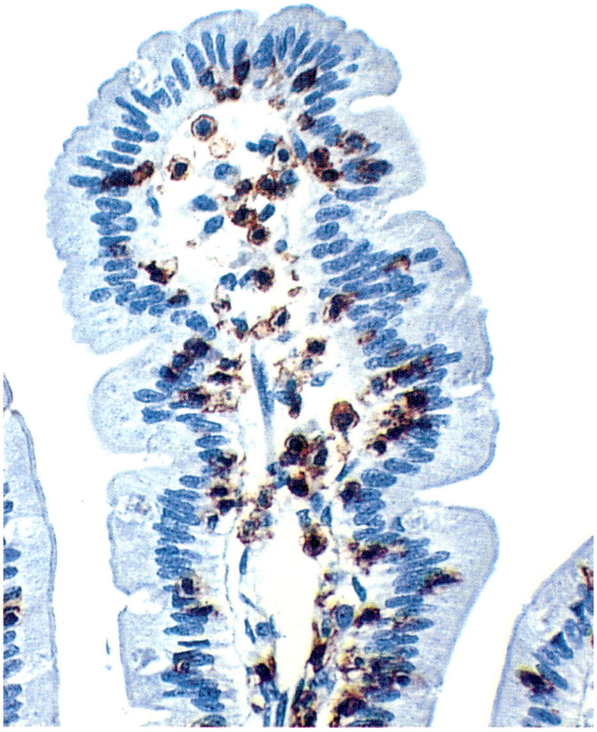

FIGURE 23.15 Small intestinal mucosa immunostained with antibody UCHL1 (CD45RO), a T-cell marker. The positive red-brown reaction highlights both the intraepithelial and lamina propria T cells. The intraepithelial cells are predominantly CD8 positive, whereas most T cells in the lamina propria are CD4 positive.

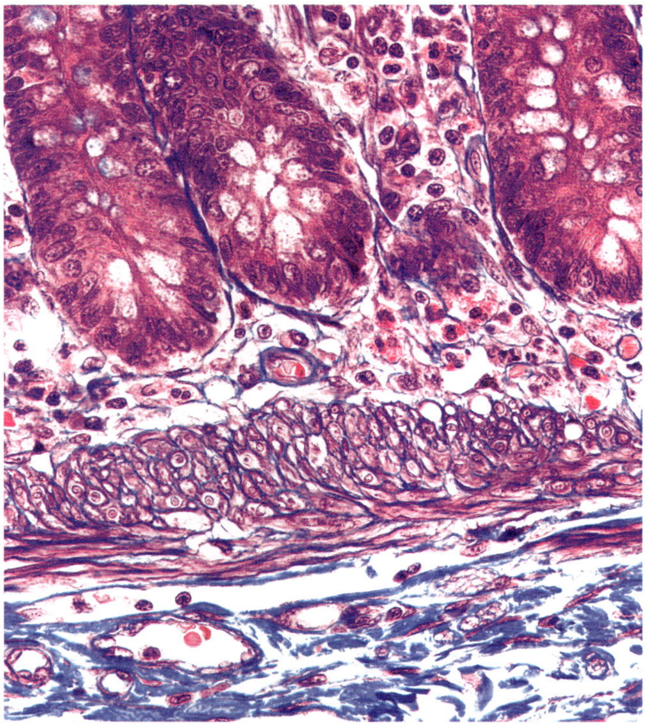

FIGURE 23.16 High-magnification photomicrograph disclosing inner circular and outer longitudinal smooth muscle bands of the muscularis mucosae; this layering is often inconspicuous on hematoxylin and eosin preparations, where the muscularis mucosae appears as a thin eosinophilic strip between the lamina propria and underlying submucosa (Masson trichrome).

› ***Muscularis Mucosae*** The muscularis mucosae, which forms the lower limit of the mucosa, is composed of a slender band of smooth muscle cell aggregates supported by elastic fibers. Smooth muscle cells are arranged in an outer longitudinal and an inner circular layer, although these layers are not well delineated on routine light microscopy (Fig. 23.16). Absent or weak immunostaining for smoothelin is characteristic of this layer (81). Bundles of smooth muscle cells radiate from the muscularis mucosae and extend into the villi. The muscularis mucosae provides an important structural foundation for the mucosa and its absence in some biopsy specimens can cause a loss of villous orientation, an artifact that may interfere with optimal evaluation (82). The muscularis mucosae is discontinuous at points where duodenal Brunner glands penetrate the mucosa and lymphoid aggregates extend from the submucosa into the mucosa.

› ***Submucosa*** The submucosa lies between the muscularis mucosae and muscularis propria. It is a loose, paucicellular layer composed of a regular, honeycomb-like arrangement (at the ultrastructural level) of collagenous and elastic fibers and a few scattered fibroblasts. The submucosa also contains adipose tissue and a sparse population of macrophages, lymphocytes, plasma cells, and mast cells. Its histologic appearance and principal role in maintaining the structural integrity of the small bowel are similar throughout the gastrointestinal tract (83). The submucosa is a major focus of vascular routing and related distribution of regional blood and lymphatic flow. Relatively large caliber arterioles, venules, and lymphatic vessels form extensive individual plexuses and networks within this layer (Fig. 23.16). From this "vascular center," numerous penetrating capillary vessels supply and drain most of the mucosa and muscularis externa. Lymphatic vessels may be distinguished from blood vessels by the thinner wall of the former and the lack of luminal erythrocytes. However, certain immunohistochemical staining patterns and electron microscopic characteristics are more helpful for definitive identification (84,85). Specifically, the endothelial cells of blood capillaries are immunopositive for PAL-E and factor VIII–related antigen, whereas lymphatic capillary endothelia typically lack these antigenic sites and remain unstained with such antibody preparations. In addition, blood capillaries, as seen by ultrastructural analysis, have a continuous basal lamina, endothelial fenestrations, and supporting pericytes. Lymphatic capillaries have a discontinuous basal lamina and lack both fenestrations and surrounding pericytes (22,23). Although small lymphatic vessels are conspicuous in the submucosa, prominent

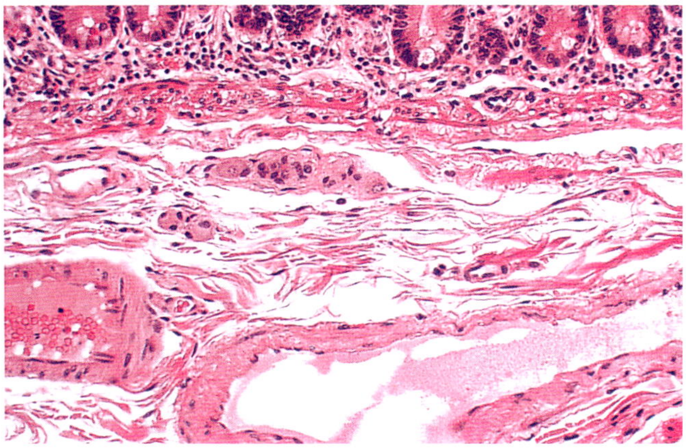

FIGURE 23.17 Normal submucosa separated from the overlying mucosa by the eosinophilic-staining muscularis mucosae. The submucosa is paucicellular, disclosing fibrocollagenous tissue and a prominent vascular component. Note the ganglion cells of the Meissner plexus just beneath the muscularis mucosae.

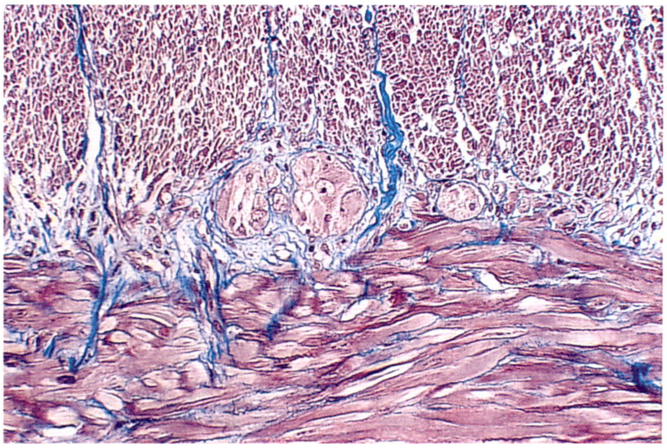

FIGURE 23.18 The Masson trichrome stain clearly delineates the inner circular (*above*) from the outer longitudinal (*below*) smooth muscle bands of the muscularis externa. The prominent muscular component (*red*) is partitioned into bundles of varying sizes by delicate collagenous fibers (*blue*). Note the ganglia of the myenteric plexus of Auerbach, characteristically located between the two muscle bands. Fibrous tissue is minimal within the muscularis externa.

dilated lymphatic structures in this layer, as well as in the mucosa, can be seen in pathologic states such as intestinal lymphangiectasia or Crohn disease (86).

The submucosa houses Meissner plexus, one of the two major integrative centers of the enteric nervous system. It consists of a network of ganglia that interconnect through neural processes (87). The ganglia contain compact aggregates of ganglion cells that can be recognized due to their characteristically large ovoid shape, abundant pink cytoplasm, vesicular nuclei, and single, often prominent nucleoli (Fig. 23.17). Abundant S100-positive Schwann cells, glial-like cells, and neural processes are also present in the Meissner plexus. The entire plexus, including the ganglia, contains no connective tissue elements or vascular structures in the normal state (88,89). The plexus is also normally devoid of inflammatory cells; therefore, if these are seen, an injury pattern specific to the neural plexus (such as an inflammatory neuropathy) or inflammatory enteritis should be considered (88). Neural interconnections exist between the Meissner plexus and the myenteric plexus of Auerbach, as well as with autonomic neural processes.

› ***Muscularis Propria*** The muscularis propria is the thick outer smooth muscle layer that surrounds the submucosa. It is covered externally by subserosal connective tissue and, in most places, by a serosa. Its two layers are arranged as outer longitudinal and inner circular muscle bands (Fig. 23.18) and stain for smooth muscle actin and smoothelin. Blood vessels, lymphatic vessels, and nerves course through the muscularis propria and slender collagenous septa surround groups of smooth muscle cells, creating characteristic bundles and packets of muscle cell aggregates (Fig. 23.18). Fibrous tissue in this layer is minimal in the normal small bowel (90) but may accumulate in cases of intermittent ischemia, irradiation, familial visceral myopathy, scleroderma, and Crohn disease (90).

The myenteric plexus of Auerbach lies between the outer longitudinal and inner circular muscle layers (Figs. 23.19 to 23.21). The Auerbach plexus is similar in composition to the submucosal plexus, although it typically contains larger ganglia, a greater number of neurons, and a more compact plexus network (87). As a consequence of these features, it is best to evaluate the myenteric plexus for specific disease processes involving the enteric nervous system, such as the various visceral neuropathies. Routine processing allows only a small portion of the plexus to be visualized

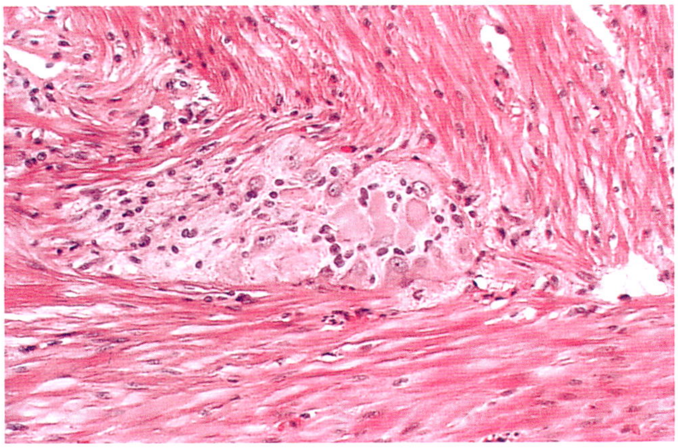

FIGURE 23.19 A single ganglion of the myenteric plexus of Auerbach located at the interface of the inner (*above*) and outer (*below*) smooth muscle layers of the muscularis externa. Ganglion cells (neuronal cell bodies) are evident and characterized by a polygonal shape, abundant pink cytoplasm, and an eccentric nucleus; spindled neural projections and Schwann cells are also intermixed.

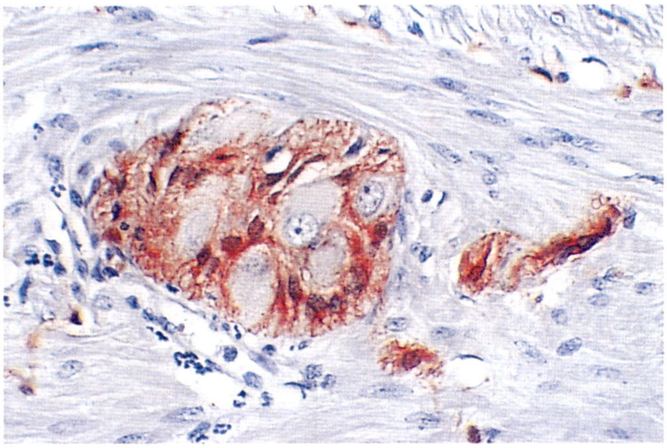

FIGURE 23.20 This S100 immunostain highlights the otherwise inconspicuous spindled Schwann cell component of the ganglion. It also marks the Schwann cells accompanying the neural projections that interconnect these ganglia to one another within the plexus system. Note that the ganglion cells show no such immunoreactivity.

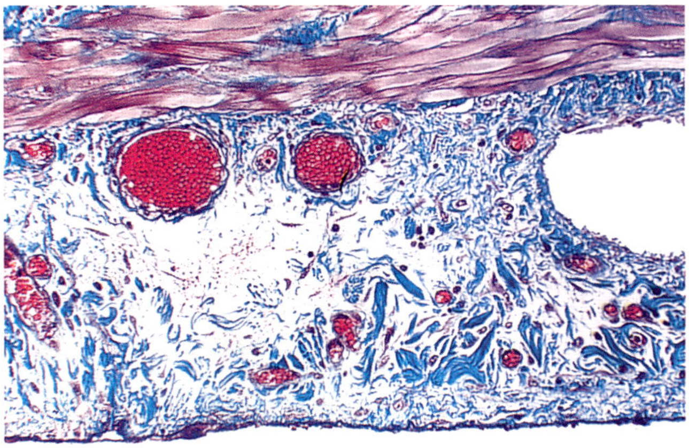

FIGURE 23.22 The subserosal region contains a delicate fibrocollagenous network, blood vessels, lymphatics, and nerves. The serosa consists of a thin fibrous layer (*blue* and *bottom*) covered by a single layer of mesothelial cells; however, the mesothelium is often denuded in surgical specimens. A portion of the outer layer of the muscularis externa is also present in this field (*top*) (Masson trichrome).

and many of these conditions cause no detectable changes on routine H&E-stained sections. Thus, special preparations of thicker, larger, and silver-stained sections cut en face are often required to diagnose many of these disorders (87). Interstitial cells of Cajal form a meshwork around the Auerbach plexus and emanate into the muscularis propria between muscle cell bundles (Fig. 23.21). They can be highlighted with CD117 immunostains (91–93).

› ***Serosa and Subserosal Region*** The serosa envelops most of the external surface of the small bowel. Its outermost layer consists of a single row of cuboidal mesothelial cells, under which lies a thin band of loose connective tissue. A subserosal zone of connective tissue lying between this mesothelial covering and the muscularis propria also contains ramifying branches of blood vessels, lymphatic vessels, and nerves (Fig. 23.22).

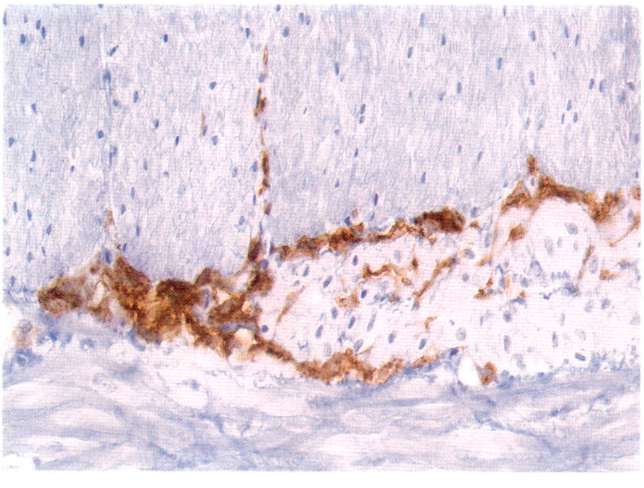

FIGURE 23.21 The CD117 (c-kit) immunostain shows ICC surrounding the myenteric plexus of Auerbach.

DISTINCTIVE REGIONAL CHARACTERISTICS OF THE SMALL BOWEL

Duodenum

The duodenum exhibits several distinctive histologic features related to its proximal location in direct continuity with the pylorus. The gastroduodenal junction, although well delineated grossly, is poorly demarcated histologically (Figs. 23.23 and 23.24) (94). A gradual transition between epithelial cell types occurs, with three comingling of

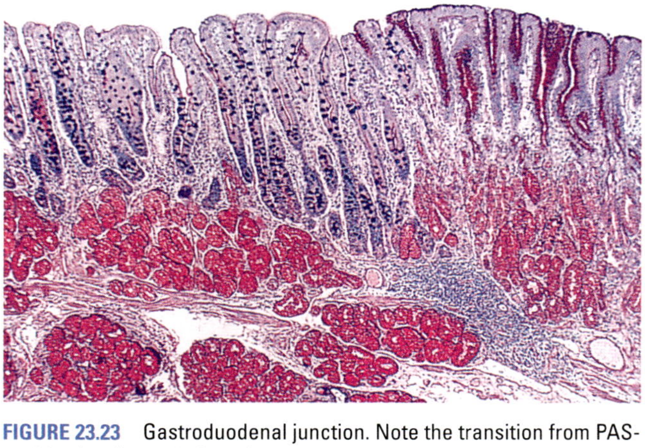

FIGURE 23.23 Gastroduodenal junction. Note the transition from PAS-positive (*red*) gastric foveolar epithelium and underlying pyloric glands (*right*) to a villous mucosal architecture of the duodenum (*left*) lined predominantly by Alcian blue/PAS-positive (*blue-purple*) goblet cells and absorptive cells. Note that both pyloric (*right*) and Brunner glands (*left*) are composed predominantly of cells containing only neutral, PAS-positive mucin. Brunner glands, however, are predominantly submucosal in location, while pyloric glands are an intramucosal structure (Alcian blue/PAS).

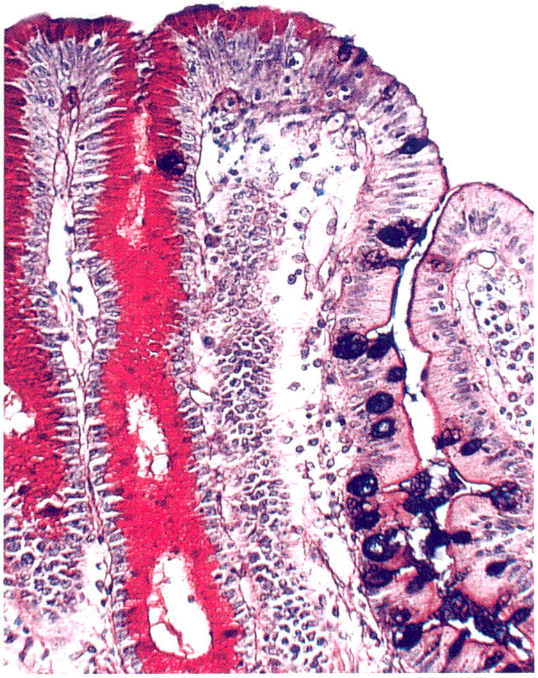

FIGURE 23.24 Several villi within the confines of the gastroduodenal junction disclosing both "usual small intestinal-type" epithelium and antral-type, PAS-positive, foveolar epithelium. This transitional-type epithelium is a characteristic hybrid found in this region. At more distal small intestinal sites, this transitional-type epithelium is termed gastric metaplasia (Alcian blue/PAS).

foveolar and intestinal-type epithelium (Fig. 23.24) (95). In the region of the gastroduodenal junction, irregular undulating strips of antral-type mucosa extend about 1 to 2 mm into the anatomic duodenum (96,97). Distal to this region, only the usual small intestinal–type mucosa is found (95). Gastric-type epithelium found in more distal aspects of the duodenum and in the rest of the small intestine is considered to be metaplastic or heterotopic in nature (94).

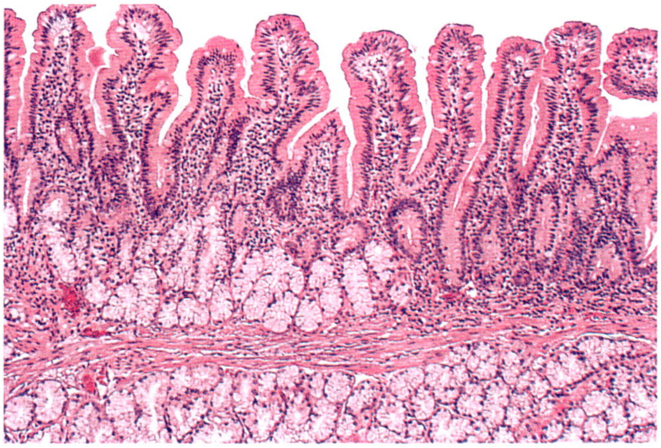

FIGURE 23.25 Short, slightly broader villi predominate in the duodenum. The underlying submucosal Brunner glands are a distinctive feature of this portion of the small bowel. Note that a fair portion of Brunner glands normally occurs above the muscularis mucosae.

Although the duodenal mucosa may demonstrate long villi with a villous-to-crypt length ratio on the order of 3:1 to 5:1, more commonly, they are shorter and broader with occasional branching extensions in the duodenal bulb (Fig. 23.25) (98). They often have a leaflike shape with few fingerlike forms when viewed under a scanning electron or dissecting microscope (Fig. 23.26) (27,94,95). Also, the number of mononuclear cells within the lamina propria is increased in the duodenum when compared with the rest of the proximal small intestine (98). This varied constellation of findings is considered normal and is probably a consequence of the effect of acidic gastric contents on this most proximal intestinal site (97).

Submucosal Brunner glands are localized to the duodenum and are typically used by the pathologist to identify a segment of small intestine as duodenum histologically. They are present just distal to the gastroduodenal junction where they are most concentrated and gradually decrease

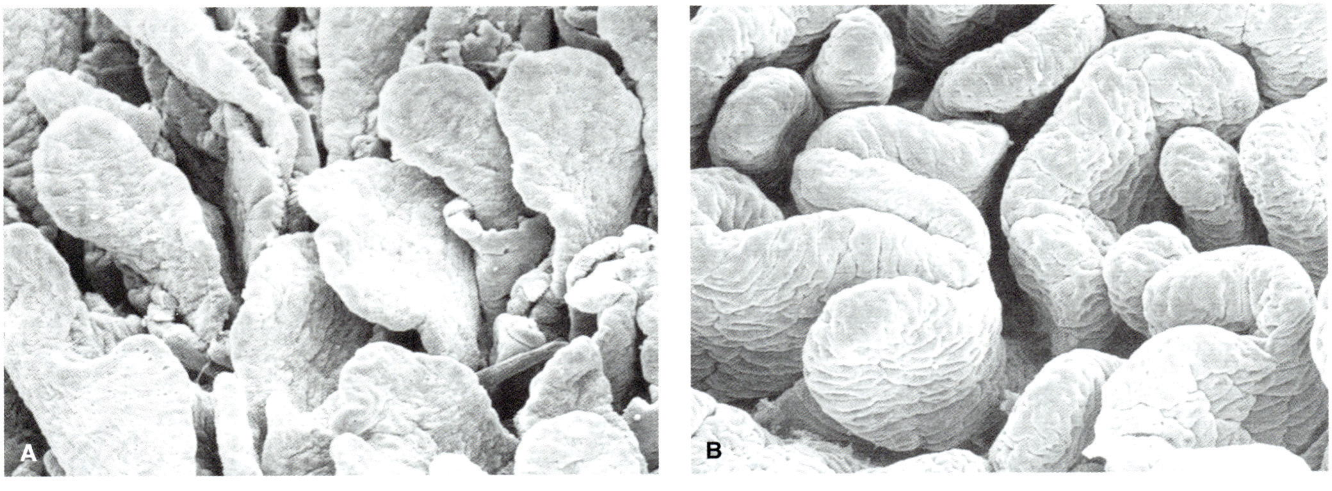

FIGURE 23.26 Two views of duodenal mucosa using scanning electron microscopy: Leaflike **(A)** and ridge-shaped **(B)** villi predominate in these normal duodenal specimens.

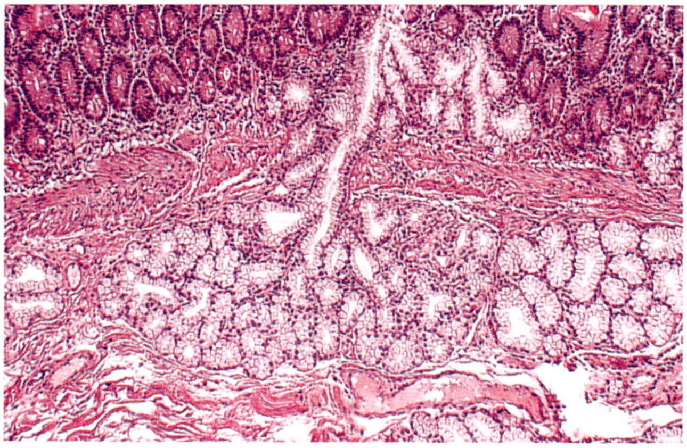

FIGURE 23.27 The submucosal Brunner gland lobule with draining duct extending through muscularis mucosae. Note the stark contrast between the crypt epithelium and that of Brunner glands and their ducts.

in quantity along the length of the duodenum (99). Beyond the entrance of the ampulla of Vater, only scattered groups can be found (99).

Brunner glands form lobular aggregates predominantly located within the submucosa; however, they extend through the muscularis mucosae into the deep portions of mucosa beneath the crypts of Lieberkühn in the first portion of the duodenum (Figs. 23.25 and 23.27). On average, about one-third of the gland population resides within the mucosa (100). Brunner glands are lined by cuboidal-to-columnar cells with pale, bubbly cytoplasm and an ovoid, basally situated nucleus. Their cytoplasm contains neutral mucins that are PAS positive and diastase resistant (Fig. 23.23). Occasionally, mucus cells with apically concentrated mucin and perinuclear vacuolization or clearing are seen. Although opinions vary, these changes are thought to represent the secretory phase of the gland after absorption of foodstuffs (101,102). The glands empty by way of ducts lined by similar epithelium, which are often seen passing through interrupted areas of muscularis mucosae (Fig. 23.27). These ducts drain into the crypts at varying levels (103). Brunner glands and their ducts can be distinguished from surrounding crypts by the absence of goblet cells and diffuse cytoplasmic PAS positivity (103).

Although most of the lining epithelial cells of Brunner glands are of the mucus type, scattered endocrine cells and Paneth cells are present as well (101). Some have been shown to contain somatostatin, gastrin, and peptide YY by immunohistochemistry (102). However, the ducts that drain Brunner glands are devoid of endocrine cells (103).

Peptidergic neural fibers, predominantly those with immunoreactivity for vasoactive intestinal peptide and substance P, course within and between individual Brunner glands. These endocrine substances are probably important in local regulation of acinar secretion, although this function has been verified only for vasoactive intestinal peptide (104). The function of Brunner glands has not been fully elucidated, but their mucus is felt to be of prime

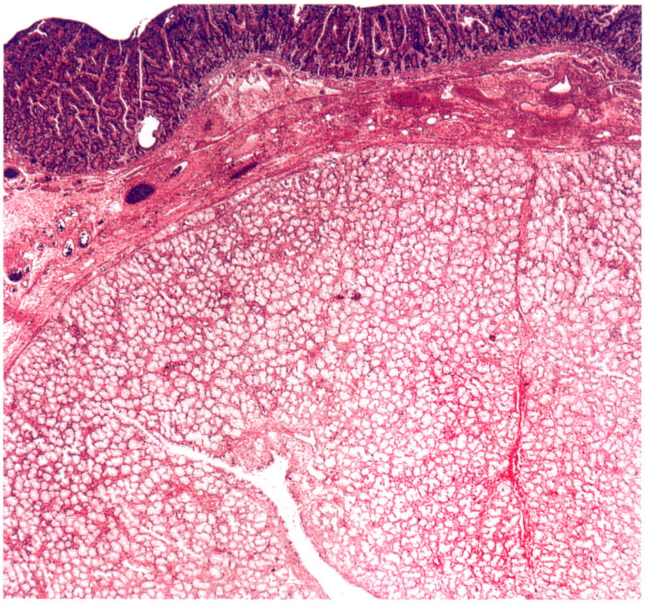

FIGURE 23.28 The Brunner gland nodule disclosing abundant normal-appearing submucosal Brunner glands intermixed with smooth muscle, underlying an unremarkable duodenal mucosal villous surface.

importance for protection of the duodenal mucosa from the potentially damaging effects of the delivered acidic gastric contents (104).

Proliferation of Brunner glands may produce endoscopically appreciable lesions and rarely, obstructive symptoms or bleeding. Endoscopic features may include diffuse or segmental mucosal nodularity or single or multiple discrete sessile or pedunculated polyps (Fig. 23.28). Histologically, these lesions feature prominent normal-appearing Brunner glands that expand the submucosa and occupy an increased proportion of the mucosal thickness. Lesions that contain cystically dilated Brunner glands, smooth muscle bundles, or adipose tissue are referred to as Brunner gland "hamartomas" to denote the presence of mesenchymal elements; however, the distinction between polypoid hyperplasia and a hamartoma is somewhat arbitrary (105). Although a size cutoff of 0.5 cm for the diagnosis of hamartoma has been suggested, evidence to support this threshold is lacking. Similarly, Brunner gland "adenoma" has been used to refer to polypoid Brunner gland proliferations and is often used interchangeably with hamartoma; no substantial evidence exists to suggest that any of these proliferations are truly neoplastic (106,107). Moreover, carcinoma arising from a population of Brunner glands has yet to be convincingly documented. Nodules or polypoid structures composed of collections of these submucosal glands in the duodenum are probably best termed Brunner gland hyperplasia/nodules (108).

Pseudomelanosis duodeni is characterized by brown-black pigment located primarily within lamina propria macrophages and rarely occurs in the proximal duodenum (Fig. 23.29). Lipomelanin, ceroid, iron, sulfide, and

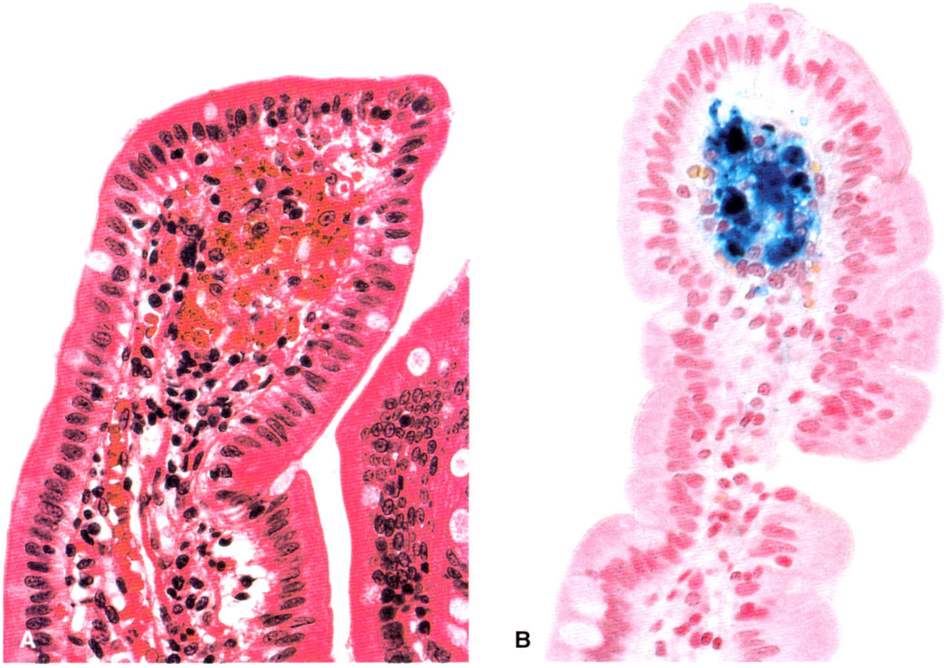

FIGURE 23.29 **A:** Macrophages containing granular brown-black pigment within the lamina propria of the duodenum, a characteristic of pseudomelanosis duodeni. **B:** Prussian blue stain disclosing the prominent iron content of the pigment.

hemosiderin have been identified in these deposits. Most patients with this finding are hypertensive and suffer from chronic renal failure or diabetes mellitus (109). Occasional cases have been reported in association with gastrointestinal hemorrhage.

Jejunum

The jejunum characteristically features prominent development of the *plicae circulares*, or permanent circular folds, also termed valves of Kerckring and *valvulae conniventes* (Fig. 23.1). The jejunal villi are taller than elsewhere in the small bowel with a villous-to-crypt ratio on the order of 3:1 to 5:1. Most jejunal villi are slender and fingerlike (Figs. 23.2 and 23.3), in contrast to the slightly shorter villi of the ileum and leaflike, occasionally branched and blunted villi of the proximal duodenum (25). These morphologic transitions are gradual, particularly in the mobile small intestine, where the separation between jejunum and ileum is arbitrarily defined.

Ileum

The ileum has a number of distinctive features, including its unique junction with the large intestine, a high concentration of lymphoid tissue, and pigment deposits. The ileum protrudes approximately 2 to 3 cm into the large intestine at the junction of the cecum and ascending colon. This nipple-like extension of the terminal ileum is encircled by large bowel mucosa and has been likened to the relationship of the uterine cervix with the vagina (110). A muscular sphincter at this site, along with the external ligamentous support, is responsible for modifying its function in order to prevent reflux and to allow forward passage of ileal contents (111). Histologically, the mucosal transition demonstrates a gradual loss of villi occurring at variable lengths along the short intracecal ileal segment (Fig. 23.30). The ileocecal region normally contains abundant fat within its submucosa, diffusely distributed and proportional to adipose content in the rest of the abdominal cavity (Fig. 23.30) (112). In fact, on rare occasions, a distinct mass of fat is evident. This benign entity, the so-called lipohyperplasia of the ileocecal region has been reported to cause abdominal pain and lower gastrointestinal bleeding in rare cases (112).

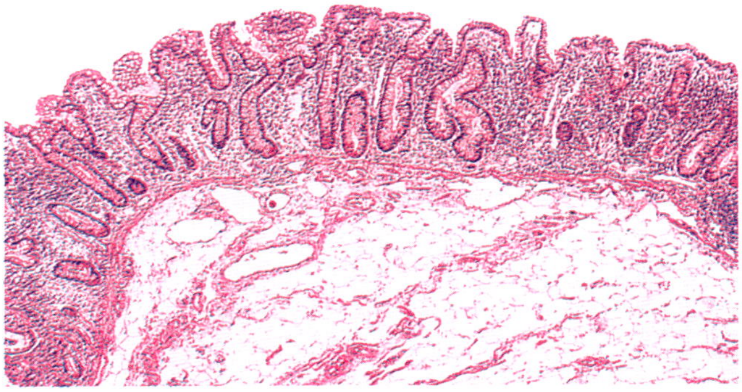

FIGURE 23.30 Transition from the villous mucosal surface of the ileum (*left*) to the flat mucosa of the large intestine (*right*) at the ileocecal junction. Note the prominent submucosal adipose tissue characteristic of this region.

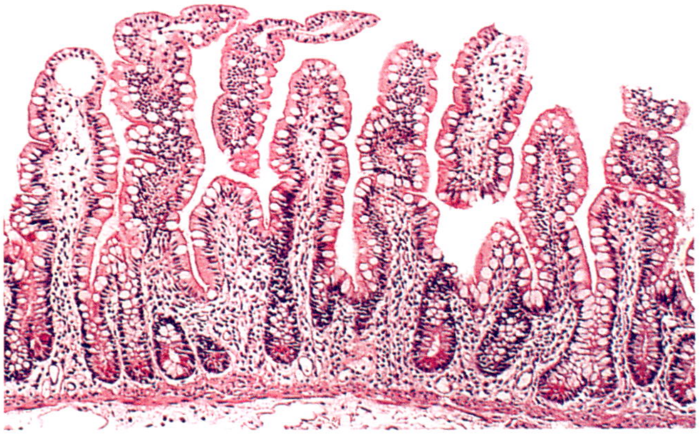

FIGURE 23.31 Characteristic ileal mucosa with slender, relatively short villi (compared with jejunal villi in Fig. 23.3) lined by abundant goblet cells with a lesser number of absorptive columnar cells.

The distinctive mucosal characteristics of the ileum, when compared with both jejunum and duodenum, include shorter and fewer plicae circulares and an increased proportion of goblet cells within the epithelium (Fig. 23.31). The villi are typically shorter and show less serration than at more proximal sites and often have a predominantly fingerlike shape (25). The villi are less straight resulting tangential appearing sections even in normal samples (113). These features become gradually more apparent along the length of the small intestine and are most evident in the distal ileum. The ileum also contains prominent organized lymphoid tissue in for form of Peyer patches, as discussed above.

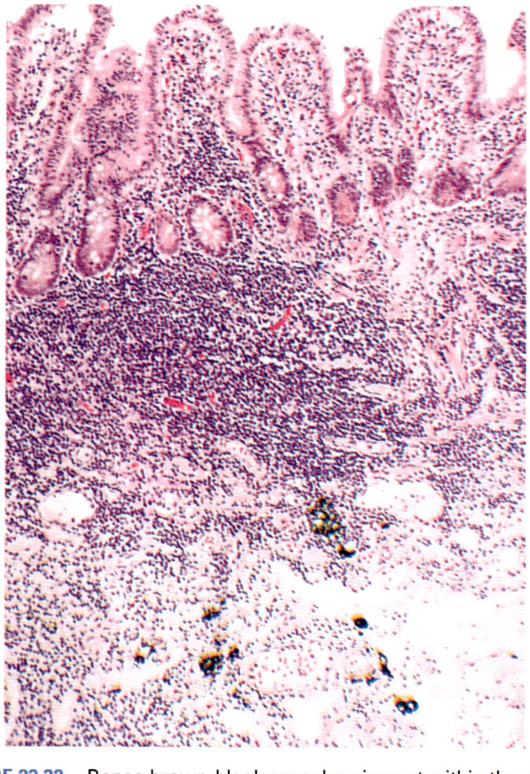

FIGURE 23.32 Dense brown-black granular pigment within the depths of the Peyer patch of the ileum. The pigment is typically confined to macrophages.

Irregularly distributed deposits of granular brown-black pigment are commonly found in the deep portions of Peyer patches in adults (Fig. 23.32) (114). Although its origin is controversial, atmospheric or dietary sources are most probable. Accumulating principally within macrophages, this pigment has been shown by x-ray spectroscopy to contain a distinct mineral composition that includes silicates, aluminum, and titanium (114). The pigment is inert and has no known clinicopathologic significance.

A final distinctive feature often seen in the ileum is the Meckel diverticulum; it is the most common intestinal congenital anomaly and is found in 1% to 2% of the general population. The Meckel diverticulum is an antimesenteric outpouching of the terminal ileum usually located approximately 20 cm from the ileocecal junction. It represents the persistence of the omphalomesenteric duct. Although the Meckel diverticulum is usually an incidental finding, it can cause lower gastrointestinal bleeding or small bowel obstruction (115). Histologically, small intestinal mucosa alone lines the diverticulum in about 50% to 70% of cases. Ectopic gastric or pancreatic tissues are found in the remainder, typically encountered at the distalmost aspect (116).

SPECIAL CONSIDERATIONS

Geographic, Age-Related, and Dietary Factors

Since geographic and local environmental factors can affect small bowel morphology, historical data regarding the residence or recent travel of an individual are essential to evaluate histologic material accurately. Specimens from individuals residing in, or visiting at length, less-developed tropical regions in Africa, southern India, and Thailand show a distinctly different villous appearance from those of individuals who live in temperate climate zones (117–119). The morphologic alterations seen in individuals from tropical areas include leaflike villi predominating over fingerlike forms in jejunal segments and an increased number of lamina propria mononuclear cells. This difference manifests as stubby villi with a pyramidal shape and occasional branched and fused villous tips. Interestingly, although the villi are shorter, the villous-to-crypt length ratio usually remains constant in all geographic settings. These alterations should be considered a normal variant because patients are typically asymptomatic and otherwise healthy. The cause of these morphologic changes is uncertain, but environmental factors, particularly regional enteric flora, presumably play a role. Similar mucosal changes are seen in normal individuals in temperate environments, but only in proximal portions of the duodenum.

Aging also modifies small bowel mucosal architecture. Although the literature is limited, specimens from elderly individuals generally have shorter and broader villi than those from younger individuals (120). Moreover, lower animal and human fetuses have been documented as having fingerlike villi exclusively, suggesting that exposure to the environment or aging itself modifies villous architecture. The functional significance of these changes is uncertain.

Diet alters villous architecture in laboratory animals. A diet high in fiber results in broad and fused villi, whereas a fiber-free diet seems to prevent the formation of leaflike forms (22). If this finding is valid in humans, it may be one factor related to the presence of stubby, leaflike villi seen in patients in less-developed countries where high-fiber diets are common.

Metaplastic and Heterotopic Tissues

Gastric-type mucosa is frequently encountered in the proximal small intestine. Gastric foveolar metaplasia characteristically consists of antral-type, PAS-positive, foveolar columnar cells lying along the surface epithelium. This change is focal and often in direct continuity with usual columnar absorptive epithelium on the same villus (121). Although discrete anatomic landmarks delineate the stomach from the duodenum, transition from gastric-antral and duodenal-type mucosa occurs in the midpylorus and is characterized by hybrid mucosa with gastric and intestinal elements (121). As such, foveolar-type surface epithelium in the duodenal bulb may be a natural distal extension of this transition zone. The presence of this type of epithelium in the normal duodenum likely facilitates *Helicobacter pylori* colonization of the proximal duodenum in infected patients. On the other hand, extensive foveolar metaplasia may represent repair or adaptive change upon exposure to increased levels of gastric acid. Scattered chief and parietal cells without any organized arrangement may be associated with this type of epithelium (121). Although some consider this finding to be part of the metaplastic process, others consider it to represent a congenital finding. There are no compelling data to support either view.

However, the presence of organized oxyntic glands likely reflects congenital rests. These gastric heterotopias form single or multiple mucosal polyps that contain all cellular elements encountered in the normal gastric oxyntic mucosa. Characteristically, mucus foveolar epithelium overlies an organized arrangement of glands lined by chief and parietal cells; this is typically well demarcated from the surrounding intestinal-type epithelium. Gastric heterotopia is fairly common, being reported in up to 2% of the population (122), although this number is likely an underestimation. Gastric heterotopia is a well-defined entity in the proximal duodenum and usually presents as a mucosal nodule on the anterior wall. Although they are usually of no clinical significance, larger heterotopias may cause obstructive symptoms or bleed (123). Gastric heterotopias distal to the ligament of Treitz are more often symptomatic due to mucosal ulceration or intussusception (123). This relationship with clinical symptoms may derive from patient

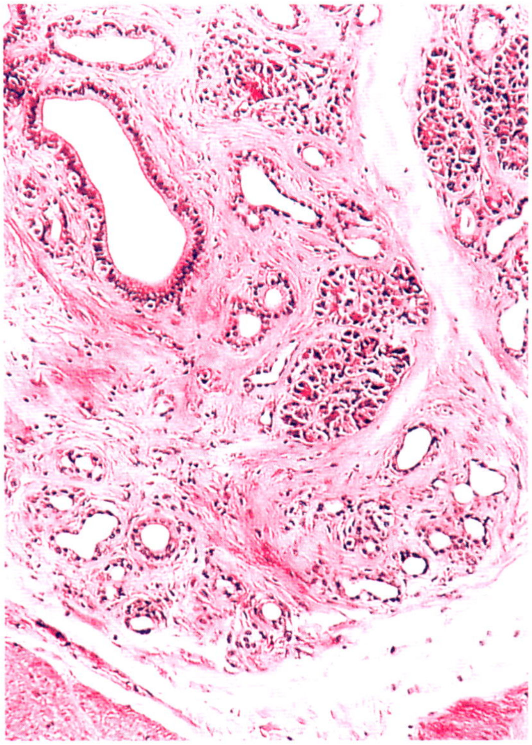

FIGURE 23.33 Heterotopic pancreas in the duodenum characterized in this instance by variably sized ducts, acini, and abundant smooth muscle.

selection bias because asymptomatic gastric heterotopias at a distal site would not be routinely detected.

Heterotopic pancreatic tissue also can be found anywhere along the small intestine but is most commonly observed in the duodenum and jejunum (124,125). It can form submucosal, intramural, or serosal nodules composed of various admixtures of pancreatic acini, ducts, and islets of Langerhans (Fig. 23.33). Isolated ductal structures admixed with smooth muscle may predominate, and in these instances the alternative term adenomyoma has been used (125). Nodules of pancreatic tissue within the small intestine are usually asymptomatic, although larger lesions with prominent mucosal involvement may become clinically significant (125). Combined submucosal pancreatic heterotopia with overlying gastric-type mucosa in the duodenal bulb has been reported (126).

Lymphoid Proliferations

The normal appearance and immunologic composition of small intestinal lymphoid populations have been detailed earlier in the chapter. All these compartments participate at some level in mucosal immune responses, but they also provide the milieu for various hyperplastic and neoplastic proliferations, as well as for certain immunodeficiency states. Some of these disorders have histologic features that deviate only slightly from a normal appearance.

Lymphoid hyperplasias are divided into two broad categories: focal and diffuse forms (55,56). Focal lymphoid hyperplasia is a localized, well-circumscribed proliferation of benign lymphoid tissue characterized by a polymorphic infiltrates within which numerous benign follicles with reactive germinal centers are dispersed. These proliferations often involve only the mucosa and submucosa, but they may extend through the entire bowel wall. Focal lymphoid hyperplasia is predominantly found in the terminal ileum. It is more common in children and adolescents and may cause ileocecal intussusception and symptoms mimicking appendicitis (127).

Diffuse or nodular lymphoid hyperplasia appears as endoscopically appreciable nodular or polypoid aggregates of benign lymphoid follicles segmentally distributed in the small intestine and/or colon (128,129). It is usually asymptomatic and incidentally encountered (127). However, this pattern may be associated with common variable immunodeficiency or selective IgA deficiency, sometimes in the setting of protozoa infection and diarrhea (127).

Low- and high-grade lymphomas of B-, T-, and NK-cell lineages develop in the small intestine (130). Low-grade B-cell lymphomas, such as mantle cell lymphoma, extranodal marginal zone lymphoma and follicular lymphoma may be particularly subtle and present as small lesions of polyps. An accurate diagnosis requires a high index of suspicion on the part of the pathologist, particularly in limited biopsy samples. In general, lymphoid proliferations that present as mass-forming or ulcerative lesions should raise the possibility of lymphoma. Large pleomorphic cells, either in sheets or in a background of mixed immune cells and solid or sheet-like growth of monotonous cells warrant further evaluation.

Morphologic Changes Associated with Ileal Diversion and Continence-Restoring Procedures

With the increasing number of diversion and continence-restoring procedures being performed after total colectomy, it is common to see biopsy and revision specimens after such operations. As a consequence, familiarity with the altered yet normal morphology within these ileal creations must be appreciated in order to evaluate them optimally. The expected mucosal changes include villous shortening and crypt lengthening (approximate 1:1 to 2:1 ratio), increased numbers of goblet cells and lymphoid follicles, and a denser mononuclear cell infiltrate within the lamina propria (131,132). Goblet cell mucin alterations have been seen in nearly 50% of pouches examined, with conversion to predominantly sulfomucins as are typically seen in colonic goblet cells (133).

MUCOSAL BIOPSY SPECIMEN EVALUATION IN SUSPECTED MALABSORPTION

Specimen Procurement and Processing

Small intestinal mucosal biopsy remains the gold standard for evaluation of malabsorptive states and other enteropathies. Clinical professional societies advocate for acquisition of

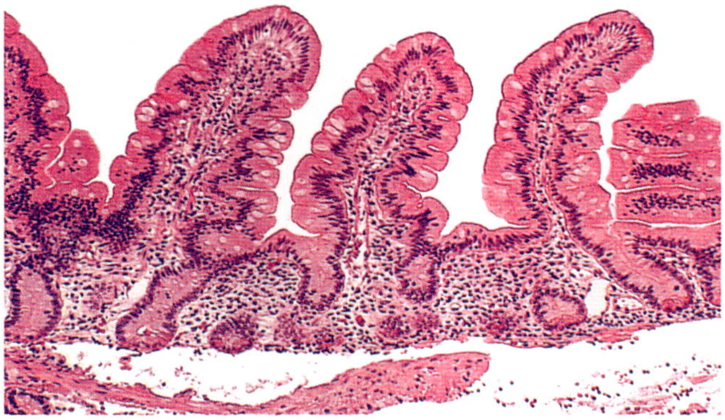

FIGURE 23.34 Absence of muscularis mucosae in a small bowel biopsy specimen, resulting in shorter- and broader-appearing villi that are widely spaced.

multiple tissue samples taken from both the duodenal bulb and the postbulbar duodenum to assess for these disorders. In the modern era, these biopsies are obtained using pinch biopsy forceps (134). Forceps trauma, both in the endoscopy suite and during specimen grossing, may distort the villi, simulating shortening or blunting (Fig. 23.34). Some evidence indicates that these artifacts can be minimized by obtaining and handling samples one tissue fragment at a time (135).

Tangential sectioning of the mucosa may produce a distorted pattern disclosing apparently short and broad villi and an expanded lamina propria. These variations should not be misinterpreted as a villous abnormality. Although some authorities have advocated for orientation of small intestinal mucosal biopsies either at the time of biopsy or during embedding, studies have failed to convincingly show that these time-consuming processes increase diagnostic yield. Interpretive challenges due to tangential sectioning are more effectively mitigated by sufficient sampling, which increases the likelihood that an adequate proportion of the villi will be sectioned to maximize the view of the villus length. In general, if four normal villi in a row are observed, the architecture the specimen is probably normal. Several features should suggest the possibility of tangential sectioning, including the presence of numerous elliptically shaped glands, a multilayered arrangement of the crypts (Fig. 23.35), and a multilayered surface epithelium (Fig. 23.35). If any of these features are present, apparent alterations in villous architecture must be interpreted with caution.

Brunner Gland–Related Artifact

Physiologic differences in the histology of the duodenal bulb compared to the postbulbar duodenum may also simulate disease states. Villi overlying Brunner glands are often short, broad, and stubby (Fig. 23.25) (135). Intraepithelial lymphocytes may be slightly increased in the duodenal bulb. The proximal duodenum is also subject to peptic injury that may cause villous blunting and increased inflammation, although neutrophils tend to be more abundant than lymphocytes in this setting. Regardless of these

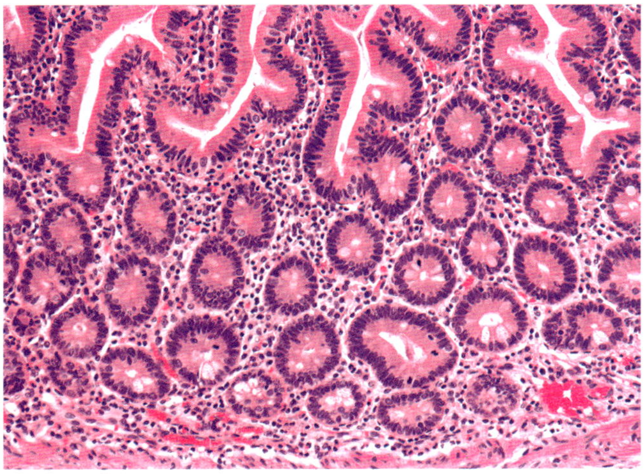

FIGURE 23.35 Another clue in identifying a tangential cut is multilayering, or "stratification," of the surface epithelium (*left portion of central villus*). The normal surface layer is one-cell thick. This broad and short villous appearance is a consequence of malorientation and should be interpreted accordingly.

potential pitfalls, samples from the duodenal bulb should still be obtained as they increase the sensitivity of enteropathy detection when interpreted in clinical context (136).

Lymphoid Aggregate–Related Artifact

Mucosal lymphoid aggregates are scattered along the small bowel and often distort the villous architecture. Villi are usually absent over lymphoid aggregates and nearby villi may be distorted, short, and stubby (Fig. 23.32).

ACKNOWLEDGMENTS

We thank James T. McMahon, PhD, for the electron micrographs.

The author wishes to thank Drs. Megan G. Lockyer and Robert E. Petras for their contributions to the prior version.

REFERENCES

1. Avisse C, Flament JB, Delattre JF. Ampulla of vater. Anatomic, embryologic, and surgical aspects. *Surg Clin North Am*. 2000;80(1):201–212.
2. Brennan PA, Standring SM, Wiseman AM. *Gray's Surgical Anatomy*. 1st ed. Elsevier; 2020.
3. Costacurta L. Anatomical and functional aspects of the human suspensory muscle of the duodenum. *Acta Anat (Basel)*. 1972;82:34–46.
4. Kahn E, Daum F. Anatomy, histology and developmental anomalies of the small intestine and colon. In: Feldman M, Friedman LS, Brandt LJ, eds. *Sleisenger and Fordtran's Gastrointestinal and Liver Disease: Pathophysiology/Diagnosis/Management*. 9th ed. Saunders Elsevier; 2010:1615–1641.
5. Parks DA, Jacobson ED. Physiology of the splanchnic circulation. *Arch Intern Med*. 1985;145:1278–1281.
6. Ohtani O, Ohtani Y. Organization and developmental aspects of lymphatic vessels. *Arch Histol Cytol*. 2008;71:1–22.
7. Phillips RJ, Powley TL. Innervation of the gastrointestinal tract: patterns of aging. *Auton Neurosci*. 2007;136(1-2):1–19.
8. Alpers DH. Digestion and absorption of carbohydrates and proteins. In: Johnson LR, ed. *Physiology of the Gastrointestinal Tract*. 2nd ed. Raven Press; 1987:1469–1487.
9. Feracci H, Bernadac A, Gorvel JP, Maroux S. Localization by immunofluorescence and histochemical labelling of a aminopeptidase N in relation to its biosynthesis in rabbit and pig enterocytes. *Gastroenterology*. 1982;82:317–324.
10. Lojda Z. The histochemical demonstration of brush border endopeptidase. *Histochemistry*. 1979;64:205–221.
11. Goodman BE. Insights into digestion and absorption of major nutrients in humans. *Adv Physiol Educ*. 2010;34:44–53.
12. Johnson LR. *Gastrointestinal Physiology*. 9th ed. Elsevier; 2019.
13. Glickman RM. Fat absorption and malabsorption. *Clin Gastroenterol*. 1983;12:323–324.
14. Farrell JJ. Digestion and absorption of nutrients and vitamins. In: Feldman M, Friedman LS, Brandt LJ, eds. *Sleisenger and Fordtran's Gastrointestinal and Liver Disease*. 9th ed. Saunders Elsevier; 2011:1695–1734.
15. Yin J, Chen JDZ. Roles of interstitial cells of Cajal in regulating gastrointestinal motility: in vitro versus in vivo studies. *J Cell Mol Med*. 2008;12(4):1118–1129.
16. Fiorenza V, Yee YS, Zfass AM. Small intestinal motility: normal and abnormal function. *Am J Gastroenterol*. 1987;82:1111–1114.
17. Elson CO, Kagnoff MF, Fiocchi C, Befus AD, Targan S. Intestinal immunity and inflammation: recent progress. *Gastroenterology*. 1986;91:746–768.
18. Rubin W. The epithelial "membrane" of the small intestine. *Am J Clin Nutr*. 1971;24:45–64.
19. Wilson JP. Surface area of the small intestine in man. *Gut*. 1967;8:618–621.
20. Holmes R, Hourihane DO, Booth CC. The mucosa of the small intestine. *Postgrad Med J*. 1961;37:717–724.
21. Toner PG, Carr KE. The use of scanning electron microscopy in the study of the intestinal villi. *J Pathol*. 1969;97:611–617.
22. Trier JS, Madara JL. Functional morphology of the mucosa of the small intestine. In: Johnson LR, ed. *Physiology of the Gastrointestinal Tract*. 2nd ed. Raven Press; 1987:1209–1249.
23. Dobbins WO III. The intestinal mucosal lymphatic in man. A light and electron microscopic study. *Gastroenterology*. 1966;51:994–1003.
24. Ensari A, Marsh MN. Exploring the villus. *Gastroenterol Hepatol Bed Bench*. 2018;11(3):181–190.
25. Neutra MR, Padykula HK. The gastrointestinal tract. In: Weiss L, ed. *Modern Concepts of Gastrointestinal Histology*. Elsevier; 1984:658–706.
26. Ferguson A, Sutherland A, MacDonald TT, Allan F. Technique for microdissection and measurement in biopsies of human small intestine. *J Clin Pathol*. 1977;30:1068–1073.
27. Kakar S, Nehra V, Murray JA, Dayharsh GA, Burgart LJ. Significance of intraepithelial lymphocytosis in small bowel biopsy samples with normal mucosal architecture. *Am J Gastroenterol*. 2003;98:2027–2033.
28. de Santa Barbara P, van den Brink GR, Roberts DJ. Development and differentiation of the intestinal epithelium. *Cell Mol Life Sci*. 2003;60:1322–1332.

29. Garrison AP, Helmrath MA, Dekaney CM. Intestinal stem cells. *J Pediatr Gastroenterol Nutr.* 2009;49:2–7.
30. Watson AJ. Necrosis and apoptosis in the gastrointestinal tract. *Gut.* 1995;37:165–167.
31. Reed JC. Mechanisms of apoptosis. *Am J Pathol.* 2000;157: 1415–1430.
32. Williamson RC. Intestinal adaptation (first of two parts). Structural, functional, and cytokinetic changes. *N Engl J Med.* 1978;298:1393–1402.
33. Ahuja V, Dieckgraefe BK, Anant S. Molecular biology of the small intestine. *Curr Opin Gastroenterol.* 2006;22:90–94.
34. Möller K, Uhlig R, Gorbokon N, et al. Comparison of INSM1 immunostaining with established neuroendocrine markers synaptophysin and chromogranin A in over 14,000 neuroendocrine and non-neuroendocrine tumors. *Mol Cell Endocrinol.* 2024;581:112106.
35. Buffa R, Rindi G, Sessa F, et al. Synaptophysin immunoreactivity and small clear vesicles in neuroendocrine cells and related tumours. *Mol Cell Probes.* 1987;1:367–381.
36. Facer P, Bishop AE, Lloyd RV, Wilson BS, Hennessy RJ, Polak JM. Chromogranin: a newly recognized marker for endocrine cells of the human gastrointestinal tract. *Gastroenterology.* 1985;89:1366–1373.
37. Sjölund K, Sandén G, Håkanson R, Sundler F. Endocrine cells in human intestine: an immunocytochemical study. *Gastroenterology.* 1983;85:1120–1130.
38. Gunawardene AR, Corfe BM, Staton CA. Classification and functions of enteroendocrine cells of the lower gastrointestinal tract. *Int J Exp Pathol.* 2011;92(4):219–231.
39. Goldman H, Antonioli DA. Mucosal biopsy of the esophagus, stomach, and proximal duodenum. *Hum Pathol.* 1982;13: 423–448.
40. Sandow MJ, Whitehead R. The paneth cell. *Gut.* 1979;20: 420–431.
41. Wehkamp J, Fellermann K, Herrlinger KR, Bevins CL, Stange EF. Mechanisms of disease: defensins in gastrointestinal diseases. *Nat Clin Pract Gastroenterol Hepatol.* 2005;2:406–415.
42. Wehkamp J, Salzman NH, Porter E, et al. Reduced Paneth cell alpha-defensins in ileal Crohn's disease. *Proc Natl Acad Sci U S A.* 2005;102:18129–18134.
43. Ermund A, Schütte A, Johansson MEV, Gustafsson JK, Hansson GC. Studies of mucus in mouse stomach, small intestine, and colon. Gastrointestinal mucus layers have different properties depending on location as well as over Peyer's patches. *Am J Physiol Gastrointest Liver Physiol.* 2013;305:G341–G347.
44. Pelaseyed T, Bergström JH, Gustafsson JK, et al. The mucus and mucins of the goblet cells and enterocytes provide the first defense line of the gastrointestinal tract and interact with the immune system. *Immunol Rev.* 2014;260:8–20.
45. Kim JJ, Khan WI. Goblet cells and mucins: role in innate defense in enteric infections. *Pathogens.* 2013;2(1):55–70.
46. Chen Z-M, Wang HL. Alteration of cytokeratin 7 and cytokeratin 20 expression profile is uniquely associated with tumorigenesis of primary adenocarcinoma of the small intestine. *Am J Surg Pathol.* 2004;28(10):1352–1359.
47. Chauhan A, Sanchez-Avila M, Manivel J, et al. Optimization of immunophenotypic panel to differentiate upper from lower gastrointestinal adenocarcinomas: analysis of new and traditional markers. *Appl Immunohistochem Mol Morphol.* 2021;29(1):13–19.
48. Isaacson P, Judd MA. Carcinoembryonic antigen (CEA) in the normal human small intestine: a light and electron microscopic study. *Gut.* 1977;18:786–791.
49. Groisman GM, Amar M, Livne E. CD10: A valuable tool for the light microscopic diagnosis of microvillus inclusion disease (familial microvillus atrophy). *Am J Surg Pathol.* 2002; 26:902–907.
50. Mayer L, Eisenhardt D, Salomon P, Bauer W, Plous R, Piccinini L. Expression of class II molecules on intestinal epithelial cells in humans. Differences between normal and inflammatory bowel disease. *Gastroenterology.* 1991;100(1):3–12.
51. Pompili S, Latella G, Gaudio E, Sferra R, Vetuschi A. The charming world of the extracellular matrix: a dynamic and protective network of the intestinal wall. *Front Med (Lausanne).* 2021;8:610189.
52. Wershil BK, Furuta GT. 4. Gastrointestinal mucosal immunity. *J Allergy Clin Immunol.* 2008;121(2 Suppl):S380–S383; quiz S415.
53. Park JI, Cho SW, Kang JH, Park TE. Intestinal Peyer's patches: structure, function, and in vitro modeling. *Tissue Eng Regen Med.* 2023;20(3):341–353.
54. Kober OI, Ahl D, Pin C, Holm L, Carding SR, Juge N. γδ T-cell-deficient mice show alterations in mucin expression, glycosylation, and goblet cells but maintain an intact mucus layer. *Am J Physiol Gastrointest Liver Physiol.* 2014;306: G582–G593.
55. Tomasi TB Jr. Mechanisms of immune regulation at mucosal surfaces. *Rev Infect Dis.* 1983;5(Suppl 4):S784–S792.
56. Ranchod M, Lewin KJ, Dorfman RF. Lymphoid hyperplasia of the gastrointestinal tract. A study of 26 cases and review of the literature. *Am J Surg Pathol.* 1978;2:383–400.
57. Powell N, Walker MM, Nicholas JT. Gastrointestinal eosinophils in health, disease, and functional disorders. *Nat Rev Gastroenterol Hepatol.* 2010;7:146–156.
58. Polydorides AD, Banner BF, Hannaway PJ, Yantiss RK. Evaluation of site-specific and seasonal variation in colonic mucosal eosinophils. *Hum Pathol.* 2008;39(6):832–836.
59. Sapp H, Ithamukkala S, Brien TP, et al. The terminal ileum is affected in patients with lymphocytic or collagenous colitis. *Am J Surg Pathol.* 2002;26:1484–1492.
60. Persson T, Andersson P, Bodelsson M, Laurell M, Malm J, Egesten A. Bactericidal activity of human eosinophilic granulocytes against *Escherichia coli*. *Infect Immun.* 2001;69: 3591–3596.
61. Conner JR, Kirsch R. The pathology and causes of tissue eosinophilia in the gastrointestinal tract. *Histopathology.* 2017; 71(2):177–199.
62. McCarthy AJ, Sheahan K. Classification of eosinophilic disorders of the small and large intestine. *Virchows Arch.* 2018;472(1):15–28.
63. Brandtzaeg P, Halstensen TS, Kett K, et al. Immunobiology and immunopathology of human gut mucosa: humoral immunity and intraepithelial lymphocytes. *Gastroenterology.* 1989;97:1562–1584.
64. Rescigno M, DiSabatino A. Dendritic cells in intestinal homeostasis and disease. *J Clin Invest.* 2009;119:2441–2450.
65. Fries PN, Giebel PJ. Mucosal dendritic cell diversity in the gastrointestinal tract. *Cell Tissue Res.* 2011;343:33–41.
66. Izcue A, Powrie F. Special regulatory T-cell review: regulatory T cells and the intestinal tract–patrolling the frontier. *Immunology.* 2008;123:6–10.

67. Schneider EN, Smoller BR, Lamps L. Histiocytic subpopulations in the gastrointestinal tract: distribution and possible relationship to function. *Appl Immunohistochem Mol Morphol.* 2004;12:356–359.
68. Kamada N, Hisamatsu T, Honda H, et al. Human CD14+ macrophages in intestinal lamina propria exhibit potent antigen-presenting ability. *J Immunol.* 2009;183:1724–1731.
69. Panarelli NC. Infectious diseases of the upper gastrointestinal tract. *Histopathology.* 2021;78(1):70–87.
70. Ribatti D. The staining of mast cells: a historical overview. *Int Arch Allergy Immunol.* 2018;176(1):55–60.
71. Hahn HP, Hornick JL. Immunoreactivity for CD25 in gastrointestinal mucosal mast cells is specific for systemic mastocytosis. *Am J Surg Pathol.* 2007;31(11):1669–1676.
72. Heib V, Becker M, Taube C, Stassen M. Advances in the understanding of mast cell function. *Br J Haematol.* 2008; 142(5):683–694.
73. Krystel-Whittemore M, Dileepan KN, Wood JG. Mast cell: a multi-functional master cell. *Front Immunol.* 2016;6:620.
74. Boeckxstaens GE. The emerging role of mast cells in irritable bowel syndrome. *Gastroenterol Hepatol (N Y).* 2018;14(4): 250–252.
75. Goldstein NS, Underhill J. Morphologic features suggestive of gluten sensitivity in architecturally normal duodenal biopsy specimens. *Am J Clin Pathol.* 2001;116(1):63–71.
76. Järvinen TT, Collin P, Rasmussen M, et al. Villous tip intraepithelial lymphocytes as markers of early-stage coeliac disease. *Scand J Gastroenterol.* 2004;39(5):428–433.
77. Cerf-Bensussan N, Schneeberger EE, Bhan AK. Immunohistologic and immunoelectron microscopic characterization of the mucosal lymphocytes of human small intestine by the use of monoclonal antibodies. *J Immunol.* 1983;130: 2615–2622.
78. Greenwood JH, Austin LL, Dobbins WO III. In vitro characterization of human intestinal intraepithelial lymphocytes. *Gastroenterology.* 1983;85:1023–1035.
79. Padmanabhan V, Callas PW, Li SC, Trainer TD. Histopathological features of the terminal ileum in lymphocytic and collagenous colitis: a study of 32 cases and review of literature. *Mod Pathol.* 2003;16(2):115–119.
80. Ribot JC, Lopes N, Silva-Santos B. γδ T cells in tissue physiology and surveillance. *Nat Rev Immunol.* 2021;21 (4):221–232.
81. Montani M, Thiesler T, Kristiansen G. Smoothelin is a specific and robust marker for distinction of muscularis propria and muscularis mucosae in the gastrointestinal tract. *Histopathology.* 2010;57:244–249.
82. Perera DR, Weinstein WM, Rubin CE. Symposium on pathology of the gastrointestinal tract-Part II. Small intestinal biopsy. *Hum Pathol.* 1975;6:157–217.
83. Lord MG, Valies P, Broughton AC. A morphologic study of the submucosa of the large intestine. *Surg Gynecol Obstet.* 1977; 145:55–60.
84. Lee AK, DeLellis RA, Silverman ML, Wolfe HJ. Lymphatic and blood vessel invasion in breast carcinoma: a useful prognostic indicator? *Hum Pathol.* 1986;17:984–987.
85. Schlingemann RO, Dingjan GM, Emeis JJ, Blok J, Warnaar SO, Ruiter DJ. Monoclonal antibody PAL-E specific for endothelium. *Lab Invest.* 1985;52:71–76.
86. Vardy PA, Lebenthal E, Shwachman H. Intestinal lymphangiectasia: a reappraisal. *Pediatrics.* 1975;55:842–851.
87. Gershon MD, Erde SM. The nervous system of the gut. *Gastroenterology.* 1981;80:1571–1594.
88. Krishnamurthy S, Schuffler MD. Pathology of neuromuscular disorders of the small intestine and colon. *Gastroenterology.* 1987;93:610–639.
89. Ferri GL, Probert L, Cocchia D, Michetti F, Marangos PJ, Polak JM. Evidence for the presence of S-100 protein in the glial component of the human enteric nervous system. *Nature.* 1982;297:409–410.
90. Gordon IO, Bettenworth D, Bokemeyer A, et al; Stenosis Therapy and Anti-Fibrotic Research (STAR) Consortium. Histopathology scoring systems of stenosis associated with small bowel Crohn's disease: a systematic review. *Gastroenterology.* 2020;158(1):137–150.e1.
91. Streutker CJ, Huizinga JD, Driman DK, Riddell RH. Interstitial cells of Cajal in health and disease. Part I: Normal ICC structure and function with associated motility disorders. *Histopathology.* 2007;50:176–189.
92. Farrugia G. Interstitial cells of Cajal in health and disease. *Neurogastroenterol Motil.* 2008;20(Suppl 1):54–63.
93. Hagger R, Finlayson C, Jeffrey I, Kumar D. Role of the interstitial cells of Cajal in the control of gut motility. *Br J Surg.* 1997;84:445–450.
94. Lawson HH. The duodenal mucosa in health and disease. A clinical and experimental study. *Surg Annu.* 1989;21: 157–180.
95. Lawson HH. Definition of the gastroduodenal junction in healthy subjects. *J Clin Pathol.* 1988;41:393–396.
96. Korn ER, Foroozan P. Endoscopic biopsies of normal duodenal mucosa. *Gastrointest Endosc.* 1974;21:51–54.
97. Kreuning J, Bosman FT, Kuiper G, Wal AM, Lindeman J. Gastric and duodenal mucosa in "healthy" individuals. An endoscopic and histopathological study of 50 volunteers. *J Clin Pathol.* 1978;31:69–77.
98. Dandalides SM, Carey WD, Petras RE, Achkar E. Endoscopic small bowel mucosal biopsy: a controlled trial evaluating forceps size and biopsy location in the diagnosis of normal and abnormal mucosal architecture. *Gastrointest Endosc.* 1989;35:197–200.
99. Robertson HE. The pathology of Brunner's glands. *Arch Pathol.* 1941;31:112–130.
100. Leeson TS, Leeson RC. The fine structure of Brunner's glands. *J Anat.* 1968;103:263–276.
101. Thompson IW, Day DW, Wright NA. Subnuclear vacuolated mucous cells: a novel abnormality of simple mucin-secreting cells of non-specialized gastric mucosa and Brunner's glands. *Histopathology.* 1987;11:1067–1081.
102. Kamiya R. Basal-granulated cells in human Brunner's glands. *Arch Histol Jpn.* 1983;46:87–101.
103. Treasure T. The ducts of Brunner's glands. *J Anat.* 1978; 127:299–304.
104. Bosshard A, Chery-Croze S, Cuber JC, Dechelette MA, Berger F, Chayvialle JA. Immunocytochemical study of peptidergic structures in Brunner's glands. *Gastroenterology.* 1989;97:1382–1388.
105. Zhu M, Li H, Wu Y, et al. Brunner's gland hamartoma of the duodenum: a literature review. *Adv Ther.* 2021;38(6): 2779–2794.
106. Lu L, Li R, Zhang G, Zhao Z, Fu W, Li W. Brunner's gland adenoma of duodenum: report of two cases. *Int J Clin Exp Pathol.* 2015;8(6):7565–7569.

107. Silverman L, Waugh JM, Huizenga KA, Harrison EG Jr. Large adenomatous polyp of Brunner's glands. *Am J Clin Pathol.* 1961;36:438–443.
108. Kamani L, Raj R, Ali R, Siddiqi F. Brunner's gland hyperplasia: a rare cause of gastrointestinal bleeding. *Clin Pract.* 2020; 10(2):1267.
109. Lopez G, D'Ercole M, Ferrero S, Croci GA. Duodenal pseudomelanosis: a literature review. *Diagnostics (Basel).* 2021; 11(11):1974.
110. Rosenberg JC, Didio LJ. Anatomic and clinical aspects of the junction of the ileum with the large intestine. *Dis Colon Rectum.* 1970;13:220–224.
111. Kumar D, Phillips SF. The contribution of external ligamentous attachments to function of the ileocecal junction. *Dis Colon Rectum.* 1987;30:410–416.
112. Axelsson C, Andersen JA. Lipohyperplasia of the ileocaecal region. *Acta Chir Scand.* 1974;140:649–654.
113. Cuvelier C, Demetter P, Mielants H, Veys EM, De Vos M. Interpretation of ileal biopsies: morphological features in normal and diseased mucosa. *Histopathology.* 2001;38: 1–12.
114. Urbanski SJ, Arsenault AL, Green FH, Haber G. Pigment resembling atmospheric dust in Peyer's patches. *Mod Pathol.* 1989;2:222–226.
115. Mackey WC, Dineen P. A fifty year experience with Meckel's diverticulum. *Surg Gynecol Obstet.* 1983;156:56–64.
116. Artigas V, Calabuig R, Badia F, Rius X, Allende L, Jover J. Meckel's diverticulum: value of ectopic tissue. *Am J Surg.* 1986;151:631–634.
117. Bennett MK, Sachdev GK, Jewell DP, et al. Jejunal mucosal morphology in healthy north Indian subjects. *J Clin Pathol.* 1985;38:368–371.
118. Lindenbaum J, Gerson CD, Kent TH. Recovery of small-intestinal structure and function after residence in the tropics. I. Studies in Peace Corps volunteers. *Ann Intern Med.* 1971; 74:218–222.
119. Gerson CD, Kent TH, Saha JR, Siddiqi N, Lindenbaum J. Recovery of small-intestinal structure and function after residence in the tropics. II. Studies in Indians and Pakistanis living in New York City. *Ann Intern Med.* 1971;75:41–48.
120. Webster SG, Leeming JT. The appearance of the small bowel mucosa in old age. *Age Ageing.* 1975;4:168–174.
121. Tsubone M, Kozuka S, Taki T, Hoshino M, Yasui A, Hachisuka K. Heterotopic gastric mucosa in the small intestine. *Acta Pathol Jpn.* 1984;34:1425–1431.
122. Van Zanten SJ, Dixon MF, Lee A. The gastric transitional zones: neglected links between gastroduodenal pathology and helicobacter ecology. *Gastroenterology.* 1999;116(5):1217–1229.
123. Lai EC, Tompkins RK. Heterotopic pancreas. Review of a 26 year experience. *Am J Surg.* 1986;151:697–700.
124. Dolan RV, ReMine WH, Dockerty MB. The fate of heterotopic pancreatic tissue. A study of 212 cases. *Arch Surg.* 1974; 109:762–765.
125. Armstrong CP, King PM, Dixon JM, Macleod IB. The clinical significance of heterotopic pancreas in the gastrointestinal tract. *Br J Surg.* 1981;68:384–387.
126. Tanemura H, Uno S, Suzuki M, et al. Heterotopic gastric mucosa accompanied by aberrant pancreas in the duodenum. *Am J Gastroenterol.* 1987;82:685–688.
127. Daniels JA, Lederman HM, Maitra A, Montgomery EA. Gastrointestinal tract pathology in patients with common variable immunodeficiency (CVID): a clinicopathologic study and review. *Am J Surg Pathol.* 2007;31:1800–1812.
128. Rambaud JC, De Saint-Louvent P, Marti R, et al. Diffuse follicular lymphoid hyperplasia of the small intestine without primary immunoglobulin deficiency. *Am J Med.* 1982; 73:125–132.
129. Matuchansky C, Touchard G, Lemaire M, et al. Malignant lymphoma of the small bowel associated with diffuse nodular lymphoid hyperplasia. *N Engl J Med.* 1985;313:166–171.
130. Kim EK, Yang WI, Ryu HJ, Ji HL, Yoon SO. Practical approach to the histologic diagnosis of gastrointestinal lymphomas through the first-line marker battery of CD20, CD3, CD30, and Epstein-Barr virus-encoded RNAs. *Adv Anat Pathol.* 2020;27(2):75–86.
131. Bechi P, Romagnoli P, Cortesini C. Ileal mucosal morphology after total colectomy in man. *Histopathology.* 1981;5: 667–678.
132. Philipson B, Brandberg A, Jagenburg R, Kock NG, Lager I, Ahrén C. Mucosal morphology, bacteriology, and absorption in intra-abdominal ileostomy reservoir. *Scand J Gastroenterol.* 1975;10:145–153.
133. Shepherd NA, Jass JR, Duval I, Moskowitz RL, Nicholls RJ, Morson BC. Restorative proctocolectomy with ileal reservoir: pathological and histochemical study of mucosal biopsy specimens. *J Clin Pathol.* 1987;40:601–607.
134. Rubio-Tapia A, Hill ID, Semrad C, et al. American College of Gastroenterology guidelines update: diagnosis and management of celiac disease. *Am J Gastroenterol.* 2023;118(1):59–76.
135. Robert ME, Crowe SE, Burgart L, et al. Statement on best practices in the use of pathology as a diagnostic tool for celiac disease: a guide for clinicians and pathologists. *Am J Surg Pathol.* 2018;42(9):e44–e58.
136. Voutilainen M, Juhola M, Färkkilä M, Sipponen P. Gastric metaplasia and chronic inflammation at the duodenal bulb mucosa. *Dig Liver Dis.* 2003;35(2):94–98.

24

Colon

Maria Westerhoff ▪ Joel K. Greenson

EMBRYOLOGY

Early in development, the gut is patterned into four asymmetrical axes—anterior–posterior, dorsoventral, left–right, and radial (1–8). The fifth axis is the functional axis of developmental immunologic programming. This axis in particular forms the environment for the gastrointestinal tract to be colonized by resident flora (9–11). Because gut epithelium is constantly differentiating from a stem cell in a progenitor pool throughout adult life, the pathways involved in embryology, axes of development, and cell–cell interactions continue to be important in the adult intestinal epithelium (7,8,12–14).

Development and differentiation along the anterior–posterior axis gives rise to the foregut, midgut, and hindgut, resulting in nuances in differentiation that are regionally specific from mouth to anus. The right colon (cecum, appendix, and ascending colon, hepatic flexure) and proximal two-thirds of the transverse colon, arise from the midgut. The hindgut gives rise to the distal one-third of the transverse colon, left colon (splenic flexure, descending colon, and sigmoid colon), and rectum (15,16). The significant variation in patterns of gene expression, physiologic function, disease distribution, and even histologic appearance between the right and left colon reflects the midgut and hindgut derivation (7,17–32).

The left–right axis is manifested in the colon by characteristic turning and looping of the gut, resulting in portions of colon with varying mesentery and fixation within the abdominal cavity (15).

The fundamental axis maintained in the adult is the radial, crypt to surface, axis. This is dependent upon colonization of the large intestine by successive consortia of bacterial species to form the microbiome, which is the complete set of genes within the microbiotica (10–12). Homeostasis of intestinal epithelium occurs throughout life along the radial axis. The epithelial and mesenchymal progenitor/proliferative cells reside deeper in the mucosa, while the differentiated functional cells are more luminal (13,33–35). Thus, the proliferative zone of the colonic mucosa is in the bases of the crypts and apoptosis is not a normal finding in this area. The mucosa-associated bacteria adherent to the cells makes up the gut microbiota. This is critical for the normal structure, development, and optimal function of not only the large intestine, but also of the mucosal immune system (10,12,36,37).

The primitive gut tube subdivides into fore-, mid-, and hindgut around the 4th developmental week. At this stage, the midgut is midline in the embryo; however, it remains open to the yolk sac (38). Around the 6th week, the midgut is pushed out into the extraembryonic coelom as the growing liver is taking up space within the abdominal cavity. It is thought that the first rotation of the gut takes place at this time, with the small intestine to the right and the colon to the left. The umbilical (extraembryonic) portion of the gut then elongates, but the growth of the intra-abdominal portion is minimal, understandably due to the space occupied by the liver. The gut returns to the intra-abdominal position around the 10th week. Traditionally, it is thought that the small bowel enters first, followed by the cecum, but recent reports observe in animal studies that the ileal loops may actually be last to enter. There is then a 180-degree rotation causing the cecum to be placed into the right upper side of the abdomen. This is followed by the descent of the cecum into the right iliac fossa. The descending colon becomes fixed and retroperitoneal when the mesentery fuses with the peritoneum of the left dorsal part of the abdominal wall (39). The sigmoid mesocolon, however, persists. The terminal aspect of the hindgut enters the cloaca (hence the anus is derived from both hindgut and the posterior part of the cloaca).

Abnormalities of any of the developmental pathways or along any axis during organogenesis may result in gross morphologic malformations. This includes diverticula, rotational malformations, atresias, duplications, and aganglionic segments (15,16,40–44). Disturbances of developmental pathways used for organ homeostasis may result in metaplasias, polyposis syndromes, and malignant transformation (12,45–49). Recent studies have implicated the microbiome as being important in a vast array of diseases both within the gastrointestinal tract and systemically. The influence of the microbiome on drug toxicity, as well as irritable bowel syndrome, inflammatory bowel disease (IBD), autoimmunity, allergy, and obesity is an area of ongoing research (50,51). Interestingly, microbiota also play a role in gut motility by having an effect on the development of enteric glial cells and the enteric nervous system as a whole (52,53).

ANATOMIC CONSIDERATIONS

The colon is the terminal 1.0 to 1.5 m segment of the gastrointestinal tract, subdivided into the cecum, ascending, transverse, descending, and sigmoid colon, with the rectum terminating at the anal canal (Fig. 24.1) (54,55). Typically, the cecum is entirely covered by peritoneum, while the ascending colon is retroperitoneal, with peritoneum on the lateral and anterior surfaces. The transverse colon has a mesentery and is completely surrounded by peritoneum. The descending colon is also retroperitoneal and thus lacks serosa on the posterior aspect. The sigmoid is intraperitoneal

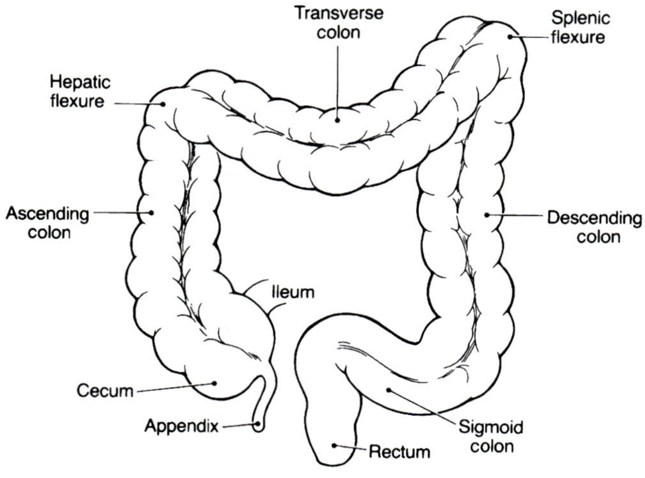

FIGURE 24.1 Major regions of the colon.

with a mesentery. The rectum is divided into thirds, with the upper having peritoneum on the anterior and lateral surfaces, middle third having peritoneum only on the anterior aspect, and the lower third having no peritoneal covering at all. These anatomic differences in peritonealization have implications on surgical margins for resections. The nonperitonealized surfaces that are cut by the surgeon represent margins, whereas the serosal aspects do not (Table 24.1).

There is considerable anatomic variation in the position of the colon segments, mesenteric coverings, and attachments to the posterior abdominal wall (55,56). Nonetheless, the vascular supply, venous drainage, and innervation pattern for the two primary (right and left) colon segments along the embryologic anterior–posterior (midgut/hindgut) axis are consistent. The right colon receives its blood supply from the superior mesenteric artery, its parasympathetic nervous innervation from the vagus nerve, and sympathetic innervation from the superior mesenteric ganglia. The left colon receives its blood supply from the inferior mesenteric artery, parasympathetic innervation from sacral nerves S2, S3, and S4 through the pelvic splanchnic nerves, and sympathetic innervation from the inferior mesenteric ganglia. Venous drainage is predominantly portal. The rectum receives blood

TABLE 24.1 Peritoneal Covering of Various Sites of Colon: Implications on Margin Status in Surgical Specimens

Site	Status of Peritoneal Covering	Implications on Surgical Margin on a Resection Case
Cecum	Entirely peritonealized	
Transverse colon	Intraperitoneal, with mesentery; mesenteric margin	Mesenteric margin
Descending colon	Anterior and lateral surfaces covered by visceral peritoneum; posterior surface lacks peritoneum	Nonperitonealized aspect will be radial margin
Sigmoid	Intraperitoneal, with mesentery	Mesenteric margin
Rectum	Upper third of rectum: anterior and lateral aspect covered by visceral peritoneum Middle third of rectum: anterior aspect covered by peritoneum Lower third of rectum: lacks peritoneal covering	Upper third: posterior aspect is radial margin Middle third: posterior and lateral aspects are radial margin Lower third: circumferential radial margin

from the middle and inferior rectal arteries, parasympathetic innervation from the pelvic splanchnic nerves, and sympathetic innervation through the hypogastric plexus through lumbar spinal segments L1, L2, and L3 (21,54–56).

Unique external features of the colon include the bands of teniae coli and saccular haustra, visible through the investing serosa and subserosal tissue. The muscular layers of the large intestine are composed of both longitudinally and circularly arranged fibers. The outer longitudinal fibers are primarily concentrated into three flat smooth muscle bands of the teniae coli, with one strip at the mesentery and the other two equidistant around the colonic circumference (54–56). Epiploic appendages are small protuberances of adipose tissue present at the two nonmesenteric taeniae coli bands. The vasculature supplying the colon penetrates the inner circular muscularis at the edges of the teniae coli. The thickness of the colonic wall varies from 0.2 cm (cecum) to 0.4 cm (sigmoid).

From the luminal perspective, various landmarks are recognized by the endoscopist. The cecum is readily identified by the ileocecal valve, the appendiceal orifice, and blind-ended, saclike appearance. It possesses the widest diameter within the colon. The subjacent portal vasculature imparts a blue hue to the mucosa at the hepatic flexure. Orientation of the teniae coli within the transverse colon results in a T-shaped lumen, ending in a slitlike orifice and acute angle of the splenic flexure. Although the descending and sigmoid colon may have thickened mucosal folds and diverticular orifices, calibration marks on the colonoscope are more reliable means of approximating the location within this region. Finally, unlike the small bowel, which possess circumferential mucosal folds (plicae circularis), the colonic mucosa exhibits partial, incomplete mucosal folds.

FUNCTION

The colon participates in several metabolic processes that include absorption, secretion, fermentation, and oxidation unique to the colonic epithelium, or in concert with the commensal bacteria. The commensal microbiota are integral in the formation of short-chain fatty acids, metabolic intermediates and vitamins, detoxification or biotransformation of bile acids, as well as phosphate and oxalate excretion (31,57–62). Fermentation of carbohydrates to form short-chain fatty acids, particularly butyrate, serves as a major source of energy for colonocytes, and butyrate plays a crucial role in colonocyte growth and differentiation (59,61,63–65). Nearly equal to the activity observed in the liver, colonocytes have the capacity to mediate biotransformation of bile salts, drugs, and xenobiotics (65,66). Many of these processes are segment-specific to either the right or the left colon.

The well-recognized function of the colon is absorption of water and storage of the feces, with a stool output of 200 to 500 g daily. The cecum receives 1.3 to 1.8 L of electrolyte-rich ileal effluent daily and is a high-capacity absorptive surface. It effectively absorbs 80% of the chylous water (and sodium ions, Na^+) during prolonged mucosal exposure to the luminal contents made possible by the retrograde peristalsis unique to the cecum (23,28,29,32). Bulk absorption of water and sodium occurs via electroneutral sodium chloride (NaCl) transport, which occurs at the surface and in the superficial portions of the crypts (23,32,67). Within the left colon, low-capacity electrogenic absorption via luminal sodium channels that are regulated by aldosterone and angiotensin, serve in further absorption of water from the fecal contents, as well as sodium preservation (23,32,68,69). In addition, another mechanism for water absorption unique to the left colon has been observed; it involves the formation of a hyperosmolar (Na^+) compartment between the colonocytes and pericrypt myofibroblasts and may be vital in extracting water from the osmotically dense feces, allowing final stool compaction (29,70,71). Lubricating the increasingly dense feces with various mucins is an additional important function of the left mucosa.

In regards to its role in immunity, the intestinal mucosa is the largest immune organ of the body (11,70–74). The plasma cells within the mucosa represent 80% of the antibody-producing cells within the entire body and produce more antibodies than any other part of the body (70–72,74). The process of antigen sampling (commensal bacterial and dietary antigens) across specialized regions of the colon results in "gut priming" and routinely confers protection from potential infection locally and in other mucosal tissues (systemic immunization) (36,37,40,66,75–83). At baseline, the gut immune system is highly activated in response to normal flora—the so-called physiologic inflammation—in which the intestinal microflora and the intestinal immunologic mechanisms influence each other locally and systemically, forming an interdependent mutualistic ecosystem, the balance of which is required for maintenance of health and prevention of disease (36,37,40,66,79,80,84,85). Locally, the colonic epithelium and commensal flora serve as important barriers to infection via tight junctions and secretion of antimicrobial substances, as well as competition for nutrient substrates (66,73,74,77,86).

LIGHT MICROSCOPY

The colon contains four histologically distinct compartments: (a) mucosa, (b) submucosa, (c) muscularis propria, and (d) serosa. The enteric nervous system spans all four compartments with ganglia and plexus in both the submucosa and muscularis propria, extending processes throughout the lamina propria, submucosa, and muscular layers.

Mucosa

The luminal colonic mucosa is the most metabolically and immunologically active compartment of the colon. The luminal surface is covered by glycocalyx (glycans, enzymes, lectins, and mucin), facilitating formation of the

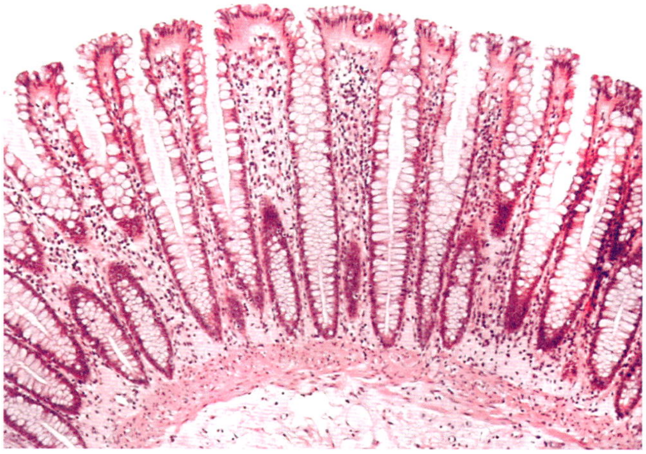

FIGURE 24.2 Normal colonic mucosa. The histologic section of this endoscopic mucosal biopsy specimen is oriented so the simple columnar surface epithelium facing the lumen is at the top of the figure and the cut surface of the specimen is at the bottom. The mucosal crypts are lined up in parallel, and they open to the lumen. The lamina propria consists of the stromal elements investing the crypts and extend from the surface epithelium to the smooth muscle cells of the muscularis mucosae at the bottom.

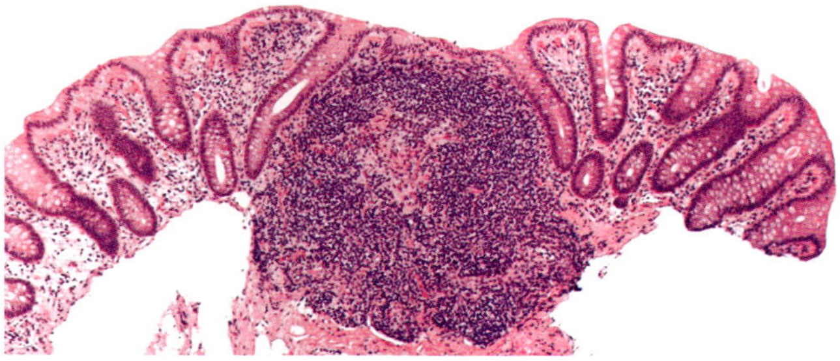

FIGURE 24.4 Colonic lymphoid aggregate. This lymphoid aggregate splays the adjacent crypts, which appear in a diagonal or near horizontal axis (rather than vertical), and mimics basal lymphoplasmacytosis at the crypt/lymphoid aggregate margin, resembling minor architectural disarray. Goblet cells are absent in the crypt epithelium adjacent to the follicle, while retained on the opposite side of the crypt. These normal features of colonic lymphoid aggregates should not be misinterpreted as architectural distortion indicative of chronic mucosal injury (i.e., inflammatory bowel disease) (see Table 24.3). These aggregates may contain well-formed germinal centers and appear as small polyps endoscopically.

commensal microbial ecosystem and serving a barrier function (87–89). Beneath this, the columnar epithelium lines regularly spaced crypts that span the depth of the lamina propria. The crypts are aligned perpendicular to and extend to the muscularis mucosae, imparting the well-known "rack of test tubes" appearance (Fig. 24.2). While some variation in space between crypts is expected in normal individuals, irregularly oriented or bifurcated crypts are generally considered abnormal (see Regional Variation in Histologic Features, below). It is important to note, however, that there are regularly occurring folds in the mucosa that lead to slight crypt distortion still considered within the range of normal. These include innominate grooves, mucosa adjacent to lymphoid follicles, epithelium within lymphoglandular complexes, and ridges created by muscularis mucosae contraction (Figs. 24.3 to 24.5). These normal variations must be distinguished from the histologic changes of chronic mucosal injury (as in IBD) (Table 24.2).

Advances in immunohistochemistry and development of new antibodies has allowed further classification of cell types, with utility in assessing both normal and pathologic histologic patterns and providing useful adjuncts to standard histochemical stains (Table 24.3) (74,90–130).

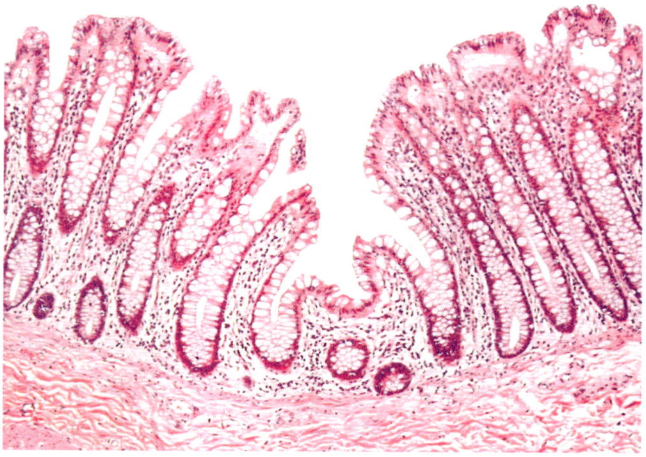

FIGURE 24.3 Innominate grooves of colonic mucosa. Multiple crypts open in a "mirror image" across a common crypt lumen at the groove, with the common crypt lumen opening to the colonic lumen. This normal finding is not a true branching of the crypts and should not be misinterpreted as architectural distortion indicative of chronic mucosal injury (i.e., inflammatory bowel disease). The innominate groove common lumen is generally within the superficial one-third of the mucosa.

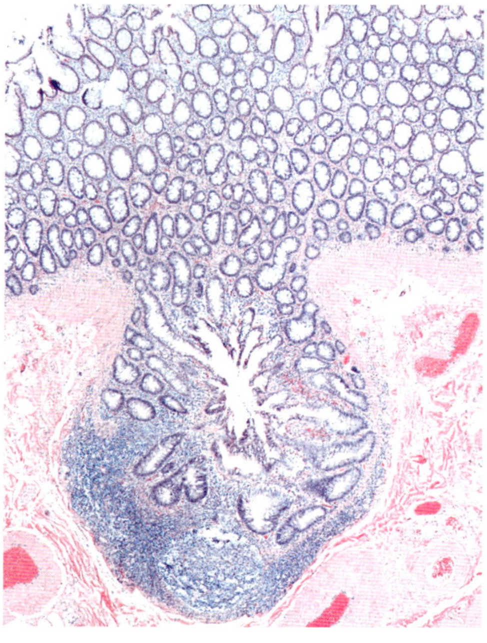

FIGURE 24.5 Lymphoglandular complex. Crypt epithelium is present within a lymphoid follicle that extends from the mucosa through the muscularis mucosae into the submucosa.

TABLE 24.2 Common Artifacts and Variants: A Guide to Evaluation and Interpretation

Histologic Feature	Common Misinterpretation	Keys to Accurate Interpretation
Mucosa with regular eosinophilic subnuclear material	Collagenous colitis ***Accurate Interpretation:*** **Normal colon, tangentially sectioned**	Assess for other features of microscopic colitis.[a] Tangential sectioning of the surface epithelium results in the cytoplasm of adjacent enterocytes appearing below the basally aligned nuclei of "in plane" enterocytes. Levels or a trichrome stain.
Normal colon—histologic features seen in association with adjacent lymphoid aggregates or follicles		
Focally increased surface intraepithelial lymphocytes with otherwise normal architecture.	Lymphocytic colitis ***Accurate Interpretation:*** **Normal colon**	Assess for other features of the microscopic colitides.[a] Level through the biopsy material to locate the adjacent lymphoid aggregate or lymphoid follicle.
Focal "architectural distortion" in association with numerous lamina propria mononuclear cells. Other biopsy fragments appear normal.	Nonspecific colitis IBD: Ulcerative colitis (UC), indeterminate colitis (IDC), Crohn disease ***Accurate Interpretation:*** **Normal colon**	Assess for other features of chronic mucosal injury.[b] Level through the biopsy material to locate the adjacent lymphoid aggregate or lymphoid follicle.
Normal colon—histologic features mimicking chronic colitis (IBD)		
Biopsy contains horizontal and unusually oriented crypts.	IBD (either UC or IDC). ***Accurate Interpretation:*** **Normal colon**	Biopsy material extracted without the muscularis mucosae frequently contains unusually oriented crypts. Assess crypt morphology only in areas in which the muscularis mucosae is present, "tethering" the crypts. Examine other tissue fragments; perform levels when necessary. Assess for other features of chronic mucosal injury.[b]
Right (or ascending) colon biopsy with numerous lamina propria mononuclear cells	Nonspecific colitis IBD (either UC or IDC) ***Accurate Interpretation:*** **Normal proximal (right) colon**	Ensure site of biopsy Determine extent of mononuclear expansion, if any. The right colon contains significantly more lamina propria inflammatory cells than the left colon. Assess for other features of chronic mucosal injury.[b]
Deeply eosinophilic granular cells at the base of crypts in the right colon (Paneth cells), could this mean chronic mucosal injury?	Nonspecific colitis UC, IDC, or Crohn's ***Accurate Interpretation:*** **Normal colon**	Assess for other features of chronic mucosal injury.[b] Endocrine cells have apical (luminal) nuclei and fine basal granules and are normal throughout the colon. Paneth cells have basal nuclei; coarse luminal granules are normal in the right colon and pathologic in the left.
Bifurcated crypts in sigmoid colon and/or rectum, without other findings	Inactive IBD ***Accurate Interpretation:*** **Normal colon or rectum**	One to two bifurcated crypts within the sigmoid colon or rectum is acceptable. Assess for other features of chronic mucosal injury.[b]
Normal colon—histologic features commonly encountered resulting from bowel preparation effects		
Widely spaced crypts, without other findings	Edema Nonspecific colitis ***Accurate Interpretation:*** **Normal colon**	Water content/edema is not a reproducible finding. Review clinical history for type of bowel preparation, sodium phosphate enemas frequently cause edema. Ensure biopsy fragment contains muscularis mucosae. If other features suggest (i.e., lamina propria pallor [reduced mononuclear cells] and crypt distortion), consider treated IBD.
Basal and/or surface epithelial apoptosis with or without reactive-appearing surface enterocytes	Resolving acute self-limited colitis Antibiotic-associated colitis ***Accurate Interpretation:*** **Normal colon**	Review the clinical history to determine bowel preparation used (oral sodium phosphate frequently causes basal apoptosis, while sodium phosphate enemas may cause surface or basal apoptosis bowel preparation effect). Alert clinical colleagues that sodium phosphate causes bowel preparation effect and is ill advised in the evaluation of diarrhea or GVHD patients.

[a]Histologic features of microscopic colitis: (a) increased intraepithelial lymphocytes; (b) accompanying damage to surface epithelium (reactive appearance, epithelial sloughing); (c) superficially dense mononuclear inflammatory infiltrate; and with or without (d) thickening or irregularity of the subepithelial basement membrane collagen table (usually entraps superficial capillaries).

[b]Histologic features of chronic mucosal injury (chronic colitis, also known as inflammatory bowel disease, or IBD): (a) mononuclear expansion of lamina propria, displacing crypts and resulting in (b) basal plasmacytosis; (c) crypt architectural distortion (bifurcated or irregularly oriented crypts; crypt "dropout"); and (d) Paneth cell (left colon only) or pyloric metaplasia.

GVHD, graft-versus-host disease; IBD, inflammatory bowel disease; IDC, indeterminate colitis; UC, ulcerative colitis.

TABLE 24.3 Predominant Immunohistochemical and Histochemical Staining Patterns of Many of the Normal Cell Types Present in the Human Colon (76,115,131)

Cell Type	Immunohistochemical Profile Positive	Negative	Histochemical Stain
Colonic enterocyte	CK20, pCEA, mCEA, villin, Cdx2, AE1/AE3, SATB2	CK7, EGFR (extremely low expression)	AB2.5—patchy apical blush
Goblet cell (*S*, surface; *C*, crypts)	MUC1 (S/C), MUC2 (S/C), MUC3 (S), MUC4 (S/C), MUC5B (C), MUC11 (S/C), MUC12 (S/C)	MUC3 (C) MUC5A (S/C) MUC5B (S)	+AB2.5, mucicarmine, PAS, PAS-D
Enteroendocrine cell	Chromo-A, Chromo-B, synaptophysin, NSE, specific peptides; AE1/AE3	—	Grimelius
Paneth cell	HL-5, HL-6 (R); AE1/AE3	—	Autofluorescent with eosin
M cell	No known definitive differentiating stains	—	No known definitive differentiating stains
Intraepithelial lymphocytes (surface)	CD3 TCRαβ, CD3 TCRγδ	CD10, CD43, CD138	—
Intraepithelial lymphocytes (M cell)	CD3, CD45 RO, CD45 RA (rare), CD20 (rare)	CD138	—
Lymphoid aggregate (follicle center)	CD19, CD20, CD10, CD68, occasional CD20, occasional S100, scattered CD45 RO/CD3	Bcl-2	—
Lymphoid aggregate (periphery/ paracortex)	CD3, CD5, CD20 (rare), S100 (IDC, occasional)	CD138	—
LP plasma cells	CD79a, CD138	CD20, CD3, CD123	—
LP lymphocytes	CD3 (CD4+/CD8+ varying ratio) CD5 (occasional)	Keratin, S100	—
Eosinophils	CD15	—	Autofluorescent with eosin
Mast cells	Tryptase, CD117 (*c-kit*)	CD34	Giemsa, toluidine blue
Macrophages	CD68, HAM56, MAC387, lysozyme, α1 antitrypsin	LCA, keratin	Iron (hemosiderin and anthracene pigment of melanosis coli)
Dendritic macrophages	CD11b (subepithelial dome), CD123	α-SMA, keratin	—
Muciphages	CD68, HAM56	—	AB2.5; PAS-D
Pericrypt myofibroblasts	Vimentin, HHF35, SMMHC, α-SMA, EMA (focal)	Desmin	MT–mixed blue and red
Subepithelial myofibroblasts	Vimentin, α-SMA	Desmin	MT–mixed blue and red
Basement membrane	Collagen IV, tenascin (minimal)	Tenascin (thick)	MT–blue; saffron–deep red; eosin–autofluorescence
Muscularis mucosa	Vimentin, HHF35, α-SMA, desmin	—	MT–red
Arterioles, capillaries, veins	*Luminal:* CD31, CD34, vimentin, vWF, factor XIII *Wall:*α-SMA	α-SMA, HHF-35	MT–red; elastin
Lymphatic vessels	CD34, vimentin, D2-40 (R)®	vWF	MT–red
Enteric glia and ganglia	Synaptophysin, PDGFR-α(R), NSE	α-SMA	—
Schwann cells	S100, vimentin	α-SMA	—
Interstitial cells of Cajal	CD34, CD117 (c-kit)	S100, CD31	—
Submucosal adipose	S100	—	Oil red O
SM lymphocytes	CD3, scattered CD138 plasma cells	—	—
Muscularis propria	α-SMA, desmin, vimentin	—	MT–deep red
Serosal mesothelium	Calretinin, vimentin, AE1/AE3, CK7	pCEA	—

—, no information; AB2.5, Alcian blue 2.5; AE1/AE3, pan-cytokeratin; C, crypts; Chromo-A, chromogranin A; Chromo-B, chromogranin B; CK20, cytokeratin 20; DGFR, epidermal growth factor; IDC, interdigitating dendritic cells; LCA, leukocyte common antigen; LP, lamina propria; mCEA, monoclonal carcinoembryonic antigen; MT, Masson trichrome; MUC, mucin gene; NSE, neuron-specific enolase; PAS, periodic acid–Schiff; PAS-D, periodic acid–Schiff with diastase digestion; pCEA, polyclonal carcinoembryonic antigen; PDGFR-α(R), platelet-derived growth factor receptor alpha; R, research; S, surface; SM, submucosa; SMMHC, smooth muscle myosin heavy chain; vWF, von Willebrand factor.

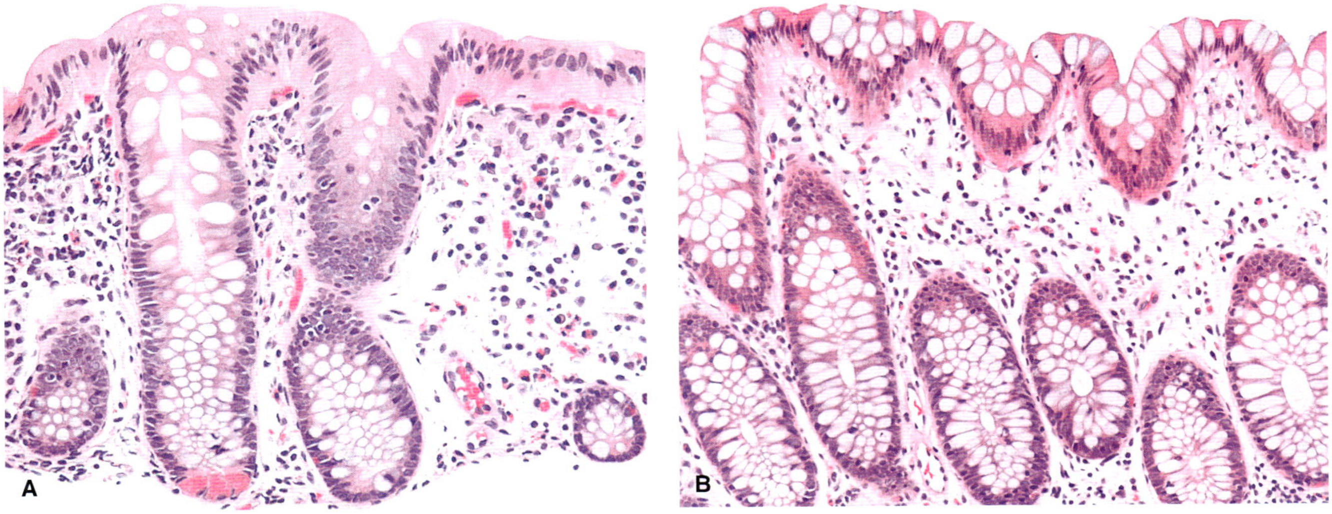

FIGURE 24.6 Normal mucosa from the cecum (**A**) and rectum (**B**). **A:** The mucosa in the cecum has more absorptive cells and fewer goblet cells compared to the rectum. The lamina propria is more cellular, with greater density of plasma cells, eosinophils, and lymphoid aggregates in the cecum as compared to the rectum. Paneth cells are normally present, residing at the base of the crypts. **B:** The rectal mucosa has a higher ratio of goblet cells to absorptive cells, with a less dense lamina propria and more easily identified muciphages. Paneth cells are not normally seen within the rectum.

Regional Variation in Histologic Features

Although the overall mucosal architecture as described above is maintained, the right and left colon do display important histologic differences. This includes the density of lamina propria mononuclear cells and sites in which the presence of Paneth cells is considered normal versus a marker of chronic injury, as will be described below. As the range of what is considered "normal" differs depending on location, we encourage gastroenterologists and surgeons to uniformly separate and appropriately label biopsy material from the right and left colon, whether it be for evaluation of diarrhea or classification of polyps (17,21,22,24,30,106).

Reflective of its absorptive and antigen processing functions, the right colon displays a higher absorptive cell (colonocyte) to goblet cell ratio (roughly 5:1) when compared to the left colon (Fig. 24.6). Proceeding distally, an increase in goblet cells is apparent, with a ratio of 3 or 4:1 for colonocyte to goblet cell. This indicates the increased formation of mucin in the descending and sigmoid colon necessary for consolidation and transit of stool (17,28,29,132). Paneth cells are normally present at the base of crypts within the midgut-derived right colon; however, importantly they are indicative of metaplasia secondary to chronic mucosal injury starting from the distal one-third of the transverse colon (Fig. 24.6).

Surface intraepithelial lymphocytes (IELs) are seen in greater concentrations in the right colon than the left and can be particularly marked overlying lymphoid aggregates (22,24,133–135). Similarly, lamina propria inflammatory cell density is also greater in the right colon than the left (descending and sigmoid colon), as are organized lymphoid aggregates, possibly related to the higher concentration of commensal microorganisms and resultant antigen sampling activities (24). Distally, as goblet cell concentration and mucin increases, lamina propria macrophages that are specifically scavenging mucin (muciphages) are increasingly observed (Table 24.4). In the sigmoid colon and rectum, most gastrointestinal pathologists will accept a few bifurcated crypts as being within the range of normal (22), although this has not been systematically studied nor reported.

Epithelium

The colonic mucosa is composed of a single layer of columnar cells making test tube–like invaginations to form millions of crypts (28,136). These are invested in basement membrane, surrounded by lamina propria (22,133,135), and separated from the submucosa by the muscularis mucosae.

The mucosal crypt architecture shows remarkable consistency despite the high epithelial turnover rate and variety

TABLE 24.4 **Low Power Impression of Right Colon vs. Left Colon**

Variation in Histologic Appearance Between Right and Left Colon on Low Magnification	Cause
Right colon appears darker	• Lamina propria contains more mononuclear cells • More absorptive cells lining crypt epithelium • Less goblet cells
Left colon appears lighter	• Fewer plasma cells in lamina propria • More goblet cells • Fewer absorptive cells

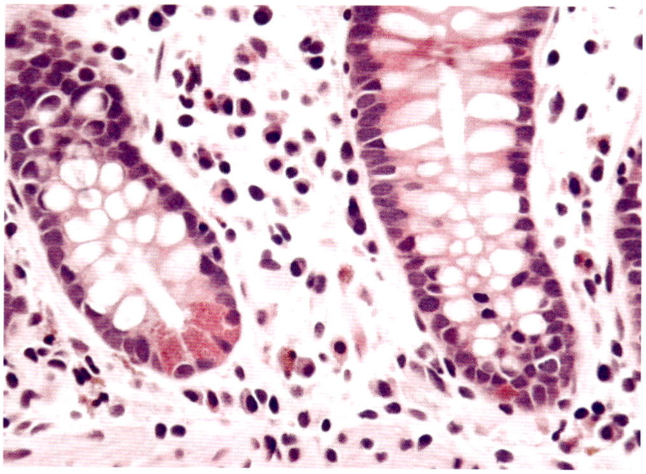

FIGURE 24.7 Normal right colon with Paneth cells and an endocrine cell. Within the crypt on the left are three Paneth cells at the base of the crypt. Note the basal nucleus and coarse, luminal-facing granules that empty into the crypt lumen. Within the crypt on the right is an endocrine cell at the base of the crypt. The endocrine cell is smaller, has a luminal nucleus and fine, basally facing granules that empty into the pericrypt myofibroblast sheath and adjacent vasculature.

of specialized cell types (137,138). Mucosal renewal is generally attributed to stem cells located near the base of the intestinal crypts and maintained within a mesenchymal niche (15,33,34,136,137,139–142). Multipotent stem cells divide and give rise to a transient population of progenitor cells in this maturation process that starts from the base of the crypt and toward the luminal surface (33,136,142). The granule-containing epithelial cell types appear to ignore the direction of luminal migration during maturation; that is, Paneth cells migrate down toward the crypt base and enteroendocrine cells home toward the mid and deeper regions of the crypts (Fig. 24.7) (15,17,143).

During migration toward the luminal surface, dividing transit cells commit and differentiate to one of five distinct epithelial cell types: absorptive colonocyte, mucus-secreting goblet cell, enteroendocrine cell, Paneth cell, or microfold (M) cell. At any given time, 75% to 80% of all colonocytes are associated with the crypts, and only 10% to 15% of colonocytes form the surface epithelium (intercrypt table) (134,138). Epithelial cell turnover is frequent. In fact, most of these cells, except for stem cells, are replaced within a week.

› ***Absorptive Colonocytes*** Absorptive colonocytes compose the majority of the surface epithelium (87,89,144). The luminal surface of the absorptive colonocyte is characterized by rigid, tightly packed apical microvilli (15,138). The tips of the microvilli contain integral membrane mucin-like glycoproteins that form a brush border glycocalyx (145). These are visible as a striate luminal border. Absorptive colonocyte cytoplasm is lightly eosinophilic, with small apical vesicles containing mucin (of different composition than goblet cell mucin) positioned for luminal release (23,82,96,106). The apical poles of columnar cells fan out over the "flask" of the goblet cells, such that only the apexes of goblet cells contact the lumen (138). Basally aligned colonocyte nuclei are oval, uniformly sized, and aligned with the long axis parallel to the long axes of the cells.

› ***Goblet Cells*** Goblet cells are dispersed throughout the surface epithelium and crypts. Although their "wine goblet" shape makes them distinctive and appear numerous, they are outnumbered by absorptive colonocytes. The large number of mucous granules takes up most of the cytoplasm and the nucleus is depressed at the base of the cell. Mucin composition varies regionally along the length of the colon due to differential synthesis of the several known secreted and membrane-bound mucins (17). This variation is reflected in the differential histochemical staining patterns commonly observed (106,146,147). Goblet cell cytoplasm is relatively clear with standard hematoxylin and eosin (H&E) stains; however, mucin granules become distinct with mucicarmine, Alcian blue pH 2.5, and periodic acid–Schiff stains (17). Goblet cell nuclei, when compared to adjacent absorptive colonocytes, appear hyperchromatic, dense, and irregular (138). Loss of nuclear polarity and a disorganized appearance to the goblet cells may be seen in conditions such as adenovirus infection and in serrated lesions (148,149).

› ***Endocrine Cells*** Endocrine cells within the gut epithelium represent the largest population of hormone-producing cells in the body (136,143,150,151). They account for approximately 1% of the individual cells lining the intestinal lumen and are predominantly located in the crypts. Rarely, they are scattered within the lamina propria (15,17,30,129,138,150,152,153). More than 30 peptide hormone genes are known to be expressed throughout the digestive tract, in a regionally and spatially distinct pattern (150,154). Enteroendocrine cells contain basally oriented, small, but distinct, deeply eosinophilic granules (143,150). The round, smoothly contoured nuclei of enteroendocrine cells are pushed towards the lumen, with opposite polarity compared to other epithelial cell types (Fig. 24.7). This reverse nuclear polarity can be a useful histologic feature. Enteroendocrine (amine precursor uptake decarboxylase [APUD]) cells can be mistaken for Paneth cells due to their eosinophilic granules. The luminal, rather than basal, location of the APUD cell nuclei is a feature that helps distinguish them from Paneth cells.

Enteroendocrine cells may be further identified by their histochemical silver staining properties and varying immunoreactivity to chromogranin A, synaptophysin, neuron-specific enolase, as well as specific antibodies to the putative peptide hormone of the cell or cell proliferation (i.e., carcinoid tumor) (Table 24.2).

› ***Paneth Cells*** Paneth cells disregard the rule of luminal migration; they are normally encountered at the base of the crypts within the midgut-derived right colon (15,73,121,136,155). Hence, they are not a sign of chronic injury when present in the cecum, ascending colon, and proximal two-thirds of the transverse colon. Nevertheless, it has been reported that Paneth cells can be present in the rectum of the pediatric population without being associated with IBD (156). These pyramidal-shaped cells have basally aligned ovoid nuclei and apical coarse, densely eosinophilic cytoplasmic granules (Fig. 24.7) (15,73). Granule and cellular contents include: α-defensins, β-defensins, NOD2, lysozyme, phospholipase A2, secretory leukocyte inhibitor, monomer IgA, TNF-α, heavy metal ions, zinc-binding protein, trypsin and trypsinogen, EGF, osteopontin, FAS ligand (CD95L), CD44v6, CD15, REG protein, and numerous others (73,136,155,157,158). The diverse Paneth cell contents, including antibacterial peptides, reflect their significant role in innate immunity. In addition putative roles in regulation of cell matrix interactions, apoptosis, and cellular immunity, as well as stem-cell niche maintenance, have been proposed (73,136,155,157,159).

In addition to characteristic granule staining with H&E stains, granules are conspicuously stained by periodic acid–Schiff (73); and, interestingly, Paneth cell autofluorescence is elicited by eosin stain (Table 24.3) (121).

› *M Cells and Follicle-Associated Epithelium* The M cell is a distinct type of enterocyte that occurs in the thin layer of epithelium covering lymphoid follicles. They are associated with both the immunologic cells and variants of absorptive colonocytes (the follicle-associated epithelium) unique to the dome region of these lymphoid follicles (131,134,136,160,161). The role of the M cell is to directly "sample" the gut microbiota, transporting the antigens to closely associated dendritic cells. These then process and present the antigen for T cells that support B-cell maturation, ultimately resulting in the secretory IgA response of the intestine. This is part of the homeostasis between the integrity of the intestinal mucosal barrier and commensal gut bacteria. Estimates of M cells in human colon vary widely, reported from "rare" to approximately 10% of surface epithelial cells (134,161–165). Light microscopy has insufficient magnification to distinguish the unique features of M cells. On electron microscopy, they show reduced and irregular microvilli compared to absorptive cells, apical microfolds, absence of thick brush borders, and the presence of a cell surface–amplifying basolateral membrane subdomain that also forms an intraepithelial pocket. This M-cell intraepithelial pocket provides the docking site for special populations of intraepithelial B and T lymphocytes and immediately overlies the dome region of lymphoid follicles (134,160–163). The M cells sample the contents of the lumen and transfer antigens to antigen-presenting cells via a specialized method of transcytosis (77,80,160,166). The follicle-associated crypts contain few or no goblet cells, enteroendocrine cells, or Paneth cells (162,164). The epithelium in this area may have the appearance of intraepithelial lymphocytosis because of the normal presence of T and B lymphocytes in proximity with the lymphoid follicles. Adenomas in this area may herniate into the lymphoid follicles and be challenging to distinguish from carcinoma. Moreover, gut-associated lymphoid tissue–associated neoplasms, also referred to as "dome carcinomas" or "GALT carcinomas" exist (167,168).

› *Intraepithelial Inflammatory Cells* IELs occur in two compartments: within the paracellular spaces of the absorptive epithelium and in highest density associated with lymphoid aggregates within M-cell pockets (Fig. 24.8) (133,135,169–171). The former are predominantly CD3+, CD8+, TCRαβ+ suppressor T cells, with between 15% and 40% TCRγδ+ T cells, while the latter are mixture of CD3+/CD45RO+ activated memory, some CD45RA+ naive T cells, and IgM-secreting B cells (93,135, 172–174). IELs are the first members of the immune system to encounter dietary antigens and commensal and pathologic microorganisms, and they likely play an integral role in oral tolerance (58,166,170,173,175,176). The IELs home toward their intraepithelial destination, migrating along various chemokine gradients produced by adjacent epithelial, inflammatory, and mesenchymal cells (131,177).

Nuclear molding and indistinct cytoplasmic contours are characteristic of IELs as they squeeze through the basement membrane to occupy paracellular spaces (Fig. 24.8). Retention of the classic lymphocyte with its round nucleus and thin rim of cytoplasm is more common in IELs overlying aggregates. Normal IEL density ranges from 1 to 5 lymphocytes per 100 colonocytes, except in follicle-associated epithelium, where M-cell–associated IELs are normally abundant (174). Therefore, lymphocytic colitis should not be diagnosed based on these areas. In general, 20 or greater lymphocytes per 100 colonocytes are considered pathologic (82,133,135,178). The number of IELs decreases from the ascending colon to rectum, with highest concentration in the lymphoid aggregate and commensal bacteria–rich cecum (133,134) (Table 24.3).

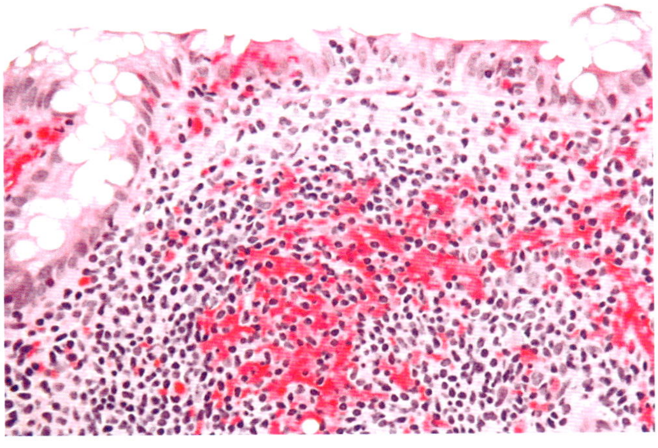

FIGURE 24.8 Intraepithelial lymphocytes (IELs) overlying a lymphoid follicle. Large numbers of IELs are typically seen overlying lymphoid aggregates. This should not be misinterpreted as lymphocytic colitis.

Intraepithelial eosinophils may occasionally be seen in the normal colon, although at much lower numbers than lymphocytes (178–180). In general, ascending colon epithelium can have more eosinophils in the epithelium than descending colon. Several studies mention that normal colons do not have aggregates of eosinophils within the colonic crypts (eosinophilic crypt abscesses) and that infiltration of eosinophils into the surface epithelium is rare (181–183).

› Stem and Dividing Transit Cells It is estimated that between four and six stem cells are present per crypt in the intestine. In addition, some dividing transit cells are apparently able to be "recruited" to serve as stem cells following injury with stem cell loss (33). These proliferative and undifferentiated cells are morphologically indistinct; however, they appear to have a large nucleus with diffuse chromatin and scant cytoplasm with few small organelles (15,136). Mitotic activity is frequently encountered in the basal one-fifth of the crypt, and apoptosis may rarely also be seen (137,139,142).

Apoptosis

The epithelial cells of the colon have remarkably short life spans (Table 24.5), during which they mature, migrate, and function (15,23,35,139,142,184,185). Programmed cell death (apoptosis) is the conclusion of the normal process of colonocyte turnover. This is manifested as apoptotic bodies predominantly affecting surface epithelium, where they are generally found in the basal aspect of the epithelium near the basement membrane (Fig. 24.9). Histologically, the apoptotic bodies consist of vacuoles containing pyknotic nuclear debris surrounded by free space (25,140,142, 186–193). Lamina propria inflammatory cells similarly undergo apoptosis, but this is frequently overlooked histologically (188,194,195). Sodium phosphate bowel preparations transiently increase the rate of apoptosis (see Bowel Preparation Effects, below); similar effects are seen with other physical and chemical agents. Increased apoptosis of both surface and/or crypt epithelium is seen in several disease states, including graft-versus-host disease, autoimmune enteropathies, systemic autoimmune disorders, viral infections, and with certain medications (79,196–200). Altered apoptosis is also a manifestation of neoplastic progression (201).

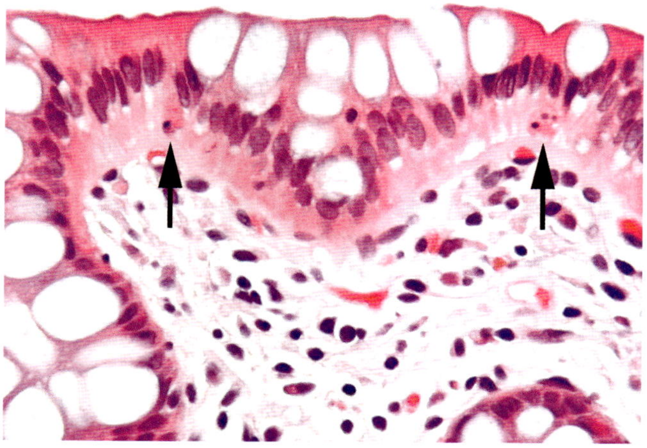

FIGURE 24.9 Apoptosis. Two apoptotic bodies are seen within the surface epithelium (*arrows*).

Basement Membrane

In well-oriented sections, the normal basement membrane is between 3 and 5 μm thick, regular, and stains with Masson trichrome, saffron, eosin von Gieson elastin, and other connective tissue stains (Fig. 24.10) (202,203). A basement membrane thickness greater than 10 μm is considered to be pathologic, particularly when it is accompanied by entrapped capillaries and inflammatory cells, as well as surface epithelial injury (122,202,204).

The basement membrane complex anchors the various epithelial cells to the underlying myofibroblast network and lamina propria. This fenestrated extracellular support matrix is produced collaboratively by epithelial and mesenchymal

TABLE 24.5 The Life Span of the Various Colon Epithelial Cells and Number of Replacements Per Average Human Life Span Vary Between Cell Types. Despite the High Rate of Turnover, Preservation of Genetic Information Is the Rule Rather Than the Exception

Cell Type	Life Span	Number of Replacements/Life
Absorptive Colonocyte	4–8 days	3,285–6,570
Goblet Cell	3–4 days	6,570–8,760
Enteroendocrine Cell	10–15 days	1,750–2,630
Paneth Cell	20 days	1,300
M Cell	Unknown	Unknown
Stem Cell Lineage/Niche	8.2 y	9–10

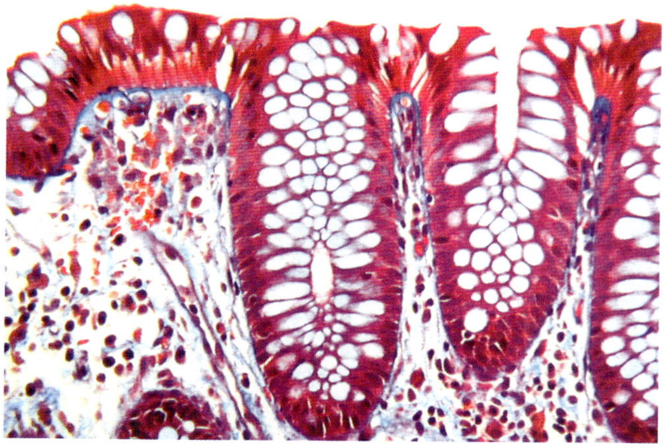

FIGURE 24.10 Normal basement membrane. The normal basement membrane is 3 to 5 μm thick and has a crisp, delicate, and regular lower border; it blends into the crypt sheath imperceptibly (trichrome stain).

cells (105,134,137,205). In addition to allowing IELs to traverse the basement membrane, the fenestrations permit epithelial, mesenchymal, and dendritic cell processes to sample and/or present antigens, and have functional implications with respect to water and ion transport (23,34,137). The basement membrane also plays an important barrier function that prevents carcinoma cells arising in adenomas from invading the lamina propria.

Lamina Propria

The lamina propria invests the colonic crypts, extending from the basement membrane complex to the muscularis mucosae. Inflammatory and mesenchymal cells of this compartment perform integral immunologic, metabolic, and proliferative functions.

› ***Lamina Propria Inflammatory Cells*** The lamina propria of the colon is home to an array of cells that comprise the gut-associated immune system. It contains more than 30,000 discrete lymphoid aggregates distributed along the length of the colon with the highest concentration within the cecum (206,207). The normal colonic lamina propria contains mature B lymphocytes, plasma cells, T lymphocytes, eosinophils, mast cells, and macrophages, functioning as antigen-sampling and processing factories. T-cell subtypes include helper, suppressor, and lymphokine-activated killer (LAK) cells, but natural killer (NK) cells are mostly absent. There is normally a decreasing inflammatory cell gradient from lumen to muscularis mucosae; there are more inflammatory cells near the luminal surface and fewer near the muscularis mucosae. As a result, the loose connective tissue of the lamina propria is obscured near the luminal aspect of the mucosa, but becomes more visible approaching the muscularis mucosae (208). The predominant cell type of the lamina propria is the IgA-secreting plasma cell, with much smaller proportions of IgM-, IgE-, and IgG-secreting plasma cells also present (194,209). Secreted IgA and IgM are transported luminally, providing humoral immune protection (144,194,209). The distinct cartwheel nucleus, perinuclear Golgi zone, and amphophilic cytoplasm characteristic of plasma cells observed in other tissues are retained in colonic plasma cells. Of the remaining lamina propria lymphocytes, more than 90% are CD3+ T cells with fewer than 50% also expressing CD8 (134,135,171). There are also CD20-positive B lymphocytes present within and adjacent to lymphoid follicles (80,85,144,206,210).

Myeloid cells that normally reside in the lamina propria include eosinophils and mast cells. The number of eosinophils is highly variable in the normal colon and is dependent upon both the region of colon sampled (178–180, 185,186,211) and the geographic residence of the patient (212). Eosinophils are generally more numerous in the lamina propria of the ascending colon compared to that of the descending colon (181). A range of normal eosinophil counts in the lamina propria has been reported as 55 per mm^2 in the right colon compared to 28 per mm^2 in the left; however, the significance of eosinophilia should be interpreted along on with the presence of other histologic factors of colonic injury and clinical features (206). Higher mean eosinophil concentrations are seen in samples obtained from patients who live in the southern United States compared to the northern United States, with a rather extreme degree of variability (212). Eosinophilia is increased in parasitic and allergic diseases, collagenous colitis, and chronic idiopathic IBD. Mast cells, or tissue-based basophils, are less numerous than eosinophils and their density is increased in the ileocecal region compared with other sites of the colon (213,214). Mast cells are difficult to identify with routine H&E stains, but are highlighted with Giemsa and toluidine blue stains as well as tryptase and CD117 immunostains (Fig. 24.11, Table 24.3) (211). The density of mast cells is highly variable in normal individuals, ranging from 11 to 55 per high-power field. Therefore, routine staining for mast cells is not necessary in normal-appearing colonic biopsies obtained to evaluate patients for chronic diarrhea, or when a diagnosis of mast cell activation syndrome is clinically considered. Mastocytic enterocolopathy, a disorder initially proposed as a cause of chronic diarrheal illness, is characterized by essentially normal histologic findings and is not a pathologic diagnosis. Neutrophils are not normally seen within the lamina propria in high numbers, although they may be seen in areas of hemorrhage and within blood vessels (215).

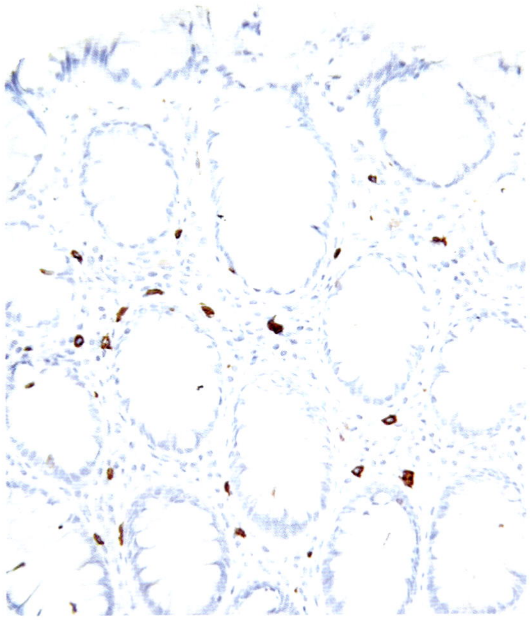

FIGURE 24.11 Mucosal mast cells. Although difficult to discern on H&E-stained sections, mucosal mast cells are easily identified with CD117. Mucosal mast cells serve well as an internal control when evaluating CD117 stains of gastrointestinal tract mesenchymal tumors. (Anti-CD117 stain.)

Macrophages are commonly seen scattered throughout the lamina propria and are occasionally concentrated at the basal aspects of the crypts (79,216–219) (Fig. 24.12).

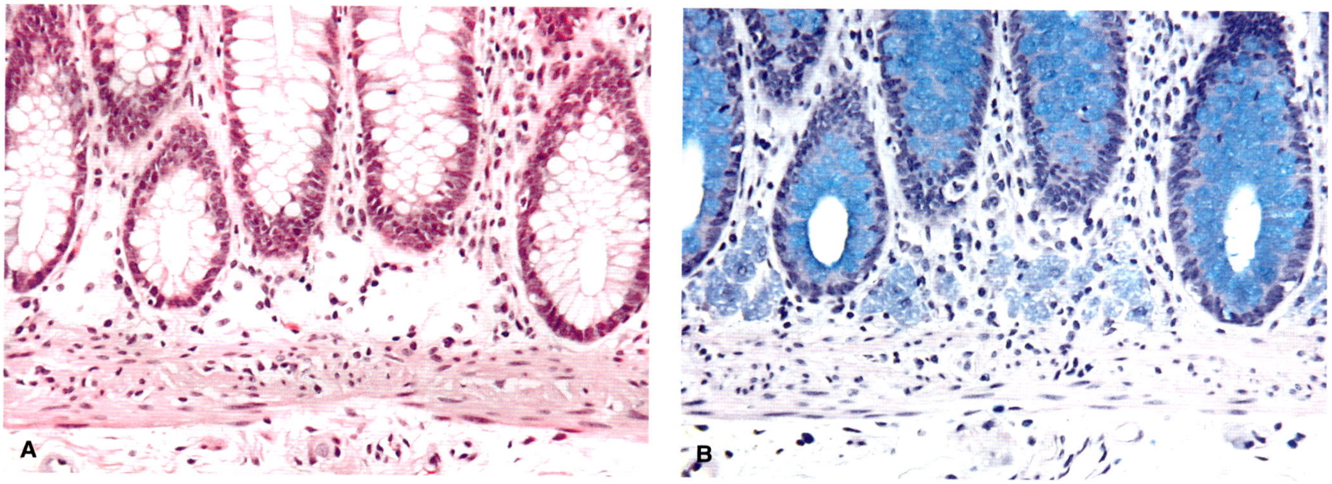

FIGURE 24.12 Lamina propria muciphages. **A:** These pale macrophages at the base of the mucosa are stuffed with mucins. This finding is not infrequent and does not generally correlate with disease. **B:** The same area stained with Alcian blue pH 2.5; the muciphages show strong cytoplasmic staining. Similar findings are seen with PAS with diastase digestion. Of note, bacteria-laden macrophages in Whipple disease are generally negative when stained with Alcian blue pH 2.5 but densely stain with PAS with diastase.

Their visualization may be enhanced by specific histochemical stains that detect the variety of materials they scavenge and store: apoptotic debris, microbes, lipofuscin, cholesterol esters, gangliosides, mucolipids, glycogen, mucopolysaccharides, and others (Fig. 24.12, Table 24.3) (217,218). Muciphages are the most commonly recognized macrophages. These cells ingest mucin and other cell products that has exuded from adjacent epithelial cells. As normal constituents of the lamina propria, muciphages have increased concentration within the left colon, in keeping with the increased number of goblet cells present in this location. Distention of the lamina propria by muciphages to the point that they appear to be replacing other lamina propria inflammatory cells is rarely normal. If encountered, this finding may indicate bacterial or fungal infection due to *Tropheryma whippelii*, *Mycobacterium avium-intracellulare* complex, *Histoplasma capsulatum*, and others, or various metabolic storage disorders. Further histochemical, PCR, or electron microscopic methodologies and additional laboratory evaluation should be considered in this situation (101,109,220–223).

Plasmacytoid dendritic cells are scattered throughout the lamina propria, while stellate dendritic cells are concentrated in the subepithelial dome space associated with lymphoid follicles (144,210,224,225). Dendritic cells are histologically indistinct and frequently require immunohistochemistry for definitive identification (Table 24.3) (79,86,85,144). Plasmacytoid dendritic cells have recently been implicated in allergic and autoimmune disorders (95), while the stellate dendritic cells are integral in antigen presentation (79,82,226).

Myofibroblasts (Pericrypt Myofibroblast Sheath and Lamina Propria Myofibroblasts)

The lamina propria contains two distinct populations of myofibroblasts: the pericryptal myofibroblast sheath and the subepithelial myofibroblast syncytia. Interacting closely with the epithelium, lamina propria inflammatory cells, and the muscularis mucosae, myofibroblasts function in absorption, ion and mucin secretion, immune regulation, and differentiation as they help maintain the stem cell niche (5,23,132,227). The rim of fusiform cells organized in close apposition to each colonic crypt was originally designated the pericryptal fibroblastic sheath but is now known to be a syncytium both anatomically and functionally (Fig. 24.13) (205,227). This specialized

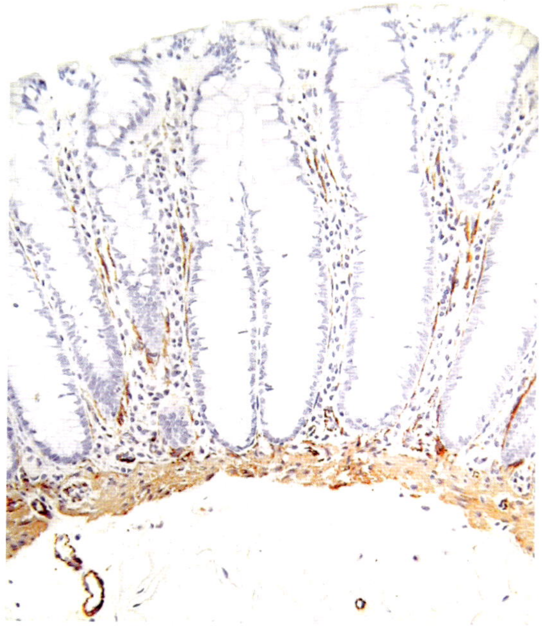

FIGURE 24.13 Lamina propria myofibroblasts, anti-smooth muscle-actin (SMA) stain. This stain highlights the pericryptal myofibroblast sheath, muscularis mucosae, and submucosal blood vessels.

population of mesenchymal cells surrounds the crypts and extends into the lamina propria, forming a reticular network within the extracellular matrix, attaching to one another with both gap and adherens junctions (90,99,205,227,228), and displaying distinct immunophenotypes (Table 24.3). These cells can be prominent in some colonic polyps, particularly those with nondysplastic serrated epithelium. In this situation, pericryptal myofibroblasts assume a whorled appearance around the crypts. This pattern, in combination with occasional staining of these cells with EMA, has led to the erroneous designation of such lesions as perineuriomas (229).

The myofibroblasts are oval and scaphoid and appear to overlap like shingles on a roof around the crypts. The subepithelial myofibroblasts exist in two distinct morphologic states: (a) the activated myofibroblast and (b) the stellate-transformed myofibroblast (205,227). Myofibroblasts are often surrounded by an incomplete basal lamina and embedded in a subepithelial sheet of reticular fibers that also contains fenestrae through which lymphocytes and macrophages traverse. Gap junctions couple some myofibroblasts to the tissue smooth muscle cells, and the cells are commonly in close apposition to terminals of nerve fibers; however, it has not been determined whether the interstitial cell of Cajal network is physically connected to the subepithelial myofibroblast network (205,227,230).

Vasculature and Lymphatic Vessels in the Lamina Propria

Capillaries and high endothelial venules are scattered throughout the lamina propria, as well as lymphatic channels that are immediately superficial to the muscularis mucosae (231,232). Capillaries are composed of a circumferential endothelial lining and may contain red blood cells as well as inflammatory cells. Irregularly shaped, distorted, and engorged capillaries frequently indicate prolapse of the mucosa. In addition to providing oxygen and nutrients to mucosal cells, vascular adhesion molecules participate in "homing" of circulating lymphocytes to their appropriate colonic microenvironment. Lymphatic tributaries may be found in small numbers near the crypt bases; they have thin walls and cross the muscularis mucosae to join the readily observed submucosal lymphatic channels (232–236). Distinction between capillaries and lymphatic vessels often requires immunohistochemical analysis (Table 24.3) (91,101,107,235).

Muscularis Mucosae

The muscularis mucosae is a thin layer of smooth muscle that forms the deep boundary of the lamina propria muscularis mucosae. This muscle layer is physically tethered to the mucosa, with occasional smooth muscle cells extending into the lamina propria or coalescing with the pericryptal myofibroblast sheath. The muscularis mucosae receives innervation via the submucosal plexus (99,236,237). Because the colonic glands are tethered to the muscularis mucosae, this structure is valuable in evaluating crypt architecture in endoscopic biopsies. Biopsies that do not contain muscularis mucosae may show artifactual architectural distortion, with glands adopting horizontal or curved configurations. This is particularly relevant in assessing samples for idiopathic IBD and sessile serrated polyps/adenomas, where crypt architectural distortions are key components of making the histologic diagnoses. Careful examination of other biopsy fragments that have muscularis mucosae, as well as assessment for other features of mucosal injury (Table 24.2), may allow an accurate diagnosis. The muscularis mucosae is normally traversed by lymphoglandular complexes (Fig. 24.5), vascular channels, and neural twigs. It participates in absorptive, secretory, proliferative, and possibly motility functions. Isolated thickening may occur with prolapse of the overlying mucosa and, in proximity to diverticula. Clear duplication of the muscularis mucosae is generally considered a feature of chronic mucosal injury.

Submucosa

The submucosa is composed of loosely arranged bundles of smooth muscle, fibroelastic tissue, and adipose tissue, in which the local enteric nervous system, vasculature, and lymphatic vessels are embedded. Lymphatic channels may be conspicuous and dilated immediately beneath the submucosa and do not contain cellular elements (232,233,235). Sparse inflammatory cells are scattered throughout, occasionally organized as submucosal lymphoid aggregates. The submucosa provides a flexible matrix that proves useful during peristalsis—it allows the mucosa to glide and move freely over the rigid muscularis propria.

Submucosal smooth muscle cells are arranged in loosely woven fascicles forming small bundles. These collections are closely apposed to interstitial cells of Cajal, which are immediately adjacent to nerve varicosities, forming the neuroeffector junctions that receive, transmit, and integrate central, parasympathetic, and sympathetic nervous system commands (230,238,239). The two submucosal neural plexuses are the submucosal plexus of Meissner located immediately beneath the muscularis mucosae and the deep submucosal plexus of Henle that lies on the inner aspect of the muscularis propria. Neural plexuses are composed of neurons, glial cells, and stromal elements (240–242). Ganglion cells contain round or ovoid nuclei, a prominent eosinophilic nucleolus, and ample basophilic cytoplasm stippled with Nissl substance (Fig. 24.14). Ganglion cells characteristically cluster together and may mimic giant cells, epithelioid cells, or granulomata. When present conspicuously in the mucosa, ganglion cells may reflect chronic damage, such as may be seen in the context of diverticular disease or idiopathic IBD. Nerve axons are fibrillar and distinguishing these axons from fibroblasts or their elastofibrotic products may require the use of histochemical or immunohistochemical stains (Fig. 24.14B, Table 24.3) (54,119,239,240,242–245).

Interstitial cells of Cajal are essentially modified myofibroblasts. These cells have a fusiform shape and large ovoid

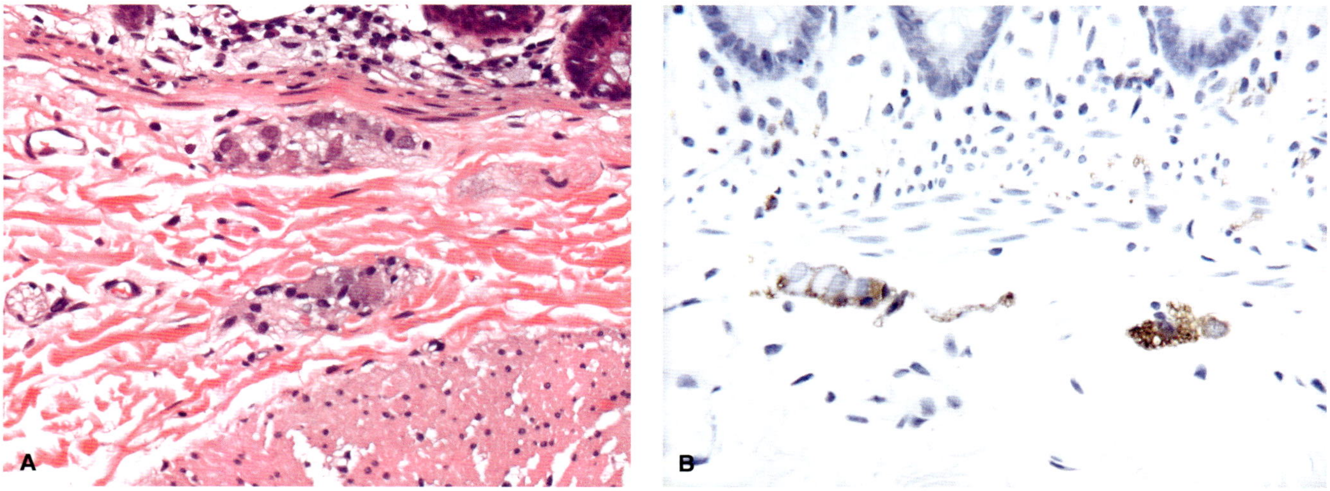

FIGURE 24.14 Ganglion cells of Meissner plexus. **A:** Submucosal nerve twigs and clusters of ganglion cells comprising Meissner plexus. (H&E stain, ×20). **B:** Same area stained with S100 (hematoxylin counterstain), highlighting the Schwann cells. The ganglion cells on the left are conspicuously negative with S100 (also ×20).

nucleus. Silver stains or immunohistochemical stains, such as anti-CD117, can distinguish these from ganglion cells or adjacent smooth muscle cells (228,239,243,246,247). These intriguing cells are thought to represent the pacemaker cells of the gut and play an important role in the control of gut motor activity (228,248,249). The normal interstitial cell of Cajal density within the submucosa is substantially less than that seen surrounding the myenteric plexus (see below) (239,248–250). The submucosa contains arterioles derived from the superior and inferior mesenteric arteries, venules, and lymphatic vessels (Fig. 24.15). These vessels are frequently distended by red blood cells and appear tortuous.

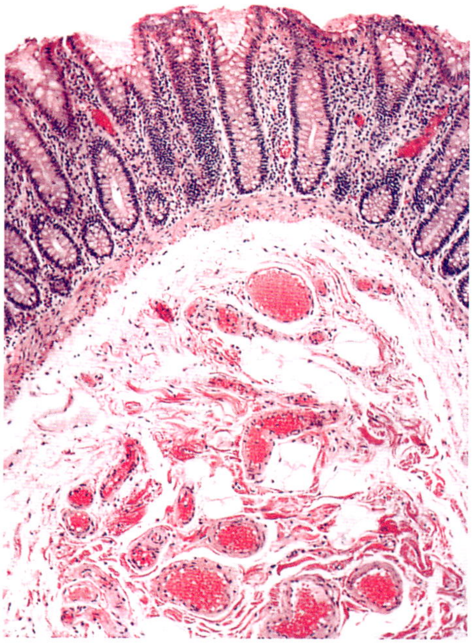

FIGURE 24.15 Colonic submucosal vasculature.

The amount of adipose within the submucosa varies substantially between the right and left colon and among patients. Of note, the ileocecal valve and cecal submucosa may appear expanded by mature adipocytes, resembling a lipoma. However, in the absence of the submucosal adipose forming a discrete, lobulated mass, this abundant adipose tissue is within the range of normal.

Muscularis Propria, Subserosal Zone, and Serosa

The muscularis propria, or external smooth muscle layers, of the colon consist of an inner circular layer and an outer longitudinal layer (Fig. 24.16) (251,252). Structural variations of the muscularis propria have been identified, which may reflect different motility and storage functions of various regions of the colon (243,246). Smoothelin, a

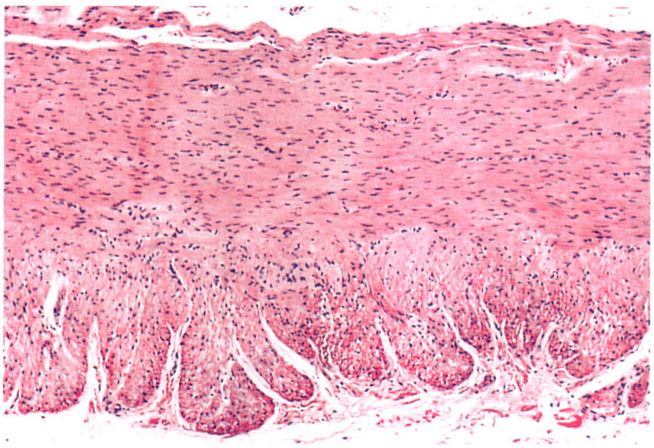

FIGURE 24.16 Muscularis propria and subserosal tissue. Both layers of the muscularis propria can be seen with the neural tissue of Auerbach plexus. Below the muscle layers is the fibrovascular adipose tissue of the subserosa.

FIGURE 24.17 Interstitial cells of Cajal (ICC). The CD117 strongly positive, dendritic-appearing cells between the muscle layers and surrounding Auerbach plexus are the interstitial cells of Cajal. These cells are considered to be the pacemaker cell of the gut and perform other functions in gut motility. (Anti-CD117 stain.)

cytoskeletal protein that is expressed in terminally differentiated contractile smooth muscle cells, may serve a role in the contractile apparatus of these cells; there is a marked reduction of expression of this marker in the outer longitudinal muscle layer of some cases of colonic inertia (253). Auerbach plexus lies between the two muscle layers and resembles Meissner plexus histologically. The interstitial cells of Cajal, the putative pacemaker cells of the gut that drive peristalsis, can be identified throughout the muscularis propria with immunostains for CD117 and CD34 (Fig. 24.17) (118,254,255).

Patients with motility disorders may have decreased numbers of interstitial cells of Cajal within the bowel wall, although the significance of this finding is not clear (249,252). Decreased numbers of pacemaker cells may contribute to decreased gut motility but it is more likely that this finding is a consequence of longstanding inertia. The muscularis is perforated by blood and lymphatic vessels and is encased in a subserosal zone of fibroadipose tissue. The serosa itself is limited to the mesothelial lining and immediately adjacent fibroelastic tissue.

EFFECTS OF PREPARATION AND ARTIFACTS

Bowel Preparation Effects

Commonly used bowel preparations for colonoscopy (sodium phosphate enemas, bisacodyl enemas and suppositories, dioctyl sodium sulfosuccinate, soapsuds enemas) can produce mucosal abnormalities that mimic or obscure inflammatory conditions and impart an edematous or hyperemic appearance to the endoscopist (256–258). Histologic features suggesting a bowel preparatory effect include flattening of the absorptive colonocytes to a cuboidal shape,

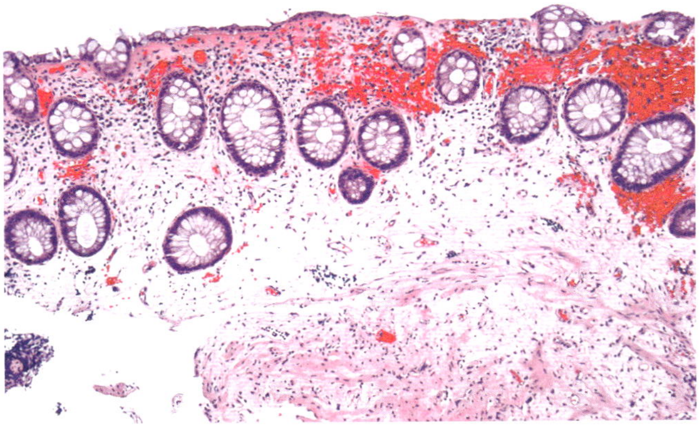

FIGURE 24.18 Enema effect. There is edema with extravasation and lysis of red blood cells (hemorrhage) within the lamina propria. The surface epithelium is largely denuded. Mucin depletion due to induced goblet cell secretion and increased apoptosis may also be seen.

reduction in goblet cell mucus resulting from increased secretion, detached surface epithelium leaving an exposed basement membrane, focal surface epithelial and crypt neutrophilic infiltrates, red blood cell extravasation within the lamina propria, and increase in crypt or surface epithelial cell apoptosis (Fig. 24.18). Oral sodium phosphate is among the most irritating of bowel preparations and, thus, tends to produce these changes most commonly. Endoscopically visible aphthous erosions and even frank ulcers have been reported due to preparatory solutions. Neutrophilic cryptitis and increased apoptotic debris in the deep crypts can also result from bowel preparations (Fig. 24.19) (256,259,260). Although the histologic changes induced by bowel preparations may not interfere with rendering a diagnosis for colorectal polyps, these inflammatory changes may be misdiagnosed as a pathologic condition. In the evaluation of patients for reasons other than colorectal cancer screening, bowel preparation with polyethylene

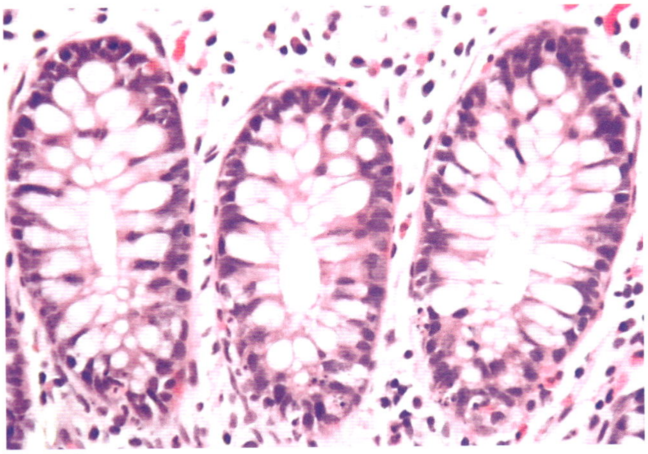

FIGURE 24.19 Oral sodium phosphate bowel preparation effect. Colonic crypts with apoptotic bodies and neutrophils are secondary to the effects of bowel preparation. Such changes could easily be interpreted as representing infectious colitis or graft-versus-host disease (GVHD) in the right clinical setting.

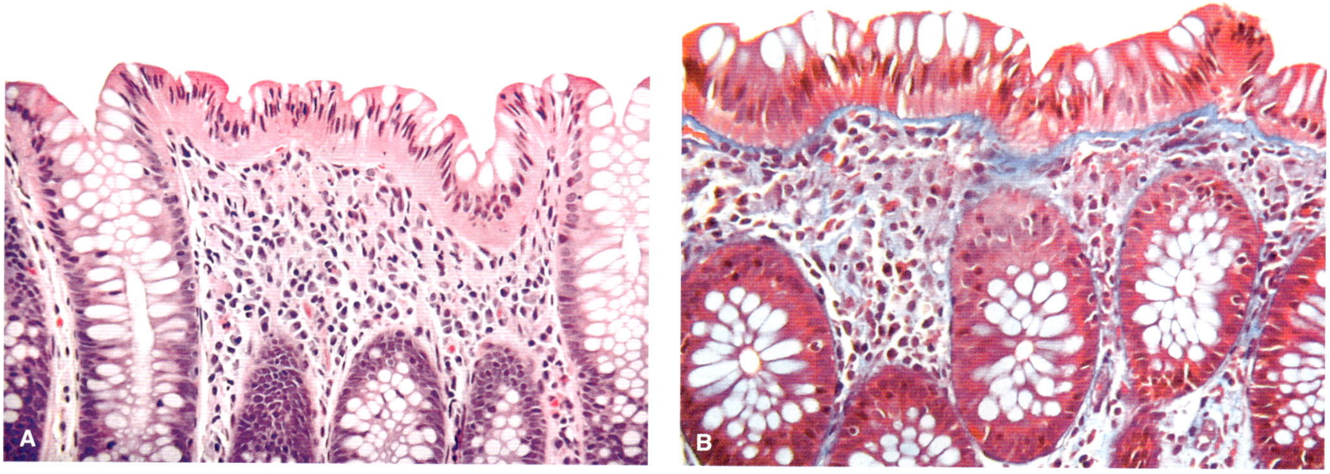

FIGURE 24.20 Normal colon mimicking collagenous colitis. **A:** This normal mucosa shows blending of the colonocyte cytoplasm with the basement membrane to give the illusion of a thickened subepithelial collagen table. Note that there is no surface damage or colitis present. **B:** This trichrome-stained section shows focal thickening in an area where the crypt sheath joins the surface tangentially. Care must be taken when evaluating tangential sections. Again note the lack of colitis or surface damage.

glycol is suggested (261,262). Although polyethylene glycol preparation has also been reported to cause superficial mucus loss, epithelial cell loss, lymphocyte and neutrophil infiltration, and rarely aphthous erosions, the changes tend to be milder than those seen in association with other agents (263).

Incorrect Tissue Orientation and Tangential Sectioning

Difficulties with proper tissue orientation are most common to endoscopically procured biopsy specimens owing to their small size. The most accurate interpretation is possible when the tissue is sectioned perpendicular to the plane of the surface epithelium, such that the long axes of the crypts are fully visualized. Evaluation of crypt architecture, inflammatory cell gradient, and thickness and regularity of the subepithelial collagen band may be substantially hindered by tangential sectioning (Table 24.2) (264). Tangentially sectioned colonic samples feature cross sections of the crypts, rather than their full lengths. Features of chronic mucosal injury may not be appreciated in this situation, especially when the sample contains only the most superficial aspects of the mucosa. Tangential sectioning through the cytoplasm of surface colonocytes may mimic a thickened subepithelial collagen layer, thereby simulating features of collagenous colitis (Fig. 24.20). However, the regular appearance of this artifact, smooth interface with the lamina propria, and lack of entrapped vessels and inflammatory cells are features that can help one avoid misinterpreting the significance of this finding. Tangential sections through the basal crypt region can simulate the appearance of an adenoma, as these areas contain less mature colonocytes with larger nuclei, less cytoplasm, and decreased goblet cells.

Tissue Trauma

Tissue trauma occurs with avulsion of the mucosa during forceps biopsy or improper handling in the pathology laboratory, resulting in a crush artifact. Irritation of the mucosa by cleansing solutions may produce endoscopically visible edema, petechiae, friability, tears, and hemorrhage when colonoscopes are not adequately rinsed (262,264). Biopsy samples may contain edematous, largely acellular lamina propria with extravasated red blood cells primarily in the luminal portion of the lamina propria. In the absence of other features of mucosal inflammation, these features should not be considered pathologic. Polyfoam pads may cause triangular artifacts in biopsy material and are not recommended (265). Crush artifacts may also occur when pressure is applied to the biopsy material with rigid forceps. Use of a plastic pipette with large bore opening for transferring biopsy material to the cassette avoids crush artifact. The mucosa of crushed tissues displays artifactual crypt crowding, tunneled epithelial cells within crypt lumina, and stripping of the surface epithelium.

Pseudolipomatosis

Pseudolipomatosis is characterized by vacuolated, unlined spaces in the lamina propria and mucosa that resemble loosely arranged adipocytes (Fig. 24.21). These lesions essentially represent trapped air that is introduced into the mucosa during colonoscopy. Mild mucosal trauma produces tiny breaks in the mucosa that allow microscopic amounts of air into the mucosa under the pressure of insufflation (266).

Electrocautery

Endoscopic removal of polyps with electrocautery ("hot" biopsy, or snare) frequently results in thermodesiccation of the tissue, compressing crypts together and altering the

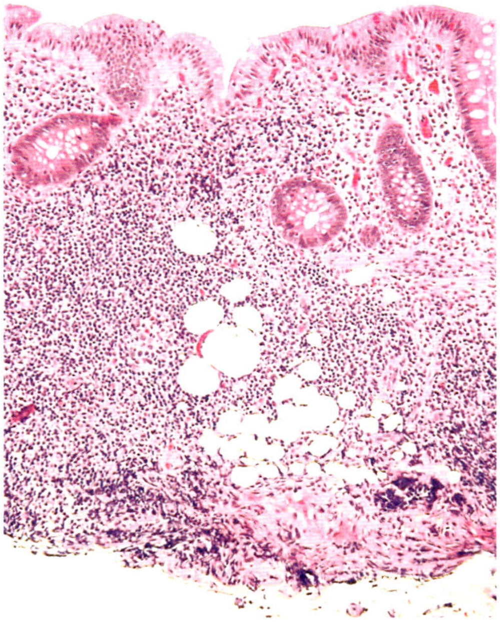

FIGURE 24.21 Pseudolipomatosis. The clear spaces within this lymphoid aggregate represent air bubbles due to insufflation during endoscopy. This artifact is frequently misinterpreted as adipose tissue/lipoma.

nuclear features. Characteristically, the crypts are closely apposed, with elongated, pyknotic, and distorted nuclei (Fig. 24.22). These features may be difficult or impossible to distinguish from a tubular adenoma. Extreme thermal effect may result in loss of both overall architecture and nuclear detail that precludes definitive assessment.

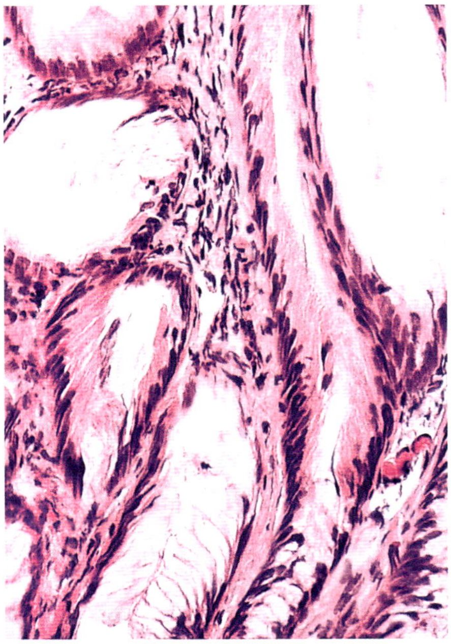

FIGURE 24.22 Electrodesiccation and compression artifact in the bases of adjacent crypts in normal colonic mucosa produced by an endoscopic electrocautery snare. Affected nuclei are pyknotic and elongated. Colonic crypts distorted by this artifact may be difficult or impossible to differentiate from the tubules of an adenoma.

ACKNOWLEDGMENTS

The authors are grateful for the continuing influence of the late Dr. Rodger C. Haggitt in the preparation of this manuscript.

REFERENCES

1. Batlle E, Henderson JT, Beghtel H, et al. Beta-catenin and TCF mediate cell positioning in the intestinal epithelium by controlling the expression of EphB/ephrinB. *Cell*. 2002; 111(2):251–263.
2. Beck F. Homeobox genes in gut development. *Gut*. 2002;51(3): 450–454.
3. Bonhomme C, Duluc I, Martin E, et al. The Cdx2 homeobox gene has a tumour suppressor function in the distal colon in addition to a homeotic role during gut development. *Gut*. 2003; 52(10):1465–1471.
4. Chailler P, Basque JR, Corriveau L, Ménard D. Functional characterization of the keratinocyte growth factor system in human fetal gastrointestinal tract. *Pediatr Res*. 2000;48(4): 504–510.
5. Haffen K, Kedinger M, Simon-Assmann P. Mesenchyme-dependent differentiation of epithelial progenitor cells in the gut. *J Pediatr Gastroenterol Nutr*. 1987;6(1):14–23.
6. Kedinger M, Simon-Assmann P, Bouziges F, Haffen K. Epithelial-mesenchymal interactions in intestinal epithelial differentiation. *Scand J Gastoenterol Suppl*. 1988;151:62–69.
7. Roberts DJ. Molecular mechanisms of development of the gastrointestinal tract. *Dev Dyn*. 2000;219(2):109–120.
8. Stutzmann J, Bellissent-Waydelich A, Fontao L, Launay JF, Simon-Assmann P. Adhesion complexes implicated in intestinal epithelial cell-matrix interactions. *Microsc Res Tech*. 2000; 51(2):179–190.
9. Kaplan JL, Shi HN, Walker WA. The role of microbes in developmental immunologic programming. *Pediatr Res*. 2011; 69(6):465–472.
10. Koenig JE, Spor A, Scalfone N, et al. Succession of microbial consortia in the developing infant gut microbiome. *Proc Natl Acad Sci U S A*. 2011;108(Suppl 1):4578–4585.
11. Shanahan F. The host-microbe interface within the gut. *Best Prac Res Clin Gastroenterol*. 2002;16(6):915–931.
12. Bajaj-Elliott M, Poulsom R, Pender SL, Wathen NC, MacDonald TT. Interactions between stromal cell–derived keratinocyte growth factor and epithelial transforming growth factor in immune-mediated crypt cell hyperplasia. *J Clin Invest*. 1998;102(8):1473–1480.
13. de Santa Barbara P, van den Brink GR, Roberts DJ. Development and differentiation of the intestinal epithelium. *Cell Mol Life Sci*. 2003;60(7):1322–1332.
14. Karam SM. Lineage commitment and maturation of epithelial cells in the gut. *Front Biosci*. 1999;4:D286–D298.
15. de Santa Barbara P, van den Brink GR, Roberts DJ. Molecular etiology of gut malformations and diseases. *Am J Med Genet*. 2002;115(4):221–230.
16. Moore KL, Persaud TVN. The digestive system. In: Moore KL, Persaud TVN, eds. *The Developing Human: Clinically Oriented Embryology*. 7th ed. WB Saunders; 2003:266–284.

17. Arai T, Kino I. Morphometrical and cell kinetic studies of normal human colorectal mucosa. Comparison between the proximal and the distal large intestine. *Acta Pathol Jpn*. 1989; 39(11):725–730.
18. Baker K, Zhang Y, Jin C, Jass JR. Proximal versus distal hyperplastic polyps of the colorectum: different lesions or a biological spectrum? *J Clin Pathol*. 2004;57(10):1089–1093.
19. Birkenkam-Demtroder K, Olesen SH, Sørensen FB, et al. Differential gene expression in colon cancer of the caecum versus the sigmoid and rectosigmoid. *Gut*. 2005;54(3): 374–384.
20. Calam J, Ghatei MA, Domin J, et al. Regional differences in concentrations of regulatory peptides in human colon mucosal biopsy. *Dig Dis Sci*. 1989;34(8):1193–1198.
21. Gervaz P, Bucher P, Morel P. Two colons-two cancers: paradigm shift and clinical implications. *J Surg Oncol*. 2004; 88(4):261–266.
22. Greenson JK, Odze RD. Inflammatory bowel disease of the large intestine. In: Odze RD, Goldblum JR, Crawford JM, eds. *Surgical Pathology of the GI Tract, Liver, Biliary Tract and Pancreas*. WB Saunders; 2004;213–214.
23. Kunzelmann K, Mall M. Electrolyte transport in the mammalian colon: mechanisms and implications for disease. *Physiol Rev*. 2002;82(1):245–289.
24. Lee E, Schiller LR, Fordtran JS. Quantification of colonic lamina propria cells by means of a morphometric point-counting method. *Gastroenterology*. 1988;94(2):409–418.
25. Liu LU, Holt PR, Krivosheyev V, Moss SF. Human right and left colon differ in epithelial cell apoptosis and in expression of Bak, a pro-apoptotic Bcl-2 homologue. *Gut*. 1999; 45(1):45–50.
26. Macfarlane GT, Gibson GR, Cummings JH. Comparison of fermentation reactions in different regions of the human colon. *J Appl Bacteriol*. 1992;72(1):57–64.
27. Moskaluk CA, Zhang H, Powell SM, Cerilli LA, Hampton GM, Frierson HF Jr. Cdx2 protein expression in normal and malignant human tissues: an immunohistochemical survey using tissue microarrays. *Mod Pathol*. 2003;16(9):913–919.
28. Naftalin RJ, Zammit PS, Pedley KC. Regional differences in rat large intestinal crypt function in relation to dehydrating capacity in vivo. *J Physiol*. 1999;514(Pt 1):201–210.
29. Naftalin RJ. The dehydrating function of the descending colon in relationship to crypt function. *Physiol Res*. 1994;43(2): 65–73.
30. Paluszkiewicz P, Berbeć H, Pawlowska-Wakowicz B, Cybulski M, Paszkowska A. p53 protein accumulation in colorectal cancer tissue has prognostic value only in left-sided colon tumours. *Cancer Detect Prev*. 2004;28(4):252–259.
31. Priebe MG, Vonk RJ, Sun X, He T, Harmsen HJM, Welling GW. The physiology of colonic metabolism. Possibilities for interventions with pre- and probiotics. *Eur J Nutr*. 2002; 41(suppl 1):I2–I10.
32. Sandle GI. Salt and water absorption in the human colon: a modern appraisal. *Gut*. 1998;43(2):294–299.
33. Booth C, Potten CS. Gut instincts: thoughts on intestinal epithelial stem cells. *J Clin Invest*. 2000;105(11):1493–1499.
34. Marshman E, Booth C, Potten CS. The intestinal epithelial stem cell. *Bioessays*. 2002;24(1):91–98.
35. Seidelin JB. Colonic epithelial cell turnover: possible implications for ulcerative colitis and cancer initiation. *Scand J Gastroenterol*. 2004;39(3):201–211.
36. Atarashi K, Tanoue T, Shima T, et al. Induction of colonic regulatory T cells by indigenous *Clostridium* species. *Science*. 2011;331(6015):337–341.
37. Ishikawa H, Tanaka K, Maeda Y, et al. Effect of intestinal microbiota on the induction of regulatory CD25+ CD4+ T cells. *Clin Exp Immunol*. 2008;153(1):127–135.
38. Metzger R, Metzger U, Fiegel HC, Kluth D. Embryology of the midgut. *Semin Pediatr Surg*. 2011;20(3):145–151.
39. Kluth D, Fiegel HC, Metzger R. Embryology of the hindgut. *Semin Pediatr Surg*. 2011;20(3):152–160.
40. Wang Y, Devkota S, Musch MW, et al. Regional mucosa-associated microbiota determine physiological expression of TLR2 and TLR4 in murine colon. *PLoS ONE*. 2010;5(10): e13607.
41. Bajpai M, Mathur M. Duplications of the alimentary tract: clues to the missing links. *J Pediatr Surg*. 1994;29(10):1361–1365.
42. Bossard P, Zaret KS. Repressive and restrictive mesodermal interactions with gut endoderm: possible relation to Meckel's diverticulum. *Development*. 2000;127(22):4915–4923.
43. Martinez-Frias ML, Bermejo E, Rodrigues-Pinilla E. Anal atresia, vertebral, genital, and urinary tract anomalies: a primary polytopic developmental field defect identified through an epidemiological analysis of associations. *Am J Med Genet*. 2000;95(2):169–173.
44. Robertson K, Mason I, Hall S. Hirschsprung's disease: genetic mutations in mice and men. *Gut*. 1997;41(4):436–441.
45. Houlston R, Bevan S, Williams A, et al. Mutations in DPC4 (SMAD4) cause juvenile polyposis syndrome, but only account for a minority of cases. *Hum Mol Genet*. 1998;7(12):1907–1912.
46. Howe JR, Roth S, Ringold JC, et al. Mutations in the SMAD4/ DPC4 gene in juvenile polyposis. *Science*. 1998;280(5366): 1086–1088.
47. Howe JR, Bair JL, Sayed MG, et al. Germline mutations of the gene encoding bone morphogenetic protein receptor 1A in juvenile polyposis. *Nat Genet*. 2001;28(2):184–187.
48. Roth S, Sistonen P, Salovaara R, et al. SMAD genes in juvenile polyposis. *Genes Chromosomes Cancer*. 1999;26(1):54–61.
49. Ruiz i Altaba A. Gli proteins and Hedgehog signaling: development and cancer. *Trends Genet*. 1999;15(10):418–425.
50. Arora T, Sharma R. Fermentation potential of the gut microbiome: implications for energy homeostasis and weight management. *Nutr Rev*. 2011;69(2):99–106.
51. Turnbaugh PJ, Bäckhed F, Fulton L, Gordon JI. Diet-induced obesity is linked to marked but reversible alterations in the mouse distal gut microbiome. *Cell Host Microbe*. 2008;3(4): 213–223.
52. Gohir W, Ratcliffe EM, Sloboda DM. Of the bugs that shape us: maternal obesity, the gut microbiome, and long-term disease risk. *Pediatr Res*. 2015;77(1–2):196–204.
53. Obata Y, Pachnis V. The effect of microbiota and the immune system on the development and organization of the enteric nervous system. *Gastroenterology*. 2016;151(5):836–844.
54. Guyton, AC. The digestive and metabolic systems. In: Guyton AC, ed. *Anatomy and Physiology*. WB Saunders; 1984: 643–700.
55. Smith ME, Morton DG. The colon. In: Smith ME, Morton DG, eds. *The Digestive System*. Churchill Livingstone; 2001: 175–186.
56. Netter FH. Abdomen. In: Netter FH, Colacino S, eds. *Atlas of Human Anatomy*. Ciba-Geigy Corp; 1989:251–256, 264–268.

57. Mai V, Morris JG Jr. Colonic bacterial flora: changing understandings in the molecular age. *J Nutr*. 2004;134(2):459–464.
58. Ouwehand A, Isolauri E, Salminen S. The role of the intestinal microflora for the development of the immune system in early childhood. *Eur J Nutr*. 2002;41(supp 1):I32–I37.
59. Pryde SE, Duncan SH, Hold GL, Stewart CS, Flint HJ. The microbiology of butyrate formation in the human colon. *FEMS Microbiol Lett*. 2002;217(2):133–139.
60. Roediger WE, Babidge W. Human colonocyte detoxification. *Gut*. 1997;41(6):731–734.
61. Macfarlane GT, Macfarlane S. Human colonic microbiota: ecology, physiology and metabolic potential of intestinal bacteria. *Scand J Gastroenterol Suppl*. 1997;222:3–9.
62. Zaharia V, Varzescu M, Djavadi I, et al. Effects of short chain fatty acids on colonic Na+ absorption and enzyme activity. *Comp Biochem Physiol A Mol Integr Phsiol*. 2001;128(2):335–347.
63. Topping DL, Clifton PM. Short-chain fatty acids and human colonic function: roles of resistant starch and nonstarch polysaccharides. *Physiol Rev*. 2001;81(3):1031–1064.
64. Grieg ER, Boot-Handford RP, Mani V, Sandle GI. Decreased expression of apical Na+ channels and basolateral Na+, K+-ATPase in ulcerative colitis. *J Pathol*. 2004;204(1):84–92.
65. Willemsen LE, Koetsier MA, van Deventer SJ, van Tol EA. Short chain fatty acids stimulate epithelial mucin 2 expression through differential effects on prostaglandin E_1 and E_2 production by intestinal myofibroblasts. *Gut*. 2003;52(10):1442–1447.
66. Lu L, Walker WA. Pathologic and physiologic interactions of bacteria with the gastrointestinal epithelium. *Am J Clin Nutr*. 2001;73(6):1124S–1130S.
67. Naftalin RJ, Pedley KC. Regional crypt function in rat large intestine in relation to fluid absorption and growth of the pericryptal sheath. *J Physiol*. 1999;514(pt 1):211–227.
68. Hirasawa K, Sato Y, Hosoda Y, Yamamoto T, Hanai H. Immunohistochemical localization of angiotensin II receptor and local renin-angiotensin system in human colonic mucosa. *J Histochem Cytochem*. 2002;50(2):275–282.
69. Thiagarajah JR, Griffiths NM, Pedley KC, Naftalin RJ. Evidence for modulation of pericryptal sheath myofibroblasts in rat descending colon by transforming growth factor beta and angiotensin II. *BMC Gastroenterol*. 2002;2:4.
70. Thiagarajah JR, Pedley KC, Naftalin RJ. Evidence of amiloride-sensitive fluid absorption in rat descending colonic crypts from fluorescence recovery of FITC-labelled dextran after photobleaching. *J Physiol (Lond)*. 2001;536(Pt 2):541–553.
71. Hopkins MJ, Sharp R, Macfarlane GT. Age and disease related changes in intestinal bacterial populations assessed by cell culture, 16S rRNA abundance, and community cellular fatty acid profiles. *Gut*. 2001;48(2):198–205.
72. Eckburg PB, Bik EM, Bernstein CN, et al. Diversity of the human instestinal microbial flora. *Science*. 2005;308(5728):1635–1638.
73. Porter EM, Bevins CL, Ghosh D, Ganz T. The multifaceted Paneth cell. *Cell Mol Life Sci*. 2002;59(1):156–170.
74. Boman HG. Antibacterial peptides: basic facts and emerging concepts. *J Intern Med*. 2003;254(3):197–215.
75. Chow J, Lee SM, Shen Y, Khosravi A, Mazmanian SK. Host-bacterial symbiosis in health and disease. *Adv Immunol*. 2010;107:243–274.
76. Kinross JM, Darzi AW, Nicholson JK. Gut microbiome-host interactions in health and disease. *Genome Med*. 2011;3(3):14.
77. Köhler H, McCormick BA, Walker WA. Bacterial-enterocyte crosstalk: cellular mechanisms in health and disease. *J Pediatr Gastroenterol Nutr*. 2003;36(2):175–185.
78. Zuercher AW, Jiang HQ, Thurnheer MC, Cuff CF, Cebra JJ. Distinct mechanisms for cross-protection of the upper versus lower respiratory tract through intestinal priming. *J Immunol*. 2002;169(7):3920–3925.
79. Demetter P, De Vos M, Van Huysse JA, et al. Colon mucosa of patients both with spondyloarthritis and Crohn's disease is enriched with macrophages expressing the scavenger receptor CD163. *Ann Rheum Dis*. 2005;64(2):321–324.
80. Didierlaurent A, Sirard JC, Kraehenbuhl JP, Neutra MR. How the gut senses its content. *Cell Microbiol*. 2002;4(2):61–72.
81. Girardin SE, Hugot JP, Sansonetti PJ. Lessons from Nod2 studies: towards a link between Crohn's disease and bacterial sensing. *Trends Immunol*. 2003;24(12):652–658.
82. Hershberg RM, Mayer LF. Antigen processing and presentation by intestinal epithelial cells—polarity and complexity. *Immunol Today*. 2000;21(3):123–128.
83. Neutra MR. Current concepts in mucosal immunity. V. Role of M cells in transepithelial transport of antigens and pathogens to the mucosal immune system. *Am J Physiol*. 1998;274(5):G785–G791.
84. Noverr MC, Huffnagle GB. Does the microbiota regulate immune responses outside the gut? *Trends Microbiol*. 2004;12(12):562–568.
85. Spahn TW, Kucharzik T. Modulating the intestinal immune system: the role of lymphotoxin and GALT organs. *Gut*. 2004;53(3):456–465.
86. Berkes J, Viswanathan VK, Savkovic SD, Hecht G. Intestinal epithelial responses to enteric pathogens: effects on the tight junction barrier, ion transport, and inflammation. *Gut*. 2003;52(3):439–451.
87. Anderson JM, Van Itallie CM. Tight junctions and the molecular basis for regulation of paracellular permeability. *Am J Physiol*. 1995;269(4 pt 1):G467–G475.
88. Kucharzik T, Walsh SV, Chen J, Parkos CA, Nusrat A. Neutrophil transmigration in inflammatory bowel disease is associated with differential expression of epithelial intercellular junction proteins. *Am J Pathol*. 2001;159(6):2001–2009.
89. Walsh SV, Hopkins AM, Nusrat A. Modulation of tight junction structure and function by cytokines. *Adv Drug Deliv Rev*. 2000;41(3):303–313.
90. Adegboyega PA, Mifflin RC, DiMari JF, Saada JI, Powell DW. Immunohistochemical study of myofibroblasts in normal colonic mucosa, hyperplastic polyps, and adenomatous colorectal polyps. *Arch Pathol Lab Med*. 2002;126(7):829–836.
91. Akishima Y, Ito K, Zhang L, et al. Immunohistochemical detection of human small lymphatic vessels under normal and pathological conditions using the LYVE-1 antibody. *Virchows Arch*. 2004;444(2):153–157.
92. Aldenborg F, Enerbäck L. The immunohistochemical demonstration of chymase and tryptase in human intestinal mast cells. *Histochem J*. 1994;26(7):587–596.
93. Brandtzaeg P, Farstad IN, Helgeland L. Phenotypes of T cells in the gut. *Chem Immunol*. 1998;71:1–26.
94. Buffa R, Marè P, Gini A, Salvadore M. Chromogranins A and B and secretogranin II in hormonally identified endocrine

cells of the gut and the pancreas. *Basic Appl Histochem*. 1988;32(4):471–484.
95. Castellaneta A, Abe M, Morelli AE, Thomson AW. Identification and characterization of intestinal Peyer's patch interferon-alpha producing (plasmacytoid) dendritic cells. *Hum Immunol*. 2004;65(2):104–113.
96. Corfield AP, Myerscough N, Longman R, Sylvester P, Arul S, Pignatelli M. Mucins and mucosal protection in the gastrointestinal tract: new prospects for mucins in the pathology of gastrointestinal disease. *Gut*. 2000;47(4):589–594.
97. Ferri GL, Probert L, Cocchia D, Michetti F, Marangos PJ, Polak JM. Evidence for the presence of S-100 protein in the glial component of the human enteric nervous system. *Nature*. 1982;297(5865):409–410.
98. Frängsmyr L, Baranov V, Hammarström S. Four carcinoembryonic antigen subfamily members, CEA, NCA, BGP and CGM2, selectively expressed in the normal human colonic epithelium are integral components of the fuzzy coat. *Tumour Biol*. 1999;20(5):277–292.
99. Fulcheri E, Cantino D, Bussolati G. Presence of intra-mucosal smooth muscle cells in normal human and rat colon. *Basic Appl Histochem*. 1985;29(4):337–344.
100. Fujisaki J, Shimoda T. Expression of cytokeratin subtypes in colorectal mucosa, adenoma, and carcinoma. *Gastroenterol Jpn*. 1993;28(5):647–656.
101. Gabbiani G, Schmid E, Winter S, et al. Vascular smooth muscle cells differ from other smooth muscle cells: predominance of vimentin filaments and specific alpha-type actin. *Proc Natl Acad Sci U S A*. 1981;78(1):298–302.
102. Galli SJ, Tsai M, Wershil BK. The c-kit receptor, stem cell factor, and mast cells. What each is teaching us about the others. *Am J Pathol*. 1993;142(4):965–974.
103. Grimelius L. Silver stains demonstrating neuroendocrine cells. *Biotech Histochem*. 2004;79(1):37–44.
104. Hamrock D, Azmi FH, O'Donnell E, Gunning WT, Philips ER, Zaher A. Infection by Rhodococcus equi in a patient with AIDS: histological appearance mimicking Whipple's disease and Mycobacterium avium-intracellulare infection. *J Clin Pathol*. 1999;52(1):68–71.
105. Higaki S, Tada M, Nishiaki M, Mitani M, Yanai H, Okita K. Immunohistological study to determine the presence of pericryptal myofibroblasts and basement membrane in colorectal epithelial tumors. *J Gastroenterol*. 1999;34(2):215–220.
106. Jass JR. Mucin core proteins as differentiation markers in the gastrointestinal tract. *Histopathology*. 2000;37(6):561–564.
107. Jones TR, Kao KJ, Pizzo SV, Bigner DD. Endothelial cell surface expression and binding of factor VIII/von Willebrand factor. *Am J Pathol*. 1981;103(2):304–308.
108. Kato H, Yamamoto T, Yamamoto H, Ohi R, So N, Iwasaki Y. Immunocytochemical characterization of supporting cells in the enteric nervous system in Hirschsprung's disease. *J Pediatr Surg*. 1990;25(5):514–519.
109. Kawana T, Nada O, Ikeda K. An immunohistochemical study of glial fibrillary acidic (GFA) protein and S-100 protein in the colon affected by Hirschsprung's disease. *Acta Neuropathol (Berl)*. 1988;76(2):159–165.
110. Kende AI, Carr NJ, Sobin LH. Expression of cytokeratins 7 and 20 in carcinomas of the gastrointestinal tract. *Histopathology*. 2003;42(2):137–140.
111. Kurki P, Virtanen I. The detection of smooth muscle antibodies reacting with intermediate filaments of desmin type. *J Immunol Methods*. 1985;76(2):329–335.
112. Lamps LW, Molina CP, West AB, Haggitt RC, Scott MA. The pathologic spectrum of gastrointestinal and hepatic histoplasmosis. *Am J Clin Pathol*. 2000;113(1):64–72.
113. Lee MJ, Lee HS, Kim WH, Choi Y, Yang M. Expression of mucins and cytokeratins in primary carcinomas of the digestive system. *Mod Pathol*. 2003;16(5):403–410.
114. Meyer T, Brinck U. Differential distribution of serotonin and tryptophan hydroxylase in the human gastrointestinal tract. *Digestion*. 1999;60(1):63–68.
115. Moll R, Lowe A, Laufer J, Franke WW. Cytokeratin 20 in human carcinomas. A new histodiagnostic marker detected by monoclonal antibodies. *Am J Pathol*. 1992;140(2):427–447.
116. O'Connell FP, Pinkus JL, Pinkus GS. CD138 (syndecan-1), a plasma cell marker immunohistochemical profile in hematopoietic and nonhematopoietic neoplasms. *Am J Clin Pathol*. 2004;121(2):254–263.
117. Ozğul C, Karaöz E, Erdoğan D, Dursun A. Expression of epidermal growth factor receptor in normal colonic mucosa and in adenocarcinomas of the colon. *Acta Physiol Hung*. 1997–1998;85(2):121–128.
118. Park HJ, Kamm MA, Abbasi AM, Talbot IC. Immunohistochemical study of the colonic muscle and innervation in idiopathic chronic constipation. *Dis Colon Rectum*. 1995;38(5):509–513.
119. Petchasuwan C, Pintong J. Immunohistochemistry for intestinal ganglion cells and nerve fibers: aid in the diagnosis of Hirschsprung's disease. *J Med Assoc Thai*. 2000;83(11):1402–1409.
120. Qualtrough D, Hinoi T, Fearon E, Paraskeva C. Expression of CDX2 in normal and neoplastic human colon tissue and during differentiation of an in vitro model system. *Gut*. 2002;51(2):184–190.
121. Rubio CA, Nesi G. A simple method to demonstrate normal and metaplastic Paneth cells in tissue sections. *In Vivo*. 2003;17(1):67–71.
122. Rubio CA, Slezak P. The subepithelial band in collagenous colitis is autofluorescent a study in H&E stained sections. *In Vivo*. 2002;16(2):123–126.
123. Sarsfield P, Rinne A, Jones DB, Johnson P, Wright DH. Accessory cells in physiological lymphoid tissue from the intestine: an immunohistochemical study. *Histopathology*. 1996;28(3):205–211.
124. Sartore S, De Marzo N, Borrione AC, et al. Myosin heavy-chain isoforms in human smooth muscle. *Eur J Biochem*. 1989;179(1):79–85.
125. Smithson JE, Warren BF, Young S, Pigott R, Jewell DP. Heterogeneous expression of carcinoembryonic antigen in the normal colon and upregulation in active ulcerative colitis. *J Pathol*. 1996;180(2):146–151.
126. Truong LD, Rangdaeng S, Cagle P, Ro JY, Hawkins H, Font RL. The diagnostic utility of desmin. A study of 584 cases and review of the literature. *Am J Clin Pathol*. 1990;93(3):305–314.
127. Werling RW, Yaziji H, Bacchi CE, Gown AM. CDX2, a highly sensitive and specific marker of adenocarcinomas of intestinal origin: an immunohistochemical survey of 476 primary and metastatic carcinomas. *Am J Surg Pathol*. 2003;27(3):303–310.
128. West AB, Isaac CA, Carboni JM, Morrow JS, Mooseker MS, Barwick KW. Localization of villin, a cytoskeletal protein specific to microvilli, in human ileum and colon and in colonic neoplasms. *Gastroenterology*. 1988;94(2):343–352.

129. Wiedenmann B, Waldherr R, Buhr H, Hille A, Rosa P, Huttner WB. Identification of gastroenteropancreatic neuroendocrine cells in normal and neoplastic human tissue with antibodies against synaptophysin, chromogranin A, secretogranin I (chromogranin B), and secretogranin II. *Gastroenterology*. 1988;95(5):1364–1374.
130. Wong NA, Herriot M, Rae F. An immunohistochemical study and review of potential markers of human intestinal M cells. *Eur J Histochem*. 2003;47(2):143–150.
131. Pickard KM, Bremner AR, Gordon JN, MacDonald TT. Microbial-gut interactions in health and disease. Immune responses. *Best Pract Res Clin Gastroenterol*. 2004;18(2):271–285.
132. Thiagarajah JR, Gourmelon P, Griffiths NM, Lebrun F, Naftalin RJ, Pedley KC. Radiation induced cytochrome c release causes loss of rat colonic fluid absorption by damage to crypts and pericryptal myofibroblasts. *Gut*. 2000;47(5):675–684.
133. Kirby JA, Bone M, Robertson H, Hudson M, Jones DEJ. The number of intraepithelial T cells decreases from ascending colon to rectum. *J Clin Pathol*. 2003;56(2):158.
134. Kraehenbuhl JP, Neutra MR. Epithelial M cells: differentiation and function. *Annu Rev Cell Dev Biol*. 2000;16:301–332.
135. Sapp H, Ithamukkala S, Brien TP, et al. The terminal ileum is affected in patients with lymphocytic or collagenous colitis. *Am J Surg Pathol*. 2002;26(11):1484–1492.
136. Brittan M, Wright NA. The gastrointestinal stem cell. *Cell Prolif*. 2004;37(1):35–53.
137. Bleuming SA, Peppelenbosch MP, Roberts, DJ, van den Brink GR. Homeostasis of the adult colonic epithelium: a role for morphogens. *Scand J Gastroenterol*. 2004;39(2):93–98.
138. Halm DR, Halm ST. Secretagogue response of goblet cells and columnar cells in human colonic crypts. *Am J Physiol Cell Physiol*. 2000;278(1):C212–C233.
139. Kim KM, Shibata D. Methylation reveals a niche: stem cell succession in human colon crypts. *Oncogene*. 2002; 21(35):5441–5449.
140. Potten CS. Epithelial cell growth and differentiation. II. Intestinal apoptosis. *Am J Physiol*. 1997;273(2 pt 1):G253–G257.
141. Potten CS, Booth C, Tudor GL, et al. Identification of a putative intestinal stem cell and early lineage marker; musashi-1. *Differentiation*. 2003;71(1):28–41.
142. Sancho E, Batlle E, Clevers H. Live and let die in the intestinal epithelium. *Curr Opin Cell Biol*. 2003;15(6):763–770.
143. Schonhoff SE, Giel-Moloney M, Leiter AB. Minireview: development and differentiation of gut endocrine cells. *Endocrinology*. 2004;145(6):2639–2644.
144. Neutra MR, Mantis NJ, Kraehenbuhl JP. Collaboration of epithelial cells with organized mucosal lymphoid tissues. *Nat Immunol*. 2001;2(11):1004–1009.
145. Corfield AP, Wiggins R, Edwards C, et al. A sweet coating—how bacteria deal with sugars. *Adv Exp Med Biol*. 2003;535:3–15.
146. Filipe MI. Mucins in the human gastrointestinal epithelium: a review. *Invest Cell Pathol*. 1979;2(3):195–216.
147. Culling CF, Reid PE, Dunn WL, Freeman HJ. The relevance of the histochemistry of colonic mucins based upon their PAS reactivity. *Histochem J*. 1981;13(6):889–903.
148. Jeon YH, Ahn JH, Chang HK. Colorectal epithelial neoplasm associated with gut-associated lymphoid tissue. *J Clin Pathol*. 2016;69(1):6–11.
149. Westerhoff M. Histologic features of colonic infections. *Pathologe*. 2022;43(1):16–30.
150. Rehfeld JF. The new biology of gastrointestinal hormones. *Physiol Rev*. 1998;78(4):1087–1108.
151. Skipper M, Lewis J. Getting to the guts of enteroendocrine differentiation. *Nat Genet*. 2000;24(1):3–4.
152. Hirschowitz L, Rode J. Changes in neurons, neuroendocrine cells and nerve fibers in the lamina propria of irradiated bowel. *Virchows Arch A Pathol Anat Histopathol*. 1991;418(2): 163–168.
153. Qian J, Hickey WF, Angeletti RH. Neuroendocrine cells in intestinal lamina propria. Detection with antibodies to chromogranin A. *J Neuroimmunol*. 1988;17(2):159–165.
154. Roth KA, Gordon JI. Spatial differentiation of the intestinal epithelium: analysis of enteroendocrine cells containing immunoreactive serotonin, secretin, and substance P in normal and transgenic mice. *Proc Natl Aca Sci U S A*. 1990;87(16): 6408–6412.
155. Ayabe T, Ashida T, Kohgo Y, Kono T. The role of Paneth cells and their antimicrobial peptides in innate host defense. *Trends Microbiol*. 2004;12(8):394–398.
156. Pezhouh MK, Cheng E, Weinberg AG, Park JY. Significance of Paneth cells in histologically unremarkable rectal mucosa. *Am J Surg Pathol*. 2016;40(7):968–971.
157. Lala S, Ogura Y, Osborne C, et al. Crohn's disease and the NOD2 gene: a role for Paneth cells. *Gastroenterology*. 2003;125(1): 47–57.
158. Ogura Y, Lala S, Xin W, et al. Expression of NOD2 in Paneth cells: a possible link to Crohn's ileitis. *Gut*. 2003;52(11): 1591–1597.
159. Lin PW, Simon PO Jr, Gerwitz AT, et al. Paneth cell cryptidins act in vitro as apical paracrine regulators of the innate inflammatory response. *J Biol Chem*. 2004;279(19):19902–19907.
160. Baranov V, Hammarstrom S. Carcinoembryonic antigen (CEA) and CEA-related cell adhesion molecule 1 (CEACAM1), apically expressed on human colonic M cells, are potential receptors for microbial adhesion. *Histochem Cell Biol*. 2004;121(2):83–89.
161. Sierro F, Pringault E, Simon-Assmann P, Kraehenbuhl JP, Debard N. Transient expression of M-cell phenotype by enterocyte-like cells of the follicle-associated epithelium of mouse Peyer's patches. *Gastroenterology*. 2000;119(3):734–743.
162. Neutra MR, Mantis NJ, Frey A, Giannasca PJ. The composition and function of M cell apical membranes: implications for microbial pathogenesis. *Semin Immunol*. 1999;11(3):171–181.
163. Cetin Y, Muller-Koppel L, Aunis D, Bader MF, Grube D. Chromogranin A (CgA) in the gastro-entero-pancreatic (GEP) endocrine system. II. CgA in mammalian entero-endocrine cells. *Histochemistry*. 1989;92(4):265–275.
164. Gebert A, Fassbender S, Werner K, Weissferdt A. The development of M cells in Peyer's patches is restricted to specialized dome-associated crypts. *Am J Pathol*. 1999;154(5):1573–1582.
165. Jepson MA, Clark MA, Hirst BH. M cell targeting by lectins: a strategy for mucosal vaccination and drug delivery. *Adv Drug Deliv Rev*. 2004;56(4):511–525.
166. Melgar S, Hammarström S, Oberg A, Danielsson A, Hammarström ML. Cytolytic capabilities of lamina propria and intraepithelial lymphocytes in normal and chronically inflamed human intestine. *Scan J Immunol*. 2004;60(1–2): 167–177.
167. Jeon YH, Ahn JH, Chang HK. Colorectal epithelial neoplasm associated with gut-associated lymphoid tissue. *J Pathol Transl Med*. 2020;54(2):135–145.
168. McCarthy AJ, Chetty R. Gut-associated lymphoid tissue or so-called "dome" carcinoma of the colon: Review. *World J Gastrointest Oncol*. 2019;11(1):59–70.

169. Farstad IN, Lundin KE. Gastrointestinal intraepithelial lymphocytes and T cell lymphomas. *Gut*. 2003;52(2):163–164.
170. Helgeland L, Dissen E, Dai KZ, Midtvedt T, Brandtzaeg P, Vaage JT. Microbial colonization induces oligoclonal expansions of intraepithelial CD8 T cells in the gut. *Eur J Immunol*. 2004;34(12):3389–3400.
171. MacDonald TT, Bajaj-Elliot M, Pender SL. T cells orchestrate intestinal mucosal shape and integrity. *Immunol Today*. 1999;20(11):505–510.
172. Brandtzaeg P. Development and basic mechanisms of human gut immunity. *Nutr Rev*. 1998;56(1 pt 2):S5–S18.
173. Ebert EC. Interleukin-12 up-regulates perforin- and Fas-mediated lymphokine-activated killer activity by intestinal intraepithelial lymphocytes. *Clin Exp Immunol*. 2004;138(2): 259–265.
174. Kagnoff MF. Current concepts in mucosal immunity. III. Ontogeny and function of gamma delta T cells in the intestine. *Am J Physiol*. 1998;274(3 Pt 1):G455–G458.
175. Chen Y, Chou K, Fuchs E, Havran WL, Boismenu R. Protection of the intestinal mucosa by intraepithelial gamma delta T cells. *Proc Natl Acad Sci U S A*. 2002;99(22):14338–14343.
176. Lin T, Yoshida H, Matsuzaki G, et al. Autospecific gammadelta thymocytes that escape negative selection find sanctuary in the intestine. *J Clin Invest*. 1999;104(9):1297–1305.
177. Shibahara T, Wilcox JN, Couse, T, Madara JL. Characterization of epithelial chemoattractants for human intestinal intraepithelial lymphocytes. *Gastroenterology*. 2001;120(1): 60–70.
178. Rothenberg ME, Mishra A, Brandt EB, Hogan SP. Gastrointestinal eosinophils. *Immunol Rev*. 2001;179:139–155.
179. Bochner BS, Schleimer RP. Mast cells, basophils, and eosinophils: distinct but overlapping pathways for recruitment. *Immunol Rev*. 2001;179:5–15.
180. Levy AM, Yamazaki K, Van Keulen VP, et al. Increased eosinophil infiltration and degranulation in colonic tissue from patients with collagenous colitis. *Am J Gastroenterol*. 2001;96(5):1522–1528.
181. Polydorides AD, Banner BF, Hannaway PJ, Yantiss RK. Evaluation of site-specific and seasonal variation in colonic mucosal eosinophils. *Hum Pathol*. 2008;39(6):832–836.
182. Turner KO, Sinkre RA, Neumann WL, Genta RM. Primary colonic eosinophilia and eosinophilic colitis in adults. *Am J Surg Pathol*. 2017;41(2):225–233.
183. Walker MM, Potter MD, Talley NJ. Eosinophilic colitis and colonic eosinophilia. *Curr Opin Gastroenterol*. 2019;35(1): 42–50.
184. Howell SJ, Wilk D, Yadav SP, Bevins CL. Antimicrobial polypeptides of the human colonic epithelium. *Peptides*. 2003;24(11): 1763–1770.
185. Harnois C, Demers MJ, Bouchard V, et al. Human intestinal epithelial crypt cell survival and death: complex modulations of Bcl-2 homologs by Fak, PI3-K/Akt-1, MEK/Erk, and p38 signaling pathways. *J Cell Physiol*. 2004;198(2):209–222.
186. Barkla DH, Gibson PR. The fate of epithelial cells in the human large intestine. *Pathology*. 1999;31(3):230–238.
187. Gomez-Angelats M, Bortner CD, Cidlowski JA. Cell volume regulation in immune cell apoptosis. *Cell Tissue Res*. 2000; 301(1):33–42.
188. Gupta S. Molecular signaling in death receptor and mitochondrial pathways of apoptosis (review). *Int J Oncol*. 2003;22(1): 15–20.
189. Huppertz B, Frank HG, Kaufmann P. The apoptosis cascade—morphological and immunohistochemical methods for its visualization. *Anat Embryol (Berl)*. 1999;200(1):1–18.
190. Luciano L, Groos S, Busche R, Von Engelhardt W, Reale E. Massive apoptosis of colonocytes induced by butyrate deprivation overloads resident macrophages and promotes the recruitment of circulating monocytes. *Cell Tissue Res*. 2002;309(3):393–407.
191. Schuster N, Krieglstein K. Mechanisms of TGF-beta-mediated apoptosis. *Cell Tissue Res*. 2002;307(1):1–14.
192. Watson AJM. Apoptosis and colorectal cancer. *Gut*. 2004; 53(11):1701–1709.
193. Xiao ZQ, Moragoda L, Jaszewski R, Hatfield JA, Fligiel SE, Majumdar AP. Aging is associated with increased proliferation and decreased apoptosis in the colonic mucosa. *Mech Ageing Dev*. 2001;122(15):1849–1864.
194. Medina F, Segundo C, Campos-Caro A, Salcedo I, García-Poley A, Brieva JA. Isolation, maturational level, and functional capacity of human colon lamina propria plasma cells. *Gut*. 2003;52(3):383–389.
195. Simon HU. Regulation of eosinophil and neutrophil apoptosis—similarities and differences. *Immunol Rev*. 2001;179: 156–162.
196. Iqbal N, Salzman D, Lazenby AJ, Wilcox CM. Diagnosis of gastrointestinal graft-versus-host disease. *Am J Gastroenterol*. 2000;95(11):3034–3038.
197. Iwamoto M, Koji T, Makiyama K, Kobayashi N, Nakane PK. Apoptosis of crypt epithelial cells in ulcerative colitis. *J Pathol*. 1996;180(2):152–159.
198. Ma C, Pai RK, Schaeffer DF, et al. Recommendations for standardizing biopsy acquisition and histological assessment of immune checkpoint inhibitor-associated colitis. *J Immunother Cancer*. 2022;10(3):e004560.
199. Ono Y, Gonzalez RS. Apoptosis, crypt dropout, and equivocal immunohistochemical staining may indicate cytomegalovirus infection in inflammatory bowel disease patients. *Am J Surg Pathol*. 2023;47(8):933–941.
200. Karamchandani DM, Chetty R. Apoptotic colopathy: a pragmatic approach to diagnosis. *J Clin Pathol*. 2018;71(12): 1033–1040.
201. Backus HH, Van Groeningen CJ, Vos W, et al. Differential expression of cell cycle and apoptosis related proteins in colorectal mucosa, primary colon tumours, and liver metastases. *J Clin Pathol*. 2002;55(3):206–211.
202. Anagnostopoulos I, Schuppan D, Riecken EO, Gross UM, Stein H. Tenascin labelling in colorectal biopsies: a useful marker in the diagnosis of collagenous colitis. *Histopathology*. 1999;34(5):425–431.
203. Gledhill A, Cole FM. Significance of basement membrane thickening in the human colon. *Gut*. 1984;25(10):1085–1088.
204. Yuan L, Wu TT, Zhang L. Microscopic colitis: lymphocytic colitis, collagenous colitis, and beyond. *Hum Pathol*. 2023;132: 89–101.
205. Powell DW, Mifflin RC, Valentich JD, Crowe SE, Saada JI, West AB. Myofibroblasts. II. Intestinal subepithelial myofibroblasts. *Am J Physiol*. 1999;277(2 pt 1):C183–C201.
206. Azzali G. Structure, lymphatic vascularization and lymphocyte migration in mucosa-associated lymphoid tissue. *Immunol Rev*. 2003;195:178–189.
207. Brandtzaeg P, Johansen FE, Baekkevold ES, Carlsen HS, Farstad IN. The traffic of mucosal lymphocytes to

extraintestinal sites. *J Pediatr Gastroenterol Nutr*. 2004; 39(suppl 3):S725–S726.

208. Goldstein NS, Bhanot P. Paucicellular and asymptomatic lymphocytic colitis. Expanding the clinicopathologic spectrum of lymphocytic colitis. *Am J Clin Pathol*. 2004;122(3):405–411.
209. Fischer M, Kuppers R. Human IgA- and IgM-secreting intestinal plasma cells carry heavily mutated VH region genes. *Eur J Immunol*. 1998;28(9):2971–2977.
210. Brandtzaeg P, Pabst R. Let's go mucosal: communication on slippery ground. *Trends Immunol*. 2004;25(11):570–577.
211. Nishida Y, Murase K, Isomoto H, et al. Different distribution of mast cells and macrophages in colonic mucosa of patients with collagenous colitis and inflammatory bowel disease. *Hepatogastroenterology*. 2002;49(45):678–682.
212. Pascal RR, Gramlich TL, Parker KM, Gansler TS. Geographic variations in eosinophil concentration in normal colonic mucosa. *Mod Pathol*. 1997;10(4):363–365.
213. Barbara G, Stanghellini V, DeGiorgio R, et al. Activated mast cells in proximity to colonic nerves correlate with abdominal pain in irritable bowel syndrome. *Gastroenterology*. 2004;126(3): 693–702.
214. Boyce JA. Mast cells: beyond IgE. *J Allergy Clin Immunol*. 2003;111(1):24–32, quiz 33.
215. Doyle LA, Sepehr GJ, Hamilton MJ, Akin C, Castells MC, Hornick JL. A clinicopathologic study of 24 cases of systemic mastocytosis involving the gastrointestinal tract and assessment of mucosal mast cell density in irritable bowel syndrome and asymptomatic patients. *Am J Surg Pathol*. 2014;38(6):832–843.
216. Mayer L. Current concepts in mucosal immunity. I. Antigen presentation in the intestine: new rules and regulations. *Am J Physiol*. 1998;274(1):G7–G9.
217. Rubio CA. Rectal muciphages are rich in lysozymes: a novel source of antimicrobial mucosal defense? *Scand J Gastroenterol*. 2002;37(6):743–744.
218. Salto-Tellez M, Price AB. What is the significance of muciphages in colorectal biopsies? The significance of muciphages in otherwise normal colorectal biopsies. *Histopathology*. 2000; 36(6):556–569.
219. Schenk M, Bouchon A, Birrer S, Colonna M, Mueller C. Macrophages expressing triggering receptor expressed on myeloid cells-1 are underrepresented in the human intestine. *J Immunol*. 2005;174(1):517–524.
220. Alkan S, Beals TF, Schnitzer B. Primary diagnosis of Whipple disease manifesting as lymphadenopathy: use of polymerase chain reaction for detection of Tropheryma whippelii. *Am J Clin Pathol*. 2001;116(6):898–904.
221. Dobbins WO III, Weinstein WM. Electron microscopy of the intestine and rectum in acquired immunodeficiency syndrome. *Gastroenterology*. 1985;88(3):738–749.
222. Lee SH, Barnes WG, Hodges GR, Dixon A. Perforated granulomatous colitis caused by Histoplasma capsulatum. *Dis Colon Rectum*. 1985;28(3):171–176.
223. Nguyen HN, Frank D, Handt S, et al. Severe gastrointestinal hemorrhage due to Mycobacterium avium complex in a patient receiving immunosuppressive therapy. *Am J Gastroenterol*. 1999;94(1):232–235.
224. Hart AL, Lammers K, Brigidi P, et al. Modulation of human dendritic cell phenotype and function by probiotic bacteria. *Gut*. 2004;53(11):1602–1609.
225. Zareie M, Singh PK, Irvine EJ, Sherman PM, McKay DM, Perdue MH. Monocyte/macrophage activation by normal bacteria and bacterial products: Implications for altered epithelial function in Crohn's disease. *Am J Pathol*. 2001;158(3): 1101–1109.
226. Bodey B, Siegel SE, Kaiser HE. Antigen presentation by dendritic cells and their significance in antineoplastic immunotherapy. *In Vivo*. 2004;18(1):81–100.
227. Powell DW, Mifflin RC, Valentich JD, Crowe SE, Saada JI, West AB. Myofibroblasts. I. Paracrine cells important in health and disease. *Am J Physiol*. 1999;277(1 pt 1):C1–C9.
228. Skalli O, Schürch W, Seemayer T, et al. Myofibroblasts from diverse pathologic settings are heterogeneous in their content of actin isoforms and intermediate filament proteins. *Lab Invest*. 1989;60(2):275–285.
229. Hissong, E, Yantiss, RK. Epithelial-stromal polyps of the colon are not perineuriomas. *Am J Clin Pathol*. 2021;156(1):109–116.
230. Ward SM, Sanders KM, Hirst GD. Role of interstitial cells of Cajal in neural control of gastrointestinal smooth muscles. *Neurogastroenterol Motil*. 2004;16(suppl 1):112–117.
231. Biberthaler P, Langer S. Comparison of the new OPS imaging technique with intravital microscopy: analysis of the colon microcirculation. *Eur Surg Res*. 2002;34(1–2):124–128.
232. Fenoglio CM, Kay GI, Lane N. Distribution of human colonic lymphatics in normal, hyperplastic, and adenomatous tissue. Its relationship to metastasis from small carcinomas in pedunculated adenomas, with two case reports. *Gastroenterology*. 1973;64(1):51–66.
233. Dobbins WO III. The intestinal mucosal lymphatics in man. A light and electron microscopic study. *Gastroenterology*. 1966;51(6):994–1003.
234. Fogt F, Zimmerman RL, Ross HM, Daly T, Gausas RE. Identification of lymphatic vessels in malignant, adenomatous and normal colonic mucosa using the novel immunostain D2-40. *Oncol Rep*. 2004;11(1):47–50.
235. Fogt F, Pascha TL, Zhang PJ, Gausas RE, Rahemtulla A, Zimmerman RL. Proliferation of D2-40-expressing intestinal lymphatic vessels in the lamina propria in inflammatory bowel disease. *Int J Mol Med*. 2004;13(2):211–214.
236. Percy WH, Fromm TH, Wangsness CE. Muscularis mucosae contraction evokes colonic secretion via prostaglandin synthesis and nerve stimulation. *Am J Physiol Gastrointest Liver Physiol*. 2003;284(2):G213–G220.
237. Percy WH, Brunz JT, Burgers RE, Fromm TH, Merkwan CL, van Dis J. Interrelationship between colonic muscularis mucosae activity and changes in transmucosal potential difference. *Am J Physiol Gastrointest Liver Physiol*. 2001;281(2): G479–G489.
238. Daniel EE, Wang YF. Gap junctions in intestinal smooth muscle and interstitial cells of Cajal. *Microsc Res Tech*. 1999; 47(5):309–320.
239. Hagger R, Gharaie S, Finlayson C, Kumar D. Regional and transmural density of interstitial cells of Cajal in human colon and rectum. *Am J Physiol*. 1998;275(6 pt 1):G1309–G1316.
240. Coerdt W, Michel JS, Rippin G, et al. Quantitative morphometric analysis of the submucous plexus in age-related control groups. *Virchows Arch*. 2004;444(3):239–246.
241. Wedel T, Spiegler J, Soellner S, et al. Enteric nerves and interstitial cells of Cajal are altered in patients with slow-transit constipation and megacolon. *Gastroenterology*. 2002; 123(5):1459–1467.
242. Wilder-Smith CH, Talbot IC, Merki HS, Meier-Ruge WA. Morphometric quantification of normal submucous plexus

in the distal rectum of adult healthy volunteers. *Eur J Gastroenterol Hepatol*. 2002;14(12):1339–1342.
243. Faussone-Pellegrini MS, Pantalone D, Cortesini C. Smooth muscle cells, interstitial cells of Cajal and myenteric plexus interrelationships in the human colon. *Acta Anat (Basel)*. 1990;139(1):31–44.
244. Tunru-Dinh V, Wu ML. Intramucosal ganglion cells in normal adult colorectal mucosa. *Int J Surg Pathol*. 2007;15(1):31–37.
245. Oh HE, Chetty R. Intramucosal ganglion cells are common in diverticular disease. *Pathology*. 2008;40(5):470–474.
246. Faussone-Pellegrini MS, Cortesini C, Pantalone D. Neuromuscular structures specific to the submucosal border of the human colonic circular muscle layer. *Can J Physiol Pharmacol*. 1990;68(11):1437–1446.
247. Mazzia C, Porcher C, Jule Y, Christen MO, Henry M. Ultrastructural study of relationships between c-kit immunoreactive interstitial cells and other cellular elements in the human colon. *Histochem Cell Biol*. 2000;113(5):401–411.
248. Rumessen JJ, Peters S, Thuneberg L. Light and electron microscopical studies of interstitial cells of Cajal and muscle cells at the submucosal border of human colon. *Lab Invest*. 1993;68(4):481–495.
249. Ward SM, Sanders KM. Interstitial cells of Cajal: primary targets of enteric motor innervation. *Anat Rec*. 2001;262(1):125–135.
250. Takayama I, Horiguchi K, Daigo Y, Mine T, Fujino MA, Ohno S. The interstitial cells of Cajal and a gastroenteric pacemaker system. *Arch Histol Cytol*. 2002;65(1):1–26.
251. Krishnamurthy S, Schuffler MD. Pathology of neuromuscular disorders of the small intestine and colon. *Gastroenterology*. 1987;93(3):610–639.
252. Fraser ID, Condon RE, Schulte WJ, DeCosse JJ, Cowles VE. Longitudinal muscle of muscularis externa in human and nonhuman primate colon. *Arch Surg*. 1981;116(1):61–63.
253. Chan OT, Chiles L, Levy M, et al. Smoothelin expression in the gastrointestinal tract: implication in colonic inertia. *Appl Immunohistochem Mol Morphol*. 2013;21(5):452–459.
254. Hirota S, Isozaki K, Moriyama Y, et al. Gain-of-function mutations of c-kit in human gastrointestinal stromal tumors. *Science*. 1998;279(5350):577–580.
255. Kindblom LG, Remotti HE, Aldenborg F, Meis-Kindblom JM. Gastrointestinal pacemaker cell tumor (GIPACT): gastrointestinal stromal tumors show phenotypic characteristics of the interstitial cells of Cajal. *Am J Pathol*. 1998;152(5):1259–1269.
256. Driman DK, Preiksaitis HG. Colorectal inflammation and increased cell proliferation associated with oral sodium phosphate bowel preparation solution. *Hum Pathol*. 1998;29(9):972–978.
257. Levine DS. Proctitis following colonoscopy. *Gastrointest Endosc*. 1988;34(3):269–272.
258. Pockros PJ, Foroozan P. Golytely lavage versus a standard colonoscopy preparation: effect on normal colonic mucosal histology. *Gastroenterology*. 1985;88(2):545–548.
259. Rejchrt S, Bures J, Siroký M, Kopácová M, Slezák L, Langr F. A prospective, observational study of colonic mucosal abnormalities associated with orally administered sodium phosphate for colon cleansing before colonoscopy. *Gastrointest Endosc*. 2004;59(6):651–654.
260. Wong NA, Penman ID, Campbell S, Lessells AM. Microscopic focal cryptitis associated with sodium phosphate bowel preparation. *Histopathology*. 2000;36(5):476–478.
261. Fa-Si-Oen PR, Penninckx F. The effect of mechanical bowel preparation on human colonic tissue in elective open colon surgery. *Dis Colon Rectum*. 2004;47(6):948–949.
262. Allen TV, Achord JL. The pickle of proper bowel biopsy orientation. *Gastroenterology*. 1977;72(4 pt 1):774–775.
263. Bucher P, Gervaz, P, Egger JF, Soravia C, Morel P. Morphologic alterations associated with mechanical bowel preparation before elective colorectal surgery: a randomized trial. *Dis Colon Rectum*. 2006;49(1):109–112.
264. Haggitt RC. Handling of gastrointestinal biopsies in the surgical pathology laboratory. *Lab Med*. 1982;13:272–278.
265. Carson FL. Polyfoam pads—a source of artifact. *J Histotechnol*. 1981;4:33–34.
266. Snover DC, Sandstad J, Hutton S. Mucosal pseudolipomatosis of the colon. *Am J Clin Pathol*. 1985;84(5):575–580.

25

Appendix

Phoenix D. Bell ■ Laura W. Lamps

GROSS ANATOMY/SURGICAL PERSPECTIVE

The vermiform appendix is a slender tubular extension of the posteromedial aspect of the cecum that originates below and within 1 to 3 cm of the ileocecal valve. The anatomy of the junction between the base of the appendix and the cecum is relatively uniform. The remainder of the appendix, however, can have several different forms, including retrocecal, subcecal, pelvic, and juxta-ileal (1–3). The retrocecal position is most common, present in nearly 70% of the population (3,4). Unusual locations including an intramural appendix buried within the cecal wall and subhepatic appendices have been documented (5,6). These unusual anatomic locations should be considered during surgical exploration for acute abdominal pain. Although the appendix itself lacks taeniae, the base of the vermiform appendix lies at the convergence of the three cecal/ascending colon taeniae. These aid in locating the appendix when it is not readily apparent; the prominent anterior taeniae is most easily traced for this purpose (1,7).

The appendix can vary remarkably in length but generally spans 7 to 10 cm (2,4) with an external diameter of 3 to 8 mm, and luminal diameter of 1 to 3 mm (8). Except in cases of an unusual anatomical location, the appendix is intraperitoneal (9). The mesoappendix is a fold of fat and peritoneum contiguous with the mesentery of the terminal ileum. It extends along the length of the appendix, terminating just proximal to the tip (1).

The appendicular artery, a derivative of the inferior branch of the ileocolic artery of the superior mesenteric trunk, provides the majority of blood to the appendix (4,9). It courses behind the terminal ileum and enters the mesoappendix near the base of the appendix. The distal portion rests against the appendiceal muscular wall (8). Blood vessels lie essentially "unprotected" on the external surface where there is no mesoappendix at the appendiceal tip (1). Variations of the appendiceal blood supply with accessory arterial contributions are not unusual (10). Venous, lymphatic, and neural components closely follow the arterial vasculature (7). Branches of the ileocolic vein drain the appendiceal venous network into the superior mesenteric vein and eventually into the portal circulation, whereas lymphatic vessels drain into the ileocolic lymph nodes (7). Innervation is derived from branches of the vagus nerve (parasympathetic) and superior mesenteric plexus (sympathetic).

The external surface of the vermiform appendix is smooth, pink-tan or gray, and glistening (Fig. 25.1) without hyperemia or injection of the vessels. Sectioning may reveal a nodular appearance due to the characteristic prominent lymphoid tissue.

Development of the Vermiform Appendix and Congenital Anomalies

The vermiform appendix originates from the primordial structure termed the cecal diverticulum, which arises from the antimesenteric border of the caudal limb of the midgut

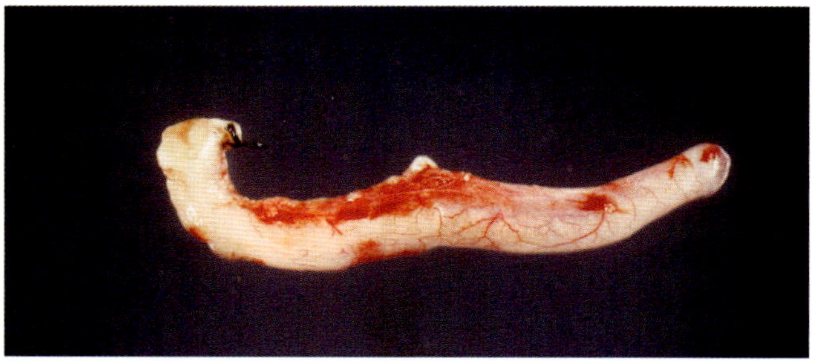

FIGURE 25.1 Gross photograph of a normal appendix. This incidentally removed appendix has a normal length and diameter, and a glistening external surface. There is focal hemorrhage due to the surgical procedure.

loop (8,11) during the 6th week of fetal life. Around the 5th month of fetal life, the most proximal portion enlarges and expands in continuity with the remainder of the large bowel, forming the cecum proper. The distal aspect or apex simply elongates, remaining narrow, and becomes the vermiform appendix (8,11). The appendix continues to grow during childhood in both length and diameter, reaching its maximum size somewhere between 3 and 6 years of age (12,13). In addition, the base of the appendix transitions from a cone shape to a rounded shape, and the appendix migrates from the cecal tip to the left posterior wall (13).

Abnormal embryologic development can result in a number of congenital anomalies, including agenesis, atresia, and various duplications or even triplication of the appendix (8,14–18). Agenesis is very rare and may be associated with other congenital anomalies (8,14). Duplications have been well described and categorized. Type A duplication is a partial duplication that features a single cecum and an appendix with a bifurcated distal appendiceal portion. Type B duplication is a complete appendiceal duplication featuring a single cecum with two separate appendices and two distinct bases. Type C duplication is a complete duplication of both cecum and appendix. It has two cecal structures, each with its own appendix (8,14–17). Some type B duplications are associated with other congenital anomalies (17), and type C duplication is always associated with other organ duplications (17). In general, appendiceal duplication is recognized by the presence of complete and separate inner circular and outer longitudinal muscle bands and the presence of a prominent lymphoid component (19). The majority of type B and virtually all type A duplications are found incidentally during surgery for imaging of suspected appendicitis. Horseshoe appendix, an extremely rare abnormality in which both ends of the appendix connect to the cecum, has been attributed to abnormal embryologic development (20).

Dissection and Processing Techniques

Routine formalin fixation is adequate for appendiceal assessment. In routine cases, description of size, appearance, and any unusual lesions should be recorded (8). The external surface should be described, including any areas of hyperemia, inflammation, perforation, or the presence of mucin. Upon sectioning, luminal patency should be assessed and the presence of ulceration, perforation, mural thickening, erythema, and localized lesions should be noted along with their focality and distribution. The tip should be closely inspected for small tumors, as well-differentiated neuroendocrine tumors commonly occur in the distal-most portion of the appendix (21). When grossly evident, they often appear as bulbous, tan-yellow expansions or nodules. On the other hand, goblet cell neoplasms produce ill-defined mural thickening with obliteration of the muscularis propria, usually in the mid appendix.

Transverse serial cross-sectioning at 2 to 5 mm intervals should be performed along the entire length of the appendix. Two cross sections, one from the middle and one from the proximal margin of resection, should be submitted. The proximal margin should be indicated clearly, as neoplasms of the appendix are often discovered incidentally during appendectomy for other reasons. A routine section of the bisected tip is also standard at many institutions.

If the appendix is dilated, a neoplasm should be suspected and it may be helpful to bisect the appendix longitudinally prior to fixation. Extra longitudinal sections involving the transition from dilated to nondilated appendix can also facilitate a diagnosis when there is only a partial dilatation (8). Careful inspection for luminal, mural, and extra-appendiceal mucin is critically important, as is documentation of perforation. In addition, it is important to try and keep the cutting board surface as clean as possible, so that contamination by spilled mucin is not interpreted as extra-appendiceal mucin in the case of a mucinous neoplasm. The entire appendix as well as all luminal and extra-appendiceal mucin should be submitted in these cases, to ensure correct staging and the detection of any neoplastic cells outside the appendix (22).

FUNCTION

The exact function of the appendix remains an unsolved medical mystery. Although some regard it as a vestigial structure without a particular purpose, the abundant quantity of organized lymphoid tissue suggests involvement in mucosal immunity (23). It has been suggested that B lymphocytes derived from the appendix migrate and populate distant sites of the gastrointestinal tract lamina propria, and evolve into functional immunoglobulin (Ig)A-secreting plasma cells (24,25). In this role, the appendix could both attenuate potentially harmful Ig responses and enhance regional mucosal immunity (25). The appendix also contains a highly concentrated biofilm, or layer of loose mucin and beneficial commensal gut bacteria adjacent to the lumen. This biofilm is not regularly sloughed as the appendix is a blind tube. Thus, the appendix may function as a haven

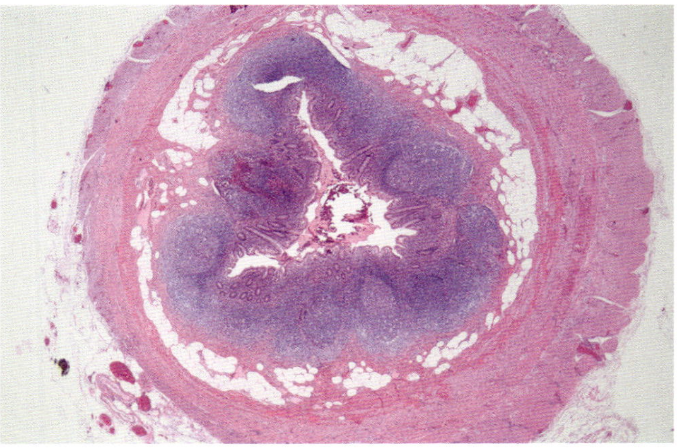

FIGURE 25.2 The histology of the appendix is very similar to the colon; it is organized into four layers, which include the mucosa, submucosa, muscularis propria, and subserosa/serosa (H&E, 2×).

for commensal bacteria, providing a reservoir of beneficial organisms to repopulate the gut in the event that its contents are purged following infection with a pathogen (23,26). It has also been postulated that the appendix and its biofilm play a crucial role in the development of a normal gastrointestinal immune system (27).

NORMAL HISTOLOGY OF THE APPENDIX

The histology of the appendix is very similar to that of the colon. It is composed of the mucosa, submucosa, muscularis propria (or externa), and serosa (Fig. 25.2). A unique feature of the appendix is the prominent lymphoid follicles surrounding the lumen.

Mucosal Architecture and Design

The appendiceal mucosa is composed of epithelium, lamina propria, and muscularis mucosae (Fig. 25.3). The appendiceal crypts are lined by a single layer of columnar epithelium.

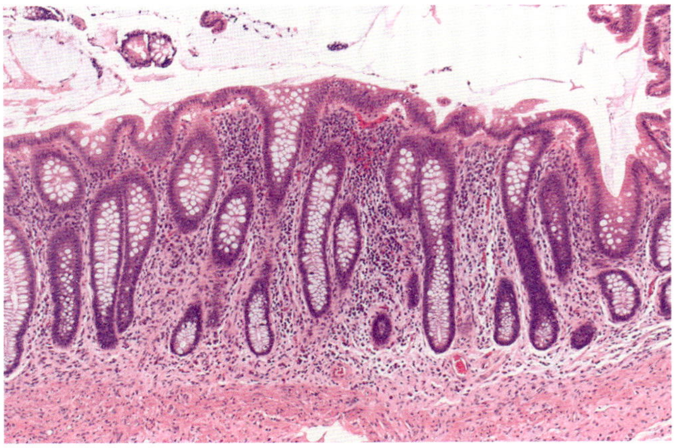

FIGURE 25.3 The appendiceal mucosa is composed of the epithelium, lamina propria, and muscularis mucosae (H&E, 10×).

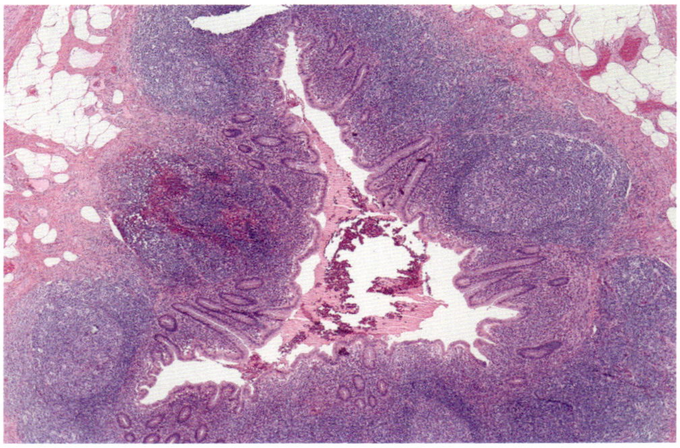

FIGURE 25.4 The distinguishing histologic feature of the appendix is the prominent lymphoid aggregates and lymphoid tissue that typically circumferentially involves, and distorts, the appendiceal lumen (H&E, 4×).

The lamina propria underlies the surface epithelium and is distributed between the crypts. It is characterized by prominent lymphoid follicles, which involve the mucosa and submucosa, often distorting the crypt architecture (Fig. 25.4) (9,20). The outermost layer of the mucosa is formed by the muscularis mucosae, which is composed of small fascicles of smooth muscle cells arranged in a thin layer. In contrast to the colon, this layer varies in thickness and may be focally absent (Fig. 25.5).

Surface Epithelium

The surface epithelium is comprised of goblet cells interspersed between nongoblet columnar cells (Fig. 25.6). The goblet cells are characterized by apical mucin droplets with pale basophilic cytoplasm and small basally located nuclei. The mucin vacuoles of goblet cells are periodic acid–Schiff (PAS) positive, reflecting the presence of neutral mucin, and they are Alcian blue positive due to the presence of acid sulfomucin. The columnar cells are tall with eosinophilic cytoplasm and round basally located nuclei. By ultrastructural examination, the columnar cells can be further

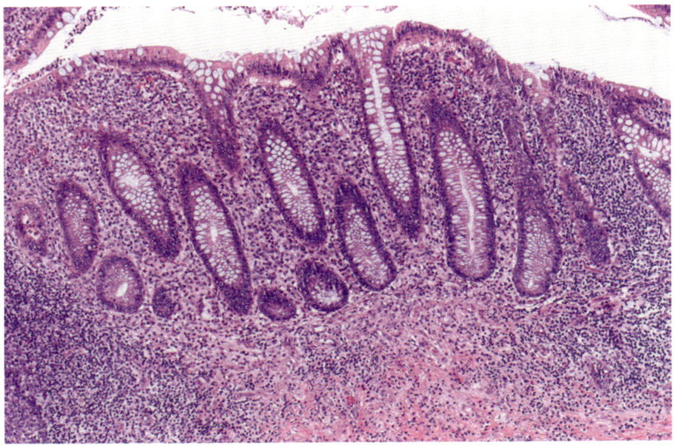

FIGURE 25.5 The lymphoid aggregates may involve both the mucosa and submucosa, interrupting the muscularis mucosae (H&E, 10×).

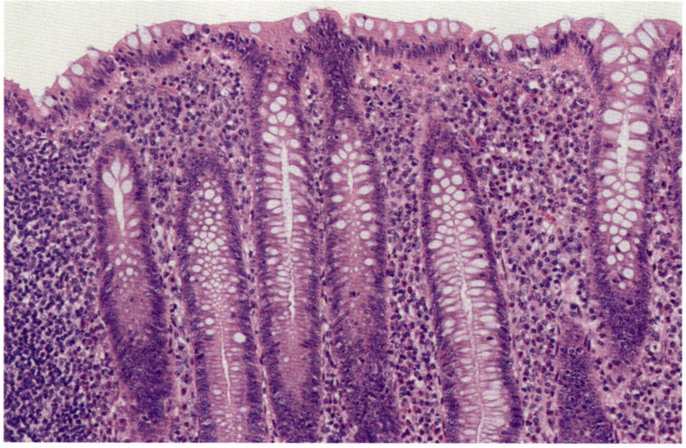

FIGURE 25.6 The luminal and crypt epithelium of the appendix is primarily composed of goblet cells interspersed between tall columnar cells with eosinophilic cytoplasm and basally located nuclei (H&E, 20×).

classified into "sensescent" mucous cells, absorptive cells, and membrane or microfold cells (M cells) (21,24).

The mucus layer overlying the surface epithelium is produced by the goblet cells and has two layers, unlike its colonic counterpart that has only one. The inner mucus layer is more compact and acts as a barrier to bacteria, while the outer mucus layer is looser and contains commensal bacteria forming the biofilm (19,26). The epithelium overlying lymphoid aggregates, as in other portions of the small and large bowel, is a specialized or follicle-associated epithelium that is different from the surrounding surface epithelium. It contains fewer goblet cells, and many of the columnar cells are of the M-cell type (Fig. 25.7) (25). The M cell, a specialized epithelial cell, assists in luminal transport of antigens into the epithelium for appropriate immunologic processing (27,28). M cells are columnar in shape with an attenuated brush border and are often associated with lymphocytes. Definite characterization of these cells relies on ultrastructural examination, which shows apical cytoplasmic vesicles and shortened microvilli or microfolds (24,27). As lymphoid aggregates are prominent in the normal appendix, this specialized follicle-associated epithelium often lines the majority of the appendiceal lumen. Thus, functionally, the surface epithelium is likely involved in antigen processing, as well as in forming a barrier to luminal contents. The luminal surface is also the site where senescent cells are sloughed into the lumen (22,23). Scattered endocrine cells can be seen within the surface epithelium but are more abundant in the underlying crypts. Migratory T and B lymphocytes can be found anywhere within the surface epithelium (29,30), but are more abundant in the follicle-associated epithelium.

Crypt Epithelium

In contrast to the colon where crypts are largely uniform, appendiceal crypts are more irregular in shape, length, and distribution (31). They are generally absent in areas with abundant lymphoid tissue or lymphoid aggregates (Fig. 25.8) (32).

Several different cell types line the crypts. The goblet cells and columnar cell variants discussed above are the most abundant. There are also scattered undifferentiated stem cells located at the crypt bases that are histologically inconspicuous but can be identified by ultrastructural means (22). Endocrine cells, either single or in clusters, are seen along the crypt epithelium. Their appearance varies from a flask-shaped cell with a narrow strip of apical cytoplasm contiguous with the surface to a spindle-shaped cell with no luminal connection (33,34). Endocrine cells contain small, dark red infranuclear granules (Fig. 25.9) (35). They can also be highlighted by immunohistochemical for chromogranin, synaptophysin, and INSM1. Ultrastructural analysis discloses secretory granules within their cytoplasm. Paneth cells can also be found in the crypt bases within the normal appendix in nearly 96% of specimens (36–38). These cells have abundant eosinophilic supranuclear granules overlying round, basally located nuclei (Fig. 25.10). Paneth cells contain lysozyme, defensins, and Igs and appear capable of

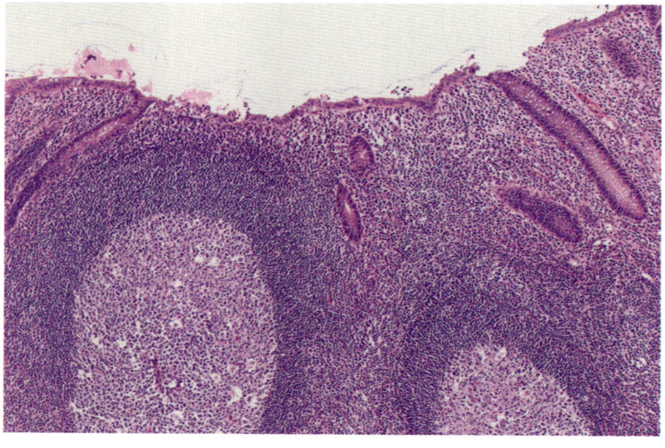

FIGURE 25.7 The follicle-associated epithelium is located above the lymphoid aggregates, is frequently devoid of goblet cells, and contains intraepithelial lymphocytes (H&E, 10×).

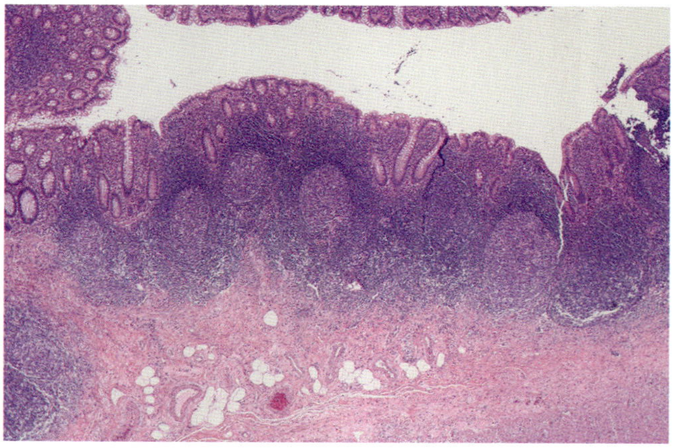

FIGURE 25.8 Crypts are deficient in the area of lymphoid aggregates. Further, the aggregates often distort the surrounding crypts (H&E, 4×).

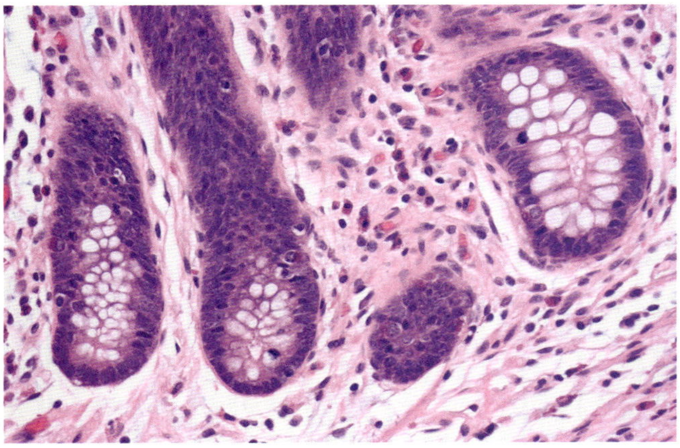

FIGURE 25.9 Endocrine cells are scattered throughout the base of appendiceal crypts. They can be identified by their infranuclear eosinophilic granules (H&E, 40×).

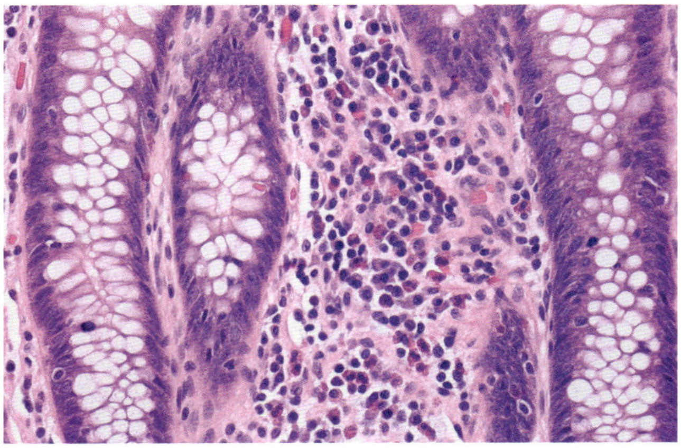

FIGURE 25.11 The lamina propria surrounding the crypts is composed of a mix of inflammatory cells, including lymphocytes, plasma cells, eosinophils, mast cells, macrophages, and dendritic cells (H&E, 40×).

phagocytosis. Their composition, in addition to their localization near stem cells, suggests that Paneth cells may protect the stem cells from damage by luminal contents and help regulate intestinal bacteria (39,40).

As above, intraepithelial lymphocytes occur within the crypt epithelium (41,42), but intraepithelial neutrophils and eosinophils are not normal constituents. Rarely, gastric, ileal, or esophageal squamous-type mucosa can be seen within the appendiceal lining, and are regarded as true heterotopias (43,44).

Subepithelial Basement Membrane

The basement membrane separates the epithelium from the lamina propria. It is composed of collagen and other matrix components (45), which help stabilize the epithelium.

Lamina Propria

The lamina propria forms a connective tissue framework around the crypts. It is composed of collagen and elastic fibers and associated fibroblasts intermingled with capillaries, lymphatic vessels, and nerve fibers (21–23). As in the colonic lymphoid tissue, its migratory cell component consists primarily of plasma cells and T cells, along with scattered B cells, macrophages, dendritic cells, eosinophils, and mast cells (21,30,46) (Fig. 25.11). Lymphoid follicles are present within, and often distort, the lamina propria architecture (Fig. 25.12). These lymphoid aggregates can extend beneath the muscularis mucosae into the underlying submucosa (Fig. 25.13), are often confluent, and appear similar in composition and function to Peyer patches of the ileum (24). As in Peyer patches, appendiceal lymphoid tissue can be compartmentalized into: (a) the lymphoid follicle; (b) the subepithelial dome; (c) the interfollicular region; and (d) follicle-associated epithelium (47–49). The follicle has, in most cases, a mitotically active germinal center containing a polymorphic population of follicular center B cells, CD4-positive Th cells, and tangible body macrophages (Fig. 25.14) (16,46–50). Immediately surrounding the germinal center is the mantle zone, a darkly staining

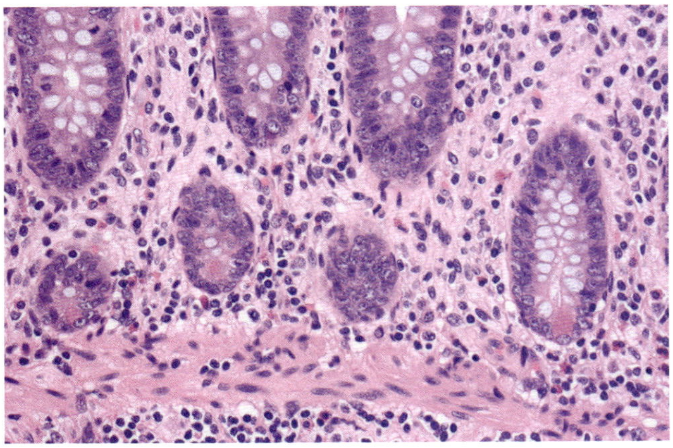

FIGURE 25.10 Paneth cells can be seen distributed throughout the appendiceal crypts, characterized by supranuclear eosinophilic granules (H&E, 40×).

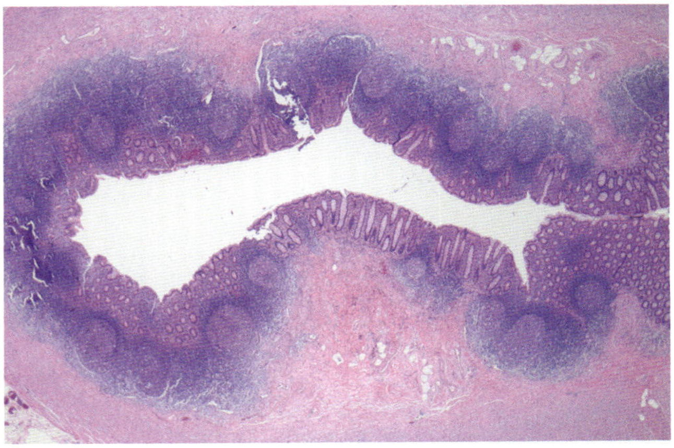

FIGURE 25.12 Lymphoid follicles are prominent throughout the appendix, often distorting the luminal architecture (H&E, 2×). These lymphoid aggregates can extend beneath the muscularis mucosae into the underlying submucosa (Fig. 25.13), are often confluent, and appear similar in composition and function to Peyer patches of the small bowel.

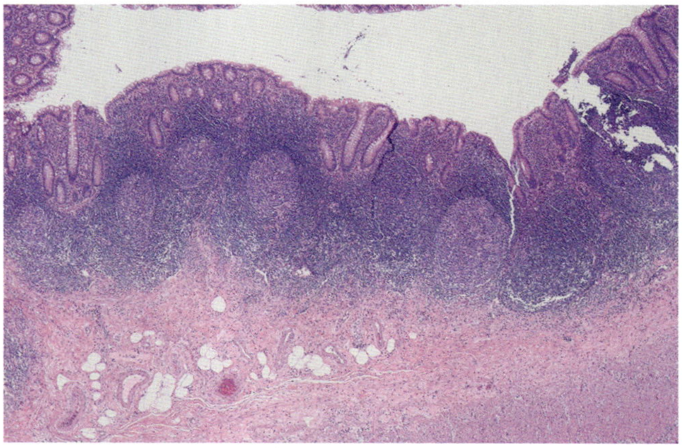

FIGURE 25.13 The lymphoid follicles may involve both the mucosa and submucosa, which results in disruption of the muscularis mucosae (H&E, 4×).

cuff of small B cells that express IgM and IgD. The subepithelial dome region overlies the lymphoid aggregate and harbors a heterogeneous population of cells including B and T cells, macrophages, dendritic cells, and occasional plasma cells (24,49). A prominent collagenous network and closely associated lymphatic vessels surround and define the lymphoid follicle (17). This collagenous/fibrous border is contiguous with the connective tissue framework of the interfollicular zones and adjacent lamina propria (17). The parafollicular region surrounding a single lymphoid follicle and the interfollicular area between confluent lymphoid follicles consist predominantly of T cells (49). Moreover, the ratio of CD4-positive T cells to CD8-positive T cells is normally about 8:1 in these areas (49). Finally, as detailed above, the follicle-associated epithelium is specialized and morphologically distinct from the adjacent absorptive-type surface epithelium.

The phenotype of the mononuclear cell population within the appendiceal mucosa is different from that of the colon. Although the quantity of B cells and plasma cells containing IgA and IgM is similar in both, IgG-containing cells are more abundant in the appendix (16,49). In fact, nearly 50% of the B cells in the subepithelial dome region are IgG immunoreactive (16). In addition, T cells express more of the integrin subunit β_7 compared to T cells throughout the gut. Integrin $\alpha_4\beta_7$ is primarily located on T cells between the lamina propria and epithelium and is involved in the tethering and rolling or homing step in attracting lymphocytes which may promote antigen processing (19).

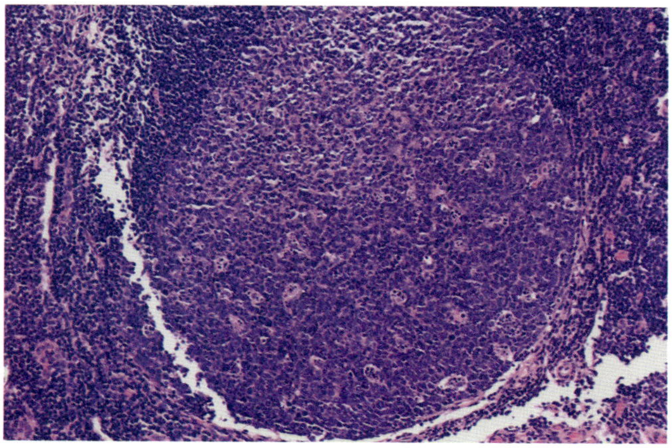

FIGURE 25.14 The lymphoid follicles usually show an activated germinal center, which contains a mixture of B and CD4-positive Th cells and tangible body macrophages (H&E, 20×).

Although a characteristic feature of the appendix, the amount of lymphoid tissue varies with age. The neonatal appendix contains scant or no lymphoid tissue. With increasing age, the lymphoid nodules accumulate, peaking in the first decade (20,51). Lymphoid aggregates then steadily diminish in quantity throughout the remainder of life. However, appendices excised incidentally from middle-aged adults can still show a prominent organized lymphoid component (52). Lymphoid nodules and associated lymphocytes may also be scant in the context of fibrous obliteration of the appendiceal lumen (52). Thus, there is great variation in the normal lymphoid content of the appendix.

The lamina propria contains a well-developed mucosal nervous plexus that is different from the more prominent submucosal (Meissner) and myenteric (Auerbach) plexuses. Although all contain neurons (ganglion cells), Schwann cells, and neural processes (axons and neuropil), only the mucosal plexus contains endocrine cells. As a consequence, this network has been termed the mucosal neuroendocrine complex (53). These complexes, located just beneath the crypts, are composed of collections of endocrine cells and can be seen on H&E-stained preparations as polygonal cells with pale granular cytoplasm, often intimately associated with spindled Schwann cells, neural processes, and occasional neurons. These collections of neuroendocrine ganglia are interconnected by neural fibers that can be highlighted by neuron-specific enolase (NSE) and, in a subset, with substance P (54). S100 protein can also delineate this network because it marks the accompanying Schwann cells. The mucosal plexus communicates with other neural networks of the enteric nervous system (53,55–57). The subepithelial endocrine cells are not always conspicuous, but can be highlighted using chromogranin, synaptophysin, and NSE, or by using electron microscopy (58,59). Most of these cells contain serotonin (55,59). The mucosal neuroendocrine complex is believed to modulate neural communication, through serotonin mediators, between the epithelium and the deeper submucosal and intermuscular plexuses (59).

Muscularis Mucosae

The muscularis mucosae is a thin band composed of fascicles of smooth muscle cells that separate the lamina propria and mucosal epithelium from the underlying submucosa. It forms a continuous layer in the colon (21), but the muscularis mucosae may be variably attenuated, poorly developed, or even focally absent in the appendix, particularly in

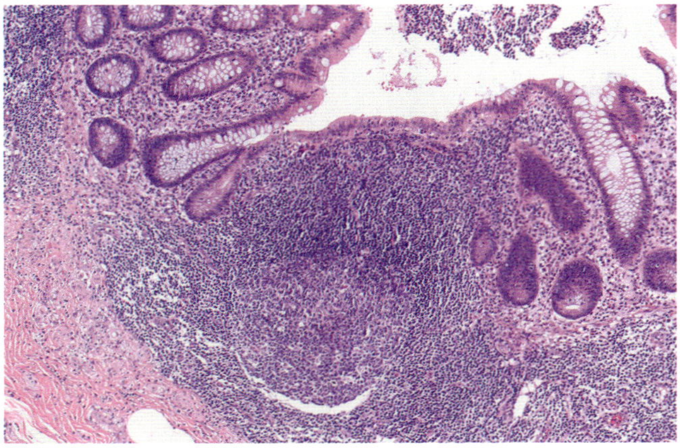

FIGURE 25.15 In contrast to other organs in the gastrointestinal tract, the muscularis mucosae of the appendix is attenuated and often focally absent, especially in regions where lymphoid aggregates involve the submucosa (H&E, 10×).

the region of penetrating lymphoid aggregates (Fig. 25.15) (32,60).

Submucosa

The submucosa separates the mucosa from the muscularis propria. It is composed of a meshwork of collagen and elastic fibers surrounding adipose tissue, arterioles, venules, capillaries, and lymphatic vessels (7,21) (Fig. 25.16). Macrophages, lymphocytes, plasma cells, and mast cells are scattered throughout (Fig. 25.17) (20,61). The submucosa also contains the submucosal Meissner plexus, which consists of collections of ganglion cells with associated neuronal processes, and Schwann cells (Fig. 25.18) (62,63). Ganglion cells are large, round to polygonal cells with abundant eosinophilic cytoplasm, eccentrically placed vesicular nuclei, and prominent nucleoli (Fig. 25.19). The surrounding spindle and wavy Schwann cells are less conspicuous on H&E-stained preparations, but can be highlighted with S100 protein immunohistochemistry.

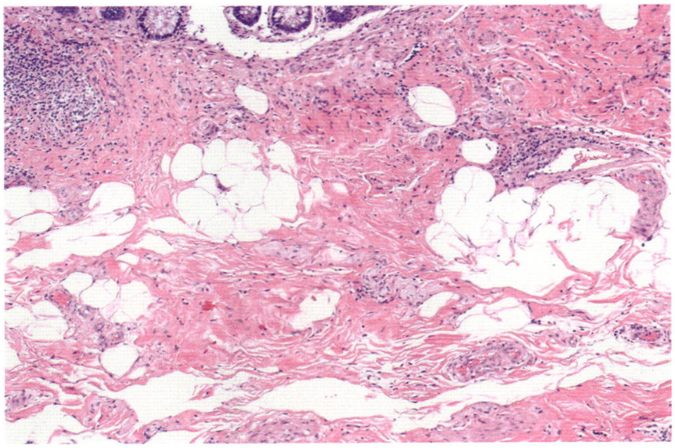

FIGURE 25.16 The submucosa is situated between the mucosa and the muscularis propria. It is composed of collagen and elastic fibers intertwined with adipocytes, blood vessels, and lymphatics (H&E, 10×).

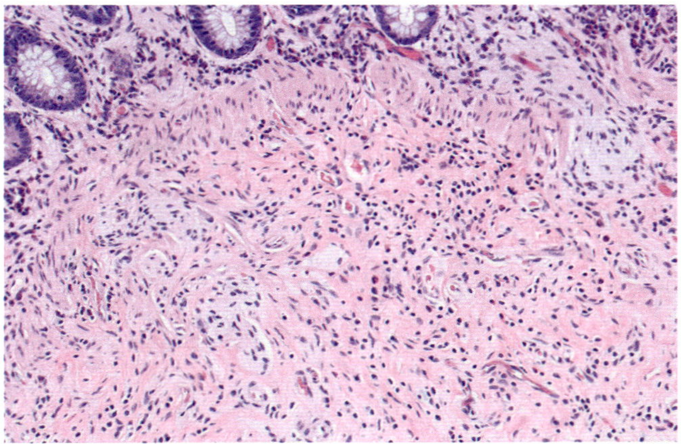

FIGURE 25.17 Inflammatory cells are also scattered throughout the submucosa and may include macrophages, lymphocytes, plasma cells, and mast cells (H&E, 20×).

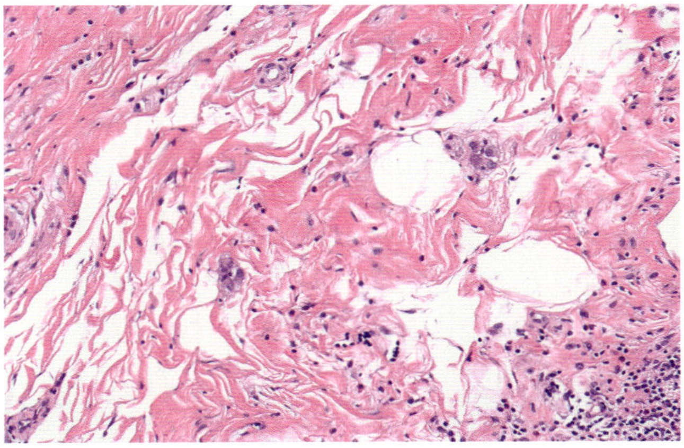

FIGURE 25.18 The submucosal, or Meissner plexus, is interspersed throughout the submucosal tissue and consists of a network of ganglia and Schwann cells (H&E, 20×).

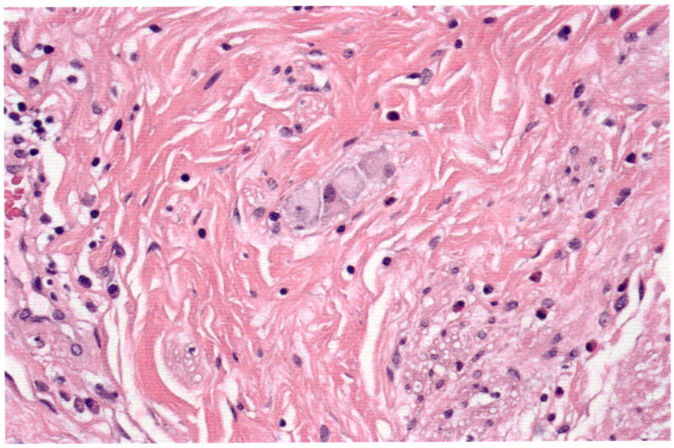

FIGURE 25.19 Ganglion cells in the submucosal plexus are large cells with abundant eosinophilic cytoplasm, eccentrically placed vesicular nuclei, and prominent nucleoli (H&E, 40×).

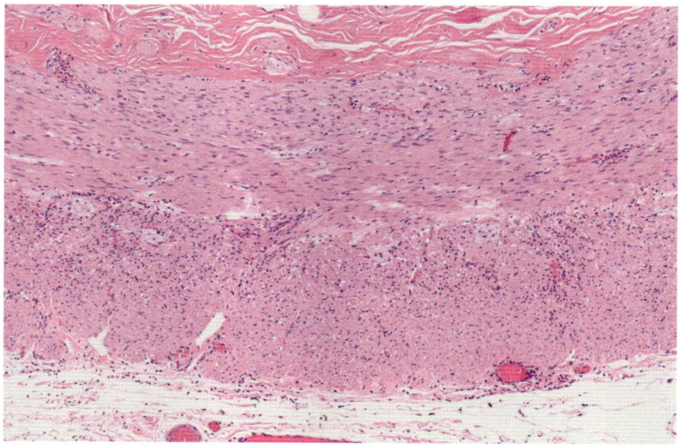

FIGURE 25.20 The muscularis propria sits external to the submucosa and adjacent to the subserosa/serosa. It is composed of an inner circular layer and an outer longitudinal layer of smooth muscle (H&E, 10×).

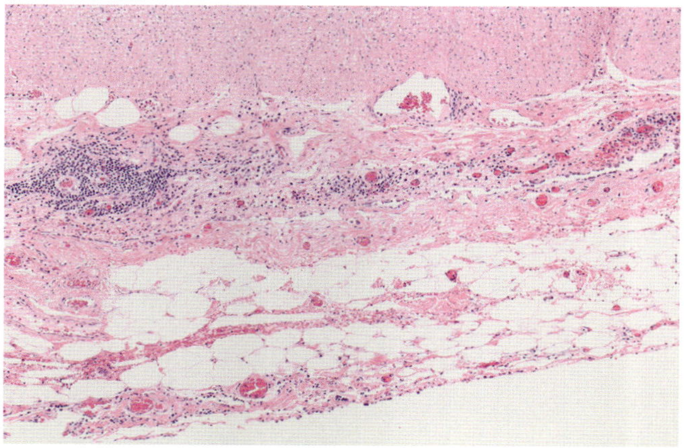

FIGURE 25.22 The subserosa consists of blood vessels, lymphatics, and nerves surrounded by loose connective tissue (H&E, 10×).

Muscularis Propria, Periappendiceal Connective Tissue, and the Serosa

The muscularis propria is the smooth muscle layer between the submucosa and serosa. It is composed of an inner circular layer and an outer longitudinal band, each of which consists of smooth muscle cells arranged in fascicles (Fig. 25.20) (32,64). The individual smooth muscle cells are ovoid with blunted ends, and they form bundles of varying sizes. Occasionally, individual or groups of smooth muscle cells show marked eosinophilic granularity, particularly within the inner circular layer (60,65). This finding is termed granular degeneration of the muscularis propria, although some authors have recently suggested that it represents a localized form of PEComatosis (66) termed "micronodular PEComa." This assumption is based on similar immunohistochemical profiles between this change and a soft tissue PEComa. Although an intriguing hypothesis, it is worth mentioning that true PEComas of the appendix are exceedingly rare; much rarer, in fact, than those affecting other parts of the gastrointestinal tract that do not show eosinophilic granularity of the muscularis propria. Between the two layers of the muscularis propria lies the myenteric (Auerbach) plexus, which is morphologically and functionally similar to the previously described submucosal plexus (Fig. 25.21) (63). However, the anatomic distribution of ganglion cells and nerves tends to be more dispersed; one can frequently identify scattered ganglion cells throughout the muscularis propria, rather than sandwiched between the muscle layers. The muscularis propria also contains blood and lymphatic vessels and nerve fibers (17,67). Just external to the outer longitudinal smooth muscle layer is the periappendiceal connective tissue or subserosal region, consisting of loose connective tissue and ramifying blood vessels, lymphatics, and nerves (Fig. 25.22). The outermost surface, or serosa, is lined by a single layer of cuboidal mesothelial cells that overlies a slender band of fibrous tissue (Fig. 25.23). Only the attachment of the fibrofatty mesoappendix lacks a serosa (1).

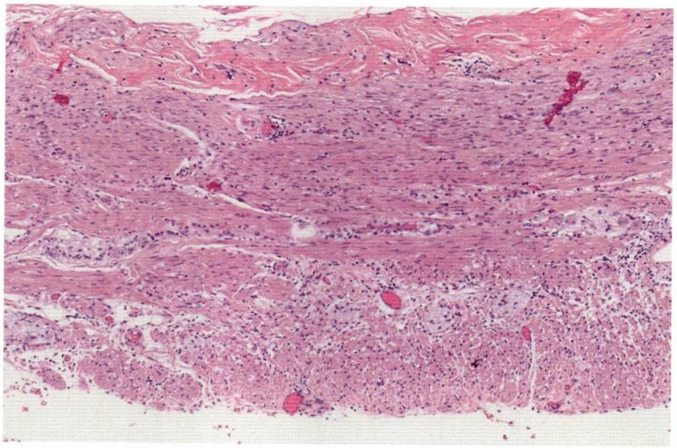

FIGURE 25.21 Similar to the submucosal plexus, the myenteric (Auerbach) plexus is composed of ganglia and Schwann cells and courses between the layers of the muscularis propria (H&E, 10×).

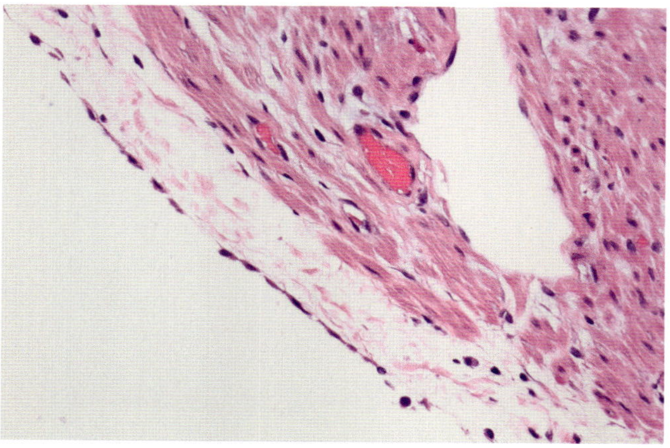

FIGURE 25.23 The outermost surface of the appendix is the serosa, which is composed of a single layer of cuboidal mesothelial cells that overlies a thin band of fibrous tissue (H&E, 40×).

SPECIAL CONSIDERATIONS

Normal Variation of Mucosal Inflammation Versus Appendicitis

Acute appendicitis has a number of definitions but is usually characterized by an abundant neutrophilic and eosinophilic infiltrate within the mucosa, submucosa, and muscularis propria with at least focal mucosal ulceration. Advanced cases show suppurative inflammation extending into and through the appendiceal wall (9,52). However, the changes seen in early appendicitis can be quite minimal, and criteria considered sufficient to diagnose early acute appendicitis are variable (9,52,68–74). Many authorities recommend submitting additional sections if acute inflammation limited to the superficial mucosa is the only finding (8). The presence of luminal neutrophils has been considered to be insufficient for a diagnosis of acute appendicitis because some incidental appendectomy specimens reportedly show similar changes, although the data supporting this approach are not robust (9,52,71–73). Sometimes luminal neutrophils may be present in the context of fecaliths but it is not clear whether this finding is a nonspecific alteration of limited clinical significance, or represents mild appendicitis. A small percentage of appendices removed due to clinical suspicion of acute appendicitis are histologically normal even after the entire appendix is removed. Often the clinical symptoms resolve following appendectomy despite an apparent absence of inflammation (75,76).

Normal Histologic Variations

The specialized features of the appendiceal microanatomy should not be confused with pathologic entities. As noted above, the muscularis mucosae is quite variable in the appendix and may even by absent; in addition, lymphoid aggregates can disrupt the muscularis mucosae. These normal variations should not be mistaken for "loss of muscularis mucosae," particularly when the diagnosis of a low-grade appendiceal mucinous neoplasm is in question. Lack of cytologic dysplasia and presence of the lamina propria may be helpful in these cases.

In contrast to the colon, where crypts are largely uniform and regularly spaced, appendiceal crypts are more irregular in shape, length, and distribution. Crypts are also frequently absent in areas with prominent lymphoid follicles. These changes should not be mistaken for the chronic architectural changes seen in appendiceal involvement by inflammatory bowel disease. Normal appendices lack active neutrophilic inflammation and prominent lamina propria lymphoplasmacytos and, thus, recognition of these features may prevent overinterpreting the significance of crypt architectural distortion.

The myenteric plexi in the appendix have much more variability in location than the myenteric plexi in the colon, and even normal appendices may lack ganglion cells (77–79). This may be particularly important when evaluating frozen sections for Hirschprung disease.

Finally, extensive involvement by the appendiceal lumen by lymphoid aggregates, particularly in children and young adults, should not be mistaken for appendiceal lymphoma. The prominent germinal centers with a mixed population of lymphoid cells and the lack of architectural effacement support a reactive/normal process.

REFERENCES

1. Williams PL, Warwick R, Dyson M, Bannister LH, eds. *Gray's Anatomy of the Human Body*. 37th ed. Churchill Livingstone; 1989.
2. Buschard K, Kjaeldgaard A. Investigation and analysis of the position, fixation, length, and embryology of the vermiform appendix. *Acta Chir Scand*. 1973;139(3):293–298.
3. Wakeley CP. The position of the vermiform appendix as ascertained by an analysis of 10,000 cases. *J Anat*. 1933;67(Pt 2): 277–283.
4. Thorek P. *Anatomy and Surgery*. 3rd ed. Springer-Verlag; 1985.
5. Chauhan S, Anand S. Intracecal appendix: an extremely rare anatomical variation. A case report and review of literature. *Surg Radiol Anat*. 2018;40(1):111–114.
6. Longani SK, Ahmed A. Classical presentation of acute appendicitis in the case of a subhepatic appendix. *Cureus*. 2019; 11(10):e6035.
7. Hollinshead WH, Rosse C. *Textbook of Anatomy*. 4th ed. Harper & Row; 1985.
8. Williams RA, Myers P. *Pathology of the Appendix and Its Surgical Treatment*. Chapman and Hall Medical; 1994.
9. Hardin DM Jr. Acute appendicitis: review and update. *Am Fam Physician*. 1999;60(7):2027–2034.
10. Savvakis S, Karamitsou P, Vardaxi C, et al. The presence and the anatomical variations of the accessory appendicular artery: a systematic review of 604 cases. *ANZ J Surg*. 2024;94(5): 903–909.
11. Moore KL. *The Developing Human: Clinically Oriented Embryology*. 3rd ed. WB Saunders; 1982.
12. Searle AR, Ismail KA, MacGregor D, Hutson JM. Changes in the length and diameter of the normal appendix throughout childhood. *J Pediatr Surg*. 2013;48(7):1535–1539.
13. Trout AT, Towin AJ, Zhang B. Journal club: the pediatric appendix: defining normal. *Am J Roentgenology*. 2014;202(5): 936–945.
14. Collins DC. Agenesis of the vermiform appendix. *Am J Surg*. 1951;82(6):689–696.
15. Woywodt A, Verhaart S, Kiss A. Atresia of the appendix. *J Pediatr Surg*. 1998;33(9):1423–1425.
16. Wallbridge PH. Double appendix. *Br J Surg*. 1962;50:346–347.
17. Nageswaran H, Khan U, Hill F, Maw A. Appendiceal duplication: a comprehensive review of published cases and clinical recommendations. *World J Surg*. 2018;42(2):574–581.
18. Tinckler LF. Triple appendix vermiformis—a unique case. *Br J Surg*. 1968;55(1):79–81.
19. Bluett MK, Halter SA, Salhany KE, O'Leary JP. Duplication of the appendix mimicking adenocarcinoma of the colon. *Arch Surg*. 1987;122(7):817–820.

20. Singh ChG, Nyuwi KT, Rangaswamy R, Ezung YS, Singh HM. Horseshoe appendix: an extremely rare appendiceal anomaly. *J Clin Diag Research*. 2016;10(3):PD25–PD26.
21. Glasser CM, Bhagavan BS. Carcinoid tumors of the appendix. *Arch Pathol Lab Med*. 1980;104(5):272–275.
22. Yantiss RK, Shia J, Kilmstra DS, Hahn HP, Odze RD, Misdraji J. Prognostic significance of localized extra-appendiceal mucin deposition in appendiceal mucinous neoplasms. *Am J Surg Pathol*. 2009;33(2):248–255.
23. Smith HF. A review of the function and evolution of the cecal appendix. *Anat Rec (Hoboken)*. 2023;306(5):972–982.
24. Bjerke K, Brandtzaeg P, Rognum TO. Distribution of immunoglobulin producing cells is different in normal human appendix and colon mucosa. *Gut*. 1986;27(6):667–674.
25. Bockman DE. Functional histology of appendix. *Arch Histol Jpn*. 1983;46(3):271–292.
26. Randal Bollinger R, Barbas AS, Bush EL, Lin SS, Parker W. Biofilms in the large bowel suggest an apparent function of the human vermiform appendix. *J Theor Biol*. 2007;249(4):826–831.
27. Kooij IA, Sahami S, Meijer SL, Buskens CJ, Te Velde AA. The immunology of the vermiform appendix: a review of the literature. *Clin Exp Immunol*. 2016;186(1):1–9.
28. Hwang JM, Krumbhaar KB. The amount of lymphoid tissue of the human appendix and its weight at different age periods. *Am J Med Sci*. 1940;199:75–83.
29. Hamilton SR. Structure of the colon. *Scand J Gastroenterol Suppl*. 1984;93:13–23.
30. Shamsuddin AM, Phelps PC, Trump BF. Human large intestinal epithelium: light microscopy, histochemistry, and ultrastructure. *Hum Pathol*. 1982;13(9):790–803.
31. Levine DS, Haggitt RC. Normal histology of the colon. *Am J Surg Pathol*. 1989;13(11):966–984.
32. Bockman DE, Cooper MD. Early lymphoepithelial relationships in human appendix. A combined light- and electron-microscopic study. *Gastroenterology*. 1975;68(5 Pt 1):1160–1168.
33. Filipe MI. Mucins in the human gastrointestinal epithelium: a review. *Invest Cell Pathol*. 1979;2(3):195–216.
34. Ermund A, Schutte A, Johansson ME, Gustafsson JK, Hansson GC. Studies of mucus in mouse stomach, small intestine and colon. I. Gastrointestinal mucus layers have different properties depending on location as well as over the Peyer's patches. *Am J Physiol Gastrointest Liver Physiol*. 2013;305(5):G341–G347.
35. Owen RL, Jones AL. Epithelial cell specialization within human Peyer's patches: an ultrastructural study of intestinal lymphoid follicles. *Gastroenterology*. 1974;66(2):189–203.
36. Dobbins WO III. Human intestinal intraepithelial lymphocytes. *Gut*. 1986;27(8):972–985.
37. Bartnik W, ReMine SG, Chiba M, Thayer WR, Shorter RG. Isolation and characterization of colonic intraepithelial and lamina proprial lymphocytes. *Gastroenterology*. 1980;78(5 pt 1):976–985.
38. Fawcett DW, Bloom W. *Bloom and Fawcett: A Textbook of Histology*. 11th ed. WB Saunders; 1986.
39. Neutra MR, Padykula HA. The gastrointestinal tract. In: Weiss L, ed. *Modern Concepts of Gastrointestinal Histology*. Elsevier; 1984:658–706.
40. Lewin KJ. The endocrine cells of the gastrointestinal tract. The normal endocrine cells and their hyperplasias. Part 1. In: Sommers SC, Rosen PP, Fechner RE, eds. *Pathology Annual*. Appleton-Century-Croft; 1986:1–27.
41. Sjolund K, Sanden G, Hakanson R, Sundler F. Endocrine cells in human intestine: an immunocytochemical study. *Gastroenterology*. 1983;85(5):1120–1130.
42. Millikin PD. Eosinophilic argentaffin cells in the human appendix. *Arch Pathol*. 1974;98(6):393–395.
43. Hofler H, Kasper M, Heitz PU. The neuroendocrine system of normal human appendix, ileum and colon, and in neurogenic appendicopathy. *Virchows Arch A Pathol Anat Histopathol*. 1983;399(2):127–140.
44. Geller SA, Thung SN. Morphologic unity of Paneth cells. *Arch Pathol Lab Med*. 1983;107(9):476–479.
45. Wehkamp J, Fellermann K, Herrlinger KR, Bevins CL, Stange EF. Mechanisms of disease: defensins in gastrointestinal diseases. *Nat Clin Pract Gastroenterol Hepatol*. 2005;2(9):406–415.
46. Wehkamp J, Salzman NH, Porter E, et al. Reduced Paneth cell alpha-defensins in ileal Crohn's disease. *Proc Natl Acad Sci U S A*. 2005;102(50):18129–18134.
47. Eastwood GL. Gastrointestinal epithelial renewal. *Gastroenterology*. 1977;72(5 pt 1):962–975.
48. Lipkin M. Proliferation and differentiation of normal and diseased gastrointestinal cells. In: Johnson LR, ed. *Physiology of the Gastrointestinal Tract*. 2nd ed. Raven Press; 1987:255–284.
49. Aubrey DA. Gastric heterotopia in the vermiform appendix. *Arch Surg*. 1970;101(5):628–629.
50. Ashley DJ. Aberrant mucosa in the vermiform appendix. *Br J Surg*. 1958;45(192):372–373.
51. Droga BW, Levine S, Baber JJ. Heterotopic gastric and esophageal tissue in the vermiform appendix. *Am J Clin Pathol*. 1963;40:190–193.
52. Parks DA, Jacobson ED. Physiology of the splanchnic circulation. *Arch Intern Med*. 1985;145(7):1278–1281.
53. Gledhill A, Cole FM. Significance of basement membrane thickening in the human colon. *Gut*. 1984;25(10):1085–1088.
54. Wolf JL, Bye WA. The membranous epithelial (M) cell and the mucosal immune system. *Annu Rev Med*. 1984;35:95–112.
55. Heatley RV. The gastrointestinal mast cell. *Scand J Gastroenterol*. 1983;18(4):449–453.
56. Tomasi TB Jr. Mechanisms of immune regulation at mucosal surfaces. *Rev Infect Dis*. 1983;5(Suppl 4):S784–S792.
57. Dotan I, Mayer L. Mucosal immunity. In: Feldman M, Friedman LS, Brandt LJ, eds. *Sleisenger and Fordtran's Gastrointestinal and Liver Disease*. 9th ed. Saunders Elsevier; 2010:21–30.
58. Spencer J, Finn T, Isaacson PG. Gut associated lymphoid tissue: a morphological and immunocytochemical study of the human appendix. *Gut*. 1985;26(7):672–679.
59. van der Valk P, Meijer CJ. The histology of reactive lymph nodes. *Am J Surg Pathol*. 1987;11(11):866–882.
60. Berry RJ, Lack LA. The vermiform appendix of man, and the structural changes therein coincident with age. *J Anat Physiol*. 1906;40(Pt 3):247–256.
61. Papadaki L, Rode J, Dhillon AP, Dische FE. Fine structure of a neuroendocrine complex in the mucosa of the appendix. *Gastroenterology*. 1983;84(3):490–497.
62. Lundqvist M, Wilander E. Subepithelial neuroendocrine cells and carcinoid tumours of the human small intestine and appendix. A comparative immunohistochemical study with regard to serotonin, neuron-specific enolase and S-100 protein reactivity. *J Pathol*. 1986;148(2):141–147.

63. Millikin PD. Extraepithelial enterochromaffin cells and Schwann cells in the human appendix. *Arch Pathol Lab Med*. 1983;107(4):189–194.
64. Walker NI, Bennett RE, Axelsen RA. Melanosis coli. A consequence of anthraquinone-induced apoptosis of colonic epithelial cells. *Am J Pathol*. 1988;131(3):465–476.
65. Stanley MW, Cherwitz D, Hagen K, Snover DC. Neuromas of the appendix. A light-microscopic, immunohistochemical and electron-microscopic study of 20 cases. *Am J Surg Pathol*. 1986;10(11):801–815.
66. Anderson WJ, Kojc N, Fletcher CDM, Hornick JL. Micronodular PEComas of the appendix. *Histopathology*. 2021;78(7): 1047–1050.
67. Facer P, Bishop AE, Lloyd RV, Wilson BS, Hennessy RJ, Polak JM. Chromogranin: a newly recognized marker of endocrine cells in the human gastrointestinal tract. *Gastroenterology*. 1985;89(6):1366–1373.
68. Rode J, Dhillon AP, Papadaki L. Serotonin-immunoreactive cells in the lamina propria plexus of the appendix. *Hum Pathol*. 1983;14(5):464–469.
69. Sobel HJ, Marquet E, Schwarz R. Granular degeneration of appendiceal smooth muscle. *Arch Pathol*. 1971;92(6): 427–432.
70. Montani M, Thiesler T, Kristiansen G. Smoothelin is a specific and robust marker for distinction of muscularis propria and muscularis mucosae in the gastrointestinal tract. *Histopathology*. 2010;57(2):244–249.
71. Lord MG, Valies P, Broughton AC. A morphologic study of the submucosa of the large intestine. *Surg Gynecol Obstet*. 1977; 145(1):55–60.
72. Gershon MD, Erde SM. The nervous system of the gut. *Gastroenterology*. 1981;80(6):1571–1594.
73. Krishnamurthy S, Schuffler MD. Pathology of neuromuscular disorders of the small intestine and colon. *Gastroenterology*. 1987;93(3):610–639.
74. Hausman R. Granular cells in musculature of the appendix. *Arch Pathol*. 1963;75:360–372.
75. Strong S, Blencowe N, Bhangu A; National Surgical Research Collaborative. How good are surgeons at identifying appendicitis? Results from a multi-centre cohort study. *Int J Surg*. 2015;15:107–112.
76. Grimes C, Chin D, Bailey C, Gergely S, Harris A. Appendiceal faecaliths are associated with right iliac fossa pain. *Ann R Coll Surg Engl*. 2010;92(1):61–64.
77. Lane VA, Levitt MA, Baker P, Minneci P, Deans K. The appendix and aganglionosis. A note of caution-how the histology can mislead the surgeon in total colonic Hirschsprung disease. *European J Pediatr Surg Rep*. 2015;3(1):3–6.
78. Hanani M. Multiple myenteric networks in the human appendix. *Auton Neurosci*. 2004;110(1):49–54.
79. Kubikova E, Sivakova I, Falougy HE, Perzelova A. Subserosal localization of myenteric ganglia in normal human appendix: immunostaining with neuronal and glial markers. *Bratisl Lek Listy*. 2018;119(12):743–746.

26

Anal Canal

Julio C. Poveda ■ Rhonda K. Yantiss

The anal canal is a small but complex structure where various muscle groups, vascular plexuses, and epithelial cell types converge and create a functional barrier between the digestive tract and the external world. The purpose of this chapter is to describe the embryologic, gross, and histologic features that are important to understanding anatomy and distinguishing normal variants from potential pathologic mimics that occur in this region.

DEFINITION AND BOUNDARIES

The anal canal can be described as an epithelium-lined cylinder that is pinched closed at both ends and surrounded by an intricate meshwork of vasculature and muscle. The anatomic structures and length included in its definition depend on the discipline queried. The embryologic anal canal measures approximately two centimeters in length and encompasses the area between the anal verge distally and the dentate line proximally (1). The surgical anal canal is the more commonly referenced and practical representation; it is defined as a nearly 4-cm segment that extends from the anal verge past the dentate line to the anorectal ring (Fig. 26.1) (2,3). In both the embryologic and surgical definitions, the anal canal is bordered posteriorly by the coccyx, laterally by the ischiorectal fossae, and anteriorly by the urethra in men and both the perineal body and posterior vaginal wall in women (1). The World Health Organization and the TNM system use the surgical definition of the anal canal for tumor classification and staging (4,5).

EMBRYOLOGY

The anatomic complexity of the anal canal and supporting structures reflects the origins of different components in the embryo: the hindgut, the cloaca, and the proctodeum (6). The primitive gut is a blind-ended tube that forms during cephalocaudal and lateral embryonic folding (7). The caudal portion of this tube is the hindgut, which is recognizable by the 4th week of development and ultimately gives rise to the distal third of the transverse colon, the descending colon, sigmoid colon, rectum, and proximal anal canal (8).

The hindgut expands into the cloaca below the level of the pubococcygeal line. The cloaca is an endoderm-lined space in continuity with the ventral urachal allantois. It has a ventrocaudal membrane that represents the blind end of the hindgut and separates the endoderm from ectoderm (8–10). During the 6th week of gestation, a layer of mesoderm known as the urorectal septum begins to divide the cloaca into a ventral urogenital cavity and a dorsal hindgut/rectal cavity. The septum is complete by the 10th week of gestation, fully dividing the rectum and urogenital cavity (Fig. 26.2). Most developmental abnormalities of the anorectum presumably arise from the anomalous growth and/or positioning of the urorectal septum at 6 to 10 weeks of gestation (7,11–13).

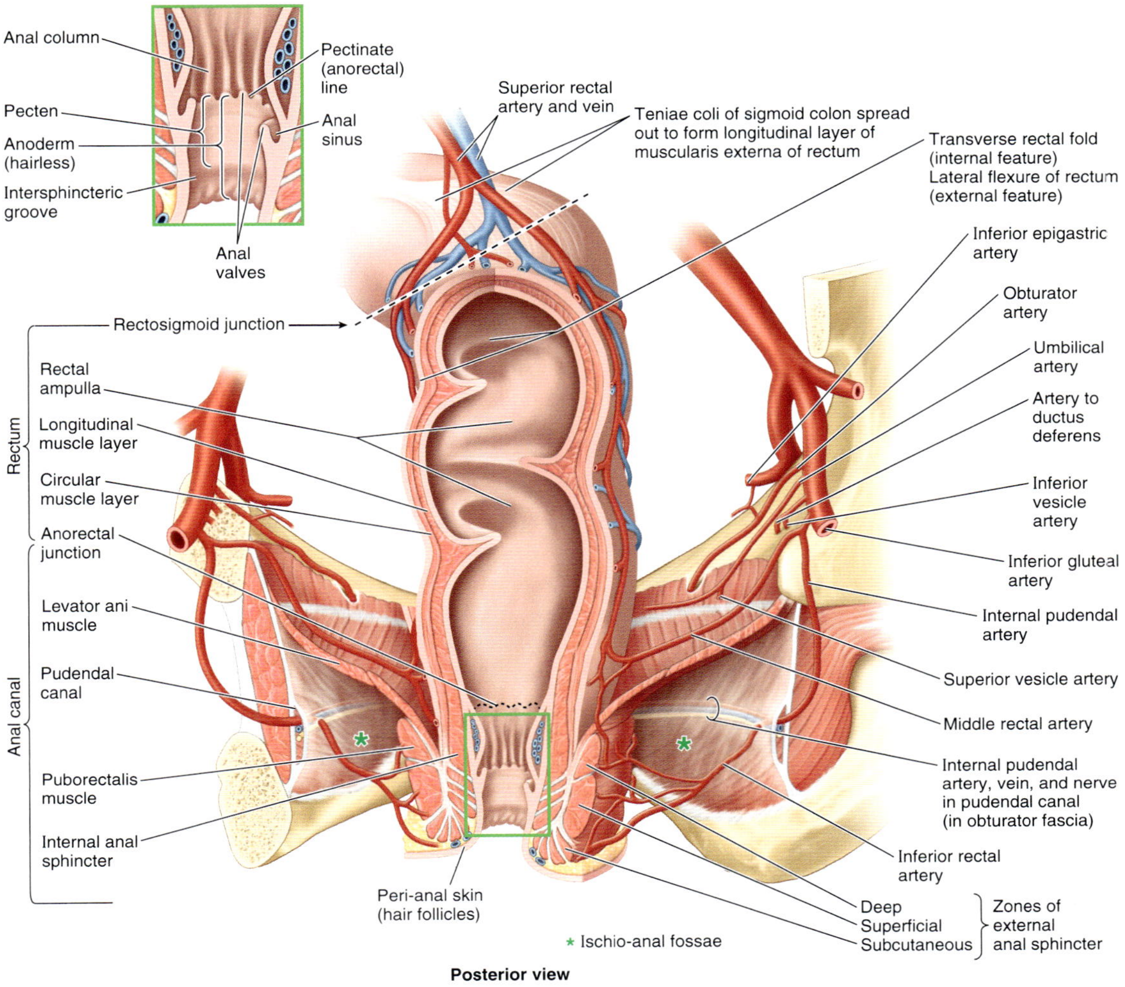

FIGURE 26.1 A coronal view of the major structures of the anal canal. The surgical anal canal (green box) extends from the anal verge to the anorectal ring, a physiologic boundary formed by the puborectalis muscle. (Dalley AF, Agur AMR. *Moore's Clinically Oriented Anatomy*. 9th ed. Wolters Kluwer; 2024.)

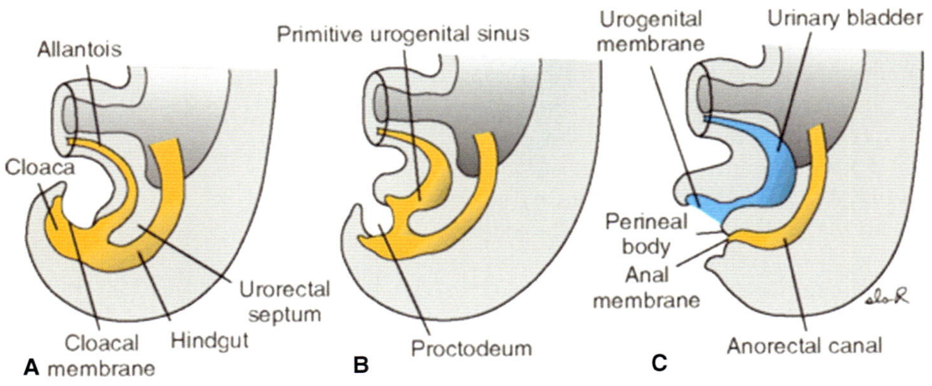

FIGURE 26.2 Embryologic development of the anal canal. The cloaca is an expansion of the primitive hindgut and the urachal allantois (**A**). The cloacal membrane is the blind ending of the hindgut. During the 6th and 7th weeks of gestation, the urorectal septum lengthens and migrates closer to the cloacal membrane (**B**). At the same time, the proctodeum invaginates and causes the cloacal membrane to move in a dorsocephalic direction. The cloacal membrane ruptures at the end of the 7th week of gestation, and by the 10th week of development, the urorectal septum has provided complete separation of the hindgut dorsally and the urogenital sinus ventrally (**C**). Reprinted with permission from Sadler TW, Langman J. *Langman's Medical Embryology*. Wolters Kluwer Health/Lippincott Williams & Wilkins; 2012.

The cloacal membrane ruptures by the end of the 7th gestational week to create an anal opening, thereby providing continuity between the endoderm and proctodeum. Proliferation and invagination of proctodeum during this time causes the cloacal membrane to migrate dorsally prior to rupture. As a result, the lower third of the anal canal is derived from the ectoderm, while the upper two-thirds are derived from the endoderm. The vasculature and innervation of the anal region reflect derivation from the endoderm and proctodeum: inferior mesenteric artery branches and autonomic nerves supply hindgut derivatives, but the internal pudendal artery and inferior rectal nerve branches supply ectoderm derivatives (6,8,10).

The external and internal anal sphincters develop during the 10th and 12th gestational weeks, respectively. Similar to other layers of the anal canal wall, they are derived from splanchnic mesenchyme (14). The anatomic orientation and development of the gut tube and associated mesenchyme are directed in part by expression of sonic hedgehog (*Shh*) genes. In the hindgut and cloaca, *Shh* signaling induces site-specific expression of *Hoxd13*, a transcription factor known to be essential for appropriate endodermal differentiation to anorectal-type epithelium (15,16).

GROSS AND FUNCTIONAL ANATOMY

The tubular anal canal consists of mucosa supported by submucosa and surrounded by a network of overlapping muscles. The internal surface has three grossly distinct regions. The proximal mucosa has a smooth, somewhat granular appearance that gives rise to 8 to 12 longitudinal mucosal folds known as the anal columns of Morgagni (3). Each longitudinal fold terminates at an anal valve created by a semilunar fold of tissue; each valve links two anal columns (17). The semilunar valves enclose small anal sinuses, or anal crypts, a subset of which contain openings to anal glands (18–20). The anal glands penetrate the submucosa and terminate within the internal anal sphincter muscle, or in the intersphincteric space proximal to the external anal sphincter (21,22). The functional role of the anal glands is unknown.

The mucosa at the level of the anal valves is pale with a circumferential undulating appearance; this boundary is the dentate, or pectinate, line. The mucosa distal to the dentate line is the pectin. It is a smooth, squamous-lined region terminating at the anal verge. Mucosa at the anal verge is wrinkled and slightly pigmented; this circumferential ring of tissue marks the end of the smooth anoderm and the beginning of perianal skin (1,3,23).

The dentate line corresponds to the site where proximal anorectal-type mucosa meets distal anodermal mucosa in fetuses and young children, but these two types of mucosae are separated by the anal transition zone in many adults (Figs. 26.3 and 26.4). The length of the anal transition zone varies from person to person but generally extends only a few millimeters above and below the dentate line. Visualization of the anal transition zone in a gross specimen requires special

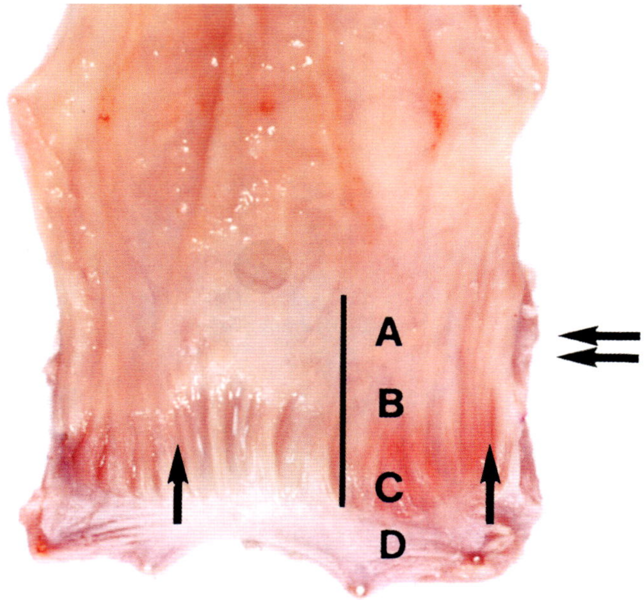

FIGURE 26.3 The simplest way to visualize the internal surface and structures of the anal canal is to cut longitudinally along one side of the cylinder and open the canal, in essence making a tubular structure into a flat rectangular piece of tissue, as seen in this surgical specimen. This autopsy specimen from the lower rectum and anal canal of an infant demonstrates well-formed anal columns (*single black arrows*). Anal valves are not apparent, and there is only a thin rim of perianal skin at the bottom. The *black vertical line* indicates the extent of the surgical anal canal and includes the anorectal zone (**A**), a small transition zone (**B**), the squamous mucosa (pecten, **C**), and the anal verge (**D**).

techniques, such as staining with Alcian blue or green to differentiate between regional epithelial mucins (Fig. 26.5) (24).

The submucosa of the anal canal contains fibroelastic connective tissue and loosely dispersed smooth muscle cells. The connective tissue is particularly rich in venules from the hemorrhoidal vessels in the right anterior, right posterior, and left lateral positions. Lush vascular structures supported by connective tissue form the anal cushions in these locations. The anal cushions provide nearly 20% of the resting anal pressure at the level of the internal anal sphincter, and are best visualized in vivo. Viewed from the external surface during anoscopy, the anal orifice is compressed by the anal cushions, resulting in a "Y-shaped" cutaneous slit oriented anteroposteriorly. It often has a corrugated appearance due to the presence of tonically contracted intersphincteric longitudinal muscle bundles that insert into the dermis of the perianal skin (25–27).

Musculature

The muscles of the anal canal can be divided into two main groups, those of the internal anal sphincter complex, and those of the external anal sphincter complex (Fig. 26.1). These muscles are necessary for maintaining continence and control of voluntary defecation.

The internal anal sphincter is the continuation of the circular layer of muscularis propria of the rectum. As the muscularis propria extends caudally, it passes through the cranial end of the puborectalis muscle, which represents

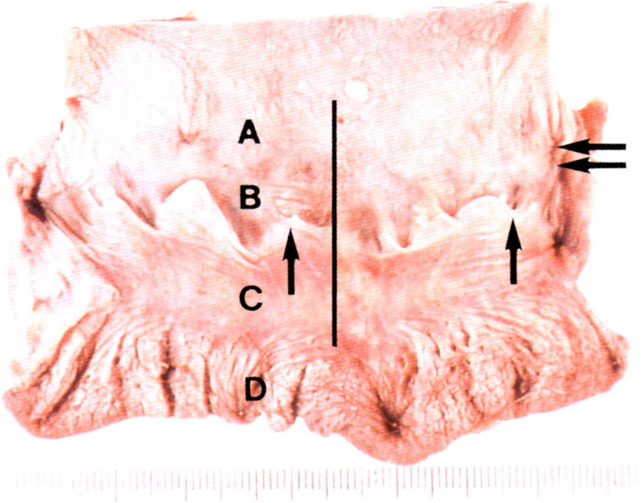

FIGURE 26.4 This autopsy specimen is from an adult. The anal columns are less apparent, but the anal valves are well formed. The *single black arrows* show the undulating dentate line, which roughly follows the base of the anal columns/valves; *double arrows* show the upper border of the anal transitional zone. The extent of the surgical anal canal is marked by the *vertical black line* and includes the anorectal zone (**A**), the transition zone (**B**), the squamous mucosa (pecten, **C**), and the wrinkled perianal skin of the anal verge (**D**).

the proximal border of the internal anal sphincter. At this transition point, the muscle enlarges due to increased wall thickness, forming a concentric aggregate of muscle cells in continuous maximal contraction. The internal anal sphincter terminates abruptly at a site proximal to the anal verge, although its precise location varies from person to person (28,29).

The external anal sphincter is anatomically complicated; it is composed of striated skeletal muscles and is vaguely elliptical in shape (1,11). It surrounds the entirety of the

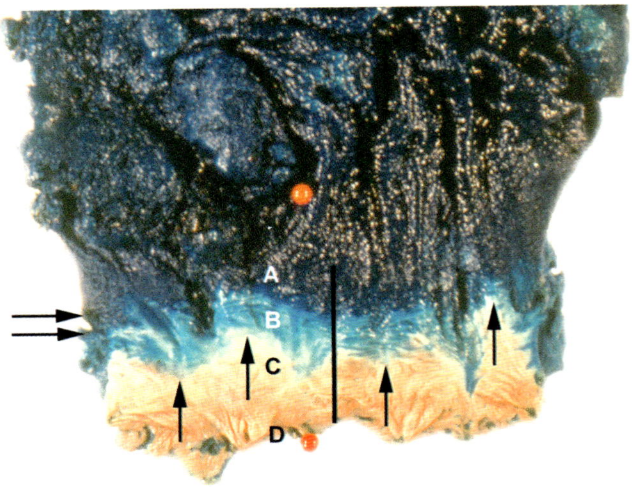

FIGURE 26.5 Surgical specimen of adult anal canal after treatment with Alcian green. The anorectal mucosa is dark blue (**A**), the transitional zone mucosa is pale blue (**B**), and the squamous mucosa (**C**) and perianal skin (**D**) do not stain. The *double arrows* represent upper border of the anal transitional zone. The undulating dentate line (*vertical arrows*) is highlighted by this stain. The extent of the surgical anal canal is depicted by the *vertical black line*.

internal anal sphincter and continues caudally to terminate approximately 1 cm distal to the internal anal sphincter (3). The muscular divisions of the external anal sphincter are not entirely understood. It is widely viewed as a tripartite structure with deep, subcutaneous, and superficial regions (19). The deep external sphincter muscle is ring-like and fused with the puborectalis muscle. The subcutaneous portion of the external sphincter is similarly ring-like in orientation. The superficial division runs between the deep and subcutaneous portions. Here, the skeletal muscle fibers are arranged with an elliptical orientation from the perineal body anteriorly to the coccyx posteriorly.

Some anatomists have suggested the external anal sphincter contains only deep and superficial layers, whereas others hypothesize that the sphincter consists of a single aggregate of fibers in continuity with the puborectalis muscle (30,31). Regardless of its precise nature, the striated muscles have increased resting tone through the tonic contractile activity of type 1 skeletal muscle fibers, thereby maintaining continence.

A potential intersphincteric space lies between the internal and external anal sphincters. This space contains the longitudinal anal muscle, which represents a continuation of the external/longitudinal smooth muscle layer of the muscularis propria (32). Skeletal muscle fibers from the levator ani, puborectalis, and pubococcygeus muscles can also be found in this area; histologic sections of the intersphincteric space contain a combination of smooth and skeletal-type muscle cells (33). At the distal end of the longitudinal anal muscle, fibroelastic septa extend through the external sphincter into the perianal dermis. These strands are known as the corrugator cutis ani muscle and cause the characteristic folding of perianal skin (34).

The levator ani, puborectalis, and pubococcygeus muscles comprise the pelvic diaphragm, the muscular structure that largely defines the pelvic floor and separates the pelvic cavity from the perineum. These striated muscles consist of predominantly slow/tonic motor units (35,36). The puborectalis muscle is especially relevant to discussions of the anal canal because it forms the anorectal ring that defines the proximal limit of the surgical anal canal (19). The puborectalis muscle is fixed anteriorly to the pubis, producing a U-shaped loop of muscle and providing the ventral kink that serves as the division between the rectum and the anal canal. Muscles of the pelvic floor play a crucial role in maintaining both fecal and urinary continence (35,37).

Innervation

Sensory and motor nerves innervate the anal canal. The sensory components are important for voluntary control of defecation, although sensation does not appear to be necessary for involuntary control (38). The upper anal canal is rich in pressure-sensitive nerve endings carried by the inferior rectal branch of the pudendal nerve (39). Sensory pain fibers are present below, but not above, the dentate line. The perianal skin of the anal verge receives somatic sensation from S4 in a dermatomal distribution (40).

Motor control is more complex, and different nerves supply various muscles of the anal sphincters. The internal anal sphincter has both intrinsic myogenic neuronal function, as well as innervation from extrinsic autonomic nerves (29). The intrinsic enteric nervous system (myenteric plexus) facilitates spontaneous, rhythmic contractions of the sphincter muscle, even in the absence of extrinsic nerve control (41). Extrinsic autonomic control is derived from the hypogastric pelvic nerve plexus, which includes nerve L5 and is predominately made up of sympathetic neurons. Sympathetic control allows for continuous excitatory signals and tonic contraction of the sphincter. Opposing parasympathetic fibers are carried from S2, S3, and S4 (42).

The external anal sphincter receives motor control from the inferior rectal branch of the pudendal nerve and the perineal branch of S4. Like the internal sphincter, these nerve fibers are predominately sympathetic in nature and provide tonic contraction. However, the skeletal component of this muscle is imparted with neural motor control, which can promote further contraction of this sphincter. When under voluntary control, the fibers of the external sphincter and the puborectalis muscle function as a unit (43).

The pelvic diaphragm, which includes the puborectalis and levator ani, is also supplied by S2–S4, which provides for sympathetic tonic contraction, as well as the ability for voluntary control. Fecal incontinence can result from damage to nerve fibers within the pelvic diaphragm, either from a specific event (e.g., childbirth) or as a consequence of aging (36,44).

Vasculature

The vascular supply of the anal canal reflects its embryologic origins. The routes described below seem straightforward in this writing; however, the actual vasculature of the anal canal consists of dense anastomosing arteriovenous networks that are important for surgical management of this area.

The anal canal above the dentate line is a derivative of the hindgut and, thus, it is supplied by branches of the inferior mesenteric artery. The inferior mesenteric artery gives rise to the left colic artery, which in turn gives rise to the superior hemorrhoidal, or rectal, artery. The superior hemorrhoidal artery bifurcates and runs through the rectal submucosa where branches ramify to supply the anal canal. Draining venules coalesce into rich, saccular venous plexuses in the right posterior, right anterior, and left lateral anal canal. These regions correspond to physiologic anal cushions and are the sites of pathologic internal hemorrhoidal varices (Fig. 26.1) (23,26,45).

The anorectum above the dentate line has venous drainage primarily to the inferior vena cava. Blood from the middle and inferior hemorrhoidal veins flows to the internal iliac vein and then to the inferior vena cava. The lymphatic vessels above the dentate line drain to inferior mesenteric and internal iliac lymph nodes (46).

Below the dentate line, the vascular and lymphatic routes correspond to those that supply the proctodeum in the embryo. The internal iliac arteries give rise to the internal pudendal artery, which gives rise to inferior hemorrhoidal arteries. The middle hemorrhoidal artery also arises from the anterior division of the internal iliac and serves this region. Superficial capillaries and veins from the internal and middle hemorrhoidal systems are the nidus for pathologic external hemorrhoids. Venous drainage from this area is predominantly to the portal venous system via the inferior mesenteric vein. Lymph flows primarily along inferior rectal lymphatic channels to the superficial inguinal lymph nodes. Because lymph also drains to perirectal and internal iliac lymph nodes, these are considered regional for the purposes of staging anal carcinomas (5,47).

LIGHT MICROSCOPY

The microscopic anatomy of the anal canal varies by anatomic region: the anorectum above the dentate line, the anal transition zone, the anal canal below the dentate line, and the anal verge (Fig. 26.6). Each region is made up of mucosa, submucosa, and muscle, which will be described below.

Mucosa

The term mucosa collectively refers to the superficial layers of epithelium, lamina propria, and muscularis mucosae of the gastrointestinal tract. The lamina propria consists of

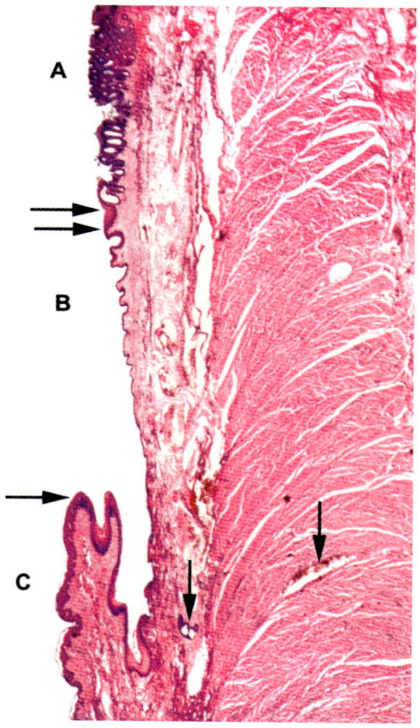

FIGURE 26.6 A longitudinal section of the anal canal demonstrates the anorectal zone (**A**), anal transition zone (**B**), and squamous mucosa (**C**). The anal transition zone extends into an anal sinus with a valve. The *vertical black arrows* indicate anal glands in the submucosa and internal sphincter. The *single horizontal arrow* indicates the dentate line; the *double horizontal arrow* marks the proximal border of the anal transition zone (H&E).

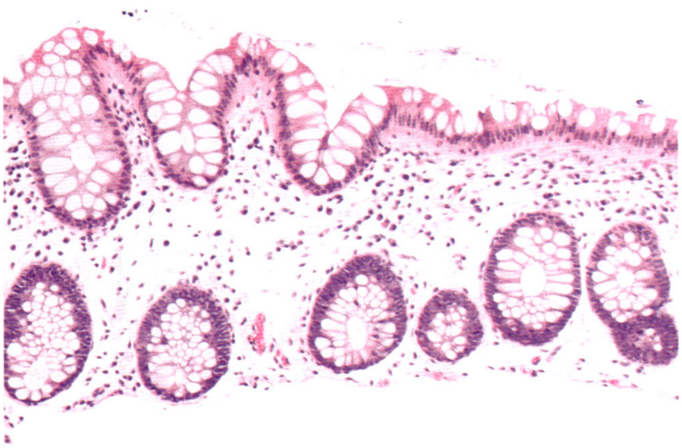

FIGURE 26.7 The mucosa in the upper anorectum is histologically similar to that of the colorectum. The epithelium contains mucin-filled goblet cells and eosinophilic absorptive colonocytes. Loose mesenchyme and occasional inflammatory cells comprise the lamina propria, which supports the epithelium.

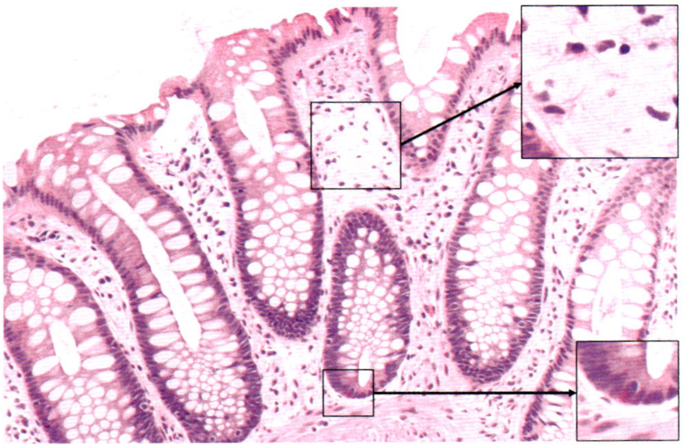

FIGURE 26.8 Scattered endocrine cells (*bottom inset*) usually contain brightly granular eosinophilic cytoplasm and round nuclei. Aggregates of muciphages (*upper inset*) are commonly present. Thin fascicles of smooth muscle cells comprise the muscularis mucosae, which runs longitudinally under the crypts.

delicate mesenchyme that supports the epithelium, and the muscularis mucosae is a thin, discontinuous band of smooth muscle cells running longitudinally between the lamina propria and submucosa.

The anorectal junction is an anatomic landmark that denotes the proximal end of the surgical anal canal; the mucosa in this region is virtually identical to that of the distal colorectum. Goblet cells, absorptive colonocytes, and endocrine cells comprise the epithelial lining in order of decreasing abundance (Fig. 26.7). Goblet cells are necessary for lubricating the mucosa. They contain abundant mucin-filled cytoplasm that appears almost clear in hematoxylin and eosin (H&E)-stained sections; these cells also contain small, flattened nuclei near the basement membrane. Colonocytes are fewer in the anorectum than the abdominal colon, and they are often inconspicuous due to high numbers of goblet cells. Similar to colonocytes elsewhere, they are tall columnar cells with faintly eosinophilic cytoplasm, microvilli, and basally located nuclei that are rounder and less hyperchromatic than those of adjacent goblet cells. Endocrine cells are small, cuboidal cells that contain densely packed, brightly eosinophilic granules and nuclei that are luminally oriented in the crypts (Fig. 26.8).

The epithelium of the anorectum grows as a single cell layer along a basement membrane; tubular crypts increase the surface area (11,48). Crypts of the abdominal colon are often described as resembling test tubes in a rack to invoke their regular architectural appearance; they are oriented parallel to one another and perpendicular to the muscularis mucosae. The crypts of the anorectum have a slightly less uniform appearance and are often irregularly dispersed in the lamina propria. They may be shortened, angulated, or occasionally branching, especially toward the surface (49). These subtle architectural changes are considered within the range of normal.

The lamina propria contains scattered lymphocytes, plasma cells, and loosely arranged collagen fibrils, as well as scattered capillaries. Mucin-filled macrophages (i.e., muciphages) can be numerous, especially under the surface epithelium (Fig. 26.8). The muscularis mucosae is thin and parallel to the luminal surface epithelium. Mucosal prolapse-type changes are common in the anorectum. In this situation, thin bundles of smooth muscle cells emanate from the muscularis mucosae vertically into the lamina propria between and parallel to the tubular crypts. Epithelial changes, such as mucin depletion, crypt serration, and erosions may also be present.

The anorectal mucosa and that of the anal transition zone merge between 3 and 20 mm proximal to the dentate line in adults, whereas transitional epithelium is absent in the fetus and newborn (50). The epithelial lining of the anal transition zone is stratified and contains four to nine cell layers. The basal layers consist of polarized cuboidal cells, whereas the surface cells display variable morphology and may be flat, cuboidal, columnar with apical mucin, or polygonal, similar to umbrella cells of the urothelium (Figs. 26.9 to 26.11) (49). The layers between the deepest and most

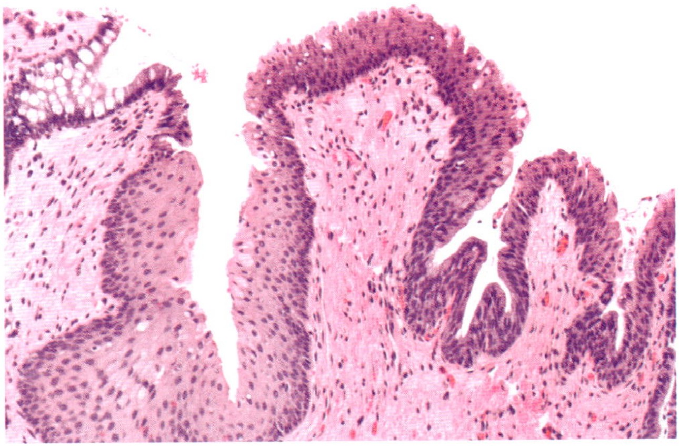

FIGURE 26.9 The anorectal mucosa and anal transition zone merge above the dentate line in adults. Goblet cell–rich epithelium gives way to transitional epithelium, which contains several layers of small basal cells with slightly hyperchromatic nuclei.

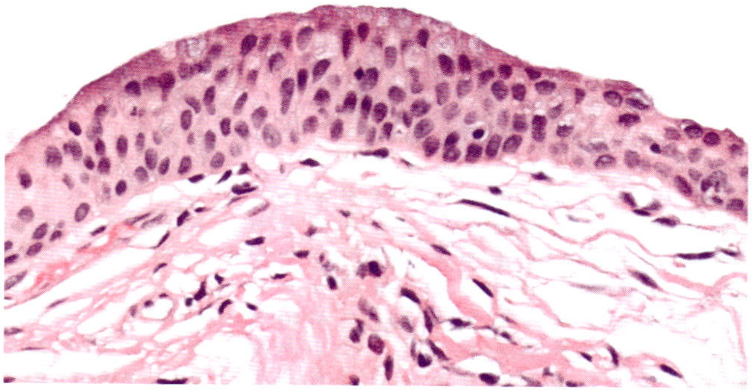

FIGURE 26.10 Superficial cells in the anal transitional zone are short columnar cells with a small amount of acid mucin that imparts blue discoloration to the cytoplasm.

superficial epithelial cell layers can be oriented perpendicular to the basement membrane with elongated nuclei that impart a streaming appearance, or their long axes may be parallel to the basement membrane, reminiscent of stratified squamous epithelium. Mucin is sparse in the anal transition zone and generally has a basophilic appearance when present; histochemical stains demonstrate a predominance of sialomucins rather than the sulphomucins more common in the colon (Table 26.1) (24,51).

Between 6 and 12 anal glands arise from the anal crypts of Morgagni. These glands extend into the submucosa, and at least half penetrate the internal sphincter. The glandular epithelium varies according to the location along the gland: squamous cells are numerous at the gland orifice on the luminal surface, the neck region is lined by transitional-type epithelium, and columnar cells are present in the deep regions where the glands extend into the submucosa (Fig. 26.12) (20). Inflamed anal glands may participate in the pathogenesis of anal fistulae (fistula-in-ano) and rarely give rise to adenocarcinomas of the anal canal (Table 26.2) (21,53).

The anal transition zone epithelium merges with the smooth, nonkeratinizing stratified squamous epithelium at, or below, the dentate line. Again, the basal cell layer is composed of uniform, cuboidal cells with basophilic cytoplasm, whereas superficial cells are flattened with eosinophilic cytoplasm and small, round, or slightly elongated nuclei with dense chromatin. Melanocytes, Langerhans cells, intraepithelial lymphocytes, and Merkel cells are present in small numbers in this region (Figs. 26.13 and 26.14) (54–56).

The stratified squamous mucosa transitions to keratinized perianal skin at the level of the anal verge. It displays a superficial granular cell layer with keratohyalin granules and a cornified layer of basket-weave keratin (Fig. 26.15). Dermal papillae are well-formed, and hair follicles and other adnexal structures are present (Fig. 26.16).

The transitions between epithelial regions of the anal canal are gradual and irregular in some cases, resulting in map-like configurations that can be appreciated with Alcian green histochemical stains performed on the gross specimen. The proximal anal canal stains a deep blue-green, corresponding to mucins of the anorectal mucosa. The distal anal canal is lined by squamous epithelium, which does not pick up the stain. The middle transitional region has a pale turquoise hue (Fig. 26.6) (49).

Submucosa

The submucosa displays regional variations along the length of the anal canal. Above the dentate line, it consists of loosely arranged collagen fibrils, fibroblasts, and extracellular matrix, in which the anal vascular cushions are embedded (Fig. 26.17) (44). The connective tissue scaffolding around the hemorrhoidal vessels loosens with advanced age, allowing veins and venules to progressively engorge under the forces of gravity, resulting in hemorrhoidal varices and/or thrombosis (Fig. 26.18) (Table 26.2) (57). Multinucleated stromal cells can be identified in the lamina propria and submucosa of the anal canal; these are benign fibroblast-type cells of no clinical consequence (Fig. 26.19) (58,59).

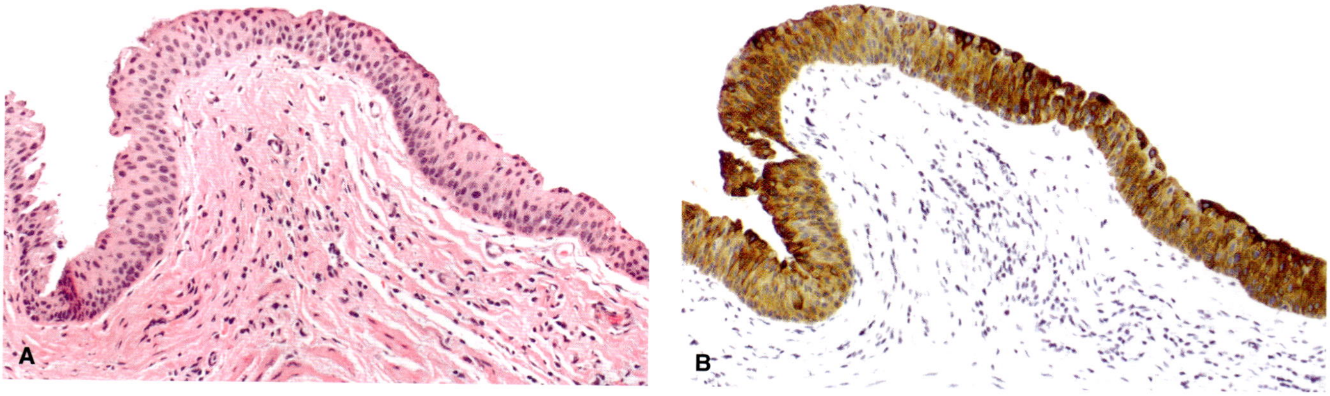

FIGURE 26.11 Surface cells in the transitional mucosa may be flattened and contain eosinophilic cytoplasm reminiscent of squamous cells (**A**). Transitional zone epithelium shows strong, diffuse immunostaining for cytokeratin 7, unlike squamous mucosa of the anus (**B**).

TABLE 26.1 Cell Types of the Anorectum and Their Staining Patterns (50,52)

	Cell Type	IHC Positive (Selected)	IHC Negative	Positive Histochemical Stains
Anorectal mucosa	Goblet cell	Pankeratin, MUC1, MUC2, MUC4, CEA	CK7	Mucicarmine, PAS, PAS-d
	Colonic absorptive cell	Pankeratin, CK20, CDX2	CK7	
	Endocrine cell	INSM1, chromogranin, synaptophysin		Grimelius
	Macrophage	CD68, CD163, CD4	S100	PAS (mucin positive)
	Plasma cell	CD19, CD38, CD138, BCMA	CD3, CD20,	
	Lymphocyte	CD3, CD20, CD45		
Anal transitional zone	Transitional epithelium	CK7, CK19	CK20, CDX2	PAS (luminal border of columnar cells)
	Anal glands	CK7, CK19, CK5/6, p63	CK20, CDX2	
	Endocrine cell	AE1/AE3, chromogranin, synaptophysin, serotonin, INSM1		Grimelius
Anal mucosa	Keratinocyte	CK5/6, p63, p40		
	Melanocyte	S100 protein, melan-A, PRAME	AE1/AE3	Fontana–Masson
	Merkel cell	CK20, NSE	CK7	
	Langerhans cell	S100 protein, CD1a, langerin		ATPase
	Lymphocyte	CD3, CD20, CD45		
Anal verge/perianal skin	Keratinocyte	CK5/6, p63, p40		
	Melanocyte	S100, melan-A	AE1/AE3	Fontana–Masson
	Adnexal structures	Variable: 34Be12, EMA, aSMA, calponin, p63		PAS, PAS-D
	Mast cell	CD117, tryptase, CD2		Toluidine blue
	Macrophage	CD68, CD163, CD4	S100	
Submucosa	Endothelial cells	CD34, CD31, ERG		
	Fibroblasts	Vimentin	Pankeratin	
	Collagen			Masson trichrome
	Adipose tissue	vimentin, S100 protein		Oil red O (frozen secton)
	Lymphatic vessels	CD31, D2-40	CD34	
	Peripheral nerves	S100 protein		

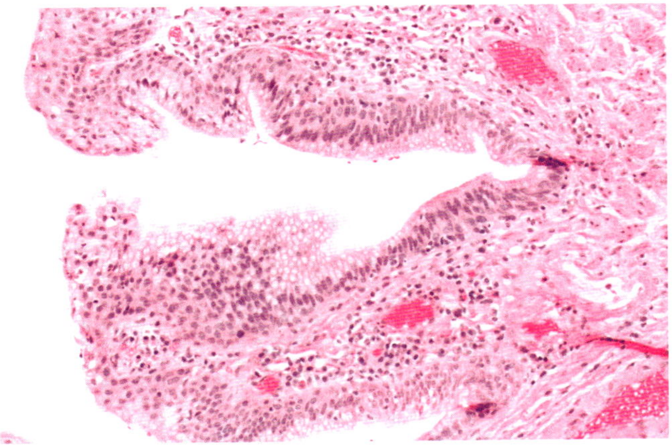

FIGURE 26.12 Anal glands are normally present in the transition zone. They are usually lined by transitional-type epithelium and scattered goblet cells. This gland opens onto the surface within an anal crypt.

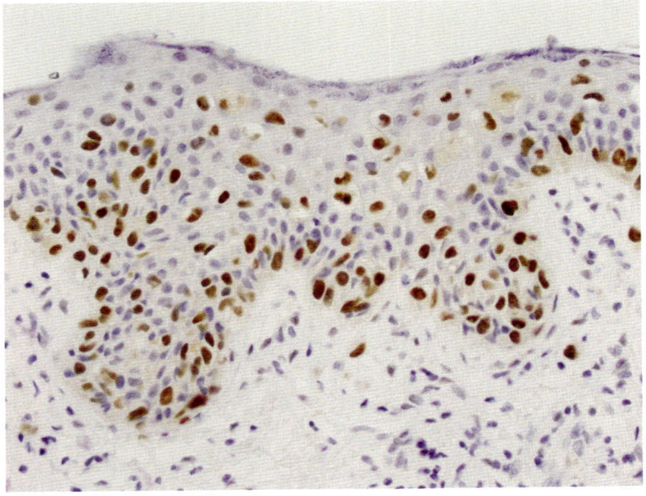

FIGURE 26.13 PRAME immunohistochemistry is a novel stain that is specific for mucosal melanomas. PRAME expression is characterized as diffuse when nuclear staining is present in over 75% on neoplastic cells, as illustrated.

TABLE 26.2 Pathologic Processes That Occur in Different Components of the Anal Canal

Normal Structures	Pathologic Disorders
Anorectal mucosa	Colorectal carcinoma Endocrine tumors
Anal transition zone	HPV-related squamous dysplasia squamous cell carcinoma
Anal glands	Anal fistulae Adenocarcinoma, anal gland subtype
Anal squamous mucosa	Squamous cell carcinoma/dysplasia Benign melanocytic nevus Malignant melanoma
Perianal skin	Squamous cell carcinoma/dysplasia Basal cell carcinoma Extramammary Paget disease Benign melanocytic nevus Malignant melanoma Hidradenoma papilliferum
Venous plexuses	Hemorrhoids Anal fissure
Muscular sphincters	Incontinence Prolapse

The pectin below the dentate line contains a submucosa rich in elastic tissue and collagen fibrils that tether the mucosa to the underlying muscle. Elastic fibers are most prominent at the anal verge underlying the transition from anal mucosa to perianal skin (Fig. 26.20) (23).

Muscles

There are two features to keep in mind when observing the muscular layers of the anal canal. First, a full-thickness

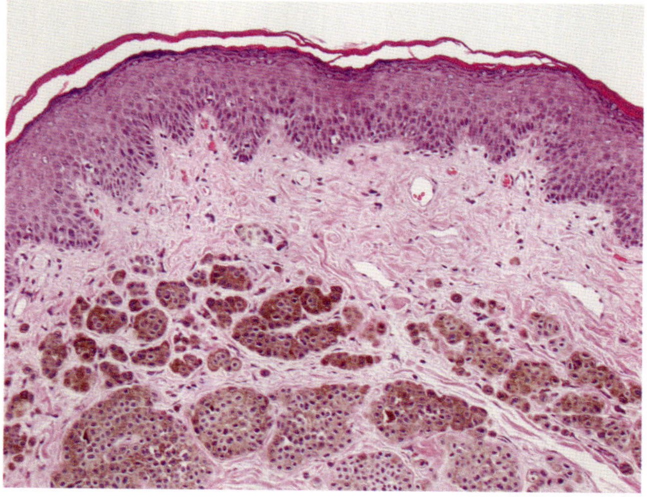

FIGURE 26.14 Congenital nevus are characterized by bland melanocytic proliferation in the dermis. The melanocytes are typically seen forming small clusters and may display variable degree of pigmentation. These lesions often present as small pigmented anal tags.

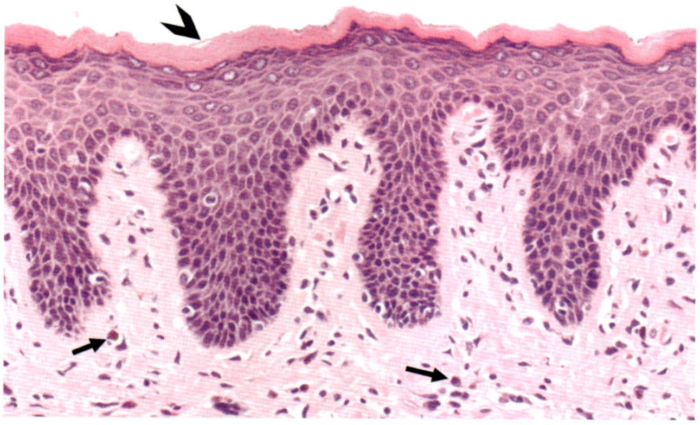

FIGURE 26.15 Perianal skin has well-formed dermal papillae, a superficial granular layer, and keratinization with a compact or basket-weave layer of brightly eosinophilic keratin at the surface (*arrowhead*). Pigmented dermal macrophages (*black arrow*) are frequently present.

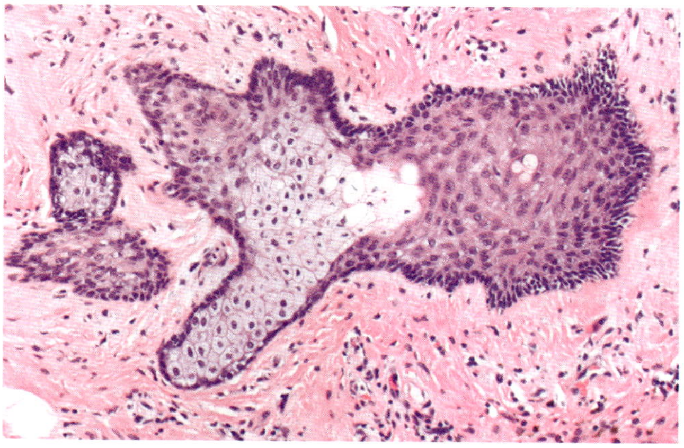

FIGURE 26.16 A sebaceous gland located in the dermis of the perianal skin.

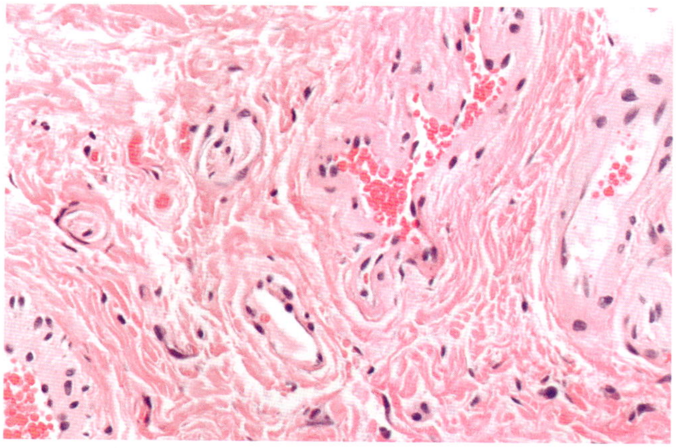

FIGURE 26.17 The vascular cushions of the anal canal represent a rich venous network within the submucosa.

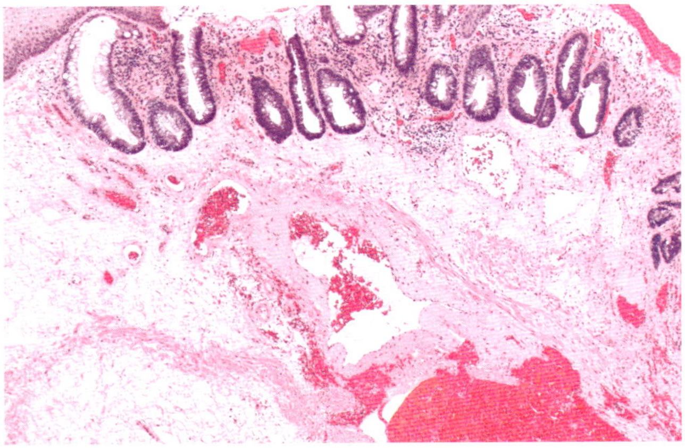

FIGURE 26.18 Pathologic hemorrhoids occur as the submucosal fibers around the plexus loosen, and veins become progressively engorged. Dilated veins fill the submucosa at the junction between anorectal and transitional epithelium in the anal canal.

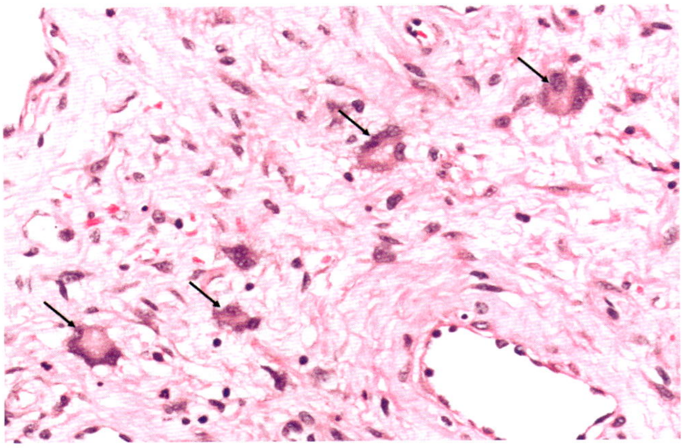

FIGURE 26.19 Multinucleated stromal cells (*black arrows*) can be found in the lamina propria and submucosa of the anal canal. They probably represent activated fibroblasts.

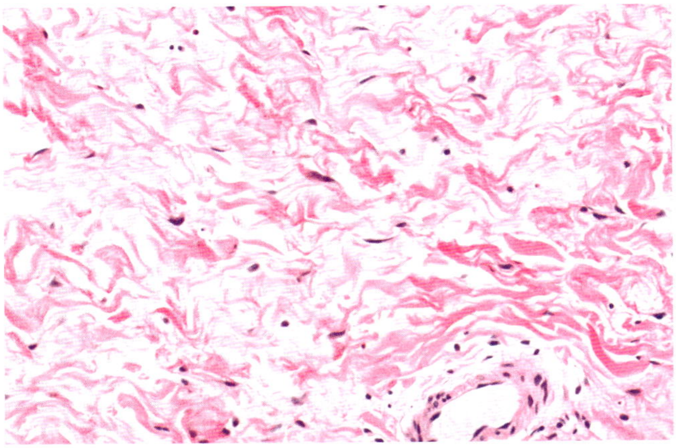

FIGURE 26.20 The submucosa of the distal anus contains increased amounts of gray-blue elastic fibers, especially near the perianal skin.

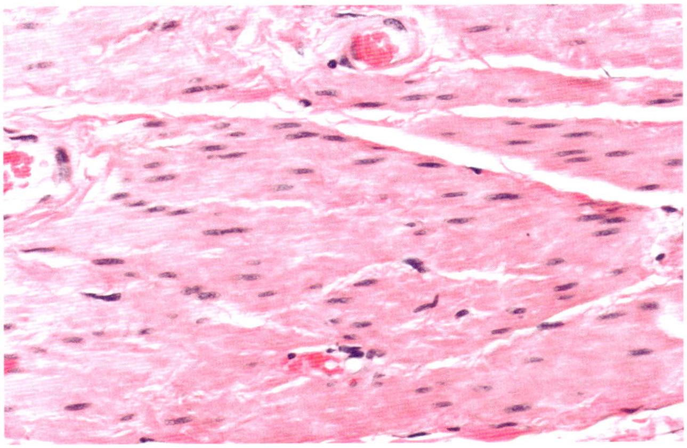

FIGURE 26.21 Tight fascicles of slender smooth muscle cells comprise the internal anal sphincter and longitudinal anal muscle.

section through the anal wall that includes sphincter muscle will show a mixture of smooth and striated muscle (35). Smooth muscle cells are thin and arranged in bundles parallel to one another (Fig. 26.21). Each smooth muscle cell contains a single nucleus. Skeletal muscle cells are polygonal and multinucleated, features that are best seen when cut in cross-section. The sarcomere pattern of striations can be appreciated in longitudinally sectioned fibers (Fig. 26.22).

Second, the muscle layers of the anal canal are intricately associated with each other. Unlike the inner and outer layers of the colonic muscularis propria, the muscles of the anal canal can be arranged in parallel, longitudinal, perpendicular, radial, or circumferential aggregates depending on their location in the sphincter and the method of tissue sampling. The net effect is a rather haphazard-appearing arrangement of irregularly intersecting fascicles. Fortunately, a clear demarcation between muscle groups is not relevant to clinical practice in most situations.

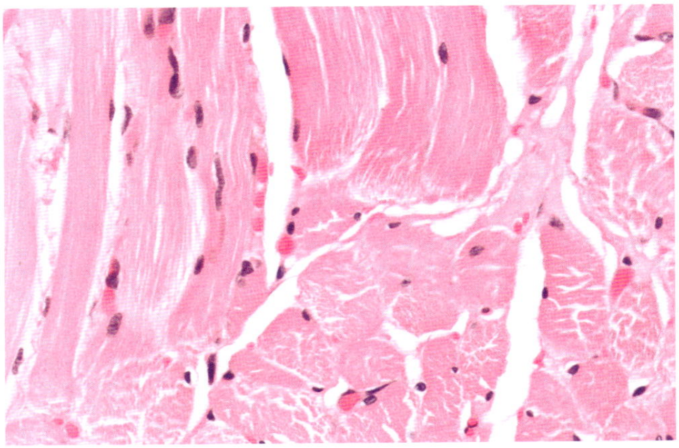

FIGURE 26.22 Brightly eosinophilic skeletal muscle fibers are multinucleated and contain cytoplasmic striations. Skeletal muscle is found in the external anal sphincter, longitudinal anal muscle, and pelvic diaphragm.

DIAGNOSTIC CONSIDERATIONS

Pathologists receive anal resection or anal biopsy specimens for evaluation of disease. Brief mention will be made here of the more common abnormalities encountered and how they differ from the normal histology described above.

Epithelial Metaplasia and Heterotopia

Epithelial cell types other than those previously described can be identified in sections taken from the anal canal. Patients who have longstanding colorectal disease, such as inflammatory bowel disease, may have metaplasia of anorectal epithelium as a result of chronic mucosal injury. Paneth cell metaplasia and pseudopyloric metaplasia are the most common altered cell types in chronic anorectal injury. Rarely, heterotopic aggregates of gastric oxyntic glands can be found in the anal canal, similar to islands of oxyntic mucosa of the proximal esophagus (inlet patches), duodenum, and intestines. This abnormality may produce the impression of a polyp, but is otherwise of no clinical significance (60,61).

Inflammatory Conditions

Anal fissures are linear erosions that can ulcerate, causing painful, nonhealing wounds. Men and women are equally affected. Potential inciting factors include constipation and hard stool, diarrhea from frequent stooling, or other underlying inflammatory or infectious conditions involving the anal canal. Fissures most commonly occur in the posterior midline, followed by the anterior midline (62). It is possible that the posterior wall is more susceptible to fissures because it is less perfused than other parts of the anus; increased anal pressure due to straining further reduces blood flow, leading to ischemia, erosion, and ultimately fissure formation (29,63,64). The microscopic findings of a sampled fissure often demonstrate inflamed and ulcerated anal mucosa usually accompanied by fibrosis and granulation tissue.

Anal fistulae are defined as inflammatory tracts leading from the epithelialized anal canal to another epithelialized surface, usually the skin. Anal fistulae are probably due to anal gland infection (21,65). The cryptoglandular theory of fistulous disease is supported by autopsy data confirming similarities between the locations and growth patterns of anal glands and those of anal fistulae. Modern classification schemes allow for more than a dozen types of fistulae; treatment is multimodal but almost always requires a surgical approach (66). Histologic features are similar to those seen in fissure formation and include inflamed epithelium, granulation tissue, submucosal fibrosis, and, in some cases, a giant cell reaction to fecal material (Fig. 26.23). Anal fistulae are commonly seen in patients with Crohn disease (67).

Infection is another cause of anal inflammatory changes. Most commonly, herpes simplex virus infection causes ulcers with classic diagnostic cytologic changes in epithelial cells at the ulcer edge. These include multinucleation, margination of chromatin, and nuclear molding, often accompanied by dense cytoplasmic eosinophilia (Fig. 26.24). Other sexually transmitted diseases, specifically chlamydia and syphilis, may manifest as dense submucosal chronic inflammation with a characteristic band-like pattern of lymphoplasmacytic inflammation (Fig. 26.25) (68). Immunohistochemical stains for spirochetes are useful, although definitive clinical diagnosis is usually confirmed with serologic testing or nucleic acid–amplification testing (69).

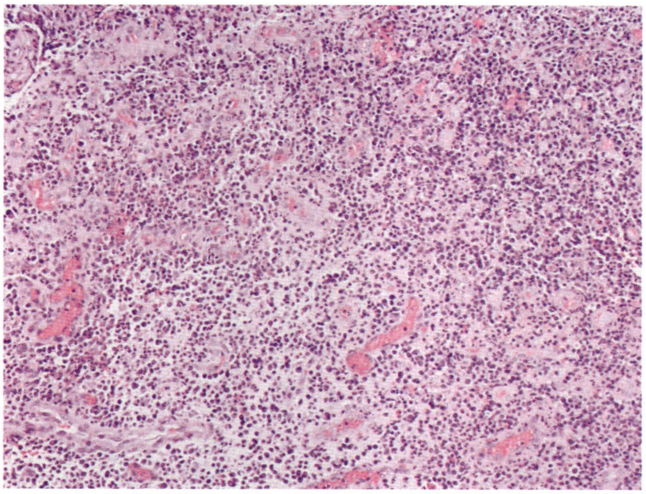

FIGURE 26.23 Granulation tissue from a fistula tract contains sheets of mature plasma cells, lymphocytes, and eosinophils with admixed small capillaries.

Neoplasia

The most common form of neoplastic disease in the anal canal is that of squamous epithelial dysplasia and/or squamous cell carcinoma (70). Similar to the transformation

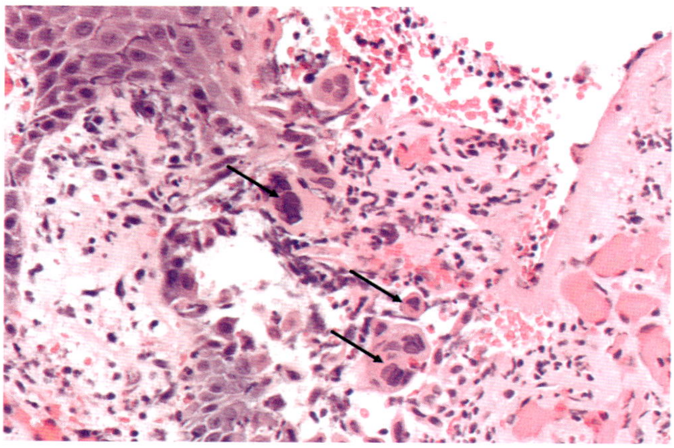

FIGURE 26.24 Herpesvirus infection induces nuclear abnormalities in squamous epithelial cells; multinucleation with hypereosinophilia (*arrows*) are accompanied by macrophage-rich ulcer debris. Infected nuclei have a "glassy" appearance with marginated chromatin and nuclear molding.

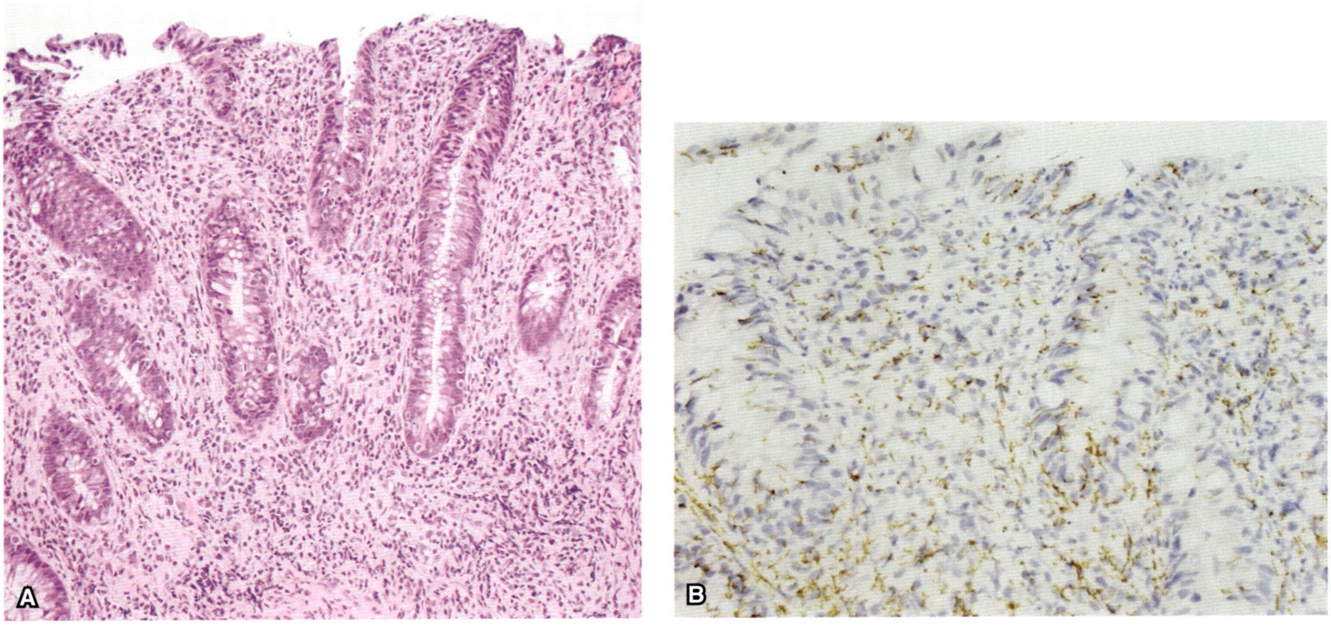

FIGURE 26.25 The hallmark of *Treponema pallidum* infection is the presence of dense subepithelial lymphoplasmacytic inflammation with characteristic "band-like" dense lymphoplasmacytic infiltrate. However, atypical presentation including active cryptitis with chronic inflammatory infiltrate within lamina propria spaces is also common (*right;* **A**). Immunohistochemistry for spirochete highlights numerous coiled spirochetes (*left;* **B**).

zone of the uterine cervix, the epithelium of the anal transitional zone is particularly susceptible to infection with human papillomavirus and, thus, is prone to squamous dysplasia. The dysplasia may be low- or high-grade and appear as a flat plaque or verrucous growth. Distinguishing dysplasia from reactive cytologic abnormalities in the transitional zone epithelium can be challenging. Features of low-grade dysplasia include dysmaturation and disorganization of squamous cells in the lower one-third of the epithelium, often accompanied by superficial koilocytosis (Fig. 26.26). High-grade dysplastic lesions show dysmaturation and mitotic figures involving more than the deep one-third of the epithelium (Fig. 26.27) (71). Immunohistochemical stains for p16 typically show diffuse block-like nuclear positivity in areas of high-grade dysplasia (72,73). Immunohistochemistry for p16 is recommended when entertaining the possibility of high-grade dysplasia (i.e., AIN 2), as this marker readily distinguishes most dysplasia cases from potential mimickers (71).

Other forms of neoplasia that can be seen in the anal area include primary extramammary Paget disease (Fig. 26.28), secondary Paget disease due to colonization of the squamous

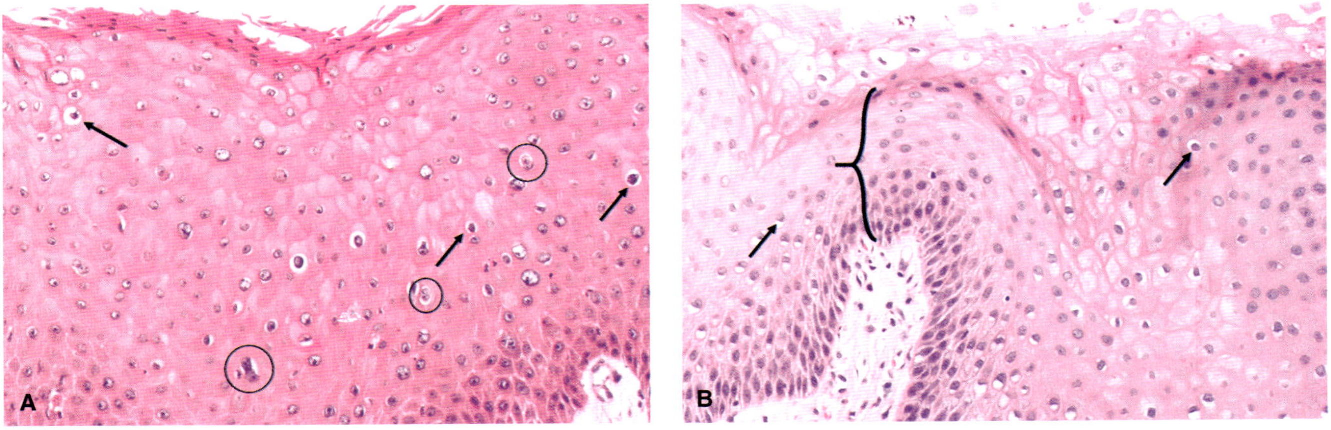

FIGURE 26.26 Human papillomavirus infection causes disorganization of squamous epithelium with binucleated keratinocytes (*black circles*) and koilocytosis (**A**, *black arrows*). Reactive changes in squamous mucosa can mimic HPV infection, but normal maturation is present (*bracket*) and cytoplasmic halos are unassociated with nuclear abnormalities (*arrow*). Superficial keratinocytes contain uniform pale cytoplasm, which may be related to surface trauma (**B**).

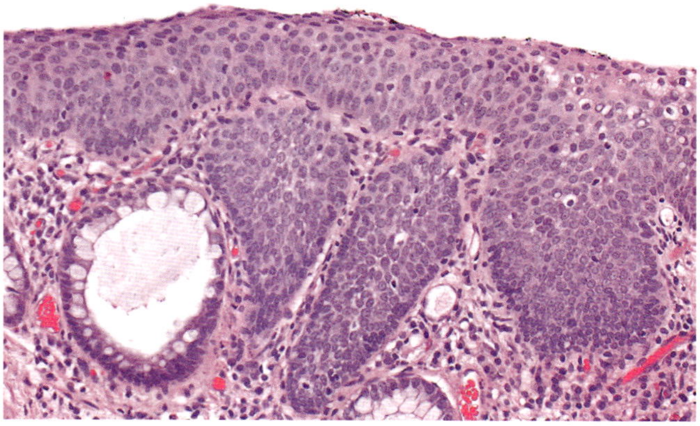

FIGURE 26.27 High-grade squamous intraepithelial neoplasia (HSIL) is a high-risk HPV-driven precancerous lesion characterized by nuclear atypia including hyperchromatic nuclei displaying nuclear membrane irregularities that show loss of maturation. Numerous mitotic figures and apoptotic squamoid cells are usually readily seen.

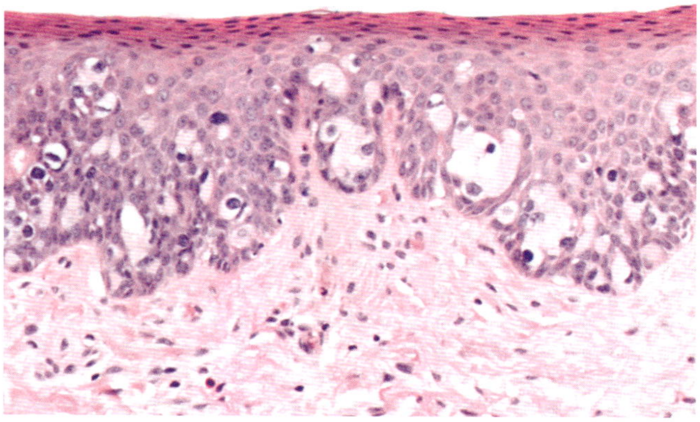

FIGURE 26.28 Extrammamary Paget disease displays clustered cells with pale cytoplasm and enlarged, hyperchromatic nuclei, predominantly in the basal third of the squamous epithelium. These malignant cells contain mucin and are immunopositive for cytokeratin 7.

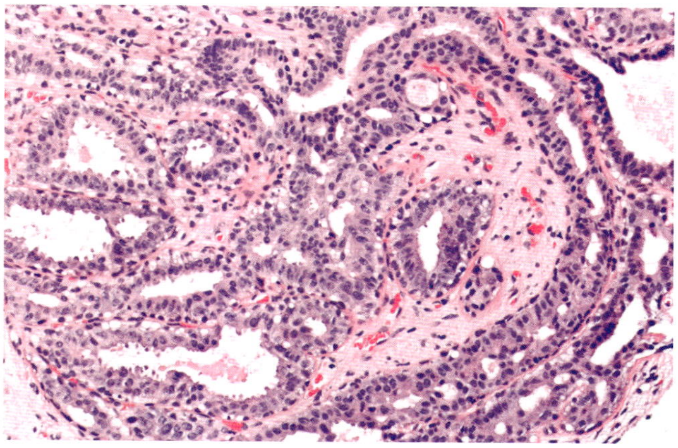

FIGURE 26.29 Hidradenoma papilliferum is a benign neoplasm that arises in perianal and vulvar skin. Its complex architecture can simulate features of adenocarcinoma. The double layer of luminal columnar cells and basal myoepithelial cells is a clue to a benign diagnosis.

epithelium by occult rectal adenocarcinoma, squamous cell carcinoma of the perianal skin, adenocarcinoma derived from anal glands, malignant melanoma, and secondary tumor involvement by adjacent genitourinary and gynecologic origin. All of these tumors can show pagetoid spread of malignant cells in the epithelium. Tumor cells are arranged singly or in clusters and typically contain abundant cytoplasm. In some cases, the clinical history may be helpful, whereas others require immunohistochemical analysis to correctly classify disease (74,75). Of note, reactive squamous epithelial cells can show cytoplasmic clearing, but their nuclear features are essentially normal (76).

Hidradenoma papilliferum (i.e., papillary hidradenoma, papillary apocrine adenoma) is the most common benign adnexal neoplasm of perianal skin. This lesion usually presents as a well-circumscribed dermal nodule (77). Tumors are composed of papillary fronds–lined luminal cuboidal cells and basal myoepithelial cells, often in densely collagenous stroma (Fig. 26.29).

REFERENCES

1. Garza A, Beart RW Jr. Anatomy and embryology of the anus, rectum, and colon. In: Corman ML, Bergamaschi RCM, Nicholls RJ, Fazio VW, eds. *Corman's Colon and Rectal Surgery*. 6th ed. Wolters Kluwer Health/Lippincott Williams & Wilkins; 2013:1–26.
2. Symington J. The rectum and anus. *J Anat Physiol*. 1888;23(Pt 1): 106–115.
3. Halligan S, Stoker J. Imaging of fistula in ano. *Radiology*. 2006; 239(1):18–33.
4. Welton ML, Lambert R, Bosman FT. Tumours of the anal canal. In: *WHO Classification of Tumours of the Digestive System*. 4th ed. IARC Press; 2010:183–194.
5. Welton ML, Steele SR, Goodman, KA, et al. Anus. In: *AJCC Cancer Staging Manual*. 8th ed. Springer; 2017:275–284.
6. Schoenwolf GC, Bleyl SB, Brauer PR, Francis-West PH. Development of the gastrointestinal tract. In: *Larsen's Human Embryology*. 5th ed. Elsevier/Churchill Livingstone; 2015: 341–374. Accessed January 18, 2018. https://www.clinicalkey.com/#!/content/book/3-s2.0-B978145570684600014X
7. Nievelstein RA, van der Werff JF, Verbeek FJ, Valk J, Vermeij-Keers C. Normal and abnormal embryonic development of the anorectum in human embryos. *Teratology*. 1998;57(2):70–78.
8. Coalson RE, Tomasek JJ. Digestive system and mesenteries. In: *Embryology*. Springer; 1992:78–85.
9. Kluth D, Fiegel HC, Metzger R. Embryology of the hindgut. *Semin Pediatr Surg*. 2011;20(3):152–160.
10. Sadler TW, Langman J. *Langman's Medical Embryology*. 12th ed. Wolters Kluwer Health/Lippincott Williams & Wilkins; 2012.
11. Schizas AMP, Williams AB. The normal anus. In: Cohen R, Windsor A, eds. *Anus*. Springer; 2014:1–12.
12. Matsumaru D, Murashima A, Fukushima J, et al. Systematic stereoscopic analyses for cloacal development: the origin of anorectal malformations. *Sci Rep*. 2015;5:13943.
13. van der Putte SC. The development of the human anorectum. *Anat Rec (Hoboken)*. 2009;292(7):951–954.

14. Moore KL, Persaud TVN, Torchia MG. Alimentary system. In: *The Developing Human*. 10th ed. Elsevier; 2016:209–240. Accessed January 18, 2018. https://www.clinicalkey.com/#!/content/book/3-s2.0-B978032331338400011X?scrollTo=%23hl0001186
15. Le Guen L, Marchal S, Faure S, de Santa Barbara P. Mesenchymal-epithelial interactions during digestive tract development and epithelial stem cell regeneration. *Cell Mol Life Sci*. 2015;72(20):3883–3896.
16. Mao J, Kim BM, Rajurkar M, Shivdasani RA, McMahon AP. Hedgehog signaling controls mesenchymal growth in the developing mammalian digestive tract. *Development*. 2010; 137(10):1721–1729.
17. Glisson F. Tractatus de ventriculo et intestinis. In: *Bibliotheca Anatomica Sive Thesaurus*. Vol 1. 2nd ed. Chouet et Ritter; 1699.
18. Morgagni GB. *Adversaria Anatomica Omnia. Advers III, Animadv. VI. Patavii*, Josephus Cominus; 1717.
19. Milligan ETC, Morgan CN. Surgical anatomy of the anal canal: with special reference to anorectal fistulæ. *Lancet*. 1934; 224:1150–1156.
20. McColl I. The comparative anatomy and pathology of anal glands. Arris and Gale lecture delivered at the Royal College of Surgeons of England on 25th February 1965. *Ann R Coll Surg Engl*. 1967;40(1):36–67.
21. Parks AG. Pathogenesis and treatment of fistula-in-ano. *Br Med J*. 1961;1(5224):463–469.
22. Seow-Choen F, Ho JM. Histoanatomy of anal glands. *Dis. Colon Rectum*. 1994;37(12):1215–1218.
23. Saraswati R, Novelli M. Surgical treatment and pathology: normal histology. In: Cohen R, Windsor A, eds. *Anus*. Springer; 2014:43–47.
24. Fenger C. The anal transitional zone. *Acta Pathol Microbiol Immunol Scand Suppl*. 1987;289:1–42.
25. Lestar B, Penninckx F, Kerremans R. The composition of anal basal pressure. An in vivo and in vitro study in man. *Int J Colorectal Dis*. 1989;4(2):118–122.
26. Thomson WH. The nature of haemorrhoids. *Br J Surg*. 1975; 62(7):542–552.
27. Thomson H. The anal cushions–a fresh concept in diagnosis. *Postgrad Med J*. 1979;55(644):403–405.
28. Goligher JC, Leacock AG, Brossy JJ. The surgical anatomy of the anal canal. *Br J Surg*. 1955;43(177):51–61.
29. Lund JN, Scholefield JH. Aetiology and treatment of anal fissure. *Br J Surg*. 1996;83(10):1335–1344.
30. Oh C, Kark AE. Anatomy of the external anal sphincter. *Br J Surg*. 1972;59(9):717–723.
31. Parks AG. Modern concepts of the anatomy of the anorectal region. *Postgrad Med J*. 1958;34(393):360–366.
32. Macchi V, Porzionato A, Stecco C, Vigato E, Parenti A, De Caro R. Histotopographic study of the longitudinal anal muscle. *Clin Anat*. 2008;21(5):447–452.
33. Lawson JO. Pelvic anatomy. II. Anal canal and associated sphincters. *Ann R Coll Surg Engl*. 1974;54(6):288–300.
34. Lunniss PJ, Phillips RK. Anatomy and function of the anal longitudinal muscle. *Br J Surg*. 1992;79(9):882–884.
35. Dickinson VA. Maintenance of anal continence: a review of pelvic floor physiology. *Gut*. 1978;19(12):1163–1174.
36. Parks AG, Swash M, Urich H. Sphincter denervation in anorectal incontinence and rectal prolapse. *Gut*. 1977;18(8): 656–665.
37. Felt-Bersma RJF. Physiology of the rectum and anus. In: Ratto C, Parello A, Donisi L, Litta F, eds. *Colon, Rectum and Anus: Anatomic, Physiologic and Diagnostic Bases for Disease Management*. Springer; 2017:55–69.
38. Read MG, Read NW. Role of anorectal sensation in preserving continence. *Gut*. 1982;23(4):345–347.
39. Duthie HL, Gairns FW. Sensory nerve-endings and sensation in the anal region of man. *Br J Surg*. 1960;47:585–595.
40. Cuming T, Bailey AC, Sashidharan PN. Perianal skin conditions. In: Cohen R, Windsor A, eds. *Anus*. Springer; 2014:253–274.
41. Penninckx F, Kerremans R, Beckers J. Pharmacological characteristics of the non-striated anorectal musculature in cats. *Gut*. 1973;14(5):393–398.
42. Frenckner B, Ihre T. Influence of autonomic nerves on the internal and sphincter in man. *Gut*. 1976;17(4):306–312.
43. Snooks SJ, Henry MM, Swash M. Anorectal incontinence and rectal prolapse: differential assessment of the innervation to puborectalis and external anal sphincter muscles. *Gut*. 1985;26(5):470–476.
44. Haas PA, Fox TA Jr. The importance of the perianal connective tissue in the surgical anatomy and function of the anus. *Dis Colon Rectum*. 1977;20(4):303–313.
45. Noorani A, Carapeti E. Haemorrhoids: anatomy, pathophysiology and presentation. In: Cohen R, Windsor A, eds. *Anus*. Springer; 2014:157–167.
46. Blair JB, Holyoke EA, Best RR. A note on the lymphatics of the middle and lower rectum and anus. *Anat Rec*. 1950;108(4): 635–644.
47. Hardy KJ. The lymphatic drainage of the anal margin. *Aust N Z J Surg*. 1971;40(4):367–369.
48. Walls EW. Observations on the microscopic anatomy of the human anal canal. *Br J Surg*. 1958;45(193):504–512.
49. Fenger C. Histology of the anal canal. *Am J Surg Pathol*. 1988; 12(1):41–55.
50. Fenger C, Filipe MI. Mucin histochemistry of the anal canal epithelium. Studies of normal anal mucosa and mucosa adjacent to carcinoma. *Histochem J*. 1981;13(6):921–930.
51. Fenger C, Filipe MI. Pathology of the anal glands with special reference to their mucin histochemistry. *Acta Pathol Microbiol Scand A*. 1977;85(3):273–285.
52. Fetissof F, Dubois MP, Assan R, et al. Endocrine cells in the anal canal. *Virchows Arch A Pathol Anat Histopathol*. 1984;404(1):39–47.
53. Behan WM, Burnett RA. Adenocarcinoma of the anal glands. *J Clin Pathol*. 1996;49(12):1009–1011.
54. Clemmensen OJ, Fenger C. Melanocytes in the anal canal epithelium. *Histopathology*. 1991;18(3):237–241.
55. Lundquist K, Kohler S, Rouse RV. Intraepidermal cytokeratin 7 expression is not restricted to Paget cells but is also seen in Toker cells and Merkel cells. *Am J Surg Pathol*. 1999;23(2):212–219.
56. Gervaz E, Dauge-Geffroy MD, Sobhani I, et al. Quantitative analysis of the immune cells in the anal mucosa. *Pathol Res Pract*. 1995;191(11):1067–1071.
57. Loder PB, Kamm MA, Nicholls RJ, Phillips RK. Haemorrhoids: pathology, pathophysiology and aetiology. *Br J Surg*. 1994;81(7):946–954.
58. Groisman GM, Amar M, Polak-Charcon S. Multinucleated stromal cells of the anal mucosa: a common finding. *Histopathology*. 2000;36(3):224–228.
59. Pitt MA, Roberts IS, Agbamu DA, Eyden BP. The nature of atypical multinucleated stromal cells: a study of 37 cases from different sites. *Histopathology*. 1993;23(2):137–145.

60. Rifat Mannan AA, Kahvic M, Bharadwaj S, Grover VK. Gastric heterotopia of the anus: report of two rare cases and review of the literature. *Indian J Pathol Microbiol*. 2008;51(2):240–241.
61. Steele SR, Mullenix PS, Martin MJ, et al. Heterotopic gastric mucosa of the anus: a case report and review of the literature. *Am Surg*. 2004;70(8):715–719.
62. Stewart DB, Gaertner W, Glasgow S, Migaly J, Feingold D, Steele SR. Clinical practice guideline for the management of anal fissures. *Dis Colon Rectum*. 2017;60(1):7–14.
63. Schouten WR, Briel JW, Auwerda JJ. Relationship between anal pressure and anodermal blood flow. The vascular pathogenesis of anal fissures. *Dis Colon Rectum*. 1994;37(7):664–669.
64. Klosterhalfen B, Vogel P, Rixen H, Mittermayer C. Topography of the inferior rectal artery: a possible cause of chronic, primary anal fissure. *Dis Colon Rectum*. 1989;32(1):43–52.
65. Robinson AM, DeNobile JW. Anorectal abscess and fistula-in-ano. *J Natl Med Assoc*. 1988;80(11):1209–1213.
66. Parks AG, Gordon PH, Hardcastle JD. A classification of fistula-in-ano. *Br J Surg*. 1976;63(1):1–12.
67. Panés J, Rimola J. Perianal fistulizing Crohn's disease: pathogenesis, diagnosis and therapy. *Nat Rev Gastroenterol Hepatol*. 2017;14(11):652–664.
68. Arnold CA, Limketkai BN, Illei PB, Montgomery E, Voltaggio L. Syphilitic and lymphogranuloma venereum (LGV) proctocolitis: clues to a frequently missed diagnosis. *Am J Surg Pathol*. 2013;37(1):38–46.
69. de Vries HJ, Zingoni A, White JA, Ross JD, Kreuter A. 2013 European Guideline on the management of proctitis, proctocolitis and enteritis caused by sexually transmissible pathogens. *Int J STD AIDS*. 2014;25(7):465–474.
70. Fléjou JF. An update on anal neoplasia. *Histopathology*. 2015;66(1):147–160.
71. Darragh TM, Colgan TJ, Cox JT, et al; Members of LAST Project Work Groups. The lower anogenital squamous terminology standardization project for HPV-associated lesions: background and consensus recommendations from the College of American Pathologists and the American Society for Colposcopy and Cervical Pathology. *Arch Pathol Lab Med*. 2012;136(10):1266–1297.
72. Pirog EC, Quint KD, Yantiss RK. P16/CDKN2A and Ki-67 enhance the detection of anal intraepithelial neoplasia and condyloma and correlate with human papillomavirus detection by polymerase chain reaction. *Am J Surg Pathol*. 2010;34(10):1449–1455.
73. Walts AE, Lechago J, Bose S. P16 and Ki67 immunostaining is a useful adjunct in the assessment of biopsies for HPV-associated anal intraepithelial neoplasia. *Am J Surg Pathol*. 2006;30(7):795–801.
74. Goldblum JR, Hart WR. Perianal Paget's disease: a histologic and immunohistochemical study of 11 cases with and without associated rectal adenocarcinoma. *Am J Surg Pathol*. 1998;22(2):170–179.
75. Dawson H, Serra S. Tumours and inflammatory lesions of the anal canal and perianal skin revisited: an update and practical approach. *J Clin Pathol*. 2015;68(12):971–981.
76. Val-Bernal JF, Pinto J. Pagetoid dyskeratosis is a frequent incidental finding in hemorrhoidal disease. *Arch Pathol Lab Med*. 2001;125(8):1058–1062.
77. Baker GM, Selim MA, Hoang MP. Vulvar adnexal lesions: a 32-year, single-institution review from Massachusetts General Hospital. *Arch Pathol Lab Med*. 2013;137(9):1237–1246.

27

Liver

Camila C. Simoes ■ Rondell P. Graham

The embryology, gross morphology, normal histology, and minor pathologic alterations of the human liver—the single largest organ in the human body—are described in this chapter. The knowledge of normal liver parenchyma provides the basis of the interpretation of liver resection and biopsy specimens, in which deviation from normal is suspected. Some morphologic changes, particularly in needle biopsy specimens, are frequently subtle but may be of diagnostic importance. The pathologist should be familiar with these histologic variations of and from the normal liver. Therefore, this chapter also discusses minor histopathologic alterations that are not readily appreciated, but the clinical findings are significant. In addition, this chapter also describes nonspecific histologic alterations that are rather frequently encountered, may even be prominent in surgical and autopsy liver specimens, but often have little clinical significance. Brown pigments that are often encountered in the interpretation of liver specimens are discussed toward the end of this chapter.

Part of this chapter is also dedicated to handling and processing of liver biopsy specimens in order to obtain optimal sections for routine histologic evaluation and immunohistochemistry, and to preserve the submitted tissue for ancillary tests including molecular tests and electron microscopy studies. Immunohistochemistry is routinely performed on liver tissue nowadays, particularly in the diagnosis of primary and metastatic liver tumors.

EMBRYOLOGY

The liver is one of the first organs to develop, and rapidly becomes the largest organ during embryogenesis. It arises as a hepatic diverticulum from the endodermal layer of the most distal portion of the foregut during the 3rd to 4th week of gestation. The endoderm also gives rise to lungs, pancreas, thyroid, and gastrointestinal tract. When the embryo reaches 4 to 5 mm in length, the hepatic diverticulum differentiates cranially into proliferating hepatic cords and caudally into the gallbladder and extrahepatic bile ducts. The anastomosing cords of hepatoblasts grow into the mesenchyme of the septum transversum. As the hepatic cords extend outward during the 5th week of gestation, they are penetrated by the inwardly growing capillary plexus, which arises from the vitelline veins in the outer margins of the septum transversum and forms the primitive hepatic sinusoids.

Scattered mesenchymal cells derived from the septum transversum lie between the endothelial walls of the sinusoids and the hepatic cords, forming the connective tissue elements of the hepatic stroma, as well as the liver capsule. Hematopoietic tissue and Kupffer cells are also derived from splanchnic mesenchyme of the septum transversum. Once these structures are established, the liver grows rapidly to fill most of the embryonal abdominal cavity and by 9 weeks of gestation accounts for approximately 10% of the total weight of the embryo. The bile canaliculi appear in the 10-mm embryo as intercellular spaces between hepatoblasts.

The extrahepatic biliary tree arises directly from the endoderm, while the epithelium of the intrahepatic bile ducts arises from the proximal part of the primitive hepatic cords. This process is largely determined by the progressive development and branching of the portal vein with its surrounding mesenchyme. First, the hepatoblasts in direct contact with the mesenchyme around the portal vein transforms into bile duct–type cells. Then a second layer transforms into bile duct epithelial cells, resulting in a circular cleft in the shape of a cylinder around the portal vein and its enveloping mesenchyme (Fig. 27.1). These cells can be identified by their positivity of cytokeratin 19 (CK19). This primitive channel duct in the 8-mm embryo (5 to 6 weeks of gestation) is referred as the ductal plate (1), which then undergoes gradual remodeling to form the normal anastomosing system of bile ducts in the portal tracts (2,3). Failure of remodeling of the ductal plate results in excess of bile duct structures retaining a fetal configuration (Fig. 27.2) (4). In patients with ductal plate malformation such as congenital hepatic fibrosis, ductal plate remnants, which should be absent at birth, are seen throughout the liver. Occasionally, ductal plate remnants can be seen embedded in fibrous stroma of the portal tracts in an otherwise normal liver.

The differentiation of intrahepatic ducts occurs in embryos of 22 to 30 mm. Despite the common ancestry of hepatocytes and bile duct cells, each cell type is structurally and functionally distinct. The walls of the terminal twigs of the biliary tree, the canals of Hering, which connect bile canaliculi to bile ducts, include both typical hepatocytes and bile duct cells, without intermediate forms.

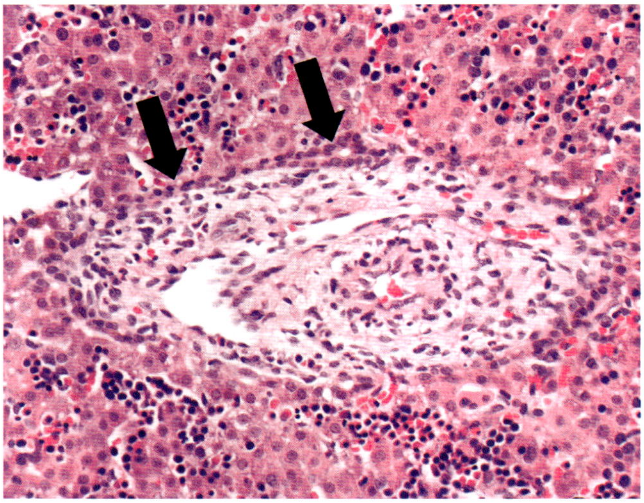

FIGURE 27.1 Ductal plate (*arrows*) developing around the portal vein mesenchyme in the liver of a 10-week-old embryo. There is extramedullary hematopoiesis in the sinusoids.

Intrahepatic hematopoiesis begins during the 6th week, hepatocyte bile formation by the 12th week, and excretion of bile into the duodenum by the 16th week. Hematopoiesis is among the most important functions of fetal liver. The third trimester marks the cessation of hematopoiesis with a concomitant decrease in liver growth so that the liver accounts for approximately 5% of the newborn's body weight. An increased extramedullary hematopoiesis or myelopoiesis in the liver beyond the third trimester suggests the presence of an active fetal response to intrauterine ascending infection (5).

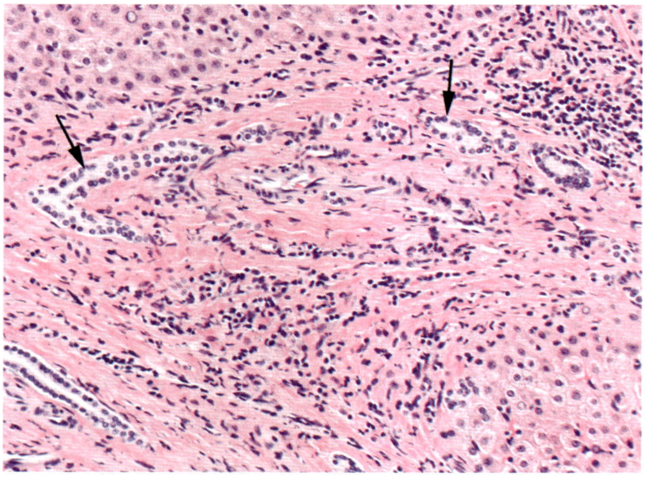

FIGURE 27.2 Ductal plate remnant (*arrows*) in portal tract stroma, similar to that seen in liver of an embryo, can occasionally be encountered in an otherwise normal liver.

GROSS MORPHOLOGY

The liver of an adult weighs 1,400 to 1,600 g, comprising 2.5% of the body weight. It is relatively larger in infancy, representing one-eighteenth of the birth weight, mainly due to a large left lobe. The liver resides predominantly in the right upper quadrant of the abdomen and is completely protected by the rib cage. It extends from the right fifth intercostal space in the midclavicular line down to the right costal margin and to the left as far as the left midclavicular line. It has the appearance of a wedge with the base to the right and measures about 10 cm in vertical span, 12 to 15 cm in thickness, and 15 to 20 cm in its greatest transverse diameter (6,7).

The superior, anterior, and lateral surfaces of the liver are smooth and almost completely covered by peritoneum, except for a small triangular area—the "bare area" below the diaphragm—which is surrounded by the reflections of the peritoneum forming the coronary ligaments. A thin layer of fibrous connective tissue, the Glisson capsule, surrounds the liver and extends into the parenchyma to form extensions that support arterial and biliary structures. Anteriorly, the falciform and round ligaments, which during fetal life conducted the left umbilical vein, connect the liver to the abdominal wall. Through the posterior surface of the liver at the base of the bare area runs the inferior vena cava, to which two to four hepatic veins connect. The fossa of the gallbladder and the round ligament separate the quadrate lobe from the right and left liver lobes, respectively. The fossa of the ductus venosus (i.e., the connection of the left umbilical vein to the inferior vena cava) and the inferior vena cava separate the caudate lobe from the left and right lobes of the liver. The horizontal portal fissure (or porta hepatis), which joins the upper ends of the gallbladder fossa and the groove of the round ligament, contains the branches of the hepatic artery, the portal vein, the hepatic nerve plexus, the hepatic ducts, and lymph vessels (6).

The liver is divided by deep grooves into two large lobes—the right (lateral to falciform ligament) and left (medial to falciform ligament)—and two smaller lobes, the caudate and quadrate lobes. This traditional division is only of topographical significance. Functionally, the division into eight segments based on either Couinaud or Bismuth segmental classification (7,8), which does not correspond to the anatomical division into lobes, is more important. Each segment is served by its own vascular inflow, outflow, and biliary drainage. The center of each segment contains branch of hepatic artery, portal vein, and bile duct, while hepatic veins are located in the periphery of each segment. This segmental division is of critical importance, particularly when dissecting small space-occupying lesions from these areas or when removing segments of liver for transplantation (9). Each segment can be resected without damaging those remaining. For the liver to remain viable, resections must be performed along vessels that define the peripheries of these segments, which mean that resection lines should be parallel to the hepatic veins. Anatomical and functional variations occur not infrequently, particularly in the right liver. Imaging studies, including magnetic resonance imaging or computed tomography, can provide detailed insight into the individual segmental anatomy (10).

HISTOLOGY

Structural Organization

The basic unit of the liver has been characterized in various different ways based on histologic landmarks, functional features, vascularization, bile flow, or susceptibility to pathologic processes. However, the most widely utilized models in histopathology are the classic (or Kiernan) lobule (11) and the acinus of Rappaport (12).

The classic hepatic lobule is based on histologic/structural landmarks and is demarcated along its perimeter by conceptual lines linking portal tracts (or actual interportal fibrous septa in some nonhuman species). Although the shape of the classic lobule is known to be highly variable, it is typically represented as a regular hexagon, with portal tracts located at its vertices and a central vein near its center. Pathologists commonly subdivide the classic lobule into three concentric regions: centrilobular or perivenular (around the central vein), midzonal, and periportal (near portal tracts and edges of lobules). These lobular regions reflect the normal gradient of various metabolic products and susceptibility to certain pathologic processes along the central vein–portal tract axis (Fig. 27.3).

The Rappaport acinus is an alternative subdivision of the liver into three zones, from the region closest to portal tracts (zone 1), an intermediate region (zone 2), to areas closest to central veins (zone 3). Unlike the lobular regions, acinar zones are curvilinear and converge toward portal tracts, forming zones that are based on the location of hepatocytes relative to the terminal portal circulation (but also hepatic artery circulation and bile flow, which show a similar distribution). While the expression of most metabolic markers and the distribution of some pathologic findings—such as steatosis—show a pattern most closely mapping to the classic lobular zones, some pathologic processes such as portal–central fibrosis or necrosis are best explained by the Rappaport acinus concept, with zones spanning from central vein to portal tracts (Fig. 27.3).

The absence of microanatomical structures along the perimeter of classic lobules in humans poses difficulties for the precise identification of its boundaries, delineation of lobular or acinar zones, and performance of zonal-based analysis of human liver tissue. The recent discovery that the classic lobular architecture of the liver follows a Voronoi diagram–like organization (13) (with central veins representing its originating points, or "seeds") has allowed us to more accurately delineate the boundaries of classic

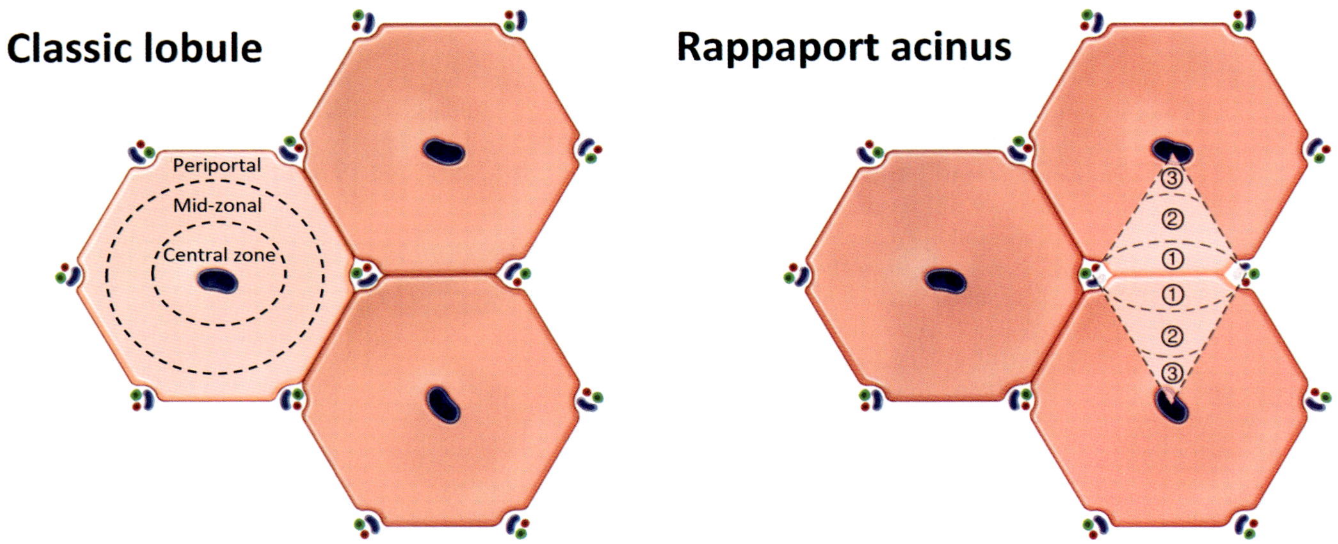

FIGURE 27.3 Traditional representation of a classic (Kiernan) lobule as a hexagon (**left**). Representation of the Rappaport acinus (**right**) occupying areas spanning two adjacent classic lobules. (Adapted from: Lau C, Kalantari B, Batts KP, et al. The Voronoi theory of the normal liver lobular architecture and its applicability in hepatic zonation. *Sci Rep.* 2021;11(1):9343, with permission.)

lobules and its zones. The newly described "Voronoi lobule" is particularly well suited for the age of computational pathology, as both lobulation and zonation can automatically be generated based on the recognition of central veins/perivenular areas using the mathematical principles behind Voronoi diagrams (Figs. 27.4 and 27.5). The Voronoi lobule is a more accurate representation of liver lobules in both human and porcine livers.

The terminal vascular branches, which bring substances for nutrition and metabolism into the acinus, run along the terminal bile ducts that drain the secretory products of the same acinus. These vessels form a vascular plexus around the bile ducts (14). Thus, as a result of the sinusoidal blood flow, structural, secretory, and functional unity is established in the acinus. The oxygen gradient, metabolic heterogeneity, and differential distribution of enzymes across the three zones of the acinus explain the zonal distribution of liver damage due to ischemia and toxic substances, with zone 3 being the most susceptible zone to ischemic and toxic injury (15,16). The periportal hepatocytes (Fig. 27.6) contain carbamoyl phosphate synthetase and glucose-6-phosphatase enzymes, whereas the predominant enzymes in perivenular hepatocytes are glutamine synthetase and NADPH-cytochrome (17). In clinical practice, one may be able to corroborate the suspicious offending agent with histologic zonal pattern of injury. Alcoholic liver disease and drug-induced injury commonly produce centrilobular or zone 3 hepatocyte injury, because the centrilobular area is populated by hepatocytes with a high level of cytochrome P-450 activity for drug metabolism (16). A long list of drugs, classically acetaminophen (15,16), can cause centrilobular hepatocyte coagulative necrosis, which may extend to midzonal areas resulting in geographic pattern of hepatocyte necrosis (Fig. 27.7). In rare instances, one may see exclusively midzonal necrosis, such as in furosemide toxicity; or periportal necrosis, such as in cocaine toxicity (15).

Hepatocytes

The hepatocytes are arranged in plates that are normally one-cell-layer thick in the adult and are separated by sinusoids along which blood flows from portal tracts to terminal hepatic venules (Fig. 27.8). Surrounding the terminal hepatic venules, the hepatocytes exhibit a more regular radial pattern. Away from the perivenular area, the liver cell plates are arranged less regularly without distinct radial arrangement. The hepatocytes in the periportal area are closely packed and smaller than other parenchymal cells with more intense nuclear staining and more basophilic cytoplasm (Fig. 27.9). The periportal area is also the regenerative compartment of the liver parenchyma. The hepatocytes bordering the portal tracts are joined together and form a distinct row called the limiting plate (Fig. 27.10). Destruction of this limiting plate by necroinflammation and/or apoptosis is a hallmark of chronic hepatitis (piecemeal necrosis or interface hepatitis) (Fig. 27.11).

In children up to 5 or 6 years of age, the liver cells are uniform and arranged in two-cell-thick plates (Fig. 27.12) (18). In adults, the presence of two-cell-thick plates, the formation of rosettes, hepatocyte buds, and the increase of mitotic activity indicate parenchymal regeneration.

The individual hepatocyte is a polygonal epithelial cell approximately 25 μm in diameter with a well-defined plasma membrane that is differentiated into three specialized regions or domains: basolateral (70% of total surface area), which faces the sinusoid; bile canalicular (15%),

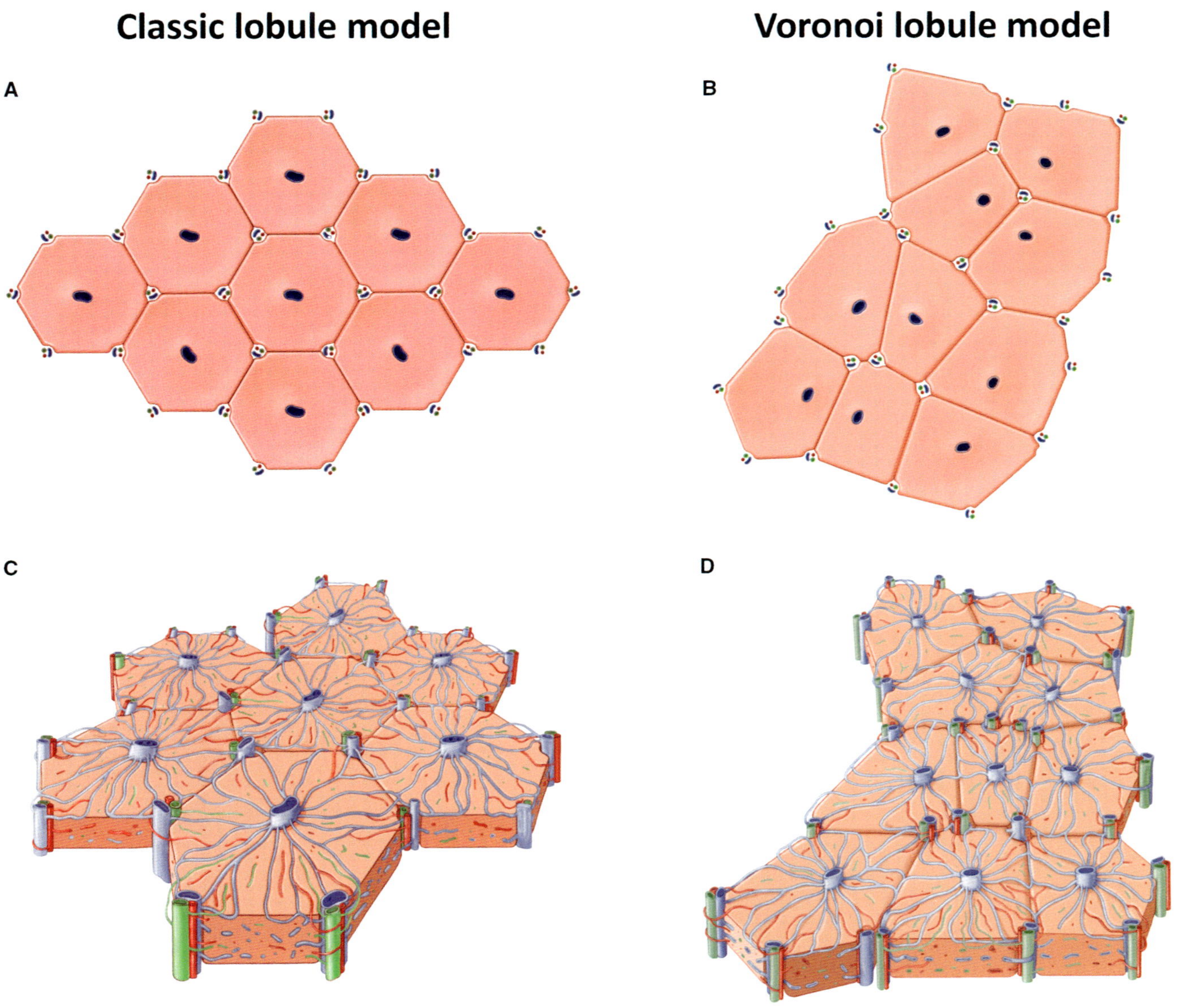

FIGURE 27.4 Comparison between the classic lobule (**A** and **C**) and Voronoi lobule (**B** and **D**). The classic lobules are represented as juxtaposed regular hexagons, while the Voronoi lobule is composed of convex polygons of various shapes, which are dictated by the position of central veins. (Adapted from: Lau C, Kalantari B, Batts KP, et al. The Voronoi theory of the normal liver lobular architecture and its applicability in hepatic zonation. *Sci Rep*. 2021;11(1):9343, with permission.)

bounding part of the intercellular space that constitutes the bile canaliculi; and lateral (15%), facing the rest of the intercellular space. Each domain has different molecular, chemical, and antigenic compositions and functions.

The nucleus is centrally located, round, and contains one or more nucleoli. At birth, all but a few hepatocytes are mononuclear. In adults, binucleate forms are not uncommon (up to 25% of cells), but mitotic activity is rare. Nuclei vary in size in the adult liver, and may contain 2, 4, 8, or more haploid chromosome sets, most are diploid (19). Some nuclei are larger than others, indicating more than two homologous sets of chromosomes, known as polyploidy, particularly in individuals over 60 years (Fig. 27.13). The significance of polyploidy is unknown and is usually more marked in the midzonal area. It may represent the mechanism to generate genetic diversity and permits adaptation of hepatocytes to xenobiotic or nutritional injury (20). Since cell size is proportional to cell ploidy, polyploidy does not provide an increased quantity of genetic material per unit volume of cytoplasm, that is, nuclear to cytoplasmic ratio remains low.

The cytoplasm of the hepatocytes is eosinophilic due to numerous mitochondria, and it also contains fine basophilic granules, which represent rough endoplasmic reticulum. The hepatocytes contain abundant cytoplasmic glycogen, which after proper fixation is stainable

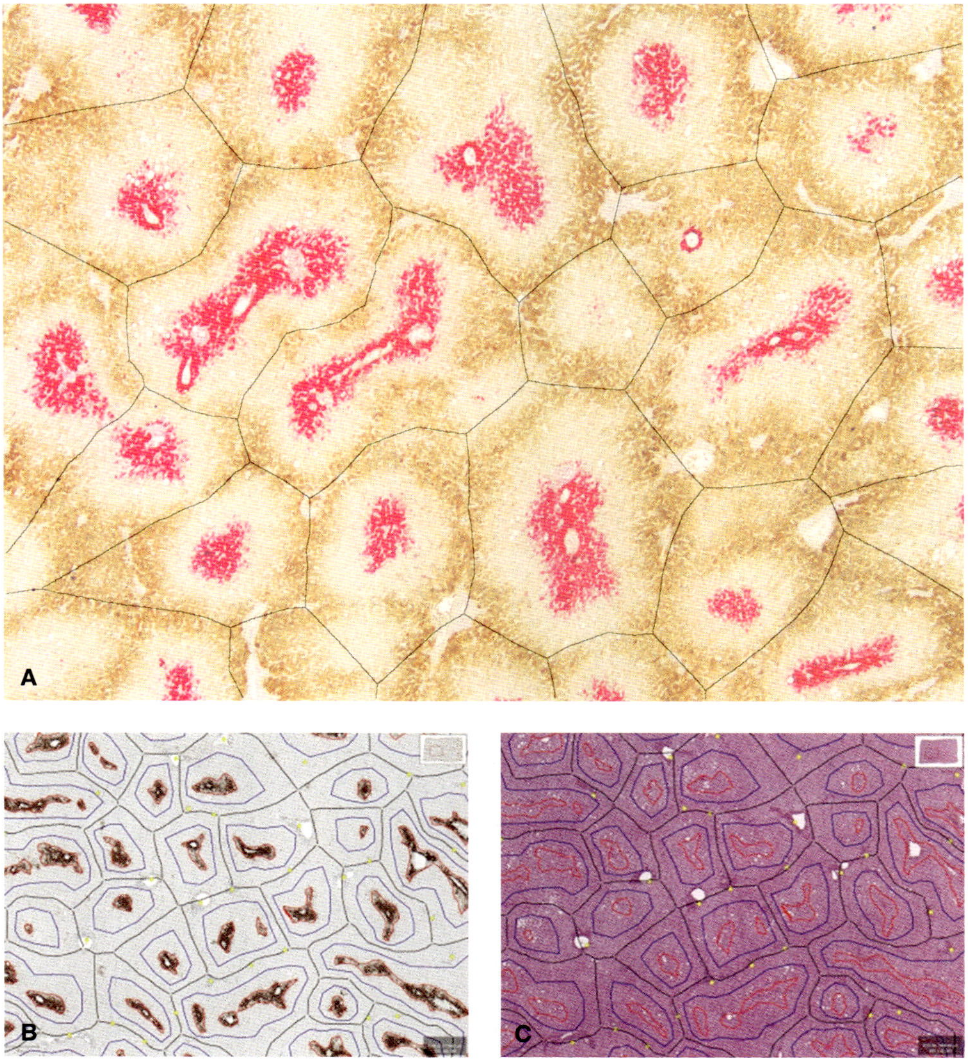

FIGURE 27.5 Voronoi lobule applied to the human liver. **A:** Computer-generated Voronoi lobulation accurately identifies the borders of lobules using centrilobular and periportal double immunostain markers glutamine synthetase (GS, *red*) and ASS1 (*brown*). Each lobule contains a central area (*red*), midzonal area (no stain), and periportal area (*brown stain*). **B:** Automated lobulation and zonation of a liver section based on the recognition of central zones (GS immunostain within *red lines*). *Blue lines* demarcate midzones and periportal areas and *black lines* demarcate the edges of Voronoi lobules. **C:** Lobulation/zonation algorithm transposed to the corresponding hematoxylin and eosin stain.

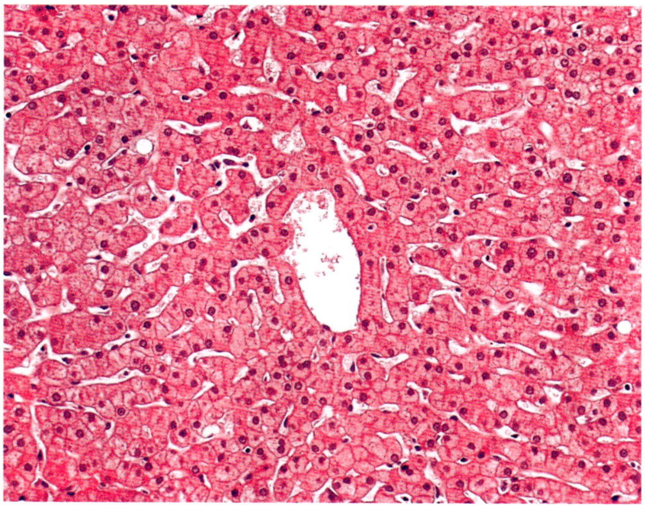

FIGURE 27.6 Terminal hepatic venule surrounded by converging hepatocyte plates and sinusoids.

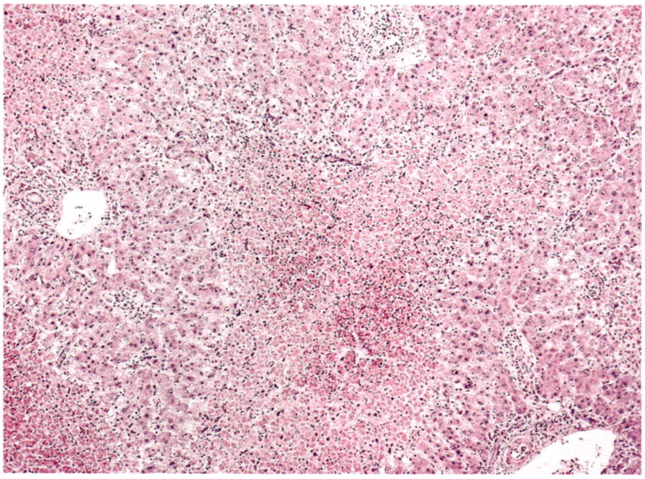

FIGURE 27.7 Centrilobular and midzonal geographic coagulative necrosis in acetaminophen-induced toxicity.

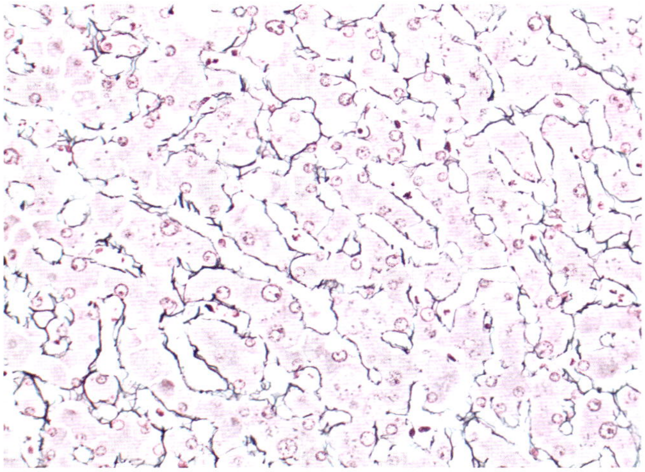

FIGURE 27.8 Normal adult liver parenchyma composed of one-cell-thick hepatocyte plates, each lined by reticulin fibers (reticulin stain).

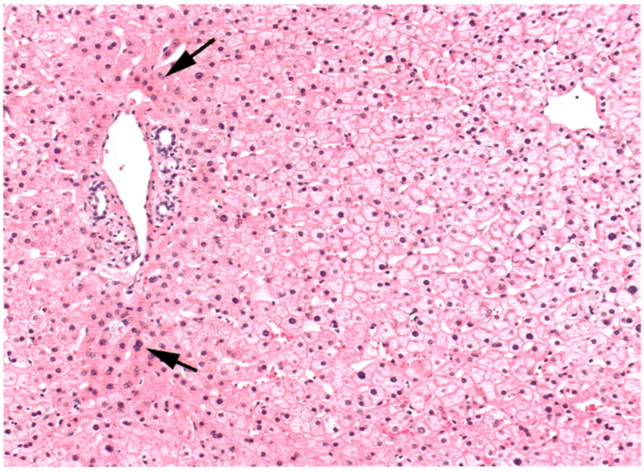

FIGURE 27.9 The periportal regenerative compartment of hepatic lobule contains smaller and more basophilic hepatocytes (*arrows*).

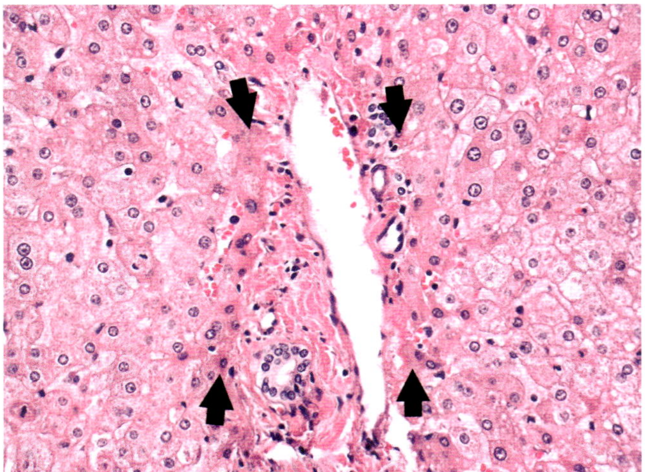

FIGURE 27.10 Normal portal tract with bile duct, hepatic arteriole, portal venule, and clearly defined limiting plate (*arrows*).

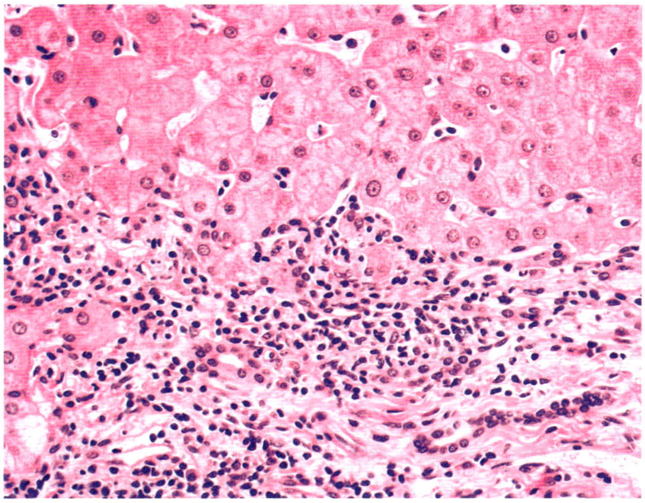

FIGURE 27.11 Interface hepatitis causing destruction of limiting plate in chronic viral hepatitis.

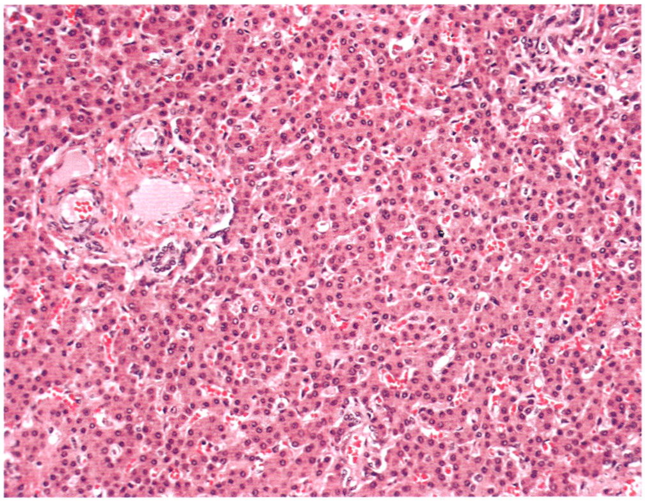

FIGURE 27.12 Liver of a child showing small uniform hepatocytes that are arranged in two-cell–thick plates.

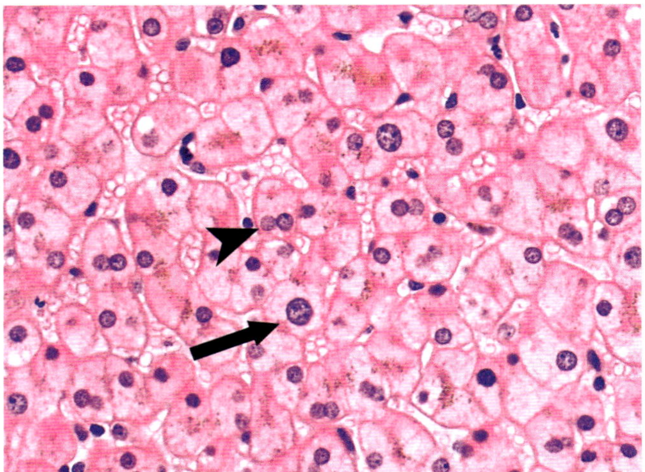

FIGURE 27.13 Liver of a 65-year-old patient showing significant polyploidy of hepatocyte nuclei (*arrow*) and binucleate forms (*arrowhead*).

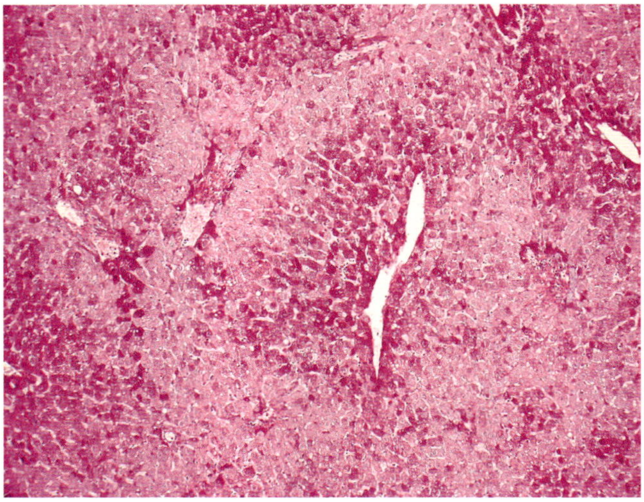

FIGURE 27.14 PAS reaction shows irregular distribution of cytoplasmic glycogen (darker color) in the hepatocytes.

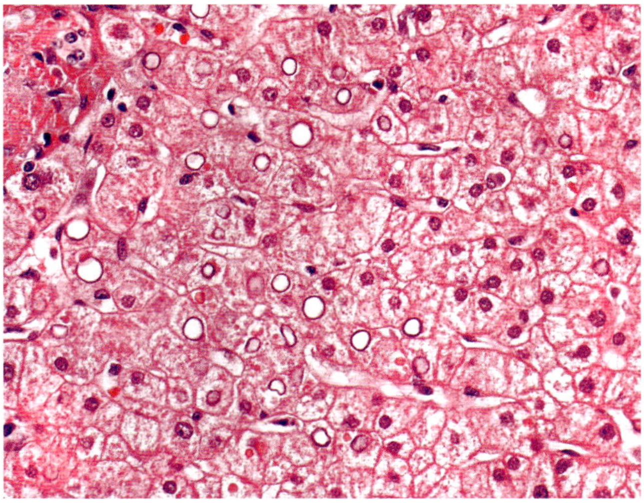

FIGURE 27.16 Glycogen accumulation in hepatocyte nuclei resulting in clear, empty appearance.

with periodic acid–Schiff (PAS) reagent (Fig. 27.14). On hematoxylin and eosin (H&E) preparations, glycogen gives a fine, reticulated, foamy appearance to the cytoplasm (Fig. 27.15). The quantity and distribution of cytoplasmic glycogen reflect diurnal and diet-related variations. An irregular distribution pattern may sometimes be found in biopsies and is not of diagnostic significance (Fig. 27.15). Glycogen accumulation in hepatocyte nuclei around portal tracts produces a vacuolated appearance and is common in adolescents and young adults (Fig. 27.16). In adults, such an appearance may be conspicuous in conditions such as glucose intolerance, diabetes mellitus, Wilson disease, and pancreatic carcinoma (21).

Acidophilic bodies (also known as apoptotic bodies) are nonspecific findings that represent an active form of programmed cell death. Isolated or rare acidophilic bodies represent normal turnover of hepatocytes (Fig. 27.17). Occasional focal necroses where chronic inflammatory cells replace a few necrotic hepatocytes are not unusual in otherwise apparently normal liver.

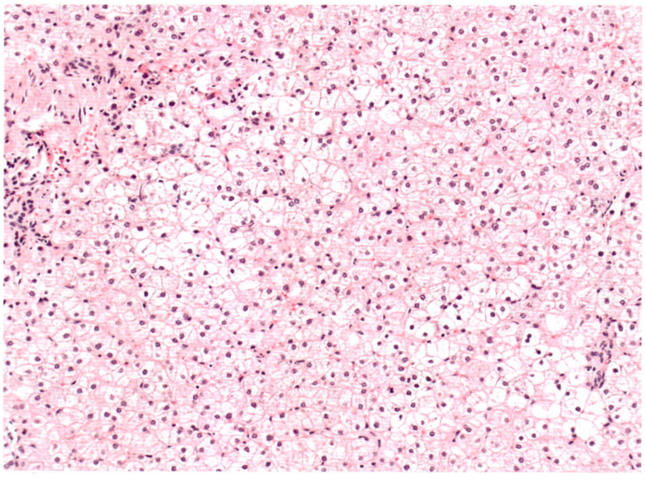

FIGURE 27.15 Irregular distribution of cytoplasmic glycogen, producing patchy collection of hepatocytes with clear cytoplasm, is of no clinical significance in the absence of other histopathologic findings.

Hepatocyte Regeneration

Liver is the only human organ that is capable of natural regeneration. Regeneration may be rapid as seen after partial hepatectomy (22). This is predominantly due to the hepatocytes reentering the cell cycle: quiescent cells stimulated by mediators including cytokines, move into a primed state (G0 to G1) when growth factors can stimulate DNA synthesis and cellular replication.

In the event of injury, regeneration is observed predominantly in the periportal hepatocytes in mild injury or throughout the hepatic parenchyma in severe injury. It is manifested by mitoses, multinucleation, and crowding of

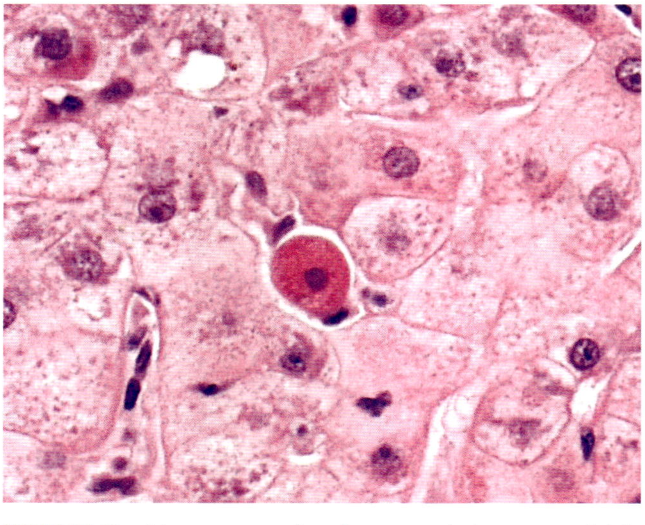

FIGURE 27.17 A hepatocyte undergoing apoptosis; hepatocyte shrinkage, nuclei fragmentation (karyorrhexis), and chromatin condensation (pyknosis).

the periportal cell plates by small, uniform, darkly stained, or basophilic hepatocytes. Liver cell plates in the periportal region become irregularly thickened and occasional hepatocellular rosettes may appear. Nuclear displacement to the sinusoidal pole with hyperchromasia is a cytologic indication of regenerative activity. All these changes impart a darkened periportal region under low magnification, which is often the only remarkable change in mild acute hepatitis or in residual hepatitis as a reaction to recent injury. In severe hepatocellular injury or hepatitis, the feature of regeneration may be overwhelmed by the amount of ballooning degeneration, cytolytic necrosis, apoptosis, bridging necrosis, and inflammatory infiltrate.

The degree of hepatocyte regeneration often correlates with the degree of necroinflammatory activity in chronic hepatitis, particularly when interface hepatitis is the predominant process.

If hepatocytes are extensively damaged, such as in massive hepatic necrosis, the conventional hepatocyte regeneration is impaired. Hepatocytes may be derived from bipotential progenitor cells, the so-called oval cells, residing in the canals of Hering (22). These cells can differentiate into either hepatocytes or cholangiocytes.

Hepatocyte Degeneration and Death

In general, there are two types of cell death that can be distinguished by morphologic features, apoptosis, and cytolytic necrosis, although it is likely that these are two ends of a spectrum with possible intermediate forms.

Apoptosis occurs at all stages during fetal growth and development of ductal plates and hepatoblasts. There is a good correlation between the proliferative and apoptotic activities in the ductal plate, depending on the remodeling process. Involution of the liver and neoplasia are also controlled by apoptosis, which is induced by transforming growth factor β1. In involution or regressing tumors, scattered apoptotic bodies, rather than massive cytolytic necrosis, are observed. In viral infections such as cytomegalovirus and herpes hepatitis, apoptosis has been proposed as a mechanism of cell death. Similarly, viral hepatitis C leads to increased apoptosis in hepatocytes (23). In viral hepatitis B and C, however, cell death may be mediated directly by the virus or by the host immune system through the release of cytokines such as tumor necrosis factor α to the infected cells (24,25). In normal liver tissue, although rare, individual apoptotic bodies may be seen, which suggests that apoptosis is a physiologic process in the liver. Apoptosis involves shrinkage, nuclear disassembly, and fragmentation of the cell into discrete bodies with intact plasma membranes, which are then rapidly phagocytosed by neighboring Kupffer cells (Fig. 27.18).

Cytolytic necrosis is manifested histologically by ballooning degeneration, and the hepatocytes become swollen and pale staining as the result mainly from dilatation of the endoplasmic reticulum (Fig. 27.19). This is a consequence of loss of mitochondrial function and resultant ATP depletion, leading to loss of ion homeostasis and plasma membrane integrity. The cytoplasm is partially rarefied, particularly along the cellular periphery, and the cytoplasmic remnants clump around the nucleus; cell membranes are frequently indistinct. Ballooned hepatocytes undergo lytic necrosis, which is not visible, but the occurrence can be inferred to small foci of stromal collapse that are accompanied by collections of lymphocytes and Kupffer cells, that is, spotty necrosis (Fig. 27.20). The degree of ballooning degeneration varies across the lobule, although classically the centrilobular region is the most severely affected. In addition to hepatocyte injury and regeneration, there are varying degrees of cholestasis, Kupffer cell activation, ductular reaction, and bile duct damage.

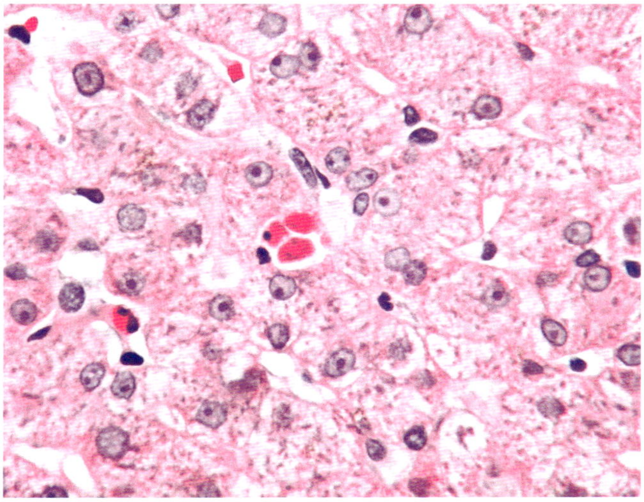

FIGURE 27.18 Acidophilic and apoptotic bodies.

Bile Canaliculi

The bile canaliculus is an intercellular space with a diameter of approximately 1 μm, formed by the apposition of the edges of gutter-like hemicanals on adjacent surfaces of two or

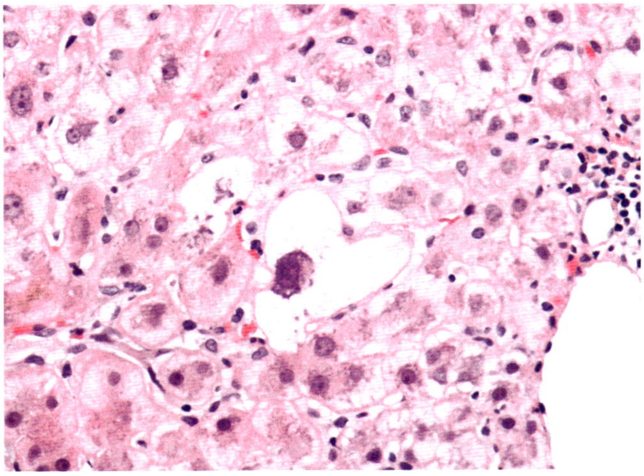

FIGURE 27.19 Ballooning degeneration of the hepatocytes.

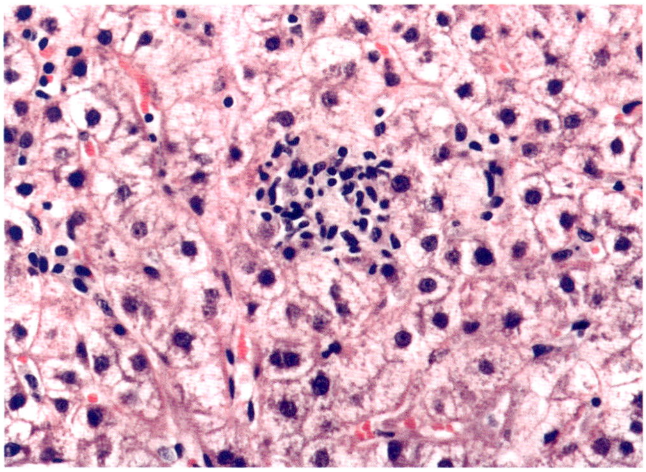

FIGURE 27.20 Focal collection of lymphocytes and macrophages surround hepatocytes with lytic necrosis.

three neighboring hepatocytes. Bile canaliculi are not readily recognized under the light microscope unless distended in conditions causing parenchymal cholestasis, which are accompanied by pseudoglandular formation of hepatocytes (Fig. 27.21). Bile canaliculi form a chicken wire–like network in the center of the hepatic plates and can be demonstrated immunohistochemically with polyclonal anticarcinoembryonic antigen (pCEA) or CD10 (Fig. 27.22) (26).

The canalicular membrane of the hepatocytes is host to several biliary proteins that are responsible for the formation and flow of bile in the liver. Inherited defect of these proteins leads to heterogeneous intrahepatic cholestatic syndromes, which are known as progressive familial intrahepatic cholestasis (27).

Sinusoidal Lining Cells

In normal liver biopsy specimens, the hepatic sinusoids are slit-like spaces that contain a few blood cells. The

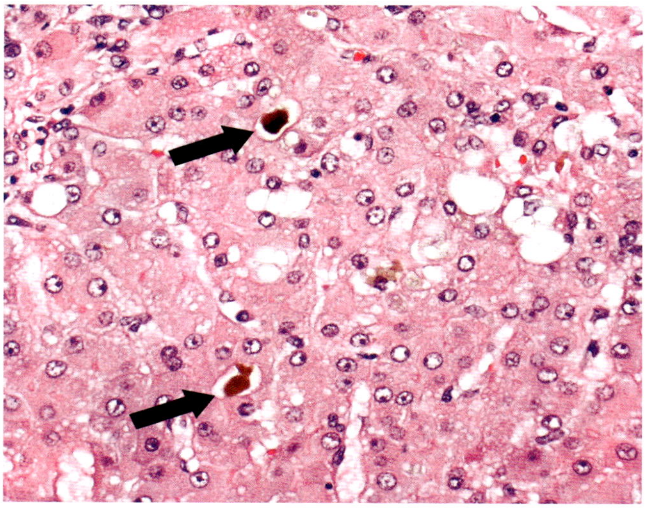

FIGURE 27.21 Canalicular bile thrombi (*arrows*) in zone 3 of the acinus.

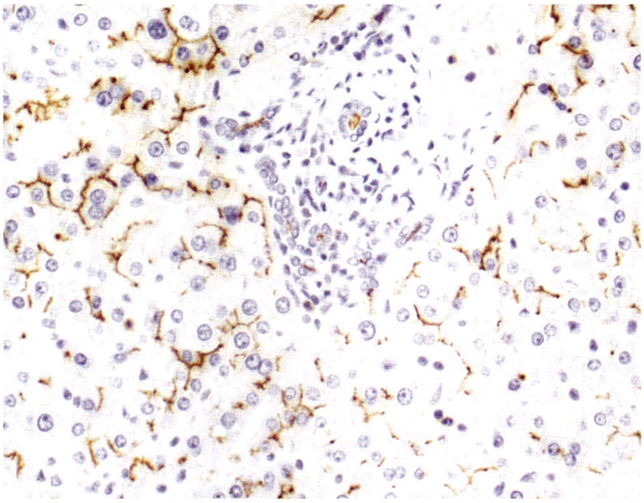

FIGURE 27.22 CD10 immunostaining delineates the twig-like structures of bile canaliculi, as well as the lumen of bile ducts and ductules.

periportal sinusoids are more tortuous than the perivenular ones. Hepatic sinusoids separate cords of hepatocytes and are lined by sinusoidal lining cells supported by reticulin fibers (Fig. 27.8). Sinusoidal lining cells, which include endothelial and Kupffer cells, constitute a coordinated defense system (28,29). They are rather inconspicuous in normal biopsy specimens. The endothelial cells have thin, indistinct cytoplasm and small, elongated, darkly stained nuclei without nucleoli. The sieve-like plates of the endothelial cytoplasm and the absence of a structurally defined basement membrane (in contrast to capillaries) facilitate exchange between blood and hepatocytes.

The Kupffer cells have a bean-shaped nucleus and plump cytoplasm with star-shaped extensions. They are more numerous near the portal tracts. They belong to the mononuclear phagocytic system and are derived in part from the bone marrow. The Kupffer cells represent up to 90% of the resident macrophages in the liver and account for approximately 15% of the total liver cell population. They contain vacuoles and, particularly in the diseased liver, many PAS diastase (PAS-D) resistant lysosomes and phagosomes, as well as acid-fast granular aggregates of ceroid pigment or bile (Figs. 27.23 and 27.24). These cells respond actively to many types of injury by proliferation and enlargement.

Between the endothelial cells and the hepatocytes lies the space of Disse, a zone of rapid intercellular exchange. It contains plasma, scanty connective tissue that constitutes the normal framework of the liver, and perisinusoidal cells such as hepatic stellate cells (Ito cells, interstitial fat-storing cells, or hepatic lipocytes) and pit cells (29). The connective tissue fibers along the sinusoids are predominantly collagen type III, which stains black in silver impregnations (reticulin) and forms a regular network radiating from the center of the lobules. Elastic fibers and basement membranes are absent from normal sinusoids. The space of Disse is not discernible in well-fixed, normal liver biopsy

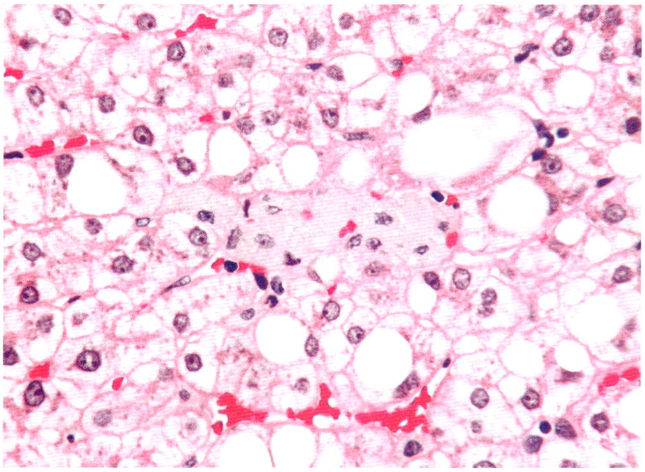

FIGURE 27.23 Collection of ceroid-containing macrophages signifying a focus of recent hepatocyte dropout.

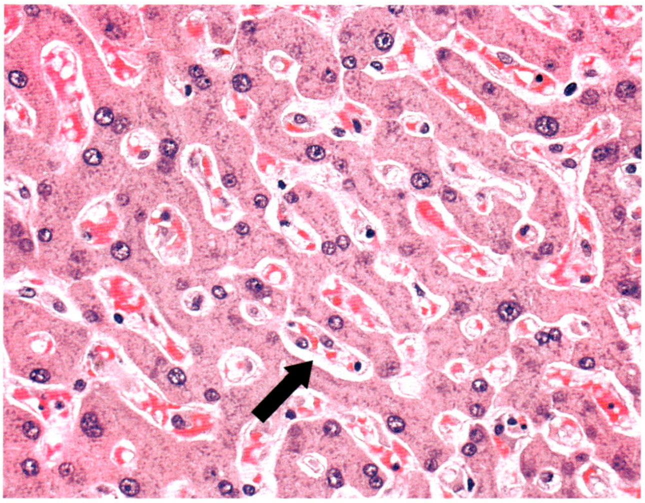

FIGURE 27.25 Autopsy liver specimen exhibiting dilatation of sinusoids and space of Disse (*arrow*) with prominent sinusoidal lining cells including endothelial and Kupffer cells.

material; but in postmortem liver, the hepatocytes shrink, pericellular edema develops, and the space becomes more conspicuous (Fig. 27.25). Pathologically, extravasation of red blood cells into the space of Disse occurs in hepatic vein outflow obstruction.

On light microscopy of normal liver, hepatic stellate cells are quiescent and difficult to differentiate from sinusoidal lining cells. They are modified resting fibroblasts that store fat and vitamin A and produce hepatocyte growth factor and collagen (30). Hepatic stellate cells are the main fibrogenic cell type in injured liver (31). When loaded with fat, such as in hypervitaminosis A (32), they may be recognized due to cytoplasmic fat droplets of rather uniform size with scalloping of the elongated nucleus (Fig. 27.26). Hepatic stellate cells are highly responsive to stimuli released during inflammation, such as oxidative stress and proinflammatory cytokines that promote their transdifferentiation into myofibroblasts. When activated, these cells contain stainable desmin and actin in their cytoplasm, justifying their designation as myofibroblasts.

Pit cells have not been characterized by light microscopy. Under the electron microscope, they have neurosecretory-like electron-dense granules and rod-cored vesicles. However, recent evidence indicates that pit cells are not endocrine cells but correspond to large granular lymphocytes and have natural killer cell activity (33,34).

Occasional inflammatory cells, lymphocytes, or polymorphonuclear leukocytes may be present in the hepatic sinusoids. During the first few weeks after birth, the presence of foci of extramedullary hematopoietic cells in the sinusoids and wall of terminal hepatic venules is a normal feature.

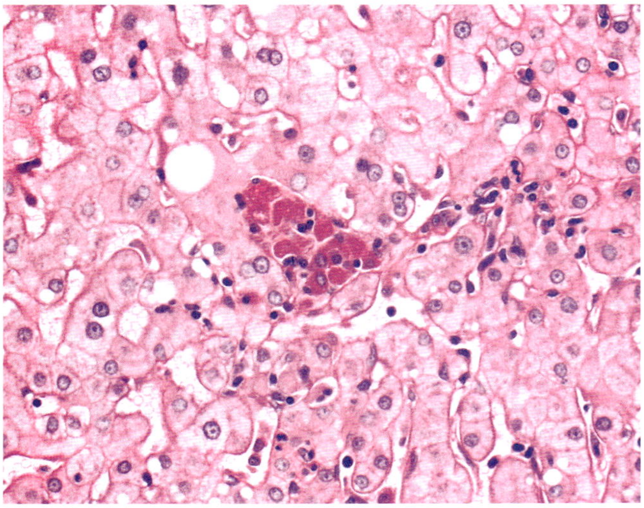

FIGURE 27.24 Ceroid pigment in macrophages as an indicator of previous hepatocyte injury/necrosis (PAS-D).

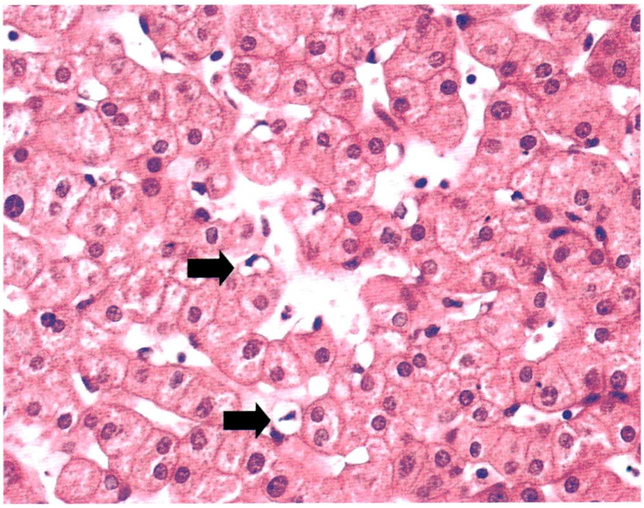

FIGURE 27.26 Prominent hepatic stellate cells (Ito cells) in the liver biopsy specimen of a patient with hypervitaminosis A. The nuclei of the hepatic stellate cells are scalloped (*arrows*) due to fat droplets in cytoplasm.

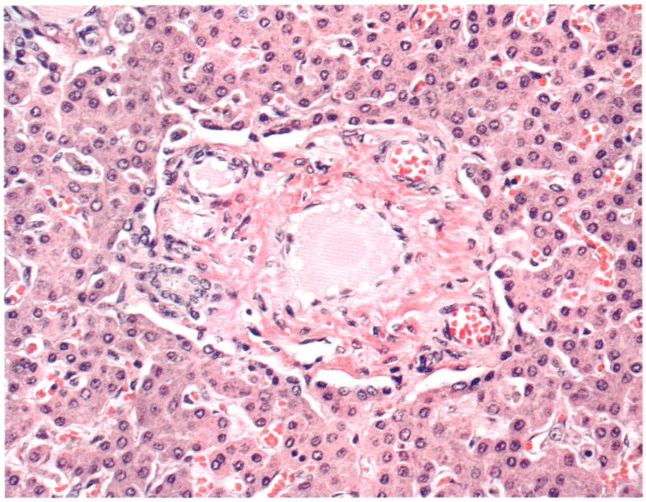

FIGURE 27.27 Normal portal tract in a newborn with bile ducts and their corresponding hepatic arteries of approximately the same diameter.

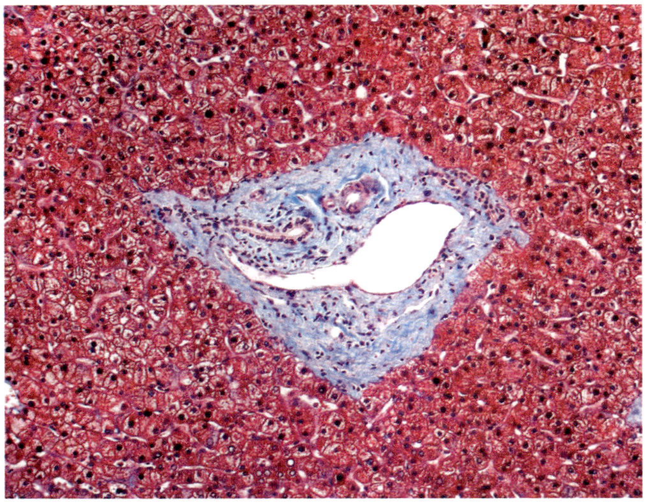

FIGURE 27.28 The connective tissue of portal tracts consists mainly collagen type I, which appears as thick, deep blue fibers on trichrome stain.

Portal Tracts

Each portal tract contains a bile duct and several bile ductules, a hepatic artery branch, a portal vein branch, and lymphatic channels embedded in connective tissue (Fig. 27.27). The amount of connective tissue and the size of the intraportal structures depend on the size of the portal tract. Nerve fibers, both sympathetic and parasympathetic for innervation of blood vessels and bile ducts, can be seen in large portal tracts. The larger portal tracts are round or triangular, the smaller ones are triangular or branching, and the smallest terminal divisions are round or oval. The size of a portal tract is approximately three to four times the diameter of the hepatic artery branch.

The portal tracts normally contain a few lymphocytes, macrophages, and mast cells, but no polymorphonuclear leukocytes or plasma cells. The number of inflammatory cells increases with age. However, their density varies from one portal tract to the next.

The portal tracts also contain portal fibroblasts. Portal fibroblasts and hepatic stellate cells are profibrogenic cells and the main resident of mesenchymal cells in normal liver. The connective tissue of the portal tracts consists mainly of collagen type I, which is seen as thick, deep blue fibers on the trichrome stain (Fig. 27.28). Newly formed collagen type III appears as fine, light blue fibers. In the subcapsular region of the liver, large portal tracts are often encountered, containing more and denser connective tissue (Fig. 27.29). Irregular extensions of fibrous tissue from the Glisson capsule into the parenchyma, sometimes connecting adjacent portal tracts, must not be interpreted as bridging fibrous septa or cirrhosis in wedge or superficial biopsy specimens of subcapsular parenchyma (Fig. 27.30) (35).

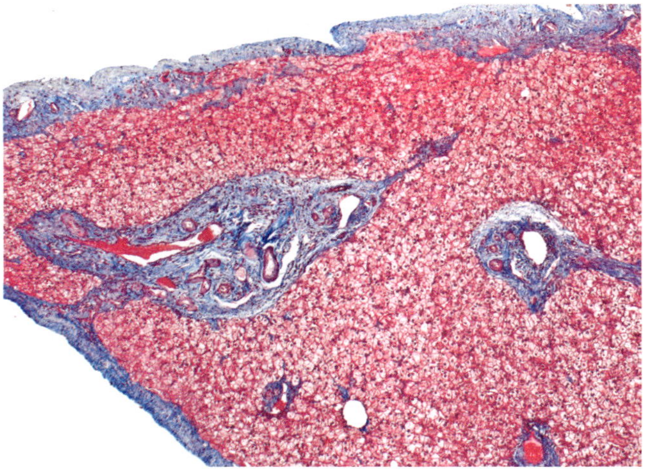

FIGURE 27.29 Subcapsular liver parenchyma containing large portal tracts (trichrome stain).

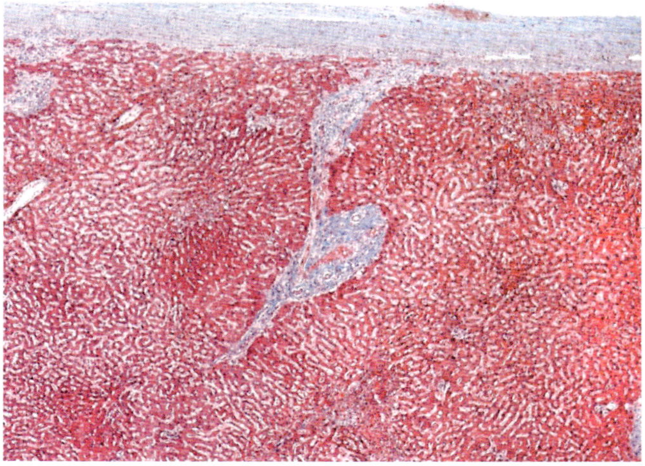

FIGURE 27.30 Subcapsular liver parenchyma with extension of the Glisson capsule into the parenchyma which may be mistaken for bridging fibrosis (trichrome stain).

Blood Supply and Drainage

The liver is nourished by a dual blood supply, approximately 75% via the portal vein and the remainder via the hepatic arteries. The portal vein carries nutrient-rich venous blood from the alimentary tract including the pancreas, whereas the hepatic artery supplies oxygen-rich arterial blood from the celiac axis for liver and biliary tree survival.

The right, middle, and left hepatic veins provide venous outflow. The right hepatic vein drains the right lobe, the middle hepatic vein drains primarily the middle portion of the left lobe and a variable portion of the right, and the left hepatic vein provides the principal drainage of the left lateral lobe. The middle and the left hepatic vein often merge to form a common trunk before entering the vena cava. In addition, there are short venous segments that drain the posterior surface of the liver directly into the inferior vena cava.

Portal Vein

Portal veins are the largest vessels in portal tracts and produce venules that empty into periportal sinusoids (Fig. 27.31). The portal vein may be absent in up to 30% of portal tracts without any clinical significance. An increase in profiles of the portal veins is commonly seen in cirrhotic liver as the result of portal hypertension.

Thrombosis of portal vein occurs secondary to cirrhosis, tumor in the hepatic hilum or pancreas, or hypercoagulable states (36,37). Portal vein thrombosis in noncirrhotic liver often causes lobar atrophy of the liver, frequently the left lobe, and portal hypertension with venous collaterals (cavernous transformation). In portal vein thrombosis, the portal vein may be absent or shows recanalization.

Obstruction confined to the small portal veins rarely causes portal hypertension, but there are conditions such as porto-sinusoidal vascular disease (PSVD) (previously termed idiopathic noncirrhotic portal hypertension, noncirrhotic portal fibrosis, hepatoportal sclerosis, or obliterative portal venopathy), congenital hepatic fibrosis, and chronic schistosomiasis that may cause clinically significant portal hypertension in absence of cirrhosis (38,39). In PSVD, the portal vein wall is thickened and hyalinized, resulting in narrowing and occasional luminal obliteration. This may alternate with some portal vein branches that are

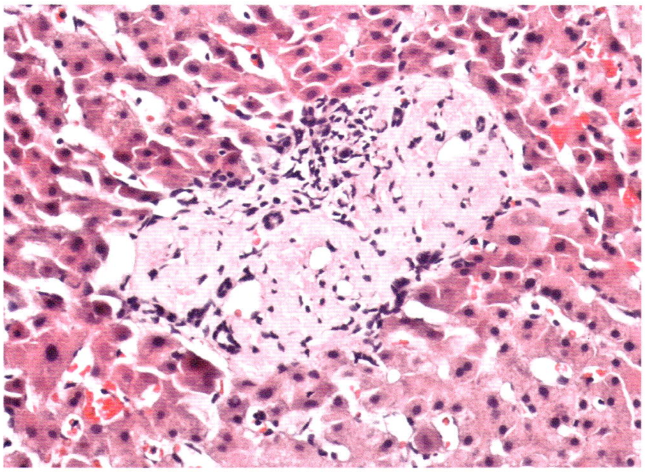

FIGURE 27.32 Luminal obliteration of portal vein in porto-sinusoidal vascular disease.

dilated and herniate into the hepatic lobule (Fig. 27.32) (38,39).

Hepatic Vein

The intrahepatic course of the valveless hepatic veins, which are embedded in a thin sheath of connective tissue, is straight to the inferior vena cava. There is a defined spatial relationship between the terminal hepatic venules and the branches of the portal vein and hepatic artery in the portal tracts, which interdigitate but do not directly connect in the three-dimensional space.

The smaller branches (or sublobular veins) and the smallest efferent veins (or terminal hepatic venules) are in direct contact with the hepatic parenchyma. The terminal hepatic venules have a very thin wall lined by endothelial cells, which is readily demonstrable after staining with trichrome for collagen or Victoria blue for elastic fibers, but they do not have an adventitia around their wall (Fig. 27.33).

Thickening of the wall of terminal hepatic venules is often part of pericellular fibrosis and central hyaline sclerosis in

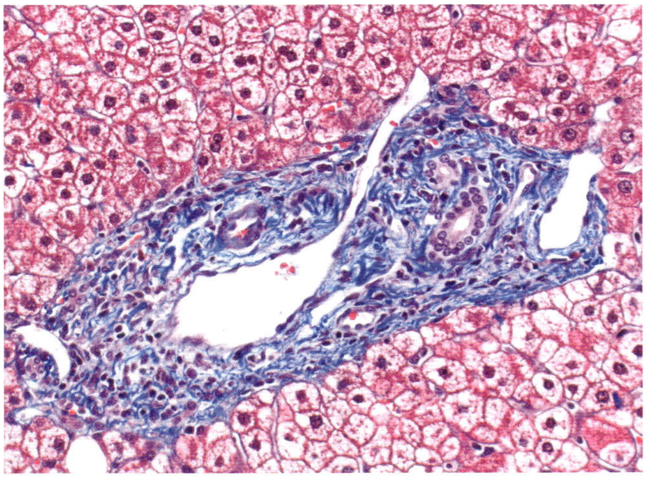

FIGURE 27.31 Portal venule empties into the periportal sinusoids (trichrome stain).

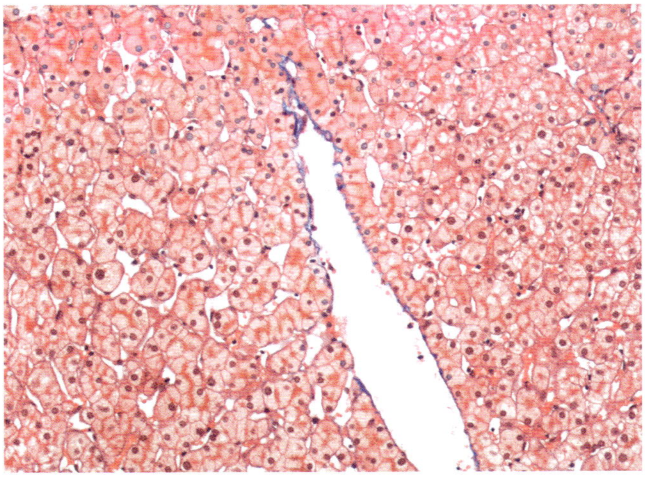

FIGURE 27.33 Normal hepatic venule has a thin fibrous wall (trichrome stain).

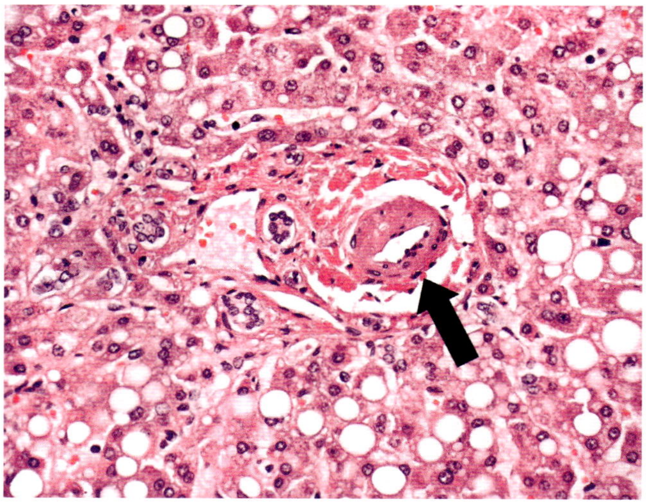

FIGURE 27.34 Thickened hepatic artery in an older individual (*arrow*).

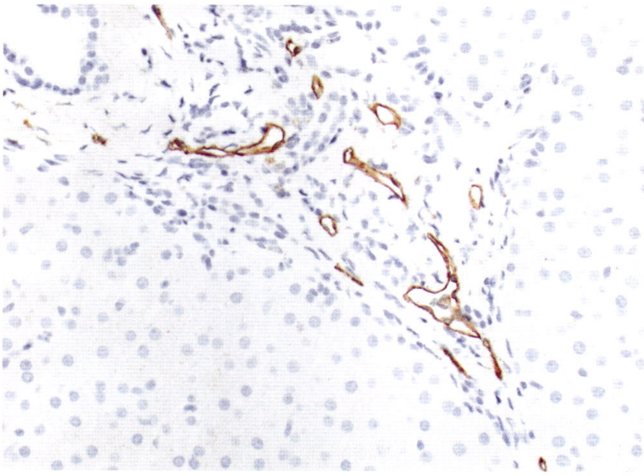

FIGURE 27.35 D2-40 immunostain highlights lymphatic channels in a cirrhotic liver.

alcoholic liver disease. In patients with a history of fatty liver disease, the findings of perivenular and pericellular fibrosis confirm previous episodes of steatohepatitis. Embedded small clusters of hepatocytes in the wall of hepatic venules are usually seen in liver with significant or higher stage of fibrosis. It should be noted, however, that mild perivenular fibrosis may be seen focally in apparently normal individuals, as well as in children up to 2 years of age.

Hepatic Artery

The hepatic artery branches are intimately related to the corresponding portal veins. They may show thickening and hyalinization of the wall in older individuals or more significantly in hypertensive individuals, although these changes are usually milder than in other organs (Fig. 27.34). The terminal hepatic arterioles regulate the parenchymal blood supply with their muscular sphincter, whereas the portal venous supply is controlled by mesenteric venous blood flow.

Insufficient arterial blood supply can significantly impact liver survival, such as in hepatic artery obstruction caused by thrombosis, arteritis, surgical ligation, or chemoembolization, resulting in biliary tree and parenchymal infarction (40). However, arteritis involving small arteries is seldom symptomatic because of the compensatory blood supply from the portal vein. Arteritis may also cause obliteration of adjacent portal vein, which then leads to nodular regenerative hyperplasia and eventually portal hypertension (41,42).

Lymphatics

The liver represents the largest single source of lymph in the body, producing 15% to 20% of the overall total volume and 25% to 50% of the thoracic duct flow (43). Furthermore, hepatic lymph has an unusually high protein content, about 85% to 95% of that in plasma, and a high content of lymphocytes. Most of the hepatic lymphatics leave the liver at the porta hepatis and drain into hepatic nodes along the hepatic artery and into the celiac nodes. Other important efferent routes are via the falciform ligament and the superior epigastric vessels to the parasternal nodes, from the bare area to posterior mediastinal nodes, and from the visceral surface to the left gastric nodes.

The capsule and the stroma of the liver are rich in lymphatic structures. The lymphatic plexus found in the capsule forms anastomoses with the intrahepatic lymphatics. The significance of these anastomoses is evident when hepatic venous pressure is increased, resulting in exudation of excess lymph from the capsular plexus that forms protein-rich ascitic fluid. The intrahepatic lymphatic system exists as a fine, valved plexus associated with branches of hepatic artery in portal tracts. Hepatic lymph is most likely formed in the interstitial space of Disse and the lymphatic channels in portal tracts drain the space of Disse. The lymph flows in the same direction as the bile, opposite to that of the blood. Lymphatic channels are readily identified in the fibrous bands, portal tracts, and the capsule of cirrhotic liver, but are less evident in the normal liver. Although rarely needed, immunohistochemical stain for D2-40 may highlight lymphatic channels in portal tracts (Fig. 27.35) (44).

Bile Ducts

Bile is formed in hepatocytes, steadily secreted into bile canaliculi, canals of Hering, and then into the intra- and extrahepatic bile ducts. The extrahepatic biliary tract consists of a gallbladder that ends in the cystic duct. The cystic duct joins the common hepatic duct to form the common bile duct, which enters the second portion of the duodenum through its muscular structure, the sphincter of Oddi.

Bile ducts accompany the hepatic artery and portal vein while coursing through the liver. They are nourished by the hepatic arteries via complex peribiliary plexus of capillaries, supplying all structures within the portal tracts. In transplanted liver, in the absence of anastomosing plexus of capillaries of the native liver, survival of the biliary tree

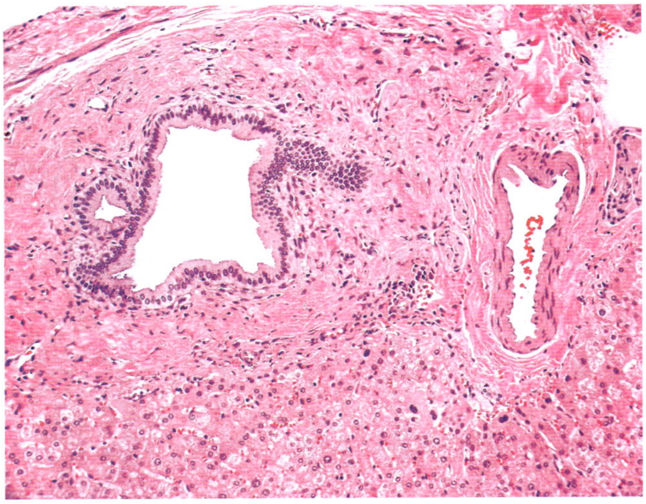

FIGURE 27.36 A large portal tract containing an artery and a bile duct lined by columnar epithelial cells.

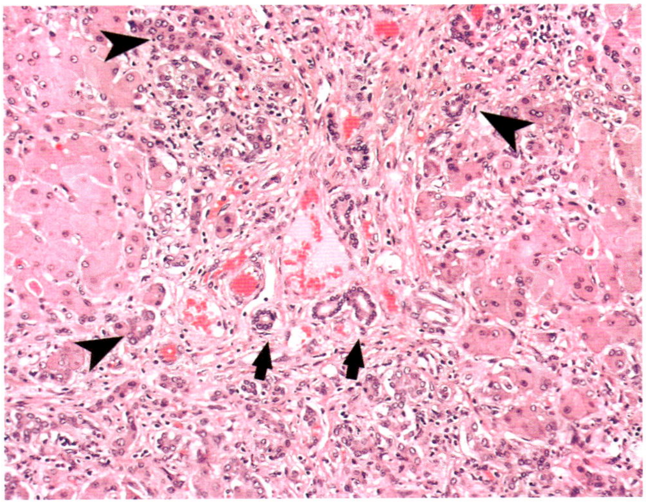

FIGURE 27.37 Bile ductules (*arrowheads*) are located in the peripheral zone of the portal tract and are smaller than the bile ducts (*arrows*).

relies solely on arterial supply from an intact hepatic artery. Thrombosis to hepatic artery leads to ischemia of the bile ducts, bile leaks, and parenchymal necrosis.

The larger intrahepatic or septal bile ducts are lined by tall columnar epithelial cells measuring about 10 mm in diameter with basally situated, pale, oval nuclei and light eosinophilic cytoplasm (Fig. 27.36). They have an internal diameter greater than 100 μm and a distinct basement membrane stainable with PAS-D. Lymphocytes may occasionally be present within the lining epithelium. The larger bile ducts are located in the central part of the portal tracts and have more periductal fibrous tissue than the smaller ones. The collagen fibers are arranged in an irregular and circumferential—but not concentric—manner as may be seen in chronic biliary tract diseases such as primary sclerosing cholangitis and as a sequel of chronic cholecystitis.

The smaller or interlobular bile ducts are lined by cuboidal or low columnar epithelium (Fig. 27.27). They have a basement membrane and a small amount of periductal connective tissue. One or more interlobular ducts may be present in a portal tract. Bile ducts are always accompanied by a hepatic artery, which has approximately the same diameter as the bile duct (external caliber ratio of bile duct to artery is 0.7:0.8) (45). In normal liver, the absence of bile duct and hepatic artery should not exceed 10% of the portal tracts. The bile ducts are connected to the bile canaliculi by bile ductules and canals of Hering.

Bile ductules are located in the peripheral zone of the portal tracts and are smaller (lumen of less than 20 μm) than the interlobular bile ducts (Fig. 27.37) (46). Bile ductules link canals of Hering and the interlobular bile ducts. Bile ductules may be observed at the edge of the portal tract stroma or may transverse the limiting plate, in which case it will have an "intralobular" as well as an "intraportal" segment. They have a basement membrane, are lined by cuboidal cholangiocytes, and are accompanied by a portal vein but not by a hepatic artery branch. Canals of Hering, which are the physiologic link between hepatocyte canaliculi and the biliary tree, are not discernible on routine sections of normal liver. They are lined partly by cholangiocytes and partly by hepatocytes. Canals of Hering can be demonstrated by staining for high–molecular-weight biliary cytokeratins (CK7, CK19), which are prominent in all cells of ductal origin (Fig. 27.38) (47). Isolated progenitor cells (equivalent to oval cells in rodent models, not visible under light microscopy examination) with CK7, CK19, and NCAM/CD56 immunophenotypes may be identified in canals of Hering (48).

In diseased liver, a unifying term "ductular reaction" is used for proliferative reaction of ductular phenotype, encompassing proliferating bile ductules, ductular

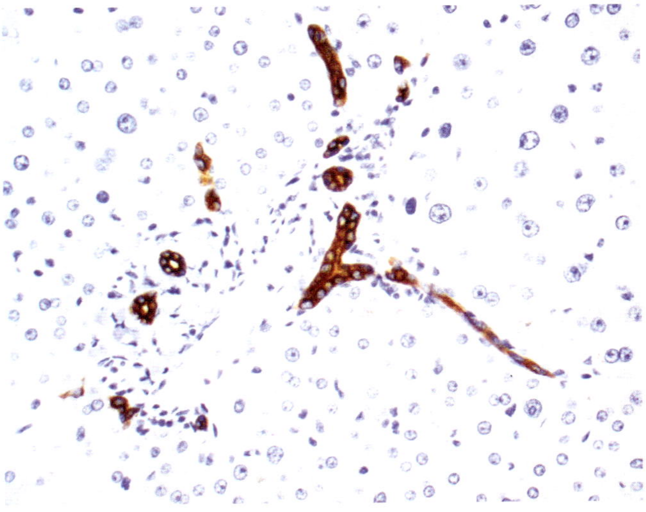

FIGURE 27.38 CK7 immunostaining of bile duct, bile ductules, and canals of Hering.

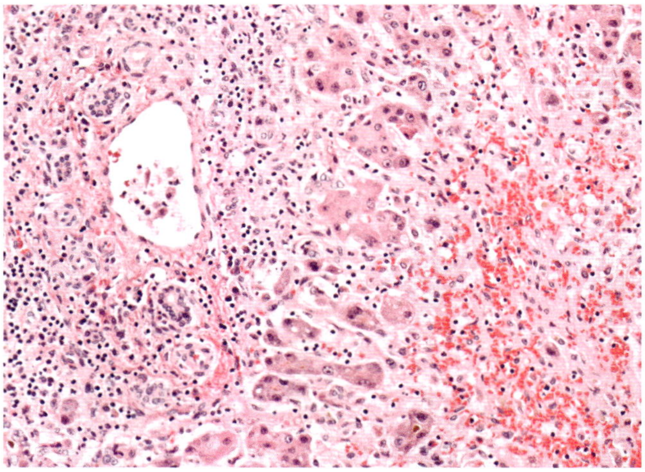

FIGURE 27.39 Ductular hepatocytes in acinar zone 1 in massive hepatic necrosis.

hepatocytes, intermediate hepatobiliary cells, and cells possibly deriving from circulation including from the bone marrow (46). Ductular reaction occurs in a variety of chronic liver diseases and can be so extensive as to raise the question of adenocarcinoma. In chronic liver diseases, ductular reaction involves not only ductular cells and stem cells, but also inflammatory cells, stroma, and other structures leading to progressive fibrosis (49,50).

Proliferating bile ductules demonstrate structural features typical of cholangiocytes, such as basement membrane formation and lumen. Proliferating bile ductules are invariably accompanied by inflammatory cell infiltrates (predominantly neutrophils). Proliferating bile ductules are typically found in large duct obstruction or chronic cholestatic diseases, such as primary biliary cholangitis and primary sclerosing cholangitis. Proliferating bile ductules can be differentiated from ductular hepatocytes that lack basement membrane and are seen abundantly in zone 1 of the liver acini in massive hepatic necrosis (Fig. 27.39).

Intermediate hepatobiliary cells (often referred to as progenitor/stem cells or oval cells) show dual characteristics of both hepatocytes and cholangiocytes, including simultaneous expression of biliary antigens (CK7, CK19, and OV-6) and hepatocyte antigens (HepPar1 and canalicular staining for polyclonal CEA), and structural features such as basement membrane formation typical of cholangiocytes and canalicular membranes typical of hepatocytes (48). Intermediate hepatobiliary cells and ductular hepatocytes are frequently found in regenerating liver after submassive or massive necrosis (51).

Nerve Supply and Innervation

The liver is innervated by two separate, but intercommunicating plexuses around the hepatic artery and portal vein (52). They include parasympathetic fibers from both vagus nerves and sympathetic fibers, which receive their preganglionic connections from spinal segments T7 to T10. The hilar plexuses also include afferent visceral and phrenic fibers. Besides their presence around vascular structures in portal tracts, nerve fibers (mostly sympathetic) are present in the parenchyma along the sinusoids. Release of neurotransmitters from the intrasinusoidal fibers modulates hepatocyte and perisinusoidal cell function. It controls in part carbohydrate and lipid metabolism and induces contraction of perisinusoidal cells, thereby regulating intrasinusoidal blood flow (26). However, neural mechanisms may have only a minor regulatory role because limited reinnervation in liver allografts does not seem to impair their function. Denervation probably explains the impaired normal response of the liver to ischemia and sinusoidal dilatation in liver allografts (53), and impaired metabolic function in cirrhosis (54).

EXTRACELLULAR MATRIX

Both interstitial and basement membrane collagens are present in the liver and play an important role, not only as structural elements but also in hepatic function. Collagen I—the main component of the dense, birefringent connective tissue fibers—is seen mainly in portal tracts and walls of hepatic veins and rarely in the normal parenchyma, whereas collagens III is present along the sinusoids. Collagen I can be demonstrated with connective tissue stains; collagen III can be seen with silver impregnation for reticulin. Collagen II, characteristic of cartilage, is absent from the liver. Collagens IV and V, the basement membrane collagens, and laminin are seen in the basement membrane of vessels, bile ducts, and bile ductules, but (except for some collagen IV) not along the sinusoids of normal human liver. Distribution of elastic fibers in the liver, as demonstrated by orcein, resorcin, or Victoria blue stains, seems to follow that of collagen I. Fibronectin, an extracellular matrix glycoprotein, is present diffusely along the sinusoidal surface of hepatocytes and in portal tracts together with the other collagens. All components of the extracellular matrix are visualized best by immunohistochemical staining using specific antibodies.

AGING CHANGES

There are several changes in the liver related to aging. These changes are more commonly seen in individuals 60 years of age and older. There is increased variation of the size of the hepatocytes and their nuclei, similar to that seen in patients on methotrexate, due to increased polyploid cells (19). With aging, the liver cell cords become atrophic and there may be apparent dilatation of sinusoids

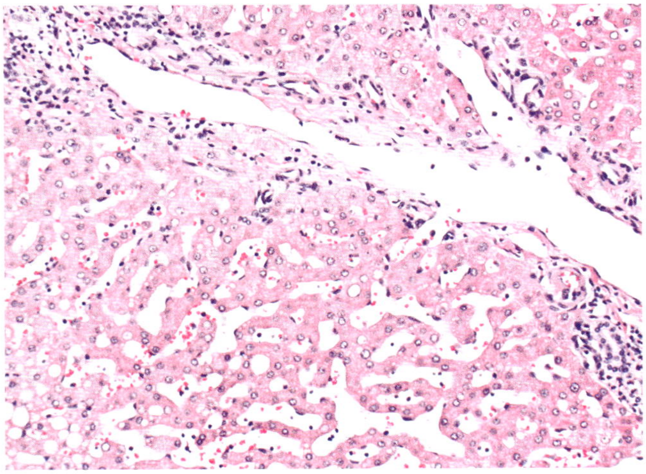

FIGURE 27.40 Hepatocyte atrophy causing sinusoidal dilatation in older individuals.

(Fig. 27.40). More abundant lipofuscin deposition is present in the centrilobular hepatocytes (Fig. 27.41), and sometimes there are some iron pigments in the periportal hepatocytes. The portal tracts contain denser collagen and may exhibit a higher number of mononuclear inflammatory cells than in younger subjects. The arteries may have thickened walls (Fig. 27.34), even in normotensive individuals. These histologic changes of aging need to be kept in mind because they are often present in viable donors for which frozen sections are requested and should not be interpreted as pathologic. These aging-related findings are accompanied by alteration in the metabolic function of the liver, including the metabolism of toxins and drugs, and therefore may increase the susceptibility of the liver to hypovolemia and drug-induced injury, decrease its capacity for regeneration and for shorter post transplantation survival (55).

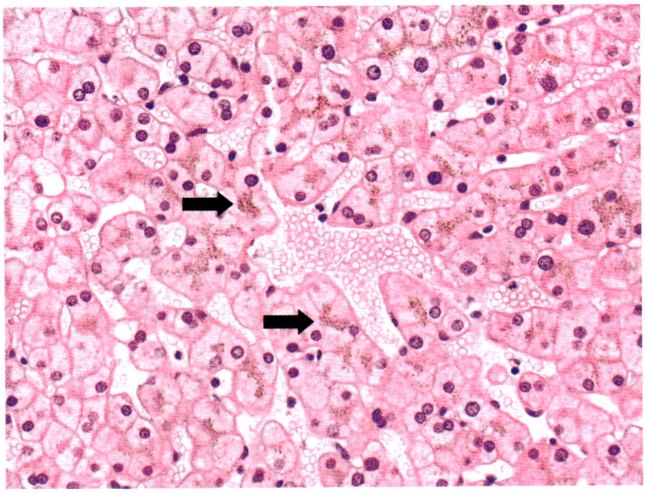

FIGURE 27.41 Marked lipofuscin pigment deposition in older individuals (*arrows*).

METHODOLOGY

Liver Biopsy

The role of liver biopsy continues to evolve over time (56,57). Historically, liver biopsy was used almost exclusively as a diagnostic tool. However, the evolution of clinical care has led the liver biopsy to take on increased importance in clinical management compared to its origins. Currently, liver biopsy has three major roles: (a) for diagnosis; (b) for assessment of prognosis (disease staging and grading); and/or (c) to assist in making therapeutic management decisions (58,59).

To justify the inherent risk in the procedure, it is essential that the resulting liver biopsy specimen is adequate to allow detailed interpretation. It should be at least 1.5 cm in total length for many liver diseases; otherwise, another pass is recommended (60). Adequate size of the specimen minimizes sampling error, improves architectural assessment, and increases accuracy of staging and grading in chronic hepatitis. Studies in patients with viral hepatitis have shown that grading and staging accuracy are reduced in biopsies less than 2.0 or 2.5 cm in length (61). It has been proposed that the adequate number of portal tracts for grading and staging of chronic liver disease should be greater than 11. Short specimens may lead to the failure to recognize cirrhosis in up to 20% of cases. It should also be noted that the size of the sample is proportional to the method and size of the needle used for sampling. Sampling error, particularly in focally or irregularly distributed disease processes, always must be taken into consideration.

Squeezing of tissue during the biopsy procedure results in distortion of cells and elongation of nuclei, which makes cytologic evaluation of the specimen difficult.

Specimen Handling

After the biopsy procedure, the liver specimen should be handled as little as possible and with utmost care to avoid squeezing and drying artifacts. If the case so indicates, small pieces of liver tissue may be frozen for ancillary studies or fixed in glutaraldehyde for electron microscopy. If infectious disease is suspected, cultures should be taken. Then the tissue should be transferred quickly into the appropriate fixative solution, usually 10% buffered formalin. Physiologic saline should not be used as it may cause distortion and dissociation of the hepatocytes on frozen section and routine histologic examination.

Needle biopsy specimens may be arranged on a piece of card to prevent distortion and fragmentation. At this stage, the gross appearance of the liver specimen is noted. Particular attention should be paid to fragmentation, which suggests cirrhosis if the fragments show rounded edges, and to the number, size, shape, and color of the fragments. Tumors or granulomas can be recognized as white areas in an otherwise reddish-brown normal liver parenchyma. Gray-black discoloration is seen in Dubin–Johnson syndrome, rusty brown in hemochromatosis, green in cholestasis, yellow in fatty liver, and brown in older individuals from the lipofuscin deposition.

Needle biopsy specimens are fixed for at least 3 hours at room temperature, whereas wedge biopsy specimens, after sectioning into 2-mm thick slices, need longer fixation. Formalin penetrates most tissues at about 0.5 mm per hour at room temperature. In order to avoid shrinkage and hardening of the tissue, it is important to process liver specimens separately from other tissues and on a more rapid schedule in the automated tissue processor. Rush liver biopsy specimens can be manually processed to shorten the time schedule to meet the needs of critically ill patients. More than 10 consecutive sections, 3 to 5 μm in thickness, can be cut without artifact from well-embedded specimens. Usually, paraffin is used for embedding but plastic embedding may be used to obtain thinner sections.

It should be emphasized that an adequate and properly processed liver specimen without artifacts is an important prerequisite for the accurate evaluation by an experienced histopathologist, who should be supplied with all relevant clinical and laboratory data.

Microscopic examination should conform to a systematic routine and include all tissue fragments and all structures of the liver (architecture, portal triads, limiting plate, hepatocytes, sinusoidal cells, and terminal hepatic venules). We usually start with careful examination of portal tracts and then move to the remainder of the parenchyma from there.

Special Stains

The tissue is routinely stained with H&E. Masson trichrome, Sirius red, or chromotrope-aniline blue stains are used for fibrous tissue; Victoria blue or orcein is used to identify elastic fibers, lipofuscin, ceroid, and copper-binding protein and less commonly hepatitis B surface antigen (HBsAg); PAS-D is used for glycoproteins, including α1-antitrypsin inclusions (Fig. 27.42), ceroid in macrophages, chronic passive congestion inclusions, basement membranes of bile ducts, cytoplasmic inclusions of cytomegalovirus, and *Mycobacterium avium-intracellulare.* Stains for reticulin and iron are also important. If it is not possible to perform all these stains, at a minimum, a special stain for connective tissue, such as Masson trichrome, and/or reticulin should be obtained in order to assess the lobular architecture and to facilitate the diagnosis of cirrhosis.

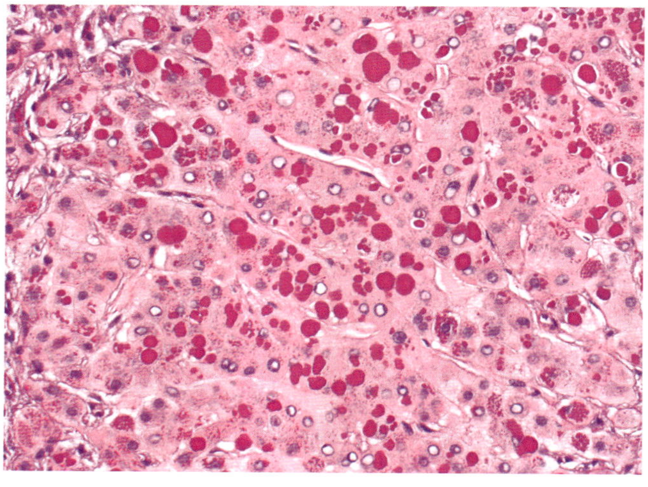

FIGURE 27.42 Alpha-1-antitrypsin granules and globules in hepatocytes (PAS-D).

Immunohistochemical Studies

The development of monoclonal antibodies and highly sensitive immunohistochemical staining procedures (peroxidase–antiperoxidase and avidin–biotin–peroxidase complex methods), has made it possible to demonstrate many antigens in routinely processed (i.e., formalin-fixed and paraffin-embedded) tissue sections. Blocking of biotin is often required prior to the application of primary antibodies because hepatocytes contain large amount of endogenous biotin, which can give a false positive reaction; such a false positive is even more pronounced with the use of antigen retrievals.

Normal Liver and Nonneoplastic Diseases

Cytokeratins (CK) are the intermediate filaments of epithelial cells and are present in hepatocytes and, in greater amounts, in the bile duct epithelium. Different as well as similar cytokeratins are expressed by hepatocytes and bile ducts. Embryonal hepatocytes contain CK8, CK18, and CK19. The expression of CK19 in hepatocytes disappears by the 10th week of gestation. Mature hepatocytes in the normal liver contain only CK8 and CK18, and therefore, stain diffusely with keratin CAM 5.2. Cytokeratin staining of hepatocytes is usually more intense in the acinar zone 1. Most hepatocytes also stain with keratin 35βH11, which reacts with CK8 only. Nonneoplastic hepatocytes do not stain with vimentin, epithelial membrane antigen (MOC31 immunohistochemical stain), CK7, and CK19. Periportal hepatocytes in cholestatic disease may, however, stain for CK7 (Fig. 27.43). Hepatocyte paraffin-1 (HepPar1) and thyroid transcription

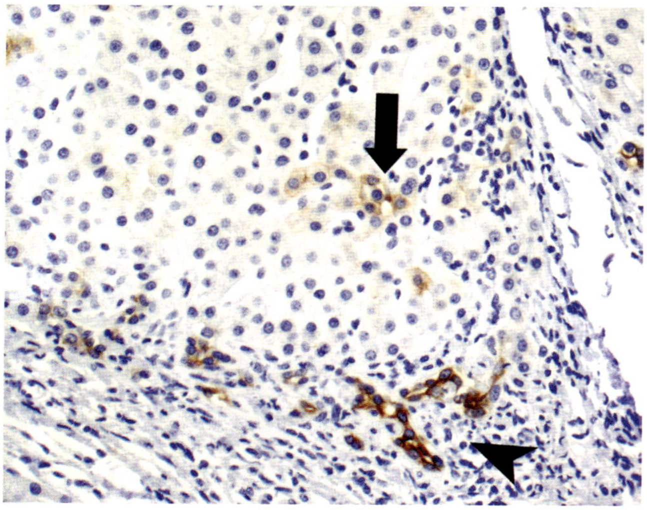

FIGURE 27.43 Proliferating bile ductules (*arrowhead*) and hepatocytes undergoing ductular metaplasia (*arrow*) in chronic cholestasis are positive for CK7.

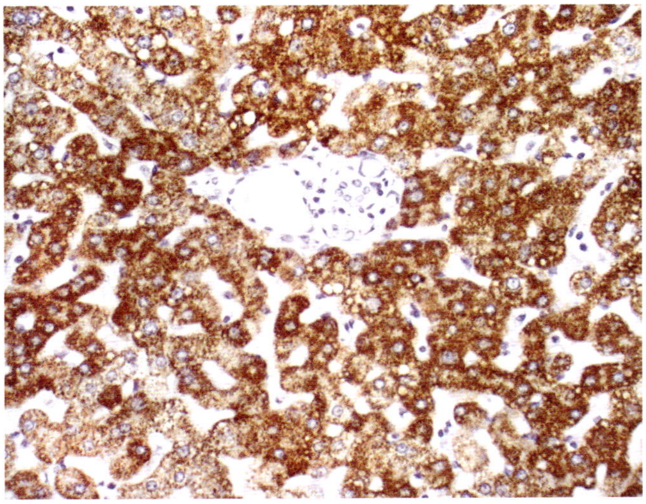

FIGURE 27.44 Granular staining of hepatocytes using HepPar1 antibody excludes other cells and all structures in the portal tract.

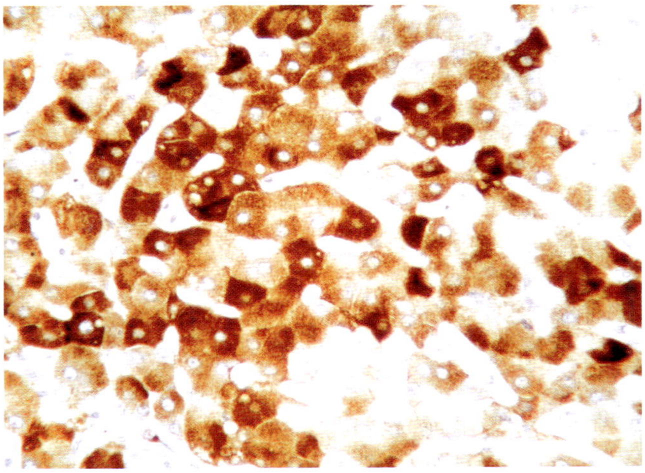

FIGURE 27.45 HBsAg immunostain demonstrates cytoplasmic hepatitis B surface antigen deposition in a patient with chronic hepatitis B.

factor-1 (TTF1) stain hepatocytes in a diffuse, granular, cytoplasmic pattern (Fig. 27.44) (62–64). Arginase-1 stains hepatocytes in cytoplasmic and nuclear pattern (62,64). Glypican-3, an oncofetal antigen, may be positive in hepatocellular carcinoma, but absent in normal hepatocytes other than in severely active hepatitis (65). Normal hepatocytes do not stain with AFP, but hepatocytes in cirrhotic nodules may occasionally show focal positive staining.

Cytokeratin polypeptides may be altered in specific liver diseases, such as alcoholic hepatitis, chronic cholestasis, and Wilson disease, with formation of Mallory–Denk hyalines. Mallory–Denk hyalines are composed of heterogeneous cytokeratin filament and usually react strongly with antibody to CK8, CK18, 34βE12, CAM 5.2, p62, and ubiquitin (66). They also occasionally react with CK7 and CK19.

Bile canaliculi can be demonstrated using pCEA and CD10 (67) (Fig. 27.22).

Sinusoidal endothelial cells show phenotypic differences with vascular endothelium. Normally, they do not bind the lectin ulex europaeus; they do not express factor VIII–related antigen or contain other molecules found in vascular endothelium, such as CD34 and CD31. They, however, assume these phenotypic properties in chronic liver diseases, in cirrhosis, and in hepatocellular carcinoma. In normal liver, positive staining for CD31 and CD34 are limited to vascular endothelium of portal tract vessels and periportal sinusoids.

Several markers of neural/neuroectodermal differentiation have been found in hepatic stellate cells. These are synaptophysin, glial fibrillary acidic protein (GFAP), and neural cell adhesion molecule (NCAM), which can be used to identify resting stellate cells. When activated, these cells show expression of vimentin, desmin, and smooth muscle actin, which suggests myofibroblastic differentiation (68).

Bile ducts and ductules are readily revealed by immunostaining for bile duct–type CK7 and CK19 (Figs. 27.38 and 27.43). Bile ducts also stain with CK8, CK18, AE1/AE3, 35βH11, and 34βE12 antibodies.

Immunohistochemistry in nonneoplastic liver diseases is commonly performed for: (a) localization of hepatotropic and nonhepatotropic viral antigens; (b) identification of biliary epithelium; and (c) identification of inclusion bodies in storage and hereditary diseases. A recent study also utilized immunohistochemistry in combination with zonal image analysis method to objectively quantify the difference in zonal distribution of hypoxia stain markers between normal and steatotic livers (69).

The presence or absence and distribution pattern of viral antigens are helpful in the diagnostic and prognostic evaluation of viral hepatitis, particularly hepatitis B surface and core antigens (Figs. 27.45 and 27.46) in hepatitis B virus (HBV) or in dual HBV and hepatitis C virus (HCV) infection. The detection of hepatitis A, C, and delta (D) antigens and of herpes virus antigens (cytomegalovirus, herpes

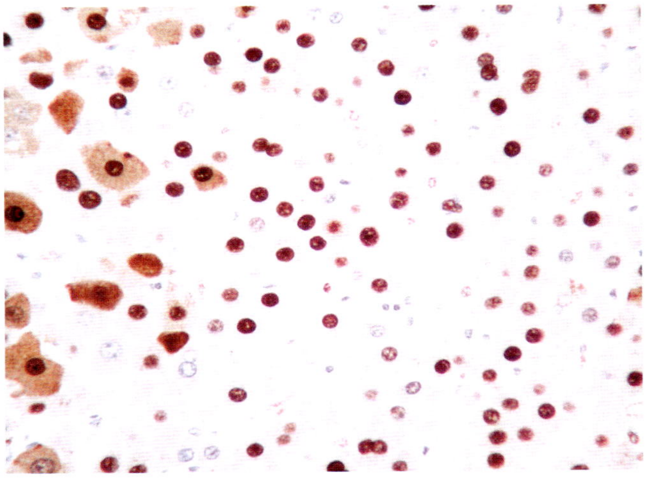

FIGURE 27.46 Immunostaining of hepatitis B core antigen (HBcAg) in numerous nuclei and cytoplasm of hepatocytes in an immunosuppressed patient with high viral replication.

simplex virus, and Epstein–Barr virus) confirms the cause of acute or chronic hepatitis. Since the cloning and sequencing of the HCV genome in 1989, there have been a number of studies for the detection of HCV antigens in the liver. However, the reports are conflicting and there is no reliable HCV antibody available yet. The detection rate of positive cases varied, which may be related to tissue sampling and differences in sensitivity of various methods and specificity and/or avidity of the antibodies. Immunohistochemical studies on frozen sections appeared to demonstrate HCV antigens more reliably than on formalin-fixed, paraffin-embedded sections.

Cytokeratin 7 or 19 immunostain is performed for identification and counting of bile duct when bile duct loss/ductopenia, graft-versus-host disease, or chronic allograft rejection is suspected. In addition, both CKs can be used to evaluate the degree of ductular reaction in biliary diseases and chronic viral hepatitis. In chronic viral hepatitis, a higher degree of ductular reaction correlates with a higher degree of interface hepatitis and fibrosis progression. CK7 staining of periportal cholestatic hepatocytes confirms long-standing cholestasis.

Alpha-1-antitrypsin and fibrinogen immunostains are performed to identify intracytoplasmic inclusion bodies in α1-antitrypsin deficiency and fibrinogen storage disease respectively.

Neoplastic Diseases

The most common use of immunohistochemistry for liver specimen is for immunophenotyping, classification, and prognostication of primary or metastatic tumors of the liver (62,64). Primary or metastatic poorly/undifferentiated tumors may lose their organ-specific antigenicity, and immunohistochemistry may fail to pinpoint organ of origin of the tumor. In these cases, clinical correlation and further imaging studies are required.

For the identification of hepatocellular carcinoma, the expected immunophenotype is cytoplasmic expression of HepPar1 (62,64), arginase-1 (62,64), glypican-3 (65), and CK8 and CK18 (64). Canalicular staining for polyclonal CEA and CD10 (67) may also be observed. Positivity for glypican-3 can be used to differentiate hepatocellular carcinoma from dysplastic nodules and cirrhotic nodules.

Cholangiocarcinoma (CCA) as well as biliary intraepithelial neoplasia is positive for biliary cytokeratins 7 and 19. Intrahepatic cholangiocarcinoma (iCCA) has two different morphologic subtypes, small- and large-duct types (70–73). By the current WHO definition, the classification of iCCA is based on morphologic features and does not require immunohistochemistry (72). However, in more difficult cases, some immunohistochemical stains can be helpful in aiding in the differentiation between small- and large-duct iCCA. C-reactive protein (more specific) and N-cadherin (less specific) are highly sensitive markers for small-duct iCCA. Tubulin β-III (TUBB3) and CD56 (NCAM) may also be positive in small-duct iCCA (73). In contrast, S100 calcium-binding protein P (S100P) is a useful marker for large-duct iCCA (73). Like pancreatic adenocarcinoma, a mutant pattern of p53 and loss of Smad4 protein expression can also be seen in CCA (70).

In benign hepatocellular tumors, often confirmed by an intact reticulin network and low proliferative rate, liver fatty acid binding protein, serum amyloid A, C-reactive protein, glutamine synthetase, and β-catenin play an important role in classifying these lesions (74). Focal nodular hyperplasia typically shows a map-like staining pattern for glutamine synthetase. Inflammatory hepatocellular adenoma shows positive staining for serum amyloid A and C-reactive protein. L-FABP–deficient hepatocellular adenoma shows loss of L-FABP while the remaining markers are within normal limits. Beta catenin-activated hepatocellular adenomas are characterized by, diffuse strong positivity for glutamine synthetase, and in cases with CTNNB1 exon3 mutations, nuclear positivity for β-catenin (Table 27.1).

Workup for metastatic tumors involves various immunohistochemical antibodies, including establishing a line of differentiation (epithelial, stromal, or melanoma) and organ of origin (75).

Electron Microscopy

Electron microscopy has a limited but a potential role in: (a) investigating hereditary and metabolic liver diseases; (b) viral infection not otherwise identified by light microscopy or serology; (c) tumors of unknown histogenesis; (d) certain drug-induced liver injuries; and (e) diseases of

TABLE 27.1 Immunohistochemical Patterns in Benign Hepatocellular Tumors

	Glutamine Synthetase	β-Catenin	L-FABP	CRP/SAA
FNH	+ (map-like)	– (membranous)	+ (retained)	–
β-Catenin–activated HCA	+ (diffuse)	+ (nuclear)	+ (retained)	–
HNF-1α–Inactivated HCA	–	– (membranous)	– (loss)	–
Inflammatory HCA	–	– (membranous)	+ (retained)	+
Unclassified HCA	–	– (membranous)	+ (retained)	–

FNH, focal nodular hyperplasia; HCA, hepatocellular adenoma; HNF-1α, hepatocyte nuclear factor 1α; L-FABP, liver fatty acid–binding protein (normally expressed in normal liver); CRP/SAA, C-reactive protein/serum amyloid A.

Modified from Burt AD, Ferrell LD, Hubscher SG.MD. *MacSween's Pathology of the Liver.* 7th ed. Elsevier; 2017, with permission from Elsevier.

unclear etiology (56). Tissue obtained for electron microscopic study should be fixed in 3% glutaraldehyde.

Molecular Studies

In regard to neoplastic diseases of the liver, the main clinical applications for molecular testing are for the diagnosis of specific tumors such as fibrolamellar carcinoma and current NIPBL::NACC1 fusion hepatic carcinoma (also called cholangioblastic variant of iCCA and solid-tubulocystic variant of iCCA) (76,77). In addition, molecular assays are currently used in investigational studies to understand the pathogenesis of benign hepatocellular tumors, preneoplastic nodules, hepatocellular carcinoma, and CCA, which can provide better surveillance, diagnosis, treatment, and prognostication of these lesions (56).

In situ hybridization (ISH) employs antigen-labeled complementary DNA or RNA sequences to localize a specific DNA or RNA sequence in tissue. ISH may be performed on formalin-fixed and paraffin-embedded tissue sections. ISH has been most commonly applied to liver tissue for the identification of hepatitis A, B, C, and D viruses, cytomegalovirus, and Epstein–Barr virus. ISH can be used to identify albumin mRNA which is highly sensitive for normal hepatocytes and hepatocellular tumors (78).

Polymerase chain reaction (PCR) is a technique to amplify exponentially a single or few copies of DNA sequence, employing DNA polymerase and generating thousands to millions of copies of the particular DNA sequence. Reverse transcription PCR allows the identification of RNA. Currently, these techniques are the most sensitive and specific method to demonstrate HBV DNA, HCV RNA, and their genotypes in the blood and liver tissue. In addition, PCR can be used to identify infectious organisms or specific genetic mutations.

Microarray analysis provides an arrayed series of thousands of DNA sequences. Each may contain a specific DNA sequence, the relative abundance of which is determined by chemiluminescence-labeled targets.

FREQUENT HISTOLOGIC CHANGES OF LITTLE SIGNIFICANCE

The Liver at Autopsy

Liver tissue obtained at autopsy often shows changes that are not usually seen in liver biopsy specimens and therefore may cause difficulties in the evaluation. Agonal loss of glycogen from hepatocytes causes increased density and eosinophilia of the cytoplasm. Poor fixation results in irregular staining of hepatocytes, particularly in the center of the specimen. This may result in striking differences in the appearance of liver cells in the peripheral versus the central part of the tissue.

Agonal necrosis, particularly of hepatocytes in zone 3 in patients with shock or heart failure, may not be reflected in elevated aminotransferase levels (Fig. 27.47). Its terminal nature is recognized from the lack of any inflammatory response. Autolysis of hepatocytes, particularly in hepatitis and cholestasis resulting in loss of cellular detail and prominent sinusoidal lining cells, is often more pronounced than in other tissues because the liver is rich in proteolytic enzymes. Loss of inflammatory cells by autolysis may make the diagnosis of hepatitis in postmortem specimens difficult. Trichrome stain assists greatly in the identification of portal tracts and central venules (and thus the appreciation of lobular architecture) and, by demonstrating the fibrous septa, the chronicity of the condition.

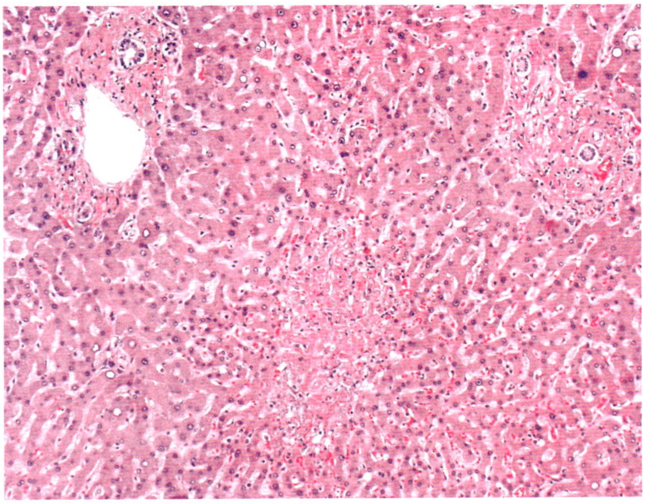

FIGURE 27.47 Agonal necrosis of centrilobular hepatocytes in autopsy liver.

Dilatation of the sinusoidal and perisinusoidal spaces of Disse is of little significance in postmortem liver tissues as opposed to similar changes in well-preserved biopsy specimens. Mild accumulation of lymphocytes in some portal tracts is frequently seen in autopsy livers and does not justify the diagnosis of chronic hepatitis. Large tissue sections obtained at autopsy often include many large triangular portal tracts with abundant connective tissue that can be distinguished from true portal fibrosis by evaluation of the size of the intraportal structures. In addition, increased fibrous tissue in portal tracts and parenchyma is a normal phenomenon in older individuals.

Surgical Liver Biopsy Specimens

Surgical biopsy specimens may have several features not seen in needle biopsy specimens that may cause diagnostic difficulties. If the surgeon removes a small, superficial wedge of liver tissue from the inferior margin, the triangular tissue fragment is covered on two sides by the Glisson capsule. The fibrous connections between the superficial portal tracts and the capsule may imitate cirrhosis (Fig. 27.30). However, these changes usually do not extend more than 2 mm into the liver parenchyma. Large portal tracts can be occasionally encountered in subcapsular liver parenchyma (Fig. 27.29).

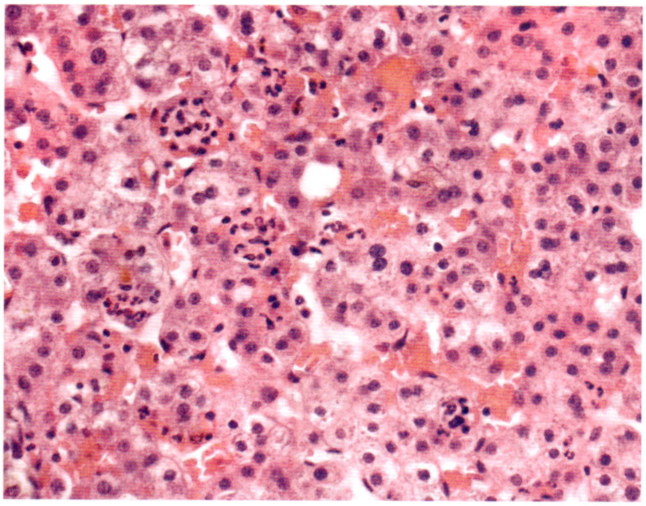

FIGURE 27.48 Surgical biopsy specimen showing clusters of polymorphonuclear neutrophils in sinusoids resembling microabscesses (surgical hepatitis).

In biopsy specimens removed at the end of a long surgical procedure, clusters of polymorphonuclear neutrophils are seen in or under the capsule, in sinusoids, around terminal hepatic venules, in portal tracts, and in areas of small focal necrosis resembling microabscesses, probably as a result of minor trauma (Fig. 27.48). This characteristic lesion must be distinguished from inflammatory liver diseases such as cholangitis. Other "innocent" hepatic lesions include focal steatosis involving small groups of hepatocytes (<5%), lipogranulomas from mineral oil deposition in perivenular areas and portal tracts in the absence of significant steatosis; and unexplained mitoses of hepatocytes that normally have a lifespan of many years.

MINOR BUT SIGNIFICANT HEPATIC INJURY PATTERNS

There are minor alterations in hepatic parenchyma that are often dismissed on histopathologic examination of the liver. These minor alterations are frequently clinically significant, may explain the abnormalities that the patients have and justify the indications for liver biopsies. They include mild inflammation in the liver parenchyma such as in nonspecific reactive hepatitis, mild acute hepatitis, and residual hepatitis; subtle alterations of hepatic parenchyma caused by abnormal vascularity such as chronic venous congestion, nodular regenerative hyperplasia, and hepatoportal sclerosis; or subtle alterations of hepatic parenchyma in the vicinity of a space-occupying lesion in a missed targeted liver biopsy. They require close inspection of the liver biopsy itself often with the aid of special stains such as reticulin stain and PAS-D stain; and most importantly the knowledge of normal liver structures and lobular architecture to identify minor and subtle alterations.

Nonspecific Reactive Hepatitis

This poorly defined histologic change represents a reaction of the liver to a variety of extrahepatic or systemic diseases, particularly infectious diseases, autoimmune diseases, connective tissue diseases, gastrointestinal diseases, and acquired immunodeficiency syndrome (AIDS) (79,80). Nonspecific reactive hepatitis must be differentiated from primary liver diseases, such as mild chronic hepatitis and residual stages of acute viral hepatitis and drug-induced injury. The alterations consist of activation of sinusoidal lining cells with prominent Kupffer cells, foci of isolated hepatocyte necrosis with accumulation of macrophages and other inflammatory cells, and mild infiltration of some portal tracts by mononuclear cells without interface hepatitis (Fig. 27.49).

Mild Acute Hepatitis and Residual/Resolving Hepatitis

Mild acute hepatitis is a mild diffuse hepatocellular injury due to hepatotropic or nonhepatotropic viruses, drugs, or other less common causes. Aminotransferase and other liver enzyme activities are mildly elevated, and the disease usually resolves in a few months. Mild acute hepatitis typically requires no specific treatment. Mild acute hepatitis is characterized by subtle, but diffuse changes, including panlobular disarray with degeneration and regeneration of the hepatocytes, accompanied by activation of sinusoidal lining cells and diffuse infiltration of sinusoids and portal tracts by inflammatory cells, primarily lymphocytes and macrophages and some plasma cells. Cholestasis or bridging necrosis is not seen in mild acute hepatitis.

The term residual/resolving hepatitis is used to describe changes in liver biopsies taken late in the course of acute hepatitis with residual abnormalities in liver enzyme activities (56). Since most of the hepatocellular injury has occurred, the prognosis of residual/resolving hepatitis is good and does not require further treatment when the

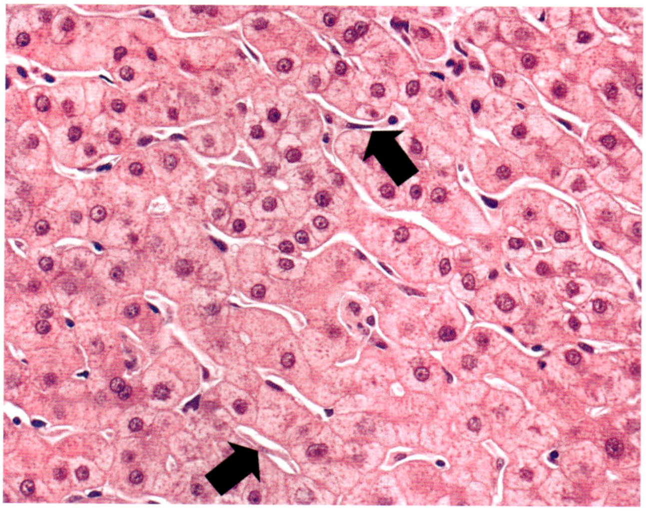

FIGURE 27.49 Nonspecific reactive hepatitis characterized by activation of sinusoidal lining cells (*arrows*).

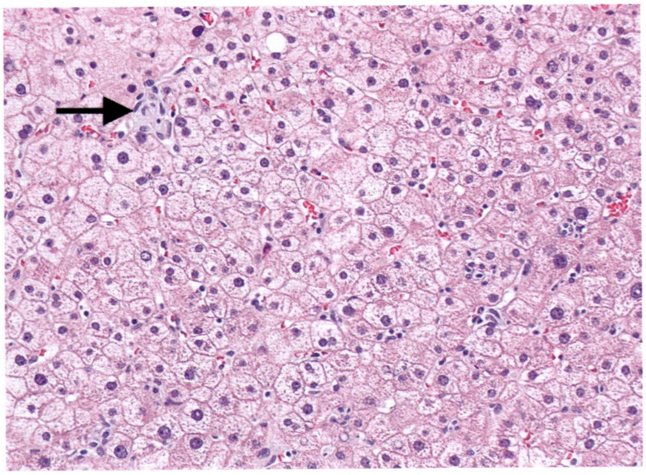

FIGURE 27.50 Cluster of macrophages (*arrow*) containing debris of necrotic hepatocyte in residual hepatitis.

cause of injury has been identified. The features of typical acute hepatitis, which include diffuse hepatocellular damage and inflammation have subsided, with mild pleomorphism of the hepatocytes, irregular cell plates with a few foci of lobular necrosis, and mild inflammation. Clusters of PAS-D–resistant macrophages, which contain debris of necrotic or apoptotic hepatocyte or ceroid pigment and may be iron positive, are observed predominantly in the centrilobular area and portal tracts (Figs. 27.24 and 27.50). Focal reticulin fiber condensation may be seen because of hepatocyte loss. The portal tracts may be expanded with mild lymphocytic infiltrate and mild fibrosis or with thin fibrous septa.

Venous Outflow Impairment Pattern (Congestive Hepatopathy)

Acute or chronic venous congestion with dilatation of sinusoids in zone 3 (pericentral) and of terminal hepatic venules is frequently a consequence of right-sided congestive heart failure, whereas irregular necrosis of hepatocytes in zone 3 is often caused by left-sided heart failure or shock.

Chronic venous congestion causes stasis and increase of hepatic venous pressure, dilatation of sinusoids in zone 3, followed by extravasation of red blood cells to the perisinusoidal spaces of Disse and atrophy of hepatocytes around the terminal hepatic venules (Fig. 27.51). In more severe chronic venous congestion cases, there is also atrophy of hepatocytes with increased lipofuscin accumulation, scattered small droplet fat vacuoles, extravasation of red blood cells to the perisinusoidal space of Disse, intracytoplasmic eosinophilic globules in centrilobular hepatocytes, enlargement of Kupffer cells containing ceroid pigment, focal cholestasis, and progressive fibrosis. Special stains for iron may demonstrate iron deposition in centrilobular Kupffer cells. Engorgement of sinusoids around terminal hepatic venules is also observed in other conditions such as Budd–Chiari syndrome, and veno-occlusive disease.

In contrast, dilatation of sinusoids in zone 1 has been observed in pregnancy, in renal transplant patients, after

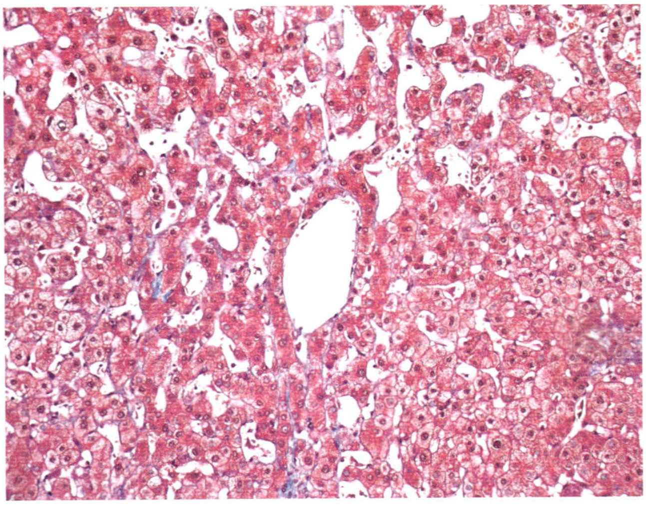

FIGURE 27.51 Dilated central venule and zone 3 sinusoids accompanied by atrophy of hepatocytes in chronic venous congestion.

exposure to anabolic/androgenic or contraceptive steroids (Fig. 27.52), and near space-occupying lesions.

Nodular Regenerative Hyperplasia

Nodular regenerative hyperplasia is a diffuse nodular regeneration of the hepatocytes, associated with various disorders, such as autoimmune diseases (rheumatoid arthritis, systemic sclerosis, systemic lupus erythematosus), primary biliary cholangitis, primary sclerosing cholangitis, hematologic malignancies, endocrine or metabolic disorders, or drugs (42,81,82). In majority of the cases, the proposed pathogenesis is either obliteration of small portal veins or arteritis.

Nodular regenerative hyperplasia occurs in both sexes at any age, but most frequently between the ages of 40 and 70. Many patients are asymptomatic, but they may develop noncirrhotic portal hypertension, and rarely hepatic failure, ascites, or rupture of liver with massive hemorrhage.

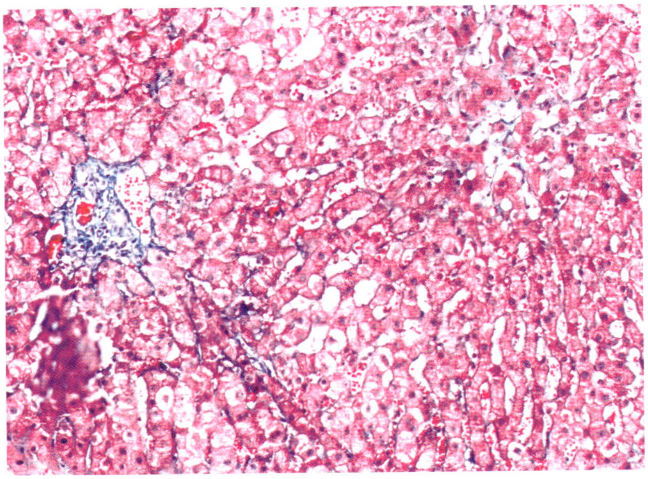

FIGURE 27.52 Dilatation of zone 1 sinusoids in contraceptive steroids injury.

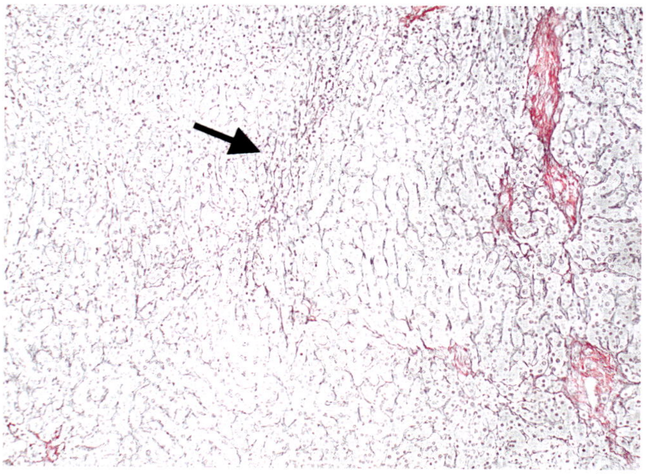

FIGURE 27.53 Compression and atrophy of centrilobular parenchyma resulting in condensation of reticulin framework (*arrow*) in nodular regenerative hyperplasia.

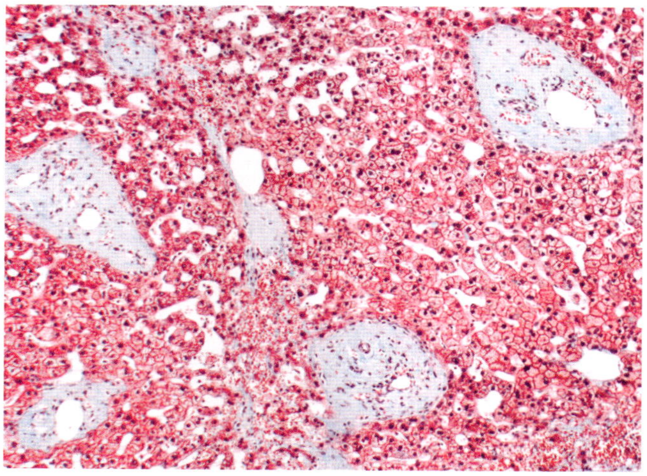

FIGURE 27.54 Porto-sinusoidal vascular disease showing atrophy of liver parenchyma resulting in close proximation of densely fibrotic portal tracts.

The liver parenchyma shows widely scattered parenchymal nodules varying in diameter from 1 to 4 mm. The capsular surface of the liver often reveals minimal shallow irregularities that may be mistaken for cirrhosis. The nodules either replace almost the entire parenchyma or are confined to one portion of the liver. The nodules in nodular regenerative hyperplasia are not surrounded by fibrous septa; therefore, they are ill-defined and soft, distinct from those found in cirrhosis. The nodules consist of hyperplastic hepatocytes arranged in plates more than one-cell-layer thick, particularly adjacent to or surrounding the portal tracts. The surrounding centrilobular parenchyma is compressed and atrophic and the sinusoids are congested (the so-called linear congestion). A reticulin stain shows condensation of reticulin framework around the expanding nodules and the irregularly thickened hepatocyte plates within the nodules, but dense septa as in cirrhosis are not seen (Fig. 27.53). The hyperplastic hepatocytes are slightly larger and more variable than normal and arranged in two-cell-thick plates with occasional binucleation. Obliteration of small portal veins is often seen.

Because of the subtle alterations of the lobular architecture, the diagnosis is often difficult in needle biopsy specimen. Alternating areas of hyperplastic hepatocytes and atrophic hepatocytes with linear congestion on needle biopsy should lead to a reticulin stain to confirm nodular regenerative hyperplasia.

Porto-Sinusoidal Vascular Disease

Noncirrhotic portal hypertension also called idiopathic portal hypertension, obliterative portal venopathy, hepatoportal sclerosis, among other terminology, is a condition consisting of portal hypertension, splenomegaly, and anemia secondary to hypersplenism characterized by dense portal tract thrombosis and obliteration of portal veins in noncirrhotic patients.

This constellation was first described by Mikkelsen et al. in 1965 (83) as hepatoportal sclerosis to describe noncirrhotic sclerosis of intrahepatic portal veins and corresponds to the clinical condition described by Banti in the early 20th century to describe portal hypertension, splenomegaly, and anemia not associated with hematologic disease (Banti disease). Also in the 1960s, other terms, noncirrhotic or idiopathic portal hypertension with histologic features of obliterative portal venopathy were also proposed, adding some complexity to the landscape of the nomenclature of this entity.

Recently, the term porto-sinusoidal vascular disorder (PSVD) has been introduced to encompass these histologic findings (38). The specific histologic lesions recognized for the diagnosis of PSVD are nodular regenerative hyperplasia, portal vein stenosis (previously known as obliterative portal venopathy lesions) (84), and incomplete septal fibrosis/cirrhosis (39).

The liver is either of normal size or more often mildly atrophic with wrinkled capsule. Atrophy may not be uniformly distributed; therefore, the liver lobes may be disproportionate in their size when compared to normal liver. Portal tracts are round and densely fibrotic with fibrous obliteration of larger portal veins and/or with numerous dilated small portal vein branches (Figs. 27.32 and 27.54). Some of the dilated small portal vein branches may dilate or herniate into the periportal liver parenchyma. Bile ducts may show periductal fibrosis. While the lobular architecture is fairly well-maintained, occasional portal-to-portal or portal-to-central bridging fibrosis may be present. The centrilobular hepatocytes become atrophic, causing dilatation of sinusoids, reverse lobulation, and close proximation of the portal tracts. The subtle histologic alterations can be underappreciated.

Perilesional Sampling (Vicinity of Space-Occupying Lesions)

Nonspecific reactive changes are also seen in patients with space-occupying lesions in the liver (55). Although

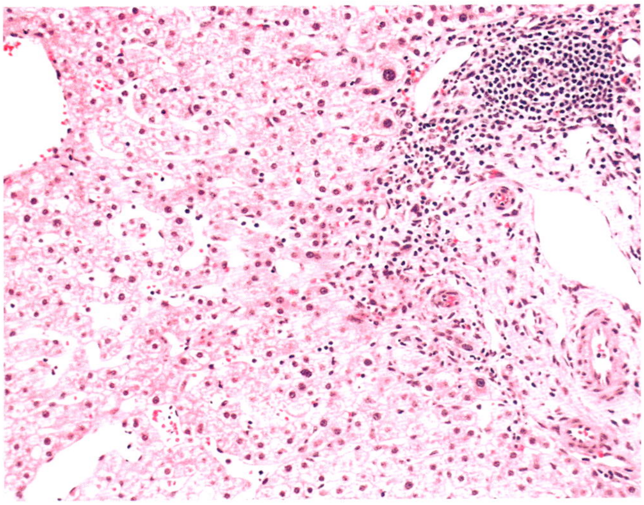

FIGURE 27.55 Liver biopsy specimen of a patient with metastatic carcinoma showing dilatation of sinusoids in zone 2, as well as ductular reaction and infiltration of portal tract by neutrophils. The specimen does not contain metastatic carcinoma, but the histologic changes are consistent with the vicinity of a space-occupying lesion.

the biopsy specimen may not include the neoplasm, the abscess, or the cyst, a characteristic histologic triad is often observed in the adjacent liver (85). The histologic changes are the result of local obstruction of blood and bile flow by the expanding mass. These changes consist of proliferating and distorted bile ductules with irregular and even atypical epithelium accompanied by neutrophils in edematous portal tracts, and focal sinusoidal dilatation and congestion (Fig. 27.55). The changes may be subtle and are usually focal, involving adjacent small portal tracts. The liver cell plates may be compressed and distorted with atrophy of hepatocytes. In contrast to large bile duct obstruction and other biliary tract diseases such as primary sclerosing cholangitis, cholestasis is usually absent. The described triad is characteristic of perilesional sampling, and its recognition in a liver specimen should lead to continued search for a mass lesion.

BROWN PIGMENTS

There are several pigments that can cause brown discoloration of the liver parenchyma. Hemosiderin and copper are abundant in the cytoplasm of hepatocytes during the first week of life, then gradually disappear and should be absent at the age of 6 to 9 months. Therefore, brown cytoplasmic pigments representing hemosiderin and copper should not be seen in normal children and adult livers. Minimal hepatocellular iron may be within normal limits, particularly in older individuals.

Lipofuscin

Lipofuscin, the "wear and tear" or "aging" pigment (86), is commonly seen in varying quantities as fine, well-delineated, light brown, PAS-D-resistant, partly acid-fast–positive granules in the cytoplasm of hepatocytes in zone 3, particularly at the canalicular pole (Fig. 27.41). Lipofuscin is produced by lysosomal oxidation of lipids. Under the electron microscope, lipofuscin granules appear as secondary lysosomes in pericanalicular location that contain irregular and lobulated electron-dense granules, lipid droplets, and a heterogeneous matrix.

Lipofuscin has no functional or clinical significance (56,57). There is a progressive increase in its amount in individual hepatocytes and in the number of cells involved in older individuals (86). An increase of lipofuscin pigment also occurs in atrophic liver, starvation, chronic wasting diseases, and malignancies, producing brown discoloration of the liver on gross examination. Lipofuscin is not found in recently regenerated hepatocytes. It may be difficult to distinguish lipofuscin from the pigment that accumulates in hepatocytes in large amounts in Dubin–Johnson syndrome.

In contrast to lipofuscin, iron and copper are coarser, birefringent, and usually deposited in periportal hepatocytes. Bile is poorly defined and less granular than the other pigments and often accumulates in bile canaliculi in zone 3.

Dubin–Johnson Pigment

Dubin–Johnson syndrome is an autosomal recessive familial disorder of bilirubin metabolism characterized by chronic or intermittent benign jaundice due to the defect in hepatocellular excretion of organic anions into bile (87). Dubin–Johnson syndrome results in black to dark-green discoloration of the liver that can be recognized macroscopically even on a needle biopsy specimen (88). Otherwise, the liver is grossly normal. Abundant dark-brown coarse melanin-like lysosomal pigment in the pericanalicular cytoplasm of centrilobular hepatocytes is noted (85). In contrast to lipofuscin, the granules are darker and more variable in size and extend into the midlobular and periportal hepatocytes. Electron microscopy demonstrates distinctive oval or irregularly shaped, pleomorphic electron-dense lysosomes. They contain very dense bodies frequently associated with lipid droplets, in a finely granular background, and lack the typical lobulation of lipofuscin granules. Other liver abnormalities, particularly cholestasis, are not seen.

Hemosiderin

Iron can accumulate in the liver in a variety of conditions, including hereditary hemochromatosis and secondary iron overload conditions associated with systemic macrophage iron accumulation (transfusions, hemolytic conditions, anemia of chronic disease, etc.), hepatitides (hepatitis C, alcoholic liver disease, porphyria cutanea tarda), and liver-specific iron accumulation of uncertain pathogenesis in cirrhosis (89).

In the early stages, hereditary hemochromatosis and secondary iron overload result in distinct pattern of hemosiderin deposition; but as the disease progresses, the histologic

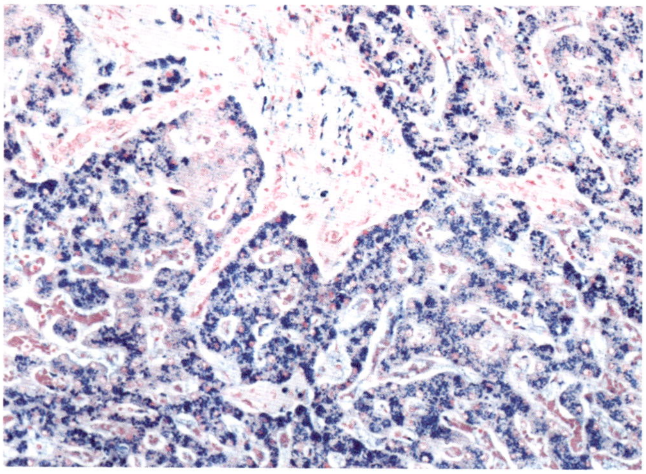

FIGURE 27.56 Severe hemosiderin deposition in hepatocytes, Kupffer cells, and portal macrophages in hereditary hemochromatosis (Perl iron stain).

distinction between them becomes less clear. Hereditary hemochromatosis causes preferential hemosiderin deposition in periportal hepatocytes of both homozygous and heterozygous C282Y patients in the early stages. This pattern may remain in heterozygous carriers throughout life. In the early stages of hereditary hemochromatosis, Kupffer cells do not contain excess iron and other liver biopsy abnormalities are not observed. With increasing age, continued iron deposition in hepatocytes leads to direct hepatocellular damage and fibrosis. Iron is deposited in all hepatocytes throughout the hepatic lobules, as well as in Kupffer cells, portal macrophages, and bile ducts and ductules (Fig. 27.56). Iron in bile ducts and ductules is typical, but not pathognomonic for hereditary hemochromatosis.

Secondary iron overload results in hemosiderin deposition primarily in Kupffer cells, distributed evenly throughout the liver lobules without any zonal preference. In severe cases of secondary iron overload, coarse uneven hemosiderin deposition is also seen in hepatocytes throughout the hepatic lobule. Fibrosis or cirrhosis is usually absent. However, in sideroblastic anemia and thalassemia major, the severe secondary iron overload mimics that of hereditary hemochromatosis. It can also cause fibrosis and cirrhosis but with more iron in clustered macrophages and Kupffer cells. Other causes of secondary iron overload include alcoholic liver disease and blood transfusion. Secondary iron overload from excessive ingestion or peripheral hemolysis (such as caused by Ribavirin) results in hemosiderin deposition in hepatocytes.

Copper-Associated Protein

Increase in copper deposition is seen in chronic cholestasis/cholate-stasis and Wilson disease (90). Copper-associated protein and free copper are usually found in periportal hepatocytes or, in cirrhosis, in hepatocytes at the periphery of nodules. In chronic cholestasis, the periportal and periseptal hepatocytes have rarefied cytoplasm, usually accompanied by ductular reaction and in biliary cirrhosis by biliary halo. In Wilson disease, copper-associated protein is diffusely distributed in hepatocytes throughout the lobule, in contrast to the periportal or periseptal deposition in chronic cholestasis. Recently, metallothionein immunohistochemistry has been proposed as a sensitive clinical biomarker for histologic diagnosis of Wilson disease (91,92).

Bile

In the normal physiologic state, bile in liver parenchyma is invisible (56). Bile can be visualized only in pathologic conditions, referred generally as cholestasis. Bile can accumulate in the cytoplasm of hepatocytes, bile canalicular spaces (Fig. 27.21), Kupffer cells, bile ducts, and ductules. Accumulation in the cytoplasm of hepatocytes is called hepatocytic or parenchymal cholestasis, in canalicular spaces is called canalicular cholestasis, and in bile ductules is called ductular or cholangiolar cholestasis.

ACKNOWLEDGMENTS

The late Dr. Hans Popper reviewed the original manuscript. The late Dr. Michael Gerber was the senior author of this chapter in the earlier editions. Dr. Swan Thung and Dr. Arief Suriawinata were authors of the prior edition. The authors thank Dr. Roger K. Moreira for his insightful contribution on the Voronoi-based algorithm of hepatic zonation.

REFERENCES

1. Roskams T, Desmet V. Embryology of extra- and intrahepatic bile ducts, the ductal plate. *Anat Rec (Hoboken)*. 2008;291(6): 628–635.
2. Carpentier R, Suñer RE, van Hul N, et al. Embryonic ductal plate cells give rise to cholangiocytes, periportal hepatocytes, and adult liver progenitor cells. *Gastroenterology*. 2011;141(4): 1432–1438, 1438.e1-4.
3. Greenbaum LE. The ductal plate: a source of progenitors and hepatocytes in the adult liver. *Gastroenterology*. 2011;141(4):1152–1155.
4. Desmet VJ. [Cystic diseases of the liver. From embryology to malformations.] *Gastroenterol Clin Biol*. 2005;29(8–9):858–860.
5. Miranda RN, Omurtag K, Castellani WJ, De las Casas LE, Quintanilla NM, Kaabipour E. Myelopoiesis in the liver of stillborns with evidence of intrauterine infection. *Arch Pathol Lab Med*. 2006;130(12):1786–1791.
6. Lau WY. *Applied Anatomy in Liver Resection and Liver Transplantation*. 1st ed. Springer; 2021.
7. Bismuth H. Surgical anatomy and anatomical surgery of the liver. *World J Surg*. 1982;6(1):3–9.
8. Couinaud C. [Dorsal sector of the liver.] *Chirurgie*. 1998;123(1): 8–15.
9. Reichert PR, Renz JF, D'Albuquerque LA, et al. Surgical anatomy of the left lateral segment as applied to living-donor and

split-liver transplantation: a clinicopathologic study. *Ann Surg*. 2000;232(5):658–664.

10. van Leeuwen MS, Noordzij J, Hennipman A, Feldberg MA. Planning of liver surgery using three dimensional imaging techniques. *Eur J Cancer*. 1995;31A(7–8):1212–1215.
11. Kiernan F. The anatomy and physiology of the liver. *Philos Trans R Soc Lond*. 1833;123:711–770.
12. Rappaport AM. The structural and functional unit in the human liver (liver acinus). *Anat Rec*. 1958;130(4):673–689.
13. Lau C, Kalantari B, Batts KP, et al. The Voronoi theory of the normal liver lobular architecture and its applicability in hepatic zonation. *Sci Rep*. 2021;11(1):9343.
14. Terada T, Ishida F, Nakanuma Y. Vascular plexus around intrahepatic bile ducts in normal livers and portal hypertension. *J Hepatol*. 1989;8(2):139–149.
15. Ramachandran R, Kakar S. Histological patterns in drug-induced liver disease. *J Clin Pathol*. 2009;62(6):481–492.
16. Fisher K, Vuppalanchi R, Saxena R. Drug-induced liver injury. *Arch Pathol Lab Med*. 2015;139(7):876–887.
17. Lamers WH, Hilberts A, Furt E, et al. Hepatic enzymic zonation: a reevaluation of the concept of the liver acinus. *Hepatology*. 1989;10(1):72–76.
18. Morgan JD, Hartroft WS. Juvenile liver. Age at which one-cell-thick plates predominate in the human liver. *Arch Pathol*. 1961;71:86–88.
19. Zhang S, Lin YH, Tarlow B, Zhu H. The origins and functions of hepatic polyploidy. *Cell Cycle*. 2019;18(12): 1302–1315.
20. Tanami S, Ben-Moshe S, Elkayam A, Mayo A, Halpern KB, Itzkovitz S. Dynamic zonation of liver polyploidy. *Cell Tissue Res*. 2017;368, 405–410.
21. Seeff LB, Zimmerman HJ. Relationship between hepatic and pancreatic disease. *Prog Liver Dis*. 1976;5:590–608.
22. Fausto N, Campbell JS. The role of hepatocytes and oval cells in liver regeneration and repopulation. *Mech Dev*. 2003;120(1): 117–130.
23. Fischer R, Baumert T, Blum HE. Hepatitis C virus infection and apoptosis. *World J Gastroenterol*. 2007;13(36):4865–4872.
24. Lin S, Zhang Y-J. Interference of apoptosis by hepatitis B virus. *Viruses*. 2017;9(8):230.
25. Ebert G, Preston S, Allison C, et al. Cellular inhibitor of apoptosis proteins prevent clearance of hepatitis B virus. *Proc Natl Acad Sci U S A*. 2015;112(18):5797–5802.
26. Loke SL, Leung CY, Chiu KY, Yau WL, Cheung KN, Ma L. Localisation of CD10 to biliary canaliculi by immunoelectron microscopical examination. *J Clin Pathol*. 1990;43(8):654–656.
27. Alam S, Lal BB. Recent updates on progressive familial intrahepatic cholestasis types 1, 2 and 3: outcome and therapeutic strategies. *World J Hepatol*. 2022;14(1):98–118.
28. Brunt EM, Gouw AS, Hubscher SG, et al. Pathology of the liver sinusoids. *Histopathology*. 2014;64(7):907–920.
29. Gibert-Ramos A, Sanfeliu-Redondo D, Aristu-Zabalza P, et al. The hepatic sinusoid in chronic liver disease: the optimal milieu for cancer. *Cancers (Basel)*. 2021;13(22):5719.
30. Kamm DR, McCommis KS. Hepatic stellate cells in physiology and pathology. *J Physiol*. 2022;600(8):1825–1837.
31. Puche JE, Saiman Y, Friedman SL. Hepatic stellate cells and liver fibrosis. *Compr Physiol*. 2013;3(4):1473–1492.
32. Nollevaux M-C, Guiot Y, Horsmans Y, et al. Hypervitaminosis A-induced liver fibrosis: stellate cell activation and daily dose consumption. *Liver Int*. 2006;26(2):182–186.
33. Nakatani K, Kaneda K, Seki S, Nakajima Y. Pit cells as liver-associated natural killer cells: morphology and function. *Med Electron Microsc*. 2004;37(1):29–36.
34. Peng H, Wisse E, Tian Z. Liver natural killer cells: subsets and roles in liver immunity. *Cell Mol Immunol*. 2016;13(3):328–336.
35. Petrelli M, Scheuer PJ. Variation in subcapsular liver structure and its significance in the interpretation of wedge biopsies. *J Clin Pathol*. 1967;20(5):743–748.
36. Lisman T, Caldwell SH, Intagliata NM. Haemostatic alterations and management of haemostasis in patients with cirrhosis. *J Hepatol*. 2022;76(6):1291–1305.
37. Okuda K. Non-cirrhotic portal hypertension versus idiopathic portal hypertension. *J Gastroenterol Hepatol*. 2002;17(Suppl 3): S204–S213.
38. De Gottardi A, Rautou PE, Schouten J, et al; VALDIG group. Porto-sinusoidal vascular disease: proposal and description of a novel entity. *Lancet Gastroenterol Hepatol*. 2019;4(5):399–411.
39. De Gottardi A, Sempoux C, Berzigotti A. Porto-sinusoidal vascular disorder. *J Hepatol*. 2022;77(4):1124–1135.
40. Hong SY, Yi NJ, Hong K, et al. Redo hepatic artery reconstruction for thrombosis without retransplantation in 1,355 adult living donor liver transplantations. *Liver Transpl*. 2023;29: 961–969.
41. Bromfield B, Tellez R, Novelli P, Duarte-Rojo A. Hepatic artery pseudoaneurysms and hepatic ischaemic injury: a rare complication of polyarteritis nodosa. *BMJ Case Rep*. 2022;15(11):e251432.
42. Parreau S, Dumonteil S, Gondran G, et al. Frequency and significance of hepatic involvement in new-onset giant cell arteritis: a study of 514 patients. *J Rheumatol*. 2023:jrheum.2023–0358.
43. Ohtani O, Ohtani Y. Lymph circulation in the liver. *Anat Rec (Hoboken)*. 2008;291(6):643–652.
44. Yokomori H, Oda M, Kaneko F, et al. Lymphatic marker podoplanin/D2-40 in human advanced cirrhotic liver–re-evaluations of microlymphatic abnormalities. *BMC Gastroenterol*. 2010;10:131.
45. Crawford AR, Lin XZ, Crawford JM. The normal adult human liver biopsy: a quantitative reference standard. *Hepatology*. 1998;28(2):323–331.
46. Roskams TA, Theise ND, Balabaud C, et al. Nomenclature of the finer branches of the biliary tree: canals, ductules, and ductular reactions in human livers. *Hepatology*. 2004;39(6): 1739–1745.
47. Saxena R, Theise N. Canals of Hering: recent insights and current knowledge. *Semin Liver Dis*. 2004;24(1):43–48.
48. Theise ND, Saxena R, Portmann BC, et al. The canals of Hering and hepatic stem cells in humans. *Hepatology*. 1999; 30(6):1425–1433.
49. Marakovits C, Francis H. Unraveling the complexities of fibrosis and ductular reaction in liver disease: pathogenesis, mechanisms, and therapeutic insights. *Am J Physiol Cell Physiol*. 2024;326(3):C698–C706.
50. Yoshizawa T, Lee JW, Hong SM, et al. Three-dimensional analysis of ductular reactions and their correlation with liver regeneration and fibrosis. *Virchows Arch*. 2023.
51. Craig CE, Quaglia A, Selden C, Lowdell M, Hodgson H, Dhillon AP. The histopathology of regeneration in massive hepatic necrosis. *Semin Liver Dis*. 2004;24(1):49–64.
52. Tiniakos DG, Lee JA, Burt AD. Innervation of the liver: morphology and function. *Liver*. 1996;16(3):151–160.
53. Dhillon AP, Sankey EA, Wang JH, et al. Immunohistochemical studies on the innervation of human transplanted liver. *J Pathol*. 1992;167(2):211–216.

54. Lee JA, Ahmed Q, Hines JE, Burt AD. Disappearance of hepatic parenchymal nerves in human liver cirrhosis. *Gut*. 1992;33(1): 87–91.
55. Schmucker DL. Age-related changes in liver structure and function: implications for disease? *Exp Gerontol*. 2005;40(8–9): 650–659.
56. Suriawinata A, Thung NS. *Liver Pathology: An Atlas and Concise Guide*. 1st ed. Demos Medical Publishing; 2011.
57. Torbenson M. *Biopsy Interpretation of the Liver*. 4th ed. Lippincott Williams & Wilkins; 2021.
58. Rockey DC, Caldwell SH, Goodman ZD, Nelson RC, Smith AD; American Association for the Study of Liver Diseases. Liver biopsy. *Hepatology*. 2009;49(3):1017–1044.
59. Tapper EB, Lok AS-F. Use of liver imaging and biopsy in clinical practice. *N Engl J Med*. 2017;377(8):756–768.
60. Cholongitas E, Senzolo M, Standish R, et al. A systematic review of the quality of liver biopsy specimens. *Am J Clin Pathol*. 2006;125:710–721.
61. Schiano TD, Azeem S, Bodian CA, et al. Importance of specimen size in accurate needle liver biopsy evaluation of patients with chronic hepatitis C. *Clin Gastroenterol Hepatol*. 2005;3(9):930–935.
62. Koehne de Gonzalez AK, Salomao MA, Lagana SM. Current concepts in the immunohistochemical evaluation of liver tumors. *World J Hepatol*. 2015;7(10):1403–1411.
63. Gu K, Shah V, Ma C, Zhang L, Yang M. Cytoplasmic immunoreactivity of thyroid transcription factor-1 (clone 8G7G3/1) in hepatocytes: true positivity or cross-reaction? *Am J Clin Pathol*. 2007;128(3):382–388.
64. Takahashi Y, Dungubat E, Kusano H, et al. Application of immunohistochemistry in the pathological diagnosis of liver tumors. *Int J Mol Sci*. 2021;22(11):5780.
65. Guo M, Zhang H, Zheng J, Liu Y. Glypican-3: a new target for diagnosis and treatment of hepatocellular carcinoma. *J Cancer*. 2020;11(8):2008–2021.
66. Strnad P, Zatloukal K, Stumptner C, Kulaksiz H, Denk H. Mallory-Denk-bodies: lessons from keratin-containing hepatic inclusion bodies. *Biochim Biophys Acta*. 2008;1782(12): 764–774.
67. Morrison C, Marsh W Jr, Frankel WL. A comparison of CD10 to pCEA, MOC-31, and hepatocyte for the distinction of malignant tumors in the liver. *Mod Pathol*. 2002;15(12):1279–1287.
68. Ijzer J, Roskams T, Molenbeek RF, et al. Morphological characterization of portal myofibroblasts and hepatic stellate cells in the normal dog liver. *Comp Hepatol*. 2006;5:7.
69. Peleman C, De Vos WH, Pintelon I, et al. Zonated quantification of immunohistochemistry in normal and steatotic livers. *Virchows Arch*. 2023;482(6):1035–1045.
70. Akita M, Fujikura K, Ajiki T, et al. Dichotomy in intrahepatic cholangiocarcinomas based on histologic similarities to hilar cholangiocarcinomas. *Mod Pathol*. 2017;30(7):986–997.
71. Balitzer D, Joseph NM, Ferrell L, et al. Immunohistochemical and molecular features of cholangiolocellular carcinoma are similar to well-differentiated intrahepatic cholangiocarcinoma. *Mod Pathol*. 2019;32, 1486–1494.
72. Nakanuma Y, Klimstra D, Komuta M, Zen Y. Intrahepatic cholangiocarcinoma. *World Health Organization Classification of Digestive System Tumours*. International Agency for Research on Cancer; 2019:254–259.
73. Zen Y. Intrahepatic cholangiocarcinoma: typical features, uncommon variants, and controversial related entities. *Hum Pathol*. 2023;132:197–207.
74. Hytiroglou P, Bioulac-Sage P, Theise ND, Sempoux C. Etiology, pathogenesis, diagnosis, and practical implications of hepatocellular neoplasms. *Cancers (Basel)*. 2022;14(15):3670.
75. Park JH, Kim JH. Pathologic differential diagnosis of metastatic carcinoma in the liver. *Clin Mol Hepatol*. 2019;25(1):12–20.
76. Alshareefy Y, Shen CY, Prekash RJ. Exploring the molecular pathogenesis, diagnosis and treatment of fibrolamellar hepatocellular carcinoma: a state of art review of the current literature. *Pathol Res Pract*. 2023;248:154655.
77. Hissong E, Al Assaad M, Bal M, et al. NIPBL::NACC1 fusion hepatic carcinoma. *Am J Surg Pathol*. 2024;48(2):183–193.
78. Nasir A, Lehrke HD, Mounajjed T, et al. Albumin in situ hybridization can be positive in adenocarcinomas and other tumors from diverse sites. *Am J Clin Pathol*. 2019;152(2):190–199.
79. Fiel MI, Schiano TD. Systemic disease and the liver-part 1: systemic lupus erythematosus, celiac disease, rheumatoid arthritis, and COVID-19. *Surg Pathol Clin*. 2023;16(3):473–484.
80. Fiel MI, Schiano TD. Systemic disease and the liver part 2: pregnancy-related liver injury, sepsis/critical illness, hypoxia, psoriasis, scleroderma/Sjogren's Syndrome, sarcoidosis, common variable immune deficiency, cystic fibrosis, inflammatory bowel disease, and hematologic disorders. *Surg Pathol Clin*. 2023; 16(3):485–498.
81. Penrice DD, Thakral N, Kezer CA, et al. Outcomes of idiopathic versus secondary nodular regenerative hyperplasia of the liver: a longitudinal study of 167 cases. *Liver Int*. 2022;42(6): 1379–1385.
82. Kounis I, Sebagh M, Evain M, et al. Nodular regenerative hyperplasia is not a rare condition after liver transplantation: incidence, predictive factors, and impact on survival. *Transplantation*. 2023;107(2):410–419.
83. Mikkelsen WP, Edmondson HA, Peters RL, Redeker AG, Reynolds TB. Extra- and intrahepatic portal hypertension without cirrhosis (hepatoportal sclerosis). *Ann Surg*. 1965;162(4): 602–620.
84. Guido M, Alves VAF, Balabaud C, et al; International Liver Pathology Study Group. Histology of portal vascular changes associated with idiopathic non-cirrhotic portal hypertension: nomenclature and definition. *Histopathology*. 2019;74(2):219–226.
85. Gerber MA, Thung SN, Bodenheimer HC Jr, Kapelman B, Schaffner F. Characteristic histologic triad in liver adjacent to metastatic neoplasm. *Liver*. 1986;6(2):85–88.
86. Hohn A, Jung T, Grimm S, Grune T. Lipofuscin-bound iron is a major intracellular source of oxidants: role in senescent cells. *Free Radic Biol Med*. 2010;48(8):1100–1108.
87. Strassburg CP. Hyperbilirubinemia syndromes (Gilbert-Meulengracht, Crigler-Najjar, Dubin-Johnson, and Rotor syndrome). *Best Pract Res Clin Gastroenterol*. 2010;24(5):555–571.
88. Sticova E, Jirsa M. New insights in bilirubin metabolism and their clinical implications. *World J Gastroenterol*. 2013; 19(38):6398–6407.
89. Batts KP. Iron overload syndromes and the liver. *Mod Pathol*. 2007;20(Suppl 1):S31–S39.
90. Ludwig J, Moyer TP, Rakela J. The liver biopsy diagnosis of Wilson's disease. Methods in pathology. *Am J Clin Pathol*. 1994;102(4):443–446.
91. Rowan DJ, Mangalaparthi KK, Singh S, et al. Metallothionein immunohistochemistry has high sensitivity and specificity for detection of Wilson disease. *Mod Pathol*. 2022;35(7):946–955.
92. Wiethoff H, Mohr I, Fichtner A, et al. Metallothionein: a game changer in histopathological diagnosis of Wilson disease. *Histopathology*. 2023;83(6):936–948.

28

Gallbladder and Extrahepatic Biliary System

Edward B. Stelow ▪ Seung-Mo Hong

The gallbladder, extrahepatic biliary system, liver, and dorsal pancreas all develop from the hepatic diverticulum of the distal foregut that appears in the 4th week of embryonic life (1). The cranial bud of the diverticulum extends outward to form the common duct and, eventually, the liver. The lumen of the common duct extends into the cystic duct and gallbladder which remains solid until the 12th week of life. Developmental anomalies are uncommon. Gallbladder agenesis results in an absence of the gallbladder, and two or more gallbladders rarely develop in the same patient (2–4).

The gallbladder is one of the most common surgical pathology specimens. It is most often resected due to clinical symptoms related to gallstones or inflammatory disease: most neoplasms are incidentally discovered in gallbladders obtained for other reasons. Normal gallbladders are commonly removed during liver transplantation, but incidental cholecystectomy specimens are otherwise uncommon. The situation is quite different with the extrahepatic bile ducts and ampullae of Vater. These sites are sampled only when neoplasia is suspected. Although the distal common bile duct and ampulla can be studied in pancreatoduodenectomy specimens but they are often abnormal due to the effects of neoplasia in close proximity to these structures. The remaining extrahepatic bile duct is rarely resected. The complete biliary system and ampulla can be only studied with autopsy material; however, autolysis is usually present in these specimens as a result of bile toxicity.

This chapter discusses the gross anatomy, physiology, histology, immunohistochemistry, and ultrastructure of the normal gallbladder, extrahepatic bile ducts, ampulla of Vater, and minor papilla.

GALLBLADDER

Gross Anatomy

The gallbladder is a piriform bladder that is attached to the extrahepatic biliary system via the cystic duct. It rests in a shallow depression located on the inferior surface of the posterior right lobe of the liver. It measures up to 10 cm long and 3 to 4 cm wide in normal adults, and its wall is 1 to 2 mm thick, varying due to the degree of muscular contraction. The gallbladder is mostly covered by serosa with the exception of a bare area where it lies in the gallbladder fossa. The gallbladder is anatomically divided into a blindly ending fundus, a large central body, and a narrow neck that joins the cystic duct. The tapered area of the body that joins the neck is considered the infundibulum. It is here that a peritoneal fold, the cholecystoduodenal ligament, attaches the gallbladder to the first portion of the duodenum. Hartmann pouch, a small bulge at the infundibulum, is probably not normal and may be the result of chronic inflammation or stone impaction (5). The neck is somewhat serpentine, measures 5 to 7 mm long, and narrows as it connects with the cystic duct (6).

Physiology

The gallbladder concentrates, stores, and releases bile. Approximately 800 to 1,000 mL of bile flow daily into the gallbladder from the liver (7). Its filling results from complex neural and hormonal stimulations that result in its relaxation and the contraction and closing of the sphincter of Oddi. When the sphincter is closed, the intraluminal pressure of the bile ducts will increase as bile is continuously produced by the liver; bile will then flow into the gallbladder. When relaxed, the gallbladder can store only 40 to 70 mL of bile and retains a constant intraluminal pressure (7). A much larger volume, however, is handled as the gallbladder concentrates bile via a sodium-coupled transport of chloride, mediated by NaK-ATPase (8). The active transport of electrolytes into the lateral intercellular space creates an osmotic gradient such that water flows through the basement membrane into the capillaries of the lamina propria, thereby concentrating the luminal contents. The gallbladder also secretes mucosubstances from the surface epithelial cells and neck mucous glands to protect the mucosal barrier from bile-related injury.

Contraction of the gallbladder is mediated by complex neural and humoral mechanisms that function both after and between meals (9). Cholecystokinin is released from the proximal duodenal mucosa after fatty meals and is the most important hormone that promotes gallbladder contraction (10). Motilin aids in interdigestive gallbladder contraction which occurs in tandem with giant migratory complexes of the intestines every 2 hours or so (9,11). Pancreatic polypeptide and somatostatin may also effect gallbladder motility (9,11–13). The vagal system plays a role both directly and indirectly in gallbladder contraction (14). The complicated balance of humoral and neural mechanisms governing gallbladder activity is sometimes disrupted, resulting in gallbladder dysmotility (15). Dysmotility may, in turn, result in gallbladder pathology (16).

Blood Supply and Lymphatic Vessel Drainage

The arterial supply of the gallbladder varies both in its anatomy and its relationship to the extrahepatic biliary system. The cystic artery supplies the gallbladder and arises from the proximal portion of the right hepatic artery in more than 70% of individuals; it arises from the superior mesenteric artery in 13% of individuals followed by the common hepatic artery, left hepatic artery, gastroduodenal artery, celiac artery, and aorta (17). Most commonly, the cystic duct artery is located superior to the cystic duct (Fig. 28.1). A double cystic artery is present in 14% to 25% of individuals based on autopsy data (17,18).

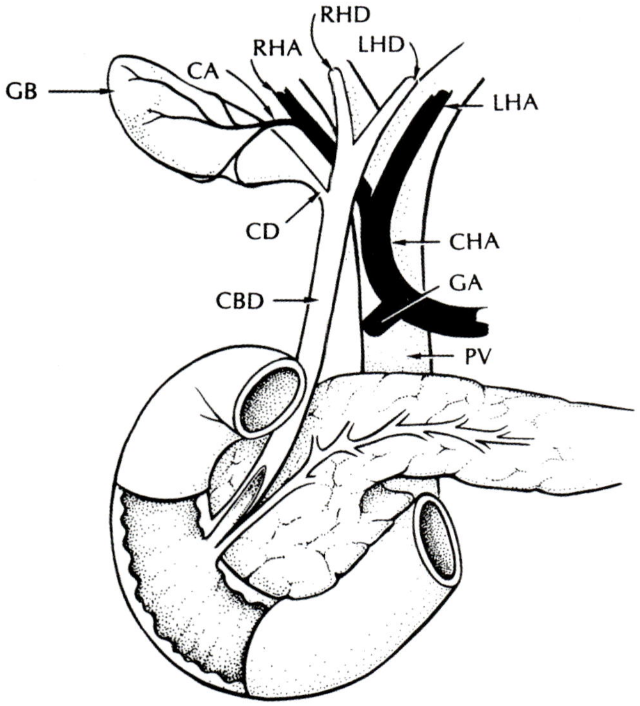

FIGURE 28.1 Although variations are common, this diagram depicts the "usual" relationships of the extrahepatic bile ducts, portal vein, and branches of the common hepatic artery. CA, cystic artery; CBD, common bile duct; CD, cystic duct; CHA, common hepatic artery; GA, gastroduodenal artery; GB, gallbladder; LHA, left hepatic artery; LHD, left hepatic duct; PV, portal vein; RHA, right hepatic artery; RHD, right hepatic duct.

The cystic artery branches to form superficial channels that course through the subserosa of the gallbladder and deep channels that lie between the gallbladder and the hepatic bed (6).

The venous drainage of the gallbladder consists of a network of small venous channels on the hepatic side of the gallbladder that lead directly into the liver. Other small veins flow toward the cystic duct and merge with channels from the common bile duct before terminating in the portal venous system.

The lymphatic vessels of the gallbladder drain first into lymph nodes at the gallbladder neck or cystic duct. Following this, drainage may proceed to retropancreatic, celiac, or mesenteric lymph nodes (19,20). These pathways converge at the abdominal aortic lymph nodes located near the superior mesenteric artery (19). Retrograde lymph flow to the hepatic hilum is unusual but may occur when the lymphatic channels are obstructed by cancer, inflammation, or surgical ligation (20).

Nerve Supply

Nerve branches from the left trunk of the vagus nerve join the hepatic plexus. The hepatic plexus then supplies the gallbladder with sympathetic, parasympathetic, and possibly afferent nerve fibers (6). Vagal stimulation, both directly and indirectly, may then stimulate interdigestive periodic contractions of gallbladder smooth muscle (14). Neuropeptide Y nerve fibers also participate in gallbladder smooth muscle contraction, and are found in all layers of the gallbladder. They form a particularly dense network in the lamina propria, running near the epithelium and paralleling the muscle bundles (21).

Histology

The layers of the gallbladder include the mucosa, muscularis propria, perimuscular subserosal connective tissue, and serosa. The gallbladder does not contain a muscularis mucosae or submucosa. The luminal folds are lined by a single layer of columnar epithelium supported by lamina propria. The height and width of the folds are variable, and branching is characteristic (Fig. 28.2). The columnar epithelial cells have lightly eosinophilic cytoplasm with occasional small apical vacuoles (Fig. 28.3). Some lining epithelial cells contain oblong supranuclear inclusions. These eosinophilic bodies were once thought to represent Cystisospora belli, although subsequent studies demonstrated that they were simply proteinaceous aggregates rather than organisms. Nuclei are aligned at the cell base or slightly more centrally. They are ovoid and uniform and have fine chromatin and smooth membranes. Nucleoli are absent or very small and inconspicuous. Occasional columnar cells are narrow with dark eosinophilic cytoplasm (5). These pencil-like cells appear to be little more than contracted columnar cells, although they reportedly have a few ultrastructural and enzymatic properties that differ from those of the usual columnar cells. Basal cells are inconspicuous and have nuclei that lie just above and parallel to the basement membrane.

Tubuloalveolar mucous glands are located only in the neck of the gallbladder (22). They have cuboid or low columnar cells with abundant clear to lightly basophilic cytoplasm and round, basally oriented nuclei (Fig. 28.4). Their lectin-binding profile is dissimilar to that of the surface epithelial cells (23). The neck mucous glands also differ morphologically and histochemically from the pyloric-type metaplastic glands found in the fundus, body, or neck of chronically inflamed gallbladders or those that contain gallstones (22). Rare endocrine cells can be found in the gallbladder mucosa (24,25). Both foveolar-type and pyloric-type metaplasia can be seen in patients with chronic cholecystitis

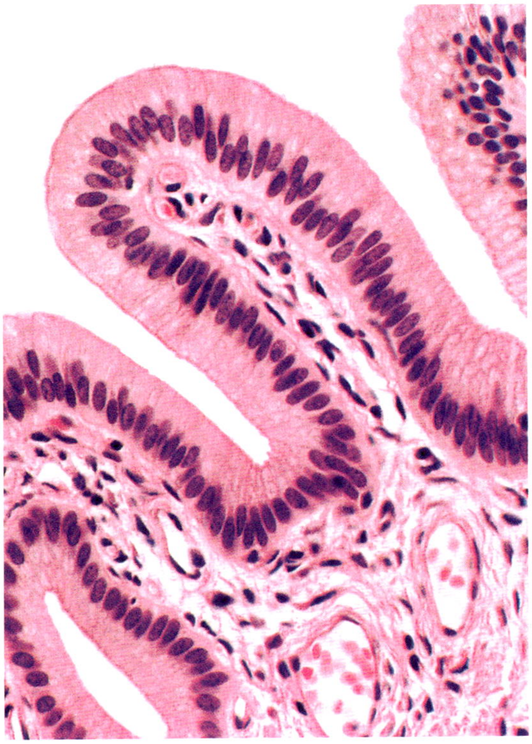

FIGURE 28.3 The gallbladder is lined by a single layer of tall columnar cells with basally oriented nuclei.

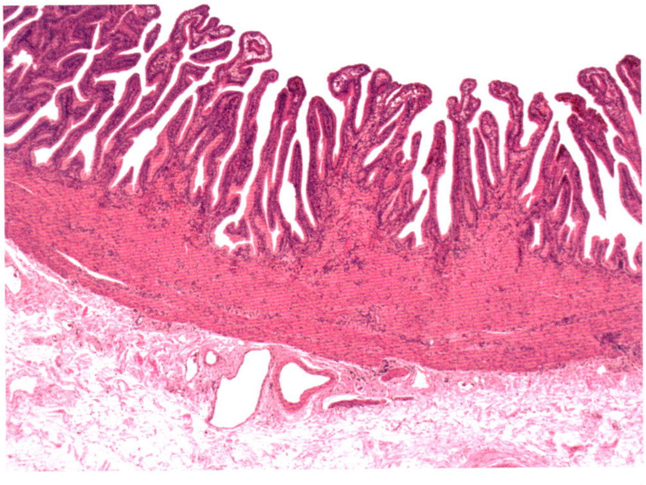

FIGURE 28.2 The luminal folds of the gallbladder vary in height and contain a delicate core of lamina propria above the bundles of smooth muscle.

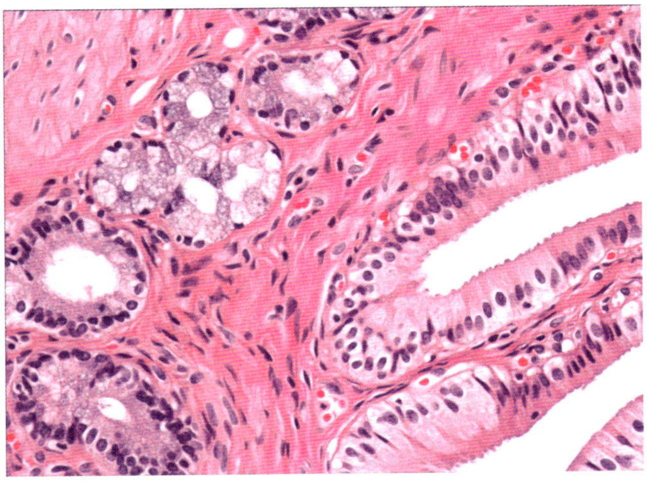

FIGURE 28.4 Mucous glands are present only in the neck of the normal gallbladder.

and cholelithiasis. Intestinal metaplasia may also be present in chronic cholecystitis and cholelithiasis, featuring goblet cells as well as a lesser component of absorptive cells, endocrine cells, and Paneth cells (5,22,26–30). Squamous epithelium is not found in the normal gallbladder and is rare in patients with cholecystitis (5). Melanocytes are not found in the normal epithelial lining. However, a few small lymphocytes often are seen between the surface columnar cells.

Gallbladder epithelial cells contain chiefly sulfated acid mucin with very small quantities of nonsulfated acid mucin (22). In contrast, metaplastic goblet cells, foveolar-type cells, and pyloric-type glands contain nonsulfated acid mucin and neutral mucin, but little sulfated acid mucin. By immunohistochemistry, the gallbladder epithelial cells express MUC5AC and MUC6 similar to gastric epithelium but they lack pepsinogens I and II (29,31). Lysozyme is also absent in the normal gallbladder columnar cells but may be found in metaplastic glands (32). Alpha-1-antitrypsin and alpha-1-antichymotrypsin are present in both normal and metaplastic epithelia (32).

Immunohistochemical staining for polyclonal carcinoembryonic antigen (CEA) shows focal weak staining along the apices of some lining cells in the normal gallbladder, although staining can be increased in inflammatory conditions (33,34). Absorption of at least one polyclonal antibody with human liver powder abolishes this immunoreactivity because it eliminates nonspecific crossreactivity with biliary glycoprotein (34). The surface epithelium and neck mucous glands are strongly immunoreactive for epithelial membrane antigen and low–molecular-weight keratin. Normal gallbladder epithelium also reacts with antibodies directed against cytokeratin (CK) 7 and carbohydrate antigen (CA) 19.9, and it does not react with antibodies directed against CK20, MUC2 or CDX2 (Fig. 28.5) (35–38). Normal gallbladder mucosa lacks immunoreactivity for estrogen receptor, although focal immunopositivity can be seen in foci of pyloric metaplasia (39).

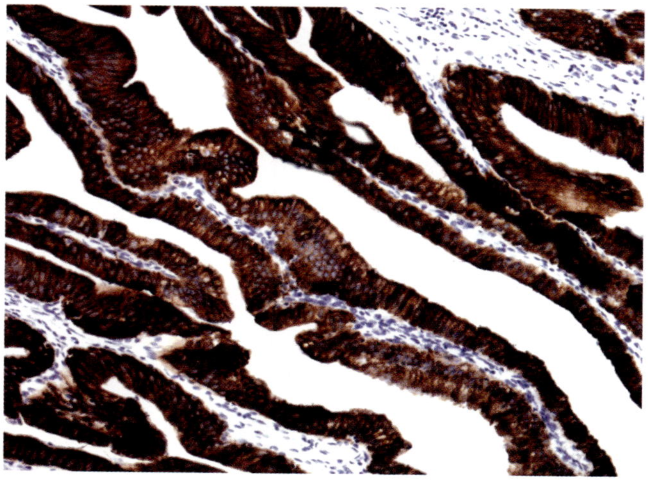

FIGURE 28.5 The epithelial lining of the gallbladder reacts strongly with anitbodies directed against cytokeratin 7 (immunoperoxidase technique).

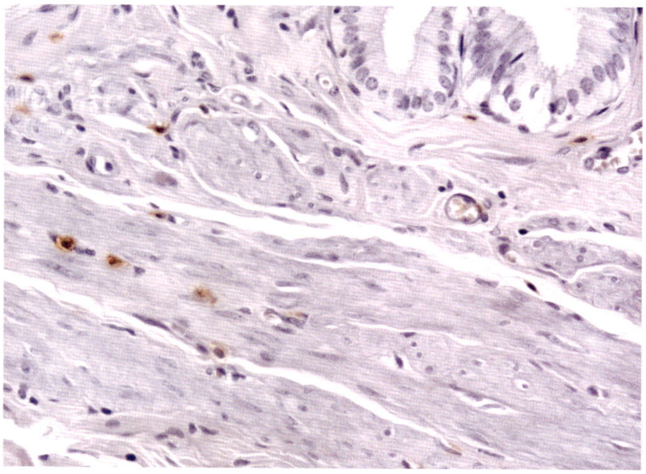

FIGURE 28.6 Occasional mast cells are identified within the gallbladder by kit-immunostaining. No interstitial cells of Cajal were identified (immunoperoxidase technique).

Adhesion molecules expressed by normal gallbladder epithelial cells include alpha-catenin, beta-catenin, gamma-catenin, CD44, CD99, and E-cadherin (40).

The lamina propria contains loose connective tissue, elastic fibers, nerve fibers, small blood vessels, and lymphatic channels. Mast cells and macrophages may be seen in small numbers, and it has been noted that these cells are more numerous in normal or minimally inflamed gallbladders than in those with overt chronic cholecystitis (Fig. 28.6) (41). Polymorphonuclear leukocytes are not normally present in the lamina propria, but small numbers of lymphocytes and plasma cells are usual. Plasma cells that contain immunoglobulin (Ig) A occur chiefly in the lamina propria, whereas IgM-containing cells are more frequent in the smooth muscle layer (42). A few IgG-containing plasma cells also may be present.

The muscularis consists of loosely arranged bundles of circular, longitudinal, and oblique fibers that do not form well-developed layers like they do in the luminal gut. Fibrovascular connective tissue focally separates the muscle bundles. Fascicles of smooth muscle cells sometimes extend high into the lamina propria to just beneath the epithelial basement membrane. The thickness of the muscle layer is quite variable which may simply reflect variable contractile states of the specimen. Ganglion cells are found in the lamina propria, between smooth muscle cell bundles, and in the subserosal connective tissue (Fig. 28.7). Smooth muscle cells are typically immunoreactive with antibodies to smoothelin and desmin consistent with a contractile muscularis propria (43). Interstitial cells of Cajal have been rarely noted and rare gastrointestinal stromal tumors of the gallbladder have been reported (5,44).

The subserosal tissue contains loose collagen fibers, fibroblasts, elastic fibers, adipocytes, blood vessels, nerves, and lymphatic vessels. Small aggregates of lymphocytes may occur around vessels and rare lymph nodes may be found in the subserosal connective tissue (45). Paraganglia are found

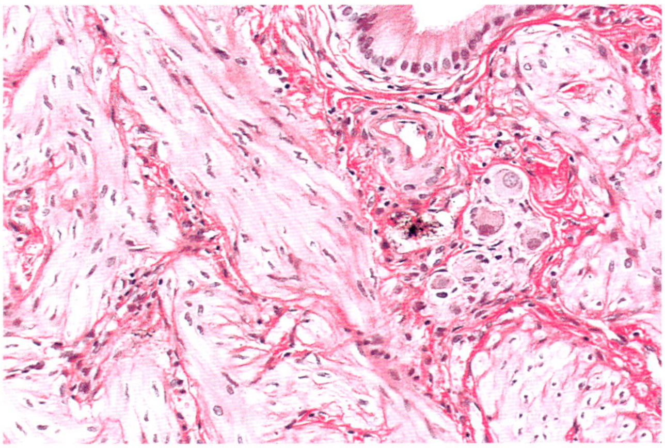

FIGURE 28.7 Ganglion cells are readily seen in the connective tissue layers of the gallbladder.

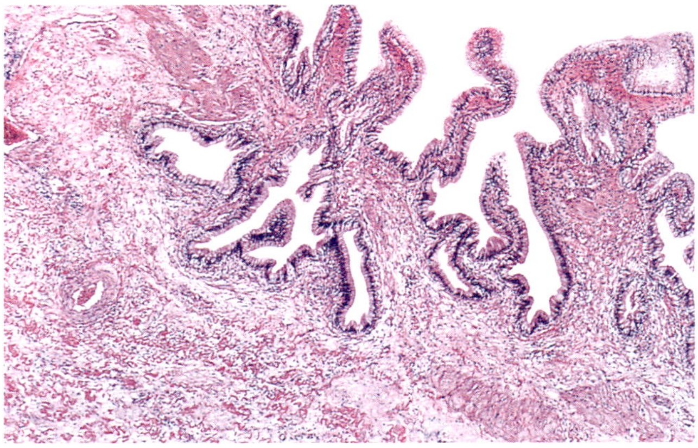

FIGURE 28.9 Rokitansky–Aschoff sinuses occur in the normal gallbladder but uncommonly penetrate through the smooth muscle bundles.

adjacent to blood vessels and small nerves but are few in number (Fig. 28.8) (46).

Rokitansky–Aschoff sinuses represent herniations of mucosa into the muscularis, or subserosal connective tissue (Fig. 28.9). These are commonly considered a feature of chronic cholecystitis, being present in nearly 90% of inflamed gallbladders muscularis (47). Superficial sinuses are observed in 42% of normal gallbladders in which case they are generally confined to the lamina propria (48). Penetration of the muscularis is infrequent in normal gallbladders, and Rokitansky–Aschoff sinuses are not found in gallbladders from fetuses and only a few superficial outpouchings are observed in organs from infants (49). The diameters of the sinuses are variable, and flask-shaped formations are usual. Although the exact mechanism for their formation is unknown, herniation of the epithelium may result from increased intraluminal pressure and extreme gallbladder contractions that propel mucosal elements through weak points in the gallbladder wall (49). The sinuses are uncommonly involved by intraepithelial neoplasia and when this happens, careful attention showed be paid to distinguish the changes from invasive adenocarcinoma (50).

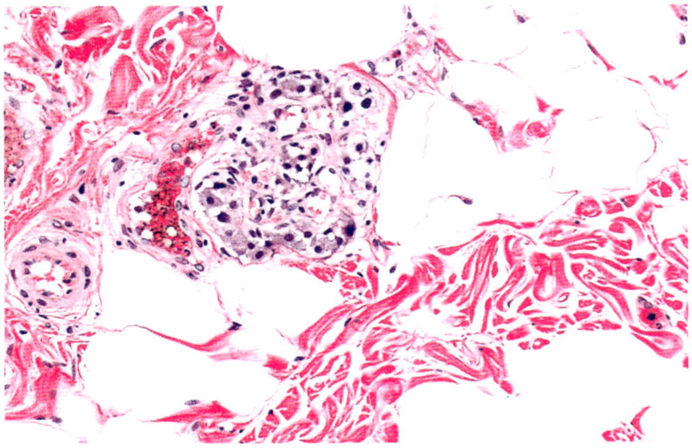

FIGURE 28.8 Paraganglia are located in the subserosal connective tissue of the normal gallbladder.

The recognition of nonneoplastic epithelium, bile within the dilatations, and lack of desmoplasia all point to noninvasive disease.

Luschka ducts are small bile ducts that lie in the connective tissue deep to the muscularis propria; they can occur in a subserosal location but are more common on the hepatic side of the gallbladder (Fig. 28.10). These ducts are present in 10% to 12% of cholecystectomy specimens, occurring in both normal and diseased organs (48,49). They have been observed in gallbladders from infants, adolescents, and adults and may represent embryonic remnants. Reports of their drainage site are varied. Some data suggest they communicate with intrahepatic bile ducts, but those beneath the serosa could drain into the peritoneal cavity or even the gallbladder lumen, especially within its neck (47,49,51).

Luschka ducts are solitary or multiple but are usually present in small groups surrounded by a distinctive ring of connective tissue. Serial sections may demonstrate a system of anastomosing channels. The diameters of their lumina vary from several microns up to a few millimeters. The ducts are lined by cells similar to those of the intrahepatic bile ducts. In some instances, small foci of hepatic parenchyma are located adjacent to the ducts (49). Luschka ducts are distinct from Rokitansky–Aschoff sinuses and the two should not communicate. It should be noted that the term Luschka duct has been used somewhat indiscriminately to refer to entities ranging from Rokitansky–Aschoff sinuses to true accessory ducts.

Larger accessory ducts are sometimes seen in the gallbladder bed. They may be mistaken grossly for small veins or thin strands of fibrous tissue (52). If these ducts are not ligated during surgery, a bile leak may develop. Although these ducts lie in the gallbladder wall, they usually do not drain into the lumen of the fundus. Rather, they communicate with the cystic or hepatic ducts (53–55).

Ectopic hepatic, pancreatic, adrenal, gastric, and thyroid tissues as well as foregut cysts have been reported in the gallbladder (56–65). Ectopic hepatic and adrenal tissues are typically incidental findings, whereas ectopic

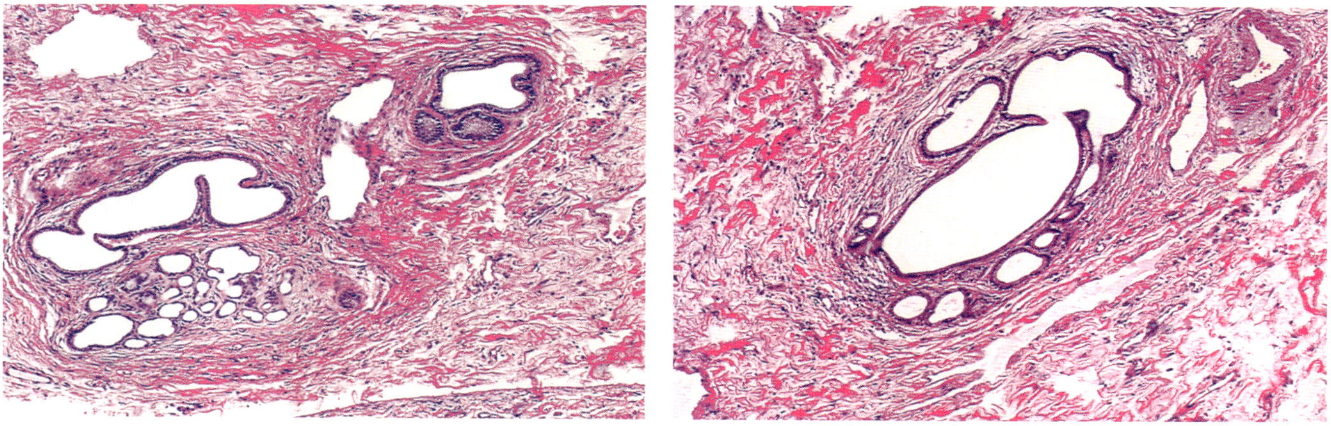

FIGURE 28.10 Luschka ducts consist of groups of small ducts having lumina of various caliber. They are surrounded by condensed connective tissue.

pancreatic or gastric tissues may cause symptoms related to their secretions (56).

Ultrastructure

The surface columnar cells measure 15 to 25 μm in height and 2.5 to 7.0 μm in width and rest on a basement membrane (66). These cells have numerous apical microvilli with a filamentous glycocalyx and core rootlets (Fig. 28.11). The microvilli are shorter and more variable in size and density than those of the intestinal epithelium. Pinocytotic vesicles are formed from the intervillous portions of the cell membrane. The lateral cell membranes are straight at the apex and connected by junctional complexes. Below this boundary, the cell membranes have complex interdigitations that surround lateral intercellular spaces (Fig. 28.11). The diameter of the intercellular space varies depending on the state of fluid transport (67). It is collapsed when there is no water transport but is distended during influx of electrolytes and water. The nuclei are ovoid, have prominent euchromatin, and occasional small nucleoli. The cytoplasm contains rough endoplasmic reticulum, mitochondria, glycogen, filaments, Golgi apparatus, mucous granules, vesicles, and lysosomes.

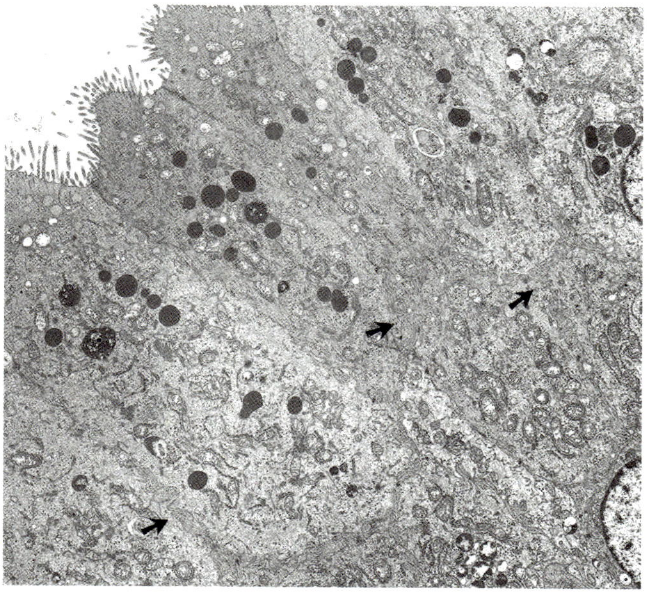

FIGURE 28.11 Ultrastructurally, the apical portion of the columnar cells of the gallbladder contains abundant microvilli with core rootlets, mitochondria, Golgi apparatus, mucous granules, lysosomes, and a few strands of rough endoplasmic reticulum. The lateral cell membranes form complex interdigitations (*arrows*).

Pencil cells have slender outlines, narrow nuclei, and dense cytoplasm that is packed with organelles. At the base of the pencil cell, cytoplasmic extensions project into the basement membrane, unlike that for the typical columnar cell (68). However, microvilli and lateral membrane interdigitations are similar for the pencil cell.

The basal cell measures 10 to 15 μm in diameter, has an irregular nucleus, and has cytoplasmic organelles that include rough and smooth endoplasmic reticulum, mitochondria, vacuoles, and ring-shaped osmiophilic inclusions (66,68). They have a cytoplasmic extension that runs parallel to the basement membrane, changes direction to run perpendicularly, and then branches toward the lumen (66). The branches are variable in length, delicate, and complex. Throughout the lining epithelium there are intraepithelial nerve endings that originate from the nerve submucosal plexus and are associated with the small basal cells (66).

Capillaries are found just below the epithelial basement membrane, and their lumina change in size according to the state of fluid transport. The epithelial cells of the glands in the gallbladder neck have a few short microvilli, relatively even lateral membranes, rare secretory granules, and round nuclei (69).

CYSTIC DUCT

The cystic duct is located at the right free edge of the lesser omentum and usually joins the right lateral portion of the common hepatic duct approximately 2 cm distal to the union of the right and left hepatic ducts. The mean length of the cystic duct is approximately 30 mm with a range from 4 to 65 mm with a collapsed diameter of 4 mm (17). Connection and drainage into the common hepatic duct varies; most cystic ducts drain laterally at an acute angle into the common bile duct (17,70). However, it may form an angular junction with either the anterior or posterior aspect of the common hepatic duct, run parallel to the common hepatic duct, or be unusually long (70). In rare instances, the cystic duct may join the right and left hepatic ducts, forming a trifurcation. The cystic duct usually passes inferiorly to the cystic artery and to the right of the right hepatic artery.

The lining of the cystic duct is pleated with short folds of varying width and height. The surface cells are identical microscopically and immunohistochemically to those of the gallbladder (71). Groups of mucous glands are embedded in the dense, collagenous lamina propria. Lectin-binding patterns of the lining cells are similar to those for the surface epithelial cells of the gallbladder body and neck, and the lectin-binding profiles for the mucous glands of the cystic duct are indistinguishable from those of the glands at the gallbladder neck (23). Enterochromaffin cells containing serotonin have been described in cystic ducts from patients with pancreaticobiliary disease (72). In this same group of patients, a few intramural gland cells have shown immunoreactivity for somatostatin.

The connective tissue of the large, oblique folds, grossly visible in the cystic duct at the junction with the gallbladder neck, contains thin groups of smooth muscle cells (i.e., spiral valve of Heister) (Fig. 28.12). The smooth muscle is believed to prevent both overdistention and collapse of the cystic duct when it is subjected to changes in pressure (6). These cells show no or only limited immunoreactivity with antibodies to smoothelin (43). Abundant collagen and some elastic fibers, nerve fibers, and ganglion cells are intermixed with the smooth muscle. Nerve fibers showing immunoreactivity for vasoactive intestinal peptide (VIP) and other peptides have been described in the wall (13,72). The loose subserosal connective tissue contains adipose tissue, nerves with occasional ganglion cells, large blood vessels, and lymphatic channels. Lymphocytes and plasma cells are sparse or absent.

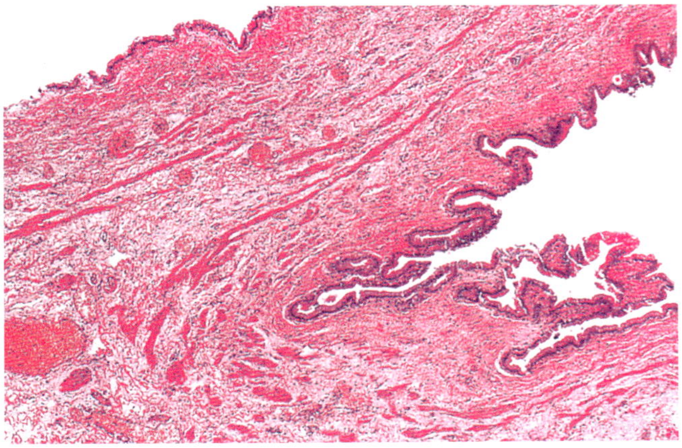

FIGURE 28.12 The stroma of the spiral valve of Heister contains thin strands of smooth muscle fibers.

RIGHT AND LEFT HEPATIC DUCTS, COMMON HEPATIC DUCT, AND COMMON BILE DUCT

Gross Anatomy

The right and left hepatic ducts, common hepatic duct, and common bile duct are embedded between the serous layers of the hepatoduodenal ligament (i.e., the right free border of the lesser omentum). The hepatic ducts emerge from the liver and, in most instances, unite in the hilum approximately 1 cm from the liver to form the common hepatic duct. In 10% to 30% of cases, two large segmental ducts drain the right hepatic lobe and join separately with the left hepatic duct, common hepatic duct, or cystic duct; it is incorrect to label one of these ducts the right hepatic duct and the other as an accessory (73). The mean length of the right hepatic duct is approximately 0.8 cm (range 0.2–2.5) and that of the left hepatic duct is 1.0 cm (range 0.2–3.5) (74). The usual diameter of each hepatic duct spans 3 to 4 mm and the length of the common hepatic duct ranges from 0.8 to 5.2 cm (mean 2.0) with a diameter ranging from 0.4 to 2.5 cm (74,75). The diameter of the common hepatic duct and its number of elastic fibers increase with age (75). The common bile duct, resulting from the union of the cystic duct and common hepatic duct, can be divided into supraduodenal, retroduodenal, pancreatic, and intraduodenal segments. It is usually about 1 mm thick and 5 cm long, but its length is quite variable (range 1.5–15.0 cm) (74,76). The diameter at its midpoint ranges from 0.4 to 1.4 cm (mean 0.66), and its lumen narrows by approximately 50% after entering the duodenal window (74,77). In an autopsy study of 100 selected subjects who ranged in age from 15 to 102 years, lacked a history of biliary tract disease, and had completely intact biliary tracts, the outer diameters of the upper portions of the common bile ducts ranged from 0.4 to 1.2 cm (mean 0.74) (78). The outer diameters increase with age but are not related to body weight or length (78). The pits in the surface epithelium (i.e., sacculi of Beale) are conspicuous in the extraduodenal portion of the common bile duct and the hepatic ducts. The wall of the common bile duct thickens approximately 2 mm from the duodenal wall because there is an increase in the amount of mural smooth muscle. This thickening corresponds to abrupt narrowing of the duct lumen.

Arterial Supply, Venous Drainage, and Relationship to Bile Ducts

The common hepatic artery arises from the celiac trunk and divides into right and left hepatic branches (Fig. 28.1).

Variations in the origins of the right and left hepatic arteries and their relationships to the extrahepatic bile ducts are typical (79). In one study, almost 42% of 200 cadavers had aberrant hepatic arteries (18). Most often, the right hepatic artery is dorsal to the common hepatic duct and right hepatic duct. The common hepatic and left hepatic arteries lie to the left of the extrahepatic bile ducts and ventral to the portal vein. The gastroduodenal artery lies to the left of the common bile duct, and a branch, the superior pancreaticoduodenal, traverses the duct either dorsally or ventrally (6).

The extrahepatic bile ducts are supplied by numerous arteries. The major arteries that supply branches to the common hepatic duct and the common bile duct include the retroduodenal, right and left hepatic, posterior superior and anterior superior pancreaticoduodenal, common hepatic, cystic, gastroduodenal, and retroportal arteries (80). The most important branches travel along the lateral borders of the common bile duct (81).

The portal vein, formed by the union of the splenic and superior mesenteric veins, lies dorsal to the bile ducts (Fig. 28.1). The mean length is 6.4 cm (range 4.8–8.8) and its mean diameter is 0.9 cm (range 0.64–1.21) (82). Venous channels that drain the superior portion of the common bile duct and common hepatic duct drain into marginal vessels that then enter the liver directly (83). Vessels from the inferior portion of the common bile duct lead to the portal vein.

Lymphatic Vessel Drainage

Lymphatic channels from the common bile duct drain into lymph nodes located along the hepatoduodenal ligament or within the posterior pancreaticoduodenal area. Drainage then proceeds to lymph nodes at the superior mesenteric artery, aorta, and common hepatic duct (84,85).

Nerve Supply

The nerve supply to the cystic and hepatic ducts derives from the anterior portion of the hepatic plexus, whereas nerves that supply the common bile duct arise from the posterior segment of the hepatic plexus. The nerve of the common bile duct, lying dorsally, is the right portion of the posterior hepatic plexus. Smaller branches from the posterior hepatic plexus travel inferiorly along the common bile duct and accompany the duct to the major duodenal papilla (6). Neuropeptide Y–containing nerve fibers have the same pattern of distribution in the common bile duct as in the gallbladder (21).

Histology

The extrahepatic bile ducts, serving as conduits for the flow of bile, are lined by a single layer of tall columnar cells surrounded by a dense connective tissue layer (Fig. 28.13). The surface of the epithelium is relatively flat or pleated. The columnar cells have basally oriented nuclei that are ovoid and uniform. Nucleoli are absent or very small. Goblet cells are absent in normal epithelium. The epithelium dips into the stroma to form shallow depressions or deeper pits (i.e., sacculi of Beale). The deeper sacculi may appear to be isolated from the surface epithelium, but deeper sections will often show their connections. Surrounding the sacculi are unevenly distributed lobules of glands that empty into the sacculi (Fig. 28.14). These glands have been termed diverticula, crypts, parietal sacculi, deep glands, biliary glands, periductal glands, and extrahepatic peribiliary glands (86). When located in the more peripheral connective tissue, the

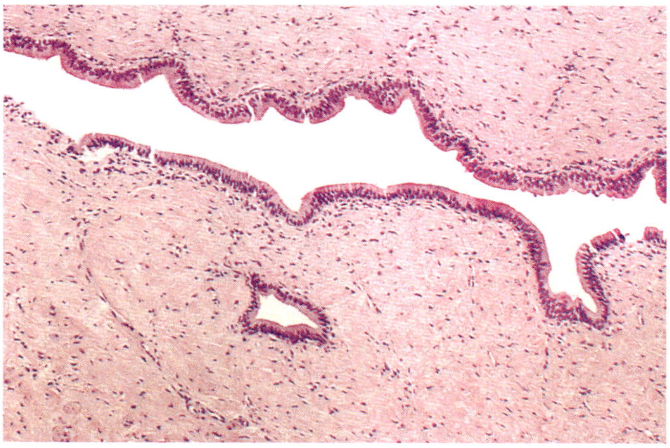

FIGURE 28.13 Intrapancreatic segment of common bile duct. The extrahepatic bile ducts are lined by a single layer of tall columnar cells overlying dense, collagenous connective tissue. In segments of the common bile duct away from the duodenum, a few small groups of smooth muscle fibers are sometimes found in the outer half of the wall.

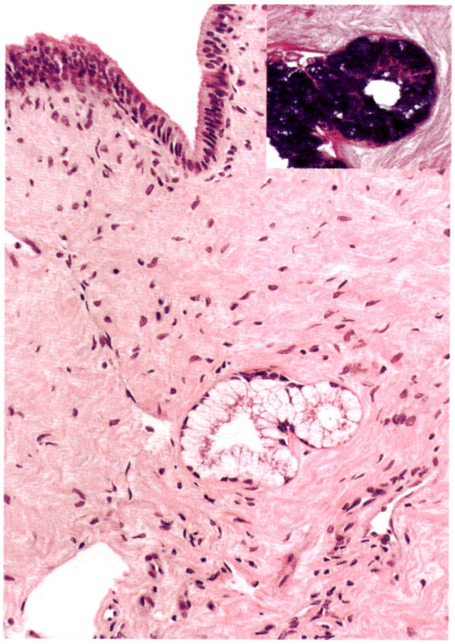

FIGURE 28.14 Glands embedded in the subepithelial collagenous stroma of the extrahepatic bile ducts typically contain cells with mucin-filled cytoplasm. (Inset: Alcian blue–periodic acid–Schiff [PAS] stain from the same field).

glands are encircled by condensed stroma. The peribiliary tubular glands are branched or, occasionally, simple and tubular (86). Although they are found in all parts of the extrahepatic bile duct system, they are less frequent in the central portion of the common bile duct and in the intrapancreatic portion than around the bile duct at the ampulla. Peribiliary tubular glands are lined by low-columnar or cuboid cells, many of which are filled with mucus (Fig. 28.14). With inflammation and fibrosis, the sacculi and peribiliary glands may become distorted, mimicking well-differentiated adenocarcinoma with desmoplastic stroma, especially in frozen sections. The lack of a lobular arrangement and the presence of marked nuclear atypia and perineural invasion are diagnostic of adenocarcinoma (87).

The surface epithelial cells contain smaller quantities of mucin than the cells that line the gallbladder (88). The former also contain sulfated acid mucin, whereas metaplastic and dysplastic cells primarily contain nonsulfated acid mucin and smaller quantities of sulfated and neutral mucins. The normal lining epithelium stains similarly to that of the gallbladder for epithelial membrane antigen and low–molecular-weight keratin (CAM 5.2 antibody). Cytokeratin 7 is consistently expressed in normal epithelium, while cytokeratin 20 expression depends on the condition of the epithelial cells. In normal cells, expression of cytokeratin 20 is usually absent, however, it may be expressed when metaplasia, hyperplasia or carcinoma are present (35,89). Polyclonal CEA immunoreactivity may be absent or appear as focal weak staining along the apices of some cells (90). Cytoplasmic staining using either polyclonal or monoclonal anti-CEA antibodies is typically absent (91). Immunoreactivity for lysozyme has been found in the cytoplasm of the cells in the glands, whereas staining of the surface epithelial cells is absent or very weak (90). In addition, cells of the peribiliary glands are usually immunoreactive for pancreatic and salivary alpha-amylase, trypsin, and lipase (86). The surface epithelium of the common bile duct also shows immunoreactivity for these enzymes.

Gastric metaplasia and intestinal metaplasia are sometimes found in inflamed and fibrotic extrahepatic bile ducts (28,88,92). Scattered endocrine cells, including cells immunoreactive for chromogranin and somatostatin, can be observed between mucin-containing cells in normal as well as diseased biliary epithelium (72,92–94).

The stroma directly beneath the surface epithelium is dense and contains abundant collagen, elastic fibers, and some small vessels (Fig. 28.13). Lymphocytes are sparse. Pancreatic acini and ducts may be seen in the wall of the intrapancreatic portion of the common bile duct (95). Small pancreatic ducts sometimes empty into this segment of the duct. The peripheral stroma of the common bile duct is less dense than the inner connective tissue and contains large blood vessels, lymphatic channels, nerves and ganglion cells, elastic fibers, and smooth muscle cell aggregates (96). This stroma merges with the connective tissue of the hepatoduodenal ligament. The distribution of smooth muscle

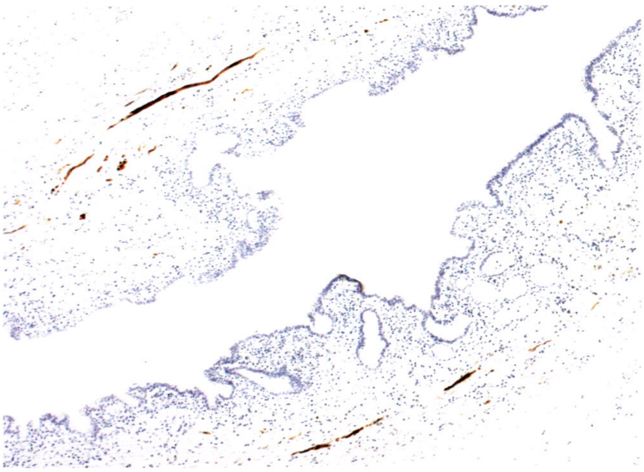

FIGURE 28.15 Occasional strands of smooth muscle are demonstrated in the upper portion of the common bile duct reacting with antibodies directed against desmin (immunoperoxidase technique).

cells varies throughout the bile duct. Scattered groups of cells are present at the upper third of the bile duct, whereas a continuous or interrupted pattern of thick smooth muscle cell bundles is present throughout the lower third of the bile duct (Fig. 28.15) (97). The smooth muscle cells are arranged longitudinally and are intermixed with collagen and elastic fibers. These smooth muscle cells are typically not immunoreactive with antibodies to smoothelin (43). Nerve fibers showing immunoreactivity for VIP are present beneath the epithelium and in smooth muscle (94).

VATERIAN SYSTEM AND MINOR PAPILLA

Gross Anatomy

The Vaterian system is composed of the segments of the common bile duct and major pancreatic duct (occurring either separately or as a common channel) at the duodenum, major papilla, and the sphincteric musculature. It also includes the extraduodenal portion of the common bile duct and major pancreatic duct that join to form a common channel outside the duodenal wall (74). It is a complex structural unit composed of a highly developed mucosa, musculature, and nerve supply that regulates the flow of bile and pancreatic secretions. Its sphincter of Oddi is a part of the overall gastrointestinal motility system and is subject to regulation by myogenic, neural, and gastrointestinal hormonal elements (13,98).

The major pancreatic duct of Wirsung drains many small channels in its course from the tail of the pancreas to the duodenal ostium. It typically inserts into the duodenal window caudal or a little lateral to the common bile duct. Its lumen narrows at the duodenal wall. The minor duct of Santorini joins the major pancreatic duct at a variety of angles and locations within the pancreas. Uncommonly, the duct of Wirsung is smaller than the duct of Santorini

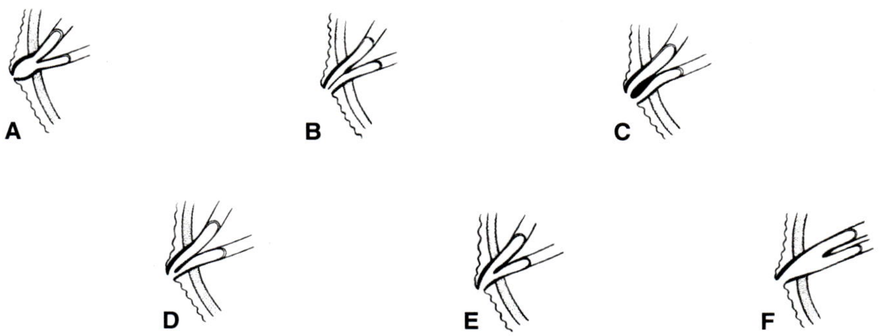

FIGURE 28.16 Relationship of the common bile duct and duct of Wirsung at the major papilla: (**A**) ampulla, (**B**) interposed septum, (**C**) separate openings, (**D**) short common channel, (**E**) long common channel, and (**F**) extended common channel.

and the latter may be the chief conduit for drainage of the pancreas (99). The duct of Santorini leads into the minor papilla in most patients but ends blindly in 10% to 20% of cases (99,100). The luminal pressure of the major pancreatic duct is nearly always higher than that for the common bile duct except when the gallbladder empties (101).

The relationship of the common bile duct and duct of Wirsung at the papilla is complex and variable. The ducts may have separate openings into the duodenum, an interposed septum, or a common channel that sometimes forms an ampulla (Fig. 28.16). The ampulla, defined strictly, is a dilated, juglike conduit resulting from the union of the common bile duct and major pancreatic duct. Data from various studies of the pancreaticoduodenal junction show the frequency for separate openings into the duodenal lumen ranges from 12% to 54% and 46% to 88% share a common channel (73,99,100,102–106). The best data suggest that approximately 74% of patients have a common channel, 19% have separate openings for the pancreatic duct and common bile duct, and 7% have an interposed septum (102). On occasion, the ducts will unite before the duodenal wall is breached, forming an extended common channel. In one study, the length of the extended common channel ranged from 0.9 to 3.3 cm (mean 2.2) (107). In another series of 390 autopsy cases, the authors found that 25% of patients had a well-defined ampulla, 18% have a long (>3 cm) common channel in the absence of an ampulla, and 31% have a short (<3 cm) common channel (102). This confluence of the pancreatic and bile ducts outside of the duodenal wall has been increasingly described in association with congenital dilatation of the bile duct, choledochal cyst, and adenocarcinoma of the extrahepatic bile ducts and gallbladder (108–110).

The major papilla, a cylindrical protuberance housing the terminations of the common bile duct and major pancreatic duct or a common channel, is situated medially at the midportion of the second part of the duodenum. It is usually completely or partially covered by a triangular fold of duodenal mucosa; a longitudinal mucosal fold projects from the caudal portion of its base, forming a frenulum, which is absent in about one quarter of the cases (111). Rarely, the major papilla is located at or just below the level of the duodenal mucosa or is absent. Mucosal reduplications (i.e., valves of Santorini) at the ostium of the major papilla are found in approximately one-third of adults, but not in fetuses. They consist of columnar-shaped protrusions and transverse leaflike flaps of ductal mucosa (112,113). The columnar-shaped projections arising from the terminal common bile duct number one to four per patient and range from 1 to 5 mm in length (113). Leaflike flaps are commonly present in the caudal wall of the common channel in fetuses and adults. The flaps are separated by small cul-de-sacs of variable size, and sometimes extend into the major pancreatic duct. They may be limited to the orifice of the duct of Wirsung when a common channel is absent. Presumably, these flaps flatten during the flow of pancreatic juice into the duodenum; when the cul-de-sacs are filled, the ostium is blocked and regurgitation is prevented (113).

The sphincter of Oddi consists of the intrinsic circular and longitudinal musculature of the Vaterian system. It is embryologically and functionally distinct from the musculature of the duodenal wall although, the muscle cell bundles from the duodenal wall aid in anchoring the Vaterian system in place. The papilla and duodenal wall form both a morphologic and a functional unit (114). Connective tissue fibers and aggregates of smooth muscle cells spread from the orifice of the papilla to the circular duodenal musculature. The arrangement of muscle cell bundles that form the sphincter is highly complex. Important components include the smooth muscle cell aggregates around the intrapancreatic and intraduodenal portions of the common bile duct, or sphincter choledochus (115). Circular bundles of smooth muscle cells extend 13 to 14 mm up the common bile duct from the pore of the papilla (111). Smooth muscle cell aggregates are also present in the wall of the common channel, around the duct of Wirsung, and near the ostium of the papilla. It is controversial whether the smooth muscle bundles around the pancreatic duct above the common channel have important sphincteric function, but the finding of a sustained pancreatic duct high-pressure zone with phasic contractions after sphincterotomy may be evidence that the sphincter of Oddi extends to include portions of the pancreatic duct (115–117). Fascicles of smooth muscle cells have been shown to extend up the pancreatic duct a mean of 7.3 mm from the papillary pore (111). The tunica muscularis of the duodenum may not have a primary role in managing the flow of bile and pancreatic juice at the choledochoduodenal junction.

The sphincter of Oddi serves to inhibit the flow of bile into the duodenum, pumps bile into the duodenum when necessary, and likely precludes the entry of duodenal contents into the common bile duct and major pancreatic duct (98). Manometric studies have shown that the control of the flow of bile during fasting results from the phasic contractions of the sphincter of Oddi that may primarily be under neural control (118,119). The flow of pancreatic juice is also regulated. The contractions are in addition to the steady basal pressure of the sphincter of Oddi, which is several mm Hg higher than that for the common bile duct and pancreatic duct (120). The high pressure zone measures 4 to 6 mm in length and the phasic contractions may be antegrade, retrograde, or simultaneous (117). Cholecystokinin inhibits the phasic contractions of the sphincter and decreases the basal pressure, allowing the flow of large quantities of bile into the duodenum (118). Glucagon-like cholecystokinin decreases sphincteric pressure, whereas gastrin and secretin elevate basal pressure (117). The phasic contractions and basal tone of the sphincter can be increased or decreased by exogenous drugs. For instance, most narcotics increase sphincteric pressure, whereas atropine decreases it (117).

The minor papilla is nearly always present but may be difficult to locate grossly (100). Its size is variable. It is usually situated 2 cm proximal to the major papilla (100,121).

Vascular and Nerve Supply and Lymphatic Vessel Drainage

The intraduodenal portion of the common bile duct is supplied by vessels from the anterior and posterior superior pancreaticoduodenal arteries (6). Venous drainage occurs via small veins that lead to the portal vein. The fine venous architecture of the major papilla has been described in detail (122). Lymphatic drainage is variable but generally lymphatic vessels from the pancreaticoduodenal junction drain into the anterior and posterior pancreaticoduodenal lymph nodes and then to the nodes near the inferior pancreaticoduodenal artery (123). The Vaterian system is innervated extrinsically by parasympathetic nerve fibers in the vagal nerve and by sympathetic nerve fibers in the splanchnic nerves (13,98). Although little is known regarding the role of these nerve fibers in regulating the sphincter of Oddi, some evidence indicates that its motility is inhibited by vagal activation (98,124). Three separate ganglia cell groups provide intrinsic innervation. These are found at the base of the papilla in the duodenal wall, within the musculature of the papilla, and within the submucosa (98). This intrinsic innervation appears to provide tonic inhibition and is similar to that for other gastrointestinal sphincters, including the lower esophageal, pyloric, and internal anal sphincters.

Histology

The epithelial lining of the duct of Wirsung is identical to that of the common bile duct. The cytoplasm of the

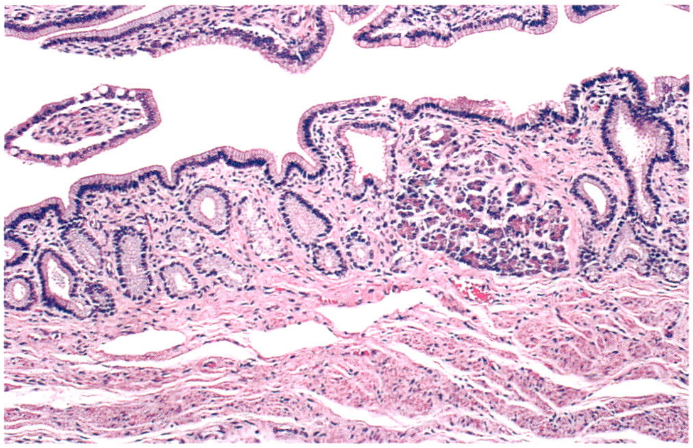

FIGURE 28.17 The duct of Wirsung at the papilla of Vater is lined by a single layer of tall columnar cells with occasional interspersed goblet cells. Accessory pancreatic ducts and acini are also observed.

columnar cells also contains sulfated acid mucin (125). The epithelium may undergo hyperplastic, metaplastic, or dysplastic changes. Surrounding the normal epithelium is a dense fibrous layer with abundant collagen and elastic fibers. A few ganglion cells may be seen in the outer half of the fibrous wall. Small pancreatic ducts draining acini traverse the dense fibrous layer. The epithelium at the orifice of Wirsung duct is folded to create mucosal reduplications that increase the surface area. Goblet cells are interspersed between the columnar lining cells within the papilla. Numerous small accessory pancreatic ducts drain into the ductal lumen near the ostium, and pancreatic acini are sometimes present just beneath the lining of the duct (Fig. 28.17). A few lymphocytes may be seen within the ductal epithelium, and lymphocytes, plasma cells, and mast cells sparsely populate the fibrovascular lamina propria. Circular smooth muscle bundles are present around the duct as it penetrates the duodenal wall (121).

The epithelium of the terminal common bile duct and common channel covers long, slender papillary fronds or valvules (Fig. 28.18). They correspond to the mucosal reduplications seen grossly. These papillary formations are

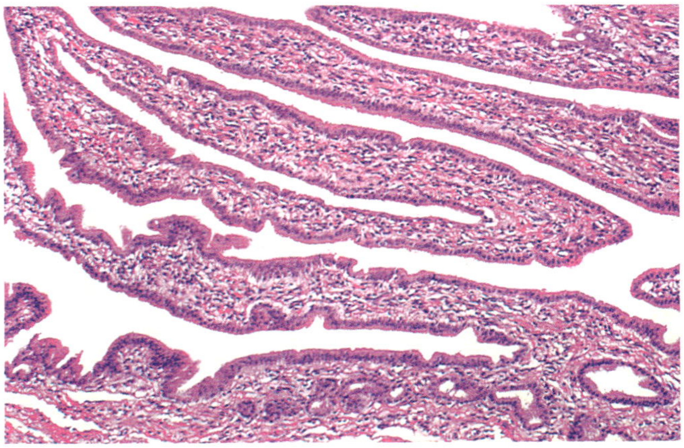

FIGURE 28.18 Near the ostium of the major papilla, the epithelium of the common bile duct lines prominent papillary fronds (valvules).

considerably larger than the duodenal villi, which are few or absent at the surface of the papilla. The valvules may branch and sometimes project beyond the ostium of the papilla; shorter fronds are more numerous at the periphery while the longer ones are increased centrally (77,126,127). The columnar lining cells contain eosinophilic cytoplasm and basal nuclei. Interspersed goblet cells are more numerous near the ostium. The stroma supporting the fronds contains a few lymphocytes, mast cells, and plasma cells. Interrupted bundles of smooth muscle cells are present at the frond bases and are occasionally found in their stroma. The smooth muscle of the sphincter choledochus becomes apparent in the duct wall several millimeters before it enters the duodenal window. About 5 mm from the duodenal wall, longitudinally arranged bundles of smooth muscle cells encircle approximately two-thirds of the common bile duct; the duct is completely encircled by smooth muscle approximately 2 mm from the duodenal musculature (121). These intrinsic smooth muscle of the duct is separated from the muscularis propria of the duodenum by connective tissue and, at times, pancreatic tissue (121). Variable numbers of smooth muscle cell bundles are arranged in a circular or longitudinal muscle fashion and surround the common channel when it is present. Before forming a common channel, however, the common bile duct is set apart from the pancreatic duct by a septum that eventually loses its muscle fibers, becoming a thin connective tissue membrane (121). When the common bile duct and duct of Wirsung are separate within the papilla, they are distinguishable by light microscopy because the common bile duct is larger, has more prominent fronds, a greater amount of enveloping smooth muscle, and bile in its lumen.

A bewildering assortment of peribiliary glands and ducts of various caliber surround the common bile duct at the papilla. Frequently, it is only possible to distinguish mucous glands from the terminations of accessory pancreatic ducts by studying serial sections (95). Mucous glands drain into the shallow or deep recesses between the papillary fronds (Fig. 28.19). The number and distribution of these glands are variable. Glands near the surface of the papilla may be distended with mucus and some may even represent dilated accessory pancreatic ducts (126). Small accessory pancreatic channels empty into the common bile duct (Fig. 28.20), duct of Wirsung, common channel, surface of papilla, or through the duodenal mucosa near the papilla (95,128). They are sometimes numerous and may cause obstruction of the common bile duct, duct of Wirsung, or common channel. In such instances, a diagnosis of accessory duct hyperplasia may be appropriate, although some may use terms such as pancreatic heterotopia or adenomyosis to describe this finding (126). Accessory pancreatic ducts are extremely common, being present in nearly all major papilla (128). Pancreatic acini are also commonly found in this location, being present in nearly 10% major papillae, whereas pancreatic islets are generally absent from this area (129). The accessory ducts drain these small lobules of pancreatic acini and appear as packets of multiple small-caliber lumens encircled by a cellular fibrovascular stroma (Fig. 28.21). Generally, a larger central duct is surrounded by smaller branches within a group of ducts. These groups are sometimes seen penetrating the duodenal smooth muscle. Small groups of heterotopic pancreatic acini and ducts also occur in the submucosa of the duodenum away from the major papilla (Fig. 28.22).

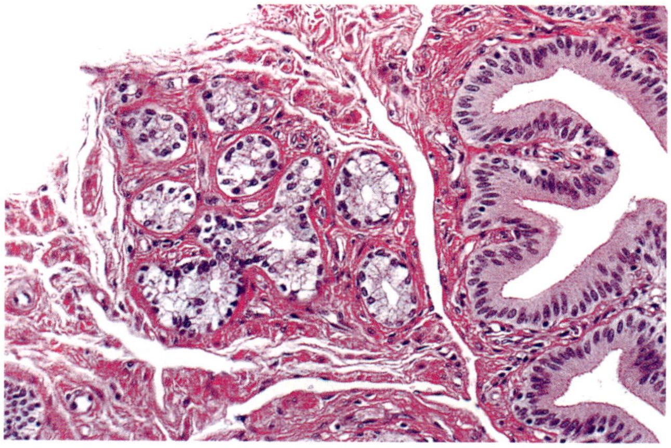

FIGURE 28.19 Mucous glands are present around the common bile duct at the papilla and drain into recesses between the papillary fronds.

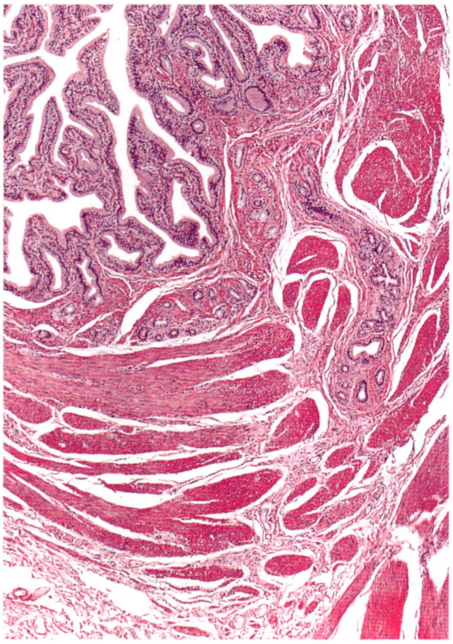

FIGURE 28.20 Accessory pancreatic ducts pierce the large smooth muscle bundles to empty into the lumen of the common bile duct.

Immunohistochemically, the cells lining the common bile duct and the duct of Wirsung at the papilla are positive for low–molecular-weight keratin (CAM 5.2 antibody), cytokeratin 7, and epithelial membrane antigen. There may be linear apical staining for CEA. The adjacent mucous glands and accessory pancreatic ducts have the same immunoreactivity for keratin and epithelial membrane

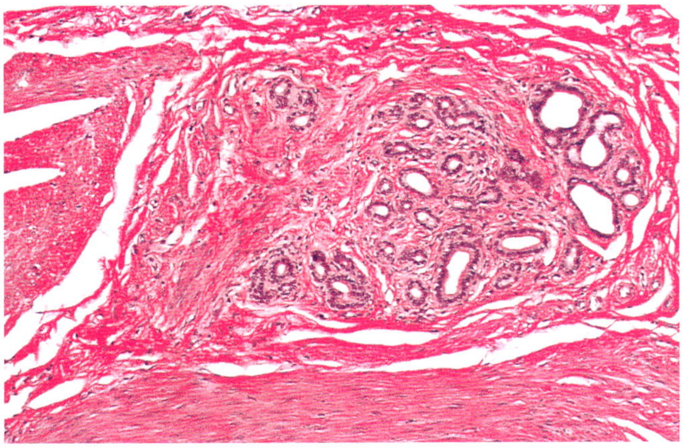

FIGURE 28.21 Accessory pancreatic ducts that penetrate the smooth muscle bundles at the choledochoduodenal junction are surrounded by a fibrovascular stroma.

antigen. A few scattered cells lining the large ducts within the pancreas are positive for neuron-specific enolase, chromogranin A, insulin, and glucagon (130). Chromogranin-positive cells are sometimes located in the lining epithelium of the duct of Wirsung and common bile duct within the papilla. Mucous glands and accessory pancreatic ducts also contain scattered cells immunoreactive for neuron-specific enolase and chromogranin A (Fig. 28.23). In patients with pancreaticobiliary disease, a few cells lining the lumen of the papilla and in adjacent mucous glands have been found to be immunoreactive for somatostatin (72). Although usually absent, endocrine cell micronests may be scattered singly or are grouped in the stroma adjacent to pancreatic ducts, ductules, or accessory glands but not around the

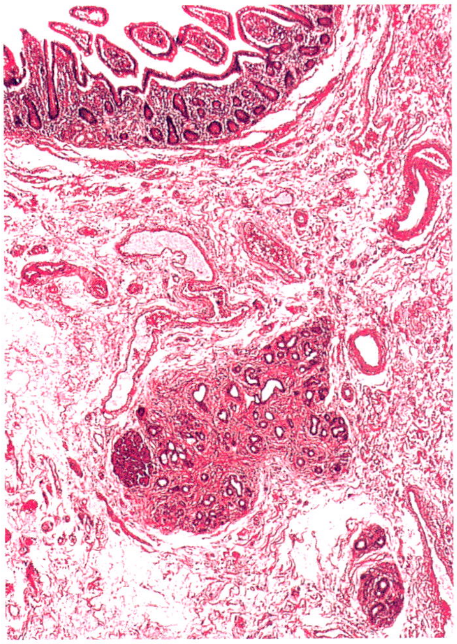

FIGURE 28.22 Groups of heterotopic pancreatic ducts and acini may be seen in the submucosa of the duodenum away from the papilla of Vater.

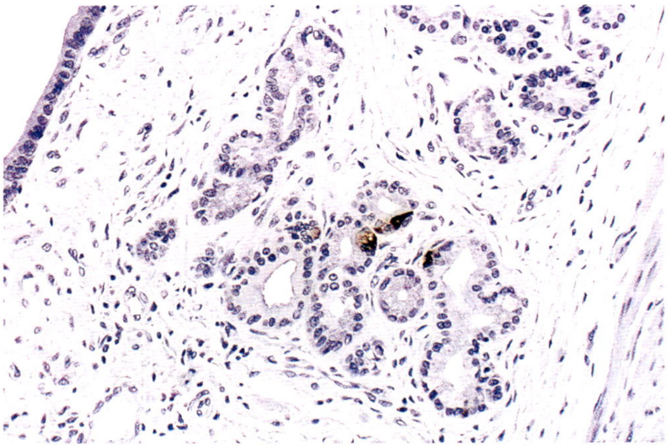

FIGURE 28.23 Some of the small ducts around the duct of Wirsung at the major papilla contain a few cells that are immunoreactive for chromogranin A (immunoperoxidase technique).

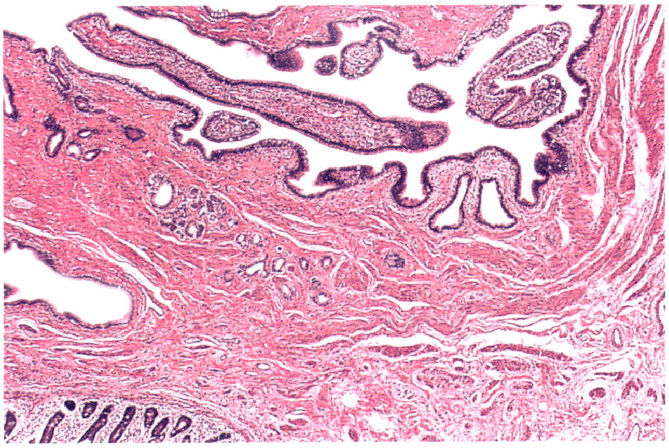

FIGURE 28.24 The duct of Santorini at the minor papilla contains papillary fronds and is surrounded by muscle bundles.

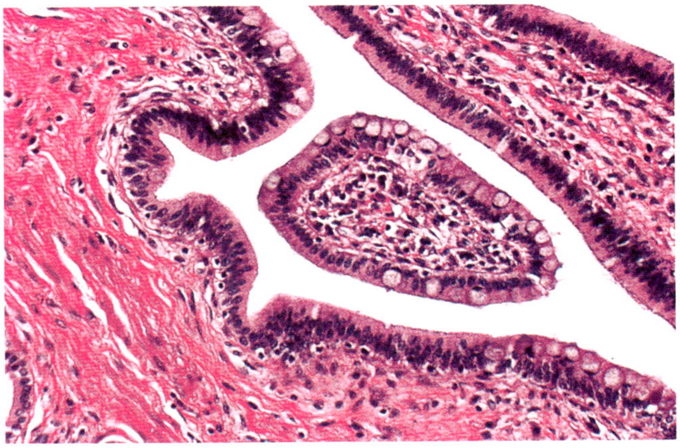

FIGURE 28.25 The duct of Santorini at the minor papilla is lined by tall columnar cells with interspersed goblet cells.

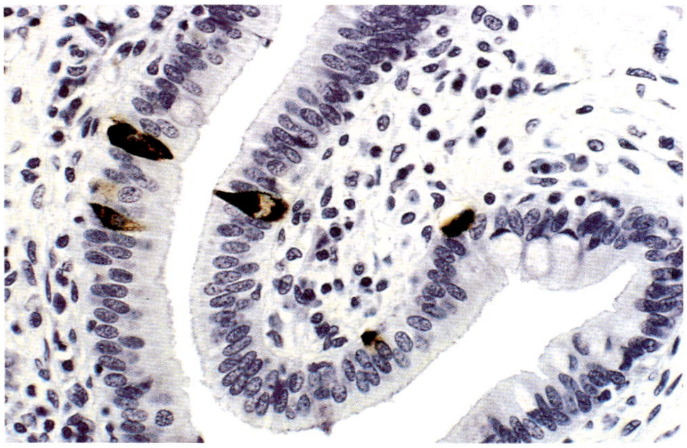

FIGURE 28.26 A few cells that line the duct of Santorini at the minor papilla are flask shaped and immunoreactive for chromogranin A (immunoperoxidase technique).

common bile duct (129). They have been found in about 3% of major papillae and consist of round, trabecular, or ribbonlike groups of cells that immunohistochemically are distinct from those of pancreatic islets. They are typically scattered, rarely nodular, and immunohistochemically stain for somatostatin and pancreatic polypepide. It is unclear whether they are a normal finding or represent a pathologic condition. The functional role of these endocrine cells in the papilla of Vater is unknown.

The pancreatic duct of Santorini contains papillary fronds that are lined by simple columnar epithelium with some goblet cells (Figs. 28.24 and 28.25). Small pancreatic ducts open into the lumen of the duct at the minor papilla or separately into the duodenum (100). Small lobules of pancreatic acini may be present within the connective tissue of the minor papilla. One author observed them in 77% of minor papillae, as well as well-formed islets of Langerhans in 14% of patients (129). Smooth muscle bundles separated by collagen, small nerves, and ganglion cells surround the duct. The bundles of muscle cells are occasionally continuous with those of the muscularis mucosae of the duodenum, but continuity between these groups of muscle cells is usually lacking (100). The lining epithelial cells and those of the small pancreatic ducts stain strongly for low–molecular-weight keratin (CAM 5.2 antibody), cytokeratin 7, and weakly for CEA at the luminal surface. A few cells within small ducts and some that line the lumen of the duct of Santorini are flask shaped and immunoreactive for neuron-specific enolase and chromogranin A (Fig. 28.26). Small groups of neuroendocrine cells may extend below the epithelial lining (Fig. 28.27). In the above-mentioned study of 167 minor papillae, 16% contained endocrine "micronests," which were predominantly scattered and rarely nodular (129). They were usually immunoreactive for somatostatin and pancreatic polypeptide and lacked staining for insulin and glucagon. It is possible that some of these micronests represent metaplasia/hyperplasia or neoplasia.

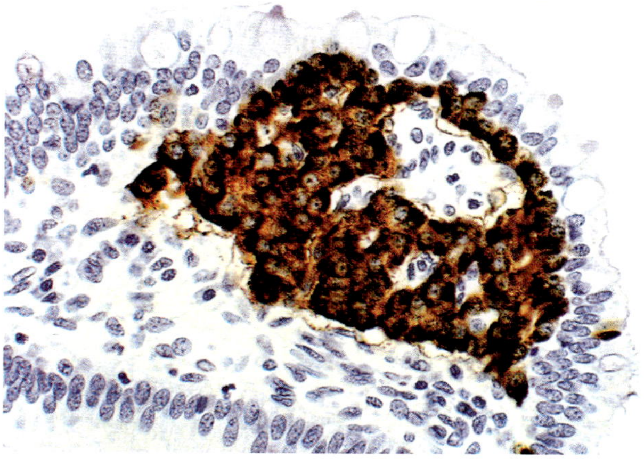

FIGURE 28.27 A group of cells below the lining epithelium of the duct of Santorini at the minor papilla is immunoreactive for chromogranin A (immunoperoxidase technique).

REFERENCES

1. Ando H. Embryology of the biliary tract. *Dig Surg*. 2010;27(2): 87–89.
2. Frey C, Bizer L, Ernst C. Agenesis of the gallbladder. *Am J Surg*. 1967;114(6):917–926.
3. Malde S. Gallbladder agenesis diagnosed intra-operatively: a case report. *J Med Case Reports*. 2010;4:285.
4. Kurzweg FT, Cole PA. Triplication of the gallbladder: review of literature and report of a case. *Am Surg*. 1979;45(6):410–412.
5. Albores-Saavedra J, Henson DE, Klimstra DS. *Normal Anatomy. Tumors of the Gallbladder, Extrahepatic Bile Ducts, and Ampulla of Vater*. 3rd ed. Armed Forces Institute of Pathology; 2000:1–16.
6. Lindner HH. Embryology and anatomy of the biliary tree. In: Way LW, Pellegrini CA, eds. *Surgery of the Gallbladder and Bile Ducts*. WB Saunders; 1987:3–22.
7. Guyton AC. The liver and biliary system. In: *Textbook of Medical Physiology*. WB Saunders; 1976:936–944.
8. Frizzell RA, Heintze K. Transport functions of the gallbladder. In: Javitt NB, ed. *Liver and Biliary Tract Physiology*. University Park Press; 1980:221–247.
9. Shaffer EA. Review article: control of gall-bladder motor function. *Aliment Pharmacol Ther*. 2000;14(Suppl 2):2–8.
10. Rehfeld JF. Clinical endocrinology and metabolism. Cholecystokinin. *Best Pract Res Clin Endocrinol Metab*. 2004;18(4): 569–586.
11. Pomeranz IS, Davison JS, Shaffer EA. In vitro effects of pancreatic polypeptide and motilin on contractility of human gallbladder. *Dig Dis Sci*. 1983;28(6):539–544.
12. Fisher RS, Rock E, Levin G, Malmud L. Effects of somatostatin on gallbladder emptying. *Gastroenterology*. 1987;92(4): 885–890.
13. Balemba OB, Salter MJ, Mawe GM. Innervation of the extrahepatic biliary tract. *Anat Rec A Discov Mol Cell Evol Biol*. 2004;280(1):836–847.
14. Magee DF, Naruse S, Pap A. Vagal control of gall-bladder contraction. *J Physiol*. 1984;355:65–70.
15. Lavoie B, Balemba OB, Godfrey C, et al. Hydrophobic bile salts inhibit gallbladder smooth muscle function via stimulation

of GPBAR1 receptors and activation of KATP channels. *J Physiol*. 2010;588(Pt 17):3295–3305.
16. Colecchia A, Sandri L, Staniscia T, et al. Gallbladder motility and functional gastrointestinal disorders. *Dig Liver Dis*. 2003;35(Suppl 3):S30–S34.
17. Moosman DA, Collier FA. Prevention of traumatic injury to the bile ducts; a study of the structures of the cystohepatic angle encountered in cholecystectomy and supraduodenal choledochostomy. *Am J Surg*. 1951;82(1):132–143.
18. Michels NA. The hepatic, cystic and retroduodenal arteries and their relations to the biliary ducts with samples of the entire celiacal blood supply. *Ann Surg*. 1951;133(4):503–524.
19. Ito M, Mishima Y, Sato T. An anatomical study of the lymphatic drainage of the gallbladder. *Surg Radiol Anat*. 1991;13(2):89–104.
20. Shirai Y, Yoshida K, Tsukada K, Ohtani T, Muto T. Identification of the regional lymphatic system of the gallbladder by vital staining. *Br J Surg*. 1992;79(7):659–662.
21. Ding WG, Fujimura M, Mori A, Tooyama I, Kimura H. Light and electron microscopy of neuropeptide Y-containing nerves in human liver, gallbladder, and pancreas. *Gastroenterology*. 1991;101(4):1054–1059.
22. Laitio M. Morphology and histochemistry of non-tumorous gallbladder epithelium. A series of 103 cases. *Pathol Res Pract*. 1980;167(2–4):335–345.
23. Karayannopoulou G, Damjanov I. Lectin binding sites in the human gallbladder and cystic duct. *Histochemistry*. 1987; 88(1):75–83.
24. Delaquerriere L, Tremblay G, Riopelle JL. Argentaffine cells in chronic cholecystitis. *Arch Pathol*. 1962;74:142–151.
25. Yamamoto M, Nakajo S, Tahara E. Endocrine cells and lysozyme immunoreactivity in the gallbladder. *Arch Pathol Lab Med*. 1986;110(10):920–927.
26. Albores-Saavedra J, Nadji M, Henson DE, Ziegels-Weissman J, Mones JM. Intestinal metaplasia of the gallbladder: a morphologic and immunocytochemical study. *Hum Pathol*. 1986;17(6):614–620.
27. Kozuka S, Hachisuka K. Incidence by age and sex of intestinal metaplasia in the gallbladder. *Hum Pathol*. 1984;15(8): 779–784.
28. Kozuka S, Kurashina M, Tsubone M, Hachisuka K, Yasui A. Significance of intestinal metaplasia for the evolution of cancer in the biliary tract. *Cancer*. 1984;54(10):2277–2285.
29. Tatematsu M, Furihata C, Miki K, et al. Complete and incomplete pyloric gland metaplasia of human gallbladder. *Acta Pathol Jpn*. 1987;37(1):39–46.
30. Tsutsumi Y, Nagura H, Osamura Y, Watanabe K, Yanaihara N. Histochemical studies of metaplastic lesions in the human gallbladder. *Arch Pathol Lab Med*. 1984;108(11):917–921.
31. Chang HJ, Kim SW, Lee BL, Hong EK, Kim WH. Phenotypic alterations of mucins and cytokeratins during gallbladder carcinogenesis. *Pathol Int*. 2004;54(8):576–584.
32. Aroni K, Kittas C, Papadimitriou CS, Papacharalampous NX. An immunocytochemical study of the distribution of lysozyme, a1-antitrypsin and a1-antichymotrypsin in the normal and pathological gall bladder. *Virchows Arch A Pathol Anat Histopathol*. 1984;403(3):281–289.
33. Albores-Saavedra J, Nadji M, Morales AR, Henson DE. Carcinoembryonic antigen in normal, preneoplastic and neoplastic gallbladder epithelium. *Cancer*. 1983;52(6):1069–1072.
34. Maxwell P, Davis RI, Sloan JM. Carcinoembryonic antigen (CEA) in benign and malignant epithelium of the gall bladder, extrahepatic bile ducts, and ampulla of Vater. *J Pathol*. 1993;170(1):73–76.
35. Cabibi D, Licata A, Barresi E, Craxi A, Aragona F. Expression of cytokeratin 7 and 20 in pathological conditions of the bile tract. *Pathol Res Pract*. 2003;199(2):65–70.
36. Sakamoto H, Mutoh H, Ido K, Satoh K, Hayakawa H, Sugano K. A close relationship between intestinal metaplasia and Cdx2 expression in human gallbladders with cholelithiasis. *Hum Pathol*. 2007;38(1):66–71.
37. Osawa H, Kita H, Satoh K, et al. Aberrant expression of CDX2 in the metaplastic epithelium and inflammatory mucosa of the gallbladder. *Am J Surg Pathol*. 2004;28(9):1253–1254.
38. Agrawal V, Goel A, Krishnani N, Pandey R, Agrawal S, Kapoor VK. p53, carcinoembryonic antigen and carbohydrate antigen 19.9 expression in gall bladder cancer, precursor epithelial lesions and xanthogranulomatous cholecystitis. *J Postgrad Med*. 2010;56(4):262–266.
39. Yamamoto M, Nakajo S, Tahara E. Immunohistochemical analysis of estrogen receptors in human gallbladder. *Acta Pathol Jpn*. 1990;40(1):14–21.
40. Choi YL, Xuan YH, Shin YK, et al. An immunohistochemical study of the expression of adhesion molecules in gallbladder lesions. *J Histochem Cytochem*. 2004;52(5):591–601.
41. Hudson I, Hopwood D. Macrophages and mast cells in chronic cholecystitis and "normal" gall bladders. *J Clin Pathol*. 1986;39(10):1082–1087.
42. Green FH, Fox H. An immunofluorescent study of the distribution of immunoglobulin-containing cells in the normal and the inflamed human gall bladder. *Gut*. 1972;13(5):379–384.
43. Raparia K, Zhai QJ, Schwartz MR, Shen SS, Ayala AG, Ro JY. Muscularis mucosae versus muscularis propria in gallbladder, cystic duct, and common bile duct: smoothelin and desmin immunohistochemical study. *Ann Diagn Pathol*. 2010; 14(6):408–412.
44. Mendoza-Marin M, Hoang MP, Albores-Saavedra J. Malignant stromal tumor of the gallbladder with interstitial cells of Cajal phenotype. *Arch Pathol Lab Med*. 2002;126(4): 481–483.
45. Weedon D. *Pathology of the Gallbladder*. Masson; 1984.
46. Fine G, Raju UB. Paraganglia in the human gallbladder. *Arch Pathol Lab Med*. 1980;104:265–268.
47. Elfving G. Crypts and ducts in the gallbladder wall. *Acta Pathol Microbiol Scand*. 1960;49(Suppl 135):1–45.
48. Robertson HE, Ferguson WJ. The diverticula (Luschka's crypts) of the gallbladder. *Arch Pathol (Chic)*. 1945;40:312–333.
49. Halpert B. Morphological studies on the gall-bladder. II. The "true Luschka ducts" and the "Rokitansky-Aschoff sinuses" of the human gallbladder. *Bull Johns Hopkins*. 1927;41:77–103.
50. Albores-Saavedra J, Shukla D, Carrick K, Henson DE. In situ and invasive adenocarcinomas of the gallbladder extending into or arising from Rokitansky-Aschoff sinuses: a clinicopathologic study of 49 cases. *Am J Surg Pathol*. 2004;28(5):621–628.
51. Beilby JO. Diverticulosis of the gall bladder. The fundal adenoma. *Br J Exp Pathol*. 1967;48(4):455–461.
52. Moosman DA. Accessory bile ducts: their significance during cholecystectomy. *J Mich State Med Soc*. 1964;63:355–358.
53. Foster JH, Wayson EE. Surgical significance of aberrant bile ducts. *Am J Surg*. 1962;104:14–19.
54. McQuillan T, Manolas SG, Hayman JA, Kune GA. Surgical significance of the bile duct of Luschka. *Br J Surg*. 1989; 76(7):696–698.

55. Goor DA, Ebert PA. Anomalies of the biliary tree. Report of a repair of an accessory bile duct and review of the literature. *Arch Surg*. 1972;104(3):302–309.
56. Tejada E, Danielson C. Ectopic or heterotopic liver (choristoma) associated with the gallbladder. *Arch Pathol Lab Med*. 1989;113(8):950–952.
57. Mutschmann PN. Aberrant pancreatic tissue in the gallbladder wall. *Am J Surg*. 1946;72:282–283.
58. Busuttil A. Ectopic adrenal within the gall-bladder wall. *J Pathol*. 1974;113(4):231–233.
59. Curtis LE, Sheahan DG. Heterotopic tissues in the gallbladder. *Arch Pathol*. 1969;88(6):677–683.
60. Bulut AS, Karayalcin K. Ciliated foregut cyst of the gallbladder: report of a case and review of literature. *Patholog Res Int*. 2010;2010:193535.
61. Cassol CA, Noria D, Asa SL. Ectopic thyroid tissue within the gall bladder: case report and brief review of the literature. *Endocr Pathol*. 2010;21(4):263–265.
62. Liang K, Liu JF, Wang YH, Tang GC, Teng LH, Li F. Ectopic thyroid presenting as a gallbladder mass. *Ann R Coll Surg Engl*. 2010;92(4):W4–W6.
63. Al-Shraim M, Rabie ME, Elhakeem H, Kandeel A, Shah MT, Jamil S. Pancreatic heterotopia in the gallbladder associated with chronic cholecystitis: a rare combination. *JOP*. 2010;11(5):464–466.
64. Hayama S, Suzuki Y, Takahashi M, et al. Heterotopic gastric mucosa in the gallbladder: report of two cases. *Surg Today*. 2010;40(8):783–787.
65. Mrak K, Eberl T, Tschmelitsch J, Langner C. Heterotopic pancreatic tissue in the cystic duct: complicating factor or coexisting pathology. *South Med J*. 2010;103(5):471–473.
66. Gilloteaux J, Pomerants B, Kelly TR. Human gallbladder mucosa ultrastructure: evidence of intraepithelial nerve structures. *Am J Anat*. 1989;184(4):321–333.
67. Kaye GI, Wheeler HO, Whitlock RT, Lane N. Fluid transport in the rabbit gallbladder. A combined physiological and electron microscopic study. *J Cell Biol*. 1966;30(2):237–268.
68. Evett RD, Higgins JA, Brown AL Jr. The fine structure of normal mucosa in human gall bladder. *Gastroenterology*. 1964;47:49–60.
69. Laitio M, Nevalainen T. Gland ultrastructure in human gall bladder. *J Anat*. 1975;120(Pt 1):105–112.
70. Berci G. Biliary ductal anatomy and anomalies. The role of intraoperative cholangiography during laparoscopic cholecystectomy. *Surg Clin North Am*. 1992;72(5):1069–1075.
71. Repassy G, Schaff Z, Lapis K, Marton T, Jakab F, Sugar I. Mucosa of the Heister valve in cholelithiasis: transmission and scanning electron microscopic study. *Arch Pathol Lab Med*. 1978;102(8):403–405.
72. Dancygier H, Klein U, Leuschner U, Hubner K, Classen M. Somatostatin-containing cells in the extrahepatic biliary tract of humans. *Gastroenterology*. 1984;86(5 Pt 1):892–896.
73. Northover JMA, Terblanche. J. Applied surgical anatomy of the biliary tree. In: Blumgart LH, ed. *The Biliary Tract Clinical Surgery International*. Vol 5. Churchill Livingstone; 1982:1–16.
74. Dowdy GS Jr, Waldron GW, Brown WG. Surgical anatomy of the pancreatobiliary ductal system. Observations. *Arch Surg*. 1962;84:229–246.
75. Takahashi Y, Takahashi T, Takahashi W, Sato T. Morphometrical evaluation of extrahepatic bile ducts in reference to their structural changes with aging. *Tohoku J Exp Med*. 1985;147(3):301–309.
76. Blidaru D, Blidaru M, Pop C, Crivii C, Seceleanu A. The common bile duct: size, course, relations. *Rom J Morphol Embryol*. 2010;51(1):141–144.
77. Baggenstoss AH. Major duodenal papilla. Variations of pathologic interest and lesions of the mucosa. *Arch Pathol Lab Med*. 1938;26:853–868.
78. Mahour GH, Wakim KG, Ferris DO. The common bile duct in man: its diameter and circumference. *Ann Surg*. 1967;165(3):415–419.
79. Benson EA, Page RE. A practical reappraisal of the anatomy of the extrahepatic bile ducts and arteries. *Br J Surg*. 1976;63(11):853–860.
80. Chen WJ, Ying DJ, Liu ZJ, He ZP. Analysis of the arterial supply of the extrahepatic bile ducts and its clinical significance. *Clin Anat*. 1999;12(4):245–249.
81. Northover JM, Terblanche J. A new look at the arterial supply of the bile duct in man and its surgical implications. *Br J Surg*. 1979;66(6):379–384.
82. Douglass BE, Baggenstoss AH, Hollinshead WH. The anatomy of the portal vein and its tributaries. *Surg Gynecol Obstet*. 1950;91(5):562–576.
83. Vellar ID. Preliminary study of the anatomy of the venous drainage of the intrahepatic and extrahepatic bile ducts and its relevance to the practice of hepatobiliary surgery. *ANZ J Surg*. 2001;71(7):418–422.
84. Yoshida T, Shibata K, Yokoyama H, et al. Patterns of lymph node metastasis in carcinoma of the distal bile duct. *Hepatogastroenterology*. 1999;46(27):1595–1598.
85. Yoshida T, Matsumoto T, Sasaki A, et al. Lymphatic spread differs according to tumor location in extrahepatic bile duct cancer. *Hepatogastroenterology*. 2003;50(49):17–20.
86. Terada T, Kida T, Nakanuma Y. Extrahepatic peribiliary glands express alpha-amylase isozymes, trypsin and pancreatic lipase: an immunohistochemical analysis. *Hepatology*. 1993;18(4):803–808.
87. Qualman SJ, Haupt HM, Bauer TW, Taxy JB. Adenocarcinoma of the hepatic duct junction. A reappraisal of the histologic criteria of malignancy. *Cancer*. 1984;53(7):1545–1551.
88. Laitio M. Carcinoma of extrahepatic bile ducts. A histopathologic study. *Pathol Res Pract*. 1983;178(1):67–72.
89. Rullier A, Le Bail B, Fawaz R, Blanc JF, Saric J, Bioulac-Sage P. Cytokeratin 7 and 20 expression in cholangiocarcinomas varies along the biliary tract but still differs from that in colorectal carcinoma metastasis. *Am J Surg Pathol*. 2000;24(6):870–876.
90. Nagura H, Tsutsumi Y, Watanabe K, et al. Immunohistochemistry of carcinoembryonic antigen, secretory component and lysozyme in benign and malignant common bile duct tissues. *Virchows Arch A Pathol Anat Histopathol*. 1984;403(3):271–280.
91. Davis RI, Sloan JM, Hood JM, Maxwell P. Carcinoma of the extrahepatic biliary tract: a clinicopathological and immunohistochemical study. *Histopathology*. 1988;12(6):623–631.
92. Hoang MP, Murakata LA, Padilla-Rodriguez AL, Albores-Saavedra J. Metaplastic lesions of the extrahepatic bile ducts: a morphologic and immunohistochemical study. *Mod Pathol*. 2001;14(11):1119–1125.
93. Yamamoto M, Nakajo S, Tahara E, Miyoshi N. Endocrine cell carcinoma of extrahepatic bile duct. *Acta Pathol Jpn*. 1986;36(4):587–593.

94. Dancygier H. Endoscopic transpapillary biopsy (ETPB) of human extrahepatic bile ducts–light and electron microscopic findings, clinical significance. *Endoscopy*. 1989;21(Suppl 1): 312–320.
95. Cross KR. Accessory pancreatic ducts; special reference to the intrapancreatic portion of the common duct. *AMA Arch Pathol*. 1956;61(5):434–440.
96. Hong SM, Presley AE, Stelow EB, Frierson HF Jr, Moskaluk CA. Reconsideration of the histologic definitions used in the pathologic staging of extrahepatic bile duct carcinoma. *Am J Surg Pathol*. 2006;30(6):744–749.
97. Hong SM, Kang GH, Lee HY, Ro JY. Smooth muscle distribution in the extrahepatic bile duct: histologic and immunohistochemical studies of 122 cases. *Am J Surg Pathol*. 2000;24(5):660–667.
98. Allescher HD. Papilla of Vater: structure and function. *Endoscopy*. 1989;21(Suppl 1):324–329.
99. Millbourn E. On the excretory ducts of the pancreas in man, with special reference to their relations to each other, to the common bile duct and to the duodenum. *Acta Anat (Basel)*. 1950;9(1–2):1–34.
100. Baldwin WM. The pancreatic ducts in man, together with a study of the microscopical structure of the minor duodenal papilla. *Anat Rec*. 1911;5:197–228.
101. Parry EW, Hallenbeck GA, Grindlay JH. Pressures in the pancreatic and common ducts; values during fasting, after various meals, and after sphincterotomy; an experimental study. *AMA Arch Surg*. 1955;70(5):757–765.
102. DiMagno EP, Shorter RG, Taylor WF, Go VL. Relationships between pancreaticobiliary ductal anatomy and pancreatic ductal and parenchymal histology. *Cancer*. 1982;49(2):361–368.
103. Howard J, Jones R. The anatomy of the pancreatic ducts; the etiology of acute pancreatitis. *Am J Med Sci*. 1947;214(6): 617–622.
104. Newman HF, Weinberg SB, Newman EB, Northup JD. The papilla of Vater and distal portions of the common bile duct and duct of Wirsung. *Surg Gynecol Obstet*. 1958; 106(6):687–694.
105. Stamm BH. Incidence and diagnostic significance of minor pathologic changes in the adult pancreas at autopsy: a systematic study of 112 autopsies in patients without known pancreatic disease. *Hum Pathol*. 1984;15(7):677–683.
106. Sterling JA. The common channel for bile and pancreatic ducts. *Surg Gynecol Obstet*. 1954;98(4):420–424.
107. Suda K, Matsumoto Y, Miyano T. An extended common channel in patients with biliary tract carcinoma and congenital biliary dilatation. *Surg Pathol*. 1988;1:65–69.
108. Okada A, Nakamura T, Higaki J, Okumura K, Kamata S, Oguchi Y. Congenital dilatation of the bile duct in 100 instances and its relationship with anomalous junction. *Surg Gynecol Obstet*. 1990;171(4):291–298.
109. Hara H, Morita S, Sako S, et al. Relationship between types of common channel and development of biliary tract cancer in pancreaticobiliary maljunction. *Hepatogastroenterology*. 2002;49(44):322–325.
110. Kamisawa T, Suyama M, Fujita N, et al. Pancreatobiliary reflux and the length of a common channel. *J Hepatobiliary Pancreat Sci*. 2010;17(6):865–870.
111. Flati G, Flati D, Porowska B, Ventura T, Catarci M, Carboni M. Surgical anatomy of the papilla of Vater and biliopancreatic ducts. *Am Surg*. 1994;60(9):712–718.
112. Suarez CV. The Santorini valves. *Mt Sanai J Med*. 1981;48(2): 149–157.
113. Brown JO, Echenberg RJ. Mucosal reduplications associated with the ampullary portion of the major duodenal papilla in humans. *Anat Rec*. 1964;150:293–302.
114. Dziwisch L, Lierse W. Three-dimensional arrangement of dense connective tissue around the human major duodenal papilla. Including the ampullary region and the distal choledochal duct. *Acta Anat (Basel)*. 1989;135(3):231–235.
115. Boyden EA. The anatomy of the choledochoduodenal junction in man. *Surg Gynecol Obstet*. 1957;104(6):641–652.
116. Suarez CV. Structure of the major duodenal papilla. *Mt Sinai J Med*. 1982;49(1):31–37.
117. Goff JS. The human sphincter of Oddi. Physiology and pathophysiology. *Arch Intern Med*. 1988;148(12):2673–2677.
118. Toouli J, Hogan WJ, Geenen JE, Dodds WJ, Arndorfer RC. Action of cholecystokinin-octapeptide on sphincter of Oddi basal pressure and phasic wave activity in humans. *Surgery*. 1982;92(3):497–503.
119. Tanaka M. Function and dysfunction of the sphincter of Oddi. *Dig Surg*. 2010;27(2):94–99.
120. Coelho JC, Moody FG. Certain aspects of normal and abnormal motility of sphincter of Oddi. *Dig Dis Sci*. 1987;32(1): 86–94.
121. Hand BH. An anatomical study of the choledochoduodenal area. *Br J Surg*. 1963;50:486–494.
122. Biazotto W. The fine venous architecture of the major duodenal papilla in human beings. *Anat Anz*. 1990;171(2): 105–108.
123. Shirai Y, Ohtani T, Tsukada K, Hatakeyama K. Patterns of lymphatic spread of carcinoma of the ampulla of Vater. *Br J Surg*. 1997;84(7):1012–1016.
124. Smirnov VM, Lychkova AE. Mechanism of synergism between sympathetic and parasympathetic autonomic nervous systems in the regulation of motility of the stomach and sphincter of Oddi. *Bull Exp Biol Med*. 2003;135(4):327–329.
125. Kozuka S, Sassa R, Taki T, et al. Relation of pancreatic duct hyperplasia to carcinoma. *Cancer*. 1979;43(4):1418–1428.
126. Edmondson HA. Tumors of the gallbladder and extrahepatic bile ducts. *Atlas of Tumor Pathology, Fascicle 26*. Armed Forces Institute of Pathology; 1967:121–167.
127. Suda K, Ootaka M, Yamasaki S, et al. Distended glands or overreplacement of ampullary mucosa at the papilla of Vater. *J Hepatobiliary Pancreat Surg*. 2004;11(4):260–265.
128. Loquvam GS, Russell WO. Accessory pancreatic ducts of the major duodenal papilla. Normal structures to be differentiated from cancer. *Am J Clin Pathol*. 1950;20(4):305–313.
129. Noda Y, Watanabe H, Iwafuchi M, et al. Carcinoids and endocrine cell micronests of the minor and major duodenal papillae. Their incidence and characteristics. *Cancer*. 1992;70(7): 1825–1833.
130. Alpert LC, Truong LD, Bossart MI, Spjut HJ. Microcystic adenoma (serous cystadenoma) of the pancreas. A study of 14 cases with immunohistochemical and electron-microscopic correlation. *Am J Surg Pathol*. 1988;12(4):251–263.

29

Pancreas

Monica Garcia-Buitrago

The pancreas is an unpaired organ located in the left superior retroperitoneum. It is principally an epithelial organ that includes both exocrine elements (acini and ducts) and endocrine elements (islets of Langerhans). The stroma is sparse in the normal gland, although areas of fibrosis commonly develop as individuals age. There are relatively few opportunities for pathologists to observe the normal histology of the pancreas because normal pancreatic tissue is rarely resected, the gland quickly autolyzes, and the non-neoplastic parenchyma adjacent to the resected neoplasms usually has secondary changes due to obstruction and peritumoral inflammation. For this reason, normal histology and minor variations may be unfamiliar to practicing pathologists.

ANATOMY

Location and Relationship to Other Structures

The pancreas is positioned posterior to the omental bursa at the level of the second and third lumbar vertebrae in the retroperitoneum. It extends from the duodenal loop right of the midline to near the hilum of the spleen on the left of the midline (Fig. 29.1) (1). Four continuous anatomic areas are defined by the neighboring organs and vessels. The head of the pancreas is nestled within the C-shaped second and third portions of the duodenum and extends to the level of the superior mesenteric vein. The left lobe of the liver also lies anterior to the head, from which the distal portion of the common bile duct enters the posterosuperior aspect of the pancreatic head to drain its contents into the duodenum at the ampulla of Vater. The neck is a slender portion of pancreas located anterior to the mesenteric vessels, just below the pylorus. The body of the pancreas extends laterally from the neck from the left border of the superior mesenteric vein to the left border of the aorta. The posterior wall of the gastric antrum usually lies superior while the proximal jejunum passes inferior to the pancreatic body. The left adrenal gland and left kidney are each located posterior to the body (2). The tail of the pancreas extends laterally from the left border of the aorta, tapering gradually to a blunt end within a few centimeters of the splenic hilum. The anatomic location of the tail relative to the hilum varies among individuals and can be located either centrally (50%), inferiorly (42%), or superiorly (8%) (3,4). The posterior peritoneal lining of the lesser sac covers the anterior aspect of the pancreas and the superior surfaces of the neck, body, and tail. The rest of the gland is located in the retroperitoneum (5). Due to the proximity of the pancreas to various organs and anatomic structures, identifying the origin of neoplasms at these sites can be challenging radiographically. Pancreatic masses and

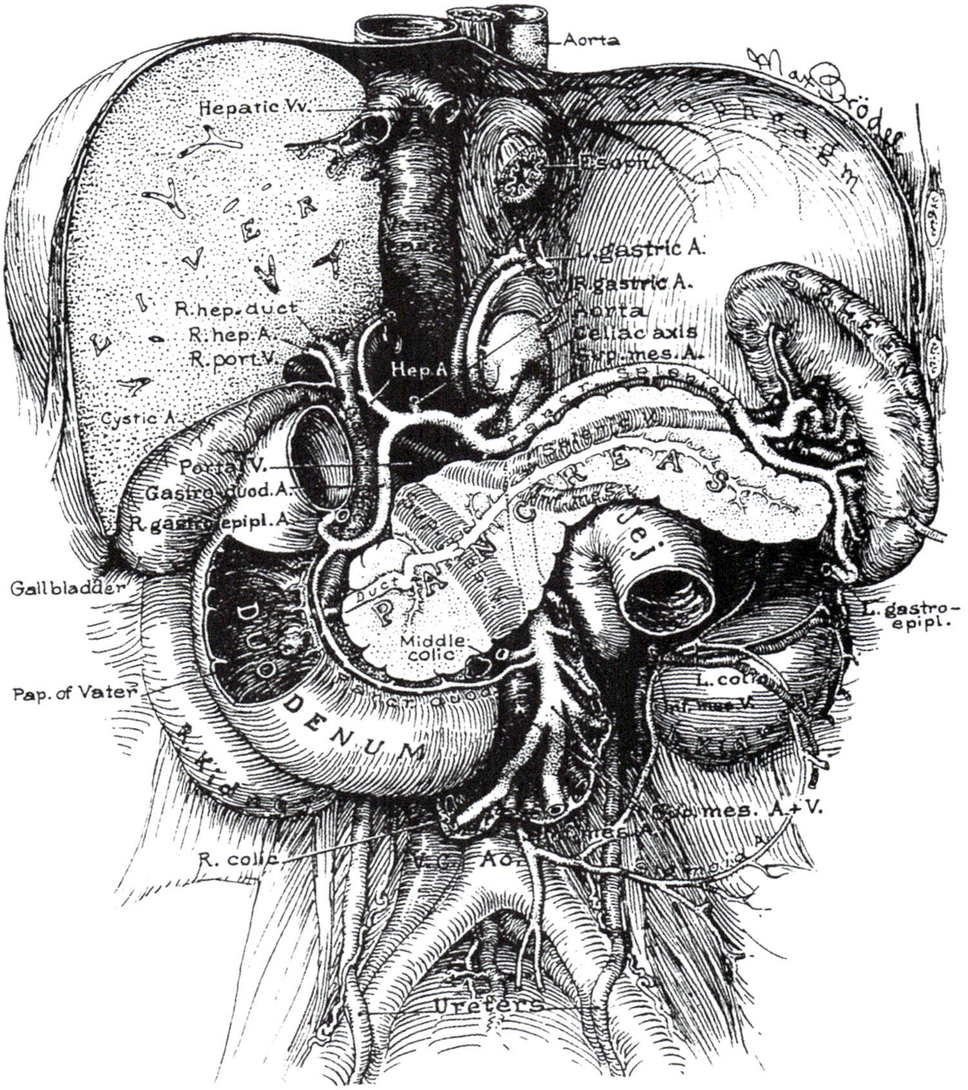

FIGURE 29.1 Anatomic relationships of the pancreas. The anterior aspect of the upper abdominal viscera is shown after removal of the stomach and omentum. Note the close relationship of the pancreas to the duodenum, jejunum, spleen, and major vessels. Illustration courtesy of the Max Brödel Archives.

cysts may be mistaken for neoplasms of the duodenum, ampulla, common bile duct, left adrenal gland, superior pole of either kidney, spleen, left lobe of the liver, greater curvature of the stomach, mesentery, and superior retroperitoneum. Recognizing these anatomic landmarks is helpful when classifying and staging pancreatic tumors.

Several major blood vessels are located close to the pancreas (5,6). The body of the pancreas rests on the aorta. The celiac trunk arises from the aorta superior to the neck of the pancreas, giving rise to the hepatic and splenic arteries, which run superior to the pancreatic tail. The superior mesenteric artery emerges posterior to the junction of the neck and body, with the adjacent superior mesenteric vein passing through a groove between the head and the neck. The superior mesenteric vessels extend around the uncinate process anterior to the aorta. This constitutes a posterior free surface of pancreatoduodenectomy specimens that is usually bluntly separated from the vessels during surgery. It is important to differentiate the surface of this vascular groove from the adjacent uncinate process margin, which is manually dissected during surgery (Fig. 29.2) (7). The head of the gland rests on the inferior vena cava, the right renal vessels, and the left renal vein, all of which are located posterior to the pancreas (8). The splenic vein accompanies the splenic artery along the superior aspect of the tail, ultimately joining the superior mesenteric vein to form the portal vein near the posterosuperior border of the head.

The pancreas receives its arterial supply from branches of the celiac trunk and the superior mesenteric artery with numerous anastomoses and variations between individuals (2,4,9–12). The neck of the pancreas is the narrowest part of the organ that lies between the head and the body. It is supplied by small branches from the splenic artery and the superior pancreaticoduodenal artery and drained by small veins that join the splenic vein. The body of the pancreas is the largest part of the organ. It crosses the midline of the body and is supplied by the dorsal, greater, and transverse pancreatic arterial branches of the splenic artery. It

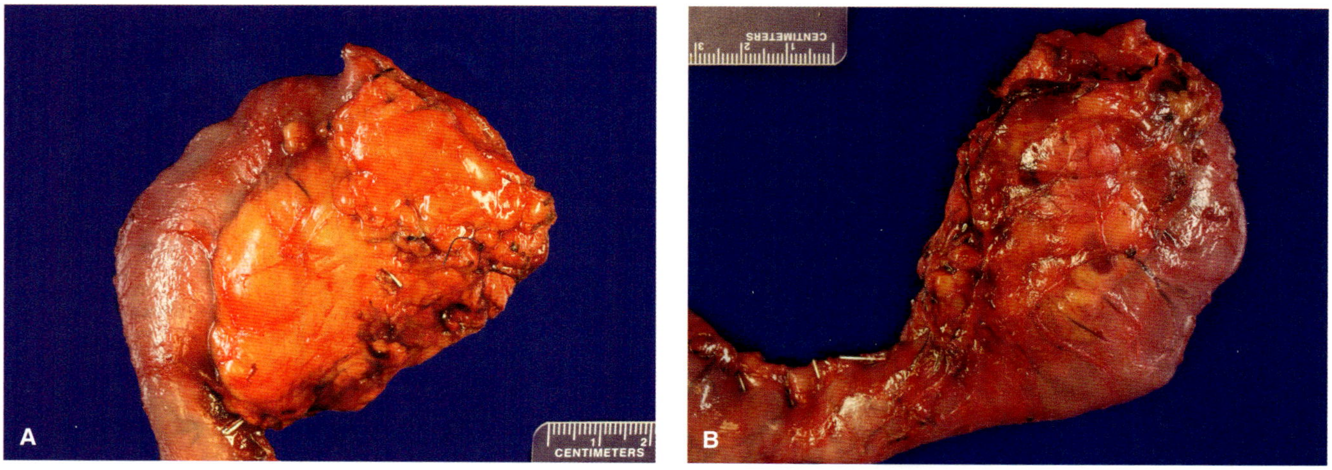

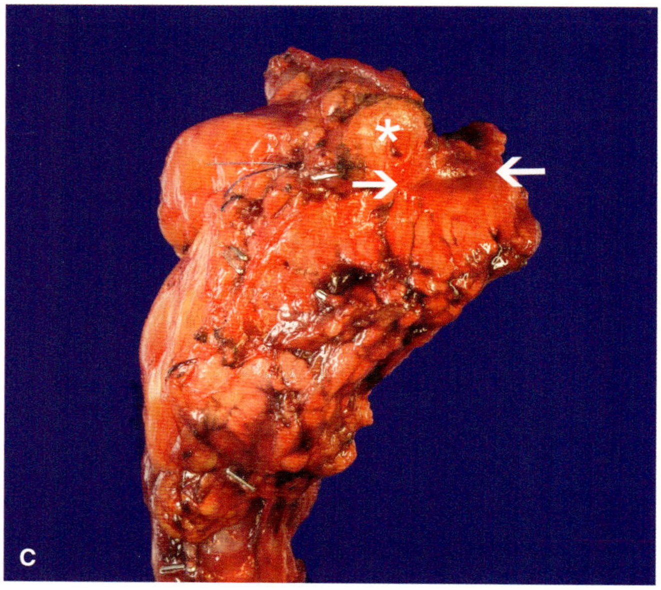

FIGURE 29.2 Gross orientation of a pancreatoduodenectomy specimen. The anterior surface of the pancreatic head (**A**) is covered by a smooth peritoneal surface. The posterior surface (**B**) is somewhat more irregular, as it has been dissected free from the retroperitoneal soft tissues. The vascular groove (**C**) is the smooth surface (*arrows*) that can be seen between the cut pancreatic neck margin (*asterisk*) and the uncinate process.

is drained by the dorsal, greater, and transverse pancreatic veins, which join the splenic vein. The tail of the pancreas is the leftmost part of the organ that extends to the hilum of the spleen. It is supplied by the caudal pancreatic artery, which is a branch of the splenic artery. It is drained by the caudal pancreatic vein, which joins the splenic vein.

The lymphatic drainage follows the pancreatic arteries and veins to drain into the pancreatic, splenic, and pyloric lymph nodes. Approximately 55% of interlobular lymphatic vessels are closely related to the accompanying artery and vein, 25% are separated from other structures by connective tissue, and 20% are closely related to acinar cells. Only 2% border the ductal system (13). Collecting channels arise on the surface or along the border of the pancreas (12).

There are two major systems of lymph nodes draining the pancreas: one encircles the pancreas and the other surrounds the aorta from the level of the celiac trunk to the origin of the superior mesenteric artery. These lymph nodes may be closely opposed to the periphery of the gland or even embedded within its parenchyma, especially along its inferior and superior borders and in the anterior and posterior pancreatoduodenal regions. Lymph nodes are also found around the celiac axis, adjacent to the common bile duct, and at the splenic hilum (14).

Several classification systems exist for these lymph nodes, including assigned station numbers and specific anatomic locations. Station numbers are seldom used in practice, while anatomic location is often used to stage lymphatic spread of pancreatic malignancies. Carcinomas of the head and neck of the pancreas spread to regional lymph nodes located along common bile duct, common hepatic artery, portal vein, pyloric, anterior and posterior pancreaticoduodenal arcades, superior mesenteric vein, and right lateral wall of superior mesenteric artery. Regional lymph nodes for carcinomas of the body and tail include nodal groups along common hepatic artery, celiac axis, splenic artery, and splenic hilum. Tumor involvement of nonregional lymph nodes is staged as distant metastasis.

FIGURE 29.3 Gross appearance of the normal pancreas. The bulbous head (**left**) is connected to the neck, body, and tail, which merge imperceptibly. The parenchyma consists of distinctly lobulated pink–tan fleshy tissue. The pancreatic duct (opened longitudinally) is thin and smooth throughout its course.

The parasympathetic and sympathetic nerves of the pancreas are supplied by the vagus nerve and the splanchnic nerves, both of which originate from the celiac and superior mesenteric plexi. Sympathetic and parasympathetic innervation play a crucial role in regulating intrapancreatic blood flow, pancreatic exocrine secretion, and pancreatic endocrine function (2,5,12). The course of the nerves accompanies the vasculature (4).

Gross Anatomy

The normal adult pancreas measures 15 to 20 cm in length and weighs 85 to 120 g (6,15). At birth, the pancreas weighs 2 to 3 g, gradually increasing to 7 g by the age of 1 year (16). Its weight gradually diminishes with age to an average of 70 g by the ninth decade of life (15,17).

The head, neck, body, and tail of the pancreas are grossly indistinct (Fig. 29.3). The bulk of the organ is composed of the head. The uncinate process has a vaguely nodular, irregular appearance on the posterior inferior surface; it lies posterior to the pancreatic neck and the superior mesenteric vessels. The vascular groove is a concave smooth surface adjacent to the uncinate process (7). The neck and body extend from the head in a somewhat triangular cross-section. The tail tapers and flattens as it approaches the splenic hilum (2).

The normal pancreas is pink-tan to yellow and uniformly lobulated. Its anterior surface is smooth and covered by peritoneum, while the remaining surfaces are covered by a thin layer of loose fibroconnective tissue. The boundary with surrounding retroperitoneal adipose tissue becomes indistinct when there is significant parenchymal lobular atrophy. Cut sections of the pancreas reveal thin-walled white ducts branching into distinct lobules. The main pancreatic duct of Wirsung averages 2 to 3 mm in diameter (range: 1.8 to 9 mm) (18), reaching a maximum of 4.5 mm average thickness near the ampulla where contents of the pancreatic ducts drain into the duodenum. The main duct is narrower in the tail and tends to widen slightly (up to 1 mm) with age. A main duct diameter exceeding 10 mm is considered pathologically dilated. Approximately 50 secondary ducts drain into the main duct (19,20), entering alternately from either side in a herringbone pattern (Fig. 29.4) (6,21). The trajectory of the major ducts varies depending on developmental fusion and atrophy patterns. In general, the main pancreatic duct of Wirsung begins in the tail, collecting tributaries as it traverses the body and neck toward the head. The duct turns inferiorly and joins the accessory duct of Santorini from the superior head and the major duct from the uncinate process, ultimately exiting through the ampulla at the major papilla. The accessory duct does not have a separate opening into the duodenum, although retention of embryonic patency through the minor duodenal papilla may occur.

The relationship between the main pancreatic duct and the distal common bile duct is variable. The most common variant is a long common channel formed by the fusion of the two ducts near the duodenum, forming the ampulla. In some individuals, the ducts join at the duodenal wall forming a short common channel of less than 3 mm in length, while in others, the ducts remain separate, entering the duodenum side by side or apart (Fig. 29.5) (4,12,15,22). Data from autopsy studies indicate 55% to 86% of individuals have a long common channel (23). Villiform mucosal projections, the ampullary valves of Santorini, line the distal ducts to prevent duodenal reflux (24,25). The intraduodenal portions of the pancreatic and bile ducts are surrounded by thin fascicles of smooth muscle cells in the sphincter of Oddi, which are continuous with both the muscularis mucosae and the muscularis propria of the duodenum.

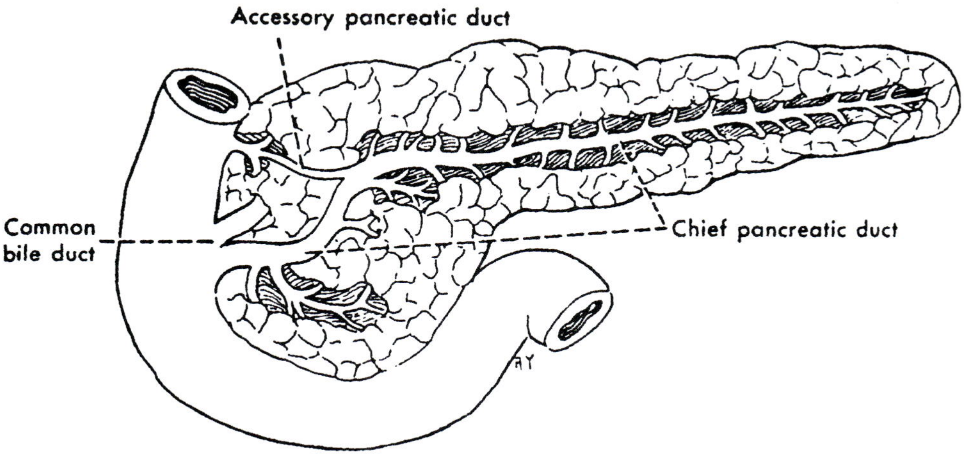

FIGURE 29.4 Schematic diagram of the pancreas showing the pattern of the major ducts and their tributaries. In this example, the accessory duct is patent at the duodenum through the minor papilla. (Reprinted with permission from Cubilla AL, Fitzgerald PJ. Tumors of the exocrine pancreas. In: Hartmann WH, Sobin LH, eds. *Atlas of Tumor Pathology.* 2nd series, fascicle 19. Armed Forces Institute of Pathology; 1984. Permission obtained from The Joint Pathology Center.)

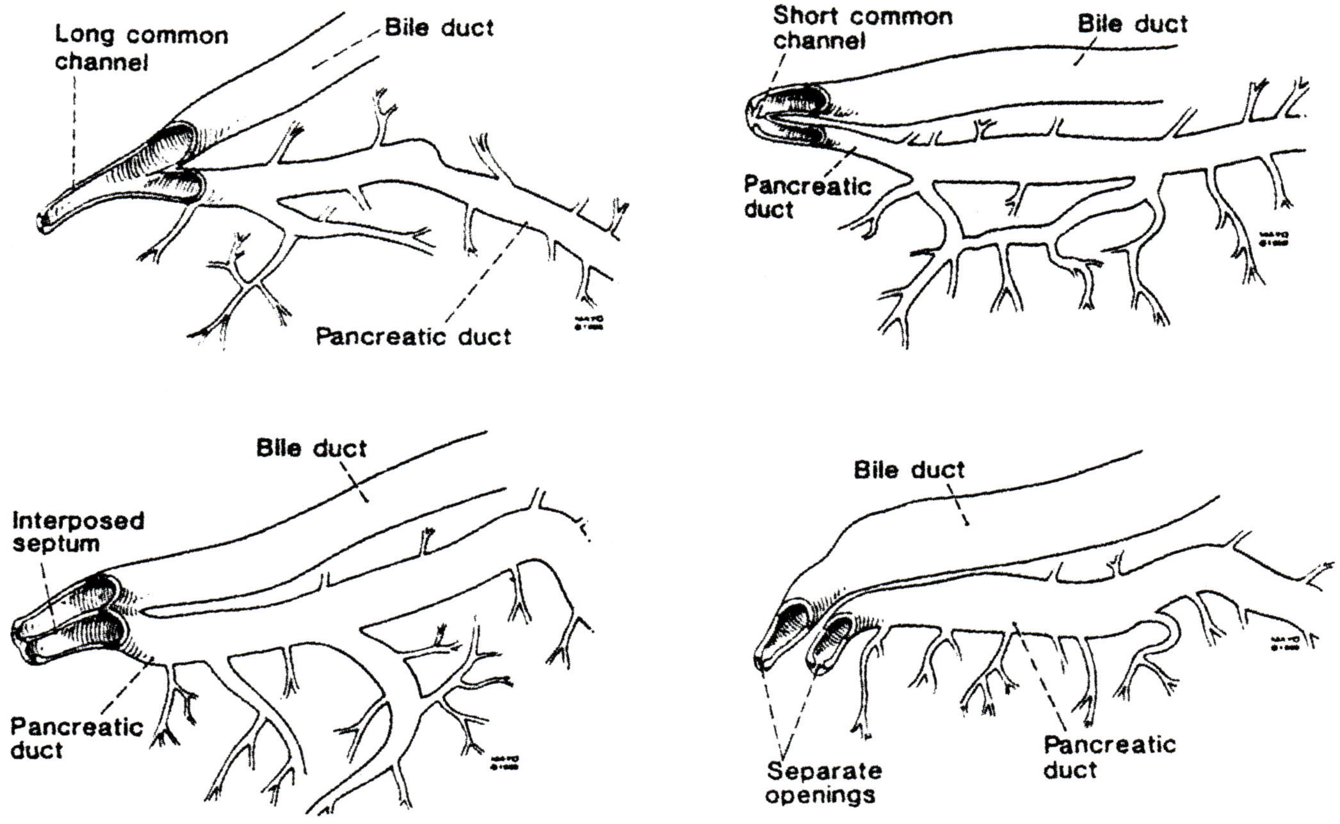

FIGURE 29.5 Anatomic variations in the paths of the pancreatic and biliary ducts at the ampulla. A long common channel (the prototypical ampulla) is only present in some individuals. In others, the ducts fuse within only a few millimeters of the duodenum, resulting in a short common channel, or the two ducts enter separately. Reprinted with permission from DiMagno EP, Shorter RG, Taylor WF, Taylor WF, Go VL. Relationships between pancreaticobiliary ductal anatomy and pancreatic ductal and parenchymal histology. *Cancer.* 1982;49(2):361–368.

DEVELOPMENT

Organogenesis

The pancreas develops during the 4th to 7th weeks of gestation. The gland is ultimately formed by fusion of the dorsal and ventral buds of septum transversum endoderm located near the junction of the embryonic foregut and midgut (26,27). The dorsal bud forms opposite the hepatic diverticulum, while the ventral bud forms adjacent to the hepatic diverticulum (20). This positions the ventral pancreatic duct in proximity to the common bile duct as it develops from the hepatic diverticulum. As the duodenum rotates during the 6th week of gestation, the ventral pancreas and common bile duct migrate around the posterior aspect of the duodenum to lie posteriorly and inferiorly to the dorsal pancreas (Fig. 29.6) (28). Fusion typically occurs in the 7th week, with the dorsal portion forming most of the adult gland including the superior head, neck, body, and tail. The ventral portion contributes to the inferior head and uncinate process (27). The ductal systems of the two lobes normally fuse. Thus, the distal two-thirds of the main pancreatic duct develop from the dorsal duct while the proximal third originates from the ventral duct. The accessory duct arises from the proximal dorsal duct. The ampulla develops during the 8th week after duct fusion is completed.

Cytogenesis

Pancreatic embryonic development is complex and occurs across three overlapping stages (29,30). First, the foregut endoderm patterns into dorsal and ventral pancreatic buds containing multipotent progenitor cells. Second, the cells within the buds commit to either endocrine or exocrine cell lineages. Third, extensive growth and branching lead to organ maturation. Multiple transcription factors cooperate to specify pancreatic organogenesis (26,30). The earliest transcription factor that initiates development is the homeodomain protein PDX1. PDX1 is first expressed in the multipotent progenitors of the budding endoderm (29) that ultimately differentiate into all three pancreatic epithelial lineages (31,32). Downstream transcription factors include SOX9 and SOX17 early in pancreatic development, followed by MNX1, GATA4 and GATA6, ISL1, and PTF1A and NR5A2 (32,33). Early fate determination between exocrine and endocrine pancreatic

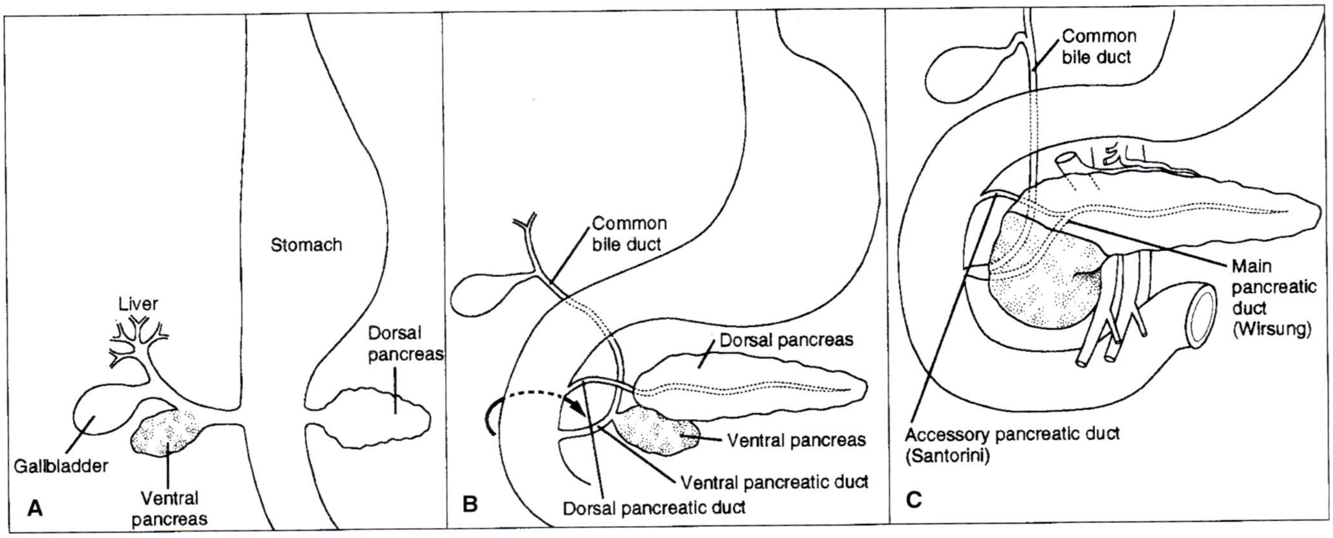

FIGURE 29.6 Development of the pancreas. The dorsal and ventral buds form on opposite sides of the duodenum (**A**). During the 6th week, the ventral pancreas migrates posteriorly around the duodenum (**B**) to lie inferior to the dorsal pancreas, where it comprises much of the head of the gland (**C**). (Reprinted with permission from Skandalakis LJ, Rowe JS Jr, Gray SW, et al. Surgical embryology and anatomy of the pancreas. *Surg Clin North Am.* 1993;73(4):661–697, with permission from Elsevier.)

cells is controlled in part by the balance between NKX6.1 and PTF1A. Transcription factors with a significant role in directing endocrine fate include, but are not limited to, NGN3, HES1, NEUROD/Beta2, PAX4, PAX6, NKX2.2, and NKX6.1 (32,34). Transcription factors vital for ductal cell types include PDX1, NKX6.1, SOX9 and FOXA2, HNF6, and HNF1β (35). Temporal Notch and Hedgehog ligand gradients regulate signal transduction pathways that coordinate developmental transcriptional programs. Notch signals are required for ductal branching and endocrine and exocrine lineage commitment (33,36,37). Sonic hedgehog (Shh) gradients in midgestational embryos specify foregut and gastrointestinal development. Later termination of Shh signaling facilitates transcriptional activation of differentiation gene subsets that guide pancreatic organ maturation (38).

The pancreatic ductal network develops by a process of branching morphogenesis off the developing pancreatic buds. Lumina are formed as the ducts branch (39). Progenitor cells destined for acinar fate segregate around the ductal tips while those destined for endocrine fates segregate along ductal trunks (39,40). Acinar cells and supportive mesenchyme orient into lobular units to facilitate zymogen secretion into the duct tips such that zymogen granules can be detected in acinar cytosol by the 4th month of gestation (16). The earliest immature granules are elongated with a fibrillary matrix. They first appear at 15 weeks of gestation and gradually condense into mature spherules by 20 weeks (41–43). Of note, zymogen granules often revert to the immature elongated structures in acinar pancreatic neoplasms (42,44–48). Pancreatic endocrine cells also differentiate from ductal elements to coalesce into islets of Langerhans within the surrounding mesenchyme at 8 to 10 weeks of gestation (49). Insulin-expressing cells appear first, followed by others expressing glucagon, somatostatin, and other secretory polypeptides (50–53).

The alpha and beta cells segregate to opposite ends of the islets by 16 weeks of gestation. These bipolar islets are replaced by mantle islets with centrally located beta cells surrounded by a rim of alpha cells at 18 to 20 weeks (52–54). Islet cell maturation occurs late in gestation and the early postnatal period (55).

The pancreatic lobules mature around branching duct structures during the 3rd to 4th months of gestation (Fig. 29.7A,B). The ductal lining cells develop specific characteristics based on their locations within the ductal system. Periductal mesenchyme is a prominent feature of early pancreatic development (Fig. 29.7C) and resembles the ovarian-like stroma seen in mucinous cystic neoplasms (56,57). As the pancreas develops, the mesenchyme becomes less abundant, ultimately constituting a minimal component of the epithelium-rich adult gland.

Developmental Anomalies and Heterotopia

Pancreatic agenesis is extremely rare; it can occur as an isolated condition or as part of a congenital syndrome such as polysplenia/heterotaxy syndrome (58). It results from germline mutation in genes encoding pancreatic developmental transcription factors *PDX1*, *PTF1A*, and *GATA6* or mutations in cis-regulatory elements required for their expression. Partial pancreatic agenesis may be asymptomatic or manifest with diabetes mellitus and/or pancreatic exocrine insufficiency depending on its extent. Pancreatic hypoplasia is more common than agenesis and may occur as part of a congenital syndrome or an isolated anomaly due to germline mutations in *HNF1B* (58). Affected individuals may

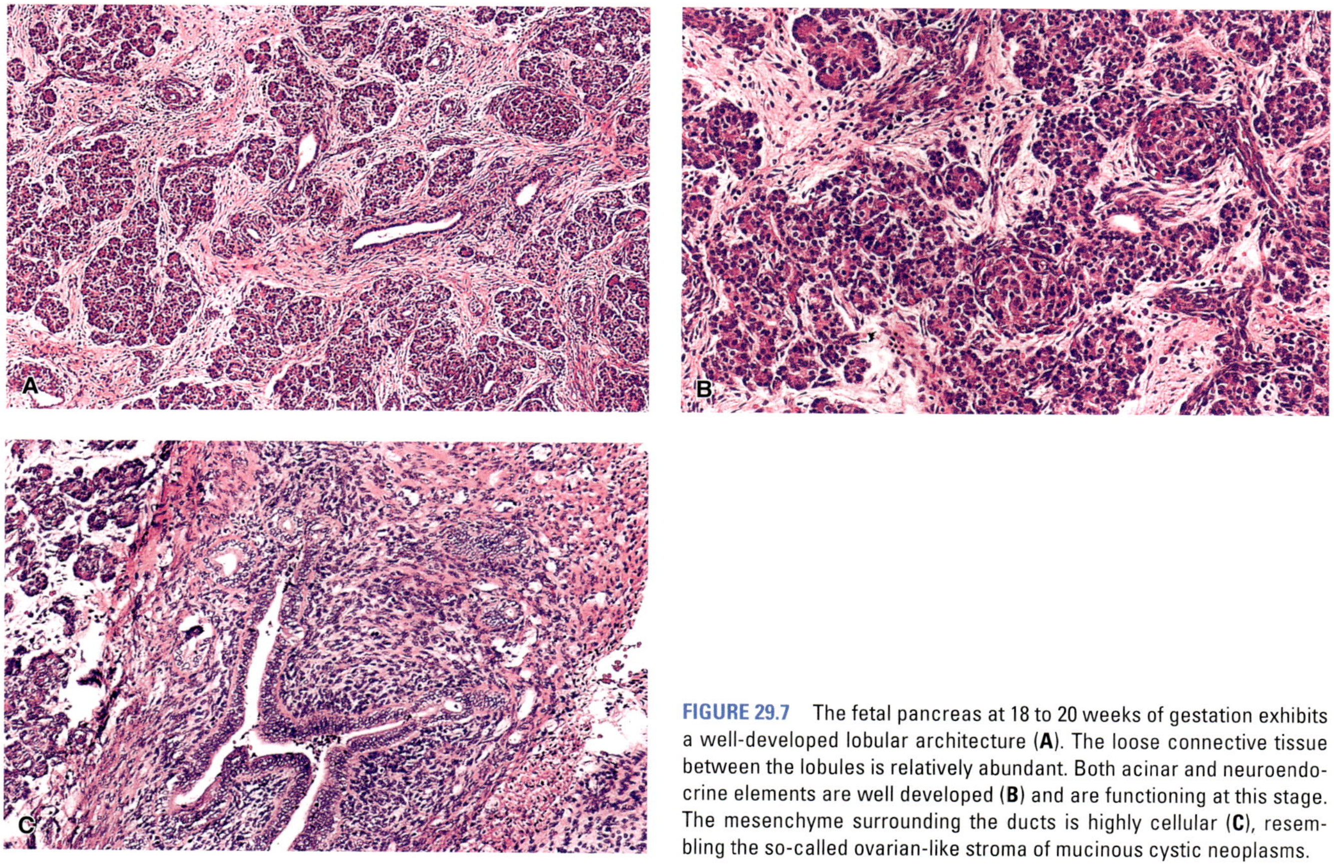

FIGURE 29.7 The fetal pancreas at 18 to 20 weeks of gestation exhibits a well-developed lobular architecture (**A**). The loose connective tissue between the lobules is relatively abundant. Both acinar and neuroendocrine elements are well developed (**B**) and are functioning at this stage. The mesenchyme surrounding the ducts is highly cellular (**C**), resembling the so-called ovarian-like stroma of mucinous cystic neoplasms.

be asymptomatic or exhibit organ dysfunction including exocrine insufficiency, mature onset diabetes of the young, type 5 (MODY5), or anomalies of the renal and genital tract (59–64).

Annular pancreas is an extremely rare developmental anomaly characterized by partial or complete encirclement of the second part of the duodenum by pancreatic tissue (Fig. 29.8). This condition affects only 0.015% of the population and is usually accompanied by pancreas divisum (65). It possibly results from the failure of one of the lobes of the ventral embryonic bud to regress, causing it to encircle the duodenum during the normal rotation of the duodenum. Annular pancreas can cause duodenal obstruction and some cases are associated with duodenal atresia (66). The band of pancreatic tissue encircling the duodenum is flattened and may be embedded within the muscularis propria. Histologically, the annular tissue contains all the normal parenchymal elements intermixed with smooth muscle. The encircling pancreatic portion originates from the ventral pancreas and is accordingly enriched with pancreatic islets (67).

The anatomy of the pancreatic duct system and its relationship with the common bile duct is highly variable among individuals (20,66,68,69). An embryonic communication between the dorsal pancreatic duct and the duodenum persists in approximately 50% of people. This results in a patent accessory duct at the minor papilla, usually 2 cm proximal to the major papilla. The dorsal duct becomes the main route of drainage for the gland in 4% of individuals, in which case it is usually larger in diameter than the ventral duct (69). This condition is more prevalent in children, suggesting that obliteration of

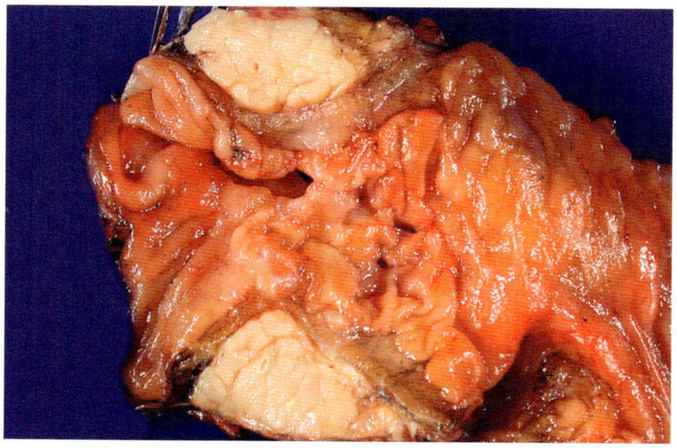

FIGURE 29.8 Annular pancreas. The pancreatic tissue completely encircles the duodenum, which is opened longitudinally to expose the pancreas around it. Reprinted with permission from Hruban RH, Pitman MB, Klimstra DS. Tumors of the pancreas. In: Silverberg SG, Sobin LH, eds. *Atlas of Tumor Pathology*. 4th ed. American Registry of Pathology; 2006.

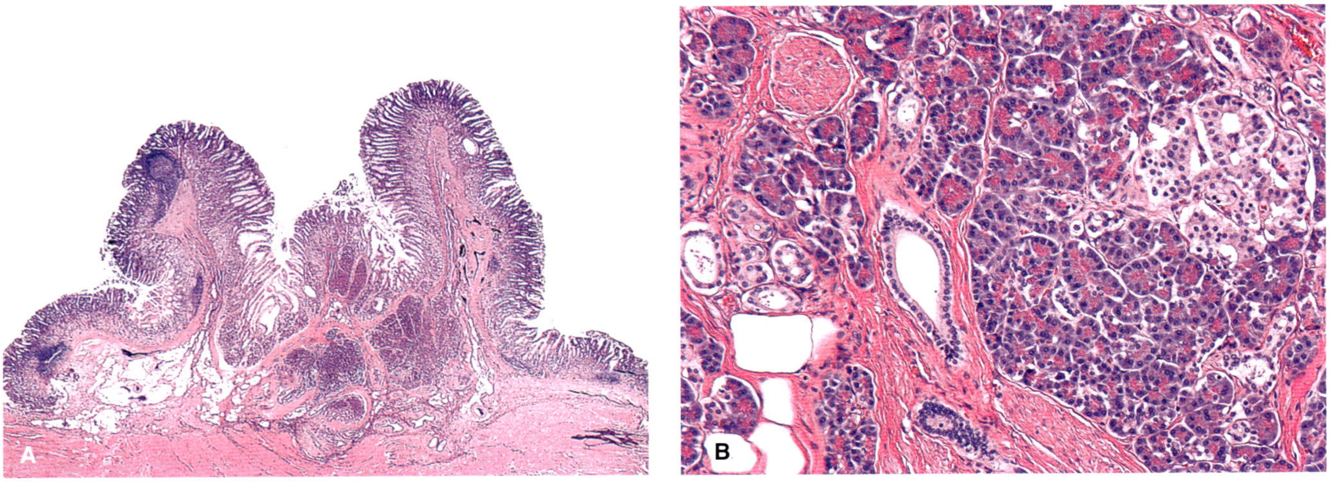

FIGURE 29.9 Heterotopic pancreatic tissue in the stomach. At low power, a submucosal nodule of pancreatic tissue results in an umbilicated appearance (**A**). In this example, there are acini, ducts, and islets arranged in a disorganized pattern, with interspersed bundles of smooth muscle (**B**).

the accessory duct opening persists in the few adults who retain this anatomy (70). Incomplete fusion of the dorsal and ventral bud duct systems creates two separate ductal systems in approximately 5% to 10% of individuals and is more commonly associated with pancreas divisum than the annular pancreas.

Pancreas divisum is classified into three types. Type 1, or classical pancreas divisum, reflects a complete failure of duct fusion and results in most pancreatic secretions draining through the accessory duct at the minor papilla, while the main duct drains only the inferior head of the pancreas through the major papilla. Type 2 results from complete regression of the ventral duct, leaving the single dorsal duct at the minor papilla as the primary outlet for exocrine secretions. Type 3 represents an incomplete pancreas divisum with a small communicating branch of the ventral duct remaining intact. It is controversial whether pancreas divisum is associated with acute or chronic pancreatitis (69,71,72). Aside from abnormal pancreatic duct fusions, anomalous junctions between the main pancreatic duct and the distal common bile duct 2 cm proximal to the duodenum can also occur (20,66). This abnormality may be associated with choledochal cysts and carcinomas of the extrahepatic bile ducts or gallbladder (20,73,74). Bifid pancreas is a rare anomaly in which the main pancreatic duct bifurcates within the body of the pancreas (75).

Pancreatic heterotopia refers to pancreatic tissue located outside of the normal anatomical position of the gland (66). While found in up to 1.5% of individuals at autopsy, the surgical or endoscopic incidence is only 0.1% to 0.2%. Of the latter, between 25% and 50% are symptomatic with typical onset in adulthood (76–78). The most common sites are the duodenum near the ampulla and the stomach (79,80). Duodenal heterotopias are usually incidental nodules resembling polyps within an otherwise normal background. Gastric pancreatic heterotopias present in similar fashion and can even form large masses. Ectopic pancreatic tissue present in an inflammatory background, as in patients with autoimmune gastritis, is best considered to represent pancreatic metaplasia. However, pancreatic tissue at the gastroesophageal junction, jejunum, ileum, Meckel diverticulum, colon, spleen, gallbladder, mesentery, and liver most likely represent true heterotopias (81,82). In the tubular gastrointestinal tract, heterotopic pancreas usually presents grossly as firm, lobulated yellow to white-tan submucosal nodule ranging millimeters to centimeters in size. The overlying mucosa may be umbilicated, ulcerated, or unremarkable (Fig. 29.9A). All three pancreatic epithelial constituents (ducts, acini, islets) are present in 12% of cases (82). More commonly, only ducts are present surrounded by interlacing smooth muscle fascicles, resembling an adenomyoma (Fig. 29.9B). Acini of pancreatic heterotopia in the duodenum are sometimes co-populated with cells resembling Brunner glands (19). It is important to distinguish benign pancreatic heterotopias from ductal adenocarcinomas, especially when there are surface erosions and reactive epithelial changes (83–85). Benign entities such as chronic pancreatitis and pseudocysts can arise in pancreatic heterotopias. Pancreatic neoplasms have also been rarely reported including pancreatic intraepithelial neoplasia, intraductal papillary mucinous neoplasm, acinar cell carcinoma, and pancreatic endocrine tumor (86–89).

Heterotopic tissues may also be found within the pancreas. Accessory splenic tissue may be found in the tail of the pancreas or, less commonly, in the head (90). In most cases, the splenic tissue is small (less than 2 cm), dark red, and spherical. Intrapancreatic adrenal cortical tissue has also been described (91). Benign developmental cysts such as foregut cysts can also occur in the pancreas and should not be confused with pancreatic cystic neoplasms (Fig. 29.10).

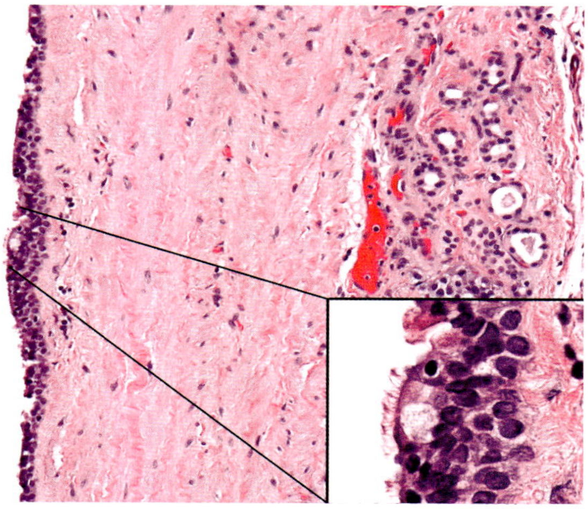

FIGURE 29.10 A rare pancreatic foregut cyst is lined by a single layer of ciliated columnar epithelial cells (*inset*).

MICROSCOPIC FEATURES

Microscopically, the pancreas consists of lobules ranging from 1 to 10 mm in size (Fig. 29.11). The parenchyma is composed of the three epithelial elements: acini, ducts, and islets of Langerhans. A sparse amount of fibroconnective tissue containing vessels, matrix, stellate cells, and nerves separates the lobules.

Acini

Acinar cells comprise 85% of the pancreatic mass and serve as the main exocrine secretory component. While histologic sections often depict acini as single-layered polygonal cells surrounding a central lumen, it must be remembered that these represent two-dimensional cross-sections of three-dimensional spherical and tubular structures (Fig. 29.12). Most acini are located at the proximal (originating) end of the ductules, although some can be found abutting the duct periphery or even be situated in between cells lining the ducts. Anastomosing loops of acini also may be found

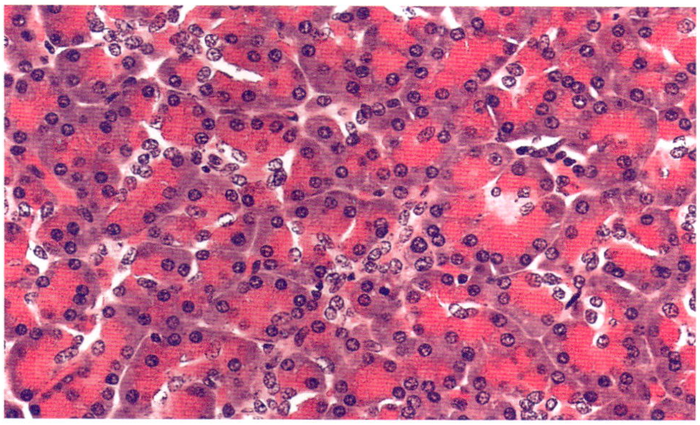

FIGURE 29.12 Acinar cells show basally located nuclei, basophilic basal cytoplasm, and abundant granular eosinophilic apical cytoplasm. Most acini consist of spherically arranged individual acinar cells.

(92) (Fig. 29.13). Thus, there are diverse pathways for acinar secretions to reach the ductal system.

Acinar cells are polarized epithelial cells with basally located round nuclei and small distinct nucleoli. Acinar cells are some of the most highly secretory cells in the human body. An abundant rough endoplasmic reticulum is required to ensure proper oxidative folding of secretory peptides. The resulting basolateral concentration of ribonucleoproteins imparts a basophilic appearance on H&E stains. The apical portion of the cell is filled with numerous zymogen granules. These impart an eosinophilic appearance (Fig. 29.12). Acinar cells adjust zymogen granule content in response to digestive hormone paracrine signals (1). The acinar cells are not surrounded by a myoepithelial cell layer. Rather, they are directly anchored to the underlying basement membrane matrix. In general, acinar cells from different pancreatic regions are morphologically and functionally similar (93,94). Zymogen granules also label with periodic acid–Schiff (PAS) stains, and this is resistant to diastase digestion (Fig. 29.14). Butyrate esterase stains also label acinar cells, reflecting the presence of enzymatically active lipase (45).

Immunohistochemical stains for pancreatic enzymes such as trypsin, chymotrypsin, lipase, amylase, and elastase also label acinar cells (Fig. 29.15) and with the exception of

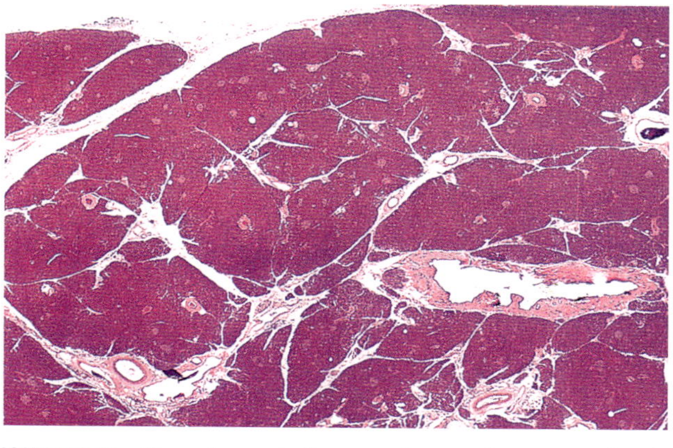

FIGURE 29.11 Normal pancreatic parenchyma at low power showing lobular architecture, highly cellular glandular tissue with scant stroma.

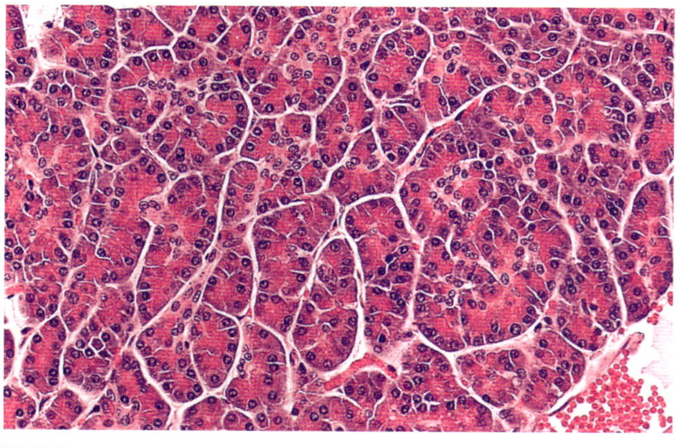

FIGURE 29.13 Pancreatic acini with tubular configuration, exhibiting interanastomosing loops when studied by serial sectioning.

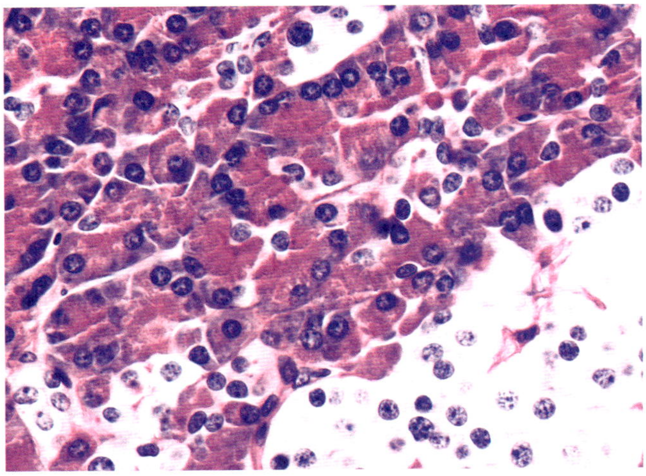

FIGURE 29.14 Zymogen granules are positive with periodic acid–Schiff stain with diastase pretreatment. Islet cells (**lower right**) are negative.

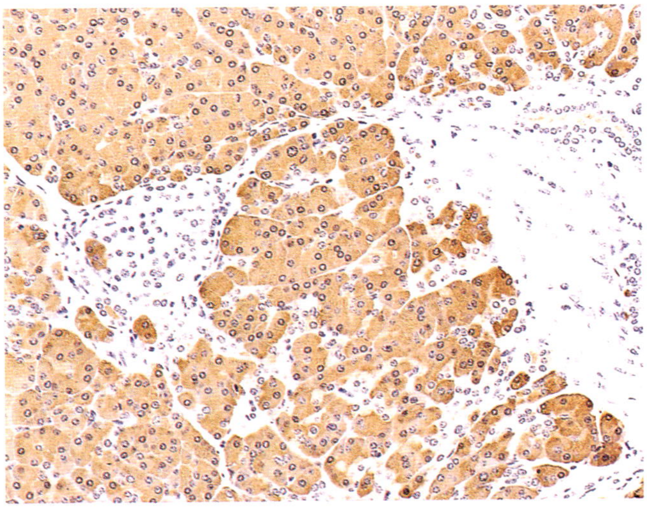

FIGURE 29.15 Immunohistochemical staining for trypsin shows intense labeling of acinar cells but not ductal or islet cells. Faint luminal labeling of ductal cells may be seen due to deposition of luminal enzymatic secretions on the apical cell surfaces.

amylase, these stains are also sensitive markers for acinar differentiation in pancreatic neoplasms (45–47,95). Antibodies targeting the C-terminal of BCL10 also detect acinar differentiation due to homology between BCL10 and the digestive enzyme carboxyl ester hydrolase (96). Immunohistochemical antibodies detect the inactive digestive proenzymes, as these are the forms that are stored within the zymogen granules (97,98). Acinar cells express epithelial cytokeratin intermediate filaments that are immunopositive for CAM5.2, cytokeratin 8, and cytokeratin 18, and negative for cytokeratins 7, 19, 20, and AE1/3 (Fig. 29.16). Acinar cells do not produce mucin and are likewise negative for mucin stains including DUPAN-2, carcinoembryonic antigen (CEA), CA19.9, and MUC proteins. Endocrine markers for chromogranin, INSM1, and synaptophysin are also negative.

Ultrastructurally, the basal cytoplasm is filled with parallel stacks of rough endoplasmic reticulum, scattered mitochondria, and interspersed free polyribosomes (Fig. 29.17) (99). Immature electron-dense zymogen granules emanate from the *trans* side of the Golgi apparatus located centrally. Beyond the Golgi, the apical cytosol is filled with large, round, homogeneous zymogen granules. These mature 0.25 to 1 µm granules contain dense secretory content and fuse with the apical plasma membrane upon stimulation to expel their contents into the lumen. The emptied vesicle membranes are then recycled to the Golgi apparatus and filled with another round of zymogens (100,101). Short apical microvilli containing actin myofilaments are connected to a terminal web of cytosolic intermediate filaments beneath the plasma membrane (99). Acinar cell adhesion is mediated by lateral junctional complexes including zonula occludens and zonula adherens. Basolateral desmosomes attach cells to the basement membrane matrices (16,99).

Pancreatic Ducts

Pancreatic ductal epithelial cells secrete approximately 2 L of fluid per day composed of water, chloride, and bicarbonate via transmembrane electrolyte and fluid transporters.

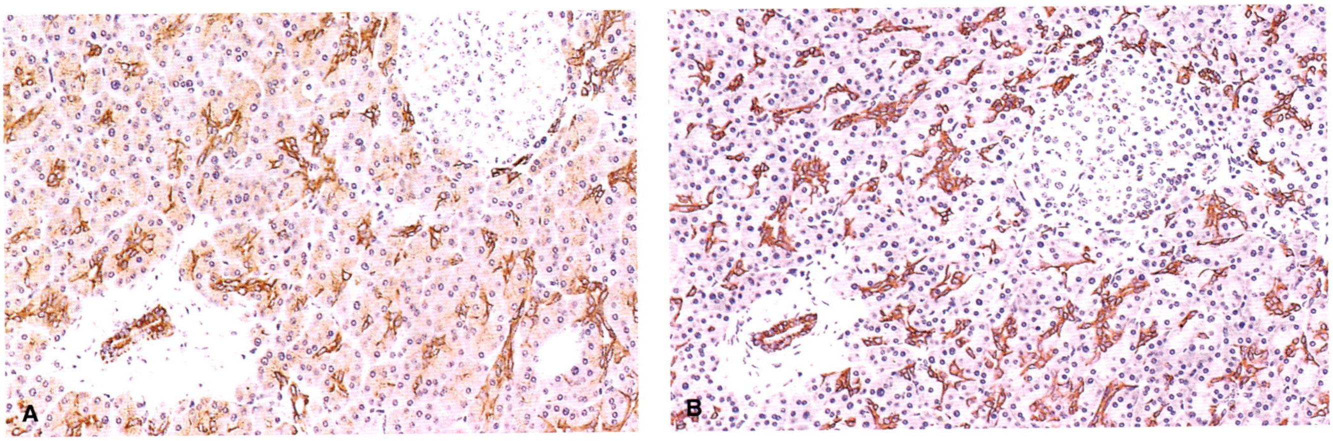

FIGURE 29.16 Immunohistochemical labeling for keratins. The CAM5.2 antibody shows diffuse labeling of acinar cells and ductal cells, the latter being more intensely positive (**A**). The AE1/AE3 antibodies label only ductal cells (**B**). Islet cells show only focal faint labeling with these antibodies.

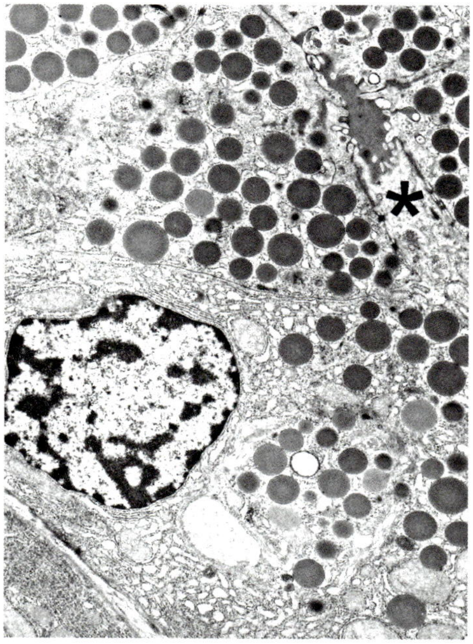

FIGURE 29.17 The ultrastructural appearance of acinar cells. These polarized cells have abundant parallel stacks of rough endoplasmic reticulum with interspersed mitochondria in the basal cytoplasm. Homogeneous electron-dense zymogen granules are concentrated in the apical cytoplasm underlying the lumina. The luminal spaces are lined by short microvilli. Adjacent acinar cells are joined to one another and to the centroacinar cells (*asterisk*) by apical junctional complexes. The centroacinar cells have lucent cytoplasm devoid of secretory granules.

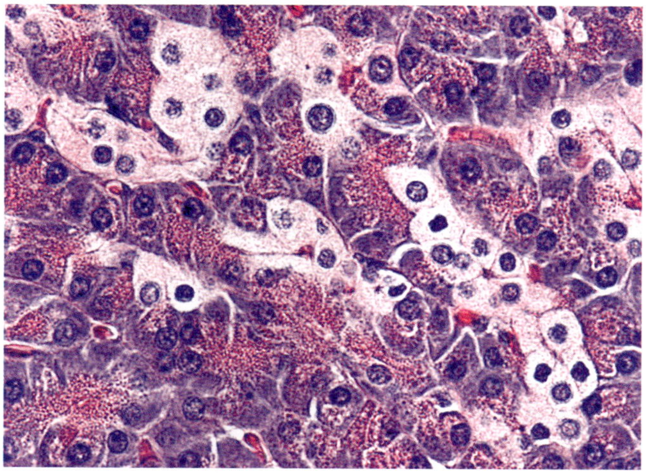

FIGURE 29.19 Some centroacinar cells are enlarged and have oncocytic cytoplasm, a variation of uncertain significance.

Ductal fluid buffers the acidity of the pancreatic juices, transports acinar secretions out of the pancreas, and stabilizes the proenzymes secreted by acini until activation in the duodenum (102). The ductal system comprises five continuous segments: centroacinar cells; intercalated ducts; intralobular ducts; interlobular ducts; and main ducts (103). The cellular arrangement of the ductular system begins with centroacinar cells located immediately adjacent to acinar cells. These are small, inconspicuous, flat to cuboidal cells with pale cytoplasm and central oval nuclei (Fig. 29.18). Centroacinar cells are located in the middle of the acini, where they are partially connected to the acinar cells by tight junctions. The cytosolic ultrastructure is largely devoid of organelles, reflecting a nonsecretory state of low metabolic activity. The apical plasma membranes contain scattered short microvilli similar to those on the adjacent acinar cells. Centroacinar cells adhere to each other with interdigitating junctional complexes. Occasional cells in the centroacinar compartment have more abundant granular oncocytic cytoplasm (Fig. 29.19) that reflects numerous mitochondria (103,104) and may represent a clinically trivial form of oncocytic metaplasia (described later).

The intercalated ducts are the smallest ducts outside the acini. These elements are responsible for secreting most of the ductal pancreatic fluid (Fig. 29.20) (105,106). They are lined by cuboidal cells with a central ovoid nuclei and indistinct nucleoli. Both centroacinar and intercalated ductal cells lack mucin and are negative for mucicarmine or

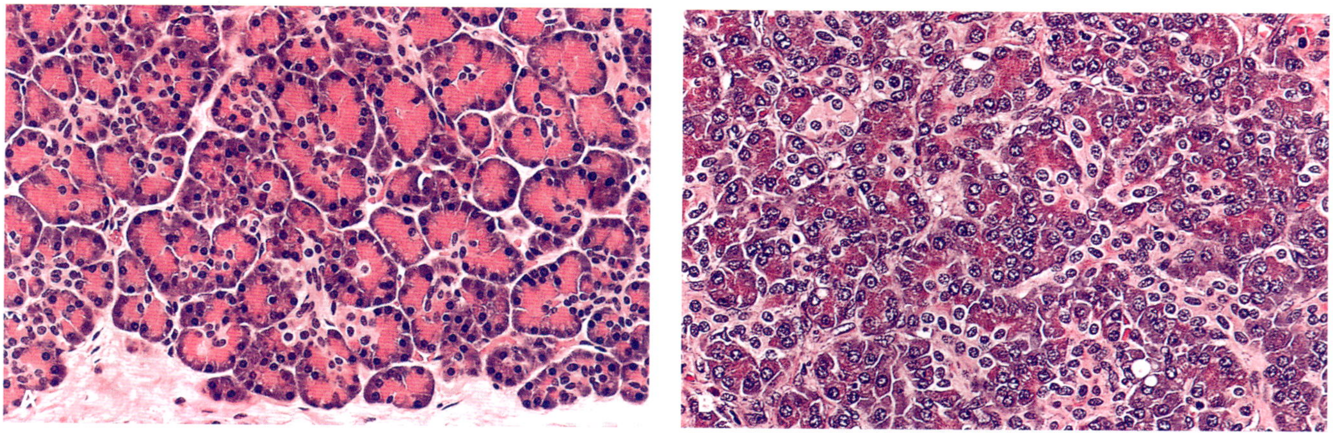

FIGURE 29.18 Centroacinar cells constitute the beginning of the ductal system, show minimal cytoplasm and oval nuclei, and are situated in the center of the acini (**A**). In some areas, centroacinar cells may be more prominent, with more abundant lightly eosinophilic cytoplasm (**B**).

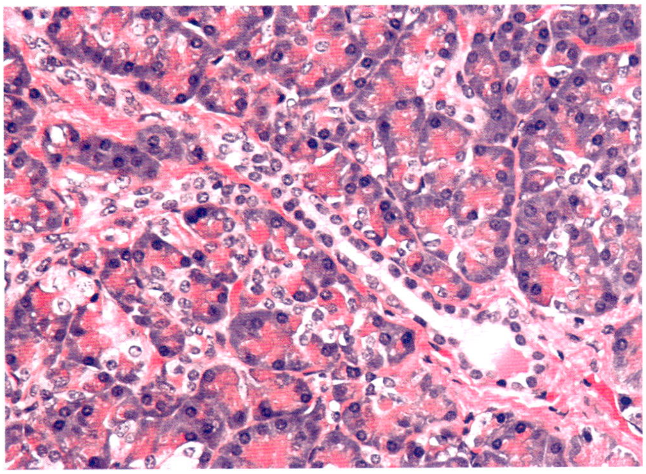

FIGURE 29.20 The ductal system consists of innumerable intercalated ducts that fuse to form these intralobular ducts. The transition from one to the next is imperceptible. The intralobular ducts show minimal collagenized stroma.

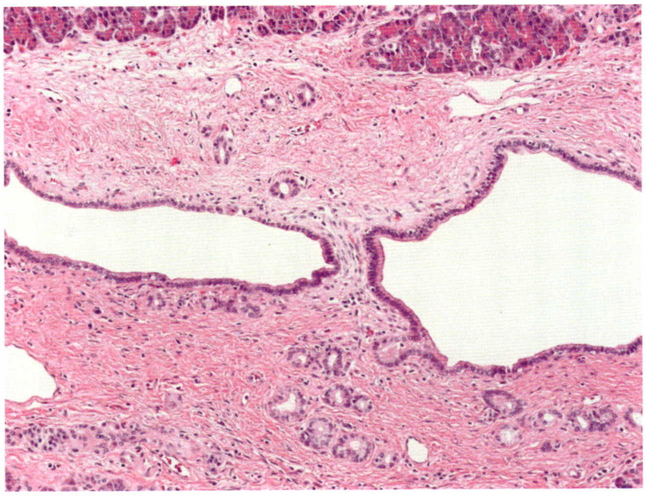

FIGURE 29.22 Interlobular ducts are surrounded by a thick rim of collagen and show small lobular aggregates of ductules within the wall.

Alcian blue mucin stains. The intercalated ducts seamlessly merge into larger intralobular ducts. The intralobular ducts are lined by cells identical to those of the intercalated ducts (Fig. 29.20). Both intercalated and intralobular ductal epithelial cells are attached to the basement membrane and lack contractile myoepithelial or basal cells.

Intralobular ducts exit the lobules to join and form interlobular ducts that are invested with variably thick collagen (Fig. 29.21). The interlobular ductal cells have slightly more cytoplasm compared to the intralobular ducts and assume a low columnar shape. Similar to the smaller ducts, they lack mucin; the presence of mucin visible by routine H&E stains usually indicates neoplastic transformation. Other more specialized stains can detect the presence of some types of mucin in normal interlobular ducts. As the interlobular ducts approach the main pancreatic ducts, they develop progressively thicker collagenous support containing lobular aggregates of small ductules reminiscent of Beale ducts that surround the major bile ducts (Fig. 29.22).

The main pancreatic duct receives numerous tributaries of interlobular ducts (Fig. 29.23). Main ducts are lined by low columnar epithelium with basal round nuclei and without papillary projections, although simple papillae can be found in the ampulla (Fig. 29.24). Apical cytoplasmic clearing reflects scant mucin production, but tall columnar cells with abundant cytosolic mucin are not normally present (103). The scant mucins of intralobular and smaller interlobular ductal cells are predominantly sulphomucins that stain with Alcian blue at pH 1.0 (Fig. 29.25) (107,108). Sulphomucin production is lower in the larger ducts, which instead produce more neutral mucins and sialomucins (107). The thick connective tissue wall of the largest ducts contains numerous periductal ductules and bundles of smooth muscle (1).

Ultrastructural features of large ductal cells include cytosolic mucigen granules, rough endoplasmic reticulum,

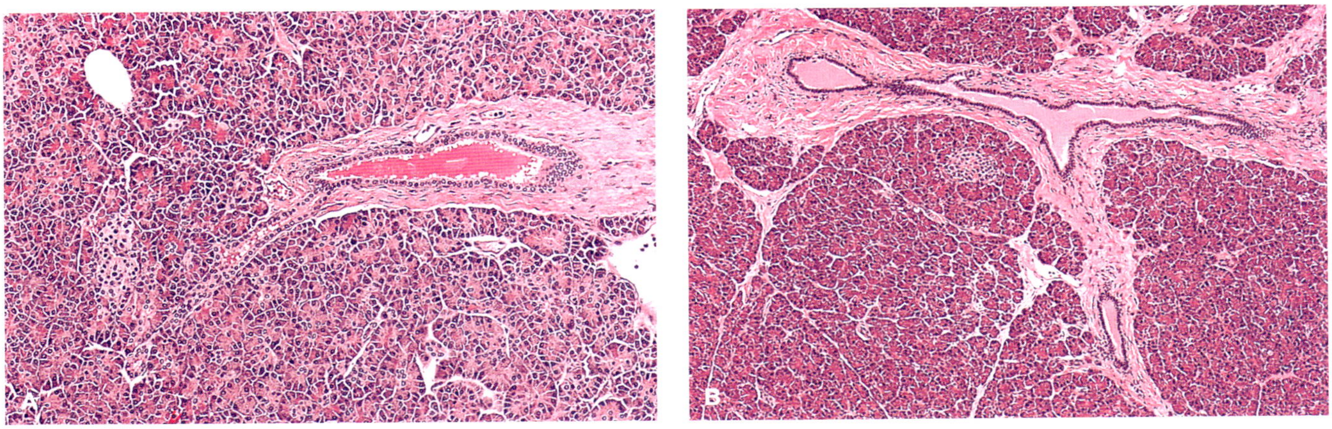

FIGURE 29.21 Intralobular ducts come together to form interlobular ducts (**A**). The interlobular ducts are surrounded by a variably thick rim of dense fibrous tissue and carry the pancreatic secretions to the major ducts, receiving tributaries of small interlobular ducts as they pass through the connective tissue septa of the gland (**B**).

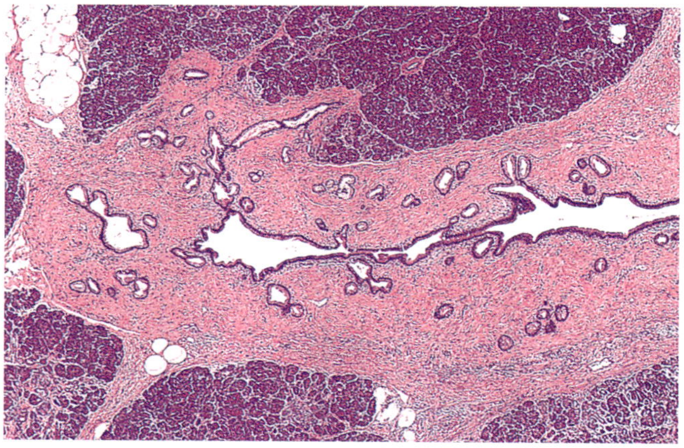

FIGURE 29.23 Numerous interlobular ducts join the main pancreatic duct along its course.

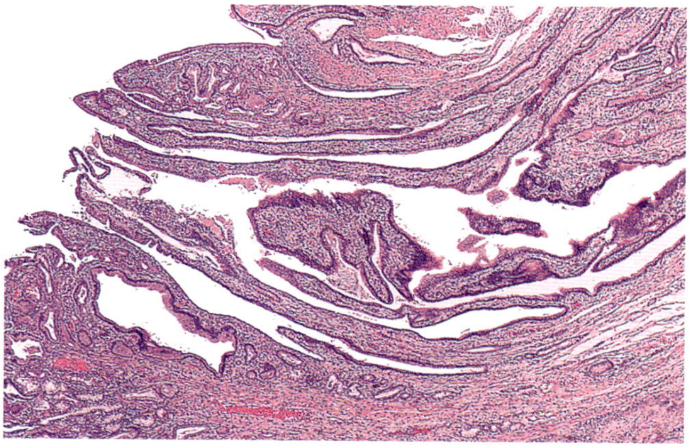

FIGURE 29.24 As the main pancreatic duct enters the ampulla (of Vater), the ductal epithelium forms broad, simple papillae known as the ampullary valves (of Santorini).

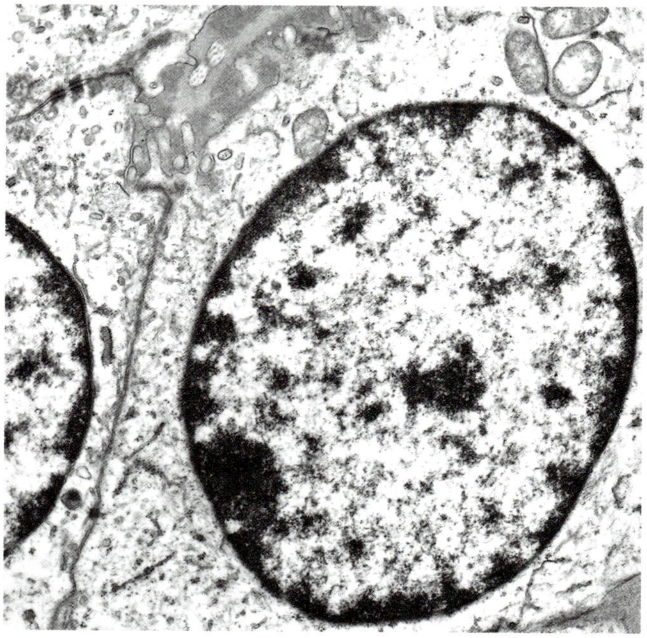

FIGURE 29.26 Ultrastructural appearance of intralobular ductal cells. The cytoplasm resembles that of centroacinar cells, with an electron-lucent appearance, scattered mitochondria, and rough endoplasmic reticulum. In the smaller ducts, mucigen granules are largely absent. The luminal surface exhibits short microvilli. In addition, scattered cilia are present, the cross-section of which may be seen within the lumen (**top**).

mitochondria, and Golgi reminiscent of acinar cells (Figs. 29.26 and 29.27). Single long cilia project from the apical surfaces of intercalated ductal cells into ductal lumens (Fig. 29.26) (99,103). The motile cilia are connected to basal bodies in the paranuclear cytoplasm and may function in mixing and propulsion of the pancreatic secretions (109).

Ductal cells express cytokeratins 7, 8, 18, and 19, AE1/3, and CAM5.2. by immunohistochemistry (Fig. 29.16). They do not normally express cytokeratin 20. Exocrine enzymes

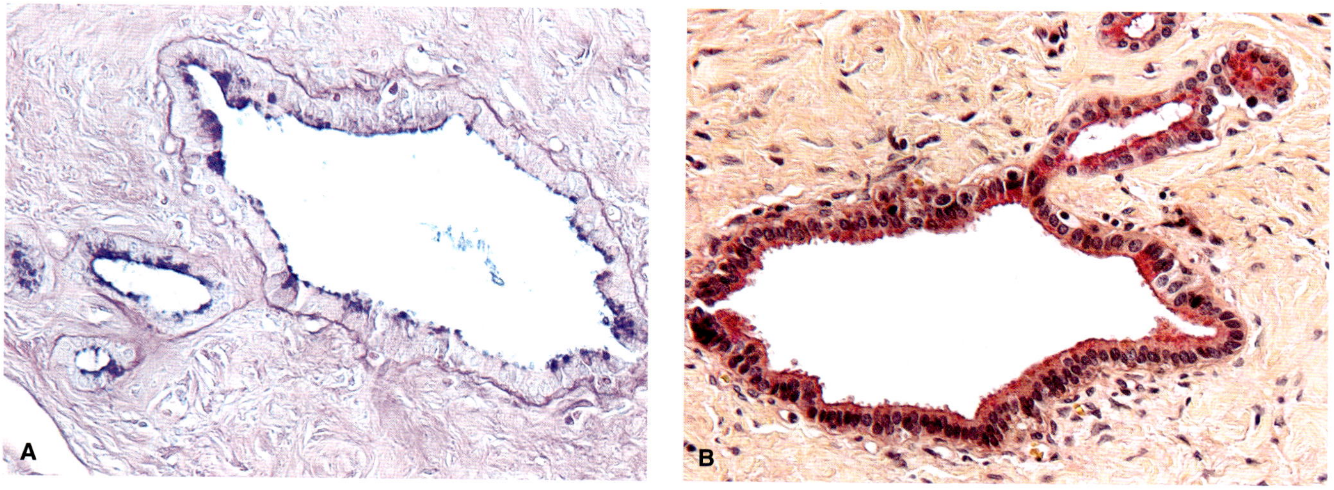

FIGURE 29.25 The cells of the intralobular and smaller interlobular ducts contain mucin in the apical cytoplasm that stains positively with Alcian blue/PAS (**A**). Staining with mucicarmine shows a similar distribution of mucin (**B**).

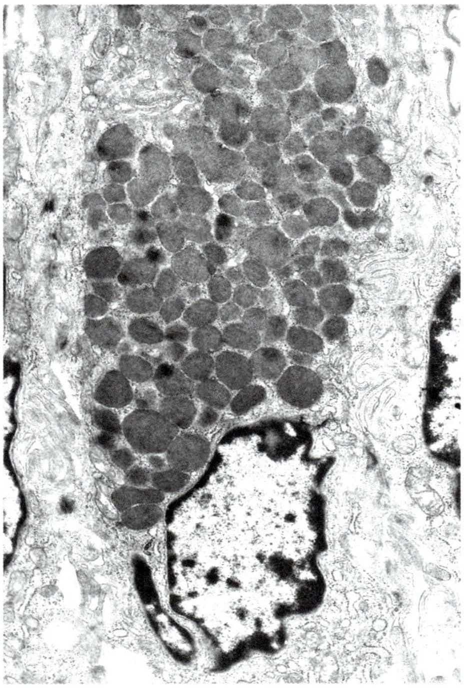

FIGURE 29.27 In the larger ducts, more abundant mucigen granules accumulate within the apical cytoplasm. These granules vary in size and have irregular contours and heterogeneous, variably electron-dense secretory contents. There are complex interdigitations of the lateral membranes between adjacent cells.

and endocrine markers are also absent. It should be noted that individual endocrine cells can be found situated within the larger ducts, and acinar enzyme markers can label the apical surfaces of ductal cells due to precipitation of luminal enzymatic contents.

Intercalated and intralobular ductal cells express carbonic anhydrase as required for fluid and ion transport (110,111). Expression is weaker in the larger interlobular ducts (Fig. 29.28). Ductal cells also express CA19-9, DUPAN-2, cystic fibrosis transmembrane conductance regulator (CFTR), N-terminal gastrin-releasing peptide (N-GRP), and transcription factors such as HNF1β, HNF6, and SOX9 (112–117). Normal ducts are negative for monoclonal CEA (118), B72.3, and CA125. The expression of different MUC glycoproteins varies: MUC1 is present in smaller intralobular and intercalated ducts, MUC6 is expressed in centroacinar cells and intercalated ducts. MUC2, MUC4, and MUC5AC are not normally expressed in pancreatic ducts (119–121).

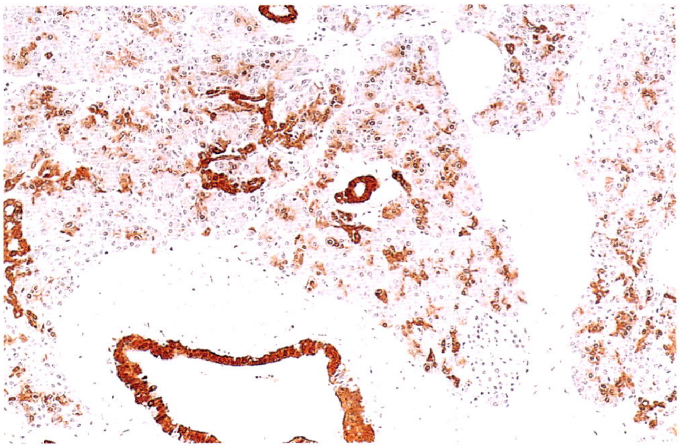

FIGURE 29.28 Immunohistochemical staining for carbonic anhydrase. In this preparation, there is staining of ductal cells of all sizes, including centroacinar, intercalated duct, as well as intralobular and interlobular ductal cells.

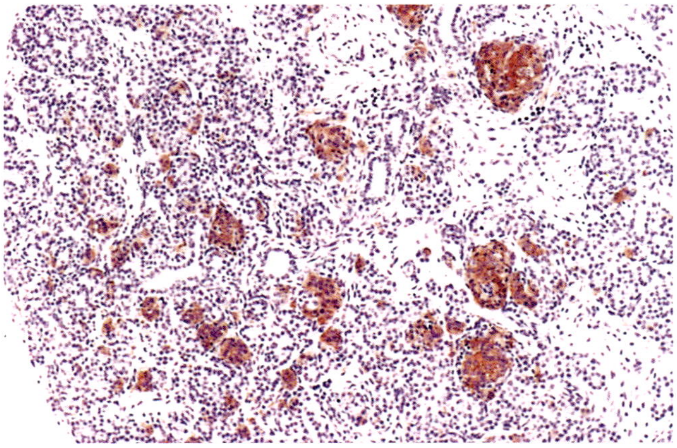

FIGURE 29.29 A fetal pancreas stained immunohistochemically for chromogranin. Note the abundance of neuroendocrine cells relative to acinar cells at this stage of development.

Islets of Langerhans

The endocrine pancreas decreases from 10% of the total organ volume in newborns to only 1% to 2% in adults (Fig. 29.29) (1,94,108,122). Over one million islets of Langerhans contain the vast majority of endocrine cells. Islets are distributed throughout the pancreas but are more numerous in the tail (123,124). Variability in islet distribution can lead to regions with abundant islets juxtaposed with areas exhibiting sparse distribution (Fig. 29.30), and this can be accentuated by endocrine cell hyperplasia, as seen in chronic pancreatitis. The apparent volume of islets varies with age and the presence of exocrine atrophy, as commonly occurs in chronic pancreatitis (125,126).

Islets are either compact or diffuse. Most (90%) are compact islets composed of sharply circumscribed round nests of endocrine cells that span 75 to 225 μm in diameter (127). Compact islets are located predominantly in the body and tail of the gland. Diffuse islets may measure up to 450 μm in diameter and are restricted to the posteroinferior head derived from the embryonic ventral lobe (128,129).

Despite the circumscribed appearance, interdigitating trabecular substructures resembling small lobules can be seen in cross-sections of compact islets (127). Most cells within compact islets exhibit uniform round nuclei with coarsely clumped chromatin and inconspicuous nucleoli (Fig. 29.31). The cytoplasm is pale and amphophilic. Occasional islet cells have nuclei two to four times larger

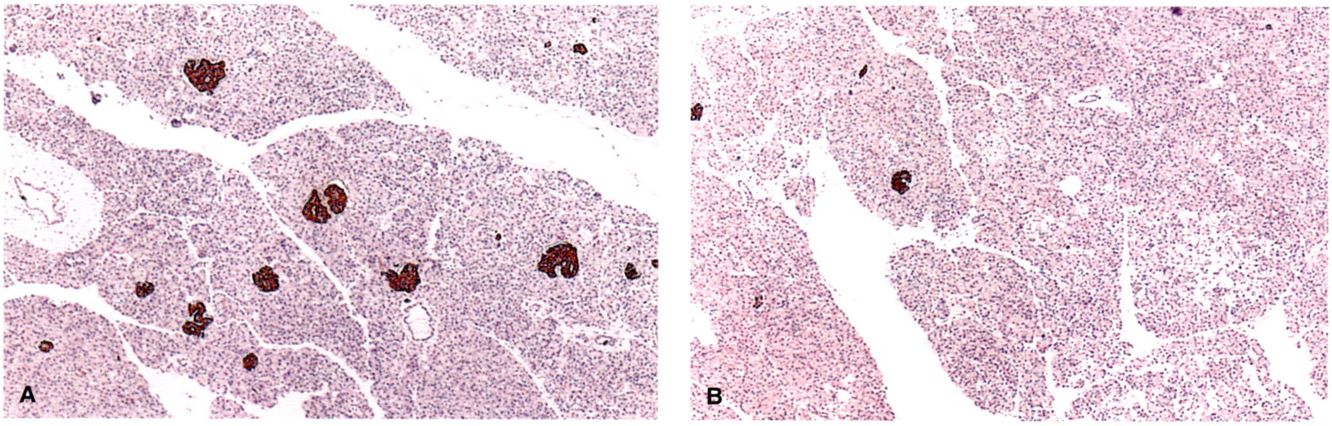

FIGURE 29.30 As highlighted by immunohistochemical labeling for chromogranin, the concentration of islets varies considerably from one lobule (**A**) to the next (**B**).

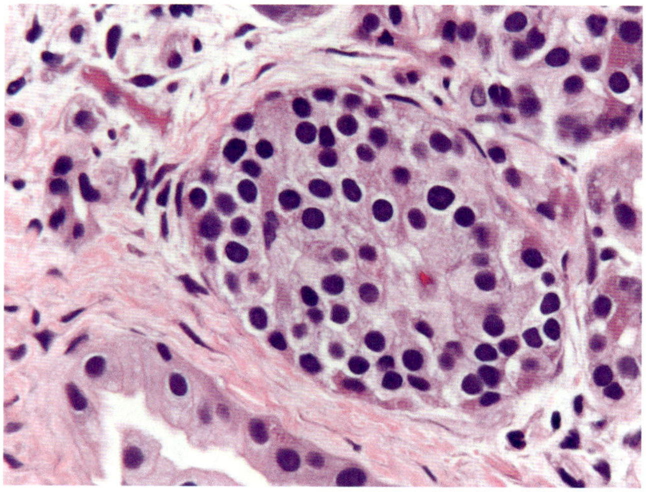

FIGURE 29.31 Compact islets that consist of sharply circumscribed round to oval nests of neuroendocrine cells. The individual cells show nuclei with a stippled chromatin pattern and amphophilic cytoplasm.

than their neighbors. These nuclei have a 4N or 8N DNA content, are exclusive to β-cells; they do not have any known pathologic significance (130). Mitotic figures are exceptionally rare in normal islets (131). Pancreatic islets are supplied with numerous capillaries lined by fenestrated endothelium that are nearly imperceptible on routine H&E light microscopy. This microvascular network is in direct contact with essentially all islet cells (132). A thin layer of connective tissue separates compact islets from the surrounding acinar tissue, but they lack true encapsulation.

Diffuse islets display an overtly trabecular appearance with winding cords of cells intermingled among acini (Fig. 29.32A). They are composed of columnar-shaped cells with basophilic cytoplasm, hyperchromatic nuclei, and prominent nucleoli (Fig. 29.32B). Given their less common occurrence and pseudoinfiltrative appearance, diffuse islets may be mistakenly interpreted as neuroendocrine tumors, particularly in cases of chronic pancreatitis with lobular atrophy and islet hyperplasia.

Each endocrine cell type that populates pancreatic islets is specialized to produce one peptide hormone. Beta

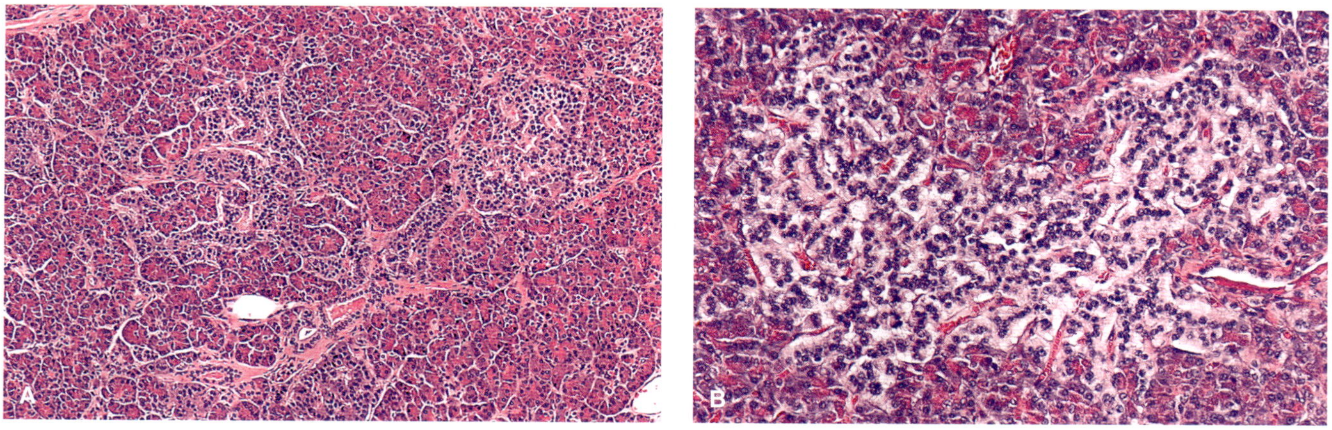

FIGURE 29.32 Diffuse islets are composed of trabeculae of neuroendocrine cells interspersed between adjacent acini (**A**). The borders of the diffuse islets are ill defined (**B**). The cells have somewhat basophilic cytoplasm.

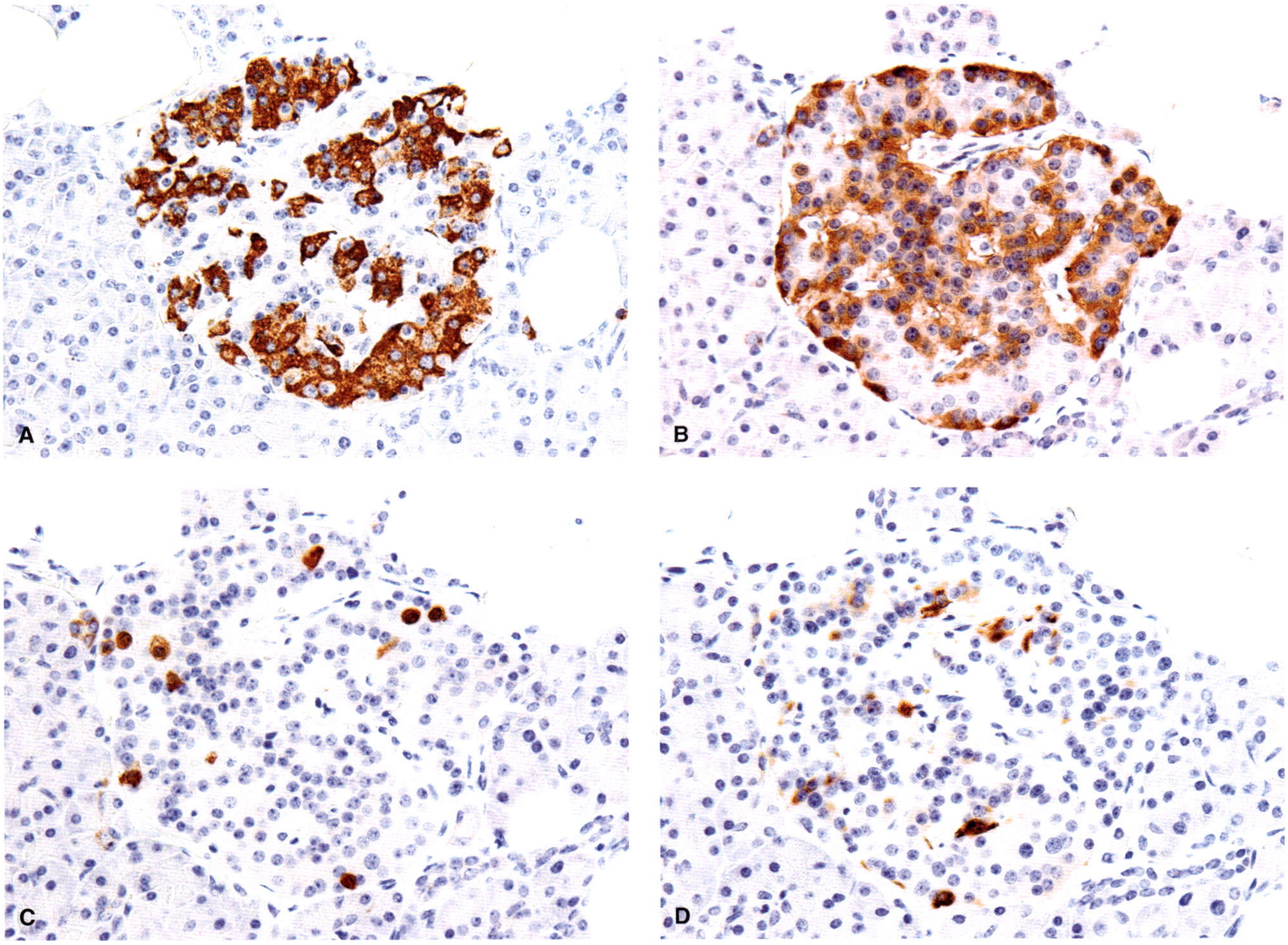

FIGURE 29.33 The distribution of the different peptide-producing cells within the islets of Langerhans. Beta cells (labeled for insulin) are the most numerous (**A**) and are situated in the central regions of the islet. Alpha cells (labeled for glucagon) are generally arranged around the periphery (**B**). Delta cells (labeled for somatostatin) (**C**) and gamma cells (labeled for pancreatic polypeptide) (**D**) are much less numerous and do not display an obvious pattern of arrangement.

cells are dedicated to producing insulin, alpha cells produce glucagon, delta cells produce somatostatin, gamma cells produce pancreatic polypeptide, and epsilon cells produce ghrelin (133,134). Immunohistochemical antibodies against the peptide hormones can be used to distinguish the different endocrine cell lineages. Compact islets are composed predominantly of centrally located beta cells (60% to 70%) and peripheral alpha cells (15% to 20%) (127,135,136) (Fig. 29.33). Delta cells (5% to 10%) and epsilon cells (~1%) constitute minor subpopulations (127,135,136). Diffuse islets are composed of gamma cells (70%), beta cells (20%), alpha cells (5%), and delta cells (5%) (Fig. 29.34) (137–139). The different proportions of endocrine cell lineages between compact and diffuse islets reflect their divergent embryologic origins (128,129). These proportions also vary with age. For example, delta cells constitute one-third of the compact islet cell population in infants (122,140).

All islet cells express general endocrine markers including synaptophysin, chromogranin, neuron-specific enolase (NSE), CD56, and insulinoma-associated protein 1 (INSM1) (Fig. 29.35) (141). Islet cells weakly express CAM5.2 cytokeratins. Other epithelial cytokeratins and acinar enzymes are not normally expressed. Normal islet

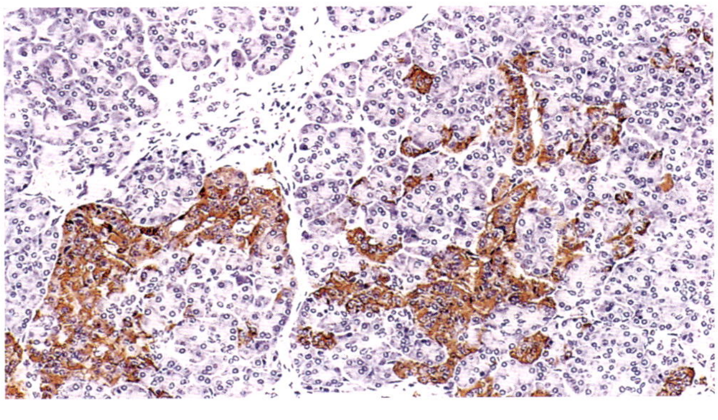

FIGURE 29.34 Immunohistochemistry for pancreatic polypeptide shows positive staining in gamma cells, which constitute most of the cells in the diffuse islets.

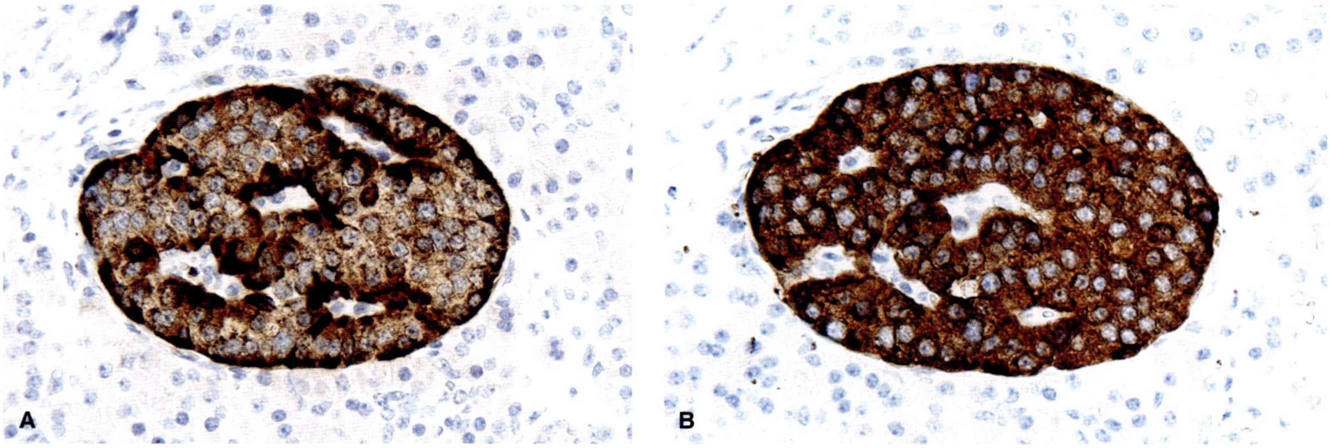

FIGURE 29.35 Alpha cells stain more intensely for chromogranin than do beta cells (**A**), whereas all the islet cells stain uniformly for synaptophysin (**B**).

cells also express CD99, progesterone receptors, PDX1, and ISL1 (142–146).

Ultrastructurally, endocrine cells are interconnected by both tight and gap junctions (127,136). Like acinar cells, islet cells are highly secretory, and their cytoplasm is accordingly filled with the essential organelles for secretory protein synthesis including rough endoplasmic reticulum, mitochondria, and Golgi (Fig. 29.36). The cytosol is likewise filled with secretory granules. Unlike apically concentrated zymogen granules in acinar cells, secretory granules are concentrated near the basolateral surface in close proximity to the capillaries into which the granules are secreted. Granule size and morphology are specific to each islet cell lineage (147,148). Alpha cell granules measure 200 to 300 nm and contain an eccentric core of dense proteinaceous material separated from the outer lipid membrane by a less dense peripheral "halo" (Fig. 29.37A). Beta cells contain 225 to 375 nm granules with either a finely granular or a crystalline core (Fig. 29.37B) and lipid inclusions (or ceroid bodies) (Fig. 29.36) (136). Delta cell granules are

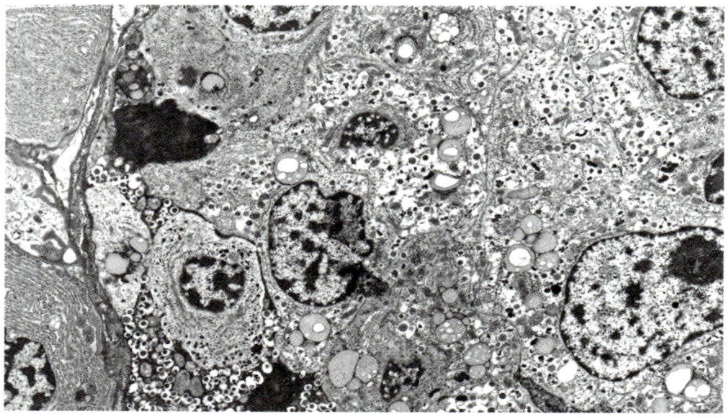

FIGURE 29.36 Ultrastructural appearance of islets of Langerhans. The islets are circumscribed and separated from the adjacent acinar cells (**left**). The peptide granules are randomly distributed within the cytoplasm. Lipid inclusions (ceroid bodies) are found within the beta cells.

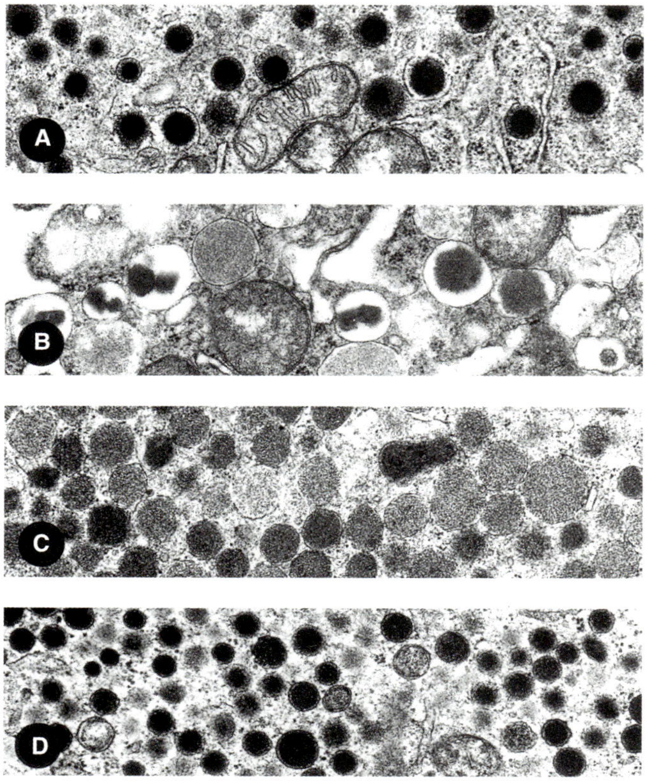

FIGURE 29.37 Ultrastructural appearance of islet cell granules. Alpha cell granules (**A**) are round and contain an eccentric electron-dense core within a less dense peripheral region. There is a thin halo beneath the limiting membrane. Beta cell granules (**B**) are polymorphous and contain crystalline cores with a wide halo beneath the limiting membrane. Delta cell granules (**C**) are round with a moderately dense core surrounded by a very thin halo. The granules of PP cells (**D**) are smaller and have homogeneous hyperdense cores. Reprinted with permission from Bommer G, Friedl U, Heitz PU, Klöppel G. Pancreatic PP cell distribution and hyperplasia. Immunocytochemical morphology in the normal human pancreas, in chronic pancreatitis and pancreatic carcinoma. *Virchows Arch A Pathol Anat Histol.* 1980;387(3):319–331.

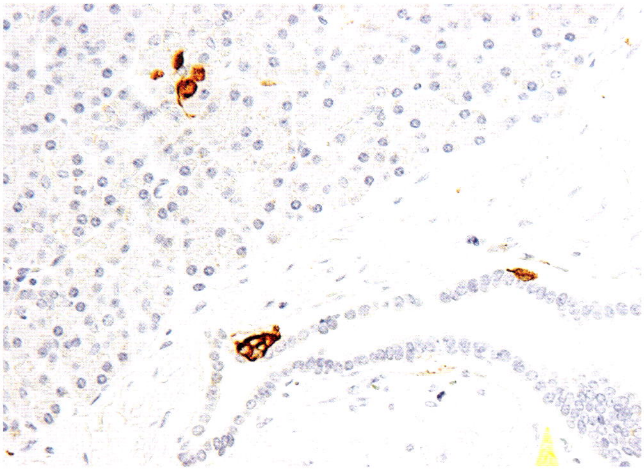

FIGURE 29.38 Extrainsular neuroendocrine cells within the ducts and between the acini are only detectable with immunohistochemical labeling for chromogranin.

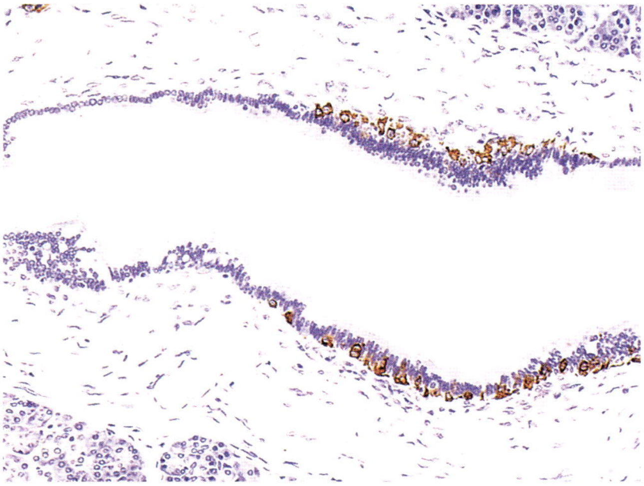

FIGURE 29.39 The number of ductal neuroendocrine cells is increased in areas showing proliferation (low-grade PanIN) of the ductal epithelium (**right**).

slightly smaller (170 to 220 nm) with a uniformly dense core (Fig. 29.37C). Cytoplasmic processes extend from delta cells toward nearby alpha and beta cells, which could facilitate paracrine somatostatin signaling in addition to systemic release into the bloodstream (127). Pancreatic polypeptide (PP) cells are filled with 180 to 220 nm granules in the ventral pancreas and smaller (120 to 150 nm) homogeneous granules in the other regions (Fig. 29.37D) (108). The cells are separated from the fenestrated endothelial cells of the capillaries by thin basement membrane material (136).

Extrainsular Endocrine Cells

In addition to the islets, a small number of individual endocrine cells are found within the pancreatic ducts and scattered among the acini (Fig. 29.38). These extrainsular endocrine cells are abundant in infants but constitute less than 10% of the total pancreatic endocrine cell population in adults (108). The ductal endocrine cells are mostly located in the main and larger interlobular ducts (103,149). Some are apically oriented toward the lumen and form tight junctions directly with ductal cells (150). Others are basally oriented between ductal cells and basement membrane. Ductal neuroendocrine cells secrete their peptides directly into pancreatic ducts. Some can be detected in pancreatic juice (150) including insulin, somatostatin, and pancreatic polypeptide (149,151). An additional subset of large duct endocrine cells produce serotonin (151). This population is also notable as it may represent the cell of origin for rare serotonin-producing pancreatic endocrine tumors that usually present as small masses obstructing the major pancreatic ducts (152–154). Ductal endocrine cell numbers increase in the presence of pancreatic intraepithelial neoplasia (Fig. 29.39) (149).

Connective Tissues

The mesenchyme comprises nearly 30% of the volume of the neonatal pancreas (Fig. 29.7A). This gradually decreases during infancy (122) to the point where very little interlobular connective tissue remains in adult pancreas. Basement membrane invests the periphery of acinar lobular units while a capillary-rich double basement membrane courses through the islets. Basement membrane matrix is composed mostly of laminin isoforms and heparin sulfate proteoglycans (Fig. 29.40).

Small vessels transport hormonal secretions from the islets directly to the exocrine elements (93,94,155,156). A complex capillary network surrounds the acini often accompanied by nerve fibers. The acini and ducts are innervated by bundles of unmyelinated nerves that travel through the interlobular connective tissues. Nerve endings are separated from the acinar or ductal cells by the basement membrane, although a minority make direct contact with the epithelial cells (103). Islets are innervated by sympathetic, parasympathetic, and peptidergic fibers from the autonomic ganglia. Periacinar nerves produce neuropeptide Y; periductal and intra-islet nerves produce vasoactive intestinal polypeptide (157). Nerve-derived substance P, cholecystokinin, and calcitonin gene–related peptide can also be detected (157). Small autonomic ganglia are located between acinar units (Fig. 29.41). Paraganglia are present in peripancreatic soft tissues, thereby accounting for the occurrence of peripancreatic paragangliomas (158). Muscular arteries course within interlobular connective tissue away from the centrally located pancreatic ducts. This separation of the ducts and muscular arteries is maintained in chronic pancreatitis. The presence of a gland in close proximity to a muscular artery therefore suggests malignancy (159). Adipocytes are also present within the pancreas (Fig. 29.23) in proportions ranging from 3% to 20% (6). Older and overweight individuals tend to have more intrapancreatic fat (160). More intraparenchymal fat is present in regions derived from the dorsal lobe compared to those from the ventral lobe (161,162).

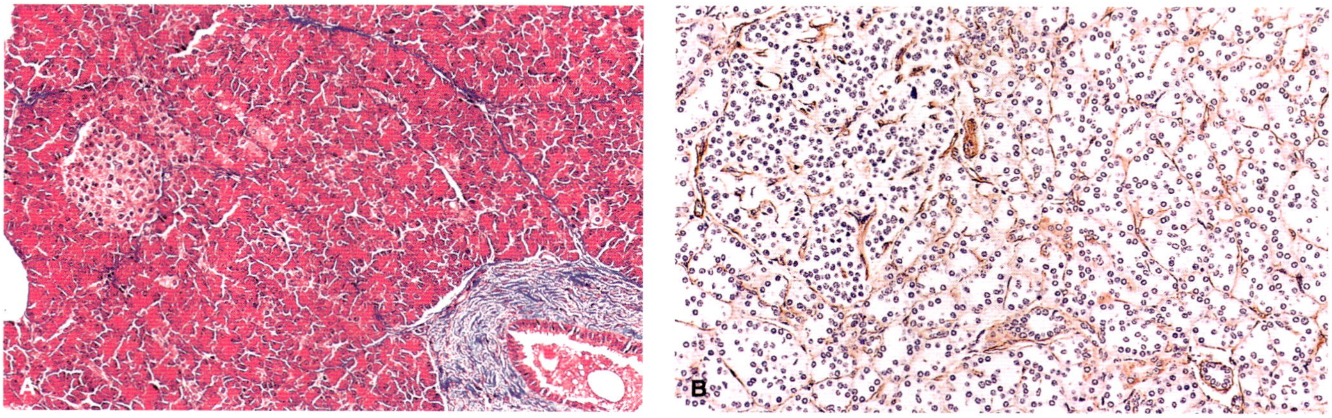

FIGURE 29.40 Connective tissues are minimal within the pancreas. Collagen is largely restricted to the tissues surrounding the interlobular ducts, with only thin bands extending into the lobules, as demonstrated with a trichrome stain (**A**). Immunohistochemical staining for type IV collagen (**B**) shows that an extensive network of basement membranes surrounds each acinus and duct and accompanies the capillaries within the islets (**upper left**).

Resident fibroblast-like stromal cells are referred to as pancreatic stellate cells (163,164). These cells are thought to be biologically analogous to hepatic stellate cells. At rest, pancreatic stellate cells are "inactivated" and filled with vitamin A storage droplets. Cytokines released from damaged acini and ducts during pancreatic injury activate resident stellate cells to synthesize and secrete extracellular matrix that provides structural scaffolds that help resolve tissue injury. If injury persists, continued stellate cell activation results in dense fibrosis that slowly replaces the lobular acinar units (163,165). In addition to stellate cells, stromal interstitial cells of Cajal resembling those of the gastrointestinal tract have been reported in the pancreas (166).

Cytologic Features

Fine-needle aspirates are typically done for pancreatic neoplasms and therefore contain very little normal pancreatic cells if the lesion is adequately sampled. Reactive cellular changes, fibrosis, and atrophy are frequently present adjacent to neoplasm, thereby posing diagnostic challenges when cellular and reactive populations mimic well-differentiated ductal adenocarcinomas. Since acinar cells comprise 85% of the gland, aspirates of normal pancreas yield abundant acinar cells, scattered ductal cells, and a few islet cells. Benign acinar cells maintain their lobular architecture in smear preparations, appearing as cohesive, well-organized grape-like clusters and two-dimensional rosette-like acini (Fig. 29.42) (167–169). Scattered single cells and occasional stripped nuclei can be seen in the background. Acinar cell nuclei are round and regular with uniform chromatin and variably prominent nucleoli. The polygonal cells have abundant finely granular cytoplasm. The appearance of the zymogen granules varies depending on the staining method. Granules are turquoise with the standard Papanicolaou stain, eosinophilic on H&E (Fig. 29.42), and light purple with the Romanowsky stain variants including Diff-Quik and Giemsa. Scattered

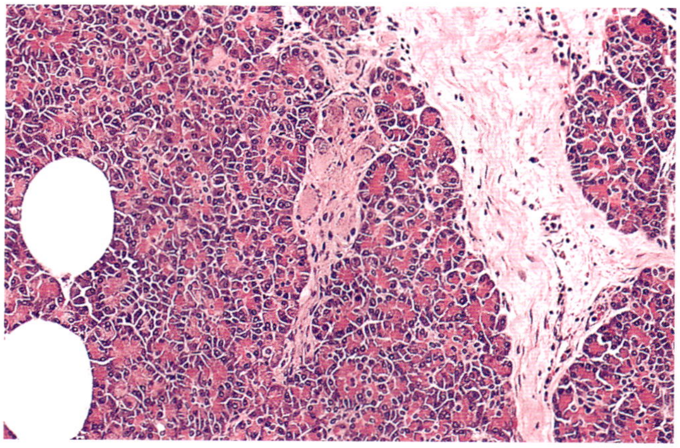

FIGURE 29.41 Small autonomic ganglia are located within lobules of acinar tissue.

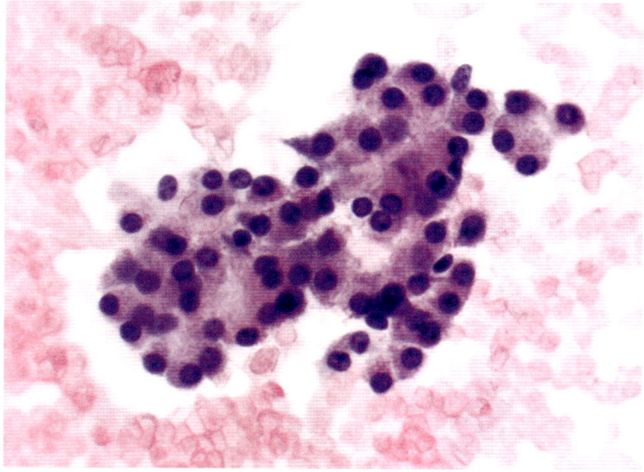

FIGURE 29.42 Cytology of normal acinar cells on hematoxylin and eosin stain. The cells are organized in acini. The cytoplasm has a pyramidal shape with round basal nuclei, small nucleoli, and apical eosinophilic zymogen granules.

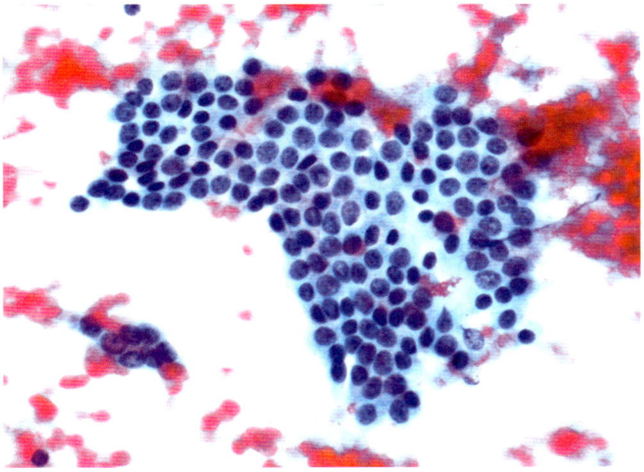

FIGURE 29.43 Cytology of normal ductal cells on hematoxylin and eosin stain. The flat sheet of cells has evenly spaced, round, and uniform nuclei.

small cytoplasmic vacuoles may be visible on Romanowsky-type stains. Acinar neoplasms form large, crowded aggregates and irregular clusters rather than the small, uniform grape-like clusters of benign acini (169).

Normal pancreatic ducts form flat two-dimensional cohesive sheets or tubules on FNA smears. The ducts are lined by bland, evenly spaced epithelial cells containing uniform nuclei. This appearance is referred to as honeycomb-like, which is synonymous with a benign pancreatic process (Fig. 29.43) (167,168). The apically oriented cytosol and basolaterally oriented nuclei demonstrate preserved polarity and are best observed in strips or at the edges of sheets. Ductal cells contain round to ovoid nuclei with uniform chromatin and inconspicuous nucleoli. Their nongranular cytoplasm stains are aqua blue with the Papanicolaou stain and indigo to purple with Romanowsky-type stains. Reactive ductal epithelial cells display mild to moderate variation in nuclear size and architecture but remain cohesive with a preserved "honeycomb" appearance. Fine uniform chromatin is also preserved, and nuclear membranes remain smooth with only minor irregularities. The presence of ductal sheets showing abundant apical mucin and/or crowded, disorganized cells with partial or complete loss of polarity is referred to as a drunken honeycomb pattern. This appearance is synonymous with neoplasia as can be seen in PanIN, mucinous neoplasms, and invasive pancreatic ductal adenocarcinomas.

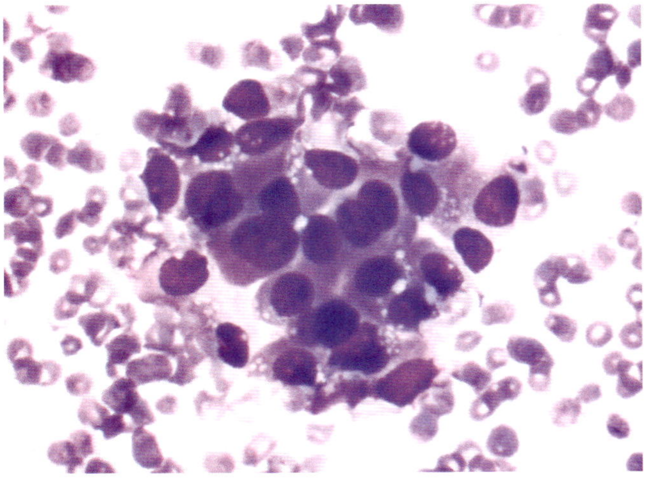

FIGURE 29.44 Cytology of normal islet cells on Diff-Quik stain. Polygonal to plasmacytoid islet cells are loosely arranged with round nuclei and amphophilic cytoplasm.

Islet cells are rarely appreciable in aspirate smears but can be distinguished from acinar cells by their relatively larger nuclei, denser chromatin, polygonal to plasmacytoid shape, and pale amphophilic cytoplasm (Fig. 29.44). Mild variation in nuclear size is not unusual.

MINOR ALTERATIONS

Physiologic changes, injury, or aging may produce minor benign alterations to some components of the normal pancreas. Some are subtle and easily overlooked. Others are more noticeable and can even mimic neoplastic processes. Recognizing the diagnostic pitfalls is essential for the accurate interpretation of biopsies and FNA smears, especially since some of these alterations can coexist with pancreatic neoplasms.

Acinar Alterations

Incidental acinar cell nodules are common and consist of well-circumscribed nests of benign acinar cells with pale cytoplasm and hyperchromatic nuclei arranged in a vaguely trabecular architecture that is clearly distinct from surrounding acini on routine H&E stains (Fig. 29.45A) (170–172). Although eosinophilic zymogen granules are often present, in some instances they are depleted and the nucleus-to-cytoplasm ratio is increased, imparting a more basophilic appearance (Fig. 29.45B). They have been referred to by a variety of terms including pale acinar nodule, atypical acinar cell nodule, focal acinar transformation, eosinophilic degeneration, and focal acinar cell dysplasia (15,108,173–176). The latter term was coined in the 1980s as it was thought at the time that these nodules might represent neoplastic precursors for acinar cell carcinomas. However, no evidence has since supported that hypothesis, and the causes of these benign nodules remain unknown. They are more prevalent in adults, suggesting they are acquired (173). Acinar cell nodules range in size from 100 to 700 μm. As such, they can be confused with islets or endocrine microadenomas if not properly recognized.

Acinar dilatation or ectasia may occur during active secretion (1) or when downstream ducts are obstructed (174). It has been associated with premortem uremia, septicemia, and dehydration at autopsy (15,177,178). The entire lobular unit(s) involved are usually ectatic and resemble

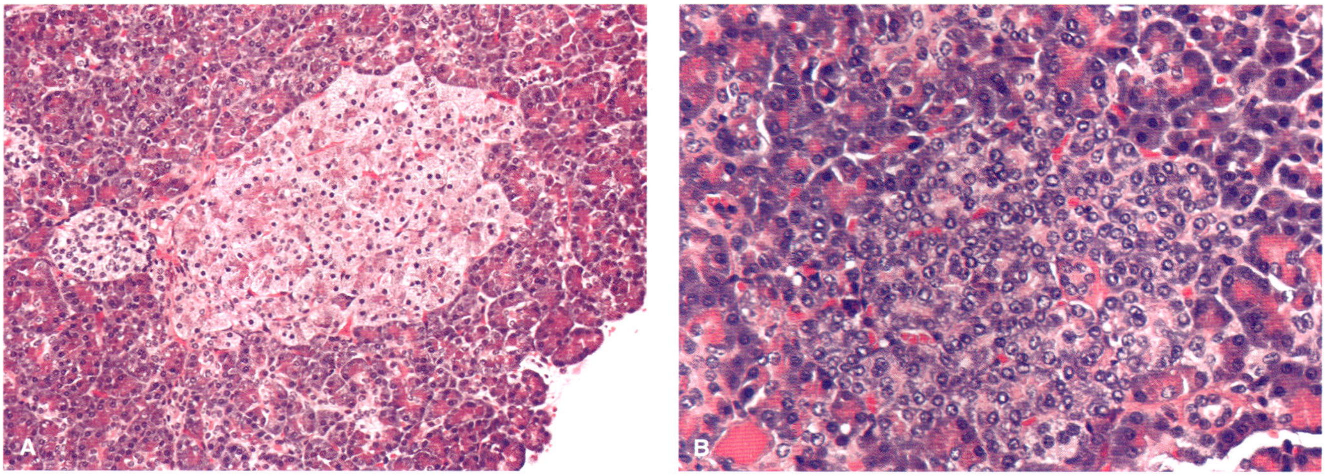

FIGURE 29.45 Atypical acinar cell nodules. The eosinophilic type (**A**) is more common and consists of a collection of acini showing loss of basophilia in the basal cytoplasm and hyperchromasia of the nuclei. The eosinophilic granules within the apical cytoplasm are maintained. An islet of Langerhans, with which these lesions may be confused, is present at the **lower right**. The basophilic type of atypical acinar cell nodule (**B**) exhibits an increased nucleus-to-cytoplasm ratio with loss of eosinophilic granularity of the apical cytoplasm. Some degree of nuclear atypia is also present.

small ductules lined by flattened epithelial cells will loss of most zymogen granules (Fig. 29.46). Adjacent centroacinar cells may proliferate and colonize the dilated acini (178). Inspissated eosinophilic secretions are found within the ectatic lumina.

Rare metaplastic changes involving acinar units include replacement of acinar cells by goblet cells or squamous cells (179). Proliferating centroacinar cells may also replace the acinar cells during early atrophy (174). A more common metaplastic change occurs during chronic pancreatitis, termed acinar to ductal metaplasia. This change was previously considered to represent a benign acinar cystadenoma but is now thought to develop when acinar cells reversibly transdifferentiate into immature ductal cells that form neoductules (Fig. 29.47) (180–184). Because acinar to ductal metaplasia is triggered by injury-induced activation of (K)RAS signaling, metaplastic ductules are proliferative and can display some features that overlap with well-differentiated ductal adenocarcinomas with irregularly shaped ducts and nuclear enlargement. Unlike adenocarcinomas, acinar to ductal metaplasias always remain confined within the lobular unit. Although benign, data from animal models suggest that chronic pancreatitis and acinar to ductal metaplasia increase the probability that native ductal or metaplastic ductal epithelial cells will acquire congenic *KRAS* mutations with subsequent neoplastic transformation (184,185).

Duct Alterations

Neoplastic transformation of the ducts appears as mucinous epithelium that grows within small to intermediate ducts (PanINs) or within large main ducts or their side

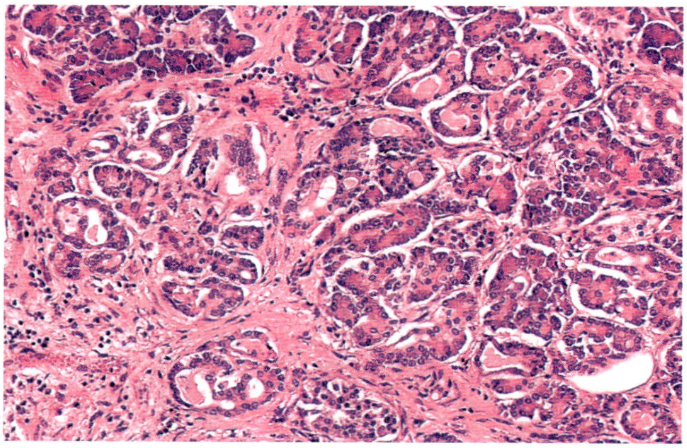

FIGURE 29.46 In acinar ectasia, the acinar lumina are dilated and filled with eosinophilic secretions. The lining cells have a flattened appearance, often resembling small ductules cells more than acinar cells.

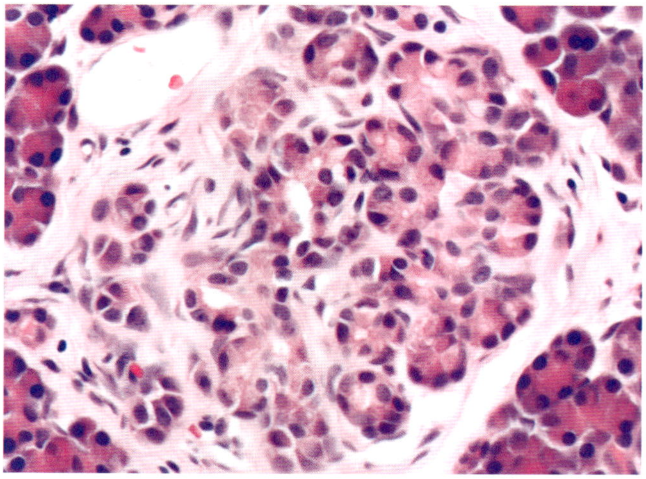

FIGURE 29.47 The apparent replacement of acinar cells by centroacinar or ductular cells has been designated "acinar to ductal metaplasia."

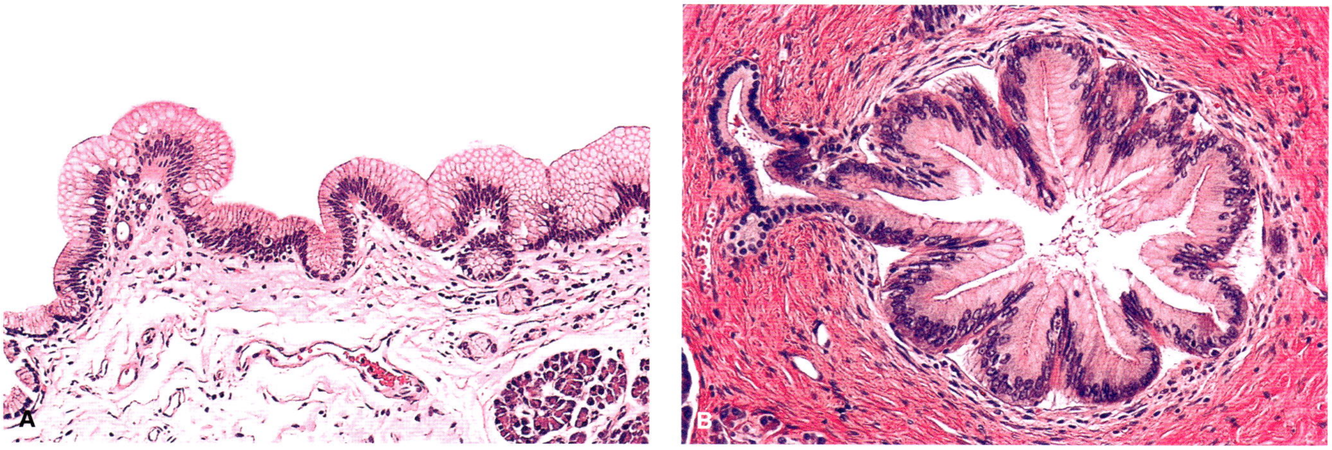

FIGURE 29.48 Low-grade pancreatic intraepithelial neoplasia (PanIN). The normal cuboidal to low columnar ductal epithelial cells are replaced by tall columnar cells containing abundant apical mucin (**A**); in this example with no papilla formation (previously termed PanIN-1A). The nuclei remain basally located and show minimal pseudostratification. Another lesion (**B**) has similar cytologic features but demonstrates papilla formation (previously termed PanIN-1B). Note the transition to normal ductal epithelium (**left**).

branches (intraductal papillary mucinous neoplasms). These lesions are distinct from normal ductal cells and acinar to ductal metaplasias, displaying an undulating or papillary intraductal growth pattern and abundant cytosolic mucins that are easily visualized (Figs. 29.48 and 29.49). Neoplastic cells within both PanINs and intraductal papillary mucinous neoplasms are capable of detaching from the site of origin into the lumen and migrating to other locations within the ductal system (186,187). Invasive adenocarcinomas may also colonize native ducts to mimic a precursor neoplasm (187,188). Oncogenic *KRAS* mutations and telomere shortening represent the earliest genetic events that facilitate neoplastic transformation with intraductal neoplasia (189–194). PanINs develop additional mutations in driver genes that accumulate over time (*CDK2NA*, *TP53*, *SMAD4*, and others) and are accompanied by progressively increasing degrees of dysplasia (186,187,195–200). Intraductal papillary mucinous neoplasms acquire a different set of driver mutations beyond *KRAS*, most commonly *RNF43* and *GNAS* (200,203–205).

Low-grade dysplasia of the ductal system (197–199) features either a well-organized flat epithelial monolayer (Fig. 29.48A) or, as cells become more crowded, a mildly disorganized monolayer with undulating papillary growths that project into the ductal lumen (Fig. 29.48B). Low-grade nuclear abnormalities include pseudostratification, enlargement, and hyperchromatic chromatin (Fig. 29.50). Mitotic figures are infrequent. Low-grade dysplastic lesions carry a very low risk of progression to invasive adenocarcinoma (201,202). Intraductal neoplasia may involve a single duct, locally extend into adjacent ducts, or be found multifocally throughout the ductular system (Fig. 29.49). The cytosol of PanINs and mucinous cysts is commonly filled with abundant neutral mucins and sialomucins with

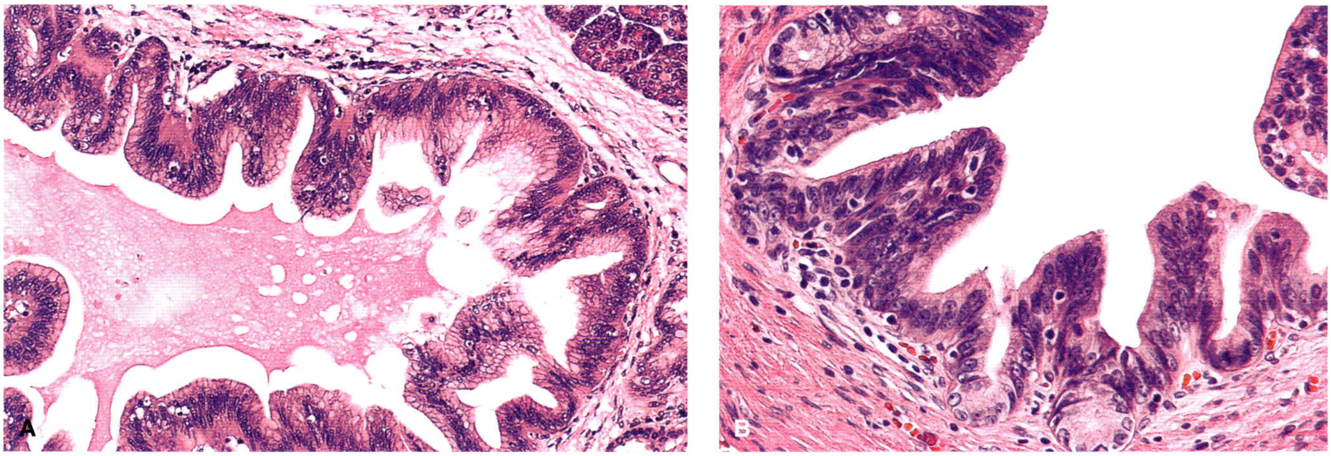

FIGURE 29.49 Low-grade PanIN may also show pseudostratified nuclei with focal loss polarity (**A**). At higher power (**B**) the nuclei are enlarged, moderately atypical, and overlapping. This moderate level of atypia was previously termed PanIN-2.

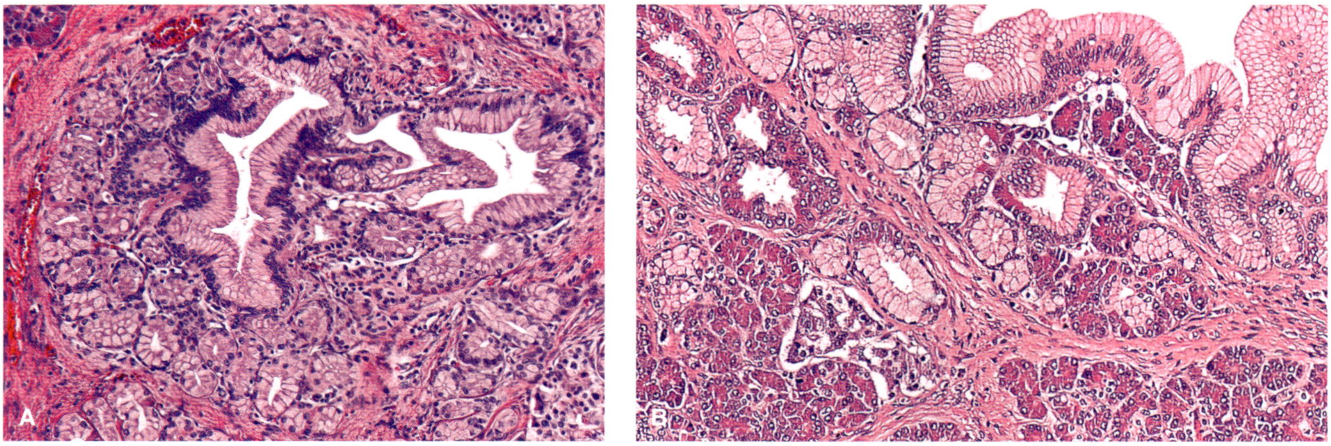

FIGURE 29.50 Low-grade PanIN may involve aggregates of small ductules around larger ducts, resulting in clusters of mucinous glands (**A**). Because the smallest components of the ductal system are involved, cells with low-grade PanIN may abut adjacent acinar cells (**B**).

less sulphomucins, thereby resembling the mucinous cells of foveolar or pyloric epithelia (108,116,188). Some intraductal neoplasms contain goblet cells, particularly intestinal-type intraductal papillary mucinous neoplasms. Low-grade intraductal neoplasms are very common and are thought to increase in incidence with age. Data from imaging, surgical, and autopsy studies (15,19,174,187,196) indicate that intraductal papillary mucinous neoplasms are present in 10% to 20% of patients over 60 years and up to 50% of nontumorous adult pancreata harbor PanINs.

High-grade dysplasia is the immediate precursor of invasive adenocarcinoma and demonstrates more significant architectural and cytologic abnormalities than low-grade dysplasia (Fig. 29.51). They contain crowded, proliferative epithelial cells forming papillary or micropapillary projections into ductal lumina. Papillae may fuse to form solid or cribriform structures. It is not uncommon for necrosis to develop at the tips of such structures as they become hypoxic. Nuclei are markedly enlarged with increased nuclear to cytosolic ratios. Complete loss of polarity is accompanied by nuclear membrane irregularities, coarse chromatin with chromocenters or prominent nucleoli, and readily identifiable mitotic activity including atypical forms (197,198,206). The morphologic progression from low- to high-grade dysplasia is associated with increasingly abnormal expression of tumor-associated glycoproteins such as CEA, B72.3, and CA125 (115).

Benign morphologic alterations to non-neoplastic ducts include squamous, oncocytic, goblet cell, and acinar metaplasia. Squamous metaplasia is reportedly present in up to one-third of adult pancreata (19,174). It is most common in the intralobular and intercalated ducts, with metaplastic cells occasionally extending to the center of the acini (Fig. 29.52). Overt keratinization or a granular cell layer is occasionally observed in a background of well-developed chronic pancreatitis (Fig. 29.53). Metaplastic ducts are lined by immature stratified squamous epithelium. The metaplastic epithelium of smaller ducts may fill and

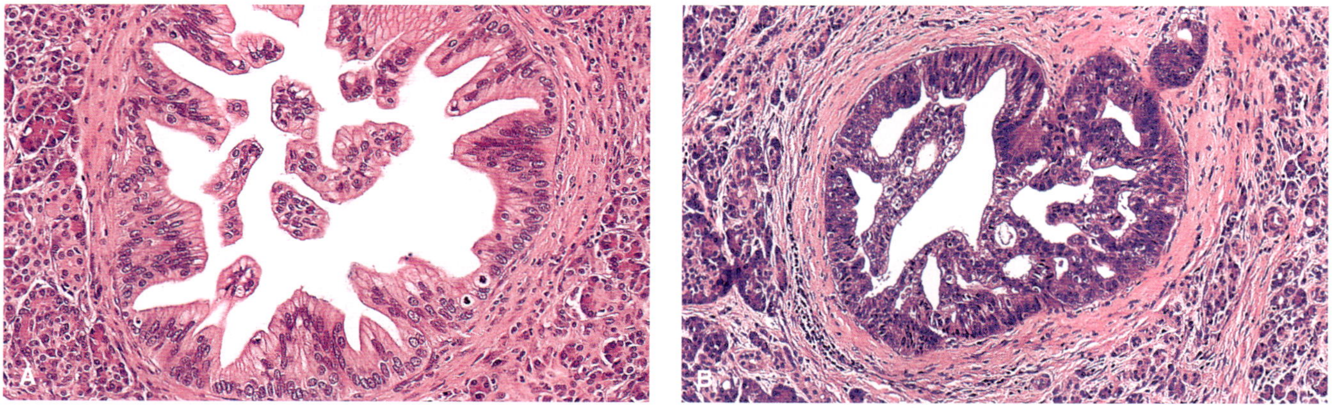

FIGURE 29.51 In high-grade PanIN there is complete loss of polarity, with budding of disorganized cellular clusters into the ductal lumen. The nuclei are markedly irregular and vary in morphology between adjacent cells. Increased mitoses are present (**A**). In another focus (**B**), architectural abnormalities include cribriforming, and there is extreme nuclear atypia.

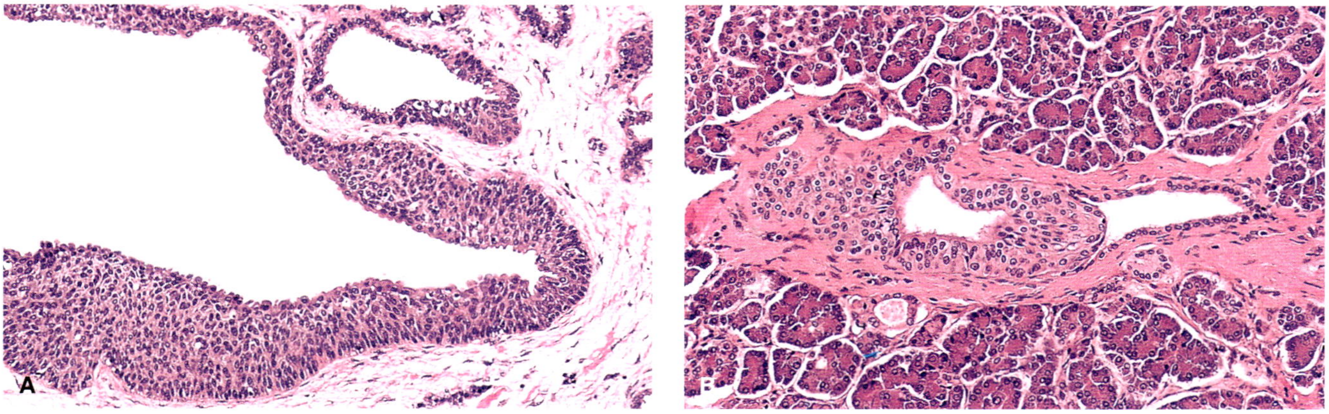

FIGURE 29.52 Squamous metaplasia of the ducts. There are multiple layers of immature-appearing squamous cells without keratinization. The process may involve larger ducts (**A**). In this example (**B**) there is partial involvement of a small interlobular duct.

obliterate ductal lumina. Squamous metaplasia is associated with chronic inflammatory conditions including pancreatitis and duodenal or bile reflux. A grossly cystic lesion of the pancreas lined by similar epithelium is termed "squamoid cyst of pancreatic ducts" (207).

Oncocytic changes are common in centroacinar cells and may extend into intercalated and intralobular ducts (104,208,209). Oncocytic ductal cells accumulate mitochondria that impart a granular eosinophilic appearance to the cytosol. Nuclei may be enlarged with prominent nucleoli (Fig. 29.54). Benign oncocytic metaplasia may occur sporadically or in a background of chronic pancreatitis (19,210). Oncocytic intraductal papillary neoplasms represent a distinct neoplastic entity with overtly dysplastic features and recurrent PRKCA/B rearrangements (211).

Goblet cells may be found intermixed with main duct epithelium near the ampulla of Vater, presumably due to the close proximity of adjacent intestinal mucosa (1,107). Benign goblet cells in smaller ducts distant from the ampulla presumably reflect a true metaplastic change (Fig. 29.55)

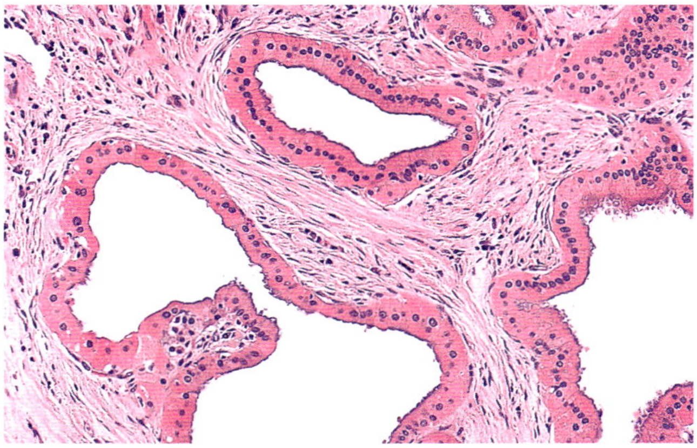

FIGURE 29.54 Oncocytic changes involving small ducts. In this example, there is associated fibrosing pancreatitis.

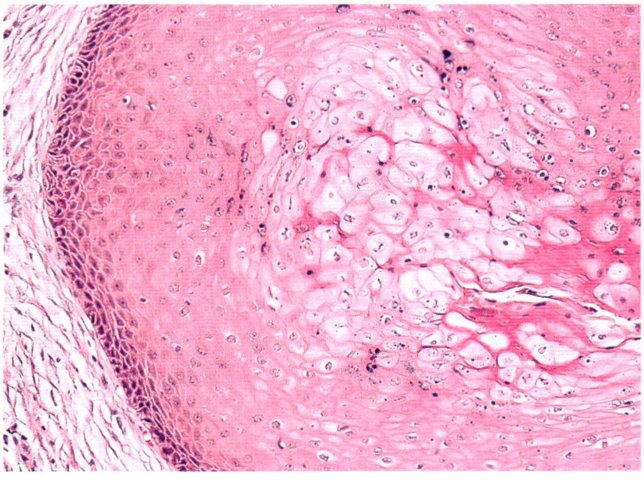

FIGURE 29.53 Squamous metaplasia of the ducts with keratinization, an uncommon finding.

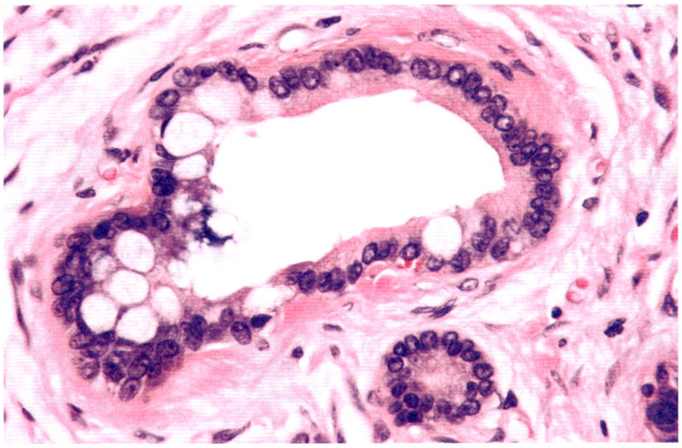

FIGURE 29.55 Goblet cell metaplasia involving a small duct. Flask-shaped goblet cells are distinct from the mucinous columnar cells of low-grade PanIN.

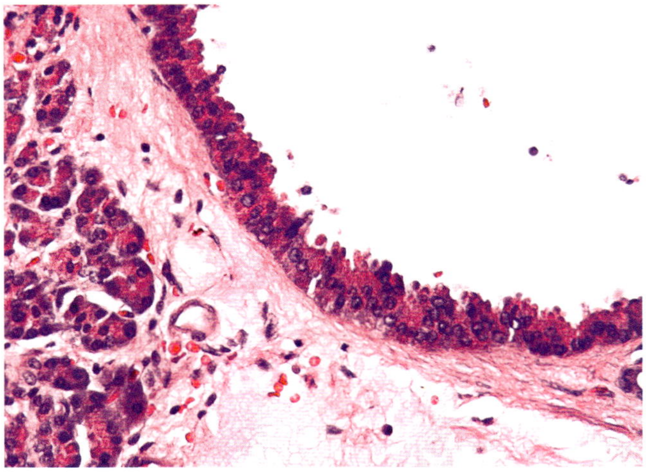

FIGURE 29.56 This small interlobular duct shows acinar metaplasia, with slightly enlarged acinar cells replacing the normal cuboidal ductal epithelium.

(174,179). In contrast to low-grade PanIN, the mucin-containing goblet cells are more flask shaped and do not grow in a confluent fashion. Metaplasia is not to be confused with intestinal-type intraductal papillary neoplasms, which are grossly visible cysts lined by dysplastic intestinal-type epithelium enriched with goblet cells that express intestinal markers CDX2 and MUC2 (212,213).

Epithelial cells lining small ducts may occasionally be partially or completely replaced with acinar cells. This phenomenon has been referred to as acinar ductal metaplasia (3). Acinar cells occur either singly or in clusters within the ducts, displaying an identical appearance to their lobular counterparts (Fig. 29.56). Like normal acinar cells, metaplastic acinar cells express trypsin, chymotrypsin, and BCL10 by immunohistochemistry. Staining is diffuse in the cytosol, unlike labeling of deposited intraluminal enzyme secretions that are restricted to the apical surfaces of ductal cells (Fig. 29.15). Acinar ductal metaplasia is a localized process that does not typically form cysts. Larger multilocular cystic lesions lined by benign acinar cells are a distinct non-neoplastic entity termed "acinar cystic transformation." These cysts are primarily located in the head of the pancreas and lined by layers of acinar epithelium (214–217).

Duct ectasia most often occurs in older adults with chronic pancreatitis (15,19,218). Ectatic ducts usually reach several millimeters in diameter, and main pancreatic duct ectasia of greater than 4 mm is present in 16% of patients at autopsy (15). Ectasias that extend into multiple ducts may coalesce or become tortuous, forming unilocular or multilocular cysts; these lesions have been termed retention cysts and can be seen on gross and histologic cross-sections (Fig. 29.57A). Careful examination demonstrates their continuity with the ductal system. Their lining epithelium may harbor mucinous cells (Fig. 29.57B). In those cases, distinguishing simple cysts with PanIN from IPMNs can be challenging (198,199,219–224). Generally, multicystic lesions with micropapillae or true papillae are classified as intraductal papillary mucinous neoplasms, whereas solitary cysts >1 cm with flat epithelium are designated as benign simple mucinous cysts. Mucinous cystic lesions with ovarian-type stroma are termed mucinous cystic neoplasms.

Islet Alterations

Islet hyperplasia is defined as an absolute increase in the size or quantity of islets compared to the normal islet volume for a given age. Individual compact islets larger than 250 μm in diameter are classified as hyperplastic (225). However, discerning whether there is a true increase in the total volume of islet tissue in the pancreas poses challenges. Any cause of pancreatic atrophy with exocrine gland loss can give the illusion of increased islet numbers due to preservation of islets in the midst of reductions to lobular acinar structures. The density of islets also varies across different regions of the pancreas. Thus, a quantitative assessment of islet volume is required to diagnose true islet hyperplasia.

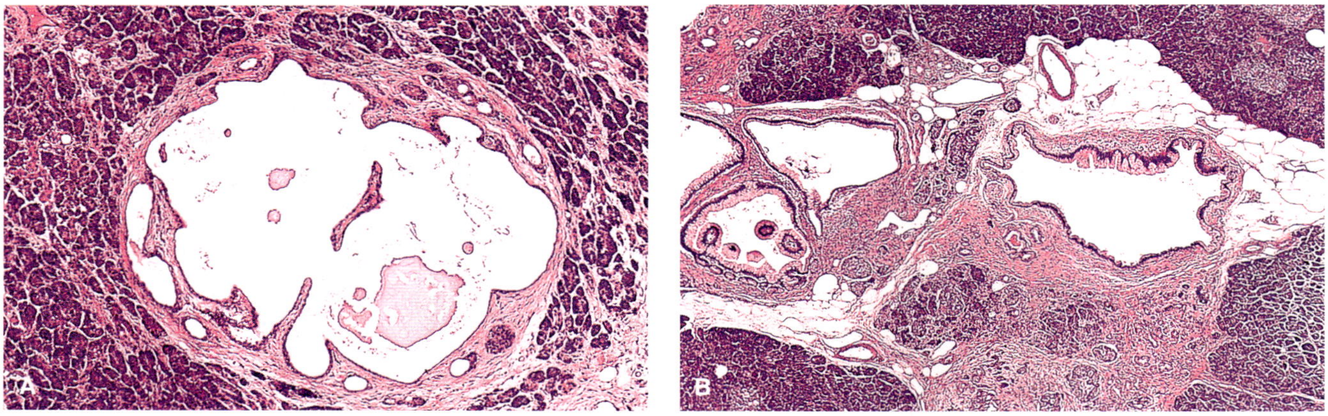

FIGURE 29.57 Duct ectasia. The dilated ducts may be lined by a flattened cuboidal epithelium resembling normal ductal epithelium (**A**) or there may be involvement by PanIN (**B**). In the latter circumstance, the lesion merges morphologically with intraductal papillary mucinous neoplasm and simple mucinous cyst.

Conditions associated with islet hyperplasia in infancy encompass Beckwith–Wiedemann syndrome, maternal diabetes mellitus, erythroblastosis fetalis, and hyperinsulinemic hypoglycemia (108). Islet cell hyperplasia can also occur in adults with hyperinsulinemia (226). Sources of islet hyperplasia include proliferation of preexisting endocrine cells or expansion of multipotent progenitors that differentiate into endocrine cells (136). The normal distribution of endocrine peptide is usually maintained.

Nesidioblastosis refers to a histomorphology finding of malformed islet aggregates containing hypertrophic beta cells that bud from adjacent pancreatic ductules ("ductuloinsular complexes"). This finding was originally thought to occur primarily in infants with hyperinsulinemic hypoglycemia, although this phenomenon has now been observed in normal infants without hyperinsulinemia and in adults with acquired hyperinsulinemia following gastric bypass procedures (227–231). Both focal and diffuse types are described (227). Focal nesidioblastosis presents as localized aggregates of malformed islets that cluster near small ductules (Fig. 29.58). The islets in the remaining pancreas are normal. Diffuse nesidioblastosis is characterized by malformed islets throughout the gland without discrete localization. In either case, the hypertrophic beta cells of nesidioblastosis display up to 40% larger nuclear volume compared with age-matched controls, although they may not necessarily be directly responsible for the hyperinsulinemia. Some cases of congenital hyperinsulinemic hypoglycemia appear to have a genetic cause, as they are associated with germline mutations genes that encode the ATP-sensitive potassium channels (KATP) present on the plasma membrane of beta cells (232–235).

The apparent increase in number of islets related to exocrine atrophy has been termed "islet aggregation" and does not reflect true islet hyperplasia. In chronic pancreatitis, most acini are lost resulting in progressive atrophy that spares native islets amidst the expanding fibrous or adipose tissue

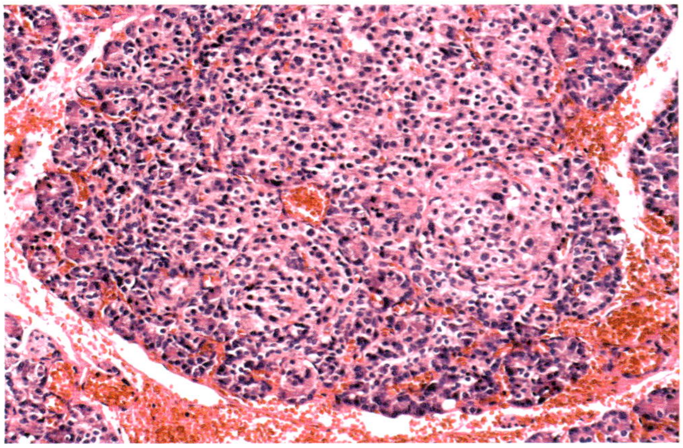

FIGURE 29.58 The pancreas from an infant with persistent neonatal hyperinsulinemic hypoglycemia shows the focal form of nesidioblastosis. There is localized aggregation of islets separated by thin bands of acini.

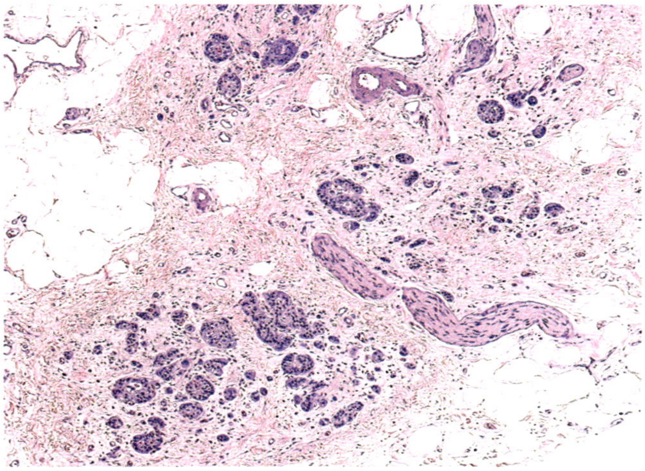

FIGURE 29.59 Islet aggregation involving compact islets. Atrophic parenchyma is characterized by complete loss of exocrine elements, leaving the clustered islets embedded in fibrous stroma.

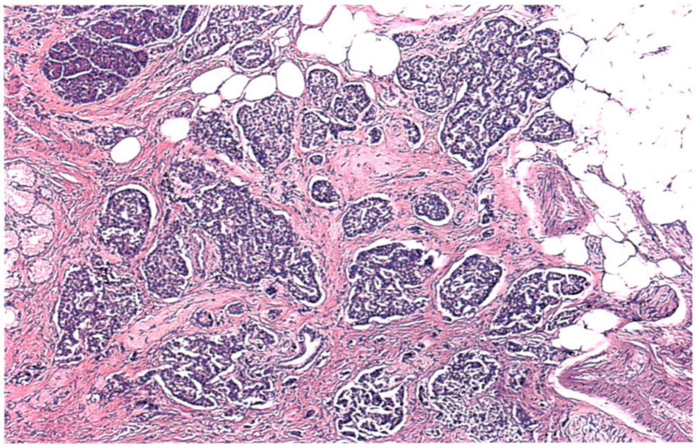

FIGURE 29.60 Atrophic parenchyma with aggregation of the remaining neuroendocrine elements that may simulate a neoplasm. The aggregates are poorly circumscribed and can extend into peripancreatic adipose tissue.

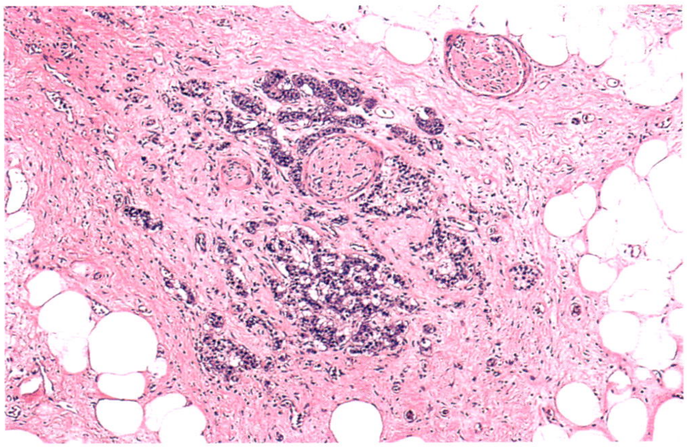

FIGURE 29.61 Exocrine atrophy in areas containing diffuse-type islets results in a pseudoinfiltrative pattern of individual cells and trabeculae.

(Fig. 29.59) (182). Thus, islet aggregation in this context should be distinguished from true islet cell hyperplasia. Extreme cases of islet aggregation result in coalescence of native islets into nests and trabeculae that can mimic pancreatic neuroendocrine tumors (Fig. 29.60).

Islets are also often present in peripancreatic adipose tissue. The appearance of infiltrative growth can be striking in the head of the pancreas since it contains diffuse islets rather than the insular arrangement of compact islets elsewhere in the gland. When exocrine elements atrophy, diffuse islets may appear as small clusters, trabeculae, or individual cells (Fig. 29.61).

Most pancreatic neuroendocrine tumors have well-defined borders, many are encapsulated, and some express abnormal endocrine peptides by immunohistochemistry (for example, vasoactive intestinal polypeptide or gastrin). In contrast, islet aggregation borders are ill-defined and abnormal peptides are not expressed (135,236). Bear in mind gamma cells are more abundant in diffuse islets than compact islets (Fig. 29.62). Coalescence of islets

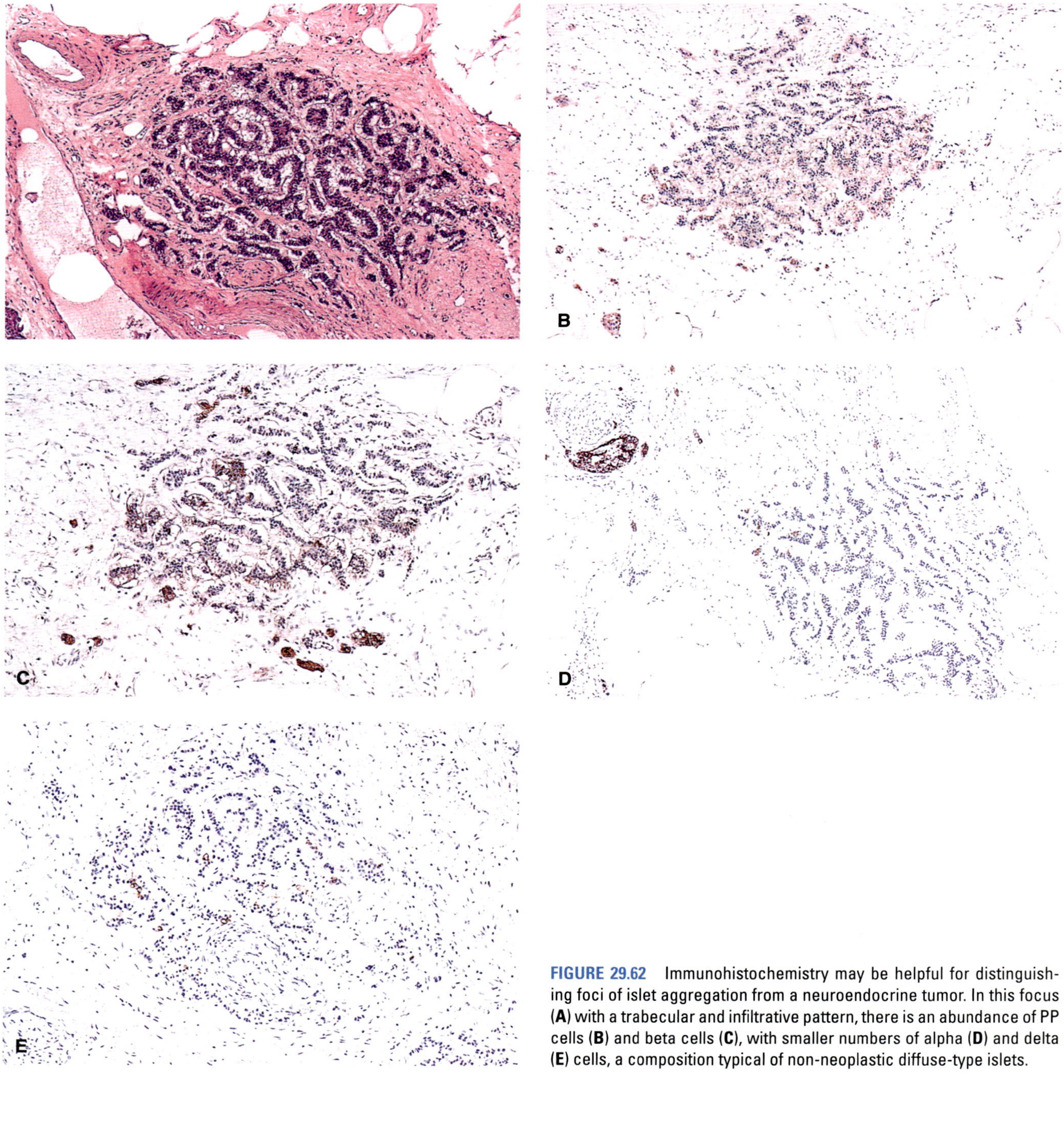

FIGURE 29.62 Immunohistochemistry may be helpful for distinguishing foci of islet aggregation from a neuroendocrine tumor. In this focus (**A**) with a trabecular and infiltrative pattern, there is an abundance of PP cells (**B**) and beta cells (**C**), with smaller numbers of alpha (**D**) and delta (**E**) cells, a composition typical of non-neoplastic diffuse-type islets.

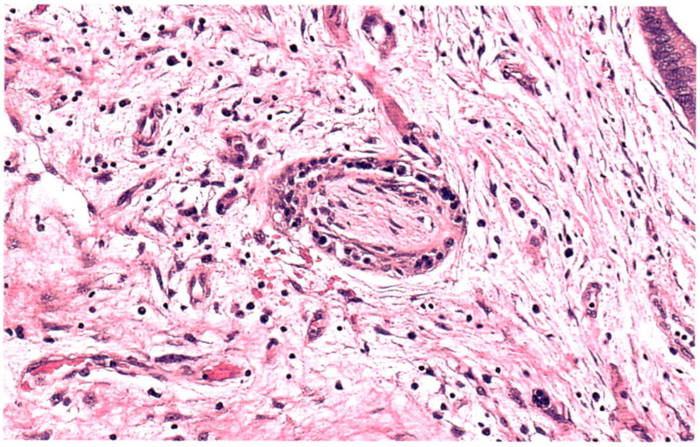

FIGURE 29.63 Perineural invasion by islet cells in chronic pancreatitis may simulate carcinoma.

around nerves can resemble perineural invasion as seen in tumors, most often pancreatic ductal adenocarcinoma (Fig. 29.63).

Peliosis insulis refers to dilated blood-filled spaces within islets (Fig. 29.64). This phenomenon can be seen in pancreata of patients with multiple endocrine neoplasia-1 (MEN-1) (237) but may also be seen in normal pancreas as well. Its etiology and significance remain unclear. The blood-filled spaces are not lined by endothelium on ultrastructural examination, indicating that they are probably not abnormal capillaries.

Islet amyloid deposition is a common finding in pancreata of older individuals (Fig. 29.65), especially those with non–insulin-dependent (type II) diabetes mellitus (238,239). Insular amyloid is restricted to islets; it is biochemically different from systemic amyloid, and is not associated with systemic amyloidosis. Conditions of generalized pancreatic fibrosis may result in amyloid-like deposits in islets. However, this finding is best characterized as insular hyalinizing fibrosis and the deposited

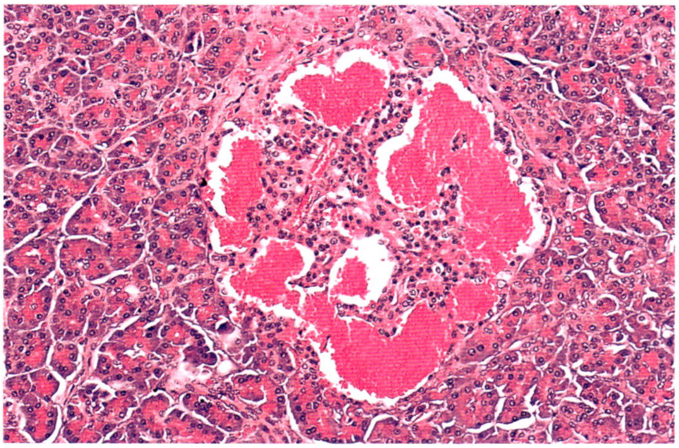

FIGURE 29.64 The appearance of dilated, blood-filled spaces within the islets has been called peliosis insulis. It is probably of no clinical significance.

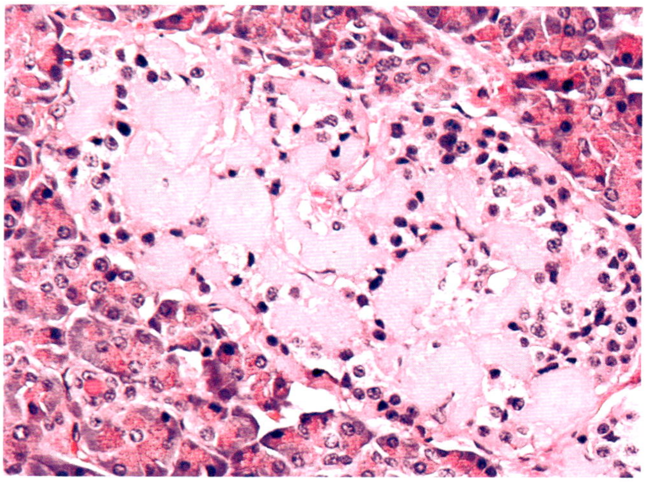

FIGURE 29.65 Amyloid-like hyalinization of the perivascular tissue may be seen in the islets, especially in older patients with type II diabetes. The surrounding acinar tissue is not fibrotic.

material lacks the ultrastructural features of true amyloid fibrils (239).

CHRONIC PANCREATITIS, ATROPHY, AND FIBROSIS

Chronic pancreatitis can be caused by alcoholism, duct obstruction, fibrosing autoimmune disorders, malnutrition, and germline mutations (240–246). Most of these conditions culminate in similar histologic features including fibrosis, mixed inflammation, and atrophy (19). A notable variant is autoimmune pancreatitis, which manifests histologically with a characteristic storiform pattern of fibrosis, obliterative venous phlebitis, and dense IgG4-positive plasma cell infiltrates in some cases. Apart from these systemic forms of pancreatitis, microscopic foci of fibrosis and atrophy are also common incidental findings in otherwise normal pancreata.

Adipose tissue may replace atrophic pancreatic parenchyma during long-standing ductal obstruction, occasionally with residual islets left between lobules of fat (Fig. 29.66). Shwachman syndrome is an extremely rare autosomal recessive disorder of fatty infiltration affecting the pancreas, bone marrow, and skeleton (247). It may result in gland enlargement with reduced parenchymal tissue and exocrine insufficiency (19). The terms "lipomatosis" or "lipomatous pseudohypertrophy" have been applied to pancreata containing over 25% adipose tissue (Fig. 29.67). Adipose distribution is not uniform and may even radiographically simulate pancreatic neoplasms when localized (248). Pancreatic lipomatosis occurs in older individuals with parenchymal atrophy (15). Other associations include adult-onset diabetes and atherosclerosis. Pancreatic lipomatosis is not necessarily associated with obesity (15).

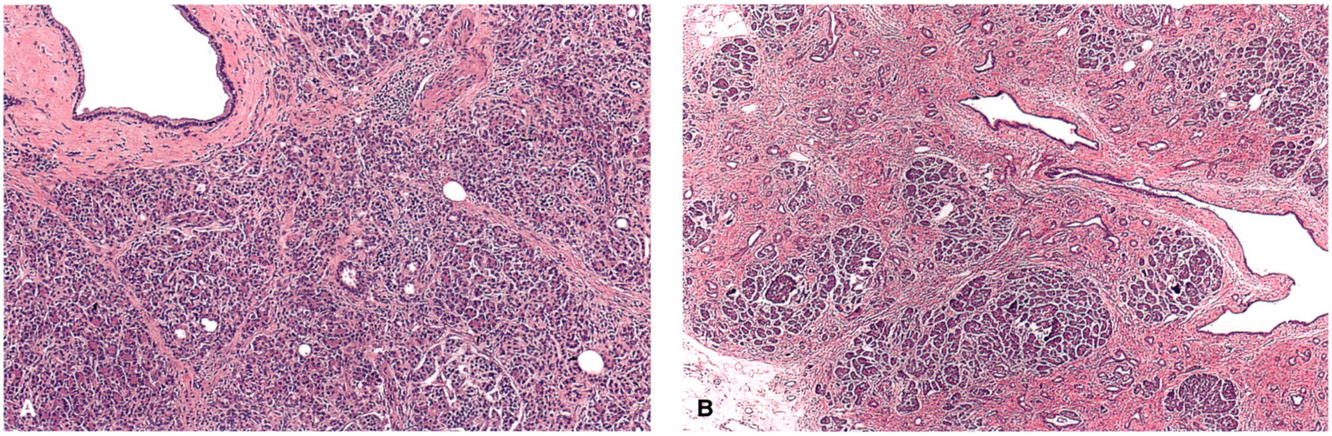

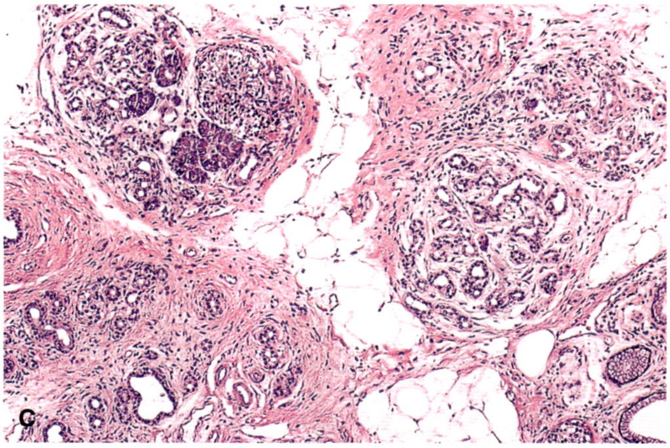

FIGURE 29.66 Progressive changes in chronic pancreatitis. In the early stages (**A**) the fibrosis is largely limited to the periductal and septal areas of the gland. The lobules show prominence of ductules. There are scattered aggregates of chronic inflammatory cells. As the pancreatitis progresses (**B**), the amount of fibrosis is increased, entrapping small lobules of residual acinar tissue. The ducts are ectatic. In the terminal stages (**C**), most of the acinar tissue is atrophic, leaving lobular aggregates of small ductules and islets within a fibrotic stroma surrounded by adipose tissue.

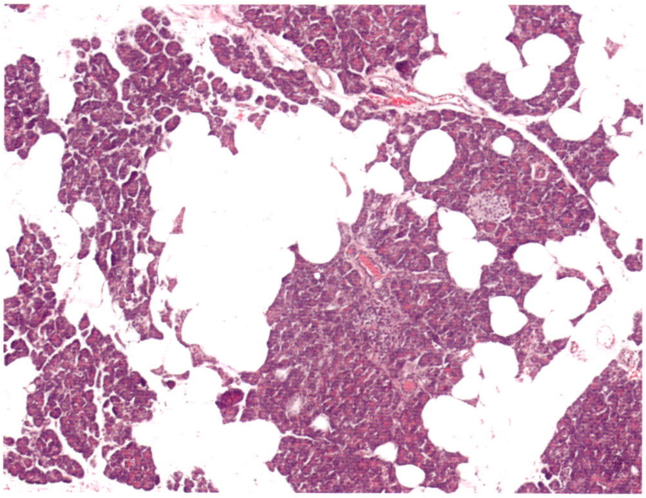

FIGURE 29.67 Pancreatic lipomatosis characterized by the replacement of parenchymal tissue with adipose tissue, constituting over 25% of the gland's volume. As in this case, the remaining parenchyma may not necessarily show changes in atrophic chronic pancreatitis.

REFERENCES

1. Fawcett DW. *Bloom and Fawcett: A Textbook of Histology*. 12th ed. Chapman and Hall; 1994.
2. Moore K, Dalley AF, Agur AMR. *Clinically Oriented Anatomy*. 8th ed. LWW; 2017.
3. Lack EE. *Pathology of the Pancreas, Gallbladder, Extrahepatic Biliary Tract, and Ampullary Region*. Oxford University Press, Inc; 2003.
4. Skandalakis LJ, Rowe JS Jr., Gray SW, Skandalakis JE. Surgical embryology and anatomy of the pancreas. *Surg Clin North Am*. 1993;73(4):661–697.
5. Pansky B. Anatomy of the pancreas. Emphasis on blood supply and lymphatic drainage. *Int J Pancreatol*. 1990;7(1–3):101–108.
6. Bockman DE. Anatomy of the pancreas. In: Go VL, Brooks FP, DiMagno EP, et al., eds. *The Exocrine Pancreas. Biology, Pathobiology, and Diseases*. Raven Press; 1986:1–7.
7. Adsay NV, Basturk O, Saka B, et al. Whipple made simple for surgical pathologists: orientation, dissection, and sampling of pancreaticoduodenectomy specimens for a more practical and accurate evaluation of pancreatic, distal common bile duct, and ampullary tumors. *Am J Surg Pathol*. 2014;38(4):480–493.
8. Ibukuro K. Vascular anatomy of the pancreas and clinical applications. *Int J Gastrointest Cancer*. 2001;30(1–2):87–104.

9. Okahara M, Mori H, Kiyosue H, Yamada Y, Sagara Y, Matsumoto S. Arterial supply to the pancreas; variations and cross-sectional anatomy. *Abdom Imaging*. 2010;35(2): 134–142.
10. Bertelli E, Di Gregorio F, Bertelli L, Orazioli D, Bastianini A. The arterial blood supply of the pancreas: a review. IV. The anterior inferior and posterior pancreaticoduodenal aa., and minor sources of blood supply for the head of the pancreas. An anatomical review and radiologic study. *Surg Radiol Anat*. 1997;19(4):203–212.
11. Bertelli E, Di Gregorio F, Bertelli L, Civeli L, Mosca S. The arterial blood supply of the pancreas: a review. II. The posterior superior pancreaticoduodenal artery. An anatomical and radiological study. *Surg Radiol Anat*. 1996;18(1):1–9.
12. Casillas J, Levi JU, Quiroz AO, et al. *Multidisciplinary Teaching Atlas of the Pancreas*. 2016.
13. Navas V, O'Morchoe PJ, O'Morchoe CC. Lymphatic system of the rat pancreas. *Lymphology*. 1995;28(1):4–20.
14. Cubilla AL, Fortner J, Fitzgerald PJ. Lymph node involvement in carcinoma of the head of the pancreas area. *Cancer*. 1978;41(3):880–887.
15. Stamm BH. Incidence and diagnostic significance of minor pathologic changes in the adult pancreas at autopsy: a systematic study of 112 autopsies in patients without known pancreatic disease. *Hum Pathol*. 1984;15(7):677–683.
16. Heitz PU, Beglinger C, Gyr K. Anatomy and physiology of the exocrine pancreas. In: Kloppel G, Heitz PU, eds. *Pancreatic Pathology*. Churchill-Livingstone; 1984:3–21.
17. Saisho Y, Butler AE, Meier JJ, et al. Pancreas volumes in humans from birth to age one hundred taking into account sex, obesity, and presence of type-2 diabetes. *Clin Anat*. 2007; 20(8):933–942.
18. Birnstingl M. A study of pancreatography. *Br J Surg*. 1959;47: 128–139.
19. Cubilla AL, Fitzgerald PJ. Tumors of the exocrine pancreas. In: Hartmann WH, Sobin LH, eds. *Atlas of Tumor Pathology. 2nd Series, Fascicle 19*. Armed Forces Institute of Pathology; 1984.
20. Kozu T, Suda K, Toki F. Pancreatic development and anatomical variation. *Gastrointest Endosc Clin N Am*. 1995;5(1): 1–30.
21. Noe M, Rezaee N, Asrani K, et al. Immunolabeling of cleared human pancreata provides insights into three-dimensional pancreatic anatomy and pathology. *Am J Pathol*. 2018;188(7):1530–1535.
22. Baggenstoss AH. Major duodenal papilla. Variations of pathologic interest and lesions of the mucosa. *Arch Pathol*. 1938;26: 853–868.
23. Horiguchi S, Kamisawa T. Major duodenal papilla and its normal anatomy. *Dig Surg*. 2010;27(2):90–93.
24. DiMagno EP, Shorter RG, Taylor WF, Taylor WF, Go VL. Relationships between pancreaticobiliary ductal anatomy and pancreatic ductal and parenchymal histology. *Cancer*. 1982;49(2):361–368.
25. Frierson HF Jr. The gross anatomy and histology of the gallbladder, extrahepatic bile ducts, Vaterian system, and minor papilla. *Am J Surg Pathol*. 1989;13(2):146–162.
26. Larsen HL, Grapin-Botton A. The molecular and morphogenetic basis of pancreas organogenesis. *Semin Cell Dev Biol*. 2017;66:51–68.
27. Jennings RE, Berry AA, Strutt JP, Gerrard DT, Hanley NA. Human pancreas development. *Development*. 2015;142(18): 3126–3137.
28. Watt AJ, Zhao R, Li J, Duncan SA. Development of the mammalian liver and ventral pancreas is dependent on GATA4. *BMC Dev Biol*. 2007;7:37.
29. Edlund H. Pancreas: how to get there from the gut? *Curr Opin Cell Biol*. 1999;11(6):663–668.
30. Dassaye R, Naidoo S, Cerf ME. Transcription factor regulation of pancreatic organogenesis, differentiation and maturation. *Islets*. 2016;8(1):13–34.
31. Burlison JS, Long Q, Fujitani Y, Wright CV, Magnuson MA. Pdx-1 and Ptf1a concurrently determine fate specification of pancreatic multipotent progenitor cells. *Dev Biol*. 2008; 316(1):74–86.
32. Conrad E, Stein R, Hunter CS. Revealing transcription factors during human pancreatic β cell development. *Trends Endocrinol Metab*. 2014;25(8):407–414.
33. Bastidas-Ponce A, Scheibner K, Lickert H, Bakhti M. Cellular and molecular mechanisms coordinating pancreas development. *Development*. 2017;144(16):2873–2888.
34. Bechard ME, Bankaitis ED, Hipkens SB, et al. Precommitment low-level Neurog3 expression defines a long-lived mitotic endocrine-biased progenitor pool that drives production of endocrine-committed cells. *Genes Dev*. 2016;30(16):1852–1865.
35. Aigha II, Abdelalim E.M. NKX6.1 transcription factor: a crucial regulator of pancreatic β cell development, identity, and proliferation. *Stem Cell Res Ther*. 2020;11:459.
36. Murtaugh LC, Stanger BZ, Kwan KM, Melton DA. Notch signaling controls multiple steps of pancreatic differentiation. *Proc Natl Acad Sci U S A*. 2003;100(25):14920–14925.
37. Afelik S, Jensen J. Notch signaling in the pancreas: patterning and cell fate specification. *Wiley Interdiscip Rev Dev Biol*. 2013;2(4):531–544.
38. Qu X, Afelik S, Jensen JN, et al. Notch-mediated post-translational control of Ngn3 protein stability regulates pancreatic patterning and cell fate commitment. *Dev Biol*. 2013; 376(1):1–12.
39. Villasenor A, Chong DC, Henkemeyer M, et al. Epithelial dynamics of pancreatic branching morphogenesis. *Development*. 2010;137(24):4295–4305.
40. Zhou Q, Law AC, Rajagopal J, Anderson WJ, Gray PA, Melton DA. A multipotent progenitor domain guides pancreatic organogenesis. *Dev Cell*. 2007;13(1):103–114.
41. Chong JM, Fukayama M, Shiozawa Y, et al. Fibrillary inclusions in neoplastic and fetal acinar cells of the pancreas. *Virchows Arch*. 1996;428(4–5):261–266.
42. Laitio M, Lev R, Orlic D. The developing human fetal pancreas: an ultrastructural and histochemical study with special reference to exocrine cells. *J Anat*. 1974;117(Pt 3):619–634.
43. Lebenthal E, Lev R, Lee PC. Prenantal and postnatal development of the human exocrine pancreas. In: Go VL, Brooks FP, DiMagno EP, et al., eds. *The Exocrine Pancreas. Biology, Pathobiology, and Diseases*. Raven Press; 1986:33–43.
44. Hassan MO, Gogate PA. Malignant mixed exocrine-endocrine tumor of the pancreas with unusual intracytoplasmic inclusions. *Ultrastruct Pathol*. 1993;17(5):483–493.
45. Klimstra DS, Heffess CS, Oertel JE, Rosai J. Acinar cell carcinoma of the pancreas. A clinicopathologic study of 28 cases. *Am J Surg Pathol*. 1992;16(9):815–837.

46. Klimstra DS, Rosai J, Heffess CS. Mixed acinar-endocrine carcinomas of the pancreas. *Am J Surg Pathol*. 1994;18(8): 765–778.
47. Klimstra DS, Wenig BM, Adair CF, Heffess CS. Pancreatoblastoma. A clinicopathologic study and review of the literature. *Am J Surg Pathol*. 1995;19(12):1371–1389.
48. Tucker JA, Shelburne JD, Benning TL, Yacoub L, Federman M. Filamentous inclusions in acinar cell carcinoma of the pancreas. *Ultrastruct Pathol*. 1994;18(1–2):279–286.
49. Gouzi M, Kim YH, Katsumoto K, Johansson K, Grapin-Botton A. Neurogenin3 initiates stepwise delamination of differentiating endocrine cells during pancreas development. *Dev Dyn*. 2011;240(3):589–604.
50. Jeon J, Correa-Medina M, Ricordi C, Edlund H, Diez JA. Endocrine cell clustering during human pancreas development. *J Histochem Cytochem*. 2009;57(9):811–824.
51. Clark A, Grant AM. Quantitative morphology of endocrine cells in human fetal pancreas. *Diabetologia*. 1983;25(1): 31–35.
52. Grasso S, Palumbo G, Fallucca F, Lanzafame S, Indelicato B, Sanfilippo S. The development and function of the endocrine pancreas of fetuses and infants born to normal and diabetic mothers. *Acta Endocrinol Suppl (Copenh)*. 1986;277: 130–135.
53. Robb P. The development of the islets of Langerhans in the human foetus. *Q J Exp Physiol Cogn Med Sci*. 1961;46: 335–343.
54. Conklin JL. Cytogenesis of the human fetal pancreas. *Am J Anat*. 1962;111:181–193.
55. Puri S, Hebrok M. Dynamics of embryonic pancreas development using real-time imaging. *Dev Biol*. 2007;306(1):82–93.
56. Albores-Saavedra J, Gould EW, Angeles-Angeles A, et al. Cystic tumors of the pancreas. *Pathol Annu*. 1990;25(Pt 2): 19–50.
57. Compagno J, Oertel JE. Mucinous cystic neoplasms of the pancreas with overt and latent malignancy (cystadenocarcinoma and cystadenoma). A clinicopathologic study of 41 cases. *Am J Clin Pathol*. 1978;69(6):573–580.
58. Khan N, Dandan W, Al Hassani N, Hadi S. A newly-discovered mutation in the RFX6 gene of the rare Mitchell–Riley syndrome. *J Clin Res Pediatr Endocrinol*. 2016;8(2):246–249.
59. Duval H, Michel-Calemard L, Gonzales M, et al. Fetal anomalies associated with HNF1B mutations: report of 20 autopsy cases. *Prenat Diagn*. 2016;36(8):744–751.
60. Haldorsen IS, Vesterhus M, Raeder H, et al. Lack of pancreatic body and tail in HNF1B mutation carriers. *Diabet Med*. 2008;25(7):782–787.
61. Balasubramanian M, Shield JPH, Acerini CL, et al. Pancreatic hypoplasia presenting with neonatal diabetes mellitus in association with congenital heart defect and developmental delay. *Am J Med Genet A*. 2010;152A(2):340–346.
62. Schnedl WJ, Piswanger-Soelkner C, Wallner SJ, et al. Agenesis of the dorsal pancreas and associated diseases. *Dig Dis Sci*. 2009;54(3):481–487.
63. Klein WA, Dabezies MA, Friedman AC, Caroline DF, Boden GH, Cohen S. Agenesis of dorsal pancreas in a patient with weight loss and diabetes mellitus. *Dig Dis Sci*. 1994;39(8): 1708–1713.
64. Wang JT, Lin JT, Chuang CN, et al. Complete agenesis of the dorsal pancreas—a case report and review of the literature. *Pancreas*. 1990;5(4):493–497.
65. England RE, Newcomer MK, Leung JW, Cotton PB. Case report: annular pancreas divisum—a report of two cases and review of the literature. *Br J Radiol*. 1995;68(807):324–328.
66. Newman BM, Lebenthal E. Congenital abnormalities of the exocrine pancreas. In: Go VL, Brooks FP, DiMagno EP, et al., eds. *The Exocrine Pancreas. Biology, Pathobiology, and Diseases*. Raven Press; 1986:773–782.
67. Dowsett JF, Rode J, Russell RC. Annular pancreas: a clinical, endoscopic, and immunohistochemical study. *Gut*. 1989;30(1): 130–135.
68. Dimitriou I, Katsourakis A, Nikolaidou E, Noussios G. The main anatomical variations of the pancreatic duct system: review of the literature and its importance in surgical practice. *J Clin Med Res*. 2018;10(5):370–375.
69. Adibelli ZH, Adatepe M, Imamoglu C, Esen OS, Erkan N, Yildirim M. Anatomic variations of the pancreatic duct and their relevance with the Cambridge classification system: MRCP findings of 1158 consecutive patients. *Radiol Oncol*. 2016;50(4):370–377.
70. Dawson W, Langman J. An anatomical-radiological study on the pancreatic duct pattern in man. *Anat Rec*. 1961;139: 59–68.
71. Pezzilli R. Pancreas divisum and acute or chronic pancreatitis. *JOP*. 2012;13(1):118–119.
72. Bertin C, Pelletier A-L, Vullierme MP, et al. Pancreas divisum is not a cause of pancreatitis by itself but acts as a partner of genetic mutations. *Am J Gastroenterol*. 2012;107(2): 311–317.
73. Kimura K, Ohto M, Saisho H, et al. Association of gallbladder carcinoma and anomalous pancreaticobiliary ductal union. *Gastroenterology*. 1985;89(6):1258–1265.
74. Kinoshita H, Nagata E, Hirohashi K, Sakai K, Kobayashi Y. Carcinoma of the gallbladder with an anomalous connection between the choledochus and the pancreatic duct. Report of 10 cases and review of the literature in Japan. *Cancer*. 1984; 54(4):762–769.
75. Krishnamurty VS, Rajendran S, Korsten MA. Bifid pancreas. An unusual anomaly associated with acute pancreatitis. *Int J Pancreatol*. 1994;16(2–3):179–181.
76. Lai EC, Tompkins RK. Heterotopic pancreas. Review of a 26-year experience. *Am J Surg*. 1986;151(6):697–700.
77. Pang LC. Pancreatic heterotopia: a reappraisal and clinicopathologic analysis of 32 cases. *South Med J*. 1988;81(10): 1264–1275.
78. Tanaka K, Tsunoda T, Eto T, et al. Diagnosis and management of heterotopic pancreas. *Int Surg*. 1993;78(1): 32–35.
79. Laughlin EH, Keown ME, Jackson JE. Heterotopic pancreas obstructing the ampulla of Vater. *Arch Surg*. 1983;118(8): 979–980.
80. Tsunoda T, Eto T, Yamada M, et al. Heterotopic pancreas: a rare cause of bile duct dilatation—report of a case and review of the literature. *Jpn J Surg*. 1990;20(2):217–220.
81. Seifert G. Congenital anomalies. In: Kloppel G, Heitz PU, eds. *Pancreatic Pathology*. Churchill-Livingstone; 1984:22–26.
82. Zhang Y, Sun X, Gold JS, et al. Heterotopic pancreas: a clinicopathological study of 184 cases from a single high-volume medical center in China. *Hum Pathol*. 2016;55:135–142.
83. Persson GE, Boiesen PT. Cancer of aberrant pancreas in jejunum. Case report. *Acta Chir Scand*. 1988;154(10): 599–601.

84. Tanimura A, Yamamoto H, Shibata H, Sano E. Carcinoma in heterotopic gastric pancreas. *Acta Pathol Jpn*. 1979;29(2): 251–257.
85. Goodarzi M, Rashid A, Maru D. Invasive ductal adenocarcinoma arising from pancreatic heterotopia in rectum: case report and review of literature. *Hum Pathol*. 2010;41(12): 1809–1813.
86. Tsapralis D, Charalabopoulos A, Karamitopoulou E, et al. Pancreatic intraductal papillary mucinous neoplasm with concomitant heterotopic pancreatic cystic neoplasia of the stomach: a case report and review of the literature. *Diagn Pathol*. 2010;5:4.
87. Ma C, Gocke CD, Hruban RH, Belchis DA. Mutational spectrum of intraepithelial neoplasia in pancreatic heterotopia. *Hum Pathol*. 2016;48:117–121.
88. Sun Y, Wasserman PG. Acinar cell carcinoma arising in the stomach: a case report with literature review. *Hum Pathol*. 2004;35(2):263–265.
89. Chetty R, Weinreb I. Gastric neuroendocrine carcinoma arising from heterotopic pancreatic tissue. *J Clin pathol*. 2004; 57(3):314–317.
90. Landry ML, Sarma DP. Accessory spleen in the head of the pancreas. *Hum Pathol*. 1989;20(5):497.
91. Albores-Saavedra J. The pseudometastasis. *Patologia*. 1994; 32:63–71.
92. Akao S, Bockman DE, Lechene de la Porte P, Sarles H. Three-dimensional pattern of ductuloacinar associations in normal and pathological human pancreas. *Gastroenterology*. 1986;90(3):661–668.
93. Henderson JR, Daniel PM, Fraser PA. The pancreas as a single organ: the influence of the endocrine upon the exocrine part of the gland. *Gut*. 1981;22(2):158–167.
94. Williams JA, Goldfine ID. The insulin–acinar relationship. In: Go VL, Brooks FP, DiMagno EP, et al., eds. *The Exocrine Pancreas. Biology, Pathobiology, and Diseases*. Raven Press; 1986:347–360.
95. Hoorens A, Lemoine NR, McLellan E, et al. Pancreatic acinar cell carcinoma. An analysis of cell lineage markers, p53 expression, and Ki-ras mutation. *Am J Pathol*. 1993;143(3):685–698.
96. La Rosa S, Franzi F, Marchet S, et al. The monoclonal anti-BCL10 antibody (clone 331.1) is a sensitive and specific marker of pancreatic acinar cell carcinoma and pancreatic metaplasia. *Virchows Arch*. 2009;454(2):133–142.
97. Bendayan M, Roth J, Perrelet A, Orci L. Quantitative immunocytochemical localization of pancreatic secretory proteins in subcellular compartments of the rat acinar cell. *J Histochem Cytochem*. 1980;28(2):149–160.
98. Kraehenbuhl JP, Racine L, Jamieson JD. Immunocytochemical localization of secretory proteins in bovine pancreatic exocrine cells. *J Cell Biol*. 1977;72(2):406–423.
99. Kern HF. Fine structure of the human exocrine pancreas. In: Go VL, Brooks FP, DiMagno EP, et al., eds. *The Exocrine Pancreas. Biology, Pathobiology, and Diseases*. Raven Press; 1986: 9–19.
100. Palade G. Intracellular aspects of the process of protein synthesis. *Science*. 1975;189(4200):347–358.
101. Romagnoli P. Increases in apical plasma membrane surface paralleling enzyme secretion from exocrine pancreatic acinar cells. *Pancreas*. 1988;3(2):189–192.
102. Pallagi P, Hegyi P, Rakonczay Z Jr. The physiology and pathophysiology of pancreatic ductal secretion: the background for clinicians. *Pancreas*. 2015;44(8):1211–1233.
103. Kodama T. A light and electron microscopic study on the pancreatic ductal system. *Acta Pathol Jpn*. 1983;33(2):297–321.
104. Greider MH. Oxyphil cells of the human pancreas. *Anat Rec*. 1967;157:251.
105. Delporte C. Aquaporins and gland secretion. *Adv Exp Med Biol*. 2017;969:63–79.
106. Burghardt B, Nielsen S, Steward MC. The role of aquaporin water channels in fluid secretion by the exocrine pancreas. *J Membr Biol*. 2006;210(2):143–153.
107. Roberts PF, Burns J. A histochemical study of mucins in normal and neoplastic human pancreatic tissue. *J Pathol*. 1972;107(2):87–94.
108. Solcia E, Capella C, Kloppel G. Tumors of the pancreas. In: Rosai J, Sobin LH, eds. *Atlas of Tumor Pathology. 3rd Series, Fascicle 20*. Armed Forces Institute of Pathology; 1996.
109. Dilorio P, Rittenhouse AR, Bortell R, Jurczyk A. Role of cilia in normal pancreas function and in diseased states. *Birth Defects Res C Embryo Today*. 2014;102(2):126–138.
110. Schulz I. Electrolyte and fluid secretion in the exocrine pancreas. In: Johnson LR, ed. *Physiology of the Gastrointestinal Tract*. Raven Press; 1981:795–819.
111. Spicer SS, Sens MA, Tashian RE. Immunocytochemical demonstration of carbonic anhydrase in human epithelial cells. *J Histochem Cytochem*. 1982;30:864–873.
112. Atkinson BF, Ernst CS, Herlyn M, Steplewski Z, Sears HF, Koprowski H. Gastrointestinal cancer-associated antigen in immunoperoxidase assay. *Cancer Res*. 1982;42(11):4820–4823.
113. Borowitz MJ, Tuck FL, Sindelar WF, Fernsten PD, Metzgar RS. Monoclonal antibodies against human pancreatic adenocarcinoma: distribution of DU-PAN-2 antigen on glandular epithelia and adenocarcinomas. *J Natl Cancer Inst*. 1984; 72(5):999–1005.
114. Haglund C, Lindgren J, Roberts PJ, Nordling S. Gastrointestinal cancer-associated antigen CA 19-9 in histological specimens of pancreatic tumours and pancreatitis. *Br J Cancer*. 1986;53(2):189–195.
115. Klimstra DS, Hameed MR, Marrero AM, et al. Ductal proliferative lesion associated with infiltrating ductal adenocarcinoma of the pancreas. *Int J Pancreatol*. 1994;16:224–225.
116. Sessa F, Bonato M, Frigerio B, et al. Ductal cancers of the pancreas frequently express markers of gastrointestinal epithelial cells. *Gastroenterology*. 1990;98(6):1655–1665.
117. Furuhata A, Minamiguchi S, Shirahase H, et al. Immunohistochemical antibody panel for the differential diagnosis of pancreatic ductal carcinoma from gastrointestinal contamination and benign pancreatic duct epithelium in endoscopic ultrasound-guided fine-needle aspiration. *Pancreas*. 2017;46(4):531–538.
118. Kim JH, Ho SB, Montgomery CK, Kim YS. Cell lineage markers in human pancreatic cancer. *Cancer*. 1990;66(10): 2134–2143.
119. Balague C, Gambus G, Carrato C, et al. Altered expression of MUC2, MUC4, and MUC5 mucin genes in pancreas tissues and cancer cell lines. *Gastroenterology*. 1994;106(4): 1054–1061.
120. Terada T, Ohta T, Sasaki M, Nakanuma Y, Kim YS. Expression of MUC apomucins in normal pancreas and pancreatic tumours. *J Pathol*. 1996;180(2):160–165.
121. Moschovis D, Bamias G, Delladetsima I. Mucins in neoplasms of pancreas, ampulla of Vater and biliary system. *World J Gastrointest Oncol*. 2016;8(10):725–734.

122. Rahier J, Wallon J, Henquin JC. Cell populations in the endocrine pancreas of human neonates and infants. *Diabetologia*. 1981;20(5):540–546.
123. Wittingen J, Frey CF. Islet concentration in the head, body, tail and uncinate process of the pancreas. *Ann Surg*. 1974;179(4):412–414.
124. Wang X, Misawa R, Zielinski MC, et al. Regional differences in islet distribution in the human pancreas–preferential beta-cell loss in the head region in patients with type 2 diabetes. *PloS One*. 2013;8(6):e67454.
125. Fowler JL, Lee SS-Y, Wesner ZC, Olehnik SK, Kron SJ, Hara M. Three-dimensional analysis of the human pancreas. *Endocrinology*. 2018;159(3):1393–1400.
126. Stefan Y, Grasso S, Perrelet A, Orci L. A quantitative immunofluorescent study of the endocrine cell populations in the developing human pancreas. *Diabetes*. 1983;32(4):293–301.
127. Grube D, Bohn R. The microanatomy of human islets of Langerhans, with special reference to somatostatin (D-) cells. *Arch Histol Jpn*. 1983;46(3):327–353.
128. Malaisse-Lagae F, Stefan Y, Cox J, Perrelet A, Orci L. Identification of a lobe in the adult human pancreas rich in pancreatic polypeptide. *Diabetologia*. 1979;17(6):361–365.
129. Stefan Y, Grasso S, Perrelet A, Orci L. The pancreatic polypeptide-rich lobe of the human pancreas: definitive identification of its derivation from the ventral pancreatic primordium. *Diabetologia*. 1982;23(2):141–142.
130. Ehrie MG, Swartz FJ. Diploid, tetraploid and octaploid beta cells in the islets of Langerhans of the normal human pancreas. *Diabetes*. 1974;23(7):583–588.
131. Lecompte PM, Merriam JC Jr. Mitotic figures and enlarged nuclei in the Islands of Langerhans in man. *Diabetes*. 1962; 11:35–39.
132. El-Gohary Y, Sims-Lucas S, Lath N, et al. Three-dimensional analysis of the islet vasculature. *Anat Rec (Hoboken)*. 2012; 295(9):1473–1481.
133. Wierup N, Sundler F, Heller RS. The islet ghrelin cell. *J Mol Endocrinol*. 2014;52(1):R35–R49.
134. Pisania A, Weir GC, O'Neil JJ, et al. Quantitative analysis of cell composition and purity of human pancreatic islet preparations. *Lab Invest*. 2010;90(11):1661–1675.
135. Bommer G, Friedl U, Heitz PU, Klöppel G. Pancreatic PP cell distribution and hyperplasia. Immunocytochemical morphology in the normal human pancreas, in chronic pancreatitis and pancreatic carcinoma. *Virchows Arch A Pathol Anat Histol*. 1980;387(3):319–331.
136. Kloppel G, Lenzen S. Anatomy and physiology of the endocrine pancreas. In: Kloppel G, Heitz PU, eds. *Pancreatic Pathology*. Churchill-Livingstone; 1984:133–153.
137. Orci L, Baetens D, Ravazzola M, Stefan Y, Malaisse-Lagae F. Pancreatic polypeptide and glucagon: non-random distribution in pancreatic islets. *Life Sci*. 1976;19(12):1811–1815.
138. Orci L, Malaisse-Lagae F, Baetens D, Perrelet A. Pancreatic-polypeptide-rich regions in human pancreas. *Lancet*. 1978;2(8101):1200–1201.
139. Wang X, Zielinski MC, Misawa R, et al. Quantitative analysis of pancreatic polypeptide cell distribution in the human pancreas. *PloS one*. 2013;8(1):e55501.
140. Orci L, Stefan Y, Malaisse-Lagae F, Perrelet A. Instability of pancreatic endocrine cell populations throughout life. *Lancet*. 1979;1(8116):615–616.
141. Lan MS, Breslin MB. Structure, expression, and biological function of INSM1 transcription factor in neuroendocrine differentiation. *FASEB J*. 2009;23(7):2024–2033.
142. Fellinger EJ, Garin-Chesa P, Triche TJ, Huvos AG, Rettig WJ. Immunohistochemical analysis of Ewing's sarcoma cell surface antigen p30/32MIC2. *Am J Pathol*. 1991;139(2): 317–325.
143. Hochwald SN, Zee S, Conlon KC, et al. Prognostic factors in pancreatic endocrine neoplasms: an analysis of 136 cases with a proposal for low-grade and intermediate-grade groups. *J Clin Oncol*. 2002;20(11):2633–2642.
144. Weidner N, Tjoe J. Immunohistochemical profile of monoclonal antibody O13: antibody that recognizes glycoprotein p30/32MIC2 and is useful in diagnosing Ewing's sarcoma and peripheral neuroepithelioma. *Am J Surg Pathol*. 1994; 18(5):486–494.
145. Park JY, Hong S-M, Klimstra DS, Goggins MG, Maitra A, Hruban RH. Pdx1 expression in pancreatic precursor lesions and neoplasms. *Appl Immunohistochem Mol Morphol*. 2011; 19(5):444–449.
146. Hermann G, Konukiewitz B, Schmitt A, Perren A, Klöppel G. Hormonally defined pancreatic and duodenal neuroendocrine tumors differ in their transcription factor signatures: expression of ISL1, PDX1, NGN3, and CDX2. *Virchows Arch*. 2011;459(2):147–154.
147. Kloppel G. Endokrines pankreas und diabetes mellitus. In: Doerr W, Seifert G, eds. *Spezielle Pathologische Anatomie*. Vol. 14. Springer; 1981.
148. Pelletier G. Identification of four cell types in the human endocrine pancreas by immunoelectron microscopy. *Diabetes*. 1977;26(8):749–756.
149. Chen J, Baithun SI, Pollock DJ, Berry CL. Argyrophilic and hormone immunoreactive cells in normal and hyperplastic pancreatic ducts and exocrine pancreatic carcinoma. *Virchows Arch A Pathol Anat Histopathol*. 1988;413(5):399–405.
150. Bendayan M. Presence of endocrine cells in pancreatic ducts. *Pancreas*. 1987;2(4):393–397.
151. Oertel JE, Heffess CS, Oertel YC. Pancreas. In: Sternberg SS, ed. *Histology for Pathologists*. Raven Press; 1992:657–668.
152. Patchefsky AS, Solit R, Phillips LD, et al. Hydroxyindole-producing tumors of the pancreas. Carcinoid-islet cell tumor and oat cell carcinoma. *Ann Intern Med*. 1972;77(1):53–61.
153. Wilson RW, Gal AA, Cohen C, DeRose PB, Millikan WJ. Serotonin immunoreactivity in pancreatic endocrine neoplasms (carcinoid tumors). *Mod Pathol*. 1991;4(6):727–732.
154. McCall CM, Shi C, Klein AP, et al. Serotonin expression in pancreatic neuroendocrine tumors correlates with a trabecular histologic pattern and large duct involvement. *Hum Pathol*. 2012;43(8):1169–1176.
155. Chey WY. Hormonal control of pancreatic exocrine secretion. In: Go VL, Brooks FP, DiMagno EP, et al., eds. *The Exocrine Pancreas. Biology, Pathobiology, and Diseases*. Raven Press; 1986:301–313.
156. Henderson JR, Daniel PM. A comparative study of the portal vessels connecting the endocrine and exocrine pancreas, with a discussion of some functional implications. *Q J Exp Physiol Cogn Med Sci*. 1979;64(4):267–275.
157. Adeghate E, Donath T. Distribution of neuropeptide Y and vasoactive intestinal polypeptide immunoreactive nerves in normal and transplanted pancreatic tissue. *Peptides*. 1990;11(6): 1087–1092.

158. Zhang Y, Nose V. Endocrine tumors as part of inherited tumor syndromes. *Adv Anat Pathol*. 2011;18(3):206–218.
159. Sharma S, Green KB. The pancreatic duct and its arteriovenous relationship: an underutilized aid in the diagnosis and distinction of pancreatic adenocarcinoma from pancreatic intraepithelial neoplasia. A study of 126 pancreatectomy specimens. *Am J Surg Pathol*. 2004;28(5):613–620.
160. Olsen TS. Lipomatosis of the pancreas in autopsy material and its relation to age and overweight. *Acta Pathol Microbiol Scand A*. 1978;86A(5):367–373.
161. Orci L, Stefan Y, Malaisse-Lagae F, Perrelet A, Patel Y. Pancreatic fat. *N Engl J Med*. 1979;301(23):1292.
162. Suda K, Mizuguchi K, Hoshino A. Differences of the ventral and dorsal anlagen of pancreas after fusion. *Acta Pathol Jpn*. 1981;31(4):583–589.
163. Jaster R. Molecular regulation of pancreatic stellate cell function. *Mol Cancer*. 2004;3:26.
164. Haqq J, Howells LM, Garcea G, Metcalfe MS, Steward WP, Dennison AR. Pancreatic stellate cells and pancreas cancer: current perspectives and future strategies. *Eur J Cancer*. 2014; 50(15):2570–2582.
165. Masamune A, Watanabe T, Kikuta K, Shimosegawa T. Roles of pancreatic stellate cells in pancreatic inflammation and fibrosis. *Clin Gastroenterol Hepatol*. 2009;7(Suppl 11): S48–S54.
166. Popescu LM, Hinescu ME, Ionescu N, Ciontea SM, Cretoiu D, Ardelean C. Interstitial cells of Cajal in pancreas. *J Cell Mol Med*. 2005;9(1):169–190.
167. Centeno BA, Pitman MB. Fine needle aspiration biopsy of the pancreas. In: Centeno BA, Pitman MB, eds. *Neoplasms of the Exocrine and Endocrine Pancreas*. Butterworth-Heinemann; 1999:109–160.
168. Hruban RH, Pitman MB, Klimstra DS. Tumors of the pancreas. In: Silverberg SG, Sobin LH, eds. *Atlas of Tumor Pathology*. 4th ed. American Registry of Pathology; 2006.
169. Sigel CS, Klimstra DS. Cytomorphologic and immunophenotypical features of acinar cell neoplasms of the pancreas. *Cancer Cytopathol*. 2013;121(8):459–470.
170. Shinozuka H, Lee RE, Dunn JL, Longnecker DS. Multiple atypical acinar cell nodules of the pancreas. *Hum Pathol*. 1980;11(4):389–391.
171. Tanaka T, Mori H, Williams GM. Atypical and neoplastic acinar cell lesions of the pancreas in an autopsy study of Japanese patients. *Cancer*. 1988;61(11):2278–2285.
172. Troxell ML, Drachenberg C. Allograft pancreas: pale acinar nodules. *Hum Pathol*. 2016;54:127–133.
173. Longnecker DS, Hashida Y, Shinozuka H. Relationship of age to prevalence of focal acinar cell dysplasia in the human pancreas. *J Natl Cancer Inst*. 1980;65(1):63–66.
174. Oertel JE. The pancreas. Nonneoplastic alterations. *Am J Surg Pathol*. 1989;13(Suppl 1):50–65.
175. Longnecker DS, Shinozuka H, Dekker A. Focal acinar cell dysplasia in human pancreas. *Cancer*. 1980;45(3):534–540.
176. Kodama T, Mori W. Atypical acinar cell nodules of the human pancreas. *Acta Pathol Jpn*. 1983;33(4):701–714.
177. Baggenstoss AH. The pancreas in uremia; a histopathologic study. *Am J Pathol*. 1948;24:1003–1017.
178. Walters MN. Studies on the exocrine pancreas. I. Nonspecific pancreatic ductular ectasia. *Am J Pathol*. 1964;44:973–981.
179. Walters MN. Goblet-cell metaplasia in ductules and acini of the exocrine pancreas. *J Pathol Bacteriol*. 1965;89:569–572.
180. Schmid RM. Acinar-to-ductal metaplasia in pancreatic cancer development. *J Clin Invest*. 2002;109(11):1403–1404.
181. Hruban RH, Adsay NV, Albores-Saavedra J, et al. Pathology of genetically engineered mouse models of pancreatic exocrine cancer: consensus report and recommendations. *Cancer Res*. 2006;66(1):95–106.
182. Bockman DE, Boydston WR, Anderson MC. Origin of tubular complexes in human chronic pancreatitis. *Am J Surg*. 1982; 144(2):243–249.
183. Aichler M, Seiler C, Tost M, et al. Origin of pancreatic ductal adenocarcinoma from atypical flat lesions: a comparative study in transgenic mice and human tissues. *J Pathol*. 2012; 226(5):723–734.
184. Hong S-M, Heaphy CM, Shi C, et al. Telomeres are shortened in acinar-to-ductal metaplasia lesions associated with pancreatic intraepithelial neoplasia but not in isolated acinar-to-ductal metaplasias. *Mod Pathol*. 2011;24(2):256–266.
185. Shi C, Hong S-M, Lim P, et al. KRAS2 mutations in human pancreatic acinar-ductal metaplastic lesions are limited to those with PanIN: implications for the human pancreatic cancer cell of origin. *Mol Cancer Res*. 2009;7(2):230–236.
186. Makohon-Moore AP, Matsukuma K, Zhang M, et al. Precancerous neoplastic cells can move through the pancreatic ductal system. *Nature*. 2018;561(7722):201–205.
187. McDonald OG. The biology of pancreatic morphology. *Pathology*. 2022;54(2):236–247.
188. Hutchings D, Waters KM, Weiss MJ, et al. Cancerization of the pancreatic ducts: demonstration of a common and under-recognized process using immunolabeling of paired duct lesions and invasive pancreatic ductal adenocarcinoma for p53 and Smad4 expression. *Am J Surg Pathol*. 2018; 42(11):1556–1561.
189. Brat DJ, Lillemoe KD, Yeo CJ, Warfield PB, Hruban RH. Progression of pancreatic intraductal neoplasias to infiltrating adenocarcinoma of the pancreas. *Am J Surg Pathol*. 1998;22(2):163–169.
190. Goggins M, Hruban RH, Kern SE. BRCA2 is inactivated late in the development of pancreatic intraepithelial neoplasia: evidence and implications. *Am J Pathol*. 2000; 156(5):1767–1771.
191. Maitra A, Adsay NV, Argani P, et al. Multicomponent analysis of the pancreatic adenocarcinoma progression model using a pancreatic intraepithelial neoplasia tissue microarray. *Mod Pathol*. 2003;16(9):902–912.
192. Moskaluk CA, Hruban RH, Kern SE. p16 and K-ras gene mutations in the intraductal precursors of human pancreatic adenocarcinoma. *Cancer Res*. 1997;57(11):2140–2143.
193. van Heek NT, Meeker AK, Kern SE, et al. Telomere shortening is nearly universal in pancreatic intraepithelial neoplasia. *Am J Pathol*. 2002;161(5):1541–1547.
194. Wilentz RE, Iacobuzio-Donahue CA, Argani P, et al. Loss of expression of Dpc4 in pancreatic intraepithelial neoplasia: evidence that DPC4 inactivation occurs late in neoplastic progression. *Cancer Res*. 2000;60(7):2002–2006.
195. Klimstra DS, Longnecker DS. K-ras mutations in pancreatic ductal proliferative lesions. *Am J Pathol*. 1994;145(6): 1547–1550.
196. Mukada T, Yamada S. Dysplasia and carcinoma in situ of the exocrine pancreas. *Tohoku J Exp Med*. 1982;137(2):115–124.
197. Hruban RH, Adsay NV, Albores-Saavedra J, et al. Pancreatic intraepithelial neoplasia: a new nomenclature and

classification system for pancreatic duct lesions. *Am J Surg Pathol*. 2001;25(5):579–586.

198. Hruban RH, Takaori K, Klimstra DS, et al. An illustrated consensus on the classification of pancreatic intraepithelial neoplasia and intraductal papillary mucinous neoplasms. *Am J Surg Pathol*. 2004;28(8):977–987.
199. Basturk O, Hong S-M, Wood LD, et al. A revised classification system and recommendations from the Baltimore consensus meeting for neoplastic precursor lesions in the pancreas. *Am J Surg Pathol*. 2015;39(12):1730–1741.
200. Hayashi A, Hong J, Iacobuzio-Donahue CA. The pancreatic cancer genome revisited. *Nat Rev Gastroenterol Hepatol*. 2021;18(7):469–481.
201. Gaujoux S, Brennan MF, Gonen M, et al. Cystic lesions of the pancreas: changes in the presentation and management of 1,424 patients at a single institution over a 15-year time period. *J Am Coll Surg*. 2011;212(4):590–600; discussion 600–603.
202. Allen PJ, D'Angelica M, Gonen M, et al. A selective approach to the resection of cystic lesions of the pancreas: results from 539 consecutive patients. *Ann Surg*. 2006;244(4):572–582.
203. Hosoda W, Chianchiano P, Griffin JF, et al. Genetic analyses of isolated high-grade pancreatic intraepithelial neoplasia (HG-PanIN) reveal paucity of alterations in TP53 and SMAD4. *J Pathol*. 2017;242(1):16–23.
204. Hong S-M, Park JY, Hruban RH, Goggins M. Molecular signatures of pancreatic cancer. *Arch Pathol Lab Med*. 2011; 135(6):716–727.
205. Scarlett CJ, Salisbury EL, Biankin AV, Kench J. Precursor lesions in pancreatic cancer: morphological and molecular pathology. *Pathology*. 2011;43(3):183–200.
206. Klein WM, Hruban RH, Klein-Szanto AJ, Wilentz RE. Direct correlation between proliferative activity and dysplasia in pancreatic intraepithelial neoplasia (PanIN): additional evidence for a recently proposed model of progression. *Mod Pathol*. 2002;15(4):441–447.
207. Othman M, Basturk O, Groisman G, Krasinskas A, Adsay NV. Squamoid cyst of pancreatic ducts: a distinct type of cystic lesion in the pancreas. *Am J Surg Pathol*. 2007; 31(2):291–297.
208. Tasso F, Picard D. Sur les oncocytes du pancreas humain. *C R Soc Biol (Paris)*. 1969;163:1855–1858.
209. Tasso F, Sarles H. Canalicular cells and oncocytes in the human pancreas. Comparative study on the normal condition and in chronic pancreatitis. *Ann Anat Pathol (Paris)*. 1973;18:277–300.
210. Frexinos J, Ribet A. Oncocytes in human chronic pancreatitis. *Digestion*. 1972;7(5):294–301.
211. Singhi AD, Wood LD, Parks E, et al. Recurrent rearrangements in PRKACA and PRKACB in intraductal oncocytic papillary neoplasms of the pancreas and bile duct. *Gastroenterology*. 2020;158(3):573–582.e2
212. Adsay NV, Merati K, Andea A, et al. The dichotomy in the preinvasive neoplasia to invasive carcinoma sequence in the pancreas: differential expression of MUC1 and MUC2 supports the existence of two separate pathways of carcinogenesis. *Mod Pathol*. 2002;15(10):1087–1095.
213. Adsay NV, Merati K, Basturk O, et al. Pathologically and biologically distinct types of epithelium in intraductal papillary mucinous neoplasms: delineation of an "intestinal" pathway of carcinogenesis in the pancreas. *Am J Surg Pathol*. 2004;28(7):839–848.
214. Zamboni G, Terris B, Scarpa A, et al. Acinar cell cystadenoma of the pancreas: a new entity? *Am J Surg Pathol*. 2002;26(6): 698–704.
215. Singhi AD, Norwood S, Liu T-C, et al. Acinar cell cystadenoma of the pancreas: a benign neoplasm or non-neoplastic ballooning of acinar and ductal epithelium? *Am J Surg Pathol*. 2013;37(9):1329–1335.
216. Khor TS, Badizadegan K, Ferrone C, et al. Acinar cystadenoma of the pancreas: a clinicopathologic study of 10 cases including multilocular lesions with mural nodules. *Am J Surg Pathol*. 2012;36(11):1579–1591.
217. Rift CV, Hasselby JP, Hansen CP, Federspiel B. Acinar cystic transformation of the pancreas: report of a case and a review of the literature. *Pathol Res Pract*. 2020;216(6):152928
218. Komatsu K. Pancreatographical and histopathological study of dilations of the pancreatic ductules with special references to cystic dilatation. *Juntendoo Med J*. 1974;19:250–269.
219. Agostini S, Choux R, Payan MJ, Sastre B, Sahel J, Clement JP. Mucinous pancreatic duct ectasia in the body of the pancreas. *Radiology*. 1989;170(3 Pt 1):815–816.
220. Nagai E, Ueki T, Chijiiwa K, Tanaka M, Tsuneyoshi M. Intraductal papillary mucinous neoplasms of the pancreas associated with so-called "mucinous ductal ectasia." Histochemical and immunohistochemical analysis of 29 cases. *Am J Surg Pathol*. 1995;19(5):576–589.
221. Nishihara K, Fukuda T, Tsuneyoshi M, Kominami T, Maeda S, Saku M. Intraductal papillary neoplasm of the pancreas. *Cancer*. 1993;72(3):689–696.
222. Sessa F, Solcia E, Capella C, et al. Intraductal papillary-mucinous tumours represent a distinct group of pancreatic neoplasms: an investigation of tumour cell differentiation and K-ras, p53 and c-erbB-2 abnormalities in 26 patients. *Virchows Arch*. 1994;425(4):357–367.
223. Tian FZ, Myles J, Howard JM. Mucinous pancreatic ductal ectasia of latent malignancy: an emerging clinicopathologic entity. *Surgery*. 1992;111(1):109–113.
224. Krasinskas AM, Oakley GJ, Bagci P, et al. "Simple mucinous cyst" of the pancreas: a clinicopathologic analysis of 39 examples of a diagnostically challenging entity distinct from intraductal papillary mucinous neoplasms and mucinous cystic neoplasms. *Am J Surg Pathol*. 2017;41(1):121–127.
225. Solicia E, Capella C, Kloppel G. Tumors of the endocrine pancreas. In: *Atlas of Tumor Pathology of the Pancreas*. Armed Forces Institute of Pathology; 1997:145–196.
226. Weidenheim KM, Hinchey WW, Campbell WG Jr. Hyperinsulinemic hypoglycemia in adults with islet-cell hyperplasia and degranulation of exocrine cells of the pancreas. *Am J Clin Pathol*. 1983;79(1):14–24.
227. Goossens A, Gepts W, Saudubray JM, et al. Diffuse and focal nesidioblastosis. A clinicopathological study of 24 patients with persistent neonatal hyperinsulinemic hypoglycemia. *Am J Surg Pathol*. 1989;13(9):766–775.
228. Stanley CA, Thornton PS, Ganguly A, et al. Preoperative evaluation of infants with focal or diffuse congenital hyperinsulinism by intravenous acute insulin response tests and selective pancreatic arterial calcium stimulation. *J Clin Endocrinol Metab*. 2004;89(1):288–296.
229. Suchi M, MacMullen C, Thornton PS, Ganguly A, Stanley CA, Ruchelli ED. Histopathology of congenital hyperinsulinism: retrospective study with genotype correlations. *Pediatr Dev Pathol*. 2003;6(4):322–333.

230. Thomas PM, Cote GJ, Wohllk N, et al. Mutations in the sulfonylurea receptor gene in familial persistent hyperinsulinemic hypoglycemia of infancy. *Science*. 1995;268(5209): 426–429.
231. Kloppel G, Anlauf M, Raffel A, Perren A, Knoefel WT. Adult diffuse nesidioblastosis: genetically or environmentally induced? *Hum Pathol*. 2008;39(1):3–8.
232. Clayton PT, Eaton S, Aynsley-Green A, et al. Hyperinsulinism in short-chain L-3-hydroxyacyl-CoA dehydrogenase deficiency reveals the importance of beta-oxidation in insulin secretion. *J Clin Invest*. 2001;108(3):457–465.
233. Glaser B, Kesavan P, Heyman M, et al. Familial hyperinsulinism caused by an activating glucokinase mutation. *N Engl J Med*. 1998;338(4):226–230.
234. Reinecke-Luthge A, Koschoreck F, Kloppel G. The molecular basis of persistent hyperinsulinemic hypoglycemia of infancy and its pathologic substrates. *Virchows Arch*. 2000;436(1): 1–5.
235. Stanley CA, Lieu YK, Hsu BY, et al. Hyperinsulinism and hyperammonemia in infants with regulatory mutations of the glutamate dehydrogenase gene. *N Engl J Med*. 1998;338(19): 1352–1357.
236. Bartow SA, Mukai K, Rosai J. Pseudoneoplastic proliferation of endocrine cells in pancreatic fibrosis. *Cancer*. 1981; 47(11):2627–2633.
237. Kovacs K, Horvath E, Asa SL, Murray D, Singer W, Reddy SS. Microscopic peliosis of pancreatic islets in a woman with MEN-1 syndrome. *Arch Pathol Lab Med*. 1986;110(7): 607–610.
238. Westermark P. Amyloid in the islets of Langerhans: thoughts and some historical aspects. *Ups J Med Sci*. 2011;116(2):81–89.
239. Kloppel G. Islet histopathology in diabetes mellitus. In: Kloppel G, Heitz PU, eds. *Pancreatic Pathology*. Churchill-Livingstone; 1984:154–192.
240. Castellani C, Bonizzato A, Rolfini R, Frulloni L, Cavallini GC, Mastella G. Increased prevalence of mutations of the cystic fibrosis gene in idiopathic chronic and recurrent pancreatitis. *Am J Gastroenterol*. 1999;94(7):1993–1995.
241. Gorry MC, Gabbaizedeh D, Furey W, et al. Mutations in the cationic trypsinogen gene are associated with recurrent acute and chronic pancreatitis. *Gastroenterology*. 1997;113(4):1063–1068.
242. Ito T, Nakano I, Koyanagi S, et al. Autoimmune pancreatitis as a new clinical entity. Three cases of autoimmune pancreatitis with effective steroid therapy. *Dig Dis Sci*. 1997;42(7): 1458–1468.
243. Kloppel G, Maillet B. Pathology of acute and chronic pancreatitis. *Pancreas*. 1993;8(6):659–670.
244. Lilja P, Evander A, Ihse I. Hereditary pancreatitis-a report on two kindreds. *Acta Chir Scand*. 1978;144(1):35–37.
245. Whitcomb DC, Gorry MC, Preston RA, et al. Hereditary pancreatitis is caused by a mutation in the cationic trypsinogen gene. *Nat Genet*. 1996;14(2):141–145.
246. Yoshida K, Toki F, Takeuchi T, Watanabe S, Shiratori K, Hayashi N. Chronic pancreatitis caused by an autoimmune abnormality. Proposal of the concept of autoimmune pancreatitis. *Dig Dis Sci*. 1995;40(7):1561–1568.
247. Seifert G. Lipomatous atrophy and other forms. In: Kloppel G, Heitz PU, eds. *Pancreatic Pathology*. Churchill-Livingstone; 1984:27–31.
248. Altinel D, Basturk O, Sarmiento JM, et al. Lipomatous pseudohypertrophy of the pancreas: a clinicopathologically distinct entity. *Pancreas*. 2010;39(3):392–397.

SECTION VIII

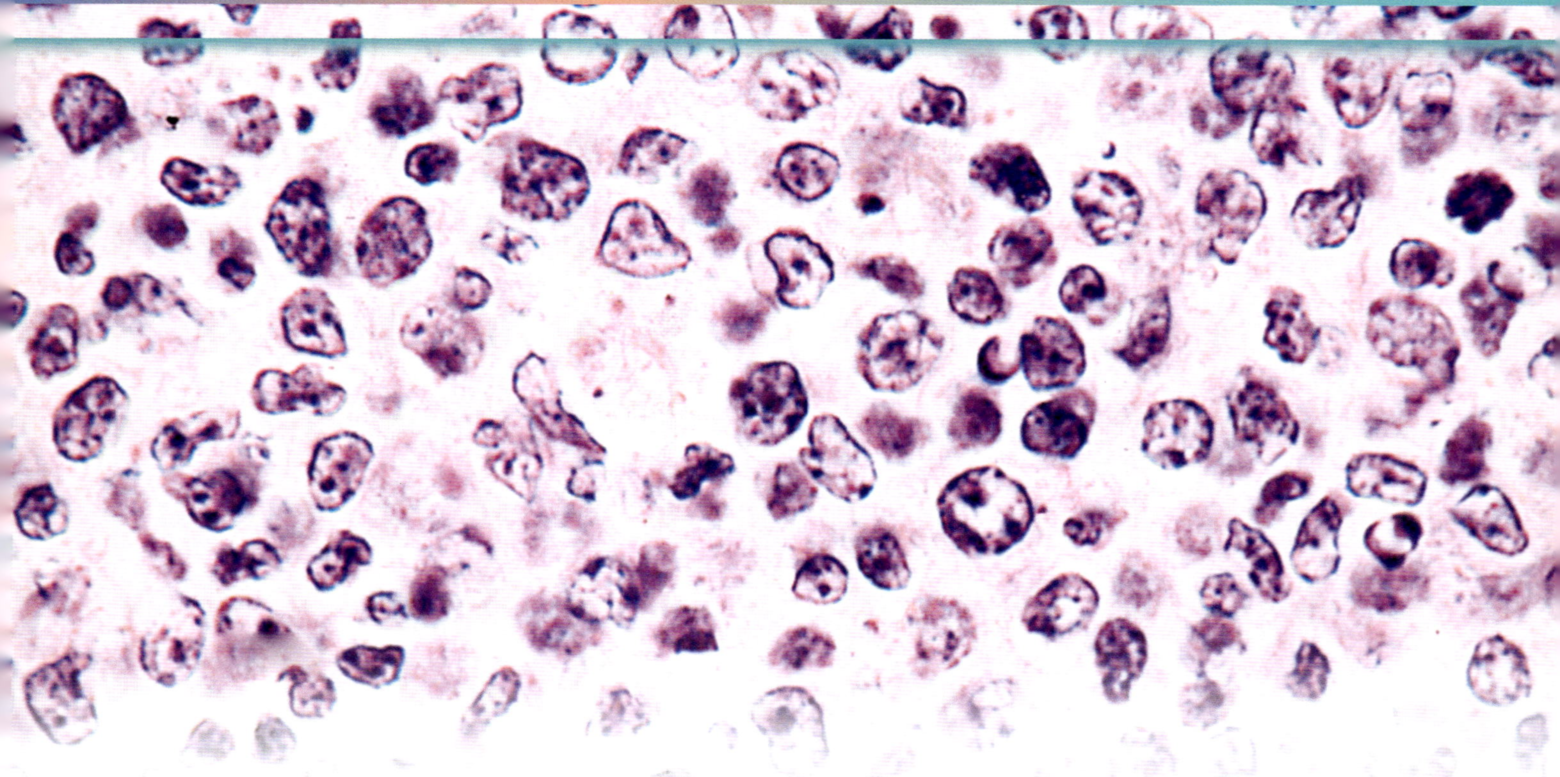

Hematopoietic System

30

Lymph Nodes

Staci Keene ■ Yuri Fedoriw

The lymph nodes are a critical component of the immune system, and central to foreign antigen presentation, B-cell maturation, and immune memory (1). Lymph nodes are concentrated throughout the body in areas where exposure to foreign antigen is most likely. The skin has numerous draining lymph nodes, partly grouped where lymphatics converge, such as the axillary, cervical, and inguinal regions. The gastrointestinal tract and the airways have small collections of lymphoid cells in their mucosal surfaces (i.e., mucosa-associated lymphoid tissue). These sites drain into the mesenteric, mediastinal, and hilar lymph nodes (2–5). Drainage from extra-abdominal areas is through the para-iliac and para-aortic nodes. All these systems converge on a single lymphatic channel, the thoracic duct, that returns the lymphatic fluid to the bloodstream. The brain is the only organ without lymphatic vessels. Its extracellular fluid drains via the cerebrospinal fluid or the perivascular Virchow–Robin space (6). Under antigen stimulation, lymph nodes can enlarge, changing their shape, size, and histology.

EMBRYOLOGY/DEVELOPMENTAL CHANGES

Relatively little is known of the development of the lymph nodes during the embryonic period. They seem to arise from the lymphatic sacs, which in turn develop from the venous system. In the first trimester, a lymphatic plexus forms from the sacs, appearing as small collections of lymphoblasts in association with the plexus (7). In the second trimester, differentiation into the cortex and medulla begins to take place, gradually forming the compartmentalized structure of the lymph node parenchyma. This process is under the influence of nonlymphoid cells, such as macrophages and interdigitating dendritic cells (IDCs), and mesenchyme-derived cells, such as follicular dendritic cells and potentially fibroblastic reticulum cells (8). After the compartmentalization is completed, no other changes take place than those that follow from antigenic stimulation.

GROSS FEATURES

Resting lymph nodes are mostly small, round or reniform structures with a thin, fibrous capsule. Resting lymph nodes span less than 1 cm, but can become much larger under antigenic stimulation. A diameter of more than 3 cm is unusual in benign lymph nodes and should raise the suspicion for a pathologic condition, particularly malignancy. The cut surface of a reactive lymph node is a pink-brown and homogeneous. A white or salmon-color or distinct nodularity is concerning for a neoplastic process.

ANATOMY

Blood Supply

The lymph node hilum contains an artery, vein, and efferent lymphatic vessel. Arterioles enter the lymph node through the hilum, branch, and rapidly form a plexus of capillaries in the parenchyma. Chemokine receptors and other homing factors, guide migration of lymphocytes to and through the vasculature of the lymphoid tissues, thereby helping to position the lymphocytes within different compartments of the lymph node (9–11).

Venous drainage accompanies the arteriolar route. The postcapillary or high endothelial venules in the lymph nodes are unique, as they are the main route of entrance of lymphocytes homing to the lymph nodes. They will be discussed in more detail later.

Lymphatic Vessels

The afferent lymphatic vessels enter the lymph node through the capsule, draining into the subcapsular sinus. This sinus is lined by endothelial cells, but the system of branching sinuses arising from it no longer have an endothelial lining. The sinuses converge on the efferent lymphatic vessel at the lymph node hilum.

LIGHT MICROSCOPY: THE DIFFERENT COMPARTMENTS, HISTOLOGY, AND FUNCTION

Examination of a sectioned lymph node at low power reveals relatively clearly defined areas or compartments (Fig. 30.1). The outermost portion of the lymph node is the capsule. The capsule is a fibrous connective tissue membrane containing delicate reticulin fibers. Branches of the capsule, called trabeculae, extend into the parenchyma and provide additional support to the blood vessels entering the lymph node. Beneath the capsule are three compartments: the cortex, the medulla, and the hilar region. The cortex can be divided into two regions: the outer cortex and paracortical region or inner cortex. Primary and/or secondary follicles are typically arranged within the outer cortex, although these structures can be found deeper in the lymph node. The follicles are composed primarily of B cells, while the paracortical areas contain predominantly T cells and the specialized postcapillary high endothelial venules.

The medullary region is the deepest portion of the lymph node. It contains two structures: the medullary cords

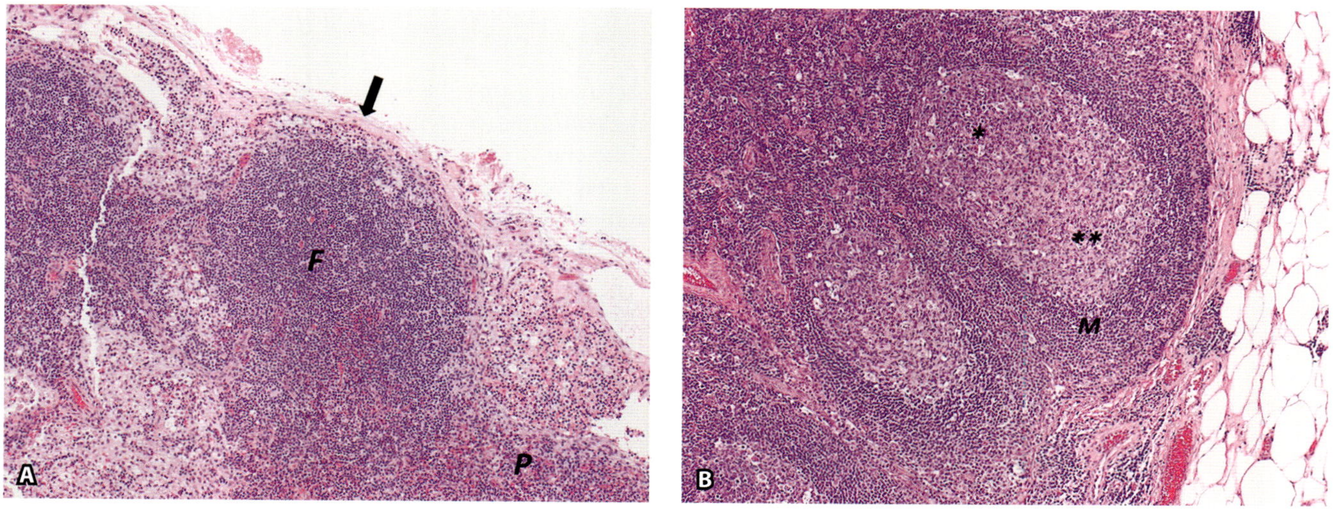

FIGURE 30.1 Resting and reactive lymph node architecture. **A:** Hematoxylin and eosin (H&E)–stained lymph node section demonstrating a resting (or primary) follicle (*F*), at the periphery of the node identified by a thin fibrous capsule (*arrow*). The pericortical area (*P*) is located between follicles and away from the capsule. **B:** These reactive (or secondary) follicles (*F*) show characteristic features of the germinal center reaction with dark (*) and light zone (**) polarization and surrounded by the mantle zone (*M*), a thin but conspicuous rim of small, mature-appearing lymphocytes.

and the medullary sinuses. The low-power view of the medullary region shows dark areas separated by lighter staining ones, often in a somewhat reticular pattern. The dark areas are the medullary cords, where plasma cells and occasional mast cells are found. The lighter areas between these sheets of dark-staining small lymphocytes are the sinuses. They are filled with macrophages that contain ample pale cytoplasm, thereby imparting the light appearance of this area. The sinuses run through the entire parenchyma but are especially visible in the medullary region.

A lymph node is virtually always stimulated by constant antigenic exposure. To some degree, activation will lead to expansion of one area, often at the cost of the volume in other compartments (Fig. 30.2). Stimulation also causes a shift in the cellular composition within the activated compartment.

This adaptability to the constantly changing antigenic challenges explains the variable histology of the normal lymph node. Almost all lymph nodes that sampled for histologic evaluation show evidence of antigenic stimulation. The follicle, paracortical area, medullary cords, and medullary sinuses are each discussed separately.

The Follicle

Follicles are classified as primary or secondary follicles. Primary follicles are aggregates of small lymphoid cells, predominantly composed of antigen-naïve B cells and follicular dendritic cells, the latter supporting the follicular architecture. After antigenic stimulation, the primary follicles will develop a germinal center, at which point they are referred to as secondary follicles. This germinal center reaction requires the cooperation of follicular dendritic cells, lymphoid cells, and tingible body macrophages (Fig. 30.3). The dark outer rim of the germinal center is the mantle zone, containing the displaced primary follicle naïve B cells. These naïve B cells migrate into the germinal center and become centroblasts. From there, the centroblasts either undergo apoptosis or mature into centrocytes with successful recombination of the B-cell receptor. Cytologic features of centroblasts and centrocytes are described below. The function of the germinal center reaction is the generation of B cells with high-affinity antigen receptors that produce an antibody with the best possible fit to the antigen and function either as a direct precursor for antibody-producing plasma cells or as long-lived memory cells. The outermost zone of the follicle

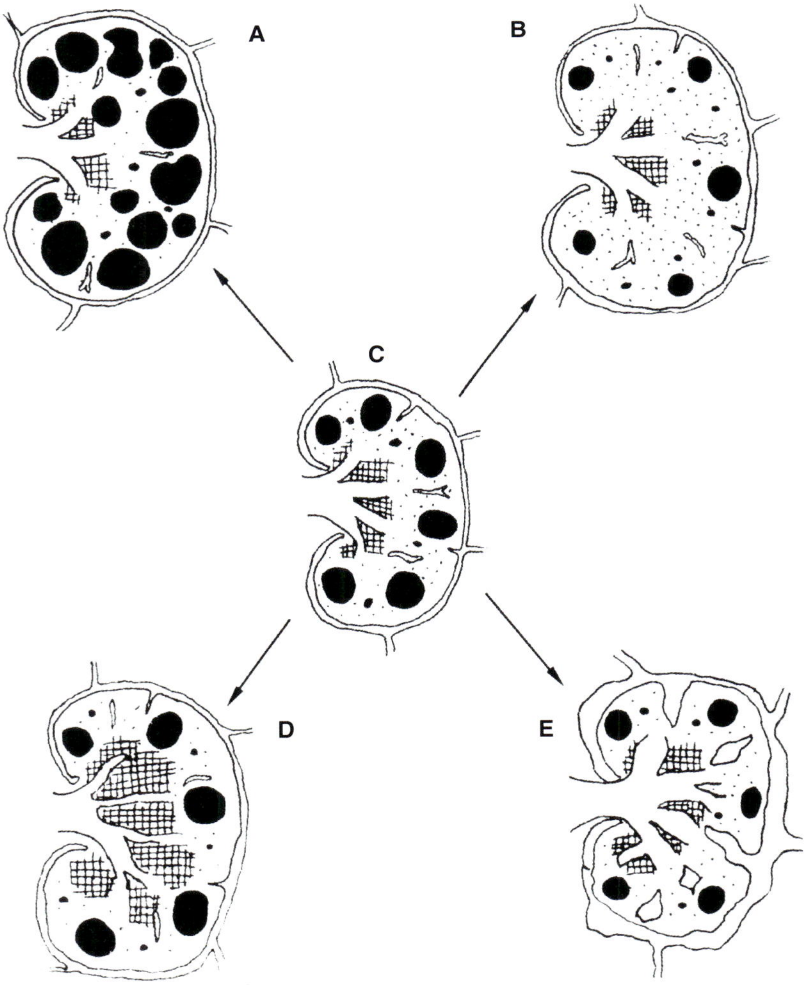

FIGURE 30.2 Schematic representation of the lymph node with enlargement of the four different compartments, respectively the follicular (**A**), the paracortical (**B**), the medullary (**D**), and the sinusoidal (**E**) compartments; (**C**) represents an "unstimulated" lymph node.

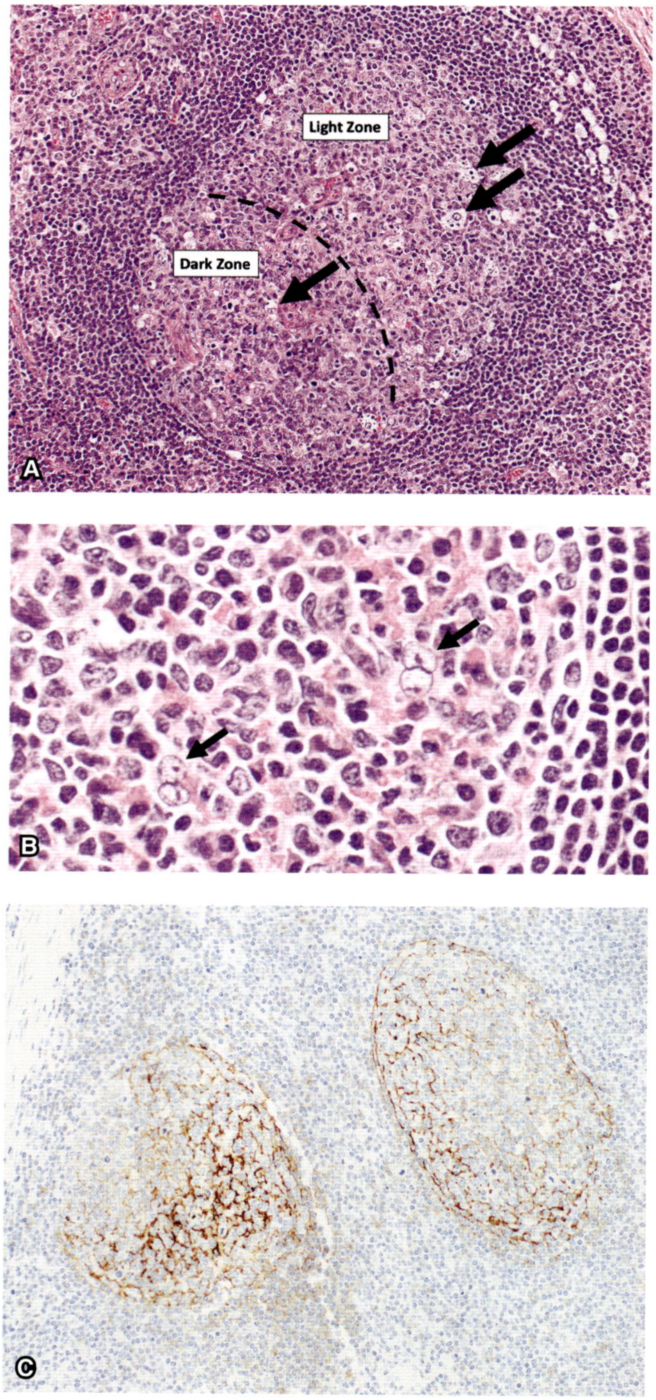

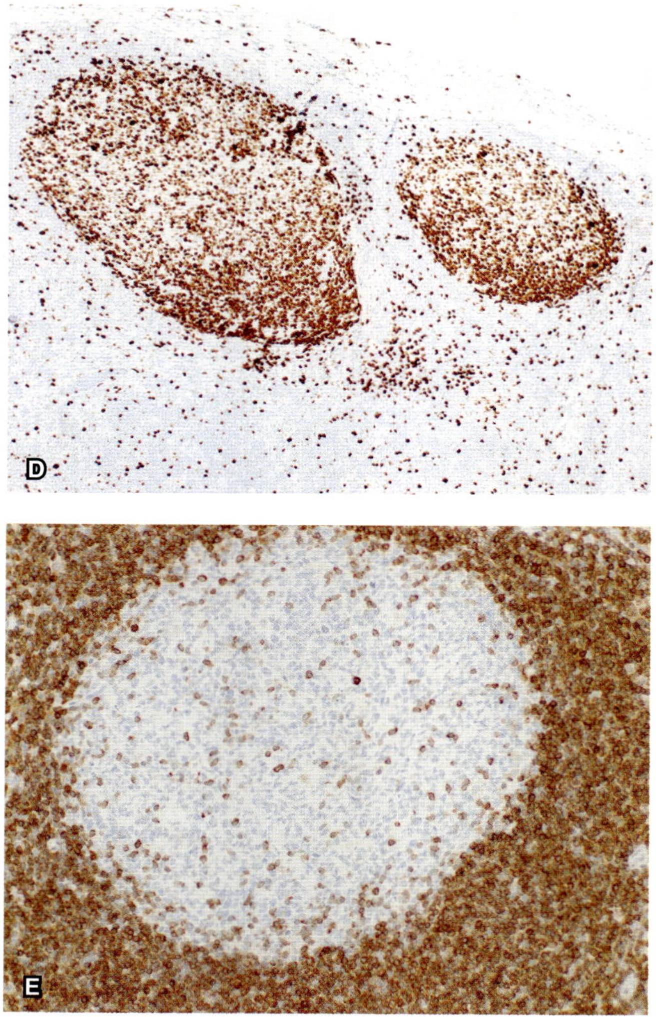

FIGURE 30.3 The follicle. **A:** H&E-stained germinal center surrounded by mantle zone B cells. The germinal center contains scattered tingible body macrophages (*arrows*) imparting the "starry-sky" appearance. Centroblasts and centrocytes can be seen, primarily populating the dark and light zones, respectively. **B:** Follicular dendritic cells are often binucleated (*arrows*). **C:** CD21 staining pattern of follicular dendritic cell meshworks seen in this reactive follicle. **D:** Polarization is evident using Ki67 immunohistochemistry which highlights the marked proliferation as part of the germinal center reaction. **E:** While T cells and mantle zone B cells express BCL2, the anti-apoptotic protein is notably downregulated in benign germinal center B cells.

is the marginal zone (12,13) because it loosely resembles the splenic marginal zone. The marginal zone is generally difficult to appreciate in the lymph node, except when it is expanded in benign or lymphomatous proliferations (14,15).

The cellular composition of both the lymphoid and nonlymphoid cells within the follicle is described below.

Follicular Dendritic Cells

Follicular dendritic cells are important for antigen presentation, trapping antigens on their surfaces, and presenting them to B cells for maturation (16–18). Follicular dendritic cells have a large, but inconspicuous nucleus with a very fine, almost vesicular chromatin and a small nucleolus. Not infrequently, they are or appear binucleated, with the nuclei pressed together (Fig. 30.3B). The cytoplasm is inconspicuous but can be better visualized with ultrathin sections or immunohistochemistry as featuring long and slender cytoplasmic protrusions. The cytoplasmic process of one cell is linked to the protrusions of other follicular dendritic cells via (hemi) desmosomes, thus forming a round network with a fingerprint-like configuration. CD21 and CD23 immunohistochemical stains highlight the follicular dendritic meshwork created by these cells (Fig. 30.3C).

Lymphoid Cells

The B cells of the primary follicle and the mantle zone have small, slightly irregular nuclei, with condensed, dark-staining chromatin. They have scanty cytoplasm and appear to be packed closely together. B cells are most commonly highlighted by immunohistochemistry using antibodies to CD20 and the transcription factor PAX5 but also express pan–B-cell markers CD19 and CD79a. The B cells express IgM, while those in the mantle zone cells express IgD simultaneously (19).

Distinctive lymphoid subsets can be found in the germinal center (20,21). Centroblasts contain large vesicular nuclei with several small peripherally located nucleoli and a small rim of basophilic cytoplasm. The areas of centroblastic proliferation constitute the "dark zone" of the secondary follicle. After antigenic stimulation and a successful germinal center reaction, the B cells gradually become smaller, lose their visible nucleoli, and are referred to as centrocytes. The centrocytes are small, mature-appearing lymphocytes with clear or faintly eosinophilic cytoplasm and make up the "light zone" of the secondary follicle. Other B-lineage cells include plasma cells and immunoblasts. Immunoblasts are large cells with big vesicular nuclei and a single centrally placed nucleolus. Immunoblasts can be highlighted with CD30. A small subset of follicular helper T cells can also be identified in the germinal center by immunohistochemistry. These cells show consistent expression for CD4, PD-1, and CXCL13, and are relevant for B-cell homing and maturation (22,23).

Tingible Body Macrophages

Tingible body macrophages are large phagocytic cells characterized by pale cytoplasm that contains phagocytized debris from lymphocytes that underwent apoptosis during maturation within the follicle. Their nuclei are large with finely dispersed chromatin. Because of their avid phagocytosis, their cytoplasm often is very clear, appearing as white spots in a sea of blue lymphocytes, a mottling feature described as the starry sky pattern. It should be noted that this highly characteristic feature is often used to distinguish between benign and malignant follicles.

Variations of the Follicular Pattern

The combination of features described above creates a characteristic morphologic pattern of the follicle: a round or ovoid area with a dark outer edge representing the mantle zone and a lighter center. The center shows mottling and upon closer inspection a very polymorphic cytologic picture with many mitotic figures. Sometimes zonation is seen, with the dark zone directed inward toward the hilus of the lymph node and the light zone directed more toward the capsule of the node (24).

Reactive conditions can cause the follicles to become very large and/or coalesce; progressive transformation of germinal centers is a good example of this feature (Fig. 30.4) (25–27). Progressive transformation of germinal centers represents infiltration of the follicle by mantle zone small B-cells and T lymphocytes, distorting follicular architecture. Progressively transformed germinal centers can be recognized by their remnant round shape highlighted by immunostains for follicular dendritic cell meshworks.

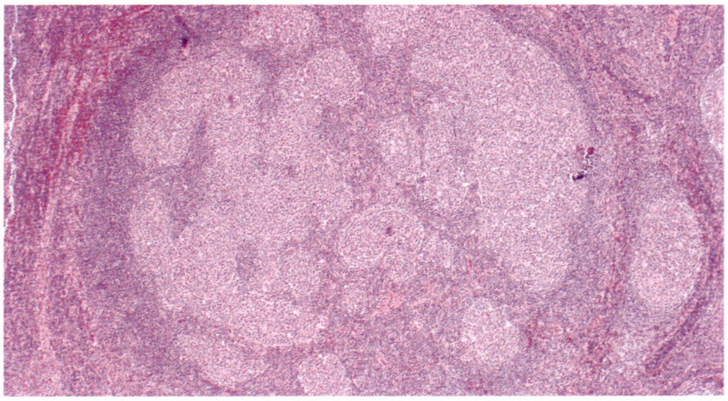

FIGURE 30.4 Progressive transformation of germinal centers (PTGC) is one of the many reactive patterns that could be identified. The follicles appear to coalesce with an associated infiltration of small, mature appearing lymphocytes, imparting a mottled appearance.

Function

The follicle is central to B-cell maturation through the germinal center reaction. Here, the naïve B lymphocytes are exposed to antigen and undergo a second round of controlled B-cell receptor rearrangement, referred to as somatic hypermutation. Survival is promoted for B cells that emerge with high-affinity receptors, or antibodies, while most B cells undergo apoptosis (28,29). Selection takes place in the reactive germinal centers and, thus, are negative for the antiapoptotic protein, BCL2 (Fig. 30.3E).

The germinal center reaction results in postgerminal center B cells that express high-avidity antibodies on their surfaces. These cells can recirculate as memory cells, spreading immune competence through the entire body and waiting for another encounter with the antigen. Other postgerminal B cells enter the medullary cords or bone marrow.

The Medullary Cords

The medullary cords are found in the hilar region of the lymph node between the sinuses (Fig. 30.5). They contain several cell populations as described below.

Lymphoid Cells

Small lymphocytes make up most of the cells in the medullary cords. They have small, round nuclei and a scant-to-moderate amount of cytoplasm. In some cells, the chromatin is clumped and peripherally distributed in the nucleus, similar to the plasma cell nucleus. Cells with this kind of chromatin tend to have more cytoplasm and sometimes even a perinuclear hof indicative of plasma cell differentiation. Thus, they are called plasmacytoid lymphocytes or lymphoplasmacytic cells. Many

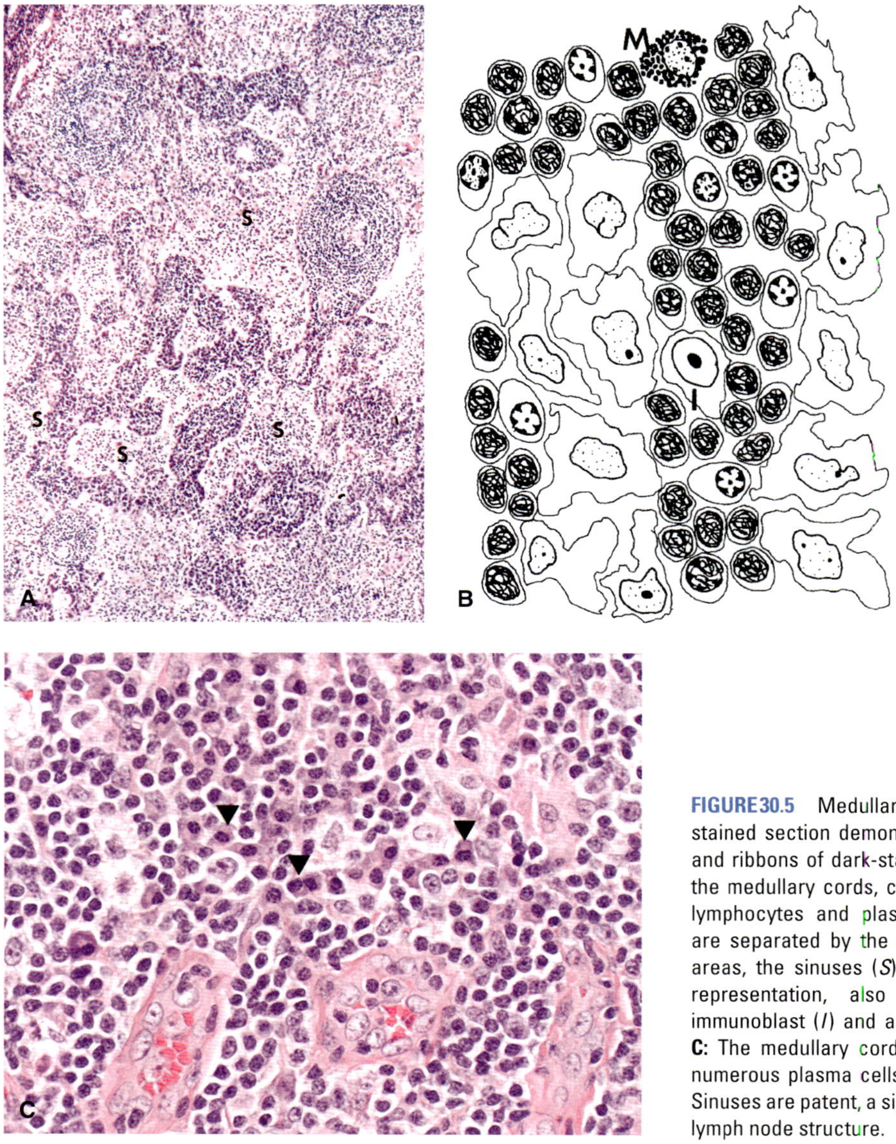

FIGURE 30.5 Medullary cords. **A:** H&E-stained section demonstrating sheets and ribbons of dark-staining cells are the medullary cords, containing small lymphocytes and plasma cells; they are separated by the lighter staining areas, the sinuses (*S*). **B:** Schematic representation, also depicting an immunoblast (*I*) and a mast cell (*M*). **C:** The medullary cords here include numerous plasma cells (*arrowheads*). Sinuses are patent, a sign of preserved lymph node structure.

of the cells with less cytoplasm and irregular nuclei are T lymphocytes and express CD2, CD3, CD5, and CD7. Their presence is necessary to modulate the process of antibody formation that takes place in the medullary cords. Plasma cells are also present, but in varying numbers (Fig. 30.5C). They are distinctive cells with small round nuclei that contain peripheral clumps of chromatin at the nuclear membrane, like the numbers on a dial or the face of a clock. The nucleus is eccentrically located in the cytoplasm and associated with a rim of cytoplasmic clearing (i.e., perinuclear hof). The outer rim of the cytoplasm is deeply basophilic. Plasma cells are CD20-negative but retain CD79a staining; they are positive for cytoplasmic immunoglobulins and CD138.

Macrophages

Macrophages are scarce in the medullary cords. They have medium-sized to large irregular nuclei and abundant cytoplasm. They are not as avidly phagocytic as tingible body macrophages from the follicle center, perhaps because they are geared more toward antigen handling and presentation than toward phagocytosis.

Other Cell Types

Mast cells can also be found in the medullary cords, but are rarely evaluated. Their granules may obscure the nucleus and they can be identified using CD117 or mast cell tryptase by immunohistochemistry.

The Paracortex

The paracortex or paracortical area contains high endothelial venules, IDCs, and T lymphocytes (Fig. 30.6).

High Endothelial Venules

These highly distinct vessels are found only in the paracortex. They are lined with plump, cuboidal, or even cylindrical endothelial cells with a large ovoid nucleus, vesicular chromatin, and indistinct nucleoli. Sometimes the lumina of these vessels appear to be obliterated by the endothelium. High endothelial venules have long been recognized as the port of entry for blood-borne lymphocytes to the lymph node parenchyma (30,31). Therefore they play a crucial role in recirculation, distribution, and homing of lymphocytes in different lymphoid organs, a process mediated by specific homing receptors on the lymphocyte surface that react with organ-specific ligands on the endothelial cell

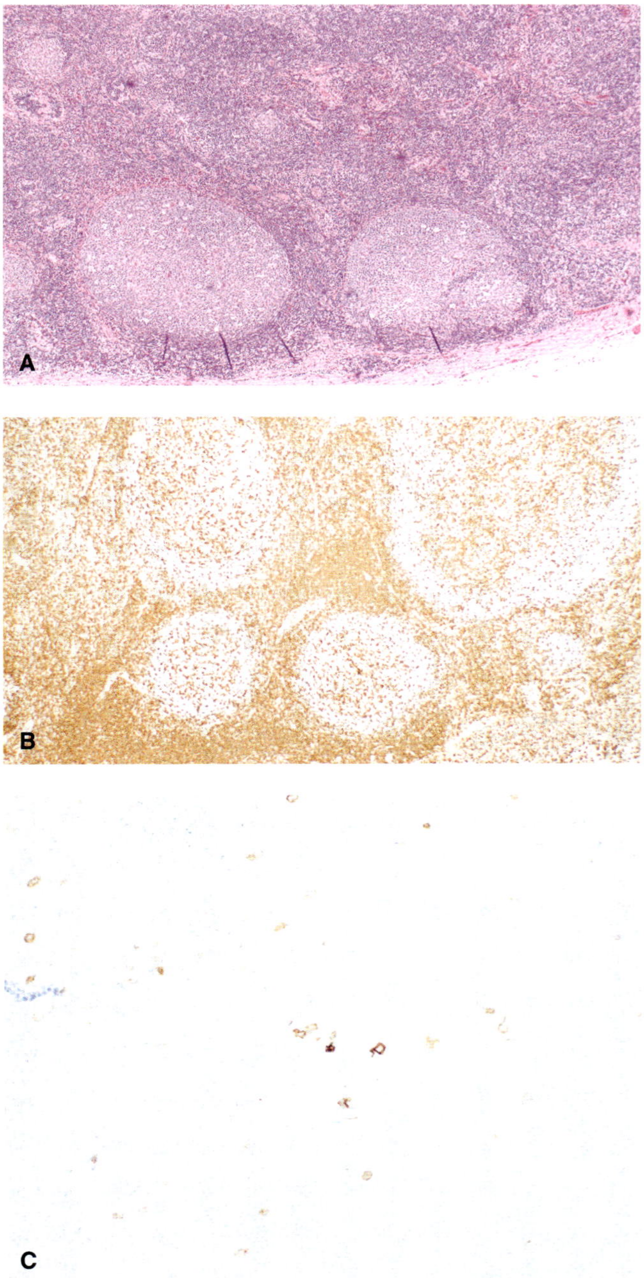

FIGURE 30.6 The paracortex. **A:** H&E-stained, low-power appearance of the paracortex (*outlined*), primarily composed of T cells highlighted by CD3 immunohistochemistry **(B)**. **C:** Scattered large immunoblasts can be identified by CD30 immunohistochemistry within the paracortex.

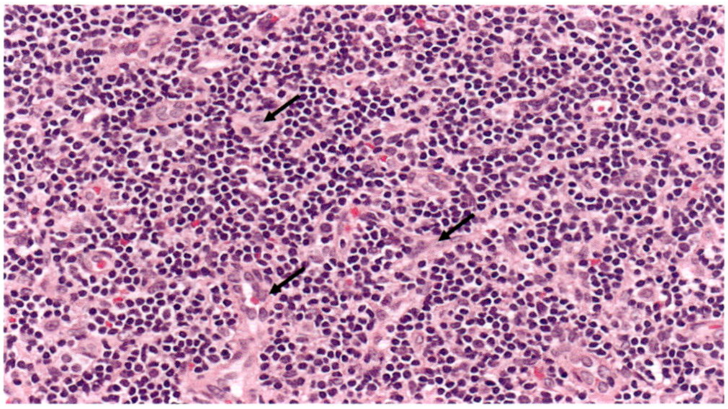

FIGURE 30.7 High endothelial venules (HEVs). H&E-stained section highlighting arborizing HEVs that are readily identified in the paracortex (*arrows*). The endothelial cells are round and plump and appear to obliterate the vascular space.

surface (Fig. 30.7) (32–35). Chemokines play an important role in this process of positioning of lymphoid cells and optimizing immune responses (36–37).

Interdigitating Dendritic Cells

IDCs are large cells with nuclei characterized by deep clefts and folds. The chromatin pattern is delicate, almost transparent, and nucleoli are inconspicuous. The cytoplasm is abundant, pale, and with ill-defined borders. When present in large numbers, IDCs impart a mottled appearance to the paracortex.

The IDC is a bone marrow–derived cell, intimately related to the Langerhans cell of the skin, which it closely resembles both morphologically and functionally (1,38–40). Both are antigen-presenting cells and are important in initiating and/or maintaining immune responses. These cells can be demonstrated with S100 protein immunohistochemistry.

Lymphoid Cells

The cell distribution within the paracortex is variable, but small T lymphocytes predominate in most cases. These T cells have small, irregular nuclei with coarse chromatin and little cytoplasm. They are demonstrated with immunostains for CD2, CD3, CD5, and CD7 and are either CD4- or CD8-positive, with the CD4-positive cells outnumbering the CD8-positive cells by approximately 2:1. Immunoblasts are present in varying numbers; they are large cells with vesicular nuclei of variable shape. Immunoblasts are typically CD30-positive by immunohistochemistry.

Other Cell Types

Finally, the fibroblastic reticulum cells are often found at the edge of the paracortex. The fibroblastic reticulum cells produce reticulin fibers that are involved in the transport of cytokines and/or antigens through the parenchyma, which is termed the fibroblastic reticulum cell conduit system This mechanism represents an effective means of spreading important activating molecules through the entire node (41–43).

The Sinuses

The sinuses are the structures carrying lymphatic fluid from the afferent lymphatics through the lymph node to the efferent lymph vessels. The afferent lymph vessels drain into the subcapsular sinus, a structure at least partly lined by endothelium. As the sinuses traverse the lymph node, they lose their endothelial lining and acquire a lining of macrophages (44–46). The macrophages in the sinuses are similar to macrophages elsewhere: they are large cells with a medium-sized to large nuclei with irregular contours and vesicular chromatin, a low nucleus-to-cytoplasm ratio, and signs of phagocytic activity. Small lymphocytes are also found in the sinuses (Fig. 30.8).

Other Cell Types

The cytokeratin-positive interstitial reticulum cell cytokeratin-positive interstitial reticulum cell is also present in lymph nodes. The cytokeratin-positive interstitial reticulum cell is an ill-defined cell type that is primarily recognized in immune stains for keratins. It is found around the subcapsular sinuses and has a coarsely dendritic morphology (47,48). Its nature is unclear, though they are probably a

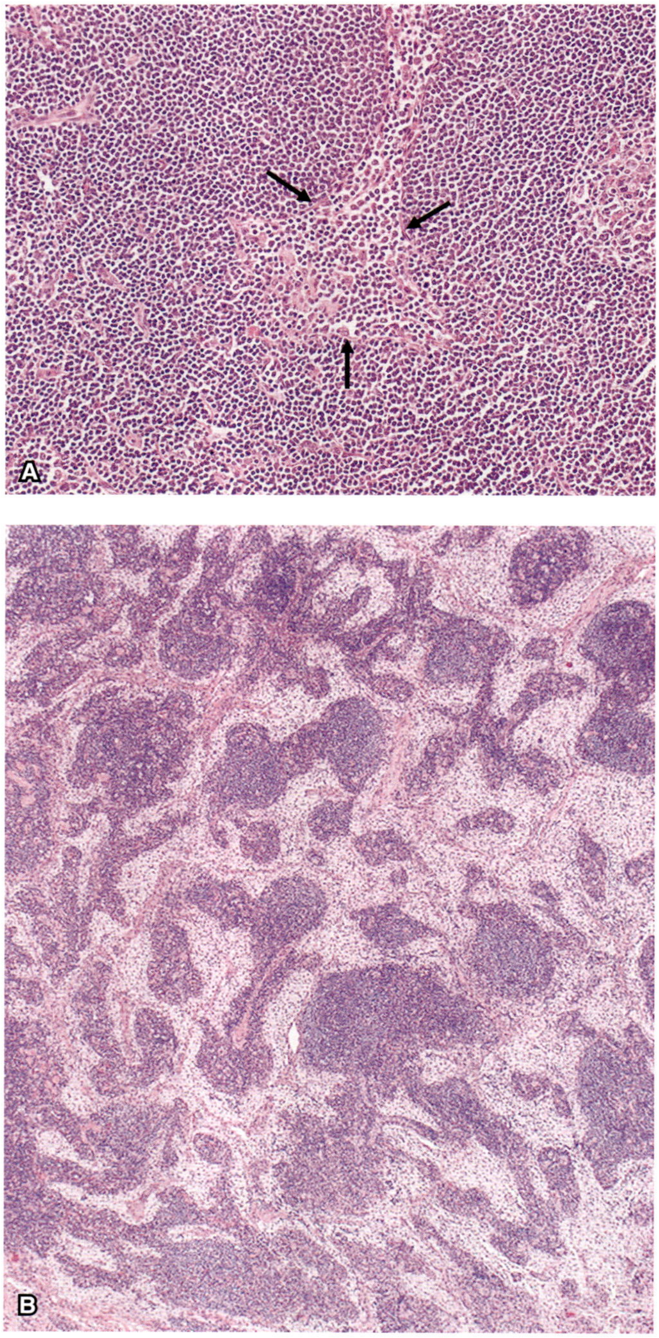

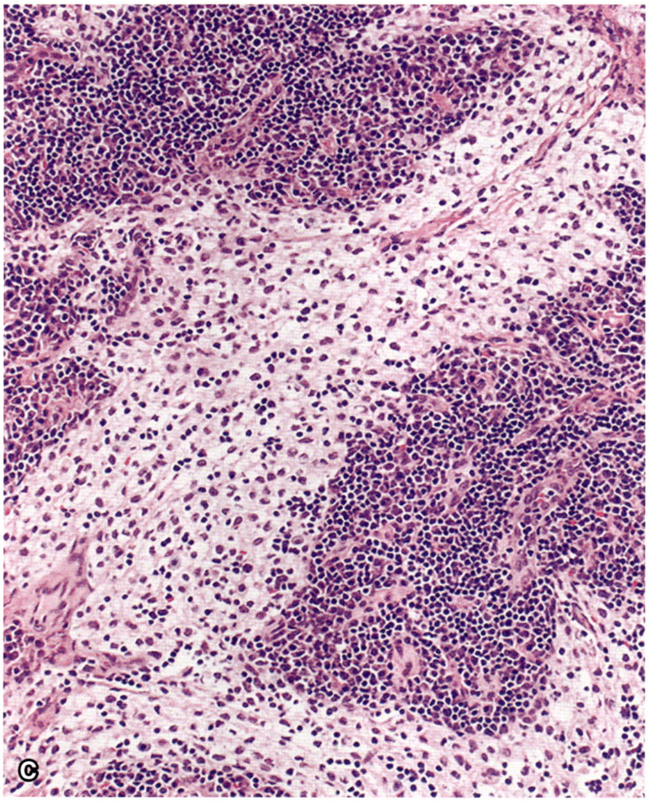

FIGURE 30.8 The sinus. **A:** High-power view of resting lymph node sinus with thin endothelial lining and numerous centrally located lymphocytes. Low-power **(B)** and high-power **(C)** florid sinus histiocytosis from a lymph node draining previous surgical site.

specialized form of fibroblastic reticulum cells characterized by the expression of low–molecular-weight keratins (1). It is important to be aware of these cells and not confuse them with metastatic tumor cells when evaluating lymph node sections. This is particularly important when evaluating sentinel node lymph nodes (49). Fortunately, the appearance of the cytokeratin-positive interstitial reticulum cell is very different from that of most carcinoma cells so they rarely pose diagnostic challenges in routinely stained sections.

HISTOLOGIC APPROACH TO EVALUATING LYMPH NODE PATHOLOGY

There are numerous causes of reactive, nonneoplastic lymphadenopathy. Unfortunately, the specific etiology is not histologically identified in most cases. Malignant proliferations are better characterized, as therapy and prognosis are more intimately tied to specific disease classification. Nonetheless, stepwise evaluation of the various nodal compartments is central to thorough histologic assessment. Table 30.1 briefly summarizes the distinctive features of benign/reactive conditions and their malignant counterparts.

Follicular Changes

Follicular hyperplasia is the most common cause of lymphadenopathy, and arguably the most common pathologic finding in a lymph node biopsy. Benign germinal centers usually reflect an exuberant reaction to an antigen. They typically show cellular pleomorphism of the follicle center with zonation of follicles and clearly distinguishable light and dark zones that can be highlighted by Ki67 staining (Fig. 30.3D). The follicles contain scattered tingible body macrophages with increased numbers of mitotic figures and a well-defined mantle zone. Other clues to a benign diagnosis include the presence of low number of follicles per surface area and a predominant cortical localization accompanied by a well-developed network of follicular dendritic cells in the follicle center and zonation of the follicles with clearly distinguishable dark and light zones.

TABLE 30.1 Benign Compartmental Enlargement and Their Malignant Counterparts

Compartment	Benign	Malignant
Follicle	Follicular hyperplasia	Follicular lymphomas
Paracortex	Paracortical hyperplasia Dermatopathic lymphadenopathy	T-cell lymphoma Mycosis fungoides
Medullary cords	Medullary hyperplasia Reactive plasmacytosis	Lymphoplasmacytic lymphoma Plasmacytoma
Sinuses	Sinus histiocytosis	Rosai–Dorfman disease

Changes in the Paracortex

Paracortical hyperplasia can take three forms, each with its own differential diagnostic considerations. The first is expansion of the paracortex by small lymphocytes, usually with increased high endothelial venules. Distinctive features suggestive of malignancy include cellular monotony and destruction of lymph node architecture by the proliferation. Features concerning malignancy include abnormal cytology or change in expected immunophenotypic pattern with marked predominance of CD4 or CD8, or loss of CD7 or CD5. Expansion of immunoblasts can be seen in viral infections, vaccinations, and in some drug reactions (25). Necrosis can occur in Kikuchi histiocytic necrotizing lymphadenitis as well as in other situations (50). On occasion, the changes can exhibit a nodular pattern (51). In such cases, a preserved architecture should raise the possibility of a benign condition, no matter the histologic picture. In short, a combination of clinical history, morphology, immunophenotyping, cytogenetic studies, and demonstration of clonality may be helpful in reaching a final diagnosis. Dermatopathic lymphadenopathy is seen in lymph nodes draining skin with pruritic skin disorders. In this situation, the paracortex is expanded by a marked increase in IDCs/Langerhans cells (52).

Sinusoidal Changes

Sinusoidal changes are very common in lymph node specimens. Lymph nodes draining tumors or inflammatory areas often show sinus histiocytosis. Typical conditions showing a sinusoidal pattern include sinus histiocytosis with massive lymphadenopathy (53) and Langerhans cell histiocytosis (54). The former condition is a peculiar clinicopathologic entity in which the histiocytes seem to engulf large numbers of lymphocytes in their cytoplasm without destroying them, a phenomenon called emperipolesis. The latter can also cause sinus histiocytosis with Langerhans cells. Langerhans cells can be recognized by the presence of their deeply grooved nuclei. An admixture of eosinophils can be a helpful clue. Immunopositivity for CD1a and S100 protein among lesional cells can facilitate a diagnosis.

A particular sinusoidal proliferation of small vessels can be seen on occasion, often as a reaction of the lymph node to ischemia or irradiation. This pattern is called vascular transformation of the sinuses and can feature a delicate vascular pattern or a more spindle-cell proliferation resembling Kaposi sarcoma (55,56).

Combined Patterns

Follicular, medullary, paracortical, and sinusoidal changes often occur simultaneously, and any combination is possible. Combined patterns are extremely rare in lymphomas, so any combination argues for a benign condition. For example, toxoplasma and Epstein–Barr virus infections often cause alterations in multiple compartments.

ARTIFACTS

Numerous extrinsic and intrinsic factors can influence lymph node histology. Though they are not specific patterns, they can cause considerable difficulties in evaluating histology and are therefore mentioned here.

Technical Artifacts

Lymph node tissue is vulnerable and easily damaged in processing. Undue pressure on a specimen during dissection can cause considerable crush artifact, to the point of obliterating morphology completely. Specimens with extensive crushing artifacts should not be evaluated.

Fixation artifacts are also common and typically occur with large specimens or insufficient fixation. If the fixation time in formalin is too short, only the outer edge of the specimen is fixed with pronounced variation evidence between the two areas (Fig. 30.9).

Intrinsic Artifacts

Though the above-described architecture is found in all nodes throughout the body, in some areas typical features can be found, mostly the result of repeated inflammation or reactions. This pattern is most often seen in inguinal nodes with fibrosis that can distort the normal architecture. Similarly, in retroperitoneal lymph nodes, hyalinization can be found.

SPECIAL TECHNIQUES AND PROCEDURES

Histologic assessment of the lymph node is central to the diagnostic process and can guide further helpful ancillary testing. While some studies require fresh tissue requested before fixation (such as flow cytometry and karyotyping), immunohistochemistry, fluorescence in situ hybridization and molecular studies have become critical to the work-up lymph node specimens and can be performed on formalin-fixed and paraffin-embedded specimens. Antibodies listed above can be interpreted only with the knowledge of expected staining patterns highlighted or discussed in this chapter. Control tissue, often tonsil resections, can be used to evaluate expected patterns of expression.

Molecular testing, such as B- and T-cell receptor gene rearrangement studies, have become commonplace in pathology laboratories, but their specificity largely relies on a morphologic correlate. This is true particularly for T cells, as T lymphocytes do not undergo somatic hypermutation, and clones can be identified in reactive, nonneoplastic, or autoimmune proliferations. In other words, a molecular finding is not always specific without a histologic pattern.

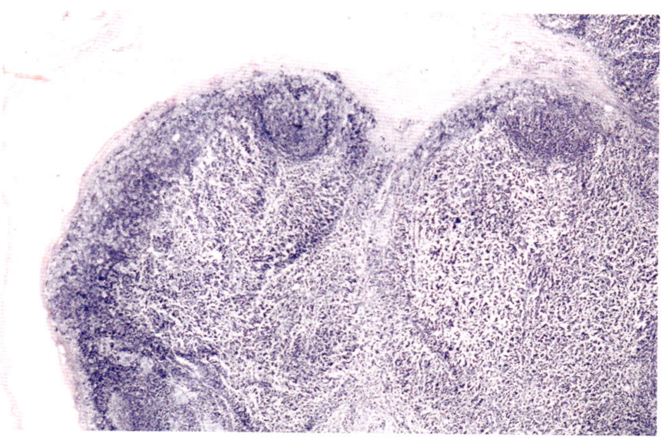

FIGURE 30.9 Fixation artifact. Edge of the specimen is properly fixed, the center, fixed in alcohol during the dehydration procedure, shows loss of tissue structure.

REFERENCES

1. Willard-Mack CL. Normal structure, function, and histology of lymph nodes. *Toxicol Pathol*. 2006;34(5):409–424
2. Cesta MF. Normal structure, function, and histology of mucosa-associated lymphoid tissue. *Toxicol Pathol*. 2006;34(5):599–608.
3. Jain S, Bemark M, Spencer J. Human gut-associated lymphoid tissue: a dynamic hub propagating modulators of inflammation. *Clin Transl Med*. 2023;13(9):e1417.
4. Vossenkämper A, Blair PA, Safinia N, et al. A role for gut-associated lymphoid tissue in shaping the human B cell repertoire. *J Exp Med*. 2013;210(9):1665–1674.
5. Azzali G. Structure, lymphatic vascularization and lymphocyte migration in mucosa-associated lymphoid tissue. *Immunol Rev*. 2003;195:178–189.
6. Dupont G, Schmidt C, Yilmaz E, et al. Our current understanding of the lymphatics of the brain and spinal cord. *Clin Anat*. 2019;32(1):117–121.
7. Hoorweg K, Cupedo T. Development of human lymph nodes and Peyer's patches. *Semin Immunol*. 2008;20:164–170.
8. Drayton D, Liao S, Mounzer R, Ruddle NH. Lymphoid organ development: from ontogeny to neogenesis. *Nat Immunol*. 2006; 7:344–353.
9. Schaerli P, Willimann K, Lang AB, Lipp M, Loetscher P, Moser B. CXC chemokine receptor 5 expression defines follicular homing T-cells with B cell helper function. *J Exp Med*. 2000; 192:1553–1562.
10. Stein JV, Nombela-Arrieta C. Chemokine control of lymphocyte trafficking: a general overview. *Immunol*. 2005;116:1–12.
11. Campbell DJ, Koch MA. Phenotypical and functional specialization of FoxP3+ regulatory T cells. *Nat Rev Immunol*. 2011; 11:119–130.
12. van den Oord JJ, de Wolf-Peeters C, Desmet VJ. The marginal zone of the human reactive lymph node. *Am J Clin Pathol*. 1986; 86:475–479.
13. Palm AKE, Kleinau S. Marginal zone B cells: from housekeeping function to autoimmunity? *J Autoimmun*. 2021;119:102627.
14. Jaffe ES, Arber DA, Campo E, eds. *Hematopathology*. 2nd ed. Elsevier; 2016.

15. Hartmann S, Kriener S, Hansmann, M. The diagnostic spectrum of reactive lymph node changes. *Pathologist*. 2008;29:253–263.
16. Rezk SA, Nathwani BN, Zhao X, Weiss LM. Follicular dendritic cells: origin, function, and different disease-associated patterns. *Hum Pathol*. 2013;44(6):937–950.
17. Heesters B, Myers R, Carroll M. Follicular dendritic cells: dynamic antigen libraries. *Nat Rev Immunol*. 2014;14:495–504.
18. Aguzzi A, Krautler NJ. Characterizing follicular dendritic cells: progress report. *Eur J Immunol*. 2010;40:2134–2138.
19. Sanz I, Wei C, Jenks SA, et al. Challenges and opportunities for consistent classification of human B cell and plasma cell populations. *Front Immunol*. 2019;10:2458.
20. Victora GD, Michel C. Germinal centers. *Ann Rev Immunol*. 2022;40(1):413–442.
21. Stebegg M, Kumar SD, Silva-Cayetano A, Fonseca VR, Linterman MA, Graca L. Regulation of the germinal center response. *Front Immunol*. 2018;9:2469.
22. Ramiscal RR, Vinuesa CG. T-cell subsets in the germinal center. *Immunol Rev*. 2013;252(1):146–155.
23. Crotty S. Follicular helper CD4 T-cells (TFH). *Annu Rev Immunol*. 2011;29:621–663.
24. Schwickert TA, Lindquist RL, Shakhar G, et al. In vivo imaging of germinal centres reveals a dynamic open structure. *Nature*. 2007;446:83–87.
25. Medeiros JL, Ioachim HL. *Ioachim's Lymph Node Pathology*. 5th ed. Wolters Kluwer; 2022.
26. Chang C-C, Osipov V, Wheaton S, Tripp S, Perkins SL. Follicular hyperplasia, follicular lysis, and progressive transformation of germinal centers. A sequential spectrum of morphologic evolution in lymphoid hyperplasia. *Am J Clin Pathol*. 2003;120:322–326.
27. Hicks J, Flaitz C. Progressive transformation of germinal centers: review of histopathologic and clinical features. *Int J Pediatr Otorhinolaryngol*. 2002;65(3):195–202.
28. Huang C. Germinal center reaction. *Adv Exp Med Biol*. 2020;1254:47–53.
29. Betzler AC, Ushmorov A, Brunner C. The transcriptional program during germinal center reaction – a close view at GC B cells, Tfh cells and Tfr cells. *Front Immunol*. 2023;14:1125503.
30. Worbs T, Föster R. T-cell migration dynamics within lymph nodes during steady state: an overview of extracellular in intracellular factors influencing the basal intranodal T-cell motility. *Curr Top Microbiol Immunol*. 2009;334:71–105.
31. Denucci CC, Mitchell JS, Shimizu Y. Integrin function in T-cell homing to lymphoid and nonlymphoid sites: getting there and staying there. *Crit Rev Immunol*. 2009;29:87–109.
32. Blanchard L, Girard JP. High endothelial venules (HEVs) in immunity, inflammation and cancer. *Angiogenesis*. 2021;24(4):719–753.
33. Arata-Kawai H, Singer MS, Bistrup A, et al. Functional contributions of N- and O-glycans to L-selectin ligands in murine and human lymphoid organs. *Am J Pathol*. 2011;178:423–433.
34. Hussain B, Kasinath V, Ashton-Rickardt GP, et al. High endothelial venules as potential gateways for therapeutics. *Trends Immunol*. 2022;43(9):728–740.
35. Wiedle G, Dunon D, Imhof BA. Current concepts in lymphocyte homing and recirculation. *Crit Rev Clin Lab Sci*. 2001;38:1–31.
36. Schaerli P, Moser B. Chemokines: control of primary and memory T-cell traffic. *Immunol Res*. 2005;31:57–74.
37. Stein JV, Nombela-Arrieta C. Chemokine control of lymphocyte trafficking: a general overview. *Immunology*. 2005;116(1):1–12.
38. Thorbecke GJ, Silberberg-Sinakin I, Flotte TH. Langerhans cells as macrophages in skin and lymphoid organs. *J Invest Dermatol*. 1980;75:32–43.
39. Takahashi K, Kenji A, Norihiro T, et al. Morphological interactions of interdigitating dendritic cells with B and T cells in human mesenteric lymph nodes. *Am J Pathol*. 2001;159(1):131–138.
40. Bousso P. T-cell activation by dendritic cells in the lymph node: lessons from the movies. *Nat Rev Immunol*. 2008;8:675–684.
41. Shklovskaya E, Roediger B, Fazekas de St Groth B. Epidermal and dermal dendritic cells display differential activation and migratory behaviour while sharing the ability to stimulate CD4+ T cell proliferation in vivo. *J Immunol*. 2008;181:418–430.
42. Roozendaal R, Mebius RE, Kraal G. The conduit system of the lymph node. *Int Immunol*. 2008;20:1483–1487.
43. Fletcher A, Acton S, Knoblich K. Lymph node fibroblastic reticular cells in health and disease. *Nat Rev Immunol*. 2015;15:350–361.
44. Forkert PG, Thliveris JA, Bertalanfy FD. Structure of sinuses in the human lymph node. *Cell Tissue Res*. 1997;183:115–130.
45. Birmingham KG, O'Melia MJ, Bordy S, et al. Lymph node subcapsular sinus microenvironment-on-a-chip modeling shear flow relevant to lymphatic metastasis and immune cell homing. *iScience*. 2020;23(11):101751.
46. Louie DAP, Liao S. Lymph node subcapsular sinus macrophages as the frontline of lymphatic immune defense. *Front Immunol*. 2019;10:347.
47. Wacker HH, Frahm SO, Heidebrecht HJ, Parwaresch R. Sinus-lining cells of the lymph nodes recognized as a dendritic cell type by the new monoclonal antibody Ki-M9. *Am J Pathol*. 1997;151:423–434.
48. Gould VE, Bloom KJ, Franke WW, Warren WH, Moll R. Increased numbers of cytokeratin-positive interstitial reticulum cells (CIRC) in reactive, inflammatory and neoplastic lymphadenopathies: hyperplasia or induced expression?. *Vichows Archiv A Pathol Anat*. 1995;425:617–629.
49. Treseler P. Pathologic examination of the sentinel lymph node: what is the best method? *Breast J*. 2006;12(5 Suppl 2):S143–S151.
50. Kuo T. Kikuchi's disease (histiocytic necrotizing lymphadenitis): a clinicopathologic study of 79 cases with an analysis of histologic subtypes, immunohistology, and DNA ploidy. *Am J Surg Pathol*. 1995;20:798–809.
51. Weiss L, O'Malley D. Benign lymphadenopathies. *Mod Pathol*. 2013;26(Suppl 1):S88–S96.
52. Garces S, Yin CC, Miranda RN, et al. Clinical, histopathologic, and immunoarchitectural features of dermatopathic lymphadenopathy: an update. *Mod Pathol*. 2020;33(6):1104–1121.
53. Ravindran A, Rech KL. How I diagnose Rosai-Dorfman disease. *Am J Clin Pathol*. 2023;160(1):1–10.
54. Callihan TR. Langerhans' cell histiocytosis (histiocytosis X). In: Jaffe ES, ed. *Surgical Pathology of the Lymph Nodes and Related Organs*. WB Saunders; 1995:537–559.
55. Samet A, Gilbey P, Talmon Y, Cohen H. Vascular transformation of the lymph node sinuses. *J Laryngol Otol*. 2001;115:760–762.
56. Chan JK, Warnke RA, Dorfman R. Vascular transformation of sinuses in lymph nodes. A study of its morphological spectrum and distinction from Kaposi's sarcoma. *Am J Surg Pathol*. 1991;15(8):732–743.

31

Spleen

J. Han J.M. van Krieken ■ Attilio Orazi

INTRODUCTION

A variety of ideas regarding the physiology and anatomy of the spleen have been developed since antiquity (1). As producer of black bile, the spleen was seen as the origin of melancholy. Galen (131–201 AD) called the human spleen an enigmatic organ, a notion that has persisted for a long time. Malpighi described the splenic lymphoid tissue as white pulp against a background of red pulp in the 17th century. In 1857, Billroth published one of the first histology studies of the human spleen in which he divided the red pulp into cord tissue and venous sinuses. Up until the second half of the 20th century, the spleen was considered a rather useless reservoir for blood cells and was hardly studied. Weiss was able to elucidate the ultrastructure of the organ using electron microscopy, thereby providing some insights into the red pulp function (2). Knowledge of the organization of the white pulp is a work in progress as of today (3–6).

Nevertheless, many pathologists still lack a clear understanding of the normal histology and functions of the human spleen. This is due to several reasons. The organ is extremely vulnerable to autolysis, which often makes histologic findings in postmortem specimens difficult to interpret and of limited teaching value. Surgically removed spleens are suitable if processed without delay. However, since the number of splenectomies performed in most institutions is relatively scarce, it is not surprising that pathologists may feel uncomfortable when interpreting splenic pathology. For this reason, the International Spleen Consortium was initiated to improve education and collaboration with respect to the spleen and its diseases (7).

A substantial source of confusion with respect to the structure and function of the human spleen lies in the terminology and definitions applied to this organ because many are based on studies of animal spleens. The human and animal spleens do not have an identical architecture and, thus, the value of comparative studies is limited.

TABLE 31.1 Summary of Splenic Histology, Function, and Relationship to Lymph Node Compartments

Spleen Compartment	Description	Function/Composition	Equivalent in Lymph Node
White Pulp			
T-cell area	Irregular area of small lymphocytes containing lymph vessels bordering arteries	Predominant CD4 lymphocytes	Paracortex
B-cell follicle	Round area of small lymphocytes surrounded by medium-sized lymphocytes (a germinal center may be present)	Production of Ig-producing cells and probably memory cells	Follicle
Perifollicular zone	Area between white and red pulp containing many erythrocytes and lacking a normal sinusoidal structure	Place of retarded blood flow with interaction of blood cells, cells, antigens, and antibodies	Medulla (?)
Red Pulp			
Sinuses/cord tissue with sheathed capillaries	Tissue containing a meshwork of sinuses (with interrupted basement membrane) and capillaries, partly sheathed	Removal of particles from blood cells. Possible place of interaction of new antigens with reticulum cells	Sinus: partly high endothelial venule Sheathed capillary: medullary sinus
Nonfiltering area	Area of red pulp tissue lacking capillaries and containing lymphocytes	Probably place of onset of immune reaction	Medulla or compartment of primary follicles
Perivascular rim	Small area along the vessel tree containing lymphocytes and plasma cells	Probably connected to lymphatics	Medulla (?)

For example, the marginal sinus as described in rodent spleens is not present in the human spleen. Furthermore, some definitions such as that of the marginal zone vary widely from author to author (6,7–11).

Conceptual difficulties also arise because the spectrum of findings in the normal spleen is widely variable. The spleen is a compartmentalized organ with each compartment performing one, or a few, functions (Table 31.1). Stimulating one of those functions can lead to morphologic changes affecting multiple elements in the compartment, or compartments, responsible for that function. It is essential to define a normal control population if one undertakes histologic studies in the spleen in specific disorders (12). For example, spleens removed incidentally during abdominal surgery performed for unrelated lesions are morphologically different from traumatically ruptured spleens. It can be difficult to distinguish physiologic from pathologic changes.

PRENATAL AND DEVELOPMENTAL CHANGES

The spleen can be recognized from about the 5th week of gestation, and blood vessels appear in it by the 9th week. Red and white pulp cannot be distinguished until the 9th month. The functional role of the spleen during prenatal development is different from that of the adult spleen, and this is reflected in the microscopic anatomy of the organ. Hematopoiesis was once believed to take place in the fetal spleen and contribute to blood cell formation in the fetus until the 6th month of gestation. It is now known, in fact, that the spleen is not a stem cell niche for hematopoiesis but functions as a site of maturation for hematopoietic precursors derived from the bone marrow (13). One may see foci of hematopoietic cells (extramedullary hematopoiesis) in the adult spleen in many reactive conditions as well as in disorders of the bone marrow associated with myelofibrosis. Extramedullary hematopoiesis in the spleen is also referred to as myeloid metaplasia.

The immune system develops during fetal growth, and this development continues after birth (14). This functional maturation is reflected in morphology: the splenic white pulp does not contain follicles and marginal zones until birth. Initially, there are immature B cells in clusters and T cells scattered throughout the organ. Their numbers increase with the developmental age of the fetus such that areas rich or B and T cell can be recognized from the end of the second semester onward (15,16). Phagocytosis can be demonstrated at the 12th week of gestation (12).

Developmental anomalies of the spleen are common (17). The presence of accessory spleens can be found in at least 25% of autopsies. These splenules are composed of splenic tissue with completely normal red and white pulp. Residual spleniculi may lead to recurrence of disease when hematopoietic disorders are treated with splenectomy.

Polysplenia is associated with immotile cilia syndrome (18). In this syndrome, left–right orientation of thoracic and abdominal organs may be abnormal and the right-sided spleen is often divided into many small pieces. This is not to be confused with acquired splenosis, in which many small fragments of spleen are present after trauma. Congenital asplenia is exceedingly rare and associated with abnormalities of the cardiovascular system. Splenogonadal fusion is a rare developmental anomaly; approximately 120 cases have been reported since the first description of this entity in 1883 by Bostroem (Fig. 31.1). Presumably, the close

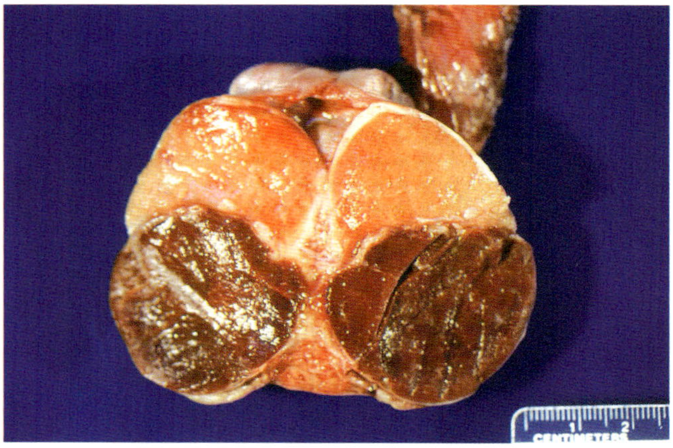

FIGURE 31.1 Gross appearance of splenogonadal fusion. Note ectopic splenic tissue within the testis.

proximity of the spleen and gonad during early embryologic development allows fusion of these seemingly unrelated organs. A testicular or scrotal mass is its usual clinical presentation (19).

APOPTOSIS

Apoptosis does not seem to play an important role in splenic development, but the lymphoid compartment shows extensive apoptosis, especially in the germinal centers of the B-cell follicles (Fig. 31.2). Extensive apoptosis results in an accumulation of macrophages that phagocytose remnants of lymphocytes, imparting a starry sky appearance to the parenchyma. Apoptosis generally occurs due to unsuccessful gene rearrangements of the antigen receptor in T cells or because immunoglobulin recognizes autoantigen. This physiologic process is important in the protection against autoimmune diseases. The Bcl-2 protein that protects against some forms of apoptosis is expressed in most B and T cells, but is lacking in germinal center B cells, thereby rendering them susceptible to apoptosis. The t(14;18) translocation of follicular lymphoma leads to aberrant expression of Bcl-2 in the tumor cells. Recognizing follicular lymphoma in the spleen can be problematic for the following reasons. Since the spleens of patients over about 20 years of age only rarely contain active germinal centers, the distinction between a primary follicle and a nodule of follicular lymphoma can be difficult. The absence of a Bcl-2 negative germinal center is not diagnostic of follicular lymphoma: the presence of the t(14;18) needs to be proven. Of note, involvement of the spleen by follicular lymphoma is often nodular but does not lead to the disturbance of the splenic architecture (20).

Apoptosis also plays an important role in maintaining normal T-cell populations. Patients with autoimmune lymphoproliferative syndrome have a decreased rate of apoptosis in T lymphocytes due to a genetic defect of FAS or Fas ligand. As a result, patients have splenomegaly due to the marked degree of lymphoid hyperplasia seen in the T-cell–rich areas of the spleen, as well as autoimmunity (21).

GROSS FEATURES/ORGAN WEIGHT

The human spleen is a bean-shaped organ surrounded by a smooth capsule covered by the peritoneum. In contrast to several species, the capsule does not contain smooth muscle fibers and, therefore, does not have the capability

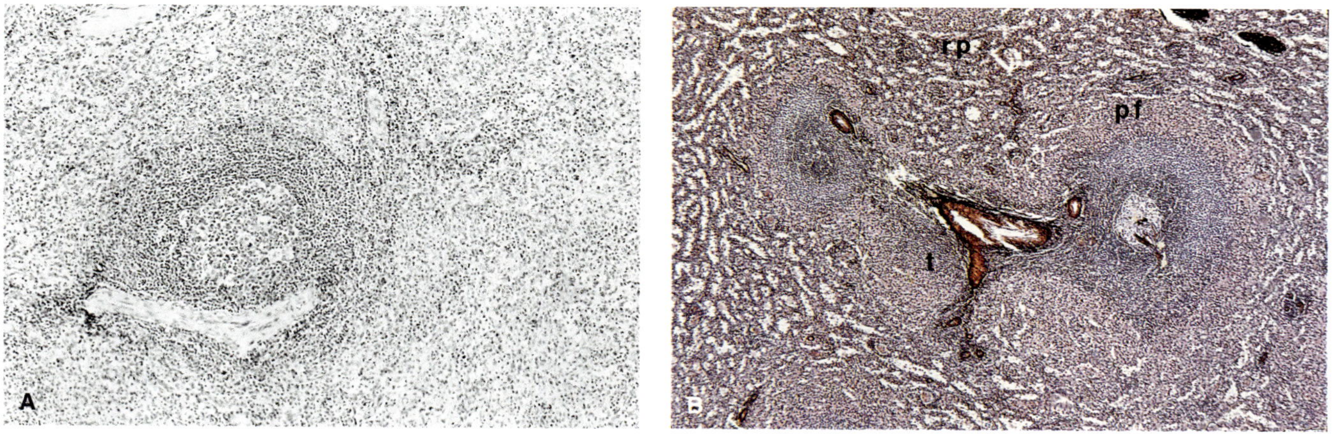

FIGURE 31.2 Spleen removed in idiopathic thrombocytopenic purpura. **A:** Formalin-fixed paraffin embedding (H&E, original magnification ×40). **B:** Methylmethacrylate embedding (methenamine-silver/H&E, original magnification ×40). Overview of red and white pulp showing central arteriole with T-cell area, a primary follicle, and a secondary follicle containing a germinal center. Note the absence of the marginal zone around the T-cell area and the presence of the erythrocyte-rich (*pink*) perifollicular zone surrounding both the T- and B-cell compartments of the white pulp. Note the lack of detail on the structure of the red pulp and the difficult discernible perifollicular zone in standard H&E section (rp, red pulp; pf, perifollicular zone; t, T-cell area).

of undergoing contraction in response to acute blood loss. It has been shown, however, that evolution in *Homo sapiens* is ongoing. Divers from Bajau, the so-called sea nomads, have acquired a mutation in *PDE10A* resulting in extraordinary breath-holding ability, thanks to splenic contraction (22).

The surface of the spleen may be covered with fibrotic or even calcified plaques, the cause of which is unknown. It is not uncommon to find several grooves at the outer surface that have no clinical significance. The weight of the spleen is highly variable (23). In adults, the spleen generally weighs 150 to 250 g; but, it is often substantially smaller in the elderly, even when there is no apparent hypofunction. Recent data suggest that the splenic volume decreases with an increase in cytokine levels (24).

On the cut surface, the red and white pulp can be discerned, the latter consisting of small (less than or equal to 2 mm) nodules. It is important to realize that involvement of the spleen in malignant lymphoma is first observed in the white pulp, which becomes enlarged in many cases.

ANATOMY

Blood Supply

Blood reaches the spleen via the splenic artery, a large branch of the celiac artery, and enters the spleen through four to six branches; their number and location are highly variable. Venous outflow occurs via four to six venous branches. These combine within the lienorenal ligament to form the splenic vein, which drains into the portal vein. For this reason, portal hypertension can produce congestive splenomegaly. The blood flow within the spleen is highly specialized and relates to the different functions of the spleen.

Nerves

The spleen is innervated by nonmyelinated fibers from the major splanchnic nerves and the celiac plexus. These nerve fibers run along the splenic artery. Innervation in human spleens is less extensive than that in cat and dog spleens, and this might be related to the more important reservoir function of the spleen in these animals. An interesting role in the immune response has also been suggested (25).

Lymphatic Vessels

No afferent lymphatic vessels are present in the spleen, so the spleen is rather a secondary lymphoid organ of the circulation (6). Its lymph drainage occurs *via* hilar lymph nodes and lymph nodes in the gastrosplenic ligament. The lymph then flows through lymphatic vessels along the splenic artery to the celiac lymph nodes along the celiac artery. The lymphatic vessels in the spleen are described below.

LIGHT MICROSCOPY

Vascular Tree

After entering the hilus, the splenic artery branches like a tree. These arterial branches, called trabecular arteries, are accompanied by veins and lymph vessels and are surrounded by collagenous fibers within the parenchyma. These vessels containing fibrous structures are usually referred to as trabeculae or septa, a term which is inappropriate to describe perivascular collagen cuffs. True septa are also present in the spleen. They are connected to the capsule, lack vessels, and only extend for a short distance into the splenic tissue. Foci of condensed reticular fibers devoid of vessels are found throughout the red pulp. The condensed reticulum is in direct continuity with the reticular meshwork of the surrounding red pulp and may represent areas of collapse or involution of the red pulp.

Trabecular arteries branch to form central arteries and arterioles that are no longer accompanied by veins or an associated collagenous cuff. Rather, they are surrounded by lymphatic tissue predominantly composed of T lymphocytes. This lymphatic compartment is usually referred to as periarterial or periarteriolar lymphoid sheath. It is present around the vessels and becomes smaller toward the capillary endings. The arterioles are usually described as branching into penicillar arterioles, which run in parallel. This phenomenon is restricted to involuted specimens in humans; the disappearance of tissue between arterioles leaves them lying close to each other.

Branching of arterioles and capillaries often occurs at right angles. Reconstructions based on serial sections have shown that the terminal end of the capillary forms a peculiar and specifically splenic structure (8) (Figs. 31.3 and 31.4). These structures are known by several names, determined partly by the species in which they have been studied. Various terms include sheathed capillaries, Hülsekapillaren, ellipsoids, or periarteriolar macrophage sheaths. In humans, they are present in the red pulp and the perifollicular zone and are generally referred to as sheathed capillaries. The sheathed capillary is surrounded by a "sheath" of mononuclear phagocytes and rare reticulum cells. Since autolysis is so rapid, visualization of the sheathed capillaries is dependent on adequate tissue processing. The endothelial lining of the capillary ends abruptly in a string of concentrically arranged macrophages. Blood cells coming from an arteriole have to pass through the sheathed capillary on their way to the lumen of the sinus, which they reach by slowly percolating through the cord macrophages and red pulp stroma (26) and then via slits in the basement membrane of the sinus (2,8). This pathway has been described as open circulation. On the other hand, a proportion of the arteriolar branches may end in close opposition to sinus walls, allowing a more rapid circulation of at least a portion of the blood flow, even in the absence

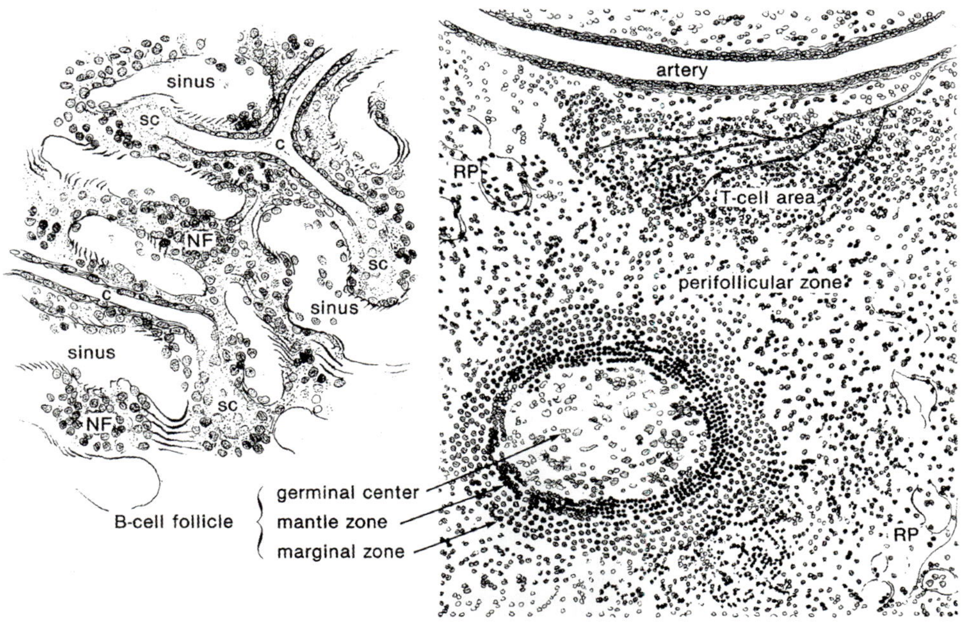

FIGURE 31.3 Schematic impression of red (**left**) and white (**right**) pulp, showing the main compartments and structures of the human spleen. The capillaries (c) end as sheathed capillaries (sc) without direct communication with the sinuses. The nonfiltering areas (NF) are bordered by sinuses and are devoid of (sheathed) capillaries. The perifollicular zone surrounds the white pulp (follicle and T-cell area) and lacks fully developed sinuses. Note the zoning in the B cell but not in the T-cell compartment. The T-cell area contains a lymphatic plexus (**Left**, original magnification ×250; **right**, original magnification ×100).

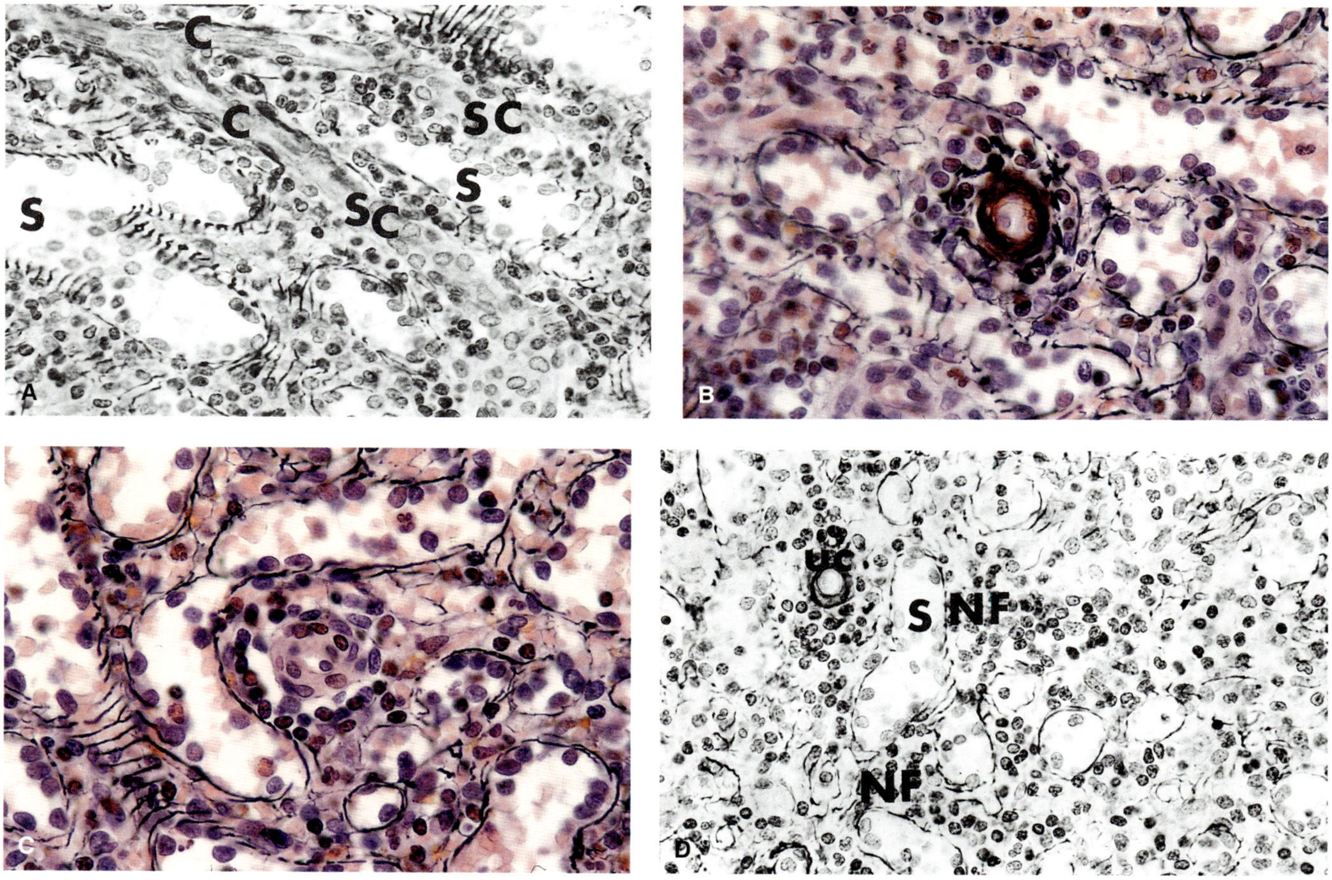

FIGURE 31.4 **A:** Traumatically ruptured spleen. Methylmethacrylate embedding (methenamine-silver/H&E, original magnification ×400). Capillary transitioning into sheathed capillary. Note the proximity to, but lack of connection with, the sinuses (*C*, capillary [unsheathed]; *SC*, sheathed capillary; *S*, sinus). **B:** Same specimen as in (**A**) (original magnification ×1,000). Detail of unsheathed capillary. **C:** Same specimen as in (**A**) (original magnification ×1,000). Detail of sheathed capillary. **D:** Same specimen as in (**A**) (original magnification ×250). Detail of the red pulp showing sinuses in cord tissue. Note the nonfiltering areas devoid of capillaries and completely surrounded by sinuses (*uc*, unsheathed capillary; *S*, sinus; *NF*, nonfiltering area).

of demonstrable direct anatomic connections. The latter pathway of flow has been termed closed circulation. The red pulp sinuses are considered the first part of the splenic venous tree. The localization of the sheathed capillaries at the end of the arterial tree seems perfect for their functioning as a filtering unit.

Within the sinusoidal meshwork, there are large sinuses that open directly into veins running along the arteries in the collagenous cuff.

Small efferent lymph vessels can be found in the T-lymphocyte compartment of the white pulp in about two-thirds of spleens. They are not seen in the surrounding perifollicular zone. A reconstruction from serial sections showed that these lymph vessels form a network around arterioles and eventually follow the arterial tree to the hilar region (8).

Red Pulp

Seventy-five percent of the volume of the spleen is made up of red pulp (8). The two-dimensional picture given by conventional histology sections suggests that the red pulp is largely composed of cordal macrophages, interconnected by their cytoplasmic processes to form a reticular meshwork that provides structural support to the venous sinuses. Serial sections have shown, however, that the red pulp also contains a loose reticular framework, is rich in capillaries, and contains the terminal ends of the penicillar arterioles (Fig. 31.4D). The sinuses account for about 30% of the red pulp (8). The sinus endothelial cells are surrounded by almost circular strands of discontinuous basement membrane that is predominantly composed of collagen IV and laminin, known as the ring fibers (Fig. 31.5). The ring fibers are both interconnected among themselves and anchored to the dendritic processes of the cordal macrophages and splenic (fibroblastic) reticulum cells. Stromal reticulum fibers and cells running throughout the red pulp cords also

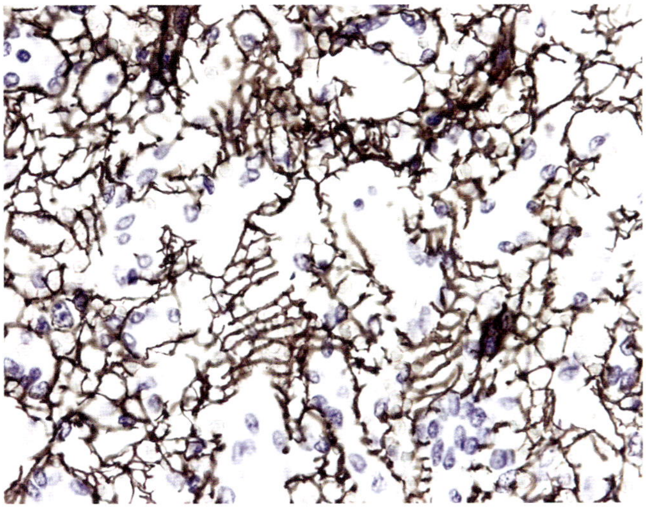

FIGURE 31.5 Normal spleen with red pulp stained with an antibody against collagen IV showing the ring fibers surrounding the sinuses (original magnification ×250).

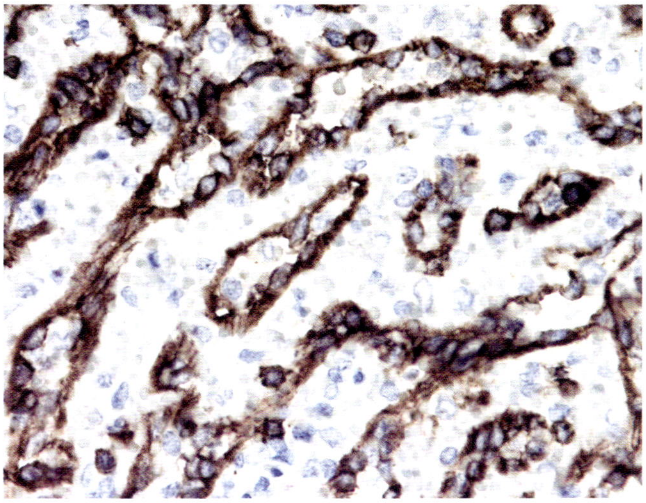

FIGURE 31.6 Normal spleen, red pulp stained with anti-CD8 showing the positive sinus endothelial cells (original magnification ×400).

contribute to provide structural support to this splenic area (the reticular meshwork of the red pulp).

A subpopulation of reticulum cells that express nerve growth factor receptor is found predominantly in the periarteriolar location (27). These cells, most likely representing adventitial reticulum cells similar to those present in the adventitia of blood vessels, also have been observed within the stroma of bone marrow and lymph nodes (28). Myoid reticulum cells express smooth muscle actin and are found scattered throughout the red pulp. These cells are, however, much more concentrated within the marginal zone of the lymphoid follicles and in the periarteriolar lymphoid sheath (29). Whether or not the SMA-positive red pulp cells correspond to fibroblastic reticulum cells that have undergone myofibroblastic differentiation, or are a truly separate population, is unclear at this time. The red pulp sinuses themselves form a complex network of their own with many interconnections and bulblike extensions with blind ends, the latter of which project into the cord tissue (see Figure 4 of van Krieken et al. (8)).

The sinuses are lined by elongated, flat endothelial cells with typical bean-shaped nuclei having a longitudinal cleft; these cells are also known as littoral cells. Immunohistochemistry has shown that these cells are positive for endothelial markers and unique among other endothelial cells for CD8 (Fig. 31.6) and often for CD68 and CD21 (30).

The preponderant function of red pulp is blood filtration. However, in serial sections, one might notice that a fair amount of the red pulp tissue does not include capillary endings, including sheathed capillaries, and that these areas are surrounded only by sinuses. Small aggregates of lymphocytes (both B and T) and mononuclear phagocytes are present (Fig. 31.4), which means that these nonfiltering areas of the red pulp should be regarded as a splenic lymphoid compartment in addition to the white pulp. Morphometrically, the size

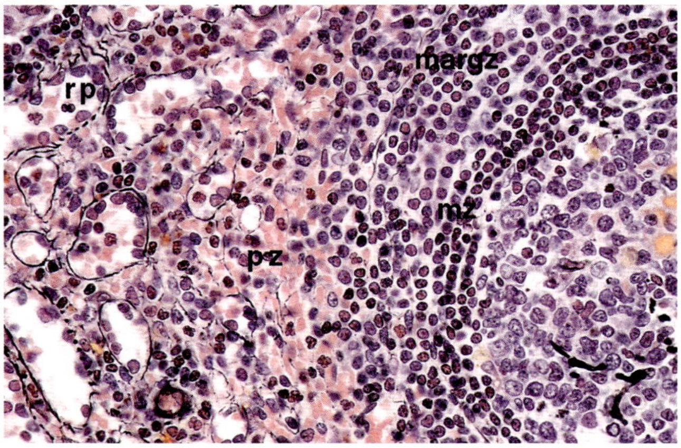

FIGURE 31.7 Same specimen as in Figure 31.3. A secondary follicle (germinal center to the right) borders the red pulp (*mz*, mantle zone; *margz*, marginal zone; *pz*, perifollicular zone; *rp*, red pulp; original magnification ×100).

of this lymphoid, nonfiltering red pulp compartment is comparable to that of the white pulp (31). Newly formed white pulp follicles might originate from the small lymphoid aggregates of these nonfiltering areas. In addition, large numbers of splenic monocytes assemble in clusters in the cords of the subcapsular red pulp. These reserve monocytes can be quickly released from the spleen are may represent an important rapid deployment force involved in the regulation of tissue inflammatory responses, including atherosclerosis (30,32,33).

Blood cells can only reach these areas by passing through large stretches of red pulp tissue or *via* influx from the sinus by passing through the sinus endothelium. Blind-ended bulblike extensions of the red pulp sinuses represent a splenic endothelial component with high endothelial venule-like characteristics. These structures exchange lymphocytes between the sinusoidal circulation and the splenic lymphoid compartment. They are surrounded by follicles and represent a distinct splenic compartment containing erythrocyte-filled vascular spaces (8,10,11). Perifollicular zone sinuses differ from the typical red pulp sinuses in their enhanced expression of CD34. This zone may represent the entry compartment for recirculating lymphocytes into the white pulp since it is capable of supporting influx and local proliferation of lymphoreticular cells, particularly CD4-positive T lymphocytes (10). It has been suggested that the entry of these cells may be dependent on the presence in the perifollicular area of specialized reticulum cells with an endothelial-like phenotype secreting lymphokines and guiding the T cells into the periarteriolar lymphoid sheath (10).

White Pulp

The white pulp consists of B- and T-cell lymphoid compartments (Fig. 31.2). The B-cell compartment mainly contains splenic lymphoid follicles. Secondary follicles are composed of a germinal center surrounded by a ring of small lymphocytes, called the mantle zone or corona, which in turn is surrounded by the marginal zone that contains medium-sized lymphocytes (Fig. 31.7). The germinal centers have similar features to those found in other lymphoid organs. They are formed by a scaffold of follicular dendritic cells that express CD21, CD23, CD35, and the low-affinity nerve growth factor receptor. The B cells of the germinal centers express CD20, CD19, CD10, and CD79a but not CD5. They have a high proliferation rate and do not express Bcl-2. The T lymphocytes present within the germinal centers are predominantly CD4-positive; tingible-body macrophages are CD68-positive. Mantle zones consist predominantly of CD5-positive small lymphocytes that are IgM positive, IgD positive, and DBA.44 positive (Fig. 31.8). The most important difference with lymphoid follicles in lymph nodes is that the splenic follicles have a remarkable unique structure surrounding the mantle zone: the splenic marginal zone. The marginal zone lymphocytes which form this anatomical structure are B lymphocytes

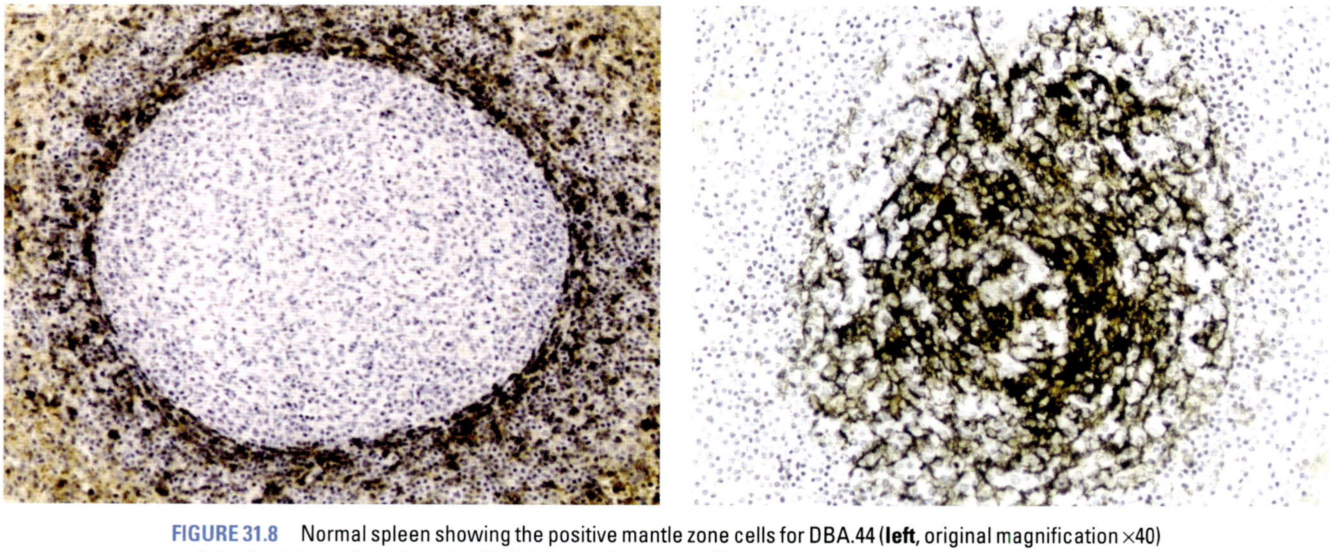

FIGURE 31.8 Normal spleen showing the positive mantle zone cells for DBA.44 (**left**, original magnification ×40) and the dendritic cells stained for CD21 (**right**, original magnification ×40).

that, in contrast to mantle cells, are negative for IgD and DBA.44 (34).

The marginal zone also contains a population of macrophages functionally distinct from the cord histiocytes of the red pulp. Marginal zone macrophages may be important in maintaining the anatomic structure of the marginal zone by attracting newly differentiated marginal zone B lymphocytes into it. These cells move into the marginal zone area from the germinal center (35) and are considered to be part of the recirculating pool of B lymphocytes (36).

The reticulin framework of the marginal zone is characterized by the presence of numerous SMA-positive reticulum cells arranged in a concentric meshwork pattern. The marginal zone SMA-positive cells continue into the T-cell zones where reticulum cells form the reticular framework of the periarteriolar lymphoid sheath. Cells with SMA positivity are also seen, although less frequently, in the periarteriolar lymphoid sheath and scattered throughout the red pulp. These cells become more prominent in the presence of red pulp congestion, such as seen in cases of fibrocongestive splenomegaly (Fig. 31.9) (9).

In the rat spleen, the mantle and marginal zones are separated by a marginal sinus that can easily be seen by light microscopic examination (37). This dividing sinus is not discernible in humans by light microscopy but marginal sinus-like structures have been described based on electron microscopy (38) but they are to be absent in active follicles (39).

The term *marginal zone* has been used with different meanings (8,10–12,40,41). Some investigators use the term to refer to the ring of medium-sized lymphocytes that surrounds the outer border of the mantle zone; few others have included the mantle zone only the bordering area between the red and the white pulp, or even the zone surrounding the T-cell areas. We prefer to reserve the term *marginal zone* for the unique splenic structure that encases

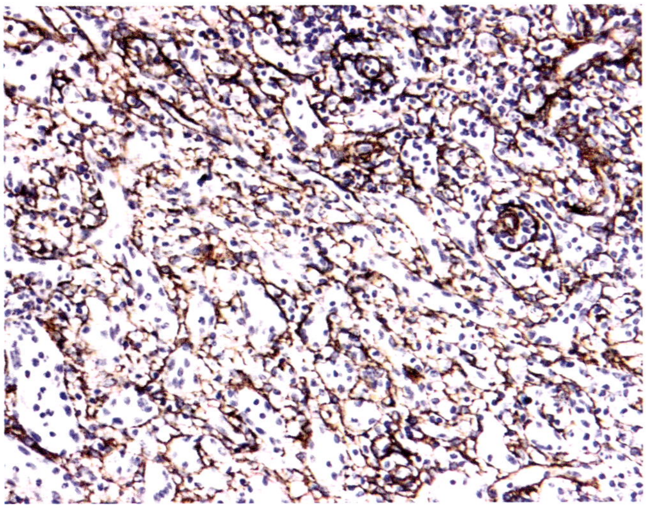

FIGURE 31.9 Spleen in fibrocongestive splenomegaly showing increased expression for smooth muscle actin (original magnification ×100).

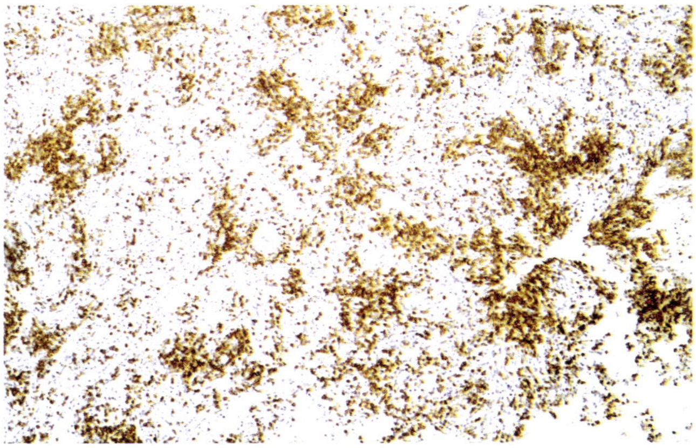

FIGURE 31.10 Normal spleen stained with anti-CD3 antibody (original magnification ×40) showing the somewhat loosely organized T-cell compartment.

the IgD-positive and IgM-positive small lymphocytes of the mantle zone. We refer to the bordering area between the red and the white pulp as the perifollicular zone.

The periarteriolar lymphoid sheath around arterioles is not regularly arranged (Fig. 31.10). The arterioles are not constantly covered by cylindrical lymphoid cuffs; they can be seen traversing follicles and even germinal centers without accompanying lymphoid tissue (12,42).

Rather, the periarteriolar lymphoid sheath is irregular aggregates of small polymorphic T lymphocytes, most of which express CD4. They represent a complex organizational structure of various subsets of T cells (43). The T-cell areas are surrounded by a perifollicular-like zone as well. The follicles sometimes border T-cell areas, with which they share a common perifollicular zone.

Perifollicular Zone

The perifollicular zone is a specialized compartment of the red pulp that is associated with its own reticular stroma. It is found both at the outside of the white pulp marginal zone (in the follicles) and at the periphery of periarteriolar lymphoid sheath. In the perifollicular zone, the reticular fibers are more widely spaced than in the rest of the red pulp (Figs. 31.3 and 31.7). In silver-stained plastic sections, the perifollicular zone can be identified by the paucity of basal membrane strands and by the presence of a vascular pattern that is different from the one seen in the rest of the red pulp. At the outer border of this area, red pulp sinuses are more widely spaced than elsewhere, and a rich network of capillaries, including sheathed capillaries, is present. The perifollicular zone contains a considerable number of erythrocytes and leukocytes (6,8,10,11).

The perifollicular zone at the outside of the white pulp stands out in silver-stained sections but may be poorly visible in routine hematoxylin and eosin (H&E)–stained sections. However, since the perifollicular zone contains a large number of erythrocytes, it can be recognized by its deeply congested appearance around the densely packed lymphocytes of the T- and B-cell areas (Fig. 31.2A). The erythrocytes are seen more regularly in perifollicular zone than in the red pulp sinuses; the red pulp sinuses outside the perifollicular zone often appear to be less filled with these cells than the cord tissue.

The perifollicular zone, which makes up about 8% of the spleen, contains a mixture of blood cells comparable with that of the peripheral blood. It has been suggested that this area is responsible for the passage of about 10% of the splenic blood, which is known to have a retarded flow. In humans, weakly expressed sialoadhesin-positive macrophages are present in the perifollicular zone and in the red pulp although, sialoadhesin is strongly expressed by dispersed perifollicular macrophages in some cases.

FLOW CYTOMETRY

The spleen has been only rarely investigated by flow cytometry, but there are reference values for lymphocyte cells in normal and reactive spleens (30,44). These are generally comparable to those seen in other lymphoid organs, but there are a few differences (Table 31.2). In contrast to the thymus and bone marrow, the spleen contains only very rare TdT-positive lymphoid precursors. Within the B-cell subsets, the spleen shows a frequency of CD19-positive/CD20-negative B cells that is higher than that in the peripheral blood or lymph nodes. This corresponds to the presence, of a sizable proportion of early plasma cells (CD138 negative), as well as more mature plasma cells. Other findings included a significant population of CD20/CD5-positive B cells, accounting for approximately 10% of the B lymphocytes; the presence of CD4/CD8 ratio of 1.2:1 (and a mean number of γ/δ-positive T cells of 6% in normal and 10% in reactive spleens). While NK/T cells account for less than 6% of the CD3-positive circulating lymphocytes in peripheral blood, there is a relatively high frequency of these cells in the spleen.

TABLE 31.2 Expression of Surface and Intracellular Markers by Human Spleen Lymphocytes

	Normal (Cadaveric) Spleen		Reactive (Nonmalignant) Spleen		
Cell Marker	N	Mean ± sd (%)	N	Mean ± sd (%)	P-value[a]
CD2	14	38±10	12	50±14	.015
CD3		31±9		43±14	.009
CD4		17±8		23±7	NS
CD5		32±8		42±13	.028
CD7		37±11		44±12	NS
CD8		14±5		19±7	.028
CD4/CD8 ratio		1.2±3		1.2±0.2	NS
CD10		1±1		1±1	NS
CD11c		28±10		25±10	NS
CD13		1±1		1±1	NS
CD16/56+CD3−		15±7		12±5	NS
CD16/56+CD3+		5±2		5±4	NS
CD19		55±11		45±14	.04
CD20		49±9		42±12	NS
CD20+CD5+		8±6		11±7	NS
CD23		34±13		35±12	NS
CD30		1±1		1±1	NS

[a]NS, not significant ($P > .05$).

ULTRASTRUCTURE

Electron microscopy and the use of microcasts from the vasculature have elucidated the functional microanatomy of the spleen. These studies have shown the routes that blood cells take through the spleen and have also clearly illustrated the pitting function of the spleen (removal of inclusions in erythrocytes) exerted by the sinusoids. In spleens examined in a diagnostic-oriented setting, however, there is hardly, if ever, the necessity of using ultrastructural studies.

FUNCTION

The human spleen has several important functions. However, splenectomy does not lead to impaired health, except for an increased risk of postsplenectomy infections caused by encapsulated bacteria (e.g., pneumococci). The reason for this is that many functions of the spleen, at least in adults, can be taken over by other organs.

The spleen is involved in the primary immune response to blood-borne antigens and polysaccharide antigens; it also acts as a regulator of immune reactions elsewhere in the body. It contains a specific environment that facilitates the binding of antibodies and antigens; cells or microorganisms covered by antibodies are trapped and destroyed in the spleen, as are erythrocytes that have decreased flexibility and lowered osmotic resistance. Each of these functions takes place in a specific splenic compartment, which is capable of undergoing rapid changes in its size and composition, even under physiologic conditions. Therefore, the main splenic functions to be considered include blood filtering, immunologic function, hematopoiesis, and reservoir.

Filter Function

The location and specialized anatomy of the spleen is especially suitable for its function as a filter of the blood.

Normal blood cells are capable of traversing the barrier of macrophages of the sheathed capillary, the red pulp cord macrophages, and the sinus endothelium (at a speed comparable to that of the blood in the capillary bed of other organs). However, in cases in which the flexibility of the red blood cell is diminished (e.g., by aging, intoxication, or congenital defects), the macrophages of the splenic filtering unit can eliminate the abnormal cell by ingesting it (45). The filtering function includes a process known as pitting, a term which is used to describe the removal of inclusions, such as nuclear remnants known as Howell–Jolly bodies, from erythrocytes without destroying the cell. The presence of such circulating erythrocytes in the peripheral blood indicates the presence of splenic hypofunction.

The macrophages of the spleen can readily take up bacteria, antigens, and immune complexes. The spleen is capable of filtering out reticulocytes, platelets, hematopoietic stem cells, lymphocytes, and dendritic cells from the blood and providing the proper microambient conditions for their further differentiation. It also sequesters monocytes from the blood and facilitates their transformation into splenic macrophages.

Immunologic Function

The spleen plays a more important role in B- and T-lymphocyte production and differentiation (4,43). The spleen receives B and T cells from the recirculating lymphocyte pool and sorts them into dedicated compartments such as the follicles and the periarteriolar lymphoid sheath where they can interact with antigens and antigen-presenting cells and become capable of mounting effective immune responses.

The marginal zone is a component of the B-cell follicle and is a remarkably larger compartment in the spleen than other organs. Although the exact physiologic function(s) of the marginal zone is still unclear, its main immunologic role relates to the thymus-independent rapid response to blood-borne microorganisms. These organisms are rapidly trapped in the spleen and brought directly into contact with numerous immunocompetent cells, the spleen is well situated for this task.

Hematopoiesis

Splenic hematopoietic function is only present in fetuses. Hematopoietic cells encountered in the adult spleen originate from circulating, marrow-derived, progenitors/early precursors that become entrapped in the spleen and are capable of undergoing further differentiation. When this physiologic phenomenon reaches pathologic relevance by causing splenomegaly, it is termed splenic myeloid metaplasia or extramedullary hematopoiesis. Although splenic myeloid metaplasia can be seen in many different conditions, the most striking examples of this condition can be observed in patients with primary myelofibrosis. In this disease condition, the splenic microenvironment is characterized by increased levels of intact, functional CXCL12, which contributes to the localization of CD34-positive cells to the spleen and the establishment of the abnormal extramedullary hematopoiesis (46). It has been hypothesized that these primitive malignant hematopoietic cells can sustain hematopoiesis for prolonged periods of time, engraft secondary recipients, and generate additional clones leading to disease progression and leukemic transformation (47).

Reservoir Function

The human spleen contains about 300 mL of blood and functions as a reservoir for factor VIII of the clotting system, platelets, granulocytes, and iron.

AGING DIFFERENCES

In infancy and childhood, the immune system is not yet fully developed, and this is also reflected in the histology of the spleen (48). The marginal zone is observed as a separate compartment only after 4 months of age; moreover, the marginal zone B cells in the spleen of infants have a different phenotype in that they lack CD21 and are positive for IgD and IgM compared with adult marginal zone B cells. An important age difference, in our experience, is the regular occurrence of germinal centers in the white pulp of normal spleens in patients younger than 20 years, whereas older patients rarely have secondary follicles (30). The often-mentioned age-dependent atrophic change has only been documented in patients in their 8th decade of life (30). However, recent evidence in mice has suggested an age-related alteration in marginal zone microarchitecture and function (49). It is unclear whether a selective loss of marginal zone function can also occur in humans.

Hyalinization of vessels in the spleen is seen frequently, even in very young children and, therefore, does not represent a pathologic finding (50).

In infants, the elastic fibers of the splenic capsule are homogeneously intermingled with collagen fibers, an arrangement that stabilizes the capsule during spleen growth and enlargement. With aging, collagen fibers predominate in the outer capsular surface over elastic fibers with the latter more evident in the deep lamina of the splenic capsule. The elastic fibers in elderly patients shorten, fragment, and thicken. The progressive decrease in the amount of elastic fibers in the splenic capsule may restrict splenic distention and contribute to involution of the spleen as one grows older (51).

DIFFERENTIAL DIAGNOSIS

Compartmentalized lymphoid tissue of the white pulp is interwoven with the filtering red pulp. Each splenic compartment reacts to external stimuli with physiologic changes in its composition and histology. As in the lymph node, the line between pathologic and physiologic reactions is vague.

As previously mentioned, normal blood cells can pass undamaged through the barrier of macrophages of the sheathed capillary and the red pulp cord tissue, as well as the sinus endothelium at a speed comparable with that of the blood flow in the capillary bed of other organs. However, when the flexibility of the blood cells is diminished by aging, intoxication, or congenital defects, the red pulp macrophages can ingest the abnormal cells. In this process, the sheathed capillaries lose their macrophages, which spread out into the surrounding red pulp or enter the sinuses to be transported to the liver. In cases characterized by chronic stimulation of the filtering function, the number and length of the capillaries increase in parallel with the hypertrophy of the red pulp, whereas the sheathed capillaries are less readily seen in the sections. In idiopathic thrombocytopenic purpura, remnants of phagocytosed thrombocytes can be seen as periodic acid–Schiff (PAS)-positive fragments in cord macrophages. If blood cells are covered by immunoglobulins or immunocomplexes, parts of the cell membrane can be removed by the sinus endothelium by pitting and culling, giving rise to a spherocyte. This happens in a fashion similar to the removal of nuclear remnants, as previously described.

In septicemia, the filtering compartment may show activation and hyperplasia of macrophages indistinguishable from cases of acute or chronic hemolysis; these changes are most likely induced by the presence of circulating immunocomplexes, fragmented cells, or antibody-coated cells. In these conditions, postmortem autolysis of the activated macrophages can lead to early disintegration of the red pulp cells and stroma. The septic spleen at autopsy probably represents an artifact that can be the result of but is not specific for, sepsis. It should not be diagnosed as splenitis. In septic spleens, there is a significant depletion of B and T areas, accompanied by a reactive germinal center hyperplasia regardless of the type of bacteria responsible. However, depletion of splenic B areas is pronounced in the setting of premortal enterococcemia in comparison with a panel of gram-negative flagellated bacteria (52).

True splenitis, in which the spleen contains an inflammatory response to a local noxious agent such as in typhoid fever or tropical diseases, is rare in the Western hemisphere. Lymphoplasmacytoid cells and plasma cells normally rim arteries and arterioles and extend along red pulp capillaries. This perivascular cellular rim also may contain some macrophages or small epithelioid granulomas, the significance of which is unclear. The perivascular presence of plasma cells is a normal finding and by no means justifies a diagnosis of splenitis, nor is the diffuse influx of granulocytes throughout the red pulp in specimens resected during prolonged surgery.

The effect of chronic cardiac congestion on the lymphoid and filtering compartments appears more likely to be the effect of concomitant infections or therapy than chronic cardiac failure. In chronic venous congestion due to portal hypertension, the sinuses are normal in size but contain fewer buds and appear rigid. The amount of cord tissue and number of capillaries are decreased; an increase of reticulin fibers (and increased expression of smooth muscle actin) in reticulum cells are seen. Infarcts in the splenic tissue are microscopically more irregularly defined and poorly demarcated than could be expected macroscopically due to the intricate distribution of the splenic vessels. In three-dimensional reconstructions, capillaries from different arterioles are seen to cross each other with overlapping territories.

Primary tumors of the spleen are rare. Metastatic carcinoma seems relatively common to occur in endocrine tumors, including small-cell carcinoma of the lung, with a conspicuous tendency for intrasinusoidal spread. Lymphomas exhibit a homing pattern to specific splenic compartments dependent on the type of lymphoma, similar to that observed in other lymphoid organs (53). The distribution of nonlymphomatous hematopoietic malignancies is similar to that observed in the bone marrow. Extramedullary erythropoiesis and megakaryopoiesis are found primarily along and within the sinuses of the red pulp, whereas myelopoiesis is found in proximity to the capillaries within the cord tissue. Blastic infiltration seen in cases of acute leukemia can be found anywhere in the spleen.

SPECIMEN HANDLING

The spleen is quite vulnerable and, due to the large numbers of macrophages and granulocytes, may undergo rapid autolysis. Proper and rapid fixation is therefore important, and this goal is not reached when the entire organ is put into formalin. For proper handling, the specimen has to be received fresh, and handling has to be rapid. An appropriate protocol involves weighing the organ and examining the surface. After that, the organ is cut into 0.5-cm slices and the cut surfaces are inspected for nodules larger than normal white pulp. Ideally, pieces should be submitted for flow cytometry, and snap frozen for cryostat section immunohistochemistry or molecular techniques. When no abnormalities are seen, at least three or four blocks are taken out randomly and processed for microscopic examination.

HISTOLOGIC TECHNIQUE

Routine paraffin embedding leads to shrinkage and loss of cellular detail. Since routine H&E staining often does not yield sufficient information, methenamine-silver/H&E stain, or at least, PAS and Gomori reticulin stains are necessary for

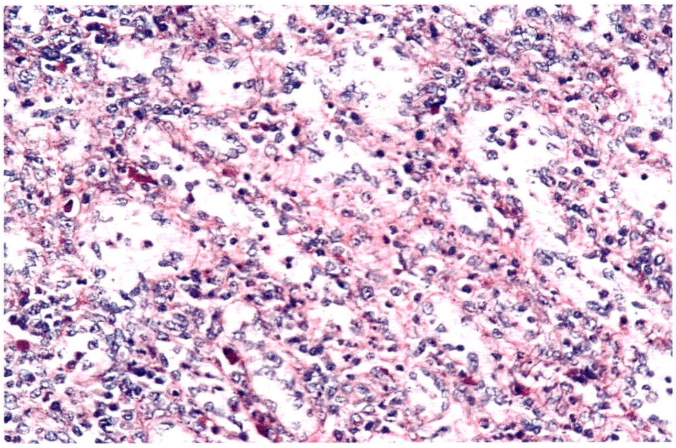

FIGURE 31.11 Same specimen as in Figure 31.1. Detail of red pulp showing with some difficulty, the structure of the sinuses (PAS stain, paraffin-embedded, original magnification ×200).

an adequate morphologic analysis of the splenic microarchitecture (Fig. 31.11).

SPECIAL PROCEDURES

The spleen is only rarely removed for diagnostic purposes. Staging laparotomy is no longer part of the required diagnostic workup of a patient with Hodgkin disease. Therefore, splenic pathology may be seen as an unsuspected incidental finding in a patient splenectomized for other reasons. Not uncommonly, splenic lymphoma may be discovered as an incidental finding. A high degree of suspicion is necessary in order to apply techniques that require fresh/frozen tissue, such as flow cytometry, frozen tissue immunohistochemistry, molecular analysis, and/or cytogenetics (Fig. 31.9). An overview of relevant panels of markers is described by Borch et al. (30).

Clonality assessment in a spleen can be used similarly as in other organs and can be used in addition to immunohistochemistry (54).

Immunohistochemistry can also be necessary to characterize nonhematopoietic tumors. Electron microscopy can be helpful in the evaluation of storage disorders such as Gaucher disease, although biochemical analysis is considered by most experts as the most practical and specific approach.

CONCLUSION

The human spleen has always been a somewhat enigmatic organ. Studies of its histology must be based on carefully selected, surgically excised normal spleens. The organ should be processed immediately and appropriately for optimal results. The spleen is a highly compartmentalized organ. Each compartment has its own structure and cell populations. The old division into red and white pulp is probably oversimplified and should be expanded.

REFERENCES

1. Paraskevas GK, Koutsouflianiotis KN, Nitsa Z, Demesticha T, Skandalakis P. Knowledge of the anatomy and physiology of the spleen throughout antiquity and the early middle ages. *Anat Sci Int*. 2016;91:43–55.
2. Chen L, Weiss L. Electron microscopy of the red pulp of human spleen. *Am J Anat*. 1972;134:425–457.
3. Neely HR, Flajnik MF. Emergence and evolution of secondary lymphoid organs. *Annu Rev Cell Dev Biol*. 2016;32:693–711.
4. Rodríguez-Perea AL, Arcia ED, Rueda CM, Velilla PA. Phenotypical characterization of regulatory T cells in humans and rodents. *Clin Exp Immunol*. 2016;185:281–291.
5. Meng W, Zhang B, Schwartz GW, et al. An atlas of B-cell clonal distribution in the human body. *Nat Biotechnol*. 2017;35: 879–884.
6. Lewis SM, Williams A, Eisenbarth SC. Structure-function of the immune system in the spleen. *Sci Immunol*. 2019;4:eaau6085.
7. O'Malley DP, Louissaint A Jr, Vasef MA, et al. International Spleen Consortium. Recommendations for gross examination and sampling of surgical specimens of the spleen. *Ann Diagn Pathol*. 2015;19(5):288–295.
8. van Krieken JH, Te Velde J, Hermans J, Welvaart K. The splenic red pulp; a histomorphometrical study in splenectomy specimens embedded in methylmethacrylate. *Histopathology*. 1985; 9:401–416.
9. Kraus MD. Splenic histology and histopathology: an update. *Semin Diagn Pathol*. 2003;20:84–93.
10. Steiniger B, Barth P, Hellinger A. The perifollicular and marginal zones of the human splenic white pulp: do fibroblasts guide lymphocyte immigration? *Am J Pathol*. 2001;159: 501–512.
11. Steiniger B, Barth P, Herbst B, Hartnell A, Crocker PR. The species-specific structure of microanatomical compartments in the human spleen: strongly sialoadhesin-positive macrophages occur in the perifollicular zone, but not in the marginal zone. *Immunology*. 1997;92:307–316.
12. van Krieken JH, te Velde J, Kleiverda K, Leenheers-Binnendijk L, van de Velde CJ. The human spleen: a histological study in splenectomy specimens embedded in methylmethacrylate. *Histopathology*. 1985;9:571–585.
13. Yamamoto K, Miwa Y, Abe-Suzuki S, et al. Extramedullary hematopoiesis: elucidating the function of the hematopoietic stem cell niche. *Mol Med Rep*. 2016;13:587–591.
14. Timens W, Rozeboom T, Poppema S. Fetal and neonatal development of human spleen: an immunohistological study. *Immunology*. 1987;60:603–609.
15. Jones JF. Development of the spleen. *Lymphology*. 1983;16: 83–89.
16. Namikawa R, Mizuno T, Matsuoka H, et al. Ontogenic development of T and B cells and non-lymphoid cells in the white pulp of human spleen. *Immunology*. 1986;57:61–69.
17. Varga I, Galfiova P, Adamkov M, et al. Congenital anomalies of the spleen from an embryological point of view. *Med Sci Monit*. 2009;12:RA269–RA276.
18. Moller JH, Nakib A, Anderson RC, Edwards JE. Congenital cardiac disease associated with polysplenia. A developmental complex of bilateral "left-sidedness." *Circulation*. 1967;36: 789–799.

19. Khairat AB, Ismail AM. Splenogonadal fusion: case presentation and literature review. *J Pediatr Surg*. 2005;40:1357–1360.
20. Howard MT, Dufresne S, Swerdlow SH, Cook JR. Follicular lymphoma of the spleen: multiparameter analysis of 16 cases. *Am J Clin Pathol*. 2009;131:656–662.
21. Oren H, Ozkal S, Gülen H, et al. Autoimmune lymphoproliferative syndrome: report of two cases and review of the literature. *Ann Hematol*. 2002;81:651–653.
22. Ilardo MA, Moltke I, Korneliussen TS, et al. Physiological and genetic adaptations to diving in sea nomads. *Cell*. 2018;173:569–580.e15.
23. Myers J, Segal RJ. Weight of the spleen. I. Range of normal in a nonhospital population. *Arch Pathol*. 1974;98:33–35.
24. Vahidy FS, Parsha KN, Rahbar MH, et al. Acute splenic responses in patients with ischemic stroke and intracerebral hemorrhage. *J Cereb Blood Flow Metab*. 2016;36:1012–1021.
25. Jung WC, Levesque JP, Ruitenberg MJ. It takes nerve to fight back: the significance of neural innervation of the bone marrow and spleen for immune function. *Semin Cell Dev Biol*. 2017;61:60–70.
26. Steiniger BS, Pfeffer H, Gaffling S, Lobachev O. The human splenic microcirculation is entirely open as shown by 3D models in virtual reality. *Sci Rep*. 2022;12:16487.
27. Cattoretti G, Schiro R, Orazi A, Soligo D, Colombo MP. Bone marrow stroma in humans: anti-nerve growth factor receptor antibodies selectively stain reticular cells in vivo and in vitro. *Blood*. 1993;81:1726–1738.
28. Orazi A, O'Malley DP, Thomas JL, et al. Stromal changes in reactive and malignant disorders of the spleen. *Mod Pathol*. 2004;17:264A.
29. Pinkus GS, Warhol MJ, O'Connor EM, Etheridge CL, Fujiwara K. Immunohistochemical localization of smooth muscle myosin in human spleen, lymph node, and other lymphoid tissues. Unique staining patterns in splenic white pulp and sinuses, lymphoid follicles, and certain vasculature, with ultrastructural correlations. *Am J Pathol*. 1986;123:440–453.
30. Borch WR, Aguilera NS, Brisette MD, O'Malley DP, Auerbach A. Practical applications in immunohistochemistry. An immunophenotypic approach to the spleen. *Arch Pathol Lab Med*. 2019;143:1093–1105.
31. van Krieken JH, te Velde J, Hermans J, Cornelisse CJ, Welvaart C, Ferrari M. The amount of white pulp in the spleen: a morphometrical study done in methacrylate-embedded splenectomy specimens. *Histopathology*. 1983;7:767–782.
32. Swirski FK, Nahrendorf M, Etzrodt M, et al. Identification of splenic reservoir monocytes and their deployment to inflammatory sites. *Science*. 2009;325:612–616.
33. Potteaux S, Ait-Oufella H, Mallat Z. Role of splenic monocytes in atherosclerosis. *Curr Opin Lipidol*. 2015;26:457–463.
34. van Krieken JH, von Schilling C, Kluin PM, Lennert K. Splenic marginal zone lymphocytes and related cells in the lymph node: a morphologic and immunohistochemical study. *Hum Pathol*. 1989;20:320–325.
35. Pillai S, Cariappa A, Moran ST. Marginal zone B lymphocytes. *Annu Rev Immunol*. 2005;23:161–196.
36. Steiniger B, Timphus EM, Barth PJ. The splenic marginal zone in humans and rodents: an enigmatic compartment and its inhabitants. *Histochem Cell Biol*. 2006;126:641–648.
37. Sasou S, Satodate R, Katsura S. The marginal sinus in the perifollicular region of the rat spleen. *Cell Tissue Res*. 1976;172:195–203.
38. Schmidt EE, MacDonald IC, Groom AC. Microcirculatory pathways in normal human spleen, demonstrated by scanning electron microscopy of corrosion casts. *Am J Anat*. 1988;181:253–266.
39. Schmidt EE, MacDonald IC, Groom AC. Changes in splenic microcirculatory pathways in chronic idiopathic thrombocytopenic purpura. *Blood*. 1991;78:1485–1489.
40. Takasaki S. Light microscopic, scanning and transmission electron microscopic, and enzyme histochemical observations on the boundary zone between the red pulp and its surroundings in human spleens. *Tokyo Yikekai Med J*. 1979;94:553–568.
41. Kraal G. Cells in the marginal zone of the spleen. *Int Rec Cytol*. 1992;132:31–74.
42. Steiniger B, Ruttinger L, Barth PJ. The three-dimensional structure of human splenic white pulp compartments. *J Histoch Cytochem*. 2003;51:655–663.
43. Woon HG, Braun A, Li J, et al. Compartmentalization of total and virus-specific tissue-resident memory CD8+ T cells in human lymphoid organs. *PLoS Pathog*. 2016;12:e1005799.
44. Colovai AI, Giatzikis C, Ho EK, et al. Flow cytometric analysis of normal and reactive spleen. *Mod Pathol*. 2004;17:918–927.
45. Pivkin IV, Peng Z, Karniadakis GE, Buffet PA, Dao M, Suresh S. Biomechanics of red blood cells in human spleen and consequences for physiology and disease. *Proc Natl Acad Sci U S A*. 2016;113:7804–7809.
46. Wang X, Cho SY, Hu CS, Chen D, Roboz J, Hoffman R. C-X-C motif chemokine 12 influences the development of extramedullary hematopoiesis in the spleens of myelofibrosis patients. *Exp Hematol*. 2015;43:100–109.e1.
47. Wang X, Prakash S, Lu M, et al. Spleens of myelofibrosis patients contain malignant hematopoietic stem cells. *J Clin Invest*. 2012;122:3888–3899.
48. Timens W, Boes A, Rozeboom-Uiterwijk T, Poppema S. Immaturity of the human splenic marginal zone in infancy. Possible contribution to the deficient infant immune response. *J Immunol*. 1989;143:3200–3206.
49. Birjandi SZ, Ippolito JA, Ramadorai AK, et al. Alterations in marginal zone macrophages and marginal zone B cells in old mice. *J Immunol*. 2011;186:3441–3451.
50. Lindley RP. Splenic arteriolar hyalin in children. *J Pathol*. 1986;148:321–325.
51. Rodrigues CJ, Sacchetti JCL, Rodrigues AJ Jr. Age-related changes in the elastic fiber network of the human splenic capsule. *Lymphology*. 1999;32:64–69.
52. Gunia S, Albrecht K, May M, Stosiek P. The white pulp in the setting of the septic spleen caused by different bacteria: a comparative morphometric study. *APMIS*. 2005;113:675–682.
53. van Krieken JH, Feller AC, te Velde J. The distribution of non-Hodgkin's lymphoma in the lymphoid compartments of the human spleen. *Am J Surg Pathol*. 1989;13:757–765.
54. van Krieken JH, Langerak AW, Macintyre EA, et al. Improved reliability of lymphoma diagnostics via PCR-based clonality testing: report of the BIOMED-2 Concerted Action BHM4-CT98-3936. *Leukemia*. 2007;21:201–206.

32

Bone Marrow

S.H. Kroft

The bone marrow is a large and complex organ that is distributed throughout the cavities of the skeleton. The bone marrow of an adult has been estimated to constitute 4% to 5% of total body mass (1,2), exceeding that of the liver. About half of this mass consists of hematopoietically inactive fatty marrow (which appears yellow) and the remainder of hematopoietically active marrow (which appears red). Although essentially hematopoietically inactive, even fatty marrow contains a few scattered microscopic foci of hematopoietic cells. The functions of hematopoietic marrow include: (a) the formation and release of various types of blood cells (hematopoiesis), mast cells, osteoclasts, and some endothelial progenitor cells; (b) the phagocytosis and degradation of circulating particulate material such as microorganisms and abnormal or senescent red cells and leukocytes; and (c) antibody production. Normal marrow produces $4–5 \times 10^{11}$ blood cells per day (3). In addition to hematopoietic stem cells, the marrow contains mesenchymal stem cells that can differentiate under appropriate conditions into adipocytes, hepatocytes, osteoblasts and osteocytes, chondrocytes, skeletal and cardiac muscle cells, kidney cells, and neural cell lineages (4). The nonhematopoietic marrow serves as a large store of reserve lipids. The various functions of hematopoietic marrow are based on a high degree of structural organization (3). However, this organization is labile, altering rapidly in response to many stimuli.

TECHNIQUES FOR STUDYING THE MARROW

Thorough microscopic evaluation of bone marrow requires examination of both aspirate smear and trephine (core) biopsy preparations. International guidelines also consider peripheral blood smears to be an essential part of a complete marrow evaluation (5); however, peripheral blood morphology is not addressed in this chapter. Aspirate smear preparations provide superior assessment of cytologic detail of hematopoietic cells, whereas trephine biopsies provide information related to architecture, cellularity, focal lesions (e.g., lymphoid aggregates, granulomas, amyloidosis, metastatic solid malignancies), fibrosis, or necrosis (5–9).

In adults, marrow is aspirated from the posterior superior iliac spine or the anterior iliac crest (5,6). Sternal aspirates may be obtained in certain situations, as well. In children, marrow is usually aspirated from the posterior superior iliac spine and, in the case of patients less than 1 year of age, also from the upper end of the medial surface of the tibia just below and medial to the tibial tuberosity. Smears of bone marrow aspirate material may be prepared in a variety of ways, including particle crush preparations, direct smears, and buffy coat smears; each technique carries certain advantages and disadvantages (5,9). The marrow smears are air dried and stained using one of several Romanowsky-type stains, including May–Grünwald–Giemsa or Wright–Giemsa (5,10). Aspirate smears will also often be stained using a Prussian blue stain for iron (Perls acid ferrocyanide method). Touch imprints of trephine biopsies may also be air dried and treated in a similar fashion as aspirate smears. In this chapter, unless otherwise stated, the descriptions of cells in marrow smears apply to smears stained by a Romanowsky method.

Trephine biopsies are generally obtained from the posterior superior iliac spine or anterior iliac crest. This provides a core of bone and associated marrow. The biopsy specimen is commonly fixed in 10% neutral buffered formalin for 6 hours but may be fixed in a variety of other solutions for varying lengths of time, including but not limited to acetic acid zinc formalin (AZF), B5 (mercuric acid and formalin), Bouin fixative (picric acid, acetic acid, and formalin), or Zenker solution (mercuric chloride, potassium dichromate, sodium sulfate, water, acetic acid). The fixed specimen is then decalcified by one of a variety of methods, including EDTA, formic acid, picric acid, nitric acid, or proprietary, commercially available decalcifying solutions, and then embedded in paraffin. Decalcification and paraffin embedding result in some shrinkage of marrow tissue, loss of activity of cellular enzymes, and variable loss of cytologic detail. In addition, certain decalcification procedures cause leaching of the iron stores (i.e., of the hemosiderin present within macrophages). Histologic studies also can be performed on clot sections of aspirated marrow. In general, this involves concentration of particles in a liquid aspirate, induction of clotting, fixation, and paraffin embedding (5,9). Clot sections, having not been decalcified, consequently exhibit better retention of immunoreactivity, and maintain better integrity of nucleic acids for molecular studies. Methods for embedding undecalcified trephine biopsies in methyl methacrylate are also available. Semithin sections of such specimens provide superior cytologic detail, but these plastic-embedding methods are technically demanding and not in wide use (5).

Sections of paraffin-embedded marrow fragments or decalcified bone cores are optimally cut to a thickness of 2 to 3 μm and are routinely stained with hematoxylin and eosin (H&E). A reticulin stain (by a silver impregnation method) and a Giemsa stain may also be routinely applied, although practice patterns vary regionally. Various other cytochemical stains may be performed on marrow sections in specific circumstances, including periodic acid–Schiff (PAS) reaction for glycogen or glycoprotein, Leder stain for chloroacetate esterase, and a trichrome stain for fibrosis.

Recent years have seen a dramatic increase in the number of commercially available immunohistochemical antibodies suitable for use in paraffin-embedded tissue sections. While the decalcification process and the wide variety of fixatives in use for bone marrow trephines have historically limited the application of immunohistochemistry, improvements in commercial reagents, antigen retrieval techniques, and decalcification processes now enable the routine use of a large number of immunohistochemical stains in bone marrow cores, although procedures often need to be modified for optimal results (11,12). Immunohistochemistry may now be used to characterize a wide variety of normal and abnormal marrow cell populations in marrow trephines and clot sections (12–17). While an exhaustive list of available antibodies used in diagnosis of pathologic processes in bone marrow sections is beyond the scope of this chapter, a list of antibodies commonly used for identifying various cell lineages is provided in Table 32.1. Certain antibodies may be applied using immunocytochemical techniques to fixed bone marrow smears although this is no longer common practice. It is important to note that, because of the wide variability in preanalytical and analytical procedures, immunohistochemical findings may differ considerably from laboratory to laboratory; recommendations for improving standardization have been published (11).

Table 32.1 Lineage Antigens Commonly Assessed in Bone Marrow Sections Using Immunohistochemistry

Cell Lineage	Antigens
Leukocytes	CD45
Immature cells (blasts)	CD34, CD117 (myeloid), TdT (lymphoid)
B cells	CD10 (immature and germinal center), CD19, CD20, CD22, CD79a, Bcl-6 (germinal center), Pax-5, MUM-1/IRF4 (postgerminal center)
Plasma cells	CD138, MUM-1/IRF4, kappa and lambda light chains
T cells	CD2, CD3, CD4, CD5, CD7, CD8
NK cells	CD2, CD7, CD56, CD57, TIA-1, granzyme-B
Maturing granulocytes	CD10 (mature), CD15, CD33, CD117 (early promyelocytes), lysozyme, myeloperoxidase
Monocytes/ macrophages	CD4, CD15, CD33, CD68, CD163, lysozyme
Erythroid precursors	CD71, CD117 (pronormoblasts), E-cadherin (pronormoblasts), CD235a (glycophorin A), hemoglobin A, spectrin
Megakaryocytes	CD31, CD41, CD42, CD61, factor VIII–related antigen (von Willebrand factor)
Mast cells	CD117, mast cell tryptase
Endothelial cells	CD31, CD34
Osteoblasts	CD56

GENERAL FEATURES OF HEMATOPOIESIS

Blood cells are produced in the embryo and fetus and throughout postnatal life. In the developing fetus and growing child, the total number of hematopoietic cells and blood cells increases progressively with time. By contrast, the hematopoietic systems of healthy adults are examples of steady-state cell renewal systems. In such systems, a relatively constant rate of loss of mature blood cells from the circulation is balanced by the production of new blood cells at the same rate. The number of hematopoietic cells and blood cells therefore remains constant.

New blood cells are eventually derived from a small number of hematopoietic stem cells, estimated at 2×10^4 total cells (18). These cells have two properties: (a) the ability to mature into all types of blood cells; and (b) an extensive capacity to generate new stem cells, and thus to maintain their own number (19). In humans, the existence of pluripotent hematopoietic stem cells with both the above properties has been demonstrated by the success of bone marrow transplantation. Hematopoietic stem cells differentiate into progenitor cells that are committed to one or more lineages; these committed progenitor cells do not have the capacity for self-renewal or to sustain long-term hematopoiesis. As these committed progenitors branch and mature, they have progressively restricted differentiation potential. In the classical, branching hierarchical model of hematopoiesis (Fig. 32.1), the pluripotent stem cells give rise to lymphoid stem cells and multipotent myeloid stem cells (20–23). The lymphoid stem cells mature into all types of lymphocytes. The myeloid stem cells mature into neutrophil, eosinophil, basophil granulocytes, monocytes, erythrocytes, platelets, mast cells, and osteoclasts. While this model is conceptually appealing, accumulated evidence over the last two decades indicates that hematopoiesis is considerably more complex (19,24).

Hematopoietic progenitor cells have been identified and characterized by their ability to form colonies containing cells of one or more hematopoietic lineages in vitro and are therefore called colony-forming units (CFUs) or colony-forming cells (CFCs). CFUs generate colonies containing a mixture of granulocytes, erythroblasts, macrophages, and megakaryocytes and are, therefore, termed CFU-GEMM. Bipotent hematopoietic progenitor cells that give rise to colonies containing granulocytes and macrophages are termed CFU-GM. There are also bipotent progenitor cell-generating colonies containing a mixture of erythroblasts and megakaryocytes (CFU-E mega). The unipotent progenitor cells that give rise to neutrophil granulocytes, eosinophil granulocytes,

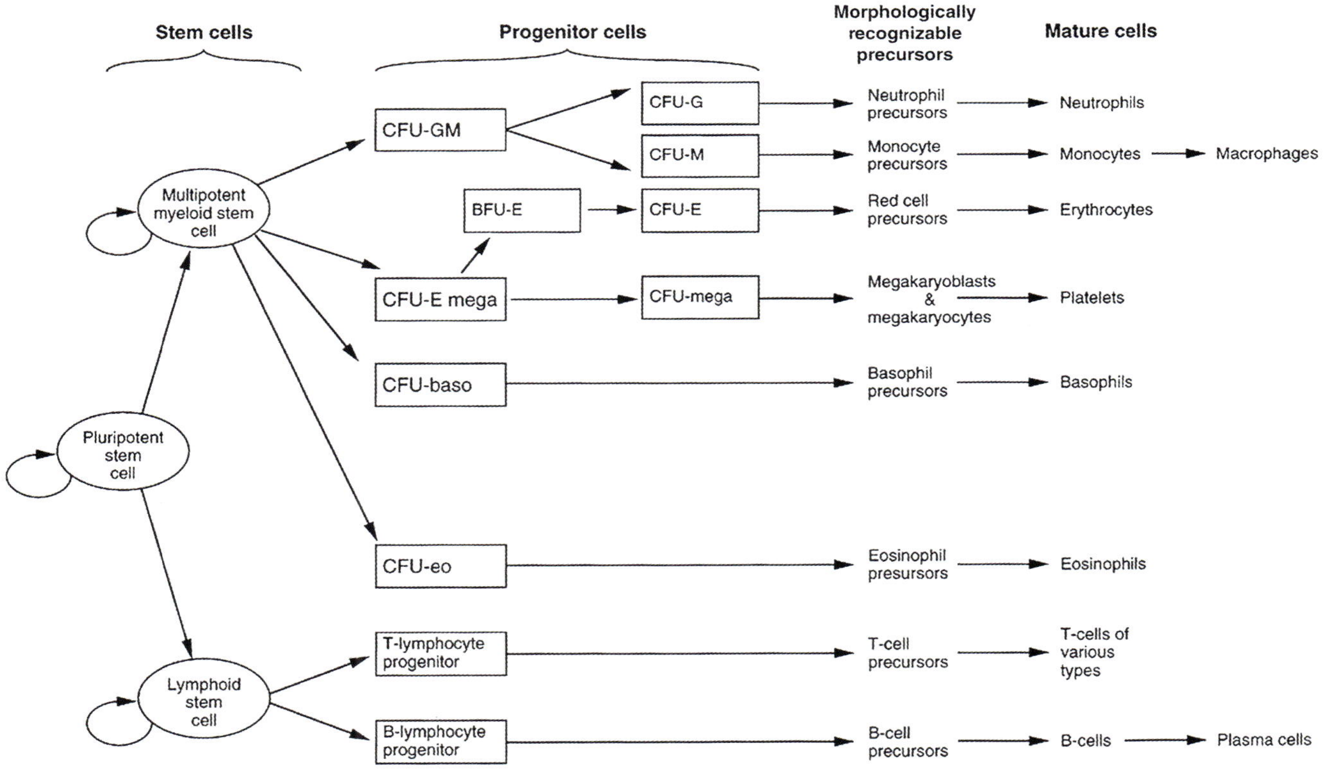

FIGURE 32.1 Classical model of hematopoiesis showing the relationships between the various types of stem cells, progenitor cells, and morphologically recognizable precursor cells. BFU-E, erythroid burst-forming units; CFU, colony-forming units; E, erythroblasts; GM, granulocytes and macrophages; eo, eosinophil granulocytes; baso, basophil granulocytes; mega, megakaryocytes; G, neutrophil granulocytes; M, macrophages.

basophil granulocytes, macrophages, erythroblasts, and megakaryocytes are described as CFU-G, CFU-eo, CFU-baso, CFU-M, CFU-E, and CFU-mega, respectively. These develop into the least mature of the morphologically recognizable blood cell precursors in the marrow. Thus, CFU-G develops into myeloblasts, CFU-eo into eosinophil promyelocytes, CFU-baso into basophil promyelocytes, CFU-M into monoblasts, CFU-E into pronormoblasts, and CFU-mega into megakaryoblasts. The stem cells and progenitor cells are found in both the blood and the marrow but cannot be distinguished morphologically. The characteristics of the various types of morphologically recognizable hematopoietic cells found in the marrow are described later in this chapter.

Two processes are involved in the formation of all types of blood cells. These are the progressive acquisition of the biochemical, functional, and morphologic characteristics of the particular cell type (i.e., differentiation) and cell proliferation. The latter results in the production of a large number of mature cells from a single cell committed to one or more differentiation pathways. Differentiation occurs at all stages of hematopoiesis, and cell proliferation occurs in the hematopoietic stem cells, progenitor cells, and, except in the megakaryocytic lineage, in the more immature morphologically recognizable precursor cells. The nearly mature blood cells seem to enter the circulation mainly by passing through the endothelial cells of the marrow sinusoids.

REGULATION OF HEMATOPOIESIS

Hematopoietic stem cells and early progenitor cells show low-level expression of transcription factors and genes specific to several hematopoietic lineages (multilineage priming). Commitment to a single lineage involves enhancement of transcription factors controlling the gene expression programs specific to that lineage and permanent silencing by those transcription factors of gene programs required for differentiation down other lineages (25).

The mechanisms underlying the commitment of a stem cell to differentiate are not yet fully understood (24,25). According to one model, the probability of a stem cell undergoing self-renewal or differentiation is a stochastic process. Environmental signals (soluble factors, cell–cell and cell–extracellular matrix interactions) mediated by specific receptor–ligand interactions operate only by influencing stem cell and progenitor cell apoptosis and proliferation. Another model proposes that all pathways toward renewal or death taken by stem cells and progenitor cells are determined by environmental signals. Bone marrow stromal cells (e.g., macrophages, nonphagocytic reticular or fibroblastoid cells, adipocytes, endothelial cells) play a major role in generating such signals; they provide niches for the attachment of stem cells and their progeny, are a source of the extracellular matrix involved in such attachment, and secrete various membrane-bound and soluble stimulatory hematopoietic growth factors and inhibitory cytokines (24). It is likely that elements of both mechanisms are operative (25).

Stem cells and early hematopoietic progenitor cells interact via specific cell surface receptors with multilineage hematopoietic growth factors (24,25). The latter include stem cell factor (steel factor, kit ligand), FLT3 ligand, interleukin (IL)-1 and IL-6 for the pluripotent stem cells, and stem cell factor, FLT3 ligand, thrombopoietin, IL-3 (multi-CSF), and granulocyte-macrophage colony–stimulating factor (GM-CSF) for the multipotent myeloid stem cells. The regulation of later progenitor cells and the morphologically recognizable hematopoietic cells is dependent both on multilineage growth factors and lineage-specific growth factors such as G-CSF, M-CSF, IL-5 (influencing CFU-eo), thrombopoietin and IL11 (influencing CFU-mega), and erythropoietin (mainly influencing late BFU-E and CFU-E). The growth factors influencing lymphocyte progenitor cells and precursors include IL-2, IL-4, IL-5, IL-6, IL-7, IL-11, and IL-21 for the B lineage and IL-2, IL-4, IL-7, IL-10, IL-15, and IL-21 for the T lineage.

Hematopoietic growth factors are glycoproteins that influence the survival, proliferation, and differentiation of their target cells via second messengers. In their absence, the target cells undergo programmed cell death (apoptosis). Some growth factors such as G-CSF and GM-CSF not only regulate hematopoiesis but also enhance the function of the mature cells. Most hematopoietic growth factors are produced by bone marrow stromal cells and T lymphocytes, either constitutively (e.g., M-CSF production by fibroblastoid cells and endothelial cells) or after their activation by various signals. Thus, fibroblastoid cells and endothelial cells that have been activated by macrophage-derived IL-1 or tumor necrosis factor (TNF) and endotoxin-stimulated macrophages produce M-CSF, GM-CSF, G-CSF, IL-6, and stem cell factor. Antigen- or IL-1–activated T cells produce IL-3, IL-5, and GM-CSF.

The kidney is the main organ of postnatal erythropoietin production, where it is synthesized by interstitial fibroblast-like cells associated with peritubular capillaries and proximal convoluted tubules (26). About 10% of erythropoietin is produced in the liver, which is the main organ of synthesis in the fetus. There is an oxygen sensor in the renal interstitial fibroblast-like cells, and the production of erythropoietin is inversely proportional to the degree of oxygenation of renal tissue. The erythropoietin receptor is upregulated at the late BFU-E and CFU-E stages, and signaling through this receptor is required to prevent apoptosis.

In addition to the stimulatory cytokines mentioned above, inhibitors (negative regulators) of hematopoiesis are produced by macrophages, fibroblastoid cells, and endothelial cells. These include transforming growth factor-β1 (TGF-β1), which inhibits multilineage progenitor cells, early erythroid progenitors, and megakaryocytes; TNF-α, which inhibits the proliferation of granulocyte precursors; interferon-α, which inhibits megakaryocyte progenitors; and macrophage inflammatory protein-1α (MIP-1α), which inhibits the proliferation of stem cells.

Recombinant forms of a number of growth factors, or alternatively agonists for their receptors, are used therapeutically to manage patients with cytopenias, and these produce predictable morphologic changes in the marrow related to their physiologic roles. Thus, recombinant G-CSF and GM-CSF produce an expansion and shift toward immaturity of the myelomonocytic lineage, as well as morphologic changes associated with granulocyte activation (toxic granulation and Dohle bodies) (27,28). Similarly, the administration of recombinant erythropoietin drives an increase in erythroid precursors and a shift toward immaturity, as well as some (usually mild) changes conventionally construed as being indicative of dyserythropoiesis, such as dyssynchrony of nuclear and cytoplasmic maturation, increased binucleation, and budding or irregularity of late precursors (29). Finally, thrombopoietin administration can cause significant expansion of marrow megakaryocytes, as well as morphologic changes mimicking myeloproliferative neoplasms (30).

HEMATOPOIESIS IN THE EMBRYO AND FETUS: DEVELOPMENT OF THE BONE MARROW

Studies in experimental animals have shown that hematopoietic stem cells responsible for embryonic (primitive) hematopoiesis develop in the yolk sac. Those responsible for fetal and postnatal (definitive) hematopoiesis are considered to arise in the aorto-gonad-mesonephros region by some investigators and the yolk sac by others (31–33). The stem cells migrate through the bloodstream to colonize the fetal liver and other fetal tissues.

In the human embryo, erythropoietic cells first appear within the blood islands of the yolk sac about 19 days after fertilization (34,35). A few megakaryocytes are found in these blood islands during the 6th and 7th weeks of gestation. Yolk sac erythropoiesis is megaloblastic and results in the production of nucleated red cells (Fig. 32.2) that contain three embryonic hemoglobins: hemoglobins Gower I ($\zeta_2\varepsilon_2$), Gower II ($\alpha_2\varepsilon_2$), and Portland I ($\zeta_2\gamma_2$), and, in later embryos, hemoglobin F ($\alpha_2\gamma_2$) (36).

Hematopoietic foci develop in the hepatic cords during the 6th week of gestation, and the liver becomes the major site of erythropoiesis in the middle trimester of pregnancy (37,38). During this period, about half the nucleated cells of the liver are erythropoietic cells (Fig. 32.3). A few granulocyte precursors and megakaryocytes also are found in this organ. Fetal hepatic erythropoiesis is normoblastic and gives rise to nonnucleated red cells containing hemoglobin F. These red cells are considerably larger than the red cells of adults. The number of erythropoietic cells in the liver decreases progressively after the 7th month of gestation; a few cells persist until the end of the 1st postnatal week.

Marrow cavities are formed as a result of the erosion of bone or calcified cartilage by blood vessels and cells from the periosteum (34). The first marrow cavity to develop is that of the clavicle (at about 2 months of gestational age). After the formation of the marrow cavities, the vascular connective tissue present within them becomes colonized by circulating hematopoietic stem cells. The latter generate erythropoietic cells during the 3rd and 4th months of gestation, the order of appearance of erythropoietic cells being the same as the order of formation of the marrow cavities. After the 6th month, the bone marrow becomes the major site of hematopoiesis (39). Erythropoiesis in fetal bone marrow is normoblastic and results in the production of nonnucleated red cells that are larger than adult red cells and contain hemoglobins F and A ($\alpha_2\beta_2$). The fetal bone marrow is the predominant site of intrauterine granulocytopoiesis and megakaryocytopoiesis. In this tissue, the myeloid/erythroid ratio (i.e., the ratio of the number of neutrophil precursors plus neutrophil granulocytes to the number of erythroblasts) remains constant at about 1:4 after 6.5 months of gestation (39).

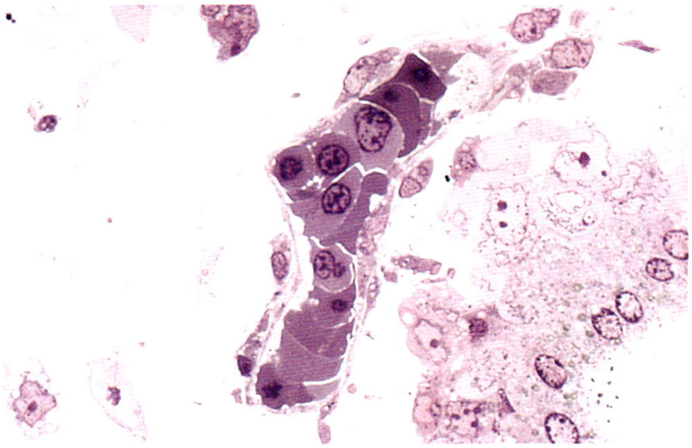

FIGURE 32.2 Semithin section of a plastic-embedded chorionic villus biopsy sample obtained at 7 weeks of gestation, showing a blood vessel containing nucleated embryonic red cells (toluidine blue).

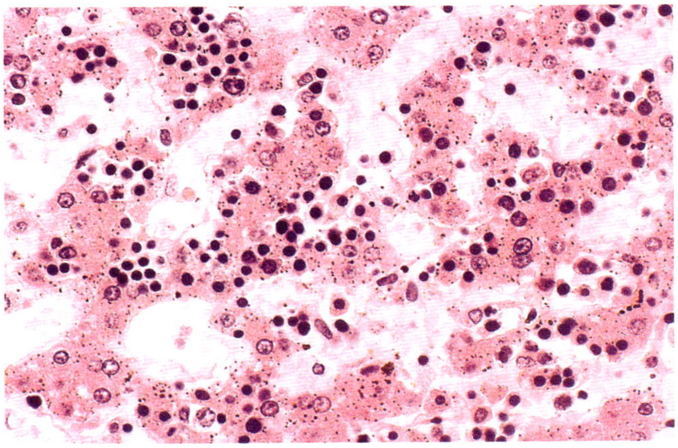

FIGURE 32.3 Fetal liver tissue obtained postmortem showing erythropoietic activity. The erythroblasts (identifiable by their darkly staining, round nuclei) are found extravascularly, both within the hepatic cords and between the cords and the sinusoidal endothelial cells. The brown material within hepatocytes is formalin pigment, a common postmortem fixation artifact (H&E).

POSTNATAL CHANGES IN THE DISTRIBUTION OF RED MARROW AND THE TYPE OF HEMOGLOBIN

At birth, all the marrow cavities contain red, hematopoietic marrow. Furthermore, the red marrow contains only a few fat cells. After the first 4 years of life, an increasing number of fat cells appear between the hematopoietic cells, particularly in certain regions of the marrow, and these regions eventually become yellow and virtually devoid of hematopoietic cells (40,41). Zones of yellow, fatty marrow are found just below the middle of the shafts of the long bones between the ages of 10 and 14 years and, subsequently, extend in both directions, distal spread being more rapid than proximal spread. By the age of about 25 years, hematopoietic marrow is confined to the proximal quarters of the shafts of the femora and humeri, the skull bones, ribs, sternum, scapulae, clavicles, vertebrae, pelvis, and the upper half of the sacrum. Although the distribution of hematopoietic marrow remains essentially unaltered throughout adult life, its fat cell content increases with increasing age.

The percentages of hemoglobins F and A in the blood of full-term neonates are 50% to 85% and 15% to 50%, respectively. The proportion of hemoglobin F decreases postnatally at different rates in different individuals, but adult levels of less than 1% are reached in nearly all children by the age of 2.5 years.

Because young children have red marrow containing few fat cells in virtually all their marrow cavities, a rapid increase in hematopoietic tissue in this age group is presumably accommodated mainly by a reduction in the proportion of marrow space occupied by sinusoids. If the increase in the rate of hematopoiesis is substantial and prolonged (e.g., in congenital hemolytic anemias), there is an increase in the total volume of the marrow cavities and the reestablishment of extramedullary hematopoiesis in organs such as the liver, spleen, and lymph nodes (42). The expansion of the marrow cavities leads to skeletal abnormalities, such as frontal and parietal bossing, dental deformities, and malocclusion of the teeth. It also causes thinning of the cortex, which may lead to fractures after minor trauma. In adults, increased hematopoiesis is initially associated with the replacement of fat cells in red marrow by hematopoietic cells and also with the spread of red marrow into marrow cavities normally containing yellow marrow (42). If the increase in hematopoiesis is marked, extramedullary hematopoiesis may develop.

STRUCTURAL ORGANIZATION OF HEMATOPOIETIC MARROW

The marrow cavities of most bones contain trabeculae of cancellous bone. The inner surface of the cortex and the outer surfaces of the trabeculae are lined by the endosteum, which consists of a single layer of cells supported on a delicate layer of reticular connective tissue. In most areas of the endosteum, the cells consist of very flat bone-lining cells (endosteal lining cells), but in some areas, they consist of osteoblasts or osteoclasts. The marrow, which is located between the trabeculae, is supplied with an extensive microvasculature and some myelinated and nonmyelinated nerve fibers. It does not have lymphatic drainage (43). The space between the small blood vessels contains a few reticulin fibers and a variety of cell types. The latter include fat cells, precursors of red cells, granulocytes, monocytes and platelets, lymphocytes, plasma cells, macrophages (phagocytic reticular cells), nonphagocytic reticular cells, and mast cells (44,45).

Blood Supply

One or more nutrient canals penetrate the shafts of the long bones obliquely. Each canal contains a nutrient artery and one or two nutrient veins. After entering the marrow, the nutrient artery divides into ascending and descending branches, which coil around the central longitudinal vein, the main venous channel of the marrow. The ascending and descending arteries give off numerous arterioles and capillaries that travel radially toward the endosteum and often open into a plexus of sinusoids (46). The sinusoids drain through a system of collecting venules and larger venous channels into the central longitudinal vein, which in turn drains mainly into the nutrient veins. In the diaphyses of long bones containing yellow fatty marrow, the nutrient artery gives off relatively few branches until it reaches the lower edge of the red marrow, where it breaks up into numerous vessels that penetrate the hematopoietic tissue. Many blood vessels of various sizes supply the marrow within flat and cuboidal bones, entering the marrow cavity via one or more large nutrient canals, as well as through numerous smaller canals.

There are interconnections between the blood supply of the bone marrow and bone through an endosteal network of blood vessels. This network communicates both with the periosteal vessels *via* fine veins passing through the bone and with branches of the nutrient artery. Furthermore, studies in experimental animals have shown that many capillaries derived from the nutrient artery enter haversian canals but swing back into the marrow and open into sinusoids or venules (47–50).

The sinusoids of human bone marrow have thin walls consisting of an inner complete layer of flattened endothelial cells with little or no underlying basement membrane and an outer incomplete layer of adventitial cells (51). Some endothelial cells show alkaline phosphatase activity. Endothelial cells contain no stainable iron except when the iron stores are increased. They produce extracellular matrix, stem cell factor, IL6, GM-CSF, IL-1α, IL-11, and G-CSF, and are intimately involved with the regulation of hematopoiesis. Endothelial cells of the sinusoids allow the bidirectional migration of progenitor cells and hematopoietic stem cells through them by a mechanism involving specific binding molecules. Adventitial cells are a type of reticular cell, forming part of the cytoplasmic network of the marrow stroma. Adventitial cells stain strongly for alkaline phosphatase.

Nerve Supply

The nerve supply of a long bone enters the bone marrow mainly via the nutrient canal but also through a number of epiphyseal and metaphyseal foramina. Bundles of nerve fibers travel together with the nutrient artery and its branches and supply the smooth muscle in such vessels or, occasionally, terminate between hematopoietic cells (52).

Extracellular Matrix

Normal marrow contains a scanty incomplete network of fine branching reticulin (type III collagen) fibers between the parenchymal cells (Fig. 32.4). A higher concentration of thicker fibers is found in and around the walls of the larger arteries and near the endosteum; such fibers are continuous with the fibers in the parenchyma. Other extracellular matrix components produced by stromal cells include fibronectin, vascular cell adhesion molecule (VCAM)-1, vitronectin, thrombospondin, and proteoglycans such as heparan sulfate and chondroitin sulfate.

Stromal Cells

The stromal cells of the marrow comprise (a) osteoblasts, adipocytes, and nonphagocytic reticular cells including myofibroblasts, all of which are derived from mesenchymal stem cells; (b) osteoclasts, macrophages, and mast cells that are derived from myeloid hematopoietic stem cells; and (c) endothelial cells that are derived from either hematopoietic stem cells or a more primitive marrow cell that also gives rise to hematopoietic stem cells (53). Some stromal cells are intimately involved in the regulation of hematopoiesis.

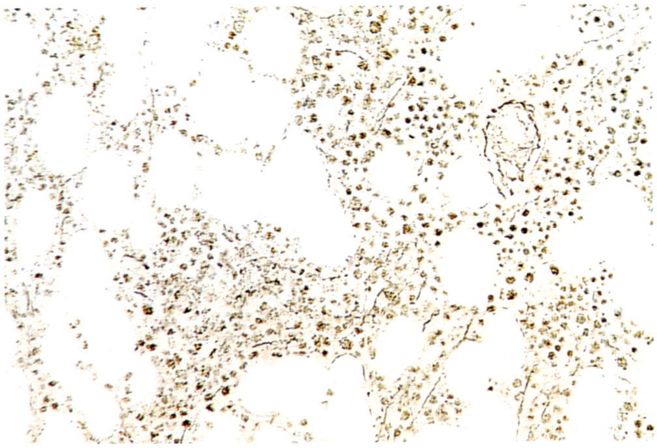

FIGURE 32.4 Section of a decalcified, paraffin-embedded trephine biopsy core from a hematologically normal adult, showing a scant network of fine reticulin fibers. The upper right-hand quadrant of the photomicrograph shows a circular arrangement of fibers associated with a blood vessel. (Silver impregnation of reticulin.)

Osteoblasts and Osteoclasts

Osteoblasts are present in the endosteum in areas of deposition of osteoid. Osteoblasts are frequently found in a continuous layer, usually one or two cells thick, and appear like an area of epithelium. They become surrounded by the osteoid they produce and eventually become osteocytes. Osteoclasts are large, multinucleate cells involved in bone resorption. Osteoblasts arise from progenitor cells closely associated with the endosteal lining cells. Although it is usually considered that osteoblast progenitor cells are not derived from hematopoietic stem cells, studies in mice indicate that osteoblasts and hematopoietic cells arise from a common primitive marrow cell (54). Osteoblasts produce cytokines such as IL6, G-CSF, and GM-CSF that influence hematopoiesis (55). Osteoclasts originate from the myeloid hematopoietic stem cells, but the specific precursor population remains poorly defined (56).

Romanowsky-stained normal marrow smears may contain groups of osteoblasts or individual osteoclasts. They are commonly seen in the marrow aspirates of children but are only seen in adults when there is increased bony remodeling activity. Osteoblasts have an ovoid or elongated shape and are 20 to 50 μm in diameter (Fig. 32.5A). They have abundant basophilic cytoplasm, often with somewhat indistinct margins, and a single eccentric nucleus. The chromatin has a reticular quality and one to three nucleoli are present. The cytoplasm contains a rounded pale area corresponding to the Golgi apparatus, which often is situated some distance from the nucleus. Osteoblasts superficially resemble plasma cells, but the latter are smaller, contain distinctly clumped chromatin, generally lack nucleoli, and have a Golgi zone situated immediately adjacent to the nucleus (Fig. 32.5B). Osteoblasts stain positively for alkaline phosphatase. In histologic sections, osteoblasts are cuboidal or pyramidal and have eccentric nuclei, distinct nucleoli, and dense, eosinophilic cytoplasm (Fig. 32.5C). Osteoclasts appear in aspirate smears as giant, multinucleate cells with abundant pale blue cytoplasm containing many azurophilic granules (Fig. 32.6A). The individual nuclei are round, uniform in size, contain a single nucleolus, and do not overlap. Osteoclasts characteristically reside in depressions or pits in the bony trabeculae, known as a Howship lacunae or resorption lacunae (Fig. 32.6B). They are strongly acid phosphatase positive and can be distinguished from megakaryocytes because the latter contain single, large lobated nuclei.

Fat Cells

The number of fat cells in hematopoietic bone marrow varies with age (57–59). In normal adults, 30% to 70% of the area of a histologic section of hematopoietic marrow consists of fat cells (Fig. 32.7). Fat cells are the largest cells in the

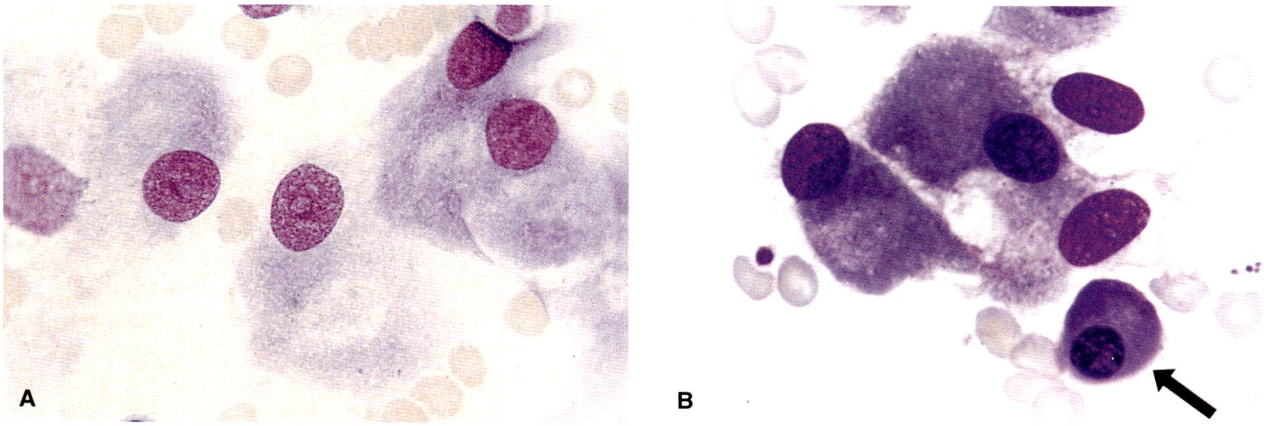

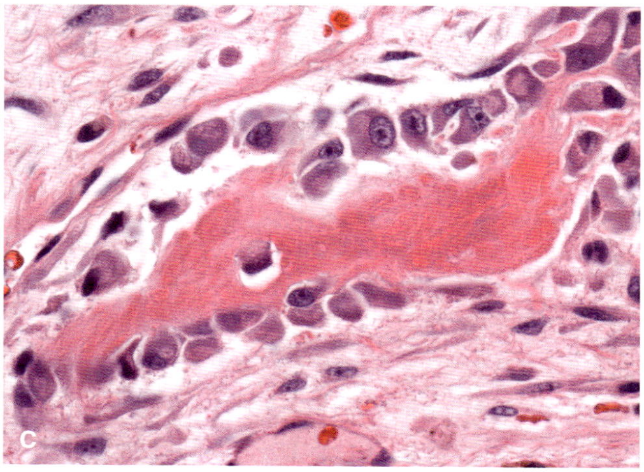

FIGURE 32.5 **A:** Group of osteoblasts from a Wright–Giemsa–stained smear of normal bone marrow. **B:** Several osteoblasts, with an adjacent plasma cell for comparison (*arrow*) (Wright–Giemsa). **C:** Exuberant osteoblastic proliferation in a marrow trephine due to repair following a previous trephine biopsy at this site (H&E).

marrow, and sections of such cells have average diameters of 85 μm. The fat cells of the bone marrow are surrounded by small quantities of reticulin and collagen fibers. They are in intimate contact with vascular channels, macrophages, and all types of hematopoietic cells. Marrow fat cells are formed by the accumulation of lipids within adventitial cells, other nonphagocytic reticular cells, and, possibly, sinus endothelial cells. Whenever there is an increase or decrease in the number of hematopoietic cells in bone marrow, there is a corresponding decrease or increase, respectively, in the number of fat cells such that the intersinusoidal space within marrow cavities is fully occupied. The mechanisms

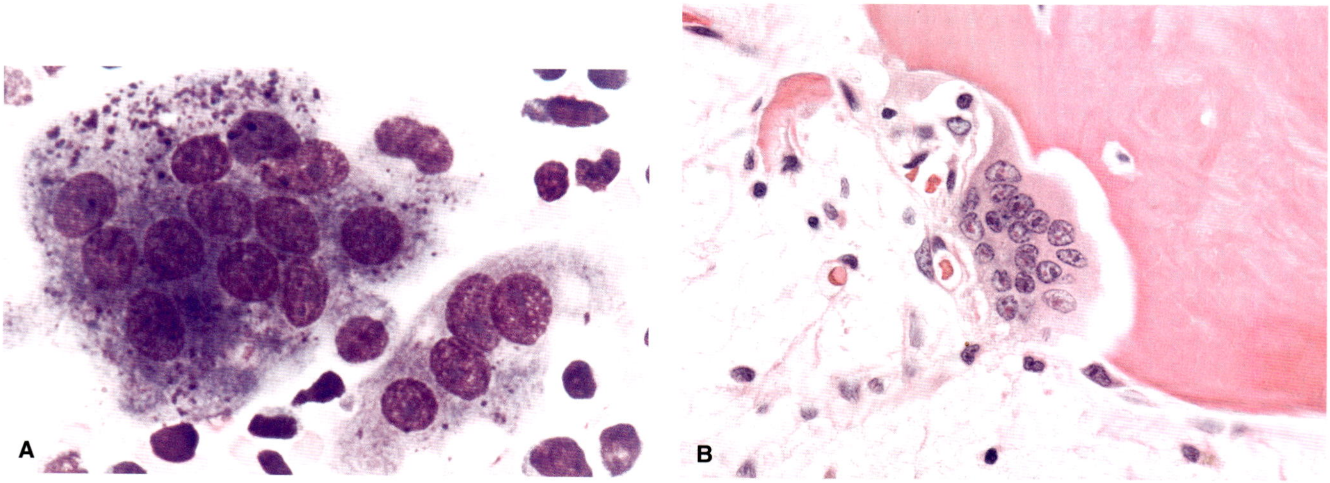

FIGURE 32.6 **A:** Two multinucleate osteoclasts from an aspirate smear of normal bone marrow with increased bony remodeling (Wright–Giemsa). **B:** A multinucleated osteoclast residing in a pit on the surface of a bone trabeculum (Howship lacuna or resorption lacuna) (H&E).

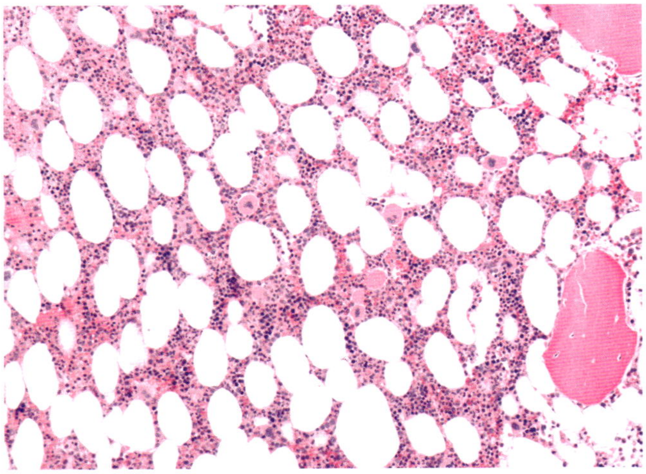

FIGURE 32.7 Section of a decalcified, paraffin-embedded trephine biopsy core from a hematologically normal adult. About 60% of the area of marrow tissue in this photomicrograph is occupied by fat cells. There may be a substantial variation in cellularity in different parts of the same section (H&E).

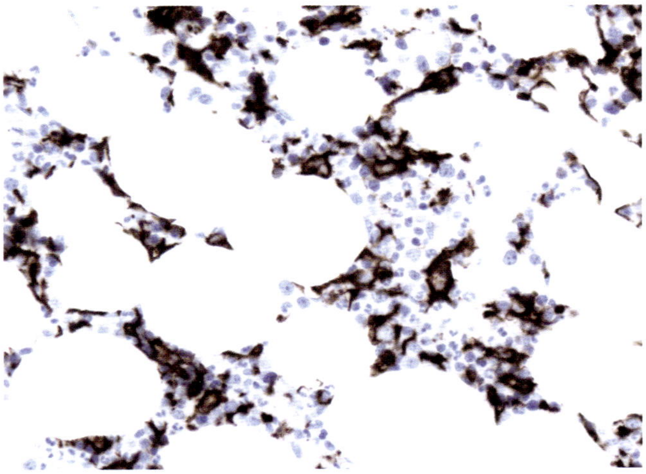

FIGURE 32.9 Immunohistochemical demonstration of macrophages in a section of a paraffin-embedded trephine biopsy core from a hematologically normal subject with anti-CD163. Note the abundant dendritic processes highlighted by the reaction.

underlying the inverse relationship between the mass of fat cells and hematopoietic cells in the marrow are uncertain. In severe anorexia nervosa or cachexia secondary to chronic disorders, there is a marked reduction in fat cells, often together with a reduction in hematopoietic tissue. In these conditions, the space normally occupied by cells is filled with a gelatinous extracellular substance composed of acid mucopolysaccharide (Fig. 32.8) (60). This "gelatinous transformation" or "serous fat atrophy" may be associated with decreased cell counts in the blood.

Macrophages (Phagocytic Reticular Cells)

The bone marrow contains many macrophages. The frequency of this cell type is best appreciated in sections of trephine biopsies stained for antigens found on macrophages such as CD68 or CD163 (Fig. 32.9) rather than in smears of aspirated bone marrow. Macrophages are moderately large cells with abundant cytoplasm. They are often irregularly shaped cells spanning 20 to 30 μm with a round or ovoid nucleus and pale, lace-like chromatin with one or more large nucleoli. The cytoplasm is voluminous, stains pale blue, and contains azurophilic granules, vacuoles, and variably sized inclusions composed of phagocytosed material (Fig. 32.10). Macrophages are derived from monocytes and, therefore originally from the hematopoietic stem cells.

Macrophages may contain refractile yellow-brown intracytoplasmic inclusions ranging from 0.5 and 4 μm in diameter (Fig. 32.11). These inclusions represent hemosiderin

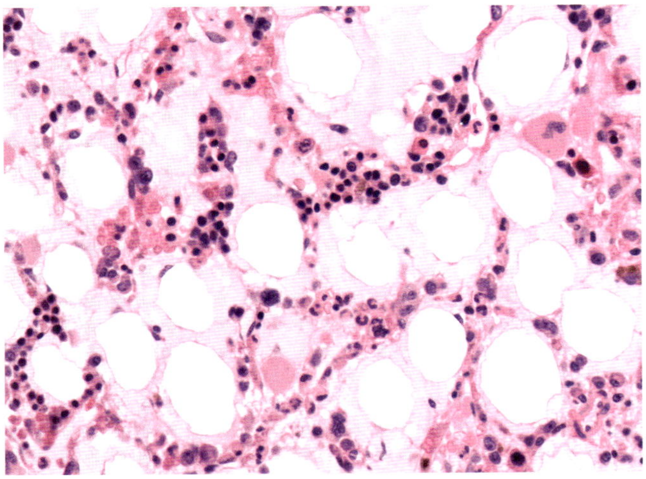

FIGURE 32.8 Bone marrow biopsy from patient with cachexia due to advanced AIDS demonstrating gelatinous transformation, also known as serous fat atrophy. There is an accumulation of homogeneous, pink mucopolysaccharide material in the interstitium, preferentially located around adipocytes.

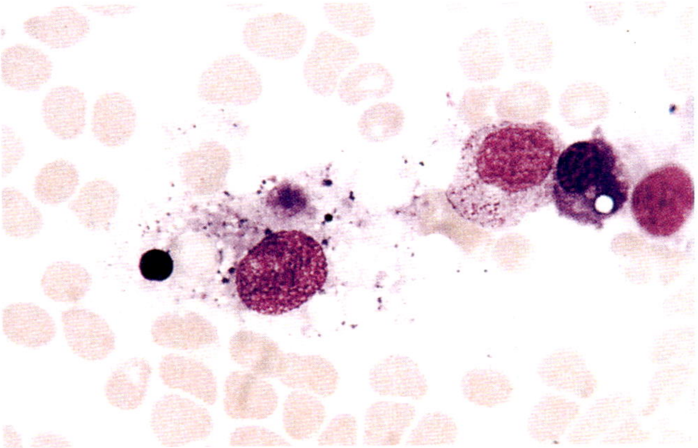

FIGURE 32.10 Macrophage from a normal marrow containing a dark extruded erythroblast nucleus and several intracytoplasmic inclusions of various shapes, sizes, and staining characteristics. The large pale rounded inclusions may represent degraded red cells (May-Grunwald Giemsa).

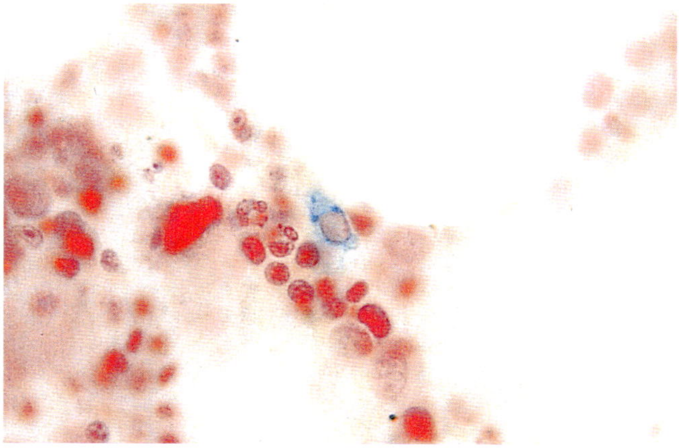

FIGURE 32.11 Section of a paraffin-embedded normal marrow fragment (*clot section*). The macrophage in the center shows blue hemosiderin-containing intracytoplasmic granules and a diffuse bluish coloration of the cytoplasm (Perls acid ferrocyanide reaction).

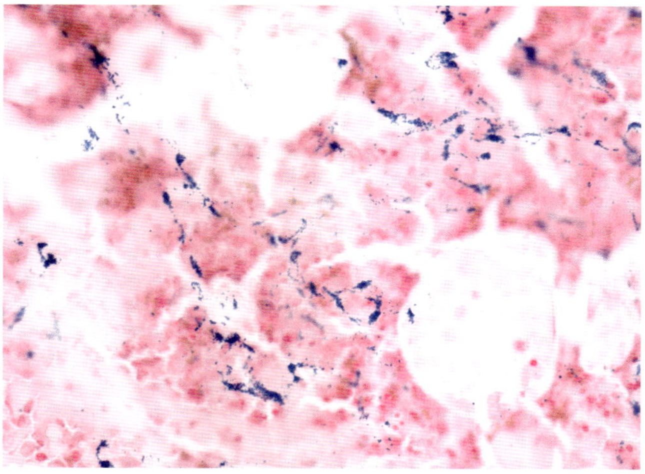

FIGURE 32.13 This iron stain demonstrates the distinct curvilinear arrays of iron granules seen following parenteral iron administration.

and appear as blue or blue-black granules when stained by Perls acid ferrocyanide method (Fig. 32.12). The amount of iron within a marrow smear or histologic section may be assessed semiquantitatively and is a useful guide to the total iron stores (61). Stainable storage iron is absent in iron deficiency and increased in hereditary hemochromatosis and transfusion-induced iron overload. Patients who have received therapeutic parenteral iron may have a unique pattern of iron deposition characterized by curvilinear arrangements of uniformly sized granules (Fig. 32.13) (62). Iron deposited in this form may not be bioavailable (63). Macrophages contain PAS-positive material and are strongly positive for α-naphthyl acetate esterase and acid phosphatase. They do not stain for α-naphthol AS-D chloroacetate esterase activity (49), and most do not stain with Sudan Black. Some macrophages appear to stain positively for alkaline phosphatase.

Macrophages are present within erythroblastic islands (Fig. 32.14), plasma cell islands, and lymphoid nodules but also may occur elsewhere in the marrow parenchyma. Some are found immediately adjacent to the endothelial cells of sinusoids, forming part of the adventitial cell layer. Bone marrow macrophages not only function as phagocytic cells but also generate various hematopoietic growth factors (e.g., c-kit ligand or stem cell factor, M-CSF, IL-1, and G-CSF) and are thus involved in short-range regulation of lymphopoiesis and myelopoiesis. They presumably also are involved in antigen processing.

Several variant macrophage forms may be encountered in routine bone marrow specimens. Hemophagocytic forms, macrophages with ingested intact or degenerating but recognizable red cells, platelets, or nucleated marrow cells, are seen in small numbers in normal marrows (Fig. 32.15) (64,65). These are increased in a group of primary and secondary disorders of histiocyte activation and proliferation,

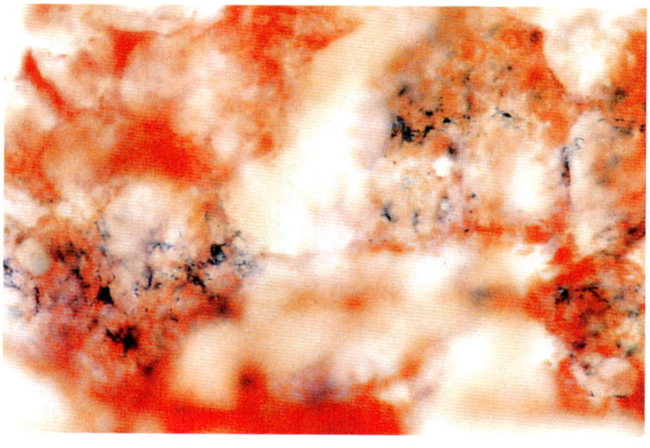

FIGURE 32.12 Marrow fragment from a normal marrow smear stained by Perls acid ferrocyanide reaction. The dark blue granular material represents hemosiderin within macrophages.

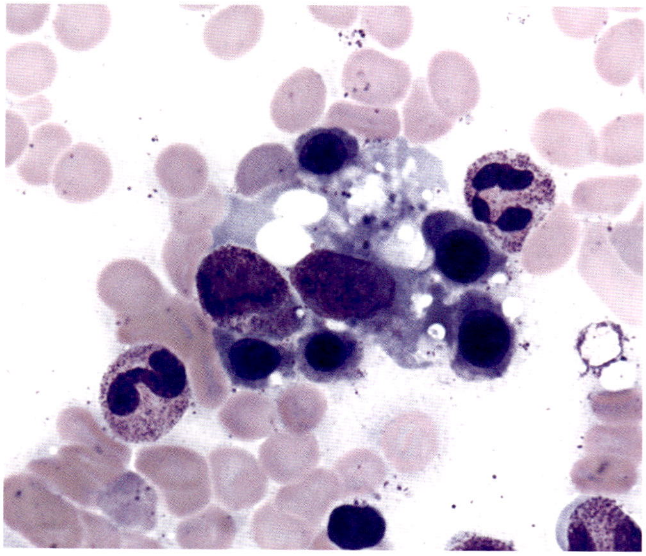

FIGURE 32.14 Several erythroid precursors intimately associated with a macrophage containing debris.

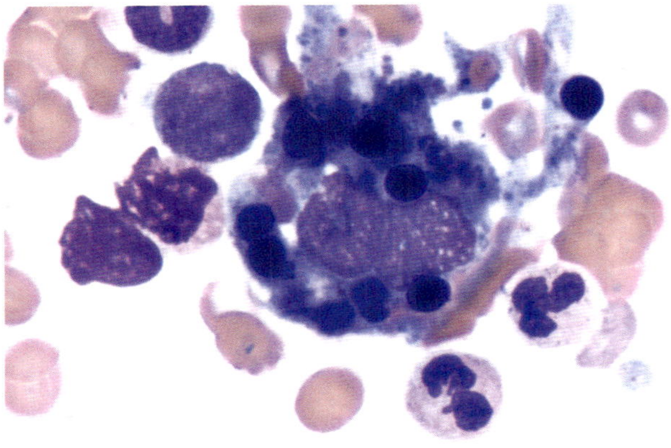

FIGURE 32.15 This hemophagocytic macrophage was encountered in an otherwise normal bone marrow. The macrophage contains an erythrocyte and multiple normoblasts in its cytoplasm.

collectively known as hemophagocytic lymphohistiocytoses (HLH). They may also be nonspecifically increased in patients with recent marrow damage of various etiologies. The presence of phagocytized nucleated cells, in particular granulocytes, and multiple ingested nucleated cells in a single macrophage are predictive of HLH as opposed to states of nonspecific increases in hemophagocytic histiocytes (66,67).

Cells morphologically resembling Gaucher cells may be seen in marrows that harbor or previously harbored conditions associated with increased cell turnover. While most commonly associated with chronic myeloid leukemia, such cells may also be encountered in the setting of other hematologic malignancies or chronic hemolytic anemia (68–74). These are terminally differentiated, long-lived cells, and can persist even after successful treatment of the underlying disease process. Notably, in the context of chronic myeloid leukemia, they are derived from the neoplastic clone and therefore can harbor *BCL/ABL1* transcripts even after successful therapy (75). Like true Gaucher cells, these "pseudo-Gaucher cells" are 20 to 100 microns in diameter, with a single, small, often eccentric nucleus and abundant cytoplasm (Fig. 32.16). The cytoplasm is pale blue to gray in Romanowsky-stained smears and eosinophilic in H&E sections. The cytoplasm has striated, fibrillary, or crinkled appearance, although the striations are typically less compact than those of true Gaucher cells. As with Gaucher cells, pseudo-Gaucher cells contain accumulated glucocerebroside, possibly as a result of liberation of membrane lipids in high turnover states, which overwhelms glucocerebrosidase in macrophages with a resultant functional enzyme deficiency. However, pseudo-Gaucher cells differ ultrastructurally from true Gaucher cells and are positive for lipid stains such as Sudan black B (76). In contrast to true Gaucher cells, pseudo-Gaucher cells lack diffuse cytoplasmic staining with Prussian blue and are only weakly positive for PAS (77).

Sea-blue histiocytes are macrophages with abundant cytoplasm that contain numerous, coarse, sea-blue granules in Wright or Giemsa-type stains (Fig. 32.17A). The nuclei are pyknotic, and either centrally or eccentrically situated. The cytoplasm is pale yellow or yellow-brown in H&E-stained sections (Fig. 32.17B). The material within sea-blue-histiocytes consists of ceroid or ceroid-like pigment. Cytochemically these cells are reactive for PAS, Sudan black B, phospholipid, and acid-fast stains. As with pseudo-Gaucher cells, sea-blue histiocytes are encountered in small numbers in states of accelerated cell turnover (78–81). Notably, both types of cells may be encountered in the same bone marrow, and histiocytes that show morphologic overlap between sea-blue histiocytes and pseudo-Gaucher cells may be seen. Sea-blue histiocytes may also be found in patients

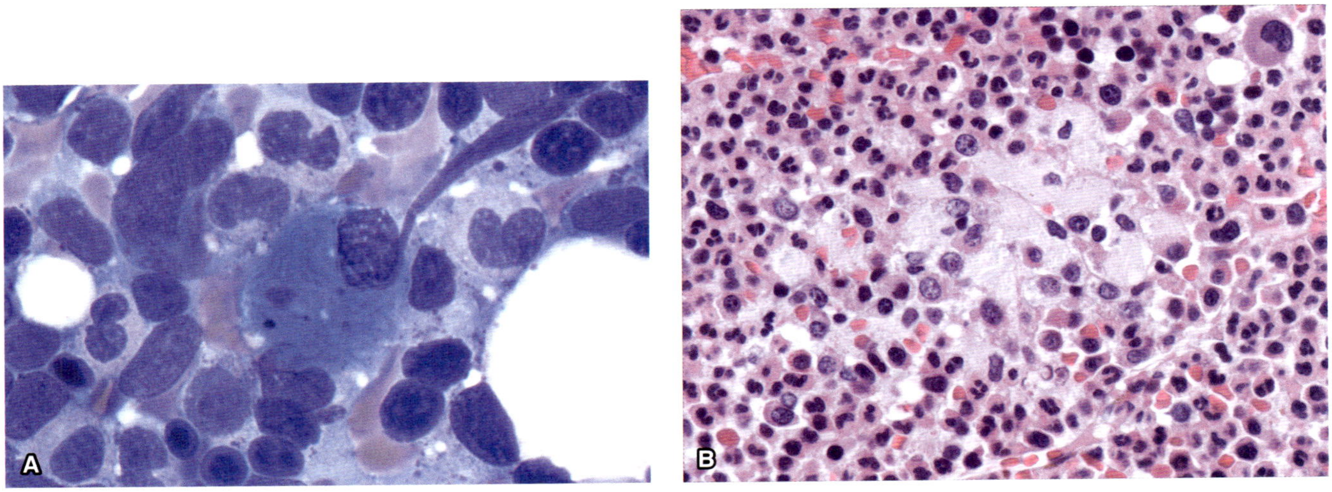

FIGURE 32.16 **A.** This pseudo-Gaucher cell was encountered in a Wright–Giemsa–stained bone marrow aspirate smear from a patient with chronic myeloid leukemia in complete remission. The cell has an eccentric nucleus and abundant basophilic cytoplasm with a distinctly striated appearance, similar to classic Gaucher cells. **B.** A cluster of pseudo-Gaucher cells in a marrow trephine from a patient with a myelodysplastic/myeloproliferative neoplasm. The cells have eccentric nuclei and abundant pink cytoplasm with a striated appearance (H&E).

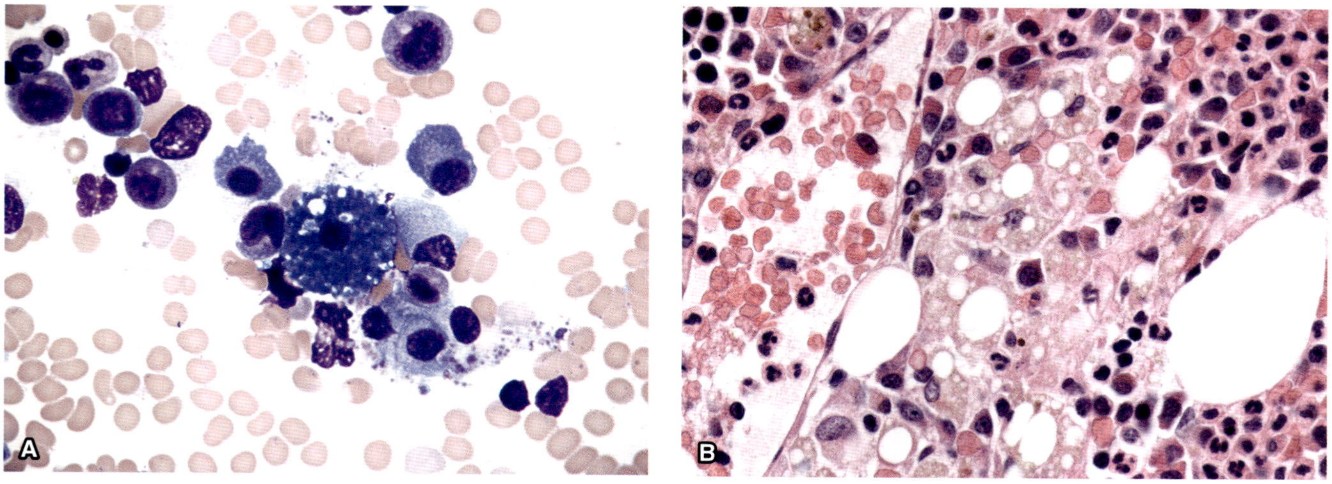

FIGURE 32.17 **A.** This sea-blue histiocyte was encountered in a Wright–Giemsa–stained bone marrow aspirate from a patient with active plasma cell myeloma. The cell has a central nucleus and abundant cytoplasm containing numerous bright blue inclusions and several vacuoles. **B.** This cluster of sea-blue histiocytes was present in the trephine biopsy from a patient with chronic myelomonocytic leukemia. In H&E sections, the abundant cytoplasm of sea-blue histiocytes stains pale yellow-brown with interspersed vacuoles.

with abnormal lipid metabolism, such as hyperlipidemia or long-term therapy with total parenteral nutrition containing lipid emulsions (82–85). In the latter setting, infiltration by sea-blue histiocytes can be extensive, resulting in cytopenias. Sea-blue histiocytes may also be a feature of some congenital storage disorders, including in adult, nonneuronopathic forms of Neimann–Pick disease (86–88).

Nonphagocytic Reticular Cells

Nonphagocytic reticular cells have an irregular or spindle shape and resemble macrophages, except that they lack large intracytoplasmic inclusions. Light microscopic cytochemical and histochemical data indicate that these cells are PAS negative, strongly positive for alkaline phosphatase, negative for acid phosphatase, negative or only weakly positive for α-naphthyl acetate esterase, and negative for stainable iron. Thus, there seems to be some overlap between the cytochemical characteristics of nonphagocytic reticular cells and macrophages (49,50). At least some of the nonphagocytic reticular cells arise from a mesenchymal stem cell capable of giving rise to colonies of fibroblast-like or myofibroblast-like cells in vitro. As mentioned earlier, nonphagocytic reticular cells play an important role in the microenvironmental regulation of hematopoiesis, both by binding to primitive hematopoietic cells (89) and producing hematopoietic growth factors both constitutively and in response to stimulation by monokines (90). Nonphagocytic reticular cells synthesize collagen (types I and III) and fibronectin.

Mast Cells

Mast cells tend to be found in association with the periphery of lymphoid aggregates, the adventitia of small arteries, adjacent to the endosteal cells of bone trabeculae, and the endothelial cells of sinusoids.

Hematopoietic stem cells generate progenitors of mast cells within the bone marrow (91), and the most mature of these cells enter the blood and migrate into tissues (92). The circulating cells, which still lack mast cell granules, migrate into the tissues, where they proliferate and mature into mast cells. Although mast cells and basophils show morphologic similarities and functional overlap, and both are derived from hematopoietic stem cells, they are distinct cell types that appear to develop through different pathways (93,94).

Unlike the water-soluble granules of basophils, those of mast cells are much less so. Nevertheless, mast cells are not easily recognized in sections of marrow stained with H&E, in which their granules are pale and refractile appearing (Fig. 32.18A). By contrast, they are readily identified in sections stained with the Giemsa stain. In such sections, mast cells have round or oval outlines and many dark purple cytoplasmic granules. The nucleus is ovoid and may be situated eccentrically. A small minority of normal mast cells have a spindled shape. Mast cells are easily highlighted in sections using immunohistochemistry for mast cell tryptase (Fig. 32.18B) or CD117, where they are seen as scattered, single cells. In marrow aspirate smears, mast cells are primarily concentrated in particles (Fig. 32.18C). In Romanowsky-stained smears, mast cells vary between 5 and 25 μm in their long axis and usually have a round or ovoid appearance (Fig. 32.18D). The cytoplasm is packed with uniform, coarse purple-black granules; in contrast to basophils, the granules in mast cells generally don't overlie the nucleus. The nucleus is small, round or oval, and either centrally or eccentrically located, with less condensed chromatin than that of basophils. The granules of mast cells are rich in heparin and stain metachromatically with toluidine blue. Mast cells are also peroxidase negative, PAS positive, acid phosphatase positive, and α-naphthol AS-D

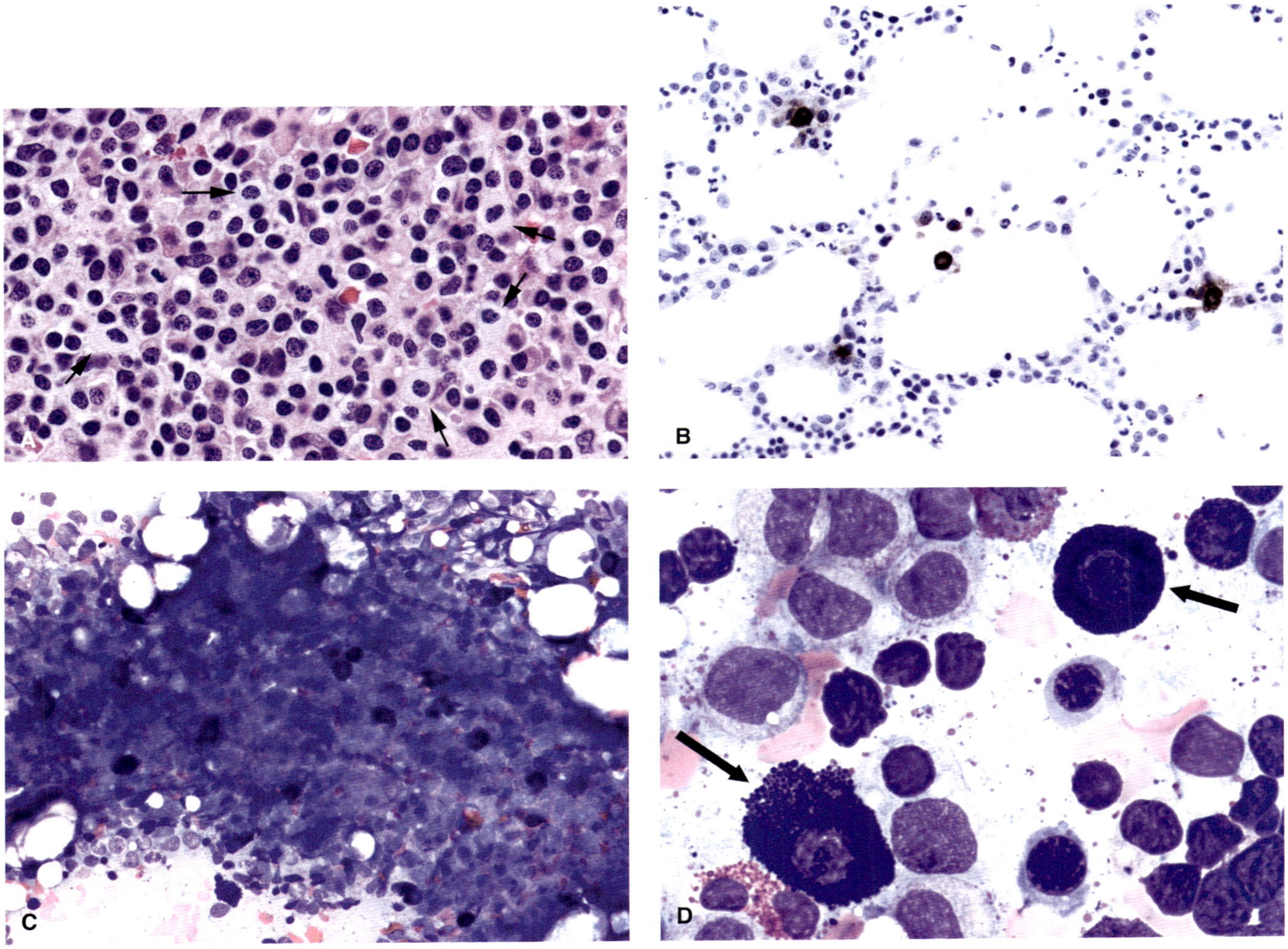

FIGURE 32.18 **A:** This lymphoid infiltrate in a bone marrow trephine biopsy contains multiple mast cells (*arrows*). These are polygonal cells with moderately abundant cytoplasm containing pale, refractile granules. **B:** Immunohistochemistry for mast cell tryptase in a normal marrow highlighting several, scattered mast cells. **C:** This marrow particle in a Wright–Giemsa–stained marrow smear with mast cell hyperplasia contains multiple, darkly staining mast cells. **D:** Two mast cells (*arrows*) in a Wright–Giemsa–stained bone marrow smear.

chloroacetate esterase positive. Unlike basophils, mast cells are capable of mitosis.

HEMATOPOIETIC CELLS

Neutrophil Precursors

Aspirate Smears

The earliest morphologically recognizable neutrophil precursor is the myeloblast. The stages through which myeloblasts mature into circulating neutrophil granulocytes are termed neutrophil promyelocytes, neutrophil myelocytes, neutrophil metamyelocytes, neutrophil bands, and segmented neutrophils (Fig. 32.19). Cell division occurs in myeloblasts, promyelocytes, and myelocytes but not in more mature cells.

A myeloblast is 10 to 20 μm in diameter. It has a large, rounded nucleus that may be flattened on one side with finely dispersed chromatin and two to five nucleoli. The nuclear-to-cytoplasmic ratio is moderately high, and the cytoplasm is moderately basophilic with no, or a few, fine granules. It is likely that only some myeloblasts mature into neutrophil promyelocytes and that others mature into eosinophil or basophil promyelocytes.

Neutrophil promyelocytes are larger than myeloblasts and have more abundant basophilic cytoplasm containing a few to many purple-red (azurophilic) primary granules. Early promyelocytes resemble myeloblasts—they are medium in size with a high nuclear-to-cytoplasmic ratio and fine chromatin—but have overtly granulated cytoplasm. More mature promyelocytes are larger, with more abundant, heavily granulated, and often eccentrically distributed cytoplasm. A paranuclear clearing corresponding to the Golgi apparatus may be evident in promyelocytes, as well. Nucleoli may be prominent. Neutrophil myelocytes have a lower nuclear-to-cytoplasmic ratio than promyelocytes as well as eccentric

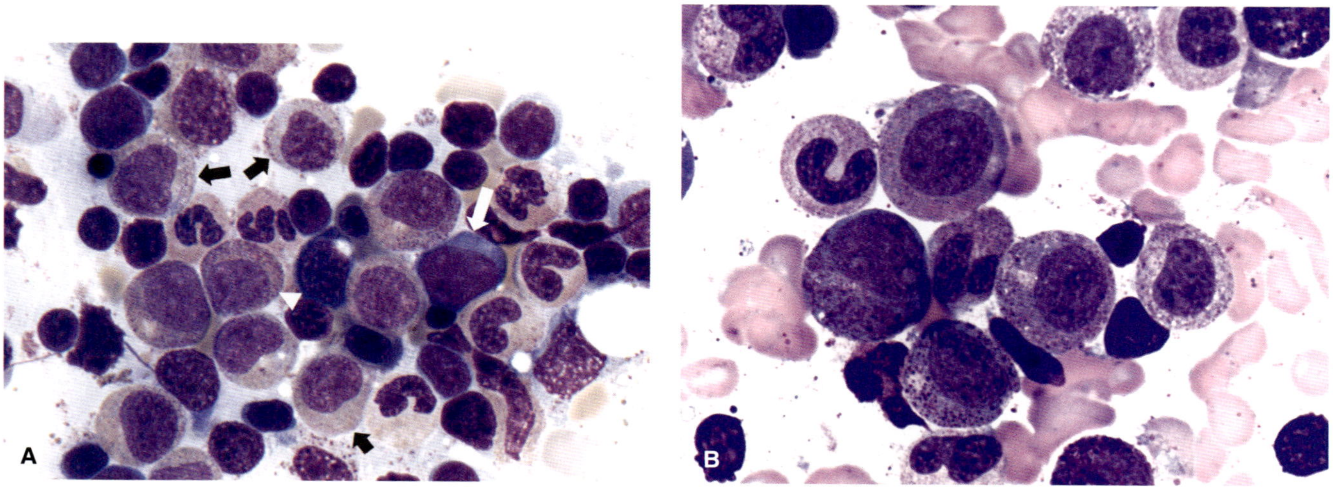

FIGURE 32.19 Neutrophil precursors from a Wright–Giemsa–stained normal marrow smear. **A:** A blast is indicated by the *white arrow*. Immediately adjacent to the blasts are five neutrophilic granulocytes. Beginning at 3 o'clock and proceeding clockwise: Band, late metamyelocyte, myelocyte, myelocyte, segmented neutrophil. A myelocyte is indicated by the *white arrowhead*. This is surrounded by four neutrophilic granulocytes. From 6 o'clock moving clockwise: Metamyelocyte, myelocyte, band, and segmented neutrophil. Several additional metamyelocytes are indicated by *black arrows*. Also note a hematogone in the upper left corner of the field. **B:** Left to right, top to bottom: Band, myelocyte, promyelocyte, late metamyelocyte, myelocyte, early metamyelocyte, myelocyte.

round or ovoid nuclei, and fine light pink or pale orange granules in the cytoplasm. The latter have been variably termed neutrophilic, secondary, or specific granules. Early myelocytes are large, have prominent residual azurophilic granules in addition to relatively few neutrophilic granules, and have relatively immature chromatin, and often distinct nucleoli. Late myelocytes are smaller, have abundant secondary granules with few or no residual azurophilic granules, have distinctly coarser chromatin, and indistinct or absent nucleoli. The neutrophil metamyelocyte has an indented nucleus, with the indentation penetrating less than half the distance from the farthest nuclear margin (95). The cytoplasm is acidophilic and contains numerous fine neutrophilic granules. Under normal conditions, no azurophilic granules are present at the metamylocyte stage and beyond, but they may be present in reactive states, such as in bacterial infections or administration of recombinant granulocyte CSF. Neutrophil bands (also called juvenile neutrophils or stab forms) have U-shaped or long, relatively narrow, band-like nuclei that may be twisted into various configurations. Early band forms are distinguished from metamyelocytes based on an indentation that penetrates greater than half the distance from the furthest nuclear margin. The nuclei contain large clumps of condensed chromatin and may show one or more partial constrictions that still contain recognizable chromatin structure. These constrictions become progressively more complete and eventually develop into fine strands or filaments that lack chromatin structure, at which point the cell is considered a segmented neutrophil. Most segmented neutrophils have two to five nuclear segments that are joined together by such filaments. Some of the neutrophil granulocytes of females have a drumstick-like nuclear appendage attached to one of the nuclear segments that represents an inactivated X chromosome.

Cytochemistry

When stained by the PAS reaction, myeloblast cytoplasm shows a diffuse, pale red-purple tinge, sometimes with fine granules of the same color. Myeloblasts either do not stain with Sudan black or show a few small sudanophilic granules near the nucleus. They are also myeloperoxidase negative and, usually, α-naphthol AS-D chloroacetate esterase negative. The cytoplasm of neutrophil promyelocytes and more mature cells of the neutrophil series stain positively with the PAS reagent, with Sudan black, and with reactions for myeloperoxidase and α-naphthol AS-D chloroacetate esterase activity. A granular staining pattern is produced with all these cytochemical reactions (Fig. 32.20). The intensity of staining increases in cell classes of increasing maturity with the PAS reaction and, to a lesser extent, with Sudan black. Promyelocytes and neutrophil myelocytes, but not neutrophil granulocytes, stain for α-naphthyl acetate esterase activity and, more weakly, for α-naphthyl butyrate esterase activity. Acid phosphatase activity is present in cells at and after the promyelocyte stage; this activity is strongest in the immature cells and weak in neutrophil granulocytes. A few neutrophil metamyelocytes stain weakly for alkaline phosphatase activity, and segmented neutrophil granulocytes stain with variable intensity (weak to strong) (96–99). Immunocytochemical studies indicate that both lysozyme (muramidase) and elastase are present in promyelocytes and all of the more mature cells of the neutrophil series, and that lactoferrin is present in neutrophil myelocytes, metamyelocytes, and granulocytes.

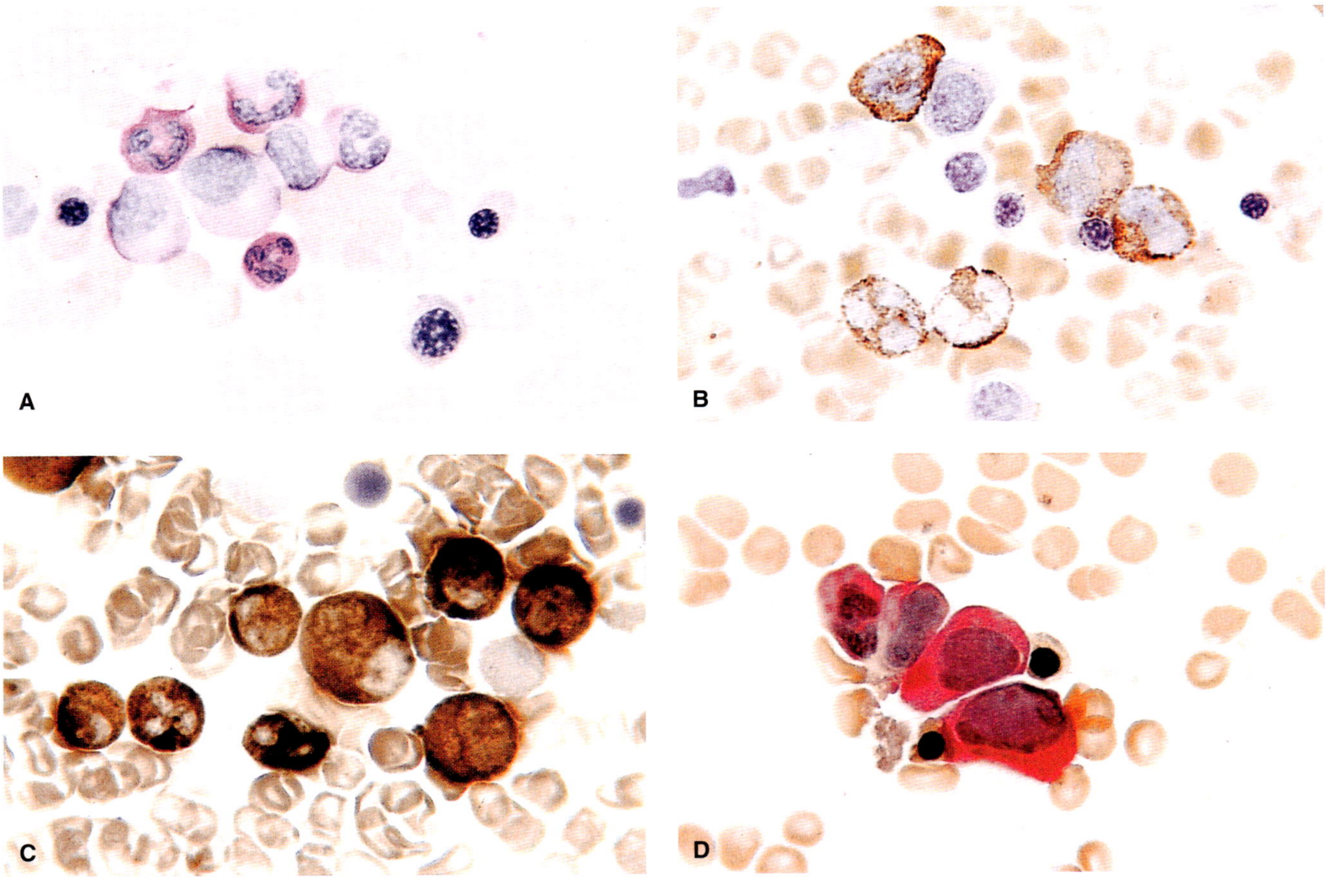

FIGURE 32.20 Cytochemical reactions of neutrophil precursors and neutrophil granulocytes. **A:** Faint PAS positivity in neutrophil myelocytes and stronger positivity in neutrophil granulocytes. The three erythroblasts are PAS negative. **B:** Sudan black positivity in two neutrophil myelocytes, one eosinophil myelocyte, a neutrophil metamyelocyte, and a neutrophil granulocyte. The lymphocytes and erythroblasts are sudanophobic. **C:** Strong peroxidase positivity in neutrophil myelocytes and granulocytes; *p*-phenylene diamine and catechol were used as the substrate. **D:** Alpha-naphthol AS-D chloroacetate esterase positivity in three neutrophil myelocytes and a neutrophil granulocyte. The two erythroblasts have not stained. The diazonium salt of fast violet-red LB was used as the capture agent.

Biopsy Sections

In marrow sections, early granulopoietic cells (myeloblasts and promyelocytes) are mainly found near the endosteum of bone trabeculae and the adventitial aspects of arterioles (Fig. 32.21A). Maturing granulocyte precursors radiate outward from these sites, and later-stage neutrophil granulocytes often are found in the center of intertrabecular areas (Fig. 32.21B), adjacent to sinusoids. A few promyelocytes and myelocytes are present singly or in small clusters at sites away from bone trabeculae and blood vessels (Fig. 32.21C).

Myeloblasts are not identifiable in H&E-stained sections of normal bone marrows. More mature granulocyte precursors are identifiable based on their moderately abundant, granular, eosinophilic cytoplasm (Fig. 32.21). Promyelocytes have round or ovoid, often eccentric nuclei, finely stippled chromatin, and one to several small, regular nucleoli. Myelocytes are cytologically similar but have more clumped chromatin and indistinct or absent nucleoli. More mature granulocytic elements are easily identifiable based on their characteristic nuclear conformation, although in histologic sections the fine chromatin strands that join the nuclear lobes of granulocytes usually are not seen. Maturing granulocytic elements may be highlighted by immunohistochemistry for myeloperoxidase (Fig. 32.22), which is contained within neutrophil primary granules. Although primary granules are present in all granule-containing cells of the neutrophil series, they normally lose their azurophilic property with maturation.

Eosinophil and Basophil Precursors

Aspirate Smears

The eosinophil and basophil granulocytes develop through stages that are essentially similar to those through which the neutrophil granulocytes develop. The earliest morphologically recognizable precursors are cells in which a few eosinophil or basophil granules have formed, that is, the eosinophil promyelocytes and basophil promyelocytes.

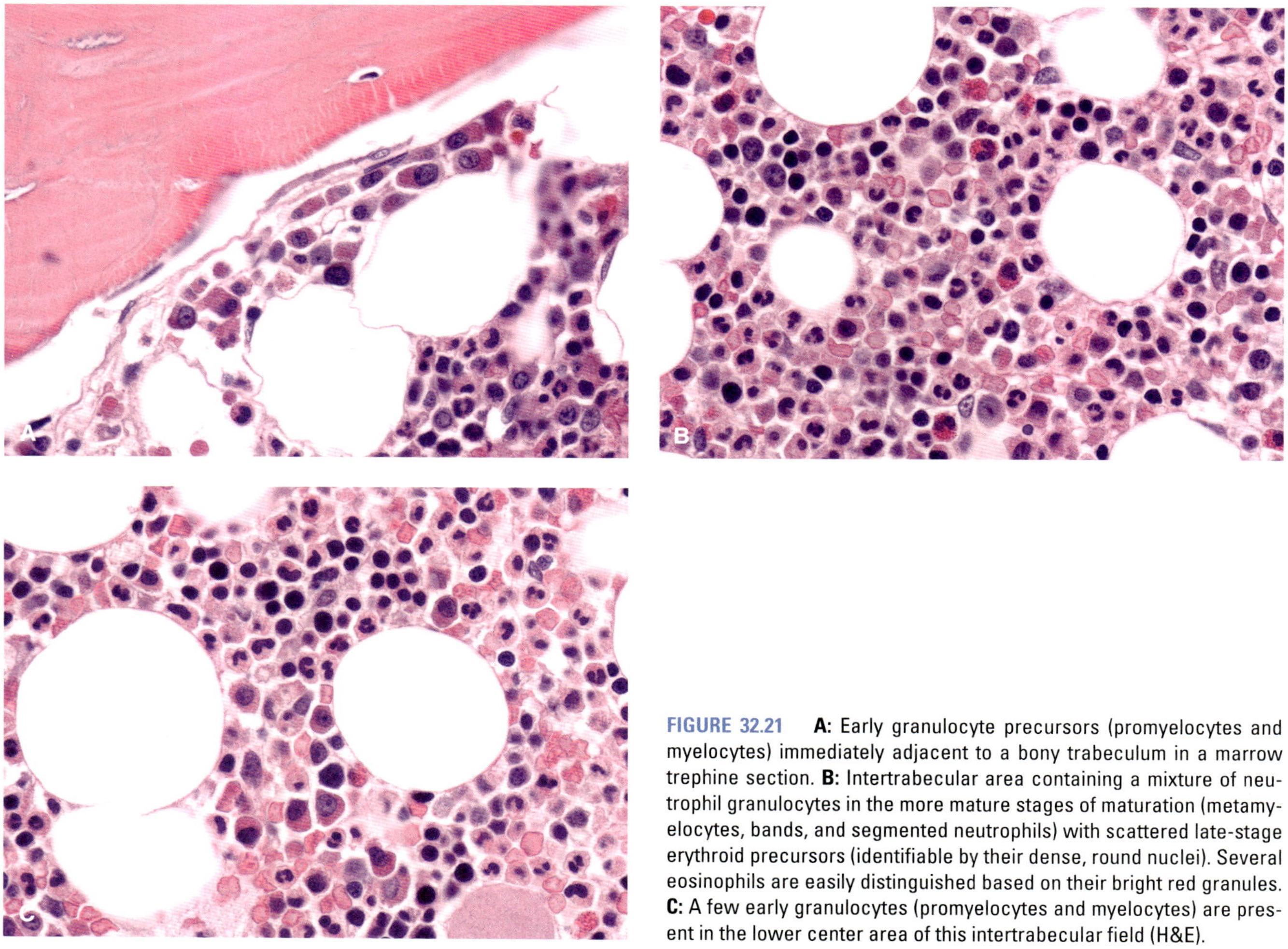

FIGURE 32.21 **A:** Early granulocyte precursors (promyelocytes and myelocytes) immediately adjacent to a bony trabeculum in a marrow trephine section. **B:** Intertrabecular area containing a mixture of neutrophil granulocytes in the more mature stages of maturation (metamyelocytes, bands, and segmented neutrophils) with scattered late-stage erythroid precursors (identifiable by their dense, round nuclei). Several eosinophils are easily distinguished based on their bright red granules. **C:** A few early granulocytes (promyelocytes and myelocytes) are present in the lower center area of this intertrabecular field (H&E).

Eosinophil promyelocytes have rounded nuclei with dispersed chromatin and nucleoli. They contain two types of granules: large red-orange granules and large, dark purple granules which may overlie the nucleus (Fig. 32.23A). Eosinophil myelocytes have smaller nuclei with more clumped chromatin, a lower nuclear-to-cytoplasmic ratio, and predominantly eosinophilic granules with few or no basophilic granules (Fig. 32.23B). Eosinophil metamyelocytes and granulocytes have only large eosinophilic granules (Fig. 32.23C,D). Basophil myelocytes, metamyelocytes, and granulocytes contain large, round, deeply basophilic granules that often overlie the nucleus (Fig. 32.24); the more mature granules stain metachromatically with toluidine blue. The majority of circulating eosinophil and basophil granulocytes have two nuclear segments.

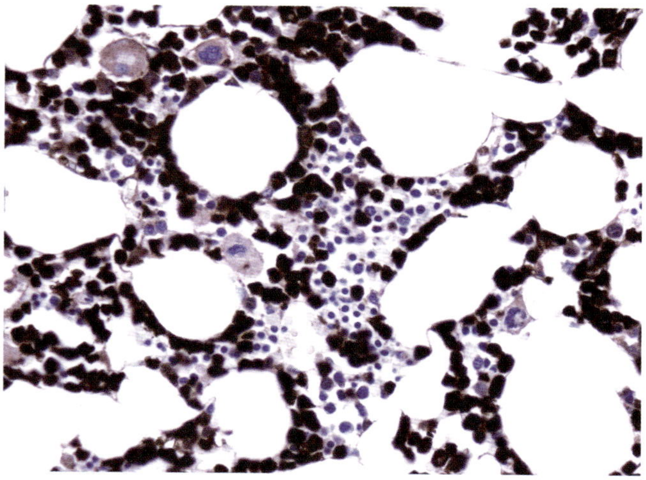

FIGURE 32.22 Immunohistochemistry for myeloperoxidase in a normal marrow section. The maturing granulocytes stain dark brown.

Cytochemistry

The granules of eosinophil and basophil granulocytes and their precursors do not stain by the PAS reaction (99,100). However, PAS-positive deposits are found between the specific granules in both cell lineages. The periphery of the eosinophil granules of all cells of the eosinophil series stains strongly with Sudan black, and the core stains weakly or not at all. Basophil granules are strongly sudanophilic in basophil promyelocytes and myelocytes, but the degree of sudanophilia decreases with increasing maturity; in mature basophils, the granules either do not stain or stain red (i.e., metachromatically). Peroxidase and acid phosphatase, but not lysozyme, are demonstrable in the eosinophil granules in

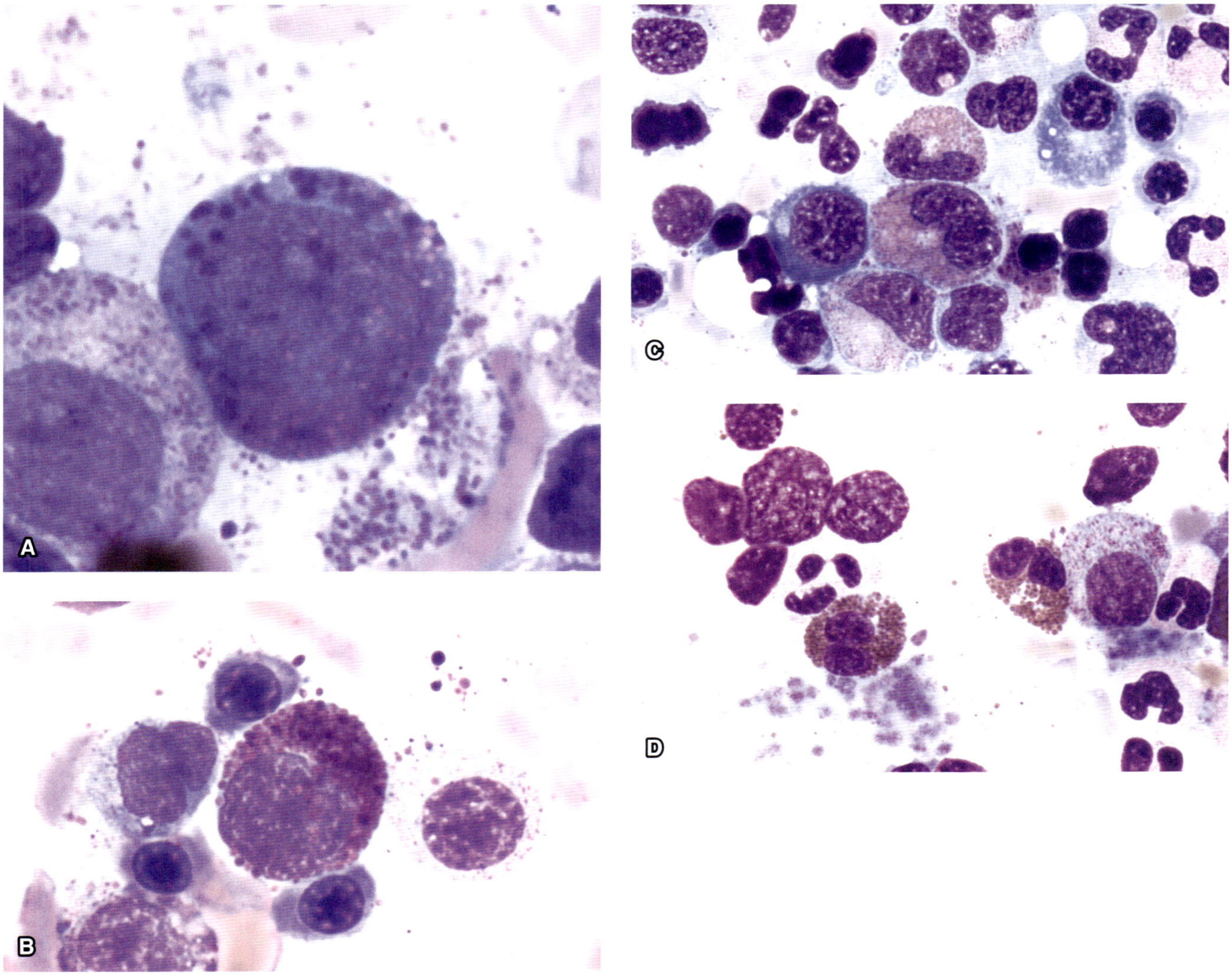

FIGURE 32.23 Eosinophil maturation sequence in Wright–Giemsa–stained bone marrow aspirates. **A:** An eosinophil promyelocyte containing both eosinophilic granules and dark basophilic granules, some of which overlie the nucleus. The nucleus is oval, with fine chromatin and a prominent nucleolus. **B:** An eosinophil myelocyte. Few residual basophilic granules remain, and the chromatin is coarser than in the promyelocyte in (**A**). **C:** An eosinophil metamyelocyte (**bottom**) and an eosinophil band (**top**). Only eosinophilic granules remain. **D:** Two segmented eosinophils.

all eosinophil precursors and eosinophils. Human eosinophil peroxidase is biochemically and immunochemically distinct from myeloperoxidase, the type of peroxidase present in the neutrophil series. In the basophil series, the granules are strongly positive for peroxidase in basophil promyelocytes and myelocytes, weakly positive in basophil metamyelocytes, and almost negative in basophil granulocytes. Basophil granules stain positively for acid phosphatase. Basophil and eosinophil granulocytes are essentially negative for α-naphthol AS-D chloroacetate esterase and α-naphthyl butyrate esterase.

Eosinophil granules contain eosinophil cationic proteins and an arginine- and zinc-rich major basic protein that are involved in the killing of metazoan parasites. The major basic protein also stimulates basophils and mast cells to release histamine. Other constituents of eosinophil granules include histaminase and arylsulfatase, which are involved in the modulation of immediate-type hypersensitivity reactions. Basophil granules contain chondroitin sulfate and heparan sulfate, which account for their property of staining metachromatically (red-violet) with toluidine blue. They also contain histamine, one of the substances released when immunoglobulin E (IgE)-coated basophils react with specific antigens.

Biopsy Sections

In H&E tissue sections, eosinophils and precursors are easily discriminated from other cellular elements based on their large, bright red, refractile granules (Fig. 32.21B). Because basophil granules are water soluble, their contents become extracted during routine fixation for histologic studies. Consequently, basophil granulocytes cannot be seen in histologic sections processed in the usual manner.

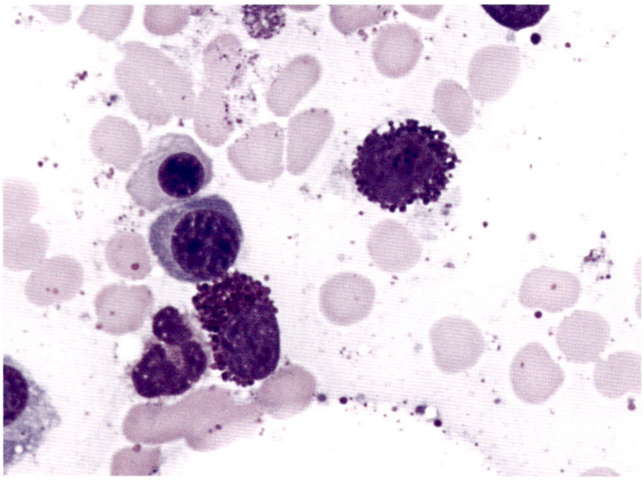

FIGURE 32.24 A basophil metamyelocyte from a Wright–Giemsa–stained normal marrow smear (**upper right**). A late eosinophil myelocyte is also present (**lower left**).

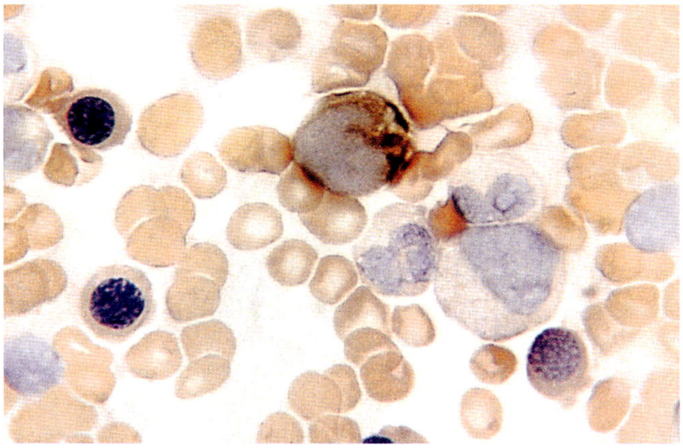

FIGURE 32.25 A cell with strong α-naphthyl acetate esterase activity from a normal marrow smear (the diazonium salt of fast blue BB was used as the capture agent). This cell has a slightly convoluted nucleus and relatively little cytoplasm and is most probably a monoblast or early promonocyte.

Monocyte Precursors

Aspirate Smears

Immature monocytic cells are often designated as "monoblasts" and "promonocytes." However, it is worth noting that the monoblasts and promonocytes seen in acute myeloid leukemias do not closely resemble any normal marrow precursor, and thus the use of these terms to describe monocytic precursors in normal marrows may be misleading. Blood monocytes are not end cells but develop further in the tissues to become macrophages. Some data suggest that macrophages and osteoclasts have a common progenitor. All these cells are considered to constitute the mononuclear phagocyte system.

In normal marrows, monocytes, including immature monocytes, are inconspicuous, but they may increase in number in reactive and regenerative states. Monoblasts are not identifiable with certainty in Romanowsky-stained marrow smears, although blastic cells exhibiting monocytic differentiation may be identified with cytochemical stains (Fig. 32.25). The most immature identifiable monocytes in marrow smears are large cells with mildly indented nuclei, slightly coarse chromatin, small or inconspicuous nucleoli, and small amounts of moderately basophilic cytoplasm containing few or no granules (Fig. 32.26). As they mature, the nuclei become more obviously indented; the cytoplasm remains basophilic but is more abundant with more obvious granularity (Fig. 32.26). Mature marrow monocytes are smaller with more deeply folded nuclei with coarser, ropy chromatin, a higher nuclear-to-cytoplasmic ratio, and more grayish cytoplasm (Fig. 32.26B,C). Note that mature marrow monocytes are often smaller and more compact-appearing than peripheral blood monocytes. The latter are larger cells with more abundant, sometimes vacuolated, pale gray-blue cytoplasm, and eccentrically placed, ovoid, kidney-shaped, or horseshoe-shaped nuclei (Fig. 32.26B). Monocytes may be hyperplastic and/or left shifted in certain reactive states (Fig. 32.26C) (101).

Cytochemistry

Some normal monocytes show several fine or moderately coarse PAS-positive granules and sudanophilic granules and a few peroxidase-positive granules scattered in their cytoplasm (96,99,102). Monocytes do not stain for alkaline phosphatase but stain strongly for acid phosphatase. They contain lysozyme.

Monocytes are α-naphthol AS-D chloroacetate esterase negative but are α-naphthyl acetate esterase positive. Alpha-naphthyl acetate esterase activity is present not only in monocytes and macrophages, but also in other myeloid cells, including neutrophil promyelocytes and myelocytes, megakaryocytes, and immature red cell precursors. Alpha-naphthyl butyrate esterase activity is stronger than α-naphthyl acetate esterase activity in monocytes and macrophages and is much weaker in the other types of myeloid cells mentioned above. Both the α-naphthyl acetate and the α-naphthyl butyrate esterase activities of monocytes are inhibited by fluoride; in granulocytes and their precursors, these enzyme activities are fluoride insensitive.

Biopsy Sections

Monocytes are not identifiable with certainty in normal biopsy sections.

Red Cell Precursors

Aspirate Smears

In this chapter, the term *erythroblast* is used to describe any nucleated red cell precursor, normal or pathologic, and the term *normoblast* to describe all cells that have the morphologic characteristics of the erythroblasts found in normal bone marrow. The terms used to describe various classes of normal red cell precursors are, in order of increasing maturity, pronormoblast, basophilic normoblast, early polychromatic normoblast, late polychromatic normoblast, marrow reticulocyte, and blood

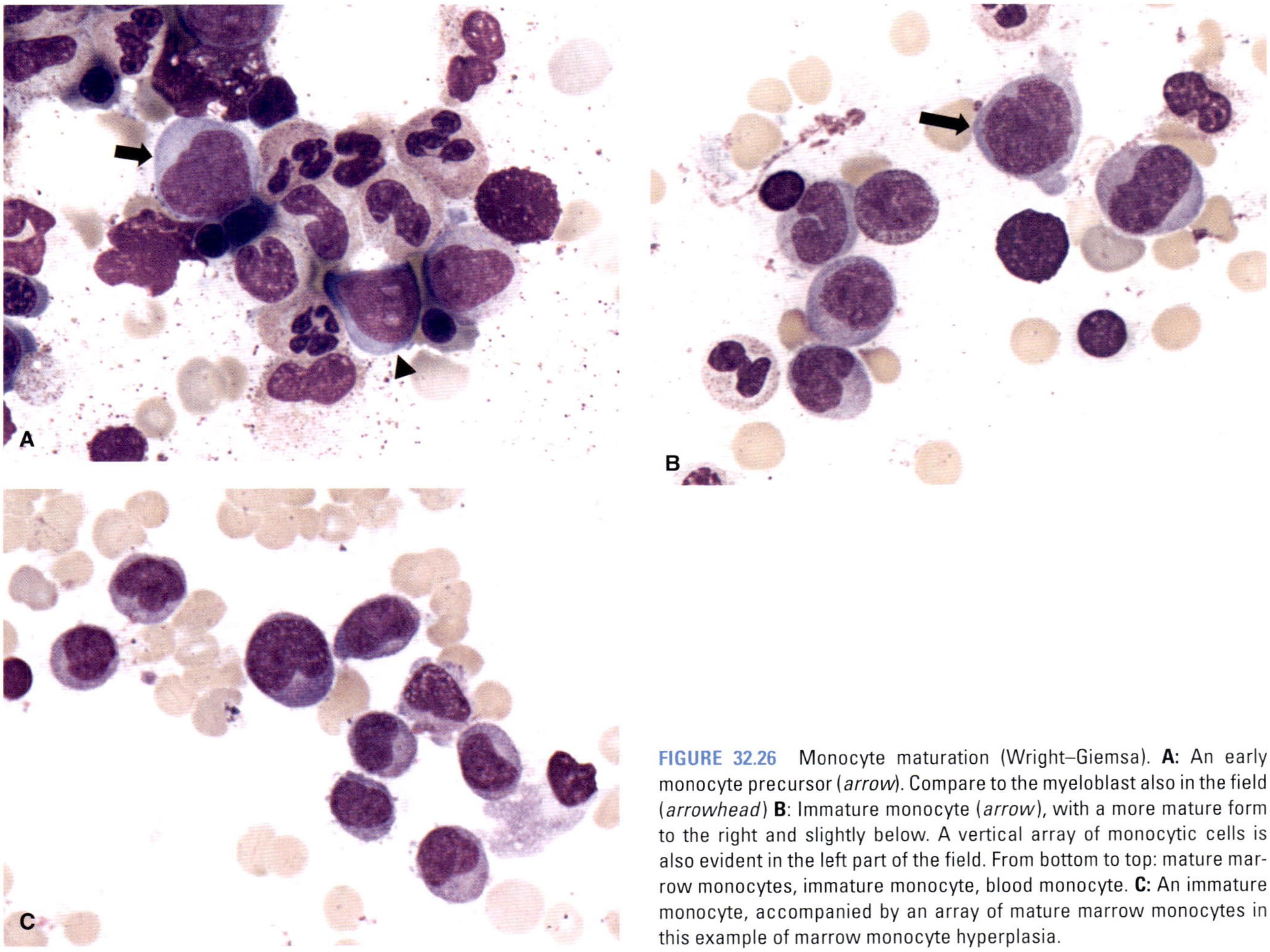

FIGURE 32.26 Monocyte maturation (Wright–Giemsa). **A:** An early monocyte precursor (*arrow*). Compare to the myeloblast also in the field (*arrowhead*) **B**: Immature monocyte (*arrow*), with a more mature form to the right and slightly below. A vertical array of monocytic cells is also evident in the left part of the field. From bottom to top: mature marrow monocytes, immature monocyte, blood monocyte. **C:** An immature monocyte, accompanied by an array of mature marrow monocytes in this example of marrow monocyte hyperplasia.

reticulocyte (Fig. 32.27). Cell division occurs only in the first three of these cytologic classes. Marrow samples containing normoblasts are said to show normoblastic erythropoiesis.

Pronormoblasts are large cells with a diameter of 12 to 20 μm. They have distinctly round nuclei and have moderate amounts of deeply basophilic, agranular cytoplasm distributed as an even rim around the nucleus. The chromatin has a finely stippled appearance, and there are one or more prominent, irregular nucleoli. The basophilic normoblasts resemble pronormoblasts except that their chromatin is slightly more condensed and consequently has a coarsely granular appearance. The early polychromatic normoblasts are smaller than basophilic normoblasts and have a smaller nucleus and a lower nuclear-to-cytoplasmic ratio. The cytoplasm is polychromatic and agranular, and the nucleus contains several medium-sized clumps of condensed chromatin, particularly adjacent to the nuclear membrane. The polychromasia results from the presence of moderate quantities of cytoplasmic RNA that stains blue and hemoglobin that stains red. Late polychromatic normoblasts are even smaller and show a further reduction in the ratio of the area of the nucleus to the area of the cytoplasm. The cytoplasm is predominantly orthochromatic but is faintly polychromatic with a grayish tinge. The nucleus is small and eccentric and contains large clumps of condensed chromatin. The nuclear diameter is less than about 6.5 μm. When mature, late polychromatic normoblasts extrude their nuclei and become marrow reticulocytes; the extruded nuclei are rapidly phagocytosed and degraded by adjacent macrophages. The marrow reticulocyte is irregular in outline and has faintly polychromatic cytoplasm. It is motile and soon enters the marrow sinusoids. When marrow and blood reticulocytes are stained supravitally with brilliant cresyl blue, the ribosomal RNA responsible for their polychromasia precipitates into a basophilic reticulum, hence the term *reticulocyte*. Reticulocytes circulate in the blood for 1 to 2 days before becoming mature red cells. The average volume of blood reticulocytes is 20% larger than that of red cells. The latter are circular, biconcave, and acidophilic (i.e., stain red) and, in dried fixed smears, have an average diameter of 7.2 μm (range: 6.7 to 7.7 μm).

Cytochemistry

Normal erythroblasts are PAS negative. They also fail to stain with Sudan black and are peroxidase negative. Most

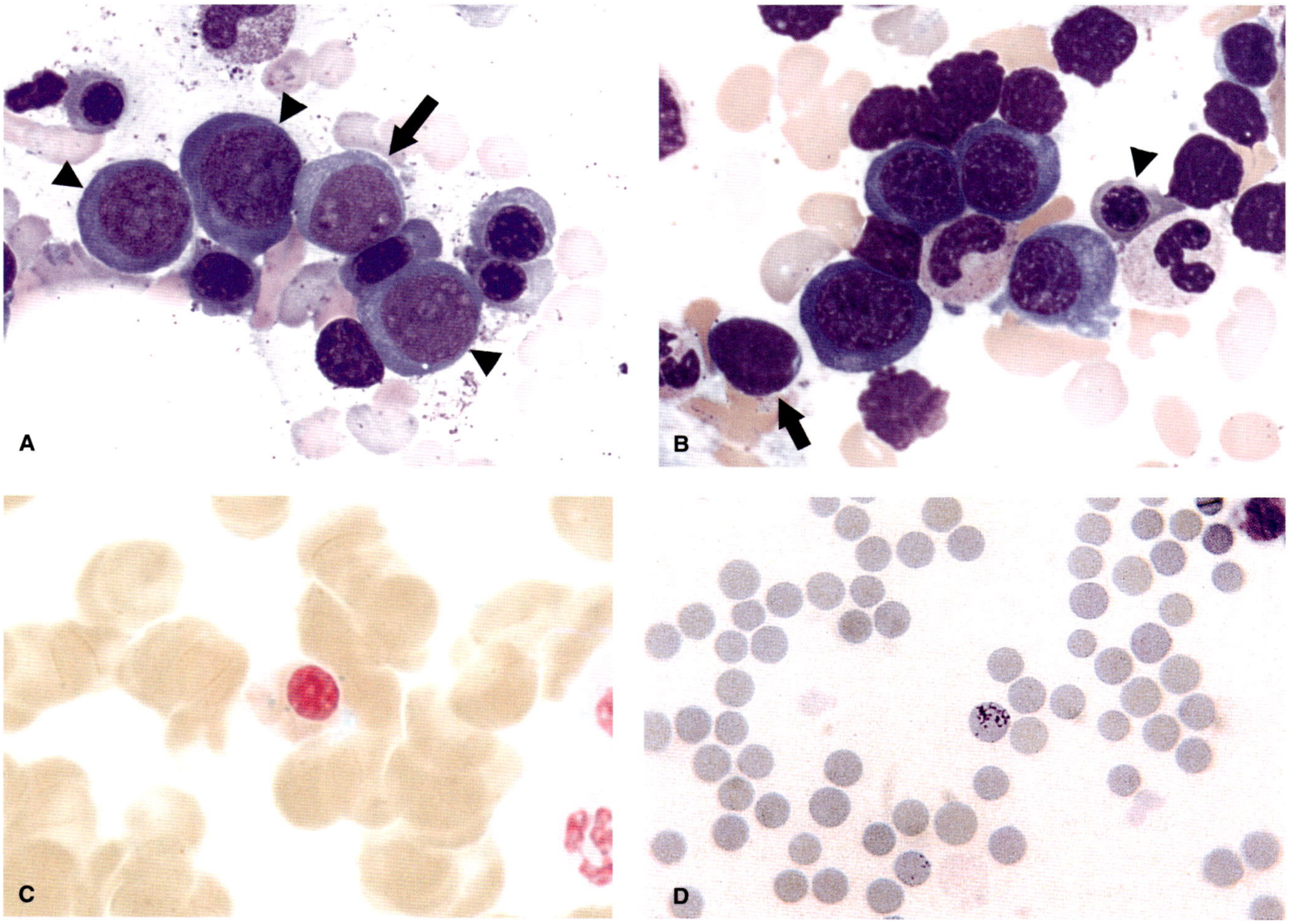

FIGURE 32.27 Red cell precursors from a normal Wright–Giemsa–stained bone marrow smear (**A–C**) and a reticulocyte from normal peripheral blood (**D**). **A:** Three pronormoblasts (*arrowheads*) and five polychromatophilic normoblasts. A myeloblast (*arrow*) is also present. **B:** Four basophilic normoblasts and a late polychromatic normoblast (*arrowhead*). Also present is a hematogone (maturing B-cell precursor, *arrow*) **C:** A sideroblast showing two fine, blue siderotic granules. (**A** and **B**, Wright–Giemsa stain; **C**, Perls acid ferrocyanide reaction; **D**, supravital staining with brilliant cresyl blue.)

nucleated red cells are α-naphthol AS-D chloroacetate esterase negative, but occasional cells show a few positive granules. A few α-naphthol butyrate esterase-positive granules are seen in some nucleated red cells of all degrees of maturity; the positive granules are sometimes seen at the nuclear margin. Coarse acid phosphatase–positive paranuclear granules are frequently present in all types of erythroblasts.

In normal bone marrow smears stained by Perls acid ferrocyanide method, 30% to 50% of the polychromatic erythroblasts contain one to five small blue granules that are usually just visible at high magnification (Fig. 32.27C). These iron-containing granules are randomly distributed within the cytoplasm. Erythroblasts containing siderotic granules are termed *sideroblasts*. In iron-deficiency anemia and anemia of chronic disease (anemia of inflammation), the percentage of sideroblasts is decreased, or sideroblasts may be altogether absent. The percentage of sideroblasts, the average number of siderotic granules per cell, and the average size of such granules are increased in hemolytic anemias and other conditions associated with an increased percentage saturation of transferrin. Normal sideroblasts must be distinguished from ring sideroblasts, a pathologic form in which iron is abnormally deposited in mitochondria. In ring sideroblasts, the siderotic granules are increased in number and closely opposed to the nucleus, rather than randomly distributed in the cytoplasm (Fig. 32.28). The most commonly employed definition of a ring sideroblast requires at least five granules extending around at least one-third of the nuclear circumference (103).

Biopsy Sections

Erythroblasts of varying degrees of maturity are found in aggregates or islands in marrow sections (Fig. 32.29). Pronormoblasts and basophilic normoblasts are large cells with round nuclei and distinct nuclear membranes,

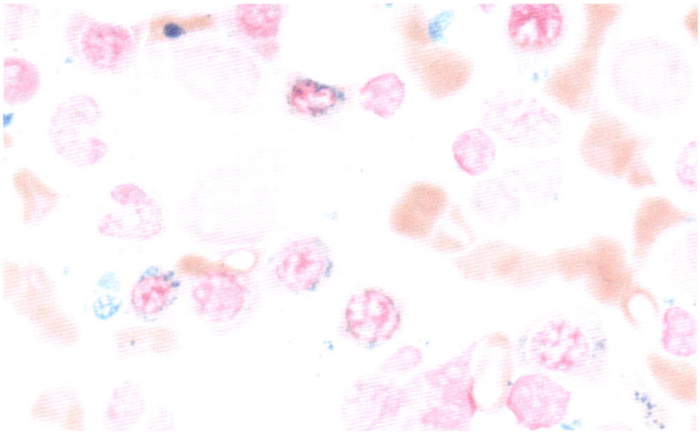

FIGURE 32.28 Multiple ring sideroblasts in a patient with myelodysplastic syndrome with ring sideroblasts and *SF3B1* mutation. The erythroblasts contain numerous siderotic granules that ring the nuclei of the cells (Perls acid ferrocyanide reaction).

stippled chromatin, and irregular nucleoli. Their cytoplasm is densely amphophilic, in contrast to the granular eosinophilic appearance of early granulocyte precursors. The late erythroblasts contain round, heavily stained nuclei showing little structural detail, and have moderate quantities of poorly staining cytoplasm, usually with a distinct cytoplasmic membrane. Intermediate-to-late erythroblasts can be reliably identified by immunohistochemical staining for glycophorin A or hemoglobin A (Table 32.1, Fig. 32.30). Normal pronormoblasts should lack both of these antigens, but are positive for E-cadherin, which is absent in more mature forms (104). CD71 (transferrin receptor) may be used to identify all stages of erythroblasts (105). Similarly, spectrin marks all stages of erythroblast maturation, with diffuse cytoplasmic staining of early forms and membranous staining of later stages and mature erythrocytes (106).

Dyserythropoiesis and Ineffective Erythropoiesis

Most of the erythroblasts in normal bone marrow are uninucleate and do not display any unusual morphologic features. However, when 400 to 1,000 consecutive erythroblasts (excluding mitoses) were studied in bone marrow smears from each of 10 healthy volunteers with stainable iron in the bone marrow, 0% to 0.57% (mean: 0.31%) were found to be binucleate, 0.7% to 4.8% (mean: 2.4%) showed intererythroblastic cytoplasmic bridges, 0% to 0.9% (mean: 0.24%) showed cytoplasmic stippling, and 0% to 0.7% (mean: 0.39%) showed cytoplasmic vacuolation.

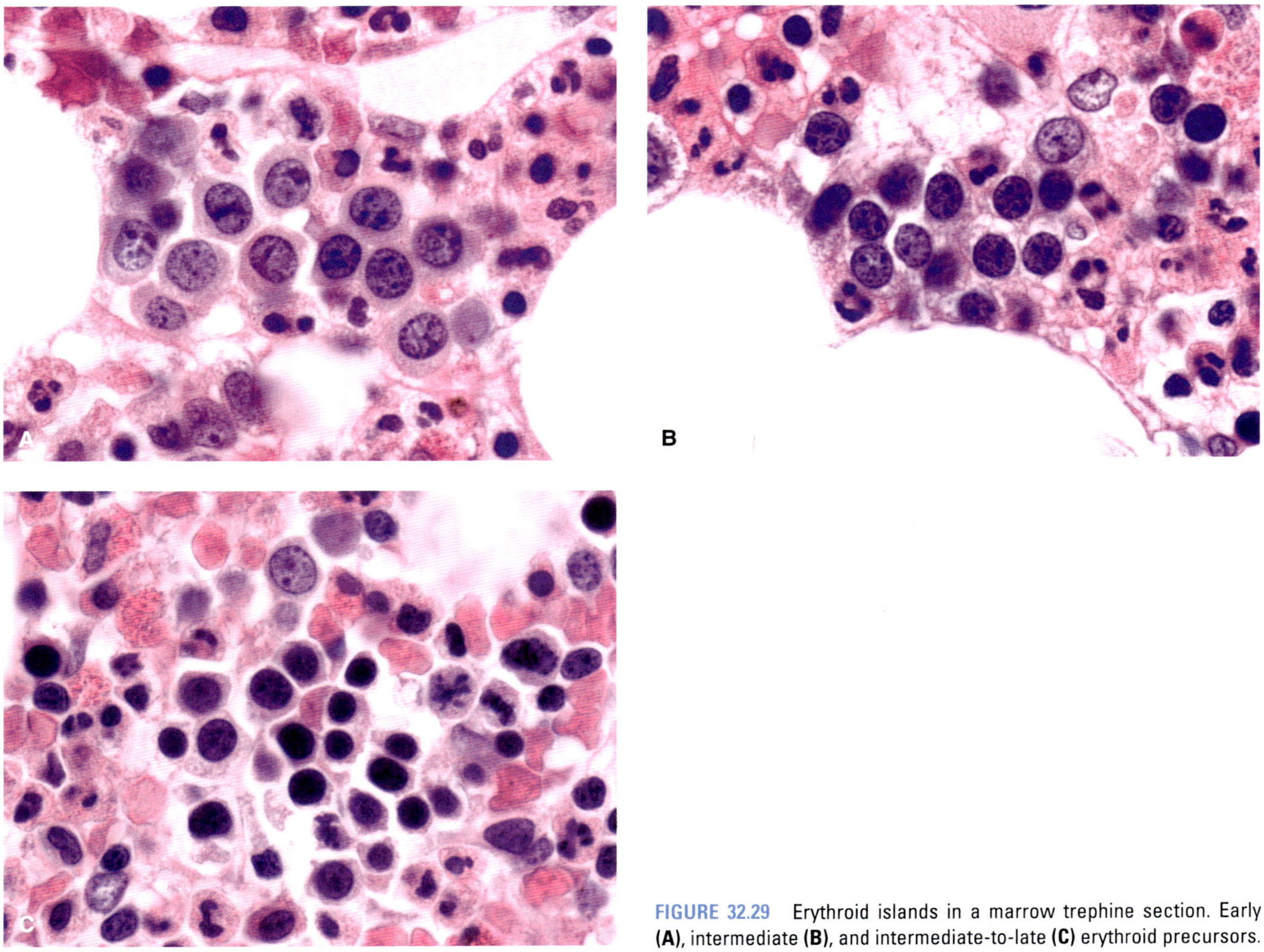

FIGURE 32.29 Erythroid islands in a marrow trephine section. Early **(A)**, intermediate **(B)**, and intermediate-to-late **(C)** erythroid precursors.

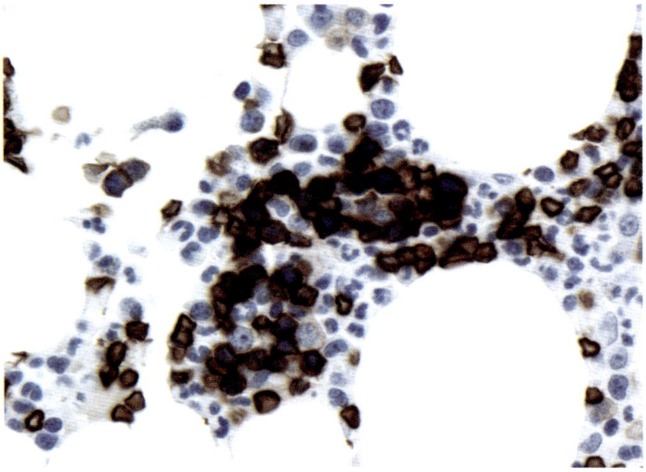

FIGURE 32.30 Immunohistochemistry for glycophorin A highlighting both maturing normoblasts and mature erythrocytes.

In addition, 0% to 0.55% (mean: 0.22%) had markedly irregular nuclear outlines or karyorrhectic nuclei, and 0% to 0.39% (mean: 0.18%) contained Howell–Jolly bodies (micronuclei), a marker of chromosome breaks (Fig. 32.31) (107). In another study of 15 healthy males in which 5,000 erythroid cells (including mitoses) were assessed per subject, 0.14% ± 0.04 (SD) were found to be binucleate or multinucleate cells or to be pluripolar mitoses (108). When there is increased erythropoietin (either endogenously or exogenously), in addition to an expansion of the number of erythroid precursors, there may be a shift toward immaturity, increased binucleation, megaloblastoid change (dyssynchrony between nuclear and cytoplasmic maturation), and nuclear irregularity of the late-stage nucleated red cells (terminal dyserythropoiesis) (Fig. 32.32).

The phrase *ineffective erythropoiesis* is used to describe the loss of potential erythrocytes due to the phagocytosis and destruction of developing erythroblasts within the bone marrow. The extent of ineffective erythropoiesis in normal bone marrow is small (46). In a number of conditions, such as homozygous β-thalassemia and megaloblastic anemia, there

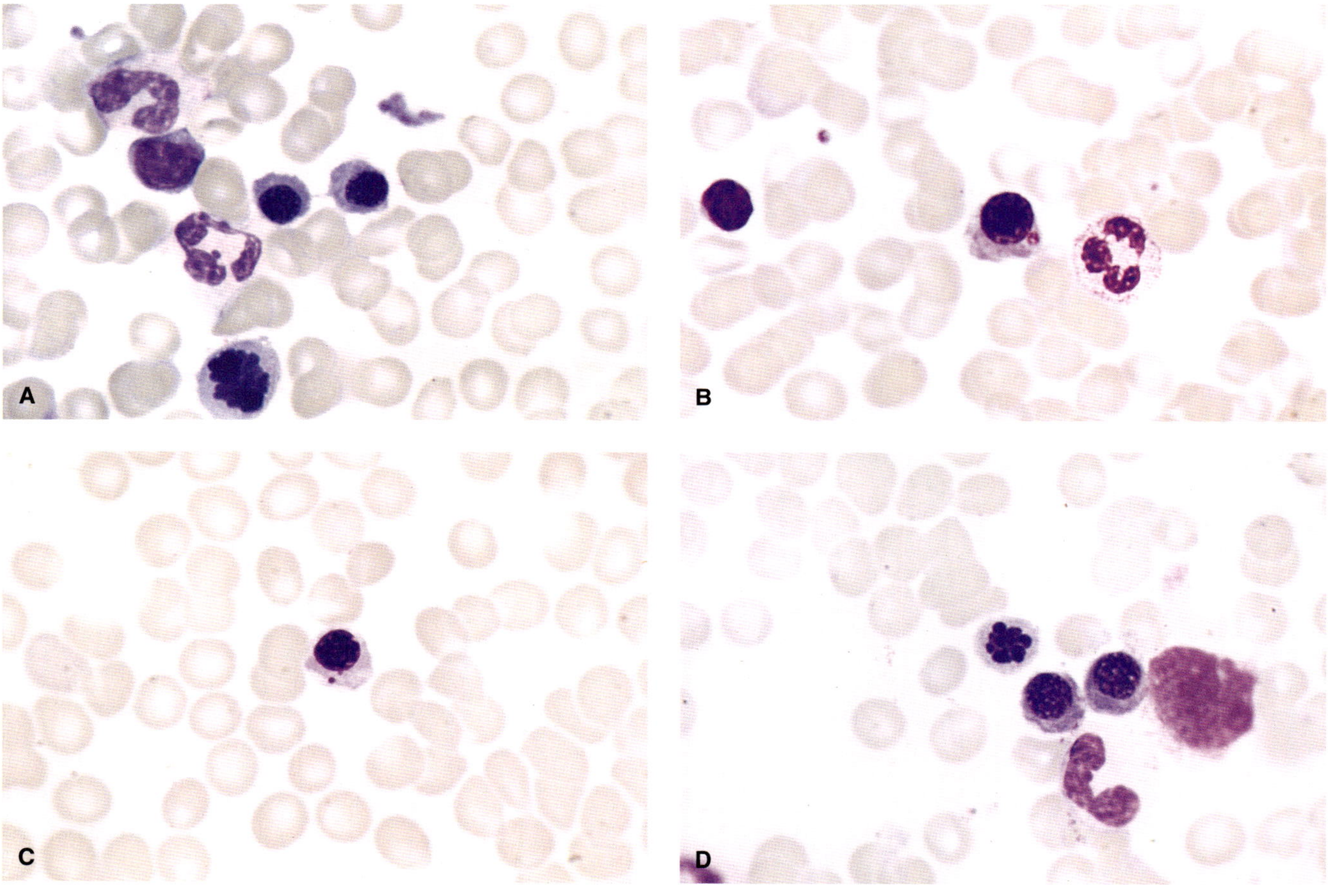

FIGURE 32.31 Morphologic evidence of dyserythropoiesis in bone marrow smears from healthy volunteers. **A:** Intererythroblastic cytoplasmic bridge. **B:** Large Howell–Jolly body in an early polychromatic erythroblast. **C:** Two smaller Howell–Jolly bodies in a late polychromatic erythroblast. **D:** Karyorrhexis in a late polychromatic erythroblast.

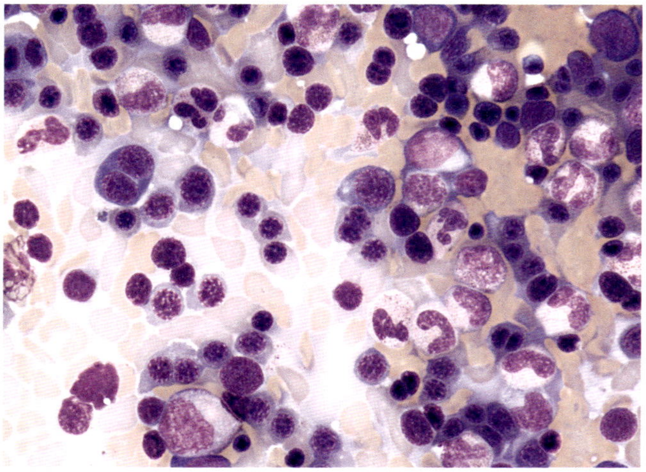

FIGURE 32.32 Erythroid hyperplasia in a patient with hemolytic anemia. There are multiple binucleate erythroid precursors and one with nuclear irregularity (**center**).

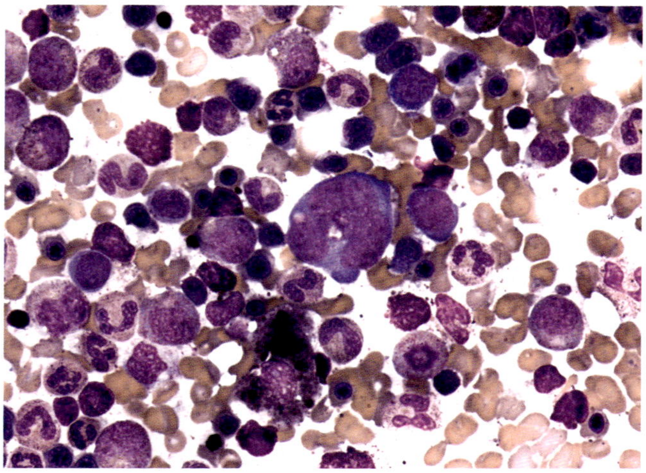

FIGURE 32.33 This promegakaryocyte in a Wright–Giemsa preparation demonstrates a high nuclear-to-cytoplasmic ratio, a C-shaped nucleus, and basophilic cytoplasm devoid of lavender granulation.

is a gross increase in the ineffectiveness of erythropoiesis; some of the abnormal erythroblasts undergo apoptosis prior to phagocytosis. In such conditions, erythroblasts at various stages of degradation may be recognized within marrow macrophages. Apoptosis at the late BFU-E and CFU-E stages is thought to be a major factor controlling the rate of erythropoiesis. When there is substantially increased ineffective erythropoiesis, the marrow displays increased erythroid precursors with a paradoxical decrease in the production of reticulocytes.

Megakaryocytes

Aspirate Smears

The majority of the cells of the megakaryocyte series are larger than other hematopoietic cells and have polyploid DNA contents. The earliest morphologically recognizable cells in this series are called megakaryoblasts. These are 20 to 30 μm in diameter and have a single large, ovoid, kidney-shaped, or lobed nucleus that is surrounded by a narrow rim of intensely basophilic, agranular cytoplasm. The nucleus contains several nucleoli. Megakaryoblasts (group I megakaryocytes) mature into promegakaryocytes (group II megakaryocytes), which in turn develop into granular megakaryocytes (group III megakaryocytes). Promegakaryocytes are larger than megakaryoblasts and have a larger volume of cytoplasm relative to that of the nucleus (Fig. 32.33). They possess a single multilobed nucleus with the overlapping lobes arranged in a C-shaped formation. The cytoplasm is less basophilic than that of megakaryoblasts and contains a few azurophilic granules that are usually grouped within the concavity formed by the overlapping nuclear lobes. The granular megakaryocytes (Fig. 32.34) are up to 100 μm in diameter and have abundant pale-staining cytoplasm containing many lavender-colored granules. The nucleus has multiple

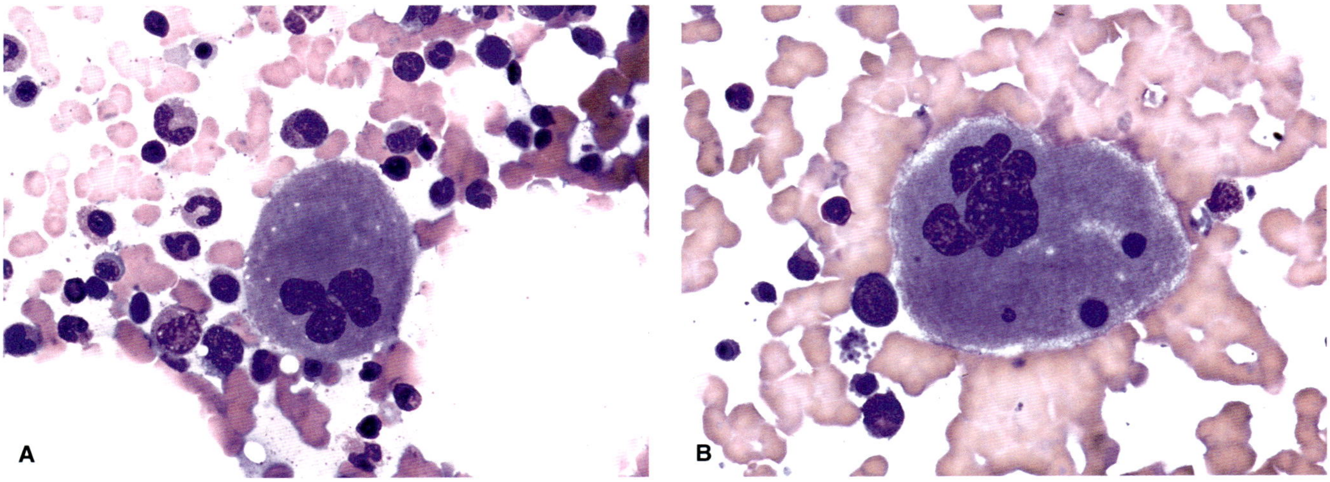

FIGURE 32.34 Two granular megakaryocytes from a Wright–Giemsa–stained bone marrow smear. The megakaryocyte on the right has two lymphocytes in its cytoplasm (emperipolesis).

lobes, and these become fairly tightly packed together before the shedding of platelets. The chromatin has a coarse, "hammered metal" appearance. Platelets are formed by the fragmentation of cytoplasmic processes of the mature granular megakaryocytes. When platelet formation is completed, a bare nucleus remains. Some normal megakaryocytes display the phenomenon of emperipolesis (109,110). This term is used to describe the movement of one cell type within the cytoplasm of another without destruction. The cytoplasm of an affected megakaryocyte may contain one or more intact cells of a number of types, including neutrophil and eosinophil granulocytes and their precursors, lymphocytes, erythroblasts, and red cells (Fig. 32.34). The physiologic relevance of megakaryocyte emperipolesis is uncertain; one suggestion has been that certain marrow cells may enter the circulation via the processes of megakaryocyte cytoplasm that protrude into marrow sinusoids.

Mature platelets are usually 2 to 3 μm in diameter and are irregular in outline. The cytoplasm stains pale blue and has a number of azurophilic granules at its center. Newly formed platelets are slightly larger than mature ones.

About 40% of megakaryoblasts, 20% of promegakaryocytes, and 2% of granular megakaryocytes synthesize DNA (111). However, cell division is probably uncommon in megakaryoblasts and is not seen in the other two cell types. The occurrence of cycles of DNA replication without cytokinesis results in the characteristic polyploidy of these cells. The total DNA content of megakaryoblasts ranges between 4c and 32c, and of promegakaryocytes and granular megakaryocytes between 8c and 64c (1c = the haploid DNA content). There is a positive correlation between the nuclear area and DNA content of megakaryocytes.

Cytochemistry

When stained by the PAS reaction, megakaryocytes show a diffuse and finely granular positivity over both the nucleus and the perinuclear and intermediate zones of the cytoplasm (96–99). A narrow peripheral zone of the cytoplasm is often PAS negative, and this may be surrounded by clumps of positive granules within attached platelets. Within platelets, PAS-positive material appears as scattered, lightly staining fine granules at the periphery and as clumps of darkly staining coarse granules at the center. Megakaryocytes and platelets are usually unstained by Sudan black, but occasional megakaryocytes may show a diffuse positivity with fine positive granules scattered both in the cytoplasm and over the nucleus. Megakaryocytes and platelets display strong acid phosphatase activity.

Peroxidase activity cannot be demonstrated in megakaryocytes by light microscopy but can be demonstrated in a characteristic distribution using the electron microscope.

Megakaryocytes show no α-naphthol AS-D chloroacetate esterase activity. However, they have substantial α-naphthyl acetate esterase activity and weaker α-naphthyl butyrate esterase activity; the latter generates many coarse or fine positive granules in the cytoplasm and over the nucleus.

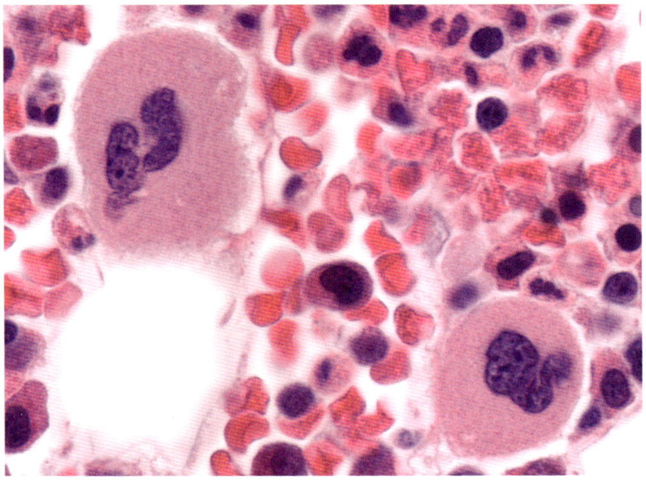

FIGURE 32.35 Two megakaryocytes in a normal bone marrow trephine section (H&E).

Biopsy Sections

Megakaryocytes are readily recognized by their large size, dense, light or dark pink cytoplasm, and lobated nuclei in sections stained either with H&E (Fig. 32.35) or Giemsa. They are normally scattered singly or in loose clusters of two to five cells, and are usually not found in a paratrabecular location. Small numbers of small megakaryocytes with tightly convoluted, hyperchromatic nuclei and minimal or no cytoplasm may also be seen (Fig. 32.36). These senescent megakaryocytes should not be confused with dysplastic megakaryocytes. Empiropolesis may be evident in trephine sections.

Megakaryocytes and megakaryoblasts may be identified immunohistochemically using antibodies directed to a variety of antigens, including CD31, CD41 (platelet glycoprotein IIb), CD42 (platelet glycoprotein Ib), CD61 (platelet glycoprotein IIIa), and factor VIII–related antigen (von Willebrand factor) (Fig. 32.37). Using a CD61 antibody,

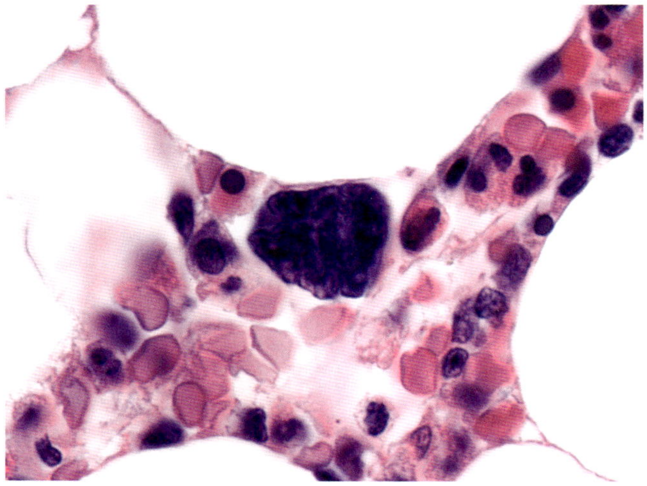

FIGURE 32.36 Hyperchromatic, hyperconvoluted megakaryocyte "nucleus" in a normal bone marrow trephine section (H&E).

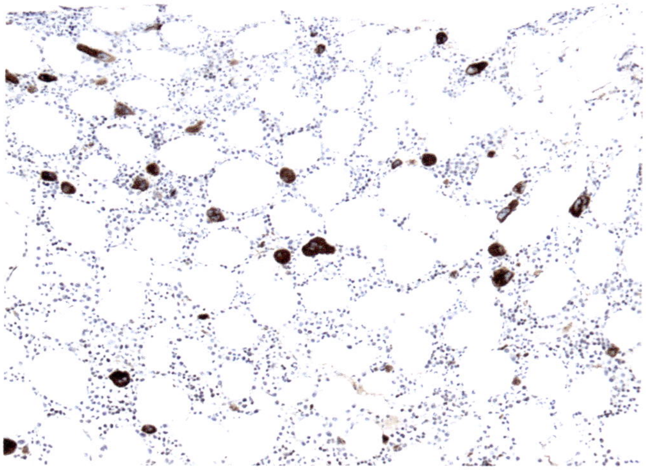

FIGURE 32.37 Normal bone marrow trephine stained with an antibody to factor VIII–related antigen (von Willebrand factor).

the mean value for the total number of megakaryocytes and megakaryoblasts in 15 normal subjects was 24/mm^2 (range: 14 to 38) and for megakaryoblasts alone, it was 2.8/mm^2 (range: 1.2 to 4.9) (112).

Lymphocytes and Plasma Cells

All lymphocytes are derived from the lymphoid stem cells present in the marrow, which are in turn derived from the pluripotent hematopoietic stem cells. The lymphoid stem cells generate both B-cell progenitors and T-cell progenitors. The former mature through a number of antigen-independent intermediate stages into B cells; this maturation occurs within the microenvironment of the marrow. The newly formed B cells travel via the blood into the B-cell zones of peripheral lymphoid tissue. Either the lymphoid stem cells or early T-cell progenitors migrate from the marrow through the blood into the thymus. Here, these cells undergo antigen-independent maturation into T cells, and those T cells that recognize self are deleted. The mature T cells then travel through the blood into the T-cell zones of the peripheral lymphoid organs. The mature B and T lymphocytes that enter the peripheral lymphoid tissue are triggered into division when they react with specific antigens in the presence of appropriate accessory cells. Their progeny develops into effector cells or memory cells. In the case of B cells, the effector cell is an antibody-secreting plasma cell. Antigen-dependent proliferation of B cells occurs in normal marrow and results in the presence of plasma cells in this tissue.

Immunohistochemical studies show that the ratio of T cells to B cells in normal adult marrow is around 3:1. The appearance of mature bone marrow lymphocytes is indistinguishable from those of other lymphocytes in the body. Some T and B lymphocytes have fine or coarse PAS-positive granules arranged in one to four (usually one or two) rings around the nucleus, and occasional cells have large clumps of PAS-positive material. Lymphocytes are peroxidase negative and α-naphthol AS-D chloroacetate esterase negative, and over 99% of cells are alkaline phosphatase negative. Some lymphocytes show a positive paranuclear dot when stained for α-naphthyl butyrate esterase; this staining is unaffected by fluoride. A substantial proportion of normal lymphocytes show either a paranuclear dot or diffuse granular positivity when stained for acid phosphatase. A paranuclear dot is found in both T cells and B cells but more frequently in T cells.

Mature lymphocytes are distributed inconspicuously through the interstitium among maturing hematopoietic elements in the normal marrow. In a minority of otherwise normal marrows, discrete lymphoid aggregates are present, and the frequency of such aggregates increases with increasing age (113). Benign lymphoid aggregates are small, only rarely exceeding 1 mm in diameter, and round. They are typically well circumscribed, with little infiltration into the surrounding interstices. They are composed predominantly of small, bland lymphocytes with regular to mildly irregular nuclei, clumped chromatin, and scanty cytoplasm (Fig. 32.38). A few admixed histiocytes are often present, and they may contain one or more small blood vessels. Benign lymphoid aggregates almost never occupy a paratrabecular location; lymphoid aggregates exhibiting an intimate association with bony trabeculae almost always represent lymphomatous infiltrates. Immunohistochemistry of focal benign lymphoid aggregates typically reveals an admixture of T cells and B cells, usually with a predominance of the former (Fig. 32.38) (114). B-cell–predominant lymphoid aggregates are more likely to represent lymphomatous infiltrates (115,116). Rarely, lymphoid aggregates containing reactive germinal centers will be observed, almost always in states of immune dysregulation. These will display the morphologic and immunophenotypic features of reactive germinal centers in extramedullary site (Fig. 32.39) (117,118).

Small numbers of maturing B-cell precursors (colloquially known as "hematogones") are detectable in small numbers (usually less than 1%) by flow cytometry in most marrows from children and adults (119). However, they may be hyperplastic in some reactive and regenerative states, particularly in younger patients, rarely comprising up to 70% of nucleated bone marrow cells. In Romanowsky-type smears, hematogones show a spectrum of maturation (Fig. 32.40). The most immature hematogones are medium to large in size, with regular or mildly indented nuclei, dense, evenly distributed chromatin, inconspicuous or absent nuclei, and narrow rim of pale basophilic, agranular cytoplasm. These may be confused with myeloblasts, but may be differentiated based on denser chromatin, less conspicuous nucleoli, and scantier cytoplasm. As hematogones mature, they become smaller and their chromatin becomes denser, although it remains evenly distributed (rather than clumped, as in mature lymphocytes). These more mature hematogones are also distinguished by their very scanty cytoplasm, with the nucleus appearing to form the cell border for a large proportion of the circumference of the cell.

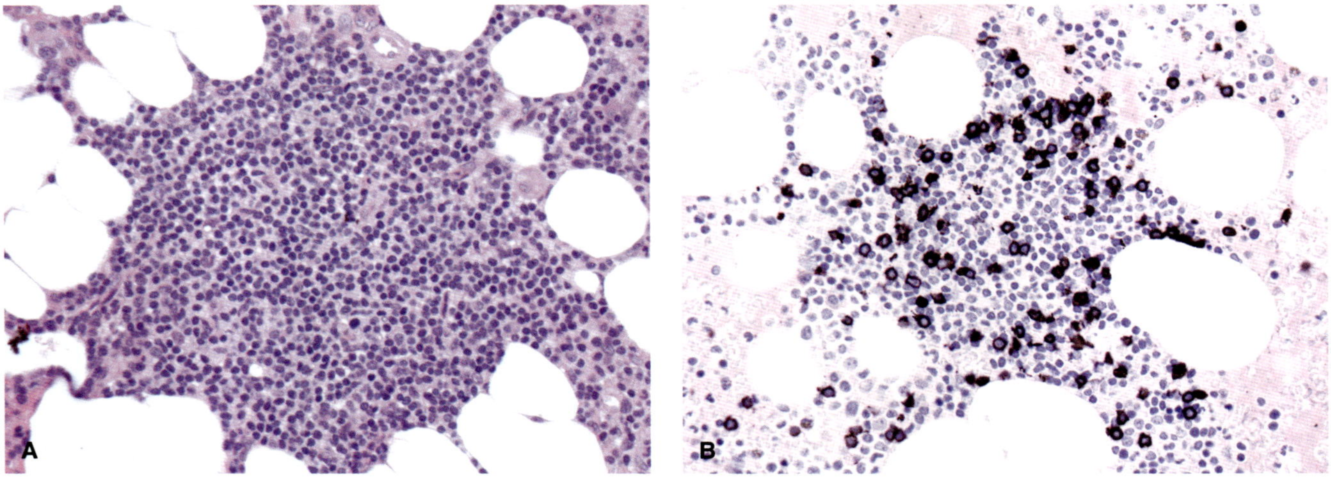

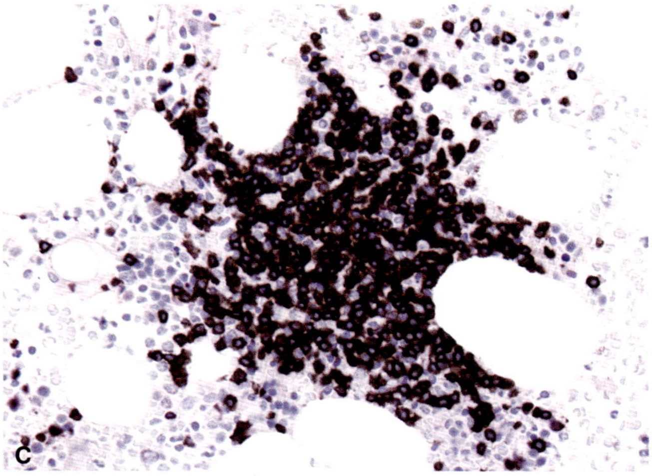

FIGURE 32.38 Focal benign lymphoid aggregate in an otherwise normal marrow. **A:** H&E appearance. **B:** Immunohistochemistry for CD20 (B-cell marker). **C:** Immunohistochemistry for CD3 (T-cell marker).

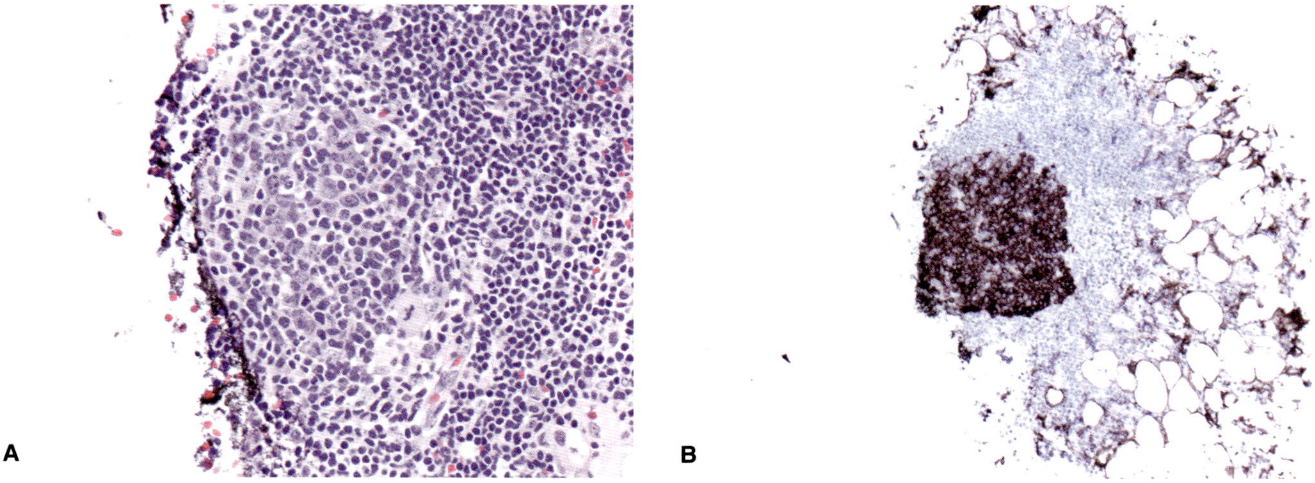

FIGURE 32.39 Reactive germinal center encountered in a trephine biopsy from a patient with aplastic anemia. **A.** On H&E, the germinal center has the same features as reactive germinal centers in extramedullary tissues. As in other sites, germinal centers in bone marrow will be CD10(+) **(B)** and Bcl-2(-) **(C)**. They will also contain follicular dendritic meshworks, highlighted by CD21, CD23, or CD35 (not illustrated).

FIGURE 32.39 *(Continued)*

In marrow sections, hyperplastic hematogones form interstitial concentrations without distorting the normal marrow architecture. Cytologically they resemble mature lymphocytes, but show slightly more open, granular chromatin.

Plasma Cells

Plasma cells in smears of normal bone marrow vary somewhat in size and appearance (Fig. 32.41). Most are 14 to 20 μm in diameter and have deep blue cytoplasm. The cytoplasm has a pale paranuclear area corresponding to the Golgi apparatus and may contain one or more vacuoles. The nucleus is small relative to the volume of cytoplasm, contains moderate quantities of condensed chromatin, and is eccentrically located. Although most plasma cells are uninucleate, a few are binucleate (Fig. 32.41B) or multinucleate. Some normal plasma cells have other features. For example, occasional cells may contain one or a few large, round acidophilic, PAS-positive cytoplasmic inclusions (i.e., Russell bodies), or multiple smaller, slightly basophilic round inclusions (i.e., Mott cells, grape cells, or morular cells) (Fig. 32.41C). Rare plasma cells have many pleomorphic cytoplasmic inclusions, and consequently appear reticulated. Others have eosinophilic cytoplasm, usually at the periphery; when the eosinophilia is confined to the periphery, it contrasts markedly with the intense basophilia of the rest of the cytoplasm, resulting in the appearance of a flaming cell. Occasional plasma cells have azurophilic rods that resemble Auer rods present in acute myeloid leukemia, but that are PAS, Sudan black, and peroxidase negative. Plasma cells show strong acid phosphatase activity, particularly around the nucleus and over the Golgi zone. They do not stain for α-naphthol AS-D chloroacetate esterase. Plasma cells normally comprise ≤2% of the cellularity in marrow aspirates.

In trephine sections, plasma cells are often located in intimate association with small blood vessels that lack a smooth muscle layer (Fig. 32.42) (120), although they may

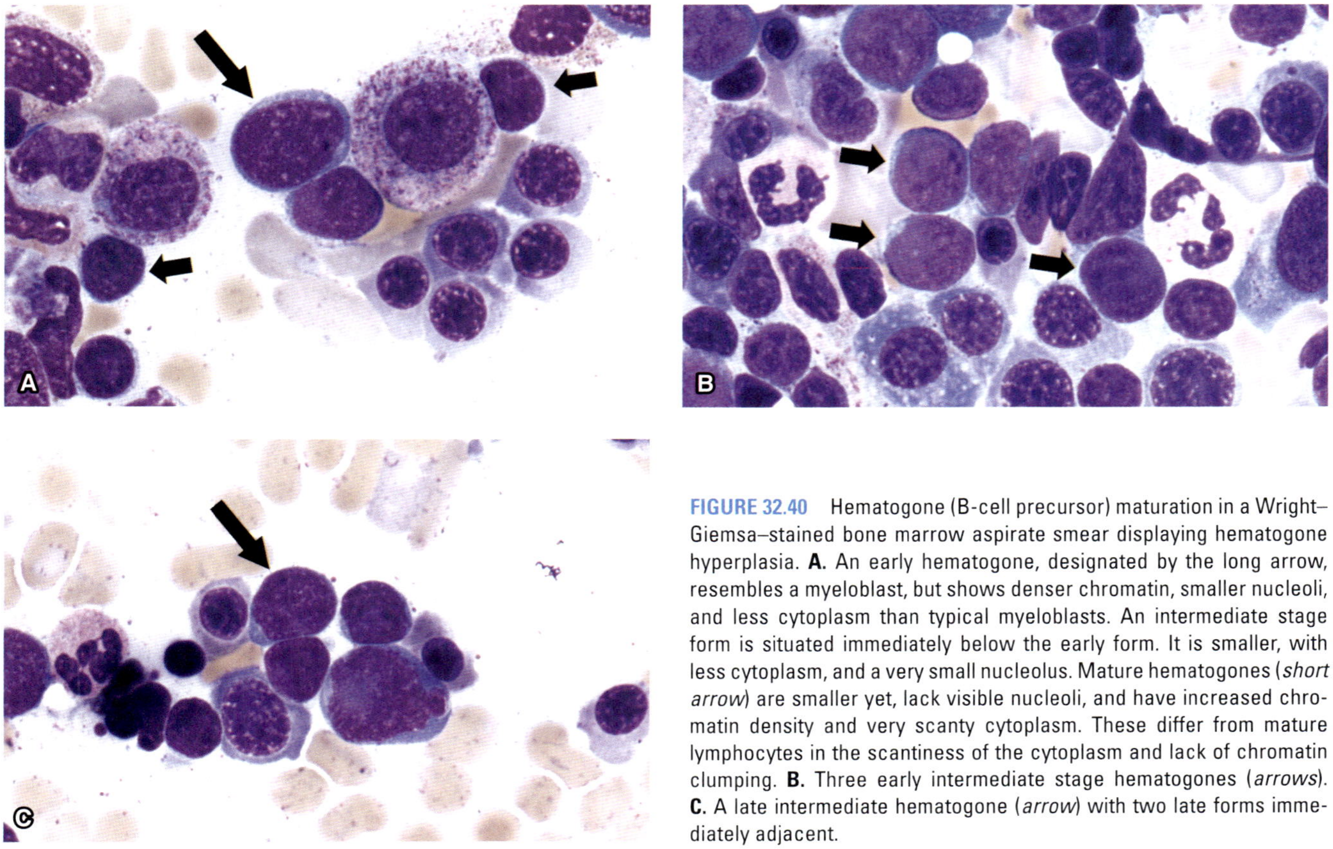

FIGURE 32.40 Hematogone (B-cell precursor) maturation in a Wright–Giemsa–stained bone marrow aspirate smear displaying hematogone hyperplasia. **A.** An early hematogone, designated by the long arrow, resembles a myeloblast, but shows denser chromatin, smaller nucleoli, and less cytoplasm than typical myeloblasts. An intermediate stage form is situated immediately below the early form. It is smaller, with less cytoplasm, and a very small nucleolus. Mature hematogones (*short arrow*) are smaller yet, lack visible nucleoli, and have increased chromatin density and very scanty cytoplasm. These differ from mature lymphocytes in the scantiness of the cytoplasm and lack of chromatin clumping. **B.** Three early intermediate stage hematogones (*arrows*). **C.** A late intermediate hematogone (*arrow*) with two late forms immediately adjacent.

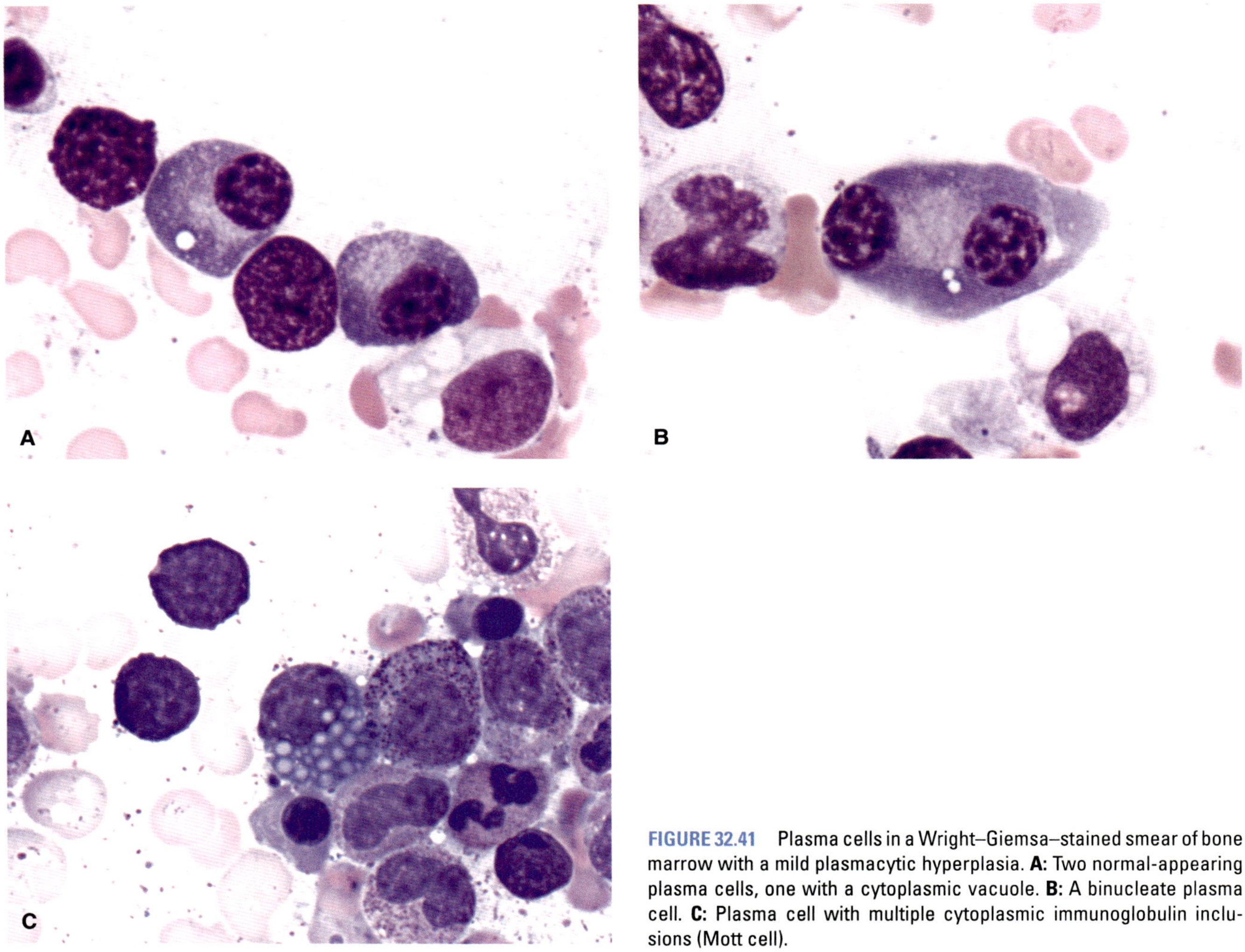

FIGURE 32.41 Plasma cells in a Wright–Giemsa–stained smear of bone marrow with a mild plasmacytic hyperplasia. **A:** Two normal-appearing plasma cells, one with a cytoplasmic vacuole. **B:** A binucleate plasma cell. **C:** Plasma cell with multiple cytoplasmic immunoglobulin inclusions (Mott cell).

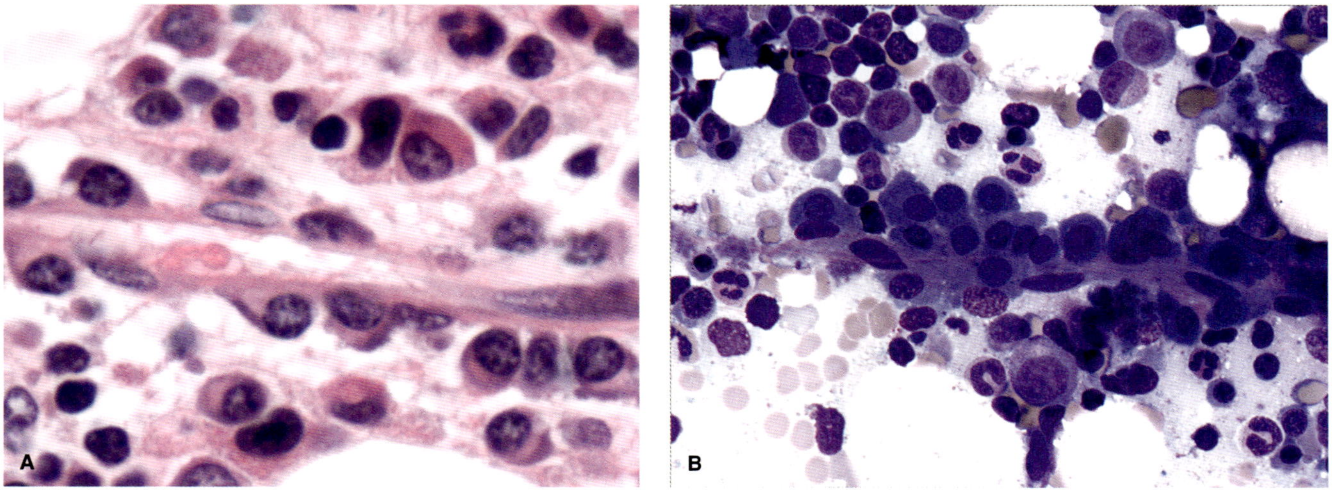

FIGURE 32.42 **A:** Normal perivascular localization of plasma cells in a marrow trephine section (H&E). **B:** An aspirated blood vessel in a Wright–Giemsa–stained aspirate smear demonstrating adhesion of plasma cells to the vessel.

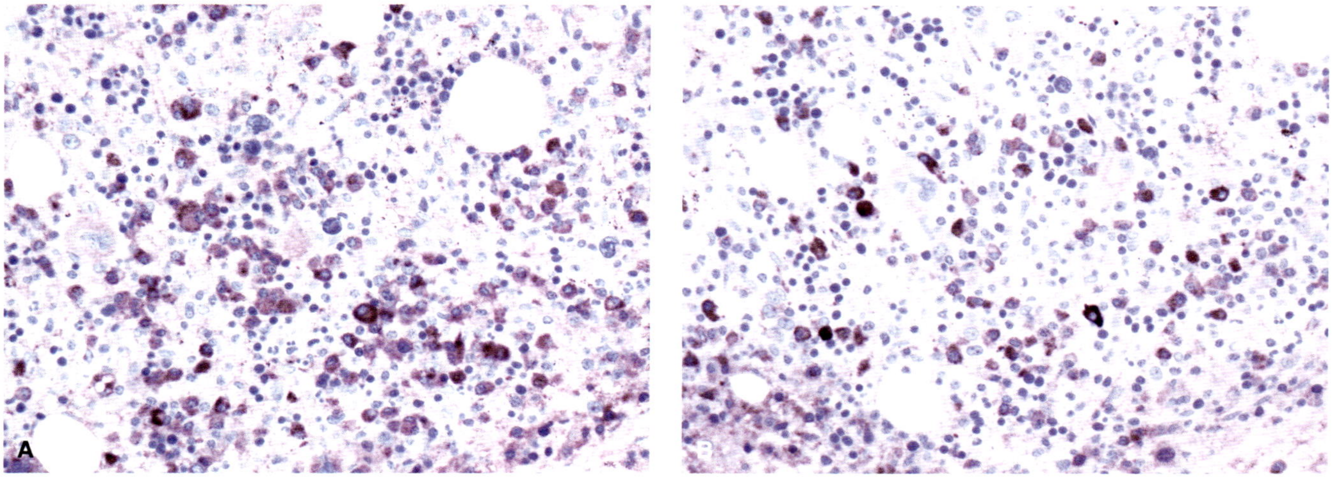

FIGURE 32.43 Light chain immunohistochemistry in an aspirate clot section from a patient with reactive plasmacytosis. **A:** Kappa light chain. **B:** Lambda light chain.

also be seen as single cells scattered throughout the interstitium. Occasionally, an intact blood vessel with adherent plasma cells will be present in bone marrow aspirate smears, as well (Fig. 32.42B). In marrow sections, plasma cells exhibit characteristic features: eccentric nucleus with distinctly, often peripherally clumped chromatin and moderately abundant, densely amphophilic cytoplasm with a paranuclear clearing. Immunohistochemistry or in situ hybridization for kappa and lambda will demonstrate an admixture of kappa- and lambda-expressing plasma cells, usually with a slight excess of the former (Fig. 32.43).

Plasma cells normally comprise ≤2% of the cellularity in marrow aspirates. This is often an underestimate compared to the core biopsy, as plasma cells are often concentrated in the particles in the aspirate smears. Plasma cells may be modestly increased in number in marrow aspirates as a nonspecific reactive finding in a wide variety of disease states (121,122). Rarely, florid reactive bone marrow plasma cell proliferations are encountered, up to 30% of aspirate cellularity. These morphologically mimic plasma cell myeloma, but can easily be distinguished through assessment of light chain expression. Reported associated disease states include autoimmune diseases, AIDS, drug reactions, multicentric Castleman disease, Hodgkin lymphoma, acute leukemias, or carcinomas (123–129). The constituent plasma cells generally lack atypia. However, for unclear reasons, plasma cells in florid plasmacytosis in the setting of AIDS can show atypical features, such as nuclear budding (123) (Fig 32.44). In sections from marrows with plasmacytic hyperplasia, the perivascular localization of plasma cells is characteristically accentuated.

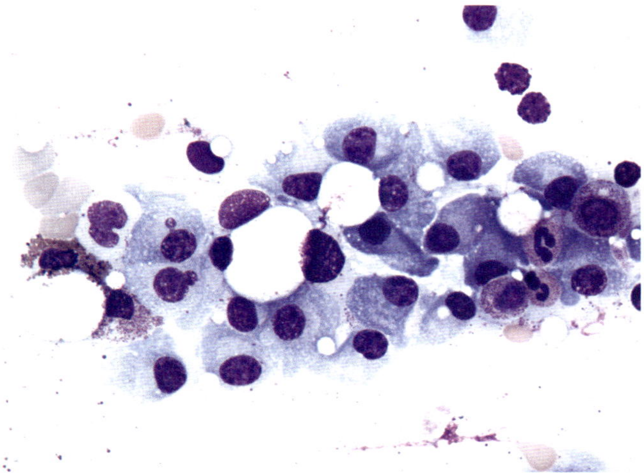

FIGURE 32.44 Florid reactive plasmacytosis in a patient with AIDS. Plasma cells comprised approximately 30% of the aspirate cellularity. While the plasma cells were predominantly normal appearing, a subset showed nuclear budding (left side of field), a finding that seems to be unique to this clinical setting (Wright–Giemsa).

CELLULARITY OF THE MARROW

The term *marrow cellularity* is usually defined as the proportion of the area of a histologic section excluding bone occupied by hematopoietic cells (by cells other than fat cells). Cellularity may be assessed by point counting using an eyepiece with a graticule (histomorphometry) or, more accurately, by computerized image analysis (130). However, crude estimation serves for clinical purposes. The shrinkage of tissue subjected to decalcification and paraffin embedding results in the cellularity of paraffin-embedded sections being about 5% lower than in plastic-embedded sections (131).

In healthy subjects, cellularity varies with age (57–59). In neonates, there are very few fat cells in the marrow, and the cellularity approaches 100%. Average cellularity drops off about 15% between childhood and early adulthood. During adulthood cellularity declines approximately 3% per decade, and then plateaus after age 70 (59).

In assessing cellularity, it should be noted that cellularity may vary considerably from one intertrabecular space to the next in a single biopsy specimen, particularly in patients who have received treatment for a hematolymphoid malignancy, so that a reliable estimate requires the examination of at least five such spaces. Furthermore, the immediate subcortical marrow of the ilium is frequently less cellular than deeper marrow. Additionally, marrows from older patients may sometimes display heterogeneity in cellularity, with lower cellularity sometimes evident immediately adjacent to bony trabeculae. Notably, pelvic radiation for solid malignancies may result in prolonged, localized bone marrow aplasia, despite preserved peripheral counts (132).

A study of postmortem biopsy samples from 100 normal subjects who died suddenly without evidence of bone or marrow disease showed only slight differences in the cellularity at different hematopoietic sites. The percentage cellularity (± SD) in biopsies from the anterior iliac crest, posterior iliac crest, lumbar vertebrae, and sternum were 60 ± 6, 62 ± 7, 64 ± 7, and 61 ± 8, respectively (133).

MARROW DIFFERENTIAL COUNT

During the first day of life, the erythroblasts account for 18.5% to 65% (mean: 40%) of the nucleated cells in a marrow smear. Over the next 8 to 10 days, this figure decreases progressively from 0% to 20.5% (mean: 8%). After a period of erythroblastopenia lasting about 3 weeks, the percentage of erythroblasts increases again, reaching values of 6.5% to 31.5% (mean: 16%) at the age of 3 months (134). These changes are caused by an increase in arterial oxygen saturation soon after birth and the consequent suppression of erythropoietin production. Erythropoietin production increases again 6 to 13 weeks later when the hemoglobin concentration in the blood decreases to about 11 g/dL. The proportion of granulocytes and their precursors ranges between 20% and 73% (mean: 46%) of the nucleated marrow cells on the first day of life (134), increases during the next 3 weeks, and then decreases again to reach a stable value of about 55% after the 2nd month. The average value for the proportion of lymphocytes in the marrow increases from 12% during the first 2 days of life, to 33% at 7 to 10 days, and 47% at 1 month. The lymphocyte percentage then remains stable until the end of the 1st year, after which it decreases slowly to 19% at 4 to 4.5 years, which is only slightly higher than the adult value of 15% (134–137). Plasma cells are infrequent in the neonate, accounting for up to 0.4% (mean: 0.016%) of nucleated marrow cells (138). They gradually increase in number to reach a mean value of 0.386% at the age of 12 to 15 years (5,000 cell differential count). Plasma cells in healthy adults comprise 0.4% to 2.0% of marrow aspirate cells.

The differential count on 2,000 consecutive nucleated cells in bone marrow smears from normal adults is given in Table 32.2 (133). The mean and range for the myeloid/erythroid ratio in healthy adults are 3.1 and 2.0:8.3, respectively (139).

Table 32.2 Differential Counts[a] on Marrow Smears From 28 Healthy Adults Aged Between 20 and 29 Years

	Percentages		
Cell Type	**Mean**	**95% Confidence Limits**	**Observed Range**
Myeloblasts	1.21	0.75–1.67	0.75–1.80
Promyelocytes	2.49	0.99–3.99	1.00–3.75
Myelocytes			
Neutrophil	17.36	11.54–23.18	12.25–22.65
Eosinophil	1.37	0–2.85	0.25–3.45
Basophil	0.08	0–0.21	0.00–0.25
Metamyelocytes			
Neutrophil	16.92	11.40–22.44	11.45–23.60
Eosinophil	0.63	0.07–1.19	0.25–1.30
Juvenile neutrophil granulocytes (stab forms)	8.70	3.58–13.82	4.85–13.95
Granulocytes			
Neutrophil	13.42	4.32–22.52	8.70–8.95
Eosinophil	0.93	0.21–1.65	0.45–1.55
Basophil	0.20	0–0.48	0.05–0.50
Monocytes	1.04	0.36–1.72	0.65–2.10
Plasma cells	0.46	0–0.96	0.10–0.95[b]
Lymphocytes	14.60	6.66–22.54	9.35–25.05
Basophilic erythropoietic cells	0.92	0.40–1.44	0.50–1.60
Early polychromatic normoblasts	6.76	2.56–10.96	3.30–12.20
Late polychromatic normoblasts	11.58	6.16–17.0	7.85–19.55
Reticular cells	0.24	0–0.54	0.05–0.65

[a]2,000 cells were studied in each individual.

[b]The observed range in 63 cases, aged 20–93 years, was 0.10–2.00%.

From Jacobsen KM. Untersuchungen uber das knochenmarkspunktat bei normalen individuen verschiedener altersklassen. *Acta Med Scand.* 1941;106:417–446.

REFERENCES

1. Ellis RE. The distribution of active bone marrow in the adult. *Phys Med Biol.* 1961;5:255–258.
2. Woodard HQ, Holodny E. A summary of the data of Mechanik on the distribution of human bone marrow. *Phys Med Biol.* 1960;5:57–59.

3. Nombela-Arrieta C, Manz MG. Quantification and three-dimensional microanatomical organization of the bone marrow. *Blood Adv.* 2017;1:407–416.
4. Pittenger MF, Discher DE, Peault BM, Phinney DG, Hare JM, Caplan AI. Mesenchymal stem cell perspective: cell biology to clinical progress. *NPJ Regen Med.* 2019;4:22.
5. Lee SH, Erber WN, Porwit A, Tomonaga M, Peterson LC. ICSH guidelines for the standardization of bone marrow specimens and reports. *Int J Lab Hematol.* 2008;30:349–364.
6. Bain BJ. Bone marrow aspiration. *J Clin Pathol.* 2001;54:657–663.
7. Brynes RK, McKenna RW, Sundberg RD. Bone marrow aspiration and trephine biopsy. An approach to a thorough study. *Am J Clin Pathol.* 1978;70:753–759.
8. Bain BJ. Bone marrow trephine biopsy. *J Clin Pathol.* 2001;54:737–742.
9. Reichard K, Foucar K. Procurement and indications for bone marrow examination. In: Foucar K, Reichard K, Czuchlewski D, eds. *Bone Marrow Pathology.* 4th ed. ASCP Press; 2020:48–61.
10. Woronzoff-Dashkoff KP. The Ehrlich-Chenzinsky-Plehn-Malachowski-Romanowsky-Nocht-Jenner-May-Grunwald-Leishman-Reuter-Wright-Giemsa-Lillie-Roe-Wilcox stain. The mystery unfolds. *Clin Lab Med.* 1993;13:759–771.
11. Torlakovic EE, Brynes RK, Hyjek E, et al. ICSH guidelines for the standardization of bone marrow immunohistochemistry. *Int J Lab Hematol.* 2015;37:431–449.
12. Fend F, Tzankov A, Bink K, et al. Modern techniques for the diagnostic evaluation of the trephine bone marrow biopsy: methodological aspects and applications. *Prog Histochem Cytochem.* 2008;42:203–252.
13. Reichard K, Czuchlewski D. Specialized testing in bone marrow evaluation. In: Foucar K, Reichard K, Czuchlewski D, eds. *Bone Marrow Pathology.* ASCP Press; 2020:62–79.
14. Olsen RJ, Chang C-C, Herrick JL, Zu Y, Ehsan A. Acute leukemia immunohistochemistry: a systematic diagnostic approach. *Arch Pathol Lab Med.* 2008;132:462–475.
15. Pileri SA, Ascani S, Milani M, et al. Acute leukaemia immunophenotyping in bone-marrow routine sections. *Br J Haematol.* 1999;105:394–401.
16. Kremer M, Quintanilla-Martinez L, Nahrig J, von Schilling C, Fend F. Immunohistochemistry in bone marrow pathology: a useful adjunct for morphologic diagnosis. *Virchows Arch.* 2005;447:920–937.
17. Pileri SA, Roncador G, Ceccarelli C, et al. Immunohistochemistry of bone-marrow biopsy. *Leuk Lymphoma.* 1997;26(Suppl 1):69–75.
18. Abkowitz JL, Catlin SN, McCallie MT, Guttorp P. Evidence that the number of hematopoietic stem cells per animal is conserved in mammals. *Blood.* 2002;100:2665–2667.
19. Cheng H, Zheng Z, Cheng T. New paradigms on hematopoietic stem cell differentiation. *Protein Cell.* 2020;11:34–44.
20. Kondo M, Weissman IL, Akashi K. Identification of clonogenic common lymphoid progenitors in mouse bone marrow. *Cell.* 1997;91:661–672.
21. Morrison SJ, Wandycz AM, Hemmati HD, Wright DE, Weissman IL. Identification of a lineage of multipotent hematopoietic progenitors. *Development.* 1997;124:1929–1939.
22. Akashi K, Traver D, Miyamoto T, Weissman IL. A clonogenic common myeloid progenitor that gives rise to all myeloid lineages. *Nature.* 2000;404:193–197.
23. Manz MG, Miyamoto T, Akashi K, Weissman IL. Prospective isolation of human clonogenic common myeloid progenitors. *Proc Natl Acad Sci U S A.* 2002;99:11872–11877.
24. Borger DK, Means RT. Origin and development of blood cells. In: Means RT, Rodgers GM, Glader B, et al., eds. *Wintrobe's Clinical Hematology.* 15th ed. Wolters Kluwer/Lippincott Williams & Wilkins; 2024:69–92.
25. Kaushansky K. Hematopoietic stem cells, progenitors, and cytokines. In: Kaushansky K, Lichtman MA, Prchal JT, et al., eds. *Williams Hematology.* 10th ed. McGraw Hill; 2021:273–294.
26. Souma T, Suzuki N, Yamamoto M. Renal erythropoietin-producing cells in health and disease. *Front Physiol.* 2015;6:167.
27. Ryder JW, Lazarus HM, Farhi DC. Bone marrow and blood findings after marrow transplantation and rhGM- CSF therapy [see comments]. *Am J Clin Pathol.* 1992;97:631–637.
28. Schmitz LL, McClure JS, Litz CE, et al. Morphologic and quantitative changes in blood and marrow cells following growth factor therapy. *Am J Clin Pathol.* 1994;101:67–75.
29. Horina JH, Schmid CR, Roob JM, et al. Bone marrow changes following treatment of renal anemia with erythropoietin. *Kidney Int.* 1991;40:917–922.
30. Douglas VK, Tallman MS, Cripe LD, Peterson LC. Thrombopoietin administered during induction chemotherapy to patients with acute myeloid leukemia induces transient morphologic changes that may resemble chronic myeloproliferative disorders. *Am J Clin Pathol.* 2002;117:844–850.
31. Palis J, Robertson S, Kennedy M, Wall C, Keller G. Development of erythroid and myeloid progenitors in the yolk sac and embryo proper of the mouse. *Development.* 1999;126:5073–5084.
32. Moore MA, Metcalf D. Ontogeny of the haemopoietic system: yolk sac origin of in vivo and in vitro colony forming cells in the developing mouse embryo. *Br J Haematol.* 1970;18:279–296.
33. Dzierzak E, Medvinsky A, de Bruijn M. Qualitative and quantitative aspects of haematopoietic cell development in the mammalian embryo. *Immunol Today.* 1998;19:228–236.
34. Kelemen E, Calvo W, Fliedner TM. *Atlas of Human Hematopoietic Development.* Springer-Verlag; 1979.
35. Bloom W, Bartelmez GW. Hematopoiesis in young human embryos. *Am J Anat.* 1940;67:21–53.
36. Steinberg MH, Nagel RL. Hemoglobins of the embryo, fetus, and adult. In: Steinberg MH, Forget BG, Higgs DR, Weatherall DJ, eds. *Disorder of Hemoglobin. Genetics, Pathophysiology, and Clinical Management.* Cambridge University Press; 2009:119–135.
37. Emura I, Sekiya M, Ohnishi Y. Two types of immature erythrocytic series in the human fetal liver. *Arch Histol Jpn.* 1983;46:631–643.
38. Gilmour JR. Normal haemopoiesis in intrauterine and neonatal life. *J Pathol Bacteriol.* 1941;52:25–55.
39. Kalpaktsoglou PK, Emery JL. Human bone marrow during the last three months of intrauterine life. A histological study. *Acta Haematol.* 1965;34:228–238.
40. Piney A. The anatomy of the bone marrow with special reference to the distribution of the red marrow. *Br Med J.* 1922;2:792–795.
41. Custer RP, Ahlfeldt FE. Studies on the structure and function of bone marrow. II. Variations in cellularity in various bones with advancing. *J Lab Clin Med.* 1932;17:960–962.

42. Wickramasinghe SN, McCullough J. *Blood and Bone Marrow Pathology*. Churchill Livingstone; 2003.
43. Munka V, Gregor A. Lymphatics and bone marrow. *Folia Morphol (Praha)*. 1965;13:404–412.
44. Bain BJ, Clark DM, Lampert IA, Wilkins BS. *Bone Marrow Pathology*. Blackwell Science; 2001.
45. Bartl R, Frisch B. Normal bone marrow: histology, histochemistry and immunohistochemistry. In: Wickramasinghe SN, McCullough J, eds. *Blood and Bone Marrow Pathology*. Churchill Livingstone; 2003.
46. Wickramasinghe SN. *Human Bone Marrow*. Blackwell Scientific; 1975.
47. De Bruyn PP, Breen PC, Thomas TB. The microcirculation of the bone marrow. *Anat Rec*. 1970;168:55–68.
48. Branemark PI. Bone marrow microvascular structure and function. *Adv Microbiol*. 1968;1:1–65.
49. Trubowitz S, Masek B. A histochemical study of the reticuloendothelial system of human marrow–its possible transport role. *Blood*. 1968;32:610–628.
50. Burgio VL, Magrini U, Ciardelli L, Pezzoni G. An enzyme-histochemical approach to the study of the human bone-marrow stroma. *Acta Haematol*. 1984;71:73–80.
51. Wickramasinghe SN. Observations on the ultrastructure of sinusoids and reticular cells in human bone marrow. *Clin Lab Haematol*. 1991;13:263–278.
52. Miller MR, Kasahara M. Observations on the innervation of human long bones. *Anat Rec*. 1963;145:13–17.
53. Loges S, Fehse B, Brockmann MA, et al. Identification of the adult human hemangioblast. *Stem Cells Dev*. 2004;13:229–242.
54. Dominici M, Pritchard C, Garlits JE, Hofmann TJ, Persons DA, Horwitz EM. Hematopoietic cells and osteoblasts are derived from a common marrow progenitor after bone marrow transplantation. *Proc Natl Acad Sci U S A*. 2004;101:11761–11766.
55. Benayahu D, Horowitz M, Zipori D, Wientroub S. Hemopoietic functions of marrow-derived osteogenic cells. *Calcif Tissue Int*. 1992;51:195–201.
56. Tsai J, Kaneko K, Suh AJ, Bockman R, Park-Min K-H. Origin of osteoclasts: osteoclast precursor cells. *J Bone Metab*. 2023;30:127–140.
57. Hartsock RJ, Smith EB, Petty CS. Normal variations with aging of the amount of hematopoietic tissue in bone marrow from the anterior iliac crest. A study made from 177 cases of sudden death examined by necropsy.. *Am J Clin Pathol*. 1965;43:326–331.
58. Sturgeon P. Volumetric and microscopic pattern of bone marrow in normal infants and children. III. Histologic pattern. *Pediatrics*. 1951;7:774–781.
59. Wong J, Jackson R, Chen L, et al. Determination of age-dependent bone marrow normocellularity. *Am J Clin Pathol*. 2024;161:170–176.
60. Bohm J. Gelatinous transformation of the bone marrow: the spectrum of underlying diseases. *Am J Surg Pathol*. 2000;24:56–65.
61. Gale E, Torrance J, Bothwell T. The quantitative estimation of total iron stores in human bone marrow. *J Clin Invest*. 1963;42:1076–1082.
62. Thomason RW, Lavelle J, Nelson D, Lin K, Uherova P. Parenteral iron therapy is associated with a characteristic pattern of iron staining on bone marrow aspirate smears. *Am J Clin Pathol*. 2007;128:590–593.
63. Thomason RW, Almiski MS. Evidence that stainable bone marrow iron following parenteral iron therapy does not correlate with serum iron studies and may not represent readily available storage iron. *Am J Clin Pathol*. 2009;131:580–585.
64. Goel S, Polski JM, Imran H. Sensitivity and specificity of bone marrow hemophagocytosis in hemophagocytic lymphohistiocytosis. *Ann Clin Lab Sci*. 2012;42:21–25.
65. Qari M, Roberts G. Improving the diagnostic precision of bone marrow examination for hemophagocytic syndromes. *Lab Hematol*. 1999;5:37–42.
66. Yan M, Beck RC, Gadde R. Hemophagocyte morphology and hypertriglyceridemia correlate with diagnosis of hemophagocytic lymphohistiocytosis in patients with bone marrow hemophagocytes. *Am J Clin Pathol*. 2024;161:177–185.
67. Gars E, Purington N, Scott G, et al. Bone marrow histomorphological criteria can accurately diagnose hemophagocytic lymphohistiocytosis. *Haematologica*. 2018;103:1635–1641.
68. Dosik H, Rosner F, Sawitsky A. Acquired lipidosis: Gaucher-like cells and "blue cells" in chronic granulocytic leukemia. *Semin Hematol*. 1972;9:309–316.
69. Van Dorpe A, Broeckaert-van O, Desmet V, Verwilghen RL. Gaucher-like cells and congenital dyserythropoietic anaemia, type II (HEMPAS). *Br J Haematol*. 1973;25:165–170.
70. Ziyeh S, Harzer K. Bone marrow cytological storage phenomena in lipidoses. *Eur J Pediatr*. 1994;153:224–233.
71. Busche G, Majewski H, Schlue J, et al. Frequency of pseudo-Gaucher cells in diagnostic bone marrow biopsies from patients with Ph-positive chronic myeloid leukaemia. *Virchows Arch*. 1997;430:139–148.
72. Carrington PA, Stevens RF, Lendon M. Pseudo-Gaucher cells. *J Clin Pathol*. 1992;45:360.
73. Scullin DC Jr., Shelburne JD, Cohen HJ. Pseudo-Gaucher cells in multiple myeloma. *Am J Med*. 1979;67:347–352.
74. Zaino EC, Rossi MB, Pham TD, Azar HA. Gaucher's cells in thalassemia. *Blood*. 1971;38:457–462.
75. Anastasi J, Musvee T, Roulston D, Domer PH, Larson RA, Vardiman JW. Pseudo-Gaucher histiocytes identified up to 1 year after transplantation for CML are BCR/ABL-positive. *Leukemia*. 1998;12:233–237.
76. Takahashi K, Terashima K, Kojima M, Yoshida H, Kimura H. Pathological histochemical and ultrastructural studies on sea-blue histiocytes and Gaucher-like cells in acquired lipidosis occurring in leukemia. *Acta Pathol Jpn*. 1977;27:775–797.
77. Sahin DG, Teke HU, Karagulle M, et al. Gaucher cells or pseudo-Gaucher cells: that's the question. *Turk J Haematol*. 2014;31:428–429.
78. Howard MR, Kesteven PJ. Sea blue histiocytosis: a common abnormality of the bone marrow in myelodysplastic syndromes. *J Clin Pathol*. 1993;46:1030–1032.
79. Hayhoe FG, Flemans RJ, Cowling DC. Acquired lipidosis of marrow macrophages: birefringent blue crystals and Gaucher-like cells, sea-blue histiocytes, and grey-green crystals. *J Clin Pathol*. 1979;32:420–428.
80. Kelsey PR, Geary CG. Sea-blue histiocytes and Gaucher cells in bone marrow of patients with chronic myeloid leukaemia. *J Clin Pathol*. 1988;41:960–962.
81. Rywlin AM, Hernandez JA, Chastain DE, Pardo V. Ceroid histiocytosis of spleen and bone marrow in idiopathic thrombocytopenic purpura (ITP): a contribution to the understanding of the sea-blue histiocyte. *Blood*. 1971;37:587–593.

82. Parker AC, Bain AD, Brydon WG, et al. Sea-blue histiocytosis associated with hyperlipidaemia. *J Clin Pathol.* 1976;29:634–638.
83. Bigorgne C, Le Tourneau A, Messing B, et al. Sea-blue histiocyte syndrome in bone marrow secondary to total parenteral nutrition including fat-emulsion sources: a clinicopathologic study of seven cases [see comments]. *Br J Haematol.* 1996;95:258–262.
84. Meiklejohn DJ, Baden H, Greaves M. Sea-blue histiocytosis and pancytopaenia associated with chronic total parenteral nutrition administration. *Clin Lab Haematol.* 1997;19:219–221.
85. Wu TT, Hoff DS. Fish oil lipid emulsion-associated sea-blue histiocyte syndrome in a pediatric patient. *J Pediatr Pharmacol Ther.* 2015;20:217–221.
86. Dewhurst N, Besley GT, Finlayson ND, Parker AC. Sea blue histiocytosis in a patient with chronic non-neuropathic Niemann-Pick disease. *J Clin Pathol.* 1979;32:1121–1127.
87. Dodhy MA, Aslam W, Nazir A, Shaheen M, Habib M, Zafar H. Sea blue histiocytosis secondary to Niemann Pick disease. *Int J Pathol.* 2011;9:77–78.
88. Landas S, Foucar K, Sando GN, Ellefson R, Hamilton HE. Adult Niemann-Pick disease masquerading as sea blue histiocyte syndrome: report of a case confirmed by lipid analysis and enzyme assays. *Am J Hematol.* 1985;20:391–400.
89. Tsai S, Patel V, Beaumont E, Lodish HF, Nathan DG, Sieff CA. Differential binding of erythroid and myeloid progenitors to fibroblasts and fibronectin. *Blood.* 1987;69:1587–1594.
90. Broudy VC, Zuckerman KS, Jetmalani S, Fitchen JH, Bagby GC Jr. Monocytes stimulate fibroblastoid bone marrow stromal cells to produce multilineage hematopoietic growth factors. *Blood.* 1986;68:530–534.
91. Kirshenbaum AS, Kessler SW, Goff JP, Metcalfe DD. Demonstration of the origin of human mast cells from CD34+ bone marrow progenitor cells. *J Immunol.* 1991;146:1410–1415.
92. Dahlin JS, Hallgren J. Mast cell progenitors: origin, development and migration to tissues. *Mol Immunol.* 2015;63:9–17.
93. Varricchi G, Raap U, Rivellese F, Marone G, Gibbs BF. Human mast cells and basophils-How are they similar how are they different? *Immunol Rev.* 2018;282:8–34.
94. Franco CB, Chen C-C, Drukker M, Weissman IL, Galli SJ. Distinguishing mast cell and granulocyte differentiation at the single-cell level. *Cell Stem Cell.* 2010;6:361–368.
95. Glassy EF. *Color Atlas of Hematology.* College of American Pathologists; 2018.
96. Rozenszajn L, Leibovich M, Shoham D, Epstein J. The esterase activity in megaloblasts, leukaemic and normal haemopoietic cells. *Br J Haematol.* 1968;14:605–610.
97. Gibb RP, Stowell RE. Glycogen in human blood cells. *Blood.* 1949;4:569–579.
98. Rheingold JJ, Wislocki GB. Histochemical methods applied to hematology. *Bull New Engl Med Cent.* 1948;10:133–137.
99. Hayhoe FGJ, Quaglino D. *Haematological Cytochemistry.* Churchill Livingstone; 1988.
100. Parwaresch MR. *The Human Blood Basophil.* Springer-Verlag; 1976.
101. Hintzke M, Harrington AM, Olteanu H, Hosking P, Kroft SH. Bone marrow monocytosis: a survey of 150 cases. *Am J Clin Pathol.* 2015;144:A150.
102. Leder LD. The origin of blood monocytes and macrophages. A review. *Blut.* 1967;16:86–98.
103. Mufti GJ, Bennett JM, Goasguen J, et al. Diagnosis and classification of myelodysplastic syndrome: International Working Group on Morphology of myelodysplastic syndrome (IWGM-MDS) consensus proposals for the definition and enumeration of myeloblasts and ring sideroblasts. *Haematologica.* 2008;93:1712–1717.
104. Ohgami RS, Chisholm KM, Ma L, Arber DA. E-cadherin is a specific marker for erythroid differentiation and has utility, in combination with CD117 and CD34, for enumerating myeloblasts in hematopoietic neoplasms. *Am J Clin Pathol.* 2014;141:656–664.
105. Marsee DK, Pinkus GS, Yu H. CD71 (transferrin receptor): an effective marker for erythroid precursors in bone marrow biopsy specimens. *Am J Clin Pathol.* 2010;134:429–435.
106. Sadahira Y, Kanzaki A, Wada H, Yawata Y. Immunohistochemical identification of erythroid precursors in paraffin embedded bone marrow sections: spectrin is a superior marker to glycophorin. *J Clin Pathol.* 1999;52:919–921.
107. Wickramasinghe SN, Lee MJ, Furukawa T, Eguchi M, Reid CD. Composition of the intra-erythroblastic precipitates in thalassaemia and congenital dyserythropoietic anaemia (CDA): identification of a new type of CDA with intra-erythroblastic precipitates not reacting with monoclonal antibodies to alpha- and beta-globin chains. *Br J Haematol.* 1996;93:576–585.
108. Nemec J, Polak H. Erythropoietic polyploidy. I. The morphology of polyploid erythroid elements and their incidence in healthy subjects. *Folia Haematol Int Mag Klin Morphol Blutforsch.* 1965;84:24–40.
109. Rozman C, Vives-Corrons JL. On the alleged diagnostic significance of megakaryocytic 'phagocytosis' (emperipolesis). *Br J Haematol.* 1981;48:510.
110. Larsen TE. Emperipolesis of granular leukocytes within megakaryocytes in human hemopoietic bone marrow. *Am J Clin Pathol.* 1970;53:485–489.
111. Queisser U, Queisser W, Spiertz B. Polyploidization of megakaryocytes in normal humans, in patients with idiopathic thrombocytopenia and with pernicious anaemia. *Br J Haematol.* 1971;20:489–501.
112. Thiele J, Holgado S, Choritz H, Georgii A. Density distribution and size of megakaryocytes in inflammatory reactions of the bone marrow (myelitis) and chronic myeloproliferative diseases. *Scand J Haematol.* 1983;31:329–341.
113. Rywlin AM, Ortega RS, Dominguez CJ. Lymphoid nodules of bone marrow: normal and abnormal. *Blood.* 1974;43:389–400.
114. Thiele J, Zirbes TK, Kvasnicka HM, Fischer R. Focal lymphoid aggregates (nodules) in bone marrow biopsies: differentiation between benign hyperplasia and malignant lymphoma–a practical guideline. *J Clin Pathol.* 1999;52:294–300.
115. Naemi K, Brynes RK, Reisian N, et al. Benign lymphoid aggregates in the bone marrow: distribution patterns of B and T lymphocytes. *Hum Pathol.* 2013;44:512–520.
116. Johnston A, Brynes RK, Naemi K, et al. Differentiating benign from malignant bone marrow B-cell lymphoid aggregates: a statistical analysis of distinguishing features. *Arch Pathol Lab Med.* 2015;139:233–240.
117. Brunning RD, McKenna RW. Bone marrow manifestations of malignant lymphoma and lymphoma-like conditions. *Pathol Annu.* 1979;14:1–59.
118. Farhi DC. Germinal centers in the bone marrow. *Hematol Pathol.* 1989;3:133–136.

119. McKenna RW, Washington LT, Aquino DB, Picker LJ, Kroft SH. Immunophenotypic analysis of hematogones (B-lymphocyte precursors) in 662 consecutive bone marrow specimens by 4-color flow cytometry. *Blood.* 2001;98:2498–2507.
120. Kass L, Kapadia IH. Perivascular plasmacytosis: a light-microscopic and immunohistochemical study of 93 bone marrow biopsies. *Acta Haematol.* 2001;105:57–63.
121. Batdorf B, Kroft S, Olteanu H, Harrington A. Reactive bone marrow plasmacytosis: an update for the modern era. *Am J Clin Pathol.* 2014;142:A102.
122. Hyun BH, Kwa D, Gabaldon H, Ashton JK. Reactive plasmacytic lesions of the bone marrow. *Am J Clin Pathol.* 1976; 65:921–928.
123. Turbat-Herrera EA, Hancock C, Cabello-Inchausti B, Herrera GA. Plasma cell hyperplasia and monoclonal paraproteinemia in human immunodeficiency virus-infected patients. *Arch Pathol Lab Med.* 1993;117:497–501.
124. Tatsuno I, Nishikawa T, Sasano H, Shizawa S, Iwase H, Satoh S. Interleukin 6-producing gastric carcinoma with fever, hypergammaglobulinemia, and plasmacytosis in bone marrow. *Gastroenterology.* 1994;107:543–547.
125. Tanvetyanon T, Leighton JC. Severe anemia and marrow plasmacytosis as presentation of Sjogren's syndrome. *Am J Hematol.* 2002;69:233.
126. Kass L, Votaw ML. Eosinophilia and plasmacytosis of the bone marrow in Hodgkin's disease. *Am J Clin Pathol.* 1975; 64:248–250.
127. Jamshidi K, Arlander T, Garcia MC, Windschitl HW, Swaim WR. Azulfidine agranulocytosis with bone marrow, megakaryocytosis, histiocytosis and plasmacytosis. *Minn Med.* 1972;55:545–548.
128. Bacon CM, Miller RF, Noursadeghi M, McNamara C, Du M-Q, Dogan A. Pathology of bone marrow in human herpes virus-8 (HHV8)-associated multicentric Castleman disease. *Br J Haematol.* 2004;127:585–591.
129. Gavarotti P, Boccadoro M, Redoglia V, Golzio F, Pileri A. Reactive plasmacytosis. Case report and review of the literature. *Acta Haematol.* 1985;73:108–110.
130. Al-Adhadh AN, Cavill I. Assessment of cellularity in bone marrow fragments. *J Clin Pathol.* 1983;36:176–179.
131. Kerndrup G, Pallesen G, Melsen F, Mosekilde L. Histomorphometrical determination of bone marrow cellularity in iliac crest biopsies. *Scand J Haematol.* 1980;24:110–114.
132. Harrington AM, Currey A, Olteanu H, Kroft SH. Persistent localized bone marrow aplasia after radiotherapy with preserved peripheral counts: a study of 8 cases. *Ann Diagn Pathol.* 2010;14:168–172.
133. Jacobsen KM. Untersuchungen uber das knochenmarkspunktat bei normalen individuen verschiedener altersklassen. *Acta Med Scand.* 1941;106:417–446.
134. Gairdner D, Marks J, Roscoe JD. Blood formation in infancy. Part I. The normal bone marrow. *Arch Dis Child.* 1952;27: 128–133.
135. Rosse C, Kraemer MJ, Dillon TL, McFarland R, Smith NJ. Bone marrow cell populations of normal infants; the predominance of lymphocytes. *J Lab Clin Med.* 1977;89:1225–1240.
136. Diwany M. Sternal marrow puncture in children. *Arch Dis Child.* 1940;15:159–170.
137. Glaser K, Poncher HG, Limarzi LR. The cellular composition of the bone marrow in normal infants and children. *J Lab Clin Med.* 1948;33:1639.
138. Steiner ML, Pearson HA. Bone marrow plasmacyte values in childhood. *J Pediatr.* 1966;68:562–568.
139. Young RH, Osgood EE. Sternal marrow aspirated during life. Cytology in health and disease. *Arch Intern Med.* 1935; 55:186–203.

SECTION IX

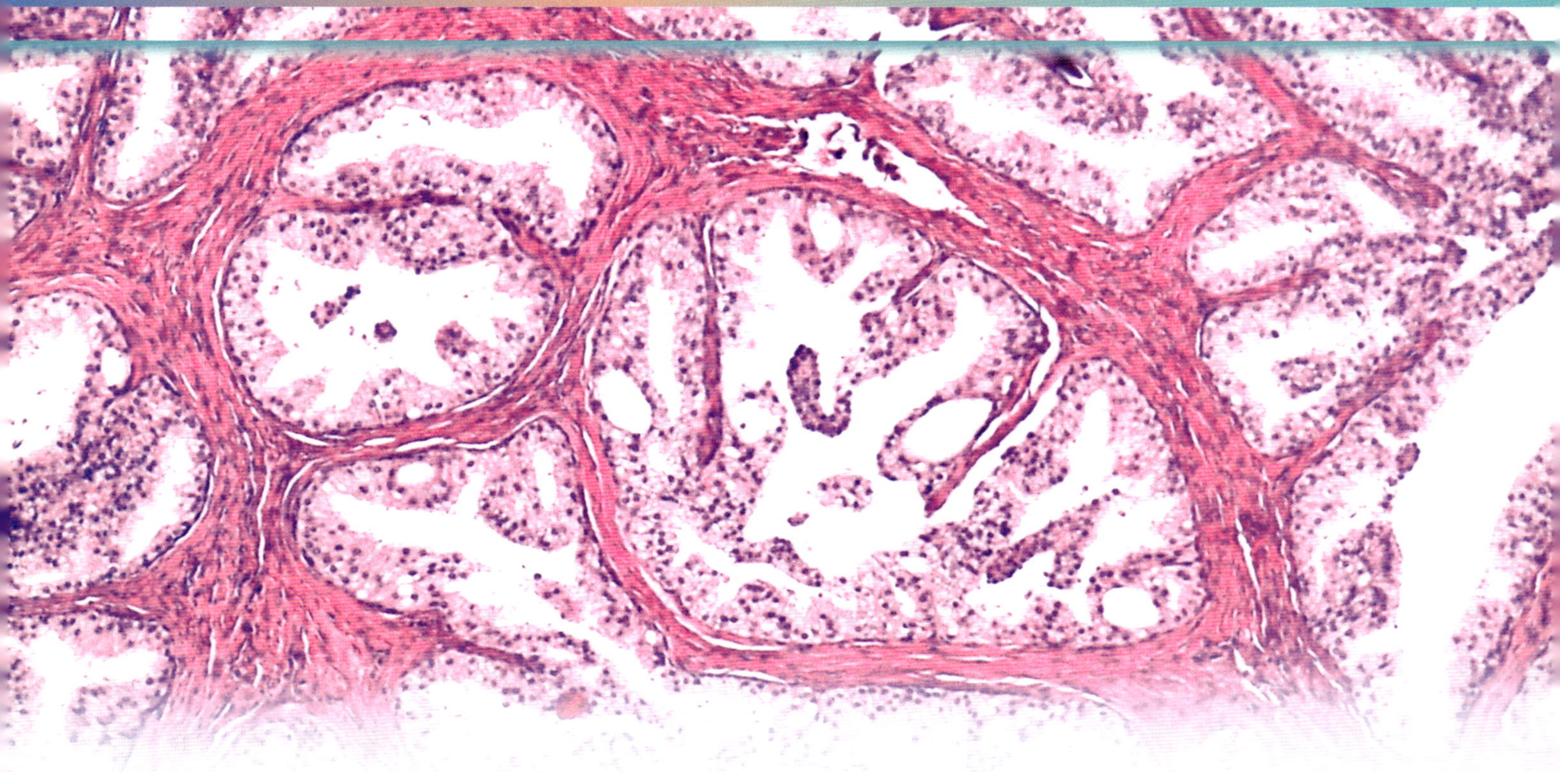

Genitourinary Tract

33

Kidney

Steven P. Salvatore

INTRODUCTION

The kidney has an intricate structure that underlies its diverse roles of excreting waste products, regulating body fluid and solute balance, regulating blood pressure, and secreting hormones. Familiarity with the basic structure of the kidney facilitates the evaluation and comprehension of diseases and functional disorders that can affect the kidney. The structure of the normal human kidney is considered in this chapter.

PEDIATRIC KIDNEY

The human kidney is structurally immature at the time of birth, and important morphologic changes occur during infancy and childhood. Pathologists unfamiliar with the histologic peculiarities of the pediatric kidney may mistake normal findings for pathologic conditions or fail to observe significant abnormalities of renal maturation.

KIDNEY DEVELOPMENT

The kidney has long been considered an excellent model system for the study of organogenesis. Detailed reviews of the morphologic and molecular aspects of kidney development are available (1–12).

EMBRYONIC KIDNEYS

Organogenesis begins during the third week of human embryogenesis with the initial formation of the central nervous and cardiovascular systems. The urogenital system represents the last organ system to develop. Kidney development goes through three successive stages: pronephros, mesonephros, and metanephros. All three systems develop from the intermediate mesoderm located between the dorsal somites and lateral plate mesoderm and extend from the cervical to the caudal regions of the embryo. The pronephros and mesonephros are transient structures; the mesonephros forms before the pronephros regresses and the metanephros develops before the mesonephros disappears (Fig. 33.1). Some genes that regulate metanephric kidney development appear to be involved in forming the earlier embryonic kidneys.

Pronephros

The pronephros develops in the cervical region at the end of the third week of human gestation. The pronephros consists of a glomerulus-like structure termed the glomus, tubules, and a duct. The glomus is not physically connected to the tubules; it projects into the coelomic cavity and filters blood. Ciliated tubules, called nephrostomes, open into the

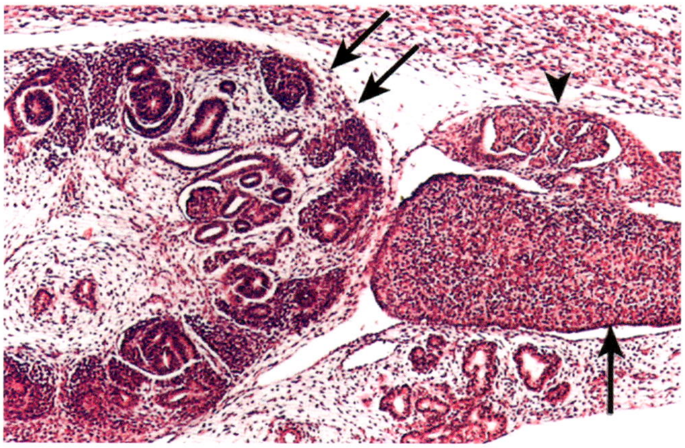

FIGURE 33.1 Mesonephros and metanephros. The mesonephros (*arrowhead*) contributes somatic cell lineages to the gonadal ridge (*single arrow*), which will develop into the gonad. Early nephron formation is present in the metanephros (*double arrows*), whose development is dependent on the presence of the mesonephros. (Reprinted with permission from Murphy WM, Grignon DJ, Perlman EJ. Tumors of the kidney, bladder, and related urinary structures. In: Silverberg SG, Sobin LH, eds. *Atlas of Tumor Pathology.* 4th series, fascicle 1. Armed Forces Institute of Pathology; 2004. Permission obtained from The Joint Pathology Center.)

coelom and collect the filtrate. The nephrostomes connect to proximal tubules which empty into a distal tubule that joins the pronephric duct. In humans, the pronephros is a rudimentary organ and does not function. As the pronephric duct extends caudally, the glomus and tubules regress. However, the pronephric duct persists and becomes the mesonephric duct.

Mesonephros

The human mesonephros develops in the middle of the fourth week of gestation as a thoracic organ caudal to the pronephros. Although 40 to 42 nephrons are formed in the human mesonephros, only 30 to 32 or less are present at any given time because the more cranial nephrons degenerate as the more caudal ones form (2). The mesonephric nephrons consist of glomeruli that are directly connected to tubules that have proximal and distal segments. Some of these directly connect to the mesonephric duct, which is also termed the wolffian duct or nephric duct. The distal mesonephric duct fuses with the cloaca, a precursor of the urinary bladder. The cranial tubules are connected to the mesonephric duct, whereas the more numerous caudal tubules never fuse with the mesonephric duct. Elegant fate mapping studies have demonstrated that the mesonephric tubules largely derive from a mesenchymal-to-epithelial transition within the mesenchyme adjacent to the mesonephric duct (13). Moreover, numerous genes involved in the formation of the definitive mammalian kidney, the metanephros, are also expressed in the mesonephric tubule (14). For example, genes expressed in the renal vesicle of metanephric kidney development are expressed in early mesonephric tubules but not retained in mature mesonephric tubules. Genes expressed in differentiated metanephric structures, such as the early proximal tubule, are expressed in more mature mesonephric tubules. These findings support the notion that the mesonephros and metanephros share some common cell and molecular pathways of tubulogenesis. The excretory function of the human mesonephros is believed to be limited. As observed with the pronephros, the mesonephros undergoes apoptosis and degenerates. In the male, some mesonephric tubules form the efferent ducts of the epididymis, whereas the mesonephric duct gives rise to the duct of the epididymis, the vas deferens, and the seminal vesicle. In females, the mesonephros undergoes dissolution, with the epoophoron, paroophoron, and Gartner duct remaining as vestigial structures. In both males and females, the mesonephric duct is involved in the formation of the paramesonephric (mullerian) duct, while it degenerates in males (15).

METANEPHROS

Overview

The metanephros is the definitive and permanent kidney and develops from a mutual inductive interaction between the mesonephric duct and a condensed area of mesenchymal cells in the caudal intermediate mesoderm, called the metanephric mesenchyme or blastema. Although the nephric duct and metanephric mesenchyme are both formed from the intermediate mesoderm, their respective epithelial and mesenchymal cell types provide for complex molecular signaling between each other. During the fourth week of gestation, the first step in human metanephric development occurs. Factors expressed by the metanephric mesenchyme induce the ureteric bud, a branch of the caudal nephric duct, to grow dorsally until it encounters the mesenchyme. The ureteric bud undergoes iterative branching to form the renal pelvis, calyces, and collecting ducts. Induced by the ureteric bud, the metanephric mesenchyme differentiates into the glomeruli, proximal and distal tubules, and Henle loops. Thus, cells of the metanephric kidney originate from two different lineages to form the collecting ducts and nephrons. The reciprocal inductive interaction between the ureteric bud and the metanephric mesenchyme is the central process of kidney development. The following outlines the morphologic features of renal organogenesis followed by the molecular aspects.

Formation of the Renal Pelvis and Calyces

Growing into the adjacent metanephric mesenchyme, the ureteric bud branches repeatedly while it also elongates. This complex three-dimensional branching pattern creates an elaborate renal architecture. The ureteric bud and its branches consist of a stalk or trunk portion and an actively growing ampullary tip. Two main categories of branching have been observed (2,3,16–19). The most common type is terminal branching of the ampullary tip with bifid terminal branching forming a "T" structure. Asymmetric bifid branching forming an "L" structure and even trifid branching have been observed. Lateral branching from the stalk segment has been noted but is far less frequent. Experimental studies have demonstrated that ureteric bud stalks that have had their existing tips removed can branch and form new tips (20). Thus, the developing collecting duct system exhibits considerable developmental plasticity. The first three to five generations of ureteric bud branches form the renal pelvis with more divisions occurring in the poles than in the midpolar region (Fig. 33.2). Urine production is accompanied by progressive dilatation and coalescence of the earlier branches to form the early pelvic–calyceal system by 11 to 12 weeks. Subsequent generations of branches form the calyces and extensive tissue remodeling of the calyceal system occurs. By 11 to 14 weeks, the calyces become compressed between the expanding renal pelvis and the aggregation of nephrons induced by collecting ducts in the developing papillae. The minor calyces convert from a bulbous configuration to their definitive cup-like shape as the papillae become conical (Fig. 33.3). The very first nephrons formed are believed to either degenerate or attach to a later generation branch that elongates eventually to reach the juxtamedullary cortex.

FIGURE 33.2 Diagram depicting early branches of ureteric bud that dilate and coalesce to form the renal pelvis. Examples of third-, fourth-, and fifth-generation branches are *circled.* (Modified with permission from Potter EL. *Normal and Abnormal Development of the Kidney.* Year Book; 1972, with permission from Elsevier.)

Formation of the Collecting System

At 8 weeks, the first nephrons can be observed attached to ureteric bud branches. The organogenetic processes of collecting duct branching and elongation and nephron formation occur simultaneously. Collecting duct morphogenesis has been divided into four periods (2). The first occurs during the 5th to the 14th week of gestation, at which time branching occurs from the ampullary tips, and individual nephrons remain attached to their ampullae. In an iterative bifurcation model of branching, one of the two new ampullae retains the old nephron whereas the other induces the formation of a new one. The second period occurs during weeks 14 to 22 and is characterized by the formation of arcades. Ampullae rarely branch but single elongating tips repeatedly induce new nephrons while carrying attached older nephrons. As new nephrons are formed, the connecting tubule of the older nephron merges its point of attachment away from the ampulla to the connective tubule of the newer nephron. Repetition of this process results in three to seven nephrons forming around a single ampulla, joined to one another in an arcade by their connective tubules. Arcades are associated with juxtamedullary nephrons in the inner cortex of the fully developed kidney.

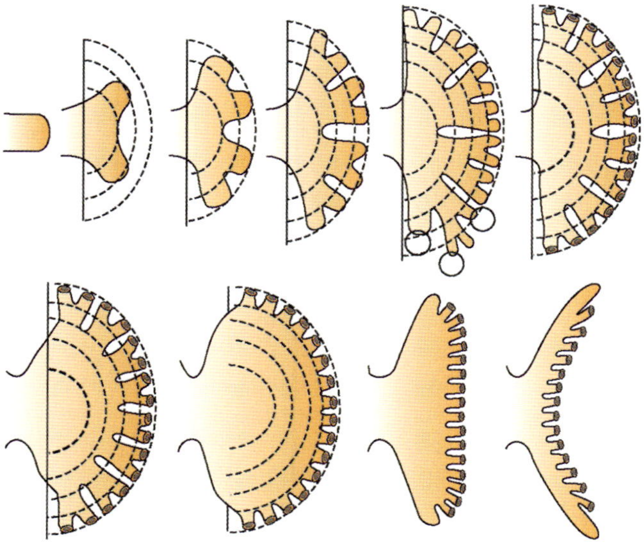

FIGURE 33.3 Diagram illustrating later branches of ureteric bud forming a minor calyx and papilla. *Circles* indicate generation branches that may expand to form part of the calyx or, if not expanding, form papillary ducts. The expanding pelvis and the peripheral zone of differentiating nephrons compress the original saccular cavity, producing the cup-like shape of the calyx and the conical configuration of the papilla. (Modified with permission from Potter EL. *Normal and Abnormal Development of the Kidney.* Year Book; 1972, with permission from Elsevier.)

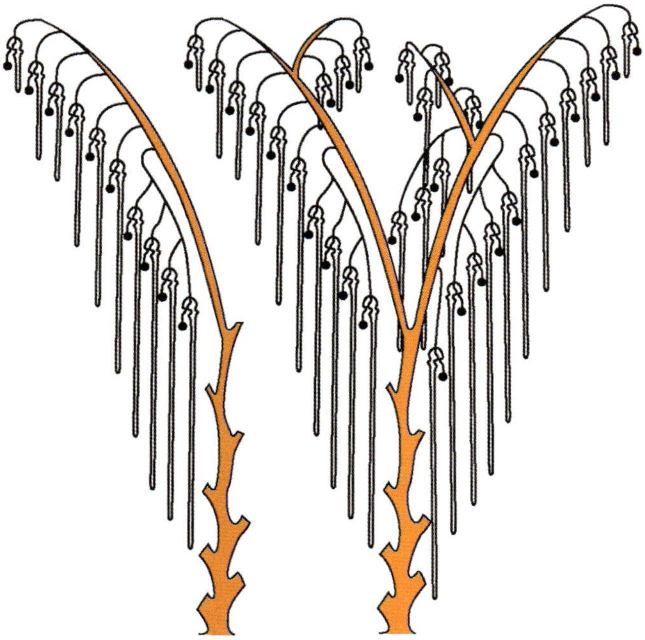

FIGURE 33.4 Diagram demonstrating the pattern of nephrons and collecting tubules at birth. **A:** The most common arrangement is for each collecting tubule to have a single arcade composed of three to five nephrons and five to seven nephrons individually attached. **B:** Depending on the division of the ampullary tips, other variations are possible. (Modified with permission from Potter EL. *Normal and Abnormal Development of the Kidney.* Year Book; 1972, with permission from Elsevier.)

In the third period during weeks 20 to 36, the ampullae advance beyond the attachment point of the arcade toward the outer surface. The ampullae do not branch but induce five to seven nephrons, each of which will have a direct connection to the developing collecting tubule. This type of nephron attachment predominates in the outer cortex of the mature kidney (Fig. 33.4). Since nephrons retain contact with their ampullae of origin, the longitudinal growth of the collecting tubules positions the attached glomeruli in the cortex. In the fourth period, beginning at 32 to 36 weeks, the ampullae disappear and no new nephrons form. Normally, nephrogenesis does not occur beyond 36 weeks of gestation. The last nephrons formed are in the outer cortex with their glomeruli near the renal capsule.

Nephron Formation

Over 100 years ago, investigations by Herring and Huber provided a fairly accurate morphologic view of human nephron development (Fig. 33.5) (21,22). They were also

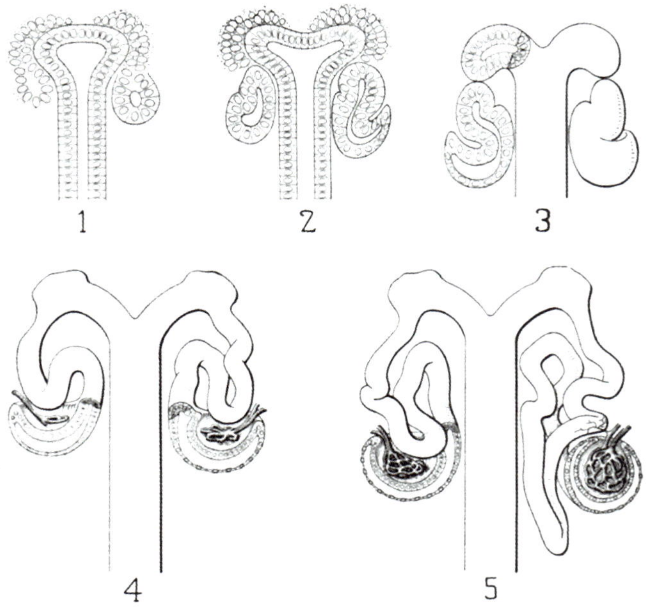

FIGURE 33.5 Huber's schematic drawings of nephron development. **1.** Condensation stage with the cap. A renal vesicle (**right**) is present. **2.** Comma-shaped body. **3.** S-shaped body. **4.** Early glomerular capillary development, Bowman capsule formation and tubule elongation. **5.** Glomerular and tubule maturation (**right**). (Reprinted from Huber GC. On the development and shape of uriniferous tubules of certain of the higher mammals. *Am J Anat.* 1905;4(suppl):1–98.)

prescient in regard to some mechanisms of nephrogenesis, for example, in the development of glomerular capillaries. From 8 weeks of gestation, the nephrons and collecting duct system develop together. The stages in individual nephron development do not vary and occur continuously throughout the periods of collecting duct formation. The formation of nephrons can be divided into two phases: the induction stage and the morphogenetic stage (4,23). In the induction stage, the mesenchyme condenses around the ampullary tips in response to inducing signals from the ureteric bud. Two types of mesenchymal condensates have been described to form in this induction stage prior to epithelial differentiation of the mesenchyme (24). The first condensate, called the cap, closely surrounds each ampullary tip. A short time later, another condensate, termed the pretubular aggregate, forms at the lateral edges of the ampullary tip, below the cap. The cap mesenchyme is believed to regulate ureteric bud branching and contains the progenitor cells of the nephron epithelia, whereas the cells of the pretubular aggregate having differentiated from the cap are believed to be committed to forming the nephron elements.

The morphogenetic stage of nephron formation involves several complex phases (Fig. 33.5). First, the cells of the pretubular aggregate undergo a mesenchyme-to-epithelium transition characterized by expression of epithelial markers and synthesis of basement membrane matrix glycoproteins. The cells develop intercellular junctions and become polarized and surrounded by a basal lamina, forming a structure termed the renal vesicle. A central cavity may be observed in the vesicle that can be difficult to distinguish between pretubular aggregates and renal vesicles in histologic sections. Soon after formation, the vesicle fuses to the ureteric duct epithelium and a continuous basal lamina surrounds both the vesicle and the duct. Opposite the area of fusion between the vesicle and the ureteric duct, a vascular cleft develops representing the site where the glomerular capillaries will emerge. The vesicle becomes a comma-shaped tubular structure. Another crevice forms near the fusion between the comma-structure and the ureteric duct. After elongation and folding, an S-shaped figure representing an early nephron forms (Fig. 33.6). At this stage, the S-shaped body is already compartmentalized into distinct cell types that are arranged into three areas. The vascular cleft lies below the upper and middle limbs of the S-shaped body, and above the lower limb. The lower limb, most distant from the ureteric bud, differentiates into the visceral

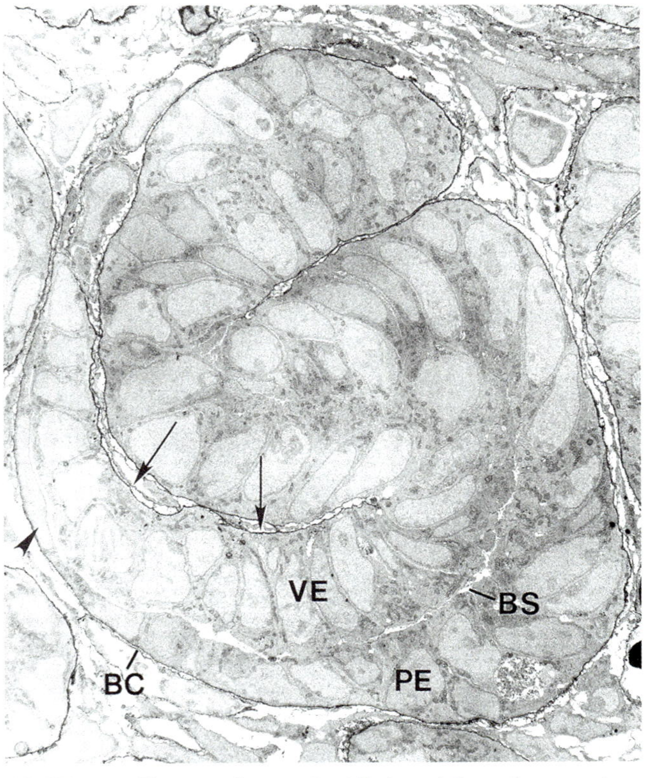

FIGURE 33.6 Electron micrograph of S-shaped figure from newborn mouse kidney. The basement membranes are rendered black by labeling with antilaminin IgG conjugated to horseradish peroxidase. The vascular cleft (*arrows*), visceral epithelial cells (*VE*), Bowman space (*BS*), parietal epithelial cells (*PE*), and Bowman capsule (*BC*) can be seen. The visceral epithelial cells will differentiate into podocytes. Some parietal epithelial cells are becoming squamous (*arrowhead*) and will line Bowman capsule. The epithelial cells above the vascular cleft will give rise to the proximal tubules, loops of Henle, and distal convoluted tubules. (Magnification ×5,000.) (Modified with permission from Clapp WL, Abrahamson DR. Development and gross anatomy of the kidney. In: Tisher CC, Brenner BM, eds. *Renal Pathology.* 2nd ed. JB Lippincott; 1994:3–59.)

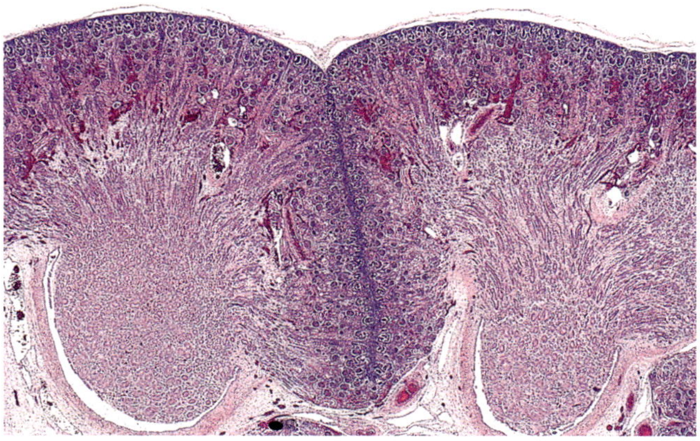

FIGURE 33.7 Developing kidney at 21 weeks of gestation showing two medullary pyramids with surrounding cortex. The nephrogenic zone represents a thin layer outlining the peripheral aspects of the lobes, both at the surface and in the midplane of the septa (column) of Bertin, between the two renal lobes.

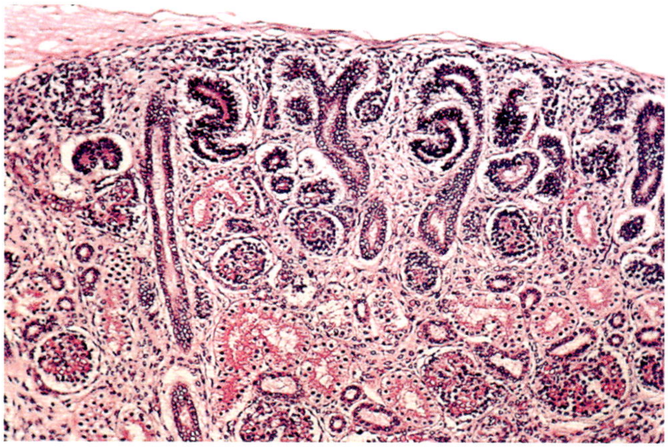

FIGURE 33.8 Nephrogenic zone from developing kidney at 26 weeks of gestation illustrating several stages of nephron formation.

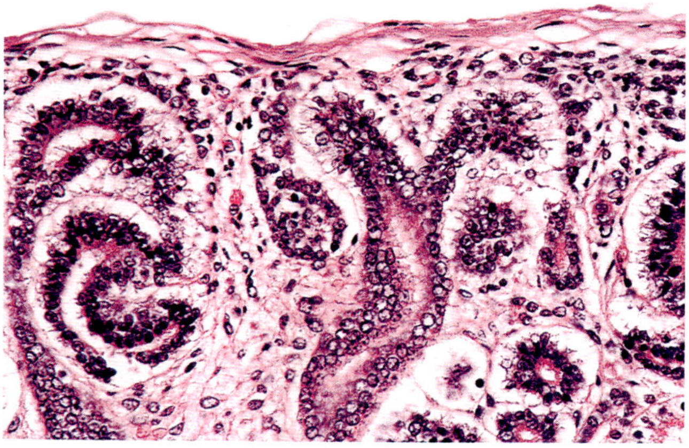

FIGURE 33.9 Higher magnification of same field as in Figure 33.7. In the center, pretubular aggregates (early renal vesicles) are present on either side of the ureteric duct. An early S-shaped body is present (**left**).

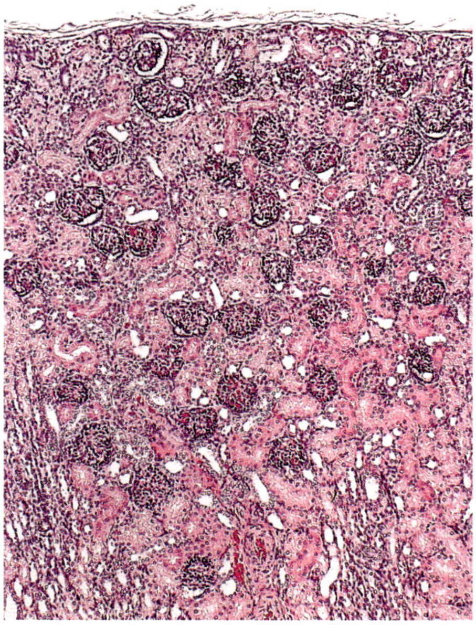

FIGURE 33.10 Newborn kidney (40 weeks of gestation). Note the absence of a nephrogenic zone. Some glomeruli are near the renal capsule.

epithelium (podocytes) and parietal epithelium (Bowman capsule) of the glomerulus. The midportion of the S-body forms the proximal tubule and the loop of Henle. The upper limb becomes the distal convoluted tubule distal convoluted tubule distal convoluted tubule and fuses with the ureteric bud branches to form the connective tubule. Active nephron formation occurs across the developing renal cortex in a band, known as the nephrogenic zone (Figs. 33.7 to 33.9). After nephron formation ceases by 36 weeks of gestation, the nephrogenic zone disappears (Fig. 33.10). The growth of the collecting ducts and the incremental formation of nephrons result in a centrifugal developmental pattern extending through the renal cortex. The earliest nephrons to form are found in the juxtamedullary zone of cortex, whereas the last nephrons to develop are in the outer cortex. This principle is fundamental to understanding postnatal structural changes in the kidney and is sometimes useful in the timing of developmental disturbances in the cortex. For example, a disturbance during the early months of development may result in an abnormality of the entire cortical thickness, whereas one that occurs in the last half of gestation may involve only the outermost layers of cortical nephrons.

MOLECULAR REGULATION OF KIDNEY DEVELOPMENT

To make a kidney requires an orchestration of numerous complex cellular and molecular events (Table 33.1).

High-throughput global gene expression studies have led to a more detailed view of kidney development. The

TABLE 33.1 Genes Involved in Kidney Development

Developmental Process Gene (Human Syndrome)	Function	Expression in Normal Developing Kidney	Mutant Kidney Phenotype (Mouse)
Specification of the Nephrogenic Mesenchyme			
Lim 1	Transcription factor	Intermediate mesoderm, nephric duct, mesonephros, ureteric bud, collecting ducts, pretubular aggregates, S-shaped bodies, podocytes	Absent pro-, meso-, and metanephros (*Lim 1* required at multiple steps of kidney development)
Eya1 (Mutations in branchio-oto-renal [BOR] syndrome)	Transcription coactivator (interacts with *Six1*)	Intermediate mesoderm, uninduced and induced metanephric mesenchyme	Intact pro- and mesonephros, but absent metanephric blastema, renal agenesis
Six1 (Mutations in branchio-oto-renal [BOR] syndrome)	Transcription factor (interacts with *Eya1*)	Uninduced and induced metanephric mesenchyme and collecting tubules	Ureteric bud forms but does not fully invade metanephric mesenchyme, which undergoes apoptosis, renal agenesis
Sall 1 (Mutations in Townes–Brocks syndrome)	Transcription factor	Mesonephros, induced metanephric mesenchyme	Ureteric bud forms but does not fully invade metanephric mesenchyme, which undergoes apoptosis, renal hypoplasia, or agenesis
Hoxa11, Hoxd11	Transcription factors	Intermediate mesoderm, metanephric mesenchyme	Impaired ureteric bud branching, renal hypoplasia in *Hoxall/Hoxdll* double mutants
Pax2	Transcription factor	Intermediate mesoderm, nephric duct, mesonephros, ureteric bud, induced metanephric mesenchyme	Intact nephric duct but mesonephric tubules and ureteric bud fail to form, renal agenesis
WT1	Transcription factor	Intermediate mesoderm, mesonephros, uninduced and induced metanephric mesenchyme, comma- and S-shaped bodies, podocytes	Absent caudal mesonephros, ureteric bud fails to form, metanephric mesenchyme undergoes apoptosis, renal agenesis
Cell Survival			
BMP7	Growth factor	Mesonephric duct, ureteric bud, induced metanephric mesenchyme, distal tubules, comma- and S-shaped bodies, podocytes	Increased apoptosis in metanephric mesenchyme, decreased ureter bud branching, hydroureter, severe renal hypoplasia
FGF-8	Growth factor	Pretubular aggregates, vesicles, tubule progenitors in S-body	Increased apoptosis in S-body precursors, truncated nephrons, severe renal hypoplasia
Bcl-2	Antiapoptotic factor	Ureteric bud, induced metanephric mesenchyme, parietal epithelium of Bowman capsule, tubules, and collecting ducts	Increased apoptosis especially in metanephric mesenchyme, cysts in tubules and collecting ducts, severe renal hypoplasia
AP-2b	Transcription factor	Distal tubules and collecting ducts	Increased apoptosis and cyst formation in distal tubules and collecting ducts
Pax2 (Heterozygous mutations in renal-coloboma syndrome)	Transcription factor	Intermediate mesoderm, nephric duct, mesonephros, ureteric bud, induced metanephric mesenchyme	Heterozygous *Pax-2* mutations result in increased apoptosis in collecting ducts, reduced ureteric branching, hypoplastic kidneys
Cell Proliferation			
FGF-7	Growth factor	Interstitial fibroblasts or stroma surrounding ureteric bud and developing collecting ducts	Decreased growth of ureteric bud and collecting ducts, decreased number of nephrons, renal hypoplasia
N-Myc	Transcription factor	Induced metanephric mesenchyme	Decreased cell proliferation, decreased ureteric bud tips and nephrons, renal hypoplasia
Glypican-3 (Mutations in Simpson–Golabi–Behmel syndrome)	Heparan sulfate proteoglycan	All metanephric mesenchyme and ureteric bud derivatives	Increased cell proliferation in cortical collecting ducts, increased apoptosis in medullary collecting ducts, renal medullary cystic dysplasia

(*continued*)

TABLE 33.1 Genes Involved in Kidney Development (*Continued*)

Developmental Process Gene (Human Syndrome)	Function	Expression in Normal Developing Kidney	Mutant Kidney Phenotype (Mouse)
Branching of Ureteric Bud			
GDNF	Growth factor	Intermediate mesoderm, mesonephros, induced metanephric mesenchyme, pretubular aggregates	Ureteric bud fails to form or has abnormal branching, renal hypoplasia, or agenesis
c-ret	GF receptor (TK) (receptor for *GDNF*)	Mesonephric duct, ureteric bud, and tips of ureteric bud	Similar to GDNF-deficient mice
GFRa1	GF co-receptor (forms signaling complex with *GDNF* and *c-ret*)	Induced metanephric mesenchyme, pretubular aggregates, mesonephric duct, ureteric bud, and tips of ureteric bud	Similar to GDNF-deficient mice
GDF11	Growth factor	Mesonephric duct, uninduced and induced metanephric mesenchyme, ureteric bud and branches	No ureteric bud formation, metanephric mesenchyme undergoes apoptosis, renal hypoplasia, or agenesis
Sprouty1	Receptor (TK) antagonist	Mesonephric duct, ureteric bud, and tips of ureteric bud	Supernumerary ureteric buds, multiple ureters, multiplex kidneys
Emx2	Transcription factor	Intermediate mesoderm, mesonephric duct, mesonephros, ureteric bud, comma- and S-shaped bodies	Ureteric bud invades metanephric mesenchyme but fails to dilate or branch, no induction of mesenchyme, renal agenesis
RARa, RARb2	Transcription factors	RARa—ureteric bud, metanephric mesenchyme, stroma RARb2—stroma only	Decreased ureteric bud branching (defective stroma signaling) in *RARa/b2* double mutants, renal hypoplasia
BMP4	Bone morphogenetic protein	Stromal mesenchymal cells around ureteric bud branches	Reduced and ectopic ureteral branching, ectopic ureterovesical junction, hydroureter, double collecting ducts, hypo/dysplastic kidneys (heterozygous *BMP4* null mice)
Foxc1/2	Transcription factor	Intermediate mesoderm, mesonephros, uninduced and induced metanephric mesenchyme	Double ureters with one a hydroureter, duplex kidneys (*Foxc1*/*Foxc2* compound heterozygous mutants similar to *Foxc1* homozygous mutants)
Slit2	Secreted protein	Mesonephric duct, anterior intermediate mesoderm, ureteric bud tips	Supernumerary ureteric buds, hydroureter, fused multiple kidneys
Robo2	Transmembrane (receptor for *Slit2*)	Intermediate mesoderm, induced metanephric mesenchyme	Supernumerary ureteric buds, hydroureter, fused multiple kidneys
Pod1	Transcription factor	Induced metanephric mesenchyme, stromal cells, and podocytes	Decreased ureteric branching, arrest in tubular and glomerular differentiation, renal hypoplasia
Heparan sulfate 2-sulfotransferase	Enzyme (synthesis of heparan sulfate, a component of HS proteoglycan)	Mesonephric duct, transient in ureteric bud, metanephric mesenchyme	Ureteric bud outgrowth but no branching, no condensation of metanephric mesenchyme, renal agenesis
Integrin $\alpha 8$	Transmembrane adhesion receptor	Intermediate mesoderm, induced metanephric mesenchyme, pretubular aggregates	Limited ureteric bud invasion of metanephric mesenchyme, decreased ureteric branching, renal hypoplasia, or agenesis
Integrin α *3b1*	Transmembrane adhesion receptor	Ureteric bud, collecting ducts, podocytes	Decreased branching of medullary collecting ducts, microcystic proximal tubules, defective glomerulogenesis
Pbxl	Transcription factor	Induced metanephric mesenchyme and stromal cells	Decreased and irregular ureteric bud branching, large mesenchymal condensates, renal hypoplasia, or unilateral agenesis

TABLE 33.1 Genes Involved in Kidney Development (*Continued*)

Developmental Process Gene (Human Syndrome)	Function	Expression in Normal Developing Kidney	Mutant Kidney Phenotype (Mouse)
Grem1	Bone morphogenetic protein (BMP) antagonist	Intermediate mesoderm, mesonephric duct, metanephric mesenchyme	Ureteric bud forms but fails to invade metanephric mesenchyme, which undergoes apoptosis, renal agenesis
Wnt11	Secreted glycoprotein	Ureteric bud tips	Loss of ureteric tips, reduced ureteric branching, renal hypoplasia
Formin (Mutations in mice cause the limb deformity syndrome)	Protein that regulates cytoskeleton function	Mesonephric duct, mesonephros, ureteric bud, metanephric mesenchyme	Decreased ureteric bud outgrowth, renal hypoplasia, unilateral, or bilateral renal agenesis
Angiotensinogen (*Agt*)	Renin substrate	Ureteric bud, stroma, S-shaped bodies, glomeruli, proximal tubules	Hypoplastic papillae, hydronephrosis, thickened blood vessels, reduced blood pressure
Renin	Enzyme (cleaves Agt to form angiotensin I)	Developing arcuate and interlobular arteries and afferent arterioles, restricted to juxtaglomerular apparatus with maturity	Hypoplastic papillae, hydronephrosis, thickened blood vessels, reduced blood pressure
Angiotensin-converting enzyme (*ACE*)	Enzyme (converts angiotensin I to angiotensin II)	Glomeruli, proximal tubules and collecting ducts, small arteries	Hypoplastic papillae, hydronephrosis, thickened blood vessels, reduced blood pressure
Angiotensin II receptor, type 1 (AT_1)	Angiotensin II receptor	Ureteric bud, stroma, S-shaped bodies, proximal tubules, collecting ducts	Hypoplastic papillae, hydronephrosis, thickened blood vessels, reduced blood pressure
Angiotensin II receptor, type 2 (AT_2)	Angiotensin II receptor	Stroma adjacent to ureteric bud stalk	3% have abnormalities, duplicated collecting system, and hydronephrotic upper pole
Mesenchymal-to-Epithelial Transition			
Wnt4	Secreted glycoprotein	Pretubular aggregates, comma-shaped body, distal S-shaped body	Ureteric bud branching occurs, no mesenchymal–epithelial transition, renal agenesis
Wnt9b	Secreted glycoprotein	Mesonephric duct, ureteric bud, and collecting ducts but not branching tips	No mesenchymal–epithelial transition, renal agenesis
Foxd1(BF2)	Transcription factor	Interstitial stroma cells	Decreased ureteric bud branching, large mesenchymal condensates, abnormal renal capsule, small fused pelvic kidneys
Pod1	Transcription factor	Induced metanephric mesenchyme, stromal cells, and podocytes	Increased size of induced mesenchymal condensates, arrest in tubular and glomerular differentiation, renal hypoplasia
Fras1 (Mutations in Fraser syndrome)	Extracellular matrix protein	Basal side of ureteric ducts	Ureteric bud invades metanephric mesenchyme but decreased induction, apoptosis of mesenchyme, renal hypoplasia, or agenesis
Grip1	Cytoplasmic protein (interacts with *Fras1*)	Basal side of ureteric ducts	Similar abnormal phenotype as in *Fras1*-deficient mice
Cadherin-6 (*K-cadherin*)	Transmembrane adhesion protein	Renal vesicle, proximal end of comma- and S-shaped bodies, developing proximal tubules and Henle loops	Delayed fusion of some comma-shaped bodies to ureteric bud leading to loss of nephrons
Glomerulogenesis			
WT1 (Mutations in WAGR, Denys-Drash, and Frasier syndromes)	Transcription factor	Intermediate mesoderm, mesonephros, uninduced and induced metanephric mesenchyme, S-shaped body, podocytes	Disturbed podocyte differentiation, glomerulosclerosis

(*continued*)

TABLE 33.1 Genes Involved in Kidney Development (*Continued*)

Developmental Process Gene (Human Syndrome)	Function	Expression in Normal Developing Kidney	Mutant Kidney Phenotype (Mouse)
PDGF-β	Growth factor	Glomerular endothelial cells and podocytes	Dilated glomerular capillaries with no mesangial cells
PDGFR-β	Growth factor receptor (receptor for *PDGF-β*)	Glomerular mesangial cells	Dilated glomerular capillaries with no mesangial cells
Notch2	Transmembrane receptor	Developing collecting ducts, comma- and S-shaped bodies, podocytes	Abnormal glomeruli arrested at capillary loop stage, with disorganized podocytes and no mesangial cells
VEGF-A	Growth factor	S-shaped bodies, podocytes, collecting ducts	Small glomeruli, lack capillary loops, few endothelial cells
Laminin α5	Basement membrane protein	Basement membranes of ureteric bud, developing tubules, and glomerular basement membrane	Abnormal glomeruli with displaced endothelial and mesangial cells, and clustered podocytes
Laminin β2	Basement membrane protein	Glomerular basement membranes beginning at capillary loop stage	Absence of podocyte foot processes, proteinuria
Laminin α3b1	Transmembrane adhesion receptor	Ureteric bud, collecting ducts, podocytes	Glomerular capillary loops dilated and fewer in number, loss of podocyte foot processes, dual GBMs (failure of fusion)
Collagen IV α5 (Mutations in Alport syndrome)	Basement membrane protein	Forms heterodimer with α3 (IV) and α5 (IV) chains in glomerular basement membrane from capillary stage onward	Thinning and "basket-weave" thickening of glomerular basement membranes
Nephrin (Mutations in congenital nephrotic syndrome of Finnish type)	Transmembrane protein	Podocyte filtration slit diaphragm	Foot process effacement, absent filtration slit diaphragms, proteinuria
CD2AP	Adapter protein (interacts with nephrin)	Podocyte filtration slit diaphragm domain	Irregular foot processes, absent filtration slit diaphragms, proteinuria
Podocin (Mutations in autosomal recessive steroid-resistant nephrotic syndrome)	Membrane protein (interacts with nephrin)	Podocyte filtration slit diaphragm domain	Irregular foot processes, absent filtration slit diaphragms, proteinuria
α-actinin (Mutations in autosomal dominant focal segmental glomerulosclerosis [FSGS])	Cross-links actin filaments	Podocyte filtration slit diaphragm domain	Foot process effacement, GBM duplication, proteinuria, and FSGS
TRPC6 (Mutations in autosomal dominant FSGS)	Cation-channel	Podocyte filtration slit diaphragm domain	Foot process effacement and FSGS (humans)
Neph1	Transmembrane protein (interacts with nephrin)	Podocyte filtration slit diaphragm	Foot process effacement, proteinuria
FAT1	Protocadherin	Podocyte filtration slit diaphragm	Foot process effacement
Lmx1b (Mutations in nail-patella syndrome)	Transcription factor	Podocytes	Abnormal foot processes, absent filtration slit diaphragms
Podocalyxin	CD34-related transmembrane protein	Podocyte apical membrane domain	Abnormal podocytes, foot process effacement, absent filtration slit diaphragms, anuria

TABLE 33.1 Genes Involved in Kidney Development (*Continued*)

Developmental Process Gene (Human Syndrome)	Function	Expression in Normal Developing Kidney	Mutant Kidney Phenotype (Mouse)
GLEPP1 (*Ptpro*)	Receptor tyrosine phosphatase	Podocyte apical membrane domain	Shortened and widened podocyte foot processes, reduced glomerular filtration rate
Kreisler (*Krml1/MafB*)	Transcription factor	Podocytes, initially at capillary loop stage	Abnormal podocyte differentiation with no foot processes
Tubular Differentiation			
PKD1 (*polycystin-1*) (Mutations in autosomal dominant polycystic kidney disease [ADPKD])	Transmembrane protein (adhesion receptor, in cilia, interacts with polycystin-2)	Developing nephron segments and collecting ducts	Renal cysts arising from developing nephron segments and collecting ducts
PKD2 (*polycystin-2*) (Mutations in autosomal dominant polycystic kidney disease [ADPKD])	Transmembrane protein (calcium channel, in cilia, interacts with polycystin-1)	Developing nephron segments and collecting ducts	Renal cysts arising from developing nephron segments and collecting ducts
Frem2 (Mutations in Fraser syndrome)	Extracellular matrix protein	Mesonephros, ureteric bud especially at tips, tubule derivatives	Cysts of collecting ducts and thick ascending limbs
Cox-2 (Cyclooxygenase-2)	Enzyme (prostaglandin synthesis)	Developing collecting ducts, S-shaped bodies, macula densa, cortical thick ascending limb, medullary interstitial cells	Progressive postnatal outer cortical dysplasia, tubular cysts, and glomerular hypoplasia
EGFR	Growth factor receptor (TK)	Mesonephric duct, ureteric bud, collecting ducts	Dilatation of collecting ducts, uremia
Brnl	Transcription factor	Renal vesicle, comma- and S-shaped bodies, developing Henle loop (HL), distal convoluted tubule (distal convoluted tubule), macula densa (MD)	Disrupted differentiation of HLs, MD, and distal convoluted tubule
Psen1/Psen2	Presinilins- (transmembrane protein with γ-secretase activity)	Developing early nephron structures	Renal vesicles and pretubular aggregates form, but no comma- or S-shaped bodies; proximal tubules and glomeruli fail to form (*Psen1/Psen2* double null mutant mice with human *PSEN1* transgene)
Tensin	Adhesion protein (phosphoprotein, binds actin)	Proximal and distal tubules (adult kidney)	Hydronephrosis, cystic dilatation of proximal tubules
R-cadherin	Transmembrane adhesion protein	Induced metanephric mesenchyme, renal vesicle, comma- and S-shaped bodies	Dilatation and cytoplasmic vacuolization of proximal tubules

Genitourinary Developmental Molecular Anatomy Project (GUDMAP), is an international consortium working to produce a high-resolution morphologic and molecular map of the developing urogenital tract (25). The GUDMAP database (http://www.gudmap.org) serves as a valuable interactive online resource for research in kidney development. For example, the GUDMAP database can be used with bioinformatic analysis to identify transcription factors that bind to conserved binding sites (*cis*-regulatory modules) in the promoters of genes highly expressed in a particular anatomic compartment (26). Recent studies are increasing the limits of resolution by evaluating gene expression at the single-cell level in the developing kidney (27). Some single-cell progenitor cells of the metanephric mesenchyme co-express markers of both nephron epithelial and stromal lineages. Moreover, single cells of the renal vesicle co-express markers of both podocyte and proximal tubule lineages. These findings are consistent with "multilineage priming," whereby promiscuous multilineage gene expression in a single progenitor cell primes the cell for subsequent specific lineage commitment. Such studies will enhance our understanding of specific cell types and general cell differentiation (28).

INTERMEDIATE MESODERM SPECIFICATION

Both the mesonephric duct and the metanephric mesenchyme form from the intermediate mesoderm. The intermediate mesoderm is a strip of tissue in the early embryo between the paraxial somatic mesoderm and the lateral plate mesoderm. Little is known about how the intermediate mesoderm becomes specified to develop into the kidney morphogenetic program of the pronephros, mesonephros, and finally the metanephros. The intermediate mesoderm is specified along both the mediolateral and anteroposterior axes. The earliest known marker genes of the intermediate mesoderm include *Osr1*, *Lhx1*, *Pax2*, and *Pax8*. Along the mediolateral axis and lateral to the intermediate mesoderm, secreted factors of the bone morphogenetic protein (BMP) family, including BMP2 and their cognate receptor ALK3, activate the intermediate mesoderm marker genes (29,30). Signals medial to the intermediate mesoderm, including activin, a member of the transforming growth factor β (TGF-β) family activate the intermediate mesoderm genes (31). It is not clear how these signals, lateral and medial to the intermediate mesoderm, are integrated.

The progression of development of the transient kidneys (pronephros and mesonephros) and the permanent kidney (metanephros) in a cranial–caudal direction along the anteroposterior axis is striking. The *Hox* genes are well known to determine the developmental fate of cells within distinct regions along the anteroposterior axis. Encoding homeodomain-containing transcription factors, the *Hox* genes are arranged into four chromosomal clusters, which are subdivided into 13 sets of paralogous genes. The *Hox* genes of paralog group 4, including *Hoxb4*, *Hoxc4*, and *Hoxd4* are involved in establishing the anterior border of the kidney morphogenetic field (32). Expression of the *Hox11* paralog group, *Hoxa11*, *Hoxc11*, and *Hoxd11*, is restricted in the intermediate mesoderm to the posterior, metanephric level (13). Deletion of the *Hox11* paralog genes results in agenesis of the metanephros but not the mesonephros (32,33). Thus, specific patterns of Hox gene expression establish the anterior and posterior borders of the intermediate mesoderm that will become the metanephric mesenchyme.

NEPHRIC DUCT

In the early embryo, the previously discussed signals along with the *Hox* gene expression domains induce the intermediate mesoderm to form the nephric duct. Several genes including *Pax2*, *Pax8*, *Lhx1*, *Gata3*, and β-catenin (*CTNNB1*) form a regulatory network essential for normal nephric duct development (34–37). As the nephric duct elongates caudally, it converts from a solid core of mesenchymal cells to an epithelial duct. It eventually contacts and fuses with the cloaca. Cellular extensions from the caudal tip of the nephric duct participate in guiding the duct toward the cloaca. Insertion of the nephric duct into the cloaca is regulated by rearranged during transfection (*Ret*) expression in the nephric duct, which is dependent upon *Gata3* and retinoic acid signaling (38). Direct contact with the nephric duct induces apoptosis in the cloaca which is necessary for fusion of the nephric duct with the cloaca (39).

URETERIC BUD FORMATION

The first step in metanephric kidney development is the outgrowth of the ureteric bud from the nephric duct. Signals from the metanephric mesenchyme induce this budding and in turn, the ureteric bud induces the metanephric mesenchyme to form the epithelia of the nephron. Failure to form a ureteric bud results in renal agenesis and incorrect positioning of the ureteric bud leads to congenital anomalies of the kidney and urinary tract (40,41). *Gdnf/Ret* signaling is a major pathway regulating ureteric bud formation (42). The glial-derived neurotrophic factor (GDNF) secreted by the metanephric mesenchyme interacts with RET, a proto-oncogene receptor tyrosine kinase and a co-receptor GFRα1, both expressed in the nephric duct, to induce ureteric bud outgrowth.

Prior to definitive ureteric bud outgrowth, a segment in the caudal nephric duct thickens. Cells with elevated RET activity migrate within the nephric duct and concentrate to form a RET-rich cellular domain that will become the initial budding tip. Simultaneously, this nephric duct segment converts from a cuboidal to a pseudostratified epithelium. Unlike the cell migration, this epithelial reorganization occurs independent of RET signaling.

The dominant view is that ureteric bud formation is mainly dependent on *Gdnf/Ret* signaling (Fig. 33.11). The ureteric bud fails to form leading to renal agenesis in most embryos without *Gdnf* or without *Ret* (40). However, several activators and inhibitors influence the *Gdnf/Ret* pathway to coordinate normal ureteric bud outgrowth (43,44). Activators of *Gdnf* expression in the metanephric mesenchyme include *Pax2*, *Eya1*, *Hox11* paralogs, *Sall1*, *Grem1*, nephronectin, and integrin $\alpha 8\beta 1$. Mutations of these genes result in defective ureteric bud formation and often renal agenesis. The opposite phenotype of supernumerary ureteric buds and ureters results when genes that limit *Gdnf* expression are mutated. Examples of inhibitor genes include *Spry1*, *Bmp4*, *Robo2*, *Slit2*, and *Foxc2*. The proteins encoded by these various genes often interact in regulatory networks. For example, *Pax2*, *Eya1*, and *Hox11* paralog proteins interact in a complex (45) and integrin $\alpha 8\beta 1$ interacts with its ligand nephronectin to activate *Gdnf* expression (46). SLIT2 in the nephric duct interacts with its receptor ROBO2 in the metanephric mesenchyme to restrict *Gdnf* expression (47). Moreover, an activator may interact with an inhibitor to facilitate ureteric bud sprouting. For example, *Grem1*, a BMP antagonist, is upregulated in the mesenchyme surrounding the nephric duct prior to initiation of ureteric bud outgrowth

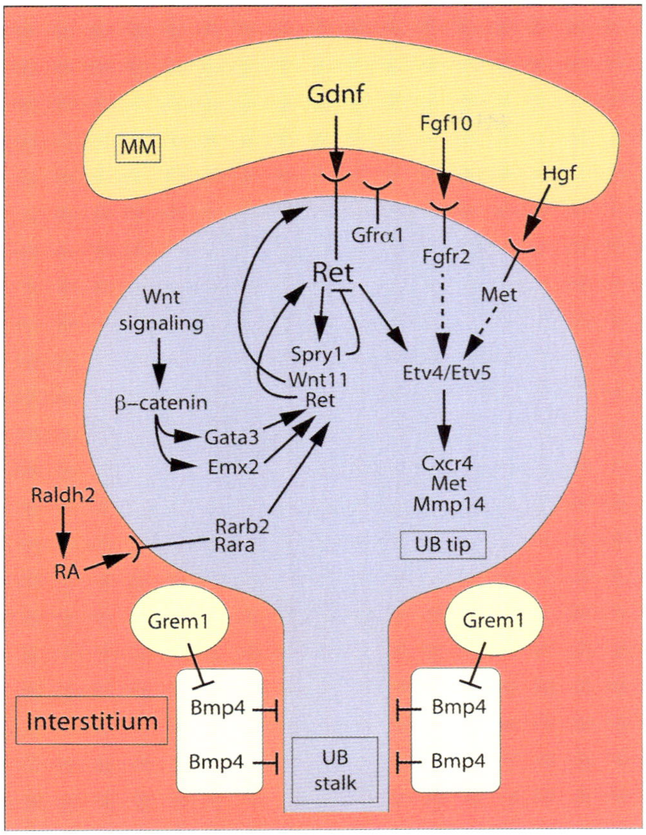

FIGURE 33.11 Schematic drawing of the genetic signaling networks controlling ureteric bud branching morphogenesis. The *Gdnf/Ret* pathway, restricted to the ureteric bud (ureteric bud) tip, involves GDNF from the metanephric mesenchyme (metanephric mesenchyme) binding to the RET receptor and GFRα1 co-receptor. *Gdnf/Ret* signaling activates the expression of *Wnt11* and *Spry1*. In feedback loops, *Spry1* negatively regulates *Ret* signaling, *Wnt11* upregulates *Gdnf*, and *Ret* positively regulates its own expression. *Ret* expression is also controlled by canonical (β-catenin-dependent) Wnt signaling through *Gata3* and *Emx2* and by retinoic acid (*RA*), generated in the interstitium from retinaldehyde dehydrogenase 2 (*Raldh2*), acting through its receptors *Rarb2* and *Rara*. *Cxcr4*, *Met*, and *Mmp14*, involved in processes such as cell migration, are activated by *Etv4/Etv5*, which are downstream targets of *Gdnf/Ret*. *Etv4/Etv5* may also be regulated by *Fgf10/Fgfr2* and *Hgf/Met* signaling (*dotted arrows*). Along the ureteric bud stalk, the inhibition of ureteral branching by *Bmp4* in the surrounding mesenchyme is suppressed by *Grem1*. (Modified from Costantini F. GDNF/Ret signaling and renal branching morphogenesis. *Organogenesis.* 2010;6:252–262.)

and reduces the activity of *Bmp4* which inhibits *Gdnf/Ret* signaling (48). In other words, the negative regulation of ureteric bud formation by *Bmp4* is suppressed by its antagonist *Grem1*.

Relatively less is known about the control of *Ret* expression in the ureteric bud. However, studies have provided support for retinoic acid, β-catenin, and *Gata3* having important roles in the regulation of *Ret* expression. *Ret* expression in ureteric bud cells is activated by paracrine retinoic acid signaling between the stromal mesenchyme and the ureteric bud (49). Specifically, retinoic acid synthesized by retinaldehyde dehydrogenase 2 (*Raldh2*) in stromal cells is secreted and binds to retinoic acid receptors in ureteric bud cells, inducing *Ret* expression. The canonical Wnt/β-catenin pathway is involved in the positive regulation of *Ret* expression. Conditional inactivation of β-catenin in the nephric duct results in decreased *Ret* expression and aberrant and ectopic ureteric bud formation causing a range of renal defects, including dysplasia, duplex kidneys, and aplasia (36). Conditional inactivation of *Gata3* in the nephric duct results in similar renal defects, thus phenocopying the β-catenin mutants (37). The expression of β-catenin is relatively unaffected in the *Gata3* mutants. These results suggest *Gata3* acts in a signaling cascade downstream of β-catenin, but upstream of *Ret*. Moreover, it is believed this cascade operates to maintain ureteric bud cells in an undifferentiated or precursor state and prevent premature differentiation and ectopic ureteric budding.

Although its expression is induced in the ureteric bud by *Gdnf/Ret* signaling, Sprouty1 (*Spry1*), an inhibitor of receptor tyrosine kinase signaling, acts in a feedback loop to negatively regulate *Gdnf/Ret* function (50). Other signaling pathways must exist to maintain ureteric bud formation in the absence of *Gdnf*, *Ret*, and *Spry1*. Fibroblast growth factor 10 (FGF10), expressed in the metanephric mesenchyme and interacting with its receptor FGFR2 in the ureteric bud, proves to be representative of one alternative pathway. A correct balance between activating and inhibitory signals from different genes is more important for normal ureteric bud outgrowth and branching than the signals from a single gene.

URETERAL BRANCHING

Gdnf/Ret Signaling

Since nephron formation begins at the ampullary tip of the ureteric bud, the number of ureteric bud branches largely determines the final nephron numbers in the adult kidney. Molecular signals that regulate the ureteric bud outgrowth continue to play important roles in ureteric bud branching. Again, GDNF from the metanephric mesenchyme interacting with RET in the ureteric bud epithelium is a key signaling pathway (Fig. 33.11). After interacting with GDNF, RET, as a tyrosine kinase receptor, activates downstream signaling, including the Ras/Erk MAP kinase, P13 kinase-Akt, and protein kinase C (PKC) (51). Inhibition of phosphatidylinositol 3-kinase (PI3K) activity blocks the ureteric bud outgrowth and branching (52). On the other hand, conditional ureteric bud deletion of *PTEN*, encoding a phosphatase that normally antagonizes the P13K pathway, results in disturbed ureteric bud branching characterized by ectopic budding (53).

Activation of RET causes autophosphorylation of several key tyrosine (Y) residues in the RET cytoplasmic domain (54). These residues serve as docking sites for intracellular adaptor molecules in the above pathways. Moreover, there are two major alternative splice isoforms of RET, RET9, and RET51, the latter being longer and containing an additional tyrosine docking site. The different tyrosines have specificity for certain intracellular adaptors which transduce the various signaling pathways. For example, recruitment of PLCγ to Y1015 (RET9 and RET51) activates the PKC pathway. RET Y1062 (RET9 and RET51) serves as a docking site for multiple adaptors. Recruitment of the GRB2-SOS complex to Y1062 activates Ras/Erk MAP kinase signaling, whereas recruitment of the GRB2-GAB1 complex to Y1062 activates the PI3K-AKT pathway. Interestingly, the tyrosine phosphatase Shp2, as part of a larger complex, binds to RET Y1062 and mediates the downstream activation of Ras/Erk MAP kinases and PI3K-AKT pathways. Conditional deletion of Shp2 in the ureteric bud results in downregulation of *Ret* target genes and severe renal hypoplasia/dysgenesis (55). These findings indicate these key tyrosine docking sites linked to different signaling pathways downstream of *Ret* activation have distinct roles in organogenesis, for example, regulating enteric nerve development and/or renal branching morphogenesis. However, the molecular basis for this organotypic specificity remains unclear.

The expression of several *Ret*-regulated genes including *Cxcr4*, *Myb*, *Met*, and *Mmp14* is severely reduced in *Etv4/Etv5* mutant kidneys suggesting they are downstream transcriptional targets of *Etv4/Etv5* (56). Cxcr4 is the receptor for the CXC chemokine ligand 12 (Cxcl 12). Inhibition of *Cxcr4* in embryonic kidney organ cultures impairs ureteric branching (57). Furthermore, an intact Cxcl 12–Cxcr4 interaction is required for the eventual development of the glomerular vasculature (58). Normal kidney development depends upon a correct balance between extracellular matrix synthesis and degradation. By degrading extracellular matrix components, mainly collagens and laminins, matrix metalloproteinases (MMP) play an important role. *Mmp14* encodes MMP-14, a membrane type. Positively regulated by *Gdnf/Ret* signaling, *Etv4/Etv5* and their downstream targets *Cxcr4*, *Met*, and *Mmp14* represent a gene network controlling ureteral branching morphogenesis. Other pathways likely intersect this gene network. For example, *Sox9*, encoding a transcriptional regulator, is required for normal ureteric branching (59). Although *Sox9* neither activates *Ret* expression nor is regulated by *Gdnf/Ret* signaling, it is required for activation of *Etv4/Etv5*, which are essential effector genes of *Gdnf/Ret* signaling.

Other Signaling Pathways

Fibroblast Growth Factors

Several other signaling pathways are involved in ureteric branching. FGF signaling is critical throughout kidney development (60). The ligands FGF7, FGF 8, and Fgf10 and the FGF receptors FGFR1, FGFR2, and FGFL1 are the most relevant to renal development. Both FGFR1 and FGFR2 are receptor tyrosine kinases and alternative splicing results in each receptor having IIIb and IIIc isoforms. Although Fgrl1 binds to some FGFs, it lacks a tyrosine kinases domain and may act as a decoy receptor, binding and sequestering FGFs away from the other Fgfrs (61). Conditional gene deletion studies have demonstrated that both *FGFR1* and *FGFR2* (mainly the IIIc isoform) in the metanephric mesenchyme are vital for full formation of the metanephric mesenchyme and ureteric branching (62,63). *FGFRs* activate several downstream signaling pathways also activated by *Ret*. The FGF receptor substrate 2α (FRS2α) is a well-characterized adaptor that binds to FGFRs and mediates the activation of these signaling pathways. However, ureteral gene deletion studies provide evidence that although FGFR2 and FRS2α have important roles in regulating ureteral branching, they appear to act independently but also in an incremental manner (64).

Renin–Angiotensin System

The renin–angiotensin system plays a critical role in kidney development. AT_1R or AT_2R are G protein-coupled receptors that represent the two major receptors for angiotensin II. Stimulation of AT_1R by angiotensin II increases tyrosine phosphorylation of the epidermal growth factor receptor (EGFR) and Ret, promoting branching of the ureteric epithelium (65,66), suggesting molecular crosstalk between G-protein and receptor tyrosine kinase signaling pathways.

Integrins

Integrins are heterodimeric transmembrane receptors consisting of associated α and β subunits (67). They are the primary cellular receptors for extracellular matrix proteins and are classified into collagen, laminin, and arginine–glycine–aspartic acid binding integrins. Some integrins play a greater role in glomerular development than in ureteral branching morphogenesis. However, conditional deletion of the integrin β1 subunit in the ureteric bud results in severe ureteral branching defects and decreased nephron formation (68,69). The integrin β1 cytoplasmic tail contains a tyrosine residue (Y783) that is important for binding to the cytosolic protein talin, which provides a link to the actin cytoskeleton. Disruption of this interaction leads to defective ureteral branching (70). Integrin-linked kinase (ILK) is a cytoplasmic protein that binds β1- and β3-integrin cytoplasmic domains and regulates the actin cytoskeleton by recruiting actin-binding proteins such as α-parvin. ILK lacks functional kinase activity and thus is a pseudokinase. Conditional deletion of ILK in the ureteric epithelium results in decreased branching but also decreased p38 MAPK activity and intraluminal obstruction of the collecting system due to continued collecting duct cell proliferation (71). ILK is required to activate p38 MAPK which induces cell cycle arrest during normal tubulogenesis.

Laminins

Laminins are trimeric molecules that consist of an α, β, and γ chain. They play key roles in several different areas of renal development. Laminin-111 and laminin-511 are expressed in ureteric bud structure basement membranes (72). Conditional deletion of the *Lamc1* gene, encoding the laminin γ1 subunit, in the ureteral epithelium causes failure of the ureteric bud outgrowth or defective ureteral branching (73). In addition, cultures of laminin γ1-deficient collecting duct cells reveal a decrease in mediators of β1-integrin signaling (73). Since several β1-integrins serve as laminin receptors, these findings indicate that a functional interaction between laminin γ1- and β1-integrin is important for ureteral morphogenesis.

Bone Morphogenetic Proteins

A fine regulatory balance exists between factors that promote and factors that inhibit ureteral branching. The BMPs, the largest family within the TGF-β superfamily of growth factors, generally inhibit ureteral branching (74). BMPs bind to a membrane heteromeric receptor complex composed of two types of serine/threonine kinase receptors, type I ALK and type II receptors. BMPs activate either the canonical Smad protein signaling pathway or the noncanonical mitogen-activated protein kinase (MAPK) family of signaling proteins (e.g., p38). The BMPs 2, 4, and 7 are the most studied in kidney development. *Bmp2* is expressed in the condensed metanephric mesenchyme adjacent to the tips of the ureteral branches. *Bmp4* is expressed in the metanephric mesenchyme surrounding the nephric duct, the nascent ureteric bud, and the mesenchyme adjacent to the ureteral trunks. Genetic evidence indicates that Bmp4 inhibits ureteral branching but promotes ureteral trunk elongation (75,76). BMP2 and BMP4 bind to the type I receptor ALK3. Conditional deletion of *Alk3* in the ureteric bud lineage reveals an inhibitory function for ALK3 during early ureteral branching (77). *Bmp7* is expressed in both the metanephric mesenchyme and the ureteric bud branches. Recent studies implicate BMP7 in inhibiting ureteric bud outgrowth and branching (78). GREM1 plays a role in this ureteric bud outgrowth and branching by antagonizing BMP7 (78) as well as BMP4 (48).

Semaphorins

Semaphorins comprise a large protein family involved in providing guidance cues in cellular processes, such as migration (79). They include secreted, transmembrane, and glycosylphosphatidylinositol (GPI)-linked proteins, many of which bind to plexin transmembrane receptors. Some secreted semaphorins bind to transmembrane co-receptors, such as neuropilin-1, which then assemble with a plexin to form a holoreceptor complex. The intracellular pathways downstream of semaphorin/plexin signaling include Ras-specific GTPase activating proteins (GAPs), PI3-Akt, Rho-GTPases, and GSK-3β. These molecules are involved in the regulation of integrin-mediated adhesion, actin dynamics, and microtubule organization (79). Both in vitro and in vivo studies have shown that both *Sema3a-neuropilin1* (*Npn1*) and *Sema4d-plexinB1* signaling inhibit ureteral branching (80,81). However, other semaphorins appear to have opposite functions. For example, *Sema4c-plexinB2* signaling appears to stimulate ureteral branching (82).

Hippo Pathway

The Hippo pathway is a conserved kinase cascade that controls tissue growth and mediates crosstalk with other signaling pathways, including Wnt, BMP, and TGF-β pathways. Activation of the Hippo kinases Mst1/2 and Lats1/2 leads to phosphorylation of the transcriptional coactivators Yap and Taz, which excludes them from the nucleus. Loss of Hippo signaling (inhibition of Mst1/2 or Lats1/2) results in nuclear accumulation of Yap and Taz which promotes proliferation and inhibits apoptosis. Yap and Taz are required for normal ureteral branching (83) and also proper Ret-dependent nephric duct insertion into the cloaca (84).

URETERIC BRANCH GROWTH

The ureteric bud undergoes branching and growth to develop an elaborate collecting system. Morphometric studies in the developing murine kidney have shown extensive ureteric bud branching in the first half of metanephric development followed by a period of ureteric bud trunk segment growth by elongation, and finally a few rounds of terminal branching prior to birth (85). However, ureteral branching displays structural and temporal discontinuity rather than a continuous, reiterative process (86,87).

Studies focusing on Wnt signaling have provided insights into the molecular mechanisms underlying ureteric bud-derived collecting duct growth. Wnt signaling pathways have been broadly categorized as canonical, with utilization of β-catenin and noncanonical, which is β-catenin independent (88). Planar cell polarity is one type of noncanonical pathway and refers to the organization of cells in a plane perpendicular to the apical–basal cellular axis (89). Genes in this pathway are involved in regulating the lengthening and narrowing of the developing collecting ducts without a change in diameter (90). Signaling by *Wnt9b* regulates this oriented cell division in the postnatal collecting duct (91).

The cells within the embryonic collecting ducts have an elongated shape arrangement, perpendicular to the longitudinal duct axis (91). Embryonic collecting ducts deficient in *Wnt9b* generally lack this elongated shape arrangement and become shortened and dilated (91). Thus, *Wnt9b* influences collecting duct growth in both embryonic and postnatal periods, cell intercalation, and oriented cell division, respectively.

Another Wnt gene, *Wnt7b*, is important for development of the renal inner medulla (92). *Wnt7b* signaling to the interstitium via the canonical β-catenin-dependent Wnt pathway regulates elongation of developing medullary collecting ducts and loops of Henle. In the absence of *Wnt7b*, these structures dilate and the renal medulla never forms. The absence of *Dkk1*, an antagonist of *Wnt7b*, leads to hypertrophy of the collecting ducts and overgrowth of the renal papilla (93). Thus, a correct balance of signaling by *Wnt7b* and its antagonist *Dkk1* is critical for collecting duct and loop of Henle morphogenesis and formation of the renal medulla. Finally, studies have implicated α3β1 integrin, a major laminin receptor, acting in concert with c-Met, the receptor for HGF, to positively regulate *Wnt7b* in the developing renal medulla (94).

COLLECTING SYSTEM DIFFERENTIATION

Ureteral Tip and Trunk

The ureteral branch consists of a trunk or stalk which elongates and an ampullary tip. It is the ureteral tip that induces the adjacent metanephric mesenchyme to form nephrons. Proliferating epithelial cells with elevated *Ret* expression driven by *Gdnf* signaling form the tips (95,96). The cellular processes underlying actual branching from the ureteral tip have not been established with certainty. On the basis of isolated ureteric bud cultures, one model proposes a purse-string mechanism, whereby contraction of actin microfilaments in the apical region changes the columnar cell shape from rectangular to more triangular (97). Subsequently, the epithelial monolayer of triangular or wedge-shaped cells with decreased apical surface area pushes outward to form smooth outpouches with a continuous lumen. This possibility is supported by the findings that a functional actin-cytoskeleton (98) and the actin depolymerizing factors, cofilin1 and destrin (99), are required for ureteral branching.

Descendents of the tip cells eventually populate much of the trunk epithelium. Thus, there is increasing cell differentiation from the tip along to the trunk. For example, a trunk epithelial marker is the water channel protein aquaporin-2, which will also be expressed in the mature collecting duct. Tip markers include *Ret*, *Vsnl1*, and *Wnt11*. *Visinin-like 1* (*Vsnl1*) encodes a calcium-sensor protein and is a new marker for tip cells (100). *Wnt11*, expressed in the ureteral tips during all stages of kidney development, is required for ureteral branching morphogenesis (101). *Wnt11* expression in the tip is dependent upon *Gdnf/Ret* signaling, and reciprocally *Gdnf* expression in the metanephric mesenchyme is dependent upon a tip *Wnt11* signal. Thus, it has been proposed that *Gdnf*, *Ret*, and *Wnt11* function in a positive, autoregulatory feedback loop to drive ureteral branching (101). A rather unusual cellular mechanism associated with cell division has been noted in the tips of ureteral branches (102). Tip cells about to divide project out and undergo mitosis in the tubular lumen while still attached to the underlying basement membrane. Subsequently, one daughter cell remains still attached and retreats back into the epithelium whereas the other daughter cell remains in the lumen and eventually reenters the epithelium a few cell diameters away. The significance of this mitosis-associated cell dispersal process is unknown.

Cell Types

A clear understanding of the process by which specific cell types are established in the developing collecting system has been complicated by the diversity of cell types within the different duct segments. The collecting system epithelium consists of principal cells, responsible for vasopressin-regulated water reabsorption and aldosterone-regulated sodium reabsorption, and intercalated cells which regulate acid–base homeostasis. The principal cells are more numerous than intercalated cells throughout the collecting system and are characterized by aquaporin-2 expression. The intercalated cells, which express carbonic anhydrase II, can be divided into three subtypes: type A, type B, and nonA–nonB cells. Type A intercalated cells, expressing apical vacuolar H^+-ATPase and the basolateral exchanger AE1, secrete H^+ into the urine. Type B intercalated cells, expressing the apical anion exchanger pendrin and basolateral H^+-ATPase, secrete HCO_3^- into the urine. The nonA–nonB intercalated cells express apical H^+-ATPase and apical pendrin. Their function remains to be established. The different intercalated cells have distinct distributions (103). Type A intercalated cells are located in the connective tubule, cortical collecting duct, outer medullary collecting duct, and initial inner medullary collecting duct, whereas type B and nonA–nonB intercalated cells are primarily present in the connective tubule and the cortical collecting duct.

During the development of the collecting system, both in the embryonic and postnatal periods, there is considerable cellular remodeling (104,105). The connective tubule cells remain by far the most numerous cell type in all segments. In the fetal kidney, intercalated cells initially appear in the connective tubule and medullary collecting duct. Type A intercalated cells remain the predominant IC subtype in these segments throughout development. After birth, the type A intercalated cells increase significantly as a percentage of the total cells in each of the collecting system segments. Although some Type B cells are present in the fetal outer medullary collecting duct, they are removed by apoptosis after birth (106) (Figs. 33.12 and 33.13). In the postnatal period, intercalated cells increase dramatically in the cortical collecting duct with the type B intercalated cells

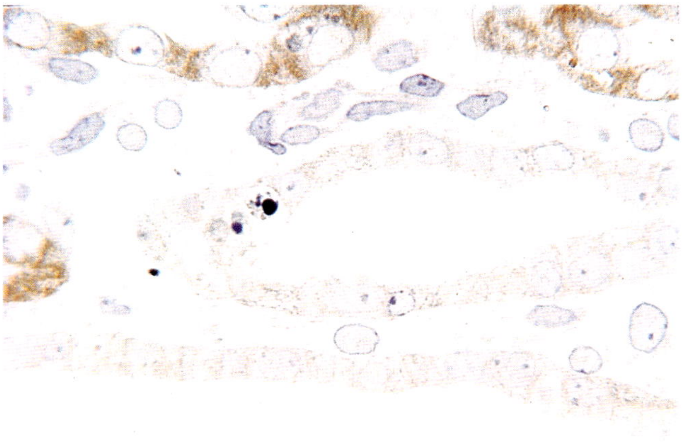

FIGURE 33.12 Developing medullary collecting duct in a postnatal kidney. After etching with sodium methoxide, toluidine blue is removed from normal nuclei but remains in the nuclear fragments of apoptotic bodies (Epon, ×300; courtesy of Dr. Jin Kim).

becoming the most abundant subtype in this duct segment. After birth, some nonA–nonB intercalated cells are located in the cortical collecting duct but more are situated in the connective tubule, where they are similar in abundance as the type B intercalated cells but still less than the type A intercalated cells.

A clear understanding of cell type differentiation in the collecting duct remains a challenge. Immunolocalization, lineage tracing, and gene deletion studies indicate that a subset of ureteric bud tip cells expressing an amino-terminus truncated p63 (ΔNp63) serve as progenitors of intercalated cells (107). The expression of ΔNp63 is transient and ceases after birth. The forkhead gene *Foxi1* encodes a transcription factor that is expressed in both type A and type B intercalated cells but not in principal cells

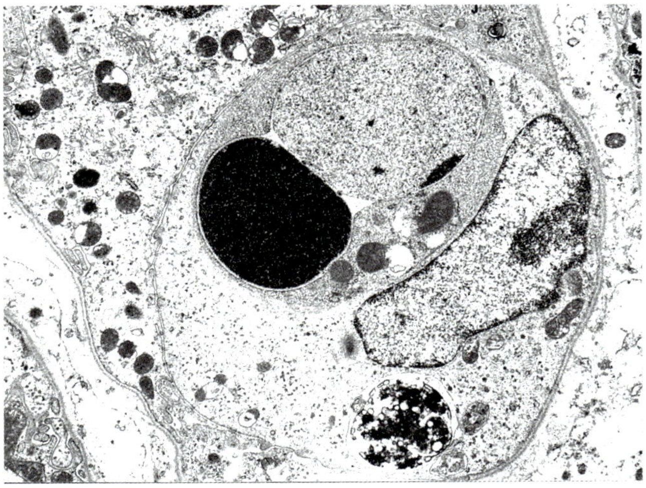

FIGURE 33.13 Electron micrograph of a postnatal medullary collecting duct illustrating a phagocytosed apoptotic body composed of a nucleus with condensed chromatin and organelle remnants (×1,200; courtesy of Dr. Jin Kim).

(108). Most data suggest principal cells and intercalated cells arise from a common progenitor cell and that activation of Foxi1 is necessary for differentiation of these precursor cells into intercalated cells. At least some precursor cells that give rise to intercalated cells have been reported to express aquaporin 2 (109). The transcription factor TFCP2L1 has been reported to induce the expression of intercalated cell–specific genes (110).

Notch signaling is a cell–cell communication pathway that is involved in cell fate determination and differentiation. After conditional inactivation of Notch signaling in the developing collecting duct, mice display increased urine production, decreased urine osmolality, sodium wasting, and a severe urinary concentrating defect compatible with nephrogenic diabetes insipidus (111). The collecting ducts show a prominent decrease in principal cells and an increase in intercalated cells, such that the latter abnormally outnumber the former. Transgenic overexpression of Notch signaling in the defective collecting ducts leads to a dramatic reversal in the cellular composition such that the entire duct is made up of principal cells. Other studies also show that increased notch signaling confers a principal cell fate (112). Thus, Notch signaling appears critical for principal cell differentiation in the collecting duct. The transcription factor Elf5 has been reported to induce the expression of principal cell–specific genes (113).

METANEPHRIC MESENCHYME

Specification

Recent studies have provided insights into the specification of early cell lineages in the metanephric mesenchyme. *Osr1*, encoding a transcriptional regulator, is broadly expressed in the intermediate mesoderm and the lateral plate mesoderm. Most cell types in the developing metanephric kidney, including the ureteral epithelium, the metanephric mesenchyme and its nephron epithelial derivatives, the interstitium, vasculature, and smooth muscle arise from progenitor cells expressing *Osr1* (114). Prior to ureteric bud invasion of the metanephric mesenchyme, *Osr1* positive cells give rise to both epithelial (*Pax2/Six2*$^+$) and interstitial cell (*Foxd1*$^+$) lineages. However, after ureteric bud invasion, *Osr1* expression becomes restricted to the cap mesenchyme nephron progenitors (114). The following genes define the early nephron progenitor cells (Fig. 33.14). *Osr1* is the earliest known marker gene of the metanephric mesenchyme. *Eya1* is another factor in the early lineage determination of the metanephric mesenchyme. *Eya1* regulates *Six1*, *Pax2*, and *Gdnf* expression and loss of *Eya1* leads to renal agenesis (115,116). Functioning as a transcriptional coactivator with phosphatase activity, *Eya1* interacts with *Six1*, converting the latter into a transcriptional activator (117). *Six1*, expressed

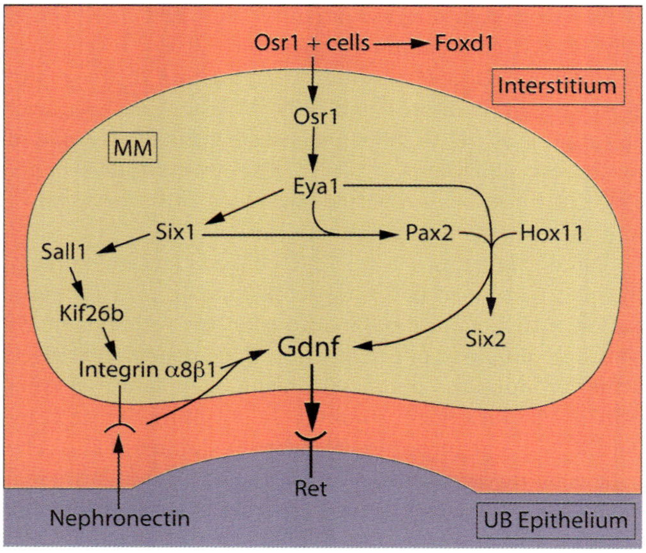

FIGURE 33.14 Schematic diagram illustrating a working model of the hierarchical cascade of transcription factors in the metanephric mesenchyme (metanephric mesenchyme) that defines the early nephron progenitor cells and leads to regulation of *Gdnf* expression. The earliest gene, *Osr1*, is expressed in cells that give rise to either nephron (*Pax2/Six2*+) or interstitial (*Foxd1*+) progenitors. *Eya1*, another early factor, regulates *Six1*, and an *Eya1/Six1* complex activates *Pax2.* A complex of *Eya1/Pax2/Hox11* upregulates *Six2,* which is required for nephron progenitor maintenance. The *Eya1/Pax2/Hox 11* complex also activates *Gdnf.* Downstream of *Six1, Sall1* upregulates *Kif26b,* which maintains expression of integrin α8β1. The binding of integrin α8β1 to nephronectin, expressed in the ureteric bud epithelium, is an interaction important for the maintenance of *Gdnf* expression.

in the metanephric mesenchyme, is required for ureteric bud invasion into the mesenchyme. Loss of *Six1* results in reduced expression of *Pax2*, *Sall1*, *Six2*, and *Gdnf* in the metanephric mesenchyme and renal agenesis (118). Studies indicate *Six1* upregulates the expression of *Grem1* in the metanephric mesenchyme, which in turn, reduces the activity of *Bmp4*, an inhibitor of ureteric bud outgrowth, leading to the initiation of ureteral branching (119).

Pax2, encoding a transcription factor, is expressed in the nephric duct but also in the metanephric mesenchyme prior to induction. Moreover, Pax2 and Eya1 form a complex with the Hox11 paralogs (Hoxa11, Hoxc11, and Hoxd11) to activate the expression of *Gdnf* and *Six2* in the metanephric mesenchyme (45). *Sall1* encodes a transcription factor that is expressed in the metanephric mesenchyme and its deletion results in renal agenesis (120). Studies have revealed that in the absence of *Sall1* in the metanephric mesenchyme, *Wnt9b*, normally specifically expressed in the ureteral stalk, becomes ectopically expressed in the ureteral tips disrupting ureteral branching (121). Thus, *Sall1* in the metanephric mesenchyme normally downregulates *Wnt9b* expression in the ureteral tips, which normally express *Wnt11*, allowing for the initiation of ureteral branching. *Sall1* is likely a transcriptional target of *Six1* (118). A kinesin family gene, *Kif26b*, is expressed in the metanephric mesenchyme and is a downstream target of *Sall1*. Deletion of *Kif26b* in mice leads to reduced expression of integrin α8β1 and *Gdnf* in the metanephric mesenchyme and renal agenesis (122). Integrin α8β1 expressed in the metanephric mesenchyme binds to its ligand nephronectin expressed in the ureteral bud epithelium. Thus, *Osr1* functions at the top of a genetic hierarchical cascade in the metanephric mesenchyme followed by *Eya1*, *Six1*, *Pax2*, *Hox11 paralogs*, *Six2*, *Sall1*, *Kif26b*, and integrin α8β1 to regulate *Gdnf* expression (Fig. 33.14).

Nephron Progenitor Population

As the ureteric bud invades the metanephric mesenchyme, it provides a signal that induces the condensation of metanephric mesenchymal cells around the ureteric bud branches. These condensates will give rise to all the nephron epithelial structures. These cellular interactions occur within the nephron niche, which includes the ureteral ampulla, cap mesenchyme, pretubular aggregate, renal vesicle, S-body, stroma, and endothelial progenitors (123). In recent years, our understanding of the nephron progenitor population in the cap mesenchyme has increased (124,125).

Although *Wnt11* is required for ureteral branching and stimulates *Gdnf* expression (101), it is the *Wnt* gene *Wnt9b*, expressed throughout the ureteric bud epithelium that primarily induces the formation of the cap mesenchyme (126). Cells expressing *Six2* and *Cited1* seem to define a stem cell population capable of self-renewal and expansion of the cap mesenchyme but that also undergoes depletion as nephrogenesis ceases.

Gene expression studies have demonstrated that the cap mesenchyme is not a homogeneous compartment but can be divided into three subdomains; the inner capping mesenchyme (*Six2*+, *Cited1*+), the outer capping mesenchyme (*Six2*+, *Cited1*+, Eya1+, *Meox1*+) and the induced mesenchyme (*Six2*+, *Eya1*+, *Wnt4*+) (Fig. 33.15) (127). However, this spatial and molecular complexity of the cap mesenchyme is certainly not appreciated by light microscopy since the cells of the cap mesenchyme resemble mesenchymal "blastemal" cells. The inner and the outer capping mesenchyme contain the *Six2*+, *Cited1*+ cell population which appears to be true stem cells (128). The survival of these progenitor cells is maintained by several other genes including *Fgf9*, *Fgf10*, *Frs2a*, *Bmp7*, *Sall1*, and *p53* (129–134). Moreover, *Osr1* and *Six2* interact synergistically to maintain the progenitor pool (135).

Metabolic programming also plays a role in nephron progenitor cell maintenance. Younger nephron progenitor cells use glycolysis to a significantly greater degree than older nephron progenitor cells and inhibition of glycolysis leads to accelerated differentiation and enhanced nephrogenesis in embryonic kidneys (136). Recent studies are exploring the role of epigenetics in kidney development

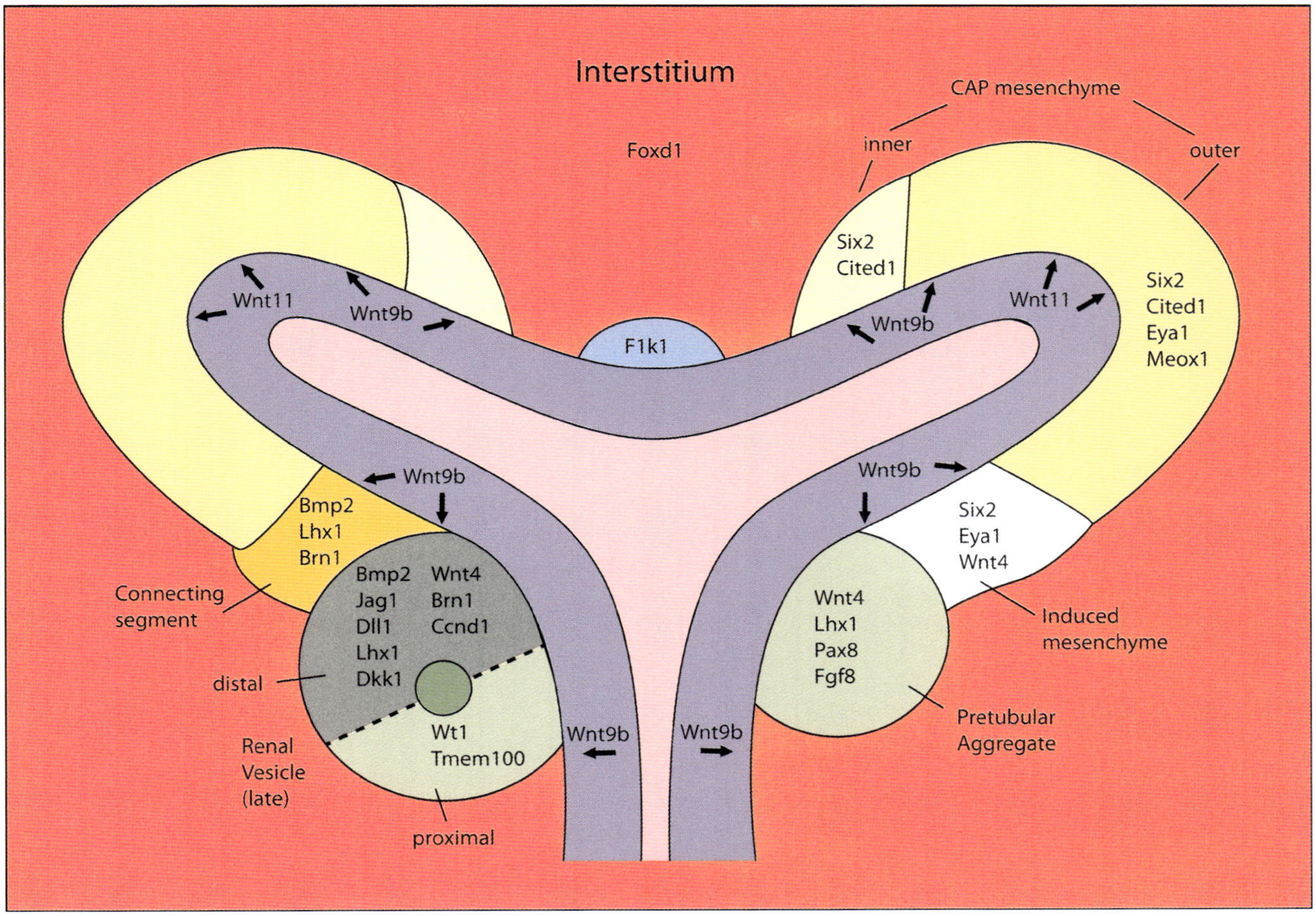

FIGURE 33.15 Schematic representation of the compartments and molecules involved in early nephrogenesis. *Wnt11*, expressed at the ureteric bud tips, is important for ureteral branching and *Gdnf* expression. However, *Wnt9b*, expressed throughout the ureteral epithelium except the very tips, primarily induces the formation of the cap mesenchyme (cap mesenchyme). The cap mesenchyme is divided into three subdomains: inner capping mesenchyme (*Six2, Cited1*⁺), outer capping mesenchyme (*Six2, Cited1, Eya1, Meox1*⁺) and induced mesenchyme (*Six2, Eya1, Wnt4*⁺). Cells positive for *Six2* and *Cited1* are the progenitors that give rise to all the epithelial elements of the nephron. Differentiation of the cap mesenchyme proceeds around and under the ureteric bud tips. *Wnt9b* is required for the formation of the pretubular aggregate (**right**) which expresses *Wnt4, Lhx1, Pax8*, and *Fgf8*. *Wnt4* is necessary to transform the mesenchymal pretubular aggregate to an epithelial renal vesicle (renal vesicle) (**left**). Within the renal vesicle, a proximal domain defined by *Wt1* and *Tmem100* expression, will give rise to the glomerular podocytes and proximal tubules. A distal domain of the renal vesicle, characterized by numerous genes (*Bmp2, Jag1, Dll1, Lhx1, Dkk1, Wnt4, Brn1, Ccnd1*) will generate the loops of Henle and the distal tubule. The connecting segment (*Bmp2, Lhx1, Brn1*⁺), which joins the late renal vesicle to the ureteric bud tip, is formed from cells in the distal renal vesicle. Interstitial cells express *Foxd1*. Vascular endothelial cell progenitors are marked by *Flk1* expression. (Modified from Mugford JW, Yu J, Kobayashi A, et al. High-resolution gene expression analysis of the developing mouse kidney defines novel cellular compartments within the nephron progenitor population. *Dev Biol*. 2009;333:312–323, with permission from Elsevier; Georgas K, Rumballe B, Valerius MT, et al. Analysis of early nephron patterning reveals a role for distal renal vesicle proliferation in fusion to the ureteric tip via a cap mesenchyme-derived connecting segment. *Dev Biol*. 2009; 332:273–286, with permission from Elsevier.)

and disease (137). For example, histone deacetylases 1 and 2 (HDAC1 and HDAC2) interact with Six2, Osr1, and Sall1 to maintain a correct balance of nephron progenitor cell self-renewal and differentiation into nephrons (138).

The cessation of nephrogenesis is associated with an accelerated wave of new nephron formation in the outer nephrogenic zone with an altered topology, such that multiple new nephrons are attached to a single ureteral tip (139,140). There is coincident loss of the cap mesenchyme. The molecular mechanisms causing the cessation of nephrogenesis are not clear. However, studies have revealed that genes associated with cell proliferation are activated prior to genetic markers of cell differentiation, which in turn, are increased prior to the downregulation of *Six2* and *Cited1* (141). Intrinsic cell aging and cell–cell contact alterations within the cap mesenchyme progenitor populations appears to play a role in cessation of nephrogenesis (142). Surprisingly, loss of one *Tsc1* allele in nephron progenitor cells results in delayed cessation of nephrogenesis and increased numbers of nephrons. Notch signaling downregulates Six2, primes nephron progenitors for differentiation, and is required for formation of all the nephron elements (143,144). Thus, there are multiple molecular pathways involved in the termination of nephrogenesis.

As cells transition through these subdomains, from the capping mesenchyme to the induced mesenchyme ($Six2^+$, $Wnt4^+$, $Cited1^-$), they progress toward epithelialization (Fig. 33.15). *Wnt9b*, expressed in the ureteric epithelium, is the primary signal via the canonical (β-catenin-dependent) pathway that induces the cap mesenchyme to initiate a mesenchymal-to-epithelial transition (126,145). *Wnt4* is expressed in the induced mesenchyme as a first molecular response and is essential for propagating the pathway toward epithelialization (146,147). There is recent evidence suggesting that noncanonical Wnt signaling (β-catenin-independent), likely the calcium/NFAT pathway, plays an important role in mediating the *Wnt4*-induced mesenchymal-to-epithelial transition (148,149).

PATTERNING OF THE NEPHRON

Early Events: Pretubular Aggregate and Renal Vesicle

A group of cells within the induced mesenchyme forms a pretubular aggregate in the curve beneath the ureteric bud tips. The pretubular aggregate undergoes a mesenchymal-to-epithelial transition to form a renal vesicle, which develops a central lumen and becomes surrounded by a basement membrane. Thus, the pretubular aggregate is the precursor to the renal vesicle. Cells transition through the compartments from the capping mesenchyme to the induced mesenchyme to the pretubular aggregate and the renal vesicle with overlapping gene expression profiles (Fig. 33.15) (150). The mesenchymal-to-epithelial transition of the cap mesenchyme to nephron epithelia is a genetic switch regulated by the balance between signals which favor stem cell maintenance and those signals driving epithelial differentiation. Deletion of *Six2* results in depletion of the progenitor cells with accompanying massive differentiation of the cap mesenchyme (151,152). *Wnt9b* derived from the ureteric bud induces the cap mesenchyme to form the pretubular aggregate, which expresses *Wnt4*. In the absence of *Wnt9b*, the pretubular aggregate does not form (126). *Wnt4* expression in the pretubular aggregate is necessary for epithelialization in forming the renal vesicle (146,147). A working model is that *Six2* blocks mesenchymal-to-epithelial transition by antagonizing the actions of *Wnt9b*. However, studies have provided evidence that canonical *Wnt9b* signaling can mediate expansion of the progenitor cell population by promoting proliferation of the $Six2^+$ cells, as well as inducing epithelial differentiation in cells with reduced levels of *Six2* (153). These results indicate that rather than opposing *Wnt9b*, *Six2* appears to coact with *Wnt9b* to promote progenitor cell proliferation. The dual roles of Wnt9b in nephron progenitor renewal and differentiation require β-catenin. Different levels of β-catenin activity in metanephric mesenchyme control these disparate cell fates, with low levels promoting progenitor renewal and high levels directing progenitor differentiation (154).

In addition to its role in ureteral branching, the Hippo pathway is important in nephron formation. Deletion of the transcriptional effector Yap from the Six2-positive progenitors impairs nephron induction and nephron formation (155). However, deletion of Lats1/2, the upstream negative regulators of Yap and Taz, disrupts the epithelialization of nephron progenitor cells and converts all the progenitor cells into myofibroblasts (156).

With downregulation of *Six2*, the cells in the induced mesenchyme activate *Wnt4*, *Lhx1*, *Pax8*, and *Fgf8* expression to form the pretubular aggregate. Large-scale microarray analysis of laser microdissected compartments coupled with dual section ISH/immunohistochemistry (IHC) and three-dimensional reconstruction revealed over 60 genes with robust expression levels in the renal vesicle compared to other compartments (150). There was remarkable polarization of gene expression profiles in the proximal and distal domains of the renal vesicle (Fig. 33.15).

The majority of polarized genes are expressed in the distal renal vesicle compared to the proximal renal vesicle. Proximal domain renal vesicle gene markers include *Wt1* and *Tmem100* which have continued restricted expression in the glomerular podocytes and proximal tubules. *Wt1* tumor suppressor gene is actually expressed throughout the cap mesenchyme prior to pretubular aggregate and renal vesicle development. The *Wt1* gene encodes a zinc finger transcription factor that activates or represses target gene transcription depending on the developmental, cellular, or even promoter context. *Wt1* is expressed in diverse developing tissues and often in cells undergoing mesenchymal-to-epithelial transition (157). Wt1 has different roles in different structures during kidney development. It mediates the mesenchymal-to-epithelial transition from nephron progenitors to nephron epithelia. Studies using chromatin immunoprecipitation (ChIP) coupled to microarray analysis (ChIP-chip) revealed that *Bmp7*, *Pax2*, and *Sall1* are *Wt1* transcriptional targets in nephron progenitor cells that likely play a role in the mesenchymal-to-epithelial transition mediated by *Wt1* (158). *Wt1* activates *Wnt4* expression in the kidney mesenchyme by binding to the *Wnt4* gene promoter and altering the state of its chromatin to an active state (159). It is possible that during nephrogenesis, *Wt1* represses *Wnt4* expression in other cell types by binding to the *Wnt4* promoter and switching the chromatin state to a repressive state.

Another Wilms tumor suppressor gene, *Wtx*, is expressed in the metanephric mesenchyme and early nephron epithelia similar to that of *Wt1*. Interestingly, *Wtx* appears to have a dual role as both an activator and an inhibitor of the Wnt signaling pathway (160,161).

The distal domain renal vesicle gene markers include those involved in BMP (*Bmp2*), Notch (*Jag1*, *Dll1*), and Wnt (*Wnt4*, *Dkk1*, *Ccnd1*) signaling pathways (Fig. 33.15) (150). The distal markers show continued restricted expression in the medial and distal limbs of the S-shaped body and later in the loop of Henle and distal tubule. They are not expressed in the developing glomeruli at any stage. The distal renal vesicle markers *Bmp2*, *Lhx1*, and *Brn1* are also expressed in the early connecting segment, which does not express the ureteric bud marker *Calb1* (calbindin). Fusion of the renal vesicle with the ureteric bud creating the connecting segment involves loss of the ureteric bud basement membrane and penetration of cells, originally derived from the cap mesenchyme, but expressing distal renal vesicle markers into the ureteric bud tip (150). Thus, the evidence favors the connecting segment originating from the cap mesenchyme and not the ureteric epithelium.

Later Events: Proximal and Distal Tubules

The Notch signaling pathway is an evolutionarily conserved pathway that enables short-range communication between adjacent cells to determine their developmental fate (162). There are four receptors, Notch1-4, and five different ligands, Delta-like 1(Dll1), Dll3, Dll4, Jagged 1 and 2. Signaling is initiated by the binding of a ligand on one cell to a receptor on an adjacent cell, followed by a series of proteolytic cleavages of the receptor. An extracellular domain of the Notch receptor is cleaved by metalloproteases and this extracellular domain bound to the ligand is endocytosed by the ligand-presenting cell. The transmembrane domain of the residual receptor is cleaved by γ-secretase and the resulting Notch intracellular domain translocates to the nucleus, where it associates with a transcription complex leading to the activation of Notch target genes.

Notch1 and *Notch2* and genes encoding their ligands *Jag1* and *Dll1* are all expressed in the renal vesicle. Previous studies implicated Notch signaling in proximal tubule differentiation and repression of other nephron segments, including distal tubules (163–165). However, more recent investigations demonstrate Notch signaling downregulates Six2 and is required for the formation of all segments of the nephron (143,144).

Brn1, a POU-domain transcription factor, is expressed in the distal domain of the renal vesicle and also in the connecting segment. At later stages, *Brn1* is expressed in the developing Henle loop, distal convoluted tubule, and macula densa but not in the glomerulus, proximal tubule, or collecting duct (166). *Brn1*-deficient mice die after birth of renal failure due to disturbed development of Henle loop accompanied by increased apoptosis and defective differentiation of the distal convoluted tubule and macula densa. Thus, *Brn1* has an essential role in distal tubule patterning.

The Hox genes are master regulators during development. Although they are most known in general for their role in determining segment identity and in the kidney for defining the anterior and posterior kidney morphogenetic fields, they likely have multiple cellular functions related to ureteral branching, replenishing nephron progenitors, and stromal development (167,168). Hox genes also have a role in directing specific cell differentiation within tubule segments.

INTERSTITIUM

Compared to other parenchymal components, there is far less known about the interstitium in kidney development (169). It is generally believed that the metanephric mesenchyme is originally derived from the intermediate mesoderm. A traditional view is that cells of the metanephric mesenchyme originally derived from the intermediate mesoderm not induced by the ureteric bud will become interstitial cells. Although *Osr1* is expressed in the early metanephric mesenchyme, there is evidence that some *Osr1*-positive cells, prior to ureteric bud invasion and induction, give rise to cells expressing *Foxd1*, the earliest known marker of interstitial cells (127). There are at least three distinct *Foxd1* stromal compartments: capsular, cortical, and medullary. Foxd1-expressing cortical stromal cells are maintained by self-renewal and as discussed before, a fraction of these cells reside in the $Six2^+$ cap mesenchyme in early nephrogenesis (170). Nephron progenitor cells lacking Pax2 have been shown to differentiate into interstitial cells, thus Pax2 has a role in maintaining nephron progenitor cells and repressing an interstitial differentiation pathway (171). There is evidence that β-catenin is essential for the differentiation of stromal progenitors to form the medullary stroma (172). Data from cell lineage tracing studies suggest that most metanephric interstitial cells derive from the paraxial mesoderm and few originate from the intermediate mesoderm (173). This may potentially explain why paraxial mesodermal derivatives, such as cartilage and muscle, may be ectopically present in renal dysplasia or Wilms tumor.

In addition to providing a structural framework around the other components, an emerging view is that the developing interstitium plays an essential role in nephron and collecting duct differentiation (169,174). The interstitial cell marker *Foxd1* encodes a forkhead transcription factor, which also plays a role in nephrogenesis (175). The loss of *Foxd1* also results in renal capsule abnormalities that include the ectopic presence of *Bmp4*-expressing cells, and fused kidneys that never ascend from the pelvis to the lumbar region (176). Studies implicate the *Hox10* genes in development of the interstitium and capsule as the defects observed in the *Foxd1* mutant kidneys are largely phenocopied in *Hox10* triple mutant kidneys (177). Foxd1 represses the expression of the proteoglycan decorin which inhibits Bmp–Smad signaling, which stimulates nephron progenitor differentiation (178). Stromal cells isolated from the cortical interstitium and capsule of human kidneys have been reported to be distinct mesenchymal stromal cell populations (179,180).

FAT4, an atypical cadherin produced in the interstitium, operates in the Hippo pathway and allows the activation of β-catenin target genes that promote nephron progenitor differentiation (181). Loss of FAT4 or its ligands DCHS1/2 (Dachsous 1 and 2) results in expansion of the nephron progenitor pool (182). Hedgehog signaling is involved in diverse areas of developmental and cell biology (183). Hedgehog binding to the Patched family of cell surface receptors results in the activation of the GLI transcriptional mediators and controls nephron formation (184).

Stromal cells express retinaldehyde dehydrogenase 2 (Raldh2) which synthesizes retinoic acid. Retinoic acid binds to its receptors in ureteral branch cells, inducing Ret expression and promoting branching (49). Moreover, stromal retinoic acid activates the expression of Ecm1, which downregulates Ret expression in the ureteral branch (185). Other stroma cell–expressed molecules include *Rara, Rarb2*, *Pod1*, *Fgf7*, *Bmp4*, and *Pbx1*. Thus, the interstitium and renal capsule provide signals that are critical for correct patterning of the nephron and ureteral components.

A loose stroma containing spindle-shaped cells surrounds the early ureteric bud branches and early nephrons and is known as the primary interstitium. As nephrogenesis proceeds, a cortical interstitium and a medullary interstitium form. In the postnatal period, the interstitial cells resemble fibroblasts, dendritic cells, or macrophages according to morphologic and immunophenotypical findings (186–188).

GLOMERULOGENESIS

To appreciate how some glomerular diseases arise or how the glomerulus responds to injury, an understanding of glomerular development is indispensable (189–192). Glomerular development proceeds through a sequence of structures described as vesicle, comma-shaped, S-shaped, capillary loop, and maturing glomerulus stages. The vesicle and comma-shaped stages were discussed previously. At the S-shaped stage, the lower limb beneath the vascular cleft separates into two layers divided by a narrow developing Bowman space (Fig. 33.6). Visceral epithelial cells that will differentiate into podocytes form the upper internal layer, while the parietal epithelial cells form the lower outer layer on the opposite side of Bowman space.

During the S-shaped stage, microvessels can often be identified within the vascular cleft of the S-figure. Emerging evidence suggests that endothelial cells of the early glomerular capillaries originate from intrinsic angioblasts derived from the metanephric mesenchyme (193,194). Release of growth factors, such as vascular endothelial growth factor (VEGF), from the immature podocytes may attract the angioblasts, expressing VEGF receptors (as Flk1) into the vascular clefts. Other signaling systems such as the angiopoietin (ligand)-Tie (receptor) axis also play a role in endothelial cell and vascular development (195). At this stage, the endothelial cells contain few fenestrae. The early podocytes are cuboidal or columnar, whereas the parietal epithelial cells are already flattening. Situated between the endothelial and podocyte layers are two basement membranes. The basement membrane beneath the podocytes is usually thicker and more continuous than the one underneath the endothelial cells.

During the capillary loop stage, the capillaries start to fill out into an expanding Bowman space. The endothelial cells flatten and develop numerous fenestrae. With maturation, the endothelial cells lose their diaphragms which bridge the fenestrae (196). This endothelial maturation depends on ADAM10, a regular of Notch signaling (197). The podocytes develop a complex cellular architecture as they become terminally differentiated and cease to undergo mitosis. They flatten and form cytoplasmic primary processes, which in turn, elaborate foot processes that interdigitate with those from adjacent podocytes and adhere to the developing glomerular basement membrane. Intercellular junctional complexes are present at the apical membranes between podocytes. With foot process development, these junctions migrate down the lateral surfaces of the emerging foot processes and disappear, when they are either replaced by or converted into the slit diaphragms. This cellular process has been elegantly demonstrated by block-face scanning electron microscopy (198). Slit diaphragms, specialized intercellular junctions, bridge the space between adjacent foot processes. The slit diaphragm is connected to the podocyte cytoskeleton as part of a multifunctional protein complex, that includes nephrin, CD2-associated protein (CD2AP), and podocin (199). Nephrin is the protein encoded by the *NPHS1* gene that is mutated in congenital nephrotic syndrome of the Finnish type, which is associated with loss of the slit diaphragm, abnormal foot processes, and massive proteinuria (200). Thus, the slit diaphragm is critical for maintaining podocyte architecture and the glomerular filtration barrier.

The dual glomerular basement membrane, synthesized by both the endothelium and podocytes, is still present but areas of fusion between the two membranes are found. Beginning during the capillary loop stage, a complex series of transitions in the glomerular basement membrane protein composition occurs. There is developmental switching of both type IV collagen and laminin isoforms in the glomerular basement membrane, events which are essential for forming normal glomerular capillaries. Whereas the immature glomerular basement membranes of comma- and S-shaped nephrons contain laminin $\alpha1\beta1\gamma1$ (laminin 111), glomerular basement membranes at later developmental stages and in adults contain laminin $\alpha5\beta2\gamma1$ (laminin 521) (190). Collagen type IV is made up of six distinct α chains (α1[IV] to α6[IV]), which form different triple helical molecules called protomers. There are three types of protomers: $\alpha1.\alpha2.\alpha1$(IV), $\alpha3.\alpha4.\alpha5$(IV), and $\alpha5.\alpha6.\alpha5$(IV). The immature glomerular basement membranes of comma-, S-shaped, and early capillary loop stage glomeruli contain collagen $\alpha1.\alpha2.\alpha1$(IV). Starting in capillary loop stages, this collagen network is replaced by $\alpha3.\alpha4.\alpha5$(IV), which persists as the only collagen IV network normally present in the mature glomerular basement membrane. The mechanisms responsible for both laminin and collagen IV isoform switching in the developing glomerular basement membrane are not known. The laminin isoform switching process precedes the one for type IV collagen isoforms (201). Ultrastructural immunolabeling studies have demonstrated that whereas endothelial cells, mesangial cells, and podocytes synthesize collagen $\alpha1.\alpha2.\alpha1$(IV), only podocytes make collagen $\alpha3.\alpha4.\alpha5$(IV) (202).

Glomeruli in the maturing stage resemble adult glomeruli by histology but are smaller in diameter. The podocytes of the maturing glomeruli may have a cuboidal appearance. A single-fused glomerular basement membrane predominates and areas of dual unfused basement membranes are rarely seen. At this time, the synthesis of components for the glomerular basement membrane is largely by the podocytes. In areas where foot process interdigitation is continuing, irregular outpouchings of basement membrane are found beneath the podocytes. These loops reflect newly synthesized glomerular basement membrane, which will be deposited into the existing glomerular basement membrane.

The development of the mesangium occurs relatively later in glomerulogenesis. Mesangial cells likely derive from the metanephric mesenchyme, although their origin is not entirely clear. The mesangial cell precursors are believed to be distinct from the VEGF-receptor–expressing angioblasts that differentiate into the glomerular endothelial cells. However, the emergence of mesangial cells in the glomerulus is dependent on platelet-derived growth factor-B (PDGF-β), produced by podocytes and endothelial cells, and its receptor, PDGF receptor-β (PDGF-Rβ), expressed on mesangial cells (203,204).

VASCULATURE

Renal vascularization appears synchronized with nephrogenesis resulting in a complex network of vessels of variable size and containing several cell types (205,206). Several major morphogenetic processes are believed to underlie renal vascular development: vasculogenesis, angiogenesis, and hemovasculogenesis. In vasculogenesis, vessels originate from intrinsic cells. An example is endothelial cell precursors forming tubes followed by coating of vascular smooth muscle cells. Angiogenesis, the sprouting of new vessels from preexisting vessels, occurs in branching from the renal artery and from juxtamedullary efferent arterioles to form the vasa recta. There is less known about hemovasculogenesis, which refers to the simultaneous formation of blood precursors and vessels. One concept is that vasculogenesis predominates during early nephrogenesis and vessel branching; elongation occurs later in kidney development and continues after birth. Whole mount three-dimensional analysis of early human embryos have shown in several tissues a phase of vasculogenesis preceding angiogenesis (207).

Vascular progenitor cells reside within the interstitium. *Foxd1*-positive stromal cells give rise to vascular smooth muscle cells, pericytes, renin-expressing cells, and mesangial cells (208). A subset of *Foxd1*-positive cells express the transcription factor Tbx18 and differentiate into vascular smooth muscle cells, pericytes, and mesangial cells (209). A hemovascular progenitor expressing the stem cell leukemia/T-cell acute lymphoblastic protein 1 (SCL/Tal1) was shown to give rise to endothelial cells and also blood precursors (210). These studies also demonstrated that sphingosine-1-phosphate (S1P) and its receptor SIPR1, expressed in SCL/Tal1-derived blood cells and endothelial cells, operate in a signaling pathway to regulate renal vascular development. Endothelial cell precursors are heterogeneous. For example, a CD146$^+$ cell population, essential for development of the renal microvasculature, differentiates into CD31$^+$ endothelial cells (211). Another population, a subset of cortical stroma *Foxd1*-positive cells, expressing Flk1 (VEGFR2) is critical for forming peritubular capillaries (212).

Detailed anatomic spatiotemporal studies suggest that renal vascularization is mostly an angiogenic process whereby blood vessels initially wrap around the ureteric bud, migrate to the nephrogenic zone, and form new polygonal vascular plexuses around ureteric bud tips and nephron progenitor cells (213). The vascular networks contain red cells, are enclosed by a basement membrane, and are connected to older blood vessels. This patterning of developing renal blood vessels has been corroborated by another study, which in addition showed prominent molecular heterogeneity of endothelial cells (214).

DEVELOPMENT OF THE JUXTAGLOMERULAR APPARATUS

Renin expression in the metanephric kidney has been detected as early as 8 weeks of gestation and renin mRNA levels are significantly higher in the developing kidney than in the adult organ (215,216). Renin-expressing cells are found in intrarenal arteries of the developing kidney, including the arcuate and interlobular arteries. As development progresses, the distribution of renin-expressing cells shifts from the larger vessels to the juxtaglomerular apparatus, primarily the terminal afferent arteriole, in the mature kidney (217,218). Within the afferent arterioles themselves, heterogeneous patterns of renin expression exist (219). The juxtaglomerular cell, as a cellular component of the mature juxtaglomerular apparatus, is located in the wall of the terminal afferent arteriole close to the glomerulus. Renin progenitor cells differentiate into juxtaglomerular apparatus cells (220). Renin cells display a unique set of genes distinct from other cell types (221). The juxtaglomerular apparatus cells have a dual endocrine-contractile phenotype, which has been demonstrated to be maintained by RBP-J, a transcriptional mediator of the Notch pathway (222). The unique gene repertoire of renin cells affords them considerable plasticity to differentiate along different pathways. They can differentiate into non–renin-expressing cells, such as vascular smooth muscle cells and glomerular mesangial cells (220).

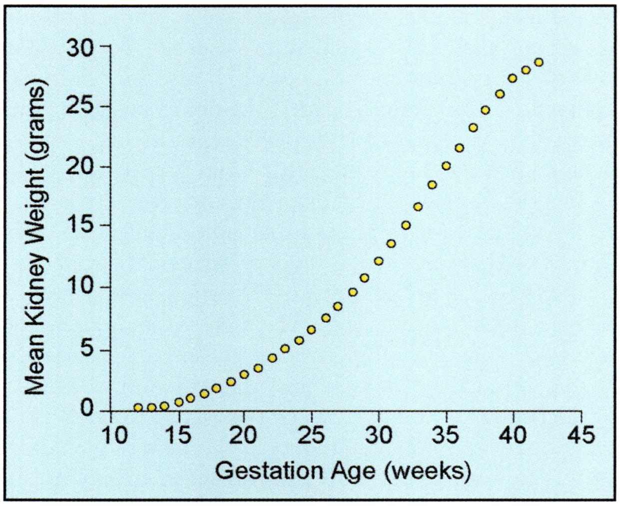

FIGURE 33.16 Mean combined **(right and left)** weight of kidneys from second and third-trimester fetuses and neonates. (Modified with permission from Hansen K, Sung CJ, Huang C, Pinar H, Singer DB, Oyer CE. Reference values for second trimester fetal and neonatal organ weights and measurements. *Pediatr Dev Pathol.* 2003;6:160–167.)

GROSS ANATOMY

Kidney Position and Blood Supply

Upon formation, the metanephric kidneys are situated close to each other in the pelvis at the level of the upper sacrum. Between the sixth and ninth weeks of gestation, the kidneys are found further apart and at higher levels in the abdomen until they reach their final upper lumbar position (223). This ascent of the kidneys is believed to result largely from differential growth of the caudal part of the embryo away from the kidneys (224,225). However, others have argued that the cephalad movement of the kidneys is active and not caused by differential growth of the vertebral column (226). With this migration, the renal hilum rotates from a ventral orientation to face anteromedially. Initially, the kidneys receive their blood supply from branches of the common iliac arteries. With their ascension, the kidneys are supplied by arteries originating from progressively higher levels of the distal aorta (223). As the ascending kidneys receive new branches from the aorta, the older, caudal branches undergo involution. Persistence of these inferior vessels may result in accessory renal arteries. The most cephalad branches arising from the abdominal aorta become the permanent main renal arteries.

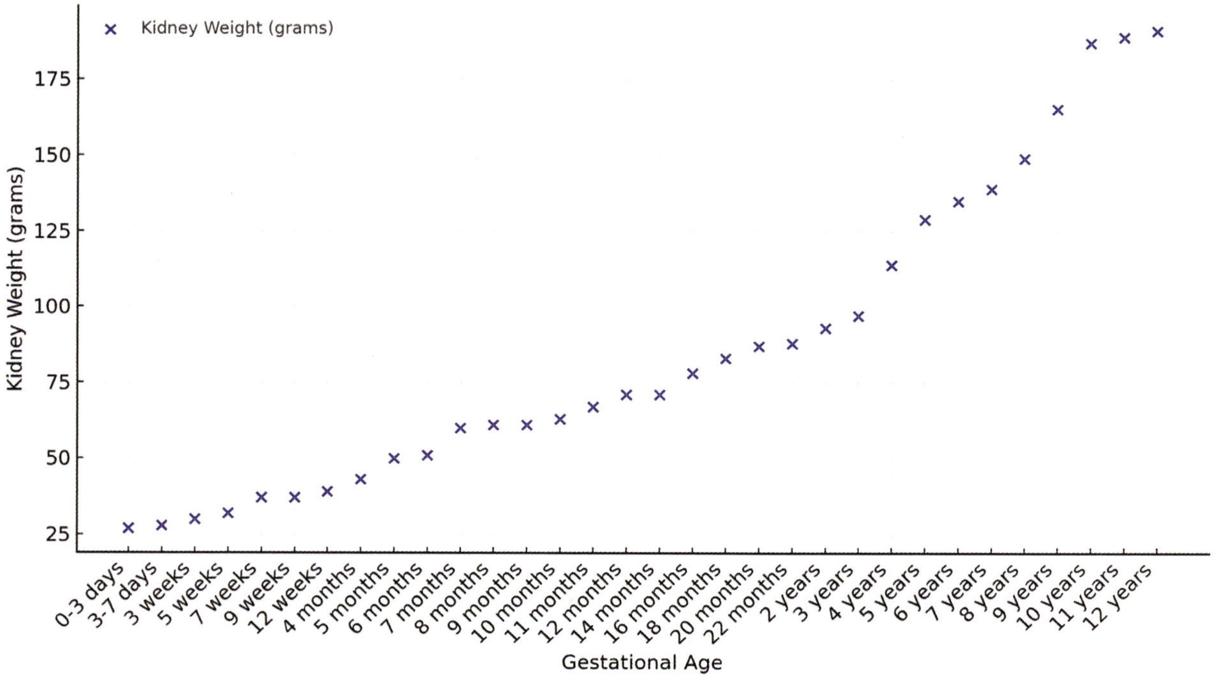

FIGURE 33.17 Mean combined **(right and left)** weight of kidneys at various postnatal ages. (Data from Emery JL, Mithal A. The weights of kidneys in late intra-uterine life and childhood. *J Clin Pathol.* 1960;13:490–493.)

Kidney Weight and Configuration

The various reference values reported for fetal and neonatal kidney weights correspond relatively closely, despite potential variability due to factors such as the social and economic status and the level of health care in a given population (227–232). Separate values from nonmacerated and macerated cases are available (232), as well as values using prefixation (206,207) and postfixation weighing (231,232) of the kidneys. Data for the combined weight (right and left) of the kidneys during the second and third trimesters are shown in Figure 33.16. Different reference values published for combined kidney weights during infancy and childhood also favorably compare (227,233). The data of Emery and Mithal (227) are illustrated in Figure 33.17.

The newborn kidney has a shorter, more rounded configuration than that of the adult. The upper and lower poles project further medially, so the renal sinus is relatively deeper in the infant (Figs. 33.18 and 33.19) (234). The renal sinus of infants contains much less fat and connective tissue than in adults, and the cortical septa of Bertin more closely approach the pelvic–calyceal system. Figure 33.19 illustrates the process of "unrolling" of the renal poles during childhood, as the kidney assumes the more elongated configuration observed in adults. This change in configuration produces a shallower renal sinus with partial exteriorization of the pelvis. As the pelvic–calyceal system assumes a more exterior position, it also becomes displaced further from the parenchyma lining the sinus. The additional space created between the pelvic–calyceal system and the cortical columns of Bertin is normally filled with fat, which increases in amount as the kidney approaches maturity.

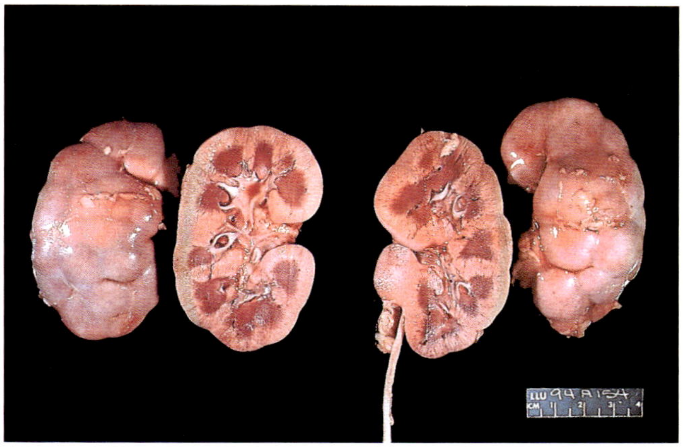

FIGURE 33.18 Gross appearance of newborn kidneys. The rounded configuration with a relatively deeper sinus, characteristic of the infantile kidney, is seen on the sectioned surface. Fetal lobations are prominent on the external surface.

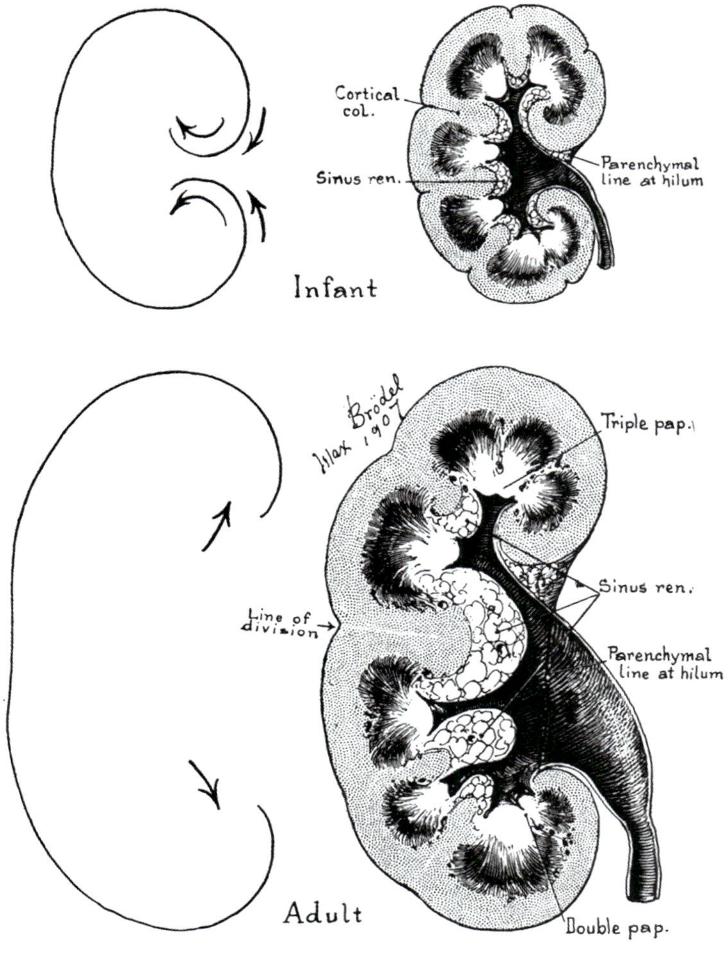

FIGURE 33.19 Max Brödel's classic illustration of the process of unrolling of the kidney in postnatal life. The rounded configuration of the infantile organ becomes elongated as the upper and lower poles diverge and pelvic–calyceal structures are partially everted from their original position within the renal sinus. The space thus created within the renal sinus is filled by fat, which is far more abundant in the adult kidney than in the infant kidney. (From Kelley HA, Burnam CF. *Diseases of the Kidneys, Ureters, and Bladder*. Vol. 1. Appleton; 1925.)

Fetal Lobations

A renal lobe consists of a medullary pyramid and its surrounding cortical parenchyma (Figs. 33.7 and 33.18). The centrilobar cortex covers the base of a pyramid, whereas the septal cortex surrounds the sides of a pyramid. Thus, a column (septum) of Bertin represents the confluence of two layers of septal cortex from two adjacent lobes. Although a lobe does not represent a functional renal unit, it may be viewed as an anatomic organizational unit (234). Lobation begins at 6 to 7 weeks and proceeds to maximum development with an average of 14 lobes at 28 weeks of gestation (235,236). At this stage, generally, 14 papillae and calyces are present corresponding with the same number of lobes. Deep clefts on the surface separate the lobes. After the 28th week, a process of variable lobar fusion decreases the number of surface fissures, papillae, and calyces. The degree of calyceal fusion is greater than papillae fusion. In full-term infant kidneys, the mean number reported for calyces is 9, and for papillae, it is 11 (236). Considerably more lobar fusion, creating compound papillae, occurs in the polar regions than in the midpolar region, where simple papillae are more likely to be retained.

The surface of the neonatal kidney is divided into polygons by prominent fissures that correspond roughly, although not precisely, to the lobar outlines (Fig. 33.18). These fetal lobations usually decrease in prominence with advancing age, persisting longer on the ventral surface than the dorsal surface of the organ. Although there is considerable individual variation in the chronology of their disappearance, they are usually inconspicuous by 4 to 5 years of age (237). The only valid generalization is that fetal lobations usually diminish in number and prominence in the first few years of life, but they remain apparent, especially on the ventral surfaces, in a significant proportion of adult kidneys. In one study, one or more interlobar fissures were detected in up to 50% of adult kidneys (236). In older children and adults, it is important to distinguish persistent fetal lobations from cortical scars. Fetal lobation is a more accurate term than "fetal lobulation" because a lobule is an architectural feature of the cortex, primarily observed on the histologic level.

CORTICAL ARCHITECTURE

The developing renal cortex has unique temporal and spatial features of organization. Each generation of nephrons, most easily observed histologically by the glomeruli, forms a "layer" over the preceding generation. The earliest formed nephrons are situated in the inner cortex (juxtamedullary) near the future medulla, whereas the nephrogenic zone in the outer cortex (superficial) contains the last nephrons formed. Between the 32nd and 36th weeks of gestation, nephron formation ceases and the nephrogenic zone disappears. The nephrogenic zone persists in kidneys of children born prematurely (238,239). A study examined the kidneys of preterm neonates who survived from a few days to over 2 months (240). The findings included a decreased nephrogenic zone width, decreased percentage of immature glomeruli, and an increased number of glomerular generations indicative of accelerated postnatal renal maturation. Compared to gestational controls, the glomeruli from the preterm kidneys had an increased diameter and a significant number in the outer cortex had shrunken tufts. These findings suggest preterm infants may have an increased vulnerability to renal disease later in life.

The histologic features of the developing renal cortex have been used as an index of fetal maturation (239). The ratio of the width of the nephrogenic zone to the width of the remaining cortex decreases in a linear fashion as the birth weight increases (229,241). This approach has been used to detect infants with reduced intrauterine growth in whom this ratio is less than expected for the birth weight. Using glomeruli as representative of nephrons, studies have employed counting of the number of layers of glomeruli from inner to outer cortex, as a method to evaluate nephrogenesis (242–244). Each successive layer is assumed to represent a new glomerular generation. These estimates are performed on well-oriented sections that are orthogonal to the cortex and display a well-defined corticomedullary junction. The studies are in fair agreement between gestations of 24 to 36 weeks in which the average number of rows of glomeruli are as follows: 5 to 7 (24 weeks), 8 to 9 (28 weeks), 9 to 10 (32 weeks), and 10 to 14 (36 weeks). A comprehensive study of 71 infants generally confirmed these results and found the number of glomerular generations was directly proportional to gestational age, body weight, and kidney weight (245). However, there was considerable variability in the final number of glomerular generations, ranging from 8 to 12 per kidney. There was also some variation in the timing of cessation of nephrogenesis. Interestingly, the study revealed an increase in glomerular size from midgestation to term in females, whereas glomerular size in males remained constant over this period. The proportional number of podocytes and endothelial cells within mature glomeruli appeared to remain constant during gestation.

After the full complement of nephrons is attained by the normal fetus, subsequent renal growth reflects hypertrophy and maturation of the nephrons. As tubules elongate and increase in diameter, they become interposed between glomeruli. The glomeruli in the outer cortex of newborns are crowded together (Fig. 33.10), whereas the older, more mature glomeruli deeper in the cortex are more widely separated. This process of tubular growth not only separates glomeruli from one another, but also tends to separate them from the cortical surface near where they originally developed. In a normal-term newborn, one observes many glomeruli very close to the renal capsule (Fig. 33.10). By about 2 months of age, the process

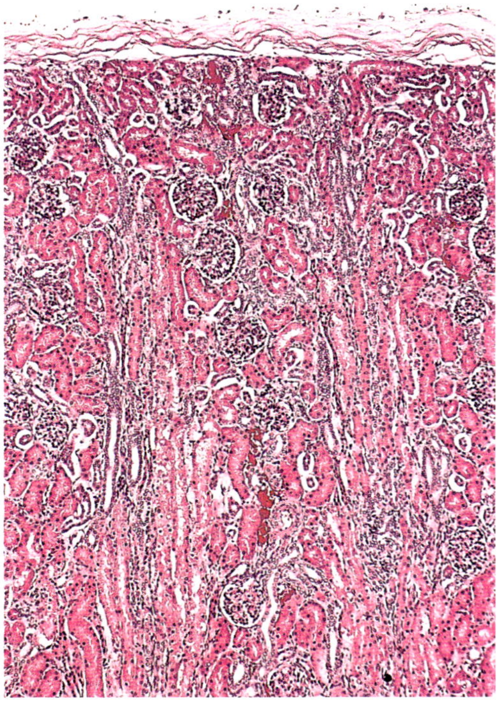

FIGURE 33.20 Renal cortex at 2 months of age. Glomeruli in the outer cortex are becoming more widely spaced due to tubular elongation. Superficial glomeruli are beginning to separate from the renal capsule, the first indication of the cortex corticis.

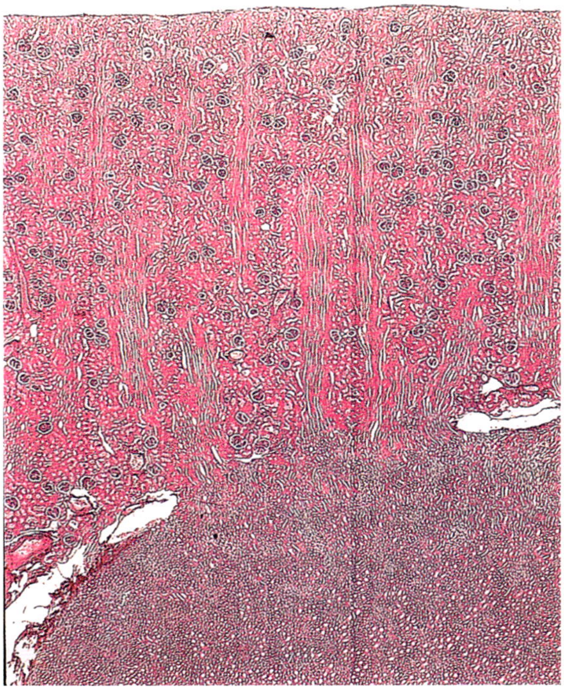

FIGURE 33.22 Kidney at 21 months of age. This perpendicularly oriented section illustrates the full length of several medullary rays that extend from the corticomedullary junction to a level near the cortex corticis. Each medullary ray marks the center of a cortical lobule.

of nephron growth has begun to separate the outermost glomeruli and a narrow zone largely devoid of glomeruli develops beneath the renal capsule (Fig. 33.20). This latter zone has been termed the cortex corticis (246). The cortex corticis becomes progressively wider during childhood (Fig. 33.21). Although it is normal to find an occasional glomerulus adjacent to the capsular surface in normal

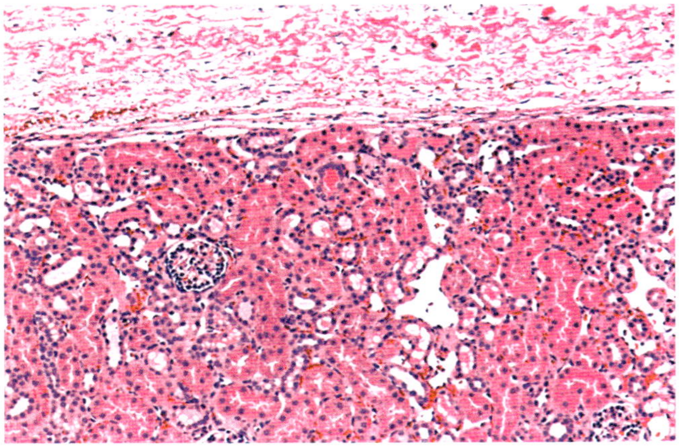

FIGURE 33.21 Micrograph of well-developed renal cortex corticis. The zone without glomeruli beneath the distinct renal capsule is evident. (Reprinted with permission from Murphy WM, Grignon DJ, Perlman EJ. Tumors of the kidney, bladder, and related urinary structures. In: Silverberg SG, Sobin LH, eds. *Atlas of Tumor Pathology*. 4th series, fascicle 1. Armed Forces Institute of Pathology; 2004. Permission obtained from The Joint Pathology Center.)

infants and children, the presence of numerous very superficial glomeruli suggests defective renal growth during late fetal or early postnatal life. Abnormally crowded glomeruli also can be an important clue to defects in nephron growth and differentiation, which may involve only the outer cortex, or the entire cortical mantle. Thus, the cortical architecture may be viewed as a record of the developmental history of the kidney.

The cortex is subdivided into distinctly demarcated lobules by radially oriented groups of tubules termed medullary rays that extend from the base of the pyramids upward into the cortex (Fig. 33.22). A lobule is defined as the cortical domain surrounding a medullary ray. Despite their name, the medullary rays (of Ferrein) actually are part of the cortex and contain the straight segments of the proximal tubule, the thick ascending limbs, and the collecting ducts. Medullary rays usually extend to near the cortical surface in infants, but not in older children. The presence of complete medullary rays is a good indicator that the plane of a given section is perpendicular and reflective of the true thickness of the cortex. Medullary ray nodules are complex tangled tubular configurations commonly seen in the medullary rays of infants during the early months of life, being most prominent between 1 and 6 months of age (Fig. 33.23) (247). These structures apparently represent a normal transitory developmental phenomenon of variable prominence, which is pathologic only when extreme.

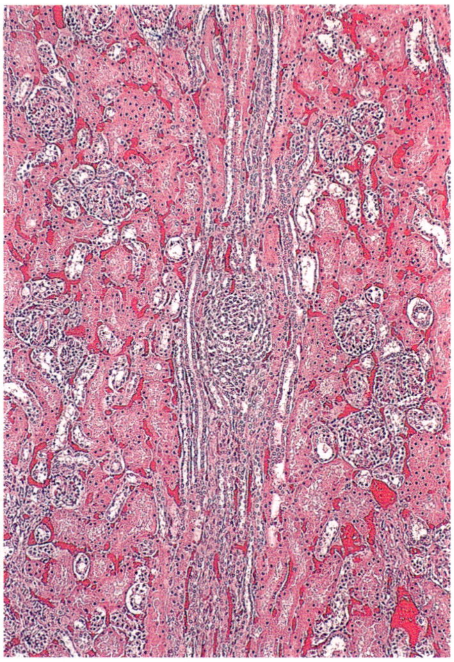

FIGURE 33.23 Medullary ray nodule. In the center of the micrograph, a tangled cluster of collecting ducts forms a nodule near the midportion of a medullary ray. This structure is usually transitory, being uncommon in the first month of life and extremely rare after 1 year of age.

NEPHRON NUMBER

The number of nephrons in a kidney is determined *in utero*. The final number is dependent on gestational age and a favorable intrauterine environment. Unbiased, precise stereologic methods have been used to count glomeruli which serve as a surrogate for nephron number. One study of human intrauterine renal growth revealed the glomerular number increased from 15,000 at 15 weeks of gestation to 740,000 by 40 weeks (247). The greatest rate of nephron induction has been observed between 15 and 17 weeks of gestation, however, approximately 60% of the total nephrons are believed to form during the third trimester (247). Nephron number, inferred from glomerular number estimates in children and adults without renal disease, is strongly correlated with birth weight (248). Intrauterine growth retardation has been shown to impair nephron formation, as measured in fetuses and in infants dying within a year of birth (249,250).

It is apparent that a kidney having a quantitative abnormality, such as a low nephron number, may not demonstrate an obvious defect of nephron spatial topography on histologic examination. However, studies have demonstrated an inverse correlation between the number of glomeruli and mean glomerular volume, suggesting glomeruli increase in size to compensate for an innate low nephron number (248,250). Large glomeruli may be susceptible to scarring. Glomerulomegaly has been used as an adverse factor to assess the risk of disease progression in childhood nephrotic syndrome (251). Thus, increased glomerular size may be useful as an indicator of nephron deficiency in individuals susceptible to renal disease.

Until a few years ago, each human kidney was believed to contain 1 million nephrons. It is now appreciated that there is a remarkably wide variation in total nephrons per kidney among normal adults, ranging from as low as 227,000 to over 2,000,000 (252–254). Nephron endowment is programmed in the perinatal period capping the nephron number in an individual's lifetime. Some time ago, Brenner et al. postulated that an inborn deficit of nephrons predisposes to acquired renal disease, including hypertension, in adults (255). A study showing that hypertensive individuals had fewer nephrons but a larger glomerular volume than age-matched normotensive controls supports this hypothesis (256). The endowment of nephrons from nephrogenesis and the developmental origins of renal disease are fertile areas for investigation.

GLOMERULAR MATURATION AND GROWTH

Newly formed glomeruli are structurally distinctive, and their evolution toward a mature form is a gradual process. Since the period of glomerular development spans a 6- to 7-month period of fetal life, a spectrum of maturational stages is normally present in infant kidneys. As mentioned before, this spectrum is organized in a temporal–spatial manner in the cortex. Familiarity with normal glomerular maturation can facilitate an assessment of the renal developmental status in infants.

Dramatic changes in glomerular structure and size occur through the early months and years of life. For convenience of study, several stages of glomerular development have been defined (256–259). Figure 33.24 shows representative glomeruli from the midcortical region of infants and children from birth to 9 years of age to illustrate the maturational changes. Figure 33.24A shows the characteristic appearance of recently formed glomeruli from a term newborn infant. In addition to their small size, the most obvious distinctions from mature glomeruli are the simple character of the tufts with relatively few capillary loops and the layer of cuboidal cells surfacing the visceral layer of the tuft. This cuboidal layer, representing developing podocytes, is often continuous and covers most of the circumference of the tufts. Ultrastructurally, these primitive podocytes are closely approximated to one another and often lack foot processes (Fig. 33.25A). The glomerular basement membrane is thin and two lamina densa structures, representing basal laminae produced by endothelial cells and podocytes prior to their fusion into a single lamina, may be focally observed. The thickness of the glomerular basement membrane increases progressively with age (Fig. 33.25). The approximate values for glomerular

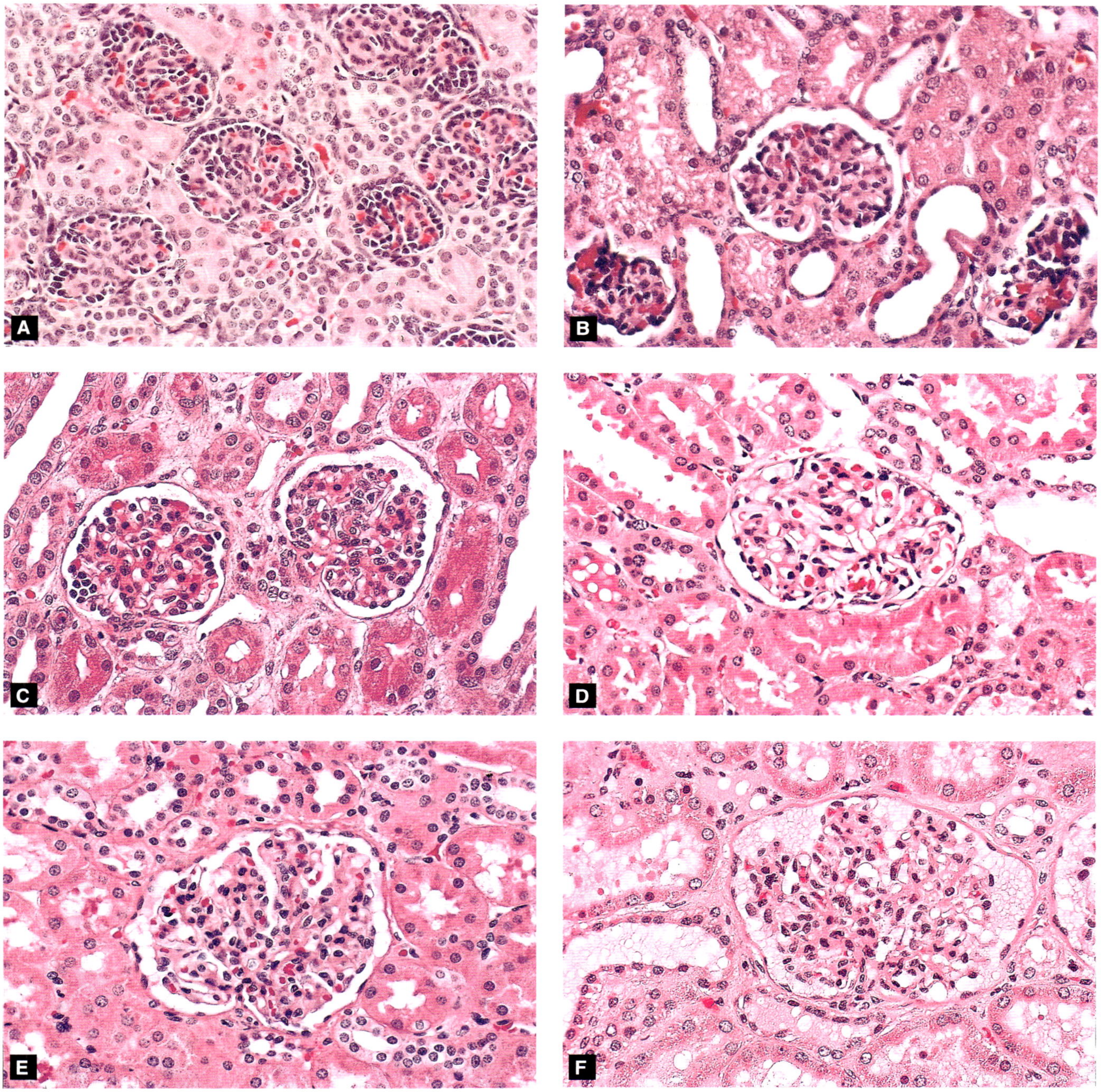

FIGURE 33.24 Normal midcortical glomeruli of six infants and children from birth to 9 years of age. All the micrographs were taken at the same magnification. **A:** Newborn. **B:** 6 months. **C:** 11 months. **D:** 21 months. **E:** 5 years. **F:** 9 years.

basement membrane thickness during childhood range as follows: 100 to 130 nm in fetal kidneys, 170 nm (±30) at birth; 208 nm (±24) at 1 year; 245 nm (±49) at 2 years; 268 nm (±43) at 6 years; and 300 nm (±42) by 10 years (258,260,261). The growth rate is greatest prior to 2 years of age. In contrast to adults, the glomerular basement membrane thickness in children appears less sex dependent (249). Further glomerular basement membrane growth in adolescence, which is less documented, must occur to reach the adult glomerular basement membrane thickness in men (373 nm ± 42) and women (326 nm ± 45) (262).

The continuous cuboidal layer of cells is transient as the maturing podocytes flatten over the surfaces of the developing capillaries. A few small clusters of cuboidal podocytes remain in infant glomeruli (Fig. 33.24B, C). After a glomerulus has been in existence for more than 12 months, remnants of the cuboidal layer are usually not seen in normally developed glomeruli. In 1940,

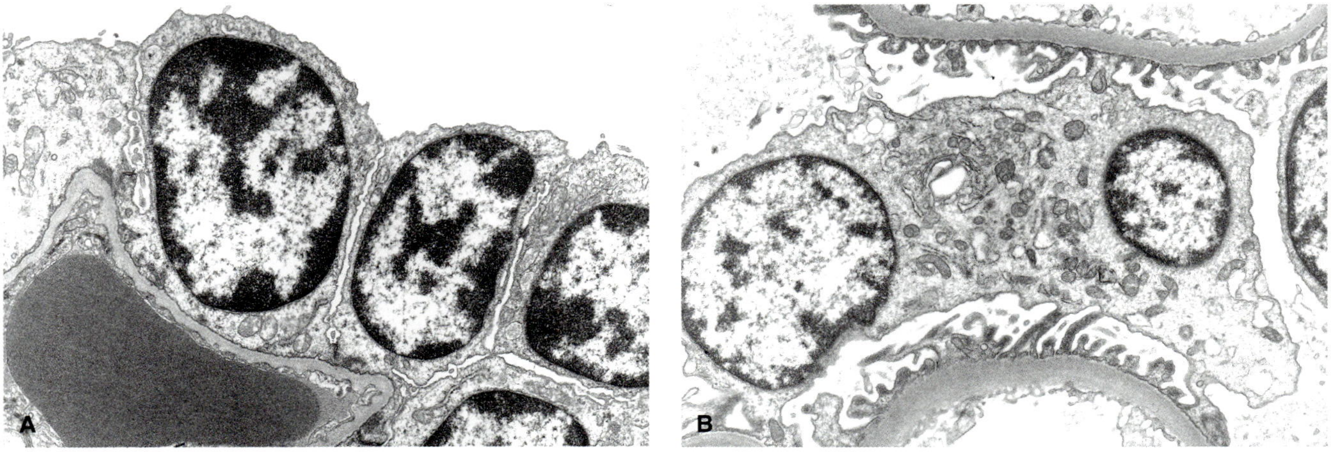

FIGURE 33.25 **A:** Electron micrograph of a normal glomerulus from an infant 2 months of age. Note the continuous layer of primitive podocytes lacking foot processes and the thin glomerular basement membrane. **B:** Mature glomerulus from a patient 16 years of age, photographed at the same magnification as in A. There is prominent foot process development, and the glomerular basement membrane is distinctly thicker than in the infant. (Courtesy of Dr. Gary W. Mierau.)

Gruenwald and Popper suggested that some podocytes may be sloughed into Bowman space as part of the maturation process (256). In fact, several studies have demonstrated that podocytes are shed from the glomerulus and excreted in the urine in various glomerular diseases as well as healthy individuals (263,264). Whether podocyturia occurs in the normal neonate and infant remains to be established. The above podocyte alterations are coincident with the emergence of more conspicuous capillary loops (Fig. 33.24C, D). Eventually, the capillary loops are arranged into lobules (Fig. 33.24E, F). However, an occasional glomerulus with the small size and immature appearance of a neonatal glomerulus may be observed in older infants, especially in the outer third of the cortex (Fig. 33.26).

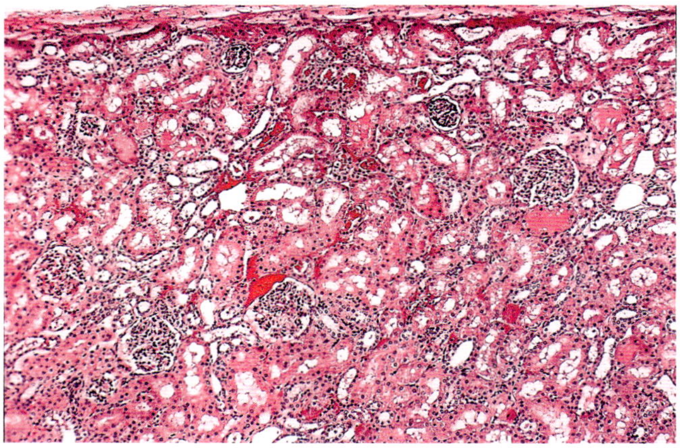

FIGURE 33.26 Persistent immature glomeruli at 12 months of age. Two miniature glomeruli with cuboidal cells at the periphery of the tuft are present near the renal capsule. Small numbers of defective glomeruli can be found in infant kidneys, but they are destined to undergo sclerosis and involution.

Glomerular growth during childhood has been evaluated in several investigations (265–269). It is well established that glomerular size increases from birth to adolescence. The midcortical and juxtamedullary mean glomerular area in fetuses actually decreased between 12 and 20 months of gestation (267). After this initial decrease, the glomeruli remain at the same size until birth, after which they steadily grow. The mean glomerular diameter in normal children increases from 112 to 167 μm between birth and 15 years of age, averaging 3.6 μm per year during this period (268). Glomerular growth occurs by an increase in the number or length of capillaries rather than by hemodynamic capillary dilatation.

Juxtamedullary glomeruli are larger than superficial glomeruli at birth and during infancy. However, some uncertainty exists regarding these regional differences in glomerular size in later childhood and young adults. Some authors have observed no size difference between juxtamedullary and superficial glomeruli by the 14th to 36th postnatal month (265,267), whereas others have found a size difference persists until at least 15 years of age (266,268). Methodologic differences in the studies or the wide variation in glomerular size existing among individuals (270) may account for these disparate findings.

Investigations using stereologic methods have provided estimates for the number of cells in glomeruli; the total number of cells per glomerulus increases along with the mean glomerular volume in normal individuals under 20 years of age compared to those greater than 20 years of age (271). The number of endothelial cells and mesangial cells increased with age, whereas the number of podocytes remained unchanged. The proliferation marker Ki-67 is expressed in podocyte precursors in comma- and S-shaped bodies but its expression is markedly reduced in podocytes of the glomerular capillary loop stage. The cyclin-dependent kinase (CDK) inhibitors p27 and p57 are absent in the comma- and S-shaped bodies but expressed in maturing podocytes of the capillary loop stage as well as mature podocytes in adult kidneys. These findings suggest that CDK inhibitors are involved with arresting the cell cycle of podocytes at the capillary loop stage and maintaining the fully mature podocytes in a quiescent differentiated state (272–274).

A stereologic approach combined with IHC and confocal microscopy has provided an evaluation of podocyte number in both children and adults (275). Glomeruli from children, including some less than 3 years of age, contain the same number of podocytes (450 to 500 podocytes per glomerulus) as small- and medium-sized glomeruli from adults. However, large adult glomeruli contained more podocytes (800 podocytes per glomerulus). The postnatal origin of these additional podocytes is not clear. The large adult glomeruli have a lower podocyte density, resulting in relative podocyte depletion, which may make them more susceptible to glomerular injury.

Early Juxtamedullary Glomeruli

The maturational process in the very early generations of glomeruli may be accelerated because even in very young fetuses these juxtamedullary glomeruli rarely possess a cuboidal layer and are considerably larger than their immediate superficial neighbors. These large juxtamedullary glomeruli subsequently disappear, suggesting that they are transient structures (276,277). It is presumed that these precociously formed glomeruli are functionally important during fetal life and likely undergo involution early in the postnatal period. Relatively little study has been made of these interesting structures.

Glomerulosclerosis in Infants

Glomerulosclerosis in infantile kidneys is commonly observed. In most cases, it is a presumably normal phenomenon that must be distinguished from the pathologic changes of glomerular disease (278–280).

Emery and Macdonald conducted thorough studies of glomerulosclerosis in children's kidneys that were considered morphologically within normal limits (277). Their series of 475 cases included kidneys from fetuses at 24 weeks of gestation to children 15 years of age. The percentage of sclerotic glomeruli in each kidney was most often in the range of 1% to 2% (65% of cases), although 3% to 10% of glomeruli were affected in 30% of cases. The proportion of infants having sclerotic glomeruli is age dependent. Scarred glomeruli occur in 25% to 40% of kidneys from late fetuses and newborns, and 70% of kidneys by 2 months of age, remaining near this level throughout the first year. Afterward, their incidence steadily declines to approximately 10% of children at 6 years of age.

Emery and Macdonald also found that sclerotic glomeruli are localized to two areas; the deep inner cortex near the arcuate vessels and the superficial outer cortex. The presence of scarred glomeruli in the juxtamedullary zone coincides with the disappearance of the large glomeruli seen in this region during the months after birth, as discussed previously. Thus, it appears the sclerosing glomeruli in the juxtamedullary zone represent the involution of the large glomeruli that localize to this zone during nephrogenesis, whereas glomerular scarring in the outer cortex likely reflects a different etiologic process. In a study of 800 infant kidneys, Thomas reported a similar distribution pattern of sclerotic glomeruli in the cortex (281). These changes should not be overinterpreted as evidence of glomerular disease, unless their number is considerably above the usual range (i.e., greater than 20%). It is noteworthy that glomerular involution primarily in the outer cortex has been described as a special form of global sclerosis associated with relapsing minimal change nephritic syndrome (282).

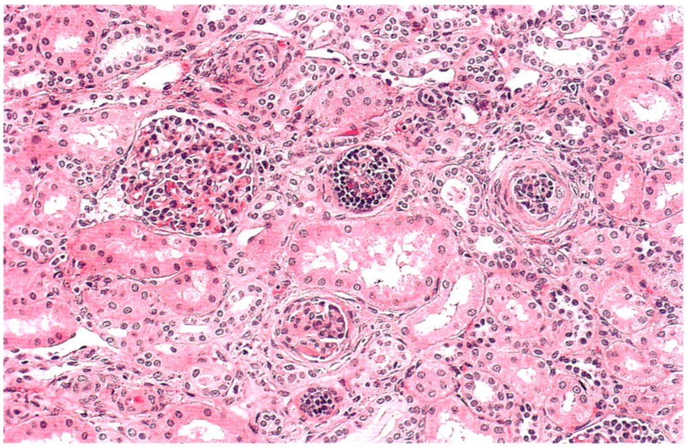

FIGURE 33.27 Infantile glomerulosclerosis at 9 months of age. One small, developmentally immature glomerulus is seen near the center, and two adjacent glomeruli are undergoing involutional sclerosis.

A typical example of infantile glomerulosclerosis is illustrated in Figure 33.27. The scarred glomeruli may occur singly or in small groups. They are usually smaller than normal, immature in appearance, and variably hyalinized. The afferent arteriole is often thickened, and periglomerular fibrosis may be present as well as some chronic inflammatory cells in the interstitium. In later stages, only a small globule of hyaline material in a small focus of sclerosis without capillary lumens may be seen. The tubules associated with the sclerotic glomeruli often contain proteinaceous material and apparently disappear along with the glomeruli.

Ectopic Glomeruli

In the kidneys of fetuses and infants, glomeruli are often found outside the confines of the renal parenchyma, either in the renal sinus (Fig. 33.28) or in the connective tissue around interlobar vessels. These ectopic glomeruli appear to degenerate during postnatal life and are not found in

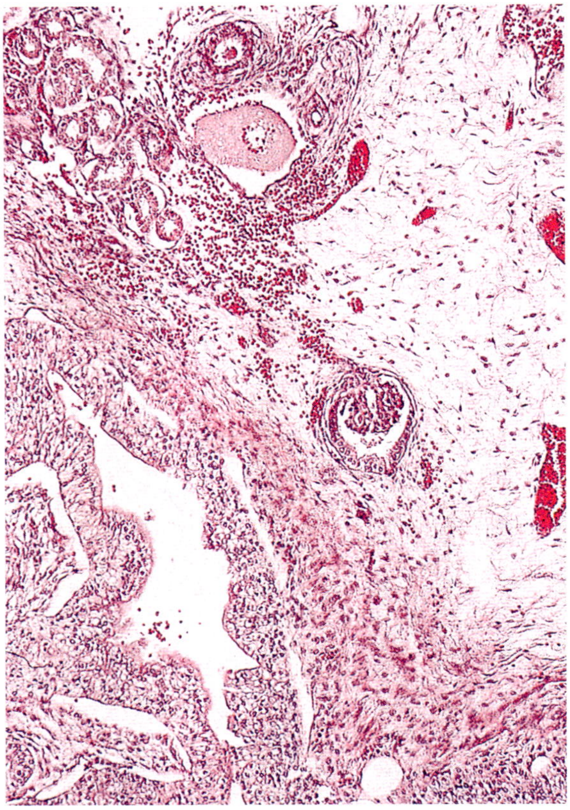

FIGURE 33.28 Ectopic glomerulus in the renal sinus.

adult kidneys. It has been suggested that some vessels supplying the pelvic mucosa and medulla may be derived from degenerated ectopic glomeruli (283,284).

TUBULAR MATURATION AND GROWTH

There is less known about the maturation and growth of tubules between birth and adulthood compared to glomeruli. Although all tubular segments increase in size during postnatal maturation, the proximal convoluted tubules undergo very prominent elongation and increased tortuosity (2). The mean lengths of proximal tubules are about 2 mm at birth, 3.5 mm at 3 months, 6.5 mm at 1 year of age, 7.7 mm at 2 years, and 12.0 mm at 12 years. The mean proximal tubular length of 20 mm in adult kidney indicates that proximal tubule elongation continues through adolescence and young adulthood.

Proximal tubules are less uniform in size than glomeruli at birth. Those in the outer cortex of the newborn kidney are the shortest, those in the midcortex intermediate in length, and those from the juxtamedullary cortex are the longest, consistent with a centrifugal pattern of development (265). These regional differences in proximal tubules decrease significantly after 1 month of age and disappear by 14 months. Whether tubular function, such as solute and volume reabsorption, of the maturing proximal tubules from different cortical regions also follows a centrifugal pattern of maturation remains uncertain (285). It is interesting that the proximal tubules from the outer cortex of adult kidneys are longer than those from the mid- and juxtamedullary cortex (265). The ratio of glomerular surface area to proximal tubular volume decreases with development, suggesting morphologic dominance of glomeruli over tubules early in life until the tubules, in effect, grow up to their glomeruli (265).

During postnatal maturation, the loops of Henle undergo striking elongation (2). The newborn kidney lacks a well-formed inner medulla and contains loops of Henle that are relatively short (2,16,22). As the kidney increases in size after birth, the loops of Henle elongate, the medullary interstitium increases, and the medulla becomes separated into outer and inner zones. From birth until full maturation, loops of Henle may increase in length as much as threefold. In the mature kidney, the location of the tips of the loops of Henle relates to the age of their associated nephrons. The loops of the last formed nephrons in the outer cortex reach the junction between the outer and inner medulla, and are known as short loops of Henle. In contrast, the loops of the earliest formed nephrons extend deep into the inner medulla near the tip of the papilla and are referred to as long loops of Henle. Prior to birth, thin portions of loops of Henle are only observed in the long loops of Henle. These thin portions are present in the descending limbs of the long loops of Henle and continue to lengthen after birth. Descending thin limbs are not seen in short loops of Henle until after birth. Only long loops of Henle develop ascending thin limbs ascending limb, which are derived from the thick ascending limbs, likely by an ascending process of apoptotic remodeling (286,287).

ADULT KIDNEY

Gross Anatomy

The kidneys lie within the retroperitoneum and extend from the 12th thoracic to the 3rd lumbar vertebrae with the right kidney usually slightly more caudad. Craniocaudal movement of the kidneys during respiration may be up to 4 cm and they may be 2.5 cm lower in the erect than in the supine position. The kidneys are situated within the perirenal space, which contains abundant fat and is traversed by fine fibrous septa (288–290). Anterior and posterior layers of the renal fascia, known as Gerota fascia, enclose the kidneys. Each kidney weighs 125 to 170 g in men and 115 to 155 g in women (291). If differences in body build are considered, kidney weight correlates best with body surface area, whereas age, sex, and race have less influence (292). Each kidney is 11 to 12 cm in length, 5 to 7.5 cm in width, and 2.5 to 3 cm in thickness. Magnetic resonance imaging has shown mean kidney lengths of

12.4 ± 0.9 cm for men and 11.6 ± 1.1 cm for women and mean kidney volumes of 202 ± 36 mL for men and 154 ± 33 mL for women (293). However, the estimated renal volume may vary with changes in blood pressure and intravascular volume. The upper poles slant somewhat toward the midline and the posterior and the hilar aspect of each kidney has an anteromedial orientation. The anterior surface of each kidney is more convex compared to the flatter posterior surfaces. The left kidney tends to be slightly larger and may demonstrate irregularities of the lateral contour from compression by the spleen in up to 10% of normal individuals (294). A glistening tough fibroelastic capsule surrounds the kidney.

The aperture on the concave medial surface of each kidney is the hilum, through which passes the vessels, nerves, and ureter. The main renal artery branches in the hilum to form anterior and posterior divisions (Fig. 33.29) that, in turn, divide into segmental arteries that supply the apical, upper, middle, lower, and posterior segmental regions of the parenchyma (295,296). Most of the segmental arteries arise from the anterior division. No collateral circulation has been demonstrated between the segmental arteries. Some of the accessory arteries actually represent normal segmental arteries with an early origin from the main-stem renal artery or aorta (296). Therefore, ligation of such a segmental artery in the belief that it is an accessory vessel results in necrosis of the corresponding parenchymal segment. The intrarenal veins do not follow a segmental distribution, and there are numerous anastomoses of the veins throughout the kidney. There are variable drainage patterns of the large extrarenal veins which join to form the main renal vein (297). A relatively common occurrence is a posterior primary venous tributary with a retropelvic position that should be remembered during renal surgical procedures.

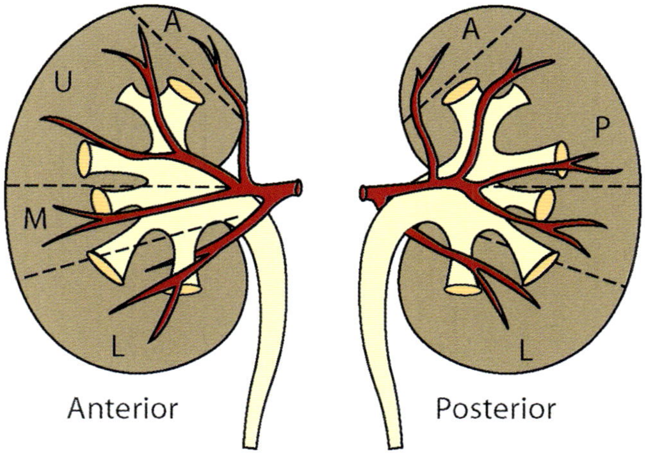

FIGURE 33.29 Diagram of the vascular supply of the human kidney. The anterior division of the renal artery divides into three segmental branches that supply the upper (*U*) and middle (*M*) segments of the anterior surface and most of the lower (*L*) segment. The small apical (*A*) segment is usually supplied by a branch from the anterior division. The posterior division of the renal artery supplies the posterior (*P*) segment, which represents more than half of the posterior surface of the kidney. (Modified with permission from Graves FT. The anatomy of the intrarenal arteries and its application to segmental resection of the kidney. *Br J Surg.* 1954;42:132–139. Copyright © 1954 British Journal of Surgery Society Ltd.)

The outer cortex, striated inner region, and medulla can be grossly distinguished. The presence of glomeruli and convoluted tubules results in the cortex having a more granular appearance. The renal medulla is divided into 8 to 18 striated conical masses called pyramids (Fig. 33.30).

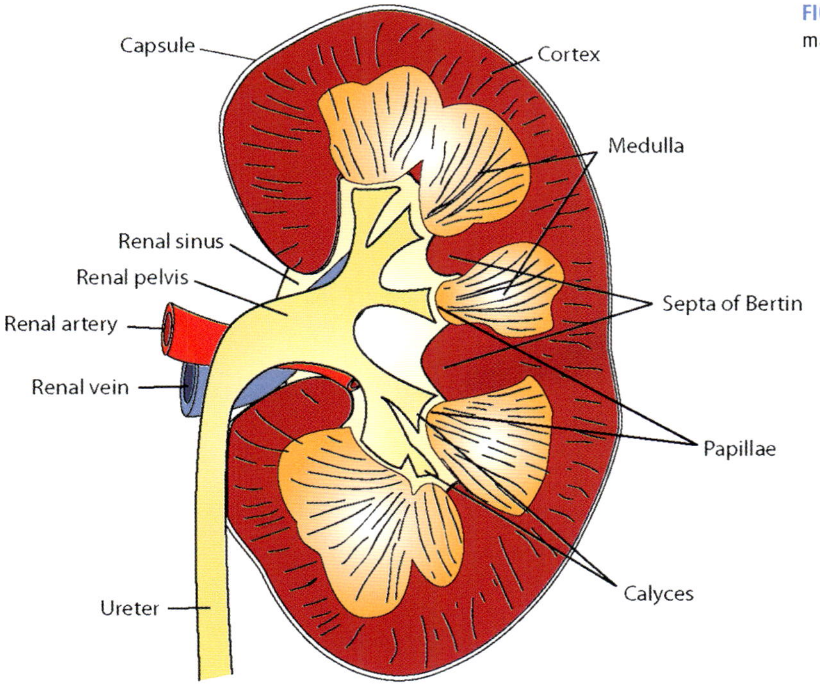

FIGURE 33.30 Diagram of a bisected kidney illustrating major anatomic structures.

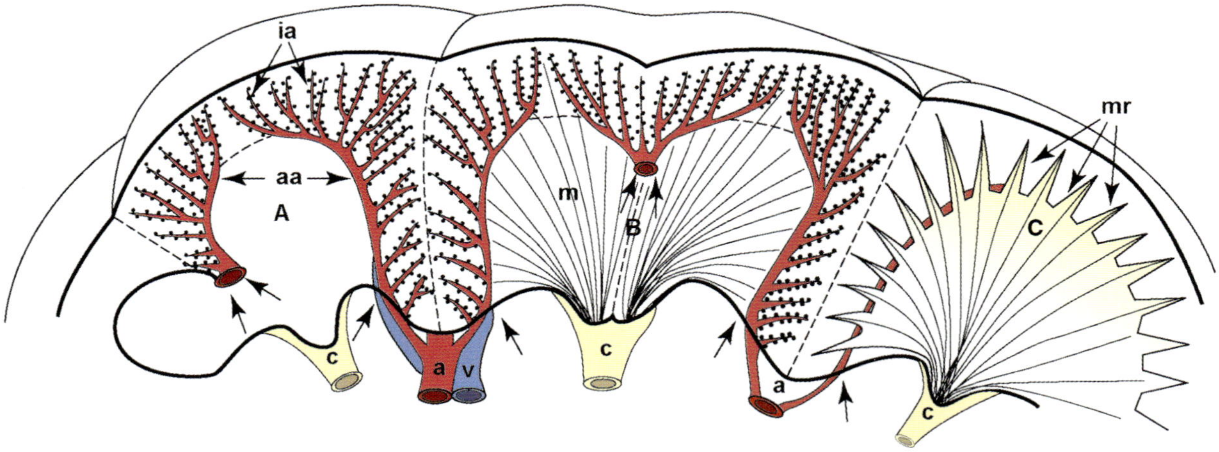

FIGURE 33.31 Diagram of three renal lobes. **A:** Arcuate (*aa*) and interlobular (*ia*) arteries. **B:** Cortex and medulla (*m*) are illustrated in a double lobe with fused double papillae. **C:** Lobe showing medullary rays (*mr*). A septum of Bertin represents the approximation of two layers of septal cortex from two adjacent lobes. (*Small double arrows* in A and B, subsidiary septal arteries; *single arrows*, location where arcuate vessels enter the renal parenchyma; *a*, interlobar arteries; *v*, interlobar vein; *c*, calyces). (Modified with permission from Hodson CJ. The renal parenchyma and its blood supply. *Curr Probl Diagn Radiol.* 1978;7:1–32. Copyright © 1978 Elsevier.)

The striated appearance reflects the parallel linear orientation of the loops of Henle and collecting ducts. The base of each pyramid is located at the corticomedullary junction, whereas the apex extends toward the renal pelvis, forming a papilla. The tip of each papilla, the area cribrosa, is perforated by 20 to 70 small openings (3) that represent the distal ends of the collecting ducts of Bellini. The cortex is about 1 cm in thickness, encircles the base of each pyramid, and extends downward between pyramids to form the columns of Bertin.

Despite well-described radiologic features (298,299), an enlarged column of Bertin has on occasion been clinically mistaken for a renal tumor. Longitudinal striations extending from the base of the pyramids out into the cortex are termed the medullary rays of Ferrein. These structures are actually part of the cortex and are formed by the straight segments of the proximal tubules, the cortical thick ascending limbs, and the cortical collecting ducts. The medullary rays may be visualized during excretory urography in conditions with tubular fluid stasis (300).

A single pyramid with its surrounding cortical parenchyma constitutes a renal lobe (Fig. 33.31) (301). The human kidney has an average of 14 lobes. During development, variable lobar fusion leads to coalescence of some papillae and remodeling of the corresponding calyces gradually reducing the number of papillae and calyces. The mean number of calyces and papillae reported is 9 and 11, respectively (236). There is a greater degree of lobar fusion in the polar regions than in the midpolar region of the kidney. Although the mature kidney eventually develops a smooth outer surface, a degree of persistent fetal lobation may be observed in some adult kidneys.

There are two main types of renal papillae (301). Simple papillae drain only one lobe and have convex tips containing small, often slit-like orifices. Compound papillae drain two or more adjacent fused lobes and have flattened, ridged, or concave tips with round, often gaping orifices. The distribution of papillae types within the kidney is related to the embryologic pattern of fusion involving the lobes, papillae, and calyces (Fig. 33.32). It is believed that the more open orifices of compound papillae are less capable of preventing intrarenal reflux (302), which may be associated with an increase in intrapelvic pressure. This concept is supported by the observation that pyelonephritic scars associated with intrarenal reflux are present more commonly in the renal poles, where the compound papillae predominantly occur.

The renal pelvis is the sac-like expansion of the upper ureter. Two or three outpouchings or major calyces, the infundibula, extend from the pelvis and divide into the minor calyces, into which the papillae protrude. In addition, elaborate leaf-like extensions, termed fornices, extend from the minor calyces into the medulla, and secondary pouches increase the pelvic surface area (303). The walls of the calyces, pelvis, and ureters contain specialized cells which serve a pacemaker function to facilitate urine movement to the bladder.

The renal sinus is located on the medial or concave aspect of each kidney (Fig. 33.33) (304,305). It contains the renal pelvis, the major renal arteries and veins, the lymphatic vessels, and neural structures that supply the kidney. The renal hilum is the entry into the sinus. Fat fills the renal sinus and is contiguous with the perirenal fat. Within the renal sinus, the renal capsule does

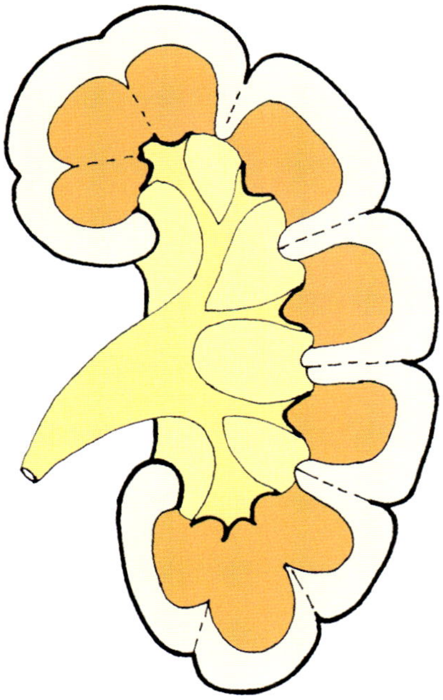

FIGURE 33.32 Schematic representation of the lobar architecture. In the polar regions, there is a greater degree of lobe fusion, resulting in the formation of compound papillae and calyces and the loss of septal cortex. The individual lobes tend to be retained in the midpolar region, and the septal cortex extends between renal pyramids, as septa of Bertin, to the renal sinus. (Modified with permission from Hodson CJ. The renal parenchyma and its blood supply. *Curr Probl Diagn Radiol.* 1978;7:1–32, with permission from Elsevier.)

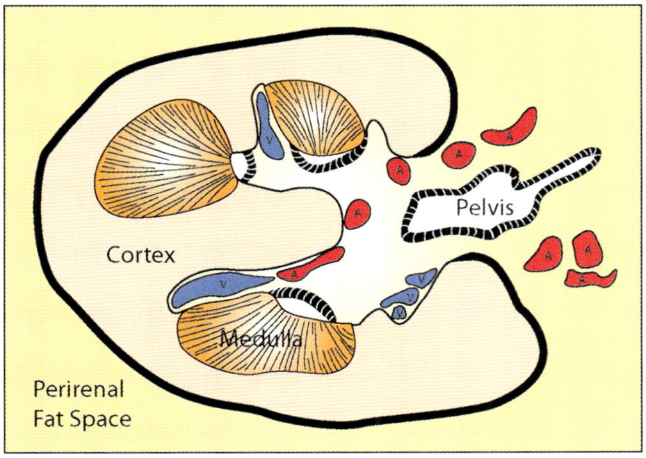

FIGURE 33.33 Diagram of kidney cross-section, showing the renal sinus which is filled with fat (*yellow*). The renal sinus contains the pelvic–calyceal system and the major renal vessels. The renal capsule (*thick black line*) surrounds the convex surface of the kidney but disappears from the cortical surface as the latter enters the renal sinus. The cortical surfaces, including the septa of Bertin, facing the renal sinus lack a capsule. The capsule surrounding the pelvic–calyceal system is indicated by a cross-hatched thick black line. The capsule covering the calyces, appears to extend over the medullary pyramids on cross-section, and it envelopes the renal pelvis as well. (Modified with permission from Murphy WM, Grignon DJ, Perlman EJ. Tumors of the kidney, bladder, and related urinary structures. In: Silverberg SG, Sobin LH, eds. *Atlas of Tumor Pathology.* 4th series, fascicle 1. Armed Forces Institute of Pathology; 2004. Permission obtained from The Joint Pathology Center.)

not enclose the cortical parenchymal surface. Beckwith called attention to the importance of the renal sinus as a pathway for tumor dissemination in Wilms tumor (305), and this has also been shown in renal cell carcinomas (306,307).

NEPHRON

The structural and functional unit of the kidney is the nephron. It consists of the renal corpuscle, which consists of the glomerulus and Bowman capsule, proximal tubule, thin limbs, and distal tubule, all of which originate from the metanephric blastema. The total number of nephrons in a human kidney varies markedly among normal individuals (308). A 10-fold variation in nephron number has been reported, from approximately 200,000 to more than 2.5 million nephrons per kidney. The usual range is about 600,000 to 1,200,000 nephrons per kidney.

Nephrons can be classified according to the position of their glomeruli in the cortex or the length of their loop of Henle (Fig. 33.34). Superficial nephrons have glomeruli located in the outer cortex and their efferent arterioles usually ascend to the cortical surface. The glomeruli of juxtamedullary nephrons are located immediately above the corticomedullary junction in the inner cortex, and their efferent arterioles form the descending vasa recta. The glomeruli of midcortical nephrons are situated in the midcortex above the juxtamedullary region but below the superficial nephrons.

Alternatively, nephrons can be classified based on the length of the loop of Henle: those with a short loop of Henle and those with a long loop. The length of the loop of Henle is generally related to the location of its parent glomerulus in the cortex. The short loops generally form their bend at various levels within the inner stripe of the outer medulla, whereas the long loops of Henle enter and turn back within the inner medulla. Although there are numerous gradations between these two main types of nephrons, there are seven times shorter than long-loop nephrons in human kidneys (309). A correlation between the urinary concentrating ability and the relative length of the medulla has been established (310).

The connecting segment or tubule joins the nephron to the collecting duct system and is believed to originate from the metanephric blastema. The collecting duct system includes the initial collecting tubule, the cortical collecting duct, the outer medullary collecting duct, and the inner medullary collecting duct (311,312). It is derived from the ureteric bud and collects tubule fluid from different nephrons, so it has not been classically considered a nephron component. Although not strictly correct in an anatomic sense, the term nephron is commonly used to include the connecting segment and entire collecting duct.

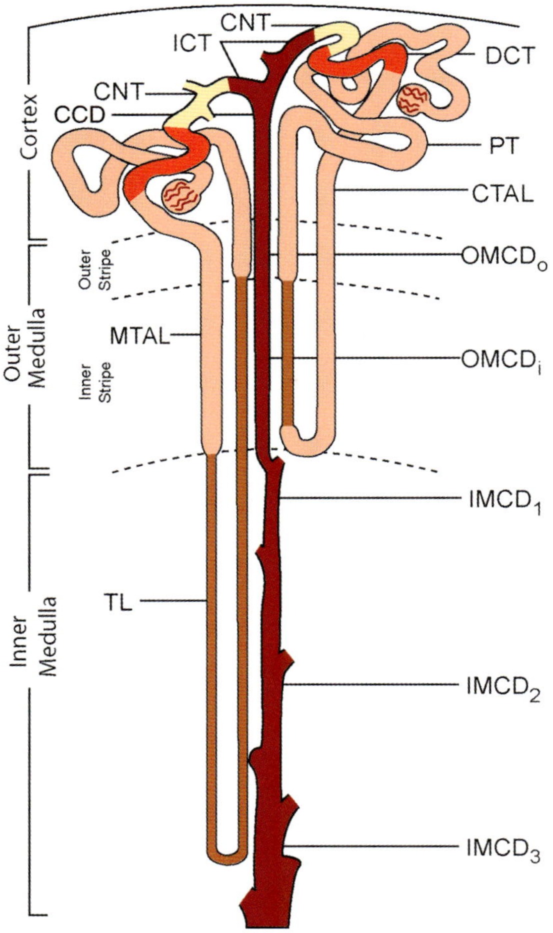

FIGURE 33.34 Diagram illustrating the segments of the nephron and the zones of the kidney. *PT*, proximal tubule; *TL*, thin limb of Henle loop; *MTAL*, medullary thick ascending limb; *CTAL*, cortical thick ascending limb; *DCT*, distal convoluted tubule; *CNT*, connecting tubule; *ICT*, initial collecting tubule; *CCD*, cortical collecting duct; *OMCD$_o$*, collecting duct in outer stripe of outer medulla; *OMCD$_i$*collecting duct in inner stripe of outer medulla; *IMCD$_1$*, outer third of inner medullary collecting duct; *IMCD$_2$*, middle third of inner medullary collecting duct; *IMCD$_3$*, inner third of inner medullary collecting duct. (Modified with permission from Madsen KM, Tisher CC. Structural-functional relationships along the distal nephron. *Am J Physiol.* 1986;250 (pt 2):F1–F15. Copyright © 1986, The American Physiological Society.)

ARCHITECTURE

The renal cortex can be divided into lobules. A renal lobule consists of a centrally positioned medullary ray and its surrounding cortical parenchyma containing all nephrons draining into the collecting ducts of the medullary ray.

The nephron segments and blood vessels in the cortex and medulla have a specific geometric arrangement (313). This intricate architecture allows for integration of complex transport functions along the length of a specific nephron segment, as well as between different nephron segments in a specific region or zone (314).

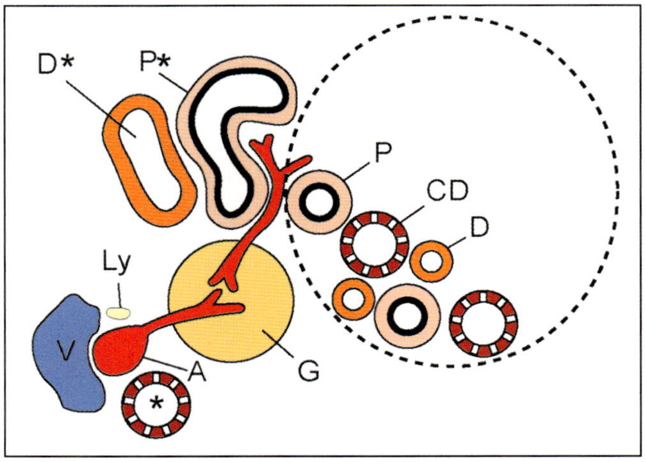

FIGURE 33.35 Diagram of architectural regions of the renal cortex. A medullary ray is encircled by the *dotted line*, and the cortical labyrinth is outside the *dotted line.* The proximal straight (*P*) and distal straight ascending limb (*D*) tubules and the collecting ducts (*CD*) are located in the medullary ray. The adjacent cortical labyrinth contains the interlobular vessels: Arteries (*A*), veins (*V*), and lymphatics (*Ly*); arcades (*) of connective tubules; glomeruli (*G*); and the proximal (*P**) and distal (*D**) convoluted tubules. (Modified with permission from Alpern RJ, Caplan MJ, Moe OW, Quaggin SE. *Seldin and Giebisch's The Kidney: Physiology and Pathophysiology.* 6th ed. Elsevier Shop; 2025.)

Two architectural regions of the renal cortex can be distinguished: the cortical labyrinth and the medullary rays (Fig. 33.35). The cortical labyrinth represents a continuous parenchymal zone that surrounds the regularly distributed medullary rays. Glomeruli, proximal and distal convoluted tubules, interlobular vessels, and a rich capillary network are situated in the cortical labyrinth. The large majority of convoluted tubular profiles are proximal tubules. Connective tubules of juxtamedullary nephrons fuse and form the arcades, which are adjacent to the interlobular vessels within the cortical labyrinth. Individual nephrons with their interlobular vessels, glomeruli, and attached tubular segments, are difficult to distinguish in this complex topography by histologic assessment. The medullary rays (Figs. 33.36 and 33.37) contain the proximal and distal straight tubules and collecting ducts, all of which enter into the medulla. The distal straight tubules are the thick ascending limbs. Within an individual medullary ray, the straight tubules of superficial nephrons are situated centrally, the straight tubules of midcortical nephrons are localized peripherally, and the collecting ducts occupy a position between the two groups. The straight tubules of juxtamedullary nephrons descend directly into the medulla, never entering the medullary rays.

The tubules in the cortex have a compact back-to-back appearance with little intervening interstitium. Rarely, a discrete cortical scar consistent with a previous biopsy site

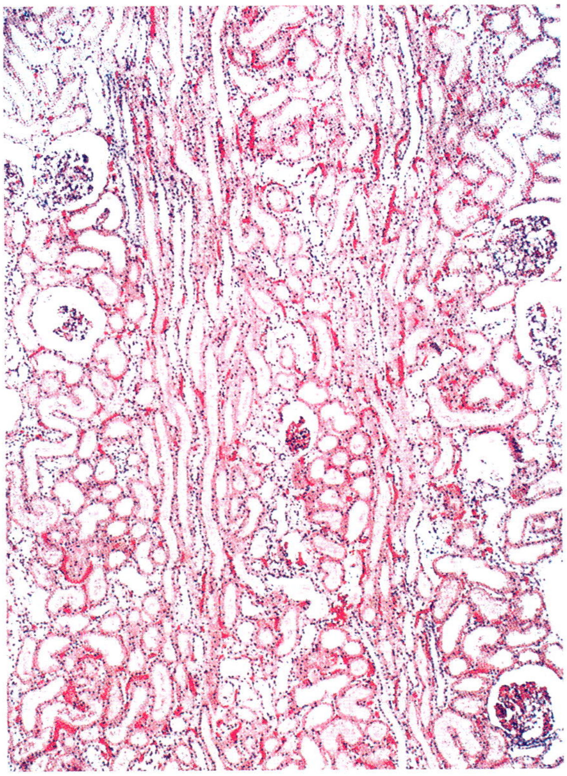

FIGURE 33.36 Longitudinal section of cortex demonstrating two linear aggregates of tubules representing medullary rays (H&E, ×50).

may be observed (Fig. 33.38). Incidental tumors may be detected in a renal biopsy (Fig. 33.39).

The localization of specific segments of the nephrons at various levels in the medulla accounts for the division of the medulla into an outer and inner zone, with the former subdivided into an inner and outer stripe (Fig. 33.34). The relative tissue volumes for the cortex and the outer and inner medulla are 70%, 27%, and 3%, respectively (312). Glomeruli are not present in the medulla.

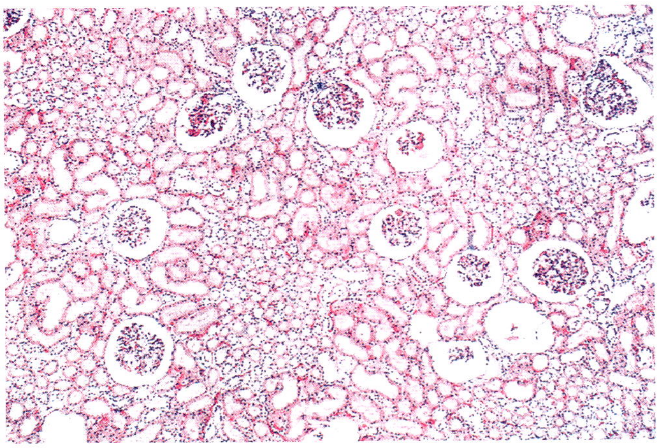

FIGURE 33.37 Cross-section of cortex illustrating medullary rays that are regularly distributed within the cortical labyrinth (H&E, ×50).

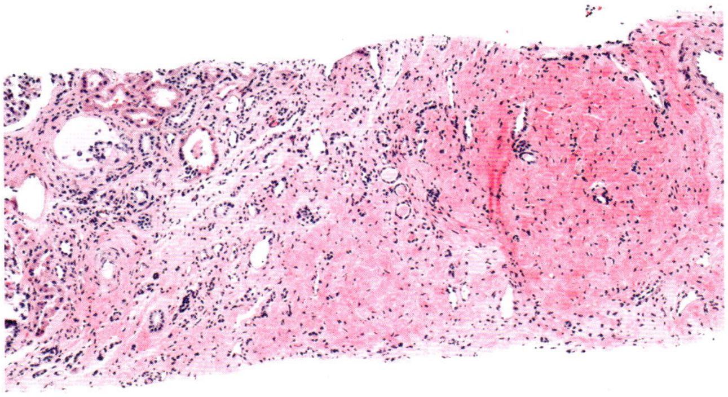

FIGURE 33.38 Focal fibrous scar, suggestive of previous biopsy site, situated in middle of renal cortex, which is otherwise intact (H&E, ×100).

The outer stripe of the outer medulla is relatively thin. It contains the terminal portions of the proximal straight tubules, the thick ascending limbs, and the collecting ducts. The outer stripe is also distinguished by the absence of thin limbs of Henle. In contrast to the outer stripe, the inner stripe of the outer medulla is thicker. It contains thin descending limbs, thick ascending limbs, and collecting ducts. It is further characterized by the absence of the proximal straight tubules. Aggregations of descending and ascending vasa recta known as vascular bundles develop in the outer stripe but are located predominantly in the inner stripe. Compared with the kidneys of some mammals with very high urine-concentrating ability, the human kidney has a simple medulla (313). In contrast to the complex medulla, the vascular bundles of the simple medulla do not fuse to form larger vascular structures, and they do not incorporate the descending thin limbs of short loops (Fig. 33.40). The inner medulla contains the thin descending and thin ascending limbs of long loops, as well as the collecting ducts. Thick ascending limbs are absent in the inner medulla.

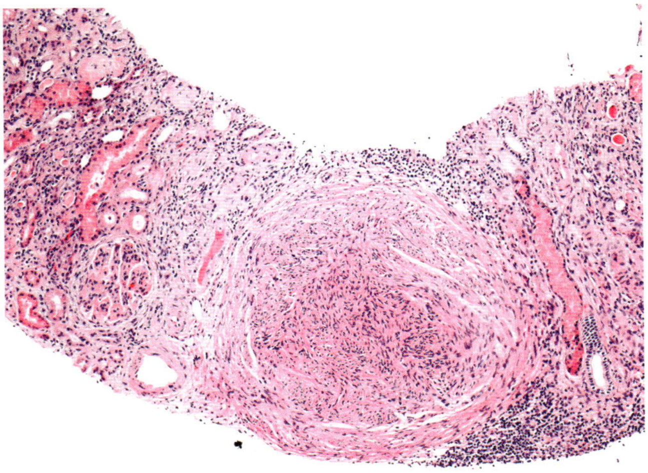

FIGURE 33.39 Circumscribed leiomyoma enclosed within a renal biopsy specimen (H&E, ×100).

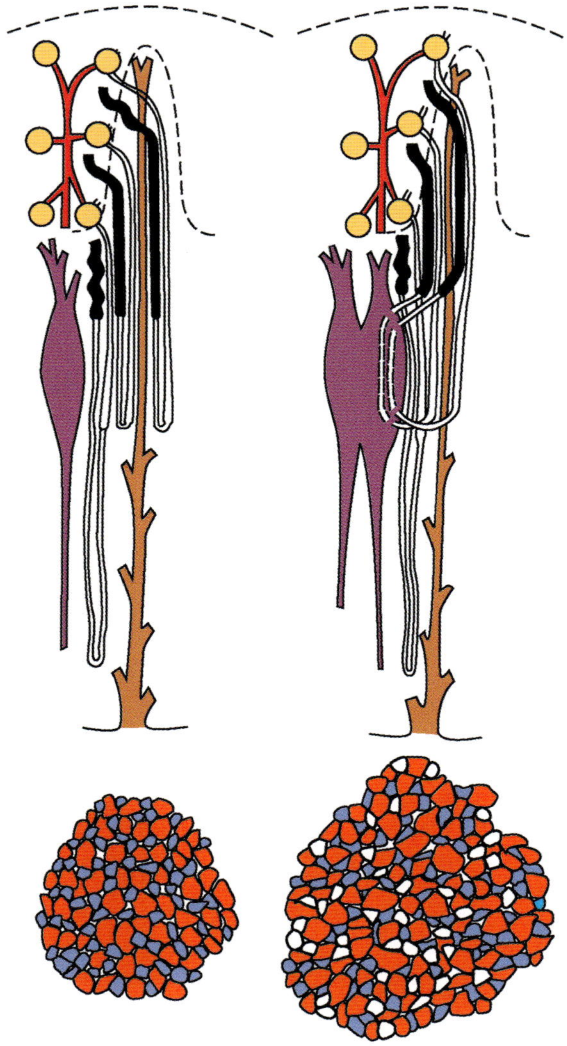

FIGURE 33.40 Schematic diagram demonstrating the simple and complex types of medulla. **Upper left:** In the simple medulla, the loops of Henle remain separate from the vascular bundle (*purple*). The vascular bundle itself (**lower left** cross-section) contains only descending (*red*) and ascending (*blue*) vasa recta. **Upper right:** In the complex medulla, the descending thin limbs of short loops of Henle descend within the vascular bundles (*purple*), which tend to fuse. Therefore, the complex bundles (**lower right** cross-section) contain the descending thin limbs of short loops (*white*) in addition to descending (*red*) and ascending (*blue*) vasa recta. (Modified with permission from Jamison RL, Kriz W. *Urinary Concentrating Mechanism: Structure and Function*. Oxford University Press; 1982.)

PARENCHYMA

An accurate morphologic evaluation of the kidney requires a detailed systematic examination of the glomeruli, tubules, interstitium, and blood vessels of the renal parenchyma. A standard nomenclature for structures of the kidney exists (315). A detailed approach to the histopathologic evaluation of the kidney has been described (316) and technical guidelines for handling the renal biopsy are available (317,318). The following discussion emphasizes normal morphologic aspects and structural–functional relationships in the kidney. The reader is directed to detailed discussions for more information (313,319).

GLOMERULUS

Overview

Malpighi first described the glomeruli and demonstrated their continuity with the renal vasculature (320). Bowman later elucidated the capillary architecture of the glomerulus and the continuity between its surrounding capsule and the proximal tubule (321,322). The renal corpuscle consists of a tuft of interconnected capillaries and an enclosing capsule named after Bowman. The term glomerulus is commonly used to refer to the glomerular capillary tuft and Bowman capsule, although the term renal corpuscle is more accurate in a strict anatomic sense. The glomerulus does not simply represent a ball of capillaries. Providing structural support for the capillary tuft is a central region of mesangium, which contains cells and their surrounding matrix material. The capillaries are lined by a thin layer of endothelial cells, contain a basement membrane, and are covered by podocytes that form the visceral layer of Bowman capsule. The parietal epithelium is continuous with the visceral epithelium at the vascular pole where the afferent arteriole enters the glomerulus and the efferent arteriole exits. The glomerulus somewhat resembles a blind-pouched extension of the proximal tubule invaginated by a tuft of capillaries (Fig. 33.41) (323). The cavity situated between the two epithelial layers of Bowman capsule is called Bowman space or the urinary space. At the urinary pole, this space and the parietal layer of Bowman capsule continue into the lumen and epithelium of the proximal tubule. The glomerular tuft originates from the afferent arteriole, which enters the glomerulus at the vascular pole and divides into several lobules. Anastomoses are believed to exist between individual capillaries within a lobule as well as between lobules (313,323,324). The efferent arteriole is formed by rejoined capillaries and leaves the glomerulus at the vascular pole. In contrast to the afferent arteriole, the efferent arteriole has a more continuous intraglomerular segment. The glomerulus is responsible for the ultrafiltration of plasma. The glomerular filtration barrier consists of the fenestrated endothelium, the peripheral glomerular basement membrane, and the slit diaphragms between the podocyte foot processes.

The glomerulus has a round configuration and an average diameter of about 200 μm (313,319). It has been reported that the glomeruli in solitary functioning kidneys are significantly larger than those in control patients with two kidneys (325). In adults 50 to 70 years of age, the volume of normal-appearing glomeruli in the superficial cortex has been reported to

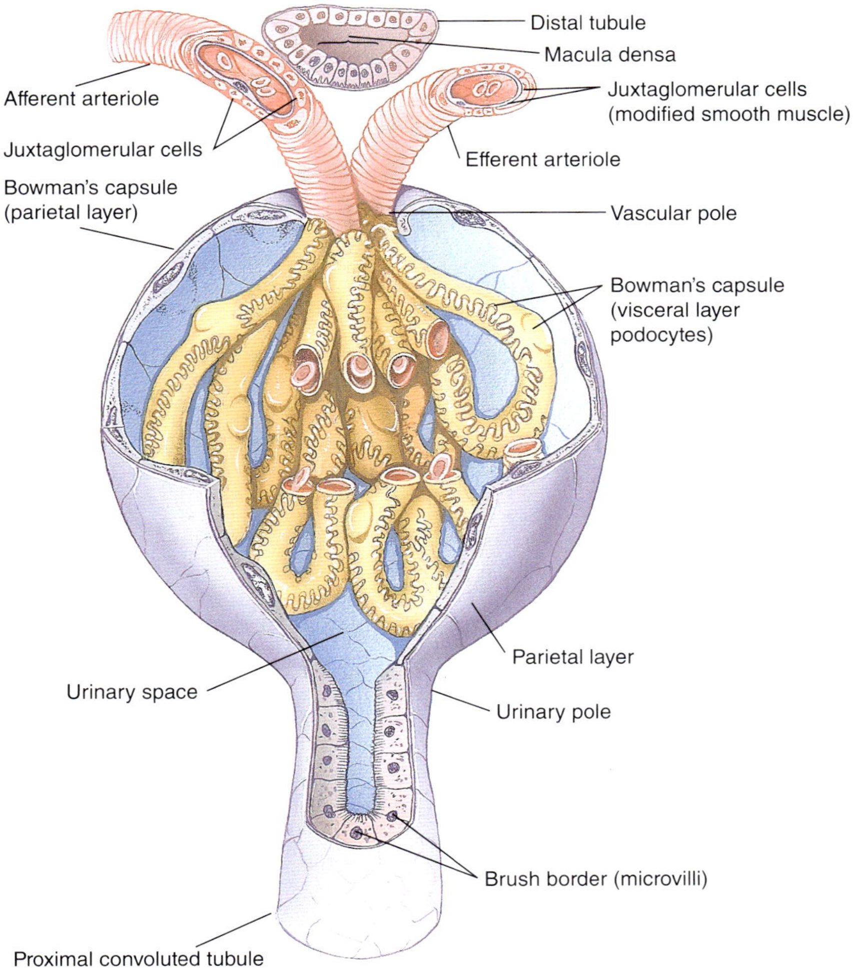

FIGURE 33.41 Schematic three-dimensional representation of the glomerulus. (Modified with permission from Gartner LP, Lee LMJ. *Gartner and Hiatt's Atlas and Text of Histology.* 8th ed. Wolters Kluwer; 2023.)

be 20% larger than those in the juxtamedullary cortex (270). This may reflect compensatory glomerular hypertrophy accompanying global glomerulosclerosis in the superficial cortex of these older patients. Although the glomerulus has a lobular architecture, the lobulation is often inconspicuous in light microscopic sections. An accentuated degree of lobulation may be more prominent in autopsy kidneys than in biopsy specimens. An apparent increase in the number of glomerular cells can be observed with an increased section thickness, therefore an accurate assessment of glomerular cellularity requires histologic sections 2 to 4 μm thick (Fig. 33.42). In general, the presence of more than three cells in a mesangial area away from the vascular pole constitutes hypercellularity. The delicate character of the glomerular capillary walls can be observed on thin histologic and frozen sections (Fig. 33.43) (326). A fortuitous tangential section may show a prominent lumen of a hilar arteriole which should not be mistaken for abnormal dilatation (Fig. 33.44). Occasionally, two glomerular tufts appearing as a bifid glomerulus with an apparent single vascular pole can be observed (Fig. 33.45). Some renal biopsies have limited numbers of glomeruli. The glomeruli are occasionally displaced from the main tissue fragment and may be detected upon careful examination as single glomeruli isolated in space or in surrounding tissue (Fig. 33.46A, B).

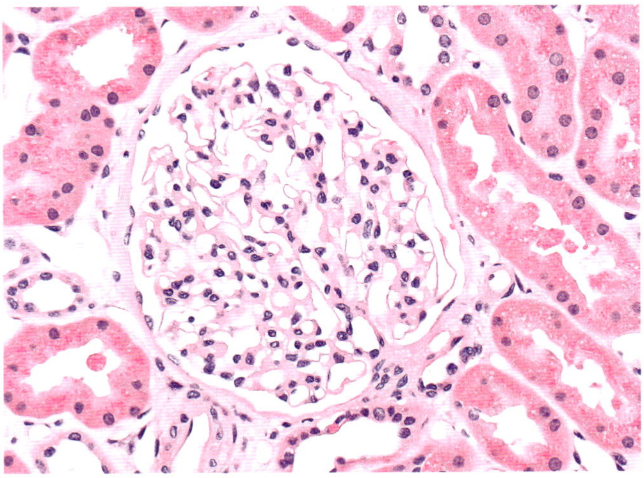

FIGURE 33.42 Glomerulus exhibiting round configuration and normal cellularity (H&E, ×400).

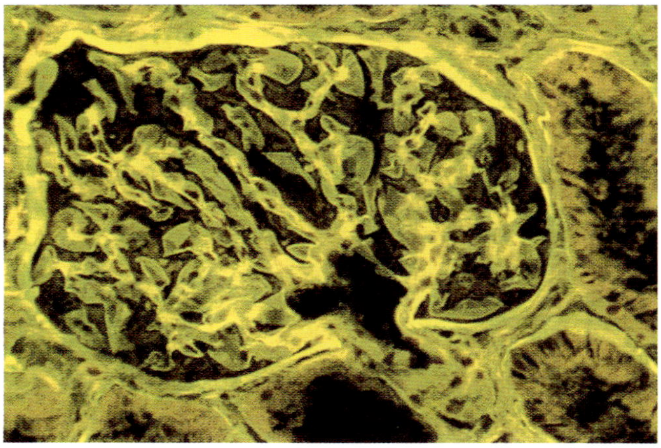

FIGURE 33.43 Fluorescent micrograph of an H&E-stained frozen section. Note the delicate character of the glomerular capillaries. (Courtesy of Dr. Stephen M. Bonsib.)

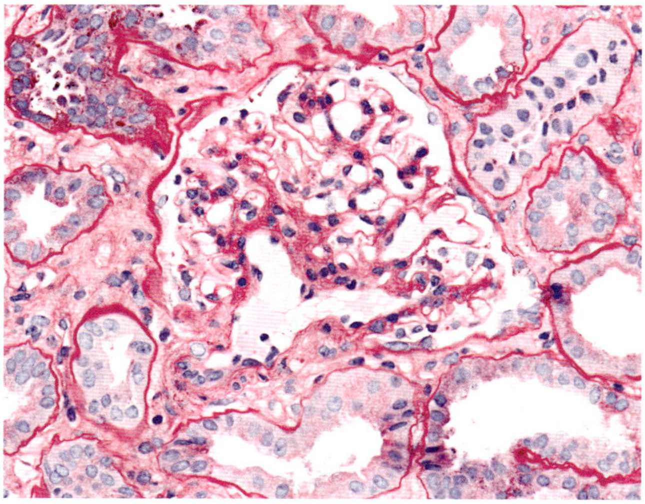

FIGURE 33.44 PAS-stained tangential section through a glomerular hilar arteriole (PAS, ×400).

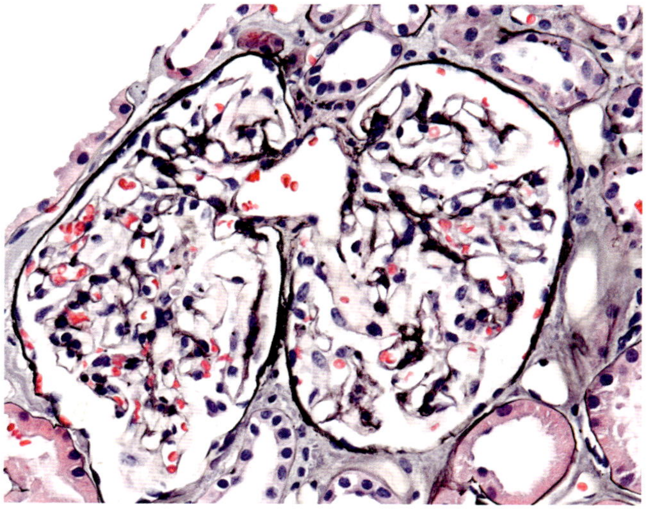

FIGURE 33.45 Bifid glomerulus with two capillary tufts sharing a hilar arteriole and even a juxtaglomerular apparatus (Jones silver stain, ×400).

Global glomerulosclerosis may occur as part of aging in patients without renal disease. The mean percentage of global glomerulosclerosis in normal kidneys is less than 1% between 1 and 20 years of age, 2% between 20 and 40 years, 7% between 40 and 60 years, and 11% between 60 and 80 years (326–328). It has been suggested that the 90th percentile for global glomerulosclerosis may be generally estimated within a given patient by subtracting 10 from half the patient's age (328). More recent studies have examined the changes in human kidneys with aging (329). Kidney volume decreases after the age of 50 years (330), and simple renal cysts increase with age (331). In one study of biopsies from over 2,000 ostensibly healthy living donor kidneys, the normal number of globally sclerotic glomeruli ranges from 1 glomerulus for a 20-year-old to 5.5 glomeruli for a 70-year old when 17 to 32 glomeruli are present in the sample. Another study of living donor kidneys showed a dramatic loss of intact glomeruli with aging (332). From the youngest age group (18 to 29 years) to the oldest age group (70 to 75 years), the number of nonsclerotic glomeruli was 48% lower, and the number of globally sclerotic glomeruli increased by only 15%.

Endothelial Cells

A thin fenestrated endothelium lines the glomerular capillaries. Endothelial cells have light eosinophilic cytoplasm and slightly ovoid nuclei The endothelial cells are extremely attenuated around the capillary lumen, and the thicker portions of the cells containing the nuclei lie adjacent to the mesangium away from the urinary space. The cytoplasm contains microtubules, microfilaments, and intermediate filaments (333). The attenuated portion of endothelial cytoplasm is perforated by fenestrae ranging from 70 to 100 nm in diameter (323). It has been argued that endothelial fenestrae do not simply arise from fusion of plasma membrane invaginations, called caveolae (334). As mentioned earlier, most investigations have indicated that adult glomerular endothelial cells lack diaphragms across the fenestrae, whereas they have been observed in the embryonic glomerulus.

The endothelial cell surface carries a negative charge because of the presence of polyanionic glycoproteins, including podocalyxin, a major sialoprotein of glomerular endothelial cells as well as of podocytes (335,336). The glomerular endothelial cells have a glycocalyx surface layer that fills the fenestrae forming sieve plugs (337,338). This glycocalyx has been best visualized with special electron microscopic techniques using cationic dyes, lipid particles, or with high-pressure freezing (339,340). The glycocalyx, consists of membrane-bound proteoglycans (syndecan and glypican), secreted glycoproteins (perlecan and versican), and secreted glycoaminoglycans (GAGs, hyaluronan). There is accumulated evidence that the glomerular endothelial glycocalyx is an important component of the glomerular filtration barrier (341,342). There is strong expression of CD34 in glomerular endothelial cells (Fig. 33.47).

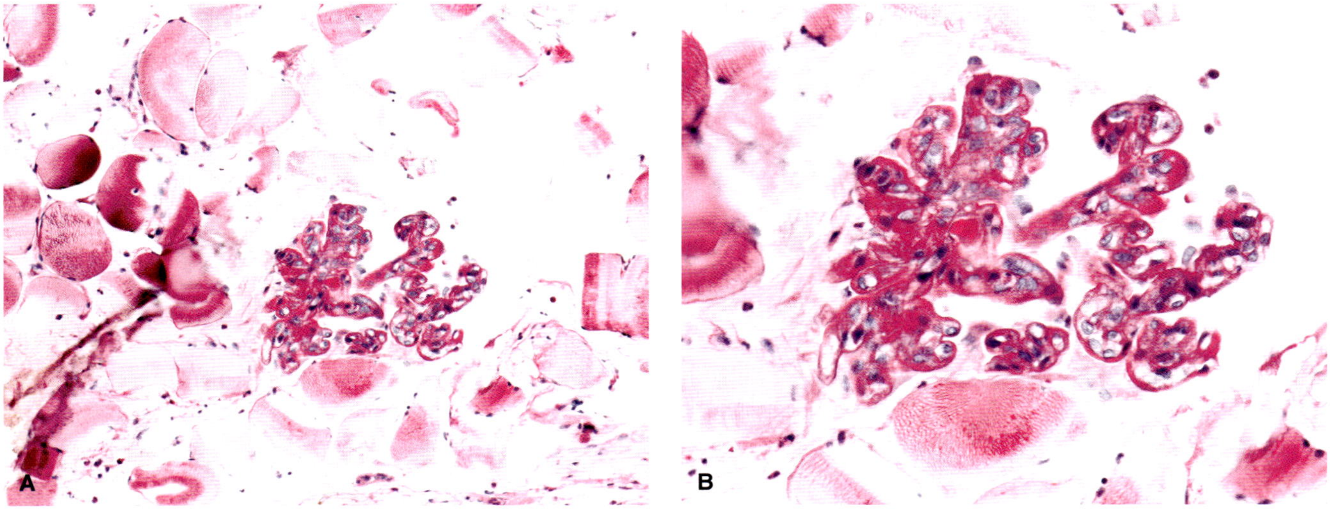

FIGURE 33.46 **A:** Isolated glomerulus, the only glomerulus present in a renal biopsy specimen, embedded within skeletal muscle (PAS, ×200). **B:** Higher magnification reveals glomerular hypercellularity and thickened capillary walls due to subendothelial deposits ("wire-loop" lesions) characteristic of lupus nephritis (PAS, ×400).

Mesangial Cells

The mesangium, composed of mesangial cells and their surrounding matrix, is observed as a periodic acid–Schiff (PAS) and methenamine silver-positive structural support for the glomerular capillary loops (Fig. 33.48). The mesangial cells usually can be distinguished by their location and dark-staining nuclei. Ultrastructurally, they are irregular in shape and have elongated thin cytoplasmic processes that may extend a short distance between the endothelium and the glomerular basement membrane. The mesangial cell processes contain microfilaments that contain actin, myosin, and α-actinin (343,344). With smooth muscle contractile properties, the mesangial cell has been proposed to be a specialized pericyte that likely modulates glomerular filtration (345). Whereas the endothelium forms a continuous layer around the inner circumference of the glomerular capillary, the basement membrane, and the visceral epithelial cell layer do not completely encircle the capillary but enclose the mesangial matrix and cells between the capillaries (Fig. 33.49).

Mesangial cells are involved in the generation of the mesangial matrix, which is similar to the glomerular basement membrane but contains several types of collagens (e.g., collagen type IV α1 and α2 chains, collagen V), various laminin isoforms, as well as fibronectin and proteoglycans (perlecan). The matrix is especially rich in microfibrils, unbranched, noncollagenous structures that contain fibrillins and microfibril-associated glycoproteins (346,347). The presence of microfibril-mediated attachments between mesangial cell processes and the glomerular basement membrane suggests that the mesangial cell and the glomerular basement membrane represent a biomechanical

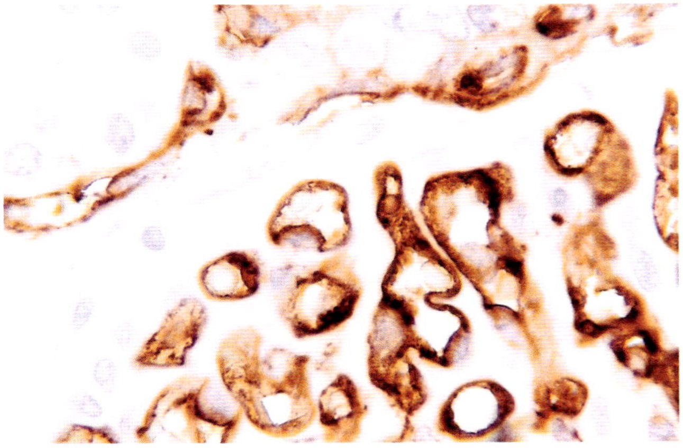

FIGURE 33.47 Glomerular endothelium showing immunoreactivity for CD34. Some peritubular capillaries show staining along the upper border. (CD34 immunohistochemistry, ×630.)

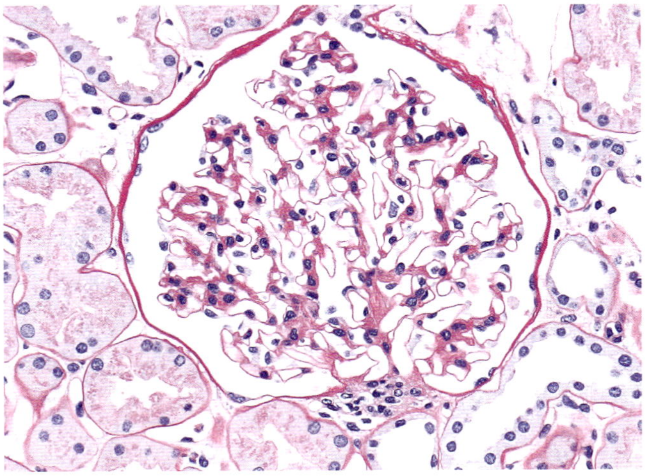

FIGURE 33.48 PAS-stained normal glomerulus illustrating the PAS-positive mesangium within the central regions of the glomerular capillaries (×400).

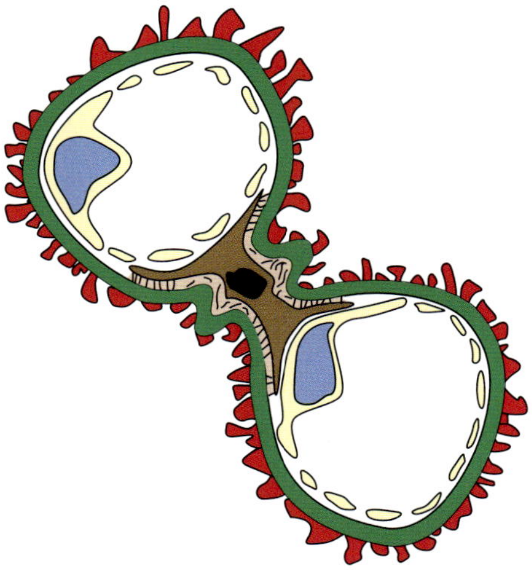

FIGURE 33.49 Schematic diagram illustrating the relationship between the mesangium and the glomerular capillaries. The visceral epithelial cell (podocyte) cytoplasm (*red*) and endothelial cell cytoplasm (*yellow*) are depicted. Note that the glomerular basement membrane (*green*) encloses the mesangium and its attached capillaries. The central mesangial cell is represented by *dark brown* cytoplasm and a *black* nucleus, and the mesangial matrix is represented by the *light brown* fibrillar texture. The cytoplasmic processes of mesangial cells are connected to the glomerular basement membrane directly or indirectly by microfibrils in the mesangial matrix.

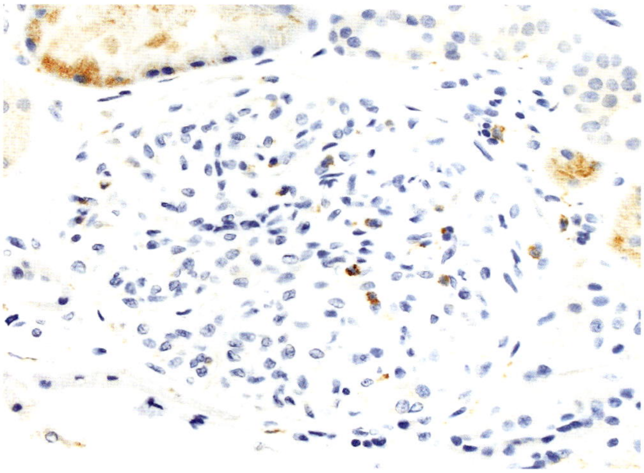

FIGURE 33.50 KP-1 immunohistochemical stain illustrating that a small percentage of cells in the mesangium label as tissue monocytes (KP-1 immunohistochemistry, ×600).

functional unit (348–350). It has been proposed that the contractile apparatus of the mesangium maintains the structure of the capillary walls by counteracting the distention caused by the intracapillary hydraulic pressure (351). Several molecules mediate interactions between mesangial cells and the surrounding matrix, other glomerular cells, and the glomerular basement membrane (352). Afadin, a F-actin–binding protein, localizes to cell contacts between mesangial cells and endothelial cells (353). Integrin $\alpha3\beta1$ and Lu/BCAM are mesangial receptors that mediate adhesion of mesangial cells to the laminin $\alpha5$ chain in the glomerular basement membrane (354). Integrin $\alpha8\beta1$ produced by mesangial cells binds to its ligand nephronectin in the glomerular basement membrane and this interaction is believed to regulate mesangial cell adhesion (355). The fibronectin receptor integrin $\alpha4\beta1$ is also expressed in mesangial cells (356).

The mesangial cell also has phagocytic capability and plays a role in the clearance of macromolecules and debris from the mesangium (357). Mesangial cells can respond to, as well as generate, a variety of molecules, including interleukin I, PDGF, and arachidonic acid metabolites, which may play a central role in the response to glomerular injury (358). PDGF-β, the main ligand for the receptor PDGFR, is a mitogen for mesangial cell proliferation and genetic deletion of PDGF-β and PDGFR leads to an absence of mesangial cells and the mesangium (359). It is recognized that a small subpopulation of cells in the mesangium are bone marrow–derived macrophages and play a role in immune responsiveness (Fig. 33.50) (360–362). Occasional leukocytes are observed in the normal glomerulus and labeled with leukocyte common antigen (CD45 and CD45RB) IHC.

The mesangium is continuous with the extraglomerular mesangium, a component of the juxtaglomerular apparatus, along the glomerular stalk. Interestingly, it has been demonstrated that renin lineage cells residing in the extraglomerular mesangial region may migrate and repopulate the intraglomerular mesangium upon glomerular injury (362,363).

Glomerular Basement Membrane

The glomerular basement membrane can be demonstrated on light microscopy by PAS and Jones silver stains (Fig. 33.51). The silver preparation is more specific for membranes and gives a distinct outline of the glomerular basement membrane. Examination of a normal peripheral capillary loop away from the vascular pole and the mesangium shows a delicate basement membrane (Fig. 33.52). Hematoxylin and eosin (H&E) and PAS preparations may stain capillary luminal contents and the cytoplasm of the endothelial and podocyte cell layers, resulting in an apparent thickening of the glomerular basement membrane. The basement membrane is situated between the endothelium and the podocytes in the glomerular capillary wall. On ultrastructural examination, the glomerular basement membrane completely surrounds the endothelium of the peripheral capillary loops and the mesangium between the loops. The glomerular basement membrane consists of a central dense layer, the lamina densa, and two surrounding thinner electron-lucent layers, the lamina rara interna and the lamina rara externa. In comparison with laboratory animals, the electron-lucent layers appear less prominent in the human glomerulus (Fig. 33.53).

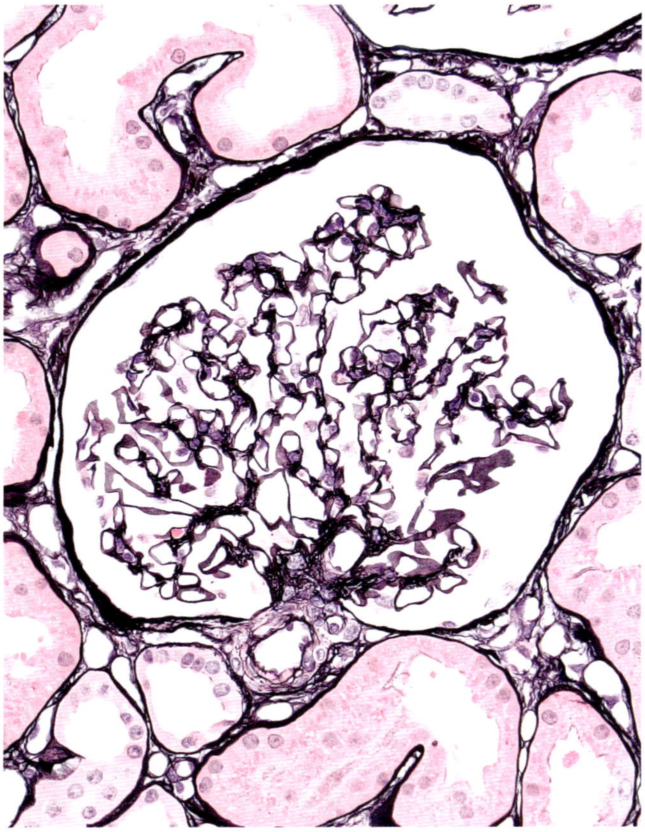

FIGURE 33.51 Jones silver-stained normal glomerulus illustrating the silver-positive glomerular basement membrane and positive basement membrane of Bowman capsule (×400).

The adult glomerular basement membrane ranges between 310 and 380 nm in mean thickness (364–366). It is significantly thicker in men (mean: 373 nm) than in women (mean: 326 nm) and increases in width until the fourth decade of life (262). Proteomic analysis has identified at least 212 proteins in the normal human glomerular extracellular matrix (367,368). However, the major components of the glomerular basement membrane include type IV collagen, laminin, nidogen (entactin), and heparan sulfate proteoglycans (369); collagen IV is the major constituent of the glomerular basement membrane. Six chains, α1(IV) to α6(IV), make up the collagen IV protein family (370). Three chains of collagen IV self-associate to form triple-helical molecules called protomers. Despite many possible combinations, the six chains of collagen IV form only three types of protomers, which are designated as α1.α2.α1(IV), α3.α4.α5(IV), and α5.α6.α5(IV). The triple helical protomers unite at the noncollagenous domain (NC1) at the carboxy terminus forming hexamers, which in turn, assemble to form a polymerized network that serves as a scaffold for integration of other glomerular basement membrane components. Three canonical sets of hexamers form three distinct networks in basement membranes: the α1.α2.α1 to α1.α2.α1 (IV), the α3.α4.α5(IV) to α3.α4.α5(IV), and the α1.α2.α1 to α5.α6.α5 (IV) networks (371). The α3.α4.α5(IV) to α3.α4.α5(IV) network predominates in the adult glomerular basement membrane and mutations of the genes encoding the α3, α4, and α5(IV) chains cause Alport syndrome (372). In Goodpasture syndrome, autoantibodies are targeted to the α3(IV) chain (372). The α1.α2.α1 to α5.α6.α5 (IV) network is found in Bowman capsule. The α1.α2.α1(IV) protomers are synthesized by both endothelial cells and podocytes but the α5.α6.α5 (IV) protomers are secreted only by podocytes (202).

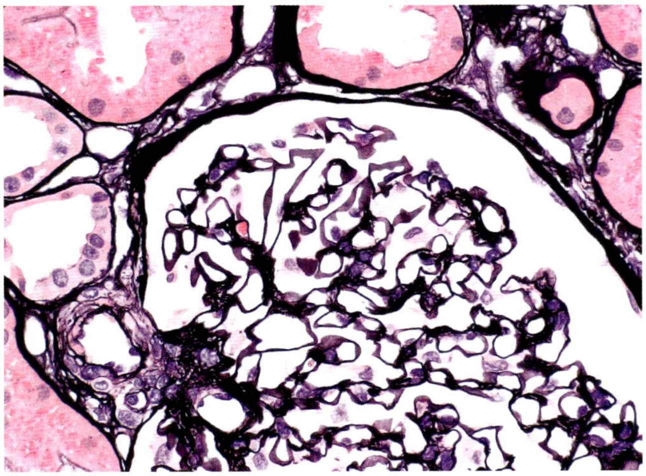

FIGURE 33.52 Higher magnification micrograph demonstrating thin regular glomerular basement membranes of peripheral capillary loops. (Jones silver stain, ×600.)

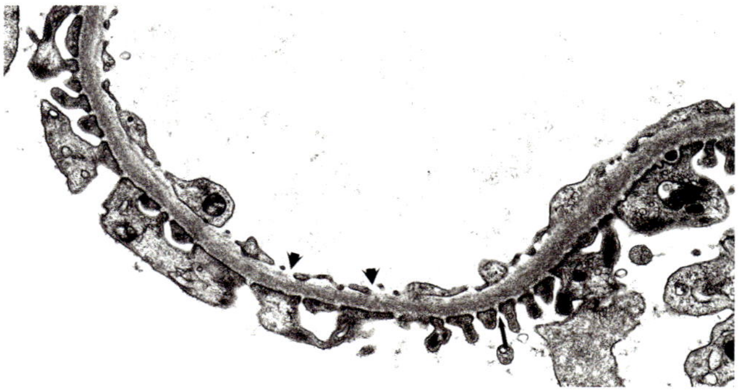

FIGURE 33.53 Transmission electron micrograph of a glomerulus. Bowman space is below and the capillary lumen is above the glomerular capillary wall. A fenestra or pore (*arrowhead*) in the endothelium is evident. Note the regular alignment of the foot processes of the podocytes. Filtration slit diaphragms (*arrows*) are present between individual foot processes. The glomerular basement membrane consists primarily of the lamina densa. The electron-lucent lamina rara interna and lamina rara externa are not prominent (×12,000).

Laminins are heterotrimers composed of three chains: α, β, and γ. Laminin-521, containing the α5, β2, and γ1 chains, is the major laminin isoform in the adult glomerular basement membrane (369). Laminin α5 and β2 are synthesized by both glomerular endothelial cells and podocytes (373). Mutations of laminin β2 result in a congenital nephrotic syndrome called Pierson syndrome in humans (374).

Nidogen is a glycoprotein that binds to both collagen IV and laminin in the glomerular basement membrane (375) but does not appear essential for glomerular basement membrane formation (376). Proteoglycans consist of glycosaminoglycan chains bound to a core protein (377). The glycosaminoglycan chains contain either heparan sulfate or chondroitin sulfate. Heparan sulfate proteoglycans include agrin, perlecan, and collagen XVIII. They are primarily responsible for the negative charge of the glomerular basement membrane (378,379). Agrin is the major heparan sulfate proteoglycan found in the glomerular basement membrane, whereas perlecan and collagen XVIII are present mainly in the mesangium matrix (380–382). Advanced microscopic methods, including stochastic optical reconstruction microscopy, have revealed a nanoscale organization of molecules within the glomerular basement membrane (383). The $\alpha 3\alpha 4\alpha 5$(IV) network is in the center of the glomerular basement membrane, whereas the $\alpha 1\alpha 2\alpha 1$(IV) network localizes near the endothelial side of the glomerular basement membrane. Laminin-521 maps to the central portion of the glomerular basement membrane but also in two layers near the endothelial and podocyte sides of the glomerular basement membrane. Agrin is situated in two layers along the endothelial and podocyte surfaces of the glomerular basement membrane with more detected near the podocytes.

Podocytes

The podocytes are the largest cells in the glomerulus. By light microscopy, they are positioned on the outside of the glomerular capillary wall, often bulge into the urinary space, and have prominent nuclei and abundant light eosinophilic cytoplasm. Scanning electron microscopy shows that the podocytes have a prominent cell body containing the nucleus and organelles and long cytoplasmic ramifications, the primary processes that surround the glomerular capillaries and divide into individual foot processes (Fig. 33.54) (384). The foot processes cover the capillary wall, contact the lamina rara externa of the glomerular basement membrane, and interdigitate with foot processes from different podocytes. Advanced microscopy methods, including serial block-face scanning electron microscopy and focused ion beam scanning electron microscopy, have revealed that some foot processes arise directly from the podocyte cell body as well as from the elongated cytoplasmic processes (384,385). These studies also demonstrated tortuous ridge-like prominences along the basal surface of the cell body and the processes, from which the proximal portions of the foot processes emerge.

Three-dimensional electron microscopy reconstructions have revealed additional complex architectural features of the podocyte layer. Neal et al. also demonstrated foot processes (termed anchoring processes) directly between the podocyte cell body and the glomerular basement membrane (386). Moreover, their study characterized

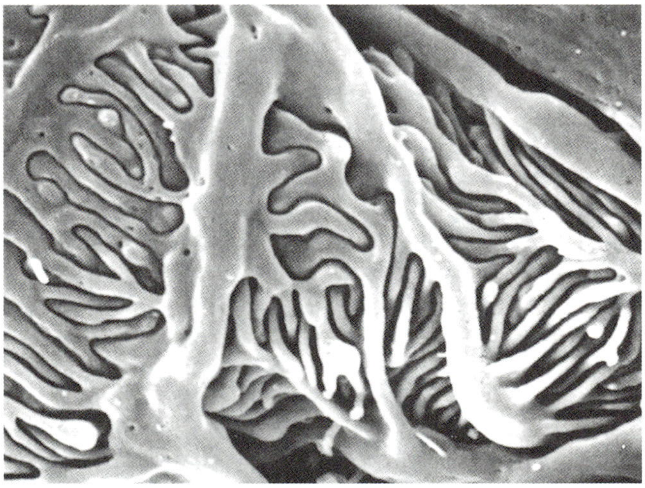

FIGURE 33.54 Scanning electron micrograph of glomerulus illustrating the primary processes of podocytes wrapping around the capillary loops. Note the interdigitation of the foot processes (×13,000; courtesy of Dr. Jill W. Verlander).

three distinct compartments of Bowman space. The subpodocyte space, first described by Gautier in 1950 (387), exists as a restricted area under the podocyte cell body and is associated with a reported 60% of the glomerular filtration surface (386). The subpodocyte space has been reported to connect via narrow passages, called subpodocyte exit pores (SEPs), with the main Bowman space. The subpodocyte space is a higher-resistance pathway than the parallel pathway across the barrier surface not covered by the subpodocyte space (388). In addition, this resistance across the subpodocyte space may be regulated by alterations of the SEPs (389). The interpodocyte space was characterized as a narrow anastomosing region interconnecting the subpodocyte space with the main peripheral Bowman space (386). Serial block-face scanning electron microscopy studies suggest that the subpodocyte space and interpodocyte space are not free-flowing urinary spaces (390). The physiologic role of these podocyte partitions of Bowman space remains to be clarified.

By transmission electron microscopy, the cells have abundant rough endoplasmic reticulum, a well-developed Golgi apparatus, and prominent lysosomes. Podocytes have an elaborate cytoskeleton that underlies their stability, shape, adhesion, and response to stress (391). There are numerous intermediate filaments, microtubules, and microfilaments in the cytoplasm (333). The intermediate filaments and microtubules predominate in the cell body and primary processes whereas, the foot processes contain a dense microfilament contractile apparatus (392–394). The latter, containing actin, myosin, α-actinin, talin, and vinculin, connects to the intermediate filaments and microtubules of the primary processes (395). Electron microscopic studies have demonstrated at least two distinct actin filament networks in podocyte foot processes

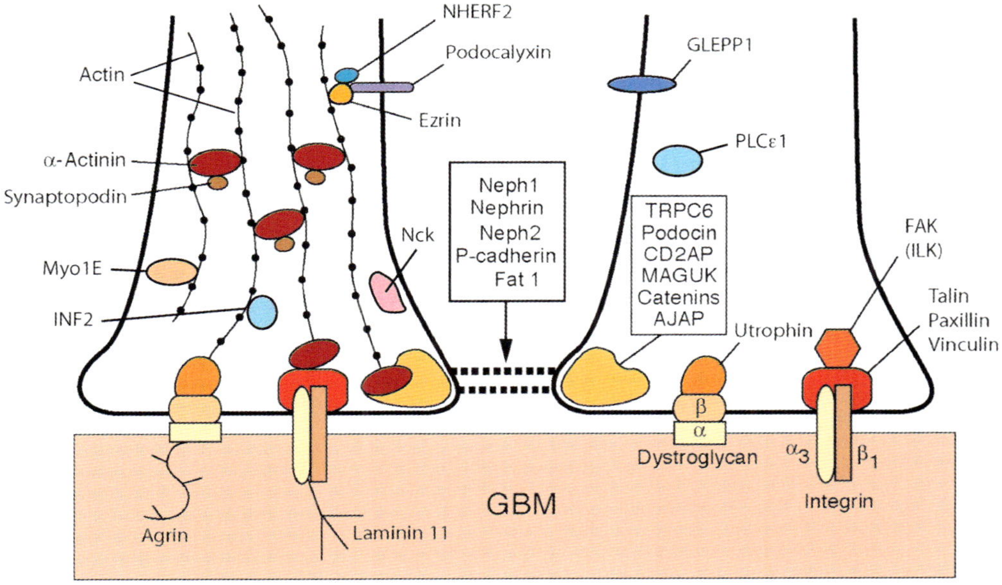

FIGURE 33.55 Schematic drawing of the membrane domains of the podocyte: The filtration slit diaphragm, basal membrane, and apical membrane domains. Molecular interactions within these domains of the podocyte foot processes are represented. See the text for further explanations. (Modified from Pavenstadt H, Kriz W, Kretzler M. Cell biology of the glomerular podocyte. *Physiol Rev.* 2003;83:253–307; Kerjaschki D. Caught flat-footed: podocyte damage and the molecular bases of focal glomerulosclerosis. *J Clin Invest.* 2001;108:1583–1587.)

(396). Bundles of actin filaments containing α-actinin and synaptopodin extend along the longitudinal axis of the foot processes. A second actin network containing cortactin is situated between the longitudinal network and the plasma membrane. In glomerular diseases with proteinuria, the podocyte cytoskeleton is disrupted leading to foot process effacement. It is known that most cells contain both contractile and noncontractile actin filaments. The former typically contain both α-actinin and myosin (397). Whereas the actin filaments in the center of the foot process contain α-actinin and synaptopodin, the actin filaments in the cell body and primary processes contain myosin IIA but lack synaptopodin (398). These results suggest the actin filaments in the foot process are noncontractile, whereas the actin filaments in the cell body and primary processes are contractile. This ultrastructural arrangement of contractile proteins may allow the podocytes to play an active role in modifying the glomerular filtration surface area.

Using serial block-face scanning electron microscopy, three-dimensional reconstruction studies have confirmed the complex geometry of podocytes and their foot processes (399). Podocyte cell body volumes range from 30% to 50% of the total cell volume and major or primary cell processes 30% to 50% of the cell volume. Although the foot processes account for only 20% of the cell volume, they constitute nearly 60% of the surface area. Complex models of podocyte actin cytoskeletal dynamics have been constructed to further investigate podocyte function and response to injury.

Two structures in the foot processes, focal adhesions, and filtration slit diaphragms, interact with and control the actin cytoskeleton (400). Focal adhesions anchor the foot processes to the glomerular basement membrane. They consist of transmembrane protein complexes containing integrins and their interacting partners which link to the actin cytoskeleton (401). For example, α3β1 integrin interconnects laminin in the glomerular basement membrane with the talin, paxillin, vinculin adaptor cytoplasmic complex, which connects to the actin cytoskeleton. Mutations of the integrin α3 subunit in humans are associated with severe proteinuria (402). GTPases, several of which regulate maturation of focal adhesions, control actin dynamics. These include the small GTPases, including RhoA, Rac1, and Cdc42 (403), and the large GTPase protein dynamin (404).

The adjacent foot processes near the glomerular basement membrane are separated by a 30- to 40-nm space termed the filtration pore or slit, which is bridged by a thin extracellular structure called the filtration slit diaphragm. In addition to functioning as a critical barrier to filtration, the slit diaphragm regulates actin cytoskeletal dynamics in the foot processes (405). On ultrastructural examination, the slit diaphragm has always attracted the attention of pathologists. The slit diaphragm consists of a network of winding strands, about 30 to 35 nm long, which merge centrally into a longitudinal density. The strands, creating a slit diaphragm thickness between 5 and 10 nm, surround pores the same size or smaller than albumin molecules. The pores are variable in shape without a central filament (406). High-resolution helium ion scanning electron microscopy studies have shown the slit diaphragm to consist of cross-bridging filaments surrounding pores, thereby forming a ladder-like structure in the middle of the filtration slit (407,408). Bridging shorter strands in the lower part of the slit diaphragm closest to the glomerular basement membrane consist of Neph1 whereas longer strands in the top toward the apical side contain nephrin. The slit diaphragm shares similarities with tight and adherens junctions (409,410). It separates the different membrane surfaces of the podocyte. Podocytes, similar to other epithelial cells, are polarized with distinct membrane domains (Fig. 33.55) (411). The basal

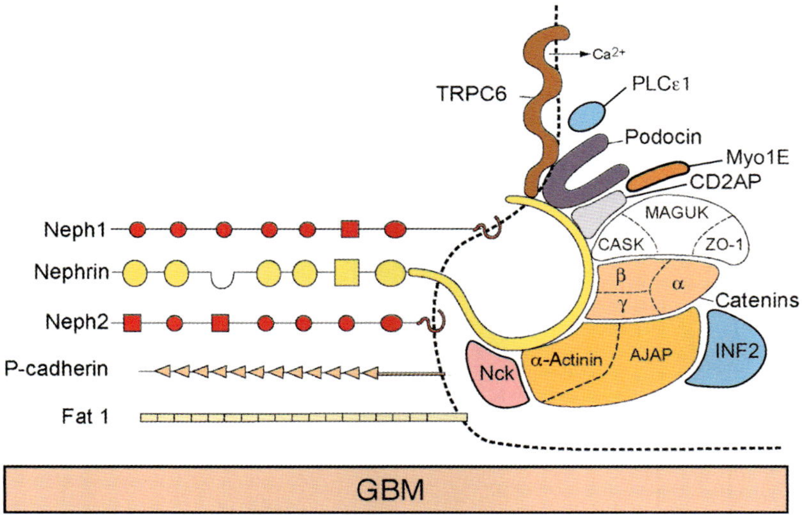

FIGURE 33.56 Schematic drawing of a working model of the podocyte filtration slit diaphragm domain. The **left** side portrays the molecular composition of the slit diaphragm itself, which spans between adjacent foot processes. The **right** side illustrates the molecular interactions of the nephrin-associated multiprotein complex within the podocyte foot process membrane and cytoplasm. See the text for further explanations. (Modified from Pavenstadt H, Kriz W, Kretzler M. Cell biology of the glomerular podocyte. *Physiol Rev.* 2003;83:253–307; Kerjaschki D. Caught flat-footed: podocyte damage and the molecular bases of focal glomerulosclerosis. *J Clin Invest.* 2001;108:1583–1587; Huber TB, Benzing T. The slit diaphragm: a signaling platform to regulate podocyte function. *Curr Opin Nephrol Hypertens.* 2005;14:211–216; and Lehtonen S, Ryan JJ, Kudlicka K, et al. Cell junction-associated proteins IQGAP1, MAGI-2, CASK, spectrins, and alpha-actinin are components of the nephrin multiprotein complex. *Proc Natl Acad Sci USA.* 2005;102:9814–9819. Copyright (2005) National Academy of Sciences, USA.)

membrane domain and the apical membrane domain are located below and above the slit diaphragm, respectively. The slit diaphragm area, including the bridging diaphragm as well as the adjacent podocyte foot process membrane and cytoplasm, may also be considered a very specialized surface domain.

A major advance in our understanding of the podocyte was the identification of the protein nephrin, encoded by *NPHS1*, the gene mutated in congenital nephrotic syndrome of the Finnish type (199,200). Nephrin, a transmembrane adhesion protein of the immunoglobulin superfamily, localizes to strands of the slit diaphragm (411). Lack of nephrin leads to the loss of the slit diaphragm, foot process effacement, and massive proteinuria (200,412). An increasing number of other proteins localize to the slit diaphragm domain, where they interact with nephrin and other partners, forming a multifunctional complex (Fig. 33.56) (199,395,413). Some proteins are present in the actual slit diaphragm itself (414).

Other proteins are present in the podocyte cytoplasm or plasma membrane adjacent to the slit diaphragm. For example, members of the membrane-associated guanylate kinase family of scaffolding proteins, including MAGI-1, MAGI-2, CASK, and ZO-1 localize adjacent to the slit diaphragm (415,416). These scaffolding proteins connect junctional membrane proteins to the actin cytoskeleton and signaling cascades. In addition, multiple adherens junction–associated proteins, including α-actinin, IQGAP1, αII spectrin, and βII spectrin are also components of this expanding nephrin-associated multiprotein complex (415,416).

Mutations or deficiencies of genes encoding many of the proteins that comprise the slit diaphragm domain complex, including nephrin (200), Neph1 (417), Fat1 (418), podocin (419), CD2AP (420), α-actinin (421,422), TRPC6 (423,424), PLCE1 (425), INF2 (426), and MYO1E (427) (Fig. 33.51) cause glomerular diseases characterized by absent or defective slit diaphragms, foot process effacement and proteinuria. Nephrin plays a central role in signaling. One crucial pathway of nephrin influencing the actin cytoskeleton involves the Nck adaptor proteins (428,429). Phosphorylation of tyrosine residues on the nephrin cytoplasmic domain results in the recruitment of the Nck adaptor proteins Nck1 and/or Nck2 to interact with nephrin. After this interaction, Nck binds and activates the neuronal Wiskott–Aldrich syndrome protein, an actin nucleation protein. This protein binds and activates the Arp2/3 multiprotein complex, which induces actin polymerization (430) and stabilizes foot processes (431–433). Moreover, nephrin plays a role in its endocytic trafficking within the podocyte and is important for turnover and maintenance of the slit diaphragm (434).

The basal membrane domain, the "sole" of the podocyte foot process, is embedded in the underlying glomerular basement membrane (Fig. 33.55). Two types of surface receptors in the basal membrane, integrins and dystroglycan, anchor the foot processes by binding to their ligands in the glomerular basement membrane (356,435). The α3β1 integrin binds to collagen type IV, laminin, and nidogen/entactin whereas, dystroglycan binds to laminin, agrin, and perlecan (395). Both integrins and dystroglycan are coupled by adapter molecules to the actin cytoskeleton. The integrins bind the talin, paxillin, and vinculin complex (393), and dystroglycan binds utrophin (435). The induction of cellular responses from integrin–ligand interactions, known as "outside-in" signaling, is believed to be mediated by focal adhesion kinase (FAK) and ILK (395,436). Similar to the filtration slit domain, an intact basal membrane domain is required to maintain foot process integrity.

The apical membrane domain above the slit diaphragm has a prominent glycocalyx surface of negatively charged glycoproteins (Fig. 33.55) (437). These include podocalyxin (437) and GLEPP1 (438). Podocalyxin interacts with a complex composed of ezrin and NHERF2 (Na^+/H^+ exchanger-regulatory factor 2), which in turn, associates

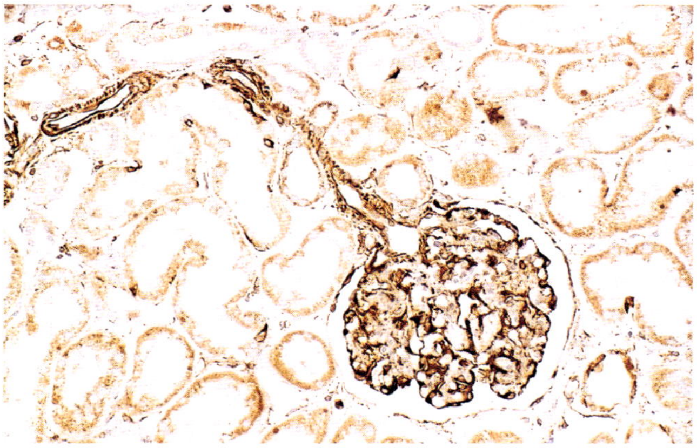

FIGURE 33.57 Vascular structures including the glomerulus are heavily labeled with antibody to vimentin. Tubular and interstitial components show weaker and sparse labeling. (Vimentin immunohistochemistry, ×100.)

with the actin cytoskeleton (439,440). Genetic evidence exists indicating podocalyxin is important for foot process stability (441). Proteins within podocyte organelles are also important for podocyte function and intact filtration.

Podocytes are positive for WT1, CD10, GLEPP1, podocalyxin, and nephrin by IHC. Mature podocytes are terminally differentiated cells and generally do not replicate (Figs. 33.57 and 33.58). They express the CDK inhibitors P27 and p57, which are involved in maintaining cell quiescence (442). The Wilms tumor suppressor gene, WT1, plays an indispensable role in the regulation of cell growth and differentiation during early nephrogenesis (157). Embryonic mice homologous for a targeted mutation of WT1 fail to develop kidneys (443). Striking evidence also exists for the importance of WT1 in glomerular podocyte differentiation. During kidney development, WT1 expression is detected in the metanephric mesenchyme and becomes stronger in the renal vesicle, but the highest levels occur during glomerulus formation within the podocyte cell layer

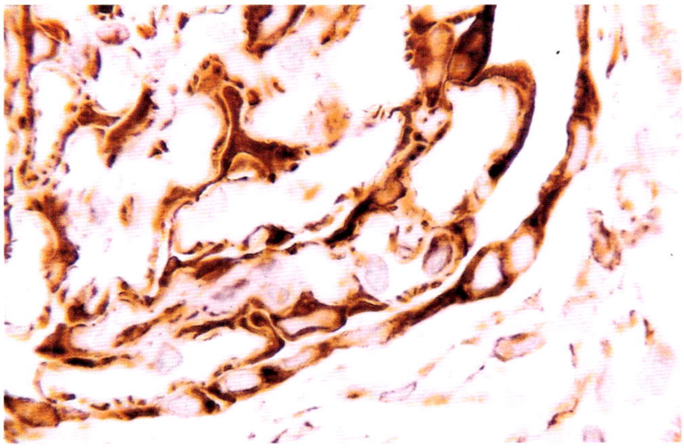

FIGURE 33.58 Prominent expression of vimentin is seen in glomerular endothelium, podocytes, and parietal epithelium (vimentin immunohistochemistry, ×630).

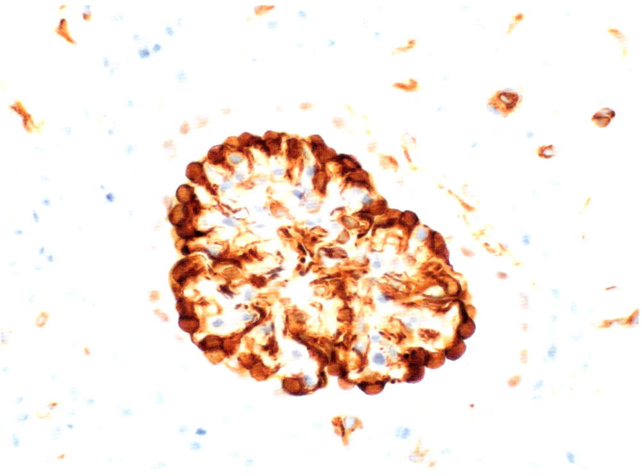

FIGURE 33.59 Light micrograph demonstrating expression of the WT1 protein in podocyte nuclei of a fetal glomerulus (WT1 immunohistochemistry, ×600).

(444,445). Expression of the WT1 protein, a transcription factor, in the nuclei of podocytes does not disappear with glomerular maturation but persists in the adult kidney (Fig. 33.59). Greater than 95% of patients with the Denys–Drash syndrome (nephrotic syndrome and genital anomalies and/or Wilms tumor), characterized by shrunken glomeruli with hypertrophied podocytes, have point mutations affecting the zinc finger DNA-binding domain of WT1 (446). These findings provide a functional link between a molecular defect of WT1 and podocyte pathology and indicate that WT1 has a major role in maintenance of podocyte structure and function in the mature kidney. As a master regulator, WT1 regulates the expression of other podocyte genes that are required for intact glomerular filtration (447,448). Studies using a multiomic integrative approach, including whole exome sequencing and mass spectrometry-based proteomics are identifying podocyte genes which will enhance our understanding of the role of podocytes in health and disease (449,450).

Glomerular Filtration Barrier

The glomerular capillary wall functions as a filter that is selective for the size and charge of molecules (451–453). To pass through the capillary wall, a molecule must journey along an extracellular pathway through the fenestrated endothelium, the glomerular basement membrane, and the filtration slit diaphragm.

The permeability of the glomerular capillary wall to water and small molecules is very high, whereas its permeability to molecules the size of albumin and larger is very low. Studies employing mathematical modeling have suggested that the glomerular basement membrane and filtration slits contribute equally to the total resistance to water filtration, although some data suggest the slit diaphragm is the most restrictive part of the barrier to the filtration of macromolecules (454). Mutations or deficiencies of genes encoding proteins of the filtration slit diaphragm domain result in massive proteinuria, providing genetic evidence for the crucial role of the slit diaphragm in glomerular permselectivity (413).

The glomerular basement membrane has been favored as the principal structure responsible for the charge-selective permeability of the glomerular capillary wall (451,452). This charge selectivity in the glomerular basement membrane has been associated with the presence of polyanionic molecules, such as heparan sulfate proteoglycans.

Agrin is the major heparan sulfate proteoglycan in the glomerular basement membrane. However, there is evidence that an intact glomerular basement membrane does serve as a major barrier to protein permeability. The "integrated view" of the glomerular filtration barrier is that the endothelium, the glomerular basement membrane, and the podocytes and their slit diaphragms do not act independently but are linked to one another in a functional unit.

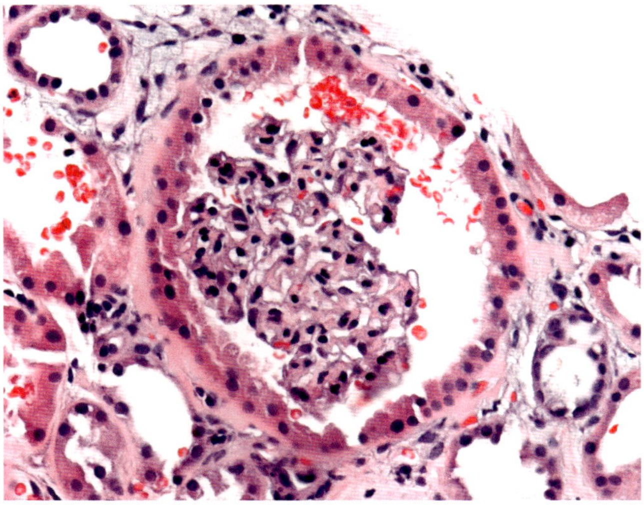

FIGURE 33.61 Glomerulus with circumferential tubular metaplasia of Bowman capsule (H&E, ×400).

Parietal Epithelial Cells

The parietal layer of Bowman capsule consists of relatively flat squamous epithelial cells with proliferative potential (455). Keratins, cadherins, the transcription factor Pax2, and claudin-1 are expressed in parietal epithelial cells (Fig. 33.60) (456–460). The cells are 0.1 to 0.3 μm in height but may increase to 2 to 3.5 μm at the nucleus. The epithelium rests on the basement membrane of Bowman capsule, which has a range from 1,200 to 1,500 nm in thickness and may have a lamellated appearance. In contrast to the glomerular basement membrane, the basement membrane of Bowman capsule expresses the α6 chain of type IV collagen, which is part of the α1.α2.α1(IV) to α5.α6.α5(IV) network (460). The parietal epithelial cell layer functions as a physical barrier for the glomerular ultrafiltrate. When this layer is experimentally compromised, macromolecules can leak into the periglomerular interstitial space (461). On occasion, the flat parietal epithelial cell layer may be replaced with proximal tubular epithelium, especially following acute tubular injury (Fig. 33.61).

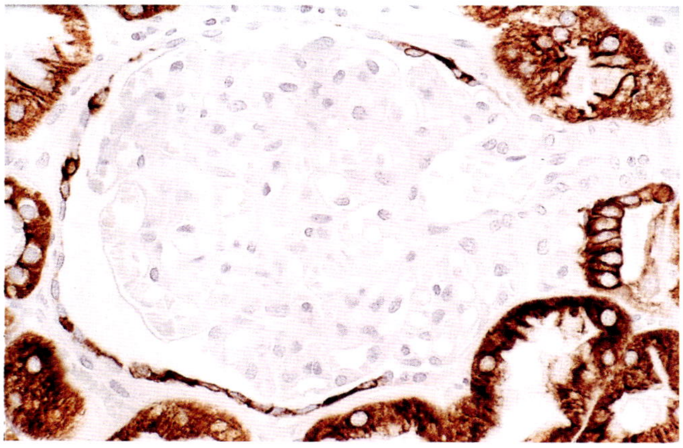

FIGURE 33.60 Keratin expression is not observed in the glomerular tuft, but immunoreactivity is seen in the parietal epithelial cells. The vascular pole is on the **right**. The macula densa (**far right center**) and other tubule segments are immunoreactive for keratin (CAM 5.2 immunohistochemistry, ×400).

Different populations of parietal epithelial cells appear to exist (462). Peripolar cells are situated at the vascular pole between parietal epithelial cells and podocytes and contain prominent cytoplasmic granules (463). They have an immunophenotype intermediate between parietal epithelial cells and podocytes. Other cells lining Bowman capsule near the vascular pole that express podocyte markers have been called parietal podocytes or ectopic podocytes (464,465). Parietal epithelial cells can become activated in some glomerular diseases, in which case they express CD44 and show increased proliferation, migration, and matrix deposition (466,467).

Parietal epithelial cells may serve as progenitor cells to renew podocytes (468). They can transdifferentiate into podocytes and repopulate the glomerular tuft (469–472) or migrate onto Bowman capsule and express parietal epithelial cell markers (473–475). The functional relevance of these apparent bidirectional movements of podocytes and parietal epithelial cells remains to be determined.

JUXTAGLOMERULAR APPARATUS

The juxtaglomerular apparatus is situated at the vascular pole of the glomerulus and includes the afferent and efferent arterioles, extraglomerular mesangial region, and macula densa (Fig. 33.62). A prominent juxtaglomerular apparatus may occasionally be observed in a normal glomerulus and should not be mistaken for segmental

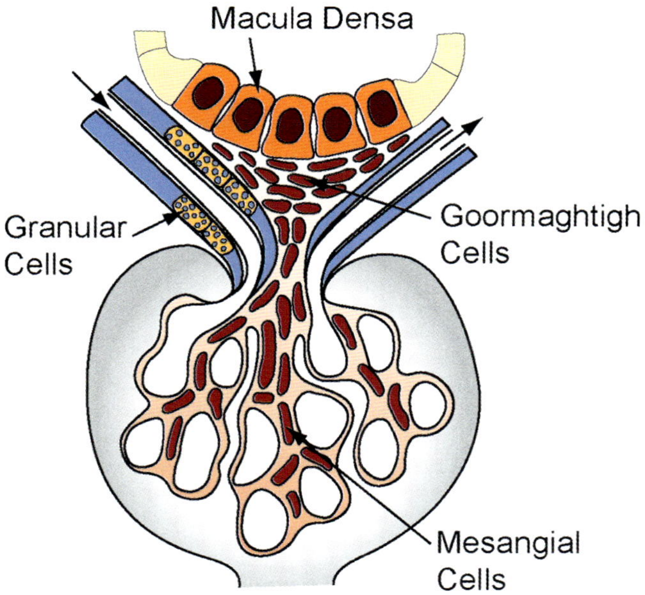

FIGURE 33.62 The basic components of the juxtaglomerular apparatus. (Modified with permission from Kriz W, Kaissling B. Structural organization of the mammalian kidney. In: Seldin DW, Giebisch D, eds. *The Kidney: Physiology and Pathophysiology.* 3rd ed. Lippincott Williams & Wilkins; 2000:587–654.)

glomerulosclerosis (Fig. 33.63). The juxtaglomerular apparatus is the major structural unit of the renin–angiotensin system. Although the general outline of this anatomical unit usually can be observed in light microscopic sections (Fig. 33.64), histochemical or immunocytochemical methods are usually required to demonstrate the distinctive juxtaglomerular granular cells. These cells tend to occur in clusters and are most abundant in the wall of the afferent arteriole but are also found in the wall of the

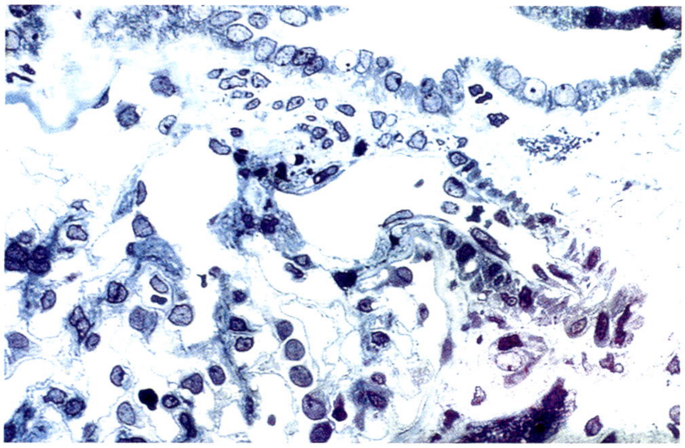

FIGURE 33.64 Light micrograph depicting the juxtaglomerular apparatus. From the top to the bottom of the micrograph are the macula densa, extraglomerular mesangium, afferent arteriole, and glomerulus (×750; courtesy of Dr. Luciano Barajas). (Reprinted with permission from Barajas L, Salido EC, Smolens P, et al. Pathology of the juxtaglomerular apparatus including Bartter's syndrome. In: Tisher CC, Brenner BM, eds. *Renal Pathology with Clinical and Functional Correlations.* 2nd ed. JB Lippincott; 1994:948–978.)

efferent arteriole and the extraglomerular mesangial region (476–478). Ultrastructural analysis shows the presence of myofilaments, attachment bodies, a well-developed endoplasmic reticulum and Golgi apparatus, and numerous membrane-bound granules (Fig. 33.65). The granules are variable in shape and size. It is believed that the smaller, often rhomboid-shaped granules with a crystalline substructure observed in the Golgi region represent mature granules. Renin and angiotensin II have been immunolocalized to the granules of these cells (479,480). Renin release occurs by exocytosis. It is believed to be modulated by adrenergic nerve activity (481,482).

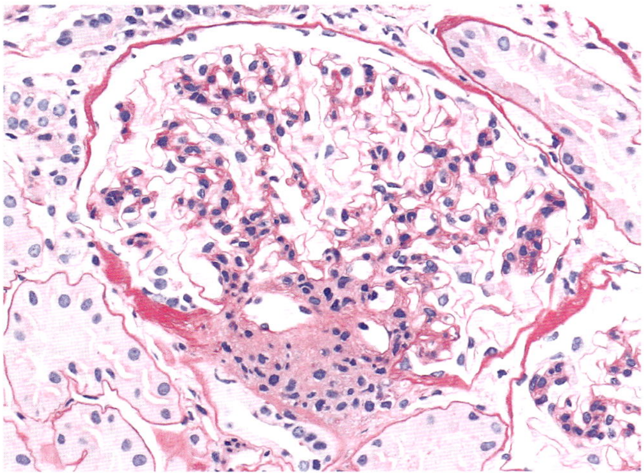

FIGURE 33.63 Glomerulus with prominent juxtaglomerular apparatus, in this case in a patient with Gitelman syndrome. Caution in interpretation is necessary for the evaluation of glomerular cellularity at the vascular pole because of the continuum of the extraglomerular and intraglomerular mesangial regions (PAS, ×400).

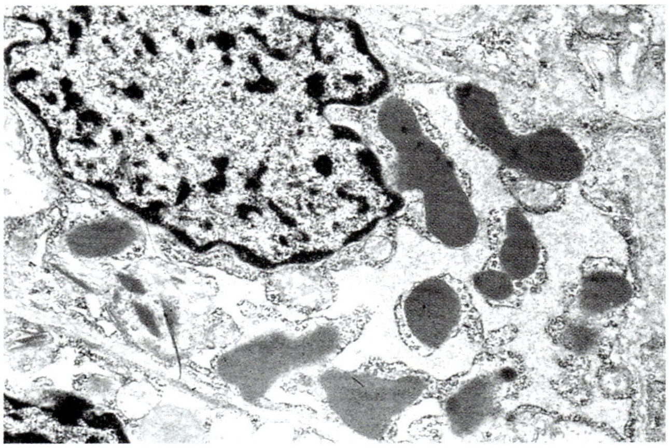

FIGURE 33.65 Electron micrograph of a juxtaglomerular granular cell. Note the prominent cytoplasmic membrane-bound granules (×19,000; courtesy of Dr. Luciano Barajas). (Reprinted with permission from Barajas L, Bloodworth JMB Jr, Hartroft PM. Endocrine pathology of the kidney. In: Bloodworth JMB Jr, ed. *Endocrine Pathology: General and Surgical.* 2nd ed. Williams & Wilkins; 1982:723–766.)

In addition to the classical role of renin cells in blood pressure regulation, recent studies have revealed that renin-expressing cells display considerable plasticity and have roles in regeneration and oxygen sensing (483). Several studies using genetic labeling techniques and experimental podocyte depletion have shown that cells of renin lineage serve as podocyte progenitors (484–488). Evidence for the endocrine plasticity of renin cells exists. Deletion of the von Hippel–Lindau (PVHL), which targets hypoxia-inducible factor-2 (HIF-2) for degradation, converts juxtaglomerular cells from a renin-expressing to erythropoietin-producing cells, likely in response to HIF-2 stabilization (489–491).

The extraglomerular mesangium, also called lacis or cells of Goormaghtigh, is located between the afferent and efferent arterioles and has extensive contact with the basal surface of the macula densa. This extraglomerular region is continuous with the intraglomerular mesangium and the Goormaghtigh cells are similar in ultrastructure to the mesangial cells. There are numerous gap junctions between the extraglomerular mesangial cells and the cells of the intraglomerular mesangium and glomerular arterioles (492,493). The gap junctions consist of connexin proteins, especially connexin 40 (494,495). These morphologic features and the central position within the juxtaglomerular apparatus suggest that the extraglomerular mesangium may represent the structural–functional link between the macula densa and the glomerular arterioles and mesangium.

The macula densa represents a plaque of specialized tubular cells within the cortical thick ascending limb of Henle adjacent to the hilum of the glomerulus. The cells are low columnar and their apically situated nuclei may protrude into the tubular lumen (Fig. 33.66). By electron microscopy, they contain cytoplasmic organelles largely lateral to and beneath the apical nuclei, and basal cellular processes that interdigitate with the extraglomerular mesangial cells. The lateral intercellular spaces between the macula densa cells vary in width but usually are more dilated compared with the lateral intercellular spaces of other nephron segments (496). In contrast with contiguous portions of the thick ascending limb, there is evidence that the macula densa lacks epidermal growth factor and Tamm–Horsfall protein but is water permeable (Fig. 33.67) (497–499). The anatomic arrangement of the juxtaglomerular apparatus is suited for functional regulation of the adjacent structures. The macula densa plays a role in tubuloglomerular feedback, a mechanism whereby luminal concentrations of sodium and/or chloride are sensed by the macula densa leading to the transfer of a signal to the glomerular arterioles to regulate the glomerular filtration rate (500,501). Studies support the following general sequence: increased sodium chloride concentration at the macula densa results in formation and release of ATP and adenosine; adenosine interacts with receptors on extraglomerular mesangial cells triggering an increase in cytosolic calcium; gap junctions transmit the calcium flux to the adjacent afferent arteriole resulting in vasoconstriction, inhibition of renin release, and decreased glomerular filtration. The neuronal isoform of nitric oxide synthase (nNOS) and the cyclooxygenase enzyme COX-2 immunolocalize to the macula densa (502–504). There is evidence that both nitric oxide and COX-2–generated prostaglandins play a role in the signaling between the macula densa and the renin-secreting cells in the afferent arteriole (505,506).

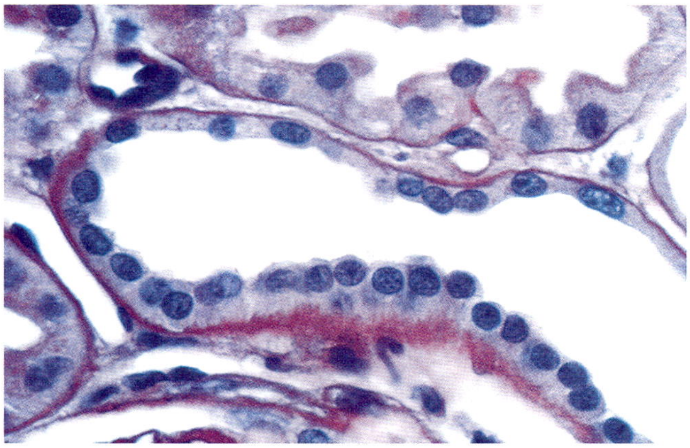

FIGURE 33.66 Macula densa characterized as a morphologically distinct plaque of low columnar cells with apically situated nuclei (PAS, ×750).

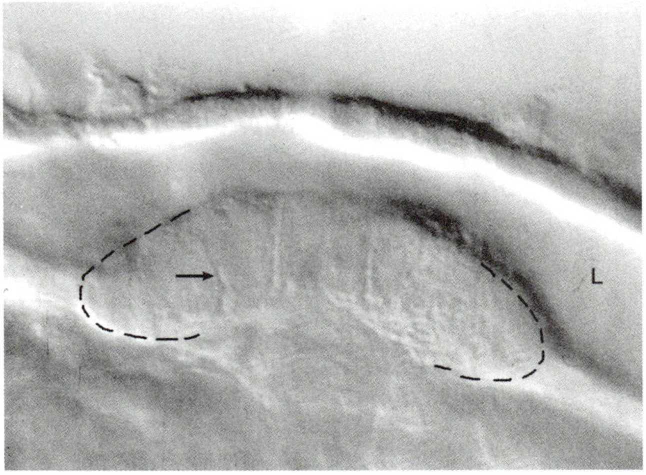

FIGURE 33.67 Differential interference contrast image of an isolated thick ascending limb segment perfused in vitro. The tubular lumen (*L*) and macula densa (*dashed lines* enclose the ends of the macula densa plaque) are observed. In response to a reduction in tubular luminal osmolality, there is dilatation of the lateral intercellular spaces (*arrow*) in the macula densa, suggesting increased water flow (×1,250). (Modified with permission from Kirk KL, Bell PD, Barfuss DW, Ribadeneira M. Direct visualization of the isolated and perfused macula densa. *Am J Physiol.* 1985;248(pt 2):F890–F894.)

PROXIMAL TUBULE

The proximal tubule is divided into an initial convoluted portion, the pars convoluta, and a straight portion, the pars recta. The convoluted portion forms several coils around its parent glomerulus in the cortex and continues into the straight portion, which is located in the medullary ray. The human proximal tubule is approximately 14 mm in length (507). In histologic sections of the cortex, sectioned profiles of proximal convoluted tubules represent the major parenchymal component. The appearance of the cortex and especially the proximal tubules varies, according to the method of fixation. A decrease in blood pressure results in decreased filtration and renal volume (508,509). After immersion fixation of excised pieces of renal tissue, the cortex has a more homogeneous compact appearance and there is collapse of the proximal tubular lumens (Fig. 33.68) (510). Free nuclei and vesicular membranous material may be observed in the proximal tubular lumens.

It is well known that renal parenchyma, especially the tubules, undergo significant postmortem autolysis and that these changes occur more rapidly than in other tissues, such as the liver, heart, and skeletal muscle (510,511). Within the kidney, proximal tubules typically show greater autolytic changes including loss of cellular adhesion and nuclei (Figs. 33.69 and 33.70). In autopsy specimens, it may be difficult to distinguish acute tubular injury from autolysis, and in a given case, both cellular lesions may be present. Most cases of ischemic-type acute tubular injury show a more patchy distribution of injured tubules which are often dilated with attenuated or sloughed epithelium, luminal casts, and occasional mitotic figures. Further studies are needed to confirm the useful of reported particular tubular epithelial lesions as a feature of acute tubular injury in postmortem kidneys (512) and the value of immunohistochemical expression of KIM-1, a proximal tubule marker upregulated in acute kidney injury (513,514).

Fixation of an experimental functioning kidney in situ by rapid freezing, dripping of fixative on the renal surface, or vascular perfusion results in more conspicuous intertubular interstitial spaces and widely open lumens of the proximal tubules. The cells of the proximal tubule are cuboidal to low columnar with eosinophilic, often granular

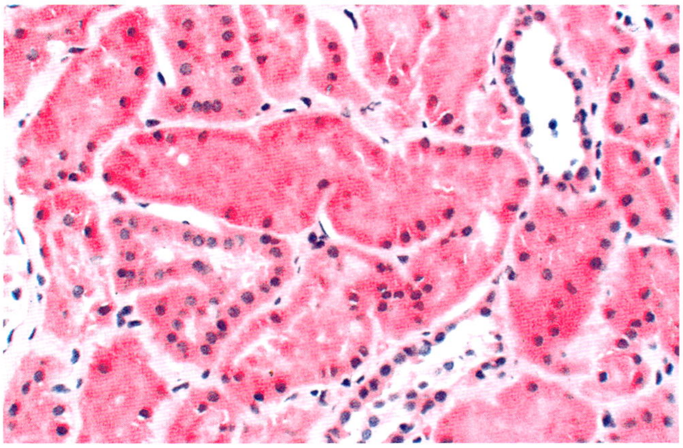

FIGURE 33.68 Immersion-fixed renal biopsy specimen demonstrating diffuse collapse of the proximal tubular lumens. Note the patent lumens of the distal nephron segments (H&E, ×250).

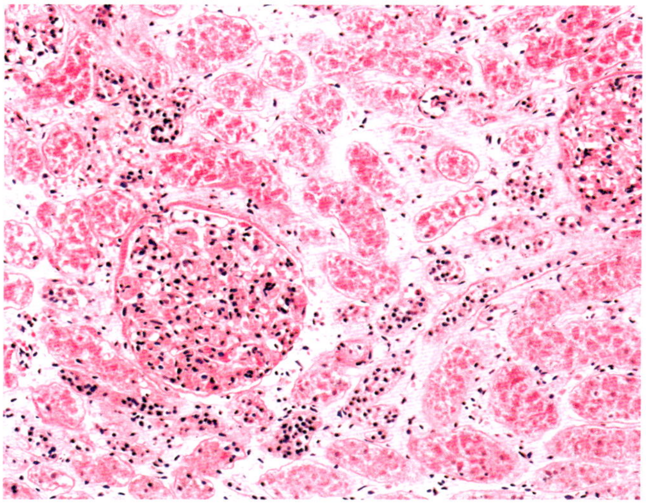

FIGURE 33.69 Autolysis showing loss of cellular features including nuclei, especially in proximal tubules. Cellular degeneration is present in the distal tubules and glomeruli but their nuclei are still apparent (H&E, ×200).

cytoplasm and round nuclei situated in the center or near the base of the cells (Fig. 33.71). *Ex vivo* perfusion of donor organs by providing oxygen and nutrients and more suitable storage time are transforming organ transplantation. Donor kidneys perfused *ex vivo* display more patent and somewhat more dilated lumens in glomeruli, tubules, and peritubular capillaries on histologic examination (Figs. 33.72 to 33.74).

The lateral cell borders of the proximal tubules are indistinct because of extensive interdigitations of lateral cellular processes from adjacent cells (Fig. 33.75). A complex intercellular space forms from these interdigitations. There are

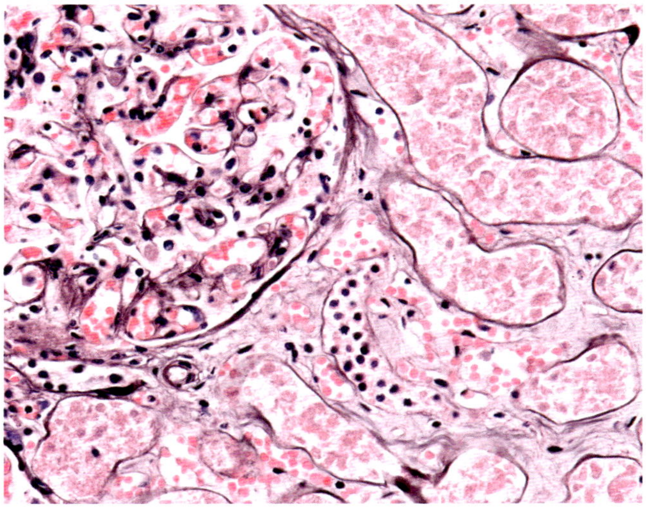

FIGURE 33.70 Silver stain illustrating autolysis. Proximal tubules are most affected. A few distal tubules are identified by their dark nuclei. The glomerular basement membranes, Bowman capsule and tubular basement membranes are still relatively intact (Jones silver stain, ×400).

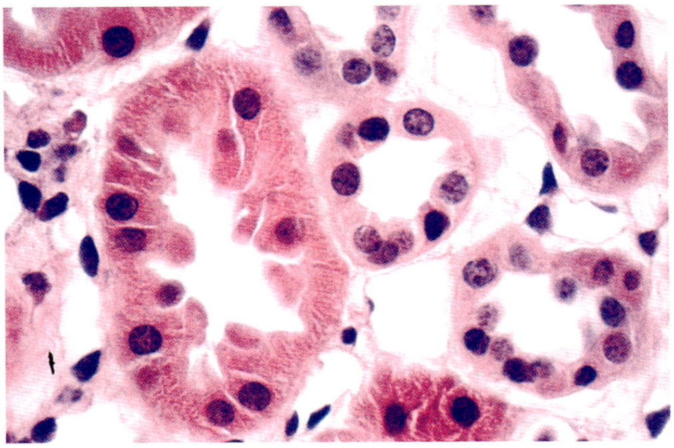

FIGURE 33.71 Cross-section of a proximal tubule (**left of center**). The proximal tubular cells are taller and more eosinophilic than the cells of the distal nephron segments (**right**) (H&E, ×630).

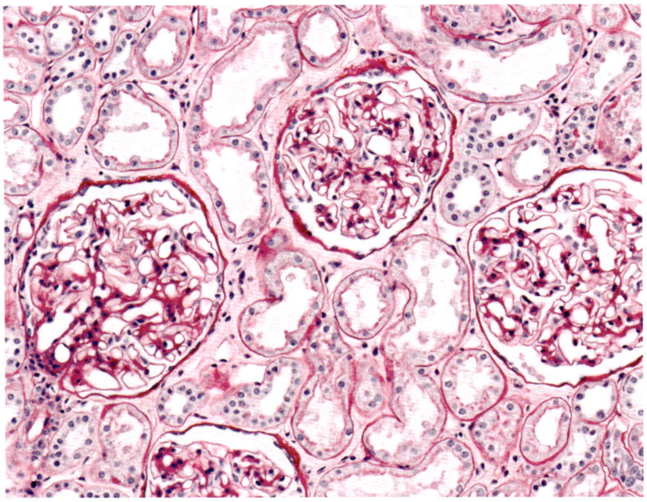

FIGURE 33.72 Ex vivo perfused kidney with intact architecture and patent and dilated lumens in glomeruli and tubules. (PAS, ×200).

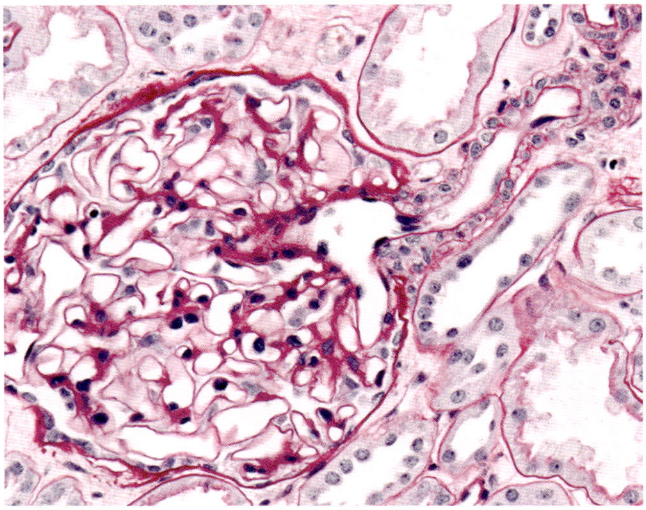

FIGURE 33.73 Ex vivo perfused kidney illustrating crisp cellular detail of a glomerulus, hilar arteriole, and proximal and distal tubules. The brush border of the proximal tubules is somewhat thinned (PAS, ×400).

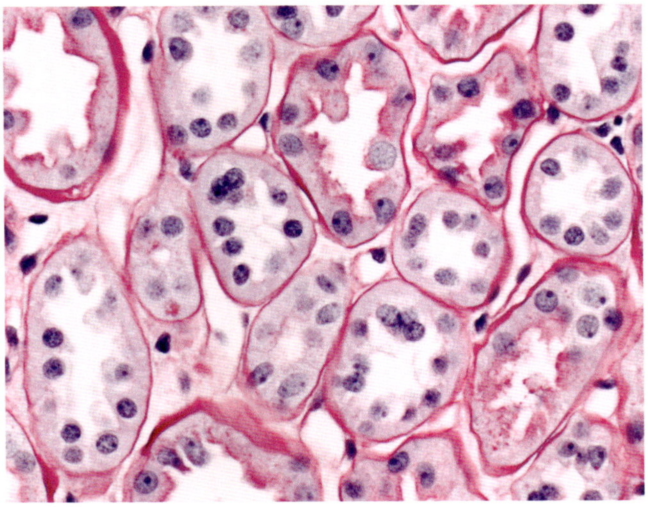

FIGURE 33.74 Ex vivo perfused kidney with proximal and distal tubules. The peritubular capillaries surrounding the tubules are easily identified and display flattened endothelial nuclei (PAS, ×630).

vertical striations that represent numerous elongated mitochondria in the basal regions of the cells. Cytoplasmic apical vacuoles and granules correspond to a well-developed endocytic-lysosomal apparatus. There is a prominent PAS-positive luminal brush border composed of the numerous densely packed long microvilli (Fig. 33.76). Thin actin filaments, 6 nm in diameter, are inside each microvillus and extend into the apical cytoplasm. The brush border, apical cytoplasmic vacuoles, and basal striations are less prominent in the pars recta. Beneath the microvilli in the apical cytoplasm is the terminal web, an arrangement of filaments containing spectrin and myosin (515). Lectins have been

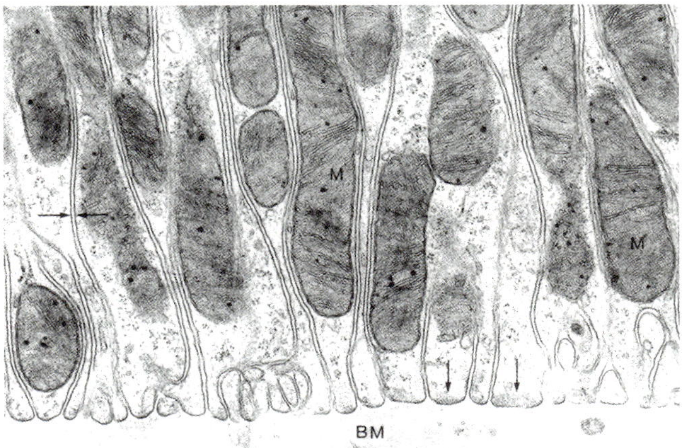

FIGURE 33.75 Electron micrograph illustrating extensive interdigitation of cellular processes in the basal region of proximal tubular cells. The mitochondria (*M*) are elongated. The width of the extracellular space (*opposing arrows*) is constant, and there are bundles of cytoplasmic filaments (*single arrows*) adjacent to the basement membrane (*BM*) (×40,000). (Reprinted with permission from Maunsbach AB, Christensen EI. Functional ultrastructure of the proximal tubule. In: Windhager EE, ed. *Handbook of Physiology. Renal Physiology.* Oxford University Press; 1992:41–107.)

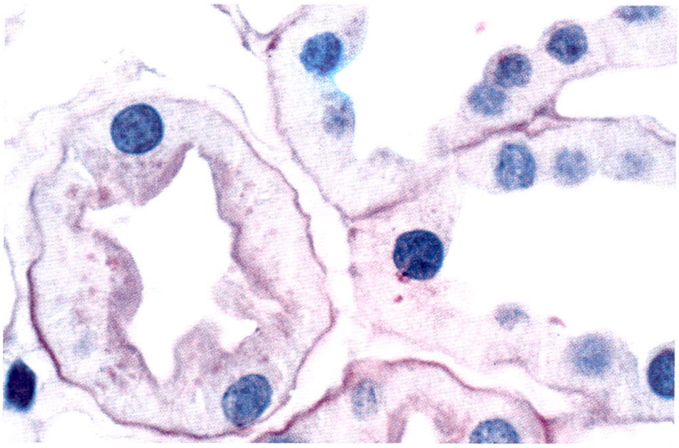

FIGURE 33.76 PAS-stained cross-section of proximal tubule to the left of the center of the micrograph. Note the prominent PAS-positive brush border (PAS, ×1,250).

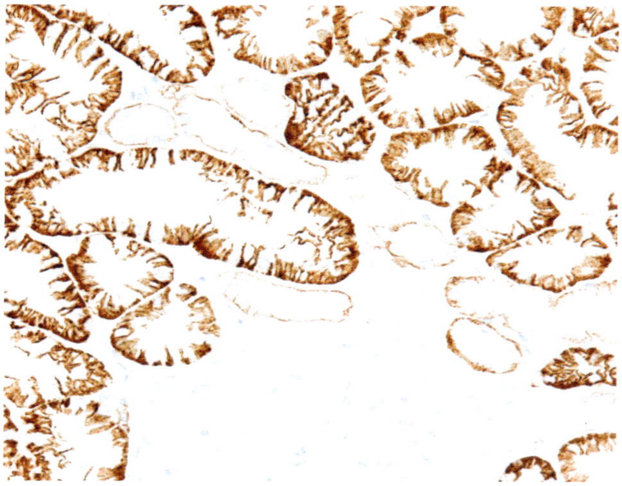

FIGURE 33.78 CD138 (syndecan-1) immunohistochemistry showing strong immunoreactivity in basolateral membranes and cytoplasm of proximal tubules and mild staining along basal aspects of distal tubules and collecting ducts. Glomeruli are negative (CD138 immunohistochemistry, ×400).

used as selective probes to delineate renal tubular segments (516–518). Although a certain degree of nonspecificity has been reported, the lectin *Lotus tetragonolobus* has been used as a marker of proximal tubular epithelium (Fig. 33.77). CD10, CD15, and CD138 (Fig. 33.78) are proximal tubule markers.

Keratins 8 and 18 are expressed in both the convoluted and straight portions of the proximal tubule, whereas keratin 19 is focally expressed in the straight portion (456,457,519). The specific expression of the cell adhesion protein cadherin-6 in the proximal tubule has been reported (520). Cytokeratins 7 and 34βE12 are absent in proximal tubules.

Distinct segments of the tubule portion of the nephron can be distinguished by structural and functional differences, mostly on an ultrastructural level (521,522). These tubule segments often can be detected on light microscopy because of their known distribution within specific zones of the kidney (Fig. 33.79). Although a pars convoluta and a pars recta have been described in the human kidney (523), the segmentation of the proximal tubule into three divisions has not been closely examined.

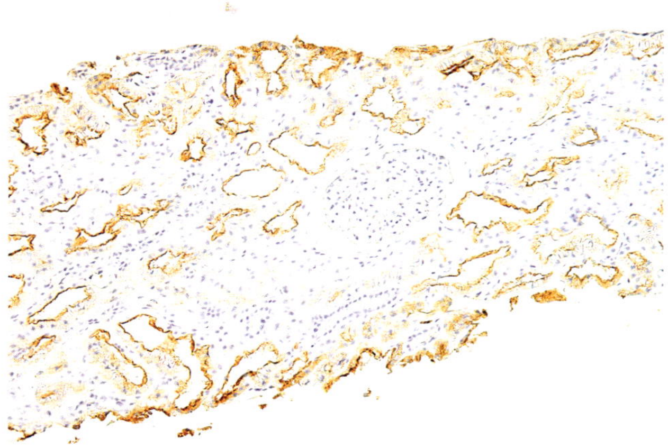

FIGURE 33.77 Staining of the brush border of the proximal tubules with lectin *Lotus tetragonolobus*. The distal nephron segments and glomeruli are negative (×200).

The cells in the S_1 segment have a tall brush border, a well-developed endocytic-lysosomal apparatus, numerous elongated mitochondria, and extensive basolateral invaginations and interdigitations. The cells in the S_2 segment are similar to those in the S_1 segment; however, the brush border is shorter, and the endocytic organelles, mitochondria, and basolateral invaginations and interdigitations are less prominent (Fig. 33.80). The cells in the S_3 segment are more cuboidal and have relatively fewer endocytic organelles, small mitochondria, and inconspicuous membrane invaginations and interdigitations. The length of the brush border in the S_3 segment varies among species, but it appears shorter in humans.

The proximal tubule is responsible for the reabsorption of about 65% of the filtered water, sodium, chloride, potassium, and other solutes. The reabsorption of chloride, bicarbonate, glucose, amino acids, and fluid is coupled to the active transport of sodium (524). An excellent correlation exists along the length of the proximal tubule between the elaborately developed basolateral membrane expressed as surface area (Fig. 33.81), the high Na^+/K^+-ATPase activities that are localized to the basolateral membrane, and the capacity to transport sodium and ions (525,526). The α1β2 heterodimer is the main Na^+/K^+-ATPase isozyme of the kidney but the α2- and α3-isoforms have been detected (527,528). This sodium pump-mediated active transport of Na^+ out of the cell across the basolateral membrane establishes a lumen-to-cell concentration gradient for Na^+. The transport of Na^+ from the lumen into the proximal tubule cell, down its concentration gradient, is mediated by the Na^+/H^+ exchanger, NHE3, expressed in the brush border (529,530). NHE3 in response to angiotensin II traffics into the brush border microvilli, where it likely increases

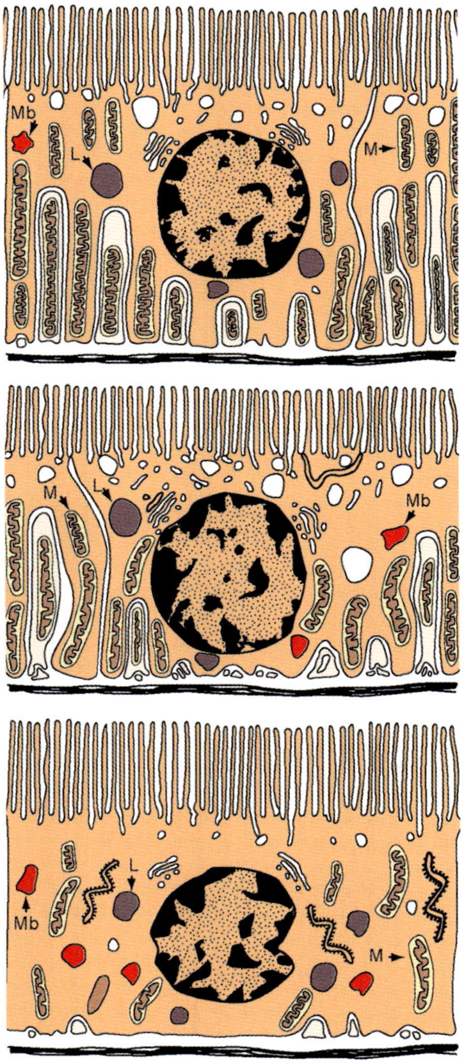

FIGURE 33.79 Schematic diagram of the three segments of the proximal tubule: **upper**, S_1; **middle**, S_2; **lower**, S_3. The prominent basolateral processes are lined with mitochondria. The interdigitating cellular processes that come from adjacent cells are shaded lighter (*Mb*, microbody; *M*, mitochondrion; *L*, lysosome). (Modified with permission from Maunsbach AB, Christensen EI. Functional ultrastructure of the proximal tubule. In: Windhager EE, ed. *Handbook of Physiology. Renal Physiology.* Oxford University Press; 1992:41–107.)

sodium intake (531). Reabsorption of chloride, bicarbonate, glucose, amino acids, and fluid is coupled to sodium transport. The numerous mitochondria located in close proximity to the plasma membrane provide a source of the cellular energy required for active transport. Although the mitochondria appear as isolated organelles, three-dimensional electron microscopic studies have demonstrated that they are actually connected with extensive branching (532). In general, the intrinsic rates at which fluid and solutes are transported decrease along the proximal tubule from S_1 to S_3. The discovery of the aquaporins, a family of water channel proteins, has enhanced our understanding of the kidney's role as the primary organ that

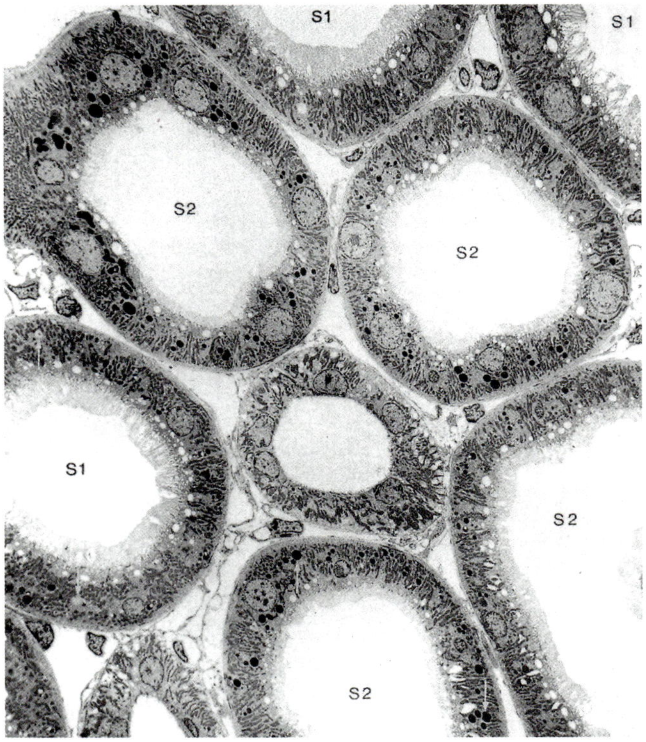

FIGURE 33.80 Electron micrograph of rat renal cortex. The cells in the S_1 segment are taller and have a more prominent brush border than the cells in the S_2 segment (×1,870). (Reprinted with permission from Maunsbach AB, Christensen EI. Functional ultrastructure of the proximal tubule. In: Windhager EE, ed. *Handbook of Physiology. Renal Physiology.* Oxford University Press; 1992:41–107.)

regulates water balance (533,534). Aquaporin-1, AQP1, abundant in both the apical and basolateral membranes of the proximal tubule, mediates osmotic water permeability in this segment (535,536).

The well-developed endocytic-lysosomal apparatus in the proximal tubule (Fig. 33.82) plays an important role in reabsorption and degradation of albumin and low–

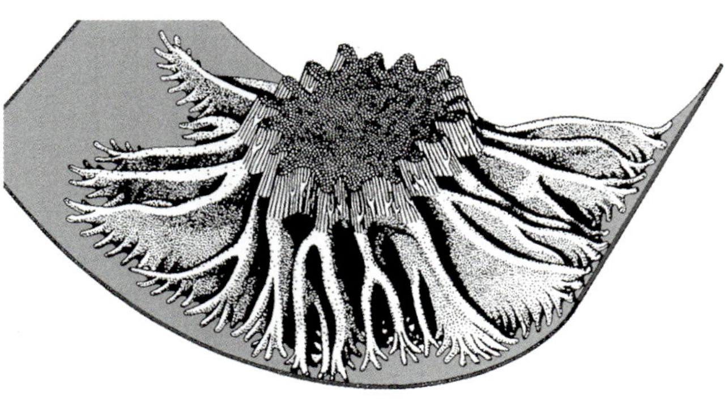

FIGURE 33.81 Three-dimensional schematic drawing of the proximal convoluted tubule illustrating the complex basal and lateral cellular processes that interdigitate with those from adjacent cells. (Modified with permission from Welling LW, Welling DJ. Shape of epithelial cells and intercellular channels in the rabbit proximal nephron. *Kidney Int.* 1976;9:385–394, with permission from Elsevier.)

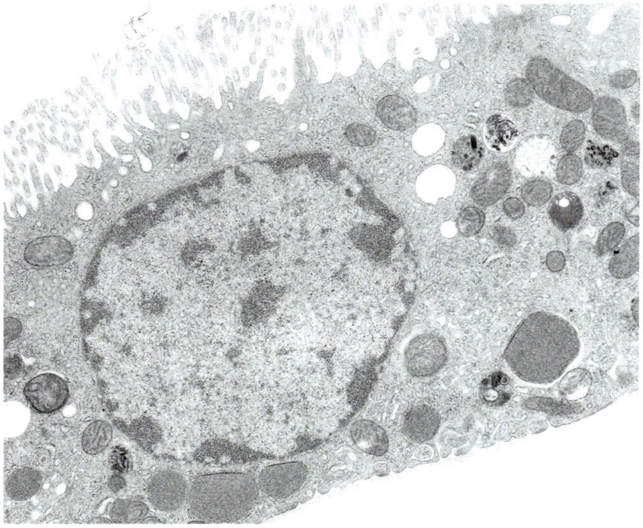

FIGURE 33.82 Electron micrograph illustrating an isolated perfused S_2 segment of the rabbit proximal tubule. Note the endocytic compartment consisting of coated pits and vesicles, apical tubules, small endocytic vesicles, and larger endocytic vacuoles. The lysosomes are heterogeneous and contain electron-dense material. The mitochondria are numerous (×15,000). (Reprinted with permission from Clapp WL, Park CH, Madsen KM, Tisher CC. Axial heterogeneity in the handling of albumin by the rabbit proximal tubule. *Lab Invest.* 1988;58:549–558.)

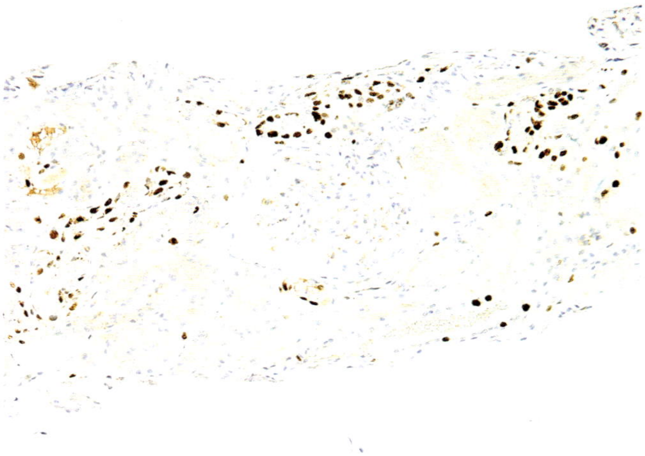

FIGURE 33.83 Biopsy of kidney allograft with delayed graft function. Normally there is sparse labeling for Ki-67, a nuclear protein expressed by proliferating cells. In this example, there is a prominent increase in labeling of the tubular nuclei (Ki-67 immunohistochemistry, ×200).

molecular-weight proteins filtered by the glomerulus (537). This apparatus includes clathrin-coated pits, small vesicles, variable-sized endosomes, apical dense tubules, and lysosomes. Proteins are absorbed by endocytosis and transferred through the endosomal compartment to the lysosomes, where they are degraded. This is a selective process dependent upon the size and charge of the protein molecule (538–540). The capacity for protein degradation decreases from the S_1 to the S_3 segment (541). Upon exposure to an acidic environment in the endosomes (542), the internalized ligand–receptor complexes are segregated and the receptors are recycled back to the luminal membrane *via* small vacuolar structures, termed apical dense tubules (543). The apical dense tubules form an intricate anastomosing network in the apical cytoplasm. Megalin and cubilin are multiligand, endocytic receptors expressed throughout the endocytic apparatus of the proximal tubule (544). The receptors function independently but also interact as a dual complex to facilitate the uptake of albumin (545), as well as numerous ligands, including low–molecular-weight proteins, vitamin-binding proteins, hormones, lipoproteins, and drugs (546). Certain aspects of proximal tubular handling of albumin are controversial. These involve studies suggesting that normally, large amounts of filtered albumin undergo transcytosis in the proximal tubule and is recycled intact to the blood versus the pathway that normally small amounts of filtered albumin undergo megalin/cubilin mediated uptake leading to lysosomal degradation of albumin within the proximal tubule cells (547–549).

In autophagy, cytoplasmic material is sequestered in a double-membrane structure, the autophagosome, which fuses with lysosomes, resulting in degradation of the enclosed material. Autophagic structures are often observed in proximal tubular cells and are important for normal tubule homeostasis. Moreover, studies indicate autophagy in proximal tubules has a protective role against acute kidney injury (550).

The normal adult kidney has a relatively low rate of cell turnover with little proliferation (551–553), although renal cell proliferation is accelerated during hypertrophy and following injury. Histologic assessment of cell proliferation can be made by determining a mitotic index or immunostaining with an antibody that detects proteins present during the cell cycle (554). During recovery after tubular injury, such as ischemia or toxin exposure, cell proliferation in tubular cells, especially in the proximal tubule, increases markedly (Fig. 33.83) (555,556). Recent studies have addressed whether reparative tubular cells arise from a pre-existing intratubular stem cell population or if fully differentiated tubular cells dedifferentiate and then proliferate to replace injured neighboring cells. A population of CD24- and CD133-positive cells have been described scattered throughout the proximal tubule (556). These scattered tubular cells are not conspicuous, are smaller, have fewer mitochondria, and have no brush border compared to surrounding cells. Experimental genetic labeling studies indicate that the scattered tubular cells are not a fixed stem cell population but represent terminally differentiated cells that undergo dedifferentiation and express putative stem cell markers (557–559). Apoptosis has been documented in the adult kidney during the repair response to various forms of tubular injury, including ischemia, toxic insults, and hydronephrosis (Fig. 33.84) (560–563).

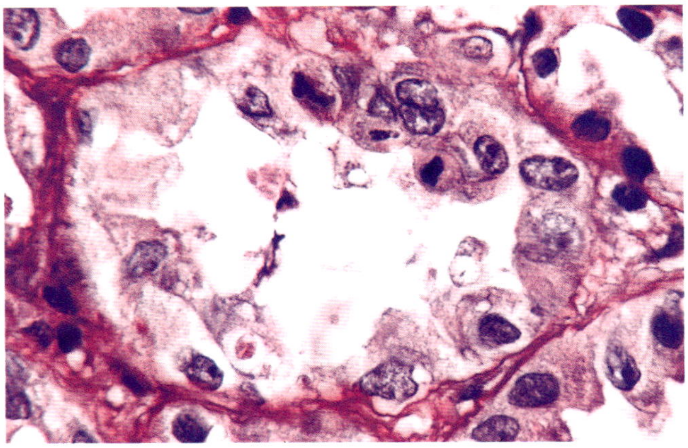

FIGURE 33.84 Micrograph from a patient with cyclosporin A toxicity. Several apoptotic nuclei are present (1 to 2 o'clock positions) in the tubular epithelium (PAS, ×400).

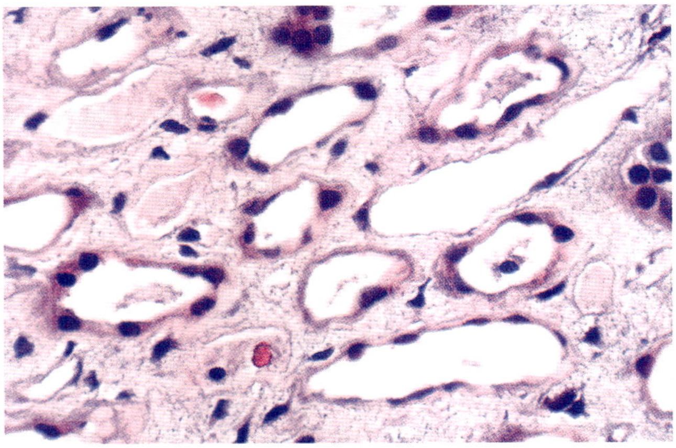

FIGURE 33.85 Several thin limbs of Henle are depicted in the center of the light micrograph. The lining epithelium is extremely attenuated, and the nuclei protrude into the lumens (H&E, ×200).

THIN LIMBS OF HENLE LOOP

The transition of the thin limbs of Henle loop with other nephron segments marks the borders between certain zones of the kidney (Fig. 33.34). Between the outer and inner stripes of the outer medulla, there is an abrupt transition from the proximal tubule to the descending thin limb of Henle loop. Short-looped nephrons have only a short descending thin limb located in the inner stripe of the outer medulla. Near the hairpin turn of the short loop, the descending thin limb continues into the thick ascending limb. Long-looped nephrons have both a long descending thin limb and a long ascending thin limb. The long descending thin limb traverses the inner stripe of the outer medulla and enters the inner medulla, whereas the long ascending thin limb resides entirely within the inner medulla. At the border between the outer and inner medulla, the long ascending thin limb continues into the thick ascending limb. Nephrons originating in the extreme outer cortex have short loops that remain in the cortex and do not reach the medulla.

By light microscopy, the thin limb is lined with a flat, simple epithelium about 1 to 2 μm thick (Fig. 33.85). The lenticularly shaped nucleus bulges slightly into the lumen. At least two different types of epithelium have been demonstrated in humans (564). Type I is present in the descending thin limb of short-looped nephrons. It is an extremely thin, simple epithelium with few cellular interdigitations and cell organelles. Type II epithelium lines the initial part of the descending thin limb of long-looped nephrons located in the outer medulla. This epithelium is characterized by taller cells, short microvilli, and more prominent cell organelles than in the other epithelial types. Type III epithelium, found in the descending thin limb of long-looped nephrons in the inner medulla, is composed of simple cells with few organelles and without lateral interdigitations. Type IV epithelium forms the bends of the long loops and lines the entire ascending thin limb in the inner medulla. It is characterized by low, flattened cells with few organelles and no microvilli, but abundant lateral interdigitations. Thin limb epithelium (Fig. 33.86) has been reported to be immunoreactive for keratins 7, 8, 18, and 19 (456,457).

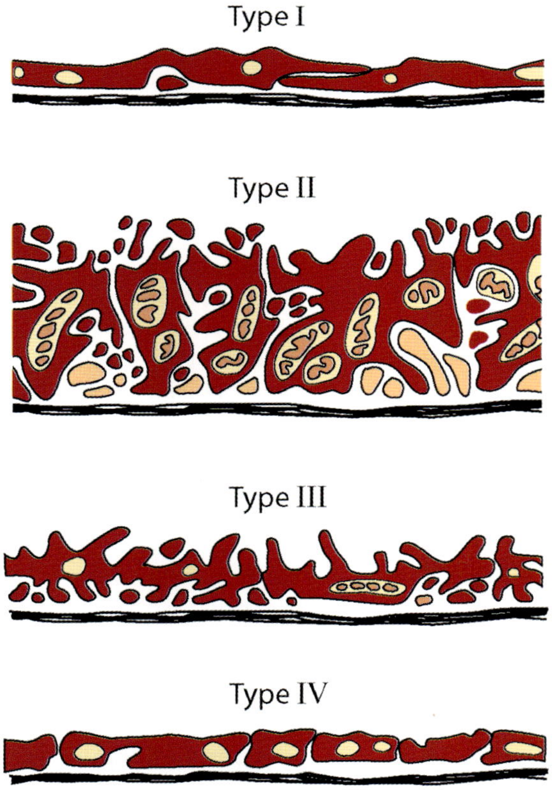

FIGURE 33.86 Schematic drawing of the four types of epithelium in the thin limbs of Henle loop. (The interdigitating cellular processes that come from adjacent cells are shaded lighter.) (Modified with permission from Madsen KM, Tisher CC. Anatomy of the kidney. In: Brenner BM, ed. *Brenner and Rector's The Kidney*. 7th ed. WB Saunders; 2004:3–72. Copyright © 2004 Elsevier.)

The thin limb of Henle loop plays an important role in urinary concentration, a complex process that is not yet fully understood. Physiologic studies have demonstrated that the descending thin limb is permeable to water but has low permeability to sodium chloride, whereas the thin ascending limb is largely impermeable to water but has a high permeability to sodium chloride (312,565). These physiologic investigations are supported by immunohistochemical studies (535,566,567). The aquaporin water channel protein AQP1 mediates water permeability and is expressed in the descending thin limb, primarily of long loops, but is absent in ascending thin limbs. The urea transporter UT-A2, which mediates urea secretion into the loop, is expressed in specific segmental patterns in descending thin limbs. The kidney-specific chloride channel C1C-K1 is expressed exclusively in the ascending thin limb (568,569). In the passive model proposed by Kokko and Rector (570) and Stephenson (571), a hypertonic medullary interstitium concentrates sodium chloride in the descending thin limb by extraction of water. The fluid that then enters the ascending thin limb has a higher sodium chloride concentration, resulting in passive salt absorption and dilution of the fluid of the ascending thin limb. The morphologic features of a simple epithelium with few organelles in the ascending thin limb are consistent with the lack of demonstrable active transport in this segment. Urea secreted into the descending thin limb is returned to the collecting ducts where it is reabsorbed. Thus, urea is recycled. Thus, the thin limb contributes to the maintenance of a hypertonic medullary interstitium and delivers a dilute fluid to more distal segments.

The inner stripe of the outer medulla and the outer aspect of the inner medulla display a complex tubulovascular organization (572–574). Both the outer medulla and the inner contain two distinct topographic regions: the vascular bundle and the collecting duct cluster regions. The vascular bundles, containing the descending vasa recta and ascending vasa recta, surround the collecting duct cluster region, which contain the collecting ducts, ascending thin limbs, some aquaporin-1 negative descending thin limbs, and a peritubular capillary bed. Some aquaporin-1 positive descending thin limbs enter the vascular bundle region of the inner medulla but not the outer medulla. A detailed understanding of the role of this complex architecture in supporting the steep corticomedullary osmolality gradient required for urinary concentration remains to be determined.

DISTAL TUBULE

The distal tubule consists of three distinct segments: The thick ascending limb of Henle loop, the macula densa, and the distal convoluted tubule (Fig. 33.34). The macula densa, as previously discussed, is a specialized plaque of cells within the thick ascending limb. The thick ascending limb in the cortex extends beyond the macula densa before it joins the distal convoluted tubule. Sometimes, the term distal tubules is used informally and not precisely to refer to all nephron segments beyond the thin limbs of Henle loop.

Thick Ascending Limb

The transition from the descending thin limb to the thick ascending limb occurs before the hairpin turns into short looped nephrons. In long-looped nephrons, the transition from the thin ascending limb to the thick ascending limb marks the border between the inner medulla and the inner stripe of the outer medulla. The thick ascending limb can be divided into a medullary and a cortical segment. The ratio of medullary to cortical thick ascending limb for an individual nephron depends on its origin. Juxtamedullary nephrons contain mainly medullary thick ascending limbs, whereas superficial cortical nephrons contain mainly cortical thick ascending limbs. The cells are eosinophilic and cuboidal, and their round nuclei tend to be located in the apical region, causing a bulge of the cell into the lumen (Fig. 33.87). Similar to the proximal tubule cells, the cells of the thick ascending limb have indistinct lateral borders because of elaborate basolateral membrane invaginations and interdigitations. They also have cytoplasmic basal striations because of elongated mitochondria. These morphologic features are characteristic of epithelial cells involved in active transport. However, in contrast to the proximal tubule, the cells are lower and less eosinophilic and there is no brush border in the thick ascending limb. As the thick ascending limb ascends into the cortex, there is a gradual decrease in cell height, basolateral membrane area, and size of the mitochondria (575). Scanning electron microscopy has shown two luminal surface configurations of cells in the thick ascending limb (576). Cells with

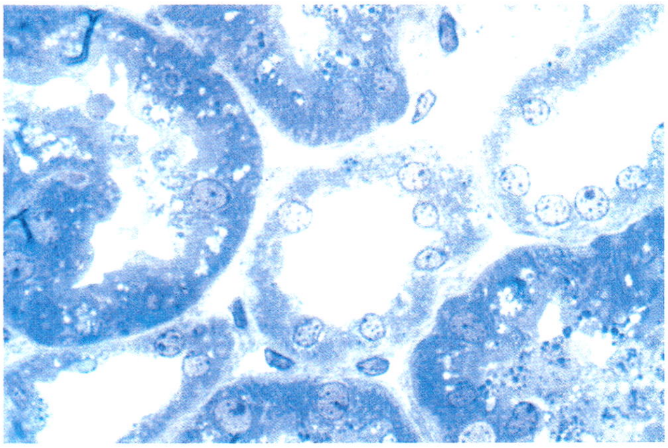

FIGURE 33.87 Light micrograph demonstrating a cross-section of a thick ascending limb in the center. There is no brush border, and the cells are lower than the adjacent proximal tubular cells (toluidine blue stained, 1-μm Epon section, ×750).

a relatively smooth surface are most commonly found in the medullary segment, whereas cells with a rough surface due to luminal microprojections and apical lateral membrane invaginations predominate in the cortical segment. The functional significance of these structural findings remains unexplained. The cortical thick ascending limb continues into the distal convoluted tubule just beyond the macula densa. The cells of the thick ascending limb synthesize Tamm–Horsfall protein and secrete it into the tubular lumen (498). This segment expresses keratins 8 and 18 (457,519,577), as well as kidney-specific (Ksp)-cadherin (578).

An important function of the thick ascending limb is the active reabsorption of sodium chloride. There is a correlation between structure and function in the ascending limb. The basolateral membrane surface area, mitochondrial density, and the Na^+/K^+-ATPase activity are all greater in the medullary segment than in the cortical segment of the thick ascending limb (524,575,579). The reabsorption of sodium chloride in both the medullary and cortical segments of the thick ascending limb is driven by the basolateral Na^+/K^+-ATPase creating a favorable electrochemical gradient. Entry of sodium chloride across the apical membrane is mediated by the $Na^+/K^+/2Cl^-$ cotransporter (NKCC2), which localizes to the thick ascending limb apical plasma membrane (580). NKCC2 is the target of loop diuretics, such as furosemide. This reabsorption of salt coupled with the water impermeability of the thick ascending limb results in a hypertonic interstitium and delivery of a hypotonic fluid to more distal tubular segments.

Distal Convoluted Tubule

The distal convoluted tubule begins just beyond the macula densa in the cortex and represents the terminal part of the distal tubule. The cells of the distal convoluted tubule are similar to those of the thick ascending limb and contain numerous mitochondria. Distal convoluted tubule cells contain prominent basolateral membrane infoldings. However, the distal convoluted tubule cells are taller, characteristically have nuclei closer to the lumen, and lack lateral interdigitations in the apical region between adjacent cells (Fig. 33.88). The cells have a single cilium and small microprojections on the luminal surface. In comparison with the proximal tubule, the cells of the distal convoluted tubule are lower and less eosinophilic, have a less prominent apical endocytic apparatus, and lack a brush border. More nuclei are observed in a cross-section than in the proximal tubule, and the lumen is normally open. The epithelium of the distal convoluted tubule shows immunoreactivity for keratins 8, 18, and 19 (456,519), and Ksp-cadherin (578).

The distal convoluted tubule has a higher level of Na^+/K^+-ATPase activity than any other tubular segment (579) and this correlates with the high mitochondrial density. Basolateral membrane Na^+/K^+-ATPase in the distal convoluted tubule drives solute and water reabsorption. Apical membrane transport of NaCl is mediated by the cotransporter NCC, which is the target of thiazide diuretics (581,582). The NCC cotransporter is distinct from the furosemide-sensitive cotransporter NKCC2, presents in the thick ascending limb. The distal convoluted tubule, similar to the thick ascending limb, is relatively impermeable to water but is responsible for the reabsorption of sodium chloride (583–585). Altered transport activity of upstream tubular segments such as the thick ascending limb results in increased NCC expression and sodium reabsorption in the distal convoluted tubule (586).

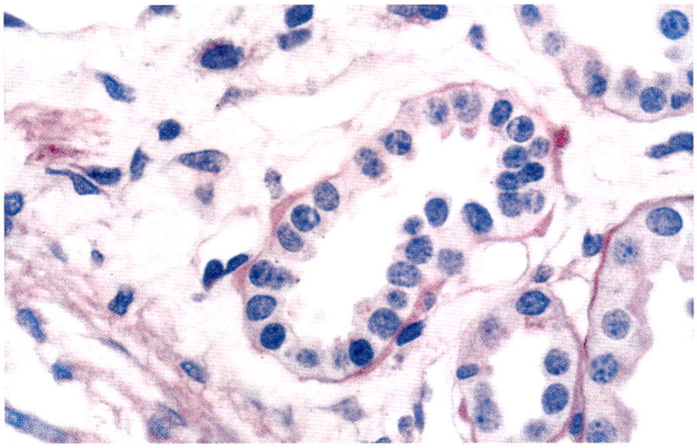

FIGURE 33.88 Light micrograph showing a distal convoluted tubule. Note the absence of a brush border and the nuclei situated close to the lumen (PAS, ×630).

CONNECTING TUBULE

The connective tubule is a transitional segment that connects the distal convoluted tubule with the collecting duct system. In superficial nephrons, the connective tubule continues directly into an initial collecting tubule (Fig. 33.89). In contrast, the connective tubules of juxtamedullary nephrons and many midcortical nephrons join to form an arcade that ascends in the cortex before draining into an initial collecting tubule. In humans, most nephrons empty individually into initial collecting tubules (312). Fourteen percent of the nephrons are connected to arcades, and each arcade consists of approximately three nephron attachments (3). Each cortical collecting tubule receives an average of 11 nephrons (3). In superficial nephrons, the connective tubule is situated adjacent to the afferent arteriole of its parent glomerulus (587). Physiologic studies have shown that increased sodium reabsorption in the connective tubule is associated with dilatation of the adjacent afferent arteriole (588). This functional link has been called connecting tubule glomerular feedback. It results in an increase in renal blood flow and glomerular filtration rate which favors sodium excretion.

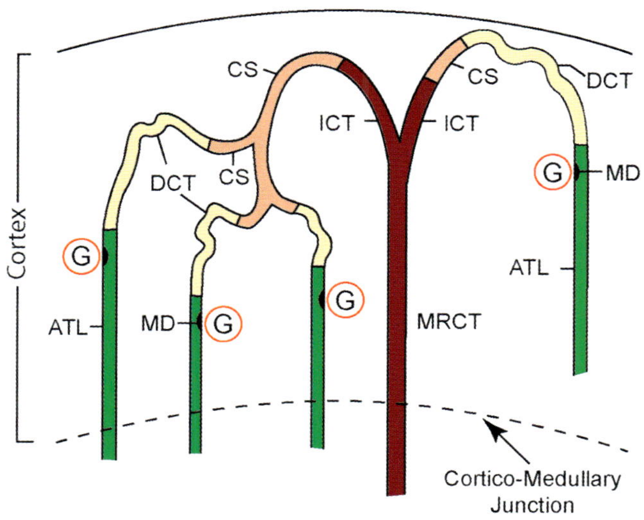

FIGURE 33.89 Diagram of the various anatomic arrangements of the distal tubule connecting to the cortical collecting duct in superficial, midcortical, and juxtamedullary nephrons. (*G*, glomerulus; *ATL*, ascending thin limb [of Henle]; *MD*, macula densa; *DCT*, distal convoluted tubule; *CS*, connecting segment; *ICT*, initial collecting tubule; *MRCT*, medullary ray collecting tubule).

In most species, including humans, the connective tubule contains different cell types resulting from an intermixing of cells from the adjacent distal convoluted tubule and cortical collecting duct (589). However, the connective tubule cell is the most characteristic cell type of this transitional segment. It occurs only in this segment. They display ultrastructural features intermediate between the distal convoluted tubule cells and the principal cells of the cortical collecting duct, and contain true infoldings of the basal cell membrane (590). Various types of intercalated cells, similar to those in the cortical collecting duct, are also present in the collecting system and are likely involved in tubular acid–base regulation. Three types of intercalated cells, type A, type B, and non A–non B are found in the connective tubule The connective tubule is an important site of sodium reabsorption, potassium secretion, calcium reabsorption, and significant amounts of water transport. Proteins in the connective tubule mediating these transport functions include the sodium channel, ENaC, potassium channel ROMK, Na^{+}/Ca^{2+} exchanger, Ca^{2+}-ATPase, calcium channel TRPV5, and aquaporin-2 (AQP2) (591–595).

COLLECTING DUCT

The collecting duct begins in the cortex and descends to the tip of the papilla, also called the area cribrosa, where the inner medullary segments terminate as the ducts of Bellini. During its course, there is an increase in diameter from the cortical portion to the terminal segments at the area cribrosa. The collecting duct can be divided into the cortical collecting duct, outer medullary collecting duct, and inner medullary collecting duct. Significant cellular heterogeneity exists along the collecting duct. Lectins *Dolichos biflorus* and *Arachis hypogaea* have been used as markers for collecting duct epithelium (Fig. 33.90) (518). The distal tubules and collecting ducts show variable but generally more intense staining for keratins than the proximal tubules (Figs. 33.91 and 33.92). Keratins 8, 18, and 19 are prominently expressed throughout the cortical collecting duct and medullary collecting duct (456,457,519). There is also staining for keratin 7. Scattered keratin 7- and 19-negative cells have been observed to be intercalated cells (456). Keratins 5/6, 17, and 20 as well as vimentin are restricted primarily to medullary collecting ducts (519). GATA3, a transcription factor, is strongly expressed throughout the entire collecting duct system (Fig. 33.93). There is less immunoreactivity for

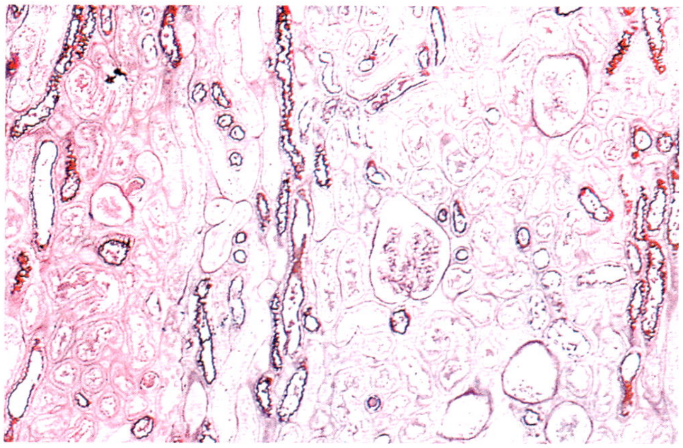

FIGURE 33.90 Staining of the collecting ducts and thick ascending limbs with lectin *Arachis hypogaea*. The proximal tubules and glomeruli are negative. (Courtesy of Dr. Randolf A. Hennigar.)

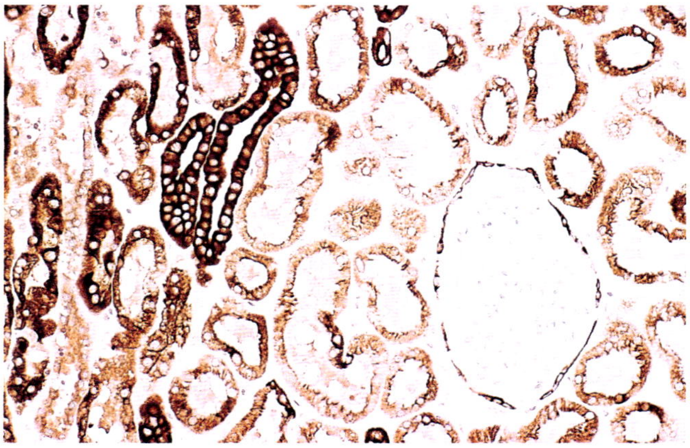

FIGURE 33.91 Distal tubules, collecting ducts, proximal tubules, and parietal epithelium lining Bowman capsule showing expression of keratins. The distal tubules and collecting ducts are labeled more intensely than the proximal tubules (CAM 5.2 immunohistochemistry, ×100).

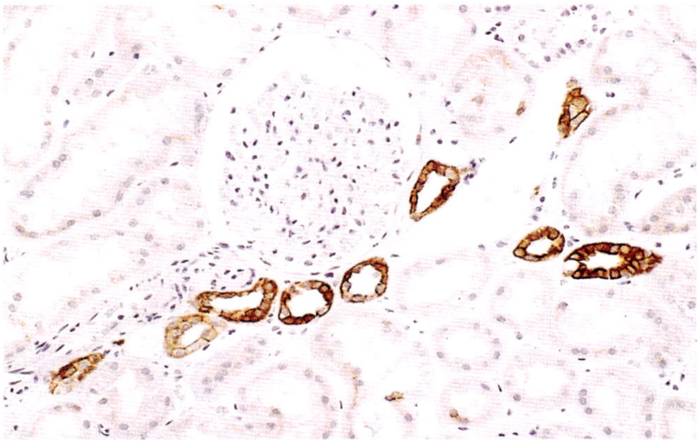

FIGURE 33.92 Detection of keratin expression varies depending on the specificity and dilution of the antibody. In this micrograph, there is prominent immunoreactivity of the distal tubules and collecting ducts (35βH11 immunoperoxidase, ×100).

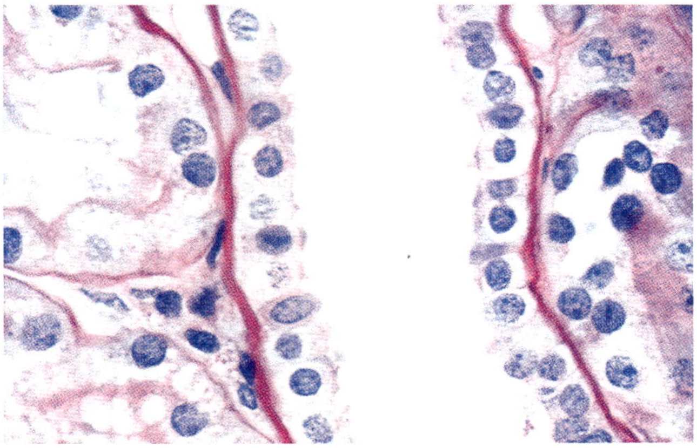

FIGURE 33.94 Micrograph illustrating a cortical collecting duct. Note the distinct lateral cell orders (PAS, ×500).

Ksp-cadherin in collecting ducts compared to the thick ascending limbs and distal convoluted tubules (578).

Cortical Collecting Duct

The cortical collecting duct can be subdivided further into the initial collecting tubule and the medullary ray portion. The latter is the main segment and it runs in parallel with the cortical pars recta and thick ascending limb in the medullary ray. By light microscopy, the epithelium of the cortical collecting duct consists of cuboidal cells with distinct lateral cell borders and central round nuclei (Fig. 33.94). The lumen is prominently open, and there is no brush border. The cortical collecting duct is composed of principal cells and intercalated cells. Principal cells are more abundant and mainly responsible for salt and water transport and the intercalated cells are involved in acid–base regulation. It is difficult to distinguish principal cells from intercalated cells, although the principal cells have their nuclei closer to the apical surface compared to intercalated cells. The principal cells have extremely light or clear cytoplasm. By electron microscopy, the principal cells have relatively few cell organelles and no interdigitation of lateral cellular processes from adjacent cells, which accounts for the distinct cell borders observed on light microscopy (Fig. 33.95). However, there are prominent infoldings of the basal plasma membrane, which gives the basal region an accentuated clear appearance (596). The principal cell has a fairly smooth luminal surface with short microvilli and a single cilium by scanning electron microscopy (see Fig. 33.104).

Principal cells are involved in sodium reabsorption and potassium secretion. Sodium reabsorption is mediated by the amiloride-sensitive sodium channel, ENaC, located in the apical membrane of principal cells throughout the entire collecting duct (597,598). Experimental conditions

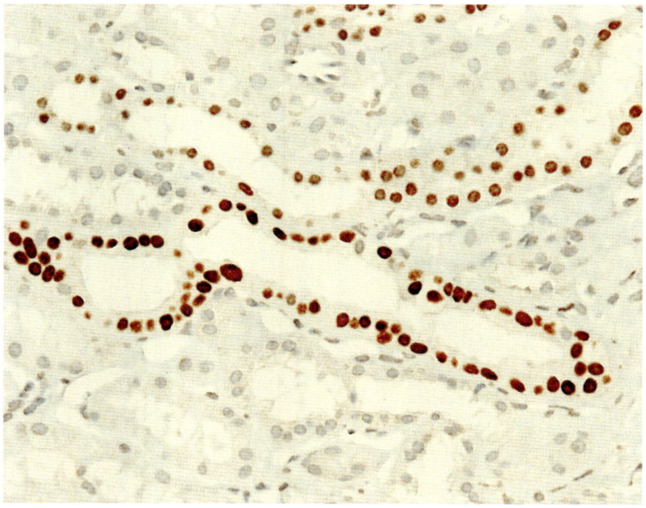

FIGURE 33.93 Elongated collecting ducts displaying positive nuclear immunoreactivity for GATA3. (GATA3 immunohistochemistry, ×400.)

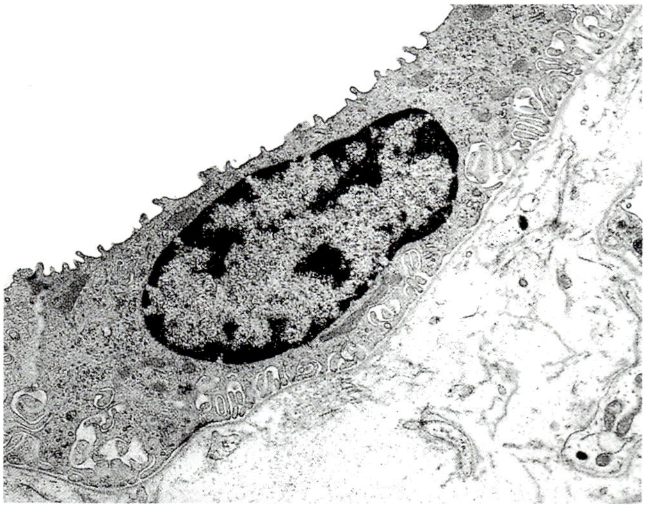

FIGURE 33.95 Electron micrograph of principal cell from the collecting duct. Note the relatively prominent infoldings of the basal plasma membrane (×12,500).

of dietary potassium loading or mineralocorticoid stimulation have shown increases in potassium secretion and Na^+/K^+-ATPase activity in the cortical collecting duct along with an increase in the surface area of the basolateral membrane of the principal cells (599–603). These findings indicate that the principal cells are involved in potassium secretion in the cortical collecting duct. Potassium secretion is mediated largely by the apical membrane potassium channel, ROMK. The entire collecting duct becomes permeable to water in the presence of the antidiuretic hormone vasopressin. After vasopressin binds to its receptor on the basolateral membrane of principal cells (604) small apical cytoplasmic tubulovesicles, called aggrephores, containing the water channel AQP2 are shuttled to the apical membrane, which markedly increases water permeability (605,606). The presence of the water channels AQP3 and AQP4 in the basolateral membranes of principal cells facilitates the final exit of water into the interstitium (607,608).

The intercalated or dark cells are interspersed in the lining epithelia of the collecting duct. Intercalated cells usually represent the minority cell type in epithelia where they are found. They are present in the connecting segment, the outer medullary collecting duct, and the initial portion of the inner medullary collecting duct. Intercalated cells may be identified on 1-μm thick toluidine blue-stained Epon sections by their densely staining cytoplasm and their often-convex luminal surfaces covered with numerous microprojections (Fig. 33.96). The darkly staining cytoplasm is due in part to the presence of relative more organelles, especially mitochondria. Three distinct populations of intercalated cells, types A and B and non A–non B have been described in the cortical collecting duct of mammals (609–612). On ultrastructural examination, the type A intercalated cells have prominent microprojections of the apical membrane and extensive tubulovesicular structures in the apical cytoplasm (Fig. 33.97). In comparison with the type A cells, the type B intercalated cells have a denser cytoplasm, more mitochondria, a smaller apical membrane area, a small number of microprojections on the apical surface, more spherical vesicular structures throughout the cytoplasm but fewer vesicles beneath the apical membrane, and a larger basolateral membrane surface area (Fig. 33.98). By scanning electron microscopy, the type A cells have a large convex luminal surface covered with numerous complex microprojections or small folds called microplicae (Fig. 33.99), whereas the type B cells display a small angular luminal surface with relatively small microvilli (Fig. 33.100) (611). The type B intercalated cells may be inconspicuous on scanning electron microscopy.

The non A–non B intercalated cells are primarily located in the connective tubule and the initial collecting tubule. They are larger than type A and type B intercalated cells and have abundant mitochondria with prominent apical microprojections similar to those of type A cells.

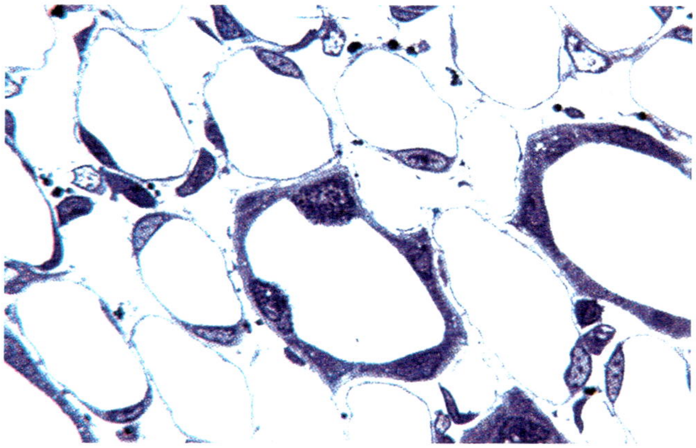

FIGURE 33.96 Light micrograph of the outer medulla showing intercalated cells in the collecting ducts. The intercalated cells exhibit a bulging apical surface covered with microprojections and dark-staining cytoplasm (1-μm toluidine blue-stained Epon section, ×160).

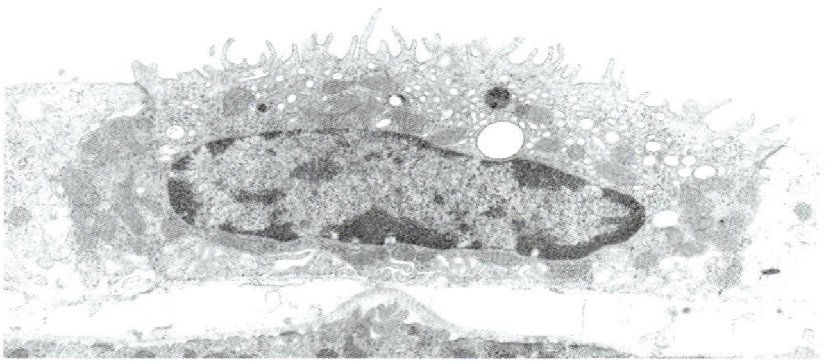

FIGURE 33.97 Electron micrograph of a type A intercalated cell in the cortical collecting duct. Note the prominent tubulovesicular membrane compartment in the apical cytoplasm and the numerous microprojections on the luminal surface (×11,800; courtesy of Dr. Jill W. Verlander).

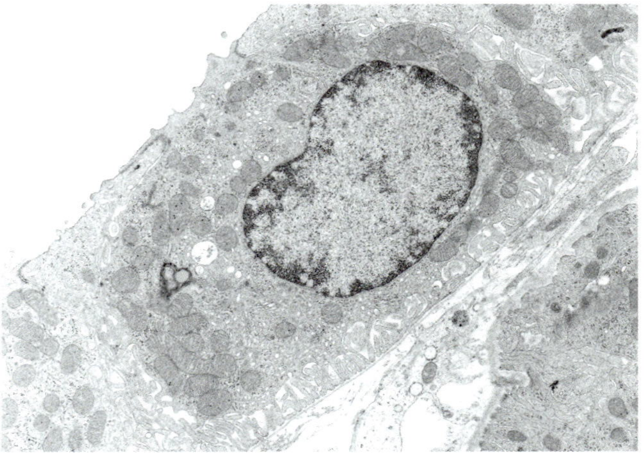

FIGURE 33.98 Electron micrograph of a type B intercalated cell in the cortical collecting duct. There are numerous vesicles throughout the cytoplasm, and the basolateral membrane is prominent. Note the paucity of microprojections on the luminal surface. Compared to the type A intercalated cell, there are fewer vesicles beneath the apical membrane (×11,800; courtesy of Dr. Jill W. Verlander).

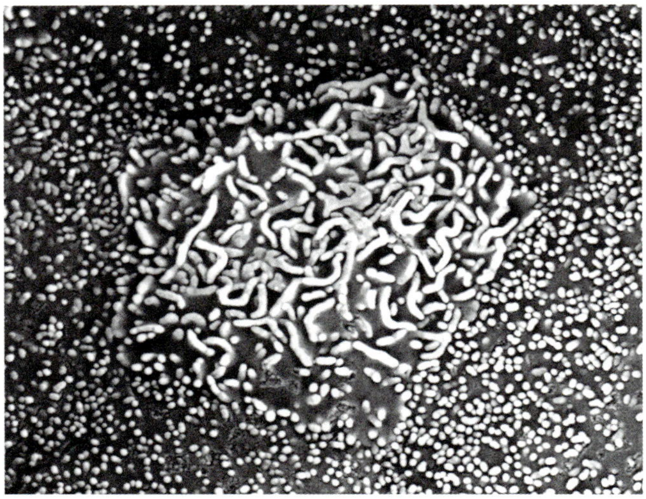

FIGURE 33.99 Scanning electron micrograph of the luminal surface of a type A intercalated cell in the cortical collecting duct. The type A cell is well demarcated and has a large luminal surface covered primarily with microplicae but also microvilli (×15,000; courtesy of Dr. Jill W. Verlander).

All intercalated cell subtypes are typified by their high levels of carbonic anhydrase type II, an enzyme that catalyzes the interconversion of CO_2 to HCO_3, consistent with their role in urine acidification (613). Physiologic studies have demonstrated that the cortical collecting duct reabsorbs bicarbonate in acid-loaded animals (614) and secretes bicarbonate in alkali-loaded animals (615).

Studies have immunolocalized the vacuolar-type proton pump H^+-ATPase in the apical membrane (616–618) and the Cl^-/HCO_3^- exchanger, AE1, in the basolateral membrane (609,618–620) of type A intercalated cells. The AE1 protein is specific for the type A intercalated cell. Type A cells are responsible for H^+ secretion in the cortical collecting duct. The immunolocalization of the H^+-ATPase to the basolateral membrane of type B intercalated cells (617,618) and the physiologic evidence for an apical Cl^-/HCO_3^- exchange in these cells (621) indicate that type B cells are involved in bicarbonate secretion. Apical Cl^-/HCO_3^- exchange in the type B cell is mediated by the protein pendrin, which immunolocalizes to the apical membrane and apical cytoplasmic vesicles of the B cell (103,622,623). Furthermore, the renal cortical expression of pendrin is increased in alkali-loaded animals and decreased in acid-loaded animals (624). Type B intercalated cells are most numerous in the cortical collecting duct. Pendrin is not in the same protein family as AE1 (625). Mutations of the gene encoding pendrin result in Pendred syndrome, a disorder mainly associated with a thyroid goiter and deafness (626). The cell-specific ultrastructural features, distribution of the transporters, and cell responses to physiologic changes have established that type A intercalated cells secrete acid whereas type B intercalated cells secrete base.

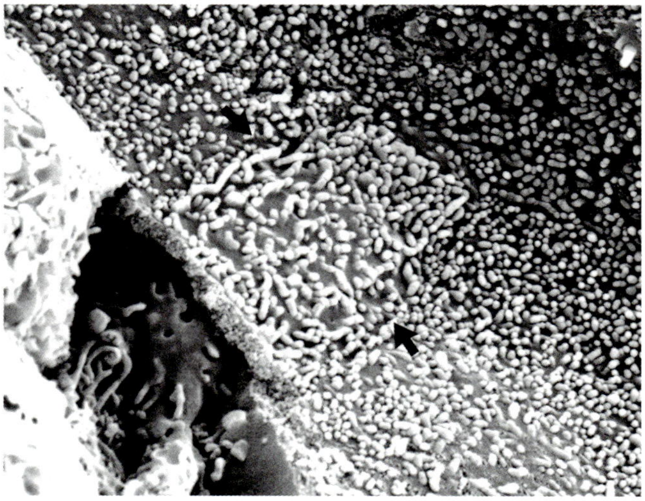

FIGURE 33.100 Scanning electron micrograph of the luminal surface of a cortical collecting duct. A type B intercalated cell (*arrows*) displays a small angular luminal surface covered with short microprojections, mainly microvilli (×15,000; courtesy of Dr. Jill W. Verlander).

The non A–non B intercalated cells express pendrin in the apical membrane and cytoplasmic vesicles like type B intercalated cells but also express the H^+-ATPase in the apical membrane like type A intercalated cells (103,627). Thus, the types of intercalated cells may be defined by their cellular distribution of the H^+-ATPase and the presence or absence of the anion exchangers, AE1 and pendrin (628). Although the function of non A–non B intercalated cells is not well understood, they appear to have an important role in chloride reabsorption via pendrin-mediated apical Cl^-/HCO_3^- exchange (629,630). Pendrin expression in the cortical collecting duct appears to be regulated by *Nedd4-2*, a E3 ubiquitin-protein ligase, which is highly expressed in type B and non A–non B cells (631).

Type A cells express CD117 by IHC (Fig. 33.101). Single-cell expression investigation has found hybrid cells

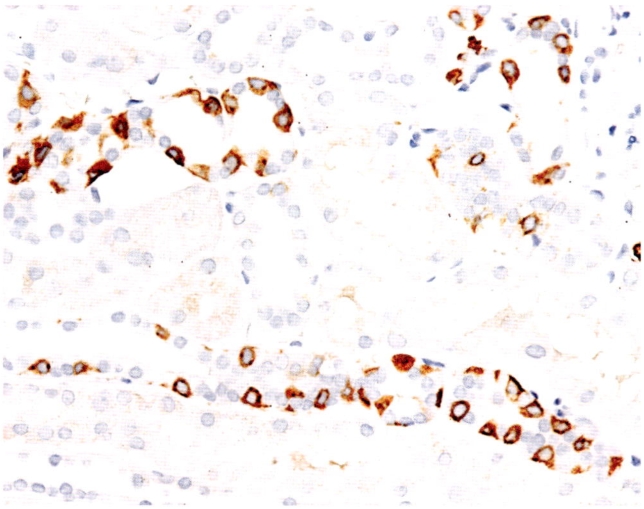

FIGURE 33.101 Type A intercalated cells along collecting ducts showing expression of c-Kit. (c-Kit immunohistochemistry ×400).

in the collecting duct to express both aquaporin-2 (principal cell marker) and either AE1 (type A intercalated cell marker) or pendrin (type B intercalated cell marker), as well as markers of both principal cells and intercalated cells (632,633).

Outer Medullary Collecting Duct

The collecting duct traverses the outer medulla without receiving tributaries. Similar to the cortical collecting duct, the outer medullary collecting duct contains principal cells and intercalated cells (Fig. 33.102). The principal cells in this segment are similar to those in the cortical collecting duct but are taller and have fewer organelles and basal membrane infoldings. They also express apical membrane ENaC and AQP2 and basolateral Na^+/K^+-ATPase consistent with sodium and water reabsorption. The intercalated cells gradually decrease along outer medullary collecting duct (634,635). The intercalated cells of this segment resemble the type A intercalated cells in the cortical collecting duct but are taller and have a less dense cytoplasm. Type B and non A–non B intercalated cells are typically absent in the outer medullary collecting duct.

The outer medullary collecting duct plays a major role in urine acidification. An increase in the surface area of the apical plasma membrane of the type A intercalated cells in this segment has been demonstrated after hydrogen ion stimulation (636,637). These type A cells with their apical and basolateral membranes containing the H^+-ATPase and the chloride/bicarbonate exchanger, AE1, respectively are responsible for hydrogen ion secretion (616,620). The outer medullary collecting duct is also an important segment for potassium reabsorption, especially during dietary potassium restriction. This is mediated by both the gastric H^+/K^+-ATPase $\alpha 1$ and colonic H^+/K^+-ATPase $\alpha 2$ isoforms in the apical membrane of the type A intercalated cells (638–642).

Ammonia metabolism plays an important role in acid–base homeostasis. Most of the ammonia secretion is handled by the collecting duct (643). Rhesus (Rh) transporters Rhbg and Rhcg, mainly expressed in type A intercalated cells throughout the collecting duct, mediate renal ammonia secretion. Type B intercalated cells do not express the Rh transporters.

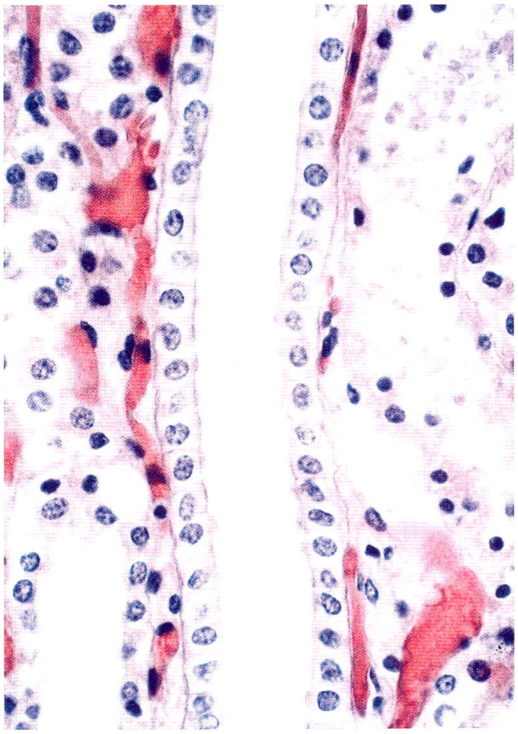

FIGURE 33.102 Light micrograph illustrating longitudinal section of an outer medullary collecting duct (H&E, ×250).

Inner Medullary Collecting Duct

The inner medullary collecting duct inner medullary collecting duct represents the terminal portion of the collecting duct. Although the inner medullary collecting duct is often called the papillary collecting duct, only the inner two-thirds of the inner medullary collecting duct is located in the papilla. Descending through the inner medulla, the collecting ducts join in successive fusions, which result in an arborescent architectural arrangement. There is a significant increase in diameter and height of the epithelium as the ducts descend (644). The height of the cells increases gradually from cuboidal to columnar (Fig. 33.103). However, in the terminal portion of the human inner medulla there is often an abrupt transition between collecting ducts lined with cuboidal cells and the ducts of Bellini, which are composed of tall columnar cells.

Structural and functional heterogeneity exists along the inner medullary collecting duct (644). It can be subdivided arbitrarily into three portions: the outer, middle, and inner thirds. However, there is physiologic evidence for the division of the inner medullary collecting

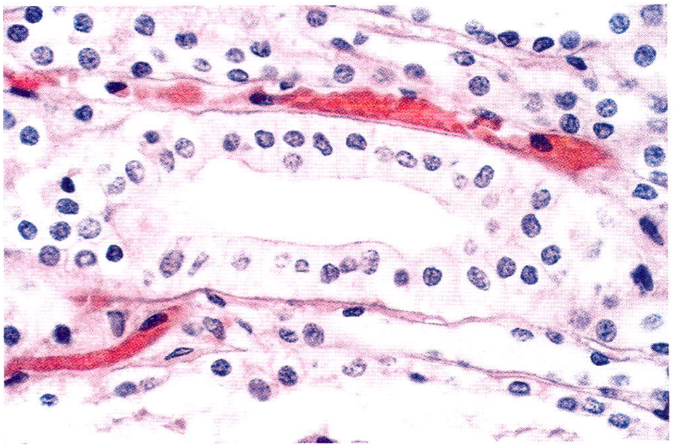

FIGURE 33.103 Micrograph illustrating columnar cells of the collecting duct in the inner medulla (H&E, ×500).

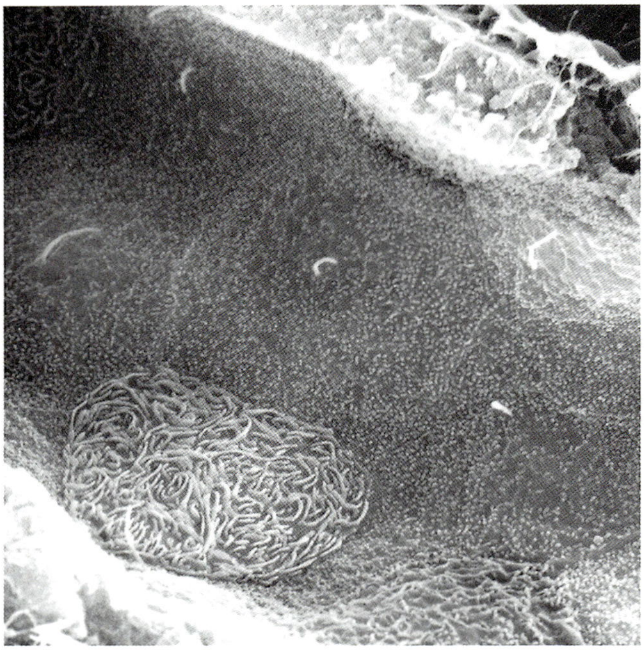

FIGURE 33.104 Scanning electron micrograph of the inner medullary collecting duct. The intercalated cell is round and exhibits a convex luminal surface covered with microplicae without cilia. The adjacent principal cells are characterized by short microvilli and a single central cilium on their luminal surface (×12,000). (Reprinted with permission from Clapp WL, Madsen KM, Verlander JW, Tisher CC. Morphologic heterogeneity along the rat inner medullary collecting duct. *Lab Invest.* 1989;60:219–230.)

duct into two functionally distinct segments, which are termed initial inner medullary collecting duct and terminal inner medullary collecting duct (Fig. 33.104) (645,646). The initial inner medullary collecting duct is the outer segment and mainly corresponds to the outer third, whereas the terminal inner medullary collecting duct includes most of the middle and inner thirds. The initial inner medullary collecting duct consists mainly of cells that are similar in structure to the principal cells in the outer medullary collecting duct. Intercalated cells are rare to absent in the initial inner medullary collecting duct. The terminal inner medullary collecting duct is composed of mainly one cell type, the inner medullary collecting duct cell. Compared with principal cells, the inner medullary collecting duct cells are taller and have small stubby apical microprojections, lighter staining cytoplasm containing numerous ribosomes, small lysosomes in the basal cytoplasm, fewer infoldings of the basal plasma membrane but prominent lateral membrane infoldings (Fig. 33.105) (647). By scanning electron microscopy, the inner medullary collecting duct cells display more numerous small microvilli and lack the central cilium characteristic of principal cells (Fig. 33.106). In fact, only two epithelial cell types in the kidney, inner medullary collecting duct cells and intercalated cells, have no central cilium.

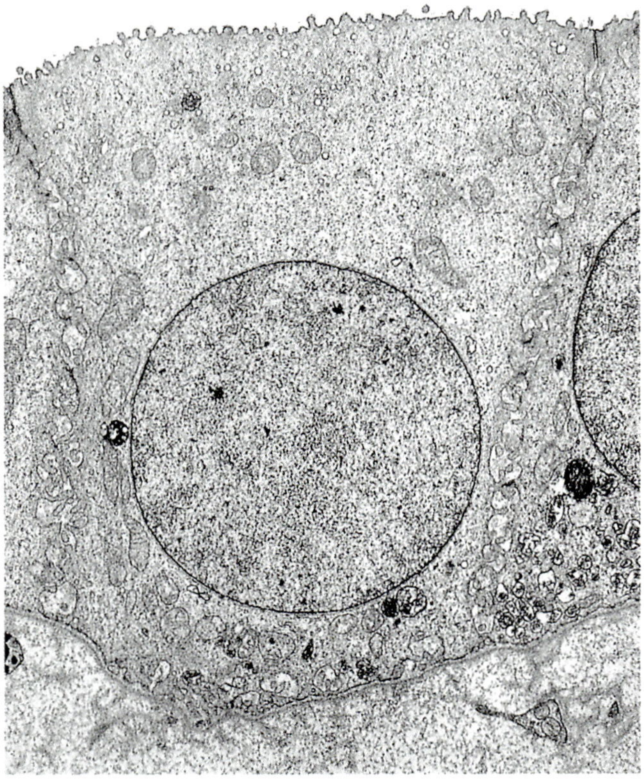

FIGURE 33.105 Electron micrograph of an inner medullary collecting duct cell. The cell is tall, has extensive lateral membranes, and exhibits small stubby microvilli. Infoldings of the basal plasma membrane are not prominent (×12,500). (Reprinted with permission from Clapp WL, Madsen KM, Verlander JW, Tisher CC. Morphologic heterogeneity along the rat inner medullary collecting duct. *Lab Invest* 1989;60:219–230.)

PAPILLARY SURFACE EPITHELIUM

A layer of cuboidal epithelium lines the external surface of the renal papilla. These cells have a relatively smooth surface but with glycocalyx, cytoplasmic vesicles, and few mitochondria (648,649). They lack the luminal asymmetric unit membrane typical of transitional epithelium. Various proteins including UT-B1, H^+/K^+-ATPase, and osteopontin have been localized to the papillary surface epithelium, which support studies indicating urea transport, acid secretion, and inhibition of crystal deposition in this epithelium (650–652). In antidiuresis, widely dilated intercellular spaces have been observed in the papillary epithelium, suggesting significant fluid movement (653).

INTERSTITIUM

The renal interstitium, a complex space between the parenchymal components of both the cortex and medulla, includes extracellular matrix, several types of interstitial

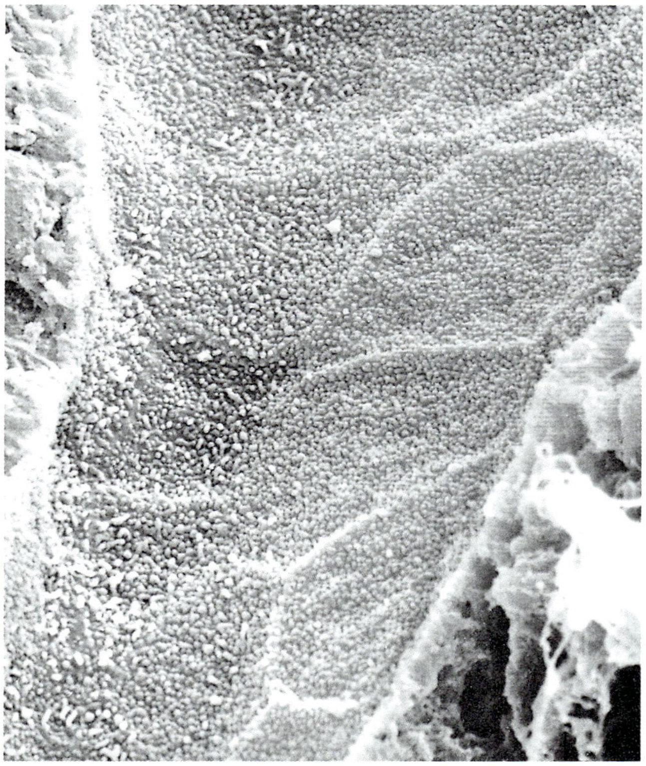

FIGURE 33.106 Scanning electron micrograph of the terminal inner medullary collecting duct. The entire luminal surface of the inner medullary collecting duct cells is covered with abundant short microvilli. There is an absence of cilia (×12,500). (Reprinted with permission from Clapp WL, Madsen KM, Verlander JW, Tisher CC. Morphologic heterogeneity along the rat inner medullary collecting duct. *Lab Invest* 1989;60:219–230.)

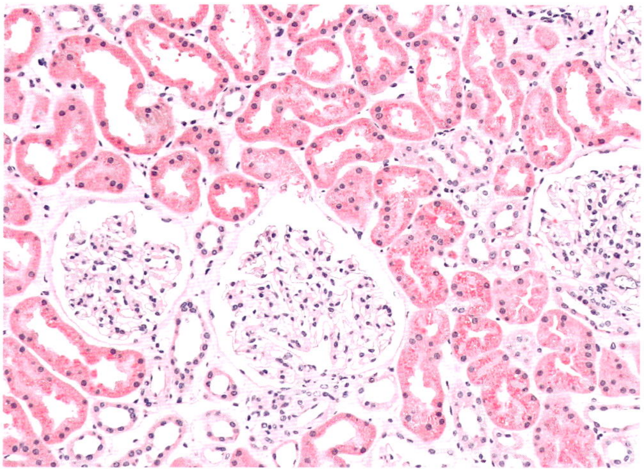

FIGURE 33.107 Biopsy specimen of the cortex of a kidney donated for transplantation. Note the compact arrangement of the tubules and the limited amount of interstitial tissue (H&E, ×200).

cells, lymphatic vessels, and nerves (654,655). In humans, estimates of the relative cortical interstitial volume range from 5% to 20%, with a mean of 12% (327,656,657). A significant increase with age has been reported (657). The relative volume of the renal interstitium increases from the cortex to the tip of the papilla.

The interstitium in both the cortex and medulla may be divided into different compartments (654). The cortical interstitium includes the periarterial connective tissue and the peritubular interstitium. The periarterial connective tissue constitutes a loose sheath around the intrarenal arteries and contains the lymphatic vessels and nerves. The sheath communicates with the peritubular interstitium. It should not be overinterpreted as representing focal interstitial fibrosis in the cortex. The peritubular interstitium includes the spaces between the glomeruli, tubules, and peritubular capillaries. In the normal cortex, it is inconspicuous on light microscopy, and the tubules and capillaries often have a back-to-back architectural appearance (Fig. 33.107). In the medulla, interstitial spaces include a narrow zone in the outer stripe of the outer medulla, the interbundle region of the inner stripe, and the expansive inner medulla. This appreciable amount of interstitium in the medulla should not be mistaken for interstitial fibrosis by the pathologist. The extracellular matrix of the interstitium consists of fibrils within a ground substance (654,655,658). It contains sulfated and nonsulfated glycosaminoglycans, fibronectin, laminin, interstitial collagens (types I, III, VI), and microfibrils.

Various types of fibroblasts, pericytes, and immune cells, including dendritic cells, macrophages, and lymphocytes are within the interstitium (659–661). Cortical fibroblasts have a spindle shape but may have a stellate appearance. They have elongated cell processes rich in actin filaments, contain prominent endoplasmic reticulum and express the enzyme ecto-5-nucleotidase (5-NT). The strongest 5-NT expression is from fibroblasts in the deep inner cortex. Fibroblasts in the renal capsule or in the periarterial connective tissue show little 5-NT expression. Fibroblasts (5-NT–positive) in the deep inner cortex produce erythropoietin (662,663). Studies have indicated that some erythropoietin-(EPO) producing interstitial cells may have originated from migrating neural crest cells (664) whereas others have reported these cells derive from FoxD1-expressing stromal cells (665).

Pericytes surround capillaries and display characteristic ultrastructural features. They have close contacts with endothelial cells through special membrane invaginations, called peg-sockets, which contain adherens junctions (666). They are present in the cortex and also in the medulla, especially surrounding the descending vasa recta, and are essential for kidney function (667). Evidence from numerous studies indicate kidney fibrosis results from a proliferation of resident renal cells transforming into myofibroblasts rather than tubular epithelial–mesenchymal transition (668). Genetic fate tracing studies indicate pericytes are progenitors of myofibroblasts in renal injury (669). A population of mesenchymal stem cell-like cells expressing Gli1 and localizing to the

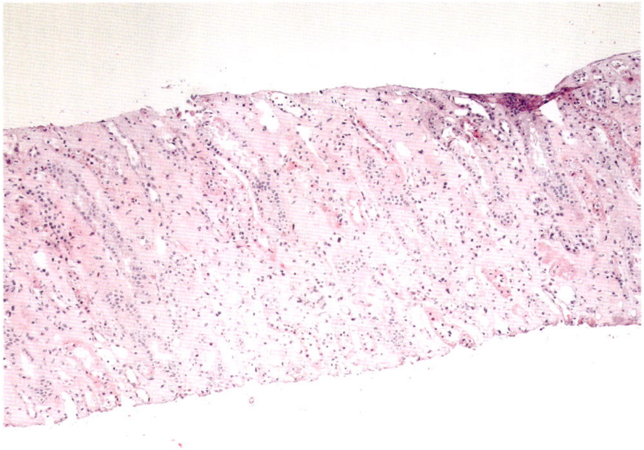

FIGURE 33.108 Renal biopsy specimen illustrating the inner medulla. Note the prominent amount of interstitium surrounding the tubules (H&E, ×200).

pericyte niche are reported to be a source of myofibroblasts in renal injury (670).

Complex networks of resident immune cells, including dendritic cells and macrophages, exist in the kidney (671). Moreover, there are different subsets of these cells. These cells have multiple immune roles and link innate and adaptive immunity. Dendritic cells are professional antigen-presenting cells. Multiple markers are necessary to distinguish renal dendritic and macrophages. In one study, CD11c$^+$, MHC-II$^+$, F4/80$^-$, CD64$^-$, ZBTB46$^+$ cells were designated as dendritic cells whereas CD11c$^+$, MHC-II$^+$, F4/80$^+$, CD64$^+$, ZBTB46$^-$ cells were called macrophages (672). Lymphocytes and granulocytes are uncommon in the kidney.

The medullary interstitium has a gelatinous appearance on light microscopy (Fig. 33.108). Several types of interstitial cells exist in the medulla, but the most distinctive are the lipid-laden cells, especially prominent in the inner medulla. These cells, called renomedullary interstitial cells, are often arranged in rows between the loop of Henle and the vasa recta, have irregular, long cytoplasmic processes, and contain lipid inclusions. These cells can be observed on 1-μm thick toluidine blue-stained sections of the inner medulla. The lipid droplets contain mainly triglycerides that are rich in unsaturated fatty acids, including arachidonic acid, phospholipids, and cholesterol (654). The renomedullary interstitial cells are believed to exert an antihypertensive function of the renal medulla, largely attributed to the production of medullipin (673,674). These cells also express cyclooxygenase (COX-2) (675).

VASCULATURE

The intricate microvasculature of the kidney underlies its complex hemodynamic functions (313). The segmental arteries, originating from the anterior and posterior

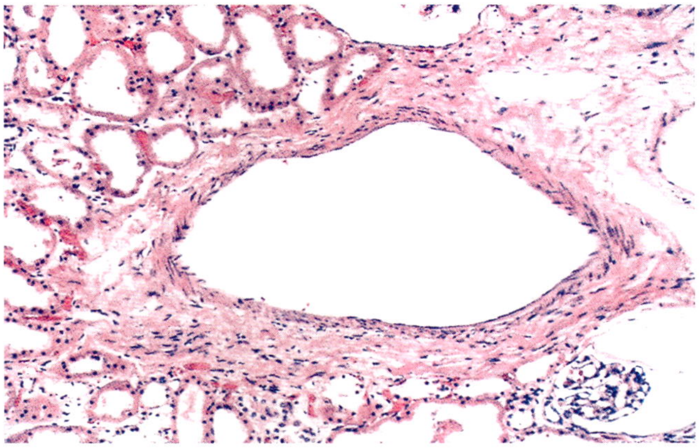

FIGURE 33.109 Micrograph of the corticomedullary junction illustrating an arcuate artery (H&E, ×125).

divisions of the main renal artery, divide to form the interlobar arteries, which course toward the cortex along the septa of Berlin between adjacent renal pyramids. At the corticomedullary junction, the interlobar arteries give rise to the arcuate arteries, which follow a gently curved course along the base of the pyramids parallel to the kidney surface (Fig. 33.109). The interlobular arteries branch sharply from the arcuate arteries and ascend in the cortex in a radial fashion toward the renal surface. Most afferent arterioles originate from the interlobular arteries, and each supplies a single glomerulus. The angle of origin of the afferent arterioles becomes less recurrent and more open as the interlobular arteries extend to the outer cortex (Fig. 33.110) (676). The length of the afferent arterioles is variable; average values of 170 to 280 μm have been reported (Fig. 33.111) (677,678). Some rare branches of the intrarenal arteries that do not terminate in glomeruli, the so-called aglomerular vessels, may result from degeneration of the connected glomeruli (676). Aglomerular arterioles near the corticomedullary junction have been observed to enter the medulla, and shunt arterioles between afferent

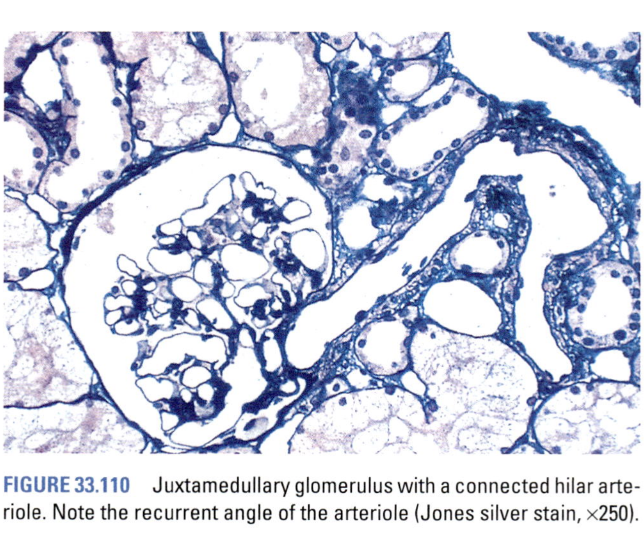

FIGURE 33.110 Juxtamedullary glomerulus with a connected hilar arteriole. Note the recurrent angle of the arteriole (Jones silver stain, ×250).

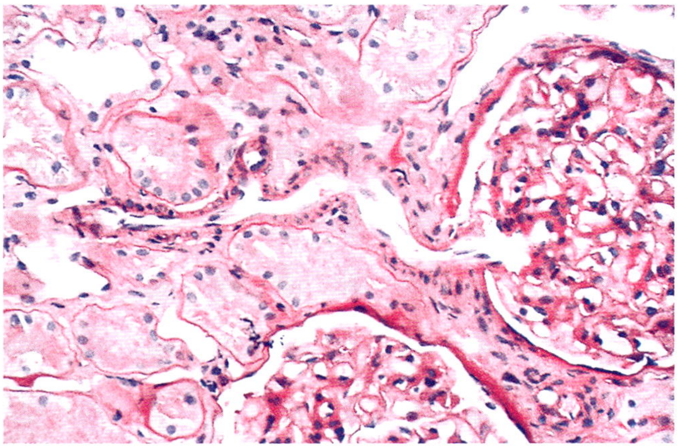

FIGURE 33.111 Micrograph depicting the transverse course of an afferent arteriole supplying a glomerulus (PAS, ×250).

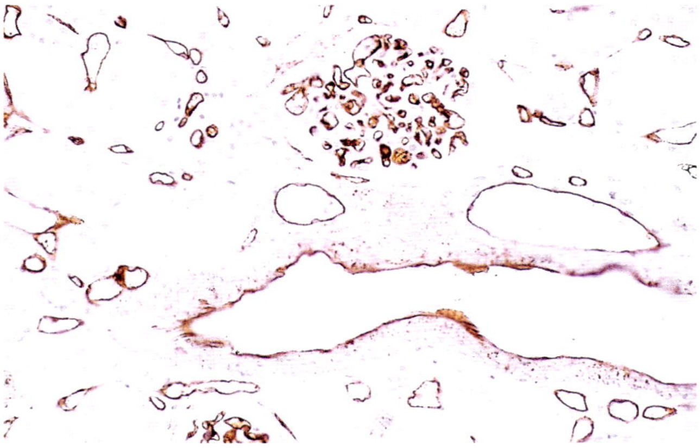

FIGURE 33.113 CD34 immunoperoxidase staining demonstrates a greater variety of vascular structures that label more intensely than with factor VIII. In the micrograph, arteries, veins, glomeruli, and peritubular capillaries are immunoreactive (CD34 immunohistochemistry, ×100).

and efferent arterioles have been reported (678–680). The wall structure of the intrarenal arteries and the proximal portion of the afferent arterioles resembles that of blood vessels of the same size elsewhere in the body. The endothelium stains for factor VIII–related antigen (Fig. 33.112) (681,682) and CD34 (Fig. 33.113) (683,684), whereas the muscularis stains for smooth muscle actin (Fig. 33.114) (685) and vimentin.

The efferent arterioles from the glomeruli branch to form a complex postglomerular microcirculation (Fig. 33.115). Although gradations exist, three basic types of efferent arterioles may be distinguished (685,686). The superficial or outer cortical efferent arterioles are fairly long and divide into extensive capillary networks that supply the convoluted tubules of the cortical labyrinth. They are readily identified by CD34 and smooth muscle actin staining (Figs. 33.114, 33.116, and 33.117).

The midcortical efferent arterioles are variable in length and supply the cortical labyrinth as well as the straight tubules of the medullary rays. With the exception of the outer cortex, there is dissociation between the tubule segments and the efferent arterioles of their parent glomeruli. In the midcortex and inner cortex, tubule segments are supplied by capillaries of efferent arterioles from other glomeruli (687,688). The efferent arterioles from juxtamedullary nephrons descend and supply the entire medulla. In contrast to the efferent arterioles of superficial and midcortical glomeruli (Fig. 33.118), those from juxtamedullary glomeruli are larger in diameter, display more layers of smooth muscle cells, and have more endothelial cells on cross-sections (313). In the outer stripe of the outer medulla, the efferent arterioles of juxtamedullary nephrons divide to form the descending vasa recta that descend in the vascular bundles but at intervals leave the bundles to form capillary plexuses.

The ascending (or venous) vasa recta drain the renal medulla. The ascending vasa recta from the inner medulla

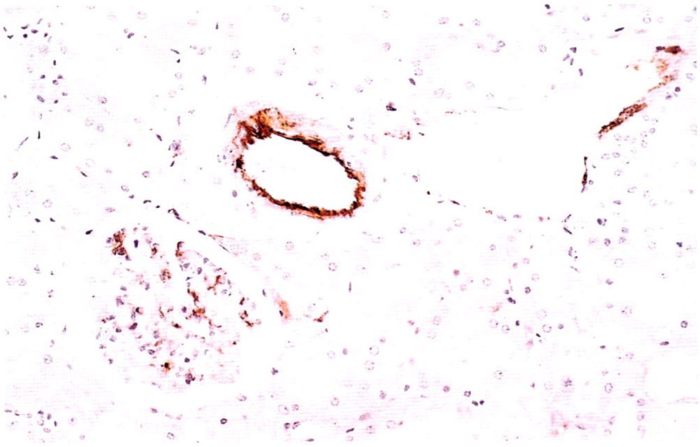

FIGURE 33.112 Factor VIII is produced by endothelium. The micrograph shows factor VIII immunoperoxidase staining of a medium-sized artery (**center**), vein (**right**), and glomerulus (**left**) (factor VIII immunohistochemistry, ×100).

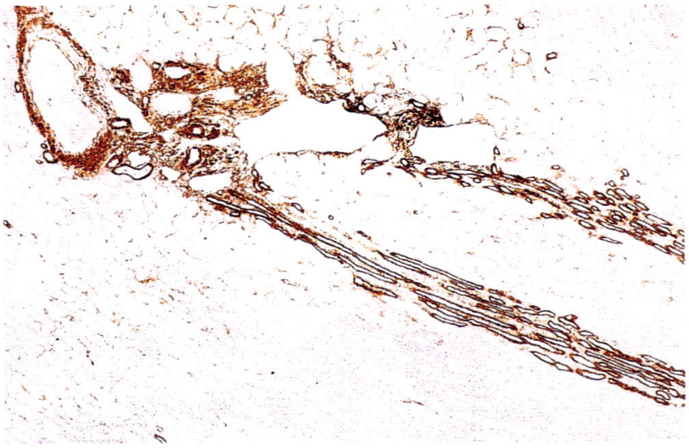

FIGURE 33.114 Smooth muscle actin immunoperoxidase stain illustrating a labeled large artery (**upper left**) and arterioles at the corticomedullary junction and two positive-stained columns of vasa recta (**midright**) penetrating the medulla. Two venous profiles (**upper center**) have minimal muscularis. (SMA immunohistochemistry, ×40)

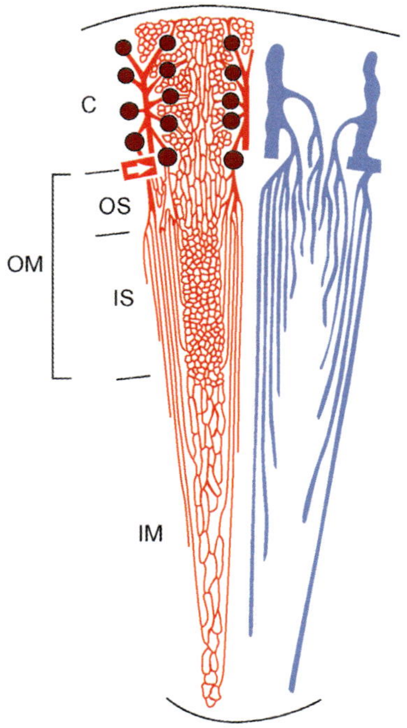

FIGURE 33.115 The renal microvasculature. The **left** side (*red*) illustrates the arterial vessels, glomeruli, and capillaries. An interlobular artery originates from an arcuate artery (*white arrow*) and gives rise to the afferent arterioles, which supply the glomeruli (*dark brown*). The efferent arterioles of the superficial and midcortical glomeruli supply the capillary plexuses of the cortical labyrinth and the medullary rays. The efferent arterioles of the juxtamedullary glomeruli descend into the medulla and form the descending vasa recta, which supply the adjacent capillary plexuses. Note the prominence of the capillary plexus in the inner stripe of the outer medulla. The **right** side (*blue*), which may be superimposed on the left side, displays the venous system. The ascending vasa recta drain the medulla and empty into the arcuate and interlobular veins, which drain the cortex. The vasa recta from the inner medulla ascends within the vascular bundles, whereas most vasa recta from the inner stripe ascend between the bundles (*C*, cortex; *OM*, outer medulla; *OS*, outer stripe; *IS*, inner stripe; *IM*, inner medulla). (Modified with permission from Kriz W, Kaissling B. Structural organization of the mammalian kidney. In: Seldin DW, Giebisch D, eds. *The Kidney: Physiology and Pathophysiology*. 3rd ed. Lippincott Williams & Wilkins; 2000:587–654.)

joins the vascular bundles, whereas most from the inner stripe of the outer medulla ascend between the bundles (313). This architectural arrangement creates a functional separation of the blood flow to the outer and the inner medulla. The close proximity of the arterial descending and venous ascending vasa recta within the vascular bundles allows for effective countercurrent exchange (313). The ascending vasa recta at the cortico–medullary junction empty into the arcuate and interlobular veins (Fig. 33.114), which do form extensive anastomoses in contrast to the arcuate arteries. The interlobular veins, which accompany the interlobular arteries, drain the cortex and empty into the arcuate veins. The intrarenal veins have less musculature than comparably sized veins in other organs (Fig. 33.114). The arcuate veins empty into the interlobar veins, which converge to form a single renal vein that exits at the hilum of the kidney.

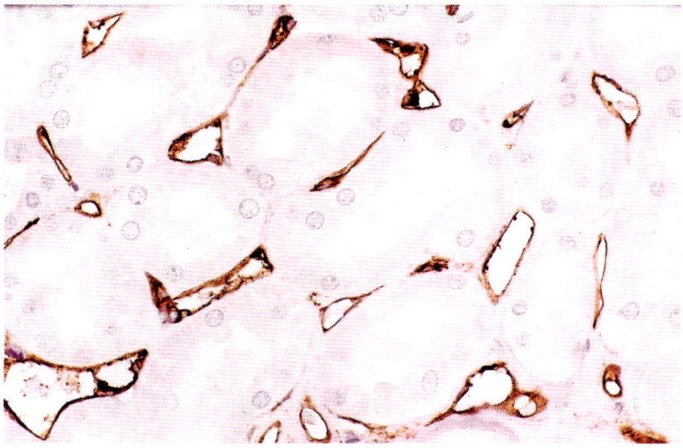

FIGURE 33.116 The extensive cortical peritubular capillary network is shown by endothelial labeling with CD34 antibody (CD34 immunohistochemistry, ×400).

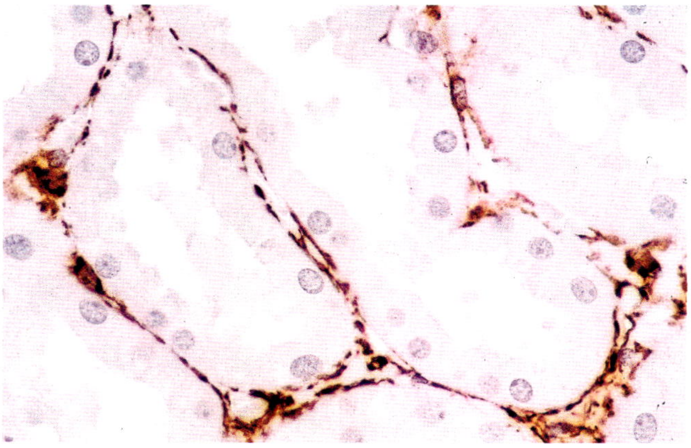

FIGURE 33.117 Smooth muscle actin expression complements and parallels the CD34 expression in documenting the cortical peritubular capillaries (SMA immunohistochemistry, ×400).

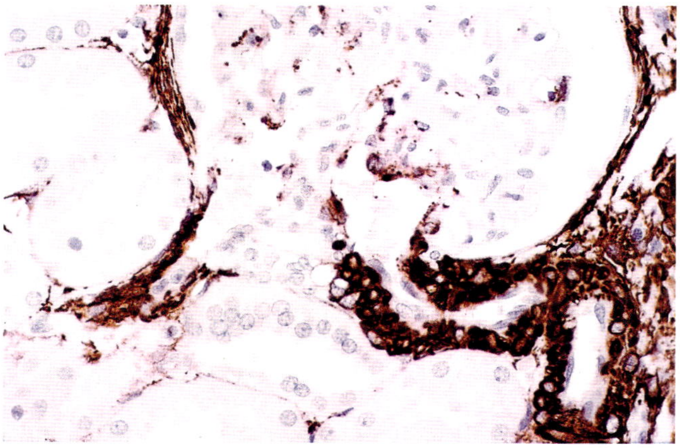

FIGURE 33.118 Smooth muscle actin immunoperoxidase of a superficial glomerulus delineating the more prominent smooth muscle investment of the afferent arteriole (**right**) compared with the efferent arteriole (**left**) (SMA immunohistochemistry, ×400).

LYMPHATIC VESSELS

Lymphatic networks in the kidney include a deeper cortical system and less extensive capsular and subcapsular networks (689–691). The less prominent lymphatic vessels within the capsule drain into the subcapsular lymphatic channels which appear to connect with the larger lymphatic vessels in the cortex. The larger channels surround the interlobular arteries and empty into arcuate and interlobal lymphatic vessels which finally drain into the larger lymph vessels at the hilum (Fig. 33.119). They are often more noticeable around interlobular veins than around interlobular arteries but are sporadic around glomeruli or between tubules (692). The interlobal and hilar lymphatic channels possess valves. Lymphatic vessels are sparse to absent in the medulla of health kidneys (691,692). In the cortex, they are embedded in the loose periarterial connective tissue but they are not conspicuous on routine histologic sections. The lymphatic vessels may allow exchange with the venous system and serve as a route for the intrarenal distribution of hormones and inflammatory cells (693). Several markers of lymphatic endothelium are now available including: VEGFR-3 (receptor for VEGF-C), LYVE-1 (hyaluronate receptor), hyakuronate receptor, Prox-1 (lymphatic transcription factor), podoplanin (a membrane glycoprotein), and D2-40 (lymphatic endothelium-specific protein) (694,695). Interestingly, fenestrated ascending vasa recta have been reported to express both endothelial and lymphatic markers, suggesting that they are specialized hybrid vessels (696).

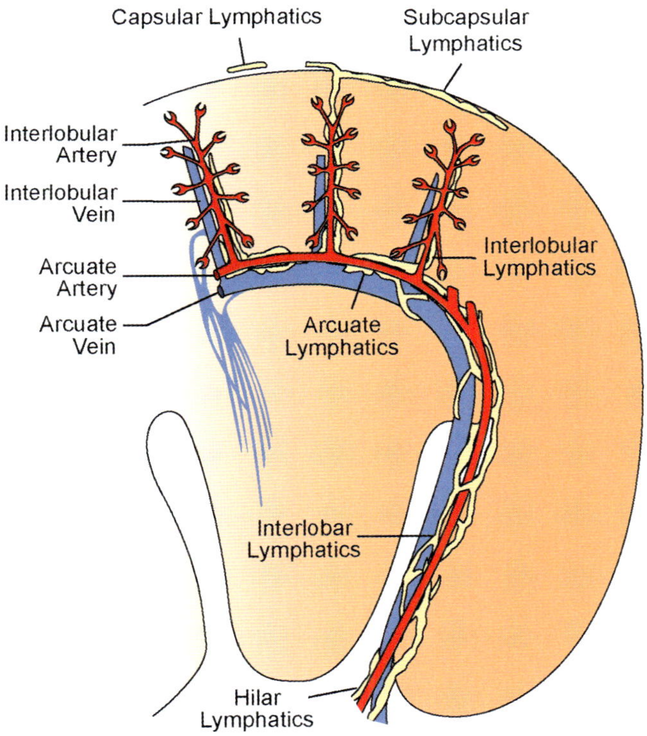

FIGURE 33.119 The lymphatic vessels of the kidney. The arteries (*red*), veins (*blue*), and lymphatics (*yellow*) are illustrated. The lymphatics are primarily distributed in the cortex, although a subcapsular network is also present. Note the absence of the lymphatics in the medulla. (Modified with permission from Madsen KM, Tisher CC. Anatomy of the kidney. In: Brenner BM, ed. *Brenner and Rector's The Kidney*. 7th ed. WB Saunders; 2004:3–72. Copyright © 2004 Elsevier.)

NERVES

The nerve supply to the kidney derives from postganglionic fibers primarily from the celiac plexus (697). The nerve fibers generally accompany the arteries and arterioles in the cortex and the outer medulla (698). They generally lie within the perivascular interstitium but penetrate vessel walls to innervate the smooth muscle. On 1-μm thick toluidine blue–stained sections, a nerve often has a small round configuration but should not be mistaken for a small glomerulus (Fig. 33.120). Staining for myelin (Fig. 33.121) or for peripheral axons with antibodies against phosphoneurofilament (pNF) (Fig. 33.122) demonstrates the nerve fibers (699–702). There is prominent innervation of the juxtaglomerular apparatus (Fig. 33.123) (703). The efferent arterioles and the descending vasa recta are accompanied by nerve fibers (703). Although there is innervation of the tubules, it is less extensive than of the vasculature (704,705). Relative to its length, the thick ascending limb receives the largest nerve supply of any tubule segment (706). Studies using the efferent nerve marker tyrosine hydroxylase and the afferent nerve marker calcitonin-related peptide showed efferent fibers predominant along the renal artery (707).

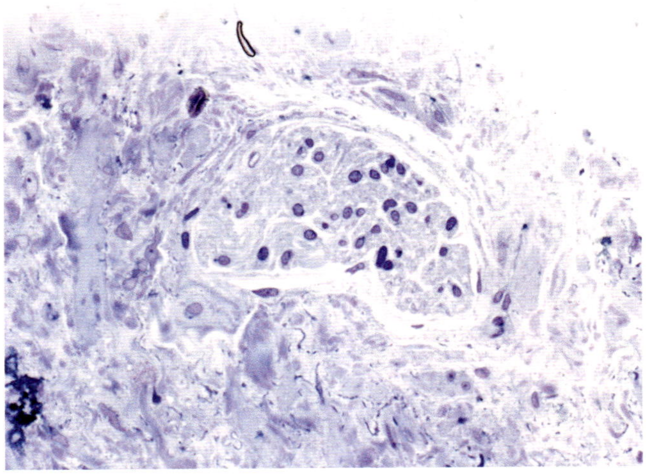

FIGURE 33.120 Light micrograph of renal cortex showing a small nerve. It has a circumscribed contour but in contrast to a glomerulus, lacks capillaries and their lumens and a Bowman capsule (1-μm thick toluidine blue–stained Epon section, ×630).

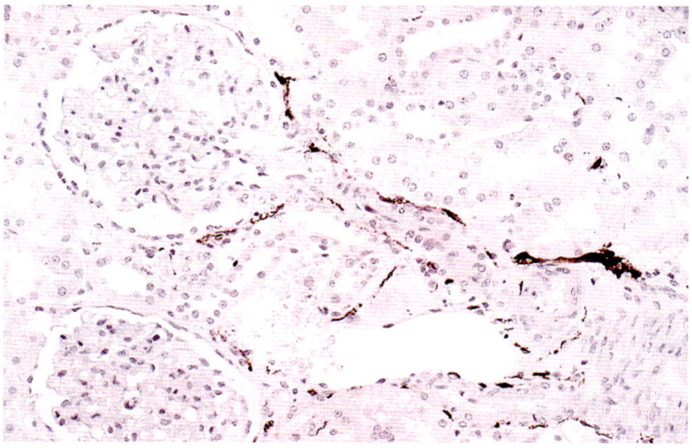

FIGURE 33.121 S100 immunohistochemical stain demonstrating nerves extending along the afferent arteriole to the vascular pole of the glomerulus (S100 immunohistochemistry, ×100).

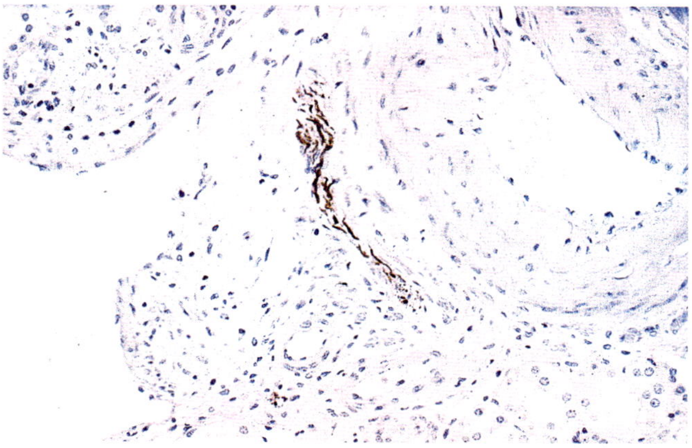

FIGURE 33.122 Phosphoneurofilament (pNF) immunohistochemical stain showing a nerve between an artery (**right**) and a vein (**left**) (pNF immunohistochemistry, ×100).

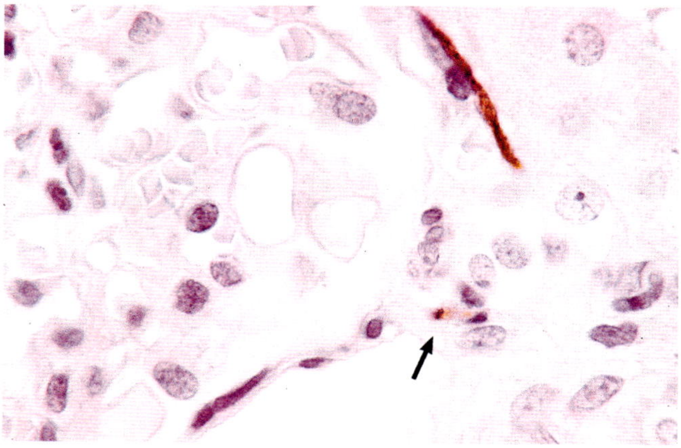

FIGURE 33.123 Two axons are demonstrated in this view of the vascular pole of a glomerulus (**center**). One axon has a longitudinal profile (**upper right**), and the other is observed in cross-section as a dot (*arrow*) (pNF immunohistochemistry, ×630).

ACKNOWLEDGMENTS

Portions of this chapter are adapted from contributions by Drs. Bruce Beckwith, Byron Croker, and William Clapp in previous editions.

REFERENCES

1. Grobstein C. Inductive interaction in the development of the mouse metanephros. *J Exp Zool.* 1955;130:319–340.
2. Potter EL. *Normal and Abnormal Development of the Kidney*. Year Book Medical Publishers, Inc; 1972.
3. Oliver J. *Nephrons and Kidneys: A Quantitative Study of Development and Evolutionary Mammalian Renal Architectonics*. Hoeber Medical Division, Harper & Row; 1968.
4. Saxen L. *Organogenesis of the Kidney*. Cambridge University Press; 1987.
5. Clapp WL, Abrahamson DR. Development and gross anatomy of the kidney. In: Tisher CC, Brenner BM, eds. *Renal Pathology*. 2nd ed. JB Lippincott; 1994;3–59.
6. Vize PD, Woolf AS, Bard JBL. *The Kidney: From Normal Development to Congenital Disease*. Academic Press; 2003.
7. Dressler GR. Advances in early kidney specification, development and patterning. *Development*. 2009;136:3863–3874.
8. Costantini F, Kopan R. Patterning a complex organ: branching morphogenesis and nephron segmentation in kidney development. *Dev Cell*. 2010;18:698–712.
9. Little MH, McMahon AP. Mammalian kidney development: principles, progress, and projections. *Cold Spring Harb Perspect Biol*. 2012;4:a008300.
10. Little MH, ed. *Kidney Development, Disease, Repair and Regeneration*. Elsevier Inc; 2016.
11. McMahon AP. Development of the mammalian kidney. *Curr Topics in Dev Biol*. 2016;117:31–64.
12. Oxburgh L, Carroll TJ, Cleaver O, et al. (Re)Building a kidney. *J Am Soc Nephrol*. 2017;28:1370–1378.
13. Mugford JW, Sipila P, Kobayashi A, Behringer RR, McMahon AP. Hoxd11 specifies a program of metanephric kidney development within the intermediate mesoderm of the mouse kidney. *Dev Biol*. 2008;319:396–405.
14. Georgas KM, Chiu HS, Rumballe BA, Lesieur E, Little MH. Expression of metanephric nephron-patterning genes in differentiating mesonephric tubule. *Dev Dyn*. 2011;240:1600–1612.
15. Masse J, Watrin T, Laurent A, Deschamps S, Guerrier D, Pellerin I. The developing female genital tract: from genetics to epigenetics. *Int J Dev Biol*. 2009;53:411–424.
16. Peter K. *Untersuchungen ueber Bau und Entwicklung der Niere*. Gustav Fischer; 1927.
17. Davies J. Development of the ureteric bud. In: Vize PD, Woolf AS, Bard JB, eds. *The Kidney: From Normal Development to Congenital Disease*. Academic Press; 2003:165–179.
18. Al-Awqati Q, Goldberg MR. Architectural patterns in branching morphogenesis in the kidney. *Kidney Int*. 1998;54:1832–1842.

19. Watanabe T, Costantini F. Real-time analysis of ureteric bud branching morphogenesis in vitro. *Dev Biol.* 2004;271: 98–108.
20. Sweeny D, Lindstrom N, Davies JA. Developmental plasticity and regenerative capacity in the renal ureteric/collecting duct system. *Development.* 2008;135:2505–2510.
21. Herring PT. The development of the Malpighian bodies of the kidney, and its relation to pathological changes which occur in them. *J Pathol Bacteriol.* 1900;6:459–496.
22. Huber GC. On the development and shape of uriniferous tubules of certain of the higher mammals. *Am J Anat.* 1905;4(Suppl):1–98.
23. Bard J. The metanephros. In: Vize PD, Woolf AS, Bard JB, eds. *The Kidney: From Normal Development to Congenital Disease*. Academic Press; 2003:139–148.
24. Sariola H, Sainio K, Bard J. Fates of the metanephric mesenchyme. In: Vize PD, Woolf AS, Bard JB, eds. *The Kidney: From Normal Development to Congenital Disease*. Academic Press; 2003:181–193.
25. Harding SD, Armit C, Armstrong J, et al. The GUDMAP database -an online resource for genitourinary research. *Development.* 2011;138:2845–2853.
26. Potter SS, Brunskill EW, Patterson LT. Defining the genetic blueprint of kidney development. *Pediatr Nephrol.* 2011;26:1469–1478.
27. Brunskill EW, Park JS, Chung E, Chen F, Magella B, Potter SS. Single cell-dissection of early kidney development: Multilineage priming. *Development.* 2014;141:3093–3101.
28. Potter SS. Single-cell RNA sequencing for the study of development, physiology and disease. *Nat Rev Nephrol.* 2018; 14:479–492.
29. James RG, Schultheiss TM. BMP signaling promotes intermediate mesoderm gene expression in a dose-dependent, cell-autonomous and translation-dependent manner. *Dev Biol.* 2005;288:113–125.
30. Giovanni VD, Alday A, Chi L, et al. ALK3 controls nephron number and androgen production via lineage-specific effects in intermediate mesoderm. *Development.* 2011;138: 2717–2727.
31. Noon PBE, Brak H, Guttmann-Raviv N, Reshef R. Interplay between activin and Hox genes determines the formation of the kidney morphogenetic field. *Development.* 2009;136:1995–2004.
32. Patterson LT, Pembaur M, Potter SS. Hoxa11 and Hoxd11 regulate branching morphogenesis of the ureteric bud in the developing kidney. *Development.* 2001;128: 2153–2161.
33. Wellik DM, Hawkes PJ, Capecchi MR. Hox11 paralogous genes are essential for metanephric kidney induction. *Genes Dev.* 2002;16:1423–1432.
34. Bouchard M, Souabni A, Mandler M, Neubüser A, Busslinger M. Nephric lineage specification by Pax2 and Pax8. *Genes Dev.* 2002;16:2958–2970.
35. Grote D, Souabni A, Busslinger M, Bouchard M. Pax2/8-regulated Gata 3 expression is necessary for morphogenesis and guidance of the nephric duct in the developing kidney. *Development.* 2006;133:53–61.
36. Marose TD, Merkel CE, McMahon AP, Carroll TJ. Beta-catenin is necessary to keep cells of ureteric bud/Wolffian duct epithelium in a precursor state. *Dev Biol.* 2008;314: 112–126.
37. Grote D, Boualia SK, Souabni A, et al. Gata3 acts downstream of beta-catenin signaling to prevent ectopic metanephric kidney induction. *PLoS Genet.* 2008;4: e1000316.
38. Chia I, Grote D, Marcotte M, Batourina E, Mendelsohn C, Bouchard M. Nephric duct insertion is a crucial step in urinary tract maturation that is regulated by a Gata3-Raldh2-Ret molecular network in mice. *Development.* 2011;138: 2089–2097.
39. Hoshi M, Reginensi A, Joens MS, Fitzpatrick JAJ, McNeill H, Jain S. Reciprocal spatiotemporally controlled apoptosis regulates Wolffian duct cloaca fusion. *J Am Soc Nephrol.* 2018;29:775–783.
40. Schedl A. Renal abnormalities and their developmental origin. *Nat Rev Genet.* 2007;8:791–802.
41. Uetani N, Bouchard M. Plumbing in the embryo: developmental defects of the urinary tract. *Clin Genet.* 2009;75: 307–317.
42. Costantini F. GDNF/Ret signaling and renal branching morphogenesis: from mesenchymal signals to epithelial cell behaviors. *Organogenesis.* 2010;6:252–262.
43. Costantini F, Shakya R. GDNF/ret signaling and the development of the kidney. *Bioessays.* 2006;28:117–127.
44. Boyle S, deCaestecker M. Role of transcriptional networks in coordinating early events during kidney development. *Am J Physiol Renal Physiol.* 2006;291:F1–F8.
45. Gong KQ, Yallowitz AR, Sun H, Dressler GR, Wellik DM. A Hox-Eya-Pax complex regulates early kidney developmental gene expression. *Mol Cell Biol.* 2007;27: 7661–7668.
46. Linton JM, Martin GR, Reichardt LF. The Ecap mesenchyme protein nephronectin promotes kidney development via integrin $\alpha 8\beta 1$-mediated stimulation of Gdnf expression. *Development.* 2007;134:2501–2509.
47. Grieshammer U, Ma L, Plump AS, Wang F, Tessier-Lavigne M, Martin GR. SLIT2-mediated ROBO2 signaling restricts kidney induction to a single site. *Dev Cell.* 2004;6: 709–717.
48. Michos O, Goncalves A, Lopez-Rios J, et al. Reduction of BMP4 activity by gremlin1 enables ureteric bud outgrowth and GDNF/WNT11 feedback signaling during kidney branching morphogenesis. *Development.* 2007;134: 2397–2405.
49. Rosselot C, Spraggon L, Chia I, et al. Non-cell-autonomous retinoid signaling is crucial for renal development. *Development.* 2010;137:283–292.
50. Basson MA, Akbulut S, Watson-Johnson J, et al. Sprouty 1 is a critical regulator of GDNF/RET-mediated kidney induction. *Dev Cell.* 2005;8:229–239.
51. Takahashi M. The GDNF/RET signaling pathway and human disease. *Cytokine Growth Factor Rev.* 2001;12: 361–373.
52. Tang MJ, Cai Y, Tsai SJ, Wang YK, Dressler GR. Ureteric bud outgrowth in response to RET activation is mediated by phophatidylinositol 3-kinase. *Dev Biol.* 2002;243: 128–136.
53. Kim D, Dressler GR. PTEN modulates GDNF/RET mediated chemotaxis and branching morphogenesis in the developing kidney. *Dev Biol.* 2007;307:290–299.
54. Jain S. The many faces of RET dysfunction in the kidney. *Organogenesis.* 2009;5:177–190.

55. Willecke R, Heuberger J, Grossmann K, et al. The tyrosine phophatase Shp2 acts downstream of GDNF/Ret in branching morphogenesis of the developing mouse kidney. *Dev Biol.* 2011;360:310–317.
56. Lu BC, Cebrian C, Chi X, et al. Etv4 and Etv5 are required downstream of GDNF and Ret for kidney branching morphogenesis. *Nat Genet.* 2009;41:1295–1302.
57. Ueland J, Yuan A, Marlier A, Gallagher AR, Karihaloo A. A novel role for the chemokine receptor Cxcr4 in kidney morphogenesis: an in vitro study. *Dev Dyn.* 2009;238:1083–1091.
58. Takabatake Y, Sugiyama T, Kohara H, et al. The CXCL12 (SDF-1)/CXCR4 axis is essential for the development of renal vasculature. *J Am Soc Nephrol.* 2009;20:1714–1723.
59. Reginensi A, Clarkson M, Neirijnck Y, et al. SOX9 controls epithelial branching by activating RET effector genes during kidney development. *Hum Mol Genet.* 2011;20:1143–1153.
60. Bates CM. Role of fibroblast growth factor receptor signaling in kidney development. *Am J Physiol Renal Physiol.* 2011;301:F245–F251.
61. Trueb B. Biology of FGFRL1, the fifth fibroblast growth factor receptor. *Cell Mol Life Sci.* 2011;68:951–964.
62. Poladia DP, Kish K, Kutay B, et al. Role of fibroblast growth factor receptors 1 and 2 in the metanephric mesenchyme. *Dev Biol.* 2006;291:325–339.
63. Sims-Lucas S, Cusack B, Baust J, et al. Fgfr1 and the IIIc isoform of Fgfr2 play critical roles in the metanephric mesenchyme mediating early inductive events in kidney development. *Dev Dyn.* 2011;240:240–249.
64. Sims-Lucas S, Cusack B, Eswarakumar VP, Zhang J, Wang F, Bates CM. Independent roles of Fgfr2 and Frs2α in ureteric epithelium. *Development.* 2011;138:1275–1280.
65. Yosypiv IV, Schroeder M, El-Dahr SS. Angiotensin II type I receptor-EGF receptor cross-talk regulates ureteric bud branching morphogenesis. *J Am Soc Nephrol.* 2006;17:1005–1014.
66. Song R, Spera M, Garrett C, Yosypiv IV. Angiotensin II-induced activation of c-Ret signaling is critical in ureteric bud branching morphogenesis. *Mech Dev.* 2010;127:21–27.
67. Matthew S, Chen X, Pozzi A, Zent R. Integrins in renal development. *Pediatr Nephrol.* 2012;27:891–900.
68. Wu W, Kitamura S, Truong DM, et al. β1-integrin is required for kidney collecting duct morphogenesis and maintenance of renal function. *Am J Physiol Renal Physiol.* 2009;297:F210–F217.
69. Zhang X, Mernaugh G, Yang DH, et al. β1 integrin is necessary for ureteric bud branching morphogenesis and maintenance of collecting duct structural integrity. *Development.* 2009;136:3357–3366.
70. Mathew S, Palamuttam RJ, Mernaugh G, et al. Talin regulates integrin β1-dependent and –independent cell functions in ureteric bud development. *Development.* 2017;144:4148–4158.
71. Smeeton J, Zhang X, Bulus N, et al. Integrin-linked kinase regulates p38 MAPK-dependent cell cycle arrest in ureteric bud development. *Development.* 2010;137:3233–3243.
72. Miner JH, Yurchenco PD. Laminin functions in tissue morphogenesis. *Annu Rev Cell Dev Biol.* 2004;20:255–284.
73. Yang DH, McKee KK, Chen ZL, et al. Renal collecting system growth and function depend upon embryonic γ1 laminin expression. *Development.* 2011;138:4535–4544.
74. Cain JE, Hartwig S, Bertram JF, Rosenblum ND. Bone morphogenetic protein signaling in the developing kidney: present and future. *Differentiation.* 2008;76:831–842.
75. Miyazaki Y, Oshima K, Fogo A, Hogan BL, Ichikawa I. Bone morphogenetic protein 4 regulates the budding site and elongation of the mouse ureter. *J Clin Invest.* 2000;105:863–873.
76. Cain JE, Bertram JF. Ureteric branching morphogenesis in BMP4 heterozygous mutant mice. *J Anat.* 2006;209:745–755.
77. Hartwig S, Bridgewater D, Di Giovanni V, Cain J, Mishina Y, Rosenblum ND. BMP receptor ALK3 controls collecting system development. *J Am Soc Nephrol.* 2008;19:117–124.
78. Goncalves A, Zeller R. Genetic analysis reveals an unexpected role of BMP7 in initiation of ureteric bud outgrowth in mouse embryos. *PLoS One* 2011;6:e19370.
79. Tran TS, Kolodkin AL, Bharadwaj R. Semaphorin regulation of cellular morphology. *Annu Rev Cell Dev Biol.* 2007;23:263–292.
80. Tufro A, Teichman J, Woda C, Villegas G. Semaphorin3a inhibits ureteric bud branching morphogenesis. *Mech Dev.* 2008;125:558–568.
81. Korostylev A, Worzfeld T, Deng S, et al. A functional role for semaphorin 4D/plexin B1 interactions in epithelial branching morphogenesis during organogenesis. *Development.* 2008;135:3333–3343.
82. Perala N, Jakobson M, Ola R, et al. Sema4C-Plexin B2 signaling modulates ureteric branching in developing kidney. *Differentiation.* 2011;81:81–91.
83. Reginensi A, Enderle L, Gregorieff A, Johnson RL, Wrana JL, McNeill H. A critical role for NF2 and the Hippo pathway in branching morphogenesis. *Nat Commun.* 2017;7:12309.
84. Reginensi A, Hoshi M, Boualia SK, et al. Yap and Taz are required for Ret-dependent urinary tract morphogenesis. *Development.* 2015;142:2696–2703.
85. Cebrian C, Borodo K, Charles N, Herzlinger DA. Morphometric index of the developing murine kidney. *Dev Dyn.* 2004;231:601–608.
86. Short KM, Combes AN, Lefevre J, et al. Global quantification of tissue dynamics in the developing mouse kidney. *Dev Cell.* 2014;29:188–202.
87. Sampogna RV, Schneider L, Al-Awqati Q. Developmental programming of branching morphogenesis in the kidney. *J Am Soc Nephrol.* 2015;26:2414–2422.
88. McNeill H, Woodgett JR. When pathways collide: collaboration and connivance among signaling proteins in development. *Nat Rev Mol Cell Biol.* 2010;11:404–413.
89. McNeill H. Planar cell polarity and the kidney. *J Am Soc Nephrol.* 2009;20:2104–2111.
90. Fischer E, Legue E, Doyen A, et al. Defective planar cell polarity in polycystic disease. *Nat Genet.* 2006;38:21–23.

91. Karner CM, Chirumamilla R, Aoki S, Igarashi P, Wallingford JB, Carroll TJ. Wnt9b signaling regulates planar cell polarity and kidney tubule morphogenesis. *Nat Genet.* 2009;41: 793–799.
92. Yu J, Carroll TJ, Rajagopal J, et al. A Wnt7b-dependent pathway regulates the orientation of epithelial cell division and establishes the cortico-medullary axis of the mammalian kidney. *Development.* 2009;136:161–171.
93. Pietila I, Ellwanger K, Railo A, et al. Secreted Wnt antagonist Dickkopf-1 controls kidney papilla development coordinated by Wnt-7b signaling. *Dev Biol.* 2011;353:50–60.
94. Liu Y, Chattopadhyay N, Qin S, et al. Coordinate integrin and c-Met signaling regulate Wnt gene expression during epithelial morphogenesis. *Development.* 2009;136: 843–853.
95. Michael L, Davies JA. Pattern and regulation of cell proliferation during murine ureteric bud development. *J Anat.* 2004;204:241–255.
96. Shakya R, Watanabe T, Costantini F. The role of GDNF/Ret signaling in ureteric bud cell fate and branching morphogenesis. *Dev Cell.* 2005;8:65–74.
97. Meyer TN, Schwesinger C, Bush KT, et al. Spatiotemporal regulation of morphogenetic molecules during in vitro branching of the isolated ureteric bud; toward a model of branching through budding in the developing kidney. *Dev Biol.* 2004;275:44–67.
98. Michael L, Sweeney DE, Davies JA. A role for microfilament-based contraction in branching morphogenesis of the ureteric bud. *Kidney Int.* 2005;68:2010–2018.
99. Kuure S, Cebrian C, Machingo Q, et al. Actin depolymerizing factors cofilin1 and destrin are required for ureteric bud branching morphogenesis. *PLoS Genet.* 2010;6: e1001176.
100. Ola R, Jakobson M, Kvist J, et al. The GDNF target Vsnl1 marks the ureteric tip. *J Am Soc Nephrol.* 2011;22: 274–284.
101. Majumdar A, Vainio S, Kispert A, McMahon J, McMahon AP. Wnt11 and Ret/Gdnf pathways cooperate in regulating ureteric branching during metanephric kidney development. *Development.* 2003;130:3175–3185.
102. Packard A, Georgas K, Michos O, et al. Luminal mitosis drives epithelial dispersal within the branching ureteric bud. *Dev Cell.* 2013;27:319–330.
103. Kim YH, Kwon TH, Frische S, et al. Immunocytochemical localization of pendrin in intercalated cell subtypes in rat and mouse kidney. *Am J Physiol Renal Physiol.* 2002; 283:F744–F754.
104. Song HK, Kim WY, Lee HW, et al. Origin and fate of pendrin-positive intercalated cells in developing mouse kidney. *J Am Soc Nephrol.* 2007;18:2672–2682.
105. Hiatt MJ, Ivanova L, Toran N, Tarantal AF, Matsell DG. Remodeling of the fetal collecting duct epithelium. *Am J Pathol.* 2010;176:630–637.
106. Kim J, Cha H, Tisher CC, Madsen KM. Role of apoptotic and nonapoptotic cell death in removal of intercalated cells from developing rat kidney. *Am J Physiol Renal Physiol.* 1996;270:F575–F592.
107. El-Dahr SS, Li Y, Gutierrez E, et al. p63+ ureteric bud tip cells are progenitors of intercalated cells. *JCI Insight.* 2017;2:e89996.
108. Blomqvist SR, Vidarsson H, Fitzgerald S, et al. Distal renal tubular acidosis in mice that lack the forkhead transcription factor Foxi1. *J Clin Invest.* 2004;113:1560–1570.
109. Wu H, Chen L, Zhou Q, et al. Aqp2-expressing cells give rise to renal intercalated cells. *J Am Soc Nephrol.* 2013;24: 243–252.
110. Werth M, Schmidtt-Ott KM, Leete T, et al. Transcription factor *TFCP2L1* patterns cells in the mouse kidney collecting duct. *Elife.* 2017;6:e24265.
111. Jeong HW, Jeon US, Koo BK, et al. Inactivation of notch signaling in the renal collecting duct causes nephrogenic diabetes insipidus in mice. *J Clin Invest.* 2009;119: 3290–3300.
112. Guo Q, Wang Y, Tripathi P, et al. Adam10 mediates the choice between principal cells and intercalated cells in the kidney. *J Am Soc Nephrol.* 2015;26:149–159.
113. Grassmeyer J, Mukherjee M, deRiso J, et al. Elf5 is a principal lineage specific transcription factor in the kidney that contributes to Aqp2 and Avpr2 gene expression. *Dev Biol.* 2017;424:77–89.
114. Mugford JW, Sipila P, McMahon JA, McMahon AP. Osr1 expression demarcates a multi-potent population of intermediate mesoderm that undergoes progressive restriction to an Osr1-dependent nephron progenitor compartment within the mammalian kidney. *Dev Biol.* 2008;324:88–98.
115. Sajithlal G, Zou D, Silvius D, Xu PX. Eya1 acta as a critical regulator for specifying the metanephric mesenchyme. *Dev Biol.* 2005;284:323–336.
116. Xu PX, Adams J, Peters H, Brown MC, Heaney S, Maas R. Eya1-deficient mice lack ears and kidneys and show abnormal apoptosis of organ primordial. *Nat Genet.* 1999;23: 113–117.
117. Li X, Oghi KA, Zhang J, et al. Eya protein phosphatase activity regulates Six1-Dach-Eya transcriptional effects in mammalian organogenesis. *Nature.* 2003;426:247–254.
118. Xu PX, Zhang W, Huang L, Maire P, Laclef C, Silvius D. Six1 is required for the early organogenesis of mammalian kidney. *Development.* 2003;130:3085–3094.
119. Nie X, Xu J, El-Hashash A, Xu PX. Six1 regulates Grem1 expression in the metanephric mesenchyme to initiate branching morphogenesis. *Dev Biol.* 2011;352:141–151.
120. Nishinakamura R, Matsumoto Y, Nakao K, et al. Murine homolog of SALL1 is essential for ureteric bud invasion in kidney development. *Development.* 2001;128:3105–3115.
121. Kiefer SM, Robbins L, Stumpff KM, Lin C, Ma L, Rauchman M. Sall1-dependent signals affect Wnt signaling and ureter tip fate to initiate kidney development. *Development.* 2010;137:3099–3106.
122. Uchiyama Y, Sakaguchi M, Terabayashi T, et al. Kif26b, a kinesin family gene, regulates adhesion of the embryonic kidney mesenchyme. *Proc Natl Acad Sci U S A.* 2010;107: 9240–9245.
123. Carroll TJ, Das A. Defining the signals that constitute the nephron progenitor niche. *J Am Soc Nephrol.* 2013;24:873–876.
124. Kopan R, Chen S, Little N. Nephron progenitor cells: shifting the balance of self-renewal and differentiation. *Curr Top Dev Biol.* 2014;107:293–331.
125. O'Brien LL, McMahon AP. Induction and patterning of the metanephric nephron. *Semin Cell Dev Biol.* 2014;36: 31–38.

126. Carroll TJ, Park JS, Hayashi S, Majumdar A, McMahon AP. Wnt9b plays a central role in the regulation of mesenchymal to epithelial transitions underlying organogenesis of the mammalian urogenital system. *Dev Cell.* 2005;9:283–292.
127. Mugford JW, Yu J, Kobayashi A, McMahon AP. High-resolution gene expression analysis of the developing mouse kidney defines novel cellular compartments within the nephron progenitor population. *Dev Biol.* 2009;333:312–323.
128. Combes AN, Lefevre JG, Wilson S, Hamilton NA, Little MH. Cap mesenchyme cell swarming during kidney development is influenced by attraction, repulsion and adhesion to the ureteric tip. *Dev Biol.* 2016;418:297–306.
129. Barak H, Huh SH, Chen S, et al. FGF9 and FGF10 maintain the stemness of nephron progenitors in mice and man. *Dev Cell.* 2012;22;1191–1207.
130. Giovanni VD, Walker KA, Bushnell D, et al. Fibroblast growth factor receptor-Frs2α signaling is critical for nephron progenitors. *Dev Biol.* 2015;400:82–93.
131. Tomita M, Asada M, Asada N, et al. Bmp7 maintains undifferentiated kidney progenitor population and determines nephron numbers at birth. *PLoS One.* 2013;8:e73554.
132. Brown AC, Muthukrishnan SD, Guay JA, et al. Role for compartmentalization in nephron progenitor differentiation. *Proc Natl Acad Sci U S A.* 2013;110:4640–4645.
133. Kanda S, Tanigawa S, Ohmori T, et al. Sall1 maintains nephron progenitors and nascent nephrons by acting as both an activator and a repressor. *J Am Soc Nephrol.* 2014;25:2584–2595.
134. Li Y, Liu J, Li W, et al. p53 enables metabolic fitness and self renewal of nephron progenitors. *Development.* 2015;142:1228–1241.
135. Xu J, Liu H, Park JS, Lan Y, Jiang R. Osr1 acts downstream of and interacts synergistically with Six2 to maintain nephron progenitor cells during kidney organogenesis. *Development.* 2014;141:1442–1452.
136. Liu J, Edgington-Giordano F, Dugas C, et al. Regulation of nephron progenitor cell self-renewal by intermediary metabolism. *J Am Soc Nephrol.* 2017;28:3323–3335.
137. Hilliard SA, El-Dahr SS. Epigenetics of renal development and disease. *Yale J Biol Med.* 2016;89:565–573.
138. Liu H, Chen S, Yao X, et al. Histone deacetylases 1 and 2 regulate the transcriptional programs of nephron progenitors and renal vesicles. *Development.* 2018;145:dev 153619.
139. Hartman HA, Lai HL, Patterson LT. Cessation of renal morphogenesis in mice. *Dev Biol.* 2007;310:379–387.
140. Rumballe BA, Georgas KM, Combes AN, Ju AL, Gilbert T, Little MH. Nephron formation adopts a novel spatial topology at cessation of nephrogenesis. *Dev Biol.* 2011;360:110–122.
141. Brunskill EW, Lai HL, Jamison DC, Potter SS, Patterson LT. Microarrays and RNA-Seq identify molecular mechanisms driving the end of nephron production. *BMC Dev Biol.* 2011;11:15.
142. Chen S, Brunskill EW, Potter SS, et al. Intrinsic age-dependent changes and cell-cell contacts regulate nephron progenitor lifespan. *Dev Cell.* 2015;35:49–62.
143. Chung E, Deacon P, Marable S, Shin J, Park JS. Notch signaling promotes nephrogenesis by downregulating Six2. *Development.* 2016;143:3907–3913.
144. Chung E, Deacon P, Park JS. Notch is required for the formation of all nephron segments and primes nephron progenitors for differentiation. *Development.* 2017;144:4530–4539.
145. Park JS, Valerius MT, McMahon AP. Wnt/(beta)-catenin signaling regulates nephron induction during mouse kidney development. *Development.* 2007;134:1103–1108.
146. Stark K, Vainio S, Vasslleva G, McMahon AP. Epithelial transformation of metanephric mesenchyme in the developing kidney regulated by Wnt-4. *Nature.* 1994;372:679–683.
147. Kispert A, Vainio S, McMahon AP. Wnt-4 is a mesenchymal signal for epithelial transformation of metanephric mesenchyme in the developing kidney. *Development.* 1998;125:4225–4234.
148. Tanigawa S, Wang H, Yang Y, et al. Wnt4 induces nephronic tubules in metanephric mesenchyme by a non-canonical mechanism. *Dev Biol.* 2011;352:58–69.
149. Burn SF, Webb A, Berry RL, et al. Calcium/NFAT signaling promotes early nephrogenesis. *Dev Biol.* 2011;352:288–298.
150. Georgas K, Rumballe B, Valerius MT, et al. Analysis of early nephron patterning reveals a role for distal renal vesicle proliferation in fusion to the ureteric tip via a cap mesenchyme-derived connecting segment. *Dev Biol.* 2009;332:273–286.
151. Self M, Lagutin OV, Bowling B, et al. Six2 is required for suppression of nephrogenesis and progenitor renewal in the developing kidney. *EMBO J.* 2006;25:5214–5228.
152. Kobayashi A, Valerius MT, Mugford JW, et al. Six2 defines and regulates a multipotent self-renewing nephron progenitor population throughout mammalian kidney development. *Cell Stem Cell.* 2008;3:169–181.
153. Karner CM, Das A, Ma Z, et al. Canonical Wnt9b signaling balances progenitor cell expansion and differentiation during kidney development. *Development.* 2011;138:1247–1257.
154. Ramalingam H, Fessler AR, Das A, et al. Disparate levels of beta-catenin activity determine nephron progenitor cell fate. *Dev Biol.* 2018;440:13–21.
155. Reginensi A, Scott RP, Gregorieff A, et al. Yap- and Cdc42-dependent nephrogenesis and morphogenesis during mouse kidney development. *PLoS Genet.* 2013;9:e1003380.
156. McNeill H, Reginensi A. Lats1/2 regulate Yap/Taz to control nephron progenitor epithelization and inhibit myofibroblast formation. *J Am Soc Nephrol.* 2017;28:852–861.
157. Hastie ND. Wilms tumor 1(WT1) in development, homeostasis and disease. *Development.* 2017;144:2862–2872.
158. Hartwig S, Ho J, Pandey P, et al. Genomic characterization of Wilms tumor suppressor 1 targets in nephron progenitor cells during kidney development. *Development.* 2010;137:1189–1203.
159. Essafi A, Webb A, Berry RL, et al. A Wt1-controlled chromatin switching mechanism underpins tissue-specific Wnt4 activation and repression. *Dev Cell.* 2011;21:559–574.
160. Tannberger K, Pfister AS, Brauburger K, et al. Amer1/WTX couples Wnt-induced formation of Ptdins (4,5)P2 to LRP6 phosphorylation. *EMBO J.* 2011;30:1433–1443.
161. Major MB, Camp ND, Berndt JD, et al. Wilms tumor suppressor WTX negatively regulates WNT/β-catenin signaling. *Science.* 2007;316:1043–1046.

162. Kopan R, Hagan MX. The canonical notch signaling pathway: unfolding the activation mechanism. *Cell.* 2009;137: 216–233.
163. Cheng HT, Miner JH, Lin M, et al. Gamma-secretase activity is dispensable for mesenchyme-to-epithelium transition but required for podocyte and proximal tubule formation in developing mouse kidney. *Development.* 2003;130: 5031–5042.
164. Cheng HT, Kim M, Valerius MT, et al. Notch2, but not Notch1, is required for proximal fate acquisition in the mammalian nephron. *Development.* 2007;134:801–811.
165. Surendran K, Botle S, Barak H, et al. The contribution of Notch1 to nephron segmentation in the developing kidney is revealed in a sensitized Notch2 background and can be augmented by reducing Mint dosage. *Dev Biol.* 2010;337:386–395.
166. Nakai S, Sugitani Y, Sato H, et al. Crucial roles of Brn1 in distal tubule formation and function in mouse kidney. *Development.* 2003;130:4751–4759.
167. Drake KA, Adam M, Mahoney R, et al. Disruption of Hox9, 10, 11 function results in cellular level lineage infidelity in the kidney. *Sci Rep.* 2018;8:6306.
168. Magella B, Mahoney R, Adam M, Potter SS. Reduced Abd-B Hox function during kidney development results in lineage infidelity. *Dev Biol.* 2018;438:84–93.
169. Li W, Hartwig S, Rosenblum ND. Developmental origins and functions of stromal cells in the normal and diseased mammalian kidney. *Dev Dyn.* 2014;243:853–863.
170. Kobayashi A, Mugford JW, Krautzberger AM, Naiman N, Liao J, McMahon AP. Identification of a multipotent self-renewing stromal progenitor population during mammalian kidney organogenesis. *Stem Cell Reports.* 2014;3: 650–662.
171. Naiman N, Fujioka K, Fujino M, et al. Repression of interstitial identity in nephron progenitor cells by Pax2 establishes the nephron-interstitium boundary during kidney development. *Dev Cell.* 2017;41:349–365.
172. Boivin FJ, Bridgewater D. β-Catenin in stromal progenitors controls medullary stromal development. *Am J Physiol Renal Physiol.* 2018;314:F1177–F1187.
173. Guillaume R, Bressan M, Herzlinger D. Paraxial mesoderm contributes stromal cells to the developing kidney. *Dev Biol.* 2009;329:169–175.
174. Levinson R, Mendelsohn C. Stromal progenitors are important for patterning epithelial and mesenchymal cell types in the embryonic kidney. *Sem Dev Biol.* 2003;14:225–231.
175. Hatini V, Huh SO, Herzlinger D, Soares VC, Lai E. Essential role of stromal mesenchyme in kidney morphogenesis revealed by targeted disruption of winged helix transcription factor BF-2. *Genes Dev.* 1996;10:1467–1478.
176. Levinson RS, Batourina E, Choi C, Vorontchikhina M, Kitajewski J, Mendelsohn CL. Foxd1-dependent signals control cellularity in the renal capsule, a structure required for normal renal development. *Development.* 2005;132: 529–539.
177. Yallowitz AR, Hrycaj SM, Short KM, Smyth IM, Wellik DM. Hox10 genes function in kidney development in the differentiation and integration of the cortical stroma. *PLoS One.* 2011;6:e23410.
178. Fetting JL, Guay JA, Karolak MJ, et al. FOXD1 promotes nephron progenitor differentiation by repressing decorin in the embryonic kidney. *Development.* 2014;141:17–27.
179. Leuning DG, Reinders ME, Li J, et al. Clinical-grade isolated human kidney perivascular stromal cells as an organotypic cell source for kidney regenerative medicine. *Stem Cells Transl Med.* 2017;6:405–418.
180. Leuning DG, Engelse MA, Lievers E, et al. The human kidney capsule contains a functionally distinct mesenchymal stromal cell population. *PLoS One.* 2017:12:e0187118.
181. Das A, Tangigawa S, Karner CM, et al. Stromal-epithelial crosstalk regulates kidney progenitor cell differentiation. *Nature Cell Biol.* 2013;15:1035–1044.
182. Bagherie-Lachidan M, Reginensi A, Pan Q, et al. Stromal Fat4 acts non-autonomously with Dchs1/2 to restrict the nephron progenitor pool. *Development.* 2015;142: 2564–2573.
183. Briscoe J, Therond PP. The mechanisms of hedgehog signaling and its roles in development and disease. *Nat Rev Mol Cell Biol.* 2013;14:416–429.
184. Rowan CJ, Li W, Martirosyan H, et al. Hedgehog-GLI signaling in *Foxd1*-positive stromal cells promotes nephrogenesis via TGFβ signaling. *Development.* 2018;145: dev159947.
185. Paroly SS, Wang F, Spraggon L, et al. Stromal protein Ecm1 regulates ureteric bud patterning and branching. *PLoS One.* 2013;9:e84155.
186. Sundelin B, Bohman SO. Postnatal development of the interstitial tissue of the rat kidney. *Anat Embryol (Berl).* 1990;182:307–317.
187. Maric C, Ryan GB, Alcorn D. Embryonic and postnatal development of the rat renal interstitium. *Anat Embryol (Berl).* 1997;195:503–514.
188. Marxer-Meier A, Hegyi I, Loffing J, Kaissling B. Postnatal maturation of renal cortical peritubular fibroblasts in the rat. *Anat Embryol (Berl).* 1998;197:143–153.
189. Abrahamson DR, Wang R. Development of the glomerular capillary and its basement membrane. In: Vize PD, Woolf AS, Bard JBL, eds. *The Kidney: From Normal Development to Congenital Disease.* Academic Press; 2003: 221–249.
190. Miner JH, Abrahamson DR. Molecular and cellular mechanisms of glomerular capillary development. In: Alpern RJ, Moe OW, Caplan M, eds. *Seldin and Giebisch's The kidney: Physiology and Pathophysiology.* 5th ed. Elsevier, Inc; 2013: 891–910.
191. Quaggin SE, Kreidberg JA. Development of the renal glomerulus:good neighbors and good fences. *Development.* 2008;135:609–620.
192. Miner JH. Organogenesis of the kidney glomerulus. *Organogenesis.* 2011;7:75–82.
193. Robert B, St. John PL, Hyink DP, Abrahamson DR. Evidence that embryonic kidney cells expressing flk-1 are intrinsic, vasculogenic angioblasts. *Am J Physiol.* 1996;271: F744–F753.
194. Robert B, St. John PL, Abrahamson DR. Direct visualization of renal vascular morphogenesis in Flk1 heterozygous mutant mice. *Am J Physiol.* 1998;275:F164–F172.
195. Woolf AS, Yuan HT. Development of kidney blood vessels. In: Vize PD, Woolf AS, Bard JBL, eds. *The Kidney: From Normal Development to Congenital Disease.* Academic Press; 2003:251–266.
196. Ichimura K, Stan RV, Kurihara H, Sakai T. Glomerular endothelial cells form diaphragms during development

and pathologic conditions. *J Am Soc Nephrol.* 2008;19: 1463–1471.
197. Farber G, Hurtado R, Loh S, et al. Glomerular endothelial maturation depends on ADAM10, a key regulator of notch signaling. *Angiogenesis.* 2018;21:335–347.
198. Ichimura K, Kakuta S, Kawasaki Y, et al. Morphological process of podocyte development revealed by block-face scanning electron microscopy. *J Cell Sci.* 2017;130:132–142.
199. Patrakka J, Tryggvason K. Nephrin—a unique structural and signaling protein of the kidney filter. *Trends Mol Med.* 2007; 13:396–403.
200. Kestila M, Lenkkeri U, Mannikko M, et al. Positionally cloned gene for a novel glomerular protein—nephrin—is mutated in congenital nephrotic syndrome. *Mol Cell.* 1998;1: 575–582.
201. Abrahamson DR, St. John PL, Stroganova L, Zelenchuk A, Steenhard BM. Laminin and type IV collagen isoform substitutions occur in temporally and spatially distinct patterns in developing kidney glomerular basement membranes. *J Histochem Cytochem.* 2013;61:706–718.
202. Abrahamson DR, Hudson BG, Stroganova L, Borza DB, St John PL. Cellular origins of type IV collagen networks in developing glomeruli. *J Am Soc Nephrol.* 2009;20: 1471–1479.
203. Leveen P, Pekny M, Gebre-Medhin S, Swolin B, Larsson E, Betsholtz C. Mice deficient for PDGF B show renal, cardiovascular, and hematological abnormalities. *Genes Dev.* 1994;8:1875–1887.
204. Soriano P. Abnormal kidney development and hematological disorders in PDGF beta-receptor mutant mice. *Genes Dev.* 1994;8:1888–1896.
205. Sequeira-Lopez MLS, Gomez RA. Development of the renal arterioles. *J Am Soc Nephrol.* 2011;22:2156–2165.
206. Mohamed T, Sequeira-Lopez MLS. Development of the renal vasculature. *Semin Cell Dev Biol.* 2018; pii:S1084-9521 (17)30450-0.
207. Belle M, Godefroy D, Couly G, et al. Tridimensional visualization and analysis of early human development. *Cell.* 2017;169:161–173.
208. Sequeira-Lopez MLS, Lin EE, Li M, Hu Y, Sigmund CD, Gomez RA. The earliest metanephric arteriolar progenitors and their role in kidney vascular development. *Am J Physiol Regul Integr Comp Physiol.* 2015;308:R138–R149.
209. Xu J, Nie X, Cai X, Cai CL, Xu PX. Tbx18 is essential for normal development of vasculature network and glomerular mesangium in the mammalian kidney. *Dev Biol.* 2014;391: 17–31.
210. Hu Y, Li M, Gothert JR, Gomez RA, Sequeira-Lopez ML. Hemovascular progenitors in the kidney require sphingosine-1-phosphate receptor for vascular development. *J Am Soc Nephrol.* 2016;27:1984–1995.
211. Halt KJ, Parssinen HE, Junttila SM, et al. CD146+ cells are essential for kidney vasculature development. *Kidney Int.* 2016;90:311–324.
212. Muherjee E, Maringer K, Papke E, et al. Endothelial marker-expressing stromal cells are critical for kidney formation. *Am J Physiol Renal Physiol.* 2017;313:F611–F620.
213. Munro ADA, Hohenstein P, Davies JA. Cycles of vascular plexus formation within the nephrogenic zone of the developing mouse kidney. *Sci Rep.* 2017;7:3273.
214. Daniel E, Azizoglu DB, Ryan AR, et al. Spatiotemporal heterogeneity and patterning of developing renal blood vessels. *Angiogenesis.* 2018;21:617–634.
215. Celio MR, Groscurth P, Inagami T. Ontogeny of renin immunoreactive cells in the human kidney. *Anat Embryol (Berl).* 1985;173:149–155.
216. Gomez RA, Lynch KR, Chevalier RL, et al. Renin and angiotensinogen gene expression in maturing rat kidney. *Am J Physiol.* 1988;254:F582–F587.
217. Minuth M, Hackenthal E, Poulsen K, Rix E, Taugner R. Renin immunocytochemistry of the differentiating juxtaglomerular apparatus. *Anat Embryol (Berl).* 1981;162: 173–181.
218. Gomez RA, Lynch KR, Sturgill BC, et al. Distribution of renin mRNA and its protein in the developing kidney. *Am J Physiol.* 1989;257:F850–F858.
219. Reddi V, Zaglul A, Pentz ES, Gomez RA. Renin-expressing cells are associated with branching of the developing kidney vasculature. *J Am Soc Nephrol.* 1998;9:63–71.
220. Sequeira Lopez ML, Pentz ES, Nomasa T, Smithies O, Gomez RA. Renin cells are precursors for multiple cell types that switch to the renin phenotype when homeostasis is threatened. *Dev Cell.* 2004;6:719–728.
221. Brunskill EW, Sequeira-Lopez MLS, Pentz ES, et al. Genes that confer the identity of the renincell. *J Am Soc Nephrol.* 2011;22:2213–2225.
222. Castellanos-Rivera RM, Pentz ES, Lin E, et al. Recombination signal binding protein for Ig-κJ region regulates juxtaglomerular cell phenotype by activating the myo-endocrine program and suppressing ectopic gene expression. *J Am Soc Nephrol.* 2015;26:67–80.
223. Moore KL, Persaud TVN. *The Developing Human; Clinically Oriented Embryology.* 7th ed. Saunders; 2003.
224. Gruenwald P. The normal changes in the position of the embryonic kidney. *Anat Rec.* 1943;85:163–176.
225. Friedland GW, De Vries P. Renal ectopia and fusion. Embryologic basis. *Urology.* 1975;5:698–706.
226. Muller F, O'Rahilly R. Somitic-vertebral correlation and vertebral levels in the human embryo. *Am J Anat.* 1986; 177:3–19.
227. Emery JL, Mithal A. The weights of kidneys in late intrauterine life and childhood. *J Clin Pathol.* 1960;13:490–493.
228. Gruenwald P, Minh HN. Evaluation of body and organ weights in perinatal pathology. *Am J Clin Pathol.* 1960;34: 247–253.
229. Singer DB, Sung CR, Wigglesworth JS. Fetal growth and maturation: with standards for body and organ development. In: Wigglesworth JS, Singer DB, eds. *Textbook of Fetal and Perinatal Pathology*. 2nd ed. Blackwell; 1998:8–40.
230. Guihard-Costa AM, Menez F, Delezoide AL. Organ weights in human fetuses after formalin fixation: standards by gestational age and body weight. *Pediatr Dev Pathol.* 2002;5: 559–578.
231. Hansen K, Sung CJ, Huang C, Pinar H, Singer DB, Oyer CE. Reference values for second trimester fetal and neonatal organ weights and measurements. *Pediatr Dev Pathol.* 2003;6:160–167.
232. Maroun LL, Graem N. Autopsy standards of body parameters and fresh organ weights in nonmacerated and macerated human fetuses. *Pediatr Dev Pathol.* 2005;8:204–217.

233. Coppoletta JM, Wolbach SB. Body length and organ weights of infants and children. *Am J Pathol.* 1933;9:55–70.
234. Hodson J. The lobar structure of the kidney. *Br J Urol.* 1972; 44:246–261.
235. Lofgren F. *Das Topographische System der Malpighischen Pyramiden der Menschenniere.* Hakan Ohlssons Boktryckeri; 1949.
236. Sykes D. The morphology of renal lobulations and calices, and their relationship to partial nephrectomy. *Br J Surg.* 1964; 51:294–304.
237. Crelin ES. *Functional Anatomy of the Newborn.* Yale University Press; 1973.
238. Campos ES. Pathological changes in the kidney in congenital syphilis. *Johns Hopkins Hosp Bull.* 1923;34:253–263.
239. Potter EL, Thierstein ST. Glomerular development in the kidney as an index of fetal maturity. *J Pediatr.* 1943;22: 695–706.
240. Sutherland MR, Gubhaju L, Moore L, et al. Accelerated maturation and abnormal morphology in the preterm neonatal kidney. *J Am Soc Nephrol.* 2011;22:1365–1374.
241. Singer DB, Klish W. Morphometric studies of the renal glomerulogenic zone. *Am J Pathol.* 1970;59:32a.
242. Tsuda S. Histologic investigation of the foetal kidney. *Jap J Obstet Gynecol.* 1934;17:337–341.
243. Dorovini-Zis K, Dolman CL. Gestational development of brain. *Arch Pathol Lab Med.* 1977;101:192–195.
244. Hinchliffe SA, Sargent PH, Chan YF, et al. "Medullary ray glomerular counting" as a method of assessment of human nephrogenesis. *Path Res Pract.* 1992;188:775–782.
245. Ryan D, Sutherland MR, Flores TJ, et al. Development of the human fetal kidney from mid to late gestation in male and female infants. *EBioMedicine.* 2018;27:275–283.
246. Peter K. Harnorgane. Organe uropoietica. In: Peter K, Wetzel G, Heiderich F, eds. *Handbuch der Anatomie des Kindes.* Vol. 2. JF Bergmann; 1938:1–41.
247. Hinchliffe SA, Sargent PH, Howard CV, Chan YF, van Velzen D. Human intrauterine renal growth expressed in absolute number of glomeruli assessed by the disector method and Cavalieri principle. *Lab Invest.* 1991;64:777–784.
248. Hughson M, Farris AB, Douglas-Denton R, Hoy WE, Bertram JF. Glomerular number and size in autopsy kidneys: the relationship to birth weight. *Kidney Int.* 2003;63: 2113–2122.
249. Hinchliffe SA, Lynch MRJ, Sargent PH, Howard CV, Van Velzen D. The effect of intrauterine growth retardation on the development of nephrons. *Br J Obstet Gynaecol.* 1992; 99:296–301.
250. Manalich R, Reyes L, Herrera M, Melendi C, Fundora I. Relationship between weight at birth and the number and size of renal glomeruli in humans: a histomorphometric study. *Kidney Int.* 2000;58:770–773.
251. Fogo A, Hawkins EP, Berry PL, et al. Glomerular hypertrophy in minimal change disease predicts subsequent progression to focal glomerulosclerosis. *Kidney Int.* 1990;38: 115–123.
252. Nyengaard JR, Bendtsen TF. Glomerular number and size in relation to age, kidney weight, and body surface in normal man. *Anat Rec.* 1992;232:194–201.
253. Merlet-Benichou C, Gilbert T, Vilar J, Moreau E, Freund N, Lelièvre-Pégorier M. Nephron number: variability is the rule. Causes and consequences. *Lab Invest.* 1999;79: 515–527.
254. Hoy WE, Hughson MD, Bertram JF, Douglas-Denton R, Amann K. Nephron number, hypertension, renal disease, and renal failure. *J Am Soc Nephrol.* 2005;16:2557–2564.
255. Brenner BM, Garcia DL, Anderson S. Glomeruli and blood pressure: less of one, more of the other? *Am J Hypertens.* 1988;1:335–347.
256. Gruenwald P, Popper H. The histogenesis and physiology of the renal glomerulus in early postnatal life: histological examinations. *J Urol.* 1940;43:452–459.
257. Macdonald MS, Emery JL. The late intrauterine and postnatal development of human renal glomeruli. *J Anat.* 1959;93:331–340.
258. Vernier RL, Birch-Andersen A. Studies of the human fetal kidney. I. Development of the glomerulus. *J Pediatr.* 1962;60:754–768.
259. Thony HC, Luethy CM, Zimmermann A, Laux-End R, Oetliker OH, Bianchetti MG. Histological features of glomerular immaturity in infants and small children with normal or altered tubular function. *Eur J Pediatr.* 1995;154(Suppl 3): S65–S68.
260. Volger C, McAdams J, Homan SM. Glomerular basement membrane and lamina densa in infants and children: an ultrastructural evaluation. *Pediatr Pathol.* 1987;7: 527–534.
261. Ramage IJ, Howatson AG, McColl JH, Maxwell H, Murphy AV, Beattie TJ. Glomerular basement membrane thickness in children: a stereologic assessment. *Kidney Int.* 2002; 62:895–900.
262. Steffes MW, Barbosa J, Basgen JM, Sutherland DE, Najarian JS, Mauer SM. Quantitative glomerular morphology of the normal human kidney. *Lab Invest.* 1983;49:82–86.
263. Vogelmann SU, Nelson WJ, Myers BD, Lemley KV. Urinary excretion of viable podocytes in health and renal disease. *Am J Physiol.* 2003;285:F40–F48.
264. Petermann AT, Krofft R, Blonski M, et al. Podocytes that detach in experimental membranous nephropathy are viable. *Kidney Int.* 2003;64:1222–1231.
265. Fetterman GH, Shuplock NA, Philipp FJ, Gregg HS. The growth and maturation of human glomeruli and proximal convolutions from term to adulthood. *Pediatrics.* 1965;35:601–619.
266. Zolnai B, Palkovits M. Glomerulometrics III. Data referring to the growth of the glomeruli in man. *Acta Biol Hung.* 1965;15:409–423.
267. Souster LP, Emery JL. The sizes of renal glomeruli in fetuses and infants. *J Anat.* 1980;130:595–602.
268. Moore L, Williams R, Staples A. Glomerular dimensions in children under 16 years of age. *J Pathol.* 1993;171:145–150.
269. Akaoka K, White RHR, Raafat F. Human glomerular growth during childhood: a morphometric study. *J Pathol.* 1994; 173:261–268.
270. Samuel T, Hoy WE, Douglas-Denton R, Hughson MD, Bertram JF. Determinants of glomerular volume in different cortical zones of the human kidney. *J Am Soc Nephrol.* 2005;16:3102–3109.
271. Steffes MW, Schmidt D, McCrery R, Basgen JM; International Diabetic Nephropathy Study Group. Glomerular cell number in normal subjects and in type 1 diabetic patients. *Kidney Int.* 2001;59:2104–2113.

272. Combs HL, Shankland SJ, Setzer SV, Hudkins KL, Alpers CE. Expression of the cyclin kinase inhibitor, p27kip1, in developing and mature human kidney. *Kidney Int.* 1998; 53:892–896.
273. Nagata M, Nakayama K, Terada Y, Hoshi S, Watanabe T. Cell cycle regulation and differentiation in the human podocyte lineage. *Am J Pathol.* 1998;153:1511–1520.
274. Hiromura K, Haseley LA, Zhang P, et al. Podocyte expression of the CDK-inhibitor p57 during development and disease. *Kidney Int.* 2001;60:2235–2246.
275. Puelles VG, Douglas-Denton RN, Cullen-McEwen LA, et al. Podocyte number in children and adults: associations with glomerular size and numbers of other glomerular resident cells. *J Am Soc Nephrol.* 2015;26:2277–2288.
276. Kampmeier OF. The metanephros or so-called permanent kidney in part provisional and vestigial. *Anat Rec.* 1926;33: 115–120.
277. Emery JL, Macdonald MS. Involuting and scarred glomeruli in the kidneys of infants. *Am J Pathol.* 1960;36:713–723.
278. Herxheimer G. Uber hyaline Glomeruli der Neugeborenen und Sauglinge. *Frankfurt Ztschr Path.* 1909;2:138–152.
279. Schwarz L. Weitere Beitrage zur Kenntnis der anatomischen Nierenveranderungen der nNeugeborenen und sauglinge. *Virchows Arch Path Anat.* 1928;267:654–689.
280. Friedman HH, Grayzel DM, Lederer M. Kidney lesions in stillborn and newborn infants. "Congenital glomerulosclerosis." *Am J Pathol.* 1942;18:699–713.
281. Thomas MA. Congenital glomerulosclerosis. *Pathology.* 1969; 1:105–112.
282. Dijkman HBPM, Wetzels JFM, Gemmink JH. Glomerular involution in children with frequently relapsing minimal change nephrotic syndrome: an unrecognized form of glomerulosclerosis? *Kidney Int.* 2007;71:44–52.
283. Moffat DB, Fourman J. Ectopic glomeruli in the human and animal kidney. *Anat Rec.* 1964;149:1–11.
284. MacCallum DB. The bearing of degenerating glomeruli on the problem of the vascular supply of the mammalian kidney. *Am J Anat.* 1939;65:69–103.
285. Evan AP, Larsson L. Morphologic development of the nephron. In: Edelmann CM Jr, Bernstein J, Meadow SR, et al., eds. *Pediatric Kidney Disease.* 2nd ed. Little Brown & Co; 1992:19–48.
286. Kim J, Lee GS, Tisher CC, Madsen KM. Role of apoptosis in development of the ascending thin limb of Henle in rat kidney. *Am J Phsiol Renal Physiol.* 1996;271:F831–F845.
287. Neiss WF. Histogenesis of the loop of Henle in the rat kidney. *Anat Embryol (Berl).* 1982;164:315–330.
288. Raptopoulos V, Kleinman PK, Mark S, Snyder M, Silverman PM. Renal fascial pathway; posterior extension of pancreatic effusions within the anterior pararenal space. *Radiology.* 1986;158:367–374.
289. Tobin CE. The renal fascia and its relation to the transversalis fascia. *Anat Rec.* 1944;89:295–311.
290. Kunin M. Bridging septa of the perinephric space: anatomic, pathologic, and diagnostic considerations. *Radiology.* 1986;158:361–365.
291. Wald H. The weight of normal adult human kidneys and its variability. *Arch Pathol Lab Med.* 1937;23:493–500.
292. Kaisiske BL, Umen AJ. The influence of age, sex, race, and body habitus on kidney weight in humans. *Arch Pathol Lab Med.* 1986;110:55–60.
293. Cheong B, Muthupillai R, Rubin MF. Normal values for renal length and volume as measured by magnetic resonance imaging. *Clin J Am Soc Nephrol.* 2007;2:38–45.
294. Frimann-Dahl J. Normal variations of the left kidney. An anatomical and radiologic study. *Acta Radiol.* 1961;55:207–216.
295. Graves FT. The anatomy of the intrarenal arteries and its application to segmental resection of the kidney. *Br J Surg.* 1954;42:132–139.
296. Graves FT. *Anatomical Studies for Renal and Intrarenal Surgery.* Wright; 1986.
297. Satyapal KS. Classification of the drainage patterns of the renal veins. *J Anat.* 1995;186:329–333.
298. Hodson CJ, Mariani S. Large cloisons. *AJR Am J Roentgenol.* 1982;139:327–332.
299. Lafortune M, Constantin A, Breton G, Vallee C. Sonography of the hypertrophied column of Bertin. *AJR Am J Roentgenol.* 1986;146:53–56.
300. Bigongiari LR, Patel SK, Appelman H, Thornbury JR. Medullary rays. Visualization during excretory urography. *AJR Am J Roentgenol.* 1975;125:795–803.
301. Ransley PG, Risdon RA. Renal papillary morphology in infants and young children. *Urol Res.* 1975;3:111–113.
302. Ransley PG. Intrarenal reflux. Anatomical, dynamic and radiologic studies–part I. *Urol Res.* 1977;5:61–69.
303. Schmidt-Nielsen B. The renal pelvis. *Kidney Int.* 1987;31: 621–628.
304. Amis ES, Cronan JJ. The renal sinus: an imaging review and proposed nomenclature for sinus cysts. *J Urol.* 1988;139: 1151–1159.
305. Beckwith JB. National Wilms Tumor Study: an update for pathologists. *Pediatr Dev Pathol.* 1998;1:79–84.
306. Bonsib SM, Gibson D, Mhoon M, Greene GF. Renal sinus involvement in renal cell carcinomas. *Am J Surg Pathol.* 2000; 24:451–458.
307. Bonsib SM. The renal sinus is the principal invasive pathway. A prospective study of 100 renal cell carcinomas. *Am J Surg Pathol.* 2004;28:1594–1600.
308. Bertram JF, Douglas-Denton RN, Diouf B, Hughson MD, Hoy WE. Human nephron number: implications for health and disease. *Pediatr Nephrol.* 2011; 26(9):1529–1533.
309. Oliver J. *Architecture of the Kidney in Chronic Bright's Disease.* Harper & Row. Hoeber Medical Division; 1939.
310. Schmidt-Nielsen B, O'Dell R. Structure and concentrating mechanism in the mammalian kidney. *Am J Physiol.* 1961; 200:1119–1124.
311. Madsen KM, Tisher CC. Structural-functional relationships along the distal nephron. *Am J Physiol.* 1986;250:F1–F15.
312. Jamison RL, Kriz W. *Urinary Concentrating Mechanism: Structure and Function.* Oxford University Press; 1982.
313. Kriz W, Kaissling B. Structural organization of the mammalian kidney. In: Alpern RJ, Caplan MJ, Moe OW, eds. *Seldin and Giebisch's The Kidney: Physiology and Pathophysiology.* 5th ed. Elsevier, Inc; 2013:595–691.
314. Knepper M, Burg M. Organization of nephron function. *Am J Physiol.* 1983;244:F579–F589.
315. Kriz W, Bankir L. A standard nomenclature for structures of the kidney. *Kidney Int.* 1988;33:1–7.
316. Zhou XJ, Laszik Z, Nadasdy T, et al., eds. Algorithmic approach to the interpretation of renal biopsy. *Silva's Diagnostic Renal Pathology.* 2nd ed. Cambridge University Press; 2017:69–91.

317. Pirani CL, Croker BP. Handling and processing of renal biopsy and nephrectomy specimens. In: Tisher CC, Brenner BM, eds. *Renal Pathology*. 2nd ed. JB Lippincott; 1994:1683–1694.
318. Walker PD, Cavallo T, Bonsib SM; Ad Hoc Committee on Renal Biopsy Guidelines of the Renal Pathology Society. Practice guidelines for the renal biopsy. *Mod Pathol*. 2004;17: 1555–1563.
319. Fenton RA, Praetorius J. Anatomy of the kidney. In: Skorecki K, Chertow GM, Marsden PA, et al., eds. *Brenner and Rector's The Kidney*. 10th ed. Elsevier; 2016:42–82.
320. Haymann JM Jr. Malpighi's "Concerning the structure of the kidneys." *Ann Med Hist*. 1925;7:242–263.
321. Bowman W. On the structure and use of the Malpighian bodies of the kidney, with observations on the circulation through that gland. *Philos Trans R Soc Lond*. 1842;132:57–80.
322. Fine LG. William Bowman's description of the glomerulus. *Am J Nephrol*. 1985;5:437–440.
323. Jorgensen F. *The Ultrastructure of the Normal Human Glomerulus*. Munksgaard; 1966.
324. Tisher CC, Brenner BM. Structure and function of the glomerulus. In: Tisher CC, Brenner BM, eds. *Renal Pathology*. 2nd ed. JB Lippincott; 1994:143–161.
325. Newbold KM, Howie AJ, Koram A, Adu D, Michael J. Assessment of glomerular size in renal biopsies including minimal change nephropathy and single kidneys. *J Pathol*. 1990;160:255–258.
326. Kaplan C, Pasternak B, Shah H, Gallo G. Age-related incidence of sclerotic glomeruli in human kidneys. *Am J Pathol*. 1975;80:227–234.
327. Kappel B, Olsen S. Cortical interstitial tissue and sclerosed glomeruli in the normal human kidney, related to age and sex. *Virchows Arch a Pathol Anat Histol*. 1980;387: 271–277.
328. Smith SM, Hoy WE, Cobb L. Low incidence of glomerulosclerosis in normal kidneys. *Arch Pathol Lab Med*. 1989;113:1253–1255.
329. Hommos MS, Glassock RJ, Rule AD. Structural and functional changes in human kidneys with healthy aging. *J Am Soc Nephrol*. 2017;28:2838–2844.
330. Wang X, Vrtiska TJ, Avula RT, et al. Age, kidney function, and risk factors associate differently with cortical and medullary volumes of the kidney. *Kidney Int*. 2014;85: 677–685.
331. Rule AD, Sasiwimonphan K, Lieske JC, Keddis MT, Torres VE, Vrtiska TJ. Characteristics of renal cystic and solid lesions based on contrast-enhanced computed tomography of potential kidney donors. *Am J Kidney Dis*. 2012;59: 611–618.
332. Denic A, Lieske JC, Chakkera HA, et al. The substantial loss of nephrons in healthy kidneys with aging. *J Am Soc Nephrol*. 2017;28:313–320.
333. Vasmant D, Maurice M, Feldmann G. Cytoskeletal ultrastructure of podocytes and glomerular endothelial cells in man and in the rat. *Anat Rec*. 1984;210:17–24.
334. Sorensson J, Fierlbeck W, Heider T, et al. Glomerular endothelial fenestrae in vivo are not formed from caveolae. *J Am Soc Nephrol*. 2002;13:2639–2647.
335. Horvat R, Hovoka A, Dekan G, Poczewski H, Kerjaschki D. Endothelial cell membranes contain podocalyxin—the major sialoprotein of visceral glomerular epithelial cells. *J Cell Biol*. 1986;102:484–491.
336. Kerjaschki D, Sharkey DJ, Farquhar MG. Identification and characterization of podocalyxin—the major sialoprotein of the renal glomerular epithelial cell. *J Cell Biol*. 1984;98:1591–1596.
337. Rostgaard J, Qvortrup K. Electron microscopic demonstrations of filamentous sieve plugs in capillary fenestrae. *Microvasc Res*. 1997;53:1–13.
338. Rostgaard J, Qvortrup K. Sieve plugs in fenestrae of glomerular capillaries—site of the filtration barrier? *Cells Tissues Organs*. 2002;170:132–138.
339. Dane MJC, van den Berg BM, Lee DH, et al. A microscopic view on the renal endothelial glycocalyx. *Am J Physiol Renal Physiol*. 2015;308:F956–F966.
340. Hegermann J, Lunsdorf H, Ochs M, Haller H. Visualization of the glomerular endothelial glycocalyx by electron microscopy using cationic colloidal thorium dioxide. *Histochem Cell Biol*. 2016;145:41–51.
341. Hjalmarsson C, Johansson BR, Haraldsson B. Electron microscopic evaluation of the endothelial surface layer of glomerular capillaries. *Microvas Res*. 2004;67:9–17.
342. Jeansson M, Haraldsson B. Morphological and functional evidence for an important role of the endothelial cell glycocalyx in the glomerular barrier. *Am J Physiol Renal Physiol*. 2006;290:F111–F116.
343. Becker CG. Demonstration of actomyosin in mesangial cells of the renal glomerulus. *Am J Pathol*. 1972;66:97–110.
344. Drenckhahn D, Schnittler H, Nobiling R, Kriz W. Ultrastructural organization of contractile proteins in rat glomerular mesangial cells. *Am J Pathol*. 1990;137:1343–1351.
345. Schlondorff D. The glomerular mesangial cell: An expanding role for a specialized pericyte. *FASEB J*. 1987;1: 272–281.
346. Sterzel RB, Hartner A, Schlotzer-Schrehardt U, et al. Elastic fiber proteins in the glomerular mesangium in vivo and in cell culture. *Kidney Int*. 2000;58:1588–1602.
347. Schaefer L, Mihalik D, Babelova A, et al. Regulation of fibrillin-1 by biglycan and decorin is important for tissue preservation in the kidney during pressure-induced injury. *Am J Pathol*. 2004;165:383–396.
348. Mundel P, Elger M, Sakai T, Kriz W. Microfibrils are a major component of the mesangial matrix in the glomerulus of the rat kidney. *Cell Tissue Res*. 1988;254:183–187.
349. Sakai T, Kriz W. The structural relationship between mesangial cells and basement membrane of the renal glomerulus. *Anat Embryol (Berl)*. 1987;176:373–386.
350. Kriz W, Elger M, Lemley K, Sakai T. Structure of the glomerular mesangium: a biomechanical interpretation. *Kidney Int*. 1990;38(suppl 30):2–9.
351. Kriz W, Elger M, Mundel P, Lemley KV. Structure-stabilizing forces in the glomerular tuft. *J Am Soc Nephol*. 1995;5:1731–1739.
352. Kurihara H, Sakai T. Cell biology of mesangial cells: the third cell that maintains the glomerular capillary. *Ana Sci Int*. 2017;92:173–186.
353. Tsurumi H, Kurihara H, Miura K, et al. Afadin is localized at cell-cell contacts in mesangial cells and regulates migratory polarity. *Lab Invest*. 2016;96:49–59.
354. Kikkawa Y, Virtanen I, Miner JH. Mesangial cells organize the glomerular capillaries by adhering to the G domain of laminin alpha5 in the glomerular basement membrane. *J Cell Biol*. 2003;161:187–196.

355. Zimmerman SE, Hiremath C, Tsunezumi J, Yang Z, Finney B, Marciano DK. Nephronectin regulates mesangial cell adhesion and behavior in glomeruli. *J Am Soc Nephrol.* 2018;29:1128–1140.
356. Kerjaschki D, Ojha PP, Susani M, et al. A beta-1-integrin receptor for fibronectin in human kidney glomeruli. *Am J Pathol.* 1989;134:481–489.
357. Michael AF, Keane WF, Raij L, Vernier RL, Mauer SM. The glomerular mesangium. *Kidney Int.* 1980;17:141–154.
358. Sterzel RB, Lovett DH. Interactions of inflammatory and glomerular cells in the response to glomerular injury. In: Wilson CB, Brenner BM, Stein JH, eds. *Immunopathology of Renal Disease*. Churchill Livingstone; 1988:137–173.
359. Floege J, Eitner F, Alpers CE. A new look at platelet-derived growth factor in renal disease. *J Am Soc Nephrol.* 2008;19:12–23.
360. Falini B, Flenghi L, Pileri S, et al. PG-M1: a new monoclonal antibody directed against a fixative-resistant epitope on the macrophage-restricted form of the CD689 molecule. *Am J Pathol.* 1993;142:1359–1372.
361. Imasawa T, Utsunomiya Y, Kawamura T, et al. The potential of bone marrow-derived cells to differentiate to glomerular mesangial cells. *J Am Soc Nephrol.* 2001;12:1401–1409.
362. Hugo C, Shankland SJ, Bowen-Pope DF, Couser WG, Johnson RJ. Extraglomerular origin of the mesangial cell after injury. *J Clin Invest.* 1997;100:786–794.
363. Starke C, Betz H, Hickmann L, et al. Renin lineage cells repopulate the glomerular mesangium after injury. *J Am Soc Nephrol.* 2015;26:48–54.
364. Jorgensen F, Bentzon MW. The ultrastructure of the normal human glomerulus. Thickness of glomerular basement membranes. *Lab Invest.* 1968;18:42–48.
365. Osawa G, Kimmelstiel P, Seling V. Thickness of glomerular basement membranes. *Am J Clin Pathol.* 1966;45:7–20.
366. Osterby R. Morphometric studies of the peripheral glomerular basement membrane in early juvenile diabetes. Development of initial basement membrane thickening. *Diabetologica.* 1972;8:84–92.
367. Lennon R, Byron A, Humphries JD, et al. Global analysis reveals the complexity of the human glomerular extracellular matrix. *J Am Soc Nephrol.* 2014;25:939–951.
368. Hobeika L, Barati MT, Caster DJ, McLeish KR, Merchant ML. Characterization of glomerular extracellular matrix by proteomic analysis of laser-captured microdissected glomeruli. *Kidney Int.* 2017;91:501–511.
369. Miner JH. The glomerular basement membrane. *Exp Cell Res.* 2012;318:973–978.
370. Hudson BG, Reeders SI, Tryggvason K. Type IV collagen: structure, gene organization and role in human diseases. *J Biol Chem.* 1993;268:26033–26036.
371. Hudson BG. The molecular basis of Goodpasture and Alport syndromes: beacons for the discovery of the collagen IV family. *J Am Soc Nephrol.* 2004;15:2514–2527.
372. Hudson BG, Tryggvason K, Sundaramoorthy M, Neilson EG. Alport's syndrome, Goodpasture's syndrome, and type IV collagen. *N Eng J Med.* 2003;348:2543–2556.
373. St. John PL, Abrahamson DR. Glomerular endothelial cells and podocytes jointly synthesize laminin-1 and -11 chains. *Kidney Int.* 2001;60:1037–1046.
374. Zenker M, Aigner T, Wendler O, et al. Human laminin beta2 deficiency causes congenital nephrosis with mesangial sclerosis and distinct eye abnormalities. *Hum Mol Genet.* 2004;13:2625–2632.
375. Katz A, Fish AJ, Kleppel MM, Michael AF, Butkowski RJ. Renal entactin (nidogen): isolation, characterization and tissue distribution. *Kidney Int.* 1991;40:643–652.
376. Murshed M, Smyth N, Miosge N, et al. The absence of nidogen 1 does not affect murine basement membrane formation. *Mol Cell Biol.* 2000;20:7007–7012.
377. Iozzo RV. Basement membrane proteoglycans: from cellar to ceiling. *Nature Rev Mol Cell Biol.* 2005;6:646–656.
378. Fraquhar MG. The glomerular basement membrane. A selective macromolecular filter. In: Hay ED, ed. *Cell Biology of Extracellular Matrix*. 2nd ed. Plenum; 1991:365–418.
379. Mahan JD, Sisson-Ross SS, Vernier RC. Anionic sites in the human kidney: ex vivo perfusion studies. *Mod Pathol.* 1989;2:117–124.
380. Groffen A, Ruegg MA, Dijkman H, et al. Agrin is a major heparan sulfate proteoglycan in the human glomerular basement membrane. *J Histochem Cytochem.* 1998;46:19–27.
381. Groffen AJ, Hop FW, Tryggvason K, et al. Evidence for the existence of multiple heparan sulfate proteoglycans in the human glomerular basement membrane and mesangial matrix. *Eur J Biochem.* 1997;247:175–182.
382. McCarthy KJ, Wassenhove-McCarthy DJ. The glomerular basement membrane as a model system to study the bioactivity of heparan sulfate glycosaminoglycans. *Microsc Microanal.* 2012;18:3–21.
383. Suleiman H, Zhang L, Roth R, et al. Nanoscale protein architecture of the kidney glomerular basement membrane. *Elife.* 2013;2:e01149.
384. Ichimura K, Miyazaki N, Sadayama S, et al. Three-dimensional architecture of podocytes revealed by block-face scanning electron microscopy. *Sci Rep.* 2015;5:8993.
385. Burghardt T, Hochapfel F, Salecker B, et al. Advanced electron microscopic techniques provide a deeper insight into the peculiar features of podocytes. *Am J Physiol Renal Physiol.* 2015;309:F1082–F1089.
386. Neal CR, Crook H, Bell E, Harper SJ, Bates DO. Three-dimensional reconstruction of glomeruli by electron microscopy reveals a distinct restrictive urinary subpodocyte space. *J Am Soc Nephrol.* 2005;16:1223–1235.
387. Gautier A, Bernhard W, Oberling C. [The existence of a pericapillary lacunar apparatus in the malpighian glomeruli revealed by electronic microscopy]. *C R Seances Soc Biol Fil.* 1950;144:1605–1607.
388. Salmon AHJ, Toma I, Sipos A, et al. Evidence for restriction of fluid and solute movement across the glomerular capillary wall by the subpodocyte space. *Am J Physiol Renal Physiol.* 2007;293:F1777–F1786.
389. Neal CR, Muston PR, Njegovan D, et al. Glomerular filtration into the subpodocyte space is highly restricted under physiological perfusion conditions. *Am J Physiol Renal Physiol.* 2007;293:F1787–F1798.
390. Arkill KP, Qvortrup K, Starborg T, et al. Resolution of the three dimensional structure of components of the glomerular filtration barrier. *BMC Nephrology.* 2014;15:24.
391. Schell C, Huber TB. The evolving complexity of the podocyte cytoskeleton. *J Am Soc Nephrol.* 2017;28:3166–3174.

392. Andrews PM, Bates SB. Filamentous actin bundles in the kidney. *Anat Rec*. 1984;210:1–9.
393. Drenckhahn D, Franke R. Ultrastructural organization of contractile and cytoskeletal proteins in glomerular podocytes of chicken, rat and man. *Lab Invest*. 1988;59:673–682.
394. Holthofer H, Miettinen A, Lehto V, Lehtonen E, Virtanen I. Expression of vimentin and cytokeratin types of intermediate filament proteins in developing and adult human kidneys. *Lab Invest*. 1984;50:552–559.
395. Pavenstadt H, Kriz W, Kretzler M. Cell biology of the glomerular podocytes. *Physiol Rev*. 2003;83:253–307.
396. Ichimura K, Kurihara H, Sakai T. Actin filament organization of foot processes in vertebrate glomerular podocytes. *Cell Tissue Res*. 2007;329:541–557.
397. Pellegrin S, Mellor H. Actin stress fibers. *J Cell Sci*. 2007; 120:3491–3499.
398. Suleiman HY, Roth R, Jain S, Heuser JE, Shaw AS, Miner JH. Injury-induced actin cytoskeleton reorganization in podocytes revealed by super-resolution microscopy. *JCI Insight*. 2017;2:e94137.
399. Falkenberg CV, Azeloglu EU, Stothers M, et al. Fargility of foot process morphology in kidney podocytes arises from chaotic spatial propagation of cytoskeletal instability. *PLoS Comput Biol*. 2017;13:e1005433.
400. Perico L, Conti S, Benigni A, Remuzzi G. Podocyte-actin dynamics in health and disease. *Nat Rev Nephrol*. 2016;12: 692–710.
401. Sever S, Schiffer M. Actin dynamics at focal adhesions: a common endpoint and putative therapeutic target for proteinuria diseases. *Kidney Int*. 2018;93:1298–1307.
402. Has C, Sparta G, Kiritsi D, et al. Integrin α3 mutations with kidney, lung and skin disease. *New Eng J Med*. 2012; 366:1508–1514.
403. Mouawad F, Tsui H, Takano T. Role of Rho-GTPases and their regulatory proteins in glomerular podocyte function. *Can J Physiol Pharmacol*. 2013;91:773–782.
404. Gu C, Lee HW, Garborcauskas G, Reiser J, Gupta V, Sever S. Dynamin autonomously regulates podocyte focal adhesion maturation. *Am J Soc Nephrol*. 2017;28:446–451.
405. Grahammer F, Schell C, Huber TB. The podocyte slit diaphragm – from a thin grey line to a complex signaling hub. *Nat Rev Nephrol*. 2013;9:587–598.
406. Gagliardini E, Conti S, Benigni A, Remuzzi G, Remuzzi A. Imaging the porous ultrastructure of the glomerular epithelial filtration slit. *J Am Soc Nephrol*. 2010;21: 2081–2089.
407. Rice WL, van Hoek AN, Paunescu TG, et al. High resolution helium ion scanning microscopy of the rat kidney. *PLoS One*. 2013;8:e57051.
408. Tsuji K, Paunescu TG, Suleiman H, et al. Re-characterization of the glomerulopathy in CD2AP deficient mice by high-resolution helium ion scanning microscopy. *Sci Rep*. 2017;7:8321.
409. Schnabel E, Anderson JM, Farquhar MG. The tight junction protein ZO-1 is concentrated along slit diaphragms of the glomerular epithelium. *J Cell Biol*. 1990;111:1255–1263.
410. Reiser J, Kriz W, Kretzler M, Mundel P. The glomerular slit diaphragm is a modified adherens junction. *J Am Soc Nephrol*. 2000;11:1–8.
411. Ruotsalainen V, Ljungberg P, Wartiovaara J, et al. Nephrin is specifically located at the slit diaphragm of glomerular podocytes. *Proc Natl Acad Sci U S A*. 1999;96: 7962–7967.
412. Putaala H, Soininen R, Kilelainen P, Wartiovaara J, Tryggvason K. The murine nephrin gene is specifically expressed in kidney, brain and pancreas:inactivation of the gene leads to massive proteinuria and neonatal death. *Hum Mol Genet*. 2001; 10:1–8.
413. Huber TB, Benzing T. The slit diaphragm: a signaling platform to regulate podocyte function. *Curr Opin Nephrol Hypertens*. 2005;14:211–216.
414. Grahammer F, Wigge C, Schell C, et al. A flexible, multilayered protein scaffold maintains the slit in between glomerular podocytes. *JCI Insight*. 2017;1:e86177.
415. Lehtonen S, Ryan JJ, Kudlicka K, Iino N, Zhou H, Farquhar MG. Cell junction-associated proteins IQGAP1, MAG1-2, CASK, spectrins, and α-actinin are components of the nephrin multiprotein complex. *Proc Natl Acad Sci U S A*. 2005;102: 9814–9819.
416. Hirabayashi S, Mori H, Kansaku A, et al. MAGI-1 is a component of the glomerular slit diaphragm that is tightly associated with nephrin. *Lab Invest*. 2005;85:1528–1543.
417. Donoviel DB, Freed DD, Vogel H, et al. Proteinuria and perinatal lethality in mice lacking NEPH1, a novel protein with homology to NEPHRIN. *Mol Cell Biol*. 2001;21: 4829–4836.
418. Ciana L, Patel A, Allen ND, Ffrench-Constant C. Mice lacking the giant protocadherin mFAT1 exhibit renal slit junction abnormalities and a partially penetrant cyclopia and anophthalmia phenotype. *Mol Cell Biol*. 2003;23: 3575–3582.
419. Roselli S, Heidet L, Sich M, et al. Early glomerular filtration defect and severe renal disease in podocin-deficient mice. *Mol Cell Biol*. 2004;24:550–560.
420. Shih NY, Li J, Karpitskii V, et al. Congenital nephrotic syndrome in mice lacking CD2-associated protein. *Science*. 1999;286:312–315.
421. Kaplan JM, Kim SH, North KN, et al. Mutations in ACTN4, encoding α-actinin-4, cause familial focal segmental glomerulosclerosis. *Nat Genet*. 2000;24:251–256.
422. Kos CH, Le TC, Sinha S, et al. Mice deficient in α-actinin-4 have severe glomerular disease. *J Clin Invest*. 2003;111:1683–1690.
423. Winn MP, Conlon PJ, Lynn KL, et al. A mutation in the TRPC6 cation channel causes familial focal segmental glomerulosclerosis. *Science*. 2005;308:1801–1804.
424. Reiser J, Polu KR, Moller CC, et al. TRPC6 is a glomerular slit diaphragm-associated channel required for normal renal function. *Nat Genet*. 2005;37:739–744.
425. Hinkes B, Wiggins RC, Gbadegesin R, et al. Positional cloning uncovers mutations in PLCE1 responsible for a nephritic syndrome variant that may be reversible. *Nat Genet*. 2006; 38:1397–1405.
426. Brown EJ, Schondorff JS, Becker DJ, et al. Mutations in the forming gene INF2 cause focal segmental glomerulosclerosis. *Nat Genet*. 2010;42:72–77.
427. Mele C, Iatropoulos P, Donadelli R, et al. MYO1E mutations and childhood familial focal segmental glomerulosclerosis. *New Eng J Med*. 2011;365:295–306.
428. Jones N, Blasutig IM, Eremina V, et al. Nck adaptor proteins link nephrin to the actin cytoskeleton of kidney podocytes. *Nature*. 2006;440:818–823.

429. Verma R, Kovari I, Soofi A, Nihalani D, Patrie K, Holzman LB. Nephrin ectodomain engagement results in Src kinase activation, nephrin phosphorylation, Nck recruitment and actin polymerization. *J Clin Invest.* 2006;116:1346–1359.
430. Okrut J, Prakash S, Wu Q, Kelly MJ, Taunton J. Allosteric N-WASP activation by an inter-SH3 domain linker in Nck. *Proc Nat Acad Sci U S A.* 2015;112:E6436–E6445.
431. New LA, Martin CE, Scott RP, et al. Nephrin tyrosine phosphorylation is required to stabilize and restore podocyte foot process architecture. *J Am Soc Nephrol.* 2016;27:2422–2435.
432. Jones N, New LA, Fortino MA, et al. Nck proteins maintain the adult glomerular filtration barrier. *J Am Soc Nephrol.* 2009;20:1533–1543.
433. Schell C, Baumhakl L, Salou S, et al. N-WASP is required for stabilization of podocyte foot processes. *J Am Soc Nephrol.* 2013;24:713–721.
434. Martin CE, Peterson KA, Aoudjit L, et al. ShcA adaptor protein promotes nephrin endocytosis and is upregulated in proteinuric nephropathies. *J Am Soc Nephrol.* 2018;29:92–103.
435. Regele HM, Fillipovic E, Langer B, et al. Glomerular expression of dystroglycans is reduced in minimal change nephrosis but not in focal segmental glomerulosclerosis. *J Am Soc Nephrol.* 2000;11:403–412.
436. Hannigan GE, Leung-Hagesteijn C, Fitz-Gibbon L, et al. Regulation of cell adhesion and anchorage-dependent growth by a new β1-integrin-linked protein kinase. *Nature.* 1996;379:91–96.
437. Sawada H, Stukenbrok H, Kerjaschki D, Farquhar MG. Epithelial polyanion (podocalyxin) is found on the sides but not the soles of the foot processes of the glomerular epithelium. *Am J Pathol.* 1986;125:309–318.
438. Wiggins RC, Wiggins JE, Goyal M, Wharram BL, Thomas PE. Molecular cloning of cDNAs encoding human GLEPP1, a membrane protein tyrosine phosphatase. *Genomics.* 1995;27:174–181.
439. Takeda T, McQuistan T, Orlando RA, Farquhar MG. Loss of glomerular foot processes is associated with uncoupling of podocalyxin from the actin cytoskeleton. *J Clin Invest.* 2001;108:289–301.
440. Orlando RA, Takeda T, Zak B, et al. The glomerular epithelial cell anti-adhesion podocalyxin associates with the actin cytoskeleton through interactions with ezrin. *J Am Soc Nephrol.* 2001;12:1589–1598.
441. Doyonnas R, Kershaw DB, Duhme C, et al. Anuria, omphalocele and perinatal lethality in mice lacking the CD34-related protein podocalyxin. *J Exp Med.* 2001;194:13–27.
442. Shankland SJ, Eitner F, Hudkins KL, Goodpaster T, D'Agati V, Alpers CE. Differential expression of cyclin-dependent kinase inhibitors in human glomerular disease: role in podocyte proliferation and maturation. *Kidney Int.* 2000;58:674–683.
443. Kreidburg JA, Sariola H, Loring JM, et al. WT-1 is required for early kidney development. *Cell.* 1993;74:679–691.
444. Pritchard-Jones K, Fleming S, Davidson D, et al. The candidate Wilms tumor gene is involved in genitourinary development. *Nature.* 1990;346:194–197.
445. Mundlos S, Pelletier J, Darveau A, Bachmann M, Winterpacht A, Zabel B. Nuclear localization of the protein encoded by the Wilms tumor gene WT1 in embryonic and adult tissues. *Development.* 1993;119:1329–1341.
446. Pelletier J, Bruening W, Kashatn CE. Germline mutations in the Wilms tumor suppressor gene are associated with abnormal urogenital development in Denys-Drash syndrome. *Cell.* 1991;67:437–447.
447. Schumacher VA, Schlotzer-Schrehardt U, Karumanchi SA, et al. WT1-dependent sulfatase expression maintains the normal glomerular filtration barrier. *J Am Soc Nephrol.* 2011;22:1286–1296.
448. Kann M, Ettou S, Jung YL, et al. Genome-wide analysis of Wilms tumor 1-controlled gene expression in podocytes reveals key regulatory mechanisms. *J Am Soc Nephrol.* 2015;26:2097–2104.
449. Warejko JK, Tan W, Dga A, et al. Whole exome sequencing of ptainets with steroid-resistant nephrotic syndrome. *Clin J Am Soc Nephrol.* 2018;13:53–62.
450. Rinschen MM, Godel M, Grahammer F, et al. A multilayered quantitative *in vivo* expression atlas of the podocyte unravels kidney disease candidate genes. *Cell Reports.* 2018;23:2495–2508.
451. Deen WM. What determines glomerular capillary permeability? *J Clin Invest.* 2004;114:1412–1414.
452. Haraldsson B, Nystrom J, Deen WM. Properties of the glomerular barrier and mechanisms of proteinuria. *Physiol Rev.* 2008;88:451–487.
453. Kanwar YS. Continuum of historical controversies regarding structural-functional relationship of the glomerular ultrafiltration unit (GUU). *Am J Physiol Renal Physiol.* 2015;308:F420–F424.
454. Edwards A, Daniels BS, Deen WM. Ultrastructural model for size selectivity in glomerular filtration. *Am J Physiol.* 1999;276:F892–F902.
455. Dijkman H, Smeets B, van der Laak J, Steenbergen E, Wetzels J. The parietal epithelial cell is crucially involved in human idiopathic focal segmental glomerulosclerosis. *Kidney Int.* 2005;68:1562–1572.
456. Moll R, Hage C, Thoenes W. Expression of intermediate filament proteins in fetal and adult human kidney: modulation of intermediate filament patterns during development and in damaged tissue. *Lab Invest.* 1991;65:74–86.
457. Oosterwijk E, van Muijen GNP, Oosterwijk-Wakka JC, Warnaar SO. Expression of intermediate-sized filaments in developing and adult human kidney and in renal cell carcinoma. *J Histochem Cytochem.* 1990;38:385–392.
458. Yaoita E, Franke WW, Yamamoto T, Kawasaki K, Kihara I. Identification of renal podocytes in multiple species: higher vertebrates are vimentin positive/lower vertebrates are desmin positive. *Histochem Cell Biol.* 1999;111:107–115.
459. Floege J, Alpers CE, Sage EH, et al. Markers of complement-dependent and complement-independent glomerular visceral epithelial injury in vivo. *Lab Invest.* 1992;67:486–497.
460. Peissel B, Geng L, Kalluri R, et al. Comparative distribution of the alpha 1(IV), alpha 5(IV), and alpha 6(IV) collagen chains in normal human adult and fetal tissues and in kidneys from X-linked Alport syndrome patients. *J Clin Invest.* 1995;96:1948–1957.
461. Ohse T, Chang AM, Pippin JW, et al. A new function for parietal epithelial cells: a second glomerular barrier. *Am J Physiol Renal Physiol.* 2009;297:F1566–F1574.

462. Shankland SJ, Smeets B, Pippin JW, Moeller MJ. The emergence of the glomerular parietal epithelial cell. *Nat Rev Nephrol.* 2014;10:158–173.
463. Alcorn D, Ryan GB. The glomerular peripolar cell. *Kidney Int Suppl.* 1993;42:S35–S39.
464. Gibson IW, Downie I, Downie TT, Han SW, More IA, Lindop GB. The parietal podocyte: a study of the vascular pole of the human glomerulus. *Kidney Int.* 1992;41:211–214.
465. Bariety J, Mandet C, Hill GS, Bruneval P. Parietal podocytes in normal human glomeruli. *J Am Soc Nephrol.* 2006; 17:2770–2780.
466. Smeets B, Stucker F, Wetzels J, et al. Detection of activated parietal epithelial cells on the glomerular tuft distinguishes early focal segmental glomerulosclerosis from minimal change disease. *Am J Pathol.* 2014;184:3239–3248.
467. Miesen L, Steenbergen E, Smeets B. Parietal cells-new perspectives in glomerular disease. *Cell Tissue Res.* 2017;369: 237–244.
468. Shankland SJ, Freedman BS, Pippin JW. Can podocytes be regenerated in adults? *Curr Opin Nephrol Hypertens.* 2017; 26:154–164.
469. Appel D, Kershaw DB, Smeets B, et al. Recruitment of podocytes from glomerular parietal epithelial cells. *J Am Soc Nephrol.* 2009;20:333–343.
470. Berger K, Schulte K, Boor P, et al. The regenerative potential of parietal epithelial cells in adult mice. *J Am Soc Nephrol.* 2014;25:693–705.
471. Wanner N, Hartleben B, Herbach N, et al. Unraveling the role of podocyte turnover in glomerular aging and injury. *J Am Soc Nephrol.* 2014;25:707–716.
472. Lasagni L, Angelotti ML, Ronconi E, et al, Podocyte regeneration driven by renal progenitors determines glomerular disease remission and can be pharmacologically enhanced. *Stem Cell Reports.* 2015;5:248–263.
473. Hackl MJ, Burford JL, Villanueva K, et al. Tracking the fate of glomerular epithelial cells in vivo using serial multiphoton imaging in new mouse models with fluorescent lineage tags. *Nat Med.* 2013;19:1661–1666.
474. Schulte K, Berger K, Boor P, et al. Origin of parietal podocytes in atubular glomeruli mapped by lineage tracing. *J Am Soc Nephrol.* 2014;25:129–141.
475. Kaverina NV, Eng DG, Schneider RRS, Pippin JW, Shankland SJ. Partial podocyte replenishment in experimental: FSGS derives from nonpodocyte sources. *Am J Physiol Renal Physiol.* 2016;310:F1397–F1413.
476. Barajas L. Anatomy of the juxtaglomerular apparatus. *Am J Physiol.* 1979;237:F333–F343.
477. Barajas L, Bloodworth JMB Jr, Hartroft PM. Endocrine pathology of the kidney. In: Bloodworth JMB Jr, ed. *Endocrine Pathology*. 2nd ed. Williams & Wilkins; 1982:723–766.
478. Barajas L, Salido EC, Smolens P, et al. Pathology of the juxtaglomerular apparatus including Bartter's syndrome. In: Tisher CC, Brenner BM, eds. *Renal Pathology*. 2nd ed. JB Lippincott; 1994;948–978.
479. Cantin M, Gutkowska J, Lacasse J, et al. Ultrastructural immunocytochemical localization of renin and angiotensin II in the juxtaglomerular cells of the ischemic kidney. *Am J Pathol.* 1984;115:212–224.
480. Taugner R, Mannek E, Nobiling R, et al. Coexistence of renin and angiotensin II in epithelioid cell secretory granules of rat kidney. *Histochemistry.* 1984;81:39–45.
481. Barajas L, Wang P. Localization of tritiated norepinephrine in the renal arteriolar nerves. *Anat Rec.* 1979;195:525–534.
482. Kopp UC, DiBona GF. Neural regulation of renin secretion. *Semin Nephrol.* 1993;13:543–551.
483. Gomez RA, Sequeira-Lopez MLS. Renin cells in homeostasis, regeneration and immune defense mechanisms. *Nat Rev Nephrol.* 2018;14:231–245
484. Pippin JW, Sparks MA, Glenn ST, et al. Cells of renin lineage are progenitors of podocytes and parietal epithelial cells in experimental glomerular disease. *Am J Pathol.* 2013; 183:542–557.
485. Pippin JW, Kaverina NV, Eng DG, et al. Cells of renin lineage are adult pluripotent progenitors in experimental glomerular disease. *Am J Physiol Renal Physiol.* 2015;309: F341–F358.
486. Lichtnekert J, Kaverina NV, Eng DG. Renin-angiotensin-aldosterone system inhibition increases podocyte derivation from cells of renin lineage. *J Am Soc Nephrol.* 2016;27: 3611–3627.
487. Kaverina NV, Kadoya H, Eng DG, et al. Tracking the stochastic fate of cells of the renin lineage after podocyte depletion using multicolor reporters and intravital imaging. *PLoS One.* 2017;12:e0173891.
488. Eng DG, Kaverina NV, Scneider RRS, et al. Detection of renin lineage cell transdifferentiation to podocytes in the kidney glomerulus with dual lineage tracing. *Kidney Int.* 2018;93:1240–1246.
489. Kurt B, Paliege A, Schwarzensteiner I, et al. Deletion of von Hippel-Lindau protein converts renin-producing cells into erythropoietin-producing cells. *J Am Soc Nephrol.* 2013;24: 433–444.
490. Kurt B, Gerl K, Karger C, Schwarzensteiner I, Kurtz A. Chronic hypoxia-inducible transcription factor-2 activation stably transforms juxtaglomerular renin cells into fibroblast-like cells in vivo. *J Am Soc Nephrol.* 2015;26:587–596.
491. Gerl K, Miqwerol L, Todorov VT, et al. Inducible glomerular erythropoietin production in the adult kidney. *Kidney Int.* 2015;88:1345–1355.
492. Pricam C, Humbert F, Perrelet A, Orci L. Gap junctions in mesangial and lacis cells. *J Cell Biol.* 1974;63:349–354.
493. Taugner R, Schiller A, Kaissling B, Kriz W. Gap junctional coupling between the JGA and the glomerular tuft. *Cell Tissue Res.* 1978;186:279–285.
494. Wagner C, Kurtz A. Distribution and functional relevance of connexins in renin-producing cells. *Pflugers Arch-Eur J Physiol.* 2013;465:71–77.
495. Yao J, Oite T, Kitamura M. Gap junctional intercellular communication in the juxtaglomerular apparatus. *Am J Physiol Renal Physiol.* 2009;296:F939–F946.
496. Kaissling B, Kriz W. Variability of intercellular spaces between macula densa cells: a transmission electron microscopic study in rabbits and rats. *Kidney Int.* 1982;22(suppl): 9–17.
497. Salido EC, Barajas L, Lechago J, Laborde NP, Fisher DA. Immunocytochemical localization of epidermal growth factor in mouse kidney. *J Histochem Cytochem.* 1986;34: 1155–1160.
498. Sikri KL, Foster CL, MacHugh N, Marshall RD. Localization of Tamm-Horsfall glycoprotein in the human kidney using immunofluorescence and immunoelectron microscopical techniques. *J Anat.* 1981;132:597–605.

499. Kirk KL, Bell PD, Barfuss DW, Ribadeneira M. Direct visualization of the isolated and perfused macula densa. *Am J Physiol.* 1985;248:F890–F894.
500. Schnermann J. The juxtaglomerular apparatus: from anatomical peculiarity to physiological relevance. *J Am Soc Nephrol.* 2003;14:1681–1694.
501. Schnermann J. Concurrent activation of multiple vasoactive signaling pathways in vasoconstriction by tubuloglomerular feedback: a quantitative assessment. *Annu Rev Physiol.* 2015; 77:301–322.
502. Wilcox CS, Welch WJ, Murad F, et al. Nitric oxide synthase in macula densa regulates glomerular capillary pressure. *Proc Natl Acad Sci U S A.* 1992;89:11993–11997.
503. Mundel P, Bachmann S, Bader M, et al. Expression of nitric oxide synthase in kidney macula densa cells. *Kidney Int.* 1992;42:1017–1019.
504. Harris RC, McKanna JA, Akai Y, Jacobson HR, Dubois RN, Breyer MD. Cyclooxygenase-2 is associated with the macula densa of rat kidney and increases with salt restriction. *J Clin Invest.* 1994;94:2504–2510.
505. Welch WJ, Wilcox CS, Thomson SC. Nitric oxide and tubuloglomerular feedback. *Semin Nephrol.* 1999;19: 251–262.
506. Harris RC, Breyer MD. Physiological regulation of cyclooxygenase-2 in the kidney. *Am J Physiol.* 2001;281:F1–F11.
507. Rouillier C. General anatomy and histology of the kidney. In: Rouillier C, Muller AF, eds. *The Kidney: Morphology, Biochemistry, Physiology*. Academic Press; 1969:61–156.
508. Swann HG. The functional distention of the kidney: a review. *Tex Rep Biol Med.* 1960;18:566–596.
509. Hodson CJ. Physiological change in size of the human kidney. *Clin Radiol.* 1961;12:91–94.
510. Genest DR, Williams MA, Greene MF. Estimating the time of death in stillborn fetuses: I. Histologic evaluation of fetal organs; an autopsy study of 150 stillborns. *Obstet Gynecol.* 1992;80:575–584.
511. Tomita Y, Nihira M, Ohno Y, Sato S. Ultrastructural changes during in situ early postmortem autolysis in kidney, pancreas, liver, heart and skeletal muscle of rats. *Leg Med.* 2004;6: 25–31.
512. Kocovski L, Duflou J. Can acute tubular necrosis be differentiated from autolysis at autopsy? *J Forensic Sc.* 2009; 54:439–442.
513. Ichimura T, Bonventre JV, Bailly V, et al. Kidney injury molecule-1 (Kintermediate mesoderm-1), a putative epithelial cell adhesion molecule containing a novel immunoglobulin domain, is upregulated in renal cells after injury. *J Biol Chem.* 1998;273:4135–4142.
514. Han WK, Bailly V, Abichandani R, Thadhani R, Bonventre JV. Kidney injury molecule-1 (Kintermediate mesoderm-1): a novel biomarker for human renal proximal tubule injury. *Kidney Int.* 2002;62:237–244.
515. Rodman JS, Mooseker M, Faruhar MG. Cytoskelatal proteins of the rat kidney proximal tubule brush border. *Eur J Cell Biol.* 1986;42:319–327.
516. Farraggiana F, Malchiodi F, Prado A, Churg J. Lectin-peroxidase conjugate reactivity in normal kidney. *J Histochem Cytochem.* 1982;30:451–458.
517. Hennigar RA, Schulte BA, Spicer SS. Heterogeneous distribution of glycoconjugates in human kidney tubules. *Anat Rec.* 1985;211:376–390.
518. Silva FG, Nadasdy T, Laszik Z. Immunohistochemical and lectin dissection of the human nephron in health and disease. *Arch Pathol Lab Med.* 1993;117:1233–1239.
519. Skinnider BF, Folpe AL, Hennigar RA, et al. Distribution of cytokeratins and vimentin in adult renal neoplasms and normal renal tissue. *Am J Surg Pathol.* 2005;29:747–754.
520. Paul R, Ewing CM, Robinson JC, et al. Cadherin-6, a cell adhesion molecule specifically expressed in the proximal renal tubule and renal cell carcinomas. *Cancer Res.* 1997;57: 2741–2748.
521. Letts RFR, Zhai XY, Bhikka C, et al. Nephron morphometry in mice and rats using tomographic microscopy. *Am J Physiol Renal Physiol.* 2017;312:F210–F229.
522. Maunsbach AB, Christensen EI. Functional ultrastructure of the proximal tubule. In: Windhager EE, ed. *Handbook of Physiology. Section 8: Renal Physiology*. Oxford University Press; 1992:41–107.
523. Tisher CC, Bulger RE, Trump BF. Human renal ultrastructure. I. Proximal tubule of healthy individuals. *Lab Invest.* 1966;15:1357–1394.
524. Mount DB, Yu ASL. Transport of sodium, chloride, potassium. In: Brenner BM, ed. *Brenner & Rector's The Kidney*. 10th ed. Elsevier; 2016:144–184.
525. Welling LW, Welling DJ. Shape of epithelial cells and intercellular channels in the rabbit proximal nephron. *Kidney Int.* 1976;9:385–394.
526. Welling LW, Welling DJ. Relationship between structure and function in renal proximal tubule. *J Electron Microsc Tech.* 1988;9:171–185.
527. Ahn KY, Madsen KM, Tisher CC, Kone BC. Differential expression and cellular distribution of mRNAs encoding α- and β-isoforms of Na^+-K^+-ATPase in rat kidney. *Am J Physiol.* 1993;265:F792–F801.
528. Clapp WL, Bowman P, Shaw GS, Patel P, Kone BC. Segmental localization of mRNAs encoding Na^+-K^+-ATPase α- and β-subunit isoforms in rat kidney using RT-PCR. *Kidney Int.* 1994;46:627–638.
529. Biemesderfer D, Pizzonia J, Abu-Alfa A, et al. NHE3: a Na^+/H^+ exchanger isoform of renal brush border. *Am J Physiol.* 1993;265:F736–F742.
530. Amemiya M, Loffing J, Lotscher M, Kaissling B, Alpern RJ, Moe OW. Expression of NHE-3 in the apical membrane of rat proximal tubule and thick ascending limb. *Kidney Int.* 1995;48:1206–1215.
531. Riquier-Brison AD, Leong PK, Pihakaski-Maunsbach K, et al. Angiotensin II stimulates trafficking of NHE3, NaPi2 and associated proteins into the proximal tubule microvilli. *Am J Physiol Renal Physiol.* 2010;298:F177–F186.
532. Bergeron M, Guerette D, Forget J, Thiery G. Three-dimensional characteristics of the mitochondrial of the rat nephron. *Kidney Int.* 1980;17:175–185.
533. Agre P, King LS, Yasui M, et al. Aquaporin water channels-from atomic structure to clinical medicine. *J Physiol.* 2002; 542:3–16.
534. King LS, Kozono D, Agre P. From structure to disease: the evolving tale of aquaporin biology. *Nat Rev Mol Cell Biol.* 2004; 5:687–698.
535. Nielsen S, Smith BL, Christensen EI, Knepper MA, Agre P. CHIP28 water channels are localized in constitutively water-permeable segments of the nephron. *J Cell Biol.* 1993; 120:371–383.

536. Maunsbach AB, Marples D, Chin E, et al. Aquaporin-1 water channel expression in human kidney. *J Am Soc Nephrol.* 1997; 8:1–14.
537. Christensen EI, Nielsen R, Birn H. Renal filtration, transport and metabolism of albumin and albuminuria. In: Alpern RJ, Moe OW, Caplan M, eds. *Seldin and Giebisch's The Kidney: Physiology and Pathophysiology*. 5th ed. Elsevier, Inc; 2013:2457–2474.
538. Christensen EI, Rennke HG, Carone FA. Renal tubular uptake of protein: effect of molecular charge. *Am J Physiol.* 1983;244:F436–F441.
539. Park CH, Maack T. Albumin absorption and catabolism by isolated perfused proximal convoluted tubules of the rabbit. *J Clin Invest.* 1984;73:767–777.
540. Park CH. Time course and vectorial nature of albumin metabolism in isolated perfused rabbit PCT. *Am J Physiol.* 1988;255:F520–F528.
541. Clapp WL, Park CH, Madsen KM, Tisher CC. Axial heterogeneity in the handling of albumin by the rabbit proximal tubule. *Lab Invest.* 1988;58:549–558.
542. Larsson L, Clapp WL, Park CH, Cannon JK, Tisher CC. Ultrastructural localization of acidic compartments in cells of isolated rabbit proximal convoluted tubule. *Am J Physiol.* 1987;253:F95–F103.
543. Christensen EI. Rapid membrane recycling in renal proximal tubule cells. *Eur J Cell Biol.* 1982;29:43–49.
544. Christensen EI, Birn H. Megalin and cubilin: Synergistic endocytic receptors in renal proximal tubule. *Am J Physiol.* 2001;280:F562–F573.
545. Birn H, Fyfe JC, Jacobsen C, et al. Cubilin is an albumin binding protein important for renal tubular albumin reabsorption. *J Clin Invest.* 2000;105:1353–1361.
546. Nielsen R, Christesen EI, Birn H. Megalin and cubilin in proximal tubule protein reabsorption: from experimental models to human disease. *Kidney Int.* 2016;89:58–67.
547. Russo LM, Sandoval RM, McKee M, et al. The normal kidney filters nephrotic levels of albumin retrieved by proximal tubules cells: retrieval is disrupted in nephrotic states. *Kidney Int.* 2007;71:504–513.
548. Sandoval RM, Wagner MC, Patel M, et al. Multiple factors influence glomerular albumin permeability in rats. *J Am Soc Nephrol.* 2012;23:447–457.
549. Weyer K, Andersen PK, Schmidt K, et al. Abolishment of proximal tubule albumin endocytosis does not affect plasma albumin during nephrotic syndrome in mice. *Kidney Int.* 2018;93:335–342.
550. Jiang M, Wei Q, Dong G, Komatsu M, Su Y, Dong Z. Autophagy in proximal tubules protects against acute tubular injury. *Kidney Int.* 2012;82:1271–1283.
551. Olsen S, Solez K. Acute tubular necrosis and toxic renal injury. In: Tisher CC, Brenner BM, eds. *Renal Pathology*. 2nd ed. JB Lippincott; 1994:769–809.
552. Nadasdy NT, Laszik Z, Blick KE, Johnson LD, Silva FG. Proliferative activity of intrinsic cell populations in the normal human kidney. *J Am Soc Nephrol.* 1994;4:2032–2039.
553. Droz D, Zachar D, Charbit L, Gogusev J, Chrétein Y, Iris L. Expression of the human nephron differentiation molecules in renal cell carcinomas. *Am J Pathol.* 1990; 137:895–905.
554. Gerdes J, Becker MHG, Key G, Cattoretti G. Immunohistochemical detection of tumor growth fraction (Ki-67 antigen) in formalin fixed and routinely processed tissues. *J Pathol.* 1992;168:85–87.
555. Kliem V, Johnson RJ, Alpers CE, et al. Mechanisms involved in the pathogenesis of tubulointerstitial fibrosis in 5/6-nephrectomized rats. *Kidney Int.* 1996;49:666–678.
556. Hansson J, Hultenby K, Cramnert C, et al. Evidence for a morphologically distinct and functionally robust cell type in the proximal tubules of human kidney. *Hum Pathol.* 2014; 45:382–393.
557. Smeets B, Boor P, Dijkman H, et al. Proximal tubular cells contain a phenotypically distinct, scattered cell population involved in tubular regeneration. *J Pathol.* 2013;229:645–659.
558. Kusaba T, Lalli M, Kramann R, Kobayashi A, Humphreys BD. Differentiated kidney epithelial cells repair injured proximal tubule. *Proc Natl Acad Sci U S A.* 2014;111:1527–1532.
559. Berger K, Bangen J-M, Hammerich L, et al. Origin of regenerating tubular cells after acute kidney injury. *Proc Natl Acad Sci U S A.* 2014;111:1533–1538.
560. Gobe GC, Axelsen RA, Searle JW. Cellular events in experimental unilateral ischemic renal atrophy and in regeneration after contralateral nephrectomy. *Lab Invest.* 1990;63: 770–779.
561. Schumer M, Colombel MC, Sawczuk IS, et al. Morphologic, biochemical and molecular evidence of apoptosis during the reperfusion phase after brief periods of renal ischemia. *Am J Pathol.* 1992;140:831–838.
562. Schimizu A, Yamanaka N. Apoptosis and cell desquamation in repair process of ischemic tubular necrosis. *Virchows Arch B Cell Pathol Incl Mol Pathol.* 1993;64:171–180.
563. Dieterich HJ, Barrett JM, Kriz W, Bülhoff JP. The ultrastructure of the thin limbs of the mouse kidney. *Anat Embryol (Berl).* 1975;147:1–13.
564. Bulger RE, Tisher CC, Myers CH, Trump BF. Human renal ultrastructure. II. The thin limb of Henle's loop and the interstitium in healthy individuals. *Lab Invest.* 1967;16: 124–141.
565. Sands JM, Layton HE, Fenton RA. Urine concentration and dilution. In: Skorecki K, Chertow GM, Marsden PA, et al., eds. *Brenner & Rector's The Kidney*. 10th ed. Elsevier; 2016:258–280.
566. Nielsen S, Pallone T, Snith BL, Christensen EI, Agre P, Maunsbach AB. Aquaporin-1 water channels in short and long loop descending thin limbs and in descending vasa recta in rat kidney. *Am J Physiol.* 1995;268:F1023–F1037.
567. Zhai XY, Fenton RA, Andeason A, Thomsen JS, Christensen EI. Aquporin-1 is not expressed in descending thin limbs of short-loop nephrons. *J Am Soc Nephrol.* 2007;18: 2937–2944.
568. Uchida S, Sasaki S, Nitta K, et al. Localization and functional characterization of rat kidney-specific chloride channel, ClC-K1. *J Clin Invest.* 1995;95:104–113.
569. Takeuchi Y, Uchida S, Marumo F, Sasaki S. Cloning, tissue distribution, and intrarenal localization of ClC chloride channels in human kidney. *Kidney Int.* 1995;48: 1497–1503.
570. Kokko JP, Rector FC Jr. Countercurrent multiplication system without active transport in inner medulla. *Kidney Int.* 1972;2:214–223.
571. Stephenson JL. Concentration of urine in a central core model of the renal counterflow system. *Kidney Int.* 1972; 2:85–94.

572. Pannbecker TL, Layton AT. Targeted delivery of solutes and oxygen in the renal medulla: role of microvessel architecture. *Am J Physiol Renal Physiol.* 2014;307:F649–F655.
573. Dantzler WH, Layton AT, Layton HE, Pannabecker TL. Urine-concentrating mechanism in the inner medulla: function of the thin limbs of the loops of Henle. *Clin J Am Soc Nephrol.* 2014;9:1781–1789.
574. Wei G, Rosen S, Dantzler WH, Pannabecker TL. Architecture of the human renal inner medulla and functional implications. *Am J Physiol Renal Physiol.* 2015;309:F626–F637.
575. Kone BC, Madsen KM, Tisher CC. Ultrastructure of the thick ascending limb of Henle in the rat kidney. *Am J Anat.* 1984;171:217–226.
576. Allen F, Tisher CC. Morphology of the ascending thick limb of Henle. *Kidney Int.* 1976;9:8–22.
577. Stamenkovic I, Skalli O, Gabliani G. Distribution of intermediate filament proteins in normal and diseased human glomeruli. *Am J Pathol.* 1986;125:465–475.
578. Shen SS, Krishna B, Chirala R, Amato RJ, Truong LD. Kidney-specific cadherin, a specific marker for the distal portion of the nephron and related renal neoplsams. *Mod Pathol.* 2005;18:933–940.
579. Garg LC, Knepper MA, Burg MB. Mineralocorticoid effects on Na-K-ATPase in individual nephron segments. *Am J Physiol.* 1981;240:F536–F544.
580. Nielsen S, Maunsbach AB, Ecelbarger CA, Knepper MA. Ultrastructural localization of Na-K-2Cl cotransporter in thick ascending limb and macula densa of rat kidney. *Am J Physiol.* 1998;275:F885–F893.
581. Bachmann S, Velazquez H, Obermuller N, Reilly RF, Moser D, Ellison DH. Expression of the thiazide-sensitive Na-Cl cotransporter by rabbit distal convoluted tubule cells. *J Clin Invest.* 1995;96:2510–2514.
582. Plotkin MD, Kaplan MR, Verlander JW, et al. Localization of the thiazide-sensitive Na-Cl cotransporter, rTSC1, in the rat kidney. *Kidney Int.* 1996;50:174–183.
583. Woodhall PB, Tisher CC. Response of the distal tubule and cortical collecting duct to vasopressin in the rat. *J Clin Invest.* 1973;52:3095–3108.
584. Gross JB, Imai M, Kokko JP. A functional comparison of the cortical collecting tubule and the distal convoluted tubule. *J Clin Invest.* 1975;55:1284–1294.
585. Kaissling B. Structural aspects of adaptive changes in renal electrolyte excretion. *Am J Physiol.* 1982;243:F211–F226.
586. Kaissling B, Bachmann S, Kriz W. Structural adaptation of the distal convoluted tubule to prolonged furosemide treatment. *Am J Physiol Renal Physiol.* 1985;248:F374–F381.
587. Dorup J, Morsing P, Rasch R. Tubule-tubule and tubule-arteriole contacts in rat distal kidney distal nephrons. A morphologic study based on computer-assisted three-dimensional reconstructions. *Lab Invest.* 1992;67:761–769.
588. Ren Y, Garvin JL, Liu R, Carretero OA. Crosstalk between the connecting tubule and the afferent arteriole regulates renal microcirculation. *Kidney Int.* 2007;71:1116–1121.
589. Myers CH, Bulger RE, Tisher CC, Trump BF. Human renal ultrastructure. IV. Collecting duct of healthy individuals. *Lab Invest.* 1966;15:1921–1950.
590. Kaissling B, Kriz W. Structural analysis of the rabbit kidney. *Adv Anat Embryol Cell Biol.* 1979;56:1–123.
591. Loffing J, Korbmacher C. Regulated sodium transport in the renal connecting tubule (connective tubule) via the epithelial sodium channel (ENaC). *Pflugers Arch.* 2009;458:111–135.
592. Wade JB, Fang L, Coleman RA, et al. Differential regulation of ROMK (Kir1.1) in distal nephron segments by dietary potassium. *Am J Physiol Renal Physiol.* 2011;300:F1385–F1393.
593. Boros S, Bindels RJ, Hoenderop JG. Active Ca(2+) reabsorption in the connecting tubule. *Pflugers Arch.* 2009;458:99–109.
594. Coleman RA, Wu DC, Liu J, Wade JB. Expression of aquaporins in the renal connecting tubule. *Am J Physiol Renal Physiol.* 2000;279:F874–F883.
595. Kortenoeven ML, Pedersen NB, Miller RL, Rojek A, Fenton RA. Genetic ablation of aquaporin-2 in the mouse connecting tubules results in defective renal water handling. *J Physiol.* 2013;591:2205–2219.
596. Welling LW, Evan AP, Welling DJ. Shape of cells and extracellular channels in rabbit cortical collecting ducts. *Kidney Int.* 1981;20:211–222.
597. Duc C, Farman N, Canessa CM, Bonvalet JP, Rossier BC. Cell-specific expression of epithelial sodium channel alpha, beta, and gamma subunits in aldosterone-responsive epithelia from the rat: localization by in situ hybridization and immunocytochemistry. *J Cell Biol.* 1994;127:1907–1921.
598. Hager H, Kwon TH, Vinnikova AK, et al. Immunocytochemical and immunoelectron microscopic localization of alpha-, bata-, and gamma-ENaC in rat kidney. *Am J Physiol.* 2001;280:F1093–F1096.
599. Stanton BA, Biemesderfer D, Wade JB, Giebisch G. Structural and functional study of the rat nephron. Effects of potassium adaptation and depletion. *Kidney Int.* 1981;19:36–48.
600. Petty KJ, Kokko JP, Marver D. Secondary effect of aldosterone on Na-K-ATPase activity in the rabbit cortical collecting tubule. *J Clin Invest.* 1981;68:1514–1521.
601. Mujais SK, Chekal MA, Jones WJ, Hayslett JP, Katz AI. Regulation of renal Na-K-ATPase in the rat: role of the natural mineralo- and glucocorticoid hormones. *J Clin Invest.* 1984;73:13–19.
602. Kaissling B, Le Hir M. Distal tubular segments of the rabbit kidney after adaptation to altered Na- and K-intake. I. Structural changes. *Cell Tissue Res.* 1982;224:469–492.
603. Wade JB, O'Neil RG, Pryor JL, Boulpaep EL. Modulation of cell membrane area in renal collecting tubules by corticosteroid hormones. *J Cell Biol.* 1979;81:439–445.
604. Kirk KL, Buku A, Eggena P. Cell specificity of vasopressin binding in renal collecting duct: computer-enhanced imaging of a fluorescent hormone analog. *Proc Natl Acad Sci U S A.* 1987;84:6000–6004.
605. Nielsen S, Digiovanni SR, Christensen EI, Knepper MA, Harris HW. Cellular and subcellular immunolocalization of vasopressin-regulated water channel in rat kidney. *Proc Natl Acad Sci U S A.* 1993;90:11663–11667.
606. Nielsen C, Chou CL, Marples D, Christensen EI, Kishore BK, Knepper MA. Vasopressin increases water permeability of kidney collecting duct by inducing translocation of aquaporin-CD water channels to plasma membrane. *Proc Natl Acad Sci U S A.* 1995;92:1013–1017.
607. Ecelbarger CA, Terriis J, Frindt G, et al. Aquaporin-3 water channel localization and regulation in rat kidney. *Am J Physiol.* 1995;269:F663–F672.

608. Hasegawa H, Ma T, Skach W, Matthay MA, Verkman AS. Molecular cloning of a mercurial-insensitive water channel expressed in selected water-transporting tissues. *J Biol Chem.* 1994;269:5497–5500.
609. Schuster VL, Bonsib SM, Jennings ML. Two types of collecting duct mitochondria-rich (intercalated) cells: lectin and band 3 cytochemistry. *Am J Physiol.* 1986;251: C347–C355.
610. Verlander JW, Madsen KM, Tisher CC. Effect of acute respiratory acidosis on two populations of intercalated cells in the rat cortical collecting duct. *Am J Physiol.* 1987;253: F1142–F1156.
611. Teng-umnuay P, Verlander JW, Yuan W, Tisher CC, Madsen KM. Identification of distinct subpopulations of intercalated cells in the mouse collecting duct. *J Am Soc Nephrol.* 1996;7:260–274.
612. Kim J, Kim YH, Cha JH, Tisher CC, Madsen KM. Intercalated cell subtypes in connecting tubule and cortical collecting duct of rat and mouse. *J Am Soc Nephrol.* 1999;10: 1–12.
613. Lonnerholm G. Histochemical demonstration of carbonic anhydrase activity in the human kidney. *Acta Physiol Scand.* 1973;88:455–468.
614. McKinney TD, Burg MB. Bicarbonate absorption by rabbit cortical collecting tubules in vitro. *Am J Physiol.* 1978; 234:F141–F145.
615. McKinney TD, Burg MB. Bicarbonate secretion by rabbit cortical collecting tubules in vitro. *J Clin Invest.* 1978;61: 1421–1427.
616. Brown D, Gluck S, Hartwig J. Structure of the novel membrane-coating material in proton-secreting epithelial cells and identification as an H^{+}ATPase. *J Cell Biol.* 1987;105:1637–1648.
617. Brown D, Hirsh S, Gluck S. An H^{+}-ATPase in opposite plasma membrane domains in kidney epithelial cell subpopulations. *Nature.* 1988;331:622–624.
618. Alper SL, Natale J, Gluck S, Lodish HF, Brown D. Subtypes of intercalated cells in rat kidney collecting duct defined by antibodies against erythroid band 3 and renal vacuolar H^{+} ATPase. *Proc Natl Acad Sci U S A.* 1989;86: 5429–5433.
619. Drenckhahn D, Schluter K, Allen DP, Bennett V. Colocalization of band 3 with ankyrin and spectrin at the basal membrane of intercalated cells in the rat kidney. *Science.* 1985;230:1287–1289.
620. Verlander JW, Madsen KM, Low PS, Allen DP, Tisher CC. Immunocytochemical localization of band 3 protein in the rat collecting duct. *Am J Physiol.* 1988;255:F115–F125.
621. Weiner ID, Hamm LL. Regulation of intracellular pH in the rabbit cortical collecting tubule. *J Clin Invest.* 1990; 85:274–281.
622. Royaux IE, Wall SM, Karniski LP, et al. Pendrin, encoded by the Pendred syndrome gene, resides in the apical region of renal intercalated cells and mediates bicarbonate secretion. *Proc Natl Acad Sci U S A.* 2001;98:4221–4226.
623. Soleimani M, Greeley T, Petrovic S, et al. Pendrin: an apical $Cl^{-}/OH^{-}/HCO_3^{-}$ exchanger in the kidney cortex. *Am J Physiol.* 2001;280:F356–F364.
624. Frische S, Kwon TH, Frokiaer J, Madsen KM, Nielsen S. Regulated expression of pendrin in rat kidney in response to chronic NH_4Cl or $NAHCO_3$ loading. *Am J Physiol.* 2003;284: F584–F593.
625. Romero MF. Molecular pathophysiology of SLC4 bicarbonate transporters. *Curr Opin Nephrol Hypertens.* 2005;14: 495–501.
626. Everett LA, Glaser B, Beck JC, et al. Pendred syndrome is caused by mutations in a putative sulphate transporter gene (PDS). *Nat Genet.* 1997;17:411–422.
627. Wall SM, Hassell KA, Royaux IE, et al. Localization of pendrin in mouse kidney. *Am J Physiol.* 2003;284:F229–F241.
628. Wall SM. Recent advances in our understanding of intercalated cells. *Curr Opin Nephrol Hypertens.* 2005;14:480–484.
629. Verlander JW, Kim YH, Shin W, et al. Dietary Cl(-) restriction upregulates pendrin expression within the apical plasma membrane of type B intercalated cells. *Am J Physiol Renal Physiol.* 2006;291:F833–F939.
630. Wall SM, Weinstein AM. Cortical distal nephron Cl transport in volume homeostasis and blood pressure regulation. *Am J Physiol Renal Physiol.* 2013;305:F427–F438.
631. Nanami M, Pham TD, Kim YH, et al. The role of intercalated cell *Nedd4-2* in BP regulation, ion transport and transporter expression. *J Am Soc Nephrol.* 2018;29:1706–1719.
632. Chen L, Lee JW, Chou CL, et al. Transcriptomes of major renal collecting duct cell types in mouse identified by single-cell RNA-seq. *Proc Natl Acad Sci U S A.* 2017; 114:E9989–E9998.
633. Park J, Shrestha R, Qiu C, et al. Single-cell transcriptomics of the mouse kidney reveals potential cellular targets of kidney disease. *Science.* 2018;360:758–763.
634. LeFurgey A, Tisher CC. Morphology of rabbit collecting duct. *Am J Anat.* 1979;115:111–124.
635. Hansen GP, Tisher CC, Robinson RR. Response of the collecting duct to disturbances of acid-base and potassium balance. *Kidney Int.* 1980;17:326–337.
636. Madsen KM, Tisher CC. Cellular response to acute respiratory acidosis in rat medullary collecting ducts. *Am J Physiol.* 1983;245:F670–F679.
637. Madsen KM, Tisher CC. Response of intercalated cells of rat outer medullary collecting duct to chronic metabolic acidosis. *Lab Invest.* 1984;51:268–276.
638. Garg LC, Narang N. Ouabain-insensitive K^{+} adenosine triphosphatase in distal nephron segments of the rabbit. *J Clin Invest.* 1988;81:1204–1208.
639. Wingo CS. Active proton secretion and potassium absorption in the rabbit outer medullary collecting duct: functional evidence of H^{+} K^{+} ATPase. *J Clin Invest.* 1989;84: 361–365.
640. Wingo CS, Madsen KM, Smolka A, Tisher CC. H^{+} K^{+} ATPase immunoreactivity in cortical and outer medullary collecting duct. *Kidney Int.* 1990;38:985–990.
641. Ahn KY, Kone BC. Expression and cellular localization of mRNA encoding the "gastric" isoform of H^{+}-K^{+}-ATPase α-subunit in rat kidney. *Am J Physiol.* 1995;268: F99–F109.
642. Campbell-Thomson ML, Verlander JW, Curran KA, et al. In situ hybridization of H-K-ATPase β-subunit mRNA in rat and rabbit kidney. *Am J Physiol.* 1995;269:F345–F354.
643. Weiner ID, Verlander JW. Ammonia transport in the kidney by Rhesus glycoproteins. *Am J Physiol Renal Physiol.* 2014; 306:F1107–F1120.
644. Madsen KM, Clapp WL, Verlander JW. Structure and function of the inner medullary collecting duct. *Kidney Int.* 1988; 34:441–454.

645. Sands JM, Nonoguchi H, Knepper MA. Vasopressin effects on urea and H_2O transport in inner medullary collecting duct subsegments. *Am J Physiol.* 1987;253:F823–F832.
646. Clapp WL, Madsen KM, Verlander JM, Tisher CC. Intercalated cells of the rat inner medullary collecting duct. *Kidney Int.* 1987;31:1080–1087.
647. Nielsen S, Terris J, Smith CP, Hediger MA, Ecelbarger CA, Knepper MA. Cellular and subcellular localization of the vasopressin-regulated urea transporter in rat kidney. *Proc Natl Acad Sci U S A.* 1996;93:5495–5500.
648. Silverblatt FJ. Ultrastructure of the renal pelvic epithelium of the rat. *Kidney Int.* 1974;5:214–220.
649. Khorshid MR, Moffat DB. The epithelia lining the renal pelvis in the rat. *J Anat.* 1974;118:561–569.
650. Lucien N, Bruneval P, Lasbennes F, et al. UT-B1 urea transporter is expressed along the urinary and gastrointestinal tracts of the mouse. *Am J Physiol Regul Integr Comp Physiol.* 2005;288:R1046–R1056.
651. Verlander JW, Moudy RM, Cambell WG, Cain BD, Wingo CS. Immunohistochemical localization of H-K-ATPase a2 subunit in rabbit kidney. *Am J Physiol Renal Physiol.* 2001;281:F357–F365.
652. Madsen KM, Zhang L, Shamat AR, Siegfried S, Cha JH. Ultrastructural localization of osteopontin in the kidney: induction by lipopolysaccharide. *J Am Soc Nephrol.* 1997;8:1043–1053.
653. Bonventre JV, Karnovsky MJ, Lechene CP. Renal papillary epithelial morphology in antidiuresis and water diuresis. *Am J Physiol Renal Fluid Electrolyr Physiol.* 1978;235:F69–F76.
654. Lemley KV, Kriz W. Anatomy of the interstitium. *Kidney Int.* 1991;39:370–381.
655. Bohman SO. The ultrastructure of the renal medulla and the interstitial cells. In: Cotran RS, ed. *Tubulo-Interstitial Nephropathies.* Churchill Livingstone; 1983:1–34.
656. Hestbech J, Hansen HE, Amdisen A, Olsen S. Chronic renal lesions following long-term treatment with lithium. *Kidney Int.* 1977;12:205–213.
657. Bohle A, Grund KE, MacKensen S, Tolon M. Correlations between renal interstitium and level of serum creatinine. *Virchows Arch a Pathol Anat Histol.* 1977;373:15–22.
658. Mounier F, Foidart JM, Gubler MC. Distribution of extracellular matrix glycoproteins during normal development of human kidney: an immunohistochemical study. *Lab Invest.* 1986;54:394–401.
659. Zeisberg M, Kalluri R. Physiology of the interstitium. *Clin J Am Soc Nephrol.* 2015;10:1831–1840.
660. Kaissling B, Hegyi I, Loffing J, Le Hir M. Morphology of interstitial cells in the healthy kidney. *Anat Embryol (Berl).* 1996;193:303–318.
661. Kaissling B, Le Hir M. The renal interstitium: morphological and functional aspects. *Histochem Cell Biol.* 2008;130:247–262.
662. Bachmann S, LeHir M, Eckardt KU. Colocalization of erythropoietin mRNA and ecto-5-nucleotidase immunoreactivity in peritubular cells of the rat renal cortex suggests that fibroblasts produce erythropoietin. *J Histochem Cytochem.* 1993;41:335–341.
663. Maxwell PH, Osmond MK, Pugh CW, et al. Identification of the renal erythropoietin-producing cells using transgenic mice. *Kidney Int.* 1993;44:1149–1162.
664. Asada N, Takase M, Nakamura J, et al. Dysfunction of fibroblasts of extrarenal origin underlies renal fibrosis and renal anemia in mice. *J Clin Invest.* 2011;121:3981–3990.
665. Kobayashi H, Liu Q, Binns TC, et al. Distinct subpopulations of FOXD1 stroma-derived cells regulate renal erythropoietin. *J Clin Invest.* 2016;126:1926–1938.
666. Sims DE. The pericyte-A review. *Tissue Cell.* 1986;18:153–174.
667. Lemos DR, Marsh G, Huang A, et al. Maintenance of vascular integrity by pericytes is essential for normal kidney function. *Am J Physiol Renal Physiol.* 2016;311:F1230–F1242.
668. Kriz W, Kaissling B, Le Hir M. Epithelial-mesenchymal transition (EMT) in kidney fibrosis: fact and fantasy? *J Clin Invest.* 2011;121:468–474.
669. Humphreys BD, Lin SL, Kobayashi A, et al. Fate tracing reveals the pericyte and not epithelial origin of myofibroblast in kidney fibrosis. *Am J Pathol.* 2010;176:85–97.
670. Kramann R, Schneider RK, DiRocco DP, et al. Perivascular Gli1 progenitors are key contributors to injury-induced organ fibrosis. *Cell Stem Cell.* 2015;16:51–66.
671. Viehmann SF, Bohner AMC, Kurts C, Brähler S. The multifaceted role of the renal mononuclear phagocyte system. *Cell Immunol.* 2018; pii:S0008-8749(18)30180-1.
672. Brahler S, Zinselmeyer BH, Raju S, et al. Opposing roles of dendritic cell subsets in experimental glomerulonephritis. *J Am Soc Nephrol.* 2018;29:138–154.
673. Muirhead EE. The medullipin system of blood pressure control. *Am J Hypertens.* 1991;4:556s–568s.
674. Folkow B. Incretory renal functions-Tigerstedt, renin and its neglected antagonist medullipin. *Acta Physiol.* 2007;190:99–102.
675. Kurtz A. Endocrine functions of the renal interstitium. *Pflgers Arch Eur J Physiol.* 2017;469:869–876.
676. Fourman J, Moffat DB. *The Blood Vessels of the Kidney.* Blackwell Scientific; 1971.
677. Edwards JG. Efferent arterioles of glomeruli in the juxtamedullary zone of the human kidney. *Anat Rec.* 1956;125:521–529.
678. Casellas D, Mimran A. Shunts in renal microvasculature of the rat. A scanning electron microscopic study of corrosion casts. *Anat Rec.* 1981;201:237–248.
679. Ljungqvist A. Ultrastructural connection between afferent and efferent arterioles in juxtamedullary glomerular units. *Kidney Int.* 1975;8:239–244.
680. Ljungqvist A. Fetal and postnatal development of the intrarenal arterial pattern in man. *Acta Paediatr.* 1963;52:443–464.
681. Sanfilippo F, Pizzo SV, Croker BP. Immunohistochemical studies of cell differentiation in a juxtaglomerular tumor. *Arch Pathol Lab Med.* 1982;106:604–607.
682. Fina L, Molgard HV, Robertson D, et al. Expression of the CD34 gene in vascular endothelial cells. *Blood.* 1990;75:2417–2425.
683. Civin CL, Trischmann TM, Fackler MJ, et al. Summary of CD34 cluster workshop section. In: Knapp W, ed. *Leucocyte Typing IV.* Academic Press; 1989:818–825.
684. Gabbiani G, Schmid E, Winter S, et al. Vascular smooth muscle cells differ from other smooth muscle cells: predominance of vimentin filaments and a specific α-type actin. *Proc Natl Acad Sci U S A.* 1981;78:298–302.
685. Rollhauser H, Kriz W, Heinke W. Das gefass–system der rattenniere. *Z Zellforsch.* 1964;64:381–403.

686. Kriz W, Barrett JM, Peter S. The renal vasculature: anatomical–functional aspects. In: Thurau K, ed. *Kidney and Urinary Tract Physiology II*. University Park Press; 1976:1–21.
687. Beeuwkes R, Bonventre JV. Tubular organization and vascular-tubular relations in the dog kidney. *Am J Physiol*. 1975; 229:695–713.
688. Beeuwkes R. Vascular-tubular relationships in the human kidney. In: Leaf A, Giebisch G, Bolis L, et al., eds. *Renal Pathophysiology*. Raven Press; 1980:155–163.
689. Pierce EC. Renal lymphatics. *Anat Rec*. 1944;90:315–335.
690. Bell RD, Keyl MJ, Shrader FR, Jones EW, Henry LP. Renal lymphatics: the internal distribution. *Nephron*. 1968;3: 454–463.
691. Kriz W, Dieterich HJ. Das lymphagefass system der niere bei einigen saugetieren: licht-und elektronenmikroskipische untersuchungen. *Z Anat Entwickl Gesch*. 1970;131:111–147.
692. Ishikawa Y, Akasaka Y, Kiguchi H, et al. The human renal lymphatics under normal and pathological conditions. *Histopathology*. 2006;49:265–273.
693. Kriz W. A periarterial pathway for intrarenal distribution of renin. *Kidney Int*. 1987;31(suppl 20):551–556.
694. Yang Y, Oliver G. Development of the mammalian lymphatic vasculature. *J Clin Invest*. 2015;124:888–897.
695. Zheng W, Aspelund A, Alitalo K. Lymphangiogenic factors, mechanisms and applications. *J Clin Invest*. 2015;124:878–887.
696. Kenig-Kozlovsky Y, Scott RP, Onay T, et al. Ascending vasa recta are angiopoietin/Tie2-dependent lymphatic-like vessels. *J Am Soc Nephrol*. 2018;29:1097–1107.
697. Mitchell GAG. The nerve supply of the kidneys. *Acta Anat (Basel)*. 1950;10:1–37.
698. Gosling JA. Observations on the distribution of intrarenal nervous tissue. *Anat Rec*. 1969;163:81–88.
699. Stefansson K, Wollmann RL, Jerkovic M. S-100 protein in soft tissue tumors derived from Schwann cells and melanocytes. *Am J Pathol*. 1982;106:261–268.
700. Nakajima T, Uatanabe S, Sato Y, Kameya T, Hirota T, Shimosato Y. An immunoperoxidase study of S-100 protein distribution in normal and neoplastic human tissues. *Am J Surg Pathol*. 1982;6:715–727.
701. Trojanowski JQ, Lee VMY, Schlaepfer WW. An immunohistochemical study of human central and peripheral nervous system tumors, using monoclonal antibodies against neurofilaments and glial filaments. *Hum Pathol*. 1984;15:248–257.
702. Lee VMY, Carden MJ, Schlaepfer WW. Structural similarities and differences between neurofilament proteins from five different species as revealed using monoclonal antibodies. *J Neurosci*. 1986;6:2179–2186.
703. Fourman J. The adrenergic innervation of the efferent arterioles and the vasa recta in the mammalian kidney. *Experientia*. 1970;26:293–294.
704. Barajas L, Powers K, Wang P. Innervation of the renal cortical tubules: a quantitative study. *Am J Physiol*. 1984;247:F50–F60.
705. Barajas L, Powers K. Innervation of the thick ascending limb of Henle. *Am J Physiol*. 1988;255:F340–F348.
706. Barajas L, Liu L, Poers K. Anatomy of the renal innervation: intrarenal aspects and ganglia of origin. *Can J Physiol Pharmacol*. 1992;70:735–749.
707. Sakakura K, Ladich E, Cheng Q, et al. Anatomic assessment of sympathetic peri-arterial renal nerves in man. *J Am Coll Cardiol*. 2014;64:635–643.

34

Urinary Bladder, Ureter, and Renal Pelvis

Oleksandr N. Kryvenko

The urinary system extends from the pelvicalyceal system in the kidneys receiving the urine and ends at the urethral orifice where the urine leaves the body during urination. The urinary bladder is an epithelium-lined muscular viscus which has the ability to distend and accommodate up to 400 to 500 mL of urine without a change in intraluminal pressure. The empty bladder lies within the pelvis minor while the distended bladder extends into the abdominal cavity where its dome may be elevated to the level of the umbilicus. Upon reaching its capacity, the urinary bladder sends neural stimuli to the central nervous system to initiate micturition. It is able to initiate and sustain a contraction until the organ is empty. Interestingly, micturition may be initiated or inhibited voluntarily despite the involuntary nature of the organ by activating skeletal muscles of the pelvis. The ureters are epithelium-lined muscular tubes designed to transport urine from the kidneys to the urinary bladder with the aid of peristalsis. The renal pelvis and calyces represent a reservoir between the renal papillae and the ureter, serving to collect the urine excreted from the kidney and transport it to the ureter proper. The urethra is the most distal part of the urinary system. Its sphincter is under voluntary control to start or stop the urination. The entire urinary system is surfaced by urothelium with some site-specific variations which are discussed below.

EMBRYOLOGY

At approximately the 6th to 7th weeks of gestation, the cloaca is divided by the urorectal septum into a dorsal rectum and a ventral urogenital sinus (1,2). It is this urogenital sinus that will give rise to the majority of the urinary bladder, aided by the caudal migration of the cloacal membrane which will close the infraumbilical portion of the abdominal wall. The caudal portions of the mesonephric ducts become dilated and eventually fuse with the urogenital sinus in the midline dorsally, contributing to the formation of the bladder trigone. While these ducts contribute initially to the formation of the mucosa of the trigone, this is subsequently entirely replaced by the endodermal epithelium of the urogenital sinus. The gradual absorption of the mesonephric ducts brings about the separate openings of the ureters into the urinary bladder in the area of the trigone. During embryologic development, the allantois regresses completely, forming a thick, epithelium-lined tube, the urachus, which extends from the umbilicus to the apex (dome) of the bladder (1). Vestiges of the urachus may be found deep in the muscularis propria of the bladder at the dome (Fig. 34.1A). Before or shortly after birth, the urachus involutes further becoming simply a fibrous cord. Incomplete obliteration of the urachus may lead to developmental abnormalities including a patent urachus, umbilicourachal sinus, vesicourachal diverticulum, and urachal cyst. Pathologists commonly refer to the fibrous cord that extends from the dome of the bladder to the umbilicus as the urachal remnant, but it should be called the median umbilical ligament since urachal remnant refers to remnants of the epithelial lining of the urachus which occasionally persists within the median umbilical ligament (3). The epithelial lining of the urachus is urothelium similar to that of the urinary bladder

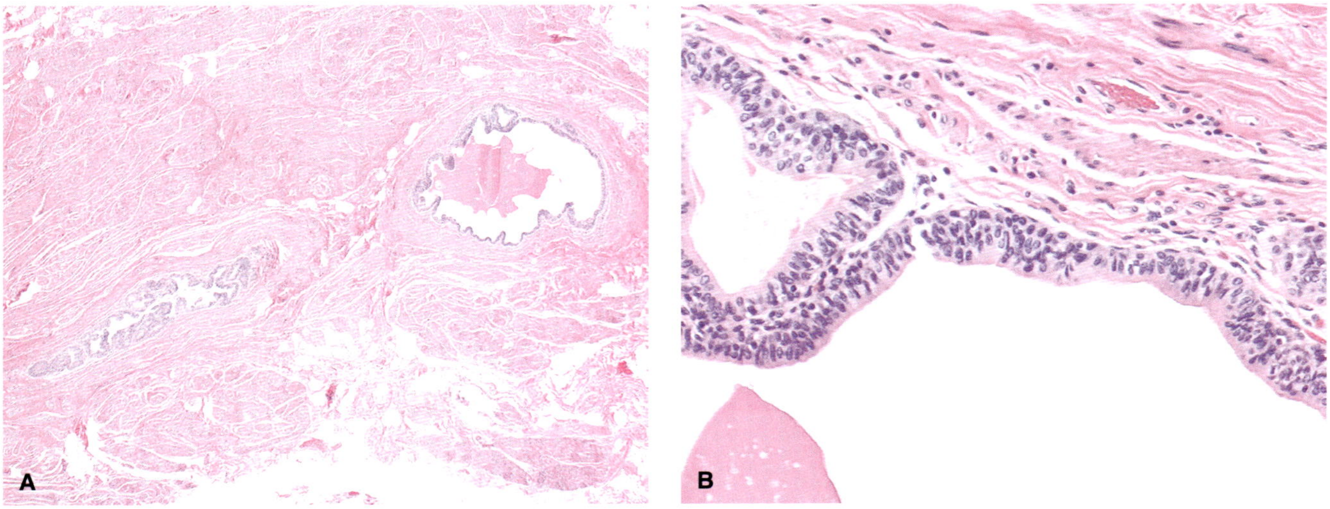

FIGURE 34.1 Urachal remnant found in section of the dome of a cystectomy specimen. **A:** Low-power magnification showing an epithelium-line cystic structure within muscularis propria. **B:** High-power magnification showing epithelial lining reminiscent of urothelium with luminal mucin and inspissated dense proteinaceous section.

and the ureter, but it may undergo metaplastic change, mostly glandular, and be associated with mucin or inspissated proteinaceous material (Fig. 34.1B). Carcinomas that develop in the urachal remnants are often adenocarcinomas with mucinous or signet ring cell differentiation.

The epithelium of the urinary bladder is endodermally derived from the cranial portion of the urogenital sinus in continuity with the allantois. The lamina propria, the muscularis propria (detrusor muscle), and the adventitia develop from the adjacent splanchnic mesenchyme. These facts are important in understanding the histogenesis and nomenclature of lesions arising from the epithelial surface, as well as the bladder wall. Glandular features within benign urothelium (e.g., cystitis cystica and glandularis) and glandular differentiation in urothelial carcinomas are not due to mesodermal or müllerian rests within the trigone but come about through a process of metaplasia and reflect the histologic plasticity of the urothelium. Since the mesonephric ducts involute totally during embryologic development, it is incorrect to refer to tumors with mixed epithelial and sarcomatoid features arising in the bladder epithelium as mesodermal mixed tumors. They are, in fact, "endodermal mixed" tumors and are usually called sarcomatoid carcinomas. Müllerian duct remnants may be found within the bladder wall (Fig. 34.2A) and appear as endometriosis, endosalpingiosis, or endocervicosis (Fig. 34.2B). These findings are collectively termed müllerianosis (4). Endometriosis is the most common histologic

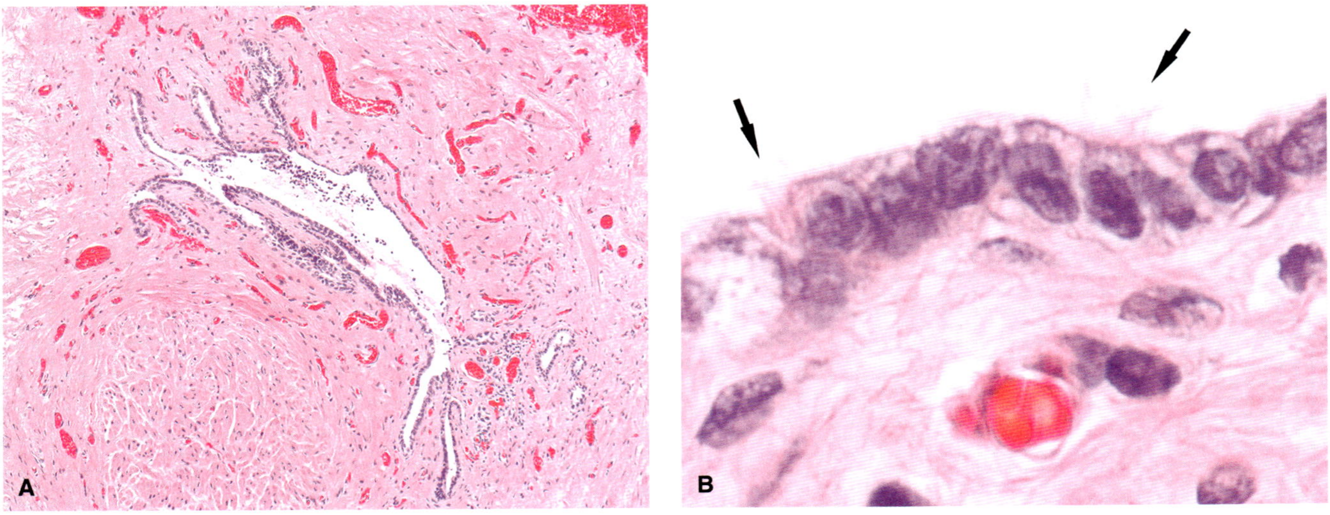

FIGURE 34.2 Müllerianosis within the bladder wall. **A:** Müllerian duct remnant showing branching cystic structure. **B:** High-power magnification shows luminal cilia (*arrows*) concluding that the structure represents an endosalpingiosis. (*continued*)

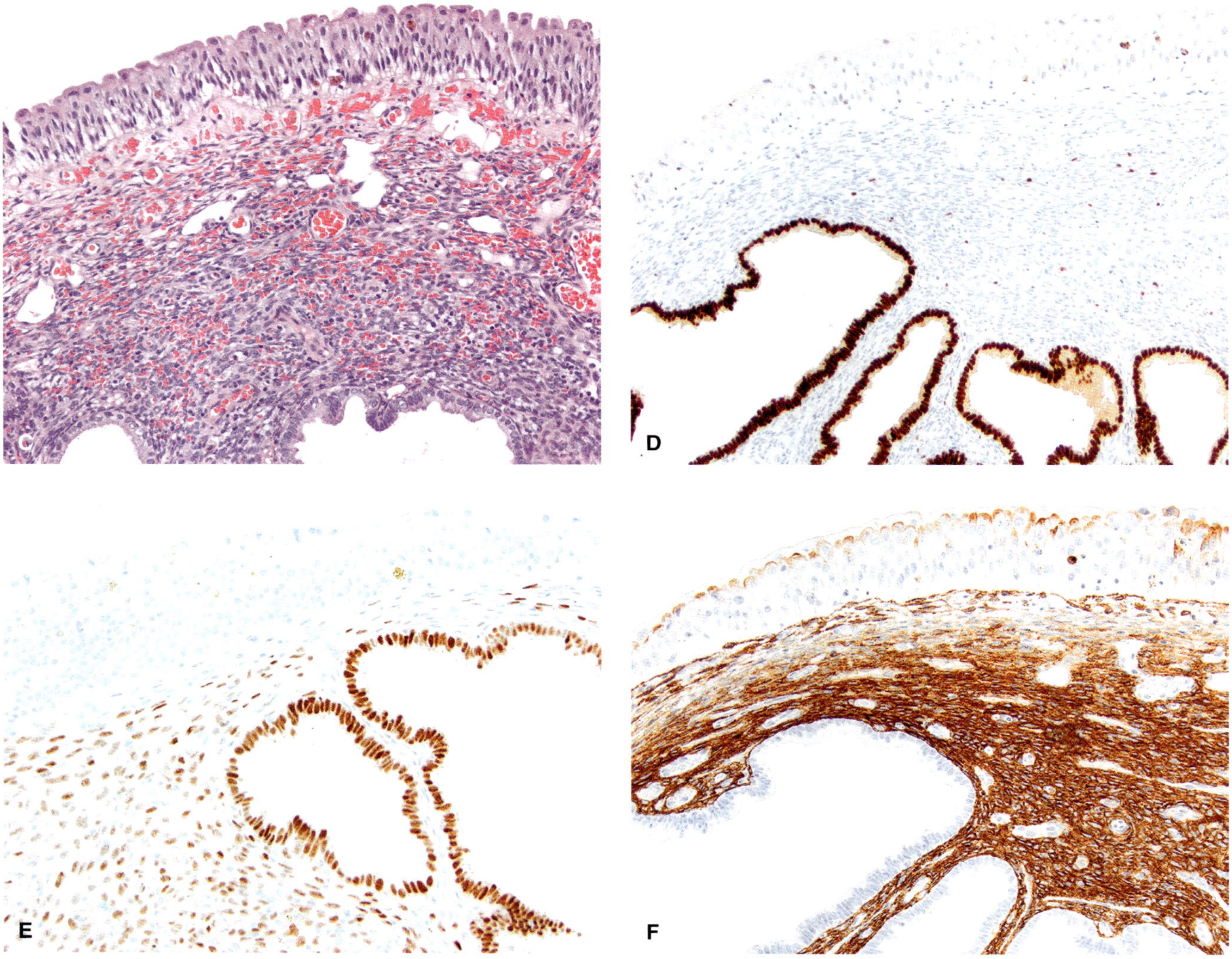

FIGURE 34.2 *(Continued)* **C:** Bladder endometriosis shows abundant congested endometrial stroma. Evidence of bleeding is seen as hemosiderin accumulation in the urothelium. Benign endometrial glands are seen at the bottom of the figure. **D:** PAX8 highlights endometrial glands but is negative in the stroma. **E:** Estrogen receptor is positive in both endometrial stromal cells and endometrial glands. **F:** Endometrial stromal is positive for CD10.

type of müllerianosis (Fig. 34.2C). All types of müllerianosis are positive for PAX8 and estrogen receptor by immunohistochemistry (Fig. 34.2D, E) (5). Unlike endosalpingiosis and endocervicosis, endometriosis is accompanied by a CD10-positive spindle cell stroma, often with fresh hemorrhage or hemosiderin deposits (Fig. 34.2F). Endometriosis may be symptomatic with hematuria and/or pelvic pain that may require surgical intervention, especially when a mass lesion is visible at cystoscopy. Other forms of müllerianosis are usually incidental findings discovered in specimens procured for unrelated reasons. Mesonephric rests may occasionally be encountered within the bladder.

The ureters develop by branching and elongation of the ureteric bud, or metanephric diverticulum, which begins as a dorsal bud from the mesonephric duct (1,2). The stalk of the ureteric bud becomes the ureter, whereas the cranial end forms the renal pelvis, as well as the calyces and the collecting tubules. The epithelium of the ureter and the renal pelvis, although histologically identical to that of the bladder, is of mesodermal derivation.

ANATOMICAL CONSIDERATIONS

Bladder

The empty bladder in an adult has the shape of a four-sided inverted pyramid and is enveloped by the vesical fascia. The superior surface is covered by the pelvic parietal peritoneum. The posterior surface, or base, lies posteroinferiorly; it is separated from the rectum by the uterine cervix and the proximal portions of the vagina in females and by the seminal vesicles and the ampulla of the vasa deferentia in males. The two inferolateral surfaces of the bladder face laterally, inferiorly, and anteriorly and are in contact with the fascia of the levator ani muscles. The most anterosuperior point of the bladder is known as the apex and it is located at the point of contact of the superior surfaces and the two inferolateral surfaces. The apex, or dome, marks the point of insertion of the median umbilical ligament and consequently is the area where urachal carcinomas are centered.

The trigone is a complex anatomic structure located at the base of the bladder and extending to the posterior bladder neck. In the proximal and lateral aspects of the trigone, the ureters enter into the bladder obliquely. The muscle in this region is a combination of smooth muscle of the longitudinal layer of the intramural ureter and the detrusor muscle (6,7). The intramural ureter is surrounded by a fibromuscular sheath (Waldeyer sheath) which is fused into the ureteral muscle. This fibromuscular tissue fans out in the area of the trigone and mixes with the detrusor muscle, thus fixing the intramural ureter to the bladder. As the bladder distends, the surrounding musculature exerts pressure on the obliquely oriented intramural ureter, producing closure of the ureteral lumen and thus avoiding reflux of urine. The most distal portion of the bladder is the bladder neck and represents the convergence of the posterior and inferolateral walls at the urethra. In men, the bladder neck merges with the prostate gland. It is important to recognize the existence of the occasional presence of prostatic ducts in this area since their involvement by urothelial carcinoma should not be mistaken with invasive carcinoma. The internal sphincter is located in the region of the bladder neck with major contributions from the middle circular layer of the detrusor muscle.

The main arterial blood supply of the bladder comes from the inferior vesical arteries that branch from the internal iliac arteries. The superior vesical, obturator, and inferior gluteal arteries also supply the bladder as do the uterine and vaginal arteries in females. The veins of the urinary bladder drain into the internal iliac veins and form the vesical venous plexus. In men, this plexus envelops the bladder base, prostate, and seminal vesicles and connects with the prostatic venous plexus. In females, it covers the bladder neck and urethra and communicates with the vaginal plexus. Lymphatic drainage is through the external and internal lymph nodes although drainage of portions of the bladder neck region may be through the sacral or common iliac nodes.

The urinary bladder is supplied by both sympathetic and parasympathetic nerves that form the vesical nerve plexus (8,9). The former are derived from T_{11} to L_2 nerves and play no role in micturition. On the other hand, the parasympathetic nerves come from S_2 to S_4 and travel to the bladder via the pelvic nerve and the inferior hypogastric plexus. These nerves are important to micturition since they contract the fibers of the muscularis propria which in turn produce traction upon the bladder neck, opening the internal sphincter of the bladder. In fact, it is believed that micturition is initiated by voluntary relaxation of the perineal muscles and the striated muscle of the external sphincter located along the urethra. This action decreases urethral resistance as well as triggers contraction of the smooth muscle of the trigone and the remaining bladder, closing the ureteral orifices and increasing the hydrostatic pressure within the viscus. The bladder also contains sensory nerves which travel along the pelvic and hypogastric nerves and account for the sensation of pain as the bladder becomes too distended.

Ureters

The ureters measure approximately 30 cm in length. The abdominal ureter takes a vertical course downward and medially along the anterior surface of the psoas muscle. It is covered by adventitia that represents an extension of Gerota fascia. The pelvic ureter can be subdivided into a longer parietal and a shorter intravesical portion. The parietal portion is intimately related to the peritoneum. It descends posterolaterally and, as it approaches the bladder base, it becomes medially directed to reach the urinary bladder. The ureters enter the base of the bladder obliquely and empty into the bladder at the ureteral orifices. The distal parietal portion and the intravesical segments are enveloped in Waldeyer sheath which aids in fixing the ureter to the bladder.

Depending on the anatomic level, the ureter receives blood from branches of the renal, abdominal aortic, gonadal, hypogastric, vesical, and uterine arteries which form a richly intercommunicating plexus of vessels surrounding the organ. Venous drainage is variable but tends to follow a pattern similar to the arterial distribution. The upper ureteral lymphatic vessels drain into the lateral aortic lymph nodes, the middle portion drains into the common iliac lymph nodes, and the inferior portion drains into either the common, external, or internal iliac lymph nodes.

Renal Pelvis

The renal pelvis lies primarily within the hilum, a space formed medially when one draws a vertical plane through the medial aspects of the upper and lower poles of the kidney. The renal sinus is located within the medial and antral portions of the kidney and is occupied by the renal pelvis, renal calyces, renal vessels and nerves, and fat. The fibrous capsule which lines the kidney passes over the lips of the hilum and lines the renal sinus, becoming continuous with the renal calyces. Within the renal sinus, the renal pelvis divides into two and rarely three major calyces, which in turn divide into 7 to 14 minor calyces. Urine from the distal collecting ducts within the renal medulla flows into the minor calyces at the tips of the renal papillae. The blood supply of the renal pelvis comes from branches of the renal artery and the venous drainage follows a similar distribution. Its lymphatic drainage is into the renal hilar lymph nodes.

Histologic Layers and Landmarks

The urinary system hollow organs have a relatively similar histologic composition with variable prominence of different layers in different anatomic sites. The lumen is lined by urothelium resting on the basement membrane. Although histologically and immunohistochemically identical, the urothelium of the bladder and proximal urethra is endodermally derived, whereas that of the pelvicalyceal systems and ureters originates from the mesoderm (10). The urothelium is supported by the lamina propria, which is composed of delicate areolar connective tissue rich in small vessels and some inflammatory cells. The lamina propria is sometimes termed the subepithelial connective tissue or submucosa. It contains a thin band of poorly organized bundles of smooth muscle cells, or muscularis mucosae, in some sites within the genitourinary system. The muscularis propria is composed of thick bundles of smooth muscle cells and is responsible for contraction and peristalsis. The muscularis propria of the bladder is termed the detrusor muscle. The adventitia represents the outermost layer of the genitourinary system in those areas that are not surfaced by the peritoneum. The superior urinary bladder is covered by the peritoneum (Fig. 34.3). To allow its superior expansion with distention, the bladder is surrounded by loose fibroadipose tissue except in the area of the bladder neck where the ligaments anchor the bladder and prevent its displacement (11). These ligaments include the pubovesical ligaments in women and the puboprostatic ligaments in men. The muscularis propria merges with the muscular stroma of the prostate in men. These histologic and anatomic landmarks are important to consider when assessing pathologic stage of bladder carcinomas (Fig. 34.3).

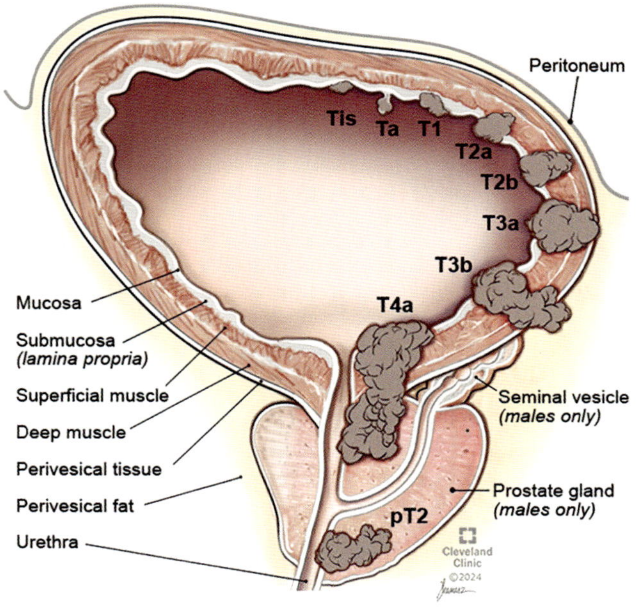

FIGURE 34.3 Bladder cancer staging diagram.

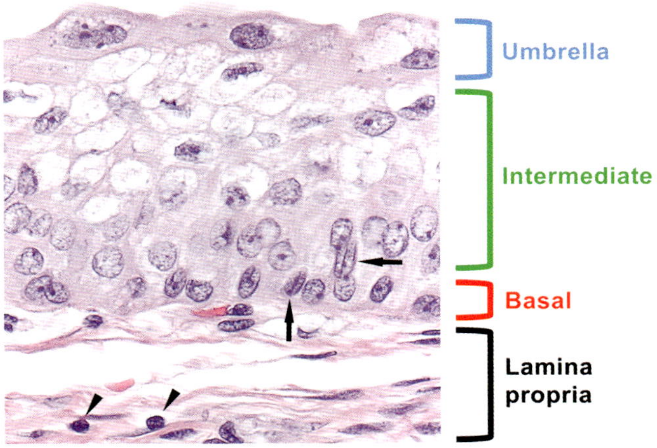

FIGURE 34.4 The urothelium is composed of 1 layer of umbrella cells, 0–5 layers of intermediate cells, and 1 layer of basal cells. Each umbrella cell overlies several intermediate cells. Groves are often seen in intermediate cells imparting them a coffee bean appearance (*arrows*). The nuclei of urothelium are euchromatic and may have small nucleoli. The size of intermediate cell nuclei is within a range of two to three times of a lymphocyte (*arrowheads*).

Urothelium

The urothelium is a specialized multilayered epithelium that lines the entire urinary system (10). It primarily functions as a barrier to prevent absorption of uric acid and urea and resist entry of the pathogens from the external (i.e. intraluminal) environment. Some data suggest that the urothelium may also modify urine composition through either secretion or reabsorption of water and electrolytes depending on hydration status and/or bladder distension (12–14).

The urothelium is a complex structure composed of three functionally and immunohistochemically distinct cell layers: the basal cell layer, intermediate cell layer, and apical, or umbrella, cell layer (Fig. 34.4).

Basal cells are the smallest urothelial cells and this layer is home to the urothelial stem cells that provide a life-long regeneration of the urothelium (15). These cuboidal cells are arranged in two to three layers, the deepest of which is attached to the basement membrane via hemidesmosomes. Benign urothelium usually does not denude and, thus, extensive denudation in bladder biopsies and transurethral resections should be documented when samples are obtained from patients without prior intravesical therapy or instrumentation. Extensive denudation raises concern for urothelial carcinoma in situ, as the lack of hemidesmosomes and desmosomes in the tumor cells results in their separation from the basement membrane and each other, correspondingly (16). Urothelial cells in the basal cell express keratin 5/6, whereas the more superficial urothelial cell layers lack expression of this marker (Fig. 34.5; Table 34.1) (10,17). Invasive urothelial carcinomas are typically negative for keratin 5/6 or demonstrate a single-cell pattern keratin 5/6 immunoreactivity at the periphery of invasive nests that corresponds to focal basal cell differentiation (17). Urothelial basal cells and prostatic

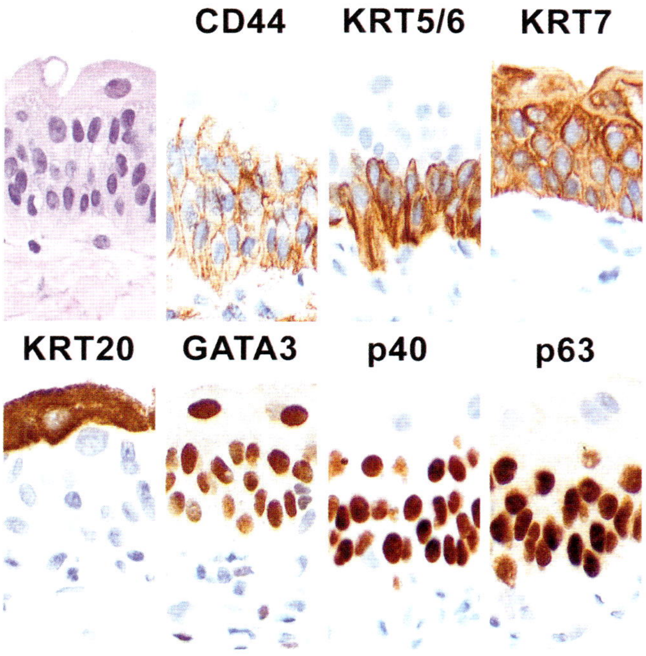

FIGURE 34.5 Immunohistochemistry of benign urothelium. See also Table 34.1.

basal cells are morphologically and immunohistochemically similar, with the exception of D2-40 staining in the latter. This feature of prostatic basal cells may be exploited to distinguish intraductal spread of urothelial carcinoma within the prostate from stromal invasion by the tumor (17).

Intermediate cells are morphologically similar to basal cells, although they contain slightly more cytoplasm and slightly more polygonal shape. The distinction is more evident in noninvasive papillary urothelial carcinomas, particularly low-grade neoplasms (Fig. 34.6). The number of layers of intermediate cells is variable and can range from none to five (Fig. 34.7). Benign urothelium virtually never undergoes hyperplasia and those areas where such a possibility is considered usually represent tangential sections of normal mucosa. The number of epithelial layers in flat urothelial lesions has limited diagnostic significance. The intermediate cells are larger than basal cells and attached to each other via desmosomes. Intermediate cell nuclei are euchromatic and roughly two to three times the size of a lymphocyte. They often have a single longitudinal nuclear groove imparting a coffee bean

TABLE 34.1 Pattern of Immunoreactivity in Benign Urothelium

	CD44	KRT5/6	KRT7	KRT20	GATA3	p40	p63
Umbrella cells	–	–	+	+	+	–	–
Intermediate cells	+	–	+	–	+	+	+
Basal cells	+	+	+	–	+	+	+

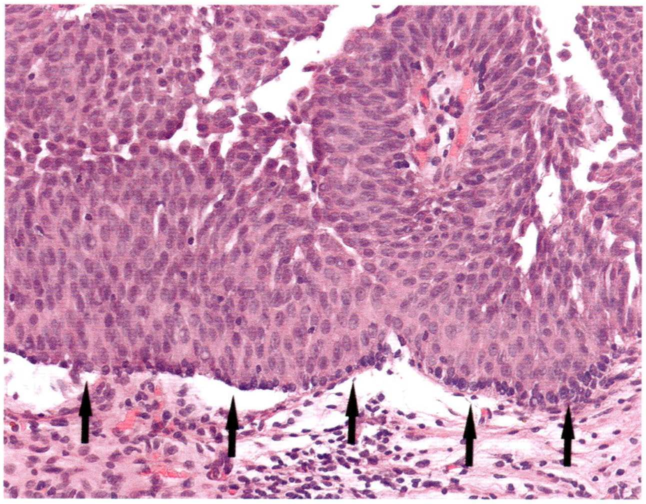

FIGURE 34.6 Prominent basal cell layer (*arrows*) in noninvasive low-grade papillary urothelial carcinoma.

appearance to the nuclei (Fig. 34.4). Small nucleoli may be seen at high-power magnification (Fig. 34.4).

Umbrella cells are arranged in a single layer at the luminal surface. Umbrella cells are larger than intermediate cells and they vary dramatically in size, often overlying several intermediate cells. The umbrella cells are dome-shaped when the bladder is empty and relaxed (Fig. 34.4 and 34.7), but become horizontally arranged and flattened when the bladder is distended (Fig. 34.8). The nuclei of umbrella cells are horizontally oriented and more widely spaced in the distended bladder (compare Fig. 34.7 vs. 34.8). Umbrella cells may be binucleated or show degenerative-type nuclear atypia. Umbrella cells are the only cells that form tight junctions to maintain a barrier to luminal contents (18). The umbrella cells of benign urothelium express keratin 20

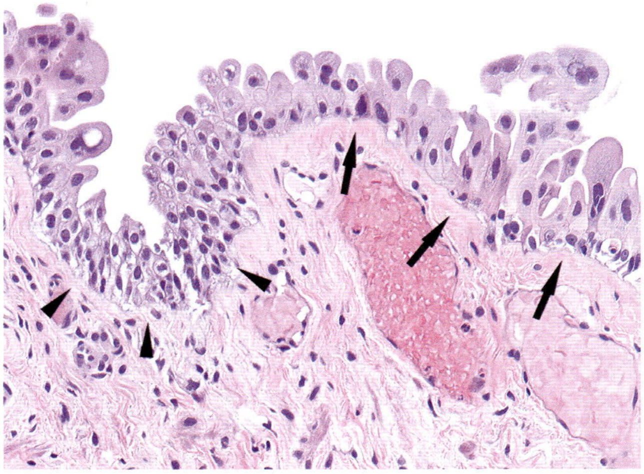

FIGURE 34.7 Preservation of the mucosal folds and the shape of umbrella cells suggest that the bladder is not distended. However, the number of intermediate cell layers varies. Right-hand side is represented only by umbrella and basal cell layers (*arrows*) whereas 2–3 layers of intermediate cells are seen on the left-hand side (*arrowheads*).

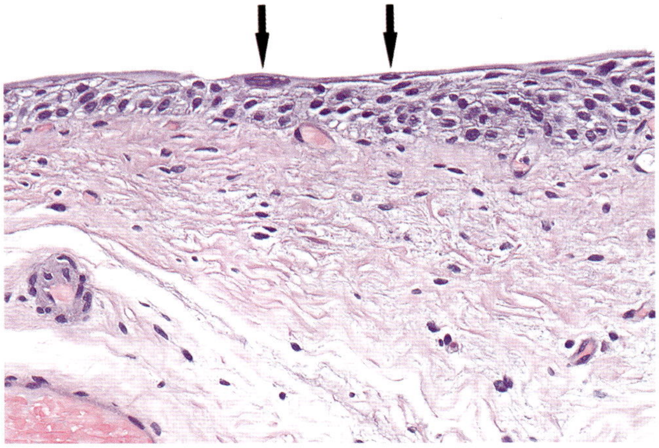

FIGURE 34.8 The umbrella cells become flattened with elongated and wider spaced nuclei (*arrows*) in distended bladder. Compare to Figures 34.4 and 34.7.

(Fig. 34.5; Table 34.1) and this can be a helpful diagnostic feature because noninvasive high-grade urothelial carcinomas usually show full-thickness immunopositivity for this marker. However, this feature is not absolute. Benign urothelium may show full-thickness immunopositivity with keratin 20 (Fig. 34.9A,B) and noninvasive high-grade cancers can be negative for keratin 20 (Fig. 34.10A,B).

Normal (Nonneoplastic) Variations of Urothelium

Most of the urinary system is lined by the urothelium with histologic and immunohistochemical features as described above. However, it is normal to encounter nonkeratinizing squamous epithelium in the trigone and urethra (Fig. 34.11A). These areas show immunostaining for GATA3 and demonstrate full-thickness keratin 5/6 immunoreactivity, but they are negative for keratin 20 due to a lack of umbrella cells. Some authors consider this finding to

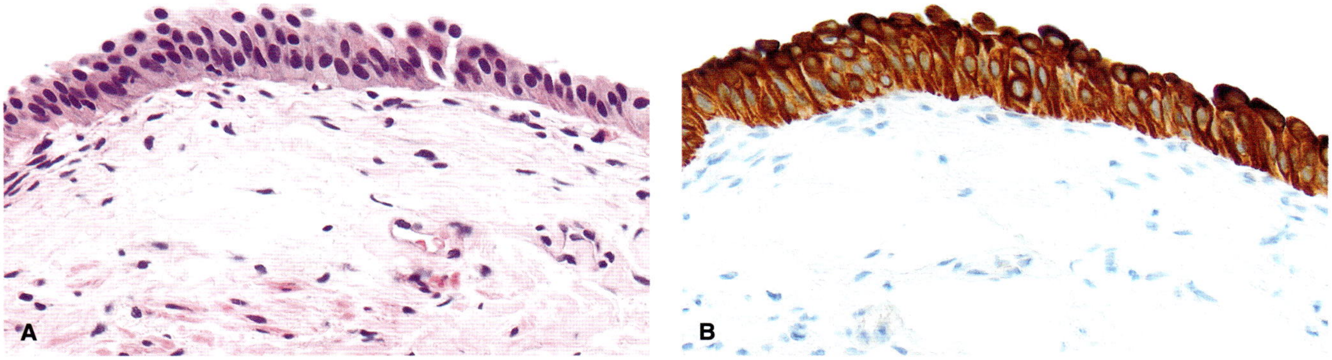

FIGURE 34.9 False-positive full-thickness staining of benign urothelium with keratin 20. **A:** Normal urothelium without any features of tumor or reactive changes. **B:** Keratin 20 shows a full-thickness staining which should not be used as evidence of tumor.

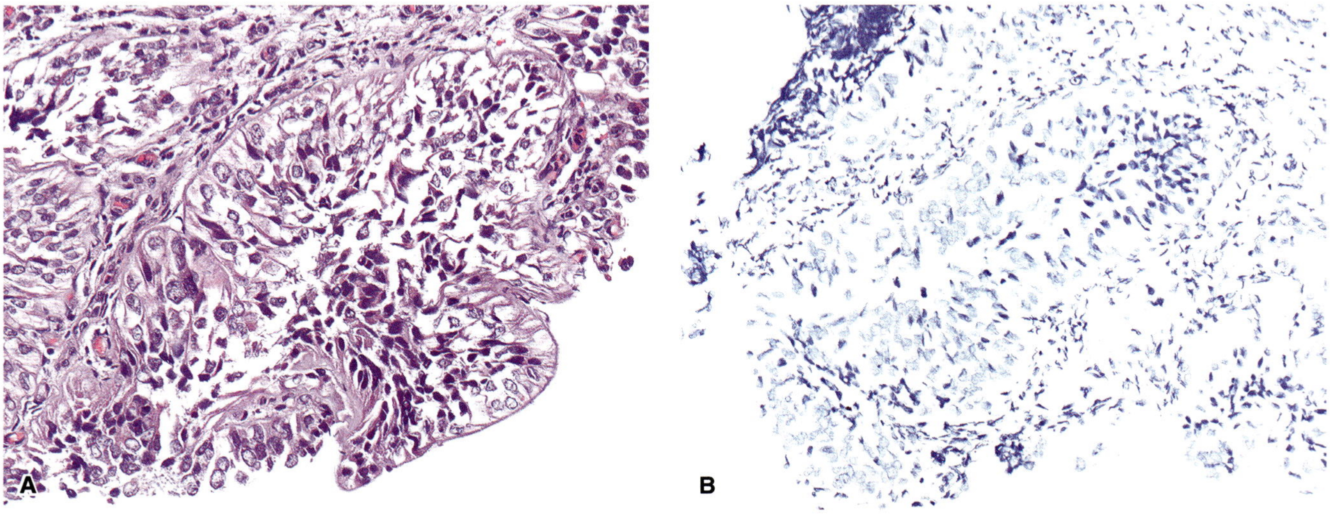

FIGURE 34.10 False-negative staining of in situ urothelial carcinoma (CIS) with keratin 20. **A:** The routine H&E sections demonstrate nuclear pleomorphism, hyperchromasia, nucleomegaly, and cytologic dyscohesion—typical features of CIS allowing classification of routine H&E preparation. **B:** Submitted keratin 20 immunostain is negative in carcinoma. In such cases, a definitive diagnosis of CIS can be issued based on H&E sections alone.

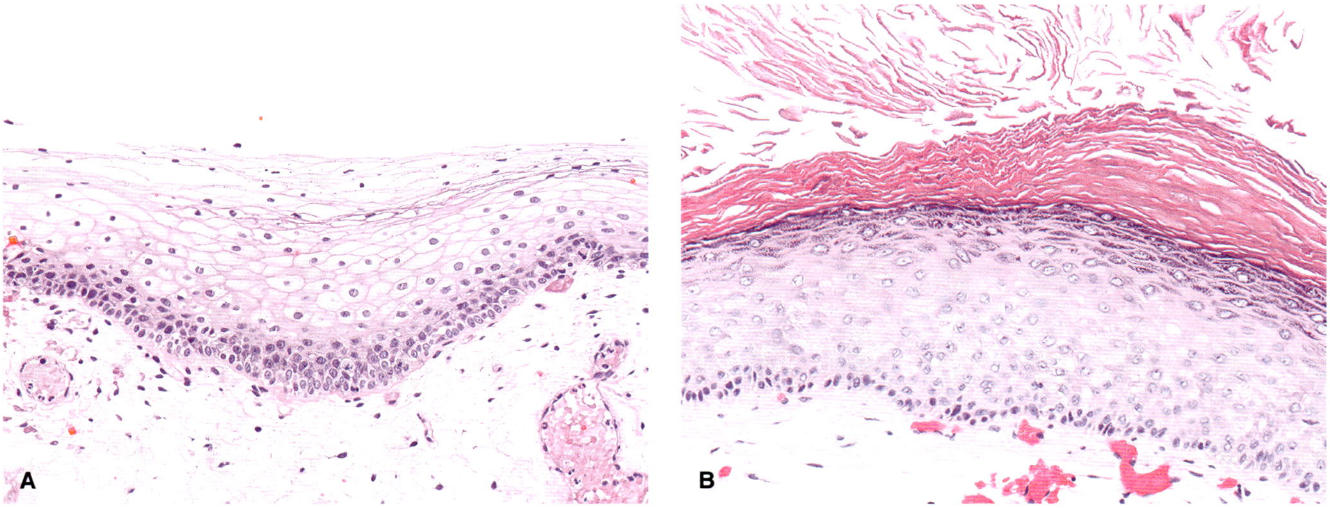

FIGURE 34.11 Squamous metaplasia of the urothelium. **A:** Nonkeratinizing squamous metaplasia. The squamous cells have prominent cell membranes and their cytoplasm contains glycogen (*pale areas*). The umbrella cell layer is absent. **B:** Keratinizing squamous metaplasia. Thick layer of keratin and granular cell layer is evident. The umbrella cell layer is absent.

represent nonkeratinizing squamous metaplasia but others consider it within the spectrum of normal findings in the urinary system. Keratinizing squamous epithelium is always abnormal and is suggested to be a preneoplastic process (Fig. 34.11B) (19). Mounds of keratin and a prominent granular cell layer are usually evident in such cases. Specimens demonstrating these findings should be reported as "benign urothelium with keratinizing squamous metaplasia."

The urothelium normally produces mucin which maintains its integrity and protects from invading bacteria (20). Mucin production can be increased in some reactive conditions and urothelial carcinomas, even in the absence of glandular differentiation. Benign urothelium may also undergo mucin cell metaplasia with variable numbers of mucin cells in the urothelium (Fig. 34.12). These mucin cells are not goblet cells and are negative for CDX2.

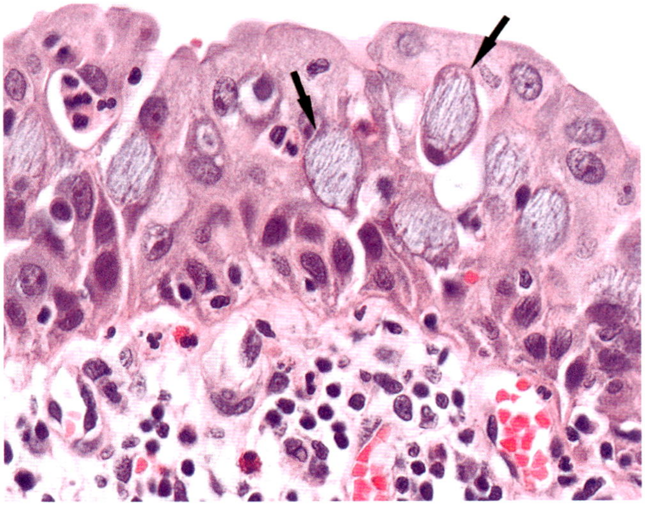

FIGURE 34.12 Mucin cell metaplasia (*arrows*) within benign urothelium.

Lamina Propria/Subepithelial Connective Tissue

The lamina propria is composed of loose connective tissue lying between the urothelium and muscularis propria. Its thickness varies among different anatomic sites of the urinary system and changes depending on the distention of the organ. With the exception of the trigone region, the lamina propria of the bladder contains a prominent muscularis mucosae (Fig. 34.13) (21). Intermediate-sized prominent muscular arteries can be seen at the same level in lamina propria as muscularis mucosae (Fig. 34.13B), and in some areas the lamina propria may be thin and these structures can be absent (Fig. 34.13C). The muscularis

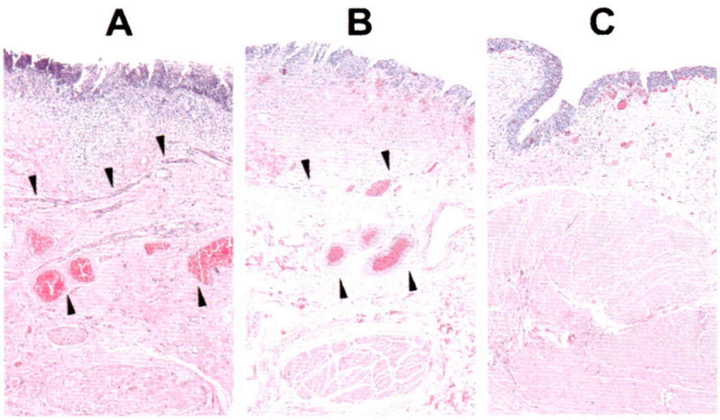

FIGURE 34.13 The thickness and composition of the lamina propria (subepithelial connective tissue) may vary. **A:** This lamina propria is thick and demonstrates both wispy muscle bundles (muscularis mucosa) and intermediate-sized vessels separating it into superficial and deeper layers (*arrowheads*). **B:** While this lamina propria is thick and muscularis mucosa is absent, intermediate-sized vessels can be used as a reference for the depth of cancer invasion (*arrowheads*). **C:** This lamina propria is thin and lacks both muscularis mucosa and intermediate-sized vessels. The substaging into more superficial or deeper invasion is not possible in such specimens.

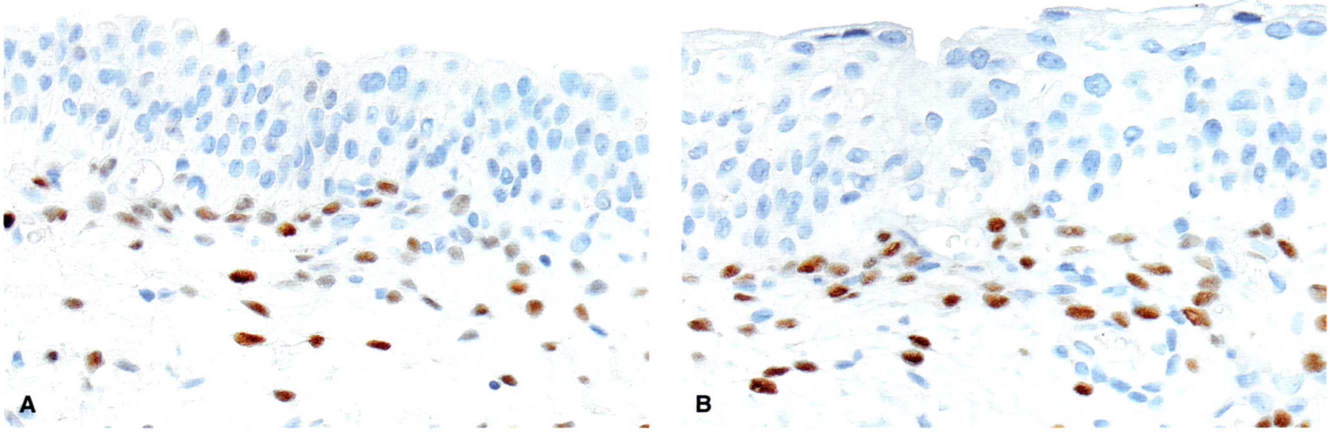

FIGURE 34.14 Stromal cells in submucosa are normally positive for estrogen (**A**) and progesterone (**B**) receptors.

mucosae and intermediate-sized muscular arteries separate lamina propria into its more superficial and deeper compartments (22). This distinction is of clinical significance because the extent of lamina propria invasion has prognostic value. In fact, the 5th edition World Health Organization Classification of the Urinary and Male Genital Tumours suggests that an extent of lamina propria invasion in transurethral resection specimens should be regularly attempted (23,24). Alternative approaches suggest reporting the width of the largest focus of carcinoma invading the lamina propria or the cumulative width of all foci.

The stromal cells of the lamina propria normally express estrogen and progesterone receptors (Fig. 34.14A,B). The loose connective tissue in the lamina propria plays a critical function in adjusting the surface of the mucosa according to the degree of bladder distention. The mucosa of the nondistended bladder has a folded architecture (Fig. 34.15A). These normal folds are short and uniform with pointy lumina tips and tapering toward the lumen. Exaggerated folds may simulate features of polypoid cystitis, although the latter usually demonstrates irregular mucosal protrusions that are broad-based and have variably edematous or fibrotic stroma (Fig. 34.15B) (25). Normal mucosal folds feature nonbranching exophytic protrusions, whereas papillomas and noninvasive papillary carcinomas are usually exophytic with branching protrusions (Fig. 34.15C).

Adipose tissue is normally present in the lamina propria and muscularis propria (Fig. 34.16). For this reason, invasion of fat by carcinoma cells is not considered in determining the depth of invasion by carcinoma in transurethral resection specimens.

Benign Urothelial Proliferations

Von Brunn Nests

Von Brunn nests are benign inverted urothelial proliferations present in the lamina propria of the renal pelvis, ureters, and bladder (26) (Fig. 34.17A). They may show slightly more prominent nuclear atypia and pleomorphism compared to the surface urothelium but nuclear hyperchromasia and mitotic activity are not prominent (5). Von Brunn nests may be distributed in clusters or form a small mass designated by urologists as a tumor (Fig. 34.17B). Florid hyperplasia of Von Brunn nests may mimic the nested subtype of invasive urothelial carcinoma. However, Von Brunn nests are never present within the muscularis propria, nor do they show retraction artifact or an associated desmoplastic stromal reaction. Von Brunn nests may undergo cystic change (Fig. 34.17C). In this situation, the luminal surface is lined by umbrella cells. The presence of umbrella cells allows distinction of cystic Von Brunn nests from cystitis cystica et glandularis.

Cystitis Cystica et Glandularis Including Intestinal Metaplasia

Cystitis cystica et glandularis has overlapping features with hyperplastic Von Brunn nests that show cystic change. However, the lumina in cystitis cystica et glandularis are lined by columnar eosinophilic cells rather than umbrella cells (Fig. 34.18). Inverted urothelial proliferations in cystitis cystica et glandularis may undergo intestinal metaplasia with intestinal-type CDX2-positive goblet cells and occasionally extend into the muscularis propria (Fig. 34.19A,B). Cystitis cystica et glandularis in the muscularis propria is not associated with a desmoplastic reaction, although it may be accompanied by large pools of dissecting mucin containing floating epithelium. This finding can cause diagnostic challenges, especially when seen in combination with low- or high-grade glandular dysplasia (Fig. 34.19C). Such cases may be very difficult to distinguish from invasive intestinal-type adenocarcinoma. Clues to a malignant diagnosis include the presence of desmoplasia, an infiltrative growth pattern, and solid epithelial areas.

Diverticulum

Diverticula are inverted invaginations of urothelium that often penetrate the muscularis propria or even adventitia

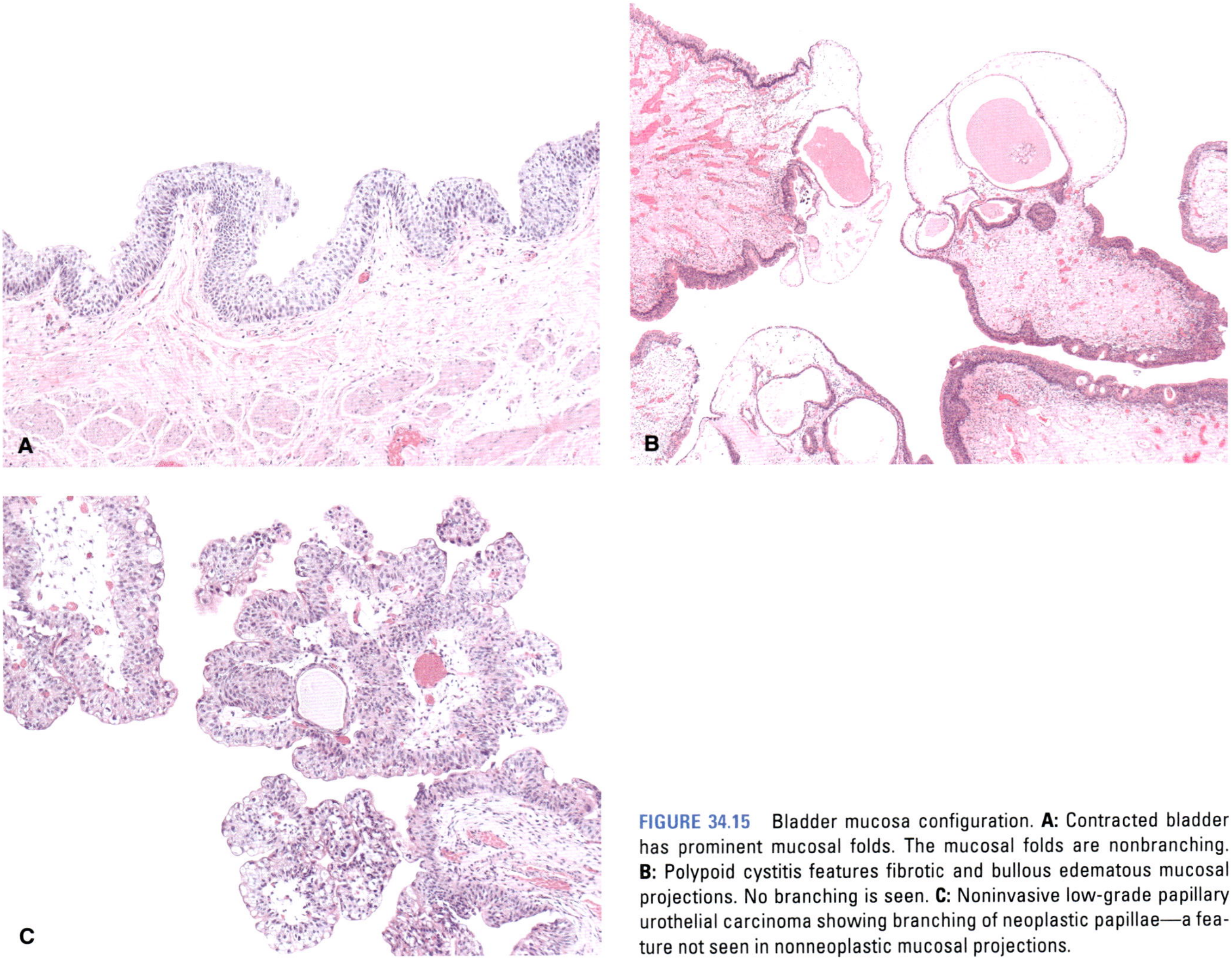

FIGURE 34.15 Bladder mucosa configuration. **A:** Contracted bladder has prominent mucosal folds. The mucosal folds are nonbranching. **B:** Polypoid cystitis features fibrotic and bullous edematous mucosal projections. No branching is seen. **C:** Noninvasive low-grade papillary urothelial carcinoma showing branching of neoplastic papillae—a feature not seen in nonneoplastic mucosal projections.

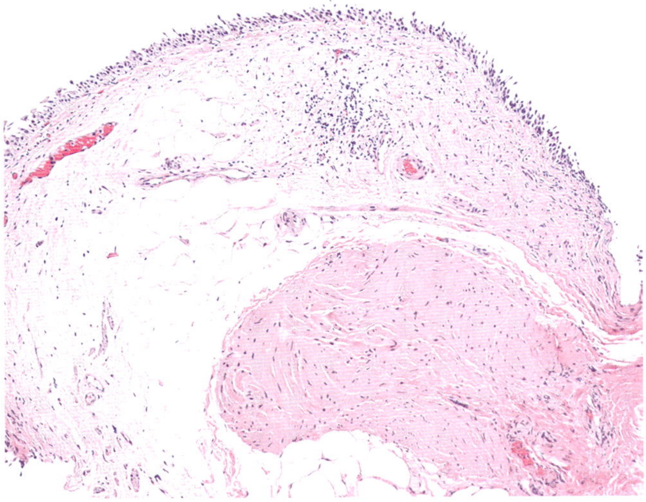

FIGURE 34.16 Except for urothelium proper, mature adipose tissue can be present at any layer of bladder wall.

(Fig. 34.20A,B). Unlike cystically dilated Von Brunn nests, diverticula occur singly. The etiology, location, and age distribution of diverticula are different among men and women. In men, diverticula usually arise in older patients in the bladder secondary to increased pressure related to benign prostatic hyperplasia. In women, the diverticula occur in the third and fourth decades of life and mostly arise in the urethra. They are most likely trauma-related following vaginal delivery. Rarely, diverticula are observed in children, in which case they likely represent developmental malformations (27). Because of urine stasis, diverticula are prone to develop urinary stones (Fig. 34.21). Colonization of diverticula by noninvasive urothelial carcinoma may mimic invasion and be misinterpreted to represent a muscle-invasive urothelial carcinoma (Fig. 34.22). In this situation, the lack of an irregular infiltrative growth pattern, prominent retraction artifact, desmoplastic reaction, and paradoxical differentiation all point to a noninvasive process. Diverticula lack a muscularis propria which is an important consideration

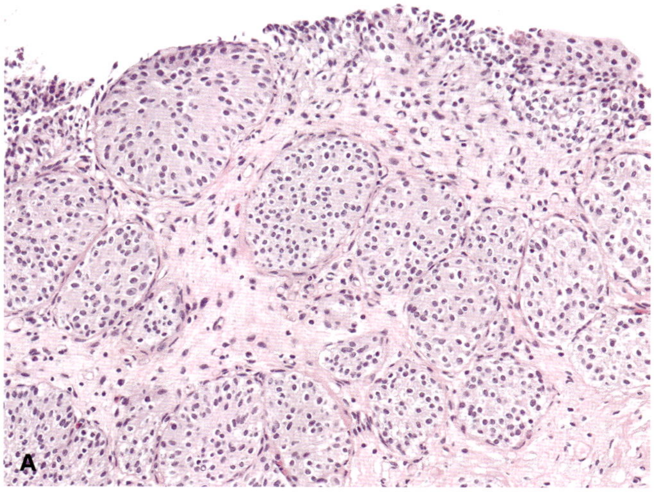

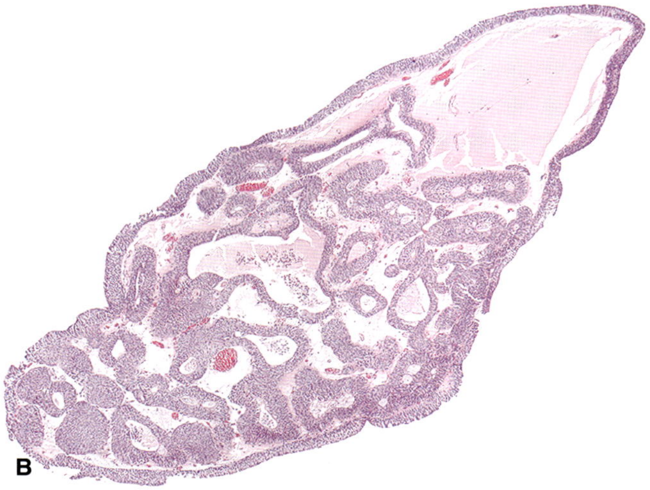

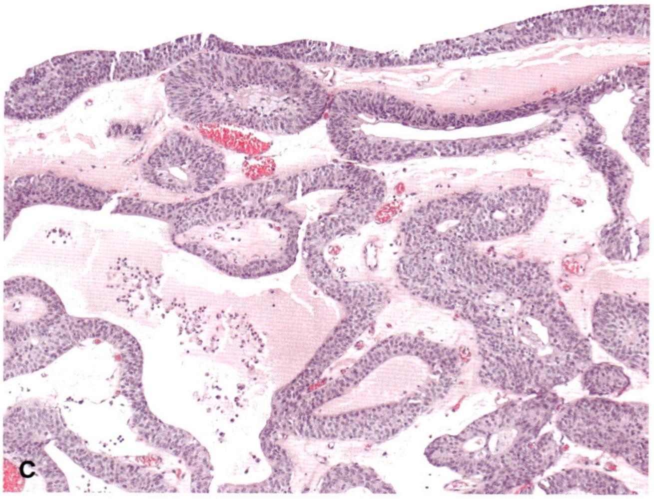

FIGURE 34.17 Von Brunn nests (VBN). **A:** Tightly packed VBNs in the lamina propria. Mild epithelial atypia could be seen which is normal for benign inverted lesions. **B:** This cluster of florid VBN hyperplasia was labeled by urologist as a bladder tumor resection. The lesion appears to be polypoid but the entire epithelial proliferation is endophytic. **C:** VBNs may have cystic change. Unlike cystitis cystica et glandularis where the lumen is lined by columnar cells, cystically dilated VBNs are lined by umbrella cells.

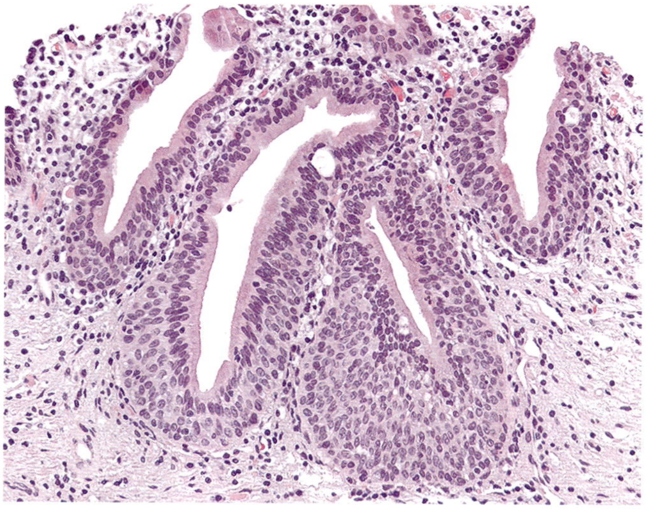

FIGURE 34.18 Cystitis cystica et glandularis. Unlike Von Brunn nests with cystic change, cystitis cystic et glandularis is lined by columnar cells. Note the presence of occasional goblet cells (**upper right**).

when assigning tumor stage to invasive carcinomas arising within diverticula.

Nephrogenic Adenoma

Nephrogenic adenoma was once considered a metaplastic process but we now know that the cells forming nephrogenic adenomas originate from the renal tubules (28). Using fluorescent in situ hybridization (FISH), Mazal *et al.* demonstrated the presence of Y chromosome in the cells of nephrogenic adenoma in female patients with orthotopic kidney transplants from male donors. Mucosal damage is a predisposing condition to nephrogenic adenoma. Presumably, shed renal tubular epithelial cells implant and proliferate in the area of urothelial denudation. Conventional nephrogenic adenomas can be papillary, tubular, or flat (Fig. 34.23A). The epithelial cells forming nephrogenic adenoma are positive for PAX8 and Napsin A, and negative for urothelial (e.g., GATA3 and p63) and intestinal (e.g., CDX2 and keratin 20) markers (Fig. 34.23B,C). Hobnail

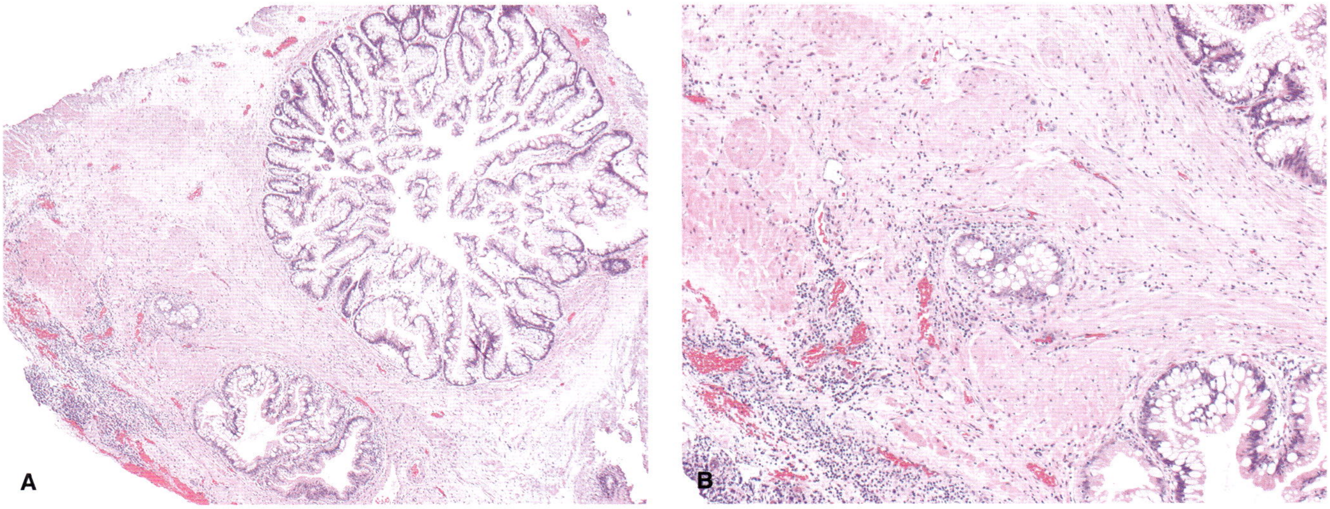

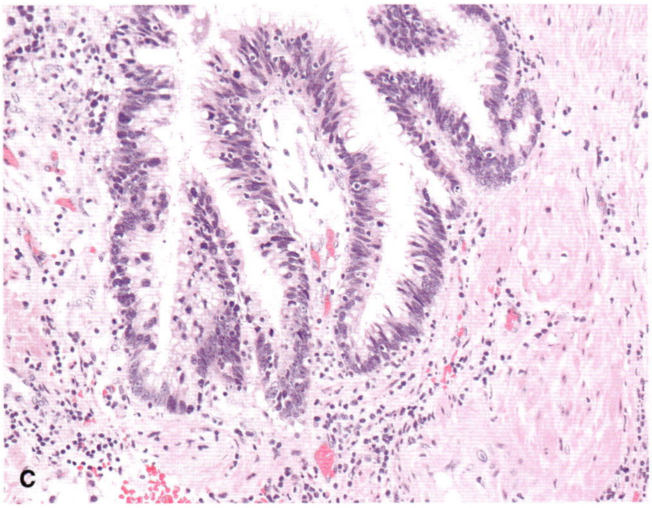

FIGURE 34.19 Cystitis cystica et glandularis with intestinal metaplasia. **A:** Low-power view showing extensive glandular proliferation with luminal mucin accumulation. **B:** Some glands are present within muscularis propria—a finding permissible for benign glandular lesions. **C:** This focus of cystitis cystica et glandularis with intestinal metaplasia within muscularis propria (thick muscle bundles on the right-hand side) developed a low-grade dysplasia. Criteria for dysplasia are the same as in colonic lesions. Lack of desmoplasia and infiltrative growth pattern help rule out an invasive intestinal-type adenocarcinoma.

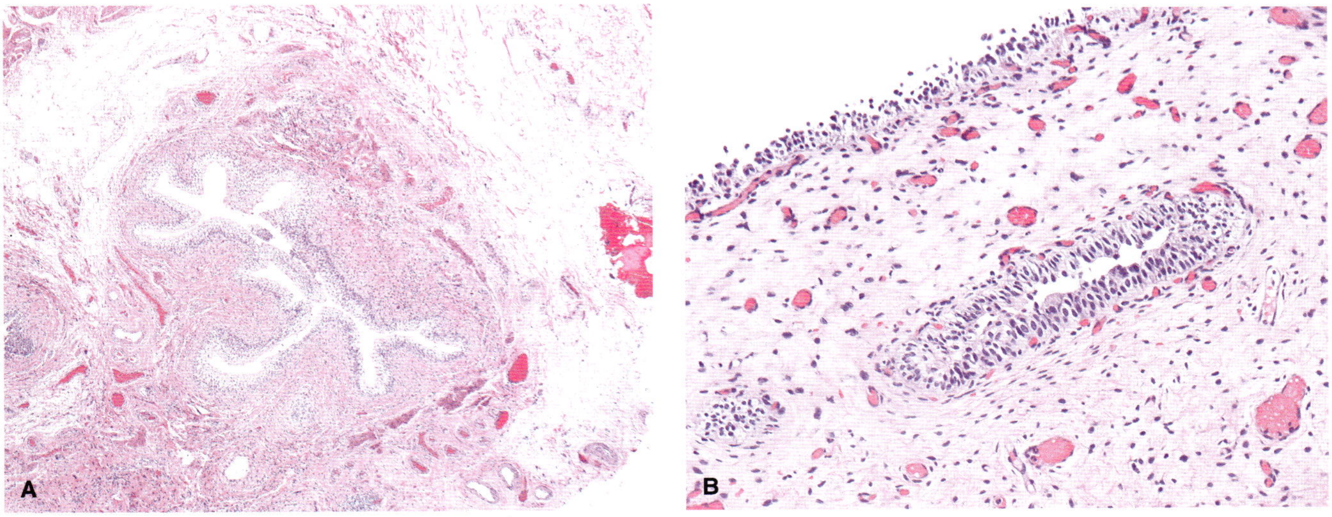

FIGURE 34.20 Bladder diverticulum. **A:** Low-power magnification of a larger diverticulum penetrating into the perivesical adipose tissue. The wall of the diverticulum lacks muscularis propria. **B:** Smaller bladder diverticulum which is present in the lamina propria.

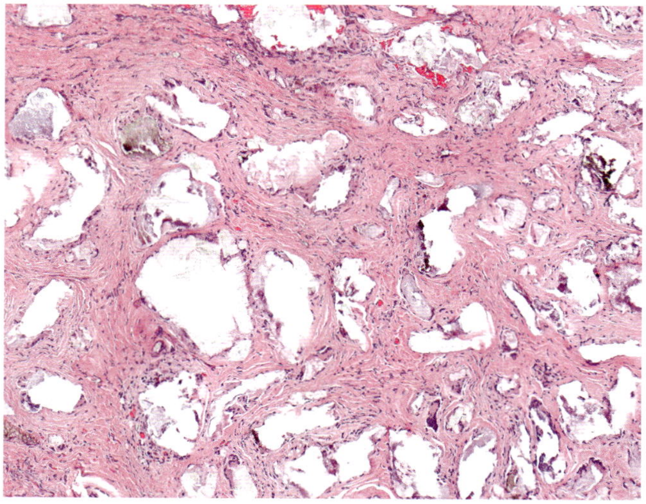

FIGURE 34.21 Impacted stone within the bladder wall is a common complication of diverticula. Urine stasis predisposes to the formation of calculi.

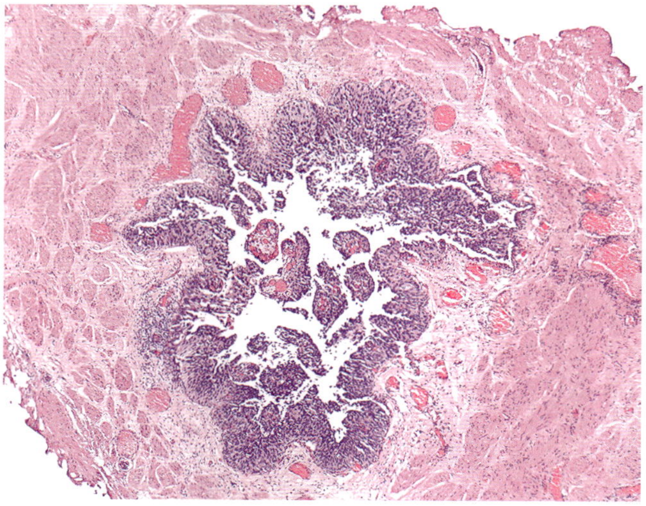

FIGURE 34.22 This bladder diverticulum is colonized by a noninvasive high-grade papillary urothelial carcinoma. Same was present on the surface of the bladder wall in other tissue chips. The cancer may primarily arise in the diverticulum. This finding may be misinterpreted as a muscle-invasive bladder cancer—a drastic mistake to make.

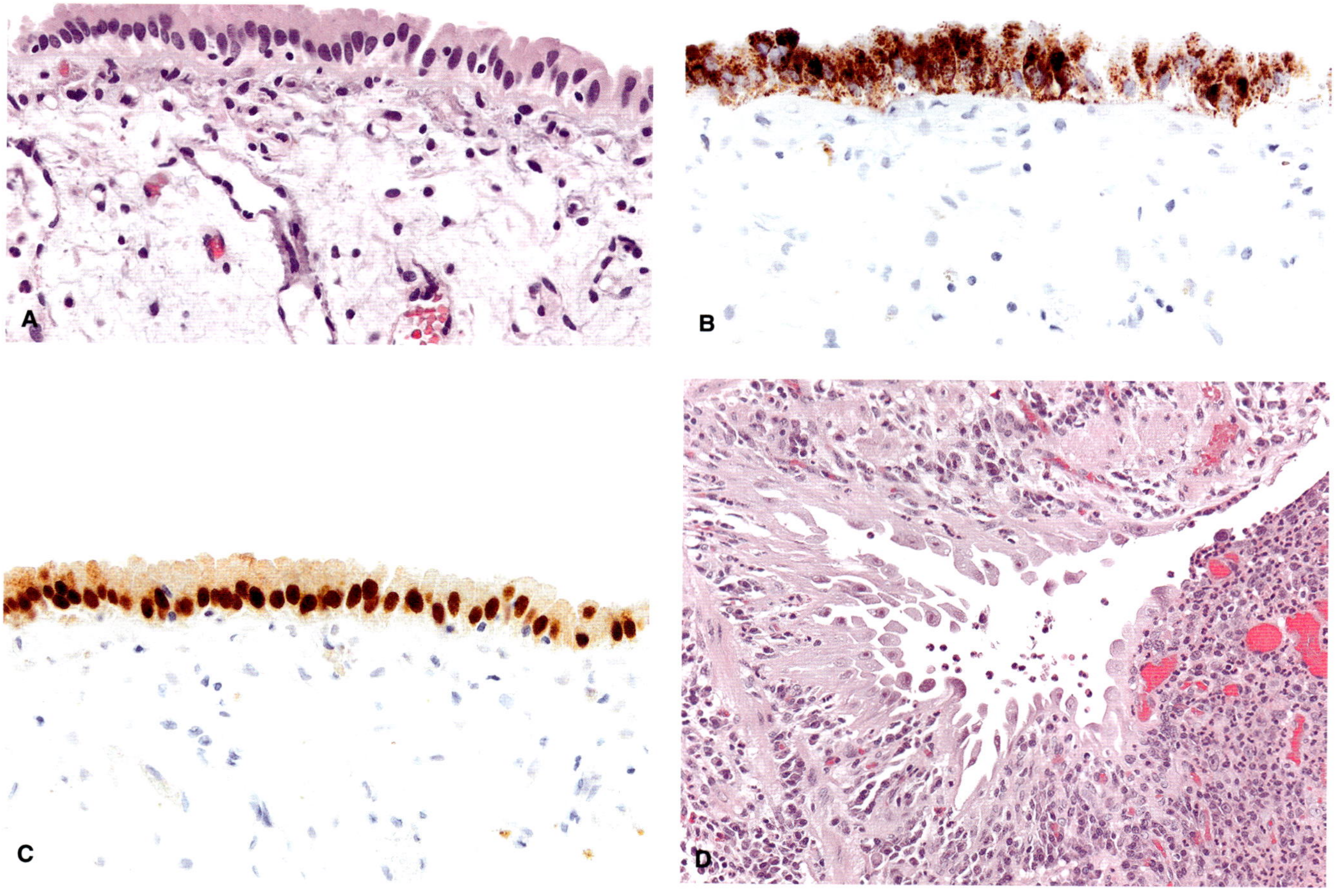

FIGURE 34.23 Nephrogenic adenoma. **A:** Flat nephrogenic adenoma composed of tall columnar cells. Napsin A (**B**) and PAX8 (**C**) are great immunostains for nephrogenic adenoma. **D:** This nephrogenic adenoma developing in a site of prior transurethral resection has hobnailing and significant cytologic atypia. Such atypical appearing nephrogenic adenomas may be misclassified as urothelial carcinoma and clear cell carcinoma.

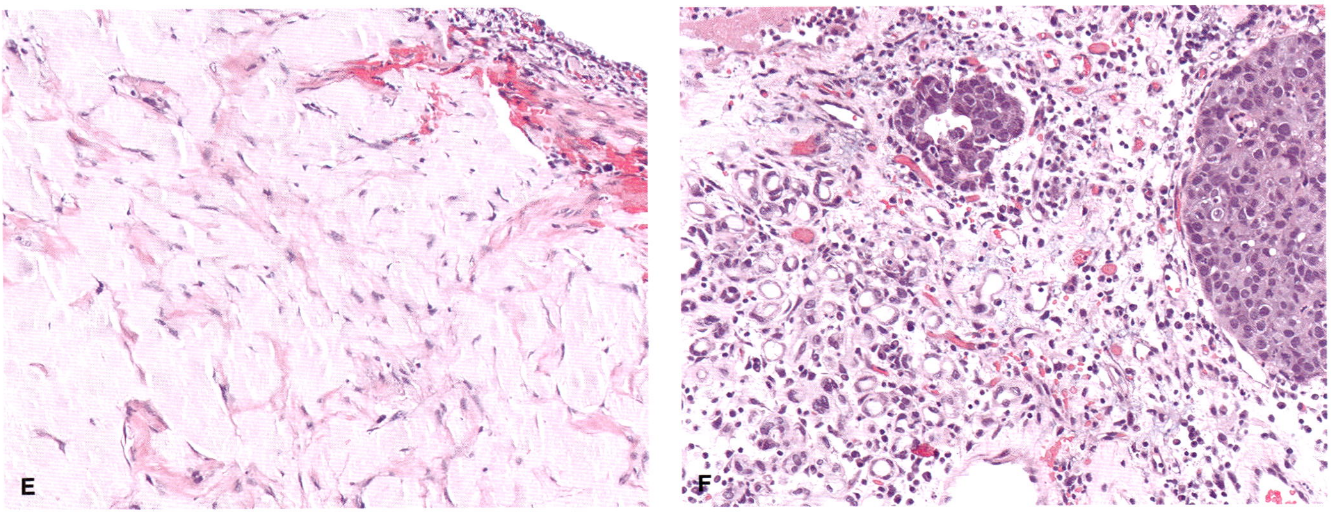

FIGURE 34.23 *(Continued)* **E:** Fibromyxoid nephrogenic adenoma. The epithelial component is barely seen and may require immunohistochemistry with keratins and PAX8 for proper classification. Fibromyxoid nephrogenic adenomas are negative for Napsin A. **F:** Tubular nephrogenic adenoma underneath in situ urothelial carcinoma (CIS). Tiny formed tubules with mucin may mimic signet ring cells, glandular differentiation, and invasion in otherwise noninvasive cancer.

cells and significant cytologic atypia may be seen in nephrogenic adenoma (Fig. 34.23D), but solid growth pattern is not permissible; the latter is a useful feature in the differential diagnosis between nephrogenic adenoma and clear cell adenocarcinoma. Fibromyxoid nephrogenic adenoma is a less frequent variant that features flattened, compressed tubular cells within myxoid tissue and a lack of staining for Napsin A (Fig. 34.23E) (29,30). Predisposing factors for both conventional and fibromyxoid nephrogenic adenoma include radiation therapy, prior surgery, and bladder stones. Nephrogenic adenoma may be intermixed with urothelial carcinoma, thereby mimicking a urothelial carcinoma with glandular differentiation (Fig. 34.23F) (31). Both conventional tubular and fibromyxoid nephrogenic adenoma may be present within muscularis propria (Fig. 34.24A,B).

Regional Variations of Urothelium

Pelvicalyceal System

The renal pelvis is continuous with the ureter distally and connects to two or three major calyces proximally. The major calyces divide into 7 to 14 minor calyces where the distal collecting tubules (i.e., ducts of Bellini) open at the tips of the renal papillae (i.e., zona cribrosa) in minor

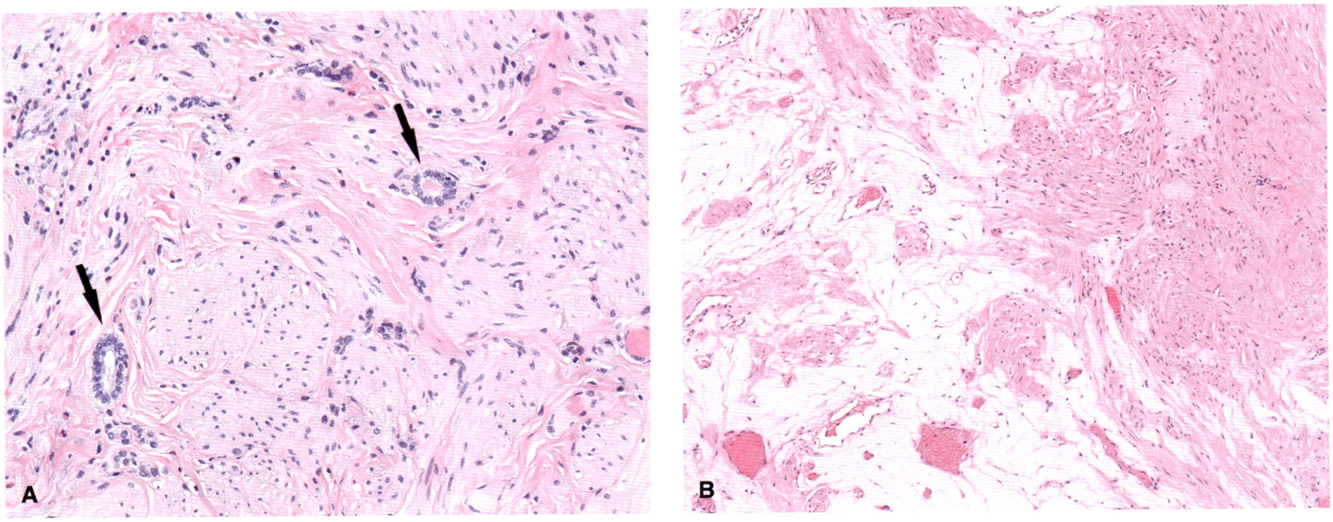

FIGURE 34.24 Nephrogenic adenoma mimicking invasive adenocarcinoma. **A:** Tubular nephrogenic adenoma within muscularis propria (*arrows* indicate tubules of nephrogenic adenoma within muscularis prporia). **B:** Fibromyxoid nephrogenic adenoma dissecting prostatic stroma.

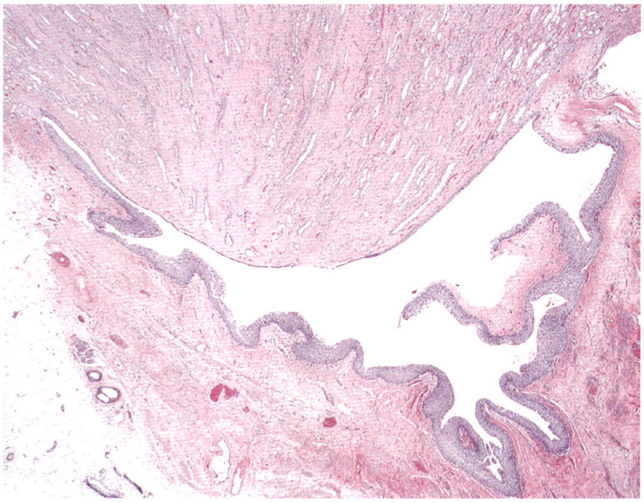

FIGURE 34.25 Renal papilla and minor calyx.

calyces (Fig. 34.25). Lamina propria is essentially absent in the renal papillae and is very thin in both the calyxes and renal pelvis (Fig. 34.26). Renal papillae are lined by a single layer of the urothelium whereas the calyces and renal pelvis have a urothelium lining featuring all three cell layers (Fig. 34.26). Muscularis mucosae is not present in the pelvicalyceal system.

Ureter

The ureters contain a sizable amount of lamina propria which allows for the formation of longitudinal mucosal folds when the organ is not distended (Fig. 34.27). There is no muscularis mucosae in the ureter (Fig. 34.28). The muscularis propria thickness increases in the distal ureter (32). Prominent hyperplasia of Von Brunn nests is often seen in the ureters and pelvicalyceal system (Fig. 34.29). Like in the bladder, the Von Brunn nests sharply stop at the level of the muscularis propria and never penetrate the latter.

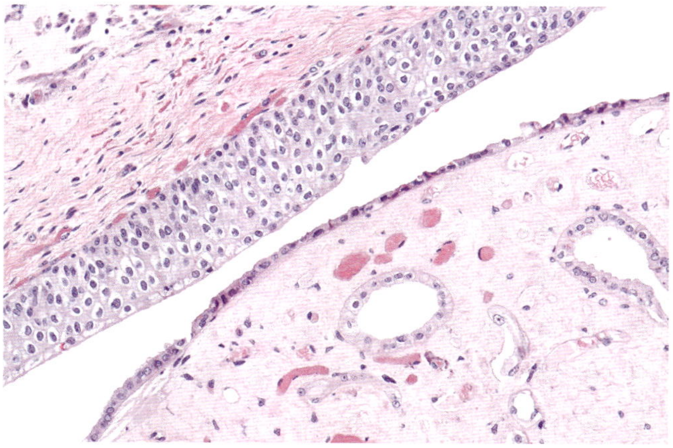

FIGURE 34.26 Urothelium lining minor calyx (**upper**) contrasts with that lining a renal papilla (**lower**). While calyx is covered by a urothelium analogous to elsewhere in the urinary system, the papilla is lined by a single layer of urothelium. Distal collecting tubules (ducts of Bellini) are seen in the distal portion of papilla under urothelium (zona cribrosa).

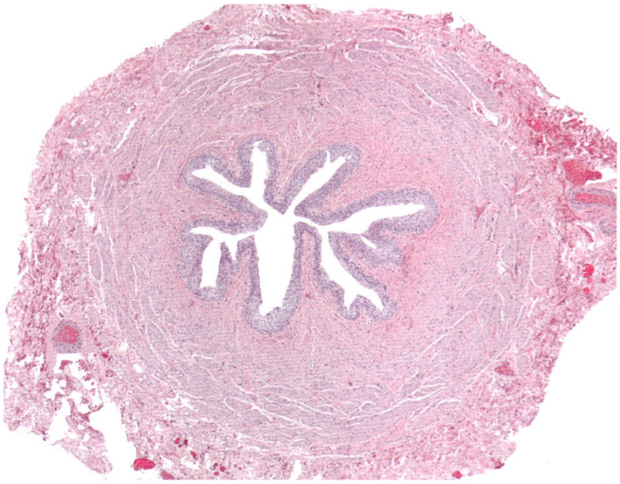

FIGURE 34.27 A cross-section of the ureter showing longitudinal folds and well-developed lamina propria and muscularis propria.

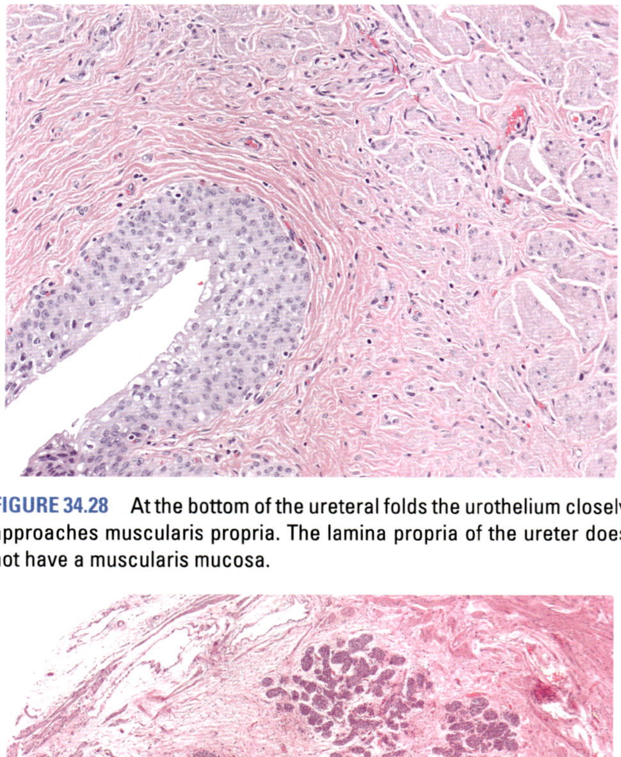

FIGURE 34.28 At the bottom of the ureteral folds the urothelium closely approaches muscularis propria. The lamina propria of the ureter does not have a muscularis mucosa.

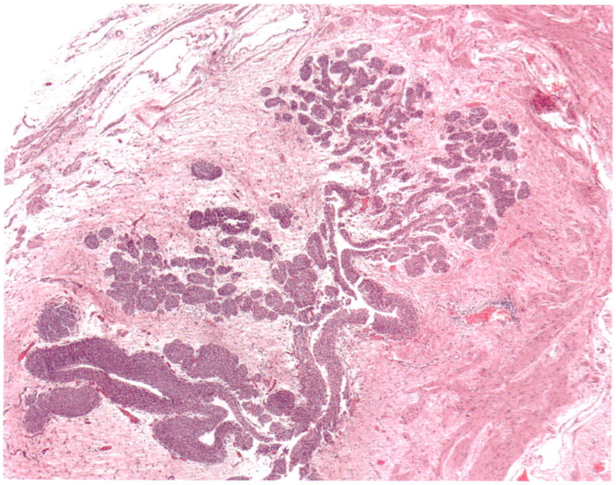

FIGURE 34.29 Florid Von Brunn nest hyperplasia is often present in the ureter. Note how Von Brunn nests abruptly stop at the same level and do not penetrate into the muscularis propria.

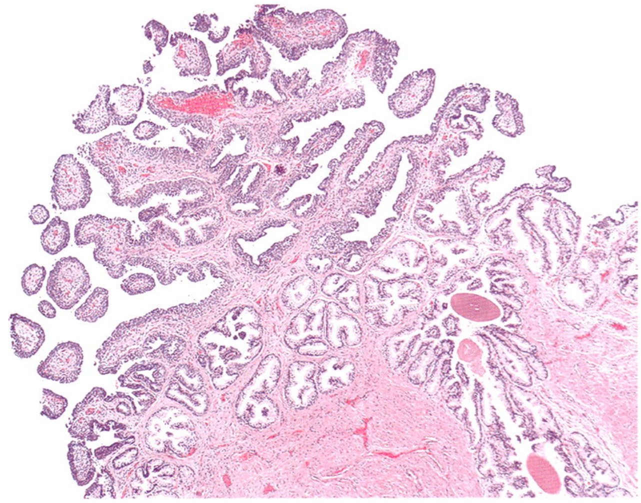

FIGURE 34.30 Urothelial mucosa of the prostatic urethra may not infrequently have a prominent papillary architecture. This should not be confused with a papillary tumor. Benign prostatic tissue is seen underneath.

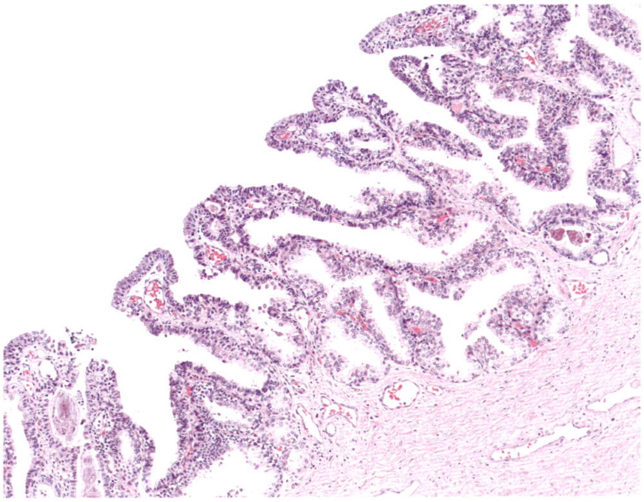

FIGURE 34.32 Prostatic urethra may be lined by prostatic glands with papillary architecture without overlying urothelium.

Prostatic Urethra

The prostatic urethra urothelium usually contains all three cell layers and is histologically comparable to that of the bladder. Nonkeratinizing squamous metaplasia is common. The mucosa of the prostatic urethra features broad, fibrotic polypoid fronds which should not be misclassified as a papillary tumor (Fig. 34.30). The prostatic urethra also contains a thin band of lamina propria devoid of muscularis mucosae. Thin-walled cavernous vessels may be seen in the lamina propria and they represent a continuation of the vasculature of the corpus spongiosum (Fig. 34.31). In some specimens, the surface urothelium is absent and the surface is represented by the prostatic tissue with a villous appearance that may mimic prostatic ductal adenocarcinoma (Fig. 34.32).

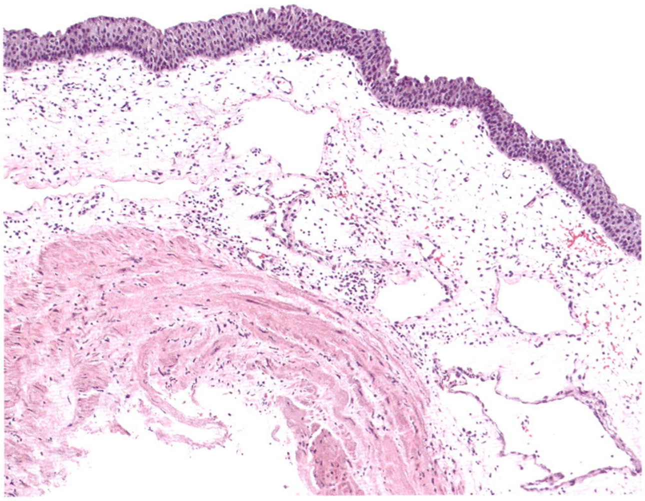

FIGURE 34.31 Prominent thin-walled cavernous vessels may be seen in the lamina propria of prostatic urethra which are the continuation of corpus spongiosum vasculature.

Reactive Urothelial Changes

Reactive epithelial changes are often seen in association with intraepithelial inflammation (Fig. 34.33). Both acute and/or chronic intraepithelial inflammation may lead to reactive epithelial changes. Inflammation in the subepithelial connective tissue usually does not lead to reactive changes in the overlying urothelium. Reactive epithelial changes include nucleomegaly with single nucleoli that may be prominent. Mitotic figures are usually frequent and the chromatin is open without hyperchromasia. A combination of keratin 20, p53, and CD44 immunostains can help separate reactive urothelial atypia from urothelial carcinoma in situ (33).

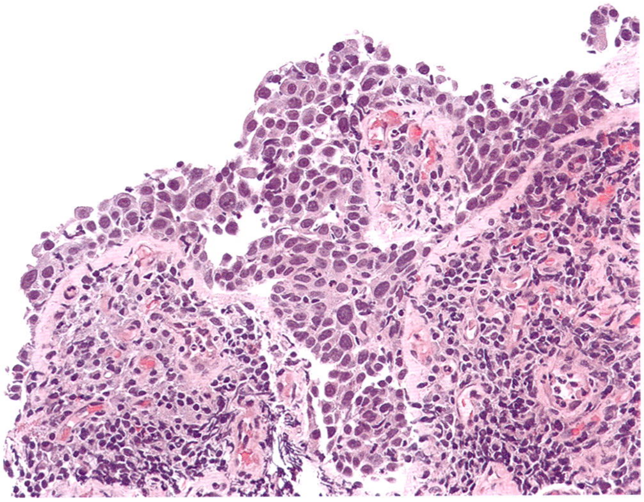

FIGURE 34.33 Reactive epithelial changes in the urothelium. Intraepithelial inflammation (inflammatory cells between urothelial cells) is noted. Although some nuclear pleomorphism is present, the nuclei have open chromatin pattern and a single nucleolus. No nuclear hyperchromasia is seen.

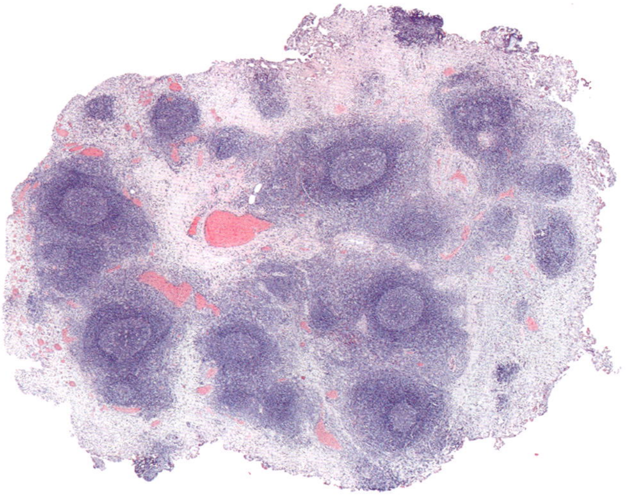

FIGURE 34.34 Follicular cystitis. Large lymphoid aggregates with prominent germinal centers are a typical finding.

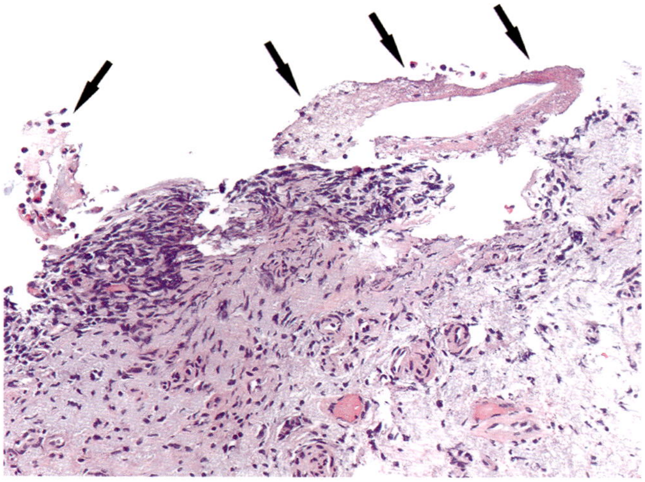

FIGURE 34.35 Hunner ulcer (Hunner lesion). The urothelial surface is denuded and covered by flakes of fibrin (*arrows*). Fibrin is usually not present in cases of the in situ urothelial carcinoma (CIS) with extensive denudation.

Follicular Cystitis

Follicular cystitis is a reactive inflammatory condition characterized by subepithelial lymphoid aggregates with germinal centers and mucosal nodularity (Fig. 34.34). Follicular cystitis is more common in women. Although bacteriuria or lower urinary tract infection is detected in some patients with follicular cystitis, most patients are asymptomatic (34). The differential diagnosis is mostly with low-grade lymphomas.

Hunner Ulcer (Hunner Lesion)

Hunner ulcers are nearly pathognomonic of interstitial cystitis, but are seen only in 5% to 10% of patients with interstitial cystitis (35,36). Interstitial cystitis is mostly seen in women (36). Hunner ulcers appear as areas of denuded urothelium with extensive flakes of fibrin on the surface and chronic inflammation in the subepithelial connective tissue (Fig. 34.35) (37). Extensive fibrin is usually lacking in cases of urothelial carcinoma in situ with extensive denudation.

Urothelial Tumors

The only two benign urothelial tumors are inverted and exophytic papilloma. Inverted papillomas have no risk of recurrence or progression once completely resected and they do not require any specific follow-up. Exophytic papillomas are followed with repeat cystoscopy and urine cytology sampling similar to noninvasive low-grade papillary urothelial carcinomas due to a risk for recurrence or progression. Urothelial carcinomas are dichotomized into low- and high-grade. Low-grade carcinomas feature scant mitotic figures, hyperchromasia of occasional nuclei, and nucleomegaly that does not exceed five times the size of a lymphocyte (Fig. 34.15C). High-grade carcinomas demonstrate nuclear hyperchromasia of most nuclei, significant nuclear pleomorphism, brisk mitotic activity, and nucleomegaly exceeding five times the size of a lymphocyte (Fig. 34.36A). As all noninvasive high-grade urothelial carcinomas are luminal type, most of them diffuse express keratin 20 by immunohistochemistry (Fig. 34.36B). Noninvasive urothelial carcinomas are divided into flat and papillary tumors (Fig. 34.36A,C). While noninvasive papillary urothelial carcinomas can be of low- and high-grade, flat lesions are considered to represent carcinoma in situ and are only considered to be high-grade. Urothelial dysplasia is no longer recognized in the WHO 5th edition of the Classification of Tumours of the Urinary and Male Genital Tumours. Flat lesions that do not meet diagnostic criteria of carcinoma in situ should be reported as urothelial atypia suspicious for but not diagnostic of carcinoma in situ. Finally, the tumors that violate the basement membrane and invade underlying tissue are regarded as invasive urothelial carcinomas, and these are usually high-grade (Fig. 34.36D). Diagnostic features of invasion include desmoplastic reaction surrounding infiltrating tumor nests (Fig. 34.36D), paradoxical differentiation of the invasive tumor nests (Fig. 34.36E), complex infiltrative growth pattern that cannot be explained by colonization of preexistent benign structures (Fig. 34.36E), prominent retraction artifact around the infiltrating tumor nests (Fig. 34.36F), and solid urothelial proliferation within the muscularis propria. Invasive tumors are classified as luminal type, basal type, and neuronal type (i.e., small cell carcinoma). While routine immunohistochemical or molecular assessment is not routinely required, different molecular types have variable immunoprofiles. For example, the luminal type lacks keratin 5/6 immunoreactivity, whereas diffuse keratin 5/6 immunoreactivity is commonly seen in basal type invasive urothelial carcinoma. In adult patients with unresectable or metastatic tumors who received prior systemic chemotherapy, Her2/neu testing should be considered in view

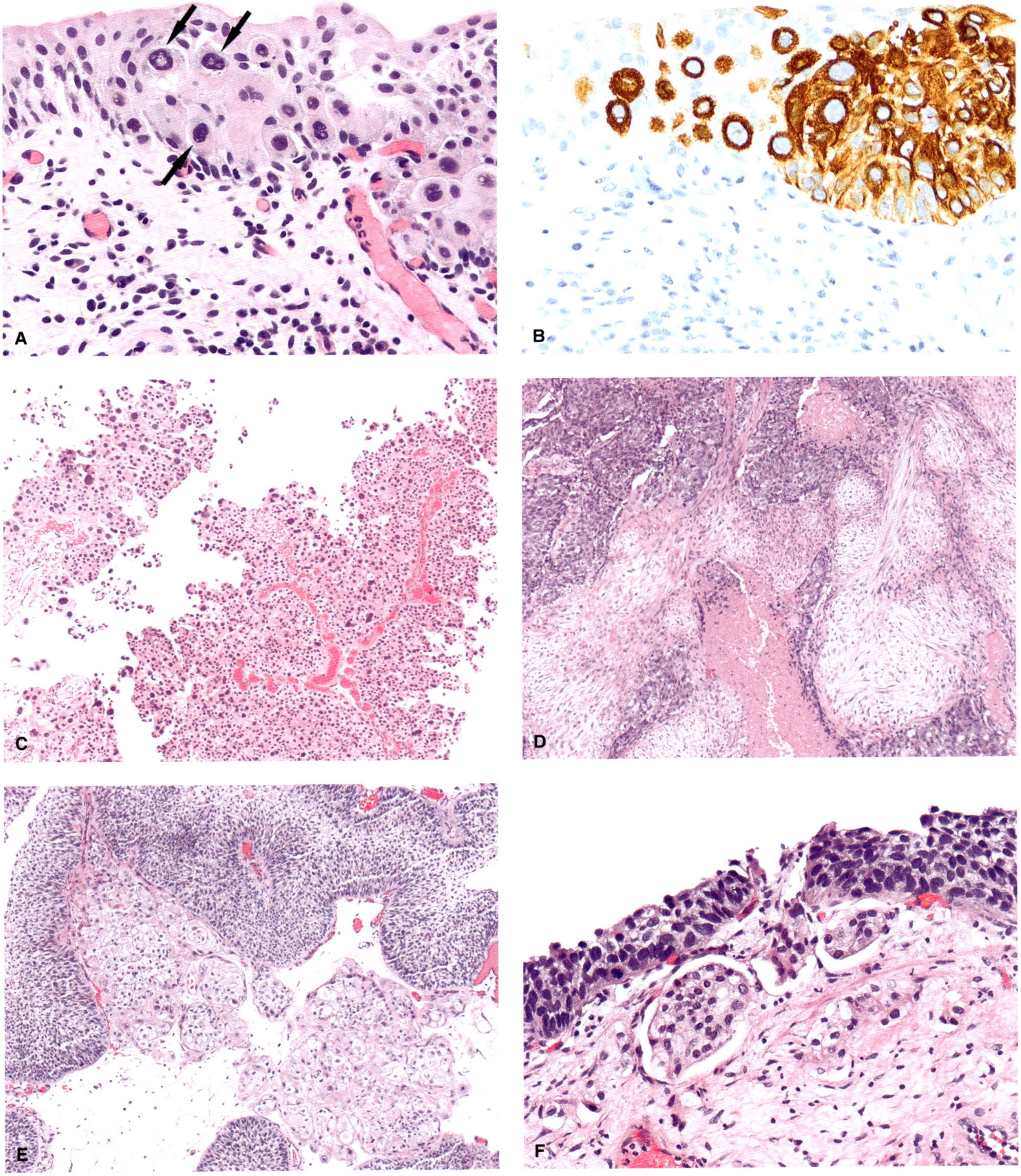

FIGURE 34.36 Urothelial neoplasia. **A:** In situ urothelial carcinoma (CIS) demonstrating cells with large hyperchromatic nuclei (*arrows*). The CIS nuclei are significantly larger than the nuclei of adjacent benign urothelium (left-hand side). The CIS nuclei also exceed the five times lymphocyte size threshold. **B:** Keratin 20 is positive in CIS and negative in benign urothelium. **C:** Noninvasive high-grade papillary urothelial demonstrating thing branching fibrovascular cores covered by thickened urothelium with significant cytologic atypia. **D:** Invasive high-grade urothelial carcinoma. Desmoplasia and complex irregular growth patterns are diagnostic of invasion. **E:** More eosinophilic nests (paradoxical differentiation/maturation) within the lamina propria with complex irregular growth patterns are diagnostic of invasion. **F:** Prominent retraction artifact around invasive tumor nests. Surface CIS is attached to the basement membrane.

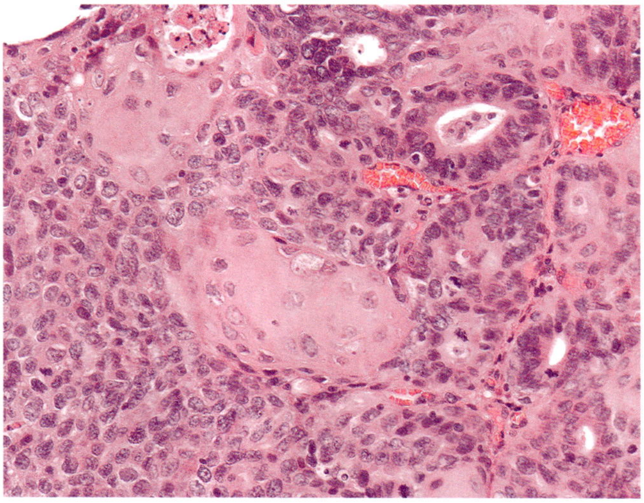

FIGURE 34.37 High-grade urothelial carcinoma demonstrating squamous and glandular divergent differentiation.

of a recent FDA approval of Enhertu® in such patients. Both noninvasive and invasive urothelial carcinomas may demonstrate divergent differentiation including squamous, glandular, trophoblastic, and müllerian differentiation (Fig. 34.37). Pure adenocarcinomas and squamous cell carcinomas may arise in the bladder without demonstration of any urothelial carcinoma component and should be classified as such only when no evidence of urothelial carcinoma component is established in a well-sampled specimen.

REFERENCES

1. Gredler ML, Larkins CE, Leal F, et al. Evolution of external genitalia: insights from reptilian development. *Sex Dev*. 2014; 8(5):311–326.
2. Kruepunga N, Hikspoors J, Mekonen HK, et al. The development of the cloaca in the human embryo. *J Anat*. 2018; 233(6):724–739.
3. Parada Villavicencio C, Adam SZ, Nikolaidis P, Yaghmai V, Miller FH. Imaging of the urachus: anomalies, complications, and mimics. *Radiographics*. 2016;36(7):2049–2063.
4. Young RH, Clement PB. Mullerianosis of the urinary bladder. *Mod Pathol*. 1996;9(7):731–737.
5. Kryvenko ON, Epstein JI. Mimickers of urothelial neoplasia. *Ann Diagn Pathol*. 2019;38:11–19.
6. Tanagho EA, Smith DR, Meyers FH. The trigone: anatomical and physiological considerations. 2. In relation to the bladder neck. *J Urol*. 1968;100:633–639.
7. Elbadawi A. Anatomy and function of the ureteral sheath. *J Urol*. 1972;107:224–229.
8. Moore KL, Dalley AF, Agur AMR. *Clinically Oriented Anatomy*. Lippincott Williams & Wilkins; 2006.
9. Weiss L. *Cell and Tissue Biology: A Textbook of Histology*. Urban & Schwarzenberg; 1988.
10. Jafari NV, Rohn JL. The urothelium: a multi-faceted barrier against a harsh environment. *Mucosal Immunol*. 2022;15(6):1127–1142.
11. Albers DD, Faulkner KK, Cheatham WN, Elledge EF, Coalson RE. Surgical anatomy of the pubovesical (puboprostatic) ligaments. *J Urol*. 1973;109(3):388–392.
12. Rubenwolf PC, Georgopoulos NT, Kirkwood LA, Baker SC, Southgate J. Aquaporin expression contributes to human transurothelial permeability in vitro and is modulated by NaCl. *PLoS One*. 2012;7(9):e45339.
13. Spector DA, Deng J, Stewart KJ. Hydration status affects sodium, potassium, and chloride transport across rat urothelia. *Am J Physiol Renal Physiol*. 2013;305(12):F1669–F1679.
14. Sugaya K, Ogawa Y, Nishizawa O, de Groat WC. Decrease in intravesical saline volume during isovolumetric cystometry in the rat. *Neurourol Urodyn*. 1997;16(2):125–132.
15. Osborn SL, Kurzrock EA. Production of urothelium from pluripotent stem cells for regenerative applications. *Curr Urol Rep*. 2015;16(1):466.
16. Levi AW, Potter SR, Schoenberg MP, Epstein JI. Clinical significance of denuded urothelium in bladder biopsy. *J Urol*. 2001;166(2):457–460.
17. Iakymenko OA, Briski LM, Delma KS, Jorda M, Kryvenko ON. Utility of D2-40, cytokeratin 5/6, and high-molecular-weight cytokeratin (Clone 34betaE12) in distinguishing intraductal spread of urothelial carcinoma from prostatic stromal invasion. *Am J Surg Pathol*. 2022;46(4):454–463.
18. Dalghi MG, Montalbetti N, Carattino MD, Apodaca G. The urothelium: life in a liquid environment. *Physiol Rev*. 2020; 100(4):1621–1705.
19. Guo CC, Fine SW, Epstein JI. Noninvasive squamous lesions in the urinary bladder: a clinicopathologic analysis of 29 cases. *Am J Surg Pathol*. 2006;30(7):883–891.
20. Balish MJ, Jensen J, Uehling DT. Bladder mucin: a scanning electron microscopy study in experimental cystitis. *J Urol*. 1982; 128(5):1060–1063.
21. Dixon JS, Gosling JA. Histology and fine structure of the muscularis mucosae of the human urinary bladder. *J Anat*. 1983; 136(Pt 2):265–271.
22. Ro JY, Ayala AG, el-Naggar A. Muscularis mucosa of urinary bladder. Importance for staging and treatment. *Am J Surg Pathol*. 1987;11(9):668–673.
23. Williamson SR, Al-Ahmadie HA, Cheng L, Downes MR, Lopez-Beltran A, McKenney JK, et al. Invasive urothelial carcinoma. In: WHO Classification of Tumours Editorial Board. *WHO Classification of Tumours of the Urinary and Male Genital Tumours*. IARC; 2022:151–191.
24. Kryvenko ON, Williamson SR, Netto GJ. *Survival Guide to Kidney and Bladder*. Innovative Science Press; 2025.
25. Lane Z, Epstein JI. Polypoid/papillary cystitis: a series of 41 cases misdiagnosed as papillary urothelial neoplasia. *Am J Surg Pathol*. 2008;32(5):758–764.
26. Wiener DP, Koss LG, Sablay B, Freed SZ. The prevalence and significance of Brunn's nests, cystitis cystica and squamous metaplasia in normal bladders. *J Urol*. 1979;122(3):317–321.
27. Barrett DM, Malek RS, Kelalis PP. Observations on vesical diverticulum in childhood. *J Urol*. 1976;116(2):234–236.
28. Mazal PR, Schaufler R, Altenhuber-Muller R, et al. Derivation of nephrogenic adenomas from renal tubular cells in kidney-transplant recipients. *N Engl J Med*. 2002;347(9):653–659.
29. Li L, Williamson SR, Castillo RP, et al. Fibromyxoid nephrogenic adenoma: a series of 43 cases reassessing predisposing conditions, clinical presentation, and morphology. *Am J Surg Pathol*. 2023;47(1):37–46.

30. Hansel DE, Nadasdy T, Epstein JI. Fibromyxoid nephrogenic adenoma: a newly recognized variant mimicking mucinous adenocarcinoma. *Am J Surg Pathol*. 2007;31(8):1231–1237.
31. Kryvenko ON, Wasco MJ, Williamson SR. Nephrogenic adenoma intermixed with urothelial carcinoma. *Arch Pathol Lab Med*. 2022;147(5):552–558.
32. Notley RG. he musculature of the human ureter. *Br J Urol*. 1970;42(6):724–727.
33. McKenney JK, Desai S, Cohen C, Amin MB. Discriminatory immunohistochemical staining of urothelial carcinoma in situ and non-neoplastic urothelium: an analysis of cytokeratin 20, p53, and CD44 antigens. *Am J Surg Pathol*. 2001;25(8):1074–1078.
34. Hansson S, Hanson E, Hjälmås K, et al. Follicular cystitis in girls with untreated asymptomatic or covert bacteriuria. *J Urol*. 1990;143(2):330–332.
35. Hunner GL. A rare type of bladder ulcer in women: report of cases. *Trans South Surg Gynecol Assoc*. 1914;27:247–317.
36. Hand JR. Interstitial cystitis; report of 223 cases (204 women and 19 men). *J Urol*. 1949;61(2):291–310.
37. Lynes WL, Flynn SD, Shortliffe LD, Stamey TA. The histology of interstitial cystitis. *Am J Surg Pathol*. 1990;14(10):969–976.

35

Prostate

Samson W. Fine ■ Jesse K. McKenney

EMBRYOLOGY AND DEVELOPMENT OF THE PROSTATE

The prostate appears in early embryonic development as a condensation of mesenchyme along the course of the pelvic urethra. By 9 weeks' gestation, the mesenchymal condensation is most dense along the posterior (rectal) and distal (apical) aspects of the urethra (Fig. 35.1), where it is in contact with the urethral lining epithelium (1). Between its midpoint and the bladder neck, the proximal urethral segment shows a sharp anterior angulation. However, the highly condensed mesenchyme does not follow this angulation, but rather continues directly proximal to a dome-shaped prostatic base, leaving a gap between condensed prostatic mesenchyme and proximal urethra. The ejaculatory ducts penetrate this mesenchyme toward the future verumontanum, which is located at the urethral midpoint. The ejaculatory ducts are wolffian duct structures, but their surrounding stroma is indistinguishable from that of the remaining prostatic mesenchyme, which is mainly derived from the urogenital sinus (2). The portion of mesenchyme that surrounds the ejaculatory ducts and expands proximally to occupy nearly the entire prostate base is distinguishable in the adult as the central zone, which is derived from the wolffian duct similar to the ejaculatory ducts and seminal vesicles (1). In this concept, the prostate is of dual embryonic derivation.

Epithelial buds begin to branch from the walls of the distal (apex to mid) urethral segment at approximately 10 weeks' gestational age. This process is most pronounced posteriorly and laterally and features protrusions of epithelium into the condensed mesenchyme in a pattern that is essentially identical to that seen in the adult.

The prostate grows at a slow rate after birth, reaching less than 2 cm in diameter by puberty. During this period, the ducts and acini are lined by epithelium similar to that seen in the neonatal period. Gland spaces are lined by cells that are crowded with multilayered dark nuclei (Fig. 35.2).

The pubertal growth acceleration and maturation of the prostate gland are not complete until at least 20 years of age. By this time, the average prostate measures about 4.5 cm in width, 3.5 to 4.0 cm in length, and 3 cm in thickness. Focal or multifocal resumption of growth can occur in older men, resulting in benign prostatic hyperplasia. This process increases the thickness of the gland, and results in disproportionate enlargement of a single region of the gland in the transition zone. The normal mass of the glandular portion of the prostate is otherwise nearly constant until 70 years of age.

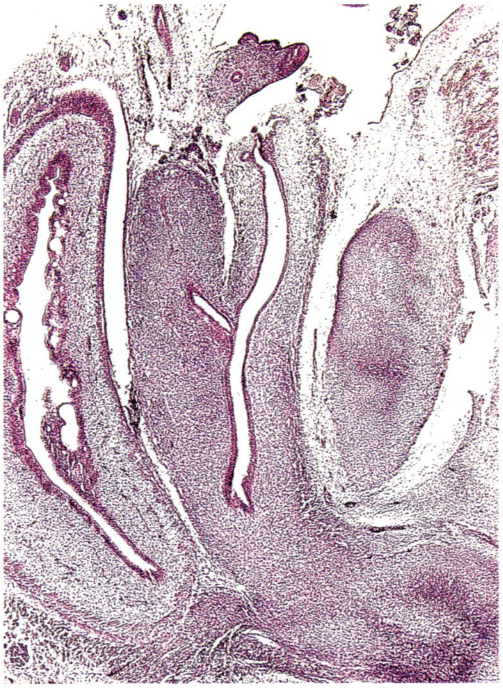

FIGURE 35.1 Embryonic prostate, age 9 weeks, in the sagittal plane of the pelvis. Urethra (narrow central lumen) is angulated to the right at the midpoint, where the ejaculatory duct approaches from above left. A vertical strip of highly condensed prostate mesenchyme contacts the posterior urethral wall only distal to the ejaculatory ducts. The prostate is flanked by the rectum (**left**) and pubis (**right**). Duct buds have not yet formed.

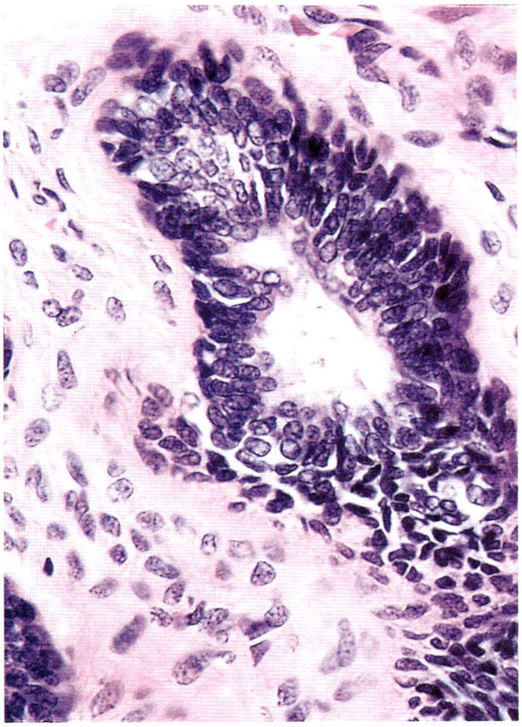

FIGURE 35.2 Prepubertal prostatic duct lined by epithelium with multiple layers of nuclei and showing no cytoplasmic differentiation.

GENERAL TOPOGRAPHIC RELATIONSHIPS: McNEAL'S ZONAL ANATOMY

The human prostate gland is a composite organ, comprised of several glandular and nonglandular components. These different "zones" are tightly fused together within a common sheath of fibromuscular tissue—the "capsule"—such that gross dissection is not possible. In a series of elegant dissections in postmortem specimens, Dr. McNeal found that anatomic relationships are best demonstrated by examination of cut sections in the sagittal, coronal, and oblique coronal planes (3,4). Two main conclusions can be drawn from these studies. First, there are three distinct glandular regions: the peripheral, central, and transition zones. Second, the main nonglandular tissue of the prostate is concentrated anteromedially and is responsible for much of the anterior convexity of the organ.

The urethra is a primary reference point for describing anatomic relationships within the prostate. Visualized in a sagittal plane of section (Fig. 35.3), the prostatic urethra is divided into proximal and distal segments of approximately equal length by an anterior angulation at the midpoint between the prostate apex distally and the bladder neck proximally (1,5). The verumontanum protrudes from the

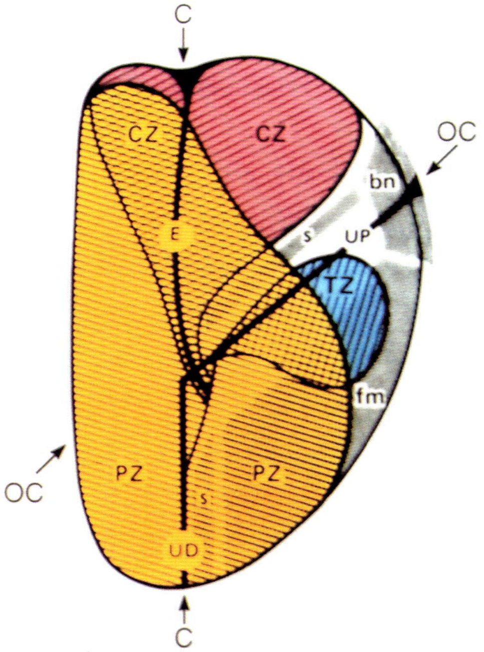

FIGURE 35.3 Sagittal diagram of distal prostatic urethral segment (*UD*), proximal urethral segment (*UP*), and ejaculatory ducts (*E*) showing their relationships to a sagittal section of the anteromedial nonglandular tissues (bladder neck [*bn*]; anterior fibromuscular stroma [*fm*]; preprostatic sphincter [*s*]; distal semicircular ["striated"] sphincter [*s*]). These structures are shown in relation to a three-dimensional representation of the glandular prostate (central zone [*CZ*]; peripheral zone [*PZ*]; transition zone [*TZ*]). Coronal plane (*C*) of Figure 35.4 and oblique coronal plane (*OC*) of Figure 35.5 are indicated by *arrows*.

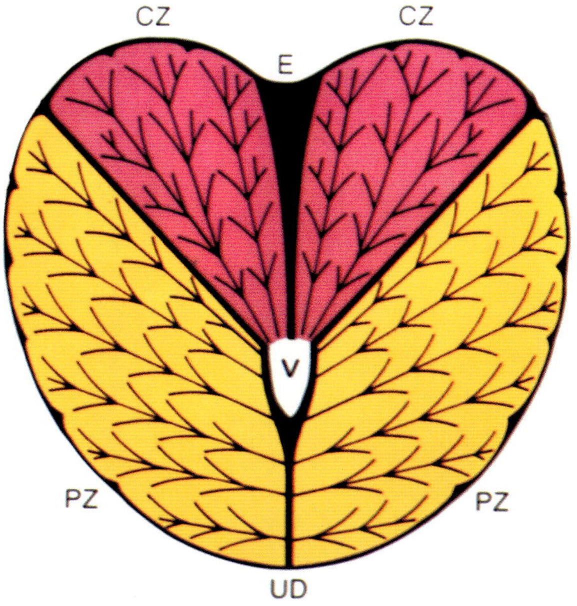

FIGURE 35.4 Coronal section diagram of prostate showing location of central zone (*CZ*) and peripheral zone (*PZ*) in relation to the distal urethral segment (*UD*), verumontanum (*V*), and ejaculatory ducts (*E*). The branching pattern of prostatic ducts is indicated; subsidiary ducts provide uniform density of acini along the entire main duct course.

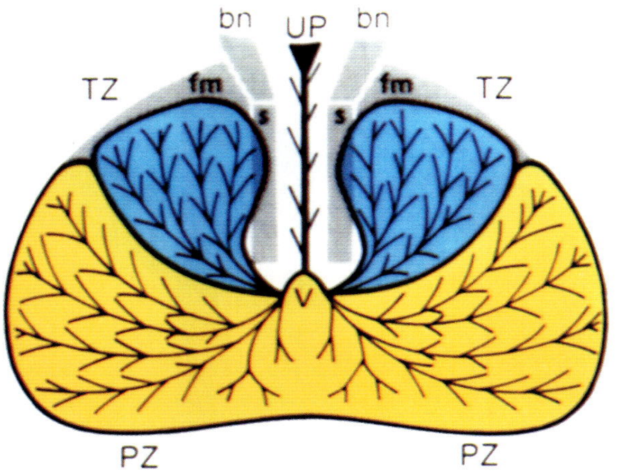

FIGURE 35.5 Oblique coronal section diagram of prostate showing location of peripheral zone (*PZ*) and transition zone (*TZ*) in relation to proximal urethral segment (*UP*), verumontanum (*V*), preprostatic sphincter (*s*), bladder neck (*bn*), anterior fibromuscular stroma (*fm*), and periurethral region with periurethral glands. Branching pattern of prostatic ducts is indicated: the medial transition zone ducts penetrate the sphincter.

posterior urethral wall at the point of angulation and is the point at which the ejaculatory ducts join the prostatic urethra. The ejaculatory ducts then extend proximally from the verumontanum (mid) to the base of the prostate, following a course that is nearly a direct extension of the long axis of the distal (apex to mid) urethral segment, although usually offset posteriorly by a few millimeters.

A coronal plane of section (Fig. 35.4) along the course of the ejaculatory ducts and distal (apex to mid) urethral segment provides the best demonstration of the anatomic relationships between the two major regions of glandular prostate, the peripheral and central zones (6). The peripheral zone comprises about 65% of the mass of the normal glandular prostate. Its ducts exit posterolateral from the urethral wall along a double row extending from the verumontanum to the prostatic apex. The ducts extend mainly laterally in the coronal plane, with branches that curve anteriorly and posteriorly.

The central zone comprises about 30% of the glandular prostate mass. Its ducts arise in a small focus on the verumontanum and immediately surround the ejaculatory duct orifices. The ducts branch directly toward the base of the prostate along the course of the ejaculatory ducts, fanning out to form an inverted conical structure that is flattened in the anterior–posterior plane. The base of this cone comprises almost the entire prostatic base. The most lateral central zone ducts run parallel to the most proximal peripheral zone ducts, separated only by a narrow band of stroma that is imperceptible in clinical specimens.

An oblique coronal section (Fig. 35.5) along the proximal segment of the prostatic urethra from verumontanum to the bladder neck best defines its glandular relationships. Normally, the proximal urethral segment is most closely related to 5% of prostatic glandular tissue, almost entirely comprised of transition zone (7). This zone is formed by two small lobes with ducts that leave the posterolateral recesses of the urethral wall at a single point. The main ducts of the transition zone extend laterally and curve sharply anteriorly, arborizing toward the bladder neck.

The main nonglandular tissue of the prostate is the anterior fibromuscular stroma which overlies the urethra in the anteromedial prostate. Its bulk and consistency vary considerably from apex to base, as described further in this chapter.

SECTIONING OF RADICAL PROSTATECTOMY SPECIMENS

Radical prostatectomy, including removal of the seminal vesicles, is the definitive surgical procedure for patients with prostate cancer. Current convention is to ink the intact prostate gland and seminal vesicles in two colors while fresh, to assess laterality. The most apical region of the gland is amputated in a transverse plane to include the opening of the distal urethra, and then sectioned parasagittally, perpendicular to the inked surface (Fig. 35.6) for paraffin embedding. This allows visualization of apical prostatic tissue in planes that are perpendicular to the true apical surface and more accurate evaluation of cancer penetration and margin status. The bladder neck margin may be handled in a similar fashion to the prostatic apex, although some institutions shave a thin section of the most superficial muscle fragment(s) surrounding the proximal prostatic urethra and submit this tissue *en face* (8).

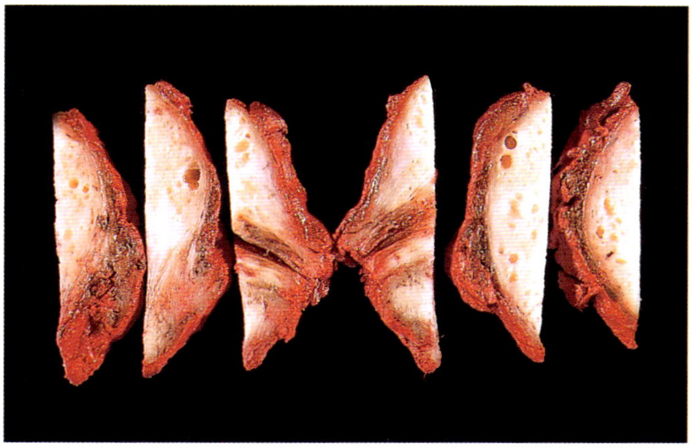

FIGURE 35.6 Apex of prostate seen grossly after 5-mm thick apical block has been subsectioned parasagittally at 3-mm intervals. Orientation of sections and localization of lesions are easily demonstrated, and cuts through the tissue are nearly perpendicular to the apical surface.

While all sample the junction of the seminal vesicles and the prostate, to what extent the remainder of the seminal vesicle is submitted is not standardized (9). A reasonable approach involves taking two sections of each seminal vesicle, one at the prostatic junction, as well as a mid–seminal vesicle section to account for both contiguous spread of tumor from the prostatic base, as well as seminal vesicle invasion that occurs in the context of extraprostatic tumor extension.

The remaining bulk of the gland is sectioned from apex to base in the anterior–posterior plane at approximately 3-mm intervals (10,11). The resulting transverse macrosections are submitted as either full whole mount sections or divided into half or quadrant sections. Although there is debate regarding extent of sampling (8), a minimum of every other intervening macrosection is recommended.

ANATOMY OF THE PROSTATE GLAND IN SURGICAL PATHOLOGY SPECIMENS

Gross Anatomy

In situ operative views of the prostate reveal a cone-shaped organ with its base surrounding the proximal urethral segment and abutting the bladder neck, while its apex narrows around the distal urethral segment as it approaches the urogenital diaphragm. Surgical manipulation and detachment of the prostate from native connective tissue leads to superior retraction of the distal urethra and yields a spherical specimen for surgical pathology evaluation. Significant tissue shrinkage due to formalin fixation and tissue processing (12,13) is common, resulting in artifactual shortening of the distance from prostatic apex to verumontanum and eversion of the posterior periurethral tissue into the urethral space. These effects create an artificial "promontory" in the apical portions of the gland (Fig. 35.7) (14).

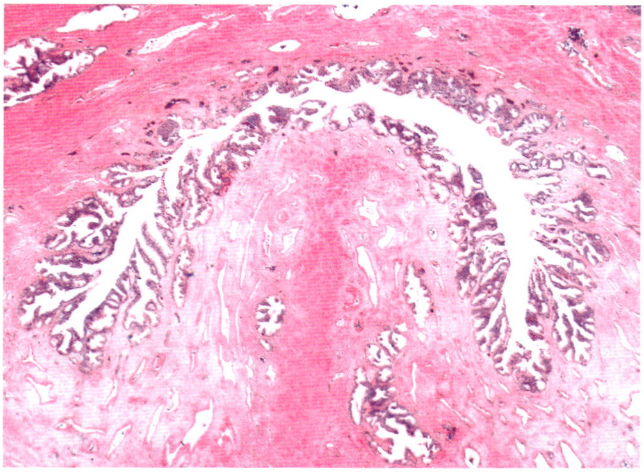

FIGURE 35.7 Distal urethra near the apex. Note the "promontory" or eversion of posterior periurethral tissue into the urethral space. The posterior portion of the semicircular sphincter (see Fig. 35.8) is present as a central muscular column.

Histo-Anatomic Variation by Region

Apical One-Third of the Prostate (Apex)

In a surgical pathology specimen sectioned in the anterior–posterior plane, the apical (distal) urethra is located near the center of the section (Fig. 35.8). The urethra is immediately surrounded by a thin layer of stroma and a variable number of periurethral glands. These periurethral glands intermingle anteriorly and medially with a semicircular band of medium-sized, compactly arranged, and vertically oriented

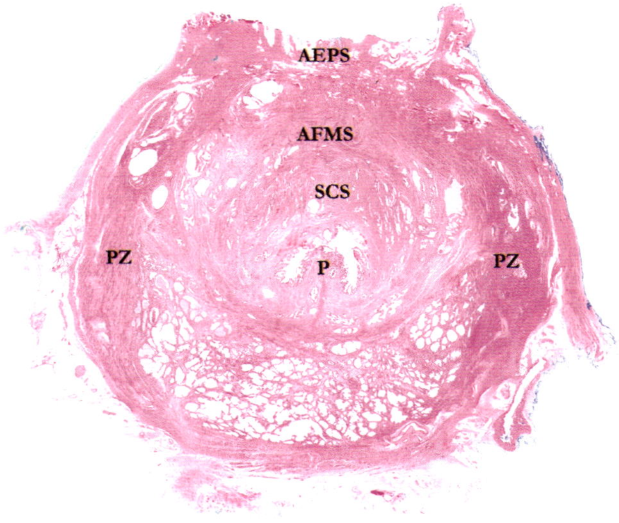

FIGURE 35.8 Whole-mount section from apex of prostate. The urethra and promontory (*P*) are central and proceeding anteriorly, the semicircular sphincter (*SCS*) and anterior fibromuscular stroma (*AFMS*) are visualized. The posterior, lateral, and anterolateral portions of the apex are composed of peripheral zone (*PZ*) tissue. Most anteriorly, the anterior extraprostatic space (*AEPS*) contains vascular and adipose remnants of the dorsal vascular complex.

muscle fibers. This band is incomplete posteriorly, appearing as a densely eosinophilic, aglandular muscular column which extends posteriorly from the urethra (Fig. 35.7). The compact morphology led to designation as the "striated sphincter" (5,10,15), yet histopathologic evaluation shows that it is composed of smooth muscles. This impression is confirmed by the lack of immunohistochemical staining for sarcomeric actin, a marker of mature skeletal muscle (14).

Anterior to the semicircular muscle, fibers of the anterior fibromuscular stroma traverse horizontally and laterally as they extend to the anterior- and apical-most aspects of the prostate. While heightened intraprostatic pressure may cause bulging of hypertrophic transition zone acini in benign prostatic hyperplasia, no normal transition zone tissue is in the prostatic apex. The bilateral peripheral zone, which composes essentially all the glandular tissue at the apex, occupies the posterior, lateral, and anterolateral prostate, abutting the anterior fibromuscular stroma medially and forming a nearly complete ring in histologic sections.

Middle One-Third of the Prostate (Midgland)

McNeal's studies revealed a 35-degree angulation of the prostatic urethra at midgland, dividing the urethra into proximal and distal segments toward the base and apex, respectively (Fig. 35.9A,B). The key anatomic landmark in the midgland is the verumontanum, an exaggerated area of glandular–stromal tissue, located subjacent to the posterior urethral wall, into which the ejaculatory ducts insert and from which the glandular zones arise (16). Histologically, the verumontanum consists of a crowded collection of prostatic glands lined by secretory epithelium with abundant corpora amylacea, directly underlying the urothelium of the prostatic urethra. When prominent, the term verumontanum mucosal gland hyperplasia is applied (Fig. 35.10) (17). The corpora amylacea in this region characteristically display a deep orange-red or rust-like color, unlike the pink hue of peripheral/transition zone corpora, which may make this region distinguishable on needle biopsy specimens. At midgland, the transition zone becomes evident as bilateral lobed tissue in the anteromedial prostate. The ducts of the transition zone appear to arise from the posterior boundary of the periurethral space and follow a course anterolaterally to serve as a boundary between transition and peripheral zones. In the normal midgland, the peripheral zone still comprises the posterior, lateral, and most of the anterolateral prostatic glandular tissue (Fig. 35.9A). In prostates with benign prostatic hyperplasia, this tissue may be significantly compressed toward the lateral most portions of the gland, forming "anterolateral horns" (Fig. 35.9B). The anterior fibromuscular stroma in this region may be less evident due to the increased density of glandular tissue in the midgland coupled with effects of organ contraction (14).

Basal One-Third of the Prostate

Progressing proximally from midgland to base, the urethra becomes invested by a thick layer of short smooth muscle fibers, constituting the "preprostatic sphincter" (Figs. 35.11

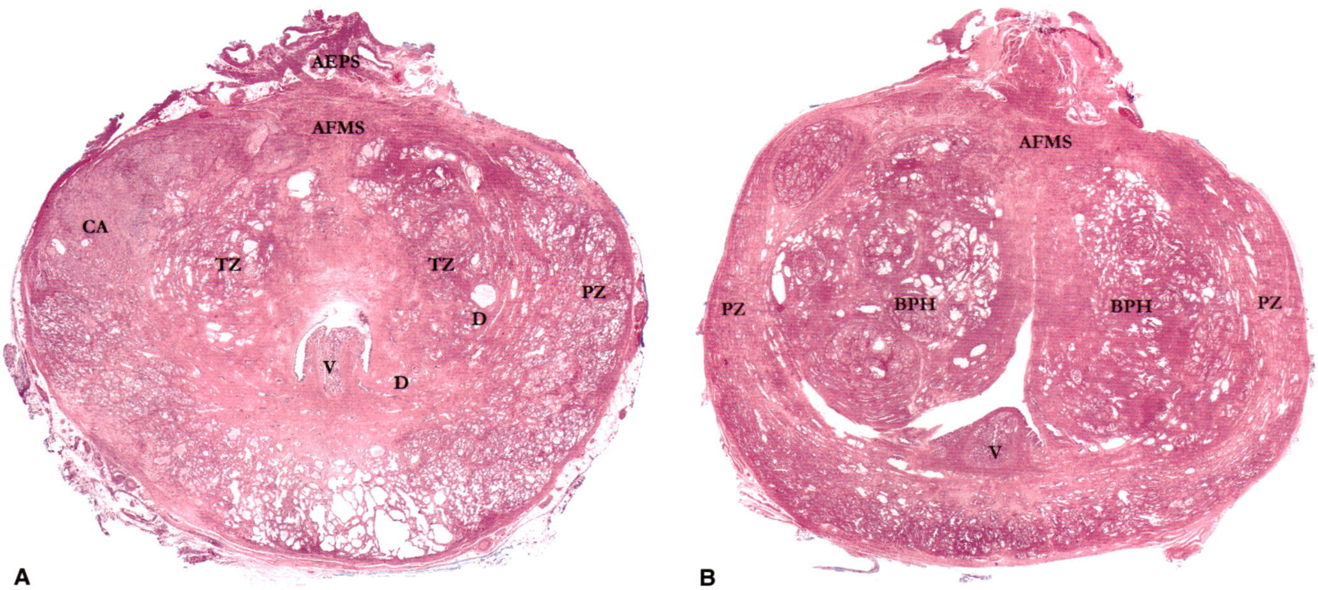

A B

FIGURE 35.9 **A:** Whole mount section from mid prostate at the level of the verumontanum (*V*). Note the bilobed transition zone (*TZ*) arising from elongated ducts (*D*) which course anterolaterally. The peripheral zone (*PZ*) still occupies the posterior, lateral, and anterolateral portions of the gland, with a cancer nodule (*CA*) evident in the right anterior peripheral zone. In the mid prostate, the anterior fibromuscular stroma (*AFMS*) is much condensed and the anterior extraprostatic space (*AEPS*) largely retains its apical consistency. **B:** Whole mount section from mid prostate at the level of the verumontanum (*V*) in a gland with extensive benign prostatic hypertrophy (*BPH*). Anterolateral "horns" of the peripheral zone (*PZ*) are compressed laterally by the expanded transition zone tissue and the *AFMS* is diminished in extent.

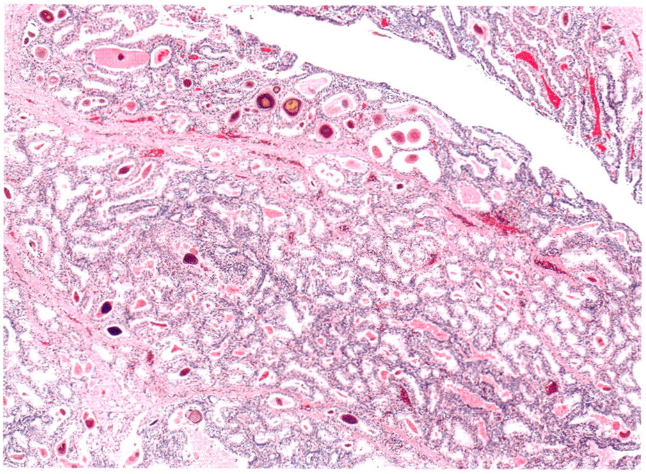

FIGURE 35.10 The verumontanum often contains densely packed prostatic glands lined by benign secretory cells, often with abundant intraluminal corpora amylacea.

and 35.12), which may be in contact with the glands and acini of the transition zone at its lateral edge (5,7). The preprostatic sphincter is thought to function during ejaculation to prevent retrograde flow of seminal fluid from the distal urethral segment and may have resting tone that maintains closure of the proximal urethral segment (18). At the base, the glands of the transition zone gradually recede, and the few remaining peripheral zone acini once again comprise the anterior glandular tissue. In contrast with its apical appearance, however, the peripheral zone rarely extends anteromedially due to the abundant stroma in this region. This stroma is an expansive

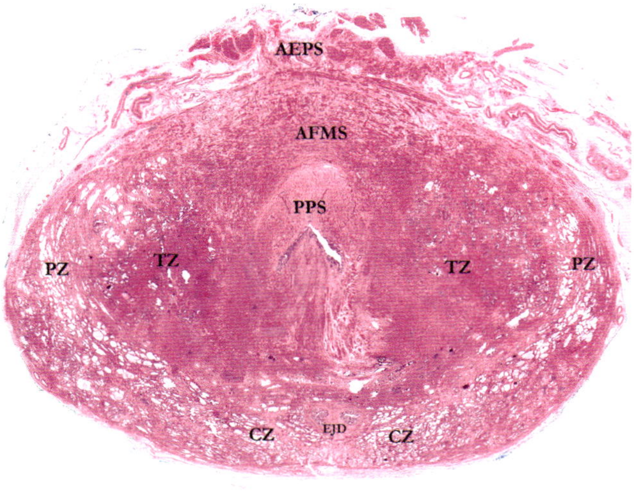

FIGURE 35.11 Whole mount section from base of prostate. The preprostatic sphincter (*PPS*) is evident as a pale area surrounding the proximal urethra. The transition zone (*TZ*) shows abortive small acini and is covered anteriorly by a vast anterior fibromuscular stroma (*AFMS*) which merges with smooth muscle bundles in the anterior extraprostatic space (*AEPS*). Posteriorly, the expansive central zone (*CZ*) surrounds the ejaculatory duct complex (*EJD*), while some peripheral zone (*PZ*) is still apparent posterolaterally.

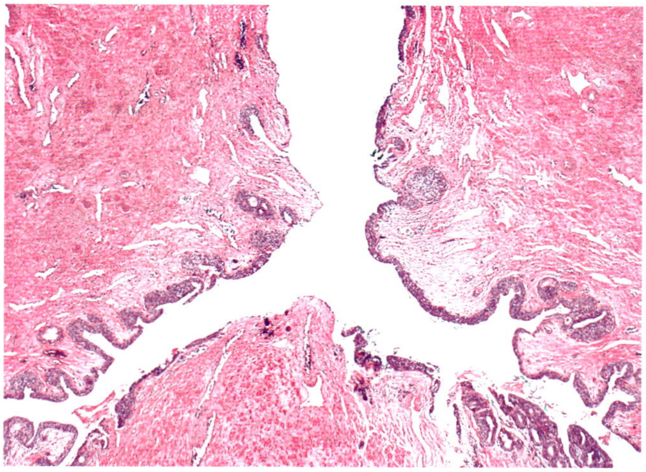

FIGURE 35.12 Preprostatic sphincter investing the proximal urethra in section from the base of prostate. The sphincter is composed of compact, short smooth muscle fibers distinct from prostatic smooth muscle and conveying a paler appearance to the periurethral zone as seen in Figure 35.11.

strip of tissue consisting of preprostatic sphincter and anterior fibromuscular stroma. The latter often merges with large smooth muscle bundles located in the anterior extraprostatic space. With increasing angulation, the prostatic urethra is identified further anteriorly in histologic sections, eventually breaching the anterior-most border of tissue sections at the level of the bladder neck. Posterior and posterolateral, the central zone becomes evident surrounding the ejaculatory ducts. The ejaculatory ducts are enveloped by a sheath of loose fibrous tissue with abundant lymphovascular spaces (Fig. 35.13). In the most basal portions of the gland, the well-formed muscular coat at the base of the seminal vesicles emerges and separates from the bulk of the prostatic tissue creating a fibroadipose tissue septum. The last vestiges of the central zone are present at the most lateral aspects of these emerging seminal vesicles (14).

Nonglandular Components of Prostatic and Extraprostatic Tissues

Prostatic Capsule and Anterior Fibromuscular Stroma

The prostatic "capsule" is a condensed fibromuscular tissue (Fig. 35.14) comprised of an inner layer of smooth muscle fibers and an outer collagenous membrane. However, the relative and absolute amounts of fibrous and muscle tissues and their arrangement vary considerably from region to region (19,20). At its inner border, transverse smooth muscle of the capsule blends with periglandular prostatic smooth muscle, and a clear separation cannot be identified (19). The distance from terminal acini of the peripheral and central zones to the prostate surface is also variable (7,21) and the proportion and arrangement of collagenous tissue is inconsistent. Consequently, the prostate capsule cannot be regarded as a well-defined anatomic structure.

While the capsule envelopes most of the external surface of the prostate, there is a defect of variable caliber at the

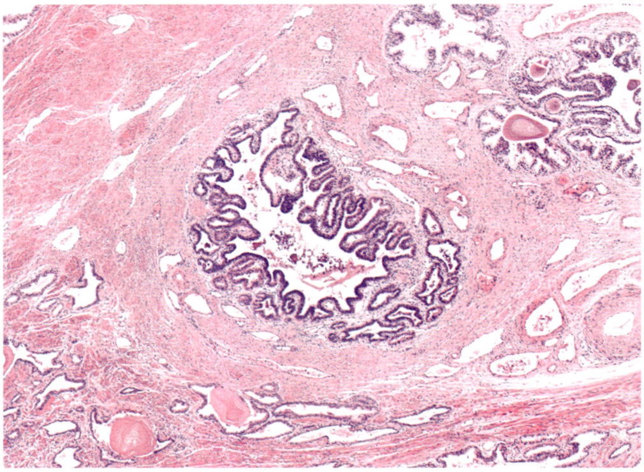

FIGURE 35.13 Ejaculatory duct encircled by a fibroconnective tissue sheath containing numerous lymphovascular spaces.

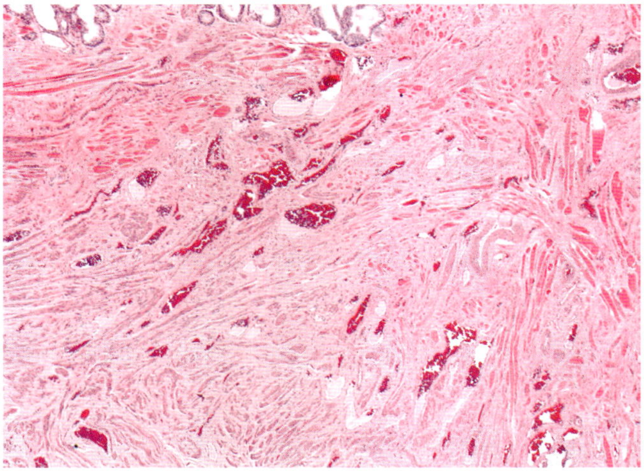

FIGURE 35.15 Anterior fibromuscular stroma showing admixture of smooth muscle bundles with skeletal muscle fibers from the urogenital diaphragm at the apex of the prostate.

prostatic apex anterior and anterolateral. Hence, if carcinoma is present in the apical third of the prostate anteriorly, it may be quite difficult to determine whether it has invaded beyond the boundary of the gland. Similar difficulty is encountered at the most proximal portion of the urethra in the bladder neck section, in which no clear capsule is evident.

The anterior fibromuscular stroma is an apron of tissue that extends downward from the bladder neck over the anteromedial surface of the prostate, narrowing to join the urethra at the prostate apex (Fig. 35.3) (7). Its deep surface is in contact with the preprostatic sphincter and the transition zone proximally toward the base, as well as the semicircular sphincter distally toward the apex. It is composed of large bundles of smooth muscle cells that are more randomly oriented than those of the bladder neck and may blend with the bladder neck muscle at its proximal edge.

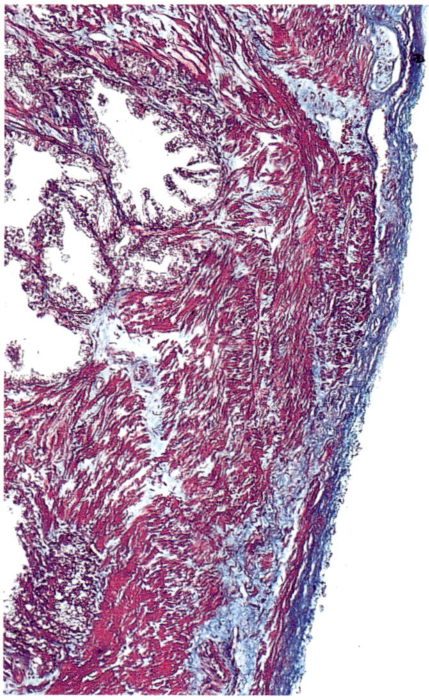

FIGURE 35.14 The prostate capsule consists of a layer of mainly transverse smooth muscle bundles (*red*), which is of variable thickness and blends with periacinar smooth muscle bundles at the capsule's poorly defined inner aspect (**left**). Collagen fibers (*blue*) are always present and usually concentrated in a thin compact membrane at the external capsular border (**right**) (trichrome stain).

Unlike the posterolateral prostate, the anterior-most region of the gland does not exhibit a distinct capsule (19). Rather, as one proceeds from apex to base, the anterior fibromuscular stroma is variably intertwined with skeletal muscle fibers emanating from the urogenital diaphragm at the apex (Fig. 35.15) or levator ani muscles in the midgland and may fuse with detrusor smooth muscle in the midgland region or base. Moreover, the anterior fibromuscular stroma also contains blood vessels that supply/drain the anterior prostate. Due to the complex tissue composition of the anterior and anterolateral prostate and lack of a definitive border, the task of separating prostatic from extraprostatic tissue is challenging.

Extraprostatic Tissues, Prostatic Innervation, and Vascular Supply

The tissue immediately anterior to the prostate is the dorsal venous or vascular complex, a group of veins and arteries set in fibroadipose tissue that runs over the anterior prostate and continues distally to supply/drain the penis (22). At radical prostatectomy, the dorsal vascular complex is ligated and divided, with a portion of blood vessels and fibroadipose tissue remaining adherent to the prostate specimen. These may be identified as the anterior extraprostatic tissue from apex through midgland. The most proximal two to three sections typically reveal medium- to large-sized smooth muscle bundles admixed with adipose tissue (Fig. 35.16). These fibers are morphologically identical to the detrusor muscle and possibly represent the inferior border of the bladder neck (23).

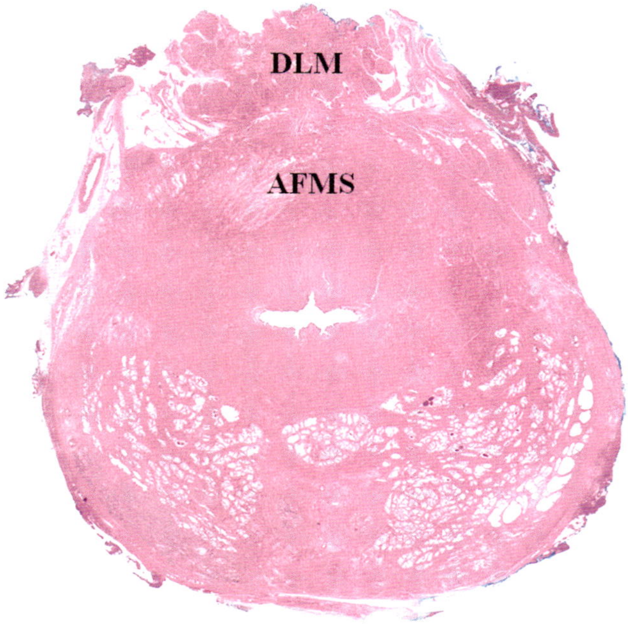

FIGURE 35.16 Prostate at mid to base of the gland—anterior extraprostatic space displays numerous medium- to large-sized discrete (detrusor-like) muscle bundles (*DLM*) admixed with adipose tissue, which merge with the anterior fibromuscular stroma (*AFMS*).

Over the medial half of the posterior surface of the prostate, the thickness of the capsule is increased by its fusion to Denonvilliers fascia (Figs. 35.17 and 35.18). This fascia is a thin, compact collagenous membrane with a smooth posterior surface that rests directly against rectal wall muscle (24). The capsule is typically fused to Denonvilliers fascia with an interposed adipose layer. In the adult, there are only scattered microscopic islands of fat along with a variable number of smooth muscle fibers.

Denonvilliers fascia extends superiorly above the prostate to cover the posterior surface of the seminal vesicles in a loosely adherent fashion (Fig. 35.17). Laterally, the fascia leaves the posterior capsule where the prostate surface begins to deviate anteriorly, and it continues in a coronal plane to anchor against the pelvic sidewalls.

Where the seminal vesicles emerge from the prostate base, there is no discrete separation between the two organs. The degree of fusion between the two muscular walls is variable, but there is frequently no boundary between the two organs medially with a minimal amount of common muscular wall separating the most basal central zone gland lumen from the seminal vesicle lumen (Figs. 35.18 and 35.19).

Where Denonvilliers fascia separates from the prostate capsule posterior and posterolateral, the space between them is filled with adipose tissue in a thick layer between the anterior aspect of the fascia and the posterolateral capsular surface of the prostate. The autonomic nerves, from the pelvic

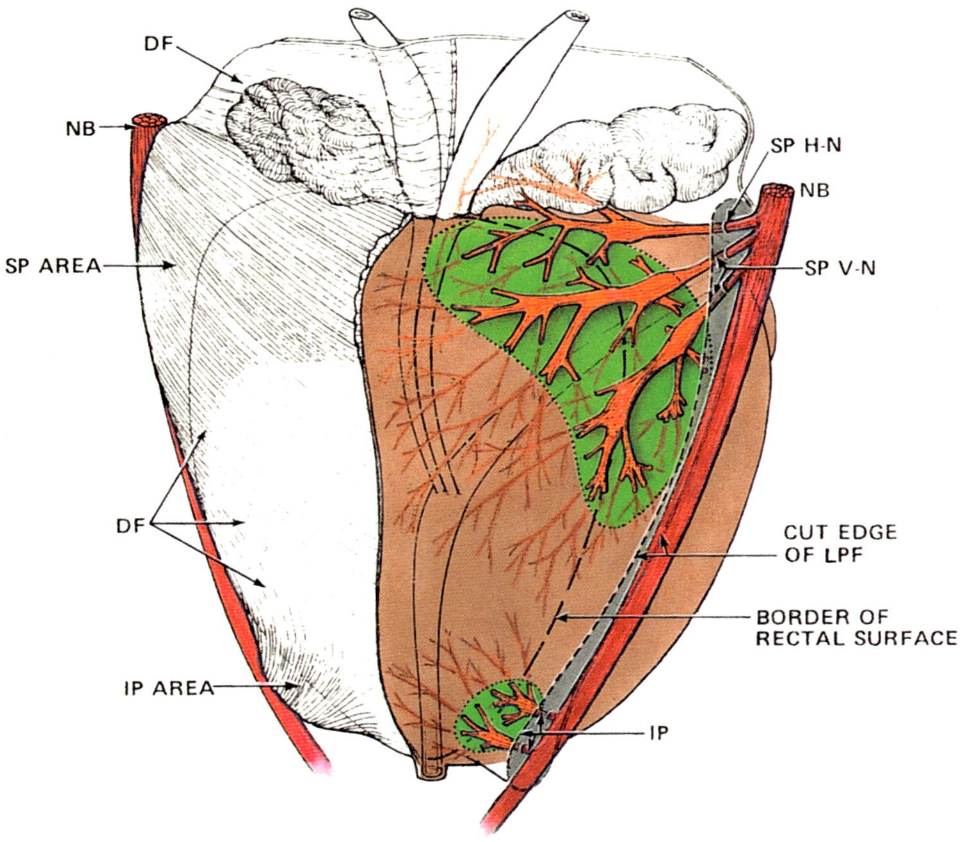

FIGURE 35.17 Distribution of nerve branches to the prostate, right posterolateral view. Nerves within the neurovascular bundle (*NB*) (*red*) branch to supply the prostate (*brown*) in a large superior pedicle (*SP*) at the prostate base and a small inferior pedicle (*IP*) at the prostate apex. Nerve branches (*orange*) leave the lateral pelvic fascia (not shown) to travel in Denonvilliers fascia (*DF*), which has been cut away from the right half of the prostate. Nerve branches from the superior pedicle fan out over a large area. A small horizontal subdivision (*H-N*) crosses the base to midline; a large vertical subdivision (*V-N*) fans out extensively over the prostate surface as far distally as mid prostate. Branches continue their course within the prostate after penetration into the capsule within a large nerve penetration area (*green*). A small inferior pedicle has a limited ramification and nerve penetration area (*green*). LPF, lateral pelvic fascia.

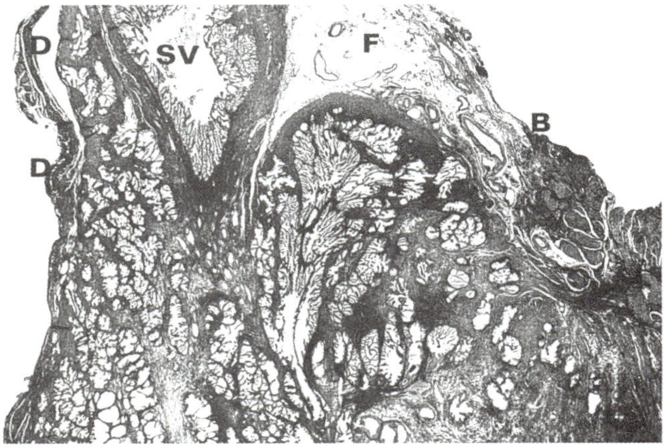

FIGURE 35.18 Parasagittal section of prostate base located almost at midline. Bladder neck smooth muscle above the level of bladder neck lumen is seen as a small dark patch (*B*) at far right. A layer of fat (*F*) covers the dome-shaped surface of the anterior central zone (**top center**). All glandular tissues within is central zone. One main duct (**center**) is seen in profile as it flares out toward the base, generating elaborate acinar structures. Behind the seminal vesicle (*SV*), the posterior central zone extends superiorly as a narrow plate. Denonvilliers fascia (*D*) is not adherent behind the seminal vesicles but blends with the capsule below.

plexus to the seminal vesicles, prostate, and corpora cavernosa of the penis, travel in this fatty layer. These nerves, along with the blood vessels to the prostate, originate from bilateral neurovascular bundles that follow a course along the pelvic sidewalls (Fig. 35.17) (25). Most of the nerve branches to the prostate leave the neurovascular bundle just superior to the prostate base and extend medially as the superior pedicle. These nerve branches fan out to penetrate the capsule, centered at the posterolateral aspect of the prostate base (25,26) and extend as far as midgland. Some nerve trunks travel medially across the prostate base, sending branches into the central zone, but the majority of nerve branches fan out distally and penetrate the capsule at an oblique angle. Small microscopic paraganglia may be seen in association with

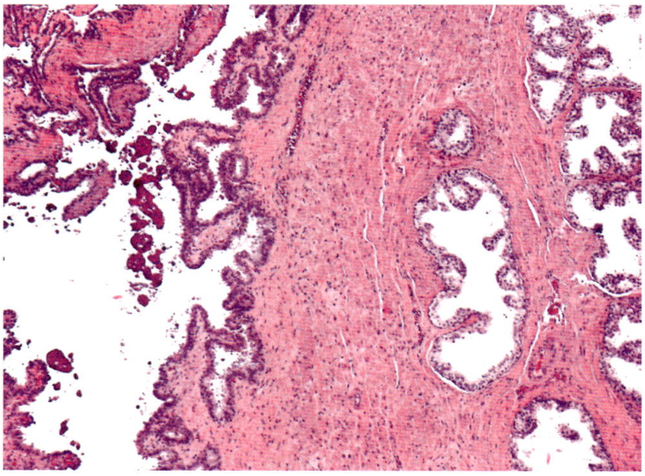

FIGURE 35.19 Minimal muscular tissue separating the prostatic central zone (**right**) from the seminal vesicles (**left**) at the base of the prostate.

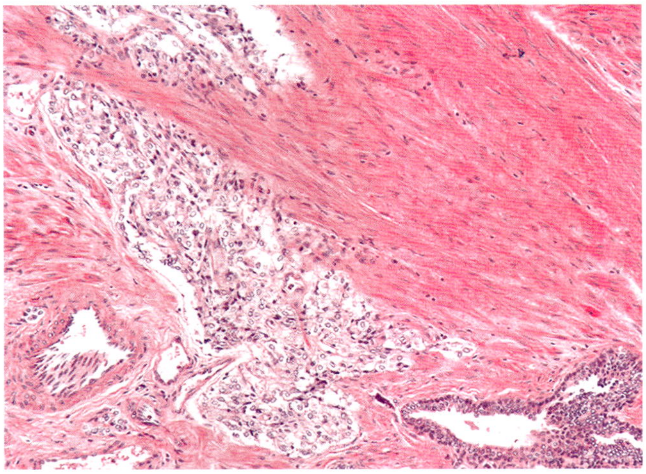

FIGURE 35.20 While paraganglia may be seen in extraprostatic tissues, often in association with peripheral nerves and ganglia, very rare examples may be intraprostatic.

extraprostatic nerves and ganglia, reportedly identified in up to 8% of radical prostatectomies (27). They are characterized by small collections of round cells with clear or amphophilic cytoplasm, and an associated small capillary vasculature (Fig. 35.20). Some variable cytologic atypia may be present in paraganglia, which could be confused with carcinoma, particularly in rare intraprostatic examples.

Before supplying the corpora cavernosa, nerve branches leave the neurovascular bundle at the prostate apex in the very small inferior pedicle and penetrate the capsule directly lateral and posterolateral (26). Here the distance from neurovascular bundle to prostate capsule is narrowed to only a few millimeters and hence sparing the nerves involved in erectile function requires dissecting very close to the prostate in this region (28,29).

Arterial branches follow the nerve branches from the neurovascular bundle; they spread over the prostate surface and penetrate the capsule to extend inward toward the distal urethral segment between the radiating duct systems of the central and peripheral zones (30,31). A major arterial branch enters the prostate at each side of the bladder neck and runs toward the verumontanum parallel to the course of the proximal urethral segment. It supplies the periurethral region and medial transition zone.

MICROSCOPIC ARCHITECTURAL AND CYTOLOGIC FEATURES OF THE GLANDULAR PROSTATE

Architectural Patterns

The biologic role of the prostate calls for the slow accumulation and occasional rapid expulsion of small volumes of fluid. These requirements are optimally met by a muscular organ having a large storage capacity and low secretory capacity. It is

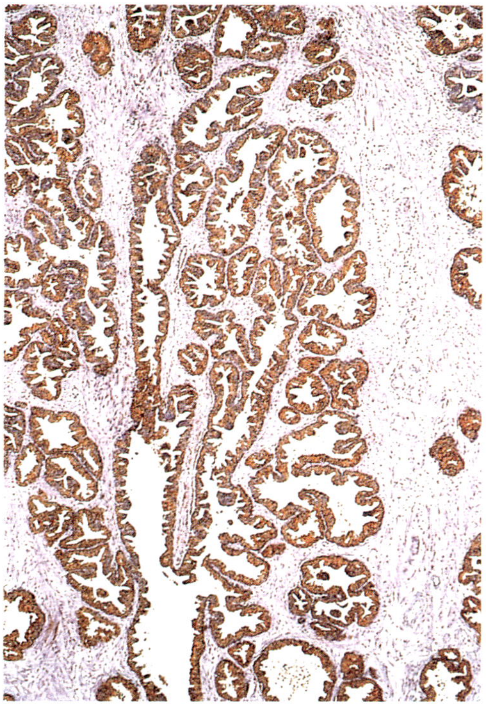

FIGURE 35.21 Ducts and acini of peripheral zone, immunohistochemically stained with anti-PSA and showing uniform distribution of protein throughout the cytoplasm of all ducts and acini.

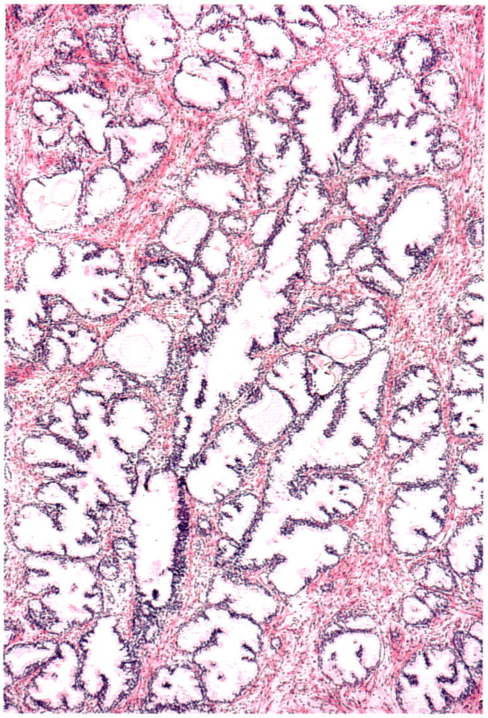

FIGURE 35.22 Subsidiary duct and branches in peripheral zone, terminating in small, rounded acini with undulating borders. Ducts and acini have similar calibers and histologic appearances.

fitting then that the prostatic ducts are morphologically identical to the acini except for their geometry, and both function as distensible secretory reservoirs. Within each prostate zone, the entire duct–acinar system is lined by columnar secretory cells of identical appearance between ducts and acini. Immunohistochemical staining for prostate-specific antigen (PSA) and prostatic acid phosphatase (PAP) shows uniform granular staining of all ductal and acinar cells (Fig. 35.21).

Except for the main transition zone ducts, which terminate at the anterior fibromuscular stroma, the main ducts of the prostate originate at the urethra and terminate near the capsule (3,6,7) (Figs. 35.4 and 35.5). Since ducts and acini within each zone have similar caliber, spacing, and histologic appearance, prostatic ducts, ductules, and acini cannot reliably be distinguished microscopically. Therefore, abnormalities of architectural pattern such as adenosis, prostatic intraepithelial neoplasia (PIN), and prostatic carcinoma are identified in routine sections as deviations from normal size and spacing of glandular units.

The main excretory duct orifices of the peripheral zone arise every 2 mm from the distal urethral segment along a double lateral line. A cluster of three or four subsidiary ducts arise about every 2 mm along each main excretory duct from urethra to capsule. These subsidiary ducts branch and extend only a short distance, rebranching and giving rise to groups of acini (Fig. 35.22). Hence, acini tend to be distributed with nearly uniform density along the course of the main duct between urethra and capsule. No acini are found immediately adjacent to the urethra.

In the peripheral and transition zones, ducts and acini have simple rounded contours with undulations of the epithelial border (4,6). The undulations presumably allow expansion of the lumens as secretory reservoirs.

Central zone ducts and acini are larger than those of the peripheral zone and transition zone (Fig. 35.23). Both ducts and acini of the central zone become progressively larger toward the capsule at the prostate base. This is reflected in the expansion of central zone cross-sectional area from

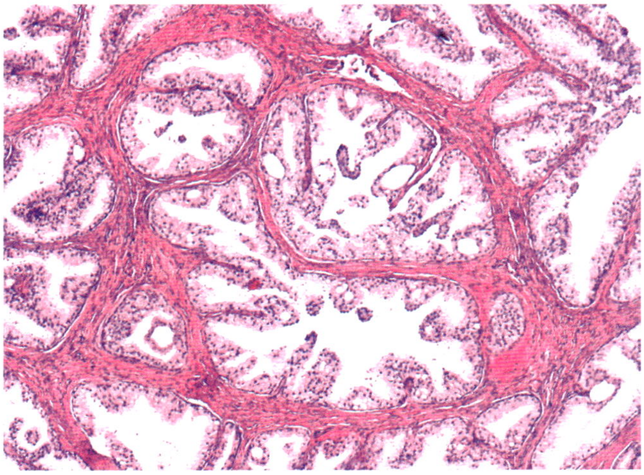

FIGURE 35.23 Low-power view of prostatic central zone architecture; large glands with complex luminal infoldings and distinct intraluminal bridges ("Roman arches").

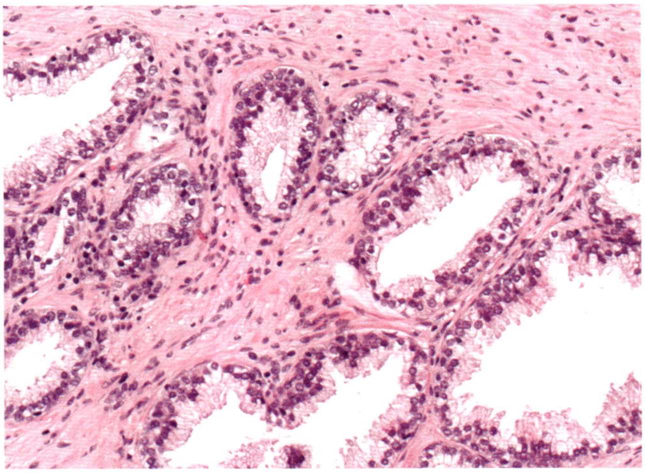

FIGURE 35.24 Prostatic acini with basal cells demonstrating little to no cytoplasm and arranged parallel to the basement membrane.

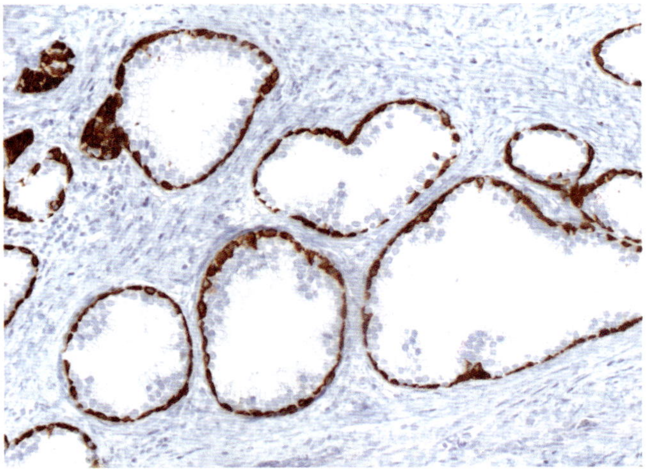

FIGURE 35.25 Same acini as in Figure 35.24 labeled with a double immunohistochemical stain for 34βE12 (cytoplasmic) and p63 (nuclear) in basal cells.

a small focus on the verumontanum to almost the entire prostate base. The corrugations in central zone duct/acinar walls are often exaggerated into distinctive intraluminal bridges reminiscent of Roman arches.

In some specimens, there is a contrast in stromal morphology that delineates the boundary between peripheral and transition zones (21). The transition zone stroma is composed of compact interlacing smooth muscle bundles. This stromal density differs from the adjacent loose peripheral zone stroma but blends with the stroma of the preprostatic sphincter and anterior fibromuscular stroma. Stromal distinctions are less evident in older prostates and may be obliterated by disease (32,33).

Cytologic Features

As with other glandular organs, the secretory cells throughout the prostate are separated from the basement membrane and stroma by a layer of basal cells. The basal cells are typically elongated and flattened parallel to the basement membrane and have slender dark nuclei with little or no discernible cytoplasm (Fig. 35.24) (34). They are typically inconspicuous and, in routine preparations, may appear incomplete or even absent around individual ducts or acini. Immunohistochemical labeling with high–molecular-weight cytokeratin and/or antibodies to p63 highlights the basal cell population (Fig. 35.25) (35–37). These stains are consistently negative in the cells of invasive malignant glands (36) because basal cells are absent. Basal cells are not myoepithelial cells analogous to those of the breast and do not contain muscle filaments (34).

The prostatic epithelium contains a small population of randomly scattered endocrine–paracrine cells (38) that are rich in serotonin-containing granules and contain neuron-specific enolase. Subpopulations of these cells also contain a variety of peptide hormones, such as somatostatin, calcitonin, and bombesin. They rest on the basal cell layer between secretory cells and often have laterally spreading dendritic processes. They are not reliably identified microscopically except with immunohistochemical and other special stains. Though their specific role in prostate biology is unknown, they presumably have paracrine function.

The secretory cells of the prostate contribute a variety of products to the seminal plasma. PSA and PAP are produced by the secretory cells of the ducts and acini of all zones, while pepsinogen II (39), tissue plasminogen activator (40), and lactoferrin are normally produced in the ducts and acini of the central zone. While PSA and PAP have historically been utilized as immunohistochemical markers of prostatic epithelium, newer antibodies against the NKX3.1 protein are also frequently employed in routine practice (41).

The cytoplasmic appearance of the normal secretory cell in all zones is similar with an abundance of small clear secretory vacuoles. Vacuoles in peripheral and transition zone cytoplasm are tightly packed (42), whereas in the central zone, a more abundant dense cytoplasm is associated with wider vacuole spacing and lower vacuole density. Since the secretory vacuoles appear empty by routine microscopy, peripheral and transition zone cells are typically pale to clear, while central zone cells are typically somewhat darker (Figs. 35.26 and 35.27).

The appearance of cell cytoplasm on tissue sections is strongly influenced by staining technique and by the type of fixative used. In the peripheral and transition zones, light hematoxylin and eosin (H&E) staining after formalin fixation shows that normal cells are "clear cells" in which a faint network of pale-staining cytoplasmic partitions between vacuoles can be visualized with scrutiny under high magnification. Only an occasional cell shows complete outlines that define numerous intact vacuoles, but immunostaining with PSA (Fig. 35.28) or PAP on the same tissue sharply outlines all the cytoplasmic vacuolar partitions.

The secretory lining of peripheral zone glands conveys an orderly appearance, with a single layer of columnar

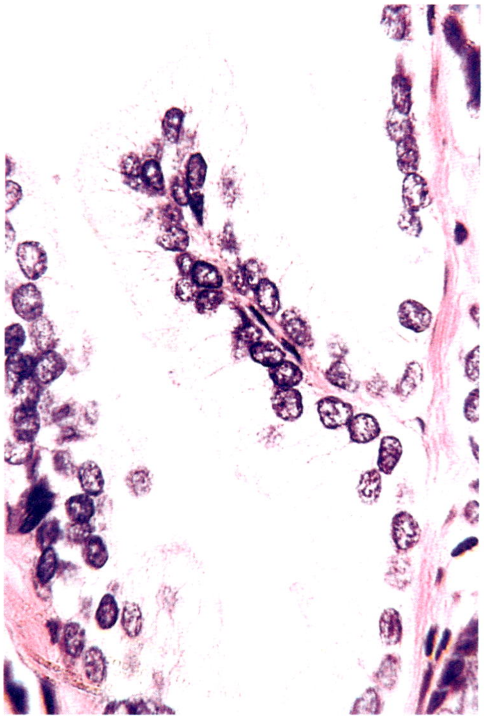

FIGURE 35.26 Peripheral zone epithelium showing clear cells in which cytoplasm is barely discernible as composed of a sheet of small empty vacuoles with delicate pale partitions.

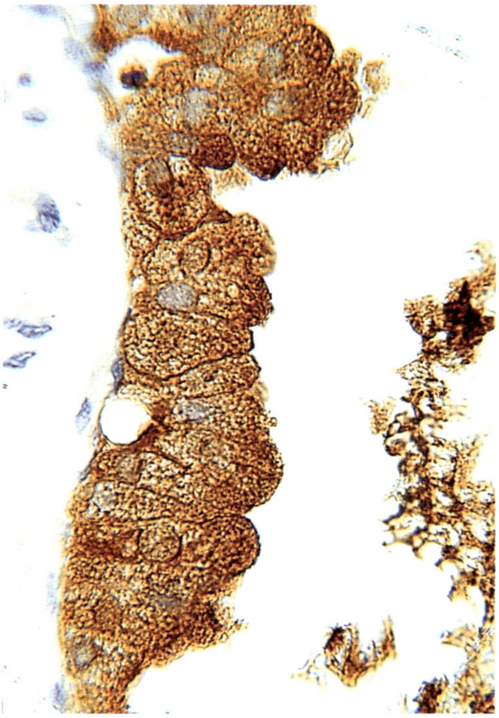

FIGURE 35.28 Peripheral zone epithelium immunostained with antibody to PSA. Protein is concentrated in a reticulated pattern that spares vacuole lumens and accentuates portions of vacuole partitions.

cells having basally oriented nuclei. In most glands, however, the epithelial row shows considerable random variation between neighboring cells in the ratio of cell height to width. Nuclear location also varies from the basal aspect to the midportion of the cell. The luminal cell border is consequently often uneven, and its irregularity is accentuated by frequent cells whose luminal aspect appears frayed.

Central zone epithelium shows an accentuation of the mild disorder of cell arrangement of the peripheral/

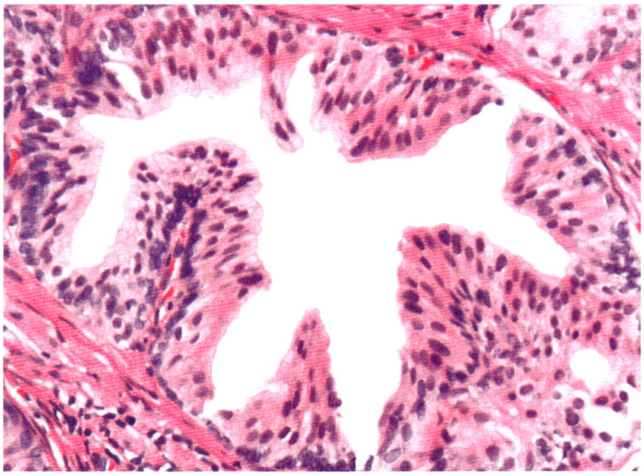

FIGURE 35.27 Central zone epithelium with eosinophilic cytoplasm and prominent basal cell layer.

transition zone (Figs. 35.23 and 35.27). Here the epithelium is variably thickened by prominent cell crowding. Nuclei, which are usually larger than in the peripheral zone, are often displaced further from the cell base than in the peripheral zone and appear pseudostratified.

The dark cytoplasm, thickened epithelium, and complex architecture in the central zone may be misinterpreted as PIN on needle biopsies. However, the distinctive histology coupled with the absence of enlarged nuclei, nucleoli, or hyperchromasia and an often-prominent basal cell layer are useful in excluding PIN.

DEVIATIONS FROM NORMAL HISTOLOGY

After 30 years of age, many prostates begin to show a variety of deviations from normal morphology (3,6,32,33). The prevalence and extent of these changes increase with age so that most prostates are quite heterogeneous in histologic composition by the seventh decade of life. Although these histologic patterns seldom have clinical significance, their distinction from adenocarcinoma is sometimes difficult.

Early morphologic studies concluded that focal atrophy in the prostate was a manifestation of aging and was seen as early as 40 years of age. However, focal atrophy in the prostate is often the consequence of previous inflammation rather than aging (3,6). The extent of atrophic foci tends to

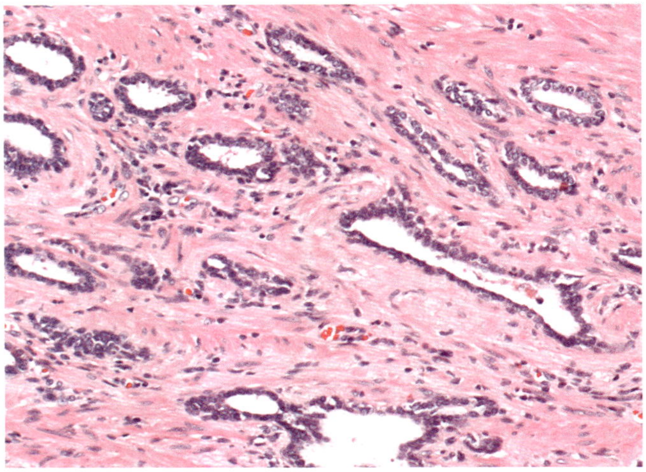

FIGURE 35.29 Simple atrophy demonstrating angulated, basophilic glands with limited to no cytoplasm and dark nuclei; scattered chronic inflammatory cells are also present.

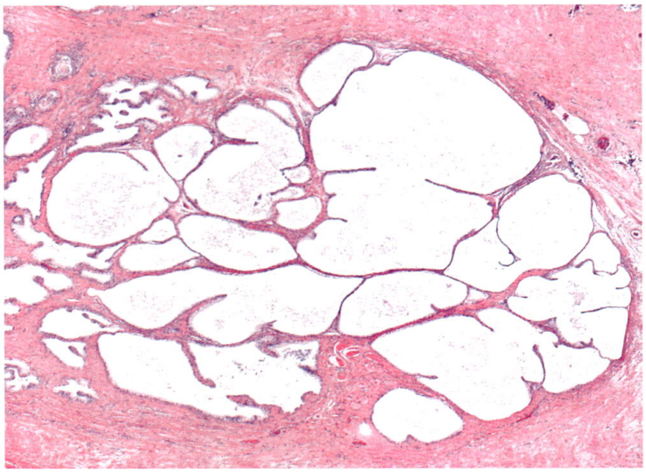

FIGURE 35.30 Back-to-back large caliber rounded acini with little to no cytoplasm characteristic of cystic atrophy.

be greater in older men, but their histologic appearance is identical to that of isolated foci found as early as 30 years of age.

Atrophy is an extremely common lesion and is mainly seen in the peripheral zone, where its distribution is typically segmental along the ramifications of a duct branch (3,6). Publication of a Working Group classification (43) highlighted four patterns of focal atrophy with distinctive histologic features. Simple atrophy is most common and features irregular or angulated, basophilic acini at low magnification. Individual acini have reduced cytoplasm, yet nuclei often retain their usual size (Fig. 35.29). Simple atrophy is commonly associated with chronic inflammation that may involve the prostatic stroma or epithelium. Atrophic glands may exhibit small nucleoli, especially in the setting of inflammation. The combination of small, angulated glands with variable architectural distortion and nucleoli may mimic cancer and cause diagnostic difficulty.

Simple atrophy with cyst formation is characterized by rounded acini of very large diameter which have a sieve-like gross appearance with numerous microscopically evident cysts. Glands show back-to-back architecture with little intervening stroma (Fig. 35.30). Cyst-like acini contain cells that have little to no apparent cytoplasm and unlike simple atrophy, are uncommonly associated with chronic inflammation.

Postatrophic hyperplasia, or lobular atrophy, has a basophilic appearance at low magnification and is composed of small round acini in a vaguely lobular arrangement. These acini often seem to surround, or radiate from, a dilated duct (Fig. 35.31). The closely packed appearance of multiple small acini conveys a histologic impression of hyperplastic glands, yet whether this is truly a hyperplastic process remains unknown. The atrophic acini are often embedded in a variable degree of fibrotic or sclerotic stroma. Postatrophic hyperplasia features low cuboidal cells with scant cytoplasm and small- to medium-sized nucleoli. Chronic inflammation is often present, and the differential diagnosis of adenocarcinoma is often raised, especially in needle biopsy material.

Partial atrophy is distinct, in that cytoplasm is attenuated, but variably present, with a nonbasophilic appearance at low magnification. Partial atrophy displays more cytoplasm lateral to the nucleus, increasing internuclear distance and imparting a pale low-power impression (Fig. 35.32). Small- to medium-sized nucleoli, as well as intraluminal dense pink sections/crystalloids may mimic carcinoma. Partial atrophy is frequently seen admixed with foci of simple atrophy suggesting that they represent a spectrum of atrophic changes.

In contrast to atrophy, the histologic hallmark of benign prostatic hyperplasia is the expansile nodule, produced by

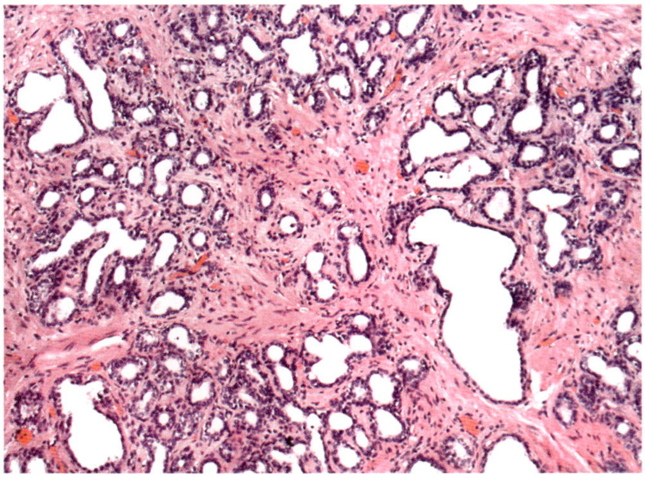

FIGURE 35.31 Postatrophic hyperplasia with atrophic acini in a lobular array surrounding a central dilated duct.

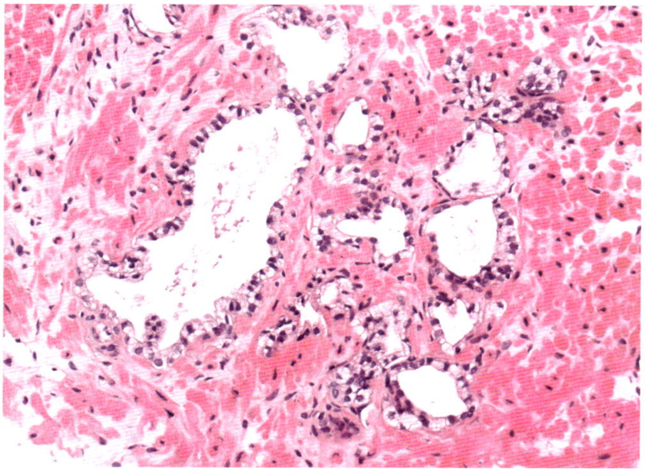

FIGURE 35.32 Focus of partial atrophy with attenuated pale cytoplasm and wispy eosinophilic intraluminal secretions. The nuclei appear relatively evenly spaced due to retention of the lateral cytoplasm.

the budding and branching of newly formed duct–acinar structures, by the focal proliferation of stroma (Fig. 35.33), or by a combination of both elements (Fig. 35.34) (3,4,7, 44). It mainly affects the transition zone.

Grossly benign prostatic hyperplasia is recognized as a globular mass replacing each transition zone, composed of numerous individual nodules. Only the nodular component is recognizable histologically as a deviation from normal pattern; internodular tissue, even when increased, is not distinguishable microscopically from normal transition zone tissue.

The enlargement of transition zone benign prostatic hyperplasia produces a characteristic progressive deformity of overall prostate contour. The expansion is chiefly anterior and toward the apex, resulting in stretching and thinning of the anterior fibromuscular stroma and producing an increase in the anteroposterior thickness of the gland.

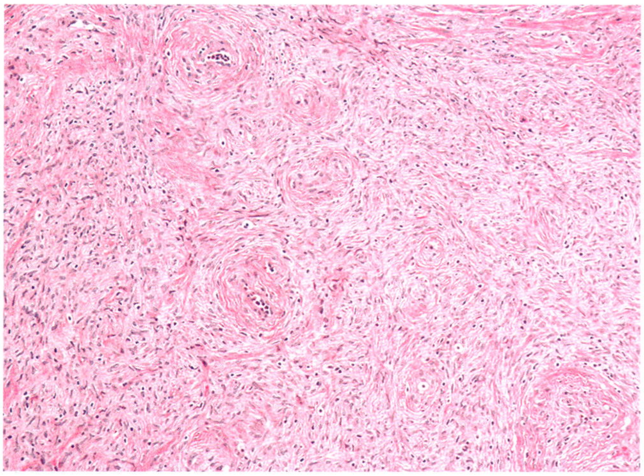

FIGURE 35.33 Prostatic stromal hyperplasia is characterized by a relatively lobular or nodular proliferation of cytologically bland spindle cells associated with prominent small round blood vessels.

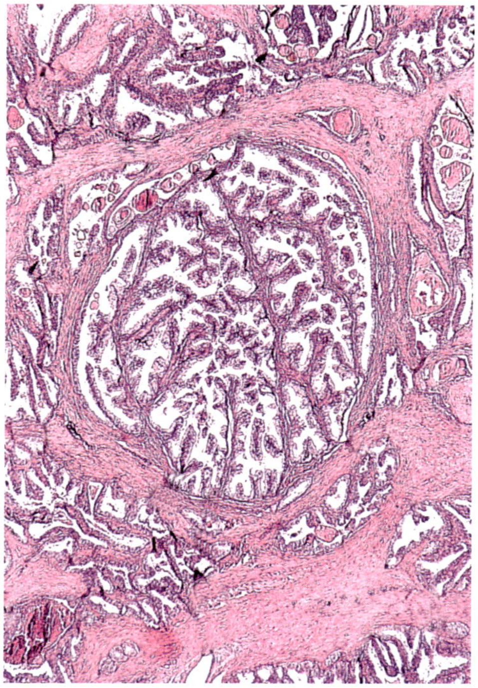

FIGURE 35.34 Nodule of glandular benign nodular hyperplasia in the prostatic transition zone.

The anterolateral horns of the peripheral zone (Fig. 35.9B) are compressed and thinned concomitant with increase of overall prostate width.

Basal cell hyperplasia is most often seen as a secondary change in benign prostatic hyperplasia nodules or inflammatory foci (45). The basal cells of ducts and acini become rounded with oval nuclei and form a multilayered lining (Fig. 35.35) that stains for basal cell–specific

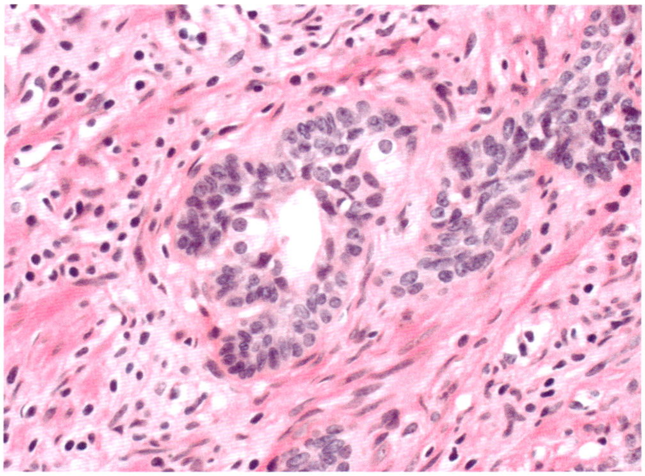

FIGURE 35.35 Focus of basal cell hyperplasia showing multiple layers of rounded basal cells with a central flattened layer of eosinophilic secretory cells. The basal cells commonly contain small prominent nucleoli.

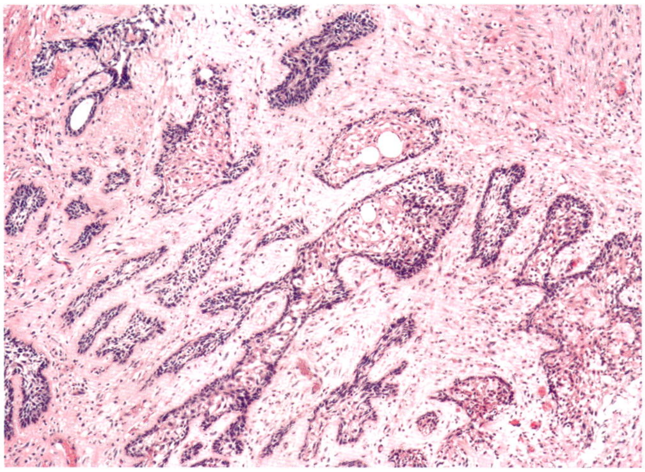

FIGURE 35.36 Prominent squamous metaplasia is often present surrounding areas of ischemia or frank infarct.

high–molecular-weight cytokeratins. There is typically a single luminal row of columnar secretory cells that stain positive for PSA. When ischemia or frank infarcts are present, squamous metaplasia may also become prominent at the periphery of the infarct. This benign metaplastic change may closely mimic urothelial carcinoma (Fig. 35.36) (46).

Because many men with prostate cancer are treated by radiation therapy, pathologists must be familiar with its effects on benign glands. After radiation, the normal glands typically become atrophic, but with cytoplasmic eosinophilia that imparts a squamoid appearance. Scattered nuclei become enlarged and hyperchromatic but display degenerative appearing, "smudgy," chromatin (Fig. 35.37). Many of the cells within these glands assume a basal cell phenotype, so expression with basal cell markers, including GATA3, is common. Because of the cytologic atypia, the latter immunophenotypic finding may cause confusion with urothelial carcinoma (47,48).

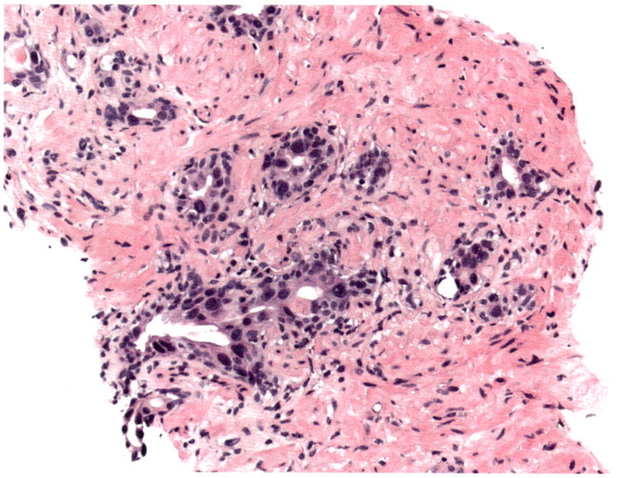

FIGURE 35.37 Radiation therapy induces atrophic changes in benign prostatic glands, often with associated cytoplasmic eosinophilia and nuclear pleomorphism.

CONSIDERATIONS IN TRANSURETHRAL RESECTION AND NEEDLE BIOPSY SPECIMENS

Tissue distortion by thermal artifact near the edges of transurethral resection (TUR) tissue fragments can create important diagnostic problems. Basal cell hyperplasia, adenomatous hyperplasia, atrophy, and benign prostatic hyperplasia with small glands may be difficult to distinguish from carcinoma without utilizing adjunctive immunohistochemical stains. Loss of nuclear detail occurs more homogeneously across the tissue fragments than obvious cell distortion. Hence, small foci of cancer may be more difficult to diagnose because of the artifactual absence of nucleoli.

The same problems are seen in needle biopsies, in which case the artifact is presumably due to compression rather than heat. The presence of artifact is usually limited to loss of nuclear detail and is more subtle because areas of severe tissue distortion are not often identified.

The regions of the prostate sampled by TUR and by needle biopsy are also different. Most needle biopsies represent posterior peripheral zone tissue. Unless a special effort is made, as in a transition zone–directed needle biopsy (49), the needle seldom reaches the more anterior portions of the gland. While this is true for transrectal ultrasound-directed needle biopsies, with the advent of transperineal biopsy techniques, the anterior portion of the prostate is now more commonly sampled (50).

In most cases, TUR specimens consist of transition zone tissue, urethral and periurethral tissues, bladder neck fragments, and anterior fibromuscular stroma (44) with variable amounts of peripheral zone tissue. On rare occasions, central zone fragments may be sampled and show the distinctive architectural and cytologic features described above.

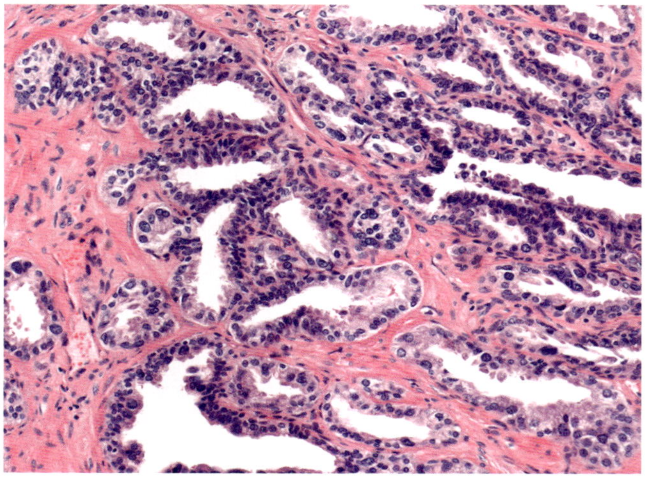

FIGURE 35.38 Benign tubular invaginations from the wall of the ejaculatory duct showing architectural (small, crowded acini) and cytologic (focal prominent nucleoli) features that suggest carcinoma. Focal yellow-brown cytoplasmic pigment, as seen in the **upper right–hand portion** of the figure, is characteristic of this epithelium.

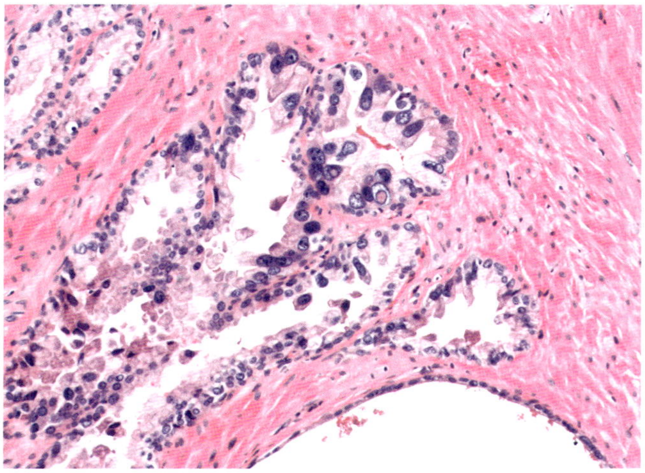

FIGURE 35.39 Benign ejaculatory duct/seminal vesicle–type epithelium demonstrating bizarre enlarged hyperchromatic nuclei.

They may be accompanied by fragments of ejaculatory duct/seminal vesicle and the tiny tubular outgrowths from the walls of these structures may be misinterpreted as adenocarcinoma when seen in tangential sections that do not reveal the main lumen (Fig. 35.38). This impression of carcinoma may be further encouraged by the frequent presence of enlarged dark nuclei with bizarre contours in the seminal vesicle and ejaculatory duct epithelium (Fig. 35.39). The presence of golden-brown cytoplasmic granules, which may be few, helps to establish the benign diagnosis.

ACKNOWLEDGMENT

The content of this chapter builds upon the prodigious work of the late Dr. John E. McNeal, who in a series of manuscripts and monographs from the 1970s through the early 1990s, defined the modern approach to prostatic anatomy. Although this and previous iterations have been reorganized to focus on anatomy relevant to evaluation of prostate cancer in routine prostatectomy specimens, they remain based on Dr. McNeal's principles.

REFERENCES

1. McNeal JE. Developmental and comparative anatomy of the prostate. In: Grayhack J, Wilson J, Scherbenske M, eds. *Benign Prostatic Hyperplasia. DHEW Publication No. (NIH) 76–1113*. Department of Health, Education and Welfare; 1975:1–10.
2. Cunha GR, Donjacour AA. Mesenchymal–epithelial interactions in the growth development of the prostate. In: Lepor H, Ratliff TL, eds. *Urologic Oncology*. Kluwer Academic; 1989:159–175.
3. McNeal JE, Stamey TA, Hodge KK. The prostate gland: morphology, pathology, ultrasound anatomy. *Monogr Urol*. 1988;9:36–54.
4. McNeal JE. Anatomy of the prostate and morphogenesis of BPH. *Prog Clin Biol Res*. 1984;145:27–53.
5. McNeal JE. The prostate and prostatic urethra: a morphologic synthesis. *J Urol*. 1972;107:1008–1016.
6. McNeal JE. Regional morphology and pathology of the prostate. *Am J Clin Pathol*. 1968;49:347–357.
7. McNeal JE. Origin and evolution of benign prostatic enlargement. *Invest Urol*. 1978;15:340–345.
8. Samaratunga H, Montironi R, True L, et al; ISUP Prostate Cancer Group. International Society of Urological Pathology (ISUP) consensus conference on handling and staging of radical prostatectomy specimens. Working group 1: specimen handling. *Mod Pathol*. 2011;24:6–15.
9. Berney D, Wheeler TM, Grignon DJ, et al; ISUP Prostate Cancer Group. International Society of Urological Pathology (ISUP) consensus conference on handling and staging of radical prostatectomy specimens. Working group 4: seminal vesicles and lymph nodes. *Mod Pathol*. 2011;24:39–47.
10. McNeal JE. Normal histology of the prostate. *Am J Surg Pathol*. 1988;12(8):619–633.
11. Wheeler TM. Anatomic considerations in carcinoma of the prostate. *Urol Clin North Am*. 1989;16:623–634.
12. McNeal JE, Bostwick DG, Kindrachuk RA, Redwine EA, Freiha FS, Stamey TA. Patterns of progression in prostate cancer. *Lancet*. 1986;1:60–63.
13. Partin AW, Epstein JI, Cho KR, Gittelsohn AM, Walsh PC. Morphometric measurement of tumor volume and percent of gland involvement as predictors of pathological stage in clinical stage B prostate cancer. *J Urol*. 1989;141:341–345.
14. Fine SW, Al-Ahmadie HA, Gopalan A, Tickoo SK, Scardino PT, Reuter VE. Anatomy of the anterior prostate and extraprostatic space: a contemporary surgical pathology analysis. *Adv Anat Pathol*. 2007;14:401–407.
15. Myers RP, Goellner JR, Cahill DR. Prostate shape, external striated urethral sphincter, and radical prostatectomy: the apical dissection. *J Urol*. 1987;138:543–550.
16. McNeal JE. The zonal anatomy of the prostate. *Prostate*. 1981;2:35–49.
17. Gagukas RJ, Brown RW, Wheeler TM. Verumontanum mucosal gland hyperplasia. *Am J Surg Pathol*. 1995;19:30–36.
18. Greene DR, Wheeler TM, Egawa S, Weaver RP, Scardino PT. Relationship between clinical stage and histological zone of origin in early prostate cancer: Morphometric analysis. *Br J Urol*. 1991;68:499–509.
19. Ayala AG, Ro JY, Babaian R, Troncoso P, Grignon DJ. The prostatic capsule: does it exist? Its importance in the staging and treatment of prostatic carcinoma. *Am J Surg Pathol*. 1989;13:21–27.
20. McNeal JE, Villers AA, Redwine EA, Freiha FS, Stamey TA. Capsular penetration in prostate cancer. Significance for natural history and treatment. *Am J Surg Pathol*. 1990;14:240–247.
21. McNeal JE, Redwine EA, Freiha FS, Stamey TA. Zonal distribution of prostatic adenocarcinoma: correlation with histologic patterns and direction of spread. *Am J Surg Pathol*. 1988;12:897–906.
22. Walsh PC. Radical retropubic prostatectomy with reduced morbidity: an anatomic approach. *NCI Monogr*. 1988;7:133–137.
23. Tanagho EA, Smith DR. The anatomy and function of the bladder neck. *Br J Urol*. 1966;38:54–71.
24. Villers A, McNeal JE, Freiha FS, Boccon-Gibod L, Stamey TA. Invasion of Denonvilliers' fascia in radical prostatectomy specimens. *J Urol*. 1993;149:793–798.
25. Lepor H, Gregerman M, Crosby R, Mostofi FK, Walsh PC. Precise localization of the autonomic nerves from the pelvic

plexus to the corpora cavernosa: a detailed anatomical study of the adult male pelvis. *J Urol.* 1985;133:207–212.
26. Villers A, McNeal JE, Redwine EA, Freiha FS, Stamey TA. The role of perineural space invasion in the local spread of prostatic adenocarcinoma. *J Urol.* 1989;142:763–768.
27. Ostrowski ML, Wheeler TM. Paraganglia of the prostate: location, frequency, and differentiation from prostatic adenocarcinoma. *Am J Surg Pathol.* 1994;18:412–420.
28. Catalona WJ, Dresner SM. Nerve-sparing radical prostatectomy: extraprostatic tumor extension and preservation of erectile function. *J Urol.* 1985;134:1149–1151.
29. Eggleston JC, Walsh PC. Radical prostatectomy with preservation of sexual function: pathological findings in the first 100 cases. *J Urol.* 1985;134:1146–1148.
30. Flocks RH. The arterial distribution within the prostate gland: its role in transurethral prostatic resection. *J Urol.* 1937;37:524–525.
31. Clegg EV. The vascular arrangements within the human prostate gland. *Br J Urol.* 1956;28:428–435.
32. McNeal JE. Age-related changes in the prostatic epithelium associated with carcinoma. In: Griffiths K, Pierrepoint CG, eds. *Some Aspects of the Aetiology and Biochemistry of Prostatic Cancer*. Tenovus; 1970:23–32.
33. McNeal JE. Aging and the prostate. In: Brocklehurst JC, ed. *Urology in the Elderly*. Churchill Livingstone; 1984:193–202.
34. Mao P, Angrist A. The fine structure of the basal cell of human prostate. *Lab Invest.* 1966;15:1768–1782.
35. Brawer MK, Peehl DM, Stamey TA, Bostwick DG. Keratin immunoreactivity in the benign and neoplastic human prostate. *Cancer Res.* 1985;45:3663–3667.
36. Hedrick L, Epstein JI. Use of keratin 903 as an adjunct in the diagnosis of prostate carcinoma. *Am J Surg Pathol.* 1989;13:389–396.
37. Weinstein MH, Signoretti S, Loda M. Diagnostic utility of immunohistochemical staining for p63, a sensitive marker of prostatic basal cells. *Mod Pathol.* 2002;15:1302–1308.
38. di Sant'Agnese PA. Neuroendocrine differentiation in prostatic carcinoma. *Cancer.* 1995;75:1850–1859.
39. Reese JH, McNeal JE, Redwine EA, Samloff IM, Stamey TA. Differential distribution of pepsinogen II between the zones of the human prostate and the seminal vesicle. *J Urol.* 1986;136:1148–1152.
40. Reese JH, McNeal JE, Redwine EA, Stamey TA, Freiha FS. Tissue type plasminogen activator as a marker for functional zones, within the human prostate gland. *Prostate.* 1988;12:47–53.
41. Gelmann EP, Bowen C, Bubendorf L. Expression of NKX3.1 in normal and malignant tissues. *Prostate* 2003;55:111–117.
42. deVries CR, McNeal JE, Bensch K. The prostatic epithelial cell in dysplasia: an ultrastructural perspective. *Prostate* 1992; 21:209–221.
43. De Marzo AM, Platz EA, Epstein JI, et al. A working group classification of focal prostate atrophy lesions. *Am J Surg Pathol.* 2006;30:1281–1291.
44. Price H, McNeal JE, Stamey TA. Evolving patterns of tissue composition in benign prostatic hyperplasia as a function of specimen size. *Hum Pathol.* 1990;21:578–585.
45. Cleary KR, Choi HY, Ayala AG. Basal cell hyperplasia of the prostate. *Am J Clin Pathol.* 1983;80:850–854.
46. Milord RA, Kahane H, Epstein JI. Infarct of the prostate gland: experience on needle biopsy specimens. *Am J Surg Pathol.* 2000;24:1378–1384.
47. Wobker SE, Khararjian A, Epstein JI. GATA3 positivity in benign radiated prostate glands: a potential diagnostic pitfall. *Am J Surg Pathol.* 2017;41:557–563.
48. Tian W, Dorn D, Wei S, et al. GATA3 expression in benign glands with radiation atypia: a diagnostic pitfall. *Histopathol.* 2017;71:150–155.
49. Haarer CF, Gopalan A, Tickoo SK, et al. Prostatic transition zone directed needle biopsies uncommonly sample clinically relevant transition zone tumors. *J Urol.* 2009;182:1337–1341.
50. Meyer AR, Mamawala M, Winoker JS, et al. Transperineal prostate biopsy improves the detection of clinically significant prostate cancer among men on active surveillance. *J Urol.* 2021; 205:1069–1074.

36

Testis and Excretory Duct System

Muhammad T. Idrees

The adult testes are paired organs that lie within the scrotum suspended by the spermatic cord (Fig. 36.1). The average weight of each is 12 to 20 g, the right usually being, on average, 10% heavier than the left (1,2). Mean testicular volume and weight strongly correlate with height, weight, body surface area, and standardized body weight (1). The left testis hangs slightly lower than the right in standing posture due to positive right-sided intra-abdominal pressure during the development and movement of abdominal viscera (3,4). The asymmetric levels help in the thermodynamics of the organ (5). The scrotal coverings are skin, dartos fascia, an external spermatic fascia, cremasteric fascia, internal spermatic fascia, and the parietal layer of the tunica vaginalis (Fig. 36.2). The dartos muscle is nonstriated and closely attached to the overlying skin, allowing it to glide freely over the underlying loose fascia layer. Colles fascia contributes to the scrotal septum.

SUPPORTING STRUCTURES

The supporting structures of the testis consist of a tough capsule (the tunica) and a number of fibrous septa that extend from the inner surface of the tunica into the parenchyma, dividing the testis into approximately 250 lobules. The posterior middle upper part of the testis is not covered by the capsule and is called the hilum. It contains blood and lymphatic vessels, nerves, and the extratesticular portion of the rete testis. The capsule has three distinctive layers: the outer serosa or visceral tunica vaginalis, the thick, collagenous tunica albuginea, and the inner tunica vasculosa. The tunica vaginalis consists of a flattened layer of mesothelial cells overlying a well-developed basement membrane. It forms a sac with two components: a visceral layer covering the testis and head of the epididymis and a parietal layer that is formed as the lining reflects posteriorly and superiorly at the hilum and the epididymis and then covers the internal spermatic fascia. The space between the two layers normally contains a small amount of straw-colored serous fluid. Infrequently, transitional, or squamous metaplasia of the surface mesothelium may be present, the former occurring as Walthard nests identical to those found more commonly on the serosa of the fallopian tubes. The tunica albuginea is composed of a layer of collagen fibers within which are embedded fibroblasts, myocytes, mast cells, and nerves.

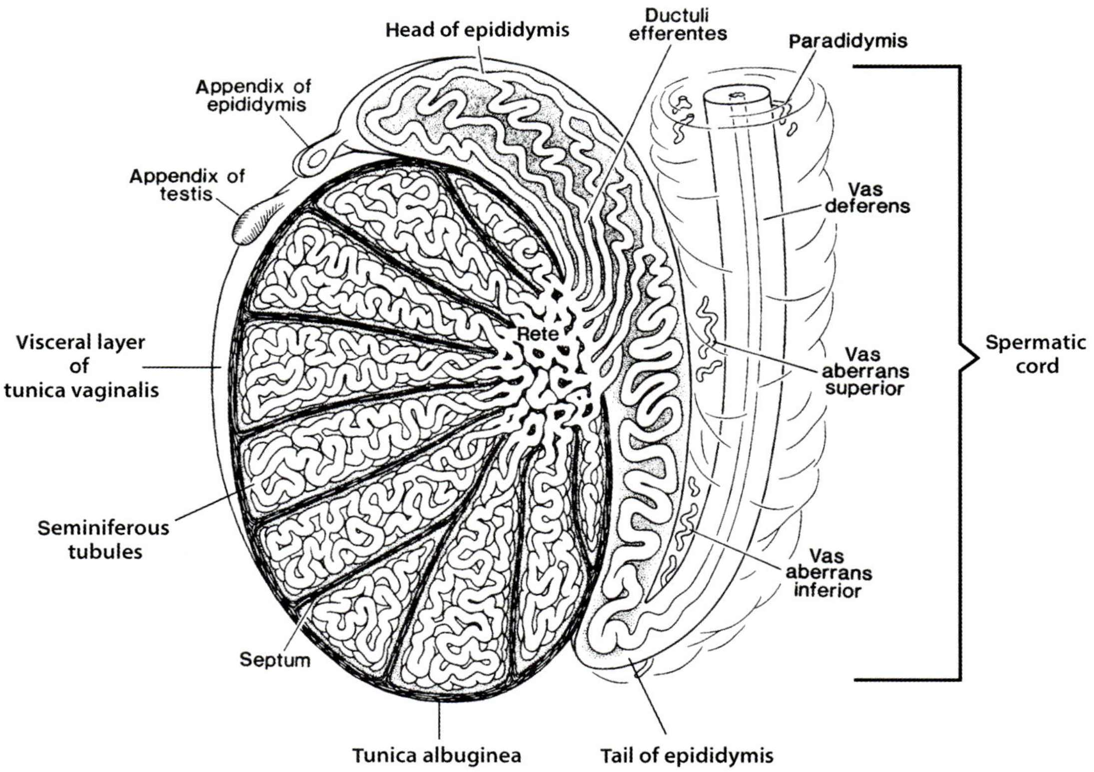

FIGURE 36.1 Diagrammatic views of testis, epididymis, and portion of ductus (vas) deferens.

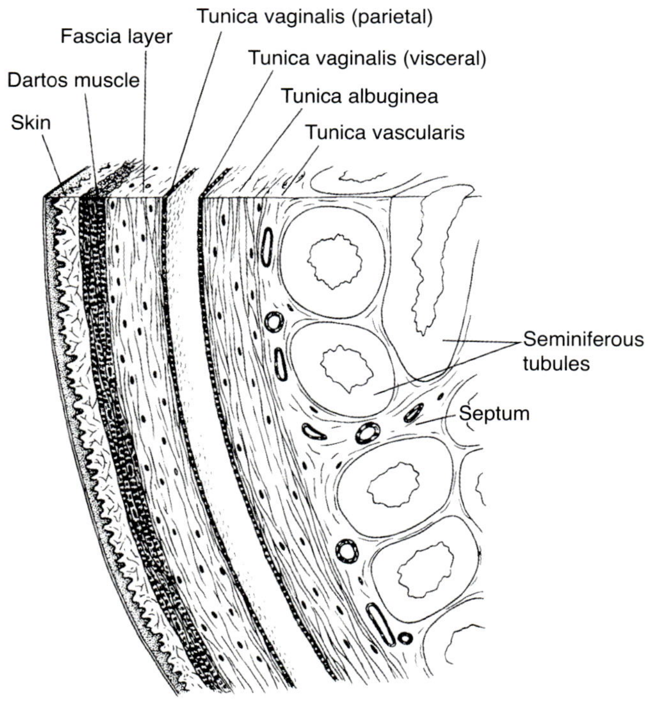

FIGURE 36.2 Scrotal covering layers and capsule of testis.

The myocytes, found mainly in the posterior region of the testis, undergo regular contractions that cause a transient increase in intratesticular pressure. The tunica vasculosa, a loose connective tissue containing blood vessels and lymphatic vessels, sends septa into the testicular parenchyma to form the individual lobules. The tunica vasculosa is a common location to identify lymphovascular space invasion by germ cell tumors. The tunica thickness varies with age, averaging 300 μm at birth, 400 to 450 μm in young adults, and 900 to 950 μm in men over 65 years of age (6).

SEMINIFEROUS TUBULES

Each lobule of the testis contains one to four seminiferous tubules (Fig. 36.1). The individual tubule is a highly convoluted, closed-loop structure with numerous communications between the arms of the loop but without any blind endings or branches. The absence of branching in the tubules can help distinguish intratubular tumor from intravascular tumor, since vessels may frequently show a branched pattern. Each arm of the looped tubule empties into the septal portion of the rete testis. The collective length of the tubules in each testis has been estimated to be between 299 and 981 m, with an average of 540 m (7). The average tubule diameter in young adults is 180 μm (± 30). The usual open testicular biopsy may encompass tubules from up to 5 lobules along with portions of the intervening

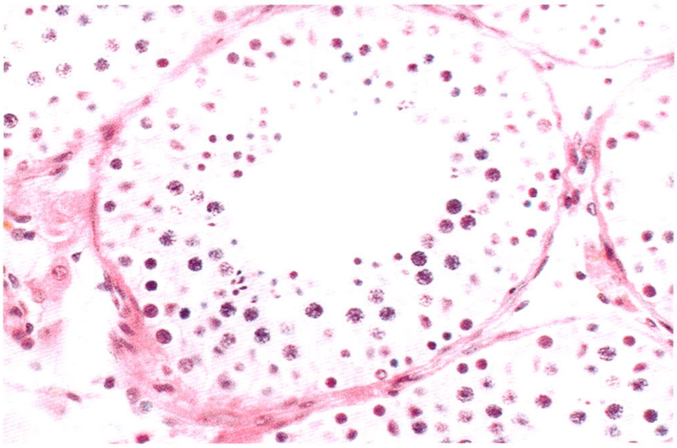

FIGURE 36.3 A cross-sectional view of seminiferous tubule and interstitium. Germ cell maturation is variable around the tubule, a normal finding.

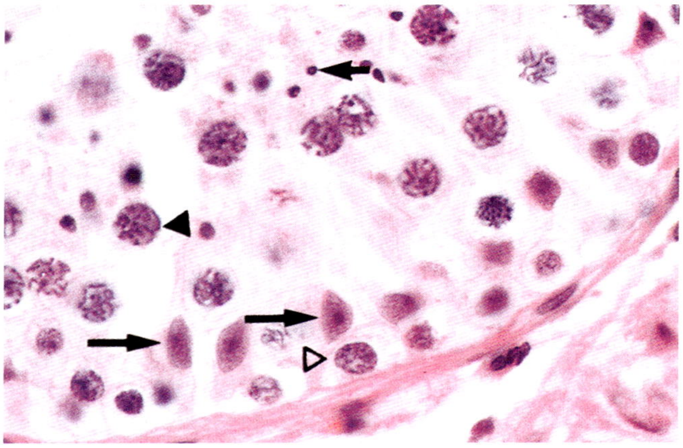

FIGURE 36.4 Seminiferous tubule with Sertoli cells (*long arrows*), spermatogonia (Δ), primary spermatocytes (▲), and spermatids (*short arrow*).

septa. It is important not to interpret the latter as foci of fibrosis. The seminiferous tubules are composed of germ cells in varying stages of differentiation and Sertoli cells. Each tubule has a distinctive basement membrane and a thin lamina propria (Fig. 36.3). At puberty, the tubule is divided into a basal compartment and an adluminal compartment via the development of a tight junction complex between adjacent Sertoli cells. The basal compartment or spermatogonial niche houses the spermatogonia and preleptotene spermatocytes and the adluminal section contains all the more developed forms. Germ cell neoplasia in situ (previously termed intratubular germ cell neoplasia unclassified) occupies the spermatogonial niche.

SERTOLI CELLS

Sertoli cells play important, yet different roles in the fetal and adult testes, and these divergent roles are reflected in their proliferative activity, cell protein markers, and the nature of cellular intermediate filaments at various periods of the cell life span. Adult Sertoli cells are nondividing cells. The role of Sertoli cells in spermatogenesis is indispensable. Specific ablation of Dicer, an RNAaseIII endonuclease from the Sertoli cells leads to loss of all germ cell types in the adult mouse testis (8). The role of Sertoli cell-derived exosomes in regulating spermatogenesis in a paracrine fashion has recently been elucidated (9). It is also becoming clear that Sertoli cells play important role in immune regulation of germ cells and favoring proliferation (10). The number of Sertoli cells is important since they have the capacity to nurture only a finite number of germ cells. Genetic, hormonal, and environmental factors play significant roles in determining the final number of Sertoli cells in the adult testis (11). These tall, irregular, columnar cells are attached to the underlying basal lamina and have abundant but relatively inconspicuous cytoplasm and an ill-defined cytoplasmic membrane. The cells send intricate cytoplasmic extensions around the germ cell elements and continuously alter their contours to accommodate the changing size and shape of the germ cells that they cradle. Adult Sertoli cell nuclei have round to slightly irregular shapes, with a folded nuclear membrane, homogeneous chromatin distribution, and a prominent, round nucleolus (Figs. 36.4 and 36.5). These features are in sharp contrast to those of the fetal and prepubertal Sertoli cells, which have an oval or elongated nucleus, a smoothly contoured nuclear membrane, and an inconspicuous nucleolus. The Sertoli cell nuclei represent about 10% of the nuclei in a normal adult tubule cross-section. They are located toward the basal side of the tubule and lie just adluminal to the spermatogonia and preleptotene spermatocytes. Sertoli cell cytoplasm may contain lipid vacuoles and/or eosinophilic granular debris. Much of this material represents phagocytosed remnants of the residual bodies of the spermatids or degenerated earlier germ cell forms. Vimentin is the predominant intermediate filament in the adult Sertoli cell, whereas embryonic Sertoli cells also contain cytokeratins

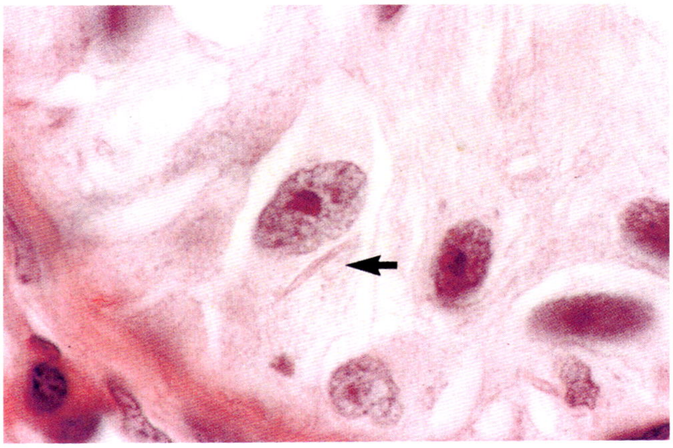

FIGURE 36.5 Sertoli cell with intracytoplasmic Charcot–Böttcher filaments (*arrow*) from a patient with germ cell aplasia. Note the prominent nucleolus and slightly wrinkled nuclear membrane.

8 and 18 (12,13). The transient appearance of cytokeratins is of interest in view of the report of malignant Sertoli cell tumors containing both cytokeratins and vimentin (14,15). Low-molecular-weight cytokeratins may also be identified in some Sertoli cells of atrophic tubules in the adult testis (16), in tubules containing germ cell neoplasia in situ (17), and in Sertoli cells of the contralateral testis in patients with germ cell neoplasms. Sertoli cells in these situations have an immature morphologic appearance and, in the latter two instances, appear to be part of the "testicular dysgenesis syndrome" (18). Adult Sertoli cells, unlike fetal or prepubertal Sertoli cells, express androgen receptors in their nuclei, but, at the same time, have lost expression of cytoplasmic anti-müllerian hormone (AMH), which is present in the immature forms (11,19). A wide variety of other molecules are produced by Sertoli cells and may be identified in immunohistochemical study, although some results are variable. These include inhibin/activin, insulin-like growth factor, platelet-derived growth factor, transforming growth factor, interleukins 1 and 6, neurofilament proteins, nuclear WT-1, and steroidogenic factor-1 (20–24). The cytoplasm of fetal Sertoli cells contains the P450 enzyme CYP26B1, which is involved in breaking down retinoic acid derived from the mesonephric tubules. Sertoli cells immunohistochemically express inhibin, calretinin, SF-1, WT1, S100, CD99, SOX9, AMH, and vimentin (25–27). Retinoic acid is a key substance in the process of directing fetal germ cells into meiosis. CYP26B1 acts as meiosis-inhibiting substance. Reduced levels of retinoic acid in the male may also play a role in directing the fetal male germ cell to mitotic arrest (28). In vitro studies have demonstrated that increasing temperatures lead to impairment of Sertoli cell function which is integral for spermatogenesis (29). The distinctive Charcot–Böttcher crystalloids are a cytoplasmic feature of adult Sertoli cells (30). These bundles of filamentous structures are located primarily in the basal portion of the cells and are best seen on ultrastructural examination but are occasionally large enough to be recognized by light microscopic studies (Fig. 36.5). Ultrastructurally, they appear to merge with vimentin-labeled intermediate filaments, both of which are increased in the cryptorchid testis and the Sertoli cell–only syndrome (17).

At puberty, the Sertoli cells develop a complex of intercellular junctions that divide the seminiferous tubule into two distinctive compartments: the basal compartment containing the various stages of spermatogonia and preleptotene spermatocytes, and the adluminal compartment housing the primary and secondary spermatocytes and spermatids. This junction structure comprises the blood–testis barrier. As preleptotene spermatocytes move from the basal compartment to the adluminal compartment, this junction complex undergoes disassembly and a new complex is formed simultaneously behind the preleptotene spermatocyte as the latter moves into the adluminal compartment. As a result, there is a transient intermediate compartment between new and old junction complexes where leptotene spermatocytes reside during transport to the adluminal compartment. These processes allow the testis to maintain an intact immunologically important blood–testis barrier (31,32). At the same time as the blood–testis barrier is being formed, Sertoli cells develop hemidesmosome-like junctions with the basal lamina and a variety of junctions (e.g., desmosome-like, adherens, and ectoplasmic specializations) with the developing germ cells. These Sertoli cell–germ cell junctions undergo assembly and disassembly as the maturing germ cells migrate toward the lumen of the seminiferous tubule (33). Still unclear is the mechanism that serves to promote the movement of germ cells from the basal compartment to the adluminal compartment. Sertoli cell senescence, reduced proliferation, and blood–testis barrier dysfunction are triggered by upregulation of miR-143-3p and targeting ubiquitin-conjugating enzymes. Targeting this pathway by downregulating miR-143-3p may improve reproductive outcomes in infertility cases (34). Several other micro-RNAs and mitochondrial uncoupling proteins are being studied for their roles in regulation of Sertoli cell functions and integrity (35,36).

GERM CELLS

Germ cells are derived originally from a subset of cells from the proximal epiblast. These primitive cells then move into extraembryonic tissue at the base of the allantois near the developing hindgut and subsequently migrate through the hindgut mesentery to the gonadal ridge. Deviations in this migratory route may account for the appearance in abnormal locations of tumors derived from these cells (Fig. 36.6), although another possibility, at least for some extragonadal germ cell tumors, is originating from stem cells. The fate of the germ cells (male or female) is determined not by their own chromosomal constitution but by the environment within which they are found. If the developing gonad is destined to be a testis, the germ cell will become a spermatogonium and, if the gonad is destined to become an ovary, the germ will become an oocyte (37). The germ cell elements comprise the majority of the cells in the adult seminiferous tubule (Figs. 36.3 and 36.4). Spermatogenesis, including proliferation of committed spermatogonia, covers a period

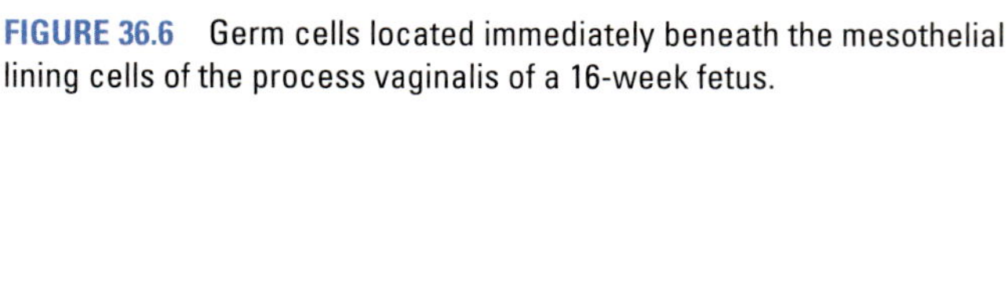

FIGURE 36.6 Germ cells located immediately beneath the mesothelial lining cells of the process vaginalis of a 16-week fetus.

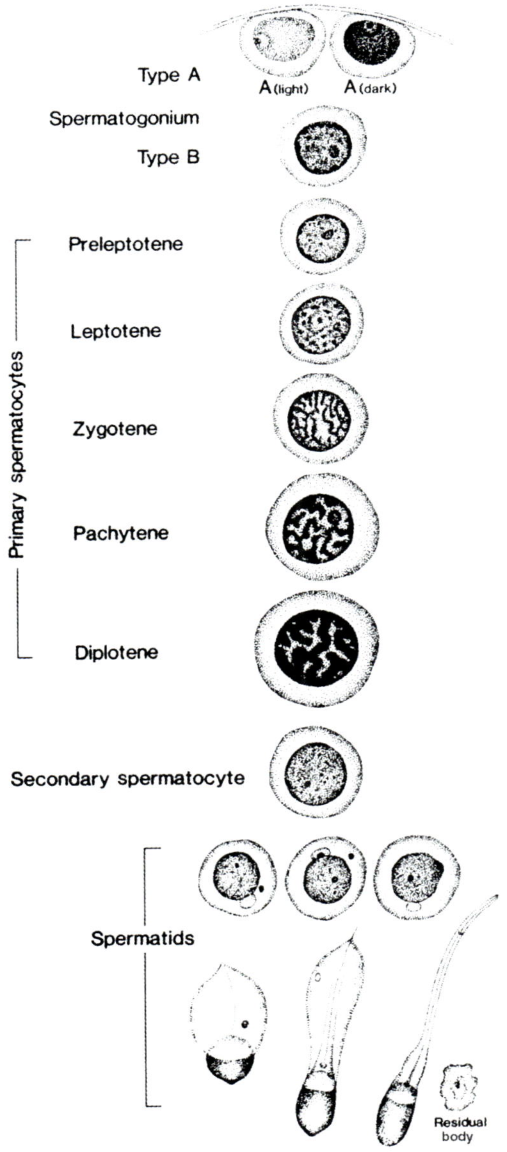

FIGURE 36.7 Steps in spermatogenesis.

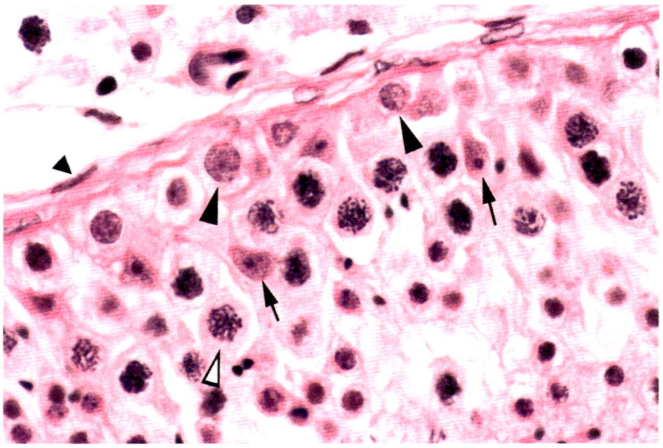

FIGURE 36.8 Portion of seminiferous tubule showing spermatogonia (*solid arrowhead*), primary spermatocytes (*open arrowhead*), Sertoli cells (*arrow*), and fibromyocyte of tunica propria (solid triangle). The smaller cells in the lower right are mainly secondary spermatocytes and early spermatids.

of ~74 days with no evidence that this time requirement is altered by age or pathologic states (38). Marked morphologic transitions occur in these cells during this time period, as shown in Figure 36.7.

The undifferentiated spermatogonia lie in the basal compartment of the adult testis. Their nuclei are ovoid to round and, depending on the plane of section, may have one or two easily identifiable nucleoli (Fig. 36.8). Crystalloids of Lubarsch are found within the cytoplasm in a perinuclear location. These structures, best seen on electron microscopic examination, measure up to 3 μm in length and represent a mixture of parallel arrays of fibrils and ribosome-like granules (39). These structures are also sometimes found in primary spermatocytes and resemble Charcot–Böttcher crystalloids of Sertoli cells (40). Utilizing special fixatives such as Zenker-formal, spermatogonia have been divided into type A_{dark} and type A_{pale} cells based on their nuclear chromatin staining pattern. It should be pointed out that these staining differences are not so evident with commonly employed fixatives such as Bouin solution, which produces coarse clumping in the nuclei of all of the spermatogonial subtypes, precluding a clear separation (41). The progenitor A_{dark} cells are considered to be the true spermatogonial stem cells or reserve cells, with only limited mitotic division.

Several molecular markers, including OCT3/4, GFRA1, PLZF, CDH1 (E-cadherin), ID4, RET, and NEUROG3, are positive in the more undifferentiated A spermatogonia. CD117 (KIT) first appears in the later A spermatogonia forms (42–48). The MAGEA4 protein is a general marker for all spermatogonia, whereas UCHL1, GFRA1, and UTF1 are heterogeneously expressed by undifferentiated spermatogonia, and CD117 is a marker for differentiated spermatogonia (49). Several other markers including SOX3, STRA8, SOHLH1, and SOHLH2 are expressed at different stages of differentiation (50). These progenitor cells are located in the spermatogonial stem cell niche. This niche is poorly defined in the testis because of the unavailability of a clear-cut marker of spermatogonial stem cells, but it is found in close relationship to the basement membrane of the tubule as well as the Leydig cells and vasculature of the interstitium (51). Some spermatogonial stem cells become committed to the process of spermatogenesis. These cells undergo further division while maintaining a narrow intercellular bridge connection among all of the offspring. This bridge allows for simultaneous maturation of interconnected cells. Of interest is the lack of intercellular connections in seminomas and germ cell neoplasia in situ, but their occasional presence in spermatocytic tumors (52). Extracellular vesicles discharged from testicular cells have been isolated and characterized, shedding light on their contribution to the maintenance of regular spermatogenesis and steroidogenesis within the testis. Within the seminiferous tubules, extracellular vesicles originating from Sertoli cells are found to facilitate the differentiation of spermatogonial stem cells, while extracellular vesicles released by undifferentiated A spermatogonia are observed to impede the proliferation of spermatogonial stem cells (53). Seminal plasma is abundant in extracellular vesicles, which are sourced from various glands including the testes, epididymis, prostate, seminal vesicles, and bulbourethral and periurethral glands. This diverse origin suggests that extracellular vesicles present in seminal plasma are derived from these glandular secretions. These extracellular vesicles are pivotal in protecting sperm and influencing key sperm attributes such as motility, morphology, capacitation, and the acrosome reaction (54). The extracellular vesicles harbor numerous proteins and miRNAs, the discussion of which extends beyond the scope of this book. The pattern of division of A_{pale} spermatogonia is still uncertain. Some of them proliferate to produce more type A_{pale} cells while others produce type B spermatogonia, which soon thereafter convert to the preleptotene form of primary spermatocytes. In humans, it appears that eight preleptotene spermatocytes are produced per each original pair of A_{pale} spermatogonia (38). The preleptotene spermatocytes then move into the adluminal compartment and start the process of the first meiotic division. How these interconnected cells move from one compartment to the other is still unclear.

Early studies divided the seminiferous epithelium cycle into six stages based on morphologic evaluation to understand the process of spermatogenesis. However, it is practically impossible to assign stages by light microscopy due to poor preservation and the frequent presence of multiple stages being represented in any given section of the tubules. High-resolution light microscopy has since improved germ cell characterization (55).

The classification of primary spermatocytes is based on altered nuclear chromatin patterns (41,55). These cells exhibit doubling of DNA content within their nuclei due to replication of each chromosome into chromatid pairs in anticipation of the initial meiotic division. The chromatin pattern of leptotene primary spermatocytes features a filamentous structure with a fine-beaded arrangement. Zygotene spermatocytes have an even coarser granularity of the chromatin filaments, with a tendency for the chromatin to gather eccentrically in the nucleus. Pachytene and diplotene spermatocytes are the most easily recognized because of their large size and prominent nuclei; they contain thick, short chromatin filaments (Fig. 36.8). The primary spermatocyte phase occupies a period of 24 days (56). After this relatively long period of gametogenesis, the first meiotic division occurs, with the formation of secondary spermatocytes.

Secondary spermatocytes have an extremely short half-life and make up only a small minority of the cells seen in a cross-section of the tubule. Their nuclei are substantially smaller than those of the primary spermatocytes and have a finely granular chromatin pattern with a haploid number of chromosomes, but a diploid amount of chromatin due to the presence of the chromatid pairs. They are located near the tubule lumen, differing only slightly in appearance from the very early spermatids with which they are closely associated (Figs. 36.4 and 36.8). They undergo the second meiotic division to produce spermatids. The spermatids are classified based on morphologic appearance with particular emphasis on the nuclear and body shape and the development of the acrosome. The 6-type classification of Heller and Clermont, with designations of Sa, Sb1, Sb2, Sc, Sd1, and Sd2, is most commonly employed (57). The earlier stages possess a spherical nucleus akin to secondary spermatocytes, albeit with a haploid chromosome count and half the usual DNA content of a normal cell (Figs. 36.7 and 36.8). The late spermatid forms are characterized by a change in the nuclear shape to an ovoid contour followed by an elongated appearance and a marked condensation of the chromatin. At the same time, excess cytoplasm is discarded by the spermatid and is phagocytosed by the Sertoli cell. Soon thereafter, the Sertoli cell–germ cell connections are removed and the intercellular bridges that connected earlier germ cell forms are dissolved, allowing the disconnected spermatozoa to enter the tubule lumen. Abnormal sloughing of immature germ cells, often with persistent intercellular connections, may be seen in patients with varicocele or other pathologic states, suggesting a failure of Sertoli cells to regulate this maturation process. Failed maturation may manifest with multiheaded spermatozoa in the seminal fluid. One must take care to separate this pathologic sloughing process from the artificial sloughing that frequently occurs in open biopsy specimens (Fig. 36.9).

Maturation of germ cells proceeds in an ordered, nonrandom fashion along the length of the seminiferous tubule. Groups of evolving germ cells of one level of development tend to be found in association with developing germ cells of another level of development at any point along the tubule. Clermont (41) described 6 such cell associations in the human testis, with 2 to 4 cell associations being present in a single tubule cross-section (38). It is important to

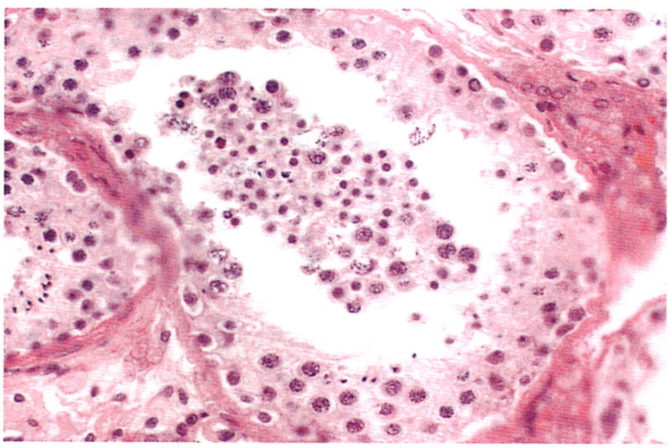

FIGURE 36.9 Seminiferous tubule, with artifactual sloughing of germ cell elements into tubule lumen.

recognize that not all stages of germ cell differentiation may be seen in any one cross-sectional view of a human seminiferous tubule. Mature spermatozoa and late spermatids may be seen in one portion while the opposite wall shows maturation only to the early spermatid level (Fig. 36.3). Furthermore, not all spermatogonia progress to become spermatozoa; apoptotic or degenerative changes in these precursor cells can be seen regularly in the seminiferous tubules (58). This normal physiologic process should not be mistaken for maturation arrest. The broad category of germ cell elements should be recognized with relative ease, and one should be able to distinguish spermatogonia, primary spermatocytes, secondary spermatocytes, and spermatids, using standard fixation/staining methods.

Advanced techniques have been developed to quantitatively evaluate germ cell maturation and to elucidate the correlation between spermatogenesis and sperm density in seminal fluid (59–62). One study proposed a semiquantitative method of analysis that includes uniformity of histologic features, tubule diameter, presence or absence of a lumen, evaluation of maturity of Sertoli cells, presence or absence of germ cell elements, degree of hypospermatogenesis, and the presence of germ cell maturation arrest and stage of arrest. The presence of late spermatids, even in a limited number of tubules, precludes the diagnosis of "germ cell arrest" (63).

Two alternative methods are helpful to the surgical pathologist. The first method (64) involves establishing a germ cell to Sertoli cell ratio by counting at least 30 tubule cross-sections. This ratio is approximately 13:1 in young healthy men. An average of 10 to 12 Sertoli cells per tubule cross-section is considered normal and approximately half of the germ cell elements within the tubule should be in the spermatid phase. An assumption is made that the Sertoli cell population is stable throughout adult life. A reasonably good assessment of the presence or absence of hypospermatogenesis or maturation arrest can be made with this technique. A second method involves counting spermatids per tubule cross-section (65). Only the more mature spermatids with ovoid or elongated nuclei and densely stained chromatin are counted. Excellent correlations have been made with seminal fluid sperm counts. A spermatid/tubule cross-section count of 45 should correspond to a seminal fluid sperm count of 85×10^6/mL. Average spermatid/tubule cross-section counts of 40, 20, or 6 to 10 correspond to sperm counts of 45, 10, or 3×10^6/mL, respectively. A minimum of 20 tubules must be counted.

INTERSTITIUM

The interstitium of the testis comprises 25% to 30% of the total testicular mass. This region can be broadly categorized into interstitial and peritubular regions. Complex cellular and molecular interactions of the interstitium provide a suitable milieu to guarantee a step-wise differentiation of spermatogonial stem cells through the process of spermatogenesis and spermiogenesis. Interstitial and peritubular cells play a pivotal role in regulating the differentiation of spermatogonial stem cells, thus maintaining and promoting male fertility (66). Each seminiferous tubule is surrounded by the lamina (tunica) propria in a sheath-like fashion; the lamina propria consists of an inner basement membrane surrounded by a thin zone of multilayered peritubular myoid cells intermingled with type 1 collagen fibrils and elastic fibers (Figs. 36.8 and 36.10). The outermost cells in this layer stain for vimentin, calponin, CD34, and actin, whereas the cells of the inner layer stain also for desmin (67). In addition to providing structural support, their contractile function plays an important role in compressing the seminiferous tubules and moving spermatozoa into the rete testis. These cells also contain androgen receptors in their nuclei (19). Androgen receptors play an important role in modulating Sertoli cell functions in both the fetus and adult (68,69). There is accumulating evidence that peritubular myoid cells contribute to self-renewal of spermatogonial stem cells through androgen and glial cell line–derived neurotrophic factor (GDNF) signaling pathways (70,71).

The basement membrane is composed of type 4 collagen fibrils intermixed with laminin, fibronectin, entactin, and heparan sulfate proteoglycans. This zone plays an important role in the function of Sertoli cells, the bases of which are apposed to the basement membrane. This zone also influences Leydig cell proliferation and testosterone production (72). Within the interstitial region are Leydig cells, blood vessels, lymphatic vessels, nerves, mast cells, and macrophages. The latter are often found in close association with the Leydig cells (73), where the two cells form complex cell–cell interactions. A recently identified cell type known as telocyte has been documented in vertebrates. Telocytes

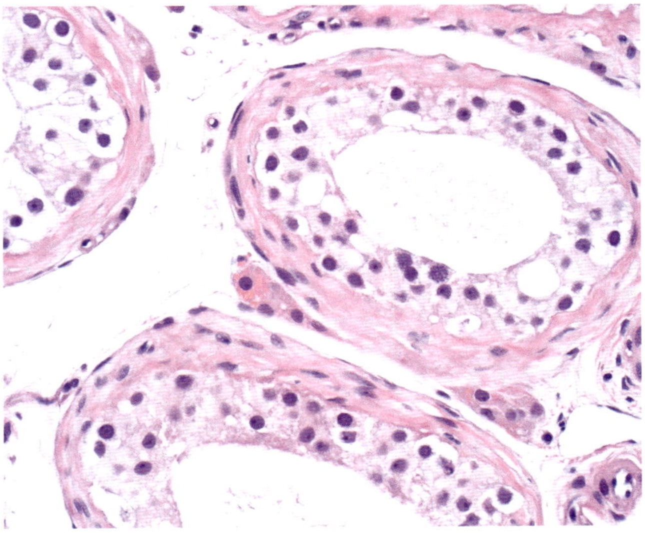

FIGURE 36.10 Seminiferous tubules with multiple layers of peritubular myoid cells. The interstitium comprise loose connective tissue with scattered Leydig cells, small vessels, and fibroblasts.

have elongated projections that allow both homo- and heterocellular junctions, thereby forming a three-dimensional network crucial for maintaining interstitial homeostasis. Telocytes presumably contribute to various physiologic functions including extracellular matrix organization and activity, structural support, paracrine and juxtacrine communication, immunomodulation, immune surveillance, cell survival, and apoptosis (74). Cytokines and reactive oxygen species from macrophages influence steroidogenesis of the Leydig cells (75) and also play a role in the function of the peritubular myoid cells. A direct interaction with spermatogonial stem cells via CSF1 and retinoic acid pathways has been proposed (76). Mast cells also have an influence on these structures (68). Elastic fibers first appear at puberty in the outermost layer of the lamina propria (77). There is a striking absence of elastic fibers in the lamina propria of the sclerotic tubules in patients with Klinefelter syndrome, in contrast to their abundance in patients with postpubertal sclerosis of other causes (78). Extracellular vesicles have been detected in diverse cell types within the testicular interstitium, such as endothelial cells, macrophages, telocytes, and Leydig cells. Nevertheless, their specific functions remain unclear in contrast to the intratubular environment, where certain novel insights have been noted (53).

A common finding in patients with oligospermia or azoospermia due to primary testicular failure is the accumulation of eosinophilic, acellular material in the lamina propria. This material is an admixture of increased collagen fibers, elastic fibrils, and basement membrane–like material (62). The peritubular tissue of patients with hypogonadotropic hypogonadism is underdeveloped, having only one or two layers of myoid cells mixed with a few collagen fibers. In contrast, there is a large accumulation of collagen in the lamina propria of some of the tubules of adult patients with cryptorchidism.

LEYDIG CELLS

Adult Leydig cells, the source of testicular androgens and insulin-like factor 3 (INSL3), only rarely undergo mitotic division (79,80). They are found singly and in clusters within the interstitium of the testis, some lying immediately adjacent to capillaries and others being located next to the peritubular fibromyocytes (Fig. 36.11). They may be seen in the lumen of sclerotic seminiferous tubules and in the tunica albuginea, the epididymis, the spermatic cord, and the hilum of the testis (81). They are often intimately associated with nerve fibers, sometimes as a large cluster adjacent to the nerve (Fig. 36.12), but more frequently they are scattered randomly throughout the nerve fiber (78,82). They may be found at these sites in the fetus as well as the adult. The single nucleus of the cell is round and vesicular with one or two eccentrically located nucleoli. An occasional binucleated cell may be seen. The nuclei may exhibit a ground glass appearance (Fig. 36.13). The cytoplasm is

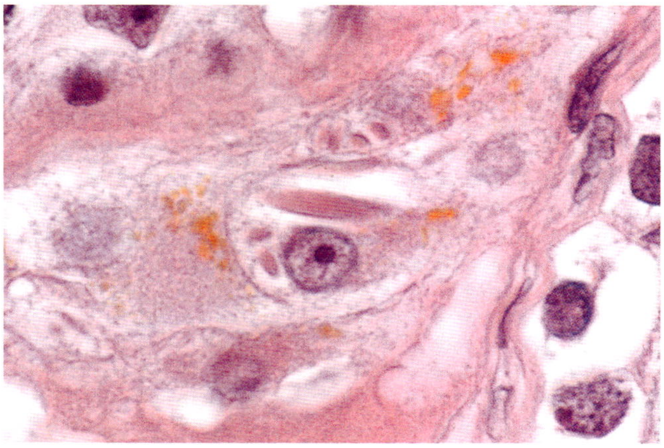

FIGURE 36.11 Leydig cells in interstitium of testis. An eosinophilic crystal of Reinke and abundant lipofuscin are prominent features in the cytoplasm of the Leydig cell in the center of this field.

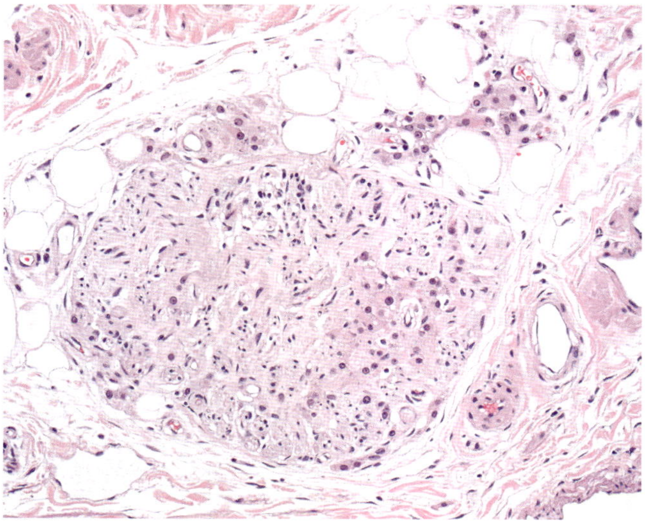

FIGURE 36.12 Leydig cells in intimate association with a nerve in the hilus of the testis.

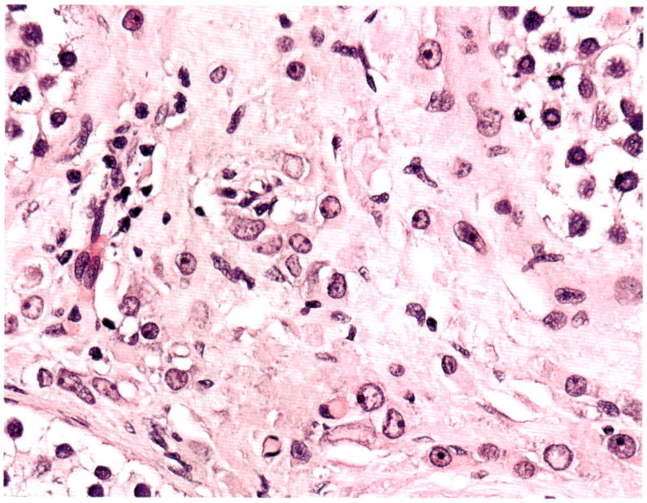

FIGURE 36.13 Leydig cells with ground glass nuclei.

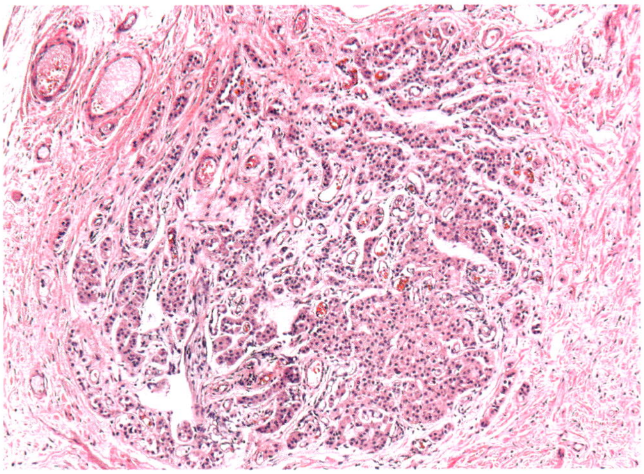

FIGURE 36.14 A nodular steroid cell nest in hilar soft tissue. These cells are different from Leydig cell and adrenal cortical cells.

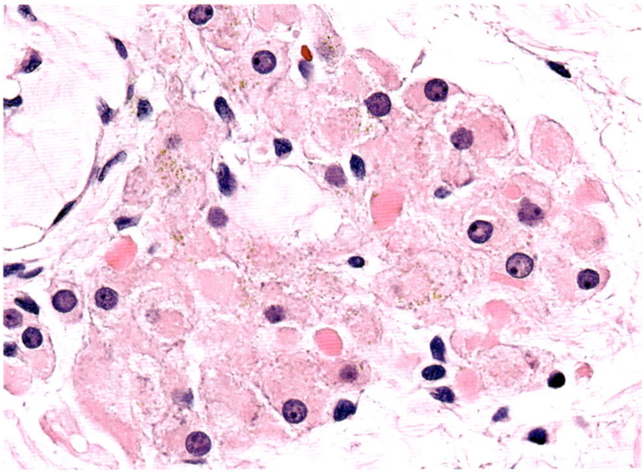

FIGURE 36.16 Leydig cells in Leydig cell tumor with Reinke crystals and globular eosinophilic intracellular material. Abundant lipofuscin is present.

usually abundant and eosinophilic. Lipid droplets and lipofuscin pigment are found in the cytoplasm, first appearing at the time of puberty and increasing in prominence in the aging testis. Sometimes the lipid accumulation is so intense that confusion of these cells with ectopic adrenal cortical cells may occur. Similar appearing cells occur in clusters designated as nodular steroid cell nests in close association with the rete testis. These cells form nonencapsulated aggregates with a sinusoidal and trabecular appearance. Although they are steroidogenic in nature and resemble Leydig cells (but lack Reinke crystals), their properties are more similar to those of adrenal cortical cells. However, they do not exhibit the compact arrangement and zonation characteristic of typical adrenal cortical nests, which are usually found in the spermatic cord rather than in the hilum (Fig. 36.14).

The characteristic Reinke crystals of the Leydig cells are present only in the postpubertal state (Fig. 36.11). Their presence is highly variable in the normal testis, and they are frequently absent or difficult to find in Leydig cell tumors, being present in approximately 30% of cases (Fig. 36.15). Crystals are often associated with globular eosinophilic intracellular material thought to be their precursors (Fig. 36.16). The nature of the material is unknown but is presumed to represent a protein product of the cell (83). The crystals stain positively with periodic acid–Schiff (PAS) and trichrome stains (red), and are immunohistochemically negative for actin, vimentin, and desmin (Fig. 36.17).

Leydig cells produce testosterone after stimulation by luteinizing hormone (LH) through interaction with LH receptors on the cell surface. This is accomplished by upregulating the expression of the steroidogenic enzyme 17-beta-hydroxysteroid dehydrogenase. Testosterone interacts with androgen receptors inducing local effects or exerting its effect by binding to androgen binding protein (ABP) and elevating testosterone levels in seminiferous tubules and epididymis (84). Androgen receptors are present in Sertoli cells, peritubular myoid cells, Leydig cells, and spermatids. Testosterone also exerts its influence by an indirect

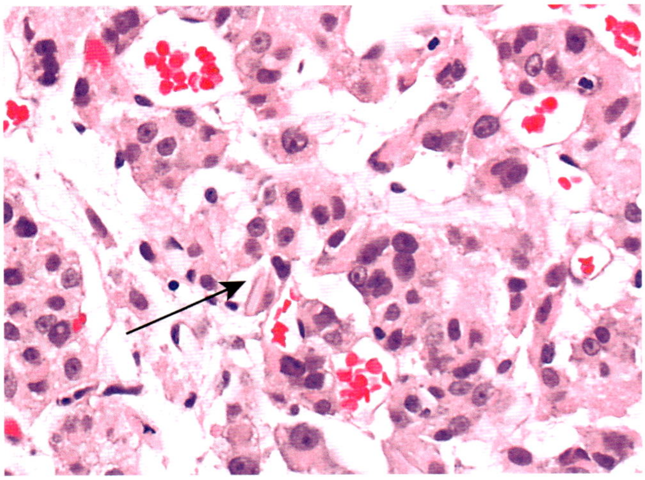

FIGURE 36.15 Leydig cell tumor with abundant Reinke crystals (*arrow*).

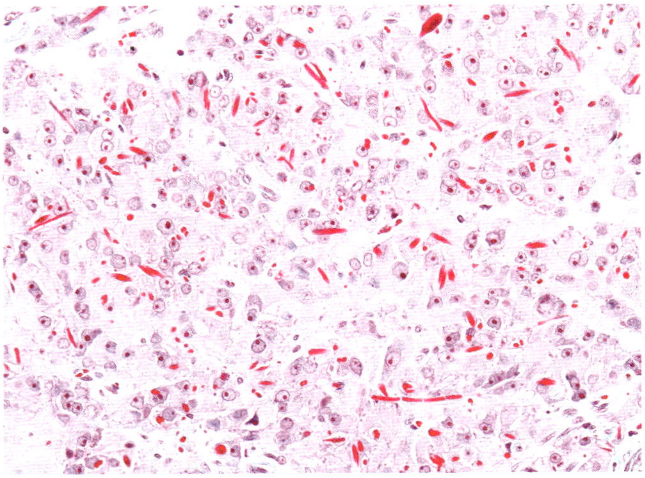

FIGURE 36.17 Trichrome display intense red staining of Reinke crystals in a Leydig cell tumor.

Sertoli cell–regulated attachment process and through peritubular myoid cells secreted factors (85,86). The roles of SF-1 steroidogenesis, spermatogenesis, and male fertility have been elucidated recently (87).

Quantitation of Leydig cells is difficult. Heller et al. (88) used a Leydig cell to Sertoli cell ratio to define a normal value of 0.39 and a range from 0.19 to 0.72 based on an assumption that the Sertoli cell population in the adult testis is stable, which was later verified by others (89). As a rule, normal adult testis tissue should have four or five Leydig cells for each tubule cross-section. The mean number of Leydig cells per one cross-section of seminiferous tubule is defined as the Leydig cell index. In both oligospermic and azoospermic groups, the Leydig cell index is significantly elevated compared to normal individuals, suggesting Leydig cell hyperplasia is a feature of the infertile testis (90). Leydig cell micronodules/hyperplasia are common in men with impaired spermatogenesis and testicular dysgenesis syndrome, and are associated with a decreased testosterone/LH ratio (91,92).

Vimentin is the predominant intermediate filament in Leydig cells, but actin filaments and neurofilament triplet proteins have been identified in both Leydig cells and Sertoli cells (24). Antibodies to inhibin α and calretinin stain Leydig cells intensely. Calretinin is also present in mesothelial cells and cells of the rete testis but not in Sertoli cells (93). INSL3 is a very specific marker for functioning fetal and adult types of Leydig cells (94). Leydig cells also stain positively for SF-1, S-100 protein, glial fibrillary acidic protein, synaptophysin, chromogranin A and B, CD99, Melan A, colony-stimulating factor 1 (CSF1) and neuron-specific enolase (95–98). The astrocyte-like markers in Leydig cells suggest a role similar to that of central nervous system astrocytes in inducing and maintaining a barrier feature of testicular capillaries (99). A subpopulation of cells stain intensely for nestin, an intermediate filament seen mainly in nerve and muscle progenitor cells (100). The presence of all of these substances undoubtedly indicates important paracrine functions that Leydig cells share with Sertoli cells, endothelial cells, peritubular cells, macrophages, and nerves.

VASCULAR SUPPLY

The blood supply of the testis is derived primarily from the internal spermatic (testicular) artery. A small contribution comes from the deferential artery, a branch of the inferior vesicle artery, and from the cremasteric artery, a branch of the inferior epigastric artery (101). Anastomoses between the testicular artery and these other two vessels occur regularly prior to the entrance of the testicular artery into the testis proper (102). A few anastomoses may be seen between the parenchymal branches of the testicular artery and the deferential artery. Each testicular artery, arising from the aorta immediately distal to the renal artery, is highly coiled and extremely long relative to its diameter. The vessel has a low pulse pressure as it enters the testis (103). The artery plays an important role in thermal regulation of the testis via countercurrent heat exchange with the veins of the pampiniform plexus. The combination of this vascular heat exchange and heat lost via the thin scrotal layer serves to maintain the testicular temperature 2 to 3 °C below body temperature. The testicular artery enters the testis posteriorly, where it localizes to the tunica vasculosa. It then courses to the inferior pole, and subsequently turns back superiorly along the anterior surface, sending branches in a centripetal fashion into the septa and then to the hilum, where they form a dense cluster. Only a few branches of these centripetal arteries enter the lobules. The small arterial segments, called recurrent arteries, pass from the hilum into the parenchyma, where they branch into arterioles and then capillaries. Each recurrent artery supplies blood to only a single lobule, as determined by the septa (104). The veins run either centripetally or centrifugally to the hilum or capsule respectively and eventually anastomose to form the pampiniform plexus of the testicular vein. Biopsy specimens of the testis from patients with varicoceles often show a striking sclerosis of vascular walls of both arteries and veins (105). The significance of these vascular alterations with respect to seminiferous tubule function in these patients is uncertain.

There is extensive development of the intratesticular microvasculature at puberty, the most notable feature being a marked coiling of the arteries and a great expansion of the peritubular capillary network. Unlike other capillaries of the body, the testicular capillary walls have a prominent basement membrane and an incomplete outer layer of pericytes (106).

The capillary network appears to have a very structured arrangement with respect to the Leydig cells and the seminiferous tubules. The capillaries arising from arterioles first intermingle with Leydig cells (arterial side inter-Leydig cell capillary). They then extend to the lamina propria of the seminiferous tubules (intramural capillary) and subsequently return to the interstitium, where they are again surrounded by Leydig cells (venous side inter-Leydig cell capillary). At one point in their journey through the tunica propria, the capillary basement membrane comes in close contact with the basement membrane of the seminiferous tubule. This site is the only location where the capillaries have defined fenestrations (107), thereby allowing exchange of a variety of substances between the microvasculature lumen and the seminiferous tubule. The capillaries in the testis have some similarities to those in the brain, supporting some type of biologic barriers. Both of these capillary beds have a high density of the GLUT-1 isoform of the glucose transporter and immunoreactivity for P-glycoprotein. Unlike the nervous tissue vascular bed, localized fenestrations can be found in the vessels of the testis (99). Vascular and perivascular cells play an important role in stem cell development as well as localization in the testis (108). VEGFA secreted by Leydig and Sertoli cells is crucial for endothelial proliferation, survival, permeability, and migration. It plays an important role in testis-specific vascular remodeling during fetal development (109).

There is a remarkable species-to-species variation in the distribution of lymphatic channels in the testis (110). In humans, there are only rare lymphatic vessels in the interstitium and a peritubular network is lacking. The lymphatic vessels drain into the septa and thence to either the capsule or the hilum and subsequently join on the posterior aspect of the testis. They then anastomose with lymphatic channels of the epididymis, enter the spermatic cord, and drain into the retroperitoneal lymph nodes.

FETAL AND PREPUBERTAL TESTIS

The fetal testis first becomes recognizable at 7 to 8 weeks of gestation, primarily through the influence of the *SRY* gene on the Y chromosome, which initiates a process by which the primitive cells derived from the coelomic surface that are destined to become either Sertoli cells or granulosa cells are directed toward the former. Clusters of Sertoli cells begin to form recognizable cords into which the germ cells migrate. The latter are derived from extraembryonic epiblast cells in the posterior wall of the yolk sac adjacent to the base of the allantois during the third week of gestation (111). These primordial germ cells then migrate into the developing hindgut and subsequently through the dorsal root mesentery during the 4th and 5th weeks to reach the gonadal ridge, where they are now called gonocytes. Signaling via CD117 and stem cell factor (SCF) interactions plays a key role in this migration (112). The gonocyte numbers increase to 1,000 cells by the time they reach the gonadal ridge and up to 30,000 cells by the 9th week of gestation (113). The vast majority of these cells develop in a unipotent direction toward the production of spermatozoa, although small numbers may remain pluripotent and potentially become precursors of embryonal germ cells in culture (114).

Early in gestation, the gonocytes are mitotically active and have a round nucleus, a prominent nucleolus, and relatively little cytoplasm. They are located in basilar and suprabasilar portions of the seminiferous tubules and are surrounded by the immature Sertoli cells (Figs. 36.18 and 36.19). During the latter half of fetal life and the first 6 months after birth, they undergo a maturation process in which they enter mitotic arrest, although a few Ki-67 positive cells may still be present in early postnatal life. They become larger, acquire more cytoplasm and a coarser nuclear chromatin pattern, and are now referred to as primary spermatogonia (Fig. 36.20). It should be noted that at any time after about 20 weeks of gestation, there may be a mixture of gonocytes in different stages of maturation along with primary spermatogonia. These maturing gonocytes are called "prospermatogonia" (115), "fetal spermatogonia" (116), "intermediate cells," or "prespermatogonia" (117). By the 6th postnatal month, virtually all of the germ cells are spermatogonia.

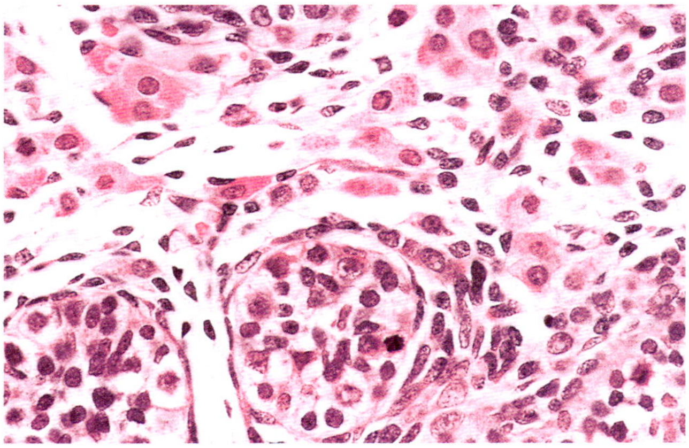

FIGURE 36.18 Fetal testis of 20 weeks of gestation, with numerous Leydig cells throughout the interstitium.

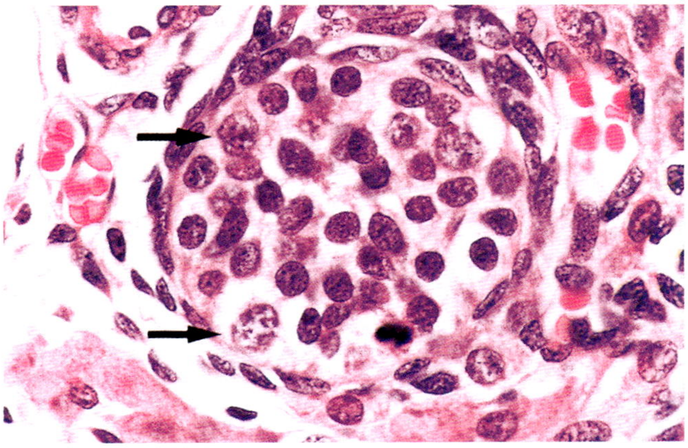

FIGURE 36.19 Same testis as shown in Figure 36.17. Note the mitotic figure, probably of a Sertoli cell. Larger cells (*arrows*) are undifferentiated germ cells. The remaining cells are immature Sertoli cells.

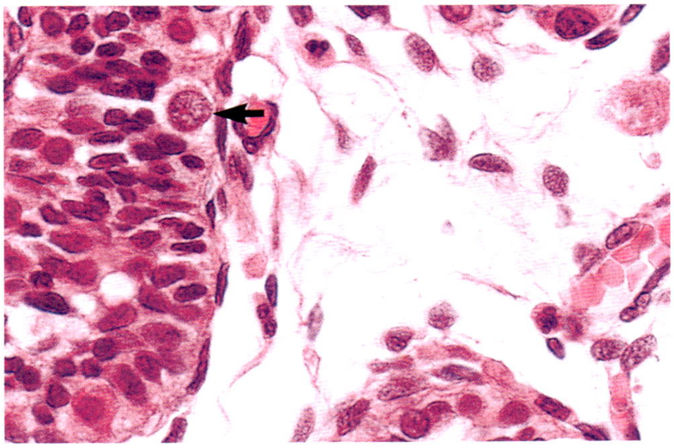

FIGURE 36.20 Testis of an 11-month-old child. A spermatogonium is presented adjacent to the basement membrane (*arrow*). The interstitium contains undifferentiated spindle cells.

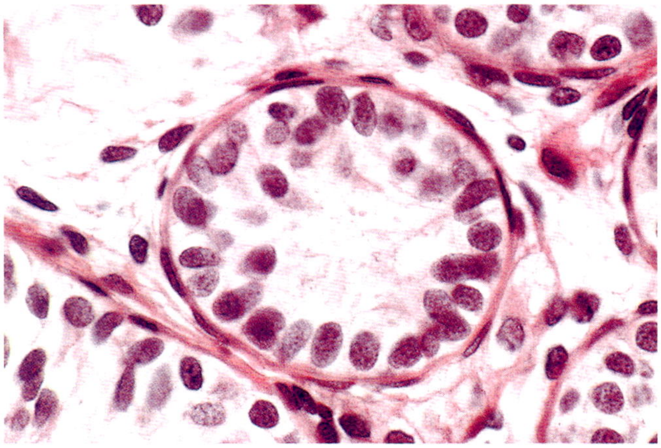

FIGURE 36.21 Thirteen-year-old prepubertal boy with bilateral cryptorchidism. Mature Leydig cells are absent in the interstitium. The tubules lack a lumen. The Sertoli cells are immature, and germ cells are rare.

This transition from the fetal stem cell pool to the adult stem cell pool is an important first step in the maturation process and appears to be defective in patients with the testicular dysgenesis syndrome and who subsequently later in life are prone to develop germ cell neoplasia (118). At birth, the germ cells average two to four per tubule cross-section. From the age of 1 to 4 years, the number of germ cells average one or two per tubule cross-section and this number doubles between the ages of 5 and 8 years (119).

Early in puberty, repeated waves of incomplete spermatogenesis take place, with an orderly process of complete maturation not appearing until the end of puberty (120). Several studies have demonstrated a marked diminution in the number of germ cells in the undescended, prepubertal testis (Fig. 36.21), and a less predictable decrease of germ cells in the truly ectopic testis (39,121). Early primordial germ cells undergo striking epigenetic reprogramming during their migration, including extensive DNA demethylation, histone modifications, RNA-associated gene expression silencing, and erasure of parental imprinted genes. Epigenetic remethylation and reestablishment of imprinted genes in a sex-specific manner occurs in an asynchronous manner in subsequent stages of gametogenesis. Although the exact timing of imprint reacquisition is still not known in humans, it generally occurs in males before meiosis begins. However, other epigenetic reprogramming events may occur in sperm cells released into the epididymis (122–124). This pattern of demethylation and loss of imprinting is also demonstrated in many of the germ cell neoplasms derived from these gonocytes (125). Epigenetic perturbations have also been demonstrated in spermatozoa of men with subfertility (126). Primordial germ cells and gonocytes acquire a number of markers that are useful in their identification and in the evaluation of germ cell neoplasms derived from these cells. These include the membranous placenta–like alkaline phosphatase (PLAP) and CD117 and the nuclear markers: OCT3/4, NANOG, SALL4, SOX2, SOX17, MAGE-A4, and VASA. PLAP was one of the earliest markers employed to identify early migrating primordial germ cells. Germ cells in the first trimester of gestation are PLAP positive but by 19 weeks of gestation only a few cells remain positive. In recent years, PLAP has been largely replaced by CD117, NANOG, and OCT3/4 as markers of gonocytes. CD117, located on the cell membrane and within the cytoplasm of gonocytes, is a tyrosine kinase receptor for the ligand SCF. The latter is normally present in Sertoli cells. CD117 appears early in the migrating germ cells and tends to persist somewhat longer than the other markers; although it is usually not easily detectable in adult testes using older immunohistochemical methodologies, newer, more sensitive techniques show CD117 in spermatogonia in many cases (127).

OCT3/4 (also known as POU5F1), SALL4, and NANOG (a homeobox domain protein, regulating self-renewal) tend to parallel each other in their staining patterns. They are located in the nucleus and stain both primordial germ cells and gonocytes. At 7 weeks postfertilization, twice as many germ cells stain positive for CD117 as there are staining for NANOG or OCT3/4. By 8 to 10 weeks, there are approximately equal numbers positive for these markers (Fig. 36.22), but by week 15, the number staining for CD117 is 16 times higher than for OCT3/4 and NANOG. At birth, only rare OCT3/4- or NANOG-positive cells can be found (113). By the 4th week after birth, NANOG is no longer detected but a rare cell staining positive for OCT3/4 may still be found at 6 months after birth (128). After the 12th postnatal month, none of the germ cells should be positive for OCT3/4. There is evidence that failure of switching off expression of OCT3/4 in gonocytes leads to coexpression of testis-specific Y-encoded protein in the spermatogonial niche. This coexpression and the concomitant expression of Kit ligand (SCF) by Sertoli cells induce neoplastic conversion of immature (delayed) gonocytes to germ cell neoplasia in situ (129–133). The retained expression of OCT3/4 beyond 12 months of age is

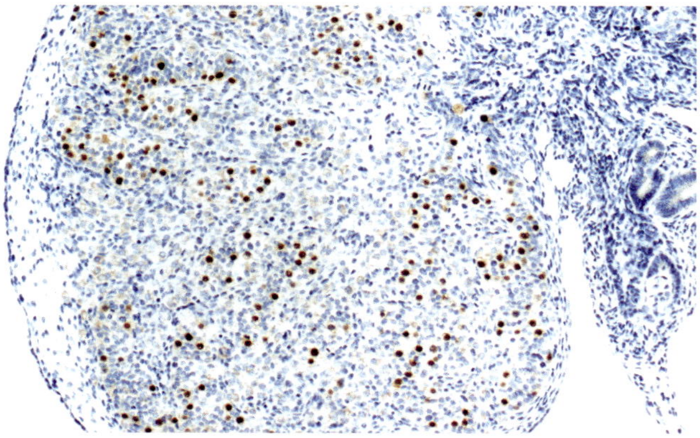

FIGURE 36.22 Fetal testis of 10 weeks of gestation. OCT 3/4 immunohistochemical stain of gonocytes. This section of testis would demonstrate an identical staining pattern of the gonocytes with CD117 (KIT).

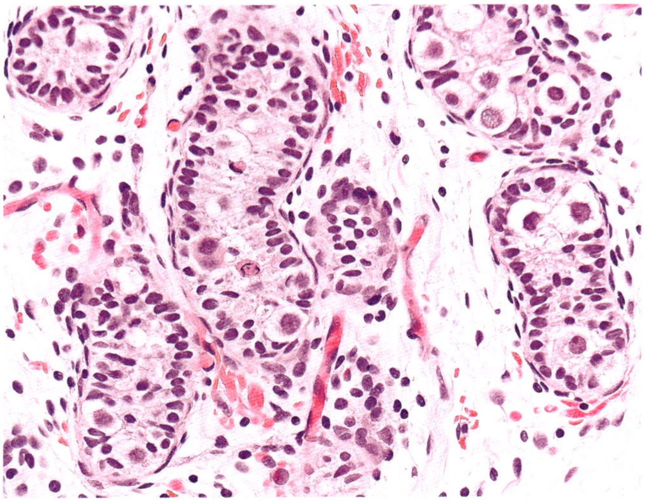

FIGURE 36.23 Delayed maturation; atypical germ cells are present within the tubules which immunohistochemically expressed OCT3/4.

considered maturation arrest and proclivity to develop germ cell neoplasia (Fig. 36.23).

VASA and MAGE-A4 stain more mature gonocytes as well as primary spermatogonia. Some costaining of gonocytes with MAGE-A4 and CD117 occurs (117) and some colocalization of VASA and OCT3/4 can be seen (134). CD30, somewhat useful in separating seminoma (−) and embryonal cell carcinoma (+), does not stain normal gonocytes (135). Corresponding to their mitotic activity and prior to their mitotic arrest soon after birth, gonocytes and spermatogonia reveal strong nuclear staining with Ki-67.

Immunohistochemical stains are critical for the diagnosis of germ cell tumors of the testis. Seminoma display expression of PLAP, c-Kit (CD117), OCT3/4, and D2-40. OCT3/4 is positive in both seminoma and embryonal carcinoma; however, CD30 expression in embryonal carcinoma is helpful in sorting out this diagnosis. Several sensitive and relatively specific markers have been useful for the identification of yolk sac tumor including alpha-fetoprotein (AFP), glypican 3, ZBTB16, and CDX2. Choriocarcinoma stains with several markers including human chorionic gonadotropin (hCG), human placental lactogen, inhibin and glypican 3 (in syncytiotrophoblasts), and GDF3, p63, and GATA3 (in cytotrophoblasts). SALL4 has emerged as a pan-germ cell marker; it is sensitive but not very specific. Cytokeratins may be variably positive in different germ cell tumors and should be carefully interpreted (136–138).

Leydig cells first appear in the fetus at 7 weeks of gestation in the first of a triphasic pattern of Leydig cell development and reach a peak level at 14 to 20 weeks of gestation. The earliest fetal Leydig cells are spindle-shaped and are gradually replaced by cells with round nuclei and abundant eosinophilic cytoplasm. By the 20th week of gestation, abundant, well-developed Leydig cells fill the interstitium (Fig. 36.13). The Leydig cells gradually undergo regression with only a few recognizable cells at birth. The first wave of Leydig cells may be derived from microvascular pericytes, epithelial components of the adjacent mesonephros, the coelomic epithelium, or even neural crest cells (139). There is strong support for a common origin of all the steroidogenic cells of the urogenital system from the adrenogenital primordium (140). Likewise, the mechanisms leading to their regression in late fetal life regression are unknown (141). The second wave of Leydig cell development occurs 2 to 3 months after birth, corresponding to the activation of the hypothalamus/pituitary/gonad axis (142). Equally unclear is whether this second population of Leydig cells arises from the same or from a separate mesenchymal precursor cell as the first wave of Leydig cells (143). These proliferations represent a mix of mature-appearing Leydig cells, spindle cells, and smaller cells with round nuclei and fairly prominent nucleoli but without the eosinophilic cytoplasm of mature Leydig cells. After this second phase of Leydig cell activity ends at about the 6th neonatal month, only immature spindle cells (Fig. 36.15) are recognized until just before puberty, when the third wave of adult Leydig cells begins to make its appearance.

Sertoli cells are the first cells to become recognizable in the fetal testis and they play a crucial role in the differentiation of the Leydig cells, suppression of entry into meiosis of the gonocytes, and production of AMH (11). The conventional viewpoint is that Sertoli cells are derived from thickened coelomic epithelium but, like the Leydig cell, some data suggest that Sertoli cells arise from steroidogenic progenitor cells (140). Fetal Sertoli cells outnumber germ cells in a ratio of 7:1 (144) and undergo active mitotic division during this period. The nucleus of the earliest fetal Sertoli cell is elliptical, but a mixture of round and ovoid nuclei is present later, and these nuclear shapes persist until puberty. They are in sharp contrast to the irregular tripartite shape and the highly folded nuclear membrane of the adult Sertoli cell.

The nucleoli of fetal and infantile Sertoli cells are inconspicuous in contrast to the prominent nucleoli of the adult forms. Inhibin-α and vimentin are reliable markers of both immature and mature Sertoli cells but are not specific. AMH is present in Sertoli cells from the 7th week of gestation until puberty but is not present in mature cells. Cytokeratin 18 is present in the first 20 weeks of gestation and completely absent thereafter. This latter marker is of interest because of its presence in some Sertoli cells in adults with testicular dysgenesis syndrome. In contrast, androgen receptor nuclear expression is a feature seen only in pubertal and adult Sertoli cells (11,145).

Sertoli cell nodules or congeries, frequently seen in cryptorchid but also in descended testes (Fig. 36.24) are composed of nonencapsulated immature Sertoli cell tubular nests, showing a prominent peripheral hyaline-like basement membrane as well as central basement membrane deposits. Laminated microliths are often present and occasional spermatogonia can frequently be found in the nodules. These Sertoli cells are proliferating and stain for AMH SOX9, only focally for cytokeratin 18, and variably for androgen receptor (145,146).

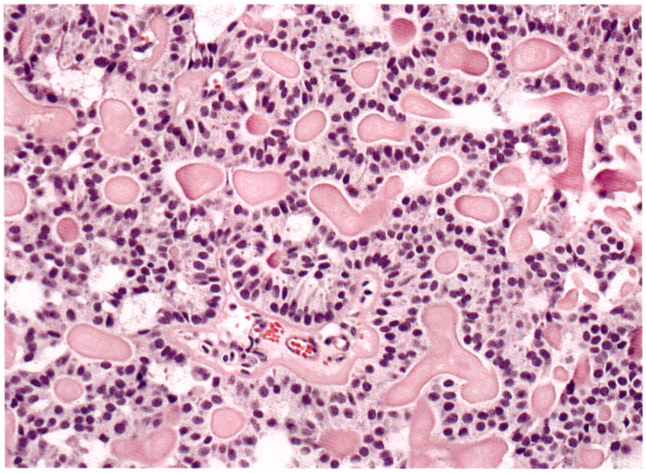

FIGURE 36.24 Sertoli cell nodule (or congeries). The tubules are composed of immature Sertoli cells, lack germ cells, and have prominent basement membrane deposits.

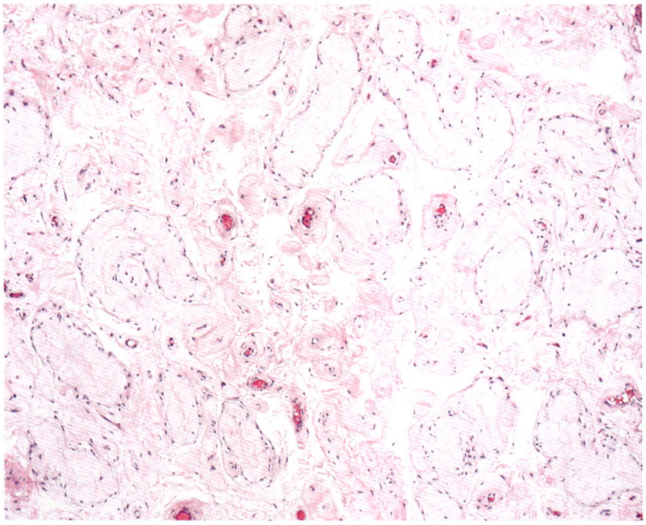

FIGURE 36.25 Area of seminiferous sclerosis with corresponding vascular hyalinization. Note absence of Leydig cells in interstitium.

Fetal Sertoli cells are mitotically active and increase in number through the first few months after birth. After the neonatal proliferative phase, the Sertoli cell population remains relatively stable until the onset of puberty when another increase in number takes place. Sertoli cells become mitotically inactive at the end of puberty and remain so throughout adult life. Stereologic studies indicate the Sertoli cell total number increases from 260 million late in fetal life to 1,500 million between 3 months and 10 years, and to 3,700 million in the adult testis (147). The testis shows a sixfold increase in size in the 1st year after birth, primarily as a result of Sertoli cell proliferation and the marked lengthening of the seminiferous tubules rather than a significant increase in tubule diameter (120). Sertoli cells average 30 per tubule cross-section in the fetal testis at 20 weeks of gestation, increase to 42 at the 4th postnatal month, and decrease thereafter to 26 at 13 years and 12 to 15 in the adult (39,148). At puberty, there is a fivefold increase in Sertoli cell volume and cytoplasmic Charcot–Böttcher crystalloids first appear. The seminiferous tubules at 20 weeks of fetal life measure 45 to 50 μm in diameter, are solidly filled with Sertoli and germ cells in a ratio of 7:1, and lack a well-defined lumen (148). The postnatal tubules slowly increase in size to reach a prepubertal diameter of 64 μm (range: 43 to 70 μm), at which time lumina begin to appear (144).

AGING TESTIS

There is a gradual and progressive decline in testicular volume with aging. Well et al. demonstrated a decrease in testicular volume from 16.5 cm^3 in males between 20 and 30 years to 14 cm^3 in men between 80 and 90 years (149). Histologic changes include a patchy pattern of hypospermatogenesis, peritubular fibrosis, and hyalinization of the seminiferous tubules, although an occasional sclerotic tubule may be found in an otherwise normal testis (150). A variable number of tubules will contain all elements of developing germ cells, including mature spermatozoa. Although the total number of peritubular cells is maintained in elderly men, there is a sharp decline in the proportion of cells that stain positively for desmin and actin (67).

Testicular arterial and arteriolar walls show thickening with hyalinization in older men; this feature is present in over 90% of testes in which there are large zones of tubular fibrosis (Fig. 36.25). The capillary bed in the aged testis becomes sparse and poorly organized (151). These vascular changes likely play a causal role in the peritubular sclerosis and tubular hyalinization (Fig. 36.25).

Substantial controversy exists regarding the Leydig cell population in the aging testis (152,153). Some authors suggest that Leydig cell numbers progressively decline from the early postpubertal state and throughout the life of the individual, with nearly 50% of that loss occurring in the first 30 years after puberty (154). The production of testosterone by Leydig cells is relatively maintained despite this loss, probably because of the large reserve of these cells in the adult testis. When the Leydig cell mass decreases below a threshold, daily sperm production does decline. The aged Leydig cell contains large amounts of lipofuscin pigment, numerous vacuoles, and increased numbers of Reinke crystals.

Abnormal sperm maturation, sloughing of germ cells into the tubule lumen, degeneration of germ cell elements, and Sertoli cell lipid accumulation with cytoplasmic vacuolization are frequent findings in the aged testis. Earlier studies suggested that the Sertoli cell population is stable throughout adulthood (155,156). However, Johnson et al. (59) found that men of ages 20 to 48 years had significantly more Sertoli cells per tubule cross-section than did men of 50 to 85 years of age, and that there was a relatively constant relationship between Sertoli cells and immature germ cells in both age groups. Sertoli cells in those tubules with hypospermatogenesis have increased amounts of vimentin and the reappearance of cytokeratins 8 and 18, suggesting that reversion of the intermediate filament pattern to that of the fetal/prepubertal testis is related to the alteration of the spermatogenic process (157).

TESTIS IN GENDER DYSPHORIA (GENDER AFFIRMATION)

Given the rising numbers of patients undergoing gender-affirming surgery, it becomes increasingly important for pathologists to discern histologic changes in these individuals (158). Various stages of maturation and regression are evident, despite all patients receiving hormonal therapy prior to surgery. The majority exhibit seminiferous tubules with thickened basement membranes, only a minority show full maturation, with no cases displaying maturation arrest. Sertoli cell–only profiles are occasionally noted. Of particular interest are easily discernible large, atypical germ cells characterized by clear cytoplasm and prominent nucleoli, along with occasional binucleated forms and mitotic figures (Fig. 36.26). Although resembling germ cell neoplasia in situ at low magnification, closer examination reveals regular nuclear borders and a lack of intensely hyperchromatic and stippled chromatin. We have identified germ cell neoplasia in situ in one individual who underwent gender-affirming surgery at our institution. Leydig cells often exhibit degenerative changes and diminished numbers. Scattered mast cells are present in the interstitium which may be edematous, fibrotic, or both. Immunohistochemistry for PLAP, CD117, and OCT4 may prove beneficial as rare instances of germ cell neoplasia in situ might be overlooked when relying solely on routine H&E staining. Rete testis, epididymal epithelial hyperplasia, and periepididymal muscular hyperplasia are observed (158).

RETE TESTIS

The rete testis, a complex network of channels situated at the hilum of the testis, collects the luminal contents originating from the seminiferous tubules (Fig. 36.1).

It comprises three segments: the septal part housing the tubuli rete, the mediastinal rete, and the extratesticular portion, alternatively referred to as the bullae retis (159,160). The tubuli rete are short tubules, 0.5 to 1.0 mm in length, that connect the two ends of the seminiferous tubule loop to the hilum testis. The terminal end of the seminiferous tubule usually consists only of Sertoli cells, forming an epithelial plug–like structure as it protrudes into the rete lumen (Fig. 36.27). There are approximately 1,500 openings of seminiferous tubules into the rete.

Some tubules directly enter the mediastinal rete without passing through intervening tubuli rete. The hilar rete is a cavernous network of interconnecting branching channels that exits from the testis to form several dilated, vesicular channels or antechamber-like structures called the bullae retis. These structures, measuring up to 3.0 mm in width, anastomose to form the ductuli efferentia. The epithelium of the rete is characterized by low columnar cells with frequent nuclear grooves, and featuring a luminal surface adorned with microvilli (Fig. 36.27). Each cell contains a single, central flagellum that is inconspicuous on light microscopic examination. The epithelium is supported by a relatively thick basal lamina, beneath which reside a small number of fibroblasts and myoid cells interspersed among collagen and elastic fibers. Traversing the hilum and the extratesticular rete are epithelium-covered columns or strands called chordae retis. These columns, often appearing as islands on a cross-section of the rete testis (Fig. 36.28), vary from 15 to 100 μm in length and 5 to 40 μm in thickness. They serve to connect opposing

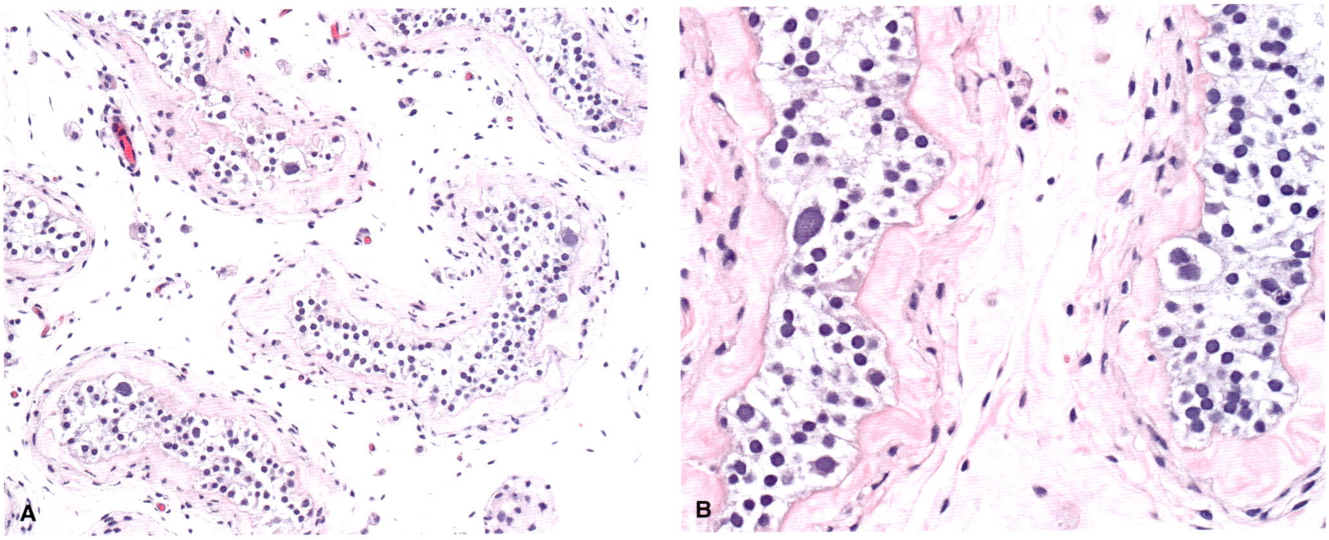

FIGURE 36.26 **A, B.** Gender dysphoria.

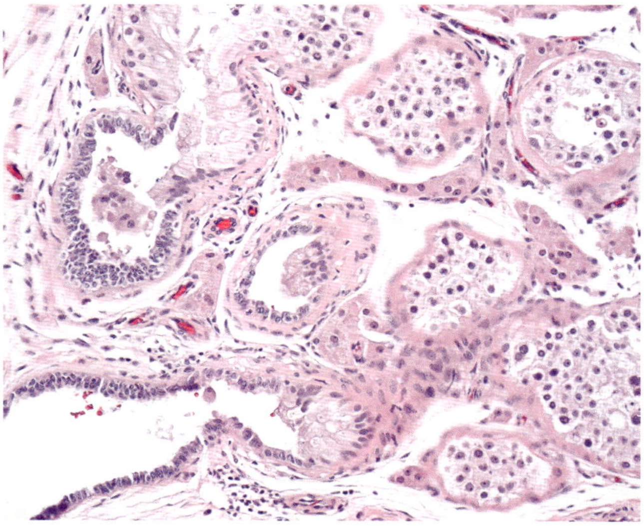

FIGURE 36.27 Junction of septal rete testis and terminal end of seminiferous tubule. Note the Sertoli cells "pouting" into the lumen of the rete. Rete epithelium is a low columnar type with frequent nuclear grooves.

walls of the chambers. The cytoplasm of the rete epithelium contains intermediate filaments composed of keratin and vimentin. Keratin is mainly situated in the apical part of the cell, while vimentin is predominantly found in the basal region. Keratins are detectable as early as the 10th week of fetal development, preceding the appearance of vimentin by 2 to 3 weeks (161). One would expect coexpression of these two intermediate filaments in the rare carcinoma of the rete or in hyperplasia of the rete. The benign rete testis can express androgen receptors, PAX2, PAX8, WT-1, CK7, and CK20, variable calretinin, and progesterone receptor; patchy expression of these markers can be seen in carcinomas of the rete testis (25,162). Calretinin, WT-1, and PAX-8 have also been reported in rete testis carcinoma

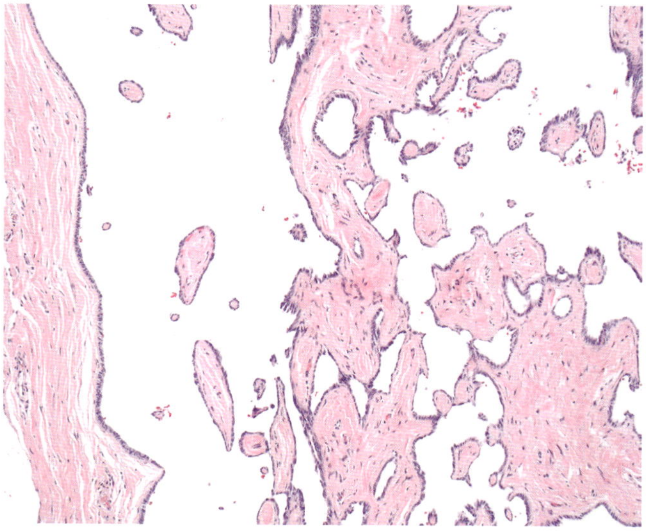

FIGURE 36.28 Rete testis, mediastinal portion, with irregular cavernous channels and cross sections of intratubular chordae (*arrows*).

(25,163). A report of nine cases of hyperplasia of the rete testis showed strong cytokeratin and EMA staining but a negative reaction for vimentin (164). Occasionally, hyaline, refractile, and eosinophilic globules, which stain positive with PAS, may be observed within the cells or lumen of the rete. These globules are sometimes positive for α1-antitrypsin but negative for α-fetoprotein (165). They are especially prominent when the rete undergoes hyperplasia, a reaction that is most frequently provoked by rete invasion by a germ cell tumor. They should not be confused with the globules produced by yolk sac tumors of the testis. The epithelium of the rete, along with that of the efferent ductules and the head of the epididymis, also possesses receptors for estrogen, progesterone, and androgen (166). The presence of estrogen receptors perhaps accounts for the hyperplasia of the rete and efferent ductules seen in patients undergoing sex-reversal procedures (167).

The rete fulfills several roles: (a) it acts as a mixing chamber for the contents of the seminiferous tubules, (b) it establishes a pressure gradient between the seminiferous tubules and the epididymis, (c) it potentially contributes unknown components to seminal fluid, and (d) it serves as a site for the reabsorption of proteins from the luminal content (168).

DUCTULI EFFERENTES

The ductuli efferentes consist of 10 to 12 tubules that arise from the extratesticular rete testis (Fig. 36.1) (169). They are involved primarily in resorption of fluid and do not appear to store spermatozoa for any length of time. These tubules aggregate to form a significant portion of the head of the epididymis proper (Fig. 36.29). Unlike the body of the epididymis, the lumina of the ductuli have an undulating border. The cells are composed of ciliated and nonciliated columnar cells, basal cells, and scattered intraepithelial lymphocytes, giving the epithelium a pseudostratified

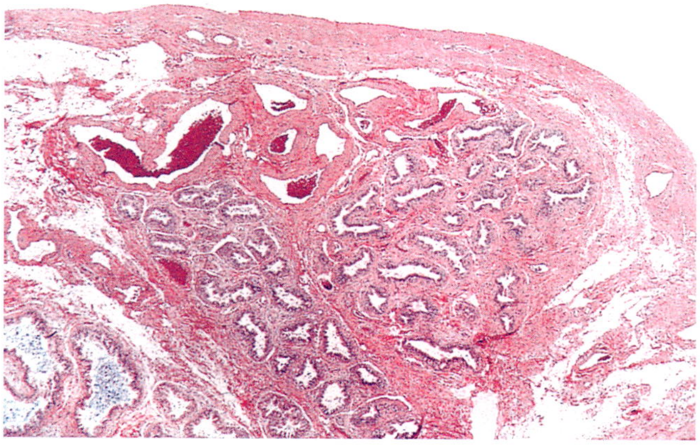

FIGURE 36.29 Head of epididymis, showing cross sections of distal portions of ductuli efferentia (**upper right**) and epididymis (**lower left**).

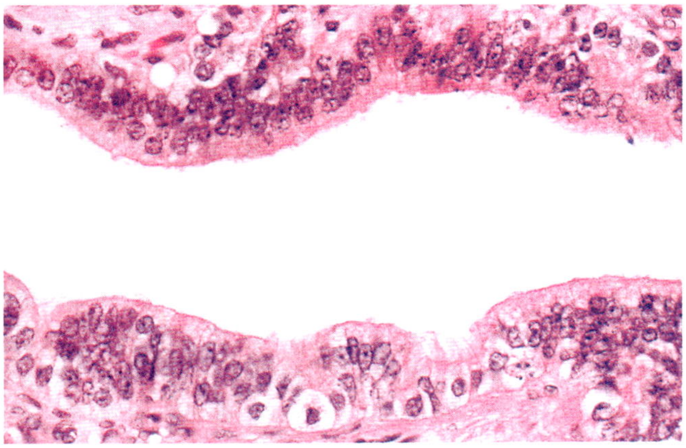

FIGURE 36.30 Epithelium of efferent ductulus. The columnar epithelial cells are mixed with basal cells and occasional intraepithelial lymphocytes, giving the epithelium a pseudostratified appearance.

appearance (Fig. 36.30). Occasional cells with a Paneth cell–like appearance are seen (Fig. 36.31), displaying abundant and intensely eosinophilic globules that are positive for PAS staining and resistant to diastase, while being negative for chromogranin A. They most likely represent prominent lysosomes, are less frequently seen in the rest of the epididymis, and are most often encountered in patients with epididymal obstruction (170). Rarely, similar findings can be seen in benign rete testis. Golden-brown lipofuscin pigment is often seen in the cytoplasm of the efferent ductules. The epithelium sits on a thick basement membrane, surrounding which is a coat of smooth muscle cells and fibroblasts, as well as a few scattered macrophages. Intraluminal macrophages that are actively phagocytizing spermatozoa are occasionally present, particularly when duct obstruction exists. Coexpression of low–molecular-weight cytokeratins, vimentin, and EMA is evident within the epithelial cytoplasm (161). Other markers including AR, ER, PR, PAX-8, CK7, and CK20 have been reported in the epithelium (25).

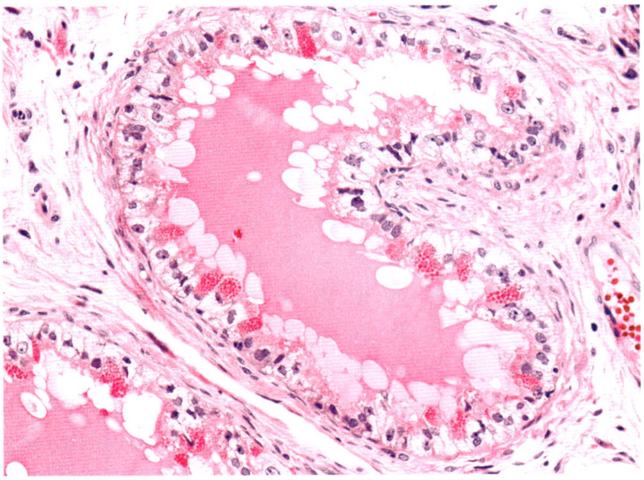

FIGURE 36.31 Epithelium of ductulus with prominent Paneth cell–like intracytoplasmic vacuoles.

EPIDIDYMIS

The epididymis, linking the ductuli efferentes to the vas deferens, is a tightly coiled tubular structure anatomically segmented into the head, body, and tail regions (Fig. 36.1). The epididymis plays an important role in sperm transport and sperm maturation, including the acquisition of motility, sperm concentration, and sperm storage. The average sperm transit time through the epididymis in humans is 12 days (171). The transport mechanism is by way of muscle contractions of the thick, muscular coat that surrounds the epididymal tubules. There is extensive reabsorption of intraluminal fluid, particularly in the head portion of the epididymis. Most of the sperm are stored in the tail segment until ejaculation occurs, and it is in this location in humans that final sperm maturation takes place (172). Many spermatozoa undergo senescence and degeneration in the tail via an unknown mechanism.

The epithelium of the epididymis consists of tall columnar or principal cells, basal cells, clear cells, tall slender or apical cells rich in mitochondria (apical mitochondria–rich cells), and scattered intraepithelial lymphocytes and macrophages (173). The principal cells form elaborate tight junctions that serve to regulate the intraluminal contents at different loci of the epididymis (174). Occasional large cells with atypical nuclei similar to those seen in the seminal vesicle are sometimes present. These have no clinical significance. The principal cells, comprising over 95% of the columnar cells, have straight stereocilia (Fig. 36.32), which are tall and nearly obliterate the lumen in the head but become progressively shorter as the tail is reached, these have no clinical significance. The columnar cell nuclei often show eosinophilic inclusions (Fig. 36.32). The epididymal epithelium secretes epididymosomes, which facilitate the transfer of certain molecules (such as CD52, GliPr1L1, MIF, P25b, P34h, ATP2B4, SPAM1, AKR1B1, SLC27A2, EDDM3B, KRT19, and WFDC8) to spermatozoa, contributing significantly to sperm maturation (175). The principal cells stain strongly for vimentin, EMA, and acid phosphatase. Both basal and principal cells stain positively for low–molecular-weight keratins, with more intense staining in the body and tail sections than the head region (176). In cryptorchid testes, the intensity of staining for low–molecular-weight keratins, particularly cytokeratin 18, is markedly diminished (177). The intensity of vimentin staining progressively declines in the tail section (176). Intense CD10 apical membranous staining is evident both in the epididymis and vas deferens, a finding that has been utilized to ascertain the possible origin of glandular structures found adjacent to the epididymis or vas deferens; positive glands are likely of wolffian origin, whereas glands of müllerian origin are negative (178). The apical mitochondria–rich cells, located primarily in the head, show intense staining for cytokeratins and acid phosphatase and less intense reactivity for EMA and vimentin. Their configuration varies from slender cells extending from the basement membrane all the way to the lumen to those that appear to be located

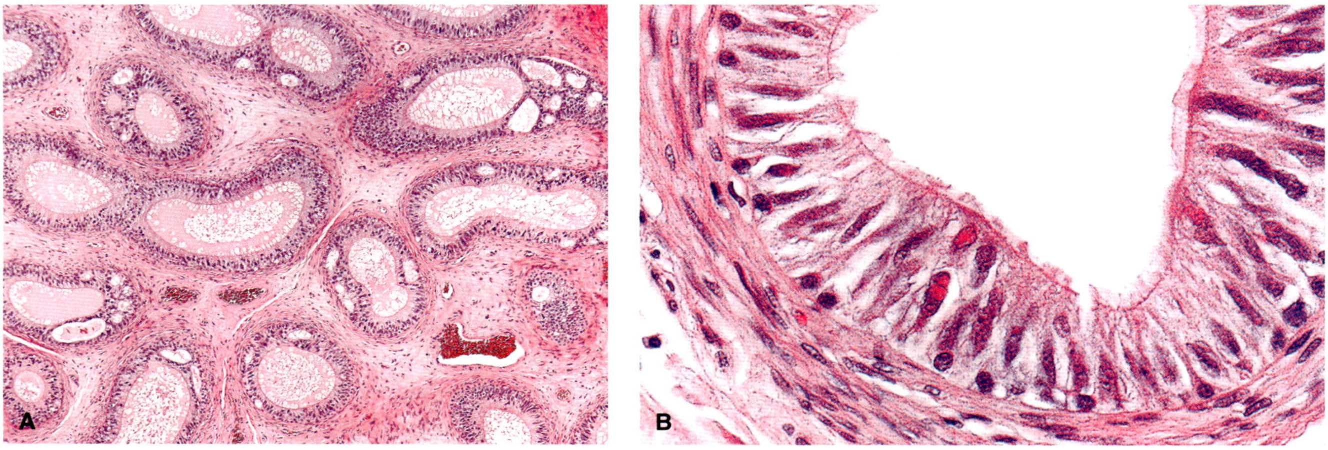

FIGURE 36.32 Epithelium of epididymis: **A.** Few ducts with cribriform architecture. **B.** High magnification of epididymal tubules shown in inset. Compare the tall columnar cells of the epididymis with the pseudostratified cells of the efferent ductulus. A few intraepithelial lymphocytes are present. The stereocilia are somewhat short, indicating the tail segment. A layer of muscle cells forms the wall. Note intranuclear eosinophilic inclusions.

only in the base of the duct (Fig. 36.33). Small foci of epithelium may rarely have the appearance of prostatic epithelium, including positive immunostaining for prostate-specific antigen. Whether this process represents metaplasia or ectopia is unclear (179). Epididymal cells may contain lipofuscin pigment, which tends to be more prominent in the head segment and is particularly evident when there is obstruction of the epididymis (180). The lumina are generally round and regular (Fig. 36.34). In up to 50% of individuals, a focal cribriform pattern is seen, which is considered a normal variation. Often this process is associated with cytoplasmic vacuolization and nuclear atypia (171,181). Importantly, these changes should not be mistaken for intraepididymal spread from a testicular germ cell tumor or a primary epididymal carcinoma. The focal distribution of the nuclear atypia and the absence of mitotic figures should be helpful features supporting a benign process.

Intranuclear, eosinophilic, PAS-positive, and diastase-resistant inclusions (Fig. 36.35) are found in the columnar cells of the adult epididymis as well as throughout the vas and seminal vesicles. These inclusions measure 1 to 14 μm and ultrastructurally represent electron-dense globules enclosed by a single membrane (182). They are most common in the distal epididymis and adjacent vas and least common in the ampulla of the vas and seminal vesicles.

The epididymis is supported by a thick basement membrane surrounded by a well-defined muscular coat. The latter plays an important role in sperm movement through the epididymis. After puberty, scattered elastic

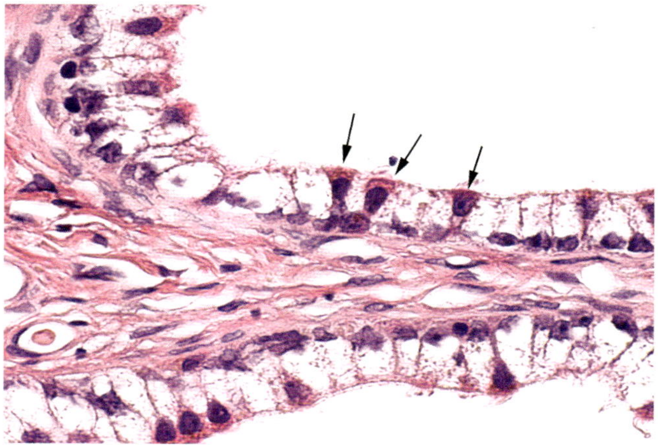

FIGURE 36.33 Epithelium of head of epididymis, demonstrating clear cells and apical mitochondria–rich cells (*arrows*).

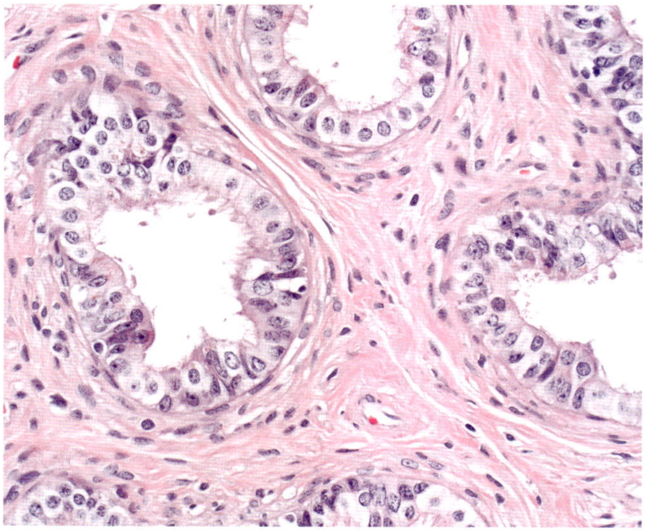

FIGURE 36.34 Epididymal cells containing lipofuscin. The lumens are rounded showing cilia projecting in the lumen. Prominent clear cells are present.

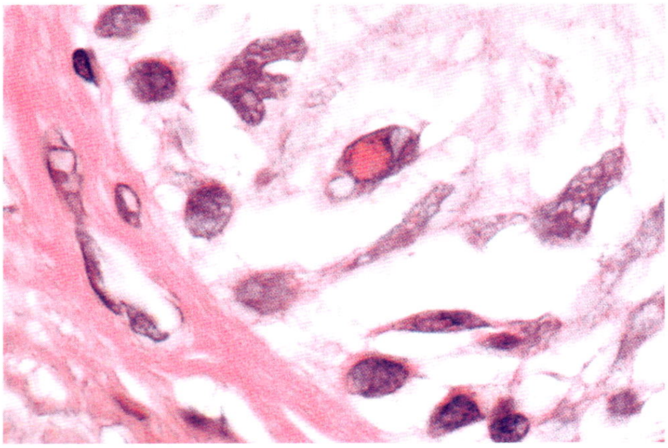

FIGURE 36.35 Intranuclear inclusions present in the epithelium of the epididymis. Similar inclusions are found in the epithelium of the ductus (vas) deferens.

fibers can be seen in both the ductuli and the epididymis. Mast cells are found throughout the connective tissue of the epididymis in a pattern similar to that seen in the tunica and interstitium of the testis (183). They are numerous in infancy, decrease in childhood, and then increase at the time of puberty. A progressive decline in numbers occurs in later adulthood. The epididymis is a rare site for carcinomas but does develop either sporadic or von Hippel–Lindau syndrome–associated papillary cystadenomas (184–186).

DUCTUS (VAS) DEFERENS

The ductus deferens, also known as the vas deferens, originates from the distal end of the epididymis and spans approximately 40 cm in length. Toward its distal end, typically within the range of 4 to 7 cm, it expands to create a structure called the ampulla. This enlarged section merges with the excretory duct of the seminal vesicle, forming the ejaculatory duct (Fig. 36.36). The adult vas deferens is lined by a pseudostratified columnar epithelium composed of columnar cells and basal cells by light microscopic examination. Ultrastructural studies show four different cell types: principal cells, pencil or peg cells, mitochondria-enriched cells, and basal cells. The luminal surface of the columnar cell is lined by tall stereocilia throughout most of the vas deferens (187). These stereocilia are substantially shorter and sparser in the ampullary region. Prominent intranuclear eosinophilic inclusions, as described above in the epididymis, may be seen (182). In addition, occasional lipid-positive vacuoles are present within the cytoplasm. The epithelium of the vas deferens is thrown into folds, which are relatively simple proximally (Fig. 36.37) but become much more complicated in the ampullary segment (Fig. 36.38). The ampulla has highly complex infoldings and many outpouchings or diverticula that extend into the muscle coat. Beneath the epithelium of the adult vas deferens is a loose connective tissue stroma that contains

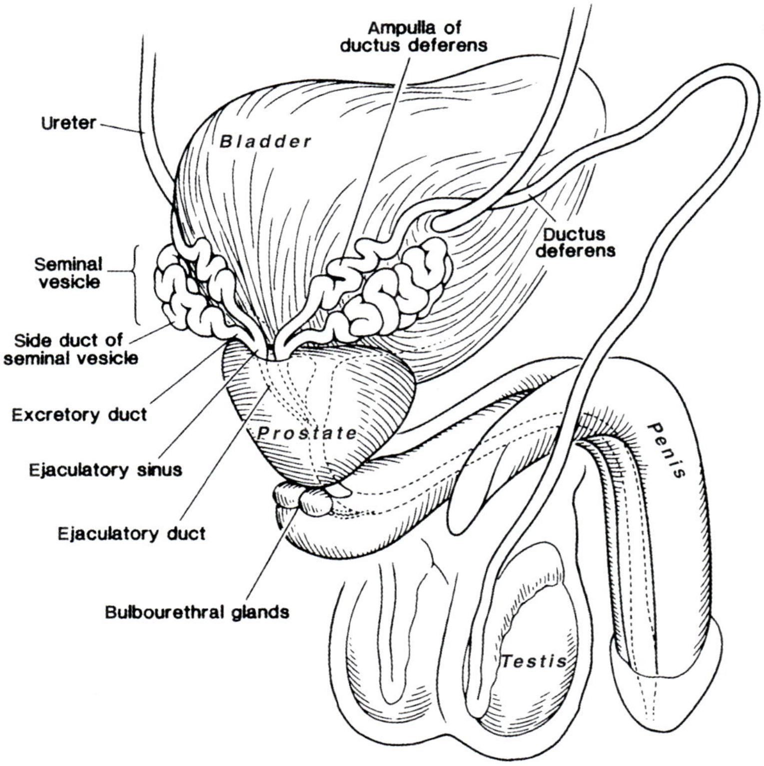

FIGURE 36.36 Diagram of excretory duct system from the vas to the ejaculatory ducts.

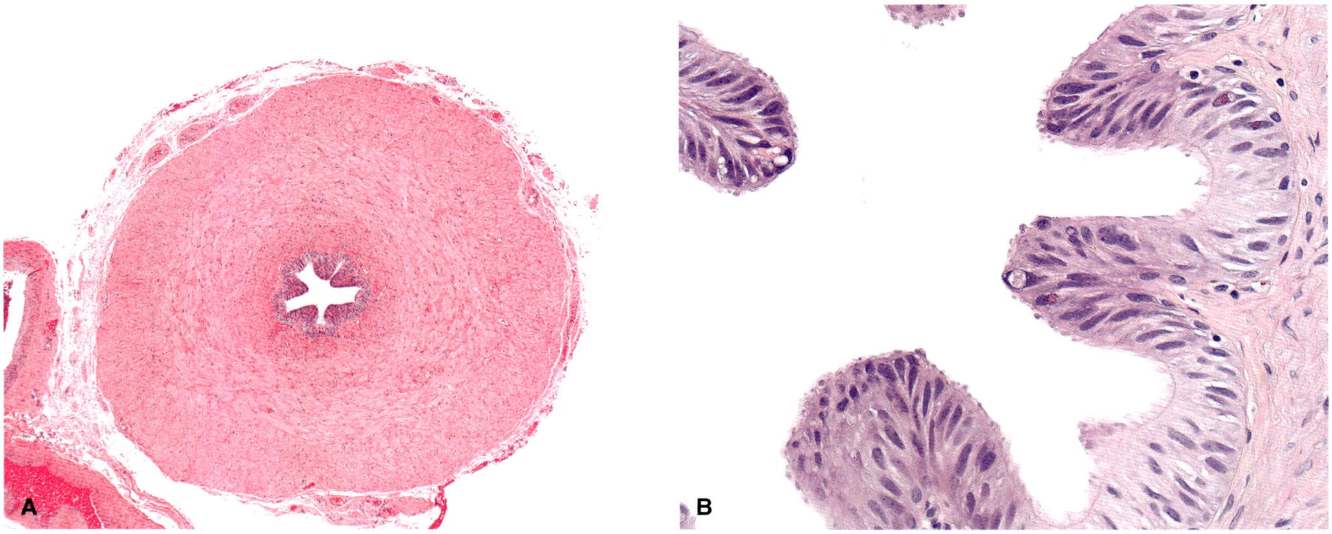

FIGURE 36.37 Proximal ductus (vas) deferens. **A:** This cross-section shows a thick muscle coat, tiny lumen, and slightly folded mucosa. **B:** Higher magnification showing pseudosratified epithelium with cilia. Some nuclei show intranuclear eosinophilic inclusions.

a well-defined, circumferentially oriented layer of elastic fibers (188). These fibers, lacking in infants and children, become frayed and fragmented in the aged vas deferens. The muscular layer is notably robust, featuring inner and outer longitudinal layers along with a central oblique or circular region. As one approaches the ampulla, the overall muscle mass diminishes gradually, albeit the inner longitudinal layer thickens somewhat toward the distal end (187). The epithelium in the ampulla contains significant amounts of lipofuscin pigment and rather closely resembles the epithelium seen in the seminal vesicles. Active phagocytosis of degenerated spermatozoa has been demonstrated in the ampullary region of a number of mammalian species (189). Vas deferens epithelium is immunohistochemically positive for AR, PAX8 and CK7, E-cadherin, and GATA3. ER and PR may occasionally be positive (25,190).

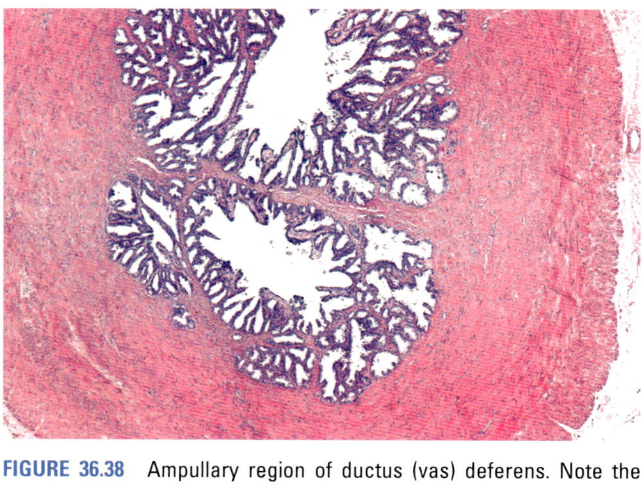

FIGURE 36.38 Ampullary region of ductus (vas) deferens. Note the complex folding and outpouching of the mucosa into the muscular coat.

SEMINAL VESICLES

The seminal vesicles are paired, intricately coiled tubular structures positioned posterolaterally to the base of the bladder, running parallel to the path of the ampulla of the vas deferens (Fig. 36.36). In adults, each vesicle typically measures between 3.5 and 7.5 cm in length and 1.2 to 2.4 cm in thickness. The main duct, which exhibits duplication in roughly 10% of individuals, extends approximately 10 to 15 cm in length when straightened. The main duct gives rise to six to eight primary side ducts, from which several secondary side ducts originate. The upper part of the main duct is bent backward in a hooklike fashion. A short excretory duct combines with the ampulla of the vas deferens to form the ejaculatory duct (Fig. 36.36). The wall of the seminal vesicle has a thin external longitudinal and a thicker internal circular muscle layer. The mucosal folds, relatively simple and shallow in infancy and childhood, become highly complex and alveolus-like in the reproductive years (Fig. 36.39) and are blunted in the aged vesicle. The lumen may contain a few sloughed epithelial cells and debris. Eosinophilic secretions, often with crystalloid structures, are commonly seen within the lumina. They usually have a plate-like arrangement but sometimes appear as smaller crystalloids similar to those seen in the gland lumina of well-differentiated prostate carcinomas. Their significance is unknown, but one should be aware of their presence in biopsies where the seminal vesicle is inadvertently sampled (191). Spermatozoa, refluxed from the ejaculatory duct, occasionally may be present, although they are not normally stored within the seminal vesicles.

The vesicle epithelium is composed of columnar and basal cells. The former have short microvilli projecting

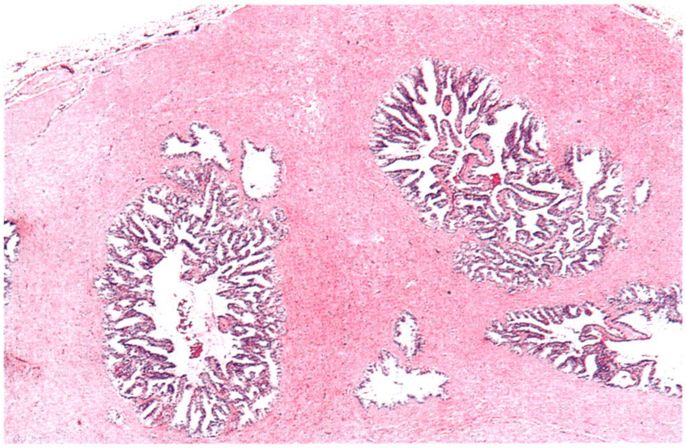

FIGURE 36.39 Seminal vesicle: Alveolus-like arrangement of mucosal folds and cross sections of side ducts.

from the surface. The cytoplasm often contains a large amount of lipofuscin pigment, a feature important in recognizing these cells obtained by needle biopsy or aspiration cytology studies of the prostate. Similar lipofuscin pigment is found in the ampulla of the ductus deferens and in the epithelium of the ejaculatory ducts. The pigment has been divided into two different types based on its appearance. Type 1 consists of coarse, highly refractile, golden-brown granules of uniform diameter (1 to 2 μg). Type 2 granules are much more variable in size (0.25 to 4 μg). Type 2 granules are further divided into type 2A and type 2B. Type 2A granules are poorly refractile to nonrefractile and yellow to gray-brown. Type 2B granules are nonrefractile and are dark to light purple or pink. Both types 1 and 2 granules are found in the seminal vesicle, vas deferens, and ejaculatory duct, whereas only the type 2 granules are found in prostatic epithelium (192).

An unusual feature of the seminal vesicle epithelium is the presence of peculiar epithelial cells with marked cytologic atypia (Fig. 36.36). Similar cells may be seen in the ampulla of the vas deferens and, less commonly, more proximally in the vas deferens or epididymis. These cells have enlarged, hyperchromatic, and often irregularly shaped nuclei and are found in approximately three-quarters of adult seminal vesicles. They often have prominent intranuclear cytoplasmic inclusions (Fig. 36.40). They are not seen in infants or children. Their genesis is unknown but may be related to endocrine influences, similar to the Arias-Stella cells seen in gestational endometrium. Since these cells may be encountered in both needle biopsy and aspiration biopsy specimens, the surgical pathologist must be alert to avoid identifying them as malignant cells (193). Degenerative-type atypia and the presence of lipochrome pigment in the cells provide helpful clues to identify them as normal seminal vesicle epithelium, and their associated basal cells provide additional assurance that they do not represent prostate cancer.

Hyaline, pink globules (Fig. 36.41) are also encountered in the muscle portion of the seminal vesicle and are thought to represent degenerating smooth muscle cells. They may be seen occasionally in the muscular coat of the vas deferens and within the prostate parenchyma (194). Deposits of amorphous pink material under the epithelium and variably replacing the muscle wall of the seminal vesicle is a normal phenomenon in aging individuals. This condition, usually referred to as senile amyloidosis, may also involve the vas deferens and ejaculatory duct, has no association with systemic amyloidosis, and is without any clinical sequelae (Fig. 36.42). It contains semenogelin 1, a major secretory product of the seminal vesicles, and lactoferrin (195,196). Immunohistochemical expression of NKX3.1 is typical in prostate tissue, aiding in the distinction between prostate tissue or carcinoma and seminal vesicles in needle biopsy samples (197).

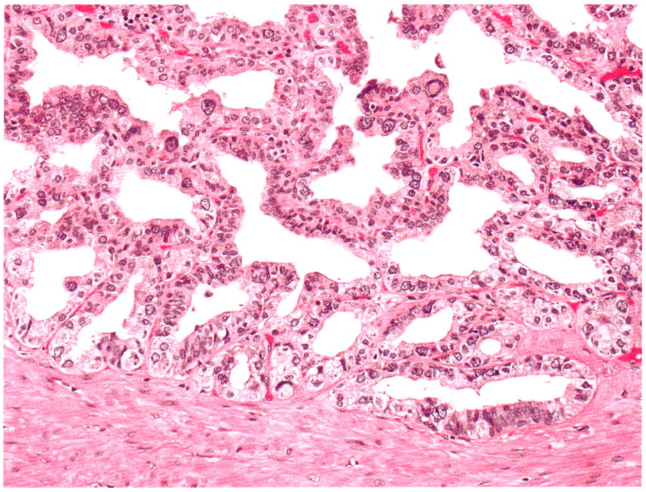

FIGURE 36.40 Seminal vesicle epithelium. Tall columnar to cuboidal cells line the lumen. Hyperchromatic "monster" cells are present. Nuclear inclusion in a large cell is present. These cells should not be mistaken for malignant cells, a distinction which is aided by the cytoplasmic lipofuscin.

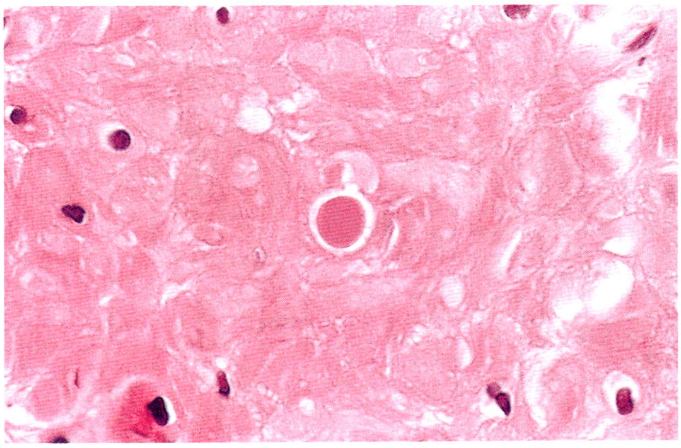

FIGURE 36.41 Muscle coat of seminal vesicle showing a hyalin globule, which probably represents a degenerated smooth muscle cell.

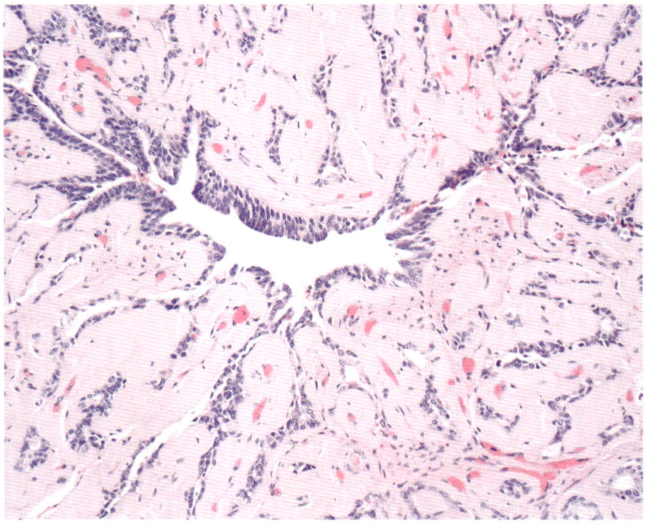

FIGURE 36.42 Senile amyloidosis in seminal vesical. Subepithelial deposits of amorphous eosinophilic amyloid.

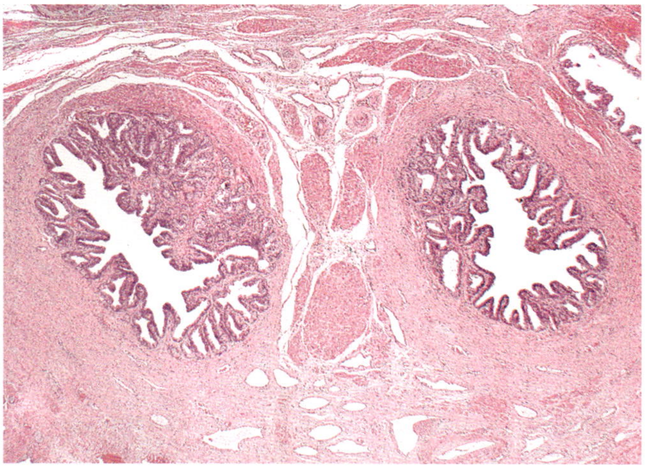

FIGURE 36.43 Paired ejaculatory ducts within the prostatic parenchyma.

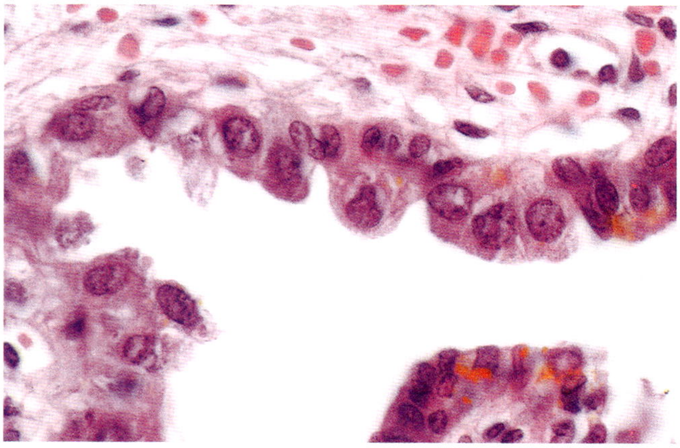

FIGURE 36.44 Epithelium of prostatic portion of ejaculatory duct. This epithelium may be encountered in a needle biopsy or aspiration biopsy of the prostate and should not be misinterpreted as malignant.

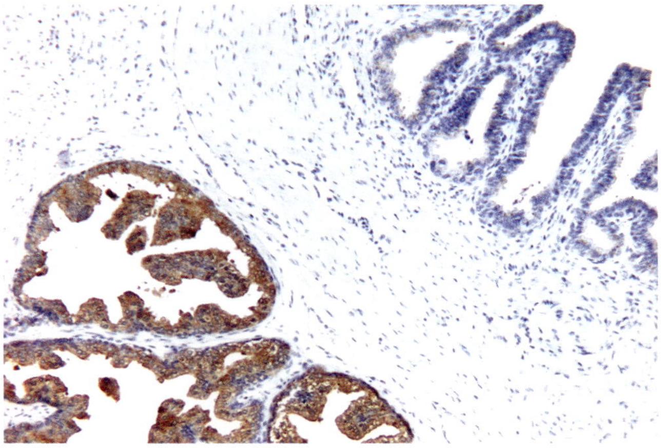

FIGURE 36.45 Intraprostatic ejaculatory duct (**right**) and adjacent prostatic tissue (**left**). Prostate-specific antigen immunoperoxidase demonstrates strongly positive cytoplasmic staining of the prostatic secretory cells and negative staining of ejaculatory duct epithelium.

EJACULATORY DUCTS

The ejaculatory ducts are short (1.5 cm) paired ducts, arising from the confluence of the excretory duct of the seminal vesicle and the ampulla of the vas deferens, that quickly converge and enter the prostate (Fig. 36.43). They run through the central zone of the prostatic parenchyma and enter the posterior aspect of the distal prostate urethra at the verumontanum (198). The outer portions of the ejaculatory ducts are invested with a thin muscular coat that progressively becomes more attenuated as the ducts pass through the prostate. The epithelium of the ejaculatory ducts resembles that of the seminal vesicle and ampulla of the vas deferens (Fig. 36.44). On occasion, a needle biopsy of the prostate will sample a portion of one of these ducts, making it imperative that the surgical pathologist be aware of the characteristics of these cells. The presence and character of the lipofuscin pigment in the cytoplasm should give a clue to the cell origin. Immunoperoxidase stains for prostate-specific antigen and NKX3.1 demonstrate a sharp contrast between the positively stained prostate epithelium and the negatively stained cells of the intraprostatic ejaculatory ducts (Fig. 36.45). An opposite staining pattern is observed with PAX8 and PAX2, which are positive in the ejaculatory duct epithelium and negative in the prostatic epithelium (162).

MESONEPHRIC AND MÜLLERIAN REMNANTS

Remnants of either the mesonephric (wolffian) duct or the paramesonephric (müllerian) duct are commonly encountered in tissues adjacent to the testis, epididymis, or vas deferens (160). These are the appendix testis (hydatid of Morgagni), appendix epididymis, vas aberrans (organ of Haller), and paradidymis (organ of Giraldes) (Fig. 36.1).

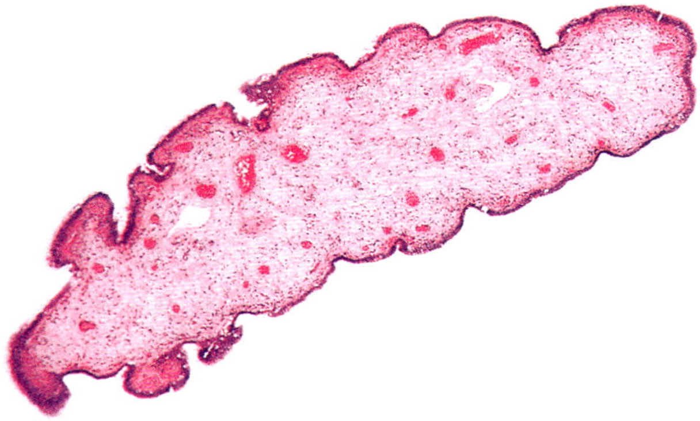

FIGURE 36.46 Appendix testis. This specimen was an incidental finding in a surgically removed testis. It was pedunculated and measures 0.9 cm in greatest length.

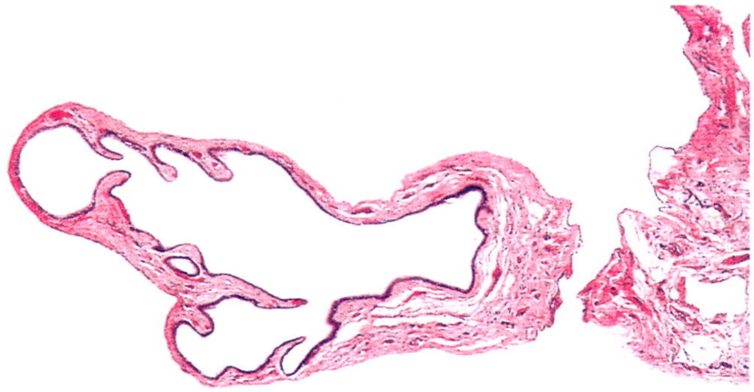

FIGURE 36.48 Appendix epididymis. A pedunculated cystic structure is attached to the head of the epididymis.

The appendix testis, a remnant of the cranial portion of the müllerian duct, is attached to the tunica vaginalis on the anterosuperior aspect of the testis just below the head of the epididymis. Occasionally, it is attached to both the testis and the epididymis. It is most often sessile and either oval or fan-shaped (90%) and, less commonly, pedunculated (10%) and measures 0.5 to 2.5 cm in greatest dimension (Fig. 36.46). It is sometimes represented by only a slight roughening or a calcified thickening of the tunica vaginalis (199,200). Approximately 80% of individuals have an appendix testis, with bilaterality in one-third of them. The appendix testis is covered by a cuboidal or columnar epithelium, which may be ciliated (Fig. 36.47). The structure has a highly vascular fibrous core containing variable numbers of smooth muscle cells. Tubular invaginations and small gland-like structures may also be found in the stroma. Rarely, a macroscopic cyst may be present. Because of its occasionally pedunculated structure, the appendix testis can become twisted, causing hemorrhagic infarction and producing severe testicular pain (201). This event occurs most often in prepubertal or pubertal boys and may be related to the presence of androgen and estrogen receptors known to be present in the appendix epithelium and the possible growth of the appendix as a result of stimulation by androgens and estrogens at this period of time (202).

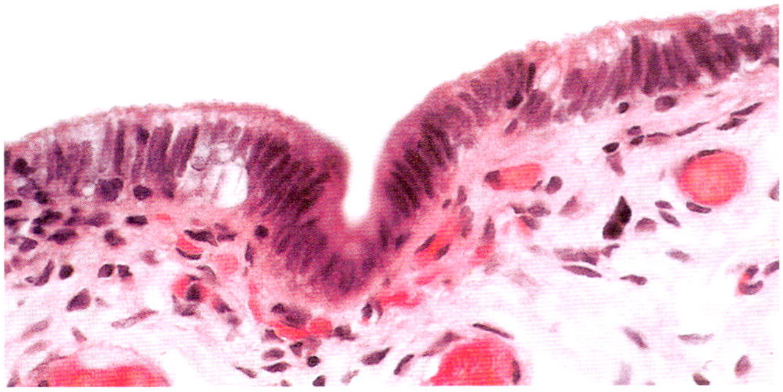

FIGURE 36.47 Appendix testis with covering of low columnar, nonciliated epithelium.

The appendix epididymis (Fig. 36.48), a remnant of the most cranial portion of the mesonephric duct, is present in ~25% of testes (200). It is almost invariably cystic; the vesicle lumen being filled with amorphous protein secretions. The epithelium lining the cyst is columnar and often ciliated (Fig. 36.49). The external surface is covered by a flattened or low cuboidal layer of cells. Since it may be pedunculated, it is also subject to torsion and infarction. Appendix testis and appendix epididymis express AR, ER, PR, PAX-8, WT1, and CK7 (25).

The other appendicular structures are derived from remnants of either the mesonephric tubules or the müllerian ducts and are variably encountered in the fat, usually as microscopic incidental findings. These are the vas aberrans inferior, the vas aberrans superior, and the paradidymis. They all have somewhat similar histologic features, with a low columnar epithelium lining a small cystic space and a thin muscular coat. Some investigators refer to these structures collectively as the paradidymis (203). The vas aberrans inferior is a tubular structure (Fig. 36.50) located near the junction of the vas deferens and the tail portion

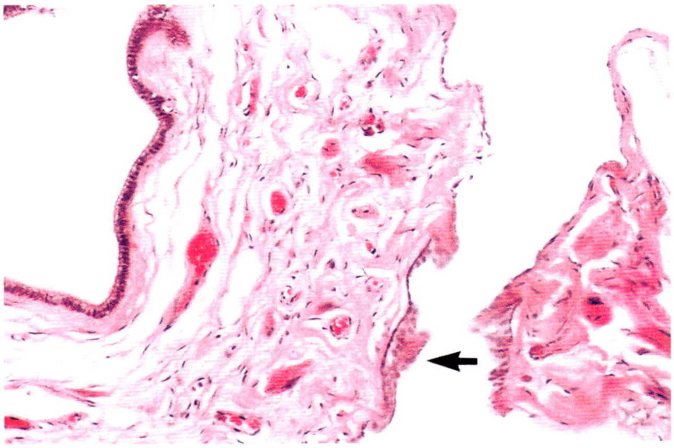

FIGURE 36.49 Appendix epididymis. Low columnar epithelium lines the cystic space. The surface is lined by focally thickened mesothelium (*arrow*).

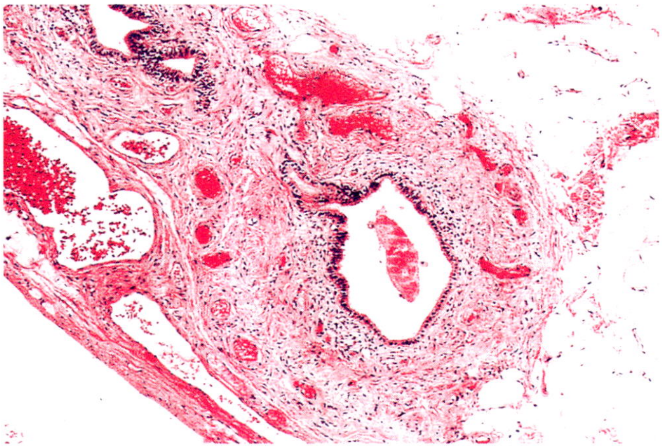

FIGURE 36.50 Vas aberrans inferior. A tubular structure is found near the tail portion of the epididymis.

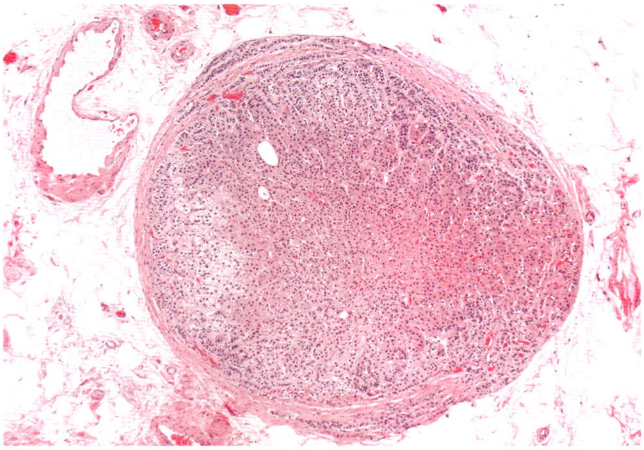

FIGURE 36.52 Adrenal cortical rest in fat adjacent to the epididymis. Note partial encapsulation and zonation similar to normal adrenal.

of the epididymis; it may or may not communicate with either structure. The vas aberrans superior is a small collection of tubules located near the head or body of the epididymis. It may communicate with the epididymis or the rete testis. Remnants of the vas aberrans may represent the origin of cord cysts, which are seen sporadically, in patients whose mothers had been treated with diethylstilbestrol (204), and in patients with von Hippel–Lindau syndrome (205).

The paradidymis is represented by one or more tubules embedded in the spermatic cord, adjacent to the ductus (vas) deferens, and near the head of the epididymis. These tubules may be encountered in a section of the wall of an inguinal hernia sac and should not be mistaken for a portion of the ductus deferens or epididymis (206). Some of these structures mimic the epididymis (Fig. 36.51) and others resemble the vas deferens. CD10 staining of the epithelial cells and the relatively thick muscle coat of the vas deferens may help distinguish the true vas deferens from vas-like tubules and, to a lesser degree, the epididymis from epididymis-like remnants (178). Rarely, macroscopic cysts may form in the spermatic cord (207).

One may find adrenal cortical rests in the fat of the spermatic cord, adjacent to the vas, the epididymis, or the rete testis. Adrenal cortical rests are usually small and display partial or complete capsulation and zonation similar to normal adrenal (Fig. 36.52). Very rarely, adrenal medullary tissue may also be present. Even less commonly seen is splenogonadal fusion, in which the splenic and gonadal anlages fuse in early embryonal life. This process is almost always on the left side and most commonly is found as an incidental finding in a cryptorchid testis (Fig. 36.53). It occasionally presents as a mass in the scrotum or inguinal canal, especially in patients with diseases associated with splenomegaly. Two forms are recognized: a continuous type that features a fibrous cord connecting the testicular splenic tissue to the eutopic spleen and a discontinuous type, in which there is no such connection. The former is associated with congenital anomalies, particularly limb defects (208).

Epididymis-like intratesticular tubular or microcystic structures are occasionally encountered, usually in elderly

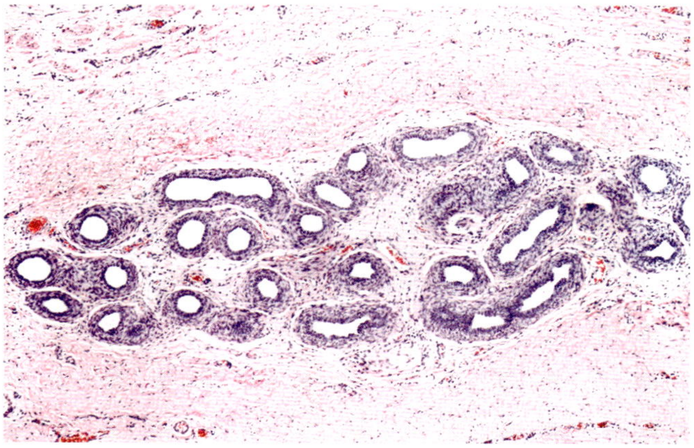

FIGURE 36.51 Epididymis-like tubules in hernia sac of a 6-year-old.

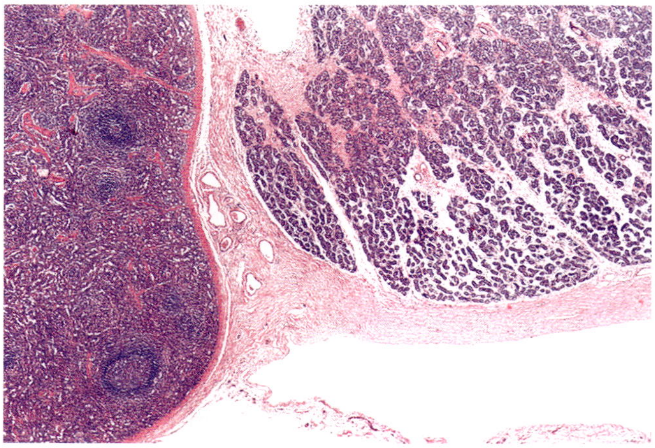

FIGURE 36.53 Splenogonadal fusion of a cryptorchid testis in a 3-year-old child. Spleen is on the **left**, and gonad, on the **right**.

men (209). They are most often seen near the mediastinal rete. The epithelium is pseudostratified and stains positively for cytokeratins 8, 18, and 19, and for CD10. Vimentin staining of the basal cells and some of the columnar cells is observed. These structures most likely are derived from wolffian duct elements.

GUBERNACULUM

The gubernaculum, also known as the tail ligament of the testis, has been the center of attention with respect to the descent of the testis since it was first described by John Hunter in 1762, and its specific role is still being debated. In the fetus, this cylindrical, gelatinous structure is attached cranially to the testis and epididymis (Fig. 36.54) and caudally to the anterior abdominal wall at the site of the inguinal canal. Just before the descent of the testis through the inguinal canal, the gubernaculum increases in net weight disproportionately to the testis, supporting the theory that this structure plays a crucial role in this phase of the passage of the testis into the scrotum.

The fetal gubernaculum is composed of a loose undifferentiated mesenchymal tissue similar to Wharton jelly. Large amounts of glycosaminoglycans fill the extracellular space and separate the individual spindle cells, most of which are fibroblasts. Fibroblasts tend to decrease with increasing gestational age. A few striated muscle cells, presumably derived from the cremasteric muscle, can be identified at the periphery of the distal gubernaculum where it attaches to the inguinal wall. The cranial portion of the early fetal gubernaculum is completely devoid of striated muscle. Smooth muscle cells are confined to the vessel walls. Collagen and elastic fibers increase in number later in gestation (210). After the testis descends into the scrotum, the gubernaculum undergoes degenerative changes, loses much of the intercellular matrix, and becomes infiltrated by blood vessels, collagen fibers, and striated muscle.

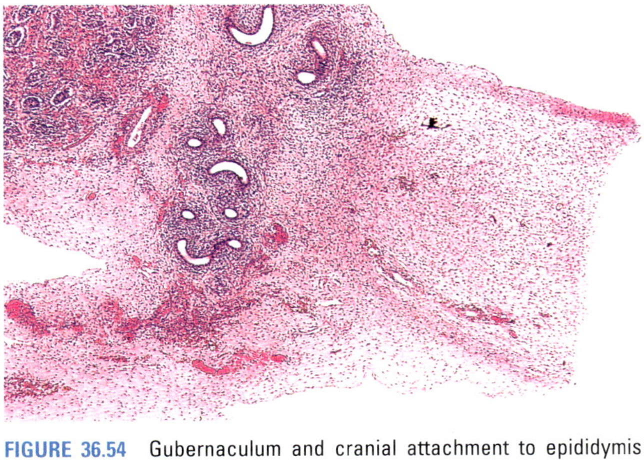

FIGURE 36.54 Gubernaculum and cranial attachment to epididymis and testis in a 26-week-old fetus.

REFERENCES

1. Handelsman DJ, Staraj S. Testicular size: the effects of aging, malnutrition, and illness. *J Androl*. 1985;6(3):144–151.
2. Prader A. Testicular size: assessment and clinical importance. *Triangle*. 1966;7(6):240–243.
3. Nayak BS. Why the left testis hangs at a lower level than the right? *Med Hypotheses*. 2009;73(2):269–270.
4. Kumar A, Kumar CJ. Swinging high and low: why do the testes hang at different levels? A theory on surface area and thermoregulation. *Med Hypotheses*. 2008;70(3):698.
5. Prasad H. Asymmetric testicular levels in the crotch: a thermodynamic perspective. *Med Hypotheses*. 2008;71(1):148–149.
6. Sosnik H. Studies on the participation of tunica albuginea and rete testis (TA and RT) in the quantitative structure of human testis. *Gegenbaurs Morphol Jahrb*. 1985;131(3):347–356.
7. Lennox B, Ahmad KN, Mack WS. A method for determining the relative total length of the tubules in the testis. *J Pathol*. 1970;102(4):229–238.
8. Papaioannou MD, Pitetti J-L, Ro S, et al. Sertoli cell Dicer is essential for spermatogenesis in mice. *Dev Biol*. 2009;326(1):250–259.
9. Tian H, Wang X, Li X, Song W, Mi J, Zou K. Regulation of spermatogonial stem cell differentiation by Sertoli cells-derived exosomes through paracrine and autocrine signaling. *J Cell Physiol*. 2024;239(4):e31202.
10. Washburn RL, Dufour JM. Complementing testicular immune regulation: the relationship between Sertoli cells, complement, and the immune response. *Int J Mol Sci*. 2023;24(4):3371.
11. Sharpe RM, Mckinnell C, Kivlin C, Fisher JS. Proliferation and functional maturation of Sertoli cells, and their relevance to disorders of testis function in adulthood. *Reproduction*. 2003;125(6):769–784.
12. Franke FE, Pauls K, Rey R, Marks A, Bergmann M, Steger K. Differentiation markers of Sertoli cells and germ cells in fetal and early postnatal human testis. *Anat Embryol (Berl)*. 2004;209(2):169–177.
13. Vogl A, Vaid K, Guttman J. The Sertoli cell cytoskeleton. In: Cheng C, ed. *Molecular Mechanisms in Spermatogenesis*. Landes Bioscience and Springer Science; 2008.
14. Nielsen K, Jacobsen GK. Malignant Sertoli cell tumour of the testis. An immunohistochemical study and a review of the literature. *APMIS*. 1988;96(8):755–760.
15. Henley JD, Young RH, Ulbright TM. Malignant Sertoli cell tumors of the testis: a study of 13 examples of a neoplasm frequently misinterpreted as seminoma. *Am J Surg Pathol*. 2002;26(5):541–550.
16. Stosiek P, Kasper M, Karsten U. Expression of cytokeratins 8 and 18 in human Sertoli cells of immature and atrophic seminiferous tubules. *Differentiation*. 1990;43(1):66–70.
17. Rogatsch H, Hittmair A, Mikuz G, Feichtinger H, Jezek D. Expression of vimentin, cytokeratin, and desmin in Sertoli cells of human fetal, cryptorchid, and tumour-adjacent testicular tissue. *Virchows Arch*. 1996;427(5):497–502.
18. Hoei-Hansen CE, Holm M, Rajpert-De Meyts E, Skakkebaek NE. Histological evidence of testicular dysgenesis in contralateral biopsies from 218 patients with testicular germ cell cancer. *J Pathol*. 2003;200(3):370–374.

19. Suárez-Quian CA, Martínez-García F, Nistal M, Regadera J. Androgen receptor distribution in adult human testis. *J Clin Endocrinol Metab.* 1999;84(1):350–358.
20. Compérat E, Tissier F, Boyé K, De Pinieux G, Vieillefond A. Non-Leydig sex-cord tumors of the testis. The place of immunohistochemistry in diagnosis and prognosis. A study of twenty cases. *Virchows Arch.* 2004;444(6):567–571.
21. Fisher JS, Macpherson S, Marchetti N, Sharpe RM. Human 'testicular dysgenesis syndrome': a possible model using in-utero exposure of the rat to dibutyl phthalate. *Hum Reprod.* 2003;18(7):1383–1394.
22. Sangoi AR, McKenney JK, Brooks JD, Higgins JP. Evaluation of SF-1 expression in testicular germ cell tumors: a tissue microarray study of 127 cases. *Appl Immunohistochem Mol Morphol.* 2013;21(4):318–321.
23. Emerich DF, Hemendinger R, Halberstadt CR. The testicular-derived Sertoli cell: cellular immunoscience to enable transplantation. *Cell Transplant.* 2003;12(4):335–349.
24. Davidoff MS, Middendorff R, Pusch W, Müller D, Wichers S, Holstein AF. Sertoli and Leydig cells of the human testis express neurofilament triplet proteins. *Histochem Cell Biol.* 1999;111(3):173–187.
25. Magers MJ, Udager AM, Chinnaiyan AM, et al. Comprehensive immunophenotypic characterization of adult and fetal testes, the excretory duct system, and testicular and epididymal appendages. *Appl Immunohistochem Mol Morphol.* 2016;24(7):e50–e68.
26. Mesa H, Gilles S, Datta MW, et al. Comparative immunomorphology of testicular Sertoli and sertoliform tumors. *Hum Pathol.* 2017;61:181–189.
27. Banco B, Palmieri C, Sironi G, et al. Immunohistochemical expression of SOX9 protein in immature, mature, and neoplastic canine Sertoli cells. *Theriogenology.* 2016;85(8):1408–1414.e1.
28. Bowles J, Koopman P. Sex determination in mammalian germ cells: extrinsic versus intrinsic factors. *Reproduction.* 2010;139(6):943–958.
29. Jorban A, Lunenfeld E, Huleihel M. Effect of temperature on the development of stages of spermatogenesis and the functionality of Sertoli cells in vitro. *Int J Mol Sci.* 2024;25(4):2160.
30. Schulze C. Sertoli cells and Leydig cells in man. *Adv Anat Embryol Cell Biol.* 1984;88:1–104.
31. Mruk DD, Cheng CY. Tight junctions in the testis: new perspectives. *Philos Trans R Soc Lond B Biol Sci.* 2010;365(1546):1621–1635.
32. Lui W-Y, Mruk D, Lee WM, Cheng CY. Sertoli cell tight junction dynamics: their regulation during spermatogenesis. *Biol Reprod.* 2003;68(4):1087–1097.
33. Kopera IA, Bilinska B, Cheng CY, Mruk DD. Sertoli-germ cell junctions in the testis: a review of recent data. *Philos Trans R Soc Lond B Biol Sci.* 2010;365(1546):1593–1605.
34. Xiao Z, Liang J, Huang R, et al. Inhibition of miR-143-3p restores blood-testis barrier function and ameliorates sertoli cell senescence. *Cells.* 2024;13(4):313.
35. Carrageta DF, Freire-Brito L, Guerra-Carvalho B, et al. Mitochondrial uncoupling proteins regulate the metabolic function of human Sertoli cells. *Reproduction.* 2024;167(2):e230229.
36. Wang Qi, Xie J-F, Yao T-T, et al. MicroRNA-30a-5p regulates cypermethrin-induced apoptosis of Sertoli cells by targeting KLF9 in vitro. *Reprod Toxicol.* 2023;119:108414.
37. Wilhelm D, Palmer S, Koopman P. Sex determination and gonadal development in mammals. *Physiol Rev.* 2007;87(1):1–28.
38. Amann RP. The cycle of the seminiferous epithelium in humans: a need to revisit? *J Androl.* 2008;29(5):469–487.
39. Hadziselimović F. Cryptorchidism. Ultrastructure of normal and cryptorchid testis development. *Adv Anat Embryol Cell Biol.* 1977;53(3):3–71.
40. Nagano T. The crystalloid of Lubarsch in the human spermatogonium. *Z Zellforsch Mikrosk Anat.* 1969;97(4):491–501.
41. Clermont Y. Kinetics of spermatogenesis in mammals: seminiferous epithelium cycle and spermatogonial renewal. *Physiol Rev.* 1972;52(1):198–236.
42. Hermann BP, Sukhwani M, Hansel MC, Orwig KE. Spermatogonial stem cells in higher primates: are there differences from those in rodents? *Reproduction.* 2010;139(3):479–493.
43. Tokuda M, Kadokawa Y, Kurahashi H, Marunouchi T. CDH1 is a specific marker for undifferentiated spermatogonia in mouse testes. *Biol Reprod.* 2007;76(1):130–141.
44. Niedenberger BA, Busada JT, Geyer CB. Marker expression reveals heterogeneity of spermatogonia in the neonatal mouse testis. *Reproduction.* 2015;149(4):329–338.
45. Oatley MJ, Kaucher AV, Racicot KE, Oatley JM. Inhibitor of DNA binding 4 is expressed selectively by single spermatogonia in the male germline and regulates the self-renewal of spermatogonial stem cells in mice. *Biol Reprod.* 2011;85(2):347–356.
46. Song H-W, Wilkinson MF. Transcriptional control of spermatogonial maintenance and differentiation. *Semin Cell Dev Biol.* 2014;30:14–26.
47. Wu X, Schmidt JA, Avarbock MR, et al. Prepubertal human spermatogonia and mouse gonocytes share conserved gene expression of germline stem cell regulatory molecules. *Proc Natl Acad Sci U S A.* 2009;106(51):21672–21677.
48. Kaucher AV, Oatley MJ, Oatley JM. NEUROG3 is a critical downstream effector for STAT3-regulated differentiation of mammalian stem and progenitor spermatogonia. *Biol Reprod.* 2012;86(5):164, 1–11.
49. Boitani C, Di Persio S, Esposito V, Vicini E. Spermatogonial cells: mouse, monkey and man comparison. *Semin Cell Dev Biol.* 2016;59:79–88.
50. Suzuki H, Ahn HW, Chu T, et al. SOHLH1 and SOHLH2 coordinate spermatogonial differentiation. *Dev Biol.* 2012;361(2):301–312.
51. Caires K, Broady J, McLean D. Maintaining the male germline: regulation of spermatogonial stem cells. *J Endocrinol.* 2010;205(2):133–145.
52. Gondos B. Ultrastructure of developing and malignant germ cells. *Eur Urol.* 1993;23(1):68–74; discussion 75.
53. Ma Y, Ma Q-W, Sun Y, Chen XF. The emerging role of extracellular vesicles in the testis. *Hum Reprod.* 2023;38(3):334–351.
54. Simon C, Greening DW, Bolumar D, Balaguer N, Salamonsen LA, Vilella F. Extracellular vesicles in human reproduction in health and disease. *Endocr Rev.* 2018;39(3):292–332.
55. Nihi F, Gomes MLM, Carvalho FAR, et al. Revisiting the human seminiferous epithelium cycle. *Hum Reprod.* 2017;32(6):1170–1182.
56. Kerr JB, de Kretser DM. The cytology of the human testis. In: Burger M, de Kretser D, eds. *The Testis.* Raven Press; 1981.
57. Heller CH, Clermont Y. Kinetics of the germinal epithelium in man. *Recent Prog Horm Res.* 1964;20:545–575.
58. Bartke A. Apoptosis of male germ cells, a generalized or a cell type-specific phenomenon? *Endocrinology.* 1995;136(1):3–4.

59. Johnson L, Zane RS, Petty CS, Neaves WB. Quantification of the human Sertoli cell population: its distribution, relation to germ cell numbers, and age-related decline. *Biol Reprod.* 1984;31(4):785–795.
60. Weissbach L, Ibach B. Quantitative parameters for light microscopic assessment of the tubuli seminiferi. *Fertil Steril.* 1976;27(7):836–847.
61. Zukerman Z, Rodriguez-Rigau LJ, Weiss DB, Chowdhury AK, Smith KD, Steinberger E. Quantitative analysis of the seminiferous epithelium in human testicular biopsies, and the relation of spermatogenesis to sperm density. *Fertil Steril.* 1978;30(4):448–455.
62. de Kretser DM, Kerr JB, Paulsen CA. The peritubular tissue in the normal and pathological human testis. An ultrastructural study. *Biol Reprod.* 1975;12(3):317–324.
63. McLachlan RI, Meyts ER-D, Hoei-Hansen CE, de Kretser DM, Skakkebaek NE. Histological evaluation of the human testis–approaches to optimizing the clinical value of the assessment: mini review. *Hum Reprod.* 2007;22(1):2–16.
64. Skakkebaek NE, Heller CG. Quantification of human seminiferous epithelium. I. Histological studies in twenty-one fertile men with normal chromosome complements. *J Reprod Fertil.* 1973;32(3):379–389.
65. Silber SJ, Rodriguez-Rigau LJ. Quantitative analysis of testicle biopsy: determination of partial obstruction and prediction of sperm count after surgery for obstruction. *Fertil Steril.* 1981;36(4):480–485.
66. Potter SJ, DeFalco T. Role of the testis interstitial compartment in spermatogonial stem cell function. *Reproduction.* 2017;153(4):R151–R162.
67. Arenas MI, Bethencourt FR, De Miguel MP, Fraile B, Romo E, Paniagua R. Immunocytochemical and quantitative study of actin, desmin and vimentin in the peritubular cells of the testes from elderly men. *J Reprod Fertil.* 1997;110(1):183–193.
68. Albrecht M. Insights into the nature of human testicular peritubular cells. *Ann Anat.* 2009;191(6):532–540.
69. Welsh M, Saunders PTK, Atanassova N, Sharpe RM, Smith LB. Androgen action via testicular peritubular myoid cells is essential for male fertility. *FASEB J.* 2009;23(12):4218–4230.
70. Spinnler K, Kohn FM, Schwarzer U, Mayerhofer A. Glial cell line-derived neurotrophic factor is constitutively produced by human testicular peritubular cells and may contribute to the spermatogonial stem cell niche in man. *Hum Reprod.* 2010;25(9):2181–2187.
71. Chen LY, Willis WD, Eddy EM. Targeting the Gdnf Gene in peritubular myoid cells disrupts undifferentiated spermatogonial cell development. *Proc Natl Acad Sci U S A.* 2016; 113(7):1829–1834.
72. Siu MKY, Cheng CY. Extracellular matrix: recent advances on its role in junction dynamics in the seminiferous epithelium during spermatogenesis. *Biol Reprod.* 2004;71(2):375–391.
73. Miller SC, Bowman BM, Rowland HG. Structure, cytochemistry, endocytic activity, and immunoglobulin (Fc) receptors of rat testicular interstitial-tissue macrophages. *Am J Anat.* 1983;168(1):1–13.
74. Pawlicki P, Yurdakok-Dikmen B, Tworzydlo W, Kotula-Balak M. Toward understanding the role of the interstitial tissue architects: possible functions of telocytes in the male gonad. *Theriogenology.* 2024;217:25–36.
75. Cheng CY, Wong EWP, Yan HHN, Mruk DD. Regulation of spermatogenesis in the microenvironment of the seminiferous epithelium: new insights and advances. *Mol Cell Endocrinol.* 2010;315(1-2):49–56.
76. Defalco T, Potter SJ, Williams AV, Waller B, Kan MJ, Capel B. Macrophages contribute to the spermatogonial niche in the adult testis. *Cell Rep.* 2015;12(7):1107–1119.
77. De Menezes AP. Elastic tissue in the limiting membrane of the human seminiferous tubule. *Am J Anat.* 1977;150(2): 349–373.
78. Nistal M, Paniagua R. Non-neoplastic disease of the testis. In: Bostwick D, Eble J, eds. *Urologic Surgical Pathology.* Mosby; 1997.
79. Amat P, Paniagua R, Nistal M, Martín A. Mitosis in adult human Leydig cells. *Cell Tissue Res.* 1986;243(1):219–221.
80. Tanaka T, Kanatsu-Shinohara M, Lei Z, Rao CV, Shinohara T. The luteinizing hormone-testosterone pathway regulates mouse spermatogonial stem cell self-renewal by suppressing WNT5A expression in Sertoli cells. *Stem Cell Reports.* 2016;7(2):279–291.
81. Jun S-Y, Ro JY, Park YW, Kim K-R, Ayala AG. Ectopic Leydig cells of testis. An immunohistochemical study on tissue microarray. *Ann Diagn Pathol.* 2008;12(1):29–32.
82. Nistal M, Paniagua R. Histogenesis of human extraparenchymal Leydig cells. *Acta Anat (Basel).* 1979;105(2):188–197.
83. Nagano T, Otsuki I. Reinvestigation of the fine structure of Reinke's crystal in the human testicular interstitial cell. *J Cell Biol.* 1971;51(1):148–161.
84. Smith LB, Walker WH. The regulation of spermatogenesis by androgens. *Semin Cell Dev Biol.* 2014;30:2–13.
85. Schlatt S, Ehmcke J. Regulation of spermatogenesis: an evolutionary biologist's perspective. *Semin Cell Dev Biol.* 2014;29:2–16.
86. Wen Q, Cheng CY, Liu Y-X. Development, function and fate of fetal Leydig cells. *Semin Cell Dev Biol.* 2016;59:89–98.
87. Smith OE, Morin F, Roussel V, Bertucci MC, Boyer A, Murphy BD. The role of steroidogenic factor 1 (SF-1) in steroidogenic cell function of the testes and ovaries of mature mice. *Reproduction.* 2023;165(1):1–17.
88. Heller CG, Lalli MF, Pearson JE, Leach DR. A method for the quantification of Leydig cells in man. *J Reprod Fertil.* 1971;25(2):177–184.
89. Weiss DB, Rodriguez-Rigau L, Smith KD, Chowdhury A, Steinberger E. Quantitation of Leydig cells in testicular biopsies of oligospermic men with varicocele. *Fertil Steril.* 1978;30(3):305–312.
90. Hashimoto J, Yamamoto M, Miyake K, Hoshino T. [A morphological study of the testis in patients with idiopathic male infertility–quantification and ultrastructure of Leydig cells.] *Hinyokika Kiyo.* 1988;34(11):1995–2011.
91. Soerensen RR, Johannsen TH, Skakkebaek NE, Rajpert-De Meyts E. Leydig cell clustering and Reinke crystal distribution in relation to hormonal function in adult patients with testicular dysgenesis syndrome (TDS) including cryptorchidism. *Hormones (Athens).* 2017;15(4):518–526.
92. Holm M, Rajpert-De Meyts E, Andersson A-M, Skakkebaek NE. Leydig cell micronodules are a common finding in testicular biopsies from men with impaired spermatogenesis and are associated with decreased testosterone/LH ratio. *J Pathol.* 2003;199(3):378–386.
93. Cao QJ, Jones JG, Li M. Expression of calretinin in human ovary, testis, and ovarian sex cord-stromal tumors. *Int J Gynecol Pathol.* 2001;20(4):346–352.

94. Anand-Ivell RJK, Relan V, Balvers M, et al. Expression of the insulin-like peptide 3 (INSL3) hormone-receptor (LGR8) system in the testis. *Biol Reprod.* 2006;74(5):945–953.
95. Davidoff MS, Middendorff R, Köfüncü E, Müller D, Jezek D, Holstein AF. Leydig cells of the human testis possess astrocyte and oligodendrocyte marker molecules. *Acta Histochem.* 2002;104(1):39–49.
96. Davidoff MS, Middendorff R, Müller D, Holstein AF. The neuroendocrine Leydig cells and their stem cell progenitors, the pericytes. *Adv Anat Embryol Cell Biol.* 2009;205:1–107.
97. Ulbright TM, Srigley JR, Hatzianastassiou DK, Young RH. Leydig cell tumors of the testis with unusual features: adipose differentiation, calcification with ossification, and spindle-shaped tumor cells. *Am J Surg Pathol.* 2002;26(11):1424–1433.
98. Lau HD, Kao C-S, Williamson SR, Cheng L, Ulbright TM, Idrees MT. Immunohistochemical characterization of 120 testicular sex cord-stromal tumors with an emphasis on the diagnostic utility of SOX9, FOXL2, and SF-1. *Am J Surg Pathol.* 2021;45(10):1303–1313.
99. Holash JA, Harik SI, Perry G, Stewart PA. Barrier properties of testis microvessels. *Proc Natl Acad Sci U S A.* 1993;90(23):11069–11073.
100. Lobo MVT, Arenas MI, Alonso FJM, et al. Nestin, a neuroectodermal stem cell marker molecule, is expressed in Leydig cells of the human testis and in some specific cell types from human testicular tumours. *Cell Tissue Res.* 2004; 316(3):369–376.
101. Mostafa T, Labib I, El-Khayat Y, El-Rahman El-Shahat A, Gadallah A. Human testicular arterial supply: gross anatomy, corrosion cast, and radiologic study. *Fertil Steril.* 2008;90(6): 2226–2230.
102. Jarow JP. Intratesticular arterial anatomy. *J Androl.* 1990;11(3): 255–259.
103. Kormano M, Suoranta H, Reijonen K. Blood supply to testis and excurrent ducts. Raspe G, ed. *Advances in Biosciences.* Vol. 10. Pergamon Press; 1972.
104. Ergün S, Stingl J, Holstein AF. Segmental angioarchitecture of the testicular lobule in man. *Andrologia.* 1994;26(3):143–150.
105. Andres TL, Trainer TD, Lapenas DJ. Small vessel alterations in the testes of infertile men with varicocele. *Am J Clin Pathol.* 1981;76(4):378–384.
106. Fawcett DW, Leak LV, Heidger PM Jr. Electron microscopic observations on the structural components of the blood-testis barrier. *J Reprod Fertil Suppl.* 1970;10:105–122.
107. Ergun S, Davidoff M, Holstein AF. Capillaries in the lamina propria of human seminiferous tubules are partly fenestrated. *Cell Tissue Res.* 1996;286(1):93–102.
108. Chan F, Oatley MJ, Kaucher AV, et al. Functional and molecular features of the Id4+ germline stem cell population in mouse testes. *Genes Dev.* 2014;28(12):1351–1362.
109. Nowak DG, Woolard J, Amin EM, et al. Expression of pro- and anti-angiogenic isoforms of VEGF is differentially regulated by splicing and growth factors. *J Cell Sci.* 2008;121(Pt 20): 3487–3495.
110. Holstein AF, Orlandini GE, Möller R. Distribution and fine structure of the lymphatic system in the human testis. *Cell Tissue Res.* 1979;200(1):15–27.
111. McLaren A. Primordial germ cells in the mouse. *Dev Biol.* 2003;262(1):1–15.
112. Figueira MI, Cardoso HJ, Correia S, Maia CJ, Socorro S. Hormonal regulation of c-KIT receptor and its ligand: implications for human infertility? *Prog Histochem Cytochem.* 2014;49(1-3):1–19.
113. Kerr CL, Hill CM, Blumenthal PD, Gearhart JD. Expression of pluripotent stem cell markers in the human fetal testis. *Stem Cells.* 2008;26(2):412–421.
114. Culty M. Gonocytes, the forgotten cells of the germ cell lineage. *Birth Defects Res C Embryo Today.* 2009;87(1):1–26.
115. Wartenberg H. Comparative cytomorphologic aspects of the male germ cells, especially of the "Gonia." *Andrologia.* 1976; 8(2):117–130.
116. Fukuda T, Hedinger C, Groscurth P. Ultrastructure of developing germ cells in the fetal human testis. *Cell Tissue Res.* 1975;161(1):55–70.
117. Gaskell TL, Esnal A, Robinson LLL, Anderson RA, Saunders PTK. Immunohistochemical profiling of germ cells within the human fetal testis: identification of three subpopulations. *Biol Reprod.* 2004;71(6):2012–2021.
118. Sharpe RM. Environmental/lifestyle effects on spermatogenesis. *Philos Trans R Soc Lond B Biol Sci.* 2010;365(1546): 1697–1712.
119. Hadziselimović F, Thommen L, Girard J, Herzog B. The significance of postnatal gonadotropin surge for testicular development in normal and cryptorchid testes. *J Urol.* 1986;136(1 Pt 2): 274–276.
120. Chemes HE. Infancy is not a quiescent period of testicular development. *Int J Androl.* 2001;24(1):2–7.
121. Nistal M, Paniagua R, Queizán A. Histologic lesions in undescended ectopic obstructed testes. *Fertil Steril.* 1985;43(3): 455–462.
122. Weaver JR, Susiarjo M, Bartolomei MS. Imprinting and epigenetic changes in the early embryo. *Mamm Genome.* 2009;20(9-10):532–543.
123. Kerjean A, Dupont JM, Vasseur C, et al. Establishment of the paternal methylation imprint of the human H19 and MEST/PEG1 genes during spermatogenesis. *Hum Mol Genet.* 2000;9(14):2183–2187.
124. Laprise SL. Implications of epigenetics and genomic imprinting in assisted reproductive technologies. *Mol Reprod Dev.* 2009;76(11):1006–1018.
125. Furukawa S, Haruta M, Arai Y, et al. Yolk sac tumor but not seminoma or teratoma is associated with abnormal epigenetic reprogramming pathway and shows frequent hypermethylation of various tumor suppressor genes. *Cancer Sci.* 2009;100(4):698–708.
126. Boissonnas CC, Abdalaoui HEl, Haelewyn V, et al. Specific epigenetic alterations of IGF2-H19 locus in spermatozoa from infertile men. *Eur J Hum Genet.* 2010;18(1):73–80.
127. Aflatoonian B, Moore H. Germ cells from mouse and human embryonic stem cells. *Reproduction.* 2006;132(5): 699–707.
128. Mitchell RT, Cowan G, Morris KD, et al. Germ cell differentiation in the marmoset (Callithrix jacchus) during fetal and neonatal life closely parallels that in the human. *Hum Reprod.* 2008;23(12):2755–2765.
129. Stoop H, Honecker F, Van De Geijn GJM, et al. Stem cell factor as a novel diagnostic marker for early malignant germ cells. *J Pathol.* 2008;216(1):43–54.
130. Kaprova-Pleskacova J, Stoop H, Brüggenwirth H, et al. Complete androgen insensitivity syndrome: factors influencing gonadal histology including germ cell pathology. *Mod Pathol.* 2014;27(5):721–730.

131. Oram SW, Liu XX, Lee T-L, Chan W-Y, Lau Y-FC. TSPY potentiates cell proliferation and tumorigenesis by promoting cell cycle progression in HeLa and NIH3T3 cells. *BMC Cancer.* 2006;6:154.
132. Oosterhuis JW, Stoop H, Dohle G, et al. A pathologist's view on the testis biopsy. *Int J Androl.* 2011;34(4 Pt 2):e14–e19; discussion e20.
133. Honecker F, Stoop H, De Krijger RR, Lau Y-FC, Bokemeyer C, Looijenga LHJ. Pathobiological implications of the expression of markers of testicular carcinoma in situ by fetal germ cells. *J Pathol.* 2004;203(3):849–857.
134. Mitchell RT, Saunders PTK, Childs AJ, et al. Xenografting of human fetal testis tissue: a new approach to study fetal testis development and germ cell differentiation. *Hum Reprod.* 2010;25(10):2405–2414.
135. Berney DM, Lee A, Randle SJ, Jordan S, Shamash J, Oliver RTD. The frequency of intratubular embryonal carcinoma: implications for the pathogenesis of germ cell tumours. *Histopathology.* 2004;45(2):155–161.
136. Emerson RE, Ulbright TM. Intratubular germ cell neoplasia of the testis and its associated cancers: the use of novel biomarkers. *Pathology.* 2010;42(4):344–355.
137. Xiao G-Q, Priemer DS, Wei C, Aron M, Yang Qi, Idrees MT. ZBTB16 is a sensitive and specific marker in detection of metastatic and extragonadal yolk sac tumour. *Histopathology.* 2017;71(4):562–569.
138. Osman H, Cheng L, Ulbright TM, Idrees MT. The utility of CDX2, GATA3, and DOG1 in the diagnosis of testicular neoplasms: an immunohistochemical study of 109 cases. *Hum Pathol.* 2016;48:18–24.
139. Svechnikov K, Landreh L, Weisser J, et al. Origin, development and regulation of human Leydig cells. *Horm Res Paediatr.* 2010;73(2):93–101.
140. Griswold SL, Behringer RR. Fetal Leydig cell origin and development. *Sex Dev.* 2009;3(1):1–15.
141. O'Shaughnessy PJ, Baker PJ, Johnston H. The foetal Leydig cell—differentiation, function and regulation. *Int J Androl.* 2006;29(1):90–95; discussion 105–108.
142. Nistal M, Paniagua R, Regadera J, Santamarìa L, Amat P. A quantitative morphological study of human Leydig cells from birth to adulthood. *Cell Tissue Res.* 1986;246(2):229–236.
143. Wu X, Wan S, Lee MM. Key factors in the regulation of fetal and postnatal Leydig cell development. *J Cell Physiol.* 2007;213(2):429–433.
144. Müller J, Skakkebaek NE. Quantification of germ cells and seminiferous tubules by stereological examination of testicles from 50 boys who suffered from sudden death. *Int J Androl.* 1983;6(2):143–156.
145. Brehm R, Rey R, Kliesch S, Steger K, Marks A, Bergmann M. Mitotic activity of Sertoli cells in adult human testis: an immunohistochemical study to characterize Sertoli cells in testicular cords from patients showing testicular dysgenesis syndrome. *Anat Embryol (Berl).* 2006;211(3):223–236.
146. Kao C-S, Idrees MT, Young RH, Ulbright TM. "Dissecting gonadoblastoma" of scully: a morphologic variant that often mimics germinoma. *Am J Surg Pathol.* 2016;40(10):1417–1423.
147. Cortes D, Müller J, Skakkebæk NE. Proliferation of Sertoli cells during development of the human testis assessed by stereological methods. *Int J Androl.* 1987;10(4):589–596.
148. Waters BL, Trainer TD. Development of the human fetal testis. *Pediatr Pathol Lab Med.* 1996;16(1):9–23.
149. Well D, Yang H, Houseni M, et al. Age-related structural and metabolic changes in the pelvic reproductive end organs. *Semin Nucl Med.* 2007;37(3):173–184.
150. Paniagua R, Nistal M, Amat P, Rodriguez MC, Martin A. Seminiferous tubule involution in elderly men. *Biol Reprod.* 1987;36(4):939–947.
151. Suoranta H. Changes in the small blood vessels of the adult human testis in relation to age and to some pathological conditions. *Virchows Arch A Pathol Pathol Anat.* 1971; 352(2):165–181.
152. Kothari LK, Gupta AS. Effect of ageing on the volume, structure and total Leydig cell content of the human testis. *Int J Fertil.* 1974;19(3):140–146.
153. Kaler LW, Neaves WB. Attrition of the human Leydig cell population with advancing age. *Anat Rec.* 1978;192(4):513–518.
154. Neaves WB, Ohnson L, Petty CS. Seminiferous tubules and daily sperm production in older adult men with varied numbers of Leydig cells. *Biol Reprod.* 1987;36(2):301–308.
155. Rowley MJ, Heller CG. Quantitation of the cells of the seminiferous epithelium of the human testis employing the Sertoli cell as a constant. *Z Zellforsch Mikrosk Anat.* 1971;115(4): 461–472.
156. Steinberger A, Steinberger E. Replication pattern of Sertoli cells in maturing rat testis in vivo and in organ culture. *Biol Reprod.* 1971;4(1):84–87.
157. de Miguel MP, Bethencourt FR, Arenas MI, Fraile B, Paniagua R. Intermediate filaments in the Sertoli cells of the ageing human testis. *Virchows Arch.* 1997;431(2):131–138.
158. Cornejo KM, Oliva E, Crotty R, et al. Clinicopathologic features and proposed grossing protocol of orchiectomy specimens performed for gender affirmation surgery. *Hum Pathol.* 2022;127:21–27.
159. Roosen-Runge EC, Holstein AF. The human rete testis. *Cell Tissue Res.* 1978;189(3):409–433.
160. Srigley JR. The paratesticular region: histoanatomic and general considerations. *Semin Diagn Pathol.* 2000;17(4): 258–269.
161. Dinges HP, Zatloukal K, Schmid C, Mair S, Wirnsberger G. Co-expression of cytokeratin and vimentin filaments in rete testis and epididymis. An immunohistochemical study. *Virchows Arch A Pathol Anat Histopathol.* 1991;418(2): 119–127.
162. Tong G-X, Memeo L, Colarossi C, et al. PAX8 and PAX2 immunostaining facilitates the diagnosis of primary epithelial neoplasms of the male genital tract. *Am J Surg Pathol.* 2011;35(10):1473–1483.
163. Rubegni P, Poggiali S, De Santi M, et al. Cutaneous metastases from adenocarcinoma of the rete testis. *J Cutan Pathol.* 2006;33(2):181–184.
164. Hartwick RW, Ro JY, Srigley JR, Ordonez NG, Ayala AG. Adenomatous hyperplasia of the rete testis. A clinicopathologic study of nine cases. *Am J Surg Pathol.* 1991;15(4):350–357.
165. Jones EC, Murray SK, Young RH. Cysts and epithelial proliferations of the testicular collecting system (including rete testis). *Semin Diagn Pathol.* 2000;17(4):270–293.
166. Hittmair A, Zelger BG, Obrist P, Dirnhofer S. Ovarian Sertoli-Leydig cell tumor: a SRY gene-independent pathway of pseudomale gonadal differentiation. *Hum Pathol.* 1997;28(10): 1206–1210.
167. Sapino A, Pagani A, Godano A, Bussolati G. Effects of estrogens on the testis of transsexuals: a pathological and

immunocytochemical study. *Virchows Arch A Pathol Anat Histopathol.* 1987;411(5):409–414.
168. Hinton BT, Keefer DA. Evidence for protein absorption from the lumen of the seminiferous tubule and rete of the rat testis. *Cell Tissue Res.* 1983;230(2):367–375.
169. Saitoh K, Terada T, Hatakeyama S. A morphological study of the efferent ducts of the human epididymis. *Int J Androl.* 1990;13(5):369–376.
170. Shah VI, Ro JY, Amin MB, Mullick S, Nazeer T, Ayala AG. Histologic variations in the epididymis: findings in 167 orchiectomy specimens. *Am J Surg Pathol.* 1998;22(8):990–996.
171. Rowley MJ, Teshima F, Heller CG. Duration of transit of spermatozoa through the human male ductular system. *Fertil Steril.* 1970;21(5):390–396.
172. Hinrichsen MJ, Blaquier JA. Evidence supporting the existence of sperm maturation in the human epididymis. *J Reprod Fertil.* 1980;60(2):291–294.
173. Regadera J, Cobo P, Paniagua R, Martínez-García F, Palacios J, Nistal M. Immunohistochemical and semiquantitative study of the apical mitochondria-rich cells of the human prepubertal and adult epididymis. *J Anat.* 1993;183 (Pt 3): 507–514.
174. Dubé É, Dufresne J, Chan PTK, Hermo L, Cyr DG. Assessing the role of claudins in maintaining the integrity of epididymal tight junctions using novel human epididymal cell lines. *Biol Reprod.* 2010;82(6):1119–1128.
175. Barrachina F, Battistone MA, Castillo J, et al. Sperm acquire epididymis-derived proteins through epididymosomes. *Hum Reprod.* 2022;37(4):651–668.
176. Kasper M, Stosiek P. Immunohistochemical investigation of different cytokeratins and vimentin in the human epididymis from the fetal period up to adulthood. *Cell Tissue Res.* 1989;257(3):661–664.
177. De Miguel MP, Mariño JM, Gonzalez-Peramato P, Nistal M, Regadera J. Epididymal growth and differentiation are altered in human cryptorchidism. *J Androl.* 2001;22(2):212–225.
178. Cerilli LA, Sotelo-Avila C, Mills SE. Glandular inclusions in inguinal hernia sacs: morphologic and immunohistochemical distinction from epididymis and vas deferens. *Am J Surg Pathol.* 2003;27(4):469–476.
179. Lee LY-J, Tzeng J, Grosman M, Unger PD. Prostate gland-like epithelium in the epididymis: a case report and review of the literature. *Arch Pathol Lab Med.* 2004;128(4):e60–e62.
180. Rajalakshmi M, Kumar BV, Kapur MM, Pal PC. Ultrastructural changes in the efferent duct and epididymis of men with obstructive infertility. *Anat Rec.* 1993;237(2):199–207.
181. Oliva E, Young RH. Paratesticular tumor-like lesions. *Semin Diagn Pathol.* 2000;17(4):340–358.
182. Madara JL, Haggitt RC, Federman M. Intranuclear inclusions of the human vas deferens. *Arch Pathol Lab Med.* 1978;102(12):648–650.
183. Nistal M, Santamaría L, Paniagua R. Mast cells in the human testis and epididymis from birth to adulthood. *Acta Anat (Basel).* 1984;119(3):155–160.
184. Nozawa T, Konda R, Ohsawa T, et al. Clear cell papillary cystadenocarcinoma of the epididymis: a case report and immunohistochemistry of markers for renal cell carcinoma. *Histol Histopathol.* 2013;28(3):321–326.
185. Yu CC, Huang JK, Chiang H, Chen MT, Chang LS. Papillary cystadenocarcinoma of the epididymis: a case report and review of the literature. *J Urol.* 1992;147(1):162–165.
186. Gondek AS,Masip VJ, Muñoz CJ, Hernández HS, Zurriaga CR, Castellarnau XT. Adolescent hydrocele carrying a surprise: a case of papillary cystadenoma of the epididymis. *Urology.* 2018;112:172–175.
187. Paniagua R, Regadera J, Nistal M, Abaurrea MA. Histological, histochemical and ultrastructural variations along the length of the human vas deferens before and after puberty. *Acta Anat (Basel).* 1981;111(3):190–203.
188. Paniagua R, Regadera J, Nistal M, Santamaría L. Elastic fibres of the human ductus deferens. *J Anat.* 1983;137 (Pt 3): 467–476.
189. Murakami M, Nishida T, Shiromoto M, Inokuchi T. Scanning and transmission electron microscopic study of the ampullary region of the dog vas deferens, with special reference to epithelial phagocytosis of spermatozoa and latex beads. *Anat Anz.* 1986;162(4):289–296.
190. Ortiz-Rey JA, Domínguez-de Dios J, Pérez-Schoch M, Miguel-Fraile PS, Gómez-de María C. [E-Cadherine and GATA-3 are useful in the confirmation of the vas deferens in vasectomy specimens.] *Rev Esp Patol.* 2020;53(4):218–225.
191. Shah RB, Lee MW, Giraldo AA, Amin MB. Histologic and histochemical characterization of seminal vesicle intraluminal secretions. *Arch Pathol Lab Med.* 2001;125(1):141–145.
192. Shidham VB, Lindholm PF, Kajdacsy-Balla A, Basir Z, George V, Garcia FU. Prostate-specific antigen expression and lipochrome pigment granules in the differential diagnosis of prostatic adenocarcinoma versus seminal vesicle-ejaculatory duct epithelium. *Arch Pathol Lab Med.* 1999; 123(11):1093–1097.
193. Kuo T, Gomez LG. Monstrous epithelial cells in human epididymis and seminal vesicles. A pseudomalignant change. *Am J Surg Pathol.* 1981;5(5):483–490.
194. Kovi J, Jackson MA, Akberzie ME. Unusual smooth muscle change in the prostate. *Arch Pathol Lab Med.* 1979;103(4): 204–205.
195. Linke RP, Joswig R, Murphy CL, et al. Senile seminal vesicle amyloid is derived from semenogelin I. *J Lab Clin Med.* 2005;145(4):187–193.
196. Rath-Wolfson L, Bubis G, Shtrasburg S, Shvero A, Koren R. Seminal tract amyloidosis: synchronous amyloidosis of the seminal vesicles, deferent ducts and ejaculatory ducts. *Pathol Oncol Res.* 2017;23(4):811–814.
197. Pitra T, Pivovarcikova K, Alaghehbandan R, et al. Utility of NKX3.1 immunohistochemistry in the differential diagnosis of seminal vesicles versus prostatic tissue in needle biopsy. *Ann Diagn Pathol.* 2020;49:151644.
198. McNeal JE. Normal histology of the prostate. *Am J Surg Pathol.* 1988;12(8):619–633.
199. Rolnick D, Kawanoue S, Szanto P, Bush IM. Anatomical incidence of testicular appendages. *J Urol.* 1968;100(6):755–756.
200. Sahni D, Jit I, Joshi K, Sanjeev. Incidence and structure of the appendices of the testis and epididymis. *J Anat.* 1996;189 (Pt 2):341–348.
201. Skoglund RW, McRoberts JW, Ragde H. Torsion of testicular appendages: presentation of 43 new cases and a collective review. *J Urol.* 1970;104(4):598–600.
202. Samnakay N, Cohen RJ, Orford J, King PA, Davies RJ. Androgen and oestrogen receptor status of the human appendix testis. *Pediatr Surg Int.* 2003;19(7):520–524.
203. Sadler T. *Langman's Medical Embryology.* 7th ed. Williams & Wilkins; 1995.

204. Whitehead ED, Leiter E. Genital abnormalities and abnormal semen analyses in male patients exposed to diethylstilbestrol in utero. *J Urol.* 1981;125(1):47–50.
205. Bernstein J, Gardner K. Renal cystic disease and renal dysplasia. In: Walsh P, ed. *Campbell's Urology*. Vol. 2. 5th ed. WB Saunders; 1986.
206. Popek EJ. Embryonal remnants in inguinal hernia sacs. *Hum Pathol.* 1990;21(3):339–349.
207. Wollin M, Marshall FF, Fink MP, Malhotra R, Diamond DA. Aberrant epididymal tissue: a significant clinical entity. *J Urol.* 1987;138(5):1247–1250.
208. Mcpherson F, Frias JL, Spicer D, Opitz JM, Gilbert-Barness EF, Splenogonadal fusion-limb defect "syndrome" and associated malformations. *Am J Med Genet A.* 2003;120A(4):518–522.
209. Nistal M, García-Cabezas MA, Castello MC, De Miguel MP, Regadera J. Age-related epididymis-like intratesticular structures: benign lesions of Wolffian origin that can be misdiagnosed as testicular tumors. *J Androl.* 2006;27(1):79–85.
210. Costa WS, Sampaio FJB, Favorito LA, Cardoso LEM. Testicular migration: remodeling of connective tissue and muscle cells in human gubernaculum testis. *J Urol.* 2002;167(5):2171–2176.

37

Penis and Distal Urethra

Elsa F. Velazquez ■ Antonio L. Cubilla

The penis is an external male sex organ, which also serves as the male urinary duct. It is composed of three cylindrical and tubular erectile structures, the two corpora cavernosa and the corpus spongiosum containing the pendulous urethra. The organ is divided into a distal portion that includes glans, coronal sulcus, and foreskin, and a proximal portion, the corpus or shaft (Fig. 37.1) (1). Most penile carcinomas arise from the distal portion of the organ (Fig. 37.2).

DISTAL PENIS

Glans

Anatomic Levels of the Glans

Epithelium
Lamina propria
Corpus spongiosum
Tunica albuginea[a]
Corpus cavernosum[a]

[a]The distal portions of the corpora cavernosa encased by the tunica albuginea are part of the glans in 77% of the cases. The tunica albuginea is part of the corpus cavernosum.

Anatomic Features

The conically shaped glans, covered by a pink smooth mucous membrane, is the most distal portion of the organ and contains the meatus urethralis in its central and ventral regions. The expanded anterior end of the corpus spongiosum, which has the shape of an obtuse cone similar to the cap of a mushroom, is the central and main tissue of the glans. It is molded over and attached to the blunt extremity of the corpus cavernosum and extends farther over their dorsal than their ventral surfaces (1). The base of this conus is an elevated rim or border, the corona, occupying 80% of the circumferential head of the glans; it is interrupted in the ventral portion by the mucosal fold of the frenulum (Fig. 37.1). The diameter of the corona is wider than the shaft and the remainder of the glans. The distal portion of the urethra that opens at the summit of the glans as a vertical, slit-like external orifice termed the urethral meatus (Figs. 37.1 and 37.2). Anatomical structures

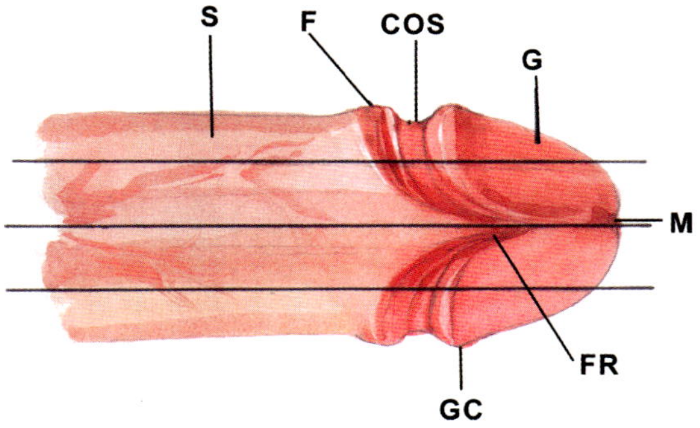

FIGURE 37.1 The penis can be subdivided into a distal portion that includes glans (*G*), coronal sulcus (*COS*), and foreskin (*F*) and a proximal portion, the corpus or shaft (*S*). *M*, urethral meatus; *GC*, glans corona; *FR*, frenulum.

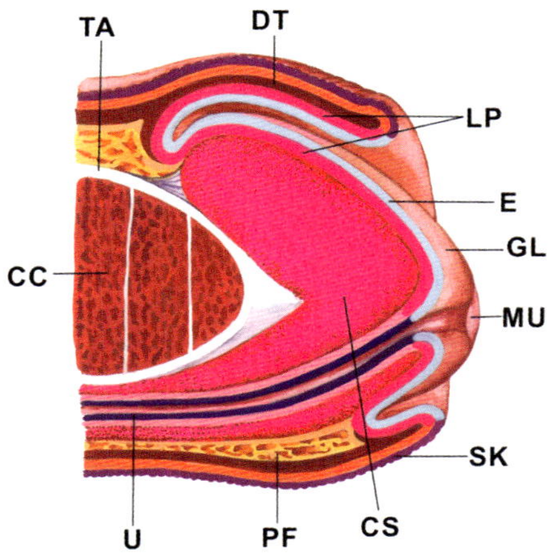

FIGURE 37.2 Diagram illustrating the distal portion of the penis, which includes glans (GL), coronal sulcus, and foreskin. *E*, epithelium; *LP*, lamina propria; *CS*, corpus spongiosum; *TA*, tunica albuginea; *CC*, corpus cavernosum; *DT*, dartos; *SK*, skin; *U*, urethra; *MU*, meatus urethralis; *PF*, penile, or buck fascia.

occasionally found in sexually active males and located in the proximal glans are called pearly penile papules, hirsutoid papillomas, or papillomatosis corona penis or glans (2–5). Grossly, they appear as 1- to 3-mm skin-colored, domed papules evenly distributed circumferentially around the corona and extending proximally on each side of the frenulum. Microscopically, they are squamous papillomas not related to human papillomaviruses (HPVs); however, they may be a source of anxiety for patients as they may be mistaken for warts (2).

Considering the glans as the tissues distal to a line passing through the coronal sulcus, the distal portions of the corpus cavernosum encased by the tunica albuginea extend out into the glans in approximately 77% of the cases (Fig. 37.3) (6). However, since they are the main constituents of the shaft, they are discussed in detail in that section.

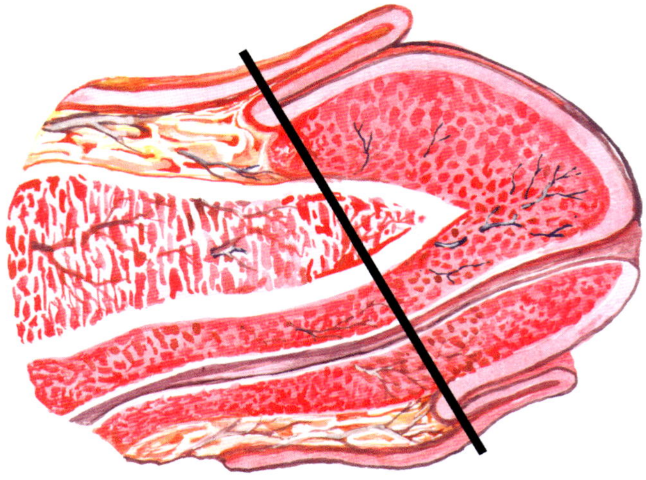

FIGURE 37.3 The distal portion of corpora cavernosa encased by the tunica albuginea is part of the glans in the majority of cases. The diagonal line passing through the coronal sulcus divides glans from shaft.

It is important for the surgical pathologists to recognize the anatomical levels of the glans upon gross examination of carcinoma cases because local extent of tumor determines both the pathologic T stage and the anatomic site of regional lymph node metastases (7,8). In fact, current TNM staging models for penile cancer are mainly based on anatomical level of tumor invasion (9). The epithelium is thin (less than 1 mm in thickness), soft, and white-gray. When hyperplastic, it is brightly white contrasting with the darker tissues below. The lamina propria is white-pink, and measures from 1 to 4 mm in thickness contrasting with the darker reddish color of the highly vascular corpus spongiosum. The tunica albuginea is a thick white fibrous tissue separating corpora spongiosa from cavernosa.

Microscopic and Immunohistochemical Features

› **EPITHELIUM** Both circumcised and uncircumcised men have partially keratinized (5 to 10 cells layer thick) stratified squamous epithelium over the glans (Fig. 37.4). Some textbooks state that the epithelium in circumcised men is thicker and more keratinized than that of uncircumcised individuals but well-controlled studies to support this view are lacking. The normal squamous epithelium is usually positive for keratins AE1/AE3 and 34βE12 and negative for CAM5.2, CK7, and CK20. Expression of p63 is seen in basal/suprabasal cells. Langerhans cells are found scattered among keratinocytes, and they are increased in number in inflammatory conditions. Langerhans cells express S100 protein, CD4, CD1a, and Langerin (Fig. 37.5). Rarely Merkel cells are also present, and they are very difficult to demonstrate by routine or immunohistochemical techniques. They are usually negative for chromogranin and positive for CK20 (10). The glans epithelium and the mucosal epithelium of the foreskin do not appear to contain melanocytes (11,12). Rarely, mucus-producing cells can be noted in the perimeatal region of the glans epithelium (13). They may be the source of the adenosquamous carcinoma of the glans penis. Intraepithelial free nerves ending close to the surface of the epithelium are noted (14). No adnexal or glandular structures are present in the glans. Histologically, the pearly penile papules appear like fibrovascular papillary projections lined by squamous epithelium (Fig. 37.6) (2). No koilocytosis is seen in these structures.

› **LAMINA PROPRIA** The lamina propria is the prolongation of the foreskin lamina propria and separates the corpus spongiosum from the glans epithelium. In the glans, the loose connective tissue of the penile fascia and the fibrous tunica albuginea are lacking so that the lamina propria adheres firmly to the underlying corpus spongiosum (Fig. 37.4) (15). The connective tissue of the lamina propria is somewhat similar to that of the corpus spongiosum, although it is more compact and contains fewer peripheral nerve bundles and elastic fibers than the erectile structure. The transition between lamina propria and corpus spongiosum is not sharply delineated and sometimes difficult to determine at

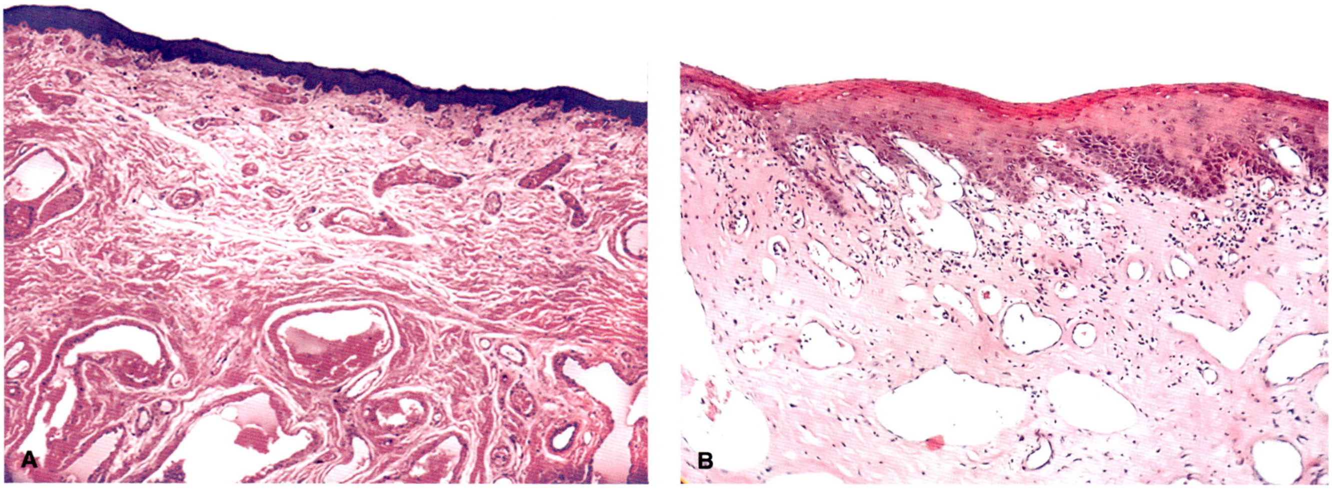

FIGURE 37.4 **A** and **B:** Low- and higher-power views of glans. The three layers of the glans are noted: partially keratinized stratified squamous epithelium, lamina propria, and corpus spongiosum.

medium or high magnifications. However, this delimitation is evident and follows a line corresponding to the geographic limit of the extension of venous sinuses of the corpus spongiosum upon gross examination. The thickness of the lamina propria varies from 1 mm at the glans corona to 2.5 mm near the meatus. Scattered specialized genital corpuscles are identified in the lamina propria underneath the squamous epithelium. These genital corpuscles are found mainly in the glans corona and frenulum. A predominance of free nerve endings over the genital corpuscles has been described in the glans (11,16). A few dermal Merkel cells have been identified at the end of the free nerves.

› ***Corpus Spongiosum*** The corpus spongiosum is the principal component of the glans and is composed of specialized venous sinuses (Fig. 37.7). In the glans, the erectile tissue has the character of a dense, venous plexus (Fig. 37.8) (15). As compared with the corpus cavernosum, the interstitial fibrous connective tissue of the corpus spongiosum is more abundant and contains more elastic fibers and less smooth muscle bundles (Figs. 37.9 and 37.10) (17,18). The

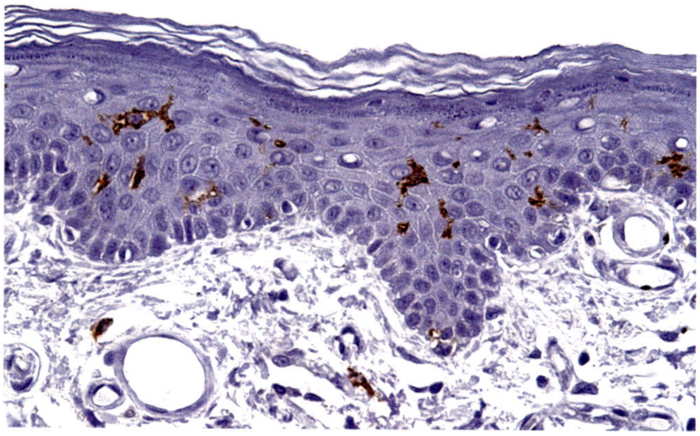

FIGURE 37.5 CD1a immunostain highlights the presence of scattered Langerhans cells in the glans epithelium.

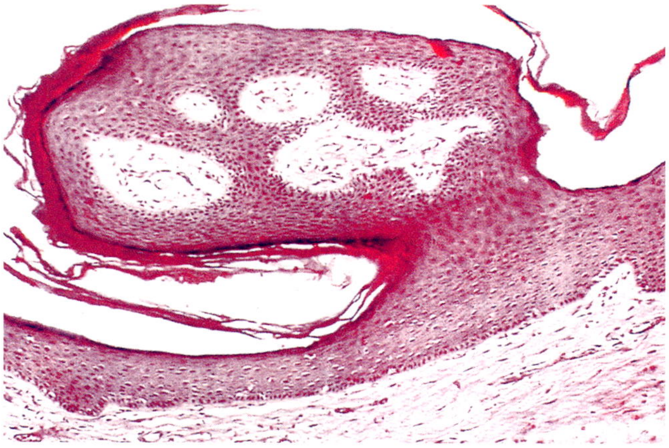

FIGURE 37.6 Pearly penile papule appears histologically as a papillary structure with a fibrovascular core lined by squamous epithelium with no evidence of koilocytosis.

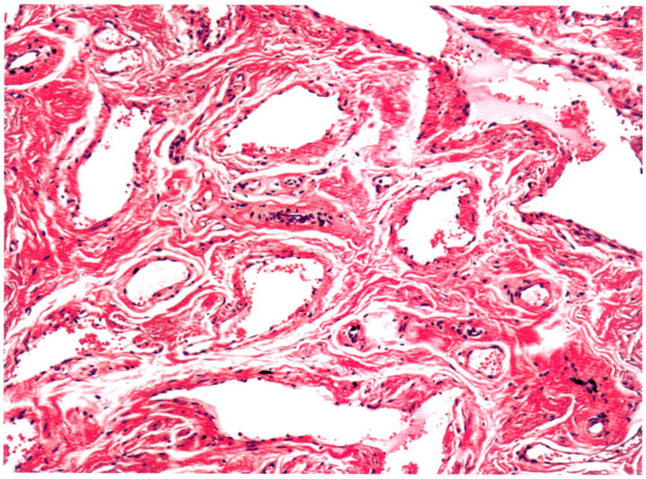

FIGURE 37.7 Glans. Higher-power view of the corpus spongiosum showing specialized venous sinuses.

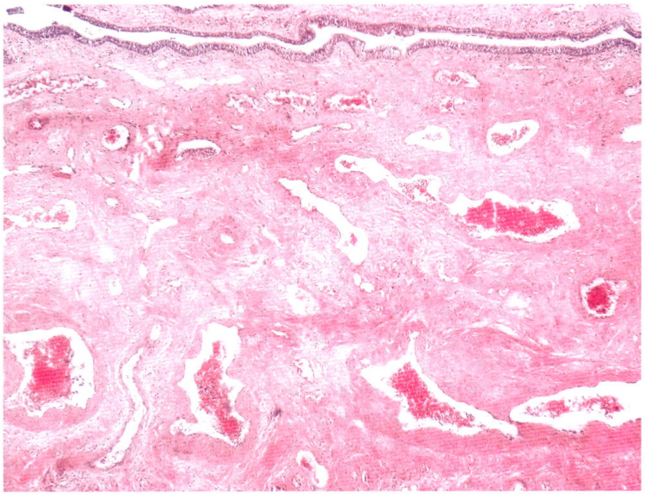

FIGURE 37.8 Corpus spongiosum. Note the dense venous plexus in corpus spongiosum periurethralis.

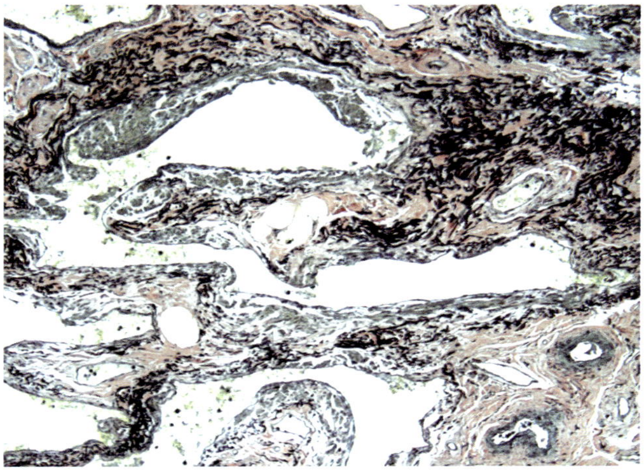

FIGURE 37.9 Corpus spongiosum. The interstitial fibrous connective tissue is more abundant and contains more elastic fibers than in the corpus cavernosum. Note the presence of nutritional veins and arteries in the interstitium of this erectile tissue (van Gieson elastic stain).

loose stroma between the vascular spaces contains some nerve endings and lymphatic vessels. In this erectile tissue, we also find nutritional veins and arteries (Fig. 37.9).

It may be difficult to distinguish corpus spongiosum from cavernosum in poorly oriented biopsies or specimens. Useful features include a wider separation of vascular lumina in the corpus spongiosum (Fig. 37.8) and little or no separation in the corpus cavernosum. In the former, the vessels are thinner and rounder, and in the latter, the vessels are thicker, irregular, or convoluted due to the prominent smooth muscle fibers (Fig. 37.11). Occasionally, especially in the elderly, erectile vessels at the corpus spongiosum are thicker and irregularly shaped with a narrowing of intervascular soft tissues, making a distinction with corpora cavernosa erectile tissues problematic.

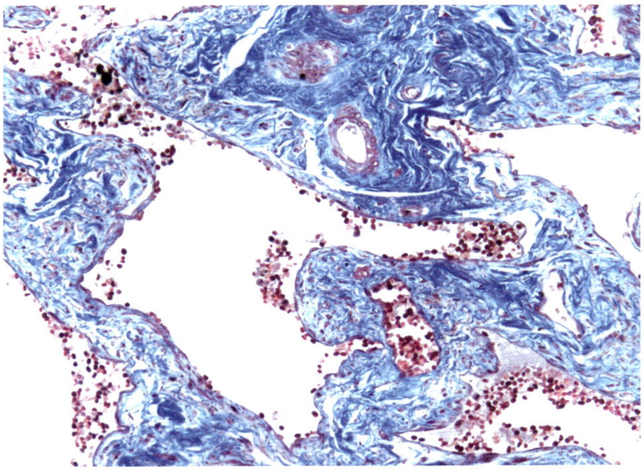

FIGURE 37.10 Corpus spongiosum. The interstitial fibrous connective tissue contains fewer smooth muscle fibers than the corpus cavernosum (Masson trichrome).

Coronal Sulcus

Anatomic Levels of the Coronal Sulcus

Epithelium
Lamina propria
Buck fascia

Anatomic Features

The coronal sulcus is a narrow and circumferential cul-de-sac located proximal to the glans corona (Figs. 37.1 and 37.2). It is found in both lateral and dorsal aspects of the penis, but not in the ventral region, which is occupied by the frenulum, a mucosal fold that fixes the foreskin to the inferior portion of the glans, just below the urethral meatus.

Microscopic and Immunohistochemical Features

Three main histologic layers are seen in the coronal sulcus: (a) the squamous epithelium, identical to the glans epithelium; (b) a thin lamina propria, which is a prolongation of the foreskin and glans lamina propria; and (c) buck fascia

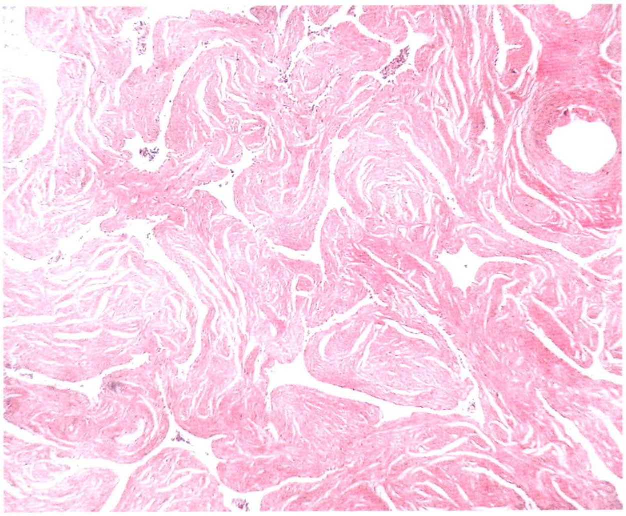

FIGURE 37.11 Corpus cavernosum: note the thick smooth muscle fibers.

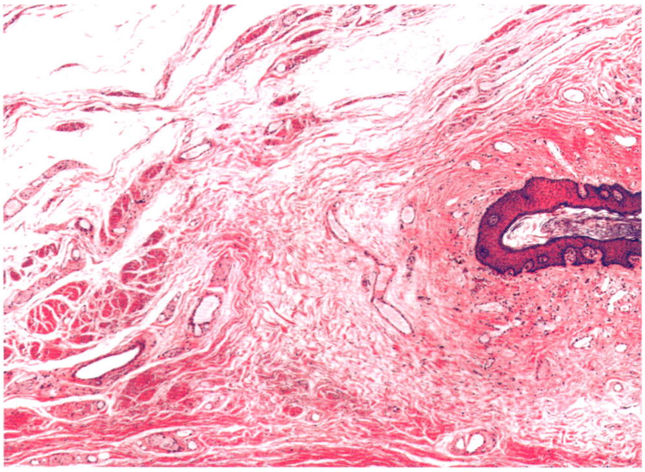

FIGURE 37.12 A section of the coronal sulcus. Histologic components of both glans and penile body are present. This specimen from an uncircumcised person (a portion of the foreskin is seen on **top**) shows the squamous epithelium (**right**), lamina propria below, then buck fascia and dartos smooth muscle bundles (**left**).

and the point of insertion of some of the smooth muscle fibers coming from the penile body dartos (Fig. 37.12).

The epithelium is a stratified squamous epithelium that is 5 to 12 cell layers thick. It is covered by a thin keratin layer and supported by lamina propria composed of loose connective tissue containing numerous capillary and lymphatic vessels, peripheral nerves, and a few Pacini bodies. Buck fascia is composed of fibrous elastic connective tissue containing small- and medium-sized blood vessels and nerves. A fourth layer can be present in more than half of specimens: the dartos (Fig. 37.13), which is a wide and loose layer (6- to 15-mm thick) surrounding the coronal sulcus. The dartos is composed of fibrous connective tissue and irregular bundles of smooth muscle cells. This layer is continuous with the corporal and preputial dartos. There are small blood and lymphatic vessels. The absence of dartos could facilitate the spread of carcinomas that involve the coronal sulcus by allowing them to extend to deeper levels such as the loose and vascular tissue of the buck fascia (19).

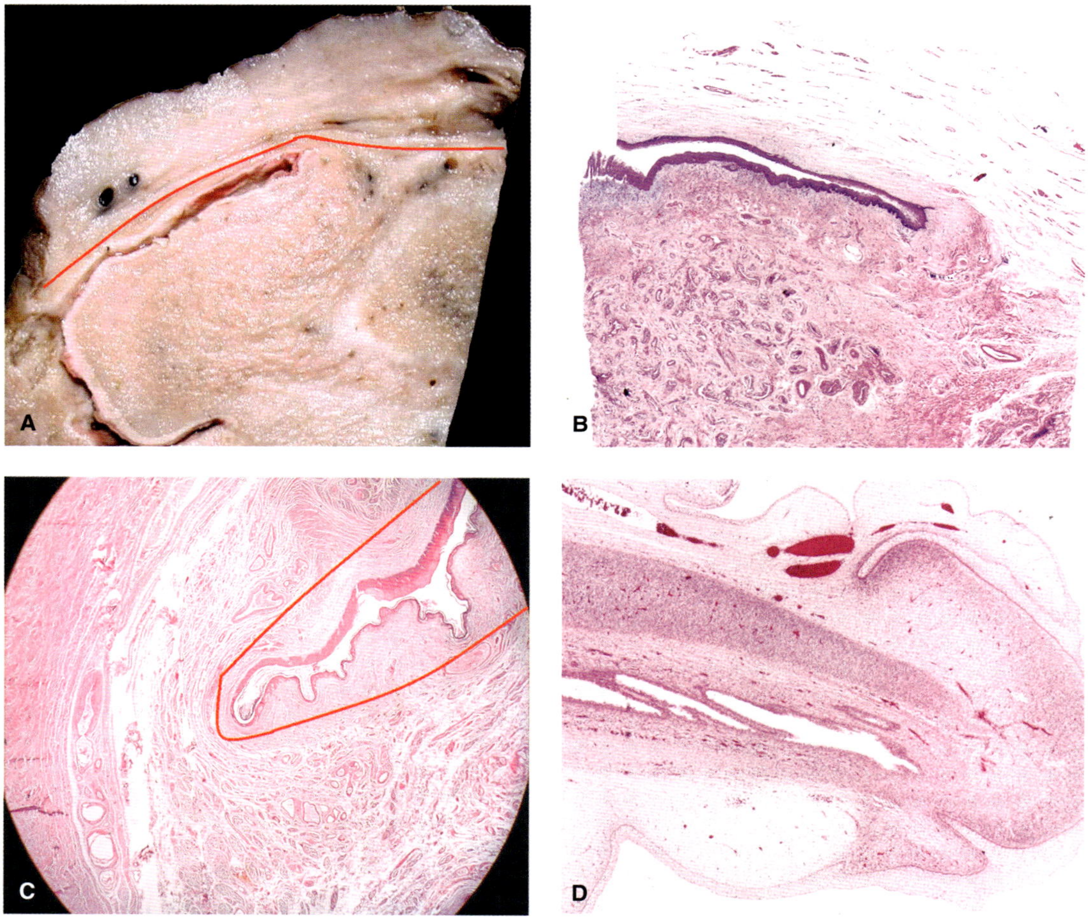

FIGURE 37.13 **A:** Cut surface of a partial penectomy specimen showing the epithelial compartments glans, foreskin, and coronal sulcus. The red line indicates the anatomical location of the dartos layer from foreskin directly to penile shaft. **B:** Low-power microscopic view of the coronal sulcus. **C:** Anatomical levels of coronal sulcus: keratinized squamous epithelium, lamina propria, and dartos. The *red line* indicates the mode in which the dartos encases the coronal sulcus. **D:** Fetal penis. Note the coronal sulcus formation in relation to the buck fascia. Glans and foreskin inner surfaces are firmly adherent.

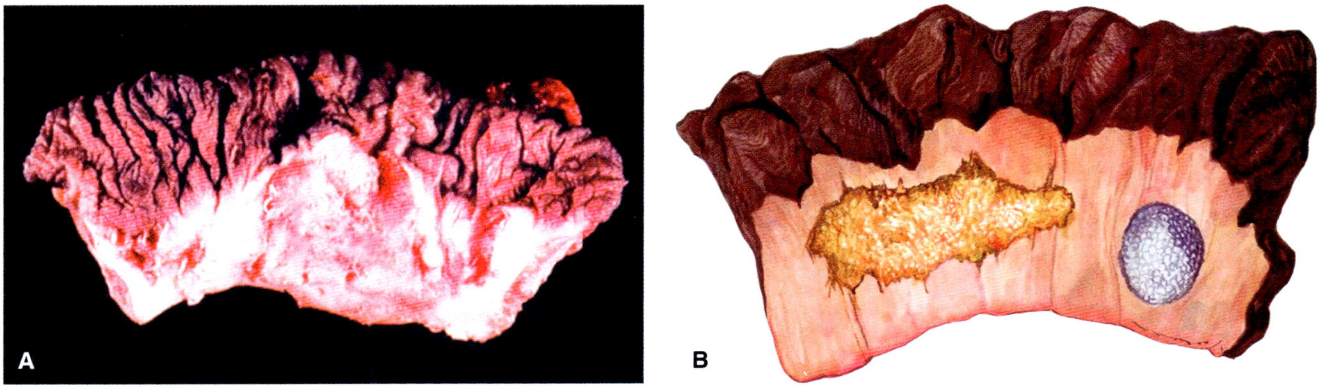

FIGURE 37.14 **A** and **B:** Gross appearance and corresponding diagram of an adult foreskin (prepuce) with multicentric carcinomas. The cutaneous surface (**top**) is darker and more wrinkled. The mucosa (**bottom**) is pale beige and slightly irregular. Two separate carcinomas (in *yellow* and *light blue* in the diagram) are present in the mucosal surface, the most common location of preputial carcinomas.

The coronal sulcus has been reported as the most frequent site of Tyson glands (1,3,17,18,20–23). These modified sebaceous glands are reported to be responsible for smegma production, which represents epithelial debris and secretions collected in this space (24). The existence of Tyson glands is somewhat controversial (11,24,25). Several studies have failed to demonstrate these glands (24,26). We could not find Tyson glands in a pathologic study of 67 totally sectioned penises removed for carcinoma. Apparently, the original descriptions by Tyson (27) were based on primate studies that could not be confirmed in humans. After circumcision, occasionally some sebaceous glands can be found in the mucosa adjacent to the skin but they are probably cutaneous sebaceous glands misplaced after surgery. Sebaceous glands associated with or without hair follicles are found in the penile shaft and cutaneous aspect of the foreskin. We have observed the presence of sebaceous glands that are unrelated to hair follicles at the mucocutaneous junction of the foreskin and adjacent mucosa, but not in the coronal sulcus. The presence of sebaceous gland hyperplasia or ectopic sebaceous glands (Fordyce condition) is more frequent on the penile shaft and foreskin but may also occur in the glans (2).

Foreskin

Anatomic Levels of the Foreskin

- Epidermis
- Dermis
- Dartos
- Lamina propria
- Mucosal epithelium

Anatomic Features and Circumcision

The foreskin or prepuce is the prolongation of the shaft's skin and normally covers most of the glans, reflecting beyond itself and transforming into a mucosal inner surface. The epithelium covering the mucosal inner surface is continuous with the epithelium of the coronal sulcus, as well as the surface of the glans (Figs. 37.1 and 37.2). Grossly, the adult foreskin shows a cutaneous surface (continuous with the skin of the shaft) that is dark and wrinkled and a mucosal surface (continuous with the coronal sulcus) that is pink to tan (Fig. 37.14).

Male circumcision is the removal of some or all the foreskin from the penis. The word "circumcision" comes from Latin *circum* (meaning "around") and *cædere* (meaning "to cut"). Male circumcision is commonly practiced for religious, cultural, and medical reasons. From very early years to the present, there have been controversies regarding the role of the foreskin and the importance of circumcision (28–33). Preputial functions are related to protection of the glans from external irritation or contamination, and it has been shown that the foreskin is a normal erogenous tissue (11,31,34). In addition, it has been demonstrated that the squamous mucosa of the glans, coronal sulcus, and foreskin are fused during the embryologic development of the penis (Fig. 37.15), and they can be considered as one tissue compartment. The fused mucosa of the glans and inner lining of the foreskin separate gradually over the years (11). Most newborn males show an unretracted foreskin at the time of delivery (35). When boys reach the age of 5 to 6 years, the foreskin can be completely retracted in most cases beyond the level of the glans corona. According to these observations, a tight preputial orifice due to an immature preputial plate does not represent an adhesion but a normal stage of penile development. Therefore, neonatal circumcision before the foreskin has naturally separated involves tearing the common prepuce/glans mucosa apart, with the possible complications of excoriation and injury to the glans and ablation of the frenular artery and meatal stenosis (11).

On the other hand, different studies suggest that circumcision reduces the risk of urinary tract infection, common sexually transmitted infections, and penile carcinoma (28,36–44). It has been proposed that the mucosal inner surface of the foreskin from newborns shows a propensity to be colonized by fimbriated bacteria, with the subsequent

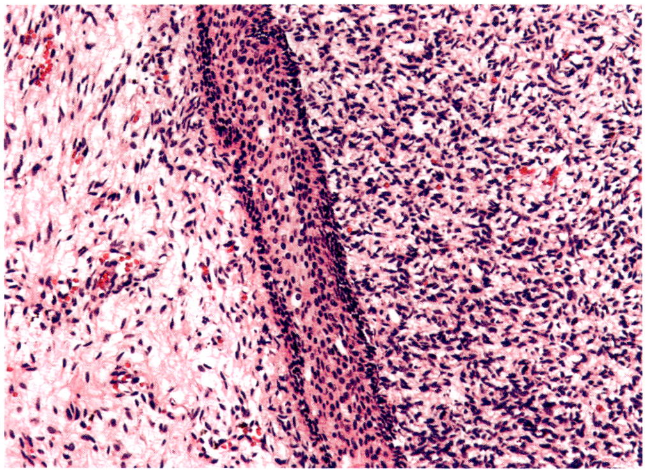

FIGURE 37.15 Fused squamous mucosa of the glans, coronal sulcus, and foreskin during the embryologic development of the penis. Note the common immature epithelial plate (**center**). The dense stromal cells of the glans will form the corpus spongiosum (**right**). The lamina propria of the glans is not yet formed. The foreskin is on the **left**.

occurrence of serious urinary tract infection (38). Studies suggest that male circumcision is associated with a reduced risk of penile HPV infection and, in the case of men with a history of multiple sexual partners, a reduced risk of cervical cancer in their current female partners (42–44). Furthermore, randomized controlled trials have shown that adult circumcision reduces the risk of acquiring human immunodeficiency virus (HIV) infection (45,46). However, it appears clear that the risk of sexually transmitted infections, including HPV and HIV, correlates with sexual behavior, and behavioral factors appear to be far more important risk factors than circumcision status. Moreover, proper hygiene and access to clean water have been shown to reduce the rate of development of squamous cell carcinoma of the penis in the uncircumcised population (44). Nevertheless, in areas where safe sexual and hygienic practices are poorly adhered to, circumcision can have a relative protective effect against the transmission of sexually transmitted infections and may lower the incidence of penile carcinoma (37). In general, complications of circumcision are minor and treatable, especially at young ages. Higher frequency of complications, and severe complications, are seen when the procedure is undertaken by inexperienced providers, in nonsterile settings, or with inadequate equipment and supplies (30). Existing scientific evidence demonstrates potential medical benefits of newborn male circumcision; however, these data are not sufficient to recommend routine neonatal circumcision. Parents should be given accurate and unbiased information to determine what is in the best interest of the child in each individual case. In a policy statement, the American Academy of Pediatrics stated that "although health benefits are not great enough to recommend routine circumcision for all male newborns, the benefits of circumcision are sufficient to justify access to this procedure for families choosing it and to warrant third-party payment for circumcision of male newborns. It is important that clinicians routinely inform parents of the health benefits and risks of male newborn circumcision in an unbiased and accurate manner." "Parents ultimately should decide whether circumcision is in the best interests of their male child. They will need to weigh medical information in the context of their own religious, ethical, and cultural beliefs and practices. The medical benefits alone, may not outweigh these other considerations for individual families" (47).

Studies related to the relation of length and amount of smegma in the balanopreputial sulcus show that phimosis is found in 4% of boys 6 to 17 years of age, but with a diminishing incidence in later years (28,48). In nonphimotic boys, where the preputial space can be inspected, smegma is present in 5% of the cases. Production of smegma appears to increase in quantity in the 16- to 17-year-old group (48). In a recent prospective study in a high-risk population, we found a variation in foreskin length (Fig. 37.16): long phimotic foreskins were significantly more frequent in patients with penile carcinoma as compared with the general population (28). Coexistence of a long foreskin and phimosis may explain the high incidence of penile cancer in some geographic regions, and circumcision in patients with long and phimotic foreskins living in high-risk areas may be indicated (28).

The foreskin is frequently involved by carcinomas, usually as a secondary invasion from tumor originating in the glans, the most common site of penile cancer. However, there are carcinomas exclusive of the foreskin. It is important to identify these primary preputial tumors since they have a better prognosis than those of other penile sites (49). Primary foreskin squamous cell carcinomas most frequently affect the mucosal surface of the prepuce, tend to be of lower grade, and are multicentric (Fig. 37.14) compared with those of the glans (13,50). Similarly to the glans, there

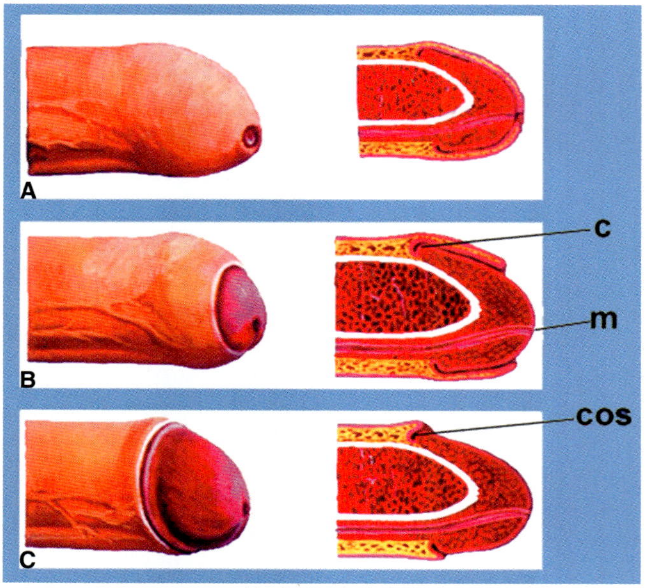

FIGURE 37.16 Foreskin length variation. **A:** The most common type, long foreskin. **B:** Intermediate. **C:** Short, unusual. *c* and *cos*, coronal sulcus; *m*, meatus urethralis.

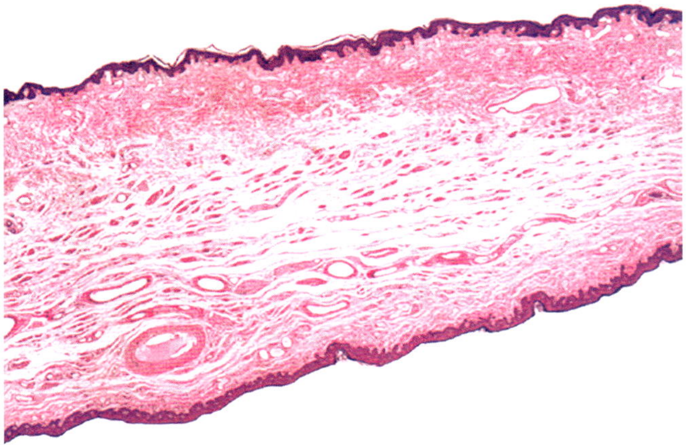

FIGURE 37.17 Full thickness of the foreskin showing all five layers: keratinized stratified squamous epithelium (**top**), dermis, dartos, submucosa, and squamous epithelium of the mucosal portion (**bottom**).

is a correlation between foreskin anatomical levels of invasion and regional cancer spread (13,50).

Microscopic and Immunohistochemical Features

The prepuce is formed by a midline collision of ectoderm, neuroectoderm, and mesenchyme, resulting in a pentalaminar structure (11). There are five layers in the histologic evaluation of the foreskin (Fig. 37.17):

1. *The epidermis* consists of a keratinized stratified squamous cell epithelium that is similar to the epidermis seen in the cutaneous tegument. Melanocytes, Langerhans cells, and Merkel cells are also present. Compared to the mucosal epithelium, the epidermis is thinner and shows better-developed rete ridges and usually a pigmented basal layer (Fig. 37.18). Vellus hairs, sebaceous, and sweat glands may be seen connected to the epidermis. Carcinomas arising in the epidermis of the foreskin are vanishingly rare.

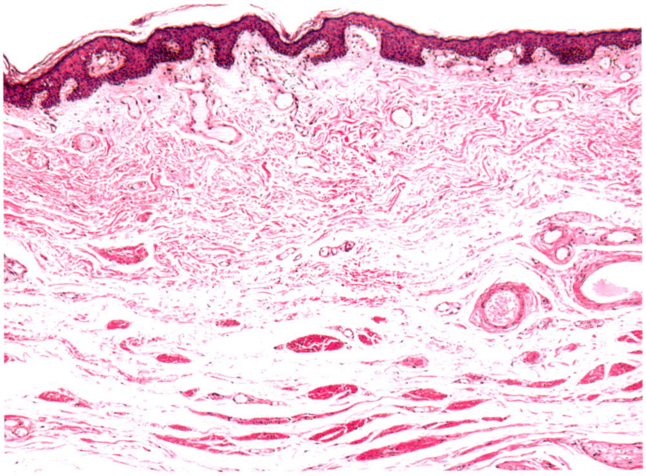

FIGURE 37.18 Cutaneous surface of the foreskin. Compared to the mucosal epithelium, the epidermis is thinner with better-developed rete ridges and usually a pigmented basal layer. Underneath the dermis the smooth muscle bundles of the dartos can be appreciated.

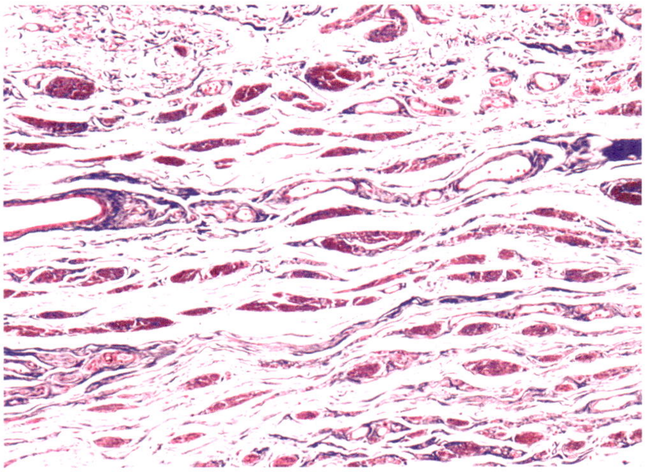

FIGURE 37.19 Smooth muscle bundles, the main component of the preputial dartos, are shown in *red* (Masson trichrome).

2. *The dermis* of the foreskin consists of connective tissue with blood vessels and nerve bundles. Meissner corpuscles are present in the dermal papillae and a few Vater–Pacini corpuscles may be found in deeper areas. Scattered vellus hairs and sebaceous and sweat glands are noted, and they are usually confined to the dermis without extension beyond the dartos. The dermis appears to have more elastic fibers than the lamina propria.
3. *The dartos* consists of smooth muscle fibers invested with elastic fibers, and it is the central axis of the foreskin (Fig. 37.19). From the foreskin, the delicate penile dartos surrounds the penile shaft and is continuous with the scrotal dartos. Similar to the penile body dartos, the smooth muscle cell bundles vary in their disposition. At the edge of the foreskin, the fibers are transversely arranged to form a sphincter to close this edge over the distal end of the glans. There are numerous nerve endings in close association with smooth muscle cells (Fig. 37.20). Nerve bundle

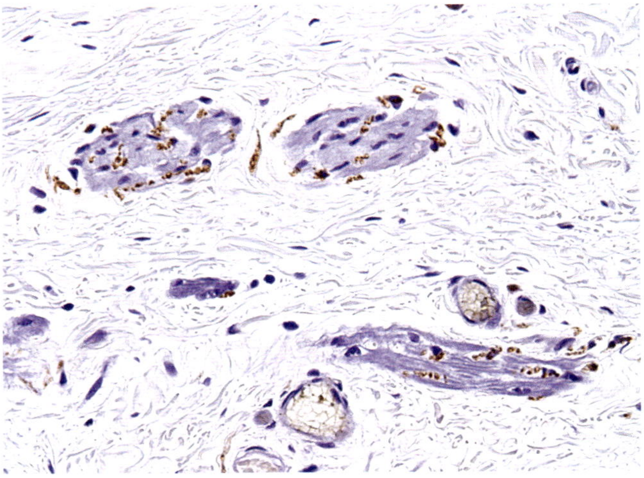

FIGURE 37.20 Neurofilament immunostain highlights numerous free nerve endings associated with smooth muscle bundles in the foreskin dartos.

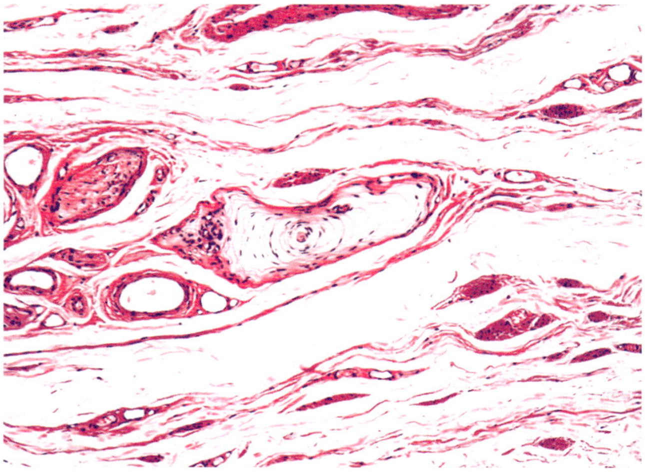

FIGURE 37.21 A Vater–Pacini corpuscle is present deep in the dartos layer of the foreskin.

density in the foreskin is highest in the ventral preputial tissue (mean: 17.9 bundles per nm) as opposed to lateral (8.6 per nm) and dorsal (6.2 per nm) tissues (51). A few Vater–Pacini corpuscles may be found scattered between these nerve bundles (Fig. 37.21).

4. *The lamina propria or chorion*, is composed of a vascular connective tissue looser than the glans lamina propria. Scattered genital corpuscles and free nerve endings are seen in the lamina propria immediately underneath the epithelium. The genital corpuscles are usually found in clusters of three to five (Fig. 37.22). Some authors believe that the corpuscular receptors are more numerous at the mucocutaneous junction of the foreskin; however, this assertion needs further study (11,31). The mucosal lamina propria is devoid of hair follicles and sweat and sebaceous glands. We have observed a few specimens with rare sebaceous glands unrelated to hair structures at the mucocutaneous junction and immediately adjacent foreskin mucosa, but it is not clear if these represent ectopic glands or variation of normal anatomy.

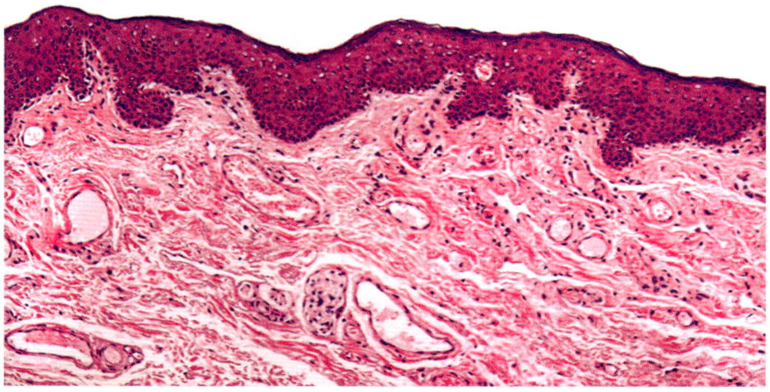

FIGURE 37.23 The squamous epithelium of the mucosal portion of the foreskin is usually thicker with more irregular and broad rete ridges than in the cutaneous portion. Adnexal structures are absent.

5. *The mucosal squamous epithelium* (Fig. 37.23) is identical to and a prolongation of the epithelia of the glans and coronal sulcus. Toward the free edge of the foreskin and approaching the mucocutaneous junction, the basal layer shows a progressive pigmentation more similar to what is seen in the epidermis of the cutaneous portion of the foreskin (Fig. 37.24). The immunohistochemical characteristics of this epithelium are identical to the glans squamous epithelium. Langerhans and Merkel cells are present but not melanocytes. Intraepithelial nerves have been described (11,14). Rarely, mucous cell metaplasia of foreskin squamous mucosal epithelia may be found.

Anatomical Features Related to Staging and Cancer Spread

The anatomical levels of the glans (epithelium, lamina propria, corpus spongiosum, and corpus cavernosum), and foreskin (mucosal epithelium, lamina propria, dartos, dermis, and epidermis) are important landmarks for cancer staging and prediction of cancer progression. It is the base on which the TNM system was constructed (9) and crucial

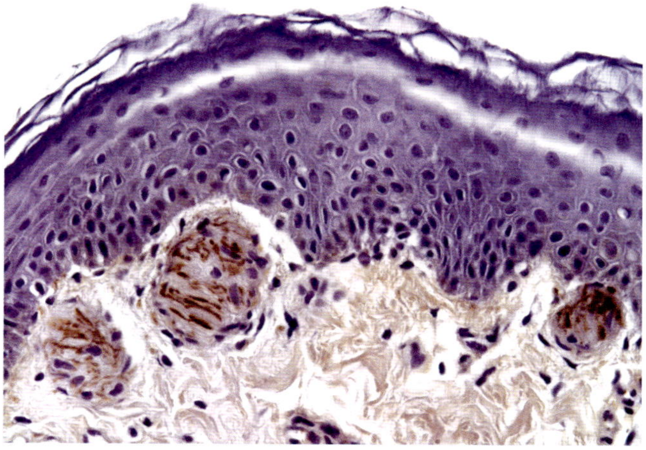

FIGURE 37.22 The genital corpuscles are usually found in clusters of three to five underneath the mucosal epithelium, in the lamina propria of the foreskin (neurofilament immunostain).

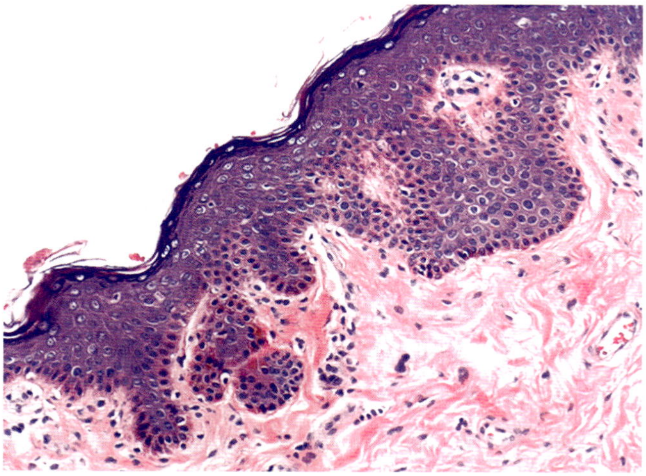

FIGURE 37.24 Foreskin showing the squamous epithelium and its pigmentation in basal layers.

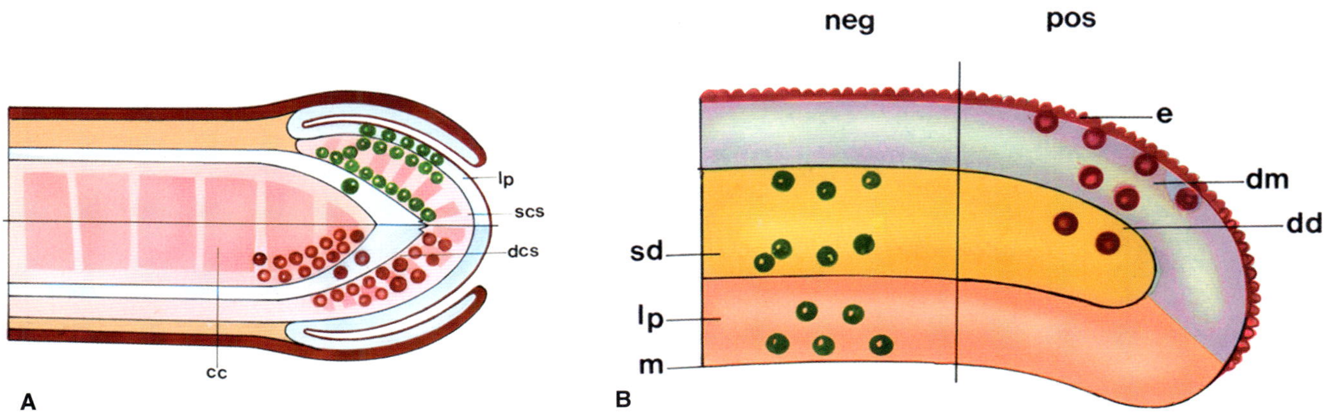

FIGURE 37.25 Anatomical features related to cancer spread. A: Diagram representing cut surface of glans. The horizontal line separates sites associated with and without nodal metastasis. Above, *green dots* represent negative nodes, and below, *red dots* represent nodal metastasis. The vast majority of tumors compromising corpora cavernosa are associated with regional spread. **B:** This diagram represents the cut surface of the foreskin. Tumors associated with nodal metastasis were those invasive of the deep dartos or outer skin. *m*, mucosal epithelium; *lp*, lamina propria; *sd*, superficial dartos; *dd*, deep dartos; *dm*, dermis; *e*, epidermis.

for the design of the prognostic index and nomograms, useful tools for cancer spread and outcome prediction (52–54). This correlation is illustrated in the diagrams. In a series of 51 patients with penile squamous cell carcinoma, a larger number of nodal metastases were found in tumors infiltrating corpus cavernosum, whereas tumors limited to lamina propria only rarely metastasized (Fig. 37.25A). Likewise, in a series of 20 carcinomas exclusive of the foreskin most of the metastatic cases were tumors infiltrating deep dartos or outer dermis (Fig. 37.25B). Occasionally, it is difficult to stage a superficially invasive glans tumor as T1 or T2 due to lack of precise anatomical delimitation or boundary between the lamina propria and the corpus spongiosum. However, the lamina propria rarely measure more than 3 mm in thickness and, thus, any tumor surpassing this measure should probably be considered to invade the corpus spongiosum and be staged as T2.

PROXIMAL PENIS (OR SHAFT)

Anatomical Levels of the Penile Shaft

- Epidermis
- Dermis
- Dartos
- Buck fascia
- Tunica albuginea
- Corpora cavernosa
- Corpus spongiosum

Anatomic Features

The penile shaft, body, or corpus, of the penis is mainly composed of three cylindrical masses of erectile tissue, the two corpora cavernosa and a corpus spongiosum with central urethra (Fig. 37.26). The posterior portions of the corpora cavernosa are two divergent and gradually tapering structures, called crura, that insert into the ischiopubic bone from where they converge to fuse at the level of the inferior portion of the pubic symphysis. The distal three-fourths of the two corpora cavernosa are intimately bound together and make up the greater part of the shaft of the penis. They retain a uniform diameter in the shaft and terminate anteriorly in a bluntly rounded extremity, being embedded in a cap formed by the corpus spongiosum of the glans (1). The erectile tissue of the

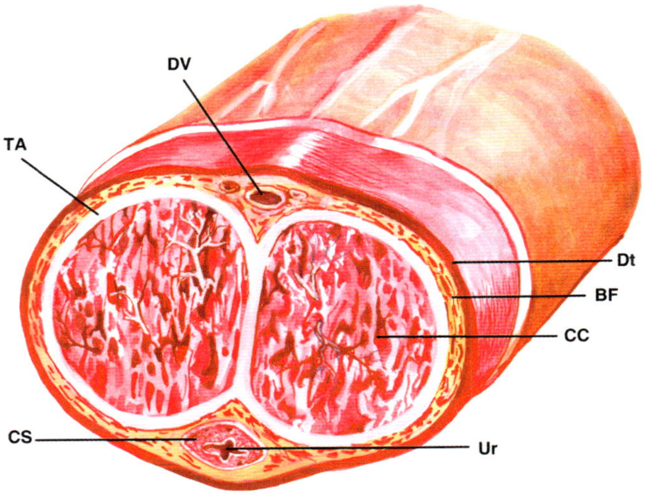

FIGURE 37.26 Cross section of the penis illustrating both corpora cavernosa (corpus cavernosum), each surrounded by the tunica albuginea (*TA*); they form the septum of the penis by their junction in the median plane. A shallow groove that marks their junction on the upper surfaces lodges the deep dorsal vein of the penis (*DV*). The dorsal arteries are located on both sides of the vein. Note the corpus spongiosum (*CS*) and central urethra (*Ur*) located in the concave space on the undersurface of both corpora cavernosa, buck fascia (*BF*), and dartos (*Dt*).

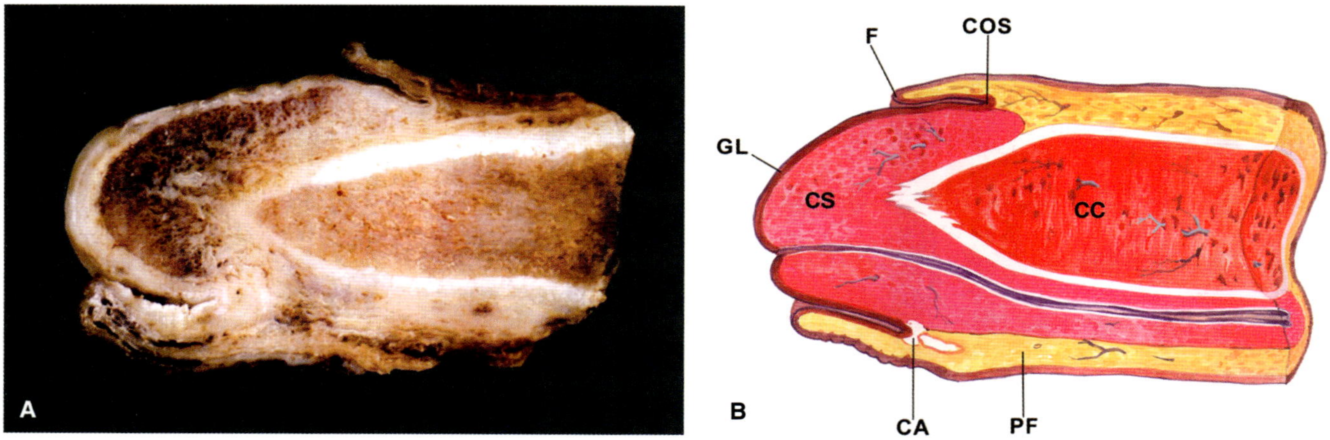

FIGURE 37.27 Gross picture (**A**) and diagram (**B**) of two longitudinal sections of a partial penectomy specimen with a small squamous cell carcinoma (*CA*) located in the coronal sulcus (*COS*). The albuginea (*in white*) surrounds the corpus cavernosum. The diagram from a parallel but more central section of the same specimen, illustrates the urethra running within the corpus spongiosum. *F*, foreskin; *GL*, glans; *PF*, buck fascia; *CS*, corpus spongiosum; *CC*, corpus cavernosum.

corpus cavernosum is a vast, sponge-like system of irregular vascular spaces fed by the afferent arteries and drained by the efferent veins. In the flaccid condition of the organ, the cavernous spaces contain little blood and appear as collapsed irregular clefts. In erection, they become large cavities engorged with blood under pressure (17).

The corpus cavernosum is surrounded by a firm, thick, fibrous envelope, the tunica albuginea. In the flaccid state, the tunica albuginea measures 2 to 3 mm in thickness and becomes thinner (about 0.5 mm) during erection. On longitudinal sections of the organ, the albuginea covering the corpus cavernosum terminates in a ">"-shaped pattern variably ending beyond or at the level of coronal sulcus or, less frequently, behind it (Fig. 37.27) (6). The tunica albuginea enveloping the corpus cavernosum is thicker and less elastic than that surrounding the corpus spongiosum (17,18). The superficial longitudinal fibers of the tunica albuginea form a single tube that encloses both corpora cavernosa whereas the deep fibers are arranged circularly around each corpus, forming the septum of the penis by their junction in the median plane (Fig. 37.26) (1). The septum is thick and complete in the proximal shaft and discontinuous distally. A shallow groove that marks their junction on the upper surfaces lodges the deep dorsal vein of the penis.

The corpus spongiosum and central urethra are located in the concave space on the undersurface of both corpora cavernosa (Fig. 37.26). The middle portion of the corpus spongiosum located in the penile shaft is a uniform cylinder somewhat smaller than the corpus cavernosum. At its ends, it expands, the distal extremity forming the glans and the proximal forming the bulb. The urethra enters the corpus spongiosum 1 to 2 cm from the posterior extremity of the bulb by piercing the dorsal surface. The bulb is just superficial to the urogenital diaphragm, and its posterior portion projects backward toward the anus beyond the entrance of the urethra.

The three cylindrical structures forming the penile shaft are covered by a thin, delicate, and elastic skin. Beneath the dermis, there is a discontinuous smooth muscle cell layer, the dartos (Fig. 37.26), embedded in a thin layer of connective tissue corresponding to the superficial fascia of the classical descriptions. Between the dartos and the albuginea, lies a highly elastic yellowish tube-like sheath encasing both corpora cavernosa and the corpus spongiosum; this is designated as buck fascia (i.e., deep penile fascia) (Figs. 37.26 and 37.27). A septum of fascia extends inward between the corpus cavernosum and the corpus spongiosum, providing separate tubular investments for these columns of erectile tissue and dividing the penis into its dorsal (corpus cavernosum) and ventral (corpus spongiosum) portions (55). When using the terms *fascia* or *penile fascia* in this chapter, we are referring to buck fascia since the superficial fascia is just part of the connective tissue surrounding the dartos. In hematoxylin and eosin (H&E)-stained sections, it is difficult to distinguish and separate superficial from deep penile fascias. For practical purposes, it is better to designate them as one fascia.

Microscopic and Immunohistochemical Features

› **SKIN** The skin covering the penile shaft is rugged and elastic. It shows a thin epidermis composed of a few cell layers and minimal keratinization. The epidermis shows well-formed rete ridges and variable pigmentation of the basal layer. The dermal papillae are thin and deep. Hair follicles are present in the dermis of the penile body and are more numerous in the proximal body. Hair follicles and other adnexa can extend to the cutaneous foreskin in some individuals. They are scanty and contain no piloerector muscle. There are a few sebaceous glands unrelated to hair follicles. Occasionally, there are also poorly developed sweat glands.

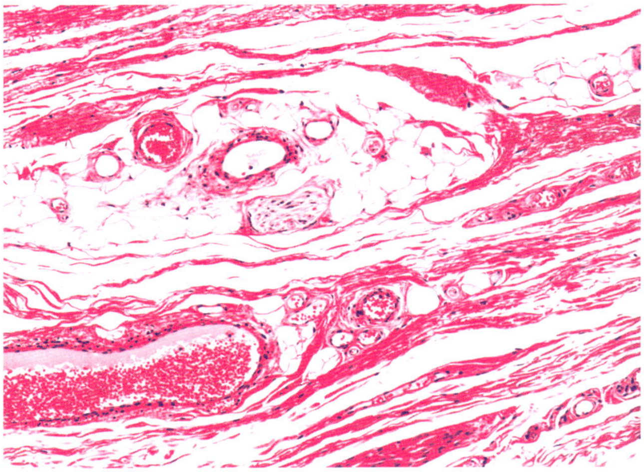

FIGURE 37.28 Buck fascia is composed of loose connective tissue with adipose cells, numerous blood vessels, and peripheral nerve bundles.

› **DARTOS** The penile dartos is composed of a discontinuous layer of smooth muscle cells that are variably arranged in transverse and longitudinal bundles. Some bundles end at the coronal sulcus whereas others run farther to become the preputial dartos. The dartos is embedded in a loose fibrovascular connective tissue with numerous nerve bundles that correspond to the superficial penile fascia, and it is the penile equivalent of the skin subcutaneous tissue or hypodermis, but without adipose tissue (1). Similarly to the scrotal smooth muscle cell component, the penile dartos produces a retraction of genital structures when the exterior temperature falls.

› **BUCK FASCIA** Buck fascia is a well-developed and continuous fibrovascular sheath that encases the corpus cavernosum and the corpus spongiosum. It is composed of loose connective tissue with numerous blood vessels and peripheral nerve bundles running within and beneath it (56). Vater–Pacini corpuscles are often seen in the penile fascia. Its yellow color is due to the presence of adipose tissue and abundant elastic fibers (Fig. 37.28). The skin and dartos slide over this fascia. Buck fascia is important from the surgical pathology point of view since it is a frequent pathway of tumor invasion in penile cancer progression (57). This is most likely due to the loose quality of this tissue and the presence of numerous lymphovascular and neural structures. Some investigators point to buck fascia as the site of origin of Peyronie disease (56).

› **TUNICA ALBUGINEA** The tunica albuginea is a thick sheath of partially hyalinized collagen fibers covering both the corpus cavernosum and corpus spongiosum. It is a poorly vascular structure, with only a few branches of circumflex vessels traversing through it, as demonstrated by factor VIII and CD31 immunostains. It is mainly composed of collagen fibers arranged in an outer longitudinal and an inner circular layer (17,18). The outer layer, which appears to determine the variation in the thickness and strength of the tunica, is absent in the ventral portion of the corpus spongiosum, transforming this portion of tunica into a vulnerable area to perforation. This anatomic aspect probably explains why most prostheses tend to extrude in this area (58). The tunica albuginea forms an incomplete fibrous septum separating both corpora cavernosa. The collagen fibers are wavy in the flaccid state and become straight during erection. The fibers are arranged in such a way so as to permit some elasticity necessary for erection. Elastic fibers are rare in the tunica albuginea of the corpus cavernosum. The tunica albuginea surrounding the corpus spongiosum is thinner and contains more elastic fibers than the one around the corpus cavernosum. In some unusual cases, an infection of the lower urinary tract can spread to the corpus spongiosum to cause Fournier gangrene (59). Eventually, the tunica albuginea may be penetrated, and, with involvement of buck fascia, the infection can rapidly spread to the dartos and directly extend to Colles scrotal fascia and Scarpa fascia of the anterior abdominal wall. The infection can spread to the buttocks, thigh, and ischiorectal space. The tunica albuginea is probably the real barrier to the infiltration of squamous cell carcinoma, contrary to the old concept that buck fascia was the barrier to the spread of cancer (57).

› **CORPORA CAVERNOSA** The corpora cavernosa are the main anatomic structures used during erection. The substance of the corpora cavernosa consists of a three-dimensional network of trabeculae. These are composed of connective tissue and smooth muscle cells and are covered by endothelium, creating a network of interanastomosing vascular spaces between them. These spaces tend to be larger in the central parts of each corpus cavernosum and smaller at their peripheries (17,18). It seems that the smooth muscle cell bundles are the main components of the trabeculae in the corpora cavernosa (Fig. 37.29). There is a highly structured criss-crossing of interconnected fibers and spaces that are

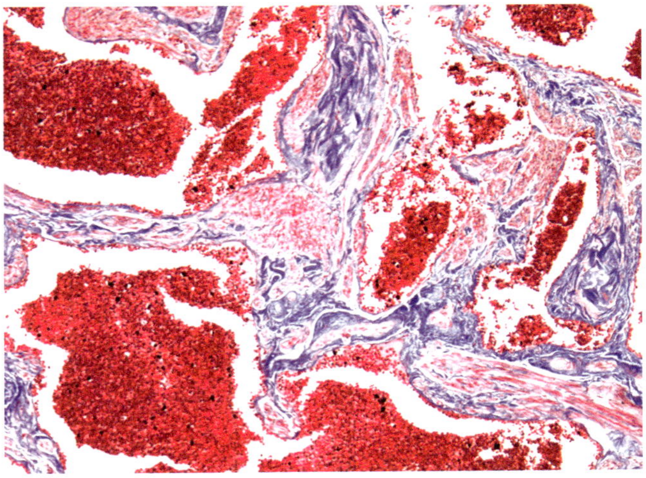

FIGURE 37.29 Corpus cavernosum. The interstitial fibrous connective tissue contains more smooth muscle fibers than in the corpus spongiosum (Masson trichrome).

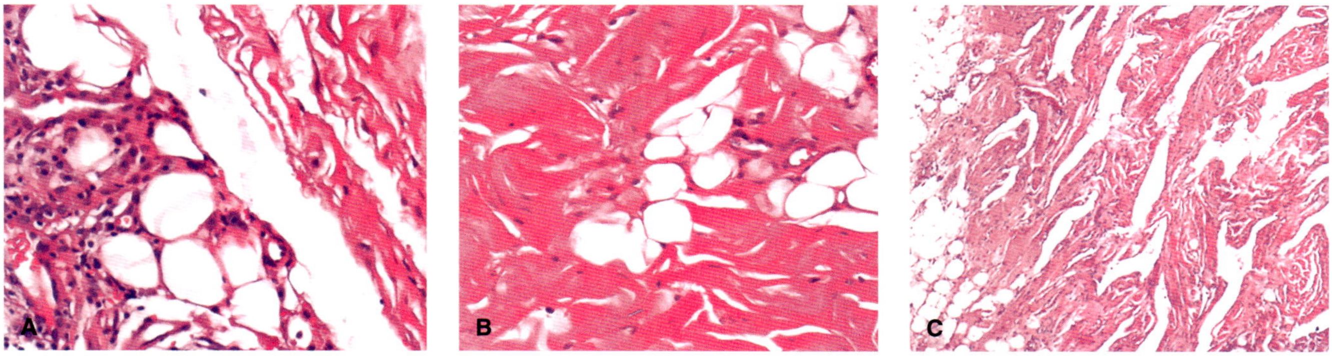

FIGURE 37.30 Adipose tissue. **A:** High-power view with adipose tissue as part of the penile fascia, which surrounds the tunica. **B:** High-power view of a cluster of adipose tissue cells within the tunica albuginea. **C:** High-power view showing adipose tissue cells within the corpus cavernosum (*CC*).

tensed as the cylinder expands during erection (60,61). This creates an internal strength and rigidity that is far greater than that possible in a hollow tube filled to equivalent pressure (61). In the flaccid state, the vascular spaces appear as 1-mm slits but they increase several times in diameter with erection. The interconnection between the venous sinuses is so wide that, if a contrast is injected at one point, both corpora cavernosa can be immediately and completely visualized. The precise nature of vascular connections between the corpus cavernosum and corpus spongiosum remains controversial. Cavernospongious arterial anastomoses have been described but their physiologic roles in erection remain unknown (62). These arterial anastomoses could explain how drugs penetrate into the corpus cavernosum via the corpus spongiosum after transurethral diffusion (62,63). No arteriovenous shunts or venous connections between the corpus cavernosum and the corpus spongiosum have been demonstrated (63).

A progressive increase of collagen fibers and decrease of smooth muscle cells and elastic fibers may be seen in the cavernous trabeculae over the course of time (64).

› *Adipose Tissue in Corpora Cavernosa* Adipose tissue, arteries, veins, and peripheral nerves are normal constituents of mesenchymal tissues present in penile fascias, which encases the corpus cavernosum at the level of penile shaft. We recently reported the presence of adipose tissue within the corpus cavernosum in a study of 63 consecutive partial penectomy specimens for squamous cell carcinoma. Fat was present within the tunica albuginea in 19% and within the corpus cavernosum in 52% of cases (65). Adipose tissue may be focal or multifocal and tends to be scant and peripherally located at the junction of the tunica albuginea with the corpora. It is possible that adipose tissue, along with small nutritional vessels and nerves, perforates from the fascia, where fat is usually present, through the tunica albuginea to reach the corpora (Fig. 37.30). In previous examination of the local routes of cancer spread, we found this pathway to be one-way carcinomas invade the penile corpora from the fascia (Fig. 37.31) (66).

› *Corpus Spongiosum* There are widely interconnected, branching vascular spaces separated by trabeculae in the corpus spongiosum of the shaft. These variably sized vascular spaces are lined by endothelial cells and surrounded by a thin layer of smooth muscle cells. Bundles of smooth muscle cells coalesce in the vascular walls to form the subendothelial cushions or polsters. The lacunae become continuous with a mucosal plexus of veins toward the urethra and they communicate with the venous network of the albuginea at the periphery (15). Compared to the corpus spongiosum of the shaft, the substance of the glans corpus spongiosum is made up of convolutions of large veins rather than spaces separated by trabeculae (18).

The main differences between the corpus spongiosum and the corpus cavernosum from the penile shaft are that the blood spaces in the corpus spongiosum are the same size in peripheral and central areas and the trabeculae between them contain more elastic fibers, whereas smooth muscle bundles are relatively scarce when compared to the trabeculae of the corpus cavernosum (Fig. 37.32) (17,18). However, there is variability and sometimes it can be difficult to distinguish corpus cavernosum from spongiosum by histology alone.

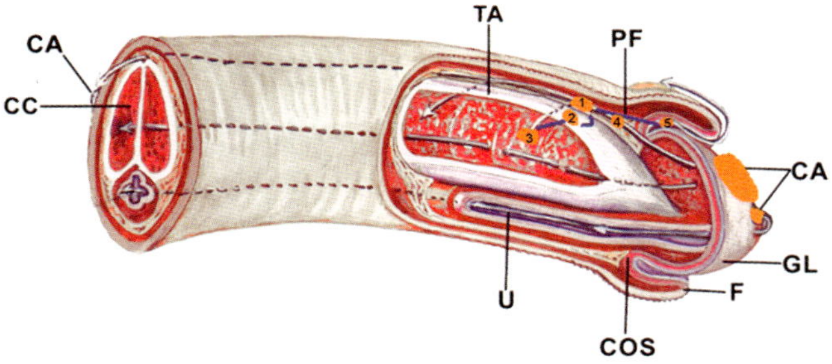

FIGURE 37.31 Diagram representing cancer spread. Cancer (*CA*) is located in the glans (*GL*). Blue line shows the route of invasion following penile fascia (*PF*) and the hypothetic way of invasion of corpora cavernosa (*CC*) through the tunica albuginea (*TA*). (1) Adipose tissue in buck fascia. (2) Adipose tissue in tunica albuginea. (3) Adipose tissue in corpora cavernosa. *F*, foreskin; *COS*, coronal sulcus; *U*, urethra.

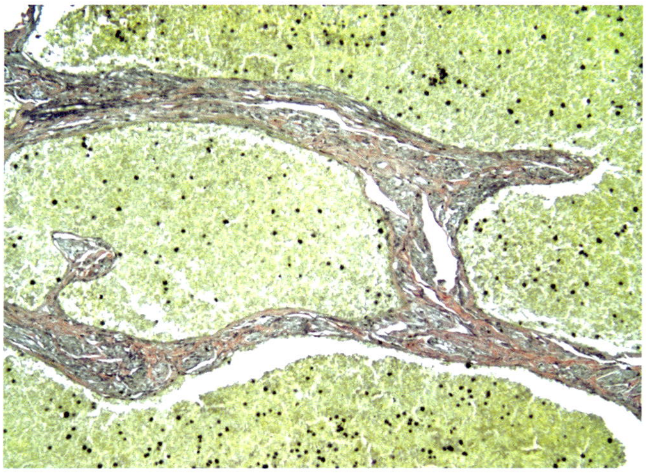

FIGURE 37.32 Corpus cavernosum. The interstitial fibrous connective tissue contains fewer elastic fibers than the corpus spongiosum (van Gieson elastic stain).

Anatomical Features Related to Staging

The new College of American Pathology protocol for the examination of specimens from patients with penile carcinomas emphasizes the importance of the site of origin or tumor involvement (67). It delineates four anatomical sites, glans, foreskin, coronal sulcus, and shaft. According to the last edition of the AJCC TNM manual, the staging of penile carcinomas is mainly based on anatomical foundations (9,50). Carcinomas are assigned a local tumor stage according to progressive anatomical levels of invasion. The model was originally designed for carcinomas of the glans and generalized to other sites, even though the histologic layers are not identical in all sites. Carcinomas originating on the shaft are unusual and may require use of a different staging system based on the sequential invasion of anatomical levels from the skin to the corpus cavernosum (68).

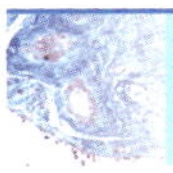

DISTAL URETHRA

Anatomic Levels of the Urethra and Periurethral Tissues

- Urethral epithelium
- Lamina propria
- Corpus spongiosum
- Tunica albuginea
- Buck fascia

Anatomic Features

The distal (anterior) urethra consists of the bulbous and penile (pendulous) segments. The bulbous urethra is 3 to 4 cm in length and is located between the inferior margin of the urogenital diaphragm and the penoscrotal junction. It courses through the root of the penis within the bulb of the corpus spongiosum. The penile urethra measures approximately 15 cm in length and extends from the penoscrotal junction to the external meatus; it is closely associated with the corpus spongiosum that forms a protective cylindrical sheath around it (1,13,23). The distal 4 to 6 mm of the penile urethra corresponds to the fossa navicularis, a distal saccular expansion that is contiguous with the urethral meatus. The penile urethra has a more central position within the corpus spongiosum, in contrast to the more dorsally positioned bulbous urethra. The mucosa of the distal urethra contains numerous recesses, the Morgagni lacunae, which extend deeply into the mucin-secreting Littré glands present in the lateral walls of the bulbous and penile urethra. A stellate-shaped lumen can be noted in cross section of the penile urethra owing to these folds of epithelium and lamina propria (Fig. 37.33A). From the surgical pathology point of view, the urethra with surrounding periurethral tissues is an important resection margin and should be carefully examined in partial penectomy specimens with penile carcinoma, as these margins are most frequently involved margin of resection (57). The anatomical levels at the urethral/periurethral margin are epithelium, lamina propria, periurethral corpus spongiosum, tunica albuginea, and buck fascia (Fig. 37.33A,B).

Microscopic and Immunohistochemical Features

› **Urethral Epithelium** There are two conflicting embryologic explanations of the differentiation of the distal urethra: The ectodermal ingrowth theory suggests that the distal urethra originates in the ectoderm and penetrates from the glans to urethra, while the endodermal differentiation theory proposes that the distal urethra is formed by differentiation of endodermal tissues from urethra to the glans (69,70). Independent of embryogenesis, there are histologic differences between the epithelium of the anterior urethra and the classical transitional urothelium of the posterior urethra and urinary tract. The fossa navicularis is lined by nonkeratinizing squamous epithelium; it is similar to and continuous with the epithelium covering the glans penis. In the pendulous and bulbous urethra, the surface cell layer is columnar with 4 to 15 stratified layers of uniform small cells and an absence of umbrella cells (Fig. 37.34). This distinctive epithelium appears to be related to the squamous rather than to the transitional urothelium, frequently showing squamous metaplasia with a predominance of squamous cell carcinomas rather than urothelial carcinomas, which are more common in the prostatic urethra. Penile urethral carcinomas are histologically similar to carcinomas arising in the glans. HPV-associated and HPV-independent tumors may occur (71).

The tumors in the membranous portion of the posterior urethra are more similar to those of the bulbous urethra, although adenocarcinomas preferentially arise in the bulbomembranous urethra (13). The frequent finding of intraepithelial precancerous lesions in the penile urethras of patients with penile squamous cell carcinoma is noteworthy

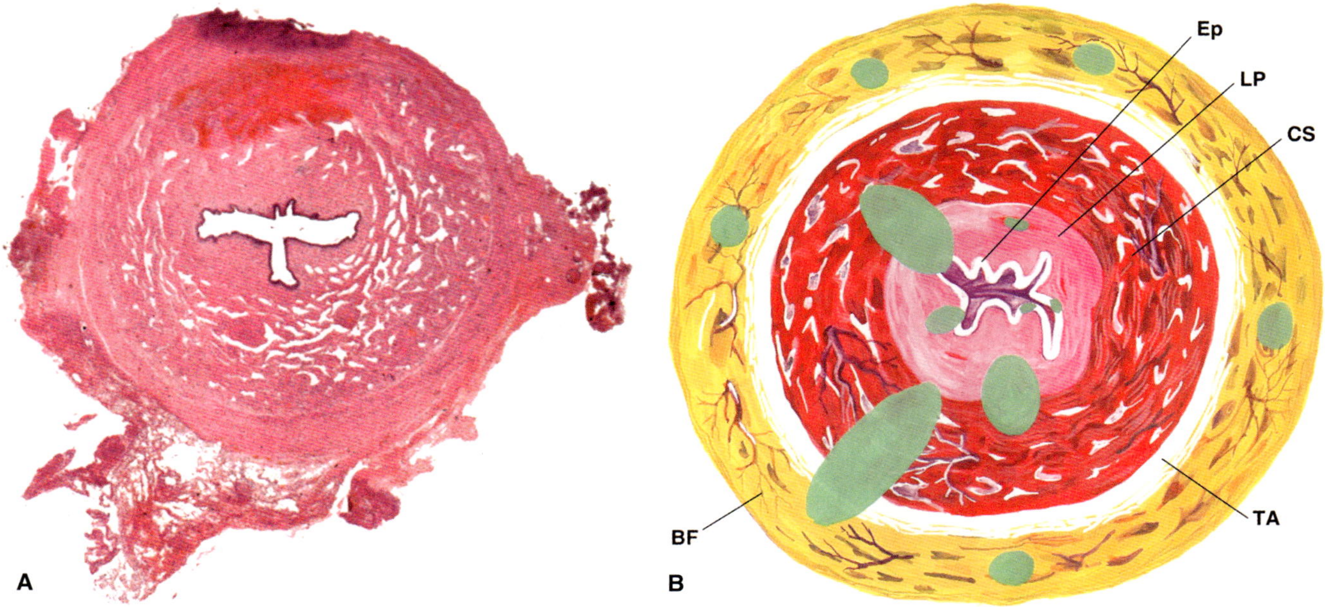

FIGURE 37.33 **A:** Cross section of the penile urethra and periurethral tissues. Note the stellate-shaped lumen at the center. **B:** Diagrammatic cross section of penile urethra and periurethral tissues. The anatomical levels at the urethral margin of resection are epithelium (*Ep*), lamina propria (*LP*), periurethral corpus spongiosum (*CS*), tunica albuginea (*TA*), and buck fascia (*BF*).

because it indicates that the urethra participates either as a mechanical pathway of penile cancer progression in a continuous manner or as an independent site of primary tumor growth in discontinuous lesions (70).

Immunohistochemical studies show that the penile urethral epithelium expresses CK7, 34βE12, and p63, but is negative for CK20. The CK7 immunostain is positive in the upper layers of the epithelium, including the columnar cells (Fig. 37.35), in contrast to the expression of p63 and 34βE12 that is mostly seen in basal and parabasal cells, but not in the superficial layer of columnar cells (Figs. 37.36 and 37.37). Occasional chromogranin-positive cells may be found close to the basal membrane by immunohistochemical stains.

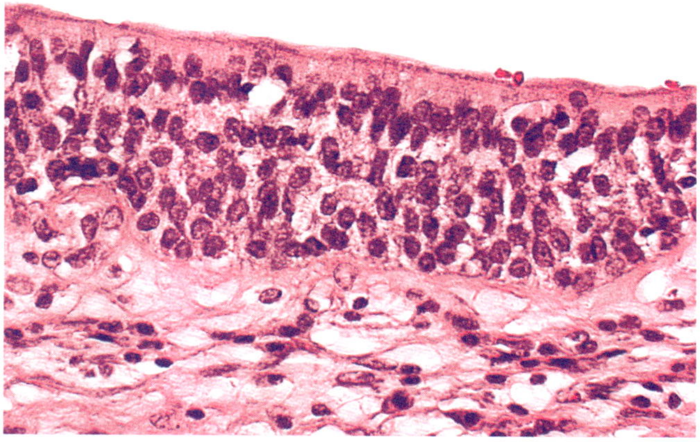

FIGURE 37.34 The epithelium of the anterior urethra is composed of columnar cells overlying 4 to 15 stratified layers of uniform small cells, usually categorized as stratified or pseudostratified columnar epithelium.

› ***Urethral and Periurethral Glands*** There are two different types of glands in the anterior urethra: The intra- or juxtaepithelial glands, with a dense eosinophilic cytoplasm and rounded basal nuclei, and the classical mucinous Littré glands, with clear mucinous cytoplasm and basally compressed nuclei resembling pyloric glands of the gastrointestinal tract (Fig. 37.38). There are histologic transitions from the intra- or juxtaepithelial glands with dense eosinophilic cytoplasm to more mucinous cells.

The recesses of the urethra (Morgagni lacunae) are lined by paraurethral mucinous Littré glands. Littré glands are tubuloacinous–mucous structures located along the full

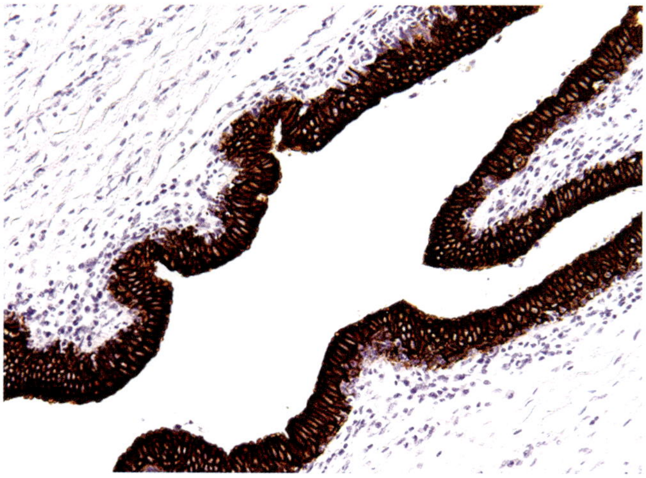

FIGURE 37.35 The expression of CK7 is seen in the upper layers of the urethral epithelium, including the columnar cells.

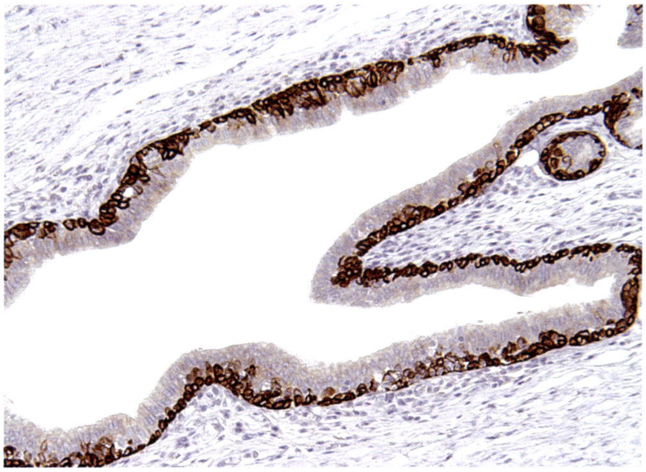

FIGURE 37.36 Urethral epithelium. Basal and parabasal cells express 34βE12. The superficial layers are negative.

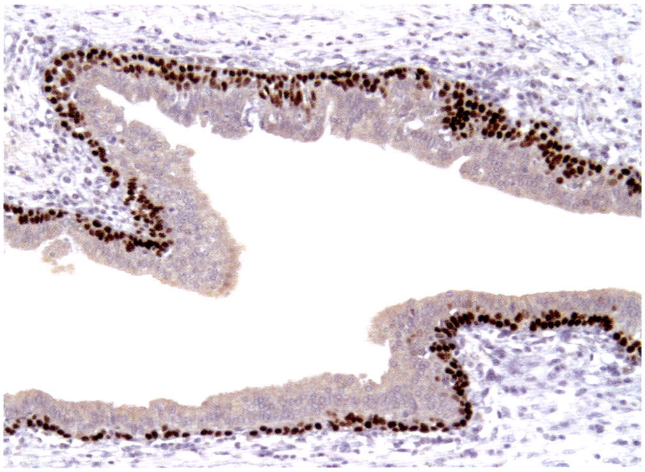

FIGURE 37.37 Urethral epithelium. In contrast to the superficial layers, basal and parabasal cells show p63 nuclear expression.

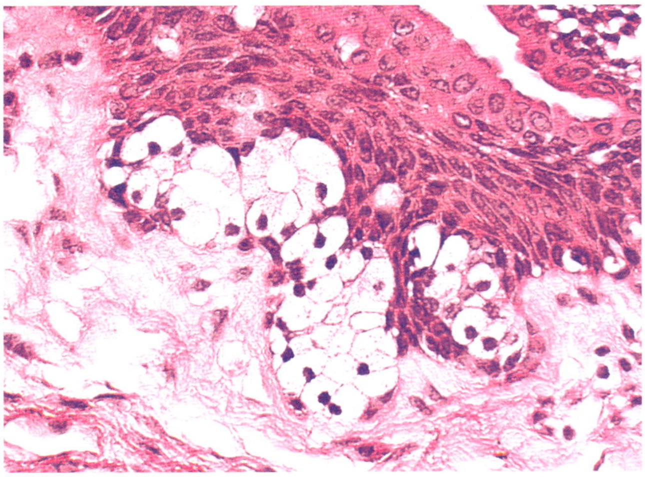

FIGURE 37.38 Penile urethra. A section illustrating intraepithelial mucinous glandular structures.

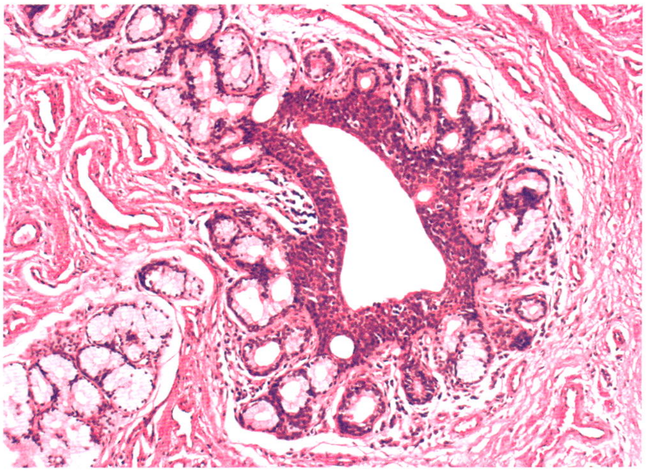

FIGURE 37.39 Penile urethra. Photomicrograph illustrating a cluster of tubular and acinar Littré mucous glands.

length of the corpus spongiosum, in close relation with erectile tissue (Fig. 37.39). Littré glands end in the urethra at the level of the intraepithelial lacunae. They can give rise to cysts or simulate a tumor when they are inflamed (72,73). Prostatic epithelial cells are normally present in periurethral glands along the penile urethra. They may be entirely composed of prostatic cells or, more commonly, mixed prostatic and mucinous epithelia (74) (Fig. 37.40). These glands may be partially responsible for the minimal, but persistently elevated levels of serum prostate-specific antigen (PSA) in some cases of successful radical prostatectomy (74).

The bulbourethral, or Cowper glands are two small structures deeply located at the level of the membranous (or bulbous) urethra where they terminate in two small ducts (75). They are mucous–acinar structures (Fig. 37.41). These glands normally contain clear cells that can be confused with prostatic carcinoma in core needle biopsy specimens.

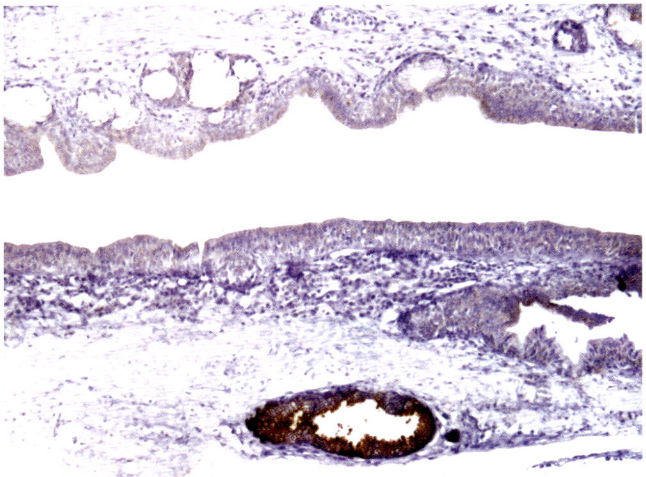

FIGURE 37.40 Some periurethral glands are positive with prostate-specific antigen (PSA) immunostain.

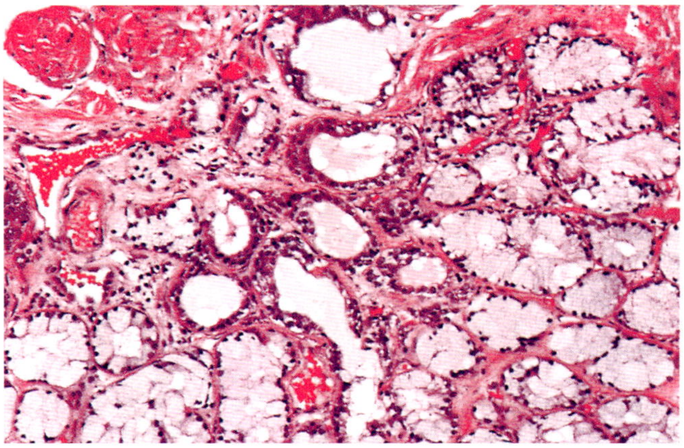

FIGURE 37.41 Bulbourethral (Cowper) gland. Specimen was taken from autopsy and shows acinous–mucous structures, which are located deeply in the membranous urethra. Courtesy of Dr. Victor Reuter, Memorial Sloan-Kettering Cancer Center.

› ***Lamina Propria*** The urethral lamina propria is a thin layer of loose fibrous and elastic tissue. Genital corpuscles are identified in the lamina propria of the most distal portion of the urethra, which is lined by squamous epithelium but are not seen in other portions of the penile urethra. We have seen cases of lichen sclerosus affecting the glans and extending into the lamina propria of the anterior urethra (70).

ARTERIES

The arteries of the penis are branches of the internal pudenda, which is a branch of the iliac artery. There are two systems: the dorsal and the cavernous arteries. The dorsal arteries are located from the base of the penis near and on both sides of the dorsal profunda vein within buck fascia and in the superior groove formed by the corpora cavernosa (Fig. 37.26). Small caliber branches or circumflex arteries, irrigate the corpus cavernosum and the periurethral corpus spongiosum. They also perforate the albuginea to reach the corpora. The terminal branches irrigate the glans, and collateral branches provide the skin nutrients. Cavernous arteries penetrate the corpus cavernosum at the site where the corpora join, and they run longitudinally near the central septum, which divides the corpora. From the cavernous arteries originate the vasa vasorum, small arteries that irrigate the erectile tissues. The helicine branches also originate from the cavernous arteries, and they are responsible for filling the vascular spaces during the process of erection; their name derives from the fact that they are coiled and twisted along the trabeculae when the penis is flaccid (76,77). These arteries have thick muscular walls; and, in addition, many possess inner thickenings of longitudinal muscle fibers that bulge into their lumina. Many of the terminal branches of the helicine arteries open directly into the spaces of the erectile tissue.

VEINS

The superficial veins are irregularly distributed and easily noted under the skin. They end in the superficial dorsal vein; this vascular structure runs straight from the foreskin to the base of the penis. It drains the foreskin venous blood and the skin and is located in the space between the dermis and buck fascia. The deeper venous system, for which the axis is the deep dorsal vein, runs along the superficial dorsal vein but in a plane separated by buck fascia (Fig. 37.26). The circumflex veins originate in the periurethral corpus spongiosum and terminate in the deep dorsal vein system. Similarly, there are veins originating in the corpus cavernosum that, after forming a small plexus at the base of the penis, terminate in the internal pudendal vein. The cavernous venous system, unlike the venous drainage from the glans penis, delays venous drainage and in doing so assists in maintaining erections (78).

Penile erection is a vascular phenomenon that results from trabecular smooth muscle relaxation, arterial dilation, and venous restriction. In further support of the concept of restriction of outflow is the observation that the walls of the circumflex veins are unusually muscular. In addition, these veins exhibit unique specializations of their lumina called *polsters*. These are local accumulations of fibroblasts and smooth muscle cells beneath the endothelium that form conspicuous longitudinal thickenings or ridges that can be followed throughout hundreds of serial sections. The pollsters likely have a role in constricting the lumen and retarding venous outflow during erection (17). There are, however, controversies regarding the real presence and significance of the polsters in the penile veins and arteries, and some authors have proposed that they represent degenerative changes (79).

The deep dorsal vein of the penis has a connection with the vertebral veins; hence it is possible for metastases to make their way to the vertebrae or even to the skull and brain without going through the heart and lungs. Pyogenic organisms may be transported by the same route (1). Invasion of veins by penile carcinoma mainly occurs in advanced stages of the disease, usually associated with systemic dissemination. Common extranodal sites of metastasis are liver, lungs, brain, bone, and heart (80).

LYMPHATIC VESSELS

The lymphatic vessels of the foreskin spring from a network that covers its internal and external surfaces; they arise from the lateral aspect and converge dorsally with the skin of the shaft lymphatic vessels to form 4 to 10 vessels that run toward the pubis. At this point, they diverge to drain into the right and left superficial inguinal lymph nodes. The lymphatics draining the glands form a rich network that, beginning in the lamina propria, course toward the frenulum, where they coalesce with two or three trunks from the distal urethra to form several collecting trunks following the coronal sulcus. A collar of lymphatic vessels entirely surrounds the corona, forming two or three trunks that run along the dorsal surface of the penis deep to the fascia and accompanied by the deep dorsal vein. At the parasymphyseal region, they form a rich anastomosing plexus draining into superficial and deep inguinal lymph nodes (81). The male urethra has a dense plexus in the mucosa. The lymphatic capillaries are especially abundant around the fossa navicularis (1). The lymphatic vessels of the urethra, corpus spongiosum, and corpus cavernosum run toward the ventral surface of the body of the penis, reaching the raphe and the dorsum, where they run with the dorsal vein and end in the superficial and deep inguinal lymph nodes. Invasion of lymphatic vessels by carcinoma in the glans is an adverse prognostic factor and affects penile cancer staging (50). Occasionally, invasion of small venules by carcinoma may be confused with lymphovascular invasion. Inmunostains with D2-4 may be of help for the distinction (82).

NERVES

The nerves originate in the sacral and lumbar plexuses. Peripheral nerves run along the arteries. Dorsal nerves are located external to the arteries, giving circumflex branches to the corpora cavernosa (83,84). The terminal branches end in the glans and the foreskin. The dorsal nerve of the penis, the principal somatosensory nerve innervating the penis, consists of two populations of axons, one to innervate the penile shaft and urethra and the other to innervate the glans. Urethral innervation by the dorsal nerve of the penis supports the view that urethral afferent impulses are a component of reflex ejaculatory activity. The pattern of glandular innervation by the dorsal nerve of the penis identifies the glans as a sensory end organ for sexual reflexes. The undulating character of the dorsal nerve of the penis is a mechanism by which the nerve can accommodate significant changes in penile length with erection (83). The dorsal nerve of the penis supplies the glans in most men, but branches of the perineal nerve can supply the ventral penis, frenulum, and periurethral area in some men. Perineural invasion by carcinoma of glans or foreskin is an adverse prognostic factor and affects the staging of penile cancer (50). It is usually located in the region of the invasive tumor front.

REFERENCES

1. Standring S, ed. *Gray's Anatomy: The Anatomical Basis of Clinical Practice*. 40th ed. Churchill Livingstone, Elsevier; 2008.
2. Bunker CB. *Male Genital Skin Disease*. Elsevier Saunders; 2004.
3. Hyman AB, Brownstein MH. Tyson's "glands." Ectopic sebaceous glands and papillomatosis penis. *Arch Dermatol.* 1969;99:31–36.
4. Tanenbaum MH, Becker SW. Papillae of the corona of the glans penis. *J Urol.* 1965;93:391–395.
5. Winer JH, Winer LH. Hirsutoid papillomas of the coronal margin of glans penis. *J Urol.* 1955;74:375–378.
6. Cubilla AL, Piris A, Pfannl R, Rodriguez I, Agüero F, Young RH. Anatomic levels: important landmarks in penectomy specimens: a detailed anatomic and histologic study based on examination of 44 cases. *Am J Surg Pathol.* 2001;25:1091–1094.
7. Ornellas AA, Nóbrega BLB, Chin EWK, et al. Prognostic factors in invasive squamous cell carcinoma of the penis: analysis of 196 patients treated at the Brazilian National Cancer Institute. *J Urol.* 2008;180(4):1354–1359.
8. Ficarra V, Martignoni G, Maffei N, et al. Predictive pathological factors of lymph nodes involvement in the squamous cell carcinoma of the penis. *Int Urol Nephrol.* 2002;34(2):245–250.
9. Amin MB, American Joint Committee on Cancer, American Cancer Society. *AJCC Cancer Staging Manual*. American College of Surgeons; 2018.
10. Moll I, Kuhn C, Moll R. Cytokeratin 20 is a general marker of cutaneous Merkel cells while certain neuronal proteins are absent. *J Invest Dermatol.* 1995;104:910–915.
11. Cold CJ, Taylor JR. The prepuce. *BJU Int.* 1999;83(Suppl 1): 34–44.
12. Tuncali D, Bingul F, Talim B, Surucu S, Sahin F, Aslan G. Histologic characteristics of the human prepuce pertaining to its clinical behavior as a dual graft. *Ann Plast Surg.* 2005;54:191–195.
13. Epstein JH, Magi-Galluci C, Zhou M. *Tumors of the Prostate Gland, Seminal Vesicles, Penis and Scrotum*. Armed Forces Institute of Pathology; 2020. ARP Press.
14. Montagna W, Kligman AM, Carlisle KS. *Atlas of Normal Human Skin*. Springer-Verlag; 1992.
15. Bailey FR, Kelly DE, Wood RL, Enders AC. *Bailey's Textbook of Microscopic Anatomy*. 18th ed. Williams & Wilkins; 1984.
16. Halata Z, Munger BL. The neuroanatomical basis for the protopathic sensibility of the human glans penis. *Brain Res.* 1986;371:205–230.
17. Fawcett DW. *Bloom and Fawcett: A Textbook of Histology*. 11th ed. WB Saunders; 1986.
18. Ham AW, Cormack DH. *Histology*. 8th ed. JB Lippincott Company; 1979.
19. Cubilla AL, Barreto J, Caballero C, Ayala G, Riveros M. Pathologic features of epidermoid carcinoma of the penis. A prospective study of 66 cases. *Am J Surg Pathol.* 1993;17(8):753–763.
20. Poirier P, Charpy A. *Traité d'Anatomie Humaine*. Masson et Cie; 1901:183.
21. Saalfeld E. Ueber die Tyson'schen drüsen. *Arch Mikr Anat.* 1899;53:212–218.
22. Tandler J, Dömeny P. Ueber Tyson'schen drüsen. *Wiener Klin Wochen.* 1898;23:555–556.
23. Testut L, Latarjet A. *Tratado de Anatomía Humana*. 9th ed. Barcelona, Salvat; 1981.
24. Parkash S, Jeyakumar S, Subramanyan K, Chaudhuri S. Human subpreputial collection: its nature and formation. *J Urol.* 1973;110:211–212.
25. Keith A, Shillitoe A. The preputial or odoriferous glands of man. *Lancet.* 1904;1:146–148.
26. Sprunk H. *Ueber Die Vermeintlichen Tyson'schen Draüsen [Dissertation]*. University of Königsberg; 1897.
27. Tyson E. *The Anatomy of a Pygmy Compared With That of A Monkey, An Ape and A Man*. University of London Press; 1699.

28. Velazquez EF, Bock A, Soskin A, Codas R, Arbo M, Cubilla AL. Preputial variability and preferential association of long phimotic foreskins with penile cancer: an anatomic comparative study of types of foreskin in a general population and cancer patients. *Am J Surg Pathol.* 2003;27:994–998.
29. Winberg J, Bollgren I, Gothefors L, et al. The prepuce: a mistake of nature? *Lancet.* 1989;1:598–599.
30. Weiss HA, Larke N, Halperin D, Schenker I. Complications of circumcision in male neonates, infants and children: a systematic review. *BMC Urol.* 2010;10:2.
31. Taylor JR, Lockwood AP, Taylor AJ. The prepuce: specialized mucosa of the penis and its loss to circumcision. *Br J Urol.* 1996;77:291–295.
32. Lukong CS. Circumcision: controversies and prospects. *J Surg Tech Case Rep.* 2011;3(2):65–66.
33. Pinto K. Circumcision controversies. *Pediatr Clin North Am.* 2012;59(4):977–986.
34. Winkelmann RK. The erogenous zones: their nerve supply and its significance. *Mayo Clin Proc.* 1959;34:39–47.
35. Ben-Ari J, Merlob P, Mimouni F, Reisner SH. Characteristics of the male genitalia in the newborn: penis. *J Urol.* 1985;134:521–522.
36. Dillner J, von Krogh G, Horenblas S, Meijer CJ. Etiology of squamous cell carcinoma of the penis. *Scand J Urol Nephrol Suppl.* 2000;205:189–193.
37. Lerman SE, Liao JC. Neonatal circumcision. *Pediatr Clin North Am.* 2001;48:1539–1557.
38. Fussell EN, Kaack MB, Cherry R, Roberts JA. Adherence of bacteria to human foreskins. *J Urol.* 1988;140:997–1001.
39. Castellsague X, Bosch FX, Munoz N, et al. International Agency for Research on Cancer Multicenter Cervical Cancer Study Group. Male circumcision, penile human papillomavirus infection, and cervical cancer in female partners. *N Engl J Med.* 2002;346:1105–1112.
40. Tobian AA, Serwadda D, Quinn TC, et al. Male circumcision for the prevention of HSV-2 and HPV infections and syphilis. *N Engl J Med.* 2009;360(13):1298–1309.
41. Wawer MJ, Tobian AAR, Kigozi G, et al. Effect of circumcision of HIV-negative men on transmission of human papillomavirus to HIV-negative women: a randomised trial in Rakai, Uganda. *Lancet.* 2011;377(9761):209–218.
42. Bailey RC, Moses S, Parker CB, et al. Male circumcision for HIV prevention in young men in Kisumu, Kenya: a randomised controlled trial. *Lancet.* 2007;369:643–656.
43. Gray RH, Kigozi G, Serwadda D, et al. Male circumcision for HIV prevention in men in Rakai, Uganda: a randomised trial. *Lancet.* 2007;369:657–666.
44. Frisch M, Friis S, Kjear SK, Melbye M. Falling incidence of penis cancer in an uncircumcised population (Denmark 1943–90). *BMJ.* 1995;311:1471.
45. Hines JZ, Ntsuape OC, Malaba K, et al. Scale-up of voluntary medical male circumcision services for HIV prevention—12 countries in Southern and Eastern Africa, 2013–2016. *MMWR Morb Mortal Wkly Rep.* 2017;66(47):1285–1290.
46. Kalichman S, Mathews C, Kalichman M, Eaton LA, Nkoko K. Male circumcision for HIV prevention: awareness, risk compensation, and risk perceptions among South African women. *Glob Public Health.* 2018;25:1–9.
47. American Academy of Pediatrics Task Force on Circumcision. Circumcision policy statement. *Pediatrics.* 2012;130(3):585–586.
48. Oster J. Further fate of the foreskin: incidence of preputial adhesions, phimosis, and smegma among Danish schoolboys. *Arch Dis Child.* 1968;43:200–203.
49. Oertell J, Caballero C, Iglesias M, et al. Differentiated precursor lesions and low-grade variants of squamous cell carcinomas are frequent findings in foreskins of patients from a region of high penile cancer incidence. *Histopathology.* 2011; 58(6):925–933.
50. Sanchez DF, Fernandez-Nestosa MJ, Canete-Portillo S, Rodriguez I, Cubilla AL. What is new in the pathologic staging of penile carcinoma in the 8th edition of AJCC TNM model: rationale for changes with practical stage-by-stage category diagnostic considerations. *Adv Anat Pathol.* 2021; 28(4):209–227.
51. Modwing R, Valderrama E. Immunohistochemical analysis of nerve distributions pattern within preputial tissues. *J Urol.* 1989;141(Suppl 1):489A.
52. Chaux A, Caballero C, Soares F, et al. The prognostic index: a useful pathologic guide for prediction of nodal metastases and survival in penile squamous cell carcinoma. *Am J Surg Pathol.* 2009;33(7):1049–1057.
53. Kattan MW, Ficarra V, Artibani W, et al. Nomogram predictive of cancer specific survival in patients undergoing partial or total amputation for squamous cell carcinoma of the penis. *J Urol.* 2006;175(6):2103–2108.
54. Ficarra V, Zattoni F, Artibani W, et al. G.U.O.N.E. Penile Cancer Project Members. Nomogram predictive of pathological inguinal lymph node involvement in patients with squamous cell carcinoma of the penis. *J Urol.* 2006;175(5):1700–1704.
55. Hricak H, Marotti M, Gilbert TJ, et al. Normal penile anatomy and abnormal penile condition: evaluation with MR imaging. *Radiology.* 1988;169:683–690.
56. Mostofi FK, Davis C Jr. Male reproductive system and prostate. In: Kissane JM, ed. *Anderson's Pathology*. Vol. 1. 8th ed. CV Mosby; 1985.
57. Velazquez EF, Soskin A, Bock A, Codas R, Barreto JE, Cubilla AL. Positive resection margins in partial penectomies: sites of involvement and proposal of local routes of spread of penile squamous cell carcinoma. *Am J Surg Pathol.* 2004;28:384–389.
58. Hsu GL, Brock G, Martínez-Pineiro L, von Heyden B, Lue TF, Tanagho EA. Anatomy and strength of the tunica albuginea: its relevance to penile prosthesis extrusion. *J Urol.* 1994;151:1205–1208.
59. Spirnack JP, Resnick MI, Hampel N, Persky L. Fournier's gangrene: report of 20 patients. *J Urol.* 1984;131:289–291.
60. Goldstein AMB, Meehan JP, Zakhary R, Buckley PA, Rogers FA. New observations on microarchitecture of corpora cavernosa in man and possible relationship to mechanism of erection. *Urology.* 1982;20:259–266.
61. Zinner NR, Sterling AM, Coleman RV, et al. The role of internal structure in human penile rigidity. *J Urol.* 1989;141:221A.
62. Droupy S, Giuliano F, Jardin A, Benôit G. Cavernospongious shunts: anatomical study of intrapenile vascular pathways. *Eur Urol.* 1999;36:123–128.
63. Vardi Y, Saenz de Tejada I. Functional and radiologic evidence of vascular communication between the spongiosal and cavernosal compartments of the penis. *Urology.* 1997;49:749–752.
64. Fontana D, Rolle L, Lacivita A, Porpiglia F, Del Noce G, Tamagnone A. Modificazioni anatomo-funzionali dei corpi cavernosi nell'anziano. *Arch Ital Urol Androl.* 1993;65:483–486.
65. Rodriguez IM, Cuevas M, Silvero A, et al. Novel histologic finding: adipose tissue is prevalent within penile tunica albuginea and corpora cavernosa: an anatomic study of 63 specimens and considerations for cancer invasion. *Am J Surg Pathol.* 2017;41(11):1542–1546.

66. WHO. *Classification of Tumours of the Urinary of the Urinary System and Male Genital Organs*. 5th ed. International Agency for Research on Cancer; 2022.
67. Velazquez EF, Amin MB, Epstein JI, Grignon DJ, et al. Members of the Cancer Committee, College of American Pathologists. Protocol for the examination of specimens from patients with carcinoma of the penis. *Arch Pathol Lab Med*. 2010;134(6):923–929.
68. Tekin B, Guo R, Cheville JC, et al. Penile squamous cell carcinoma exclusive to the shaft, with a proposal for a novel staging system. *Hum Pathol*. 2023;134:92–101.
69. Kurzrock EA, Baskin LS, Cunha GR. Ontogeny of the male urethra: theory of endodermal differentiation. *Differentiation*. 1999;64:115–122.
70. Velazquez EF, Soskin A, Bock A, et al. Epithelial abnormalities and precancerous lesions of anterior urethra in patients with penile carcinoma: a report of 89 cases. *Mod Pathol*. 2005; 18:917–923.
71. Chaux A, Han JS, Lee S, et al. Immunohistochemical profile of the penile urethra and differential expression of GATA3 in urothelial versus squamous cell carcinomas of the penile urethra. *Hum Pathol*. 2013;12:2760–2767.
72. Shiraki IW. Parameatal cysts of the glans penis: a report of 9 cases. *J Urol*. 1975;114:544–548.
73. Krawitt LN, Schechterman L. Inflammation of the periurethral glands of Littre simulating tumor. *J Urol*. 1977;118:685.
74. Cohen RJ, Garrett K, Golding JL, Thomas RB, McNeal JE. Epithelial differentiation of the lower urinary tract with recognition of the minor prostatic glands. *Hum Pathol*. 2002;33:905–909.
75. Bourne CW, Kilcoyne RF, Kraenzler EJ. Prominent lateral mucosal folds in the bulbous urethra. *J Urol*. 1981;126: 326–330.
76. Breza J, Aboseif SR, Orvis BR, Lue TF, Tanagho EA. Detailed anatomy of penile neurovascular structures: surgical significance. *J Urol*. 1989;141:437–443.
77. Krane RJ. Sexual function and dysfunction. In: Walsh PC, Gittes R, Perlmutter AD, Stamey TA, eds. *Campbell's Urology*. Vol. 1. 5th ed. WB Saunders; 1986:700–735.
78. Fitzpatrick T. The corpus cavernosum intercommunicating venous drainage system. *J Urol*. 1975;113:494–496.
79. Benson GS, McConnell JA, Schmidt WA. Penile polsters: functional structures or atherosclerotic changes? *J Urol*. 1981; 125:800–803.
80. Chaux A, Reuter V, Lezcano C, Velazquez E, Codas R, Cubilla AL. Autopsy findings in 14 patients with penile squamous cell carcinoma. *Int J Surg Pathol*. 2011;19(2):164–169.
81. Cunéo B, Marcille M. Note sur les lymphatiques du gland. *Bull Soc Anat Paris*. 1901;76:671–674.
82. Minardi D, d'Anzeo G, Lucarini G, et al. D2-40 immunoreactivity in penile squamous cell carcinoma: a marker of aggressiveness. *Hum Pathol*. 2011;11:1596–1602.
83. Yang CC, Bradley WE. Peripheral distribution of the human dorsal nerve of the penis. *J Urol*. 1998;159:1912–1916.
84. Lepor H, Gregerman M, Crosby R, Mostofi FK, Walsh PC. Precise localization of the autonomic nerves from the pelvic plexus to the corpora cavernosa: a detailed anatomical study of the adult male pelvis. *J Urol*. 1985;133:207–212.

SECTION X

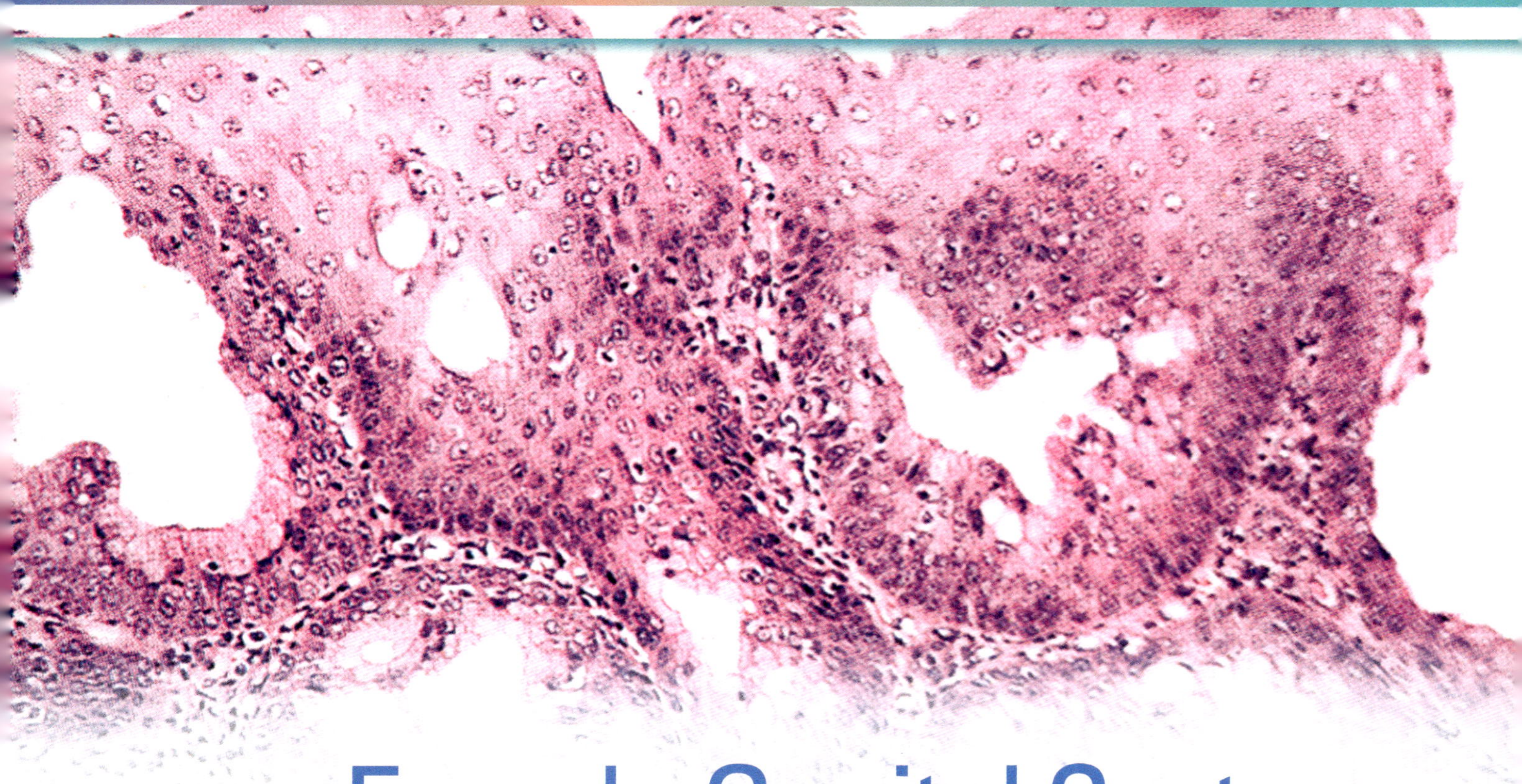

Female Genital System

38

Vulva

Krisztina Z. Hanley

CLINICAL PERSPECTIVE

Vulvar symptoms are a common cause of clinical visits to gynecologists and family practitioners. Studies have shown that 8% to 15% of reproductive age women in the general population have symptoms related to vulvar pain (1–3). Complaints may include pruritus, burning, pain, external dyspareunia, and a visible or palpable mass (4). Vulvodynia refers to vulvar pain or burning sensation for at least 3 months, without identifiable cause (5). In postmenopausal women, vulvar pain is often attributed to atrophy of the vulvovaginal epithelium due to low levels of estrogen. Chronic vulvar pain can impact quality of life. Women often seek multiple health care providers and receive several different diagnoses and treatment modalities before experiencing any improvement in symptoms.

The majority of sexually transmitted infections, as well as granulomatous, neoplastic dermatologic diseases, may involve the vulva (6–10). Graft-versus-host disease (11) and contact dermatitis (12) may create symptoms requiring evaluation of the vulva and the vagina. The vulva (pudendum femininum) is also a crucial area for detailed examination in cases of reported rape, sexual abuse, or female circumcision (13). Ambiguous genitalia and genital anomalies challenge the clinician and demand critical examination of the patient and the external genitalia (14,15). Clitoral enlargement in the newborn resulting from adrenal genital syndrome, maternal exposure to exogenous androgens, maldevelopment of the clitoris, benign tumors, and other conditions, such as the Lawrence–Seip syndrome (8,16), may result in ambiguous-appearing external genitalia. Clitoromegaly may occur in adulthood due to benign tumors, malignancy, or endocrinopathy (14,16,17).

Surgical approaches to vulvar diseases are evolving as a result of better understanding of vulvar anatomy and sexual function, as well as tumor biology. Clitoral specimens are submitted less frequently in cases of infant genital ambiguity now that follow-up studies have indicated that surgeries may result in permanent loss of function and sensation (15,18–20). In extended surgery for carcinoma of the vulva, the clitoris may be surgically spared with partial deep vulvectomy, if clinical examination does not indicate tumor involvement of that structure (15). The techniques for sentinel node evaluation used in other cancers are now applied to carcinomas of the vulva, allowing assessment of lymph nodes that receive direct lymphatic vessel drainage from the tumor and reducing morbidity associated with inguinofemoral lymph node dissection (21–23).

Cytologic evaluation of the vulva may complement biopsy and distinction from disorders that may clinically resemble extramammary Paget disease (24).

Gender reassignment surgery in case of male-to-female gender dysphoria includes bilateral orchiectomy, penectomy, labiaplasty, clitoroplasty, and vaginoplasty; these procedures are carried out to create a cosmetically acceptable and functional vagina. The neoclitoris is created from the glans penis, as it is homologous to the clitoral tissue, while labia majora and minor are constructed from scrotal skin or bowel segment, depending on how much tissue is used for vaginoplasty (25).

SPECIAL TECHNIQUES IN CLINICAL EVALUATION

Direct examination of the vulva requires adequate illumination and is enhanced by the use of a ring light or a magnifying glass (26). Some practitioners routinely use colposcopy or vulvoscopy to enhance identification of areas of hypopigmentation, scarring, fissures, small condyloma acuminatum, vestibular papillae, and vulvar intraepithelial neoplasia (VIN) (26). In cases in which condyloma acuminatum or VIN is suspected, the use of 3% acetic acid (white vinegar) is of value. Gauze sponges soaked in 3% acetic acid are applied for approximately 5 minutes, followed by prompt examination. The principle of this technique is that abnormal epithelium, especially condyloma acuminatum and VIN, becomes white (acetowhite) immediately after exposure to acetic acid reflecting differences between normal epithelium and human papillomavirus (HPV)–associated lesions. The color change to white after application of the dilute acetic acid is referred to as aceto-whitening. Although this procedure has gained wide acceptance in the evaluation of the cervical transformation zone during colposcopic examination, its use on the vulva has two limitations. First, when ulcers or fissures are present on the vulva, the application of 3% acetic acid may be associated with pain and thus is unacceptable to the patient. Second, the vulvar vestibular epithelium is normally somewhat acetowhite. The inexperienced clinician may misinterpret this finding as abnormal or as condyloma acuminatum. A biopsy is then performed of the vestibular epithelium and submitted as condyloma acuminatum. The vestibular epithelium in women of reproductive age is normally glycogenated and can be misinterpreted by the unwary pathologist as koilocytosis suggestive of condyloma acuminatum, resulting in both improper diagnosis and therapy for the patient. Inflammation within the vestibule may be associated with epithelial spongiosis, which also may resemble koilocytosis.

The diagnostic workup of vulvodynia often includes a sensory examination. The "cotton swab test" is performed with a moist cotton swab to identify painful areas of the vestibule by applying a gentle rolling motion on the entire vestibule (27). Pelvic examination of the pelvic floor muscles, uterus, and adnexa is often performed, as causes of vulvodynia may be related to endometriosis, pelvic inflammatory disease, or increased pelvic floor muscle tone.

Clinical evaluation of the vulva to assess for trauma after reported sexual assault is important and challenging. Colposcopy and vulvovaginoscopy are of value to identify related injuries (26). Documentation of findings should occur as soon as possible after the assault since most genital injuries heal rapidly without grossly evident scarring, such that there is little or no physical evidence for prosecution of sexual offenders (28). Ecchymosis resolves in 2 to 18 days, submucosal hemorrhages resolve in 2 to 14 days, and petechiae are gone within 24 hours. Superficial lacerations heal within a few days and only deep lacerations leave a scar (29).

Commonly used outpatient laboratory techniques include wet mount, potassium hydroxide preparation, and assessment of vaginal pH to exclude infectious etiologies, such as candidiasis, trichomonas, or bacterial vaginosis.

In addition to direct visualization techniques, imaging studies are increasingly applied in the vulvar assessment of female sexual response and have increased our understanding of the three-dimensional anatomy of that region. These include duplex Doppler ultrasound (30) and magnetic resonance imaging (MRI) (31,32).

From the pathologist's perspective, the majority of vulvar specimens that are examined are either diagnostic biopsies, excisional biopsies, or partial superficial or deep, or total superficial or deep vulvectomy specimens submitted as treatment for VIN, extramammary Paget disease, carcinoma, melanoma, and other diseases (9,10). An understanding of the normal histology of the vulva enhances interpretive skills and assists in arriving at an appropriate diagnosis.

ANATOMY

The female external genitalia can be defined as that portion of the female anatomy external to the hymen, extending anteriorly from the mons pubis, posteriorly to the anus, and laterally to the inguinal–gluteal folds. Included are the mons pubis, clitoris, labia minora, labia majora, vulvar vestibule and vestibulovaginal bulbs, urethral meatus, hymen, Bartholin and Skene glands and ducts, and vaginal introitus (Fig. 38.1). The superior portion of the clitoris includes the prepuce, which represents the anterior fusion of the labia minora and overlays the clitoris anteriorly, and the frenulum, which passes posteriorly and ends in its attachment to the flattened posterior aspect of the clitoris. Posteriorly, the labia minora end in the fourchette, or frenulum, of the labia. The labia majora lie lateral to the intralabial sulcus and medial to the inguinal–gluteal fold. Anteriorly, the hair-bearing lateral aspects of the labia majora blend with the mons pubis, and posteriorly, the labia majora end in the perineal body. The hair follicles of the labia majora are absent in its medial portion; however, the sebaceous gland elements are retained medial and posterior to the labia minora at the junction with the vulvar vestibule at Hart line (Fig. 38.2) (6,7,10,33). These sebaceous gland elements open directly to the epithelial surface in this portion of the vulva and can be observed as small, slightly pale to yellow elevations of the epithelium, known as Fordyce spots.

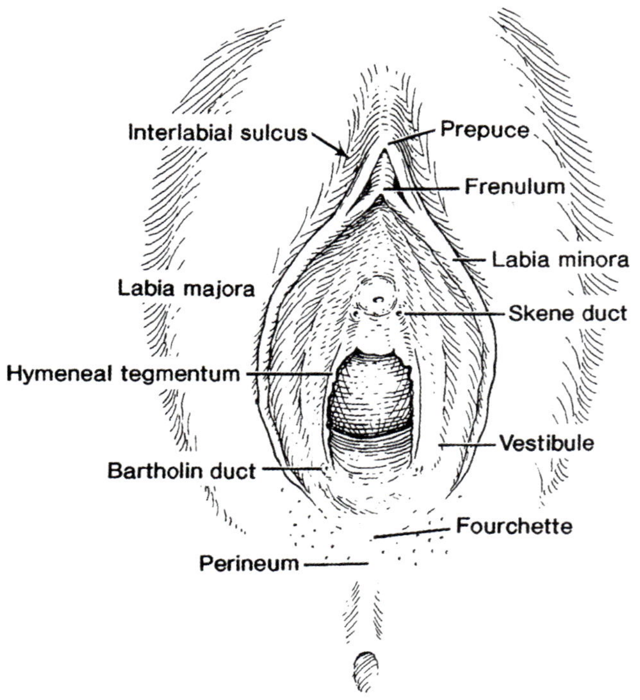

FIGURE 38.1 Topography of the vulva.

Advances in imaging techniques including MRI and digital technology have led researchers to revisit the anatomy of the vulva, especially the distal vagina, unifying the findings of prior research. MRI is often used in the diagnostic workup of suspected pelvic pathology, staging of vulvar malignancies, and characterization of müllerian duct anomalies (32). The vulva shows low- to intermediate signal intensity on T1-weighted imaging and slightly higher-signal intensity on T2-weighted imaging (32).

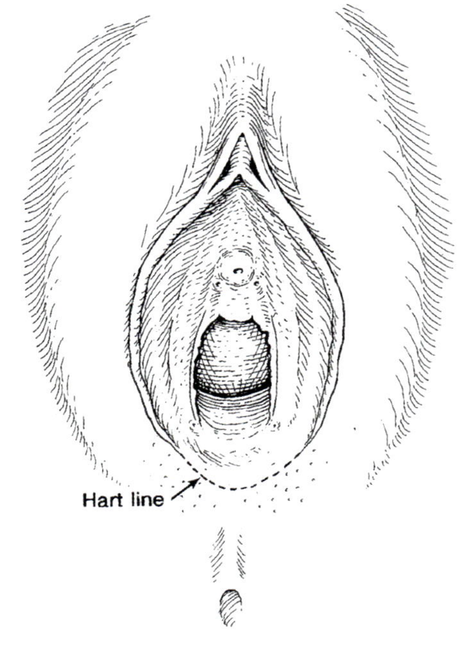

FIGURE 38.2 The vulvar vestibule and the position of Hart line. Hart line can be found on the medial aspect of the labia majora, extending in a curvilinear manner from the most inferior posterior portion of the labia minora to the vaginal fourchette.

Vulvar Biopsy

The type of lesion and its anatomic location determine which type of surgical sample is performed for histologic evaluation (34). Punch biopsies are excellent tools for evaluating melanocytic lesions, ulcers, tumors, or inflammatory processes. The sample should extend below the deepest portion of tumor in order to determine stage, prognosis, and management. Special techniques may be required when sampling modified mucous membranes (e.g., labia minora) and mucous membranes (e.g., introitus) as these surfaces are difficult to grasp with instruments. Furthermore, the surface epithelium often "slips off" due to lack of skin adnexal structures that serve to hold epithelium and subjacent connective tissue together. This is particularly problematic when evaluating immunobullous disorders. If an immunobullous disorder is suspected, tissue samples should be obtained by a broad, long technique rather than a narrow deep biopsy to ensure tissue intactness (33,34).

Shave biopsy is best performed on hair-bearing, keratinized skin (e.g., labia majora). Superficial and exophytic lesions are often sampled with shave biopsies (34). Tissue samples should be placed in 10% buffered formalin with a volume of media 10 to 20 times the amount of the specimen. Specimens requiring direct immunofluorescence (DIF) are placed in Zeus or Michel media and material for cultures is placed in a sterile container. Clinical history can be extremely helpful in the diagnosis of vulvar lesions, especially in the context of inflammatory and nonneoplastic diseases. A minimum specimen requisition should include the following: distribution, size, appearance, clinical differential diagnosis, and which part of the lesion was biopsied. Digital photographs of the lesion can be particularly helpful. Histologic evaluation of vulvar biopsies follows a systematic approach of step-by-step assessment of all tissue layers, identification, localization, and distribution of histopathologic changes, differential diagnosis, and clinicopathologic correlation. Familiarity with normal histology is a key reference point.

Sampling of the normal-appearing skin and mucosae from the mons pubis, labia, vestibule, and perineum shows site-specific differences in stratum corneum morphology, distribution of parakeratosis, and appearance of the mucocutaneous junction (35). Samples from the mons pubis show basket-weave stratum corneum with clear spaces, while the labia display a solid and compact band of eosinophilic keratin that may have an intermediate or basket-weave appearance. Basket-weave keratin is not generally seen in biopsies from the perineum.

The normal vulvar epithelium varies in thickness 0.27 ± 0.14 mm. The epithelial thickness decreases from medial to lateral, coupled with changes in stratum corneum morphology (35,36). When involved by VIN, the thickness of the epithelium is increased to approximately 0.52 ± 0.23 mm (36). The vulvar epithelium, especially that of the lateral labia minora, labia majora, and perineal body, contains melanocytes distributed among the basal cells in a ratio from 1:5 to 1:10 basal keratinocytes (37). Increased pigmentation of these areas occurs during pregnancy due to increased melanin production secondary to the effects of gestational hormones.

Four major classes of keratins have been identified in the epidermis by immunohistochemistry. The keratin pair 48 and 56 recognized by AE1/AE3 is present in proliferative epithelium (38). Keratin expression in squamous epithelium correlates with the prevalence of intermediate filament distribution within keratinocytes, and while it is retained in neoplastic process, there are some significant changes in amount and distribution (39,40). Cytokeratin AE1 generally shows a weak uniform and discontinuous staining of the basal keratinocytes in the vulva, while AE2 shows strong and diffuse staining of the granular layer and AE3 intensely stains all cell layers (41). On the other hand, CAM 5.2 is negative in normal vulvar epithelium. The pattern and distribution of keratins recognized by cytokeratin antibodies AE1, AE2, AE3, and CAM 5.2 are different in normal vulvar epithelium, hyperplasia, intraepithelial neoplasia, and squamous cell carcinoma. Thus, these antibodies can aid early identification of preneoplastic and neoplastic lesions. More specific immunohistochemical markers are also available for identification of precursor lesions of vulvar carcinoma. For example, p16 is widely used as a surrogate marker for high-risk HPV infection. The antibody targets the cyclin-dependent kinase inhibitor 2A, also known as $p16^{INK4}$, and is negative in normal vulvar squamous epithelium and squamous hyperplasia. Inflammatory lesions such as lichen sclerosus can show variable p16 expression from totally absent to heterogeneous expression in up to 42% of cases (42). Most importantly, p16 is one of the key ancillary stains used to distinguish high-grade dysplasia from its mimics as well as to separate HPV-associated VIN (p16 positive) from differentiated VIN (p16 negative).

The expression pattern and prognostic value of p53 immunostaining in vulvar lesions have gained recent attention. The extent of p53 staining in the basal and parabasal layers of the vulvar epidermis is increased in differentiated VIN compared to benign and hyperplastic squamous epithelium as well as lichen sclerosus (43). However, high-power scrutiny of the basal layer is important when assessing p53 labeling in the vulva (44). Wild-type p53 labeling patterns include: (1) scattered heterogeneous nuclear positivity at variable intensity in the basal and parabasal layer, and (2) heterogeneous strong staining in midepithelial cells with sparing of the basal cells, sometimes even the parabasal cells. On the other hand, differentiated VIN is characterized by abnormal p53 expression with a sharp interface with normal skin.

Langerhans cells are abundant in the vulvar epithelium and more prevalent in this location than in the vagina or cervix. They are predominantly located in the suprabasal layers of the epithelium with a median number of 18.7 per 100 basal squamous cells (45). They are present in keratinized and nonkeratinized epithelium, as well as within skin appendages. These dendritic cells are bone marrow-derived and can express HLA-DR antigens, as well as Fc and C3 receptors. They are capable of activating T lymphocytes as an afferent component of the immune response (46). Langerhans cells may control keratinocyte maturation; Langerhans cells are present in reduced numbers in patients with vulvar squamous cell carcinoma, adding support to this hypothesis (47).

Merkel cells are neuroendocrine cells that are present in the vulvar epithelium, as well as most other skin sites. These cells are involved in paracrine regulation of skin function (48).

Lymphocytes are commonly found in the dermis and submucosal areas of the vulva in small numbers, primarily in a perivascular distribution in the lamina propria. These sparse lymphocytes are thought to represent normal components of skin and mucosa–associated lymphoid tissue. Intraepithelial lymphocytes are infrequently seen in normal vulvar epithelium (35).

Vulvar Vestibule

The vulvar vestibule extends from the exterior surface of the hymen to the frenulum of the clitoris anteriorly, the fourchette posteriorly, the labia minora anterolaterally, and to Hart line posterolaterally on the medial aspects of the labia majora (Figs. 38.1 and 38.2) (6,7,10,15). The vestibular fossa (i.e., fossa navicularis) is the posterior portion of the vestibule that extends from the hymen to the fourchette. It is somewhat concave compared with the remainder of the vestibule. Unlike the remainder of the ectodermally derived vulvar epithelium, the epithelium of the vulvar vestibule is mostly of endodermal origin, although some authors believe the vulvar vestibular epithelium anterior to the urethra is of ectodermal origin. The vulvar vestibule is predominantly surfaced by nonkeratinized stratified squamous epithelium that blends with the thinly keratinized squamous epithelium of the labia minora, the medial labia majora at Hart line, the prepuce, and the fourchette. Although the vestibular epithelium has an embryonic origin similar to that of the distal male urethra, the epithelium is not of a typical transitional type with surface umbrella cells. Rather, it is a stratified squamous epithelium similar to the mucosa of the vagina and the ectocervix, and is glycogen rich in women of reproductive age (Fig. 38.3).

The vaginal opening and urethral orifice both lie within the vestibule, as do gland openings from the major and minor vestibular glands, as well as the paired openings of the periurethral Skene ducts. Skene ducts are the homologs of the male prostate gland. The major vestibular glands, or Bartholin glands (i.e., glandula vestibularis major), are located symmetrically posterolateral to the vaginal opening beneath the hymen, labia minora, and labia majora. They produce mucus for vaginal lubrication. Bartholin glands are

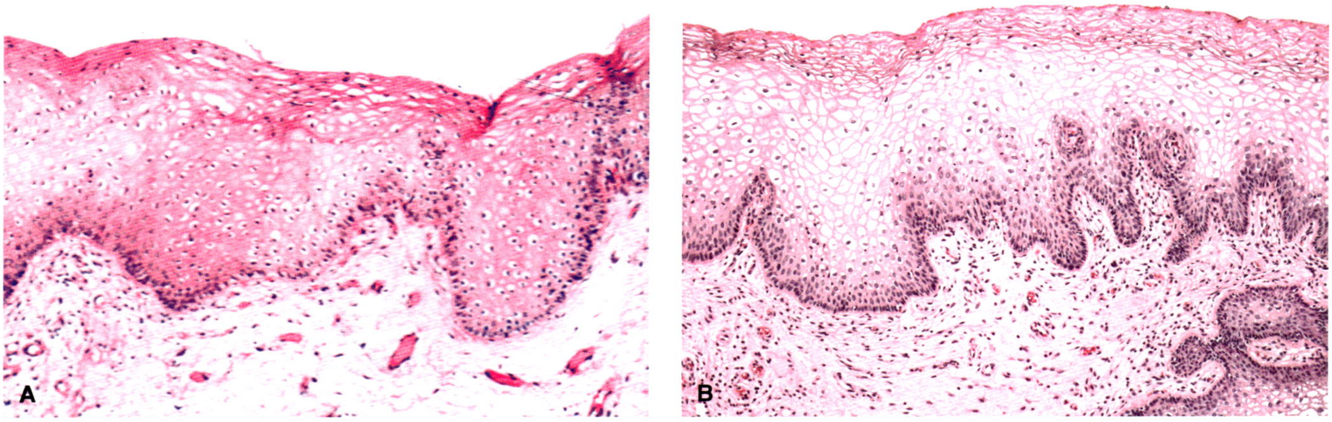

FIGURE 38.3 **A:** Vulvar vestibule with adjacent prominent vascular submucosa. Superficial thin-walled vessels are prominent and are found within the delicate fibrous stroma. A few lymphocytes are seen scattered in the superficial submucosa. **B:** Epithelium of the vulvar vestibule of a 27-year-old woman. Note that the epithelium is stratified squamous and that the superficial cells have cytoplasmic clearing, reflecting the glycogen-rich epithelium.

of ectodermal origin and consist of bilateral tubuloalveolar glands that correspond to the male bulbourethral glands, or Cowper glands. They contain several different cell types. The acini consist of mucus-secreting columnar cells (Fig. 38.4) (6,7,34). The acini drain into Bartholin duct, which measures approximately 2.5 cm in length and enters the vestibule immediately exterior (distal) and adjacent to the hymen in a posterolateral location. Bartholin duct is lined by transitional epithelium but may contain squamous or mucinous epithelial cells (Fig. 38.5). The orifice of the duct is lined by squamous epithelium to adjoin the nonkeratinized stratified squamous epithelium of the vestibule (Fig. 38.6) (6,10). Argentaffin cells can also be identified in the ductal area of the Bartholin duct epithelium and are absent in the secretory glands (49). Cysts that arise from Bartholin glands primarily result from duct obstruction and dilation (6). Carcinomas that develop in Bartholin glands can show differentiation that simulates any of these native cell types.

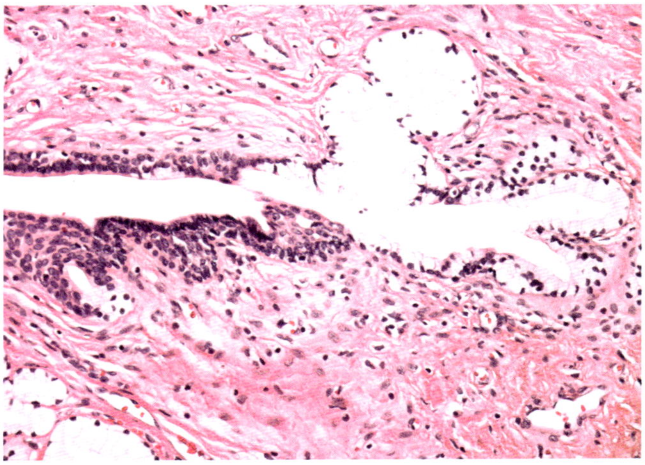

FIGURE 38.5 Bartholin duct near the gland. Bartholin duct has a transitional-like epithelial lining, with columnar cells near the surface, similar to the columnar cells lining the gland acini.

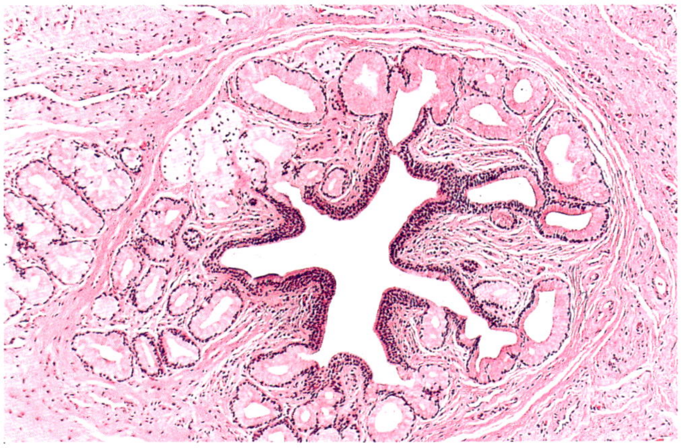

FIGURE 38.4 Bartholin gland acini are lined with a columnar epithelium. The adjacent branching Bartholin duct is present adjacent to the gland.

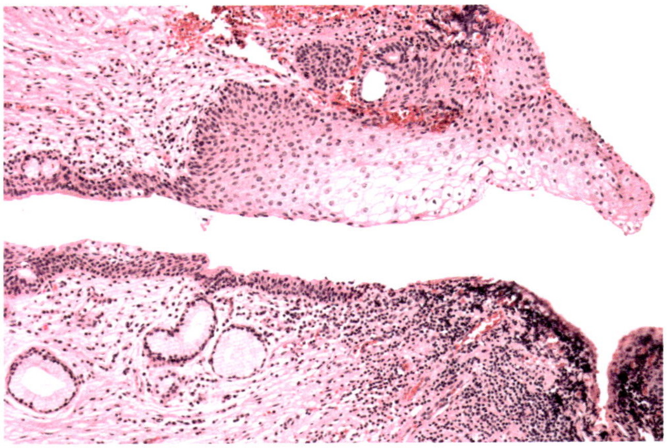

FIGURE 38.6 Bartholin duct near its exit to the vulvar vestibule. At this location the duct is lined by stratified squamous epithelium, without surface columnar cells.

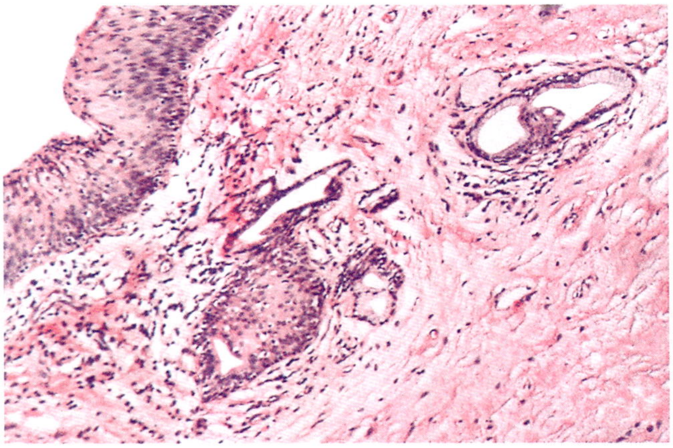

FIGURE 38.7 Minor vestibular glands of the vulvar vestibule. The vulvar vestibular glands are simple glands with a mucus-secreting columnar epithelial lining. Near their exit at the vestibular surface the glands have a stratified squamous epithelium. Vascular vestibular stroma surrounds the superficial gland elements.

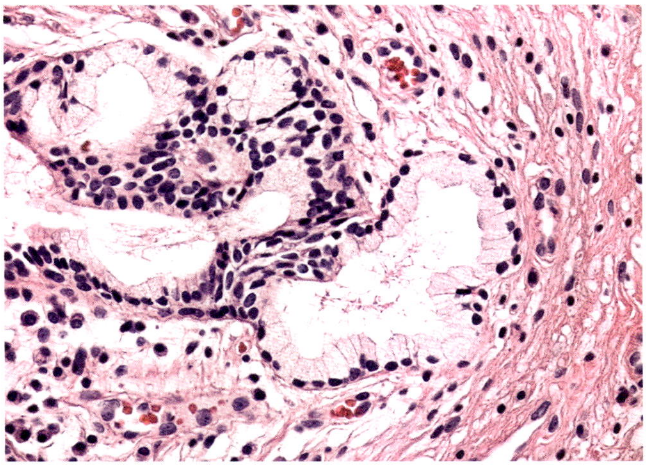

FIGURE 38.8 Vestibular gland with squamous metaplasia near the vestibular surface. Moderate chronic inflammation, consisting predominantly of lymphocytes, is seen adjacent to the gland and is consistent with vulvar vestibulitis.

The periurethral Skene glands enter the vulvar vestibule as paired gland openings immediately adjacent and posterolateral to the urethra. These glands and their ducts are generally not more than 1.5 cm in length. The periurethral glands are analogous to the male prostate gland and are lined by a pseudostratified mucus-secreting columnar epithelium. The ducts of Skene glands are lined with transitional-type epithelium that joins the stratified squamous epithelium of the vestibule. A cyst of Skene duct may result from obstruction of the duct.

The minor vestibular glands (i.e., glandulae vestibulares minores) consist of simple tubular glands that open directly to the vestibular mucosal surface (Fig. 38.7). They are analogous to the glands of Littré of the male urethra. Minor vestibular glands are small and shallow with a maximum depth of 2.27 mm (50). They are lined with a mucus-secreting columnar epithelium that merges with the stratified squamous epithelium of the vestibule (50,51). Minor vestibular glands are found in 66% of vulvar vestibulectomy specimens performed for vestibulitis and 42% of women in autopsy series (50,51). When present, they range from 1 to over 100 in number, with most women having 2 to 10 minor vestibular glands. These glands are found in greater numbers in the posterior vestibule anterior to the fourchette. The ducts of minor vestibular glands are lined by stratified squamous epithelium without surface umbrella cells. Several types of neuroendocrine cells are also present in these glands and those expressing serotonin and CXCR2 appear to increase in vestibulitis (53). Minor vestibular glands may undergo squamous metaplasia similar to that seen within the endocervix where mucus-secreting epithelial cells are replaced by stratified squamous epithelium (Fig. 38.8). This metaplastic epithelium may completely replace the glandular epithelium, resulting in the formation of a vestibular cleft (Fig. 38.9) (50). Obstruction of a minor vestibular gland associated with this metaplastic process may result in accumulation of mucous secretion within the simple tubular gland, leading to the formation of a vulvar mucous cyst (50,52,54). Vestibular adenomas have been described as arising from these minor vestibular glands (54).

Small glandopreputial glands have been described by van der Putte within the glandopreputial sulcus that is found bilaterally and immediately adjacent and lateral to the clitoris. Although these glands are quite small, they were consistently found at the base of the clitoris in female fetuses beyond 17 weeks gestational age and in adults. These eccrine glands have an eccrine cribriform secretory coil. The excretory duct of these glands is tortuous and empties into the cranial portion of the open glandopreputial sulcus to provide moisture (55,56).

Severe vulvar sensitivity and tenderness, often with dyspareunia but without associated inflammation of the vestibule, is termed vulvodynia (4). It is a relatively common

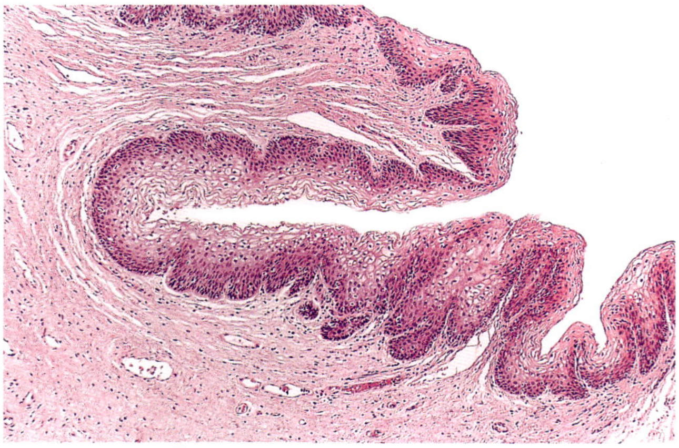

FIGURE 38.9 Vulvar vestibular cleft. The vulvar vestibular cleft has a stratified squamous epithelial lining, similar to the vulvar vestibule. These clefts appear to arise as a result of squamous metaplasia of vestibular glands.

but poorly understood condition that is a common cause of visits to vulvar clinics (4). Histologic examination of vestibular biopsies obtained from women with vestibulitis syndrome often shows proliferation of peripheral nerve bundles accompanied by neuroendocrine cell hyperplasia (57).

Urethral Orifice (Meatus Urinarius)

The urethra has a transitional epithelial lining that merges with the stratified squamous epithelium at the urethral orifice. The periurethral glands of Huffman enter the urethra throughout most of its length (10,58). Obstruction or inflammation of these periurethral glands may result in a urethral diverticulum or periurethral abscess. Partial prolapse of the urethra results in a polypoid mass, termed a caruncle. These lesions are often ulcerated and inflamed with vascular dilation and engorgement.

Hymen

The hymen marks the distal-most extent of the vagina and the most proximal boundary of the vulvar vestibule. The hymen may be imperforate, round, annular, septate, cribriform, or porous. The vaginal surface of the hymen has a nonkeratinized stratified squamous epithelium, which is glycogenated in women of reproductive age, newborn female infants, and postmenopausal women receiving estrogen therapy. The vulvar surface of the vestibular epithelium is similar to the vaginal epithelium in women of reproductive age (Figs. 38.3 and 38.10). The hymenal ring contains some Merkel tactile discs for touch and moderate numbers of free nerve endings, which are pain receptors; the hymenal ring lacks other types of receptors present in the labia majora (58).

The hymen lacks its normal opening when it is imperforate. This leads to accumulation of menstrual exodus in the vagina, resulting in vaginal distension, a condition referred to as hematocolpos. Coitus, or the routine use of intravaginal tampons, tears the hymen and may result in small soft tags referred to as carunculae hymenales or carunculae myrtiformes. There may be small vestibular papillae on the external hymen and on the vulvar vestibule. Multiple papillae are seen in the condition known as vestibular papillomatosis (6,7). They are usually linearly and symmetrically distributed, soft, delicate, and easily separated from each other in contrast to condyloma accuminatum. Such papillae are generally considered variants of normal anatomy and are not considered associated with HPV (59). Solitary, asymptomatic papillae on the hymen usually represent a normal anatomic variant (Fig. 38.11) (59).

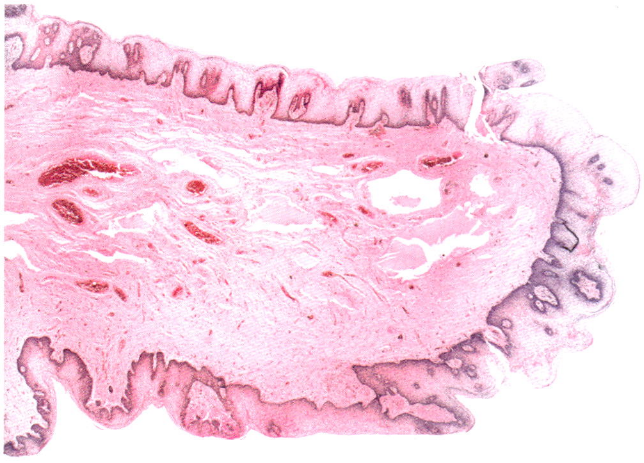

FIGURE 38.10 Cross section of the hymen of a 26-year-old woman. The epithelium of the vaginal (**upper**) and vestibular (**lower**) surfaces of the hymen is a stratified squamous epithelium, which is nonkeratinized and glycogen rich. The fibrovascular component of the hymen supports the epithelium.

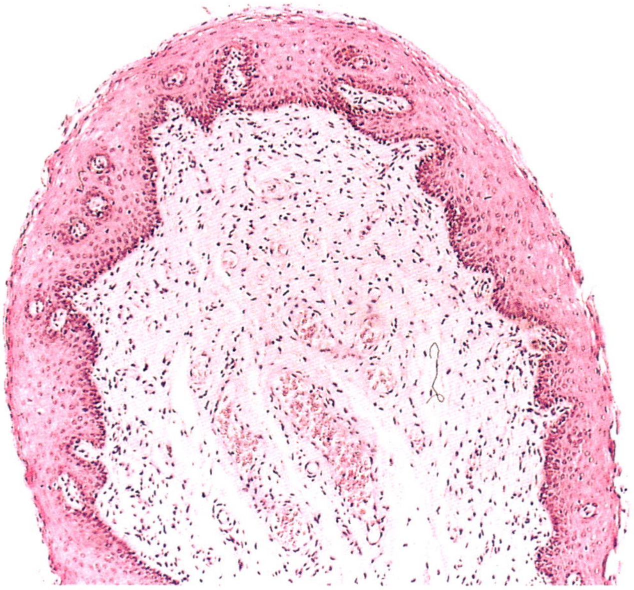

FIGURE 38.11 Vulvar vestibular papillae. The papillae have a stratified squamous epithelial surface and fibrovascular stalks.

During the prepubertal period, mucosal redundancy in the lower vagina can appear near the hymen as a fold. Histologically, these mucosal redundancies have no fibrovascular cores, but may cause confusion on physical examination after sexual assault. These redundancies disappear as the vagina enlarges with the onset of puberty (60).

Clitoris

The clitoris is the descendant of the embryonic phallus, homologous to the corpus cavernosum of the male penis. The human fetus begins sexual differentiation at approximately 8 weeks of gestation. The genital tubercle in males develops into a penis under the influence of testis-determining factor protein, a product of sex-determining gene of the Y chromosome. The size of the clitoris may be highly variable, although it measures approximately 2 cm in adults.

The external components of clitoris consist of a nonerectile tip and glans, while deeper structures include the body, two crura, and vestibular bulbs. The crura are composed of

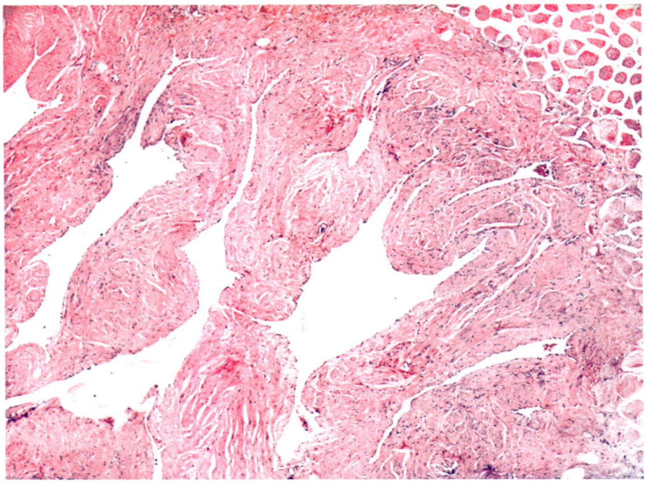

FIGURE 38.12 Erectile tissue of the labia minora.

erectile tissues similar to that of the male corpora cavernosa (10,34). They contain cavernous veins surrounded by longitudinal bundles of smooth muscle cells, as well as small centrally placed muscular arteries, enveloped by the tunica albuginea. The tunica albuginea is composed of wavy collagen fibers and straight elastic fibers. Loose connective tissue surrounds the tunica albuginea and supports the nerves and receptors of this area. The glans clitoridis is covered with squamous mucosa without glands, rete, or papillae (61). The cavernous tissue of the male corpus spongiosum does not have its counterpart in the clitoris; it is found instead in the vascular erectile tissue of the labia minora (Fig. 38.12). The clitoris contains fewer nerve endings than the labia majora, although pacinian corpuscles are more abundant. Peritrichous nerve endings for touch reception are absent. The other receptors are present with a variable distribution (58). Meissner corpuscles and Merkel tactile disks are present in reduced numbers in the clitoris compared with the labia majora or mons pubis. Pacinian corpuscles are present in large numbers (58). The free nerve endings for pain reception are found throughout the vulva and in relatively high concentrations in the clitoris, labia majora, and mons pubis (58). Ruffini and Dogiel–Krause corpuscles, which may be associated with temperature or sexual sensation, are found throughout the vulva but not in the hymenal ring (58). The rich blood supply of the clitoris is derived from the external pudendal artery, dorsal clitoral artery, and perineal arteries.

Labia Minora

The bilateral labia minora derive from the embryonic medial folds (genital folds) and lie lateral to the vulvar vestibule and medial to the labia majora, bounded by the intralabial sulcus. Development of the external genitalia in both sexes is a complex process, driven by sequential expression of regulatory genes. Hedgehog, *Wnt*, and fibroblast growth factor-mediated signaling pathways play key roles in the cross-talk between epithelium and adjacent mesenchyme during the development of external genitalia (62). The labia minora have their male embryologic counterparts in the penile corpus spongiosum (61). The minora measure approximately 5 cm in length and 0.5 cm in thickness in adults but can vary considerably between individuals or be asymmetric in an individual (46). Labia minora enlargement could be considered a variation of the normal anatomy and it is very subjective to define the "normal" vulva. Enlarged labia minora can cause functional, aesthetic, and psychosocial problems (63). The epithelium of the labia minora is of ectodermal origin, a nonkeratinizing stratified squamous type on its vestibular surface but has a thin keratin layer lateral from Hart line.

Most of the labia minora do not contain skin appendages, although the lateral labia minora may contain sweat and/or sebaceous glands (62). The epithelium can be pigmented, especially in lateral and posterior areas (Fig. 38.13). The epithelium is supported by highly vascular, loose connective tissue rich in elastic fibers. The vestibular bulbs (bulbi vestibuli) are composed of erectile tissue and invested by the bulbocavernous muscles; they lie posterior and deep to the labia minora. The labia minora lack adipose tissue. The vessels and erectile tissue are supported by a rich elastic fiber component. The nerve endings within the labia majora are similar to those found within the clitoris, although Meissner corpuscles and Merkel tactile disks are fewer in number (58). Free nerve endings, Meissner corpuscles, and pacinian corpuscles are seen in stratum basale, spinosum, and granulosum of the epidermis (64).

Developmental anomalies of labia minor rare and are usually accompanied by congenital adrenal hyperplasia or other malformation syndromes. Genital hypoplasia has been reported in association with CHARGE syndrome and Cenani–Lenz syndrome (16). Congenital enlargement of the labia minora may occur and may be asymmetrical.

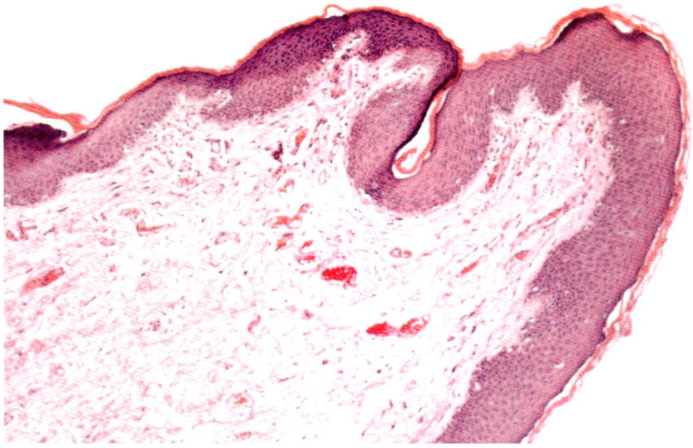

FIGURE 38.13 Lateral labia minora biopsy from a 27-year-old White woman. Within this area, the labia minora contain no skin appendages. The epithelium is pigmented, and melanocytes and pigmented basal epithelial cells are seen in the basal layer. The stratified squamous epithelium has a thinly keratinized surface. Beneath the epithelium, there is an elastic fiber–rich stroma without fat or skin appendages. Moderate numbers of small vessels can be seen. Deeper tissue is demonstrated in Figure 38.12.

Enlargement can be secondary to irritation, edema, or minor trauma. Surgical reduction of the labia minora or local excision for therapeutic reasons does not impede normal sexual function or response, although excision of the labia minora for female circumcision is associated with introital stenosis, vulvar keratinous cysts, and sexual and urinary dysfunction (13).

Labia Majora

The labia majora arise from the embryonic lateral folds (i.e., genital folds, labial folds), which arise lateral to the cloacal plate and do not fuse (10). The endodermally derived epithelium of the vestibule joins with the ectodermally derived epithelium of the medial labial majora that develops from the urogenital sinus. This junction is at Hart line where the epithelium of the medial (inner) labia majora joins the nonkeratinized squamous epithelium of the vestibule (6,7,30). The labia majora merge with the mons pubis anteriorly and with the perineal body posteriorly. The labia majora lie immediately lateral and parallel to the labia minora, separated by the interlabial sulcus. The labia majora are bounded by the vulvar vestibule in the medial posterior positions, and merge with the inguinal–gluteal folds laterally. The latter separate the labia from the medial aspect of the thighs. The length of the labia majora can vary considerably between individuals (30). Although some asymmetry of the labia majora is normal, marked asymmetry may be early evidence of type 1 neurofibromatosis (65). Chronic inflammation, varicosities, edema, Bartholin cysts, and benign or malignant tumors may cause asymmetry of the labia majora.

Age-related changes include an increase in size of the labia with puberty secondary to increased fat within the labia. There are dramatic changes in hair growth during puberty followed by progressive loss of hair follicles and labial hair after menopause (66,67). The postmenopausal labia may shrink due to loss of fat within the labia. The labia majora can also undergo changes related to parity. Progesterone-induced vascular dilation and stasis during gestation can result in development of vulvar venous varicosities (68).

Each hair follicle of the vulva contains a hair root surrounded by the dermal root sheath, which invests the root sheath of the hair follicle. The inner root sheath is composed of an external clear epithelial cell layer (Henle layer) and an inner granular epithelial cell layer (Huxley layer). The hair matrix matures to the formed hair that has an outer cuticle with a cortex and medulla. The hair papilla is found at the base of the hair root and is supported by the dermal root sheath (67). Hair follicles comprise a portion of the pilosebaceous unit, which includes sebaceous glands (Fig. 38.14).

Sebaceous glands of the labia majora can be found with and without associated hair follicles. The sebaceous glands are arranged in a lobular manner and invested by collagen fibers. The cells of the sebaceous glands secrete in a holocrine manner, with the more mature cells accumulating sebaceous secretion (sebum) within their cytoplasm. The secretions are released as the cells undergo necrosis. The secretion may be released adjacent to the hair shaft in the pilosebaceous unit or directly to the surface when no hair shaft is present. There are two types of sweat glands: apocrine and merocrine (58). Apocrine glands are tubular with a columnar secretory epithelium characterized by a prominent eosinophilic granular cytoplasm (Fig. 38.15A,B). They release cytoplasmic secretions and are associated with scent production. The scent associated with these sudoriferous glands is related to bacterial growth supported by the secretory products (58). Myoepithelial cells can be seen beneath the epithelium and are arranged at the periphery of the gland; their contractions promote expression of the secretory contents from the gland lumen. The ducts of the apocrine glands are similar to those of the merocrine glands but may secrete into the upper hair follicle rather than to the skin surface.

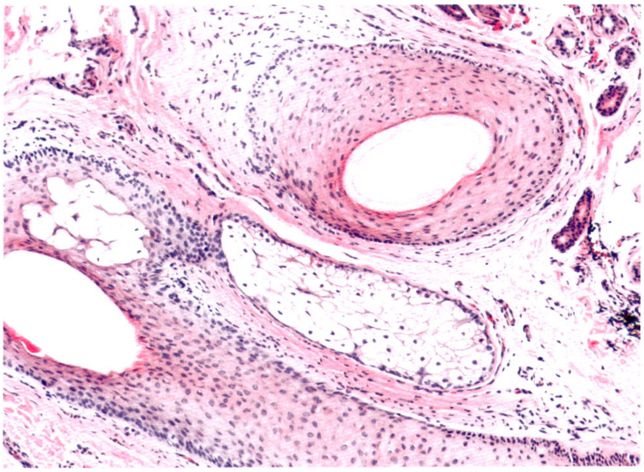

FIGURE 38.14 Pilosebaceous unit with hair follicle and adjacent sebaceous gland.

The merocrine glands are eccrine glands that produce clear, watery sweat. The secretory cells have a pale, slightly granular cytoplasm and an outer layer of myoepithelial cells. The glands are simple and coiled and located deep to the reticular dermis. The sweat duct is lined by a dual layer of cuboidal epithelium that joins with the stratified squamous epithelial surface. Unlike sebaceous and apocrine glands, merocrine glands are not significantly stimulated by the sex hormones (58).

The epithelium of the labia majora peripheral to Hart line is thinly keratinized and pigmented (Fig. 38.16). At the posterior fourchette, the retia are relatively deeper than those in the posterior lateral area (Fig. 38.17). Pigmented cells and small cells with clear cytoplasm (Toker cells) are present in the basal epithelium (Fig. 38.18). A granular layer may be present immediately beneath the keratinized surface (stratum corneum). This granular layer arises from the underlying prickle cell (spinous cell) layer of the stratified squamous epithelium while the stratum malpighii overlying the basal layer. The basal layer (stratum germinativum) is present immediately adjacent to the basement membrane (69). The medial hairless surfaces of the labia majora contain an abundance of sebaceous glands. These glands are not

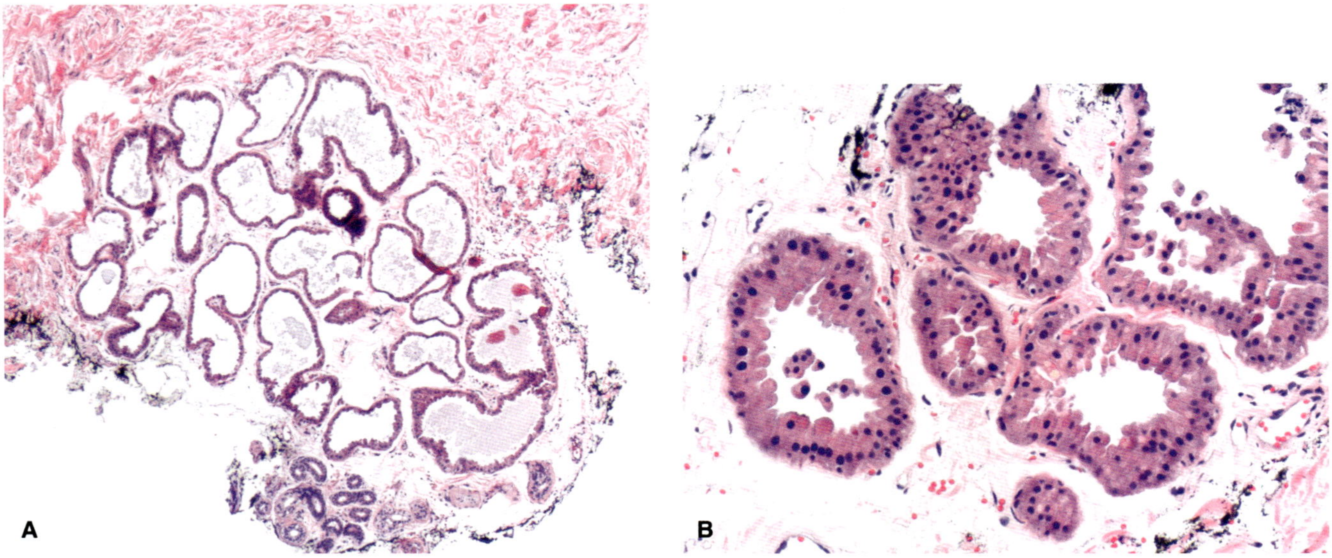

FIGURE 38.15 **A:** Apocrine tubular glands (***low power***). **B:** Glands are lined by columnar cells containing eosinophilic granular cytoplasm. Apical snouts are often seen (***high power***).

associated with hair-bearing pilosebaceous units and open directly onto the epithelial surface where a short segment of the duct lined by nonkeratinized squamous epithelium joins the keratinized squamous epithelial surface (Fig. 38.19). Sebaceous glands have a depth of approximately 2 mm (69). Keratinous (epithelial) cysts may be associated with these sebaceous gland elements (6). Hair follicles in the midportion of the labia majora are associated with the sebaceous gland elements and may be as deep as 2.38 mm (Figs. 38.14 and 38.20) (69,70). Apocrine and eccrine sweat glands are present on the hair-bearing areas of the vulva but are generally absent in the vestibule and medial nonhair–bearing areas of the medial labia majora (Fig. 38.21). A delicate muscle layer (tunica dartos labialis) is present deep within the dermis of the labia majora, and is supported by a fascial layer with prominent elastic fibers (58). Adipose tissue may be abundant in women of reproductive age.

The round ligament joins with the deep longitudinal smooth muscle layer (cremaster muscle) of the labia majora immediately adjacent to the inguinal canal (58). The round ligament may include entrapped peritoneum (processus vaginalis), which can be dilated and result in a cyst of the canal of Nuck (6). Peritoneum-lined cysts are typically encountered in the anterior portion of the labia majora adjacent to or within the inguinal canal.

The skin of the labia majora is rich in nerve endings and contains touch receptors, including Meissner corpuscles, Merkel tactile disks, and peritrichous nerve endings (64). Pacinian corpuscles for pressure sensation are present

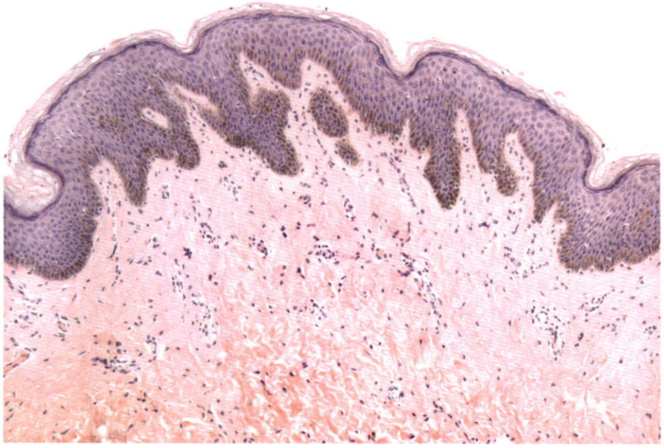

FIGURE 38.16 Posterior medial labia majora of a 27-year-old White woman peripheral to Hart line. This pigmented portion of the labia majora has a stratified squamous epithelium with a thin keratinized surface. The epithelium has deeper rete ridges than those seen in the minora. The dermis is elastic fiber rich and moderately vascular. Sebaceous gland–bearing skin was immediately adjacent to this area and has a moderately vascular dermis.

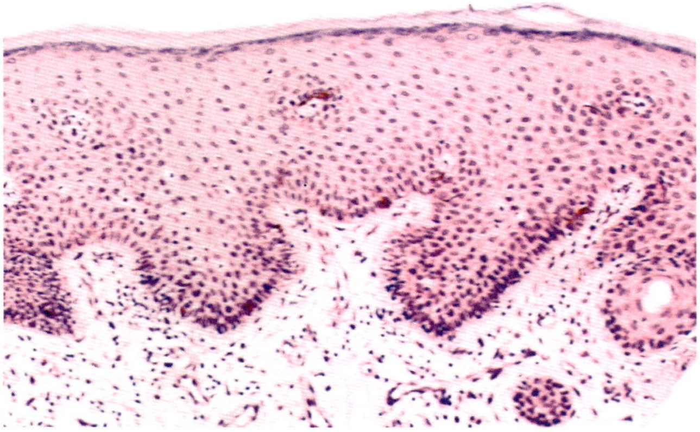

FIGURE 38.17 Posterior vulvar fourchette. The epithelium has a thinly keratinized surface, moderately deep rete, and some melanin pigmentation within the basal cells.

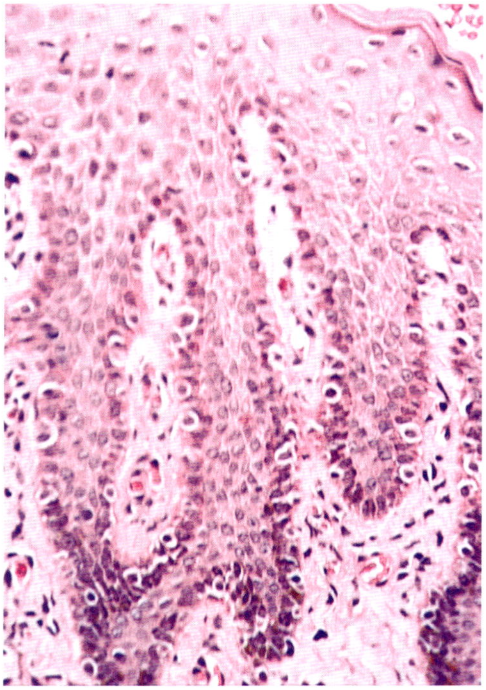

FIGURE 38.18 Posterior perineal body of a 27-year-old White woman. The skin at the perineal body is pigmented, and melanocytes and pigmented keratinocytes are present within the basal layer. The epithelium is stratified squamous epithelium, which has a thin keratin layer. The perinuclear halos present within the epithelial cells are normally seen and should not be confused with koilocytosis. Small clear cells are seen in the epithelial stromal junction within many of the retia.

within the fatty layer of the labia majora, as well as within the labia minora, clitoris, and mons pubis. Free nerve endings for pain reception are present within the labia majora, as well as within the associated muscle cells and blood vessels (64). Ruffini corpuscles are seen throughout the

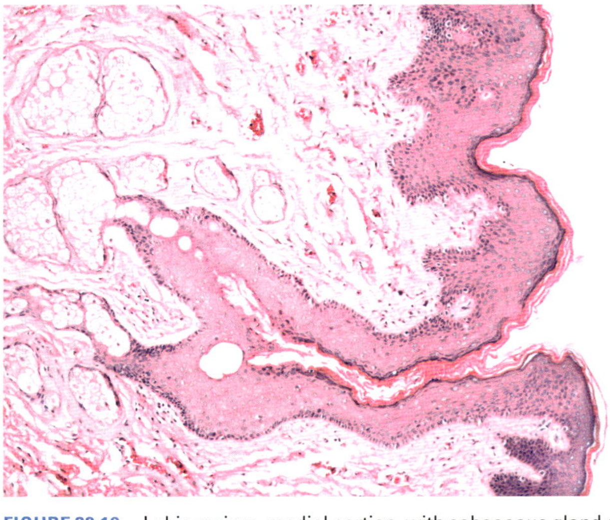

FIGURE 38.19 Labia majora, medial portion, with sebaceous gland elements exiting directly to the skin surface. The epithelium of the medial labia majora has a thin keratin and granular layer. The sebaceous glands may be seen clinically as Fordyce spots.

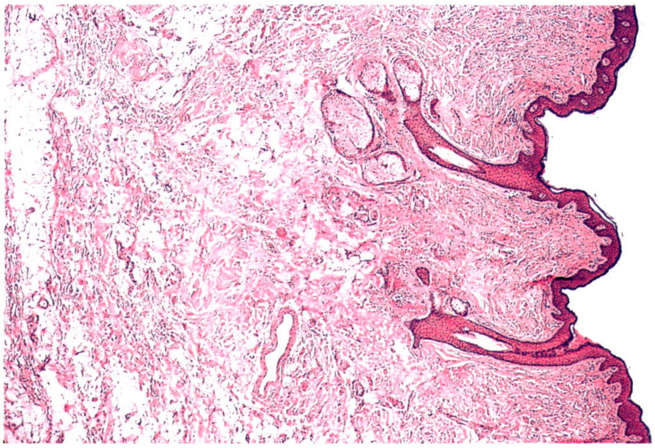

FIGURE 38.20 Labia majora, midportion, with underlying dermis and deep fatty tissue. The thickness of the dermis can be seen in this section of the labia majora. A few deep hair follicles can be seen within the elastic fiber–rich dermis. The dermal junction with the deep fatty tissue is irregular.

subcutaneous tissues of the labia majora, labia minora, clitoris, and mons pubis. They are absent in the hymen. Their exact function in the vulva is uncertain; they may be temperature receptors and/or receptors for sexual stimuli (64). Dogiel–Krause receptors have a distribution similar to that of Ruffini corpuscles but are present in a relatively smaller concentration in the mons pubis and labia majora (64).

Anogenital mammary-like glands are present within the sulcus between the labia majora and the labia minora (sulcus interlabialis). These glands were first described by Hartung in 1872 and were thought to represent ectopic breast tissue from the caudal remnant of the milk line. Van der Putte later proposed that they are normal components of the anogenital region. Current evidence supports the latter interpretation; the milk line does not extend to the vulva in the human, and breast-like tissue in this location is composed of histologically distinct anogenital mammary-like

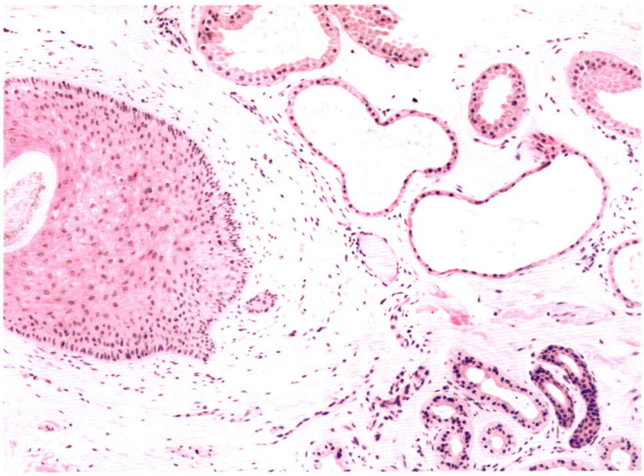

FIGURE 38.21 Labia majora with apocrine glands and sweat ducts adjacent to a hair follicle. Moderate vascularity of the collagen-rich dermis of the labia majora surrounds these sweat gland elements.

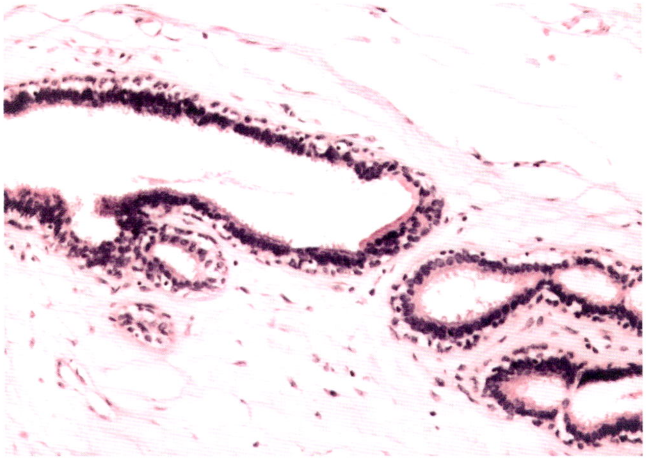

FIGURE 38.22 Mammary-like anogenital gland with duct and small acini. The epithelial lining is composed of a two-layered epithelium with an underlying myoepithelial cell layer and a low columnar epithelial luminal epithelium.

glands (71,72). These glands are morphologically heterogeneous and range from small simple glands to complex lobular units similar to breast lobules. The stroma adjacent to lobules may be loose or fibrotic. The excretory duct system is lined by columnar epithelium containing apocrine secretion and underlying myoepithelium. The duct lining transitions into squamous epithelium and myoepithelial layer is lost as the excretory duct enters the skin surface. Rare Toker cells may be detected near the opening of the duct. The myoepithelial cells are immunoreactive for smooth muscle actin and S100-protein, as well as low–molecular-weight keratin. The superficial luminal epithelial cells have an apocrine appearance with visible “snouts” (Fig. 38.22). These cells are immunoreactive for low–molecular-weight keratin and human milk fat globule antigen. Individual cells are immunoreactive for carcinoembryonic antigen and S100-protein, as well as estrogen and progesterone receptors. Mucus-containing and ciliated cells are not present, distinguishing the cysts of anogenital mammary-like glands from vestibular mucous cysts, müllerian -related cysts, or cysts derived from Bartholin gland. The lack of a stratified squamous epithelium and transitional epithelium also distinguishes them from Bartholin duct cysts, keratinous cysts, and vestibular glandular cysts that have undergone squamous metaplasia (72). A recent study on comparing molecular pathways involved in pathogenesis of histologically similar lesions of the anogenital mammary-like glands and breast tissue found mutations of *PIK3CA, AKT1, MET, ABL1* and *TP53* genes. It appears that the *PI3K-AKT* cascade plays a role in the pathogenesis of tumors arising at both sites (73).

Vulvar Toker Cells

Toker cells were first described in the lower half of the epidermis of nipple, where they show predilection to the ostia of lactiferous sinuses. Similar clear cells were also identified in the epidermis of the anogenital region. They are often seen as single cells, clusters, or small tubules between the epidermal keratinocytes. Toker cell precursors can be identified as early as 15 to 16.5 weeks of gestation at the interlabial sulci of the vulva (74). They steadily increase in number such that by 24 and 39 weeks of gestation, Toker cells are readily identified as variably distributed clusters of clear cells with round nuclei and scant cytoplasm within the lower half of the epidermis. They are most numerous around the ostia of mammary-type glands. Toker cells are the presumed precursors of primary extramammary Paget disease (75). Similar to the tumor cells of extramammary Paget disease and Merkel cells, Toker cells are strongly positive for CK7 and negative for CK20. They express estrogen receptor (ER) and lack HER2 expression (Table 38.1) (76,77).

TABLE 38.1 Immunohistochemical Staining Characteristics of Various Cell Types and Extramammary Paget Disease in the Vulva (79)

Stain	Toker Cells	Melanocytes	Squamous Cells	Paget Cells
LMWCK	+	–	+/–	+
HMWCK	–	–	+	–
EMA	+/–	–	+	+
S100 Protein	+/–	+	–	+/–
mCEA	–	–	–	+/–
HER2	–	–	–	+/–
Melan-A	–	+	–	–
GCDFP-15	–	–	–	+/–
Mucin	+/–	–	–	+/–

LMWCK, low–molecular-weight cytokeratin (CK7, CAM5.2); HMWCK, high–molecular-weight cytokeratin (CK5/6, 34βE12, CK14, CK17); EMA, epithelial membrane antigen; mCEA, monoclonal carcinoembrionic antigen; HER2, human epidermal growth factor receptor 2; Melan-A, melanoma antigen; GCDFP-15, gross cystic disease fluid; +, positive; –, negative; +/–, may be positive.

Mons Pubis (Mons Veneris)

The mons pubis has its origin in the embryonic genital medial cranial swellings. The subcutaneous tissue of mons pubis becomes more prominent with the onset of puberty when there is a progressive increase in fat tissue beneath the mons. There is also a dramatic increase in hair growth of the mons pubis and labia majora.

Age-related changes of the mons pubis have been summarized and staged by Tanner in the following sequence (66). Stage 1 shows no visible pubic hair growth. Stage 2 features a small amount of pubic-type hair on the midportion of the mons pubis with or without a similar amount of hair on the labia majora. Mons pubis hair growth is more prominent in Stage 3, both in terms of the amount of hair and its coarseness. Stage 4 hair growth over the mons pubis is similar to that of the adult, with the exception that hair growth is lacking over the upper lateral corners. Stage 5 characterizes the adult pubic hair pattern and occurs between the ages of 12 and 17 years (66). There can be substantial variability in the amount and character of the pubic hair (escutcheon) related to racial and genetic factors. However, pubic-type hair growth generally does not extend above a horizontal line drawn between, and 2 cm above, the uppermost limits of the genitofemoral folds (66,78). Hair follicle depth within the vulva is greatest in the mons pubis, where hair follicle depth has been measured up to 2.72 mm (67). The mons pubis is richly endowed with nerve receptor types similar to those of the labia majora (64). Some experts believe that the mons pubis functions to cushion the pubic bone from impact during intercourse. In addition, it is rich in oil-secreting (sebaceous) glands that release substances involved in sexual attraction (pheromones).

Lymphatic Vessel Drainage

The lymphatic vessel system of the female external genitalia is a complex network that drains the entire labia minora, fourchette, prepuce, and distal vagina below the hymen. Large lymphoid channels run lateral to the clitoris and coalesce with lymphatic vessels draining the mons pubis and labia majora. Radiolabeled tracer studies reveal that the lymphatic vessel drainage of the perineum, clitoris, and anterior labia minora is bilateral, whereas lymph flow from well-lateralized sites in the vulva is mostly to the ipsilateral groin. Secondary lymphatic vessel drainage from the superficial inguinal nodes is through the cribriform fascia to the deep inguinal or femoral nodes, with subsequent tertiary flow under the inguinal ligaments to the deep pelvic (i.e., external iliac and obturator) lymph nodes. The lymph node of Cloquet, also called Rosenmüller lymph node, is the most cephalad of the femoral lymph nodes, often lying in the femoral canal below Poupart (inguinal) ligament (79,80).

Lymphatic channels from a lateral site may sometimes drain to the contralateral lymph node group, which has clinical relevance in treatment planning of patients with vulvar malignancies. The most common sites of metastasis from vulvar malignancies are the superficial inguinal lymph nodes. In general, 8 to 10 nodes are found in this area, with superior oblique (above the ligament of Poupart) and inferior ventral (between the ligament of Poupart and the saphenous vein and fascia lata) divisions. Midline structures, such as the clitoris and the midline perineum, drain bilaterally. A second path of lymphatic vessel drainage from the clitoris involves urethral lymphatics and lymphatics draining the dorsal vein of the clitoris. These channels lead inferior to the symphysis pubis through the anogenital diaphragm to join the lymphatic vessel plexus of the anterior bladder surface. Ultimately, these channels terminate in the interiliac and obturator nodes or course superiorly to the femoral and internal iliac nodes. Deep pelvic nodes are not generally involved unless the superficial inguinal nodes are involved.

Sentinel lymph node mapping with intraoperative lymphoscintigraphy (technetium-99m–labeled nanocolloid) is gaining application to assess inguinal lymph node status among patients with vulvar carcinoma and melanoma. Azulfidine blue vital dye (blue dye) has been used either alone or with the colloid, although a few clinicians have discontinued use of the blue dye due to rare untoward reactions. Surgical excision with histopathologic assessment of the lymph nodes picking up the radioisotope-labeled colloid/blue dye (the sentinel nodes) is performed to plan appropriate lymph node resection related to the vulvar surgery. When these nodes are found to be free of tumor by this technique, the patient may be spared radical inguinal–femoral lymphadenectomy (21–23).

Obstruction of vulvar lymphatic vessels related to prior surgical interruption, radiation therapy, or chronic inflammation (such as secondary to Crohn disease, hidradenitis suppurativa, etc.) may result in lymphangiectasia of the vulva. Obstruction of vulvar lymphatic vessels may manifest with leakage of clear fluid from the vulva and result in lymphangioma circumscriptum of the vulva; this condition is characterized by the presence of multiple small, glistening, superficial clustered vesicles on the vulvar skin resembling frog spawn (81,82). Vulvar lymphatic vessel obstruction is typically associated with some degree of epithelial and dermal edema (83). Massive vulvar edema may occur in immobilized and obese women; bilateral labial enlargement can be related to chronic lymphatic obstruction (84).

Arterial Supply

The major arterial supply of the vulva originates from the branches of the internal pudendal artery, a division of the internal iliac (hypogastric) artery. These branches supply arterial blood to the labia (labial artery) via the perineal artery, and the corpora cavernosa and vestibule via deep artery of the clitoris, and to the glands and prepuce via dorsal and deep arteries of the clitoris. Additional arterial supply originates from the superficial and deep external pudendal arteries and branches from the femoral artery; these anastomose with anterior and posterior labial branches of the perineal artery. The anterior vaginal artery supplies the vestibule and Bartholin gland areas (6,10,84).

Venous Supply

Major venous drainage of the vulva is primarily from the bilateral internal iliac veins that drain into the external iliac venous system. The internal iliac veins drain both parietal and visceral venous systems. The parietal tributaries of the internal iliac vein include the internal obturator, internal pudendal veins, superior and inferior gluteal veins, sciatic vein, and ascending lumbar veins. The visceral branches drain pelvic organs, including the uterine, ovarian, and vaginal venous systems. In a study of this drainage in 79 specimens, a single internal iliac vein was present on the side studied in 73% of the cases; in 29% of the cases, two separate iliac veins drained into the external iliac vein; and, in one case, the internal iliac vein drained directly to the inferior vena cava (10,84). Such varices correlate with insufficiency of the internal iliac venous system and also involve the tributaries of the internal iliac vein, as well as the saphenous vein (85). Varices are identified in the vulva and perivulvar area in approximately 4% of women, and are more commonly identified in pregnant women. They may also be related to arteriovenous malformations involving the vulva in Klippel–Trenaunary–Weber syndrome and Parkes syndrome (68). Although they are usually evident by their clinical appearance, vulvar varices may occasionally present as a "cyst" or "nodule" in the superficial dermis. Biopsy findings include a dilated venous vessel with an associated organized thrombus and occasional perivenous neutrophilic infiltrate (68).

Nerve Supply

The major nerves of the vulva are derived from the anterior and posterior labial nerves. The anterior nerve is a branch of the ilioinguinal nerve, and the posterior labial branch is from the pudendal nerve. The clitoral nerve supply is from the dorsal nerve of the clitoris and the cavernous nerves of the clitoris. Branches of the cavernous nerves, arising from the vaginal nerve plexus, join the clitoral dorsal nerve at the hilum of the clitoral bodies (85). The dorsal nerve of the clitoral bundle branches from the pudendal nerve. The two clitoral bodies separate to form the two clitoral crura beneath the pubic arch. The dorsal nerves form two bundles that are extensively distributed along the lateral aspects of the clitoral bodies at the 11 and 1 o'clock positions but are sparse at the 12 o'clock position. These join distally to form a single clitoral body. The densest nerve groups that enter the glans clitoris are found on the dorsal aspect of the clitoris, with a concentration of nerves under the epithelium of the glans clitoris (85).

REFERENCES

1. Reed BD, Harlow SD, Sen A, et al. Prevalence and demographic characteristics of vulvodynia in a population based sample. *Am J. Obstet Gynecol.* 2012;206(2):170.e1–170.e9.
2. Harlow BL, Stewart EG. Population-based assessment of chronic unexplained vulvar pain: have we underestimated the prevalence of vulvodynia? *J Am Med Womens Assoc.* 2003; 58(2):82–88.
3. Arnold LD, Bachman GA, Rosen R, Rhoads GG. Assessment of vulvodynia symptoms in a sample of US women: a prevalence survey with nested case control study. *Am J Obstet Gynecol.* 2007;196(2):128.e1–128.e6.
4. Haefner HK, Collins ME, Davis GD, et al. The vulvodynia guideline. *J Low Genit Tract Dis.* 2005;9:40–51.
5. Vieira-Baptista P, Donders G, Margesson L, Edwards L, Haefner HK, Pérez-López FR. Diagnosis and management of vulvodynia in postmenopausal women. *Maturitas.* 2018;108:84–94.
6. Wilkinson EJ, Massoll N. Benign diseases of the vulva. In: Kurman RJ, ed. *Blaustein's Pathology of the Female Genital Tract.* 6th ed. Springer-Verlag; 2011:3–46.
7. Wilkinson EJ, Stone IK. *Atlas of Vulvar Disease.* 3rd ed. Wolters Kluwer/Lippincott Williams & Wilkins; 2012.
8. Neill SM, Lewis FM. *Ridley's The Vulva.* 3rd ed. Wiley-Blackwell; 2009.
9. Wilkinson EJ. Premalignant and malignant tumors of the vulva. In: Kurman RJ, Ellenson LH, Ronnett BM, eds. *Blaustein's Pathology of the Female Genital Tract.* 6th ed. Springer-Verlag; 2011:56–103.
10. Kurman RJ, Ronnett BM, Sherman ME, et al. Tumors of the cervix, vagina, and vulva. In: Rosai J, ed. *AFIP Atlas of Tumor Pathology, Series 4.* American Registry of Pathology; 2010:1–22.
11. Spiryda LB, Laufer MR, Soiffer RJ, Antin JA. Graft-versus-host disease of the vulva and/or vagina: diagnosis and treatment. *Biol Blood Marrow Transplant.* 2003;9:760–765.
12. Nardelli A, Degreef H, Goossens A. Contact allergic reactions of the vulva: a 14-year review. *Dermatitis.* 2004;15:131–136.
13. Thabet SM, Thabet AS. Defective sexuality and female circumcision: the cause and the possible management. *J Obstet Gynaecol Res.* 2003;29:12–19.
14. Creighton SM, Minto CL, Steele SJ. Objective cosmetic and anatomical outcomes at adolescence of feminizing surgery for ambiguous genitalia done in childhood. *Lancet.* 2001; 358:124–125.
15. Baskin LS. Anatomical studies of the female genitalia: surgical reconstructive implications. *J Pediatr Endocrinol Metab.* 2004;17:581–587.
16. Seely JR, Seely BL, Bley R Jr, Altmiller CJ. Localized chromosomal mosaicism as a cause of dysmorphic development. *Am J Hum Genet.* 1984;36:899–903.
17. Hanna SJ, Kaiser L, Muneer A, Nottingham JF, Kunkler RB. Squamous cell carcinoma of the bladder presenting as vulvitis and cliteromegaly. *Gynecol Oncol.* 2004;95:722–723.
18. Lee PA, Witchel SF. Genital surgery among females with congenital adrenal hyperplasia: changes over the past five decades. *J Pediatr Endocrinol Metab.* 2002;15:1473–1477.
19. Crouch NS, Minto CL, Laio LM, Woodhouse CR, Creighton SM. Genital sensation after feminizing genitoplasty for congenital adrenal hyperplasia: a pilot study. *BJU Int.* 2004;93: 135–138.
20. Minto CL, Liao LM, Woodhouse CR, Ransley PG, Creighton SM. The effect of clitoral surgery on sexual outcome in individuals who have intersex conditions with ambiguous genitalia: a cross-sectional study. *Lancet.* 2003;361:1252–1257.
21. Hakam A, Nasir A, Raghuwanshi R, et al. Value of multilevel sectioning for improved detection of micrometastases in sentinel lymph nodes in invasive squamous cell carcinoma of the vulva. *Anticancer Res.* 2004;24:1281–1286.
22. Moore RG, Granai CO, Gajewski W, Gordinier M, Steinhoff MM. Pathologic evaluation of inguinal sentinel lymph nodes in vulvar cancer patients: a comparison of immunohistochemical staining versus ultrastaging with hematoxylin and eosin staining. *Gynecol Oncol.* 2003;91:378–382.
23. Moore RG, DePasquale SE, Steinhoff MM, et al. Sentinel node identification and the ability to detect metastatic tumor to inguinal lymph nodes in squamous cell cancer of the vulva. *Gynecol Oncol.* 2003;89:475–479.
24. Brown HM, Wilkinson EJ. Cytology of secondary vulvar Paget's disease of urothelial origin: a case report. *Acta Cytol.* 2005;49:71–74.
25. Selvaggi G, Bellringer J. Gender reassignment surgery: an overview. *Nat Rev Urol.* 2011;8(5):274–282.
26. Mancino P, Parlavecchio E, Melluso J, Monti M, Russo P. Introducing colposcopy and vulvovaginoscopy as routine

examinations for victims of sexual assault. *Clin Exp Obstet Gynecol.* 2003;30:40–42.
27. Stenson AL. Vulvodynia: diagnosis and management. *Obstet Gynecol Clin N Am.* 2017;44:493–508.
28. Pillai M. Genital findings in prepubertal girls: what can be concluded from an examination? *J Pediatr Adolesc Gynecol.* 2008;21(4):177–185.
29. McCann J, Miyamoto S, Boyle C, Rogers K. Healing of non-hymenal genital injuries in prepubertal and adolescent girls: a descriptive study. *Pediatrics.* 2007;120(5):1000–1011.
30. Bechara A, Bertolino MV, Casabé A, et al. Duplex Doppler ultrasound assessment of clitoral hemodynamics after topical administration of alprostadil in women with arousal and orgasmic disorders. *J Sex Marital Ther.* 2003;29(Suppl 1):1–10.
31. Grant LA, Sala E, Griffin N. Congenital and acquired conditions of the vulva and vagina on magnetic resonance imaging: a pictorial review. *Semin Ultrasound CT MR.* 2010;31(5):347–362.
32. Griffin N, Grant LA, Sala E. Magnetic resonance imaging of vaginal and vulval pathology. *Eur Radiol.* 2008;18(6):1269–1280.
33. Reuter JC. High-yield vulvar histopathology for the clinician. *Obstet Gynecol Clin North Am.* 2017;44(3):329–338.
34. Day T, Holland SM, Scurry J. Normal vulvar histology: variation by site. *J Low Genit Tract Dis.* 2016;20(1):64–69.
35. Jones IS. A histological assessment of normal vulval skin. *Clin Exp Dermatol.* 1983;8(5):513–521.
36. Benedet JL, Wilson PS, Matisic J. Epidermal thickness and skin appendage involvement in vulvar intraepithelial neoplasia. *J Reprod Med.* 1991;36:608–612.
37. Hu F. Melanocyte cytology in normal skin. In: Ackerman AB, ed. *Masson Monographs in Dermatology-1.* Masson; 1981.
38. Moll R, Franke WW, Schiller DL, Geiger B, Krepler R. The catalog of human cytokeratins: patterns of expression in normal epithelia, tumors and cultured cells. *Cell.* 1982:31:11–24.
39. Eichner R, Bonitz P, Sun TT. Classification of epidermal keratins according to their immunoreactivity, isoelectric point and mode of expression. *J Cell Biol.* 1984;98:1388–1396.
40. Cooper D, Schermer A, Sun TT. Classification of human epithelia and their neoplasms using monocolonal antibodies to keratin; strategies, applications, and limitations. *Lab Invest.* 1985;52:243–256.
41. Esquius J, Brisigotti M, Matias-Guiu X, Prat J. Keratin expression in normal vulva, non-neoplastic epithelial disorders, vulvar intraepithelial neoplasia, and invasive squamous cell carcinoma. *Int J Gynecol Pathol.* 1991;4:341–355.
42. Riethdorf S, Neffen EF, Cviko A, Löning T, Crum CP, Riethdorf L. p16 INK4A expression as a biomarker for HPV16-related vulvar neoplasia. *Human Pathol.* 2004;35:1477–1483.
43. Yang B, Hart WR. Vulvar intraepithelial neoplasia of the simplex (differentiated) type: a clinicopathologic study including analysis of HPV and p53 expression. *Am J Surg Pathol.* 2000;24:429–441.
44. Tessier-Cloutier B, Kortekaas KE, Thompson E, et al. Major p53 immunohistochemical patterns in in situ and invasive squamous cell carcinomas of the vulva and correlation with TP53 mutation status. *Mod Pathol.* 2020;33:1595–1605.
45. Edwards JN, Morris HB. Langerhans' cells and lymphocyte subsets in the female genital tract. *Br J Obstet Gynaecol.* 1985; 92:974–982.
46. McLean JM. Anatomy and physiology of the vulva. In: Ridley CM, ed. *The Vulva.* Churchill Livingstone; 1988:39–65.
47. Rotsztejn H, Trznadel-Budzko E, Jesionek-Kupnicka D. Do Langerhans cells play a role in vulvar epithelium resistance to squamous cell carcinoma? *Arch Immunol Ther Exp (Warsz).* 2007;55(2):127–130.
48. Gould VE, Moll R, Moll I, Lee I, Franke WW. Biology of disease. Neuroendocrine (Merkel) cells of the skin: hyperplasias, dysplasias, and neoplasms. *Lab Invest.* 1985;52:334–353.
49. Fetissof F, Berger G, Dubois MP, et al. Endocrine cells in the female genital tract. *Histopathology.* 1985;9:133–145.
50. Pyka RE, Wilkinson EJ, Friedrich EG Jr, Croker BP. The histopathology of vulvar vestibulitis syndrome. *Int J Gynecol Oncol Pathol.* 1988;7:249–257.
51. Friedrich EG Jr, Wilkinson EJ. Mucous cysts of the vulvar vestibule. *Obstet Gynecol.* 1973;42:407–414.
52. Robboy SJ, Ross JS, Prat J, Keh PC, Welch WR. Urogenital sinus origin of mucinous and ciliated cysts of the vulva. *Obstet Gynecol.* 1978;51:347–351.
53. Slone S, Reynolds L, Gall S, et al. Localization of chromogranin, synaptophysin, serotonin, and CXCR2 in neuroendocrine cells of the minor vestibular glands: an immunohistochemical study. *Int J Gynecol Pathol.* 1999;18(4):360–365.
54. Axe S, Parmley T, Woodruff JD, Hlopak B. Adenomas in minor vestibular glands. *Obstet Gynecol.* 1986;68:16–18.
55. van der Putte SCJ, Sie-Go DM. Development and structure of glandopreputial sulcus of the human clitoris with a special reference to glandopreputial glands. *Anat Rec.* 2011;294(1):156–164.
56. Huffman JW. The detailed anatomy of the paraurethral ducts in the adult human female. *Am J Obstet Gynecol.* 1948;55:86–101.
57. Halperin R, Zehavi S, Vaknin Z, Ben-Ami I, Pansky M, Schneider D. The major histopathologic characteristics in the vulvar vestibulitis syndrome. *Gynecol Obstet Invest.* 2005;59(2):75–79.
58. Krantz KE. The anatomy and physiology of the vulva and vagina and the anatomy of the urethra and bladder. In: Philipp EE, Barnes J, Newton M, eds. *Scientific Foundations of Obstetrics and Gynaecology.* Year Book; 1977:65–78.
59. Bergeron C, Ferenczy A, Richart RM, Guralnick M. Micropapillomatosis labialis appears unrelated to human papillomavirus. *Obstet Gynecol.* 1990;76:281–286.
60. Altcheck A, Wasserman B, Deligdisch L. Prepubertal distal longitudinal vaginal folds. *J Pediatr Adolesc Gynecol.* 2008;21(6): 351–354.
61. Amenta PS. *Elias-Pauly's Histology and Human Microanatomy.* 5th ed. John Wiley & Sons; 1987:502–503.
62. Miyagawa S, Monn A, Haraguchi R, et al. Dosage-dependent hedgehog signals integrated with *Wnt*/β;-catenin signaling regulate external genitalia formation as an appendicular program. *Development.* 2009;136:3969–3978.
63. Clerico C, Lari A, Mojallal A, Boucher F. Anatomy and aesthetics of the labia minora: the ideal vulva? *Aesthetic Plast Surg.* 2017;41(3):714–719.
64. Schober J, Aardsma N, Mayoglou L, Pfaff D, Martín-Alguacil N. Terminal innervation of female genital, cutaneous sensory receptors of the epithelium of the labia minora. *Clin Anat.* 2015;28(3):392–398.
65. Friedrich EG Jr, Wilkinson EJ. Vulvar surgery for neurofibromatosis. *Obstet Gynecol.* 1985;65:135–138.
66. Tanner JM. *Growth at Adolescence.* 2nd ed. Blackwell; 1962.
67. Barman JM, Astore J, Pecoraro V. The normal trichogram of people over 50 years. In: Montagna W, Dobson RL, eds. *Advances in Biology of Skin.* Vol. IX. Hair Growth. Pergamon Press; 1969.
68. Bell D, Kane PB, Liang S, Conway C, Tornos C. Vulvar varicies: an uncommon entity in surgical pathology. *Int J Gynecol Pathol.* 2006;26:99–101.

69. Zelickson AS. *Electron Microscopy of Skin and Mucous Membranes*. Charles C Thomas; 1963.
70. Shatz P, Bergeron C, Wilkinson EJ, Arseneau J, Ferenczy A. Vulvar intraepithelial neoplasia and skin appendage involvement. *Obstet Gynecol*. 1989;74(5):769–774.
71. van der Putte SC, van Gorp LH. Cysts of mammarylike glands in the vulva. *Int J Gynecol Pathol*. 1995;14(2):184–188.
72. van der Putte SC. Mammary-like glands of the vulva and their disorders. *Int J Gynecol Pathol*. 1994;13:150–160.
73. Konstantinova AM, Vanecek T, Martinek P, et al: Molecular alterations in lesions of anogenital mammary-like glands and their mammary counterparts including hidradenoma papilliferum, intraductal papilloma, fibroadenoma and phyllodes tumor. *Ann Diagn Pathol*. 2017;28:12–18.
74. van der Putte SC. Clear cells of Toker in the developing anogenital region of male and female fetuses. *Am J Dermatopathol*. 2011;33(8):811–818.
75. Willman JH, Golitz LE, Fitzpatrick JE. Vulvar clear cells of Toker: precursors of extramammary Paget's disease. *Am J Dermatopathol*. 2005;27(3):185–188.
76. Park S, Suh YL. Useful immunohistochemical markers for distinguishing Paget cells from Toker cells. *Pathology*. 2009; 41(7):640–644.
77. Lundquist K, Kohler S, Rouse RV. Intraedermal cytokeratin 7 expression is not restricted to Paget cells but is also seen in Toker cells and Merkel cells. *Am J Surg Pathol*. 1999;23(2):212–219.
78. Lunde O. A study of body hair density and distribution in normal women. *Am J Phys Anthropol*. 1984;64:179–184.
79. Parry-Jones E. Lymphatics of the vulva. *J Obstet Gynaecol Br Commonw*. 1963;70:751–765.
80. Russel AH, Duska LR. *Cancer of Vulva in Liebel and Phillip: Textbook on Radiation Oncology*. 3rd ed. Elsevier Saunders; 2010, Chapter 52:1085.
81. Sims SM, McLean FW, Davis JD, Morgan LS, Wilkinson EJ. Vulvar lymphangioma circumscriptum: a report of 3 cases, 2 associated with squamous cell carcinoma and 1 with hidradenitis suppprativa. *J Low Genit Tract Dis*. 2010;14(3):234–238.
82. Handfield-Jones SE, Prendiville WJ, Norman S. Vulval lymphangiectasia. *Genitourin Med*. 1989;65(5):335–337.
83. McCluggage WG, Nielsen GP, Young RH. Massive vulval edema secondary to obesity and immobilization: a potential mimic of aggressive angiomyxoma. *Int J Gynecol Pathol*. 2008;27(3):447–452.
84. LaPage PA, Villavicencio JL, Gomez ER, Sheridan MN, Rich NM. The valvular anatomy of the iliac venous system and its clinical implications. *J Vasc Surg*. 1991;14:678–683.
85. Yucel S, De Souza A Jr, Baskin LS. Neuroanatomy of the human female lower urogenital tract. *J Urol*. 2004;172(1): 191–195.

Vagina

Baris Boyraz ▪ Jennifer A. Bennett

As primary diseases of the vagina are relatively uncommon, tissue from the vagina is infrequently examined histologically. The primary clinical indications for vaginal sampling include assessment of abnormal cytology/positive HPV test, evaluation of abnormal bleeding following hysterectomy, and identification of benign cysts/nodules incidentally detected on routine examination. In these situations, a biopsy is usually performed as partial or complete vaginectomies are rare.

This chapter will describe the development, anatomy, and microscopy of the normal vagina, as well as address distinctive, but uncommonly sampled anatomical structures or metaplastic changes that may pose a diagnostic challenge.

EMBRYOLOGY

The paired müllerian (paramesonephric) ducts appear during week 6 of gestation as funnel-shaped openings of the coelomic epithelium. These develop into paired, undifferentiated tubes that grow caudally, using the previously formed wolffian (mesonephric) ducts as a guide to reach the urogenital sinus (Fig. 39.1).

In week 8, the paired müllerian ducts fuse caudally, forming a uterovaginal canal lined by simple columnar epithelium (Fig. 39.2). A temporary midline septum is initially present, but this largely dissolves during week 9. The point of contact with the urogenital sinus is termed the "müllerian tubercle." The müllerian epithelium at this site proliferates and occludes the lumen forming a solid vaginal plate. At 12 weeks of gestation, the plate is composed of PAX2-positive müllerian epithelium cranially and FOXA1-positive urogenital sinus epithelium caudally (Fig. 39.3) (1). The FOXA1-positive urogenital sinus epithelium then grows cranially to completely replace the PAX2-positive müllerian epithelium to the external cervical os.

The transition to squamous epithelium occurs around the time estrogen receptors appear in the vaginal stroma (2,3). In week 16, the squamous epithelium lining the vagina thickens and becomes glycogenated due to an increase in endogenous estrogens (4). As epithelial cells mature, they lose cellular adhesion and desquamate, resulting in canalization of the vaginal lumen. By week 22, vaginal development is complete (1). Nonetheless, various congenital abnormalities can result from failed development at any step (Table 39.1) (5).

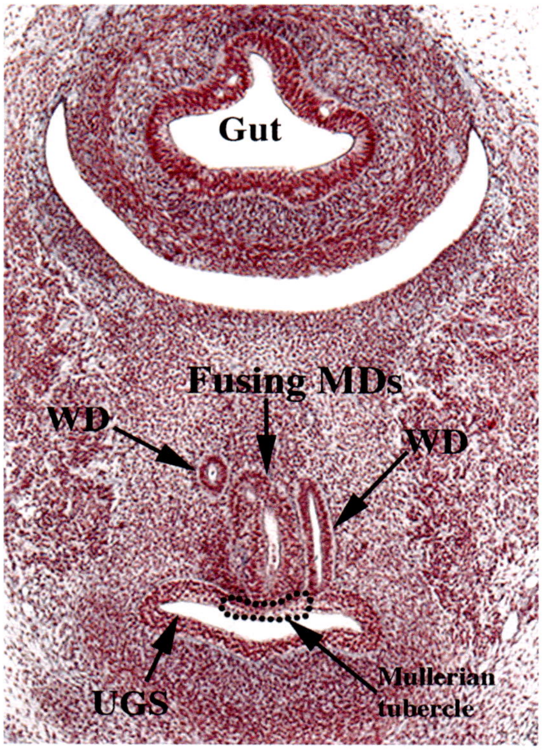

FIGURE 39.1 Pelvic section of a stage 23 Carnegie embryo (56 days) showing contact of the fusing müllerian ducts (MD) with the urogenital sinus (UGS). The point of contact is called the müllerian tubercle. Wolffian ducts (WD) also join the urogenital sinus just lateral to the müllerian tubercle. (From the Virtual Human Embryo Project (http://virtualhumanembryo.lsuhsc.edu) with permission.)

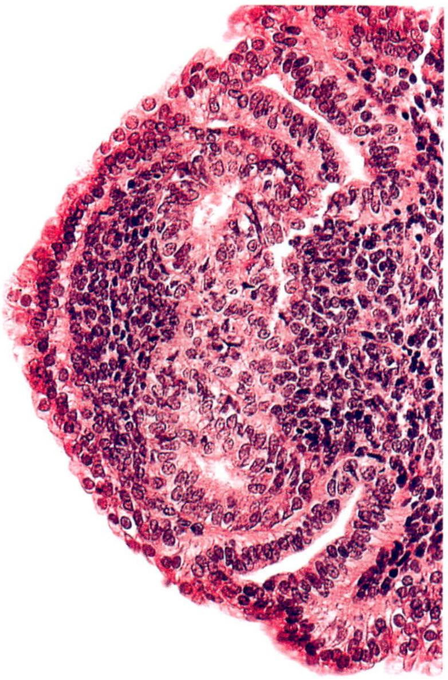

FIGURE 39.2 Region of urogenital sinus disclosing the tips of two central müllerian ducts that have grown down the (outer) paired wolffian ducts (circa day 54). The cytologic features of the cells comprising both types of ducts are indistinguishable on light microscopy at early stages of development. (From Robboy SJ, Ellington KS. *Pathology of the Female Genital Tract in Kodachrome Slides*. Robboy Associates, LLC; 1996.)

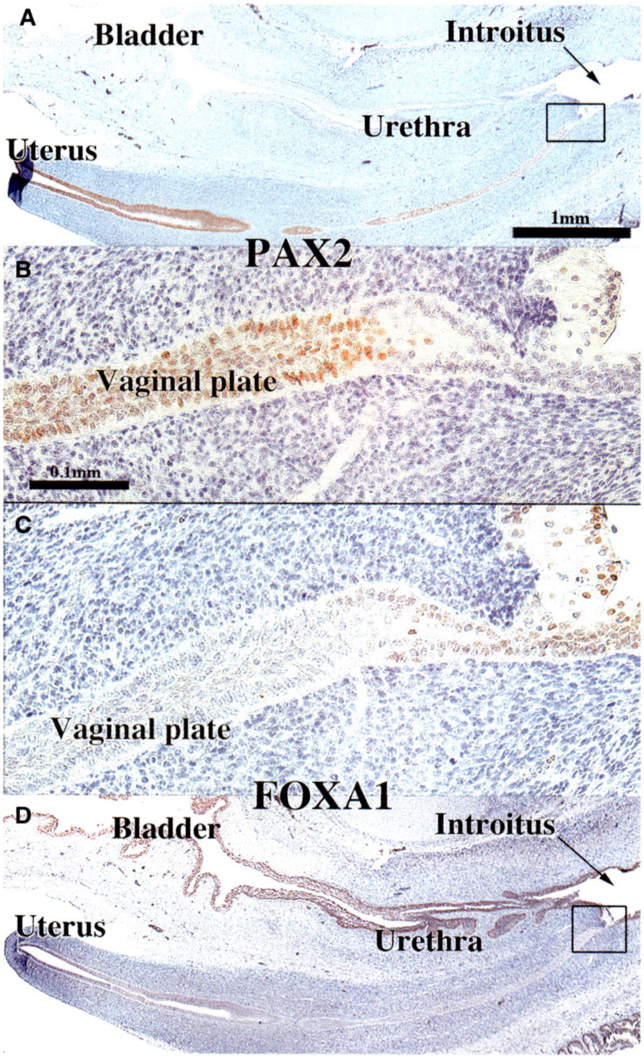

FIGURE 39.3 Sagittal sections of a 12-week human female fetal reproductive tract show PAX2-reactive epithelial cells extending near the junction with the introitus/urethra (**A,B**) and FOXA1-reactive epithelial cells extending only a short distance into the solid vaginal plate (**C,D**). (From Robboy SJ, Kurita T, Baskin L, et al. New insights into human female reproductive tract development. *Differentiation*. 2017;97:9–22, with permission from Elsevier.)

GROSS ANATOMY

The vagina extends from the vulvar vestibule to the uterine cervix, lying posterior to the urinary bladder and anterior to the rectum (Fig. 39.4). When standing, the lower vaginal axis is vertical, but the upper axis becomes more horizontal at the level of the pelvic diaphragm (6). Overall, vaginal length is variable, ranging from 6.5 to 12.5 cm (7). The anterior wall averages 6 cm in length and the posterior wall 10 cm, with the cervix filling the difference (8). The length is often slightly decreased in women with a prior hysterectomy or pelvic reconstruction; moreover, every additional 10 years of age also slightly decreases total length (9). The vaginal apex surrounds the exocervix and forms vault-like fornices

TABLE 39.1 Congenital Anomalies Due to Failures in Development

Congenital Anomaly	Etiology
Transverse Vaginal Septum	• Failed fusion between müllerian ducts and urogenital sinus • Failed canalization of vaginal plate
Longitudinal Vaginal Septum	• Defective lateral fusion of müllerian ducts • Incomplete reabsorption of caudal müllerian ducts • Often occurs with partial or complete duplication of uterine cervix and corpus
Segmental Müllerian Hypoplasia or Agenesis	• Failed development of the lower portion of müllerian ducts • Usually results in absence of uterus and upper vagina
Vaginal Atresia	• Failure of the urogenital sinus to form the caudal vagina
Müllerian Agenesis (Mayer–Rokitansky–Küster–Hauser syndrome)	• Incomplete development of müllerian ducts • Absence of uterus and vagina

between its cervical attachment and the vaginal wall. Thus, the vagina is widest at the apex and narrowest at the introitus. In the adult, the anterior and posterior walls are relaxed and remain in contact with each other, whereas the lateral walls remain fairly rigid and separated. This is thought to give an "H" appearance to the vaginal canal on cross section (10), although three-dimensional imaging with magnetic resonance imaging has also recognized a "W" shape (11).

The upper one-fourth of the vagina is related to the rectouterine space (cul-de-sac or pouch of Douglas) posteriorly and covered with peritoneum. The middle half of the vagina is

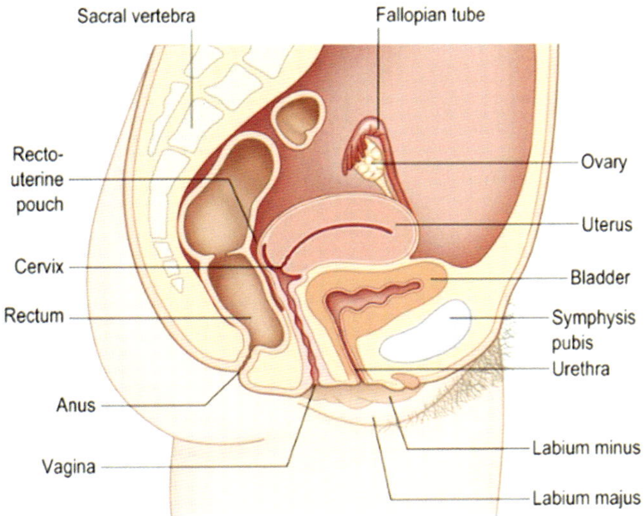

FIGURE 39.4 Structural relationships of the vagina. (From Bean S, Prat J, Robboy SJ. Vagina. In: Mutter GL, Prat J, eds. *Pathology of the Female Reproductive Tract.* Churchill Livingstone/Elsevier; 2014:132–159, with permission from Elsevier.)

closely apposed to the rectum (rectovaginal space) but lacks a discrete fascial layer (12). The lower one-fourth of the vagina is separated from the anal canal by the anal and rectal sphincters, as well as the perineal body, from which the bulbocavernosus and superficial transverse perineal muscles originate.

The urinary bladder and urethra lie anterior to the vagina. No discrete fascial layer is present between the urethra and the vagina (13). The urethra usually terminates at the vulvar vestibule, just anterior to the vaginal introitus.

The size and shape of the vaginal introitus are related to the state of the hymen (14). When the inner edges of the hymen are apposed, the vaginal opening resembles a cleft. When stretched, the hymen may persist in the form of a ring-like structure around the vaginal orifice (see Chapter 38 for anatomy of the hymenal region).

Ligaments

The vaginal structural supports (ligaments) are intimately related to the uterus, urethra, bladder, and rectum, and are best described in three levels: uterosacral/cardinal ligament complex (upper vaginal support), paravaginal attachments (mid vaginal support), and surrounding structures (lower vaginal support) (15). The lateral supports are the cardinal ligaments, the posterior supports, uterosacral ligaments. They originate where the uterine cervix and corpus meet and course outward to the lateral and posterior pelvic walls, suspending the uterus and upper vagina and maintaining vaginal length. The paravaginal attachments include the arcus tendineus fascia pelvis anteriorly and arcus tendineus levator ani posteriorly. The lower vagina is directly attached to its surrounding structures, including the urethra anteriorly, perineal membrane and pubovaginalis muscle of the levator ani laterally, and perineal body posteriorly (5,10). Failure of support at any of these levels can result in prolapse of one or more pelvic organs.

Blood Supply

The blood supply to the vagina is complex with extensive anastomoses maintaining an adequate blood supply to all areas. The internal iliac (hypogastric) artery is the principal source of blood supply to the vagina. The upper vagina is supplied by the descending/cervical branch of the uterine arteries and the vaginal arteries, while the lower vagina receives its supply from the middle rectal and internal pudendal arteries. These arteries divide to send rami to the anterior and posterior vaginal walls, which form a plexus around the vagina. A rich venous plexus also surrounds the vagina and communicates with the vesicle, pudendal, and rectal venous plexuses, which drain into the internal iliac veins.

Nerves

The upper vagina is innervated by the pelvic splanchnic nerves (S2, S3, ±S4) and the lower vagina by the pudendal nerve (S2, S3, S4) (16). The autonomic nervous system of the pelvis functions to coordinate sympathetic, parasympathetic, and somatic innervation activities that regulate clitoral erection, vaginal secretions, smooth muscle contractions of the vagina, and somatic pelvic muscles that accompany orgasm.

Lymphatic Vessels

The lymphatic vessels begin as a delicate plexus of small channels involving the entire mucosa and lamina propria, most of which lie within 5 mm of the vaginal surface (17), and then drain into a deep muscular network. They terminate in a perivaginal plexus from which collecting trunks arise; the latter coalesce into several larger channels. The lymphatic drainage follows patterns that reflect the functionally diverse geographic regions. The lymphatic vessels of the upper vagina follow the uterine artery to drain to the internal and external iliac lymph nodes, intermediate lymphatic vessels accompany the vaginal artery to the internal iliac lymph nodes, and the lower lymphatic vessels drain to the superficial inguinal lymph nodes (16).

HISTOLOGY

Epithelium

The vaginal wall consists of three principal layers: mucosa (epithelium and lamina propria), muscularis, and adventitia. The epithelium is approximately 0.4 mm thick and, on gross examination, exhibits a characteristic pattern of folds or rugae separated by furrows of variable depth. There are two longitudinal (anterior and posterior) and multiple transverse furrows. The rugal pattern of the vaginal mucosa, which contributes to the organ's elasticity, produces an undulating appearance on microscopic examination in contrast to the flat surface of the cervix. The rugae, which are more prominent in nulliparous than multiparous women, reinforce the gripping effect of the levator ani and vaginal constrictor muscles during intercourse. The epithelium consists of nonkeratinizing glycogenated squamous epithelium similar to the cervix. Although the normal vaginal mucosa lacks glands, its surface is lubricated by fluids that pass directly through the mucosa and cervical mucus.

The mature, stratified squamous epithelium can be subdivided into several layers, typical of squamous epithelia elsewhere in the body: deep, intermediate, and superficial zones (Fig. 39.5). The deep zone contains the basal cell layer and, above this, the parabasal layer. The basal layer consists of a single row of columnar cells with their long axes arranged perpendicular to the surface. The cells contain basophilic cytoplasm and relatively large ovoid nuclei; mitotic figures may be present and occasional melanocytes can be found. The parabasal layer is poorly demarcated from the overlying cell layers. It usually consists of two layers of small polygonal

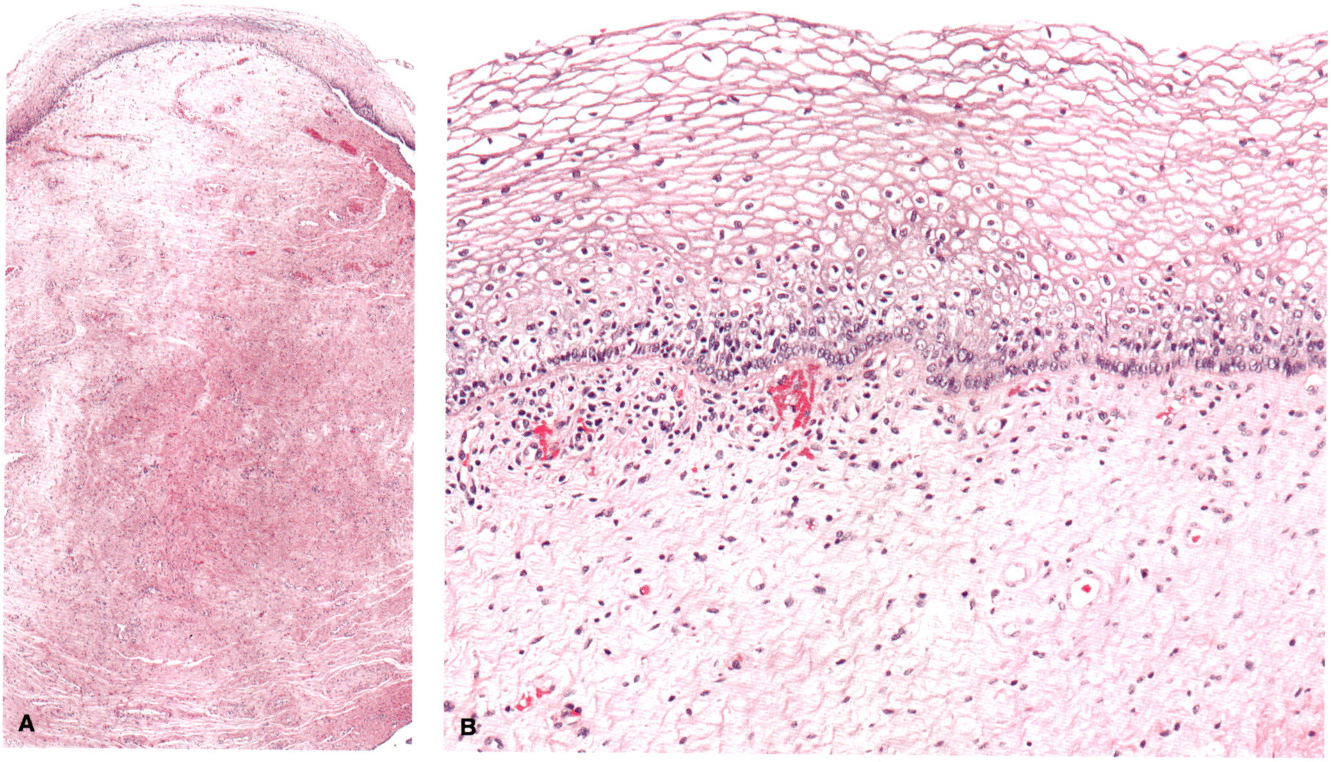

FIGURE 39.5 Vaginal biopsy demonstrating the epithelium, lamina propria, and muscularis (**A**). The epithelium is predominantly composed of glycogen-rich squamous cells with pyknotic nuclei (**B**). There is a single layer of basal cells, two to four layers of parabasal cells, and several layers of intermediate cells. Beneath the epithelium lies the lamina propria.

cells, often with intercellular bridges. The cells contain basophilic cytoplasm, a relatively large, centrally placed, round nucleus, and occasional mitotic figures. The intermediate layer is of variable thickness, but usually around 10 layers. The cells have prominent intercellular bridges and a navicular configuration, with their long axes parallel to the surface. Their cytoplasm is basophilic, although some glycogen may be present. The nuclei are round, oval, or irregular, with finely granular chromatin. The superficial layer is also of variable thickness, commonly composed of approximately 10 layers. The cells are polygonal when viewed from above and flattened on cross section. The cytoplasm is glycogenated and may contain keratohyalin granules. The nuclei are centrally located, small, round, and pyknotic.

Dendritic processes of Langerhans cells are distributed throughout the mucosa (31). Both CD8-positive and, to a lesser degree, CD4-positive T lymphocytes are also found, whereas macrophages and B lymphocytes are relatively uncommon.

Lamina Propria

The lamina propria lies directly beneath the basal layer of the epithelium. It contains elastic fibers and collagen that show considerable differences in biochemical and cellular composition in disease states such as pelvic organ prolapse (18–20). The lamina propria also contains a rich venous and lymphatic vessel network.

Occasionally, a band-like zone of loose connective tissue containing atypical polygonal to stellate stromal cells with scant cytoplasm is observed in the superficial lamina propria (Fig. 39.6). Many cells are multinucleated or have multilobulated hyperchromatic nuclei, but mitotic figures are absent. These atypical stromal cells are of fibroblastic origin and may be prominent in fibroepithelial stromal polyps occurring in the cervix, vagina, and vulva (21).

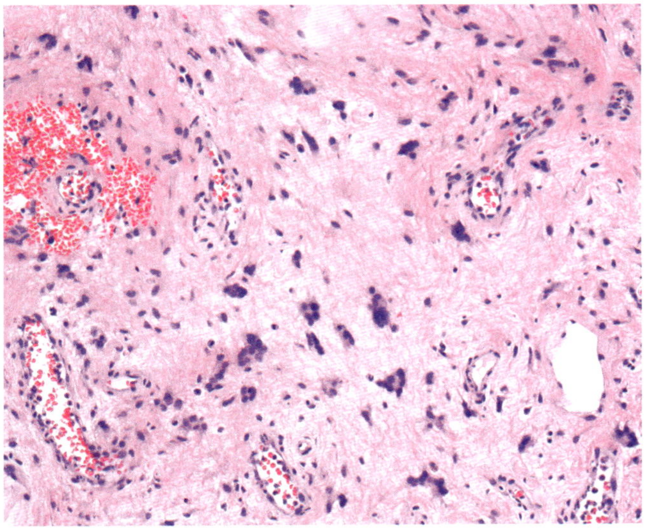

FIGURE 39.6 Bizarre stromal cells can occasionally be seen in lamina propria. Mitotic figures are absent.

Muscularis

The muscularis is composed of smooth muscle cell bundles admixed with fibrous tissue, blood vessels, and nerves (12). The smooth muscle cells are arranged in outer longitudinal and inner circular layers. Proximally, the longitudinal layer is continuous with the smooth muscle of the uterus, while distally it courses to the hymen and gradually disappears in the connective tissue. Anteriorly, however, it blends into the levator ani or obturator internus muscles (13).

Adventitia

The adventitia is composed of a thin layer of fibroadipose tissue admixed with blood vessels, nerves, and ganglia (12). It merges with the loose connective tissue of the rectovaginal space, vesicovaginal space, and endopelvic fascia.

HISTOLOGIC CHANGES DURING LIFE

Vaginal epithelial cells proliferate and mature in response to estrogen and are inhibited by progesterone. Thus, at birth, the epithelium appears mature due to circulating maternal estrogens in utero (Fig. 39.7), but shortly postpartum, the epithelium becomes atrophic, remaining this way until puberty. Quantitative studies measuring the rate of change in the maturation index in the infant's vagina from birth to the postnatal atrophic state indicate that vaginal cells replace themselves in less than 2 weeks; that is, the time required for basal cells to become desquamated superficial cells.

During the reproductive years, superficial cells with abundant intracytoplasmic glycogen dominate at the peak of estrogenic activity just before ovulation (Fig. 39.8). In contrast, intermediate cells predominate when progesterone levels are high, such as during the postovulatory phase

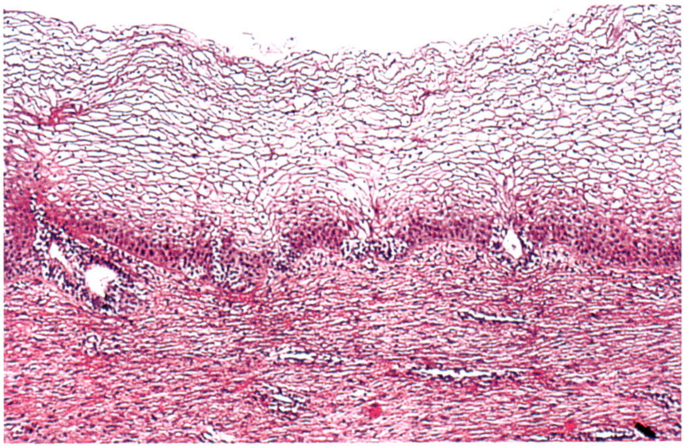

FIGURE 39.7 Vaginal mucosa of a near-term fetus shows a predominance of mature cells and cannot be distinguished from those of the adult. In addition, two glands of the embryonic type are present. (From Robboy SJ, Ellington KS. *Pathology of the Female Genital Tract in Kodachrome Slides*. Robboy Associates, LLC; 1996.)

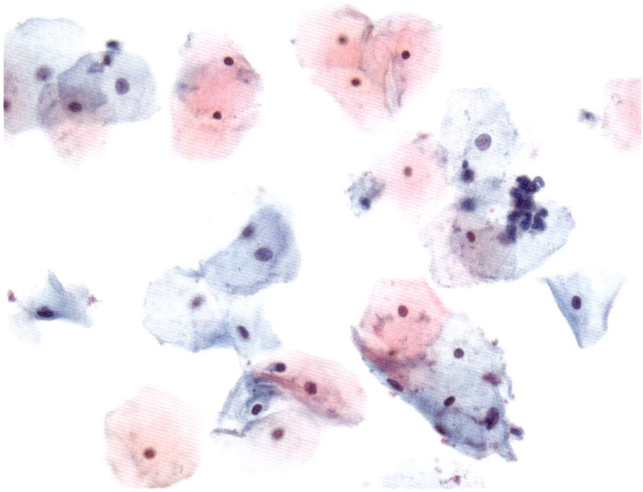

FIGURE 39.8 Vaginal smear from a premenopausal woman demonstrating superficial and intermediate squamous cells as well as occasional neutrophils. (Courtesy of Dr. Tatjana Antic, University of Chicago.)

of the menstrual cycle or pregnancy. However, whether the total number of squamous cell layers change during the normal menstrual cycle is unclear. In an early report (22), the epithelium was shown to build slowly during the proliferative phase, reaching its thickest at ovulation (average 45 layers). After ovulation, the number receded to 33 on day 19 and to 23 on day 24. In another study, the number of layers decreased slightly from about 28 during the proliferative phase to 26 during the secretory phase (23).

In menopause, lack of hormonal stimulation results in atrophy of the vaginal epithelium, with loss of the superficial and intermediate cell layers (Fig. 39.9). Hence, the epithelium is predominantly comprised of a few layers of parabasal cells, and can mimic a high-grade squamous dysplasia in histologic sections and vaginal smears. Careful attention to nuclear details, as well as use of p16 immunohistochemistry may be helpful in challenging situations. Specific therapies, including tamoxifen (24) and hormone replacement therapy (25), have a maturation effect on the vaginal epithelium in postmenopausal women.

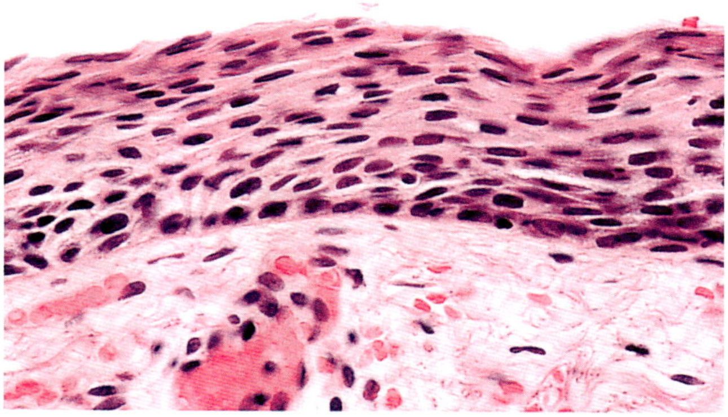

FIGURE 39.9 Atrophic vagina with loss of the superficial and intermediate cell layers. (From Robboy SJ, Ellington KS. *Pathology of the Female Genital Tract in Kodachrome Slides*. Robboy Associates, LLC; 1996.)

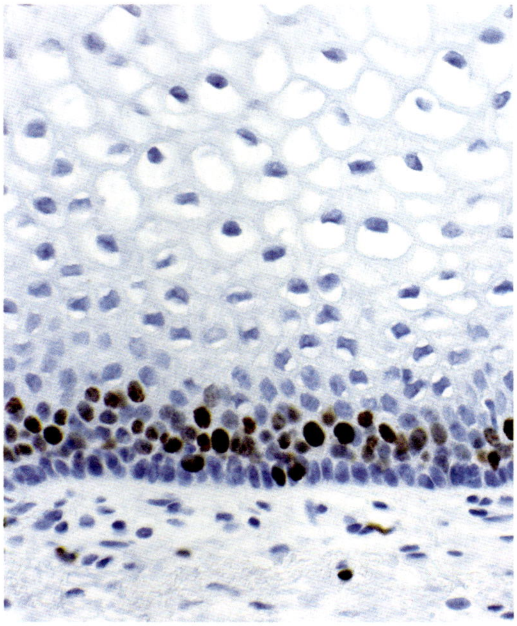

FIGURE 39.10 Ki-67 highlights proliferating basal and parabasal cells during the late G1, G2, and M phases of the cell cycle. (From Robboy SJ, Ellington KS. *Pathology of the Female Genital Tract in Kodachrome Slides*. Robboy Associates, LLC; 1996.)

IMMUNOHISTOCHEMISTRY AND ULTRASTRUCTURE

Similar to other sites with squamous epithelium, the vaginal epithelium is positive for squamous cell markers including p40, p63, and high–molecular-weight keratins (e.g., CK5/6 and CK903). It is also positive for broad-spectrum keratins, such as AE1/AE3 or MNF116. Basal cells are positive for cytokeratin 14 and parabasal cells for cytokeratin 13. Ki-67 highlights proliferating basal and parabasal cells during late G1, G2, and M phases of the cell cycle (Fig. 39.10).

On electron microscopy, the most superficial epithelial cells are large (50 µm in greatest dimension) and polygonal (26). Their surface is composed of a complex network of microridges, which are interanastomotic longitudinal elevations of the plasma membrane (27). Arranged in dense convolutions, they tie one cell to another, operating in a zipper-fastener principle.

SPECIAL ANATOMY/METAPLASTIC CHANGES

Mesonephric Remnants

Mesonephric remnants may persist along the length of the vestigial wolffian (mesonephric) duct, including the lateral

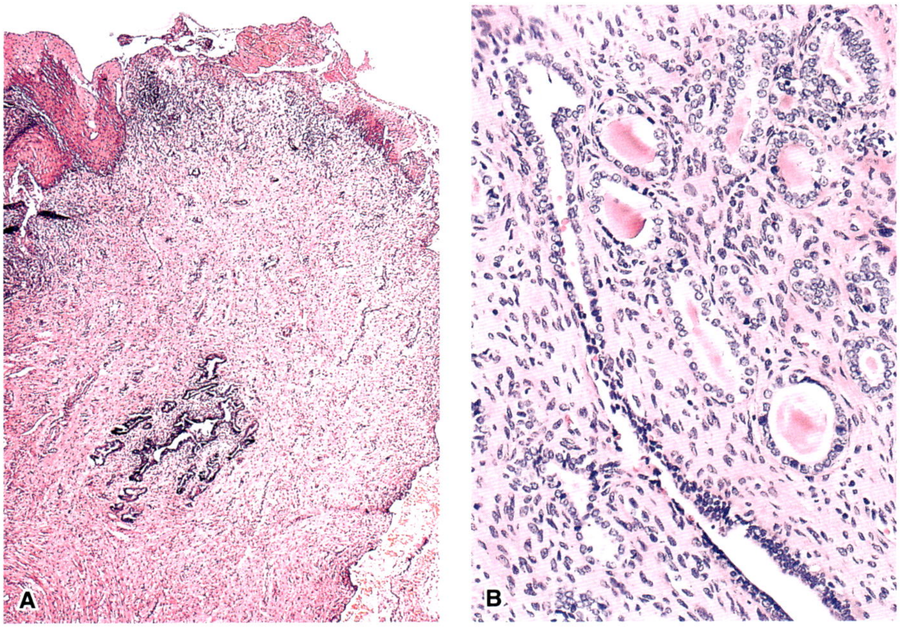

FIGURE 39.11 Vestigial mesonephric remnants present deep in the vaginal wall (**A**). Central duct surrounded by small tubules containing eosinophilic secretions (**B**). (From Robboy SJ, Ellington KS. *Pathology of the Female Genital Tract in Kodachrome Slides*. Robboy Associates, LLC; 1996.)

walls of the vagina. They are composed of groups of small-to-medium tubules lined by bland cuboidal cells with scant cytoplasm; a dilated duct can be present (Fig. 39.11). Overlapping nuclei and nuclear grooves may be observed, but cilia are absent, and mitoses are exceedingly rare. The lumen is frequently filled with a deeply eosinophilic, colloid-like secretion. In the lower genital tract, mesonephric remnants are usually positive for GATA3 and PAX8 (28). CD10 shows a distinctive luminal (apical) pattern of staining, while ER and PR are negative. Uncommonly, lesions may arise from vaginal mesonephric remnants, including Gartner duct cysts and very rarely, mesonephric adenocarcinoma (29–31).

Paraurethral Glands (Skene Glands)

Paraurethral glands, also referred to as Skene glands, lesser vestibular glands, and periurethral glands, lie on the anterior vaginal wall at the distal aspect of the urethra, and are homologous with the prostate gland in males (32). Although they drain into the urethra, they may occasionally be encountered in vaginal biopsies or resection specimens. The glands are lined by pseudostratified mucinous columnar epithelium that express prostate markers (e.g., NKX3.1, PSA, PSAP) and drain into transitional-lined ducts (Fig. 39.12) (33,34). Obstruction of the duct opening may result in the formation of cysts (Skene gland cyst) or abscesses (31,33,35). Other lesions that may develop include tubulosquamous polyps (36,37), and very rarely, Skene gland adenocarcinoma, which morphologically and immunohistochemically resembles prostatic adenocarcinoma (38–40).

G-spot (Gräfenberg Spot)

The G-spot, also known as the Gräfenberg spot, is of debatable existence (41,42). Described as an erogenous area located in the anterior vaginal wall, a reproducible morphologic entity has yet to be identified (43–45). Some believe it may represent neurovascular tissue (46), while others favor it consists of several structures, and should more aptly be designated the "G-zone" (47).

Remnants of Müllerian Duct Epithelium (Adenosis)

Adenosis results when embryonic müllerian tissue is not replaced by squamous epithelium during development. The exceedingly rare embryonic type is composed of small glands at the epithelial–stromal junction, with basal nuclei and copious PAS-negative, mucicarmine-negative cytoplasm. Adenosis likely takes on its adult forms, mucinous and tuboendometrial

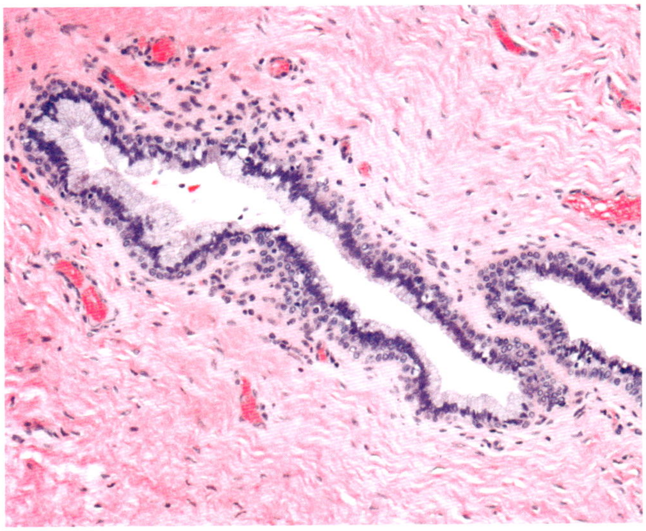

FIGURE 39.12 Skene gland with pseudostratified mucinous epithelium.

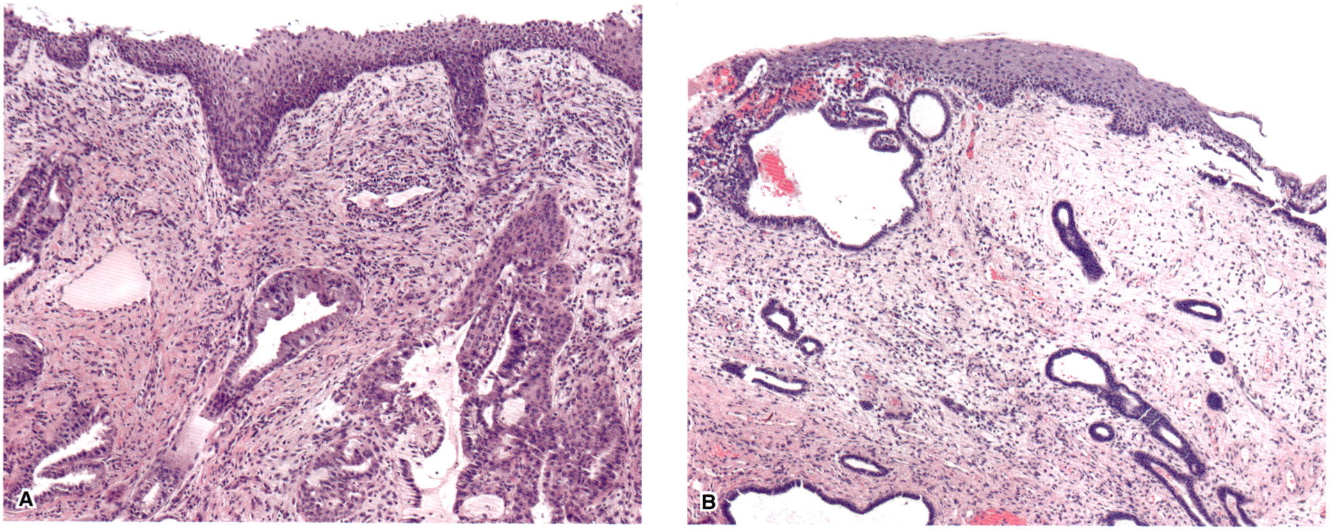

FIGURE 39.13 Mucinous adenosis composed of mucin-rich glands in the lamina propria (**A**). Tuboendometrial adenosis with ciliated glands in the lamina propria (**B**). Note the absence of glycogen in the overlying epithelium.

(Fig. 39.13), during puberty (48). The mucinous type is characterized by a single layer of columnar mucinous cells resembling endocervix, which lines glands in the lamina propria and/or replaces the surface epithelium (49). This form is the most frequently encountered (62% of biopsies with vaginal adenosis) and the most likely to show colposcopic changes (50). The tuboendometrial type is composed of a simple or pseudostratified layer of cuboidal to columnar cells; cilia are often present. The cells are usually found in glands in the lamina propria and not on the vaginal surface. This type likely represents the precursor to clear cell carcinoma, possibly through atypical adenosis, a transitional form (51,52).

In adenosis, the typical glycogen-rich surface squamous epithelium is replaced by glycogen-poor cells, which may also replace underlying glands, forming squamous pegs (53,54). Mucinous droplets can occur in both mucinous and tuboendometrial types but are more frequently associated with the former (50). Droplets may also be identified in metaplastic squamous cells in the absence of discrete gland formation.

Much of our knowledge of vaginal adenosis is due to the nonsteroidal estrogen, diethylstilbestrol (DES), a drug first introduced in the 1940s to prevent pregnancy complications (55). By 1971, up to 2 million women had taken the drug, at which time it was linked to the extremely rare development of vaginal clear cell carcinoma in young female offspring (56). Several long-term follow-up studies later confirmed this association (57–59). Thirty-five to 45% of patients with in utero DES exposure develop vaginal adenosis (50,53), but it may also occur in the absence of exposure. The precise incidence of this finding is unclear but it is estimated to affect less than 4% of women (48,53,60). Vaginal clear cell carcinoma has also rarely been reported in women unexposed to DES, most of whom have vaginal adenosis as well (60,61).

Prostatic Metaplasia

Vaginectomy specimens from patients receiving testosterone treatment for gender dysphoria may show prostatic metaplasia of the surface epithelium (Fig. 39.14) (62). Similar

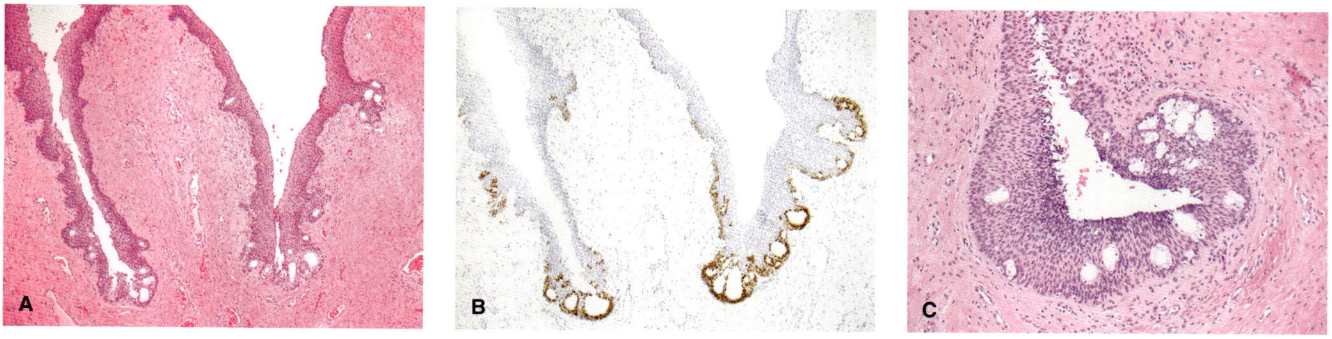

FIGURE 39.14 Vaginectomy specimen from a patient receiving testosterone for gender dysphoria shows prostatic metaplasia of the surface epithelium (**A**) highlighted by NKX3.1 (**B**). High-power magnification of the prostatic-type glands within vaginal epithelium (**C**). (Courtesy of Dr. Miranda Machacek, Massachusetts General Hospital.)

findings can be observed in patients with congenital disorders associated with endogenous androgen excess. Prostatic metaplasia is positive for NKX3.1 and androgen receptor.

MICROBIOME

The vagina plays other roles than simply serving as a conduit for intercourse and birth. As an interface between host and environment, its microbiome effectively defends against invasive microbial infections (63). It fosters vaginal health by maintaining a microenvironment of endogenous *Lactobacilli* and retains critical mediators of acquired and innate immunity (64). *Lactobacilli* metabolize glycogen to lactic acid, which maintains an acidic vaginal pH (<4.5), protecting it against pathogens and nonindigenous microbes (65).

REFERENCES

1. Robboy SJ, Kurita T, Baskin L, Cunha GR. New insights into human female reproductive tract development. *Differentiation*. 2017;97:9–22.
2. Cunha GR, Kurita T, Cao M, Shen J, Robboy S, Baskin L. Molecular mechanisms of development of the human fetal female reproductive tract. *Differentiation*. 2017;97:54–72.
3. Taguchi O, Cunha GR, Robboy SJ. Expression of nuclear estrogen-binding sites within developing human fetal vagina and urogenital sinus. *Am J Anat*. 1986;177(4):473–480.
4. Bulmer D. The development of the human vagina. *J Anat*. 1957;91(4):490–509.
5. Hoffman BL, Schorge JO, Halvorson LM, Hamid CA, Corton MM, Schaffer JI. *Williams Gynecology*. 4th ed. McGraw-Hill Education; 2020.
6. Barnhart KT, Izquierdo A, Pretorius ES, Shera DM, Shabbout M, Shaunik A. Baseline dimensions of the human vagina. *Hum Reprod*. 2006;21(6):1618–1622.
7. Lloyd J, Crouch NS, Minto CL, Liao LM, Creighton SM. Female genital appearance: "normality" unfolds. *BJOG*. 2005; 112(5):643–646.
8. Luo J, Betschart C, Ashton-Miller JA, DeLancey JO. Quantitative analyses of variability in normal vaginal shape and dimension on MR images. *Int Urogynecol J*. 2016;27(7):1087–1095.
9. Tan JS, Lukacz ES, Menefee SA, Luber KM, Albo ME, Nager CW. Determinants of vaginal length. *Am J Obstet Gynecol*. 2006;195(6):1846–1850.
10. Barber MD. Contemporary views on female pelvic anatomy. *Cleve Clin J Med*. 2005;72(Suppl 4):S3–S11.
11. Barnhart KT, Pretorius ES, Malamud D. Lesson learned and dispelled myths: three-dimensional imaging of the human vagina. *Fertil Steril*. 2004;81(5):1383–1384.
12. Maldonado PA, Carrick KS, Montoya TI, Corton MM. Posterior vaginal compartment anatomy: implications for surgical repair. *Female Pelvic Med Reconstr Surg*. 2020;26(12): 751–757.
13. Hamner JJ, Carrick KS, Ramirez DMO, Corton MM. Gross and histologic relationships of the retropubic urethra to lateral pelvic sidewall and anterior vaginal wall in female cadavers: clinical applications to retropubic surgery. *Am J Obstet Gynecol*. 2018;219(6):597 e1–e8.
14. O'Connell HE, Eizenberg N, Rahman M, Cleeve J. The anatomy of the distal vagina: towards unity. *J Sex Med*. 2008;5(8):1883–1891.
15. DeLancey JO. Anatomic aspects of vaginal eversion after hysterectomy. *Am J Obstet Gynecol*. 1992;166(6 Pt 1):1717–1724; discussion 24–28.
16. Strandring S. *Gray's Anatomy: The Anatomical Basis of Clinical Practice*. Elsevier; 2020.
17. Choo JJ, Scudiere J, Bitterman P, Dickler A, Gown AM, Zusag TW. Vaginal lymphatic channel location and its implication for intracavitary brachytherapy radiation treatment. *Brachytherapy*. 2005;4(3):236–240.
18. Kerkhof MH, Ruiz-Zapata AM, Bril H, et al. Changes in tissue composition of the vaginal wall of premenopausal women with prolapse. *Am J Obstet Gynecol*. 2014;210(2):168e1–e9.
19. Meijerink AM, van Rijssel RH, van der Linden PJ. Tissue composition of the vaginal wall in women with pelvic organ prolapse. *Gynecol Obstet Invest*. 2013;75(1):21–27.
20. Sridharan I, Ma Y, Kim T, Kobak W, Rotmensch J, Wang R. Structural and mechanical profiles of native collagen fibers in vaginal wall connective tissues. *Biomaterials*. 2012;33(5): 1520–1527.
21. Chapel DB, Cipriani NA, Bennett JA. Mesenchymal lesions of the vulva. *Semin Diagn Pathol*. 2021;38(1):85–98.
22. Hafez ESE, Evans TN. *The Human Vagina*. Elsevier/North Holland Biomedical Press; 1978.
23. Patton DL, Thwin SS, Meier A, Hooton TM, Stapleton AE, Eschenbach DA. Epithelial cell layer thickness and immune cell populations in the normal human vagina at different stages of the menstrual cycle. *Am J Obstet Gynecol*. 2000;183(4):967–973.
24. Love RR, Kurtycz DF, Dumesic DA, Laube DW, Yang Y. The effects of tamoxifen on the vaginal epithelium in postmenopausal women. *J Womens Health Gend Based Med*. 2000;9(5):559–563.
25. Binder RL, Freedman MA, Sharma KB, et al. Histological and gene expression analysis of the effects of menopause status and hormone therapy on the vaginal introitus and labia majora. *J Clin Med Res*. 2019;11(11):745–759.
26. Johannessen JV. *Electron Microscopy in Human Medicine: Urogenital System and Breast*: McGraw-Hill International Book Company; 1978.
27. Ferenczy AR, Richart M. *Female Reproductive System: Dynamics of Scan and Transmission Electron Microscopy*: John Wiley; 1974.
28. Howitt BE, Nucci MR. Mesonephric proliferations of the female genital tract. *Pathology*. 2018;50(2):141–150.
29. Jacox N, Yao HH, Baverstock R, Trpkov K, Carlson K. Periurethral and anterior vaginal wall masses: etiology, presentation, and treatment outcomes. *Obstet Gynecol*. 2022;140(5): 778–783.
30. Shoeir S, Balachandran AA, Wang J, Sultan AH. Mesonephric adenocarcinoma of the vagina masquerading as a suburethral cyst. *BMJ Case Rep*. 2018;2018.
31. Zuo SW, Napoe GS. Evaluation and management of urethral and periurethral masses in women. *Curr Opin Obstet Gynecol*. 2023;35(6):517–524.
32. Tomalty D, Giovannetti O, Hannan J, et al. Should we call it a prostate? A review of the female periurethral glandular tissue

morphology, histochemistry, nomenclature, and role in iatrogenic sexual dysfunction. *Sex Med Rev*. 2022;10(2):183–194.
33. Heller DS. Lesions of Skene glands and periurethral region: a review. *J Low Genit Tract Dis*. 2015;19(2):170–174.
34. Roma AA. Tubulosquamous polyps in the vagina. Immunohistochemical comparison with ectopic prostatic tissue and Skene glands. *Ann Diagn Pathol*. 2016;22:63–66.
35. Kelly P, McBride HA, Kennedy K, Connolly LE, McCluggage WG. Misplaced Skene's glands: glandular elements in the lower female genital tract that are variably immunoreactive with prostate markers and that encompass vaginal tubulosquamous polyp and cervical ectopic prostatic tissue. *Int J Gynecol Pathol*. 2011;30(6):605–612.
36. McCluggage WG, Young RH. Tubulo-squamous polyp: a report of ten cases of a distinctive hitherto uncharacterized vaginal polyp. *Am J Surg Pathol*. 2007;31(7):1013–1019.
37. Roy SF, Comeau L, Provencher D, Rahimi K. Tubulosquamous polyp of the vagina: misplaced Skene's glands analogous to the male prostate. *Int J Surg Pathol*. 2021;29(5):510–512.
38. Li XH, Zhang YT. Skene gland adenocarcinoma of the urethra: A systematic review. *Asian J Surg*. 2024;47(3):1587–1588.
39. Pongtippan A, Malpica A, Levenback C, Deavers MT, Silva EG. Skene's gland adenocarcinoma resembling prostatic adenocarcinoma. *Int J Gynecol Pathol*. 2004;23(1):71–74.
40. Tregnago AC, Epstein JI. Skene's glands adenocarcinoma: a series of 4 cases. *Am J Surg Pathol*. 2018;42(11):1513–1521.
41. Jannini EA, Buisson O, Rubio-Casillas A. Beyond the G-spot: clitourethrovaginal complex anatomy in female orgasm. *Nat Rev Urol*. 2014;11(9):531–538.
42. Vieira-Baptista P, Lima-Silva J, Preti M, Xavier J, Vendeira P, Stockdale CK. G-spot: fact or fiction?: a systematic review. *Sex Med*. 2021;9(5):100435.
43. Hoag N, Keast JR, O'Connell HE. The "G-Spot" is not a structure evident on macroscopic anatomic dissection of the vaginal wall. *J Sex Med*. 2017;14(12):1524–1532.
44. Ostrzenski A, Krajewski P, Ganjei-Azar P, et al. Verification of the anatomy and newly discovered histology of the G-spot complex. *BJOG*. 2014;121(11):1333–1339.
45. Ostrzenski A. G-spot anatomy: a new discovery. *J Sex Med*. 2012;9(5):1355–1359.
46. Ostrzenski A. G-spot anatomy and its clinical significance: a systematic review. *Clin Anat*. 2019;32(8):1094–1101.
47. Goldstein I, Goldstein SW, Komisaruk BR. Should we call it (G-spot) a G-zone? *Sex Med Rev*. 2022;10(2):181–182.
48. Robboy SJ. A hypothetic mechanism of diethylstilbestrol (DES)-induced anomalies in exposed progeny. *Hum Pathol*. 1983; 14(10):831–833.
49. Herbst AL, Kurman RJ, Scully RE. Vaginal and cervical abnormalities after exposure to stilbestrol in utero. *Obstet Gynecol*. 1972;40(3):287–298.
50. Robboy SJ, Kaufman RH, Prat J, et al. Pathologic findings in young women enrolled in the National Cooperative Diethylstilbestrol Adenosis (DESAD) project. *Obstet Gynecol*. 1979;53(3):309–317.
51. Robboy SJ, Welch WR, Young RH, Truslow GY, Herbst AL, Scully RE. Topographic relation of cervical ectropion and vaginal adenosis to clear cell adenocarcinoma. *Obstet Gynecol*. 1982;60(5):546–551.
52. Robboy SJ, Young RH, Welch WR, et al. Atypical vaginal adenosis and cervical ectropion. Association with clear cell adenocarcinoma in diethylstilbestrol-exposed offspring. *Cancer*. 1984;54(5):869–875.
53. Herbst AL, Poskanzer DC, Robboy SJ, Friedlander L, Scully RE. Prenatal exposure to stilbestrol. A prospective comparison of exposed female offspring with unexposed controls. *N Engl J Med*. 1975;292(7):334–339.
54. Welch WR, Robboy SJ, Townsend DE, Barnes AB, Scully RE, Herbst AL. Comparison of histologic and colposcopic findings in DES-exposed females. *Obstet Gynecol*. 1978;52(4):457–461.
55. Laronda MM, Unno K, Butler LM, Kurita T. The development of cervical and vaginal adenosis as a result of diethylstilbestrol exposure in utero. *Differentiation*. 2012;84(3):252–260.
56. Herbst AL, Ulfelder H, Poskanzer DC. Adenocarcinoma of the vagina. Association of maternal stilbestrol therapy with tumor appearance in young women. *N Engl J Med*. 1971; 284(15):878–881.
57. Hoover RN, Hyer M, Pfeiffer RM, et al. Adverse health outcomes in women exposed in utero to diethylstilbestrol. *N Engl J Med*. 2011;365(14):1304–1314.
58. Huo D, Anderson D, Palmer JR, Herbst AL. Incidence rates and risks of diethylstilbestrol-related clear-cell adenocarcinoma of the vagina and cervix: Update after 40-year follow-up. *Gynecol Oncol*. 2017;146(3):566–571.
59. Troisi R, Hatch EE, Titus L, et al. Prenatal diethylstilbestrol exposure and cancer risk in women. *Environ Mol Mutagen*. 2019;60(5):395–403.
60. Han T, Jin Y, Li Y, Bi Y, Pan L. Clinicopathologic features and outcomes of primary vaginal adenosis as a dermatologic and gynecologic burden: A retrospective study. *Medicine (Baltimore)*. 2018;97(49):e13470.
61. Pang L, Li L, Zhu L, Lang J, Bi Y. Malignant transformation of vaginal adenosis to clear cell carcinoma without prenatal diethylstilbestrol exposure: a case report and literature review. *BMC Cancer*. 2019;19(1):798.
62. Anderson WJ, Kolin DL, Neville G, et al. Prostatic metaplasia of the vagina and uterine cervix: an androgen-associated glandular lesion of surface squamous epithelium. *Am J Surg Pathol*. 2020;44(8):1040–1049.
63. Amabebe E, Anumba DOC. The vaginal microenvironment: the physiologic role of lactobacilli. *Front Med (Lausanne)*. 2018;5:181.
64. Anderson DJ, Marathe J, Pudney J. The structure of the human vaginal stratum corneum and its role in immune defense. *Am J Reprod Immunol*. 2014;71(6):618–623.
65. Smith SB, Ravel J. The vaginal microbiota, host defence and reproductive physiology. *J Physiol*. 2017;595(2):451–463.

40

Uterus and Fallopian Tubes

Susanne K. Jeffus ▪ Charles M. Quick

EMBRYOLOGY

The fallopian tube, uterus, and ovarian surface epithelium comprise the extended müllerian system (1,2), which gives rise to a common set of neoplasms and metaplastic epithelial changes.

The uterus and fallopian tubes have a complex developmental sequence (3–11). Both male and female internal genitalia are laid down early in embryogenesis in the indifferent stage of genital development. Upon completion of this indifferent stage, female differentiation is accompanied by regression of the male anlage and, conversely, male differentiation is accompanied by regression of the female anlage. Topographically, both systems are intimately related to the developing urinary tract, and, not surprisingly, anomalous development of the internal genitalia is often accompanied by anomalies of the urinary tract. Fetal sexual differentiation is completed during the first half of gestation; the last half is marked primarily by growth of the newly established genitalia. Relevant milestones have been summarized by Ramsey (Fig. 40.1) (12).

THE INDIFFERENT STAGE

The urogenital sinus and the mesonephric (wolffian) ducts are well established by the 6th week of fetal life. At this time, the paired müllerian (paramesonephric) ducts begin their development as an invagination of the coelomic epithelium adjacent to that investing each developing ovary. The müllerian ducts are intimately related to the mesonephric ducts, and their normal formation appears to be dependent on the presence of the mesonephros.

As the müllerian ducts grow caudally, they approach the midline where the distal portions fuse. Shortly after this fusion, the apposed medial duct walls disappear, bringing the two lumina into continuity to form a single cavity. Further downward growth of the fused müllerian structures (now termed the uterovaginal primordium)

AGE	GLANDS	URINARY TRACT	♂ DUCTS ♀		EXTERNAL GENITALIA
3-4 weeks	PRIMORDIAL GERM CELLS	PRONEPHROS (nonfunctional) Tubules and Ducts	PRONEPHRIC		
4-9 weeks		MESONEPHROS or WOLFFIAN BODY (temporary function) Tubules and Ducts	MESONEPHRIC or WOLFFIAN		CLOACA
5th week	UROGENITAL RIDGE				
6th week	INDIFFERENT GONAD: GERMINAL AND CORE EPITHELIUM	METANEPHROS or KIDNEY (permanent) Tubules and Ducts	PARAMESONEPHRIC or MÜLLERIAN		CLOACA SUBDIVIDES GENITAL TUBERCLE
7th week	MALE TYPE CORDS				ANAL AND URETHRAL MEMBRANES RUPTURE
8th week	TESTIS AND OVARY				URETHRAL AND LABIOSCROTAL FOLDS, PHALLUS AND GLANS
9th week			MÜLLERIAN DUCTS FUSE AT TUBERCLE		
10th week			MÜLLERIAN DUCTS DEGENERATE	WOLFFIAN DUCTS DEGENERATE	
11th week			SEMINAL VESICLES, EPIDIDYMIS, VAS DEFERENS		
12th week	OVARY DESCENT COMPLETE			WALLS FORM	SEX DISTINGUISHABLE
5 months	TESTIS AT INGUINAL RING			SINUS EPITHELIUM GROWS IN VAGINAL CLEFT	
8 months TERM	TESTIS DESCENT COMPLETE			RAPID UTERINE GROWTH	

FIGURE 40.1 Chart showing interrelations and time sequence of events in development of genitourinary system. (From Ramsey E. Embryology and developmental defects of the female reproductive tract. In: Danforth D, Scott J, eds. *Obstetrics and Gynecology*. JB Lippincott; 1986:106–119.)

brings them into contact with the urogenital sinus. The fetus has both mesonephric ducts and müllerian ducts at this stage.

Female Differentiation

The differentiation of the indifferent internal genitalia into male or female structures depends on whether the fetus possesses ovaries or testes. Leydig cells and the Sertoli cells in the developing testes secrete testosterone and a nonsteroidal müllerian inhibiting substance, respectively (13). The net effect of this secretory activity is to ensure the persistence, differentiation, and growth of the mesonephric ducts to form the male genital system and the regression of the müllerian system. In the absence of a secreting testis (i.e., in a normal female fetus with ovaries or in a fetus with nonfunctioning gonads) the müllerian structures persist, whereas the mesonephric ducts regress. The nonfused portions of the müllerian ducts form the fallopian tubes; the fused segments develop into the uterus and upper third of the vagina. Incomplete fusion of the caudal portion of the müllerian ducts results in a spectrum of uterovaginal abnormalities (14).

By the 21st week, the uterus and vagina are well formed. In contrast to the adult cervix, the cervix of the prenatal uterus is disproportionately large and makes up two-thirds of the length of the organ. The second half of gestation is marked by uterine growth; the fetal uterus doubles in size from the 28th gestational week to birth. However, the earlier cervicocorpus disproportion is maintained into childhood.

The events described above are driven, at least in part, by the expression of secreted ligands of the wingless (WNT) gene family and transcriptional regulators of the homeobox (HOX) gene family (13,15).

GROSS ANATOMY

Premenarchal Uterus and Fallopian Tubes

Neonatal Period

At birth, the uterus averages about 4 cm in length, and its bulk and shape are dominated by its disproportionately large cervix (the cervicofundal ratio is approximately 3 to 5:1). The maternal hormonal environment results in a markedly thickened rugal vaginal mucosa and a proliferative or weakly secretory endometrium. Maternal estrogen also results in cervical squamous cell maturation with glycogen storage. These mucosal changes regress shortly after birth (16–18).

Infancy

Uterine growth continues into the second year of life, at which time it reaches a plateau that persists until the premenarchal growth spurt at approximately 9 years of age. The cervix continues to account for greater than half of the uterine length until approximately the 13th year of life.

Adult Uterus and Fallopian Tubes

General Relations and Attachments

The uterus is anterior to the rectum and posterior to the bladder (Fig. 40.2). It is covered anteriorly and posteriorly by a reflection of pelvic peritoneum that continues laterally to form the anterior and posterior leaves of the broad ligament. The posterior peritoneal reflection forms the uterine wall of the pouch of Douglas and covers a longer segment of the uterine isthmus than does the anterior peritoneal reflection. The tent-like broad ligaments house the major uterine vessels, the efferent lymphatic trunks, and the apical portions of the fallopian tubes. Each ovary is attached to the ipsilateral uterine cornu by the utero-ovarian ligament, which is situated posterolateral and inferior to the uterine attachment of the fallopian tubes. The round ligaments arise anterolateral and inferior to the attachment of the fallopian tubes and pass anteriorly to insert into the canal of Nuck. These anatomic relations enable proper orientation of the hysterectomy specimen. The anterior surface of the uterus is distinguished by its longer "bare" region (i.e., lacking peritoneum) and the anteriorly directed stump of the round ligament. The posterior surface is more extensively covered by peritoneum, and the utero-ovarian ligament is attached to the posterior cornual aspect of the uterus. The uterus is anchored to its surroundings by a number of connective tissue bands; notable among them are the cardinal, uterosacral, and pubocervical ligaments (11,18–20).

Gross Anatomic Features of the Uterus

The adult nulliparous uterus is a hollow, pear-shaped muscular organ weighing 40 to 80 g and measuring approximately 7.0 to 8.0 cm along its long axis, 5.0 cm at its broadest extent (cornu to cornu), and 2.5 cm in anteroposterior dimension. These measurements vary considerably as a function of age, phase of the menstrual cycle, and parity. In general, high parity and youth correlate with larger uterine size (21). The adult uterus consists of an expanded body, the corpus, and a smaller cervix. That portion of the corpus cephalad to a line connecting the origin of the two fallopian tubes is called the fundus. The cornua are the two lateral regions of the fundus associated with the intramural portion of the fallopian tubes. The remainder of the corpus tapers from the fundus into the

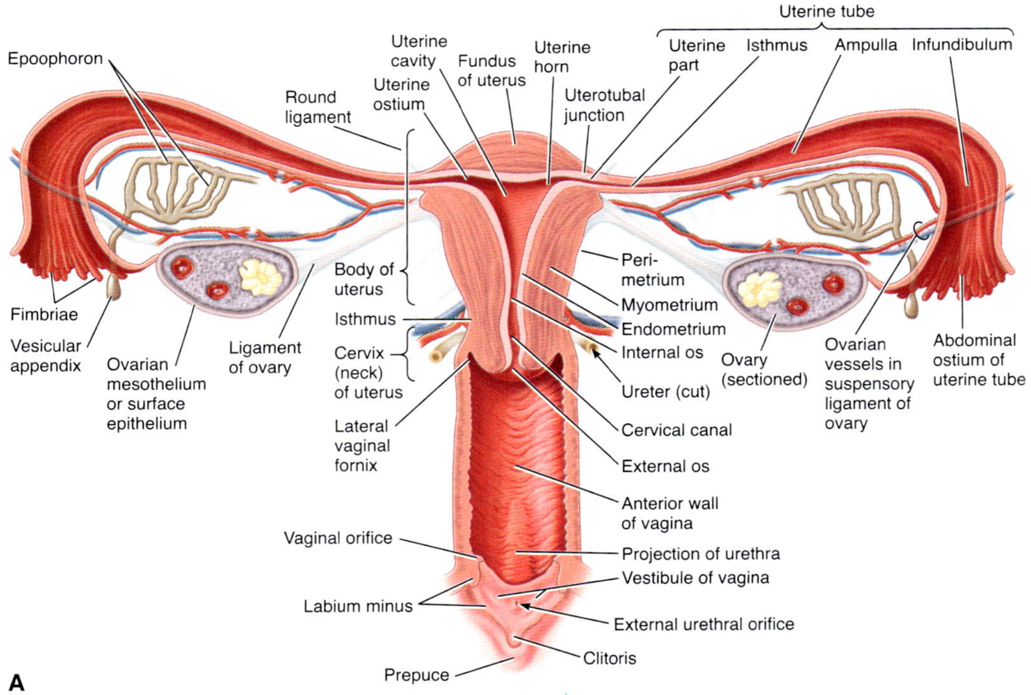

FIGURE 40.2 The normal internal female genitalia showing their relationship to other pelvic structures (**A**) and in cross-section (**B**). (**A:** Reprinted with permission from Dudek RW, Louis TM. *High-Yield Gross Anatomy*. Wolters Kluwer Health; 2015. **B:** Reprinted with permission from *Female Reproductive System Anatomical Chart*. Wolters Kluwer Health; 2000.) (*continued*)

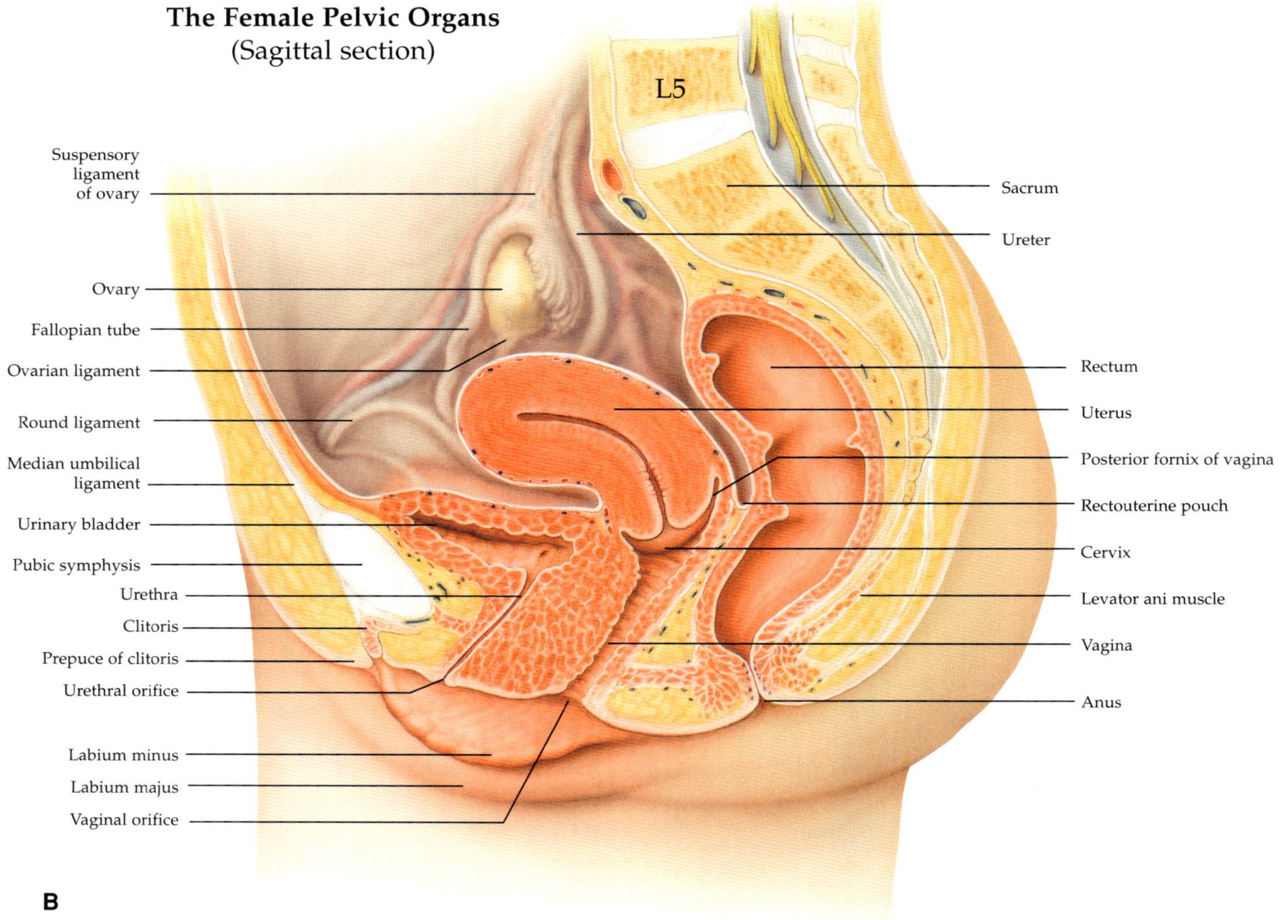

FIGURE 40.2 (*Continued*)

isthmus or the lower uterine segment, which shares histologic features with both of the uterine segments that it bridges: the uterine corpus and the endocervix. The existence of an anatomically and functionally significant lower uterine segment has been disputed by some authorities (22). The uterine cavity has the approximate configuration of the uterus, but its internal dimensions are much smaller, reflecting the substantial thickness of the uterine wall. The cavity is triangular with apices that are continuous with the lumina of the fallopian tubes at the cornua and with the endocervical canal at the internal os. The length of the cavity is approximately 6 cm, although it varies considerably with the age and parity of the individual (23). The cervix and internal endocervical canal are roughly cylindrical and normally measure 3 to 4 cm in length (24). Traditionally, the endocervical canal has been described as having an external os that opens onto the exocervix and an internal os that separates the endocervical canal from the endometrial cavity. Although the former is a reasonable anatomic landmark, the latter is not; the transition from endometrial cavity to endocervix is gradual both grossly and histologically. The mucosal surface of the endocervical canal is deeply clefted to form the plicae palmatae. The parametria are the lateral connective tissue attachments of the uterus, which contain vessels, nerves, lymphatics, and lymph nodes.

The normal myometrium consists of two strata: an outer longitudinal muscle layer covering the fundus and an inner circular submucosal muscle layer extending to surround the internal os and the tubal ostia. There is an interposed thick middle layer, richly populated by vessels and composed of randomly interdigitating fibers (25). The magnetic resonance imaging (MRI) correlate of these layers is the outer zone and the submucosal low-intensity halo "junctional zone" (26,27). Functionally, the junctional zone appears to be more involved with menstruation while the outer zone assumes a prominent role in gestation and parturition.

Toth has described two lateral subserosal longitudinal bands of muscle fibers termed the fasciculus cervicoangularis (28,29). On occasion, epithelium that is immunohistochemically and histologically similar to cervical mesonephric remnants is present within these bundles, suggesting that they represent the vestiges of the wolffian (mesonephric) duct. These remnants are more commonly encountered in the cervical stroma ("mesonephric rests") and lateral vagina (i.e., Gardner duct and derivative cysts).

Substantial deviations from the nulliparous adult uterus naturally occur throughout adult life. The uterus undergoes small-amplitude changes in size during the menstrual cycle, attaining its greatest volume during the secretory phase (27). During pregnancy, of course, the uterus enlarges much more dramatically to accommodate the growing conceptus. This growth is largely due to myocyte hypertrophy and hyperplasia, increased uterine vasculature, and deposition of extracellular matrix; the net weight increases 10-fold during pregnancy. After delivery, uterine size rapidly decreases, and a striking resorption of connective tissue occurs over the ensuing weeks that is associated with a decrease in the size of individual myocytes (30). However, the uterus does not completely return to its nulliparous size and weight. Prior pregnancy (parity) can be deduced from several gross features. The multiparous nongravid uterus tends to weigh more due to its thicker and more prominently layered muscular walls; the increase in weight is proportional to the patient's parity (21). The vasculature of the multiparous uterus tends to be more prominent. The most suggestive changes of previous pregnancy, however, are seen in the cervix. The nulliparous external os is small and circular but is transformed after pregnancy into a slit with prominent anterior and posterior lips. In addition, healed cervical lacerations may be pronounced, and enough endocervical tissue may reside on the exocervix to give it a red granular appearance near the os. With the waning of ovarian hormone synthesis during the menopausal years, the uterus involutes and becomes atrophic with decreasing weight and dimensions. On occasion, the endocervical canal is almost completely obliterated. Exogenous estrogens administered during this period can sometimes maintain uterine weight despite the loss of ovarian hormonal support.

Gross Anatomic Features of the Fallopian Tubes

The fallopian tubes are hollow epithelium-lined muscular structures 11 to 12 cm in length that run through the apex of the broad ligament to span the uterine cornu medially and the ovary laterally. Each tube is divided into four anatomic segments. The intramural segment begins at the funnel-like uppermost recess of the uterine cornu and ends where the tube emerges from the uterine wall. The segment has an 8 mm, pinpoint lumen and its course varies from straight to highly convoluted (31). The proximal tube continues for 2 to 3 cm beyond the uterine wall as the isthmus, which is a thick-walled segment with a narrow caliber lumen. The isthmus transitions into a comparatively thin-walled area with an expanded lumen termed the ampulla. The distal tube ends in the trumpet-shaped infundibulum that opens into the peritoneal cavity and is fringed by approximately 25 fimbriae. One of these, the ovarian fimbrium, attaches to the ovary. The infundibulum forms a cap over the ovarian surface at the time of ovulation, thereby creating the ovarian bursa. The tubal mucosa and underlying endosalpingeal stroma are thrown up into longitudinal, branching folds (the plicae) that become increasingly complex from the isthmus to the infundibulum. The plicae terminate in the fimbria. The fimbria sweep over the surface of the ovary to facilitate egg capture at the time of ovulation (13,32–35).

Uterine and Tubal Vasculature

The major arterial supply of the uterus derives from the right and left uterine arteries, which arise from the corresponding internal iliac arteries. The uterine arteries divide into ascending and descending branches laterally at the level of the uterine isthmus. The ascending uterine artery anastomoses freely with the ovarian artery in the mesosalpinx, whereas the descending branch anastomoses with the vaginal arterial supply. Both the ascending and descending uterine arteries give rise to a complex network of circumferentially arranged arcuate arteries in the subserosa that, in turn, give rise to a series of radial arteries that penetrate the myometrium. Each radial vessel branches in the inner third of the myometrium, giving rise to straight arteries supplying the basalis and spiral arteries of the endometrium (36–38).

A striking characteristic of the adult intramyometrial uterine arteries is their marked tortuosity, which may be related to variation in uterine size during reproductive life. In the postmenopausal years, striking degenerative changes may be seen in the uterine arteries, including intimal proliferation, fibrosis, and medial calcification. The severity of these changes is typically out of proportion to degenerative changes in nonuterine arteries. The venous drainage of the uterus parallels its arterial supply.

Uterine and Tubal Lymphatic Vessels

Lymphatic vessels are present in both the cervix and the corpus. In the endometrium, these vessels are intimately associated with the glands of the functionalis. The myometrium and cervical stroma contain a complex labyrinth of lymphatics that course toward the subserosal plexus. The channels forming the latter ramify over the entire surface of the uterus, and the confluence of these channels forms the major efferent lymphatic trunks of the uterus. The chief interest in lymphatic drainage for the pathologist is as a guide to the dissemination of carcinoma. The major lymph node groups draining cervical and endometrial carcinoma are indicated in Figure 40.3.

Lymphatic vessel anastomoses exist between the cervical and corpus systems in both the mucosal and muscular layers, and, on occasion, cervical carcinomas may take advantage of this route to spread to the corpus. However, the converse is not necessarily true, nor is it clear whether carcinomas of the corpus metastasize in a similar distribution to that of cervical carcinomas once they invade the cervix. Indeed, involvement of cervical draining nodes by endometrial carcinoma does not necessarily imply cervical involvement. For further detail, the reader is referred to specialty works and textbooks of gynecologic oncology (39–41).

Tubal lymphatic vessels accompany the ovarian vessels and drain into lymph nodes near the right and left renal

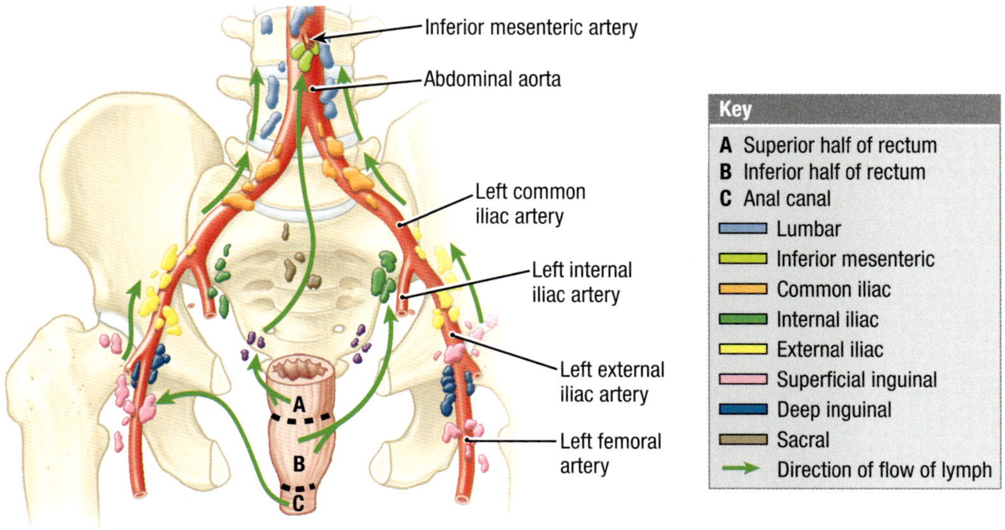

FIGURE 40.3 Lymphatic flow for the pelvis. *Green arrows* show the direction of lymphatic flow. This can be helpful in directing the pathologist to the most likely involved lymph nodes (and possible sentinel lymph nodes) by metastatic carcinoma depending on the location of the tumor. (From Aqur A, Dalley A. *Grant's Atlas of Anatomy.* 12th ed. Lippincott Williams and Wilkins; 2013:217.)

veins, as well as the presacral and common iliac nodes. Lymphatic spread of tubal malignancy may reach extrapelvic sites early in its dissemination (42,43).

UTERINE CERVIX

The uterine cervix or "neck" is the elongated fibromuscular portion of the uterus that measures 2.5 to 3.0 cm. Part of this structure protrudes into the upper part of the vagina and is termed the portio vaginalis, whereas the remainder lies above the vaginal vault and is considered to supravaginal portion. The outer surface of the vaginal portion is termed the ectocervix or exocervix. It is mostly covered by stratified squamous epithelium that is continuous with, and histologically identical to, the mucosae of the vaginal fornices. The segment of cervix in relation to the endocervical canal is known as the anatomic endocervix. The endocervical canal is mostly lined by mucin-secreting epithelium that merges with the squamous epithelium of the exocervix distally and with the epithelium of the lower uterine segment proximally, thereby bringing the vagina into communication with the endometrial cavity. The anatomic opening of the endocervical canal onto the exocervix is the external os. In parous women, this most often takes on a slit-like configuration that serves to divide the exocervix into anterior and posterior lips (18,44). This configuration is thought to be important in uterine function during gestation (45). The internal os forms the upper limit of the endocervical canal. It is more of a gradual funnel-shaped widening of the endocervical canal than a discrete orifice and represents a transition from endocervical to endometrial epithelium of the lower uterine segment. The squamocolumnar junction is defined as the point at which the endocervical glandular mucosa merges with the squamous epithelium of the exocervix. This junction is typically located on the exocervix, where it can easily be inspected with the colposcope; it does not always lie at the external os. This is further discussed in the section devoted to the transformation zone.

The uterine cervix obviously plays an important role in the anatomic support of the internal genitalia and labor and delivery, but its primary role is the production of cervical mucus. Cervical mucus acts as a functional gate that prevents vaginal microorganisms from gaining access to the upper genital tract and, with the exception of a small midcycle window before ovulation, denies sperm access to the uterus and fallopian tubes. At midcycle, however, the chemical composition of the cervical mucus changes and its viscosity decreases, thereby allowing the passage of sperm into the upper genital tract. These changes can be clinically manifest as Spinnbarkeit and fern tests. The cervical mucus also prevents sperm phagocytosis by removing seminal plasma constituents and provides a suitable environment for sperm storage, capacitation, and migration (46,47).

The following discussion first focuses on the epithelium of the exocervix, the endocervix, and the transformation zone. It then turns to the stroma of the cervix and the changes that occur in the cervix during pregnancy.

Epithelium of the Exocervix

The squamous epithelium covering the exocervix is normally noncornified. It grows, matures, and accumulates glycogen in its upper layers in response to circulating estrogens, most notably estradiol (Fig. 40.4). Because low blood

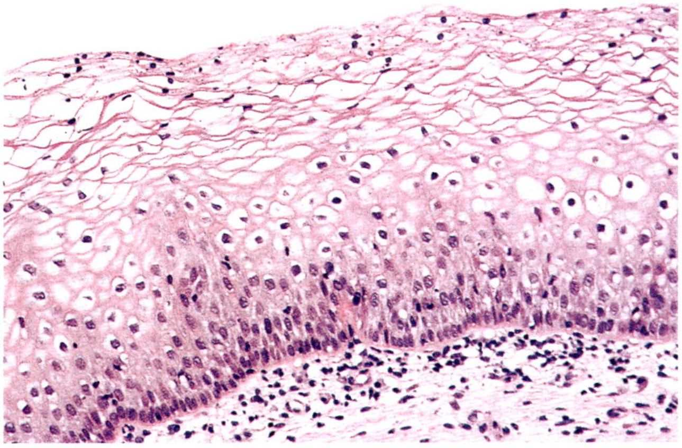

FIGURE 40.4 Mature squamous epithelium of the exocervix demonstrating a normal maturation sequence from basal cells to superficial cells. The cleared cytoplasm indicating glycogen storage should not be confused with koilocytosis.

levels of estrogen are the rule during childhood and the postmenopausal years, the squamous cells of the cervix do not proliferate or mature, and glycogen is not stored in the upper layers of the epithelium unless exogenous estrogen is made available, or the patient has a functional ovarian tumor (18). The squamous epithelium of the newborn cervix is fully mature due to maternal estrogen, but the epithelium quickly becomes atrophic and glycogen disappears as maternal estrogen levels decrease.

The estrogenically stimulated cervical squamous epithelium of the sexually mature woman can be divided into three layers: the basal/parabasal cell layer, the midzone layer (or stratum spongiosum), and the superficial layer (Fig. 40.4). The basal cell layer is composed of cells with scant cytoplasm and oval to cuboidal nuclei with dense chromatin. These cells are usually mitotically inactive and do not mark immunohistochemically with proliferation markers (e.g., Ki-67 and PCNA [proliferating cell nuclear antigen]) (48). The cells immediately above the basal layer comprise the lower portion of the midzone layer and are known as parabasal cells, a term often used in cytopathology. The parabasal cells are somewhat larger than the basal cells due to their increased cytoplasm, and the nuclei have slightly less dense chromatin. In contrast to the basal layer, mitotic figures are usually present but are not morphologically abnormal or particularly numerous. This layer also displays increased expression of proliferation markers (48). The midzone layer is composed of intermediate cells with even more abundant cytoplasm and somewhat smaller, vesicular nuclei. Glycogen accumulates in most intermediate cells, imparting a finely granular or clear appearance to the cytoplasm. The superficial cells contain small round and regular pyknotic nuclei, and abundant clear cytoplasm reflecting even greater glycogen accumulation. Keratinization occurs in both the superficial and intermediate cells and renders them flat and plate-like when they are spread on a slide. The cytoplasmic clearing characteristic of normal intermediate and superficial cells is often perinuclear. Because perinuclear clearing is also a feature of cells infected by human papillomavirus (HPV), there is a potential for misinterpreting normal epithelial cells containing glycogen as abnormal. However, normal squamous cells contain shrunken nuclei with evenly dispersed chromatin, whereas koilocytes typically show nuclear enlargement, nuclear membrane irregularities, and contain chromatin with a ropy texture. The cervical squamous mucosa undergoes cyclic changes during the menstrual cycle similar to the estrogen-progesterone–induced changes in the vaginal mucosa, although the cells composing the latter are a more reliable index of hormonal status. There is a predominance of intermediate cells during the luteal phase and pregnancy when progesterone levels are high.

The exocervical epithelium in postmenopausal women is composed mainly of basal and parabasal cells with scant cytoplasm and little or no cytoplasmic glycogen (Fig. 40.5). The cells may have an increased nucleus-to-cytoplasm ratio approaching that of squamous dysplasia. Consequently, atrophic epithelium is included in the differential diagnosis of high-grade squamous dysplasia, and care should be taken whenever a diagnosis of high-grade dysplasia is contemplated in a postmenopausal woman. However, the basal and parabasal cells in atrophic epithelia do not demonstrate the nuclear abnormalities and high mitotic rate usually seen in areas of high-grade dysplasia. Immunohistochemistry for p16 can be helpful in problematic cases. Atrophic epithelium is generally negative for p16 or shows only patchy nuclear staining, whereas high-grade dysplasia demonstrates block-like positivity of nuclei involving at least the bottom third of the epithelium with or without concomitant cytoplasmic staining (49). In-situ hybridization for high-risk HPV can also be used for direct visualization of HPV in high-grade dysplasia, whereas atrophy is negative for this marker (50).

Endocrine cells can be identified in the squamous epithelium of the exocervix by immunohistochemical techniques. Their function is unknown, but cervical carcinoid tumors do rarely occur, potentially recapitulating the phenotype of this cell type (49–56). Langerhans cells are present in the ectocervical epithelium and transformation zone

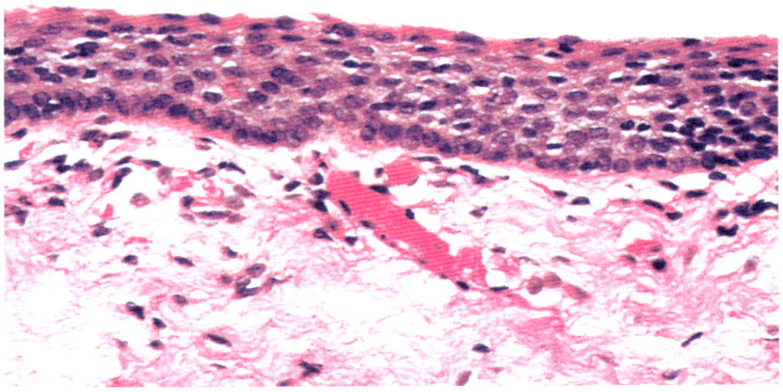

FIGURE 40.5 Postmenopausal atrophy of the cervical squamous epithelium. The immature cells can resemble the cells in high-grade SIL.

(57–60). They are involved in antigen presentation to T lymphocytes. Melanin-containing cells have been reported in the cervical epithelium and provide a plausible cell of origin for the uncommon cervical melanoma and blue nevus (61).

Epithelium of the Endocervix

The anatomic endocervix extends from the external os to the internal os, but endocervical glandular epithelium is not exclusively limited to this anatomic area, particularly during the reproductive years. Rather, endocervical epithelium occupies significant regions of the anatomic exocervix during childhood and after the menarche. The shift of the endocervical epithelium out of the canal onto the exocervix is discussed in more detail below in the section devoted to the transformation zone.

The endocervix is lined by a single layer of mucin-secreting epithelium composed of cells with small, often basilar, nuclei (Fig. 40.6). Goblet cells are sometimes encountered (Fig. 40.7). The nuclei are generally small and elongate with dense chromatin. When the endocervical epithelium has been damaged and is regenerating, the nuclei may become larger and more rounded, but mitotic figures are difficult to find in nonneoplastic endocervical cells (62). If mitotic figures are easily found in endocervical cells, consideration should be given to well-differentiated adenocarcinoma or carcinoma in situ, particularly if the nuclei are enlarged and nucleoli are prominent. Nucleoli are inconspicuous in resting endocervical cells, but they may become prominent during regeneration, pregnancy, and neoplastic transformation. Mitotic figures may be found in the constituent glandular cells of cervical endometriosis.

Other types of cells may be identified in the endocervical epithelium. Ciliated cells are almost always present and, when numerous, have been termed "ciliary (or tubal) metaplasia" (Fig. 40.8A–C) (62–68). Ciliated cells can be a useful marker of a benign process when the appearance of the endocervical glandular epithelium raises concerns for well-differentiated adenocarcinoma (63). However, they can contain enlarged dense nuclei that simulate the appearance of neoplastic cells. As a result, care should be taken to look for cilia before diagnosing in-situ neoplasia in the endocervix. Immunohistochemistry for vimentin and CEA may aid this distinction: adenocarcinoma in situ (AIS) is usually positive for CEA and negative for vimentin, while the opposite is true of tubal metaplasia (69). Immunostains for p16 and high-risk HPV in-situ hybridization can be useful in distinguishing neoplasia from ciliary metaplasia since most AIS is HPV related (49,50).

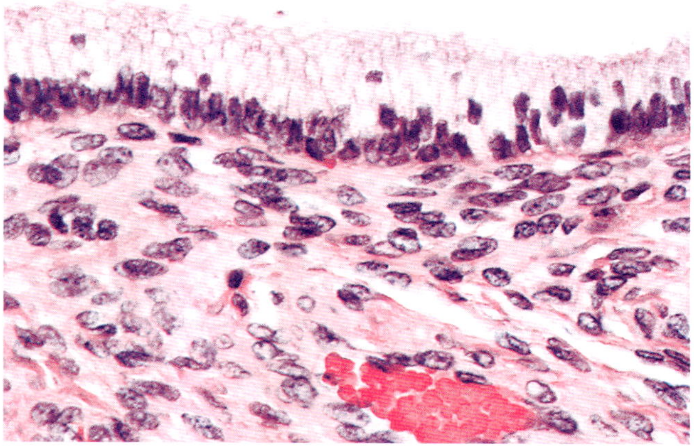

FIGURE 40.6 Normal endocervical mucosa with most nuclei in the characteristic basilar location. Enlargement of these nuclei and loss of apical mucin are features that should cause a closer inspection of the endocervical glands to ensure that neoplastic transformation is not present.

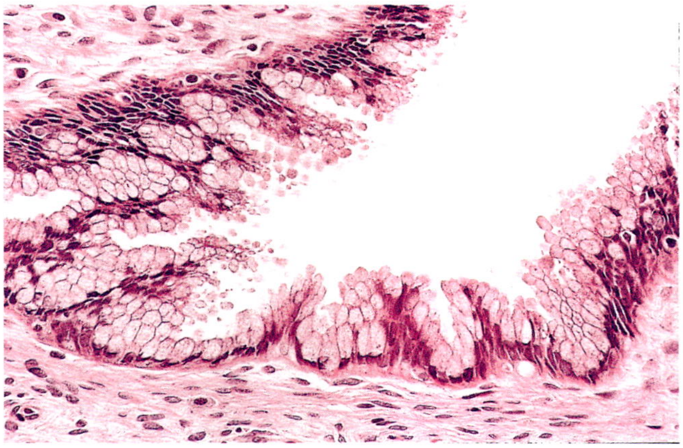

FIGURE 40.7 Goblet cells in the endocervix. Not infrequently, the nuclei of mucin-containing cells are displaced to the base of the cell and compressed by cytoplasmic mucin to produce a goblet cell. The presence of goblet cells and neuroendocrine cells in the normal endocervical mucosa tends to destabilize the conventional distinction in ovarian pathology between müllerian (i.e., cervical) mucinous and intestinal mucinous differentiation.

Subcolumnar reserve cells with potential to differentiate into ciliated and mucous secretory cells have been reported to populate the endocervix, although some evidence suggests that differentiated mucous cells are capable of division without the intercession of reserve cells (62,63). It is easy to confuse the lymphocytes that have populated the glandular epithelium with epithelial reserve cells (70).

Endocrine cells are also present within the endocervical epithelium. Their normal function is unclear, but it is generally held that they give rise to the endocrine neoplasms such as carcinoids and neuroendocrine carcinomas that occasionally occur in the cervix (51,56).

The endocervical epithelium not only lines the surface of the endocervical canal, but it also dips, to a variable degree, into the underlying stroma to form elongate clefts (Fig. 40.9A). These clefts are typically cut transversely in histologic sections, imparting the false impression that true endocervical glands are present within the stroma. However, true glands have different types of epithelia lining their ductal and secretory portions, whereas the endocervical mucosa is more or less uniform in appearance at all levels. Further evidence that these are not true glands was provided in a

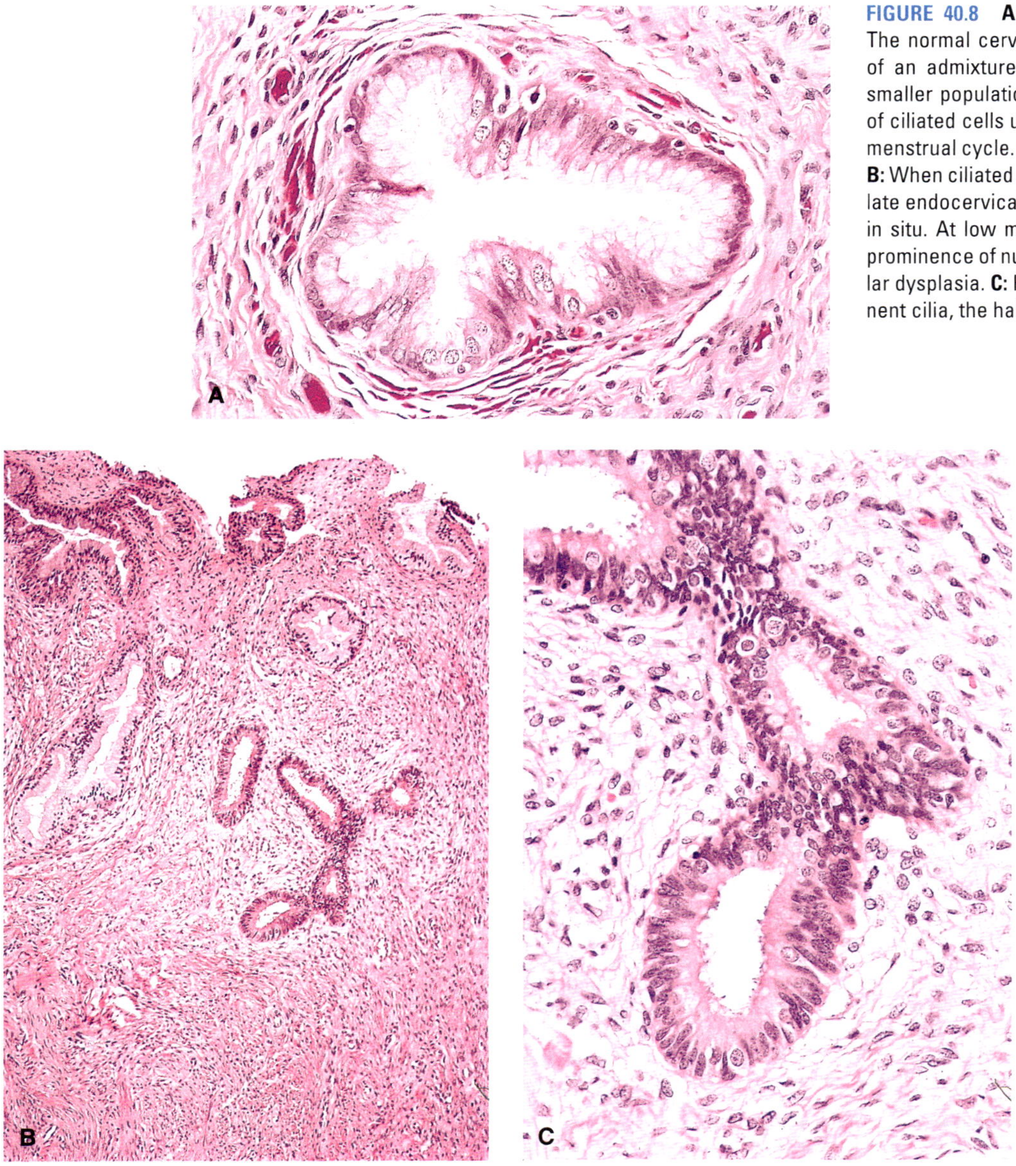

FIGURE 40.8 **A:** Ciliated cells in the endocervix. The normal cervical mucinous epithelium consists of an admixture of mucin-containing cells and a smaller population of ciliated cells. The population of ciliated cells undergoes cyclic variation with the menstrual cycle. **B** and **C:** Cervical tubal metaplasia. **B:** When ciliated cells are prominent they may simulate endocervical glandular dysplasia or carcinoma in situ. At low magnification, the glands feature a prominence of nuclei, a feature shared with glandular dysplasia. **C:** Higher magnification shows prominent cilia, the hallmark of ciliated cell metaplasia.

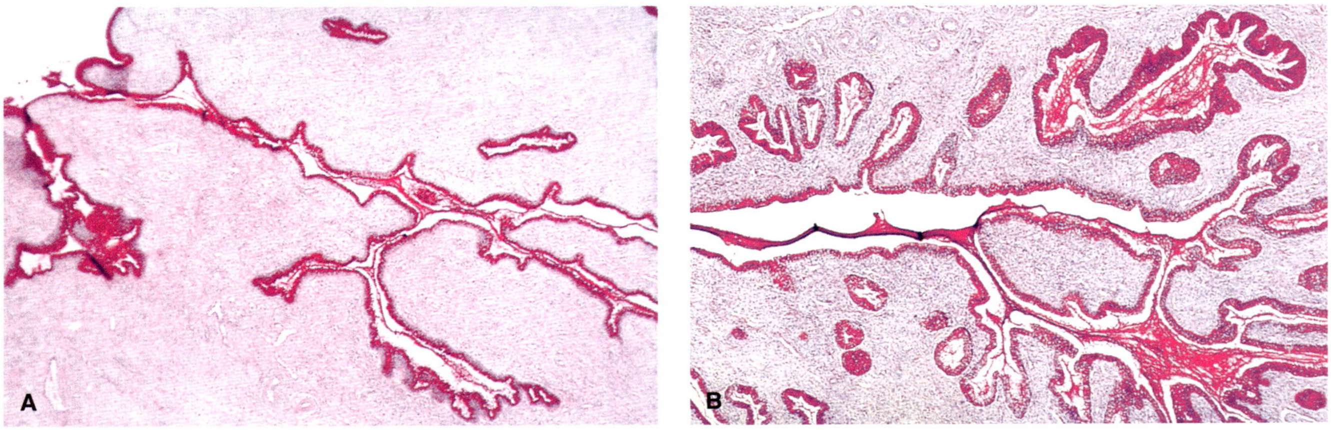

FIGURE 40.9 **A:** Tangential section of the endocervix stained with PAS to show how the gland clefts extend into the stroma and branch to form channels. **B:** When the endocervical mucosa undergoes hyperplasia and increases its surface area, as in pregnancy, the branches of the clefts proliferate and form even more collaterals ("tunnel clusters").

study conducted over 60 years ago by Fluhmann (71,72). He demonstrated by means of serial sections and three-dimensional reconstructions that what appeared to be endocervical glands within the stroma are actually complex protrusions of the endocervical lining that form clefts into the underlying stroma. When the endocervical epithelium lining the stromal clefts proliferates, side channels grow out from the clefts, giving rise to a histologic pattern that even more closely suggests acini of glands (Fig. 40.9B). Fluhmann labeled these side channels "tunnel clusters," although they are sometimes given the euphonious designation of "Fluhmann lumens." When secretions inspissate in tunnel clusters due to obstruction or viscosity, it appears as bright eosinophilic material resembling colloid of the thyroid (see Fig. 40.14). Having now discharged our obligation to anatomic accuracy, we shall continue to use the terms endocervical "gland(s)" and "cleft(s)" interchangeably.

The depth to which benign endocervical glands can extend in the cervical stroma varies from cervix to cervix. They can be found as deep as 1 cm but usually are found at a depth of less than 5 mm (73–75). This anatomic variation becomes important when considering a diagnosis of "minimal deviation adenocarcinoma (MDA)" (76,77). In this form of adenocarcinoma, the cytologic features differ only minimally from normal endocervical epithelium, and the diagnosis depends to a large extent on the identification of abnormally shaped glands at an inappropriate depth within the cervical stroma. Comparing the depth of the glands in question with that of clearly benign glands nearby can help resolve the issue. Glands around nerves or vessels, glands with an irregular "lobster claw" glandular configuration, and a granulation tissue stromal host response around glands can all be helpful clues to a cancer diagnosis. Ciliated cells are extremely uncommon in malignant endocervical glandular proliferations; this finding argues strongly against a diagnosis of adenocarcinoma.

Endocervical cells show only minimal morphologic changes during the menstrual cycle, and even this amounts only to a shifting of the basally situated nuclei to a mid-cell position at the height of the proliferative phase. These minor cytologic changes are in contrast to the dramatic biochemical changes that occur within the endometrial cells during the menstrual cycle (46). Throughout the proliferative phase of the cycle, the endocervical cells secrete mucus of lower viscosity than at other times of the cycle. This is thought to aid penetration of the cervical canal by spermatozoa (46). When progesterone levels attain their zenith during the luteal phase, the endocervical glandular secretion becomes thick and scant. It is at this stage that the secretion may become inspissated and more visible in histologic sections. During pregnancy, the number of tunnel clusters increases, and when this phenomenon is extreme, the term "cervical glandular hyperplasia" is often used (78). Pregnancy also causes the secretions of the endocervical cells to thicken and form a mucus plug that blocks the endocervical canal (47,79).

Epithelium of the Transformation Zone

The endocervical mucosa shifts over the various divisions of the anatomic cervix throughout life (18,44,80). At birth, the endocervical mucosa resides on the exocervix in two-thirds of infants, but it quickly moves back into the anatomic endocervical canal where it usually remains until near the menarche. After the onset of puberty, the endocervical mucosa again moves out onto the exocervix, where it tends to be more prominent on the anterior than posterior part (Fig. 40.10). The mechanism whereby endocervical mucosa changes location is apparently a mechanical one caused by swelling of the cervical stromal tissue in response to hormonal stimulation. As the lips of the cervix swell, they roll anteriorly and posteriorly, pulling the endocervical mucosa out of the canal onto the exocervix. The exposed endocervical tissue is often referred to as ectropion. Because the exposed endocervical mucosa appears red and ulcerated to the naked eye, it also has been interpreted as an erosion. There is, in fact, no erosion of the mucosa, rather the process is one of physiologic ectopy. After menarchial ectropion occurs, the endocervical tissue is gradually replaced by squamous epithelium throughout the reproductive years. The area where the glandular tissue is being replaced by squamous epithelium is known as the transformation zone. The junction between the two types of epithelium is labeled the squamocolumnar junction (44,81). Two squamocolumnar junctions are usually recognized (Fig. 40.10B). The original squamocolumnar junction is the point where the native (original) exocervical squamous epithelium joins the endocervical glandular epithelium and is out on the exocervix during the reproductive years (80). This junction is usually sharply defined, and it is anatomically fixed. After squamous metaplasia has replaced endocervical tissue, the original squamocolumnar junction is the fusion point between the new squamous epithelium laid down in the transformation zone and the native squamous epithelium (Fig. 40.11). The functional squamocolumnar junction is the point of active replacement of columnar endocervical epithelium by squamous cells. This junction is often irregular and patchy, and it changes its contours and its locations during reproductive life. The functional squamocolumnar junction is usually implied when the term "squamocolumnar junction" is used without a modifier and the area between the two squamocolumnar junctions is the transformation zone. During pregnancy, particularly the first pregnancy, even more endocervical tissue moves out onto the exocervix, enlarging the area of ectopic endocervical epithelium. This phenomenon also can occur during progestogen therapy.

Because endocervical glandular epithelium is present on the exocervix, the transformation zone can be visualized with the aid of a colposcope. This is fortunate because neoplastic change begins most commonly in the transformation zone, and neoplastic transformation is accompanied by structural alterations that can be recognized using the colposcope. The combination of papillomavirus detection, cytologic preparations, colposcopic examination, biopsy,

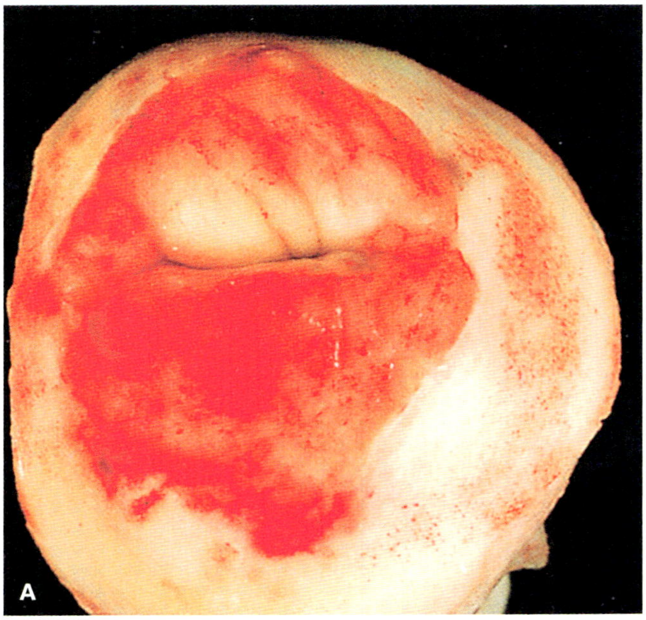

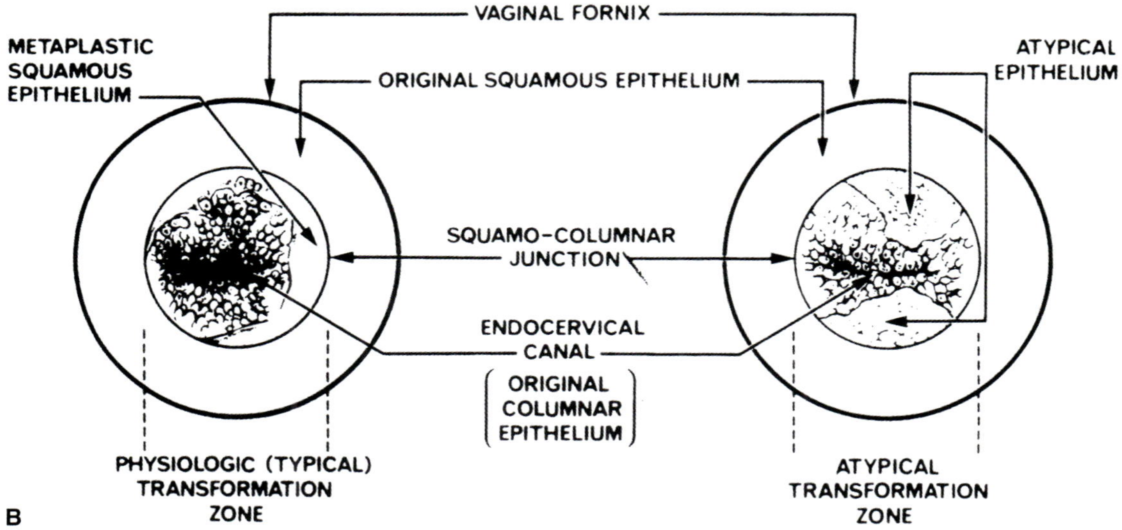

FIGURE 40.10 **A:** Multiparous cervix during the reproductive years. Note the slit-like configuration of the external os and the erythematous endocervical tissue out on the anatomic exocervix. This endocervical tissue undergoes conversion to squamous epithelium throughout the reproductive years. The squamocolumnar junction is visible as a sharp line between the white squamous epithelium and the erythematous glandular tissue. **B:** Diagram of the cervix demonstrating the transformation zone. On the left is a normal transformation zone in which metaplastic squamous epithelium is replacing endocervical columnar epithelium. "Squamocolumnar junction" refers to the original squamocolumnar junction. On the right, the metaplastic process is composed of dysplastic squamous cells and hence the process is cervical intraepithelial neoplasia. (Reprinted with permission from Fox H. *Haines and Taylor Obstetrical and Gynaecological Pathology*. 3rd ed. WB Saunders; 1987, with permission from Elsevier.)

and excision of intraepithelial abnormalities in the transformation zone under colposcopic visualization is a powerful tool for the early detection and successful treatment of in-situ neoplasia involving the cervix.

In the latter years of reproductive life the functional squamocolumnar junction reaches the area near the anatomic external os and, reversing its menarchal journey, begins to move up the anatomic endocervical canal. By the perimenopausal years, the squamocolumnar junction is usually concealed within the endocervical canal above the external os.

Two mechanisms are thought to be operative in transforming endocervical mucinous epithelium to squamous epithelium: (a) squamous epithelialization and (b) squamous metaplasia (3). The first involves the direct ingrowth of mature native squamous epithelium from the exocervix.

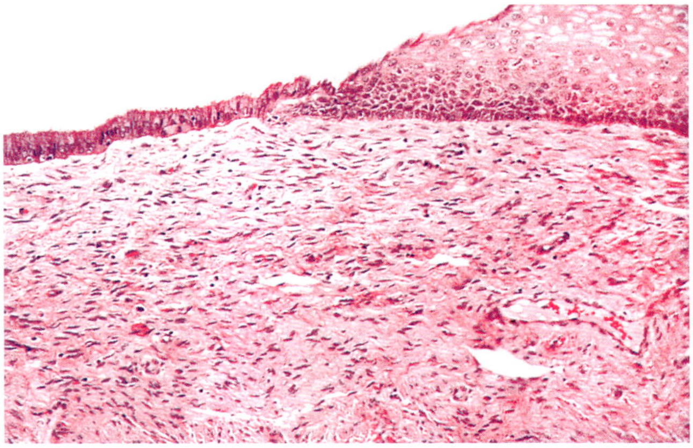

FIGURE 40.11 Squamocolumnar junction with a distinct transition from mature squamous epithelium on the right to endocervical glandular tissue on the left. Such a sharp change can be seen at the original squamocolumnar junction, as well as the junction formed by squamous epithelium with endocervical tissue in the transformation zone when squamous epithelium is mature.

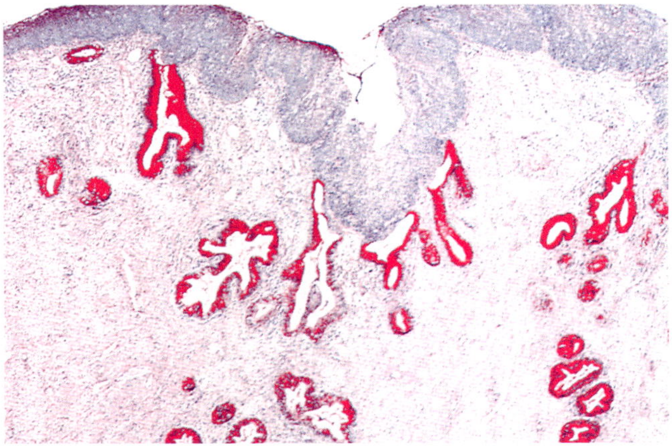

FIGURE 40.13 Conversion to squamous epithelium in the cervix may occur more rapidly on the surface than in the clefts, causing squamous epithelium to overlie endocervical gland clefts. When the newly laid down squamous epithelium blocks the orifices of the clefts, nabothian cysts, or "mucification" of tunnel clusters, as seen in Figure 40.14, may result.

This process is usually labeled "squamous epithelialization." During squamous epithelialization, mature squamous cells come to lie beneath the endocervical glandular cells. They push the endocervical cells off the basement membrane, and gradually the columnar cells degenerate and are sloughed. Squamous epithelialization initially spares the openings of the underlying endocervical glands, and at this stage, the openings to the glands have the appearance of pores when examined with the colposcope. Eventually, the ingrowth of squamous epithelium involves the orifices of the glandular clefts, and then it can extend for varying distances down into the cleft spaces (Figs. 40.12 and 40.13). When this process involves the orifice, it may plug the opening, and if the mucinous epithelium below continues to secrete, a mucin-filled cyst (nabothian cyst) or tunnel clusters filled with eosinophilic secretion result (Fig. 40.14). If squamous epithelialization involves the cleft and its ramifying tunnels, squamous epithelium will be surrounded by endocervical stroma. Consequently, histologic sections taken in an area of squamous epithelialization may show nabothian cysts, mucification of tunnel clusters, and/or islands of benign squamous epithelium in the stroma beneath the surface epithelium.

When the endocervical clefts undergo squamous epithelialization, care must be taken not to confuse the deep-lying benign squamous cells with invasive carcinoma. Although the cells in squamous epithelialization may have enlarged nuclei and prominent nucleoli, they do not demonstrate anaplasia, pleomorphism, chromatin abnormalities, or abnormal mitotic figures characteristic of invasive carcinoma. Moreover, the benign cells conform to the rounded configuration of the preexisting cleft and do not infiltrate

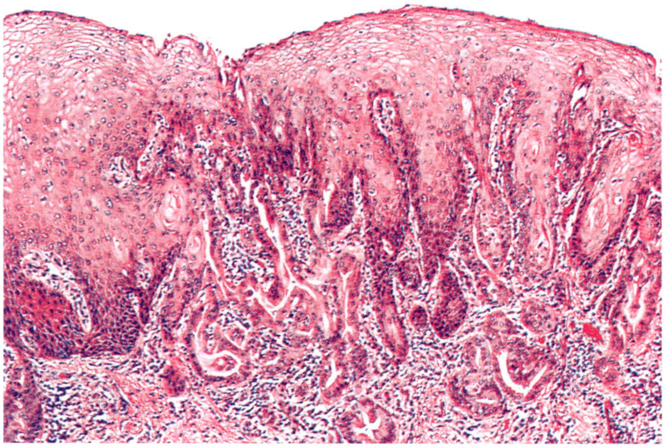

FIGURE 40.12 Squamous epithelialization of the endocervix. Note mature squamous epithelium extending into endocervical gland clefts. This process can mimic invasive carcinoma, especially when paired with epithelial dysplasia and tangential sectioning.

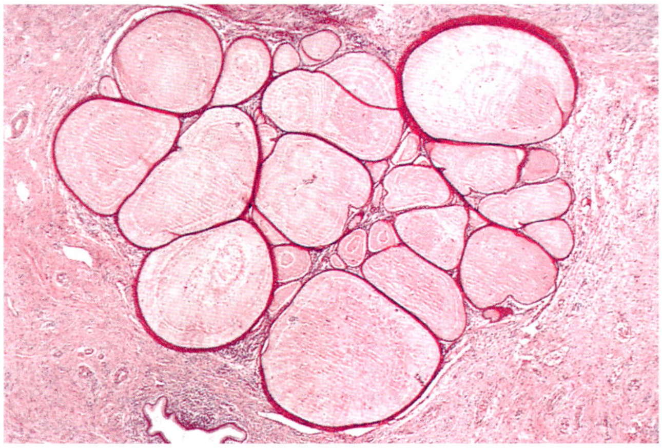

FIGURE 40.14 When the newly formed squamous epithelium in the transformation zone covers the endocervical gland cleft orifices and secretion continues, the tunnel clusters fill up with secretion that may become inspissated.

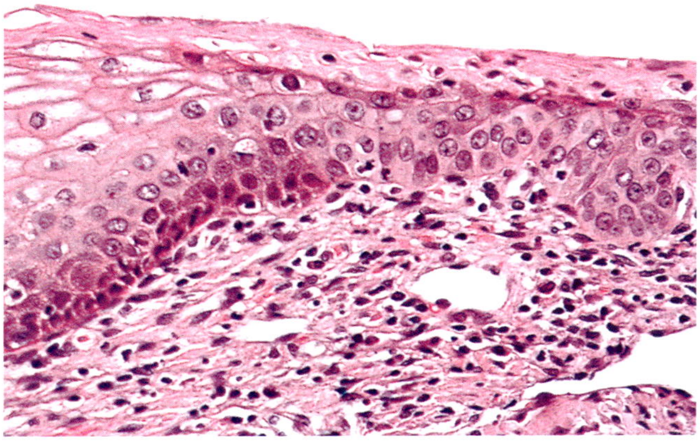

FIGURE 40.15 Functional squamocolumnar junction with metaplastic epithelium on the right. Note that in this example maturation has proceeded to the parabasal cell stage with abrupt keratinization rather than the normal maturation sequence to superficial cells as seen on the left.

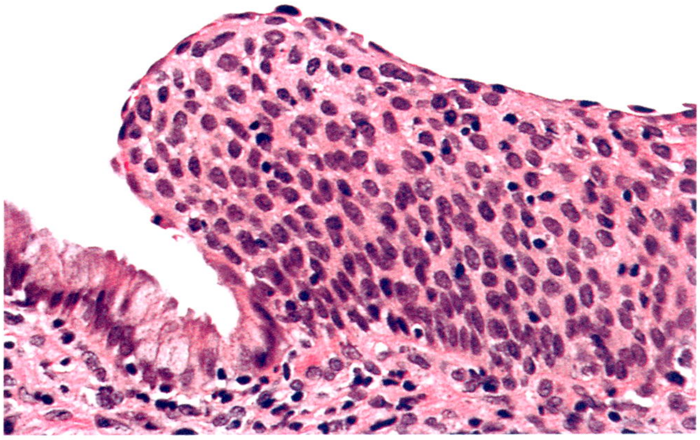

FIGURE 40.16 Immature squamous metaplasia on the right. The constituent cells do not demonstrate evidence of maturation and are similar to cells normally present in the basal layer of squamous epithelium. This type of metaplasia can be confused with a high-grade squamous intraepithelial lesion.

the stroma irregularly. Typically, there is no granulation tissue host response to squamous epithelialization, although chronic inflammation may be present. If squamous epithelialization involves tunnel clusters, small groups of squamous cells come to lie deep within the cervical stroma, imparting an architectural pattern that even more resembles infiltrating squamous cell carcinoma. Squamous epithelialization seems to be stimulated by chronic inflammation and local trauma, including cauterization or laser surgery.

The second mechanism thought to contribute to the conversion of endocervical mucinous epithelium to squamous epithelium entails first the proliferation of endocervical "reserve cells," and then the differentiation of these cells into squamous cells rather than mucin-producing cells (82). This process, known as squamous metaplasia or prosoplasia, can be distinguished from squamous epithelialization because, unlike the cells in squamous epithelialization, the reserve cells initially do not have squamous characteristics; rather, they appear as cuboidal cells with round nuclei growing beneath the mucinous epithelium (Fig. 40.15). In fact, these cuboidal cells are identical in appearance to the basal or parabasal cells of the squamous epithelium. After the reserve cells proliferate and stratify, they differentiate into squamous cells that initially have only slightly increased amounts of cytoplasm (immature squamous metaplasia). Later the cells may fully mature into glycogen-containing squamous cells indistinguishable from the superficial cells of the exocervix (Fig. 40.11). Confusingly, "squamous metaplasia" is commonly used as a generic term for both metaplasia and squamous epithelialization.

Immature squamous metaplastic cells without fully developed squamous characteristics or glycogen accumulation can come to occupy most or all of the thickness of an epithelium (Fig. 40.16) (83,84). Immature squamous metaplasia can bear a close resemblance to a high-grade squamous intraepithelial lesion because it lacks fully mature squamous cells toward the surface and the cytoplasm of the immature cells is relatively scant with elongated nuclei. However, the nuclei in immature squamous metaplasia are uniform with minimal chromatin abnormalities and smooth nuclear contours. Although mitotic figures may be present in immature squamous metaplasia, abnormal forms are not found. The possibility of immature squamous metaplasia should be considered whenever a diagnosis of high-grade squamous dysplasia is contemplated.

Squamous metaplasia is usually patchy, giving rise to the characteristic irregularity of the functional squamocolumnar junction (Fig. 40.17). As squamous metaplasia proceeds, the islands of squamous cells form bridges to other centers of metaplasia, ultimately producing a solid area of squamous epithelium.

Whatever the mechanism—either squamous epithelialization or squamous metaplasia—squamous replacement of mucinous epithelium on the exocervix is a normal process that must be distinguished from in-situ and invasive neoplasia. Features that are often found in neoplasia include moderate to marked pleomorphism, lack of maturation, irregular nuclear outlines, suprabasal mitotic figures, and abnormal mitotic figures. Nucleoli are usually inconspicuous in high-grade squamous dysplasia but are often prominent in metaplasia, epithelialization, and reactive changes in response to cervicitis.

A population of embryonic squamocolumnar junctional cells in the transformation zone has been recently identified. These cells are cytologically different from the endocervical and squamous cells and have a unique gene-expression profile (85). They are strongly cytokeratin 7-positive and able to undergo squamous metaplasia. With overlapping immunophenotypic profiles to squamous and adenocarcinomas of the cervix, it has been postulated that these squamocolumnar junctional cells might be the targets for HPV infection and give rise to HPV-related cervical neoplasia (86).

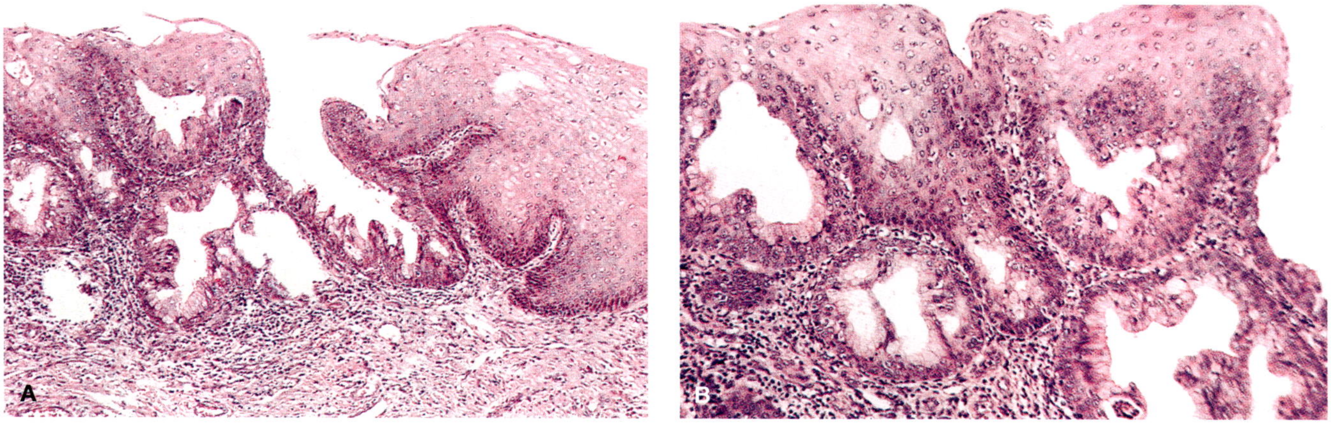

FIGURE 40.17 **A:** Islands of metaplastic squamous epithelium in the transformation zone at the functional squamocolumnar junction. These islands will eventually coalesce. **B:** Higher-power photomicrograph of the area of squamous metaplasia demonstrated in **A**.

Cervical Stroma

In contrast to the wall of the uterine corpus, which is predominately muscular, the stroma of the exocervix is mainly fibrous tissue admixed with elastin through which run infrequent aggregates of smooth muscle cells (87–90). A large number of vessels course through the stroma and a rich capillary network interfaces with the epithelium at the stromal–epithelial junction. This interface is irregular and features fingers of connective tissue containing vessels overlain by a squamous cell mucosa of variable thickness. The endocervix contains mostly fibroelastic tissue with the exception of its upper end where the superficial fibrous stroma blends imperceptibly into the endometrial stroma of the lower uterine segment. Consequently, the superficial stroma of the upper endocervix and the stroma of the lower uterine segment has a hybrid endometrial–cervical appearance that may cause localization problems when determining whether a neoplastic process involves the endometrium, the endocervix, or both. Of course, the presence of endocervical or endometrial glands can suggest the origin of the tissue in some cases, but both types of glands or even hybrid glands may be present in the transition area between the endocervix and lower uterine segment. The endocervix contains a greater number of smooth muscle cell fascicles in its deeper stroma than does the exocervix, and in the lower uterine segment, these blend into the myometrium.

The cervix bridges the sterile environment of the uterine cavity and the microbiologic milieu of the lower genital tract. It is not surprising that this important immunologic role would be marked by a conspicuous lymphoid presence (44). Thus, large numbers of T lymphocytes normally populate the endocervical stroma, whereas B-lineage lymphoid cells manifest as either plasma cells or germinal centers (91). Lymphocytes also may migrate into the endocervical epithelium, where they may assume the appearance of "cleared cells." Such cells have been misconstrued as "reserve" cells in the past (70). Endocervical plasma cells can be helpful in determining whether the sampled tissue originated from the endocervix or the endometrium, as the latter is typically devoid of plasma cells.

In addition, dendritic cells are numerous in the cervix. A subset of these are Langerhans cells that express MHC class II antigens and the CD4 receptor on their surfaces. These cells are involved in internalizing antigens and presenting them to T lymphocytes in the regional lymph nodes (92–95).

The relevance of these observations to the surgical pathologist is chiefly to discourage the overuse of "chronic cervicitis"; the presence of lymphoid tissue in the cervix is as normal as its presence in the small intestine. A diagnosis of chronic cervicitis should be withheld unless the lymphoid infiltrate is very heavy and/or lymphoid nodules are numerous. Particularly important for the diagnosis of chronic cervicitis is the presence of large numbers of plasma cells. Scattered plasma cells are normal in the cervix. Acute cervicitis is not uncommon, but true inflammatory erosion or microabscesses are rare in the cervix.

Remnants of the wolffian duct—commonly known as mesonephric rests—can be found in the endocervical stroma of the lateral portions of the cervix in about one-third of women (96) (Fig. 40.18). These structures are usually present in the deep stroma, but they are occasionally found near the surface or even blend with the endocervical gland clefts. Mesonephric rests are tubular structures lined by a single row of cuboidal cells with a central round, cytologically bland nucleus. Their tubules form lumens that contain hyaline, eosinophilic secretions and are often peripherally arranged around a central elongate duct. The combination of deep stromal location, the eosinophilic secretions, and the cuboidal cells usually serves to make identification of mesonephric rests straightforward. Even though tunnel clusters may ramify from a central cleft and contain eosinophilic secretions, they are lined by mucin-producing cells. The importance of this wolffian remnants

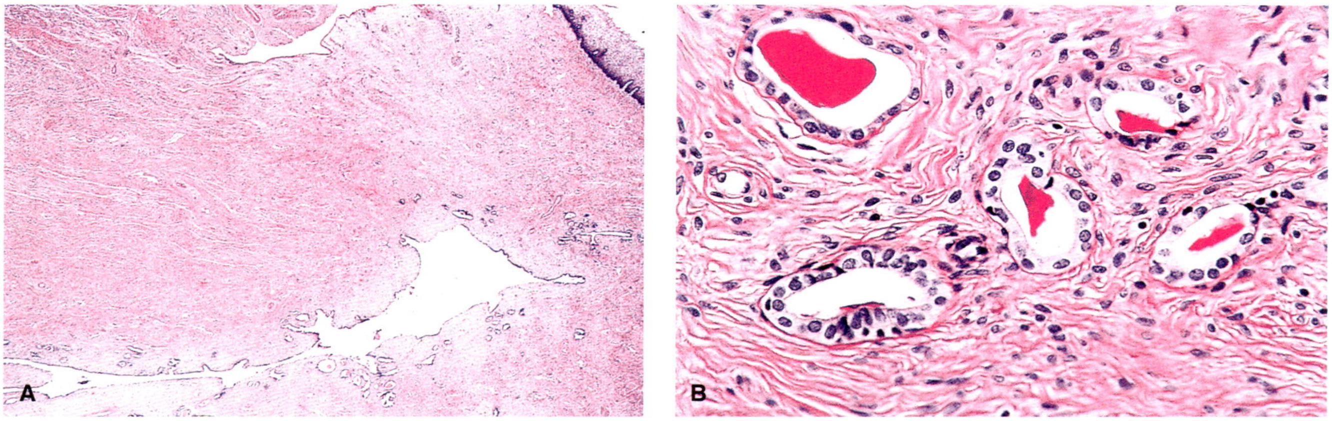

FIGURE 40.18 **A:** Mesonephric remnants in the cervix. A long cleft-like space deep in the stroma surrounded by tubules is the characteristic architectural finding. **B:** Ducts lined by bland cuboidal cells containing luminal PAS-positive eosinophilic secretion are key features of mesonephric remnants. The blandness of the constituent cells, low-power lobular architecture, and the organization around a central cleft are the most helpful features in distinguishing this from well-differentiated adenocarcinoma.

lies in their mimicry of well-differentiated adenocarcinoma. Mitotic figures are usually absent in mesonephric rests, and the chromatin of the cells is bland. Moreover, mesonephric rests do not exhibit the raggedly infiltrative growth of carcinoma even though they are located deep in the stroma. Rarely, atypical hyperplastic and neoplastic processes may involve mesonephric remnants (78,96,97). Benign and malignant mesonephric proliferations often express CD10 (98,99).

Multinucleated giant cells rarely are found in the normal superficial endocervical stroma. These cells have enlarged and sometimes bizarre-shaped nuclei with smudged chromatin similar to those seen in fibroepithelial stromal polyps (100). They should not be mistaken for a neoplasm (101–103).

Cervix During Pregnancy

The endocervical epithelium proliferates during pregnancy such that its mucus-secreting surface area increases. This proliferation leads to formation of polypoid protrusions of endocervical epithelium into the endocervical canal and an increased number of tunnel clusters budding off preexisting clefts within the cervical stroma. The overall impression is one of an increase in the amount of endocervical tissue, and consequently, this normal process is often termed "endocervical glandular hyperplasia" or "microglandular hyperplasia" when numerous small glands are packed together. Identical changes can be produced by artificial progestogens. The endocervical mucus during pregnancy is thick and functions as a plug to seal off the endometrial cavity from the vagina (22,104). Arias-Stella reaction may be seen in the endocervical glandular cells (105). As in the endometrium, the large cells with prominent nucleoli characteristic of the Arias-Stella reaction can raise concern about clear cell carcinoma, but the absence of mitotic figures and the gestational setting should quickly eliminate this possibility.

The stroma of the cervix undergoes a complex series of biochemical and biomechanical changes during pregnancy and parturition collectively known as cervical "ripening" (105). The initial change seems to be extensive destruction of collagen fibers by various collagenases accompanied by the accumulation of gel-like acid mucopolysaccharides. This process causes the cervix to soften and reaches its zenith immediately before parturition such that the cervix is easily effaced by the presenting part of the emerging infant. Thus, the usually cylindrical cervix is transformed into a thin saccular structure. The increased fluid in the cervical stroma during pregnancy causes the cervical lips to roll further out into the vagina, everting more of the endocervical mucosa beyond the external os. Squamous epithelialization and metaplasia rapidly ensue, and at the time of delivery, there is often considerable immature squamous epithelium in the transformation zone. As noted previously, the cells in immature squamous metaplasia can closely resemble those found in squamous dysplasia so caution should be exercised when examining cervical specimens taken from pregnant women.

The cervical stromal cells, particularly those near the surface of the endocervical canal, may undergo decidual change during pregnancy (Fig. 40.19). Cervical decidual reaction is typically patchy and manifests with focal replacement of the cervical stroma by aggregates of epithelioid cells. Awareness of this physiologic process during pregnancy and close attention to the cytologic features of the suspect cells should avoid misdiagnosis as invasive large cell nonkeratinizing carcinoma or other entities (106).

Normal findings in the cervix that have relevance to histopathologic differential diagnosis are presented in Table 40.1.

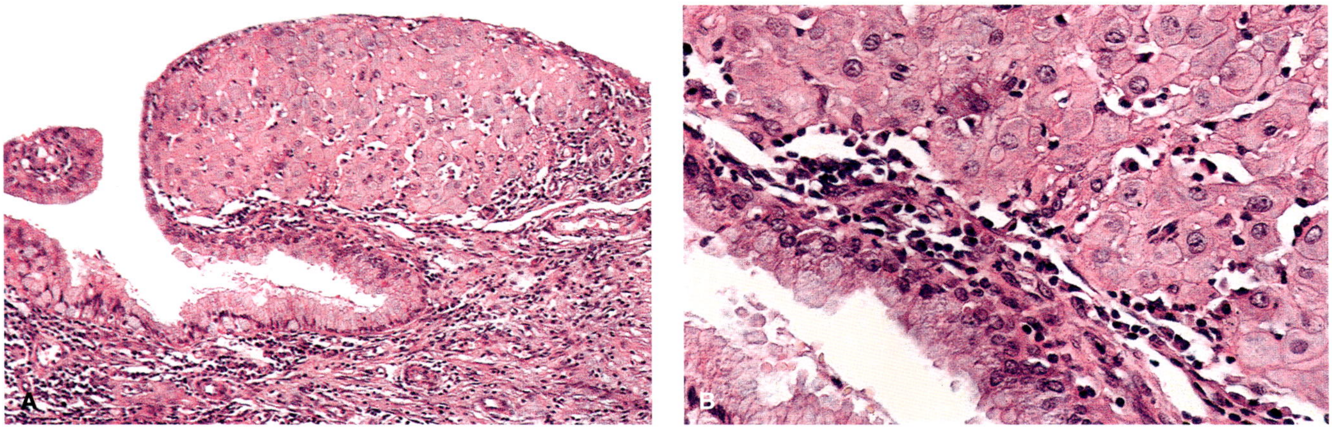

FIGURE 40.19 **A, B:** Decidual reaction in the cervix. The sheet-like arrangement of the cells can mimic squamous cell carcinoma, but the nuclei are bland.

TABLE 40.1 Normal Findings in the Cervix That Have Relevance to Histopathologic Differential Diagnosis

Finding	Diagnostic Confusion	Suggestions for Resolution	References When Relevant
Deeply situated normal endocervical gland clefts or Nabothian cysts	Minimal deviation adenocarcinoma (MDA)	• Lobster claw configuration in MDA • Ciliated cells in benign proliferations are almost always absent in malignant cervical glandular proliferations	(69,76,77)
Easily found mitotic figures in endocervical epithelium	Adenocarcinoma, invasive or in situ (AIS)	• Compare mitotically active epithelium with background normal epithelium Note: In normal endocervical epithelium mitotic figures are rare. In squamous metaplasia and regenerating cervical epithelium mitotic figures may be numerous but abnormal forms are not present • p16 immunohistochemistry and/or high-risk HPV in-situ hybridization	(49,50)
Mesonephric remnants	MDA or mesonephric carcinoma	• MDA often has an associated superficial component of AIS • The glands of mesonephric remnants are rounded and smooth-contoured; those of carcinoma are usually jagged and infiltrative • The glands of mesonephric remnants branch of a central, elongate, ductal structure; those of invasive adenocarcinoma are haphazard and lack a central originating duct	(96)
Decidual reaction	Large cell nonkeratinizing squamous cell carcinoma	• Mitotic figures, cytologic atypia in squamous cell carcinoma • Pancytokeratin negative in decidual cells and positive in squamous cell carcinoma	(107)
Arias-Stella (AS) reaction	Clear cell carcinoma of the cervix/vagina	• AS reaction usually does not feature mitotic figures that are easily found in clear cell carcinoma • Low Ki-67 in AS reaction	(108,109)
Lower uterine segment vs. endocervical fragments in endometrial biopsy	Determine endometrial vs. cervical primary if malignancy is present	• Follow-up with endometrial curettage • Imaging studies • Immunohistochemistry for CEA, p16, ER, PR, and vimentin can be helpful	(107)
Squamous metaplasia and epithelialization involving cervical glands/clefts	Invasive carcinoma	• Assess nuclear features, mitotic activity, and evaluate for the presence or absence of infiltration	
Endometriosis: benign endometrial glands and stroma with associated hemosiderin-laden macrophages and extravasated red cells	Adenocarcinoma: when stroma inconspicuous Adenosarcoma: when stromal mitotic figures conspicuous	• Additional/deeper levels • Cytologic atypia is usually minimal in endometriosis but mitotic figures may be present • History of endometriosis	(100)

TABLE 40.1 **Normal Findings in the Cervix That Have Relevance to Histopathologic Differential Diagnosis (*Continued*)**

Finding	Diagnostic Confusion	Suggestions for Resolution	References When Relevant
Microglandular hyperplasia and endocervical glandular hyperplasia	Adenocarcinoma	• Prominent nucleoli, nuclear atypia, and abnormal mitotic figures are features of carcinoma. Easily found mitotic figures should prompt consideration of carcinoma	(69,75,110)
Immature squamous metaplasia	High-grade SIL; may mimic because nuclei of the cells are large and cytoplasm is relatively scant	• Nucleoli-often present in metaplasia, but often inconspicuous in SIL • Abnormal mitotic figures and abnormal chromatin patterns in SIL • p16 immunohistochemistry or high-risk HPV in-situ hybridization	(49,50)
Glycogen storage in superficial squamous cells	Koilocytosis (HPV infection)	• Koilocytic nuclei: (1) enlarged, (2) irregular nuclear membranes, (3) dense ropy chromatin • High- and low-risk HPV in-situ hybridization	(49,50)
Tubal (ciliated cell) metaplasia	Adenocarcinoma in situ (AIS)	• Luminal mitotic figures, abnormal mitotic figures, apoptotic bodies, and stratification of nuclei are features of AIS • Look for ciliated cells; if numerous the process is almost surely benign • p16 immunohistochemistry or high-risk HPV in-situ hybridization	(66–69,73,110)
Multinucleated stromal giant cells: These may be a normal finding; probably myofibroblastic cells	Neoplasm, particularly sarcoma. Granulomatous inflammation.	• Abnormal mitotic figures, high cellularity, and necrosis all favor sarcoma • Granulomas have a syncytial appearance, lack mitotic activity, and are composed of CD68/CD163 positive histiocytes	
Postbiopsy epithelial displacement	Invasive carcinoma	• History of recent biopsy • Biopsy site changes • Chronic inflammation and giant cells may be present • Bland cytologic features and lack of mitotic activity	
Thermal (cautery) artifact	Dysplasia/SIL	• p16 immunohistochemistry or high-risk HPV in-situ hybridization	

See text for additional differential diagnostic clues (67).

ENDOMETRIUM

Tissue Sampling and Associated Problems

A variety of endometrial tissue sampling techniques are available to the clinician. These techniques differ with respect to their indications, their limitations, and their associated complications (104–106,108,111–119). Cervical dilation and endometrial curettage entail the removal of most of the uterine lining by scraping with a sharp curette. Under ideal circumstances, the excision is complete or nearly complete. Endometrial biopsy involves the removal of a more limited sample of tissue than does the complete curettage and is performed with a smaller curette. Single strips of endometrium are usually taken from both the anterior and the posterior fundal surfaces. Even though the sample is limited, the accuracy of diagnosis approximates that of the dilation and curettage. The chief advantage of endometrial biopsy is that it does not require cervical dilation or anesthesia. Endometrial biopsy thus combines convenience and low cost, with little sacrifice of diagnostic accuracy. The major limitation of endometrial biopsy lies in its inherent potential to miss focal lesions, including polyps and localized carcinomas. Accordingly, when carcinoma is suspected clinically, a complete dilatation and curettage must follow a negative biopsy because only this technique ensures the absence of carcinoma. Hysteroscopy in combination with endometrial sampling is thought by some to increase the detection rate of uterine abnormalities (120), although not all authorities agree on this point (121). Samples taken during hysteroscopy may demonstrate superficial reactive epithelial changes, hob-nailing, shedding of epithelial cells, and hemorrhage that have been collectively termed "exfoliation artifact." If endometrial biopsy is performed as part of an infertility workup, tissue should be obtained well into the presumed secretory phase 2 to 3 days prior to the next anticipated menstrual period as estimated by clinical and laboratory findings. Although biopsy in the late luteal phase might, in principle, destroy an early gestation, this seems not to be the case in clinical practice (122). Additional discussion of this topic is present in the section on Relevance of Endometrial Dating to Diagnostic Surgical Pathologists.

Three artifacts of sectioning and tissue preparation can hamper interpretation of endometrial samples. A frequent finding in endometrial curettings is the "telescoped" gland, which is characterized by an "inside-out" gland within the lumen of a gland with a normal configuration. This artifact is seen when an intussuscepted or telescoped gland (produced by the traumatic removal of the tissue) is cross-sectioned, and it occurs most frequently in straight glands. Another artifact results from tangential sectioning of a gland to produce a "pseudo-gland-within-gland" pattern that simulates malignancy. Confusion with adenocarcinoma can be avoided by attention to cytologic detail, comparison with surrounding glands, the focality of the alteration in benign endometrium, and awareness of this topologic problem. A third artifact involves tangential sectioning of the endometrial surface to produce pseudocystic and pseudobudded glands. Poor fixation can sometimes result in the retraction of endometrial glands from their surrounding stromal envelope. Cytoplasmic vacuolization resulting from autolysis can simulate early secretory epithelium.

Histology of the Normal Endometrium

The normal endometrium has a multiplicity of constantly changing normal patterns that depend on the nature and intensity of ovarian hormonal stimulation. The purpose of this section is to analyze the morphology of the normal nongravid endometrium in some detail from three points of view. We discuss regional variations, the individual components of the endometrium, and the temporal variations in the histology of the endometrium that occur throughout life.

Regional Variations

The uterine lining can be divided into two regions on the basis of its morphology: the mucosa of the lower uterine segment and the mucosa of the corpus proper. The mucosa of the lower uterine segment (isthmus) is thinner than the fundal mucosa. The glands and stroma are sluggishly responsive to hormonal stimulation and, thus, this part of the endometrium often lags behind the rest of the endometrium in development. The morphologic transition from endocervical mucosa to lower uterine segment mucosa is gradual; the hybrid endocervical–endometrial appearance of both glands and stroma serves to identify the lower uterine segment in endometrial curettings (Fig. 40.20).

The major portion of the uterine lining, the corpus mucosa proper, is normally fully responsive to hormonal stimulation. Two layers can be readily identified within the endometrium throughout this region: the lowermost is the basalis and the overlying one is the functionalis. The basalis is a zone of weakly proliferative glands and associated dense-spindled stroma immediately adjacent to the myometrium (Fig. 40.21). Characteristically, the junction of the basalis and myometrium is irregular with comingling of smooth muscle and endometrial stroma that, when florid, may give the false impression that endometrial tissue is pathologically isolated within the myometrium (Fig. 40.22). This deception is particularly important when evaluating the presence or absence of superficial myometrial invasion in patients with endometrial adenocarcinoma. Of less importance is the confusion it creates in the diagnosis of adenomyosis. The basalis, despite its inactive and undifferentiated appearance, plays a crucial role in the endometrial economy because it constitutes the "reserve cell layer" of the endometrium. After the bulk of the overlying functionalis is shed during menstruation, or after the functionalis is removed by curettage, the basalis and the residual deep functionalis are responsible for regenerating the endometrium. The remaining surface epithelium of the lower uterine segment also participates in this regeneration (123).

The appearance of the basalis is relatively constant throughout the menstrual cycle. The glands usually appear weakly proliferative with pseudostratified elongated nuclei, rare mitotic figures, and dense basophilic chromatin. They generally lack secretory change, and the stroma is spindled and nondecidualized with the exception of during the latter half of pregnancy when secretory glandular changes and stromal decidualization are often present. The importance of recognizing the basalis of the endometrium lies in not mistaking it for the functionalis in a curettage specimen. This confusion would result in an erroneous impression that this weakly proliferative appearance represented the fully developed state of the functionalis.

The functionalis exhibits the protean changes characteristic of the normal endometrium. This layer has been traditionally divided into two strata—the compactum and the

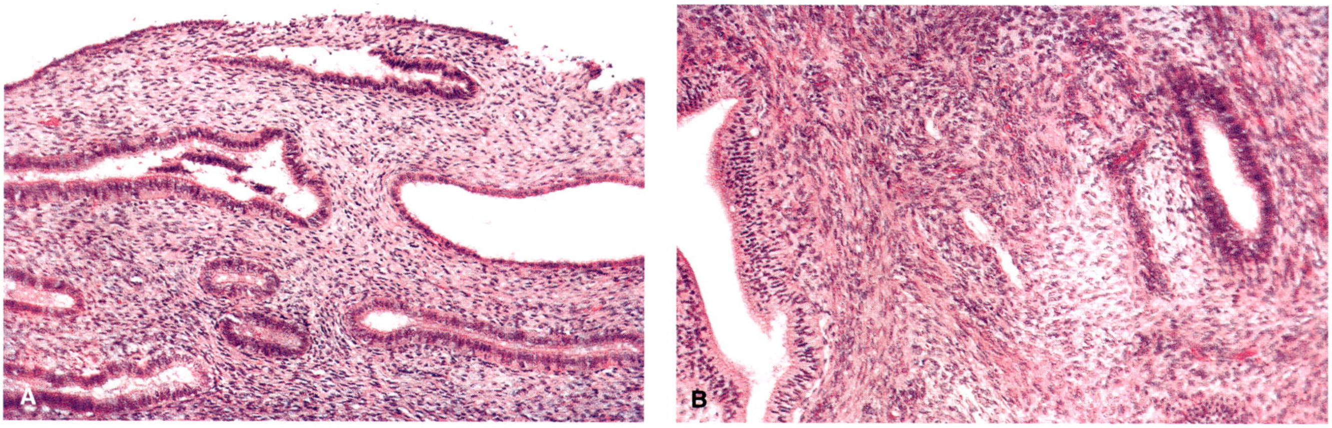

FIGURE 40.20 The lower uterine segment contains stroma and glands that are either hybrid between those seen in the endocervix and those in the fundus or a mixture of endometrial and endocervical glands and stroma. **A:** The stroma is fibrous appearing but more cellular than that typically found in the endocervix. **B:** An endometrial gland and an endocervical gland are found next to each other in this area of the lower uterine cervix.

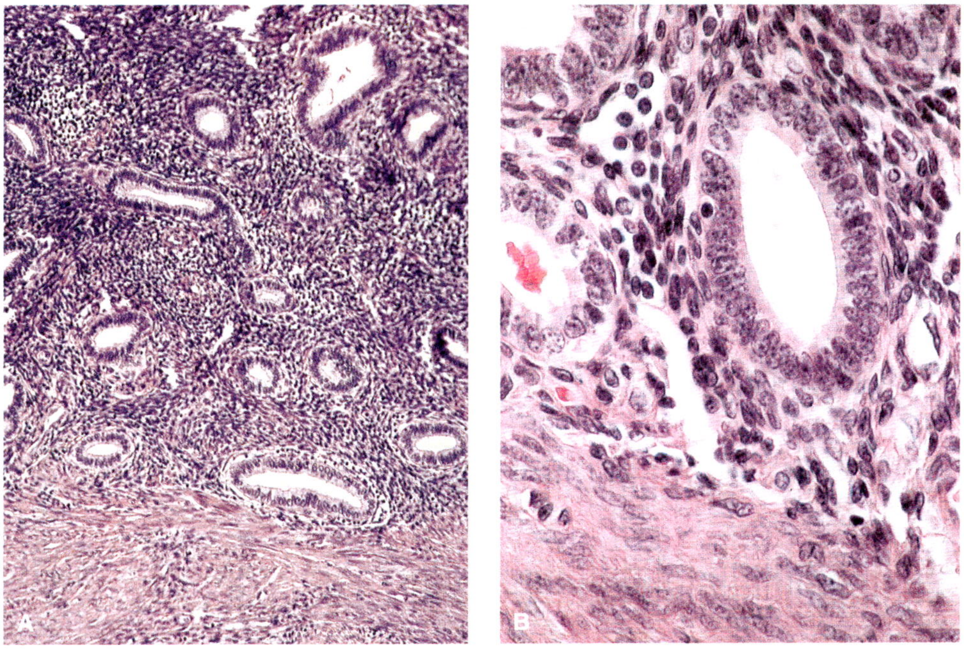

FIGURE 40.21 **A, B:** Endometrial basalis. Throughout the menstrual cycle the basalis maintains a weakly proliferative appearance. As a result, dating of endometrium should be performed on fragments containing surface epithelium.

spongiosum—based on the morphologic appearances of each during the late secretory phase of the menstrual cycle and during pregnancy. Unless otherwise specified, the term "endometrium" refers to the functionalis in the subsequent discussion.

Individual Components of the Endometrium

The normal endometrium consists of both epithelial (surface and glandular) and mesenchymal (stromal and vascular) elements that first synchronously proliferate, then differentiate, and finally disintegrate at roughly monthly intervals during the reproductive years.

› ***Epithelial Elements*** The basalis-type cells and the proliferative cells of the functionalis are morphologically similar. Both have high nucleus-to-cytoplasm ratios and elongate sausage-shaped nuclei with dense chromatin and inconspicuous nucleoli. The cytoplasm is scant and generally basophilic to amphophilic (see Figs. 40.21 and 40.33). Mitotic figures are common in the cells that compose the glands of the functionalis during the proliferative phase. When proliferative cells are the predominant cell type composing the epithelium, the nuclei appear pseudostratified.

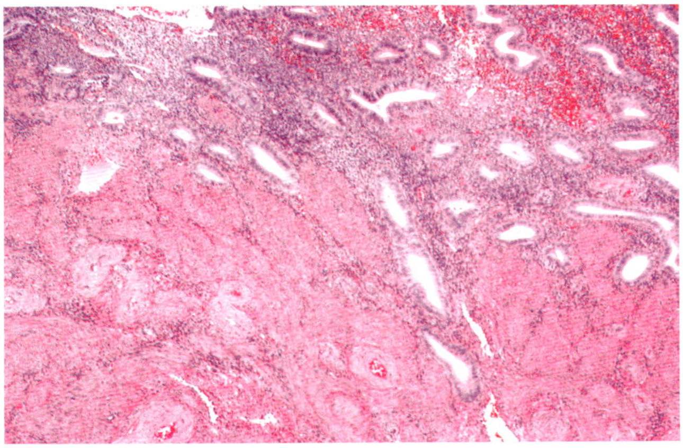

FIGURE 40.22 This is an example of an irregular endometrial–myometrial junction. This phenomenon is important to think about when determining whether or not adenocarcinoma is superficially invading the myometrium.

The characteristic cytoplasmic differentiation of the endometrial epithelial cell is nonmucinous secretion. Shortly after ovulation, secretory products accumulate in a subnuclear location in the proliferative cells; these products gradually shift to a supranuclear position and are ultimately released into the glandular lumens. This sequence of changes results in two easily recognizable secretory cell types: vacuolated and nonvacuolated secretory cells (see Figs. 40.33B and 40.34C). Although vacuolated cells may have a nucleus similar to those of proliferative phase cells, the nonvacuolated secretory cells possess rounded, vesicular nuclei with uniformly dispersed chromatin and occasionally nucleoli. The nonvacuolated secretory cells have uniform, moderately dense eosinophilic cytoplasm and often a frayed luminal border (see Fig. 40.34C).

Another type of secretory cell closely resembles the secretory cell of the fallopian tube. This cell has an elongated nucleus with coarse chromatin, a moderate amount of densely eosinophilic cytoplasm, and a rounded luminal bleb similar to those found in apocrine glands. These cells are common in the surface epithelium and occasionally line an entire endometrial gland. Some of these cells may represent ciliated cells.

The ciliated cells of the endometrium are consistently present in endometrial specimens and presumably represent one line of differentiation open to the basalis-type cell. These cells are more prominent near the uterine isthmus and during the proliferative phase (124,125). They have round, smoothly contoured nuclei containing finely stippled chromatin (Fig. 40.23). Although the nuclear features remain relatively unchanged throughout cell development,

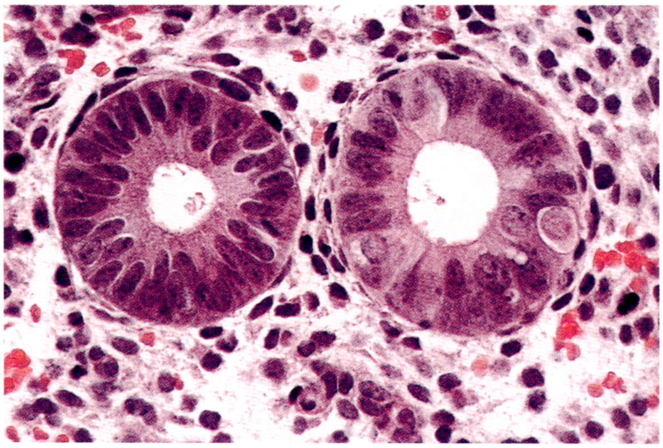

FIGURE 40.23 Proliferative phase glands with ciliated cells in the gland at the right. The round cell with clear cytoplasm at the 3:00 o'clock position has the characteristic appearance of ciliated cells before they have extruded their cilia into the glandular lumen. The other ciliated cells have a pyramidal shape.

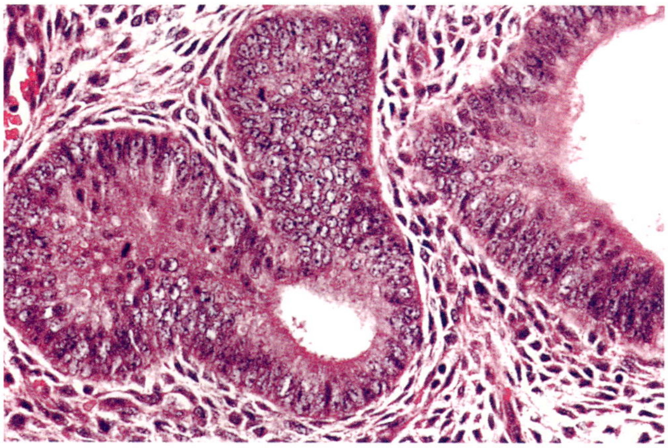

FIGURE 40.24 Proliferative phase glands and stroma. Note the elongate rather than rounded shape of the stromal cell nuclei. However, it is not uncommon for stromal cell nuclei to be elongated.

the configuration and location of ciliated cells vary as a function of the stage of ciliogenesis. The earliest identifiable ciliated cells are situated adjacent to the basal lamina of the gland and are roughly pyramidal in shape. They possess distinctively clear cytoplasm with central round nuclei. A rounded cytoplasmic zone containing eosinophilic fibrillary material can be identified with routine stains. This zone corresponds to the intracytoplasmic cilia seen with the electron microscope. When the growing ciliated cells reach the luminal surface, the cilia are exposed to the glandular lumen. Initially, the luminal surface of the ciliated cell is concave, but as the cell continues its development, this surface becomes convex, and ultimately the cilia pinch off as a merocrine secretion. The cell at this stage has a characteristic fusiform or pear shape. Ciliated cells can come to predominate the cellular population of glands.

› ***The Gland as a Whole*** The normal endometrial gland is lined by the aforementioned cells arranged in a nonstratified cuboidal-to-columnar epithelium that is pseudostratified during the proliferative phase. The glands are straight with narrow lumens in the early proliferative phase (Fig. 40.24), and exhibit increasing degrees of coiling, but not branching beginning in the midproliferative period and lasting throughout the rest of the cycle. By the late secretory phase, the glands have a serrated saw-toothed appearance. The surface epithelium is composed predominantly of apocrine-like secretory cells and ciliated cells and has a relatively constant appearance throughout the cycle.

Mesenchymal and Hematopoietic Elements

› ***Endometrial Stroma*** The endometrial stromal cell is the predominant cellular component of the stroma, and its appearance varies greatly with the menstrual cycle. The early proliferative phase features stromal cells that are small due to the scant, indistinct cytoplasm and dense ovoid-to-fusiform nuclei (Fig. 40.25). The stromal cells become progressively elongated and acquire more cytoplasm in the first part of the menstrual cycle. During the late proliferative phase and well into the secretory phase, electron microscopy shows increasing amounts of rough endoplasmic reticulum and both intra- and extracytoplasmic collagen. Toward the end of the secretory phase, the perivascular stromal cells become rounded, acquire more cytoplasm, and develop vesicular nuclei with occasionally prominent nucleoli. Cytoplasmic borders become more fully developed and the entire endometrial stroma is gradually transformed into sheets of polygonal cells with distinct cytoplasmic borders, abundant cytoplasm, and centrally placed vesicular nuclei (Fig. 40.26).

This unique müllerian stromal transformation is called decidualization when fully developed during pregnancy and predecidualization when partially developed during the late secretory phase of the menstrual cycle (126). Decidualization of the endometrial stroma is also characteristically seen in response to exogenous hormone administration, particularly high-dose progestins. The abundant cytoplasm of the decidual cell is filled with dilated rough endoplasmic reticulum,

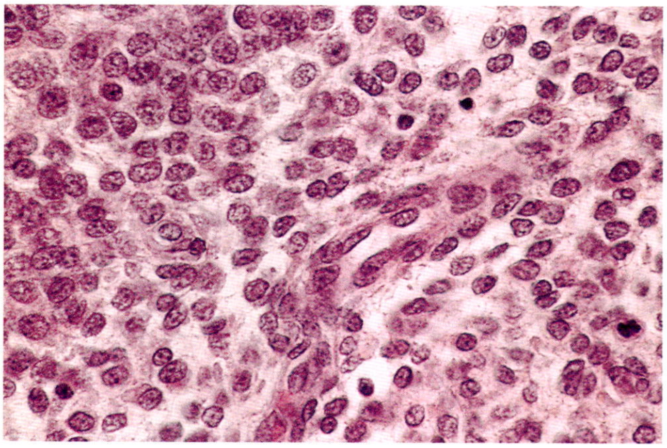

FIGURE 40.25 The proliferative phase endometrial stromal cells have scant, hard-to-discern cytoplasm and usually round nuclei (compare to Fig. 40.26). Thin-walled blood vessels populate the endometrial stroma.

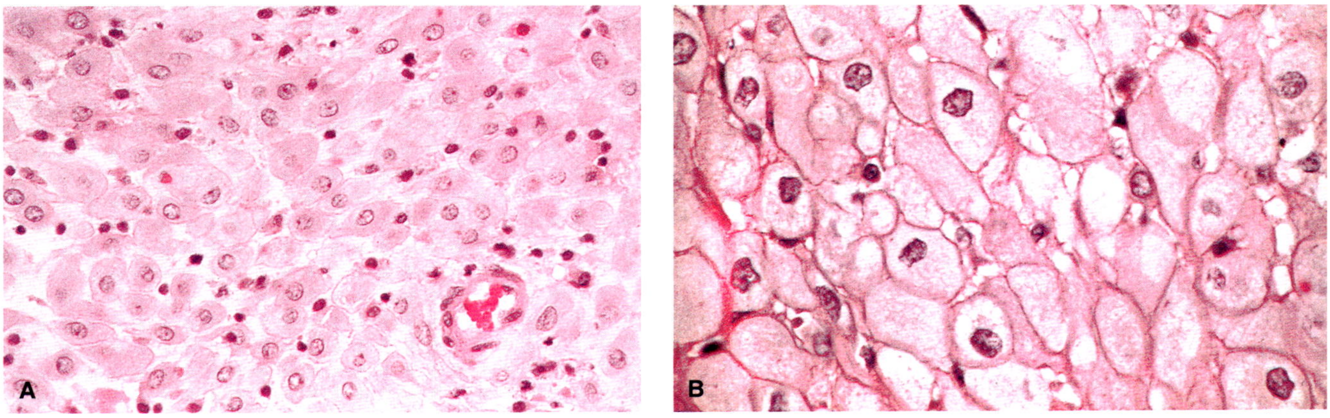

FIGURE 40.26 **A, B:** The decidual reaction during pregnancy is typified by cells with abundant pink cytoplasm and sharp cell borders.

Golgi apparatus, and mitochondria. Decidual cells form basal lamina and have complex intercellular interdigitations and tight junctions. The prominent intercellular borders are due to the accumulation of pericellular matrix (127).

The decidua is the specialized endometrium of pregnancy and plays an active role in implantation and in mediating the relationship between the fetoplacental unit and the mother. The decidua functions as an endocrine organ, secreting a host of products (e.g., prolactin, relaxin, renin, insulin-like growth factors [IGFs], IGF-binding proteins) involved in the paracrine and autocrine regulation of the fetomaternal interface (13,128). Decidual cells are capable of phagocytosis and are thought to play a role in dismantling the collagen scaffolding at the implantation site (112).

› ***Hematopoietic Cells*** A second prominent cellular constituent, particularly in the late secretory phase and during pregnancy, is what historically has been referred to as the "stromal granulocyte" but is now known to be a uterine natural killer (uNK) cell (129). These are rounded cells with bilobed nuclei and pale cytoplasm that contain eosinophilic granules. Their immunoprofile differs from that of blood NK cells. Their number appears to be positively correlated with the degree of predecidualization or decidualization in the surrounding endometrium; indeed, the number of such cells has been used as a dating criterion. Some workers hypothesize that uNKs play a role in the control of trophoblast invasion and spiral artery remodeling in human pregnancy and suggest that uNK cells may be important in initiating and maintaining decidualization. Alternatively, the death of uNK cells might be an early event in the onset of endometrial breakdown at menstruation (130–133).

The endometrium contains other leukocyte subsets that vary in number in a menstrual cycle–dependent fashion. These subsets include neutrophils and eosinophils, both of which are rare until the premenstrual phase; macrophages; mast cells and T-lymphocyte populations (13,133,134). Lymphoid follicles are commonly seen in the basalis of normal endometria (Fig. 40.27). Plasma cells are routinely seen in postpartum endometrial samplings and as an associated feature in a variety of pathologic settings (endometrial polyps, endometrial carcinoma, etc.). It has traditionally been held that, outside these settings, plasma cells are normally not present in the endometrium and when identified suggest endometritis. Although this view is plausible when large numbers are present, data from some studies have raised questions regarding the clinical relevance of scattered plasma cells (135,136). In fact, small numbers of B lymphocytes and plasma cells can be detected in normal endometrial samples from healthy patients using flow cytometry (133).

Occasionally, cells with bean-shaped nuclei and abundant vacuolated lipid-containing cytoplasm are present in the stroma of endometria stimulated by estrogen. These are termed stromal foam cells and probably represent modified endometrial stromal cells (Fig. 40.28). Similar appearing foam cells are seen as a component of an inflammatory infiltrate, particularly one produced by foreign material (e.g., keratin).

› ***Reticulin Framework*** The endometrial stromal cells elaborate a reticulin framework that becomes progressively denser as the endometrium develops during the menstrual

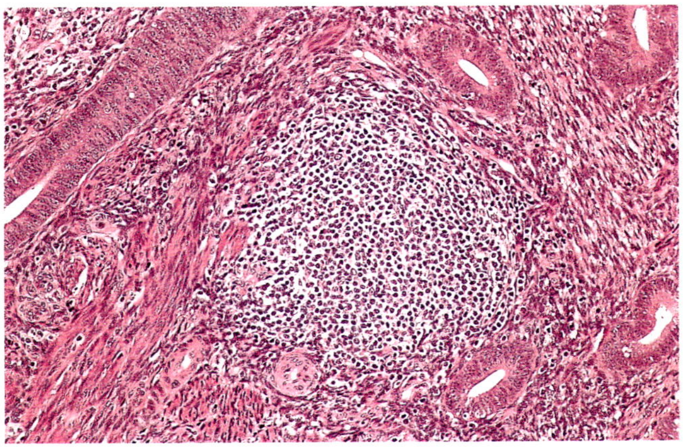

FIGURE 40.27 Lymphoid follicle. Scattered lymphoid follicles may be encountered in clinically normal women with otherwise unremarkable endometrium. When present in large numbers and when associated with plasma cells, this finding is pathologic.

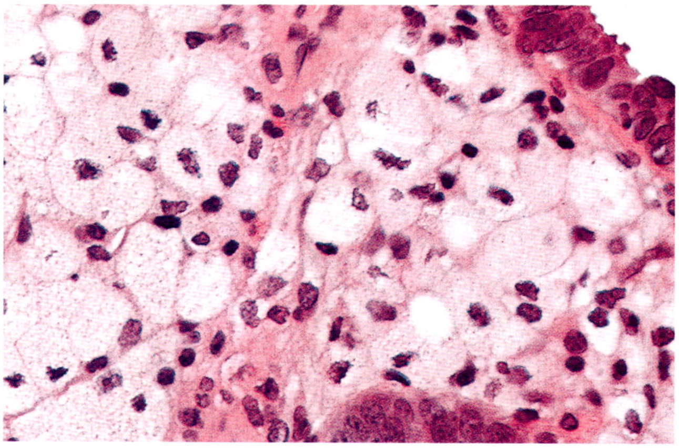

FIGURE 40.28 Stromal foam cells. Most likely these are modified endometrial stromal cells. In an inflammatory setting, stromal foam cells are probably macrophages.

cycle, such that each stromal cell is enmeshed in reticulin by the late secretory phase. This framework undergoes dissolution during menstruation. The stromal intercellular space is also rich in high–molecular-weight mucopolysaccharides during the midproliferative and late secretory phase.

› ***Vascular Elements*** The radial arteries of the endometrium derive from the myometrial arcuate system. As the radial arteries course toward the uterine cavity, they give off basal branches and then continue as endometrial spiral arteries. The basal arteries are unresponsive to steroid hormones, whereas the spiral arteries respond to varying hormone levels both by proliferation and, during the luteal phase of the menstrual cycle, by intermittent contraction (137,138).

Angiogenesis is central to the menstrual cycle, implantation, and subsequent gestation. Vessels ruptured during the menstrual phase must be repaired, vascular ingrowth must rapidly occur during the proliferative phase, spiral arteries must develop during the secretory phase and, to come full circle, the dismantling of the vasculature is required as the endometrium enters the menstrual phase (13). All these stages are closely monitored by activators and inhibitors (139–142). Angiogenesis is again pivotal in negotiating the connection of the fetoplacental unit to the maternal circulation during implantation and gestation. This entails an extensive remodeling of the placental bed spiral arteries to form the utero-placental vessels (142–144).

› ***Ultrastructural Features and Immunohistochemistry*** Both transmission and scanning electron microscopic study of the endometrium have produced an immense body of literature (115,145,146). Mention should be made of two distinctive ultrastructural features found in the early secretory glandular cell: the giant mitochondrion complex (142) and the nucleolar channel system (147). These two features seem to be the earliest postovulatory morphologic change in the endometrial glandular cell.

Normal endometrial and endocervical epithelium have some differences with respect to their immunohistochemical profiles. Endometrial glands are positive for low- and high–molecular-weight cytokeratins, vimentin, and occasionally CEA, while endocervical epithelium expresses CEA but is negative for vimentin and low–molecular-weight cytokeratin (99). The endometrial stroma and mesonephric remnants are strongly positive for CD10 while the endometrial and endocervical glands are not. Normal smooth muscle cells occasionally express CD10 and about 40% of leiomyomas contain CD10-positive cells at least focally. Both the endometrial stromal cells and smooth muscle cells express vimentin, muscle-specific actin, smooth muscle actin, and Bcl-2. Smooth muscle cells typically express desmin and caldesmon whereas endometrial stromal cells do not.

The endometrium produces many secreted proteins that serve local cell signaling functions important for the developing endometrium and embryo. There is a large and growing body of literature concerning these products of endometrium, a topic beyond the scope of this chapter. However, the reader is directed toward a number of excellent recent reviews of this subject (13,113,148–151).

Temporal Variations

Unlike morphologically unchanging epithelia that have a constant appearance throughout the cell's lifetime, the endometrium undergoes dramatic temporal morphologic changes. The changes are cyclic and particularly striking during the reproductive years. They can be conveniently considered under six headings: Newborn, premenarchal, perimenarchal, reproductive, perimenopausal, and postmenopausal years (113,152,153).

› ***Newborn*** The genitalia of the newborn girl responds to the high levels of circulating maternal and placental steroids by a transient burst of precocious development. The endometrium may be well developed and have either a proliferative or, less commonly, a secretory appearance. These changes regress within 2 weeks, and the long hormonal quiescence of the premenarchal years begins. This quiescent period is characterized by a thin endometrium populated by inactive glands set within a spindled inactive stroma. Rarely, estrogen-secreting lesions of the ovary (follicular cysts, sex cord gonadal stromal tumors) may cause abnormal endometrial growth, with resulting abnormal bleeding as part of the syndrome of precocious pseudopuberty. The endometrium under these circumstances is proliferative or hyperplastic. This phenomenon shows that the inactive appearance of the endometrium during this period is secondary to a lack of hormonal stimulation (10).

› ***Menarche*** The onset of uterine bleeding (menarche) is one of the many changes that signal the maturation of the reproductive system. This usually occurs between 12 and 15 years of age in the United States. Characteristically, the perimenarchal period is marked by greater variability in the length of individual cycles than is seen in the reproductive years and by the occurrence of anovulatory cycles (153). Disordered proliferative endometria are commonly encountered in this setting (see section Disordered Proliferative Endometrium).

› ***Reproductive Years*** The reproductive years are characterized by regularly occurring, roughly monthly, cycles that end with menstrual bleeding (154,155). The dominant ovarian steroid hormone secreted during the first half of the menstrual cycle is estradiol (E2), which induces endometrial proliferation. Immediately following ovulation and for the duration of the cycle both progesterone and estradiol are present, and this combination of hormones induces endometrial glandular secretion and stromal predecidualization. Upon withdrawal of corpus luteum steroidal support, the endometrium is shed, setting the stage for the next cycle. These regularly recurring cycles may be interrupted by pregnancies, but cycling is soon restored after the termination of pregnancy.

The biochemistry of the steroid molecules and their receptors have been the subject of intense study (156–158). Steroid molecules are hydrophobic and easily diffuse through cell membranes, such that they freely enter all cells. The endometrium, vaginal mucosa, and other steroid-sensitive tissues are target organs by virtue of the presence of high-affinity, high-specificity, low-capacity saturable receptors for E2 and progesterone. These nucleus-based receptors are absent in nonresponsive cells. In addition to being highly responsive to circulating hormones, the endometrium also synthesizes glycoproteins that affect the hypothalamic–pituitary–ovary axis and the endometrium itself. A detailed description of hormone regulation is beyond the scope of this chapter, but several excellent reviews on the topic have been published and can be found in the general references listed at the end of the chapter. In broad terms, the steroid molecule combines with the appropriate receptor, and the steroid–receptor complex becomes associated with a nonhistone nucleoprotein. The net effect of this linkage is to qualitatively and quantitatively alter DNA-dependent RNA transcription. The consequence is an altered profile of protein biosynthesis. Furthermore, the unique response of a particular target cell type depends on what specific growth and differentiation program is initiated by the steroidal signal. The endometrium is a major target organ for this continual barrage of steroidal information. It responds by undergoing the dramatic morphologic alterations that constitute the normal menstrual cycle, as well as producing proteins important for hypothalamic feedback, modulation of placental hormone secretion, regulation of macrophages, and regeneration after menses.

› ***Morphology of the Normal Menstrual Cycle Endometrium*** The first day of the menstrual cycle has conventionally been identified as the first day of menstrual flow. Menses usually lasts for fewer than 5 days and is followed by the endometrial proliferative phase, the length of which exhibits great variation (9 to 20 days) but lasts for 10 days on average. After ovulation, the coordinated and highly predictable series of stromal and glandular changes characteristic of the secretory (luteal) phase takes place. The traditional view is that the length of this phase is constant (14 days), and it is this constancy that provides the basis for endometrial dating. Due to the sensitivity and specificity of serum hormonal studies, it is exceedingly rare for the surgical pathologist to be called upon to date the endometrium for infertility purposes. However, given the prevalence of polycystic ovarian disease, dysfunctional uterine bleeding, and iatrogenic hormonal use it remains important for the surgical pathologist to recognize when ovulation has occurred and the normal responses of the endometrium to various hormonal stimuli. The following discussion is a concise description of the normal menstrual cycle with more detailed information in the accompanying figures (Figs. 40.29 and 40.30, Table 40.2) (112,113,123,159–161).

TABLE 40.2 Decision Tree for Endometrial Dating

What type of gland is present?
- A. Proliferative gland (early proliferative, midproliferative, late proliferative, interval)
  - Is the gland straight or coiled?
    - *Straight:* Early proliferative
    - *Coiled:* Midproliferative, late proliferative, interval
  - Is there stromal edema?
    - *Yes:* Midproliferative
    - *No:* Late proliferative, interval
  - Are there scattered subnuclear vacuoles present, but with less than 50% of the glands exhibiting uniform subnuclear vacuolization?
    - *No:* Late proliferative
    - *Yes:* Interval—consistent with but not diagnostic of POD 1
- B. Secretory gland-vacuolated (early secretory)
  - POD 2: Subnuclear vacuolization uniformly present, leading to exaggerated nuclear pseudostratification: (greater than 50% of the glands exhibit uniform subnuclear vacuolization); mitotic figures frequent
  - POD 3: Subnuclear vacuoles and nuclei uniformly aligned; scattered mitotic figures
  - POD 4: Vacuoles assume luminal position; mitotic figures rare
  - POD 5: Vacuoles infrequent; secretion in lumen of gland, cells have a nonvacuolated secretory appearance
- C. Secretory gland-nonvacuolated (midsecretory, late secretory, menstrual)
  - Is there stromal predecidualization?
    - *No:* Midsecretory
      - POD 6: Secretion prominent
      - POD 7: Beginning stromal edema
      - POD 8: Maximal stromal edema
    - *Yes:* Late secretory, menstrual
  - Is there crumbling of the stroma?
    - *No:* Late secretory
      - POD 9: Spiral arteries first prominent
      - POD 10: Thick periarterial cuffs of predecidual
      - POD 11: Islands of predecidual in superficial compactum
      - POD 12: Beginning coalescence of islands of predecidual
      - POD 13: Confluence of surface islands; stromal granulocytes prominent
      - POD 14: Extravasation of red cells in stroma; prominence of stromal granulocytes
    - *Yes:* Menstrual
      - Crumbling stroma, hemorrhage
      - Intravascular fibrin thrombi
      - Stromal granulocytes prominent
      - Polymorphs present
      - Late menstrual: Regenerative changes prominent

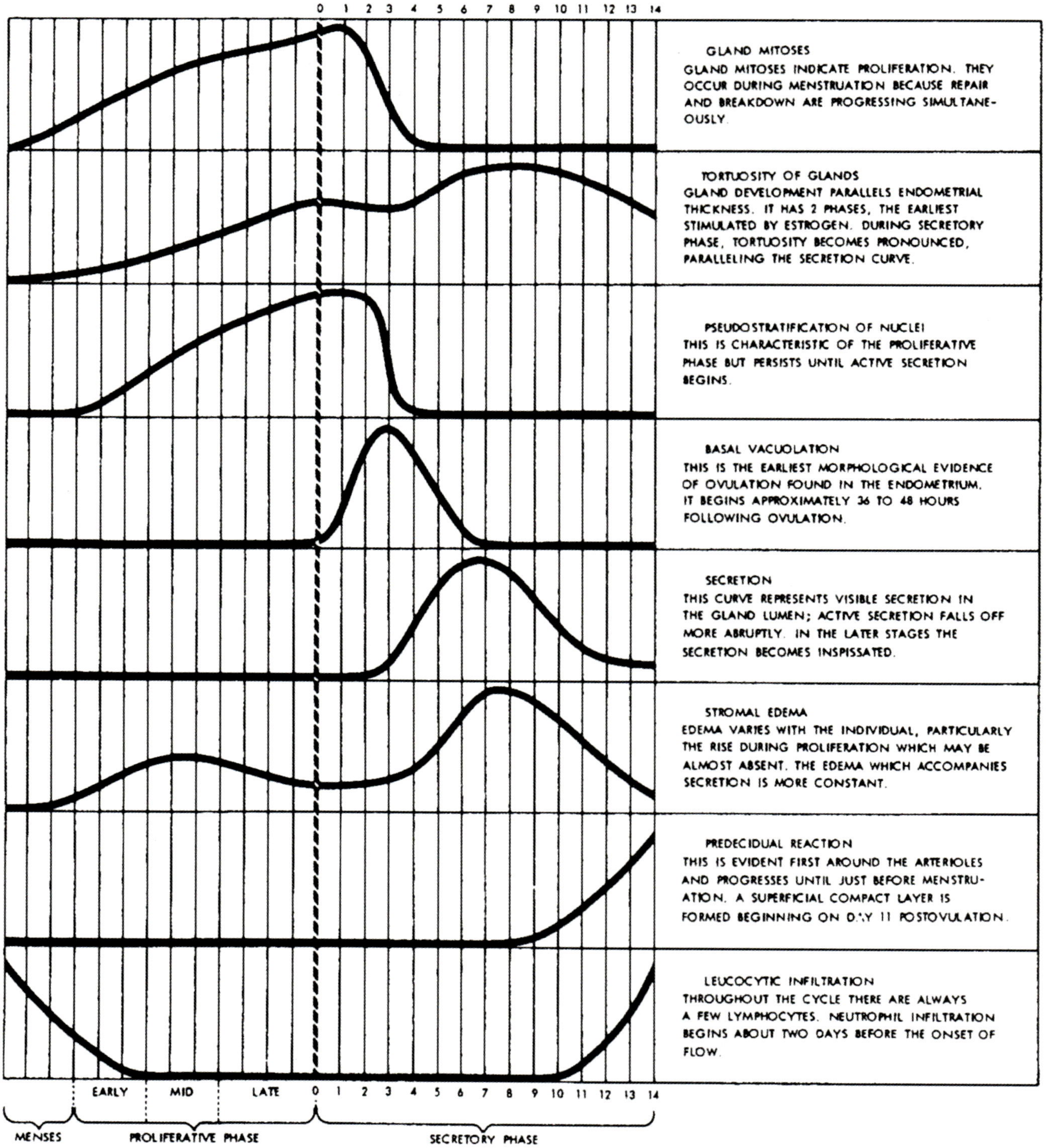

FIGURE 40.29 Approximate qualitative changes in eight morphologic criteria found to be most useful in dating human endometrium. (From Marshall J. The physiology of the myometrium. In: Hertig A, Norris H, Abell M, eds. *The Uterus*. Williams & Wilkins; 1973:89–109.)

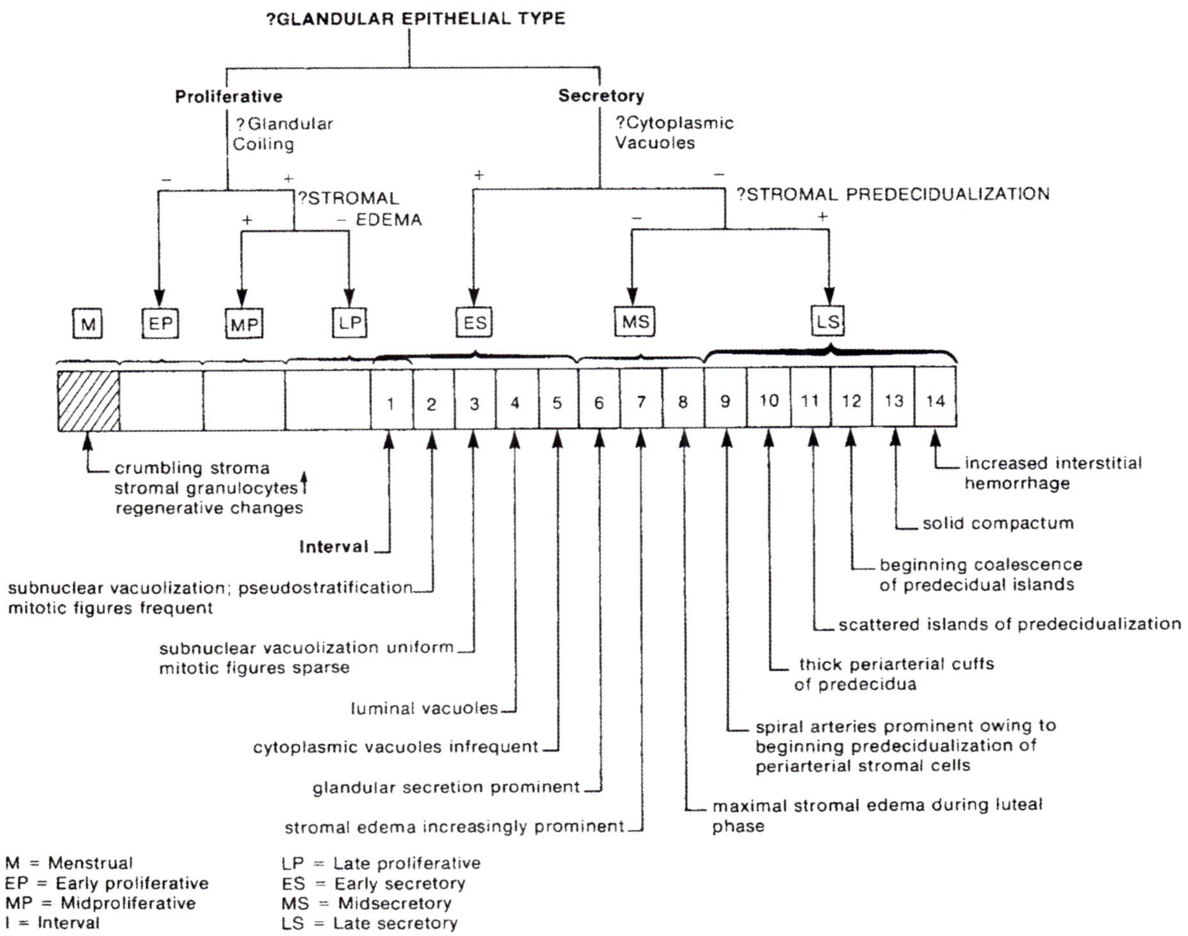

FIGURE 40.30 Decision tree for endometrial dating. (From Marshall J. The physiology of the myometrium. In: Hertig A, Norris H, Abell M, eds. *The Uterus*. Williams & Wilkins; 1973:89–109.)

Proliferative Endometrial Phase (Ovarian Follicular Phase)

The proliferative phase spans the time from the previous menstrual period to ovulation. The endometrium responds to rising estrogen levels by synchronous proliferation of glands, stroma, and vessels. During the first third of the proliferative phase (early proliferative), the rate of growth of all three of these elements is coordinated, and, as a consequence, both vessels and glands are noncoiled. After a few days, the growth of both glands and vessels outstrips that of the stroma resulting in coiling of these tubular structures in the mid- and late proliferative endometrium (Fig. 40.31).

The glands are lined by mitotically active pseudostratified columnar cells with high nuclear-to-cytoplasmic ratios and dense chromatin (Fig. 40.32). Mitotic figures are almost always easy to find. These cells are present throughout the proliferative phase and even into early secretion. After 10 to 11 days, irregular subnuclear vacuoles begin to appear. During the last 2 days, mitotic activity decreases, glandular coiling becomes more prominent, and vacuoles are easily found.

The interval period is the 48 hours between ovulation and the presence of uniformly vacuolated cells indicative of postovulatory day (POD) 2. Mitotic figures are present

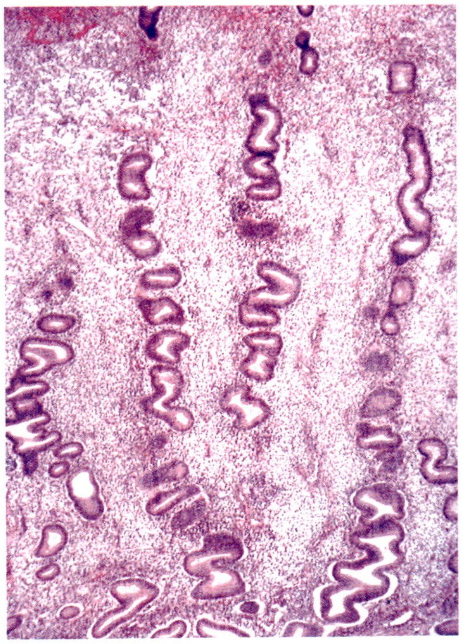

FIGURE 40.31 Midproliferative endometrium. Note the early coiling and synchronously developed glands.

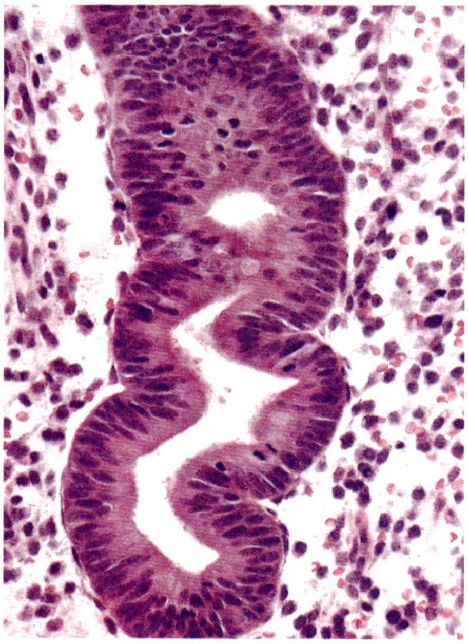

FIGURE 40.32 High-power view of a proliferative phase gland. The constituent cells are pseudostratified, and the characteristically elongated glandular nuclei have dense chromatin.

during this period, and the cells retain their proliferative nuclear features.

Endometrial Secretory Phase (Ovarian Luteal Phase)

Ovulation is mediated by the luteinizing hormone (LH) surge, a synchronous burst of LH and follicle-stimulating hormone secretion that peaks on the 14th day of a 28-day cycle. Ovulation occurs roughly 10 to 12 hours after this peak. The length of time from the last menstrual period to the day of ovulation for an individual woman is the length of her follicular phase. The surge of pituitary hormones initiates a complex series of events that results in release of the oocyte from the developed tertiary follicle and the transformation of the follicle into a corpus luteum, which now secretes estradiol and large quantities of progesterone. The biosynthetic lifetime of the corpus luteum defines the ovarian luteal phase that corresponds to endometrial secretory development. Both ovarian and endometrial luteal phases last 14 days on average, but this can be highly variable (see section on Relevance of Endometrial Dating to Diagnostic Surgical Pathologists below).

The endometrium, which has proliferated and been primed by estrogen, responds to the simultaneous stimulation of estrogen and progesterone by differentiating in a distinctive fashion. The morphologic changes can be divided into four periods: interval, early secretory, midsecretory, and late secretory (Fig. 40.30). The first 24 to 36 hours of the secretory phase are morphologically silent because the endometrium mostly retains its late proliferative appearance, although scattered nonuniform subnuclear vacuoles may appear. This morphologically indeterminate endometrium is termed "interval." The first unequivocal light microscopic indication that ovulation has occurred is the presence of uniform subnuclear vacuoles involving more than 50% of the endometrial glands. Recognizing this feature can aid the pathologist in excluding anovulatory cycling in an endometrial biopsy. Over the next few days, these vacuoles shift from a subnuclear to a supranuclear location. By the fifth POD, most of the secretions have been discharged into the gland lumen. The morphologic hallmark of the early secretory phase (PODs 2 to 5) is the vacuolated gland (Fig. 40.33).

The midsecretory phase lasts from PODs 5 to 9 and is characterized by nonvacuolated, prominently coiled secretory glands set within a spindled edematous stroma. Luminal secretion is most prominent during this period, and the overall appearance is one of glandular crowding (162). The gland-to-stroma ratio should not be mistaken for simple hyperplasia. The secretory cells usually have round, somewhat vesicular nuclei, which serves to separate them from the nuclei found in early secretory phase cells (Fig. 40.34). The distinctive feature of the late secretory endometrium (PODs 10 to 14) is stromal predecidualization. This diagnostic stromal change is

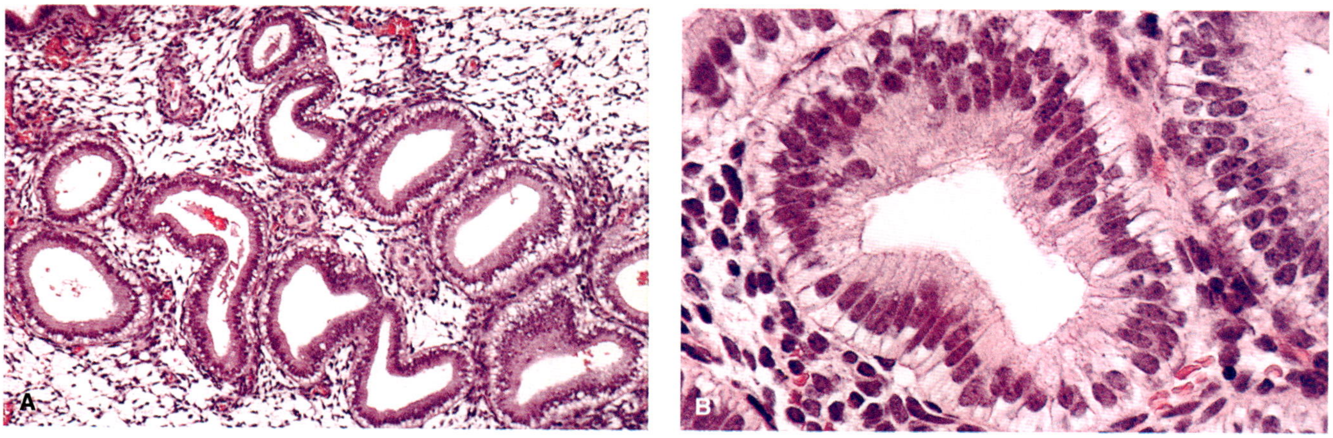

FIGURE 40.33 **A, B:** Early secretory endometrium with subnuclear vacuoles. The nuclei retain the dense chromatin of the proliferative phase. Uniformly present cytoplasmic vacuoles are the most useful marker of the first third of the secretory phase. Rare glandular mitotic figures may be seen.

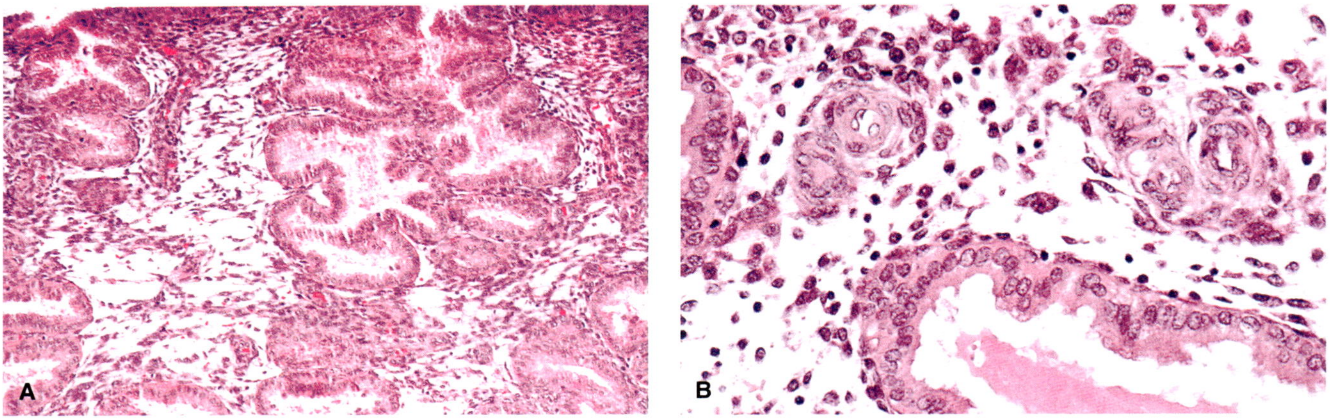

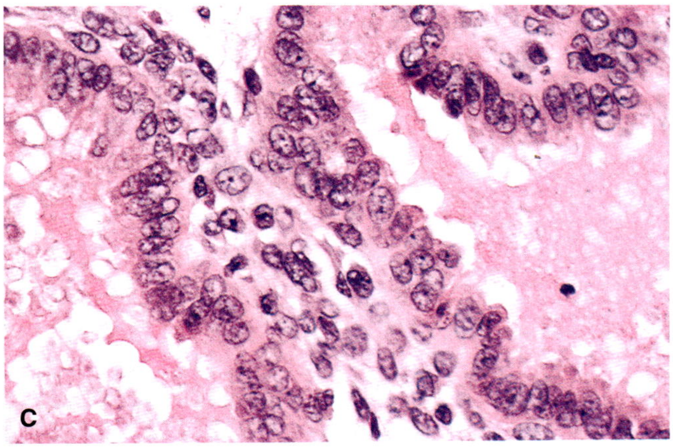

FIGURE 40.34 Midsecretory endometrium. The hallmarks of this period can be seen in these three photomicrographs. At low magnification (**A**) extreme glandular coiling and stromal edema are apparent. At a somewhat higher magnification (**B**), the coiled spiral arteries are seen within an edematous stroma. Perivascular predecidual reaction has not occurred. At yet higher magnification (**C**), the characteristically round vesicular nuclei of the midsecretory endometrium are apparent (contrast with proliferative phase nuclei in Fig. 40.23).

heralded by an increased prominence of the spiral arteries. By the 10th POD cuffs of predecidual cells are present around these arteries, initially involving the part of the vessel adjacent to the surface of the endometrium (Fig. 40.35). Subsequently, islands of predecidual cells appear in the superficial compactum. By POD 13, these islands become confluent. The extent of predecidualization is roughly paralleled by the degree of stromal infiltration by stromal granulocytes, although some investigators have suggested that the intensity of this infiltration is more closely correlated with the time of onset of menses (163). The appearance of the glands during the late secretory phase is not significantly different from that during the midsecretory phase. They are lined by nonvacuolated secretory cells with round vesicular nuclei. During the later days of this period, the glands typically have a saw-toothed appearance sometimes referred to as secretory exhaustion.

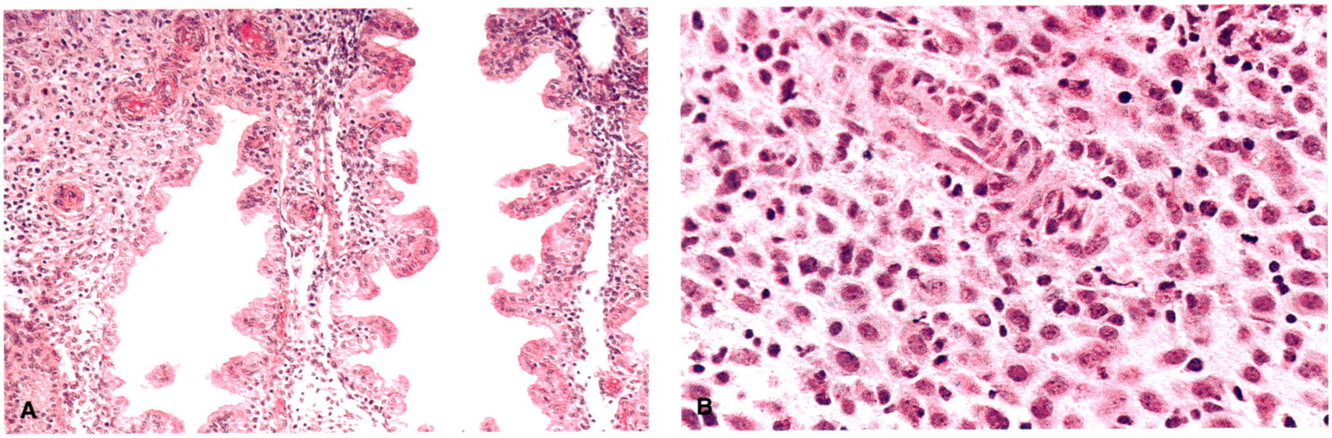

FIGURE 40.35 Late secretory endometrium. At low magnification (**A**), the serrated appearance of the gland reflects their coiled state. The stroma cells have undergone predecidual reaction. **B:** Predecidual reaction begins around the spiral arteries, and this reaction serves to distinguish midsecretory endometria from late secretory endometria.

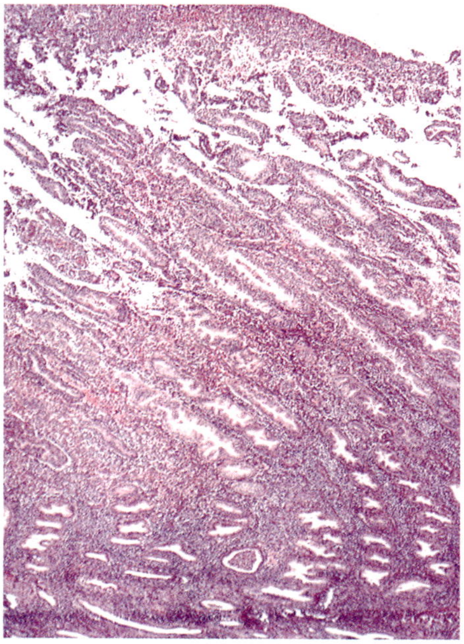

FIGURE 40.36 A low-power view of a menstrual endometrium showing the unresponsive basalis in the lower part of the photomicrograph and the functionalis disintegrating near the surface.

As the late secretory phase progresses, the cells become apoptotic and apoptotic bodies accumulate within stromal macrophages.

Menstrual Phase (Cycle Days 1 to 4)

The abrupt withdrawal of both estrogen and progesterone accompanying the demise of the corpus luteum initiates menstrual bleeding. The molecular events initiating and guiding the controlled bleeding of menstruation are complex in contrast to the relatively straightforward light microscopic picture (164). The endometrium on the first day of menstrual bleeding (cycle day 1) is thin and compact. It is composed of the basalis and a substantially shrunken, dense functionalis. The basalis maintains a relatively constant histologic appearance throughout the endometrial cycle. The functionalis thins mainly due to the withdrawal of interstitial fluid. The constituent glands and stroma of the functionalis fragment and crumble, often referred to as stromal breakdown. Fibrin thrombi appear in vessels and within the stroma. As the stroma disintegrates, the endometrial glands artificially crowd together, giving the appearance of hyperplasia. In addition, the degenerative atypia of these glands and the necrotic background may suggest a diagnosis of malignancy (Fig. 40.36). The general strategy of not making a diagnosis of malignancy when well-preserved glands and stroma are absent should prevent such a mistake (Fig. 40.37).

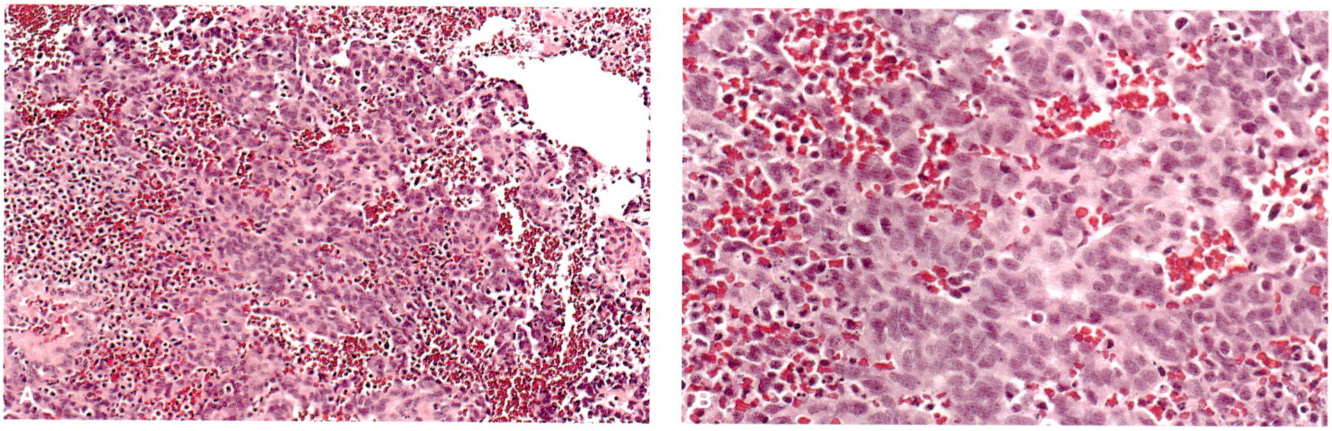

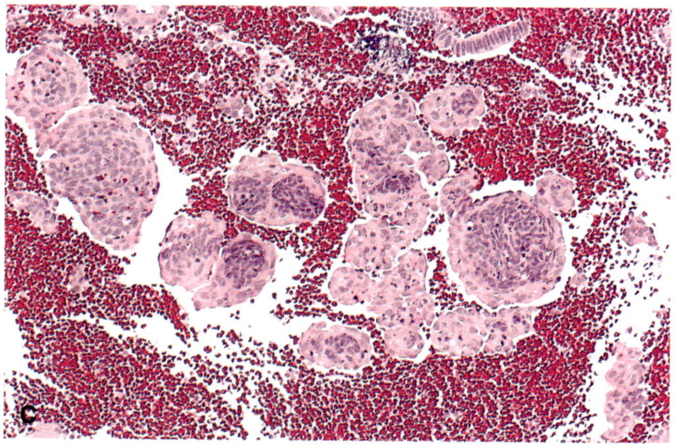

FIGURE 40.37 Stromal breakdown. Disintegrating endometrium may simulate endometrial malignancy. **A:** Sheets of disintegrating stromal cells may simulate an endometrial stromal neoplasm. **B:** Higher magnification shows individual cell necrosis and inflammation, findings that should raise the possibility of disintegrating nonneoplastic endometrium. **C:** Confirmation of this possibility is made by finding inflamed epithelium associated with degenerating stromal cells. In this case, characteristic epithelium-covered spherules are produced.

By the second or third day, the endometrium begins to repair and is completed by the fourth or fifth day. This early repair phase is thought by some to be estrogen independent as is the initial regrowth of the vasculature (165,166). Almost 50% of menstrual blood loss occurs on the first day in the majority of women (108,161,162).

Endometrial Morphology During the Luteal Phase of the Cycle of Conception

If implantation of a blastocyst occurs, it will be during the midsecretory phase (PODs 6 to 8), and this event is associated with a resurgence of glandular secretion and a persistence of stromal edema. After this time, biopsy findings will feature prominent glandular secretion, stromal edema, and stromal predecidualization (167,168).

The developing predecidual is gradually converted to decidua after POD 14 of the luteal phase of conception. This transformation is complete by the end of the first month of gestation. The fully developed decidual reaction is distinctive. Almost all of the endometrial stroma is converted into pavement-like sheets of epithelioid cells with prominent cytoplasmic margins and central vesicular nuclei (Fig. 40.26). The glands are compressed and their lining becomes flattened and endothelium-like in the superficial portion of the functionalis. This compact sheet-like zone, the zona compactum, overlies saw-toothed scalloped glands, the zona spongiosum, which continue to exhibit secretory features. Many of the glands in the spongiosum are lined by cells with enlarged nuclei with atypical nuclear features approaching those of the Arias-Stella reaction (Fig. 40.38). Scattered uterine NK cells are present. Decidual cells may exhibit substantial nuclear pleomorphism and cytologic atypia. This is particularly prominent in the region of the implantation site, which is also infiltrated by intermediate trophoblastic cells that normally have a bizarre cytologic appearance. These cells are positive for both keratin and human placental lactogen (hPL). The infiltration of decidua and the underlying myometrium by trophoblasts has been termed "syncytial metritis," not a particularly felicitous label because the process has little to do with inflammation (Fig. 40.39). This "implantation site reaction" is diagnostic of intrauterine pregnancy (169). Both placental site reaction and decidual atypia may incorrectly suggest malignancy, but confusion with adenocarcinoma is avoided by noting the secretory setting of these findings, as well as the clinical history.

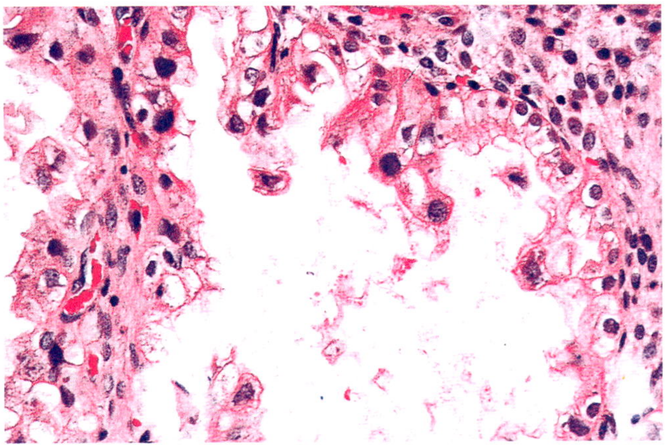

FIGURE 40.38 Gestational endometrium. This gland is lined by cells with enlarged dense nuclei characteristic of the Arias-Stella reaction.

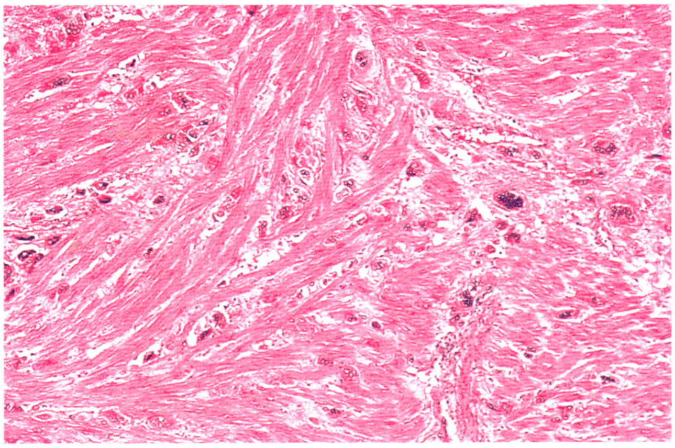

FIGURE 40.39 Myometrium beneath an implantation site containing infiltrating trophoblasts. This must not be misinterpreted as evidence of gestational trophoblastic disease.

The decidual reaction occurs in the setting of high progesterone stimulation and can also be seen in extrauterine (ectopic) pregnancy, iatrogenic use of progestational agents, or with persistence of the corpus luteum unassociated with pregnancy (i.e., corpus luteum cyst). For this reason, intrauterine implantation should only be diagnosed in the presence of chorionic villi or the presence of intermediate trophoblasts associated with enlarged vessels replaced by hyaline, or with fragments of fibrinoid matrix (169).

On rare occasions, the finding of trophoblastic tissue in a curetting does not exclude an ectopic gestation. Thus, a high clinical suspicion for ectopic pregnancy should not be negated by the presence of either chorionic villi or an implantation site in a curettage specimen (170).

Previous intrauterine pregnancy is strongly suggested in the postpartum endometrium by the presence of cuffs of hyalinized decidua around sclerotic ectatic spiral arteries. The decidual cells composing this cuff often have hyperchromatic smudged and degenerated nuclei, and the vessels are usually thrombosed. These findings have been referred to as subinvolution of the placental site and can be responsible for postpartum hemorrhage presumably because these severely altered vessels are incapable of contraction (171). The foci of hyalinization shrink over time and the nuclei may disappear, leaving a small pink scar resembling an ovarian corpus atreticum. These distinctive foci have been referred to as pregnancy plaques and may persist in the basalis for many years. Similar lesions featuring intermediate trophoblasts have been termed placental site nodule and placental site plaque (171). Occasionally, a placental site nodule or exaggerated placental site can be confused for a gestational trophoblastic tumor such as placental site trophoblastic tumor or epithelioid trophoblastic tumor. While conventional morphology is still very useful in distinguishing placental site nodules and plaques, immunohistochemical profiles can also be helpful. Studies over the last two decades have shown that there are a variety of intermediate trophoblasts involved in implantation and establishing vascular connections to the maternal blood flow. These trophoblasts seem to have distinct immunohistochemical properties, some of which are shared with the cells of gestational trophoblastic tumors. An exaggerated placental site, which is likely a normal variant of placental implantation, is characterized by infiltration of individual cells with preservation of normal structures and essentially no mitotic figures. Placental site trophoblastic tumors are characterized by mitotically active confluent masses of cells that disrupt normal architecture. All trophoblasts express keratins, so while expression may highlight the trophoblasts, it does little to predict the aggressive potential of the proliferation. Proposed algorithms for evaluating placental site nodules and trophoblastic tumors utilize a combination of hPL, p63, and Ki-67. Briefly, the intermediate trophoblasts of an exaggerated placental site are hPL-positive, p63-negative, and have a Ki-67 labeling index of less than 1%. A placental site trophoblastic tumor will have the same immunohistochemical profile with a Ki-67 approximating 10%. The intermediate trophoblasts in a placental site nodule are hPL-negative, p63-positive, and have a Ki-67 labeling index of less than 10%. Epithelioid trophoblastic tumor can sometimes be confused with a placental site nodule but has a Ki-67 labeling index greater than 10%, higher cellularity, and frequent calcifications. Occasionally, the diagnostic dilemma includes placental site nodule and squamous cell carcinoma; inhibin and cytokeratin 18 can be useful as placental site nodule is positive for both of these antibodies while squamous cell carcinomas are negative (172–174).

Hyperprogestational states are sometimes associated with the distinctive glandular change to which Arias-Stella first drew attention in 1954 and that now bears his name (175,176). This change is usually encountered in endometrial glands, but may be present in foci of endometriosis or adenomyosis, in endocervical glands, in fallopian tube epithelium, or in the glands within polyps. The Arias-Stella phenomenon characteristically involves a focus on tightly packed glands whose extreme coiling and collapse throw the lining epithelium into prominent papillary folds. This epithelium is composed of cells exhibiting marked nuclear pleomorphism and hyperchromatism. The nuclei typically have a smudged appearance. The cell cytoplasm may be strikingly hypervacuolated and cleared (clear cells) or densely eosinophilic (dark cells) (Fig. 40.38). One or the other cell type may predominate from area to area. Occasionally, the eosinophilic cells line the glands in a hobnail fashion. Mitotic figures are only rarely present. Elsewhere, the endometrium exhibits the other changes one would anticipate with progestational stimulation, such as secretory glands and a stromal decidual reaction.

An endometrial Arias-Stella change may be present in a variety of clinical settings, including normal intrauterine pregnancy, extrauterine pregnancy, gestational trophoblastic disease, and persistent corpus luteum. It also may be produced by the administration of ovulation-inducing drugs or progestational agents (175).

Because Arias-Stella phenomenon often features marked nuclear atypia of the epithelium lining the closely packed glands, it can be confused with adenocarcinoma, particularly clear cell carcinoma. This difficulty largely can be avoided by remembering that the Arias-Stella phenomenon occurs in a secretory setting; that is, more conventional secretory glands and decidualized or predecidualized stroma are usually found elsewhere in the specimen and the patient is premenopausal. Moreover, glandular mitotic figures are a prominent feature, although they may occasionally be encountered (177). Clear cell carcinoma is fundamentally a proliferative process, and the constituent cells not only possess malignant features but also exhibit mitotic activity. Most importantly, clear cell carcinoma of the endometrium develops almost exclusively in postmenopausal women. Immunochemistry can be useful in separating the two as Ki-67 staining is absent in Arias-Stella phenomenon and present in clear cell carcinoma; p53 is wild-type in Arias-Stella phenomenon but is abnormal in nearly half of all clear cell carcinoma cases (178).

Glandular nuclei in a gestational setting may occasionally exhibit marked nuclear clearing reminiscent of herpetic infections (Fig. 40.40) (179).

Endometrial Vasculature

The transition from the endometrium of early gestation to that of fully developed pregnancy is marked by the accelerated development of the endometrial vasculature, resulting in increased thickness of the spiral arteries, as reflected in their mean cross-sectional diameter (180,181).

› *PERIMENOPAUSAL AND POSTMENOPAUSAL YEARS* With the waning of hormonal function in the fifth decade of life, a woman enters the perimenopause, during which time

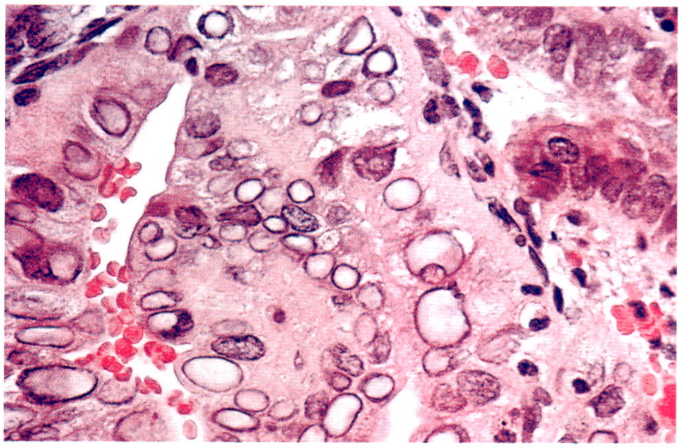

FIGURE 40.40 Nuclear clearing of the glandular cells in the endometrium may be seen in a gestational setting. This change should not be misconstrued as evidence of herpes virus infection.

uterine bleeding characteristically again becomes erratic and the length of time between bleeding episodes lengthens. Thus, both the perimenopause and the perimenarche may be marked by erratic ovarian function and consequent dysfunctional uterine bleeding (182).

The end of ovarian follicular development and ovulation results in cessation of the menstrual periods and the menopause. Thereafter, the uterus enters a second inactive period, and the endometrial glandular epithelium, as in the premenarchal years, is typically atrophic. However, the glandular architecture and thickness of the endometrium may vary considerably. Several different patterns may be seen in the peri- and postmenopausal endometrium, which are described in the following sections.

Atrophic Endometrium

Atrophic epithelium is nonstratified and composed of a single layer of flattened to cuboidal cells. Mitotic figures are not present, the nucleus-to-cytoplasm ratio is high, and there is usually no specific cytoplasmic differentiation, although cilia may be present. The defining and only constant feature of atrophic endometria is the atrophic epithelial lining of its constituent glands, which may have any configuration, including cystic dilatation and glandular crowding. The stroma is spindled. The nuclei may be densely pyknotic or plump. Stromal disintegration may be present (Fig. 40.41) (183–185).

Atrophic endometria can occur in a variety of clinical settings. It is the normal state during the premenarchal and later postmenopausal years. Atrophic patterns during the reproductive years may be seen in association with premature ovarian failure or, more commonly, in patients taking hormonal contraceptives.

Weakly Proliferative Endometrium

Weakly proliferative epithelium is nonstratified, although some degree of nuclear pseudostratification may be present, and its constituent cells are thin. In contrast to normal proliferative endometrium, atrophic endometrial cells have a paucity of mitotic figures, denser nuclear chromatin, and more disorganized glands. The glands may be of any configuration, but the glands-to-stroma ratio is almost always near unity or there can be a slight stromal predominance. The stroma is spindled; stromal cell nuclei may be densely pyknotic or plump.

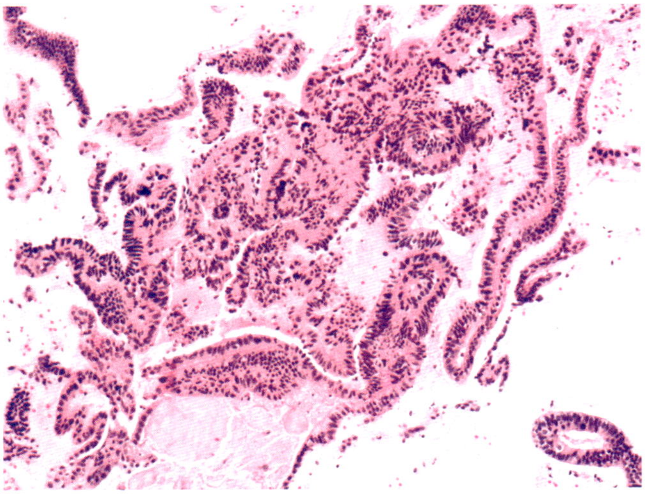

FIGURE 40.41 Endometrial atrophy as seen in an endometrial sampling. On occasion, atrophic surface epithelium may be removed in coiled masses and simulate hyperplasia because the glands are closely approximated.

The morphologic variability of the weakly proliferative endometrium parallels that of the atrophic endometrium. The difference between them is solely based on the appearance of the epithelial cells: Weakly proliferative rather than flattened or cuboidal. The clinical settings in which these two endometria occur overlap considerably and the weakly proliferative patterns are a histologic transition between normal proliferation and atrophy. Weakly proliferative endometria is most often encountered in patients in the peri- or postmenopausal years whose endometria appears to be weakly supported by low levels of endogenous or exogenous estrogen. This pattern is normal in the hormonally hyporesponsive lower uterine segment of the normally cycling premenopausal women.

During the perimenopausal period, some women are prone to chronic stromal breakdown due to the inconsistent hormonal milieu and anovulatory cycles. This can result in surface metaplasias such as eosinophilic or papillary metaplasia. The former is characterized by enlarged cells with abundant pink cytoplasm with conspicuous nucleoli. The latter often forms small papillary aggregates which layer the surface epithelium but can be detached and free-floating in an endometrial sampling. When atypia is present in these cells, this benign process can be mistaken for serous (papillary) carcinoma or its precursor serous endometrial intraepithelial carcinoma. In this setting, Ki-67 and p53 can be useful adjuvant tests as metaplasia has a low Ki-67 labeling index and is usually p53 wild-type while serous carcinoma demonstrates an elevated Ki-67 proliferation index accompanied by an p53 aberrant immunophenotype. While this panel will not be confirmatory in all cases, it can prove helpful in confirming the histologic impression (186).

Anovulatory/"Disordered Proliferative" Endometrium

"Disordered" proliferation is the morphologic bridge spanning normally proliferating endometrium and endometrial hyperplasia. It is the endometrial pattern encountered in women experiencing sporadic anovulatory cycles and thus is most commonly encountered in the perimenopausal and perimenarchal years and may be seen in women receiving estrogen therapy. It should be noted that "anovulatory" does not mean a complete cessation of ovulatory activity, but instead a relative decrease in ovulatory frequency. Anovulatory endometrium differs from normal by virtue of a loss of synchrony of glandular development so that some glands are tubular, whereas others are cystically dilated or have complex shapes. Epithelial budding and continued epithelial growth may lead to a shift in the gland-to-stroma ratio in

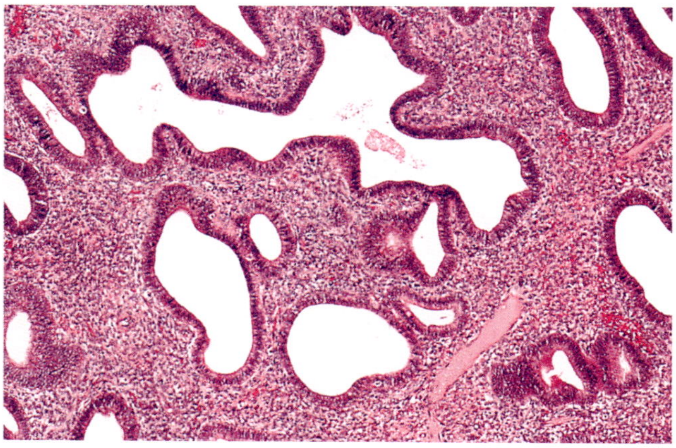

FIGURE 40.42 The anovulatory pattern of endometrium is normal during the perimenopausal years and is characterized by nonsynchronous growth of the glands including budding, but the glands-to-stroma ratio is unity or with a slight glandular predominance. This pattern of endometrial growth serves as a bridge between normal proliferation and hyperplasia.

favor of the glands, but this shift is usually focal and never more than 2:1. It is common in the anovulatory pattern to have focal glandular crowding but when the glandular predominance is so marked that diffusely there is a shift in the glands to stroma ratio (3:1 in favor of the glands), a diagnosis of hyperplasia is warranted (Fig. 40.42).

Histologically, the variously shaped glands are lined by normal proliferative cells with elongated dense nuclei that are most often pseudostratified. Mitotic figures are usually present and may be numerous. Tubal metaplasia is usually present and most pronounced in the cystically dilated glands. Stromal cells are spindled with plump nuclei. There is no evidence that this pattern is a marker for an increased risk for the subsequent development of endometrial carcinoma.

› *Relevance of Endometrial Dating to Diagnostic Surgical Pathologists* Due to the very sophisticated ability to measure serum steroid and gonadotropin levels, the endometrial biopsy no longer plays the significant role it used to in the evaluation of the infertility patient. Setting aside organic causes of infertility, the endometrial biopsy, functioning as a bioassay, is unreliable in separating fertile from infertile women. This is due largely to the substantial morphologic heterogeneity of the normally functioning endometrium (187–192).

On the very rare occasion, an endometrial biopsy is obtained for fertility evaluation, its chief function is to determine whether ovulation has occurred and document anovulatory cycles (which would point toward an ovarian factor in the patient's infertility). Organic reproductive tract disease may be detected on the endometrial biopsy such as endometrial polyps, leiomyomas, endometritis, hyperplasia, or carcinoma (192–194).

What follows is an abbreviated version of endometrial dating for a reference and for those rare occasions a pathologist is asked to provide a fine-tuned morphologic date (i.e., more than proliferative, early, mid, or late secretory). The precise details of the morphologic patterns corresponding to standard cycle dates are presented in Table 40.2 and Figures 40.29 and 40.30.

Some general points are important to keep in mind when using these aids.

Morphologic Dates and Chronologic Dates

In evaluating the endometrium, it is important to carefully distinguish between the morphologic POD assigned to a morphologically normal endometrium and the chronologic POD (112,152,195,196). The morphologic date is a summary characterization of the histologic development of the endometrium based on an assessment of glandular and stromal features. The morphologic findings may be summarized either in terms of PODs, cycle days, or phases. For example, the morphologic pattern associated with a particular "standard" POD is assigned the number of that day; for example, POD 12 refers to the pattern seen on POD 12 of the "standard" cycle. Equivalently, this morphologic information can be conveyed using cycle day or phase (e.g., late secretory).

Endometrial Dating Should Be Reported as a Range of 2 Days

It is now known that both the follicular and luteal phases vary among women and not all women have a standard 28-day cycle (197). Adding to this biologic heterogeneity is the poor interobserver reproducibility of the conventional Noyes dating.

Noyes and Haman and, subsequently, others reported poor agreement on single-day morphologic date designation of an endometrium and significantly better agreement for a 2-day designation (154,198–201). As a consequence, the results of endometrial dating should be reported as a range of 2 days to reflect this variability.

Determining Whether the Patient Has Ovulated

During the late luteal phase, the presence of a late secretory endometrium means ovulation. Deciding whether ovulation has occurred in the early luteal phase is more problematic. The characteristic light microscopic postovulatory secretory changes in the endometrium lag behind actual ovulation by at least 1 day. A biopsy on POD 2 usually reveals subnuclear vacuoles in the vast majority of glandular cells in the superficial functionalis but these may be spotty (interval pattern). Such vacuolization may be seen in late proliferative endometria, as well as in other types of nonsecretory endometria, and is, for this reason, not diagnostic of ovulation. Because of the ambiguous morphologic picture in the early postovulatory period, endometrial biopsies are usually obtained well into the presumed luteal phase of the cycle, preferably on PODs 11 to 13. We require uniform subnuclear vacuolization in 50% of the glands before diagnosing the earliest morphologic evidence of ovulation, which is usually present by POD 3.

Some Practical Details

Accurate endometrial dating requires attention to a number of details (195). Dating features should be sought only in fragments of endometrial functionalis, which are lined by

TABLE 40.3 The Contents of the Pathology Report for an Infertility Endometrial Biopsy

Statement of the adequacy of the biopsy
Suggest re-biopsy if:
Sample is inadequate
The endometrium is disintegrating and not clearly menstrual (i.e., postovulatory)
The endometrium is disintegrating, postovulatory but otherwise undatable and the main concern is with the adequacy of the luteal phase
Statement about evidence of ovulation
Statement about pattern uniformity
An estimate of the endometrium's morphologic date (2-day interval)
Correlation with information about the chronologic date
Statement about presence or absence of "organic disease"
Inflammation
Endometrial polyps
Submucosal leiomyomas

surface epithelium; fragments of basalis and lower uterine segment should be ignored. The assigned date should be of the most developed area and should be based on features near the surface epithelium (195).

The preconditions for assigning a morphologic date to an endometrium are described as follows. Assigning a date to an endometrial biopsy is obviously not possible in the absence of an adequate specimen, in a noncycling endometrium (e.g., nonsecretory pattern other than proliferative), in a patient being treated with medication that alters the morphology of the endometrium, or in an endometrium that is inflamed or that houses an intrauterine device. Thus, a precondition for applying the dating criteria is that one is dealing with a roughly normal endometrial pattern. Scanning power examination should establish the presence of pattern uniformity (apart from the expected variation due to sectioning randomly oriented fragments and normal out-of-phase curettage inhabitants such as fragments of the lower uterine segment, cervical epithelium, and stratum basalis), the absence of significant budding and branching of glands, and the absence of necrosis and inflammation. Examination at higher power should exclude the presence of significant epithelial nuclear atypia and the absence of significant numbers of stromal plasma cells.

A checklist for the contents of the pathology report of endometrial biopsies is provided in Table 40.3.

Endometrial–Myometrial Junction

The endometrial–myometrial junction is normally irregular, a phenomenon that must be kept in mind when assessing the presence or absence of myoinvasion by endometrial adenocarcinoma and in diagnosing superficial adenomyosis (Fig. 40.22).

Apoptosis and the Endometrium

The role of programmed cell death or apoptosis in the remodeling of the endometrium was mentioned earlier in this chapter. Apoptosis and its hormonal control in both human and animal endometrium is an area of active investigation by a number of groups (202–205).

Withdrawal of progesterone in the rabbit endometrium correlates with the development of apoptosis (206–209).

Bcl-2 is a proto-oncogene initially described in the (14,18) translocation in follicular lymphoma. It has been shown to prolong cell survival by preventing apoptosis. Gompel et al. studied Bcl-2 expression immunohistochemically in the normal endometrium (210). They found that Bcl-2 predominated in glandular cells and reached a maximum at the end of the follicular phase but disappeared at the onset of secretory activity. Maia et al. found a precipitous drop in Bcl-2 after ovulation, further supporting an increase in Bcl-2 in response to estrogen and a decrease in response to progesterone (211). In addition, this group found an elevated p53 expression in the proliferative phase but a drop in the late luteal phase. These results strongly suggest hormone-dependent regulation of Bcl-2 expression and cell cycling.

The role of tumor necrosis factor-alpha (TNF-α) in the induction of apoptosis was explored by Tabibzadeh et al. (212). They found that the TNF receptor as well as Fas protein were expressed in endometrial epithelium throughout the entire menstrual cycle and were most prominent in the basalis. They concluded that endometrial epithelium, by expressing receptors for TNF-α and Fas protein can respond to ligands that regulate apoptosis.

Normal findings in the endometrium that have relevance to histopathologic differential diagnosis are shown in Table 40.4.

MYOMETRIUM

The bulk of the myometrium comprises smooth muscle cells, but an important contribution is made by extracellular components such as collagen and elastin. The smooth muscle within the corpus is more concentrated relative to collagen and elastin than the muscle in either the cervix or the lower uterine segment. This distribution of muscle is consistent with the passive role of the cervix during parturition; the uterine contents propelled by fundal contractions are thought to passively dilate a cervix previously softened by the action of collagenase. The uterine smooth muscle cells are spindled, with blunt-ended fusiform nuclei. Their cytoplasmic volume depends on the cycle and whether the patient is pregnant (214–217). Scattered normal mitotic figures may be encountered, particularly during the secretory phase of the endometrial cycle (217,218). Characteristic ultrastructural features of smooth muscle include (a) numerous dense, 60- to 80-A myofilaments without cross striations, which almost fill the cytoplasm; (b) small round dense bodies along the trajectory of the filaments; (c) dense plaques arranged along the inner aspect of the plasma membrane; and (d) plasma membrane–related vesicles that

TABLE 40.4 Normal Findings in the Endometrium That Have Relevance to Histopathologic Differential Diagnosis

Findings	Diagnostic Confusion	Suggestions for Resolution	References When Relevant
"Glands within glands" this can occur from trauma of D&C (telescoping) and tangential cutting	**Endometrial hyperplasia:** Rolled-up carpet effect may simulate the architectural complexity of hyperplasia or carcinoma	This artifact is usually only focal and often at the periphery or otherwise normal fragments of functionalis. Check that epithelium is identical to architecturally normal glands in the sample.	
Strips of normal surface epithelium (commonly seen with atrophic endometrium)	**Endometrial Hyperplasia/ Carcinoma:** Rolled-up carpet effect may simulate the architectural complexity of hyperplasia or carcinoma	In contrast to hyperplasia and carcinoma, the rolled-up atrophic endometrium is cytologically bland and mitotically inactive.	(165,185)
Lower uterine segment	**Dating problem:** The glands respond weakly at best to hormones and the stroma is often fibrous	Look for mucinous epithelial component. Unless the sample is inadequate for dating the endometrium, there will be other fragments around with an appearance more typical of functionalis.	(189)
Normal basalis	**Dating problem:** The glands respond weakly at best to hormones and the stroma is inactive	When assigning a morphologic date to an endometrial fragment, use only glands in close approximation to surface epithelium	
Ciliated cells	**Adenocarcinoma:** Intraepithelial ciliated cells have rounded pear-shaped contours and open rounded nuclei with nucleoli	Look for cilia. Endometria-containing ciliated cells have a polymorphous appearance; the cytologic features are uniform within each subpopulation of cells.	
Foam cells	**Endometritis:** Foam cells are commonly encountered in hypoestrogenic settings and are thought to be one differentiated form of endometrial stromal cell	Look for plasma cells to establish the diagnosis of chronic endometritis. Macrophages (xanthoma cells) with foamy cytoplasm may occur in the endometrium as a component of a chronic inflammatory process. Commonly seen as a reaction to a foreign body or to keratin (e.g., in setting of proliferations shedding keratin).	(135,213)
Arias-Stella reaction	**Adenocarcinoma:** Particularly clear cell carcinoma	Clear cell carcinoma is most frequently encountered in postmenopausal women. Arias-Stella reaction usually does not feature mitotic figures, which are almost always easily found in clear cell carcinoma. A more difficult problem is distinguishing clear cell carcinoma from Arias-Stella reaction in a postmenopausal woman who is taking progestational medication.	(174,176,177)
Implantation site	**Gestational trophoblastic disease:** Early gestations may mimic choriocarcinoma. The often bizarre shapes of trophoblastic cells may raise the possibility of choriocarcinoma, a molar gestation, or PSTT	Clinical history to establish time course of gestation. Insistence on a bilaminar pattern of cytotrophoblast alternating with syncytiotrophoblast for choriocarcinoma; villi are not present in choriocarcinoma. PSTT features.	
	Clinically relevant endometritis: A mixed plasmacytic–lymphocytic infiltrate is a normal finding at the implantation site	Clinically relevant gestational endomyometritis can occur but is only rarely diagnosed initially on endometrial samplings. The vast majority of cases of "chronic endometritis" seen in practice represent a clinically insignificant "physiologic" finding.	(135,136,213)
Irregular endometrial–myometrial junction	**Adenomyosis:** Glands and stroma surrounded by smooth muscle	Associated smooth muscle hypertrophy is a common feature of adenomyosis and not irregular junctions. Superficial myometrial location of suspected foci favors irregular junctions.	
	Myoinvasive adenocarcinoma: The irregular junction, when expanded by hyperplasia or carcinoma, often simulates myoinvasive carcinoma. Tangential sectioning often reinforces this impression	Absence of stromal host response that is usual in myoinvasive carcinoma. Presence of endometrial stroma or normal glands means irregular junction (or adenomyosis).	

TABLE 40.4 Normal Findings in the Endometrium That Have Relevance to Histopathologic Differential Diagnosis (*Continued*)

Findings	Diagnostic Confusion	Suggestions for Resolution	References When Relevant
Menstrual endometrium	**Carcinoma:** "Cytologic atypia" produced by degenerating, poorly preserved glandular elements **Endometrial stromal neoplasm:** Degenerating stroma may produce a pattern resembling endometrial stromal sarcoma with or without sex cord elements	Obtain history; menstrual fragments have degenerating not neoplastic nuclei.	
Lymphoid nodules	**Endometritis**	The presence of scattered lymphoid nodules does not correlate with significant clinical disease. Large numbers of lymphoid nodules and germinal centers are usually associated with the conventional sine qua non of "chronic endometritis," the plasma cell.	
Stromal "granulocytes"	**Acute or chronic endometritis**	"Stromal granulocytes" or uterine NK cells are a normal feature of the late secretory endometrium and serve as one of the original Noyes dating criteria (see text). Clinically significant inflammation features easily found plasma cells and is usually accompanied by acute inflammation and necrosis of the surface epithelium. Endometrial stromal cells or normal lymphoid constituents may have somewhat eccentric nuclei and amphophilic cytoplasm that mimics that of a plasma cell. This is usually a problem with one or two scattered cells in a fragment. Plasma cells have to be easily found.	(135)
Ground-glass appearance of nuclei of epithelial cells in gestational setting and with progestogen therapy	**Herpes endometritis**	Herpes endometritis is rare; obtain history. Immunohistochemistry may be helpful.	(179)
Appearance of gestational endometrial stroma	**Decidualized stroma of progestogen therapy**	Clinical history (intrauterine device or medication vs. pregnancy or postpartum state).	

See text for additional differential diagnostic clues.

may play a role in ionized calcium movement across the plasma membrane during contraction. In addition, there is the usual complement of cytoplasmic organelles, including smooth and rough endoplasmic reticulum, mitochondria, and a Golgi apparatus. Typically, these organelles arrange themselves around the nucleus, which has an irregular shape. These ultrastructural features reflect the dual function of the uterine myocyte: muscular contraction and collagen and elastin synthesis. The ultrastructural appearance varies with the levels of circulating steroidal hormones. In particular, estrogen appears to sharply increase myocyte protein synthesis. This correlates morphologically with an increased volume of rough endoplasmic reticulum and increased numbers of cytoplasmic contractile elements. The biochemistry and electrophysiology of the myometrium have been extensively reviewed (217,219,220). The histologic and ultrastructural appearance of smooth muscle cells differ substantially from those of the endometrial stromal cell. These differences are set out in Table 40.5. However, it should be noted that cells with a hybrid smooth muscle–stromal phenotype occur normally at the endometrial–myometrial junction and that this phenotypic ambiguity is sometimes expressed by spindle cell neoplasms of the uterine corpus (221). Some uterine smooth muscle cells have been shown to express some classes of keratins (222–225). The immunohistochemical profile of myometrial cells is presented in Table 40.5.

Pregnancy-Related Changes

To accommodate the growing fetus and to prepare for its role in fetal expulsion, the uterus undergoes a 10-fold increase in size and weight during pregnancy both by hypertrophy and, to a much lesser extent, hyperplasia. Normal mitotic figures are often increased and may be present in large numbers. Uterine growth during pregnancy appears to be largely promoted by estradiol, whereas progesterone probably functions to inhibit uterine contractions during gestation. The light microscopic appearance of the hypertrophied uterine smooth muscle cells of pregnancy is distinctive. They are enlarged and have abundant, rather glassy cytoplasm and vesicular elongated nuclei with occasionally prominent

TABLE 40.5 Comparison of Differentiated Features for Endometrial Stromal Cells and Smooth Muscle Cells

Technique	Endometrial Stromal Cell	Usual Smooth Muscle Cell	Epithelioid Smooth Muscle Cells	Endometrial Stroma With Epithelioid and/or Glandular Areas
Light microscopy Architectural features	Haphazardly arranged cells resembling normal proliferative phase endometrial stromal cells	Cells arranged in looping intersecting fascicles	Rounded or polygonal cells with moderate amount of cytoplasm	Biphasic pattern Stromal component (featuring cellular endometrial stroma or fibroblastic stroma) + Epithelioid component • Trabecular epithelial pattern or • nests or • insular pattern or • Sertoli-like tubular pattern or • glands
	Complex plexiform vascular pattern Hyaline sometimes abundant with a tendency to be glassy	Vascular component not complex	Foci of neoplasm often exhibit standard smooth muscle features	When the entire neoplasm has this biphasic appearance the term "uterine tumor resembling ovarian sex cord tumors" has been used When the change is focal within what is otherwise an endometrial stromal neoplasm (either stromal nodule or endometrial stromal sarcoma), the endometrial stromal diagnosis is supplemented with "epithelioid or glandular areas"
Nuclear cytologic features	Blunt, round-to-fusiform, uniform, bland	Elongate, cigar-shaped	Round, crumpled	Small, round, and regular nuclei • Minimal nuclear pleomorphism • Only rare mitotic figures
Cytoplasm	Scanty (on H&E and trichrome)	Moderate amount (on H&E and trichrome), typically fibrillar	Cytoplasm may be clear around the nucleus, clear at the periphery of the cell, or entirely clear	Scanty cytoplasm Or May have abundant eosinophilic or foamy (lipid-rich) cytoplasm
Immunohistochemistry	Normal endometrial stromal cells express CD10 and may express desmin; they are negative for cytokeratin and EMA	Uterine smooth muscle cells express desmin and caldesmon. They also express CD10 (approximately 40%), CD34 (about a third), keratin (about 20%), and EMA (roughly half)	AE1 is positive in 40% of cases, desmin is positive in 80%, and CD34 in 10%	Epithelioid areas Muscle-specific actin (HHF-35) positive Vimentin positive Cytokeratin positive
References	(221,223)	(222,225)	(223,225)	(219)

nucleoli. Changes occur in the ultrastructural appearance of the smooth muscle cells as well. In addition to an increase in size and the number of myofilaments, there is a striking increase in the number of gap junctions (109,226). These establish the contact between cells required for the coordinated uterine contractions that expel the term infant (227). These myometrial changes are closely coordinated with the dramatic structural changes of the cervix required for cervical effacement. In the postpartum period, the uterus undergoes an extraordinary 85% reduction in weight within 3 weeks of delivery (25). This weight loss is primarily due to a reduction in individual cell volume rather than a reduction in cell number. In addition, a large amount of collagen is degraded over this brief period. Complete return of the uterus to the nulliparous weight does not occur if gestation has proceeded beyond the second trimester.

THE FALLOPIAN TUBE

Histology of the Fallopian Tube

The fallopian tube is lined by a nonstratified epithelium that is separated from the endosalpingeal stroma by a basement membrane. Each of the tubal segments is lined by three basic cell types: ciliated cells, secretory cells, and intercalated (peg) cells (Figs. 40.43 and 40.44). In recent years it has become apparent that the peg cell is, in reality, a stage in the cyclic variation during the menstrual cycle of the secretory cell (228). The relative number of these cells differs in each of the anatomic regions of the tube and the variable numbers of each cell type account for the varied histology. In addition, many investigators believe that the numbers of the three types of cells within each of these anatomic regions undergo regular variations throughout the menstrual cycle (109,225–234). Ciliated cells are most prominent at the ovarian (distal) end of the tube—particularly in the fimbrial mucosa—and predominate during midcycle; their numbers diminish progressively to achieve a nadir at the time of menstruation (Fig. 40.44). The number of cilia continues to decrease in a gestational cycle. Ciliary movement, rather than muscular contractions, is chiefly responsible for the movement of the egg toward the site of fertilization at the ampulloisthmic junction.

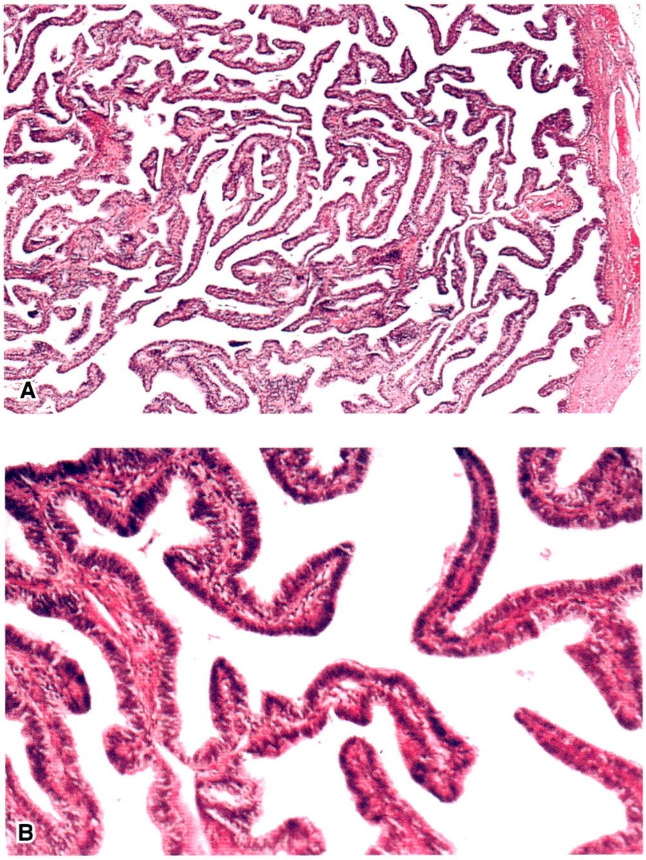

FIGURE 40.43 **A, B:** Ampulla of the fallopian tube. Note the long slender plicae or folds resting on the muscularis.

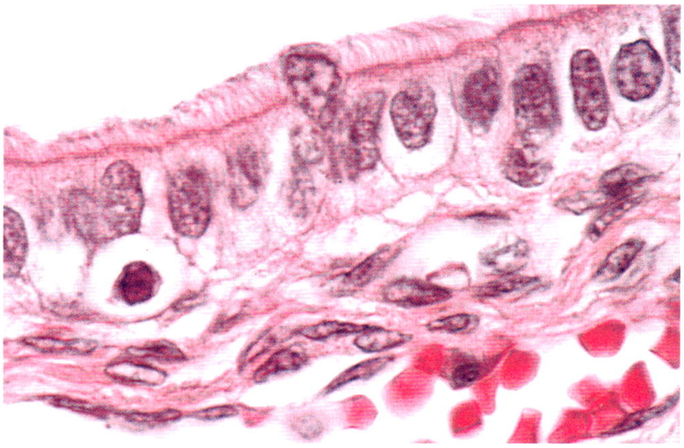

FIGURE 40.44 This high-power photomicrograph of tubal epithelium shows numerous ciliated cells with a compressed secretory cell nucleus above the level of the ciliated cells. The cell with clear cytoplasm is probably a lymphocyte.

Secretory cells are most prominent toward the uterine end of the tube and undergo cyclic changes in cell height and appearance, reflecting their elaboration, accumulation, and discharge of oviduct secretions as the menstrual cycle proceeds. Most often they have ovoid somewhat dense nuclei, and they may contain an apical vacuole (Fig. 40.45). The oviduct fluid secreted by these cells serves many important functions and has been the subject of a review (235).

Intercalated cells have been thought to represent either effete secretory cells or some type of reserve cells. They have a thin dense nucleus and little cytoplasm. Endocrine cells have been noted in the fallopian tube; their function is as mysterious here as it is in the uterus (236). The "basal cells" reported in the early literature have been shown to be lymphocytes, which may represent a tubal component of a mucosa-associated lymphoid system (237–241). Scattered lymphocytes and occasional lymphoid follicles should be considered within normal limits and constitute part of the mucosa-associated lymphoid system (242).

Ciliogenesis is promoted by estradiol and deciliation by progesterone. Prolonged exposure to progestogens (whether endogenously as in pregnancy or exogenously) or withdrawal of estrogen (as in the postmenopausal years) leads to epithelial atrophy. Postmenopausal estrogen administration leads to regrowth of cilia.

Mitotic figures are rarely seen in the fallopian tube epithelium, so no cyclic regeneration occurs as in the endometrium. Both the transmission and scanning electron microscopic appearance of the normal tubal mucosa have been extensively documented over the past three decades. The plica becomes blunted in the postmenopausal state secondary to contraction of the lining and decreases in epithelial cells.

Many metaplasias including mucinous, endometrioid, and transitional (Walthard rests) variants can arise in the fallopian tube and are benign findings.

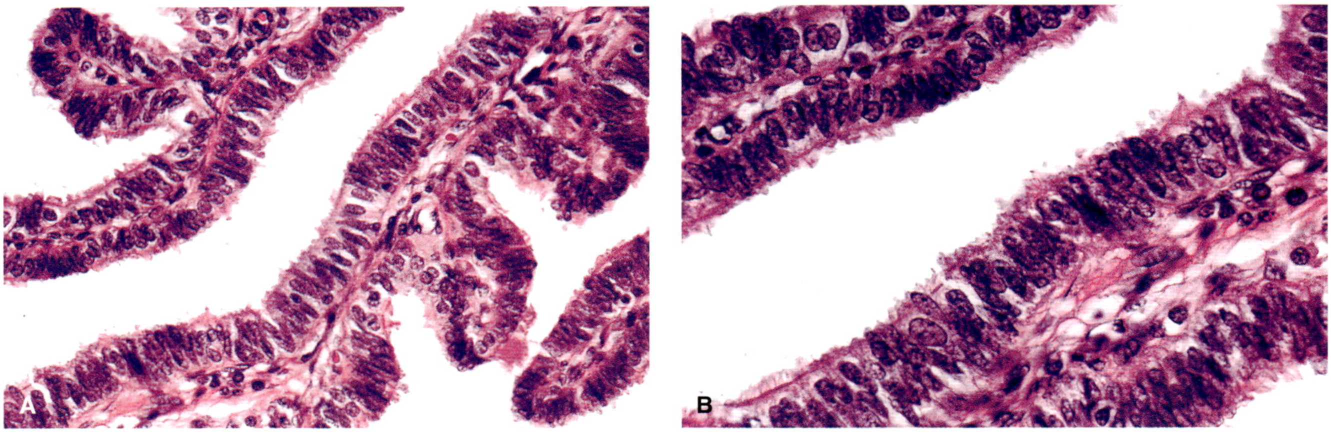

FIGURE 40.45 **A, B:** In this area, the tubal epithelial cells are crowded, a pattern that is common in the fallopian tube when ciliated cells are not numerous.

Of interest to the diagnostic pathologist are the abnormalities of ciliogenesis found in patients with Kartagener syndrome (243,244).

With the advent of genetic testing for *BRCA* 1 and 2 mutations and an increased understanding of the fallopian tube as a nidus for serous carcinoma of the ovary and peritoneum, prophylactic salpingo-oophorectomies are being performed. Although there are no inherent morphologic changes in the epithelium of a fallopian tube from a woman who harbors a *BRCA* 1 or 2 mutation, she is at an increased risk of having an in-situ or invasive carcinoma. Sectioning and extensively examining the fimbriated end, via the "SEE-FIM" (Sectioning and Extensively Examining the Fimbriated End) protocol, is recommended for any patient with a family history of ovarian cancer or known genetic predisposition such as *BRCA-1* or *BRCA-2* mutation. Accordingly, the fallopian tube is entirely submitted in multiple cross-sections at 3-mm intervals except for the distal 2 cm of the fimbriated end, which is cut radially/longitudinally (245). Serous tubal intraepithelial carcinoma (STIC) and serous tubal epithelial proliferations of uncertain significance, will arise in otherwise normal fallopian tubes and have some overlapping features with reactive atypia; recognizing this process is critical to proper patient management. For detailed evidence and categorization, the reader is directed to several excellent review articles (Fig. 40.46) (246,247).

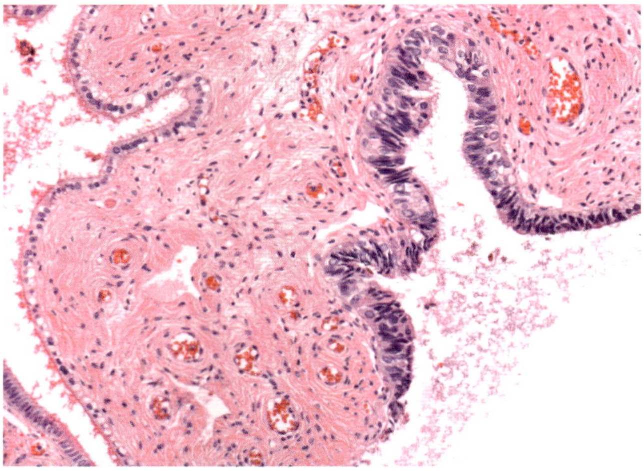

FIGURE 40.46 Fallopian tube with an area of atypia. Evaluation of p53 and Ki-67 should be considered in lesions such as this. The significance of p53 wild-type tubal atypia is currently uncertain.

Myosalpinx

The myosalpinx is composed of an inner circular layer and an outer longitudinal layer. The isthmus near the uterotubal junction also possesses an inner longitudinal layer. The presence of the muscular layer can be particularly helpful to the pathologist when trying to distinguish a tubo-ovarian complex from an ovarian serous neoplasm.

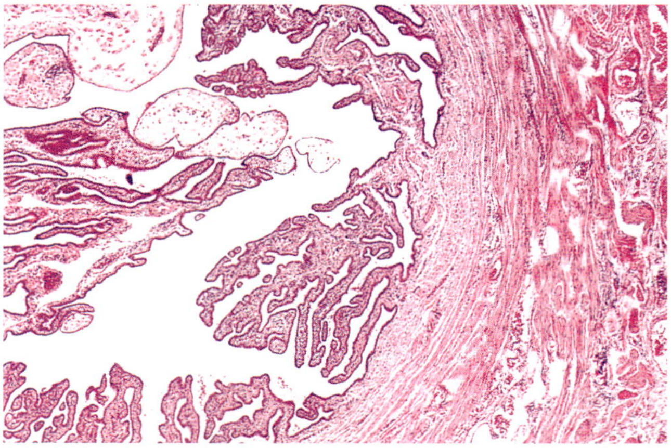

FIGURE 40.47 Fallopian tube containing decidual cells. The stromal cells of the plicae frequently undergo decidual change during pregnancy.

Fallopian Tube in Pregnancy

The fallopian tube has already played its part when the fertilized ovum implants in the endometrium. It is now inactive

TABLE 40.6 Normal Findings in the Fallopian Tube and Uterine Serosa That Have Relevance to Histopathologic Differential Diagnosis

Finding	Diagnostic Confusion	Suggestions for Resolution	References When Relevant
Decidual reaction (usually encountered in postpartum tubal ligation specimens but may also be seen in the tubes of patients on progestational agents)	Do not misinterpret as carcinoma	Nuclei are bland in decidua.	(251,252)
Endosalpingiosis	Distinguish from endometriosis and ovarian serous tumor of low malignant potential (borderline tumors)	Look for endometrial stroma to confirm diagnosis of endometriosis; look for complex papillae with micropapillae to confirm diagnosis of serous borderline tumor.	
Crowded cells and dense nuclei	STIC	STIC features crowding, nuclear pleomorphism, prominent nucleoli, and (abnormal) mitotic figures; p53 is aberrant (mutant) and Ki-67 is elevated. Nuclear atypia and glandular complexity can also be significant in chronic salpingitis but p53 is wild type.	
Mucinous and eosinophilic metaplasia (may be associated with Peutz–Jeghers syndrome)	STIC	Metaplasia lacks the architectural complexity, the nuclear atypia, and the mitotic activity of STIC; immunohistochemistry for p53 and Ki-67 can help	
Squamous metaplasia Walthard rests	STIC	Metaplasia lacks the architectural complexity, the nuclear atypia, and the mitotic activity of STIC; immunohistochemistry for p53 and Ki-67 can help	

See text for additional differential diagnostic clues.

throughout the gestational period. A muted version of endometrial decidual change often occurs in the endosalpingeal stroma during the latter part of pregnancy, while the epithelium of the fallopian tube undergoes atrophy (Fig. 40.47) (243). Arias-Stella reaction can occasionally occur. The fallopian tube is the most common site of ectopic gestation. A review of physiologic factors in its development has been presented (248). Morphologic changes in the fallopian tube can be produced by birth control pills and of course tubal ligation (249–251).

Normal findings in the fallopian tubes that have relevance to histopathologic diagnosis can be found in Table 40.6.

Paraovarian and Paratubal Structures

The broad ligament, round ligament, and mesosalpinx form a network of support to the pelvic organs. These environs are populated by a variety of tubular and cystic structures with the propensity to form clinically or surgically noticeable cysts (38,253). Many are lined by müllerian-type epithelium. Walthard rests are universal findings over the serosal surface of the fallopian tubes, and are lined by transitional-type epithelium. A more or less constant finding in sections that include the peritubal soft tissue is the tortuous remnants of the mesonephric ducts. These are lined by cuboidal epithelium and possess a fibromyovascular cuff.

REFERENCES

1. Lauchlan SC. Metaplasias and neoplasias of Müllerian epithelium. *Histopathology*. 1984;8:543–557.
2. Lauchlan SC. The secondary müllerian system revisited. *Int J Gynecol Pathol*. 1994;13:73–79.
3. Moore K. *The Developing Human. Clinically Oriented Embryology*. W.B. Saunders Company; 2003;207–221.
4. Acién P. Embryological observations on the female genital tract. *Hum Reprod*. 1992;7:437–445.
5. O'Rahilly R. *Prenatal Human Development*. Plenum Medical Book Company; 1989:35–56.
6. Jost A, Vigier B, Prépin J, Perchellet JP. Studies on sex differentiation in mammals. *Recent Prog Horm Res*. 1973;29:1–41.
7. Gondos B. Development of the reproductive organs. *Ann Clin Lab Sci*. 1985;15:363–373.
8. Szamborski J, Laskowska H. Some observations on the developmental histology of the human foetal uterus. *Biol Neonat*. 1968;13:298–314.
9. Gray CA, Bartol FF, Tarleton BJ, et al. Developmental biology of uterine glands. *Biol Reprod*. 2001;65:1311–1323.
10. Ramsey ED. Development of the human uterus and relevance to the adult condition. In: Chard T, Grudzinskas J, eds. *The Uterus*. Cambridge; 1994:41–53.
11. McLean J. Embryology and anatomy of the female genital tract. In: Fox H, Wells M, eds. *Haines and Taylor Obstetrical and Gynaecological Pathology*. Churchill Livingstone; 2002:1–40.

12. Ramsey E. Embryology and developmental defects of the female reproductive tract. In: Danforth D, Scott J, eds. *Obstetrics and Gynecology*. JB Lippincott; 1986: 106–119.
13. Strauss JF III, Barbieri RL. The structure, function, and evaluation of the female reproductive tract. In: Strauss JF III, Barbieri RL, eds. *Yen and Jaffe's Reproductive Endocrinology. Physiology, Pathophysiology, and Clinical Management*. Elsevier Saunders; 2004:255–306.
14. Patton G, Kistner R. *Atlas of Infertility Surgery*. Little, Brown and Company; 1984.
15. Tulac S, Nayak NR, Kao LC, et al. Identification, characterization, and regulation of the canonical Wnt signaling pathway in human endometrium. *J Clin Endocrinol Metab*. 2003;88:3860–3866.
16. Haber HP, Mayer EI. Ultrasound evaluation of uterine and ovarian size from birth to puberty. *Pediatr Radiol*. 1994; 24:11–13.
17. Nussbaum A, Sanders R, Jones M. Neonatal uterine morphology as seen on real-time US. *Radiology*. 1986;160:641–643.
18. Singer A, Chow C. Anatomy of the cervix and physiological changes in cervical epithelium. In: Fox H, Wells M, eds. *Haines and Taylor Obstetrical and Gynaecological Pathology*. Churchill Livingstone; 2002:247–272.
19. Eddy CA, Pauerstein CJ. Anatomy and physiology of the fallopian tube. *Clin Obstet Gynecol*. 1980;23:1177–1193.
20. Gray H, Williams PL, Warwick R, et al. *Gray's Anatomy*. Churchill Livingstone; 1989.
21. Langlois P. The size of the normal uterus. *J Reprod Med*. 1970;4:220–228.
22. Calder A. The cervix during pregnancy. In: Chard T, Grudzinskas J, eds. *The Uterus*. Cambridge; 1994:288–307.
23. Kurz K, Tadesse E, Haspels A. In vivo measurements of uterine cavities in 795 women of fertile age. *Contraception*. 1984;29:495–510.
24. Zemlyn S. The length of the uterine cervix and its significance. *Clin Ultrasound*. 1981;9:267–269.
25. Finn C, Porter D. *The Uterus*. Publishing Sciences Group; 1975.
26. Togashi K, Nakai A, Sugimura K. Anatomy and physiology of the female pelvis. MR imaging revisited. *J Magn Reson Imaging*. 2001;13:842–849.
27. Hoad CL, Raine-Fenning NJ, Fulford J, Campbell BK, Johnson IR, Gowland PA. Uterine tissue development in healthy women during the normal menstrual cycle and investigations with magnetic resonance imaging. *Am J Obstet Gynecol*. 2005; 192:648–654.
28. Toth A. Studies on the muscular structure of the human uterus. II. Fasciculi cervicoangulares: vestigial or functional remnant of the mesonephric duct? *Obstet Gynecol*. 1977;49:190–196.
29. Toth S, Toth A. Undescribed muscle bundle of the human uterus: fasciculus cervicoangularis. *Am J Obstet Gynecol*. 1974;118:979–984.
30. Huszar G, Naftolin F. The myometrium and uterine cervix in normal and preterm labor. *N Engl J Med*. 1984;311:571–581.
31. Merchant RN, Prabhu SR, Chougale A. Uterotubal junction–morphology and clinical aspects. *Int J Fertil*. 1983;28:199–205.
32. Vizza E, Correr S, Muglia U, Marchiolli F, Motta PM. The three-dimensional organization of the smooth musculature in the ampulla of the human fallopian tube: a new morpho-functional model. *Hum Reprod*. 1995;10:2400–2405.
33. Croxatto HB. Physiology of gamete and embryo transport through the fallopian tube. *Reprod Biomed Online*. 2002;4: 160–169.
34. Talbot P, Geiske C, Knoll M. Oocyte pickup by the mammalian oviduct. *Mol Biol Cell*. 1999;10:5–8.
35. Gordts S, Campo R, Rombauts L, Brosens I. Endoscopic visualization of the process of fimbrial ovum retrieval in the human. *Hum Reprod*. 1998;13:1425–1428.
36. Greiss FJ Jr, Rose JC. Vascular physiology of the nonpregnant uterus. In: Wynn R, Jollie W, eds. *Biology of the Uterus*. Plenum; 1989:69–88.
37. Ramsey E. Vascular anatomy. In: Wynn R, Jollie W, eds. *Biology of the Uterus*. Plenum; 1989:58–68.
38. Agur AMR, Dalley AF. *Grant's Atlas of Anatomy*. 12th ed. Lippincott Williams and Wilkins; 2009:244.
39. DiSaia PJ, Creasman WT. *Clinical Gynecologic Oncology*. Mosby; 1997:viii, 657.
40. Major FJ, Blessing JA, Silverberg SG, et al. Prognostic factors in early-stage uterine sarcoma. A Gynecologic Oncology Group study. *Cancer*. 1992;71(4 suppl):1702–1709.
41. Plentl A, Friedman EA. Lymphatic system of the female genitalia. *The Morphologic Basis of Oncologic Diagnosis and Therapy*. WB Saunders; 1971.
42. Klein M, Rosen A, Lahousen M, et al. Lymphogenous metastasis in the primary carcinoma of the fallopian tube. *Gynecol Oncol*. 1994;55:336–338.
43. Klein M, Rosen AC, Lahousen M, Graf AH, Rainer A. Lymphadenectomy in primary carcinoma of the Fallopian tube. *Cancer Lett*. 1999;147:63–66.
44. Wright TC, Ferenczy A. Anatomy and histology of the uterine cervix. In: Kurman RJ, ed. *Blaustein's Pathology of the Female Genital Tract*. Springer-Verlag; 2002:207–224.
45. Aspden RM. The importance of a slit-like lumen cross-section for the mechanical function of the cervix. *Br J Obstet Gynaecol*. 1987;94:915–916.
46. Gorodeski G. The cervical cycle. In: Adashi E, Rock J, Rosenwaks Z, eds. *Reproductive Endocrinology, Surgery, and Technology*. Lippincott-Raven; 1996:302–324.
47. Gipson IK. Mucins of the human endocervix. *Front Biosci*. 2001;6:D1245–D1255.
48. Konishi I, Fujii S, Nonogaki H, Nanbu Y, Iwai T, Mori T. Immunohistochemical analysis of estrogen receptors, progesterone receptors, Ki-67 antigen, and human papillomavirus DNA in normal and neoplastic epithelium of the uterine cervix. *Cancer*. 1991;68:1340–1350.
49. Darragh TM, Colgan TJ, Cox TJ, et al. The lower anogenital squamous terminology standardization project for HPV-associated lesions: background and consensus recommendations from the College of American Pathologists and the American Society for Colposcopy and Cervical Pathology. *Int J Gynecol Pathol*. 2013;32(1):76–115.
50. Mills AM, Dirks DC, Poulter MD, Mills SE, Stoler MH. HR-HPV E6/E7 mRNA in situ hybridization: validation against PCR, DNA in situ hybridization and p16 immunohistochemistry in 102 samples of cervical vulvar, anal, and head and neck neoplasia. *Am J Surg Pathol*. 2017;41(5):607–615.
51. Albores-Saavedra J, Gersell D, Gilks CB, et al. Terminology of endocrine tumors of the uterine cervix: results of a workshop sponsored by the College of American Pathologists and the National Cancer Institute. *Arch Pathol Lab Med*. 1997;121:34–39.

52. Fetissof F, Arbeille B, Boivin F, Sam-Giao M, Henrion C, Lansac J. Endocrine cells in ectocervical epithelium. An immunohistochemical and ultrastructural analysis. *Virchows Arch A Pathol Anat Histopathol*. 1987;411:293–298.
53. Fetissof F, Berger G, Dubois MP, et al. Endocrine cells in the female genital tract. *Histopathology*. 1985;9:133–145.
54. Fetissof F, Dubois MP, Heitz PU, Lansac J, Arbeille-Brassart B, Jobard P. Endocrine cells in the female genital tract. *Int J Gynecol Pathol*. 1986;5:75–87.
55. Fetissof F, Heitzman A, Machet MC, Lansac J. Unusual endocervical lesions with endocrine cells. *Pathol Res Pract*. 1993;189:928–939.
56. Fetissof F, Serres G, Arbeille B, de Muret A, Sam-Giao M, Lansac J. Argyrophilic cells and ectocervical epithelium. *Int J Gynecol Pathol*. 1991;10:177–190.
57. Scully R, Aguirre P, DeLellis R. Argyrophilia, serotonin, and peptide hormones in the female genital tract and its tumors. *Int J Gynecol Pathol*. 1984;3:51–70.
58. Chan JKC, Tsui WMS, Tung SY, Ching RC. Endocrine cell hyperplasia of the uterine cervix. A precursor of neuroendocrine carcinoma of the cervix? *Am J Clin Pathol*. 1989;92:825–830.
59. Hussain LA, Kelly CG, Fellowes R, et al. Expression and gene transcript of Fc receptors for IgG, HLA class II antigens and Langerhans cells in human cervico-vaginal epithelium. *Clin Exp Immunol*. 1992;90:530–538.
60. Morelli AE, Di Paola G, Fainboim L. Density and distribution of Langerhans cells in the human uterine cervix. *Arch Gynecol Obstet*. 1992;252:65–71.
61. Osamura RY, Watanabe K, Oh M. Melanin-containing cells in the uterine cervix: histochemical and electron-microscopic studies of two cases. *Am J Clin Pathol*. 1980;74:239–242.
62. Hiersche HD, Nagl W. Regeneration of secretory epithelium in the human endocervix. *Arch Gynecol*. 1980;229:83–90.
63. Gould PR, Barter RA, Papadimitriou JM. An ultrastructural, cytochemical, and autoradiographic study of the mucous membrane of the human cervical canal with reference to subcolumnar basal cells. *Am J Pathol*. 1979;95:1–16.
64. Ismail SM. Cone biopsy causes cervical endometriosis and tubo-endometrioid metaplasia. *Histopathology*. 1991;18:107–114.
65. Jonasson JG, Wang HH, Antonioli DA, Ducatman BS. Tubal metaplasia of the uterine cervix: a prevalence study in patients with gynecologic pathologic findings. *Int J Gynecol Pathol*. 1992;11:89–95.
66. Novotny DB, Maygarden SJ, Johnson DE, Frable WJ. Tubal metaplasia. A frequent potential pitfall in the cytologic diagnosis of endocervical glandular dysplasia on cervical smears. *Acta Cytol*. 1992;36:1–10.
67. Pacey F, Ayer B, Greenberg M. The cytologic diagnosis of adenocarcinoma in situ of the cervix uteri and related lesions. III. Pitfalls in diagnosis. *Acta Cytol*. 1988;32:325–330.
68. Suh KS, Silverberg SG. Tubal metaplasia of the uterine cervix. *Int J Gynecol Pathol*. 1990;9:122–128.
69. Marques T, Andrade LA, Vassallo J. Endocervical tubal metaplasia and adenocarcinoma in situ: role of immunohistochemistry for carcinoembryonic antigen and vimentin in differential diagnosis. *Histopathology*. 1996;28:549–550.
70. Peters WM. Nature of "basal" and "reserve" cells in oviductal and cervical epithelium in man. *J Clin Pathol*. 1986;39:306–312.
71. Fluhmann C. *The Cervix Uteri and Its Diseases*. WB Saunders; 1961.
72. Fluhmann CF. The nature and development of the so-called glands of the cervix uteri. *Am J Obstet Gynecol*. 1957;74:753–766; discussion 766–768.
73. Young RH, Clement PB. Pseudoneoplastic glandular lesions of the uterine cervix. *Semin Diagn Pathol*. 1991;8:234–249.
74. Clement PB, Young RH. Deep nabothian cysts of the uterine cervix. A possible source of confusion with minimal-deviation adenocarcinoma (adenoma malignum). *Int J Gynecol Pathol*. 1989;8:340–348.
75. Teshima S, Shimosato Y, Kishi K, Kasamatsu T, Ohmi K, Uei Y. Early stage adenocarcinoma of the uterine cervix. Histopathologic analysis with consideration of histogenesis. *Cancer*. 1985;56:167–172.
76. Bertrand M, Lickrish GM, Colgan TJ. The anatomic distribution of cervical adenocarcinoma in situ: implications for treatment. *Am J Obstet Gynecol*. 1987;157:21–25.
77. Gilks CB, Reid PE, Clement PB, Owen DA. Histochemical changes in cervical mucus-secreting epithelium during the normal menstrual cycle. *Fertil Steril*. 1989;51:286–291.
78. Nucci MR. Symposium part III: tumor-like glandular lesions of the uterine cervix. *Int J Gynecol Pathol*. 2002;21:347–359.
79. Lagow E, DeSouza MM, Carson DD. Mammalian reproductive tract mucins. *Hum Reprod Update*. 1999;5:280–292.
80. McDonnell JM, Emens JM, Jordan JA. The congenital cervicovaginal transformation zone in sexually active young women. *Br J Obstet Gynaecol*. 1984;91:580–584.
81. Burch DJ, Spowart KJ, Jesinger DK, Randall S, Smith SK. A dose-ranging study of the use of cyclical dydrogesterone with continuous 17 beta oestradiol. *Br J Obstet Gynaecol*. 1995;102:243–248.
82. Forsberg J. Cervicovaginal epithelium: its origin and development. *Am J Obstet Gynecol*. 1973;115:1025–1043.
83. Crum CP, Egawa K, Fu YS, et al. Atypical immature metaplasia (AIM). A subset of human papilloma virus infection of the cervix. *Cancer*. 1983;51:2214–2219.
84. Duggan MA. Cytologic and histologic diagnosis and significance of controversial squamous lesions of the uterine cervix. *Mod Pathol*. 2000;13:252–260.
85. Herfs M, Yamamoto Y, Laury Y, et al. A discrete population of squamocolumnar junction cells implicated in the pathogenesis of cervical cancer. *Proc Natl Acad Sci U S A*. 2012;109(26):10516–10521.
86. Yang E, Quick M, Hanamornroongruang S, et al. Microanatomy of the cervical and anorectal squamocolumnar junctions; a proposed model for anatomical differences in HPV-related cancer risk. *Mod Pathol*. 2015;28(7):994–1000.
87. Aspden RM. Collagen organisation in the cervix and its relation to mechanical function. *Coll Relat Res*. 1988;8:103–112.
88. Kiwi R, Neuman MR, Merkatz IR, Selim MA, Lysikiewicz A. Determination of the elastic properties of the cervix. *Obstet Gynecol*. 1988;71:568–574.
89. Leppert PC, Cerreta JM, Mandl I. Orientation of elastic fibers in the human cervix. *Am J Obstet Gynecol*. 1986;155:219–224.
90. Leppert PC, Yu SY. Three-dimensional structures of uterine elastic fibers: scanning electron microscopic studies. *Connect Tissue Res*. 1991;27:15–31.
91. Johansson EL, Rudin A, Wassen L, Holmgren J. Distribution of lymphocytes and adhesion molecules in human cervix and vagina. *Immunology*. 1999;96:272–277.

92. Miller CJ, McChesney M, Moore PF. Langerhans cells, macrophages and lymphocyte subsets in the cervix and vagina of rhesus macaques. *Lab Invest*. 1992;67:628–634.
93. Edwards JN, Morris HB. Langerhans' cells and lymphocyte subsets in the female genital tract. *Br J Obstet Gynaecol*. 1985;92:974–982.
94. Hughes RG, Norval M, Howie SE. Expression of major histocompatibility class II antigens by Langerhans' cells in cervical intraepithelial neoplasia. *J Clin Pathol*. 1988;41:253–259.
95. Roncalli M, Sideri M, Gie P, Servida E. Immunophenotypic analysis of the transformation zone of human cervix. *Lab Invest*. 1988;58:141–149.
96. Ferry JA, Scully RE. Mesonephric remnants, hyperplasia, and neoplasia in the uterine cervix: a study of 49 cases. *Am J Surg Pathol*. 1990;14:1100–1111.
97. Seidman JD, Tavassoli FA. Mesonephric hyperplasia of the uterine cervix: a clinicopathologic study of 51 cases. *Int J Gynecol Pathol*. 1995;14:293–299.
98. Oliva E. CD10 expression in the female genital tract: does it have useful diagnostic applications? *Adv Anat Pathol*. 2004;11:310–315.
99. McCluggage WG, Oliva E, Herrington CS, McBride H, Young RH. CD10 and calretinin staining of endocervical glandular lesions, endocervical stroma and endometrioid adenocarcinomas of the uterine corpus: CD10 positivity is characteristic of, but not specific for, mesonephric lesions and is not specific for endometrial stroma. *Histopathology*. 2003;43:144–150.
100. Nucci MR, Young RH, Fletcher CD. Cellular pseudosarcomatous fibroepithelial stromal polyps of the lower female genital tract: an underrecognized lesion often misdiagnosed as sarcoma. *Am J Surg Pathol*. 2000;24:231–240.
101. Abdul-Karim FW, Cohen RE. Atypical stromal cells of lower female genital tract. *Histopathology*. 1990;17:249–253.
102. Clement PB. Multinucleated stromal giant cells of the uterine cervix. *Arch Pathol Lab Med*. 1985;109:200–202.
103. Metze K, Andrade LA. Atypical stromal giant cells of cervix uteri–evidence of Schwann cell origin. *Pathol Res Pract*. 1991;187:1031–1035; discussion 1036–1038.
104. Ledger WL, Anderson ABM. The influence of steroid hormones on the uterine cervix during pregnancy. *J Steroid Biochem*. 1987;27:1029–1034.
105. Leppert PC. Anatomy and physiology of cervical ripening. *Clin Obstet Gynecol*. 1995;38:267–279.
106. Pisharodi LR, Jovanoska S. Spectrum of cytologic changes in pregnancy. A review of 100 abnormal cervicovaginal smears, with emphasis on diagnostic pitfalls. *Acta Cytol*. 1995;39:905–908.
107. Kaspar HG, Crum CP. The utility of immunohistochemistry in the differential diagnosis of gynecologic disorders. *Arch Pathol Lab Med*. 2015;139:39–54.
108. Nucci MR, Young RH. Arias-Stella reaction of the endocervix: a report of 18 cases with emphasis on its varied histology and differential diagnosis. *Am J Surg Pathol*. 2004;28:608–612.
109. Clement PB, Young RH, Scully RE. Nontrophoblastic pathology of the female genital tract and peritoneum associated with pregnancy. *Semin Diagn Pathol*. 1989;6:372–406.
110. Oliva E, Clement PB, Young RH. Tubal and tubo-endometrioid metaplasia of the uterine cervix: unemphasized features that may cause problems in differential diagnosis: a report of 25 cases. *Am J Clin Pathol*. 1995;103:618–623.
111. Manganiello PD, Burrows LJ, Dain BJ, Gonzalez J. Vabra aspirator and pipelle endometrial suction curette. A comparison. *J Reprod Med*. 1998;43:889–892.
112. Mutter GL, Ferenczy A. Anatomy and histology of the uterine corpus. In: Kurman RJ, ed. *Blaustein's Pathology of the Female Genital Tract*. Springer-Verlag; 2002:383–420.
113. Giudice LC, Ferenczy A. The endometrial cycle. In: Adashi EY, Rock JA, Rosenwaks Z, eds. *Reproductive Endocrinology, Surgery, and Technology*. Lippincott-Raven; 1996:272–300.
114. Buckley CH. Normal endometrium and non-proliferative conditions of the endometrium. In: Fox H, Wells M, eds. *Haines and Taylor Obstetrical and Gynaecological Pathology*. Churchill Livingstone; 2002:391–442.
115. Warren M, Li T, Klentzeris L. Cell biology of the endometrium: histology, cell types and menstrual changes. In: Chard T, Grudzinskas J, eds. *The Uterus*. Cambridge; 1994:94–124.
116. Cooper JM, Erickson ML. Endometrial sampling techniques in the diagnosis of abnormal uterine bleeding. *Obstet Gynecol Clin North Am*. 2000;27:235–244.
117. Chambers JT, Chambers SK. Endometrial sampling: when? where? why? with what? *Clin Obstet Gynecol*. 1992;35:28–39.
118. Mihm LM, Quick VA, Brumfield JA, Connors AF Jr, Finnerty JJ. The accuracy of endometrial biopsy and saline sonohysterography in the determination of the cause of abnormal uterine bleeding. *Am J Obstet Gynecol*. 2002;186:858–860.
119. Revel A, Shushan A. Investigation of the infertile couple: hysteroscopy with endometrial biopsy is the gold standard investigation for abnormal uterine bleeding. *Hum Reprod*. 2002;17:1947–1949.
120. Tahir MM, Bigrigg MA, Browning JJ, Brookes ST, Smith PA. A randomised controlled trial comparing transvaginal ultrasound, outpatient hysteroscopy and endometrial biopsy with inpatient hysteroscopy and curettage. *Br J Obstet Gynaecol*. 1999;106:1259–1264.
121. Ben-Yehuda OM, Kim YB, Leuchter RS. Does hysteroscopy improve upon the sensitivity of dilatation and curettage in the diagnosis of endometrial hyperplasia or carcinoma? *Gynecol Oncol*. 1998;68:4–7.
122. Hill GA, Herbert CM III, Parker RA, Wentz AC. Comparison of late luteal phase endometrial biopsies using the Novak curette or PIPELLE endometrial suction curette. *Obstet Gynecol*. 1989;73:443–445.
123. Ferenczy A, Bergeron C. Histology of the human endometrium: from birth to senescence. *Ann N Y Acad Sci*. 1991;622:6–27.
124. Denholm R, More I. Atypical cilia of the human endometrial epithelium. *J Anat*. 1980;131:309–315.
125. Comer MT, Andrew AC, Leese HJ, Trejdosiewicz LK, Southgate J. Application of a marker of ciliated epithelial cells to gynaecological pathology. *J Clin Pathol*. 1999;52:355–357.
126. Kearns M, Lala P. Life history of decidual cells: a review. *Am J Reprod Immunol*. 1983;3:78–82.
127. Iwahashi M, Muragaki Y, Ooshima A, Yamoto M, Nakano R. Alterations in distribution and composition of the extracellular matrix during decidualization of the human endometrium. *J Reprod Fertil*. 1996;108:147–155.
128. Speroff L, Fritz MA. The uterus. In: Speroff L, Fritz MA, eds. *Clinical Gynecologic Endocrinology and Infertility*. Lippincott Williams & Wilkins; 2004:113–144.
129. Whitelaw PF, Croy BA. Granulated lymphocytes of pregnancy. *Placenta*. 1996;17:533–543.

130. Bulmer JN, Lash GE. Human uterine natural killer cells: a reappraisal. *Mol Immunol*. 2005;42:511–521.
131. King A. Uterine leukocytes and decidualization. *Hum Reprod Update*. 2000;6:28–36.
132. Kayisli UA, Guzeloglu-Kayisli O, Arici A. Endocrine-immune interactions in human endometrium. *Ann N Y Acad Sci*. 2004;1034:50–63.
133. Gaynor LM, Colucci F. Uterine natural killer cells: functional distinctions and influence on pregnancy in humans and mice. *Front Immunol*. 2017;8:467.
134. Givan AL, White HD, Stern JE, et al. Flow cytometric analysis of leukocytes in the human female reproductive tract: comparison of fallopian tube, uterus, cervix, and vagina. *Am J Reprod Immunol*. 1997;38:350–359.
135. Tabibzadeh S. Proliferative activity of lymphoid cells in human endometrium throughout the menstrual cycle. *J Clin Endocrinol Metab*. 1990;70:437–443.
136. Kiviat N, Wolner-Hanssen P, Eschenbach D, et al. Endometrial histopathology in patients with culture-proved upper genital tract infection and laparoscopically diagnosed acute salpingitis. *Am J Surg Pathol*. 1990;14:167–175.
137. Ramsey E. *Vascular Anatomy*. Plenum Press; 1977:59–76.
138. Ramsey E. Anatomy of the uterus. In: Chard T, Grudzinskas J, eds. *The Uterus*. Cambridge University Press; 1994: 18–40.
139. Gargett CE, Rogers PAW. Human endometrial angiogenesis. *Reproduction*. 2001;121:181–186.
140. Taylor RN, Lebovic DI, Hornung D, Mueller MD. Endocrine and paracrine regulation of endometrial angiogenesis. *Ann N Y Acad Sci*. 2001;943:109–121.
141. Rees MC, Bicknell R. Angiogenesis in the endometrium. *Angiogenesis*. 1998;2:29–35.
142. Albrecht ED, Pepe GJ. Steroid hormone regulation of angiogenesis in the primate endometrium. *Front Biosci*. 2003;8:416–429.
143. Anin SA, Vince G, Quenby S. Trophoblast invasion. *Hum Fertil Camb*. 2004;7:169–174.
144. Lyall F. Priming and remodelling of human placental bed spiral arteries during pregnancy–a review. *Placenta*. 2005; 26(suppl A):S31–S36.
145. Spornitz UM. The functional morphology of the human endometrium and decidua. *Adv Anat Embryol Cell Biol*. 1992;124:1–99.
146. Cornillie FJ, Lauweryns JM, Brosens IA. Normal human endometrium: an ultrastructural survey. *Gynecol Obstet Invest*. 1985;20:113–129.
147. Dockery P, Pritchard K, Warren MA, Li TC, Cooke ID. Changes in nuclear morphology in the human endometrial glandular epithelium in women with unexplained infertility. *Hum Reprod*. 1996;11:2251–2256.
148. Paria BC, Reese J, Das SK, Dey SK. Deciphering the cross-talk of implantation: advances and challenges. *Science*. 2002;296:2185–2188.
149. Tazuke SI, Giudice LC. Growth factors and cytokines in endometrium, embryonic development, and maternal: embryonic interactions. *Semin Reprod Endocrinol*. 1996;14: 231–245.
150. Mylonas I, Jeschke U, Wiest I, et al. Inhibin/activin subunits alpha, beta-A and beta-B are differentially expressed in normal human endometrium throughout the menstrual cycle. *Histochem Cell Biol*. 2004;122:461–471.
151. Stavreus-Evers A, Koraen L, Scott JE, Zhang P, Westlund P. Distribution of cyclooxygenase-1, cyclooxygenase-2, and cytosolic phospholipase A2 in the luteal phase human endometrium and ovary. *Fertil Steril*. 2005;83:156–162.
152. Speroff L, Fritz MA. *Clinical Gynecologic Endocrinology and Infertility*. Lippincott Williams & Wilkins; 2004.
153. Treloar A, Boynton R, Behn B, Brown BW. Variation of the human menstrual cycle through reproductive life. *Int J Fertil*. 1970;12:77–126.
154. Hall J. Neuroendocrine control of the menstrual cycle. Yen and Jaffe's reproductive endocrinology. In: Strauss J, Barbieri R, eds. *Physiology, Pathophysiology, and Clinical Management*. Elsevier Saunders; 2004:195–212.
155. Hodgen G. Neuroendocrinology of the normal menstrual cycle. *J Reprod Med*. 1989;34:68–75.
156. Koehler KF, Helguero LA, Haldosen LA, Warner M, Gustafsson JA. Reflections on the discovery and significance of estrogen receptor beta. *Endocr Rev*. 2005;26:465–478.
157. Alberts B, Johnson A, Lewis J, et al. Cell communication. In: Alberts B, Johnson A, Lewis J, et al., eds. *Molecular Biology of the Cell*. Garland Science; 2002:831–906.
158. Rhen T, Cidlowski JA. Steroid hormone action. In: Strauss J, Barbieri R, eds. *Yen and Jaffe's Reproductive Endocrinology. Physiology, Pathophysiology, and Clinical Management*. Elsevier Saunders; 2004:155–174.
159. Li T, Dockery P, Rogers A, Cooke ID. How precise is histologic dating of endometrium using the standard dating criteria? *Fertil Steril*. 1989a;51:759–763.
160. Strauss JF III, Gurpide E. The endometrium: regulation and dysfunction (Chapter 9). In: Yen S, Jaffe R, eds. *Reproductive Endocrinology: Physiology, Pathophysiology and Clinical Management*. WB Saunders Company; 1991:309–356.
161. Wynn RM. The human endometrium: cyclic and gestational changes. In: Wynn R, Jollie W, eds. *Biology of the Uterus*. Plenum Medical Book Company; 1989:289–332.
162. Milwidsky A, Palti ZVI, Gutman A. Glycogen metabolism of the human endometrium. *J Clin Endocrinol Metab*. 1980; 51:765–770.
163. Daly D, Tohan N, Doney T, Maslar IA, Riddick DH. The significance of lymphocytic-leukocytic infiltrates in interpreting late luteal phase endometrial biopsies. *Fertil Steril*. 1982;37:786–791.
164. Critchley HO, Kelly RW, Brenner RM, Baird DT. The endocrinology of menstruation—a role for the immune system. *Clin Endocrinol Oxf*. 2001;55:701–710.
165. Rogers PAW, Lederman F, Taylor N. Endometrial microvascular growth in normal and dysfunctional states. *Hum Reprod Update*. 1998;4:503–508.
166. Salamonsen LA. Tissue injury and repair in the female human reproductive tract. *Reproduction*. 2003;125:301–311.
167. Hertig A. Gestational hyperplasia of endometrium. A morphologic correlation of ova, endometrium, and corpora lutea during early pregnancy. *Lab Invest*. 1964;13:1153–1191.
168. Parr M, Parr E. *The Implantation Reaction*. Plenum Medical Book Company; 1989:233–288.
169. O'Connor D, Kurman R. Intermediate trophoblast in uterine curettings in the diagnosis of ectopic pregnancy. *Obstet Gynecol*. 1988;72:665–670.
170. Gruber K, Gelven PL, Austin RM. Chorionic villi or trophoblastic tissue in uterine samples of four women with ectopic pregnancies. *Int J Gynecol Pathol*. 1997;16:28–32.

171. Young RH, Kurman RJ, Scully RE. Placental site nodules and plaques. A clinicopathologic analysis of 20 cases. *Am J Surg Pathol*. 1990;14:1001–1009.
172. Shih IM, Seidman JD, Kurman RJ. Placental site nodule and characterization of distinctive types of intermediate trophoblast. *Hum Pathol*. 1999;30:687–694.
173. Shih IM, Kurman RJ. Ki-67 labeling index in the differential diagnosis of exaggerated placental site, placental site trophoblastic tumor, and choriocarcinoma: a double immunohistochemical staining technique using Ki-67 and Mel-CAM antibodies. *Hum Pathol*. 1998;29:27–33.
174. Shih IM, Kurman RJ. p63 expression is useful in the distinction of epithelioid trophoblastic and placental site trophoblastic tumors by profiling trophoblastic subpopulations. *Am J Surg Pathol*. 2004;28:1177–1183.
175. Arias-Stella J. The Arias-Stella reaction: facts and fancies four decades after. *Adv Anat Pathol*. 2002;9:12–23.
176. Huettner PC, Gersell DJ. Arias-Stella reaction in nonpregnant women: a clinicopathologic study of nine cases. *Int J Gynecol Pathol*. 1994;13:241–247.
177. Arias-Stella J Jr, Arias-Velasquez A, Arias-Stella J. Normal and abnormal mitoses in the atypical endometrial change associated with chorionic tissue effect [corrected]. *Am J Surg Pathol*. 1994;18:694–701.
178. Vang R, Barner R, Wheeler DT, Strauss BL. Immunohistochemical staining for Ki-67 and p53 helps distinguish endometrial Arias-Stella reaction from high-grade carcinoma, including clear cell carcinoma. *Int J Gynecol Pathol*. 2004;23:223–233.
179. Mazur M, Hendrickson M, Kempson R. Optically clear nuclei: an alteration of endometrial epithelium in the presence of trophoblast. *Am J Surg Pathol*. 1983;7:415–423.
180. Lichtig C, Deutch M, Brandes J. Vascular changes of endometrium in early pregnancy. *Am J Clin Pathol*. 1984;81:702–707.
181. Hustin J, Wells M. Pathology of the pregnant uterus. In: Fox H, Wells M, eds. *Haines and Taylor Obstetrical and Gynaecological Pathology*. Churchill Livingstone; 2002:1327–1357.
182. Taffe JR, Dennerstein L. Menstrual patterns leading to the final menstrual period. *Menopause*. 2002;9:32–40.
183. Archer D, Mcintyreseltman K, Wilborn WW Jr, et al. Endometrial morphology in asymptomatic postmenopausal women. *Am J Obstet Gynecol*. 1991;165.
184. Choo YC, Mak KC, Hsu C, Wong TS, Ma HK. Postmenopausal uterine bleeding of nonorganic cause. *Obstet Gynecol*. 1985;66:225–228.
185. Moodley M, Roberts C. Clinical pathway for the evaluation of postmenopausal bleeding with an emphasis on endometrial cancer detection. *J Obstet Gynaecol*. 2004;24:736–741.
186. Quddus MR, Sung CJ, Zheng W, Lauchlan SC. p53 immunoreactivity in endometrial metaplasia with dysfunctional uterine bleeding. *Histopathology*. 1999;35:44–49.
187. Haney AF. Endometrial biopsy: a test whose time has come and gone. *Fertil Steril*. 2004;82:1295–1296; discussion 301–302.
188. Coutifaris C, Myers ER, Guzick DS, et al. Histological dating of timed endometrial biopsy tissue is not related to fertility status. *Fertil Steril*. 2004;82:1264–1272.
189. Murray MJ, Meyer WR, Zaino RJ, et al. A critical analysis of the accuracy, reproducibility, and clinical utility of histologic endometrial dating in fertile women. *Fertil Steril*. 2004;81:1333–1343.
190. Myers ER, Silva S, Barnhart K, et al. Interobserver and intraobserver variability in the histological dating of the endometrium in fertile and infertile women. *Fertil Steril*. 2004;82:1278–1282.
191. Glatstein IZ, Harlow BL, Hornstein MD. Practice patterns among reproductive endocrinologists: the infertility evaluation. *Fertil Steril*. 1997;67:443–451.
192. Balasch J. Investigation of the infertile couple: investigation of the infertile couple in the era of assisted reproductive technology: a time for reappraisal. *Hum Reprod*. 2000;15:2251–2257.
193. Balasch J, Fabregues F, Creus M, Vanrell JA. The usefulness of endometrial biopsy for luteal phase evaluation in infertility. *Hum Reprod*. 1992;7:973–977.
194. Peters AJ, Lloyd RP, Coulam CB. Prevalence of out-of-phase endometrial biopsy specimens. *Am J Obstet Gynecol*. 1992;166:1738–1745; discussion 1745–1746.
195. Noyes R. Normal phases of the endometrium. In: Hertig A, Norris H, Abell M, eds. *The Uterus*. Williams & Wilkins; 1973:110–135.
196. Noyes RW, Hertig AT, Rock J. Dating the endometrial biopsy. *Fertil Steril*. 1950;1:3–25.
197. McNeely MJ, Soules MR. The diagnosis of luteal phase deficiency: a critical review [see comments]. *Fertil Steril*. 1988;50:1–15.
198. Noyes RW, Haman JO. Accuracy of endometrial dating. *Fertil Steril*. 1953;4:504–517.
199. Li TC, Rogers AW, Lenton EA, Dockery P, Cooke I. A comparison between two methods of chronological dating of human endometrial biopsies during the luteal phase, and their correlation with histologic dating. *Fertil Steril*. 1987;48:928–932.
200. Gibson M, Badger GJ, Byrn F, Lee KR, Korson R, Trainer TD. Error in histologic dating of secretory endometrium: variance component analysis. *Fertil Steril*. 1991;56:242–247.
201. Scott RT, Snyder RR, Strickland DM, et al. The effect of interobserver variation in dating endometrial histology on the diagnosis of luteal phase defects. *Fertil Steril*. 1988;50:888–892.
202. Otsuki Y. Apoptosis in human endometrium: apoptotic detection methods and signaling. *Med Electron Microsc*. 2001;34:166–173.
203. Kokawa K, Shikone T, Nakano R. Apoptosis in the human uterine endometrium during the menstrual cycle. *J Clin Endocrinol Metab*. 1996;81:4144–4147.
204. Konno R, Igarashi T, Okamoto S, et al. Apoptosis of human endometrium mediated by perforin and granzyme B of NK cells and cytotoxic T lymphocytes. *Tohoku J Exp Med*. 1999;187:149–155.
205. Sivridis E, Giatromanolaki A. New insights into the normal menstrual cycle-regulatory molecules. *Histol Histopathol*. 2004;19:511–516.
206. Rotello RJ, Lieberman RC, Purchio AF, Gerschenson LE. Coordinated regulation of apoptosis and cell proliferation by transforming growth factor beta 1 in cultured uterine epithelial cells. *Proc Natl Acad Sci U S A*. 1991;88:3412–3415.
207. Gerschenson LE, Rotello RJ. Apoptosis: a different type of cell death. *FASEB J*. 1992;6:2450–2455.
208. Rotello RJ, Hocker MB, Gerschenson LE. Biochemical evidence for programmed cell death in rabbit uterine epithelium. *Am J Pathol*. 1989;134:491–495.
209. Rotello RJ, Lieberman RC, Lepoff RB, Gerschenson LE. Characterization of uterine epithelium apoptotic cell death

kinetics and regulation by progesterone and RU 486. *Am J Pathol*. 1992;140:449–456.
210. Gompel A, Sabourin JC, Martin A, et al. Bcl-2 expression in normal endometrium during the menstrual cycle. *Am J Pathol*. 1994;144:1195–1202.
211. Maia H Jr, Maltez A, Studart E, Athayde C, Coutinho EM. Ki-67, Bcl-2 and p53 expression in endometrial polyps and in the normal endometrium during the menstrual cycle. *BJOG*. 2004;111:1242–1247.
212. Tabibzadeh S, Zupi E, Babaknia A, Liu R, Marconi D, Romanini C. Site and menstrual cycle-dependent expression of proteins of the tumour necrosis factor (TNF) receptor family, and BCL-2 oncoprotein and phase-specific production of TNF alpha in human endometrium. *Hum Reprod*. 1995;10:277–286.
213. Achilles SL, Amortegui AJ, Wiesenfeld HC. Endometrial plasma cells: do they indicate subclinical pelvic inflammatory disease? *Sex Transm Dis*. 2005;32:185–188.
214. Cole W, Garfield R. Ultrastructure of the myometrium. In: Wynn R, Jollie W, eds. *Biology of the Uterus*. Plenum; 1989:455–504.
215. Garfield R, Yallampalli C. Structure and function of uterine muscle. In: Chard T, Grudzinskas J, eds. *The Uterus*. Cambridge; 1994:54–93.
216. Huszar G, Walsh MP. Biochemistry of the myometrium and cervix. In: Wynn R, Jollie W, eds. *Biology of the Uterus*. Plenum; 1989:355–402.
217. Kao C. Electrophysiological properties of uterine smooth muscle. In: Wynn R, Jollie W, eds. *Biology of the Uterus*. Plenum; 1989:403–454.
218. Kawaguchi K, Fujii S, Konishi I, Nanbu Y, Nonogaki H, Mori T. Mitotic activity in uterine leiomyomas during the menstrual cycle. *Am J Obstet Gynecol*. 1989;160:637–641.
219. Pradhan N, Mohatny SK. Uterine tumors resembling ovarian sex cord tumors. *Arch Pathol Lab Med*. 2013:137(12);1832–1836.
220. Marshall J. The physiology of the myometrium. In: Hertig A, Norris H, Abell M, eds. *The Uterus*. Williams & Wilkins; 1973:89–109.
221. Oliva E, Clement PB, Young RH, Scully RE. Mixed endometrial stromal and smooth muscle tumors of the uterus: a clinicopathologic study of 15 cases. *Am J Surg Pathol*. 1998;22:997–1005.
222. Azumi N, Ben-Ezra J, Battifora H. Immunophenotypic diagnosis of leiomyosarcomas and rhabdomyosarcomas with monoclonal antibodies to muscle-specific actin and desmin in formalin-fixed tissue. *Mod Pathol*. 1988;1:469–474.
223. Brown D, Theaker J, Banks P, Gatter KC, Mason DY. Cytokeratin expression in smooth muscle and smooth muscle tumours. *Histopathology*. 1987;11:477–486.
224. Gown AM, Boyd HC, Chang Y, Ferguson M, Reichler B, Tippens D. Smooth muscle cells can express cytokeratins of "simple" epithelium. *Am J Pathol*. 1988;132(2):223–232.
225. Norton AJ, Thomas JA, Isaacson PG. Cytokeratin-specific monoclonal antibodies are reactive with tumours of smooth muscle derivation. An immunocytochemical and biochemical study using antibodies to intermediate filament cytoskeletal proteins. *Histopathology*. 1987;11:487–499.
226. Silverberg S, Kurman RJ. *Atlas of Tumor Pathology: Tumors of the Uterine Corpus and Gestational Trophoblastic Disease*. AFIP; 1992.
227. Garfield RE, Hayashi RH. Appearance of gap junctions in the myometrium of women during labor. *Am J Obstet Gynecol*. 1981;140:254–260.
228. Brenner RM, Slayden OD. The fallopian tube cycle. In: Adashi E, Rock J, Rosenwaks Z, eds. *Reproductive Endocrinology, Surgery, and Technology*. Lippincott-Raven; 1996: 326–339.
229. Bonilla-Musoles F, Ferrer-Barriendos J, Pellicer A. Cyclical changes in the epithelium of the fallopian tube. Studies with scanner electron microscopy (SEM). *Clin Exp Obstet Gynecol*. 1983;10:79–86.
230. Donnez J, Casanas-Roux F, Caprasse J, Ferin J, Thomas K. Cyclic changes in ciliation, cell height, and mitotic activity in human tubal epithelium during reproductive life. *Fertil Steril*. 1985;43:554–559.
231. Jansen RPS. Endocrine response in the fallopian tube. *Endocr Rev*. 1984;5:525–551.
232. Lindenbaum ES, Peretz BA, Beach D. Menstrual-cycle-dependent and -independent features of the human Fallopian tube fimbrial epithelium: an ultrastructural and cytochemical study. *Gynecol Obstet Invest*. 1983;16: 76–85.
233. Verhage HG, Bareither ML, Jaffe RC, Akbar M. Cyclic changes in ciliation, secretion and cell height of the oviductal epithelium in women. *Am J Anat*. 1979;156:505–521.
234. Menezo Y, Guerin P. The mammalian oviduct: biochemistry and physiology. *Eur J Obstet Gynecol Reprod Biol*. 1997;73:99–104.
235. Leese HJ. The formation and function of oviduct fluid. *J Reprod Fertil*. 1988;82:843–856.
236. Sivridis E, Buckley C, Fox H. Argyrophil cells in normal, hyperplastic, and neoplastic endometrium. *J Clin Pathol*. 1984;37:378–381.
237. Constant O, Cooke J, Parsons CA. Reformatted computed tomography of the female pelvis: normal anatomy. *Br J Obstet Gynaecol*. 1989;96:1047–1053.
238. de Castro A, Yebra C, Aznar F, et al. Measurement of the endometrial cavity length using Wing Sound I. *Adv Contracept*. 1987;3:133–137.
239. Hricak H. MRI of the female pelvis: a review. *AJR Am J Roentgenol*. 1986;146:1115–1122.
240. Morris H, Emms M, Visser T, Timme A. Lymphoid tissue of the normal fallopian tube–a form of mucosal-associated lymphoid tissue (MALT)? *Int J Gynecol Pathol*. 1986;5: 11–22.
241. Boehme M, Donat H. Identification of lymphocyte subsets in the human fallopian tube. *Am J Reprod Immunol*. 1992;28:81–84.
242. Kutteh WH, Blackwell RE, Gore H, Kutteh CC, Carr BR, Mestecky J. Secretory immune system of the female reproductive tract. II. Local immune system in normal and infected fallopian tube. *Fertil Steril*. 1990;54:51–55.
243. Lurie M, Tur-Kaspa I, Weill S, Rabinovici J, Goldenberg S. Ciliary ultrastructure of respiratory and fallopian tube epithelium in a sterile woman with Kartagener's syndrome: a quantitative estimation. *Chest*. 1989;95:578–581.
244. Halbert SA, Patton DL, Zarutskie PW, Soules MR. Function and structure of cilia in the fallopian tube of an infertile woman with Kartagener's syndrome. *Hum Reprod*. 1997;12:55–58.
245. Gwin K, Wilcox R, Montag A. Insights into selected genetic diseases affecting female reproductive tract and their implication for pathologic evaluation of gynecologic specimens. *Arch Pathol Lab Med*. 2009;133(7):1041–1052.

246. Meserve EEK, Brouwer J, Crum CP. Serous tubal intraepithelial neoplasia: the concept and its application. *Mod Pathol*. 2017;30(5):710–721.
247. Bogaerts JMA, van Bommel MHD, Hermens RPMG, et al. Consensus based recommendations for the diagnosis of serous tubal intraepithelial carcinoma: an international Delphi study. *Histopathology*. 2023;83(1):67–79.
248. Pulkkinen MO, Talo A. Tubal physiologic consideration in ectopic pregnancy. *Clin Obstet Gynecol*. 1987;30:164–172.
249. Donnez J, Casanas-Roux F, Ferin J. Macroscopic and microscopic studies of fallopian tube after laparoscopic sterilization. *Contraception*. 1979;20:497–509.
250. Donnez J, Casanas-Roux F, Ferin J, Thomas K. Tubal polyps, epithelial inclusions, and endometriosis after tubal sterilization. *Fertil Steril*. 1984;41:564–568.
251. Mills SE, Fechner RE. Stromal and epithelial changes in the fallopian tube following hormonal therapy. *Hum Pathol*. 1980;11:583–585.
252. Lindblom B, Wilhelmsson L, Wikland M, Hamberger L, Wiqvist N. Prostaglandins and oviductal function. *Acta Obstet Gynecol Scand Suppl*. 1983;113:43–46.
253. Moyle P, Kataoka M, Nakai A, Takahata A, Reinhold C, Sala E. Nonovarian cystic lesions of the pelvis. *Radiographics*. 2010;30(4):921–938.

41

Ovary

C. Blake Gilks ■ Jutta Huvila

EMBRYOLOGY

Approximately 5 weeks after fertilization, a thickening of the coelomic epithelium (mesothelium) along the medial and ventral borders of the mesonephros leads to the formation of the genital ridge. The gonadal anlage forms as a result of continued proliferation of this epithelium and the subjacent mesenchyme (1). Simultaneously, primordial germ cells migrate to the gonad from the yolk sac endoderm, reaching the genital ridge during the 5th and 6th weeks of embryonic life (2). These cells (oogonia) undergo mitotic activity and become most numerous at midgestation; two-thirds of them will undergo atresia by term (1,3). At 12 to 15 weeks of gestation, the oogonia begin meiosis and arrest in meiotic prophase and are now referred to as primary oocytes (3–5).

The primitive gonad is recognizable as an ovary at 2 months of gestational age because, in contrast to the testis, it remains basically unaltered thereafter. By 7 to 9 weeks of gestation, the outer zone of the ovary has enlarged to form the definitive cortex, which consists of confluent sheets of primitive germ cells admixed haphazardly with a smaller number of pregranulosa cells (4,6). At 12 to 15 weeks, vascular connective tissue septa begin to radiate from the medullary mesenchyme into the inner portion of the cortex and extend into the superficial part of the cortex by 20 weeks (5,6). Thus, the cortex becomes divided into cellular groups composed of oocytes and pregranulosa cells (sex cords) while pregranulosa cells simultaneously begin to surround individual germ cells to form primordial follicles. Folliculogenesis begins in the inner part of the cortex at 14 to 20 weeks of gestation (2,3,5,7) and gradually extends to the outer cortex by the early neonatal period (8). The occasional follicles that mature into preantral and antral follicles in late gestation are eventually surrounded by a condensation of mesenchymal cells that becomes the theca interna (4,7). The rete ovarii is present in the hilus as early as 12 weeks (5).

The origin of gonadal sex cords (pregranulosa cells in the ovary and Sertoli cells in the testis) is controversial (2,4,8–12), but is likely of mesonephric origin (13–16). Hummitzsch et al. have proposed that granulosa cells are derived from gonadal-ridge epithelial–like (GREL) cells of the genital ridge primordium; these cells are also suggested to be the precursor of ovarian surface epithelial cells (16).

GROSS ANATOMY

The ovaries are paired pelvic organs that lie on either side of the uterus close to the lateral pelvic wall, behind the broad ligament and anterior to the rectum. Each ovary is attached along its anterior (hilar) margin to the posterior aspect of the broad ligament by a double fold of peritoneum, the mesovarium. The medial pole of the ovary is connected to the ipsilateral uterine cornu by the ovarian (or utero-ovarian) ligament and the superior aspect of the lateral pole is supported by the infundibulopelvic (or suspensory) ligament which extends to the lateral pelvic wall. The location of the ovary posterior to the broad ligament and a similar relationship of the ovarian ligament to the ipsilateral uterine (fallopian) tube aids in determining the laterality of a salpingo-oophorectomy specimen.

Prepubertal Ovaries

The ovary in the newborn is a tan, elongated, and flattened structure that lies above the true pelvis. It sometimes has a lobulated appearance with irregular edges (Fig. 41.1A). It has approximate dimensions of 1.3 cm × 0.5 cm × 0.3 cm and a weight of less than 0.3 g (17–19). Throughout infancy and childhood, the ovary increases in weight 30-fold and changes in shape, so that by the time of puberty it has reached the size, weight, and shape of the adult ovary and lies within the true pelvis (18,19). Inspection of the external and cut surfaces, particularly during the first few months of life and at puberty, may reveal prominent cystic follicles (20) similar to those seen in polycystic ovary disease (Fig. 41.1B).

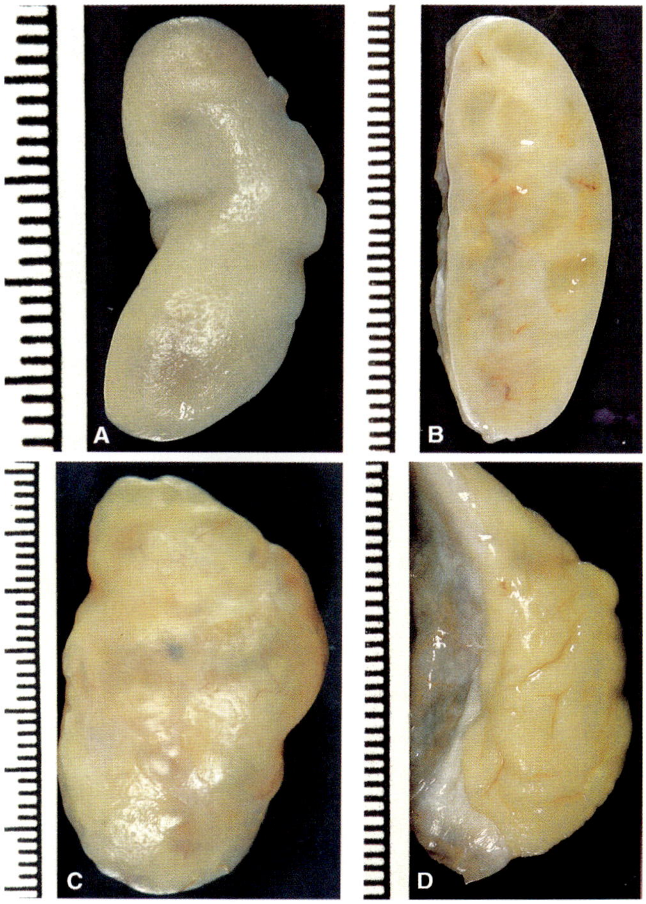

FIGURE 41.1 Gross appearance of ovary. **A:** Newborn, external aspect. **B:** Pubertal (age 15 years), sectioned surface. Note elongate shape and multiple cystic follicles. **C:** Adult (age 30 years), external aspect. **D:** Postmenopausal, external aspect. Note shrunken, gyriform appearance.

Adult Ovaries

Adult ovaries are ovoid with dimensions of approximately 3.0 to 5.0 cm × 1.5 to 3.0 cm × 0.6 to 1.5 cm and a weight of 5 to 8 g. Their size and weight, however, vary considerably depending on their content of follicular derivatives. They have a pink-white exterior that is smooth in early reproductive life (Fig. 41.1C) but become increasingly convoluted thereafter. Three ill-defined zones are discernible on the cut surface: an outer cortex, an inner medulla, and the hilus. Cystic follicles, yellow corpora lutea, and white corpora albicantia are typically visible in the cortex and medulla.

Postmenopausal Ovaries

After menopause, the ovaries typically shrink to approximately one-half of their size in the reproductive era (21). Their size varies considerably, however, with the number of ovarian stromal cells and unresorbed corpora albicantia (22). Most postmenopausal ovaries have a shrunken, gyriform, external appearance (Fig. 41.1D), whereas some are smooth and uniform. They have a firm consistency and a predominantly solid, pale cut surface, although occasional inclusion cysts measuring several millimeters in diameter may be discernible within the cortex. Small white corpora albicantia represent scars and are typically present within the medulla. Thick-walled blood vessels may be appreciable within the medulla and the hilus.

BLOOD SUPPLY

The ovarian artery, a branch of the aorta, courses along the infundibulopelvic ligament and the mesovarial border of the ovary where it anastomoses with the ovarian branch of the uterine artery. Approximately 10 arterial branches from this arcade penetrate the ovarian hilus, becoming coiled and branched as they course through the medulla (23). These helicine arteries possess longitudinal ridges of intimal smooth muscle along their length. The medullary arteries and arterioles form a plexus at the corticomedullary junction where they give rise to smaller, straight cortical arterioles that penetrate the cortex in a radial fashion, perpendicular to the ovarian surface. The cortical arterioles branch and anastomose several times, forming sets of interconnected vascular arcades (23). These arcades give rise to capillaries that form dense networks within the theca layers of the ovarian follicles. The intraovarian veins accompany the arteries, becoming large and tortuous in the medulla and forming a hilar plexus that drains into the ovarian veins; the latter traverse the mesovarium and course along the infundibulopelvic ligament (23). The ovarian veins also anastomose with tributaries of the uterine veins. The left and right ovarian veins drain into the left renal vein and the inferior vena cava, respectively.

In postmenopausal women, the medullary blood vessels may appear to be numerous and closely packed (Fig. 41.2), but they should not be mistaken for a hemangioma or vascular anomaly. In addition, many of the same vessels may be calcified or have thickened walls and narrowed lumina due to medial deposition of a hyaline, amyloid-like material.

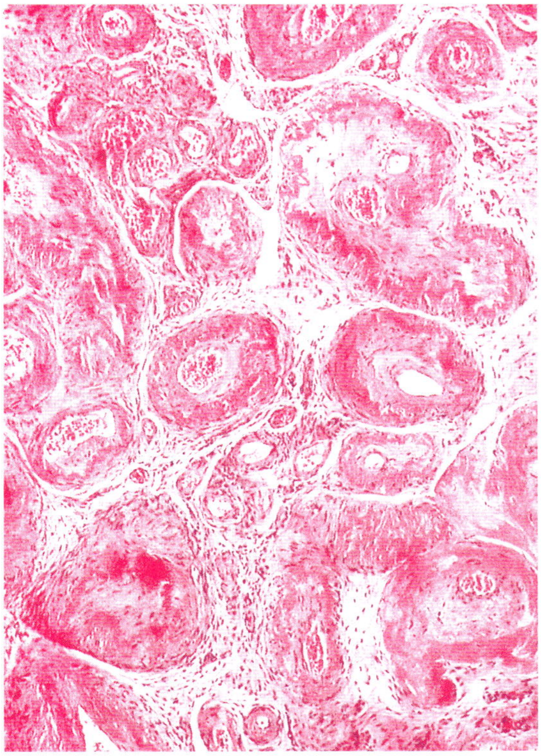

FIGURE 41.2 Numerous crowded thick-walled blood vessels within ovarian medulla of a postmenopausal woman. Some of the vessels have an eosinophilic amyloid-like material within their walls.

LYMPHATIC VESSELS

The lymphatic vessels of the ovary originate predominantly within the thecal layers of the follicles. The granulosa layer of a maturing follicle is devoid of lymphatic vessels in contrast to its counterpart within the corpus luteum, which possesses a rich supply (24). The lymphatic vessels pass through the ovarian stroma, independent of blood vessels, to drain into larger trunks that form a plexus at the hilus. The lymphatic vessels and blood vessels converge in the hilus, where the former coil around veins in a helicoid fashion. Four to eight efferent channels pass into the mesovarium where they converge to form the subovarian plexus, which, in turn, is joined by branches from the uterine (fallopian) tube and uterine fundus (24). Leaving the plexus, the drainage trunks diminish in number and size, passing along the free border of the infundibulopelvic ligament enmeshed with the ovarian veins. From there they accompany the ovarian vessels, juxtaposed to the psoas muscle, and drain into the upper para-aortic lymph nodes at the level of the lower pole of the kidney (24,25). The major lymphatic drainage of the ovary is therefore in a cephalad direction toward the para-aortic lymph nodes. Accessory channels, however, may bypass the subovarian plexus, passing through the broad ligament to the internal iliac, external iliac, and interaortic lymph nodes or, in some females, via the round ligament to the iliac and inguinal lymph nodes (24,25). When the pelvic and para-aortic lymph nodes are extensively replaced by tumor, retrograde lymphatic flow may represent a rare mechanism of tumor spread to the ovaries.

NERVE SUPPLY

The nerve supply of the ovary arises from a sympathetic plexus that is enmeshed with the ovarian vessels in the infundibulopelvic ligament (26). Nerve fibers, which are predominantly nonmyelinated, accompany the ovarian artery, entering the ovary at the hilus. Delicate terminal fibers, many surrounding small arteries and arterioles, penetrate the medulla and cortex to terminate as plexuses surrounding the follicles (26,27). Adrenergic nerve fibers and terminals are in close contact with smooth muscle cells in the cortical stroma and theca externa. The physiologic significance of ovarian sympathetic innervation is not clear, although it has been suggested that it may play a role in follicular maturation, follicular rupture, or both (26,28–29). In addition, catecholamines can stimulate progesterone production by the ovarian follicles and androgen production by the ovarian stroma in vitro (30).

SURFACE EPITHELIUM

Histology

The surface epithelium of the ovary consists of a single, focally pseudostratified layer of modified peritoneal cells. The cells vary from flat to cuboidal to columnar and several types may be seen in different areas of the same ovary (Fig. 41.3). The surface cells are separated from the underlying stroma by a basement membrane. This epithelium is extremely fragile and is almost always denuded in oophorectomy specimens because of rubbing of the surface by the surgeon and the pathologist, as well as lack of prompt fixation that results in drying. Preserved epithelium is often confined to areas protected by surface adhesions or the lining of sulci.

Histochemical studies have demonstrated glycogen, as well as acid and neutral mucopolysaccharides, within surface epithelial cells (31,32). Seventeen-beta hydroxysteroid dehydrogenase activity, absent in extraovarian mesothelial cells, has also been demonstrated (31).

Epithelial inclusion glands are thought to arise from either the surface epithelium or detached fimbrial epithelium (i.e., endosalpingiosis). The former arise from cortical invaginations of the surface epithelium that have lost their connections with the surface. They often become cystic, resulting in epithelial inclusion cysts that may be recognized on macroscopic examination; a diameter of 1 cm has been suggested as a dividing line between an epithelial inclusion cyst and the smallest cystadenoma. Alternatively, epithelial inclusion cysts are thought to be derived from detached fimbrial epithelium. These cysts are typically multiple, scattered singly or in small clusters throughout the superficial cortex (Fig. 41.4); less commonly, extension into the deeper cortical or medullary stroma may occur. Inclusion glands and cysts are typically lined by a single layer of ciliated tubal-type columnar cells; psammoma bodies are occasionally present within their lumina or the adjacent stroma. Similar glands, with or without associated psammoma bodies, are designated "endosalpingiosis" when encountered on the ovarian surface, within periovarian adhesions, or on the extraovarian peritoneum and omentum (33). Epithelial inclusion glands have been identified in ovaries from all age groups, including fetuses, infants, and adolescents (34,35). Epithelial inclusion glands and cysts become more numerous with age and are common incidental findings in late reproductive and postmenopausal age groups. Less

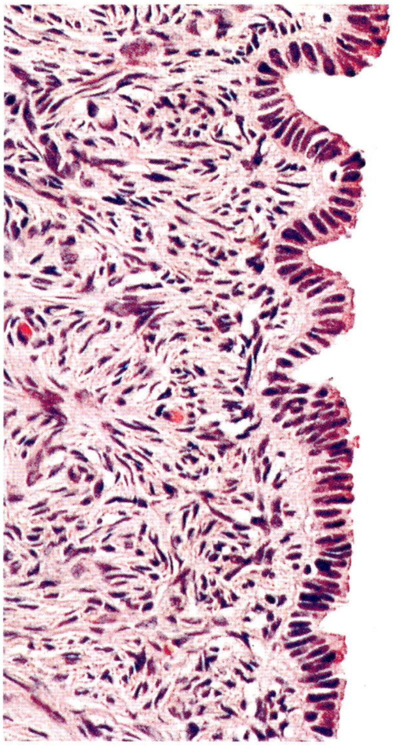

FIGURE 41.3 Ovarian surface epithelium composed of a single layer of columnar cells.

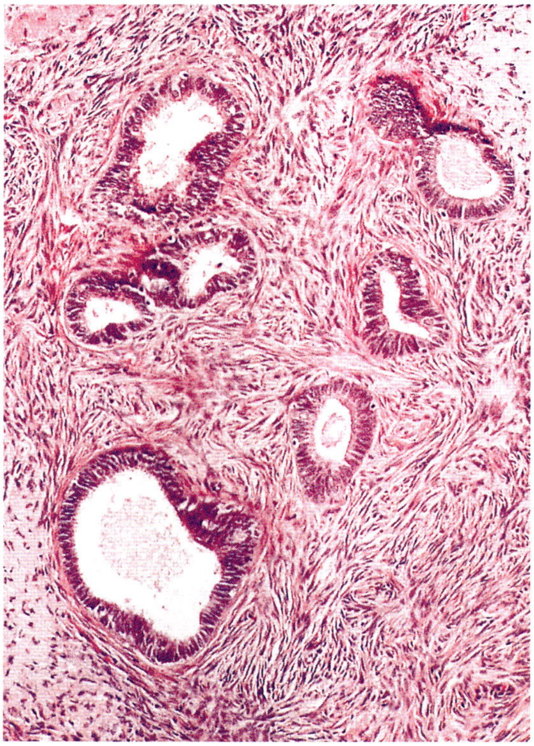

FIGURE 41.4 Epithelial inclusion glands within ovarian cortex.

frequently, epithelial inclusion glands and cysts may be lined by endometrioid or mucinous epithelia, or nonspecific columnar or flattened cells (36,37).

Epithelial inclusion glands and cysts were previously believed to be the site of origin of most surface epithelial tumors (38), but recent evidence indicates that most high-grade serous carcinomas arise from the fallopian tube, whereas most clear cell and endometrioid carcinomas arise from endometriosis (39–41).

The immunostaining profiles of the ovarian surface epithelium and that lining epithelial inclusion glands and cysts vary depending on whether the epithelium is flattened and/or cuboidal, or columnar with tubal metaplasia. The former cells stain for mesothelial markers such as calretinin and mesothelin, whereas the latter columnar-ciliated cells show either focal immunoreactivity or complete absence of staining for these markers (Fig. 41.5A) (42). Cells with a columnar shape and tubal metaplasia, whether on the ovarian surface or lining epithelial inclusion glands and cysts, stain positively for epithelial markers including oviduct-specific glycoprotein-1 and E-cadherin, whereas the flat ovarian surface epithelial cells do not express these markers (Fig. 41.5C–F). PAX8 is expressed by all cells, flat and columnar, lining epithelial inclusion glands and cysts but not by most ovarian surface epithelial cells on the ovarian surface (42,43).

Urothelial or transitional-type differentiation is also within the metaplastic potential of the ovarian surface epithelium and pelvic peritoneum. Such differentiation typically takes the form of Walthard nests of transitional cells, a common microscopic finding within the serosa or the immediately subjacent stroma of the fallopian tube, mesosalpinx, and mesovarium, or less commonly, the ovarian hilus (Fig. 41.6) (44–47). The larger nests are often cystic and may be lined by

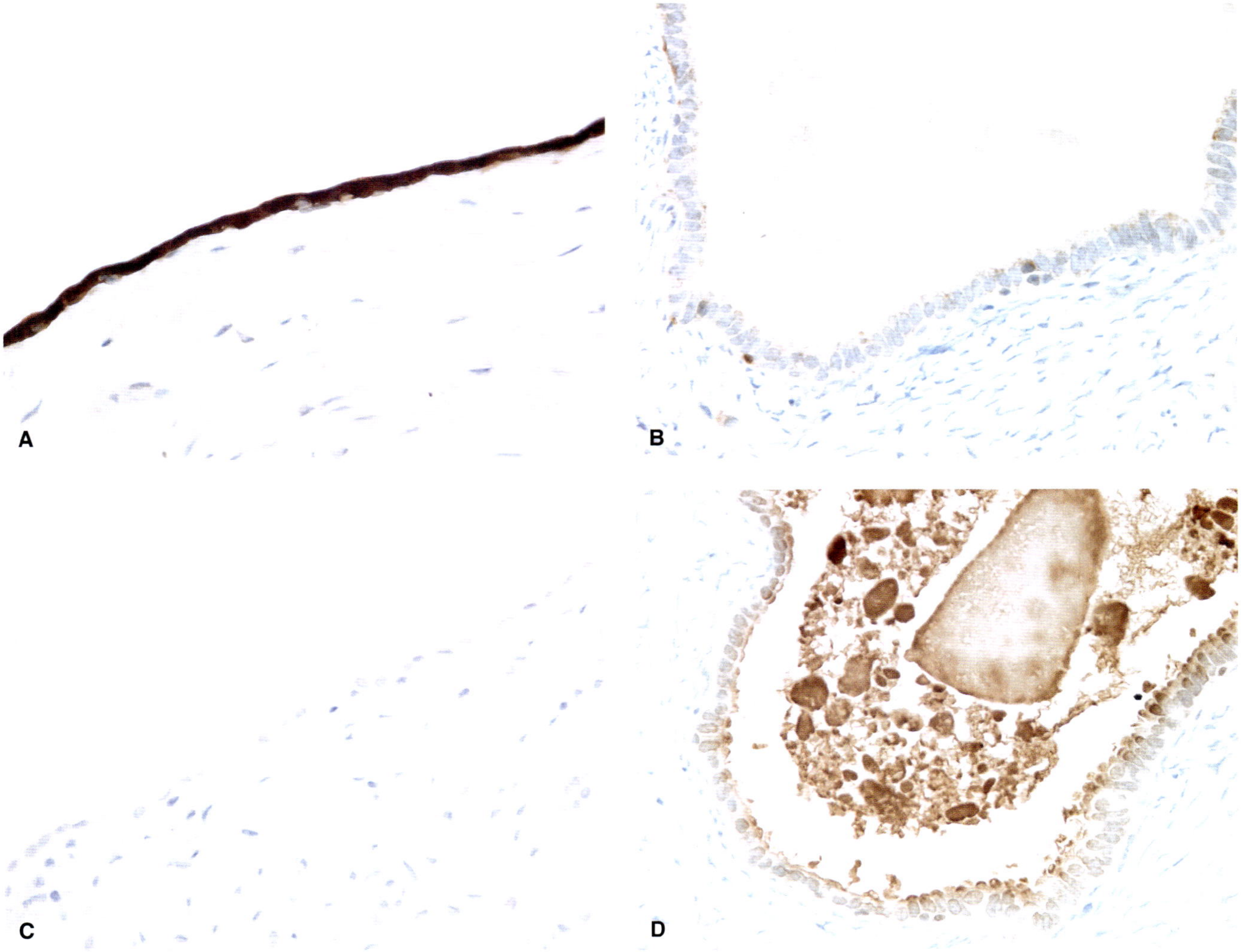

FIGURE 41.5 Immunostaining profile of normal ovarian surface epithelium and cells lining epithelial inclusion glands. The flattened and cuboidal ovarian surface epithelium cells stain positively for the mesothelial marker calretinin (**A**), and are negative for oviduct-specific glycoprotein-1 (**C**) and E-cadherin (**E**), while the columnar cells with tubal metaplasia, lining the epithelial inclusion gland, are negative for calretinin (**B**), and express both oviduct-specific glycoprotein-1 (**D**) and E-cadherin (**F**). (*continued*)

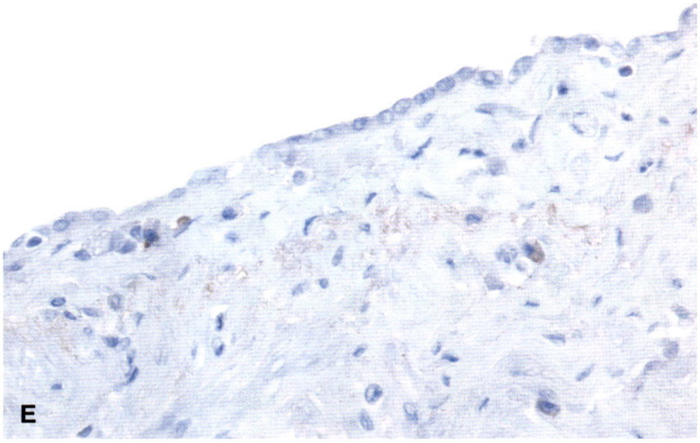

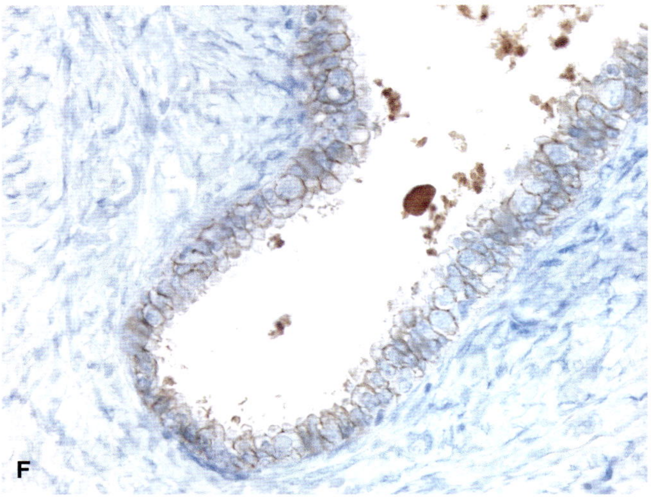

FIGURE 41.5 (*Continued*)

columnar mucinous cells. Brenner tumors are also characterized by transitional differentiation. The frequent coexistence of Brenner tumors and ovarian mucinous tumors has led some to suggest that both arise from Walthard nests (48).

Hyperplastic mesothelial cells can develop as a response to chronic pelvic inflammation, in which case they involve the surface of the ovary and focally replace the ovarian surface epithelium. Florid examples exhibit tubulopapillary (Fig. 41.7) and pseudoinfiltrative patterns, as well as varying degrees of nuclear atypia; these must be distinguished from malignant mesothelioma and carcinoma derived from the ovary or a metastasis.

Ultrastructure

The ultrastructural appearance of the ovarian surface epithelium is similar to that of the extraovarian peritoneum (49–51). The cell surfaces by scanning and transmission electron microscopy have dome-shaped apices covered by numerous, often branching, microvilli, occasional single cilia, and

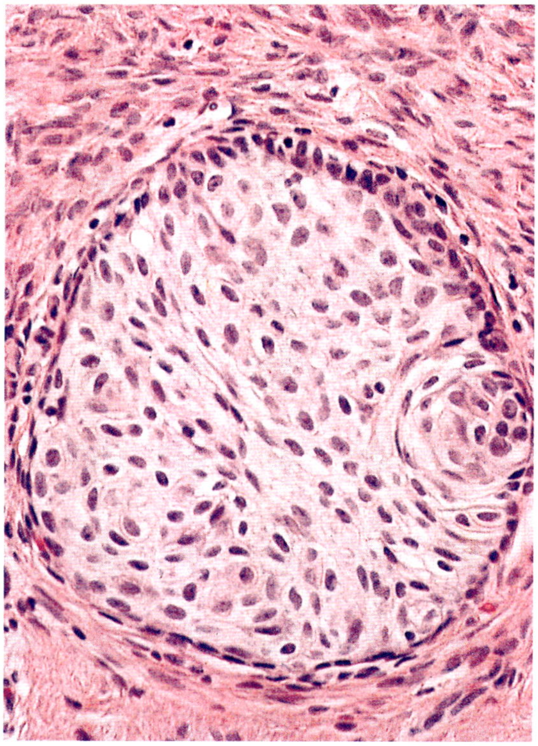

FIGURE 41.6 Walthard nest within ovarian hilus abutting medullary stroma.

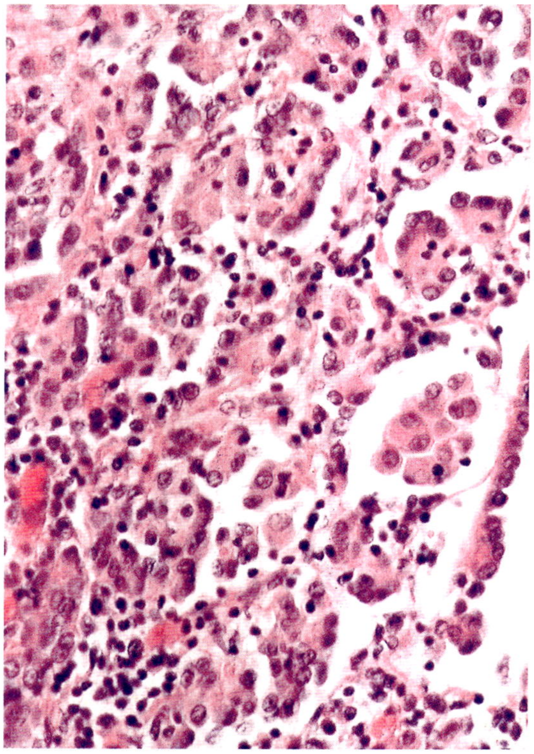

FIGURE 41.7 Hyperplastic mesothelial cells on ovarian surface. Note admixed inflammatory cells.

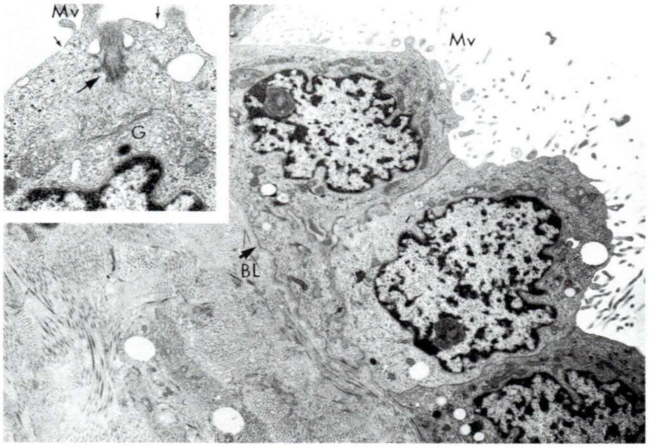

FIGURE 41.8 Electron micrograph of ovarian surface epithelium. The cells have numerous microvilli (*Mv*) and well-developed organelles in a perinuclear location. The nuclei have indented membranes and peripheral nucleoli. The lateral plasma membranes are reinforced by luminal junctional complexes and scattered desmosomes but are occasionally widely separated producing dilated intercellular spaces. A well-defined basal lamina (*BL*) separates the cells from the underlying stroma (original magnification ×6,400). *Inset:* The surface microvilli are associated with micropinocytotic vesicles (*short arrows*) and occasional single cilia (*long arrows*). Note Golgi complex (*G*) (original magnification ×22,000). (Reprinted with permission from Ferenczy A, Richart RM. *Female Reproductive System: Dynamics of Scan and Transmission Electron Microscopy*. John Wiley & Sons; 1974.)

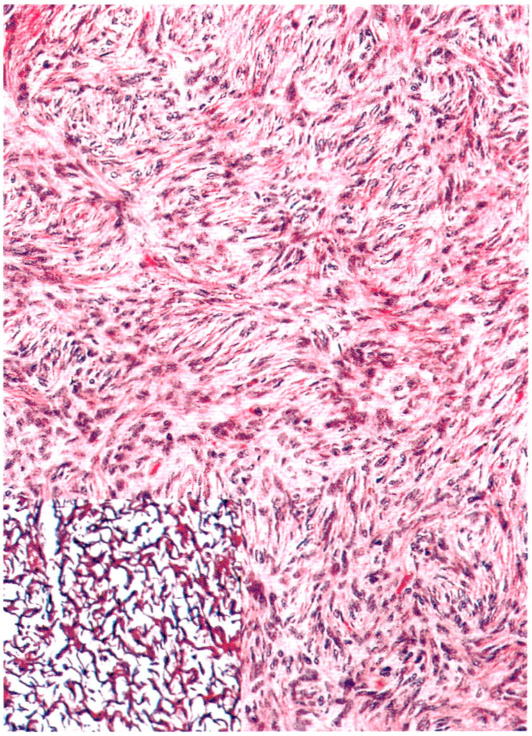

FIGURE 41.9 Ovarian stroma composed of whorls of plump spindle cells of fibroblastic type. *Inset:* Note dense reticulin network (reticulin stain).

pinocytotic vesicles (Fig. 41.8). The cytoplasm contains abundant polysomes, free ribosomes, numerous mitochondria, and bundles of intermediate filaments and tonofilaments. Lipid droplets are sometimes present in the basal cytoplasm. The nuclei have indented nuclear membranes and peripheral nucleoli. Straight or convoluted lateral plasma membranes are reinforced by luminal junctional complexes, scattered desmosomes, and desmosomal–tonofilament complexes. The membranes may be widely separated in areas, creating dilated intercellular spaces (50). A well-developed basal lamina separates the surface epithelium from the underlying stroma.

STROMA

Histology

As the cortical and medullary stroma is continuous and similar in appearance, the boundary between these two zones is ill-defined and arbitrary. The spindle-shaped stromal cells have scant cytoplasm and are arranged in whorls or a storiform pattern (Fig. 41.9). Fine cytoplasmic lipid droplets may be appreciable with special stains, especially in the late reproductive and postmenopausal age groups (52). Immunohistochemical stains reveal cytoplasmic vimentin expression (53–57). Stromal cells are separated by a dense reticulin network (Fig. 41.9) and a variable amount of collagen that is most abundant in the superficial cortex. Although the latter is frequently referred to as the tunica albuginea, it lacks the densely collagenous, almost acellular appearance and sharp delineation of the tunica albuginea of the testis.

A variety of other cells may be found within the ovarian stroma, most of which are probably derived from the cells of fibroblastic type. Luteinized stromal cells lie in the stroma at a distance from the follicles and are found singly or in small nests, most often in the medulla. These polygonal cells contain abundant eosinophilic to clear cytoplasm with variable amounts of lipid, a central round nucleus, and a prominent nucleolus (Fig. 41.10). Luteinized stromal cells are typically immunoreactive for inhibin (58–60), calretinin (61), melan-A (62), CD10 (63), and occasionally, testosterone (64). The numbers of luteinized stromal cells increase during pregnancy and after the menopause, likely secondary to elevated levels of circulating gonadotropins during these periods (22,52). In one autopsy study, luteinized stromal cells were demonstrated after diligent searching in 13% of the women under the age of 55 years and in one-third of the women over that age; the frequency of their detection increased with increasing degrees of stromal proliferation (22). More exhaustive sampling might indicate that luteinized stromal cells are a normal finding in the ovary, particularly in later life. In this age group, the presence of luteinized cells is not usually associated with clinical evidence of a hormonal disturbance. In some older women, but more often in younger patients, however, more striking

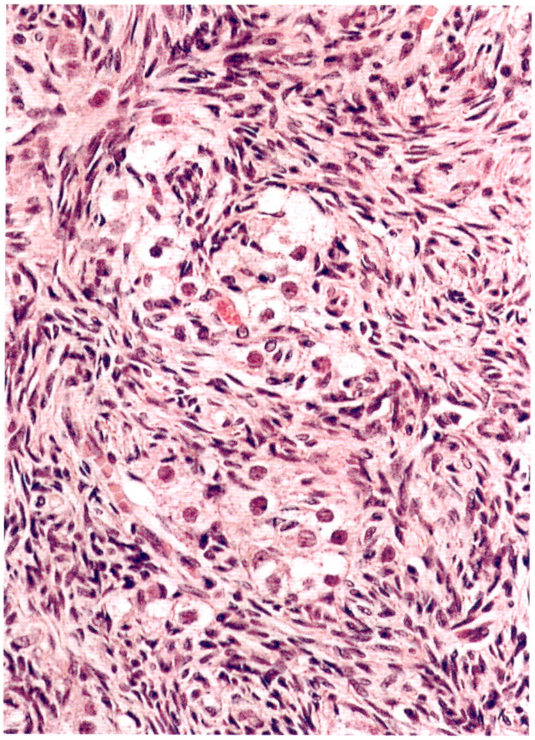

FIGURE 41.10 Luteinized stromal cells.

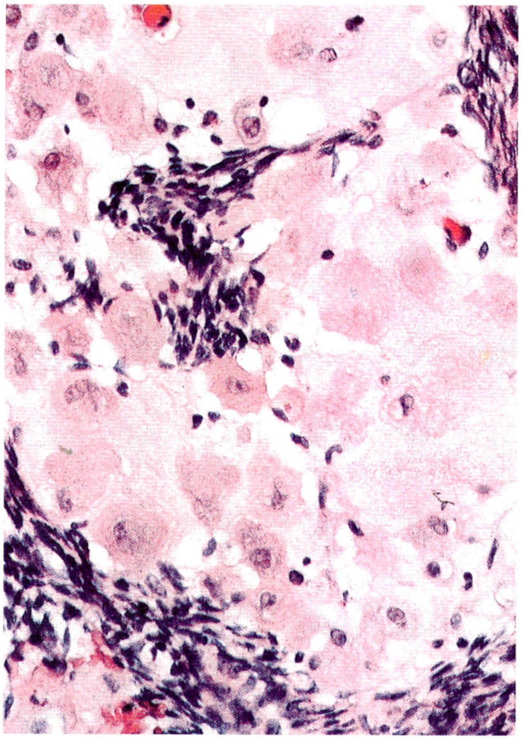

FIGURE 41.11 A nest of decidual cells within the ovarian stroma.

degrees of stromal luteinization (stromal hyperthecosis) are frequently associated with androgenic and estrogenic manifestations. In such cases, nodules of luteinized stromal cells may be appreciable on low-power microscopic examination (i.e., nodular hyperthecosis).

Enzymatically active stromal cells are characterized by their oxidative and other enzymatic activities (52,65,66). The frequency of their detection and their numbers increase with age, occurring in over 80% of postmenopausal women, typically in the medulla (65,66). Some enzymatically active stromal cells correspond to luteinized stromal cells, but most cannot be distinguished from neighboring, nonreactive stromal cells in routine histologic preparations (65).

Decidual cells may occur singly, as small nodules, or as confluent sheets within the stroma of the superficial cortex or within periovarian adhesions (Fig. 41.11). The appearance of the decidual cells is usually identical to eutopic decidua, but occasional examples may exhibit cytologic atypia potentially mimicking metastatic carcinoma on histologic examination (67–72). A network of capillaries and a sprinkling of lymphocytes are typically present within the decidual foci. A decidual reaction within the ovary is almost always a response of the ovarian stromal cells to elevated circulating or local levels of progesterone; the process is seen most commonly in pregnancy and occurs as early as the 9th week of gestation. Less commonly, it may occur in association with trophoblastic disease, in patients treated with progestins, in the vicinity of a corpus luteum, or in association with hormonally active, hyperplastic or neoplastic ovarian lesions (22,67,69). Foci of ovarian decidua have been occasionally described in both pre- and postmenopausal women with no obvious cause (22,69).

Foci of *smooth muscle* may be seen within the ovarian stroma (Fig. 41.12), most commonly in peri- or postmenopausal

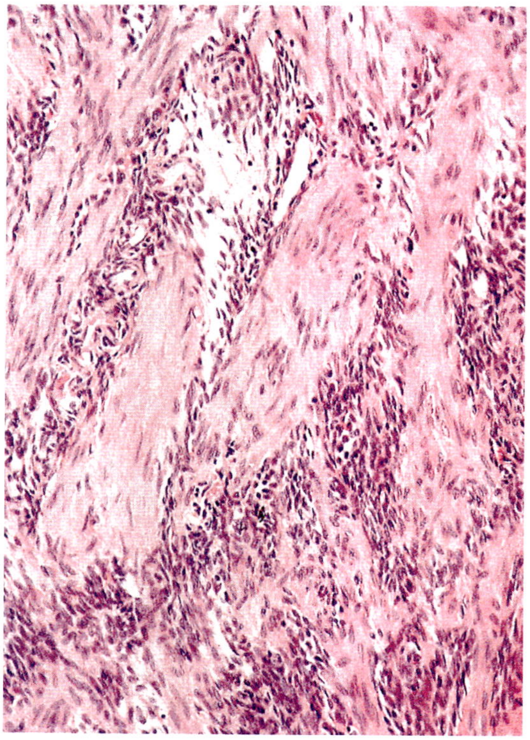

FIGURE 41.12 Smooth muscle cells within ovarian stroma.

women (73). The smooth muscle is bilateral in about 25% of cases and usually is confined to a few microscopic fields. It is often associated with other findings in the ovary, occurring with the hyperplastic ovarian stroma associated with stromal hyperthecosis or sclerocystic ovaries (74), and within the stroma surrounding nonneoplastic and neoplastic cysts, including endometriotic cysts. Rare endometriotic cysts may show endomyometriosis with prominent amounts of smooth muscle (75). One study (73) found that almost 90% of women with ovarian smooth muscle metaplasia had uterine leiomyomas.

Nests of cells resembling *endometrial stromal cells* ("stromal endometriosis") occur within the ovarian stroma, usually in the absence of typical endometriosis (Fig. 41.13) (76,77). Foci of mature *fat cells* may be encountered as an incidental histologic finding within the subcapsular ovarian stroma (78,79). Reinke crystal–containing Leydig cells, presumably representing transformed stromal cells, may occur rarely, and are typically associated with stromal hyperthecosis or within the nonneoplastic stroma in or adjacent to an ovarian neoplasm (80–82). *Neuroendocrine-type cells* have been demonstrated within the ovarian stroma in approximately 6% of the normal women (83). These cells occur in small groups in the corticomedullary stromal junction and are argyrophilic and argentaffinic. Their clinical significance and hormonal function, if any, is unknown, but it has been suggested that they may represent the cell of origin of rare primary ovarian carcinoid tumors unaccompanied by teratomatous or mucinous elements.

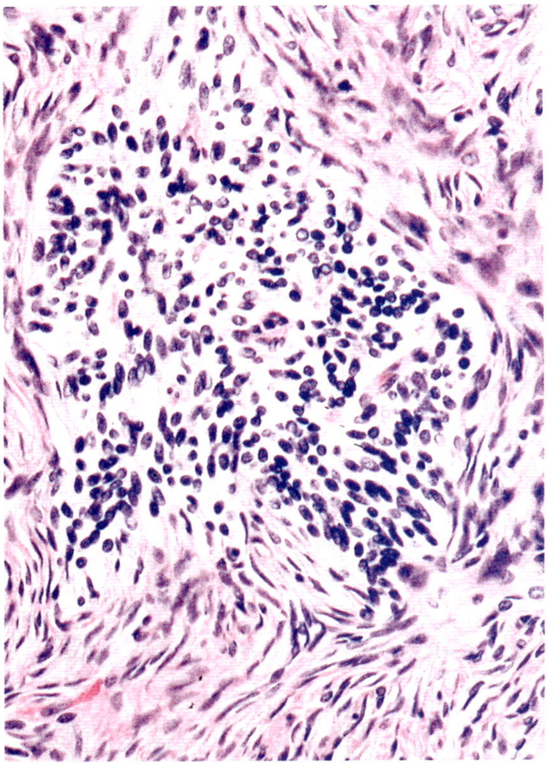

FIGURE 41.13 Focus of endometrial stromal cells ("stromal endometriosis") within ovarian cortex.

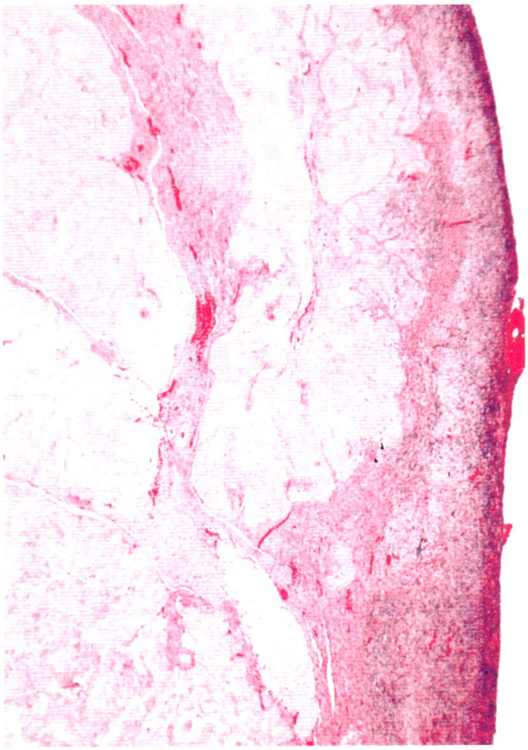

FIGURE 41.14 Atrophic postmenopausal ovary. The cortex is thin and multiple corpora albicantia are present within the medulla.

Aging Changes

Although there is typically a gradual increase in its volume from the fourth to the seventh decades (21,84), the ovarian stroma in postmenopausal women exhibits a wide spectrum of appearances (22,50,68). At one extreme, there is stromal atrophy manifested by a thin cortex and minimal amounts of medullary stroma (Fig. 41.14). At the other extreme, there is marked stromal proliferation warranting the designation "stromal hyperplasia." Most postmenopausal subjects, however, exhibit varying degrees of nodular or diffuse proliferation of the cortical and medullary stromal cells that lie between these two extremes (Fig. 41.15) (22,65), making the "normal" appearance difficult to define. Broad irregular areas of cortical fibrosis may be encountered in peri- and postmenopausal ovaries (22). When well-circumscribed, these foci resemble a small fibroma, but this designation is applied to lesions 1 cm or greater in diameter. A similar size limit could be used to distinguish between the foci of surface stromal papillarity commonly encountered in this age group (Fig. 41.16) and serous surface papillomas. Cortical "granulomata" are common incidental microscopic findings in the late reproductive and postmenopausal age groups, having been demonstrated in up to 45% of women over the age of 40 years (22,76,77,85–87). They consist of spherical circumscribed aggregates of epithelioid cells, lymphocytes, multinucleated giant cells, and anisotropic fat crystals (Fig. 41.17). Cortical granulomata and the spherical, cloud-like, hyaline scars (Fig. 41.18) present within the superficial

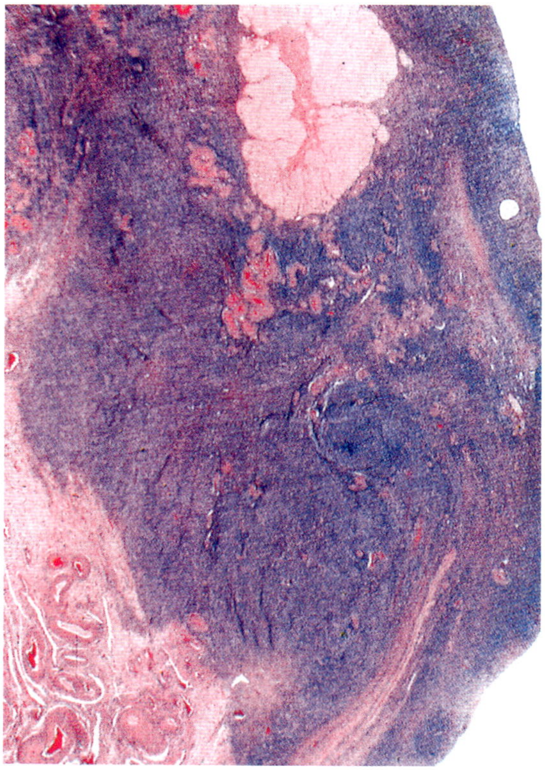

FIGURE 41.15 Postmenopausal ovary with a moderate degree of stromal proliferation.

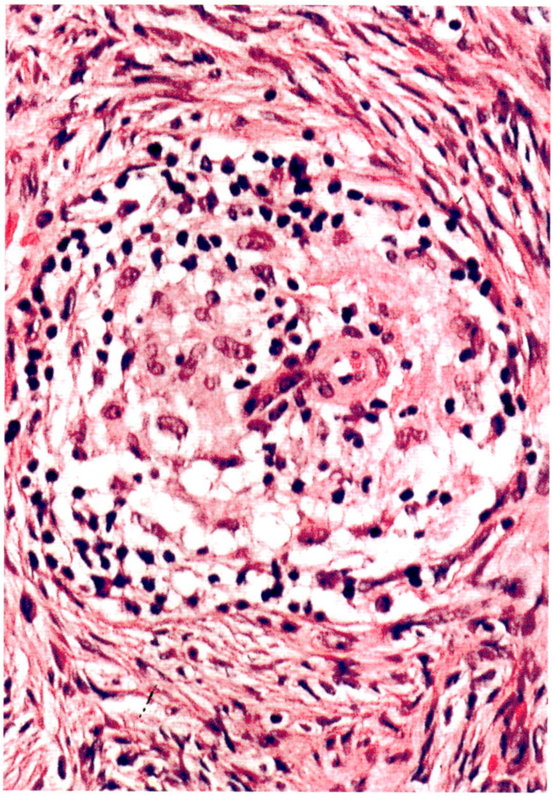

FIGURE 41.17 Cortical granuloma.

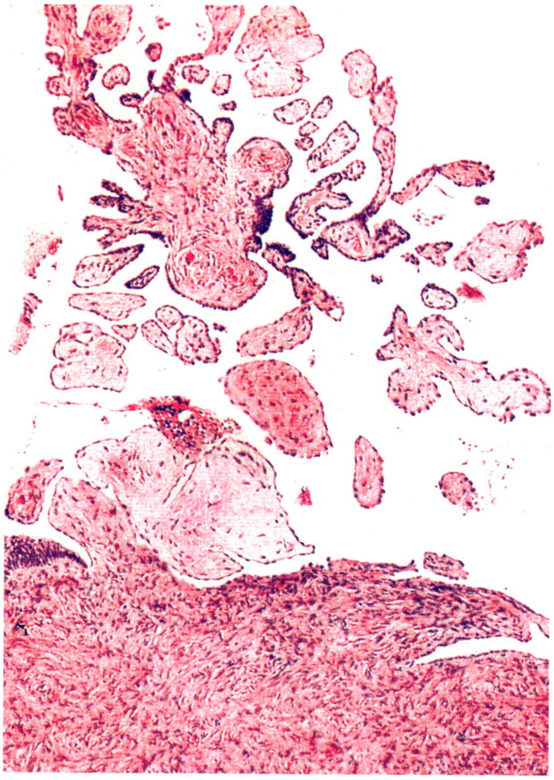

FIGURE 41.16 Papillary stromal projections from ovarian surface.

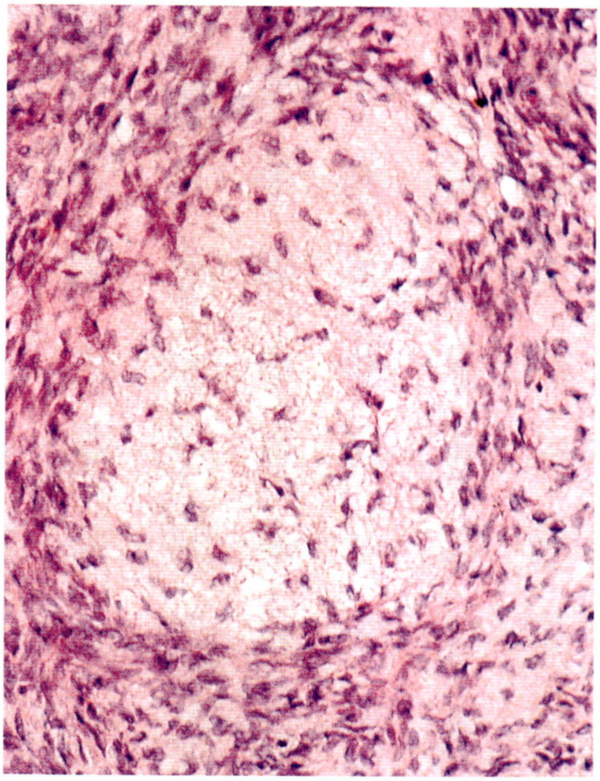

FIGURE 41.18 Hyaline scar.

cortical stroma of almost all postmenopausal ovaries are of uncertain histogenesis. It has been suggested that they may represent regressed foci of stromal endometriosis, ectopic decidua, or luteinized stromal cells.

Ultrastructure

Typical ovarian stromal cells have slender, spindle-shaped nuclei and complex cytoplasmic processes (50,65). Their scant cytoplasm is rich in organelles required for collagen synthesis, including free ribosomes and mitochondria. Tropocollagen, concentrated at the periphery of the cytoplasm, is deposited in the extracellular space and eventually converted to collagen. Rows of micropinocytotic vesicles occur along the plasma membrane and desmosome-like attachments may be found between the cells (65). Luteinized stromal cells have abundant cytoplasm containing lipid droplets and steroidogenic organelles, including smooth endoplasmic reticulum, mitochondria with tubular cristae, and Golgi (50,65,66,88).

Hormonal Aspects

Numerous studies have demonstrated the steroidogenic potential and the gonadotropin-responsiveness of the ovarian stroma in both pre- and postmenopausal women (89–101). In vitro incubation of ovarian stromal tissue indicates that its principal steroid product is androstenedione, in addition to smaller quantities of testosterone and dehydroepiandrosterone (102). In vitro production of androgens is enhanced by human chorionic gonadotropin (hCG), pituitary gonadotropins, and insulin, consistent with the presence of receptors for these hormones within the stromal cells (63,66,103). To what extent the ovarian stroma contributes to the androgen pool in normal premenopausal women is unknown, but it is likely that it is the source of small amounts of testosterone. With cessation of follicular activity at the time of the menopause, the ovarian stroma becomes, together with the adrenal glands, the major source of androgens. Testosterone and androstenedione are the major androgens secreted by the ovarian stroma in postmenopausal women (89–91,93,101), and both in vitro and in vivo studies have shown that ovaries with stromal hyperplasia secrete more androstenedione, estrone, and estradiol than normal ovaries (94,96). Approximately 80% of the circulating androstenedione in postmenopausal women, however, is of adrenal origin (90). Despite a cessation of follicular synthesis of estradiol (E2) in postmenopausal subjects, small amounts of this hormone are present in the circulation (probably derived from the adrenal glands) by peripheral conversion of estrone (90,104), and from the ovarian stroma itself (89,91,105). Estrone, however, becomes the major circulating estrogen after the menopause, derived predominantly from the peripheral aromatization of androstenedione that occurs in fat, muscle, liver, kidney, brain, and adrenals (90,105,106). Increased aromatization in postmenopausal women, likely due to high endogenous luteinizing hormone (LH) levels in these subjects, leads to a twofold increase in the daily production rate of estrone compared to that in premenopausal women; aromatization is also higher in obese subjects. In some postmenopausal women, sufficient estrogen is elaborated by this mechanism to prevent the clinical manifestations of estrogen withdrawal and to play a role in the genesis of endometrial carcinoma (84,90). An association between the degree of stromal proliferation and postmenopausal endometrial adenocarcinoma has been noted (84), and the ovarian stroma in postmenopausal women with endometrial adenocarcinoma produces more androgens in vitro than that of control subjects without endometrial cancer (107). The variations that exist in the ovarian steroid hormone output from one postmenopausal woman to another may correspond to similar variations in the morphologic appearance of the stroma in this age group, although no correlative functional and structural studies have been performed.

PRIMORDIAL FOLLICLES

Histology

Approximately 400,000 primordial follicles present at the time of birth fill the ovarian cortex (Fig. 41.19). After this period, their numbers decrease progressively through the processes of atresia and folliculogenesis until their

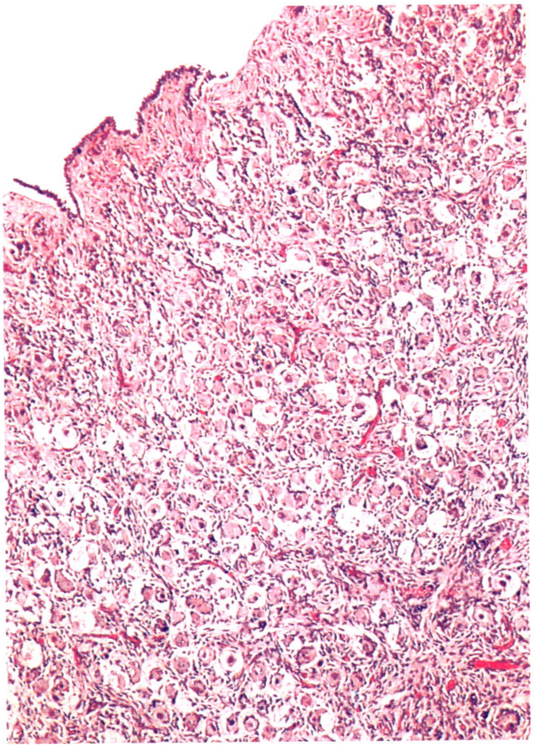

FIGURE 41.19 Newborn ovary. Multiple primordial follicles fill the ovarian cortex.

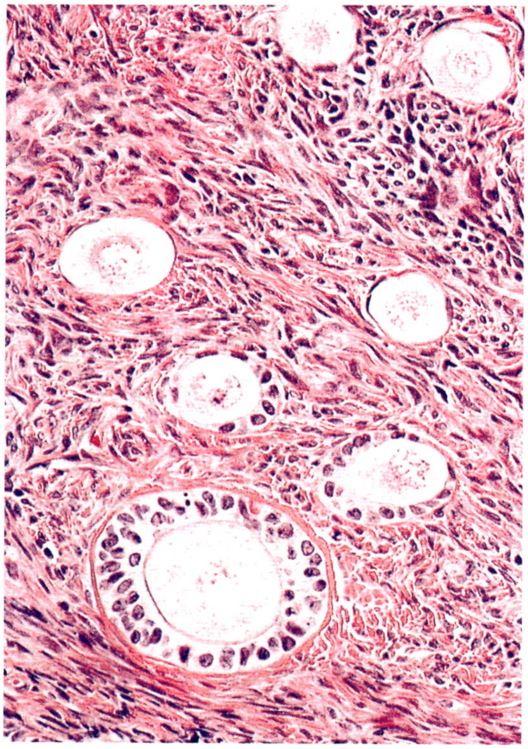

FIGURE 41.20 Primordial follicles (four at the top of the figure) and primary follicles (three at the bottom of the figure).

eventual disappearance marks the end of the menopause. Rare follicles may persist for several years after the cessation of menses, accounting for sporadic ovulation and occasional episodes of postmenopausal bleeding (108). In the reproductive era, primordial follicles are found scattered irregularly in clusters throughout a narrow band in the superficial cortex. They consist of a primary oocyte, measuring 40 to 70 μm in diameter, surrounded by a single layer of flattened, mitotically inactive, granulosa cells resting on a thin basal lamina (Fig. 41.20). Rare primordial and maturing follicles may contain multiple oocytes, particularly in individuals who are less than 20 years of age (19,109–111). The oocyte is arrested at the dictyate stage of meiotic prophase at the time of birth, enters an interphase period until follicular maturation prior to ovulation, or undergoes degeneration during atresia (112). The large spherical nucleus of the oocyte has finely granular, uniformly dispersed chromatin and one or more dense, thread-like nucleoli (112); rare oocytes may contain multiple nuclei (110,111). The cytoplasm of the oocyte contains a paranuclear, eosinophilic, crescent-shaped zone consisting of a complex of interrelated organelles, the so-called Balbiani vitelline body (113,114). Within the vitelline body is a dark spot (the centrosome) surrounded by a halo, which in turn is flanked by darker, periodic acid–Schiff (PAS)-positive, granular zones rich in mitochondria (113,114). The cytoplasm of the oocyte lacks the abundant glycogen and the high alkaline phosphatase activity characteristic of the primordial germ cells and the oogonia of the embryonic gonad.

Ultrastructure

The granulosa cells of the primordial follicle have sparse organelles, occasional desmosomal attachments with each other, and microvillus projections that attach to the oocyte by tight apposition (50). Within the oocyte, the juxtanuclear centrosome of the Balbiani vitelline body (Fig. 41.21A) consists of dense granules, closely packed vesicles, and dense fibers that form a basket-like structure at the periphery of the centrosome (Fig. 41.21B) (113,114). The centrosome is surrounded by a zone of smooth endoplasmic reticulum that represents the halo seen by light microscopy. More peripheral and constituting the rest of the Balbiani vitelline body are a concentration of most of the oocyte's organelles, including multiple Golgi complexes, prominent compound aggregates, numerous mitochondria intimately associated with sparsely granular endoplasmic reticulum, and annulate lamellae (Fig. 41.21B) (113,114). The latter structures, which may be attached or immediately adjacent to the nucleus or free within the Balbiani vitelline body, are constantly present in primary oocytes and other rapidly growing embryonal or neoplastic cells. They are arranged in stacks or concentric arrangements of up to 100 parallel, smooth, paired membranes that delineate greatly flattened cisternal spaces, 30 to 50 μm wide.

MATURING FOLLICLES

Histology and Ultrastructure

Folliculogenesis

Folliculogenesis refers to the continuous process occurring throughout reproductive life whereby cohorts of primordial follicles undergo maturation during each menstrual cycle. Follicular maturation begins during the luteal phase and continues throughout the follicular phase of the next cycle. Each month only one such follicle, the preovulatory (or dominant) follicle, achieves complete maturation, culminating in the release of the oocyte (ovulation). The other follicles that have begun the maturational process undergo atresia at earlier stages of their development. Folliculogenesis and atresia also occur prenatally, throughout childhood and during pregnancy, although maturing follicles rarely reach the preovulatory follicle stage during these periods (19,115–123).

The first morphologic evidence of follicular maturation is the assumption of a cuboidal to columnar shape of the granulosa cells accompanied by enlargement of the oocyte (*primary follicle*) (Fig. 41.20). Mitotic activity in the granulosa cells results in their stratification and three to five concentric layers around the oocyte (*secondary or preantral follicle*) (Fig. 41.22). At this stage an eosinophilic, PAS-positive, homogeneous, acellular layer, known as the zona pellucida, appears, encasing the oocyte. Its formation is usually attributed to the granulosa cells, but the oocyte may also play a role. At the end of its development, the zona pellucida is a 20 to 25 μm thick membrane rich in acid mucopolysaccharides and glycoprotein (Figs. 41.22 to 41.25) (50).

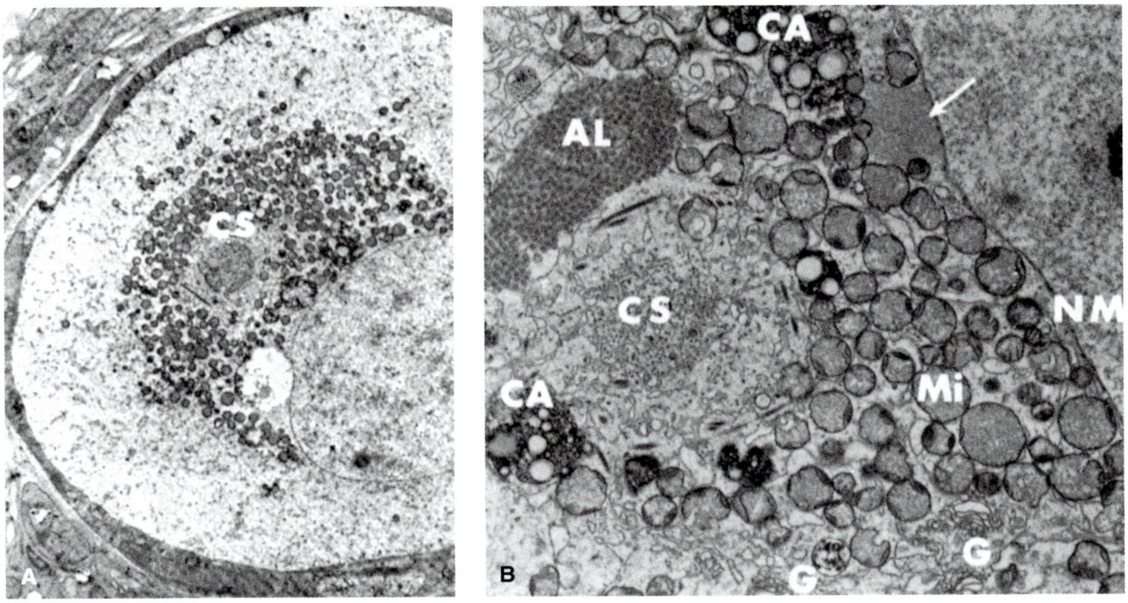

FIGURE 41.21 Electron micrograph of a primordial follicle. **A:** Balbiani vitelline body consists of a centrosome (*CS*) surrounded by a condensation of mitochondria, Golgi complexes (*G*), endoplasmic reticulum, and lysosomes (original magnification ×2,400). **B:** Detailed view of Balbiani vitelline body. A cluster of closely packed spiral fibrils (*arrow*) is attached to the nuclear envelope (*NM*). The centrosome (*CS*) is composed of dense granules, some arranged periodically on fine fibers, and small vesicles, with a peripheral zone of endoplasmic reticulum and dense fibers. Surrounding the centrosome are masses of mitochondria (*Mi*) and compound aggregates (*CAs*). A stack of annulate lamellae (*AL*) is seen tangentially. Note the prominent endoplasmic reticulum in close association with multiple Golgi complexes at the periphery of the vitelline body. (Reprinted with permission from Hertig AT. The primary human oocyte: Some observations on the fine structure of Balbiani's vitelline body and the origin of the annulate lamellae. *Am J Anat.* 1968;122:107–137.)

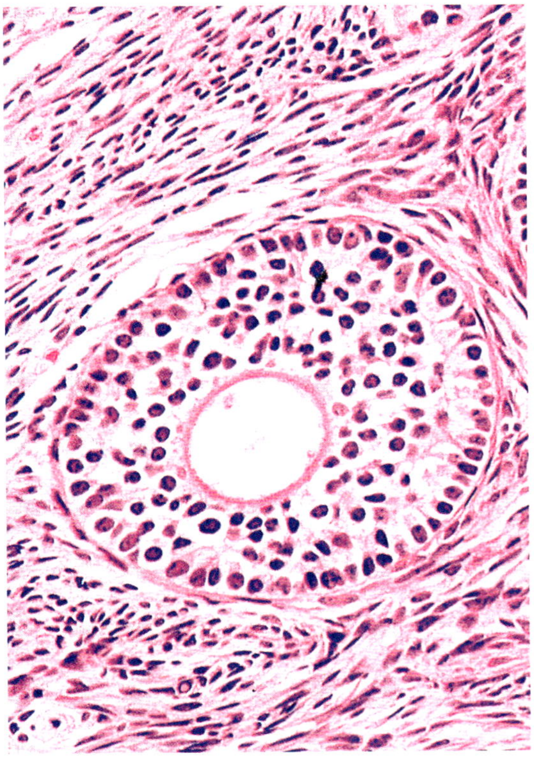

FIGURE 41.22 Preantral follicle. Several layers of granulosa cells surround the oocyte. A theca interna layer is not yet apparent.

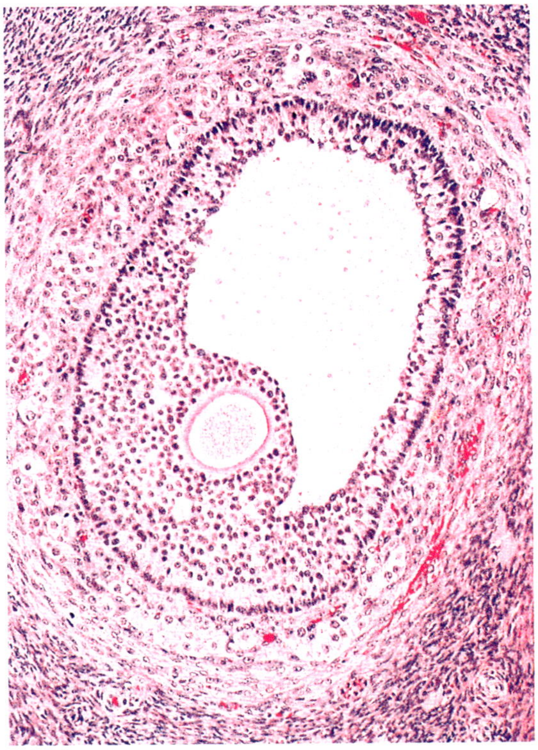

FIGURE 41.23 Mature follicle. Oocyte within cumulus oophorus projects into antrum. The theca layers are well developed.

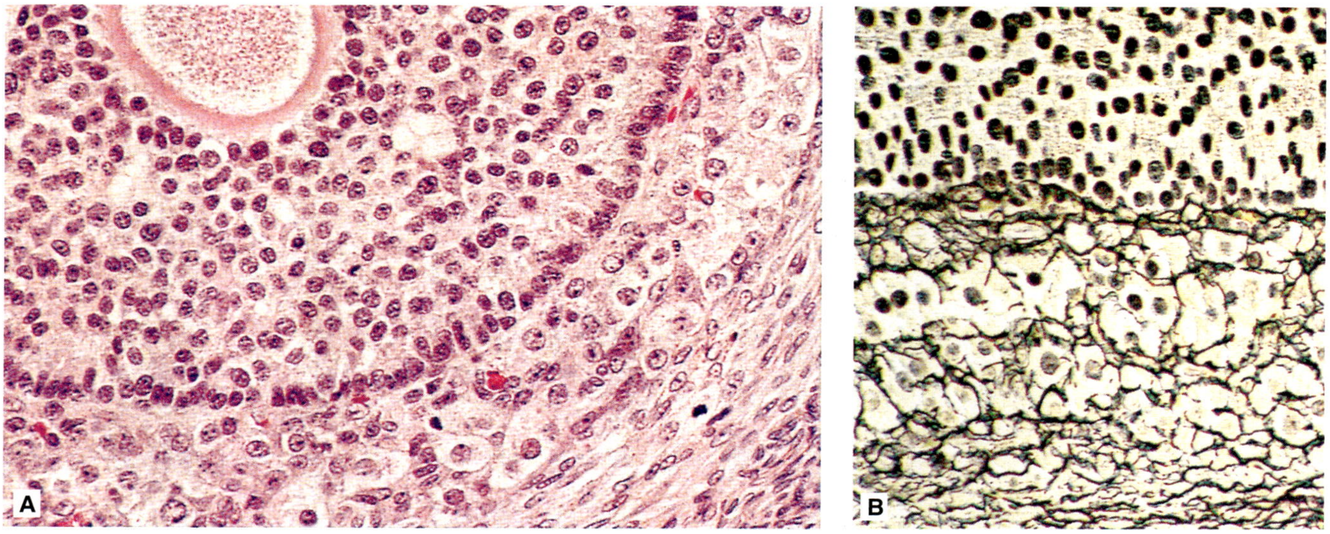

FIGURE 41.24 **A:** Mature follicle, high-power view. The granulosa layer, which contains several Call–Exner bodies, abuts the zona pellucida of the oocyte. The granulosa layer is surrounded by a layer of luteinized theca interna cells. Note mitotic figures in granulosa and theca cells. **B:** There is a reticulin network in the theca interna layer, but an absence of reticulin in the granulosa layer.

Preantral follicles measure from 50 to 400 μm in diameter, and as they increase in size, they migrate into the deeper cortex and medulla. Simultaneously, the surrounding ovarian stromal cells become specialized into several layers of theca interna cells and an outer, ill-defined layer of theca externa cells. Secretion of mucopolysaccharide-rich fluid by the granulosa cells results in their separation by fluid-filled clefts that eventually coalesce to form a single large cavity or antrum lined by several layers of granulosa cells (*tertiary, antral, or vesicular follicle*). The first evidence of antrum formation occurs in follicles that are from 200 to 400 μm in diameter, after which the follicles progressively enlarge due to continued fluid secretion into the antrum. Concurrently, the oocyte enlarges to its definitive size and assumes an eccentric position at one pole of the follicle. At this site, the granulosa cells proliferate to form the cumulus oophorus which, containing the oocyte in its center, protrudes into the antrum (*mature or graafian follicle*) (Fig. 41.23).

Ovulation

During each cycle, only a small number of mature follicles (<4 per ovary) reach a diameter of 4 to 5 mm by the mid- to late luteal phase; one of them will become the preovulatory follicle of the subsequent cycle (124,125). Late in follicular growth, the oocyte, its surrounding zona pellucida, and a single layer of radially disposed, columnar granulosa cells (the corona radiata) detach from the cumulus oophorus and float in the antral fluid. The preovulatory follicle, shortly before ovulation, reaches a diameter of 15 to 25 mm (28,125) and partially protrudes from the ovarian surface at the eventual rupture point, or stigma. Here the overlying surface epithelial cells exhibit progressive flattening, degeneration, and desquamation. The stroma in this area becomes attenuated and almost avascular, with degeneration of the stromal cells, fragmentation of collagen fibers, and an accumulation of intercellular fluid (28). These surface epithelial and stromal changes immediately preceding ovulation may be secondary to local ischemia and the release of proteolytic enzymes and prostaglandins into the stroma. The preovulatory follicle then ruptures with liberation of the follicular fluid and oocyte (with its surrounding granulosa cell layers, the corona radiata) into the peritoneal cavity. Following ovulation, the stigma is occluded by a mass of coagulated

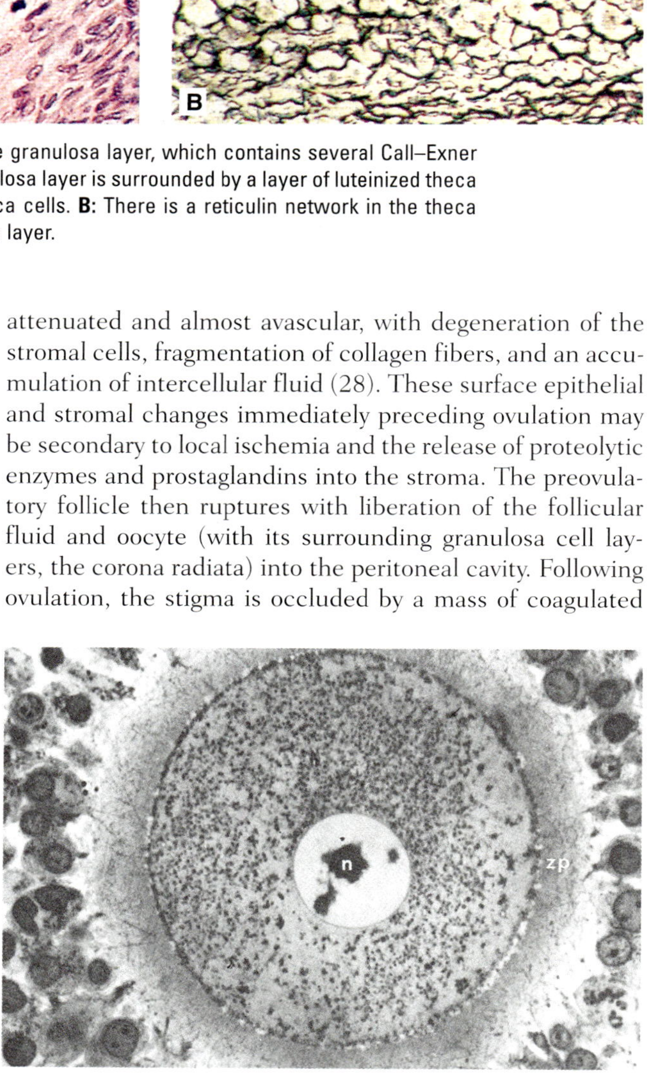

FIGURE 41.25 Maturing oocyte. Note the uniform distribution of the organelles and the row of dense granules in the cytoplasm immediately subjacent to the plasma membrane of the oocyte. A continuous zona pellucida (*zp*) surrounds the oocyte and separates it from the granulosa cells. Numerous cytoplasmic processes of the granulosa cells are visible within the zona pellucida. *n*, nucleolus. Thick section, OsO_4 fixed, Epon-embedded, toluidine blue stain. (Reprinted with permission from Baca M, Zamboni L. The fine structure of the human follicular oocytes. *J Ultrastruct Res*. 1967;19:354–381, with permission from Elsevier.)

follicular fluid, fibrin, blood, granulosa, and connective tissue cells; it is eventually converted to scar tissue.

Shortly before ovulation, the oocyte within the ovulatory follicle enters telophase of the first meiotic division. Chromosomal reduction occurs by the migration of one-half of the oocyte chromosomes into a portion of the oocyte cytoplasm that separates from the cell as the first polar body. The first meiotic division begun in fetal life is now complete, and the oocyte is now designated the secondary oocyte. Immediately after expulsion of the first polar body, the secondary oocyte enters the second meiotic division, arresting at metaphase until fertilization occurs.

Granulosa Layer

Granulosa cells are almost entirely formed from their embryonic precursors by the time of birth (19). Those within maturing and mature follicles are polyhedral cells from 5 to 7 μm in diameter; the cells resting on the basement membrane are often columnar. The granulosa cells have pale, scant cytoplasm, indistinct cell borders, and small, round to ovoid, hyperchromatic nuclei that typically lack nuclear grooves (Fig. 41.24A) (126). Mitotic figures within granulosa cells are usually numerous in maturing follicles but decrease in number prior to ovulation. Until the onset of luteinization several hours prior to ovulation, cytoplasmic lipid is absent (or sparse) as are the steroidogenic histochemical patterns (127–128). The cytoplasm of granulosa cells of primary, secondary, and mature follicles is immunoreactive for cytokeratin, vimentin, inhibin, CD99, melan-A, müllerian inhibiting substance, and desmoplakin, whereas they show nuclear immunoreactivity for WT1 and FOXL2 (53–55,58,62,129).

The granulosa cells typically surround small cavities, referred to as Call–Exner bodies (Fig. 41.24A), which have a distinctive appearance, representing one of the most specific features of granulosa cells, both normal and neoplastic. Call–Exner bodies are delimited from the granulosa cells by a basal lamina, and typically contain a deeply eosinophilic, PAS-positive, filamentous material consisting of excess basal lamina (50). Unlike the theca layers, the granulosa layer of the maturing and graafian follicles is avascular and devoid of a reticulin framework (Fig. 41.24B).

Mitochondria with lamelliform cristae, granular endoplasmic reticulum, free ribosomes, and Golgi gradually increase in abundance within the granulosa cells of maturing follicles. These ultrastructural features suggest active protein synthesis. Histochemical and ultrastructural features (abundant smooth endoplasmic reticulum and mitochondria with tubular cristae) indicative of steroid biosynthesis are absent until shortly before ovulation (52,65,127,128,130). The granulosa cells of follicles of varying stages contain adherens junctions, gap junctions, and desmosomes between adjacent granulosa cells (50,54,112). The slender cytoplasmic extensions of the granulosa cells of the corona radiata that traverse the zona pellucida have gap junctions and puncta adhaerentia with the plasma membrane of the oocyte (Fig. 41.25).

Theca Layers

In contrast to granulosa cells, theca cells differentiate continuously from the stromal cells at the periphery of developing follicles from fetal life until the end of menopause. The thecal component of the antral follicle is characterized by a well-developed theca interna and a less well-defined theca externa. The theca interna layer is three or four cells thick and lies external to the granulosa layer (Fig. 41.24A) from which it is separated by a basement membrane. Unlike the granulosa cells of the developing and mature follicles, the theca interna cells typically have a luteinized or partially luteinized appearance (Fig. 41.24A) and exhibit steroidogenic histochemical patterns (52,127,128). Luteinization of the theca interna of maturing follicles is particularly prominent during pregnancy. The round to polygonal cells range from 12 to 20 microns in diameter and have abundant, eosinophilic to clear, vacuolated cytoplasm containing variable amounts of lipid; a central, round, vesicular nucleus typically contains a single, prominent nucleolus (Fig. 41.24A). The cells differ from granulosa cells but resemble stromal cells in that they are immunoreactive for vimentin but not cytokeratin (55); theca cells are also immunoreactive for inhibin, calretinin, melan-A, and FOXL2 (62,129). Mitotic figures are typically present in the theca cells of maturing follicles. The layer contains a rich vascular plexus consisting of dilated capillaries, as well as a dense reticulin network that surrounds each cell (Fig. 41.24B). A tangential section through the theca interna may result in seemingly isolated nodules of luteinized theca cells that may occasionally be misinterpreted as foci of stromal luteinization.

The theca externa is an ill-defined layer of variable thickness that surrounds the theca interna and merges almost imperceptibly with the adjacent ovarian stroma. It is composed of circumferentially arranged collagen bundles, blood and lymphatic vessels, and plump spindle cells that lack steroidogenic histochemical features (131). The spindle cells of the theca externa are typically highly mitotic and may be misinterpreted as fibrosarcoma, particularly when only the edge of the follicle is seen microscopically (Fig. 41.26).

Ultrastructural examination of theca interna cells reveals the organelles associated with steroidogenesis, similar to those within granulosa-lutein cells. The theca externa cells, some of which exhibit smooth muscle differentiation, lack such organelles (132).

Hormonal Aspects

The initiation of folliculogenesis and early preantral follicular development is independent of gonadotropin influence, whereas the later stages of follicular maturation are under gonadotropin control. As a small antral follicle develops into a preovulatory follicle, the sequence of endocrine events within its antral fluid differs from most, if not all, other antral follicles in the same ovary (133,134). The early stages of this development are associated with an increase in follicle-stimulating hormone (FSH) receptors and

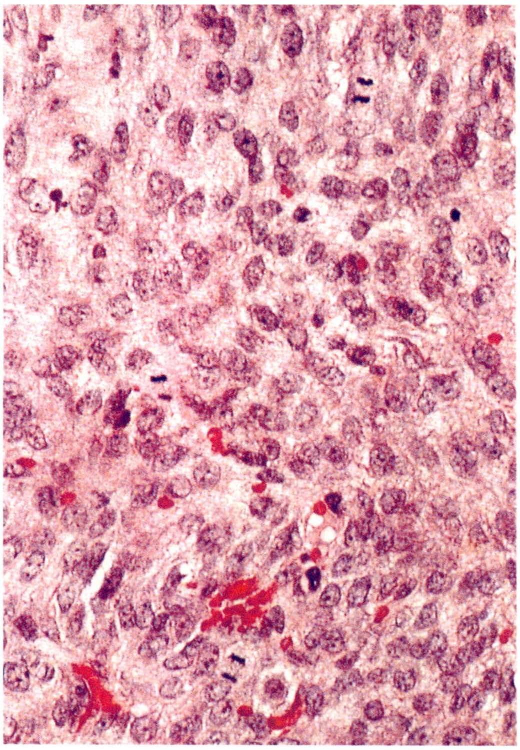

FIGURE 41.26 Theca externa of mature follicle composed of plump spindle cells. Note mitotic figures.

intrafollicular FSH within the preovulatory follicle (133–136). There is a concomitant increase in estradiol (E2) receptors within the granulosa cells and the E2 level within the follicular fluid. The latter reaches peak concentration approximately 10,000 times the circulating level during the mid- to late proliferative phase when plasma FSH levels fall to a basal level. At this stage, the preovulatory follicle is self-sustaining, continuing to mature under the influence of intrafollicular FSH and E2 (98). During the late proliferative phase, plasma LH rises and LH receptors within the granulosa cells of the preovulatory follicle, but not other follicles, become apparent (136). In contrast, LH receptors are present within the theca cells of all follicles throughout the follicular phase.

While circulating E2 is likely derived from both the granulosa cells and the LH-stimulated theca cells, intrafollicular E2 is derived almost exclusively from the granulosa cells by both de novo synthesis and FSH-dependent aromatization of theca-derived androstenedione (98,137). Aromatase activity is highest in the preovulatory follicle, thereby maintaining a high E2:androstenedione ratio (133,134,138). In contrast, follicles destined to undergo atresia are FSH- and aromatase-deficient and have high androstenedione:E2 ratios within their intrafollicular fluid. High circulating estrogen levels initiate a preovulatory surge of plasma LH (139,140) that induces luteinization of the granulosa cells, an increase in intrafollicular progesterone concentration, and a small preovulatory rise in circulating progesterone (133,134,141). The rising plasma progesterone level and the peaking estrogen level further augment the LH surge, as well as initiating a smaller increase in FSH, triggering ovulation.

The ovarian follicles also produce nonsteroidal hormones. Inhibin, a glycoprotein synthesized by the granulosa cells, is secreted into the follicular fluid and ovarian venous effluent in amounts that correlate with steroid levels (141–143). Inhibin, which is predominantly under the control of LH (144), reduces FSH secretion from the hypothalamic–pituitary unit through negative feedback mechanisms. High concentrations of prorenin are present within the fluid of mature follicles (145); their granulosa cells, theca, and stromal cells are immunoreactive for renin and angiotensin II (146). The renin–angiotensin system within the ovary has been linked to circulatory support during pregnancy (147).

CORPUS LUTEUM OF MENSTRUATION

Following ovulation on the 14th day of the typical 28-day menstrual cycle, and in the absence of fertilization, the collapsed ovulatory follicle becomes the corpus luteum of menstruation. When mature, the corpus luteum of menstruation is a 1.5 to 2.5 cm, round, yellow structure with festooned contours and a cystic center filled with a gray, focally hemorrhagic coagulum.

Histology

During the 14 days following ovulation, the corpus luteum of menstruation undergoes an orderly sequence of histologic changes that, in combination with endometrial histology and menstrual data, allow an approximate estimation of its age (148,149). However, retrospective timing of ovulation based on assessment of the corpus luteum of menstruation is subject to an error of variable magnitude due to variability in the length of each stage, as well as considerable variation in corpus luteum evolution among individuals (150).

In contrast to the granulosa cells of the maturing and preovulatory follicles, the luteinized granulosa cells of the mature corpus luteum of menstruation (granulosa-lutein cells) are large, 30 to 35 μm, polygonal cells with abundant, pale, eosinophilic cytoplasm that may contain numerous small lipid droplets (Figs. 41.27 and 41.28) (131). The spherical nucleus contains one or two large nucleoli. The histochemical pattern of these cells varies with the age of the corpus luteum of menstruation but is generally typical of steroid hormone–producing cells (52,128,151). The cytoplasm of luteinized granulosa cells contains vimentin but in contrast to granulosa cells of maturing and mature follicles, little or no cytokeratin (54). The luteinized granulosa cells are also immunoreactive for inhibin and calretinin (152,153).

The theca interna forms an irregular and often interrupted layer, several cells in thickness around the

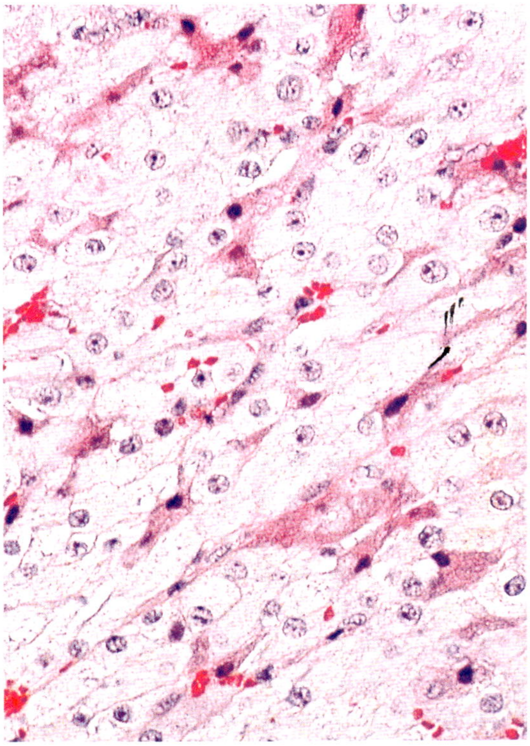

FIGURE 41.27 Mature corpus luteum of menstruation. K cells with darkly staining cytoplasm and pyknotic nuclei are interspersed between granulosa-lutein cells.

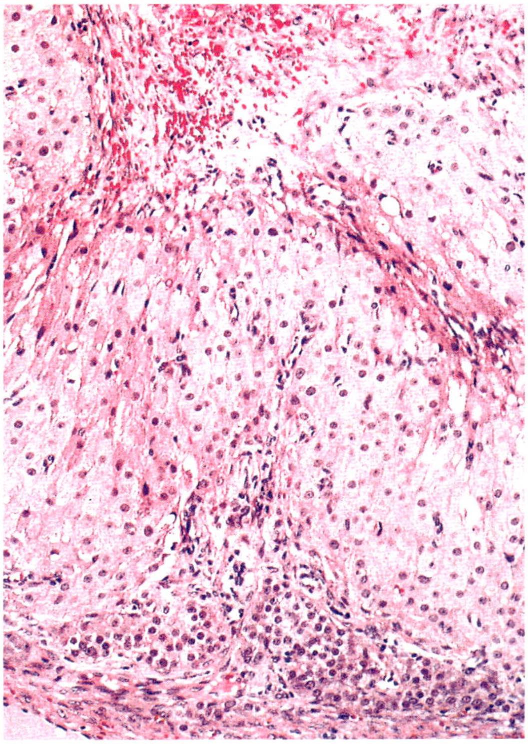

FIGURE 41.29 Mature corpus luteum of menstruation. The lining is composed of a thick layer of large granulosa-lutein cells and an outer, thinner layer of smaller theca-lutein cells. The cavity (top of the figure) contains erythrocytes and fibrin.

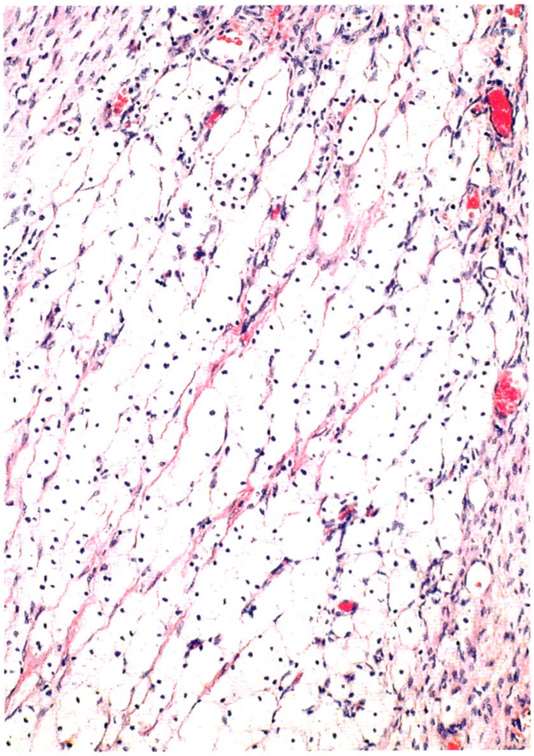

FIGURE 41.28 Degenerating corpus luteum of menstruation. Granulosa-lutein cells have pyknotic nuclei and abundant cytoplasmic lipid.

circumference of the corpus luteum of menstruation (Fig. 41.29), and ensheaths the vascular septa that extend into its center (131). When these septa are cut in cross-section, triangular-shaped nests of theca cells appear at intervals throughout the granulosa layer. In all but the earliest stages of the corpus luteum of menstruation, the theca lutein cells are approximately half the size of granulosa-lutein cells. They contain a round to oval nucleus with a single prominent nucleolus. Their less abundant, more darkly staining cytoplasm contains lipid droplets, which are usually larger than those in granulosa-lutein cells, and exhibits steroidogenic histochemical patterns (128), and immunoreactivity for inhibin, calretinin, and melan-A (62,152,153).

A third type of cell, the so-called "K" cell, occurs in small numbers within the theca interna of the mature follicle and appears in greater numbers within the granulosa layer of the early corpus luteum of menstruation (126). K cells persist until menstruation at which time they degenerate. They are characterized by a stellate shape, a deeply eosinophilic cytoplasm, and an irregular, hyperchromatic or pyknotic nucleus (Fig. 41.27). The cytoplasm is uniformly sudanophilic due to the presence of phospholipid (126). K cells lack the histochemical patterns of steroidogenic cells and have been shown to be T lymphocytes (154).

During the maturation of the corpus luteum of menstruation, capillaries originating from the theca interna layer penetrate the granulosa layer and reach the central cavity.

Fibroblasts that accompany the vessels form an increasingly dense reticulin network within the granulosa layer, as well as an inner fibrous layer that lines the central cavity (Fig. 41.28) (32).

Involutional changes begin on the 8th or 9th day following ovulation (148). The granulosa-lutein cells decrease in size, develop pyknotic nuclei, and accumulate abundant cytoplasmic lipid (Fig. 41.28). There is a decrease in histochemical staining of enzymes associated with steroid biosynthesis and an increase in hydrolytic enzymes (129). Eventually, the cells undergo dissolution and are phagocytosed (155). There is progressive fibrosis and shrinkage over a period of several months and eventual conversion to a corpus albicans.

Ultrastructure

At the ultrastructural level, luteinization is characterized by a gradually increasing content of steroidogenic organelles, specifically smooth endoplasmic reticulum and abundant mitochondria with tubular cristae (Fig. 41.30) (131,155–158). The smooth endoplasmic reticulum exhibits a characteristic regional modification in the form of a folded-membrane complex consisting of highly folded, radiating, tubular cisternae that communicate and interdigitate with adjacent cisternae (157). Well-developed, dispersed and perinuclear Golgi, free and bound ribosomes, lipid droplets, and lipofuscin pigment are also seen (Fig. 41.30) (131,155,156,157). Many irregular microvillus cytoplasmic extensions project into these pericapillary as well as the intercellular spaces (50,131,155,157). Theca-lutein cells are ultrastructurally similar to granulosa-lutein cells except for the presence of localized perinuclear Golgi and the absence of folded-membrane complexes, microvilli, and a network of fine filaments (131,157). The lutein cells of the degenerating corpus luteum of menstruation exhibit disorganization and fragmentation of the smooth endoplasmic reticulum, alterations of the mitochondria, and an increase in cytolysosomes (50). Lipid droplets are increased and irregular in size and show increased osmiophilia (155).

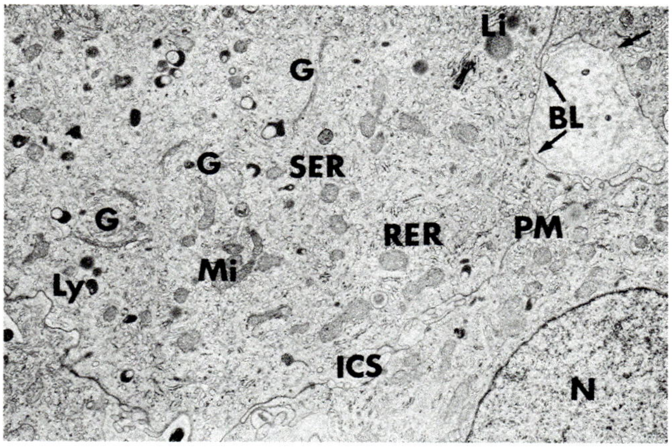

FIGURE 41.30 Electron micrograph of granulosa-lutein cell of a mature corpus luteum of menstruation. Note abundant smooth endoplasmic reticulum (*SER*), mitochondria (*Mi*), Golgi complex (*G*), rough endoplasmic reticulum (*RER*), lipid droplets (*Li*), and intercellular space (*ICS*). *BL*, basal lamina; *N*, nucleus of granulosa-lutein cell; *Ly*, lysosomes; *PM*, plasma membrane; *arrows*, micropinocytotic vesicle (original magnification ×3,600). (Reprinted with permission from Ferenczy A, Richart RM. *Female Reproductive System: Dynamics of Scan and Transmission Electron Microscopy*. John Wiley & Sons; 1974.)

Hormonal Aspects

The formation and function of the corpus luteum of menstruation are under the control of LH, reflected by the high content of LH receptors within the granulosa-lutein cells (136,142). Receptors for FSH (120) and growth hormone (138) have also been identified in the corpus luteum, although their roles in luteal function are unknown. Although progesterone is the major steroid hormone formed by the corpus luteum of menstruation in vivo and in vitro, it also synthesizes estrone and E2, as well as androgens, mostly androstenedione (159).

After ovulation, LH, FSH, and E2 levels fall, but the LH concentration is sufficient to maintain the corpus luteum of menstruation, producing a midluteal peak in progesterone and E2. If fertilization does not occur, the increased levels of progesterone and estrogen through negative feedback loops result in a fall of LH and FSH to basal levels, a reduction in LH and FSH receptors within the corpus luteum of menstruation, and a marked decline in progesterone and E2 synthesis after the 22nd day of the cycle (135,136,142,160,161). These changes are reflected by the morphologic involution of the corpus luteum of menstruation and the onset of menses. Luteolysis appears to be estrogen-related, possibly secondary to an estrogen-induced reduction in LH receptors or by enhancement of the luteolytic action of prostaglandins synthesized by the corpus luteum of menstruation (142,162). A nonsteroidal LH-receptor–binding inhibitor, which increases in concentration during the luteal phase, may also play a role (142).

CORPUS LUTEUM OF PREGNANCY

Gross Appearance

On gross inspection, the corpus luteum of pregnancy may be indistinguishable from the corpus luteum of menstruation but is usually larger and bright yellow in contrast to the orange-yellow of the late corpus luteum of menstruation (163). The larger size, which may account for up to half the ovarian volume, is due primarily to the presence of a central cystic cavity that is filled with fluid or a coagulum composed of fibrin and blood (71,121,164). The cavity size, however, can be highly variable. If the central cyst results in a corpus luteum that is over 3 cm in diameter, the corpus luteum of pregnancy (or less commonly a corpus luteum of menstruation) is designated a corpus luteum cyst; if less than this size, it is considered to represent a cystic corpus luteum. When the cavity of a corpus luteum of pregnancy is large, typically in the first trimester, the wall may lose its convolutions, becoming stretched and attenuated to the extent that it may consist focally of only the inner fibrous layer. Obliteration of the cavity usually begins by the 5th month and is typically completed by term (122). The corpus luteum of pregnancy thus gradually decreases in size, and by the last trimester, it is not a conspicuous structure. During the puerperium, the corpus luteum of pregnancy undergoes involution and conversion to a corpus albicans.

Histology

The corpus luteum of pregnancy, in contrast to the corpus luteum of menstruation, does not mature in an orderly sequence that allows an estimation of its age; however, early and late stages are recognizable on histologic examination.

Granulosa Layer

The first morphologic evidence within the corpus luteum that conception has occurred is the absence of the regressive changes that normally appear in the corpus luteum of menstruation on the 8th or 9th days. Instead, the granulosa lutein cells enlarge, reaching their maximum size of 50 to 60 μm by 8 to 9 weeks of gestation. They assume a round or polyhedral shape with abundant eosinophilic cytoplasm, round to ovoid vesicular nuclei, and one or two prominent nucleoli (122) (Fig. 41.31A). The granulosa cells of the early corpus luteum of pregnancy are characterized by cytoplasmic vacuoles that are initially tiny but eventually enlarge to occupy almost the entire cell, often with displacement and flattening of the nucleus (Fig. 41.31A). The vacuoles tend to diminish in number and size as gestation progresses and usually disappear after the 4th month. Fine, diffusely scattered, cytoplasmic lipid droplets are also commonly seen within the cells, particularly in early corpus luteum of pregnancy. With increasing age of the corpus luteum, the droplets become fewer and larger (122).

Eosinophilic colloid or hyaline droplets within the granulosa cells of a corpus luteum of pregnancy, which can be identified as early as 15 days after ovulation, are almost diagnostic of pregnancy; they may occur very rarely, however, within a corpus luteum of menstruation (122). These inclusions initially appear as small, round or irregular, often multiple, droplets that enlarge, possibly by fusion of smaller droplets into one or several large bodies that may fill the entire cell (Fig. 41.31A). They become more numerous as gestation progresses (164), although by term their numbers decrease as they undergo calcification, which continues into the puerperium (Fig. 41.31B). It is likely that these calcified bodies eventually are resorbed, as they are not a feature of corpora albicantia.

K cells identical to those within corpus luteum of menstruation are also found in the granulosa layer of the early corpus luteum of pregnancy. They are most numerous in the 2nd, 3rd, and 4th months of gestation after which time they are rarely encountered (121,122,164).

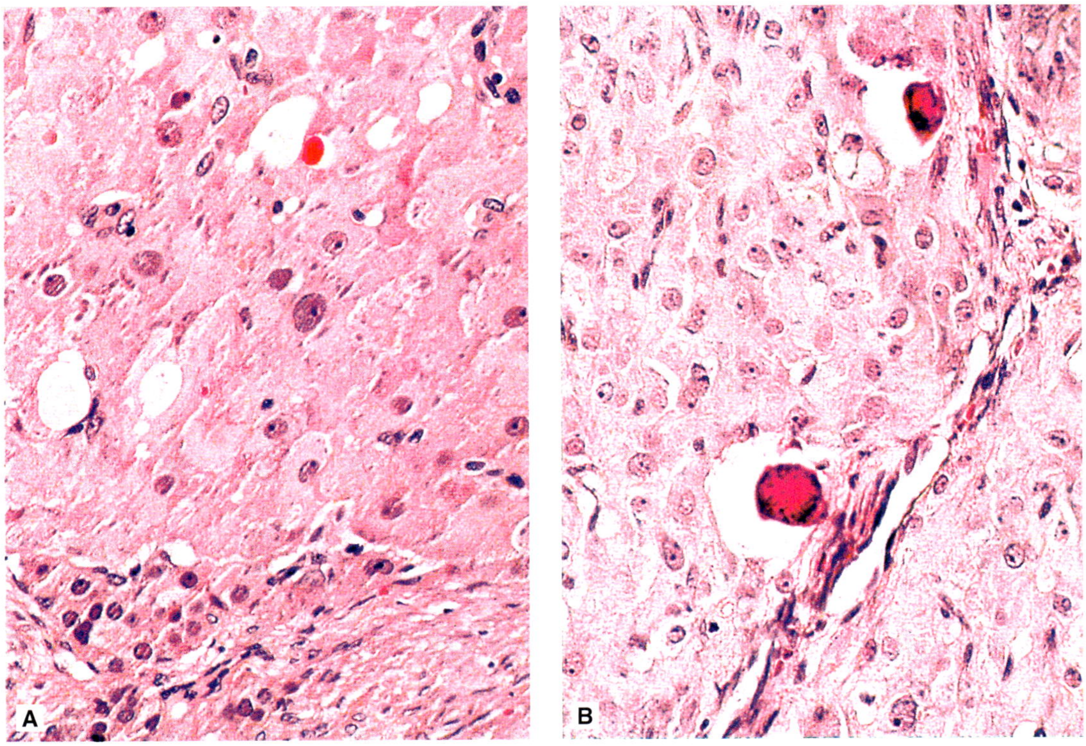

FIGURE 41.31 Corpus luteum of pregnancy. **A:** Note granulosa-lutein cells with large irregular vacuoles and densely eosinophilic hyaline body. Nests of theca cells are seen at bottom left. **B:** Focal calcification within a late CLP.

Theca Layer

The theca interna is thickest in the early corpus luteum of pregnancy at which time it resembles its counterpart in the corpus luteum of menstruation, surrounding the granulosa-lutein layer and forming triangular-shaped, vascular septa that extend into the latter. In the corpus luteum of pregnancy, the theca cells are polyhedral or round and approximately one-fourth the size of the granulosa-lutein cells (Fig. 41.31A). The cytoplasm of these cells is more darkly staining and granular than in the latter and is typically not vacuolated. The cell nuclei are central, round, and more hyperchromatic than those of the granulosa cells; one or two prominent nucleoli are usually present. The characteristic colloid inclusions seen within the granulosa cells are absent or very rare within the theca cells. Occasional K cells may be seen in early pregnancy, but in smaller numbers than in the granulosa layer (126,164). After the 4th month, the theca interna and its trabeculae become much thinner as the theca cells become smaller and fewer in number, with darker, more irregular, oblong to spindle-shaped nuclei, so that they resemble fibroblasts (164). By term, the theca interna layer has almost completely disappeared.

Connective Tissue

As in the mature corpus luteum of menstruation, the central cystic cavity is typically lined by a layer of fibrous tissue, composed of variable numbers of fibroblasts, collagen and reticulin fibers, and blood vessels (106). Its thickness is highly variable, not only within the same corpus luteum of pregnancy but also from one corpus luteum of pregnancy to another and from one phase of pregnancy to another (164). As noted, in some corpus luteum of pregnancy with large cystic cavities, the granulosa layer is focally absent, and its wall is formed entirely by this fibrous layer. As gestation advances, the central cyst or coagulum is eventually obliterated by connective tissue that may exhibit focal hyalinization and calcification (121).

Reticulin staining reveals a pattern similar to that of the mature corpus luteum of menstruation; that is, a dense pattern within the theca interna and inner fibrous layer, and a sparser framework within the granulosa layer (122). In the early corpus luteum of pregnancy, many, often large, vessels are present in the theca externa and interna, from which emanate smaller vessels that penetrate the granulosa and inner fibrous layers. In the late corpus luteum of pregnancy, the vessels develop sclerotic walls with luminal narrowing or obliteration (122,164). The amount of connective tissue around the vessels increases in proportion to the decreasing vascularization and regression of the theca interna layer.

Ultrastructure

The ultrastructural appearance of the corpus luteum of pregnancy is similar to that of the corpus luteum of menstruation and remains intact throughout pregnancy despite a reduction in its metabolic activity (165–167). The colloid or hyaline inclusions consist of homogeneous, electron-opaque material that may surround occasional needle-shaped crystals. They typically have no relationship to any organelle, although occasional smaller hyaline bodies are surrounded by rough endoplasmic reticulum.

Hormonal Aspects

Following fertilization, placental hCG stimulates progesterone production by the granulosa-lutein cells. Progesterone concentration within the postovulatory corpus luteum increases sixfold, whereas the E2 level drops to 10% of that within the preovulatory follicle (133,134). hCG alone cannot maintain progesterone secretion from the corpus luteum of pregnancy for more than a few days, and the regulation of progesterone secretion beyond that time is unknown. Progesterone production by the corpus luteum of pregnancy begins to decline by the end of the 2nd month of gestation with a concomitant increase of placental progesterone production. However, in vivo and in vitro studies indicate that the corpus luteum of pregnancy continues to produce progesterone throughout the remainder of gestation, albeit in reduced amounts, consistent with the maintenance of its structural integrity until term (120,165,166,168,169). There is a rapid decline in function during the puerperium, reflecting falling hCG levels during this period.

Relaxin, a polypeptide hormone, is also produced during gestation and the puerperium by the corpus luteum of pregnancy, probably under the control of hCG (170–173). The concentration of relaxin in ovarian vein plasma during pregnancy correlates with progesterone levels. The placenta and uterus have also been suggested as additional, but less important, sources for this hormone. Its reported actions include cervical dilatation and softening, inhibition of uterine contractions, and relaxation of the pubic symphysis and other pelvic joints (170–173). Immunoreactivity for renin and angiotensin II, similar to that noted within the preovulatory follicle (see above), has been demonstrated within the corpus luteum of pregnancy (147), consistent with the observation that prorenin, likely of ovarian origin, increases 10-fold in pregnant women soon after conception (146).

CORPUS ALBICANS

The regressing corpus luteum of menstruation is invaded by connective tissue that gradually converts it to a scar, the corpus albicans. The degenerating corpus luteum and the young corpus albicans may contain macrophages laden with ceroid and hemosiderin pigment (174,175). The mature corpus albicans is a well-circumscribed structure with convoluted borders composed almost entirely of densely packed collagen fibers with occasional admixed fibroblasts (Figs. 41.14 and 41.32). Focal calcification and ossification may be occasionally encountered. Most corpora albicantia are eventually resorbed and replaced by ovarian stroma. Persistent corpora albicantia are typically found in the medulla of postmenopausal women (Fig. 41.14) suggesting

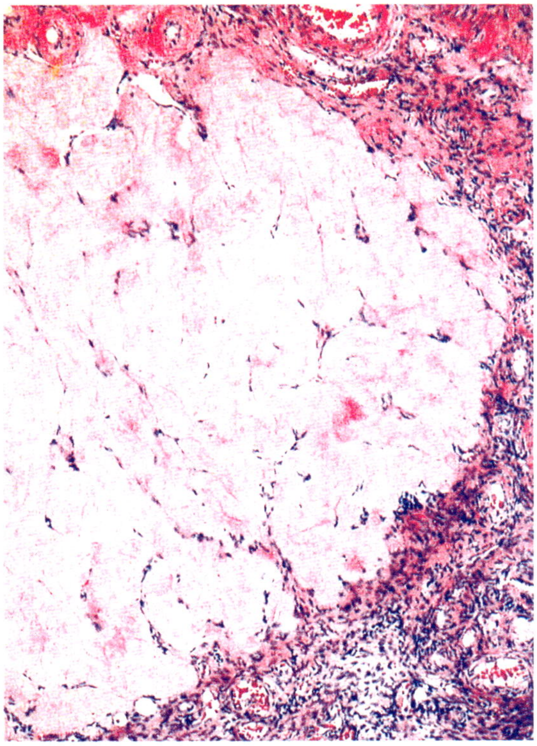

FIGURE 41.32 Corpus albicans.

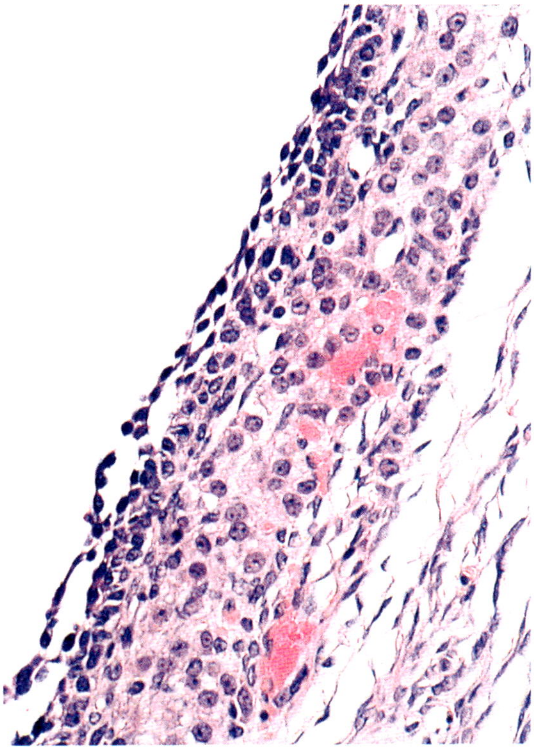

FIGURE 41.33 Lining of atretic cystic follicle composed of a thin inner layer of small, exfoliating granulosa cells and an outer, luteinized theca interna layer.

that this resorption process decelerates or terminates prior to the menopause.

ATRETIC FOLLICLES

Histology

Of the original 400,000 primordial follicles present at birth, approximately 400 mature to ovulation. The remaining 99.9% undergo atresia, a process that begins before birth and continues throughout the reproductive life but is most intense immediately after birth and during puberty and pregnancy (115–119,121,123). Factors that initiate atresia and determine which follicles will ultimately undergo atresia are unknown. The atretic process varies with the stage of follicular maturation that has been reached. Atresia of early follicles (primordial and preantral) begins with degeneration of the oocyte manifested by nuclear changes (chromatin condensation, pyknosis, and fragmentation) and cytoplasmic vacuolation. Degeneration of the granulosa cells soon follows, and the follicle disappears without a trace. In contrast, atresia of follicles that have reached the antral stage of development is more complex and variable but ultimately leads to obliterative atresia and the formation of a scar, the corpus fibrosum. The earliest evidence of this process is the mitotic inactivity of the granulosa cells and a decrease in their numbers, manifested by thinning and focal exfoliation of the granulosa layer. Some follicles may persist for an indefinite period of time at this stage as atretic cystic follicles (Fig. 41.33); those that exceed 3 cm are designated follicular cysts. Atretic cystic follicles and follicular cysts may persist for a number of years after the menopause (176,177). Atretic follicles are ultimately invaded by vascular connective tissue that eventually fills the central cavity (Fig. 41.34). The oocyte may persist for an indefinite period of time but eventually degenerates. Concurrent with these changes, the basement membrane between the granulosa and theca interna layers becomes transformed into a thick, wavy, eosinophilic, hyalinized band, the so-called "glassy membrane" (Figs. 41.34 and 41.35). The theca interna layer typically persists, often with prominent luteinization (Figs. 41.34 and 41.35), until the late stages of atresia at which time cords and nests of theca cells become surrounded by proliferating connective tissue (Fig. 41.35). Luteinization of both theca and granulosa layers is particularly striking in atretic follicles during infancy, childhood (178), and pregnancy (Fig. 41.36) (122).

Microscopic proliferations of persistent granulosa cells within the centers of atretic follicles of pregnant, and less commonly nonpregnant women, may mimic small granulosa cell tumors (Fig. 41.36), or rarely, Sertoli cell tumors (179). Similarly, structures resembling microscopic gonadoblastomas and sex cord tumors with annular tubules have been identified within atretic follicles in up to 35% of normal fetuses and infants (111,180,181). There is no evidence to suggest that any of these tumor-like proliferations represent the early stages of neoplasia.

Continued shrinkage and hyalinization of an atretic follicle produces a serpiginous strand of hyaline tissue, the corpus fibrosum or atreticum (Fig. 41.37). Like corpora

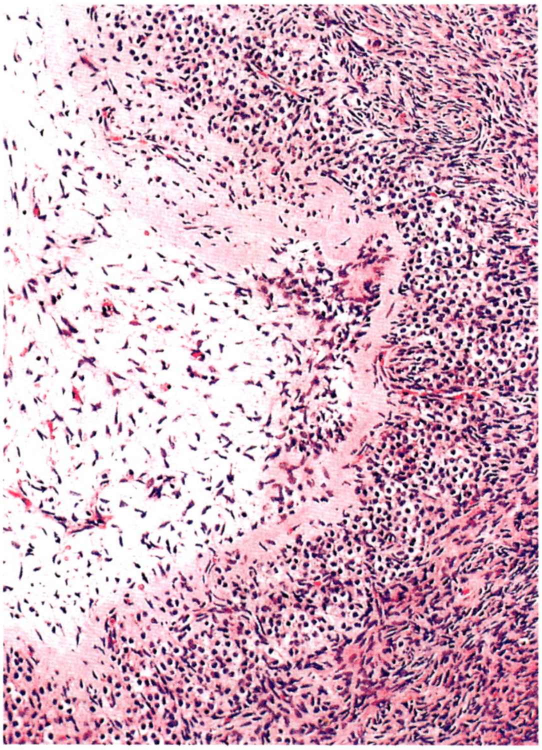

FIGURE 41.34 Atretic cystic follicle undergoing obliterative atresia. Loose connective tissue is replacing the central cavity. The wavy basement membrane ("glassy membrane") is thickened and hyalinized. A prominent layer of luteinized theca interna is evident.

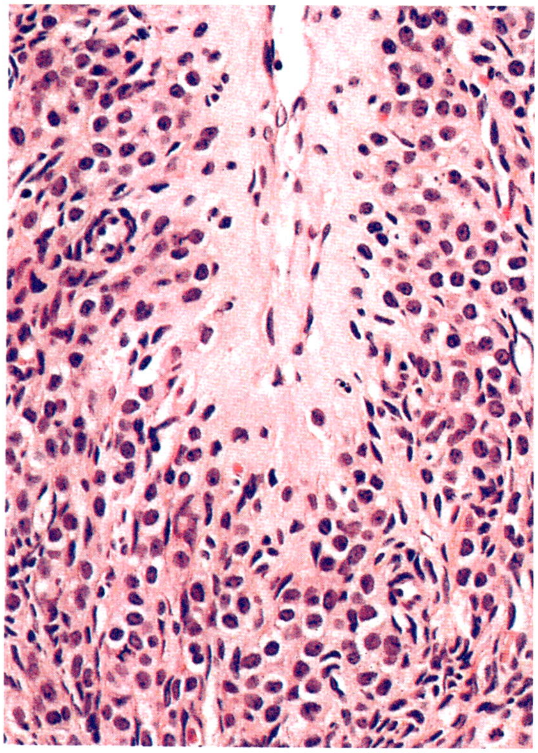

FIGURE 41.35 Edge of follicle in late stage of obliterative atresia. Hyalinized fibrous tissue occupies the central cavity and extends into the persistent luteinized theca interna layer.

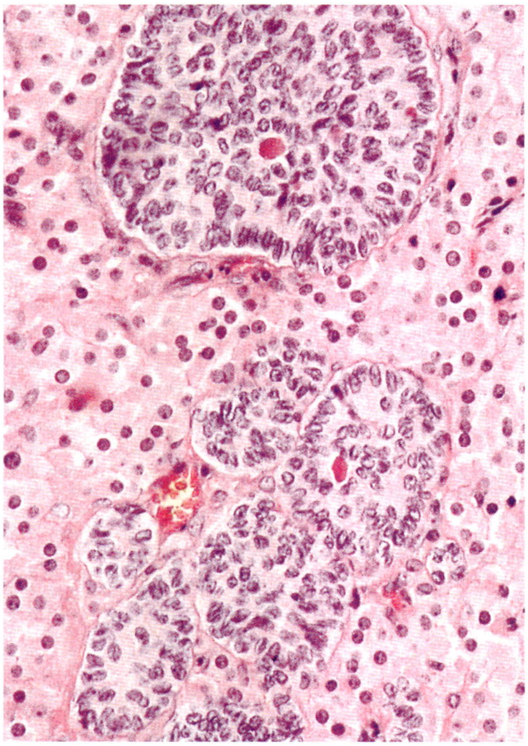

FIGURE 41.36 Atretic follicle in pregnancy. Within the center of the follicle is a proliferation of persistent granulosa cells surrounded by luteinized theca interna cells.

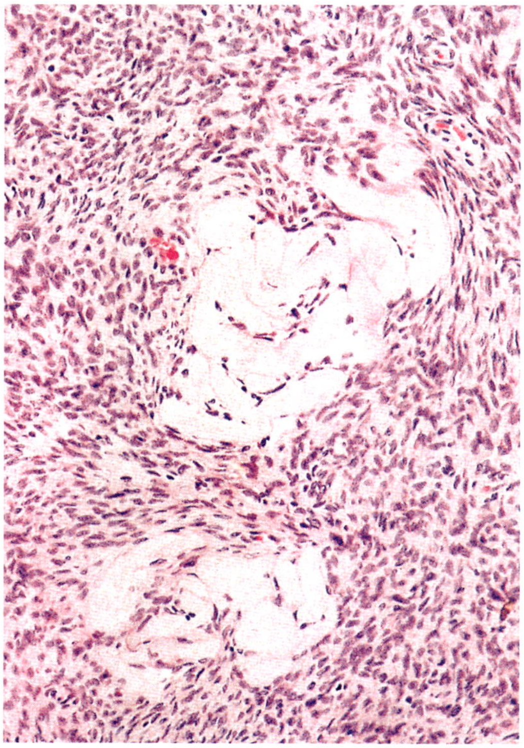

FIGURE 41.37 Two corpora fibrosa.

albicantia, most corpora fibrosa are probably resorbed by the ovarian stroma.

Hormonal Aspects

In contrast to preovulatory follicles, the microenvironment of follicles undergoing atresia is predominantly androgenic, with high concentrations of intrafollicular androstenedione and low concentrations of FSH and E2 (17,99,133,134,182). As noted, these follicles are deficient in granulosa cells, and the residual granulosa cells do not respond to FSH in vitro (125); both FSH and LH receptors are lower than in nonatretic follicles (136). Oocytes from atretic follicles are unable to complete the first meiotic division (125). It is likely that an androgenic intrafollicular milieu is the major factor that halts follicular growth and initiates atresia of that follicle.

HILUS CELLS

Histology

Ovarian hilus cells, morphologically identical to testicular Leydig cells (with the exception of a female chromatin pattern), are present during fetal life but not during childhood. They reappear at the time of puberty and are demonstrable in most postmenopausal women (183–185). Their number and location can be highly variable, and their numbers increase during pregnancy, with increasing age after the menopause and increasing degrees of ovarian stromal proliferation and luteinization (22). Mild hilus cell hyperplasia is a relatively common incidental histologic finding in postmenopausal women (66).

Hilus cell aggregates of variable size and shape are typically found in the ovarian hilus and adjacent mesovarium (Figs. 41.38 and 41.39). They are more numerous in the lateral and medial poles of the hilus and near the junction of the ovarian ligament with the ovary, typically lying close to the junction of the hilus with the medullary stroma (Fig. 41.39) (183). The aggregates are closely associated with large hilar veins and lymphatic sinusoids, and may form nodular protrusions into their lumina. Hilus cells characteristically ensheath, or less commonly lie within, nonmedullated nerves (Fig. 41.40), and occasionally surround the rete ovarii (183). Nests may also be present within the medullary stroma near the hilus, probably representing extensions of the hilus into the medulla. Also, as previously noted, cells of hilus type also rarely occur within the ovarian stroma at a distance from the hilus (stromal Leydig cells). Hilus cells may also be encountered in the perisalpinx and fimbrial endosalpinx (186).

Nests of Hilus cells are unencapsulated, typically lying within loose connective tissue, or rarely ovarian-type stroma, within the hilus (81). The cells are 15 to 25 μm in diameter, round to oval, less commonly elongate, with abundant eosinophilic cytoplasm and a spherical vesicular nucleus with one or two prominent nucleoli (Fig. 41.41).

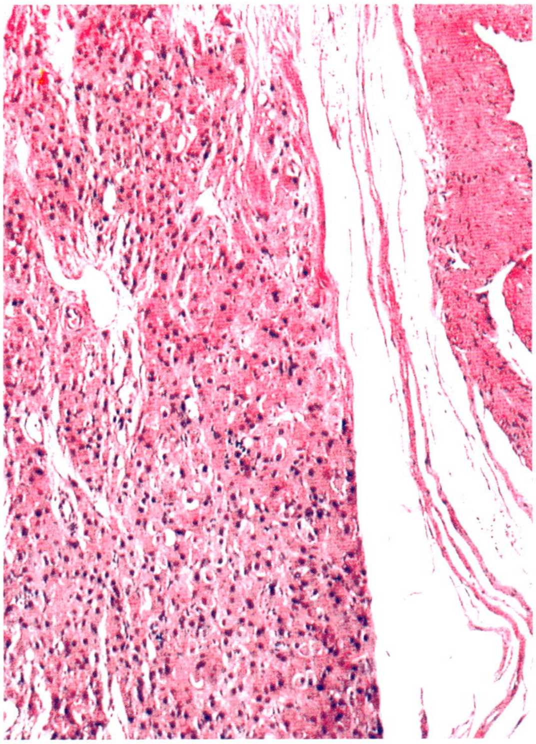

FIGURE 41.38 Nest of hilus cells adjacent to large vessel within the ovarian hilus.

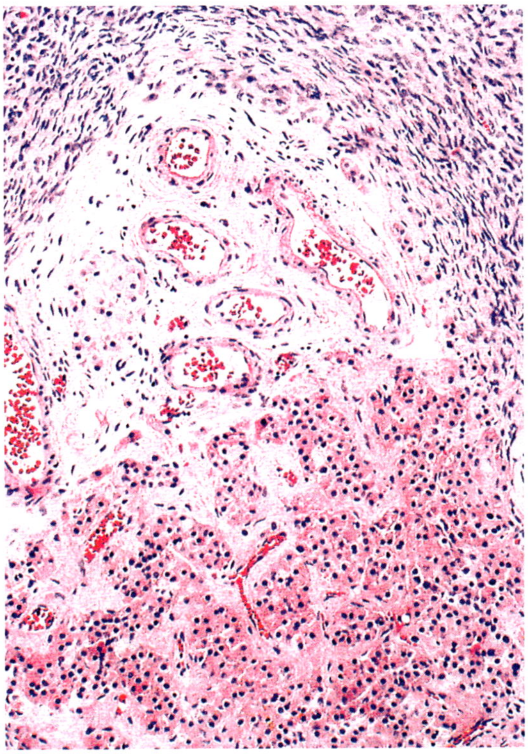

FIGURE 41.39 Nest of hilus cells with admixed small blood vessels abutting medullary stroma (top of the figure).

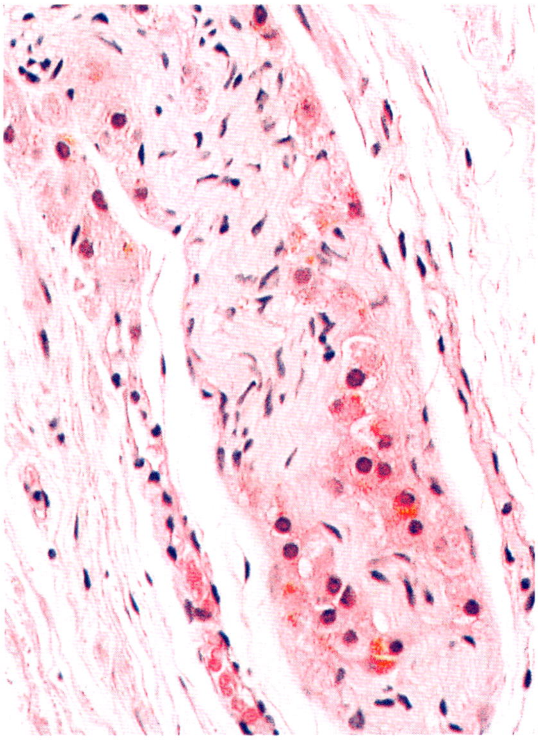

FIGURE 41.40 Perineural and intraneural hilus cells. Note fine brown lipochrome pigment within hilus cells.

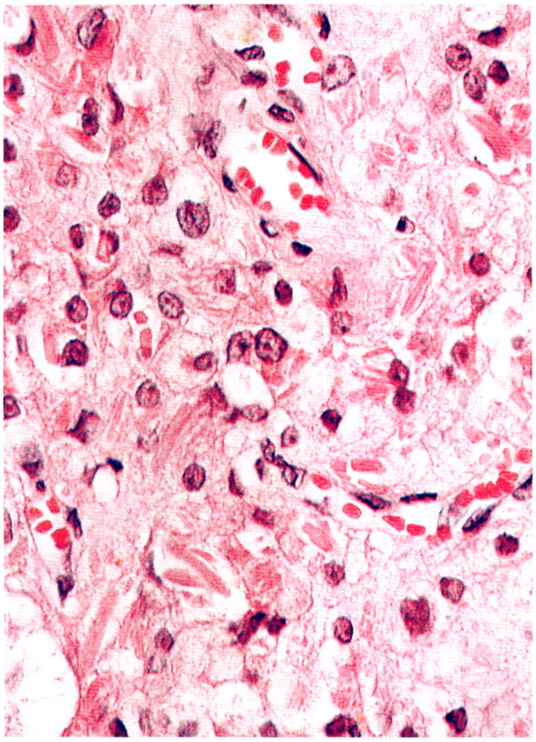

FIGURE 41.41 Hilus cells with Reinke crystals.

The nuclei, particularly in postmenopausal women, may have hyperchromatic, bizarre nuclei. Hilus cells are typically strongly immunoreactive for inhibin, calretinin, and melan-A (62,153).

Hilus cells contain crystals of Reinke, which are homogeneous, eosinophilic, nonrefractile, rod-shaped structures, 10 to 35 μm in length, with blunt, occasionally tapered, ends (Fig. 41.41). The crystals typically lie in a parallel or stacked arrangement within a cell and are often surrounded by a clear halo; they sometimes appear to extend through or overlie cell membranes. The crystals are unevenly distributed and are typically present in only a minority of cells; frequently they cannot be identified (187). Their visualization may be facilitated by the use of Masson trichrome and iron hematoxylin methods that stain them magenta and black, respectively. The crystals fluoresce yellow when hematoxylin and eosin (H&E)-stained sections are viewed by ultraviolet light (188). Spherical or ellipsoidal hyaline structures that resemble crystals of Reinke are also present within hilus cells, often in greater numbers than crystals. Elongated erythrocytes compressed within capillaries should not be confused with crystals and crystal precursors. The cytoplasm of Leydig cells may also contain perinuclear eosinophilic granules, peripheral lipid vacuoles, and golden-brown lipochrome pigment (Fig. 41.40). Delicate collagen fibrils surround each cell. Typically admixed with the hilus cells are fibroblasts and cells intermediate in appearance between the two cell types (189). The hilus cells and intermediate cells have intimate attachments to nerves, including true synaptic connections, suggesting that hilus cells may originate from hilar fibroblasts, possibly under the inductive influence of hilar nerves (183,189).

Hilus cells should be distinguished from adrenal cortical rests. The latter are extremely rare in the ovary (190) but can be found in the mesovarium or ovarian hilus in approximately one-quarter of women (191). Their histologic appearance mimics that of the normal adrenal cortex, with most of the cells containing numerous lipid vacuoles.

Ultrastructure

Hilus cells have a steroidogenic ultrastructure consisting of prominent smooth endoplasmic reticulum and mitochondria with tubular cristae, as well as well-developed Golgi, large lysosomes, and osmiophilic lipid inclusions (189). Reinke crystals have a true crystalline appearance composed of dense parallel hexagonal microtubules with a mean thickness of 12 nm, separated by clear spaces 15 nm wide producing a "woven fabric" appearance (Fig. 41.42) (189).

Hormonal Aspects

The light and electron microscopic morphology and enzyme content of hilus cells are those of steroid hormone–producing cells, although to what extent hilus cells contribute to the steroid hormone pool in normal females is unknown (32,183). In vitro incubation studies indicate that the major steroid produced by ovarian hilus cells is androstenedione and that it is produced in amounts higher than those of ovarian stroma

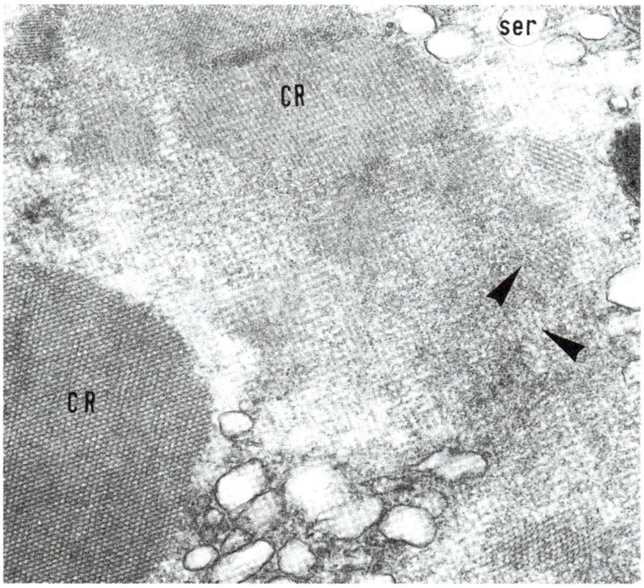

FIGURE 41.42 Reinke crystals with hexagonal internal pattern (*CR*) formed by association of precrystalline units (*arrowheads*). *ser*, smooth endoplasmic reticulum (original magnification ×25,000). (Reprinted with permission from Laffargue P, Benkoel L, Laffargue F, et al. Ultrastructural and enzyme histochemical study of ovarian hilar cells in women and their relationships with sympathetic nerves. *Hum Pathol.* 1978;9:649–659, with permission from Elsevier.)

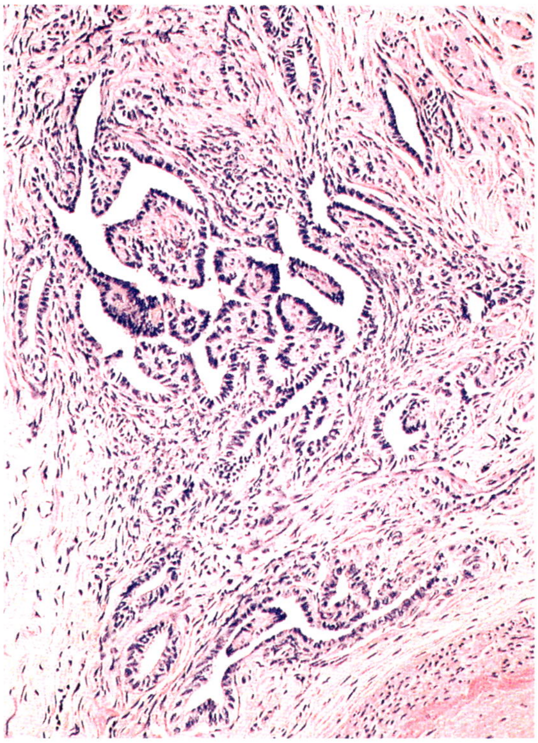

FIGURE 41.43 Rete ovarii.

(192). Lesser amounts of E2 and progesterone are also produced in vitro. Hilus cells are responsive in vivo to both exogenous and endogenous hCG stimulation, manifest by an increase in their numbers, cell size, and mitotic activity (193).

RETE OVARII

The rete ovarii, the ovarian analog of the rete testis, is present in the hilus of all ovaries. It consists of a network of irregular clefts, tubules, cysts, and intraluminal papillae, lined by an epithelium that varies from flat to cuboidal to columnar (Fig. 41.43) (193,194). Solid cords of similar cells may also be seen. Characteristically, the rete ovarii is surrounded by a cuff of spindle-cell stroma similar to, but discontinuous from, the ovarian stroma (Fig. 41.43).

The cytoplasm of the rete cells is immunoreactive for cytokeratin, epithelial membrane antigen (EMA), vimentin, and desmoplakin (54,55,195). Ultrastructural examination has revealed two types of cells, one ciliated and the other nonciliated with apical microvilli (54). The cytoplasm contains many mitochondria, a moderate amount of rough endoplasmic reticulum, many free polyribosomes, and some glycogen. Numerous desmosomes with associated tonofilament bundles connect adjacent cells. The basal lamina is well defined.

The rete ovarii juxtaposes and may communicate with mesonephric tubules within the mesovarium (193). Rare hilar cysts originate from the rete ovarii, and small tumor-like proliferations of the rete have been referred to as "rete adenomas" (193,196). The cells of the rete ovarii are negative for GATA3, which is expressed in cervical mesonephric remnants (197).

REFERENCES

1. Baker TG, Sum W. Development of the ovary and oogenesis. *Clin Obstet Gynecol*. 1976;3:3–26.
2. Minh H-N, Smadja A, Herve de Sigalony JP, et al. Etude histologique de la gonade à différenciation ovarienne au cours de l'organogenèse. *Arch Anat Cytol Pathol*. 1989;37:201–207.
3. Gondos B, Bhiraleus P, Hobel CJ. Ultrastructural observations on germ cells in human fetal ovaries. *Am J Obstet Gynecol*. 1971;110:644–652.
4. Gondos B. Cellular interrelationships in the human fetal ovary and testis. In: Federoff S, ed. *Prog Clin Biol Res. Volume 59B. Eleventh International Congress of Anatomy: Advances in the Morphology of Cells and Tissues*. Alan R. Liss Inc; 1981:373–381.
5. Konishi I, Fujii S, Okamura H, Parmley T, Mori T. Development of interstitial cells and ovigerous cords in the human fetal ovary: an ultrastructural study. *J Anat*. 1986;148:121–135.
6. Gondos B. Surface epithelium of the developing ovary. Possible correlation with ovarian neoplasia. *Am J Pathol*. 1975;81:303–321.
7. Rabinovici J, Jaffe RB. Development and regulation of growth and differentiated function in human and subhuman primate fetal gonads. *Endocr Rev*. 1990;11:532–557.

8. Van Wagenen G, Simpson ME. *Embryology of the Ovary and Testis in Homo Sapiens and Macaca Mulatta*. Yale University Press; 1965.
9. Fukuda O, Miyayama Y, Fujimoto T, Okamura H. Electron microscopic study of the gonadal development in early human embryos. *Prog Clin Biol Res*. 1989;296:23–29.
10. Pinkerton JH, McKay DG, Adams EC, Hertig AT. Development of the human ovary—a study using histochemical techniques. *Obstet Gynecol*. 1961;18:152–181.
11. Gruenwald P. The development of the sex cords in the gonads of man and mammals. *Am J Anat*. 1942;70:359–389.
12. Jirasek JE. Development of the genital system in human embryos and fetuses. In: Jirasek J, ed. *Development of the Genital System and Male Pseudohermaphroditism*. Johns Hopkins Press; 1971:3–41.
13. Byskov AG. Differentiation of mammalian embryonic gonad. *Physiol Rev*. 1986;66:71–117.
14. Satoh M. Histogenesis and organogenesis of the gonad in human embryos. *J Anat*. 1991;177:85–107.
15. Wartenberg H. Development of the early human ovary and role of the mesonephros in the differentiation of the cortex. *Anat Embryol (Berl)*. 1982;165:253–280.
16. Hummitzsch K, Anderson RA, Wilhelm D, et al. Stem cells, progenitor cells, and lineage decisions in the ovary. *Endocr Rev*. 2015;36:65–91.
17. Nicosia SV. Morphological changes in the human ovary throughout life. In: Serra GB, ed. *The Ovary*. Raven Press; 1983:57–81.
18. Pryse-Davies J. The development, structure and function of the female pelvic organs in childhood. *Clin Obstet Gynecol*. 1974;1:483–508.
19. Valdes-Dapena MA. The normal ovary of childhood. *Ann N Y Acad Sci*. 1967;142:597–613.
20. Merrill JA. The morphology of the prepubertal ovary: relationship to the polycystic ovary syndrome. *South Med J*. 1963;56:225–231.
21. Pavlik EJ, DePriest PD, Gallion HH, et al. Ovarian volume related to age. *Gynecol Oncol*. 2000;77:410–412.
22. Boss JH, Scully RE, Wegner KH, Cohen RB. Structural variations in the adult ovary. Clinical significance. *Obstet Gynecol*. 1965;25:747–764.
23. Reeves G. Specific stroma in the cortex and medulla of the ovary. Cell types and vascular supply in relation to follicular apparatus and ovulation. *Obstet Gynecol*. 1971;37:832–844.
24. Plentl AA, Friedman EA. *Lymphatic System of the Female Genitalia*. WB Saunders; 1971.
25. Eichner E, Bove ER. In vivo studies on the lymphatic drainage of the human ovary. *Obstet Gynecol*. 1954;3:287–297.
26. Jacobowitz D, Wallach EE. Histochemical and chemical studies of the autonomic innervation of the ovary. *Endocrinology*. 1967;81:1132–1139.
27. Owman C, Rosenbren E, Sjöberg N. Adrenergic innervation of the human female reproductive organs: a histochemical and chemical investigation. *Obstet Gynecol*. 1967;30:763–773.
28. Balboni GC. Structural changes: ovulation and luteal phase. In: Serra GB, ed. *The Ovary*. Raven Press; 1983:123–141.
29. Mohsin S, Pennefather JN. The sympathetic innervation of the mammalian ovary. A review of pharmacological and histochemical studies. *Clin Exp Pharmacol Physiol*. 1979;6:335–354.
30. Dyer CA, Erickson GF. Norepinephrine amplifies human chorionic gonadotropin-stimulated androgen biosynthesis by ovarian theca-interstitial cells. *Endocrinology*. 1985;116:1645–1652.
31. Blaustein A, Lee H. Surface cells of the ovary and pelvic peritoneum: a histochemical and ultrastructural comparison. *Gynecol Oncol*. 1979;8:34–43.
32. McKay DG, Pinkerton JH, Hertig AT, Danziger S. The adult human ovary: a histochemical study. *Obstet Gynecol*. 1961;18:13–39.
33. Zinsser KR, Wheeler JE. Endosalpingiosis in the omentum: a study of autopsy and surgical material. *Am J Surg Pathol*. 1982;6:109–117.
34. Blaustein A. Surface cells and inclusion cysts in fetal ovaries. *Gynecol Oncol*. 1981;12(2 Pt 1):222–233.
35. Blaustein A, Kantius M, Kaganowicz A, Pervez N, Wells J. Inclusions in ovaries of females aged day 1–30 years. *Int J Gynecol Pathol*. 1982;1:145–153.
36. Mulligan RM. A survey of epithelial inclusions in the ovarian cortex of 470 patients. *J Surg Oncol*. 1976;8:61–66.
37. Von Numers C. Observations on metaplastic changes in the germinal epithelium of the ovary and on the aetiology of ovarian endometriosis. *Acta Obstet Gynecol Scand*. 1965;44:107–116.
38. Scully RE. Ovary. In: Henson DE, Albores-Saavedra J, eds. *The Pathology of Incipient Neoplasia. Major Problems in Pathology Series*. Vol 28. 2nd ed. WB Saunders; 1993:279–300.
39. Piek JMJ, Kenemans P, Zweemer RP, van Diest PJ, Verheijen RH. Ovarian carcinogenesis: an alternative theory. *Gynecol Oncol*. 2007;107:355.
40. Crum CP, Drapkin R, Miron A, et al. The distal fallopian tube: a new model for pelvic serous carcinogenesis. *Curr Opin Obstet Gynecol*. 2007;19:3–9.
41. Singh N, Gilks CB, Wilkinson N, McCluggage WG. The secondary Müllerian system, field effect, BRCA, and tubal fimbria: our evolving understanding of the origin of tubo-ovarian high-grade serous carcinoma and why assignment of primary site matters. *Pathology*. 2015;47:423–431.
42. Auersperg N. The origin of ovarian carcinomas: a unifying hypothesis. *Int J Gynecol Pathol*. 2011;30:12–21.
43. Bowen NJ, Logani S, Dickerson EB, et al. Emerging roles for Pax8 in ovarian cancer and endosalpingeal development. *Gynecol Oncol*. 2007;104:331–337.
44. Bransilver BR, Ferenczy A, Richart RM. Brenner tumors and Walthard cell nests. *Arch Pathol*. 1974;98:76–86.
45. Danforth DN. Cytologic relationship of Walthard cell rest to Brenner tumor of ovary and the pseudomucinous cystadenoma. *Am J Obstet Gynecol*. 1942;43:984–996.
46. Roth LM. The Brenner tumor and the Walthard cell nest. An electron microscopic study. *Lab Invest*. 1974;31:15–23.
47. Teoh TB. The structure and development of Walthard nests. *J Pathol Bacteriol*. 1953;66:433–439.
48. Seidman JD, Khedmati F. Exploring the histogenesis of ovarian mucinous and transitional cell (Brenner) neoplasms, and their relationship with Walthard cell nests: a study of 120 tumors. *Arch Pathol Lab Med*. 2008;132:1753–1760.
49. Papadaki L, Beilby JO. The fine structure of the surface epithelium of the human ovary. *J Cell Sci*. 1971;8:445–465.
50. Ferenczy A, Richart RM. *Female Reproductive System: Dynamics of Scan and Transmission Electron Microscopy*. John Wiley & Sons; 1974.

51. Blaustein A. Peritoneal mesothelium and ovarian surface cells–shared characteristics. *Int J Gynecol Pathol*. 1984;3: 361–375.
52. Fienberg R, Cohen RB. A comparative histochemical study of the ovarian stromal lipid band, stromal theca cell, and normal ovarian follicular apparatus. *Am J Obstet Gynecol*. 1965;92:958–969.
53. Miettinen M, Lehto V, Virtanen I. Expression of intermediate filaments in normal ovaries and ovarian epithelial, sex cord-stromal, and germinal tumors. *Int J Gynecol Pathol*. 1983;2:64–71.
54. Czernobilsky B, Moll R, Levy R, Franke WW. Co-expression of cytokeratin and vimentin filaments in mesothelial, granulosa and rete ovarii cells of the human ovary. *Eur J Cell Biol*. 1985;37:175–190.
55. Benjamin E, Law S, Bobrow LG. Intermediate filaments cytokeratin and vimentin in ovarian sex cord-stromal tumours with correlative studies in adult and fetal ovaries. *J Pathol*. 1987;152:253–263.
56. Czernobilsky B, Shezen E, Lifschitz-Mercer B, et al. Alpha smooth muscle actin (alpha-SM actin) in normal human ovaries, in ovarian stromal hyperplasia and in ovarian neoplasms. *Virchows Arch B Cell Pathol Incl Mol Pathol*. 1989;57:55–61.
57. Lastarria D, Sachdev RK, Babury RA, Yu HM, Nuovo GJ. Immunohistochemical analysis for desmin in normal and neoplastic ovarian stromal tissue. *Arch Pathol Lab Med*. 1990; 114:502–505.
58. Matias-Guiu X, Pons C, Prat J. Müllerian inhibiting substance, alpha-inhibin, and CD99 expression in sex cord-stromal tumors and endometrioid ovarian carcinomas resembling sex cord-stromal tumors. *Hum Pathol*. 1998;29:840–845.
59. Rishi M, Howard LN, Bratthauer GL, Tavassoli FA. Use of monoclonal antibody against human inhibin as a marker for sex cord-stromal tumors of the ovary. *Am J Surg Pathol*. 1997;21:583–589.
60. Hildebrandt RH, Rouse RV, Longacre TA. Value of inhibin in the identification of granulosa cell tumors of the ovary. *Hum Pathol*. 1997;28:1387–1395.
61. McCluggage WG, Maxwell P. Immunohistochemical staining for calretinin is useful in the diagnosis of ovarian sex cord-stromal tumours. *Histopathology*. 2001;38:403–408.
62. Jungbluth AA, Busam KJ, Gerald WL, et al. A103: an anti-melan-a monoclonal antibody for the detection of malignant melanoma in paraffin-embedded tissues. *Am J Surg Pathol*. 1998;22:595–602.
63. Oliva E, Vu Q, Young RH. CD10 expression in sex cord-stromal tumors (SCTs) and steroid cell tumors (StCTs) of the ovary. *Mod Pathol*. 2002;15:204A.
64. Nagamani M, Hannigan EV, Dinh TV, Stuart CA. Hyperinsulinemia and stromal luteinization of the ovaries in postmenopausal women with endometrial cancer. *J Clin Endocrinol Metab*. 1988;67:144–148.
65. Scully RE, Cohen RB. Oxidative-enzyme activity in normal and pathologic human ovaries. *Obstet Gynecol*. 1964; 24:667–681.
66. Loubet R, Loubet A, Leboutet MJ. The ovarian stroma after the menopause: activity and ageing. In: de Brux J, Gautray JP, eds. *Clinical Pathology of the Ovary*. MTP Press Ltd; 1984:119–141.
67. Bassis ML. Pseudodeciduosis. *Am J Obstet Gynecol*. 1956;72: 1029–1037.
68. Israel SL, Rubenstone A, Meranze DR. The ovary at term. I. Decidua-like reaction and surface cell proliferation. *Obstet Gynecol*. 1954;3:399–407.
69. Ober WB, Grady HG, Schoenbucher AK. Ectopic ovarian decidua without pregnancy. *Am J Pathol*. 1957;33:199–217.
70. Bersch W, Alexy E, Heuser HP, Staemmler HJ. Ectopic decidua formation in the ovary (so-called deciduoma). *Virchows Arch A Pathol Pathol Anat*. 1973;360:173–177.
71. Starup J, Visfeldt J. Ovarian morphology in early and late human pregnancy. *Acta Obstet Gynecol Scand*. 1974;53:211–218.
72. Herr JC, Heidger PM Jr, Scott JR, Anderson JW, Curet LB, Mossman HW. Decidual cells in the human ovary at term. I. Incidence, gross anatomy and ultrastructural features of merocrine secretion. *Am J Anat*. 1978;152:7–27.
73. Doss BJ, Wanek SM, Jacques SM, Qureshi F, Ramirez NC, Lawrence WD. Ovarian smooth muscle metaplasia: an uncommon and possibly underrecognized entity. *Int J Gynecol Pathol*. 1999;18:58–62.
74. Hughesdon PE. Morphology and morphogenesis of the Stein–Leventhal ovary and of so-called "hyperthecosis." *Obstet Gynecol Surv*. 1982;37:59–77.
75. Scully RE. Smooth-muscle differentiation in genital tract disorders. *Arch Pathol Lab Med*. 1981;105:505–507.
76. Hughesdon PE. The origin and development of benign stromatosis of the ovary. *J Obstet Gynaecol Br Commonw*. 1972;79:348–359.
77. Hughesdon PE. The endometrial identity of benign stromatosis of the ovary and its relation to other forms of endometriosis. *J Pathol*. 1976;119:201–209.
78. Hart WR, Abell MR. Adipose prosoplasia of ovary. *Am J Obstet Gynecol*. 1970;106:929–931.
79. Honoré LH, O'Hara KE. Subcapsular adipocytic infiltration of the human ovary: a clinicopathological study of eight cases. *Eur J Obstet Gynecol Reprod Biol*. 1980;10:13–20.
80. Sternberg WH, Roth LM. Ovarian stromal tumors containing Leydig cells. I. Stromal–Leydig cell tumor and non-neoplastic transformation of ovarian stroma to Leydig cells. *Cancer*. 1973;32:940–951.
81. Zhang J, Young RH, Arseneau J, Scully RE. Ovarian stromal tumors containing lutein or Leydig cells (luteinized thecomas and stromal Leydig cell tumors)–a clinicopathological analysis of fifty cases. *Int J Gynecol Pathol*. 1982;1:270–285.
82. Rutgers JL, Scully RE. Functioning ovarian tumors with peripheral steroid cell proliferation: a report of twenty-four cases. *Int J Gynecol Pathol*. 1986;5:319–337.
83. Hidvegi D, Cibils LA, Sorensen K, Hidvegi I. Ultrastructural and histochemical observations of neuroendocrine granules in nonneoplastic ovaries. *Am J Obstet Gynecol*. 1982;143:590–594.
84. Snowden JA, Harkin PJR, Thornton JG, Wells M. Morphometric assessment of ovarian stromal proliferation–a clinicopathological study. *Histopathology*. 1989;14:369–379.
85. Bigelow B. Comparison of ovarian and endometrial morphology spanning the menopause. *Obstet Gynecol*. 1958;11:487–513.
86. Roddick JW Jr, Greene RR. Relation of ovarian stromal hyperplasia to endometrial carcinoma. *Am J Obstet Gynecol*. 1957;73:843–852.
87. Woll E, Hertig AT, Smith GVS, Johnson LC. The ovary in endometrial carcinoma: with notes on the morphological history of the aging ovary. *Am J Obstet Gynecol*. 1948;56: 617–633.

88. Laffargue P, Adechy-Benkoel L, Valette C. Ultrastructure of the ovarian stroma. (Functional significance). *Ann Anat Pathol (Paris)*. 1968;13:381–402.
89. Aiman J, Forney JP, Parker CR Jr. Secretion of androgens and estrogens by normal and neoplastic ovaries in postmenopausal women. *Obstet Gynecol*. 1986;68:1–5.
90. Chang RJ, Judd HL. The ovary after menopause. *Clin Obstet Gynacol*. 1981;24:181–191.
91. Dennefors BL, Janson PO, Knutson F, Hamberger L. Steroid production and responsiveness to gonadotropin in isolated stromal tissue of human postmenopausal ovaries. *Am J Obstet Gynecol*. 1980;136:997–1002.
92. Greenblatt RB, Colle ML, Mahesh VB. Ovarian and adrenal steroid production in the postmenopausal woman. *Obstet Gynecol*. 1976;47:383–387.
93. Judd HL, Judd GE, Lucas WE, Yen SS. Endocrine function of the postmenopausal ovary: concentration of androgens and estrogens in ovarian and peripheral vein blood. *J Clin Endocrinol Metab*. 1974;39:1020–1024.
94. Judd HL, Lucas WE, Yen SS. Effect of oophorectomy on circulating testosterone and androstenedione levels in patients with endometrial cancer. *Am J Obstet Gynecol*. 1974;118:793–798.
95. Longcope C, Hunter R, Franz C. Steroid secretion by the postmenopausal ovary. *Am J Obstet Gynecol*. 1980;138:564–568.
96. Lucisano A, Russo N, Acampora MG, et al. Ovarian and peripheral androgen and oestrogen levels in post-menopausal women: correlations with ovarian histology. *Maturitas*. 1986;8:57–65.
97. Mattingly RF, Huang WY. Steroidogenesis of the menopausal and postmenopausal ovary. *Am J Obstet Gynecol*. 1969;103:679–693.
98. McNatty KP, Makris A, DeGrazia C, Osathanondh R, Ryan KJ. The production of progesterone, androgens, and estrogens by granulosa cells, thecal tissue, and stromal tissue from human ovaries in vitro. *J Clin Endocrinol Metab*. 1979;49:687–699.
99. McNatty KP, Smith DM, Makris A, et al. The intraovarian sites of androgen and estrogen formation in women with normal and hyperandrogenic ovaries as judged by in vitro experiments. *J Clin Endocrinol Metab*. 1980;50:755–763.
100. Plotz EJ, Wiener M, Stein AA, Hahn BD. Enzymatic activities related to steroidogenesis in postmenopausal ovaries of patients with and without endometrial carcinoma. *Am J Obstet Gynecol*. 1967;99:182–197.
101. Vermeulen A. The hormonal activity of the postmenopausal ovary. *J Clin Endocrinol Metab*. 1976;42:247–253.
102. Rice BF, Savard K. Steroid hormone formation in the human ovary. IV. Ovarian stromal compartment; formation of radioactive steroids from acetate-1–14C and action of gonadotropins. *J Clin Endocrinol Metab*. 1966;26:593–609.
103. Barbieri RL, Makris A, Randall RW, Daniels G, Kistner RW, Ryan KJ. Insulin stimulates androgen accumulation in incubations of ovarian stroma obtained from women with hyperandrogenism. *J Clin Endocrinol Metab*. 1986;62:904–910.
104. Reed MJ, Beranek PA, Ghilchik MW, James VH. Conversion of estrone to estradiol and estradiol to estrone in postmenopausal women. *Obstet Gynecol*. 1985;66:361–365.
105. Longcope C. Metabolic clearance and blood production rates of estrogens in postmenopausal women. *Am J Obstet Gynecol*. 1971;111:778–781.
106. Grodin JM, Siiteri PK, MacDonald PC. Source of estrogen production in postmenopausal women. *J Clin Endocrinol Metab*. 1973;36:207–214.
107. Nagamani M, Stuart CA, Doherty MG. Increased steroid production by the ovarian stromal tissue of postmenopausal women with endometrial cancer. *J Clin Endocrinol Metab*. 1992;74:172–176.
108. Dawood MY, Strongin M, Kramer EE, Wieche R. Recent ovulation in a postmenopausal woman. *Int J Gynaecol Obstet*. 1980;18:192–194.
109. Sherrer C, Gerson B, Woodruff JD. The incidence and significance of polynuclear follicles. *Am J Obstet Gynecol*. 1977;128:6–12.
110. Gougeon A. Frequent occurrence of multiovular follicles and multinuclear oocytes in the adult human ovary. *Fertil Steril*. 1981;35:417–422.
111. Manivel JC, Dehner LP, Burke B. Ovarian tumorlike structures, biovular follicles, and binucleated oocytes in children: their frequency and possible pathologic significance. *Pediatr Pathol*. 1988;8:283–292.
112. Baca M, Zamboni L. The fine structure of the human follicular oocytes. *J Ultrastruct Res*. 1967;19:354–381.
113. Hertig AT. The primary human oocyte: some observations on the fine structure of Balbiani's vitelline body and the origin of the annulate lamellae. *Am J Anat*. 1968;122:107–137.
114. Hertig AT, Adams EC. Studies on the human oocyte and its follicle. I. ultrastructural and histochemical observations on the primordial follicle stage. *J Cell Biol*. 1967;34:647–675.
115. Curtis EM. Normal ovarian histology in infancy and childhood. *Obstet Gynecol*. 1962;19:444–454.
116. Dekel N, David MP, Yedwab GA, Kraicer PF. Follicular development during late pregnancy. *Int J Fertil*. 1977;22:24–29.
117. Govan AD. Ovarian follicular activity in late pregnancy. *J Endocrinol*. 1970;48:235–241.
118. Himelstein-Braw R, Byskov AG, Peters H, Faber M. Follicular atresia in the infant human ovary. *J Reprod Fertil*. 1976;46:55–59.
119. Maqueo M, Goldzieher JW. Hormone-induced alterations of ovarian morphology. *Fertil Steril*. 1966;17:676–683.
120. Mikhail G, Allen WM. Ovarian function in human pregnancy. *Am J Obstet Gynecol*. 1967;99:308–312.
121. Nelson WW, Greene RR. The human ovary in pregnancy. *Int Abstr Surg*. 1953;97:1–22.
122. Nelson WW, Greene RR. Some observations on the histology of the human ovary during pregnancy. *Am J Obstet Gynecol*. 1958;76:66–90.
123. Peters H, Himelstein-Braw R, Faber M. The normal development of the ovary in childhood. *Acta Endocrinol (Copenh)*. 1976;82:617–630.
124. McNatty KP, Hillier SG, van den Boogaard AM, Trimbos-Kemper TC, Reichert LE Jr, van Hall EV. Follicular development during the luteal phase of the human menstrual cycle. *J Clin Endocrinol Metab*. 1983;56:1022–1031.
125. McNatty KP, Smith DM, Makris A, Osathanondh R, Ryan KJ. The microenvironment of the human antral follicle: interrelationships among the steroid levels in antral fluid, the population of granulosa cells, and the status of the oocyte in vivo and in vitro. *J Clin Endocrinol Metab*. 1979;49:851–860.
126. White RF, Hertig AT, Rock J, Adams E. Histological and histochemical observations on the corpus luteum of human pregnancy with special reference to corpora lutea associated

with early normal and abnormal ova. *Contrib Embryol*. 1951; 34:55–74.

127. Jones GE, Goldberg B, Woodruff JD. Histochemistry as a guide for interpretation of cell function. *Am J Obstet Gynecol*. 1968;100:76–83.
128. Sasano H, Mori T, Sasano N, Nagura H, Mason JI. Immunolocalization of 3 beta-hydroxysteroid dehydrogenase in human ovary. *J Reprod Fertil*. 1990;89:743–751.
129. Shah SP, Köbel M, Senz J, et al. Mutation of FOXL2 in granulose-cell tumors of the ovary. *N Engl J Med*. 2009; 360:2719–2729.
130. Mestwerdt W, Muller O, Brandau H. Structural analysis of granulosa cells from human ovaries in correlation with function. In: Channing CP, Marsh JM, Sadler WA, eds. *Ovarian Follicular and Corpus Luteum Function: Advances in Experimental Medicine and Biology*. Vol 112. Plenum Press; 1978.
131. Gillim SW, Christensen AK, McLennan CE. Fine structure of the human menstrual corpus luteum at its stage of maximum secretory activity. *Am J Anat*. 1969;126:409–427.
132. Okamura H, Virutamasen P, Wright KH, Wallach EE. Ovarian smooth muscle in the human being, rabbit, and cat. Histochemical and electron microscopic study. *Am J Obstet Gynecol*. 1972;112:183–191.
133. McNatty KP. Follicular determinants of corpus luteum function in the human ovary. In: Channing CP, Marsh JM, Sadler WA, eds. *Ovarian Follicular and Corpus Luteum Function: Advances in Experimental Medicine and Biology*. Vol 112. Plenum Press; 1978:465–477.
134. McNatty KP. Cyclic changes in antral fluid hormone concentrations in humans. *Clin Endocrinol Metab*. 1978;7:577–600.
135. Erickson GF. Normal ovarian function. *Clin Obstet Gynecol*. 1978;21:31–52.
136. Shima K, Kitayama S, Nakano R. Gonadotropin binding sites in human ovarian follicles and corpora lutea during the menstrual cycle. *Obstet Gynecol*. 1987;69:800–806.
137. McNatty KP, Makris A, De Grazia C, Osathanondh R, Ryan KJ. The production of progesterone, androgens and oestrogens by human granulosa cells in vitro and in vivo. *J Steroid Biochem*. 1979;11:775–779.
138. Hillier SG. Intrafollicular paracrine function of ovarian androgen. *J Steroid Biochem*. 1987;27:351–357.
139. Pauerstein CJ, Eddy CA, Croxatto HD, Hess R, Siler-Khodr TM, Croxatto HB. Temporal relationships of estrogen, progesterone, and luteinizing hormone levels to ovulation in women and infrahuman primates. *Am J Obstet Gynecol*. 1978;130:876–886.
140. Yussman MA, Taymor ML. Serum levels of follicle stimulating hormone and luteinizing hormone and of plasma progesterone related to ovulation by corpus luteum biopsy. *J Clin Endocrinol Metab*. 1970;30:396–399.
141. Futterweit W. *Polycystic Ovarian Disease. In Clinical Perspectives in Obstetrics and Gynecology*. Springer-Verlag; 1985.
142. Tanabe K, Gagliano P, Channing CP, et al. Levels of inhibin-F activity and steroids in human follicular fluid from normal women and women with polycystic ovarian disease. *J Clin Endocrinol Metab*. 1983;57:24–31.
143. Tsonis CG, Messinis IE, Templeton AA, McNeilly AS, Baird DT. Gonadotropic stimulation of inhibin secretion by the human ovary during the follicular and early luteal phase of the cycle. *J Clin Endocrinol Metab*. 1988;66:915–921.
144. McLachlan RI, Cohen NL, Vale WW, et al. The importance of luteinizing hormone in the control of inhibin and progesterone secretion by the human corpus luteum. *J Clin Endocrinol Metab*. 1989;68:1078–1085.
145. Sealey JE, Glorioso N, Itskovitz J, Laragh JH. Prorenin as a reproductive hormone. New form of the renin system. *Am J Med*. 1986;81:1041–1046.
146. Palumbo A, Jones C, Lightman A, Carcangiu ML, DeCherney AH, Naftolin F. Immunohistochemical localization of renin and angiotensin II in human ovaries. *Am J Obstet Gynecol*. 1989;160:8–14.
147. Lumbers ER, Pringle KG. Roles of the circulating renin-angiotensin-aldosterone system in human pregnancy. *Am J Physiol Regul Integr Comp Physiol*. 2014;306:R91–R101.
148. Corner GW Jr. The histological dating of the human corpus luteum of menstruation. *Am J Anat*. 1956;98:377–401.
149. Visfeldt J, Starup J. Dating of the human corpus luteum of menstruation using histological parameters. *Acta Pathol Microbiol Scand A*. 1974;82:137–144.
150. Croxatto HD, Ortiz ME, Croxatto HB. Correlation between histologic dating of human corpus luteum and the luteinizing hormone peak—biopsy interval. *Am J Obstet Gynecol*. 1980;136:667–670.
151. Wiley CA, Esterly JR. Observations on the human corpus luteum: histochemical changes during development and involution. *Am J Obstet Gynecol*. 1976;125:514–519.
152. Cao QJ, Jones JG, Li M. Expression of calretinin in human ovary, testis, and ovarian sex cord-stromal tumors. *Int J Gynecol Pathol*. 2001;20:346–352.
153. Pelkey TJ, Frierson HF Jr, Mills SE, Stoler MH. The diagnostic utility of inhibin staining in ovarian neoplasms. *Int J Gynecol Pathol*. 1998;17:97–105.
154. Hameed A, Fox WM, Kurman RJ, Hruban RH, Podack ER. Perforin expression in human cell-mediated luteolysis. *Int J Gynecol Pathol*. 1995;14:151–157.
155. Adams EC, Hertig AT. Studies on the human corpus luteum. I. Observations on the ultrastructure of development and regression of the luteal cells during the menstrual cycle. *J Cell Biol*. 1969;41:696–715.
156. Tamura M, Sasano H, Suzuki T, et al. Immunohistochemical localization of growth hormone receptor in cyclic human ovaries. *Hum Reprod*. 1994;9:2259–2262.
157. Crisp TM, Dessouky DA, Denys FR. The fine structure of the human corpus luteum of early pregnancy and during the progestational phase of the menstrual cycle. *Am J Anat*. 1970;127:37–69.
158. Green JA, Maqueo M. Ultrastructure of the human ovary. I. The luteal cell during the menstrual cycle. *Am J Obstet Gynecol*. 1965;92:946–957.
159. LeMaire WJ, Conly PW, Moffett A, Spellacy WN, Cleveland WW, Savard K. Function of the human corpus luteum during the puerperium: its maintenance by exogenous human chorionic gonadotropin. *Am J Obstet Gynecol*. 1971;110: 612–618.
160. Centola GM. Structural changes: follicular development and hormonal requirements. In: Serra GB, ed. *The Ovary*. Raven Press; 1983:95–111.
161. Rao CV. Receptors for gonadotropins in human ovaries. In: Muldoon T, Mahesh V, Perez-Ballester B, eds. *Recent Advances in Fertility Research Part A: Developments in Reproductive Endocrinology*. Alan R. Liss; 1982:123–135.
162. Vijayakumar R, Walters WA. Ovarian stromal and luteal tissue prostaglandins, 17 beta-estradiol and progesterone in

relation to the phases of the menstrual cycle in woman. *Am J Obstet Gynecol*. 1987;156:947–951.
163. Hertig AT. Gestational hyperplasia of endometrium; a morphologic correlation ova, endometrium, and corpora lutea during early pregnancy. *Lab Invest*. 1964;13:1153–1191.
164. Visfeldt J, Starup J. Histology of the human corpus luteum of early and late pregnancy. *Acta Pathol Microbiol Scand A*. 1975;83:669–677.
165. Adams EC, Hertig AT. Studies on the human corpus luteum. II. Observations on the ultrastructure of luteal cells during pregnancy. *J Cell Biol*. 1969;41:716–735.
166. Green JA, Garcilazo JA, Maqueo M. Ultrastructure of the human ovary. II. The luteal cell at term. *Am J Obstet Gynecol*. 1967;99:855–863.
167. Pedersen PH, Larsen JF. The ultrastructure of the human granulosa lutein cell of the first trimester of gestation. *Acta Endocrinol (Copenh)*. 1968;58:481–496.
168. Le Maire WJ, Rice BF, Savard K. Steroid hormone formation in the human ovary: V. Synthesis of progesterone in vitro in corpora lutea during the reproductive cycle. *J Clin Endocrinol Metab*. 1968;28:1249–1256.
169. Weiss G, Rifkin I. Progesterone and estrogen secretion by puerperal human ovaries. *Obstet Gynecol*. 1975;46:557–559.
170. Weiss G, O'Byrne M, Hochman JA, Goldsmith LT, Rifkin I, Steinetz BG. Secretion of progesterone and relaxin by the human corpus luteum at midpregnancy and at term. *Obstet Gynecol*. 1977;50:679–681.
171. Weiss G, O'Byrne EM, Steinetz BG. Relaxin: a product of human corpus luteum of pregnancy. *Science*. 1976;194: 948–949.
172. Schmidt CL, Black VH, Sarosi P, Weiss G. Progesterone and relaxin secretion in relation to the ultrastructure of human luteal cells in culture: effects of human chorionic gonadotropin. *Am J Obstet Gynecol*. 1986;155:1209–1219.
173. Quagliarello J, Goldsmith L, Steinetz B, Lustig DS, Weiss G. Induction of relaxin secretion in nonpregnant women by human chorionic gonadotropin. *J Clin Endocrinol Metab*. 1980;51:74–77.
174. Joel RV, Foraker AG. Fate of the corpus albicans: a morphologic approach. *Am J Obstet Gynecol*. 1960;80:314–316.
175. Reagan JW. Ceroid pigment in the human ovary. *Am J Obstet Gynecol*. 1950;59:433–436.
176. Centola GM. Structural changes: atresia. In: Serra GB, ed. *The Ovary*. Raven Press; 1983:113–122.
177. Strickler RC, Kelly RW, Askin FB. Postmenopausal ovarian follicle cyst: an unusual cause of estrogen excess. *Int J Gynecol Pathol*. 1984;3:318–322.
178. Kraus FT, Neubecker RD. Luteinization of the ovarian theca in infants and children. *Am J Clin Pathol*. 1962;37:389–397.
179. Clement PB, Young RH, Scully RE. Ovarian granulosa cell proliferations of pregnancy: a report of nine cases. *Hum Pathol*. 1988;19:657–662.
180. Kedzia H. Gonadoblastoma: structures and background of development. *Am J Obstet Gynecol*. 1983;147:81–85.
181. Safneck JR, deSa DJ. Structures mimicking sex cord-stromal tumours and gonadoblastomas in the ovaries of normal infants and children. *Histopathology*. 1986;10:909–920.
182. Bomsel-Helmreich O, Gougeon A, Thebault A, et al. Healthy and atretic human follicles in the preovulatory phase: differences in evolution of follicular morphology and steroid content of follicular fluid. *J Clin Endocrinol Metab*. 1979;48:686–694.
183. Sternberg WH. The morphology, androgenic function, hyperplasia, and tumors of the human ovarian hilus cells. *Am J Pathol*. 1949;25:493–521.
184. Sternberg WH, Segaloff A, Gaskill CJ. Influence of chorionic gonadotropin on human ovarian hilus cells (Leydig-like cells). *J Clin Endocrinol Metab*. 1953;13:139–153.
185. Merrill JA. Ovarian hilus cells. *Am J Obstet Gynecol*. 1959; 78:1258–1271.
186. Honoré LH, O'Hara KE. Ovarian hilus cell heterotopia. *Obstet Gynecol*. 1979;53:461–464.
187. Janko AB, Sandberg EC. Histochemical evidence for the protein nature of the Reinke crystalloid. *Obstet Gynecol*. 1970;35:493–503.
188. Schmidt WA. Eosin-induced fluorescence of Reinke crystals. *Int J Gynecol Pathol*. 1986;5:88–89.
189. Laffargue P, Benkoel L, Laffargue F, Casanova P, Chamlian A. Ultrastructural and enzyme histochemical study of ovarian hilar cells in women and their relationships with sympathetic nerves. *Hum Pathol*. 1978;9:649–659.
190. Symonds DA, Driscoll SG. An adrenal cortical rest within the fetal ovary: report of a case. *Am J Clin Pathol*. 1973;60:562–564.
191. Falls JL. Accessory adrenal cortex in the broad ligament: incidence and functional significance. *Cancer*. 1955;8:143–150.
192. Dennefors BL, Janson PO, Hamberger L, Knutsson F. Hilus cells from human postmenopausal ovaries: gonadotrophin sensitivity, steroid and cyclic AMP production. *Acta Obstet Gynecol Scand*. 1982;61:413–416.
193. Rutgers JL, Scully RE. Cysts (cystadenomas) and tumors of the rete ovarii. *Int J Gynecol Pathol*. 1988;7:330–342.
194. Sauramo H. Development, occurrence, function and pathology of the rete ovarii. *Acta Obstet Gynecol Scand Suppl*. 1954;33:29–46.
195. Woolnough E, Russo L, Khan MS, Heatley MK. An immunohistochemical study of the rete ovarii and epoophoron. *Pathology*. 2000;32:77–83.
196. Gardner GH, Greene RR, Peckham B. Tumors of the broad ligament. *Am J Obstet Gynecol*. 1957;73:536–555.
197. Siatecka H, Massand RP. Brenner tumor associated with rete ovarii: a histologic and immunohistochemical analysis of six cases exploring the relationship between these entities. *Human Pathol*. 2021;113:84–91.

42

Placenta

Drucilla J. Roberts ■ Steven H. Lewis ■ Bradley J. Quade

The placenta is an often overlooked and neglected specimen in pathology for many reasons. It is poorly understood and poorly taught, if at all, in medical schools and pathology residencies, there are relatively few diagnoses with immediate clinical relevance, the specimens tend to be big and bloody, and there is little neoplastic pathology. However, the placenta is a uniquely important organ with significance to the offspring and the birthing person. It has properties that are not seen in any other organ. For example, it is a semiallograft with respect to the biologic mother, yet it is usually tolerated without being rejected as foreign. The placenta also can cause systemic disease in the birthing person and cannot be cured until removed from the uterus. Finally, there are a multitude of findings that predict pediatric and adult diseases.

Thus, somewhat unique in pathology, the placental specimen provides data about two patients and has three anatomic sources. It is fetal, yet mostly extraembryonic in its differentiation, and it also has maternal attachments. Thus, it stands on its own as an organ. It is the first tissue to differentiate during embryogenesis, growing and maturing over time with the fetus before becoming dispensable.

Unquestionably an objective, thorough, well-documented analysis of the placenta can provide data important for patient care. Acknowledging the histologic and pathologic variants, its complicated derivation, and its importance in patient care, the pathologist should become accustomed to obtaining information from both pediatric and obstetrical providers to better contextualize the relevance of variants and alterations seen in the placenta. We find that neonatologists, pediatric neurologists, maternal–fetal medicine specialists, and other obstetric providers are eager to know the pathologic findings of placentas from their patients as they potentially direct clinical care, explain an untoward outcome, and predict future diseases in both the offspring and birthing person. Diagnostic placental pathology also can inform public health efforts, particularly with respect to infectious diseases as well as risk factors predisposing to future conditions. Another important aspect of placental pathology is its contribution to legal proceedings. Placental pathology can often determine the cause of a bad outcome and estimate the duration of hypoxic insult or retention of a stillbirth for correlation with the clinical timeline. For all these reasons, placental pathology needs to be accurately diagnosed and carefully reported.

That said, it is not possible, practical, or recommended that all placentas be pathologically examined. We recommend that all placentas receive a targeted gross examination at delivery, and if feasible, those that are abnormal should be examined thoroughly both macroscopically and microscopically. Further, obstetric providers can be given guidelines to help them efficiently triage placentas for examination. Recommendations vetted by stakeholders in obstetric, pediatric, and pathologic societies listing the various conditions or findings that should prompt pathologic assessment of the placenta are summarized in Table 42.1 (1).

TABLE 42.1 Indications for Pathologic Examination of the Placenta[a]

- **Antepartum Conditions**
  - Maternal indications
    - Maternal death
    - Diabetes mellitus—pregestational or poorly controlled
    - Severe hypertensive disorder
    - Infection during pregnancy or the peripartum that is associated with congenital infection (e.g., intrauterine infection—chorioamnionitis, CMV, syphilis, Zika, COVID-19, etc.)
    - Maternal systemic autoimmune disease including antiphospholipid antibody syndrome
    - Metastatic malignancy
  - Obstetrical indications
    - Antenatal bleeding in the second and third trimester
    - History of placenta with pathology known to recur (e.g., abruption, chronic histiocytic intervillositis, massive perivillous fibrin deposition/maternal floor infarct, villitis of unknown etiology)
  - Fetal indications
    - Pregnancy losses: miscarriage or stillbirth/intrauterine fetal demise
    - Complications associated with multiple gestation
    - Hydrops fetalis
    - Severe fetal/placental anomaly
- **Intrapartum Indications**
  - Preterm delivery at <37 wks
  - Suspected chorioamnionitis
  - Acute abruption
  - NRFHT requiring urgent or STAT delivery
  - Retained placenta, rule out PAS
  - Postpartum hemorrhage
  - Termination of pregnancy for obstetrical or maternal indications
- **Postpartum Indications**
  - Placental indications
    - Any unusual findings (e.g., anatomy, color, mass) in any aspect of the placenta gross examination by an experienced examiner
  - Neonatal indications
    - Compromised clinical condition at birth, defined as any of the following:
      - 5-min Apgar score <7
      - Cord blood pH <7
      - Resuscitation >10 min
      - Severe anemia
      - Persistent hypoglycemia
      - Need for ventilatory assistance
      - Neurologic compromise (e.g., seizure, suspected HIE)
    - Suspected neonatal sepsis
    - Anomalies not diagnosed antepartum
    - Suspected meconium aspiration
    - Neonate with known or suspected malignancy
    - Neonatal death

[a]Modified from reference (1).

CMV, cytomegalovirus; HIE, hypoxic–ischemic encephalopathy; NRFHT, non-reassuring fetal heart testing; PAS, placenta accreta spectrum disorder; STAT, without delay.

The reader is directed to many excellent placental pathology texts for more detailed information regarding placental pathology (2–5). Numerous other examples including consensus and scientific papers have been published recently as well.

The American College of Obstetrics and Gynecology (ACOG) Committee Opinion regarding placental evaluation no longer exists. It should be noted, however, that numerous "ACOG Clinical Opinions" covering clinical conditions such as stillbirth, abruption, accreta, and other pregnancy-related conditions include information regarding placental assessment as a component of clinical care and follow-up (6–11). An expert review is now a part of ACOG's educational materials (12). The "Amsterdam" consensus panel text (13) and book (5) should be compulsory reading for any pathologists who renders diagnoses from placental pathology slides.

One important aspect of placental histology is the evolving terminology used to describe normal and abnormal findings. Dramatic advances in the science of pathology necessitate newer terminology. However, such conceptual and terminologic advances may leave clinicians with difficulty understanding how to translate those advances into the best management for their patients. Clinicians have difficulty interpreting the clinical significance of histologic diagnoses in approximately 35% of placental pathology reports, and only 21% of clinicians fully understand the nomenclature used in placenta reports (14,15). As a result, some obstetric providers fail to see the utility of placental pathology (16) and need to be reminded of the value of competent diagnostic placental pathology in lectures and manuscripts published in journals targeting the obstetric and neonatology communities (17,18).

Although much has been written and is now practiced regarding the importance of placental pathology in patient care, these complex considerations cannot be addressed without a thorough understanding of the placenta's normal structure, and it is to this end that this chapter is devoted.

Normal histology and relevant histopathology are presented for each anatomical compartment of the placenta (e.g., cord, membranes, parenchyma, and decidua). Compartmentalizing the functional units of the placenta aids understanding of this complex organ and makes reporting more concise and systematic.

ROUTINE STORAGE, EXAMINATION, AND PROCESSING

Most hospitals, do not send all placentas to pathology for assessment. It is common for academic institutions to perform histologic examination on 10% to 20% of all delivered placentas, even though far more placentas meet the criteria for assessment (19,20). Placentas that are not submitted for pathologic examination are typically discarded, thereby precluding assessment in the event that problems associated with pregnancy, labor and delivery, or the neonatal period develop. Because problems may often arise in neonates within the first few days of life, storing placentas following delivery for later examination may be helpful.

There are several strategies for storing placentas for delayed pathologic study. Refrigerated placental storage significantly slows decay. Placentas can be stored at 4 °C for up to 1 week before histologic sections are obtained but storage times exceeding 2 weeks are not recommended. Placentas should not be frozen before evaluation, as freezing renders gross examination difficult, and freeze/thaw-associated artifacts obscure many important histologic features. Immediate fixation of the placenta in 10% buffered formalin permits later examination as needed, although these large specimens require sufficient formalin and container/storage space for adequate fixation. It should be noted that fixation increases placental weights by approximately 10% (21). Formalin fixation highlights some pathologies, such as fibrin, certain kinds of infarcts, and other mass-like lesions. However, color and consistency are far easier to ascertain in fresh specimens (19).

Although assessment of the placenta is considered in the pathologist's domain educated obstetric providers may note germane clinical and gross pathology findings, empowering them to make more informed decisions about submission for pathologic examination. If the placenta is not sent to the pathology laboratory, a gross evaluation by the provider can document essential data as part of a delivery note. This information may be helpful in the event that the specimen escapes pathologic examination because it was discarded after delivery. Published guidelines are available for a simple and directed gross examination of the placenta (1,19,22).

A common procedure for gross evaluation and sectioning of placentas is outlined in Table 42.2. The gross examination directs sampling of the tissue to be inspected microscopically and is important for recognizing some entities that can only be diagnosed macroscopically.

The placenta is usually discoid, but unusual shapes or additional lobes may be present. The location of the cord insertion site should be recorded in centimeters from the margin of the placenta. Some portions may be compressed following clamping, cutting, or even collection of blood gas samples and cord blood retrieval. Pertinent pathologic parameters of the umbilical cord include length in toto, diameter, number and direction of coils, irregularities in contour and texture, surface aberrancies, color, and number of vessels contained within the cord. Deviations from normal ranges may reflect decreased umbilical cord flow or correlate with poor fetal activity or oligohydramnios. The cord is then amputated at its base on the placental disk and representative sections taken from each end. Any gross abnormalities should be sampled and immersed in fixative for subsequent histologic assessment.

TABLE 42.2 Gross Examination of the Placenta at Delivery[a,b]

- Umbilical Cord
  - Length: if <35 cm (short) or >70 cm (long), normal term length is 50–70 cm
  - Color: white is normal, yellow is associated with chorioamnionitis, green with meconium, and brown–red with hemolysis associated with intrauterine fetal demise ≥ 6 h
  - Coils: normal is two coils per 10 cm; average of <1 or >4 is abnormal. Flat or hypercoiled cords are typically easily identified without precise measurements.
  - Ulcerations[a] or plaques (e.g., abscesses[a], amnion nodosum, squamous metaplasia) on the surface
  - True knot(s) should be noted if present and described with regard to type (simple or complex) and tightness
  - Number of umbilical vessels
  - Velamentous, marginal, or furcate insertion of the umbilical cord
- Membranes
  - Membrane insertion should be predominantly marginal
  - Membranes should be thin and translucent, free of vessels or masses
  - Membranes color and clarity should be noted
- Disk
  - Size of the placenta <15 cm in greatest diameter at term is likely small, and >25 cm is likely large; weight with attached membranes and umbilical cord <400 g or >700 g is likely abnormal; placental thickness >4 cm or <1.5 cm is likely abnormal (most placental weight standards are for trimmed placenta, but we discourage clinical providers from trimming in anticipation of potential full pathologic examination)
  - Placental completeness (evidence of disruption, possibly incomplete) should be noted
  - Note any adherent or parenchymal deforming blood clot(s)
  - Visible for palpable masses should be noted

[a]Denotes critical values.

[b]Modified from reference (1).

The extraplacental membranes (i.e., amnion and chorion) should be inspected for completeness. Rupture of the gestational sac during a vaginal delivery may result in inversion of the disc and membranes such that amnion is on the exterior surface and the surface with the maternal-facing decidua is on the inside. The membranes can be manually reflected to their normal anatomic position and assessed for color, transparency, sheen, and surface irregularities, as well as for the presence of membranous vessels or accessory lobes. A membrane roll can be created by grasping the point of rupture with forceps (Fig. 42.1) and rolling the membrane in a concentric fashion to include the periphery of the placental disc. The roll is then immersed in a fixative to facilitate tissue sectioning and placement in the processing cassette. Submission of a membrane roll increases the surface area available for microscopic inspection. The outside terminus of the membrane roll as placed on the glass slide should represent the point of rupture, which corresponds to the area overlying the cervicovaginal canal where ascending bacteria first involve the placental membranes. The remaining membranes attached to the periphery of the disc may then be trimmed and set aside before focusing on the disc.

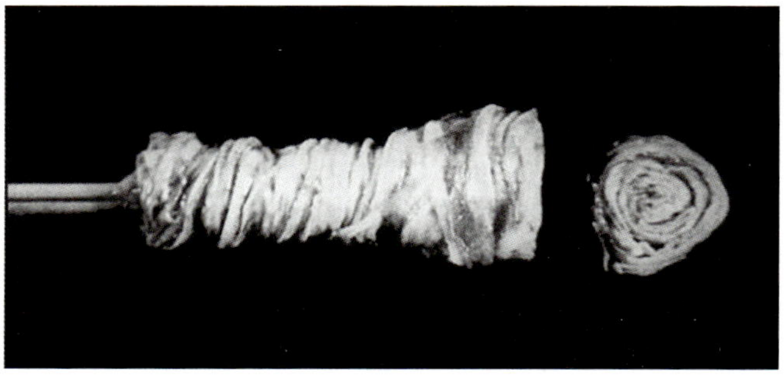

FIGURE 42.1 Proper sampling of the fetal membranes, a technique that provides the most surface area for microscopic evaluation.

The fetal surface of the placenta is measured in three dimensions and examined. Thrombi are important when considering sources of fetal vascular malperfusion. They are very fragile and easily made unrecognizable. Thus, care should be taken to examine the chorionic surface gently. In the maternal surface of the placenta any blood clot that has settled in the storage container should also be examined (Fig. 42.2). In particular, adherent or discolored brown clots suggest chronic hemorrhage, while those that depress the maternal surface are indicative of retroplacental hemorrhage and may correlate with clinically recognized abruption. The dimensions or percentage of the maternal surface involved by hematoma should be recorded. Once the membranes and cord have been removed, the placenta can be weighed: the average term placenta weighs 400 to 600 g. Placental weight varies with neonatal weight and normal weights have been reviewed for all gestational ages (4). Small, lightweight placentas are associated with fetal growth restriction, whereas, heavy, large placentas may be

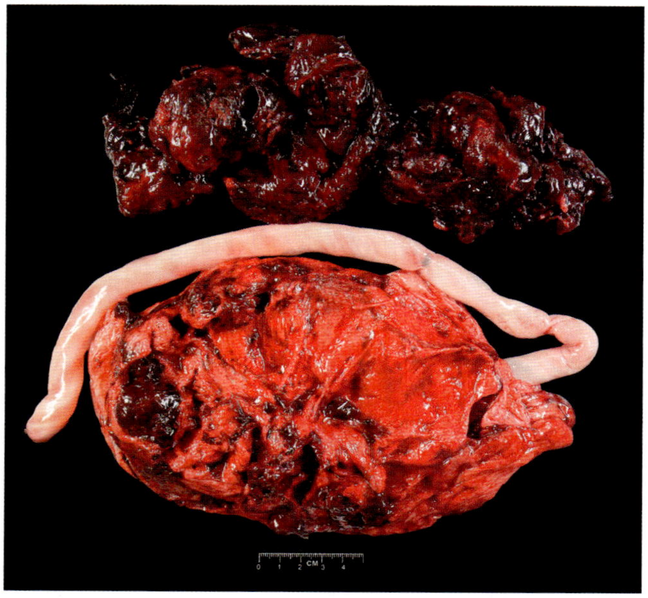

FIGURE 42.2 Loose blood clot and adherent retroplacental hematomas should be described carefully in the gross description, as some may be due to placental abruption.

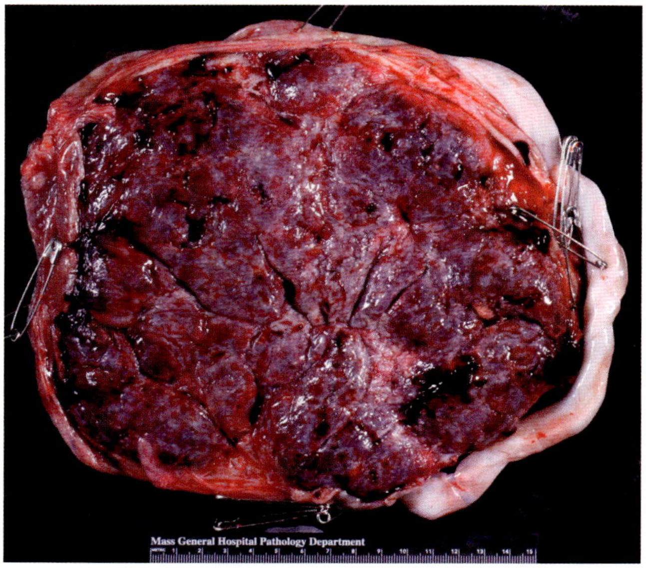

FIGURE 42.3 Maternal surface of a term placenta. Note the irregularly sized lobules or cotyledons and the relative absence of fibrin or decidua.

hydropic or reflect the presence of Beckwith–Weideman syndrome.

The parenchyma is inspected by sectioning the placenta at 1 to 2 cm intervals looking for masses and other irregularities that indicate infarction, thrombi, or other pathologic entities. The placental disc is composed of 16 to 20 cotyledons (Fig. 42.3), which represent the lobular anatomic unit of the placental disc. Partial or complete absence or disruption of a cotyledon may indicate that a portion of the placenta has been retained in utero. Representative sections of abnormal areas are blocked out, and areas of normal-appearing parenchyma (usually three) are placed into a fixative along with the already sectioned membrane roll and umbilical cord. It is important to obtain full-thickness sections spanning the placental disc from chorionic to basal plates (i.e., from maternal to fetal–maternal surfaces).

The sections taken at gross examination are placed in a specimen jar and allowed to fix for at least 4 hours before trimming and placement in cassettes for further processing. For routine examination, tissue staining with hematoxylin and eosin (H&E) is sufficient. Other standard special stains may be used for the detection of specific infectious agents, secretory activity, or structural composition (silver stains, periodic acid–Schiff [PAS], Masson trichrome, etc.). In addition, a host of immunohistochemical stains have been used to elucidate functions of specific placental cell types (illustrated in subsequent section), possible malignancies, and malperfusion. For example, rests of heterotopic fetal cells, notable for their clear cytoplasm, may be found incidentally within the placental disc (Fig. 42.4). Although these rests resemble adrenal tissue by H&E staining, they more likely represent heterotopic hepatic tissue based on their immunostaining for arginase 1 and HepPar1 (23).

The next sections will use an anatomical approach to the normal and abnormal pathology of the placenta.

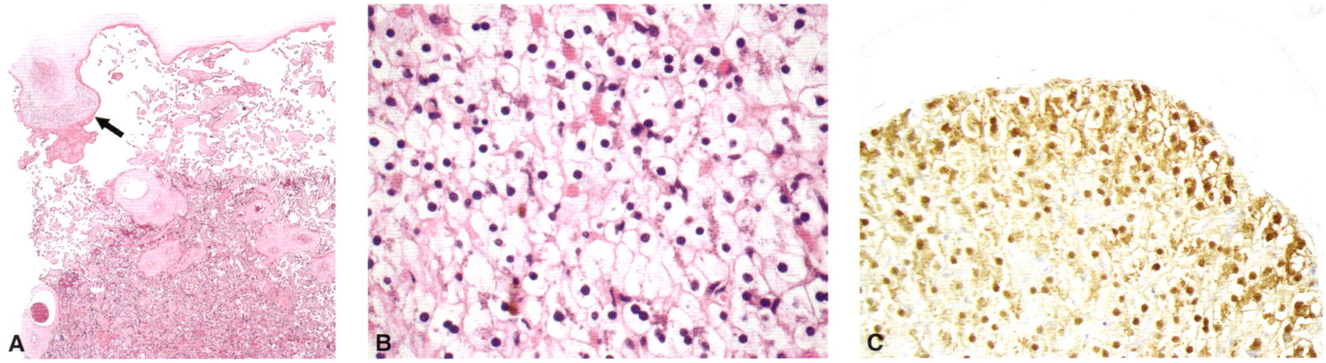

FIGURE 42.4 **A:** Hepatic heterotopia (*arrow*) in the chorionic plate of a term placenta (H&E). **B:** The uniform cells typically have clear cytoplasm, often giving the impression that they are adrenal cortical rests (H&E). **C:** Expression of immunohistochemical stains detecting arginase 1, however, indicates that these cortical nodules manifest hepatic differentiation.

UMBILICAL CORD

Embryology

Specific embryologic considerations are germane to the understanding of the normal umbilical cord structure, including its frequent possession of embryologic remnants. The open region on the ventral surface of the developing embryo diminishes in size and then forms the early umbilicus. The yolk stalk, body stalk, and allantois extend through this structure. This cylindrical primordium elongates, and its surface becomes invested by the expanding amnion, which consists of a single layer of epithelium anchored on a thin connective tissue substrate. Therefore, the developing umbilical cord is covered by amnionic epithelium and contains the yolk stalk, a pair of vitelline blood vessels, the allantois, and the allantoic blood vessels (two arteries and one vein). These anatomic relationships explain the presence of the omphalomesenteric duct, which connects the developing endoderm and the yolk sac, and the allantoic duct, which has its communication in early gestation with the urachus, within the proximal umbilical cord (4,19).

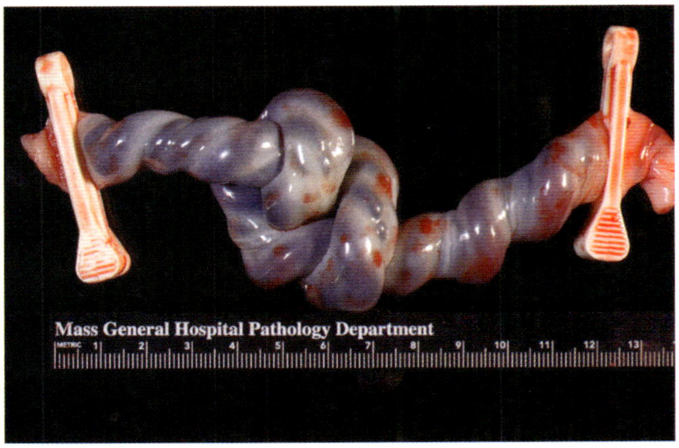

FIGURE 42.5 True knot in a hypercoiled umbilical cord. Umbilical knots can be simple or complex but only if they are tight enough to obstruct blood flow do they cause morbidity or mortality. They are more common with long umbilical cords and polyhydramnios.

Gross Morphology

Important gross anatomic features of the umbilical cord include the location of its insertion in the placental disc, its length, and the number of vessels. The presence of true knots (Fig. 42.5) may be considered normal when they are loose and there is no adverse outcome, yet this occurrence may lead to fetal demise when the knot is tight. The presence of vascular tortuosities (false knots) is common and rarely of clinical significance (Fig. 42.6). Meconium staining and surface plaques are abnormal and are described below.

The normal umbilical cord is pearly white and somewhat translucent. Its length is significant, especially when it is excessively long or short. Cord length varies with gestational age and elongation occurs as gestation proceeds. The average cord length is 32 cm at approximately 20 weeks gestational age and its normal term length is 55 to 65 cm (5,19). Cord diameter also is dependent on gestational age (24). One should not diagnose a short umbilical cord unless the placenta and cord are submitted as one specimen; a detached cord may imply that some portion of the cord was removed prior to submission to the laboratory. However, a diagnosis of a long umbilical cord (i.e., >70 cm at term) is possible even when completeness cannot be verified. The literature contains many articles that relate the significance of abnormal cord lengths with both in utero fetal activity and neonatal outcome (24–26).

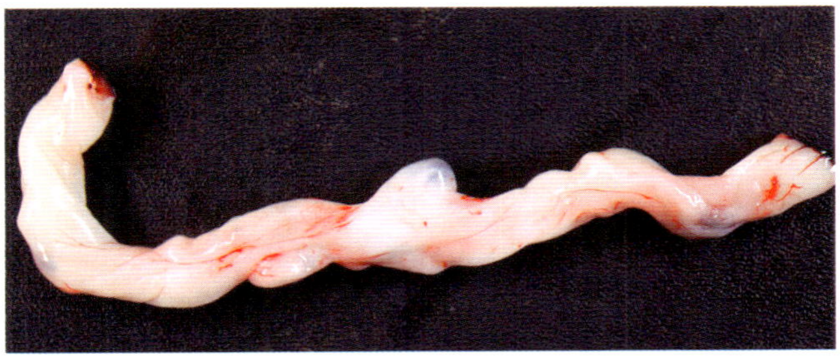

FIGURE 42.6 False knot, visible as and asymmetric dilation of the umbilical cord and representing vascular tortuosity.

Histology

Histologic examination of the umbilical cord shows reveals a well-defined single layer of amnionic epithelium on its outer surface. The epithelium is squamoid and, in the region of fetal cord insertion, often becomes multilayered. The cord amnionic epithelial cells are responsible for fluid equilibrium activities (19,20). True squamous metaplasia of the umbilical cord is morphologically similar to the fetal epidermis and considered a normal variant (27).

The amnionic epithelium is supported by Wharton jelly. This material is largely composed of hyaluronic acid, chondroitin sulfate, and other mucopolysaccharides supported by delicate interlacing microfibrils and sparse collagen. Mast cells may be prominent and their frequency is increased near the periphery of the cord vasculature (28). Macrophages are rarely identified in this area as well.

The umbilical vessels are embedded within the substance of Wharton jelly. Although the vasculature of the umbilical

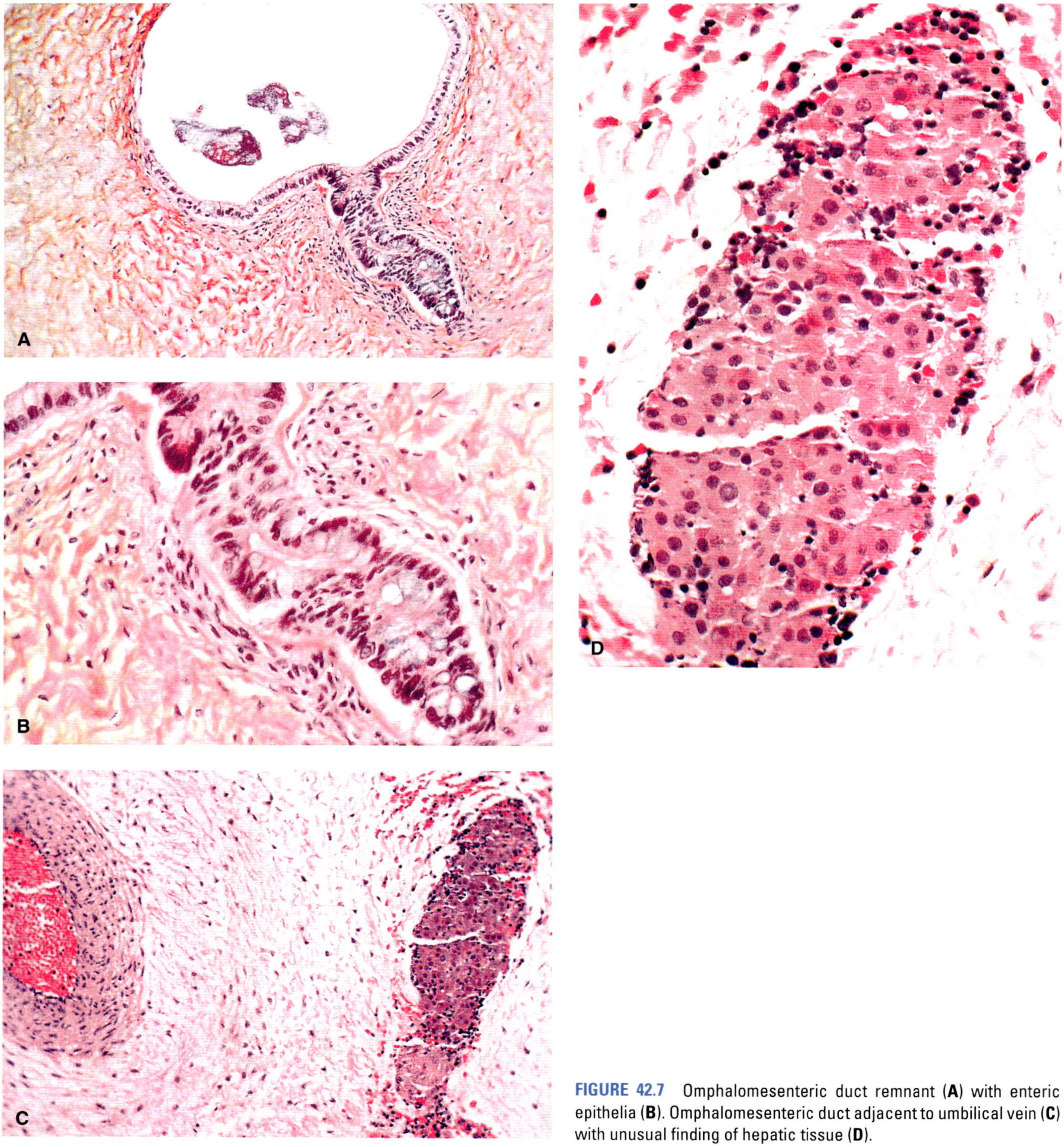

FIGURE 42.7 Omphalomesenteric duct remnant (**A**) with enteric epithelia (**B**). Omphalomesenteric duct adjacent to umbilical vein (**C**) with unusual finding of hepatic tissue (**D**).

cord is of considerable caliber, there are no vasa vasorum or lymphatic channels present in this structure. Studies investigating vascular innervation have also concluded that no nerves are present within the umbilical cord (29). Occasionally, however, autonomic nerves are identified using acetylcholinesterase thiocholine techniques in the fetal end of the cord (24,29,30). Such findings are compatible with the persistence of peripheral vagal neuronal elements associated with the ductus venosus that are entrapped in the proximal portion of the umbilical cord. These neuronal vestiges cord are best considered as remnants; there has been no demonstration of their functional significance to date (19,30).

Because the yolk sac connects to the primitive midgut through the body stalk in early development, vestiges of this epithelium-lined duct are common in the umbilical cord. The persistence of the omphalomesenteric duct is characterized by a tubular structure present within Wharton jelly and lined by a single layer of low cuboidal to columnar, mucin-secreting epithelium (Fig. 42.7). Remnants of the duct may form cystic structures that contain a variety of endodermally derived epithelia, including pancreatic, intestinal, and gastric components. Such findings are rarely of any clinical significance, although acid secretion from gastric epithelia may rarely result in umbilical vascular ulceration, hemorrhage, and fetal death (31).

The allantois differentiates as a protuberance from the yolk sac into the body stalk and is essential for the development of the umbilical vessels. This structure is incorporated into the anterior aspect of the hindgut, where it communicates with the urachus. Remnants of the allantoic duct are often found in sections of the proximal umbilical cord. Its intimate relationship with the formation of umbilical vessels explains its presence between the two umbilical arteries, when it is identified. These remnants rarely have clinical significance, unless they remain patent and reverse micturition occurs, thus expanding the umbilical cord and potentially obstructing umbilical vessels. Allantoic remnants consist of aggregates of epithelial cells with a squamoid to transitional appearance and they typically lack lumens (Fig. 42.8).

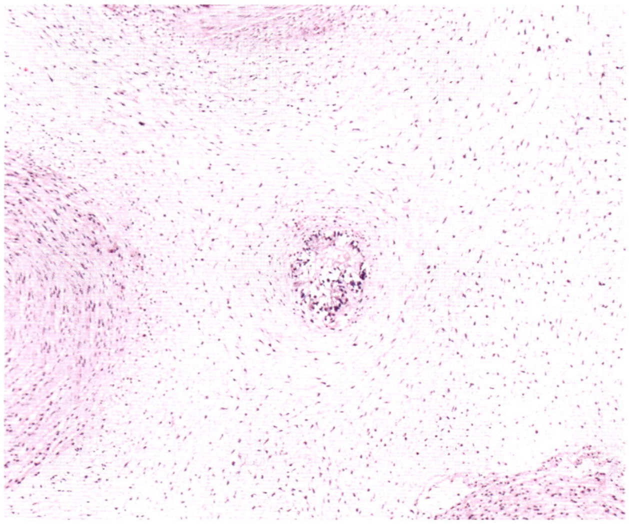

FIGURE 42.8 Allantoic duct remnant. Usually located in the center of the umbilical cord, the epithelium can be transitional or squamous. This type of remnant usually lacks a lumen (H&E).

The vasculature of the umbilical cord consists of two arteries and one vein. The arteries possess no internal elastic lamina, whereas the vein has an inner elastic lamina. The arteries feature a double-layered muscular wall composed of a network of interlacing smooth muscle cell bundles. The umbilical vein also generally has a larger diameter and possesses a thinner muscular coat with a single circular layer of smooth muscle cells (Fig. 42.9). Remnants of the vitelline vasculature (Fig. 42.10) may be observed in sections taken from the proximal cord in the form of a single vessel, a plexiform vascular network, or both. This finding should not be confused with a true hemangioma of the cord.

Another feature distinguishing the umbilical vasculature from other systemic vessels is the lack of vascular adventitia.

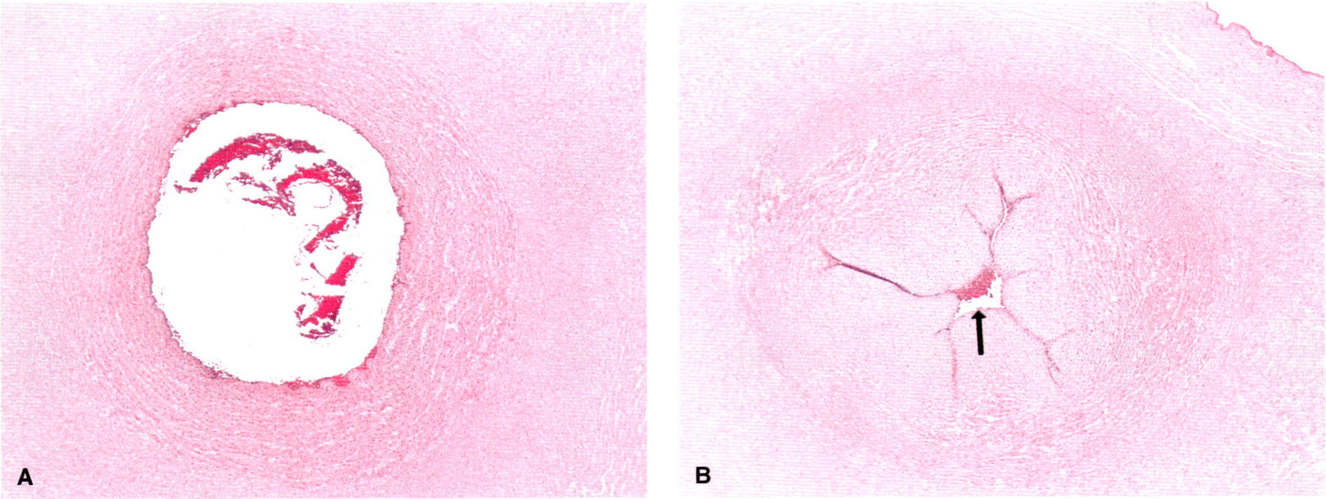

FIGURE 42.9 **A**: Umbilical vein. Note the open lumen and the relatively thin muscle wall. **B**: Umbilical artery. Thick muscular wall and opening of the lumen. (H&E).

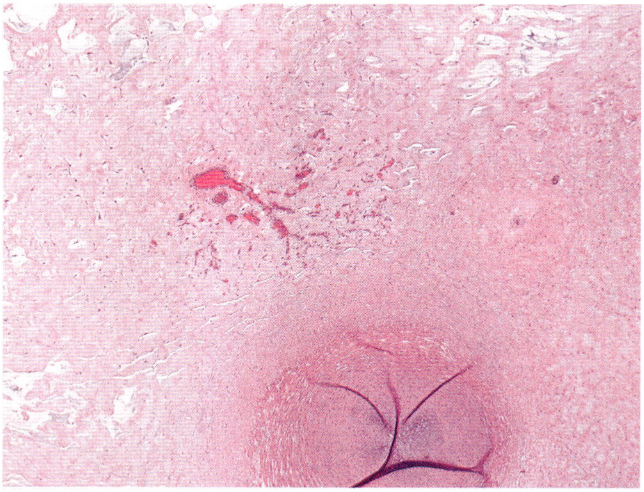

FIGURE 42.10 Vitelline vessels. Small cluster of vessels usually towards the periphery of the cord often associated with a vitelline duct. Note the feeding vessel does not come from an umbilical vessel. These are incidental benign remnants and not a hemangioma.

It is common to identify anastomotic channels between the two umbilical arteries near the placental insertion (i.e., Hyrtl anastomosis) that simulate the appearance of supernumerary vessels or a single umbilical artery (Fig. 42.11) (24,32,33). These misaligned anastomoses are of no clinical significance and may be regarded as anatomical variants.

Histopathology

Most pathology of the cord is macroscopic and, thus, histology is confirmatory. For example, a tight knot with notching that leaves an impression mark in Wharton jelly indicates a stricture associated with proximal vascular dilatation. A single umbilical artery due to either agenesis or hypoplasia of the second artery occurs in approximately 1% of neonates. This finding is associated with other congenital anomalies, especially those of the urinary tract or cardiovascular systems (19). A persistent second umbilical vein is unusual, however, it is not uncommon to identify more than three vessels in an umbilical cord. This finding is related to vascular tortuosities, or false knots (Fig. 42.6) that usually have little clinical significance, although they may be prone to thrombosis (19).

The presence of thrombotic material in the vasculature of the umbilical cord is abnormal. The process may be related to the genesis of a single umbilical artery when it occurs in early gestation. Abnormal umbilical insertions may also render vessels prone to thrombosis. Velamentous cord insertions are characterized by implantation of the umbilical vasculature in the placental membranes rather than the placental disc (Fig. 42.12). These vessels course independently within the chorion and are unguarded by Wharton jelly. Compression and thrombosis can't result from pressure on these unprotected vessels, and they are subject to injury during spontaneous or intentional membrane rupture. Other important abnormalities of the umbilical cord include vascular rupture, complete absence of Wharton jelly, and neoplasms of the umbilical cord (Fig. 42.13). Of note, perivascular umbilical hemorrhage is a common artifact of placental delivery following umbilical cord traction or clamping and should not be interpreted to represent vascular rupture.

Umbilical torsion is associated with excessive fetal movement, cocaine use, abnormal fetal heart rate tracings, and preterm deliveries (34). It may be accompanied by hypercoiling of the umbilical cord, especially long cords. Umbilical cord stricture is associated with the focal absence of Wharton jelly. Strictures may be seen in cords with torsion, kinking, or hypercoiling (Fig. 42.14). Although scant data define the dimensions of normal cords, some authors have found that abnormalities are associated with fat and thin cords (19,24,25). Both torsion and strictures are associated with adverse outcomes (19).

Leukocytes within the cord substance are indicative of fetal inflammatory response and are the result of infectious

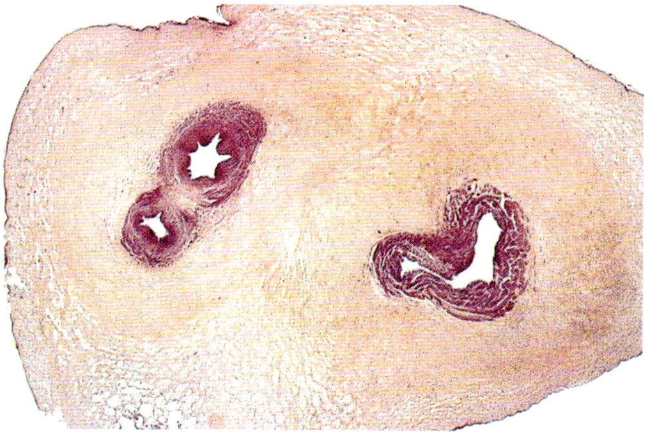

FIGURE 42.11 Hyrtl anastomosis of the umbilical arteries at the insertion to the placenta. Note the two juxtaposed arteries on the left (HPS stain).

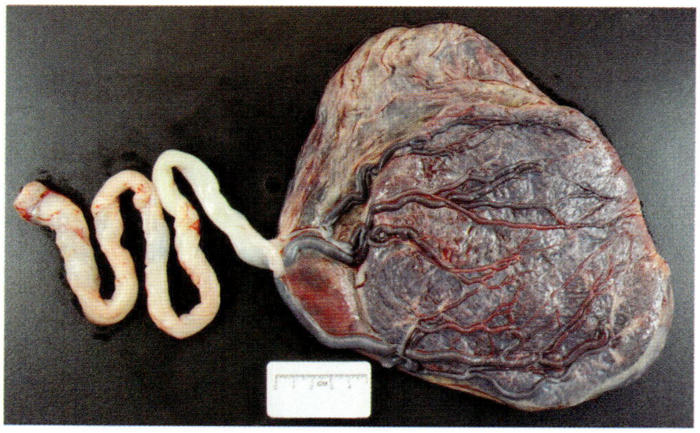

FIGURE 42.12 Velamentous insertion of the umbilical cord. The umbilical cord inserts into the membranes away from the margin of the placenta leaving its vascular branches unprotected from Wharton Jelly and prone to damage (compression, thromboses, necrosis, or inflammation). This anomaly could also be situated over the internal cervical os (vasa previa), which is a potentially dangerous condition as the vessels may rupture or be lacerated prior to delivery and result in exsanguination of the fetus.

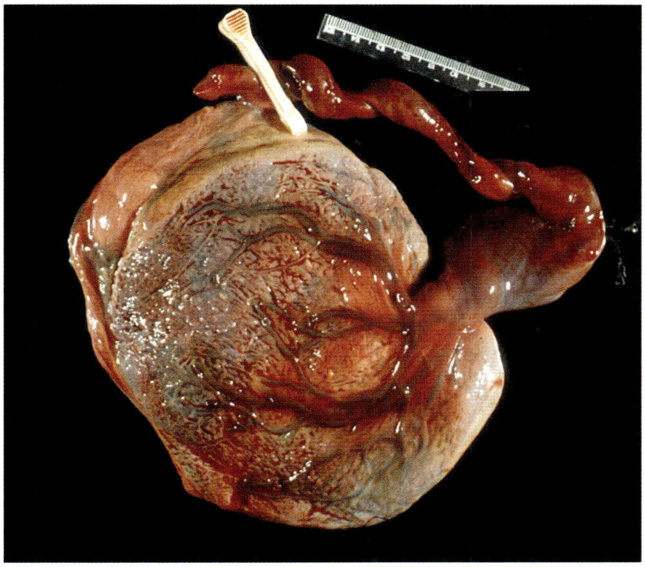

FIGURE 42.13 True hemangioma of the umbilical cord. Note marked increase in diameter of the umbilical cord at the insertion to the placenta. These are very rare and usually lethal. (Used with permission from American Registry of Pathology. Published as Figure 5-2, in Medeiros, L. Jeffrey et al. *Tumors of the Lymph Node and Spleen, AFIP Atlas of Tumor Pathology. Series 4, Fascicle 25.* American Registry of Pathology, 2017.)

antigens. Cord inflammation is vascular, perivascular, or in Wharton jelly, and is usually accompanied by maternal inflammation in the membranes (i.e., acute chorioamnionitis). These fetal and maternal inflammatory responses are staged and graded as these have associations with fetal and maternal morbidity. High-stage and high-grade fetal inflammatory cell infiltrates can be associated with

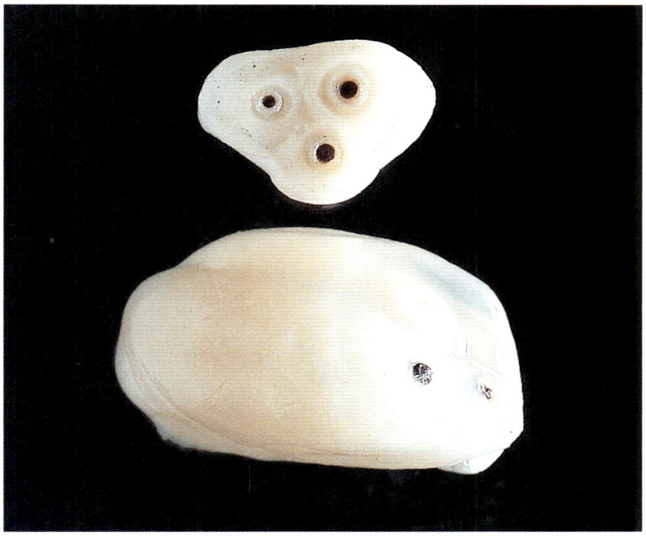

FIGURE 42.15 Necrotizing funisitis. This serious condition of fetal inflammatory response has a marked perivascular "halo" or band of degenerating and often calcified neutrophils. It is associated with significant fetal morbidity and mortality. Although it used to be a sign of syphilitic placentitis, it now is most often presumed to be due to a myriad of amniotic fluid infections.

calcifications, in which case it is classified as necrotizing funisitis (Fig. 42.15). Such severe pathology can be indicative of chronic infection due to syphilis and other pathogens (Fig. 42.16) (4,19,35). Infection by *Candida albicans* may appear as white surface plaques and wedge-shaped aggregates of acute inflammatory cells on the cord surface surrounding fungal elements, the latter of which can be highlighted with special stains for fungi (Fig. 42.17). Most candidal infections are self-limited but candidal sepsis is

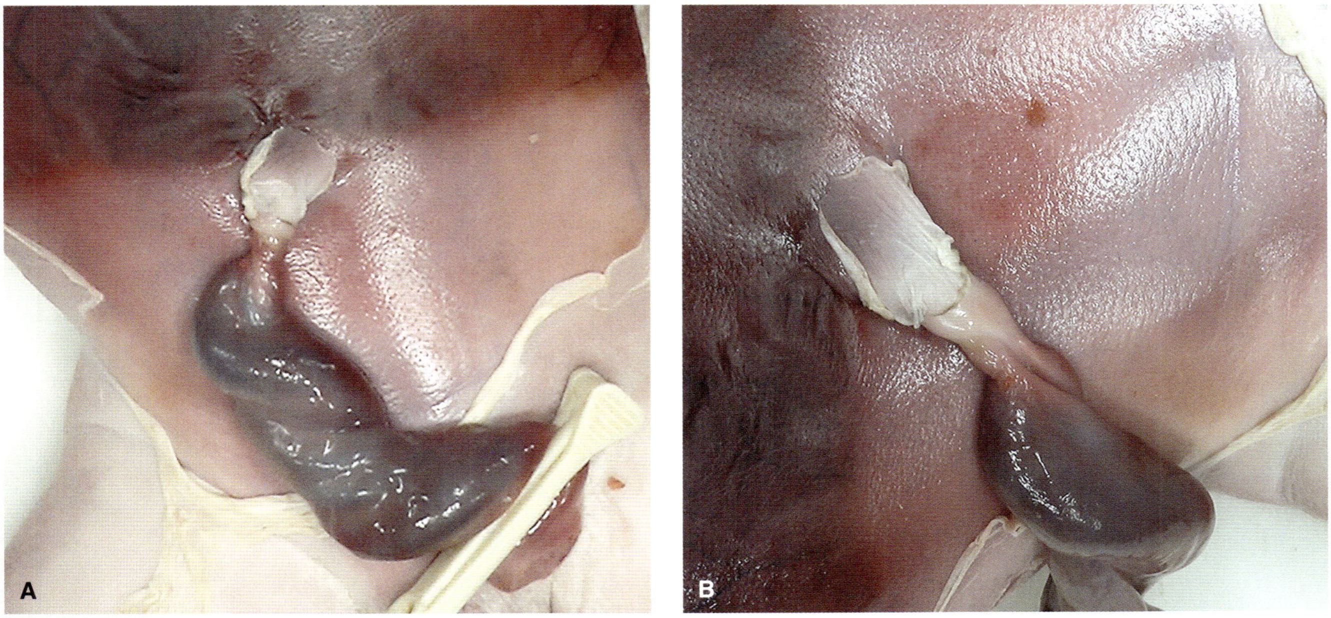

FIGURE 42.14 Focal loss of Wharton jelly in the umbilical cord segment attached to fetal abdominal wall in a case of fetal death. **A**: Focal hypercoiling (kinking) was found at delivery. **B**: One half rotation relaxed the hypercoiling but unmasks the deficiency of Wharton jelly.

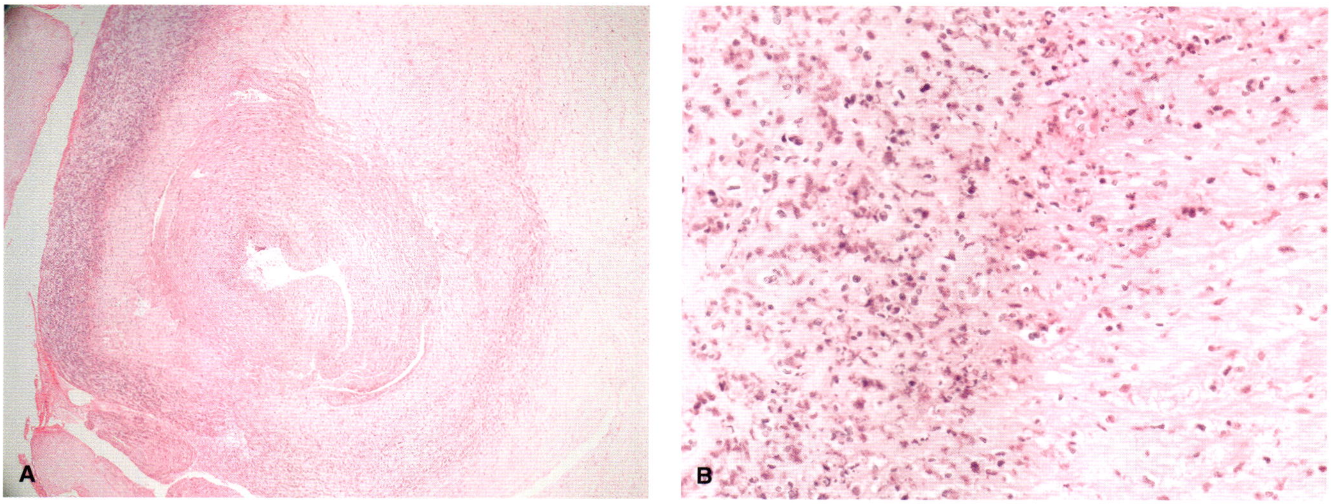

FIGURE 42.16 Necrotizing funisitis (**A**, low magnification) and (**B**, high magnification). Note the dense and necrotic and often calcified "halo" of acute inflammatory cells around the umbilical vessels.

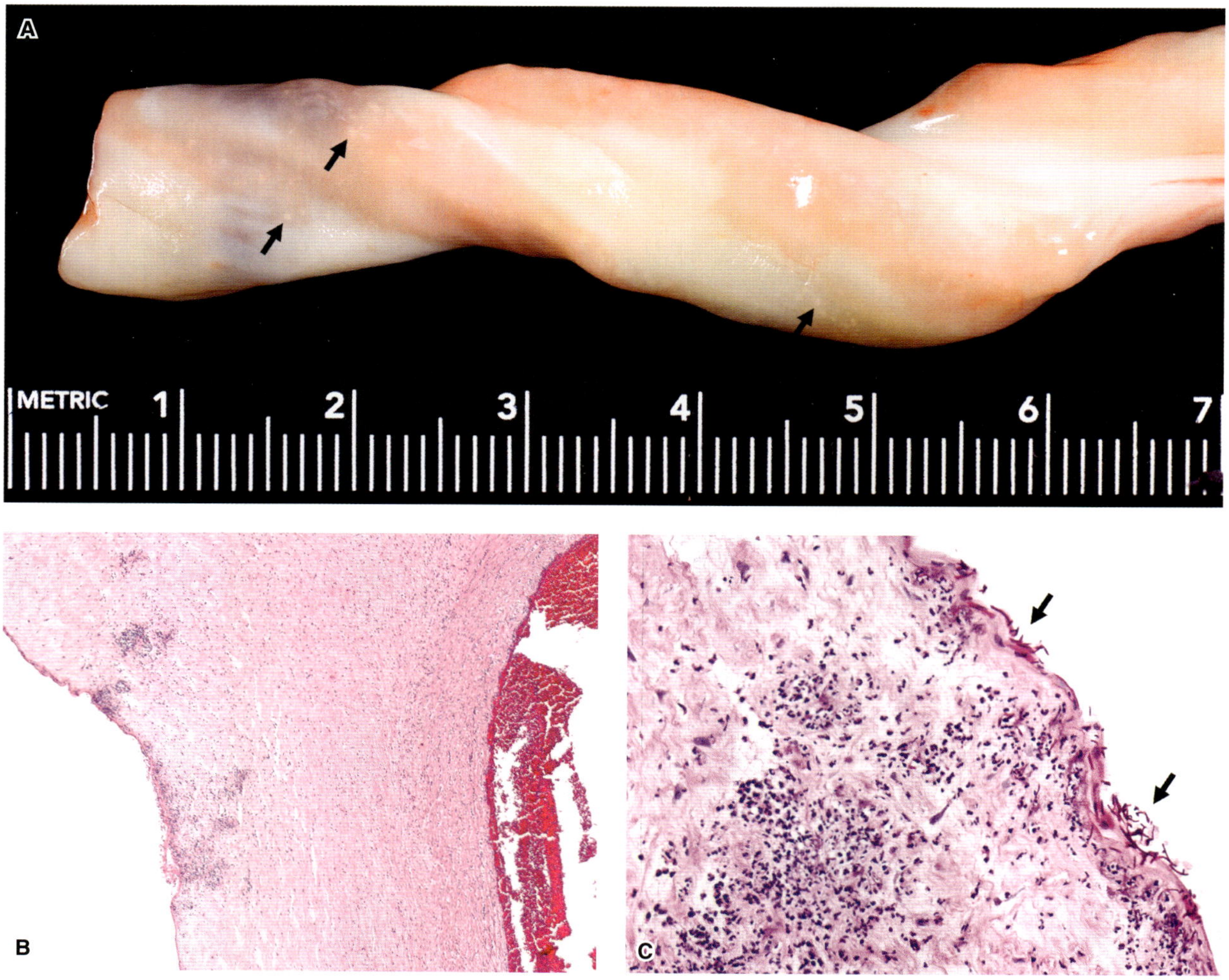

FIGURE 42.17 Candidal funisitis. **A**: Characteristic of candidal infections of the placental is the involvement of the umbilical cord by small surface abscesses visible grossly as small white/yellow plaques (*arrows*). **B**: Numerous yeast forms can be identified often by H&E stains alone, but highlighted in (**C**) with a PAS stain (*arrows*). Although most cases of candida funisitis are benign, candidal sepsis does occur and is a critical value for the neonatologist especially if suspected on gross examination.

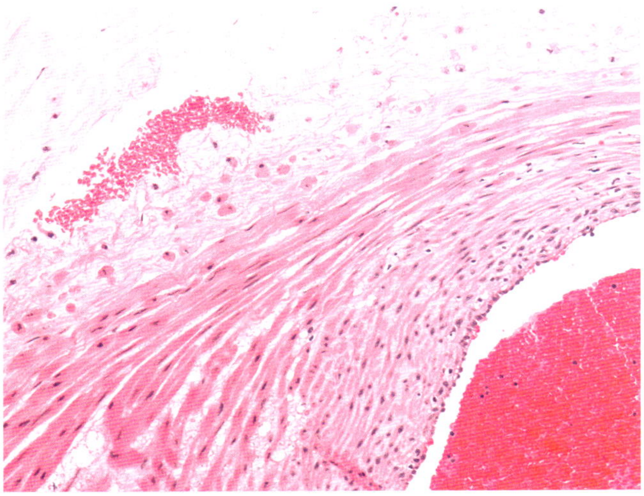

FIGURE 42.18 Meconium-associated myonecrosis of an umbilical vessel in a heavily meconium-exposed placenta. Notice the apoptotic myocytes on the outside of the vessel wall contrasting with the normal ones on the inside. This condition is rare but has associated significant morbidity to the fetus (H&E).

associated with a high neonatal mortality rate. Thus, candida plaques represent a critical value that must be reported to the neonatologist.

Meconium-induced myonecrosis is an important conditions that results from direct meconium toxicity and apoptosis of vascular media (Fig. 42.18) (19,36–38). This finding occurs as pyknotic, hypereosinophilic cells at the periphery of the affected vessel wall, typically in that aspect closest to the umbilical cord surface and in close approximation to meconium-laden macrophages (Fig. 42.19). There is often associated "vasculitis" of the involved vessel(s). Associated vascular spasms and medial degeneration may adversely affect hemodynamics in the cord and chorionic vasculature.

RAMIFICATION OF CHORIONIC VASCULATURE

The umbilical cord normally inserts into the disc in a central or eccentric fashion. Abnormal insertion at the margin and velamentous insertion in the membranes comprise a small portion of cord insertions, and both should be considered pathologic because they have the potential to cause adverse outcomes (4). The pattern of vascular ramification within the chorion is described as either magistral or dispersed. Magistral ramification is characterized by large-diameter vessels that emanate radially and diminish in caliber at the periphery of the placenta. Dispersed vessels are multiple small vessels emanating directly from the cord insertion site. The branches of the umbilical vein and umbilical arteries are histologically similar and can only be distinguished based on their gross anatomic distribution; arteries always cross over veins when observed from the fetal surface (Fig. 42.20). The notation of such vascular relationships is of extreme significance when considering vascular anastomoses in twin pregnancies (4,19,39). The primary branches of the umbilical vasculature that course through the chorionic plate periodically dive beneath this stratum to feed the stem chorionic villi. These structures establish the circulation of primary vascular ramifications ending in the terminal villi.

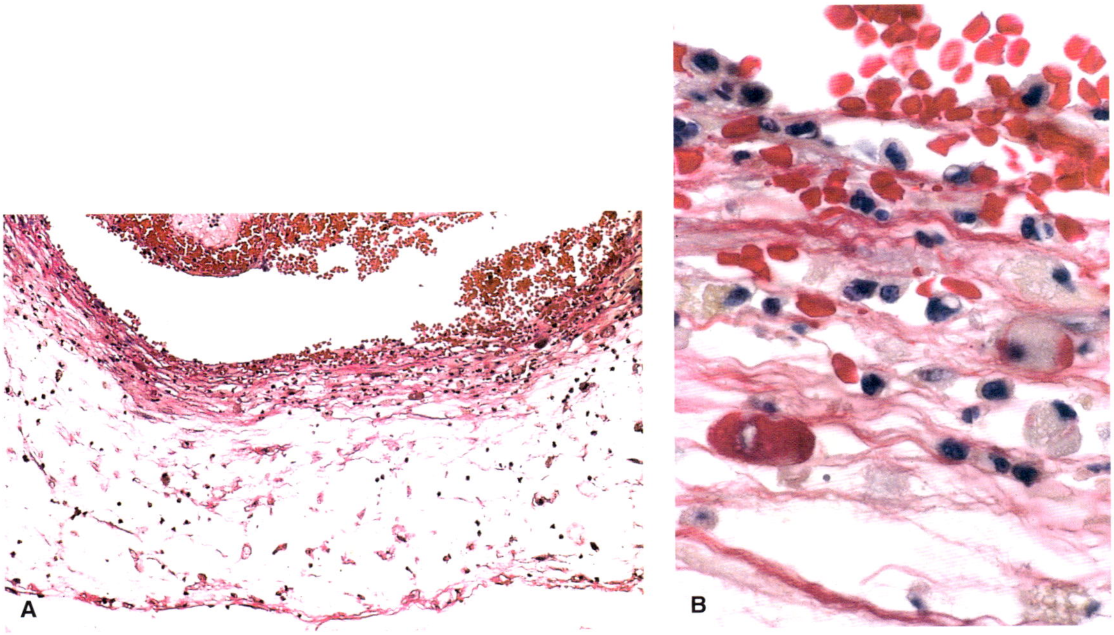

FIGURE 42.19 Globular degenerated necrobiotic medial cells secondary to meconium exposure. The process is focal and contrasts with adjacent normal myocytes (Luna Ishak stain) (**A**). Higher magnification demonstrating pigmented macrophages (**B**). (Reprinted with permission from Rana J, Ebert GA, Kappy KA. Adverse perinatal outcome with an abnormal umbilical coiling index. *Obstet Gynecol.* 1995;85:573–578.)

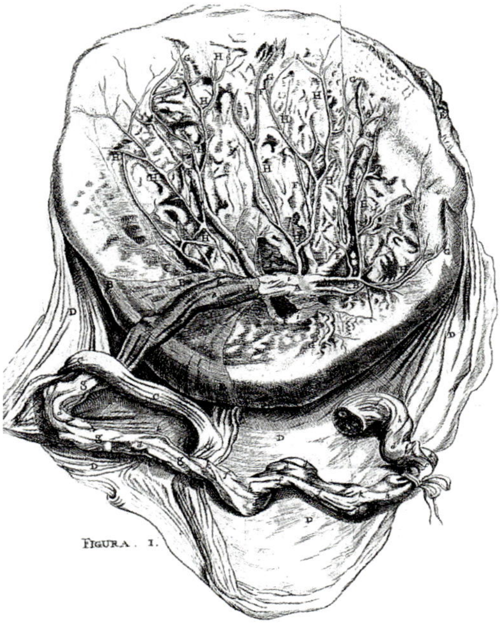

FIGURE 42.20 As early as the 1600s, Nicholas Hoboken recognized that chorionic arteries overlie veins, and this was beautifully depicted in his painstaking drawings. These are the earliest accurate drawings of the human placenta known to exist.

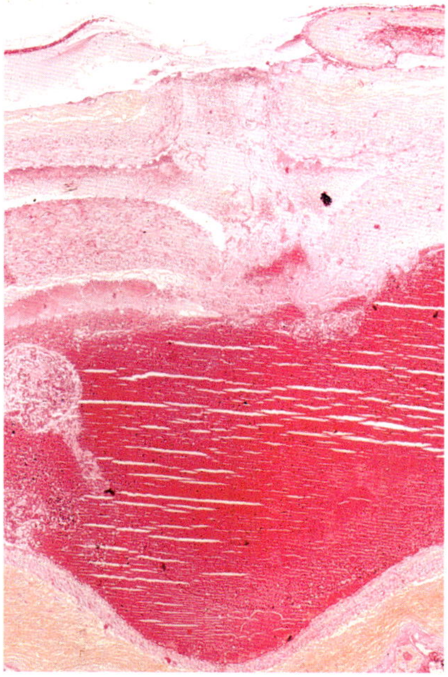

FIGURE 42.21 Iatrogenic chorionic vasculature rupture following amniocentesis for fetal lung maturation studies. The not-unusual accessory lobe was several centimeters from the main body of the placenta, obscuring its presence and vasculature. Emergent cesarean section followed a "bloody tap" and fetal heart rate abnormalities.

PATHOLOGIC ALTERATIONS OF THE CHORIONIC VASCULATURE

Velamentous and membranous vessels that connect multilobed placentas are at risk for rupture and bleeding from fetal excoriation or iatrogenesis (Fig. 42.21). Such vessels are also at increased risk of fetal part compression with subsequent occlusion and/or thrombosis because they are not cushioned by the placental parenchyma (4). Vascular thromboses may readily be identified by noting dilated vessels containing thrombi on the fetal surface. More subtle features include faintly highlighted linear white streaks that parallel the peripheral margin of the affected vessels (Fig. 42.22A). These findings can be confirmed histologically (Fig. 42.22B).

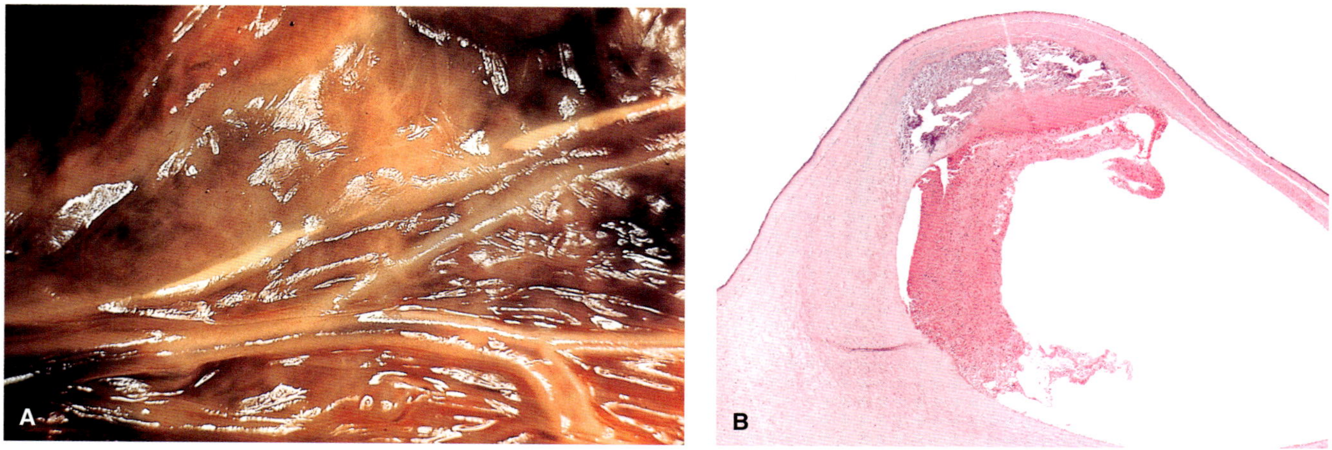

FIGURE 42.22 **A**: Chorionic vascular thrombosis characterized by a linear streak paralleling the vascular course (in this case a chorionic vein). **B**: Collection of fibrin and blood adherent to the chorionic plate vessel wall. These thrombi can be occlusive or (as in this case) nonocclusive and older lesions may calcify (H&E). (Reprinted with permission from Benirschke K, Kaufmann P. *Pathology of the Human Placenta*. 3rd ed. Springer-Verlag; 1995.)

MEMBRANES

Embryology

The placental membranes consist of the amnion and the chorion. The amnion constitutes the innermost aspect of the embryonic cavity and develops from the margin of the embryonic disc. As the embryonic disc begins to take the form of a tube, the amnionic periphery also folds inward and its attachment to the ventral body is defined. The amnionic cavity subsequently develops by the process of cavitation. Elongation of the body stalk coincides with embryonic prolapse into the amnionic compartment. As development proceeds, the resultant cavity expands, and by 12 weeks from the last menstrual period, the gossamer amniotic cavity completely occupies the chorionic sac. There is a loose connection between the amnion and chorion that can be ruptured in utero or peeled off ex utero. The amnionic cavity remains filled with amnionic fluid that amounts to approximately 1 L at term (19).

The chorion forms the base for the peripherally radiating villi and serves to encapsulate the embryo and developing amnion. As the early embryo develops, the trophoblast and its mesodermal investments continue to expand in a spherical fashion. The inner aspect of a mesodermal condensation forms an inner capsular structure deep to the peripheral trophoblast, and is termed chorion that becomes the placental disc. Chorionic villi continue to develop beneath these structures, defining the placenta proper (i.e., the chorion frondosum). The region of the chorion that covers the expanded amnionic cavity beyond the placental disc is termed chorionic laeve. As will be further seen in the discussion of placental neoplasia (below), there are significant cellular phenotypic differences between the trophoblasts forming the chorions frondosum and laeve. Here, we focus on architectural differences that develop from the trophoblastic shell after implantation. Trophoblast growth tightly coupled with hierarchical branching morphogenesis is confined to the chorion frondosum. In contrast, primordial placental villi emanating from the chorion laeve undergo atrophy, leaving just the sclerotic remnants scattered randomly throughout the placental membranes. Unlike extravillous trophoblasts at the tips of anchoring villi beneath the parenchymal disc, trophoblasts in the chorion laeve have lost their ability to invade or support branching morphogenesis. In rare cases, however, villous growth in the chorion laeve persists, resulting in unusual diffuse, spheroidal placenta membranacea (Fig. 42.23). This rare and abnormal morphologic variant is associated with fetal growth restriction, abruption, and hemorrhage, from the always present placenta previa. The fetal blood vessels invest the chorionic plate and invaginate into the branching villi in the region of the chorion frondosum. When one considers the tumorous proliferations of fetal capillaries (i.e., chorangiomas), it is tempting to speculate that these lesions represent uncoupling of branching morphogenesis in the villous trophoblasts and stroma relative to vascularization. As further

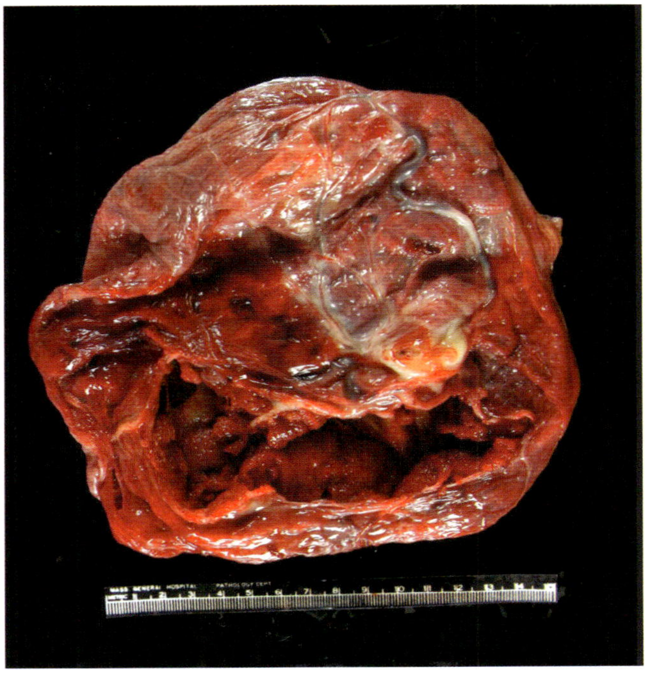

FIGURE 42.23 Placenta membranacea. Note the parenchyma covers the entire placental, there are no free membranes. This rare human anomaly is associated with growth restriction in the fetus and complications with delivery.

evidence of the tight regulation of vascularization, one should take note of how fetal vessels are confined to the chorion of the disc; the amnion is an avascular structure.

Amnion and Chorion

Gross Morphology

The fetal membranes have a particular and characteristic appearance in normal deliveries. The sac, when viewed from the fetal surface, is clear and often has a bluish hue, and the amnion is devoid of vasculature. Remnants of atrophied vasculature may be seen in the overlying chorion and appear as filamentous streaks. The chorionic plate also has a characteristic blue sheen with radiating hierarchical branches of chorionic vessels. The membranes of the chorionic plate are distinguished from the laeve as described above. It is not infrequent to find a peripheral nodule on the disc membranes that represents the remnant of the fetal yolk sac (Fig. 42.24).

Gross Morphologic Alterations

Although chorionic vessels are normal in the chorion of the chorionic plate overlying the disc, the persistence of functional vasculature in the chorion laeve is aberrant and equates to membranous vessels. These vessels may connect lobes of placenta, connect to the membranous insertion of the umbilical cord (velamentous insertion as described above), or simply loop out from and back to the disc edge.

On occasion, the chorionic plate may possess a ring of fibrin that forms a concentric ridge between the insertion of the cord and the margin of the placental disk. This

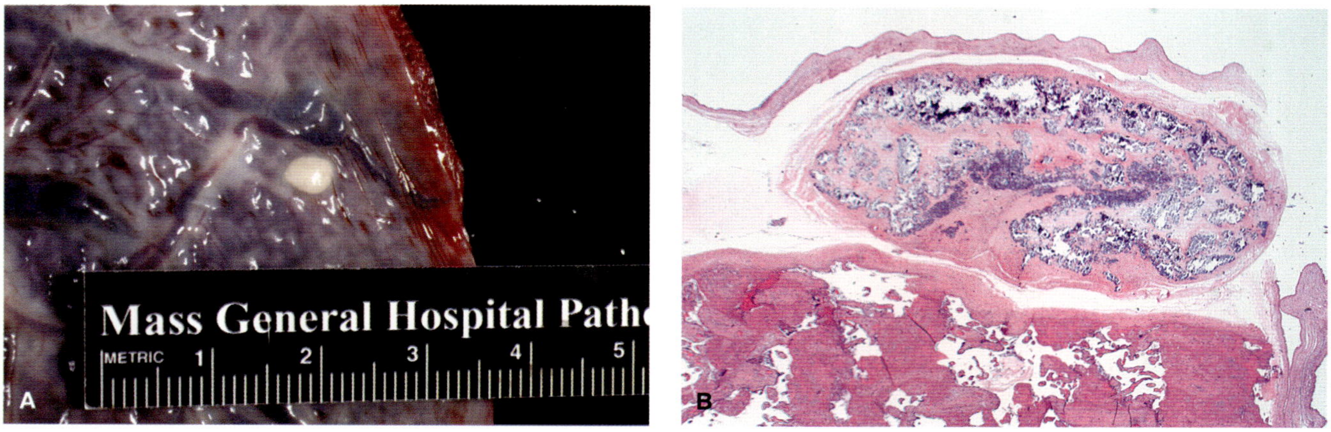

FIGURE 42.24 **A**: Gross example of a yolk sac remnant (yellow–white well circumscribed small nodule in the membranes). **B**: Histology of the remnant (H&E).

fibrin ring, which lies deep into the amnion, is indicative of extrachorial placentation. Such placentation is characterized by two forms: circumvallate and circummarginate. In the former, the membranes are reflected upon themselves at the ridge of the fibrin deposition. The membranes then cover the remaining margin of the placental disk in a loose fashion (Fig. 42.25). In the circummarginate placenta, the ring of fibrin is present over the chorion, and the overlying amnion is not reflected upon itself at this fibrinous ring (Fig. 42.26). Thus, the amnion extends to the margin of the placental disc, and its departure to form the amnionic sac occurs at this margin. It has been suggested that circummargination and circumvallation results from hemorrhage at the margin early in implantation. Such hemorrhage forms a fibrous ring around the margin (i.e., circummargination) or, if larger, results in circumvallation. Another theory for these abnormal membrane insertions is that the fibrinous ring represents placental migration in conjunction with an enlarging uterus during the second trimester (the so-called "trophotropism") (19). Of import, circummargination is clinically inconsequential but circumvallation is associated with fetal growth restriction and risk for placental abruption.

Squamous metaplasia of the amniocytes is common and usually of no clinical consequence unless extensive. Immersion of the placental membranes in water generally highlights squamous metaplasia by its resistance to wetting, in contrast to the normally hydrophilic amnion. Metaplasia may be pronounced and large plaques and nodules

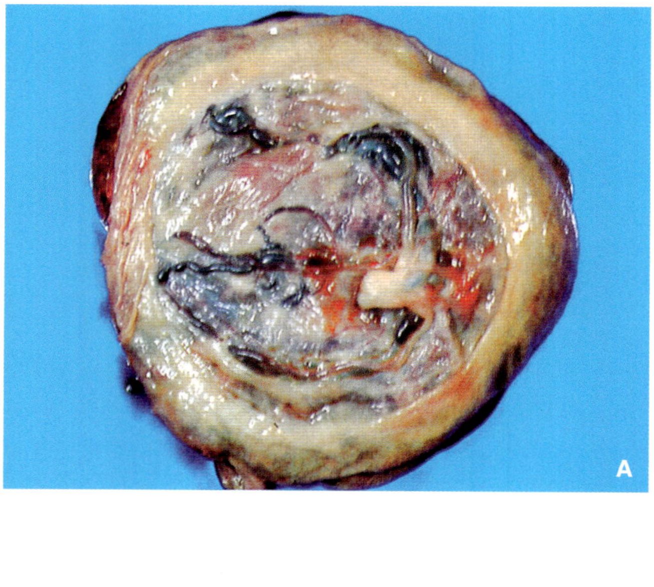

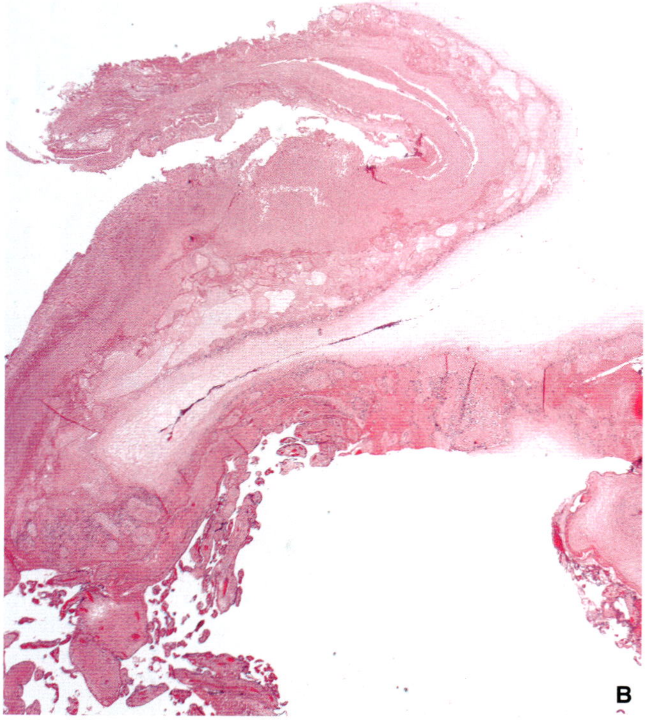

FIGURE 42.25 **A:** Circumvallate placenta. (**A.** Reprinted with permission from Benirschke K, Kaufmann P. *Pathology of the Human Placenta*. 3rd ed. Springer-Verlag; 1995.) **B:** Note loose association of amnion peripheral to marginal subamnionic fibrin deposition.

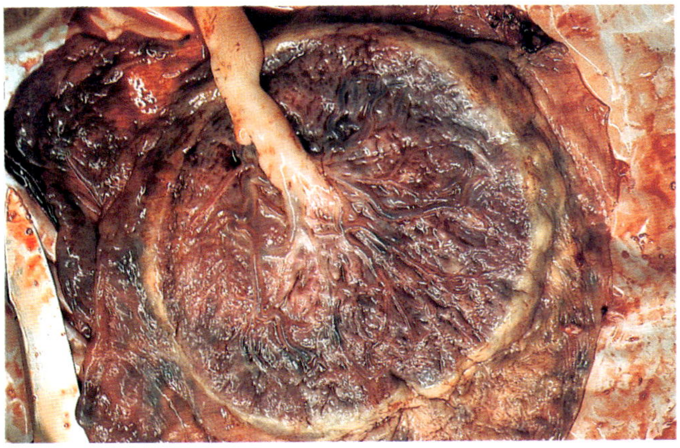

FIGURE 42.26 Circummarginate placenta. Note the close association of amnion to the disk peripheral to the fibrin ring.

may form (Fig. 42.27). It is important to distinguish squamous metaplasia from amnion nodosum, a condition in which multiple nodules form on the amnionic surface, reflecting aggregation of fetal squames and other debris that precipitate out of the amniotic fluid onto the amnion. The latter can be easily denuded from the amnion surface by slight mechanical pressure (Fig. 42.28). The nodules of cellular debris are derived from the fetal epidermis, and reflect apposition of this membrane and fetal skin in conditions where there is diminished amnionic fluid. Amnion nodosum is a significant finding as it is evidence of chronic injury to the placenta and is diagnostic of severe oligohydramnios or anhydramnios in the second or third trimester, often foretelling severe and lethal pulmonary hypoplasia.

Amnionic bands are presumably formed when loosely bound amniotic connective tissue delaminates from the chorion (40). This rare condition is responsible for in utero fetal part amputation and trauma and is a phenomenon that occurs in approximately one in 10,000 births. The precise mechanism is not known in most cases, but early rupture of the

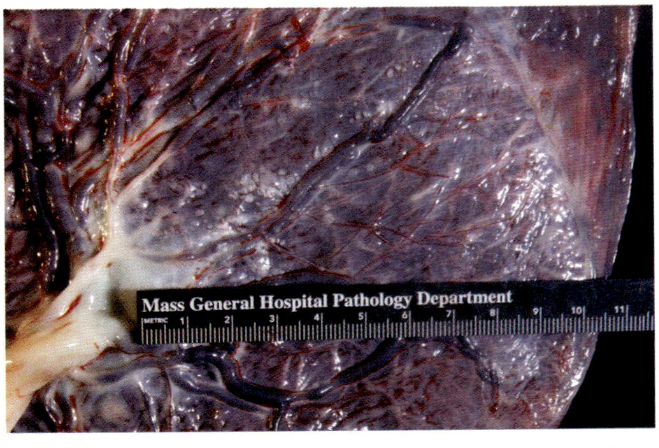

FIGURE 42.27 Gross example of extensive squamous metaplasia on the amnionic surface from a fetus with an encephalocele. It is believed that irritation from the encephalocele in this region produced extensive metaplasia.

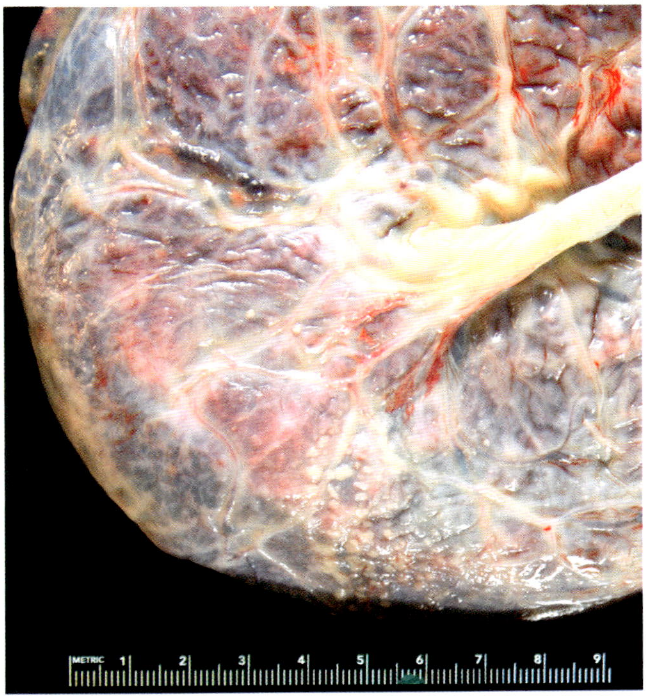

FIGURE 42.28 Gross example of amnion nodosum. Note the variable-sized yellowish nodules on the chorionic plate.

amnion allows the fetus to enter the chorionic sac. Detached amniotic remnants thus form the substance of resulting amnionic bands. At term, these placentas have highly opaque chorionic surfaces that reflect hyperplasia of this uncovered layer. Only small amounts of amnion are present, which distinguishes this condition from artifactual disruption of the amnion from the chorionic plate during the delivery process. Occasionally, an abnormal "web" (amniotic sail) will be present at the base of the cord insertion, and such redundant amnion may limit normal cord movement but is not an amniotic band (Fig. 42.29). Early amniotic rupture may cause a feature similar to amnion nodosum, called chorion nodosum, reflecting deposition of debris as the fetus injures the fibrous chorion. It is usually only a histologic diagnosis (Fig. 42.30).

Amnion Histology

The amnion, the gossamer innermost layer of the amnionic cavity, is lined by a single layer of epithelial cells that resides on a basement membrane. The basement membrane is attached to an underlying thin layer of connective tissue (Fig. 42.31) (41,42). The amnion, although adjacent to the chorion, is not truly fused to it. As the amnion develops, it is separated from the chorion by the so-called magma reticulare, which is a viscous and thixotropic gelatinous fluid (19,43). Stellate mesenchymal cells may be found within this substance. These cells also have epithelial characteristics and express cytokeratin and vimentin by immunohistochemistry.

The epithelial cell layer of the amnion is composed of one distinct cytologic type (44). It is composed of a single layer, squamoid to cuboidal epithelial cells devoid of

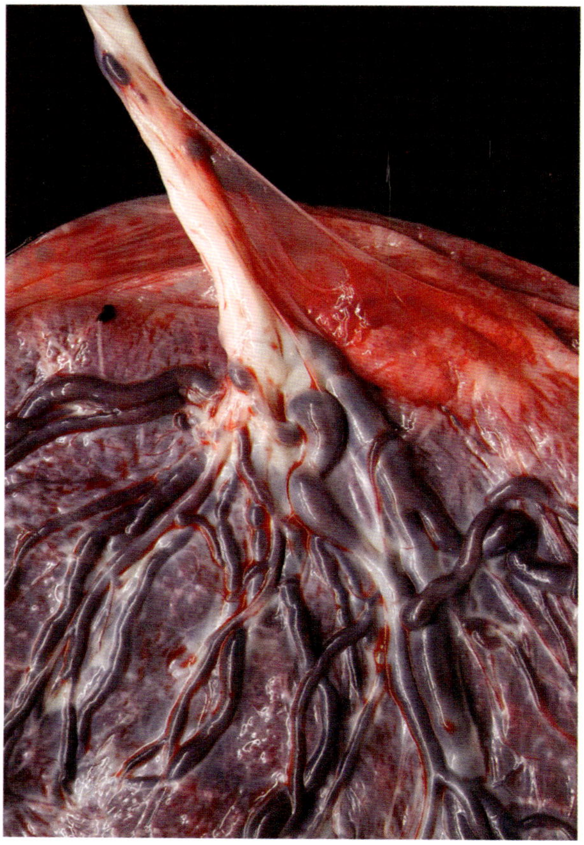

FIGURE 42.29 Amniotic sail attaching to and restricting movement of the umbilical cord.

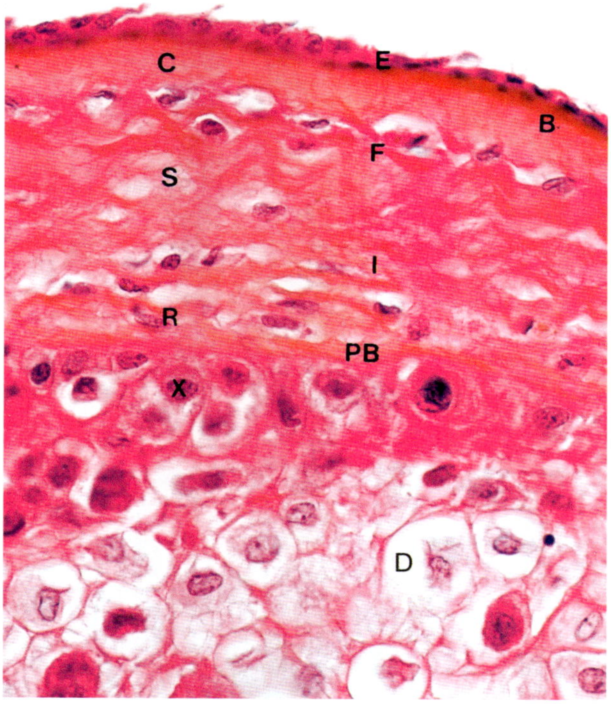

FIGURE 42.31 Flattened to cuboidal amnionic epithelial cells (*E*) adhere to their basement membrane (*B*). Beneath this is the compact layer of the amnion (*C*), which is acellular and may form a barrier to PMNs. The compact layer is rarely affected by edema and is probably the strongest amnionic layer. A fibroblastic layer (*F*) lies beneath the compact layer, and macrophages may be found. A spongy layer (*S*), relatively devoid of fibroblasts, separates the amnion from the chorion, although the two may merge imperceptibly. Often, an artifactual separation may be present near the plane of true fusion. The amnion usually measures from 0.2 to 0.5 mm in thickness (*I*). The most superficial layer of chorion is usually an incomplete cellular zone (*I*) that overlies a thick reticular layer (*R*). This layer is composed of fibroblasts and macrophages. Beneath the reticular layer is a pseudobasement membrane (*PB*) overlying extravillous trophoblast (*X*) and then maternal decidua (*D*) (HPS stain).

secretory activity. Ultrastructural studies show extensive microvillus projections with multiple vesicular structures at the base of these epithelia (19,41,43). It has been postulated that these vesicles represent pinocytotic activity (45). Presumably, the cytologic components of this layer gain their nutrition from adjacent amnionic fluid, which in turn is rich in nutrients from transudation (from fetal vasculature) and fetal excretory products. In early gestation, this nutrition is derived from the magma reticulare.

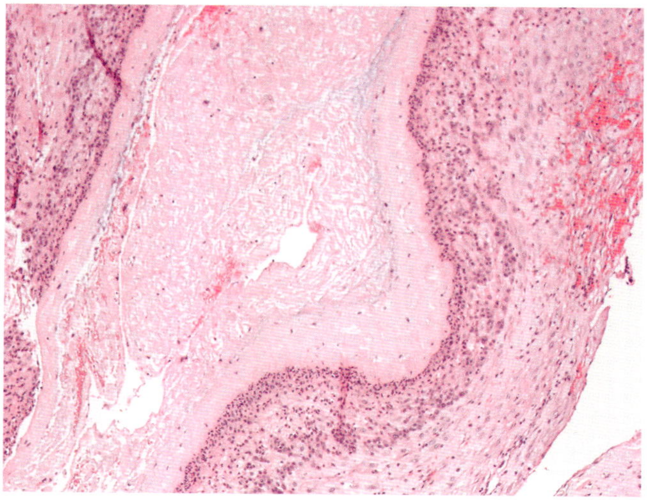

FIGURE 42.30 Chorion nodosum. Debris on the surface of the fibrous chorion for early amniotic rupture (H&E).

Amnionic epithelial cells are attached to each other by desmosomes in freeze-fracture experiments and to the underlying basement membrane by hemidesmosomes (46,47). Amnionic epithelial cells divide by mitosis. On occasion, multinucleated cells are identified. Morphometry studies have demonstrated that polyploid cells exist in this layer (48) and other karyotypic anomalies can occur, leading to potential false-positive results for chromosomal defects when amnionic cells contaminate amniocentesis preparations (19,49–51). Although the amniotic epithelium does not engage in active secretion, lipid droplets have been noted within their cytoplasm, an observation that correlates with increasing gestational age (52). Glycogen also has been found within amnionic cells (53).

Squamous metaplasia is a common occurrence in the amnionic epithelium, especially near the insertion of the umbilical cord (Fig. 42.32). This epithelium may become

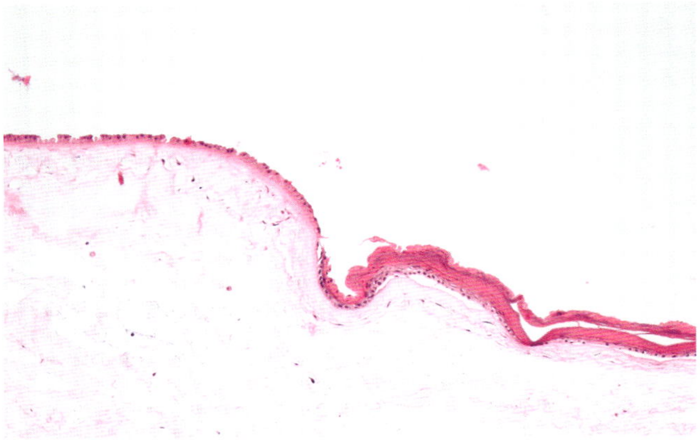

FIGURE 42.32 Squamous metaplasia of the amnion (H&E).

keratinized, and keratohyalin granules can be identified. Although this appears to result from irritation of the amnionic epithelial surface, these changes can be found in more than half of all term placentas and are inconsequential (19,47).

An additional component of the amnion is identified beneath the basement membrane. This layer principally is divided into a compact and a fibroblastic region. The connective tissue within this region may harbor macrophages, as early as the first trimester of pregnancy (54).

Amnion Histopathology

Histologic abnormalities of the amnion are heralded by an abnormal gross appearance. Membranes stained green or brown typically reflect meconium pigment staining the amnion or engulfed meconium within resident macrophages. Amnionic membranes that are white, yellow, or cloudy may indicate acute chorioamnionitis.

Abnormalities of amnionic epithelial cells that reflect degenerative or reactive changes are characterized by the presence of vacuolated cytoplasm and elongation to columnar forms. A rather rare and unusual finding associated with gastroschisis carries a pathognomonic histologic aberrancy of the amnion whereby amnionic epithelial cells contain innumerable vacuoles (Fig. 42.33) (19).

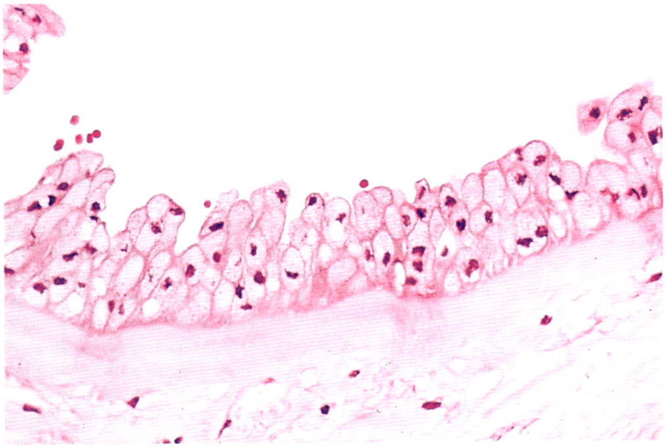

FIGURE 42.33 Unusual vacuolated elongated amnionic epithelial cells pathognomonic of gastroschisis. The pathophysiology is uncertain (H&E).

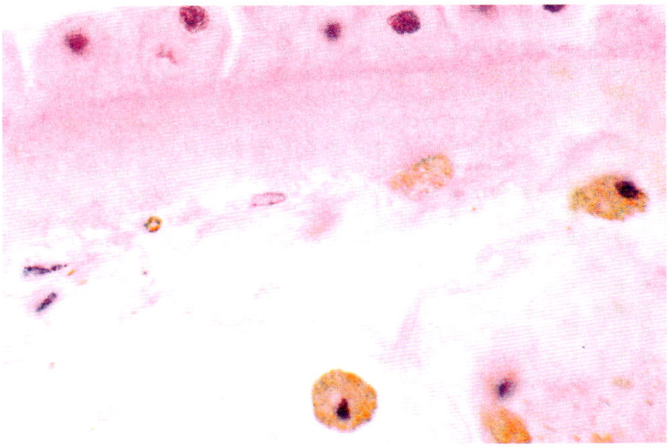

FIGURE 42.34 Amnionic epithelial degeneration overlies pigment-laden macrophages containing meconium (HPS stain).

Amnionic epithelial degeneration (Fig. 42.34) occurs with prolonged exposure to meconium. The most common finding is vacuolated amniocytes accompanied by macrophages that contain brown pigment that is opaque, waxy, or fluffy (Fig. 42.34). Meconium-laden macrophages should be distinguished from hemosiderin deposition with the use of iron stains (i.e., Prussian blue). Prior abruption leads to diffusion of heme breakdown products from bloody amniotic fluid into the amnionic membranes. Hemosiderin-laden macrophages can also be abundant in the chorionic plate when associated with massive subchorial hematomas (i.e., Breus mole), which are associated with significant pulmonary morbidity to the fetus (Fig. 42.35).

Although there are no true tumors of the amnion, occasional cysts representing edema may be identified, and these are of no clinical consequence. Occasional cysts of ectodermal and mesodermal tissue have been identified deep to the amnionic layer, but such findings are not considered to represent neoplasms; teratomas have been described but some of these lesions likely represent degenerative acardiac twins (4,55).

The papules of amnion nodosum are clearly pathologic and reflect, in most cases, significantly decreased or absent amnionic fluid. In such instances, vernix (desquamated epithelia from the fetus) and mucin form minute nodular aggregates on the ulcerated amnionic surface in this characteristic fashion (Fig. 42.36). The lesions are composed of acellular debris and cellular remnants (56,57). The material within these papules is PAS-positive and Alcian blue–positive.

Chorion Histology

The chorion is composed of a connective tissue membrane that carries the fetal vasculature (Fig. 42.37). Its inner aspect is bounded by the outer layer of the amnion, and the outer aspect is directly associated with the trophoblastic villi that sprout from its surface. There are two distinct regions of the chorion: the chorion frondosum and the chorion laeve. The chorion laeve is composed of an inner cellular layer, a "reticular" layer, a "pseudo-basement membrane," and an outer trophoblastic layer (19,42,44). The precise origin of

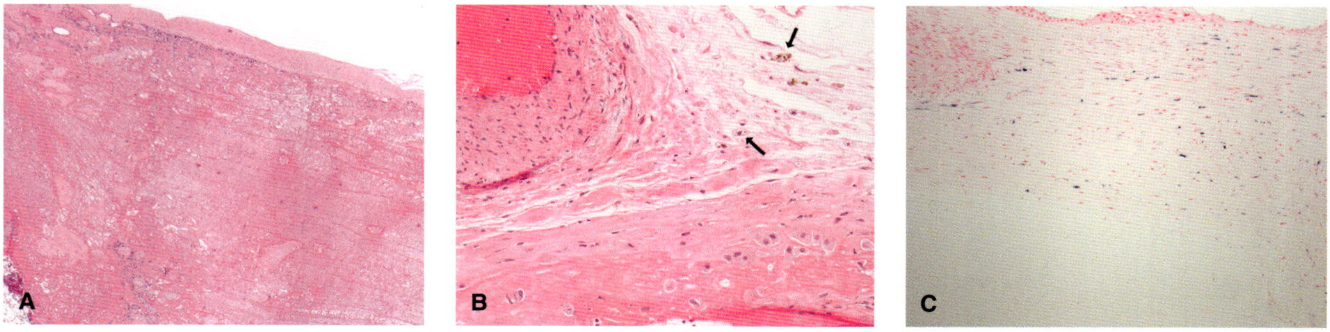

FIGURE 42.35 Large subchorionic hematoma of the fetal surface (Breus mole) (**A**, H&E) with hemosiderosis in the chorionic plate (**B**, H&E, and **C**, iron stain) with hemosiderin-laden macrophages (*arrows*).

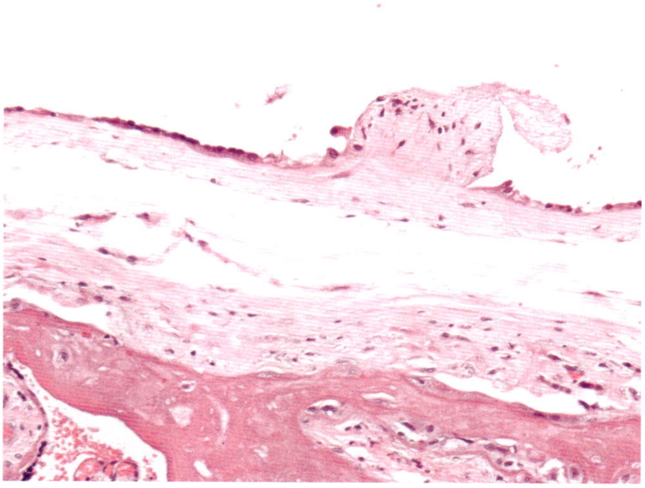

FIGURE 42.36 Amnion nodosum. Note ulceration of the amniotic epithelium with deposition of debris on the surface (H&E).

the mesenchymal component of the chorion is not clear, but it is believed that this connective tissue is derived from the primitive streak and not the trophoblasts (58). Electron microscopy studies have shown the connective tissue cells adjacent to the amnion to be rich in endoplasmic reticulum (43). Macrophages and degenerative endothelial cells have been described as part of the cytologic makeup of this layer. Acid mucopolysaccharides are prominent within the connective tissue matrix of the chorion (53). Although present in other regions of the placenta, type VI collagen is a prominent constituent of the chorionic layer (59). The chorion frondosum is similarly constituted but contains functional chorionic vessels that serve to anchor the first branch of functional chorionic villi.

Chorion Histopathology

Pathology of the chorionic vasculature includes a fetal inflammatory response in the chorionic membranes (Fig. 42.38A),

FIGURE 42.37 **A:** The separation between amnion and chorion is more apparent in early gestations as seen in this section from the chorionic plate of a 10- to 12-week placenta (upper right). **B:** The amnion is readily distinguished from the underlying chorion, which contains the easily identifiable chorionic vessels. Mesenchymal components are prominent (H&E).

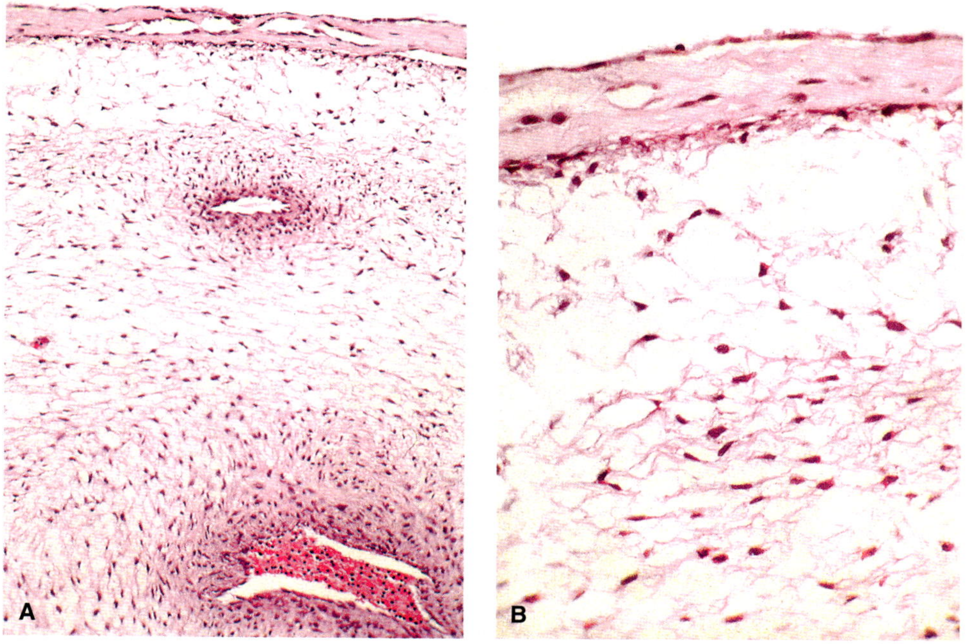

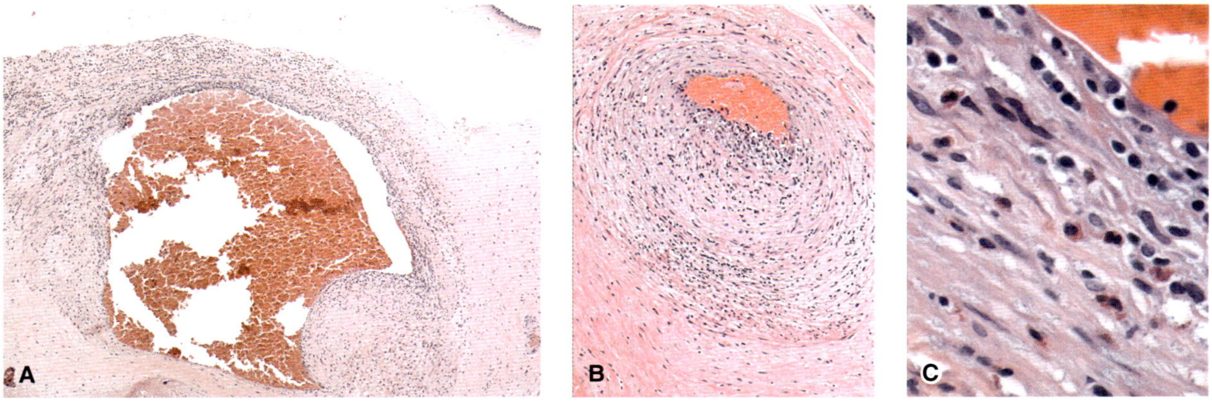

FIGURE 42.38 Patterns of inflammation involving fetal chorionic plate vessels. **A**: Severe fetal inflammatory response to amniotic fluid infection involving a chorionic plate vessel. Note the polarized migration of fetal neutrophils (and occasionally a few eosinophils migrating towards the amniotic cavity (top). **B** and **C**: T cell/ Eosinophilic vasculitis. Note the characteristic polarized migration of inflammatory cells away from the amniotic cavity (**top**) and toward the maternal blood space (**B**). At higher magnification (**C**), note that the fetal inflammatory infiltrate is composed of lymphocytes (mostly T cells) when phenotyped by immunohistochemistry (not shown) and eosinophils (H&E).

thromboses, and meconium-associated myonecrosis. Eosinophilic/T-cell chorionic plate vasculitis (Fig. 42.38B, C) is a rare condition characterized by small numbers of chorionic plate vessels infiltrated by T cells and eosinophils (19,60,61). These inflammatory cell infiltrates may injure the endothelium and lead to fetal vascular malperfusion by predisposing to thrombus formation. Fetal chorionic vessels allow permeation of polymorphonuclear cells in response to intra-amnionic bacterial antigens. Maternal neutrophils may also be seen in the chorion laeve. The amnion may become inflamed either by fetal or maternally derived neutrophilic inflammation (Fig. 42.39).

Chorionic cysts are considered normal variant gross and histologic findings, even when pronounced. Chorionic cysts are lined with extravillous CD146/MelCam-positive trophoblasts and are associated with major basic protein (MBP) production, a substance also found in eosinophilic granulocytes. These cysts are commonly found in the chorionic plate as well as placental septa. Multiple chorionic cysts are associated with massive perivillous fibrin deposition. Large chorionic cysts adjacent to the umbilical cord insertion may compromise fetal blood flow and result in downstream villous pathology.

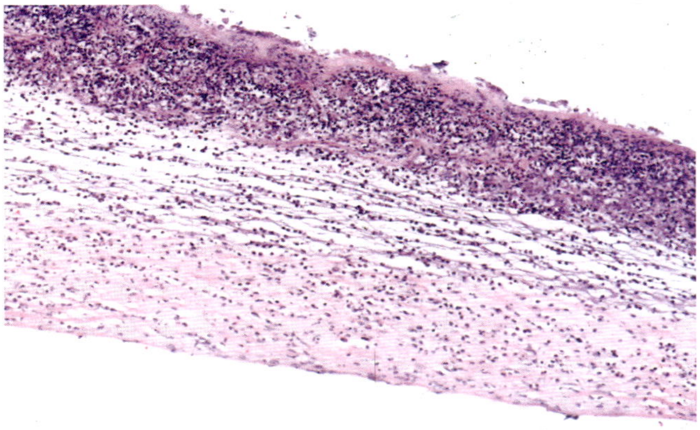

FIGURE 42.39 Acute chorioamnionitis, maternal inflammatory response, Amsterdam Stage 3 Grade 2 (H&E).

MULTIPLE GESTATION

The normal relationships of the placental membranes are germane to the understanding of twin or multiple gestations. Most twin placentas are dizygotic, although the incidences show a dependence on geographic location and ethnic background. The dizygotic twin gestation can present with separate or fused placental disks. In the latter situation, the intervening membrane can be evaluated to distinguish between dichorionic and monochorionic twin placentas. When the intervening membrane is devoid of chorion, the placenta is classified as diamniotic–monochorionic and nearly all of these placentas are monozygotic (Fig. 42.40). Rare diamnionic–monochorionic

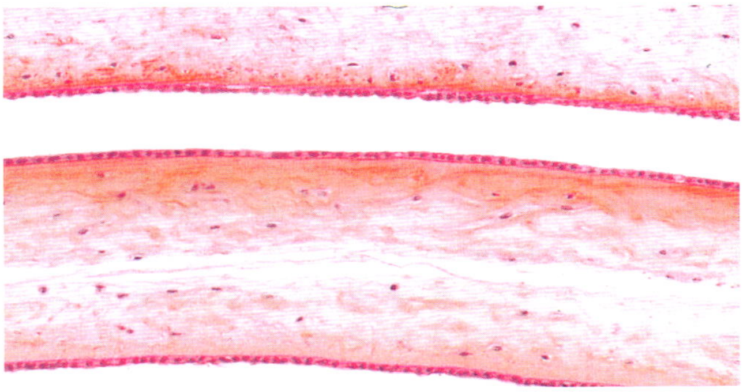

FIGURE 42.40 Intervening membrane from diamnionic monochorionic (DiMo) twin placenta. Note the absence of an intervening chorionic layer (HPS stain).

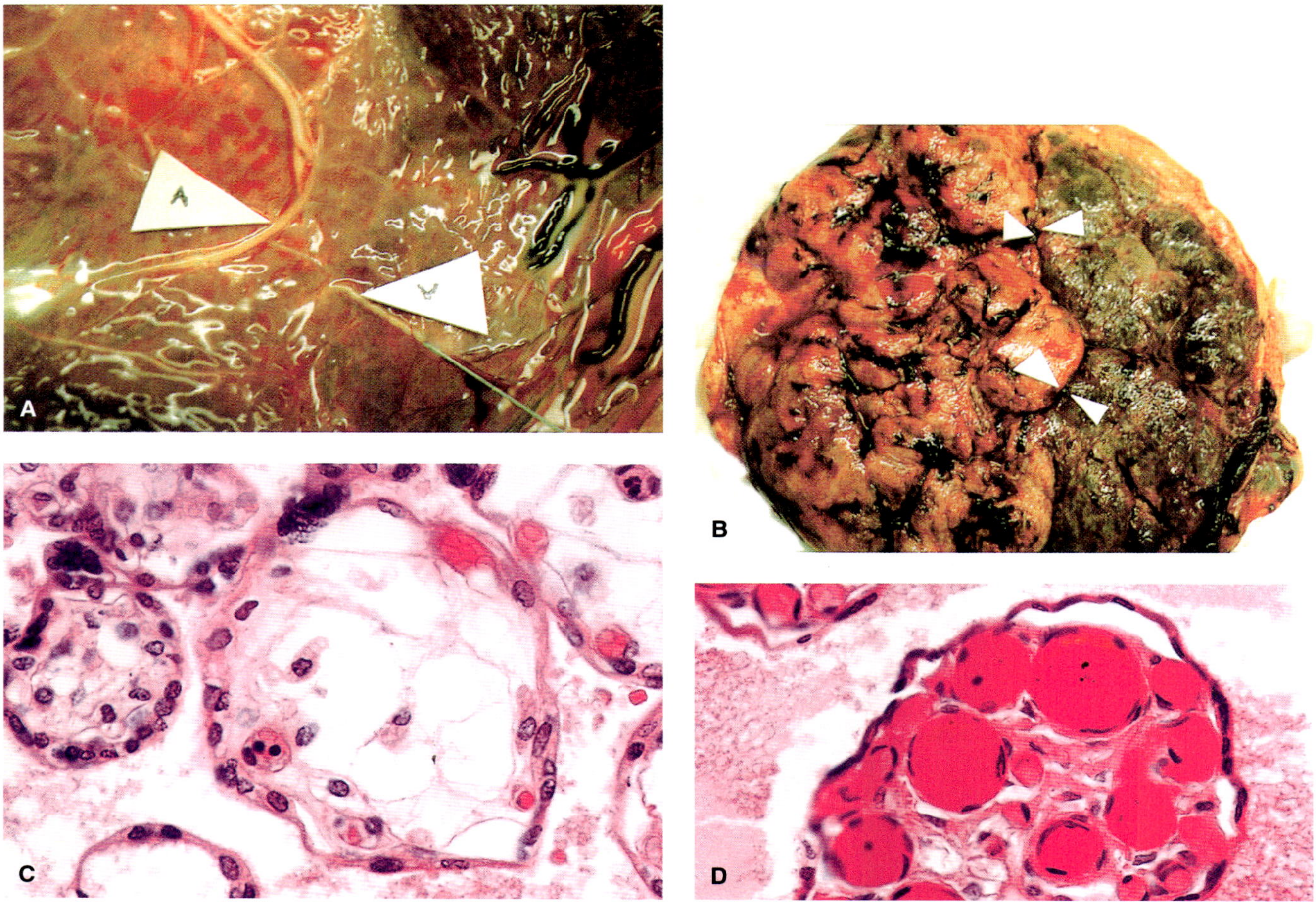

FIGURE 42.41 **A**: Twin–twin anastomosis characterized by artery-to-vein transfusion. **B**: Note pale anemic and edematous parenchyma of the donor (*left*) and dark congested parenchyma of recipient (*right*). The *arrowheads* mark the vascular equator along the maternal surface. **C**: Villi from the anemic twin are edematous with abundant macrophages, and vascular spaces contain nucleated hematologic precursors denoting high-output failure and increased red cell production, respectively. **D**: Villi from plethoric twin are markedly congested (HPS stain). Characteristic "classic" twin–twin transfusion outcomes may not always occur. Although one twin may be smaller, hemoglobin may be increased in a paradoxical fashion suggesting shifts in flow before analysis.

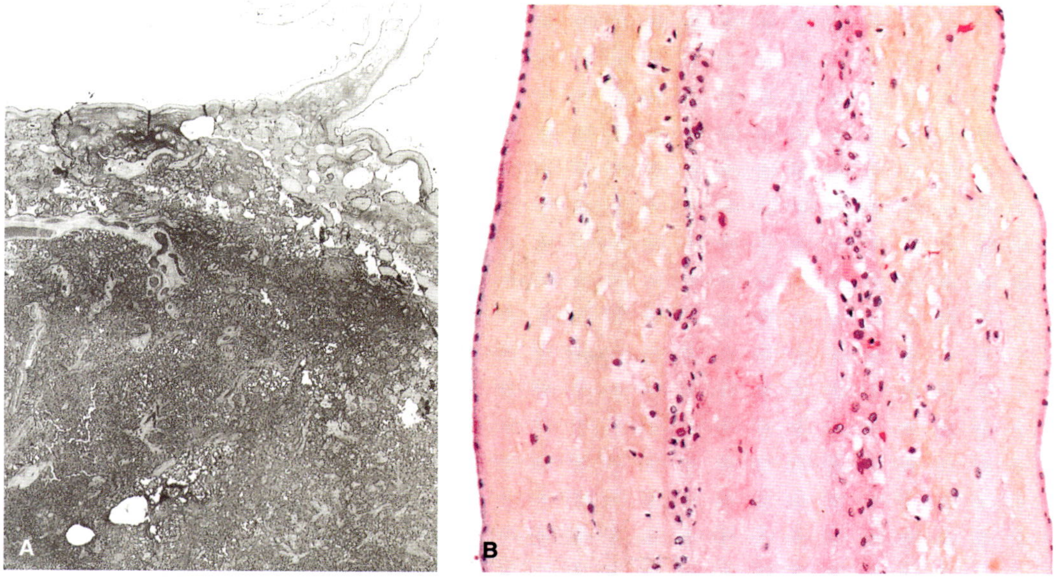

FIGURE 42.42 **A**: Diamnionic dichorionic (DiDi) intervening membrane, site of fusion at chorionic plate ("T zone") (H&E). **B**: Note the more cellular intervening chorion separating the two layers of amnion (HPS stain). (Reprinted with permission from Benirschke K, Kaufmann P. *Pathology of the Human Placenta*. 3rd ed. Springer-Verlag; 1995.)

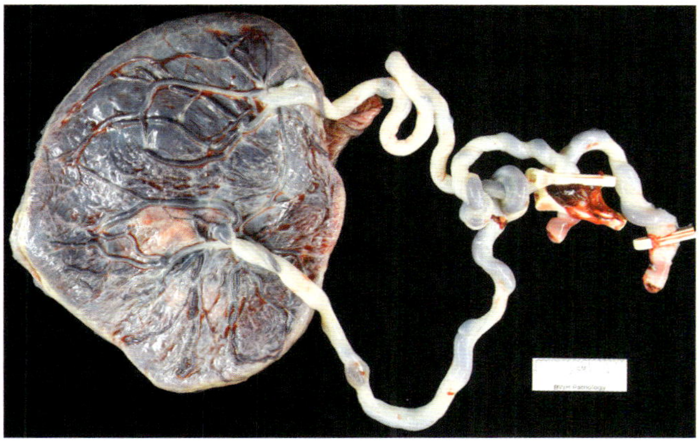

FIGURE 42.43 Monoamniotic monochorionic twin gestation with a single amnionic cavity. Note how the two umbilical cords have become entangled and knotted together.

placentas are dizygotic gestations, and they typically come to attention when sex discordance between the co-twins is appreciated. Of note, such twins are blood chimeras because hematopoietic stem cells are shared across their fused placental disks. The shared chorion of diamnionic–monochorionic placentas invests only in the chorionic plate and is not present in the intervening membrane. Diamnionic–monochorionic placentas contain shared vascular districts due to vein-to-vein and artery-to-artery anastomoses. Artery-to-vein anastomoses, having asymmetrical pressure loads, are relatively infrequent and constitute the etiology of the twin–twin transfusion syndrome (Fig. 42.41).

The diamniotic–dichorionic placenta is distinguished morphologically by examining the intervening membrane and noting the presence of two fused chorions beneath the two amnionic layers (Fig. 42.42). Most of these placentas are dizygotic; however, approximately 30% of diamniotic–dichorionic twin placentas result from monozygotic twin implantations that split within 3 days of fertilization. Vascular anastomoses are reportable in this situation (4).

The complete absence of an intervening membrane (i.e., monoamniotic–monochorionic) in a twin gestation is diagnostic of monozygotic twins (Fig. 42.43). In this situation, splitting of the embryo occurs at approximately 7 days of gestational age. Although twin–twin transfusions can occur, these are less common than in diaminionic–monochorionic placentation. The significant pathologic problem for these gestations results from cord entanglements, which result in a very high frequency of fetal death by fratricide (19). Separations occurring even later in development can result in conjoined twin fetuses (Fig. 42.44).

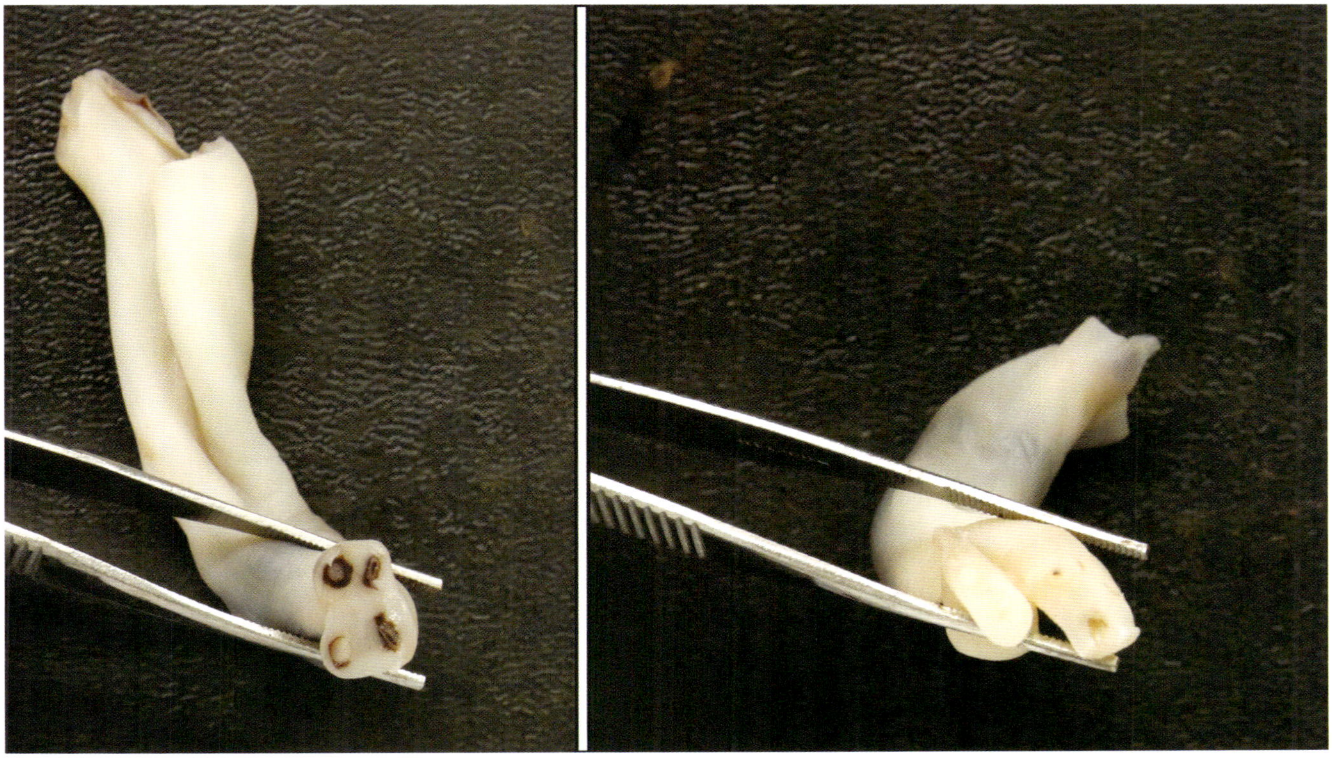

FIGURE 42.44 Abnormal furcate umbilical cord from a conjoined twin gestation. **Left**: Four vessel (two arteries and two veins) are visible in most of the cord, including the end inserting into the conjoined twins (not shown). **Right**: Each end of the furcate cord inserting into the placenta had a single umbilical artery and vein.

VILLI

Embryology

After formation of the blastocyst, the trophoblast adheres to the endometrial surface and gives rise to the placental villi and the chorion. The inner cell mass, or embryoblast, will give rise to the embryo proper, the umbilical cord, and the amnion. The villous is the functional unit of the placenta and is initially formed by invasion of cytotrophoblast into syncytiotrophoblastic trabeculae. Throughout the first trimester, trophoblastic villi are composed of an outer β-human chorionic gonadotrophin (hCG)-positive, nonproliferating syncytiotrophoblastic layer and an inner p63-positive, proliferating cytotrophoblastic layer encompassing villous mesenchyme in which the fetal vasculature differentiates. The majority of the villous is surrounded by both trophoblastic layers, although the polarity to the villi can be determined based on the distribution of cellular components. Their basal implantation regions are composed of caps of extravillous trophoblast that make up the trophoblastic cell columns at the tips of the branched villous tree, and are competent to invade into the myometrium (Fig. 42.45). Formerly, the term "X-cells" was used to describe a population of extravillous trophoblast now classified as intermediate trophoblast. Neither term adequately captures the biologic development, so the term extravillous trophoblast is currently favored by placentologists.

The placenta is comprised of a network of hierarchical branching and budding villi containing stroma, macrophages (i.e., Hofbauer cells), and vessels surrounded by cytotrophoblast and wrapped with syncytiotrophoblast. These villi derive from the chorionic plate by 4 weeks gestational age and pass through different stages of development from cell columns surrounded by syncytiotrophoblast to vascularized villi surrounded by a dual layer of cytotrophoblast and syncytiotrophoblast. The vascular system of

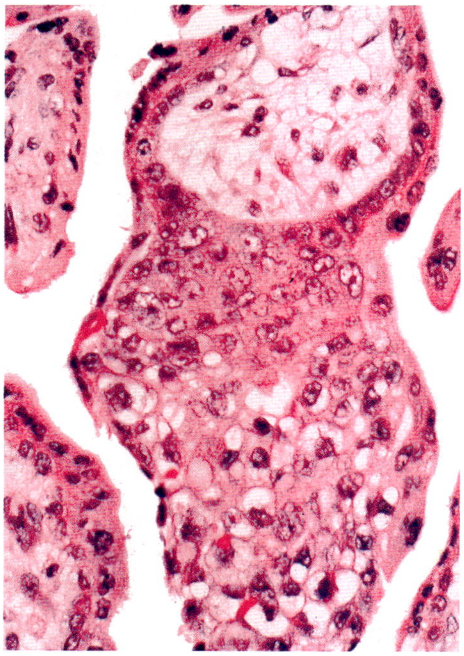

FIGURE 42.45 Polar trophoblastic proliferation comprising trophoblastic cell column. Many vacuolated extravillous trophoblasts are identified extending toward the decidual implantation site (H&E).

the placenta forms by both angiogenesis and vasculogenesis from umbilical cord arteries branching across the chorionic plate and diving into chorionic villi as they form. A fully functioning placenta capable of transporting vital materials from the mother to the fetus is present at 8 to 12 weeks of developmental age.

Placental villous morphology matures throughout gestation (Fig. 42.46). The immature placenta of less than 30 weeks gestation contains large, uniform villi with open stroma that is not vacuolated, centrally placed vessels, and two layers of trophoblast around the villi; proliferating cytotrophoblast adjacent to the stroma covered by the

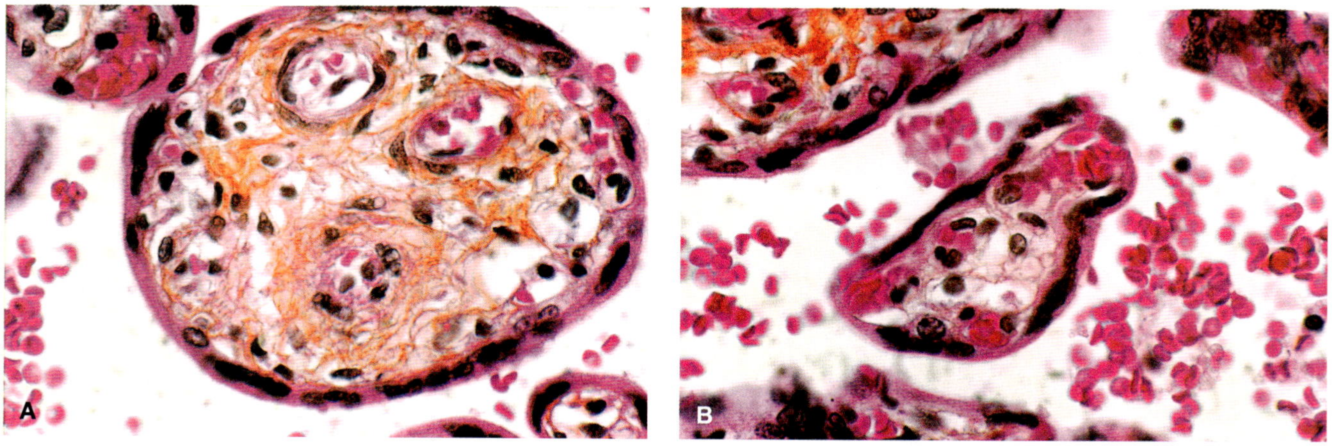

FIGURE 42.46 Third-trimester villi. Secondary villous (**A**) and tertiary villous (**B**). Note that capillary lumina are more peripherally located in the tertiary villous. Furthermore, there is less prominent villous mesenchymal substance (HPS stain).

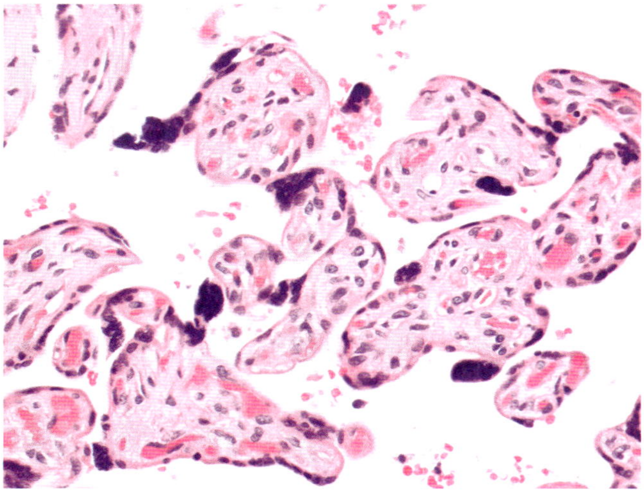

FIGURE 42.47 Mature villi. Note small size with frequent syncytiotrophoblastic knots.

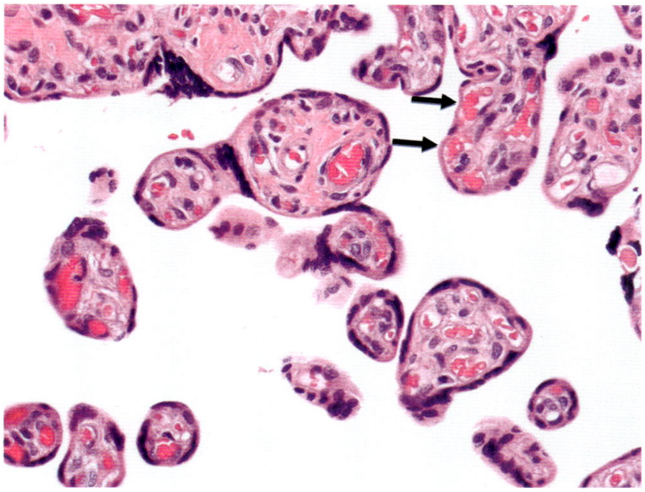

FIGURE 42.48 Mature villi demonstrating vasculosyncytial membranes (*arrows*) decreasing the distance for diffusion of gasses and transport of nutrients and waste between the maternal vascular space and the fetal vascular space (H&E).

syncytiotrophoblast immersed in the maternal space/blood. Villous syncytial trophoblast forms trophoblastic knots or sprouts that appear as attached balls of nuclei on the villous (Fig. 42.47). This is a feature of maturation and is consequently rare in immature placentas.

Several architectural and cellular specializations occur after the completion of villous growth and branching. These changes are collectively known as maturation and improve placental functional efficiency. These specializations include reduction in the size of terminal villi and compaction of villous stroma. The villous trophoblastic layer becomes progressively more attenuated with fewer p63-positive cytotrophoblast cells and thinning of syncytiotrophoblast cytoplasm concurrent with trophoblastic knots formation (usually one knot every three terminal villi). The most important feature of a mature villus is the vasculosyncytial membrane. In many foci, nuclear aggregation, cytoplasmic thinning, and peripheral displacement of fetal capillaries leave only a thin layer between maternal and fetal red blood cells, thus minimizing the diffusion distance for metabolite exchange (Fig. 42.48). In a mature placenta, these vasculosyncytial membranes should be easily found.

The course of development also has an impact on the mesenchyme. Early in development, the mesenchyme is prominent and capillary lumina are not readily identified (Fig. 42.49). After this point, the development of vascularization within villous tissue becomes more pronounced. The vasculature is derived from branches of the stem vessels that connect with the vasculature of the chorionic plate. By about 8 weeks gestational age, only nucleated hematologic precursors are evident within these villous capillary spaces. As gestation progresses, there is a decrease in the number of nucleated hematologic precursors, such that only 10% of blood cells are nucleated between 10 and 12 weeks gestational age and

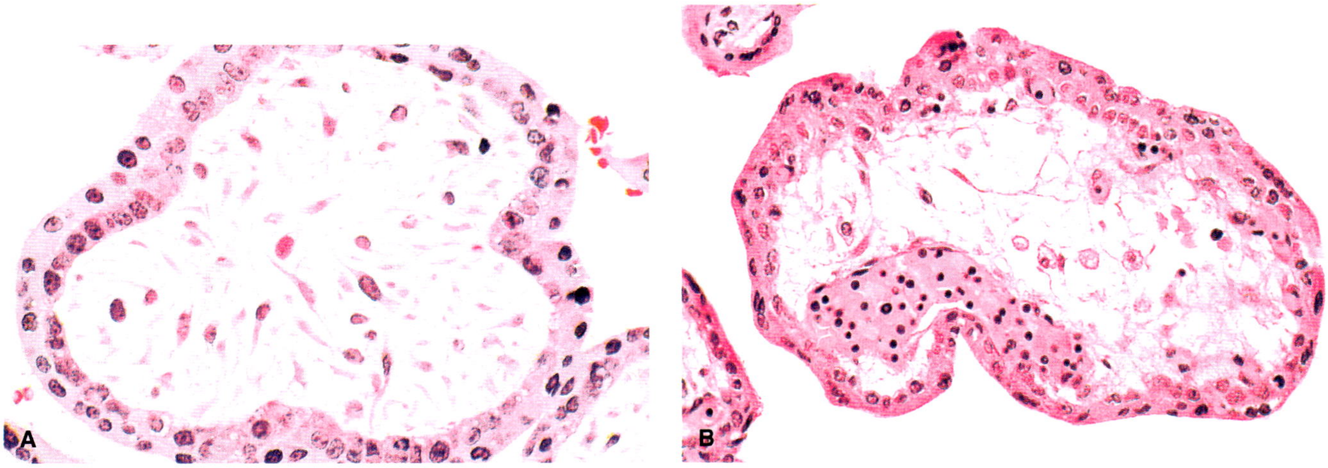

FIGURE 42.49 **A**: Near absence of nucleated hematologic precursors within villous capillaries after 12 to 20 weeks of gestation. **B**: Chorionic villus at 8 weeks of gestation. Note the nucleated hematopoietic precursors present with villous capillary spaces (H&E).

nucleated precursors are absent by 20 to 24 weeks gestational age. The remainder of the villous mesenchyme consists of primitive fibroblastic cells and macrophages. These cells bear antigens that characterize them (e.g., CD4, LeuM3, and a variety of macrophage markers) (19,54).

The villous macrophages (Hofbauer cells) that express PU.1 (SPI1) and CD68 lose their prominence as gestation proceeds. Fewer and fewer of them are present within mature villi. The function of Hofbauer cells is important in water regulation activities, transport of various nutrients and waste, and villous homeostasis via paracrine interactions with the stromal fibroblasts. Hofbauer cells may be important to immune regulatory functions as an intermediate in the processing of infectious agents that are blood borne (19,54,62). In contrast, villous macrophages become engorged by accumulated metabolites in inherited lysosomal storage diseases, yielding a stunning foamy or xanthomatous quality to an often-expanded villous stromal core.

Chorionic villi may be useful in prenatal diagnosis. Karyotype analysis may be assessed at 10 to 12 weeks gestational age through chorionic villous sampling. It should be noted that although early diagnosis is often preferable to later amniocentesis, several concerns exist. Some data suggest limb reduction abnormalities, but most of the evidence does not confirm this as an adverse event. Chorionic villous sampling is not useful in the diagnosis of neural tube defects and fragile X syndrome, both of which require amniotic fluid for evaluation. Advances in isolation of fetal cells or in silico fetal genomics based on cell-free DNA sequencing from maternal peripheral blood have enhanced prenatal diagnosis.

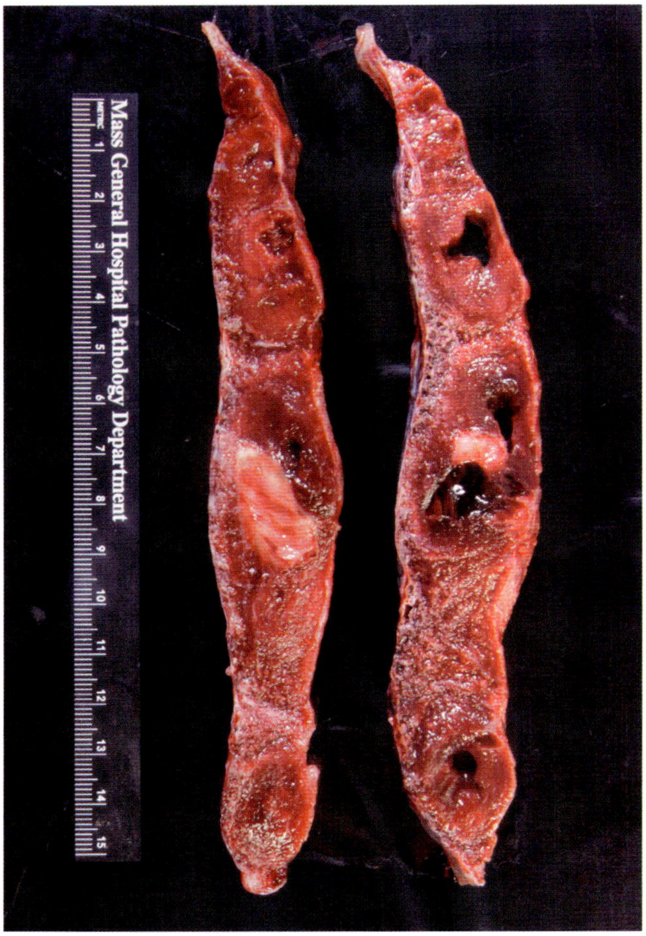

FIGURE 42.50 Gross of an infarction hematoma. Large interparenchymal hematoma with peripheral infarction.

Gross Morphology

The villous parenchyma occupies the discoid space beneath the chorionic and plates. The substance of the parenchyma is red and spongy, and is homogeneous in contour and texture. Many abnormalities are best regarded as normal variants when focal or present to a minimal extent. Examples include small infarcts and perivillous fibrin deposition. Larger or more numerous lesions, or presentations at earlier gestational ages, however, merit closer attention, as such instances are more likely to represent clinically significant pathologies.

In addition to perivillous fibrin deposition, other fibrinous depositions are common in the third trimester and are generally considered normal within the placenta. Those located beneath the chorionic plate in the maternal blood space, variously termed Langhan stria or subchorionic fibrin, are probably related to altered maternal intervillous blood flow. By virtue of distance from the decidual vascular interface, maternal blood in the subchorionic space tends to be more static and less oxygenated than maternal blood at the opposite face of the placental disc. Nitabuch fibrin is present between the floor of the placenta and the maternal decidua. This layer was once believed to prevent allograft rejection, but the precise nature and functional significance of this fibrin deposition are less certain.

On occasion, the placental parenchyma has spherical defects (usually large and grossly recognizable) (Fig. 42.50). Histologically, these lesions consist of large hematomas surrounded by a corona of infarction. Their configuration suggests they represent jet lesions or intraplacental abruptions, justifying labeling them as "infarction hematomas." Most placentas with infarction hematomas have many features of maternal vascular malperfusion and those pregnancies are very often complicated by severe hypertension.

Calcification is a common phenomenon in the mature placenta but can be a feature of maternal vascular malperfusion in immature placentas (63). Calcium deposition is seen as fine, pinhead-sized deposits of yellow–white, gritty material. Larger and more extensive precipitates may be readily visualized in utero by sonography and may be interpreted as representing gestational maturity. Extending this generalization to earlier gestational ages, however, would be inappropriate because abnormalities of maternal blood flow within the placenta may cause accelerated placental maturation. Moreover, inference of placental maturity based on calcification does not necessarily denote fetal maturity.

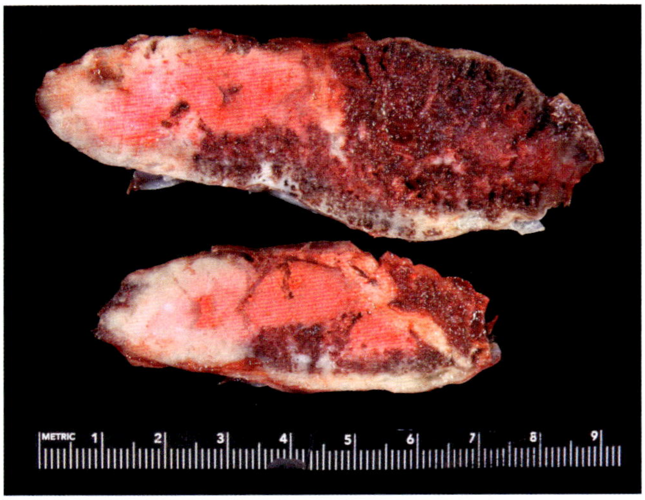

FIGURE 42.51 Old placental infarct(s) from a hypertensive pregnancy.

Gross Morphologic Alterations

Many placentas normally have some degree of infarction. Infarction is only considered pathologic when it exceeds 10% to 15% of the placental volume, is located more centrally than near the periphery, is multifocal, temporally heterogeneous, or present in a small placenta (<10th percentile for gestational age). Infarcts are characterized grossly as either recent or remote. New infarcts are pale, poorly demarcated, and slightly granular on palpation. Remote infarcts are white, often triangular in shape, and similarly granular (Fig. 42.51). There are many types of infarctions: usual type, hypertensive type, infarction hematomas, and those associated with increased perivillous fibrin. They each have characteristic gross features that are discussed more fully in the section on maternal vascular malperfusion.

Infarcts can be distinguished from perivillous fibrin deposition because they are firm with a granular cut surface. Perivillous fibrin, on the other hand, tends to be nodular and smooth.

Intervillous thrombi are triangular- or diamond-shaped lesions with a soft gelatinous red to white appearance depending on the age of the lesion (Fig. 42.52). These lamellations of fibrin (i.e., lines of Zahn) can be grossly visible and distinguish these lesions from infarcts and perivillous fibrin deposition. Intervillous thrombi are thought to be associated with fetal villous capillary leakage into flowing maternal blood (35). Most small intervillous thrombi are of no clinical consequence, although they can, in theory, result in maternal isoimmunization as a result of mixing fetal and maternal blood (4).

Histology

The histologic variations in placental architecture are largely dependent on the developmental state at which observations are made. The histology of the individual cell types involved directly in villous implantation is now described.

Extravillous trophoblasts are major constituents of the cell columns that form the deepest structural components of the implantation site. They are round to polygonal cells present singly or in groups and are usually associated with a fibrinous extracellular matrix. They tend to have pleomorphic and hyperchromatic nuclei with amphophilic to eosinophilic cytoplasm. Although predominantly mononuclear, they may also have multiple nuclei. They are epithelial derivatives and express cytokeratin, in contrast to the surrounding decidual cells. They produce human placental lactogen and MBP. They are ultrastructurally distinct insofar as they contain large numbers of mitochondria with tubular cristae. Extravillous trophoblasts can be conceptually divided into at least two types: invasive and endothelial. Invasive extravillous trophoblasts invade through the decidua into the superficial myometrium. There is almost no tissue reaction to this process, which requires cross talk between the trophoblasts and resident maternal inflammatory cells. Endothelial extravillous trophoblasts invade and transiently occlude muscularized spiral arteries in decidualized endometrial stroma. By a cell biologic process known as "transformation," vascular trophoblast invasion remodels maternal spiral arteries, converting dynamic high-pressure vessels into static low-pressure high-capacitance vessels. This transformation ensures that constant, unimpeded maternal blood flow at volumes and pressures that are optimal for gestational growth and survival.

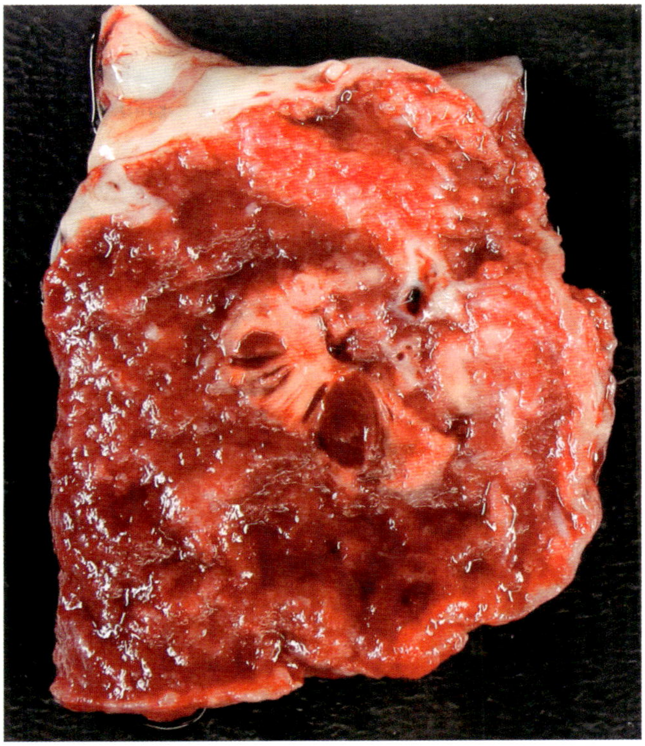

FIGURE 42.52 Intervillous thrombus. Gross example of a fresh intervillous thrombus with visible lines of Zahn.

Difficulty in distinguishing implantation site trophoblast from decidua may pose challenges in diagnosing intrauterine pregnancy in curettage specimens, especially in sparse samples without companion chorionic villi early in pregnancy. In the absence of chorionic villi, most pathologists confirm their diagnosis of intrauterine pregnancy (and hence exclude the possibility of an ectopic tubal pregnancy) based on the presence of trophoblastic derivatives in the decidua, which are distinguished from the surrounding decidua by greater nuclear pleomorphism, hyperchromasia, and multinucleation. Should difficulty arise with morphologic assessment of mononuclear cells, immunohistochemical expression of trophoblast markers can be helpful. These markers include human placental lactogen and HLA-G, though we prefer GATA3 based on its relative specificity in context, generally good performance, and ready availability (64,65). Extravillous trophoblasts are of important immunologic interest because they produce factors that regulate maternal immune cells within the decidua and vice versa (66).

An interesting normal finding within the placenta is the septal cyst. These cysts occur in placental septae and are composed largely of decidua and are lined by extravillous trophoblasts. Chorionic and septal cysts are positive for CD146 and MelCam, and negative for p63, similar to extravillous trophoblasts found in the implantation site beneath the placenta. On sectioning, gelatinous fluid may be present, and the major constituent of this fluid is one of the substances produced by extravillous trophoblast: MBP. The function of these cysts is not known. They should not be considered pathologic when present (35).

Cytotrophoblasts are stem cells of trophoblast lineage. These cells are mitotically active with round and open nuclei, and are p63-positive. Ultrastructurally, the cytotrophoblast has fewer organelles than the terminally differentiated syncytiotrophoblast, and they contain large, often numerous mitochondria. Cytotrophoblasts are located subjacent to the villous syncytiotrophoblasts and can be found proliferating at the tips of trophoblast cell columns and in the chorion (67).

Syncytiotrophoblast cells form the outer cellular layer surrounding villi and form through fusion of cytotrophoblasts. They display a number of cellular specializations. They possess a brush border composed of microvilli that are likely involved in pinocytotic activity. Vacuoles within the cytoplasm reflect the absorptive and secretory activities of these cells. The syncytiotrophoblastic cytoplasm is rich in endoplasmic reticulum, mitochondria, lipid droplets, and Golgi bodies (68). The syncytiotrophoblastic nuclei may be numerous and hyperchromatic; some pyknotic nuclei may be present reflecting a terminally differentiated state. Cellular fragments, some of which contain condensed nuclei, are continually shed from the villi into maternal circulation; fetal DNA can be extracted from these cells using maternal blood, thereby facilitating assessment for genetic disorders. The syncytiotrophoblast is the trophoblast chiefly responsible for pregnancy hormone production and secretion, including the well know hormone chorionic gonadotrophin and GCD15, the hormone recently implicated in the pathogenesis of morning sickness and hyperemesis gravidarum (69). Syncytiotrophoblasts transfer metabolites between mother and fetus through their attenuated cytoplasm, replace the anticoagulant function of endothelial cells interacting with maternal extravascular blood within the placental disc, and provide a barrier that prevents maternal bloodborne infections from reaching the fetus. In sum, syncytiotrophoblast cell specializations are integral to the physiologic function of the placenta.

Different antigen expression patterns help distinguish the different cell types, which can be particularly useful when confronted with a trophoblastic tumor or an atypical placental implantation site. Human placental lactogen may be localized within extravillous trophoblast and syncytiotrophoblast. Human chorionic gonadotropin is identified within syncytiotrophoblasts, but not within cytotrophoblasts. Nuclear β-catenin is expressed in early cytotrophoblast, whereas extravillous trophoblasts show a membranous pattern of staining. On the other hand, CD146 (MelCAM) is expressed exclusively in extravillous trophoblasts. HLA-G expression is unique to extravillous trophoblasts and is an excellent marker for these cells. Many other antigens are expressed in trophoblasts, including GATA-3, keratin 5.2, PD-L1, p63, and α-inhibin.

Histopathology

Placental maturation is a dynamic process, and its pathologic import is derived from the difference between the observed state and what would be expected based on gestational age. A 25-week placenta can be mature, in which case it is a pathologic feature of maternal vascular malperfusion. Conversely, a term placenta can manifest delayed villous maturation as a result of maternal diabetes mellitus and other factors.

Placental pathology can be broadly classified in four major categories based on the Amsterdam consensus statement (13,70). The four mechanistic disease categories are: (1) maternal vascular malperfusion, (2) fetal vascular malperfusion, (3) acute inflammatory reactions, principally due to bacterial infections, and (4) chronic inflammatory reactions due to specific infections or villitis of unknown etiology. To comprehensively cover the conceptual approach to interpreting placental histology in health and disease, we propose adding two additional categories: (5) features of hypoxia, and (6) pathologies with a significant recurrence risk. What follows are brief descriptions of the pathology and short discourses of the proposed pathobiology to illustrate the normal structure-function relationships of the placenta.

Maternal Vascular Malperfusion

Maternal vascular malperfusion is a relatively common pathology, correlating with common obstetrical problems like maternal hypertension and fetal growth restriction (71). Presumably, this condition results from poor perfusion leading to hypoxia/ischemia of the placenta due to maternal hypertension or vascular disease. Some placentas will show the pathologic features of maternal vascular malperfusion without a maternal history of hypertension, and the biologic basis for phenotypic variability is a mystery. The histopathologic stigmata of maternal vascular malperfusion include low placental weight, distal villous hypoplasia, accelerated villous maturation, decidual arteriopathy, parenchymal infarcts, excessive intervillous fibrin deposition, and retroplacental hematoma. The relative weighting of these various features remains to be determined, but the likelihood of a diagnosis increases when multiple features are present. Maternal vascular malperfusion is most predictive of adverse pregnancy outcomes when decidual arteriopathy or accelerated villous maturation is present in conjunction with any of pathologic features noted above (72).

Placental hypoplasia is arguably the most objective feature of maternal vascular malperfusion. A small placenta (<10th percentile for any given gestational age) is the most common pathologic finding. In fact, diagnosing maternal vascular malperfusion in a placenta within the normal weight range for gestational age should be made with care. Placental hypoplasia likely reflects a more chronic and prolonged underlying pathologic process.

Placental infarction represents loss of functional placental parenchyma and is associated with adverse pregnancy outcome in multivariate analysis (72). Infarcts result from either deficient maternal or fetal perfusion. Infarcts resulting from maternal underperfusion have different histologic appearances depending on their age. Acute infarcts show faint staining villi that are aggregated, compressed, or agglutinated and accompanied by lysis of intervillous maternal blood with interspersed polymorphonuclear leukocytes in the intervillous space (Fig. 42.53). The density and duration of the acute inflammatory component can be variable when reperfusion causes cytokine washout. Intervillous inflammation responding to infarcted tissue can be distinguished from acute villitis because the latter has an intravillous component in response to hematogenous disseminated pathogens. Remote infarcts show loss of nuclear basophilia that reflects DNA degradation, cytoplasmic eosinophilia

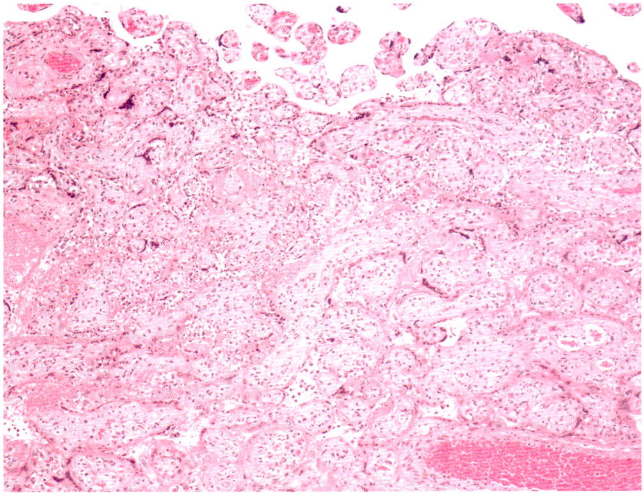

FIGURE 42.53 Acute placental infarct. Note acute inflammatory debris in the intervillous space and blood in the villous vessels (H&E).

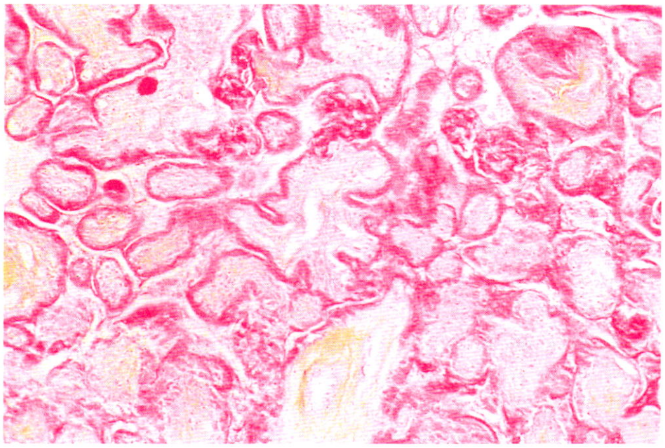

FIGURE 42.54 "Old" placental infarct illustrating complete loss of nuclear basophilia and cytoplasmic hypereosinophilia.

due to protein denaturation, and loss of villous architecture characterized by fuzzy outlines of remaining ghost villi (Fig. 42.54). The placental infarcts rarely undergo organization and fibrosis because fibroblasts are generally lacking in the placenta. The compacted villi of placental infarcts give their cut surfaces their granular texture.

There are three different morphologic patterns of maternal side infarction, each with its own etiology. The usual type or classic infarct results from cessation of maternal blood flow into a specific region of the placenta due to abruption or maternal thrombosis, creating a parabasal ischemic zone. The maternal space empties of blood and the villi of that region collapse upon one another to form a triangular/pyramidal mass of dying villi with one side adjacent to the basal plate/maternal floor. Hypertensive-type infarcts small, centrally located infarcts that are round. They are thought to represent watershed infarcts that occur in a relatively ischemic zone between the perfusion areas of two spiral arterioles. Maternal spiral arteries affected by hypertension have reduced blood flow and thinner jets that enter the maternal blood space, resulting in reduced flow to the furthest extent of watershed zones between adjacent spiral arterioles. Infarction hematomas, or rounded intraplacental hematomas, are large and multiple. They appear as poorly organized hematomas surrounded by placental infarct. This type of infarct is also associated with maternal hypertension as well as a significant risk of fetal/neonatal morbidity and death (73).

Small areas of villous infarction can result from localized fibrin deposition in the intervillous space (Fig. 42.55). These lesions are secondary to increased intervillous fibrin deposition to such an extent that it smothers the villi, thus depriving them of a free-flowing source of oxygenated maternal blood. The enmeshed villi die due to the limitations of diffusion. In contrast to the regular types of infarcts having aggregated villi and collapsed intervillous space as described above, this pattern of secondary infarction is easily discerned because the necrotic villi are separated by fibrin. The process(es) leading to localized and diffuse intervillous fibrin deposition are not well understood.

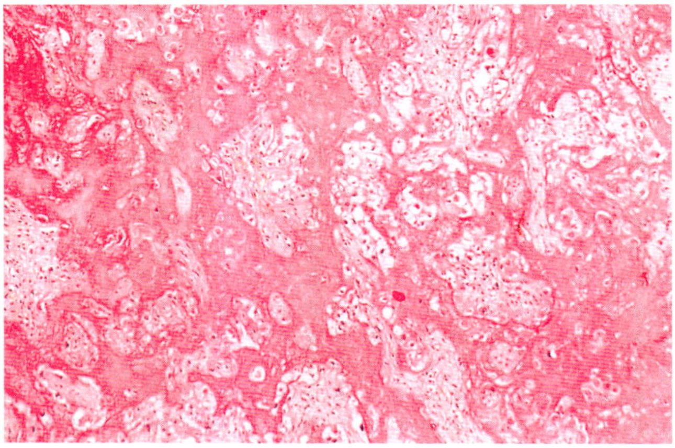

FIGURE 42.55 Increased perivillous fibrin. Note trophoblastic nuclear remnants encased in fibrin (HPS stain).

In all cases of maternal-side placental infarction, one can see continued fetal perfusion in the villous capillaries early in the infarction process. With time, however, fetal vascular involution and villous sclerosis are typical.

It is often found as a localized halo-like zone surrounding infarcts, although the whole placenta can show ischemic features in severe cases. In general, chronic ischemia causes accelerated placental maturation that is well out of proportion to the gestational age. Consequently, one should not judge overall placental maturation based on tissues in the vicinity of recognizable infarcts. One prominent feature of chronic placental ischemia near infarcts is larger or more frequent syncytial trophoblastic knots, especially in the preterm placenta. Syncytial knots in this context are best considered maternal vascular malperfusion; a maladaptive response to low oxygen tension within the intervillous space. Of note, syncytial knots shed into the maternal systemic circulation are transported to the maternal lung, where they undergo apoptosis and their nuclear DNA becomes the extracellular (cfDNA) of the maternal blood. It disappears approximately 2 days after delivery.

Depending on the time course in any given placenta, other findings associated with maternal vascular malperfusion include other features of accelerated villous maturation, markedly increased perivillous fibrin in the absence of villous necrosis, abruption, distal villous hypoplasia, and maternal vascular pathology as described subsequently.

Maternal vascular malperfusion is staged and graded according to the Amsterdam consensus statement (Table 42.3) (13). High-grade high-stage disease is associated with the most morbidity.

TABLE 42.3 Maternal Vascular Malperfusion—Patterns and Grading

Patterns
Global/Partial—exemplified by accelerated villous maturation and distal villous hypoplasia
Segmental/Complete—exemplified by placental infarcts
Grading (71)
Grade 1 (mild to moderate)—normal placental weights, <30% placental involvement, nor more than one marginal infarct
Grade 2 (severe)—placental weight <10th percentile for gestational age, multiple nonmarginal infarcts, more parenchymal involvement

Fetal Vascular Malperfusion (74)

Fetal vascular malperfusion is much less common than maternal vascular malperfusion. Evidence of fetal vascular malperfusion may be found throughout the fetal vasculature, potentially affecting the umbilical cord, chorionic plate, villous stem, intermediate, and terminal vessels. Its etiology is not clear but seems to reflect diminished blood flow between the fetus and placenta due to obstruction or increased resistance. Possible causes include tight true umbilical cord knots, excessive cord length and hypercoiling, vascular injury secondary to inflammation or mechanical compression, obstructive thrombus formation, and rare heritable fetal thrombophilia. Etiologies that affect larger vessels have the potential to impact a greater fraction of the fetal circulatory output. Variation in the morphologies of fetal vascular malperfusion depends on the duration of injury (Fig. 42.56). The early stages of fetal vascular malperfusion include thrombi and intramural fibrin deposition without organization. With time, thrombosed fetal vessels undergo organization and may recanalize. Villous stromal–vascular karyorrhexis represents an intermediate downstream lesion of fetal vascular malperfusion, and can involve both intermediate and terminal villi. Regional areas of sclerotic avascular villi corresponding to the distal vascular tree can be seen in the later stages of fetal vascular malperfusion. Fetal vascular malperfusion can also be associated with intramural fibrin deposition and prominent endothelial protuberance into the fetal vascular lumina, as well as surface or intraendothelial fibrin. Presumably, these findings reflect a remote mural thrombus that has been re-endothelialized. Similar vascular changes can take place after intrauterine demise, although they are generally diffuse throughout the placenta because there is a global cessation of fetal blood flow.

Fetal vascular malperfusion-related morbidities include fetal growth restriction, oligohydramnios, preterm labor and delivery, neurologic sequelae, and others. These morbidities are only associated with high-grade fetal vascular malperfusion as defined by the Amsterdam Criteria (Table 42.4) (13).

Acute Inflammation and Infection

Acute inflammation, commonly termed chorioamnionitis, is a common finding, especially in spontaneous preterm deliveries. As it is a membranous and umbilical cord pathology, it will be discussed in the section on that part of the placental anatomy.

Villitis of Unknown Etiology

Chronic villitis can be divided into infectious and apparently noninfectious etiologies. We say "apparently" because

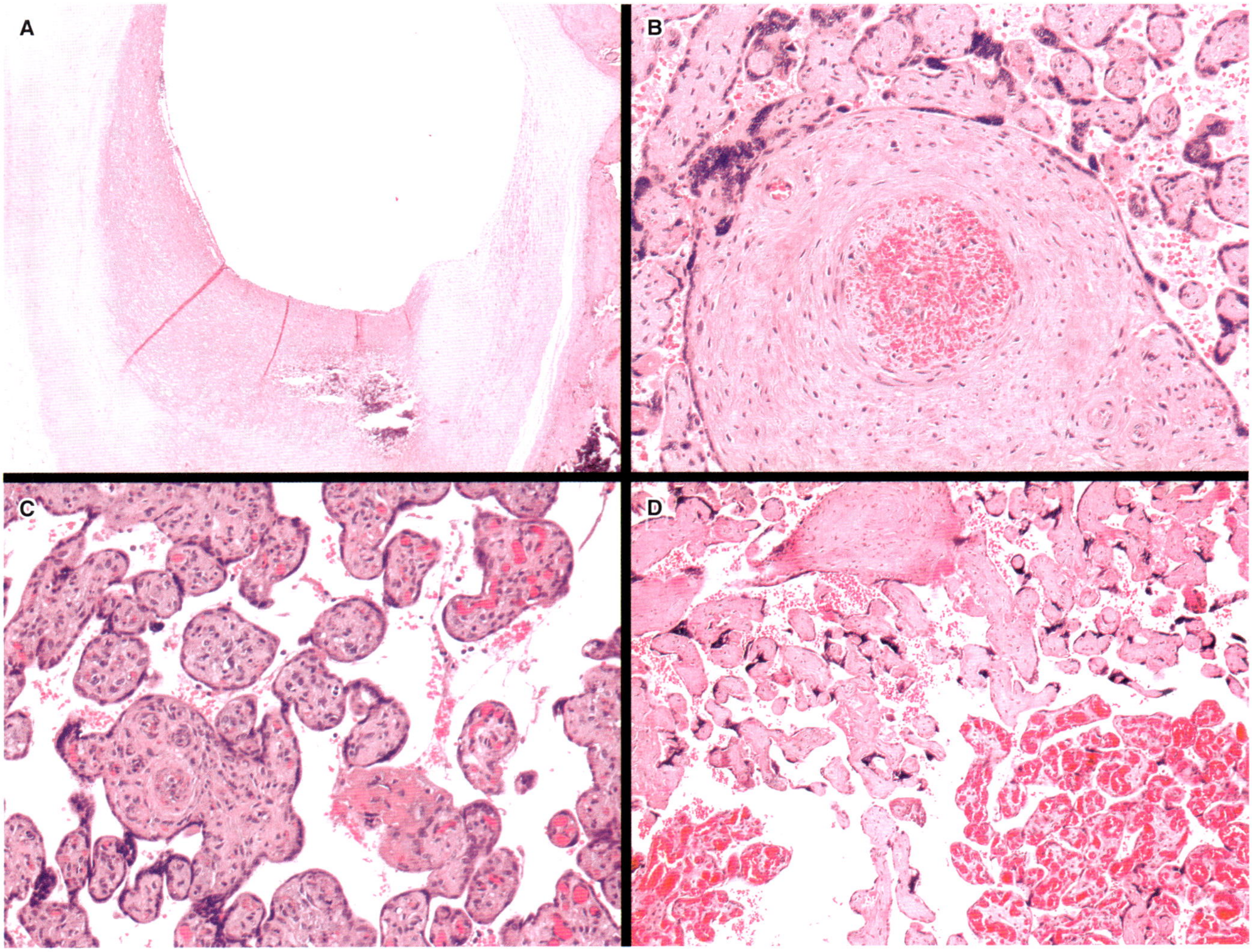

FIGURE 42.56 Examples of fetal vascular malperfusion. **A**: Thrombus in a chorionic plate vessel. **B**: Stem villous vascular obliteration. **C**: Villous stromal—vascular karyorrhexis. **D**: Avascular villi (H&E).

TABLE 42.4 Fetal Vascular Malperfusion—Patterns and Grading

Patterns
Segmental—thrombotic occlusion of chorionic or stem villous vessels or stem vessel obliteration leading to avascular villi downstream
Global—venous ectasia, intramural fibrin deposition, small foci of avascular villi or villous stromal-vascular karyorrhexis
Grading (13)
High grade
More than one focus of avascular villi (a cumulative assessment of ≥45 avascular villi over 3 sections examined or an average of >15 villi per section) with or without thrombus, or
Two or more occlusive or nonocclusive thrombi in chorionic plate or major stem villi, or
Multiple nonocclusive thrombi
Low grade
Satisfies none of the criteria for high grade

the fraction of cases designated as villitis of unknown etiology may ultimately prove attributable to a yet-to-be-defined infectious pathogen. Many examples of villitis of unknown etiology reflect an immune-mediated reaction characterized by a maternal inflammatory response to fetus-derived placental tissue grafted to the endometrium, akin to a transplant rejection to a semiallograft. Both forms of villitis share similar histologic features, although infectious villitis can be present in preterm placentas and feature plasma cells or giant cells as part of the infiltrate (Fig. 42.57). Infectious villitis usually results from one of the TORCH infections acquired hematogenously from the birthing person, and is associated with congenital infection. In the United States, cytomegalovirus (CMV) is by far the most common etiologic agent for specific chronic villitis (Fig. 42.58). However, some congenital infections are not associated with chronic villitis. For example, SARS-CoV-2 placentitis is defined by histiocytic intervillositis, massive perivillous fibrin deposition, and a distinctive necrosis of the villous syncytial trophoblast (Fig. 42.59) (75).

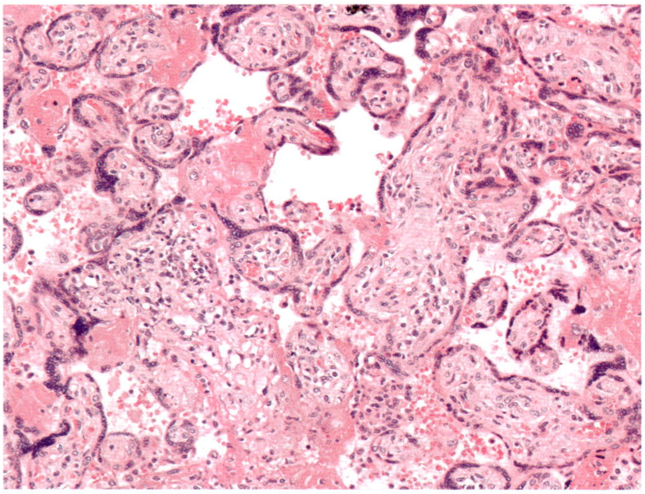

FIGURE 42.57 Chronic villitis of unknown etiology (CVUE). Note increased stromal cellularity (mononuclear cells and Hofbauer cells) and decreased villous vascularity (H&E).

Villitis of unknown etiology is seen in about 10% to 15% of all placentas examined (76). It typically occurs in late third trimester placentas and is grossly inapparent. It features maternally derived T cells with a minor component of fetal Hofbauer cells (Fig. 42.57). Not uncommonly, patches or small clusters of sclerotic villi may be present. The underlying mechanism for villous sclerosis in villitis of unknown etiology is inflammatory destruction of villous capillaries (Fig. 42.60). The presence of concurrent chronic inflammation is a useful diagnostic clue for distinguishing villitis of unknown etiology from fetal vascular malperfusion, although the differential diagnosis can be challenging. Finally, the inflammation of villitis of unknown etiology primarily affects terminal villi, although intermediate and even stem villi become inflamed. Villitis of unknown etiology may be accompanied by plasma cell deciduitis at the basal plate and/or chronic intervillositis/perivillositis. The chorionic plate is infrequently involved.

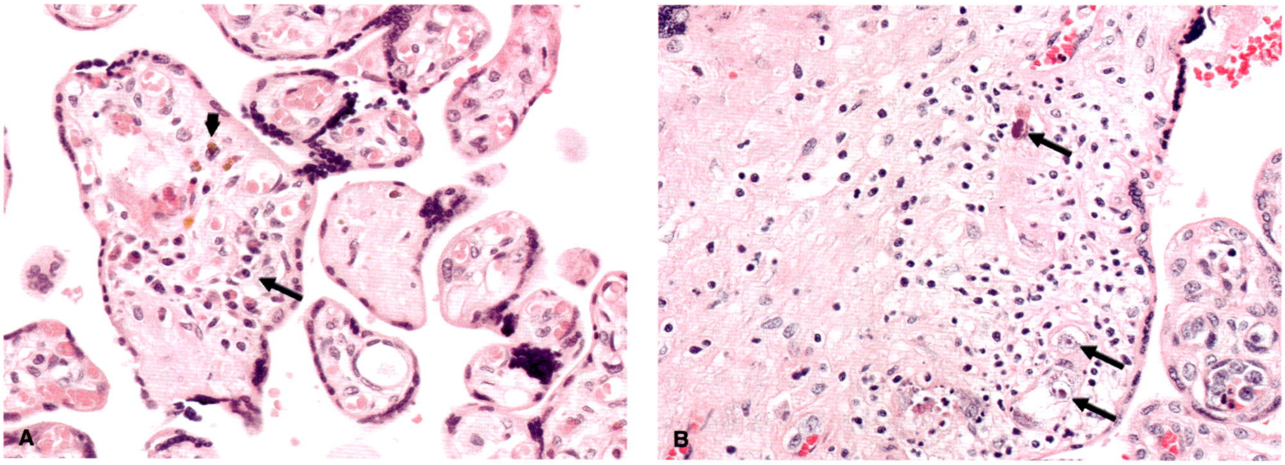

FIGURE 42.58 CMV placentitis. Plasma cell villitis, stromal hemosiderin deposition (**A**) and CMV nuclear inclusions (**B**) (H&E).

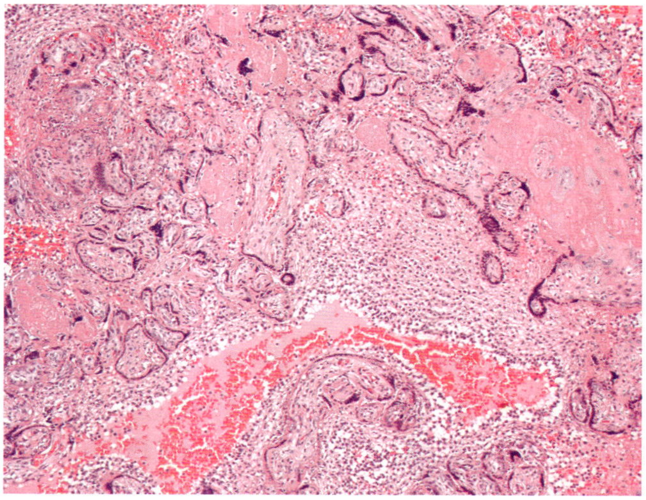

FIGURE 42.59 SARS-CoV-2 (COVID-19) placentitis. Note the increased intervillous fibrin admixed with histiocytes and the necrotic villous syncytiotrophoblast characteristic of SARS-CoV-2 infection of the placenta (H&E).

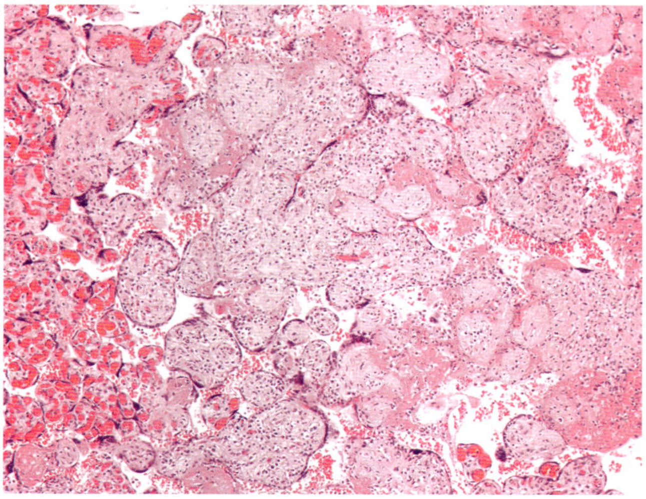

FIGURE 42.60 Cluster of severe chronic villitis of unknown etiology (H&E).

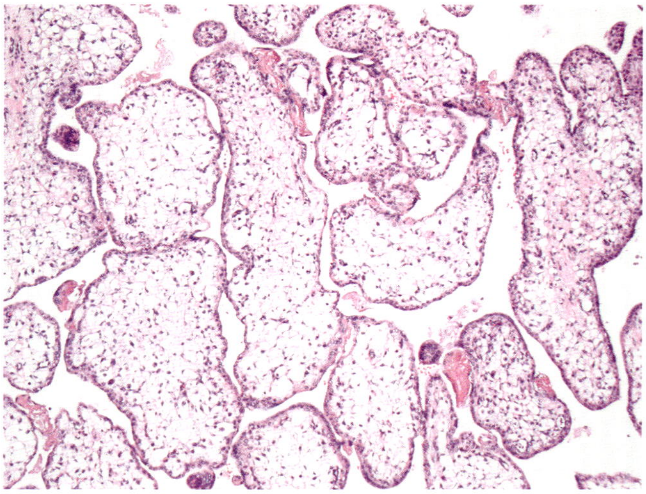

FIGURE 42.61 A 36-week gestational age placenta with marked villous edema (H&E).

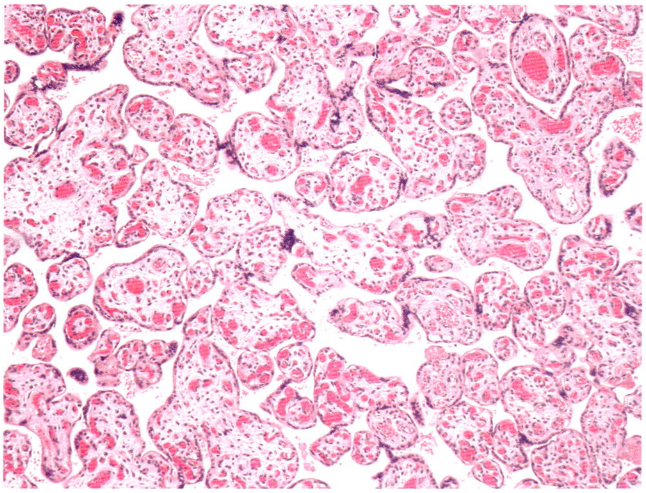

FIGURE 42.62 Chorangiosis. Note excessive numbers of villous capillaries.

Villitis of unknown etiology is also graded (Table 42.4). Diffuse or high-grade villitis is associated with fetal growth restriction and neurologic sequelae including cerebral palsy (4). Villitis of unknown etiology can recur in future pregnancies and carries a significant risk for fetal morbidity.

Placental Features of Significant Acute Hypoxia

Villous edema can displace intravillous cytologic architecture. It is considered pathologic, especially in premature gestations (Fig. 42.61) and has been reported as a cause or sequelae of fetal hypoxia (4).

Meconium exposure and meconium capture by macrophages residing in the placental membranes can be either a cause or sequelae of fetal hypoxia, although it is usually an incidental physiologic phenomenon very late in gestation. It is more likely to be a sequelae of a hypoxic event when there is significant meconium pigment deposition or when deposition is detected in preterm placentas. Meconium in the membranes and umbilical cord is described and discussed further in their respective sections.

Increased numbers of villous capillaries, termed chorangiosis, are considered to represent a fetal adaptation to prolonged hypoxia. The precise histologic definition requires finding increased capillary numbers (≥10 vessels) in multiple villi (≥10 villi) in multiple areas (≥10 fields at 100×), thus ensuring there has been diffusely increased capillary growth (Fig. 42.62) (77).

Nucleated immature red blood cells (orthochromatic normoblasts) produced by yolk sac or hepatic hematopoiesis are commonly found in fetal systemic circulation of very early gestations, but normoblastemia should not be observed in the third trimester after the shift to bone marrow–based hematopoiesis. When present, fetal anemia with increased erythroid production should be suspected. Hemolysis due to blood group incompatibility and fetal-to-maternal hemorrhage should be clinically excluded by laboratory testing, before pursuing other etiologies. Recent investigations have addressed normoblastemia as a finding in advanced-gestation placentas and correlated it with fetal erythropoietin secretion and concomitant fetal hypoxia or malperfusion (Fig. 42.63) (13,19).

Miscellaneous Pathologies

Intervillous thrombi appear as lamellated thrombotic material displacing neighboring villi (Fig. 42.64). Although the precipitating event(s) are not well understood, the layered accumulation of fibrin and maternal red blood cells suggests a relatively slow natural history requiring flowing maternal blood. There may be a small crescent of infarcted or avascular villi, but usually collateral damage is not visited upon more peripheral villi by the intervillous thrombus unless it is very large. Of note, intervillous thrombi are more common in placentas of hypertensive and diabetic pregnant people.

Mass-forming proliferations of villous vessels are termed placental chorangiomas (Fig. 42.65). Lacking the granular villous texture of infarcts and laminations of intervillous thrombi, the cut surfaces of chorangiomas are often smooth,

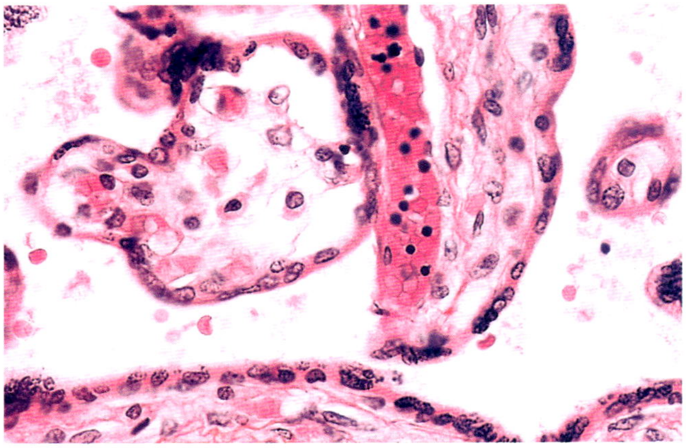

FIGURE 42.63 Fetal nucleated red blood cells are abnormal at term. Villous edema also is present (H&E).

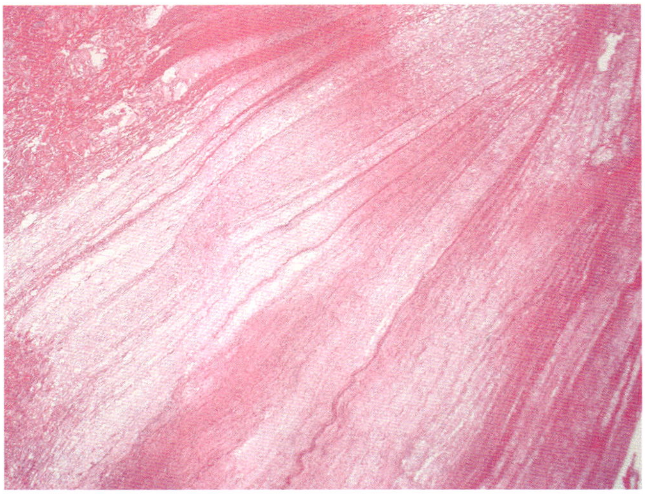

FIGURE 42.64 Histologic example of an intervillous thrombus. Note the layered fibrin deposition, forming the so-called lines of Zahn (H&E).

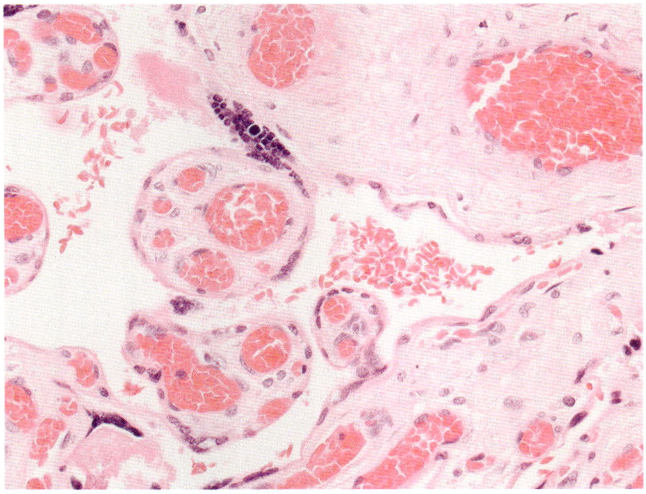

FIGURE 42.66 Maternal sickle cell trait. Note sickled red blood cells in the maternal vascular space. The fetal red blood cells won't sickle even if the fetus has trait or disease due to Hgb F production (H&E).

unless infarcted or calcified. These histologically hemangiomatous nodules only take on pathologic significance when very large lesions result in high-output cardiac failure or microangiopathic hemolysis in the fetus.

Dysmaturity of villi is considered pathologic. Villi that are immature tend to be large with open stroma and show the characteristic two-cell layer reminiscent of the first-trimester implantation. This change has been noted in diabetic mothers or in association with excessively long umbilical cords (4).

Maternal sickle cell disease or trait can be associated with sickling of maternal erythrocytes in intervillous spaces. This cytopathology is promoted by diminished oxygen tension in the setting of abnormal maternal hemoglobin S. It is wise to correlate histopathologic findings with clinical status because hypoxia during and after placental separation can cause erythrocyte sickling (Fig. 42.66) (4).

Acute placental infarcts can be accompanied by neutrophil-rich inflammatory infiltrates surrounding villi. However, an acute inflammatory infiltrate within the villi may suggest the presence of a hematogenously spread maternal bacterial infection. Specifically, *Listeria* infection causes microscopic or macroscopic abscess formation throughout the placenta. Infection with this organism can produce an odor that has been characterized as sweet or fruity (4,19,35).

Acute Villitis

The absence of inflammation does not always imply an absence of infection. Parvovirus B19 infection elicits hydropic villi in combination with nucleated red blood cells in the villous vessels, some of which contain "smudged" amphophilic intranuclear inclusion bodies (Fig. 42.67) (35). Infection can be detected by immunohistochemistry or polymerase chain reaction (PCR) for parvovirus DNA (78). Evaluation for other infections can be performed using PCR examination of amniotic fluid (79).

Placentas infected with HIV-1 or any of the hepatotrophic viruses may elicit no specific pathologic findings and only

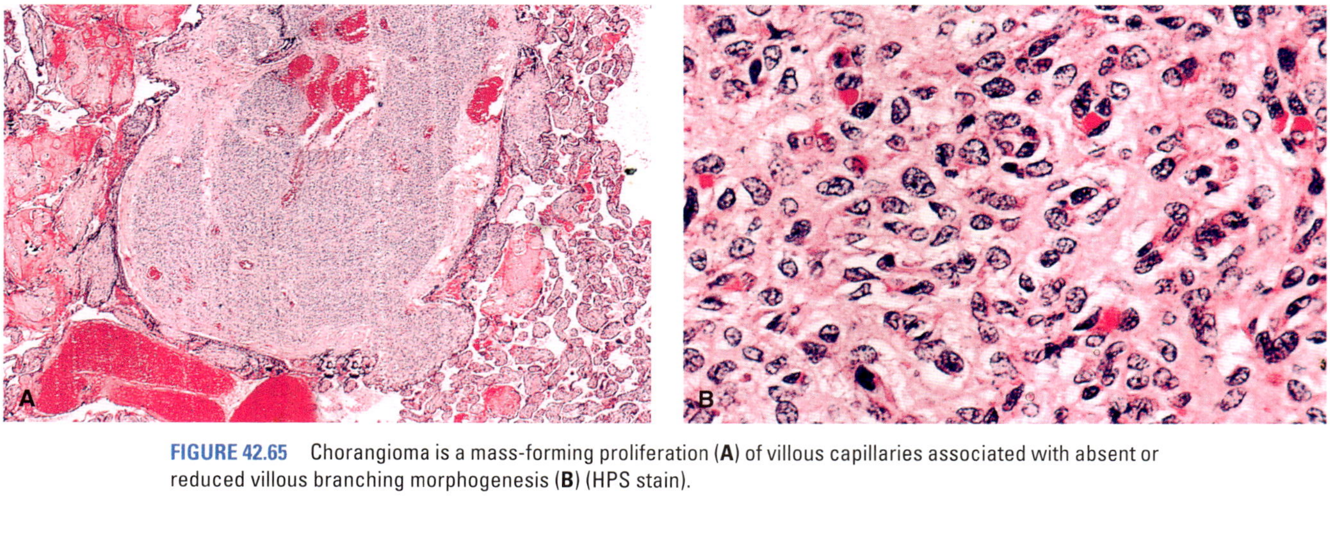

FIGURE 42.65 Chorangioma is a mass-forming proliferation (**A**) of villous capillaries associated with absent or reduced villous branching morphogenesis (**B**) (HPS stain).

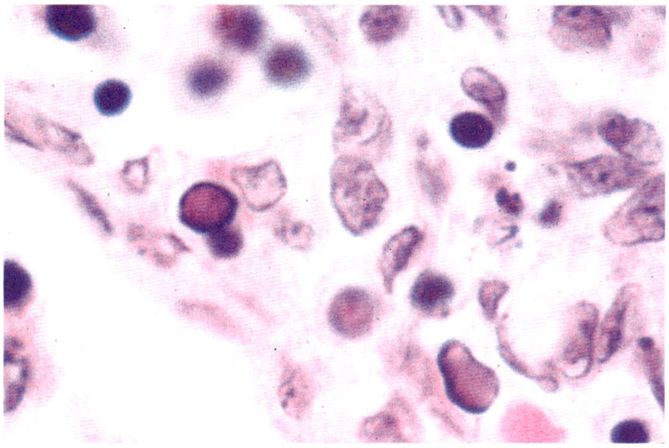

FIGURE 42.67 Parvovirus B19 infection of red blood cell progenitors is recognized by the characteristic violaceous intranuclear inclusion. Severe infections, like other causes of severe fetal anemia, may result in hydrops fetalis and extreme placental edema.

special molecular studies will detect infectious materials (80). Molecular techniques should be considered as an adjunct to histopathologic placental and fetal examination in cases of otherwise unexplained intrauterine or neonatal demise (81).

DECIDUA

Histology

The hypersecretory glandular epithelium of the endometrium, under the influence of pregestational agents, affords the proper substrate for implantation. While extrinsic to the placenta in the strictest sense, the intimate interdigitation of the gestational endomyometrium and implanting and angioinvasive trophoblast merits consideration with the placenta. At times, these hypersecretory glands may exhibit the Arias–Stella reaction in which cytologic atypia is noted and may even resemble clear cell carcinoma (Fig. 42.68). Under continued progestational influence, endometrial glands undergo

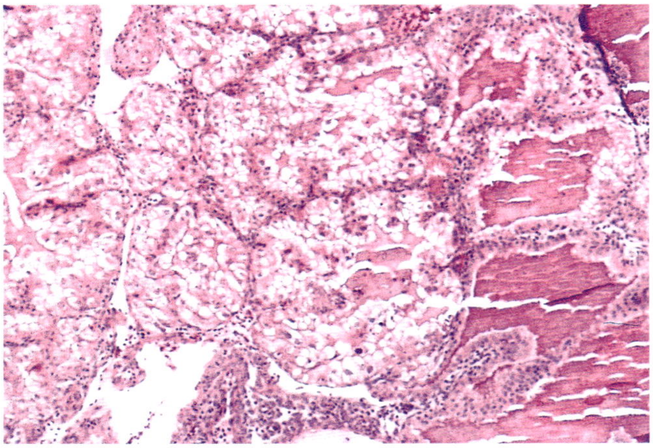

FIGURE 42.68 Arias–Stella reaction (H&E).

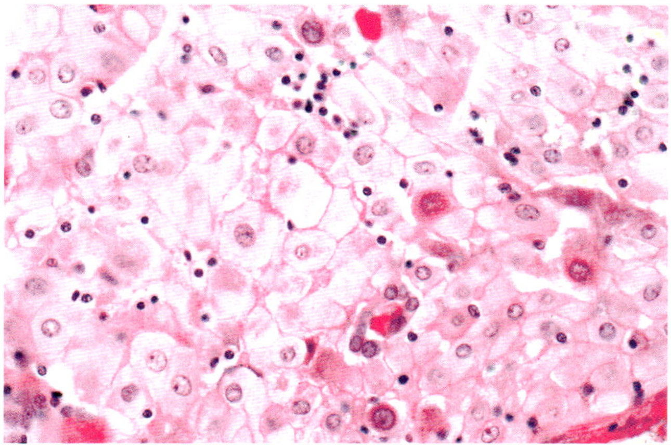

FIGURE 42.69 Decidua with stromal cells having centrally placed nuclei with clear chromatin and abundant pale cytoplasm with prominent cell borders. Near the placental disk and membranes, occasional extravillous trophoblast with darker cytoplasm and hyperchromatic nuclei may be evident.

secretory exhaustion. The decidual cells of the endometrium are epithelioid and polygonal. Their small, rounded nuclei are generally situated centrally in abundant pale eosinophilic, often vacuolated, cytoplasm with sharp cell borders (Fig. 42.69). The cytoplasm is rich in glycogen and glycoproteins. Ultrastructural examination of the decidua shows that tight junctions separate these cells (82). A basic understanding of decidual histology also is important in recognizing early implantation site and excluding ectopic pregnancy.

In addition to the decidual cells, an admixture of fibroblasts and numerous inflammatory cells (e.g., uNK cells, T lymphocytes, and neutrophils) are present in the stroma, as are relaxin-producing granular cells (83). The intercellular matrix contains abundant type IV collagen and laminin, fibronectin, and heparin sulfate proteins (84). Other collagens are also present throughout the decidual matrix, and these include types I, III, and V (19).

Prolactin and human placental lactogen are produced by invading extravillous trophoblast (85). The erroneous idea that hormonal production occurs in the decidua has resulted from the inability of many investigators to distinguish decidual components from invading trophoblastic contaminants (19).

Vascularization of the decidual component of the implantation site is critical to the developing gestation. The major branches of the uterine arteries extend deeply into the myometrial substance, resulting in arcuate arteries that branch to form radial arteries. These become the spiral arterioles, the terminal components of the endometrial vasculature. Spiral vessels are responsible for the intervillous blood flow within cotyledonary units (86). The precise number of spiral arterioles that serve to perfuse the placenta is not clear; estimates range from 25 to 300 vascular openings (19,87,88).

Endothelial extravillous trophoblastic cells replace the endothelium of the spiral arterioles early in implantation, causing fibrinoid necrosis of the vascular smooth muscle

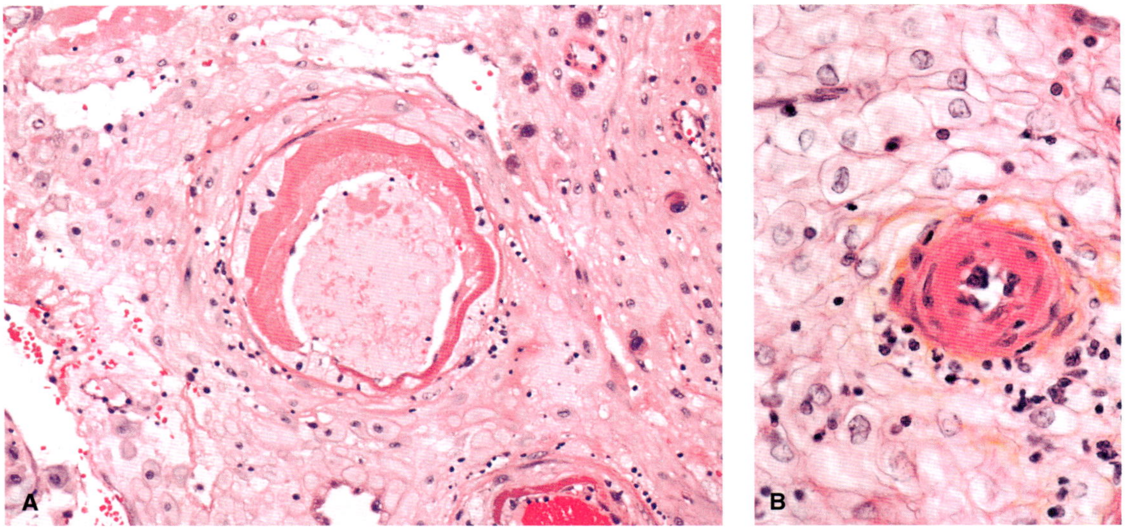

FIGURE 42.70 The histologic manifestations of severe maternal decidual arteriopathy include fibrinoid necrosis and foamy cell (atherosis) (**A**–H&E). Another finding associated with maternal hypertension is concentric medial hypertrophy (so called "non-transformation") of decidual spiral arterioles (**B**–HPS stain).

and plugging the lumina, thereby preventing blood flow to the early gestation. During this time, secretions from endometrial glands are thought to nourish the early gestation. These plugs release at approximately 10 weeks gestation and maternal blood flow to the placenta ensues. It is important to distinguish these decidual vascular changes from those that indicate pathologic conditions.

Gross inspection of the decidua is generally unrevealing, and important attributes and diagnoses are identified histologically. In full-thickness sections of the placenta disks, the underlying decidua is often denuded from the placental implantation site during the delivery process. Therefore, it is not uncommon to find sections of parenchyma devoid of decidual tissue. Decidual biopsy is mostly of historic interest, as it is a rarely submitted specimen in the current era. One region in which decidua is often prominent is on the chorion laeve. Sections of "membrane rolls," that include amnion and chorion often include sections of adherent decidua, which represent fusion of the decidua capsularis and the decidua vera during development. Decidual abnormalities may be identified in these sections. Aberrant vascular relationships of the more proximal vascular tree can be seen.

Histopathology

Principal among decidual pathology is acute atherosis of decidual vessels, which is known to accompany preeclampsia and hypertensive conditions of gestation. These lesions are characterized by fibrinoid necrosis and hyalinization of the vascular wall along with the deposition of foamy macrophages (Fig. 42.70A). Another abnormality that may characterize hypertensive pregnancies is concentric arteriolar mural hypertrophy (Fig. 42.70B) (13). Both of these pathologies are identified in the decidua capsularis/vera attached to the membrane roll.

Retroplacental bleeding is responsible for premature placental separation or abruption placenta. This has been related to two maternal conditions: increased maternal blood pressure (Fig. 42.71) and decidual vascular necrosis, both of which may indicate malperfusion (13).

When trophoblastic villi implant directly on myometrial tissue because the intervening decidua is highly attenuated or absent, placenta accreta spectrum is diagnosed (89). Placenta accreta spectrum FIGO grade 1 lacks myometrial invasion and is more common in low implantations of the placenta (i.e., placenta previa), especially when prior cesarean sections have been performed (Fig. 42.72). Placenta accreta spectrum FIGO grade 2 shows chorionic villi that penetrate the myometrium (i.e., placenta increta). Placenta accreta spectrum FIGO grade 3 is defined as implantation that results in villous tissue protruding through the uterine serosal surface (i.e., placenta percreta). Distinction between various grades, especially grades 1 and 2, is based largely on macroscopic context, as the highly magnified interface between the variations shows the same pushing

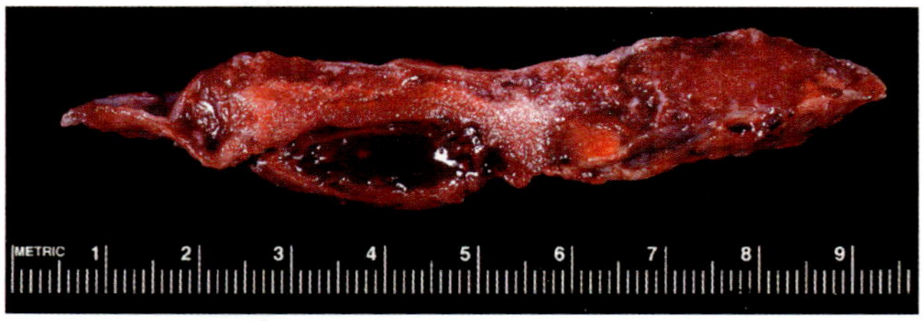

FIGURE 42.71 Gross example of an acute abruption. Note retroplacental hematoma and indentation of the parenchyma.

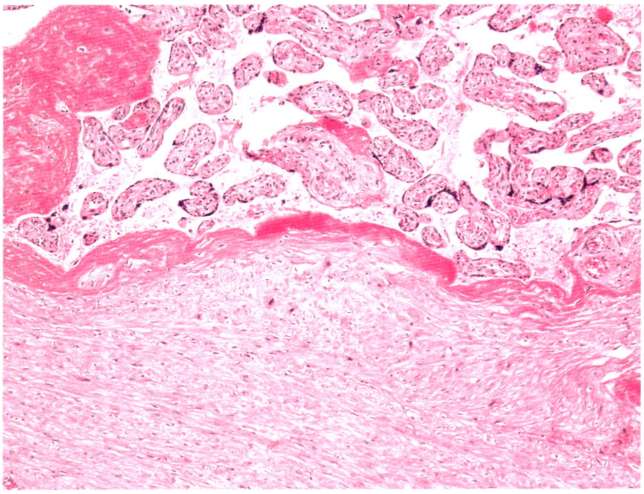

FIGURE 42.72 Placental accreta spectrum disorder FIGO grade 1 (placental accreta). Note villi directly on fibrin and muscle without intervening decidua.

border of villi. In general, if complete placental removal from the uterus at delivery cannot be achieved in patients with high-grade disease, life-threatening hemorrhage often ensues. Heroic obstetrical, surgical, and pharmacologic measures may save the uterus, but persistent hemorrhage and subsequent bacterial infection often increase the risk of morbidity and delay of definitive therapy, which is hysterectomy. Interestingly, a potential predecessor to placenta accreta spectrum may be recognized on the decidua basalis of delivered placentas (19,90,91). These microscopic lesions have adherent myometrial fibers on the placental basal plate surface. Some pathologists stage these lesions based on the presence of residual decidual cells. Stage 1 basal plate lesions retain residual decidual cells between the myometrial fibers and placenta, whereas stage 2 basal plate lesions lack decidual cells. Stage 2 disease has a higher risk of true placenta accreta spectrum in future pregnancies and must be followed.

Rare leukocytes may be seen within the decidua attached to the delivered placenta. If these cytologic constituents include plasma cells, syphilis or CMV infection should be considered. These rare lymphocytes within the decidua are likely physiologic or nonspecific in the absence of pathologic alterations. More distinctive inflammatory processes denote deciduitis and can be associated with inflammation elsewhere. Acute chorioamnionitis and acute or chronic villitis should be sought in the further examination of the placenta.

GESTATIONAL TROPHOBLASTIC DISEASE

Although an in-depth consideration of gestational trophoblastic disease and neoplasia is beyond the scope of this chapter, their brief review through the prism of basic placental histology is worthwhile as it reinforces several basic concepts covered in this chapter. Readers seeking more information about this group of placental diseases are directed toward several references (92–94).

The nosology of abnormal gestational trophoblastic proliferation has two faces. The more recent and mechanistic approach is genetic based. At the root, most gestational trophoblastic disease and neoplasia, especially hydatidiform moles and their mimics arise from errors of chromosomal division during meiosis. Almost all gestational trophoblastic disease and neoplasia will display DNA markers distinct from the genetic mother because some, if not all, of the nuclear genome is of paternal origin. Partial moles generally can be distinguished from complete moles because, with notable exceptions, they are triploid. The underlying pathobiology is an imbalance of genetic contributions from key genes under genetic regulation known as imprinting. The practical import of this knowledge is that the absent expression of one imprinted gene, p57, in villous cytotrophoblasts and stromal cells is a useful immunohistochemical tool for pathologic diagnosis of typical complete moles. Complete moles in the context of twin gestations and fascinating semimolar gestations formed by embryonic mosaicism or chimerism (95–98). An example of the normal pattern of p57 immunostain is shown in Figure 42.73. This pathobiology also explains predisposition to molar pregnancy (i.e., biallelic recurrent moles) due to maternal mutations in genes (e.g., *NLRP7*) presumably involved in imprinting and oogenesis (99–101).

The original face of gestational trophoblastic disease and neoplasia nosology is histologic. Thus, having a clear sense of normal placental histology and trophoblast cytology is a critical anchor for understanding and diagnosing gestational trophoblastic disease and neoplasia. It begins with the recognition of various histologic parts in any given gestational sample. Complete and partial hydatidiform moles are rich with placental villi, whereas placental site and epithelioid trophoblastic tumors lack villi, as does choriocarcinoma. The development of embryonic derivatives is another important distinguishing feature. Partial mole, nonmolar mimics, and molar twin gestations may show at least some embryonic development, as well as the formation of the chorion, amnion, and yolk sac. In sharp contrast, complete molar gestations do not form embryos. Choriocarcinoma, placental site, and epithelioid trophoblastic tumors represent exceptional cases with respect to embryonic development. Choriocarcinoma may be found both associated with hydatidiform moles and in situ, involving villi of spontaneous abortions and otherwise normal-term placentas (Fig. 42.74). Placental site and epithelioid trophoblastic tumors do not have embryonic development per se but arise from antecedent pregnancies ranging from months to years before subsequent tumor presentation. Thus, cataloging the presence of placental villi, membranes, and embryonic or fetal tissues is an important first step towards histologic classification of gestational trophoblastic disease and neoplasia. Another key to classification of gestational trophoblastic disease and neoplasia is proper identification of trophoblast cell types, either by cytologic or immunohistochemical observations. The molar and neoplastic trophoblasts bear a clear resemblance to their respective benign counterparts. Consequently, having a clear conceptualization of normal cytology of villous cytotrophoblasts, syncytiotrophoblasts, as well as extravillous trophoblast residing beneath the placental disc (i.e., implantation site) or membranes (i.e., chorion leave), as discussed in previous sections and illustrated in Figures 42.73, 42.75, and 42.76, facilitates accurate diagnosis.

In this closing section, it may be helpful to focus on the immunophenotypes of the various trophoblast cell types and the lessons learned from this different perspective about normal trophoblast biology, gestational trophoblastic disease, and neoplasia pathobiology.

First, all normal, molar, and neoplastic trophoblasts express cytokeratins (Figs. 42.73, 42.75, and 42.76). They also express GATA-3, though the staining frequency and intensity may fall in mature placentas (Fig. 42.75). These markers can be helpful when trophoblastic lesions are found in unusual contexts.

Villous cytotrophoblasts are the proliferating progenitor cells, giving rise to syncytiotrophoblasts and the polarized tips capable of anchoring distal villi to decidua (Fig. 42.73). It expresses p63 but does not express SALL4, unlike choriocarcinoma cells. In fact, SALL4 staining can be used as evidence in the diagnosis of choriocarcinoma (102). In addition, their well-regulated proliferation distinguishes them from the circumferential proliferation associated with hydatidiform moles, particularly the more exuberant variety found in complete moles.

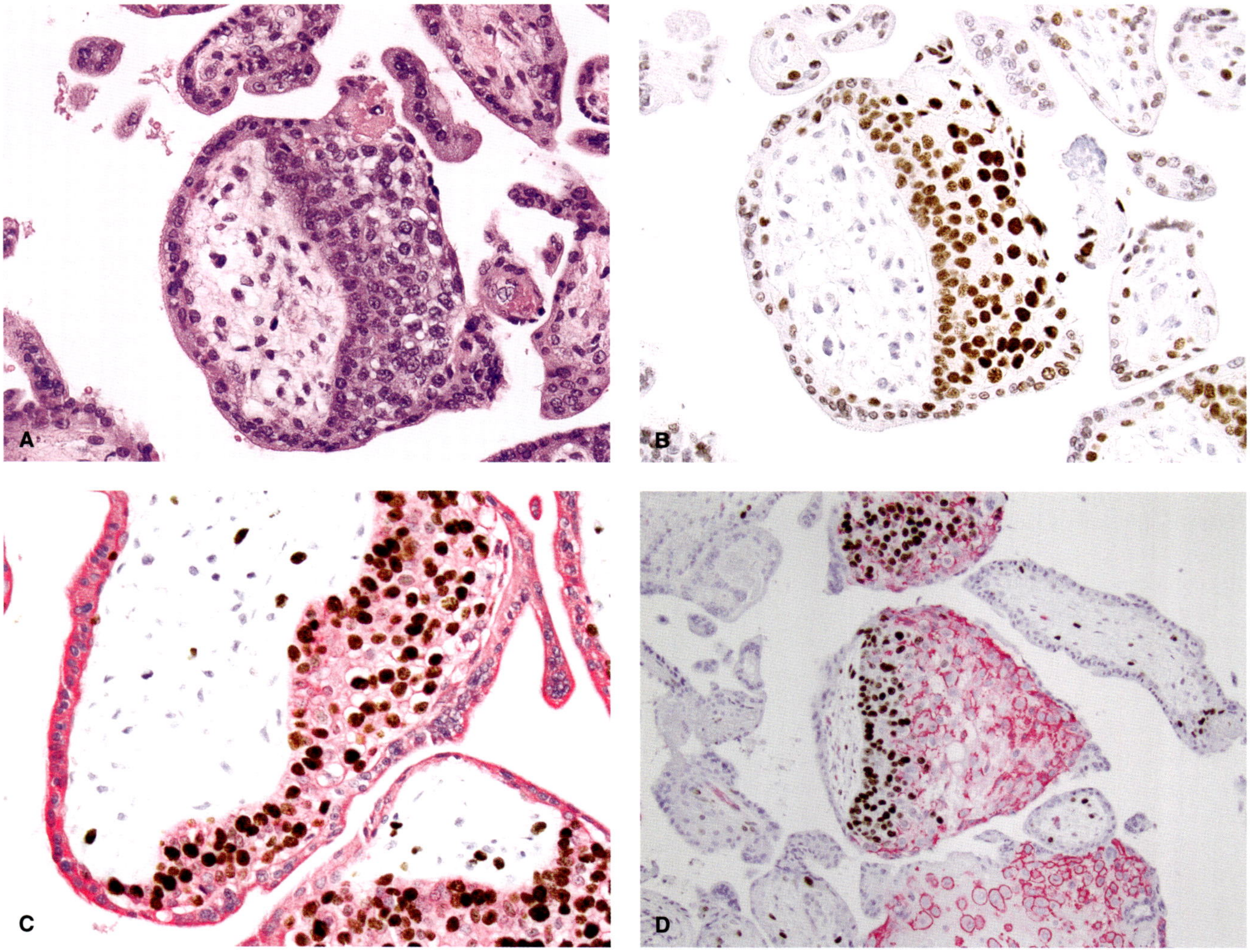

FIGURE 42.73 Immature chorionic villi—correlation of histology with common trophoblastic immunohistochemical markers. One key histologic feature of normal villi early in gestation **(A)** is polarized proliferation of mononuclear trophoblasts that will anchor the placenta and populate the implantation site. At this gestational age, all trophoblast nuclei stain strongly with GATA-3 **(B)** and cytokeratins (AE1/AE3 staining red in **(C)**. Brisk proliferation (MIB-1 staining brown in C and D) is found in the trophoblasts closest to the villi but decreases with differentiation **(D)** and concomitant CD146 expression (red staining in D). CD146 on their cell surfaces facilitates adhesion to the smooth muscle of blood vessels and myometrium (not shown).

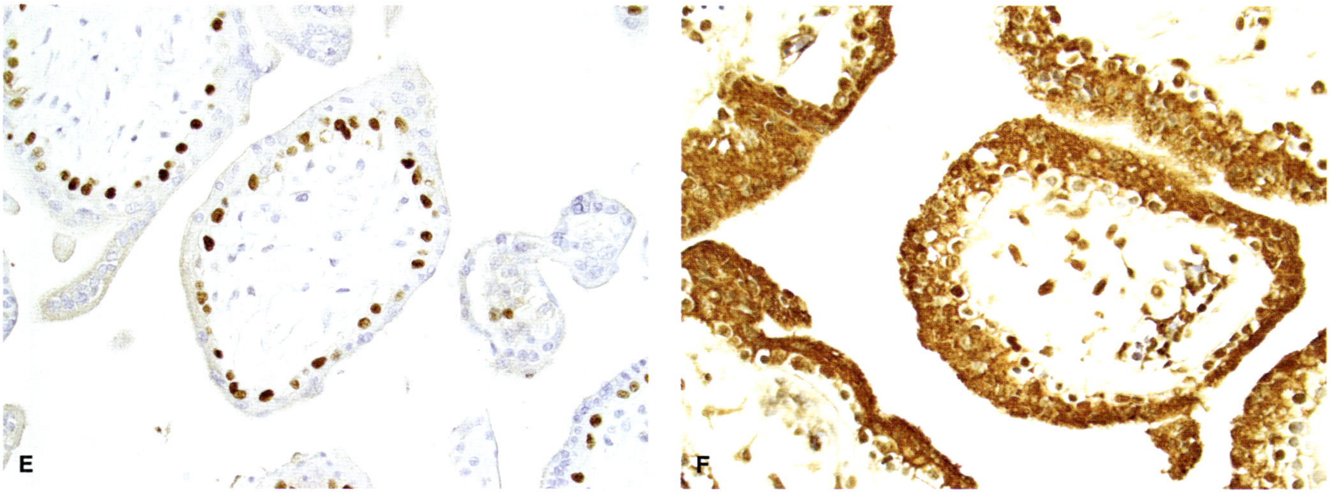

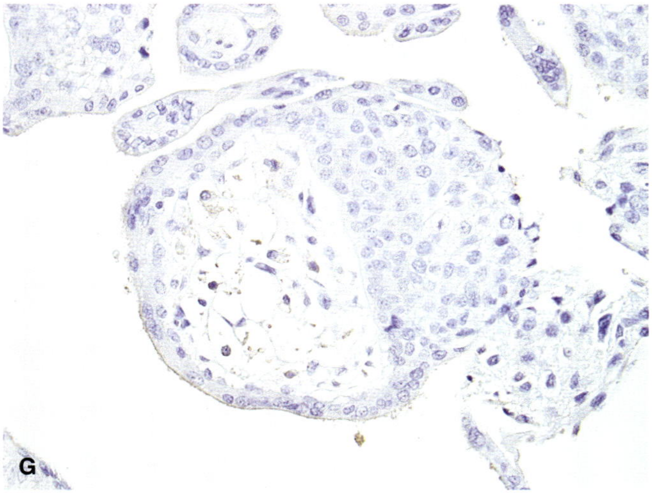

FIGURE 42.73 (*Continued*) The villous stem cell population are p63-positive cytotrophoblasts **(E)**. Differentiation in syncytiotrophoblasts includes β-hCG expression **(F)**, but the cytotrophoblasts lack such staining. Unlike the neoplastic cytotrophoblasts and syncytiotrophoblasts of choriocarcinoma (not shown), normal trophoblasts do not express SALL4 **(G)**. Understanding this immunophenotype facilitates the distinction of complete moles and placental site trophoblastic tumor.

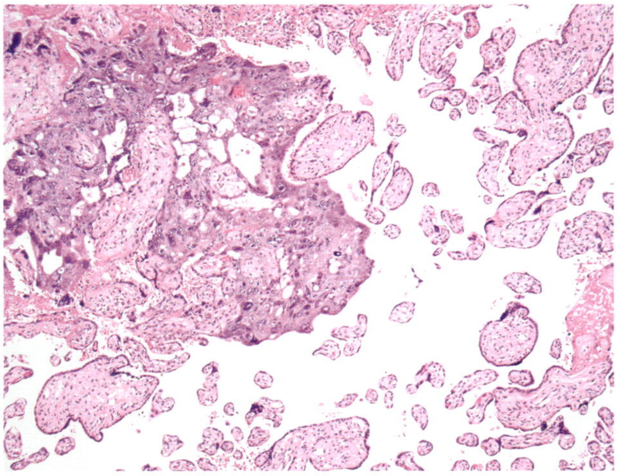

FIGURE 42.74 Intraplacental choriocarcinoma in a term placenta. Like the full-fledged counterpart of this lesion, this lesion consists of a sheet-like proliferation of highly atypical cytotrophoblasts and syncytiotrophoblasts, often associated with variable amounts of hemorrhage. In addition, intraplacental choriocarcinoma has significant metastatic potential. Unlike its counterpart, intraplacental choriocarcinoma is associated with chorionic villi. In this case, note the mature chorionic villi surrounding the small neoplastic lesion.

Syncytiotrophoblasts are nonproliferating, but functionally specialized cells producing several placenta-specific gene products, including Growth/Differentiation Factor 15 (GDF15) and hCG (Figs. 42.73, 42.75, and 42.76). hCG is well known, but GDF15 mutations have recently been implicated as the cause of pregnancy-related nausea and hyperemesis gravidarum (69). Substantial numbers of syncytiotrophoblasts are present in molar gestations and choriocarcinoma, accounting for high serum β-hCG levels, but scattered syncytiotrophoblast cells also may be found in placental site and epithelioid trophoblastic tumors, especially when using β-hCG immunohistochemistry. In those neoplasms, the smaller number of syncytiotrophoblasts correlates with the lower serum β-hCG concentration and a clinical presentation mimicking spontaneous abortions.

The polarized proliferation of cytotrophoblasts at the tips of chorionic villi is also of interest because it represents the first stage toward differentiation into extravillous trophoblasts of the implantation site. Notably, that transition is spatially associated with the onset of CD146 (i.e., MelCam and Melanoma Adhesion Molecule) expression

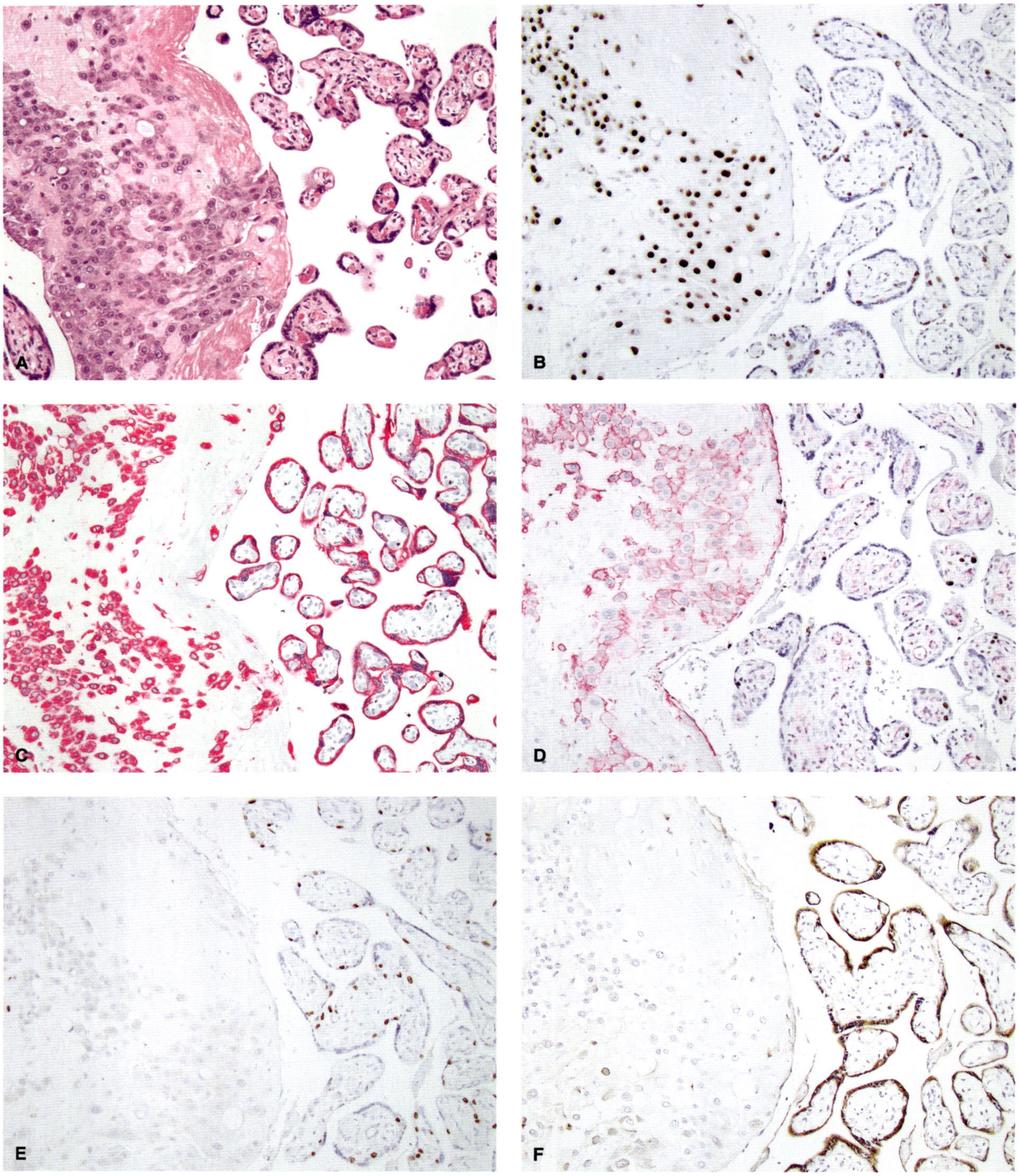

FIGURE 42.75 Mature placental parenchyma—correlation of histology with common trophoblastic immunohistochemical markers. The trophoblast population of mature placental parenchyma includes villous cytotrophoblasts and syncytiotrophoblasts, as well as implantation site-like extravillous trophoblasts in placental septa and chorionic cysts (**A**). GATA-3 expression, a useful trophoblast-selective marker, remains strong in extravillous trophoblasts and villous cytotrophoblasts, but is often decreased in syncytiotrophoblasts (**B**). That loss generally is not significant because diagnostic cell identification is not required in normal-term placenta. Cell proliferation, as highlighted by nuclear MIB-1 staining (brown in **C** and **D**) is very low, reflecting complete differentiation. Note that all trophoblasts continue to express cytokeratins (AE1/AE3 stained red in C), but only extravillous trophoblasts in the septa and implantation site express CD146 (stained red in **D**). p63 immunostaining (**E**) shows a reduction of cytotrophoblasts, and syncytiotrophoblasts continue to express β-hCG (**F**), albeit at an intensity lower than in early gestation. This immunophenotype is a yardstick from which abnormal trophoblastic proliferations can be compared.

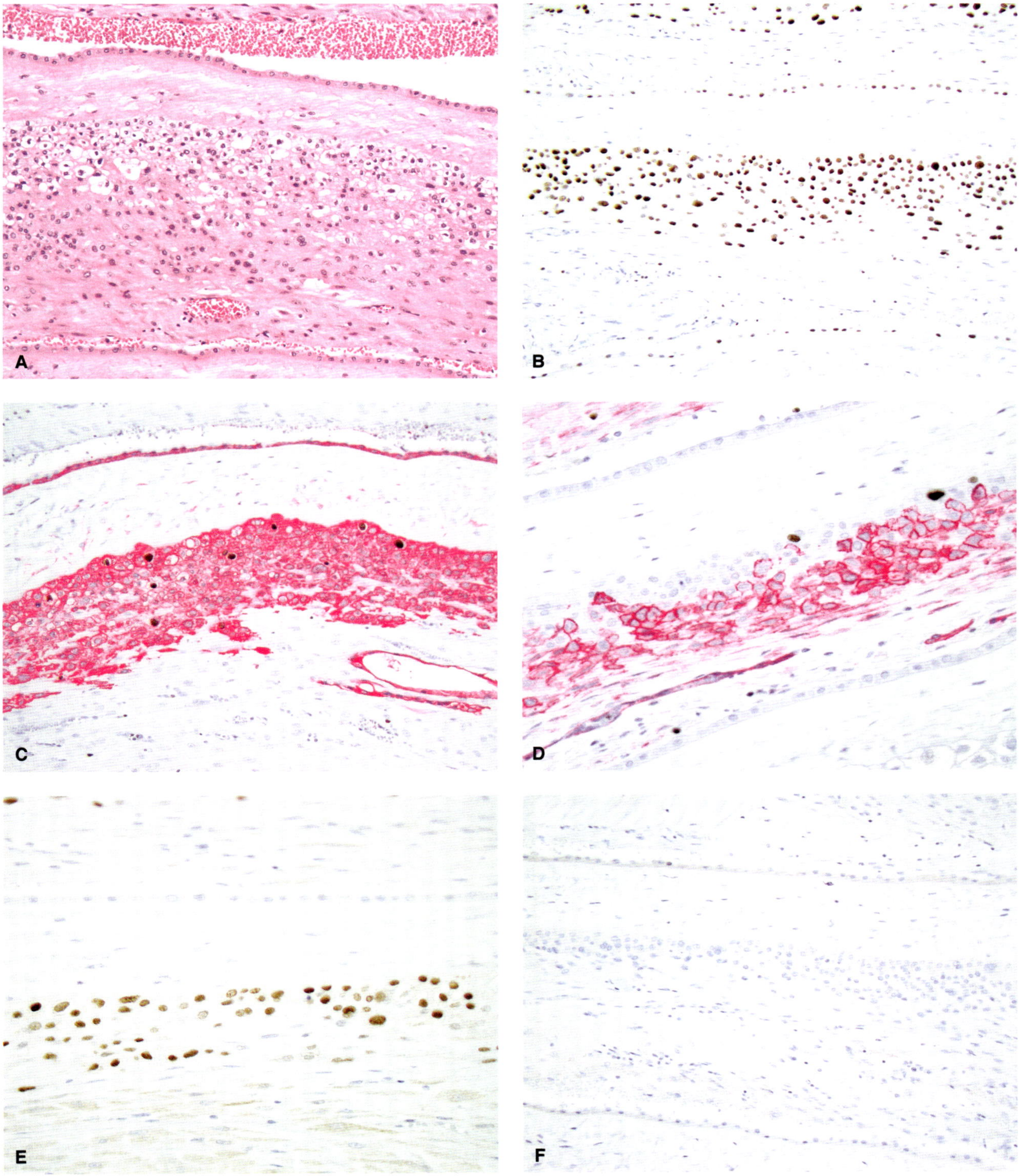

FIGURE 42.76 Mature placental membranes—correlation of histology with common trophoblastic immunohistochemical markers. Extravillous trophoblasts in the placental membranes have distinctive difference from those of the disk's implantation. Cytologically, membranous trophoblasts are a mixture of epithelioid cells with either clear vacuolated or eosinophilic cytoplasm **(A)**. Like other trophoblasts, they uniformly express GATA-3 (brown staining in **B**) and cytokeratins (AE1/AE3 with red staining in **C**). Patchy, if not highly variable, staining for CD146 (red in **D**) distinguishes these cells from implantation site extravillous trophoblasts, as does the lack of angioinvasion (note the nearby blood vessels). Further, persistent p63 staining **(E)** differentiates them from implantation site trophoblasts. In contrast to their malignant counterpart (viz., epithelioid trophoblastic tumor), the proliferation rate in membranous trophoblasts is low or absent, as demonstrated by MIB-1 staining (brown in **C** and **D**). In their differentiated state, membranous trophoblasts do not produce β-hCG **(F)**.

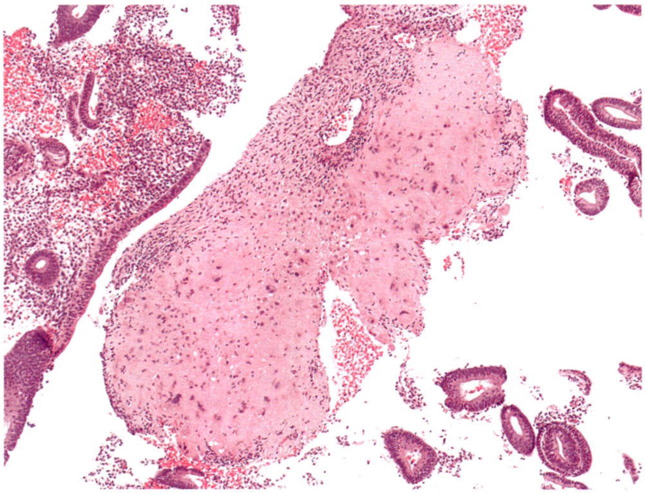

FIGURE 42.77 Placental site nodules are retained products of conception. The non-proliferating trophoblasts reside in nodular or plaque-like bodies of hyalinized extracellular matrix and may have immunophenotypes similar to extravillous trophoblasts in the placental implantation site or membranes. (H&E).

(Figs. 42.75 and 42.76). Smooth muscle cells in the decidualized endometrium and superficial myometrium also express CD146, thereby allowing intercellular adhesion and migration between implantation site trophoblasts and uterine smooth cells, that ultimately result in structural anchorage of placental villi, a cohesive interdigitation of the placental basal plate and decidua, and infiltration of maternal blood vessels. Implantation site trophoblasts migrating away from their original points of villous contact have very little or no mitotic activity. The trophoblasts of placental site trophoblastic tumor, however, retain their proliferative capacity, and CD146/MIB-1 double immunostains facilitates determination of proliferative fraction in the precise cell of interest in placental site trophoblastic tumor. In fact, one diagnostic criterion for placental site trophoblastic tumor is satisfied when the proliferative fraction is 10% or more. Greater use of immunohistochemical staining has facilitated more careful observations of atypical lesions of CD146-positive/p63-negative trophoblasts having an intermediate proliferative fraction (5% to 10%). Perhaps 10% of such lesions ultimately are diagnosed as placental site trophoblastic tumors (103).

The other type of extravillous trophoblast, which is found in the placental membranes, has a different immunophenotype. Membranous extravillous trophoblasts resemble the tumor cells of epithelioid trophoblastic tumors. Both the normal membranous extravillous trophoblast and epithelioid trophoblastic tumor cells are p63-positive. In further contrast to implantation site extravillous trophoblast and placental site trophoblastic tumor, CD146 expression is variable, ranging from patchy to absent (Fig. 42.76). Membranous extravillous trophoblasts are similar to their implantation site counterparts in the sense that their respective mitotic activities are both very low or absent. Also, analogous to the case of placental site trophoblastic tumor, normal placental site nodules (Fig. 42.77), atypical placental site nodules, and epithelioid trophoblastic tumors have minimal, intermediate, and high mitotic activities, respectively.

Hopefully, a deeper understanding of the normal histologic anatomy of the placenta and the spectrum of trophoblast cellular phenotypes will help the reader better understand the range of neoplastic and nonneoplastic conditions that occur in the placenta.

REFERENCES

1. Roberts DJ, Baergen RN, Boyd TK, et al. Criteria for placental examination for obstetrical and neonatal providers. *Am J Obstet Gynecol*. 2023;228(5):497–508.e4.
2. Heerema-McKenney A, Popek EJ, De Paepe M. Diagnostic pathology: Placenta. *Diagnostic Pathology*. 2nd ed. Elsevier; 2018.
3. Redline RW, Boyd TK, Roberts DJ, eds. *Placental and Gestational Pathology*. 1st ed. *Diagnostic Pediatric Pathology*. M. C. Cohen, I. Scheimberg, eds. Cambridge University Press; 2018:344.
4. Roberts DJ, Polizzano C. *Atlas of Placental Pathology. AFIP Atlases of Tumor and Non-Tumor Pathology*. ARP Press; 2021;6:510.
5. Khong TY, Mooney EE, Nikkels PG, et al., eds. *Pathology of the Placenta. A Practical Guide*. Springer; 2019.
6. ACOG committee opinion. Prevention of early-onset group B streptococcal disease in newborns. Number 173–June 1996. Committee on Obstetric Practice. American College of Obstetrics and Gynecologists. *Int J Gynaecol Obstet*. 1996; 54(2):197–205.
7. Bulletins ACoP. Intrauterine growth restriction. *ACOG Practice Bulletin*. 2000;12:1–3.
8. ACOG Committee on Practice Bulletins–Obstetrics. Diagnosis and management of preecclmapsia and eclampsia. *Obstet Gynecol*. 2002;99(1):159–167.
9. Committee on Obstetric Practice. ACOG committee opinion. Placenta accreta. Number 266, January 2002. American College of Obstetricians and Gynecologists. *Int J Gynaecol Obstet*. 2002;77(1):77–78.
10. Committee on Genetics. ACOG Committee Opinion No. 383: Evaluation of stillbirths and neonatal deaths. *Obstet Gynecol*. 2007;110(4):963–966.
11. American College of Obstetricians and Gynecologists. ACOG Practice bulletin no.134: Fetal growth restriction. *Obstet Gynecol*. 2013;121(5):1122–1133.
12. Redline RW. Classification of placental lesions. *Am J Obstet Gynecol*. 2015;213(4 Suppl):S21–28.
13. Khong TY, Mooney EE, Ariel I, et al. Sampling and definitions of placental lesions: Amsterdam Placental Workshop Group Consensus Statement. *Arch Pathol Lab Med*. 2016;140(7):698–713.
14. Lindley SW, Gillies EM, Hassell LA. Communicating diagnostic uncertainty in surgical pathology reports: Disparities between sender and receiver. *Pathol Res Pract*. 2014;210(10):628–633.
15. Odibo I, Gehlot A, Ounpraseuth ST, Magann EF. Pathologic examination of the placenta and its clinical utility: A survey

of obstetrics and gynecology providers. *J Matern Fetal Neonatal Med*. 2016;29(2):197–201.
16. Polnaszek BE, Clark SL, Rouse DJ. Pathologic assessment of the placenta: Evidence compared with tradition. *Obstet Gynecol*. 2022;139(4):660–667.
17. Redline RW, Roberts DJ, Parast MM, et al. Placental pathology is necessary to understand common pregnancy complications and achieve an improved taxonomy of obstetrical disease. *Am J Obstet Gynecol*. 2023;228(2):187–202.
18. Pinar H, Gordijn SJ, Grynspan D, et al. Pathologic assessment of the placenta: Evidence compared with tradition. *Obstet Gynecol*. 2022;140(2):341–342.
19. Baergen RN, Burton GJ, Kaplan CG. *Benirschke's Pathology of the Human Placenta*. 7th ed. Springer; 2022:939.
20. Langston C, Kaplan C, Macpherson T, et al. Practice guideline for examination of the placenta: developed by the Placental Pathology Practice Guideline Development Task Force of the College of American Pathologists. *Arch Pathol Lab Med*. 1997;121(5):449–476.
21. Tambouret R, Jeck WR, Roberts DJ. The difference between unfixed and postfixation placental weight. *Am J Clin Pathol*. 2019;152(2):217–220.
22. Lester SC. *Manual of Surgical Pathology: Expert Consult*. 3rd ed. Elsevier; 2010:592.
23. Heerema-McKenney A, Rabinowitz L, Bendon R, Bruehl F, Blank A, Pinar H, et al. Heterotopic nodules in the placenta, an immunohistochemical re-evaluation of the diagnosis of adrenocortical heterotopia. *Pediatr Dev Pathol*. 2021;24(1):27–33.
24. Lewis SH, Gilbert-Barness E. The placenta and its significance in neonatal outcome. In: Barness LA, ed. *Advances in Pediatrics*. Vol. 45. 1998; Mosby.
25. Naeye RL. Umbilical cord length: Clinical significance. *J Pediatr*. 1985;107(2):278–281.
26. Gardiner JP. The umbilical cord: Normal length, length in cord complications; etiology and frequency of coiling. *Surg Gynecol Obstet*. 1922;34:252–256.
27. Hoyes AD. Ultrastructure of the epithelium of the human umbilical cord. *J Anat*. 1969;105:145–162.
28. Moore RD. Mast cells of the human umbilical cord. *Am J Pathol*. 1956;32:1179–1183.
29. Nadkarni BB. Innervation of the human umbilical cord. *Am J Obstet Gynecol*. 1970;107:303–312.
30. Lauweryns JM, De Bryn M, Peuskens J, Bourgeois N. Abscence of intrinsic innervation of the human placenta. *Experientia*. 1969;25:432.
31. Blanc WA, Allan GW. Intrafunicular ulceration of persistent omphalomesenteric duct with intra-amniotic hemorrhage and fetal death. *Am J Obstet Gynecol*. 1961;82:1392–1396.
32. Priman JA. A note on the anastamosis of the umbilical arteries. *Anat Rec*. 1959;134:1–5.
33. Arts NF. Investigations on the vascular system of the placenta. I. General introduction and the fetal vascular system. *Am J Obstet Gynecol*. 1961;82:147–158.
34. Rana J, Ebert GA, Kappy KA. Adverse perinatal outcome in patients with an abnormal umbilical coiling index. *Obstet Gynecol*. 1995;85(4):573–577.
35. Benirschke K. Examination of the placenta. *Obstet Gynecol*. 1961;18:309–333.
36. Altshuler G, Arizawa M, Molnar-Nadasdy G. Meconium-induced umbilical cord vascular necrosis and ulceration: A potential link between the placenta and poor pregnancy outcome. *Obstet Gynecol*. 1992;79(5 (Pt 1)):760–766.
37. Altshuler G, Hyde S. Meconium-induced vasocontraction: A potential cause of cerebral and other fetal hypoperfusion and of poor pregnancy outcome. *J Child Neurol*. 1989;4(2):137–142.
38. King EL, Redline RW, Smith SD, Kraus FT, Sadovsky Y, Nelson DM. Myocytes of chorionic vessels from placentas with meconium-associated vascular necrosis exhibit apoptotic markers. *Hum Pathol*. 2004;35(4):412–417.
39. Hubinont C, Lewi L, Bernard P, Marbaix E, Debiève F, Jauniaux E. Anomalies of the placenta and umbilical cord in twin gestations. *Am J Obstet Gynecol*. 2015;213(4 Suppl):S91–S102.
40. Garza A, Cordero JF, Mulinare J. Epidemiology of the early amnion rupture spectrum of defects. *Am J Dis Child*. 1988;142(5):541–544.
41. Danforth D, Hull RW. The microscopic anatomy of the fetal membranes with particular reference to the detailed structure of the amnion. *Am J Obstet Gynecol*. 1958;75(3):536–547; discussion 548–550.
42. Hoyes AD. Ultrastructure of the mesenchymal layers of the human amnion in early pregnancy. *Am J Obstet Gynecol*. 1970;106(4):557–566.
43. Hoyes AD. Ultrastructure of the mesenchymal layers of the human chorion laeve. *J Anat*. 1971;109(Pt 1):17–30.
44. Bourne GL. *The Human Amnion and Chorion*. 1962; Lloyd-Luke.
45. Wynn RM. Cytotrophoblastic specializations: An ultrastructural study of the human placenta. *Am J Obstet Gynecol*. 1972;114(3):339–355.
46. Bartels H, Wang T. Intercellular junctions in the human fetal membranes. A freeze-fracture study. *Anat Embryol (Berl)*. 1983;166(1):103–120.
47. Robinson HN, Anhalt GJ, Patel HP, Takahashi Y, Labib RS, Diaz LA. Pemphigus and pemphigoid antigens are expressed in human amnion epithelium. *J Invest Dermatol*. 1984;83(3):234–237.
48. Bautzmann H, Schroder FA. [Incidence and significance of Hofbauer cells in the myometrium and decidua of pregnant white rats]. *Z Zellforsch Mikrosk Anat*. 1957;47(1):1–6.
49. Kalousek DK, Dill FJ. Chromosomal mosaicism confined to the placenta in human conceptions. *Science*. 1983;221(4611):665–667.
50. Schindler PD. Nuclear deoxyribonucleic acid (DNA) content, nuclear size and cell size in the human amnion epithelium. *Acta Anat (Basel)*. 1961;44:273–285.
51. Rossant J, Croy BA. Genetic identification of tissue of origin of cellular populations within the mouse placenta. *J Embryol Exp Morphol*. 1985;86:177–189.
52. Bautzmann H, Hertenstein C. [Histogenesis and histology of the human fetal and newborn amnion]. *Z Zellforsch Mikrosk Anat*. 1957;45(5):589–611.
53. Sala MA, Matheus M. Histochemical study of the fetal membranes in the human term pregnancy. *Gegenbaurs Morphol Jahrb*. 1984;130(5):699–705.
54. Goldstein J, Braverman M, Salafia C, Buckley P. The phenotype of human placental macrophages and its variation with gestational age. *Am J Pathol*. 1988;133(3):648–659.
55. Wang L, Du X, Li M. Placental teratoma. A case report and review of the literature. *Pathol Res Pract*. 1995;191(12):1267–1270; discussion 1270.

56. Salazar H, Kanbour AI. Amnion nodosum: Ultrastructure and histopathogenesis. *Arch Pathol*. 1974;98(1):39–46.
57. Adeniran AJ, Stanek J, Amnion nodosum revisited: clinicopathologic and placental correlations. *Arch Pathol Lab Med*. 2007;131(12):1829–1833.
58. Farkas K, Ferretti E. Derivation of human extraembryonic mesoderm-like cells from primitive endoderm. *Int J Mol Sci*. 2023;24(14):11366.
59. Hessle H, Engvall E. Type VI collagen. Studies on its localization, structure, and biosynthetic form with monoclonal antibodies. *J Biol Chem*. 1984;259(6):3955–3961.
60. Cheek B, Heinrich S, Ward K, Craver R. Eosinophilic/T-cell chorionic vasculitis: Histological and clinical correlations. *Fetal Pediatr Pathol*. 2015;34(2):73–79.
61. Katzman PJ, Li L, Wang N. Identification of fetal inflammatory cells in eosinophilic/T-cell chorionic vasculitis using fluorescent in situ hybridization. *Pediatr Dev Pathol*. 2015; 18(4):305–309.
62. Ingman K, Cookson VJKW, Jones CJP, Aplin JD. Characterisation of Hofbauer cells in first and second trimester placenta: Incidence, phenotype, survival in vitro and motility. *Placenta*. 2010;31(6):535–544.
63. Pitkin RM, Reynolds WA, Williams GA, Hargis GK. Calcium metabolism in normal pregnancy: A longitudinal study. *Am J Obstet Gynecol*. 1979;133(7):781–790.
64. Mirkovic J, Elias K, Drapkin R, Barletta JA, Quade B, Hirsch MS. GATA3 expression in gestational trophoblastic tissues and tumours. *Histopathology*. 2015;67(5):636–644.
65. Banet N, Gown AM, Shih IM, et al. GATA-3 expression in trophoblastic tissues: An immunohistochemical study of 445 cases, including diagnostic utility. *Am J Surg Pathol*. 2015; 39(1):101–108.
66. Warning JC, McCracken SA, Morris JM. A balancing act: Mechanisms by which the fetus avoids rejection by the maternal immune system. *Reproduction*. 2011;141(6):715–724.
67. Knofler M, Haider S, Saleh L, Pollheimer J, Gamage TKJB, James J. Human placenta and trophoblast development: Key molecular mechanisms and model systems. *Cell Mol Life Sci*. 2019;76(18):3479–3496.
68. Wislocki GB, Dempsey EW. Electron microscopy of the human placenta. *Anat Rec*. 1955;123(2):133–167.
69. Fejzo MS, MacGibbon KW, First O, Quan C, Mullin PM. Whole-exome sequencing uncovers new variants in GDF15 associated with hyperemesis gravidarum. *BJOG*. 2022;129(11):1845–1852.
70. Redline RW, Ravishankar S, Bagby CM, Saab S, Zarei S. Four major patterns of placental injury: A stepwise guide for understanding and implementing the 2016 Amsterdam consensus. *Mod Pathol*. 2021;34(6):1074–1092.
71. Ernst LM. Maternal vascular malperfusion of the placental bed. *APMIS*. 2018;126(7):551–560.
72. Davis DL, Lechner AC, Chapel DB, et al. Outcome-based risk stratification model for the diagnosis of placental maternal vascular malperfusion. *Mod Pathol*. 2024;37(1):100370.
73. Aurioles-Garibay A, Hernandez-Andrade E, Romero R, et al. Prenatal diagnosis of a placental infarction hematoma associated with fetal growth restriction, preeclampsia and fetal death: clinicopathological correlation. *Fetal Diagn Ther*. 2014;36(2):154–161.
74. Redline RW, Ravishankar S. Fetal vascular malperfusion, an update. *APMIS*. 2018;126(7):561–569.
75. Watkins JC, Torous VF, Roberts DJ. Defining Severe Acute Respiratory Syndrome Coronavirus 2 (SARS-CoV-2) Placentitis. *Arch Pathol Lab Med*. 2021;145(11):1341–1349.
76. Redline RW. Villitis of unknown etiology: Noninfectious chronic villitis in the placenta. *Hum Pathol*. 2007;38(10): 1439–1446.
77. Altshuler G. Chorangiosis. An important placental sign of neonatal morbidity and mortality. *Arch Pathol Lab Med*. 1984;108(1):71–74.
78. Al-Adnani M, Sebire NJ. The role of perinatal pathological examination in subclinical infection in obstetrics. *Best Pract Res Clin Obstet Gynaecol*. 2007;21(3):505–521.
79. Dong ZW, Zhou SY, Li Y, Liu RM. Detection of a human parvovirus intrauterine infection with the polymerase chain reaction. *J Reprod Med*. 2000;45(5):410–412.
80. Lewis SH, Reynolds-Kohler C, Fox HE, Nelson JA. HIV-1 in trophoblastic and villous Hofbauer cells, and haematological precursors in eight-week fetuses. *Lancet*. 1990;335(8689): 565–568.
81. Romero R, Miranda J, Kusanovic JP, et al. Clinical chorioamnionitis at term I: Microbiology of the amniotic cavity using cultivation and molecular techniques. *J Perinat Med*. 2015;43(1):19–36.
82. Lawn AM, Wilson EW, Finn CA. The ultrastructure of human decidual and predecidual cells. *J Reprod Fertil*. 1971;26(1):85–90.
83. Dallenbach FD, Dallenbach-Hellweg G. Immunohistological studies on the localization of relaxin in the human placenta and decidua. *Virchows Arch Pathol Anat Physiol Klin Med*. 1964;337:301–316.
84. Wewer UM, Faber M, Liotta LA, Albrechtsen R. Immunochemical and ultrastructural assessment of the nature of the pericellular basement membrane of human decidual cells. *Lab Invest*. 1985;53(6):624–633.
85. Kurman RJ. The morphology, biology, and pathology of intermediate trophoblast: A look back to the present. *Hum Pathol*. 1991;22(9):847–855.
86. Freese UE. The uteroplacental vascular relationship in the human. *Am J Obstet Gynecol*. 1968;101(1):8–16.
87. Haller U. [Contribution on the morphology of the utero-placental vessels]. *Arch Gynakol*. 1968;205(3):185–202.
88. Borell U, Fernstroem I, Ohlson L, Wiqvist N. Influence of uterine contractions on the uteroplacental blood flow at term. *Am J Obstet Gynecol*. 1965;93:44–57.
89. Jauniaux E, Ayres-de-Campos D, Diagnosis FPA P. Management Expert Consensus, FIGO consensus guidelines on placenta accreta spectrum disorders: Introduction. *Int J Gynaecol Obstet*. 2018;140(3):261–264.
90. Linn RL, Miller ES, Lim G, Ernst LM. Adherent basal plate myometrial fibers in the delivered placenta as a risk factor for development of subsequent placenta accreta. *Placenta*. 2015;36(12):1419–1424.
91. Hecht JL, Baergen R, Ernst LM, et al. Classification and reporting guidelines for the pathology diagnosis of placenta accreta spectrum (PAS) disorders: recommendations from an expert panel. *Mod Pathol*. 2020;33(12):2382–2396.
92. Quade BJ. Gestational trophoblastic disease. In: Redline R, Boyd TK, Roberts DJ, eds. *Placental and Gestational Pathology*. Cambridge University Press; 2018.
93. Quade BJ. Gestational trophoblastic disease. *Pathology and Genetics*. UpToDate [cited 2024].

94. Bartosch C, Nadal A, Braga AC, et al. Practical guidelines of the EOTTD for pathological and genetic diagnosis of hydatidiform moles. *Virchows Arch*. 2024;484(3):401–422.
95. Castrillon DH, Sun D, Weremowicz S, Fisher RA, Crum CP, Genest DR. Discrimination of complete hydatidiform mole from its mimics by immunohistochemistry of the paternally imprinted gene product p57KIP2. *Am J Surg Pathol*. 2001;25(10):1225–1230.
96. Lu B, Ma Y, Shao Y, Xu E. Twin pregnancy with complete hydatidiform mole and co-existing fetus: A report of 15 cases with a clinicopathological analysis and DNA genotyping. *Pathol Res Pract*. 2022;238:154116.
97. Murphy KM, Carrick K, Gwin K, et al. Rare complete hydatidiform mole with p57 expression in villous mesenchyme: Case report and review of discordant p57 expression in hydatidiform moles. *Int J Gynecol Pathol*. 2022;41(1):45–50.
98. Gaillot-Durand L, Patrier S, Aziza J, et al. p57-discordant villi in hydropic products of conception: A clinicopathological study of 70 cases. *Hum Pathol*. 2020;101:18–30.
99. Hui P. Molecular diagnosis of gestational trophoblastic disease. *Expert Rev Mol Diagn*. 2010;10(8):1023–1034.
100. Slim R, Fisher R, Milhavet F, et al. Biallelic NLRP7 variants in patients with recurrent hydatidiform mole: A review and expert consensus. *Hum Mutat*. 2022;43(12):1732–1744.
101. Nguyen NMP, Ge ZJ, Reddy R, et al. Causative mutations and mechanism of androgenetic hydatidiform moles. *Am J Hum Genet*. 2018;103(5):740–751.
102. Stichelbout M, Devisme L, Franquet-Ansart H, et al. SALL4 expression in gestational trophoblastic tumors: a useful tool to distinguish choriocarcinoma from placental site trophoblastic tumor and epithelioid trophoblastic tumor. *Hum Pathol*. 2016;54:121–126.
103. Kaur B, Short D, Fisher RA, Savage PM, Seckl MJ, Sebire NJ. Atypical placental site nodule (APSN) and association with malignant gestational trophoblastic disease; a clinicopathologic study of 21 cases. *Int J Gynecol Pathol*. 2015;34(2):152–158.

SECTION XI

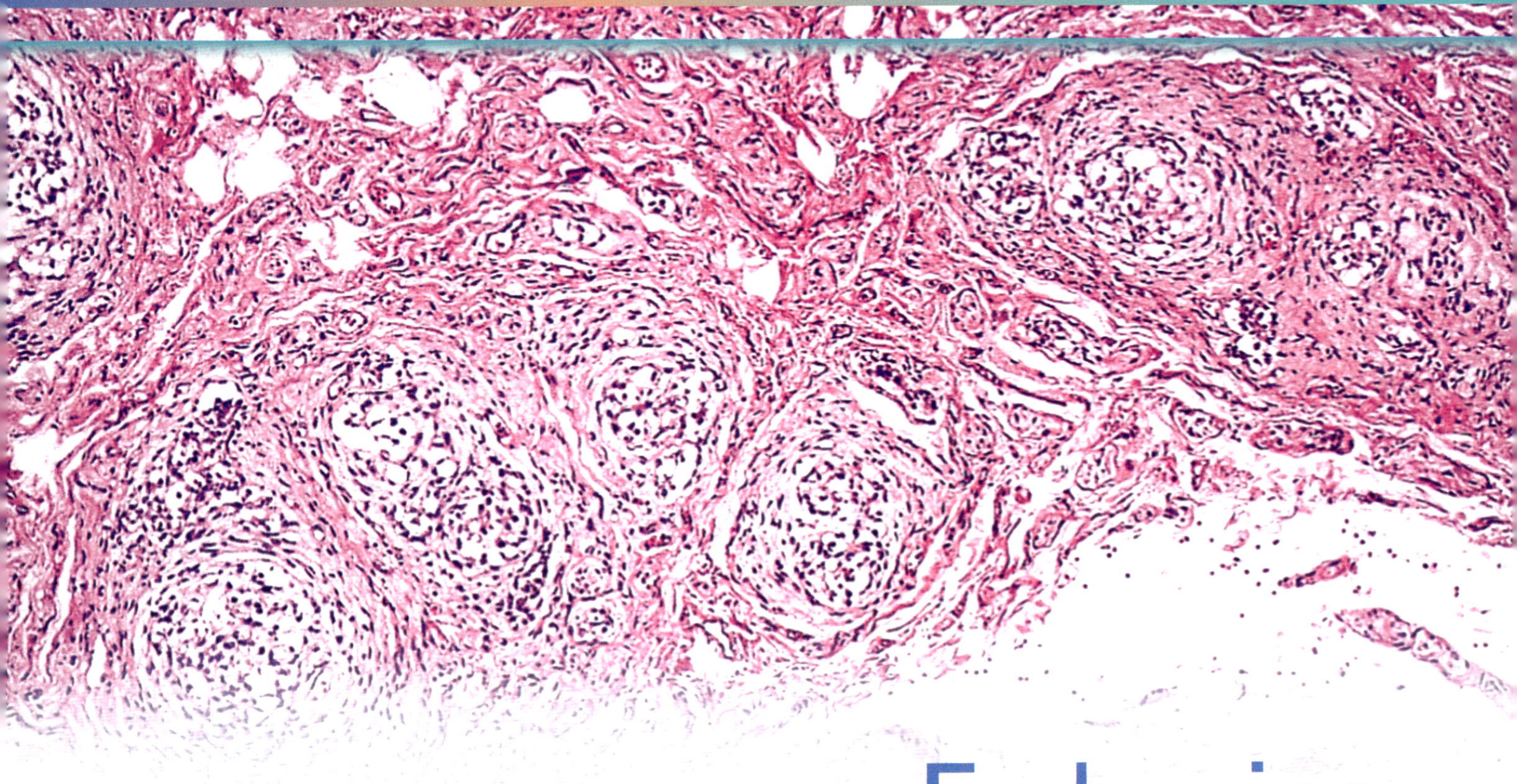

Endocrine

43

Thyroid

Theresa Scognamiglio

EMBRYOLOGY

The human thyroid first appears as a median anlage (the main anlage), and two lateral anlagens. The median anlage develops in the floor of the primitive pharynx at the foramen cecum (a dimple-like depression at the base of the tongue) from a median duct–like invagination that grows caudally, the *thyroglossal duct*. This formation contains at its base the developing thyroid gland, which is spherical at first but becomes bilobed after involution of the thyroglossal duct at about 7 weeks of gestation when it assumes its definitive position in front of the trachea (1).

During this downward migration, the thyroglossal duct undergoes atrophy, leaving a vestigial pyramidal lobe in about 40% of individuals. Faulty downward migration of the thyroid medial anlage, or persistence of parts of the thyroglossal duct, can give rise to ectopic thyroid tissue in the neck, thyroglossal duct cysts, and cervical fistulae. Microscopically, the initially solid thyroid *anlage* begins to form cords and plates of follicular cells during the 9th week of gestation. Small follicles appear by the 10th week. Inside these primitive follicles, a finely granular material begins to collect and acquires the morphologic features of colloid by the 20th week. By week 14, there are well-developed follicles containing colloid (Fig. 43.1A). Secondary follicles arise by budding from the primary follicles; they increase in number until the embryo reaches a length of about 160 mm. After this time, the follicles increase in size, but their number remains the same; the adult thyroid only forms new follicles under intense stimulation. Both the cytoplasm of the follicular cells and the intraluminal colloid contain thyroglobulin (Fig. 43.1B). Labeled amino acid studies have shown that thyroglobulin synthesis actually begins at an early stage when the thyroid is still a solid mass at the base of the tongue and long before follicle formation and colloid secretion can be identified morphologically (2–5). The fetal thyroid gland develops rapidly until the 4th month of intrauterine growth (crown–rump length 18 mm). After birth, the thyroid growth rate parallels that of the body, reaching the normal adult weight at around 15 years of age.

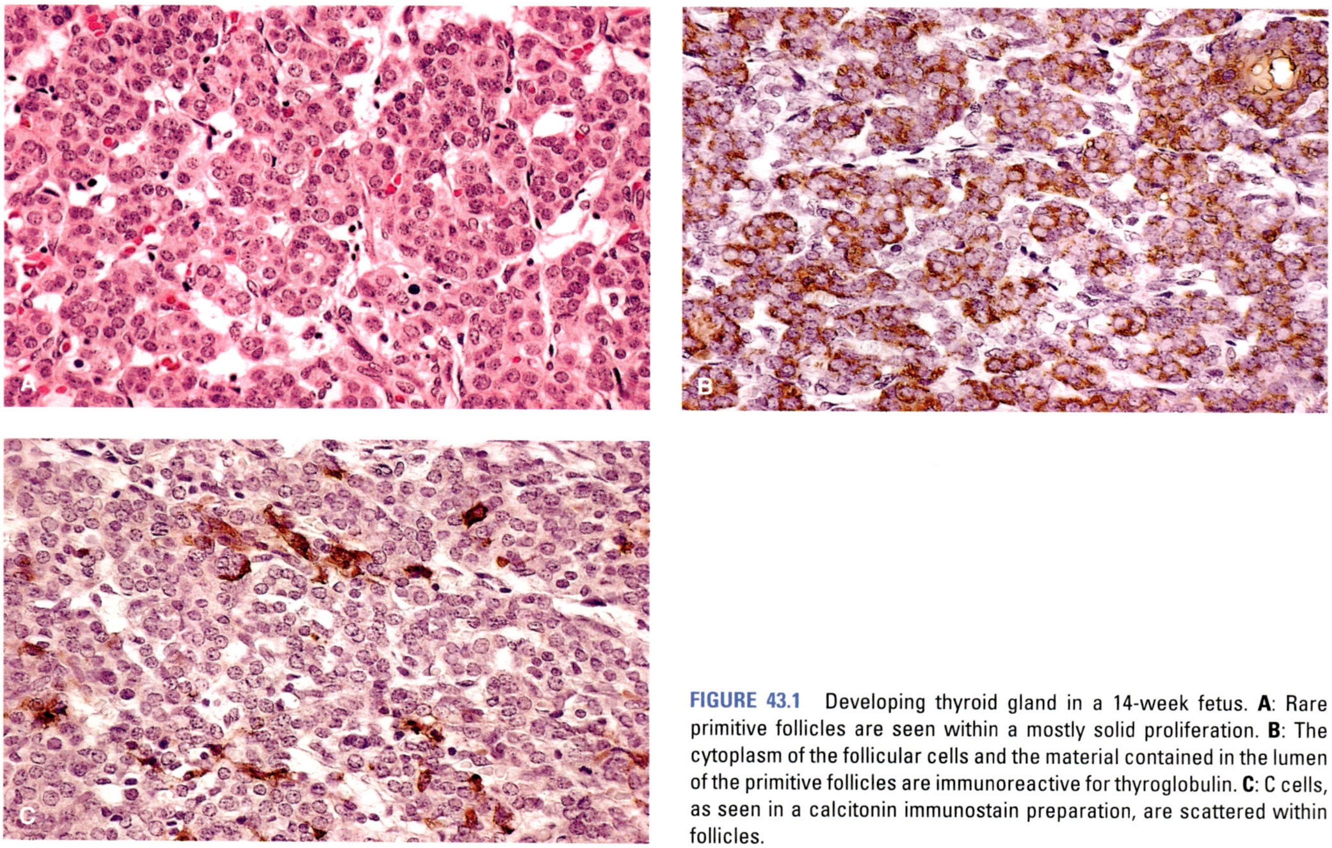

FIGURE 43.1 Developing thyroid gland in a 14-week fetus. **A**: Rare primitive follicles are seen within a mostly solid proliferation. **B**: The cytoplasm of the follicular cells and the material contained in the lumen of the primitive follicles are immunoreactive for thyroglobulin. **C**: C cells, as seen in a calcitonin immunostain preparation, are scattered within follicles.

A series of intrinsic or cell-autonomous transcription factors contribute to thyroid gland morphogenesis and follicular cell differentiation (6,7), while thyroid-stimulating hormone (TSH) influences thyroid differentiation only after the anatomic outline of the thyroid gland is well established. The most important of these transcription factors are Hhex, thyroid transcription factor 1 (TTF-1) (Nkx2-1), Pax8, and TTF-2 (Foxe1) which may be considered, collectively, as a thyroid signature within the anterior foregut endoderm (8).

Although these factors are also expressed and influence differentiation in other developing tissues, all four are coexpressed and involved in a transcriptional network of interactions of mutual dependence only in the thyroid anlage (9). Given that they regulate the expression of thyroid-specific genes, such as those responsible for the production of thyroid peroxidase and thyroglobulin, they are important not only for organogenesis but also for the functional differentiation of the thyroid gland in later stages of prenatal development, as well as postnatally (4,10–12).

Pax8 has been demonstrated in the human median thyroid anlage, in the thyroglossal duct, and in the ultimobranchial bodies (13). TTF-1 is also expressed in the median thyroid anlage (14). *Ghrelin*, a growth hormone–releasing hormone, is expressed in the follicular cells of fetal thyroid gland by immunohistochemistry (from 8 to 38 weeks of gestational age) (15). Fetal thyroid stromal tissue is immunoreactive for galectin 1 but not for galectin 3 (16).

The two lateral anlagens of the thyroid derive from the ultimobranchial bodies which in turn originate from the IVth to the Vth branchial pouch complex before the incorporation of the latter in the thyroid. The ultimobranchial bodies, while still connected to the pharynx, start their migration downward on each side of the neck together with the parathyroid IV anlage (branchial pouch stage, 3 to 12 mm, 5 to 7 weeks); at 7 to 8 weeks, they separate from the pharynx and the parathyroid (separation stage, 13 to 17 mm, 7 to 8 weeks); their lumens become obliterated by proliferating cells, so that at 8 to 9 weeks they appear as solid masses that fuse with the dorsolateral aspects of the median thyroid anlage and become incorporated into the developing lateral lobes (incorporation stage, 18 to 27 mm, 8 to 9 weeks). After fusion with the medial thyroid (at 9 weeks to term), the ultimobranchial bodies enter the dissolution phase and divides into a central thick-walled stratified epithelial cyst and a peripheral component composed of cell groups dispersed among the follicles, the C cells (28 to 520 mm, 9 weeks to term) (Fig. 43.1C). The central portion is represented by a stratified epithelial-lined cyst, whereas the peripheral portion is dispersed into a few cell groups that eventually become cystic. In postnatal life, the central epithelial cyst largely disappears, its occasional remnants corresponding to the solid cell nests (17,18).

Although data from plentiful embryologic, histochemical, and immunohistochemical studies suggest that mammalian and avian C cells are derived from the neural crest and that they migrate to the ultimobranchial bodies before their incorporation into the thyroid (19–23) the very concept of a neural crest origin of thyroid C cells is currently undergoing a major reconsideration. Recent lineage tracing data from genetic studies in mice postulate that the true progenitors to C cells arise in the endodermal germ layer, pointing to a unifying origin of all cells of the ultimobranchial bodies. According to this hypothesis, the ultimobranchial bodies may turn out to be the embryonic source of C-cell precursors and not only their carriers, not differently from other organs such as lung, gall bladder, and pancreas, all of which contain neuroendocrine cells of endodermal origin. Several transcription factors and signaling molecules involved in the development of C cells have been identified, and genes expressed in the pharyngeal pouch endoderm, neural crest-derived mesenchyme in the pharyngeal arches, and ultimobranchial bodies have been demonstrated to play critical roles for the development of C cells. Immunohistochemically, E-cadherin, an epithelial cell marker which is expressed in thyroid C cells, has been demonstrated to be colocalized with calcitonin in C cells and to be present in C-cell precursors, like the fourth pharyngeal pouch and the ultimobranchial bodies (24–28).

It has been pointed out that a unifying origin of thyroid follicular cells and C cells, even though from different endoderm domains, might contribute to the histogenesis of mixed thyroid tumors (28). Furthermore, the discovery that thyroid C cells developed from endoderm may open up new directions in the search for potential targets to treat C cell–derived tumors (29).

GROSS ANATOMY

The normal adult thyroid gland is located in the midportion of the neck, immediately in front of the larynx and trachea, where it is attached to the anterior trachea by loose connective tissue termed the lateral suspensory ligament or ligament of Berry.

The two lateral lobes surround the ventral and lateral aspects of the larynx and trachea, reaching the lower halves of the thyroid cartilage and covering the second, third, and fourth tracheal rings.

The parathyroid glands are usually located adjacent to the posterior surfaces of the thyroid lobes. The recurrent laryngeal nerves run in the cleft between the trachea and esophagus just medial to the thyroid lobes.

The normal thyroid has a shape reminiscent of a butterfly, with two bulky lateral lobes connected by a thin isthmus. Each lateral lobe is 2 to 2.5 cm wide, 5 to 6 cm long, and 2 cm deep. Their upper and lower aspects are referred to as upper and lower thyroid poles, respectively. The upper poles have a pointed shape whereas the lower poles have blunt contours. One lobe, usually the right, may be larger than the other and the isthmus may be exceptionally wide. The pyramidal lobe, a vestige of the thyroglossal duct, is found in about 40% of thyroids; it appears as a narrow projection of thyroid tissue that extends upward from the isthmus to lie on the surface of the thyroid cartilage. A gross enlargement of the pyramidal lobe may result from a diffuse pathologic process such as hyperplasia or Hashimoto thyroiditis.

The normal weight of the adult thyroid is 15 to 25 g in nongoitrous areas. However, there are significant individual variations, most of which are related to gender, age, corporal weight, hormonal status, functional status of the gland, and iodine intake (30). For instance, the thyroid gland is larger and heavier in women than in men, and it becomes even larger during pregnancy and in the secretory phase of the menstrual cycle (31).

A thin fibrous capsule invests the thyroid. Connected to this capsule are numerous fibrous septa that penetrate the thyroid parenchyma and divide it into lobules (i.e., thyromeres). Although the thyroid capsule is grossly complete, it may be microscopically incomplete in up to 62% of adults (32). Thyroid follicles may also be found within the capsule in 14% of healthy patients, and in the pericapsular connective tissue in 88%, often as nodular aggregates.

The normal thyroid is red-brown. A phenomenon exceptionally seen in otherwise normal thyroid glands of elderly individuals is the accumulation in the follicular cells of a melanin-like pigment that imparts to the gland a characteristic coal-black stain, easily apparent on gross examination. The terms *melanosis thyroidi* and *black thyroid* are used to refer to this phenomenon. These changes are qualitatively identical to those seen in more florid form in thyroids of patients on chronic minocycline therapy (33). Ultrastructurally some of the granules contain lipofuscin-type pigment. Pigment is also often present in the colloid in these cases, thus becoming ambilysosomes (34). An association between black thyroid and thyroid carcinoma has been reported. Kandil found thyroid carcinoma, predominantly of papillary type, in 65% of patients with black thyroid (35,36). The papillary carcinomas arising within black thyroids typically show decreased pigmentation of the malignant cells compared to the surrounding thyroid tissue. Abnormal thyroid peroxidase within the tumor cells has been suggested as a possible cause of this phenomenon (37).

Nodularity of thyroid parenchyma is identified grossly in about 10% of the glands of endocrinologically normal individuals and its prevalence is increased in older patients (38).

The blood supply of the thyroid gland derives primarily from the inferior thyroid artery which originates from the thyrocervical trunk of the subclavian artery, as well as the superior thyroid artery derived from the external carotid artery. A thyroid middle artery also may be variably present, ranging from an inconspicuous vessel to one the size of the inferior thyroid artery. The superior and medial thyroidal veins and the inferior vein drain via a venous plexus in the thyroid capsule into the internal jugular and the brachiocephalic veins, respectively (39,40).

An intricate lymphatic network permeates the thyroid gland, encircling the follicles and connecting the two lateral lobes through the isthmus. In neonates and children, these lymph vessels may appear as empty, elongated, tortuous spaces that simulate a retraction artifact, yet they are lined by lymphatic endothelial cells that are immunopositive for D2-40 (41). This network empties into subcapsular channels, which in turn give rise to collecting trunks within the thyroid capsule in close proximity to the veins. They empty into the regional lymph nodes, as described below:

- The *pericapsular* nodes. Whole organ sections of the thyroid have shown that the intraglandular lymph vessels penetrate the capsule and merge with the pericapsular lymph nodes, forming a plexus around the gland (42).
- The *internal jugular chain* nodes (including the *subdigastric* nodes). These lymph nodes receive lymph from vessels draining the superior portion of the thyroid lobes and isthmus.
- The *pretracheal, paratracheal*, and *prelaryngeal nodes.* These lymph nodes drain the collecting lymph vessels from the inferior portion of the gland. The prelaryngeal node located near the thyroid isthmus is sometimes referred to as the *Delphian node* (43).
- The *recurrent laryngeal nerve chain* nodes.
- The *retropharyngeal* and *retroesophageal* nodes. The *anterosuperior mediastinal* nodes. These lymph nodes are secondary to the recurrent laryngeal nerve chain and pretracheal groups. However, studies have shown that dye injected into the thyroid isthmus can also drain directly into them (44).

Some correlations exist between the site of a thyroid tumor and the location of the initial lymph node metastasis. However, the overlap in drainage to these various nodal groups is such that metastases may be found in any of them regardless of the precise location of the primary tumor. The most frequent pattern of lymph node spread from papillary thyroid carcinoma is initially to the central lymph node compartment (Level VI) followed by the lateral neck. In a retrospective study of patients who presented with clinically positive neck lymph nodes, 95% of patients had Level VI lymph node involvement, while between 54% and 68% of the patients had Levels II–IV involvement (45).

Vasomotor nonmedullated postganglionic neural fibers originating from the superior and midline cervical sympathetic ganglia influence indirectly the secretory activity of the thyroid gland through their actions on the blood vessels. In addition, adrenergic receptors in follicular cells and a network of adrenergic fibers ending near the follicular basement membrane have been demonstrated (46). It has been hypothesized, therefore, that thyroid secretion is regulated both by direct neural signals and by indirect vascular nerve signals (47–49). A role for direct neural influences in the secretion of calcitonin and other C cell–derived hormones is supported by the demonstration of a rich cholinergic network encircling the C cells in chickens (50).

Small *paraganglia* are normally present close to the thyroid and are occasionally found beneath the thyroid capsule (51). Their presence explains the rare occurrence of peri- and intrathyroidal paragangliomas (52,53).

MICROSCOPIC ANATOMY

The fundamental unit of the thyroid is the follicle, a round to slightly oval structure lined by a single layer of epithelial cells resting on a basement membrane. The lumen of the follicle contains colloid, a viscous material that is mostly composed by proteins secreted by the follicular cells, including thyroglobulin. The follicles, which are separated from each other by a loose fibroconnective tissue, have an average diameter of 200 μm. Their size may vary even within the same gland depending on the functional status of the thyroid and the age of the individual. Variations in the shape of follicles exist, but elongated follicles are a feature of hyperplastic or neoplastic conditions or are the result of compression adjacent to an expansile mass. A characteristic structure present in the normal thyroid but more often seen in hyperplastic conditions is the *Sanderson pollster*, as discussed subsequently.

The colloid, which is pale eosinophilic in the actively secreting gland, acquires a deeply eosinophilic staining quality in resting follicles. Often, numerous clumps with various shapes appear within the colloid of resting follicles, suggesting an artifactual coagulation-type phenomenon (Fig. 43.2). In some follicles, the colloid may have an amphophilic or basophilic staining quality, probably the result of an increase in the amount of acidic groups in the thyroglobulin molecule (Fig. 43.3). In the most advanced expression of this phenomenon, the intraluminal material acquires a distinct mucin–like appearance (Fig. 43.4). The glycoprotein material present within the follicles stains for periodic acid–Schiff (PAS) and Alcian blue and is

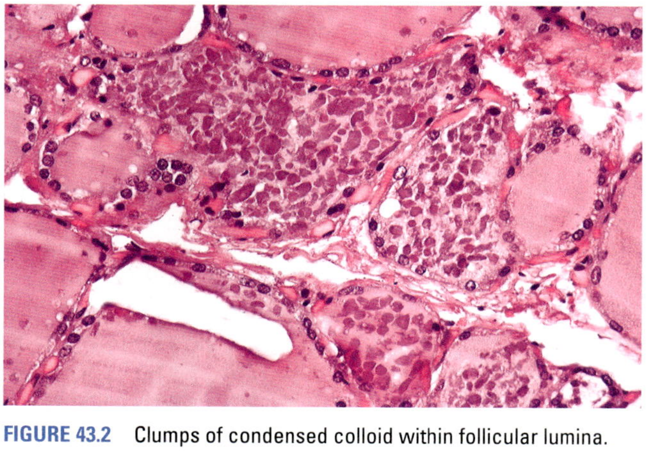

FIGURE 43.2 Clumps of condensed colloid within follicular lumina.

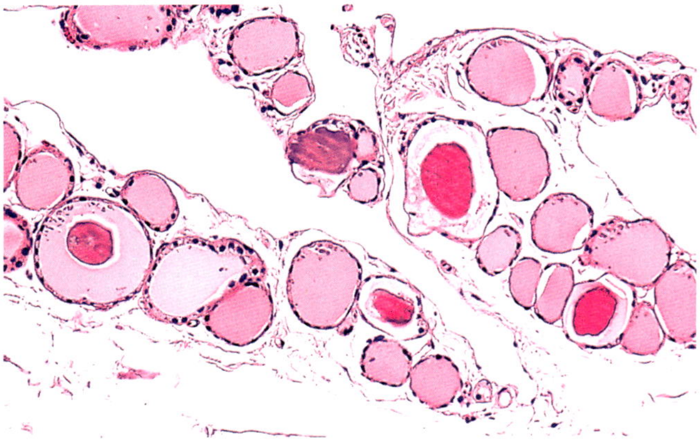

FIGURE 43.3 The colloid accumulated in these follicles exhibits different densities and tinctorial qualities, the latter ranging from acidophilic to basophilic.

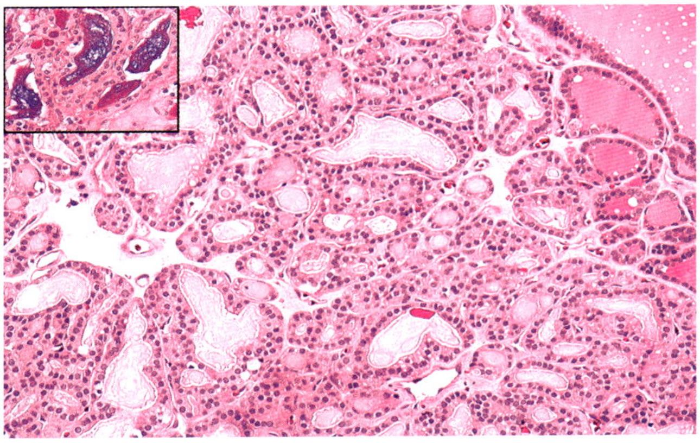

FIGURE 43.4 The basophilic colloid present in most of these follicles contrasts with the more typical red colloid present in the follicles on the right upper corner. *Inset:* The Alcian blue–PAS stain highlights the mucinous character of the basophilic colloid.

immunoreactive for thyroglobulin. A row of small vacuoles is seen at the interface between the follicular epithelium and the colloid in actively functioning glands; these are referred to as reabsorbing vacuoles. In addition, it is not unusual to find a large round or oval clear space within the colloid; this often appears empty, but it may contain *birefringent calcium oxalate crystals* (see page 1155).

Another morphologic variation of the colloid is represented by collections of round basophilic acellular corpuscles clustering at one pole of the follicle.

The epithelial cells lining the follicle are known as follicular cells or thyrocytes; among them, there is a second cellular component of C cells.

FOLLICULAR CELLS

The cells lining the follicles show variations in their shape and size according to the functional status of the gland. Three major types, expressions of a morphologic continuum, are described: flattened, cuboidal, and columnar (Fig. 43.5A). Flattened, or endothelioid, cells are relatively inactive. Cuboidal cells have a height equaling their width and are the most numerous. Their major function is to secrete colloid. The rarer columnar, or cylindrical, cells reabsorb the thyroglobulin-containing colloid, liberate the active hormones, and excrete these hormones into blood vessels. They may feature an apical cuticle, apical lipid droplets, and one or more basilar vacuoles known as the vacuoles of Bensley.

Functional polarity is apparent at the level of the follicle and the follicular cell. A single follicle may have flattened cells on one side and cuboidal or low columnar cells on the other (Fig. 43.5B), the best expression of this phenomenon has been termed the *Sanderson polster*. At the cellular level, all follicular cells manifest a definite polarity, resting with their bases on the basement membrane and having the apexes directed toward the lumen of the follicle. Size and position of the nucleus and some components of the

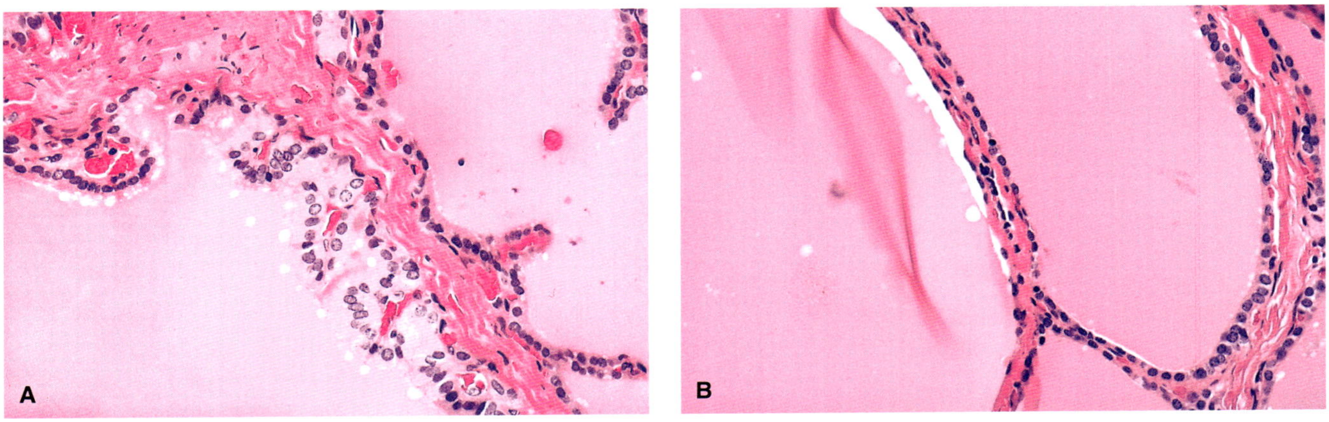

FIGURE 43.5 **A**: The epithelium of one follicle is low cuboidal and relatively inactive. The adjacent follicle shows a taller epithelium and reabsorption vacuoles. **B**: The epithelium of the same follicle is flattened on one side and cuboidal on the other, as an expression of functional polarization.

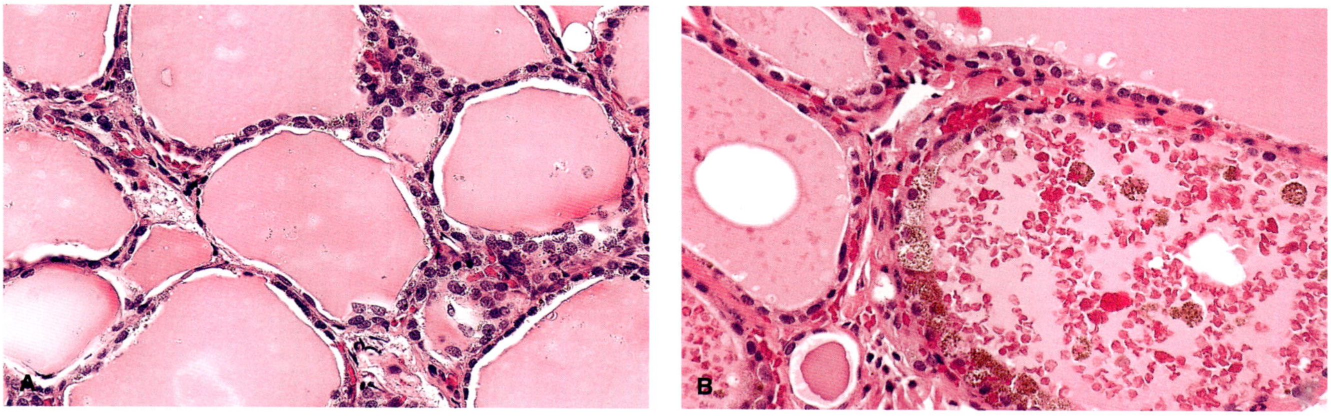

FIGURE 43.6 **A**: Lipofuscin in the cytoplasm of the follicular cells. **B**: Granular black pigment in the follicular epithelium and colloid of the thyroid of a 73-year-old patient who was not on minocycline therapy.

cytoplasm may vary considerably. In the resting thyroid, the nucleus is round or ovoid, located toward the center of the cell, and usually contains one eccentrically located nucleolus. Its chromatin may be finely granular or clumped. The nuclei of actively secreting cells are enlarged and basally located with an increase in the amount of apical cytoplasm. The cytoplasm is usually weakly eosinophilic; only exceptionally in an otherwise normal thyroid does it appear granular and, oncocytic (i.e., Hürthle cells). In contrast to parathyroid cells, little or no intracytoplasmic glycogen is present in them. Occasionally, the follicular cell cytoplasm contains a golden–brown pigment of lipofuscin type (Fig. 43.6A), which should be distinguished from the melanin-like pigment (Fig. 43.6B).

Ultrastructurally, the follicular cells are arranged in a single layer around the colloid and rest on a basement membrane that separates them from the microvilli emanating from the cell surfaces. Cell membranes of adjacent cells interdigitate in a complex fashion and are joined by junctional complexes toward the apex (54,55). The cytoplasm contains variable amounts of endoplasmic reticulum, mitochondria, and lysosomes. When the number of mitochondria is highly increased, the cytoplasm acquires an intensely eosinophilic granular appearance corresponding to Hürthle cells.

Immunohistochemistry

A wide variety of markers with various degrees of specificity and diagnostic significance are expressed by normal adult follicular cells.

Thyroglobulin (TGB)

This is probably the most specific immunohistochemical marker for follicular cells and the tumors composed of them. It can be demonstrated with either monoclonal or polyclonal antibodies and the reactivity is both in the cytoplasm and in the colloid (56,57). Oncocytes show a much lesser degree of positivity. Despite its high specificity, thyroxin-binding globulin can give rise to a common pitfall. Because of its tendency to leak out from the cytoplasm of the follicular cells and to diffuse into the adjacent tissues, it can be incorporated into the cytoplasm of other types of cells (e.g., metastatic carcinoma) and cause a false positivity (58).

Thyroid Transcription Factor 1 (TTF-1)

This is another very useful marker for thyroid follicular cells and tumors composed of them. This nuclear transcription factor, first identified in thyroid follicular cells and later in lung epithelium (pneumocytes), together with TTF-2 and Pax8, is necessary for thyroid organogenesis and differentiation (6–8). In the normal adult thyroid, its distribution is related to that of thyroglobulin and thyroperoxidase (6–8). It is consistently expressed in all types of primary and metastatic thyroid carcinomas including medullary carcinoma, although expression is often lost in anaplastic carcinomas. In fact, less than 20% of anaplastic carcinomas show retained expression of this marker (59).

Paired Box Gene 8 (PAX-8)

Along with TTF-1 and thyroglobulin, PAX8 can help confirm thyroid follicular cell differentiation. PAX-8 is a 48 kD member of the paired box family of transcription factors. While expression has been reported in medullary tumors, intrathyroidal thymic carcinomas, and parathyroid lesions, the use of a monoclonal antibody reduces the nonspecific staining and increases the specificity for thyroid follicular cell origin. In contrast to TTF-1, the expression of PAX-8 is usually maintained in anaplastic carcinomas and other high-grade tumors (59).

Keratins

The normal thyroid gland strongly expresses simple epithelial-type CKs 7 and 18 and, to a lesser degree, CKs 8 and 19, but not stratified epithelial-type CKs (60). Papillary and follicular carcinomas share the expression of simple epithelial-type CKs 7, 8, 18, and 19. CK19 is expressed particularly strongly by papillary carcinoma. The latter also usually express stratified-type CKs 5/6 and 13. This reactivity pattern of papillary carcinoma is useful for identifying small tumor foci not easily identified by conventional histologic examination and is maintained in the metastatic foci. Unfortunately, this rather distinct pattern of CK distribution in papillary carcinoma is often shared, at least focally, by the thyroid follicles of Hashimoto thyroiditis (61).

Vimentin

Some normal follicular cells occasionally express this intermediate filament in conjunction with keratin, although less commonly than in neoplastic conditions (62,63).

Epithelial Membrane Antigen (EMA)

Follicular cells are variably positive, with accentuation of the cell membranes.

Galectin

The galectins constitute a family of lactose-binding soluble lectins sharing affinity for β-galactoside residues and a significant sequence similarity in their carbohydrate-binding site. They have been implicated in cell growth and differentiation, intercellular recognition, and adhesion. The most extensively studied galectins are galectin-1 and galectin-3. The latter has been studied in great detail in the thyroid gland, where it stains carcinomas (especially those of papillary type) in a stronger and more consistent fashion than adenomas (64). It is also often focally expressed in the follicles of lymphocytic/Hashimoto thyroiditis, nodular hyperplasia, and follicular adenoma (65).

Triiodothyronine (T3) and Thyroxine (T4)

These hormones can be detected immunohistochemically both in the cytoplasm of the follicular cells and in the intraluminal colloid, but they are rarely used for diagnostic purposes, having been replaced for this purpose with thyroglobulin.

Estrogen and Progesterone Receptors

Estrogen and progesterone receptor positivity, restricted to the follicular cell nuclei, shows some correlation with the age and sex of the individual (66,67).

S100 Protein

This marker, which is mainly detected in inflammatory/hyperplastic and neoplastic thyroid conditions, is only focally and weakly expressed by normal follicular cells (68).

Claudins

Claudins and occludin are integral constituents of tight junctions and are deregulated in a variety of malignancies. They are present in the normal thyroid and in most thyroid neoplasms with different levels of expression (69).

Epidermal Growth Factor Receptor (EGFR)

The location of this receptor in normal follicular cells is mainly basal or basolateral reflecting functional polarization (70).

Thyroid Peroxidase

This enzyme is responsible for the oxidation of iodide to iodine. Immunohistochemically, it shows a pattern of staining correlated with the age of the individual. A cytoplasmic pattern of staining with apical membrane accentuation is seen in children and young adults, and a perinuclear ring distribution is seen in older individuals (71).

Sodium Iodide Symporter

At the immunohistochemical level, this molecule, responsible for the active iodide intake into the follicular epithelium, is localized mainly to the lateral basal portion of the cells (72,73).

Physiology

The main function of the thyroid gland is the production of thyroid hormones, the most important being thyroxin (T4) and triiodothyronine (T3). These hormones regulate metabolism, increase protein synthesis in every tissue of the body, and increase O_2 consumption. Thyroid hormones are particularly important for body development and for the normal maturation of the central and peripheral nervous systems.

Steps in thyroid hormone biosynthesis include ingestion of iodine ions from water and food, their absorption and transport as iodide into the extracellular fluid, and their concentration within the thyroid, where their intracellular levels are 30 times higher than in peripheral blood. The active iodide uptake across the basement membrane is mediated by human sodium iodide symporter (hNIS) in a process coupled with the flow of sodium (72). The intrathyroidal iodide is then oxidized to iodine. This last step is dependent on the action of iodide peroxidase, which oxidates the iodine ion to a highly reactive form of iodine, which in turn binds to tyrosine. The results are monoiodotyrosine when one iodine molecule is attached, and diiodotyrosine when two iodine molecules are attached. The iodotyrosine residues are condensed to form the biologically active thyroid hormones *thyroxin (T4)* and *triiodothyronine (T3)*. Thyroxin results from the coupling of two molecules of diiodotyrosine, and triiodothyronine from the coupling of one molecule of monoiodotyrosine with a molecule of diiodotyrosine (73). Thyroid hormones with numerous iodinated tyrosine residues, including biologically active T4 and T3 are stored in thyroglobulin which plays an essential role as a carrier protein of these hormones.

Thyroglobulin, a large protein with a 19 S sedimentation coefficient and a molecular weight of 670,000, is formed by two identical subunits with a 12 S sedimentation coefficient to which many oligosaccharides are linked. Variations in the sugar chains of the thyroglobulin molecule have been evaluated by the analysis of reactivity to various lectins and found to differ between the normal gland and various pathologic states, including neoplasms (74).

Thyroglobulin is encoded by a gene spreading over more than 200 kilobases in the bovine genome (75). The molecular mechanisms involved in the tissue-specific and hormone-dependent expression of the thyroglobulin gene have been studied in follicular cells in primary cultures and cell lines (76,77). Thyroglobulin is collected at the center of the thyroid follicles and is the main constituent of colloid.

Ultrastructural studies have correlated the morphologic changes that accompany thyroid hormone production and secretion. The synthesis of thyroglobulin begins in the endoplasmic reticulum and continues in the Golgi apparatus, where the end sugars of the carbohydrate site are incorporated; it is then packaged in small apical microvesicles, the contents of which are discharged into the follicular lumen after fusion of the vesicle membranes with the luminal side of the plasma membrane.

Resorption of thyroglobulin takes place through cytoplasmic pseudopodia that engulf minute portions of colloid, which are then drawn into the cell in the form of membrane-bound colloid droplets. These subsequently fuse with lysosomes, and their contents are digested by the lysosomal enzymes. The breakdown products, including T3 and T4, diffuse through the cell membrane and the basement membrane into the adjacent capillaries and most of the molecules become bound to a specific carrier protein known as thyroxin-binding globulin (77–79). Thyroxin-binding globulin normally transports more than 70% of thyroid hormones (78–80). Approximately 20% of circulating thyroid hormones are carried by transthyretin (prealbumin) and albumin (81).

Only a small portion of circulating thyroid hormones (approximately 0.05% of T3 and 0.015% of T4) is unbound and, therefore, biologically active. Free, circulating, biologically active T3 and T4 are in equilibrium with the hormones bound to the carrier proteins. The amount of circulating T4 is much larger than that of T3; however, T3 is about four times more active biologically; as a result, the final contribution of T3 to the biologic activity of thyroid hormones equals that of T4 (80).

Thyroid hormones act by binding to specific thyroid hormone receptors present in nearly all tissues. Thyroid hormones stimulate metabolism, increase oxygen consumption, and cause a rise in heat production, cardiac output, and heart rate. They are essential for normal development, growth, and maturation. The acceleration of growth may result from a direct action on the cells to increase their rate of division, by acting permissively for other hormones, or by inducing the synthesis of a variety of growth-promoting hormones (78,80,82–87).

Thyroid biosynthetic and secretory activities are controlled by the blood level of TSH, a glycoprotein synthesized and secreted by the anterior pituitary gland (88). TSH binds to thyrotropin receptors located on the basolateral surface of the follicular cell membrane, and by activating the adenylate–cyclase pathway regulates the complex mechanism responsible for T3 and T4 synthesis (81,89,90).

Stimulation of the thyroid gland by thyrotropin increases its secretory activity and vascularity and results in both hypertrophy and hyperplasia of follicular cells, accompanied by reduction of colloid storage. At the functional level, this is reflected by an increase in iodide concentration and organic binding, hormone synthesis, and hormone secretion (80,91,92).

TSH release is in turn regulated by a tripeptide secreted by the hypothalamus, thyrotropin-releasing hormone (TRH which is produced by the neurons of the medial-basal hypothalamus and carried into the pituitary gland via the hypophyseal portal vessels.). TSH and TRH releases are regulated by the circulating levels of free T3 and T4, via negative feedback on the pituitary gland and hypothalamus (low levels of free T3 and T4 stimulate the release of TSH and TRH). In contrast, TSH and TRH releases are inhibited by high levels of circulating free T3 and T4 (88,91,92).

Microscopic Variations

Sanderson Polster

A characteristic structure, present in the normal thyroid and accentuated in hyperplastic conditions, is the *Sanderson polster* (Fig. 43.7). This term refers to an aggregate of small follicles lined by flattened epithelium and covered by an undulating layer of columnar epithelium that is seen as bulging into the lumina of larger follicles. This perfectly benign and to some extent physiologic change, most likely represents the morphologic expression of the functional polarization of the thyroid follicle, and should be distinguished from intracystic papillary microcarcinoma.

Granulomata

Granulomata are relatively common in otherwise normal surgically resected thyroids and in autopsy specimens. Both foreign material and colloid may elicit this process. Suture material is the most frequent cause of granuloma formation in completion thyroidectomy specimens. Larger foreign-body granulomata sometimes clinically simulate a thyroid nodule in patients with prior laryngeal injection of Polytef (polytetrafluoroethylene) (93,94). This material may migrate through the lymphatic vessels into adjacent tissues, where it may elicit an inflammatory process.

Rarely, interstitial granulomata are seen as a reaction to oxalate crystals that have been released by broken follicles (95).

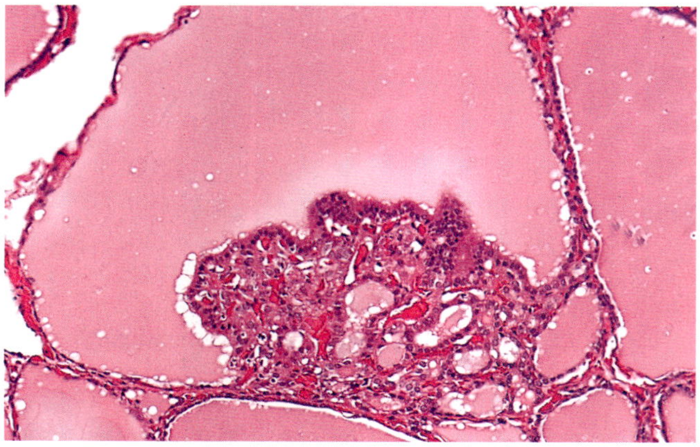

FIGURE 43.7 Sanderson polster protruding into the lumen of a follicle.

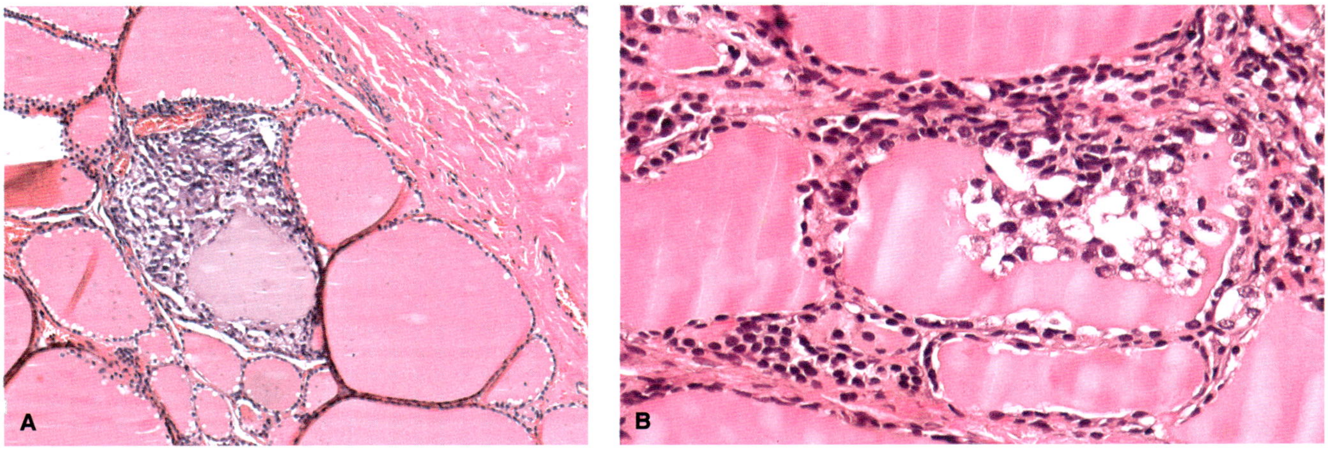

FIGURE 43.8 Palpation thyroiditis. **A**: The thyroid follicle in the center is packed with histiocytes and other inflammatory cells, with a clump of residual colloid in the center. The follicular epithelium is barely discernible. **B**: In this case, the follicle is only partially involved. Inflammatory cells and desquamated follicular cells protrude into the lumen.

Granulomatous lesions originating by the rupture of follicles and their invasion by macrophages and leukocytes as a reaction to the extruded colloid are a common incidental finding in surgically resected thyroids. Carney et al. (96) referred to this process as *multifocal granulomatous folliculitis* or *palpation thyroiditis*, and attributed it to the minor trauma resulting from physical examination of the gland. Support for this interpretation comes from the observation that the number and size of the granulomata is related to the intensity of the palpation and the fact that similar changes have been described in individuals engaged in martial arts (96,97).

Grossly, a gland affected by palpation thyroiditis appears normal or shows tiny foci of hemorrhage. Histologically, multiple small granulomata centered in disrupted follicles and composed of macrophages, lymphocytes, and plasma cells are scattered in the thyroid gland (Fig. 43.8A). Some of the macrophages are foamy, while others have the appearance of multinucleated giant cells. The appearance depends on the stage of the process, a common picture being a cluster of foamy macrophages hanging from the follicular epithelium into the lumen (Fig. 43.8B). Necrosis, hemosiderin, and iron deposition are seen only rarely. Sometimes up to four or five follicles are involved in a single granuloma.

Immunohistochemically, most of the lymphocytes are T cells; among the plasma cells, Kappa-positive cells predominate (98).

Palpation thyroiditis seems to represent a variation in the theme of *colloidophagy*, a process described many years ago and characterized by a granulomatous reaction to colloid in follicles allegedly undergoing spontaneous rupture in thyroids affected by goiter or thyroiditis (99).

Palpation thyroiditis should be distinguished from *interstitial giant cell thyroiditis* in which the granulomata are centered in the interstitium rather than in the follicles, *necrotizing granulomata* following surgical procedures characterized by a central area of necrosis surrounded by a palisading of epithelioid cells, and aggregates of C cells that are immunoreactive for calcitonin (100,101).

Crystals

Anisotropic crystals of calcium oxalate may be present within the colloid in normal adult thyroid glands, particularly in older and/or less active glands. They are best visualized under polarized light (Fig. 43.9). Their shape varies from rhomboid to irregular plaques and their size shows wide variations (102).

Intraluminal crystals are normally present in 88% of nodular goiters, 60% of follicular adenomas, 33% of follicular carcinomas, and only 5% of papillary carcinomas (103). Their overall prevalence is approximately 69% in benign nodules and 8% in malignant nodules, and heavy deposits are almost exclusively seen in benign conditions (103). Their number appears to increase with age and this, together with the observation that the crystals have been found more frequently in colloid with low positivity for thyroglobulin, has prompted some to suggest they result from variations in

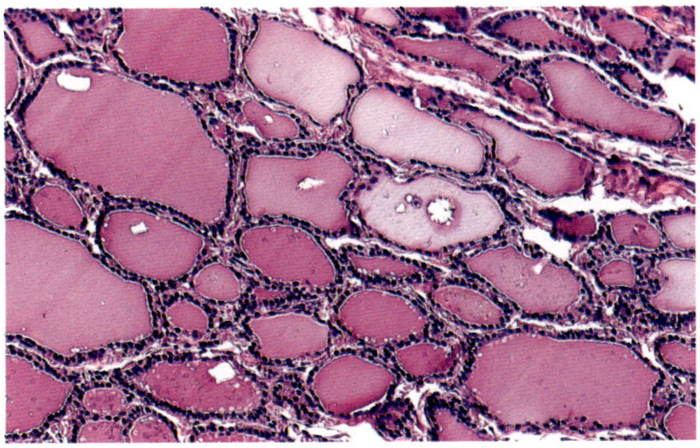

FIGURE 43.9 Multiple birefringent calcium oxalate crystals are seen in the lumina of normal thyroid follicles (polarized light).

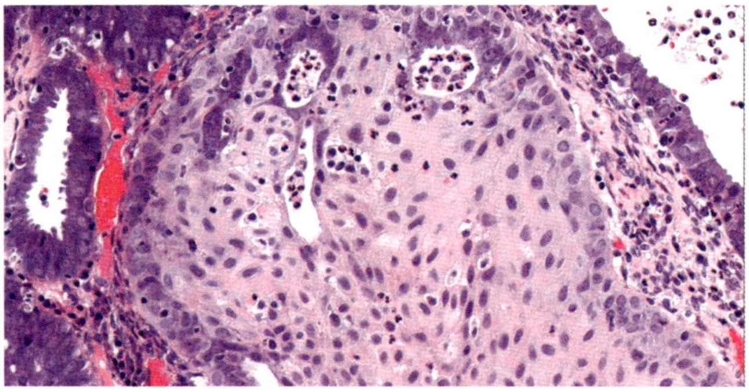

FIGURE 43.10 Squamous metaplasia in a benign follicular proliferation.

colloid and calcium concentrations in the gland secondary to a low functional state (102–104). Crystals can also be found in giant cells, remnants of colloid, and the thyroid stroma in patients with subacute thyroiditis, whereas they are infrequently found in the thyroid glands of patients with chronic thyroiditis or glandular hyperplasia (105).

Oxalate crystals are also seen in large numbers in the thyroid glands of patients undergoing dialysis for renal failure. In this setting, the thyroid is just another site of oxalate deposition, together with the kidney, myocardium, and other organs (106). Rarely, crystals released by follicular breakdown may elicit a granulomatous reaction in the nearby stroma (95). At the time of frozen section, their identification within a follicular structure can be useful in distinguishing thyroid from parathyroid gland tissue (107,108).

Squamous Metaplasia

Benign squamous cells occur as an expression of squamous metaplasia of follicular cells in various benign and malignant thyroid lesions and, under exceptional circumstances, in an otherwise normal thyroid gland (Fig. 43.10) (109–111). They should be distinguished from transversally cut follicles and the solid cell nests of ultimobranchial bodies derivation. Squamous epithelium is regularly observed as a component of the epithelium of thyroglossal duct cysts. Squamous metaplasia may represent a diagnostic pitfall in fine needle aspiration biopsy samples of thyroid nodules, particularly those with potentially benign cystic changes (112).

C CELLS

C cells (parafollicular cells) represent a minor component of the thyroid gland. They comprise not more than 0.1% of the glandular mass and are responsible for the production of the peptide hormone calcitonin. The term "C cells" was introduced by Pearse (113) to underscore their role in secreting and storing this hormone.

Immunohistochemistry for calcitonin is the most reliable method for demonstrating C cells. These polygonal cells have granular, weakly eosinophilic cytoplasm that is paler than that of follicular cells, and their nuclei are round to ovoid with a centrally located nucleolus. C cells can be difficult to identify in hematoxylin and eosin-stained sections.

C cells are located individually or in small groups within thyroid follicles. Most are found at the bases of the follicular cells without contact with the follicular lumen. Electron microscopy has shown that C cells occupy an intrafollicular, rather than interfollicular, position, and are separated from the thyroid interstitium by the follicular basal lamina. The presence of C cells in the interfollicular stroma has not been convincingly demonstrated (114).

Occasional C cells have prominent cytoplasmic processes that extend beyond the adjacent follicular cells. In normal adults and neonates, C cells are restricted to the mid-upper and upper thirds of the lateral lobes of the thyroid, in the area where ultimobranchial bodies fuse with the thyroid median anlage. The number of C cells varies with the development of the gland, being more numerous in younger patients. In one study, up to 100 C cells per low-power field were demonstrated in neonates and children, whereas in adults only a maximum of 10 cells per low-power field were counted (115). In another study, no difference in the number of C cells was found between young and middle-aged patients, while their numbers in the elderly population are variable and often increased with groups of up to 20 or more cells sometimes being observed (116). Other studies have since confirmed that normal adult thyroid glands may contain numerous C cells, sometimes in the form of large nodules (Fig. 43.11) (117). Nodules of C cells tend to be more common in men, particularly in patients over 50 years and, in the absence of disturbances in calcium metabolism and/or a family history of medullary carcinoma, do not increase the risk for medullary carcinoma. Rather, they may represent either a partial failure of embryonic C-cell migration and dispersion within the gland or age-related hyperplasia (117). As already mentioned, C cells tend to aggregate in the vicinity of solid cell nests.

C-cell hyperplasia occurs in patients with MEN2A and MEN2B syndromes and in association with a variety of other disorders. Hyperplasia in the context of a heritable

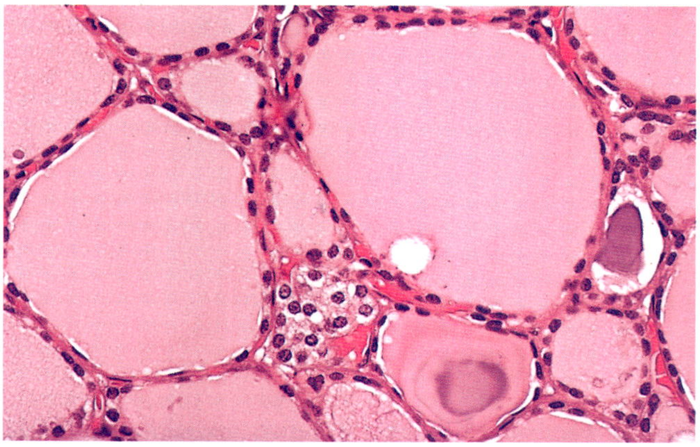

FIGURE 43.11 Clusters of C cells in the thyroid of an elderly individual with no known clinical or laboratory evidence of calcium disturbance.

syndrome is termed primary C-cell hyperplasia and is caused by a germline mutation associated with specific mutations in *RET* proto-oncogene (exon 10,11,16). Hyperplasia associated with other conditions is termed secondary or physiologic C-cell hyperplasia.

Secondary C-cell hyperplasia has been observed in the immediate periphery of nonmedullary thyroid neoplasms (e.g., papillary carcinoma, follicular carcinoma, lymphoma), and in association with a variety of conditions including sporadic medullary carcinoma, Hashimoto thyroiditis, secondary hyperparathyroidism, hypercalcemic states with hypergastrinemia, goitrous hypothyroidism, PTEN-associated tumor syndromes, medications such as cimetidine and estrogens, and after partial removal of the thyroid (118–123). Secondary C-cell hyperplasia, including cases of microdissected C-cell hyperplasia adjacent to sporadic medullary carcinoma, is unassociated with *RET* point mutations (124). Physiologic C-cell hyperplasia is characterized by the presence of ≥50 C cells per low-power (100× magnification) microscopic field; the C cells should be cytologically normal and identified only on the basis of immunostains for calcitonin (125,126).

Patients with primary C-cell hyperplasia have elevated serum levels of calcitonin and carcinoembryonic antigen (CEA). The typical location of the C-cell hyperplasia is in the upper two-thirds of the lateral lobes, although it may be found in other areas of the gland. According to the most recent criteria, primary C-cell hyperplasia is defined by the presence of >6 to 8 C cells per cluster in several foci with >50 C cells per low-power field (126). The hyperplasia may be diffuse or nodular. The C cells appear as groups of large cells, with round to ovoid centrally located nuclei and clear cytoplasm. With further progression, the follicles are filled with expansile foci of proliferating C cells (*nodular C-cell hyperplasia*). Ultrastructurally basal lamina defects are also seen more commonly in nodular hyperplasia.

The term "neoplastic" hyperplasia was proposed by Perry on the basis of the atypia of the proliferating C cells (125). Other authors have interpreted C-cell hyperplasia associated with MEN2 as an authentic preinvasive neoplasm that essentially represents carcinoma in situ of the thyroid gland parafollicular cells (127). Data from molecular studies assessing microdissected foci from thyroid glands from patients with MEN2A provide further evidence in favor of the neoplastic nature of C-cell hyperplasia in the setting of MEN2A. These studies have shown that foci of C-cell hyperplasia are monoclonal with inactivation of the same allele in both thyroid lobes and that they have different secondary alterations involving *TP53*, *RB1*, *WT1*, and *NF1* (128).

The distinction between a benign clonal proliferation of C cells and small medullary carcinoma is based on a combination of features: A nesting, expansile growth pattern, destruction of the follicular basement membrane as evidenced by PAS stain and/or collagen IV immunohistochemistry, fibrosis between the infiltrating C cells in the thyroid interstitium, and diminished intensity of calcitonin immunostaining.

Foci of nodular C-cell hyperplasia may be difficult to distinguish from squamous metaplasia, solid cell nests, intrathyroidal thymic or parathyroid rests, palpation thyroiditis, tangential cuts of follicles, and foci of metastatic carcinoma.

The main ultrastructural characteristic of C cells is the presence of endocrine-type secretory granules that range from 60 to 550 nm in diameter (129). Two main types of granules have been identified. Type I granules have an average diameter of 280 nm and a moderately electron-dense, finely granular content that is closely applied to the limiting membranes of the granules. Type II granules are smaller with an average diameter of 130 nm and more electron-dense contents separated from the limiting membranes by a distinct electron-lucent space. Most normal C cells contain mostly type I secretory granules with no, or few, type II granules. Immunocytochemical studies performed at the ultrastructural level have shown that both type I and II secretory granules contain immunoreactive calcitonin (129).

Histochemistry

Normal C cells are characterized by argyrophilia (130); lead hematoxylin (131); toluidine blue, coriophosphine O (131), and lectin Ulex Europaeus agglutinin (132). These methods, widely used in the past for the identification of C cells, have been largely replaced by immunohistochemical techniques.

Immunohistochemistry

C cells are reactive with antibodies to calcitonin (Fig. 43.12) calcitonin gene–related peptide, katacalcin, somatostatin, substance P, helodermin, gastrin-releasing peptide, TRH, serotonin, low–molecular-weight keratin, chromogranin A and synaptophysin, CEA, vimentin, and TTF-1 (130–148). Normal, hyperplastic, and neoplastic C cells are variably positive for TTF-1, whereas follicular cells are more uniformly positive (147,148), Hyperplastic C cells are generally negative for galectin 3 while medullary carcinomas

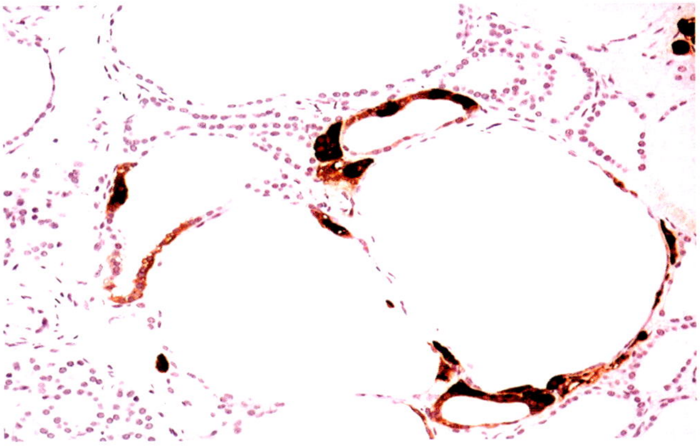

FIGURE 43.12 Immunostain for calcitonin demonstrates C cells within follicles, arranged either individually or in small groups.

associated with MEN2A and 2B are usually positive for this marker (149).

It is possible that endocrine cells other than C cells exist in the thyroid and that they represent the cells of origin of the rare thyroid "(neuro)endocrine carcinomas" having histologic and immunohistochemical features different from those of medullary carcinoma.

Physiology

Calcitonin is a 32–amino acid peptide that primarily functions to regulate the level of calcium in the plasma by a feedback mechanism. This is brought about by the inhibition of osteoclastic activity. Calcitonin is released from the thyroid gland when calcium plasma levels are increased. Calcitonin also acts in the kidney to enhance the production of vitamin D.

The major physiologic role of calcitonin is most likely the protection of the skeleton during periods of calcium stress such as growth, pregnancy, and lactation (150). However, the absence of calcitonin is not associated with hypercalcemia, nor does a marked excess of the hormone produce hypocalcemia. In addition to calcium, both gastrin and cholecystokinin induce secretion of calcitonin, as does the chronic administration of estrogenic hormones.

The calcitonin gene is located on the short arm of chromosome 11 and consists of six axons that encode katacalcin at the C-terminal flanking peptide and calcitonin gene–related peptide (136,150,151). The primary transcript of the calcitonin gene gives rise to two different mRNAs by tissue-specific alternative splicing events, leading to the production of calcitonin and calcitonin gene–related peptide mRNAs. Katacalcin, the C-terminal flanking peptide of calcitonin, is a 21–amino acid peptide that is cosecreted with calcitonin in equimolar amounts (136). Its function, however, is unknown. Calcitonin gene–related peptide is a 37–amino acid peptide that is an extremely potent vasodilator and also serves a neuromodulator or neurotransmitter function (150). Although the gene is expressed both in thyroid and nervous tissues, calcitonin is produced in large quantities only in the thyroid gland.

Basal calcitonin levels range from 3 to 36 pg/mL (0.9 to 10.5 pmol/L) in normal adult males. Plasma levels in females range from 3 to 17 pg/mL (0.9 to 5.0 pmol/L). Normal values after pentagastrin stimulation are less than 106 pg/mL (30.9 pmol/L) for males and less than 29 pg/mL (8.5 pmol/L) for females.

STROMA

Lymphocytes

It is not uncommon to observe a few collections of lymphocytes and rare plasma cells in the interstitium of the normal thyroid gland. *Simple chronic thyroiditis* and *focal lymphocytic thyroiditis* are the names given to this process, which is more common in females and most likely represents the epiphenomenon of several etiologically different conditions rather than a nosologic entity. Similar changes may be seen in the proximity of neoplasms, in thyroid glands of patients taking lithium, and in individuals who have received low-dose external radiotherapy (152).

Fibrous Tissue

The usually thin fibrous septa that separate the thyroid lobules may exhibit microscopic variations. In an autopsy study of normal thyroid glands collected from young adults, Komorowski and Hanson (32) found that 8% of the thyroid glands showed extensive fibrosis. According to their description, dense and largely acellular collagen fibers divided the thyroid into small nodules, giving it an appearance akin to micronodular cirrhosis of liver.

Another change that may occur in the thyroid interstitium, albeit rarely, is *multifocal fibrosing (sclerosing) thyroiditis* (153). This condition is characterized histologically by numerous microscopic stellate-shaped foci of cellular

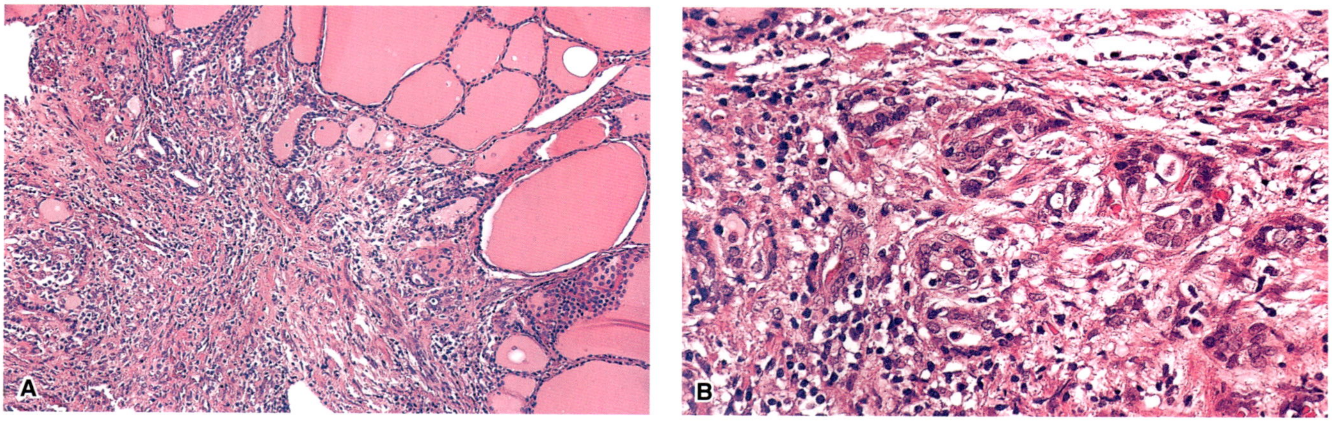

FIGURE 43.13 **A**: Multifocal sclerosing thyroiditis. On low power, the appearance resembles that of a papillary microcarcinoma. **B**: At higher power, the follicles entrapped in the fibrosis are irregularly shaped but do not show any of the cytologic features of papillary carcinoma.

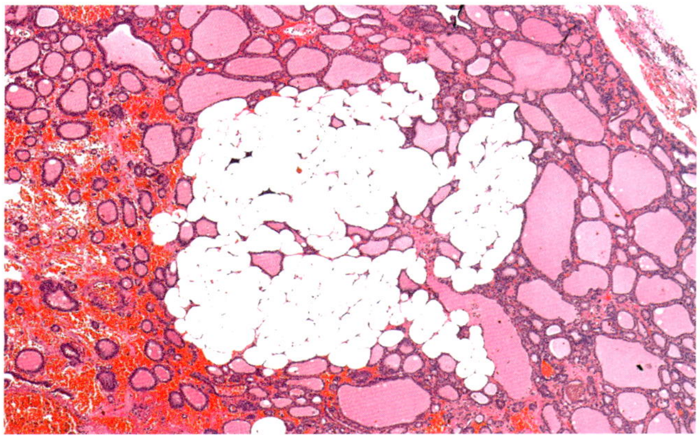

FIGURE 43.14 Adipose metaplasia of thyroid stroma. Mature adipocytes are seen between follicles.

fibroblastic tissue entrapping a few thyroid follicles in their centers. The individual lesions appear similar to papillary microcarcinomas, yet the epithelial components of such lesions lack the cytoarchitectural features of a papillary neoplasm (Fig. 43.13A,B) (153). Furthermore, the number of lesions in multifocal sclerosing thyroiditis greatly exceeds that seen in the usual case of papillary microcarcinoma. The etiology and pathogenesis of this process are not known. Because papillary microcarcinomas may occasionally be present at the edge of one of these lesions, some authors have suggested that multifocal fibrosing thyroiditis is a precursor of papillary thyroid carcinoma. However, in a series of papillary thyroid carcinomas associated with multifocal fibrosing thyroiditis, *BRAF* analysis showed that lesions of multifocal thyroiditis lack *BRAF* mutations (154).

Adipose Tissue and Skeletal Muscle

Thyroid stroma may undergo adipose metaplasia, resulting in the presence of islands of mature adipose tissue between follicles (Fig. 43.14). Mature fat may also be seen in proximity to the thyroid gland capsule, its presence in this location most likely results from the close relationship of fat

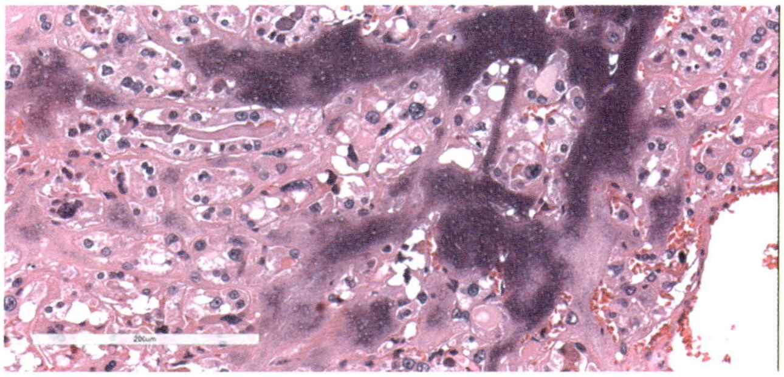

FIGURE 43.16 Dystrophic calcification.

and thyroid tissue during fetal life (155). Rarely, this change results in a clinically evident mass. This type of localized adiposity should be distinguished from lipoadenoma (156).

Other tissues that grow in close proximity to the thyroid gland during their development can be found within the capsule of adults. Striated muscle can be found within the thyroidal parenchyma, usually in the region of the isthmus or in the pyramidal lobe of the gland. Thyroid follicles can also be found in fascicles of strap muscle from the same areas (Fig. 43.15) (32).

Pancreatic tissue, a representative of foregut remnants, has been described in the wall of a perithyroidal epithelial cyst by Langlois et al. (157).

Calcifications

Dystrophic calcifications may be seen in normal thyroid glands of elderly patients, particularly in relation to vessels. They can easily be distinguished from psammoma bodies because they lack laminations and have irregular contours (Fig. 43.16).

Psammoma bodies have been described in benign thyroid lesions and in the normal thyroid gland but this is a rare occurrence (158–160). Detecting psammoma bodies in an otherwise normal thyroid gland or in a cervical lymph node should always prompt a careful search for occult thyroid papillary carcinoma (Fig. 43.17).

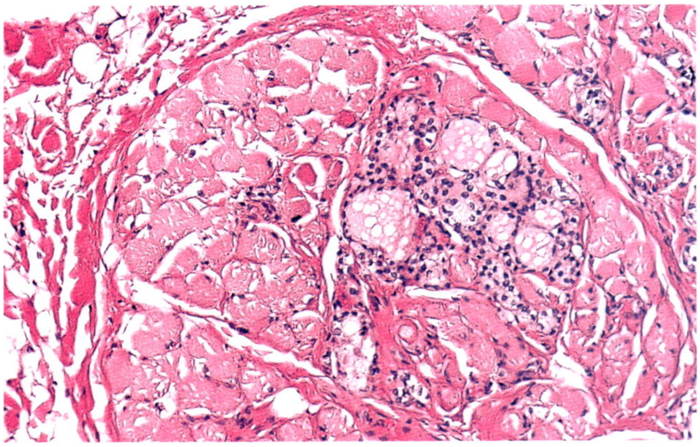

FIGURE 43.15 Clusters of thyroid tissue intimately admixed with bundles of skeletal muscle adjacent to the thyroid gland.

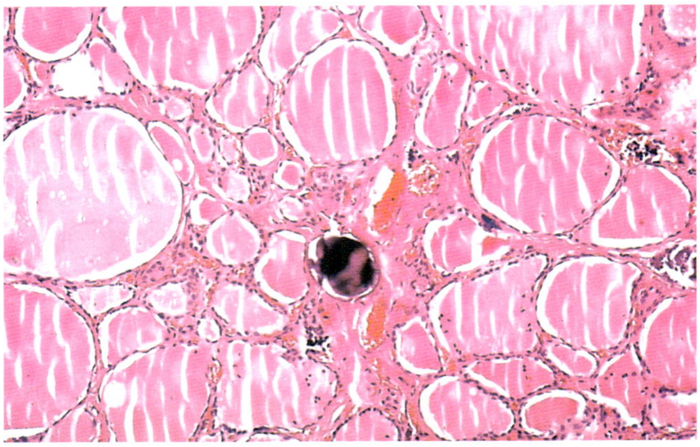

FIGURE 43.17 Psammoma body in nonneoplastic thyroid tissue adjacent to a papillary carcinoma (not shown in the picture).

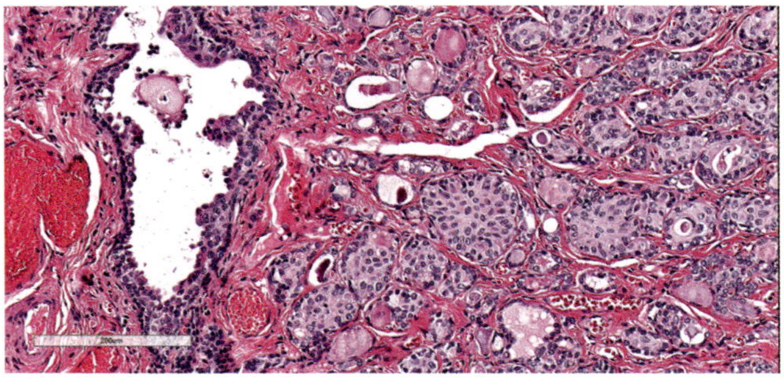

FIGURE 43.18 Hyperplastic solid cell nests adjacent to a cystic structure.

BRANCHIAL POUCH–DERIVED AND OTHER RELATED ECTOPIC TISSUES

Branchial pouch–related structures are found within the thyroid gland in various forms: solid cell nests, epithelium-lined cysts, parathyroid glands, thymic tissue, salivary gland–type tissue, and heterotopic cartilage. Solid cell nests (rests) are embryonic remnants of the ultimobranchial bodies. They appear as clusters of epithelial cells composed of main cells and C cells interspersed among the follicles.

The relationship between solid cell nests and ultimobranchial bodies was suggested by Erdheim in 1904 and Getzowa in 1907 following their demonstration of clusters of epithelial cells with solid or rarely cystic appearance in individuals with thyroid aplasia (161,162). Additional evidence along these lines was provided by the demonstration of marked similarities of human solid cell nests with the normal ultimobranchial bodies of the rat and the hyperplastic or neoplastic ultimobranchial bodies remnants in bulls (163–165). Solid cell nests are relatively common in the normal thyroid and can be detected in almost 90% of neonatal thyroid glands; the probability of finding them increases with the number of sections examined. In one study, solid cell nests were found in only 3% of routinely examined thyroids but in as many as 61% of specimens when the gland was blocked serially at 2- to 3-mm intervals (166). For unknown reasons solid cell nests are more common in males than in females. Most solid cell nests measure an average 0.1 mm in diameter, but they can occasionally reach a large size (Fig. 43.18) (167). They may be single or multiple and are more or less demarcated by the adjacent thyroid follicles by surrounding stroma. Adipose tissue, cartilage, or salivary gland tissue may be present in their vicinity (Fig. 43.19A–C).

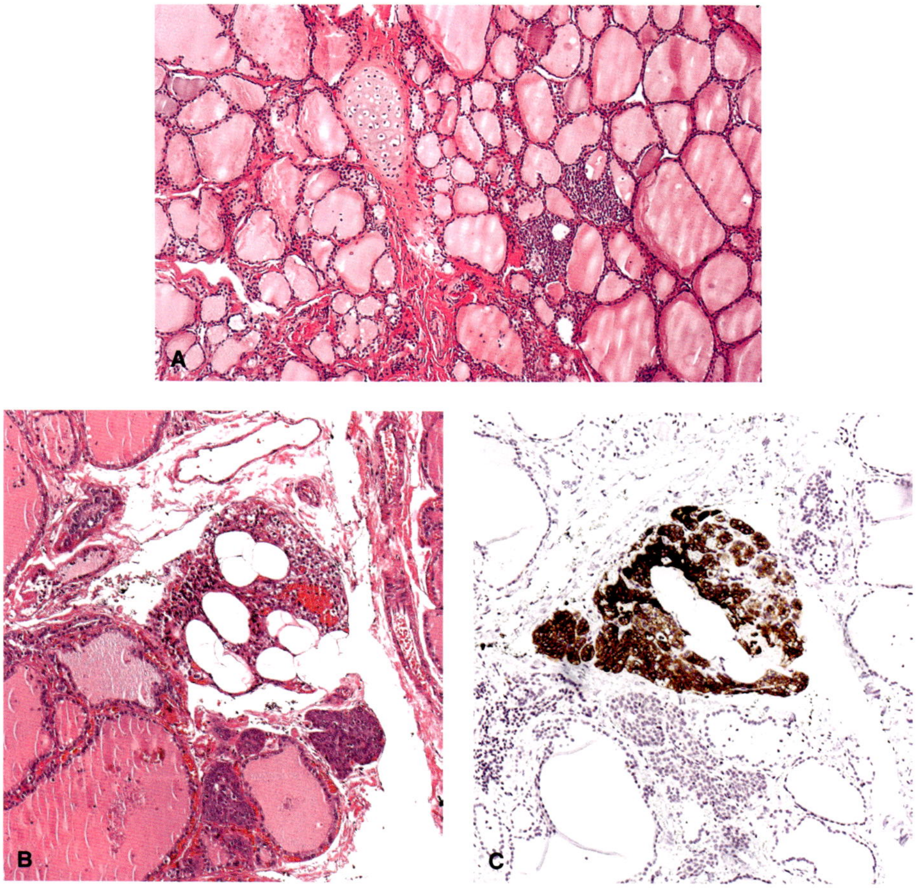

FIGURE 43.19 **A**: Cartilage island is seen in the proximity of solid cell nests. **B**: Solid cell nests with adjacent parathyroid gland. **C**: Strong PTH staining of the parathyroid gland.

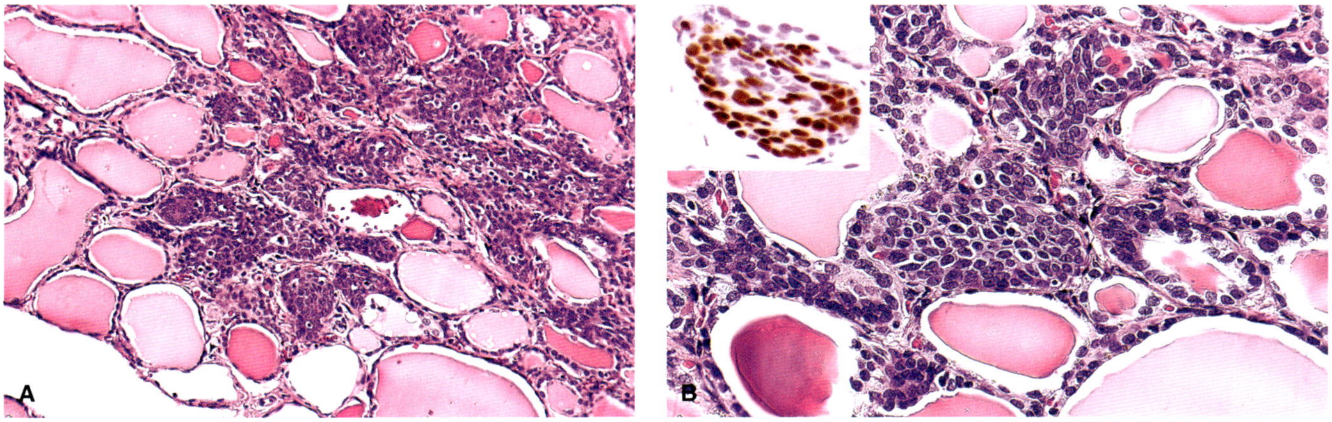

FIGURE 43.20 **A**: Low-power view shows the multilobed shape often exhibited by groups of solid cell nests. **B**: Solid cell nest in normal thyroid. Note the uniform appearance of the epithelial cells. *Inset:* Strong nuclear immunoreactivity for p63.

Most solid cell nests are found along the central axis of the middle and upper third of the lateral lobes in the same area where C cells are most numerous (168,169). Solid cell nests are often grouped in clusters featuring a multilobed shape within the interstitium (Fig. 43.20A). The main component is made up of polygonal epithelial cells with acidophilic cytoplasm and elongated nuclei that contain finely granular chromatin with frequent grooves. Since they may exhibit squamous differentiation, solid cell nests are sometimes misinterpreted as foci of squamous metaplasia in follicles. Ultrastructurally, these cells contain tonofilaments, desmosomes, and intraluminal cytoplasmic projections. Microfollicular structures lined by ciliated cells also occur (170). Some solid cell nests are connected to thyroidal follicular cells to form mixed follicles. The solid cell nests may contain small cysts with intraluminal accumulations of acidic mucins demonstrable with alcian blue histochemical stains.

The main cells of solid cell nests are positive for high- and low–molecular-weight keratins, TTF-1 (SPT24), p63, and p40; they are negative for monoclonal PAX-8 (Fig. 43.20B) (171,172). Positivity for GATA3 and monoclonal CEA was identified in 41 (73.2%) and 36 (64.3%) of solid cell nests. In addition, TTF-1 positivity has been described in rare cells forming mixed follicles (171,173). The positivity for p63 together with the expression of basal cell–type keratins (e.g., 34betaE12), telomerase, and bcl-2, is compatible with a basal/stem cell phenotype for this cellular component (174,175).

The second cell population, numerically less conspicuous, is composed by calcitonin-containing clear cells (22). They are characterized at the light microscopic level by clear cytoplasm and round nuclei, at the ultrastructural level by dense-core secretory granules, and at the immunohistochemical level by immunoreactivity to calcitonin, calcitonin gene-related peptide (CGRP), and chromogranin (168,170,173,176).

Mixed follicles consist of solid cell nests with groups of small follicles lined by low cuboidal thyroglobulin-immunoreactive epithelium (Fig. 43.21). A similar admixture of cell types is seen in mixed medullary–follicular carcinomas, leading some investigators to suggest that these rare tumors arise from uncommitted stem cells of the ultimobranchial bodies that have the potential to differentiate into C cells, follicular cells, or both (177). Solid cell nests need to be distinguished from follicles with squamous metaplasia, nodular C-cell hyperplasia, papillary microcarcinoma, and tangential sections of normal follicles (Fig. 43.22). A case of thyroid-type solid cell nests associated with struma ovarii has been reported suggesting a histogenetic link between the main cells of solid cell nests and thyroid tissue (178).

Remnants of the ultimobranchial bodies may also take the form of cysts in the soft tissues of the neck adjacent to the thyroid gland. Indeed, it is possible that some clinically evident branchial pouch cysts located in close proximity to the thyroid gland are of ultimobranchial bodies origin. Cysts derived from these remnants may also develop within the thyroid gland with or without adjacent solid cell nests (Fig. 43.23) (171). These cysts are lined by a flattened multilayered epithelium of squamous type or ciliated columnar epithelium; they often contain clumps of eosinophilic

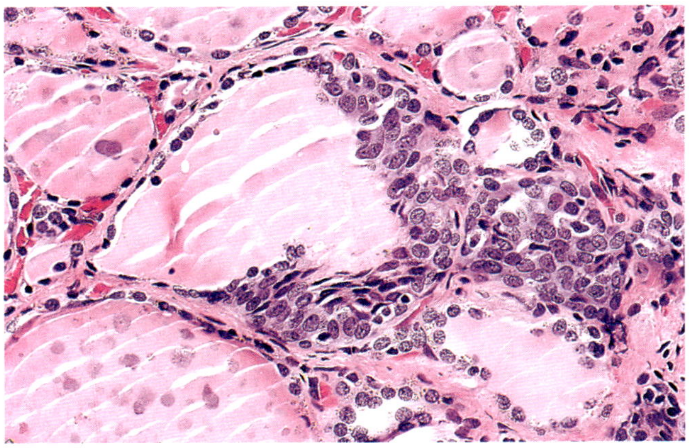

FIGURE 43.21 The so-called mixed follicle. A solid cell nest merges with a follicle lined by a flattened epithelium with colloid in the lumen.

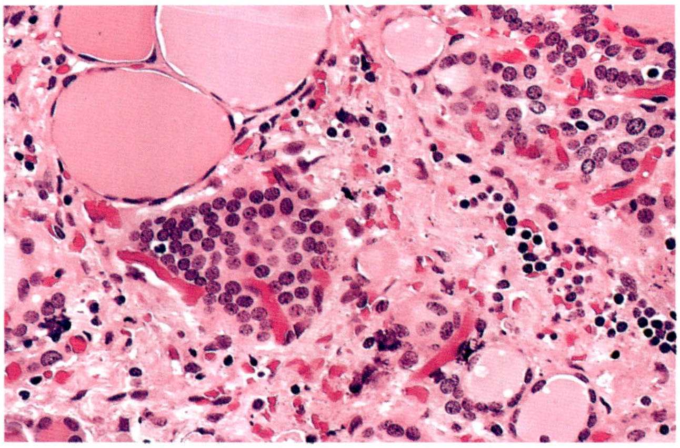

FIGURE 43.22 Tangential cut of a follicle. This should not be misinterpreted as a solid cell nest.

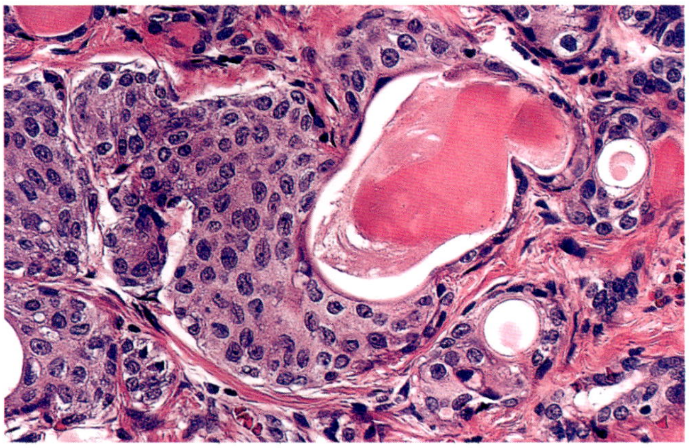

FIGURE 43.23 Solid cell nest with associated cystic formation. A dense eosinophilic material fills the lumen of the cyst.

material in their lumina (Fig. 43.24A,B). They are especially common in neonates. Cystic remnants of the ultimobranchial bodies may have an associated lymphoid component and are more commonly seen in glands with Hashimoto thyroiditis (179–182).

Parathyroid Tissue

The development of the parathyroid glands and the thymus from the branchial pouches in close proximity to the thyroid gland explains why these organs may be found adjacent to the thyroid capsule or even within the thyroid gland itself.

True intrathyroidal parathyroid glands in adults are rare. However, in an autopsy study of 58 fetal thyroid glands, intrathyroidal parathyroid tissue was found in 13 thyroid lobes from 12 fetuses (22.4%). These foci of parathyroid tissue were usually subcapsular (69%); a location deep in the thyroid gland was much less common (31%) (183). Importantly, intra- and perithyroidal parathyroid tissue can develop adenomas, primary or secondary chief cell hyperplasia, or carcinoma, and represent an often overlooked cause of surgical failure in primary hyperparathyroidism (184,185).

Thymic Tissue

Most of the thymus derives embryologically from the third branchial pouch together with the lower pair of parathyroid glands. There is also a small and inconstant portion of thymus that derives from the fourth branchial pouch together with the upper pair of parathyroid glands and the ultimobranchial bodies, which form the lateral thyroid anlage. It is the latter source that presumably gives rise to the islands of thymic tissue occasionally found in or around the thyroid gland (Fig. 43.25) (186). The fact that ectopic thymic tissue is observed more frequently in neonates and infants supports this hypothesis. Harach and Vujanic searched systematically for the presence of intrathymic tissue in 58 thyroid glands obtained at autopsy from fetuses with proven retrosternal thymus (187). Subcapsular thymic tissue was found in two cases (3.4%) and intrathyroidal thymic tissue in one (1.7%). An entire

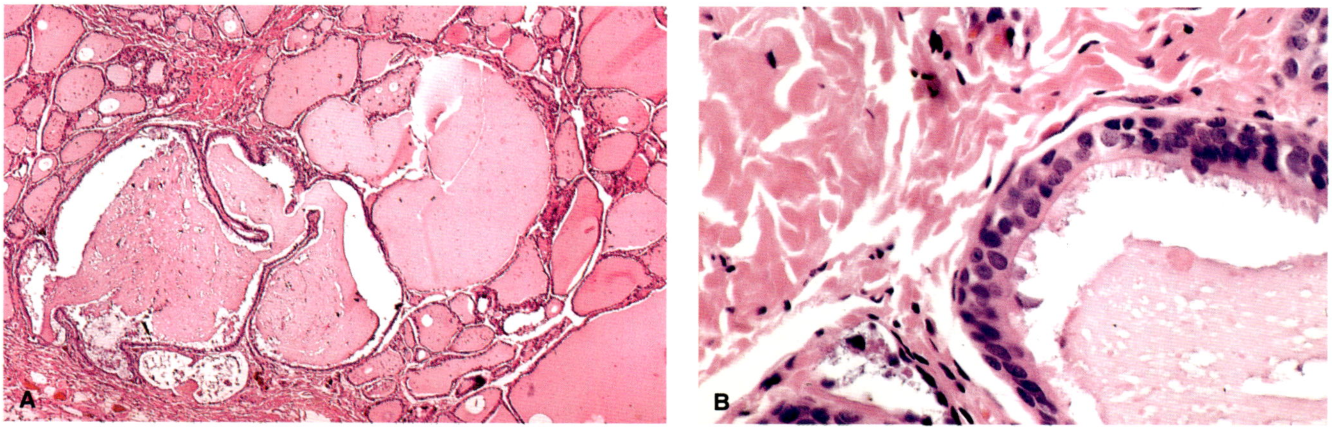

FIGURE 43.24 **A**: Intrathyroidal cyst of probable branchial pouch derivation. **B**: Higher-power view showing ciliated epithelium.

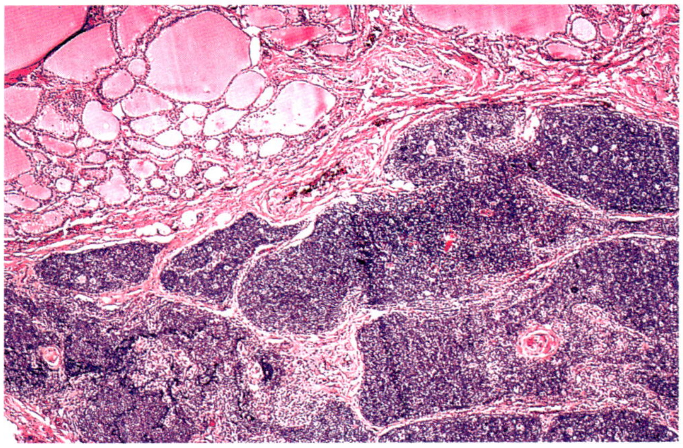

FIGURE 43.25 Intrathyroidal thymic tissue.

thymic gland within the thyroid of an infant has been described (188). Thymic rests have been reported in 1.4% of adult thyroid glands (189).

Ectopic thymic tissue may show cystic changes and present clinically as a cystic neck mass. It may also be the source of peri- and intrathyroidal thymomas (190,191).

Salivary Gland–Type Tissue

Rarely salivary gland–type tissue has been found within the thyroid gland. Most reported cases have been seen in association with a benign thyroid condition, such as multinodular goiter (192).

Ectopic Cartilage

Most intrathyroidal islands of mature cartilage probably represent remnants of the branchial pouch apparatus (193,194).

BENIGN THYROID TISSUE IN ABNORMAL LOCATIONS

The presence of nonneoplastic thyroid tissue outside the normal anatomical confines of the gland, and anatomically separate from the thyroid gland may be caused by a variety of mechanisms ranging from congenital abnormalities to acquired processes. Their main practical interest resides in the fact that a lack of knowledge of their occurrence may lead to a mistaken diagnosis of metastatic thyroid carcinoma.

Midline Structures

Most cases of ectopic thyroid are derived from abnormalities in migration patterns of the medial anlage. Thus, they are more commonly found in the neck in a midline position at any point in the normal pathway of descent of the thyroglossal duct from the foramen cecum at the base of the tongue and the normal site of the gland (194–196). The estimated frequency of ectopic thyroid is 0.17 per 1,000 patients, with lingual thyroid tissue accounting for 90% of cases (197,198). In most cases, the ectopy is partial, clinically insignificant, and discovered accidentally. The most common sites are (1) at the base of the tongue, (2) beneath the tongue, and (3) in or around the hyoid bone as a component of thyroglossal duct cyst.

Exaggerated descent of the median anlage into the mediastinum may lead to the development of substernal thyroid tissue in the preaortic area, pericardial cavity, or in the substance of the heart (199–201). Most mediastinal goiters represent a dislocation downward of originally orthotopic (cervical) glands that have been pulled down by the hyperplastic changes that occurred in them, rather than an embryonic abnormality.

Lingual thyroid is unusual as a clinical issue but relatively common as an incidental microscopic finding. In one series, almost 10% of tongues examined at autopsy had remnants of thyroid tissue in them (202). When of large size, lingual thyroid tissue may cause dysphagia, bleeding, and dyspnea (196,203). In most instances, the diagnosis is made during adolescence, and a preponderance in women has been noted (202,204). Complete failure of migration may result in an absence of thyroid tissue in its normal location (205–207). In this situation, the ectopic glands may be functionally inadequate or show compensatory hyperplasia that causes dyspnea or dysphagia. Acute hypothyroidism may follow the removal of this ectopic tissue.

The follicles of ectopic thyroid tissue are normal but because of their intimate relationship with the surrounding skeletal muscle, they may raise the differential diagnosis with carcinoma (208). Malignant tumors arising in the lingual thyroid gland are rare, with an estimated incidence of less than 1%, which is analogous to that observed in orthotopic thyroid tissue. Papillary carcinoma is the predominant histotype (209). A single case of poorly differentiated Hürthle cell carcinoma has also been reported (210).

Ectopic thyroid tissue is commonly found in the wall of thyroglossal duct cysts. It appears in the form of small group of follicles and is present in 25% to 65% of cysts; the frequency of detection is related to the number of sections submitted for histologic examination (211). The medial location and the presence of thyroid tissue in the wall distinguish thyroglossal duct cysts from the rarer bronchial pouch cysts.

Thyroglossal cyst is the most common congenital anomaly in the neck region. Most patients present with a painless mass in the midline of the neck, only rarely accompanied by dysphagia or dyspnea. Being the thyroglossal duct cyst is practically always connected with the hyoid bone upward movement of the mass on swallowing is characteristic of this condition (212). Most cysts measure from 1 to 2 cm in diameter. The original lining epithelium of the duct, cuboidal ("transitional") or columnar and often ciliated tends to become squamous or to disappear as a result of secondary inflammatory changes (213). Immunoreactivity for TTF-1 (but not thyroglobulin) has been described in the lining epithelium (214). Ectopic thyroid tissue is found commonly in the wall of *thyroglossal duct cysts.* It appears in the form of irregular small groups of follicles and is present in 25% to 65% of cysts examined histologically, its frequency being related to the number of sections submitted for histologic examination (211).

Pericapsular Soft Tissues and Skeletal Muscle

The presence of thyroid tissue in these locations is not a rare event. It most likely results from the intimate relationship of the thyroid gland with the mesodermal structures of the neck during development.

Lateral Neck

Lateral aberrant thyroid may result from implantation of thyroid tissue in the lateral neck following trauma or surgery. In this situation, a few nodules of normal-appearing thyroid tissue are seen in the lateral neck close to the cervical lymph nodes (215–217). History of previous trauma or surgery on the neck, the presence of suture material, and the benign appearance of the thyroid tissue are useful in distinguishing them from metastatic carcinoma. Spontaneous separation of thyroid tissue with subsequent implantation in the lateral neck may occur in nodular goiter or Hashimoto thyroiditis (218,219). In both of these conditions, nodules of thyroid tissue extrude and separate from the surface of the gland, depositing in the extrathyroidal soft tissue where they may acquire an autonomous blood supply. These foci have been termed parasitic nodules and can pose diagnostic challenges. Recognition of lymph node architecture which includes a fibrous capsule and subcapsular sinus will help distinguish true metastases from a parasitic nodule.

Thyroid Inclusions in Cervical Lymph Nodes

While many authors argue that the presence of normal-appearing thyroid tissue in cervical lymph nodes, particularly lateral neck nodes, always represents metastatic disease, some authors argue that this thyroid tissue can be benign and the result of a developmental anomaly (220). Developmental anomalies generally consist of a few small follicles located immediately beneath the nodal capsule (Fig. 43.26). The follicular cells that compose them lack all the cytologic features typical of papillary carcinoma

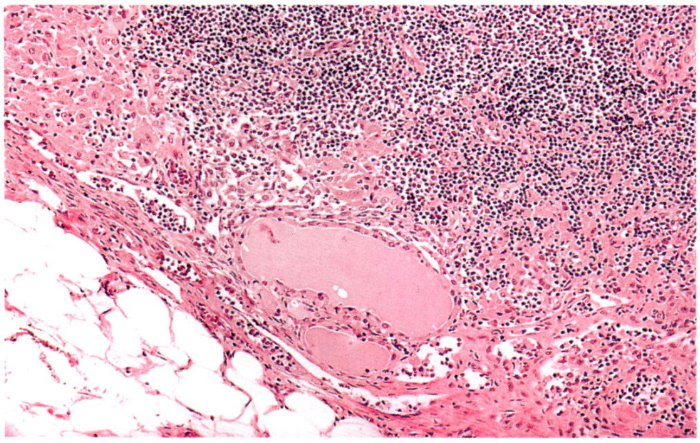

FIGURE 43.26 A group of benign-appearing thyroid follicles is seen close to the marginal sinus of a cervical lymph node. The patient did not have a carcinoma in the thyroid gland.

(221). Numerous sections are sometimes needed to rule out a metastasis from a papillary microcarcinoma, which is by far the most frequent cause of thyroid tissue in cervical nodes, even when it closely resembles nonneoplastic thyroid tissue. Criteria for diagnosing benign thyroid inclusions are very rigorous. A diagnosis of metastatic disease should be strongly considered in cases with involvement of the nodal parenchyma rather than capsular/subcapsular region, cases in which thyroid tissue has replaced one third or more of the node, and cases in which several nodes are affected. This is also true whenever the intranodal thyroid tissue shows any of the cytoarchitectural features of papillary carcinoma or psammoma bodies are present. In doubtful cases, immunohistochemical and/or molecular analysis of BRAF can be of help (221).

Other Sites

Rarely, one can find thyroid tissue in other locations including the larynx, trachea, submandibular region, aortic arch, heart and pericardium, lung, mediastinal esophagus, stomach, duodenum, diaphragm, region of gallbladder/common bile duct/porta hepatis, pancreas, liver, spleen, adrenal gland, retroperitoneum, vagina, uterus, and sella turcica (222–243).

REFERENCES

1. Hoyes AD, Kershaw DR. Anatomy and development of the thyroid gland. *Ear Nose Throat J.* 1985;64:318–333.
2. Shepard TH. Onset of function in the human fetal thyroid: Biochemical and radioautographic studies from organ culture. *J Clin Endocrinol Metab.* 1967;27:945–958.
3. Gitlin D, Biasucci A. Ontogenesis of immunoreactive thyroglobulin in the human conceptus. *J Clin Endocrinol Metab.* 1969;29:849–853.

4. Fagman H, Nilsson M. Morphogenesis of the thyroid gland. *Mol Cell Endocrinol.* 2010;323:35–54.
5. Nilsson M, Fagman H. Mechanisms of thyroid development and dysgenesis: An analysis based on developmental stages and concurrent embryonic anatomy. *Curr Top Dev Biol.* 2013; 106:123–170.
6. Trueba SS, Auge J, Mattei G, et al. PAX8, TITF1, and FOXE1 gene expression patterns during human development: New insights into human thyroid development and thyroid dysgenesis-associated malformations. *J Clin Endocrinol Metab.* 2005;90:455–462.
7. Fernandez LP, Lopez-Marquez A, Santisteban P. Thyroid transcription factors in development, differentiation and disease. *Nat Rev Endocrinol.* 2015;11:29–42.
8. Fernandez LP, Lopez-Marquez A, Gomez-Lopez G, Gómez-López G, Santisteban P. New insights into FoxE1 functions: Identification of direct FoxE1 targets in thyroid cells. *PLoS One* 2013;8:e62849.
9. Parlato R, Rosica A, Rodriguez-Mallon A, et al. An integrated regulatory network controlling survival and migration in thyroid organogenesis. *Dev Biol.* 2004;276:464–475.
10. Meeus L, Gilbert B, Rydlewski C, et al. Characterization of a novel loss of function mutation of PAX8 in a familial case of congenital hypothyroidism with in-place, normal-sized thyroid. *J Clin Endocrinol Metab.* 2004;89:4285–4291.
11. De Felice M, Di Lauro R. Thyroid development and its disorders: Genetics and molecular mechanisms. *Endocr Rev.* 2004;25:722–746.
12. Nilsson M, Fagman H. Development of the thyroid gland. *Development.* 2017;144:2123–2140.
13. Pasca di Magliano M, Di Lauro R, Zannini M. Pax8 has a key role in thyroid cell differentiation. *Proc Natl Acad Sci U S A.* 2000;97:13144–13149.
14. Lau SK, Luthringer DJ, Eisen RN. Thyroid transcription factor-1: A review. *Appl Immunohistochem Mol Morphol.* 2002;10:97–102.
15. Volante M, Allia E, Fulcheri E, et al. Ghrelin in fetal thyroid and follicular tumors and cell lines: Expression and effects on tumor growth. *Am J Pathol.* 2003;162:645–654.
16. Savin SB, Cvejic DS, Jankovic MM. Expression of galectin-1 and galectin-3 in human fetal thyroid gland. *J Histochem Cytochem.* 2003;51:479–483.
17. Norris EH. The parathyroid glands and the lateral thyroid in man: Their morphogenesis, histogenesis, topographic anatomy and prenatal growth. *Contrib Embryol Carnegie Inst* 1937;159:249–294.
18. Sugiyama S. The embryology of the human thyroid gland including ultimobranchial body and others related. *Ergeb Anat Entwicklungsgesch.* 1971;44:3–111.
19. Pearse AG, Polak JM. Cytochemical evidence for the neural crest origin of mammalian ultimobranchial C cells. *Histochemie.* 1971;27:96–102.
20. Le Douarin NM, Teillet MA. The migration of neural crest cells to the wall of the digestive tract in avian embryo. *J Embryol Exp Morphol.* 1973;30:31–48.
21. Le Douarin N, Fontaine J, Le Lievre C. New studies on the neural crest origin of the avian ultimobranchial glandular cells–interspecific combinations and cytochemical characterization of C cells based on the uptake of biogenic amine precursors. *Histochemistry.* 1974;38:297–305.
22. Nadig J, Weber E, Hedinger C. C-cell in vestiges of the ultimobranchial body in human thyroid glands. *Virchows Arch B Cell Pathol.* 1978;27:189–191.
23. Ito M, Kameda Y, Tagawa T. An ultrastructural study of the cysts in chicken ultimobranchial glands, with special reference to C-cells. *Cell Tissue Res.* 1986;246:39–44.
24. Kameda Y, Nishimaki T, Chisaka O, Iseki S, Sucov HM. Expression of the epithelial marker E-cadherin by thyroid C cells and their precursors during murine development. *J Histochem Cytochem.* 2007;55:1075–1088.
25. Johansson E, Andersson L, Örnros J, et al. Revising the embryonic origin of thyroid C cells in mice and humans. *Development.* 2015;142:3519–3528.
26. Schmid KW. Histopathology of C Cells and medullary thyroid carcinoma. *Recent Results Cancer Res.* 2015;204:41–60.
27. Kameda Y. Cellular and molecular events on the development of mammalian thyroid C cells. *Dev Dyn.* 2016;245:323–341.
28. Nilsson M, Williams D. On the origin of cells and derivation of thyroid cancer: C cell story revisited. *Eur Thyroid J.* 2016;5:79–93.
29. Andersson L, Westerlund J, Liang S, et al. Role of EphA4 receptor signaling in thyroid development: Regulation of folliculogenesis and propagation of the C-cell lineage. *Endocrinology.* 2011;152:1154–1164.
30. Hegedus L, Perrild H, Poulsen LR, et al. The determination of thyroid volume by ultrasound and its relationship to body weight, age, and sex in normal subjects. *J Clin Endocrinol Metab.* 1983;56:260–263.
31. Hegedus L, Karstrup S, Rasmussen N. Evidence of cyclic alterations of thyroid size during the menstrual cycle in healthy women. *Am J Obstet Gynecol.* 1986;155:142–145.
32. Komorowski RA, Hanson GA. Occult thyroid pathology in the young adult: An autopsy study of 138 patients without clinical thyroid disease. *Hum Pathol.* 1988;19:689–696.
33. Bell CD, Kovacs K, Horvath E, Rotondo F. Histologic, immunohistochemical, and ultrastructural findings in a case of minocycline-associated "black thyroid." *Endocr Pathol.* 2001; 12:443–451.
34. Veinot JP, Ghadially FN. Melanosis thyroidi. *Ultrastruct Pathol.* 1998;22:401–406.
35. Kandil E, Khalek MA, Ibrahim WG, et al. Papillary thyroid carcinoma in black thyroids. *Head Neck.* 2011;33:1735–1738.
36. Bann DV, Goyal N, Crist H, Goldenberg D. Black thyroid. *Ear Nose Throat J.* 2014;93:E54–E55.
37. Thompson AD, Pasieka JL, Kneafsey P, DiFrancesco LM. Hypopigmentation of a papillary carcinoma arising in a black thyroid. *Mod Pathol.* 1999;12:1181–1185.
38. Brown RA, Al-Moussa M, Beck J. Histometry of normal thyroid in man. *J Clin Pathol.* 1986;39:475–482.
39. Imada M, Kurosumi M, Fujita H. Three-dimensional imaging of blood vessels in thyroids from normal and levothyroxine sodium-treated rats. *Arch Histol Jpn.* 1986;49:359–367.
40. Imada M, Kurosumi M, Fujita H. Three-dimensional aspects of blood vessels in thyroids from normal, low iodine diet-treated, TSH-treated and PTU-treated rats. *Cell Tissue Res.* 1986;245:291–296.
41. Drut R, Altamirano E, Ollano AM. Lymphatic vessels in the thyroid gland of children. *Rev Esp Patol.* 2009;42:159–160.
42. Russell WO, Ibanez ML, Clark RL, White EC. Thyroid carcinoma. Classification, intraglandular dissemination, and

clinicopathological study based upon whole organ sections of 80 glands. *Cancer.* 1963;16:1425–1460.
43. Feind C. The head and neck. In: Haagensen CD, Feind C, Herter FP, Slanetz CA, Weinberg JA, eds. *The Lymphatics in Cancer*. WB Saunders; 1972:59–222.
44. Crile G Jr. The fallacy of the conventional radical neck dissection for papillary carcinoma of the thyroid. *Ann Surg.* 1957;145:317–320.
45. Yanir Y, Doweck I. Regional metastases in well-differentiated thyroid carcinoma: Pattern of spread. *Laryngoscope.* 2008;118:433–436.
46. Uchiyama Y, Murakami G, Ohno Y. The fine structure of nerve endings on rat thyroid follicular cells. *Cell Tissue Res.* 1985;242:457–460.
47. Melander A, Ericson LD, Sundler F, Ingbar SH. Sympathetic innervation of the mouse thyroid and its significance in thyroid hormone secretion. *Endocrinology.* 1974;94:959–966.
48. Tice LW, Creveling CR. Electron microscopic identification of adrenergic nerve endings on thyroid epithelial cells. *Endocrinology.* 1975;97:1123–1129.
49. Ingbar SH. The thyroid gland. In: Wilson JD, Foster DW, eds. *Williams Textbook of Endocrinology*. 7th ed. WB Saunders; 1985:682–815.
50. Kameda Y, Okamoto K, Ito M, Tagawa T. Innervation of the C cells of chicken ultimobranchial glands studied by immunohistochemistry, fluorescence microscopy, and electron microscopy. *Am J Anat.* 1988;182:353–368.
51. Zak FG, Lawson W. Glomic (paraganglionic) tissue in the larynx and capsule of the thyroid gland. *Mt Sinai J Med.* 1972;39:82–90.
52. LaGuette J, Matias-Guiu X, Rosai J. Thyroid paraganglioma: A clinicopathologic and immunohistochemical study of three cases. *Am J Surg Pathol.* 1997;21:748–753.
53. Lee SM, Policarpio-Nicolas MLC. Thyroid Paraganglioma. *Arch Pathol Lab Med.* 2015;139:1062–1067.
54. Heimann P. Ultrastructure of human thyroid. A study of normal thyroid, untreated and treated diffuse goiter. *Acta Endocrinol (Copenh).* 1966;53(110): 1+.
55. Klinck GH, Oertel JE, Winship T. Ultrastructure of normal human thyroid. *Lab Invest.* 1970;22:2–22.
56. Kurata A, Ohta K, Mine M, et al. Monoclonal antihuman thyroglobulin antibodies. *J Clin Endocrinol Metab.* 1984;59:573–579.
57. Stanta G, Carcangiu ML, Rosai J. The biochemical and immunohistochemical profile of thyroid neoplasia. *Pathol Annu.* 1988;23(Pt 1):129–157.
58. Rosai J, Carcangiu ML. Pitfalls in the diagnosis of thyroid neoplasms. *Pathol Res Pract.* 1987;182:169–179.
59. Baloch Z, Mete O, Asa SL. Immunohistochemical biomarkers in thyroid pathology. *Endocr Pathol.* 2018:29(2):91–112.
60. Henzen-Logmans SC, Mullink H, Ramaekers FC, Tadema T, Meijer CJ. Expression of cytokeratins and vimentin in epithelial cells of normal and pathologic thyroid tissue. *Virchows Arch a Pathol Anat Histopathol.* 1987;410:347–354.
61. Fonseca E, Nesland JM, Hoie J, Sobrinho-Simões M. Pattern of expression of intermediate cytokeratin filaments in the thyroid gland: An immunohistochemical study of simple and stratified epithelial-type cytokeratins. *Virchows Arch.* 1997;430:239–245.
62. Coclet J, Lamy F, Rickaert F, Dumont JE, Roger PP. Intermediate filaments in normal thyrocytes: Modulation of vimentin expression in primary cultures. *Mol Cell Endocrinol.* 1991;76:135–148.
63. Viale G, Dell'Orto P, Coggi G, Gambacorta M. Coexpression of cytokeratins and vimentin in normal and diseased thyroid glands. Lack of diagnostic utility of vimentin immunostaining. *Am J Surg Pathol.* 1989;13:1034–1040.
64. Xu XC, El-Naggar AK, Lotan R. Differential expression of galectin-1 and galectin-3 in thyroid tumors. Potential diagnostic implications. *Am J Pathol.* 1995;147:815–822.
65. Herrmann ME, LiVolsi VA, Pasha TL, Roberts SA, Wojcik EM, Baloch ZW. Immunohistochemical expression of galectin-3 in benign and malignant thyroid lesions. *Arch Pathol Lab Med.* 2002;126:710–713.
66. Bur M, Shiraki W, Masood S. Estrogen and progesterone receptor detection in neoplastic and non-neoplastic thyroid tissues. *Mod Pathol.* 1993;6:469–472.
67. Kawabata W, Suzuki T, Moriya T, et al. Estrogen receptors (alpha and beta) and 17beta-hydroxysteroid dehydrogenase type 1 and 2 in thyroid disorders: Possible in situ estrogen synthesis and actions. *Mod Pathol.* 2003;16:437–444.
68. Nishimura R, Yokose T, Mukai K. S-100 protein is a differentiation marker in thyroid carcinoma of follicular cell origin: An immunohistochemical study. *Pathol Int.* 1997;47:673–679.
69. Tzelepi VN, Tsamandas AC, Vlotinou HD, Vagianos CE, Scopa CD. Tight junctions in thyroid carcinogenesis: Diverse expression of claudin-1, claudin-4, claudin-7 and occludin in thyroid neoplasms. *Mod Pathol.* 2008;21:22–30.
70. Westermark K, Lundqvist M, Wallin G. EGF-receptors in human normal and pathological thyroid tissue. *Histopathology.* 1996;28:221–227.
71. Lima MA, Gontijo VA, Schmitt FCL. Thyroid peroxidase and thyroglobulin expression in normal human thyroid glands. *Endocr Pathol.* 1998;9:333–338.
72. Lin JD, Hsueh C, Chao TC, Weng HF. Expression of sodium iodide symporter in benign and malignant human thyroid tissues. *Endocr Pathol.* 2001;12:15–21.
73. Ringel MD, Anderson J, Souza SL, et al. Expression of the sodium iodide symporter and thyroglobulin genes are reduced in papillary thyroid cancer. *Mod Pathol.* 2001;14:289–296.
74. Maruyama M, Kato R, Kobayashi S, Kasuga Y. A method to differentiate between thyroglobulin derived from normal thyroid tissue and from thyroid carcinoma based on analysis of reactivity to lectins. *Arch Pathol Lab Med.* 1998;122:715–720.
75. de Martynoff G, Pohl V, Mercken L, van Ommen GJ, Vassart G. Structural organization of the bovine thyroglobulin gene and of its 5′-flanking region. *Eur J Biochem.* 1987; 164:591–599.
76. Christophe D, Gerard C, Juvenal G, et al. Identification of a cAMP-responsive region in thyroglobulin gene promoter. *Mol Cell Endocrinol.* 1989;64:5–18.
77. Lee NT, Nayfeh SN, Chae CB. Induction of nuclear protein factors specific for hormone-responsive region during activation of thyroglobulin gene by thyrotropin in rat thyroid FRTL-5 cells. *J Biol Chem.* 1989;264:7523–7530.
78. Deiss WP, Peake RL. The mechanism of thyroid hormone secretion. *Ann Intern Med.* 1968;69:881–90.
79. Brent GA. Mechanisms of thyroid hormone action. *J Clin Invest.* 2012;122:3035–3043.
80. Liddle GW, Liddle RA. Endocrinology. In: Smith LH, Thier SO, eds. *Pathophysiology: The Biological Principles of Disease*. WB Saunders; 1981.

81. Sterling K. Thyroid hormone action at the cell level (first of two parts). *N Engl J Med.* 1979;300:117–123.
82. Bernal J, Liewendahl K, Lamberg BA. Thyroid hormone receptors in fetal and hormone resistant tissues. *Scand J Clin Lab Invest.* 1985;45:577–583.
83. Müller MJ, Seitz HJ. Thyroid hormone action on intermediary metabolism. Part I. Respiration, thermogenesis and carbohydrate metabolism. *Klin Wochenschr.* 1984;62: 11–18.
84. Müller MJ, Seitz HJ. Thyroid hormone action on intermediary metabolism. II. Lipid metabolism in hyper- and hypothyroidism. *Klin Wochenschr.* 1984;62:49–55.
85. Müller MJ, Seitz HJ. Thyroid hormone action on intermediary metabolism. Part III. Protein metabolism in hyper- and hypothyroidism. *Klin Wochenschr.* 1984;62:97–102.
86. Oppenheimer JH. Thyroid hormone action at the nuclear level. *Ann Intern Med.* 1985;102:374–384.
87. Oppenheimer JH, Samuels HH, eds. *Molecular Basis of Thyroid Hormone Action.* Academic Press; 1983.
88. Larsen PR. Thyroid–pituitary interaction: Feedback regulation of thyrotropin secretion by thyroid hormones. *N Engl J Med.* 1982;306:23–32.
89. Davies T, Marians R, Latif R. The TSH receptor reveals itself. *J Clin Invest.* 2002;110:161–164.
90. Atassi MZ, Manshouri T, Sakata S. Localization and synthesis of the hormone-binding regions of the human thyrotropin receptor. *Proc Natl Acad Sci U S A.* 1991;88:3613–3617.
91. Pittman JA Jr. Thyrotropin-releasing hormone. *Adv Intern Med.* 1974;19:303–325.
92. Wilber JF. Thyrotropin releasing hormone: Secretion and actions. *Annu Rev Med.* 1973;24:353–364.
93. Walsh FM, Castelli JB. Polytef granuloma clinically simulating carcinoma of the thyroid. *Arch Otolaryngol.* 1975; 101:262–263.
94. Sanfilippo F, Shelburne J, Ingram P. Analysis of a polytef granuloma mimicking a cold thyroid nodule 17 months after laryngeal injection. *Ultrastruct Pathol.* 1980;1:471–475.
95. Chaplin AJ. Histopathological occurrence and characterization of calcium oxalate: A review. *J Clin Pathol.* 1977; 30:800–811.
96. Carney JA, Moore SB, Northcutt RC, Woolner LB, Stillwell GK. Palpation thyroiditis (multifocal granulomatous folliculitis). *Am J Clin Pathol.* 1975;64:639–647.
97. Blum M, Schloss MF. Martial-arts thyroiditis. *N Engl J Med.* 1984;311:199–200.
98. Harach HR, Jasani B. Thyroid multifocal granulomatous folliculitis (palpation thyroiditis): An immunocytochemical study. *Endocr Pathol.* 1993;4:105–109.
99. Hellwig CA. Colloidophagy in the human thyroid gland. *Science.* 1951;113:725–726.
100. Manson C, Cross P, De Sousa B. Post-operative necrotizing granulomata of the thyroid. *Histopathology.* 1992;21: 392–393.
101. Harach HR. Palpation thyroiditis resembling C cell hyperplasia. Usefulness of immunohistochemistry in their differential diagnosis. *Pathol Res Pract.* 1993;189:488–490.
102. Richter MN, McCarty KS. Anisotropic crystals in the human thyroid gland. *Am J Pathol.* 1954;30:545–553.
103. Katoh R, Suzuki K, Hemmi A, Kawaoi A. Nature and significance of calcium oxalate crystals in normal human thyroid gland. A clinicopathological and immunohistochemical study. *Virchows Arch A Pathol Anat Histopathol.* 1993;422:301–306.
104. Reid JD, Choi CH, Oldroyd NO. Calcium oxalate crystals in the thyroid. Their identification, prevalence, origin, and possible significance. *Am J Clin Pathol.* 1987;87:443–454.
105. Gross S. Granulomatous thyroiditis with anisotropic crystalline material. *AMA Arch Pathol.* 1955;59:412–418.
106. Fayemi AO, Ali M, Braun EV. Oxalosis in hemodialysis patients: A pathologic study of 80 cases. *Arch Pathol Lab Med.* 1979;103:58–62.
107. Isotalo PA, Lloyd RV. Presence of birefringent crystals is useful in distinguishing thyroid from parathyroid gland tissues. *Am J Surg Pathol.* 2002;26:813–814.
108. Wong KS, Lewis JS Jr, Gottipati S, Chernock RD. Utility of birefringent crystal identification by polarized light microscopy in distinguishing thyroid from parathyroid tissue on intraoperative frozen sections. *Am J Surg Pathol.* 2014;38:1212–1219.
109. Klinck G, Menk K. Squamous cells in the human thyroid. *Mil Surgeon.* 1951;109:406–414.
110. Harcourt-Webster JN. Squamous epithelium in the human thyroid gland. *J Clin Pathol.* 1966;19:384–388.
111. LiVolsi VA, Merino MJ. Squamous cells in the human thyroid gland. *Am J Surg Pathol.* 1978;2:133–140.
112. Pellicer DL, Sadow PM, Stephen A, Faquin WC. Atypical squamous metaplasia in a benign cystic thyroid nodule mimicking high-grade carcinoma. *Diagn Cytopathol.* 2013; 41:706–709.
113. Pearse AG. The cytochemistry of the thyroid C cells and their relationship to calcitonin. *Proc R Soc Lond B Biol Sci.* 1966;164:478–487.
114. Teitlebaum SL, Moore KE, Shieber W. Parafollicular cells in the normal human thyroid. *Nature.* 1971;230:334–335.
115. Wolfe HJ, DeLellis RA, Voelkel EF, Tashjian AH Jr. Distribution of calcitonin-containing cells in the normal neonatal human thyroid gland: A correlation of morphology with peptide content. *J Clin Endocrinol Metab.* 1975;41:1076–1081.
116. O'Toole K, Fenoglio-Preiser C, Pushparaj N. Endocrine changes associated with the human aging process. III. Effect of age on the number of calcitonin immunoreactive cells in the thyroid gland. *Hum Pathol.* 1985;16:991–1000.
117. Gibson WG, Peng TC, Croker BP. Age-associated C-cell hyperplasia in the human thyroid. *Am J Pathol.* 1982;106: 388–393.
118. Guyétant S, Wion-Barbot N, Rousselet M-C, Franc B, Bigorgne JC, Saint-Andre JP. C-cell hyperplasia associated with chronic lymphocytic thyroiditis: A retrospective quantitative study of 112 cases. *Hum Pathol.* 1994;25:514–521.
119. Scopsi L, Di Palma S, Ferrari C, Holst JJ, Rehfeld JF, Rilke F. C-cell hyperplasia accompanying thyroid diseases other than medullary carcinoma: An immunocytochemical study by means of antibodies to calcitonin and somatostatin. *Mod Pathol.* 1991;4:297–304.
120. Libbey NP, Nowakowski KJ, Tucci JR. C-cell hyperplasia of the thyroid in a patient with goitrous hypothyroidism and Hashimoto's thyroiditis. *Am J Surg Pathol.* 1989;13: 71–77.
121. Biddinger PW, Brennan MF, Rosen PP. Symptomatic C-cell hyperplasia associated with chronic lymphocytic thyroiditis. *Am J Surg Pathol.* 1991;15:599–604.

122. Tomita T, Millard DM. C-cell hyperplasia in secondary hyperparathyroidism. *Histopathology.* 1992;21:469–474.
123. Zambrano E, Holm F, Glickman J, et al. Abnormal distribution and hyperplasia of thyroid C-cells in PTEN associated tumor syndromes. *Endocr Pathol.* 2004;15:55–64.
124. Saggiorato E, Rapa I, Garino F, et al. Absence of RET gene point mutations in sporadic thyroid C-cell hyperplasia. *J Mol Diagn.* 2007;9:214–219.
125. Perry A, Molberg K, Albores-Saavedra J. Physiologic versus neoplastic C cell hyperplasia of the thyroid. Separation of distinct histologic and biologic entities. *Cancer.* 1996;77:750–756.
126. Lloyd RV, Osamura RY, Kloppel G, Rosai J. *WHO classification of Tumours of Endocrine Organs.* International Agency for Research on Cancer Lyon. 2017.
127. Carney JA, Sizemore GW, Hayles AB. Multiple endocrine neoplasia, type 2b. *Pathobiol Annu.* 1978;8:105–153.
128. Diaz-Cano SJ, de Miguel M, Blanes A, Tashjian R, Wolfe HJ. Germline RET 634 mutation positive MEN 2A-related C-cell hyperplasias have genetic features consistent with intraepithelial neoplasia. *J Clin Endocrinol Metab.* 2001;86:3948–3957.
129. DeLellis RA, Nunnemacher G, Wolfe HJ. C-cell hyperplasia. An ultrastructural analysis. *Lab Invest.* 1977;36:237–248.
130. de Lellis RA, Wolfe HJ. The pathobiology of the human calcitonin (C)-cell: A review. *Pathol Annu.* 1981;16(2):25–52.
131. Pearse AG. Common cytochemical and ultrastructural characteristics of cells producing polypeptide hormones (the APUD series) and their relevance to thyroid and ultimobranchial C cells and calcitonin. *Proc R Soc Lond B Biol Sci.* 1968;170:71–80.
132. Gonzalez-Cámpora R, Sanchez Gallego F, Martin Lacave I, Mora Marin J, Montero Linares C, Galera-Davidson H. Lectin histochemistry of the thyroid gland. *Cancer.* 1988;62:2354–2362.
133. Bussolati G, Pearse AG. Immunofluorescent localization of calcitonin in the 'C' cells of pig and dog thyroid. *J Endocrinol.* 1967;37:205–209.
134. McMillan PJ, Hooker WM, Deptos LJ. Distribution of calcitonin-containing cells in the human thyroid. *Am J Anat.* 1974;140:73–79.
135. Schmid KW, Kirchmair R, Ladurner D, Fischer-Colbrie R, Böcker W. Immunohistochemical comparison of chromogranins A and B and secretogranin II with calcitonin and calcitonin gene-related peptide expression in normal, hyperplastic and neoplastic C-cells of the human thyroid. *Histopathology.* 1992;21:225–232.
136. Ali-Rachedi A, Varndell IM, Facer P, et al. Immunocytochemical localization of katacalcin, a calcium-lowering hormone cleaved from the human calcitonin precursor. *J Clin Endocrinol Metab.* 1983;57:680–682.
137. Van Noorden S, Polak JM, Pearse AG. Single cellular origin of somatostatin and calcitonin in the rat thyroid gland. *Histochemistry.* 1977;53:243–247.
138. Yamada Y, Ito S, Matsubara Y, Kobayashi S. Immunohistochemical demonstration of somatostatin-containing cells in the human, dog and rat thyroids. *Tohoku J Exp Med.* 1977;122:87–92.
139. Kusumoto Y. Calcitonin and somatostatin are localized in different cells in the canine thyroid gland. *Biomed Res.* 1980;1:237–241.
140. Kakudo K, Vacca LL. Immunohistochemical study of substance P-like immunoreactivity in human thyroid and medullary carcinoma of the thyroid. *J Submicrosc Cytol.* 1983;15:563–568.
141. Sundler F, Christophe J, Robberecht P, et al. Is helodermin produced by medullary thyroid carcinoma cells and normal C-cells? Immunocytochemical evidence. *Regul Pept.* 1988;20:83–89.
142. Kameya T, Bessho T, Tsumuraya M, et al. Production of gastrin releasing peptide by medullary carcinoma of the thyroid. An immunohistochemical study. *Virchows Arch A Pathol Anat Histopathol.* 1983;401:99–107.
143. Sunday ME, Wolfe HJ, Roos BA, Chin WW, Spindel ER. Gastrin-releasing peptide gene expression in developing hyperplastic, and neoplastic human thyroid C-cells. *Endocrinology.* 1988;122:1551–1558.
144. Gkonos PJ, Tavianini MA, Roos BA. Thyrotropin-releasing hormone gene expression in normal thyroid parafollicular cells. *Mol Endocrinol.* 1989;3:2101–2109.
145. Nunez EA, Gershon MD. Thyrotropin-induced thyroidal release of 5-hydroxytryptamine and accompanying ultrastructural changes in parafollicular cells. *Endocrinology.* 1983;113:309–317.
146. DeLellis RA, Shin SJ, Treaba D. Immunohistochemistry of endocrine tumors (Chapter 10). In: Dabbs DJ, ed. *Diagnostic Immunohistochemistry. Theranostic and Genomic Applications.* Saunders; 2010:291–330.
147. Katoh R, Miyagi E, Nakamura N, et al. Expression of thyroid transcription factor-1 in human C-cells and medullary thyroid carcinoma. *Hum Pathol.* 2000;31:386–393.
148. Bejarano PA, Nikiforov YE, Swenson ES, Biddinger PW. Thyroid transcription factor-1, thyroglobulin, cytokeratin 7 and cytokeratin 20 in thyroid neoplasms. *Appl Immunohistochem Mol Morphol.* 2000;8:189–194.
149. Faggiano A, Talbot M, Lacroix L. et al. Differential expression of galectin-3 in medullary thyroid carcinoma and C-cell hyperplasia. *Clin Endocrinol (Oxf).* 2002;57:813–819.
150. MacIntyre I. Calcitonin: Physiology, biosynthesis, secretion, metabolism and mode of action. In: DeGroot LJ, ed. *Endocrinology*. 2nd ed. Vol 2. WB Saunders; 1989:892–901.
151. Amara SG, Jonas V, Rosenfeld MG, Ong ES, Evans RM. Alternative RNA processing in calcitonin gene expression generates mRNAs encoding different polypeptide products. *Nature.* 1982;298:240–244.
152. Kontozoglou T, Mambo N. The histopathologic features of lithium-associated thyroiditis. *Hum Pathol.* 1983;14:737–739.
153. Fellegara G, Rosai J. Multifocal fibrosing thyroiditis: report of 55 cases of a poorly recognized entity. *Am J Surg Pathol.* 2015;39:416–424.
154. Frank R, Baloch ZW, Gentile C, Watt CD, LiVolsi VA. Multifocal fibrosing thyroiditis and its association with papillary thyroid carcinoma using BRAF pyrosequencing. *Endocr Pathol.* 2014;25:236–240.
155. Schroder S, Bocker W. Lipomatous lesions of the thyroid gland: A review. *Appl Pathol.* 1985;3:140–149.
156. Morizumi H, Sano T, Tsuyuguchi M, Yoshinarl T, Morimoto S. Localized adiposity of the thyroid, clinically mimicking an adenoma. *Endocr Pathol.* 1991;2:226–229.
157. Langlois NE, Krukowski ZH, Miller ID. Pancreatic tissue in a lateral cervical cyst attached to the thyroid gland–a presumed foregut remnant. *Histopathology.* 1997;31:378–380.

158. Hunt JL, Barnes EL. Non-tumor-associated psammoma bodies in the thyroid. *Am J Clin Pathol.* 2003 Jan;119(1):90–4.
159. Batsakis JG, Nishiyama RH, Rich CR. Microlithiasis (calcospherites) and carcinoma of the thyroid gland. *Arch Pathol.* 1960;69:493–498.
160. Dugan JM, Atkinson BF, Avitabile A, Schimmel M, LiVolsi VA. Psammoma bodies in fine needle aspirate of the thyroid in lymphocytic thyroiditis. *Acta Cytol.* 1987;31:330–334.
161. Erdheim JI. Uber Schilddrusenaplasie. II. Geschwulste des Ductus Thyreoglossus. III. Uber einige menschliche Kiemenderivate. *Beitr Pathol Anat.* 1904;35:366–433.
162. Getzowa S. Zur Kenntnis des postbranchialen Korpers und der branchialen Kanalchen des Menschen. *Virchows Arch.* 1907;88:181–235.
163. Calvert R, Isler H. Fine structure of a third epithelial component of the thyroid gland of the rat. *Anat Rec.* 1970;168:23–41.
164. Black HE, Capen CC, Young DM. Ultimobranchial thyroid neoplasms in bulls. A syndrome resembling medullary thyroid carcinoma in man. *Cancer.* 1973;32:865–878.
165. Ljungberg O, Nilsson PO. Hyperplastic and neoplastic changes in ultimobranchial remnants and in parafollicular (C) cells in bulls: A histologic and immunohistochemical study. *Vet Pathol.* 1985;22:95–103.
166. Harach HR. Solid cell nests of the human thyroid in early stages of postnatal life. Systematic autopsy study. *Acta Anat (Basel).* 1986;127:262–264.
167. Fellegara G, Dorji T, Bajimeta MR, Rosai J.. "Giant" solid cell nest of the thyroid: A hyperplastic change? *Int J Surg Pathol.* 2009;17:268–269.
168. Janzer RC, Weber E, Hedinger C. The relation between solid cell nests and C cells of the thyroid gland: An immunohistochemical and morphometric investigation. *Cell Tissue Res.* 1979;197:295–312.
169. Chan JK, Tse CC. Solid cell nest–associated C-cells: Another possible explanation for "C-cell hyperplasia" adjacent to follicular cell tumors. *Hum Pathol.* 1989;20:498–499.
170. Martin V, Martin L, Viennet G, Hergel M, Carbillet JP, Fellmann D. Ultrastructural features of "solid cell nest" of the human thyroid gland: A study of 8 cases. *Ultrastruct Pathol.* 2000;24:1–8.
171. Ríos Moreno MJ, Galera-Ruiz H, De Miguel M, López MI, Illanes M, Galera-Davidson H. Inmunohistochemical profile of solid cell nest of thyroid gland. *Endocr Pathol.* 2011;22:35–39.
172. Gucer H, Mete O. Positivity for GATA3 and TTF-1 (SPT24), and negativity for monoclonal PAX8 expand the biomarker profile of the solid cell nests of the thyroid gland. *Endocr Pathol.* 2018;29:49–58.
173. Cameselle-Teijeiro J, Varela-Durán J, Sambade C, Villanueva JP, Varela-Núñez R, Sobrinho-Simoes M. Solid cell nests of the thyroid: Light microscopy and immunohistochemical profile. *Hum Pathol.* 1994;25:684–693.
174. Preto A, Cameselle-Teijeiro J, Moldes-Boullosa J, et al. Telomerase expression and proliferative activity suggest a stem cell role for thyroid solid cell nests. *Mod Pathol.* 2004;17:819–826.
175. Reis-Filho JS, Preto A, Soares P, Ricardo S, Cameselle-Teijeiro J, Sobrinho-Simões M. p63 expression in solid cell nests of the thyroid: Further evidence for a stem cell origin. *Mod Pathol.* 2003;16:43–48.
176. Yamaoka Y. Solid cell nest (SCN) of the human thyroid gland. *Acta Pathol Jpn.* 1973;23:493–506.
177. Ljungberg O, Nilsson PO. Intermediate thyroid carcinoma in humans and ultimobranchial tumors in bulls: A comparative morphological and immunohistochemical study. *Endocr Pathol.* 1991;2:24–39.
178. Cameselle-Teijeiro J, Caramés N, Romero-Rojas A, et al. Thyroid-type solid cell nests in struma ovarii. *Int J Surg Pathol.* 2011;19:627–631.
179. Louis DN, Vickery AL Jr, Rosai J, Wang CA. Multiple branchial cleft–like cysts in Hashimoto's thyroiditis. *Am J Surg Pathol.* 1989;13:45–49.
180. Apel RL, Asa SL, Chalvardjian A, LiVolsi VA. Intrathyroidal lymphoepithelial cysts of probable branchial origin. *Hum Pathol.* 1994;25:1238–1242.
181. Streutker CJ, Murray D, Kovacs K, Higgins HP. Epithelial cyst of thyroid. *Endocr Pathol.* 1997;8:75–80.
182. Manzoni M, Roversi G, Di Bella C, et al. Solid cell nests of the thyroid gland: Morphological, immunohistochemical and genetic features. *Histopathology.* 2016;68:866–874.
183. Harach HR, Vujanic GM. Intrathyroidal parathyroid. *Pediatr Pathol.* 1993;13:71–74.
184. Spiegel AM, Marx SJ, Doppman JL, et al. Intrathyroidal parathyroid adenoma or hyperplasia. An occasionally overlooked cause of surgical failure in primary hyperparathyroidism. *JAMA.* 1975;234:1029–1033.
185. Chen CL, Lin SH, Yu JC, Shih ML. Persistent renal hyperparathyroidism caused by intrathyroidal parathyroid glands. *J Chin Med Assoc.* 2014;77:492–495.
186. LiVolsi V. Branchial and thymic remnants in the thyroid and cervical region: An explanation for unusual tumors and microscopic curiosities. *Endocr Pathol.* 1993;4:115–119.
187. Harach HR, Vujanic GM. Intrathyroidal thymic tissue: An autopsy study in fetuses with some emphasis on pathological implications. *Pediatr Pathol.* 1993;13:431–434.
188. Neill J. Intrathyroid thymoma. *Am J Surg Pathol.* 1986;10:660–661.
189. Damiani S, Filotico M, Eusebi V. Carcinoma of the thyroid showing thymoma-like features. *Virchows Arch A Pathol Anat Histopathol.* 1991;418:463–466.
190. Miyauchi A, Kuma K, Matsuzuka F, et al. Intrathyroidal epithelial thymoma: An entity distinct from squamous cell carcinoma of the thyroid. *World J Surg.* 1985;9:128–135.
191. Weissferdt A, Moran CA. Ectopic primary intrathyroidal thymoma: A clinicopathological and immunohistochemical analysis of 3 cases. *Hum Pathol.* 2016;49:71–76.
192. Cameselle-Teijeiro J, Varela-Duran J. Intrathyroid salivary gland-type tissue in multinodular goiter. *Virchows Arch.* 1994;425:331--334.
193. Finkle HI, Goldman RL. Heterotopic cartilage in the thyroid. *Arch Pathol.* 1973;95:48–49.
194. Ellis PD, van Nostrand AW. The applied anatomy of thyroglossal tract remnants. *Laryngoscope.* 1977;87(5 Pt 1):765–770.
195. Guimaraes SB, Uceda JE, Lynn HB. Thyroglossal duct remnants in infants and children. *Mayo Clin Proc.* 1972;47:117–120.
196. Larochelle D, Arcand P, Belzile M, Gagnon NB. Ectopic thyroid tissue–a review of the literature. *J Otolaryngol.* 1979;8:523–530.

197. Williams ED, Toyn CE, Harach HR. The ultimobranchial gland and congenital thyroid abnormalities in man. *J Pathol.* 1989;159:135–141.
198. Ibrahim NA, Fadeyibi IO. Ectopic thyroid: Etiology, pathology and management. *Hormones (Athens)* 2011;10:261–269.
199. De Andrade MA. A review of 128 cases of posterior mediastinal goiter. *World J Surg.* 1977;1:789–797.
200. de Souza FM, Smith PE. Retrosternal goiter. *J Otolaryngol.* 1983;12:393–396.
201. Kantelip B, Lusson JR, De Riberolles C, Lamaison D, Bailly P. Intracardiac ectopic thyroid. *Hum Pathol.* 1986; 17:1293–1296.
202. Baughman RA. Lingual thyroid and lingual thyroglossal tract remnants. A clinical and histopathologic study with review of the literature. *Oral Surg Oral Med Oral Pathol.* 1972;34:781–799.
203. Reaume CE, Sofie VL. Lingual thyroid. Review of the literature and report of a case. *Oral Surg Oral Med Oral Pathol.* 1978;45:841–845.
204. Kansal P, Sakati N, Rifai A, Woodhouse N. Lingual thyroid. Diagnosis and treatment. *Arch Intern Med.* 1987;147: 2046–2048.
205. Nienas FW, Gorman CA, Devine KD, Woolner LB. Lingual thyroid. Clinical characteristics of 15 cases. *Ann Intern Med.* 1973;79:205–210.
206. Strickland AL, Macfie JA, Van Wyk JJ, French FS. Ectopic thyroid glands simulating thyroglossal duct cysts. *JAMA.* 1969;208:307–310.
207. Talib H. Lingual thyroid. *Br J Clin Pract.* 1966;20:322–323.
208. Wapshaw H. Lingual thyroid. A report of a case with unusual histology. *Br J Surg.* 1974;30:160–165.
209. Sturniolo G, Violi MA, Galletti B, et al. Differentiated thyroid carcinoma in lingual thyroid. *Endocrine.* 2016;5:189–198.
210. Seoane JM, Cameselle-Teijeiro J, Romero MA. Poorly differentiated oxyphilic (Hürthle cell) carcinoma arising in lingual thyroid: A case report and review of the literature. *Endocr Pathol.* 2002;13:353–360.
211. LiVolsi VA, Perzin KH, Savetsy L. Carcinoma arising in median ectopic thyroid (including thyroglossal duct tissue). *Cancer.* 1974;34:1303–1315.
212. Allard RH. The thyroglossal cyst. *Head Neck Surg.* 1982; 5:134–146.
213. Soucy P, Penning J. The clinical relevance of certain observations on the histology of the thyroglossal tract. *J Pediatr Surg.* 1984;19:506–509.
214. Kreft A, Hansen T, Kirkpatrick CJ. Thyroid transcription factor 1 expression in cystic lesions of the neck: An immunohistochemical investigation of thyroglossal duct cysts, branchial cleft cysts and metastatic papillary thyroid cancer. *Virchows Arch.* 2005;447:9–11.
215. Block MA, Wylie JH, Patton RB, Miller JM. Does benign thyroid tissue occur in the lateral part of the neck? *Am J Surg.* 1966;112:476–481.
216. Klopp CT, Kirson SM. Therapeutic problems with ectopic non-cancerous follicular thyroid tissue in the neck: 18 case reports according to etiological factors. *Ann Surg.* 1966;163:653–664.
217. Moses DC, Thompson NW, Nishiyama RH, Sisson JC. Ectopic thyroid tissue in the neck. Benign or malignant? *Cancer.* 1976;38:361–365.
218. Hathaway BM. Innocuous accessory thyroid nodules. *Arch Surg.* 1965;90:222–227.
219. Sisson JC, Schmidt RW, Beierwaltes WH. Sequestered nodular goiter. *N Engl J Med.* 1964;270:927–932.
220. Leon X, Sancho FJ, Garcia J, Sanudo JR, Orus C, Quer M. Incidence and significance of clinically unsuspected thyroid tissue in lymph nodes found during neck dissection in head and neck carcinoma patients. *Laryngoscope.* 2005;115(3):470–474.
221. Wallace A, Angelos P, Cipriani NA. The controversy of lateral aberrant thyroid and the use of BRAF immunostain. *Int J Surg Pathol.* 2022;30:217–220.
222. Bone RC, Biller HF, Irwin TM. Intralaryngotracheal thyroid. *Ann Otol Rhinol Laryngol.* 1972;81:424–428.
223. Donegan JO, Wood MD. Intratracheal thyroid—familial occurrence. *Laryngoscope.* 1985;95:6–8.
224. Babazade F, Mortazavi H, Jalalian H, Shahvali E. Thyroid tissue as a submandibular mass: A case report. *J Oral Sci.* 2009;51:655–657.
225. Williams RJ, Lindop G, Butler J. Ectopic thyroid tissue on the ascending aorta: An operative finding. *Ann Thorac Surg.* 2002;73:1642–1643.
226. Pollice L, Caruso G. Struma cordis. Ectopic thyroid goiter in the right ventricle. *Arch Pathol Lab Med.* 1986;110:452–453.
227. Simon M, Baczako K. Thyroid inclusion in the lung. Metastasis of an occult papillary carcinoma or ectopia? *Pathol Res Pract.* 1989;184:263–267; discussion 268–270.
228. Di Mari N, Barbagli L, Mourmouras V, Miracco C. Ectopic thyroid of the lung. An additional case. *Pathologica* 2010;102:102–103.
229. Postlethwait RW, Detmer DE. Ectopic thyroid nodule in the esophagus. *Ann Thorac Surg.* 1975;19:98–100.
230. Takahashi T, Ishikura H, Kato H, Tanabe T, Yoshiki T. Ectopic thyroid follicles in the submucosa of the duodenum. *Virchows Arch A Pathol Anat Histopathol.* 1991;418:547–550.
231. Hammers YA, Kelly DR, Muensterer OJ, Hardin WD Jr, Saeed SA, Mroczek-Musulman EC. Giant polypoid gastric heterotopia with ectopic thyroid tissue: Unusual cause of jejuno-jejunal intussusception. *J Pediatr Gastroenterol Nutr.* 2007;45:484–487.
232. Cassol CA, Noria D, Asa SL. Ectopic thyroid tissue within the gall bladder: Case report and brief review of the literature. *Endocr Pathol.* 2010;21:263–265.
233. Sekine S, Nagata M, Hamada H, Watanabe T. Heterotopic thyroid tissue at the porta hepatis in a fetus with trisomy 18. *Virchows Arch.* 2000;436:498–501.
234. Fushimi H, Kotoh K, Nakamura H, Tachibana T, Yutani C. Ectopic thyroid tissue adjacent to the gallbladder. *Histopathology.* 1998;32:90–91.
235. Eyuboglu E, Kapan M, Ipek T, Ersan Y, Oz F. Ectopic thyroid in the abdomen: Report of a case. *Surg Today.* 1999;29:472–474.
236. Seelig MH, Schonleben K. Intra-abdominal ectopic thyroid presenting as a pancreatic tumour. *Eur J Surg.* 1997;163: 549–551.
237. Salam M, Mohideen A, Stravitz RT. Ectopic thyroid presenting as a liver mass. *Clin Gastroenterol Hepatol.* 2012;10.
238. Cicek Y, Tasci H, Gokdogan C, Oneş S, Göksel S. Intra-abdominal ectopic thyroid. *Br J Surg.* 1993;80:316.

239. Shiraishi T, Imai H, Fukutome K, et al. Ectopic thyroid in the adrenal gland. *Hum Pathol.* 1999;30:105–108.
240. Tamaki S, Miyakura Y, Someya S, et al. Laparoscopic resection of retroperitoneal ectopic thyroid tissue. *Asian J Endosc Surg.* 2017;10:331–333.
241. Kurman RJ, Prabha AC. Thyroid and parathyroid glands in the vaginal wall. *Am J Clin Pathol.* 1973;59:503–507.
242. Yilmaz F, Uzunlar AK, Sogutcu N. Ectopic thyroid tissue in the uterus. *Acta Obstet Gynecol Scand.* 2005;84: 201–202.
243. Ruchti C, Balli-Antunes H, Gerber HA. Follicular tumor in the sellar region without primary cancer of the thyroid. Heterotopic carcinoma? *Am J Clin Pathol.* 1987;87: 776–780.

44

Parathyroids

Sylvia L. Asa ■ Ozgur Mete

The parathyroid glands are endocrine organs composed of small clusters of neuroendocrine cells that are present in the neck adjacent to the thyroid, embedded within the thyroid or in the upper mediastinum. They are the principal regulators of calcium homeostasis through the synthesis and secretion of parathyroid hormone.

These small glands are the source of common pathology, since asymptomatic hyperparathyroidism is now recognized as a frequent event with the advent of routine serum calcium testing (1). In addition, parathyroids are routinely removed during thyroid surgery, making them a common finding in surgical pathology.

The interpretation of parathyroid structure is highly dependent on understanding the normal anatomy, histology, and development of parathyroid glands. As with all endocrine tissues, the pathologist must also understand the function of these glands to appreciate the context of structure-function correlations.

In this chapter, we review the embryology, anatomy, and microscopic morphology of parathyroid glands to provide a clear understanding of the variability and factors that influence the structure of these glands in normal and abnormal circumstances.

HISTORY AND NOMENCLATURE

The first documented description of a parathyroid gland occurred in 1850 when Professor Richard Owen reported to the Royal College in London the presence of "a small, compact yellow glandular body attached to the thyroid at the point where the veins emerge" in the Indian rhinoceros. The paper was published 12 years later (2) and in the interim, Remak described similar glands in the cat (3). In 1855, Virchow identified the human counterparts (4). The name "glandulae parathyroidae" was coined by Ivar Sandström in 1880; as a medical student, he studied the parathyroid glands of several species (5). These studies attributed a relationship of the parathyroid glands to the thyroid, both structurally and embryologically, and it was only when Kohn identified an independent origin of parathyroids and proposed the name "Epithelkörperchen" (6,7) that they were recognized to be distinct and separate from the thyroid.

The function of the parathyroid glands was described in careful studies by Gley and Erdheim who showed an association between lack of these glands and death by tetany (8–10). The mechanism of action through regulation of calcium by parathyroid hormone was debated for many years until proven by the work of Hanson (11) and Collip (12) and ultimately by the discovery of parathyroid hormone by Aurbach in 1959 (13).

DISTRIBUTION OF PARATHYROID GLANDS

The normal adult has four parathyroid glands, but up to 13% of people have at least one supernumerary gland (14,15).

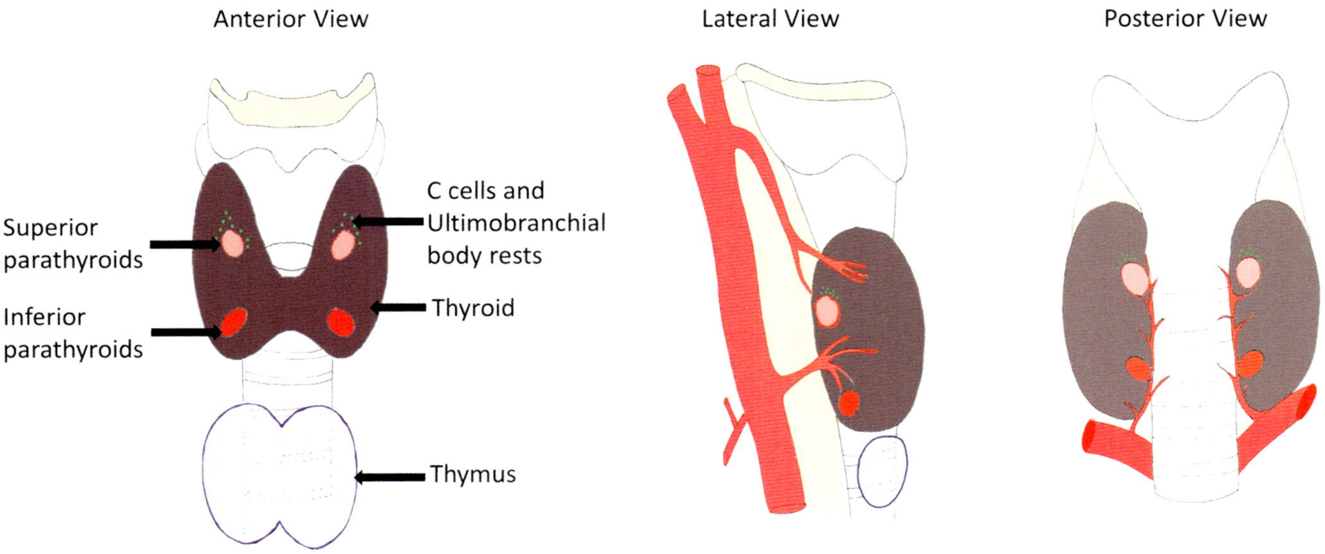

FIGURE 44.1 **Location of parathyroid glands**. There are four parathyroid glands, two on each side of the neck, behind the thyroid gland. The superior glands are at the junction of the upper and middle third of each lateral lobe; the inferior glands are usually near the inferior pole of the thyroid as illustrated, but are often lower, associated with the thymus, and may be found anywhere between the hyoid bone and the mediastinum. (Drawings courtesy of Paolo Batoni, PhD and Zoya Volynskaya, PhD.)

These data are derived from autopsy studies but also from clinical studies of patients with hyperparathyroidism (1). Reports of fewer than four glands in normal adults are attributable to failure to identify these very small structures that can be confused with lymph nodes and ganglia in the neck, or can be hidden within the thyroid gland itself.

The superior glands are usually located near the cricothyroid junction (Fig. 44.1), deep to the recurrent laryngeal nerve and superior to its junction with the inferior thyroid artery (16) (Fig. 44.2). They are usually intimately associated with the posterior thyroid at the junction of the upper and middle third of each lateral lobe; they may even be in the perithyroidal fascia or completely intrathyroidal (Fig. 44.3), and therefore easily missed until the thyroid is dissected. Less frequently, they may be located behind the pharynx or esophagus.

The inferior glands are more anterior and ventral to the recurrent laryngeal nerve (16). There is significant variation in their anatomical position, and they may be near the inferior pole of the thyroid, within the thyrothymic ligament, higher in the neck, as high as the hyoid bone, at the carotid bulb, or associated with thymus (Fig. 44.4) within the mediastinum; they have been reported anywhere from the jaw to the pericardium. The distribution tends to be symmetrical in any given patient.

Ectopic parathyroid glands have been reported in up to 22% of patients undergoing parathyroid surgery but these reports include glands that are found within the thymus, in the retroesophageal space and mediastinum as well as intrathyroidal glands that are not truly "ectopic" (17). This term should be restricted to undescended glands in a high cervical position and those trapped within the carotid sheath.

The importance of supernumerary glands is evident when surgery is performed for hyperparathyroidism and indeed the most common reason for failed parathyroidectomy is the inability to find a supernumerary or ectopic parathyroid gland (1).

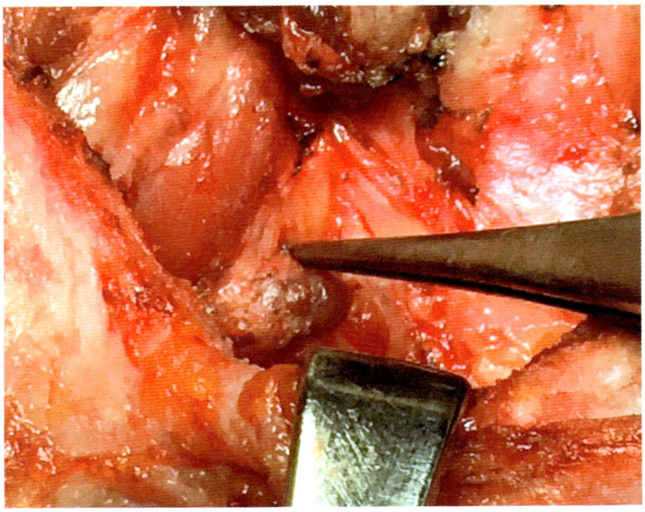

FIGURE 44.2 **Localization of a parathyroid gland**. A normal right superior parathyroid gland is embedded in perithyroidal fat posterior to the thyroid that has been removed in this intraoperative image taken post hemithyroidectomy. (Photo provided by Dr. Lorne Rotstein.)

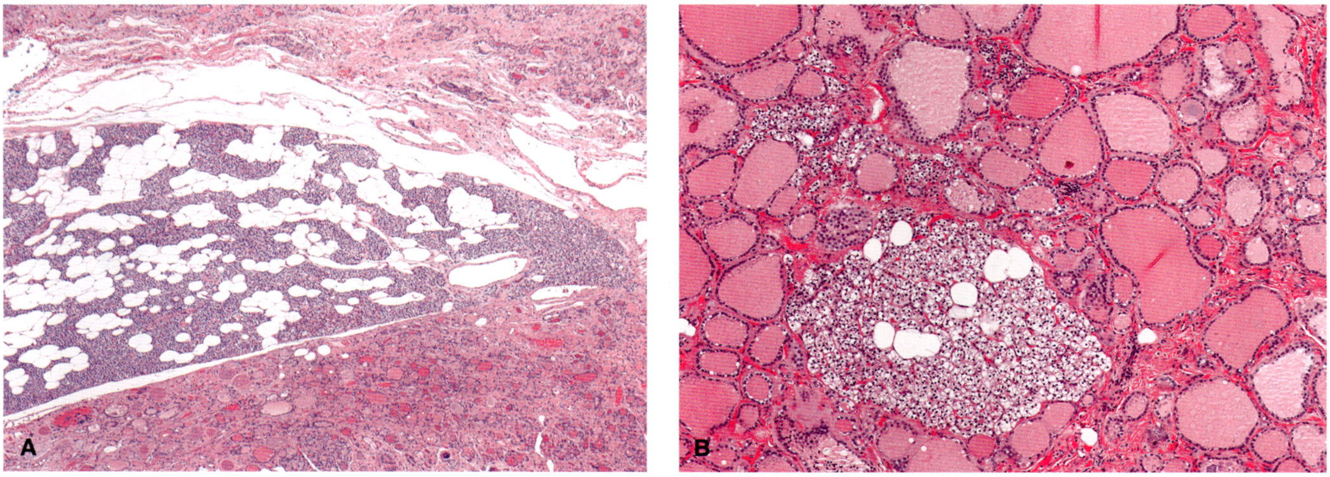

FIGURE 44.3 Intrathyroidal parathyroid glands. Parathyroids can be found within the thyroid parenchyma as in these examples (**A, B**) where parathyroids are completely embedded within the thyroid and surrounded by thyroid parenchyma.

EMBRYOLOGY

The parathyroid glands arise from the third and fourth pharyngeal pouches (also known as branchial pouches) of the endoderm between the 5th and 12th week of gestation (18–20) (Fig. 44.5). The embryology can be confusing because of the differences in migration of these structures that ultimately bring the third pouch to lie lower than or caudal to the fourth pouch.

The superior parathyroids are derived from the fourth pharyngeal pouch (parathyroid IV) that has a short embryonic migration, explaining their relative consistency in position. The third pharyngeal pouch (parathyroid III) has a more complex pattern of descent; the midline structures move cephalad faster than the lateral ones, resulting in a more variable location of the inferior glands. The parathyroid glands migrate with the thymus that attaches to the pericardium and comes to lie in the mediastinum. In the 18-mm embryo, parathyroid III is at the level of the lower pole of the thyroid, and at this stage it usually separates from the thymus, forming the lower parathyroid glands (21). However, variation in the level at which they separate accounts for the anatomic variations in the position of the inferior glands.

During migration of parathyroid III, small fragments of parathyroid tissue may become separated from the main pharyngeal pouch resulting in supernumerary glands. They occur most commonly in the thymus but have also been described in the carotid sheath (22), carotid bifurcation (23), para-aortic region (23), esophagus, hypopharynx, vagus nerve, and other sites (24). Parathyroid hormone production has been demonstrated by immunohistochemistry as early as $8^3/_7$ weeks gestational age (3.1 cm crown–rump length) (21) and is readily identified at 10 weeks (25). By 17 to 20 weeks, immunoreactivity is abundant.

The differentiation and growth of parathyroid glands is dependent on a number of genes (Table 44.1) that have been identified from studies of patients with hypoparathyroidism, particularly the DiGeorge syndrome/velocardiofacial

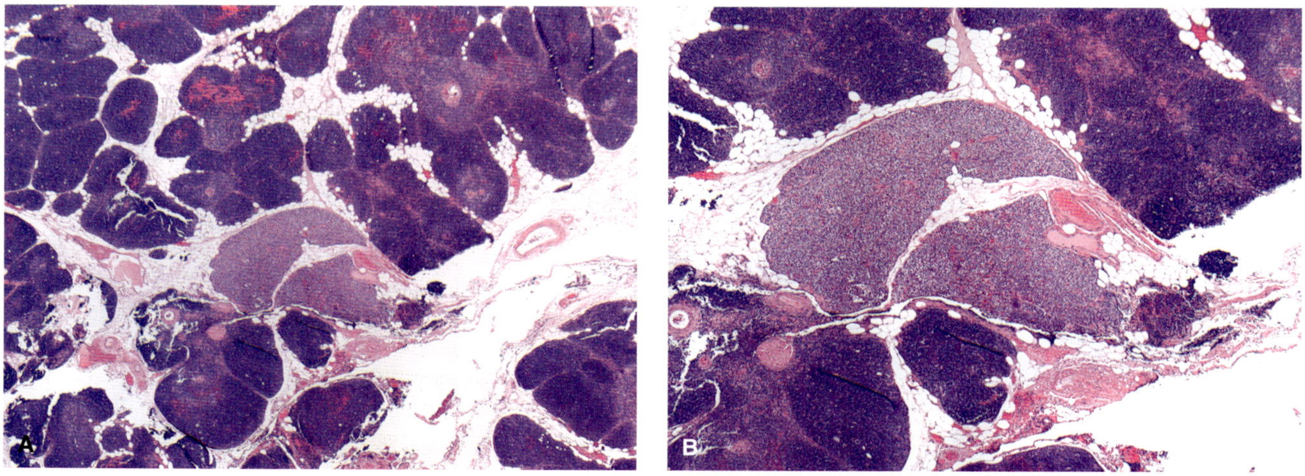

FIGURE 44.4 Intrathymic parathyroid glands. The lower or inferior parathyroid glands can sometimes be found within the thymus (**A, B**).

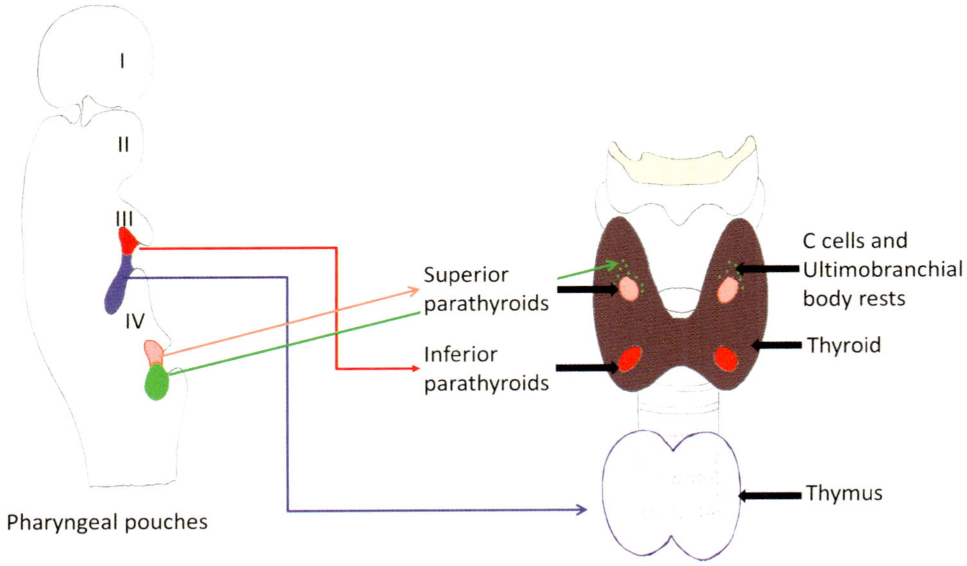

FIGURE 44.5 Embryology of parathyroid glands. The parathyroid glands derived from the third and fourth pharyngeal pouches. The superior glands originate from the fourth pouch and migrate almost directly to the junction of the upper third and lower two thirds of the thyroid lobes along with the ultimobranchial body that gives rise to calcitonin-producing C cells of the thyroid. The lower parathyroids originate higher with the thymus in the third pouch and migrate further caudally, explaining their more variable location and frequent association with the thymus. (Drawings courtesy of Paolo Batoni, PhD and Zoya Volynskaya, PhD.)

syndrome (22q11.2 deletion syndrome), and by studies of mouse models of development.

A number of transcription factors are required for parathyroid differentiation. *HOXA3,* a member of the *HOX* (homeotic genes) family, is required for the development of the organs arising from the third and fourth pharyngeal pouches (26,27). *PAX9* (paired box 9) deficiency also results in a lack of thymus, parathyroid glands, and ultimobranchial bodies (28), while *PAX1* deficiency is associated with reduced parathyroid development (29). *TBX1* (T-box gene 1)-knockout mice exhibit hypoplasia of the thymus and parathyroid glands, cardiac outflow tract abnormalities, abnormal facial structures, abnormal vertebrae, and cleft palate (30–32). Mammalian *GCM2* (Glial cells missing, *GCMB* in humans) is specifically restricted to the parathyroid primordium (33) and *GCM2*-deficient mice lack parathyroid glands (34). However, they have an expression of parathyroid hormone in the thymus driven by *GCM1*, another homolog of the Drosophila *GCM* family that functions in placental development (34). Expression of *GCM* is absent in *HOXA3*-mutant mice, indicating that it acts downstream of *HOXA3* (29). Heterozygous mutations of *GATA3* cause hypoparathyroidism with sensorineural deafness and renal dysplasia (35); *GATA3*-deficient mice fail to develop parathyroids and lack *GCM2* expression, indicating that this transcription factor is also upstream of *GCM2* (36). *Eyes absent (EYA)1* is required for the initiation of thymus and parathyroid gland formation downstream of the *HOXA3* and *PAX* genes (37) and *SIX1/4* (sineoculis homeobox homolog) also regulates patterning of the third pharyngeal pouch (38). *SOX3* (Sry-related HMG box) and *AIRE1* (autoimmune regulator) have also been implicated in parathyroid development in humans (35).

Growth factors and their signaling pathways modulate migration of the pharyngeal pouch endoderm, its interaction with the mesenchyme, and the balance between proliferation and apoptosis. The critical factors include expression of *fibroblast growth factor (FGF8)* (39,40), *bone morphogenetic proteins (BMPs)* (41), *Chordin (Chrd)* (42), and *transforming growth factor β (TGFβ)* (43), as well as silencing of *Sonic hedgehog (SHH)* (44). Mutations in tubulin chaperone E (*TBCE*) have been found in patients with the human hypoparathyroidism–retardation–dysmorphism (HRD) syndrome (45,46) but its role in parathyroid development and function remains to be proven in a mouse model.

TABLE 44.1 Transcription and Growth Factors Involved in the Development of Parathyroid Glands

Transcription Factors	Growth Factors
HOXA3	FGF8
PAX9	BMPs
PAX1	Chordin
TBX1	TGFβ
GCM2	SHH
EYA1	TBCE
GATA3	
SOX3	
AIRE1	

ANATOMY—GROSS FEATURES

Parathyroid glands are smooth, soft, flattened ovoid or nodular structures (Fig. 44.6) that vary in shape based on their relationship with surrounding structures; they can be bilobed or multilobulated in some individuals. The lower parathyroid glands may be larger than the upper glands. The capsule is smooth, shiny, and gray with a fine network of small vessels. The parenchyma is yellow to orange–tan depending on the amount of stromal fat, percent of oncocytic cells, and vascularity (5).

The adult glands usually measure up to 6 mm in length, up to 4 mm in width, and up to 2 mm in thickness, although there is significant variation. The 95% upper limit for gland

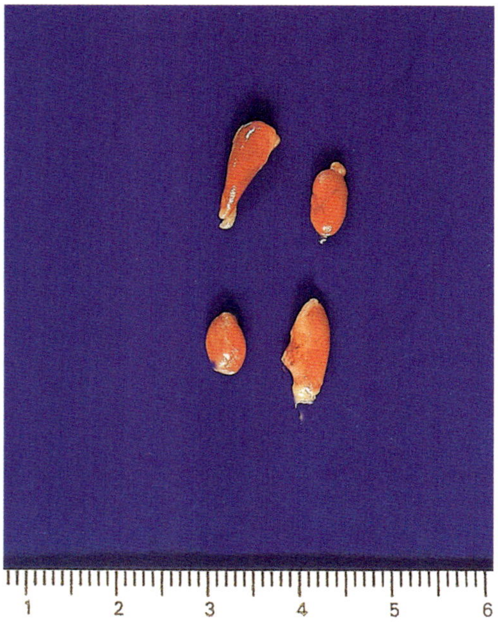

FIGURE 44.6 Gross appearance of parathyroid glands. Four normal parathyroid glands were removed at autopsy from a 53-year-old man. The glands are variable in size, shape, and weight. At the tip of the right lower gland is a small area of yellow thymus. (Courtesy of Dr. S. I. Roth.)

length has been reported as 9 mm for healthy individuals and 10 mm for hospitalized patients (47); anything larger than 1 cm is considered unequivocally abnormal. The weights of these glands are also extremely variable. The total parathyroid weight of all glands gradually increases during development, reaching 5 to 9 mg at 3 months of age (48) and a mean of 120 ± 3.5 mg in men and 142 ± 5.2 mg in women in adults (48,49). Mean weight per gland has been reported to be approximately 32 mg (49–51), has been reported to be lower in Caucasians than in blacks (51), and is higher in hospitalized patients (mean weight of 46.2 g) than in those who died suddenly (47). Pathologists should consider 60 mg to be the upper limit of normal weight for a parathyroid gland (49,52).

Because these glands have a significant amount of interspersed fat, the parenchymal content of the glands is extremely variable and difficult to evaluate. It has been reported that the parenchyma averages 74% of the weight of the gland in adults (53,54). Parenchymal content is a better indicator of gland function than gland weight alone but requires careful morphometric and densitometric analysis (55). The average parenchymal weight per gland has been calculated as 21.6 mg for men and 18.2 mg for women and mean total parenchymal weight for all four glands has been estimated at 82.0 ± 2.6 mg for men and 88.9 ± 3.9 mg for women (51).

Importantly, the amount of stromal fat is known to be dependent on body composition and constitutional fat. In young patients, parathyroid glands have scant fat and parenchymal weight is almost the same as gland weight (49). In adulthood, parenchymal weight stays relatively constant but stromal adipose tissue increases. There may be variation in stromal fat content of different glands in any given individual (49,56).

A number of studies of patients with hyperparathyroidism have examined the nontumorous glands to determine normal parameters (57). These studies are not entirely relevant for normal size, weight, and histology since some glands may be enlarged due to germline abnormalities and others may be reduced in size due to feedback inhibition. Parathyroid weight is likely to be inversely related to serum calcium concentration and direct related to serum phosphorus and renal function.

The arterial supply to the parathyroid glands is derived from the superior and inferior thyroidal arteries, sometimes with anastomosing branches from both arteries. Venous drainage of the upper glands flows to the superior and/or

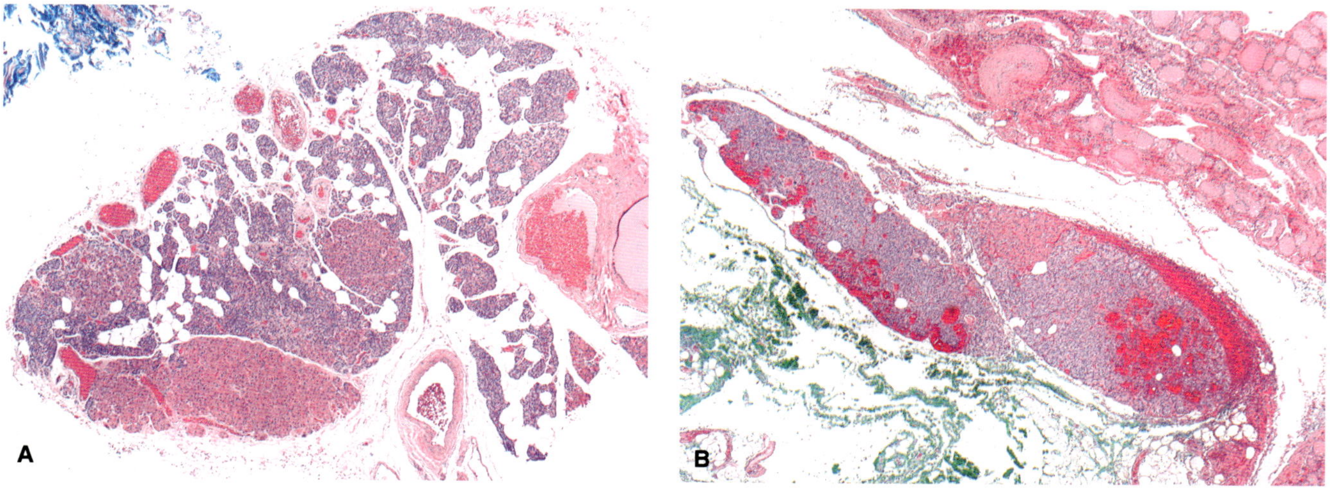

FIGURE 44.7 Vascular hilum of parathyroid glands. Parathyroid glands are well delineated but lack a true capsule; there is an incomplete thin fibrous pseudocapsule that carries the vasculature around the periphery of lobules. Histology allows clear identification of the vascular hilum of parathyroid glands. This is the region most likely to harbor nontumorous parathyroid tissue in glands with neoplasms.

lateral thyroid vein and that of the lower glands to the lateral and/or inferior thyroid vein. Lymphatic drainage is to the superior and inferior deep cervical, pretracheal, paratracheal, and retropharyngeal lymph nodes (49). The site of entry/exit of the main vascular supply forms a hilum of the gland (Fig. 44.7); this is important to identify when examining abnormal glands, since it is usually the site where nontumorous parenchyma (usually present as an atrophic rim around the lesion) can be identified in the case of neoplasia and allows the distinction of parathyroid neoplasia from hyperplasia.

HISTOLOGY AND ULTRASTRUCTURE

The embryonic parathyroid gland is a highly vascular sheet of pure chief cells with clear amphophilic cytoplasm and peripheral nuclear palisading. No oxyphil cells are present. The stroma is composed of prominent vessels but lacks fibrous tissue and adipose cells.

The normal mature parathyroid gland has a thin but incomplete, fibrous pseudocapsule (Fig. 44.7). The vascular pole, or hilum, contains an artery and a vein surrounded by fibrous tissue (Fig. 44.7). The arteries branch into a complex network of smaller arteries and veins that mainly follow the capsule; these capsular arteries and veins are connected by arterioles, capillaries, and venules located in the fibrous septa between the parenchymal cells. As in all endocrine tissues, the capillary endothelium is fenestrated (21,49) and the endothelial cells have dense bodies, pinocytotic vesicles, Weibel–Palade bodies, and tight junctions. Due to the rich capillary network, the cut surface bleeds readily, providing an easy way to distinguish parathyroid glands from lymph nodes, adipose tissue, thymus, and thyroid, which do not bleed as profusely from their cut surfaces (21). The capsule also contains two interconnecting plexuses of lymphatic capillaries: loops of the inner plexus dip into the gland parenchyma, whereas the outer plexus forms efferent drainage associated with the thyroid (58). The interstitium is composed of collagen, elastic fibers and the basement membranes of chief cells and capillaries with pericytes, scattered fibroblasts and mast cells, and a few lymphocytes. Nerve bundles in close proximity to chief cells suggest autonomic innervation that is thought to originate in the vagus nerve (59–61).

The parenchymal cells are found in lobules composed of sheets, solid nests, and trabeculae with interspersed adipocytes (Figs. 44.7 and 44.8). There are usually scattered

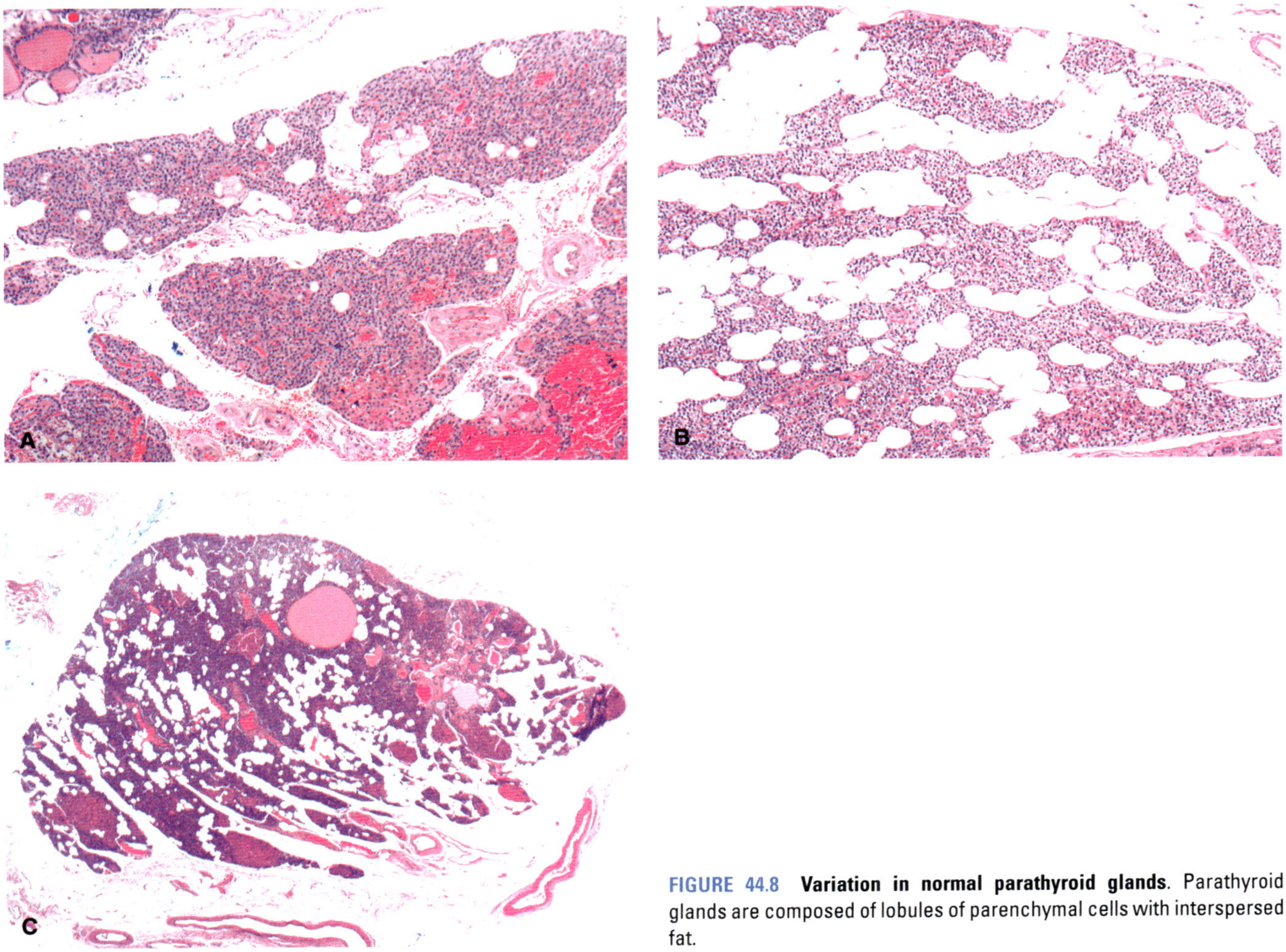

FIGURE 44.8 Variation in normal parathyroid glands. Parathyroid glands are composed of lobules of parenchymal cells with interspersed fat.

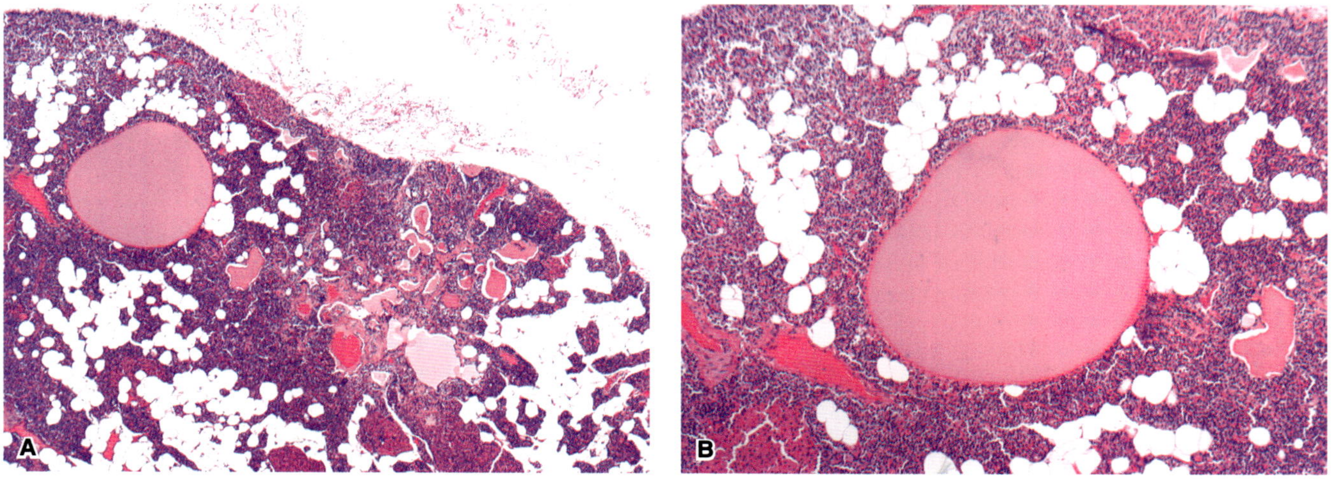

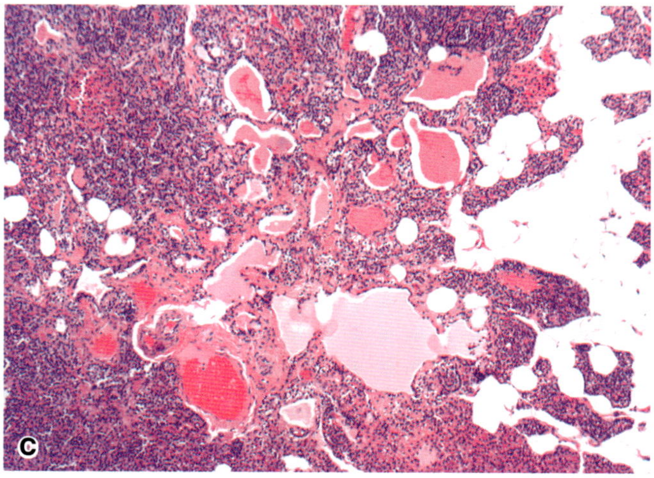

FIGURE 44.9 Colloid-like material in normal parathyroid glands. Parathyroid glands may have cysts and follicles filled with proteinaceous material that resembles colloid of the thyroid gland; proliferations with this colloid-like material can resemble thyroid neoplasms and may require special techniques to verify as of parathyroid origin. Note: Figures 44.9A, B and C are all higher magnifications of areas in the gland shown in 44.8C.

cysts (Figs. 44.8 and 44.9) and small gland-like follicles filled with pink eosinophilic homogeneous material resembling the colloid of thyroid follicles. However, thyroid follicles often contain birefringent calcium oxalate crystals that are not seen in the colloid of parathyroid glands and, thus, this feature can help distinguish thyroid from parathyroid tissue at the time of intraoperative consultation. The colloid-like material of the parathyroid gland contains glycoproteins and stains with periodic acid-Schiff (PAS) (62) and parathyroid hormone by immunohistochemistry, and has been described to have amyloid-like features (63). There is a single cell type responsible for synthesis and secretion of parathyroid hormone; this cell can exhibit variability in morphology due to functional alterations and can undergo oncocytic change. In fact, the parenchyma has been described as normally containing chief cells, oxyphil cells, and transitional cells, although all of these cell types are variants of chief cells with differing degrees of oncocytic change (49). Clear cells, or "water-clear cells," are usually found only in pathologic states (49).

Normal chief cells are round to polygonal and measure 8 to 12 μm in diameter. The nuclei are round, centrally located within the cell, and have a well-defined nuclear membrane with evenly distributed chromatin and small nucleoli. The cells have well-defined borders with cytoplasm that varies from clear to amphophilic or faintly eosinophilic (Fig. 44.10). The clear vacuolated appearance of the chief cell cytoplasm is due to accumulation of glycogen and lipid, resulting in a positive staining reaction with PAS and fat stains such as oil red O. The intracellular content of fat is thought to be inversely correlated with endocrine activity; increased cytoplasmic lipid is a feature of inactive or suppressed chief cells whereas fat is minimal in hormonally active cells.

The ultrastructural features of parathyroid chief cells (Fig. 44.11) are similar to those of other neuroendocrine cells and vary with functional activity (64,65). The organelles for hormone synthesis include rough endoplasmic reticulum where peptide hormone is translated from mRNA, the Golgi apparatus where packaging occurs, and the membrane-bound secretory granules that store hormone and carry it to the cell surface. Resting cells have poorly developed synthetic organelles, accumulation of glycogen and lysosomes, as well as large lipid bodies that correspond to the lipid seen by light microscopy; variable numbers of small dense-core secretory granules accumulate at the cell

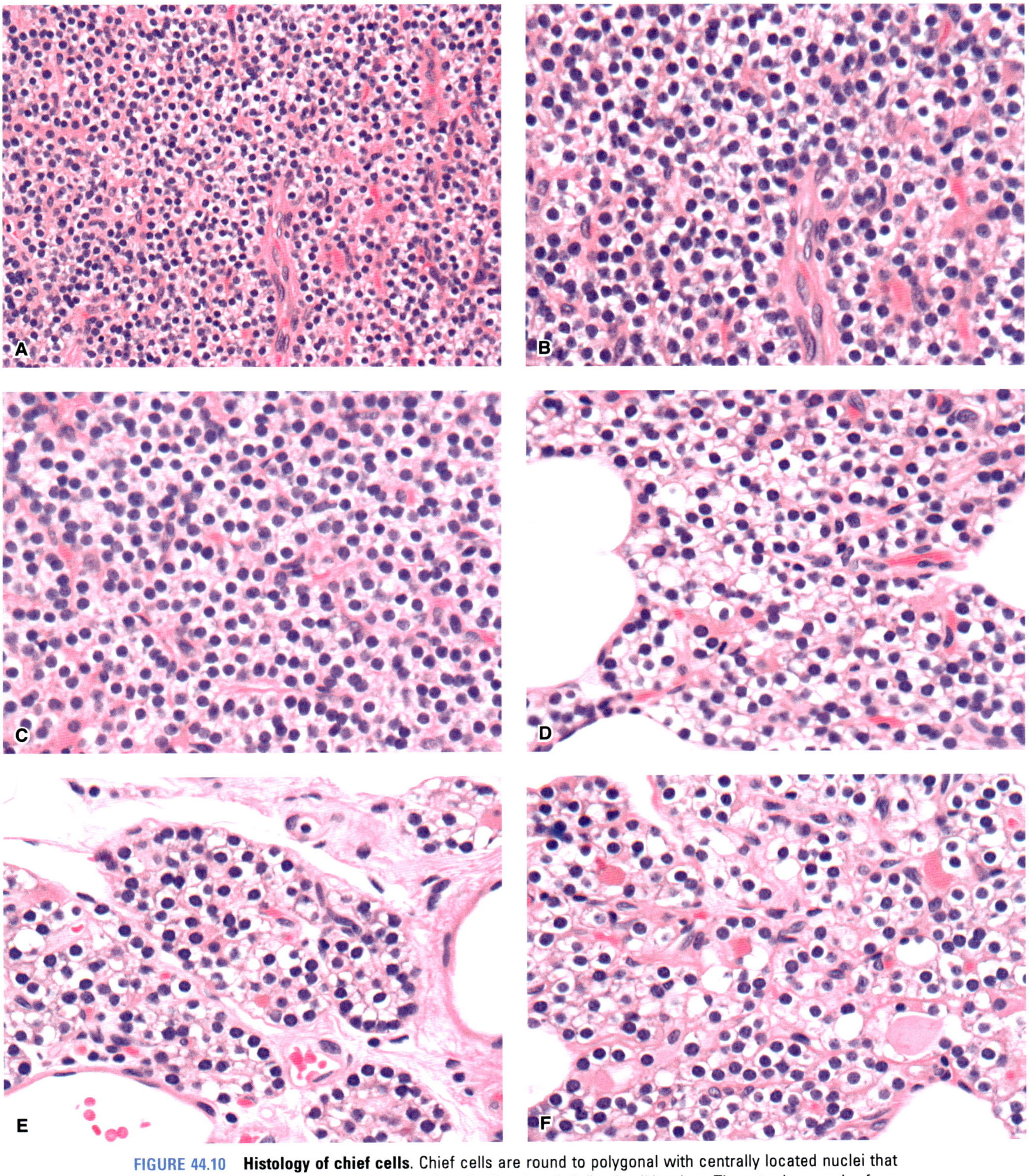

FIGURE 44.10 **Histology of chief cells**. Chief cells are round to polygonal with centrally located nuclei that have evenly distributed chromatin and small nucleoli and well-defined cell borders. The cytoplasm varies from clear to amphophilic or faintly eosinophilic. The clear vacuolated appearance is due to the accumulation of glycogen and/or lipid. Note: Figure **B** is a higher magnification of a portion of Figure **A**.

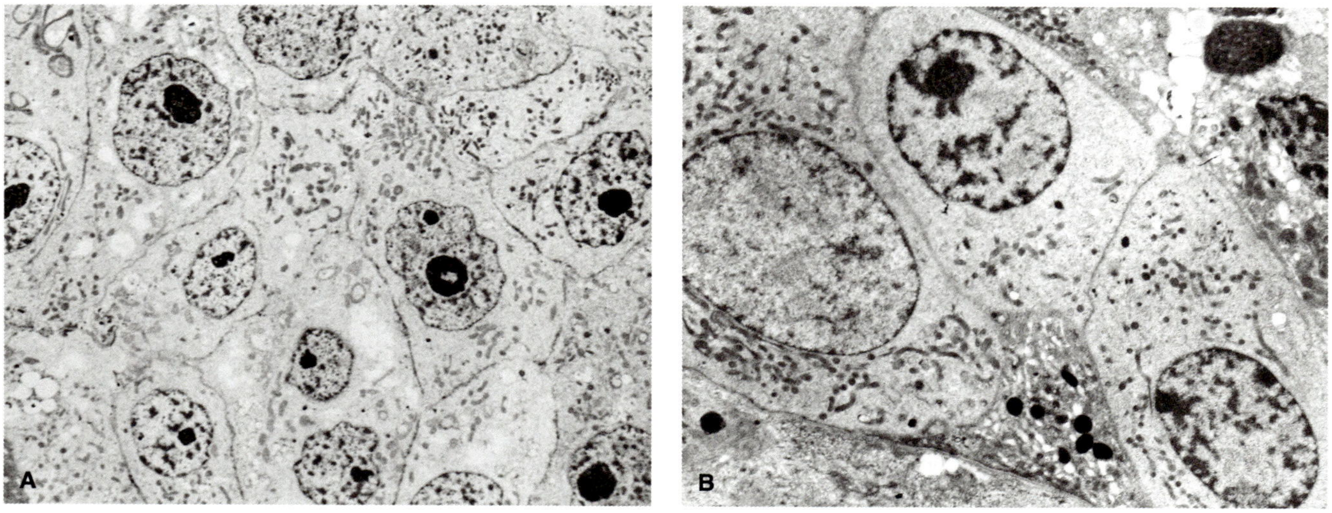

FIGURE 44.11 Ultrastructure of chief cells. By electron microscopy, parathyroid chief cells have short profiles of rough endoplasmic reticulum, variably developed Golgi complexes, scattered small secretory granules, mitochondria and vacuoles that may represent lysosomes with lipid content.

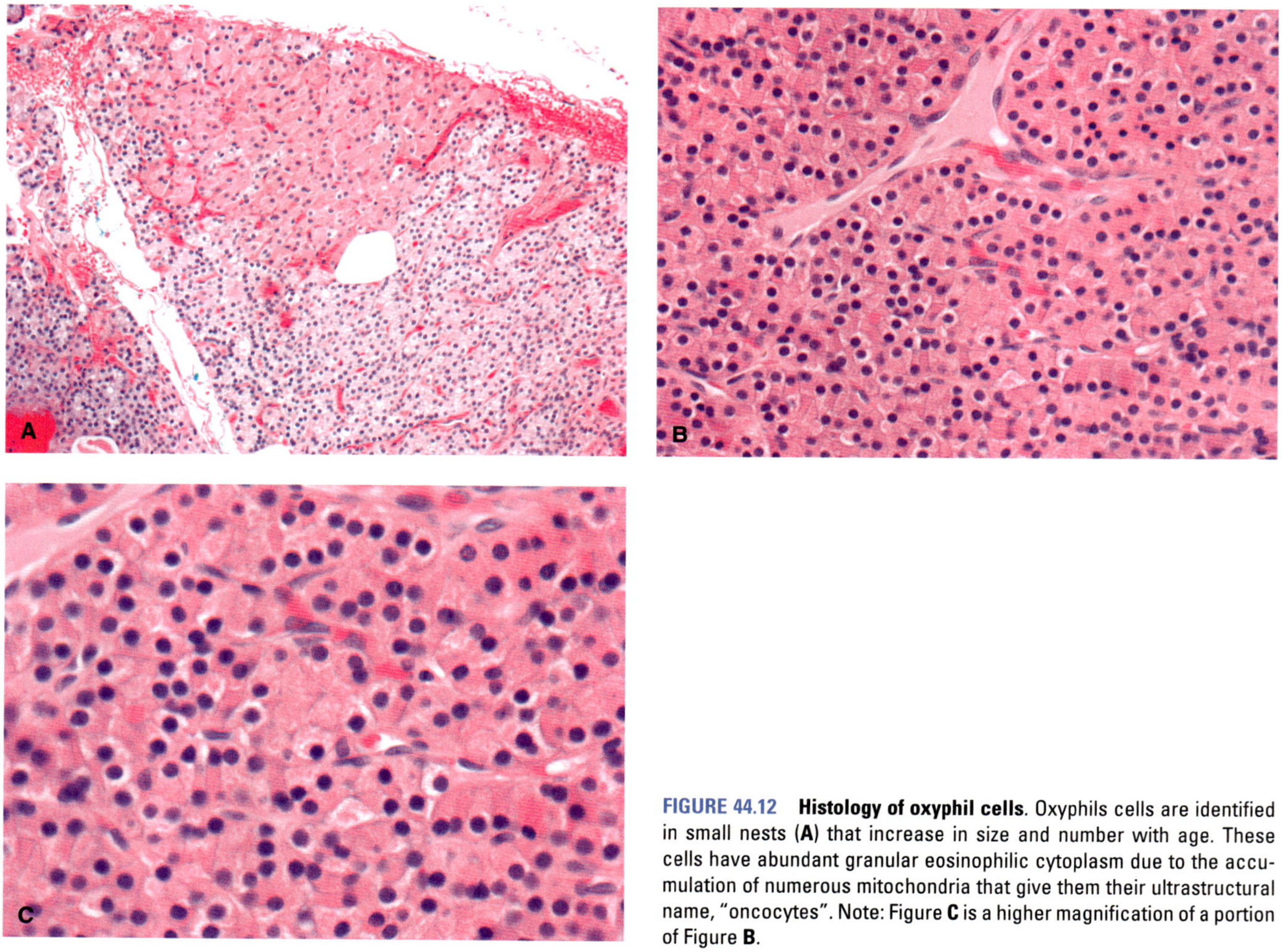

FIGURE 44.12 Histology of oxyphil cells. Oxyphils cells are identified in small nests (**A**) that increase in size and number with age. These cells have abundant granular eosinophilic cytoplasm due to the accumulation of numerous mitochondria that give them their ultrastructural name, "oncocytes". Note: Figure **C** is a higher magnification of a portion of Figure **B**.

periphery. The cell membranes are relatively straight with few interdigitations and desmosomes connecting adjacent cells. In contrast, actively synthesizing and secreting cells have well-developed parallel arrays of rough endoplasmic reticulum, usually in a perinuclear location, prominent Golgi complexes with immature secretory granules of different electron densities and electron-lucent halos, and larger dense-core secretory granules throughout the cytoplasm depleted of glycogen and lipid bodies. The cell borders are more complex with interdigitations and occasionally one can see fusion of the secretory granule membrane with the plasma membrane and emptying of secretory product into the extracellular space.

The oxyphil cell is a chief cell that has undergone oncocytic change due to accumulation of mitochondria that can be increased in both number and size. The stimulus for oncocytic change is not known but these cells increase gradually with age in the parathyroid as they do in other organs including pituitary and thyroid (66). Oxyphil cells are found singly, in small clusters, or as large nodules and sheets (Fig. 44.12). These cells are larger than chief cells, measuring 12 to 20 μm in diameter, due to the abundant eosinophilic granular cytoplasm that is filled with mitochondria.

Oxyphil cells, known as oncocytes in ultrastructural terminology, are characterized by the presence of numerous mitochondria that can be irregular in size and shape, filling the cell cytoplasm (Fig. 44.13); other organelles are scant, including short profiles of rough endoplasmic reticulum, involuted Golgi complexes, sparse secretory granules, occasional lysosomes, and lipofuscin granules, however some lack synthetic organelles entirely (64). Transitional cells are intermediate forms that have features of chief cells and oxyphilic cells with fewer mitochondria than fully developed oncocytic cells but more than chief cells (Fig. 44.14).

Clear cells are found in the fetal gland but are not present in normal adult glands; they are found in hyperplasias and adenomas (Fig. 44.15). These cells contain abundant cytoplasmic glycogen. A distinct variant of clear cells, known as "water-clear cells" (i.e., Wasserhelle cells), are very large, ranging from 15 to 40 μm in diameter, and have abundant vacuolated cytoplasm filled with glycogen and large membrane-bound vacuoles (Fig. 44.15). The granular component of the cytoplasm of the water-clear cells is typically PAS-positive but not after diastase pretreatment, proving that this diastase-sensitive material is glycogen (67). The vacuoles are generally thought to be related to the Golgi complex given their resemblance to the Golgi vesicles (68,69), although some authors have proposed a link to dilated cisternae of the granular endoplasmic reticulum given the identification of ribosomes on the surfaces of the vacuoles (70) and others have questioned a possible link to secretory granules (71).

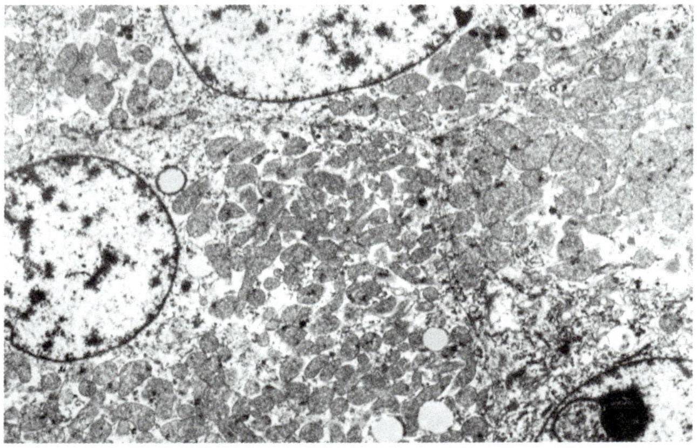

FIGURE 44.13 Ultrastructure of oxyphil cells. By electron microscopy, the cytoplasm of parathyroid oncocytic cells is almost completely replaced by the numerous mitochondria that can be enlarged and dilated; the residual rough endoplasmic reticulum, small Golgi complexes, few secretory granules, and lipid vacuoles are trapped within the accumulation of mitochondria or pushed to the perinuclear or peripheral areas.

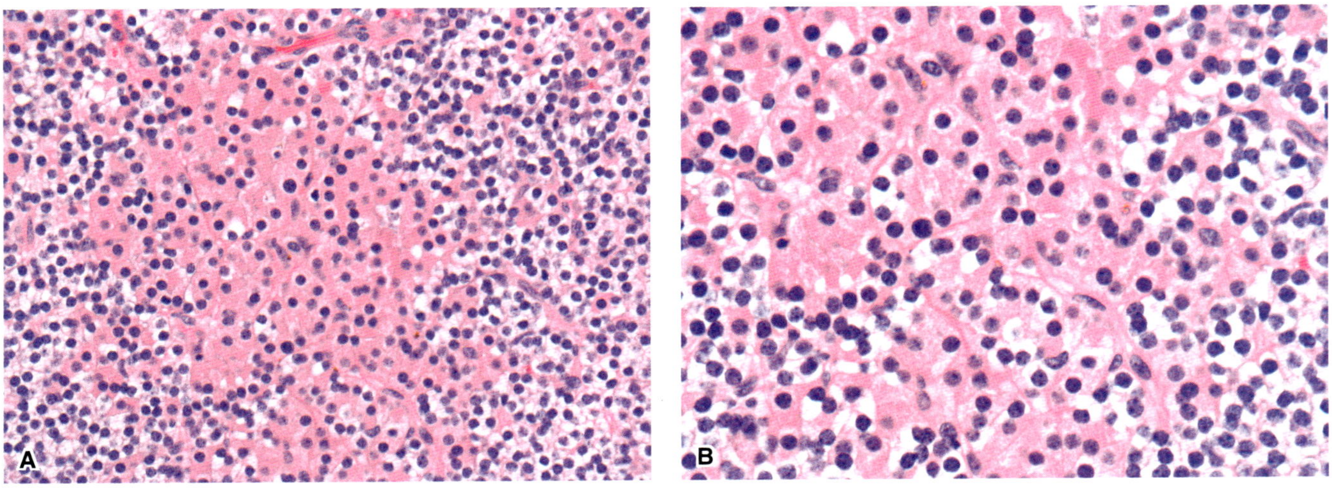

FIGURE 44.14 Histology of transitional cells. Transitional cells are chief cells with partial oncocytic change. Note: Figure **B** is a higher-power image of a portion of Figure **A**.

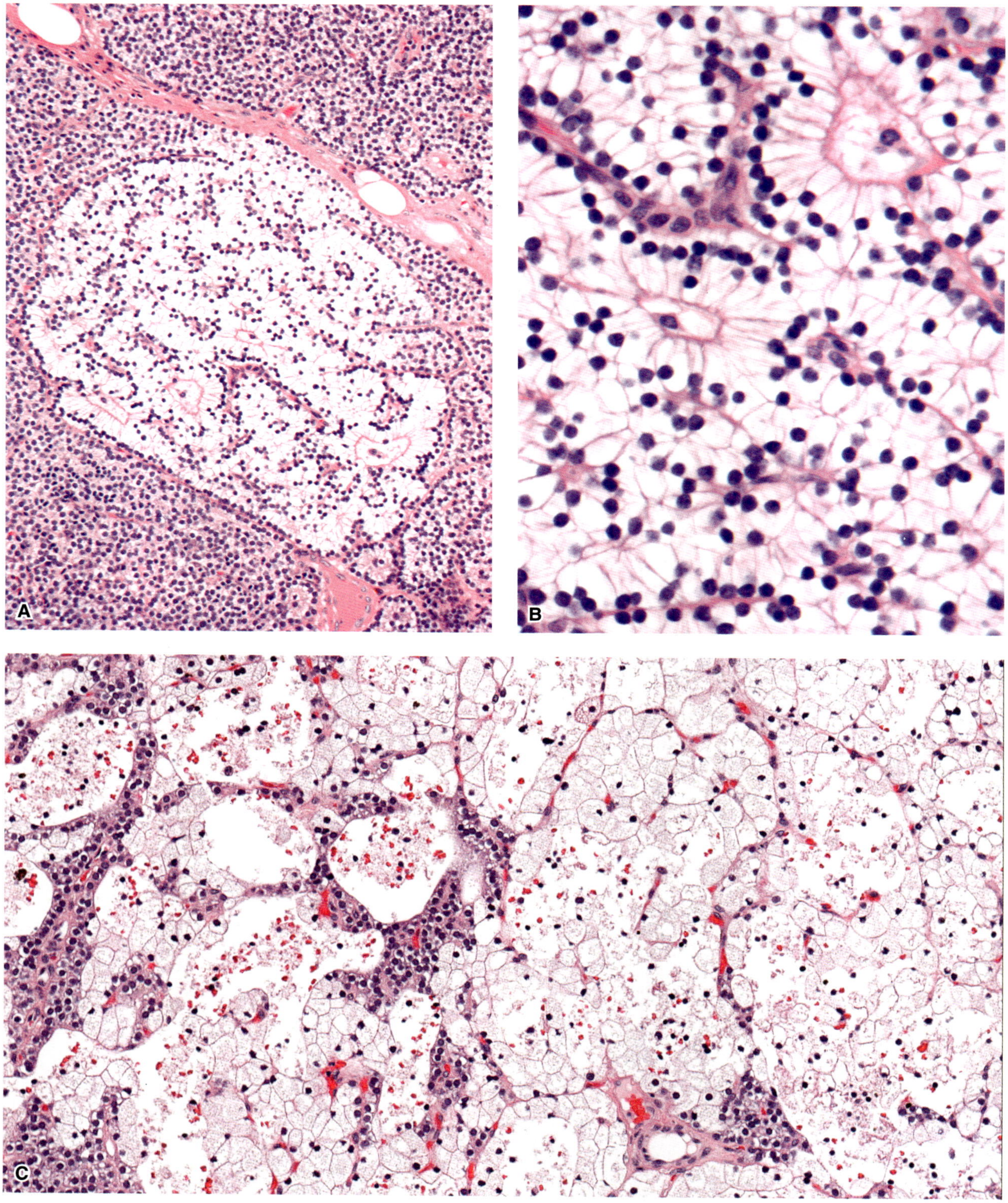

FIGURE 44.15 **The spectrum of clear-cell change in parathyroids**. Clear cells are not usually found in normal parathyroids but are present in hyperplasias (**A**, **B**) and adenomas. A distinct form of clear-cell change due to massive accumulation of glycogen and lipids gives rise to "water-clear cell" morphology (**C**). Note: Figure **B** is a high-power image from a portion of Figure **A**.

VARIATIONS WITH GENDER AND AGE

In fetuses, infants, children, and young adults the interstitium has minimal collagen and consists mainly of the capillary network. With age, there is an increasing accumulation of collagen with the formation of fibrous septa that give rise to the lobulation of the mature gland. Stromal fat cells begin to appear late in the first decade of life and increase with age, reaching a peak in the fourth decade. The amount and distribution of stromal fat varies among individuals and even between glands in the same individual. In most adults, fat cells comprise approximately 50% of the stromal volume (48). Women tend to have more stromal fat than men but this may reflect the fact that stromal fat content is proportional to total body fat. Intrathyroidal parathyroid glands often contain less stromal fat than those outside the gland.

Clear cells predominate in fetal glands but they disappear after birth. In infants and young children, only one type of cell is present, the chief cell. Transitional cells and oxyphil cells increase with age. There is less intracellular fat in the chief cells of children than adults.

PHYSIOLOGY AND PATHOPHYSIOLOGY

The main function of the parathyroid glands is the synthesis and secretion of parathyroid hormone, an 84 amino acid 9.4 kDa peptide that is encoded on the short arm of chromosome 11 (72). The gene product is initially a 115 amino acid pre-pro-hormone that is cleaved in the endoplasmic reticulum to produce a separate N-terminal 25 amino acid product and proparathyroid hormone; subsequent cleavage of a further 6 N-terminal amino acids in the Golgi produces the 84 amino acid parathyroid hormone that should also be classified as a prohormone since only the 34 N-terminal amino acids are required for the hormone-receptor interaction that is responsible for calcium homeostasis. Cleavage of the C-terminal inactive and N-terminal active fragments occurs in the liver and other sites. The half-life of the active fragment is much shorter than the complete parathyroid hormone or the C-terminal inactive fragment.

The main function of parathyroid hormone is regulation of serum calcium levels. Calcium is essential for signaling pathways that control ubiquitous cellular functions including membrane integrity, protein secretion, glycogen metabolism, division, and adhesion, as well as specialized functions such as muscle contraction, neuron excitability, and coagulation. Accordingly, its levels are closely regulated by three main hormones: parathyroid hormone, calcitriol (1,25-dihydroxycholecalciferol, also known as 1,25-dihydroxyvitamin D or 1,25[OH]$_2$D3), and calcitonin. Approximately 99% of calcium is present in bones as hydroxyapatite and only 1% can be found in extracellular fluids and soft tissues. Almost half (48%) of the total extraosseous calcium is present as ionized calcium, 46% is bound to protein and the remainder is associated with diffusible ion complexes (49). The level of ionized calcium [Ca^{2+}] is the main regulator of parathyroid hormone synthesis and secretion; increased ionized calcium inhibits transcription, translation, and secretion of parathyroid hormone by chief cells and decreased ionized calcium results in enhanced synthesis and secretion. This is in direct contrast to other endocrine cells that respond to increased calcium with increased hormone synthesis and secretion.

The regulation of parathyroid hormone by calcium is mediated by the calcium-sensing receptor (CaSR), a member of the G protein-coupled receptor (GPCR) superfamily (73–75) encoded by the *CaSR* gene on chromosome 3 at 3q13.3 to 21 (75). Hypocalcemia activates CaSR, inducing increased parathyroid hormone synthesis and secretion. The CaSR is also expressed in the renal tubules where it plays a role in calcium resorption (74,75) as well as in stomach, small intestine, and skin (76). Calcitriol indirectly alters parathyroid hormone secretion by modulating the expression of the CaSR. Inactivation of CaSR signaling is the cause of familial hypocalciuric hypercalcemia (FHH, also known as familial benign hypercalcemia) characterized by inappropriately normal parathyroid hormone levels in the face of mild hypercalcemia and hypocalcuria due to heterozygous inactivating mutation in either *CaSR*, *GNA11*, or *AP2S1* that mediates signaling (77); neonatal severe primary hyperparathyroidism, a more severe and potentially fatal disorder, is due to homozygous inactivation or heterozygous dominant-negative CaSR mutations (76,78). Gain-of-function mutations of *CaSR* result in autosomal dominant hypocalcemia, sporadic idiopathic hypoparathyroidism, and Bartter syndrome type V, characterized by hypocalcemia and hypercalcuria (76). Activating antibodies to CaSR have been described in autoimmune hypoparathyroidism including patients with polyendocrine autoimmune syndromes (76).

Rare patients with familial isolated hyperparathyroidism lack mutations in *CaSR*. A subset of these have been shown to harbor GCM2 variants that are thought to enhance the transcriptional activating capacity of the GCM2 protein (77,79).

Parathyroid hormone acts by directly and indirectly promoting calcium resorption into blood from bone, kidney, and the gut. The parathyroid hormone receptor (PTHR1) is another 7 transmembrane GPCR that increases cyclic AMP (80). When parathyroid hormone acts on this receptor, it stimulates renal tubular reabsorption of calcium and enhances calcitriol formation in kidney by stimulating 1α-vitamin D hydroxylase to convert 25(OH) vitamin D to 1α-25(OH)$_2$ vitamin D, resulting in increased gut absorption of calcium.

PTHR1 is expressed by osteoblasts and its activation is thought to prolong osteoblast life and increase activity, leading to bone formation. However, osteoblast signaling activates osteoclasts, inducing bone resorption indirectly, since they do not express PTHR1. Parathyroid hormone also activates osteocytes, causing lysis of perilacunar bone.

Paradoxically, intermittent high parathyroid hormone administration is anabolic for bone, while continuous elevation is catabolic. It is postulated that the transient stimulation can stimulate the osteoblast anabolic activity without triggering the catabolic response through osteoclasts. For this reason, parathyroid hormone is paradoxically used in the treatment of osteoporosis (72).

Mobilization of calcium from bone results in increased phosphate as well. Phosphate has no direct role in parathyroid hormone regulation but parathyroid hormone promotes excretion of phosphate by blocking its resorption in the kidney. Fibroblast growth factor 23 (FGF23) is a critical regulator of serum phosphorus; it depresses 1α-hydroxylase in the renal tubules. FGF23 produced by osteocytes and osteoblasts down regulates the type II sodium–phosphate cotransporters (NPTi2a and NPTi2c) causing phosphate wasting. FGF23 also down regulates the activity of the renal $25(OH)_2$ vitamin D 1α–hydroxylase, decreasing the serum level of $1\alpha,25(OH)_2$ vitamin D.

The PTHR1 signals through the Gsα encoded by *GNAS*. Inactivating mutations of *GNAS* result in pseudohypoparathyroidism types IA, IB, IC, and pseudopseudohypoparathyroidism. The complex *GNAS* locus on chromosome 20q13.3 undergoes parent-specific methylation at several sites (81). Heterozygous inactivating mutations of maternal *GNAS* cause pseudohypoparathyroidism type 1A with parathyroid hormone-resistant hypocalcemia and hyperphosphatemia; reduced paternal *GNAS* expression in proximal renal tubules results in little or no Gsα protein, leading to parathyroid hormone-resistant hypocalcemia and hyperphosphatemia. The same or similar *GNAS* mutations in the paternal allele are the cause of pseudopseudohypoparathyroidism (81). Autosomal dominant pseudohypoparathyroidism type IB is caused by heterozygous maternal deletions within *GNAS* or *STX16 (syntaxin 16)*, associated with loss-of-methylation (LOM) reducing Gsα expression. Epigenetic changes are also observed in sporadic pseudohypoparathyroidism 1B, and rare cases have been reported to have paternal uniparental isodisomy or heterodisomy of chromosome 20q (81).

SPECIAL PROCEDURES

A number of special procedures can be used in the evaluation of parathyroid glands. These include histologic stains and immunohistochemistry as well as molecular testing.

Histologic special stains include the Grimelius silver technique that identified the parathyroid chief cells as neuroendocrine. This stain is of historical interest and has been largely supplanted by immunohistochemistry.

The PAS stain is widely available in most laboratories and can serve to assess the amount of glycogen in chief cells, a feature that is helpful in assessing function as well as the significance of cytoplasmic clearing. Glycogen stains strongly with the PAS stain (Fig. 44.16) but is not diastase resistant, therefore is not identified in tissue stained with PAS-D (PAS after diastase digestion). The colloid material in cysts and follicles of the parathyroid is not composed of glycogen (Fig. 44.16).

Fat stains have been applied to evaluate the intracytoplasmic lipid content of chief cells that is higher in resting cells and lower in active cells. This has been applied to distinguish hyperplastic and adenomatous cells from normal cells (Fig. 44.17). A significant limitation of this tool is the need for fresh and/or frozen tissue; lipid is lost in formalin-fixed, paraffin-embedded tissues and, thus, the oil red O stain cannot be used on permanent sections. Its main application is at the time of intraoperative consultation when frozen sections or imprints can be studied with this tool. However, the use of intraoperative parathyroid hormone assays to guide the extent of surgery has revolutionized the operative approach to patients with hyperparathyroidism (1) and the need for classification of a gland as abnormal is less critical than simply the identification of tissue as parathyroid rather than lymph node or other neck structures.

In contrast to the declining role of histochemistry, immunohistochemistry has emerged as a valuable method of improving diagnosis and pathologist input into clinical care. The applications fall into three main categories: confirmation of parathyroid differentiation, distinction of benign from malignant proliferations, and determination of genetic predisposition to parathyroid disease (77).

Parathyroid lesions may be obvious and within parathyroid glands, but given the common intrathyroidal and mediastinal location of parathyroids and their tendency to form follicles, they can be easily misdiagnosed (82). Immunostains including insulinoma-associated protein-1 (INSM1), synaptophysin and chromogranin, somatostatin receptor 2 (SSTR2), as well as vesicle-associated membrane protein (VAMP), synaptosomal-associated protein 23 (SNAP-23) that is part of a soluble N-ethylmaleimide–sensitive fusion attachment protein receptor (SNARE), are all characteristic of neuroendocrine lineage. GATA3 (83), GCM2 (84), and parathyroid hormone define parathyroid cells and can distinguish them from other neuroendocrine cells that give rise to tumors in this region (85), including C cells that develop into medullary thyroid carcinomas, thymic neuroendocrine tumors, and paragangliomas, or in the case of clear cells, renal carcinoma and other large clear-cell tumors (Table 44.2, Fig. 44.18). Paragangliomas are also negative for keratins whereas the other neuroendocrine tumors including parathyroid tumors express various cytokeratins including those identified by the AE1/AE3 cocktail and cytokeratins 8/18 identified by the CAM 5.2 antibody. Chief cells are also positive for cytokeratin 19, and are negative for TTF1, PGP 9.5, and neurofilaments. The use of multiple biomarkers is encouraged, since some parathyroid lesions may have reduced staining for GCM2 (84) and parathyroid hormone expression can be reduced in some lesions.

The distinction of parathyroid hyperplasia and neoplasia from normal parathyroid can be difficult since it relies on

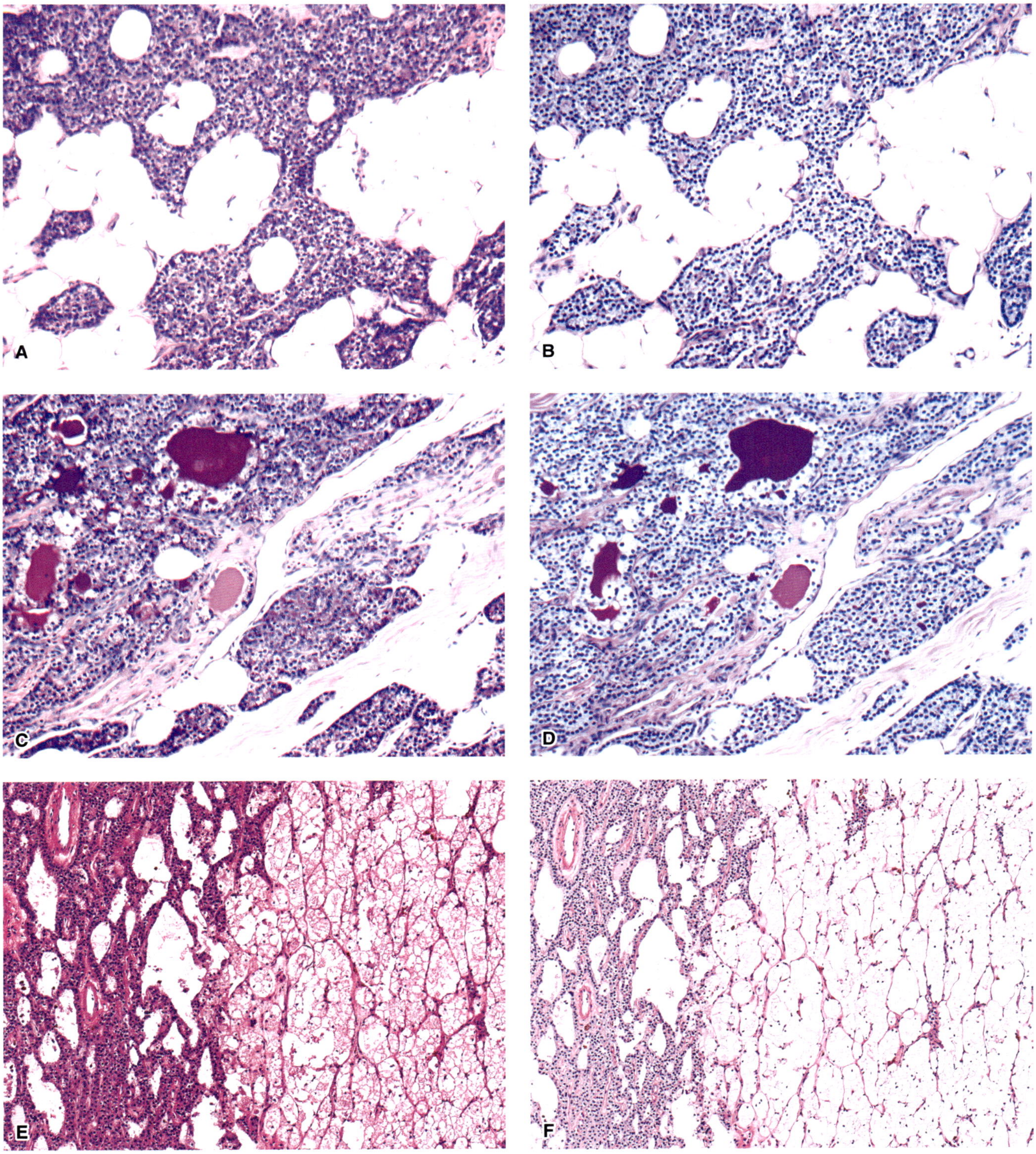

FIGURE 44.16 **Histochemistry of parathyroid: PAS and PAS-D**. Normal parathyroid chief cells contain abundant glycogen that stains with PAS (**A**) and is not diastase resistant (**B**). The colloid-like material in cysts and follicles stains intensely with PAS (**C**) and is diastase resistant, showing strong positivity in the PAS-D stain (**D**). Water-clear cells have clear vacuoles that are not filled with glycogen but there is abundant PAS-positive glycogen in the cytoplasm surrounding them (**E**) that is lost after diastase treatment (**F**).

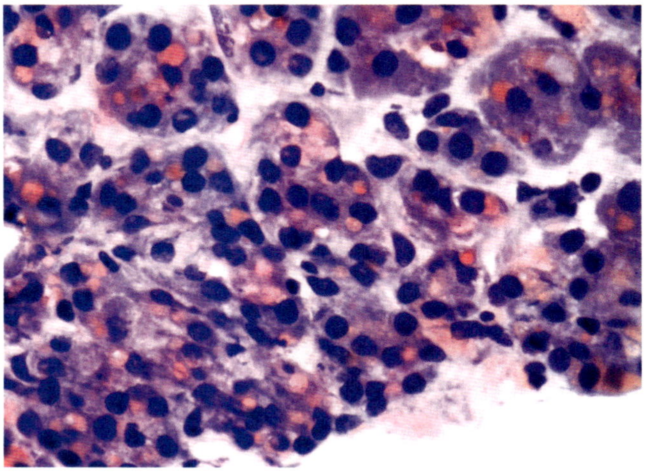

FIGURE 44.17 Histochemistry of parathyroid: Oil red O. Normal chief cells with intracytoplasmic lipid droplets in approximately 90% of the cells (frozen section Oil-Red-O stain with hematoxylin counterstain). (Courtesy of Dr. S. I Roth.)

the identification of increased cellularity that is usually visualized by assessing the amount of adipose stroma. However, the degree of stromal and even cytoplasmic fat is variable in normal glands. The recognition of a neoplasm is easy when the adjacent nontumorous gland is present to show the difference between the cellular neoplasm and the lobulated parenchyma with associated adipocytes. In the absence of a nontumorous rim, it may be impossible to distinguish hyperplasia from adenoma. Moreover, there are rare adenomas that have abundant stromal fat; these lipoadenomas can be a source of confusion. From a clinical perspective, the distinction of hyperplasia from adenoma, which was a major problem in the era of parathyroid dissections to identify all

TABLE 44.2 Immunohistochemical Biomarkers Used to Confirm Parathyroid Differentiation

Biomarkers	Reactivity Pattern
INSM1	Nuclear
Chromogranin A	Cytoplasmic[b]
Synaptophysin	Cytoplasmic[b]
SSTR2	Membranous
PTH[a]	Cytoplasmic
GATA3	Nuclear
GCM2	Nuclear
Polyclonal PAX8	Nuclear
Monoclonal PAX8	Negative
TTF1	Negative
Keratins (AE1/AE3, 7, 8/18, 19)	Cytoplasmic[b]

[a]Parathyroid cysts can be negative for PTH.
[b]May be negative in some tumors.

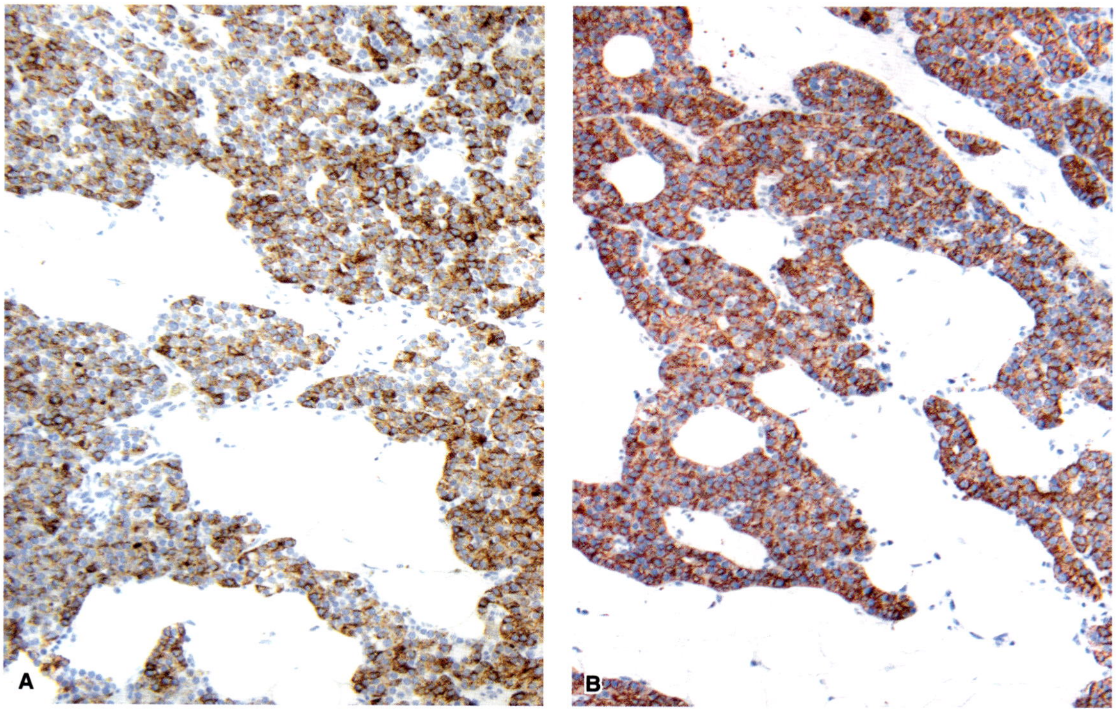

FIGURE 44.18 Immunohistochemistry: Markers of parathyroid differentiation. Normal parathyroid chief cells stain for chromogranin A (**A**) and parathyroid hormone (**B**) as well as GATA3 (**C**). Clear-cell tumors can be confirmed to be of parathyroid differentiation when they stain for GATA3 (**D**) and parathyroid hormone (**E**). Water-clear cells also express nuclear GATA3 (**F**).

FIGURE 44.18 (*Continued*)

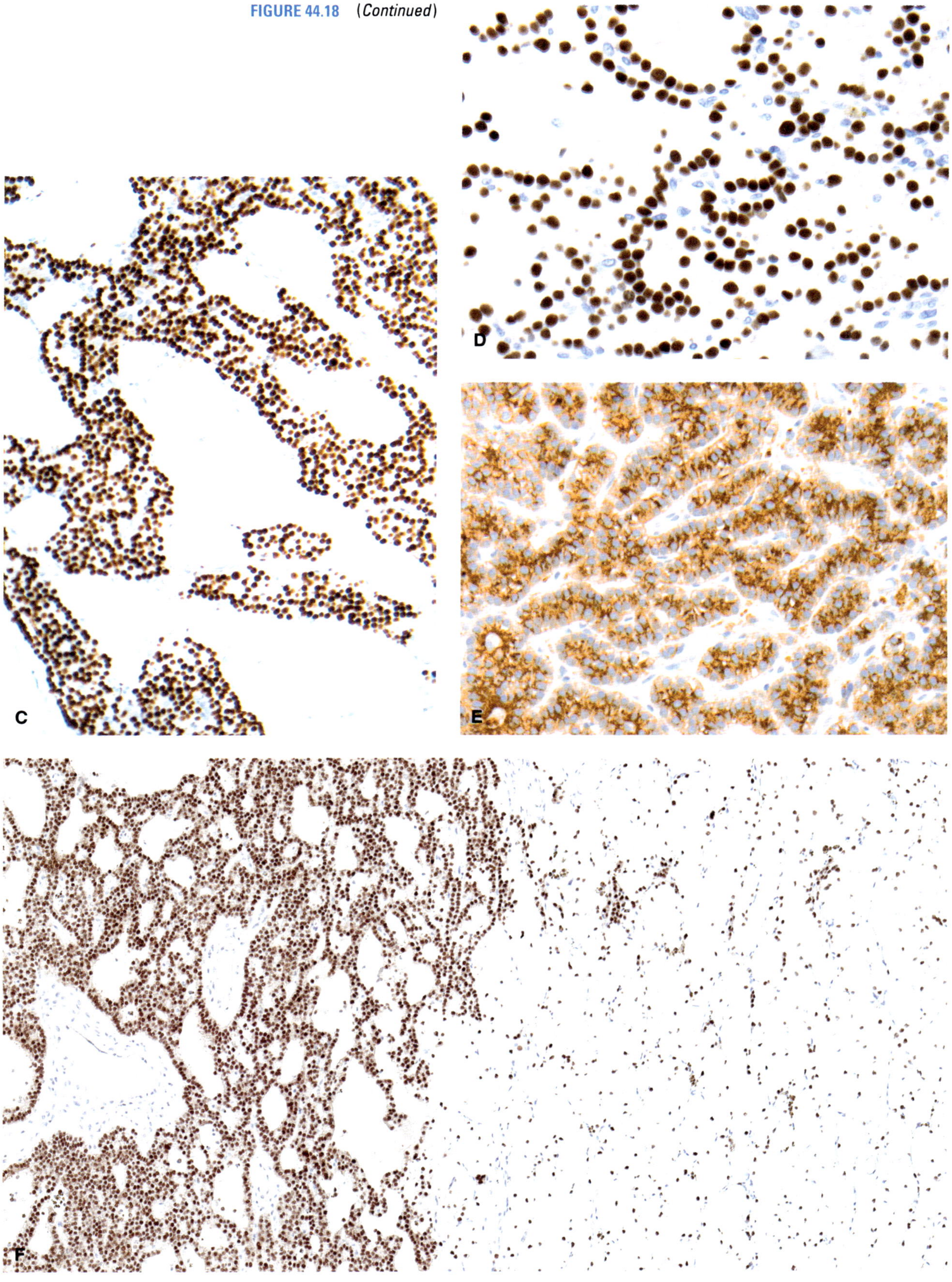

glands, is no longer critical; the use of limited parathyroidectomy based on imaging and intraoperative parathyroid hormone assays has reduced the need for accurate histologic classification of the enlarged cellular parathyroid gland (86).

Clonality studies have shown that most adenomas and carcinomas are monoclonal, whereas secondary hyperplasia is not (87–89). The concept of clonality in multiple endocrine neoplasia (MEN) is more complex and it appears that nodules in the glands of patents with MEN1 syndrome are indeed clonal proliferations (90), consistent with the concept that each cell at risk of loss of the intact tumor suppressor gives rise to a neoplastic proliferation. Therefore, affected parathyroid glands are considered to display multiple multiglandular parathyroid adenomas (91). Tertiary hyperparathyroidism is attributed to emergence of a clonal neoplasm from the background of hyperplastic tissue in secondary hyperparathyroidism.

The distinction between benign and malignant parathyroid neoplasms can be straightforward when a lesion is either completely bland and surrounded by normal tissue, clearly an adenoma, or highly infiltrative and angioinvasive (Fig. 44.19) with perineural invasion, as an obvious carcinoma. However, it is not uncommon that a patient has a fine needle biopsy of a neck mass and in that scenario, benign tumors can develop reactive changes that look worrisome: they have fibrosis, appear infiltrative, and even become trapped in surrounding tissues (92,93). In this setting, it is helpful to have biomarkers of normal and benign neoplastic cells and biomarkers of malignancy (Table 44.3).

The gene initially called *Hyperparathyroidism 2* (*HRPT2*), now known as *CDC73*, encodes a protein called parafibromin, a tumor suppressor that was initially associated with hyperparathyroidism–jaw tumor syndrome (94). Mutations inactivate this gene in that disorder and in sporadic parathyroid carcinomas. Parafibromin is immunolocalized to the nuclei and nucleoli of normal parathyroid glands (Fig. 44.20). Decreased nuclear and/or nucleolar parafibromin staining provides evidence of the malignant potential of parathyroid proliferation (Fig. 44.20) (95,96). Some studies have also

TABLE 44.3 Immunohistochemical Biomarkers Used in the Distinction of Parathyroid Adenoma and Carcinoma

Biomarker	Adenoma	Carcinoma
Parafibromin	Positive	Negative[a]
APC	Positive	Negative[b]
BCL2	Positive	Negative[b]
MDM2	Positive	Negative
RB	Positive	Negative[b]
p53 abnormal[c]	Absent	Rare
CyclinD1	Positive	Negative
Ki67	Often <5%	Often >5%
Galectin-3	Negative	Positive
PGP 9.5	Negative	Positive

[a]Nuclear expression may be retained in some parathyroid carcinomas.

[b]Can be reduced in some carcinomas.

[c]Can be global loss or diffuse overexpression.

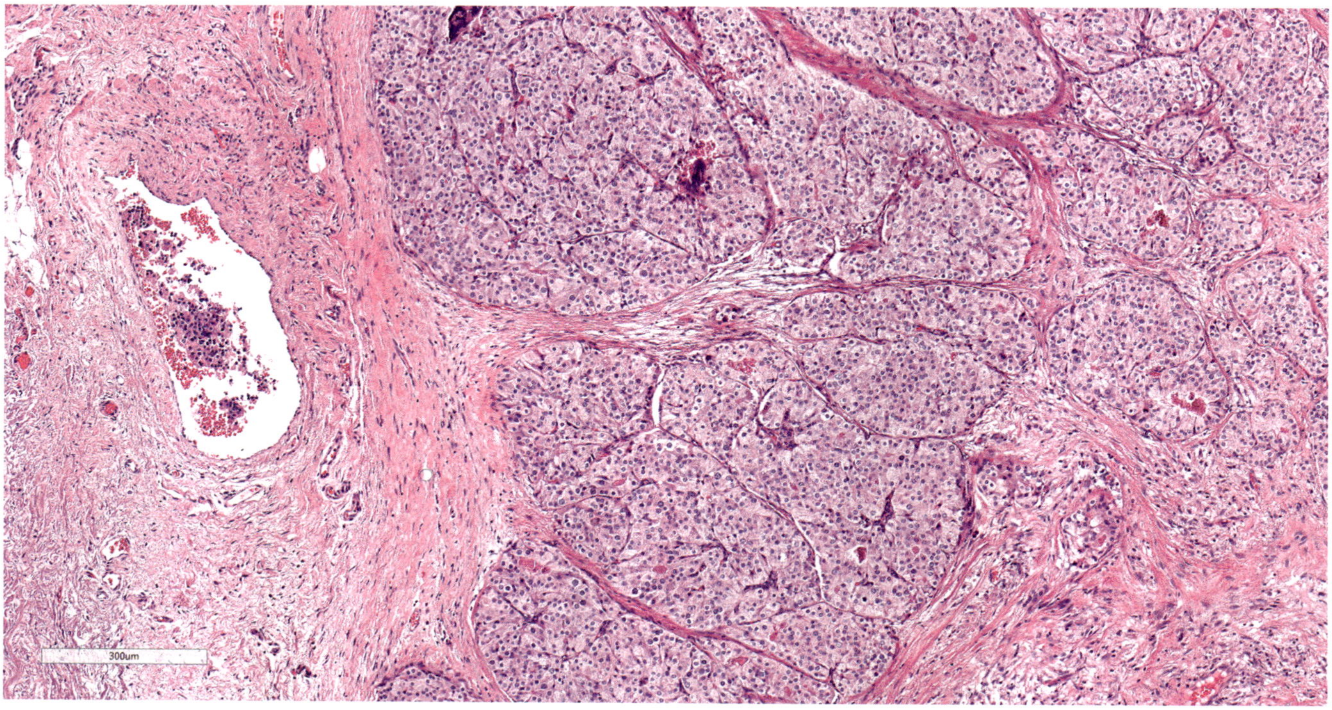

FIGURE 44.19 Vascular invasion in parathyroid carcinoma. Parathyroid carcinoma can be diagnosed on histology by the identification of unequivocal angioinvasion defined as tumor cells in a vascular lumen with associated thrombus.

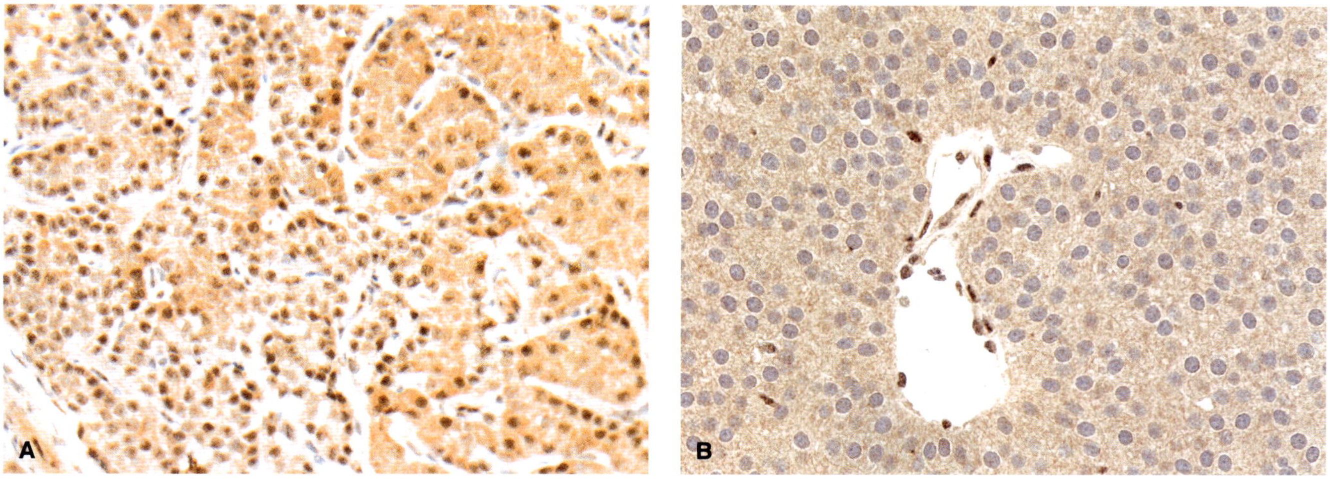

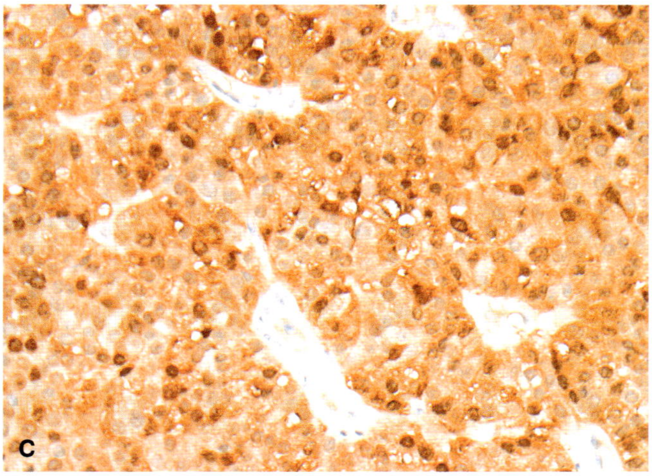

FIGURE 44.20 Immunohistochemistry for parafibromin and PGP 9.5. Parafibromin is expressed in normal, hyperplastic benign parathyroid cells (**A**); decreased nuclear and/or nucleolar staining (**B**) provides evidence of the malignant potential of a parathyroid neoplasm. The reverse pattern is seen for PGP 9.5 which is negative in normal and benign parathyroid proliferations and is upregulated in parathyroid carcinoma (**C**).

shown the reverse pattern of staining for PGP 9.5 that is not expressed in normal parathyroid and is upregulated in parathyroid carcinoma (Fig. 44.20).

Alterations in other tumor suppressor genes, including *CDKN1B* that encodes p27, *RB,* and *TP53*, have been reported in parathyroid carcinomas (97–101); loss of these tumor suppressors is seen in some carcinomas but not in hyperplasia or adenoma (Fig. 44.21). In contrast, activation of cyclic AMP by *GNAS* mutation (102) or overexpression of *cyclin D1,* the latter through the *CCND1/PRAD1*

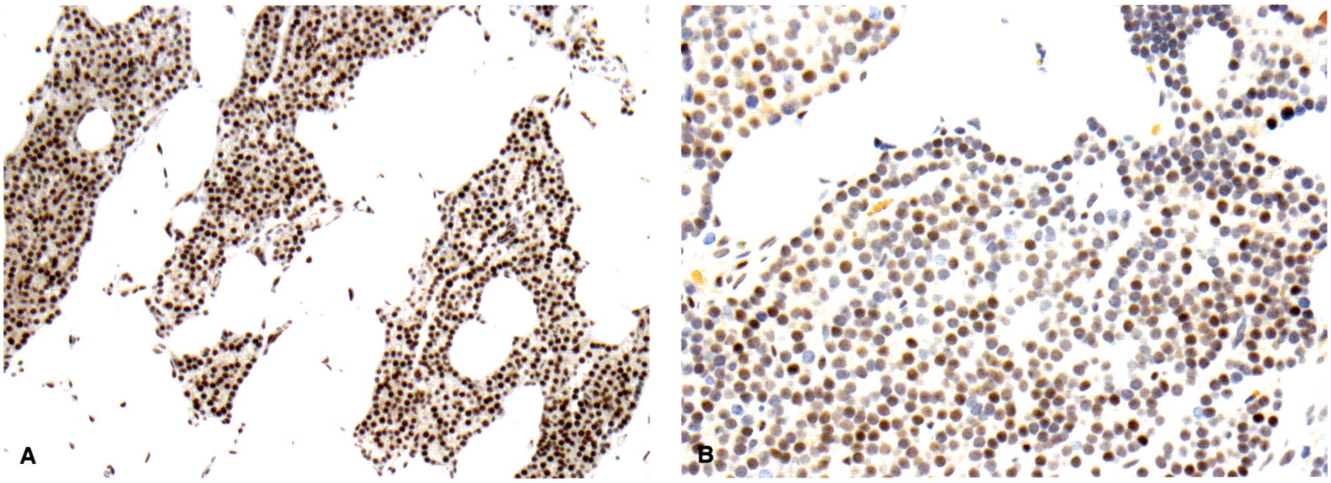

FIGURE 44.21 Dysregulation of p27, RB, and p53 in parathyroid proliferations. Normal parathyroid stains strongly for p27 (**A**) and while there may be a reduction, staining is preserved in hyperplasia (**B**) and adenomas. (*continued*)

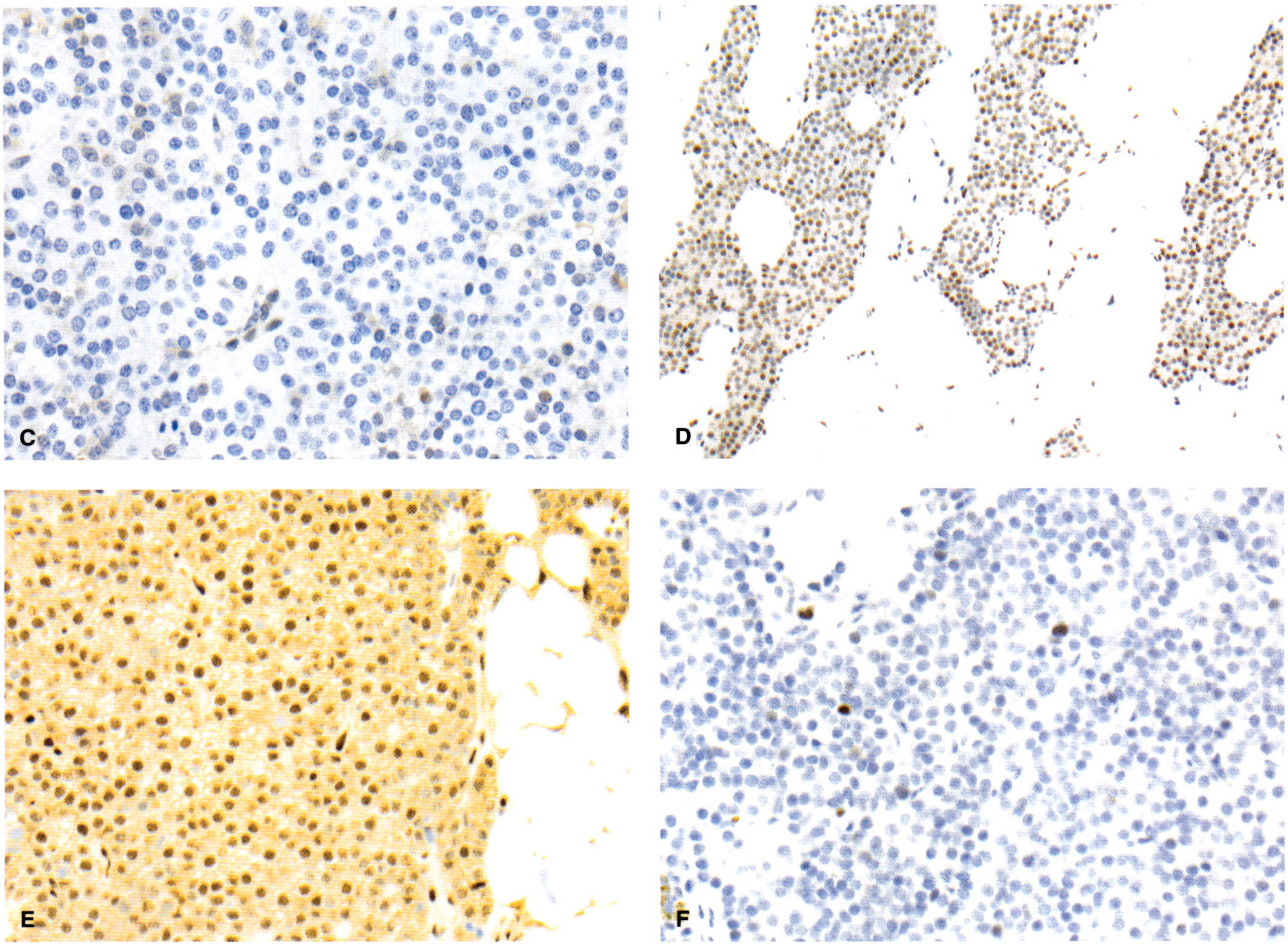

FIGURE 44.21 (*Continued*) There can be global loss in carcinoma (**C**) but this may also be a feature of benign lesions in patients with MEN4. Rb is also expressed in normal parathyroid (**D**) and hyperplasia (**E**); p53 can be focally positive in hyperplasia (**F**) and adenoma.

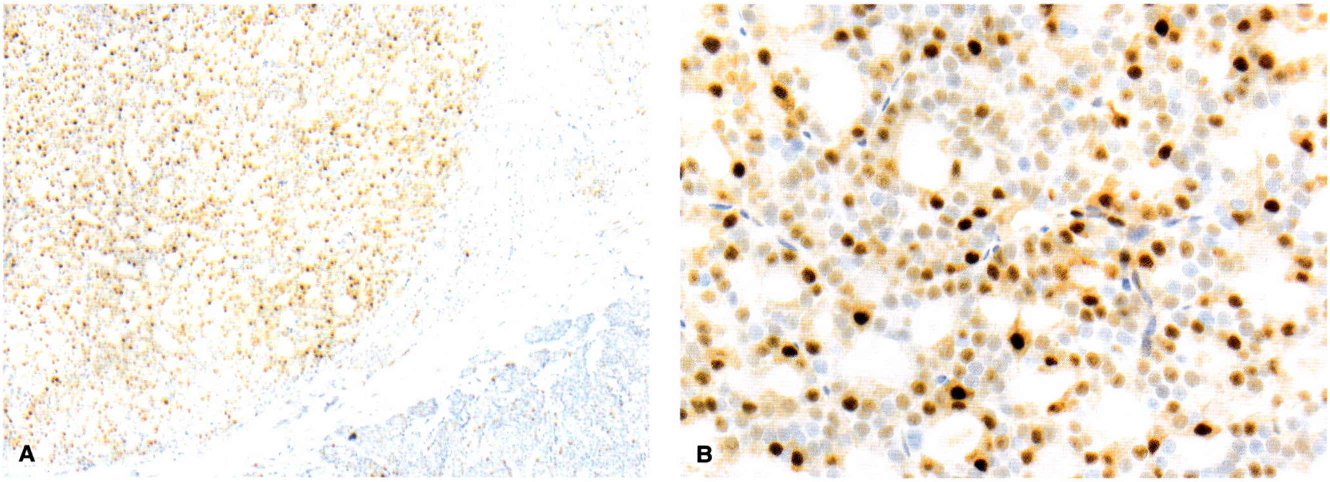

FIGURE 44.22 **Cyclin D1 upregulation in parathyroid adenoma**. Cyclin D1 upregulation is a feature of adenomas with PRAD1 rearrangements.

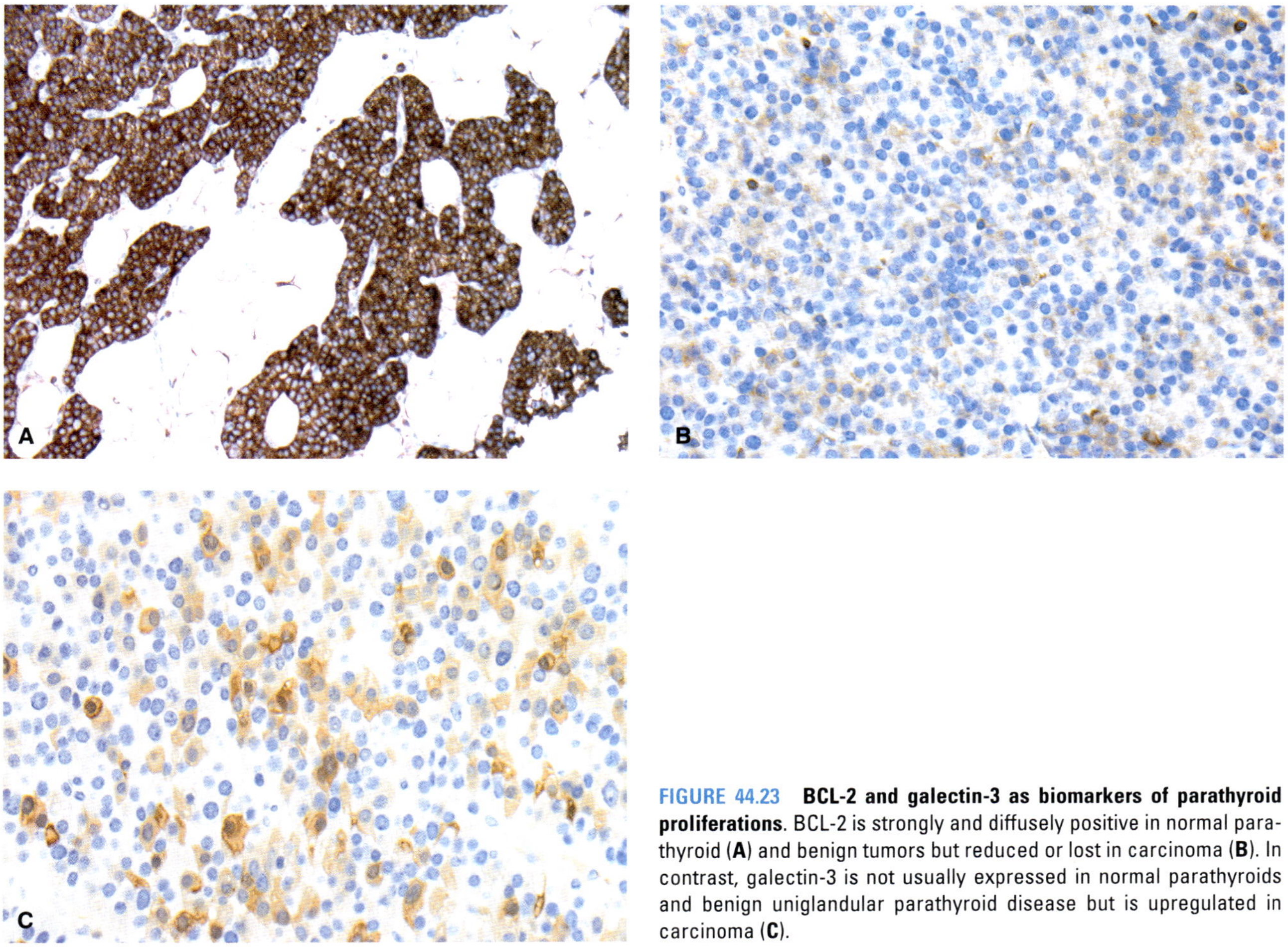

FIGURE 44.23 BCL-2 and galectin-3 as biomarkers of parathyroid proliferations. BCL-2 is strongly and diffusely positive in normal parathyroid (**A**) and benign tumors but reduced or lost in carcinoma (**B**). In contrast, galectin-3 is not usually expressed in normal parathyroids and benign uniglandular parathyroid disease but is upregulated in carcinoma (**C**).

rearrangement involving 1p15.3 to 15.1 where the *parathyroid hormone* promoter upregulates cyclin D1 (89,103), are features of adenomas (Fig. 44.22). Other biomarkers of malignancy include BCL-2 loss (Fig. 44.23) and galectin-3 overexpression (Fig. 44.23) (104), as well as loss of expression of MDM2 (105) and APC (95,106). While some authors reported that parathyroid adenomas are typically associated with a Ki67 labeling index less than 5% (106), the diagnostic role of Ki67 as a biomarker of parathyroid tumor classification remains unclear (107) (Fig. 44.24). Another ancillary

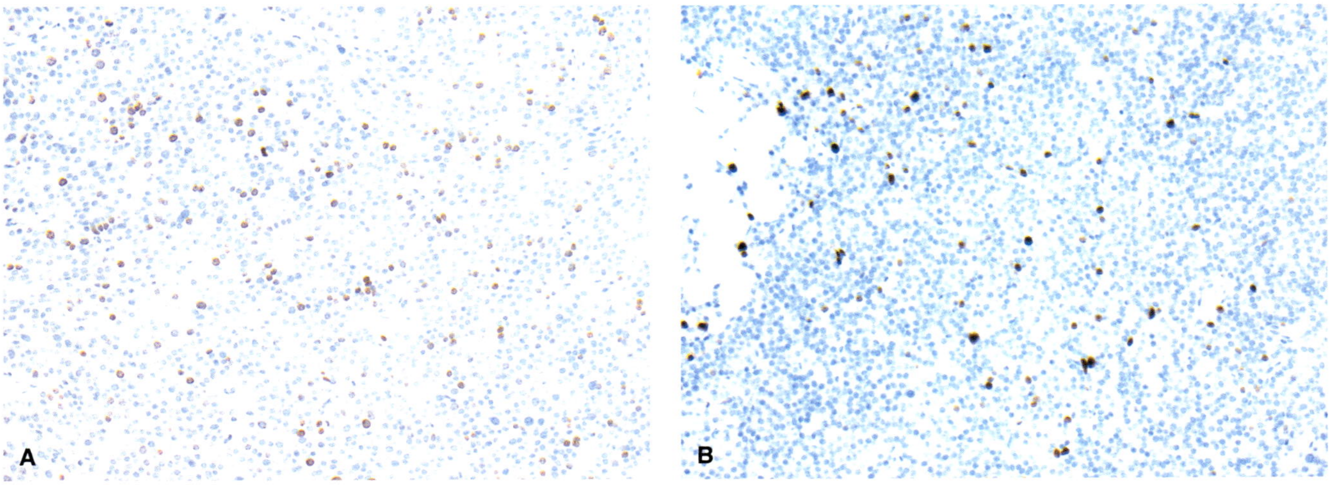

FIGURE 44.24 Ki67 labeling in hyperplasia and carcinoma. A high Ki67 labeling index is characteristic of carcinoma (**A**) but can also be seen in hyperplasia (**B**) and is therefore not specific.

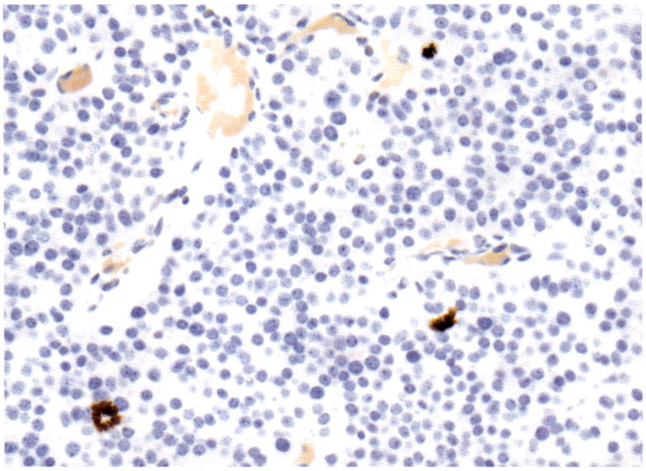

FIGURE 44.25 **PhosphoHistone-H3 (pHH3)-assisted mitotic counts.** The use of immunohistochemistry for pHH3 facilitates identification and quantification of mitoses.

tool that is helpful in identifying and counting mitoses is phosphoHistone-H3 immunohistochemistry (Fig. 44.25).

Genetic predisposition to parathyroid proliferative lesions is largely associated with MEN type 1, type 2, type 4, and type 5. MEN1 can be proposed as a likely etiologic agent when there is loss of expression of the gene product of the *MEN1* gene, menin, identified by immunohistochemistry (Fig. 44.26), and MEN4 should be considered when there is global loss of the cyclin-dependent kinase inhibitor *p27*kip1 (108,109). Loss of MAX protein is seen in parathyroid tumors of patient with MEN5 (110). As discussed above, parafibromin loss may indicate germline predisposition, especially when found in a young patient. A recent series expanded genotype–phenotype correlations of parafibromin-deficient parathyroid neoplasms (111) that are more frequent in younger individuals with larger tumors and may display certain morphologic characteristics including microcystic change, thick capsule, arborizing vasculature, extensive sheet-like growth pattern, eosinophilic cytoplasm, nuclei with coarse chromatin, and perinuclear clearing (111).

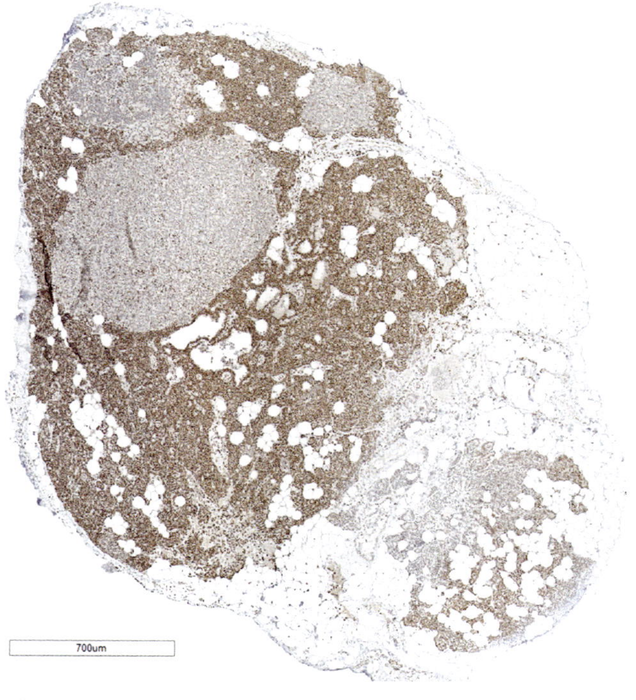

FIGURE 44.26 **Menin loss in parathyroid adenoma of MEN1 patient.** The loss of menin nuclear reactivity identifies multiple small neoplasms in the parathyroid of a patient with MEN1.

The classification of parathyroid neoplasms as members of the larger family of neuroendocrine neoplasms is important when considering therapeutic options for rare malignancies that can be lethal (49,112–114). These tumors are candidates for the targeted use of peptide receptor radiotherapy (PRRT) that is dependent on expression of somatostatin receptors, a ubiquitous feature of neuroendocrine cells (115,116).

ACKNOWLEDGMENTS

The authors gratefully acknowledge Paolo Batoni, PhD and Zoya Volynskaya, PhD, for contribution of the artwork in Figures 44.1 and 44.5, Lorne Rotstein, MD for contribution of Figure 44.2, and Dr. S. I. Roth who prepared Figures 44.6 and 44.17 for the previous edition of this book.

REFERENCES

1. Wilhelm SM, Wang TS, Ruan DT, et al. The American Association of Endocrine Surgeons Guidelines for definitive management of primary hyperparathyroidism. *JAMA Surg.* 2016;151(10):959–968.
2. Owen R. On the anatomy of the Indian rhinoceros (Rh. unicornus, L.). *Trans Zool Soc (London).* 1862;4:31–58.
3. Remak R. *Untersuchungen über die Entwickelung der Wirbelthiere.* G. Reimer; 1855.
4. Virchow R. *Die krankhaften Geschewülste.* Hirschwald; 1863.
5. Sanderström I. Om en ny Körtel hos menniskan och ätskilliga däggdjur. *Upsala Läkareförenings Förhandlingar.* 1880;15:441–471.
6. Kohn A. Studien über die Schilddrüse. *Arch Mikr Anat.* 1895;44:366–422.
7. Kohn A. Die Epithelkörperchen. *Ergeb Anat Entwicklungsgeshch.* 1899;9:194–252.
8. Gley E. Sur la toxicité des urines des chiens thyroidectomisés: contribution à l'étude des fonctions du corps thyroide. *Comptes Rendus de la Société de Biologie.* 1891;3:366–368.
9. Erdheim J. Ueber tetania parathyreopriva. *Weiner Klinishe Wachenschrift.* 1906;19:716–717.
10. Erdheim J. Beiträge zur pathologischen anatomie der menschlichen epithel-körperchen. *Zeitschrift für Heilkunst.* 1904;25:1–15.
11. Hanson AM. The hydrochloric X sicca: A parathyroid preparation for intramuscular injection. *Military Surgery.* 1924;259:218–219.

12. Collip JB. The extraction of a parathyroid hormone which will prevent or control parathyroid tetany and which regulates the level of blood calcium. *J Biol Chem.* 1925;63:395–438.
13. Aurbach GD. Isolation of parathyroid hormone after extraction with phenol. *J Biol Chem.* 1959;234:3179–3181.
14. Wang C. The anatomic basis of parathyroid surgery. *Ann Surg.* 1976;183(3):271–275.
15. Akerstrom G, Malmaeus J, Bergstrom R. Surgical anatomy of human parathyroid glands. *Surgery.* 1984;95(1):14–21.
16. Rodgers SE, Hunter GJ, Hamberg LM, et al. Improved preoperative planning for directed parathyroidectomy with 4-dimensional computed tomography. *Surgery.* 2006;140(6):932–940.
17. Roy M, Mazeh H, Chen H, Sippel RS. Incidence and localization of ectopic parathyroid adenomas in previously unexplored patients. *World J Surg.* 2013;37(1):102–106.
18. Weller GLJ. Development of the thyroid, parathyroid and thymus glands in man. *Contrib Embryol.* 1933;24:95–138.
19. Gilmour JR. The embryology of the parathyroid glands, the thymus and certain associated rudiments. *J Pathol Bacteriol.* 1937;45:507–522.
20. Norris EH. The parathyroid glands and the lateral thyroid in man: Their morphogenesis, histogenesis, topographhic anatomy and prenatal growth. *Contrib Embryol.* 1938;26:247.
21. Roth SI, Sadow PM, Johnson NB, Abu-Jawdeh GM. Parathyroid. In: Mills SE, ed. *Histology for Pathologists.* Lippincott Williams and Wilkins; 2012.
22. Sanders CD, Kirkland JD, Wolin EA. Ectopic parathyroid adenoma in the carotid sheath. *J Nucl Med Technol.* 2016;44(3):201–202.
23. Okuda I, Nakajima Y, Miura D, Maruno H, Kohno T, Hirata K. Diagnostic localization of ectopic parathyroid lesions: developmental consideration. *Jpn J Radiol.* 2010;28(10):707–713.
24. Lack EE, Delay S, Linnoila RI. Ectopic parathyroid tissue within the vagus nerve. Incidence and possible clinical significance. *Arch Pathol Lab Med.* 1988;112(3):304–306.
25. Leroyer-Alizon E, David L, Anast CS, Dubois PM. Immunocytological evidence for parathyroid hormone in human fetal parathyroid glands. *J Clin Endocrinol Metab.* 1981;52(3):513–516.
26. Manley NR, Capecchi MR. Hox group 3 paralogs regulate the development and migration of the thymus, thyroid, and parathyroid glands. *Dev Biol.* 1998;195(1):1–15.
27. Kameda Y, Arai Y, Nishimaki T, Chisaka O. The role of Hoxa3 gene in parathyroid gland organogenesis of the mouse. *J Histochem Cytochem.* 2004;52(5):641–651.
28. Peters H, Neubuser A, Kratochwil K, Balling R. Pax9-deficient mice lack pharyngeal pouch derivatives and teeth and exhibit craniofacial and limb abnormalities. *Genes Dev.* 1998;12(17):2735–2747.
29. Su D, Ellis S, Napier A, Lee K, Manley NR. Hoxa3 and pax1 regulate epithelial cell death and proliferation during thymus and parathyroid organogenesis. *Dev Biol.* 2001;236(2):316–329.
30. Jerome LA, Papaioannou VE. DiGeorge syndrome phenotype in mice mutant for the T-box gene, Tbx1. *Nat Genet.* 2001; 27(3):286–291.
31. Lindsay EA, Vitelli F, Su H, et al. Tbx1 haploinsufficieny in the DiGeorge syndrome region causes aortic arch defects in mice. *Nature.* 2001;410(6824):97–101.
32. Merscher S, Funke B, Epstein JA, et al. TBX1 is responsible for cardiovascular defects in velo-cardio-facial/DiGeorge syndrome. *Cell.* 2001;104(4):619–629.
33. Gordon J, Bennett AR, Blackburn CC, Manley NR. Gcm2 and Foxn1 mark early parathyroid- and thymus-specific domains in the developing third pharyngeal pouch. *Mech Dev.* 2001;103(1-2):141–143.
34. Gunther T, Chen ZF, Kim J, et al. Genetic ablation of parathyroid glands reveals another source of parathyroid hormone. *Nature.* 2000;406(6792):199–203.
35. Grigorieva IV, Thakker RV. Transcription factors in parathyroid development: Lessons from hypoparathyroid disorders. *Ann N Y Acad Sci.* 2011;1237:24–38.
36. Grigorieva IV, Mirczuk S, Gaynor KU, et al. Gata3-deficient mice develop parathyroid abnormalities due to dysregulation of the parathyroid-specific transcription factor Gcm2. *J Clin Invest.* 2010;120(6):2144–2155.
37. Xu PX, Zheng W, Laclef C, et al. Eya1 is required for the morphogenesis of mammalian thymus, parathyroid and thyroid. *Development.* 2002;129(13):3033–3044.
38. Zou D, Silvius D, Davenport J, Grifone R, Maire P, Xu PX. Patterning of the third pharyngeal pouch into thymus/parathyroid by Six and Eya1. *Dev Biol.* 2006;293(2):499–512.
39. Frank DU, Fotheringham LK, Brewer JA, et al. An Fgf8 mouse mutant phenocopies human 22q11 deletion syndrome. *Development.* 2002;129(19):4591–4603.
40. Gardiner JR, Jackson AL, Gordon J, Lickert H, Manley NR, Basson MA. Localised inhibition of FGF signalling in the third pharyngeal pouch is required for normal thymus and parathyroid organogenesis. *Development.* 2012;139(18):3456–3466.
41. Gordon J, Patel SR, Mishina Y, Manley NR. Evidence for an early role for BMP4 signaling in thymus and parathyroid morphogenesis. *Dev Biol.* 2010;339(1):141–154.
42. Bachiller D, Klingensmith J, Shneyder N, et al. The role of chordin/Bmp signals in mammalian pharyngeal development and DiGeorge syndrome. *Development.* 2003;130(15): 3567–3578.
43. Wurdak H, Ittner LM, Lang KS, et al. Inactivation of TGF-beta signaling in neural crest stem cells leads to multiple defects reminiscent of DiGeorge syndrome. *Genes Dev.* 2005; 19(5):530–535.
44. Moore-Scott BA, Manley NR. Differential expression of Sonic hedgehog along the anterior-posterior axis regulates patterning of pharyngeal pouch endoderm and pharyngeal endoderm-derived organs. *Dev Biol.* 2005;278(2):323–335.
45. Parvari R, Hershkovitz E, Grossman N, et al. Mutation of TBCE causes hypoparathyroidism-retardation-dysmorphism and autosomal recessive Kenny-Caffey syndrome. *Nat Genet.* 2002;32(3):448–452.
46. Parvari R, Diaz GA, Hershkovitz E. Parathyroid development and the role of tubulin chaperone E. *Horm Res.* 2007;67(1):12–21.
47. Ghandur-Mnaymneh L, Cassady J, Hajianpour MA, Paz J, Reiss E. The parathyroid gland in health and disease. *Am J Pathol.* 1986;125(2):292–299.
48. Gilmour JR, Martin WJ. The weight of the parathyroid glands. *J Pathol Bacteriol.* 1937;44:431–462.
49. DeLellis RA. *Atlas of Tumor Pathology-Tumors of the Parathyroid Gland.* 3 ed. Armed Forces Institute of Pathology, 1993.
50. Dufour DR, Wilkerson SY. The normal parathyroid revisited: percentage of stromal fat. *Hum Pathol.* 1982;13(8):717–721.
51. Dufour DR, Wilkerson SY. Factors related to parathyroid weight in normal persons. *Arch Pathol Lab Med.* 1983; 107(4):167–172.

52. Grimelius L, Akerström G, Bondeson L, et al. The role of the pathologist in diagnosis and surgical decision making in hyperparathyroidism. *World J Surg.* 1991;15:698–705.
53. Grimelius L, Akerstrom G, Johansson H, Lundqvist H. Estimation of parenchymal cell content of human parathyroid glands using the image analyzing computer technique. *Am J Pathol.* 1978;93(3):793–800.
54. Akerstrom G, Grimelius L, Johansson H, Lundqvist H, Pertoft H, Bergstrom R. The parenchymal cell mass in normal human parathyroid glands. *Acta Pathol Microbiol Scand A.* 1981;89(5):367–375.
55. Akerstrom G, Grimelius L, Johansson H, Pertoft H, Lundqvist H. Estimation of the parathyroid parenchymal cell mass by density gradients. *Am J Pathol.* 1980;99(3):685–694.
56. Akerström G, Malmaeus J, Grimelius L, Ljunghall S, Bergström R. Histological changes in parathyroid glands in subclinical and clinical renal disease. An autopsy investigation. *Scand J Urol Nephrol.* 1984;18:75–84.
57. Yao K, Singer FR, Roth SI, Sassoon A, Ye C, Giuliano AE. Weight of normal parathyroid glands in patients with parathyroid adenomas. *J Clin Endocrinol Metab.* 2004;89(7):3208–3213.
58. Balashev VN, Ignashkina MS. Lymphatic system of parathyroid glands in man. *Fed Proc Transl Suppl.* 1965;24(4):603–604.
59. Altenahr E. Electron microscopical evidence for innervation of chief cells in human parathyroid gland. *Experientia* 1971;27(9):1077.
60. Yeghiayan E, Rojo-Ortega JM, Genest J. Parathyroid vessel innervation: an ultrastructural study. *J Anat.* 1972;112(Pt 1): 137–142.
61. Isono H, Shoumura S. Effects of vagotomy on the ultrastructure of the parathyroid gland of the rabbit. *Acta Anat (Basel).* 1980;108(3):273–280.
62. Cinti S, Balercia G, Zingaretti MC, Amati S, Osculati F. The normal human parathyroid gland. A histochemical and ultrastructural study with particular reference to follicular structures. *J Submicrosc Cytol.* 1983;15(3):661–679.
63. Lieberman A, DeLellis RA. Intrafollicular amyloid in normal parathyroid glands. *Arch Pathol.* 1973;95(6):422–423.
64. Roth SI, Capen CC. Ultrastructural and functional correlations of the parathyroid gland. *Int Rev Exp Pathol.* 1974; 13(0):161–221.
65. Roth SI, Au WY, Kunin AS, Krane SM, Raisz LG. Effect of dietary deficiency in vitamin D, calcium, and phosphorus on the ultrastructure of the rat parathyroid gland. *Am J Pathol.* 1968;53(4):631–650.
66. Hamperl H. Über das Vorkommen von Onkocyten in verschiedenen Organen und ihren Geschwülsten (Mundspeicheldrüsen, Bauschpeicheldrüse, Epithelkörperchen, Hypophyse, Schilddrüse, Eileiter). *Virchows Arch.* 1936;25:327–375.
67. Bai S, LiVolsi VA, Fraker DL, Bing Z. Water-clear parathyroid adenoma: report of two cases and literature review. *Endocr Pathol.* 2012;23(3):196–200.
68. Grenko RT, Anderson KM, Kauffman G, Abt AB. Water-clear cell adenoma of the parathyroid. A case report with immunohistochemistry and electron microscopy. *Arch Pathol Lab Med.* 1995;119(11):1072–1074.
69. Roth SI. The ultrastructure of primary water-clear cell hyperplasia of the parathyroid glands. *Am J Pathol.* 1970;61(2):233–248.
70. Emura S, Shoumura S, Utsumi M, et al. Origin of the water-clear cell in the parathyroid gland of the golden hamster. *Acta Anat (Basel).* 1991;140(4):357–361.
71. Cinti S, Sbarbati A. Ultrastructure of human parathyroid cells in health and disease. *Microsc Res Tech.* 1995;32(2):164–179.
72. Potts JT. Parathyroid hormone: past and present. *J Endocrinol.* 2005;187(3):311–325.
73. Brown EM, Gamba G, Riccardi D, et al. Cloning and characterization of an extracellular Ca(2+)-sensing receptor from bovine parathyroid. *Nature.* 1993;366(6455):575–580.
74. Riccardi D, Park J, Lee WS, Gamba G, Brown EM, Hebert SC. Cloning and functional expression of a rat kidney extracellular calcium/polyvalent cation-sensing receptor. *Proc Natl Acad Sci U S A.* 1995;92(1):131–135.
75. Aida K, Koishi S, Tawata M, Onaya T. Molecular cloning of a putative Ca(2+)-sensing receptor cDNA from human kidney. *Biochem Biophys Res Commun.* 1995;214(2):524–529.
76. Alfadda TI, Saleh AM, Houillier P, Geibel JP. Calcium-sensing receptor 20 years later. *Am J Physiol Cell Physiol.* 2014;307(3):C221–C231.
77. English KA, Lines KE, Thakker RV. Genetics of hereditary forms of primary hyperparathyroidism. *Hormones (Athens).* 2023.
78. Chou YH, Pollak MR, Brandi ML, et al. Mutations in the human Ca(2+)-sensing-receptor gene that cause familial hypocalciuric hypercalcemia. *Am J Hum Genet.* 1995;56(5): 1075–1079.
79. Guan B, Welch JM, Sapp JC, et al. GCM2-activating mutations in familial isolated hyperparathyroidism. *Am J Hum Genet.* 2016;99(5):1034–1044.
80. Juppner H, Abou-Samra AB, Freeman M, et al. A G protein-linked receptor for parathyroid hormone and parathyroid hormone-related peptide. *Science.* 1991;254(5034):1024–1026.
81. Tafaj O, Juppner H. Pseudohypoparathyroidism: One gene, several syndromes. *J Endocrinol Invest.* 2017;40(4):347–356.
82. Alharbi N, Asa SL, Ezzat S. Intrathyroidal parathyroid carcinoma: An atypical thyroid lesion. *Front Endocrinol (Lausanne).* In press.
83. Ordonez NG. Value of GATA3 immunostaining in the diagnosis of parathyroid tumors. *Appl Immunohistochem Mol Morphol.* 2014;22(10):756–761.
84. Nonaka D. Study of parathyroid transcription factor Gcm2 expression in parathyroid lesions. *Am J Surg Pathol.* 2011;35(1):145–151.
85. Yan M, Roncin KL, Wilhelm S, Wasman JK, Asa SL. Images in endocrine pathology: high-grade intrathyroidal parathyroid carcinoma with Crooke's hyalinization. *Endocr Pathol.* 2020;31(2):190–194.
86. Cheung CC, Rotstein L, Asa SL. Parathyroid pathology: A practical approach. *Pathology Case Reviews.* 2003;8:34–41.
87. Arnold A, Brown MF, Urena P, Gaz RD, Sarfati E, Drueke TB. Monoclonality of parathyroid tumors in chronic renal failure and in primary parathyroid hyperplasia. *J Clin Invest.* 1995;95(5):2047–2053.
88. Arnold A, Kim HG. Clonal loss of one chromosome 11 in a parathyroid adenoma. *J Clin Endocrinol Metab.* 1989;69:496–499.
89. Arnold A, Staunton CE, Kim HG, Gaz RD, Kronenberg HM. Monoclonality and abnormal parathyroid hormone genes in parathyroid adenomas. *N Engl J Med.* 1988;218:658–652.
90. Friedman E, Sakaguchi K, Bale AE, et al. Clonality of parathyroid tumors in familial multiple endocrine neoplasia type I. *N Engl J Med.* 1989;321:213–218.
91. Erickson LA, Mete O, Juhlin CC, Perren A, Gill AJ. Overview of the 2022 WHO Classification of Parathyroid Tumors. *Endocr Pathol.* 2022;33(1):64–89.

92. Alwaheeb S, Rambaldini G, Boerner S, Coire C, Fiser J, Asa SL. Worrisome histologic alterations following fine-needle aspiration of the parathyroid. *J Clin Pathol.* 2006;59(10): 1094–1096.
93. Kim J, Horowitz G, Hong M, Orsini M, Asa SL, Higgins K. The dangers of parathyroid biopsy. *J Otolaryngol Head Neck Surg.* 2017;46(1):4.
94. Carpten JD, Robbins CM, Villablanca A, et al. HRPT2, encoding parafibromin, is mutated in hyperparathyroidism-jaw tumor syndrome. *Nat Genet.* 2002;32(4):676–680.
95. Juhlin CC, Nilsson IL, Johansson K, et al. Parafibromin and APC as screening markers for malignant potential in atypical parathyroid adenomas. *Endocr Pathol.* 2010;21(3): 166–177.
96. Gill AJ, Clarkson A, Gimm O, et al. Loss of nuclear expression of parafibromin distinguishes parathyroid carcinomas and hyperparathyroidism-jaw tumor (HPT-JT) syndrome-related adenomas from sporadic parathyroid adenomas and hyperplasias. *Am J Surg Pathol.* 2006;30(9):1140–1149.
97. Erickson LA, Jin L, Wollan P, Thompson GB, van Heerden JA, Lloyd RV. Parathyroid hyperplasia, adenomas, and carcinomas: differential expression of p27Kip1 protein. *Am J Surg Pathol.* 1999;23(3):288–295.
98. Arnold A. Molecular mechanisms of parathyroid neoplasia. *Endocrinol Metab Clin North Am.* 1994;23:93–107.
99. Cryns VL, Rubio M-P, Thor AD, Louis DN, Arnold A. p53 Abnormalities in human parathyroid carcinoma. *J Clin Endocrinol Metab.* 1994;78:1320–1324.
100. Cryns VL, Thor A, Zu H-J, et al. Loss of the retinoblastoma tumor-suppressor gene in parathyroid carcinoma. *N Engl J Med.* 1994;330:757–761.
101. Arnold A. Molecular genetics of parathyroid gland neoplasia. *J Clin Endocrinol Metab.* 1993;77:1108–1112.
102. Arnold A, Staunton CE, Kim GH, et al. Monoclonality and abnormal parathyroid hormone genes in parathyroid adenomas. *N Engl J Med.* 1988;318:658–662.
103. Arnold A, Kim HG, Gaz RD, et al. Molecular cloning and chromosomal mapping of DNA rearranged with the parathyroid hormone gene in a parathyroid adenoma. *J Clin Invest.* 1989;83:2034–2040.
104. Erovic BM, Harris L, Jamali M, et al. Biomarkers of parathyroid carcinoma. *Endocr Pathol.* 2012;23(4):221–231.
105. Stojadinovic A, Hoos A, Nissan A, et al. Parathyroid neoplasms: Clinical, histopathological, and tissue microarray-based molecular analysis. *Hum Pathol.* 2003;34(1):54–64.
106. Hosny MK, Siddiqui MT, Willis BC, et al. Parafibromin, APC, and MIB-1 are useful markers for distinguishing parathyroid carcinomas from adenomas. *Appl Immunohistochem Mol Morphol.* 2017;25(10):731–735.
107. Abbona GC, Papotti M, Gasparri G, Bussolati G. Proliferative activity in parathyroid tumors as detected by Ki-67 immunostaining. *Hum Pathol.* 1995;26(2):135–138.
108. Georgitsi M. MEN-4 and other multiple endocrine neoplasias due to cyclin-dependent kinase inhibitors (p27(Kip1) and p18(INK4C)) mutations. *Best Pract Res Clin Endocrinol Metab.* 2010;24(3):425–437.
109. Georgitsi M, Raitila A, Karhu A, et al. Germline CDKN1B/p27Kip1 mutation in multiple endocrine neoplasia. *J Clin Endocrinol Metab.* 2007;92(8):3321–3325.
110. Seabrook AJ, Harris JE, Velosa SB, et al. Multiple endocrine tumors associated with germline MAX mutations: Multiple endocrine neoplasia type 5? *J Clin Endocrinol Metab.* 2021;106(4):1163–1182.
111. Gill AJ, Lim G, Cheung VKY, et al. Parafibromin-deficient (HPT-JT Type, CDC73 Mutated) Parathyroid Tumors Demonstrate Distinctive Morphologic Features. *Am J Surg Pathol.* 2018.
112. Erovic BM, Goldstein DP, Kim D, et al. Parathyroid cancer: Outcome analysis of 16 patients treated at the princess margaret hospital. *Head Neck.* 2012.
113. Duan K, Mete O. Parathyroid Carcinoma: Diagnosis and Clinical Implications. *Turk Patoloji Derg.* 2015;31 Suppl 1:80–97.
114. Akirov A, Asa SL, Larouche V, et al. The Clinicopathological Spectrum of Parathyroid Carcinoma. *Front Endocrinol (Lausanne).* 2019;10:731.
115. Gulenchyn KY, Yao X, Asa SL, Singh S, Law C. Radionuclide therapy in neuroendocrine tumours: A systematic review. *Clin Oncol (R Coll Radiol).* 2012.
116. Opalinska M, Hubalewska-Dydejczyk A, Sowa-Staszczak A. Radiolabeled peptides: Current and new perspectives. *Q J Nucl Med Mol Imaging.* 2017;61(2):153–167.

45

Adrenal

Ozgur Mete ■ Lori Erickson

The adrenal glands are composite of endocrine organs located in the retroperitoneum, superomedial to the kidneys. They consist of the cortex and medulla; these components have a different embryonic origin, histology, and function. The adrenal cortex is composed of steroidogenic cells while the medulla consists of nests of chromaffin cells arranged in highly vascularized stroma, essentially representing intra-adrenal sympathetic paraganglia.

ANATOMY

The main portions of the adrenal glands are easily recognized due to the presence of a relatively thick yellow layer external to a narrow dark brown band that abuts a solid, pearly gray interior (Fig. 45.1). The former two zones correspond histologically to the zona fasciculata and zona reticularis of the cortex, respectively, and the latter corresponds to the medulla.

The location of the adrenal glands between several organs is responsible for their particular shape: the right adrenal gland is pyramidal in shape while the left has a crescentic shape. The depressions and ridges on the posterior surfaces (Fig. 45.2) result from their close relationships to the kidneys. When a kidney is congenitally absent, the corresponding adrenal gland is round and the characteristic longitudinal ridge on its posterior surface is absent.

EVOLUTION

The distribution of the cortex and medulla in the human adrenal gland varies based on the anatomic section of the gland. The recognition of this feature is of interest when

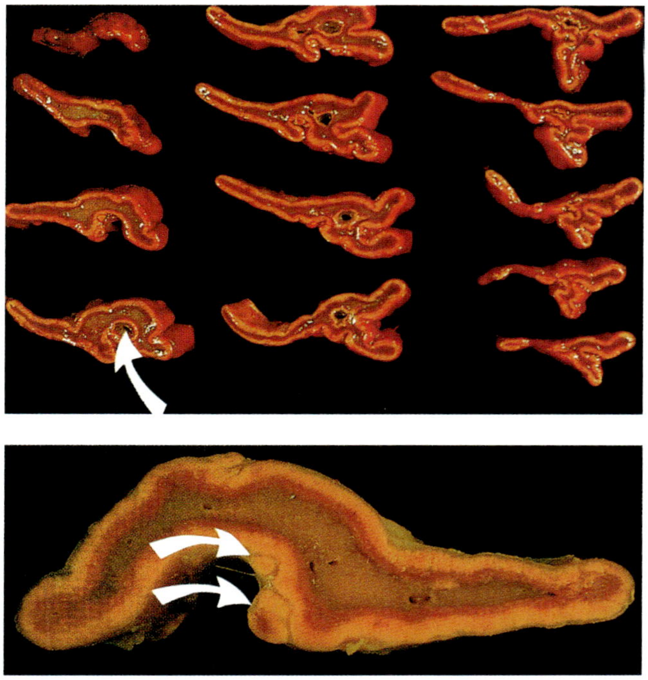

FIGURE 45.1 Normal adrenal gland. **Top:** Fresh gland sliced from head (*upper left*) through body (*center*) to tail (*lower right*). The *yellow* cortex and *pearly gray* medulla are visible in the head (*left*). *Yellow* zona fasciculata surrounds *dark brown* zona reticularis in the tail (*right*), where medulla is absent. The gland can be identified as the left adrenal because on this side the adrenal vein runs in a well-developed groove on the surface of the gland at the junction of the head and body (*arrow*). Invaginated cortex surrounds the central vein in the interior of the body. **Bottom:** Slice of formalin-fixed adrenal gland showing, from exterior inward, cortical zona fasciculata (*yellow*), zona reticularis (*brown*), and medulla (*gray*). The cortex is about 2 mm in thickness. Dilated tributaries of the central vein are seen in the medulla. The adrenal vein has been removed from its groove. Nodules of accessory cortex are present (*arrows*).

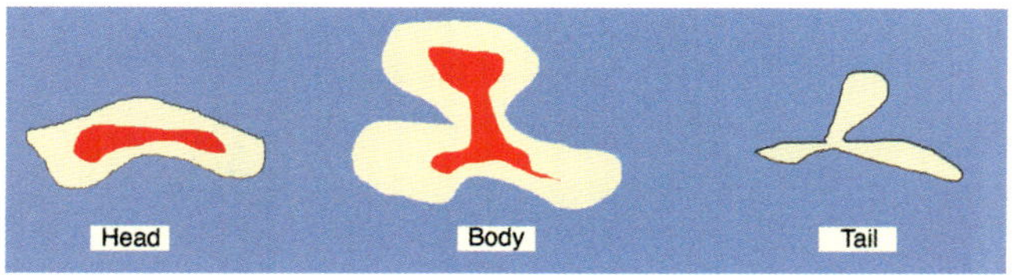

FIGURE 45.3 Diagrammatic illustration of the distribution of normal medulla in the head, body, and tail of the adrenal gland.

assessing pathologic states of adrenal medulla, particularly in the context of adrenal medullary hyperplasia (Fig. 45.3). Most of the medulla is located in the head of the gland medially, some occurs in the body, and there is usually none in the tail laterally (1). Two bands of adrenal cortex form the alae of the glands. Each cortical zone accounts for one-third of the overall thickness of the head and body of the gland, while the remaining one-third of the glandular thickness is typically occupied by the medulla.

DEVELOPMENT

Adrenal Cortex

The adrenal cortex originates from intermediate mesoderm similar to the urogenital structures. Its complex embryologic development follows five major steps starting with the formation of adrenogenital primordium, followed by adrenogenital primordium separation, invasion of precursors of cells of the medulla, encapsulation, and zonation (2).

Several transcription factors, hormone receptors, and cell signaling pathways (Table 45.1) are required for the formation of adrenogential primordium and subsequent specification of the adrenal primordium. Among the

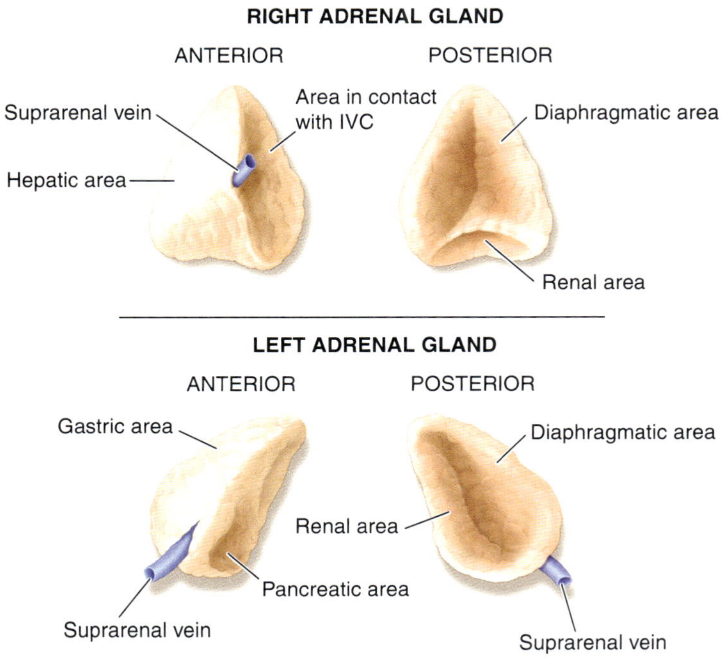

FIGURE 45.2 Topographic anatomy. There is a difference in the surface anatomy of right and left adrenal glands. IVC, inferior vena cava. (Reprinted with permission from Kawamura D, Nolan T. *Abdomen and Superficial Structures*. Wolters Kluwer Health; 2017.)

TABLE 45.1 Essential Developmental Factors and Signaling Pathways Involved in the Adrenal Cortex

Developmental Factors and Pathways	Major Function
SF1 (NR5A1)	Main regulator of steroidogenic cell differentiation and development of the adrenogenital primordium
WT1	Directly controls SF1 expression in the adrenogenital primordia; expressed in mesenchymal cells that encapsule the gland
DAX1 (NROB1)	Alters SF1 expression and regarded as a key mediator of SF1-dependent development
GATA factors (GATA6, GATA4)	Both are expressed in the adrenogenital primordium, and essential regulators of developing cortex, facilitate adrenal versus gonadal fate in developing primordia; GATA6 is carried over postpartum stages
Growth factor signaling pathways	Support the survival and progression of the development
Wnt/b-catenin pathway	Active wnt/b-catenin pathway is essential since the early developing gland
Sonic Hedgehog pathway	Support the development via paracrine signaling and likely contribute specification process

developmental factors, steroidogenic factor 1 (SF1), which is encoded by *NR5A1*, is the main regulator of the adrenal cortical development. Other critical factors of the developing cortex include DAX1 (NR0B1), WT1, GATA6 and GATA4 (2–4) (Table 45.1). DAX1 is a target gene of SF1, the latter of which is regulated by WT1. WT1 is expressed in the adrenogondal primordium and regulates mesenchymal cells during the phase of encapsulation.

The primordia appear at the 9-mm embryo stage during the 6th week of gestation as bilateral cellular aggregations (Fig. 45.4) at the mesenteric root, medial to the developing gonad and anterior to the mesonephros (5,6). These primordia are composed of two groups of mesenchymal cells: one destined to be the precursor of the fetal cortex and the other to become the adrenal capsule and its supporting connective tissue framework (6). The primordia become more defined by the 7th gestational week, at which time they separate from the coelomic lining. The primordia consist of polyhedral cells with eosinophilic, lipid-poor cytoplasm.

The primordial cells increase in size and proliferate rapidly, forming a series of parallel columns and cords that ultimately compose the bulk of the fetal cortex. The inner eosinophilic core of the gland is composed of moderately large cells arranged in tightly packed cords toward the outer aspect of the zone but more widely spaced in its innermost regions where vascular sinusoidal spaces are prominent. A thin subcapsular rim of smaller cells develops external to the fetal cortex, and ultimately develops into the adult cortex

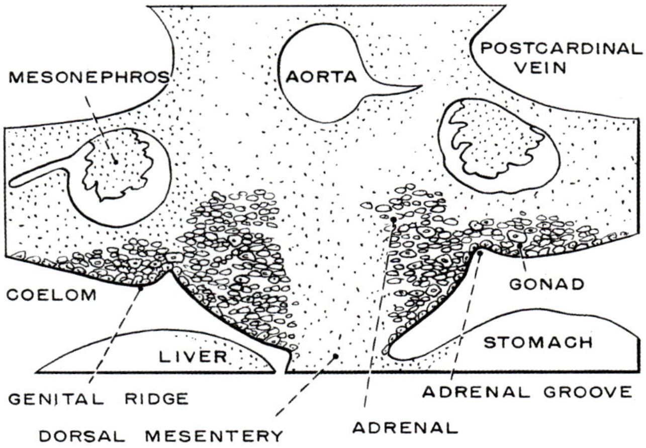

FIGURE 45.4 Diagrammatic representation of a human embryo at 6 weeks gestation showing the anatomic relationship of developing adrenal gland to coelomic cavity, gonad, and kidney (mesonephros). (Reprinted with permission from Dahl EV, Bahn RC. Aberrant adrenal cortical tissue near the testis in human infants. *Am J Pathol.* 1962;40:587–598, with permission from Elsevier.)

(Fig. 45.5). These cells are arranged in nests and arches that cap the columns of deeper cells. They have hyperchromatic, closely packed, and overlapping nuclei while the cells of the cords contain larger, more vesicular nuclei. Continuous, spottily distributed degeneration of the cells is found in the cords, and apoptotic cells are continuously replaced by the proliferation of cells in the narrow subcapsular band. Growth of the developing cortex is centripetal from the outside inward.

The adrenal glands enlarge during the second trimester in direct proportion to the total increase in body weight. Each gland weighs approximately 1.5 g by the 6th month of gestation and increases in weight thereafter, largely due to the expansion of the inner fetal zone. Each gland weighs slightly more than 4 g at birth mostly due to the presence of an abundant provisional cortex (Fig. 45.6). There are no sex differences with respect to size, shape, or weight during development.

The cut surface of the newborn adrenal gland features a relatively narrow yellow rim of permanent cortex overlying a deep brown—red provisional zone. Within hours of birth, however, the latter becomes acutely congested and begins to degenerate such that the provisional cortex is largely disorganized and necrotic by 7 to 10 days of life. Involution is substantial at 2 weeks and well established by the end of the first month, at which point the gland weight is diminished by approximately 50%. Loss of provisional cortex volume results in the conversion of the plump neonatal gland into a relatively shrunken, leaf-like organ weighing only approximately 1 g at the end of the first year of life. The gland grows slowly thereafter to reach approximately 2 g by the 10th year, after which it doubles in weight during puberty and adolescence to reach its full weight by 15 to 18 years. The key events in prepubertal cortex development include involution of the fetal zone, establishment of zonation in the outer cortex, and emergence of a recognizable zona reticularis.

The zona glomerulosa remains diffuse during childhood and adolescence. During early adult life, it assumes the

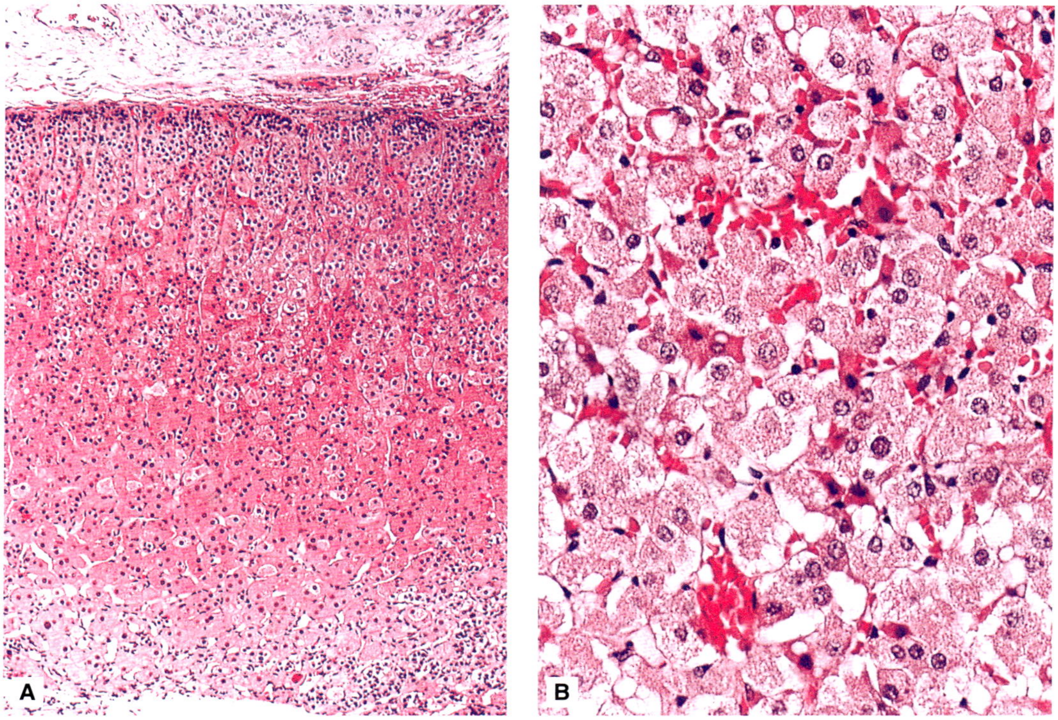

FIGURE 45.5 **A:** Provisional (fetal) cortex (29 weeks gestation, stillborn infant). Cortex is dominated by large cells with eosinophilic cytoplasm in vague columns and solid sheets with prominent capillaries. Just beneath the tenuous capsule, there is a rim of smaller cells, the source of the permanent cortex. A sympathetic ganglion and a small nerve are present in the periadrenal connective tissue (*top*). **B:** The cortex contains large cells with voluminous granular cytoplasm. Nuclei are vesicular with a single nucleolus. Vascularity is prominent.

scattered distribution characteristic of adult glands, presumably because it is displaced by the outward extension of columns of lipid-rich cells of the zona fasciculata.

The final stage in the development of the adult cortex involves the evolution of the zona reticularis. The zona reticularis is seldom seen in adrenal glands until the age of 3 years. By the age of 5 years, most glands contain scattered aggregates of compact cells on the innermost aspect of the adrenal cortex. The zona reticularis continues to expand such that virtually all glands possess a continuous concentric zona reticularis by the age of 14 years. Lipofuscin pigment increases with age.

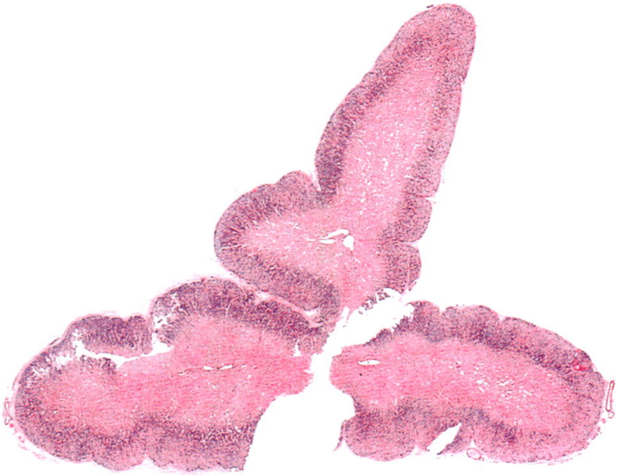

FIGURE 45.6 Adrenal cortex at birth (35 weeks gestation; infant died at 2 days). The central, degenerating, eosinophilic provisional cortex is surrounded by the developing, darkly staining outer rim of permanent cortex.

Adrenal Medulla

The adrenal medulla is an intra-adrenal sympathetic paraganglion. It is composed of chromaffin cells surrounded by steroidogenic cells of the adrenal cortex. Recent developmental evidence suggests that only a fraction of the medulla derives directly from the migratory neural crest cells. Pluripotent Schwann cell precursors associated with preganglionic nerve fibers of the adrenal anlage provide the major path for the development of chromaffin cells in the medulla (2,7).

Sympathogonia cells (i.e., those from the sympathetic anlagen) migrate further into nerves derived from the sympathetic chain and then they move alongside large blood vessels that penetrate the unencapsulated fetal adrenal cortex, primarily at its caudal pole, thereby explaining the nonuniform distribution of medulla in the adult adrenal gland. The precursor cells enter the adrenal primordium as fingerlike processes, passing among the fetal cortical cells. In this manner, sympathogonia and a plexus of nerves are initially scattered among fetal cortical cells, ultimately emerging as precursors of the medulla (Fig. 45.7).

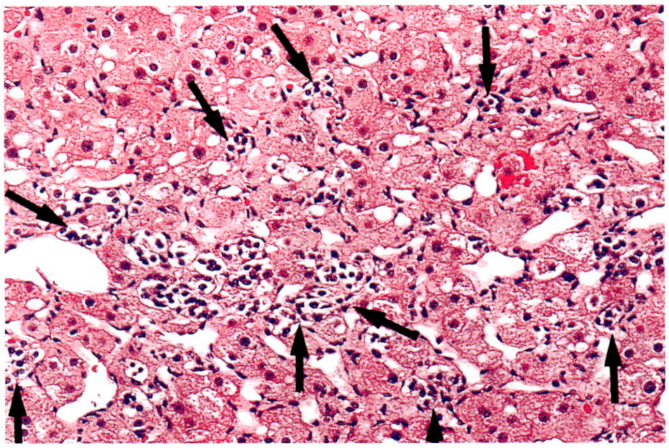

FIGURE 45.7 Developing adrenal medulla (29 weeks gestation, stillborn infant). Clusters of small medullary cells with deeply staining nuclei (*arrows*) irregularly distributed in very vascular provisional cortex. When the latter degenerates, the clusters of medullary cells survive and, lacking the support of the cortical cells, aggregate together.

At birth, the adrenal medulla comprises a central, very thin core of sympathogonia with offshoots that stretch a short distance into the periphery of the degenerating provisional cortex. The medullary cells are arranged in irregularly sized clumps containing both cell types, although the larger cells predominate. The postnatal collapse of the provisional cortex and its stroma removes the framework that supported the medullary offshoots and their associated nerve plexus in the cortex. Loss of this scaffolding leads to coalescence of these structures around the central veins.

Several transcription factors including GATA3, PHOX2B, and HAND2 play important roles in the development of chromaffin cells in the sympathetic paraganglia and chief cells in the parasympathetic paraganglia (2,7–10). Immunohistochemical stains against these biomarkers may be used in clinical practice when a distinction between paraganglioma, pheochromocytoma, and neuroblastic tumors is required.

Developmental Abnormalities

› **AGENESIS AND HYPOPLASIA** Adrenal agenesis refers to a complete absence of the adrenal gland whereas adrenal hypoplasia refers to incomplete development of the adrenal cortex. Both aplasia and hypoplasia can occur in patients with anencephaly or pathogenic *NR5A1* variants (11). Adrenal hypoplasia is also a feature of Pollister–Hall syndrome due to *GLI3* variants (12,13). *DAX1* (*NR0B1*) variants result in X-linked adrenal hypoplasia congenita with hypogonadotropic hypogonadism or Xp21 deletion (14). Hypoplastic adrenal glands may show cytomegaly with lipid inversion pattern. They often appear to have prominent medulla given the reduced thickness of the adrenal cortex but this should not be considered to reflect adrenal medullary hyperplasia.

› **HETEROTOPIA** Heterotopic (accessory) adrenal tissue can occur either as an isolated cortical rest or in association

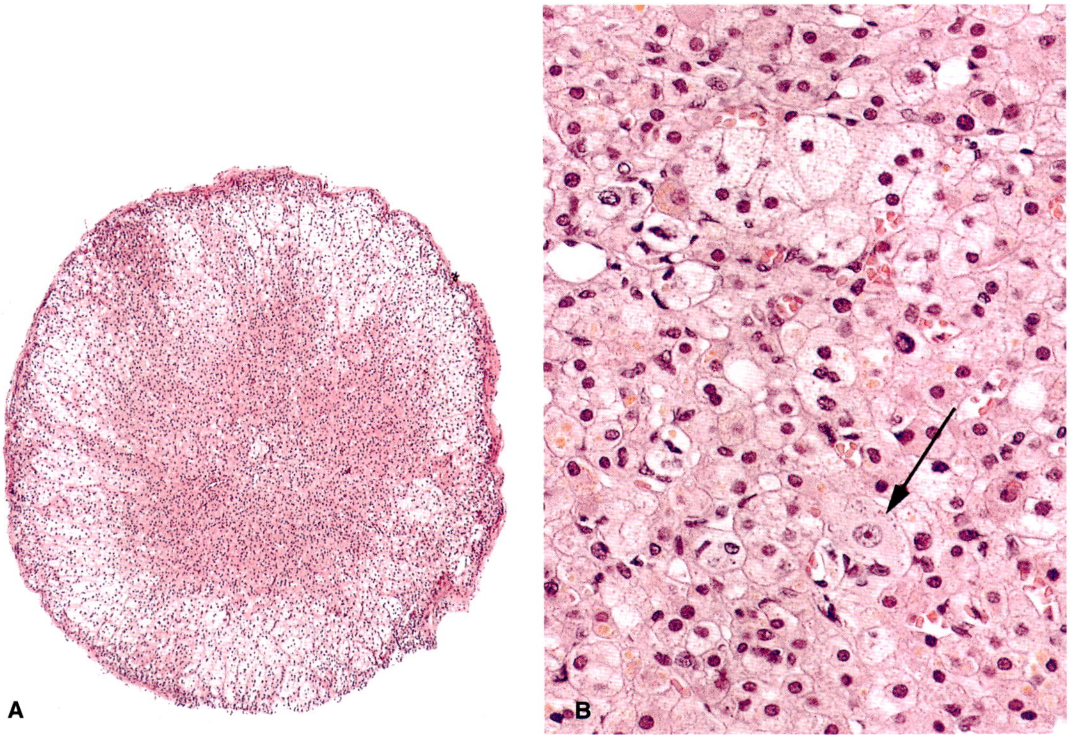

FIGURE 45.8 Accessory adrenal cortex in retroperitoneal fat. **A:** Normal zonation is suggested by the narrow rim of cells with clear cytoplasm that surrounds the main mass of cells with light eosinophilic cytoplasm. **B:** Ganglion cell (*arrow*) indicating the presence of medulla among cortical "clear" and "compact" cells. The latter contain lipofuscin.

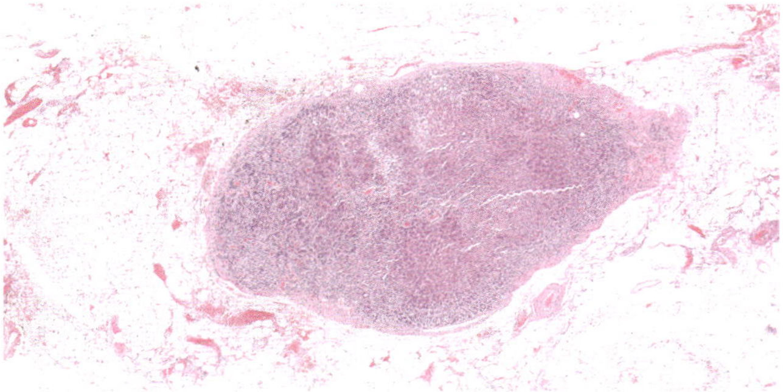

FIGURE 45.9 Ectopic adrenal cortical tissue identified as a small microscopic nodule in the paratesticular area.

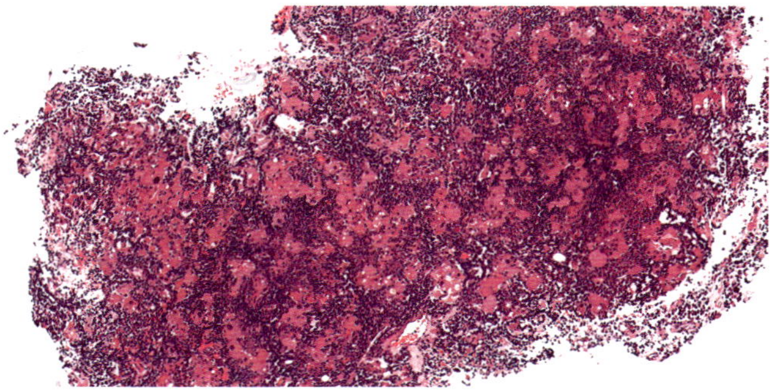

FIGURE 45.10 Adrenocortical choristoma identified in association with a pituitary corticotroph tumor. Large compact cells represent adrenocortical cells whereas smaller basophilic cells represent densely granulated corticotroph tumor.

with chromaffin cells. The adrenocortical primordium initially is unencapsulated and develops close to the gonad. Failed encapsulation of the developing adrenal cortex may explain the distribution of rests in sites aligned with the developing gonads and other intra-abdominal locations along the celiac axis, kidney, spleen, mesentery, and retroperitoneum (2,15–17). Heterotopic adrenocortical tissue is most often encountered around the adrenal glands themselves (Fig. 45.8); it also occurs in the inguinal region and around the ovary, fallopian tube, epididymis, and rete testis (Fig. 45.9). Accessory cortex shows normal zonation and responds to stimulation by adrenocorticotropic hormone.

› *Adrenal Choristoma* Adrenal choristoma is a rare abnormality that refers to the occurrence of benign adrenal cortical cells identified in unexpected locations such as lung, liver, pancreas, spinal nerves, brain, and pituitary (Fig. 45.10) (18,19).

› *Adrenal Fusions and Adhesions* Adrenal gland fusions can occur due to altered separation of developmental mesenchyme (Fig. 45.11). Most fusions occur between the liver or kidney and adrenal gland or between the adrenal glands (15). In this situation, the cellular elements of both organs are intermingled without intervening connective tissue planes. This should be remembered when dealing with tumor staging or in limited biopsy material from any organs that are subject to fusions (2).

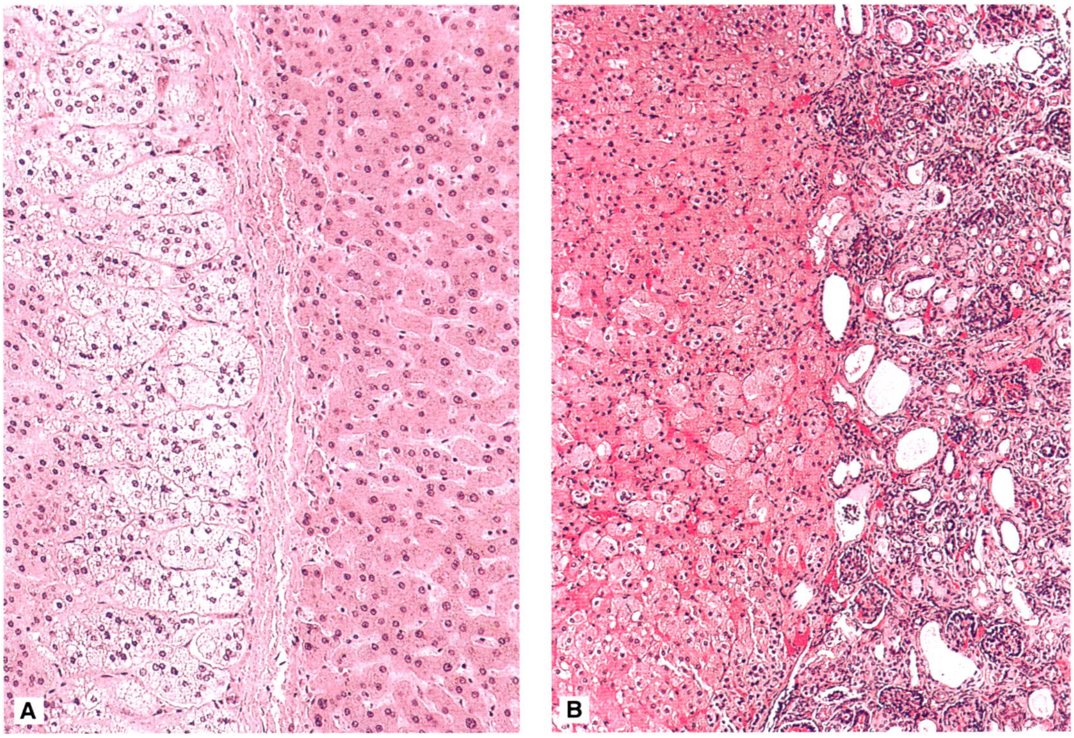

FIGURE 45.11 **A:** Common capsule of adrenal (*left*) and liver (*right*). Note that a zona glomerulosa is not evident in the adrenal cortex. **B:** Absence of an adrenal and kidney capsule results in direct contact of adrenal cortex (provisional) and renal parenchyma.

GLAND WEIGHT AND CORTICAL THICKNESS

The weight of the glands is important because assessment of the adrenal normalcy takes this feature into account. Data regarding the truly normal adrenal gland weight are difficult to obtain because the organ, particularly the cortex, responds rapidly to stress by increasing in mass. Therefore, accurate normal adrenal weight can be determined based on selected autopsy material obtained from healthy individuals who die suddenly. The combined gland weight in these circumstances ranges from 8 to 9 g and is not affected by formalin fixation (1). Gender differences are not apparent.

The adrenal glands are largest relative to body weight at the 4th month of gestation (20). The mean weight of each gland is 4.0 to 4.2 g among adults who die of sudden death without prior illness. The adrenal cortex accounts for 90% of the total gland weight. There is no correlation between absolute body weight and adrenal weight and no significant differences in the mean weight of the normal left and right glands. Neither pregnancy nor menopause appears to result in significant adrenal weight changes. If death is preceded by a prolonged illness, there is a significant increase in mean adrenal weight, with mean gland weight ranging from 5.8 to 6.2 g each. The thickness of the normal adult adrenal cortex is at least 2 mm and is best assessed microscopically.

ADRENAL GLANDS FOR HISTOLOGIC STUDY

Ideal Adrenal Glands

The adrenal gland, particularly the zona reticularis, quickly begins to show the effects of anoxia upon removal of the organ. However, glands that are less than optimally preserved are satisfactory for assessing general microanatomy.

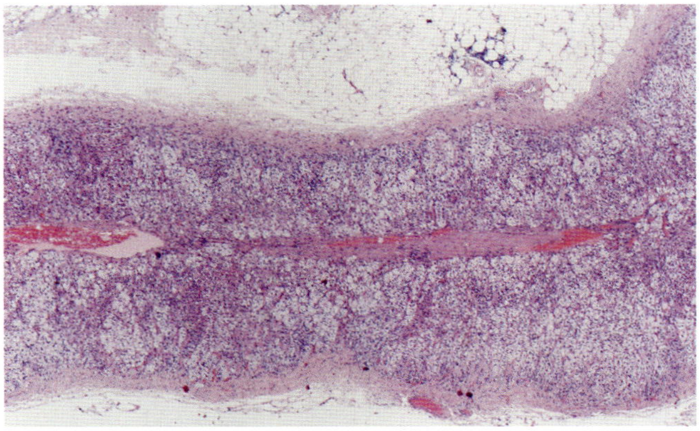

FIGURE 45.12 Adrenal cortical atrophy. Adrenal cortical atrophy is typically defined by reduced adrenal cortical thickness (<2 mm) in association with absent or reduced zona reticularis. In this photomicrograph, there is complete absence of zona reticularis in association with pathologically reduced cortical thickness.

Adrenal Glands Studied and Nontumorous Glands

Pathologists routinely examine the nontumorous cortex and medulla (7,21–23). The presence of cortical atrophy in patients with tumors of the adrenal gland (Fig. 45.11) suggests the tumor is a glucocorticoid-secreting cortical neoplasm (7,24). Such cases show reduced adrenal cortical thickness (<2 mm) in association with an absent or reduced zona reticularis (Figs. 45.12 and 45.13) (7). One of the early signs of Cushing syndrome (i.e., autonomous cortisol secretion) is the presence of a discontinuous or reduced zona reticularis layer unaccompanied by a pathologic reduction in the adrenal cortical thickness. It is important for pathologists to assess for the possibility of cortical atrophy since this information may help prevent adrenal cortical insufficiency (i.e., Addisonian crisis) in the event that the medical team is unaware of subclinical Cushing syndrome (24–26).

The adrenal medulla is normally present only in the head and body of the gland and should not exceed more than

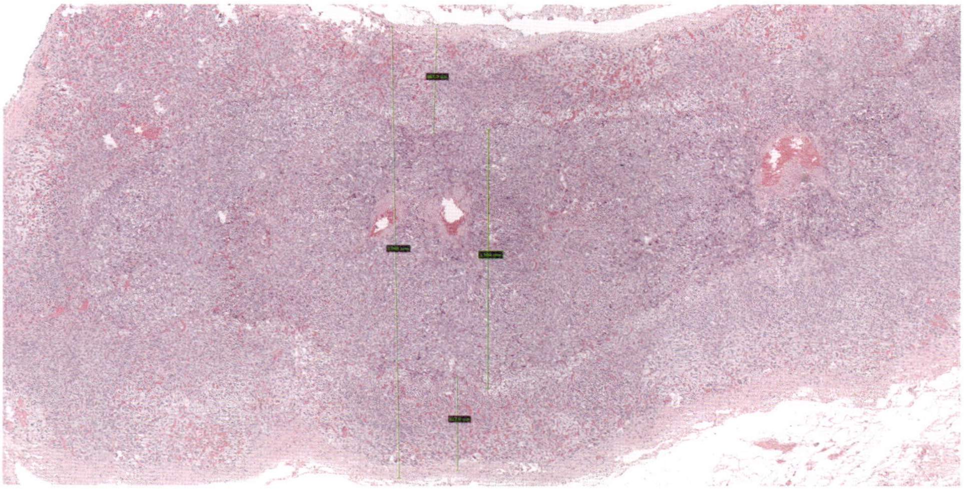

FIGURE 45.13 Normal medulla in adrenal cortical atrophy. In this case of subclinical Cushing syndrome, the lack of zona reticularis warrants atrophy and the medulla looks prominent due to atrophy of the cortex.

one-third of the gland thickness (7,23). Non-nodular expansion beyond the normal thickness warrants the diagnosis of adrenal medullary hyperplasia (7,23). Non-nodular prominence of the medulla in the tail should also be considered to represent medullary hyperplasia. Adrenal medullary hyperplasia, either alone or in association with a pheochromocytoma, is often considered a morphologic harbinger of pathogenic germline variants in *RET, MAX, TMEM127, NF1,* and *SDHB* (7,23,27). It is important to note that the adrenal medulla often appears expanded at sites of cortical atrophy (Fig. 45.13). A subcentimeter nodular proliferation almost always represents a microscopic pheochromocytoma (7).

HISTOLOGY

Capsule

The capsule of the adrenal gland is composed of coarse, hyalinized collagen bundles and elastic fibers (Fig. 45.14). It is usually thin, but it varies considerably in thickness from gland to gland and even within the same gland. It is tough and hard to cut but tears easily and does not support the unfixed gland, which is limp and readily bends. The soft consistency of the fresh glands makes them difficult to section; cooling them for 15 minutes in a refrigerator facilitates their sectioning. The adrenal capsule is surrounded by white fat in adults and brown fat in infants. Adipose tissue contains small arteries, veins, nerves, accessory cortex, excrescences of cortex, sympathetic ganglia, and an occasional paraganglion (Fig. 45.15). These points should be remembered as histologic pitfalls when assessing tumor staging or invasion in adrenal tumors.

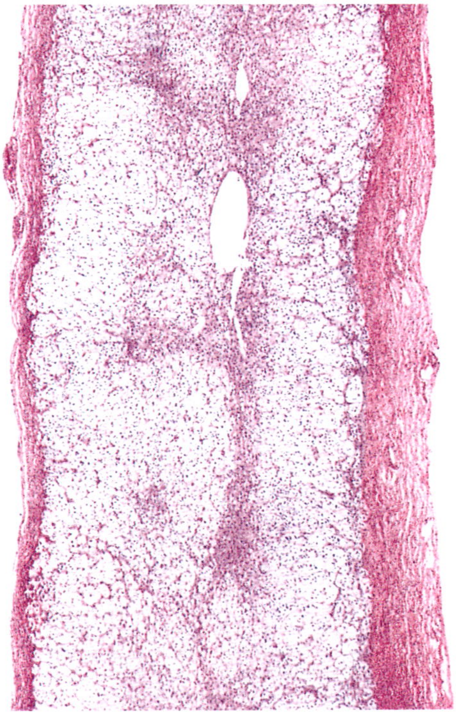

FIGURE 45.14 Adrenal capsule. Fourfold variation in the thickness of the capsule, which at its maximum thickness measures 0.3 mm.

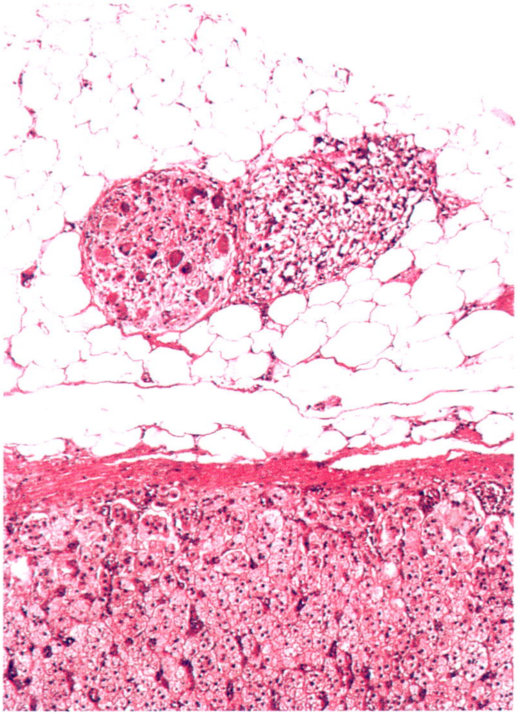

FIGURE 45.15 Normal adrenal gland. Juxtaposed sympathetic ganglion (*above left*) and paraganglion (*above right*) in periadrenal adipose tissue. The adrenal cortex features cells with clear cytoplasm (zona fasciculata). Zona glomerulosa is not evident.

The capsule is penetrated by blood vessels supplying and draining the glands, nerves to the medulla, and lymphatic channels, although only the site of exit of the adrenal vein is regularly encountered in routine sections. Capsular defects are occasionally encountered and may show protrusion of cortical tissue into the capsule or periadrenal fat to form small nodules (Fig. 45.16). These excrescences are composed predominantly of epithelial cells with a normal zonation pattern. The protrusions may contain a connective tissue component, and sometimes there is an equal mixture of cellular cords and fibrous or fibroadipose tissues (Fig. 45.17). Single rows and groups of cortical cells are commonly found here and there in capsular "pockets" (Fig. 45.18). Larger ovoid aggregates cause a slight depression in, and thinning of, the underlying cortex (Fig. 45.18).

Occasionally, wedge-shaped foci of small, plump spindle cells with hyperchromatic nuclei are attached to the capsule. These protrude into the cortex to varying depths and may be present bilaterally (Fig. 45.19). The cells are arranged in interlacing bundles and whorls. These aggregates have been termed ovarian thecal metaplasia due to their resemblance to ovarian cortical stroma, although they could also represent adrenocortical blastema that failed to mature (28–30). These foci undergo scar-like fibrosis, hyalinization, and sometimes calcification (31). Nests

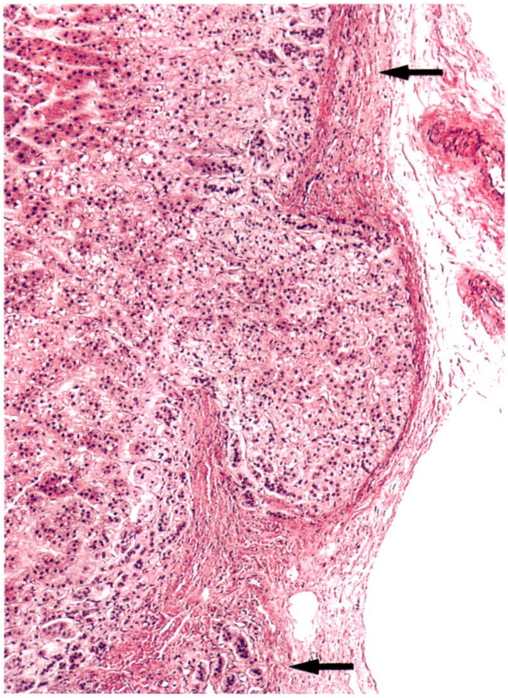

FIGURE 45.16 Normal adrenal gland. Protrusion of cortical cells surrounded by attenuated capsule through a "wide" defect in the adrenal capsule. A few cortical cells in rows and small aggregates are present in the capsule (*arrows*). A suggestive zona glomerulosa is present deep to the adrenal capsule (*top*).

of cortical cells are occasionally found in the spindle cell (Fig. 45.20) proliferation of myofibroblast origin (31), presumably due to entrapment.

These spindle cell proliferations rarely penetrate into the medulla as increasingly narrow tongues of tissue. Ovarian thecal metaplasia is said to occur in postmenopausal women, occasionally in premenopausal women, and exceptionally in older men, although we have noted them in perimenopausal or postmenopausal females. They are encountered fairly commonly in the extratumoral cortex associated with a range of functioning adrenocortical adenomas and in adrenal glands removed for other reasons. The "lesions" are generally incidental microscopic findings that are not recognized grossly.

Adrenal Cortex

The adult adrenal cortex consists of three functionally and histologically readily recognizable parenchymal cell types arranged in concentric zones or layers: the outer zona glomerulosa, the inner zona reticularis, and between them, the zona fasciculata (Figs. 45.21 and 45.22). The appearances of the different zones reflect their respective cellular arrangements, differential lipid distribution, and pigment accumulation. The normal zona glomerulosa occupies approximately 5% of the cortex whereas the fasciculata accounts for 70% and the reticularis spans 25% of the cortical thickness.

FIGURE 45.17 Normal adrenal gland. The normal cortical cells (*arrows*) can show extension/protrusion through the capsule (**A–D**). This finding should not be considered as an invasive growth.

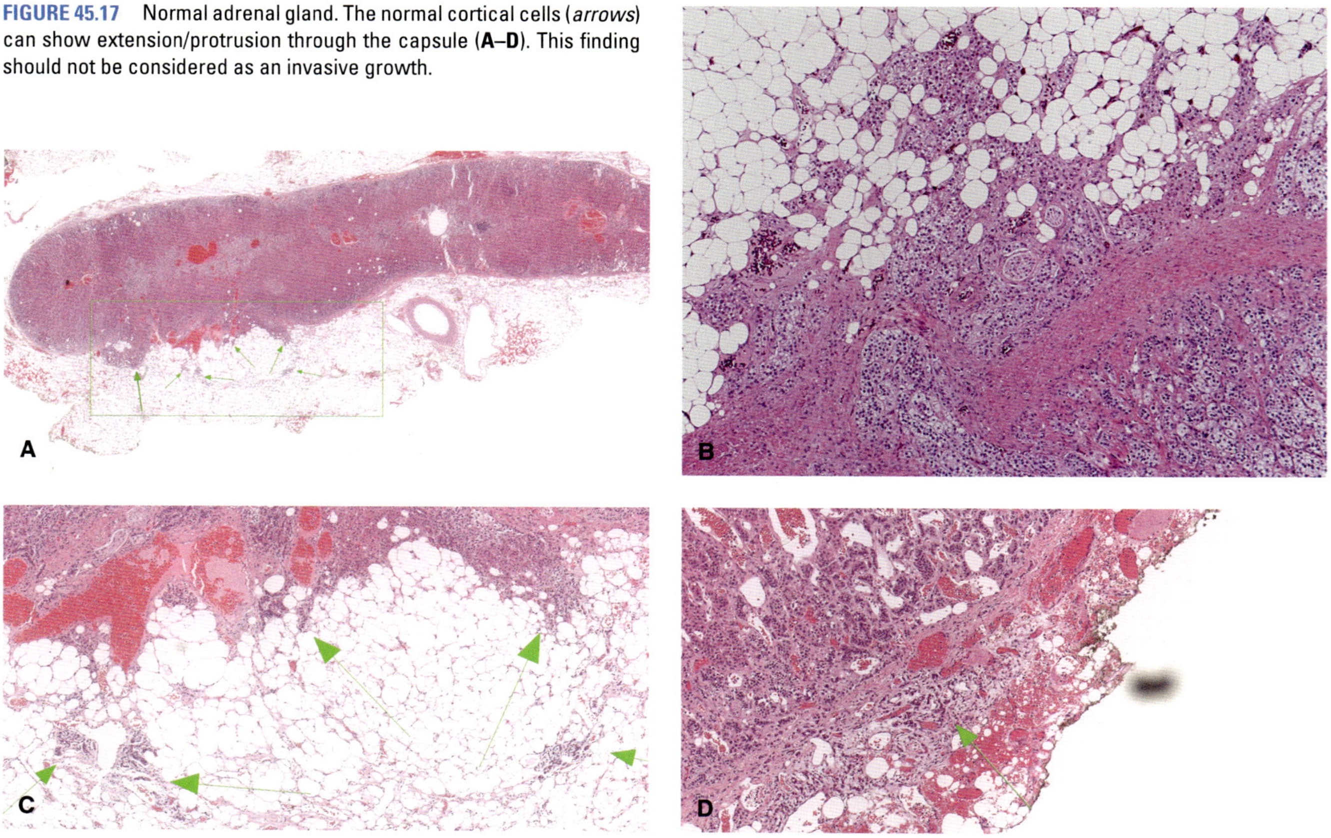

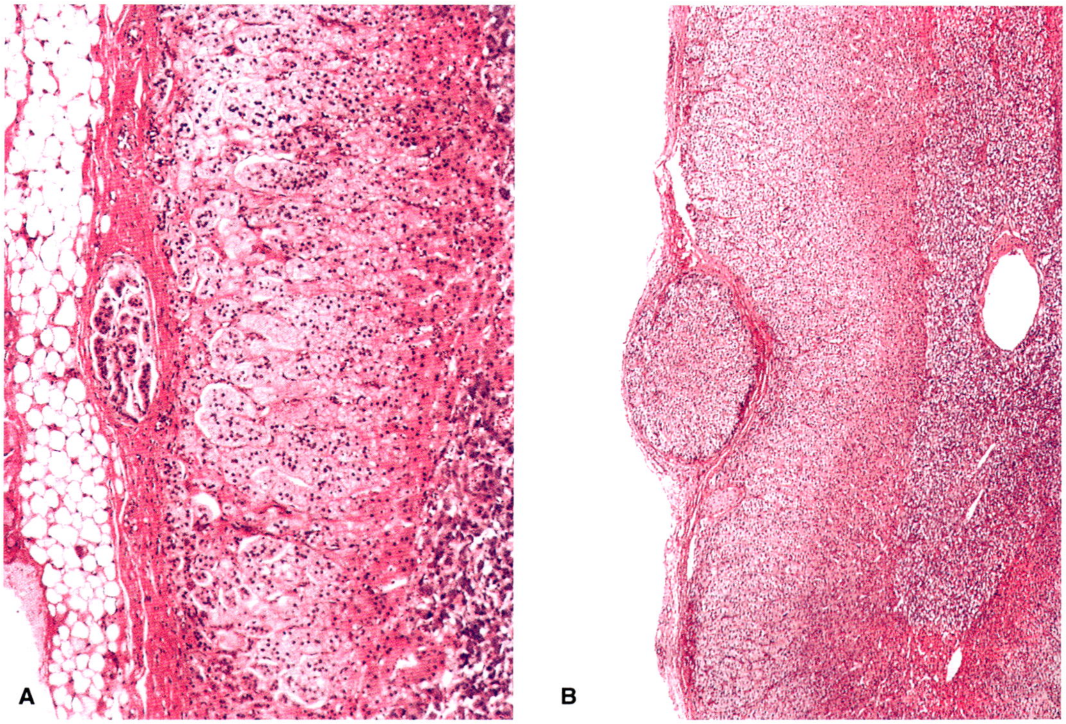

FIGURE 45.18 Normal adrenal gland. **A:** Aggregate of cells with features of zona glomerulosa type is present in a "pocket" in the capsule. **B:** Cortex featuring zona fasciculata (clear cells) and zona reticularis (compact cells) abuts medulla. A larger aggregate of cortical cells with clear (peripheral) and compact cytoplasm (central) is present in a pocket in the capsule. The cortex deep to the pocket is slightly attenuated. The zona glomerulosa is not clearly visible.

Adrenal cortical zonation becomes increasingly irregular with advancing years, although the zona reticularis is always well-developed zone regardless of age. Irregularity of cortical zonation is associated with subcentimeter adrenal cortical nodules in older patients and sporadic nodular adrenocortical disease (21).

The zona glomerulosa is the site of aldosterone production and responds to angiotensin and potassium. The zona fasciculata and zona reticularis synthesize glucocorticoids and sex hormones. Cells of all zones respond to adrenocorticotropic hormone. Mitotic figures are rare in the normal adult cortex. They may be seen in the zona fasciculata and zona reticularis, indicating that cells in the deeper areas of the cortex are capable of proliferating.

A number of modern techniques for studying cell proliferation and programmed cell death have been applied to the study of the human adrenal cortex (32). Cell proliferation as indicated by Ki-67 immunoreactivity occurs principally in

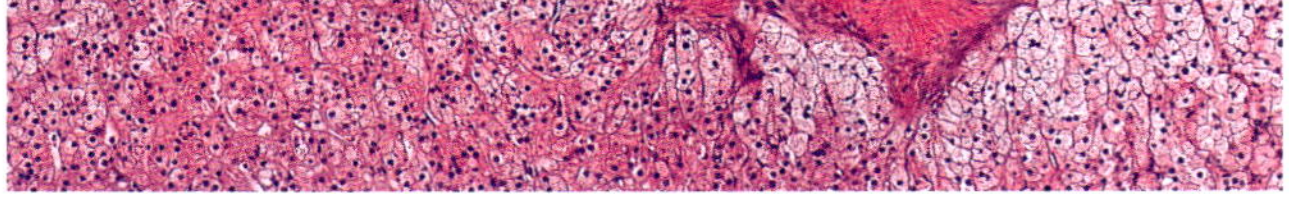

FIGURE 45.19 Ovarian thecal metaplasia (63-year-old woman with a 2-cm aldosterone-producing adrenal cortical adenoma). A group of packed spindle cells are attached to the adrenal capsule. Nests of cortical cells are trapped by hyalinized fibrous tissue. A poorly defined zona glomerulosa is present.

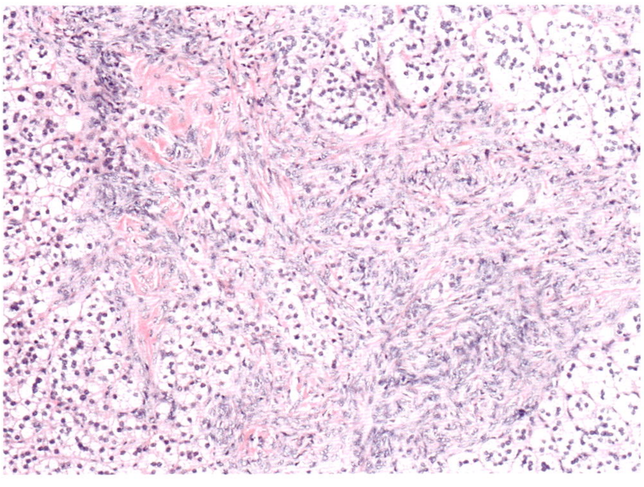

FIGURE 45.20 Ovarian thecal metaplasia. Higher magnification shows spindle cells which are often of myofibroblastic origin.

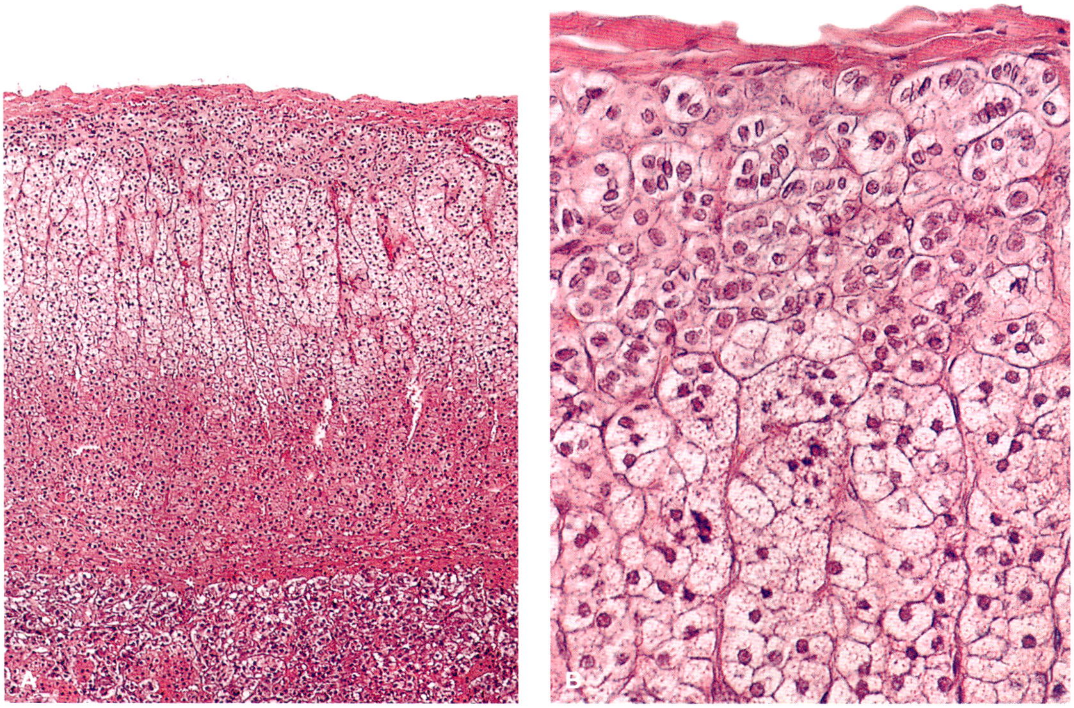

FIGURE 45.21 Normal adrenal gland. **A:** The normal pattern of zonation of the cortex is seen (clusters of cells with stainable cytoplasm in the zona glomerulosa, columns of cells with clear cytoplasm in the zona fasciculata, and cells with acidophilic cytoplasm in the zona reticularis). There is a sharp interface between the cortex (zona reticularis) and medulla (clusters of cells with basophilic cytoplasm). **B:** Zona glomerulosa composed of packed clusters and short trabeculae of cells beneath the adrenal capsule and superficial to the columns of vacuolated cells of the zona fasciculata. The zona glomerulosa nuclei tend to be oval; those of the zona fasciculata are round.

the zona fasciculata. 3-OH nick-end labeling can be used as a marker of apoptosis; labeled cells are uniformly present in the zona reticularis and in the zona glomerulosa in one-third of patients. These findings suggest that adrenal cortical cells may disperse centripetally or centrifugally from the zona fasciculata to the zona reticularis and from the zona fasciculata to the zona glomerulosa. Adrenal cortical cells undergoing apoptosis are believed to be phagocytosed by macrophages and cells lining the sinusoids.

Zona Glomerulosa

The zona glomerulosa is the narrow, irregular band of cortex situated immediately beneath the capsule and superficial to the zona fasciculata (Fig. 45.21). It remains diffuse during childhood and adolescence but is discontinuous in adults. The zona fasciculata extends to the capsule when the zona glomerulosa is atrophic (Fig. 45.23).

The zona glomerulosa cells are arranged in small clusters or rosettes supported by a minimal amount of fibrovascular stroma (Fig. 45.21). The clusters occasionally merge into short trabeculae that are straight, bent, or hairpin in shape. Some columnar cells also occur in short cords or one-cell rows parallel to the capsule. They contain faintly acidophilic or amphophilic cytoplasm that shows variably vacuolated. The round nuclei often appear slightly smaller and more deeply staining than those of the zona reticularis and zona fasciculata. They are often elongated and display a longitudinal groove.

The past decade has seen significant progress in functional immunohistochemistry of the adrenal cortex that stems from understanding of the steroidogenesis pathway of the cortex (Fig. 45.24) (33). The use of CYP11B2 immunohistochemistry to assess for aldosterone synthase has advanced our knowledge of the microscopic anatomy of the zona glomerulosa layer (33,34). It appears that small foci of CYP11B2-positive aldosterone-producing cell clusters develop with aging but have not been reported in individuals of <11 years (35–37). These cell clusters can cause primary aldosteronism and be seen in association with aldosterone-producing adenomas. Aldosterone-producing cell clusters have also been identified in adrenal glands from patients with no evidence of primary aldosteronism, raising the possibility that they reflect subclinical or a preclinical stage of primary aldosteronism (37,38).

Zona Fasciculata

The zona fasciculata occupies more than half the thickness of the cortex and appears as a broad band that lies between the zona glomerulosa and the zona reticularis (Figs. 45.21 to 45.23). The cells of the zona fasciculata are large and have distinct cell membranes. They are arranged in two-cell

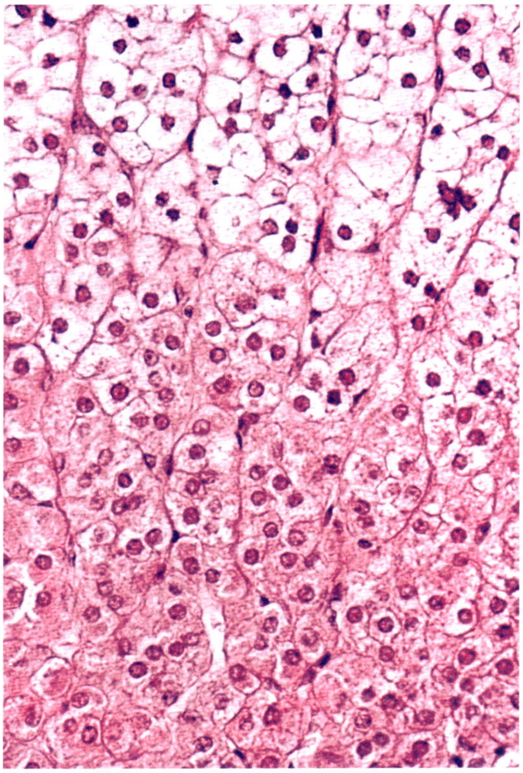

FIGURE 45.22 Normal adrenal cortex. Zona fasciculata (*upper*) features two-cell wide columns of cells with clear cytoplasm, and zona reticularis (*lower*) consists of cells having acidophilic granular cytoplasm that do not form a distinct pattern. Nuclei are vesicular, and nucleoli are small.

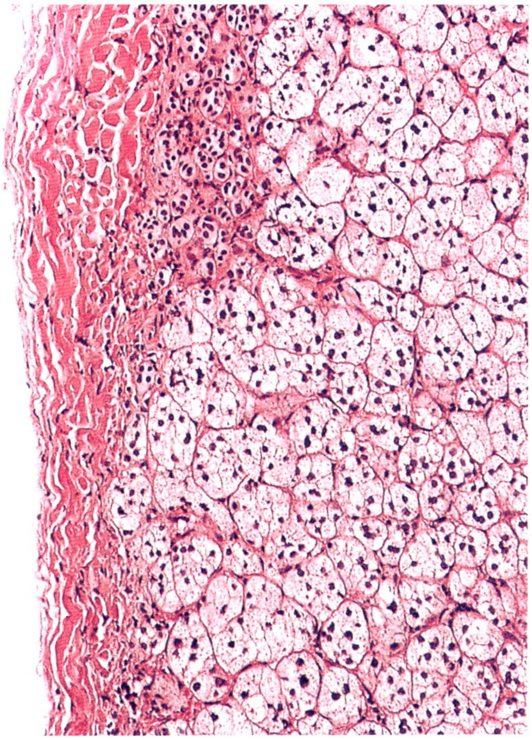

FIGURE 45.23 Normal adrenal cortex with discontinuity of zona glomerulosa. Zona glomerulosa (*upper half*) composed of clusters of cells with amphophilic cytoplasm forms a distinct band beneath the capsule and is sharply demarcated from the deeper zona fasciculata with its clustered cells having clear cytoplasm. Where the zona glomerulosa is absent (*lower half*), the zona fasciculata extends to the capsule.

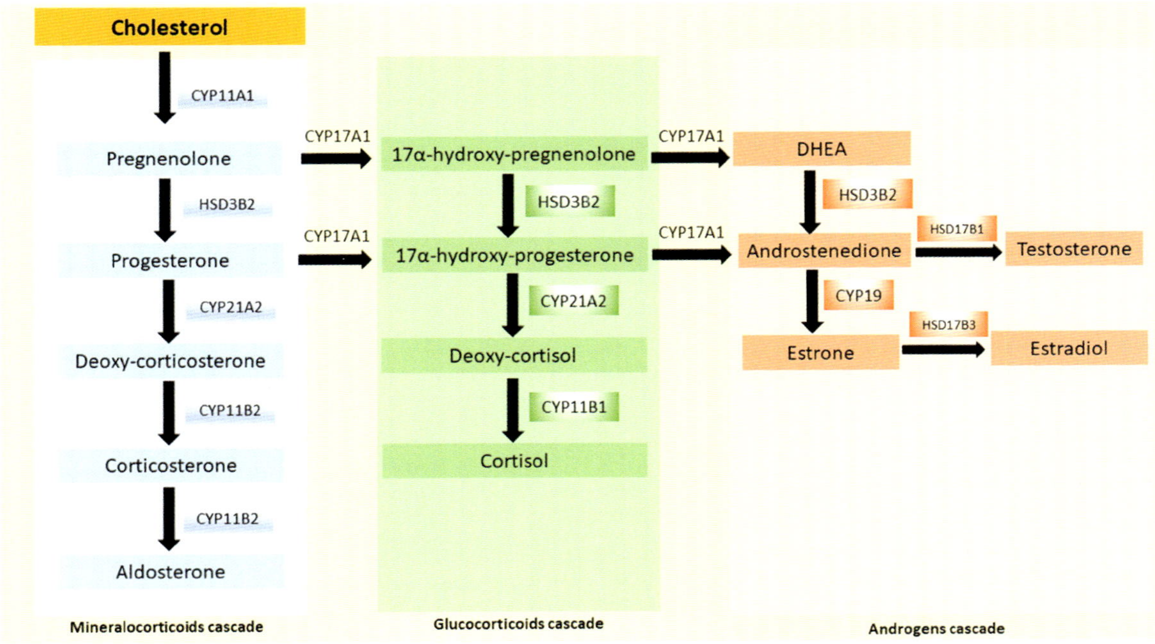

FIGURE 45.24 Steroidogenic pathway in the adrenal cortex.

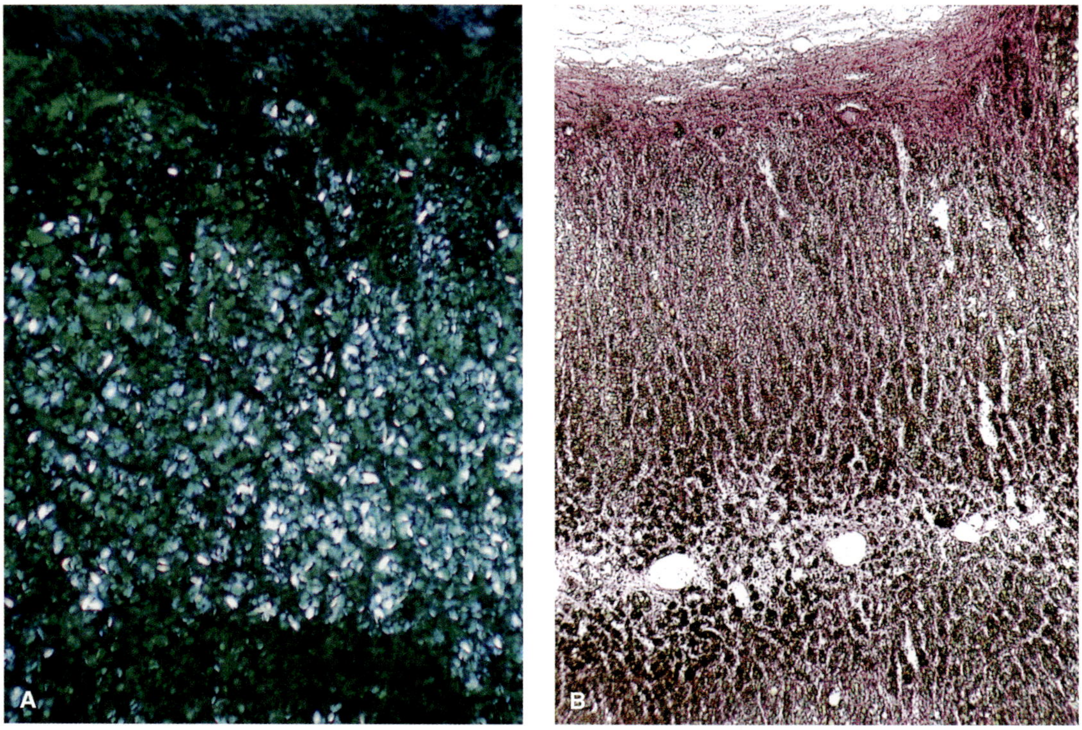

FIGURE 45.25 Normal adrenal cortex. **A:** Partial polarization shows high lipid content of zona fasciculata and low lipid content of zona glomerulosa (*above*) and zona reticularis (*below*). **B:** Fresh-frozen section of adrenal cortex stains with polychrome methylene blue. One complete band of cortex (*above*) and portion of another (*below*) is seen; the junction between the two is located by the dilated sinusoidal tributaries of the central adrenal vein. Zona fasciculata cells are packed with lipid globules. There are fewer globules in the zona glomerulosa (beneath the adrenal capsule) and in the zona reticularis (on either side of the sinusoidal vessels).

wide cords with the cord axes arranged perpendicularly to the capsule, and are bounded laterally by parallel-running fenestrated endothelium-lined capillaries. The nuclei are centrally located and more vesicular than those of the zona glomerulosa, featuring a single small nucleolus. The nuclear to cytoplasmic ratio is low. The cells are filled with cholesterol, fatty acids, and neutral fat that imparts a vacuolated, clear appearance, particularly in the outer two thirds of this region (Fig. 45.25).

Adrenocorticotropic hormone typically binds the MC2R receptor to induce cortisol production from lipid-rich polygonal cells of the zona fasciculata. Cortisol production also requires transcription of two key steroidogenic enzymes, CYP17A1 and CYP11B1 (Fig. 45.24) (39). Immunohistochemical stains for CYP11B1 and CYP17A1 can be used to aid detection of cortisol-producing nodular proliferations (33).

Zona Reticularis

The zona reticularis lies deep into the zona fasciculata and abuts the medulla (Fig. 45.21). The zona reticularis is in contact with zona reticularis and forms a lateral raphe in the tail of the gland where there is no medulla. Zona reticularis cells are arranged in a sponge-like meshwork of gently buckled anastomosing rows of cells separated by dilated capillaries. The cells are smaller than those of the zona fasciculata and contain granular or acidophilic cytoplasm with relatively scant lipids. The cytoplasm is sometimes referred to as "compact" and the reticularis cells as "compact cells." The deepest cells adjacent to the medulla usually contain yellow lipochrome pigment diffusely distributed as coarse granules in the cytoplasm or localized in a single body (Fig. 45.26). The solid granular eosinophilic cytoplasm and lipochrome pigment combine to produce the dark brown coloration of the zone seen on the cut surface of a fresh or formalin-fixed gland.

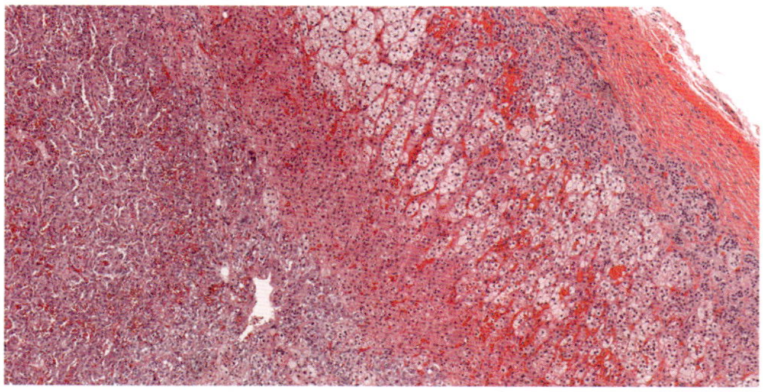

FIGURE 45.26 Zona reticularis and corticomedullary junction. The zona reticularis of the cortex is often sharply demarcated from the medulla. The deepest cells of the zona reticularis contain granular *yellowish* pigment (lipofuscin).

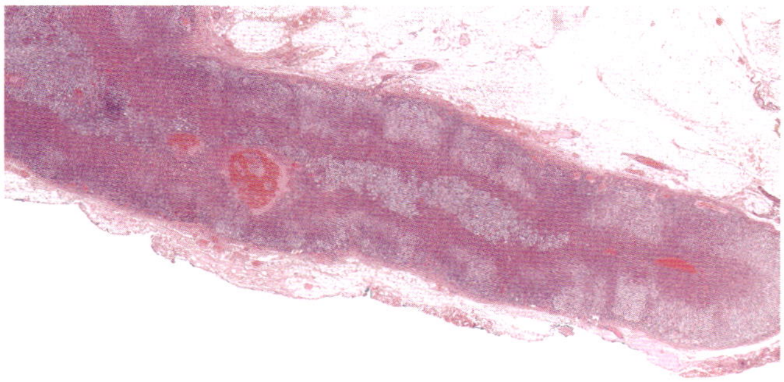

FIGURE 45.27 Normal medulla. Medulla is situated deep to the zona reticularis.

The functional output of zona reticularis, which consists of DHEA (dehydroepiandrosterone) and DHEA-S (dehydroepiandrosterone sulfate) production (Fig. 45.24), is largely mediated by CYP17A1 lyase activity in association with high CYB5 and P450 oxidoreductase expression (39). SULT2A1, a specific marker of the zona reticularis, mediates the production of DHEA-S (39). Immunohistochemical stains for these biomarkers can be used to facilitate the distinction of DHEA and/or DHEA-S–producing adrenal cortical proliferations.

Adrenal Medulla

The medulla is situated in the interior of the organ in the head and body of the gland, deep to the zona reticularis (Figs. 45.21, 45.27 to 45.30). It accounts for approximately 10% of the adrenal gland weight and rarely measures more than 2 mm in thickness (1,40). The interface between the cortex and medulla is readily visible at low-magnification owing to the acidophilia in zona reticularis and basophilia in the medulla. The junction is sharp, with no or minimal intervening connective tissue (Figs. 45.26 and 45.27).

The adrenal medulla extends to a variable degree into the crest of the gland and one or both alae (Fig. 45.31). Areas of the medulla in the alae are not necessarily in direct continuity with the main component of the medulla around the central veins. Rarely, a narrow tongue of medulla accompanied by a vessel or nerve or unaccompanied extends through the cortex to contact the capsule of the gland.

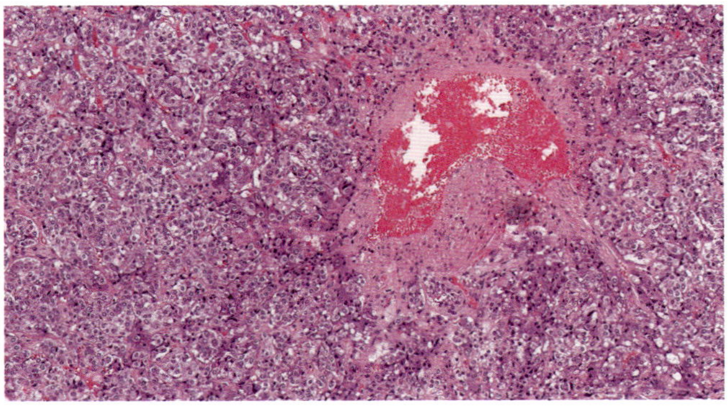

FIGURE 45.28 Normal medulla. Chromaffin cells surround central vein.

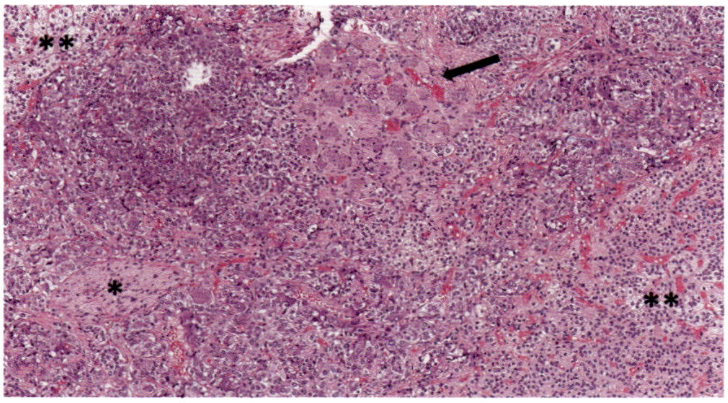

FIGURE 45.29 Normal medulla. Chromaffin cells can be admixed with clusters of ganglion cells (*arrow*) and nerves (*asterisk*). This photomicrograph is taken from the nontumorous gland of a patient with Cushing syndrome. For this reason, zona reticularis is absent. Chromaffin cells are surrounded by zona fasciculate (*double asterisk*) cells.

The adrenal medulla is composed of chromaffin cells as well as scattered small groups of cortical cells and individual ganglion cells (Fig. 45.32). The ganglion cells may contain round, hyalin bodies with a concentrically arranged fibrillar appearance. These cytoplasmic inclusions span up to 30 μm in diameter and may appear to be external to ganglion cells (Figs. 45.29 and 45.33). Immunohistochemical stains for vimentin and S100 protein demonstrate these hyaline bodies to be separated from the cells by a small amount of intercellular substance. The chromaffin cells are arranged in tight clusters and short trabeculae, supported

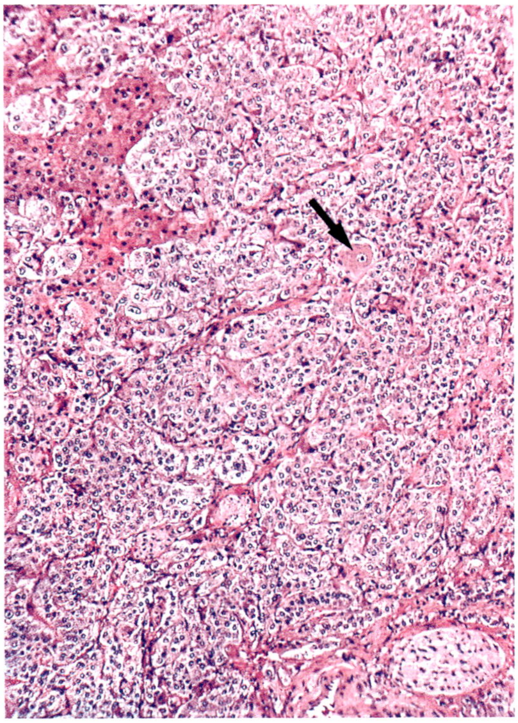

FIGURE 45.30 Norma adrenal medulla. Chromaffin cells are arranged in poorly delineated clusters. An irregularly shaped group of cortical cells (*upper left*) and an isolated ganglion cell (*arrow*) are present.

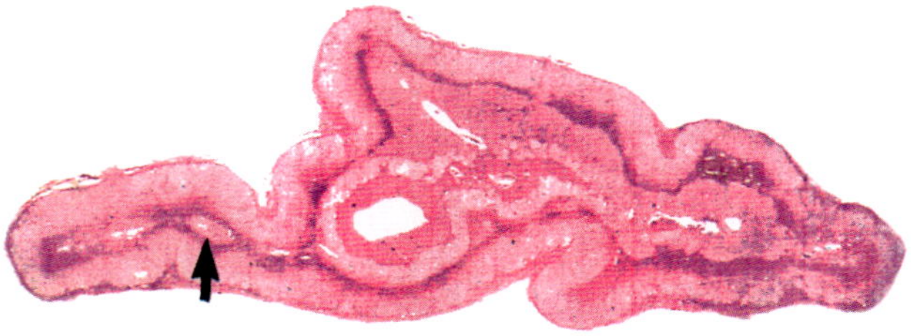

FIGURE 45.31 Normal adrenal gland. A thick capsule surrounds the adrenal cortex (outer clear zona fasciculata and inner eosinophilic zona reticularis) that encloses the basophilic adrenal medulla. An area of medulla in the ala (*arrow*) is not in continuity with the main mass of medulla. The adrenal vein is surrounded by a cuff of invaginated cortex.

by delicate fibrovascular stroma (Fig. 45.32). Sustentacular cells at the peripheries of the clusters and trabeculae are readily demonstrated with S100 protein and SOX10 immunohistochemical stains (Fig. 45.34).

The chromaffin cells are moderately large cells, polygonal to columnar, and slightly to considerably larger than cortical cells. Poorly outlined, their complete cell borders are visible only occasionally. Although the cytoplasm of most medullary cells is basophilic, it can be amphophilic or slightly acidophilic in some cases. The resulting variability of cytoplasmic staining and vacuolization often imparts an overall mottled light and dark appearance at intermediate magnification. Cytoplasmic colloid droplets are occasionally detected and are PASD-positive (Fig. 45.35). Most of

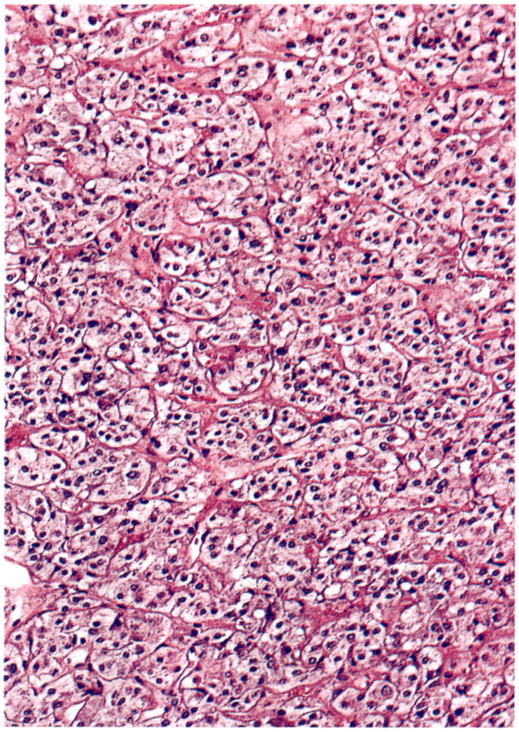

FIGURE 45.32 Normal adrenal medulla. Clusters of poorly outlined cells with basophilic cytoplasm are separated by a vascularized supporting stroma. There is some variation in nuclear size and shape.

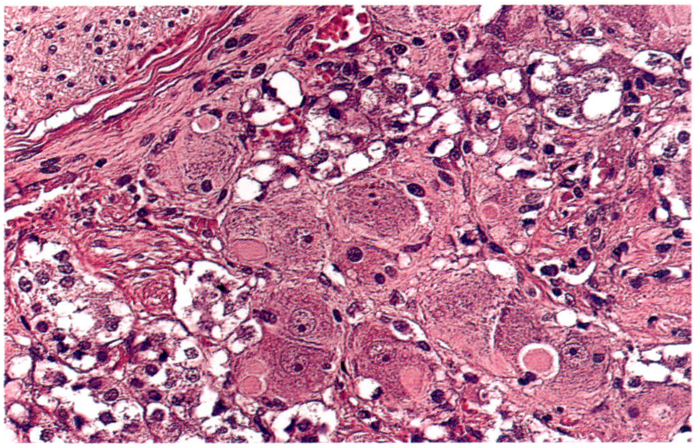

FIGURE 45.33 Normal adrenal medulla. A group of ganglion cells are demarcated by pheochromocytes (*upper right* and *lower left*) and a nerve (*upper left*). A number of the ganglion cells have cytoplasmic acidophilic bodies, some outlined by a rim of retracted cytoplasm (*arrows*). There are two such bodies in one cell.

the medullary cells are similar in size, but groups of small or large cells can be found.

The nuclei of most medullary cells are slightly larger than those of cortical cells. However, scattered cells often contain nuclei that are larger with prominent nucleoli. The usual nucleus has a finely or coarsely clumped chromatin pattern with a relatively clear nuclear background. The chromatin tends to be peripherally disposed and separated into irregular clumps. Rare cells have two or more nuclei.

Most of the nuclei are round or ovoid and centrally located. Large, intensely hyperchromatic and sometimes pleomorphic nuclei are commonly seen in singly scattered cells, especially near the corticomedullary junction (Fig. 45.36). The positions of the nuclei in the cells are not fixed; most are central, but some tend to be eccentric, located away from the vascular pole.

The adrenal medulla produces and releases catecholamines and have several distinctive histochemical reactions

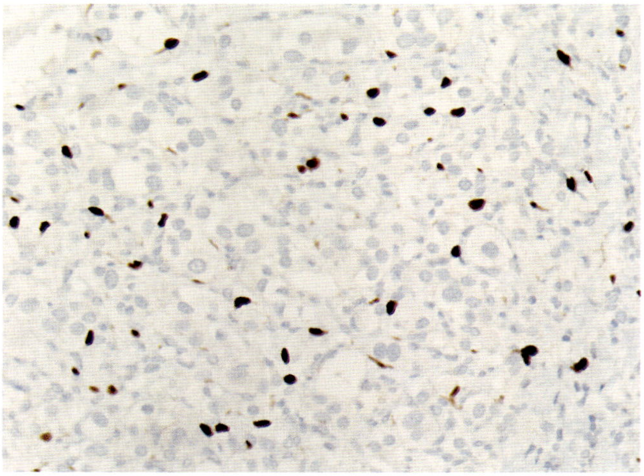

FIGURE 45.34 Sustentacular cells in normal medulla. Sustentacular cells, which surround chromaffin cells, are highlighted using SOX10.

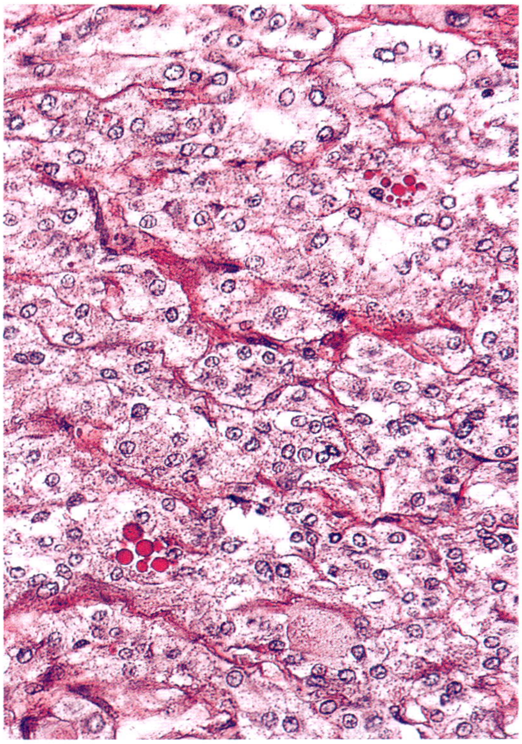

FIGURE 45.35 Normal adrenal medulla. Cytoplasmic globules, a rare finding in normal pheochromocytes, stains with periodic acid–Schiff. Condensation of the nuclear chromatin at the nuclear membrane is well demonstrated.

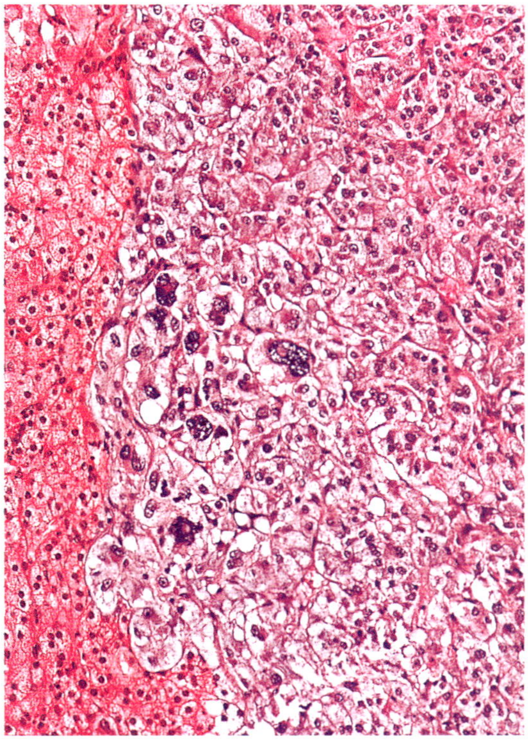

FIGURE 45.36 Normal adrenal medulla. Variability of pheochromocyte nuclei at the corticomedullary junction. The number of atypical pheochromocyte nuclei crowded together here is unusual; ordinarily, such nuclei are seen one to a medium-power field. The pheochromocytes have basophilic granular cytoplasm. The zona reticularis (*left*) features cells with granular eosinophilic cytoplasm and lipofuscin.

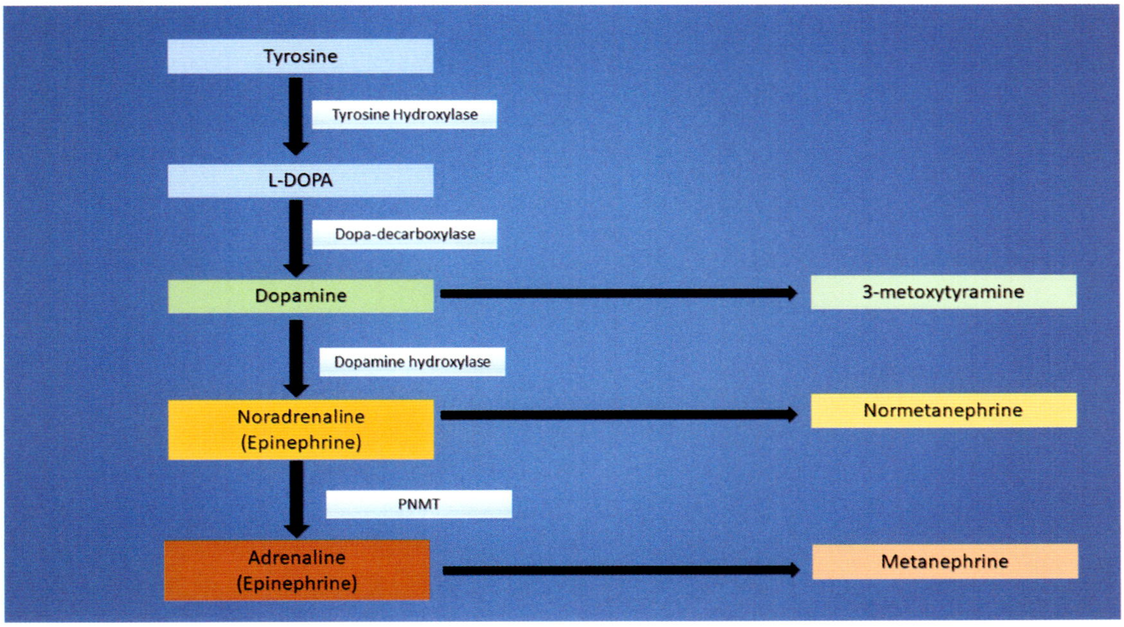

FIGURE 45.37 Catecholamine biosynthesis pathway. Parent catecholamines include dopamine, norepinephrine, and epinephrine. They are metabolized to 3-metoxytramine, normetanephrine, and metanephrine. Tyrosine hydroxylase is the rate limiting step in this biosynthetic cascade.

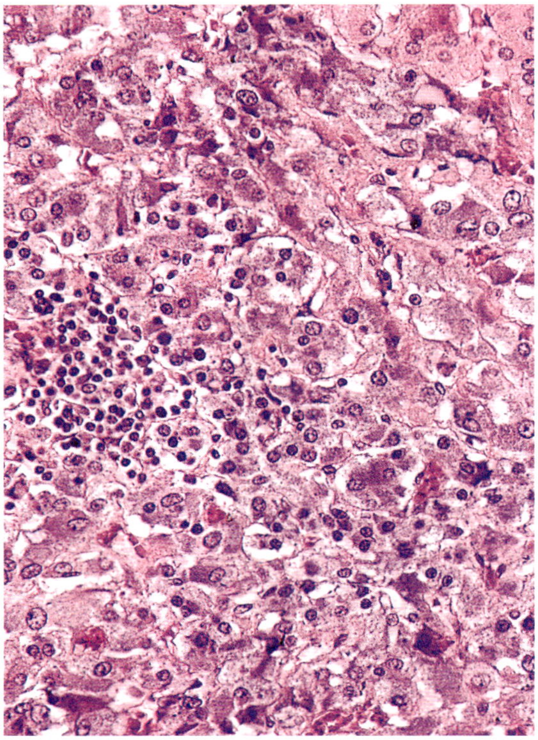

FIGURE 45.38 Normal adrenal medulla. Aggregates of lymphocytes and plasma cells are often encountered.

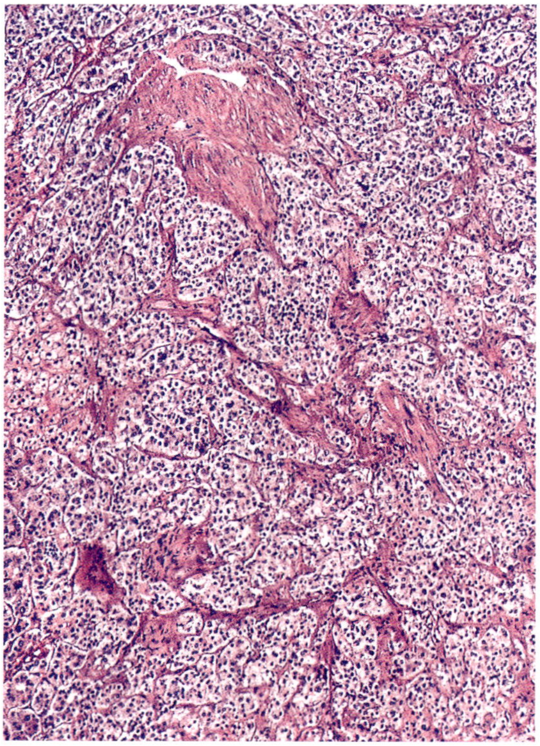

FIGURE 45.39 Normal adrenal medulla. Bundles and strands of smooth muscles are derived from the smooth muscles of central vein.

related to their secretory granule content. Catecholamines are dihydroxy derivatives of tyrosine amino acid that can be converted to colored polymers by oxidizing agents such as potassium dichromate, ferric chloride, ammoniacal silver nitrate, and osmium tetroxide. The oxidized and polymerized derivatives are termed adrenochromes and the staining has been called the chromaffin reaction. However, this assessment is no longer used since the preoperative diagnosis of catecholamine (Fig. 45.37) excess involves measuring 24-hour free serum or urine catecholamine and N-metabolites of catecholamines and the pathologic evaluation of tumors is facilitated by immunohistochemistry (7).

Single or multiple small-to-large accumulations of round cells, plasma cells, and lymphocytes are commonly seen in the normal medulla but have no known significance (Fig. 45.38). The delicate vascular stroma of the medulla is not conspicuous. Sometimes it is augmented focally by prolongations of the musculature of the central veins that separate groups of medullary cells (Fig. 45.39).

IMMUNOHISTOCHEMISTRY

Adrenal Cortex

The results of immunostaining of the prenatal and postnatal adrenal cortex and postnatal medulla are presented in Table 45.2 and illustrated in Figures 45.40 and 45.41. Staining with the antibodies varies in intensity and character. Some antibodies stain the cortex diffusely, while others show more or less staining of the outer and/or inner portions. The characteristics of adrenal cortical cells are summarized in Table 45.3.

The best marker to confirm adrenal cortical origin is SF1, which stains adrenal cortical cells in a strong, diffuse fashion (Fig. 45.42) (21,33). Cortical cells are also immunoreactive against a cytokeratin cocktail and AE1, although they are often negative or show only minimal staining for AE1/AE3 and OSCAR (41). Cells of the outer zona fasciculata and the zona reticularis are positive for Melan-A and alpha-inhibin, respectively (Fig. 45.41). Cells of the three zones were variably synaptophysin positive. Cortical cells lack reactivity for INSM1, chromogranin-A, and S100 protein antibodies.

Adrenal Medulla

The characteristics of adrenal medullary cells are summarized in Table 45.3. Chromaffin cells label with antibodies to general neuroendocrine differentiation markers (Fig. 45.43) developmental transcription factors (GATA3, PHOX2B, and HAND2) and catecholamine biosynthesis-related biomarkers including the rate-limiting enzyme, tyrosine hydroxylase (Figs. 45.32, 45.33, 45.41, 45.44 to 45.46) (7–10). Chromaffin cells are negative for keratins, neurofilament, and SOX10. Some chromaffin cells react with antibodies to S100 protein. The sustentacular cells around the clusters and trabeculae of chromaffin cells are positive

TABLE 45.2 Biomarkers at Various Stages of Developmental Prenatal and Postnatal Adrenal Cortex and Medulla

	Cortex					Medulla	
	Prenatal		Postnatal				
Immunostain	Fetal (provisional)	Adult (permanent)	Neonatal	Teenage	Adult	Prenatal	Postnatal
Vimentin	Negative (neuroblasts positive)	Negative	Negative	Strongly positive, outer one-third	Patchy positivity, strongest outer one third	Positive[a]	Sustentacular cells positive
Synaptophysin	Positive, paranuclear globular	Positive, finely granular	Weakly positive, paranuclear globular, and cytoplasmic finely granular	Rare positivity, subcapsular, paranuclear globular, and finely granular	Negative	Positive[a]	Strongly positive
Alpha-inhibin	Strongly positive, coarsely granular	Weakly positive, finely granular	Positive, inner one-half, finely granular	Strongly positive, inner one-third, finely granular	Strongly positive, inner one-third, finely granular	Negative	Negative
Melan-A	Strongly positive, very coarsely granular	Strongly positive, coarsely granular	Weakly positive, outer half, finely granular	Positive, outer one-third, coarsely granular	Positive, outer rim, coarsely granular	Negative	Negative
CD56	Intracapsular cells positive	Negative	Strongly positive outer one-half, cell membrane	Strongly positive, outer one-third cell membrane	Discontinuous positivity outer rim, cell membrane	Weakly positive[a]	Strongly positive, membranous
MIBI	Many subcapsular nuclei stained	Scattered nuclei stained	Scattered nuclei stained, subcapsular	Scattered nuclei stained, outer one-third	Scattered nuclei stained, outer one-half	Negative[a]	Negative
Chromogranin-A	Negative	Negative	Negative	Negative	Negative	Negative	Strongly positive
S100	Negative	Negative	Negative	Negative	Negative	Positive[a]	Sustentacular cells positive
Keratin AE1/AE3	Negative	Negative	Negative	Negative	Negative	Negative[a]	Negative

[a]Neuroblasts.

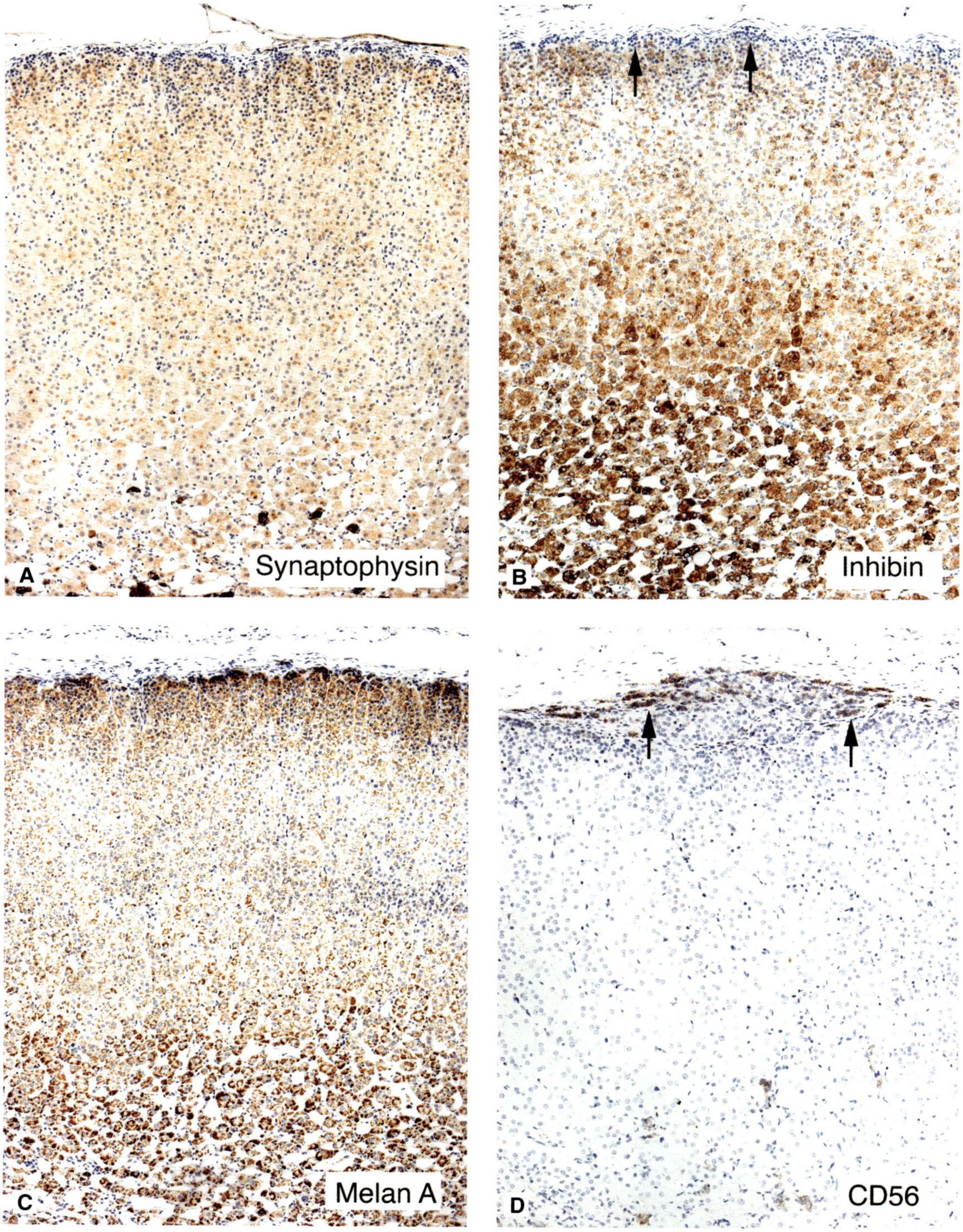

FIGURE 45.40 Immunostaining of the prenatal adrenal cortex. **A:** Synaptophysin stains almost the entire cortex leaving a narrow band of subcapsular cells unstained. **B:** Alpha-inhibin stains most of the cortex to a variable degree but do not stain a narrow subcapsular band (*arrows*). **C:** Melan-A stains the entire cortex to a variable degree. **D:** CD56 labeled a peripheral band of elongated cells, some in the capsule (*arrows*).

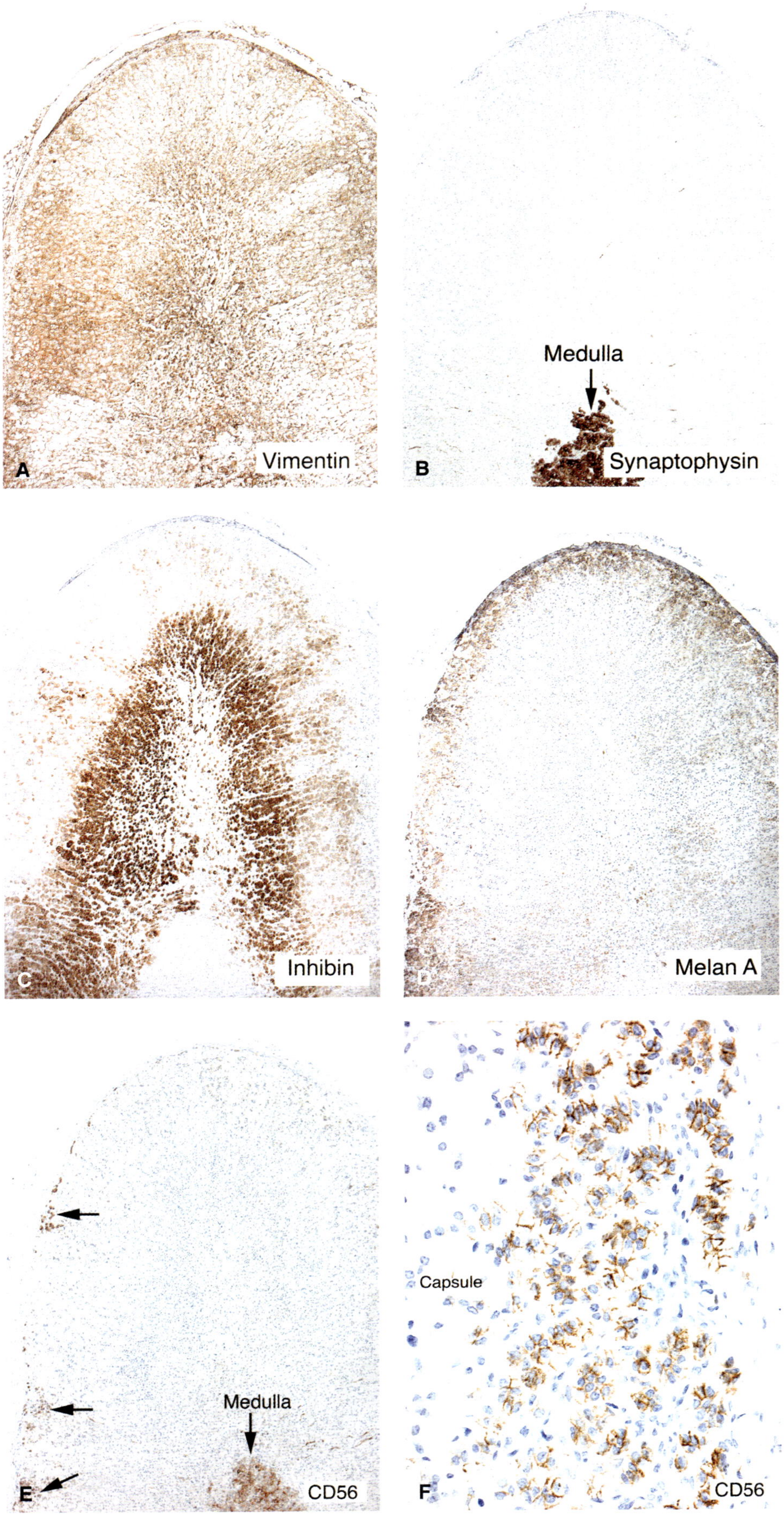

FIGURE 45.41 Immunostaining of the adult adrenal cortex. **A:** The cortex diffusely and unevenly stains with vimentin. **B:** There is weak variable staining in the cortex with synaptophysin. The medulla stains heavily. **C:** Alpha-inhibin stains the inner cortex heavily and the outer half less heavily. **D:** Melan-A stains a subcapsular band of cortical cells. **E:** CD56 scatter aggregates of subcapsular cells. The medulla stains. **F:** A subcapsular aggregate shows a membranous pattern of staining.

TABLE 45.3 Biomarkers to Confirm Adrenal Cortical and Medullary Origin in Pathology Practice

Immunohistochemical Biomarkers	Adrenal Cortex	Adrenal Medulla
SF1	Positive (diffuse)	Negative
Melan-A	Positive (variable)	Negative
Calretinin	Positive (variable)	Negative
Alpha-inhibin	Positive (variable)	Negative[a]
INSM1	Negative	Positive
Chromogranin-A	Negative	Positive
Synaptophysin	Positive (variable)	Positive
GATA3	Negative	Positive
HAND2	Negative	Positive
PHOX2B	Negative	Positive
Tyrosine hydroxylase	Negative	Positive
S100	Negative	Positive[b]
SOX10	Negative	Positive[b]

[a]A subset of pheochromocytomas can be positive for alpha-inhibin.

[b]Both S100 and SOX10 are positive in sustentacular cells of the adrenal medulla. Chromaffin cells are typically negative but pheochromocytomas may express S100.

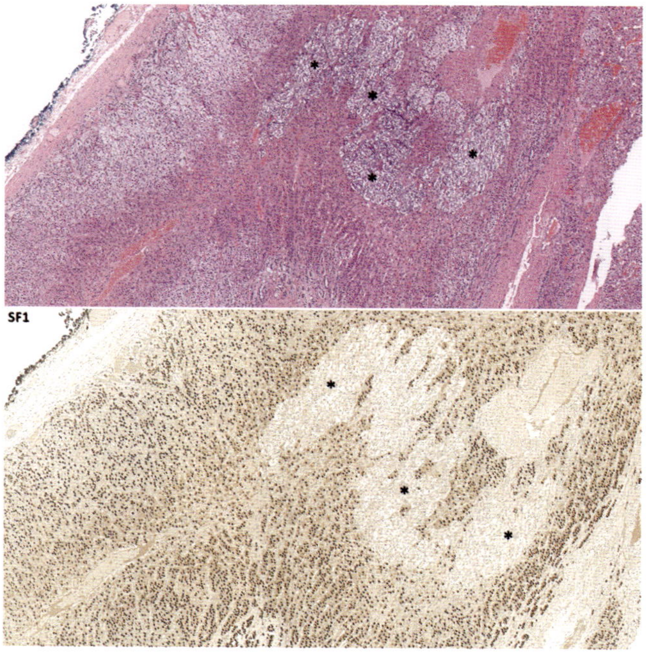

FIGURE 45.42 SF1 (steroidogenic factor 1). Unlike chromaffin cells (*asterisk*) in the medulla, adrenal cortical cells are almost always diffusely positive for SF1 (nuclear reactivity).

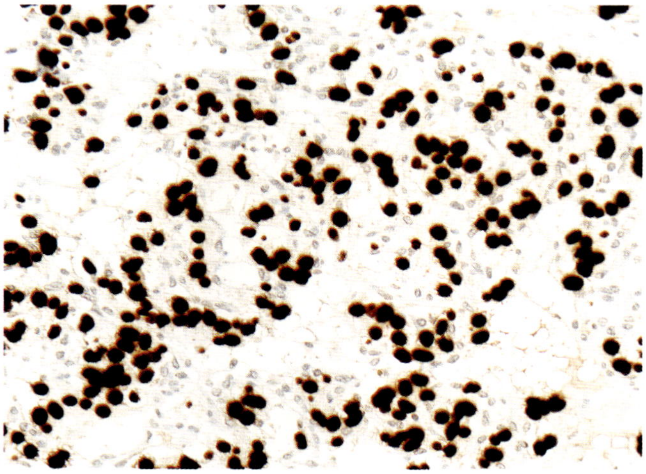

FIGURE 45.43 INSM1 (insulinoma-associated protein 1). Unlike adrenal cortical cells, chromaffin cells are diffusely positive for INSM1.

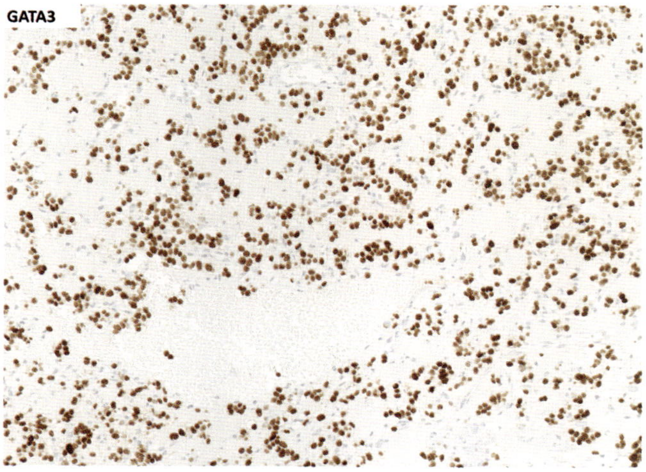

FIGURE 45.44 GATA3 (GATA-binding protein 3). Unlike adrenal cortical cells, chromaffin cells are typically positive for GATA3.

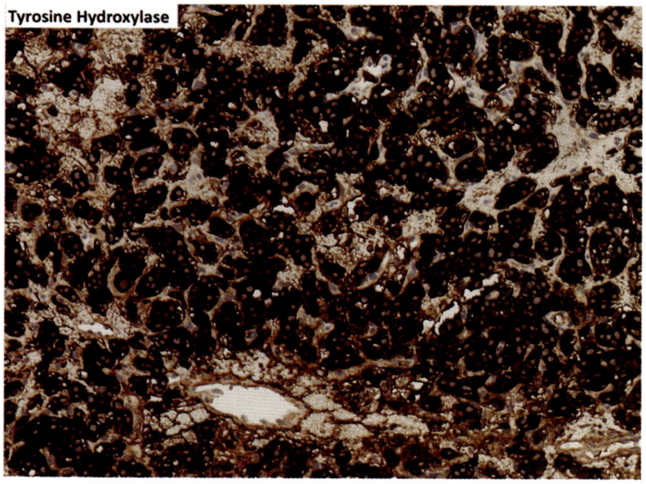

FIGURE 45.45 Tyrosine hydroxylase. Chromaffin cells typically show diffuse reactivity for tyrosine hydroxylase.

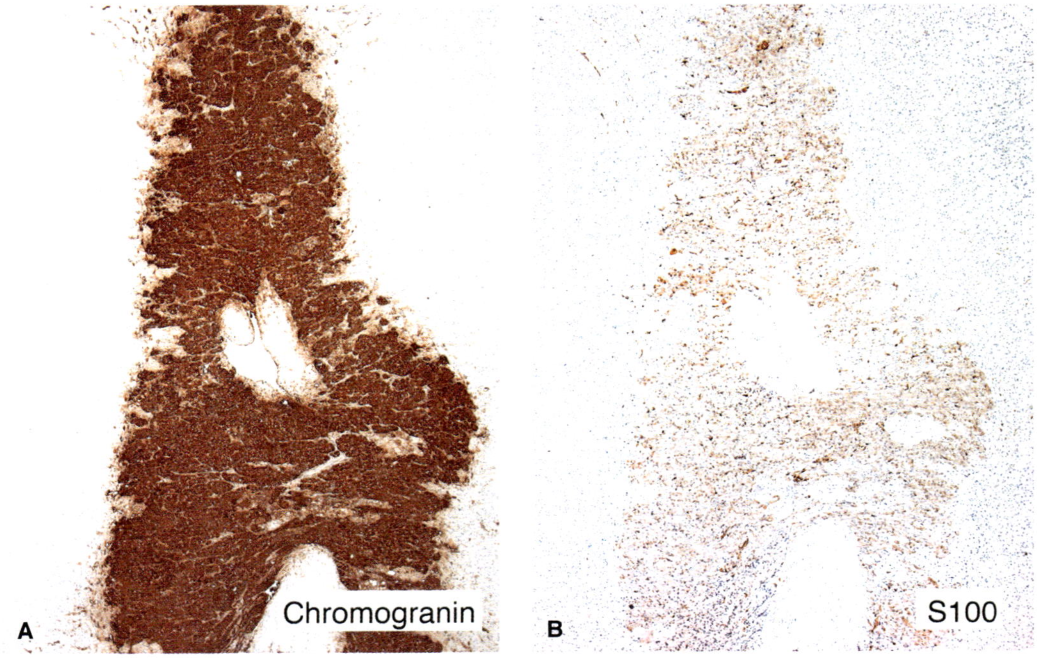

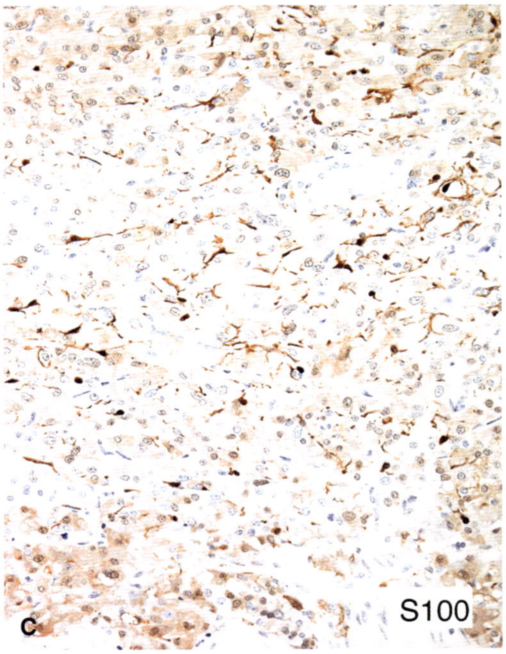

FIGURE 45.46 Immunostaining of the adrenal medulla. **A:** The medulla is heavily labeled by antibodies to chromogranin-A, the cortex is not labeled. **B:** Antibodies to S100 protein label the medulla and do not label the cortex. **C:** Sustentacular cells mantling groups of pheochromocytes show a linear discontinuous pattern.

for SOX10 (Fig. 45.34) and S100 protein (Fig. 45.46), as are nerves in the medulla.

ULTRASTRUCTURE

Adrenal Cortex

There are ultrastructural features shared by the three layers of the cortex relating to their common function of hormone synthesis. The cells feature voluminous endoplasmic reticulum, stacks of rough endoplasmic reticulum, a well-developed Golgi apparatus, lysosomes, and many mitochondria. The distribution and internal structure of some of the organelles vary from zone to zone. The usually distinct transition seen between the zones is not apparent ultrastructurally. Rather, there is a gradual alteration from one organelle distribution and type to another. Mitochondria in the zona glomerulosa are round, ovoid, or elongated with lamellar cristae, resulting in a ladder-like internal structure similar to that found in many other tissues (Fig. 45.47). Mitochondria in cells of the zona fasciculata are large, spherical, and

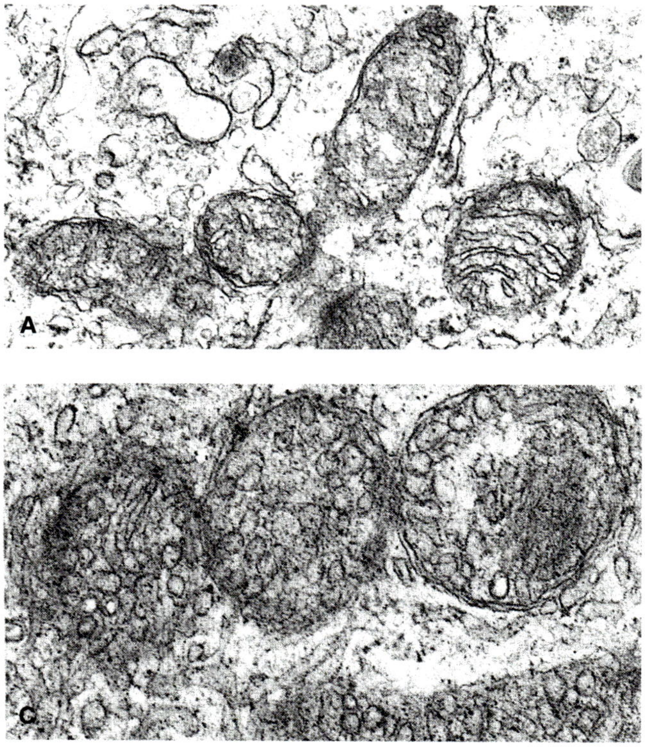

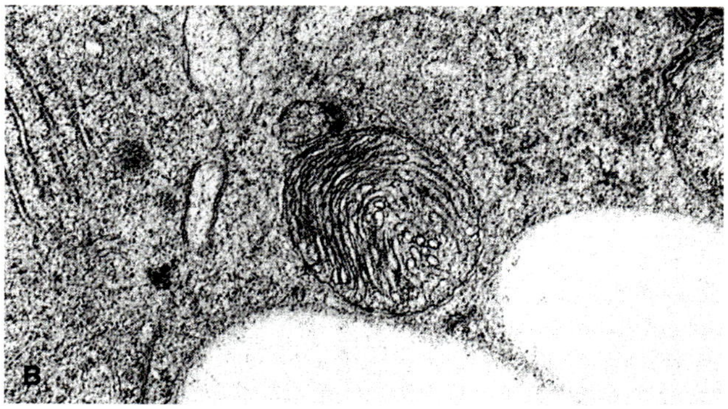

FIGURE 45.47 Normal adrenal cortex. **A:** Mitochondria in the zona glomerulosa have a lamellar pattern. **B:** Mitochondria in the zona fasciculata have a tubular and vesicular pattern. **C:** Mitochondria in the zona reticularis are elongated and have a vesicular appearance.

feature tubular cristae (Fig. 45.47). Lipid droplets are large and numerous. The mitochondria in the zona reticularis (Fig. 45.47) tend to be more elongated and exhibit tubular and vesicular cristae that are a feature of steroid-producing cells. Lipofuscin granules, membrane-bound organelles with a moderately dense matrix that contains dense granules, and clear lipid globules, are prominent. Glycogen is present.

Adrenal Medulla

Catecholamine-secreting cells dominate the medulla. Two cell types, epinephrine and norepinephrine, distinguished

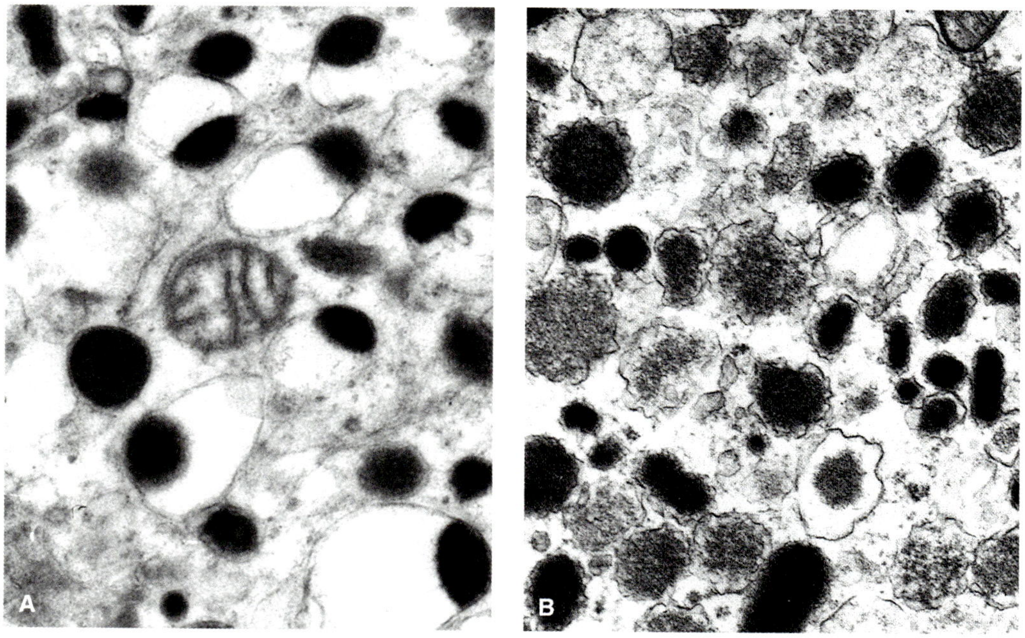

FIGURE 45.48 Normal adrenal medulla. Electron micrographs of medullary secretory granules. **A:** Content of norepinephrine granules is electron dense and often eccentrically located. **B:** Epinephrine granules with a variable electron-dense content that fill most of the sacs.

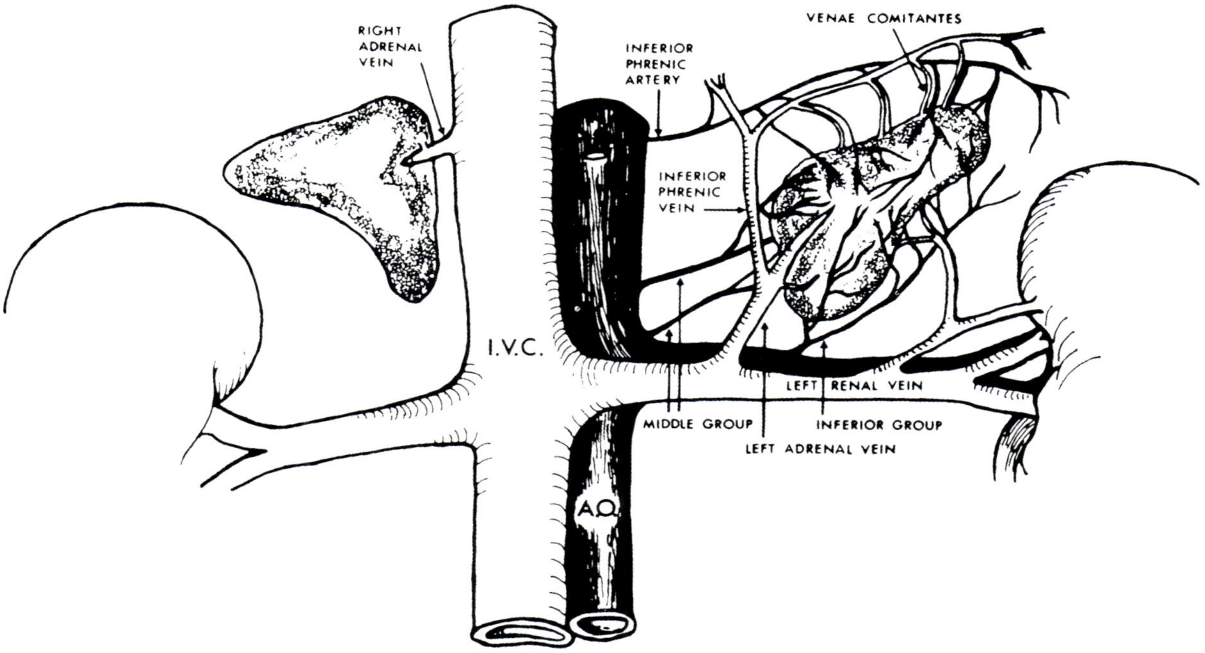

FIGURE 45.49 Diagrammatic representation of the arterial supply (*black*) and venous drainage (*light and hatched*) of the left adrenal gland. (Reprinted with permission from Symington T. *Functional Pathology of the Human Adrenal Gland*. Williams & Wilkins; 1969, with permission from Elsevier.)

by granule type are present. In tissue fixed in glutaraldehyde, cells that contain epinephrine feature granules (Fig. 45.48) measuring about 190 μm in diameter, with a moderately dense but not opaque finely granular texture fills the enclosing membrane. Norepinephrine-secreting cells contain granules (Fig. 45.49) that are electron opaque, often located eccentrically within a dilated sac, and measure about 250 μm in diameter.

OTHER ANATOMICAL STRUCTURES

Blood Vessels

The blood supply of the adrenal glands has been studied mostly from the anatomic point of view, often by observing the distribution of injected material in the vasculature. The tone of the subcapsular vascular plexus controls circulation through the organ. The histologic appearance of the vessels distal to the plexus suggests that the intravascular pressure in the organ is low.

Arteries

Each adrenal gland is supplied by three separate groups of arteries: the superior, middle, and inferior arising from the inferior phrenic artery, the aorta, and the renal artery, respectively (Fig. 45.49). The main vessels divide into 50 to 60 small feeder vessels that penetrate the anterior and posterior surfaces of the glands and form a plexus beneath the capsule. The former are commonly encountered close to the capsule and frequently exhibit atherosclerotic changes in older patients. The subcapsular plexus, important in the regulation of the circulation in the gland, is not conspicuous in routine histologic preparations.

Intraglandular Vasculature

Capillary loops from the subcapsular plexus surround the cells of the zona glomerulosa, then extend toward the interior of the organ between the columns of cells of the zona fasciculata, and ultimately open into wide interconnecting channels in the zona reticularis to form a second vascular plexus. This ends abruptly in a vascular dam at the corticomedullary junction that finally drains into the sinusoids of the medulla by relatively fewer channels. The marked vascular congestion commonly seen at the corticomedullary junction of adrenal glands obtained at autopsy may reflect this vascular barrier. Although the medulla receives some arterial blood supply, most of its vascular supply has already nourished the cortex.

The venous drainage from each gland occurs via a single vein that emerges from the anterior surface (Fig. 45.2). Inside the glands, the central adrenal vein and its tributaries have a unique muscle coat composed of two to six longitudinally running bundles of smooth muscle cells, varying in size and eccentrically situated around the vein lumen (Fig. 45.50). The bundles are heavily laden with elastic fibers that extend into tributaries of the larger central veins

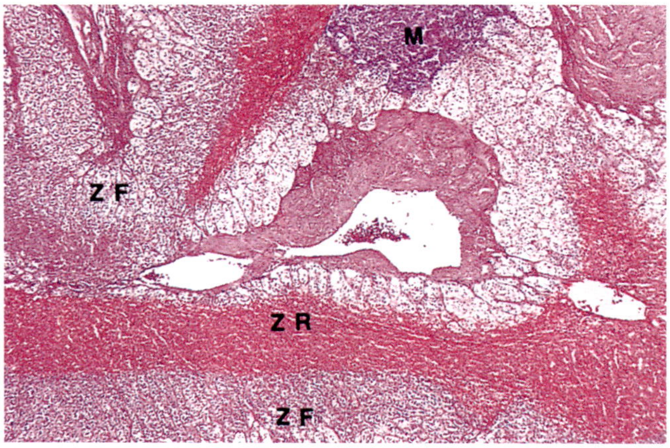

FIGURE 45.50 Central adrenal vein. Low-power view of the vein surrounded by an invaginated cuff of cortical cells with clear cytoplasm (zona fasciculata type). Distribution of smooth muscle in the vein wall is uneven. The zona fasciculata (*ZF*) and zona reticularis (*ZR*) of the cortex are evident, and there is a small amount of basophilic medulla (*M*).

and in some instances outline clusters and trabeculae of medullary cells.

The eccentricity of the muscle bundles results in a vein wall of variable thickness. Bundles of smooth muscle cells are deficient in some areas and these may be extensive such that medullary and cortical cells are separated from the bloodstream by intima and a minimal amount of subintimal connective tissue only. This peculiar anatomic structure permits medullary cells or cortical cells to occasionally form polypoid endothelium–covered projections into the lumen (Figs. 45.51 and 45.52). This ready access of chromaffin cells to the venous lumen explains the occasional finding of an intravenous tumor protrusion of pheochromocytoma, but this finding does not represent vascular invasion. A thick cuff of invaginated cortical cells surrounds the intramedullary central vein and its larger tributaries (Figs. 45.31 and 45.50). The development of a neoplasm in this "displaced" cortex probably explains the occasional cortical neoplasm that appears to have developed in the medulla of the organ.

Nerves and Ganglia

The innervation of the adrenal gland, specifically of the medulla, emanates from the lower thoracic segments of the spinal cord and passes through the greater splanchnic nerve and the upper lumbar sympathetic ganglia via the celiac plexus. This nerve supply forms a plexus of myelinated and nonmyelinated nerves on the capsule of the gland, primarily on its posterior aspect. Thus, the largely preganglionic nerve fibers pass into the medulla following either the course of emerging or penetrating vessels or connective tissue trabeculae. Occasionally, a large nerve penetrates directly into the medulla. The number of nerves visible in the medulla varies greatly from case to case; some feature a perineurium, whereas others do not (Fig. 45.53). Ganglion cells are also commonly seen singly or in clusters among the chromaffin

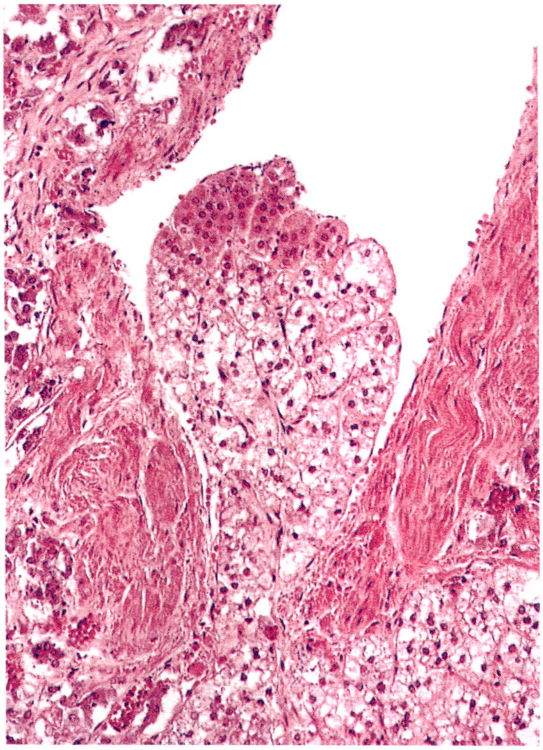

FIGURE 45.51 Central adrenal vein. Intraluminal protrusion of endothelium-covered adrenocortical cells between the discontinuous muscle bundles of a tributary of the central adrenal vein.

cells (Figs. 45.30 and 45.33). The cells of the cortex do not have a nerve supply.

Lymphatics

Injection studies demonstrate a rich plexus of lymphatic channels in the capsules of the adrenal glands. Lymphatic

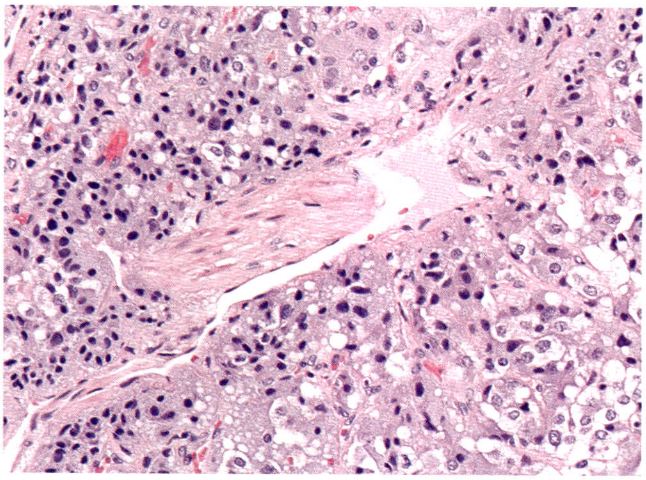

FIGURE 45.52 Adrenal medulla. Since endocrine organs typically show fenestrated endothelium, and adrenal vein has discontinuous area, it is not abnormal to identify chromaffin cells lying under the intact endothelium or even projection into the lumen.

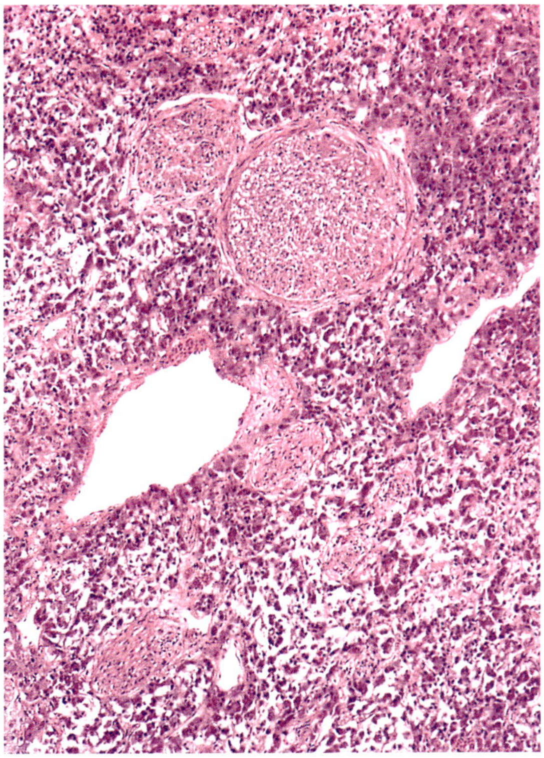

FIGURE 45.53 Adrenal medulla. Concentration of nerves, some with perineurium and others without.

vessels are distributed to the adventitia of the central vein and its main tributaries. The lymphatic vessels drain into aortic lymph nodes. There is no lymphatic vessel supply to the cortex.

ADRENOCORTICAL NODULES

Functional subcentimeter adrenocortical nodules are generally referred to as adrenal cortical neoplasms, although the HISTALDO classification applies various nomenclature for CYP11B2-expressing subcentimeter proliferations that occur in patients with primary aldosteronism (34). Sporadic nodular adrenocortical disease is defined by the 2022 WHO classification of adrenal pathologies as the presence of subcentimeter benign adrenal cortical nodules unassociated with cortical hyperfunction. This condition features unencapsulated nodular proliferations of cortical cells that demonstrate clonality (21,23). They are uncommon before the fourth decade of life and are encountered with increasing frequency among older adults (42). The true incidence and etiology of these nodules is largely unknown. They appear as multiple nodules composed of large, lipid-laden clear cells, clear and compact cells, and a minority consist only of reticularis-type cells (Fig. 45.54). The smallest nodules may be found at any level of the cortex, but they are usually located in the zona fasciculata. The smallest ill-defined nodules have the cord-like organization similar to the rest of the adrenal gland, although this appearance may be lost as the nodules enlarge.

FIGURE 45.54 Cortical nodules (36-year-old man). **A:** Suggestive nodule in the mid-cortex. **B:** Distinct nodule composed of clear cells in outer cortex.

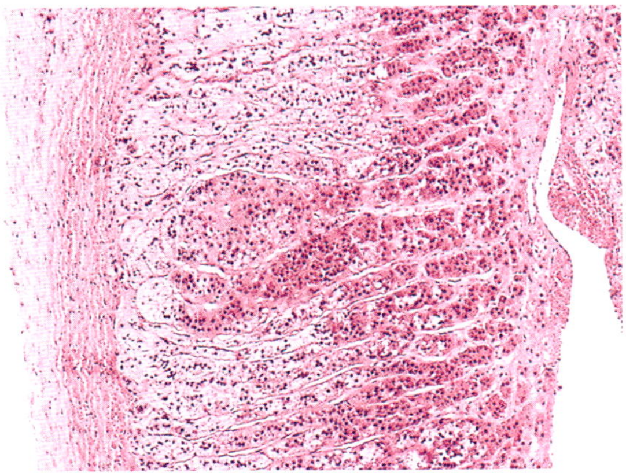

A

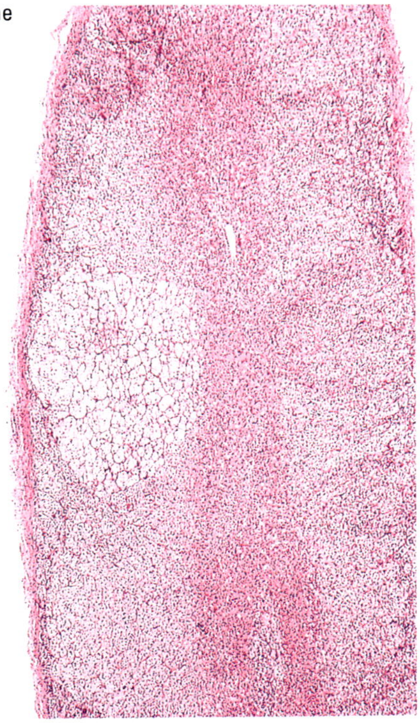

B

ACKNOWLEDGMENT

The authors of this chapter in the current edition greatly acknowledge Dr. J. Aiden Carney who previously authored this chapter and on whose detailed work the current updates have been made.

REFERENCES

1. Symington T. *Functional Pathology of the Human Adrenal Gland*. Williams & Wilkins; 1969.
2. Mete O, Erickson LA. Adrenal Pathology. In: Robertson PR, Giudice LC, Grossman A, et al., eds. *DeGroot's Endocrinology. Basic Science and Clinical Practice*. 8th ed. volume 2, Elsevier; 2023:1509–1529.
3. Kiiveri S, Liu J, Westerholm-Ormio M, et al. Transcription factors GATA-4 and GATA-6 during mouse and human adrenocortical development. *Endocr Res*. 2002;28(4):647–650.
4. Pihlajoki M, Färkkilä A, Soini T, Heikinheimo M, Wilson DB. GATA factors in endocrine neoplasia. *Mol Cell Endocrinol*. 2016;421:2–17.
5. Keene MF. Observations on the development of the human suprarenal gland. *J Anat*. 1927;61(Pt 3):302–324.
6. Crowder RE. The development of the adrenal gland in man, with special reference to origin and ultimate location of cell types and evidence in favour of the "cell migration" theory. *Carnegie Inst Contrib Embryol*. 1957;26:193–210.
7. Mete O, Asa SL, Gill AJ, Kimura N, de Krijger RR, Tischler A. Overview of the 2022 WHO classification of paragangliomas and pheochromocytomas. *Endocr Pathol*. 2022;33(1):90–114.
8. Nonaka D, Wang BY, Edmondson D, Beckett E, Sun CC. A study of gata3 and phox2b expression in tumors of the autonomic nervous system. *Am J Surg Pathol*. 2013;37(8):1236–1241.
9. Miyauchi M, Akashi T, Furukawa A, et al. PHOX2B is a sensitive and specific marker for the histopathological diagnosis of pheochromocytoma and paraganglioma. *Endocr Pathol*. 2022;33(4):506–518.
10. Soukup J, Manethova M, Stejskal V, et al. Hand2 immunohistochemistry in the diagnosis of paragangliomas and other neuroendocrine neoplasms. *Endocr Pathol*. 2024;35(1):14–24.
11. Folligan K, Roume J, Razavi F, Sepaniak S, Bouvier R, Morel Y, Trouillas J. L'hypoplasie surrénale congénitale de type anencéphalique: deux cas avec anomalies hypophysaires, sans mutation de DAX-1 et SF-1 et revue de la littérature [Hypoplasia adrenal congenita of anencephalic type: two cases with pituitary abnormalities and review of literature]. *Morphologie*. 2011;95(308):26–33. French.
12. Kempná P, Flück CE. Adrenal gland development and defects. *Best Pract Res Clin Endocrinol Metab*. 2008;22(1):77–93.
13. Biason-Lauber A, Schoenle EJ. Apparently normal ovarian differentiation in a prepubertal girl with transcriptionally inactive steroidogenic factor 1 (NR5A1/SF-1) and adrenocortical insufficiency. *Am J Hum Genet*. 2000;67:1563–1568.
14. Achermann JC, Meeks JJ, Jameson JL. Phenotypic spectrum of mutations in DAX-1 and SF-1. *Mol Cell Endocrinol*. 2001;185(1–2):17–25.
15. Lloyd RV, Douglas BR, Young WF. *Endocrine Diseases. Atlas of Nontumor Pathology. First Series, Fascicle 1*. American Registry of Pathology and the Armed Forces Institute of Pathology; 2002.
16. Duan K, Giordano TJ, Mete O. Adrenal cortex. In: Mete O, Asa SL. eds. *Endocrine Pathology*. Cambridge University Press; 2016:588–627.
17. Falco EC, Daniele L, Metovic J, et al. Adrenal rests in the uro-genital tract of an adult population. *Endocr Pathol*. 2021;32(3):375–384.
18. Mete O, Ng T, Christie-David D, McMaster J, Asa SL. Silent corticotroph adenoma with adrenal cortical choristoma: a rare but distinct morphological entity. *Endocr Pathol*. 2013;24(3):162–166.
19. Turan H, Tarçın G, Mete Ö, et al. Silent corticotroph tumor with adrenocortical choristoma in an eleven-year-old boy. *J Clin Res Pediatr Endocrinol*. 2022;14(1):126–130.
20. Ekholm E, Niemineva K. On prenatal changes in the relative weights of the human adrenals, the thymus and the thyroid gland. *Acta Paediatr*. 1950;39(1–2):67–86.
21. Mete O, Erickson LA, Juhlin CC, et al. Overview of the 2022 WHO classification of adrenal cortical tumors. *Endocr Pathol*. 2022;33(1):155–196.
22. Tischler AS, LiVolsi VA, Asa SL. Insights obtained from the nontumorous glandular tissue in patients with endocrine tumors. *Endocr Pathol*. 2023;34(4):393–405.
23. Mete O, Asa SL. Precursor lesions of endocrine system neoplasms. *Pathology*. 2013;45(3):316–330.
24. Mete O, Asa SL. Morphological distinction of cortisol-producing and aldosterone-producing adrenal cortical adenomas: not only possible but a critical clinical responsibility. *Histopathology*. 2012;60(6):1015–1016.
25. Duan K, Gomez Hernandez K, Mete O. Clinicopathological correlates of adrenal Cushing's syndrome. *J Clin Pathol*. 2015;68(3):175–186.
26. Mete O, Duan K. The many faces of primary aldosteronism and cushing syndrome: A reflection of adrenocortical tumor heterogeneity. *Front Med (Lausanne)*. 2018;5:54.
27. Romanet P, Guerin C, Pedini P, Essamet W, et al. Pathological and genetic characterization of bilateral adrenomedullary hyperplasia in a patient with germline MAX mutation. *Endocr Pathol*. 2017;28(4):302–307.
28. Reed RJ, Patrick JT. Nodular hyperplasia of the adrenal cortical blastema. *Bull Tulane Univ Med Fac*. 1967;26:151–157.
29. Wong TW, Warner NE. Ovarian thecal metaplasia in the adrenal gland. *Arch Pathol*. 1971;92(5):319–328.
30. Fidler WJ. Ovarian thecal metaplasia in adrenal glands. *Am J Clin Pathol*. 1977;67(4):318–323.
31. Mete O, Raphael S, Pirzada A, Asa SL. Is adrenal ovarian thecal metaplasia a misnomer? Report of three cases of radial scar-like spindle cell myofibroblastic nodule of the adrenal gland. *Endocr Pathol*. 2011;22(4):222–225.
32. Sasano H, Imatani A, Shizawa S, Suzuki T, Nagura H. Cell proliferation and apoptosis in normal and pathologic human adrenal. *Mod Pathol*. 1995;8(1):11–17.
33. Mete O, Asa SL, Giordano TJ, Papotti M, Sasano H, Volante M. Immunohistochemical biomarkers of adrenal cortical neoplasms. *Endocr Pathol*. 2018;29(2):137–149.
34. Williams TA, Gomez-Sanchez CE, Rainey WE, et al. International histopathology consensus for unilateral primary aldosteronism. *J Clin Endocrinol Metab*. 2021;106(1):42–54.
35. Hayashi T, Zhang Z, Al-Eyd G, et al. Expression of aldosterone synthase CYP11B2 was inversely correlated with longevity. *J Steroid Biochem Mol Biol*. 2019;191:105361.

36. Lim JS, Rainey WE. The potential role of aldosterone-producing cell clusters in adrenal disease. *Horm Metab Res.* 2020;52(6):427–434.
37. Nishimoto K, Seki T, Hayashi Y, et al. Human adrenocortical remodeling leading to aldosterone-producing cell cluster generation. *Int J Endocrinol.* 2016;2016:7834356.
38. Iwahashi N, Umakoshi H, Seki T, et al. Characterization of aldosterone-producing cell cluster (APCC) at single-cell resolution. *J Clin Endocrinol Metab.* 2022;107(9):2439–2448.
39. Basham KJ, Val P. Adrenal development and homeostasis. In: Robertson PR, Giudice LC, Grossman A, et al., eds. *DeGroot's Endocrinology. Basic Science and Clinical Practice*. 8th ed. volume 2, Elsevier; 2023:1408–1429.
40. Quinan C, Berger AA. Observations on human adrenals with especial reference to the relative weight of the normal medulla. *Ann Intern Med.* 1933;6:1180–1192.
41. Gaffey MJ, Traweek ST, Mills SE, et al. Cytokeratin expression in adrenocortical neoplasia: An immunohistochemical and biochemical study with implications for the differential diagnosis of adrenocortical, hepatocellular, and renal cell carcinoma. *Hum Pathol.* 1992;23(2):144–153.
42. Dobbie JW. Adrenocortical nodular hyperplasia: The ageing adrenal. *J Pathol.* 1969;99(1):1–18.

46

The Diffuse Neuroendocrine System

Evgeny Yakirevich ■ Ronald A. DeLellis ■ Shamlal Mangray

Several organs (pituitary gland, parathyroid glands, adrenal medulla, and paraganglia) and pancreatic islets are almost entirely composed of neuroendocrine cells, while single cells or small groups of neuroendocrine cells are diffusely distributed through the alimentary system, lung, thyroid (C-cells), thymus, urogenital tract, and skin where they are intermixed with the different predominant cell types of the respective tissues. Although neuroendocrine cells are discussed in the context of different tissues and organs in other chapters of this volume, this chapter will provide a more focused overview of these fascinating cell types that constitute the diffuse neuroendocrine system.

Numerous studies have established that there are many striking similarities between neurons and peptide hormone-producing neuroendocrine cells. Both cell types have polarized membrane orientations, separately regulated secretory pathways, secretory granules and vesicles that serve as storage sites for various peptides and amines, neurotransmitter-synthesizing enzymes, neural cell adhesion molecules (NCAMs) and, in many cases, the necessary molecular machinery for the uptake and release of neurotransmitters and neuropeptide hormones (1). Detailed biochemical and molecular studies have demonstrated a commonality of biosynthetic products that may act as classical hormones, neurotransmitters, and paracrine or autocrine factors. Accordingly, concepts of the endocrine system have been expanded to include not only the traditional endocrine glands but also the peptidergic neurons, paraganglia (specialized neurosecretory cells associated with autonomic nervous system) and the system of neuroendocrine cells that is dispersed throughout many tissues of the body.

HISTORICAL PERSPECTIVES AND NOMENCLATURE

Current concepts of the neuroendocrine system evolved from a series of observations that were initiated more than a century ago. Heidenhain first demonstrated a population of chromate positive cells (brown to yellow coloration after fixation in potassium chromate or chromic acid) in the gastric mucosa of rabbits and dogs and suggested that they might have an endocrine function (2,3). Subsequently, Kulchitsky identified similar cells in the crypts of Lieberkühn of the small intestine and noted their presence at the basal aspects of the epithelial cells, adjacent to small blood vessels (4). Based on the latter observation, Kulchitsky also suggested a possible endocrine role for these cells, and Ciaccio proposed that the term "enterochromaffin" be adopted to describe these cells based on staining with chromium salts (5). Gosset and Masson (6) demonstrated that the intestinal chromaffin cells were also argentaffin positive with silver techniques, and subsequent studies by Hamperl using argyrophilic staining led to the identification of a second and more extensive population of putative endocrine cells within the intestine and a variety of extraintestinal sites (7). Feyrter suggested that the clear cells (helle Zelle) of the gastrointestinal tract corresponded, in part, to the chromaffin and argentaffin/argyrophil positive cells and that they formed a diffuse epithelial endocrine system ("diffuse epitheliale endokrine Organe") (8,9). He also hypothesized that some of these cells might have a paracrine or local hormonal action (8,9). Similar groups of clear cells were illustrated by Frolich within the bronchial tree, and Feyrter also considered them to be a part of the diffuse epithelial endocrine system (9,10). Ultimately, the argentaffin, argyrophil, and clear cells were recognized as components of a diffusely distributed system of endocrine cells. These observations formed the basis of the concept of a diffuse regulatory network as opposed to the then current doctrine of maintenance of homeostasis by discrete endocrine glands.

The modern view of the neuroendocrine cell and neurosecretory neuron was based on observations that oxytocin and antidiuretic hormone were synthesized by hypothalamic neurons and were stored within neuronal processes in the posterior pituitary, prior to their release into the circulation (11). Additionally, the discovery that hormone-releasing and -inhibiting factors were synthesized by hypothalamic neurons, transported via axonal transport to the median eminence, and secreted into the pituitary portal system for interactions with specific adenohypophyseal cell types established without doubt that neurons could function as endocrine cells (11–13) (Fig. 46.1). These cells essentially could serve as neuroendocrine transducers by converting electrical input directly into chemical or hormonal signals (14).

The discovery that the argyrophil/argentaffin cells and the cells of Feyrter's diffuse epithelial endocrine system did, indeed, have an endocrine function originated from studies conducted in the early to mid-1960s on the source of the hormone calcitonin (15,16). The thyroid glands of many species were known to contain parafollicular cells that occasionally appeared clear in hematoxylin and eosin (H&E)–stained sections and which also had varying degrees of argyrophilia or argentaffinity (17,18). The parafollicular cells were ultimately shown to be the source of calcitonin, for which they were subsequently renamed C-cells (19,20). These studies also led to the discovery that certain endocrine cells shared a series of remarkable functional and morphologic similarities with neurons (19).

In addition to the presence of calcitonin, C-cells had the ability to synthesize and store catecholamines or indolylethylamines after uptake and decarboxylation of precursors of these substances (19). The latter property led to the introduction of the descriptive acronym APUD (Amine

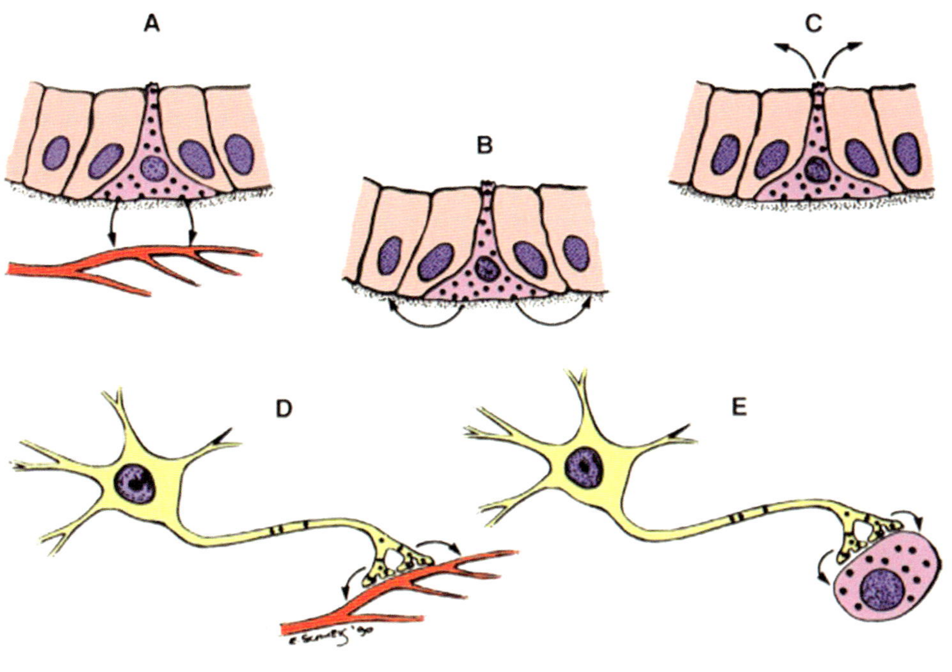

FIGURE 46.1 Secretory activities of neuroendocrine cells and neurons. **A:** Neuroendocrine cells may secrete their products through the basement membranes into adjacent capillaries for interactions with target tissues at distant sites (endocrine function). **B:** Neuroendocrine cells may secrete their products locally to influence the activities of adjacent epithelial cells (paracrine function). **C:** Neuroendocrine cells may secrete their products within a glandular lumen (luminal secretion). **D:** Neurons may secrete their products into the circulation for interactions with target tissues at distant sites (neuroendocrine function). **E:** Neurons also may secrete products that serve as neurotransmitters or neuromodulators. (Adapted with permission Fujita T. Kobayashi S. Current reviews on the paraneuron concept. *Trends Neurosci.* 1979;2:27–30, with permission from Elsevier; Dockray GJ. Evolutionary relationships of the gut hormones. *Fed Proc.* 1979; 38:2295–2301.)

TABLE 46.1 Markers of Neuroendocrine Cells Neurogenic Amine Content

1. **Amine precursor (5-hydroxytryptophan and DOPA) uptake**
2. **Aromatic amino acid decarboxylase**
3. **Nonspecific esterase or cholinesterase**
4. **Alpha glycerophosphate dehydrogenase**
5. **Peptide hormone content**
6. **Cytosolic Enzymes**
   Neuron-specific enolase (NSE), protein gene product 9.5 (PGP 9.5), histaminase, some enzymes involved in amine synthesis
7. **Secretory Granule/Membrane Proteins**
   Chromogranins/secretogranins, prohormone convertases, peptidyl glycine alpha-amidating monooxygenase (PAM) and related enzymes
8. **Synaptic Vesicle Membrane and Vesicle Fusion/Docking/Release Proteins**
   Synaptophysin, synaptic vesicle protein 2 (SV2) *N*-ethylmaleimide sensitive factor (NSF); soluble NSF attachment protein (SNAP); soluble NSF receptor (SNARE); vesicular monoamine transporters (VMAT-1, VMAT-2); synaptosomal protein-25 (SNAP-25); vesicle-associated membrane protein/synaptobrevin (VAMP 1–3); synaptotagmin; syntaxin; Rab3a; Sec1/Munc18 family; Munc13s; Sec17SNAP
9. **CD56 (NCAM) and CD57**
10. **Transcription factors**
    Mammalian homolog of achete-scute complex (mASH), human homolog of achete-scute complex (hASH), thyroid transcription factor-1 (TTF-1), forkhead box protein A1 (FoxA1), insulinoma associated protein 1 (INSM1), insulin gene enhancer binding protein 1 (islet 1), caudal type homeobox 2 (CDX2), Paired Box (PAX) family PAX2, PAX5, PAX6, PAX8, GATA3, pituitary transcription factor-1 (pit-1), T-box transcription factor (T-pit), steroidogenic factor-1 (SF-1)
11. **Somatostatin receptors**

The first six markers in this listing were described by Pearse in the original formulation of the APUD concept; however, endogenous amine content and the capacity for amine precursor uptake and decarboxylation are present in only some members of the dispersed neuroendocrine cell system.

Highlighted in blue are neuroendocrine cell markers most commonly used in diagnostic immunohistochemistry.

Precursor Uptake and Decarboxylation) (17). The APUD mechanism was subsequently identified in certain cells of the anterior pituitary and pancreatic islets. Cholinesterase, nonspecific esterases, α-glycerophosphate dehydrogenase, and certain endogenous amines were also present variably across diverse animal species and among different neuroendocrine cell types (18) (Table 46.1).

In comparing the APUD cells of the thyroid, pancreas, and pituitary to cells of known neural ancestry, Pearse concluded that "the amine storing mechanism and presence of cholinesterase together pointed towards a common ancestral cell of neural origin, perhaps coming from the neural crest" (19). The list of APUD cells was then expanded to include almost all the peptide- and amine-producing cells throughout the body, including the adrenal medulla, extra-adrenal paraganglia, and parathyroid glands.

As the numbers of candidate APUD cells increased (21), it was recognized that the synthesis of regulatory peptides was a more consistent functional parameter than was synthesis of amines, and amine synthesis was ultimately eliminated from the definition of these cells. In view of the many similarities between APUD cells and neurons, the essentially synonymous term, "paraneuron" was introduced by Fujita and Kobayashi (12). Paraneurons, according to Fujita, were endocrine and sensory cells that shared structural, functional, and metabolic features with neurons and that produced substances identical to or related to neurohormones and neurotransmitters (22). The paraneurons also possessed neurosecretory-like granules and synapse-like vesicles, and they recognized stimuli on specific receptors and released their products via the secretory portion of the cell. Since most cells of the diffuse neuroendocrine system lack APUD properties and the neural crest hypothesis of the APUD family is untrue, APUD was replaced with the term "diffuse neuroendocrine system" (21,23).

EMBRYOLOGY

Embryologic data using the chick-quail chimera system have refuted the neural crest or epiblastic origin of the majority of the dispersed neuroendocrine cells, including those of the lung, pancreas, gastrointestinal tract, and a variety of other sites (24,25). Studies of normal, chimeric, and transgenic mice have established that all gut epithelial cells, including endocrine cells, originate from multipotential stem cells present within the base of the intestinal crypts (26), whereas pancreatic endocrine cells originate from the ductal epithelium. It is now recognized that genotypic switches are present in cells previously thought to be irrevocably differentiated as a result of cellular plasticity (27). Examples of this phenomenon include the production of neuropeptides by immune cells and cardiomyocytes (28). For example, the biosynthesis of chromogranins, enkephalins, or proconvertases can be upregulated during an immune challenge.

Studies based on the chick-quail chimera system led to the concept that C-cells represented one of few remaining examples of neural crest–derived dispersed neuroendocrine cells, and this conclusion was incorrectly extrapolated to mammalian species. However, more recent investigations have challenged the neural crest origin of C-cells in mammals (29,30). These studies have demonstrated that Wnt+ neural crest cells contribute to the thyroid connective tissue but are not the source of C-cells in embryonic mouse thyroid (29). Further lineage tracing experiments have demonstrated that mouse thyroid C-cells are derived from SOX17+ progenitors originating in the anterior (pharyngeal pouch) endoderm. The development of C-cells in mammals is a multistep process that involves: (1) patterning of the 4th pharyngeal pouch; (2) budding of the ultimobranchial anlage from the 4th pharyngeal pouch; (3) migration of the ultimobranchial body; (4) fusion of the ultimobranchial body within the thyroid primordium; (5) C-cell differentiation and proliferation within the thyroid (31). Given these observations, the only neuroendocrine cells of proven neural crest origin in mammals are those of the adrenal medulla, extra-adrenal paraganglia, ganglion cells of the submucosal and myenteric plexus, and sympathetic ganglia (24,25).

In contrast, the ultimobranchial anlage of nonmammalian vertebrates and monotremes does not merge with the thyroid gland and remains as a separate organ throughout adulthood. Moreover, avian ultimobranchial C-cells may have dual origins from neural progenitors and endodermal epithelium (32).

In its current context, the term "neuroendocrine" does not imply an embryologic origin from the neuroectoderm but rather implies a shared phenotype characterized by the concurrent expression of multiple genes encoding a wide variety of neuronal and endocrine traits. In fact, many authors argue that it is more correct to refer to the diffuse neuroendocrine system as the diffuse endocrine system given the current state of our knowledge regarding its embryology and physiology.

MOLECULAR ASPECTS OF NEUROENDOCRINE CELL DEVELOPMENT

The mechanisms for the acquisition of the neuroendocrine phenotype are not fully understood; however, data from recent studies suggest important roles for both positively and negatively acting transcription factors. An important class of regulatory proteins includes those with common DNA binding and dimerization domains, the basic helix–loop–helix (b-HLH) region. The genes encoding these proteins are analogous to the *achaete–scute complex like1 (ASCL1)* which has been identified during neuronal differentiation in *Drosophila* (33) and is a master regulator in the formation of neuroendocrine cells. *ASCL1* is highly expressed in developing C-cells, pulmonary, and other endocrine cell types, but it is lost when these cells reach maturity (31,34) (Fig. 46.2). The homologous mammalian genes have been referred to as *mammalian achaete–scute homologues (mASH)* while the homologous human genes have been termed *hASH*. In Drosophila, one group of b-HLH factors encoded by genes such as *ASCL1* activates neural/endocrine differentiation, while another group of b-HLH factors encoded by *Hes1* (hairy enhancer of split) represses neuroendocrine differentiation (35–37).

The Notch pathway, which plays a crucial role in the fate of neuroendocrine cells, is mediated by ASCL1 and is upregulated by YAP1. Notch receptors are transmembrane proteins that regulate cell survival, proliferation, and differentiation after binding to the appropriate ligand(s). Hes1 and REST (RE1-silencing transcription) factor, also known as neuron-restrictive silencer factor (NRSF) are downstream targets of Notch activation. HES1 represses the transcription of ASCL1, while REST directly represses the transcription of a distinct part of the NE program (38).

Notch signaling in the embryo is important in the determination of neuroendocrine differentiation pathways of precursor cells. Notch expression directs developing cells to a nonendocrine pathway while the absence of such signaling and subsequent increased expression of ASL1 directs precursor cells to the endocrine phenotype. In addition to the Notch pathways, other pathways such as sonic hedgehog also play important roles in the differentiation of neuroendocrine cells (34,37).

Similar to many gastrointestinal tract endodermal derivatives, embryonic C-cells coexpress pioneer factors forkhead box (Fox) a1 and Foxa2 before endocrine differentiation takes place. The Hes1/mASH1 signaling pathway plays a key role in the differentiation of C-cells (31,39). In addition, a variety of other transcription factors and signaling molecules, including: PAX8 (paired box 8), TBX1 (T-box transcription factor 1), PBX1 (pre–B-cell leukemia homeobox 1), nrx2.1 (neurexin-1)/TTF-1 (thyroid transcription factor-1), Shh (sonic hedgehog), FRS2alpha (fibroblast growth factor receptor), Ripply3 (Ripply transcription receptors) and Hox3 (homeobox3 paralogs), Eya1 (eyes absent homolog), EphA (ephrin receptor gene family) are involved with

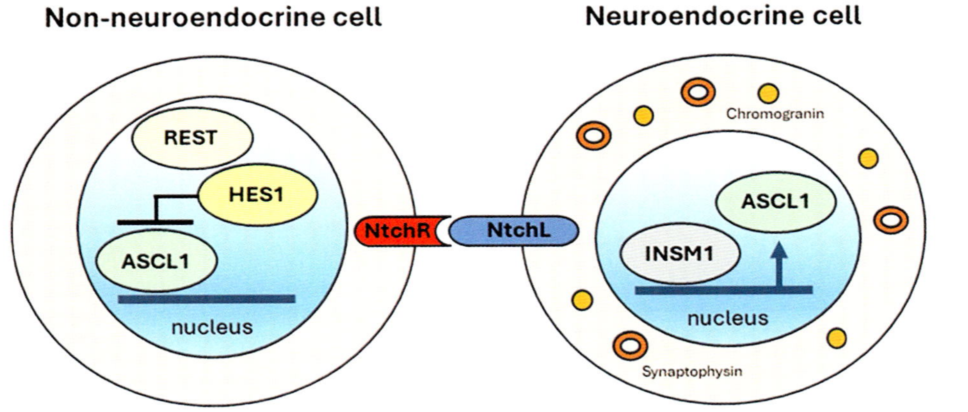

FIGURE 46.2 Neuroendocrine cell differentiation is driven by Notch signaling. INSM1 is an upstream regulator of achaete-scute complex like 1(ASCL1) transcription factor, which is a master regulator in formation of neuroendocrine cells. ASCL1 is expressed in pulmonary, thyroid C-cells, and other neuroendocrine cells. It enhances Notch ligand (NtchL) expression which activates transmembrane Notch receptors (NtchR) on neighboring progenitor cells. Hairy enhancer of split 1 (HES1) and RE1-silencing transcription (REST) factor are downstream targets of Notch activation. HES1 suppresses ASCL1, while REST suppresses different parts of neuroendocrine program in nonneuroendocrine cells.

the formation of the 4th pharyngeal pouch and the migration, differentiation, survival, and dispersion of C-cells (31,32). SOX2 (sex determining region Y-box 2) and SOX21 transcription factors function in initiation, migration, and maturation of pulmonary neuroendocrine cells (40).

Knockout *ASCL1*(−/−) mice have markedly reduced numbers of C-cells as determined by staining for calcitonin. Similar observations have been made with other neuroendocrine and neural cell types, consistent with the hypothesis that the development of this phenotype is dependent on ASCL1. Knockout of *ASCL1* also results in reduction of the numbers of pulmonary neuroendocrine cells, while knockout of *Hes1* results in a ten-fold increase in neuroendocrine cells with a concomitant decrease in Clara cells (37,41). Increased numbers of gastrointestinal neuroendocrine cells have also been observed in *Hes1* (−/−) mice.

The STAT-Ser/Hes3 signaling axis was first described as a major regulator of neural stem cells and later, as a regulator of cancer stem cells. This signaling axis also regulates several cell types with stem cell properties in the adrenal gland and pancreatic islets, consistent with a potentially important role in the endocrine and neuroendocrine cell systems (42).

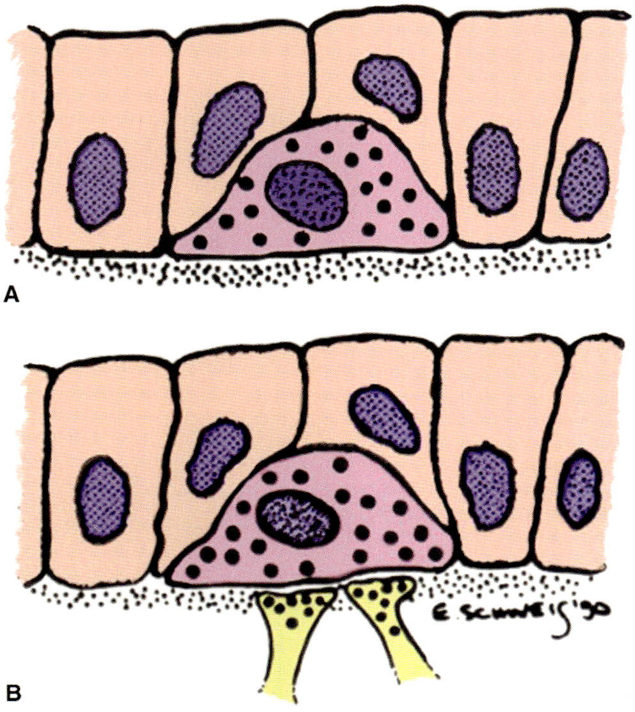

FIGURE 46.4 Closed cells are also widely distributed. **A:** The thyroid C cells are typically of the closed type. **B:** The Merkel cells of the skin are innervated closed-type neuroendocrine cells.

LIGHT MICROSCOPY AND HISTOCHEMISTRY

Neuroendocrine and endocrine cells may form grossly visible structures (e.g., adenohypophysis, parathyroid glands, adrenal medulla, extra-adrenal paraganglia) or distinct microscopic structures, such as the islets of Langerhans and pulmonary neuroepithelial bodies. However, in many organ systems, the neuroendocrine cells are dispersed among other cell types as single cells or as aggregates of three to four cells. The basal aspects of the cells are separated from the adjacent capillaries by the subjacent epithelial basement membrane (Fig. 46.3). Processes often extend from the cytoplasm to surround adjacent epithelial cells, and such cells are referred to as "paracrine cells" (13,43). The products of paracrine cells are thought to be released locally where they modulate the activities of adjacent endocrine and neuroendocrine cells (43). The apex of the endocrine cell may extend directly to the glandular lumen (opened-type cell), often with apical microvilli, or may be covered by the cytoplasm of adjacent epithelial cells and are referred to as closed endocrine cells (44–46) (Fig. 46.4). The products of opened endocrine cells may be secreted directly into the lumen of a hollow viscus. In addition, such cells may have a receptor function. Although the majority of endocrine cells are not directly innervated, some, such as those of the skin and bronchial tree, maybe innervated (Fig. 46.4B).

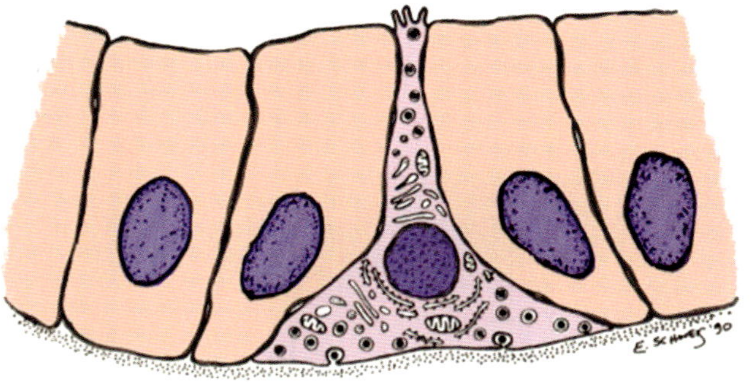

FIGURE 46.3 Diagram of typical "opened"-type neuroendocrine cell. Secretory granules are at the basal pole of the cell. Stimulation of such a cell leads to the release of hormonal products by the process of exocytosis. The basal lamina is indicated by the stippled area. Secretory granules are also present in the apical extension of the cell.

In the gastrointestinal tract, scattered neuroendocrine cells are also found within the lamina propria (47,48) without attachment to the overlying epithelium. Such cells are typically surrounded by Schwann cells and unmyelinated nerve fibers to form an enterochromaffin cell (EC)–nerve fiber complex. The EC–nerve complexes are especially prominent in appendices with chronic inflammation and neural hyperplasia (48). Stromal endocrine cells also have been identified in the prostate gland (49).

Phylogenetic and ontogenetic studies have suggested that neurons are the earliest component of the neuroendocrine system since they are present in the most primitive organisms (i.e., coelenterates) (50). The next most likely evolutionary step is the appearance of opened-type endocrine cells in the gut, which are present in the most highly developed invertebrates. Such cells become extensively diversified in vertebrates. The presence of microscopic gastroenteropancreatic endocrine cell aggregates, such as the pancreatic islets, on the other hand, is a feature that is restricted to true vertebrates (50).

These so-called dispersed neuroendocrine cells are often difficult to recognize in routinely prepared H&E-stained

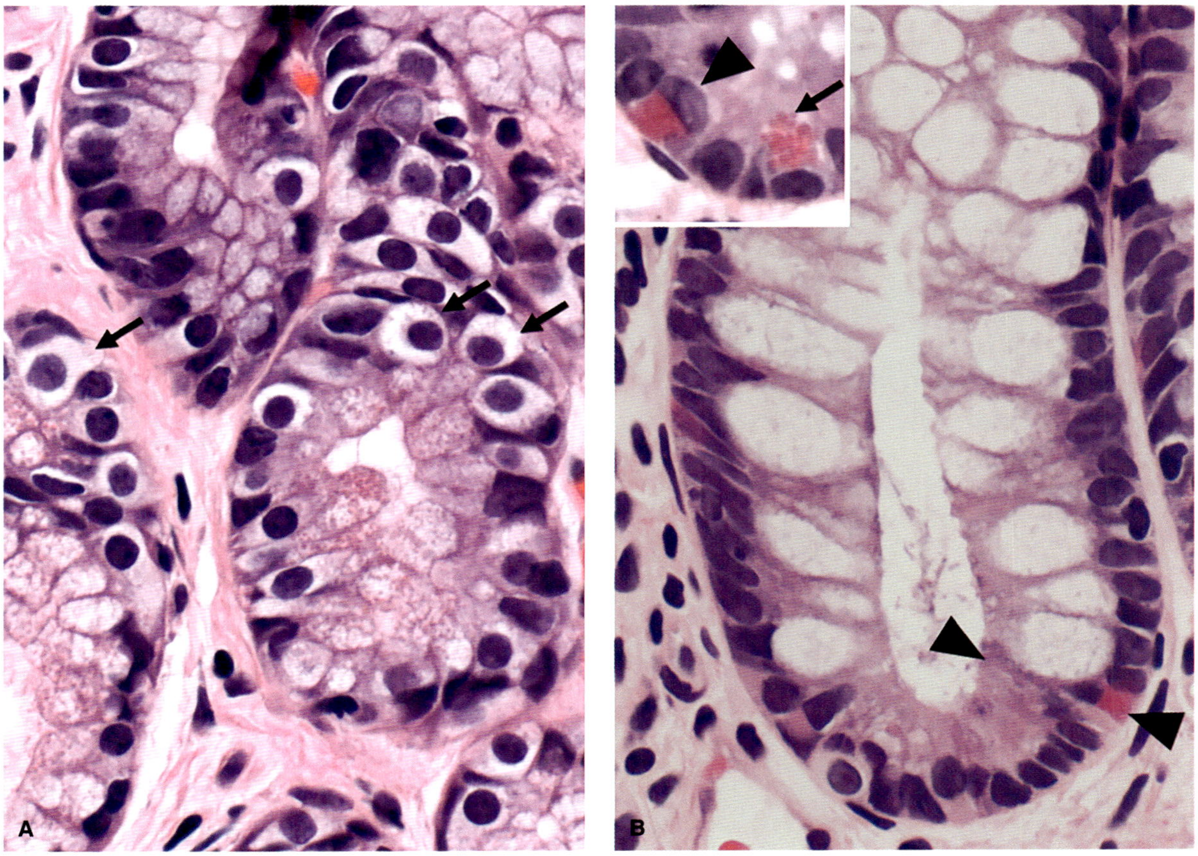

FIGURE 46.5 **A:** Neuroendocrine clear cells (helle Zelle) of gastric gland (*arrows*). **B:** Eosinophilic neuroendocrine cells of colon (*arrowheads*) that resemble Paneth cells (*arrow*) but are triangular (open type) or dome-shaped (closed type) rather than columnar and the granules are smaller than those seen in Paneth cells (*inset*).

sections, where they may appear as ovoid, pyramidal, or flask shaped, with variably clear cytoplasm (Figs. 46.5 and 46.6). In some instances, the cytoplasm may contain fine eosinophilic granules (Fig. 46.4B) that are usually visible on routine H&E sections. Some neuroendocrine cell types, such as those of the intra- and extra-adrenal paraganglia and gastrointestinal tract (ECs) develop a characteristic brown to yellow coloration after primary fixation in potassium dichromate or chromic acid. This pigment results from oxidation of cellular stores of catecholamines, serotonin, or histamine.

Some neuroendocrine cells exhibit a characteristic yellow–green fluorescence after fixation in formaldehyde and other aldehyde fixatives (42) (Fig. 46.7). In some instances, the cells may become fluorescent only after administration of L-dihydroxyphenylalanine (DOPA) or 5-hydroxytryptophan. Formaldehyde forms highly fluorescent tetrahydroisoquinoline condensation products with catecholamines and β-carboline derivatives with tryptamines such as serotonin. In some cases, freeze-dried tissues or fresh-frozen sections must be used for the demonstration of cellular stores of amines (51).

Subsets of neuroendocrine cells, including those of the gastrointestinal tract, can directly reduce ammoniacal silver to the metallic state and have been referred to as "argentaffin cells" (Fig. 46.8) (6,7). In many other neuroendocrine

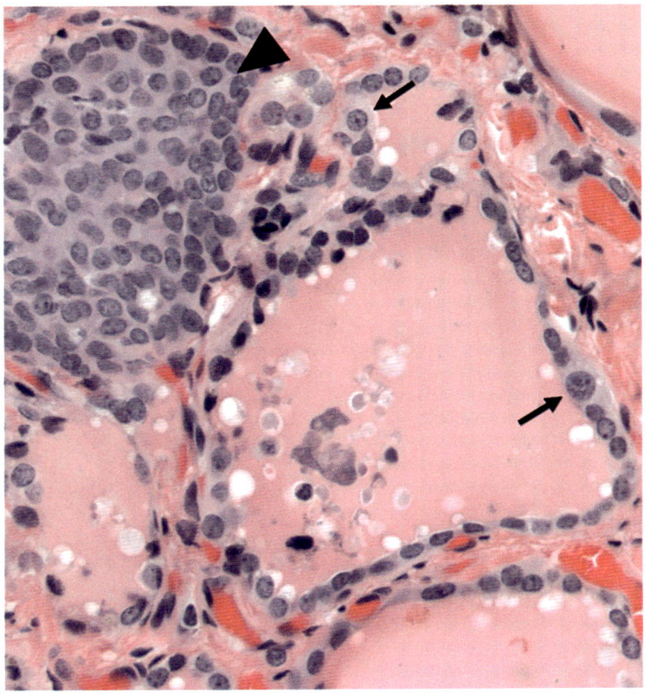

FIGURE 46.6 C-cells of the thyroid (*arrows*) that are pale and are considered as clear cells and are easily identified in association with the solid cell nests (upper left; *arrowhead*).

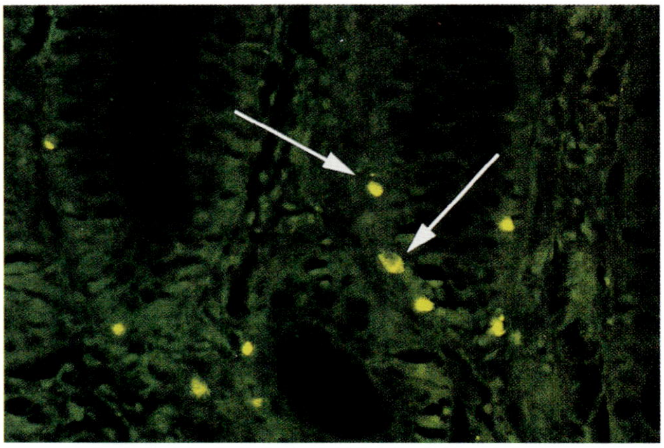

FIGURE 46.7 Formalin-fixed rectal mucosa photographed in ultraviolet light. The strongly fluorescent cells (*arrows*) correspond to the serotonin-containing enterochromaffin-type cells.

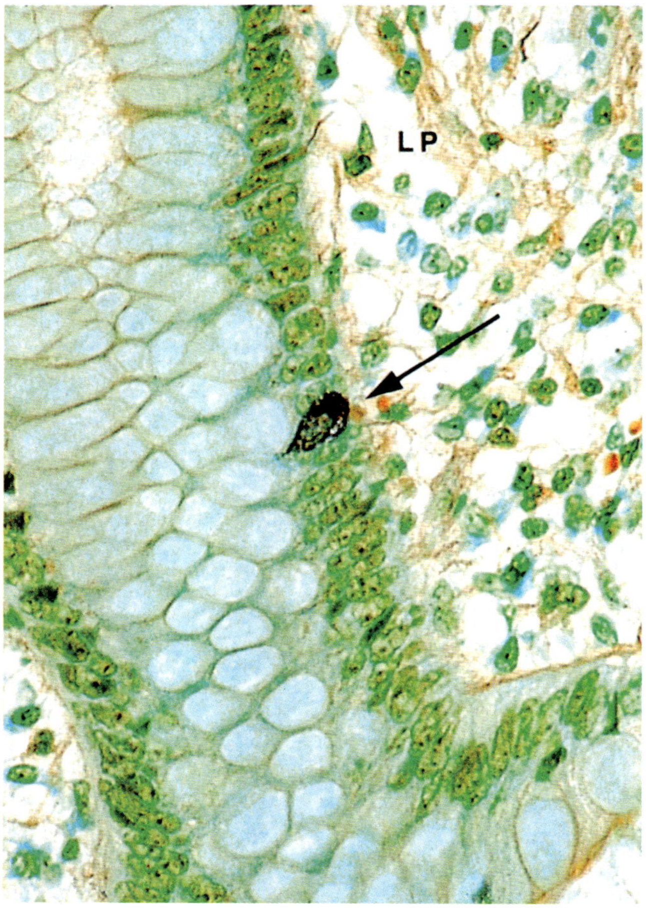

FIGURE 46.8 Colonic mucosa stained for argentaffin cells with the Masson–Fontana technique and methyl green counterstain. The argentaffin cell (*arrow*) illustrated in this field is characterized by the presence of black cytoplasmic granules. LP, lamina propria.

cells, silver positivity is evident only after the addition of an exogenous reducing agent to the staining solution, and such cells are said to be argyrophilic. The chromaffin and argentaffin reactions of endocrine cells in the gastrointestinal tract are due primarily to the presence of serotonin. While argentaffin cells are also argyrophilic, only a subset of argyrophil cells is argentaffin positive (52). The chemical basis of the argyrophil reaction is unknown, but it is apparent that reduced silver salts have an affinity for a nonamine constituent of neuroendocrine secretory granules, most likely chromogranin proteins (53).

It should be recognized that argyrophil stains are nonspecific. Cellular products such as lipofuscin, glycogen, and certain proteins including α-lactalbumin may be argyrophilic (54). Most neuroendocrine cells stain metachromatically with toluidine blue and coriophosphine O after acid hydrolysis of tissue sections (55). Lead hematoxylin also has been used for the demonstration of neuroendocrine cells (56). However, none of the histochemical stains discussed in this section is used currently in clinical practice, having been replaced by immunohistochemical methods.

ULTRASTRUCTURE

The most characteristic ultrastructural feature of neuroendocrine cells is the presence of membrane-bound secretory granules, which may vary from 50 to 500 nm in diameter (Fig. 46.9). In addition, the cells are well endowed with granular endoplasmic reticulum and prominent Golgi regions, which are characteristic of secretory cells. Because of their relatively large size, the secretory granules have also been referred to large dense core vesicles or granules (1). Immunoelectron microscopic studies have shown that the granules represent storage sites of peptide and amine hormones. Granules storing different types of hormones are characterized in some instances by differences in size, density of contents, and substructure (57). Although most neuroendocrine secretory granules are round, others, such as those of the gastrointestinal enterochromaffin and enterochromaffin-like cells, are pleomorphic with elongated, reniform, round, ovoid, or pear-shaped forms.

Secretory granules tend to be concentrated at the basal aspects of the cells in relatively close proximity to the basement membrane (Fig. 46.3). They are also prominent in cytoplasmic processes and in the apical extensions of the "opened" cells (Fig. 46.3). In addition to secretory granules, many neuroendocrine cells contain synaptic-type vesicles which have been referred to as small synaptic vesicle analogs (1). The synaptic-type vesicles are responsible for the release of amino acid neurotransmitters (gamma amino butyric acid, glutamate, glycine) and various biogenic amines in a regulated fashion in response to a variety of different stimuli (1).

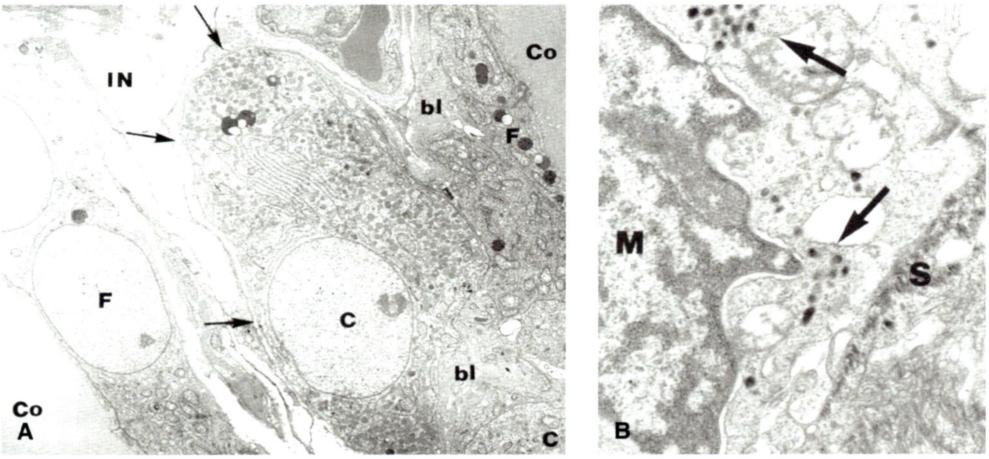

FIGURE 46.9 Ultrastructure of neuroendocrine cells. **A:** Electron micrograph of a C-cell from a patient with mild C-cell hyperplasia associated with the type 2A MEN syndrome. The C-cell is present at the base of the follicle, where it is in direct contact with the basal cytoplasm of the overlying follicular cell. The basal lamina (bl) is focally thickened at the junction of the C-cell and overlying follicular cells. The C-cell is separated from the interstitium by the follicular basal lamina (*arrows*). C, C-cell; Co, colloid; F, follicular cell; IN, interstitium (original magnification ×14,000). **B:** Electron micrograph of Merkel cell. Clusters of secretory granules (*arrows*) are present within the Merkel (M) cells. S, squamous cell (original magnification ×27,000).

FUNCTION OF NEUROENDOCRINE CELLS

The function of neuroendocrine cells has been established using immunohistochemistry and in situ hybridization techniques for the localization of specific hormones and their corresponding messenger RNAs. In many instances, the use of region-specific antisera also permits the localization of hormone precursors as well as mature hormones (58). Initial studies suggested that single endocrine cells were responsible for the production of a unique hormonal product (one-cell, one-hormone hypothesis), although newer data indicate that these cells are multimessenger units (59). Peptide hormones are synthesized within the granular endoplasmic reticulum and are packaged into secretory granules by way of the Golgi region. Multiple different peptide products may be synthesized via this route in single endocrine cells. Other nonpeptide hormone constituents such as biologically active amines are synthesized within the cytosol and are then taken up into secretory granules and small synaptic vesicles (60). Any individual endocrine cell can, therefore, vary the biosynthesis and secretion of its products in response to different signals in normal and pathologic states.

Immunohistochemistry and molecular biologic studies have led to many interesting insights into the functional interrelationships of the various components of the diffuse neuroendocrine system. For example, peptide hormones first isolated from the gastroenteropancreatic axis have been found subsequently in neurons of the central and peripheral nervous systems, where they may function as neurotransmitters or neuromodulators (14). Other peptides initially isolated from the brain have been localized to the endocrine cells of the gut, pancreas, and lung, where they may have a paracrine or autocrine function (61). Furthermore, such studies have shown that the microarchitecture of endocrine organs, which may appear homogeneous in H&E–stained sections, is often organized in a manner that permits paracrine interactions. For example, somatostatin cells of the pancreatic islets are located between the insulin and glucagon cells and typically extend short branching processes, which are in apposition to both cell types. Regulation of the secretion of insulin and glucagon may, therefore, be mediated by the local paracrine effects of somatostatin and by the endocrine effects of somatostatin reaching the islets by circulation (62,63). In the stomach, branching processes extending from somatostatin cells interdigitate with the adjacent gastrin cells, thereby regulating the secretion of gastrin in response to food intake and other stimuli (64).

Neuroendocrine cells in different tissues may produce identical peptides. Somatostatin, for example, is present in certain hypothalamic neurons (Fig. 46.10A), pancreatic D cells, gastrointestinal D cells (Fig. 46.10B), bronchopulmonary endocrine cells, thymic endocrine cells, and a subset of thyroid C cells, where it is colocalized with calcitonin (62–65). Calcitonin is present in thyroid C cells, bronchopulmonary and thymic endocrine cells, and certain urogenital endocrine cells. Gastrin-releasing peptide, a 27–amino acid peptide that is the mammalian homolog of bombesin, is present in thyroid C cells, small intensely fluorescent cells of sympathetic ganglia, neuronal cells of the gastrointestinal myenteric plexus, and bronchopulmonary endocrine cells (66,67).

Neuroendocrine cells may produce multiple distinct peptides from a common precursor molecule. For example,

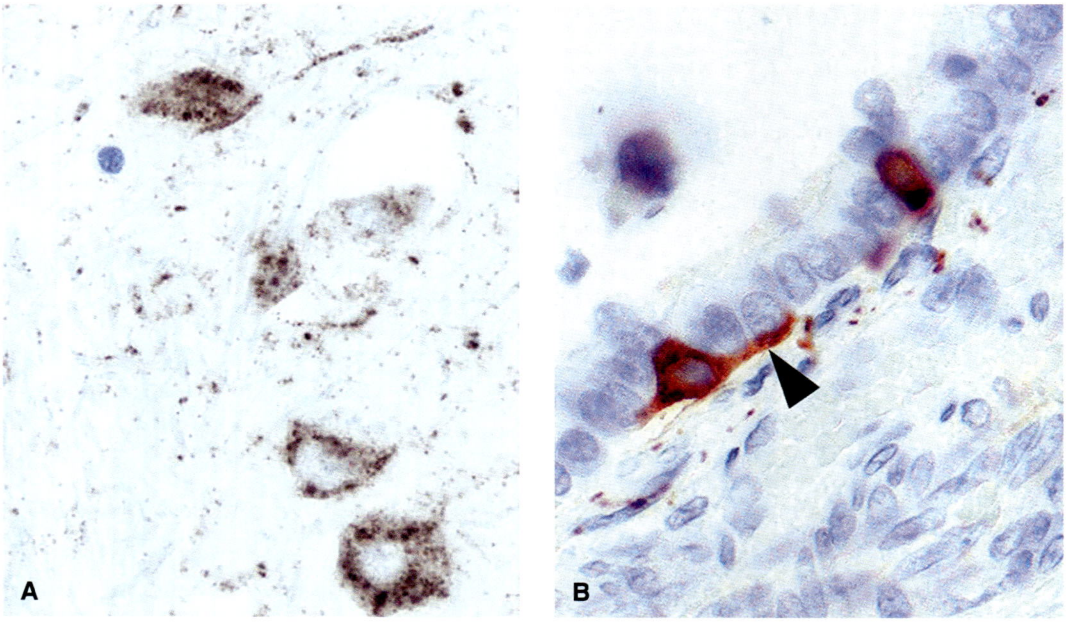

FIGURE 46.10 Somatostatin staining of hypothalamus and small intestine. **A:** Median eminence of the hypothalamus immunostained for somatostatin demonstrates positive staining of the neurons and their processes. **B:** A closed somatostatin-positive cell in the small intestine harbors the basal process (*arrowhead*) beneath the epithelium.

adrenocorticotropin (ACTH) is synthesized from the large precursor molecule pro-opiomelanocortin (68). In the adenohypophysis, pro-opiomelanocortin is processed to yield ACTH, β-lipotropin, and a 16KD N-terminal fragment. In the intermediate lobe, ACTH and β-lipotropin are processed to yield α-MSH and β-endorphin–related peptides, respectively. Hormonal diversity in endocrine cells also may result from alternative splicing pathways that produce different messenger RNAs from a single gene. Both calcitonin and the calcitonin gene–related peptide are produced from a primary RNA transcript that is spliced to produce two different forms of mature messenger RNA (69). More than one gene may also encode closely related peptides (70).

Molecular studies have led to the identification of multiple related or associated novel peptide hormones encoded by DNA sequences of known or novel hormones (71). The latest reported peptide hormones are collected in databases, such as HORDB and Hmbrase2 which include 5,729 and 7,406 peptide hormones (72,73). The individual hormone gene frequently has multiple phenotypes, because of alternative splicing, tandem organization, or differentiated maturation of the prohormone (74). As a result of these mechanisms, more than 100 different hormonally active peptides are released from the gut. For instance, the identification of several high-molecular–weight proteins that also contained the glucagon sequence was established by the molecular identification of glicentin, oxyntomodulin, glucagon-like peptide (GLP)-1 and GLP-2 in addition to two intervening peptides, all of which were generated by posttranslational processing in the L-cell (71,74).

MARKERS OF NEUROENDOCRINE CELLS

Neuroendocrine cells can be classified into those of neural (neurons, paraganglia cells) and epithelial types. The former contain neurofilaments as their major intermediate filament type while the latter contain cytokeratins with or without neurofilaments. Both cell groups can be identified based on their contents of specific hormones and neurotransmitter substances (75–77), as discussed in other chapters in this volume and by the presence of a variety nonhormonal products. The nonhormonal constituents of neuroendocrine cells include a wide array of cytosolic, secretory granule and vesicle membrane, and plasma membrane constituents (Table 46.1). These products can be identified effectively via immunohistochemistry with polyclonal antisera or monoclonal antibodies. This approach is of particular importance when evaluating tissues for the presence of neuroendocrine cells when the specific hormonal product is unknown.

Cytosolic Constituents

A variety of different enzymes can be demonstrated by immunohistochemistry in neuroendocrine cells. Although some of the enzymes are present in most neuroendocrine cells, others have a more restricted distribution. In the past, neuron-specific enolase (NSE) was used widely as a generic neuroendocrine marker for neurons and neuroendocrine cells (78). Although the sensitivity of NSE for the detection of neuroendocrine cells is high, its specificity is low.

Because of its low specificity, this marker is no longer recommended as a generic neuroendocrine marker.

The protein gene product 9.5 (PGP 9.5) is a soluble protein with a molecular weight of 27,000 Daltons that was isolated originally from the brain (79). It is a ubiquitin carboxyterminal hydrolase (UCH-L1) that plays a role in the catalytic degradation of abnormal denatured proteins (80). Immunohistochemistry studies have demonstrated that it is present in neurons and nerve fibers at all levels of the central and peripheral nervous system. It is also present in a variety of neuroendocrine cells, except for those in the normal gastrointestinal tract (79). Although the expression of this marker was originally thought to be restricted to neurons and neuroendocrine cells, it has also been identified in distal renal tubular epithelium, Leydig cells, prostatic and mammary epithelium, and other cell types (81). Additional enzymes that have a predominant cytosolic distribution include those involved in catecholamine biosynthesis (77).

Secretory Granule Constituents

Chromogranins/Secretogranins

The chromogranins/secretogranins (Cg/Sg) represent a widely distributed family of soluble proteins that represent the major constituent by weight of neurosecretory granules or large dense core vesicles (Table 46.1) (82–85). Three major granins have been identified and designated chromogranin A, chromogranin B, and to a lesser extent, secretogranin II (chromogranin C). Additional members of the granin family (numbers in parentheses refer to antibodies used for their detection) include secretogranins III (1B1075), IV (HISL-19), V (7B2), VI (NESP55), and VII (VGF) (86). The chromogranins are calcium-binding proteins that play important roles in the packaging and processing of regulatory peptides. These proteins contain multiple dibasic residues that are potential sites for proteolytic cleavage and processing producing a wide range of smaller peptides (87,88). Chromogranin A is processed by the action of proteases, in addition to posttranslational modifications, to vasostatins, chromacin, chromofungin, prochromacin, pancreastatin, catestatin, chromostatin, parastatin, serpinin, and the vasoconstrictive inhibitory factor (89). There are numerous functional roles for the granins and their smaller derivative peptides, including stabilization of the intragranular core complex, regulatory effects on the secretion of hormones and metabolites, effects on innate immunity, antimicrobial effects, vascular homeostasis, angiogenesis, tissue repair, inflammation, and heart physiology (89). Interestingly, increased plasma levels of chromogranin A have emerged as established or potential biomarkers for a variety of neoplastic, cardiovascular, and inflammatory diseases (90). Both pancreastatin and chromostatin are present in many neuroendocrine cells, in the adrenal medulla and anterior pituitary (91,92), while derivatives of chromogranin B (GAWK protein) have been localized to neuroendocrine cells in the pituitary, gastrointestinal tract, pancreas, and adrenal medulla (93).

The chromogranins are widely distributed throughout the entire system of neuroendocrine cells and have distinctive patterns of tissue and cellular distribution (94). The chromogranin proteins, particularly CgA, are among the most useful markers for the identification of normal neuroendocrine cells. Although many neuroendocrine cells contain CgA, CgB, and SgII, others contain only one or two of these proteins. For example, thyroid C cells contain CgA and SgII but lack CgB. Parathyroid chief cells, on the other hand, are positive for CgA but lack SgII. The distribution of this family of proteins is reviewed in detail by Huttner et al. (94). The chromogranins are cosecreted with other granule contents, but their replenishment is regulated differentially.

Prohormone Convertases and Peptidylglycine Alpha Amidating Monooxygenase

A variety of endopeptidases and carboxypeptidases are required for the formation of biologically active peptides from precursor molecules and are present in the trans-Golgi region and secretory granules of neuroendocrine cells. They include the prohormone convertases, PC1/PC3 (proprotein convertase subtilisin/kexin type; PCSK1), PC2 (proprotein convertase subtilisin/kexin type 2; PCSK2), and carboxypeptidases H and E (95–97). The proconvertases are widely distributed in neuroendocrine cells, while other types of endocrine cells (e.g., thyroid follicular cells, parathyroid chief cells, adrenal cortical cells, and testis) are negative. Neuroendocrine cells with a neural phenotype (e.g., adrenal medullary cells) contain a predominance of PC2 (PCSK2) while epithelial cells contain a predominance of PC1/PC3 (PCSK1). Except for parathyroid cells, the presence of PC2 (PCSK2) and P1/PC3 (PCSK1) correlates with the presence of chromogranins and secretogranins. PC2 (PCSK2) and PC1/PC3 (PCSK1) are found in normal pituitaries and adenomas, with ACTH-producing adenomas containing a predominance of PC1/PC3 (PCSK1) and other adenomas expressing a predominance of PC2 (PCSK2).

Peptidylglycine alpha-amidating monooxygenase (PAM), peptidylglycine alpha-hydroxylating monooxygenase (PHM), and peptidylamidoglycolate lyase (PAL) are present in neuroendocrine secretory granules (98,99). These enzymes are responsible for the alpha amidation of the C-terminal regions of peptide hormones, a function that is critical for the biologic function of peptides. These enzymes are not restricted in their distribution to neuroendocrine cells. For example, they have also been found in the lung in cells of the airway epithelium and submucosal glands, vascular endothelium, some chondrocytes of bronchial cartilage, alveolar macrophages, and smooth muscle cells.

Synaptic Vesicle and Vesicle Fusion/ Release Constituents

Synaptophysin (Mol. wt 38,000) was one of the earliest markers developed to visualize small synaptic vesicle analogs in neurons and neuroendocrine cells (100–102). This protein is widely distributed in nerve terminals in the central and

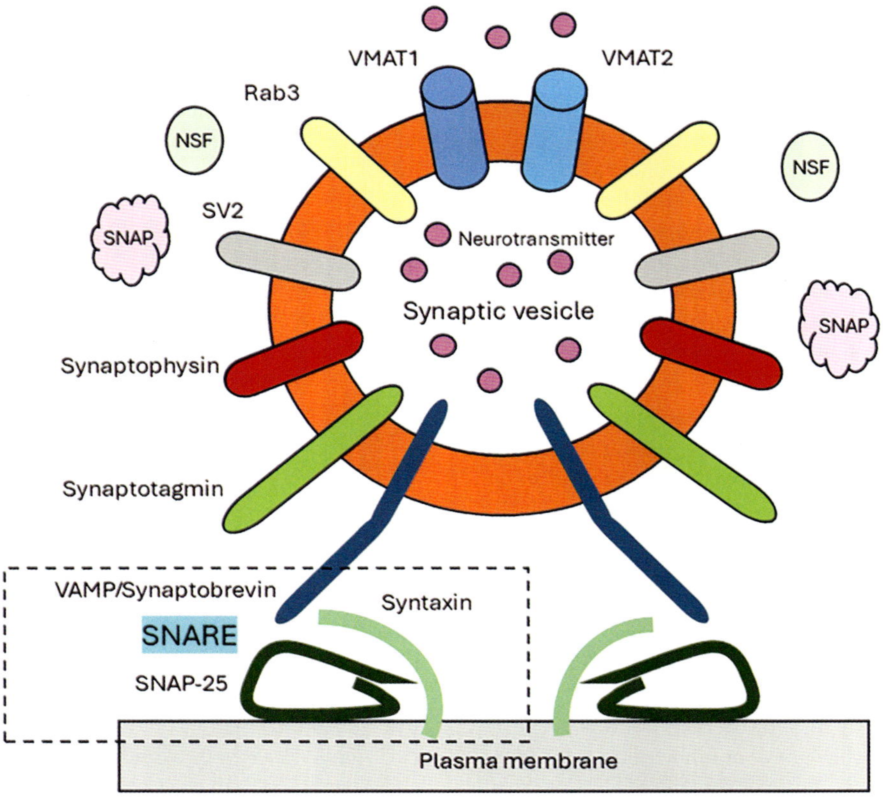

FIGURE 46.11 Synaptic vesicle constituents. Synaptic vesicle membrane contains synaptophysin, vesicle associated membrane protein (VAMP)/synaptobrevin, synaptotagmin, synaptic vesicle protein 2 (SV2), Ras-associated binding 3 (Rab3), vesicular monoamine transporter 1 (VMAT1), and vesicular monoamine transporter 2 (VMAT2). Plasma membrane carries Syntaxin and soluble N-ethymaleimide sensitive factor (NSF) attachment protein-25 (SNAP-25). The soluble cytoplasmic proteins include NSF and SNAPs. Three families of SNAP receptor (SNARE) proteins include VAMP/synaptobrevin, syntaxin, and SNAP-25 (dashed square). The process of regulated secretion of neurotransmitters is complex and is discussed in detail in references 114–117.

peripheral nervous system and is also present in neuroendocrine cells that are specialized for the regulated secretion of peptide hormones. Synaptophysin is the most abundant integral membrane protein of neuronal vesicles (Fig. 46.11). It is localized in a punctate pattern in synaptic regions of neurons and has a diffuse cytoplasmic distribution in neuroendocrine cells. Ultrastructurally, synaptophysin is present predominantly in smooth-surface synaptic-type vesicles. Although synaptophysin was originally thought to be specific for neuroendocrine cells, it is also expressed in other cell types, including the steroidogenic cells of adrenal cortex (103,104).

Synaptic vesicle protein 2 (SV2), an integral membrane protein that mediates calcium-stimulated neurotransmitter release, is present in the central and peripheral nervous system and in a wide variety of neuroendocrine cell types (105). This glycoprotein regulates the expression and trafficking of the calcium sensor protein, synaptotagmin, in addition to other functions related to neurotransmitter release and homeostasis (106,107). Immunoreactivity for SV2 is present in neuroendocrine cells in the gastrointestinal tract, pancreas, anterior pituitary, thyroid (C-cells), parathyroid, and adrenal medulla.

The vesicular monoamine transporters (VMAT1 and VMAT2) are integral membrane proteins that mediate the transport of amines into vesicles of neurons and neuroendocrine cells (108,109). These two isoforms show broad selectivity for different amines and are distributed differently in various cell types. They play critical roles in the sorting, storing, and release of neurotransmitters and in regulating the neuronal and endocrine informational output (110). VMAT2 facilitates the uptake of dopamine, norepinephrine, epinephrine, histamine, and serotonin into neurons and endocrine cells of the enterochromaffin-like type. VMAT1 does not recognize histamine as a substrate, it is absent from neurons, and its major functional activity is in endocrine cells, particularly ECs (111). In the pancreas, VMAT 1 is present in endocrine cells of the ductal epithelium, while VMAT2 is present primarily in the beta cells of the islets (112). VMAT2, on the other hand, is present in histamine-producing enterochromaffin-like cells and in central and peripheral neurons. VMAT1 and VMAT2 are both expressed by adrenal medullary cells to varying extents. In some studies, the VMATs have been used as surrogate markers for other types of neuroendocrine cells, such as those in the breast (113).

The process of regulated secretion in neurons and neuroendocrine cells is highly complex and involves a large number of molecules (1,11,114–117). Factors that govern neurotransmitter release include *N*-ethylmaleimide sensitive factor (NSF), soluble NSF attachment proteins (SNAPs), the SNAP receptors (SNAREs), Munc18-1, Munc13s, and Rab3 (Ras-associated binding 3). According to the SNARE hypothesis, the selective docking of a transport vesicle with the appropriate target membrane occurs via the formation of a complex between a vesicle membrane protein (v-SNARE) and the corresponding target membrane protein (t-SNARE) (114). The resulting SNARE complex ultimately leads to membrane fusion. Three families of SNARE proteins are currently recognized. They include the VAMP (vesicle-associated membrane protein)/synaptobrevin family of v-SNAREs and two families of t-SNAREs, the syntaxin family, and the synaptosomal-associated protein of 25kDa (SNAP-25) family. In the initial phases of exocytosis, NSF and SNAP act on synaptobrevin, syntaxin, and SNAP-25. This leads to dissociation of the SNARE complexes, activation of the SNARE proteins, and removal of the negative regulators of exocytosis. Subsequently, the vesicle protein Rab3 promotes reversible vesicle attachment (tethering) to the presynaptic membrane. Tethering permits the formation of the SNARE complex which consists of synaptobrevin, syntaxin, and SNAP-25. This series of events brings the vesicle into a docked position, immediately adjacent to the plasma membrane and calcium channels. Docking is an irreversible step in which there is some degree of membrane fusion. At some time during docking, synaptotagmin is recruited to the SNARE complex (116). The SNAREs play a key role in membrane fusion and other proteins are considered to be SNARE regulators. However, some of the regulator proteins may have direct roles in membrane fusion and neurotransmitter release. Additional proteins involved in this process include Sec1-Munc18 family and the Sec17/SNAP proteins (117).

Some of the proteins involved in the process of regulated secretion can be visualized by immunohistochemistry and have been utilized as neuroendocrine cell markers. While some of these proteins are localized within the plasma membranes (SNAP-25 and syntaxin), others (synaptobrevin, synaptophysin, Rab3, and synaptotagmin) are present in the membranes of vesicles. The soluble proteins involved in this process include NSF and SNAPs.

The VAMP/synaptobrevin plays an important role in docking and/or fusion of secretory vesicles with their target membranes. VAMP occurs in three isoforms which are designated VAMP-1, VAMP-2, and VAMP-3 (cellubrevin). VAMP-2 and VAMP-3 are expressed in pancreatic islets (118) and are involved in calcium-mediated insulin secretion. VAMP-1 is present primarily in pancreatic acinar cells (119).

The synaptotagmins (p65) include a large family of calcium-binding proteins that are constituents of the membranes of synaptic vesicles in neurons and neuroendocrine cells (120). In the normal pancreatic islets, synaptotagmins are colocalized with insulin in beta cells and are involved with calcium-induced insulin secretion (121).

Rab proteins are low molecular weight members of the Ras superfamily of monomeric G-proteins. The Rab3 isoforms are involved in the exocytosis of synaptic vesicles and secretory granules in the central nervous system (CNS) and anterior pituitary. In normal human pituitary, Rab3 isoforms are present primarily within the cytoplasm of growth hormone-producing cells with rare expression in other cell types. Among pituitary adenomas, Rab3 is most commonly expressed in growth hormone-producing adenomas but also occurs in adenomas of other types (122,123).

SNAP-25 has been studied most extensively in the pituitary gland. This protein is localized predominantly to the plasma membranes of both normal and neoplastic adenohypophyseal cells (124–126). Similar patterns of localization have been documented in the adrenal and pancreatic islets (127,128). Additionally, SNAP-25 is present in enterochromaffin-like cells, together with syntaxin and synaptobrevin (129).

CD56 and CD57

The CD56, although known as a neural cell adhesion molecule (NCAM) represents a family of glycoproteins that play key roles in cell binding, migration, differentiation, and proliferation (130,131). The NCAM family includes several major peptides that are generated by alternative splicing of RNA from a gene that is a member of the immunoglobulin supergene family. The peptide sequences that are external to the plasma membrane contain five regions that are similar to those present in immunoglobulins. The molecules are modified posttranslationally by phosphorylation, sulfation, and glycosylation. The homophilic-binding properties of NCAMs are modulated by differential expression of homopolymers of α2, 8-linked *N*-acetylneuraminic acid (polysialic acid).

CD56 is expressed on resting and activated NK cells and a subset of CD3+ cells, the brain, neuroendocrine cells of the pancreatic islets, adenohypophysis, and adrenal medulla, as well as in a variety of nonneuroendocrine cells. Studies using a monoclonal antibody reactive with a long chain of α-2,8-linked polysialic acid present on NCAMs demonstrated positive staining in cases of familial medullary thyroid carcinoma, both in the neoplastic cells and in hyperplastic C-cells adjacent to the tumor foci (132).

CD57 antigen is expressed in natural killer cells, and is associated with myelin-associated glycoprotein (MAG), NCAMs, and a granule matrix constituent of chromaffin cells (133–135). The largest of the MAGs (MAG-72) is related to the immunoglobulin supergene family proteins as well as NCAMs. CD57 also reacts with a subset of neuroendocrine cells in the anterior pituitary, lung, adrenal gland, and alimentary system (135). In addition, CD57 immunoreactivity has been identified in Schwann cells and sustentacular cells (136).

Transcription Factors

Both the mammalian (mASH1) and human (hASH1) homologs of the ASCL1, are expressed in a subset of thymic epithelial cells, thyroidal C-cells, and fetal pulmonary neuroendocrine cells (33,36,137,138). FoxA1, which regulates the development of a variety of different tissues, is expressed in adult C-cells and solid cell nests in the adult thyroid gland (139). INSM1 (insulinoma-associated protein 1) is an upstream regulator of ASCL1 and generic neuroendocrine marker (140). In fetal tissues, INSM1 is present in gastrointestinal epithelium and the enteric nervous system, pancreas, thyroid, respiratory epithelium, thymus, and cerebellum. In adult tissues, INSM1 immunoreactivity is present in adrenal medulla, pancreatic islets, and endocrine cells of the gastrointestinal tract. Positive staining is present in scattered cells in bronchial epithelium and prostate gland (140). An additional transcription factor that is present in gastrointestinal and pancreatic endocrine cells is Histone HIx (141). Insulin gene enhancer binding protein-1 (Islet-1) is a homeobox gene that is involved in the embryogenesis of the pancreatic islets as well as endocrine neoplasms of the duodenum and colon/rectum (142). CDX2 is a marker of intestinal differentiation and is present in serotonin-producing cells, as well as a handful of other gastrointestinal endocrine cells (143). TTF-1 is present in endocrine and nonendocrine cells of the thyroid and lung (144–146).

The Paired Box (PAX) genes are a family of nine nuclear transcription factors that play a crucial role during human embryogenesis and development (147). PAX2, known to control the development of the CNS, thyroid gland, kidney, eye, and female genital tract, has been also identified in glucagon-secreting pancreatic cells (148). PAX8 is a highly sensitive marker for thyroid carcinoma and is also expressed in pancreatic neuroendocrine tumors (149).

Somatostatin Receptors

Somatostatin receptors (sst1-sst5) have attracted considerable attention because of clinical applications related to their overexpression in certain tumors, including those of the neuroendocrine system (150). Because of the metabolic instability of natural somatostatin, a number of synthetic analogs/agonists (e.g., octreotide, lanreotide, pasireotide) have been developed (151). More recently, somatostatin receptor antagonists have also been developed and used both for imaging and therapy (151). The ability of tumors to express somatostatin receptors can be assessed with antibodies to the specific receptors. Positive staining of normal gastrin producing cells for sst-2A has been documented in gastric antrum, duodenum, jejunum, and ileum (152). In addition, occasional sst-2A–positive cells have been recognized in the basal cell component of bronchi. A comprehensive study of the distribution of somatostatin receptor subtypes has revealed their presence in a wide variety of neuroendocrine and nonneuroendocrine cell types, including, adrenal cortex, myocardium, skeletal muscle, ovary, and testis (153). Different neuroendocrine cells have distinctive expression patterns of somatostatin receptors. For example, C-cells are positive for sst5, while the pancreatic islets are positive for sst1-3 and sst5, albeit in different distributions (153).

DISTRIBUTION OF NEUROENDOCRINE CELLS

Bronchopulmonary and Upper Respiratory System

The endocrine components of the lung are rare epithelial cells that constitute <1% the airway cell population. They occur singly as solitary cells (Fig. 46.12) historically termed Kulchitsky (K) cells and as small aggregates composed of 4 to 10 cells that have been designated neuroepithelial bodies (36,154) (Fig. 46.12A). Solitary endocrine cells are predominantly located in bronchi and may be of the opened or closed type (Fig. 46.12B). Neuroepithelial bodies are extensively innervated and are composed of clear to faintly eosinophilic cells adjacent to the bronchial basement membranes. The neuroepithelial bodies tend to occur adjacent to sites of airway bifurcation (155). Although the functions of the two endocrine components are not known, the proper development of pulmonary endocrine cells in neuroepithelial bodies is essential for restricting the number of immune cells in the naive lung (156). Neuroepithelial bodies most likely act as modulators of lung growth and development during early phases of organogenesis and as intrapulmonary tactile- and/or chemo-receptors later in fetal growth and in the postnatal period (36).

The secretory granules of pulmonary endocrine cells vary in size and density (154,157,158). On the basis of granule size, they have been divided into three types. The P1 cells have granules that measure 40 to 50 nm in diameter, P2 cells have granules that measure 120 to 130 nm in diameter, and P3 granules measure 180 to 200 nm. The granules of Pa cells, which are found in the adult lung, measure 100 to 120 nm in diameter. Both neuroepithelial bodies and solitary endocrine cells are positive for chromogranin, synaptophysin, INSM1, and SV2 and also contain serotonin, gastrin-releasing peptide, and calcitonin, while the solitary endocrine cells also contain leu-enkephalin (67,157,159). Severely hyperplastic and dysplastic cells of the neuroepithelial bodies may also produce ACTH, vasoactive intestinal peptide, and somatostatin (57). The neuroepithelial bodies are particularly conspicuous in fetal lung tissue but are sparse in the adult (35,159). The endocrine components of the lung are also prominent in adults with hypoxic states, including chronic pulmonary diseases such as chronic obstructive pulmonary disease (36,157,159). Lungs from asthma patients also show increased pulmonary endocrine cells, which amplify allergic asthma responses (160).

Neuroendocrine cells are the earliest differentiated cells in 5 weeks postconception of human fetal lung (161). They express hASH1, the human counterpart of ASCL1

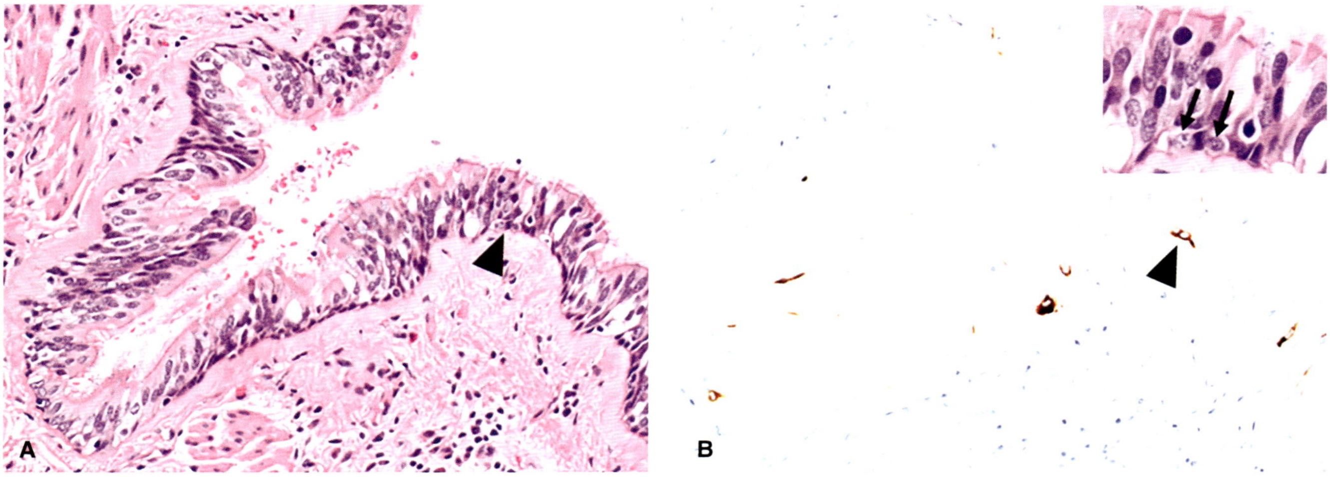

FIGURE 46.12 Neuroendocrine cells of the lung. On the hematoxylin and eosin (H&E) sections identifying neuroendocrine (neuroendocrine) cells is difficult. There are a number of cells with light or clear cytoplasm that are not neuroendocrine cells as can be seen on the chromogranin (CHR) immunohistochemical (immunohistochemistry) stain. The *arrowheads* on the H&E stain (**A**) and in the CHR immunohistochemistry (**B**) demonstrate neuroendocrine cells that are part of a neuroepithelial body with closed-type neuroendocrine cells. Adjacent closed-type neuroendocrine cells are to the left. The cells forming part of the neuroepithelial body are shown in the H&E section (*inset, arrows*). The CHR immunohistochemistry (**C**) demonstrates an open-type neuroendocrine cell in which the cytoplasm extends to the lumen in the left panel. In the right panel processes of the neuroendocrine cell are interdigitating between the epithelial cells.

which is essential for the development of the pulmonary endocrine cell system (159). In early phases of development, neuroendocrine markers are coexpressed with epithelial markers, consistent with a common cellular origin (35). Both in situ hybridization and immunohistochemistry studies have shown gastrin-releasing peptide and its corresponding messenger RNA (mRNA) as early as 8 weeks of gestation in solitary endocrine cells and neuroepithelial bodies. The numbers of cells reach a peak by 16 to 30 weeks of gestation and decline at about 6 months of age. These findings suggest that gastrin-releasing peptide may be involved in the growth and development of normal lung (159). Recent studies combining single-cell RNA sequencing and spatial transcriptomics uncovered two types of pulmonary endocrine cells in the fetal lung: gastrin-releasing peptide and ghrelin-positive cells, the latter of which are located exclusively in the most proximal airways (161,162).

Endocrine cells are rare in the larynx (163,164). They are present in the basal to middle layer of the respiratory epithelium of the ventricle and subglottic regions (165,166).

Thyroid and Thymus

The parafollicular cells, or C-cells in adult and neonatal thyroid glands are concentrated in a zone corresponding to the junctions of the upper to middle thirds of the lobes along a hypothetical central axis (167). They occupy an exclusively intrafollicular position both in neonates and adults (Fig. 46.13). The extreme upper and lower poles, as well as the isthmus, are devoid of C-cells. C-cells are larger than follicular cells and have pale cytoplasm, but, can be difficult to identify without immunohistochemical stains (calcitonin, INSM1, chromogranin and synaptophysin). They tend to be less numerous in adults than in neonates and appear flattened or spindle shaped. Typically, adult thyroid glands contain fewer than 50 C-cells per single low-power field, although occasional normal adult glands may have a higher density of these C-cells. Rarely, nodules of C-cells may be found in normal adult glands as discussed in Chapter 43.

Although endocrine cells are found commonly in thymus glands of many animal species, they are sparse in human thymic tissue. Endocrine cells may be found within the perivascular connective tissue and in association with Hassall corpuscles (168) (Fig. 46.14). Thymic cells may regulate

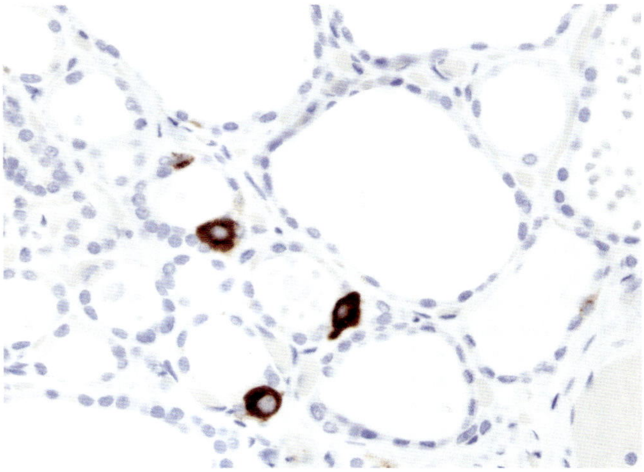

FIGURE 46.13 Adult thyroid immunostained for chromogranin with hematoxylin as the counterstain. C-cells are present within the follicle as closed-type endocrine cells.

early T-cell differentiation by the transcription of neuroendocrine genes in the stromal network and expression of cognitive receptors by immature T-cells (169).

Skin

Merkel cells (MCs) represent the neuroendocrine components of the skin (Figs. 46.4B and 46.15). These cells occur singly and in small clusters throughout the epidermis, particularly in the basal layer. The clusters are prominent in foci of specialized epithelial differentiation, such as the touch domes (170,171). Merkel cells are also present in the matrix of the proximal nail fold (172). Merkel cells are innervated by long type I myelinated fibers (Fig. 46.4B). Secretory granules are abundant and range from 80 to 130 nm in diameter. Aggregates of intermediate filament proteins are predominantly of the cytokeratin 20 (Fig. 46.15C) and 8/18 types, but neurofilament proteins, predominantly of low to intermediate molecular weights, may also be present (173). The most frequently encountered hormones include met-enkephalin, vasoactive intestinal peptide, and gastrin-releasing peptide (174,175). Interestingly, normal Merkel cells and Merkel cell carcinomas have been shown to express the transcription factor ISL-1 by immunohistochemistry (176,177).

Alimentary System

The tubular gastrointestinal tract, from the stomach to the anal canal, is richly populated by a heterogeneous collection of peptide hormone and amine-producing endocrine cells that are also referred to as enteroendocrine cells (45–47). The gut endocrine cells are responsible for the production of a wide array of different hormones (Figs. 46.16 and 46.17). In the pancreas, noninsular endocrine cells are dispersed within pancreatic ducts (Fig. 46.18) and acini, as well as within the islets (178). Endocrine cells are also present in the mucous glands of the neck of the gallbladder, extrahepatic biliary ducts, and duodenal papillae (179). Although

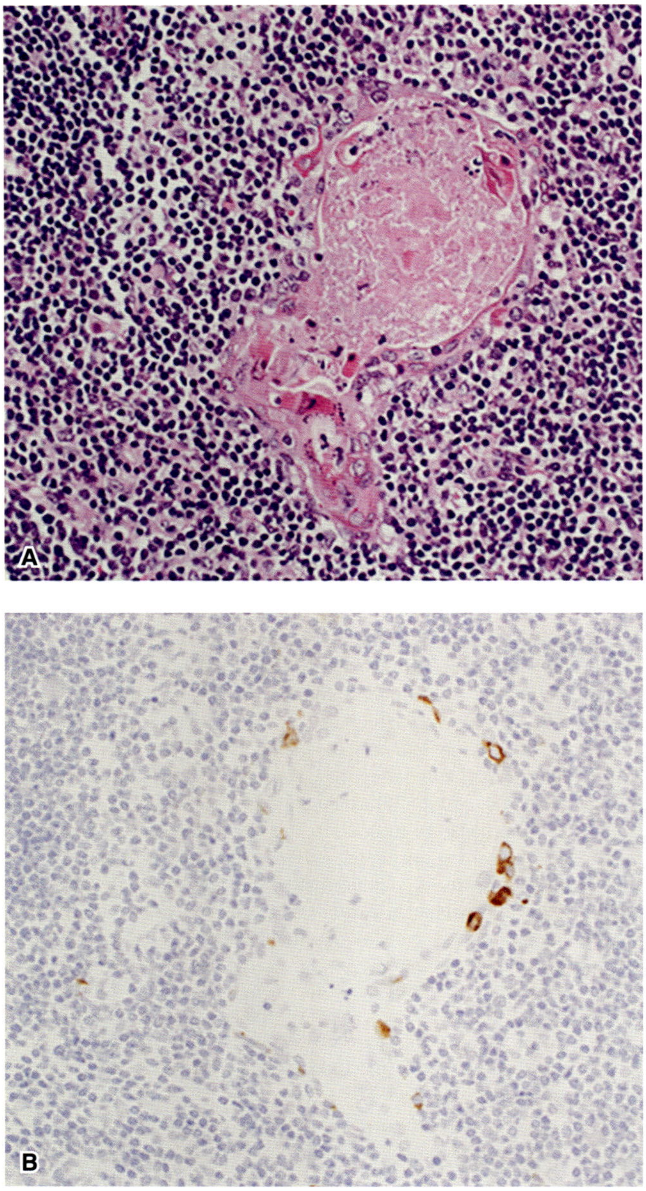

FIGURE 46.14 Thymic neuroendocrine (neuroendocrine) cells. **A:** On the hematoxylin and eosin section distinct neuroendocrine cells are not readily identified as associated with this Hassall corpuscle. **B:** Synaptophysin immunohistochemical stain demonstrates multiple neuroendocrine cells.

endocrine cells have not been described in the normal biliary epithelium of the liver or associated with the liver lobules, proliferating cholangiocytes in the diseased liver have been reported to acquire the phenotype of endocrine cells in the disease liver (180). More than 30 hormone genes are expressed in the alimentary system, making it the largest hormone producing organ in the body (74,181).

Intestinal endocrine cells originate from stem cells at the crypt bases or in the deep glands. Differentiation of the migrating pluripotent stem cells into the endocrine, absorptive goblet, and Paneth lineages occurs in the transitional zone of the crypt or gland (182). Endocrine cells of the duodenum

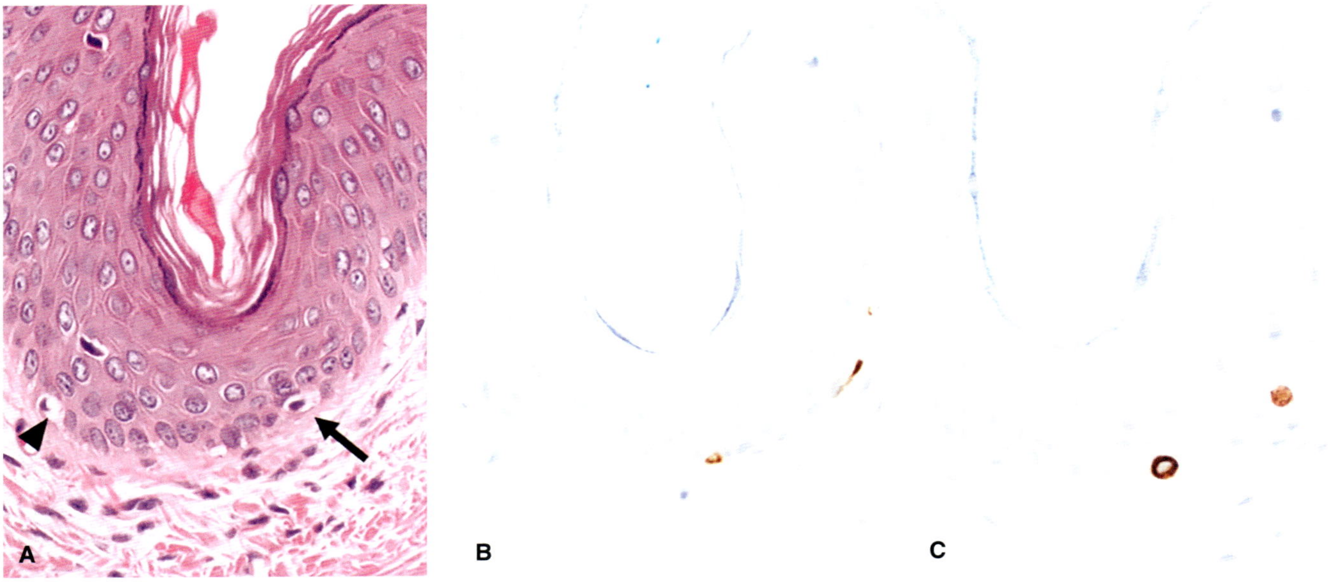

FIGURE 46.15 **A.** On H&E section the Merkel cell (*arrow*) and melanocyte (*arrowhead*) have clear cytoplasm. The Merkel cells are highlighted by synaptophysin (**B**) and cytokeratin 20 (**C**) immunohistochemical stains. Note the oval and triangular appearance of the Merkel cells.

and jejunum have a lifespan of 3 to 20 days, while longer lifetimes (up to 60 days) have been reported for ileal and colonic endocrine cells (64).

There is considerable variation in the distribution of peptide hormones and amines throughout different segments of the gastrointestinal tract. For example, gastrin predominates in the antrum, with smaller amounts present in the cells of the upper duodenum. Somatostatin-producing cells are present throughout the gastrointestinal tract, but the highest concentrations are in the stomach. Pancreatic polypeptide (PP) is produced by endocrine cells in the distal colon and rectum, in addition to the pancreatic islets. Serotonin or 5-hydroxytryptamine is produced by cells throughout the gastrointestinal tract.

Individual endocrine cells are generally classified according to their major secretory product(s) and by letter designations. In the stomach, the enterochromaffin cells produce serotonin, while the enterochromaffin-like cells produce histamine, and other populations secrete somatostatin (D), gastrin (G), and ghrelin (P/D1). Large numbers of duodenal endocrine cells produce glucose-dependent insulinotropic polypeptide, commonly known as gastric inhibitory polypeptide (K-cells), cholecystokinin (I-cells), and secretin (S-cells),

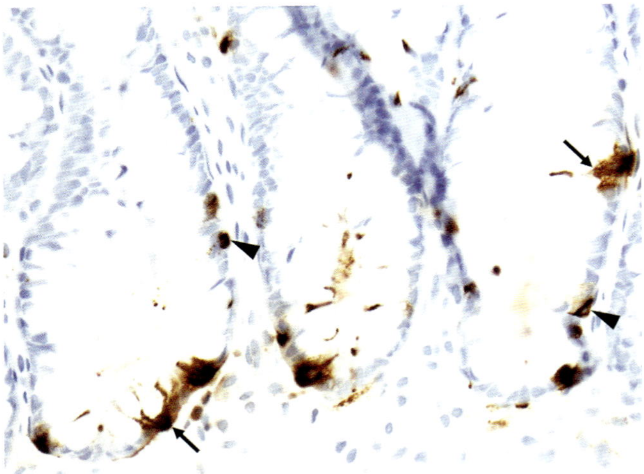

FIGURE 46.16 Colonic mucosa immunostained for chromogranin A. Open type neuroendocrine (neuroendocrine) cells (*arrows*) and closed type neuroendocrine cells (*arrowheads*) are present.

FIGURE 46.17 Gastric antrum immunostained for gastrin. The G-cells are primarily in the lower third of the gastric glands. Both open-type (*arrows*) and closed-type (*arrowhead*) G-cells are present. The G-cell on the left (*arrow*) is also an open type as seen by the triangular shape, but the full length of the cytoplasm is not demonstrated on this image.

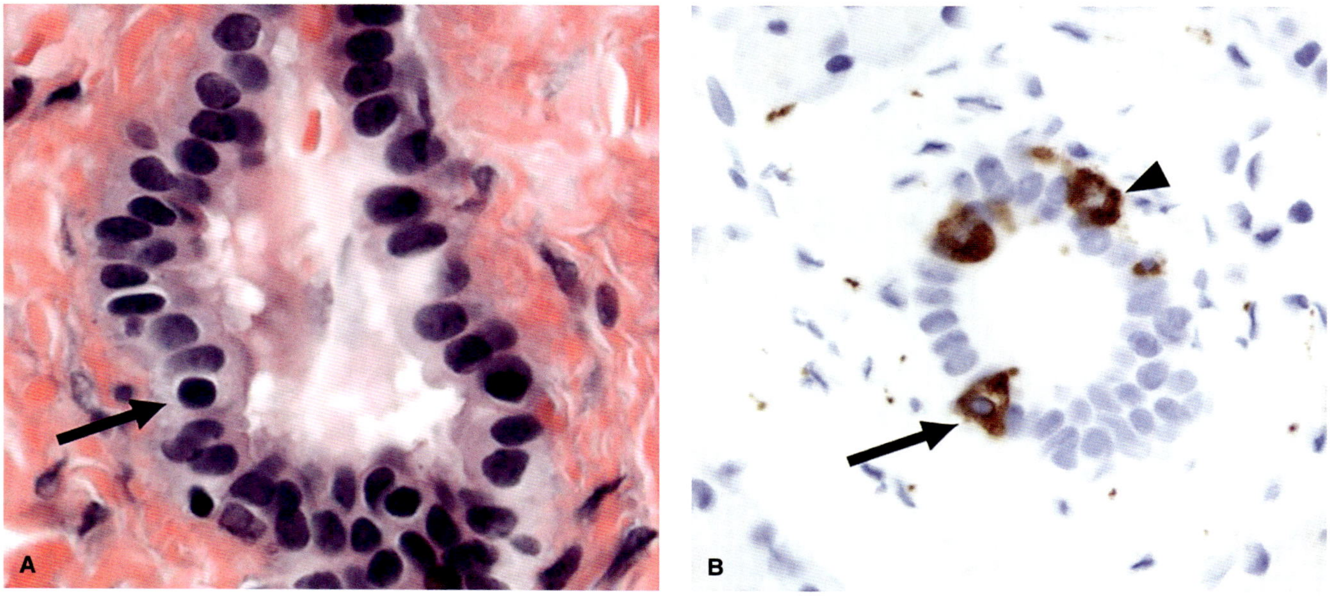

FIGURE 46.18 Neuroendocrine cells of pancreatic duct. **A:** H&E section of pancreatic duct with neuroendocrine clear cell (*arrow*). **B:** Immunostain for synaptophysin demonstrates open-type (*arrow*) and closed-type neuroendocrine cells in the epithelium (*arrowhead*).

while varying numbers of P/D1, D-, and M- (motilin) producing cells are also present. More distally, there are increasing numbers of neurotensin (N)-cells and L-cells that produce GLP-1 and GLP-2, and peptide YY (PYY).

Peptide hormones are also present within cells of submucosal glands. Brunner glands contain endocrine cells positive for somatostatin, gastrin–cholecystokinin, and PYY. Peptidergic nerve structures containing vasoactive intestinal peptide, peptide histidine methionine, substance P, neuropeptide Y, and gastrin-releasing peptide also have been identified around Brunner glands in nerve cell bodies of the submucosal ganglia (47). These findings reflect the fact that multiple peptides are involved in the control of secretion from these glands (182,183). The major peptides produced by pancreatic islet cells are discussed in detail in Chapter 29 and summarized in the Table 46.2.

Urogenital System

Although argyrophil cells are not present in the adult renal parenchyma, rare argyrophilic cells have been reported in the renal pelvis. These cells may be particularly prominent in areas of glandular metaplasia. Endocrine cells of the urothelium are predominantly of the closed type (184). These cells are positive for serotonin but do not contain peptide hormones such as ACTH, gastrin, glucagon, or somatostatin.

The endocrine cells of the prostate are present in all zones, predominantly the transition zone, and include both opened and closed types (185–188). Endocrine cells tend to be more prominent in normal or atrophic prostate than in hyperplastic foci. Most contain serotonin, and some contain somatostatin, calcitonin, or gastrin-releasing peptide.

Endocrine cells of both opened and closed types are present in the endocervical glandular epithelium and in the exocervical squamous epithelium but they are rare (189). Endocrine cells have also been described in the minor vestibular glands of the vulva (190,191).

TABLE 46.2 Main Neuroendocrine Cells and Their Products in the Alimentary System

Cell Type	Product	Anatomic Site
D	Somatostatin	Stomach antrum > corpus > duodenum
EC	Serotonin	Stomach, small intestine, colon, bile ducts, pancreatic ducts
ECL	Histamine	Stomach corpus
G	Gastrin	Stomach antrum > duodenum
I	Cholecystokinin	Small intestine
K	GIP	Duodenum, jejunum
L	GLP-2, GLP-2 Peptide YY	Ileum, colon
Mo	Motilin	Small intestine
N	Neurotensin	Ileum
P/D1	Ghrelin	Stomach corpus
PP	Pancreatic polypeptide	Distal colon/rectum
S	Secretin	Small intestine
Alpha	Glucagon	Pancreas
Beta	Insulin	Pancreas
Gamma/ PP	Pancreatic polypeptide	Pancreas
Delta	Somatostatin	Pancreas
Epsilon	Ghrelin	Pancreas (fetal>>adult)

EC, enterochromaffin; ECL, enterochromaffin-like cells; GIP, Glucose-dependent insulinotropic polypeptide (commonly known as gastric inhibitory polypeptide); GLP, glucagon-like peptide.

Other Systems

In the head and neck, individual and clusters of Merkel cells can be found in the basal layer of the oral epithelium in close contact with the tongue (192,193).

SPECIAL PROCEDURES

Molecular methodologies including in situ hybridization provide important approaches for analyzing the distribution and function of neuroendocrine cells (194,195). In situ hybridization techniques permit the identification of cells on the content of specific mRNAs. Extensive posttranslational processing and intracellular degradation of peptide products also may lead to positive hybridization signals with negative immunohistochemical reactions for the corresponding peptides. Additionally, in situ hybridization methods are of particular value for demonstrating hormone receptor mRNAs in target cells and for distinguishing de novo synthesis from uptake of hormonal peptides (194–196). The combination of in situ hybridization and immunohistochemistry has the potential for providing the maximal amount of information on the highly dynamic processes of gene transcription and translation.

In situ hybridization can combined with polymerase chain reaction (PCR) methods for demonstration of low copy number DNAs and RNAs (197,198). Detection of intracellular PCR products may be achieved indirectly by in situ hybridization using PCR product-specific probes (indirect in situ PCR) or without in situ hybridization through direct incorporation of labeled nucleotides into the PCR amplificants (direct in situ PCR). Although most protocols are designed for the demonstration of DNA, low-copy RNA sequences have been demonstrated by the addition of a reverse transcriptase (RT) step to generate cDNA from RNA templates before in situ PCR. This technique, which has been called in situ RT-PCR, may be of particular value when there are fewer than 20 copies of mRNA per cell. This technique is of great potential value for the identification of cells with low levels of mRNA-encoding hormones, hormone receptors, cytokines, growth factors, and growth factor receptors. The technical details and potential pitfalls of these methods are discussed in detail in several publications (197,199–201).

Recent advancement of single-cell RNA sequencing and spatial transcriptomics enabled mapping of the transcripts of individual cells within their tissue context (202). Based on these technologies, the Human Cell Atlas (HCA) project has emerged as a comprehensive cell reference map of human fetal and adult healthy and diseased tissues (203). Neuroendocrine cells have been profiled among millions of other cells in human lung, gut, and many other organs and systems (204).

ARTIFACTS

Since neuroendocrine cells often have a clear appearance, they must be distinguished from a variety of other cell types that also may appear clear in H&E-stained, formalin-fixed, paraffin-embedded sections. Cytoplasmic clearing may result from intracellular accumulations of lipids or glycogen; alternatively, this change may represent a shrinkage artefact. Clear cells in the intestinal epithelium may represent lymphocytes or epithelial cells with retraction of the cytoplasm from the nucleus. Neuroendocrine cells can be distinguished conclusively from other clear cells by the presence of neuroendocrine markers, including chromogranins or synaptophysin.

Significant artifacts may be associated with the use of immunohistochemical procedures for the demonstration of peptide hormones and nonhormonal markers. Appropriate positive and negative controls must be used in conjunction with these procedures (201). Nonspecific binding of immunoglobulins to secretory granules also may result from ionic interactions that may be suppressed to some extent by the use of buffers containing high concentrations of salt (205). Endogenous biotin-like activity may also contribute to nonspecific staining, particularly following microwave-induced antigen retrieval (206,207). This problem can be circumvented by the use of biotin blocking steps or by the use of biotin-free polymer-based detection systems. Diffusion of antigens from neuroendocrine cells to adjacent cells may also occur (Fig. 46.19), particularly in specimens that have not been fixed promptly.

Artifacts also may occur in in situ hybridization procedures. For example, Pagani et al. have demonstrated that oligonucleotides used for in situ hybridization procedures

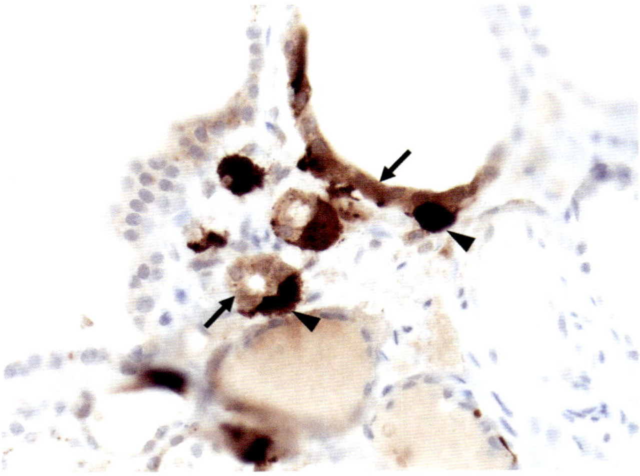

FIGURE 46.19 Immunostain for calcitonin highlighting C-cells. There is artefactual diffusion of the antibody into the follicular cells (*arrow*) and colloid which can give rise to problems with interpretation. The C-cells are the darker staining cells (*arrowheads*).

bind to neuroendocrine cells as a result of the presence of endogenous NH_2 groups (208). This type of nonspecific interaction can be blocked effectively by treating the sections with acetic anhydride. Controls for standard in situ hybridization and PCR-based *in situ* hybridization are discussed in detail in several reviews (197–200).

DIFFERENTIAL DIAGNOSIS

The differential diagnosis of various neuroendocrine cell populations is discussed in the chapters dealing with the specific organ systems in this volume.

SPECIMEN HANDLING

Most histochemical and immunohistochemical procedures for the demonstration of hormones and nonhormonal constituents of neuroendocrine cells can be performed in formalin-fixed and paraffin-embedded tissues. Other fixatives, including carbodiimide, acrolein, and diethyl pyrocarbonate, also have been used in place of formalin, and these fixatives have been reported to achieve optimal fixation of low concentrations of regulatory peptides such as those occurring in peptidergic nerve fibers (209–212).

The tissue preparative techniques for in situ hybridization studies are discussed in several review articles (194,200). In general, these methods may be performed on frozen samples that are postfixed in paraformaldehyde or in formalin-fixed samples that have been embedded in paraffin.

REFERENCES

1. Wiedenmann B, John M, Ahnert-Hilger G, Riecken E-O. Molecular and cell biological aspects of neuroendocrine tumors of the gastroenteropancreatic system. *J Mol Med.* 1998;76:637–647.
2. Heidenhain R. Untersuchangen über den Bau der Labüdusen. *Arch Mikrosk Anat Entwicklungsmech.* 1870;6:368–406.
3. Modlin IM, Champaneria MC, Bornschein J, Kidd M. Evolution of the diffuse neuroendocrine system–clear cells and cloudy origins. *Neuroendocrinol.* 2006;84:69–82.
4. Kulchitsky N. Zur Frage über den Bau der Darmkanals. *Archiv für mikroskopische Anatomie.* 1897;49:7–35.
5. Ciaccio M. Sur une nouvelle espece cellulaire dans les glandes de Lieber. *CR Seances Soc Biol Fil (Paris).* 1906;60:76–77.
6. Gosset A, Masson P. Tumeurs endocrines de l'appendice. *Press Med.* 1914;25:237–240.
7. Hamperl H. Was sind argentaffine Zellen? *Virchows Arch Path Anat.* 1932;286:811–833.
8. Feyrter F. *Uber Diffuse Endokrine Epithaliale Organe.* Barth; 1938.
9. Feyrter F. *Uber Die Periphheren Endokrine (Parakrine) Drusen Des Menshen.* Maudrich; 1954.
10. Frolich F. Die "Helle Zelle" der bronchialschleinhaut und ihre beziehungen zum problem der chemoreceptoren. *Frankfurter Z Pathol.* 1949;60:517–558.
11. Scharrer B. The neurosecretory neuron in neuroendocrine regulatory mechanisms. *Am Zool.* 1967;7:161–169.
12. Fujita T, Kobayashi S. Current reviews on the paraneuron concept. *Trends Neurosci.* 1979;2:27–30.
13. Dockray GJ. Evolutionary relationships of the gut hormones. *Fed Proc.* 1979;38:2295–2301.
14. Snyder SH. Brain peptides as neurotransmitters. *Science.* 1980;209:976–983.
15. Copp DH, Cameron EC, Cheney BA, Davidson AG, Henze KG. Evidence for calcitonin–a new hormone from the parathyroid that lowers blood calcium. *Endocrinology.* 1962;70:638–649.
16. Pearse AG. The cytochemistry of the thyroid C-cells and their relationship to calcitonin. *Proc R Soc Lond [Biol].* 1966;164:478–487.
17. Pearse AG. The cytochemistry and ultrastructure of polypeptide hormone producing cells of the APUD series and the embryologic, physiologic and pathologic implications of the concept. *J Histochem Cytochem.* 1969;17:303–313.
18. Pearse AG. 5-Hydroxytryptophan uptake by the dog thyroid C-cells and its possible significance in polypeptide hormone production. *Nature.* 1966;211:598–600.
19. Pearse AG. Common cytochemical properties of cells producing polypeptide hormones with particular reference to calcitonin and the thyroid C-cells. *Vet Rec.* 1966;79:587–590.
20. Bussolati G, Pearse AG. Immunofluorescent localization of calcitonin in the C-cells of the pig and dog thyroid. *J Endocrinol.* 1967;37:205–209.
21. Pearse AG. The diffuse neuroendocrine system and the APUD concept: Related "endocrine" peptides in brain, intestine, pituitary, placenta and anuran cutaneous glands. *Med Biol.* 1977;55:115–125.
22. Fujita T. Present status of the paraneuron concept. *Arch Cytol Histol.* 1989;52(suppl):1–8.
23. Pearse AG, Takor T. Embryology of the diffuse neuroendocrine system and its relationship to the common peptides. *Fed Proc.* 1979;38(9):2288–2294.
24. LeDouarin N, Teillet MA. The migration of neural crest cells to the wall of the digestive tract in the avian embryo. *J Embryol Exp Morphol.* 1973;30:31–48.
25. LeDouarin N. *The Neural Crest.* Cambridge University Press; 1982.
26. Thompson M, Fleming KA, Evans DJ, Fundele R, Surani MA, Wright NA. Gastric endocrine cells share a clonal origin with other gut cell lineage. *Development.* 1990;110:477–481.
27. Day R, Salzet M. The neuroendocrine phenotype, cellular plasticity, and the search for genetic switches; redefining the diffuse neuroendocrine system. *Neuor Endocrinol Lett.* 2002;23:447–451.
28. Gonzalez-Rey E, Ganea D, Delgado M. Neuropeptides: Keeping the balance between pathogen immunity and immune tolerance. *Curr Opin Pharmacol.* 2010;10(4):473–481.
29. Johansson E, Andersson L, Ornos J, et al. Revising the embryonic origin of thyroid C-cells in mice and humans. *Development.* 2015;142:3519–3528.

30. Nilsson M, Williams D. On the origin of cells and derivation of thyroid cancer: C-cell story revisited. *Eur Thyroid J.* 2016;5:79–93.
31. Kameda Y. Cellular and molecular events on the development of mammalian thyroid C-cells. *Develop Dyn.* 2016; 245:323–341.
32. Kameda Y. Morphological and molecular evolution of the ultimobranchial gland of nonmammalian vertebrates, with special reference to the chicken C-cells. *Dev Dyn.* 2017;246(10):719–739.
33. Johnson JE, Birren SJ, Anderson DJ. Two rat homologues of Drosophila achaete–scute specifically expressed in neuronal precursors. *Nature.* 1990;346:858–861.
34. Cook M, Yu X-M, Chen H. Notch in the development of thyroid C-cells and the treatment of medullary thyroid carcinoma. *Am J Transl Res.* 2010;2:119–125.
35. Ball DW, Achaete-scute *h*omolog-1 and Notch in lung neuroendocrine development and cancer. *Cancer Lett.* 2004; 204:159–169.
36. Ito T, Udaka N, Yazawa T, et al. Basic helix-loop-helix transcription factors regulate the neuroendocrine differentiation in fetal mouse pulmonary epithelium. *Development.* 2000;127:3913–3921.
37. Linniola RI. Functional facets of the pulmonary neuroendocrine system. *Lab Invest.* 2006;86:425–444.
38. Shue YT, Drainas AP, Li NY, Pearsall SM, et al. A conserved YAP/Notch/REST network controls the neuroendocrine cell fate in the lungs. *Nat Commun.* 2022;13(1):2690.
39. Lanigan TM, DeRaad SK, Russo AF. Requirement of the MASH-1 transcription factor for neuroendocrine differentiation of thyroid C-cells. *J Neurobiol.* 1998;34(2): 126–134.
40. Eenjes E, Benthem F, Boerema-de Munck A, et al. Distinct roles for SOX2 and SOX21 in differentiation, distribution and maturation of pulmonary neuroendocrine cells. *Cell Mol Life Sci.* 2023;80(3):79.
41. Ito T, Kudoh S, Fujino K, et al. Pulmonary neuroendocrine cells and small cell lung carcinoma: Immunohistochemical study focusing on mechanisms of neuroendocrine differentiation. *Acta Histochem Cytochem.* 2022;55(3):75–83.
42. Nikolakopoulou P, Poser SW, Masjkur J, et al. STAST3-Ser/Hes 3 signaling: A New molecular component of the neuroendocrine system? *Horm Metab Res.* 2016;48:77–82.
43. Larsson LI, Goltermann N, De Magistris L, Rehfeld JF, Schwartz TW. Somatostatin cell processes as pathways for paracrine secretion. *Science.* 1979;205(4413):1393–1395.
44. Fujita T, Kobayashi S. The cells and hormones of the GEP endocrine system. In: Fujita T, ed. *Gastroenteropancreatic Cell System*. Igaku-Shoin; 1973:1–16.
45. Dayal Y. Endocrine cells of the gut and their neoplasms. In: Norris HT, ed. *Pathology of the Colon, Small Intestine and Anus*. Churchill Livingstone; 1983:267–300.
46. Lechago J. The endocrine cells of the digestive and respiratory systems and their pathology. In: Bloodworth JMB Jr, ed. *Endocrine Pathology General and Surgical*. 2nd ed. Williams & Wilkins; 1982:513–555.
47. Bosshard A, Chery-Croze S, Cuber JC. Immunohistochemical study of peptidergic structures in Brunner's glands. *Gastroenterology.* 1989;97:1382–1388.
48. Aubock L, Ratzenhofer M. "Extraepithelial enterochromaffin cell complexes" in the normal human appendix and neurogenic appendicopathy. *J Pathol.* 1982;136:217–226.
49. Cohen RJ, Glezerson G, Taylor LF, Grundle HAJ, Naude JH. The neuroendocrine cell population of the human prostate gland. *J Urol.* 1993;150:365–368.
50. Falkmer S. Phylogeny and ontogeny of the neuroendocrine cells of the gastrointestinal tract. *Endocrinol Metab Clin North Am.* 1993;22:731–752.
51. Falck B, Owman CA. A detailed methodological description of the fluorescence method for the cellular distribution of biogenic monoamines. *Acta Univ Lund.* 1965;7:5–23.
52. Grimelius L. A silver nitrate stain for A2 cells of human pancreatic islets. *Acta Soc Med Upsal.* 1968;73(5–6):243–270.
53. Grimelius L, Wilander E. Silver stains in the study of endocrine cells of the gut and pancreas. *Invest Cell Pathol.* 1980;3(1):3–12.
54. Aguirre P, Scully RE, Wolfe HJ, DeLellis RA. Endometrial carcinomas with argyrophil cells. A histochemical and immunohistochemical study. *Hum Pathol.* 1984;15(3):210–217.
55. Cecilia M, Rost M, Rost FW. An improved method for staining cells of the endocrine polypeptide (APUD) series by masked metachromasia: Application of the principle of 'fixation by excluded volume'. *Histochem J.* 1976;8(1): 93–98.
56. Rode J, Dhillon AP, Papadaki L. Serotonin immunoreactive cells in the lamina propria plexus of the appendix. *Hum Pathol.* 1983;14:464–469.
57. Gould VE, DeLellis RA. The neuroendocrine cell system: its tumors, hyperplasias and dysplasias. In: Silverberg S, ed. *Principles and Practice of Surgical Pathology*. Wiley; 1983.
58. Roth J, Kasper M, Stamm B, et al. Localization of proinsulin and insulin in human insulinoma: preliminary immunohistochemical results. *Virchows Arch B Cell Pathol Incl Mol Pathol.* 1989;56:287–292.
59. Hakanson R, Sundler F. The design of the neuroendocrine system: A unifying concept and its consequences. *Trends Pharmacol Sci.* 1983;4:41–44.
60. Tischler AS. The dispersed neuroendocrine cells: The structure, function, regulation and effects of xenobiotics on this system. *Toxicol Pathol.* 1989;17:307–316.
61. Pearse AG, Polak JM, Bloom SR. The newer gut hormones. Cellular sources, physiology, pathology and clinical aspects. *Gastroenterology.* 1977;72:746–761.
62. Reichlin S. Somatostatin (part 1). *N Engl J Med.* 1983;309: 1495–1501.
63. Reichlin S. Somatostatin (part 2). *N Engl J Med.* 1983; 309:1556–1563.
64. Gribble FM, Reimann F. Enteroendocrine cells: Chemosensors in the intestinal epithelium. *Ann Rev Physio.* 2016;78:277–299.
65. Kameda Y, Oyama H, Endoh M, Horino M. Somatostatin immunoreactive C-cells in thyroid glands from various mammalian species. *Anat Rec.* 1982;204:161–170.
66. Tsutsumi Y, Osamura Y, Watanabe K, Yanaihara N. Immunohistochemical localization of gastrin releasing peptide and adrenocorticotropin releasing cells in the human lung. *Lab Invest.* 1983;48:623–632.

67. Sunday ME, Kaplan LM, Motoyama E, et al. Gastrin releasing peptide (mammalian bombesin) gene expression in health and disease. *Lab Invest.* 1988;59:5–24.
68. Kruger DT. Pituitary ACTH hyperfunction: pathophysiology and clinical aspects. In: Commani F, Mueller EE, eds. *Pituitary Hyperfunction: Pathophysiology and Clinical Aspects.* Raven Press; 1984:221–234.
69. Rosenfeld MG, Mermod JJ, Amara SJ, et al. Production of a novel neuropeptide encoded by the calcitonin gene via tissue specific RNA processing. *Nature.* 1983;304:129–135.
70. Warren TG, Shields D. Cell free biosynthesis of somatostatin precursors: Evidence for multiple forms of preprosomatostatin. *Proc Natl Acad Sci U S A.* 1982;79:3729–3733.
71. Bouillon R, Drucker DJ, Ferrennini E, Grinspoon S, Rosen CJ, Zimmet P. The past 10 years-new hormones, new functions, new endocrine organs. *Nat Rev Endocrinology.* 2015;11:681–686.
72. Zhu N, Dong F, Shi G, Lao X, Zheng H. HORDB a comprehensive database of peptide hormones. *Sci Data.* 2022; 9(1):187.
73. Kaur D, Arora A, Patiyal S, Raghava GPS. Hmrbase2: A comprehensive database of hormones and their receptors. *Hormones (Athens).* 2023;22(3):359–366.
74. Rehfeld JF. Gastrointestinal hormones and their targets. *Adv Exp Med Biol.* 2014;817:157–175.
75. Polak JM, Bloom SR. Immunocytochemistry of regulatory peptides. In: Polak JM, Vaan Noordern S, eds. *Immunocytochemistry. Practical Applications in pathology and Biology.* Wright PSG; 1983:184–211.
76. Verhofstad AAJ, Steinbusch HWM, Joosten HWJ, et al. Immunocytochemical localization of noradrenaline, adrenaline and serotonin. In: Polak JM, Van Noorden S, eds. *Immunocytochemistry. Practical Applications in Pathology and Medicine.* Wright PSG; 1983:143–168.
77. Lloyd RV. Immunohistochemical localization of catecholamines, catecholamine synthesizing enzymes and chromogranins in neuroendocrine cells and tumors. In: DeLellis RA, ed. *Advances in Immunohistochemistry.* Raven Press; 1988:317–340.
78. Schmechel D, Marangos PJ, Brightman M. Neuron specific enolase is a molecular marker for peripheral and central neuroendocrine cells. *Nature.* 1978;17:834–836.
79. Thompson RJ, Doran JF, Jackson P, Dhillon AP, Rode J. PGP 9.5—a new marker for vertebrate neurons and neuroendocrine cells. *Brain Res.* 1983;278:224–228.
80. Li GL, Fardoque M, Holtz A. Olsson Y. Expression of the ubiquitin carboxylterminal hydrolase PGP 9.5 in axons following spinal cord compression trauma. *APMIS.* 1997; 105:384–390.
81. Campbell LK, Thomas JR, Lamps LW, Smoller BR, Folpe AL. Protein gene product 9.5 (PGP 9.5) is not a specific marker of neural and nerve sheath tumors: An immunohistochemical study of 95 mesenchymal neoplasms. *Mod Pathol.* 2003;16:963–969.
82. Blaschko H, Comline RS, Schneider FH, Silver M, Smith AD. Secretion of a chromaffin protein, chromogranin, from the adrenal medulla after splanchnic nerve stimulation. *Nature.* 1967;215(5096):58–59.
83. Schober M, Fischer-Colbrie R, Schmidt KW, Bussetati G, O'Connor DI, Winkler H. Comparison of chromogranins A, B and secretogranin II in human adrenal medulla and pheochromocytoma. *Lab Invest.* 1987;57:385–391.
84. Troger J, Theurl M, Kirchmair R, et al. Granin-derived peptides. *Prog Neurobiol.* 2017;154:37–61.
85. Lloyd RV, Jin L, Kulig E, Fields K. Molecular approaches for the analysis of chromogranins and secretogranins. *Diagn Mol Pathol.* 1992;1:2–15.
86. Borges R, Díaz-Vera J, Dominguez N, Arnau MR, Machado JD. Chromogranins as regulators of exocytosis. *J of Neurochem.* 2010;114:335–343.
87. Portela-Gomes G, Stridsberg M. Selective processing of chromogranin A in the different islet cells in human pancreas. *J Histochem Cytochem.* 2001;494:483–490.
88. Portela-Gomes GM, Hacker GW, Weitgassser R. Neuroendocrine cell makers for pancreatic islets and tumors. *Appl Immunohistochem Mol Morphol.* 2004;12:183–192.
89. Helle KB, Metz-Boutigue M-H, Cerra MC, Angelone T. Chromogranins: From discovery to current times. *Pflugers Arch.* 2018;470:143–154.
90. Corti A, Marcucci F, Bachetti T. Circulating chromogranin A and its fragments as diagnostic and prognostic disease markers. *Pflugers Arch.* 2018;470:199–210.
91. Kimura N, Funakoshi A, Aunis D, Tateishi K, Miura W, Nagura H. Immunohistochemical localization of chromostatin and pancreastatin, chromogranin A derived bioactive peptides, in normal and neoplastic neuroendocrine tissues. *Endocrine Pathol.* 1995;6:35–44.
92. Schmidt WE, Siegel EG, Lamberts E, Gallwitz B, Creutzfeldt W. Pancreastatin: molecular and immunocytochemical characterization of a novel peptide in porcine and human tissues. *Endocrinology.* 1988;123:1395–1404.
93. Bishop AE, Sekiya K, Salahuddin MJ, et al. The distribution of GAWK-like immunoreactivity in neuroendocrine cells of the human gut, pancreas, adrenal and pituitary glands and its colocalization with chromogranin B. *Histochemistry.* 1989;90:475–483.
94. Huttner WB, Gerdes H-H, Rosa P. Chromogranins/secretogranins—widespread constituents of the secretory granule matrix in endocrine cells and neurons. In: Gratzl M, Langley K, eds. *Markers for Neural and Endocrine Cells. Molecular and Cell Biology, Diagnostic Applications.* VCH; 1991:93–131.
95. Lloyd RV, Jin L, Qian X, Scheithauer BW, Young WF Jr, Davis DH. Analysis of the chromogranin A post-translational cleavage product pancreastatin and the prohormone convertases PC2 and PC3 in normal and neoplastic human pituitaries. *Am J Pathol.* 1995;146:1188–1198.
96. Scopsi L, Gullo M, Rilke F, Martin S, Steiner DF. Proprotein convertases (PC1/PC3 and PC2) in normal and neoplastic tissues: Their use as markers of neuroendocrine differentiation. *J Clin Endocrinol Metab.* 1995;80:294–301.
97. Morash MG, MacDonald AB, Croll RP, Anini Y. Molecular cloning, ontogeny and tissue distribution of zebrafish (Danio rerio) prohormone convertases: pcsk1 and pcsk2. *Gen Comp Endocrinol.* 2009;162(2):179–187.
98. Scopsi L, Lee R, Gullo M, et al. Peptidylglycine alpha amidating monooxygenase in neuroendocrine tumors: Its identification, characterization, quantification and relation to the grade of morphological differentiation, amidated peptide content and granin immunocytochemistry. *Appl Immunohistochem.* 1998;6:120–132.

99. Saldise L, Martínez A, Montuenga LM, et al. Distribution of peptidyl-glycine alpha-amidating monooxygenase (PAM) enzymes in normal human lung and in lung epithelial tumors. *J Histochem Cytochem.* 1996;44:3–12.
100. Weidenmann B, Franke WW, Kuhn C, Moll R, Gould VE. Synaptophysin: A marker protein for neuroendocrine cells and neoplasms. *Proc Natl Acad Sci U S A.* 1986;83:3500–3504.
101. Gould VE, Lee I, Wiedenmann B, Moll R, Chejfec G, Franke WW. Synaptophysin: A novel marker for neurons, certain neuroendocrine cells and their neoplasms. *Hum Pathol.* 1986;17:979–983.
102. Jahn R, DeCamilli P. Membrane proteins of synaptic vesicles: markers for neurons and endocrine cells: tools for the study of neurosecretion. In: Gratzl M, Langley K, eds. *Markers for Neurons And Endocrine Cells. Molecular, Cell Biology and Diagnostic Applications.* VCH; 1991:25–91.
103. DeLellis RA. The neuroendocrine system and its tumors. An overview. *Am J Clin Pathol.* 2001;115(Suppl):S5–S16.
104. Lloyd RV. Practical markers used in the diagnosis of neuroendocrine tumors. *Endocrin Pathol.* 2003;14:293–301.
105. Portela-Gomes GM, Lukinius A, Grimelius L. Synaptic vesicle protein 2, a new neuroendocrine marker. *Am J Pathol.* 2000;157:1299–1309.
106. Nowack A, Yao J, Custer KL, Bajjakieh SM. SV2 regulates neurotransmitter release via multiple mechanisms. *Am J Physiol Cell Physiol.* 2010;299:C960–C967.
107. Dunn AR, Stout KA, Ozawa M, et al. Synaptic vesicle glycoprotein 2C (SV2C) modulates dopamine release and is disrupted in Parkinson disease. *Proc Natl Acad Sci U S A.* 2017;114:E2253–E2262.
108. Jakobsen AM, Andersson P, Saglik G, et al. Differential expression of vesicular monoamine transporter (VMAT) 1 and 2 in gastrointestinal endocrine tumors. *J Pathol.* 2001;195:463–471.
109. Rindi G, Paolotti D, Fiocca R, Wiedenmann B, Henry JP, Solcia E. Vesicular monoamine transporter 2 as a marker of gastric enterochromaffin-like cell tumors. *Virchows Arch.* 2000;436:217–223.
110. Wimalasena K. Vesicular monoamine transporters: Structure-function, pharmacology, and medicinal chemistry. *Med Res Rev.* 2011;31:483–519.
111. Schafer MKH, Weihe E, Eiden LE. Localization and expression of VMAT2 across mammalian species: A translational guide for its visualization and targeting in health and disease. *Adv Pharmacol.* 2013;68:319–334.
112. Anlauf M, Eissele R, Schäfer MK, et al. Expression of the two isoforms of the vesicular monoamine transporters (VMAT1 and VMAT2) in the endocrine pancreas and pancreatic endocrine tumors. *J Histochem Cytochem.* 2003;51:1027–1040.
113. Grönberg M, Amini RM, Stridsberg M, Janson ET, Saras J. Neuroendocrine markers are expressed in human mammary glands. *Regul Pept.* 2010;160:68–74.
114. Sollner T, Bennett MK, Whiteheart SW, Scheller RH, Rothman JE. A protein assembly–disassembly pathway in vitro that may correspond to sequential steps of synaptic vesicle docking, activation and fusion. *Cell.* 1993;75:409–418.
115. Elferink LA, Scheller RH. Synaptic vesicle proteins and regulated exocytosis. *J Cell Sci Suppl.* 1993;17:75–79.
116. Rizo J, Xu J. The synaptic vesicle release machinery. *Ann Rev Biophys.* 2015;44:339–367.
117. Wickner W, Rizo J. A cascade of multiple proteins and lipids catalyzes membrane fusion. *Mol Biol Cell.* 2017;28:707–711.
118. Regazzi R, Wolheim CB, Lang J, et al. VAMP2 and cellubrevin are expressed in pancreatic beta cells and are essential for Ca(2+) but not for GTP gamma S-induced insulin secretion. *EMBO J.* 1995;14:2723–2730.
119. Braun JE, Fritz BA, Wong SM, Lowe AW. Identification of a vesicle associated membrane protein (VAMP)-like membrane protein in zymogen granules of the rat exocrine pancreas. *J Biol Chem.* 1994;269:5328–5335.
120. Moghadam PK, Jackson MB. The functional significance of synaptotagmin diversity in neuroendocrine secretion. *Front Endocrinol (Laussane).* 2013;18:124.
121. Brown H, Meister B, Deeney J, et al. Synaptotagmin III isoform is compartmentalized in pancreatic beta cells and has a functional role in exocytosis. *Diabetes.* 2000;49:383–391.
122. Tahara S, Sanno N, Teramoto A, Osamura RY. Expression of Rab3, a Ras-related GTP-binding protein in human nontumorous pituitaries and pituitary adenomas. *Mod Pathol.* 1999;12:627–634.
123. Rotondo F, Scheithauer BW, Kovacs K, Bell DC. Rab3B immunoexpression in human pituitary adenomas. *Appl Immunohistochem Mol Morphol.* 2009;17:185–188.
124. Matsuno A, Mizutani A, Okinaga H, et al. Functional molecular morphology of anterior pituitary cells, from hormone production to intracellular transport and secretion. *Med Mol Morphol.* 2011;44:63–70.
125. Majó G, Ferrer I, Marsal J, Blasi J, Aguado F. Immunocytochemical analysis of the synaptic proteins SNAP-25 and Rab3A in human pituitary adenomas. Overexpression of SNAP-25 in the mammosomatotroph lineages. *J Pathol.* 1997;183:440–446.
126. Nishioka H, Haraoka J. Significance of immunohistochemical expression of Rab3B and SNAP-25 in growth hormone producing pituitary adenomas. *Acta Neuropathol.* 2005;109:598–602.
127. Roth D, Burgoyne D. SNAP-25 is present in a SNARE complex in adrenal chromaffin cells. *FEBS Lett.* 1994;351:207–210.
128. Sadoul K, Lang J, Montecucco C, et al. SNAP-25 is expressed in islets of Langerhans and is involved in insulin release. *J Cell Biol.* 1995;128:1019–1028.
129. Hocker M, John M, Anagnostopoulos J, et al Molecular dissection of regulated secretory pathways in human enterochromaffin-like cells; an immunohistochemical analysis. *Histochem Cell Biol.* 1999;112:205–214.
130. Heitz PU, Roth J, Zuber C, Komminoth P. Markers for neural and endocrine cells in pathology. In: Gratzl M, Langley K, eds. *Markers for Neural and Endocrine Cells. Molecular and Cell Biology, Diagnostic Applications.* VCH; 1991:203–215.
131. Jin L, Hemperly JJ, Lloyd RV. Expression of neural cell adhesion molecule in normal and neoplastic human neuroendocrine tissues. *Am J Pathol.* 1991;138:961–969.
132. Komminoth P, Roth J, Saremaslani P, Matias-Guiu X, Wolfe HJ, Heitz PU. Polysialic acid of the neural cell adhesion molecule in the human thyroid: A marker for medullary thyroid carcinoma and primary C cell hyperplasia. An immunohistochemical study on 79 thyroid lesions. *Am J Surg Pathol.* 1994;18:399–411.
133. Seeger RC, Danon YL, Rayner SA, Hoover F. Definition of thy-1 on human neuroblastoma, glioma, sarcoma and teratoma cells with a monoclonal antibody. *J Immunol.* 1982;128:983–989.

134. Lipinski M, Braham K, Cailland JM, Carlu C, Tursz T. HNK-1 antibody detects an antigen expressed on neuroectodermal cells. *J Exp Med.* 1983;158:1775–1780.
135. Tischler AS, Mobtaker H, Mann K, et al. Anti-lymphocyte antibody leu-7 (HNK-1) recognizes a constituent of neuroendocrine granule matrix. *J Histochem Cytochem.* 1986; 34:1213–1216.
136. Lloyd RV, Blaivas M, Wilson BS. Distribution of chromogranin and S100 protein in normal and abnormal adrenal medullary tissue. *Arch Pathol Lab Med.* 1985;109:633–635.
137. Jiang S-X, Kameya T, Asamura H, et al. hASH1 expression is closely correlated with endocrine phenotype and differentiation extent in pulmonary neuroendocrine tumors. *Mod Pathol.* 2004;17:222–229.
138. Altree-Tacha D, Tyrrell J, Li F. mASH1 is highly specific for neuroendocrine carcinomas: An immunohistochemical evaluation on normal and various neoplastic tissues. *Arch Pathol Lab Med.* 2017;141(2):288–292.
139. Nonaka D. A study of FoxA1 expression in thyroid tumors. *Hum Pathol.* 2017;65:217–224.
140. Rosenbaum JN, Guo Z, Baus RM, Werner H, Rehrauer WM, Lloyd RV. INSM1: A novel immunohistochemical and molecular marker for neuroendocrine and neuroepithelial neoplasms. *Am J Clin Pathol.* 2015;144:579–591.
141. Warneboldt J, Haller F, Horstmann O, et al Histone H1x is expressed in human neuroendocrine cells and tumours. *BMC Cancer.* 2008;8:388–407.
142. Graham RP, Shrestha B, Caron BL, et al. Islet-1 is a sensitive but not entirely specific marker for pancreatic neuroendocrine neoplasms and their metastases. *Am J Surg Pathol.* 2013;37:399–405.
143. La Rosa S, Rigoli E, Uccella S, Chiaravalli AM, Capella C. CDX2 as a marker of intestinal EC-cells and related well differentiated endocrine tumors. *Virchows Arch.* 2004; 445:248–254.
144. Hösgör M, Ijzendoorn Y, Mooi WJ, Tibboel D, De Krijger RR. Thyroid transcription factor-1 expression during normal human lung development and in patients with congenital diaphragmatic hernia. *J Pediatr Surg.* 2002;37:1258–1262.
145. La Rosa S, Chiaravalli AM, Placidi C, Papanikolaou N, Cerati M, Capella C. TTF1 expression in normal lung neuroendocrine cells and related tumors: immunohistochemical study comparing two different monoclonal antibodies. *Virchows Arch.* 2010;457:497–507.
146. Miskovic J, Brekalo Z, Vukojevic K, et al. Co-expression of TTF-1 and neuroendocrine markers in the human fetal lung and pulmonary neuroendocrine tumors. *Acta Hisochem.* 2015;117:451–459.
147. Robson E, He SJ, Eccles M, A PANorama of PAX genes in cancer and development. *Nat Rev Cancer.* 2006;6:52–62.
148. Ritz-Laser B, Estreicher A, Gauthier B, Philippe J. The paired homeodomain transcription factor Pax-2 is expressed in the endocrine pancreas and transactivates the glucagon gene promoter. *J Biol Chem.* 2000;275:32708–32715.
149. Liau JY, Tsai JH, Jeng YM, et al. The diagnostic utility of PAX8 for neuroendocrine tumors: An immunohistochemical reappraisal. *Appl Immunohistochem Mol Morphol.* 2016; 24:57–63.
150. Reubi JC. Somatostatin and other peptide receptors as tools for tumor diagnosis and treatment. *Neuroendocrinol.* 2004;80(Suppl 1):51–56.
151. Fani M, Nicolas GP, Wild D. Somatostatin receptor antagonists for imaging and therapy. *J Nucle Med.* 2017;58(Suppl 2): 61S–66S.
152. Gugger M, Wasser B, Kappeler A, Schonbrunn A, Reubi JC. Immunohistochemical localization of somatostatin receptor sst2A in human gut and lung tissue. Possible implications for physiology and carcinogenesis. *Ann N Y Acad Sci.* 2004;1014:132–136.
153. Unger N, Ueberberg B, Schulz S, Saeger W, Mann K, Petersenn S. Differential expression of somatostatin receptor subtype 1–5 proteins in numerous human normal tissues. *Exp Clin Endocrinol Diabetes.* 2012;120:482–489.
154. Bensch KG, Gordon GB, Miller LR. Studies on the bronchial counterpart of the Kultschitzky (argentaffin) cells and innervation of the bronchial glands. *J Ultrastruct Res.* 1965;12:668–686.
155. Noguchi M, Sumiyama K, Morimoto M. Directed migration of pulmonary neuroendocrine cells toward airway branches organizes the stereotypic location of neuroepithelial bodies. *Cell Rep.* 2015;13(12):2679–2686.
156. Branchfield K, Nantie L, Verheyden JM, Sui P, Wienhold MD, Sun X. Pulmonary neuroendocrine cells function as airway sensors to control lung immune response. *Science.* 2016;351(6274):707–710.
157. Cutz E. Neuroendocrine cells of the lung. *Exp Lung Res.* 1982;3:185–208.
158. Lauweryns JM, Peuskens JC. Neuroepithelial bodies (neuroreceptor or secretion organs?) in human infant bronchial and bronchiolar epithelium. *Anat Rec.* 1972;172:471–481.
159. Cutz E. Hyperplasia of pulmonary neuroendocrine cells in infancy and childhood. *Sem Diagn Pathol.* 2015;32:420–437.
160. Sui P, Wiesner DL, Xu J, Zhang Y, et al. Pulmonary neuroendocrine cells amplify allergic asthma responses. *Science.* 2018;360(6393):eaan8546.
161. He P, Lim K, Sun D, Pett JP, et al. A human fetal lung cell atlas uncovers proximal-distal gradients of differentiation and key regulators of epithelial fates. *Cell.* 2022;185(25): 4841–4860.
162. Sountoulidis A, Marco Salas S, Braun E, Avenel C, et al. A topographic atlas defines developmental origins of cell heterogeneity in the human embryonic lung. *Nat Cell Biol.* 2023;25(2):351–365.
163. Pesce C, Tobia-Gallelli F, Toncini C. APUD cells of the larynx. *Acta Otolaryngol.* 1984;98:158–162.
164. Torre-Rendon FE, Cisneros-Bernal E, Ochoa-Salas JA. Carcinoma indifferenciadio de cellular pequenas de la laringe. *Patologica.* 1979;17:47–57.
165. Chung JH, Lee SS, Shim YS, et al. A study of moderately differentiated neuroendocrine carcinomas of the larynx and an examination of non-neoplastic larynz tissue for neuroendocrine cells. *Laryngoscope.* 2004;114:1264–1270.
166. Triantafyllou A, Devaney KO, Hunt JL, Rinaldo A, Ferlito A. Structural biology of intraepithelial neuroendocrine cells in the larynx: Literature review. *Pathol Res Pract.* 2019; 215(1):1–4.
167. De Lellis RA, Wolfe HJ. The pathobiology of the human calcitonin (C)-cell. A review. *Pathol Annu.* 1981;16:25–52.
168. Bearman RM, Levine GD, Bensch KG. The ultrastructure of the normal human thymus. A study of 36 cases. *Anat Rec.* 1978;190:755–781.

169. Varga I, Mikusova R, Pospislova V, et al. Morphologic heterogeneity of human thymic non-lymphocytic cells. *Neuro Endocrine Lett*. 2009;30:275–283.
170. Gould VE, Moll R, Moll I, Lee I, Franke WW. Neuroendocrine (Merkel) cells of the skin: Hyperplasias, dysplasias and neoplasms. *Lab Invest*. 1985;52:334–353.
171. Munde PB, Khandekar S, Dive AM, Sharma A. Pathophysiology of the Merkel cell. *J Oral Maxillofac Pathol*. 2013;17:408–412.
172. Conejo-Mir J, Dominhuez-Cruz J, Sendin-Martin M. Nail. In: Mills SC, ed. *Histology for Pathologists*. Wolters Kluwer; 2020:43.
173. Kanik A. Skin. In: Mills SE, ed. *Histology for Pathologists*. Wolters Kluwer; 2020:9.
174. Eispert AC, Fuchs F, Brandner JM, Houdek P, Wladykowski E, Moll I. Evidence for distinct populations of Merkel cells. *Histochem Cell Biol*. 2009;132:83–93.
175. Tachibana T, Nawa T. Immunohistochemical reactions of receptors to met-enkephalin, VIP, substance P, and C gastrin-releasing peptide located on Merkel cells in the rat sinus hair follicle. *Arch Histol Cytol*. 2005;68:383–391.
176. Agaimy A, Erlenbach-Wünsch K, Konukiewitz B, et al. ISL1 expression is not restricted to pancreatic well-differentiated neuroendocrine neoplasms, but is also commonly found in well and poorly differentiated neuroendocrine neoplasms of extrapancreatic origin. *Mod Pathol*. 2013;26(7):995–1003.
177. Xiang Y, Malik F, Zhang PJ. Islet-1 is differentially expressed among neuroendocrine and non-neuroendocrine tumors and its potential diagnostic implication. *Int J Surg Pathol*. 2023;31(7):1294–1301.
178. Siegel CS, Hrubran RH, Kloppel G, Klimstra DS. Pancreas. In: Mills SE, ed. *Histology for Pathologists*. Wolters Kluwer; 2020:757.
179. Stelow EB, Hong SM. Gallbladder and extrahepatic biliary system. In: Mills SE, ed. *Histology for Pathologists*. Wolters Kluwer; 2020:719–734.
180. Alvaro D, Mancino MG, Glaser S, et al. Proliferating cholangiocytes: A neuroendocrine compartment in the diseased liver. *Gastroenterology*. 2007;132(1):415–431.
181. Gunawardene AR, Corfe BM, Staton CA. Classification and functions of enteroendocrine cells of the lower gastrointestinal tract. *Int J Exp Path*. 2011;92:219–231.
182. Busslinger GA, Weusten BLA, Bogte A, Begthel H, Brosens LAA, Clevers H. Human gastrointestinal epithelia of the esophagus, stomach, and duodenum resolved at single-cell resolution. *Cell Rep*. 2021;34(10):108819.
183. Burclaff J, Bliton RJ, Breau KA, et al. A proximal-to-distal survey of healthy adult human small intestine and colon epithelium by single-cell transcriptomics. *Cell Mol Gastroenterol Hepatol*. 2022;13(5):1554–1589.
184. Fetissof F, Dubois MP, Arbeille-Brassart B, et al. Endocrine cells in the prostate gland, urothelium and Brenner tumors. Immunohistological and ultrastructural studies. *Virchows Arch B Cell Pathol Incl Mol Pathol*. 1985;42:53–64.
185. di Sant'Agnese PA, Jensen KD. Endocrine paracrine cells of the prostate and prostatic urethra. An ultrastructural study. *Hum Pathol*. 1984;15:1034–1041.
186. di Sant'Agnese PA, Jensen KD. Somatostatin and/or somatostatin-like immunoreactive endocrine–paracrine cells in the human prostate gland. *Arch Pathol Lab Med*. 1984;108: 693–696.
187. di Sant'Agnese PA. Calcitonin-like immunoreactive and bombesin-like immunoreactive endocrine paracrine cells of the human prostate. *Arch Pathol Lab Med*. 1986;110: 412–415.
188. Santamaria L, Martin R, Martgin JJ, Alonso L. Stereologic estimation of the number of neuroendocrine cells in normal human prostate detected by immunohistochemistry. *Appl Immunohistochem and Mol Morphol*. 2001;10:275–281.
189. Scully RE, Aguirre P, DeLellis RA. Argyrophilia, serotonin and peptide hormones in the female genital tract and its tumors. *Int Rev Gynecol Pathol*. 1984;3:51–70.
190. Slone S, Reynolds L, Gall S, et al. Localization of chromogranin, synaptophysin, serotonin, and CXCR2 in neuroendocrine cells of the minor vestibular glands: An immunohistochemical study. *Int J Gynecol Pathol*. 1999;18(4): 360–365.
191. Hanley KZ. Vulva. In: Mills SE, ed. *Histology for Pathologists*. Wolters Kluwer; 2020:1036.
192. Hashimoto K. Fine structure of Merkel cell in human oral mucosa. *J Invest Dermatol*. 1972;58:381–387.
193. Pantanowitz L, Balogh K. Mouth, nose, and paranasal sinuses. In: Mills SE, ed. *Histology for Pathologists*. Wolters Kluwer; 2020:413.
194. DeLellis RA, Wolfe HJ. Analysis of gene expression in endocrine cells. In: Fenoglio-Preiser CM, Wilman CL, eds. *Molecular Diagnostics in Pathology*. Williams & Wilkins; 1991: 299–322.
195. Lloyd RV. Introduction to molecular endocrine pathology. *Endocr Pathol*. 1993;4:64–78.
196. Speel EJM, Ramaekers FCS, Hopman AHN. Cytochemical detection systems for in situ hybridization and the combination with immunohistochemistry. "Who is still afraid of red, green and blue?" *Histochem J*. 1995;27:833–858.
197. Komminoth P, Long AA. In-situ polymerase chain reaction. An overview of methods, applications and limitations of a new molecular technique. *Virchows Arch B Cell Pathol Incl Mol Pathol*. 1993;64:67–73.
198. Komminoth P, Long AA. In situ polymerase chain reaction and its application to the study of endocrine diseases. *Endocr Pathol*. 1995;6:167–171.
199. Sällström JF, Alemi M, Spets H, Zehbe I. Nonspecific amplification in in situ PCR by direct incorporation of reporter molecules. *Cell Vision*. 1994;1:243–251.
200. Nuovo G. *PCR in Situ Hybridization*. Raven Press; 1992.
201. Taylor CR, Cote RJ. *Immunomicroscopy: A Diagnostic Tool for the Surgical Pathologist*. 2nd ed. WB Saunders; 1994: 23–28.
202. Matsumoto R, Yamamoto T. Single-cell and spatial transcriptomics in endocrine research. *Endocr J*. 2024;71(2):101–118.
203. Lindeboom RGH, Regev A, Teichmann SA. Towards a human cell atlas: Taking notes from the past. *Trends Genet*. 2021;37(7):625–630.
204. Ye F, Wang J, Li J, Mei Y, Guo G. Mapping cell atlases at the single-cell level. *Adv Sci (Weinh)*. 2024;11(8):e2305449.
205. Grube D. Immunoreactivities of gastric (G-) cells. II. Nonspecific binding of immunoglobulins to G cells by ionic interactions. *Histochemistry*. 1980;66:149–167.
206. Bussolati G, Gugliotta P, Volante M, Pace M, Papotti M. Retrieved endogenous biotin: A novel marker and potential pitfall in diagnostic immunohistochemistry. *Histopathol*. 1997;31:400–407.

207. Srivastava A, Tischler AS, DeLellis RA. Endogenous biotin staining as an artifact of antigen retrieval with automated immunostaining. *Endocrine Pathol.* 2004;15:175–177.
208. Pagani A, Cerrato M, Bussolati G. Nonspecific in situ hybridization reaction in neuroendocrine cells and tumors of the gastrointestinal tract using oligonucleotide probes. *Diagn Mol Pathol.* 1993;2:125–130.
209. Kendall PA, Polak M, Pearse AG. Carbodiimide fixation for immunohistochemistry. Observations on the fixation of polypeptide hormones. *Experimentia.* 1971;27:1104–1106.
210. King JC, Lechan RM, Kugel G, Anthony ELP. Acrolein: a fixative for immunohistochemical localization of peptides in the central nervous system. *J Histochem Cytochem.* 1983; 31:62–68.
211. Pearse AG, Polak JN, Adams C, Kendall PA. Diethyl pyrocarbonate, a vapor phase fixative for immunofluorescence studies on polypeptide hormones. *Histochem J.* 1974;6:347–352.
212. Pearse AG, Polak JM. Bifunctional reagent as vapor and liquid phase fixatives for immunohistochemistry. *Histochem J.* 1975;7:179–186.

47

Paraganglia

Arthur S. Tischler ▪ Sylvia L. Asa

Paraganglia are anatomically dispersed neuroendocrine organs associated with the autonomic nervous system that are characterized by morphologically and cytochemically similar neurosecretory cells derived from neural crest precursors. For many physiologic and pathophysiologic purposes, they may be considered to comprise two groups associated with either sympathetic or parasympathetic nerves. Sympathetic paraganglia are distributed along the prevertebral and paravertebral sympathetic chains: the preaortic plexuses and sympathetic nerve branches that innervate the organs of the pelvis and retroperitoneum. The adrenal medulla is the most extensively studied and best understood example of a sympathetic paraganglion. Parasympathetic paraganglia are predominantly distributed along cervical and thoracic branches of the glossopharyngeal and vagus nerves. The prototypical parasympathetic paraganglion is the carotid body.

HISTORY AND NOMENCLATURE

The interesting and controversial history of the paraganglia is addressed in detail in several excellent reviews (1–4). The concept of a unitary paraganglionic system was first proposed by Alfred Kohn at the beginning of the 20th century (5). Several earlier investigators developed histochemical reactions demonstrating that the adrenal medulla contained substances chemically different from those in the adrenal cortex. The reaction that proved most significant from a historical perspective, development of brown coloration in the presence of chromate salts, was first discovered by Bertholdus Werner in 1857 (3). Kohn coined the terms "chromaffin reaction" for the color change and "chromaffin cells" for the reactive cells, which he described in several extra-adrenal locations in the retroperitoneum. He further noted that some cells in the carotid body exhibited a chromaffin reaction, confirming an earlier report by Stilling (3). Kohn believed that the reactive carotid body cells were derived from precursors of sympathetic ganglia and were innervated by sympathetic axons, and he suggested that they were, therefore, embryologically, histochemically, and functionally comparable to retroperitoneal chromaffin cells. He proposed a new term to encompass all the tissues composed of cells that were analogous to neurons, but not neuronal: "Since the chromaffin tissue complexes form ganglion-like bodies, since their elements are derived from ganglion anlagen, since they are connected to the sympathetic nervous system and still are not genuine ganglia, I have called them paraganglia" (5) (translated from German by Dr. Miguel Stadecker).

Obstacles to acceptance of Kohn's concept soon arose from DeCastro's finding that the innervation of the carotid body is primarily derived from the glossopharyngeal nerve (4) and from observations by many investigators that carotid body cells are usually nonchromaffin. Consequently, Watzka (4) divided the paraganglion system into chromaffin and nonchromaffin paraganglia, associated respectively with the sympathetic or parasympathetic nervous systems, and paraganglia of mixed type. Discovery of the chemoreceptor function of the carotid bodies created further difficulties because it implied that the nonchromaffin paraganglia served physiologically in a sensory role, in contrast to the endocrine function of the adrenal medulla. The suggestion was, therefore, made by Kjaergaard that the parasympathetic paraganglia be referred to by the term "chemodecton" (from the Greek *dechesthai*, to receive) (6). This name was never widely accepted, despite the earlier application by Mulligan of its counterpart, "chemodectoma," to paraganglionic tumors (4). An additional synonym for the parasympathetic paraganglia is "glomus" (from the Latin *glomus*, ball). This term is a vestige of a 19th-century hypothesis that the carotid body is of vascular origin (4). Although it aptly describes the microscopic *Zellballen* characteristic of paraganglia, it has caused confusion because it is also applied to thermoregulatory structures in the skin and other locations (e.g., glomera cutanea and glomus coccygeum) and to their corresponding tumors (glomus tumors or glomangiomas). Those structures are modified arteriovenous anastomoses unrelated to paraganglia developmentally or functionally (7).

It is now possible to return to a unitary concept of paraganglia with a synthesis of new and old information. Paraganglia are composed of a very similar basic type of neuroendocrine cell that may be used differently in various anatomic locations (8). All of these cells are probably derived from neural crest precursors (9), although the origin of those precursors now seems to be less straightforward than previously believed (10,11). All produce catecholamines detectable by more sensitive methods than the chromaffin reaction, and all express multiple additional neuroendocrine markers, including both generic markers and regulatory peptides (12). Further, chemosensory properties previously considered specific to parasympathetic paraganglia have also been documented in cells derived from sympathetic paraganglia (13). The term "paraganglion" connotes a constellation of generic characteristics of this type of cell without being dependent on a single histochemical reaction. Because it was intended by Kohn to imply analogy, rather than merely proximity, to autonomic ganglia, it continues in this context to be both conceptually helpful and literally correct.

While the chromaffin reaction is now obsolete as a basis for the classification of paraganglia, reference to the reaction persists for historical reasons. "Chromaffin cell" is the name generally accepted for the neuroendocrine cells of the normal adrenal medulla and sometimes still applied to their extra-adrenal counterparts associated with the sympathetic nervous system. Similarly, "pheochromocytoma" (from the Greek *phaios*, dusky + *chroma*, color), refers to the color change imparted by the chromaffin reaction. The current World Health Organization classification of endocrine tumors reserves the term pheochromocytoma for intra-adrenal sympathetic paragangliomas (14,15). By arbitrary convention, even functionally similar extra-adrenal tumors are classified as paragangliomas.

SYMPATHETIC VERSUS PARASYMPATHETIC PARAGANGLIOMA: A CLINICOPATHOLOGIC PERSPECTIVE

Sympathetic and parasympathetic paraganglia differ from a clinicopathologic standpoint despite similarities at the cellular level. This contrast might result from predetermination of cell fates early in development or differences in the type, timing, or intensity of signals to which the two classes of paraganglia are exposed developmentally or in adult life (11).

The only pathologic changes known to be clinically important in paraganglia are hyperplasia and neoplasia. The clinicopathologic features of these lesions reflect the innate features of their normal counterparts together with effects of tumorigenic molecular abnormalities. Several generalizations concerning these proliferative lesions underscore the differences between normal sympathetic and parasympathetic paraganglia. Although normal paraganglia of both types can produce catecholamines, proliferative lesions that produce sufficient quantities of catecholamines to cause clinical signs and symptoms usually arise in sympathetic paraganglia. This may relate to the developmental function of sympathetic paraganglia as endocrine glands which secrete catecholamines as a direct response to hypoxia before mechanisms for central neutrally mediated responses to hypoxia are established (16). In contrast, catecholamines are produced in small quantities by parasympathetic paraganglia to serve a local paracrine function within the central regulatory signaling pathways (17). In a further parallel between normal and neoplastic paraganglia, the normal adrenal medulla is essentially the body's sole source of circulating epinephrine, and tumors that produce significant amounts of epinephrine are almost invariably pheochromocytomas. Normal parasympathetic paraganglia produce predominantly dopamine (17), and dopamine excess is present in a considerable percentage of patients with parasympathetic paragangliomas (18). Further, paragangliomas that occur in patients with prolonged hypoxemia or hypercapnia typically arise in parasympathetic paraganglia and most often the carotid bodies, which play a major role in oxygen sensing (17). Examples in which genotype may be more important than tumor location include epinephrine-producing (adrenergic) paragangliomas recently reported in association with somatic *HRAS* and *FGFR1* mutations in Chinese patients (19), and norepinephrine-producing (noradrenergic) "pseudohypoxic" pheochromocytoma caused by somatic or hereditary mutations of genes that activate hypoxia-driven signaling pathways under normoxic conditions. The latter are discussed in further detail below. A still-unexplained observation is that sympathetic paraganglia give rise to both neuronal tumors (neuroblastomas, ganglioneuroblastomas, and ganglioneuromas) and paragangliomas, while parasympathetic paraganglia give rise only to paragangliomas.

A need for deeper understanding of the differences between normal paraganglia in different locations has become apparent as a result of advances in the study of hereditary pheochromocytomas and extra-adrenal paragangliomas. Germline mutations of at least 19 genes are now known to lead to the development of these tumors (20). At least 40% of pheochromocytomas and paragangliomas are associated with inherited cancer susceptibility syndromes, the highest among all tumor types (21). The major hereditary disorders are multiple endocrine neoplasia 2A and 2B (MEN2A, MEN2B), von Hippel–Lindau (VHL) disease, neurofibromatosis type 1 (NF1) and familial paraganglioma syndromes caused by mutations of genes encoding subunits of succinate dehydrogenase (*SDHA, SDHB, SDHD, SDHC*). Rarer syndromes are caused by mutations in transmembrane protein 127 (*TMEM127*), fumarate hydratase (*FH*), the tumor suppressor MYC-associated factor X (*MAX*) gene, hypoxia-inducible factor 2 alpha (*HIF2A* or *EPAS1*), prolyl hydroxylase (*PHD1*), *EGLN1* (formerly known as *PHD2*), and the Succinate Dehydrogenase Assembly Factors (*SDHAF1, SDHAF2*) (22). Tumors that arise in each of these familial syndromes show distinctive distribution, function, and metastatic potential (20). For example, VHL-associated tumors are usually noradrenergic even when intra-adrenal, while MEN2 and NF1 tumors in the adrenal gland typically produce both epinephrine and norepinephrine (23). Pheochromocytomas associated with MEN2 often arise in a background of adrenal medullary hyperplasia, whereas other syndromes are unassociated with this finding. The metastatic rate of pheochromocytomas associated with *SDHB* mutations is estimated to be 25% to 50% although the highest estimates result from incorrect interpretation of multifocal primary tumors as metastases (20,24).

Transcriptional and multiomic analyses show different clusters of markers in tumors with specific genetic backgrounds among paragangliomas from different sites (25–27). A molecular signature associated with *VHL* or *SDH* mutations shows a "pseudohypoxic" phenotype characterized by increased activity of hypoxia-driven signaling pathways under normoxic conditions, while the signatures of tumors with *RET* or *NF1* mutations show increased activity of the RAS-mediated MAPK pathway (26,27). Some genotype–phenotype correlations may be accounted for by characteristics of the predominant anatomic sites of tumor origin. For example, epinephrine production is normally confined to the adrenal medulla where there is an environment rich in adrenal cortical steroids (28). However, the basis for different gene expression clusters in tumors at any given location is, for the most, part poorly understood. An attractive but speculative theory proposes that mutations of different susceptibility genes act at different times in embryogenesis to cause defective "developmental culling" of paraganglionic progenitors that would normally undergo programmed death, and that those surviving cells later give rise to tumors (29). The theory is consistent with the fact that the adrenal medulla matures much later than extra-adrenal paraganglia, potentially providing different developmental windows (29) in which tumorigenic events can occur. Other hypotheses include differential sensitivities to hypoxia (30) and tissue-specific effects of individual mutated genes (31).

DISTRIBUTION OF PARAGANGLIA

Sympathetic paraganglia are found predominantly in the para-axial regions of the trunk along the prevertebral and paravertebral sympathetic chains and preaortic plexuses and in connective tissue in or near the walls of the pelvic organs. In adult humans, they are especially numerous along the fibers of the inferior hypogastric plexuses leading to and entering the urogenital organs, in the wall of the urinary bladder, and among the nerve fibers of the sacral plexus (Fig. 47.1) (26–29,32–34). They are not generally known by individual names, and their precise locations are variable. Exceptions are the adrenal medulla and the organ of Zuckerkandl, located at the origin of the inferior mesenteric artery (Figs. 47.1 and 47.2) (35). The distinctive characteristic of the organ of Zuckerkandl is that it is the only extra-adrenal sympathetic paraganglion that is macroscopic. Historically, it is said to have initially been shown to Alfred Kohn as an unusual lymph node by his pupil, Emil Zuckerkandl (4,35). In its most frequent anatomic configuration, it is divided into a set of paired organs (Fig. 47.2), and is therefore sometimes referred to by the plural, "organs of Zuckerkandl" (1). Because its fragmentation and proximity

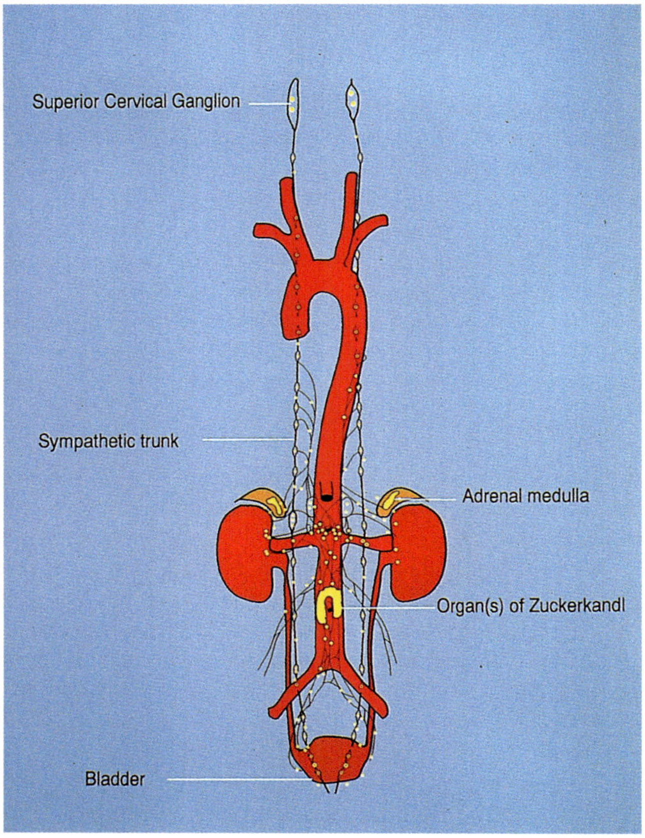

FIGURE 47.1 The distribution of sympathetic paraganglia in the human fetus. (Adapted with permission from Coupland RE. *The Natural History of the Chromaffin Cell.* Longmans Green; 1965; Glenner GG, Grimley PM. *Tumors of the Extra-adrenal Paraganglion System (Including Chemoreceptors).* Armed Forces Institute of Pathology; 1974, Permission obtained from The Joint Pathology Center.)

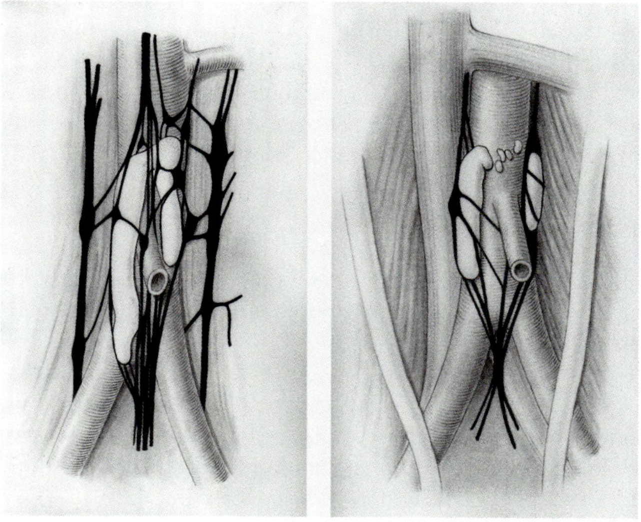

FIGURE 47.2 Modified renditions of original illustrations by Zuckerkandl representing the anatomic structure that bears his name. The bilobed configuration with a fragmented isthmus (**right**) is the most frequent variation. (Courtesy of Dr. E. E. Lack.)

to numerous smaller paraganglia can make it difficult to precisely identify, some investigators have used the plural to encompass all preaortic paraganglia between the inferior mesenteric artery and the aortic bifurcation (36). This chapter maintains the traditional, more specific, usage. In addition to these separate organs, neuroendocrine cells are present both within and adjacent to the ganglia of human sympathetic chains. The former have been referred to in neurobiology literature as small intensely fluorescent cells (37), intraganglionic chromaffin cells (38), or small granule-containing cells (39), depending on the particular technique used to detect them. In pathology literature, small intensely fluorescent cells are often regarded as intraganglionic paraganglia (40). In anatomy literature, on the other hand, some investigators reserve the term "paraganglia" for extraganglionic sites (33,41).

In contrast to sympathetic paraganglia, their parasympathetic counterparts are distributed almost exclusively along the cranial and thoracic branches of the glossopharyngeal and vagus nerves (Fig. 47.3). With the exception of the carotid bodies, which are located between the carotid arteries just above the carotid bifurcation (Fig. 47.4), parasympathetic paraganglia are highly variable in both number and location (4). Their names refer to general locations, rather than to specific structures. The middle ear, for example, contains 0 to 12 jugular and tympanic paraganglia, with an average of 2.8 (41). The principal paraganglia of the glossopharyngeal

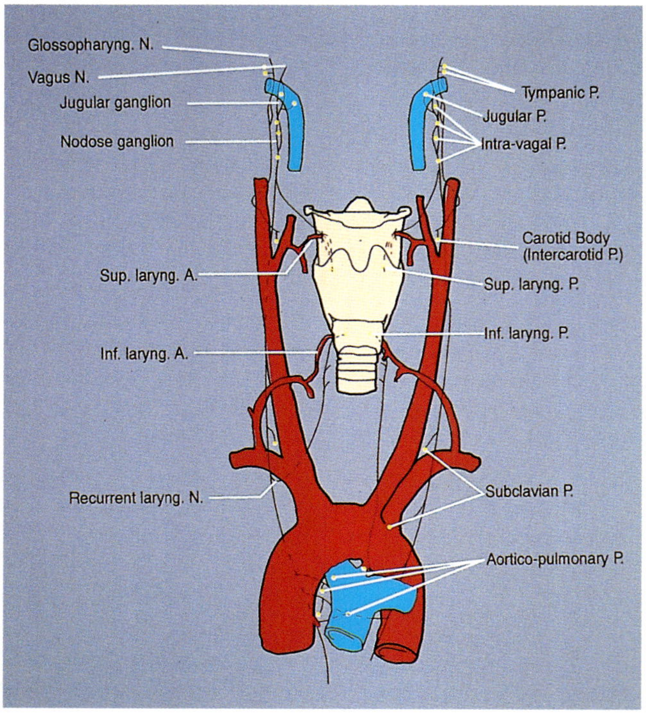

FIGURE 47.3 The distribution of the principal parasympathetic paraganglia. P, paraganglion; A, artery; N, nerve. (Adapted with permission from Glenner GG, Grimley PM. *Tumors of the Extra-adrenal Paraganglion System (Including Chemoreceptors).* Armed Forces Institute of Pathology; 1974. Permission obtained from The Joint Pathology Center.)

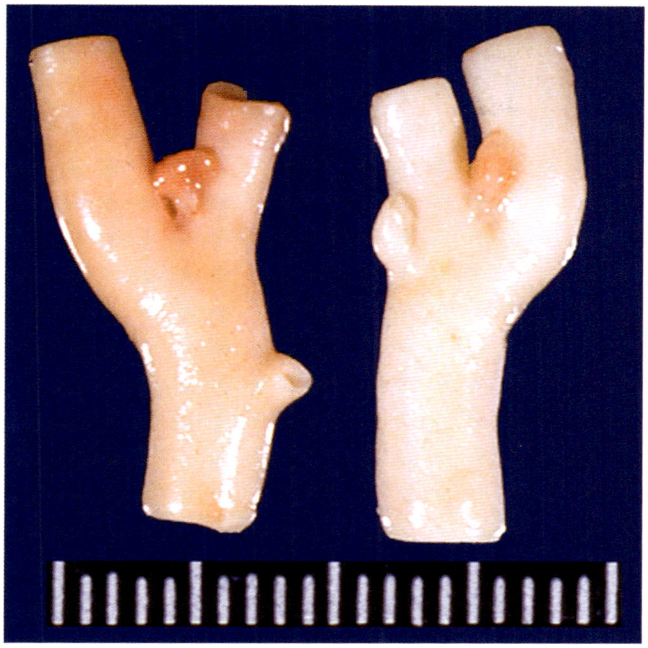

FIGURE 47.4 Gross specimens from a 5-year-old girl, illustrating normal carotid bodies and their relationship to the carotid arteries. (Courtesy of Dr. E. E. Lack.)

nerve are the tympanic paraganglia in the wall of the middle ear and the carotid bodies (4,41). Those of the vagus nerve include the jugular paraganglia in the floor of the middle ear (4,41), the superior and inferior laryngeal paraganglia (4,42), and the subclavian and aorticopulmonary or cardioaortic paraganglia near the bases of the great vessels of the heart. They sometimes also may be found in the interatrial septum (43). In addition, "intravagal" paraganglia are located within or adjacent to the vagal trunk in or near the nodose and jugular ganglia (4). These two vagal ganglia are the only sensory ganglia known to contain neuroendocrine cells comparable to the small intensely fluorescent cells of sympathetic ganglia. These cells are described as small intensely fluorescent cells in some publications (44).

Knowledge of the distribution of normal paraganglionic tissue is important because of its value in predicting the sites of origin of paragangliomas. These tumors have been reported at virtually all locations where normal paraganglia are found during fetal or adult life and tend to be most frequent in areas where paraganglionic tissue is most abundant. For example, in early childhood, paraganglionic tissue is largely extra-adrenal (see "Embryology" and "Postnatal and Developmental Changes"). Approximately 30% to 60% of paragangliomas in children are also extra-adrenal (45), most frequently arising in the vicinity of the organ of Zuckerkandl, while closer to 10% are extra-adrenal in adults. The carotid body is the most frequent site of parasympathetic paragangliomas, followed by the middle ear. However, it is also important to note that paraganglia may occur in locations outside the well-established sympathetic and parasympathetic distributions, perhaps explaining the existence of paragangliomas in unusual locations (24). Although intravagal paraganglia in humans have been identified only in the cervical and thoracic portions of the vagus nerve (46), in rodents they are also present within the abdominal portions (47). Some abdominal paraganglia in the gallbladder (48) and hilar area of the liver, may be associated with small abdominal vagus nerve branches. In scattered reports, paraganglia have been described in various sites including the orbit, mandible, paranasal sinuses, and sellar region. The validity of reports of paraganglia in the extremities is questionable.

The anatomic relationships of paraganglia in the floor or the wall of the middle ear are of particular clinical interest because of the complex clinical signs and symptoms that depend on the location of their corresponding tumors ("glomus jugulare" and "glomus tympanicum" paragangliomas). The paraganglia in the human temporal bone are distributed along the auricular branch of the vagus nerve (Arnold nerve), and the tympanic branch of the glossopharyngeal nerve (Jacobson nerve) (6,41). About 70% of paraganglia related to the Arnold nerve occur on the jugular bulb. The rest follow the nerve through the mastoid canaliculus toward the vertical portion of the facial nerve. Paraganglia along the Jacobson nerve occur anywhere from the origin of the nerve at the petrosal ganglion (10%) to the jugular bulb (28%), tympanic canaliculus (40%), promontory of the middle ear (20%), and beyond (2%). Glomus jugulare tumors may, therefore, be associated with either Arnold or Jacobson nerve, although the former is more likely. Glomus tympanicum tumors are almost always associated with Jacobson nerve (4).

From a clinical perspective, a take-home message is that paragangliomas can develop anywhere that paraganglionic neuroendocrine cells can exist, including their variable and transient locations during development. Paraganglia can be found adjacent to the thyroid gland, in the heart, within the lungs, in the hilar area of the liver, in the pancreas and mesentery, and in the prostate. In all these locations they can give rise to primary paragangliomas (24). This is an important consideration, as patients with germline predisposition to paragangliomas can develop multiple primary tumors, and in these locations, such tumors may be misclassified as metastases.

EMBRYOLOGY

Understanding of the histogenesis of paraganglia has changed greatly over the past half-decade and is still in flux. The 1970s idea of a single pluripotent sympathoadrenal progenitor migrating from the neural crest first gave way to recognition that the developmental fates of cells destined to become chromaffin cells or sympathetic neurons are largely determined before they reach their destinations in the adrenal gland or ganglia. This concept has now further evolved with the recognition of two largely separate routes of migration. A seminal 2017 lineage tracing study in mice showed that sympathetic ganglia were derived directly from migrating neural crest cells, while the adrenal medulla originated from a different wave of cells that migrated first to the dorsal root ganglia. Derivatives of those cells termed "Schwann cell precursors" subsequently migrated along the dorsal root ganglia axons to preganglionic sympathetic axons emerging from the spinal cord, and along those axons to the adrenal. In essence, this model posits that adrenal chromaffin cells originate from peripheral glial stem cells, and that peripheral nerves may serve as a stem cell niche for chromaffin cell development (10). "Bridge calls" may serve to explain transitions between neuroblast and chromaffin cell lineages. Subsequent mouse studies showed a predominant role of Schwann cell precursors in development of carotid body chief cells (49) and contributions to both chromaffin cells and neurons in the organ of Zuckerkandl (50). Single-cell transcriptomic and lineage tracing studies of human embryos have shown that Schwann cell precursors play a similar role in adrenal development, contributing to both chromaffin cells and intra-adrenal but not extra-adrenal neurons (51). These findings raise many questions that remain unanswered concerning cell fate determination (11), the nature of sustentacular cells (52), mechanisms of phenotype plasticity, and relationships to the histogenesis of pheochromocytomas, paragangliomas, neuroblastomas, and composite tumors that contain both chromaffin cells and neurons.

The new model is consistent with sometimes conflicting historic anatomic studies, some showing that primitive medullary cells appear to penetrate the adrenal cortex along preganglionic nerve fibers (1,2,53), and others reporting that the medulla can form in adrenal primordia from 4- or 5-day chick embryos explanted to chorioallantoic membranes before the onset of innervation (54). Early studies also describe primitive cells apparently migrating to developing parasympathetic paraganglia along branches of the glossopharyngeal and vagus nerves (6), and later studies employing formaldehyde-induced fluorescence of catecholamines suggest an origin of carotid body neuroendocrine cells from the sympathetic progenitors of the superior cervical ganglion (55). These observations, including the "neuronal émigré" hypothesis, are now at least partly supported by modern lineage-tracing techniques showing multiple contributions to carotid body chief cells (56). While most histologic studies of normal paraganglia date from the early to late 20th century, it is possible that revisiting classic anatomical findings will further complement modern embryology (11), providing new insights into the development of normal paraganglia and corresponding tumors.

During embryogenesis, the paraganglia are first populated by small, primitive cells that include precursors of neuroendocrine, neural, and glial cell lineages in adult paraganglia (49,56,57) (see "Light Microscopy"). They appear to be able to produce some catecholamines at the earliest stages of paraganglionic development (49,55,57) (see "Function"), and express biomarkers of immature neurons or glia (56) (see "Special Procedures"). They are readily recognized in the paraganglia at about 7 weeks of gestation, although they first arrive somewhat earlier (1,2), and they are progressively superseded by larger differentiated cells.

Studies focused on sympathoadrenal paraganglia (53,58) first show primitive sympathetic cells in large numbers around the spinal nerves and branches of the developing sympathetic trunks before the formation of the paraganglia and along the renal and spermatic arteries (53). The adrenal medulla is apparently colonized by the invasion of medullary progenitor cells into the cortex through the medial aspect of the capsule (Fig. 47.5). The invading cells initially form aggregates in the cortex (Fig. 47.5) and gradually coalesce around the central vein. They may form rosettes or pseudorosettes early in gestation. Chromaffin cells are identifiable among these aggregates from about week 8 on (1,2) and gradually increase in number. The centripetal pattern of migration may result in subcapsular and intracortical chromaffin cell rests. Nodular aggregates of primitive medullary cells, up to 400 μm in greatest dimension, can be demonstrated in the adrenal glands of all human fetuses between the ages of 10 and 30 weeks if the glands are thoroughly sectioned (58). Aggregates with a diameter of over 1 mm may occasionally be observed (58). Nerve fibers may connect intra- and extra-adrenal aggregates (58). The nodules peak in size and number between the ages of 17 and 20 weeks and then decline. Intranodular cystic degeneration is common from the age of 20 weeks onward. Occasional postnatal persistence of the nodules, apparently first described by Wiesel in 1902,

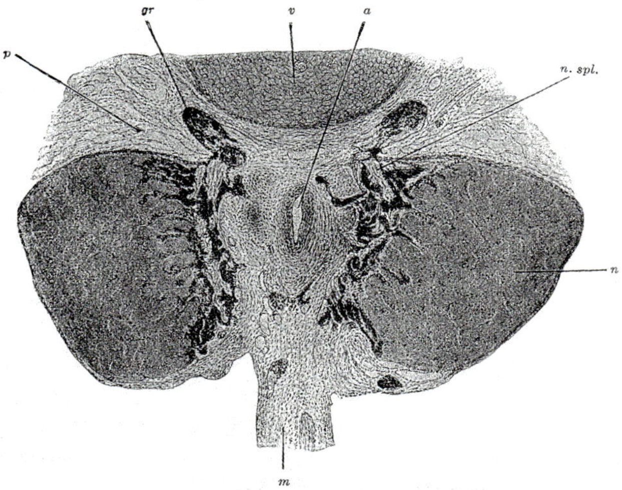

FIGURE 47.5 Original illustration by Zuckerkandl of a transverse section of a human embryo (17 mm crown–heel length) showing presumed migration of primitive sympathetic cells into the adrenal glands. The darkly stippled masses traversed by nerves are the primitive sympathetic cells. a, aorta; gr, developing sympathetic ganglia; n. Spl., splanchnic nerve; v, vertebra; p, peritoneal cushion; m, mesentery; n, adrenals. (Reprinted with permission from Zuckerkandl E. The development of the chromaffin organs and of the supra-renal glands. In: Keibel F, Mall FP, eds. *Manual of Human Embryology.* J B Lippincott; 1912.)

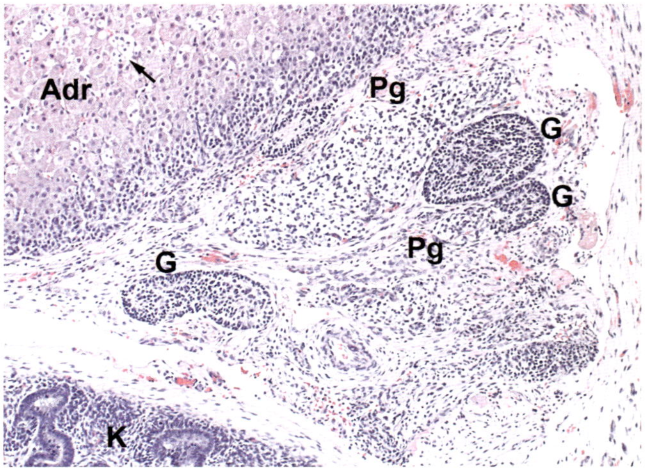

FIGURE 47.6 Section of adrenal gland from an approximately 19-week human fetus showing differences in maturation between paraganglia (Pg), sympathetic ganglia (G), and adrenal (Adr). The paraganglia are fully mature and show nests of uniform chief cells with clear to amphophilic cytoplasm. In contrast, the adrenal only shows heterogeneous aggregates of medullary progenitors (*arrow*) interspersed between the acidophilic cells of the fetal cortex, and the ganglia consist almost entirely of small blue cells with sparse cytoplasm. Pyknotic nuclei and nuclei with changes consistent with apoptosis are present within the aggregates.

may account for some erroneous diagnoses of "in situ neuroblastoma" half a century later (59–61) (see "Differential Diagnosis"). Extra-adrenal sympathetic and parasympathetic paraganglia mature cytologically earlier in development than the adrenal medulla (Fig. 47.6). Primitive cells usually disappear from these locations by week 25 but may persist in small numbers in the adrenal medulla until after birth (1,2).

POSTNATAL AND DEVELOPMENTAL CHANGES

The amount or distribution of paraganglionic tissue is known to change during development and aging. Sympathetic paraganglionic tissue in fetuses and neonates is primarily extra-adrenal, with the greatest volume in the organ of Zuckerkandl. The organ of Zuckerkandl develops maximally in humans by the age of about 3 years, when its greatest dimension may be more than 20 mm. Thereafter, the organ apparently involutes (1,34,62) while the adrenal medulla enlarges until maturity. Similarly, small intensely fluorescent cells are present in all human sympathetic ganglia at birth but are rare in adults. Parasympathetic paraganglionic tissue also appears to decrease in some locations and to increase in others. Subclavian and intrapulmonary paraganglia, for example, have been identified as prominent in human fetuses but not in adults (4), while the number of jugular and tympanic paraganglia apparently increases after birth (41). The carotid bodies increase in size between infancy and adult life, when they are normally about 3 mm in greatest dimension (4) (Fig. 47.4).

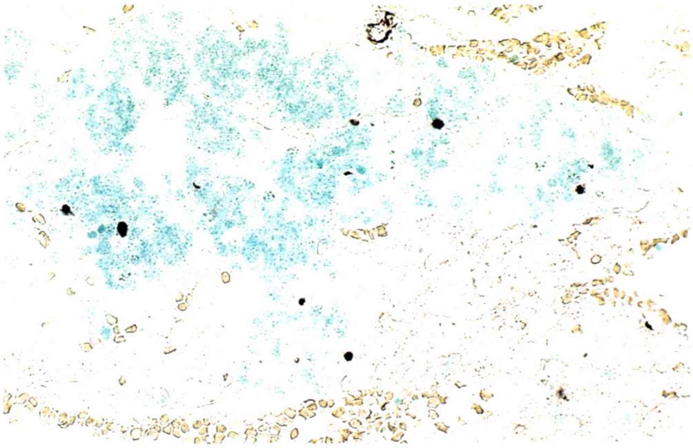

FIGURE 47.7 Section of a fetal adrenal at 16 weeks of gestation showing terminal deoxynucleotidyl transferase–mediated end-labeling of nuclei within a neuroblastic aggregate (black nuclei) consistent with developmental culling. This method, which detects fragmented DNA, can be helpful in locating apoptotic cells. (Courtesy of Dr. Salvador Diaz-Cano.)

The mechanisms involved in the developmental remodeling of the paraganglionic system may hold a number of clues to the pathobiology of paraganglionic tumors. It is generally accepted that apoptosis plays a critical role in the development of both the central and peripheral nervous systems, where excess neural progenitor cells undergo apoptotic death after failing to establish functional contacts or receive appropriate trophic substances from target tissues (63,64). There have been few studies of cell death pathways specifically focused on the paraganglia. However, apoptotic bodies can be identified within the aggregates of primitive sympathetic cells in the adrenal medulla (Fig. 47.7). Autophagy may be involved in the involution of the Organ of Zuckerkandl (62).

Although postnatal involution of the organ of Zuckerkandl is widely accepted as fact, this has only been convincingly documented in mice (1,62), and apparent disappearance of the organ in humans may in part be attributable to remodeling or fragmentation (1). While the issue remains unresolved, it is clear that microscopic sympathetic paraganglia can persist throughout life, particularly in the vicinity of the superior preaortic plexuses (1). Meticulous sketches of the distribution of sympathetic paraganglia from individual autopsies by Rex Coupland, who may be considered the father of postwar chromaffin cell biology, highlight the variable distribution and temporal changes in these structures (1). Persistent paraganglia in adults are usually encountered incidentally and can present diagnostic pitfalls by mimicking deposits of cancers, for example, in or near the prostate (65,66), bladder (67), thyroid (68), or larynx (69).

PHENOTYPIC PLASTICITY

Adult human adrenal chromaffin cells are able to "transdifferentiate" into cells closely resembling sympathetic

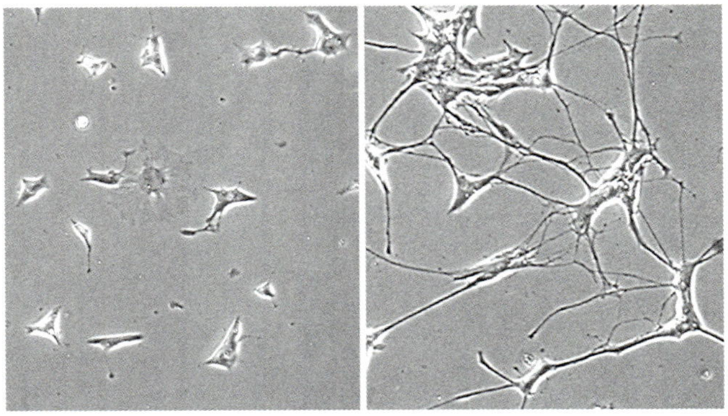

FIGURE 47.8 Phenotypic plasticity of normal adult human adrenal chromaffin cells demonstrated by acquisition of neuron-like morphology in cell culture. Cells on **left** were maintained in control medium for 2 weeks and those at **right** were in medium supplemented with nerve growth factor (50). Capacity to undergo this "transdifferentiation" is exhibited to varying degrees by chromaffin cells from other species and from other normal or neoplastic paraganglia (51). (Courtesy of Dr. James F. Powers.)

neurons when removed from their in vivo environment and exposed to appropriate neurotrophic signals (70) (Fig. 47.8). Capacity to undergo this metamorphosis is retained to varying degrees by developing or adult chromaffin cells from other species (71). The extent to which it is expressed in various extra-adrenal paraganglia, and its relationship to the development of composite tumors, have not been fully explored. However, transient transitions between neuronal and neuroendocrine states have been described in early sympathoadrenal development (11).

A somewhat different type of plasticity has been ascribed to sustentacular cells (see Cell Types), which are reported to be the progenitors of newly derived chief cells in hypoxia-induced carotid body hyperplasia (52,72). However, this finding is controversial and other studies show that already existing chief cells proliferate (73). In either case, the adult carotid body is one of the rare neural structures that can grow in response to physiologic stimuli (74).

GROSS FEATURES AND ORGAN WEIGHTS

The paraganglionic tissue of the grossly identifiable paraganglia is gray or gray–pink. Recognition of this feature is particularly important in the adrenal gland because the medulla is normally confined to the head and body of the gland but extends into the tail and alae in adrenal medullary hyperplasia (75,76). An accurate gross examination requires that the brown tissue of the cortical zona reticularis not be misidentified as medulla.

Because of the anatomic variability of the microscopic paraganglia and the organ of Zuckerkandl, meaningful weights can be ascribed only to the adrenal medulla and carotid bodies. Extensive morphometric studies (75,77) have shown that the neonatal adrenal medulla accounts for approximately 0.4% of the total volume of the gland and weighs approximately 0.012 g. These values increase to 4.2% and 0.08 g at 2 years of age, 7.0% and 0.28 g between the ages of 10 and 13 years, and 9.9% and 0.46 g in adults up to the age of 40 years. After the age of 40 years, there is a small decline in medullary weight and volume (77). The weight of the carotid bodies appears to correlate more closely with body weight than with age. Lack (78) proposed an equation to estimate carotid body weight from body weight for any age group: Combined weight of carotid bodies (mg) = 0.29 × body weight (kg) + 3.0. Standard deviations for any age group are large (75), but in normal adults, the combined weight is usually less than 30 mg (79).

ANATOMY

All paraganglia are highly vascular, a characteristic that permits them to be localized by leakage of systemically injected dye in animal studies (80). However, the details of their blood supply are highly varied according to their location and function. For example, the adrenal medulla receives arterial blood from three arteries—the inferior phrenic artery, aorta, and renal arteries—and drains via a single adrenal vein that empties into the renal vein on the left and the aorta on the right. In contrast, the carotid body receives arterial blood from one or occasionally two small arteries arising from the vicinity of the carotid bifurcation and drains via several small veins into the pharyngeal, superior laryngeal, and lingual veins (4).

The innervation of paraganglia is comparably site specific. In general, sympathetic paraganglia receive preganglionic cholinergic sympathetic innervation and variable amounts of noradrenergic and/or peptidergic innervation from intrinsic neurons, nearby sympathetic ganglia, and other sources. Most of the neuroendocrine cells in the adrenal medulla and abdominal paraganglia are innervated (81,82), although it has been suggested that the ability of paraganglia to attract or maintain innervation might determine the extent to which they persist or involute at different sites (39). Parasympathetic paraganglia generally receive their innervation from branches of either the vagus or glossopharyngeal nerves but also may receive some functional and/or vasomotor input from the nearby superior cervical ganglion and a small number of intrinsic neurons (83). Multiple neurotransmitters have been identified in both afferent and efferent nerve endings in the carotid body, and dynamic alterations of innervation occur in response to hypoxic stress (84). Multiple neurotransmitters are also present in adrenal medullary nerve endings and play a variety of roles in regulating both the development and function of adrenal chromaffin cells (84). These include acetylcholine as the predominant transmitter in splanchnic nerve endings that synapse on chromaffin cells and a variety of peptides including vasoactive intestinal peptide and substance P in intrinsic and extrinsic neurons (85).

LIGHT MICROSCOPY

Histoarchitecture and Cell Types

With the exception of the adrenal medulla and carotid bodies, paraganglia encountered by pathologists are usually microscopic organs found incidentally in fetal, pediatric, or adult specimens (Figs. 47.9 and 47.10).

The number and distribution of these structures and their proximity to visceral organs is best appreciated in fetal tissues, serving to emphasize the many locations in which paragangliomas can arise (Figs. 47.11 and 47.12). Paragangliomas contain two characteristic cell types: neuroendocrine cells and "sustentacular" cells (from Latin *sustinere*, to hold up, to support) arranged in cells in nests and cords described as *Zellballen* and *Zellsträngen* by Alfred Kohn (5). The neuroendocrine cells of sympathetic paraganglia are often referred to as "chromaffin cells" or "chromaffin-like cells." Additional terms for the subset of these cells located within sympathetic ganglia are "small granule-containing cells" and "small intensely fluorescent cells. In humans, there is only one type of small intensely fluorescent cell resembling chromaffin cells in the adrenal gland (2). A second type, present in rodents but not in humans, has features intermediate between neuroendocrine cells and neurons and is possibly a type of interneuron (2). In parasympathetic paraganglia the neuroendocrine cells are called "glomus cells," "type I cells," or "chief cells", and the sustentacular cells "type II cells." In addition to the two major cell types, there are variable numbers of connective tissue cells, vascular cells, Schwann cells, myelinated or unmyelinated nerve fibers, and intrinsic neurons (85) in both sympathetic and parasympathetic paraganglia. An additional commonly encountered cell type is the mast cell, which may be abundant in both ganglia and paraganglia (4,86).

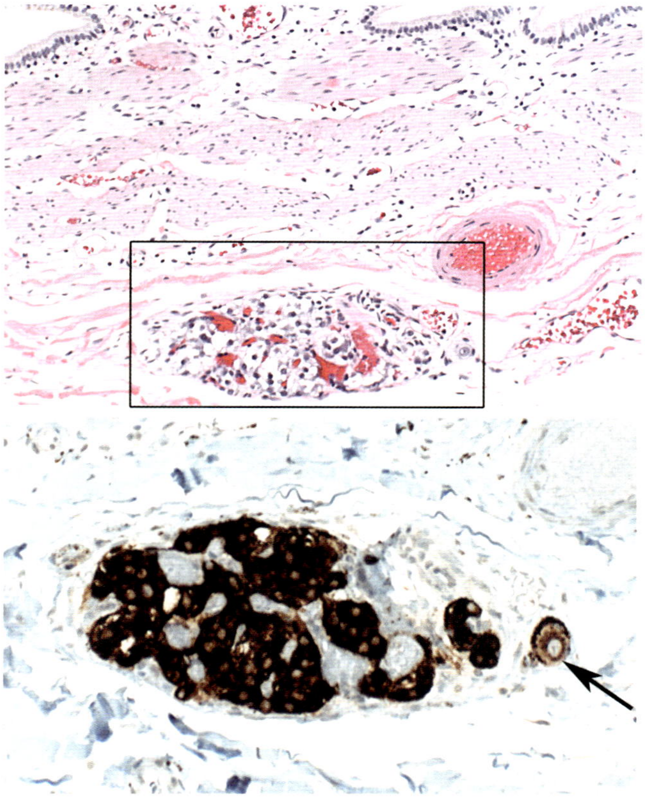

FIGURE 47.10 A microscopic paraganglion (box in **top** panel) discovered as an incidental finding in the gallbladder of a 53-year-old woman. The paraganglion is characterized by somewhat clear cells and prominent capillaries, and contains a single neuron at right. The bottom panel, represents the same tissue shown at higher magnification, decolorized and stained immunohistochemically for synaptophysin. The neuroendocrine cells show intense immunoreactivity while the neuron (*arrow*) stains weakly because it contains sparser secretory vesicles.

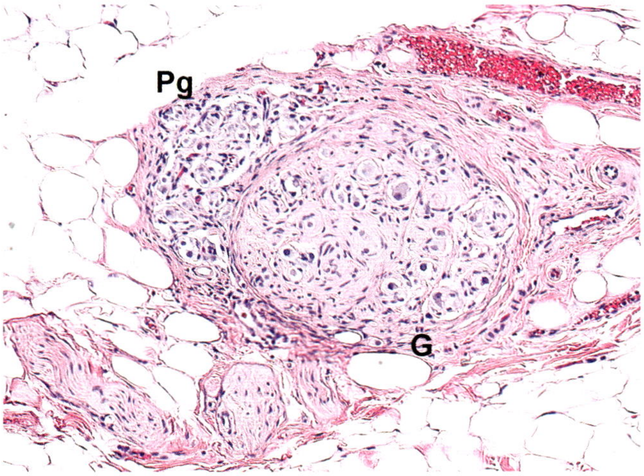

FIGURE 47.9 High magnification of an adult PG and mature ganglion found incidentally adjacent to the thyroid in a thyroidectomy specimen. The paraganglia is composed of mature chief cells and barely perceptible sustentacular cells surrounding small nests. The associated ganglion has prominent ganglion cells in a spindle cell stroma.

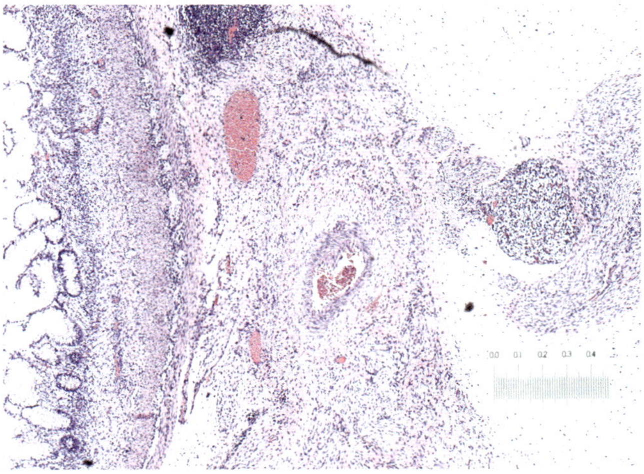

FIGURE 47.11 A freestanding abdominal paraganglion in the mesentery adjacent to the mesenteric artery in a fetus of approximately 19 weeks of gestation.

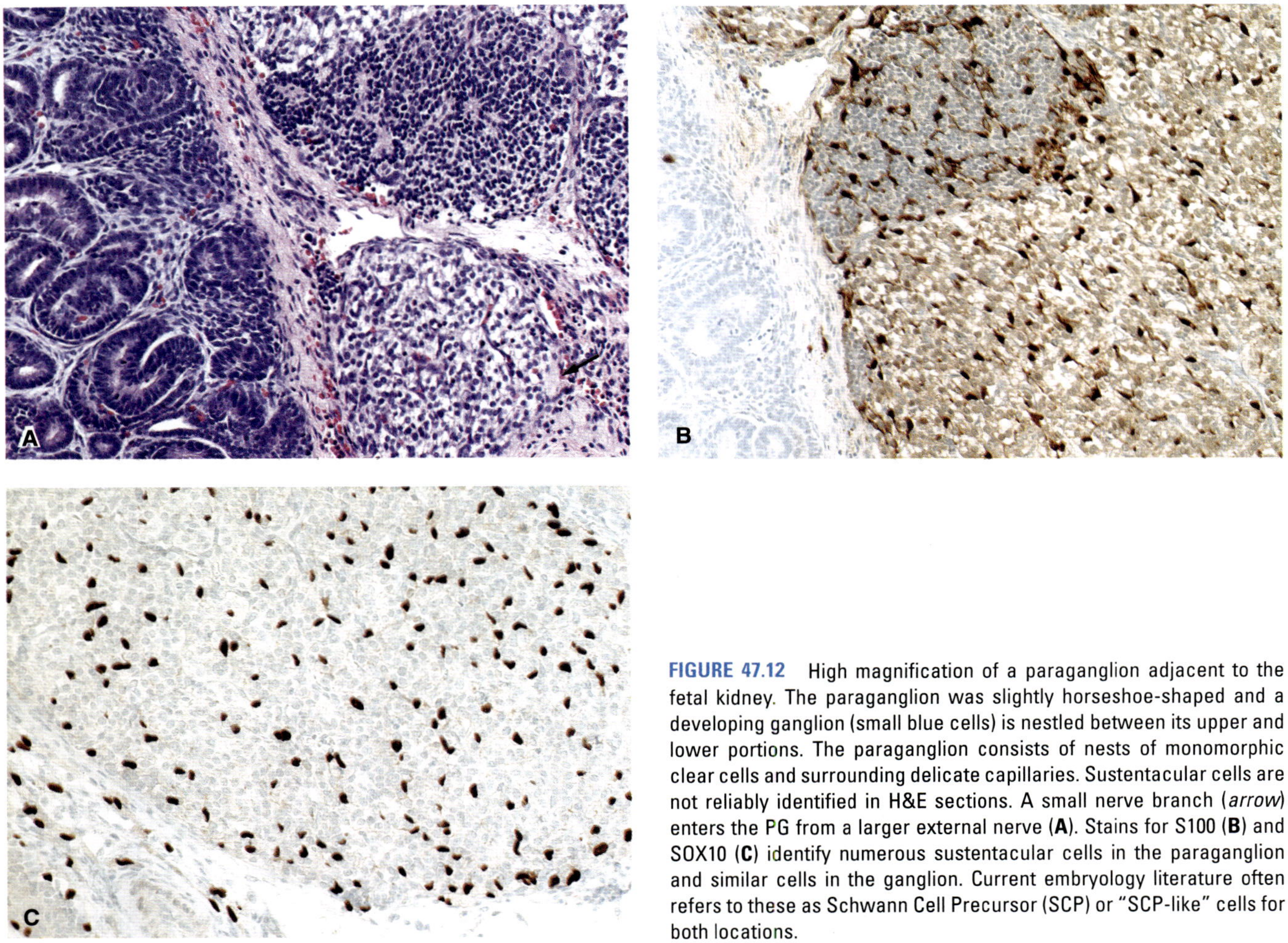

FIGURE 47.12 High magnification of a paraganglion adjacent to the fetal kidney. The paraganglion was slightly horseshoe-shaped and a developing ganglion (small blue cells) is nestled between its upper and lower portions. The paraganglion consists of nests of monomorphic clear cells and surrounding delicate capillaries. Sustentacular cells are not reliably identified in H&E sections. A small nerve branch (*arrow*) enters the PG from a larger external nerve (**A**). Stains for S100 (**B**) and SOX10 (**C**) identify numerous sustentacular cells in the paraganglion and similar cells in the ganglion. Current embryology literature often refers to these as Schwann Cell Precursor (SCP) or "SCP-like" cells for both locations.

In hematoxylin and eosin (H&E)–stained sections, paraganglionic neuroendocrine cells are round or polygonal cells with clear or amphophilic cytoplasm and small, spherical or ovoid, pale-staining nuclei (Figs. 47.9 to 47.14). Immunocytochemical stains for neuroendocrine markers such as chromogranin A (Fig. 47.10) can easily confirm their identity (see "Special Procedures"). Electron microscopy or argyrophil-type silver stains to demonstrate secretory granules or fluorescence methods to demonstrate catecholamines were employed in older publications. The sustentacular cells are glial cells, possibly related to non–myelin-forming Schwann cells elsewhere in the peripheral nervous system. They are usually flattened, with less conspicuous cytoplasm than chromaffin cells and more deeply basophilic nuclei with coarsely clumped chromatin. Immunohistochemically they can be identified by staining for S100 protein (87) and SOX10 (Figs. 47.12 to 47.14) (88) (see "Special Procedures"). A subset will also stain for glial fibrillary acidic protein (89). Sustentacular cells are present in both parasympathetic (90) and sympathetic (87) paraganglia but are more numerous in the former, where they may cause the *Zellballen* to appear more pronounced. Although SOX10 and S100 are also markers of Schwann cell precursors, it is not known whether sustentacular cells retain any of the stem cell characteristics of Schwann cell precursors in normal or neoplastic paraganglia. Consequently, they have been called "Schwann cell-like" in a single-cell transcriptome study (27).

Lobular Architecture of the Carotid Body

The carotid body is architecturally distinctive in that it consists of lobules separated by connective tissue septa (Figs. 47.14 to 47.16). Each lobule is individually reminiscent of the microscopic paraganglia that occur in other sites and is composed of nests of chief cells surrounded by other cell types. The amount of connective tissue between lobules tends to increase with age. This lobular arrangement is important to pathologists because carotid body hyperplasia is generally defined as an increase in mean lobule diameter (78,91). The carotid body increases in size and weight in individuals living at high altitudes (92) and in patients with hypoxemia due to various ailments (93).

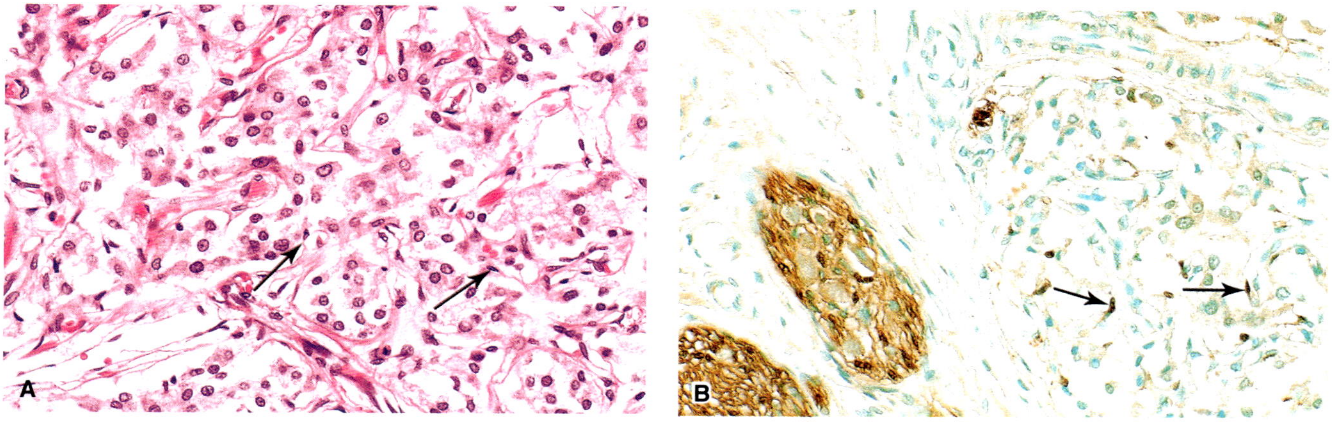

FIGURE 47.13 Section of organ of Zuckerkandl from a midtrimester human fetus, demonstrating typical cords and nests of chief cells with rounded or oval nuclei and amphophilic cytoplasm, and occasional interspersed sustentacular cells with flattened nuclei and inconspicuous cytoplasm (**A**, *arrows*). Note that the organ of Zuckerkandl is cytologically mature while the adrenal medulla in the same developmental period is still at an early formative stage. Staining for S100 protein (**B**) identifies immunoreactivity for this antigen in both nuclei and cytoplasm. Scattered sustentacular cells are stained in the organ of Zuckerkandl (*right, arrows*), where they tend to be located at the periphery of cell nests. Schwann cells are stained in the ganglion (*left*).

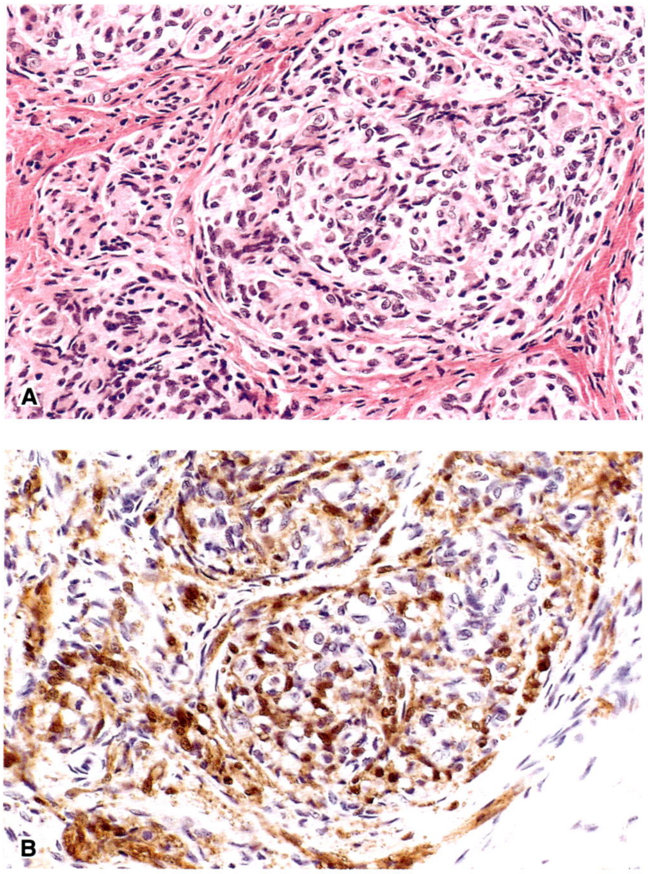

FIGURE 47.14 Section of carotid body from a 6-day-old infant demonstrating a characteristically heterogeneous cell population (**A**); small nests of chief cells are highlighted by surrounding sustentacular cells and other cell types. There are numerous sustentacular cells within the lobule that stain for S100 (**B**), accentuating the *Zellballen*, and numerous stained Schwann cells both within and adjacent to the lobule.

Some conditions are reported to result in the proliferation of sustentacular cells rather than chief cells, suggesting different mechanisms (93). While chief cell hyperplasia is associated with high altitude (92), some studies have reported that sustentacular cell proliferation is the pathognomonic feature of lobular hyperplasia in elderly patients with emphysema or hypertension (93) (Fig. 47.16). Still, others studying specimens predominantly from patients with congenital heart disease have reported proportional proliferation of sustentacular cells and chief cells (78). Schwann cell proliferation and axonal sprouting also may occur at the periphery of lobules (94). Lobular architecture similar to that of the carotid body is occasionally observed in other parasympathetic paraganglia, particularly if they are enlarged (78).

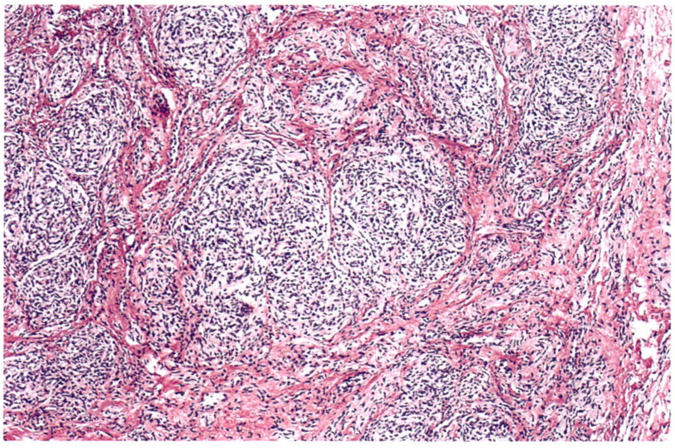

FIGURE 47.15 Carotid body from a 10-day-old infant, illustrating lobules separated by connective tissue septa.

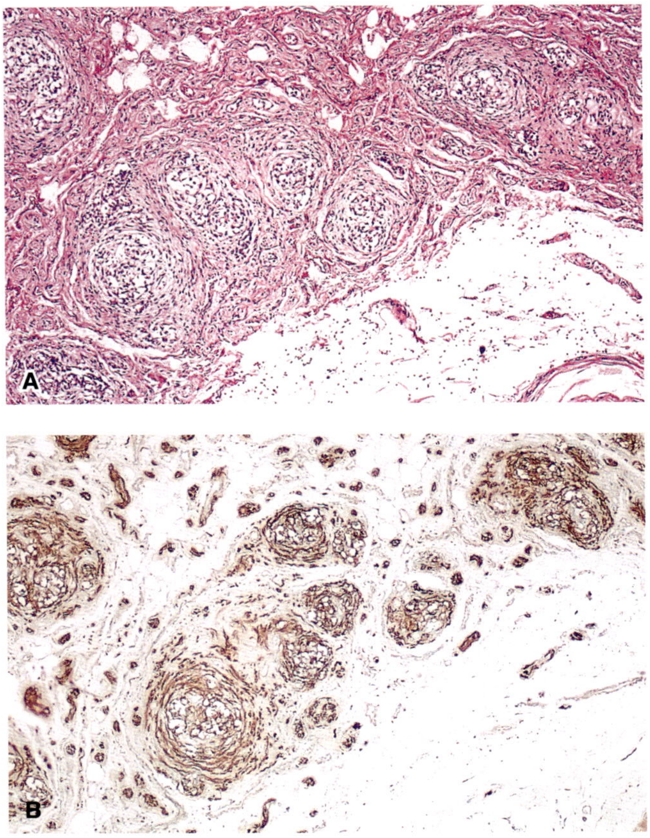

FIGURE 47.16 **A**: Carotid body from a 55-year-old woman with hypertension and emphysema. Lobules are separated by abundant connective tissue. In addition, there is a circumlobular proliferation of Schwann cells, demonstrable by staining for S100 protein. **B**: The latter change has been reported by some investigators to be characteristic of lobular hyperplasia in patients with hypertension.

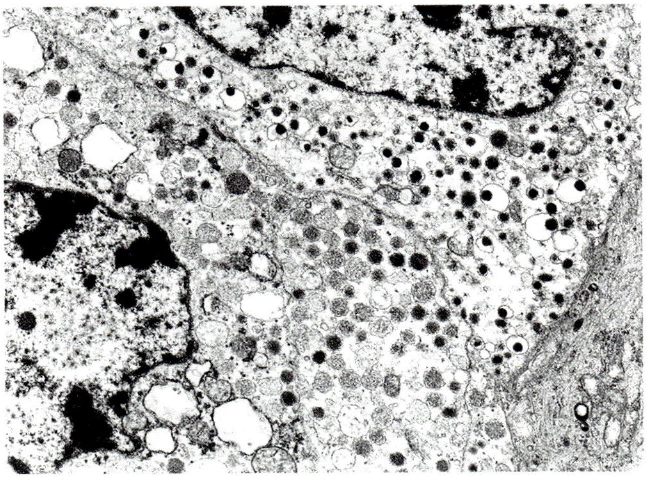

FIGURE 47.17 Electron micrograph of normal human adrenal medulla fixed in glutaraldehyde and postfixed in osmium tetroxide. Portions of cells on the left contain predominantly light, finely particulate epinephrine-type granules, whereas cells on the right contain predominantly dark, homogeneously electron-dense norepinephrine-type granules. The eccentric location of the granule cores within their surrounding membranes is a fixation artifact most commonly observed with granules of the norepinephrine type (original magnification ×9,677). (Reprinted with permission from Tischler AS. The adrenal medulla and extra-adrenal paraganglia. In: Kovacs K and Asa SL, eds. *Functional Endocrine Pathology*. Blackwell; 1990.)

ULTRASTRUCTURE

At the ultrastructural level, paraganglionic neuroendocrine cells are characterized by numerous membrane-bound granules or "dense-core vesicles" approximately 60 to 400 nm in greatest dimension. They may also contain small synaptic-like vesicles that may accumulate in clusters near the plasma membrane (95). Neuroendocrine secretory granules may vary in size, shape, and electron density, reflecting differences in the secretory products stored, the functional state of individual cells, and fixation conditions. In the rodent adrenal medulla where epinephrine and norepinephrine are mostly stored in separate cells, fixation in glutaraldehyde and postfixation in osmium tetroxide cause granules in norepinephrine cells to appear homogeneously electron dense, whereas those in epinephrine cells are lighter and finely particulate. The mechanism for these differences involves formation of an insoluble reaction product between glutaraldehyde and norepinephrine, which is subsequently darkened by osmium (96). Because epinephrine does not similarly react with glutaraldehyde, it diffuses out of the granules, leaving behind other granule constituents that are less osmiophilic. To be successful, this method requires adequate fixation of fresh tissue. Although some human adrenal medullary cells exhibit homogeneous populations of epinephrine- or norepinephrine-type granules (2) (Fig. 47.17), most have mixed granule

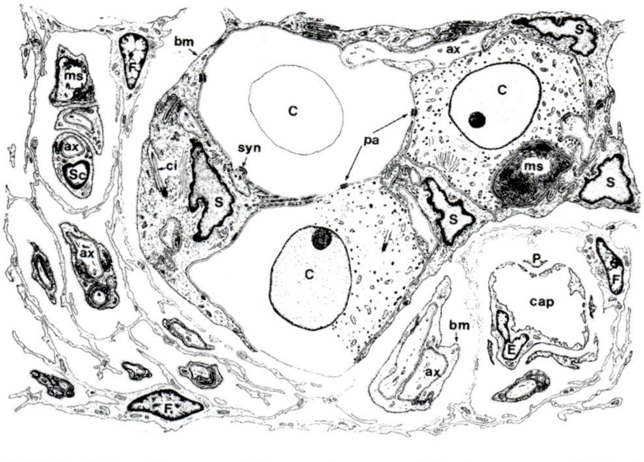

FIGURE 47.18 Diagram of the architecture of the human carotid body at the periphery of a lobule. Chief cells (C) in a small nest are insulated from the lumen of a nearby capillary (cap) by sustentacular cells (S), fibroblasts (F), and pericytes (P) and form synapses (syn) with parasympathetic axons (ax). They are also joined to each other by simple "puncta adherentia" type junctions (pa). Axons surrounded by Schwann cells (Sc) are present at the periphery of the lobule. Other illustrated structures are basement membrane (bm), endothelial cells (E), cilia (ci), and mitochondrion-rich axonal dilations termed "mitochondrial sacs" (ms). (Reprinted with permission from Böck P, Stockinger L, Vyslonzil E. The fine structure of the human carotid body. *Z Zellforsch Mikrosk Anat*. 1970;105:543–568.)

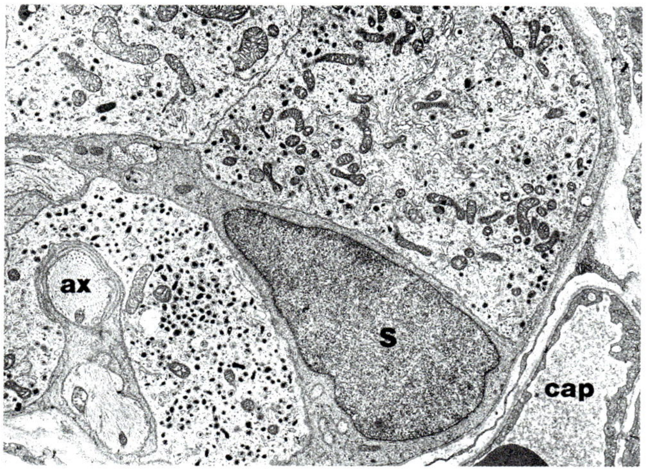

FIGURE 47.19 Electron micrograph of the human carotid body. S, sustentacular cell; cap, capillary; ax, axon (original magnification ×8,000). (Adapted with permission from Böck P, Stockinger L, Vyslonzil E. The fine structure of the human carotid body. *Z Zellforsch Mikrosk Anat.* 1970;105:543–568.)

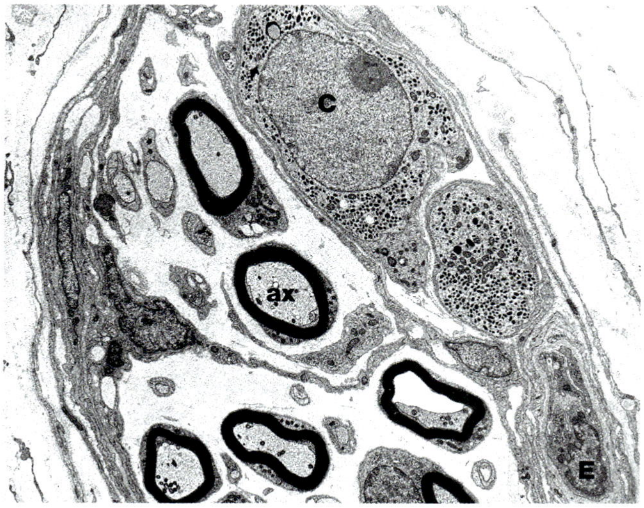

FIGURE 47.20 Electron micrograph of two chief cells enclosed within the perineurium of a small myelinated nerve in the vicinity of a human carotid body. This configuration has been described for intravagal paraganglia (56). The chief cells are separated from the lumen of a nearby capillary (original magnification ×5,400). C, chief cells; E, endothelial cell; ax, axons. (Courtesy of Professor P. Böck.)

populations and synthesize both epinephrine and norepinephrine. The electron density of most granules in extra-adrenal paraganglia is comparable to that of norepinephrine-type granules in the adrenal.

The ultrastructural organization of paraganglia and the proportions of their constituent cell types vary in different sites, apparently to suit different physiologic needs. Both sympathetic and parasympathetic paraganglia contain numerous small capillaries. In the former, portions of the surfaces of neuroendocrine cells closest to these vessels are usually separated from the capillary endothelium only by basal laminae and occasional collagen fibrils, suggesting that sympathetic paraganglia in most instances function as endocrine glands (81). In some locations, their secretory products also appear to be provided for local use (8). In contrast, the neuroendocrine cells in parasympathetic paraganglia tend to be separated from the capillary lumina by sustentacular cells, pericytes, or both (Figs. 47.18 to 47.20), consistent with a major role of their secretory products being to act directly on sensory parasympathetic nerve endings rather than to enter circulating blood (97,98).

FUNCTION

Physiologic Roles

The neuroendocrine cells in paraganglia release secretory products in response to neural or chemical stimuli. These products may be used for endocrine, paracrine, neurotransmitter, or neuromodulatory functions, depending on their anatomic context. Although their secretory products are similar, sympathetic and parasympathetic paraganglia show some differences and some overlap in the major types of stimuli to which they respond. Responses to different types of stimuli also may delineate subsets of paraganglia within the sympathetic and parasympathetic groups.

In adults the sympathetic paraganglia as represented by the adrenal medulla are considered to be essentially motor organs that respond principally to cholinergic signals from spinal cord neurons via transsynaptic stimulation, while parasympathetic paraganglia as represented by the carotid body are sensory organs that respond to low pO_2, low pH, high pCO_2, and other signals as portions of reflex loops involving the central nervous system that stimulate breathing and metabolic responses. Complex interactions in these sensory responses include synapses with afferent and efferent axons and multidirectional intracarotid transfer of information between chief cells, sustentacular cells, and neurons (17,84,99). Functional overlap between sympathetic and parasympathetic paraganglia is evident in the oxygen-sensing role of sympathetic paraganglia and immature chromaffin cells in early development (13,16). It is not known whether the persistent microscopic paraganglia in adults retain oxygen-sensing or other sensory functions.

Secretory responses to various signals are accompanied by ancillary effects including transcriptional activation of multiple genes and activation of multiple metabolic and signaling pathways to restore or maintain physiologic homeostasis. Specific responses vary according to both the cell type and aspects of the experimental model studied, including species and age. Normal and neoplastic adrenal chromaffin cells have been the major models in signaling studies because of their easy availability. Cellular responses to nerve-derived signals include activation of enzymes involved in replenishing granule constituents (100,101) and stimulation of proliferation by chromaffin cells from

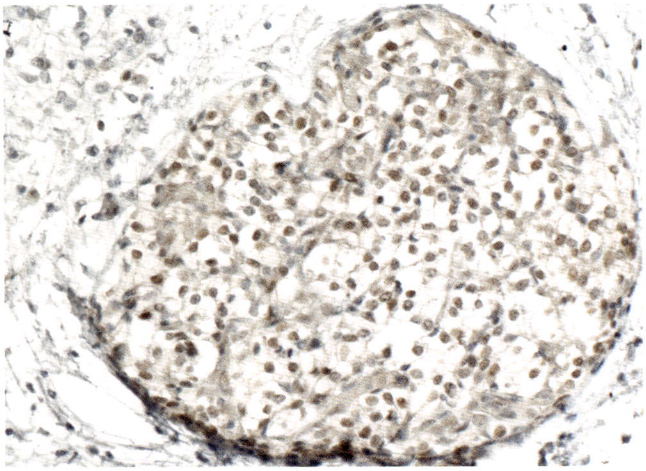

FIGURE 47.21 Immunohistochemical stain for HIF2a in a preaortic paraganglion from a human fetus at approximately 19 weeks of gestation.

rats but not humans (102). Responses may be modulated by chemical signals including corticosteroids and other hormones (100), growth factors (103), and secreted chromogranin fragments (101). Studies of the rat adrenal medulla suggest that nerve-derived signals can increase the expression of receptors that regulate chromaffin cell function, including the receptor tyrosine kinase RET (104), which is expressed at very low levels in the normal adult adrenal.

An important characteristic of oxygen-sensing paraganglia is the ability to express hypoxia-inducible transcription factor HIF2a (EPAS1) (16), which enables cellular responses that may include increased cell proliferation or survival. HIF2a is expressed in normal human sympathetic paraganglia and carotid body during development (Fig. 47.21), and the development of the carotid body requires HIF2a expression (105). Pseudohypoxic paragangliomas also express HIF2a, which has been called the Achilles heel of these tumors (106) and is being successfully targeted in clinical trials of selective inhibitors (107). It is of interest that pseudohypoxic carotid body and vagal paragangliomas can express HIF2a but normal vagal paraganglia only sometimes increase in number or size under conditions that cause carotid body hyperplasia (78). Although a tenfold increase in the prevalence of carotid body paragangliomas is reported at high altitudes (108), this strong association has not been made for paragangliomas at other sites.

Secretory Products

Although studies using the chromaffin reaction suggested that catecholamines were produced by sympathetic and not by parasympathetic paraganglia (4), more sensitive methods indicate that they are produced by paraganglia of both classes. Those methods include immunocytochemistry to demonstrate both catecholamines themselves and their biosynthetic enzymes (109,110), as well as older fluorescence techniques to demonstrate catecholamine stores (111) (see "Special Procedures"). Most of the body's epinephrine production is in the adrenal medulla, where the epinephrine-to-norepinephrine ratio is approximately 4:1. In contrast, over 90% of the catecholamine content of extra-adrenal sympathetic paraganglia is norepinephrine. Parasympathetic paraganglia produce almost no epinephrine but may produce significant quantities of dopamine. These differences in catecholamine profiles are reflected in paragangliomas that arise at different sites, with an important exception. Although most pheochromocytomas produce either epinephrine alone or epinephrine plus norepinephrine, approximately 1/3 produce almost exclusively norepinephrine. The noradrenergic phenotype, which is often associated with VHL disease and other hypoxic pathway pheochromocytomas, might arise from the small population of noradrenergic-type cells that can be found in the human adrenal either by electron microscopy or immunohistochemistry. However, tumor genotype might also influence biochemical phenotype independent of cell of origin (26,31).

In addition to producing catecholamines, both sympathetic and parasympathetic paraganglia synthesize regulatory peptides (12), the most prevalent of which are enkephalins (112). Others include vasoactive intestinal peptide (mostly in intrinsic neurons and preganglionic nerve endings), substance P, neuropeptide Y, somatostatin, and others (12,85). Pro-opiomelanocortin has been reported in the adrenal gland (113). These findings indicate that presumed "ectopic" expression of some hormones in paragangliomas and pheochromocytomas may in fact be hyperexpression of a normal product. Because regulatory peptides are not routinely searched for in these tumors it is unknown to what extent their occult expression might contribute to signs and symptoms. Regulatory peptides and amines usually coexist in the same cells and in the same secretory granules, which also contain granin proteins, adenine nucleotides, peptide-cleaving and -amidating enzymes, dopamine beta-hydroxylase, and numerous other constituents of both known and unknown functions. Peptide growth factors that might exert autocrine, paracrine, and neurotrophic effects also may be present (114). Together these granule constituents comprise a "secretory cocktail," the composition of which can be varied in different physiologic and pathologic states.

Serotonin has been reported in addition to catecholamines in some sympathetic and parasympathetic human paraganglia, but it is not clear in these mostly older studies whether the presence of serotonin is due to synthesis or uptake (115).

GENDER DIFFERENCES

Significant gender differences in the histology of paraganglia have not been reported. However, there is some evidence for functional differences. For example, women in general appear to have slightly increased susceptibility to carotid body paragangliomas. This difference is accentuated by life at high altitude, where the tumors have a female-to-male ratio of approximately 8:1 (116).

AGING CHANGES

Aging changes described in human paraganglia are limited to the topographic and involutional changes described in the sections on "Distribution of Paraganglia," "Gross Features and Organ Weights," and "Lobular Architecture of the Carotid Body." More dramatic changes in the form of hyperplasia and neoplasia occur in the adrenal medulla of aging laboratory rats (117).

SPECIAL PROCEDURES

Immunohistochemistry

Immunohistochemistry now replaces previously useful but more cumbersome or less specific techniques such as electron microscopy, catecholamine fluorescence, or silver stains for most studies of normal and abnormal paraganglia. In normal paraganglia, immunohistochemistry can reveal subtleties such as differential expression of hormones or other markers (112) that contribute to understanding pathologic conditions. For pathologists, the major application of immunohistochemistry to the paraganglionic system has until recently been the diagnosis and functional characterization of pheochromocytomas and paragangliomas. The identification of predisposition genes underlying familial paraganglioma syndromes has paved the way for immunohistochemical identification of genetic alterations (15). Loss of immunoreactivity for SDHB is a feature of any SDH-related (SDHx) tumor, and additional loss of staining for SDHA can identify tumors related to *SDHA* mutations (118). Similarly, staining for MAX (119) and FH (120) has been shown to identify tumors with mutations of the respective genes. *VHL* gene mutations can be predicted or validated by immunohistochemical staining for carbonic anhydrase IX (CAIX) (121). Expression of inhibin by paragangliomas of the pseudohypoxic type can be a potential pitfall as inhibin is often considered a marker of adrenal cortex but it can also be expressed in paragangliomas including pheochromocytoma (122). Potential future applications include the identification of markers that may be applicable to targeted therapy, such as specific somatostatin receptor subtypes (123).

Paraganglia express a plethora of markers shared to varying degrees with other neural and endocrine tissues. A partial categorization includes insulinoma associate protein 1 (INSM1), amines, regulatory peptides, granins including chromogranins A and B and other constituents of the secretory granule matrix, secretory granule membrane and cell membrane components, such as synaptophysin, and cytoskeletal proteins. Proteins known as "SNAPs" and "SNAREs" (SNAP receptors) that are involved in docking of secretory granules at the cell membrane in preparation for exocytosis also comprise an important new class of markers. Those proteins include synaptobrevin, synaptotagmin, syntaxin, and SNAP-25 (124). Developmental biologists using revolutionary new techniques for tracing cell lineage have many new markers to offer (10,125).

For diagnostic purposes, care must be taken to select from the large number of available markers those that are the most specific. For example, immunoreactivity for synaptophysin, a secretory vesicle membrane protein, is characteristically present in paraganglia and can be very helpful in confirming a neuroendocrine phenotype in appropriate contexts. However, it is also present in adrenal cortex and can, therefore, not be relied on to discriminate adrenal cortical carcinoma from pheochromocytoma. Similarly, certain SNAP and SNARE proteins are present in normal lymphoid cells. Markers that have been particularly valuable in histopathology of paraganglia are chromogranin A, catecholamine biosynthetic enzymes, S100 protein, and GATA3 (126). However, each of these has its own diagnostic caveats when extrapolating from normal to pathologic tissue and the expression of these biomarkers in other cell and tumor types must be considered.

Chromogranin A is an acidic protein that constitutes more than half the weight of many types of neuroendocrine secretory granules. It appears to be present in most neuroendocrine cells and is, therefore, a useful generic marker that can, in most instances, serve to establish the neuroendocrine nature of particular cells or tissues in the paraganglionic system (87,127,128) (Figs. 47.22 and 47.23). Since it is concentrated mostly in secretory granules, it may fail to stain cells that are degranulated due to low rates of synthesis, high rates of secretion, or low storage capacity. In sympathetic ganglia, it can be used to conveniently discriminate small intensely fluorescent cells from principal sympathetic neurons, which produce chromogranin A but have few perikaryal secretory granules (129). Chromogranin A was the first described member of the granin family of proteins, which now includes seven members (chromogranin A and chromogranin B, secretogranin II and III, 7B2, NESP55, and VGF) (130). The roles of granins include sorting of proteins to the regulated secretory pathway and directing secretory granule biogenesis. In addition, they serve as multifunctional prohormones, giving rise to cleavage fragments that exert a variety of autocrine, paracrine, and systemic effects. The various granins are differentially expressed in neuroendocrine tissues. Although most are present in the adrenal medulla, their distributions in other paraganglia are incompletely mapped. This is important for pathologists because some head and neck paragangliomas preferentially express chromogranin B and are negative or only focally positive for chromogranin A (129).

Antibodies against the catecholamine-synthesizing enzymes tyrosine hydroxylase, dopamine beta-hydroxylase and

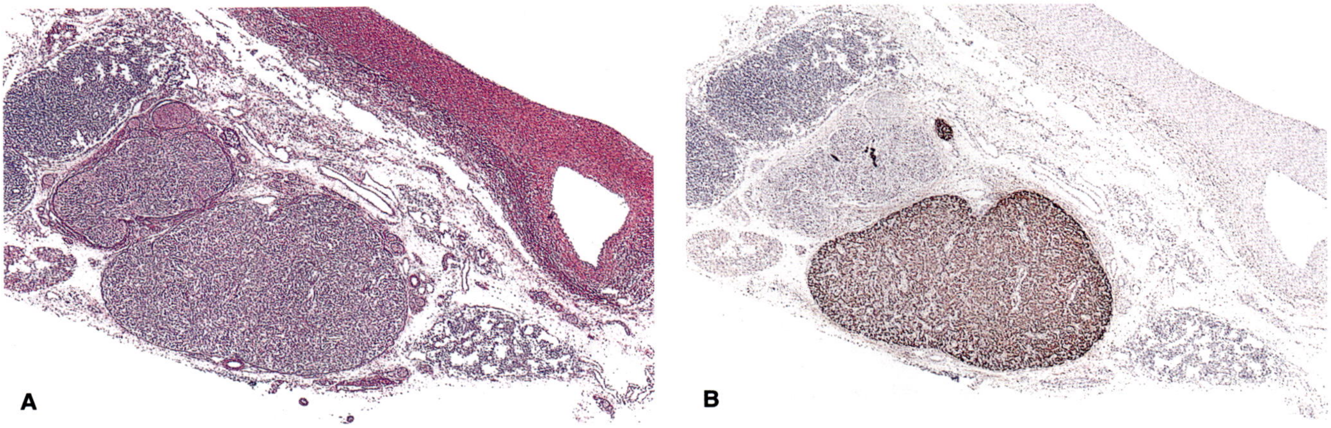

FIGURE 47.22 Transverse section through the aorta of a midtrimester human fetus at the level of the inferior mesenteric artery, demonstrating the organ of Zuckerkandl and adjacent sympathetic ganglion (**A**). Section adjacent to that shown in **A**, stained for CgA (**B**). Staining is intense in the organ of Zuckerkandl and in small nests of neuroendocrine cells within and adjacent to the ganglion, but not in neurons.

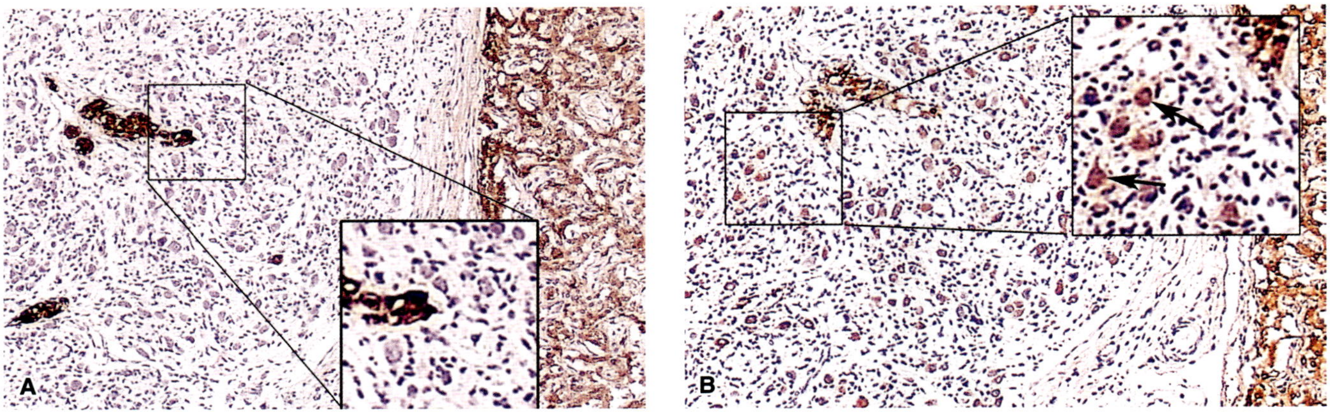

FIGURE 47.23 **A**: Higher magnification of the area shown in Figure 47.22B. The organ of Zuckerkandl (right) stains strongly for CgA. Neuroendocrine cells are also identified within the ganglion. The neurons are negative for CgA (inset). **B**: A section adjacent to that in A, stained for tyrosine hydroxylase, shows reactivity in both neuroendocrine cells and neurons (inset, *arrows*).

phenylethanolamine-N-transferase are important tools that can be used to infer from a paraffin section not only whether a tumor was producing catecholamines, but also what catecholamines were produced. Tyrosine hydroxylase is the rate-limiting enzyme in catecholamine synthesis and is therefore found in all catecholamine-producing cells (Figs. 47.24 and 47.25), whereas dopamine beta-hydroxylase is found only in cells that produce norepinephrine, and phenylethanolamine-N-transferase is found only in cells that can convert norepinephrine to epinephrine. This immunocytochemical approach provides a cellular correlate to biochemical data by demonstrating that only rarely do extra-adrenal paraganglionic cells stain for phenylethanolamine-N-transferase (109), in contrast to the adrenal medulla where the great majority of cells are stained (Figs. 47.26 and 47.27). It also has been useful in demonstrating catecholamine-synthesizing ability in pheochromocytomas and extra-adrenal paragangliomas (110). Because tyrosine hydroxylase and phenylethanolamine-N-transferase are cytosolic enzymes, staining is not dependent

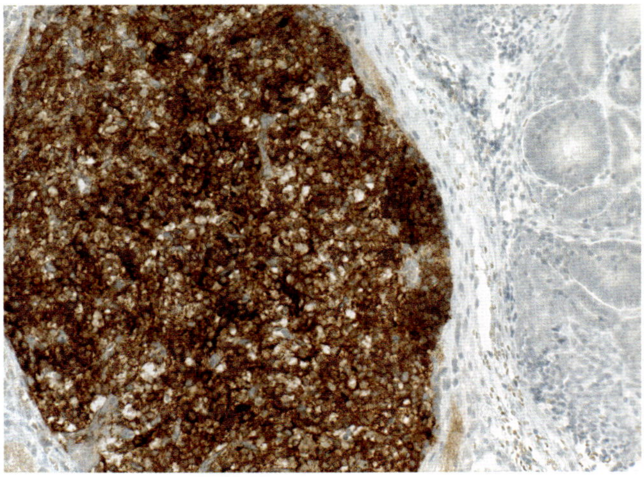

FIGURE 47.24 Immunohistochemical stain for tyrosine hydroxylase (TH) in a preaortic paraganglion from a midtrimester human fetus. Note that almost all of the cells at TH-positive and the histoarchitecture is fairly simple.

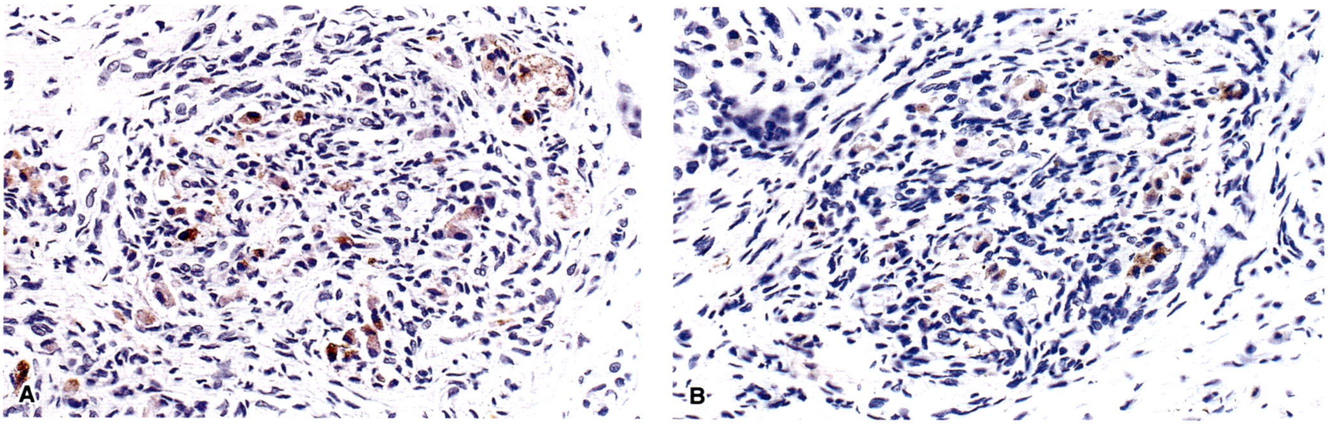

FIGURE 47.25 Lobule of a carotid body stained for CgA, demonstrating complex architecture with nests of immunoreactive chief cells surrounded by unstained cells of other types (**A**). The same structure, stained for TH, shows focal positive staining (**B**), inferring the catecholamine-synthesizing ability of chief cells.

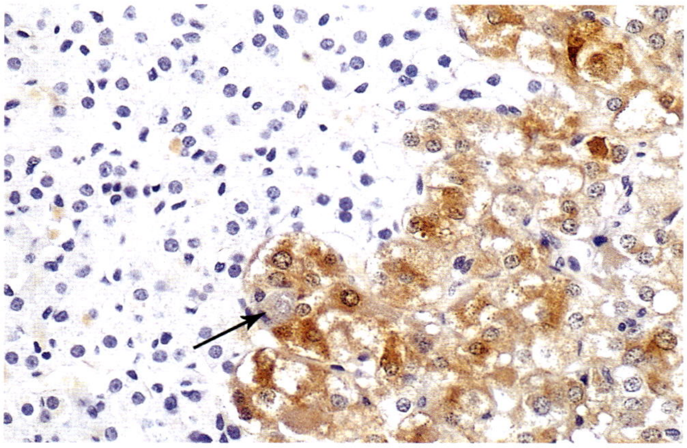

FIGURE 47.26 Adult adrenal gland stained for phenylethanolamine N-methyltransferase (PNMT). The ability of almost all neuroendocrine cells in the adrenal medulla to synthesize epinephrine is inferred from their positive staining. Occasional cells are unstained (*arrow*) as previously reported and as suggested by the electron microscopy.

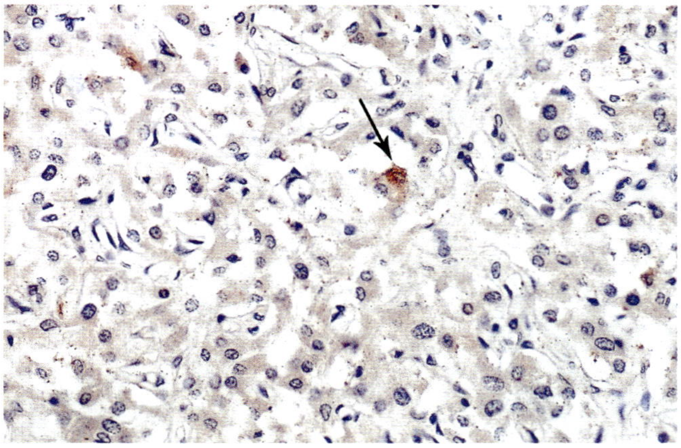

FIGURE 47.27 Organ of Zuckerkandl stained for PNMT. Only rare cells (*arrow*) contain immunoreactive PNMT; this finding is consistent with the limited ability of extra-adrenal paraganglia to synthesize epinephrine, the final step in the catecholamine biosynthetic pathway.

on storage of secretory granules. Sympathetic neurons, for example, stain strongly for tyrosine hydroxylase, in contrast to their weak or absent staining for chromogranin A. In contrast to chromogranins, which are present in many types of neuroendocrine cells, catecholamine biosynthetic enzymes in adult humans are normally present only in paraganglia and neurons (109). In addition to their biosynthetic enzymes, catecholamines themselves may be localized by immunohistochemistry. However, the presence of catecholamines without biosynthetic enzymes under some circumstances (110) suggests that synthesis cannot be distinguished from uptake by this approach. As with chromogranin A, parasympathetic paragangliomas in the head and neck can be negative for tyrosine hydroxylase (131) and, therefore, truly nonfunctional in terms of catecholamine production. A recent study suggests that they express choline acetyltransferase, an enzyme involved in acetylcholine synthesis (132).

S100 protein was initially described as a calcium-binding dimer consisting of alpha–, alpha–beta, or beta–beta chains and was initially believed to be specific for central and peripheral nervous system glial cells. Subsequent studies showed a wider distribution of immunoreactivity, and the S100 protein family has now expanded beyond the original alpha and beta subunits to include at least 17 members postulated to play various roles in different cell types (133). Nevertheless, immunostaining for S100 protein provides a useful marker in appropriate contexts. Because of their nondescript cytologic characteristics, the sustentacular cells in paraganglia are difficult to identify with certainty in sections stained with H&E. The intense nuclear immunoreactivity of these cells for S100 protein permits them to be identified in sympathetic and parasympathetic paraganglia. Both sustentacular cells and Schwann cells contain predominantly the beta subunit of S100 protein (134). These cells also express nuclear SOX10 (88).

Since sustentacular cells are frequently present in paragangliomas (87,135), immunohistochemical staining for S100 protein can sometimes be helpful in differential diagnosis. A caveat is that sustentacular cells must be distinguished from

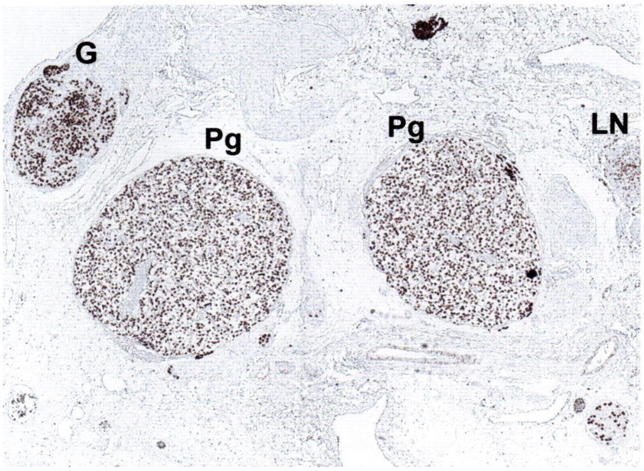

FIGURE 47.28 Immunohistochemical stain for GATA3 a in a preaortic paraganglion from human fetus at approximately 19 weeks of gestation (same case as Figure 47.21). Staining is present in two adjacent paraganglia (Pg) as well as in a developing ganglion (G) and lymph node (LN).

Langerhans cells and interdigitating reticulum cells of the immune system, which also express S100 protein. Further, other types of neuroendocrine tumors can contain bona fide sustentacular cells (136). Genetic studies of isolated sustentacular cells from paragangliomas suggest that these cells are not neoplastic (137), and therefore not an indication of bidirectional differentiation. This finding has been supported by double staining of SDH-deficient paragangliomas showing loss of cytoplasmic immunoreactivity for SDHB in neoplastic cells but retained in sustentacular cells identified by nuclear SOX10 (88). A 2023 single-cell sequencing study of paragangliomas and normal adrenal medulla provides further molecular evidence that sustentacular cells are normal (27).

GATA-3 is a transcription factor that plays a role in the development of multiple tissues, including breast, urothelium, lymphocytes and parathyroid glands as well as sympathoadrenal cells (138). It is also expressed in paraganglia (Figs. 47.28 and 47.29) and their tumors and is very useful to distinguish paragangliomas from other types of neuroendocrine neoplasms (15,126). PHOX2B, another transcription factor that is expressed during sympathetic neuronal development, has also proven to be valuable in the identification of paraganglia and their tumors (139).

Production of peptide hormones can also be detected by immunohistochemistry. As noted above, paraganglia can express enkephalins, Pro-opiomelanocortin and its derivatives, and somatostatin (Fig. 47.30).

The innervation of paraganglia can be demonstrated somewhat crudely by staining for calretinin (Fig. 47.31) or neurofilament and characterized more specifically by staining for specific secretory products (85) and transporters.

Immunohistochemical Artifacts

Important artifacts that must be borne in mind in immunohistochemical studies of paraganglia are the nonspecific

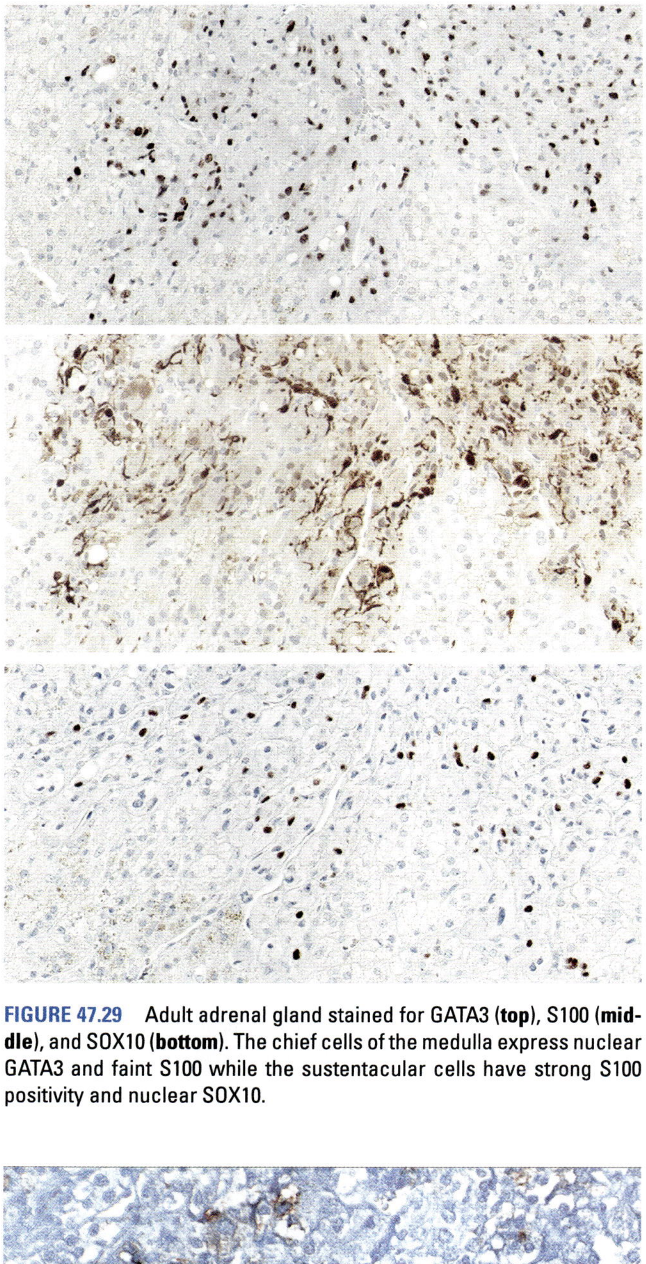

FIGURE 47.29 Adult adrenal gland stained for GATA3 (**top**), S100 (**middle**), and SOX10 (**bottom**). The chief cells of the medulla express nuclear GATA3 and faint S100 while the sustentacular cells have strong S100 positivity and nuclear SOX10.

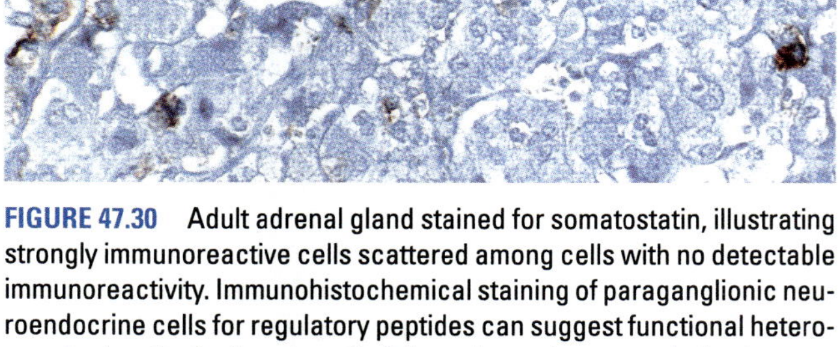

FIGURE 47.30 Adult adrenal gland stained for somatostatin, illustrating strongly immunoreactive cells scattered among cells with no detectable immunoreactivity. Immunohistochemical staining of paraganglionic neuroendocrine cells for regulatory peptides can suggest functional heterogeneity despite the fact that all of the cells produce catecholamines.

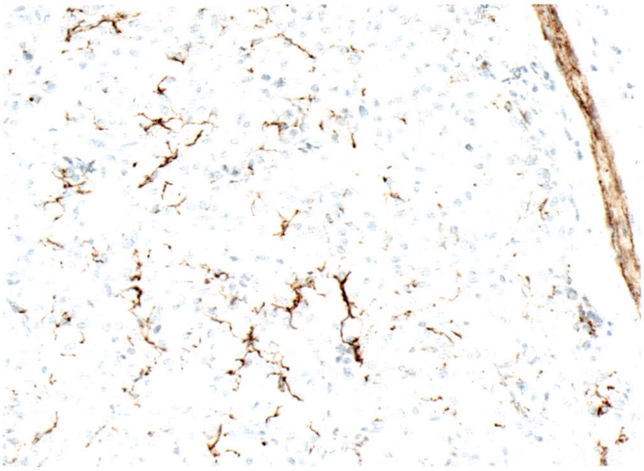

FIGURE 47.31 Immunohistochemical stain for calretinin highlighting small nerve fibers in a preaortic paraganglion from a human fetus at approximately 19 weeks of gestation. A nerve adjacent to the paraganglion is present at right.

interactions of some, but not all, antibodies with the secretory granules of mast cells (140,141) and of certain neuroendocrine cells (142). These artifacts may be particularly troublesome because, due to their inconstancy, negative controls consisting of irrelevant antibodies or normal sera are not adequate. The mast cell artifact has undoubtedly resulted in erroneous published reports of neuroendocrine secretory products in nonendocrine tissues, and also could produce incorrect results in studies of paraganglia because of their sometimes-high mast cell content. The mechanism of the mast cell artifact is not known, but in some cases the staining may be eliminated by dilution in the presence of normal serum proteins or commercial blocking kits. Nonspecific binding of immunoglobulins to neuroendocrine secretory granules appears to result from ionic interactions and may be reduced by high concentrations of salt in the buffer (142).

Other artifacts not specific to neuroendocrine cells may also be encountered in studies of paraganglia. In immunohistochemical protocols that employ a biotin bridge, such as the "ABC" technique, artifactual staining may result from the presence of endogenous biotin. This problem may be exacerbated by heat-based antigen retrieval, which can unmask endogenous biotin in addition to specific antigens (143). Commercial biotin-blocking kits offer some remedy, but a preferable solution may be to discontinue the use of biotin-based systems and switch to newer methods that employ secondary antibodies conjugated to polymer-bound reporter enzymes. Various types of artifactual staining with less clear mechanisms are also sometimes encountered. In the adrenal gland, for example, some antibodies inexplicably produce spurious staining in the adrenal cortex (144). This particularly tends to occur in the inner portion of the cortex, and might, therefore, result from nonspecific interactions of some antibodies with lipochrome pigments. Cells that are rich in mitochondria occasionally also show weak nonspecific staining. In addition to spurious staining caused by these and other artifacts, the possibility of actual immunologic cross-reactivity of an antibody with different proteins must always be considered.

In general, immunohistochemical studies should be performed with optimally diluted antibodies and verified, when practical, with antibodies from more than one source. Ideally, adsorption controls and immunoblots should also be employed for validation, particularly in research studies or when new markers or antibodies are introduced. Controls consisting solely of primary antibody omission should not be considered adequate even for routine studies. Buffer composition, blocking proteins, and controls are now becoming standardized in automated staining procedures, but should still be optimized for each new antibody as discussed in many textbooks and reviews.

Other Special Procedures

Fluorescence methods for detection of catecholamines or other biogenic amines now have very little use in routine pathology. However, literature searches of PubMed or other databases continue to demonstrate the value of these techniques in occasional, sometimes novel research applications (145). Catecholamines can be demonstrated in freeze-dried or frozen sections or in touch preparations either by formaldehyde vapor– (111) or glyoxylic acid–induced (146) fluorescence. The glyoxylic acid method is usually preferable because it produces nondiffusing fluorophores.

DIFFERENTIAL DIAGNOSIS

Paraganglia must be discriminated from normal but similar-appearing nonparaganglionic structures and from a variety of malignant tumors. The ampulloglomerular organ (147), glomus coccygeum (7), and glomera cutanea (148) are thermoregulatory structures respectively located in the suboccipital and coccygeal regions and in the skin, resembling but unrelated to paraganglia. Lobules of fetal fat may at times appear reminiscent of paraganglia. All of these structures can be readily distinguished from paraganglia by their absence of staining for chromogranin A, tyrosine hydroxylase, or other neuroendocrine markers. Reports of paraganglia in anomalous locations are also now amenable to immunohistochemical verification.

Normal paraganglia can present diagnostic pitfalls as mimics of unrelated malignant tumors. Prostatic paraganglia and paragangliomas can be misinterpreted as prostatic adenocarcinoma (65,66), bladder paraganglia can be confused with transitional cell carcinoma (67), and retroperitoneal paraganglia can be confused with metastatic clear cell carcinoma (149). Numerous mitoses, glandular or squamous differentiation, or stromal reaction are not features of paraganglia or paragangliomas. Questionable cases can be readily resolved by immunohistochemistry. For very small paraganglia that might be depleted by cutting sections, an H&E slide can be decolorized and restained for either chromogranin A or synaptophysin. The stability and abundance of those antigens make them particularly suitable for such procedures.

Cases in which paraganglia or paragangliomas must be distinguished from other normal or neoplastic neuroendocrine tissues that express many of the same markers are more problematic. Knowledge of the distribution and morphology of paraganglia is therefore essential (15). In addition, the presence of PHOX2A, GATA-3, and tyrosine hydroxylase in paragangliomas or of distinguishing markers in other types of tumors, for example, keratins, transcription factors and hormones in neuroendocrine tumors, may be helpful. Tumors showing glandular or squamous differentiation are almost certainly not paraganglionic.

In the adrenal gland, developmental neuroblastic nests (see "Embryology") must be distinguished from in situ neuroblastoma (60,61). Cortical invasion, mitoses, and necrosis are all characteristic of normal cells in this instance. In addition, the nuclei of normal adrenal medullary progenitors are reported to be smaller on average than those of neuroblastoma cells (59).

ACKNOWLEDGMENTS

We thank Drs. Peter Böck, Ernest Lack and James Powers for contributing illustrative material and Dr. Harold Kozakewich for contributing tissue blocks.

REFERENCES

1. Coupland RE. *The Natural History of the Chromaffin Cell*. Longmans Green; 1965.
2. Coupland RE. The natural history of the chromaffin cell–twenty-five years on the beginning. *Arch Histol Cytol.* 1989;52(Suppl):331–341.
3. Carmichael SW, Rochester. The history of the adrenal medulla. *Rev Neurosci.* 1989;2(2):83–100.
4. Zak FG, Lawson W. *The Paraganglionic Chemoreceptor System. Physiology, Pathology and Clinical Medicine*. Springer-Verlag; 1982.
5. Kohn A. Die paraganglien. *Arch Mikr Anat.* 1903;52:262–365.
6. Kjaergaard J. *Anatomy of the Carotid Glands and Carotid Glomus-Like Bodies (Non-Chromaffin Paraganglia)*. F.A.D.L.; 1973.
7. Rahemtullah A, Szyfelbein K, Zembowicz A. Glomus coccygeum: Report of a case and review of the literature. *Am J Dermatopathol.* 2005;27(6):497–499.
8. Furness JB, Sobels G. The ultrastructure of paraganglia associated with the inferior mesenteric ganglia in the guinea-pig. *Cell Tissue Res.* 1976;171(1):123–139.
9. Pearse AG, Polak JM, Rost FW, Fontaine J, Le LC, Le DN. Demonstration of the neural crest origin of type I (APUD) cells in the avian carotid body, using a cytochemical marker system. *Histochemie.* 1973;34(3):191–203.
10. Furlan A, Dyachuk V, Kastriti ME, et al. Multipotent peripheral glial cells generate neuroendocrine cells of the adrenal medulla. *Science.* 2017;357(6346):eaal3753.
11. Erickson AG, Kameneva P, Adameyko I. The transcriptional portraits of the neural crest at the individual cell level. *Semin Cell Dev Biol.* 2023;138:68–80.
12. Heym C, Kummer W. Regulatory peptides in paraganglia. *Prog Histochem Cytochem.* 1988;18(2):1–95.
13. Nurse CA, Salman S, Scott AL. Hypoxia-regulated catecholamine secretion in chromaffin cells. *Cell Tissue Res.* 2018;372(2):433–441.
14. WHO Classification of Tumours Editorial Board. Endocrine and neuroendocrine tumours. Lyon (France): International Agency for Research on Cancer; 2025. (WHO classification of tumours series, 5th ed.; vol. 10). https://publications.iarc.who.int/645.
15. Mete O, Asa SL, Gill AJ, Kimura N, de Krijger RR, Tischler A. Overview of the 2022 WHO classification of paragangliomas and pheochromocytomas. *Endocr Pathol.* 2022;33(1):90–114.
16. Tian H, Hammer RE, Matsumoto AM, Russell DW, McKnight SL. The hypoxia-responsive transcription factor EPAS1 is essential for catecholamine homeostasis and protection against heart failure during embryonic development. *Genes Dev.* 1998;12(21):3320–3324.
17. Iturriaga R, Alcayaga J, Chapleau MW, Somers VK. Carotid body chemoreceptors: Physiology, pathology, and implications for health and disease. *Physiol Rev.* 2021;101(3):1177–1235.
18. van der Horst-Schrivers ANA, Osinga TE, Kema IP, van der Laan BF, Dullaart RPF. Dopamine excess in patients with head and neck paragangliomas. *Anticancer Res.* 2010;30(12):5153–5158.
19. Jiang J, Zhang J, Pang Y, et al. Sino-European differences in the genetic landscape and clinical presentation of pheochromocytoma and paraganglioma. *J Clin Endocrinol Metab.* 2020;105(10):dgaa502.
20. Turin CG, Crenshaw MM, Fishbein L. Pheochromocytoma and paraganglioma: Germline genetics and hereditary syndromes. *Endocr Oncol.* 2022;2(1):R65–R77.
21. Toledo RA, Burnichon N, Cascon A, et al. Consensus Statement on next-generation-sequencing-based diagnostic testing of hereditary phaeochromocytomas and paragangliomas. *Nat Rev Endocrinol.* 2017;13(4):233–247.
22. Turchini J, Cheung VKY, Tischler AS, de Krijger RR, Gill AJ. Pathology and genetics of phaeochromocytoma and paraganglioma. *Histopathology.* 2018;72(1):97–105.
23. Eisenhofer G, Lenders JWM, Timmers H, et al. Measurements of plasma methoxytyramine, normetanephrine, and metanephrine as discriminators of different hereditary forms of pheochromocytoma. *Clin Chem.* 2011;57(3):411–420.
24. Asa SL, Ezzat S, Mete O. The diagnosis and clinical significance of paragangliomas in unusual locations. *J Clin Med.* 2018;7(9):280.
25. Brouwers FM, Gläsker S, Nave AF, et al. Proteomic profiling of von Hippel-Lindau syndrome and multiple endocrine neoplasia type 2 pheochromocytomas reveals different expression of chromogranin B. *Endocr Relat Cancer.* 2007;14(2):463–471.
26. Fishbein L, Leshchiner I, Walter V, et al. Comprehensive molecular characterization of pheochromocytoma and paraganglioma. *Cancer Cell.* 2017;31(2):181–193.

27. Zethoven M, Martelotto L, Pattison A, et al. Single-nuclei and bulk-tissue gene-expression analysis of pheochromocytoma and paraganglioma links disease subtypes with tumor microenvironment. *Nat Commun.* 2022;13(1):6262.
28. Cole TJ, Blendy JA, Monaghan AP, et al. Targeted disruption of the glucocorticoid receptor gene blocks adrenergic chromaffin cell development and severely retards lung maturation. *Genes Dev.* 1995;9(13):1608–1621.
29. Lee S, Nakamura E, Yang H, et al. Neuronal apoptosis linked to EglN3 prolyl hydroxylase and familial pheochromocytoma genes: Developmental culling and cancer. *Cancer Cell.* 2005; 8(2):155–167.
30. Fliedner SMJ, Brabant G, Lehnert H. Pheochromocytoma and paraganglioma: Genotype versus anatomic location as determinants of tumor phenotype. *Cell Tissue Res.* 2018; 372(2):347–365.
31. Fishbein L, Wilkerson MD. Chromaffin cell biology: Inferences from The Cancer Genome Atlas. *Cell Tissue Res.* 2018;372(2):339–346.
32. Hervonen A, Partanen S, Vaalasti A, Partanen M, Kanerva L, Alho H. The distribution and endocrine nature of the abdominal paraganglia of adult man. *Am J Anat.* 1978; 153(4):563–572.
33. Hervonen A, Vaalasti A, Partanen M, Kanerva L, Hervonen H. Effects of ageing on the histochemically demonstrable catecholamines and acetylcholinesterase of human sympathetic ganglia. *J Neurocytol.* 1978;7(1):11–23.
34. Baljet B, Boekelaar AB, Groen GJ. Retroperitoneal paraganglia and the peripheral autonomic nervous system in the human fetus. *Acta Morphol Neerl Scand.* 1985;23(2):137–149.
35. Ober WB. Emil Zuckerkandl and his delightful little organ. *Pathol Annu.* 1983;18 Pt 1:103–119.
36. Lack EE, Cubilla AL, Woodruff JM, Lieberman PH. Extra-adrenal paragangliomas of the retroperitoneum: A clinicopathologic study of 12 tumors. *Am J Surg Pathol.* 1980; 4(2):109–120.
37. Helen P, Alho H, Hervonen A. Ultrastructure and histochemistry of human SIF cells and paraganglia. *Adv Biochem Psychopharmacol.* 1980;25:149–152.
38. Kohn A. Die chromaffinen Zellen des sympathicus. *Anat Anz.* 1899;15(21):399–400.
39. Matthews MR. Ultrastructural studies relevant to the possible functions of small granule-containing cells in the rat superior cervical ganglion. *Adv Biochem Psychopharmacol.* 1980;25:77–86.
40. Glenner G, Grimley PM. *Tumors of the Extra-adrenal Paraganglion System (Including Chemoreceptors). Atlas of Tumor Pathology.* Armed Forces Institute of Pathology; 1974.
41. Guild SR. The glomus jugulare, a nonchromaffin paraganglion, in man. *Ann Otol Rhinol Laryngol.* 1953;62(4):1045–1071.
42. Dahlqvist A, Carlsoo B, Hellstrom S. Paraganglia of the human recurrent laryngeal nerve. *Am J Otolaryngol.* 1986; 7(5):366–369.
43. Gobbi H, Barbosa AJ, Teixeira VP, Almeida HO. Immunocytochemical identification of neuroendocrine markers in human cardiac paraganglion-like structures. *Histochemistry.* 1991;95(4):337–340.
44. Grillo MA, Jacobs L, Comroe JH Jr. A combined fluorescence histochemical and electron microscopic method for studying special monoamine-containing cells (SIF cells). *J Comp Neurol.* 1974;153(1):1–14.
45. Sarathi V. Characteristics of pediatric pheochromocytoma/paraganglioma. *Indian J Endocrinol Metab.* 2017;21(3): 470–474.
46. Plenat F, Leroux P, Floquet J, Floquet A. Intra and juxtavagal paraganglia: A topographical, histochemical, and ultrastructural study in the human. *Anat Rec.* 1988;221(3):743–753.
47. Goormaghtigh N. On the existence of abdominal vagal paraganglia in the adult mouse. *J Anat.* 1936;71(Pt 1):77–90.
48. Kuo T, Anderson CB, Rosai J. Normal paraganglia in the human gallbladder. *Arch Pathol.* 1974;97(1):46–47.
49. Hockman D, Adameyko I, Kaucka M, et al. Striking parallels between carotid body glomus cell and adrenal chromaffin cell development. *Dev Biol.* 2018;444 Suppl 1(Suppl 1):S308–S324.
50. Kastriti ME, Kameneva P, Kamenev D, et al. Schwann cell precursors generate the majority of chromaffin cells in Zuckerkandl organ and some sympathetic neurons in paraganglia. *Front Mol Neurosci.* 2019;12:6.
51. Kameneva P, Artemov AV, Kastriti ME, et al. Single-cell transcriptomics of human embryos identifies multiple sympathoblast lineages with potential implications for neuroblastoma origin. *Nat Genet.* 2021;53(5):694–706.
52. Lopez-Barneo J. Oxygen sensing and stem cell activation in the hypoxic carotid body. *Cell Tissue Res.* 2018;372(2):417–425.
53. Zuckerkandl E. The development of the chromaffin organs and of the supra-renal glands. In: Keibel F, Mall FP, eds. *Manual of Human Embryology,* Vol.1 Chapter 15. J. B. Lippincott; 1910.
54. Willier BH. A study of the origin and differentiation of the suprarenal gland in the chick embryo by chorio-allantoic grafting. *Phys Zool.* 1930;3:201–225.
55. Korkala O, Hervonen A. Origin and development of the catecholamine-storing cells of the human fetal carotid body. *Histochemie.* 1973;37(4):287–297.
56. Cooper MJ, Hutchins GM, Israel MA. Histogenesis of the human adrenal medulla. An evaluation of the ontogeny of chromaffin and nonchromaffin lineages. *Am J Pathol.* 1990; 137(3):605–615.
57. Molenaar WM, Lee VM, Trojanowski JQ. Early fetal acquisition of the chromaffin and neuronal immunophenotype by human adrenal medullary cells. An immunohistological study using monoclonal antibodies to chromogranin A, synaptophysin, tyrosine hydroxylase, and neuronal cytoskeletal proteins. *Exp Neurol.* 1990;108(1):1–9.
58. Kuntz A. The development of the sympathetic nervous system in man. *J Comp Neurol.* 1920;32:173–229.
59. Ikeda Y, Lister J, Bouton JM, Buyukpamukcu M. Congenital neuroblastoma, neuroblastoma in situ, and the normal fetal development of the adrenal. *J Pediatr Surg.* 1981;16(4 Suppl 1): 636–644.
60. Turkel SB, Itabashi HH. The natural history of neuroblastic cells in the fetal adrenal gland. *Am J Pathol.* 1974;76(2):225–244.
61. Beckwith JB, Perrin EV. In situ neuroblastomas: A contribution to the natural history of neural crest tumors. *Am J Pathol.* 1963;43(6):1089–1104.
62. Schober A, Parlato R, Huber K, et al. Cell loss and autophagy in the extra-adrenal chromaffin organ of Zuckerkandl are regulated by glucocorticoid signalling. *J Neuroendocrinol.* 2013;25(1):34–47.
63. Vogel KS. Development of trophic interactions in the vertebrate peripheral nervous system. *Mol Neurobiol.* 1993; 7(3–4):363–382.

64. Garcia I, Martinou I, Tsujimoto Y, Martinou JC. Prevention of programmed cell death of sympathetic neurons by the bcl-2 proto-oncogene. *Science.* 1992;258(5080):302–304.
65. Ostrowski ML, Wheeler TM. Paraganglia of the prostate. Location, frequency, and differentiation from prostatic adenocarcinoma. *Am J Surg Pathol.* 1994;18(4):412–420.
66. Szuman G, Amer L, Hamilton R, et al. Succinate dehydrogenase-deficient paraganglioma of the prostate. *AIM Clinical Cases.* 2023;2:e230361.
67. Rode J, Bentley A, Parkinson C. Paraganglial cells of urinary bladder and prostate: Potential diagnostic problem. *J Clin Pathol.* 1990;43(1):13–16.
68. Zak FG, Lawson W. Glomic (paraganglionic) tissue in the larynx and capsule of the thyroid gland. *Mt Sinai J Med.* 1972;39(1):82–90.
69. Hinojar AG, Prieto JR, Munoz E, Hinojar AA. Relapsing paraganglioma of the inferior laryngeal paraganglion: Case report and review of the literature. *Head Neck.* 2002;24(1):95–102.
70. Tischler AS, DeLellis RA, Biales B, Nunnemacher G, Carabba V, Wolfe HJ. Nerve growth factor-induced neurite outgrowth from normal human chromaffin cells. *Lab Invest.* 1980;43(5):399–409.
71. Anderson DJ. Cellular 'neoteny': A possible developmental basis for chromaffin cell plasticity. *Trends Genet.* 1989;5(6):174–178.
72. Camparo P, Egevad L, Algaba F, et al. Utility of whole slide imaging and virtual microscopy in prostate pathology. *APMIS.* 2012;120(4):298–304.
73. Fielding JW, Hodson EJ, Cheng X, et al. PHD2 inactivation in Type I cells drives HIF-2alpha-dependent multilineage hyperplasia and the formation of paraganglioma-like carotid bodies. *J Physiol.* 2018;596(18):4393–4412.
74. Sobrino V, Platero-Luengo A, Annese V, et al. Neurotransmitter Modulation of Carotid Body Germinal Niche. *Int J Mol Sci.* 2020;21(21):8231.
75. DeLellis RA, Wolfe HJ, Gagel RF, et al. Adrenal medullary hyperplasia. A morphometric analysis in patients with familial medullary thyroid carcinoma. *Am J Pathol.* 1976;83:177–196.
76. Carney JA, Sizemore GW, Sheps SG. Adrenal medullary disease in multiple endocrine neoplasia, type 2: pheochromocytoma and its precursors. *Am J Clin Pathol.* 1976;66(2):279–290.
77. Kreiner E. Weight and shape of the human adrenal medulla in various age groups. *Virchows Arch A Pathol Anat Histol.* 1982;397(1):7–15.
78. Lack EE. Hyperplasia of vagal and carotid body paraganglia in patients with chronic hypoxemia. *Am J Pathol.* 1978;91(3):497–516.
79. Smith P, Jago R, Heath D. Anatomical variation and quantitative histology of the normal and enlarged carotid body. *J Pathol.* 1982;137(4):287–304.
80. McDonald DM, Blewett RW. Location and size of carotid body-like organs (paraganglia) revealed in rats by the permeability of blood vessels to Evans blue dye. *J Neurocytol.* 1981;10(4):607–643.
81. Parker TL, Kesse WK, Mohamed AA, Afework M. The innervation of the mammalian adrenal gland. *J Anat.* 1993;183(Pt 2): 265–276.
82. Mascorro JA, Yates RD. Innervation of abdominal paraganglia: An ultrastructural study. *J Morphol.* 1974;142(2):153–163.
83. Case CP, Matthews MR. A quantitative study of structural features, synapses and nearest-neighbour relationships of small, granule-containing cells in the rat superior cervical sympathetic ganglion at various adult stages. *Neuroscience.* 1985;15(1):237–282.
84. Leonard EM, Salman S, Nurse CA. Sensory processing and integration at the carotid body tripartite synapse: neurotransmitter functions and effects of chronic hypoxia. *Front Physiol.* 2018;9:225.
85. Colombo-Benkmann M, Klimaschewski L, Heym C. Immunohistochemical heterogeneity of nerve cells in the human adrenal gland with special reference to substance P. *J Histochem Cytochem.* 1996;44(4):369–375.
86. Kraus R, Bezdicek P. The incidence of mastocytes in paraganglia. *Folia Morphol (Praha).* 1988;36(2):211–213.
87. Lloyd RV, blaivas M, Wilson BS. Distribution of chromogranin and S100 protein in normal and abnormal adrenal medullary tissues. *Arch Pathol Lab Med.* 1985;109(7):633–635.
88. Powers JF, Tischler AS. Immunohistochemical staining for SOX10 and SDHB in SDH-deficient paragangliomas indicates that sustentacular cells are not neoplastic. *Endocr Pathol.* 2020;31(3):307–309.
89. Achilles E, Padberg BC, Holl K, Kloppel G, Schroder S. Immunocytochemistry of paragangliomas–value of staining for S-100 protein and glial fibrillary acid protein in diagnosis and prognosis. *Histopathology.* 1991;18(5):453–458.
90. Habeck JO, Kummer W. Neuronal and neuroendocrine markers in the human carotid body in health and disease. *Adv Exp Med Biol.* 1993;337:31–35.
91. Lack EE, Perez-Atayde AR, Young JB. Carotid body hyperplasia in cystic fibrosis and cyanotic heart disease. A combined morphometric, ultrastructural, and biochemical study. *Am J Pathol.* 1985;119(2):301–314.
92. Arias-Stella J, Valcarcel J. Chief cell hyperplasia in the human carotid body at high altitudes; physiologic and pathologic significance. *Hum Pathol.* 1976;7(4):361–373.
93. Heath D, Smith P, Jago R. Hyperplasia of the carotid body. *J Pathol.* 1982;138(2):115–127.
94. Fitch R, Smith P, Heath D. Nerve axons in carotid body hyperplasia. A quantitative study. *Arch Pathol Lab Med.* 1985;109(3):234–237.
95. Verna A. Ultrastructure of the carotid body in the mammals. *Int Rev Cytol.* 1979;60:271–330.
96. Coupland RE, Hopwood D. The mechanism of the differential staining reaction for adrenaline-and noradrenaline-storing granules in tissues fixed in glutaraldehyde. *J Anat.* 1966;100(Pt 2):227–243.
97. Bock P, Stockinger L, Vyslonzil E. [The fine structure of the human carotid body]. *Z Zellforsch Mikrosk Anat.* 1970; 105(4):543–568.
98. Hervonen A, Korkala O. Fine structure of the carotid body of the midterm human fetus. *Z Anat Entwicklungsgesch.* 1972; 138(2):135–144.
99. Argent LP, Bose A, Paton JFR. Intra-carotid body inter-cellular communication. *J R Soc N Z.* 2023;53(3):332–361.
100. Sietzen M, Schober M, Fischer-Colbrie R, Scherman D, Sperk G, Winkler H. Rat adrenal medulla: Levels of chromogranins, enkephalins, dopamine beta-hydroxylase and of the amine transporter are changed by nervous activity and hypophysectomy. *Neuroscience.* 1987;22(1): 131–139.

101. Mahata SK, Mahapatra NR, Mahata M, et al. Catecholamine secretory vesicle stimulus-transcription coupling in vivo. Demonstration by a novel transgenic promoter/photoprotein reporter and inhibition of secretion and transcription by the chromogranin A fragment catestatin. *J Biol Chem.* 2003;278(34):32058–32067.
102. Tischler AS, Riseberg JC, Cherington V. Multiple mitogenic signalling pathways in chromaffin cells: A model for cell cycle regulation in the nervous system. *Neurosci Lett.* 1994;168(1-2):181–184.
103. Penberthy WT, Dahmer MK. Insulin-like growth factor-I-enhanced secretion is abolished in protein kinase C-deficient chromaffin cells. *J Neurochem.* 1994;62(5):1707–1715.
104. Powers JF, Brachold JM, Ehsani SA, Tischler AS. Up-regulation of ret by reserpine in the adult rat adrenal medulla. *Neuroscience.* 2005;132(3):605–612.
105. Macias D, Cowburn AS, Torres-Torrelo H, Ortega-Saenz P, Lopez-Barneo J, Johnson RS. HIF-2alpha is essential for carotid body development and function. *Elife.* 2018;7:e34681.
106. Tella SH, Taieb D, Pacak K. HIF-2alpha: Achilles' heel of pseudohypoxic subtype paraganglioma and other related conditions. *Eur J Cancer.* 2017;86:1–4.
107. Toledo RA, Jimenez C, Armaiz-Pena G, Arenillas C, Capdevila J, Dahia PLM. Hypoxia-inducible factor 2 Alpha (HIF2alpha) inhibitors: Targeting genetically driven tumor hypoxia. *Endocr Rev.* 2023;44(2):312–322.
108. Saldana MJ, Salem LE, Travezan R. High altitude hypoxia and chemodectomas. *Hum Pathol.* 1973;4(2):251–263.
109. Hervonen A, Pickel VM, Joh TH, et al. Immunocytochemical demonstration of the catecholamine-synthesizing enzymes and neuropeptides in the catecholamine-storing cells of human fetal sympathetic nervous system. *Adv Biochem Psychopharmacol.* 1980;25:373–378.
110. Lloyd RV, Sisson JC, Shapiro B, Verhofstad AA. Immunohistochemical localization of epinephrine, norepinephrine, catecholamine-synthesizing enzymes, and chromogranin in neuroendocrine cells and tumors. *Am J Pathol.* 1986;125(1):45–54.
111. Falck B, Bjorklund A, Lindvall O. Recent progress in aldehyde fluorescence histochemistry. *Brain Res Bull.* 1982;9(1–6):3–10.
112. Lundberg JM, Hamberger B, Schultzberg M, et al. Enkephalin- and somatostatin-like immunoreactivities in human adrenal medulla and pheochromocytoma. *Proc Natl Acad Sci U S A.* 1979;76(8):4079–4083.
113. Harno E, Gali RT, Coll AP, White A. POMC: The physiological power of hormone processing. *Physiol Rev.* 2018;98(4):2381–2430.
114. Kantorovich V, Pacak K. Pheochromocytoma and paraganglioma. *Prog Brain Res.* 2010;182:343–373.
115. Kent C, Coupland RE. On the uptake and storage of 5-hydroxytryptamine, 5-hydroxytryptophan and catecholamines by adrenal chromaffin cells and nerve endings. *Cell Tissue Res.* 1984;236(1):189–195.
116. Rodriguez-Cuevas S, Lopez-Garza J, Labastida-Almendaro S. Carotid body tumors in inhabitants of altitudes higher than 2000 meters above sea level. *Head Neck.* 1998;20(5):374–378.
117. Tischler AS, Powers JF, Alroy J. Animal models of pheochromocytoma. *Histol Histopathol.* 2004;19(3):883–895.
118. Papathomas TG, Oudijk L, Persu A, et al. SDHB/SDHA immunohistochemistry in pheochromocytomas and paragangliomas: a multicenter interobserver variation analysis using virtual microscopy: a Multinational Study of the European Network for the Study of Adrenal Tumors (ENS@T). *Mod Pathol.* 2015;28(6):807–821.
119. Roszko KL, Blouch E, Blake M, et al. Case Report of a Prolactinoma in a Patient With a Novel MAX Mutation and Bilateral Pheochromocytomas. *J Endocr Soc.* 2017;1(11):1401–1407.
120. Udager AM, Magers MJ, Goerke DM, et al. The utility of SDHB and FH immunohistochemistry in patients evaluated for hereditary paraganglioma-pheochromocytoma syndromes. *Hum Pathol.* 2018;71:47–54.
121. Favier J, Meatchi T, Robidel E, et al. Carbonic anhydrase 9 immunohistochemistry as a tool to predict or validate germline and somatic VHL mutations in pheochromocytoma and paraganglioma-a retrospective and prospective study. *Mod Pathol.* 2020;33(1):57–64.
122. Mete O, Pakbaz S, Lerario AM, Giordano TJ, Asa SL. Significance of alpha-inhibin expression in pheochromocytomas and paragangliomas. *Am J Surg Pathol.* 2021;45(9):1264–1273.
123. Korner M, Waser B, Schonbrunn A, Perren A, Reubi JC. Somatostatin receptor subtype 2A immunohistochemistry using a new monoclonal antibody selects tumors suitable for in vivo somatostatin receptor targeting. *Am J Surg Pathol.* 2012;36(2):242–252.
124. Rizo J. Mechanism of neurotransmitter release coming into focus. *Protein Sci.* 2018;27(8):1364–1391.
125. Furlan A, Adameyko I. Schwann cell precursor: a neural crest cell in disguise? *Dev Biol.* 2018;444(Suppl 1):S25–S35.
126. Miettinen M, McCue PA, Sarlomo-Rikala M, et al. GATA3: a multispecific but potentially useful marker in surgical pathology: a systematic analysis of 2500 epithelial and nonepithelial tumors. *Am J Surg Pathol.* 2014;38(1):13–22.
127. Lloyd RV, Wilson BS. Specific endocrine tissue marker defined by a monoclonal antibody. *Science.* 1983;222(4624):628–630.
128. O'Connor DT. Chromogranin: widespread immunoreactivity in polypeptide hormone producing tissues and in serum. *Regul Pept.* 1983;6(3):263–280.
129. Schmid KW, Schröder S, Dockhorn-Dworniczak B, et al. Immunohistochemical demonstration of chromogranin A, chromogranin B, and secretogranin II in extra-adrenal paragangliomas. *Mod Pathol.* 1994;7(3):347–353.
130. Helle KB, Metz-Boutigue MH, Cerra MC, Angelone T. Chromogranins: from discovery to current times. *Pflugers Arch.* 2018;470(1):143–154.
131. Tischler AS. Pheochromocytoma and extra-adrenal paraganglioma: updates. *Arch Pathol Lab Med.* 2008;132(8):1272–1284.
132. Kimura N, Shiga K, Kaneko KI, et al. Immunohistochemical expression of choline acetyltransferase and catecholamine-synthesizing enzymes in head-and-neck and thoracoabdominal paragangliomas and pheochromocytomas. *Endocr Pathol.* 2021;32(4):442–451.
133. Schafer BW, Heizmann CW. The S100 family of EF-hand calcium-binding proteins: functions and pathology. *Trends Biochem Sci.* 1996;21(4):134–140.
134. Iwanaga T, Takahashi Y, Fujita T. Immunohistochemistry of neuron-specific and glia-specific proteins. *Arch Histol Cytol.* 1989;52(Suppl):13–24.

135. Schroder HD, Johannsen L. Demonstration of S-100 protein in sustentacular cells of phaeochromocytomas and paragangliomas. *Histopathology.* 1986;10(10):1023–1033.
136. Gosney JR, Denley H, Resl M. Sustentacular cells in pulmonary neuroendocrine tumours. *Histopathology.* 1999;34(3):211–215.
137. Douwes Dekker PB, Corver WE, Hogendoorn PC, van der Mey AG, Cornelisse CJ. Multiparameter DNA flow-sorting demonstrates diploidy and SDHD wild-type gene retention in the sustentacular cell compartment of head and neck paragangliomas: chief cells are the only neoplastic component. *J Pathol.* 2004;202(4):456–462.
138. Moriguchi T, Takako N, Hamada M, et al. Gata3 participates in a complex transcriptional feedback network to regulate sympathoadrenal differentiation. *Development.* 2006; 133(19):3871–3881.
139. Miyauchi M, Akashi T, Furukawa A, et al. PHOX2B is a sensitive and specific marker for the histopathological diagnosis of pheochromocytoma and paraganglioma. *Endocr Pathol.* 2022;33(4):506–518.
140. Simson JA, Hintz DS, Munster AM, Spicer SS. Immunocytochemical evidence for antibody binding to mast cell granules. *Exp Mol Pathol.* 1977;26(1):85–91.
141. Spicer SS, Spivey MA, Ito M, Schulte BA. Some ascites monoclonal antibody preparations contain contaminants that bind to selected Golgi zones or mast cells. *J Histochem Cytochem.* 1994;42(2):213–221.
142. Grube D. Immunoreactivities of gastrin (G-) cells. II. Nonspecific binding of immunoglobulins to G-cells by ionic interactions. *Histochemistry.* 1980;66(2):149–167.
143. Srivastava A, Tischler AS, DeLellis RA. Endogenous biotin staining as an artifact of antigen retrieval with automated immunostaining. *Endocr Pathol.* 2004;15(2):175–178.
144. Powers JF, Brachold JM, Tischler AS. Ret protein expression in adrenal medullary hyperplasia and pheochromocytoma. *Endocr Pathol.* 2003;14(4):351–361.
145. Roshchina VV. The Fluorescence Methods to Study Neurotransmitters (Biomediators) in Plant Cells. *J Fluoresc.* 2016; 26(3):1029–1043.
146. de la Torre JC. Standardization of the sucrose-potassium phosphate-glyoxylic acid histofluorescence method for tissue monoamines. *Neurosci Lett.* 1980;17(3):339–340.
147. Parke WW, Valsamis MP. The ampulloglomerular organ: an unusual neurovascular complex in the suboccipital region. *Anat Rec.* 1967;159(2):193–198.
148. Bailey OT. The cutaneous glomus and its tumors-glomangiomas. *Am J Pathol.* 1935;11(6):915–936.
149. Makinen J, Nickels J. Paraganglion cells mimicking metastatic clear cell carcinoma. *Histopathology.* 1979;3(6):459–465.

Index

Note: Page number followed by f and t indicates figure and table respectively.

B

H

I

M

O